Springer Handbook

of Nanotechnology

Bharat Bhushan (Ed.)

2nd revised and extended edition
With CD-ROM, 1593 Figures and 109 Tables

Springer

Editor:

Professor Bharat Bhushan
Nanotribology Laboratory
for Information Storage and MEMS/NEMS (NLIM)
Ohio State University
W 390 Scott Laboratory
201 W. 19th AVenue
Columbus, Ohio 43210-1142
USA

Library of Congress Control Number: 2005934668

2nd edition
ISBN-10: 3-540-29855-X   e-ISBN: 3-540-29857-6
ISBN-13: 978-3-540-29855-7   Printed on acid free paper

1st edition
ISBN: 3-540-01218-4   Springer Berlin Heidelberg New York
e-ISBN: 978-3-540-29838-0

Production and typesetting: LE-TeX GbR, Leipzig
Handbook coordination: Dr. W. Skolaut, Heidelberg
Typography and layout: schreiberVIS, Seeheim
Illustrations:
schreiberVIS, Seeheim & Hippmann GbR, Schwarzenbruck
Cover design: eStudio Calamar Steinen, Barcelona
Cover production: WMXDesign GmbH, Heidelberg
Printing and binding: Stürtz AG, Würzburg

SPIN 11511250   62/3141/YL 5 4 3 2 1 0

Foreword by Neal Lane

In a January 2000 speech at the California Institute of Technology, former President W. J. Clinton talked about the exciting promise of "nanotechnology" and the importance of expanding research in nanoscale science and engineering and, more broadly, in the physical sciences. Later that month, he announced in his State of the Union Address an ambitious $497 million federal, multiagency national nanotechnology initiative (NNI) in the fiscal year 2001 budget; and he made the NNI a top science and technology priority within a budget that emphasized increased investment in U. S. scientific research. With strong bipartisan support in Congress, most of this request was appropriated, and the NNI was born. Often, federal budget initiatives only last a year or so. It is most encouraging that the NNI has remained a high priority of the G.W. Bush Administration and Congress, reflecting enormous progress in the field and continued strong interest and support by industry.

Nanotechnology is the ability to manipulate individual atoms and molecules to produce nanostructured materials and submicron objects that have applications in the real world. Nanotechnology involves the production and application of physical, chemical and biological systems at scales ranging from individual atoms or molecules to about 100 nanometers, as well as the integration of the resulting nanostructures into larger systems. Nanotechnology is likely to have a profound impact on our economy and society in the early 21st century, perhaps comparable to that of information technology or cellular and molecular biology. Science and engineering research in nanotechnology promises breakthroughs in areas such as materials and manufacturing, electronics, medicine and healthcare, energy and the environment, biotechnology, information technology and national security. It is widely felt that nanotechnology will be the next industrial revolution.

Nanometer-scale features are built up from their elemental constituents. Micro- and nanosystems components are fabricated using batch-processing techniques that are compatible with integrated circuits and range in size from micro- to nanometers. Micro- and nanosystems include micro/nanoelectro-mechanical systems (MEMS/NEMS), micromechatronics, optoelectronics, microfluidics and systems integration. These systems can sense, control, and activate on the micro/nanoscale and can function individually or in arrays to generate effects on the macroscale. Due to the enabling nature of these systems and the significant impact they can have on both the commercial and defense applications, industry as well as the federal government have taken special interest in seeing growth nurtured in this field. Micro- and nanosystems are the next logical step in the "silicon revolution."

Prof. Neal Lane

Malcolm Gillis University Professor, Department of Physics and Astronomy, and James A. Baker III Institute for Public Policy Rice University Houston, Texas, USA

Served in the Clinton Administration as Assistant to the President for Science and Technology and Director of the White House Office of Science and Technology Policy (1998–2001) and, prior to that, as Director of the National Science Foundation (1993–1998). While at the White House, he was a key figure in the creation of the NNI.

The discovery of novel materials, processes, and phenomena at the nanoscale and the development of new experimental and theoretical techniques for research provide fresh opportunities for the development of innovative nanosystems and nanostructured materials. There is an increasing need for a multidisciplinary, systems-oriented approach to manufacturing micro/nanodevices which function reliably. This can only be achieved through the cross-fertilization of ideas from different disciplines and the systematic flow of information and people among research groups.

Nanotechnology is a broad, highly interdisciplinary, and still evolving field. Covering even the most important aspects of nanotechnology in a single book that reaches readers ranging from students to active researchers in academia and industry is an enormous challenge. To prepare such a wide-ranging book on nanotechnology, Prof. Bhushan has harnessed his own knowledge and experience, gained in several industries and universities, and, for the second edition, has assembled 154 internationally recognized authors from four continents to write 58 chapters. The authors come from both academia and industry. The topics include major advances in many fields where nanoscale science and engineering is being pursued and illustrate how the field of nanotechnology has continued to emerge and blossom.

Professor Bharat Bhushan's comprehensive book is intended to serve both as a textbook for university courses as well as a reference for researchers. The first edition was a timely addition to the literature on nanotechnology, which I anticipate has stimulated further interest in this important new field and served as an invaluable resource to members of the international scientific and industrial community. It is increasingly important that scientists and engineers, whatever their specialty, have a solid grounding in the fundamentals and potential applications of nanotechnology. This important book has addressed that need. The fact that the second edition is published within two years of the first suggests that this handbook has been found to be valuable.

The editor and his team are to be warmly congratulated for bringing together this exclusive, timely, and useful nanotechnology handbook.

Foreword by James R. Heath

Nanotechnology has become an increasingly popular buzzword over the past five years or so, a trend that has been fueled by a global set of publicly funded nanotechnology initiatives. Even as researchers have been struggling to demonstrate some of the most fundamental and simple aspects of this field, the term nanotechnology has entered into the public consciousness through articles in the popular press and popular fiction. As a consequence, the expectations of the public are high for nanotechnology, even while the actual public definition of nanotechnology remains a bit fuzzy.

Why shouldn't those expectations be high? The late 1990s witnessed a major information technology (IT) revolution and a minor biotechnology revolution. The IT revolution impacted virtually every aspect of life in the western world. I am sitting on an airplane at 30 000 feet at the moment, working on my laptop, as are about half of the other passengers on this plane. The plane itself is riddled with computational and communications equipment. As soon as we land, many of us will pull out cell phones, others will check e-mail via wireless modem, some will do both. This picture would be the same if I was landing in Los Angeles, Beijing, or Capetown. I will probably never actually print this text, but will instead submit it electronically. All of this was unthinkable a dozen years ago. It is therefore no wonder that the public expects marvelous things to happen quickly. However, the science that laid the groundwork for the IT revolution dates back 60 years or more, with its origins in fundamental solid-state physics.

By contrast, the biotech revolution was relatively minor and, at least to date, not particularly effective. The major diseases that plagued mankind a quarter century ago are still here. In some third-world countries, the average lifespan of individuals has actually decreased from where it was a full century ago. While the costs of electronics technologies have plummeted, health care costs have continued to rise. The biotech revolution may have a profound impact, but the task at hand is substantially more difficult than what was required for the IT revolution. In effect, the IT revolution was based on the advanced engineering of two-dimensional digital circuits constructed from relatively simple components – extended solids. The biotech revolution is really dependent upon the ability to reverse engineer three-dimensional analog systems constructed from quite complex components – proteins. Given that the basic science behind biotech is substantially younger than the science that has supported IT, it is perhaps not surprising that the biotech revolution has not really been a proper revolution yet, and it likely needs at least another decade or so to come into fruition.

Where does nanotechnology fit into this picture? In many ways, nanotechnology depends upon the ability to engineer two- and three-dimensional systems constructed from complex components such as macromolecules, biomolecules, nanostructured solids, etc. Furthermore, in terms of patents, publications, and other metrics that can be used to gauge the birth and evolution of a field, nanotech lags some 15–20 years behind biotech. Thus, now is the time that the fundamental science behind nanotechnology is being explored and developed. Nevertheless, progress with that science is moving forward at a dramatic pace. If the scientific community can keep up this pace and if the public sector will continue to support this science, then it is possible, and even perhaps likely, that in 20 years we may be speaking of the nanotech revolution.

The first edition of the Springer Handbook of Nanotechnology was timely to assemble chapters in the broad field of nanotechnology. Given the fact that the second edition is in press two years after the publication of the first edition in April 2004, it is clear that the handbook has proven to be a valuable reference for experienced researchers as well as for novices in the field. The second edition has about one and a half times the number of chapters as compared to the first edition and has an expanded scope, giving this edition a much wider appeal.

Prof. James R. Heath

Department of Chemistry
Mail Code: 127–72
California Institute of Technology
Pasadena, CA 91125, USA

Worked in the group of Nobel Laureate Richard E. Smalley at Rice University (1984–88) and co-invented Fullerene molecules which led to a revolution in Chemistry including the realization of nanotubes. The work on Fullerene molecules was cited for the 1996 Nobel Prize in Chemistry. Later he joined the University of California at Los Angeles (1994–2002), and co-founded and served as a Scientific Director of The California Nanosystems Institute.

Preface to the 2nd Edition

On 29 December 1959 at the California Institute of Technology, Nobel Laureate Richard P. Feynman gave at talk at the Annual meeting of the American Physical Society that has become one of the 20th century classic science lectures, titled "There's Plenty of Room at the Bottom." He presented a technological vision of extreme miniaturization in 1959, several years before the word "chip" became part of the lexicon. He talked about the problem of manipulating and controlling things on a small scale. Extrapolating from known physical laws, Feynman envisioned a technology using the ultimate toolbox of nature, building nanoobjects atom by atom or molecule by molecule. Since the 1980s, many inventions and discoveries in the fabrication of nanoobjects have been a testament to his vision. In recognition of this reality, the National Science and Technology Council (NSTC) of the White House created the Interagency Working Group on Nanoscience, Engineering and Technology (IWGN) in 1998. In a January 2000 speech at the same institute, former President W. J. Clinton talked about the exciting promise of "nanotechnology" and the importance of expanding research in nanoscale science and, more broadly, technology. Later that month, he announced in his State of the Union Address an ambitious $497 million federal, multiagency national nanotechnology initiative (NNI) in the fiscal year 2001 budget, and made the NNI a top science and technology priority. The objective of this initiative was to form a broad-based coalition in which the academe, the private sector, and local, state, and federal governments work together to push the envelope of nanoscience and nanoengineering to reap nanotechnology's potential social and economic benefits.

The funding in the U.S. has continued to increase. In January 2003, the U. S. senate introduced a bill to establish a National Nanotechnology Program. On 3 December 2003, President George W. Bush signed into law the 21st Century Nanotechnology Research and Development Act. The legislation put into law programs and activities supported by the National Nanotechnology Initiative. The bill gave nanotechnology a permanent home in the federal government and authorized $3.7 billion to be spent in the four year period beginning in October 2005, for nanotechnology initiatives at five federal agencies. The funds would provide grants to researchers, coordinate R&D across five federal agencies (National Science Foundation (NSF), Department of Energy (DOE), NASA, National Institute of Standards and Technology (NIST), and Environmental Protection Agency (EPA)), establish interdisciplinary research centers, and accelerate technology transfer into the private sector. In addition, Department of Defense (DOD), Homeland Security, Agriculture and Justice as well as the National Institutes of Health (NIH) would also fund large R&D activities. They currently account for more than one-third of the federal budget for nanotechnology.

The European Union made nanosciences and nanotechnologies a priority in the Sixth Framework Program (FP6) in 2002 for the period of 2003-2006. They had dedicated small funds in FP4 and FP5 before. FP6 was tailored to help better structure European research and to cope with the strategic objectives set out in Lisbon in 2000. Japan identified nanotechnology as one of its main research priorities in 2001. The funding levels increased sharply from $400 million in 2001 to around $950 million in 2004. In 2003, South Korea embarked upon a ten-year program with around $2 billion of public funding, and Taiwan has committed around $600 million of public funding over six years. Singapore and China are also investing on a large scale. Russia is well funded as well.

Nanotechnology literally means any technology done on a nanoscale that has applications in the real world. Nanotechnology encompasses production and application of physical, chemical and biological systems at scales, ranging from individual atoms or molecules to submicron dimensions, as well as the integration of the resulting nanostructures into larger systems. Nanotechnology is likely to have a profound impact on our economy and society in the early 21st century, comparable to that of semiconductor technology, information technology, or cellular and molecular biology. Science and technology research in nanotechnology promises breakthroughs in areas such as materials and manufacturing, nanoelectronics, medicine and healthcare, energy, biotechnology, information technology and national security. It is widely felt that nanotechnology will be the next industrial revolution.

There is an increasing need for a multidisciplinary, system-oriented approach to design and manufacturing of micro/nanodevices that function reliably. This can only be achieved through the cross-fertilization of ideas from different disciplines and the systematic flow of information and people among research groups. Reliability is a critical technology for many micro- and nanosystems and nanostructured materials. A broad-based handbook was needed, and thus the first edition of Springer Handbook of Nanotechnology was published in April 2004. It presented an overview of nanomaterial synthesis, micro/nanofabrication, micro- and nanocomponents and systems, scanning probe microscopy, reliability issues (including nanotribology and nanomechanics) for nanotechnology, and industrial applications. When the handbook went for sale in Europe, it sold out in ten days. Reviews on the handbook were very flattering.

Given the explosive growth in nanoscience and nanotechnology, the publisher and the editor decided to develop a second edition merely six months after publication of the first edition. This edition of the handbook integrates the knowledge from the nanostructure, fabrication, materials science, devices, and reliability point of view. It covers various industrial applications. It also addresses social, ethical, and political issues. Given the significant interest in biomedical applications, a number of chapters in this arena have been added. The second edition consists of 59 chapters (new: 23; revised: 27; unchanged: 9). The chapters have been written by 154 internationally recognized experts in the field, from academia, national research labs, and industry.

This book is intended for three types of readers: graduate students of nanotechnology, researchers in academia and industry who are active or intend to become active in this field, and practicing engineers and scientists who have encountered a problem and hope to solve it as expeditiously as possible. The handbook should serve as an excellent text for one or two semester graduate courses in nanotechnology in mechanical engineering, materials science, applied physics, or applied chemistry.

We embarked on the development of the second edition in October 2004, and we worked very hard to get all the chapters to the publisher in a record time of about 7 months. I wish to sincerely thank the authors for offering to write comprehensive chapters on a tight schedule. This is generally an added responsibility to the hectic work schedules of researchers today. I depended on a large number of reviewers who provided critical reviews. I would like to thank Dr. Phillip J. Bond, Chief of Staff and Under Secretary for Technology, US Department of Commerce, Washington, D.C. for chapter suggestions as well as authors in the handbook. I would also like to thank my colleague, Dr. Zhenhua Tao, whose efforts during the preparation of this handbook were very useful. Last but not the least, I would like to thank my secretary Caterina Runyon-Spears for various administrative duties; her tireless efforts are highly appreciated.

I hope that this handbook will stimulate further interest in this important new field, and the readers of this handbook will find it useful.

May 2005

Bharat Bhushan
Editor

Preface to the 1st Edition

On December 29, 1959 at the California Institute of Technology, Nobel Laureate Richard P. Feynman gave a talk at the Annual meeting of the American Physical Society that has become one classic science lecture of the 20th century, titled "There's Plenty of Room at the Bottom." He presented a technological vision of extreme miniaturization in 1959, several years before the word "chip" became part of the lexicon. He talked about the problem of manipulating and controlling things on a small scale. Extrapolating from known physical laws, Feynman envisioned a technology using the ultimate toolbox of nature, building nanoobjects atom by atom or molecule by molecule. Since the 1980s, many inventions and discoveries in fabrication of nanoobjects have been a testament to his vision. In recognition of this reality, in a January 2000 speech at the same institute, former President W. J. Clinton talked about the exciting promise of "nanotechnology" and the importance of expanding research in nanoscale science and engineering. Later that month, he announced in his State of the Union Address an ambitious $497 million federal, multi-agency national nanotechnology initiative (NNI) in the fiscal year 2001 budget, and made the NNI a top science and technology priority. Nanotechnology literally means any technology done on a nanoscale that has applications in the real world. Nanotechnology encompasses production and application of physical, chemical and biological systems at size scales, ranging from individual atoms or molecules to submicron dimensions as well as the integration of the resulting nanostructures into larger systems. Nanofabrication methods include the manipulation or self-assembly of individual atoms, molecules, or molecular structures to produce nanostructured materials and sub-micron devices. Micro- and nanosystems components are fabricated using top-down lithographic and nonlithographic fabrication techniques. Nanotechnology will have a profound impact on our economy and society in the early 21st century, comparable to that of semiconductor technology, information technology, or advances in cellular and molecular biology. The research and development in nanotechnology will lead to potential breakthroughs in areas such as materials and manufacturing, nanoelectronics, medicine and healthcare, energy, biotechnology, information technology and national security. It is widely felt that nanotechnology will lead to the next industrial revolution.

Reliability is a critical technology for many micro- and nanosystems and nanostructured materials. No book exists on this emerging field. A broad based handbook is needed. The purpose of this handbook is to present an overview of nanomaterial synthesis, micro/nanofabrication, micro- and nanocomponents and systems, reliability issues (including nanotribology and nanomechanics) for nanotechnology, and industrial applications. The chapters have been written by internationally recognized experts in the field, from academia, national research labs and industry from all over the world.

The handbook integrates knowledge from the fabrication, mechanics, materials science and reliability points of view. This book is intended for three types of readers: graduate students of nanotechnology, researchers in academia and industry who are active or intend to become active in this field, and practicing engineers and scientists who have encountered a problem and hope to solve it as expeditiously as possible. The handbook should serve as an excellent text for one or two semester graduate courses in nanotechnology in mechanical engineering, materials science, applied physics, or applied chemistry.

We embarked on this project in February 2002, and we worked very hard to get all the chapters to the publisher in a record time of about 1 year. I wish to sincerely thank the authors for offering to write comprehensive chapters on a tight schedule. This is generally an added responsibility in the hectic work schedules of researchers today. I depended on a large number of reviewers who provided critical reviews. I would like to thank Dr. Phillip J. Bond, Chief of Staff and Under Secretary for Technology, US Department of Commerce, Washington, D.C. for suggestions for chapters as well as authors in the handbook. I would also like to thank my colleague, Dr. Huiwen Liu, whose efforts during the preparation of this handbook were very useful.

I hope that this handbook will stimulate further interest in this important new field, and the readers of this handbook will find it useful.

September 2003   Bharat Bhushan
                                                              Editor

Editors Vita

Dr. Bharat Bhushan received an M.S. in mechanical engineering from the Massachusetts Institute of Technology in 1971, an M.S. in mechanics and a Ph.D. in mechanical engineering from the University of Colorado at Boulder in 1973 and 1976, respectively, an MBA from Rensselaer Polytechnic Institute at Troy, NY in 1980, Doctor Technicae from the University of Trondheim at Trondheim, Norway in 1990, a Doctor of Technical Sciences from the Warsaw University of Technology at Warsaw, Poland in 1996, and Doctor Honouris Causa from the National Academy of Sciences at Gomel, Belarus in 2000. He is a registered professional engineer (mechanical). He is presently an Ohio Eminent Scholar and The Howard D. Winbigler Professor in the Department of Mechanical Engineering, Graduate Research Faculty Advisor in the Department of Materials Science and Engineering, and the Director of the Nanotribology Laboratory for Information Storage & MEMS/NEMS (NLIM) at the Ohio State University, Columbus, Ohio. He is an internationally recognized expert of tribology and mechanics on the macro- to nanoscales, and is one of the most prolific authors. He is considered by some a pioneer of the tribology and mechanics of magnetic storage devices and a leading researcher in the fields of nanotribology and nanomechanics using scanning probe microscopy and applications to micro- and nanotechnology. He has authored 5 technical books, more than 70 handbook chapters, more than 550 technical papers in referred journals, and more than 60 technical reports, edited more than 25 books, and holds 16 U.S. patents. He is coeditor of Springer NanoScience and Technology Series and coeditor of Microsystem Technologies – Micro- and Nanosystems, Information Storage and Processing Systems (formerly called the Journal of Information Storage and Processing Systems). He has given more than 250 invited presentations on five

continents and more than 110 keynote/plenary addresses at major international conferences.

Dr. Bhushan is an accomplished organizer. He organized the first symposium on Tribology and Mechanics of Magnetic Storage Systems in 1984 and the first international symposium on Advances in Information Storage Systems in 1990, both of which are now held annually. He is the founder of an ASME Information Storage and Processing Systems Division founded in 1993 and served as the founding chair during 1993–1998. His biography has been listed in over two dozen Who's Who books including Who's Who in the World and has received more than two dozen awards for his contributions to science and technology from professional societies, industry, and U.S. government agencies. He is also the recipient of various international fellowships including the Alexander von Humboldt Research Prize for Senior Scientists, Max Planck Foundation Research Award for Outstanding Foreign Scientists, and the Fulbright Senior Scholar Award. He is a foreign member of the International Academy of Engineering (Russia), Byelorussian Academy of Engineering and Technology and the Academy of Triboengineering of Ukraine, an honorary member of the Society of Tribologists of Belarus, a fellow of ASME, IEEE, STLE, and the New York Academy of Sciences, and a member of ASEE, Sigma Xi and Tau Beta Pi.

Dr. Bhushan has previously worked for the R&D Division of Mechanical Technology Inc., Latham, NY; the Technology Services Division of SKF Industries Inc., King of Prussia, PA; the General Products Division Laboratory of IBM Corporation, Tucson, AZ; and the Almaden Research Center of IBM Corporation, San Jose, CA.

List of Authors

Chong H. Ahn
University of Cincinnati
Department of Electrical and Computer
Engineering and Computer Science
814 Rhodes Hall
Cincinnati, OH 45221-0030, USA
e-mail: *chong.ahn@uc.edu*

Boris Anczykowski
nanoAnalytics GmbH
Gievenbecker Weg 11
48149 Münster, Germany
e-mail: *anczykowski@nanoanalytics.com*

Massood Z. Atashbar
Western Michigan University
Department of Electrical
and Computer Engineering
Kalamazoo, MI 49008, USA
e-mail: *massood.atashbar@wmich.ed*

Wolfgang Bacsa
University of Toulouse III (Paul Sabatier)
Laboratoire de Physique des Solides (LPST), UMR
5477 CNRS
118 Route de Narbonne
31062 Toulouse Cedex, France
e-mail: *bacsa@ramansco.ups-tlse.fr;*
bacsa@lpst.ups-tlse.fr

William Sims Bainbridge
Division of Information, Science and Engineering
National Science Foundation
Arlington, VA , USA
e-mail: *wbainbri@nsf.gov*

Antonio Baldi
Centro National Microelectrónica
Institut de Microelectronica de Barcelona (IMB)
(CNM-CSIC)
Barcelona, Spain
e-mail: *antoni.baldi@cnm.es*

Philip D. Barnes
c/o Bradley Clymer,
205 Dreese laboratory, 2015 Neil Avenue
Columbus, OH 43201-7596, USA
e-mail: *barnes.156@osu.edu*

James Batteas
Texas A&M University
Department of Chemistry
PO Box 30012
College Station, TX 77842, USA
e-mail: *batteas@mail.chem.tamu.edu*

Roland Bennewitz
McGill University
Physics Department
3600 rue University
Montreal, QC H3A 2T8, Canada
e-mail: *roland@physics.mcgill.ca*

Bharat Bhushan
The Ohio State University
Nanotribology Laboratory for Information Storage
and MEMS/NEMS
W 390 Scott Laboratory, 201 W. 19th Avenue
Columbus, OH 43202-1142, USA
e-mail: *bhushan.2@osu.edu*

Gerd K. Binnig
IBM Zurich Research Laboratory
Micro-/Nanomechanics
Säumerstraße 4
8803 Rüschlikon, Switzerland
e-mail: *gbi@zurich.ibm.com*

Marcie R. Black
Massachusetts Institute of Technology
Department of Electrical Engineering
and Computer Science
77 Massachusetts Avenue
Cambridge, MA 02139, USA
e-mail: *marcie@alum.mit.edu*

Maarten P. de Boer
Sandia National Laboratories
MEMS Devices and Reliability Physics Department
P.O. Box 5800
Albuquerque, NM 87185-1069, USA
e-mail: *mpdebo@sandia.gov*

Donald W. Brenner
North Carolina State University
Department of Materials Science and Engineering
Raleigh, NC 27695-7909, USA
e-mail: *brenner@ncsu.edu*

Jean-Marc Broto
Institut National des Sciences Appliquées
of Toulouse
Laboratoire National des Champs Magnétiques
Pulsés (LNCMP)
143 Avenue de Rangueil
31432 Toulouse Cedex 4, France
e-mail: *broto@insa-tlse.fr; broto@lncmp.or*

Guozhong Cao
University of Washington
302M Roberts Hall, Box 352120
Seattle, WA 98195-2120, USA
e-mail: *gzcao@u.washington.edu*

Robert W. Carpick
University of Wisconsin-Madison
Department of Engineering Physics
1500 Engineering Drive
Madison, WI 53706-1687, USA
e-mail: *carpick@engr.wisc.edu*

Tsung-Lin Chen
National Chiao Tung University
Department of Mechanical Engineering
HsinChu, Taiwan
e-mail: *tsunglin@mail.nctu.edu.tw*

Yu-Ting Cheng
National Chiao Tung University
Electronics Engineering Department
HsinChu, Taiwan
e-mail: *ytcheng@faculty.nctu.edu.tw*

Giovanni Cherubini
IBM Zurich Research Laboratory
Storage Technologies
Säumerstraße 4
8803 Rüschlikon, Switzerland
e-mail: *cbi@zurich.ibm.com*

Mu Chiao
The University of British Columbia
Mechanical Engineering Department
2054-6250 Applied Science Lane
Vancouver, BC V6T 1Z4, Canada
e-mail: *muchiao@mech.ubc.ca*

Jin-Woo Choi
Louisiana State University
Department of Electrical
and Computer Engineering
102 South Campus Drive
Baton Rouge, LA 70803-5901, USA
e-mail: *choi@ece.lsu.edu*

Shawn J. Cunningham
WiSpry, Inc.
Colorado Springs Design Center
7150 Campus Drive, Suite 255
Colorado Springs, CO 80920, USA
e-mail: *shawn.cunningham@wispry.com*

Dietrich Dehlinger
University of California, San Diego
Dept. of Electrical and Computer Engineering
9500 Gilman Drive, Mail Code 0407
La Jolla, CA 92093-0407, USA
e-mail: *ddehling@ucsd.edu*

Michel Despont
IBM Zurich Research Laboratory
Micro-/Nanomechanics
Säumerstraße 4
8803 Rüschlikon, Switzerland
e-mail: *dpt@zurich.ibm.com*

Lixin Dong
Swiss Federal Institute of Technology (ETH), Zürich
Institute of Robotics and Intelligent Systems (IRIS)
8092 Zürich, Switzerland
e-mail: *ldong@ethz.ch*

Gene Dresselhaus
Massachusetts Institute of Technology
Francis Bitter Magnet Laboratory
77 Massachusetts Avenue
Cambridge, MA 02139, USA
e-mail: *gene@mgm.mit.edu*

Mildred S. Dresselhaus
Massachusetts Institute of Technology
Department of Electrical Engineering
and Computer Science and Department of Physics
77 Massachusetts Avenue
Cambridge, MA 02139, USA
e-mail: *millie@mgm.mit.edu*

Martin L. Dunn
University of Colorado at Boulder
Department of Mechanical Engineering
Campus Box 427
Boulder, CO 80309, USA
e-mail: *martin.dunn@colorado.edu*

Urs T. Dürig
IBM Zurich Research Laboratory
Micro-/Nanomechanics
Säumerstraße 4
8803 Rüschlikon, Switzerland
e-mail: *drg@zurich.ibm.com*

Evangelos Eleftheriou
IBM Zurich Research Laboratory
Storage Technologies
Säumerstraße 4
8803 Rüschlikon, Switzerland
e-mail: *ele@zurich.ibm.com*

Mauro Ferrari
The University of Texas
Health Science Center at Houston
7000 Fannin, Ste 1565
Houston, TX 77030, USA
e-mail: *Mauro.Ferrari@uth.tmc.edu*

Emmanuel Flahaut
Université Paul Sabatier
CIRIMAT (Centre Interuniversitaire de Recherche
et d'Ingénierie des Matériaux)
118 Route de Narbonne
31062 Toulouse Cedex 04, France
e-mail: *flahaut@chimie.ups-tlse.fr*

László Forró
Swiss Federal Institute of Technology (EPFL)
Institute of Physics of Complex Matter
Ecublens
1015 Lausanne, Switzerland
e-mail: *laszlo.forro@epfl.ch*

Jane Frommer
IBM Almaden Research Center
Department of Science and Technology
650 Harry Road
San Jose, CA 95120, USA
e-mail: *frommer@Almaden.ibm.com*

Harald Fuchs
Universität Münster
Physikalisches Institut
Wilhelm-Klemm-Straße 10
48149 Münster, Germany
e-mail: *fuchsh@uni-muenster.de*

Christoph Gerber
Institute of Physics, University of Basel
National Competence Center for Research
in Nanoscale Science (NCCR) Basel
Klingelbergstrasse 82
4056 Basel, Switzerland
e-mail: *Christoph.Gerber@unibas.ch*

Franz J. Giessibl
Universität Regensburg
Naturwissenschaftliche Fakultät II – Physik
Universitätsstr. 31
93053 Regensburg, Germany
e-mail: *Franz.Giessibl@physik.uni-regensburg.de*

Enrico Gnecco
University of Basel
Department of Physics
Klingelbergstraße 82
4056 Basel, Switzerland
e-mail: *Enrico.Gnecco@unibas.ch*

Steve Granick
University of Illinois at Urbana-Champaign
Department of Chemistry
505 South Mathews Avenue
Urbana, IL 61801, USA
e-mail: *sgranick@uiuc.edu*

Gérard Gremaud
Swiss Federal Institute of Technology (EPFL)
Institute of Physics of Complex Matter
Ecublens
1015 Lausanne, Switzerland
e-mail: *gremaud@epfl.ch*

Jason H. Hafner
Rice University
Department of Physics & Astronomy
PO BOX 1892
Houston, TX 77251-1892, USA
e-mail: *hafner@rice.edu*

Judith A. Harrison
U.S. Naval Academy
Department of Chemistry
Annapolis, MD 21402, USA
e-mail: *jah@usna.edu*

Peter G. Hartwell
Hewlett Packard Laboratories
1501 Page Mill Road
Palo Alto, CA 94304, USA
e-mail: *peter.hartwell@hp.com*

Martin Hegner
Institute of Physics, University of Basel
National Competence Center for Research
in Nanoscale Science (NCCR) Basel
Klingelbergstrasse 82
4056 Basel, Switzerland
e-mail: *Martin.Hegner@unibas.ch*

Michael J. Heller
University of California San Diego
Dept. of Bioengineering,
Dept. of Electrical and Computer Engineering
PFBE Rm 429, 9500 Gilman Dr.
La Jolla, CA 92093-0412, USA
e-mail: *mheller@bioeng.ucsd.edu*

Stefan Hengsberger
Ecole d'ingénieur de Fribourg
Fribourg, Switzerland
e-mail: *stefan.hengsberger@eif.ch*

Seong-Jun Heo
Univeristy of Florida
Dept. of Materials Science and Engineering
100 Rhines Hall, P.O. Box 116400
Gainesville, FL 32611-6400, USA
e-mail: *heogyver@ufl.edu*

Joseph P. Heremans
The Ohio State University
Dept. of Physics,
Dept. of Mechanical Engineering
650 Ackerman Rd
Columbus, OH 43202, USA
e-mail: *heremans.1@osu.edu*

Peter Hinterdorfer
Johannes Kepler University of Linz
Institute for Biophysics
Altenbergerstraße 69
Linz, 4040, Austria
e-mail: *peter.hinterdorfer@jku.at*

Dean Ho
University of California Los Angeles
Department of Bioengineering
420 Westwood Plaza, 7523 Boelter Hall
Los Angeles, CA 90095, USA
e-mail: *deanh@seas.ucla.edu*

Dalibor Hodko
Nanogen, Inc.
10498 Pacific Center Court
San Diego, CA 92121, USA
e-mail: *dhodko@nanogen.com*

Roberto Horowitz
University of California at Berkeley
Department of Mechanical Engineering
5121 Etcheverry Hall
Berkeley, CA 94720-1742, USA
e-mail: *horowitz@me.berkeley.edu*

Hirotaka Hosoi
Japan Science and Technology Corporation
Innovation Plaza, Hokkaido
060-0819 Sapporo, Japan
e-mail: *hosoi@sapporo.jst-plaza.jp*

Xinghui Huang
University of California at Berkeley
Department of Mechanical Engineering
Berkeley, CA 94720, USA
e-mail: *xhhuang@me.berkeley.edu*

Seung-Hyun Hur
University of Illinois at Urbana-Champaign
Department of Materials Science and Engineering
1304 W. Green St.
Urbana, IL 61801, USA
e-mail: *shur@uiuc.edu*

Jacob N. Israelachvili
University of California
Department of Chemical Engineering
and Materials Department
Santa Barbara, CA 93106, USA
e-mail: *Jacob@engineering.ucsb.edu*

Ghassan E. Jabbour
University of Arizona
Optical Sciences Center
1630 East University Boulevard
Tucson, AZ 85721, USA
e-mail: *gej@optics.arizona.edu*

Guangyao Jia
University of California, Irvine
Dept. of Mechanical and Aerospace Engineering
4200 Engineering Gateway
Irvine, CA 92697, USA
e-mail: *gjia@uci.edu*

Anne Jourdain
Interuniversity Microelectronics Center (IMEC)
Kapeldreef, 75
3001 Leuven, Belgium
e-mail: *jourdain@imec.be*

Harold Kahn
Case Western Reserve University
Department of Materials Science and Engineering
10900 Euclid Avenue
Cleveland, OH 44106-7204, USA
e-mail: *kahn@cwru.edu*

Horacio Kido
University of California at Irvine
Mechanical and Aerospace Engineering
4200 Engineering Gateway
Irvine, CA 92697-3975, USA
e-mail: *hkido@uci.edu*

Jitae Kim
University of California, Irvine
Dept. of Mechanical and Aerospace Engineering
4200 Engineering Gateway
Irvine, CA 92697, USA
e-mail: *jitaekim@uci.edu*

Jongbaeg Kim
Dicon fiber Optics, Inc.
1689 Regatta Blvd.
Richmond, CA 94804-3727, USA
e-mail: *jongbaeg@gmail.com*

Nahui Kim
University of California, Irvine
Department of Electrical Engineering
and Computer Science
4200 Engineering Gateway
Irvine, CA 92697, USA
e-mail: *nahuik@uci.edu*

Andras Kis
University of California
Physics Department
Zettl Group, 366 LeConte Hall
Berkeley, CA 94720, USA
e-mail: *akis@lbl.gov*

Shih-Chung Kon
University of California at Berkeley
Department of Mechanical Engineering
Berkeley, CA 94720, USA
e-mail: *stankon@me.berkeley.edu*

Jané Kondev
Brandeis University
Physics Department
Waltham, MA 02454, USA
e-mail: *kondev@brandeis.edu*

Jing Kong
Massachusetts Institute of Technology
Department of Electrical Engineering
and Computer Science
77 Massachusetts Avenue
Cambridge, MA 02139, USA
e-mail: *jingkong@mit.edu*

Anders Kristensen
Technical University of Denmark (DTU)
MIC – Department of Micro and Nanotechnology
Building 345 East
2800 Kgs. Lyngby, Denmark
e-mail: *ak@mic.dtu.dk*

Andrzej J. Kulik
Swiss Federal Institute of Technology (EPFL)
Institute of Physics of Complex Matter
1015 Lausanne, Switzerland
e-mail: *andrzej.kulik@epfl.ch*

Hans Peter Lang
National Competence Center for Research
in Nanoscale Science (NCCR) Basel
Institute of Physics, University of Basel
Klingelbergstrasse 82
4056 Basel, Switzerland
e-mail: *Hans-Peter.Lang@unibas.ch*

Carmen LaTorre
Insulating Systems Business Owens Corning
2790 Columbus Road
Granville, OH 43023, USA
e-mail: *carmen.latorre@owenscorning.com*

Christophe Laurent
Université Paul Sabatier
CIRIMAT (Centre Interuniversitaire de Recherche
et d'Ingénierie des Matériaux)
118 Route de Narbonne
31062 Toulouse Cedex 04, France
e-mail: *laurent@chimie.ups-tlse.fr*

Abraham Lee
University of California at Irvine
Department of Biomedical Engineering
204 Rockwell Engineering Center
Irvine, CA 92697-2715, USA
e-mail: *aplee@uci.edu*

Stephen C. Lee
Ohio State University
Biomedical Engineering Center
1080 Carmack Road
Columbus, OH 43210-1002, USA
e-mail: *Lee@bme.ohio-state.edu*

Wayne R. Leifert
CSIRO Commonwealth Scientific
and Industrial Research Organisation
Health Sciences and Nutrition
PO Box 10041
Adelaide, BC South Australia 5000, Australia
e-mail: *Wayne.Leifert@csiro.au*

Yunfeng Li
University of California at Berkeley
Department of Mechanical Engineering
5121 Etcheverry Hall
Berkeley, CA 94720-1740, USA
e-mail: *yunfeng@me.Berkeley.edu*

Liwei Lin
University of California at Berkeley
Mechanical Engineering Department
5126 Etcheverry Hall, Mailstop 1740
Berkeley, CA 94720-1740, USA
e-mail: *lwlin@me.berkeley.edu*

Yu-Ming Lin
Massachusetts Institute of Technology
Department of Electrical Engineering
and Computer Science
77 Massachusetts Avenue
Cambridge, MA 02139, USA
e-mail: *yming@mgm.mit.edu*

Huiwen Liu
19060 Twilight Tr.
Eden Prairie, MN 55346, USA
e-mail: *Huiwen.Liu@seagate.com*

Gustavo S. Luengo
Department of Physics Knowledge
L'Oréal Recherche 1
Avenue Eugène Schueller
93601 Aulnay sous Bois, France
e-mail: *gluengo@rd.loreal.com*

Marc J. Madou
University of California
Dept. of Mechanical and Aerospace Engineering
4200 Engineering Gateway
Irvine, CA 92697, USA
e-mail: *mmadou@uci.edu*

Adrian B. Mann
Rutgers University
Department of Ceramics and Materials Engineering
607 Taylor Road
Piscataway, NJ 08854, USA
e-mail: *abmann@rci.rutgers.edu*

Othmar Marti
University of Ulm
Department of Experimental Physics
Albert-Einstein-Allee 11
89069 Ulm, Germany
e-mail: *Othmar.Marti@physik.uni-ulm.de*

Jack Martin
Analog Devices, Inc.
Micromachined Products Division
21 Osborn Street
Cambridge, MA 02139, USA
e-mail: *jack.martin@analog.com*

Shinji Matsui
University of Hyogo
Laboratory of Advanced Science
and Technology for Industry
3-1-2 Koto, Kamigori Ako
678-1205 Hyogo, Japan
e-mail: *matsui@lasti.u-hyogo.ac.jp*

Brendan McCarthy
University of Arizona
Optical Sciences Center
1630 East University Boulevard
Tucson, AZ 85721, USA
e-mail: *bmccarthy@optics.arizona.edu*

Edward J. McMurchie
CSIRO Commonwealth Scientific and Industrial
Research Organisation Health Sciences
and Nutrition
PO Box 10041
Adelaide BC, SA 5000, Australia
e-mail: *Ted.McMurchie@csiro.au*

Mehran Mehregany
Case Western Reserve University
Department of Electrical Engineering
and Computer Science
10900 Euclid Avenue
Cleveland, OH 44106, USA
e-mail: *mxm31@cwru.edu*

Etienne Menard
3714 Beckman Institute
Department of Materials Science and Engineering
1304 W. Green St.
Urbana, IL 61801, USA
e-mail: *emenard@uiuc.edu*

Ernst Meyer
University of Basel
Institute of Physics
Klingelbergstraße 82
4056 Basel, Switzerland
e-mail: *Ernst.Meyer@unibas.ch*

Robert Modlinski
Interuniversity Microelectronics Center (IMEC)
MCP/REMO
Kapeldreef 75
3001 Leuven, Belgium
e-mail: *modlinsk@imec.be*

Carlo D. Montemagno
University of California Los Angeles
School of Engineering and Applied Science
7523 Boelter Hall, Box 951600
Los Angeles, CA 90095–1600, USA
e-mail: *cdm@seas.ucla.edu*

Marc Monthioux
UPR A-8011 CNRS
Centre d'Elaboration des Matériaux
et d'Etudes Structurales (CEMES)
29 Rue Jeanne Marvig
31055 Toulouse Cedex 4, France
e-mail: *monthiou@cemes.fr*

Markus Morgenstern
University of Hamburg
Institute of Applied Physics
Jungiusstraße 11
20355 Hamburg, Germany
e-mail: *mmorgens@physnet.uni-hamburg.de*

Seizo Morita
Osaka University
Department of Electronic Engineering
Yamada-Oka 2-1
565–0871 Suita-City, Osaka, Japan
e-mail: *smorita@ele.eng.osaka-u.ac.jp*

Koichi Mukasa
Hokkaido University
Nanoelectronics Laboratory
Nishi-8, Kita-13, Kita-ku
060–8628 Sapporo, Japan
e-mail: *mukasa@nano.eng.hokudai.ac.jp*

Ashis Mukhopadhyay
Wayne State University
Department of Physics
Detroit, MI 48201, USA
e-mail: *ashis@physics.wayne.edu*

Ryozo Nagamune
University of California at Berkeley
Department of Mechanical Engineering
Berkeley, CA , USA
e-mail: *ryozo@me.berkeley.edu*

Bradley J. Nelson
Swiss Federal Institute of Technology (ETH)
Institute of Robotics and Intelligent Systems (IRIS)
8092 Zürich, Switzerland
e-mail: *bnelson@ethz.ch*

Michael Nosonovsky
National Institute of Standards and Technology
100 Bureau Dr., Stop 8520
Gaithersburg, MD 20899–8520, USA
e-mail: *Michael.Nosonovsky@nist.gov*

Kenn Oldham
University of California at Berkeley
Department of Mechanical Engineering
5121 Etcheverry Hall
Berkeley, CA 94720–1740, USA
e-mail: *oldham@newton.berkeley.edu*

Hiroshi Onishi
Kanagawa Academy of Science and Technology
Surface Chemistry Laboratory
KSP East 404, 3-2-1 Sakado, Takatsu-ku,
Kawasaki-shi
213–0012 Kanagawa, Japan
e-mail: *oni@net.ksp.or.jp*

René M. Overney
University of Washington
Department of Chemical Engineering
Seattle, WA 98195–1750, USA
e-mail: *roverney@u.Washington.edu*

Alain Peigney
Maître de Conférences at University of Toulouse III
(Paul Sabatier)
Centre Inter-universitaire de Recherche sur
l'Industrialisation des Matériaux (CIRIMAT)
UMR 5085 CNRS, Bâtiment 2R1
118 Route de Narbonne
31062 Toulouse Cedex 04, France
e-mail: *peigney@chimie.ups-tlse.fr*

Oliver Pfeiffer
University of Basel
Institute of Physics
Klingelbergstraß 82
4056 Basel, Switzerland
e-mail: *Oliver.Pfeiffer@stud.unibas.ch*

Rob Phillips
California Institute of Technology
Mechanical Engineering and Applied Physics
1200 California Boulevard
Pasadena, CA 91125, USA
e-mail: *phillips@aero.caltech.edu*

Haralampos Pozidis
IBM Zurich Research Laboratory
Storage Technologies
Säumerstraße 4
8803 Rüschlikon, Switzerland
e-mail: *hap@zurich.ibm.com*

Robert Puers
Katholieke Universiteit Leuven
ESAT/MICAS
Kasteelpark, Arenberg 10
3001 Leuven, Belgium
e-mail: *bob.puers@esat.kuleuven.ac.be*

Prashant K. Purohit
California Institute of Technology
Mechanical Engineering
1200 California Boulevard
Pasadena, CA 91125, USA
e-mail: *prashant@caltech.edu*

Calvin F. Quate
Stanford University
Ginzton Laboratory
Via Ortega
Stanford, CA 94305-4085, USA
e-mail: *quate@ee.stanford.edu*

Oded Rabin
Massachusetts Institute of Technology
Department of Chemistry
77 Massachusetts Avenue
Cambridge, MA 02139, USA
e-mail: *oded@mgm.mit.edu*

Françisco M. Raymo
University of Miami
Department of Chemistry
1301 Memorial Drive
Coral Gables, FL 33146-0431, USA
e-mail: *fraymo@miami.edu*

Manitra Razafinimanana
University of Toulouse III (Paul Sabatier)
Centre de Physique des Plasmas et leurs
Applications (CPPAT)
UMR 5002 CNRS, 118 Route de Narbonne
31062 Toulouse Cedex, France
e-mail: *razafinimanana@cpat.ups-tlse.fr*

Ziv Reich
Weizmann Institute of Science
Department of Biological Chemistry
Rehovot 76100, Israel
e-mail: *ziv.reich@weizmann.ac.il*

John A. Rogers
University of Illinois
Department of Materials Science and Engineering
1304 W. Green St.
Urbana, IL 61801, USA
e-mail: *jrogers@uiuc.edu*

Mark Ruegsegger
Ohio State University
Biomedical Engineering Center
1080 Carmack Road
Columbus, OH 43210, USA
e-mail: *mark@bme.ohio-state.edu*

Marina Ruths
University of Massachusetts Lowell
Department of Chemistry
Lowell, MA 01854, USA
e-mail: *Marina_Ruths@uml.edu*

Ozgur Sahin
Harvard University
Rowland Institute at Harvard
Nanomechanical Sensing
100 Edwin H. Land Blvd.
Cambridge, MA 02142, USA
e-mail: *sahin@rowland.harvard.edu*

Dror Sarid
University of Arizona
Optical Sciences Center
1630 East University Boulevard
Tucson, AZ 85721, USA
e-mail: *sarid@optics.arizona.edu*

Akira Sasahara
Kanagawa Academy of Science and Technology
Surface Chemistry Laboratory
KSP East 404, 3-2-1 Sakado, Takatsu-ku,
Kawasaki-shi
213-0012 Kanagawa, Japan
e-mail: *ryo@net.ksp.or.jp*

Helmut Schift
Paul Scherrer Institute
Laboratory for Micro- and Nanotechnology
5232 Villigen PSI, Switzerland
e-mail: *helmut.schift@psi.ch*

André Schirmeisen
University of Münster
Institute of Physics
Whilhem-Klemm-Straße 10
48149 Münster, Germany
e-mail: *schira@uni-muenster.de*

Alexander Schwarz
University of Hamburg
Institute of Applied Physics
Jungiusstraße 11
20355 Hamburg, Germany
e-mail: *aschwarz@physnet.uni-hamburg.de*

Udo D. Schwarz
Yale University
Department of Mechanical Engineering
15 Prospect Street
New Haven, CT 06510, USA
e-mail: *udo.schwarz@yale.edu*

Philippe Serp
Maître de Conférences, Institut National
Polytechnique de Toulouse
Laboratoire de Catalyse, Chimie Fine et Polymères
(LCCFP), Ecole Nationale Supérieure d'Ingénieurs
en Arts Chimiques et Technologiques
118 route de Narbonne
31077 Toulouse, France
e-mail: *pserp@ensct.fr; Philippe.Serp@ensiacet.fr*

Huamei (Mary) Shang
University of Washington
302M Roberts Hall, Box 352120
Seattle, WA 98195-2120, USA
e-mail: *hmshang@u.washington.edu*

Susan B. Sinnott
University of Florida
Department of Materials Science
and Engineering
154 Rhines Hall, P.O. Box 116400
Gainesville, FL 32611-6400, USA
e-mail: *sinnott@mse.ufl.edu*

Bryan R. Smith
Ohio State University
Biomedical Engineering Center
1080 Carmack Road
Columbus, OH 43210, USA
e-mail: *bryan@bme.ohio-state.edu*

Anisoara Socoliuc
University of Basel
Institute of Physics
Klingelbergstraße 82
4056 Basel, Switzerland
e-mail: *A.Socoliuc@unibas.ch*

Olav Solgaard
E.L. Ginzton Laboratory
Stanford, CA 94305-4088, USA
e-mail: *solgaard@stanford.edu*

W. Merlijn van Spengen
Leiden University
Kamerlingh Onnes Laboratory
Niels Bohrweg 2
2333, CA Leiden, The Netherlands
e-mail: *spengen@physics.leidenuniv.nl*

Kazuhisa Sueoka
Graduate School of Information Science
and Technology, Hokkaido University
Nanoelectronics Laboratory
kita-14, Nishi-9, Kita-ku
060-0814 Sapporo, Japan
e-mail: *sueoka@nano.isthokudai.ac.jp*

Yasuhiro Sugawara
Osaka University
Department of Applied Physics
Yamada-Oka 2-1
565-0871 Suita, Japan
e-mail: *sugawara@ap.eng.osaka-u.ac.jp*

Benjamin Sullivan
University of California, San Diego
Department of Bioengineering
9500 Gilman Drive
La Jolla, CA 92093, USA
e-mail: *bdsulliv@ucsd.edu*

Yugang Sun
University of Illinois at Urbana Champaign
Department of Materials Science and Engineering
1304 W. Green St.
Urbana, IL 61801, USA
e-mail: *ygsun@uiuc.edu*

Paul Swanson
Nanogen, Inc.
Department Advanced Technology
10398 Pacific Center Court
San Diego, CA 92121, USA
e-mail: *pswanson@nanogen.com*

Nikhil S. Tambe
GE Global Research, Bangalore # 122 EPIP
Material Systems Technologies
Phase 2, Hoodi Village, Whitefield Road
Bangalore, 560066, India
e-mail: *nikhil.tambe@ge.com*

Mike Yung-Chieh Tan
University of California at Irvine
Department of Biomedical Engineering
204 Rockwell Engineering Center
Irvine, CA 92697-2715, USA
e-mail: *ytan@uci.edu*

Kimberly L. Turner
University of California, Santa Barbara
Department of Mechanical and Environmental
Engineering
2355 Engineering Building II
Santa Barbara, CA 93106, USA
e-mail: *turner@engineering.ucsb.edu*

George W. Tyndall
IBM Almaden Research Center
Science and Technology
6369 Didion Ct.
San Jose, CA 95123, USA
e-mail: *tyndallgw@netscape.net*

Peter Vettiger
IBM Zurich Research Laboratory
Manager Micro-/Nanomechanics
Säumerstraße 4
8803 Rüschlikon, Switzerland
e-mail: *pv@zurich.ibm.com*

Guohua Wei
Ohio State University
Department of Mechanical Engineering
Nanotribology Laboratory for Information Storage
and MEMS/NEMS
Columbus, OH 43210, USA
e-mail: *guohua_wei1113@yahoo.com*

David Wendell
University of California Los Angeles
Department of Bioengineering
7523 Boelter Hall, 420 Westwood Plaza
Los Angeles, CA 90095, USA
e-mail: *dwendell@ucla.edu*

Darrin J. Young
Case Western Reserve University
Electrical Engineering and Computer Science
10900 Euclid Avenue
Cleveland, OH 44106, USA
e-mail: *djy@po.cwru.edu*

Y. Elaine Zhu
University of Notre Dame
Dept. of Chemical and Biomolecular Engineering
182 Fitzpatrick Hall
Notre Dame, IN 46556, USA
e-mail: *yzhu3@nd.edu*

Babak Ziaie
Purdue University
Department of Electrical and Computer
Engineering
465 Northwestern Avenue
West Lafayette, IN 47907-2035, USA
e-mail: *bziaie@purdue.edu*

Christian A. Zorman
Case Western Reserve University
Department of Electrical Engineering
and Computer Science
10900 Euclid Avenue
Cleveland, OH 44106, USA
e-mail: *caz@po.cwru.edu*

Jim Zoval
University of California Irvine
Department of Mechanical and Aerospace
Engineering
4200 Engineering Gateway
Irvine, CA 92697, USA
e-mail: *jzoval@uci.edu*

Philippe K. Zysset
Vienna University of Technology
Institute of Lightweight Design
and Structural Biomechanics
Gußhausstraße 27–29
Vienna, 1040, Austria
e-mail: *philippe.zysset@ilsb.tuwien.ac.at*

Contents

Part C Scanning Probe Microscopy

Part E Molecularly Thick Films for Lubrication

Part F Industrial Applications

Part G   Micro/Nanodevice Reliability

Part H Technological Convergence and Governing Nanotechnology

List of Abbreviations

μCP	microcontact printing
μTAS	micro-total analysis system
2-DEG	two-dimensional electron gas

A

ABS	air-bearing surface
AC	alternating current
ADEPT	antibody directed enzyme-prodrug therapy
AFAM	atomic force acoustic microscopy
AFM	atomic force microscopy
AIDCN	2-amino-4,5-imidazoledicarbonitrile
AM	amplitude modulation
ASA	anti-stiction agent
ASR	analyte-specific reagents
ATP	adenosine triphosphate

B

BE	boundary element
BFP	biomembrane force probe
BP	bit pitch
BPI	bits per inch
bpsi	bits per square inch
BSA	bovine serum albumin
BW	bonded washed
bioMEMS	biomedical microelectromechanical systems

C

CCVD	catalytic chemical vapor deposition
CD	compact disc
CD	critical dimension
CDS	correlated double sampling
CDW	charge density wave
CE	capillary electrophoresis
CG	controlled geometry
CMOS	complementary metal oxide semiconductor
CNT	carbon nanotube
COC	cyclic olefin copolymer
COF	chip-on-flex
CSM	continuous stiffness measurement
CTE	coefficient of thermal expansion
CVD	chemical vapor deposition
CW	continuous wave

D

DBR	distributed Bragg reflector
DC	direct current
DDT	dichlorodiphenyltrichloroethane
DEP	dielectrophoresis
DFB	distributed feedback
DFM	dynamic force microscopy
DFS	dynamic force spectroscopy
DI	deionized
DLC	diamond like carbon
DLP	digital light processing
DLVO	Derjaguin–Landau–Verwey–Overbeek
DMD	digital micromirror device
DMSO	dimethyl sulfoxide
DMT	Derjaguin–Muller–Toporov
DNA	deoxyribonucleic acid
DOF	degree-of-freedom
DOS	density of state
DPN	dip-pen nanolithography
DRIE	deep reactive ion etching
DSC	differential scanning calorimetry
DSP	digital signal processor
DT	diphtheria toxin
DVD	digital versatile disk

E

EAM	embedded atom method
EBD	electron beam deposition
ECR-CVD	electron cyclotron resonance chemical vapor deposition
EDC	1-ethyl-3-(3-dimethylaminopropyl
EDP	ethylene diamine pyrocatechol
EELS	electron energy loss spectroscopy
EFM	electrostatic force microscopy
EHD	electrohydrodynamic
EMF	electromotive force
EO	electroosmosis
EOF	electroosmotic flow
EPR	enhanced permeability and retention
ESD	electrostatic discharge

F

FACS	fluorescence-activated cell sorting
FAD	flavin adenine dinucleotide
FC	flip-chip
FCA	filtered cathodic arc

FCP	force calibration plot
FD	finite difference
FE	finite element
FEM	finite element modeling
FESEM	field-emission SEM
FESP	force modulation etched Si probe
FET	field effect transistor
FFM	friction force microscope
FIB	focused ion beam
FID	free induction decay
FKT	Frenkel–Kontorova–Tomlinson
FM	frequency modulation
FM-AFM	force modulation mode atomic force microscopy
FM-SFM	force modulation mode scanning force microscopy
FMEA	failure mode effect analysis
FMM	force modulation mode
FP	fluorescence polarization
FS	force spectroscopy

G

GFP	green fluorescent protein
GIO	grazing impact oscillator
GMR	giant magnetoresistance
GOD	glucose oxidase
GPCR	G-protein coupled receptor
Gox	glucose oxidase

H

HDD	hard disk drive
HDT	hexadecanethiol
HF	hydrofluoric acid
HOP	highly oriented pyrolytic
HOPG	highly oriented pyrolytic graphite
HPMA	hydroxyl polymethacrylamide
HRTEM	high-resolution transmission electron microscope
HTS	high throughput screening
HTSC	high-temperature superconductivity
HtBDC	hexa-tert-butyl-decacyclene

I

IBD	ion beam deposition
IC	integrated circuit
ICAM	intercellular adhesion molecule
ISE	indentation size effect
ITO	indium tin oxide

J

JKR	Johnson–Kendall–Roberts

K

KPFM	Kelvin probe force microscopy

L

LB	Langmuir–Blodgett
LCC	leadless chip carrier
LDOS	local density of state
LDV	laser Doppler velocimeter
LDV	laser Doppler vibrometry
LFA	leukocyte function-associated antigen
LFM	lateral force microscopy
LJ	Lennard-Jones
LMI	linear matrix inequalities
LN	liquid-nitrogen
LPCVD	low-pressure chemical vapor deposition
LQG	linear quadratic Gaussian
LTR	loop-transfer recovery
LVDT	linear variable differential transformer
MAH	tetramethyl-aluminum hydroxide
MALDI	matrix assisted laser desorption ionization
MAP	manifold absolute pressure

M

MD	molecular dynamics
ME	metal-evaporated
MEMS	microelectromechanical system
MFM	magnetic force microscopy
MHC	major histocompatibility complex
MHD	magnetohydrodynamic
MIM	metal-insulator-metal
MIMO	multi-input multi-output
MLE	maximum likelihood estimator
MP	metal particle
MPTMS	3-mercaptopropyltrimethoxysilane
MRAM	magnetoresistive RAM
MRFM	magnetic resonance force microscopy
MRFM	molecular recognition force microscopy
MRI	magnetic resonance imaging
MTTF	mean time to failure
MUMP	multi-user MEMS processes
MWCNT	multiwall carbon nanotube
MWNT	multiwall nanotube

N

NA	nucleic acid
NBMN	3-nitrobenzal malonitrile
NC-AFM	noncontact AFM
NCS	neocarzinostatin
NEMS	nanoelectromechanical system
NMP	no-moving-part
NNI	National Nanotechnology Initiatives
NP	silicium nitride probe

NSOM	near-field scanning optical microscopy
NTA	nitrilotriacetate–hexahistidine 6
nTP	nanotransfer printing
NVRAM	nonvolatile random-access memory

O

ODD	optical disk drive
OM	optical microscope
OTS	octadecyltrichlorosilane
OUM	ovonyx unified memory

P

PA	plasminogen activator
PAMAM	poly(amido) amine
PAPP	p-aminophenyl phosphate
PBC	periodic boundary condition
PC	polycarbonate
PC-RAM	phase change RAM
PD	proportional-derivative
pDA	1,4-phenylenediamine
PDMS	polydimethylsiloxane
PDP	2-pyridyldithiopropionyl
PE	polyethylene
PECVD	plasma enhanced chemical vapor deposition
PEEK	poly-etheretherketone
PEG	poly(ethylene glycol)
PES	photoemission spectroscopy
PES	position-error signal
PET	poly(ethylene terephthalate)
PFPE	perfluoropolyether
PMMA	polymethylmethacrylate
PS	polystyrene
PSA	prostate-specific antigen
PSD	position-sensitive detector
PSGL	P-selectin glycoprotein ligand
PTFE	polytetrafluoroethylene
PVC	polyvinychloride
PVDF	polyvinyledene fluoride
PW	power to weight ratio
PWR	plasmon waveguide resonance
PZT	lead zirconium titanate

Q

QCM	quartz crystal microbalance

R

RES	reticuloendothelial system
RF	radio-frequency
RH	relative humidity

RICM	reflection interference contrast microscopy
RIE	reactive ion etching
RNA	ribonucleic acid
RPC	reverse phase column
RPES	relative position error signal
RPM	revolutions per minute
RTP	rapid thermal processing

S

SAM	scanning acoustic microscopy
SAM	self-assembled monolayer
SCM	scanning capacitance microscopy
SCPM	scanning chemical potential microscopy
SEFM	scanning electrostatic force microscopy
SEM	scanning electron microscopy
SEcM	scanning electrochemical microscopy
SFA	surface force apparatus
SFAM	scanning force acoustic microscopy
SFD	shear flow detachment
SFIL	step-and-flash imprint lithography
SFM	scanning force microscopy
SICM	scanning ion conductance microscopy
SIMO	single-input multi-output
SISO	single-input single-output
SKPM	scanning Kelvin probe microscopy
SLAM	scanning local-acceleration microscopy
SMA	shape memory alloy
SMANCS	S-methacryl-neocarzinostatin
SMM	scanning magnetic microscopy
SN	signal-to-noise
SNOM	scanning near field optical microscopy
SNP	single nucleotide polymorphism
SOG	spin-on glass
SOI	silicon-on-insulator
SPM	scanning probe microscopy
SPR	surface plasmon resonance
sPROM	structurally programmable microfluidic system
SRAM	static random access memory
ssDNA	single-stranded DNA
SSNA	single-stranded nucleic acid
STM	scanning tunneling microscopy
SThM	scanning thermal microscopy
SWCNT	single-wall carbon nanotube
SWNT	single-wall nanotube

T

T-SLAM	variable temperature SLAM
TEM	transmission electron microscopy
TESP	tapping-mode etched silicon probe

TIRF	total internal reflection fluorescence
TIRM	total internal reflection microscopy
TMR	track misregistration
TP	track pitch
TPI	tracks per inch
TTF	tetrathiafulvalene
TV	television

U

UHV	ultrahigh vacuum

V

VCO	voltage-controlled oscillator
VCSEL	vertical-cavity surface-emitting laser
vdW	van der Waals
VLSI	very large-scale integration

Z

ZPETFFC	zero-phase-error tracking feed-forward controller

1. Introduction to Nanotechnology

A biological system can be exceedingly small. Many of the cells are very tiny, but they are very active; they manufacture various substances; they walk around; they wiggle; and they do all kinds of marvelous things—all on a very small scale. Also, they store information. Consider the possibility that we too can make a thing very small that does what we want—that we can manufacture an object that maneuvers at that level.

(From the talk "There's Plenty of Room at the Bottom," delivered by Richard P. Feynman at the annual meeting of the American Physical Society at the California Institute of Technology, Pasadena, CA, on December 29, 1959.)

1.1 Nanotechnology – Definition and Examples

Nanotechnology literally means any technology done on a nanoscale that has applications in the real world. Nanotechnology encompasses the production and application of physical, chemical, and biological systems at scales ranging from individual atoms or molecules to submicron dimensions, as well as the integration of the resulting nanostructures into larger systems. Nanotechnology is likely to have a profound impact on our economy and society in the early 21st century, comparable to that of semiconductor technology, information technology, or cellular and molecular biology. Science and technology research in nanotechnology promises breakthroughs in areas such as materials and manufacturing, nanoelectronics, medicine and healthcare, energy, biotechnology, information technology, and national security. It is widely felt that nanotechnology will be the next industrial revolution.

Nanometer-scale features are mainly built up from their elemental constituents. Examples include chemical synthesis, the spontaneous self-assembly of molecular clusters (molecular self-assembly) from simple reagents in solution, biological molecules (e.g., DNA) used as building blocks for the production of three-dimensional nanostructures, or quantum dots (nanocrystals) of arbitrary diameter (about 10 to 10^5 atoms). The definition of a nanoparticle is an aggregate of atoms bonded together with a radius between 1 and 100 nm. It typically consists of 10 to 10^5 atoms. A variety of vacuum deposition and nonequilibrium plasma chemistry techniques are used to produce layered nanocomposites and nanotubes. Atomically controlled structures are produced using molecular beam epitaxy and organometallic vapor phase epitaxy. Micro- and nanosystem components are fabricated using top-down lithographic and nonlithographic fabrication techniques and range in size from micro- to nanometers. Continued improvements in lithography for use in the production of nanocomponents have resulted in line widths as small as 10 nm in experimental prototypes. The nanotechnology field, in addition to fabrication of nanosystems, provides impetus to develop experimental and computational tools.

The discovery of novel materials, processes, and phenomena at the nanoscale and the development of new experimental and theoretical techniques for research provide fresh opportunities for the development of innovative nanosystems and nanostructured materials. The properties of materials at the nanoscale can be very different from those at a larger scale. When the dimension of a material is reduced from a large size, the properties remain the same at first, then small changes occur, until finally, when the size drops below 100 nm, dramatic

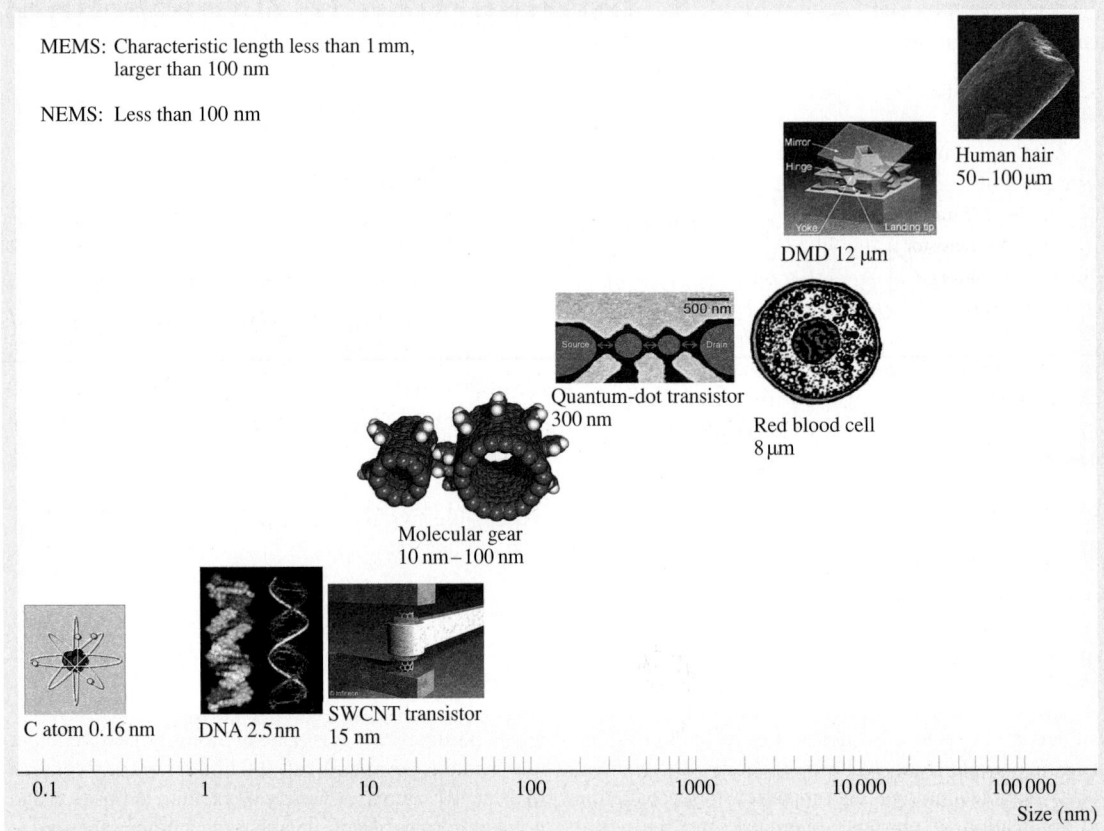

MEMS: Characteristic length less than 1 mm,
        larger than 100 nm

NEMS: Less than 100 nm

Human hair
50–100 μm

DMD 12 μm

500 nm

Quantum-dot transistor
300 nm

Red blood cell
8 μm

Molecular gear
10 nm–100 nm

C atom 0.16 nm

DNA 2.5 nm

SWCNT transistor
15 nm

| 0.1 | 1 | 10 | 100 | 1000 | 10 000 | 100 000 |

Size (nm)

Fig. 1.1 Dimensions of MEMS and NEMS in perspective. MEMS/NEMS examples shown are of a vertical single-walled carbon nanotube (SWCNT) transistor (5 nm wide and 15 nm high) [1.1], of molecular dynamic simulations of a carbon-nanotube-based gear [1.2], quantum-dot transistor obtained from [1.3], and DMD obtained from www.dlp.com

changes in properties can occur. If only one length of a three-dimensional nanostructure is of nanodimension, the structure is referred to as a quantum well; if two sides are of nanometer length, the structure is referred to as a quantum wire. A quantum dot has all three dimensions in the nanorange. The word quantum is associated with these three types of nanostructures because changes in properties arise from the quantum-mechanical nature of physics in the domain of the ultra small. Materials can be nanostructured for new properties and novel performance. This field is opening new venues in science and technology.

Micro- and nanosystems include micro/nanoelectromechanical systems. MEMS refers to microscopic devices that have a characteristic length of less than 1 mm but more than 100 nm and combine electrical and mechanical components. NEMS refers to nanoscopic devices that have a characteristic length

of less than 100 nm and combine electrical and mechanical components. In mesoscale devices, if the functional components are on a micro- or nanoscale, they may be referred to as MEMS or NEMS, respectively. These are referred to as intelligent miniaturized systems comprising sensing, processing, and/or actuating functions and combine electrical and mechanical components. The acronym MEMS originated in the USA. The term commonly used in Europe is microsystem technology (MST), and in Japan it is micromachines. Another term generally used is micro/nanodevices. MEMS/NEMS terms are also now used in a broad sense and include electrical, mechanical, fluidic, optical, and/or biological functions. MEMS/NEMS for optical applications are referred to as micro/nanooptoelectromechanical systems (MOEMS/NOEMS). MEMS/NEMS for electronic applications are referred to as radio-

Table 1.1 Dimensions and masses in perspective. (**a**) Dimensions in perspective

NEMS characteristic length	< 100 nm
MEMS characteristic length	1 mm and > 100 nm
Molecular gear	≈ 10 nm
Vertical SWCNT transistor	≈ 15 nm
Quantum-dots transistor	300 nm
Digital Micromirror	12 000 nm
Individual atoms	Typically fraction of a nm in diameter
DNA molecules	≈ 2.5 nm wide
Biological cells	In the range of thousands of nm in diameter
Human hair	≈ 75 000 nm in diameter

Table 1.2 (cont.) (**b**) Masses in perspective

NEMS built with cross sections of about 10 nm	As low as 10^{-20} N
Micromachines silicon structure	As low as 1 nN
Water droplet	≈ 10 μN
Eyelash	≈ 100 nN

frequency-MEMS/NEMS or RF-MEMS/RF-NEMS. MEMS/NEMS for biological applications are referred to as BioMEMS/BioNEMS.

To put the dimensions of MEMS and NEMS in perspective, see Fig. 1.1 and Table 1.1. Individual atoms are typically a fraction of a nanometer in diameter, DNA molecules are about 2.5 nm wide, biological cells are in the rage of thousands of nanometers in diameter, and human hair is about 75 μm in diameter. The NEMS shown in the figure range in size from 15 to 300 nm and MEMS is 12 000 nm. The mass of a micromachined silicon structure can be as low as 1 nN, and NEMS can be built with a mass as low as 10^{-20} N with cross sections of about 10 nm. In comparison, the mass of a drop of water is about 10 μN, and the mass of an eyelash is about 100 nN.

MEMS and emerging NEMS are expected to have a major impact on our lives, comparable to that of semiconductor technology, information technology, or cellular and molecular biology [1.4, 5]. MEMS/NEMS are used in electromechanical, electronics, information/communication, chemical, and biological applications. The MEMS industry in 2004 was worth about $4.5 billion and with a projected annual growth rate of 17% (Fig. 1.2) [1.6]. Growth of Si-based

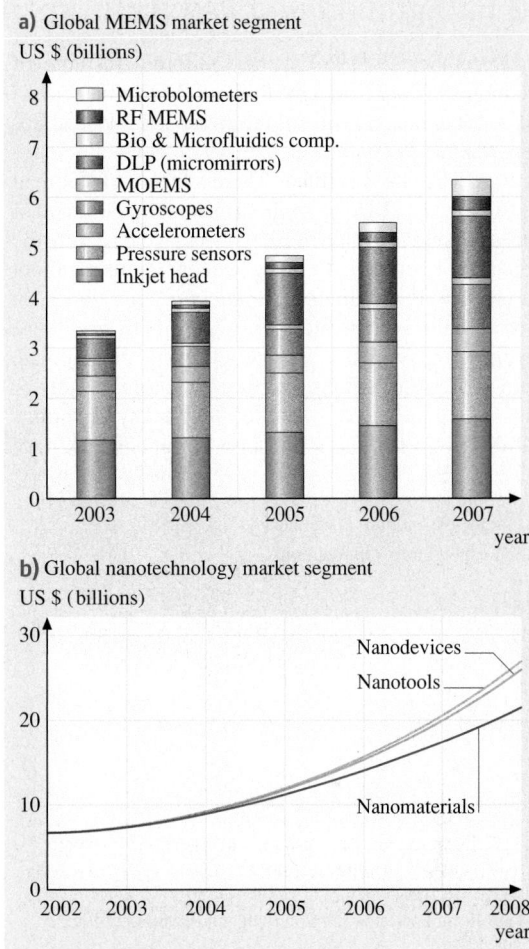

a) Global MEMS market segment
US $ (billions)

Legend:
- Microbolometers
- RF MEMS
- Bio & Microfluidics comp.
- DLP (micromirrors)
- MOEMS
- Gyroscopes
- Accelerometers
- Pressure sensors
- Inkjet head

b) Global nanotechnology market segment
US $ (billions)

Nanodevices
Nanotools
Nanomaterials

Fig. 1.2 Global MEMS and nanotechnology market segments

MEMS may slow down, and nonsilicon MEMS may pick up during the next decade. The NEMS industry was worth about $10 billion dollars in 2004, mostly in nanomaterials (Fig. 1.2) [1.7]. It is expected to expand in this decade, in nanomaterials, biomedical applications, and nanoelectronics or molecular electronics. For example, miniaturized diagnostics could be implanted for early diagnosis of illness. Targeted drug delivery devices are under development. Due to the enabling nature of these systems and because of the significant impact they can have on both commercial and defense applications, industry as well as federal governments have taken a special interest in seeing growth nurtured in this field. MEMS/NEMS are the next logical step in the "silicon revolution."

1.2 Background and Research Expenditures

On December 29, 1959 at the California Institute of Technology, Nobel Laureate Richard P. Feynman gave a talk at the annual meeting of the American Physical Society that has become a classic in 20th-century science lectures. The talk was titled "There's Plenty of Room at the Bottom" [1.8]. He presented a technological vision of extreme miniaturization in 1959, several years before the word "chip" became part of the lexicon. He talked about the problem of manipulating and controlling things on a small scale. Extrapolating from known physical laws, Feynman envisioned a technology using the ultimate toolbox of nature, building nanoobjects atom by atom or molecule by molecule. Since the 1980s, many inventions and discoveries in the fabrication of nanoobjects have

been testament to his vision. In recognition of this reality, in 1998 the White House National Science and Technology Council (NSTC) created the Interagency Working Group on Nanoscience, Engineering, and Technology (IWGN). In a January 2000 speech at the same institute, former President Bill Clinton talked about the exciting promise of "nanotechnology" and, more broadly, the importance of expanding research in nanoscale science and technology. Later that month, he announced in his State of the Union Address an ambitious $497 million federal, multiagency national nanotechnology initiative (NNI) in the 2001 fiscal year budget and made the NNI a top science and technology priority [1.9, 10]. The objective of this initiative was to form a broad-based coalition in which academia, the private sector, and local, state, and federal governments would work together to push the envelope of nanoscience and nanoengineering to reap nanotechnology's potential social and economic benefits.

Funding for this initiative in the US has continued to rise. In January 2003, the US Senate introduced a bill to establish a National Nanotechnology Program. On

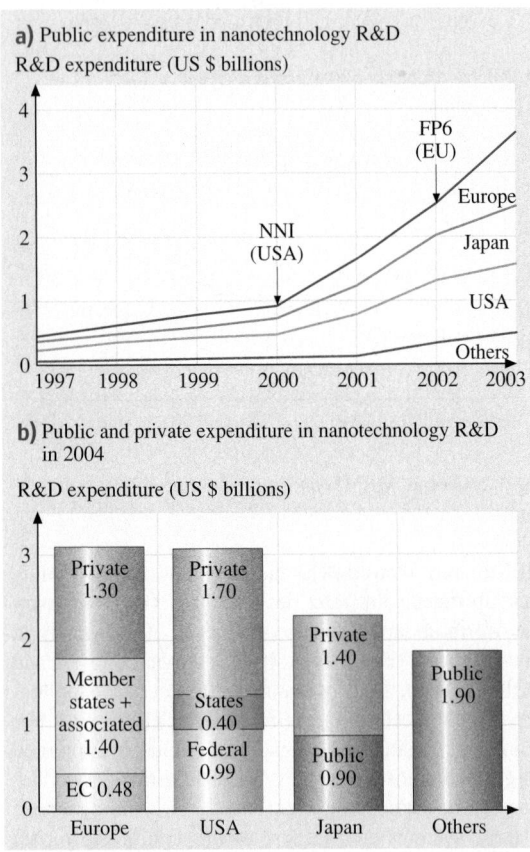

Fig. 1.3a,b Breakdown of public expenditure in nanotechnology R&D (**a**) around the world (source: European Commission 2003) and (**b**) by public and private resources in 2004 (source: European Commission 2005; private figures based upon Lux Research)

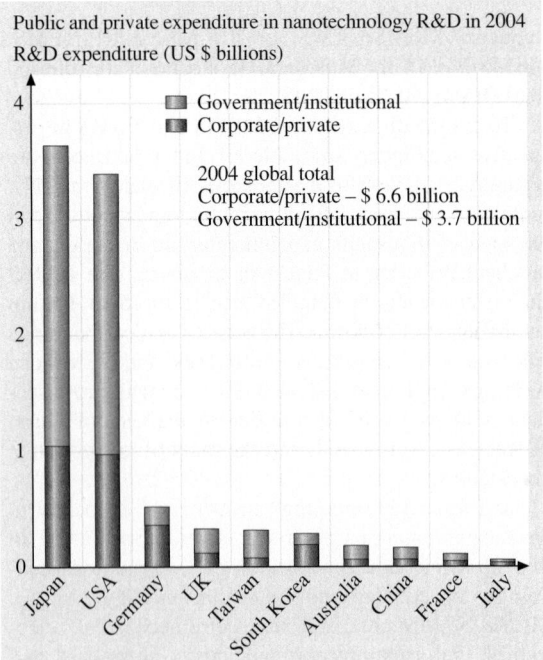

Fig. 1.4 Breakdown of public and private expenditures in nanotechnology R&D in 2004 in various countries (Lawrence 2005)

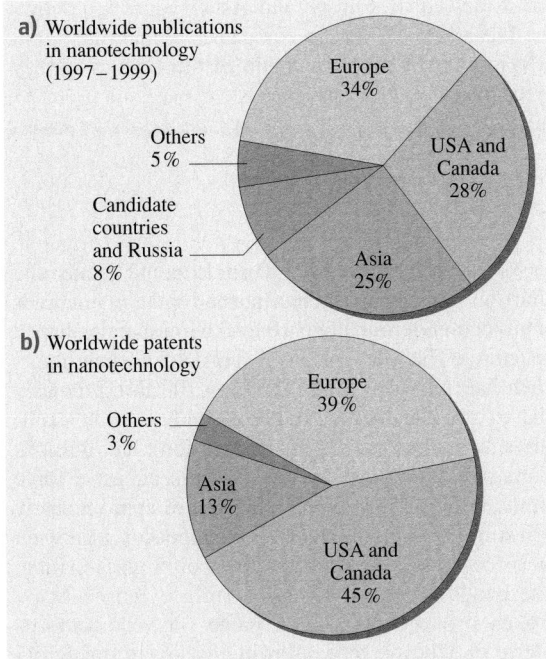

a) Worldwide publications in nanotechnology (1997–1999)
- Europe 34%
- USA and Canada 28%
- Asia 25%
- Others 5%
- Candidate countries and Russia 8%

b) Worldwide patents in nanotechnology
- Europe 39%
- USA and Canada 45%
- Asia 13%
- Others 3%

Fig. 1.5a,b Breakdown of (a) worldwide publications and (b) worldwide patents (source: European Commission 2003)

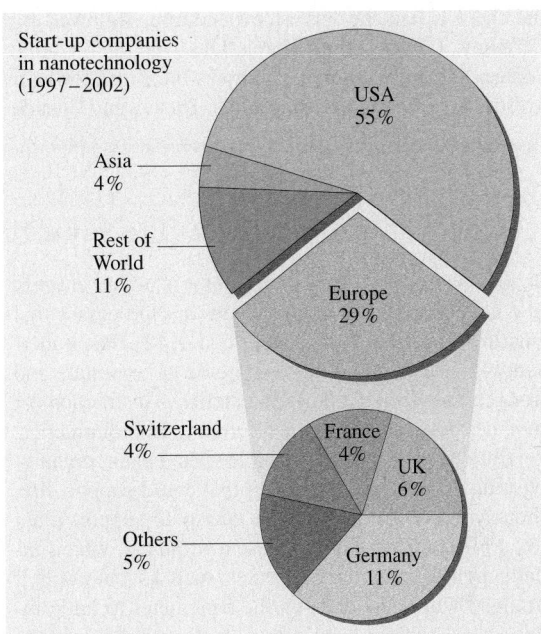

Start-up companies in nanotechnology (1997–2002)
- USA 55%
- Europe 29%
- Rest of World 11%
- Asia 4%
- Germany 11%
- UK 6%
- France 4%
- Switzerland 4%
- Others 5%

Fig. 1.6 Breakdown of startup companies around the world (1997–2002) (source: CEA, Bureau d'Etude Marketing)

December 3, 2003, President George W. Bush signed into law the 21st Century Nanotechnology Research and Development Act. The legislation put into law programs and activities supported by the National Nanotechnology Initiative. The bill gave nanotechnology a permanent home in the federal government and authorized $3.7 billion to be spent in the 4-year period beginning in October 2005 on nanotechnology initiatives at five federal agencies. The funds would provide grants to researchers, coordinate R&D across five federal agencies [National Science Foundation (NSF), Department of Energy (DOE), NASA, National Institute of Standards and Technology (NIST), and Environmental Protection Agency (EPA)], establish interdisciplinary research centers, and accelerate technology transfer into the private sector. In addition, the departments of Defense (DOD), Homeland Security, Agriculture, and Justice, as well as the National Institutes of Health (NIH), also fund large R&D activities. They currently account for more than one third of the federal nanotechnology budget.

The European Union (EU) made nanosciences and nanotechnologies a priority in the Sixth Framework Program (FP6) in 2002 for the period 2003–2006. They had dedicated modest funds in FP4 and FP5. FP6 was tailored to better help structure European research and to cope with the strategic objectives set out in Lisbon in 2000. Japan identified nanotechnology as one of its main research priorities in 2001. The funding levels increased sharply from $400 million in 2001 to around $950 million in 2004. In 2003, South Korea embarked on a 10-year program with around $2 billion of public funding, and Taiwan has committed around $600 million of public funding over 6 years. Singapore and China are also investing on a large scale. Russia is well funded as well.

Figure 1.3a shows the public expenditure breakdown in nanotechnology R&D around the world, with about US$5 billion in 2004, being about equal by USA, Japan, and Europe. Next we compare public expenditure on a per-capita basis. The average expenditures per capita for the US, EU-25, and Japan were about $3.7, $2.4, and $6.2, respectively [1.11]. Figure 1.3b shows the breakdown of expenditures in 2004 by public and private sources with more than $10 billion spent on nanotechnology research. Two thirds of this came from corporate and private funding. The private expenditure in the United States and Japan was slightly larger than that of the public, whereas in Europe it was about one third. Figure 1.4 shows the public and private expenditure breakdown in 2004 in various countries. Japan and the US had the largest expenditure, followed by Germany, Taiwan, South Korea, UK, Australia, China, France, and Italy. Figure 1.5 shows the breakdown of worldwide publications and patents. The US and Canada led, followed by Europe and Asia. Figure 1.6 shows the breakdown in startup companies around the world (1997–2002). Entrepreneurship in the USA is clearly evident followed by Europe.

1.3 Lessons from Nature (Biomimetics)

Nanotechnology is a new word, but it is not an entirely new field. Nature has many objects and processes that function on a micro- to nanoscale [1.9, 12]. The understanding of these functions can guide us to imitate and produce nanodevices and nanomaterials. Abstractions of good design from nature are referred to as biomimetics.

Billions of years ago, molecules began organizing into the complex structures that could support life. Photosynthesis harnesses solar energy to support plant life. Molecular ensembles present in plants, which include light-harvesting molecules such as chlorophyll arranged within the cells on the nanometer to micrometer scales, capture light energy and convert it into the chemical energy that drives the biochemical machinery of plant cells. Live organs use chemical energy in the body. The flagellum, a type of bacteria, rotates at over 10 000 RPM [1.13]. This is an example of a biological molecular machine. The flagellum motor is driven by the proton flow caused by the electrochemical potential differences across the membrane. The diameter of the bearing is about 20–30 nm with an estimated clearance of about 1 nm.

In the context of tribology, some biological systems have antiadhesion surfaces. First, many plant leaves (such as the lotus leaf) are covered by a hydrophobic cuticle, which is composed of a mixture of large hydrocarbon molecules that have a strong aversion to water. Second, the surface has a unique roughness distribution [1.14, 15]. It has been reported that for some leaf surfaces, the roughness of the hydrophobic leaf surface decreases wettability, which is reflected in a greater contact angle of water droplets on such surfaces.

"Geckos" (a family of lizards) are known for their amazing climbing ability. They can run up any wall, run across the ceiling, and stick to ceilings. They rely on the extreme miniaturization and multiplication of contact elements. Soles of geckos are covered with about half a million submicrometer keratin hairs, called spatulae, which are what make their feet, known as "gecko feet", so sticky. Each hair is 30–130 μm long and is only one tenth the diameter of a human hair and contains hundreds of projections terminating in 0.2–0.5 μm spatula-shaped structures. The foot typically has about 5000 hair/mm^2. Each hair produces a tiny force (≈ 100 nN), primarily due to van der Waals attraction, and possibly capillary interactions (meniscus contribution), and millions of hairs acting together create a large adhesive force on the order of 10 N with a pad area of approximately 100 mm^2 [1.16], sufficient to keep geckos firmly on their feet, even when upside down on a glass ceiling. The bonds between hair and a surface can be easily broken by "peeling," in the same way one removes a strip of adhesive tape, allowing geckos to run across ceilings.

Spiders, a family of anthropods (spiders, insects, and crustaceans) can stick to smooth, overhanging surfaces also because of a large number of hairs and the microstructure of the hairs on their feet. Spiders use claws to attach to rough surfaces but have scopulae (tufts of hairs) on their feet to adhere to smooth surfaces. The scopulae hairs of the jumping spider, *Evarcha arcuata*, branch into a very large number of smaller hairs or setulae, whose broadened ends have a contact area of about 2×10^5 nm^2 [1.17]. The number of setulae per foot is about 80, 000 giving a total of about 7 00, 000 contact points for the spider's eight legs [1.17]. This provides a large amount of adhesive force because of van der Waals attraction added on all legs. The hair surface (cuticle) is sealed with a topographically microconfigured wax layer. These surfaces are reportedly nonwettable, so capillary interactions are not expected to be significant.

Scientists are attempting to create a new type of adhesive tape by mimicking the structure of gecko or spider feet. *Geim* et al. [1.18] reported the fabrication of a "gecko" tape made by microfabrication of dense arrays of flexible plastic pillars that are little more than 2 μm tall with a pitch on a similar scale.

1.4 Applications in Different Fields

Science and technology continue to move forward in making the fabrication of micro/nanodevices and systems possible for a variety of industrial, consumer, and biomedical applications (see, e.g., [1.19, 20]). A variety of MEMS devices have been produced, and some are commercially used [1.12, 21–29]. Several types of sensors are used in industrial, consumer, defense, and biomedical applications. Various micro/nanostructures or micro/nanocomponents are used in microinstruments and other industrial applications such as micromirror arrays. The largest "killer" MEMS applications include accelerometers (some 90 million units installed in vehicles in 2004), silicon-based piezoresistive pressure sensors for manifold absolute pressure sensing for engines and for disposable blood pressure sensors (about 30 million units and about 25 million units, respectively), capacitive pressure sensors for tire pressure measurements (about 37 million units in 2005), thermal inkjet printheads (about 500 million units in 2004), and digital micromirror arrays (about $700 million in revenue in 2004). Other applications of MEMS devices include chemical/biosensors and gas sensors, microresonators, infrared detectors, and focal plane arrays for earth observations, space science, and missile defense applications, picosatellites for space applications, fuel cells, and many hydraulic, pneumatic, and other consumer products. MEMS devices are also being explored for use in magnetic storage systems (*Bhushan* [1.30]), where they are being developed for supercompact and ultrahigh-recording-density magnetic disk drives.

NEMS are produced by nanomachining in a typical top-down and bottom-up approach, largely relying on nanochemistry [1.31–37]. Examples of NEMS include microcantilevers with integrated sharp nanotips for scanning tunneling microscopy (STM) and atomic force microscopy (AFM), quantum corral formed using STM by placing atoms one by one, AFM cantilever array (Millipede) for data storage, AFM tips for nanolithography, dip-pen lithography for printing molecules, nanowires, carbon nanotubes, quantum wires, quantum boxes, quantum-dot transistors, nanotube-based sensors, biological (DNA) motors, molecular gears by attaching benzene molecules to the outer walls of carbon nanotubes, devices incorporating nanometer-thick films (e.g., in giant magnetoresistive or GMR read/write magnetic heads and magnetic media) for magnetic rigid disk drives and magnetic tape drives, nanopatterned magnetic rigid disks, and nanoparticles (e.g., nanoparticles in magnetic

tape substrates and magnetic particles in magnetic tape coatings).

Nanoelectronics can be used to build computer memory using individual molecules or nanotubes to store bits of information, molecular switches, molecular or nanotube transistors, nanotube flat-panel displays, nanotube integrated circuits, fast logic gates, switches, nanoscopic lasers, and nanotubes as electrodes in fuel cells.

BioMEMS/BioNEMS are increasingly used in commercial and defense applications; see, e.g., [1.38–44]. They are used for chemical and biochemical analyses (biosensors) in medical diagnostics (e.g., DNA, RNA, proteins, cells, blood pressure and assays, and toxin identification) [1.44, 45], tissue engineering [1.46], and implantable pharmaceutical drug delivery [1.47, 48]. Biosensors, also referred to as biochips, deal with liquids and gases. There are two types. A large variety of biosensors are based on micro/nanofluidics. Micro/nanofluidic devices offer the ability to work with smaller reagent volumes and shorter reaction times, and perform analyses multiple times at once. The second type of biosensors includes micro/nanoarrays that perform one type of analysis thousands of times. Micro/nanoarrays are a tool used in biotechnology research to analyze DNA or proteins to diagnose diseases or discover new drugs. Also called DNA arrays, they can identify thousands of genes simultaneously [1.41]. They include a microarray of silicon nanowires, roughly a few nanometers in size, to selectively bind and detect even a single biological molecule, such as DNA or protein, using nanoelectronics to detect the slight electrical charge caused by such binding, or a microarray of carbon nanotubes to electrically detect glucose.

After the tragedy of September 11, 2001, concern over biological and chemical warfare has led to the development of handheld units with bio- and chemical sensors for the detection of biological germs, chemical or nerve agents, and mustard agents and to chemical precursors to protect subways, airports, water supply, and the population at large [1.49].

BioMEMS/BioNEMS are also being developed for minimal invasive surgery including endoscopic surgery, laser angioplasty, and microscopic surgery. Other applications include implantable drug-delivery devices—micro/nanoparticles with drug molecules encapsulated in functionalized shells for site-specific targeting applications and a silicon capsule with a nanoporous membrane filled with drugs for long-term delivery.

1.5 Various Issues

There is an increasing need for a multidisciplinary, system-oriented approach to the manufacture of micro/nanodevices that function reliably. This can only be achieved through the cross-fertilization of ideas from different disciplines and the systematic flow of information and people among research groups. Common potential failure mechanisms for MEMS/NEMS requiring relative motion that need to be addressed in order to increase their reliability are adhesion, friction, wear, fracture, fatigue, and contamination [1.50–53]. Surface micro/nanomachined structures often include smooth and chemically active surfaces. Due to a large surface area to volume ratio in MEMS/NEMS, they are particularly prone to stiction (high static friction) as part of normal operation. Fracture occurs when the load on a microdevice is greater than the strength of the material. Fracture is a serious reliability concern, particularly for brittle materials used in the construction of these components, since it can immediately or would eventually lead to catastrophic failures. Additionally, debris can be formed from the fracturing of microstructures, leading to other failure processes. For less brittle materials, repeated loading over a long period of time causes fatigue that would also lead to the breaking and fracturing of the device. In principle, this failure mode is relatively easy to observe and simple to predict. However, the material properties of thin films are often not known, making fatigue predictions error prone.

Many MEMS/NEMS devices operate near their thermal dissipation limit. They may encounter hot spots that may cause failures, particularly in weak structures, such as diaphragms or cantilevers. Thermal stressing and relaxation caused by thermal variations can create material delamination and fatigue in cantilevers. In large temperature changes, as experienced in outer space, bimetallic beams will also experience warping due to mismatched coefficients of thermal expansion. Packaging has been a big problem. The contamination, which probably happens in packaging and during storage, also can strongly influence the reliability of MEMS/NEMS. For example,

a particulate dust that lands on one of the electrodes of a comb drive can cause catastrophic failure. There are no MEMS/NEMS fabrication standards, which makes it difficult to transfer fabrication steps in MEMS/NEMS between foundries.

Obviously, studies of determination and suppression of active failure mechanisms affecting this new and promising technology are critical to a high reliability of MEMS/NEMS and are determining factors in successful practical application.

Adhesion between a biological molecular layer and the substrate, referred to as "bioadhesion," and reduction of friction and wear of biological layers, biocompatibility, and biofouling for BioMEMS/BioNEMS are important.

Mechanical properties are known to exhibit a dependence on specimen size. Mechanical property evaluation of nanometer-scaled structures is carried out to help design reliable systems since good mechanical properties are of critical importance in such applications. Some of the properties of interest are Young's modulus of elasticity, hardness, bending strength, fracture toughness, and fatigue life. Finite element modeling is carried out to study the effects of surface roughness and scratches on stresses in nanostructures. When nanostructures are smaller than a fundamental physical length scale, conventional theory may no longer apply, and new phenomena emerge. Molecular mechanics is used to simulate the behavior of a nanoobject.

The societal, ethical, political, and health/safety implications are receiving considerable attention [1.11]. One of the prime reasons is to avoid some of the public skepticism that surrounded the debate over biotech advances such as genetically modified foods, while at the same time dispelling some of the misconceptions the public may already have about nanotechnology. Health/safety issues need to be addressed as well. For example, one key question is what happens to nanoparticles (such as buckyballs or nanotubes) in the environment and whether they are toxic in the human body if ingested.

1.6 Research Training

With a decreasing number of people in western countries going into science and engineering and with the rapid progress being made in nanoscience and nanotechnology, the problem of a trained work force is expected to be acute. Education and training is essential

to produce a new generation of scientists, engineers, and skilled workers with the flexible and interdisciplinary R&D approach necessary for rapid progress in nanosciences and nanotechnology [1.54]. The question is being asked: Is the traditional separation of academic disciplines into physics, chemistry, biology, and various engineering disciplines meaningful at the nano level? Generic skills and entrepreneurship are needed to translate scientific knowledge into products. Scientists and engineers in cooperation with relevant experts should address the societal, ethical, political, and health/safety implications of their work for society at large.

To increase the pool of students interested in science and technology, science needs to be projected as exciting at the high school level. Interdisciplinary curricula relevant for nanoscience and nanotechnology need to be developed. This requires revamping the education, developing new courses and course materials including textbooks [1.28, 36, 53, 55–57] and instruction manuals, and training new instructors.

1.7 Organization of Handbook

The handbook integrates knowledge from the point of view of fabrication, mechanics, materials science, and reliability. Organization of the book is straightforward. The handbook is divided into eight parts. The first part of the book includes an introduction to nanostructures, micro/nanofabrication, methods, and materials. The second part introduces various MEMS/NEMS and BioMEMS/BioNEMS devices. The third part introduces scanning probe microscopy. The fourth part provides an

overview of nanotribology and nanomechanics, which will prepare the reader for understanding the tribology and mechanics of industrial applications. The fifth part provides an overview of molecularly thick films for lubrication. The sixth part focuses on industrial applications, and the seventh part focuses on microdevice reliability. Finally, the last part focuses on technological convergence from the nanoscale as well as the social, ethical, and political implications of nanotechnology.

References

1.1   A. P. Graham, G. S. Duesberg, R. Seidel, M. Liebau, E. Unger, F. Kruepl, W. Hoenlein: Towards the Integration of Carbon Nanotubes in Microelectronics, Diamond Related Mat. **13**, 1296–1300 (2004)

1.2   D. Srivastava: Computational Nanotechnology of Carbon Nanotubes. In: *Carbon Nanotubes: Science and Applications*, ed. by M. Meyyappan (CRC Press, Boca Raton, Florida 2004) pp. 25–36

1.3   W. G. van der Wiel, S. De Franceschi, J. M. Elzerman, T. Fujisawa, S. Tarucha, L. P. Kauwenhoven: Electron Transport Through Double Quantum Dots, Rev. Modern Phys. **75**, 1–22 (2003)

1.4   Anonymous: *Microelectromechanical Systems: Advanced Materials and Fabrication Methods*, NMAB-483 (National Academy Press, Washington, D.C. 1997)

1.5   M. Roukes: Nanoelectromechanical Systems Face the Future, Physics World, 25–31 (2001)

1.6   J. C. Eloy: *Status of the MEMS Industry* (Report Yole Developpement, France 2005) presented at SPIE Photonics West, San Jose, California, January 2005

1.7   S. Lawrence: Nanotech Grows Up, Technol. Rev. **108**(Issue 6), 31 (2005)

1.8   R. P. Feynman: There's Plenty of Room at the Bottom, Eng. Sci. **23**, 22–36 (1960) www.zyvex.com/nanotech/feynman.html

1.9   I. Amato, *Nanotechnology* (2000), www.ostp.gov/nstc/html/iwgn/iwgn.public.brochure/welcome.htm or www.nsf.gov/home/crssprgm/nano/nsfnnireports.htm

1.10   Anonymous (2000): *National Nanotechnology Initiative*, www.ostp.gov/nstc/html/iwgn.fy01budsuppl/nni.pdf or www.nsf.gov/home/crssprgm/nano/nsfnnireports.htm

1.11   Anonymous: *Towards a European Strategy for Nanotechnology* (European Commission, Research Directorate, Brussels 2004)

1.12   I. Fujimasa: *Micromachines: A New Era in Mechanical Engineering* (Oxford University Press, Oxford, UK 1996)

1.13   C. J. Jones, S. Aizawa: The Bacterial Flagellum and Flagellar Motor: Structure, Assembly, and Functions, Adv. Microb. Physiol. **32**, 109–172 (1991)

1.14   M. Scherge, S. Gorb: *Biological Micro- and Nanotribology* (Springer, Berlin Heidelberg 2001)

1.15   M. Nosonovsky, B. Bhushan: Roughness Optimization for Biomimetic Superhydrophobic Surfaces, Microsyst. Technol **11**, 535–549 (2005)

1.16   K. Autumn, Y. A. Liang, S. T. Hsieh, W. Zesch, W. P. Chan, T. W. Kenny, R. Fearing, R. J. Full: Adhesive Force of a Single Gecko Foot-Hair, Nature **405**, 681–685 (2000)

1.17   A. B. Kesel, A. Martin, T. Seidl: Getting a Grip on Spider Attachment: An AFM Approach to Microstructure Adhesion in Arthropods, Smart Mater. Struct. **13**, 512–518 (2004)

1.18   A. K. Geim, S. V. Dubonos, I. V. Grigorieva, K. S. Novoselov, A. A. Zhukov, S. Y. Shapoval: Microfabricated

Adhesive Mimicking Gecko Foot-Hair, Nat. Mater. **2**, 461–463 (2003)

1.19   Anonymous: *Small Tech 101 – An Introduction to Micro and Nanotechnology* (Small Times, 2003)

1.20   M. Schulenburg: *Nanotechnology – Innovation for Tomorrow's World* (European Commission, Research DG, Brussels 2004)

1.21   R. S. Muller, R. T. Howe, S. D. Senturia, R. L. Smith, R. M. White: *Microsensors* (IEEE Press, New York 1990)

1.22   W. S. Trimmer (ed.): *Micromachines and MEMS, Classical and Seminal Papers to 1990* (IEEE Press, New York 1997)

1.23   B. Bhushan: *Tribology Issues and Opportunities in MEMS* (Kluwer, Dordrecht, Netherlands 1998)

1.24   G. T. A. Kovacs: *Micromachined Transducers Sourcebook* (WCB McGraw-Hill, Boston 1998)

1.25   M. Elwenspoek, R. Wiegerink: *Mechanical Microsensors* (Springer, Berlin Heidelberg 2001)

1.26   S. D. Senturia: *Microsystem Design* (Kluwer, Boston, Massachusetts 2000)

1.27   T. R. Hsu: *MEMS and Microsystems: Design and Manufacture* (McGraw-Hill, Boston 2002)

1.28   M. Madou: *Fundamentals of Microfabrication: The Science of Miniaturization*, 2nd edn. (CRC Press, Boca Raton, Florida 2002)

1.29   A. Hierlemann: *Integrated Chemical Microsensor Systems in CMOS Technology* (Springer, Berlin Heidelberg 2005)

1.30   B. Bhushan: *Tribology and Mechanics of Magnetic Storage Devices*, 2nd edn. (Springer, Berlin Heidelberg 1996)

1.31   K. E. Drexler: *Nanosystems: Molecular Machinery, Manufacturing and Computation* (Wiley, New York 1992)

1.32   G. Timp (ed.): *Nanotechnology* (Springer, New York 1999)

1.33   M. S. Dresselhaus, G. Dresselhaus, Ph. Avouris: *Carbon Nanotubes – Synthesis, Structure, Properties, and Applications* (Springer, Berlin Heidelberg 2001)

1.34   W. A. Goddard, D. W. Brenner, S. E. Lyshevski, G. J. Iafrate (eds.): *Handbook of Nanoscience, Engineering, and Technology* (CRC, Boca Raton, Florida 2002)

1.35   H. S. Nalwa (ed.): *Nanostructured Materials and Nanotechnology* (Academic, San Diego, California 2002)

1.36   C. P. Poole, F. J. Owens: *Introduction to Nanotechnology* (Wiley, New York 2003)

1.37   E. A. Rietman: *Molecular Engineering of Nanosystems* (Springer, Berlin Heidelberg 2001)

1.38   A. Manz, H. Becker (eds.): *Microsystem Technology in Chemistry and Life Sciences*, Topics in Current Chemistry 194 (Springer, Berlin Heidelberg 1998)

1.39   J. Cheng, L. J. Kricka (eds.): *Biochip Technology* (Harwood, Philadephia 2001)

1.40   M. J. Heller, A. Guttman (eds.): *Integrated Microfabricated Biodevices* (Marcel Dekker, New York 2001)

1.41   C. Lai Poh San, E. P. H. Yap (eds.): *Frontiers in Human Genetics* (World Scientific, Singapore 2001)

1.42   C. H. Mastrangelo, H. Becker (eds.): *Microfluidics and BioMEMS*, Proc. of SPIE, Vol. 4560 (SPIE, Bellingham, Washington 2001)

1.43   H. Becker, L. E. Locascio: Polymer Microfluidic Devices, Talanta **56**, 267–287 (2002)

1.44   A. van der Berg (ed.): *Lab-on-a-Chip: Chemistry in Miniaturized Synthesis and Analysis Systems* (Elsevier, Amsterdam 2003)

1.45   P. Gravesen, J. Branebjerg, O. S. Jensen: Microfluidics – A Review, J. Micromech. Microeng. **3**, 168–182 (1993)

1.46   R. P. Lanza, R. Langer, J. Vacanti (eds.): *Principles of Tissue Engineering*, 2nd edn. (Academic Press, San Diego 2000)

1.47   K. Park (ed.): *Controlled Drug Delivery: Challenges and Strategies* (American Chemical Society, Washington, D.C. 1997)

1.48   P. A. Oeberg, T. Togawa, F. A. Spelman: *Sensors in Medicine and Health Care* (Wiley, New York 2004)

1.49   M. Scott: MEMS and MOEMS for National Security Applications. In: *Reliability, Testing, and Characterization of MEMS/MOEMS II*, Proc. of SPIE, Vol. 4980 (SPIE, Bellingham, Washington 2003) pp. xxxvii–xliv

1.50   B. Bhushan: *Principles and Applications of Tribology* (Wiley, New York 1999)

1.51   B. Bhushan: *Handbook of Micro/Nanotribology* (CRC Press, Boca Raton, Florida 1999)

1.52   B. Bhushan: *Introduction to Tribology* (Wiley, New York 2002)

1.53   B. Bhushan: *Nanotribology and Nanomechanics – An Introduction* (Springer, Berlin Heidelberg 2005)

1.54   Anonymous: *Proc. of the Workshop on Research Training in Nanosciences and Nanotechnologies: Current Status and Future Needs* (European Commission, Research Directorate General, Brussels 2005)

1.55   M. DiVentra, S. Evoy, J. R. Heflin: *Introduction to Nanoscale Science and Technology* (Springer, Berlin Heidelberg 2004)

1.56   A. Hett: *Nanotechnology – Small Matter, Many Unknowns* (Swiss Reinsurance Company, Zurich, Switzerland 2004)

1.57   M. Koehler, W. Fritzsche: *Nanotechnology* (Wiley, New York 2004)

Part A Nanostructures, Micro/Nanofabrication and Materials

2. Nanomaterials Synthesis and Applications: Molecule-Based Devices

The constituent components of conventional devices are carved out of larger materials relying on physical methods. This top-down approach to engineered building blocks becomes increasingly challenging as the dimensions of the target structures approach the nanoscale. Nature, on the other hand, relies on chemical strategies to assemble nanoscaled biomolecules. Small molecular building blocks are joined to produce nanostructures with defined geometries and specific functions. It is becoming apparent that nature's bottom-up approach to functional nanostructures can be mimicked to produce artificial molecules with nanoscaled dimensions and engineered properties. Indeed, examples of artificial nanohelices, nanotubes, and molecular motors are starting to be developed. Some of these fascinating chemical systems have intriguing electrochemical and photochemical properties that can be exploited to manipulate chemical, electrical, and optical signals at the molecular level. This tremendous opportunity has led to the development of the molecular equivalent of conventional logic gates. Simple logic operations, for example, can be reproduced with collections of molecules operating in solution. Most of these chemical systems, however, rely on bulk addressing to execute combinational and sequential logic operations. It is essential to devise methods to reproduce these useful functions in solid-state configurations and, eventually, with single molecules. These challenging objectives are stimulating the design of clever devices that interface small assemblies of organic molecules with macroscaled and nanoscaled electrodes. These strategies have already produced rudimentary examples of diodes, switches, and transistors based on functional molecular

components. The rapid and continuous progress of this exploratory research will, we hope, lead to an entire generation of molecule-based devices that might ultimately find useful applications in a variety of fields, ranging from biomedical research to information technology.

2.1 Chemical Approaches to Nanostructured Materials

The fabrication of conventional devices relies on the assembly of macroscopic building blocks with specific configurations. The shapes of these components

are carved out of larger materials by exploiting physical methods. This top-down approach to engineered building blocks is extremely powerful and can deliver effectively and reproducibly microscaled objects. This strategy becomes increasingly challenging, however, as the dimensions of the target structures approach the nanoscale. Indeed, the physical fabrication of nanosized features with subnanometer precision is a formidable technological challenge.

2.1.1 From Molecular Building Blocks to Nanostructures

Nature efficiently builds nanostructures by relying on chemical approaches. Tiny molecular building blocks are assembled with a remarkable degree of structural control in a variety of nanoscaled materials with defined shapes, properties, and functions. In contrast to the top-down physical methods, small components are connected to produce larger objects in these bottom-up chemical strategies. It is becoming apparent that the limitations of the top-down approach to artificial nanostructures can be overcome by mimicking nature's bottom-up processes. Indeed, we are starting to see emerge beautiful and ingenious examples of molecule-based strategies to fabricate chemically nanoscaled building blocks for functional materials and innovative devices.

2.1.2 Nanoscaled Biomolecules: Nucleic Acids and Proteins

Nanoscaled macromolecules play a fundamental role in biological processes [2.1]. Nucleic acids, for example, ensure the transmission and expression of genetic information. These particular biomolecules are linear polymers incorporating nucleotide repeating units (Fig. 2.1a). Each nucleotide has a phosphate bridge and a sugar residue. Chemical bonds between the phosphate of one nucleotide and the sugar of the next ensures the propagation of a polynucleotide strand from the $5'$ to the $3'$ end. Along the sequence of alternating sugar and phosphate fragments, an extended chain of robust covalent bonds involving carbon, oxygen, and phosphorous atoms forms the main backbone of the polymeric strand.

Every single nucleotide of a polynucleotide strand carries one of the four heterocyclic bases shown in Fig. 2.1b. For a strand incorporating 100 nucleotide repeating units, a total of 4^{100} unique polynucleotide sequences are possible. It follows that nature can fabricate a huge number of closely related nanostructures relying only on four building blocks. The heterocyclic bases appended to the main backbone of alternating phos-

phate and sugar units can sustain hydrogen bonding and $[\pi \cdots \pi]$ stacking interactions. Hydrogen bonds, formed between [N−H] donors and either N or O acceptors, encourage the pairing of adenine (A) with thymine (T) and of guanine (G) with cytosine (C). The stacking interactions involve attractive contacts between the extended π-surfaces of heterocyclic bases.

In the B conformation of deoxyribonucleic acid (DNA), the synergism of hydrogen bonds and $[\pi \cdots \pi]$ stacking glues pairs of complementary polynucleotide strands in fascinating double helical supermolecules (Fig. 2.1c) with precise structural control at the subnanometer level. The two polynucleotide strands wrap around a common axis to form a right-handed double helix with a diameter of ca. 2 nm. The hydrogen bonded and $[\pi \cdots \pi]$ stacked base pairs lie at the core of the helix with their π-planes perpendicular to the main axis of the helix. The alternating phosphate and sugar units define the outer surface of the double helix. In B-DNA, approximately ten base pairs define each helical turn corresponding to a rise per turn or helical pitch of ca. 3 nm. Considering that these molecules can incorporate up to approximately 10^{11} base pairs, extended end-to-end lengths spanning from only few nanometers to hundreds of meters are possible.

Nature's operating principles to fabricate nanostructures are not limited to nucleic acids. Proteins are also built joining simple molecular building blocks, the amino acids, by strong covalent bonds [2.1]. More precisely, nature relies on 20 amino acids differing in their side chains to assemble linear polymers, called polypeptides, incorporating an extended backbone of robust [C−N] and [C−C] bonds (Fig. 2.2a). For a single polymer strand of 100 repeating amino acid units, a total of 20^{100} unique combinations of polypeptide sequences are possible. Considering that proteins can incorporate more than one polypeptide chain with over 4000 amino acid residues each, it is obvious that nature can assemble an enormous number of different biomolecules relying on the same fabrication strategy and a relatively small pool of building blocks.

The covalent backbones of the polypeptide strands form the main skeleton of a protein molecule. In addition, a myriad of secondary interactions, involving noncovalent contacts between portions of the amino acid residues, control the arrangement of the individual polypeptide chains. Intrastrand hydrogen bonds curl single polypeptide chains around a longitudinal axis in a helical fashion to form tubular nanostructures ca. 0.5 nm wide and ca. 2 nm long (Fig. 2.2b). Similarly, in-

Nanomaterials Synthesis and Applications: Molecule–Based Devices | 2.1 Chemical Approaches to Nanostructured Materials | 15

Part A | 2.1

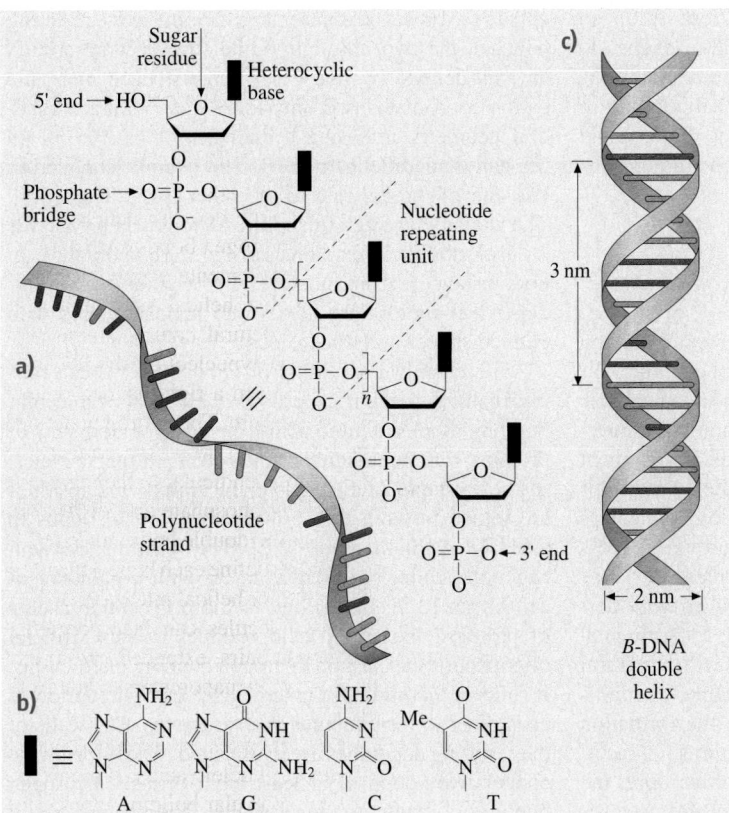

Fig. 2.1a–c A polynucleotide strand (**a**) incorporates alternating phosphate and sugar residues joined by covalent bonds. Each sugar carries one of four heterocyclic bases (**b**). Noncovalent interactions between complementary bases in two independent polynucleotide strands encourage the formation of nanoscaled double helixes (**c**)

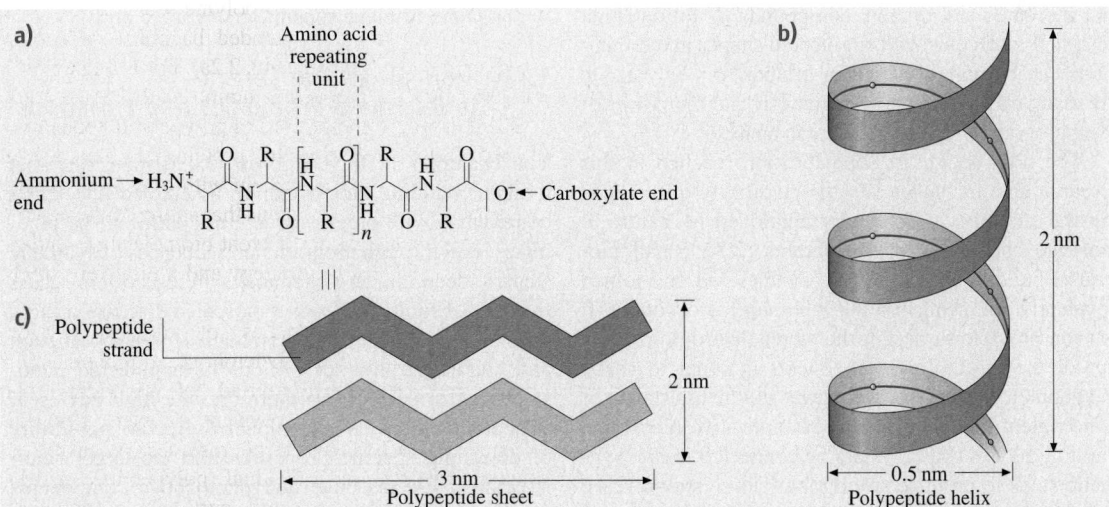

Fig. 2.2a–c A polypeptide strand (**a**) incorporates amino acid residues differing in their side chains and joined by covalent bonds. Hydrogen bonding interactions curl a single polypeptide strand into a helical arrangement (**b**) or lock pairs of strands into nanoscaled sheets (**c**)

terstrand hydrogen bonds can align from 2 up to 15 parallel or antiparallel polypeptide chains to form nanoscaled sheets with average dimensions of 2×3 nm (Fig. 2.2c). Multiple nanohelices and/or nanosheets combine into a unique three-dimensional arrangement dictating the overall shape and dimensions of a protein.

2.1.3 Chemical Synthesis of Artificial Nanostructures

Nature fabricates complex nanostructures relying on simple criteria and a relatively small pool of molecular building blocks. Robust chemical bonds join the basic components into covalent scaffolds. Noncovalent interactions determine the three-dimensional arrangement and overall shape of the resulting assemblies. The multitude of unique combinations possible for long sequences of chemically connected building blocks provides access to huge libraries of nanoscaled biomolecules.

Modern chemical synthesis has evolved considerably over the past few decades [2.2]. Experimental procedures to join molecular components with structural control at the picometer level are available. A multitude of synthetic schemes to encourage the formation of chemical bonds between selected atoms in reacting molecules have been developed. Furthermore, the tremendous progress of crystallographic and spectroscopic techniques has provided efficient and reliable tools to probe directly the structural features of artificial inorganic and organic compounds. It follows that designed molecules with engineered shapes and dimensions can be now prepared in a laboratory relying on the many tricks of chemical synthesis and the power of crystallographic and spectroscopic analyses.

The high degree of sophistication reached in this research area translates into the possibility of mimicking the strategies successfully employed by nature to fabricate chemically nanostructures [2.3]. Small molecular building blocks can be synthesized and joined covalently following routine laboratory procedures. It is even possible to design the stereoelectronic properties of the assembling components in order to shape the geometry of the final product with the assistance of noncovalent interactions. For example, five bipyridine building blocks (Fig. 2.3) can be connected in five synthetic steps to produce an oligobipyridine strand [2.4]. The five repeating units are bridged by [C−O] bonds and can chelate metal cations in the bay regions defined by their two nitrogen atoms. The spontaneous assembly of two organic strands in a double helical arrangement oc-

curs in the presence of inorganic cations. In the resulting helicate, the two oligobipyridine strands wrap around an axis defined by five Cu(I) centers. Each inorganic cation coordinates two bipyridine units with a tetrahedral geometry imposing a diameter of ca. 0.6 nm on the nanoscaled helicate [2.5]. The overall length from one end of the helicate to the other is ca. 3 nm [2.6]. The analogy between this artificial double helix and the *B*-DNA double helix shown in Fig. 2.1c is obvious. In both instances, a supramolecular glue combines two independent molecular strands into nanostructures with defined shapes and dimensions.

The chemical synthesis of nanostructures can borrow nature's design criteria as well as its molecular building blocks. Amino acids, the basic components of proteins, can be assembled into artificial macrocycles. In the example of Fig. 2.4, eight amino acid residues are joined through the formation of [C−N] bonds in multiple synthetic steps [2.7]. The resulting covalent backbone defines a circular cavity with a diameter of ca. 0.8 nm [2.8]. In analogy to the polypeptide chains of proteins, the amino acid residues of this artificial oligopeptide can sustain hydrogen bonding interactions. It follows that multiple macrocycles can pile on top of each other to form tubular nanostructures. The walls of the resulting nanotubes are maintained in position by the cooperative action of at least eight primary hydrogen bonding contacts per macrocycle. These noncovalent interactions maintain the mean planes of independent macrocycles in an approximately parallel arrangement with a plane-to-plane separation of ca. 0.5 nm.

2.1.4 From Structural Control to Designed Properties and Functions

The examples in Figs. 2.3 and 2.4 demonstrate that modular building blocks can be assembled into target compounds with precise structural control at the picometer level through programmed sequences of synthetic steps. Indeed, modern chemical synthesis offers access to complex molecules with nanoscaled dimensions and, thus, provides cost-effective strategies for the production and characterization of billions of engineered nanostructures in parallel. Furthermore, the high degree of structural control is accompanied by the possibility of designing specific properties into the target nanostructures. Electroactive and photoactive components can be integrated chemically into functional molecular machines [2.9]. Extensive electrochemical investigations have demonstrated that inorganic and organic compounds can exchange electrons with macroscopic

Nanomaterials Synthesis and Applications: Molecule–Based Devices | 2.1 Chemical Approaches to Nanostructured Materials 17

Part A | 2.1

electrodes [2.10]. These studies have unraveled the processes responsible for the oxidation and reduction of numerous functional groups and indicated viable design criteria to adjust the ability of molecules to accept or donate electrons [2.11]. Similarly, detailed photochemical and photophysical investigations have elucidated the mechanisms responsible for the absorption and emission of photons at the molecular level [2.12]. The vast knowledge established on the interactions between light and molecules offers the opportunity to engineer chromophoric and fluorophoric functional groups with defined absorption and emission properties [2.11, 13].

The power of chemical synthesis to deliver functional molecules is, perhaps, better illustrated by the molecular motor shown in Fig. 2.5. The preparation of this [2]rotaxane requires 12 synthetic steps starting from known precursors [2.14]. This complex molecule incorporates a Ru(II)-trisbipyridine stopper bridged to a linear tetracationic fragment by a rigid triaryl spacer. The other end of the tetracationic portion is terminated by a bulky tetraarylmethane stopper. The bipyridinium unit of this dumbbell-shaped compound is encircled by a macrocyclic polyether. No covalent bonds join the macrocyclic and linear components. Rather, hydrogen bonding and $[\pi \cdots \pi]$ stacking interactions maintain the macrocyclic polyether around the bipyridinium unit. In addition, mechanical constrains associated with the bulk of the two terminal stoppers prevent the macrocycle to slip off the thread. The approximate end-to-end distance for this [2]rotaxane is ca. 5 nm.

The bipyridinium and the 3,3′-dimethyl bipyridinium units within the dumbbell-shaped component undergo two consecutive and reversible monoelectronic

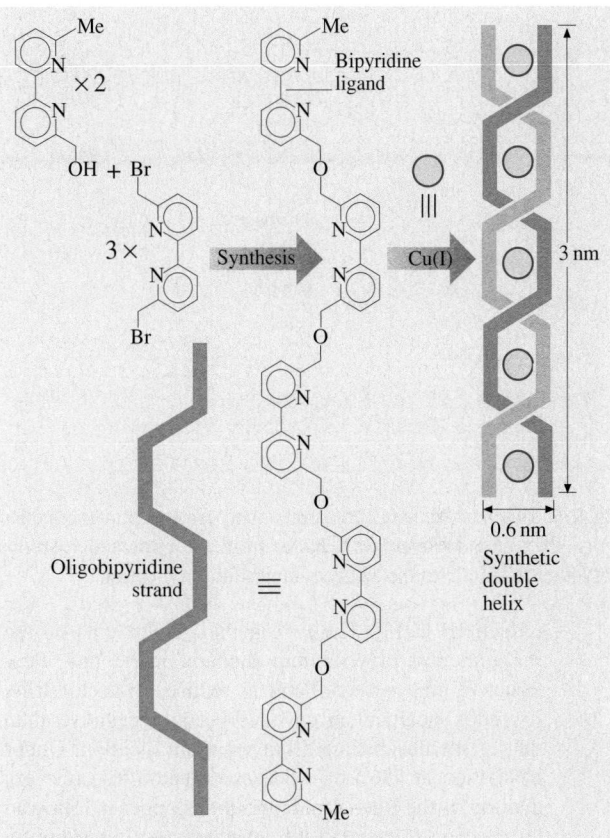

Fig. 2.3 An oligobipyridine strand can be synthesized joining five bipyridine subunits by covalent bonds. The tetrahedral coordination of pairs of bipyridine ligands by Cu(I) ions encourages the assembly two oligobipyridine strands into a double helical arrangement

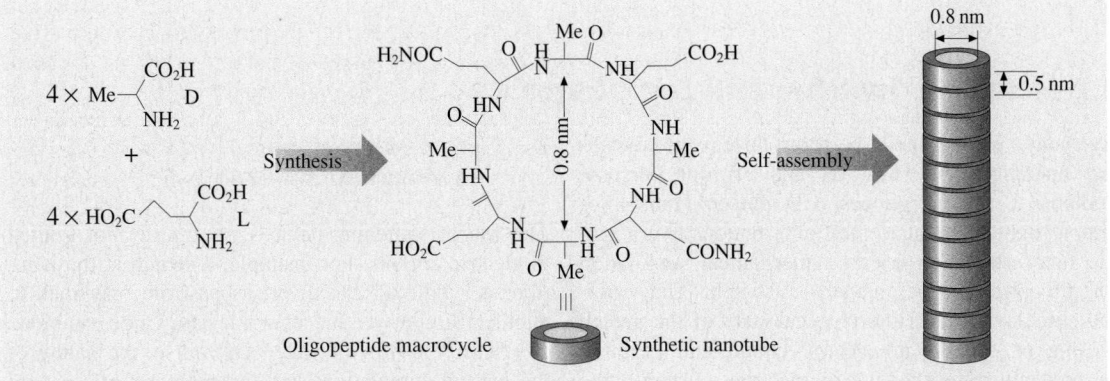

Fig. 2.4 Cyclic oligopeptides can be synthesized joining eight amino acid residues by covalent bonds. The resulting macrocycles self-assemble into nanoscaled tube-like arrays

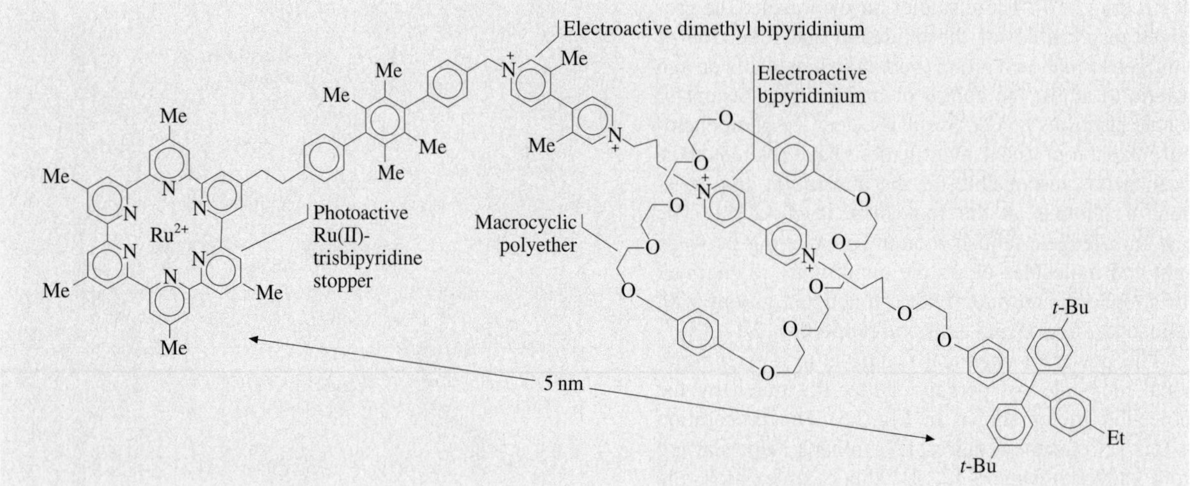

Fig. 2.5 This nanoscaled [2]rotaxane incorporates a photoactive Ru(II)-trisbipyridine stopper and two electroactive bipyridinium units. Photoinduced electron transfer from the photoactive stopper to the encircled electroactive unit forces the macrocyclic polyether to shuttle to the adjacent bipyridinium dication

reductions [2.14]. The two methyl substituents on the 3,3′-dimethyl bipyridinium dication make this electroactive unit more difficult to reduce. In acetonitrile, its redox potential is ca. 0.29 V more negative than that of the unsubstituted bipyridinium dication. Under irradiation at 436 nm in degassed acetonitrile, the excitation of the Ru(II)-trisbipyridine stopper is followed by electron transfer to the unsubstituted bipyridinium unit. In the presence of a sacrificial electron donor (triethanolamine) in solution, the photogenerated hole in the photoactive stopper is filled, and undesired back electron transfer is suppressed. The permanent and light-induced reduction of the dicationic bipyridinium unit to a radical cation depresses significantly the magnitude of the noncovalent interactions holding the macrocyclic polyether in position. As a result, the macrocycle shuttles from the reduced unit to the adjacent dicationic 3,3′-dimethyl bipyridinum. After the diffusion of molecular oxygen into the acetonitrile solution, oxidation occurs restoring the dicationic form of the bipyridinium unit and its ability to sustain strong noncovalent bonds. As a result, the macrocyclic polyether shuttles back to its original position. This amazing example of a molecular shuttle reveals that dynamic processes can be controlled reversibly at the molecular level relying on the clever integration of electroactive and photoactive fragments into functional and nanoscaled molecules.

2.2 Molecular Switches and Logic Gates

Everyday, we routinely perform dozens of switching operations. We turn on and off our personal computers, cellular phones, CD players, radios, or simple light bulbs at a click of a button. Every single time, our finger exerts a mechanical stimulation on a control device, namely a switch. The external stimulus changes the physical state of the switch closing or opening an electric circuit and enabling or preventing the passage of electrons. Overall, the switch transduces a mechanical input into an electrical output.

2.2.1 From Macroscopic to Molecular Switches

The use of switching devices is certainly not limited to electric circuits. For example, a switch at the junction of a railroad can divert trains from one track to another. Similarly, a faucet in a lavatory pipe can block or release the flow of water. Of course, the nature of the control stimulations and the character of the final outcome vary significantly from case to case, but the operating principle behind each switching device is the same. In all cases, input stimulations reach the switch changing its physical state and producing a specific output.

The development of nanoscaled counterparts to conventional switches is expected to have fundamental scientific and technological implications. For instance, one can envisage practical applications for ultraminiaturized switches in areas ranging from biomedical research to information technology. The major challenge in the quest for nanoswitches, however, is the identification of reliable design criteria and operating principles for these innovative and fascinating devices. Chemical approaches to implement molecule-sized switches appear to be extremely promising. The intrinsically small dimensions of organic molecules coupled with the power of chemical synthesis are the main driving forces behind these exploratory investigations.

Certain organic molecules adjust their structural and electronic properties when stimulated with chemical, electrical, or optical inputs. Generally, the change is accompanied by an electrochemical or spectroscopic response. Overall, these nanostructures transduce input stimulations into detectable outputs and, appropriately, are called molecular switches [2.15, 16]. The chemical transformations associated with these switching processes are often reversible. The chemical system returns to the original state when the input signal is turned off. The interconverting states of a molecular switch can be isomers, an acid and its conjugated base, the oxidized and reduced forms of a redox active molecule, or even the complexed and uncomplexed forms of a receptor [2.9, 13, 15, 16]. The output of a molecular switch can be a chemical, electrical, and/or optical signal that varies in intensity with the interconversion process. For example, changes in absorbance, fluorescence, pH, or redox potential can accompany the reversible transformation of a molecular switch.

2.2.2 Digital Processing and Molecular Logic Gates

In present computer networks, data are elaborated electronically by microprocessor systems [2.17] and are exchanged optically between remote locations [2.18]. Data processing and communication require the encoding of information in electrical and optical signals in the form of binary digits. Using arbitrary assumptions, logic thresholds can be established for each signal and, then, 0 and 1 digits can be encoded following simple conventions. Sequences of electronic devices manipulate the encoded bits executing logic functions as a result of basic switching operations.

The three basic AND, NOT, and OR operators combine binary inputs into binary outputs following precise logic protocols [2.17]. The NOT operator converts an input signal into an output signal. When the input is 0, the output is 1. When the input is 1, the output is 0. Because of the inverse relationship between the input and output values, the NOT gate is often called "inverter" [2.19]. The OR operator combines two input signals into a single output signal. When one or both inputs are 1, the output is 1. When both inputs are 0, the output is 0. The AND gate also combines two input signals into one output signal. In this instance, however, the output is 1 only when both inputs are 1. When at least one input is 0, the output is 0.

The output of one gate can be connected to one of the inputs of another operator. A NAND gate, for example, is assembled connecting the output of an AND operator to the input of a NOT gate. Now the two input signals are converted into the final output after two consecutive logic operations. In a similar fashion, a NOR gate can be assembled connecting the output of an OR operator to the input of a NOT gate. Once again, two consecutive logic operations determine the relation between two input signals and a single output. The NAND and NOR operations are termed universal functions because any conceivable logic operation can be implemented relying only on one of these two gates [2.17]. In fact, digital circuits are fabricated routinely interconnecting exclusively NAND or exclusively NOR operators [2.19].

The logic gates of conventional microprocessors are assembled interconnecting transistors, and their input and output signals are electrical [2.19]. But the concepts of binary logic can be extended to chemical, mechanical, optical, pneumatic, or any other type of signal. First it is necessary to design devices that can respond to these stimulations in the same way transistors respond to electrical signals. Molecular switches respond to a variety of input stimulations producing specific outputs and can, therefore, be exploited to implement logic functions [2.13, 20, 21].

2.2.3 Molecular AND, NOT, and OR Gates

More than a decade ago, researchers proposed a potential strategy to execute logic operations at the molecular level [2.22]. Later, the analogy between molecular switches and logic gates was recognized in a seminal article [2.23], in which it was demonstrated that AND, NOT, and OR operations can be reproduced with fluorescent molecules. The pyrazole derivative **1** (Fig. 2.6) is a molecular NOT gate. It imposes an inverse relation between a chemical input (concentration of H^+) and an optical output (emission intensity). In a mixture of methanol and water, the fluorescence quantum yield of **1** is 0.13 in the presence of only 0.1 equivalents of H^+ [2.23]. The quantum yield drops to 0.003 when the equivalents of H^+ are 1000. Photoinduced electron transfer from the central pyrazoline unit to the pendant benzoic acid quenches the fluorescence of the protonated form. Thus, a change in H^+ concentration (I) from a low to a high value switches the emission intensity (O) from a high to a low value. The inverse relationship between the chemical input I and the optical output O translates into the truth table of a NOT operation if a positive logic convention (low = 0, high = 1) is applied to both signals. The emission intensity is high (O = 1) when the concentration of H^+ is low (I = 0). The emission intensity is low (O = 0) when the concentration of H^+ is high (I = 1).

The anthracene derivative **2** (Fig. 2.6) is a molecular OR gate. It transduces two chemical inputs (concentrations of Na^+ and K^+) into an optical output (emission intensity). In methanol, the fluorescence quantum yield is only 0.003 in the absence of metal cations [2.23]. Photoinduced electron transfer from the nitrogen atom of the azacrown fragment to the anthracene fluorophore quenches the emission. After the addition of 1000 equivalents of either Na^+ or K^+, the quantum yield raises to 0.053 and 0.14, respectively. Similarly, the quantum yield is 0.14 when both metal cations are present in solution. The complexation of one of the two metal cations inside the azacrown receptor depresses the efficiency of the photoinduced electron transfer enhancing the fluorescence. Thus, changes in the concentrations of Na^+ (I1) and/or

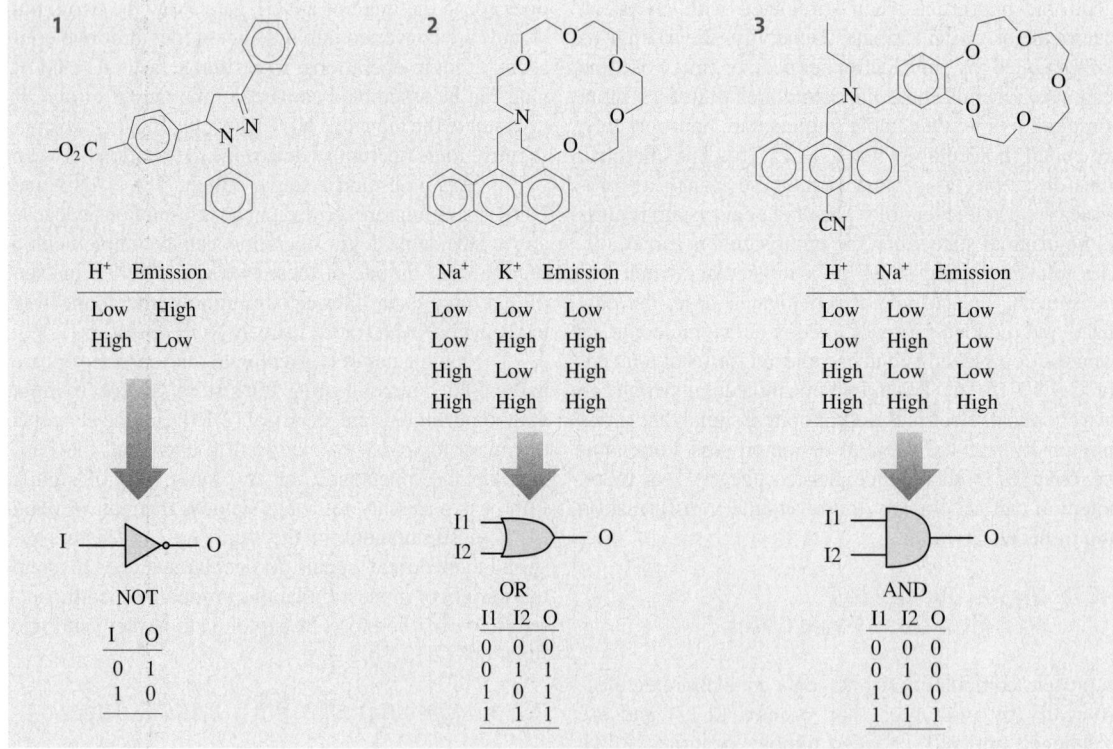

Fig. 2.6 The fluorescence intensity of the pyrazoline derivative **1** is high when the concentration of H^+ is low, and vice versa. The fluorescence intensity of the anthracene derivative **2** is high when the concentration of Na^+ and/or K^+ is high. The emission is low when both concentrations are low. The fluorescence intensity of the anthracene **3** is high only when the concentrations of H^+ and Na^+ are high. The emission is low in the other three cases. The signal transductions of the molecular switches **1**, **2**, and **3** translate into the truth tables of NOT, OR, and AND gates, respectively, if a positive logic convention is applied to all inputs and outputs (low = 0, high = 1)

K^+ (I2) from low to high values switch the emission intensity (O) from a low to a high value. The relationship between the chemical inputs I1 and I2 and the optical output O translates into the truth table of an OR operation if a positive logic convention (low = 0, high = 1) is applied to all signals. The emission intensity is low (O = 0) only when the concentration of Na^+ and K^+ are low (I1 = 0, I2 = 0). The emission intensity is high (O = 1) for the other three input combinations.

The anthracene derivative **3** (Fig. 2.6) is a molecular AND gate. It transduces two chemical inputs (concentrations of H^+ and Na^+) into an optical output (emission intensity). In a mixture of methanol and *iso*-propanol, the fluorescence quantum yield is only 0.011 in the absence of H^+ or Na^+ [2.23]. Photoinduced electron transfer from either the tertiary amino group or the catechol fragment to the anthracene fluorophore quenches the emission. After the addition of either 100 equivalents of H^+ or 1000 equivalents of Na^+, a modest change of the quantum yield to 0.020 and 0.011, respectively, is observed. Instead, the quantum yield increases to 0.068 when both species are present in solution. The protonation of the amino group and the insertion of the metal cation in the benzocrown ether receptor depress the efficiency of the photoinduced electron transfer processes enhancing the fluorescence. Thus, changes in the concentrations of H^+ (I1) and Na^+ (I2) from low to high values switch the emission intensity (O) from a low to a high value. The relationship between the chemical inputs I1 and I2 and the optical output O translates into the truth table of an AND operation if a positive logic convention (low = 0, high = 1) is applied to all signals. The emission intensity is high (O = 1) only when the concentration of H^+ and Na^+ are high (I1 = 1, I2 = 1). The emission intensity is low (O = 0) for the other three input combinations.

2.2.4 Combinational Logic at the Molecular Level

The fascinating molecular AND, NOT, and OR gates illustrated in Fig. 2.6 have stimulated the design of related chemical systems able to execute the three basic logic operations and simple combinations of them [2.13, 20, 21]. Most of these molecular switches convert chemical inputs into optical outputs. But the implementation of logic operations at the molecular level is not limited to the use of chemical inputs. For example, electrical signals and reversible redox processes can be exploited to modulate the output of a molecular

switch [2.24]. The supramolecular assembly **4** (Fig. 2.7) executes a XNOR function relying on these operating principles. The π-electron rich tetrathiafulvalene (TTF) guest threads the cavity of a π-electron deficient bipyridinium (BIPY) host. In acetonitrile, an absorption band associated with the charge-transfer interactions between the complementary π-surfaces is observed at 830 nm. Electrical stimulations alter the redox state of either the TTF or the BIPY units encouraging the separation of the two components of the complex and the disappearance of the charge-transfer band. Electrolysis at a potential of $+0.5$ V oxidizes the neutral TTF unit to a monocationic state. The now cationic guest is expelled from the cavity of the tetracationic host as a result of electrostatic repulsion. Consistently, the absorption band at 830 nm disappears. The charge-transfer band, however, is restored after the exhaustive back reduction of the TTF unit at a potential of 0 V. Similar changes in the absorption properties can be induced addressing the BIPY units. Electrolysis at -0.3 V reduces the dicationic BIPY units to their monocationic forms encouraging the separation of the two components of the complex and the disappearance of the absorption band. The original absorption spectrum is restored after the exhaustive back oxidation of the BIPY units at a potential of 0 V. Thus, this supramolecular system responds to electrical stimulations producing an optical output. One of the electrical inputs (I1) controls the redox state of the TTF unit switching between 0 and $+0.5$ V. The other (I2) determines the redox state of the bipyridinium units switching between -0.3 and 0 V. The optical output (O) is the absorbance of the charge-transfer band. A positive logic convention (low = 0, high = 1) can be applied to the input I1 and output O. A negative logic convention (low = 1, high = 0) can be applied to the input I2. The resulting truth table corresponds to that of a XNOR circuit (Fig. 2.7). The charge-transfer absorbance is high (O = 1) only when one voltage input is low and the other is high (I1 = 0, I2 = 0) or vice versa (I1 = 1, I2 = 1). It is important to note that the input string with both I1 and I2 equal to 1 implies that input potentials of $+0.5$ and -0.3 V are applied simultaneously to a solution containing the supramolecular assembly **4** and not to an individual complex. Of course, the concomitant oxidation of the TTF guest and reduction of the BIPY units in the very same complex would be unrealistic. In bulk solution, instead, some complexes are oxidized while others are reduced, leaving the average solution composition unaffected. Thus, the XNOR operation executed by this supramolecular system is a consequence

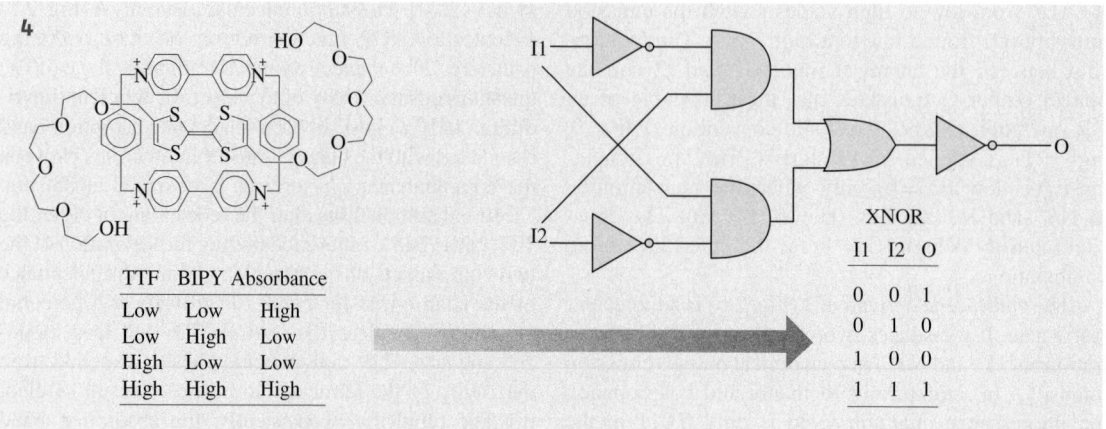

Fig. 2.7 The charge-transfer absorbance of the complex **4** is high when the voltage input addressing the tetrathiafulvalene (TTF) unit is low and that stimulating the bipyridinium (BIPY) units is high and vice versa. If a positive logic convention is applied to the TTF input and to the absorbance output (low = 0, high = 1) while a negative logic convention is applied to the BIPY input (low = 0, high = 1), the signal transduction of **4** translates into the truth table of a XNOR circuit

of bulk properties and not a result of unimolecular signal transduction.

Optical inputs can be employed to operate the three-state molecular switch of Fig. 2.8 in acetonitrile solution [2.25]. This chemical system responds to three inputs producing two outputs. The three input stimulations are ultraviolet light (I1), visible light (I2), and the concentration of H^+ (I3). One of the two optical outputs is the absorbance at 401 nm (O1), which is high when the molecular switch is in the yellow-green state **6** and low in the other two cases. The other optical output is the absorbance at 563 nm (O2), which is high when the molecular switch is in the purple state **7** and low in the other two cases. The colorless spiropyran state **5** switches to the merocyanine form **7** upon irradiation with ultraviolet light. It switches to the protonated merocyanine from **6** when treated with H^+. The colored state **7** isomerizes back to **5** in the dark or upon irradiation with visible light. Alternatively, **7** switches to **6** when treated with H^+. The colored state **6** switches to **5**, when irradiated with visible light, and to **7**, after the removal of H^+. In summary, this three-state molecular switch responds to two optical inputs (I1 and I2) and one chemical input (I3) producing two optical outputs (O1 and O2). Binary digits can be encoded on each signal applying positive logic conventions (low = 0, high = 1). It follows that the three-state molecular switch converts input strings of three binary digits into output strings of two binary digits. The corresponding truth table (Fig. 2.8) reveals that the optical output O1 is high (O1 = 1) when only the input I3 is applied (I1 = 0, I2 = 0, I3 = 1), when only the input I2 is

not applied (I1 = 1, I2 = 0, I3 = 0), or when all three inputs are applied (I1 = 1, I2 = 0, I3 = 0). The optical output O2 is high (O2 = 1) when only the input I1 is applied (I1 = 1, I2 = 0, I3 = 0) or when only the input I3 is not applied (I1 = 1, I2 = 0, I3 = 0). The combinational logic circuit (Fig. 2.8) equivalent to this truth table shows that all three inputs determine the output O1, while only I1 and I3 control the value of O2.

2.2.5 Intermolecular Communication

The combinational logic circuits in Figs. 2.7 and 2.8 are arrays of interconnected AND, NOT, and OR operators. The digital communication between these basic logic elements ensures the execution of a sequence of simple logic operations that results in the complex logic function processed by the entire circuit. It follows that the logic function of a given circuit can be adjusted altering the number and type of basic gates and their interconnection protocol [2.17]. This modular approach to combinational logic circuits is extremely powerful. Any logic function can be implemented connecting the appropriate combination of simple AND, NOT, and OR gates.

The strategies followed so far to implement complex logic functions with molecular switches are based on the careful design of the chemical system and on the judicious choice of the inputs and outputs [2.13, 20, 21]. A specific sequence of AND, NOT, and OR operations is programmed in a single molecular switch. No digital communication between distinct gates is needed since

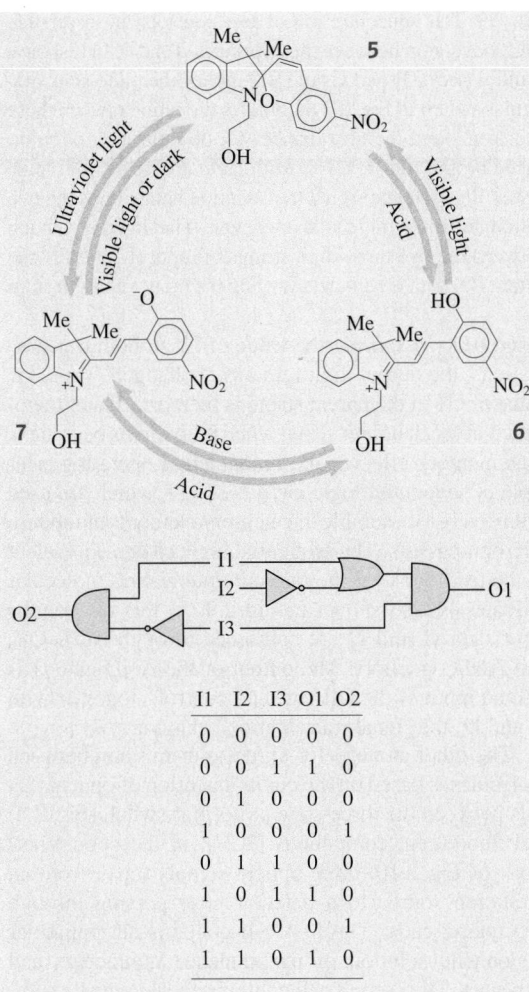

Fig. 2.8 Ultraviolet light (I1), visible light (I2), and H^+ (I3) inputs induce the interconversion between the three states **5**, **6**, and **7**. The colorless state **5** does not absorb in the visible region. The yellow-green state **6** absorbs at 401 nm (O1). The purple state **7** absorbs at 563 nm (O2). The truth table illustrates the conversion of input strings of three binary digits (I1, I2, and I3) into output strings of two binary digits (O1 and O2) operated by this three-state molecular switch. A combinational logic circuit incorporating nine AND, NOT, and OR operators correspond to this particular truth table

I1	I2	I3	O1	O2
0	0	0	0	0
0	0	1	1	0
0	1	0	0	0
1	0	0	0	1
0	1	1	0	0
1	0	1	1	0
1	1	0	0	1
1	1	0	1	0

ized. In addition, the degree of complexity that can be achieved with only one molecular switch is fairly limited. The connection of the input and output terminals of independent molecular AND, NOT, and OR operators, instead, would offer the possibility of assembling any combinational logic circuit from three basic building blocks.

In digital electronics, the communication between two logic gates can be realized connecting their terminals with a wire [2.19]. Methods to transmit binary data between distinct molecular switches are not so obvious and must be identified. Recently we developed two strategies to communicate signals between compatible molecular components. In one instance, a chemical signal is communicated between two distinct molecular switches [2.26]. They are the three-state switch illustrated in Fig. 2.8 and the two-state switch of Fig. 2.9. The merocyanine form **7** is a photogenerated base. Its *p*-nitrophenolate fragment, produced upon irradiation of the colorless state **5** with ultraviolet light, can abstract a proton from an acid present in the same solution. The resulting protonated form **6** is a photoacid. It releases a proton upon irradiation with visible light and can protonate a base co-dissolved in the same medium. The orange azopyridine **8** switches to the red-purple azopyridinium **9** upon protonation. This process is reversible, and the addition of a base restores the orange state **8**. It follows that photoinduced proton transfer can be exploited to communicate a chemical signal from **6** to **8** and from **9** to **7**. The two colored states **8** and **9** have different absorption properties in the visible region. In acetonitrile, the orange state **8** absorbs at 422 nm, and the red-purple state **9** absorbs at 556 nm. The changes in absorbance of these two bands can be exploited to monitor the photoinduced exchange of protons between the two communicating molecular switches.

The three-state molecular switch and the two-state molecular switch can be operated sequentially when dissolved in the same acetonitrile solution. In the presence of one equivalent of H^+, the two-state molecular switch is in state **9** and the absorbance at 556 nm is high (O = 1). Upon irradiation with ultraviolet light (I1 = 0), **5** switches to **7**. The photogenerated base deprotonates **9** producing **8** and **6**. As a result, the absorbance at 556 nm decreases (O = 0). Upon irradiation with visible light (I2 = 1), **6** switches to **5** releasing H^+. The result is the protonation of **8** to form **9** and restore the high absorbance at 556 nm (O = 1). In summary, the three-state molecular switch transduces two optical inputs (I1 = ultraviolet light, I2 = visible light) into a chemical signal (proton transfer) that is communicated to the

they are built in the same molecular entity. Though extremely elegant, this strategy does not have the same versatility of a modular approach. A different molecule has to be designed, synthesized, and analyzed every single time a different logic function has to be real-

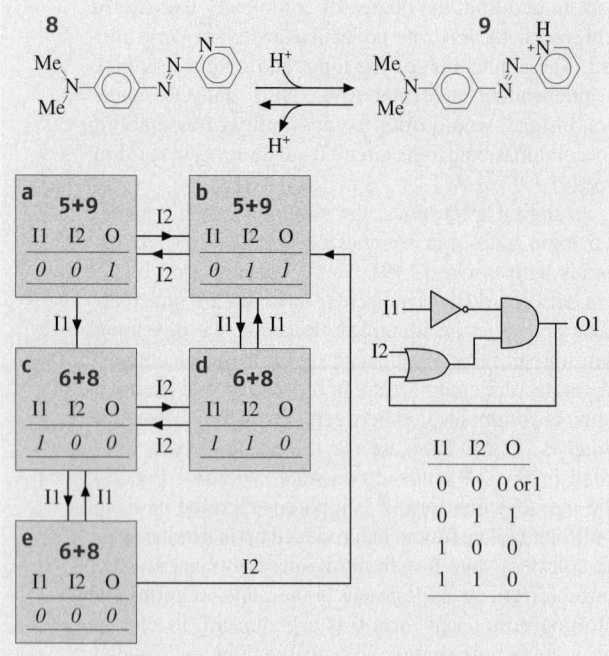

Fig. 2.9 The concentration of H$^+$ controls the reversible interconversion between the two states **8** and **9**. In response to ultraviolet (I1) and visible (I2) inputs, the three-state molecular switch in Fig. 2.7 modulates the ratio between these two forms and the absorbance (O) of **9** through photoinduced proton transfer. The truth table and sequential logic circuit illustrate the signal transduction behavior of the two communicating molecular switches. The interconversion between the five three-digit strings of input (I1 and I2) and output (O) data is achieved varying the input values in steps

string 011 (**b**) varying the value of I2. In both transformations, the output digit remains unchanged. Thus, the value of O1 in the parent string is memorized and maintained in the daughter string when both inputs become 0. This memory effect is the fundamental operating principle of sequential logic circuits [2.17], which are used extensively to assemble the memory elements of modern microprocessors. The sequential logic circuit equivalent to the truth table of the two communicating molecular switches is also shown in Fig. 2.9. In this circuit, the input data I1 and I2 are combined through NOT, OR, and AND operators. The output of the AND gate O is also an input of the OR gate and controls, together with I1 and I2, the signal transduction behavior.

The other strategy for digital transmission between molecules is based on the communication of optical signals between the three-state molecular switch (Fig. 2.8) and fluorescent compounds [2.27]. In the optical network of Fig. 2.10, three optical signals travel from an excitation source to a detector after passing through two quartz cells. The first cell contains an equimolar acetonitrile solution of naphthalene, anthracene, and tetracene. The second cell contains an acetonitrile solution of the three-state molecular switch. The excitation source sends three consecutive monochromatic light beams to the first cell stimulating the emission of the three fluorophores. The light emitted in the direction perpendicular to the exciting beam reaches the second cell. When the molecular switch is in state **5**, the naphthalene emission at 335 nm is absorbed and a low intensity output (O1) reaches the detector. Instead, the anthracene and tetracene emissions at 401 and 544 nm, respectively, pass unaffected and high intensity outputs (O2 and O3) reach the detector. When the molecular switch is in state **6**, the naphthalene and anthracene emissions are absorbed and only the tetracene emission reaches the detector (O1 = 0, O2 = 0, O3 = 1). When the molecular switch is state **7**, the emission of all three fluorophores is absorbed (O1 = 0, O2 = 0, O3 = 0). The interconversion of the molecular switch between the three states is

two-state molecular switch and converted into a final optical output (O = absorbance at 556 nm).

The logic behavior of the two communicating molecular switches is significantly different from those of the chemical systems illustrated in Figs. 2.6, 2.7, and 2.8 [2.26]. The truth table in Fig. 2.9 lists the four possible combinations of two-digit input strings and the corresponding one-digit output. The output digit O for the input strings 01, 10, and 11 can take only one value. In fact, the input string 01 is transduced into a 1, and the input strings 10 and 11 are converted into 0. Instead, the output digit O for the input string 00 can be either 0 or 1. The sequence of events leading to the input string 00 determines the value of the output. The boxes **a–e** in Fig. 2.9 illustrates this effect. They correspond to the five three-digit input/output strings. The transformation of one box into any of the other four is achieved in one or two steps by changing the values of I1 and/or I2. In two instances (**a** and **b**), the two-state molecular switch is in state **9**, and the output signal is high (O = 1). In the other three cases (**c**, **d**, and **e**), the two-state molecular switch is in state **8**, and the output signal is low (O = 0). The strings 000 (**e**) and 001 (**a**) correspond to the first entry of the truth table. They share the same input digits but differ in the output value. The string 000 (**e**) can be obtained only from the string 100 (**c**) varying the value of I1. Similarly, the string 001 (**a**) can be accessed only from the

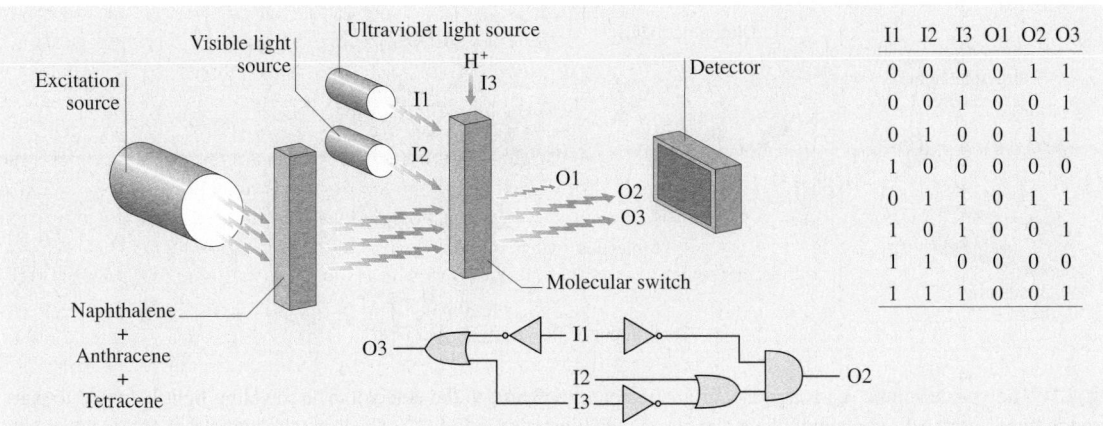

I1	I2	I3	O1	O2	O3
0	0	0	0	1	1
0	0	1	0	0	1
0	1	0	0	1	1
1	0	0	0	0	0
0	1	1	0	1	1
1	0	1	0	0	1
1	1	0	0	0	0
1	1	1	0	0	1

Fig. 2.10 The excitation source sends three monochromatic light beams (275, 357, and 441 nm) to a quartz cell containing an equimolar acetonitrile solution of naphthalene, anthracene and tetracene. The three fluorophores absorb the exciting beams and reemit at 305, 401, and 544 nm, respectively. The light emitted in the direction perpendicular to the exciting beams passes through another quartz cell containing an acetonitrile solution of the three-state molecular switch shown in Fig. 2.7. Ultraviolet (I1), visible (I2), and H$^+$ (I3) inputs control the interconversion between the three states of the molecular switch. They determine the intensity of the optical outputs reaching the detector and correspond to the naphthalene (O1), anthracene (O2), and tetracene (O3) emissions. The truth table and equivalent combinational logic illustrate the relation between the three inputs and the three outputs. The output O1 is always 0, and it is not influenced by the three inputs. Only two inputs determine the value of O3, while all of them control the output O2

induced addressing the second cell with ultraviolet (I1), visible (I2) and H$^+$ (I3) inputs. Thus, three independent optical outputs (O1, O2 and O3) can be modulated stimulating the molecular switch with two optical and one chemical input. The truth table in Fig. 2.10 illustrates the relation between the three inputs (I1, I2 and I3) and the three outputs (O1, O2 and O3), when positive logic conventions are applied to all signals. The equivalent logic circuit shows that all three inputs control the anthracene channel O2, but only I1 and I3 influence the tetracene channel O3. Instead, the intensity of the naphthalene channel O1 is always low, and it is not affected by the three inputs.

The operating principles of the optical network in Fig. 2.10 can be simplified to implement all-optical logic gates. The chemical input inducing the formation of the protonated form **6** of the molecular switch can be eliminated. The interconversion between the remaining two states **5** and **7** can be controlled relying exclusively on ultraviolet inputs. Indeed, ultraviolet irradiation induces the isomerization of the colorless form **5** to the colored species **7**, which reisomerizes to the original state in the dark. Thus, a single ultraviolet source is sufficient to control the switching from **5** to **7** and vice versa. On the basis of these considerations, all-optical NAND, NOR, and NOT gates can be imple-

mented operating sequentially or in parallel from one to three independent switching elements [2.28]. For example, the all-optical network illustrated in Fig. 2.11 is a three-input NOR gate. A monochromatic optical signal travels from a visible source to a detector. Three switching elements are aligned along the path of the traveling light. They are quartz cells containing an acetonitrile solution of the molecular switch shown in Fig. 2.8. The interconversion of the colorless form **5** into the purple isomer **7** is induced stimulating the cell with an ultraviolet input. The reisomerization from **7** to **5** occurs spontaneously, as the ultraviolet sources is turned off. Using three distinct ultraviolet sources, the three switching elements can be controlled independently.

The colorless form **5** does not absorb in the visible region, while the purple isomer **7** has a strong absorption band at 563 nm. Thus, a 563 nm optical signal leaving the visible source can reach the detector unaffected only if all three switching elements are in the nonabsorbing state **5**. If one of the three ultraviolet inputs I1, I2, or I3 is turned on, the intensity of the optical output O drops to 3–4% of its original value. If two or three ultraviolet inputs are turned on simultaneously, the optical output drops to 0%. Indeed, the photogenerated state **7** absorbs and blocks the traveling light. Applying positive logic

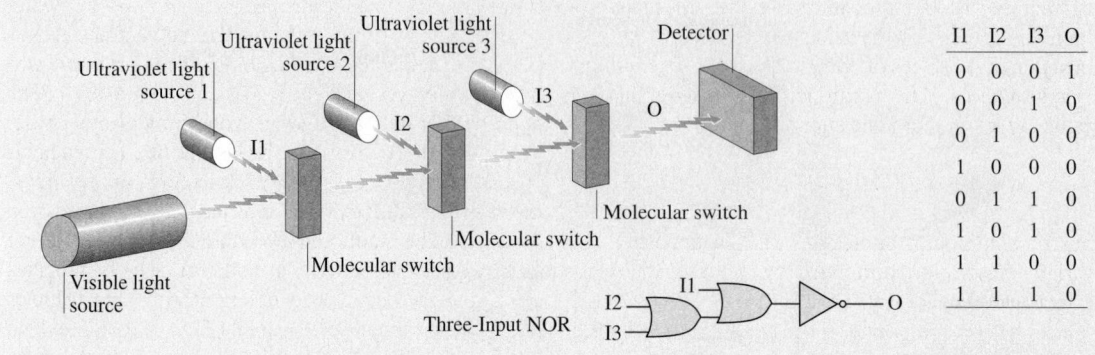

I1	I2	I3	O
0	0	0	1
0	0	1	0
0	1	0	0
1	0	0	0
0	1	1	0
1	0	1	0
1	1	0	0
1	1	1	0

Fig. 2.11 The visible source sends a monochromatic beam (563 nm) to the detector. The traveling light is forced to pass through three quartz cells containing the molecular switch illustrated in Fig. 2.7. The three switching elements are operated by independent ultraviolet inputs. When at least one of them is on, the associated molecular switch is in the purple form **7**, which can absorb and block the traveling light. The truth table and equivalent logic circuit illustrate the relation between the three inputs I1, I2, and I3 and the optical output O

conventions to all signals, binary digits can be encoded in the three optical inputs and in the optical output. The resulting truth table is illustrated in Fig. 2.11. The output O is 1 only if all three inputs I1, I2, or I3 are 0.

The output O is 0 if at least one of the three inputs I1, I2, or I3 is 1. This signal transduction corresponds to that executed by a three-input NOR gate, which is a combination of one NOT and two OR operators.

2.3 Solid State Devices

The fascinating chemical systems illustrated in Figs. 2.6–2.11 demonstrate that logic functions can be implemented relying on the interplay between designed molecules and chemical, electrical and/or optical signals [2.13, 20, 21].

2.3.1 From Functional Solutions to Electroactive and Photoactive Solids

These molecular switches, however, are operated exclusively in solution and remain far from potential applications in information technology at this stage. The integration of liquid components and volatile organic solvents in practical digital devices is hard to envisage. Furthermore, the logic operations executed by these chemical systems rely on bulk addressing. Although the individual molecular components have nanoscaled dimensions, macroscopic collections of them are employed for digital processing. In some instances, the operating principles cannot even be scaled down to the unimolecular level. Often bulk properties are responsible for signal transduction. For example,

a single fluorescent compound **2** cannot execute an OR operation. Its azacrown appendage can accommodate only one metal cation. As a result, an individual molecular switch can respond to only one of the two chemical inputs. It is a collection of numerous molecular switches dissolved in an organic solvent that responds to both inputs enabling an OR operation.

The development of miniaturized molecule-based devices requires the identification of methods to transfer the switching mechanisms developed in solution to the solid state [2.29]. Borrowing designs and fabrication strategies from conventional electronics, researchers are starting to explore the integration of molecular components into functional circuits and devices [2.30–33]. Generally, these strategies combine lithography and surface chemistry to assemble nanometer-thick organic films on the surfaces of microscaled or nanoscaled electrodes. Two main approaches for the deposition of organized molecular arrays on inorganic supports have emerged so far. In one instance, amphiphilic molecular building blocks are compressed into organized monolayers at air/water interfaces. The resulting films can be transferred on supporting solids employing the Langmuir–Blodgett technique [2.34]. Alternatively, certain molecules can be designed to adsorb spontaneously on the surfaces of compatible solids from liquid or vapor phases. The result is the self-assembly of organic layers on inorganic supports [2.35].

2.3.2 Langmuir–Blodgett Films

Films of amphiphilic molecules can be deposited on a variety of solid supports employing the Langmuir–Blodgett technique [2.34]. This method can be extended to electroactive compounds incorporating hydrophilic and hydrophobic groups. For example, the amphiphile **10** (Fig. 2.12) has a hydrophobic hexadecyl tail attached to a hydrophilic bipyridinium dication [2.36, 37]. This compound dissolves in mixtures of chloroform and methanol, but it is not soluble in moderately concentrated aqueous solutions of sodium perchlorate. Thus the spreading of an organic solution of **10** on an aqueous sodium perchlorate subphase affords a collec-

tion of disorganized amphiphiles floating on the water surface (Fig. 2.12), after the organic solvent has evaporated. The molecular building blocks can be compressed into a monolayer with the aid of a moving barrier. The hydrophobic tails align away from the aqueous phase. The hydrophilic dicationic heads and the accompanying perchlorate counterions pack to form an organized monolayer at the air/water interface. The compression process can be monitored recording the surface pressure (π)-area per molecule (A) isotherm, which indicates a limiting molecular area of ca. $50\,\text{Å}^2$. This value is larger than the projected area of an oligomethylene chain. It correlates reasonably, however, with the overall area of a bipyridinium dication plus two perchlorate anions.

The monolayer prepared at the air/water interface (Fig. 2.12) can be transferred on the surface of a indium-tin oxide electrode pre-immersed in the aqueous phase. The slow lifting of the solid support drags the monolayer away from the aqueous subphase. The final result

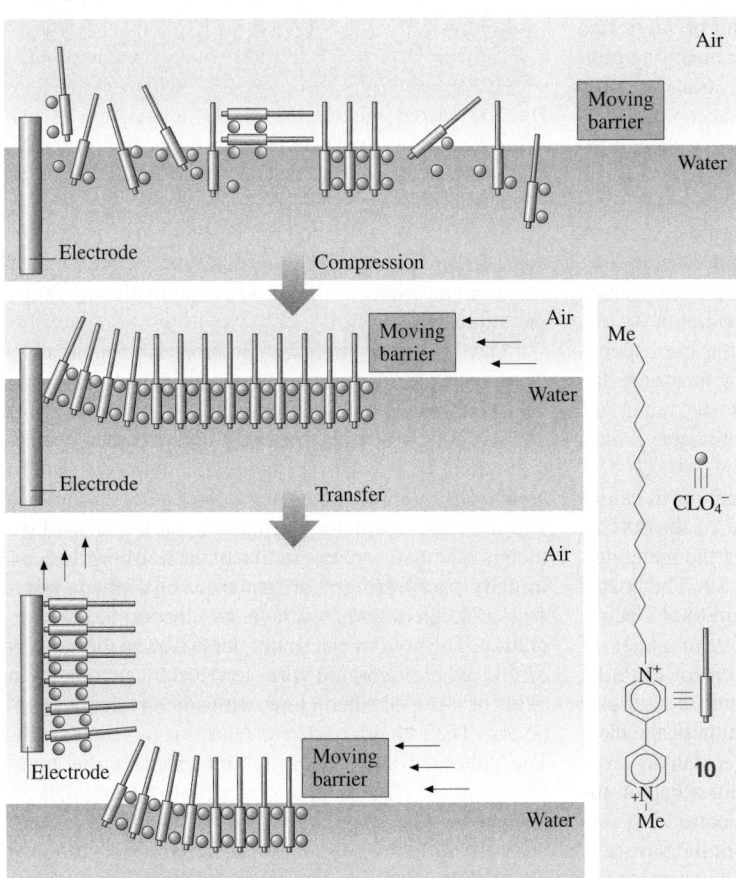

Fig. 2.12 The compression of the amphiphilic dication **10** with a moving barrier results in the formation of a packed monolayer at the air/water interface. The lifting of an electrode pre-immersed in the aqueous subphase encourages the transfer of part of the monolayer on the solid support

is the coating of the electrode with an organic film containing electroactive bipyridinium building blocks. The modified electrode can be integrated in a conventional electrochemical cell to probe the redox response of the electroactive layer. The resulting cyclic voltammograms reveal the characteristic waves for the first reduction process of the bipyridinium dications, confirming the successful transfer of the electroactive amphiphiles from the air/water interface to the electrode surface. The integration of the redox waves indicates a surface coverage of ca. 4×10^{10} mol cm^{-2}. This value corresponds to a molecular area of ca. $40 \text{\AA}^2$ and is in excellent agreement with the limiting molecular area of the $\pi-A$ isotherm.

These seminal experiments demonstrate that electroactive amphiphiles can be organized into uniform monolayers at the air/water interface and then transferred efficiently on the surface of appropriate substrates to produce electrode/monolayer junctions. The resulting electroactive materials can become the functional components of molecule-based devices. For example, bipyridinium-based photodiodes can be fabricated following this approach [2.38, 39]. Their operating principles rely on photoinduced electron transfer from chromophoric units to bipyridinium acceptors. The electroactive and photoactive amphiphile **11** (Fig. 2.13) incorporates hydrophobic ferrocene and pyrene tails and a hydrophilic bipyridinium head. Chloroform solutions of **11** containing ten equivalents of arachidic acid can be spread on an aqueous calcium chloride subphase in a Langmuir trough. The amphiphiles can be compressed into a mixed monolayer, after the evaporation of the organic solvent. Pronounced steps in the corresponding $\pi-A$ isotherm suggest that the bulky ferrocene and pyrene groups are squeezed away from the water surface. In the final arrangement, both photoactive groups align above the hydrophobic dication.

A mixed monolayer of **11** and arachidic acid can be transferred from the air/water interface to the surface of a transparent gold electrode following the methodology illustrated for the system in Fig. 2.12. The coated electrode can be integrated in a conventional electrochemical cell. Upon irradiation at 330 nm under an inert atmosphere, an anodic photocurrent of ca. 2 nA develops at a potential of 0 V relative to a saturated calomel electrode. Indeed, the illumination of the electroactive monolayer induces the electron transfer from the pyrene appendage to the bipyridinium acceptor and then from the reduced acceptor to the electrode. A second intramolecular electron transfer from the ferrocene donor to the oxidized pyrene fills its photogenerated

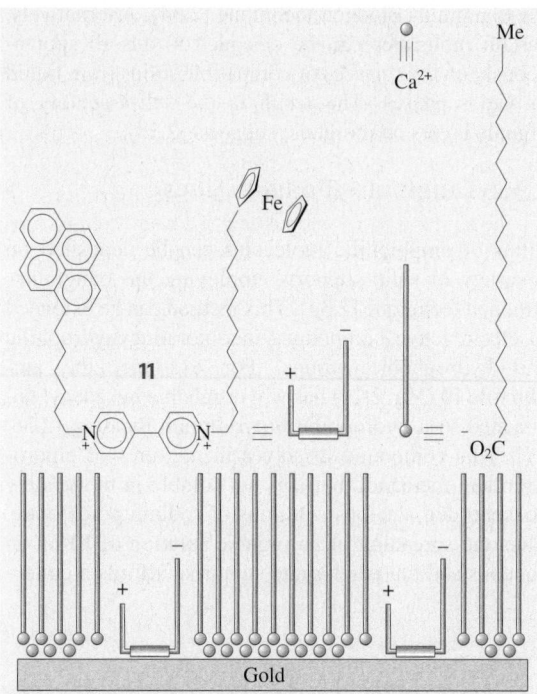

Fig. 2.13 Mixed monolayers of the amphiphile **11** and arachidic acid can be transferred from the air/water interface to the surface of an electrode to generate a molecule-based photodiode

hole. Overall, a unidirectional flow of electrons across the monolayer/electrode junction is established under the influence of light.

The ability to transfer electroactive monolayers from air/water interfaces to electrode surfaces can be exploited to fabricate molecule-based electronic devices. In particular, arrays of interconnected electrode/monolayer/electrode tunneling junctions can be assembled combining the Langmuir–Blodgett technique with electron beam evaporation [2.33]. Fig. 2.14 illustrates a schematic representation of the resulting devices. Initially, parallel fingers are patterned on a silicon wafer with a silicon dioxide overlayer by electron beam evaporation. The bottom electrodes deposited on the support can be either aluminum wires covered by an aluminum oxide or n-doped silicon lines with silicon dioxide overlayers. Their widths are ca. 6 or 7 μm, respectively. The patterned silicon chip is immersed in the aqueous subphase of a Langmuir trough prior to monolayer formation. After the compression of electroactive amphiphiles at the air/water interface, the substrate is pulled out of the aqueous phase to encourage the transfer of the molecular layer on the parallel bottom electrodes as well as on the gaps between them. Then, a second set of electrodes orthogonal to the first is deposited through a mask by electron beam evaporation. They consist of a titanium underlayer plus an aluminum overlayer. Their thicknesses are ca. 0.05 and 1 μm, respectively, and their width is ca. 10 μm. In the final assembly, portions of the molecular layer become sandwiched between the bottom and top electrodes. The active areas of these electrode/monolayer/electrode junctions are ca. 60–70 μm^2 and correspond to ca. 10^6 molecules.

The [2]rotaxane **12** (Fig. 2.14) incorporates a macrocyclic polyether threaded onto a bipyridinium-based backbone [2.40, 41]. The two bipyridinium dications are bridged by a *m*-phenylene spacer and terminated by tetraarylmethane appendages. These two bulky groups trap mechanically the macrocycle preventing its dissociation from the tetracationic backbone. In addition, their hydrophobicity complements the hydrophilicity of the two bipyridinium dications imposing amphiphilic character on the overall molecular assembly. This compound does not dissolve in aqueous solutions and can be compressed into organized monolayers at air/water interfaces. The corresponding π–A isotherm reveals a limiting molecular area of ca. 130 Å^2. This large value is a consequence of the bulk associated with the hydrophobic tetraarylmethane tails and the macrocycle encircling the tetracationic backbone.

Monolayers of the [2]rotaxane **12** can be transferred from the air/water interface to the surfaces of the bottom aluminum/aluminum oxide electrodes of a patterned silicon chip with the hydrophobic tetraarylmethane groups pointing away from the supporting substrate. The subsequent assembly of a top titanium/aluminum electrode affords electrode/monolayer/electrode junctions. Their current/voltage signature can be recorded grounding the top electrode and scanning the potential of the bottom electrode. A pronounced increase in current is observed when the potential is lowered below −0.7 V. Under these conditions, the bipyridinium-centered LUMOs mediate the tunneling of electrons from the bottom to the top electrode leading to a current enhancement.

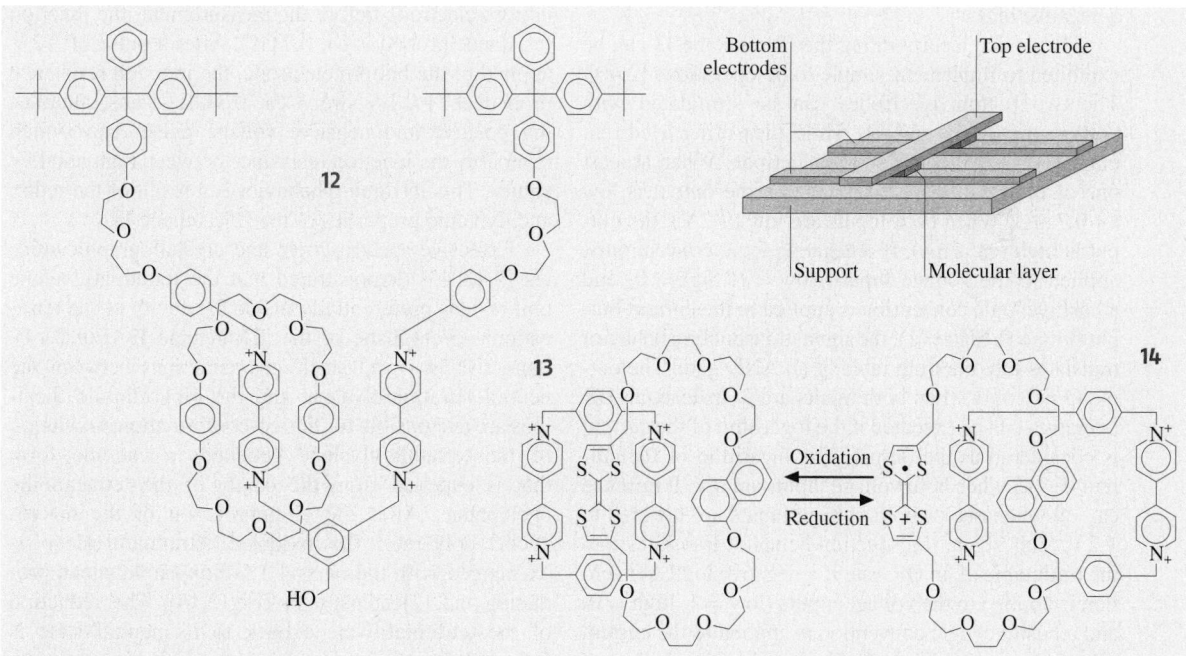

Fig. 2.14 The [2]rotaxane **12** and the [2]catenane **13** can be compressed into organized monolayers at air/water interfaces. The resulting monolayers can be transferred on the bottom electrodes of a patterned silicon support. After the deposition of a top electrode, electrode/monolayer/electrode junctions can be assembled. Note that only the portion of the monolayer sandwiched between the top and bottom electrodes is shown in the diagram. The oxidation of the tetrathiafulvalene unit of the [2]catenane **13** is followed by the circumrotation of the macrocyclic polyether to afford the [2]catenane **14**. The process is reversible, and the reduction of the cationic tetrathiafulvalene unit restores the original state

A similar current profile is observed if the potential is returned to 0 and then back to -2 V. Instead, a modest increase in current in the opposite direction is observed when the potential is raised above $+0.7$ V. Presumably, this trend is a result of the participation of the phenoxy-centered HOMOs in the tunneling process. After a single positive voltage pulse, however, no current can be detected if the potential is returned to negative values. In summary, the positive potential scan suppresses irreversibly the conducting ability of the electrode/molecule/electrode junction. The behavior of this device correlates with the redox response of the [2]rotaxane **12** in solution. Cyclic voltammograms reveal reversible monoelectronic reductions of the bipyridinium dications. But they also show two irreversible oxidations associated, presumably, with the phenoxy rings of the macrocycle and tetraarylmethane groups. These observations suggest that a positive voltage pulse applied to the electrode/monolayer/electrode junction oxidizes irreversibly the sandwiched molecules suppressing their ability to mediate the transfer of electrons from the bottom to the top electrode under a negative bias.

The device incorporating the [2]rotaxane **13** can be exploited to implement simple logic operations [2.40]. The two bottom electrodes can be stimulated with voltage inputs (I1 and I2) while measuring a current output (O) at the common top electrode. When at least one of the two inputs is high (0 V), the output is low (< 0.7 nA). When both inputs are low (-2 V), the output is high (ca. 4 nA). If a negative logic convention is applied to the voltage inputs (low = 1, high = 0) and a positive logic convention is applied to the current output (low = 0, high = 1), the signal transduction behavior translates into the truth table of an AND gate. The output O is 1 only when both inputs are 1. Instead, an OR operation can be executed if the logarithm of the current is considered as the output. The logarithm of the current is -12 when both voltage inputs are 0 V. It raises to ca. -9 when one or both voltage inputs are lowered to -2 V. This signal transduction behavior translates into the truth table of an OR gate if a negative logic convention is applied to the voltage inputs (low = 1, high = 0) and a positive logic convention is applied to the current output (low = 0, high = 1). The output O is 1 when at least one of the two inputs is 1.

The [2]catenane **13** (Fig. 2.14) incorporates a macrocyclic polyether interlocked with a tetracationic cyclophane [2.42, 43]. Organic solutions of the hexafluorophosphate salt of this [2]catenane and six equivalents of the sodium salt of dimyristoylphosphatidic acid

can be co-spread on the water surface of a Langmuir trough [2.44]. The sodium hexafluorophosphate formed dissolves in the supporting aqueous phase, while the hydrophilic bipyridinium cations and the amphiphilic anions remain at the interface. Upon compression, the anions align their hydrophobic tails away from water surface forming a compact monolayer above the cationic bipyridinium derivatives. The corresponding π–A isotherm indicates limiting molecular areas of ca. $125\,\text{Å}^2$. This large value is a consequence of the bulk associated with the two interlocking macrocycles.

Monolayers of the [2]catenane **13** can be transferred from the air/water interface to the surfaces of the bottom n-doped silicon/silicon dioxide electrodes of a patterned silicon chip with the hydrophobic tails of the amphiphilic anions pointing away from the supporting substrate [2.45, 46]. The subsequent assembly of a top titanium/aluminum electrode affords electrode/monolayer/electrode arrays. Their junction resistance can be probed grounding the top electrode and maintaining the potential of the bottom electrode at $+0.1$ V. If a voltage pulse of $+2$ V is applied to the bottom electrode before the measurement, the junction resistance probed is ca. $0.7\,\text{G}\Omega$. After a pulse of -2 V applied to the bottom electrode, the junction resistance probed at $+0.1$ V drops ca. $0.3\,\text{G}\Omega$. Thus, alternating positive and negative voltage pulses can switch reversibly the junction resistance between high and low values. This intriguing behavior is a result of the redox and dynamic properties of the [2]catenane **13**.

Extensive spectroscopic and crystallographic studies [2.42, 43] demonstrated that the tetrathiafulvalene unit resides preferentially inside the cavity of the tetracationic cyclophane of the [2]catenane **13** (Fig. 2.14). Attractive $[\pi \cdots \pi]$ stacking interactions between the neutral tetrathiafulvalene and the bipyridinium dications are responsible for this co-conformation. Oxidation of the tetrathiafulvalene generates a cationic form that is expelled from the cavity of the tetracationic cyclophane. After the circumrotation of the macrocyclic polyether, the oxidized tetrathiafulvalene is exchanged with the neutral 1,5-dioxynaphthalene producing the [2]catenane **14** (Fig. 2.14). The reduction of the tetrathiafulvalene back to its neutral state is followed by the circumrotation of the macrocyclic polyether, which restores the original state **14**. The voltage pulses applied to the bottom electrode of the electrode/monolayer/junction oxidize and reduce the tetrathiafulvalene unit inducing the interconversion between the forms **13** and **14**. The difference in the stereoelectronic properties of these two states translates into distinct current/voltage signatures. Indeed, their ability to mediate the tunneling of electrons across the junction differs significantly. As a result, the junction resistance probed at a low voltage after an oxidizing pulse is significantly different from that determined under the same conditions after a reducing pulse.

2.3.3 Self–Assembled Monolayers

In the examples illustrated in Figs. 2.12–2.14, monolayers of amphiphilic and electroactive derivatives are assembled at air/water interfaces and then transferred on the surfaces of appropriate substrates. An alternative strategy to coat electrodes with molecular layers relies on the ability of certain compounds to adsorb spontaneously on solid supports from liquid or vapor phases [2.35]. In particular, the affinity of certain sulfurated functional groups for gold can be exploited to encourage the self-assembly of organic molecules on microscaled and nanoscaled electrodes.

The electrode/monolayer/electrode junction in Fig. 2.15 incorporates a molecular layer between two gold electrodes mounted on a silicon nitride support. This device can be fabricated combining chemical vapor deposition, lithography, anisotropic etching, and self-assembly [2.47]. Initially, a silicon wafer is coated with a 50 nm thick layer of silicon nitride by low pressure chemical vapor deposition. Then, a square of $400 \times 400 \, \mu$m is patterned on one side of the coated wafer by optical lithography and reactive ion etching. Anisotropic etching of the exposed silicon up to the other side of the wafer leaves a suspended silicon nitride membrane of $40 \times 40 \, \mu$m. Electron beam lithography and reactive ion etching can be used to carve a bowl-shaped hole (diameter = 30–50 nm) in the membrane. Evaporation of gold on the membrane fills the pore producing a bowl-shaped electrode. Immersion of the substrate in a solution of the thiol **15** results in the self-assembly of a molecular layer on the narrow part of the bowl-shaped electrode. The subsequent evaporation of a gold film on the organic monolayer produces an electrode/monolayer/electrode junction (Fig. 2.15) with a contact area of less than 2000 nm^2 and approximately 1000 molecules.

Under the influence of voltage pulses applied to one of the two gold electrodes in Fig. 2.15, the conductivity of the sandwiched monolayer switches reversibly between low and high values [2.48]. In the initial state, the monolayer is in a low conducting mode. A current output of only 30 pA is detected, when a probing voltage of $+ 0.25$ V is applied to the bowl-shaped electrode. If the

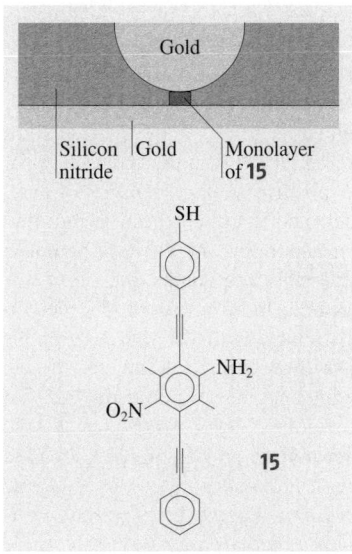

Fig. 2.15 A monolayer of the thiol **15** is embedded between two gold electrodes maintained in position by a silicon nitride support

same electrode is stimulated with a short voltage pulse of $+5$ V, the monolayer switches to a high conducting mode. Now a current output of 150 pA is measured at the same probing voltage of $+ 0.25$ V. Repeated probing of the current output at various intervals of time indicates that the high conducting state is memorized by the molecule-based device, and it is retained for more than 15 min. The low conducting mode is restored after either a relatively long period of time or the stimulation of the bowl-shaped electrode with a reverse voltage pulse of -5 V. Thus the current output switches from a low to a high value, if a high voltage input is applied. It switches from a high to a low value, under the influence of a low voltage pulse. This behavior offers the opportunity to store and erase binary data in analogy to a conventional random access memory [2.17]. Binary digits can be encoded on the current output of the molecule-based device applying a positive logic convention (low = 0, high = 1). It follows that a binary 1 can be stored in the molecule-based device applying a high voltage input, and it can be erased applying a low voltage input [2.48].

The ability of thiols to self-assemble on the surface of gold can be exploited to fabricate nanocomposite materials integrating organic and inorganic components. For example, the bisthiol **16** forms monolayers (Fig. 2.16a) on gold electrodes with surface coverages of ca. 4.1×10^{10} mol cm^2 [2.49, 50]. The formation of a thiolate–gold bond at one of the two thiol ends of **16** is responsible for adsorption. The remaining thiol group points away from the supporting surface and can be

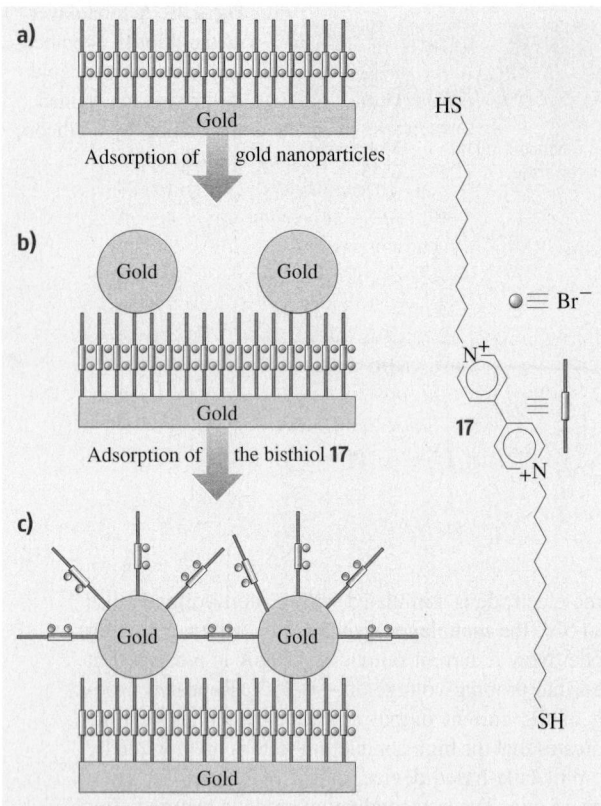

Fig. 2.16 (a) The bisthiol **16** self-assembles on gold electrodes as a result of thiolate–gold bond formation. (b) Gold nanoparticles adsorb spontaneously on the molecular layer. (c) Exposure of the composite assembly to a solution of **16** results in the formation of an additional molecular layer on the surface of the gold nanoparticles

appears. The pronounced potential shift indicates that $[Ru(NH_3)_6]^{3+}$ accepts electrons only after the surface-confined bipyridinium dications have been reduced. The lack of reversibility indicates that the back oxidation to the bipyridinium dications inhibits the transfer of electrons from the $[Ru(NH_3)_6]^{2+}$ to the electrode. Thus the electroactive multilayer allows the flow of electrons in one direction only in analogy to conventional diodes.

The current/voltage behavior of individual nanoparticles in Fig. 2.16b can be probed by scanning tunneling spectroscopy in an aqueous electrolyte under an inert atmosphere [2.51]. The platinum-iridium tip of a scanning tunneling microscope is positioned above one of the gold particles. The voltage of the gold substrate relative to the tip is maintained at -0.2 V while that relative to a reference electrode immersed in the same electrolyte is varied to control the redox state of the electroactive units. Indeed, the bipyridinium dications in the molecular layer can be reduced reversibly to a monocationic state. The resulting monocations can be reduced further and, once again, reversibly to a neutral form. Finally, the current flowing from the gold support to the tip of the scanning tunneling microscope is monitored as the tip–particle distance increases. From the distance dependence of the current, inverse length decays of ca. 16 and 7 nm^{-1} for the dicationic and monocationic states, respectively, of the molecular spacer can be determined. The dramatic decrease indicates that the reduction of the electroactive unit facilitates the tunneling of electrons through the gold/molecule/nanoparticle/tip junction. In summary, a change in the redox state of the bipyridinium components can be exploited to gate reversibly the current flowing through this nanoscaled device.

Similar nanostructured materials, combining molecular and nanoparticles layers, can be prepared on layers on indium-tin oxide electrodes following multistep procedures [2.52]. The hydroxylated surfaces of indium-tin oxide supports can be functionalized with 3-ammoniumpropylysilyl groups and then exposed to gold nanoparticles having a diameter of ca. 13 nm [2.53, 54]. Electrostatic interactions promote the adsorption of the nanoparticles on the organic layer (Fig. 2.17a). The treatment of the composite film with the bipyridinium cyclophane **17** produces an organic layer on the gold nanoparticles (Fig. 2.18b). Following this approach, alternating layers of inorganic nanoparticles and organic building blocks can be assembled on the indium-tin oxide support. Cyclic voltammograms of the resulting materials show the oxidation of the gold nanoparticles and the reduction of the bipyridinium units. The peak current for both processes increases with

exploited for further functionalization. Gold nanoparticles adsorb on the molecular layer (Fig. 2.16b), once again, as a result of thiolate–gold bond formation. The immersion of the resulting material in a methanol solution of **16** encourages the adsorption of an additional organic layer (Fig. 2.16c) on the composite material. Following these procedures, up to ten alternating organic and inorganic layers can be deposited on the electrode surface. The resulting assembly can mediate the unidirectional electron transfer from the supporting electrode to redox active species in solution. For example, the cyclic voltammogram of the $[Ru(NH_3)_6]^{3+/2+}$ couple recorded with a bare gold electrode reveals a reversible reduction process. In the presence of ten alternating molecular and nanoparticle layers on the electrode surface, the reduction potential shifts by ca. -0.2 V and the back oxidation wave dis-

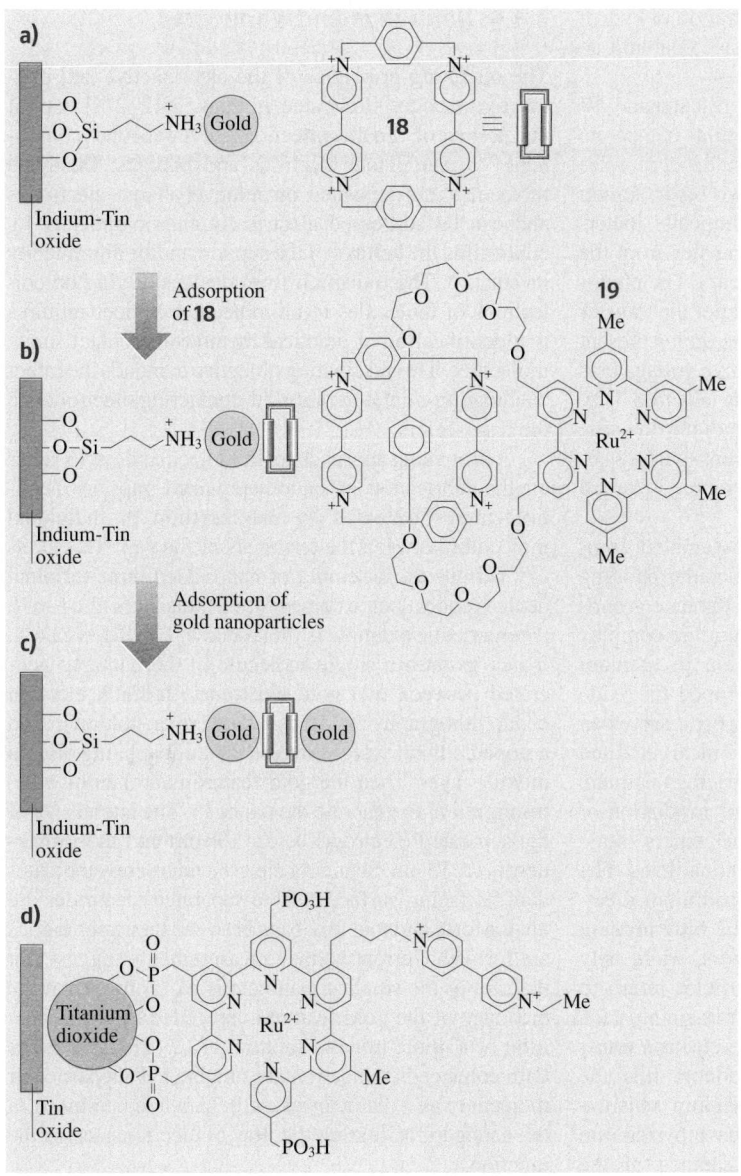

Fig. 2.17 (a) Gold nanoparticles assemble spontaneously on pre-functionalized indium-tin oxide electrodes. **(b)** Electrostatic interactions encourage the adsorption of the tetracationic cyclophane **17** on the surface-confined nanoparticles. **(c)** An additional layer of nano-particles assembles on the cationic organic coating. Similar composite films can be prepared using the tetra-cationic [2]catenane **18** instead of the cyclophane **17**. **(d)** Phosphonate groups can be used to anchor molecular building blocks to titanium dioxide nanoparticles

the number of alternating layers. Comparison of these values indicates that the ratio between the number of tetracationic cyclophanes and that of the nanoparticles is ca. 100 : 1.

The tetracationic cyclophane **17** binds dioxyarenes in solution [2.55, 56]. Attractive supramolecular forces between the electron deficient bipyridinium units and the electron rich guests are responsible for complexation. This recognition motif can be exploited to probe the ability of the composite films in Fig. 2.17b,c to sense

electron rich analytes. In particular, hydroquinone is expected to enter the electron deficient cavities of the surface-confined cyclophanes. Cyclic voltammograms consistently reveal the redox waves associated with the reversible oxidation of hydroquinone even when very small amounts of the guest (ca. 1×10^5 M) are added to the electrolyte solution [2.53, 54]. No redox response can be detected with a bare indium–tin oxide electrode under otherwise identical conditions. The supramolecular association of the guest and the surface confined cyclophanes increases the local concentration of hydroquinone at the electrode/solution interface enabling its electrochemical detection.

Following a related strategy, the [2]catenane **18** (Fig. 2.17) can be incorporated into similar composite arrays [2.57, 58]. This interlocked molecule incorporates a Ru(II)/trisbipyridine sensitizer and two bipyridinium acceptors. Upon irradiation of the composite material at 440 nm, photoinduced electron transfer from the sensitizer to the appended acceptors occurs. The photogenerated hole in the sensitizer is filled after the transfer of an electron from a sacrificial electron donor present in the electrolyte solution. Under a positive voltage bias applied to the supporting electrode, an electron flow from the bipyridinium acceptors to the indium-tin oxide support is established. The resulting current switches between high and low values as the light source is turned on and off.

Another photoresponsive device, assembled combining inorganic nanoparticles with molecular building blocks, is illustrated in Fig. 2.17d. Phosphonate groups can be used to anchor a Ru(II)/trisbipyridine complex with an appended bipyridinium dication to titanium dioxide nanoparticles deposited on a doped tin oxide electrode [2.59, 60]. The resulting composite array can be integrated in a conventional electrochemical cell filled with an aqueous electrolyte containing triethanolamine. Under a bias voltage of -0.45 V and irradiation at 532 nm, 95% of the excited ruthenium centers transfer electrons to the titanium dioxide nanoparticles. The other 5% donate electrons to the bipyridinium dications. All the electrons transferred to the bipyridinium acceptors return to the ruthenium centers, while only 80% of those accepted by the nanoparticles return to the transition metal complexes. The remaining 15% reach the bipyridinium acceptors, while electron transfer from sacrificial triethanolamine donors fills the photogenerated holes left in the ruthenium sensitizers. The photoinduced reduction of the bipyridinium dication is accompanied by the appearance of the characteristic band of the radical cation in the absorption spectrum. This band persists for hours under open circuit conditions. But it fades in ca. 15 s under a voltage bias of $+1$ V, as the radical cation is oxidized back to the dicationic form. In summary, an optical stimulation accompanied by a negative voltage bias reduces the bipyridinium building block. The state of the photogenerated form can be read optically, recording the absorption spectrum in the visible region, and erased electrically, applying a positive voltage pulse.

2.3.4 Nanogaps and Nanowires

The operating principles of the electroactive and photoactive devices illustrated in Figs. 2.12–2.17 exploit the ability of small collections of molecular components to manipulate electrons and photons. Designed molecules are deposited on relatively large electrodes and can be addressed electrically and/or optically by controlling the voltage of the support and/or illuminating its surface. The transition from devices relying on collections of molecules to unimolecular devices requires the identification of practical methods to contact single molecules. This fascinating objective demands the rather challenging miniaturization of contacting electrodes to the nanoscale.

A promising approach to unimolecular devices relies on the fabrication of nanometer-sized gaps in metallic features followed by the insertion of individual molecules between the terminals of the gap. This strategy permits the assembly of nanoscaled three-terminal devices equivalent to conventional transistors [2.61–63]. A remarkable example is illustrated in Fig. 2.18a [2.61]. It incorporates a single molecule in the nanogap generated between two gold electrodes. Initially electron beam lithography is used to pattern a gold wire on a doped silicon wafer covered by an insulating silicon dioxide layer. Then the gold feature is broken by electromigration to generate the nanogap. The lateral size of the separated electrodes is ca. 100 nm and their thickness is ca. 15 nm. Scanning electron microcopy indicates that the facing surfaces of the separated electrodes are not uniform and that tiny gaps between their protrusions are formed. Current/voltage measurements suggest that the size of the smallest nanogap is ca. 1 nm. When the breakage of the gold feature is preceded by the deposition of a dilute toluene solution of C_{60} (**19**), junctions with enhanced conduction are obtained. This particular molecule has a diameter of ca. 0.7 nm and can insert in the nanogap facilitating the flow of electrons across the junction.

The unique configuration of the molecule-based device in Fig. 2.18a can reproduce the functions of a conventional transistor [2.19] at the nanoscale. The two gold terminals of the junction are the drain and source of this nanotransistor, and the underlying silicon wafer is the gate. At a temperature of 1.5 K, the junction conductance is very small, when the gate bias is low, and increases in steps at higher voltages [2.61]. The conductance gap is a consequence of the finite energy required to oxidize/reduce the single C_{60} positioned in the junction. It is interesting that the zero-conductance window

a)

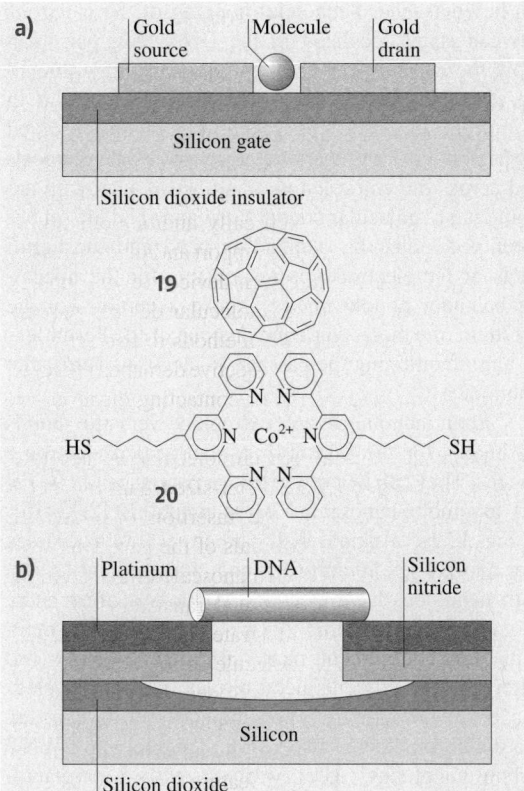

b)

Fig. 2.18 (a) Nanoscaled transistors can be fabricated inserting a single molecule (**19** or **20**) between source and drain electrodes mounted on a silicon/silicon dioxide support. **(b)** A DNA nanowire can bridge nanoelectrodes suspended above a silicon dioxide support

also changes with the gate voltage and can be opened and closed reversibly adjusting the gate bias.

A similar strategy can be employed to fabricate a nanoscaled transistor incorporating the Co(II) complex **20** shown in Fig. 2.18 [2.63]. In this instance, a silicon dioxide layer with a thickness of ca. 30 nm is grown thermally on a doped silicon substrate. Then a gold wire with a width of ca. 200 nm and a thicknesses of ca. 10–15 nm is patterned on the silicon dioxide overlayer by electron beam lithography. After extensive washing of the substrate with acetone and methylene chloride and cleaning with oxygen plasma, the gold wire is exposed to a solution of the bisthiol **20**. The formation of thiolate–gold bonds promotes the self-assembly of the molecular building block on the gold surface. At this point, electromigration-induced breakage produces a gap of 1–2 nm in

the gold wire. The surface-confined bisthiol **20** is only 0.24 nm long and, therefore, it can insert in the nanogap producing an electrode/molecule/electrode junction.

The cobalt center in **20** can be oxidized/reduced reversibly between Co(II) and Co(III) [2.63]. When this electroactive molecule is inserted in a nanogap (Fig. 2.18a), its ability to accept and donate electrons dictates the current/voltage profile of the resulting electrode/molecule/electrode junction. More precisely, no current flows across the junction below a certain voltage threshold. As the source voltage is raised above this particular value, the drain current increases in steps. The threshold associated with the source voltage varies in magnitude with the gate voltage. This intriguing behavior is a consequence of the finite energy necessary to oxidize/reduce the cobalt center and of a change in the relative stabilities of the oxidized and reduced forms Co(II) and Co(III) with the gate voltage. In summary, the conduction of the electrode/molecule/electrode junction can be tuned adjusting the voltage of the silicon support. The behavior of this molecule-based nanoelectronic device is equivalent to that of a conventional transistor [2.19]. In both instances, the gate voltage regulates the current flowing from the source to the drain.

The electromigration-induced breakage of preformed metallic features successfully produces nanogaps by moving apart two fragments of the same wire. Alternatively, nanogaps can be fabricated reducing the separation of the two terminals of much larger gaps. For example, gold electrodes separated by a distance of 20–80 nm can be patterned on a silicon/silicon dioxide substrate by electron beam lithography [2.64]. The relatively large gap between them can be reduced significantly by the electrochemical deposition of gold on the surfaces of both electrodes. The final result is the fabrication of two nanoelectrodes separated by ca. 1 nm and with a radius of curvature of 5–15 nm. The two terminals of this nanogap can be "contacted" by organic nanowires grown between them [2.65]. In particular, the electropolymerization of aniline produces polyaniline bridges between the gold nanoelectrodes. The conductance of the resulting junction can be probed immersing the overall assembly in an electrolyte solution. Employing a bipotentiostat, the bias voltage of the two terminals of the junction can be maintained at 20 mV, while their potentials are scanned relative to that of a silver/silver chloride reference electrode. Below ca. 0.15 V, the polymer wire is in an insulating state and the current flowing across the junction is less than 0.05 nA. At this voltage threshold, however, the current raises abruptly to ca. 30 nA. This value corresponds to a conductivity for the polymer nanojunction of $10-100\,S\,cm^{-1}$. When the potential is lowered again below the threshold, the current returns back to very low values. The abrupt decrease in current in the backward scan is observed at a potential that is slightly more negative than that causing the abrupt current increase in the forward scan. In summary, the conductance of this nanoscaled junction switches on and off as a potential input is switched above and below a voltage threshold.

It is interesting to note that the influence of organic bridges on the junction conductance can be exploited for chemical sensing. Nanogaps fabricated following a similar strategy but lacking the polyaniline bridge alter their conduction after exposure to dilute solutions of small organic molecules [2.66]. Indeed, the organic analytes dock into the nanogaps producing a marked decrease in the junction conductance. The magnitude of the conductance drop happens to be proportional to the analyte–nanoelectrode binding strength. Thus the presence of the analyte in solution can be detected probing the current/voltage characteristics of the nanogaps.

Nanogaps between electrodes patterned on silicon/silicon dioxide supports can be bridged also by DNA double strands [2.67, 68]. The device in Fig. 2.18b has a 10.4 nm long poly(G)–poly(C) DNA oligomer suspended between two nanoelectrodes. It can be fabricated patterning a 30 nm wide slit in a silicon nitride overlayer covering a silicon/silicon dioxide support by electron beam evaporation. Underetching the silicon dioxide layer leaves a silicon nitride finger, which can be sputtered with a platinum layer and chopped to leave a nanogap of 8 nm. At this point, a microdroplet of a dilute solution of DNA is deposited on the device and a bias of 5 V is applied between the two electrodes. Electrostatic forces encourage the deposition of a single DNA wire on top of the nanogap. As soon as the nanowire is in position, current starts to flow across the junction. The current/voltage signature of the electrode/DNA/electrode junction shows currents below 1 pA at low voltage biases. Under these conditions, the DNA nanowire is an insulator. Above a certain voltage threshold, however, the nanowire becomes conducting and currents up to 100 nA can flow across the junction through a single nanowire. Assuming that direct tunneling from electrode to electrode is extremely unlikely for a relatively large gap of 8 nm, the intriguing current/voltage behavior has to be a consequence of the participation of the molecular states in the electron transport process. Two possible mechanisms

can be envisaged. Sequential hopping of the electrons between states localized in the DNA base pairs can allow the current flow above a certain voltage threshold. But this mechanism would presumably result in a Coulomb blockade voltage gap that is not observed experimentally. More likely, electronic states delocalized across the entire length of the DNA nanowire are producing a molecular conduction band. The off-set between the molecular conduction band and the Fermi levels of the electrodes is responsible for the insulating behavior at low biases. Above a certain voltage threshold, the molecular band and one of the Fermi levels align facilitating the passage of electrons across the junction.

Carbon nanotubes are extremely versatile building blocks for the assembly of nanoscaled electronic devices. They can be used to bridge nanogaps [2.69–72] and assemble nanoscaled cross junctions [2.73–75]. In Fig. 2.19a, a single-wall carbon nanotube crosses over another one in an orthogonal arrangement [2.73]. Both nanotubes have electrical contacts at their ends. The fabrication of this device involves three main steps. First, alignment marks for the electrodes are patterned on a silicon/silicon dioxide support by electron beam lithography. Then the substrate is exposed to a dichloromethane suspension of single-wall SWNT carbon nanotubes. After washing with isopropanol, crosses of carbon nanotubes in an appropriate alignment relative to the electrode marks are identified by tapping mode atomic force microscopy. Finally chromium/gold electrodes are fabricated on top of the nanotube ends, again, by electron beam lithography. The conductance of individual nanotubes can be probed by exploiting the two electric contacts at their ends. These two-terminal measurements reveal that certain nanotubes have metallic behavior, while others are semiconducting. It follows that three distinct types of cross junctions differing in the nature of their constituent nanotubes can be identified on the silicon/silicon dioxide support. Four terminal current/voltage measurements indicate that junctions formed by two metallic nanotubes have high conductance and ohmic behavior. Similarly, high junction conductance and ohmic behavior is observed when two semiconducting nanotubes cross. The current/voltage signature of junctions formed when a metallic nanotube crosses a semiconducting one are, instead, completely different. The metallic nanotube depletes the semiconducting one at the junction region producing a nanoscaled Schottky barrier with a pronounced rectifying behavior.

Similar fabrication strategies can be exploited to assemble nanoscaled counterparts of conventional transistors. The device in Fig. 2.19b is assembled patterning an aluminum finger on a silicon/silicon dioxide substrate by electron beam lithography [2.75]. After exposure to air, an insulating aluminum oxide layer forms on the aluminum finger. Then a dichloromethane suspension of single-wall carbon nanotubes is deposited on the resulting substrate. Atomic force microscopy can be used to select carbon nanotubes with a diameter of ca. 1 nm positioned on the aluminum finger. After registering their coordinates relative to alignment markers, gold contacts can be evaporated on their ends by electron beam lithography. The final assembly is a nanoscaled three-terminal device equivalent to a conventional field effect transistor [2.19]. The two gold contacts are the source and drain terminals, while the underlying aluminum finger reproduces the function of the gate. At a source to drain bias of ca. − 1.3 V, the drain current jumps from ca. 0 to ca. 50 nA when the gate voltage is lowered from − 1.0 to − 1.3 V. Thus moderate changes in the gate voltage vary significantly the current flowing through the nanotube-based device in analogy to a conventional enhancement mode *p*-type field effect transistor [2.19].

The nanoscaled transistor in Fig. 2.18a has a microscaled silicon gate that extends under the entire chip [2.61, 63]. The configuration in Fig. 2.19b, instead, has nanoscaled aluminum gates for every single carbon nanotube transistor fabricated on the same support [2.75]. It follows that multiple nanoscaled transistors can be fabricated on the same chip and operated independently following this strategy. This unique feature offers the possibility of fabricating nanoscaled digital circuits by interconnecting the terminals of independent nanotube transistors. The examples in Fig. 2.19c,d illustrate the configurations of nanoscaled NOT and NOR gates implemented using one or two nanotube transistors. In Fig. 2.19c, an off-chip bias resistor is connected to the drain terminal of a single transistor while the source is grounded. A voltage input applied to the gate modulates the nanotube conductance altering the voltage output probed at the drain terminal. In particular, a voltage input of − 1.5 V lowers the nanotube resistance (26 MΩ) below that of the bias resistor (100 MΩ). As a result, the voltage output drops to 0 V. When the voltage input is raised to 0 V, the nanotube resistance increases above that of the bias resistor and the voltage output becomes − 1.5 V. Thus the output of this nanoelectronic device switches from a high (0 V) and to a low (− 1.5 V) level as the input shifts from a low (− 1.5 V) to a high (0 V) value. The inverse relation

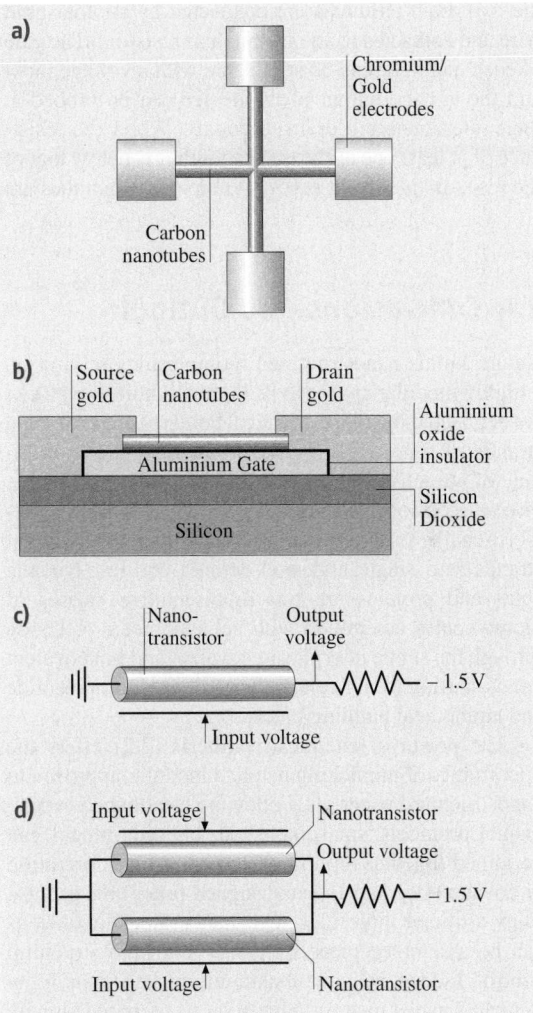

Fig. 2.19 (a) Nanoscaled junctions can be assembled on silicon/silicon dioxide supports crossing pairs of orthogonally arranged single-wall carbon nanotubes with chromium/gold electrical contacts at their ends. **(b)** Nanotransistors can be fabricated contacting the two ends of a single-wall carbon nanotube deposited on an aluminum/aluminum oxide gate with gold sources and drain. One or two nanotube transistors can be integrated into nanoscaled NOT **(c)** and NOR **(d)** logic gates

between input and output translates into a NOT operation if a negative logic convention (low = 1, high = 0) is applied to both signals.

In Fig. 2.15d, the source terminals of two independent nanotube transistors fabricated on the same chip are connected by a gold wire and grounded [2.75]. Similarly, the two drain terminals are connected by another gold wire and contacted to an off-chip bias resistors. The gate of each nanotube can be stimulated with a voltage input and the voltage output of the device can be probed at their interconnected drain terminals. When the resistance of at least one of the two nanotubes is below that of the resistor, the output is 0 V. When both nanotubes are

in a nonconducting mode, the output voltage is -1.5 V. Thus if a low voltage input -1.5 V is applied to one or both transistors, the output is high (0 V). When both voltage inputs are high (0 V), the output is low (-1.5 V). If a negative logic convention (low $= 1$, high $= 0$) is applied to all signals, the signal transduction behavior translates in to a NOR operation.

2.4 Conclusions and Outlook

Nature builds nanostructured biomolecules relying on a highly modular approach [2.1]. Small building blocks are connected by robust chemical bonds to generate long strands of repeating units. The synergism of a multitude of attractive supramolecular forces determines the three-dimensional arrangement of the resulting polymeric chains and controls the association of independent strands into single and well-defined entities. Nucleic acids and proteins are two representative classes of biomolecules assembled with subnanometer precision through the subtle interplay of covalent and noncovalent bonds starting from a relatively small pool of nucleotide and amino acid building blocks.

The power of chemical synthesis [2.2] offers the opportunity of mimicking nature's modular approach to nanostructured materials. Following established experimental protocols, small molecular building blocks can be joined together relying on the controlled formation of covalent bonds between designed functional groups. Thus artificial molecules with nanoscaled dimensions can be assembled piece by piece with high structural control. Indeed, helical, tubular, interlocked, and highly branched nanostructures have been all prepared already exploiting this general strategy and the synergism of covalent and noncovalent bonds [2.3].

The chemical construction of nanoscaled molecules from modular building blocks also offers the opportunity for engineering specific properties in the resulting assemblies. In particular, electroactive and photoactive fragments can be integrated into single molecules. The ability of these functional subunits to accept/donate electrons and photons can be exploited to design nanoscaled electronic and photonic devices. Indeed, molecules that respond to electrical and optical stimulations producing detectable outputs have been designed already [2.16]. These chemical systems can be employed to control the interplay of input and output signals at the molecular level. Their conceptual analogy with the signal transduction operated by conventional logic gates in digital

circuits is evident. In fact, electroactive and photoactive molecules able to reproduce AND, NOT, and OR operations as well as simple combinational of these basic logic functions are already a reality [2.13, 20, 21].

Most of the molecular switches for digital processing developed so far rely on bulk addressing. In general, relatively large collections of functional molecules are addressed simultaneously in solution. The realization of molecule-based devices with reduced dimensions as well as practical limitations associated with liquid phases in potential applications are encouraging a transition from the solution to the solid state. The general strategy followed so far relies on the deposition of functional molecules on the surfaces of appropriate electrodes following either the Langmuir–Blodgett methodology [2.34] or self-assembly processes [2.35]. The combination of these techniques with the nanofabrication of insulating, metallic, and semiconducting features on appropriate supports has already allowed the realization of fascinating molecule-based devices [2.30–33, 52]. The resulting assemblies integrate inorganic and organic components and, in some instances, even biomolecules to execute specific functions. They can convert optical stimulations into electrical signals. They can execute irreversible and reversible switching operations. They can sense qualitatively and quantitatively specific analytes. They can reproduce the functions of conventional rectifiers and transistors. They can be integrated within functioning nanoelectronic devices capable of simple logic operations.

The remarkable examples of molecule-based materials and devices now available demonstrate the great potential and promise for this research area. At this stage, the only limit left to the design of functional molecules is the imagination of the synthetic chemist. All sort of molecular building blocks with tailored dimensions, shapes, and properties are more or less accessible with the assistance of modern chemical synthesis. Now, the major challenges are (1) to master the operating principles of the molecule-based devices that have been and continue to be assembled and (2) to expand and improve the fabrication strategies available to incorporate molecules into reliable device architectures. As we continue to gather further insights in these directions, design criteria for a wide diversity of molecule-based devices will emerge. It is not unrealistic to foresee the evolution of an entire generation of nanoscaled devices, based on engineered molecular components, that will find applications in a variety of fields ranging from biomedical research to information technology. Perhaps nature can once again illuminate our path, teaching us not only how to synthesize nanostructured molecules but also how to use them. After all, nature is replete with examples of extremely sophisticated molecule-based devices. From tiny bacteria to higher animals, we are all a collection of molecule-based devices.

References

2.1   D. Voet, J. G. Voet: *Biochemistry* (Wiley, New York 1995)

2.2   K. C. Nicolau, E. C. Sorensen: *Classics in Total Synthesis* (VCH, Weinheim 1996)

2.3   J.-M. Lehn: *Supramolecular Chemistry: Concepts and Perspectives* (VCH, Weinheim 1995)

2.4   M. M. Harding, U. Koert, J.-M. Lehn, A. Marquis-Rigault, C. Piguet, J. Siegel: Synthesis of unsubstituted and 4,4'-substituted oligobipyridines as ligand strands for helicate self-assembly, Helv. Chim. Acta **74**, 594–610 (1991)

2.5   J.-M. Lehn, A. Rigault, J. Siegel, B. Harrowfield, B. Chevrier, D. Moras: Spontaneous assembly of double-stranded helicates from oligobipyridine ligands and copper(I) cations: Structure of an inorganic double helix, Proc. Natl. Acad. Sci. USA **84**, 2565–2569 (1987)

2.6   J.-M. Lehn, A. Rigault: Helicates: Tetra- and pentanuclear double helix complexes of Cu(I) and poly(bipyridine) strands, Angew. Chem. Int. Ed. Engl. **27**, 1095–1097 (1988)

2.7   J. D. Hartgerink, J. R. Granja, R. A. Milligan, M. R. Ghadiri: Self-assembling peptide nanotubes, J. Am. Chem. Soc. **118**, 43–50 (1996)

2.8   M. R. Ghadiri, J. R. Granja, R. A. Milligan, D. E. McRee, N. Khazanovich: Self-assembling organic nanotubes based on a cyclic peptide architecture, Nature **366**, 324–327 (1993)

2.9   V. Balzani, A. Credi, F. M. Raymo, J. F. Stoddart: Artificial molecular machines, Angew. Chem. Int. Ed. **39**, 3348–3391 (2000)

2.10   A. J. Bard, L. R. Faulkner: *Electrochemical Methods: Fundamentals and Applications* (Wiley, New York 2000)

2.11   V. Balzani (Ed.): *Electron Transfer in Chemistry* (Wiley-VCH, Weinheim 2001)

2.12   J. D. Coyle: *Principles and Applications of Photochemistry* (Wiley, New York 1988)

2.13   V. Balzani, M. Venturi, A. Credi: *Molecular Devices and Machines* (Wiley-VCH, Weinheim 2003)

2.14   P. R. Ashton, R. Ballardini, V. Balzani, A. Credi, K. R. Dress, E. Ishow, C. J. Kleverlaan, O. Kocian, J. A. Preece, N. Spencer, J. F. Stoddart, M. Venturi, S. Wenger: A photochemically driven molecular-level abacus, Chem. Eur. J. **6**, 3558–3574 (2000)

2.15   M. Irié (Ed.): Photochromism: memories and switches, Chem. Rev. **100**, 1683–1890 (2000)

2.16   B. L. Feringa (Ed.): *Molecular Switches* (Wiley-VCH, Weinheim 2001)

2.17   R. J. Mitchell: *Microprocessor Systems: An Introduction* (Macmillan, London 1995)

2.18   D. R. Smith: *Digital Transmission Systems* (Van Nostrand Reinhold, New York 1993)

2.19   S. Madhu: *Electronics: Circuits and Systems* (SAMS, Indianapolis 1985)

2.20   F. M. Raymo: Digital processing and communication with molecular switches, Adv. Mater. **14**, 401–414 (2002)

2.21   A. P. de Silva: Molecular computation – Molecular logic gets loaded, Nature Mater. **4**, 15–16 (2005)

2.22   A. Aviram: Molecules for memory, logic and amplification, J. Am. Chem. Soc. **110**, 5687–5692 (1988)

2.23   A. P. de Silva, H. Q. N. Gunaratne, C. P. McCoy: A molecular photoionic AND gate based on fluorescent signaling, Nature **364**, 42–44 (1993)

2.24   M. Asakawa, P. R. Ashton, V. Balzani, A. Credi, G. Mattersteig, O. A. Matthews, M. Montalti, N. Spencer, J. F. Stoddart, M. Venturi: Electrochemically induced molecular motions in pseudorotaxanes: A case of dual-mode (oxidative and reductive) dethreading, Chem. Eur. J. **3**, 1992–1996 (1997)

2.25   F. M. Raymo, S. Giordani, A. J. P. White, D. J. Williams: Digital processing with a three-state molecular switch, J. Org. Chem. **68**, 4158–4169 (2003)

2.26   F. M. Raymo, S. Giordani: Signal communication between molecular switches, Org. Lett. **3**, 3475–3478 (2001)

2.27   F. M. Raymo, S. Giordani: Multichannel Digital Transmission in an Optical Network of Communicating Molecules, J. Am. Chem. Soc. **124**, 2004–2007 (2002)

2.28   F. M. Raymo, S. Giordani: All-optical processing with molecular switches, Proc. Natl. Acad. Sci. USA **99**, 4941–4944 (2002)

2.29   A. J. Bard: *Integrated Chemical Systems: A Chemical Approach to Nanotechnology* (Wiley, New York 1994)

2.30   C. Joachim, J. K. Gimzewski, A. Aviram: Electronics using hybrid-molecular and mono-molecular devices, Nature **408**, 541–548 (2000)

2.31   J. M. Tour: Molecular electronics. Synthesis and testing of components, Acc. Chem. Res. **33**, 791–804 (2000)

2.32   A. R. Pease, J. O. Jeppesen, J. F. Stoddart, Y. Luo, C. P. Collier, J. R. Heath: Switching devices based on interlocked molecules, Acc. Chem. Res. **34**, 433–444 (2001)

2.33   R. M. Metzger: Unimolecular electrical rectifiers, Chem. Rev. **103**, 3803–3834 (2003)

2.34   M. C. Petty: *Langmuir–Blodgett Films: An Introduction* (Cambridge Univ. Press, Cambridge 1996)

2.35   A. Ulman: *An Introduction to Ultrathin Organic Films* (Academic, Boston 1991)

2.36   C. Lee, A. J. Bard: Comparative electrochemical studies of *N*-Methyl-*N*′-hexadecyl viologen monomolecular films formed by irreversible adsorption and the Langmuir–Blodgett method, J. Electroanal. Chem. **239**, 441–446 (1988)

2.37   C. Lee, A. J. Bard: Cyclic voltammetry and Langmuir film isotherms of mixed monolayers of *N*-docosoyl-*N*′-methyl viologen with arachidic acid, Chem. Phys. Lett. **170**, 57–60 (1990)

2.38   M. Fujihira, K. Nishiyama, H. Yamada: Photoelectrochemical responses of optically transparent electrodes modified with Langmuir–Blodgett films consisting of surfactant derivatives of electron donor, acceptor and sensitizer molecules, Thin Solid Films **132**, 77–82 (1985)

2.39   M. Fujihira: Photoelectric conversion with Langmuir–Blodgett films. In: *Nanostructures Based on Molecular Materials*, ed. by W. Göpel, C. Ziegler (VCH, Weinheim 1992) pp. 27–46

2.40   C. P. Collier, E. W. Wong, M. Belohradsky, F. M. Raymo, J. F. Stoddart, P. J. Kuekes, R. S. Williams, J. R. Heath: Electronically configurable molecular-based logic gates, Science **285**, 391–394 (1999)

2.41   E. W. Wong, C. P. Collier, M. Belohradsky, F. M. Raymo, J. F. Stoddart, J. R. Heath: Fabrication and transport properties of single-molecule-thick electrochemical junctions, J. Am. Chem. Soc. **122**, 5831–5840 (2000)

2.42   M. Asakawa, P. R. Ashton, V. Balzani, A. Credi, C. Hamers, G. Mattersteig, M. Montalti, A. N. Shipway, N. Spencer, J. F. Stoddart, M. S. Tolley, M. Venturi, A. J. P. White, D. J. Williams: A chemically and electrochemically switchable [2]catenane incorporating a tetrathiafulvalene unit, Angew. Chem. Int. Ed. **37**, 333–337 (1998)

2.43   V. Balzani, A. Credi, G. Mattersteig, O. A. Matthews, F. M. Raymo, J. F. Stoddart, M. Venturi, A. J. P. White, D. J. Williams: Switching of pseudorotaxanes and catenanes incorporating a tetrathiafulvalene unit by redox and chemical inputs, J. Org. Chem. **65**, 1924–1936 (2000)

2.44   M. Asakawa, M. Higuchi, G. Mattersteig, T. Nakamura, A. R. Pease, F. M. Raymo, T. Shimizu, J. F. Stoddart: Current/Voltage characteristics of monolayers of redox-switchable [2]catenanes on gold, Adv. Mater. **12**, 1099–1102 (2000)

2.45   C. P. Collier, G. Mattersteig, E. W. Wong, Y. Luo, K. Beverly, J. Sampaio, F. M. Raymo, J. F. Stoddart, J. R. Heath: A [2]catenane based solid-state electronically reconfigurable switch, Science **289**, 1172–1175 (2000)

2.46   C. P. Collier, J. O. Jeppesen, Y. Luo, J. Perkins, E. W. Wong, J. R. Heath, J. F. Stoddart: Molecular-based electronically switchable tunnel junction devices, J. Am. Chem. Soc. **123**, 12632–12641 (2001)

2.47   J. Chen, M. A. Reed, A. M. Rawlett, J. M. Tour: Large on-off ratios and negative differential resistance in a molecular electronic device, Science **286**, 1550–1552 (1999)

2.48   M. A. Reed, J. Chen, A. M. Rawlett, D. W. Price, J. M. Tour: Molecular random access memory cell, Appl. Phys. Lett. **78**, 3735–3737 (2001)

2.49   D. I. Gittins, D. Bethell, R. J. Nichols, D. J. Schiffrin: Redox-controlled multilayers of discrete gold particles: A novel electroactive nanomaterial, Adv. Mater. **9**, 737–740 (1999)

2.50   D. I. Gittins, D. Bethell, R. J. Nichols, D. J. Schiffrin: Diode-like electron transfer across nanostructured films containing a redox ligand, J. Mater. Chem. **10**, 79–83 (2000)

2.51   D. I. Gittins, D. Bethell, D. J. Schiffrin, R. J. Nichols: A nanometer-scale electronic switch consisting of a metal cluster and redox-addressable groups, Nature **408**, 67–69 (2000)

2.52   A. N. Shipway, M. Lahav, I. Willner: Nanostructured gold colloid electrodes, Adv. Mater. **12**, 993–998 (2000)

2.53   A. N. Shipway, M. Lahav, R. Blonder, I. Willner: Bis-bipyridinium cyclophane receptor–Au nanoparticle superstructure for electrochemical sensing applications, Chem. Mater. **11**, 13–15 (1999)

2.54   M. Lahav, A. N. Shipway, I. Willner, M. B. Nielsen, J. F. Stoddart: An enlarged bis-bipyridinum cyclophane–Au nanoparticle superstructure for selective electrochemical sensing applications, J. Electroanal. Chem. **482**, 217–221 (2000)

2.55   R. E. Gillard, F. M. Raymo, J. F. Stoddart: Controlling self-assembly, Chem. Eur. J. **3**, 1933–1940 (1997)

2.56   F. M. Raymo, J. F. Stoddart: From supramolecular complexes to interlocked molecular compounds, Chemtracts – Organic Chemistry **11**, 491–511 (1998)

2.57   M. Lahav, T. Gabriel, A. N. Shipway, I. Willner: Assembly of a Zn(II)-porphyrin-bipyridinium dyad and Au-nanoparticle superstructures on conductive surfaces, J. Am. Chem. Soc. **121**, 258–259 (1999)

2.58   M. Lahav, V. Heleg-Shabtai, J. Wasserman, E. Katz, I. Willner, H. Durr, Y. Hu, S. H. Bossmann: Photoelectrochemistry with integrated photosensitizerelectron acceptor Au-nanoparticle arrays, J. Am. Chem. Soc. **122**, 11480–11487 (2000)

2.59 G. Will, S. N. Rao, D. Fitzmaurice: Heterosupramolecular optical write-read-erase device, J. Mater. Chem. **9**, 2297–2299 (1999)

2.60 A. Merrins, C. Kleverlann, G. Will, S. N. Rao, F. Scandola, D. Fitzmaurice: Time-resolved optical spectroscopy of heterosupramolecular assemblies based on nanostructured TiO_2 films modified by chemisorption of covalently linked ruthenium and viologen complex components, J. Phys. Chem. B **105**, 2998–3004 (2001)

2.61 H. Park, J. Park, A. K. L. Lim, E. H. Anderson, A. P. Alivisatos, P. L. McEuen: Nanomechanical oscillations in a single C_{60} transistor, Nature **407**, 57–60 (2000)

2.62 W. Liang, M. P. Shores, M. Bockrath, J. R. Long, H. Park: Kondo resonance in a single-molecule transistor, Nature **417**, 725–729 (2002)

2.63 J. Park, A. N. Pasupathy, J. I. Goldsmith, C. Chang, Y. Yaish, J. R. Petta, M. Rinkoski, J. P. Sethna, H. D. Abruna, P. L. McEuen, D. C. Ralph: Coulomb blockade and the Kondo effect in single-atom transistors, Nature **417**, 722–725 (2002)

2.64 C. Z. Li, H. X. He, N. J. Tao: Quantized tunneling current in the metallic nanogaps formed by electrodeposition and etching, Appl. Phys. Lett. **77**, 3995–3997 (2000)

2.65 H. He, J. Zhu, N. J. Tao, L. A. Nagahara, I. Amlani, R. Tsui: A conducting polymer nanojunction switch, J. Am. Chem. Soc. **123**, 7730–7731 (2001)

2.66 A. Bogozi, O. Lam, H. He, C. Li, N. J. Tao, L. A. Nagahara, I. Amlani, R. Tsui: Molecular adsorption onto metallic quantum wires, J. Am. Chem. Soc. **123**, 4585–4590 (2001)

2.67 A. Bezryadin, C. N. Lau, M. Tinkham: Quantum suppression of superconductivity in ultrathin nanowires, Nature **404**, 971–974 (2000)

2.68 D. Porath, A. Bezryadin, S. de Vries, C. Dekker: Direct measurement of electrical transport through DNA molecules, Nature **403**, 635–638 (2000)

2.69 S. J. Tans, M. H. Devoret, H. Dai, A. Thess, E. E. Smalley, L. J. Geerligs, C. Dekker: Individual single-wall carbon nanotubes as quantum wires, Nature **386**, 474–477 (1997)

2.70 A. F. Morpurgo, J. Kong, C. M. Marcus, H. Dai: Gate-controlled superconducting proximity effect in carbon nanotubes, Nature **286**, 263–265 (1999)

2.71 J. Nygård, D. H. Cobden, P. E. Lindelof: Kondo physics in carbon nanotubes, Nature **408**, 342–346 (2000)

2.72 W. Liang, M. Bockrath, D. Bozovic, J. H. Hafner, M. Tinkham, H. Park: Fabry–Perot interference in a nanotube electron waveguide, Nature **411**, 665–669 (2001)

2.73 M. S. Fuhrer, J. Nygård, L. Shih, M. Forero, Y.-G. Yoon, M. S. C. Mazzoni, H. J. Choi, J. Ihm, S. G. Louie, A. Zettl, P. L. McEuen: Crossed nanotube junctions, Science **288**, 494–497 (2000)

2.74 T. Rueckes, K. Kim, E. Joselevich, G. Y. Tseng, C.-L. Cheung, C. M. Lieber: Carbon nanotube-based nonvolatile random access memory for molecular computing, Science **289**, 94–97 (2000)

2.75 A. Bachtold, P. Hadley, T. Nakanishi, C. Dekker: Logic circuits with carbon nanotube transistors, Science **294**, 1317–1320 (2001)

3. Introduction to Carbon Nanotubes

Carbon nanotubes are remarkable objects that look set to revolutionize the technological landscape in the near future. Tomorrow's society will be shaped by nanotube applications, just as silicon-based technologies dominate society today. Space elevators tethered by the strongest of cables; hydrogen-powered vehicles; artificial muscles: these are just a few of the technological marvels that may be made possible by the emerging science of carbon nanotubes.

Of course, this prediction is still some way from becoming reality; we are still at the stage of evaluating possibilities and potential. Consider the recent example of fullerenes – molecules closely related to nanotubes. The anticipation surrounding these molecules, first reported in 1985, resulted in the bestowment of a Nobel Prize for their discovery in 1996. However, a decade later, few applications of fullerenes have reached the market, suggesting that similarly enthusiastic predictions about nanotubes should be approached with caution.

There is no denying, however, that the expectations surrounding carbon nanotubes are very high. One of the main reasons for this is the anticipated application of nanotubes to electronics. Many believe that current techniques for miniaturizing microchips are about to reach their lowest limits, and that nanotube-based technologies are the best hope for further miniaturization. Carbon nanotubes may therefore provide the building blocks for further technological progress, enhancing our standards of living.

In this chapter, we first describe the structures, syntheses, growth mechanisms and properties of carbon nanotubes. Then we discuss nanotube-related nano-objects, including those formed by reactions and associations of all-carbon nanotubes with foreign atoms, molecules and compounds, which may provide the path to hybrid materials with even better properties than "pristine" nanotubes. Finally, we will describe the most

important current and potential applications of carbon nanotubes, which suggest that the future for the carbon nanotube industry looks very promising indeed.

Carbon nanotubes have long been synthesized as products of the action of a catalyst on the gaseous species originating from the thermal decomposition of hydrocarbons (see Sect. 3.2). The first evidence that the nanofilaments produced in this way were actually nanotubes – that they exhibited an inner cavity – can be found in the transmission electron microscope micrographs published by *Radushkevich* et al. in 1952 [3.1]. This was of course related to and made possible by the progress in transmission electron microscopy. It is then likely that the carbon filaments prepared by *Hughes* and *Chambers* in 1889 [3.2], which is probably the first patent ever deposited in the field, and whose preparation method was also based on the catalytically enhanced thermal cracking of hydrocarbons, were already carbon nanotube-related morphologies. The preparation of vapor-grown carbon fibers was actually reported over a century ago [3.3, 4]. Since then, the interest in carbon nanofilaments/nanotubes has been recurrent, though within a scientific area almost limited to the carbon material scientist community. The reader is invited to consult the review published by *Baker* et al. [3.5] regarding the early works. Worldwide enthusiasm came unexpectedly in 1991, after the catalyst-free formation of nearly perfect concentric multiwall carbon nanotubes (c-MWNTs, see Sect. 3.1) was reported [3.6] as by-products of the formation of fullerenes via the electric-arc technique. But the real breakthrough occurred two years later, when attempts to fill the nanotubes in situ with various metals (see

Sect. 3.5) led to the discovery – again unexpected – of single-wall carbon nanotubes (SWNTs) simultaneously by *Iijima* et al. [3.7] and *Bethune* et al. [3.8]. Single-wall carbon nanotubes were really new nano-objects with properties and behaviors that are often quite specific (see Sect. 3.4). They are also beautiful objects for fundamental physics as well as unique molecules for experimental chemistry, although they are still somewhat mysterious since their formation mechanisms are the subject of controversy and are still debated (see Sect. 3.3). Potential applications seem countless, although few have reached marketable status so far (see Sect. 3.6). Consequently, about five papers a day are currently published by research teams from around the world with carbon nanotubes as the main topic, an illustration of how extraordinarily active – and highly competitive – this field of research is. It is an unusual situation, similar to that for fullerenes, which, by the way, are again carbon nano-objects structurally closely related to nanotubes.

This is not, however, only about scientific exaltation. Economic aspects are leading the game to a greater and greater extent. According to experts, the world market was estimated to be more than 430 million dollars in 2004 and it is predicted to grow to several billion dollars before 2009. That is serious business, and it will be closely related to how scientists and engineers deal with the many challenges found on the path from the beautiful, ideal molecule to the reliable – and it is hoped, cheap – manufactured product.

3.1 Structure of Carbon Nanotubes

It is relatively easy to imagine a single-wall carbon nanotube (SWNT). Ideally, it is enough to consider a perfect graphene sheet (graphene is a polyaromatic monoatomic layer consisting of sp^2-hybridized carbon atoms arranged in hexagons; genuine graphite consists of layers of this graphene) and to roll it into a cylinder (Fig. 3.1), making sure that the hexagonal rings placed in contact join coherently. Then the tips of the tube are sealed by two caps, each cap being a hemi-fullerene of the appropriate diameter (Fig. 3.2a–c).

3.1.1 Single-Wall Nanotubes

Geometrically, there is no restriction on the tube diameter. However, calculations have shown that collapsing the single-wall tube into a flattened two-layer rib-

bon is energetically more favorable than maintaining the tubular morphology beyond a diameter value of ≈ 2.5 nm [3.9]. On the other hand, it is easy to grasp intuitively that the shorter the radius of curvature, the higher the stress and the energetic cost, although SWNTs with diameters as low as 0.4 nm have been synthesized successfully [3.10]. A suitable energetic compromise is therefore reached for ≈ 1.4 nm, the most frequent diameter encountered regardless of the synthesis technique (at least for those based on solid carbon sources) when conditions ensuring high SWNT yields are used. There is no such restriction on the nanotube length, which only depends on the limitations of the preparation method and the specific conditions used for the synthesis (thermal gradients, residence time, and so on). Experimental data are consistent with these statements, since SWNTs

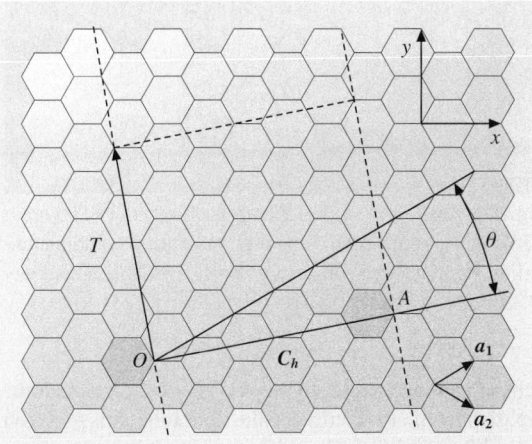

Fig. 3.1 Sketch of the way to make a single-wall carbon nanotube, starting from a graphene sheet (adapted from [3.11])

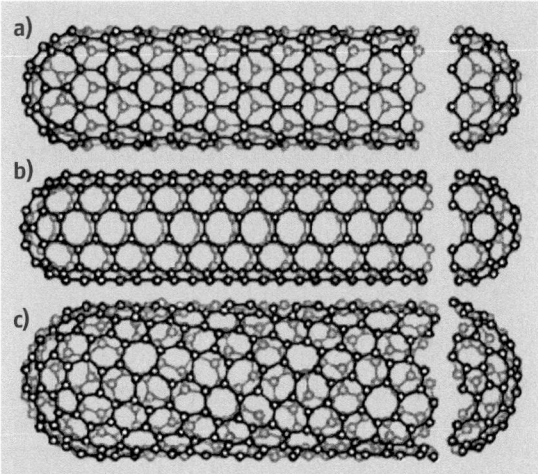

Fig. 3.2a–c Sketches of three different SWNT structures that are examples of (**a**) a zig-zag-type nanotube, (**b**) an armchair-type nanotube, (**c**) a helical nanotube (adapted from [3.12])

wider than 2.5 nm are only rarely reported in the literature, whatever the preparation method, while the length of the SWNTs can be in the micrometer or the millimeter range. These features make single-wall carbon nanotubes a unique example of single molecules with huge aspect ratios.

Two important consequences derive from the SWNT structure as described above:

1. All carbon atoms are involved in hexagonal aromatic rings only and are therefore in equivalent positions, except at each nanotube tip, where $6 \times 5 = 30$ atoms are involved in pentagonal rings (considering that adjacent pentagons are unlikely) – though not more, not less, as a consequence of Euler's rule that also governs the fullerene structure. For ideal SWNTs, chemical reactivity will therefore be highly favored at the tube tips, at the locations of the pentagonal rings.
2. Although carbon atoms are involved in aromatic rings, the C=C bond angles are not planar. This means that the hybridization of carbon atoms is not pure sp^2; it has some degree of the sp^3 character, in a proportion that increases as the tube radius of curvature decreases. The effect is the same as for the C_{60} fullerene molecules, whose radius of curvature is 0.35 nm, and whose bonds therefore have 10% sp^3 character[3.13]. On the one hand, this is believed to make the SWNT surface a bit more reactive than regular, planar graphene, even though it still consists of aromatic ring faces. On the other hand, this some-

how induces variable overlapping of energy bands, resulting in unique and versatile electronic behavior (see Sect. 3.4).

As illustrated by Fig. 3.2, there are many ways to roll a graphene into a single-wall nanotube, with some of the resulting nanotubes possessing planes of symmetry both parallel and perpendicular to the nanotube axis (such as the SWNTs from Fig. 3.2a and 3.2b), while others do not (such as the SWNT from Fig. 3.2c). Similar to the terms used for molecules, the latter are commonly called "chiral" nanotubes, since they are unable to be superimposed on their own image in a mirror. "Helical" is however sometimes preferred (see below). The various ways to roll graphene into tubes are therefore mathematically defined by the vector of helicity C_h, and the angle of helicity θ, as follows (referring to Fig. 3.1):

$$OA = C_h = na_1 + ma_2$$

with

$$a_1 = \frac{a\sqrt{3}}{2}x + \frac{a}{2}y \quad \text{and} \quad a_2 = \frac{a\sqrt{3}}{2}x - \frac{a}{2}y$$

where $a = 2.46 \text{ Å}$

and

$$\cos\theta = \frac{2n+m}{2\sqrt{n^2+m^2+nm}}$$

where n and m are the integers of the vector OA considering the unit vectors a_1 and a_2.

The vector of helicity $C_h(= OA)$ is perpendicular to the tube axis, while the angle of helicity θ is taken with respect to the so-called zig-zag axis: the vector of helicity that results in nanotubes of the "zig-zag" type (see below). The diameter D of the corresponding nanotube is related to C_h by the relation:

$$D = \frac{|C_h|}{\pi} = \frac{a_{CC}\sqrt{3(n^2 + m^2 + nm)}}{\pi} \,,$$

where

$$\underset{\text{(graphite)}}{1.41\,\text{Å}} \leq a_{C=C} \leq \underset{(C_{60})}{1.44\,\text{Å}} \,.$$

The C–C bond length is actually elongated by the curvature imposed by the structure; the average bond length in the C_{60} fullerene molecule is a reasonable upper limit, while the bond length in flat graphene in genuine graphite is the lower limit (corresponding to an infinite radius of curvature). Since C_h, θ, and D are all expressed as a function of the integers n and m, they are sufficient to define any particular SWNT by denoting them (n, m). The values of n and m for a given SWNT can be simply obtained by counting the number of hexagons that separate the extremities of the C_h vector following the unit vector a_1 first and then a_2 [3.11]. In the example of Fig. 3.1, the SWNT that is obtained by rolling the graphene so that the two shaded aromatic cycles can be superimposed exactly is a (4,2) chiral nanotube. Similarly, SWNTs from Fig. 3.2a to 3.2c are (9,0), (5,5), and (10,5) nanotubes respectively, thereby providing examples of zig-zag-type SWNT (with an angle of helicity = 0°), armchair-type SWNT (with an angle of helicity of 30°) and a chiral SWNT, respectively. This also illustrates why the term "chiral" is sometimes inappropriate and should preferably be replaced with "helical". Armchair (n, n) nanotubes, although definitely achiral from the standpoint of symmetry, exhibit a nonzero "chiral angle". "Zig-zag" and "armchair" qualifications for

Fig. 3.3 Image of two neighboring chiral SWNTs within a SWNT bundle as seen using high-resolution scanning tunneling microscopy (courtesy of Prof. Yazdani, University of Illinois at Urbana, USA)

achiral nanotubes refer to the way that the carbon atoms are displayed at the edge of the nanotube cross-section (Fig. 3.2a and 3.2b). Generally speaking, it is clear from Figs. 3.1 and 3.2a that having the vector of helicity perpendicular to any of the three overall C=C bond directions will provide zig-zag-type SWNTs, denoted $(n, 0)$, while having the vector of helicity parallel to one of the three C=C bond directions will provide armchair-type SWNTs, denoted (n, n). On the other hand, because of the sixfold symmetry of the graphene sheet, the angle of helicity θ for the chiral (n, m) nanotubes is such that $0 < \theta < 30°$. Figure 3.3 provides two examples of what chiral SWNTs look like, as seen via atomic force microscopy.

The graphenes in graphite have π electrons which are accommodated by the stacking of graphenes, allowing van der Waals forces to develop. Similar reasons make fullerenes gather and order into fullerite crystals and SWNTs into SWNT ropes (Fig. 3.4a). Provided the SWNT diameter distribution is narrow, the SWNTs in ropes tend to spontaneously arrange into hexagonal arrays, which correspond to the highest compactness achievable (Fig. 3.4b). This feature brings new periodicities with respect to graphite or turbostratic polyaromatic carbon crystals. Turbostratic structure corresponds to

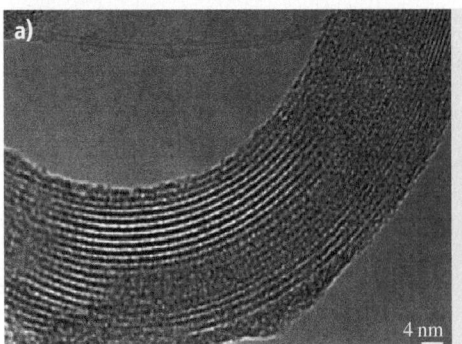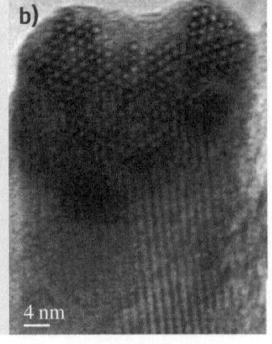

4 nm   4 nm

Fig. 3.4a,b High-resolution transmission electron microscopy images of a SWNT rope. (**a**) Longitudinal view. An isolated single SWNT also appears at the top of the image. (**b**) Cross-sectional view (from [3.14])

graphenes that are stacked with random rotations or translations instead of being piled up following sequential ABAB positions, as in graphite structure. This implies that no lattice atom plane exists other than the graphene planes themselves (corresponding to the (001) atom plane family). These new periodicities give specific diffraction patterns that are quite different to those of other sp^2-carbon-based crystals, although hk reflections, which account for the hexagonal symmetry of the graphene plane, are still present. On the other hand, $00l$ reflections, which account for the stacking sequence of graphenes in regular, "multilayered" polyaromatic crystals (which do not exist in SWNT ropes) are absent. This hexagonal packing of SWNTs within the ropes requires that SWNTs exhibit similar diameters, which is the usual case for SWNTs prepared by electric arc or laser vaporization processes. SWNTs prepared using these methods are actually about 1.35 nm wide (diameter of a (10,10) tube, among others), for reasons that are still unclear but are related to the growth mechanisms specific to the conditions provided by these techniques (see Sect. 3.3).

3.1.2 Multiwall Nanotubes

Building multiwall carbon nanotubes is a little bit more complex, since it involves the various ways graphenes can be displayed and mutually arranged within filamentary morphology. A similar versatility can be expected to the usual textural versatility of polyaromatic solids. Likewise, their diffraction patterns are difficult to differentiate from those of anisotropic polyaromatic solids. The easiest MWNT to imagine is the concentric type (c-MWNT), in which SWNTs with regularly increasing diameters are coaxially arranged (according to a Russian-doll model) into a multiwall nanotube (Fig. 3.5). Such nanotubes are generally formed either by the electric arc technique (without the need for a catalyst), by catalyst-enhanced thermal cracking of gaseous hydrocarbons, or by CO disproportionation (see Sect. 3.2). There can be any number of walls (or coaxial tubes), from two upwards. The intertube distance is approximately the same as the intergraphene distance in turbostratic, polyaromatic solids, 0.34 nm (as opposed to 0.335 nm in genuine graphite), since the increasing radius of curvature imposed on the concentric graphenes prevents the carbon atoms from being arranged as in graphite, with each of the carbon atoms from a graphene facing either a ring center or a carbon atom from the neighboring graphene. However, two cases allow a nanotube to reach – totally or partially – the 3-D crystal periodicity of graphite. One is to consider a high num-

ber of concentric graphenes: concentric graphenes with a long radius of curvature. In this case, the shift in the relative positions of carbon atoms from superimposed graphenes is so small with respect to that in graphite that some commensurability is possible. This may result in MWNTs where both structures are associated; in other words they have turbostratic cores and graphitic outer parts [3.15]. The other case occurs for c-MWNTs exhibiting faceted morphologies, originating either from the synthesis process or more likely from subsequent heat treatment at high temperature (such as 2500 °C) in inert atmosphere. Facets allow the graphenes to resume a flat arrangement of atoms (except at the junction between neighboring facets) which allows the specific stacking sequence of graphite to develop.

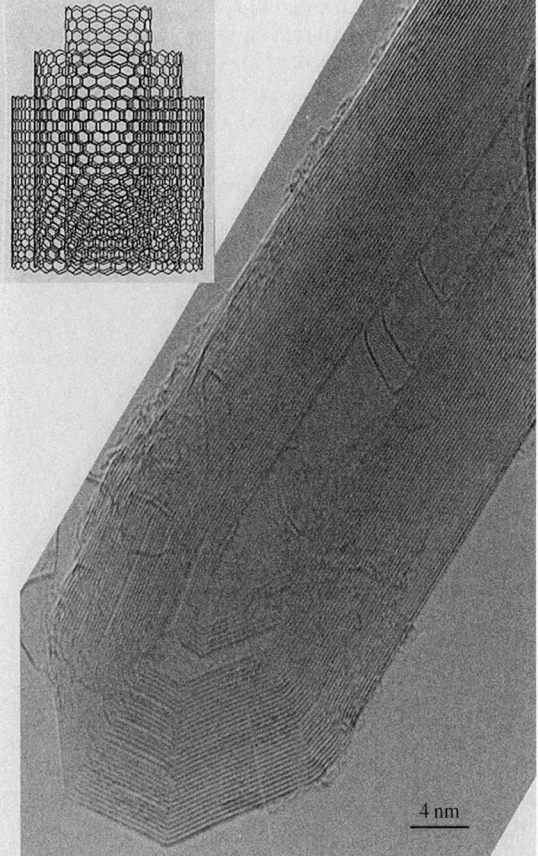

Fig. 3.5 High-resolution transmission electron microscopy image (longitudinal view) of a concentric multiwall carbon nanotube (c-MWNT) prepared using an electric arc. The *insert* shows a sketch of the Russian doll-like arrangement of graphenes

Another frequent inner texture for multiwall carbon nanotubes is the so-called herringbone texture (h-MWNTs), in which the graphenes make an angle with respect to the nanotube axis (Fig. 3.6). The angle value varies upon the processing conditions (such as the catalyst morphology or the composition of the atmosphere), from 0 (in which case the texture becomes that of a c-MWNT) to 90° (in which case the filament is no longer a tube, see below), and the inner diameter varies so that the tubular arrangement can be lost [3.16], meaning that the latter are more accurately called nanofibers rather than nanotubes. h-MWNTs are exclusively obtained by processes involving catalysts, generally catalyst-enhanced thermal cracking of hydrocarbons or CO disproportionation. One unresolved question is whether the herringbone texture, which actually describes the texture projection rather than the overall three-dimensional texture, originates from the scroll-like spiral arrangement of a single graphene ribbon or from the stacking of independent truncated cone-like graphenes in what is also called a "cup-stack" texture.

Fig. 3.7a,b Transmission electron microscopy images from bamboo multiwall nanotubes (longitudinal views). (**a**) Low magnification of a bamboo-herringbone multiwall nanotube (bh-MWNT) showing the nearly periodic nature of the texture, which occurs very frequently. (from [3.18]); (**b**) High-resolution image of a bamboo-concentric multiwall nanotube (bc-MWNT) (modified from [3.19])

Another common feature is the occurrence, to some degree, of a limited amount of graphenes oriented perpendicular to the nanotube axis, thus forming a "bamboo" texture. This is not a texture that can exist on its own; it affect either the c-MWNT (bc-MWNT) or the h-MWNT (bh-MWNT) textures (Figs. 3.6 and 3.7). The question is whether such filaments, although hollow, should still be called nanotubes, since the inner cavity is no longer open all the way along the filament as it is for a genuine tube. These are therefore sometimes referred as "nanofibers" in the literature too.

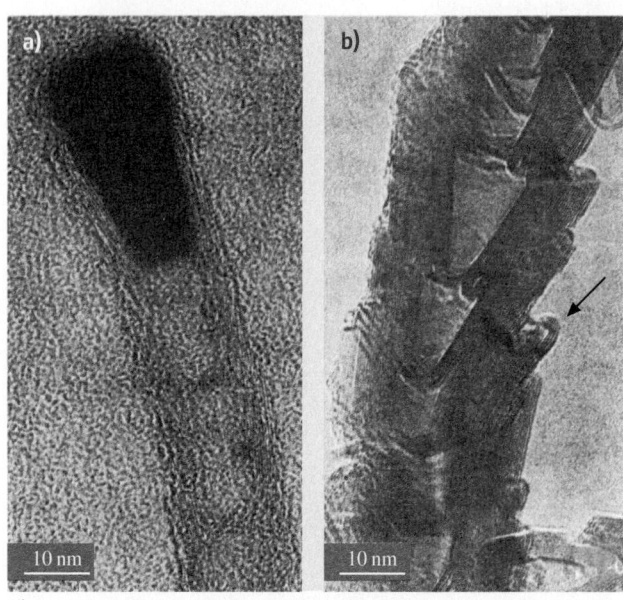

Fig. 3.6a,b Some of the earliest high-resolution transmission electron microscopy images of a herringbone (and bamboo) multiwall nanotube (bh-MWNT, longitudinal view) prepared by CO disproportionation on Fe-Co catalyst. (**a**) As-grown. The nanotube surface is made of free graphene edges. (**b**) After 2900 °C heat treatment. Both the herringbone and the bamboo textures have become obvious. Graphene edges from the surface have buckled with their neighbors (*arrow*), closing off access to the intergraphene space (adapted from [3.17])

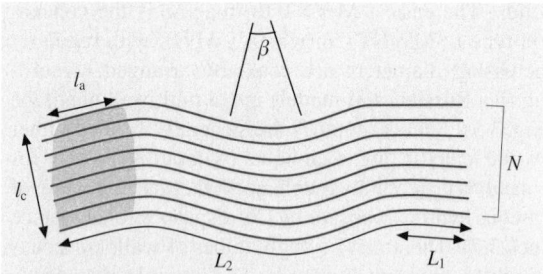

Fig. 3.8 Sketch explaining the various parameters obtained from high-resolution (lattice fringe mode) transmission electron microscopy, used to quantify nanotexture: L_1 is the average length of perfect (distortion-free) graphenes of coherent areas; N is the number of piled-up graphenes in coherent (distortion-free) areas; L_2 is the average length of continuous though distorted graphenes within graphene stacks; β is the average distortion angle. L_1 and N are related to the L_a and L_c values obtained from X-ray diffraction

One nanofilament that definitely cannot be called a nanotube is built from graphenes oriented perpendicular to the filament axis and stacked as piled-up plates. Although these nanofilaments actually correspond to h-MWNTs with a graphene/MWNT axis angle of 90°, an inner cavity is no longer possible, and such filaments are therefore often referred to as "platelet nanofibers" in the literature [3.16].

Unlike SWNTs, whose aspect ratios are so high that it is almost impossible to find the tube tips, the aspect ratios for MWNTs (and carbon nanofibers) are generally lower and often allow one to image tube ends by transmission electron microscopy. Aside from c-MWNTs derived from electric arc (see Fig. 3.5), which grow in a catalyst-free process, nanotube tips are frequently found to be associated with the catalyst crystals from which they were formed.

The properties of the MWNT (see Sect. 3.4) will obviously largely depend on the perfection and the orientation of the graphenes in the tube (for example, the spiral angles of the nanotubes constituting c-MWNTs has little importance). Graphene orientation is a matter of texture, as described above. Graphene perfection is a matter of nanotexture, which is commonly used to describe other polyaromatic carbon materials, and which is quantified by several parameters preferably obtained from high-resolution transmission electron microscopy (Fig. 3.8). Both texture and nanotexture depend on the processing conditions. While the texture type is a permanent, intrinsic feature which can only be completely altered upon a severe degradation treatment (such as oxidation), the nanotexture can be improved by subsequent thermal treatments at high temperatures (such as $> 2000\,°C$) and potentially degraded by chemical treatments (such as slightly oxidizing conditions).

3.2 Synthesis of Carbon Nanotubes

Producing carbon nanotubes so that the currently planned applications currently planned become marketable will require solving some problems that are more or less restrictive depending on the case. Examples include specifically controlling the configuration (chirality), the purity, or the structural quality of SWNTs, and adapting the production capacity to the application. One objective would be to understand the mechanism of nanotube nucleation and growth perfectly, and this remains a controversial subject despite an intense, worldwide experimental effort. This problem is partly due to our lack of knowledge regarding several parameters controlling the conditions during synthesis. For instance, the exact and accurate role of the catalysts in nanotube growth is often unknown. Given the large number of experimental parameters and considering the large range of conditions that the synthesis techniques correspond to, it is quite legitimate to think of more than one mechanism intervening during nanotube formation.

3.2.1 Solid Carbon Source–Based Production Techniques for Carbon Nanotubes

Among the different SWNT production techniques, the four processes (laser ablation, solar energy, dc electric arc, and three-phase ac arc plasma) presented in this section have at least two points in common: a high-temperature ($1000\,\mathrm{K} < T < 6000\,\mathrm{K}$) medium and the fact that the carbon source originates from the erosion of solid graphite. Despite these common points, the morphologies of the carbon nanostructures and the SWNT yields can differ notably with respect to the experimental conditions.

Before being utilized for carbon nanotube synthesis, these techniques permitted the production of fullerenes. Laser vaporization of graphite was actually the very first method to demonstrate the existence of fullerenes, including the most common one (because it is the most stable and therefore the most abundant), C_{60} [3.20]. On the other hand, the electric arc technique was (and still is) the first method of producing fullerenes in relatively large quantities [3.21–23]. Unlike fullerene formation, which requires the presence of carbon atoms in high-temperature media and the absence of oxygen, the utilization of these techniques for the synthesis of nanotubes (of SWNT type at least) requires an additional condition: the presence of catalysts in either the electrode or the target.

The different mechanisms (such as carbon molecule dissociation and atom recombination processes) involved in these high-temperature techniques take place at different time scales, from nanoseconds to microseconds and even milliseconds. The formation of nanotubes and other graphene-based products occurs afterward with a relatively long delay.

The methods of laser ablation, solar energy, and electric arc are all based on one essential mechanism: the energy transfer resulting from the interaction between either the target material and an external radiation source (a laser beam or radiation emanating from solar energy) or the electrode and the plasma (in case of an electric arc). This interaction causes target or anode erosion, leading to the formation of a plasma: an electrically neutral ionized gas, composed of neutral atoms, charged particles (molecules and ionized species) and electrons. The ionization degree of this plasma, defined by the ratio $(n_e/n_e + n_o)$, where n_e and n_o are the electron and that of neutral atom densities respectively, highlights the importance of energy transfer between the plasma and the material. The characteristics of this plasma and notably the ranges in temperature and concentrations of the various species present in the plasma thereby depend not only on the nature and composition of the target or the electrode but also on the energy transferred.

One of the advantages of these synthesis techniques is the ability to vary a large number of parameters that modify the composition of the high-temperature medium and consequently allow the most relevant parameters to be determined so that the optimal conditions for the control of carbon nanotube formation can be obtained. However, a major drawback of these techniques – and of any other technique used to produce SWNTs – is that the SWNTs formed are not pure: they are associated with other carbon phases and remnants of the catalyst. Although purification processes have been proposed in the literature and by some commercial companies for removing these undesirable phases, they are all based on oxidation (such as acid-based) processes that are likely to significantly affect the SWNT structure [3.14]. Subsequent thermal treatments at $\approx 1200\,°C$ under inert atmosphere, however, succeed in recovering structural quality somewhat [3.26].

Laser Ablation

After the first laser was built in 1960, physicists immediately made use of it as a means of concentrating a large quantity of energy inside a very small volume within a relatively short time. The consequence of this energy input naturally depends upon the characteristics of the device employed. During the interaction between the laser beam and the material, numerous phenomena occur at the same time and/or follow each other within the a certain time period, and each of these processes are sensitive to different parameters such as the characteristics of the laser beam, the incoming power density (also

termed the "fluence"), the nature of the target, and the environment surrounding it. For instance, the solid target can merely heat up, melt or vaporize depending on the power provided.

While this technique was successfully used to synthesize fullerene-related structures for the very first time [3.20], the synthesis of SWNTs by laser ablation took another ten years of research [3.24].

Laser Ablation – Experimental Devices

Two types of laser devices are currently utilized for carbon nanotube production: lasers operating in pulsed mode and lasers operating in continuous mode, with the latter generally providing a smaller fluence.

An example of the layout of a laser ablation device is given in Fig. 3.9. A graphite pellet containing the catalyst is placed in the middle of a quartz tube filled with inert gas and placed in an oven maintained at a temperature of $1200\,°C$ [3.24, 25]. The energy of the laser beam focused on the pellet permits it to vaporize and sublime the graphite by uniformly bombarding its surface. The carbon species, swept along by a flow of neutral gas, are then deposited as soot in different regions: on the conical water-cooled copper collector, on the quartz tube walls, and on the backside of the pellet.

Various improvements have been made to this device in order to increase the production efficiency. For example, *Thess* et al. [3.27] employed a second pulsed laser that follows the initial impulsion but at a different frequency in order to ensure a more complete and efficient irradiation of the pellet. This second impulsion vaporizes the coarse aggregates issued from the first ablation, causing them to participate in the active carbon feedstock involved in nanotube growth. Other modifications were suggested by *Rinzler* et al. [3.26], who inserted a second quartz tube of a smaller diameter coaxially inside the first one. This second tube reduces the vaporization zone and so permits an in-

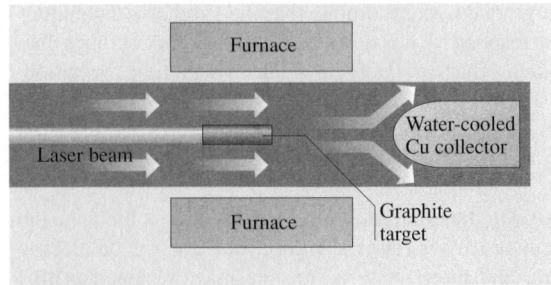

Fig. 3.9 Sketch of an early laser vaporization apparatus (adapted from [3.24, 25])

creased amounts of sublimed carbon to be obtained. They also arranged the graphite pellet on a revolving system so that the laser beam uniformly scans its whole surface.

Other groups have realized that, where the target contains both the catalyst and the graphite, the latter evaporates first and the pellet surface becomes more and more metal-rich, resulting in a decrease in the efficiency of nanotube formation during the course of the process. To solve this problem, *Yudasaka* et al. [3.29] utilized two pellets facing each other, one made entirely from the graphite powder and the other from an alloy of transition metals (catalysts), and irradiated them simultaneously.

A sketch of a synthesis reactor based on the vaporization of a target at a fixed temperature by a continuous CO_2 laser beam ($\lambda = 10.6\,\mu m$) is shown in Fig. 3.10 [3.28]. The power can be varied from 100 W to 1600 W. The temperature of the target is measured with an optical pyrometer, and these measurements are used to regulate the laser power to maintain a constant vaporization temperature. The gas, heated by contact with the target, acts as a local furnace and creates an extended hot zone, making an external furnace unnecessary. The gas is extracted through a silica pipe, and the solid products formed are carried away by the gas flow through the

pipe and then collected on a filter. The synthesis yield is controlled by three parameters: the cooling rate of the medium where the active, secondary catalyst particles are formed, the residence time, and the temperature (in the range 1000–2100 K) at which SWNTs nucleate and grow [3.30].

However, devices equipped with facilities to gather data such as the target temperature in situ are scarce and, generally speaking, this is one of the numerous variables of the laser ablation synthesis technique. The parameters that have been studied the most are the nature of the target, the nature and concentration of the catalyst, the nature of the neutral gas flow, and the temperature of the outer oven.

Laser Ablation – Results

In the absence of catalysts in the target, the soot collected mainly contains multiwall nanotubes (c-MWNTs). Their lengths can reach 300 nm. Their quantity and structural quality are dependent on the oven temperature. The best quality is obtained for an oven temperature set at 1200 °C. At lower oven temperatures, the structural quality decreases, and the nanotubes start presenting many defects [3.24]. As soon as small quantities (a few percent or less) of transition metal (Ni, Co) catalysts are incorporated into the graphite pellet, the products yielded undergo significant modifications, and SWNTs are formed instead of MWNTs. The yield of SWNTs strongly depends on the type of metal catalyst used and is seen to increase with the furnace temperature, among

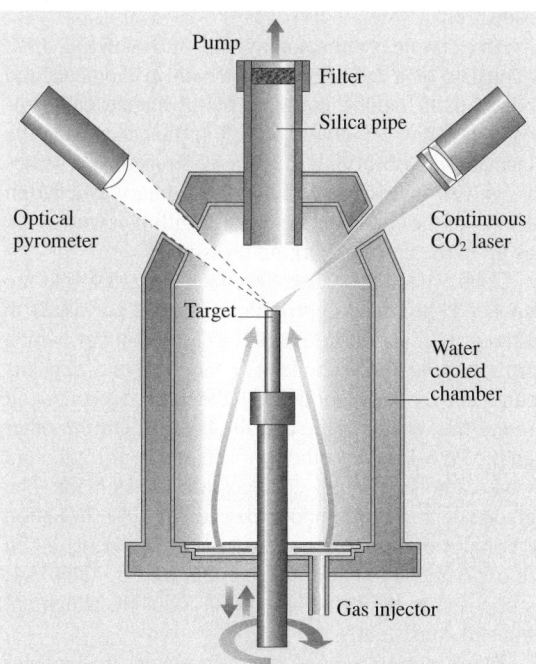

Fig. 3.10 Sketch of a synthesis reactor with a continuous CO_2 laser device (adapted from [3.28])

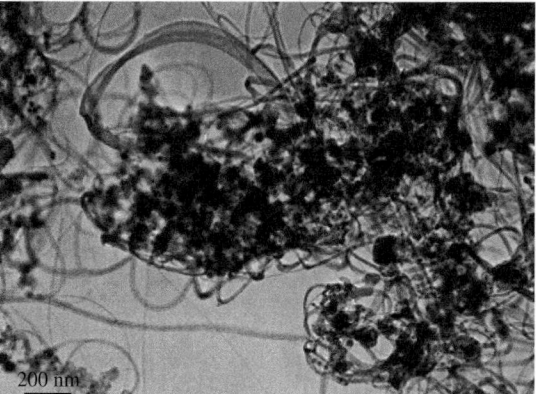

Fig. 3.11 Low-magnification TEM images of a typical raw SWNT material obtained using the laser vaporization technique. The fibrous structures are SWNT bundles, and the dark particles are remnants of the catalyst. Raw SWNT materials obtained from an electric arc exhibit similar features (from [3.14])

other factors. The SWNTs have remarkably uniform diameters and they self-organize into rope-like crystallites 5–20 nm in diameter and tens to hundreds of micrometers in length (Fig. 3.11). The ends of all of the SWNTs appear to be perfectly closed with hemispherical end-caps that show no evidence of any associated metal catalyst particle, although, as pointed out in Sect. 3.1, finding the two tips of a SWNT is rather challenging, considering the huge aspect ratio of the nanotube and their entangled nature. Another feature of the SWNTs produced with this technique is that they are supposedly "cleaner" than those produced using other techniques; in other words they associated with smaller amounts of the amorphous carbon that either coats the SWNTs or is gathered into nanoparticles. This advantage, however, only occurs for synthesis conditions designed to ensure high-quality SWNTs. It is not true when high-yield conditions are preferred; in this case SWNTs from an electric arc may appear cleaner than SWNTs from laser vaporization [3.14].

The laser vaporization technique is one of the three methods currently used to prepare SWNTs as commercial products. SWNTs prepared this way were first marketed by Carbon Nanotechnologies Inc. (Houston, TX, USA), with prices as high as $ 1000/g (raw materials) until December 2002. Probably because lowering the amount of impurities in the raw materials using this technique is impossible, they have recently decided to focus on fabricating SWNTs using the HiPCo technique (see Sect. 3.2.2). Laser-based methods are generally not considered to be competitive in the long term for the low-cost production of SWNTs compared to CCVD-based methods (see Sect. 3.2.2). However, prices as low as $ 0.03 per gram of raw high concentration have been estimated possible from a pre-industrial project study (Acolt S.A., Yverdon, Switzerland).

Electric Arc Method

Electric arcs between carbon electrodes have been studied as light sources and radiation standards for a very long time. They have however received renewed attention more recently due to their use in the production of new fullerene-related molecular carbon nanostructures, such as genuine fullerenes or nanotubes. This technique was first brought to light by *Krätschmer* et al. [3.21] who utilized it to achieve the production of fullerenes in macroscopic quantities. In the course of investigating other carbon nanostructures formed along with the fullerenes, and more particularly the solid carbon deposit that formed on the cathode, *Iijima* [3.6] discovered the catalyst-free formation of perfect c-MWNT-type carbon nanotubes. Then, as mentioned in the *Introduction*, the catalyst-promoted formation of SWNTs was accidentally discovered after some amounts of transition metals were introduced into the anode in an attempt to fill the c-MWNTs with metals during growth [3.7, 8]. Since then, a lot of work has been carried out by many groups using this technique in order to understand the mechanisms of nanotube growth as well as the role played by the catalysts (if any) in the synthesis of MWNTs and/or SWNTs [3.31–43].

Electric Arc Method – Experimental Devices

The principle of this technique is to vaporize carbon in the presence of catalysts (iron, nickel, cobalt, yttrium, boron, gadolinium, cerium, and so forth) in a reduced atmosphere of inert gas (argon or helium). After triggering an arc between two electrodes, a plasma is formed consisting of the mixture of carbon vapor, the rare gas (helium or argon), and the catalyst vapors. The vaporization is the consequence of energy transfer from the arc to the anode made of graphite doped with catalysts. The importance of the anode erosion rate depends on the power of the arc and also on other experimental conditions. It is worth noting that a high anode erosion does not necessarily lead to a high carbon nanotube production.

An example of a reactor layout is shown in Fig. 3.12. It consists of a cylinder about 30 cm in diameter and about 1 m in height, equipped with diametrically opposed sapphire windows located so that they face the plasma zone, observing the arc. The reactor possesses two valves, one for performing the primary evacuation (0.1 Pa) of the chamber, the other for filling it with a rare gas up to the desired working pressure.

Contrary to the solar energy technique, SWNTs are deposited (provided appropriate catalysts are used) in different regions of the reactor: (1) the collaret, which forms around the cathode; (2) the web-like deposits found above the cathode; (3) the soot deposited all around the reactor walls and the bottom. On the other hand, MWNTs are formed in a hard deposit adherent to the cathode whether catalysts are used or not. The cathode deposits form under the cathode. The formation of collaret and web is not systematic and depends on the experimental conditions, as indicated in Table 3.1, as opposed to the cathode deposit and soot, which are obtained consistently.

Two graphite rods of few millimeters in diameter constitute the electrodes between which a potential difference is applied. The dimensions of these electrodes vary according to the authors. In certain cases, the cathode has a greater diameter than the anode in order to facilitate their alignment [3.34, 44]. Other authors utilize electrodes of the same diameter [3.43]. The whole device can be designed horizontally [3.35, 43] or vertically [3.36, 38–40]. The advantage of the latter is the symmetry brought by the verticality with respect to gravity, which facilitates computer modeling (regarding convection flows, for instance).

Two types of anode can be utilized when catalysts are introduced: (1) a graphite anode containing a coaxial hole several centimeters in length into which a mixture of the catalyst and the graphite powder is placed; (2) a graphite anode within which the catalysts are homogeneously dispersed [3.45]. The former are by far the most popular, due to their ease of fabrication.

Optimizing the process in terms of the nanotube yield and quality is achieved by studying the roles of various parameters such as the type of doped anode (homogeneous or heterogeneous catalyst dispersion), the nature as well as the concentration of the catalyst, the nature of the plasmagen gas, the buffer gas pressure, the arc current intensity, and the distance between electrodes. Investigating the influences of these parameters on the type and amount of carbon nanostructures formed is, of course, the preliminary work that has been done. Although electric arc reactors equipped with the facilities to perform such investigations are scarce (see Fig. 3.12), investigating the missing link (the effect of

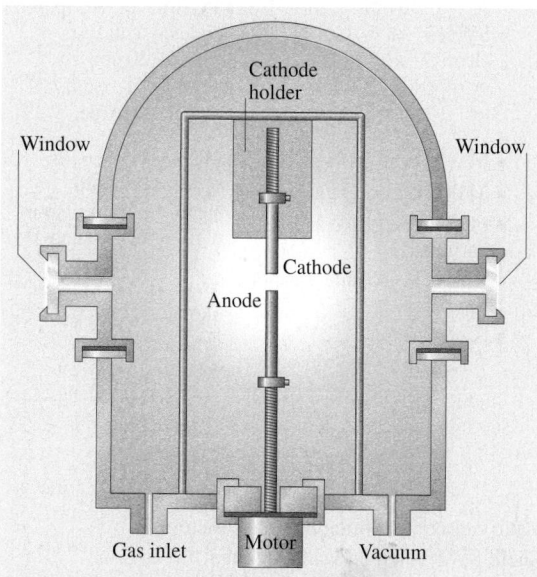

Fig. 3.12 Sketch of an electric arc reactor

varying the parameters on the plasma characteristics – the species concentrations and temperature) is likely to provide a more comprehensive understanding of the phenomena involved during nanotube formation. This has been recently performed using atomic and molecular optical emission spectroscopy [3.36, 38–41, 43].

Finally, we should mention attempts to create an electric arc in liquid media, such as liquid nitrogen [3.46] or water [3.47, 48]. The goal here is to make processing easier, since such systems should not require pumping devices or a closed volume and so they are more likely to allow continuous synthesis. This adaptation has not, however, reached the stage of mass production.

Electric Arc Method – Results

In view of the numerous results obtained with this electric arc technique, it is clear that both the morphology and the production efficiency of nanotubes strongly depends upon the experimental conditions used and, in particular, upon the nature of the catalysts. It is worth noting that the products obtained do not consist solely of carbon nanotubes. Nontubular forms of carbon, such as nanoparticles, fullerene-like structures including C_{60}, poorly organized polyaromatic carbons, nearly amorphous nanofibers, multiwall shells, single-wall nanocapsules, and amorphous carbon have all been obtained, as reported in Table 3.1 [3.37, 39, 40]. In addition, remnants of the catalyst are found all over the place – in the soot, the collaret, the web and the cathode deposit – in various concentrations. Generally, at a helium pressure of about 600 mbar, for an arc current of 80 A and for an electrode gap of 1 mm, the synthesis of SWNTs is favored by the use of Ni/Y as coupled catalysts [3.7, 35, 49]. In these conditions, which give high SWNT yields, SWNT concentrations are highest in the collaret (50–70%), then in the web ($\approx$ 50% or less) and then in the soot. On the other hand, c-MWNTs are found in the cathode deposit. SWNT lengths are micrometric and, typical outer diameters are around 1.4 nm. Using the latter conditions (Table 3.1, column 4), Table 3.1 illustrates the consequence of changing the parameters. For instance (Table 3.1, column 3), using Ni/Co instead of Ni/Y as catalysts prevents the formation of SWNTs. But when the Ni/Co catalysts are homogeneously dispersed in the anode (Table 3.1, column 1), the formation of nanotubes is promoted again, but MWNTs with two or three walls prevail over SWNTs, among which DWNTs (double-wall nanotubes) dominate. However, decreasing the ambient pressure from 60 to 40 kPa (Table 3.1, column 2) again suppresses nanotube formation.

Part A | 3.2

Table 3.1 Different carbon morphologies obtained by changing the type of anode, the type of catalyst and the pressure in a series of arc discharge experiments (electrode gap = 1 mm)

Catalyst (atom%) Arc conditions	0.6Ni + 0.6Co (homogeneous anode) $P \sim 60$ kPa $I \sim 80$ A	0.6Ni + 0.6Co (homogeneous anode) $P \sim 40$ kPa $I \sim 80$ A	0.5Ni + 0.5Co $P \sim 60$ kPa $I \sim 80$ A	4.2Ni + 1Y $P \sim 60$ kPa $I \sim 80$ A
Soot	• **MWNT + MWS + POPAC** or **Cn** ± catalysts $\phi \sim 3$–35 nm • NANF + catalysts • AC particles + catalysts • [*DWNT*], [*SWNT*], ropes or isolated, + POPAC	• **POPAC** and **AC** particles + **catalysts** $\phi \sim 2$–20 nm • NANF + catalysts $\phi \sim 5$–20 nm + MWS • [*SWNT*] $\phi \sim 1$–1.4 nm, distorted or damaged, isolated or ropes + Cn	• **AC** and POPAC particles + catalysts $\phi \sim 3$–35 nm • **NANF** + catalysts $\phi \sim 4$–15 nm • [*SWNT*] $\phi \sim 1.2$ nm, isolated or ropes	• **POPAC** and **AC** + particles + catalysts $\phi \leq 30$ nm • SWNT $\phi \sim 1.4$ nm, clean + Cn, short with tips, [*damaged*], isolated or ropes $\phi \leq 25$ nm • [*SWNC*] particles
Web	• [*MWNT*], **DWNT**, ϕ 2.7 – **4** – 5.7 nm SWNT ϕ 1.2–1.8 nm, isolated or **ropes** $\phi < 15$ nm, + POPAC ± Cn • AC particles + catalysts $\phi \sim 3$–40 nm + MWS • [*NANF*]	None	None	• SWNT, $\phi \sim 1.4$ nm, isolated or **ropes** $\phi \leq 20$ nm, + **AC** • POPAC and AC particles + catalysts $\phi \sim 3$ – **10** – 40 nm + MWS
Collaret	• **POPAC** and **SWNC** particles • Catalysts $\phi \sim 3$–250 nm, < **50** nm + MWS • SWNT ϕ 1–1.2 nm, [*opened*], **distorted**, isolated or **ropes** $\phi < 15$ nm, + **Cn** • [*AC*] particles	• **AC** and **POPAC** particles + catalysts $\phi \sim 3$–25 nm • SWNT $\phi \sim 1$–1.4 nm clean + Cn, [*isolated*] or ropes $\phi < 25$ nm • Catalysts $\phi \sim 5$–50 nm + MWS, • [*SWNC*]	• **Catalysts** $\phi \sim 3$–170 nm + MWS • AC or POPAC particles + **catalysts** $\phi \sim 3$–50 nm • SWNT $\phi \sim 1.4$ nm clean + Cn isolated or ropes $\phi < 20$ nm	• **SWNT** $\phi \sim 1.4$–2.5 nm, clean + Cn, [*damaged*], isolated or **ropes** $\phi < 30$ nm • POPAC or AC particles + catalysts $\phi \sim 3$–30 nm • [*MWS*] + catalysts or catalyst-free.
Cathode deposit	• **POPAC** and SWNC particles • **Catalysts** $\phi \sim$ 5–300 nm MWS • MWNT $\phi < 50$ nm • [*SWNT*] $\phi \sim 1.6$ nm clean + Cn, isolated or ropes	• **POPAC** and **SWNC** particles + Cn • Catalysts $\phi \sim 20$–100 nm + MWS	• **MWS**, catalyst-free • **MWNT** $\phi < 35$ nm • POPAC and PSWNC particles • [*SWNT*], isolated or ropes • [*Catalysts*] $\phi \sim 3$–30 nm	•**SWNT** $\phi \sim 1.4$–4.1 nm, clean + **Cn**, short with tips, isolated or ropes $\phi \leq 20$ nm. • **POPAC** or AC particles + catalysts $\phi \sim 3$–30 nm • MWS + catalysts $\phi < 40$ nm or catalyst-free • [*MWNT*]

Abundant – Present – [*Rare*]

Glossary: AC: amorphous carbon; POPAC: poorly organized polyaromatic carbon; Cn: fullerene-like structure, including C_{60}; NANF: nearly amorphous nanofiber; MWS: multiwall shell; SWNT: single-wall nanotube; DWNT: double-wall nanotube, MWNT: multiwall nanotube; SWNC: single-wall nanocapsule.

Recent works have attempted to replace graphite powder (sp^2 hybridized carbon) by diamond powder (sp^3 hybridized carbon). This is mixed with the catalyst powder and placed in hollowed-type graphite anodes. The result is an unexpected but quite significant increase (up to +230%) in the SWNT yield [3.41, 42]. Such experiments have revealed, as in the comparison between the results from using homogeneous instead of heterogeneous anodes, that the physical phenomena (charge and heat transfers) that occur in the anode during the arc are of the utmost importance, a factor which was neglected before this.

It is clear that while the use of a rare earth element (such as Y) as a single catalyst does not provide the right conditions to grow SWNTs, associating it with a transition metal (Ni/Y for instance) seems to lead to the best combinations that give the highest SWNT yields [3.44]. On the other hand, using a single rare earth element may lead to unexpected results, such as the closure of graphene edges from a c-MWNT wall with the neighboring graphene edges from the same wall side, leading to the preferred formation of telescope-like and open c-MWNTs that are able to contain nested Gd crystals [3.38, 40]. The effectiveness of bimetallic catalysts is believed to be due to the transitory formation of nickel particles coated with yttrium carbide, which has a lattice constant that is somewhat commensurable with that of graphene [3.50].

Figure 3.13 illustrates the kind of information provided by analyzing the plasma using emission spectroscopy, including profiles of radial temperature (Fig. 3.13a) or C_2 species concentration (Fig. 3.13b) in the plasma. A common feature is that a huge vertical gradient (≈ 500 K/mm) rapidly establishes (≈ 0.5 mm from the center in the radial direction) from the bottom to the top of the plasma, probably due to convection phenomena (Fig. 3.13a). The zone of actual SWNT formation is beyond the limit of the volume analyzable in the radial direction, corresponding to colder areas. The C_2 concentration increases dramatically from the anode to the cathode and decreases dramatically in the radial direction (Fig. 3.13b). This demonstrates that C_2 moieties are secondary products resulting from the recombination of primary species formed from the anode. It also suggests that C_2 moieties may be the building blocks for MWNTs (formed at the cathode) but not for SWNTs [3.40, 42].

Although many aspects of it still need to be understood, the electric arc method is one of the three methods currently used to produce SWNTs as commercial products. Though not selling bare nanotubes

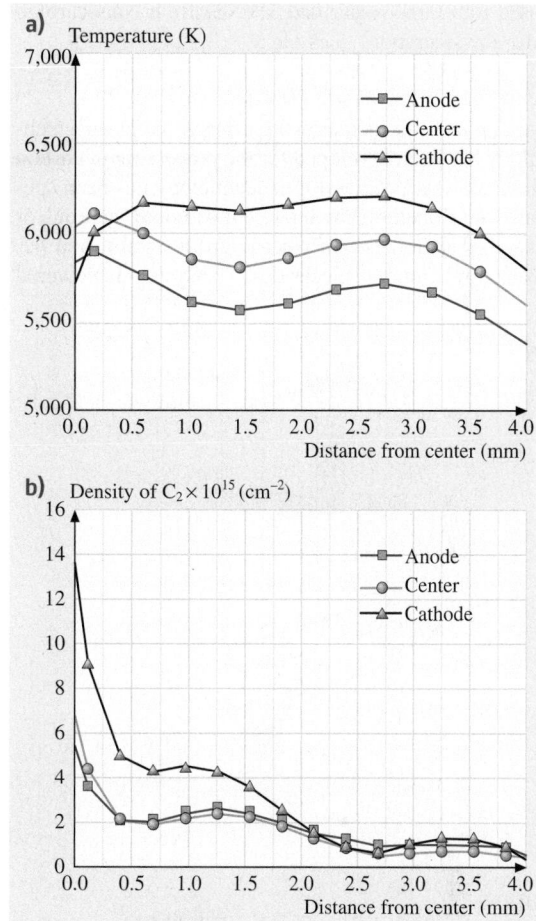

Fig. 3.13a,b Typical temperature (**a**) and C_2 concentration (**b**) profiles for plasma at the anode surface (*squares*), at the center of the plasma (*dots*), and at the cathode surface (*triangles*) at "standard" conditions (see text). Gradients are similar whichever catalyst is used, although absolute values may vary

anymore. Nanoledge S.A. (Montpellier, France), for instance, had a current production that reached several tens of kilograms per year (raw SWNTs, in other words unpurified), with a market price of ≈ 65 Euros/g in 2005, which was much cheaper than any other production method. It is however likely that the drop of prices for raw SWNTs down to 2–5 Euros/g which was anticipated for 2007 will not be possible. Actually, Bucky USA (Houston, Texas, USA) are still supplying raw SWNTs derived from electric arcs at a market price of 250 $/g in 2006 (which is, however, a 75% decrease in two years), which is barely lower than the ≈ 350 $/g proposed for 70–90%-purified SWNTs from Nanocarblab (Moscow, Russia).

Three-Phase AC Arc Plasma

An original semi-industrial three-phase AC plasma technology has been developed for the processing of carbon nanomaterials [3.52, 53]. The technology has been specially developed for the treatment of liquid, gaseous or dispersed materials. An electric arc is established between three graphite electrodes. The system is powered by a three-phase AC power supply operated at 600 Hz and at arc currents of 250–400 A. Carbon precursors, gaseous, liquid or solid, are injected at the desired (variable) position into the plasma zone. The reactive mixture can be extracted from the reaction chamber at different predetermined positions. After cooling down to room temperature, the aerosol passes through a filtering system. The main operating parameters, which are freely adjustable, include the arc current, the flow rate and the nature of the plasma gas (N_2, Ar, H_2, He, and so on), the carbon precursor (gaseous, liquid, solid, up to 3 kg/h), the injection and extraction positions, and the quenching rate. This plasma technology has shown very high versatility and it has been demonstrated that it can be

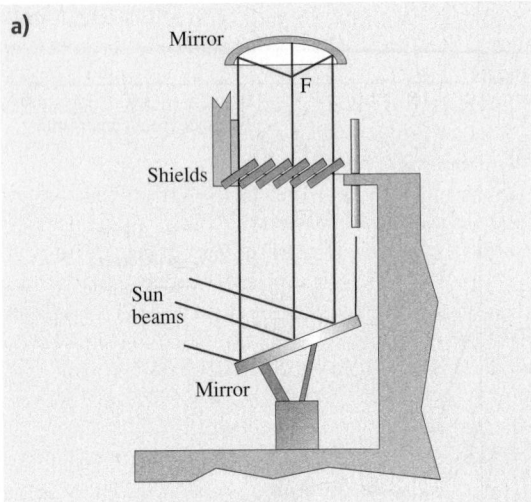

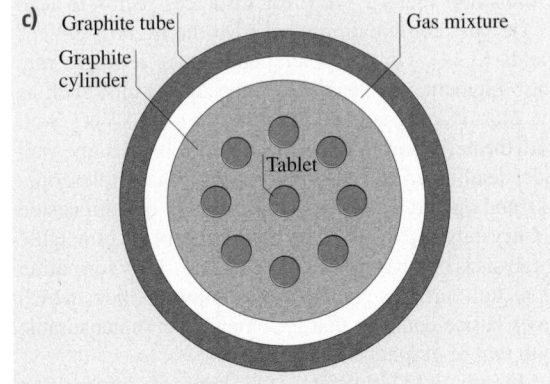

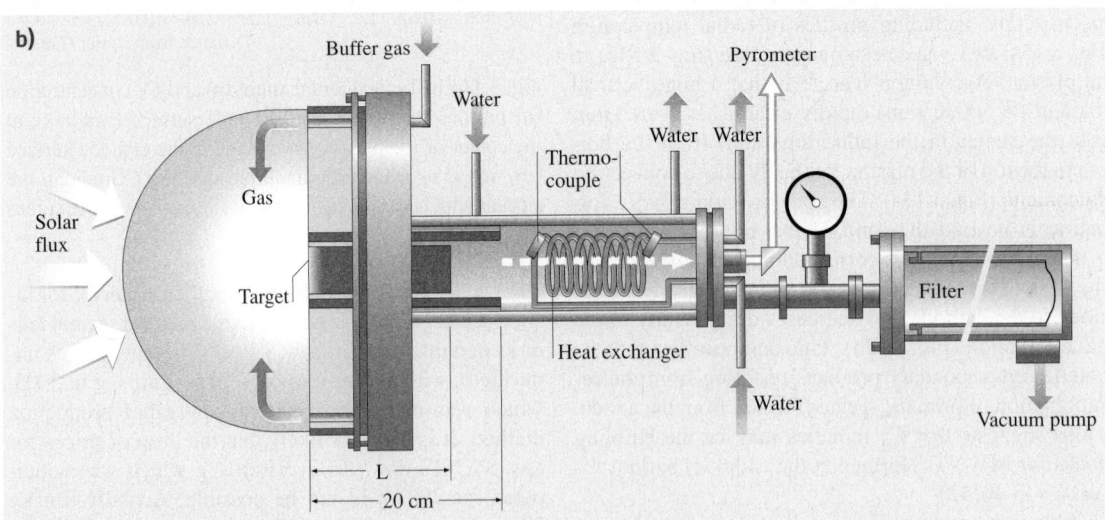

Fig. 3.14a–c Sketch of a solar energy reactor in use in the PROMES-CNRS Laboratory, Odeilho (France). (**a**) Gathering of sun rays, focused at F; (**b**) Side view of the experimental set-up at the focus of the 1 MW solar furnace; (**c**) top view of the target graphite rod. (Adapted from [3.51])

used to produce a wide range of carbon nanostructures ranging from carbon blacks to carbon nanotubes over fullerenes with a high product selectivity.

Solar Furnace

Solar furnace devices were originally utilized by several groups to produce fullerenes [3.54–56]. *Heben* et al. [3.57] and *Laplaze* et al. [3.58] later modified their original devices to achieve carbon nanotube production. This modification consisted mainly of using more powerful ovens [3.59, 60].

Solar Furnace – Experimental Devices

The principle of this technique is again based on the sublimation of a mixture of graphite powder and catalysts placed in a crucible in an inert gas. An example of such a device is shown in Fig. 3.14. The solar rays are collected by a plain mirror and reflected toward a parabolic mirror that focuses them directly onto a graphite pellet in a controlled atmosphere (Fig. 3.14a). The high temperature of about $4000\,K$ causes both the carbon and the catalysts to vaporize. The vapors are then dragged by the neutral gas and condense onto the cold walls of the thermal screen. The reactor consists of a brass support cooled by water circulation, upon which Pyrex chambers of various shapes can be fixed (Fig. 3.14b). This support contains a watertight passage permitting the introduction of the neutral gas and a copper rod onto which the target is mounted. The target is a graphite rod that includes pellets containing the catalysts, which is surrounded by a graphite tube (Fig. 3.14c) that acts as both a thermal screen to reduce radiation losses (very important in the case of graphite) and a duct to lead carbon vapors to a filter, which stops soot from being deposited on the Pyrex chamber wall. The graphite rod target replaces the graphite crucible filled with powdered graphite (for fullerene synthesis) or the mixture of graphite and catalysts (for nanotube synthesis) that were used in the techniques we have discussed previously.

These studies primarily investigated the target composition, the type and concentration of catalyst, the flow-rate, the composition and pressure of the plasmagenic gas inside the chamber, and the oven power. The objectives were similar to those of the works associated with the other solid carbon source-based processes. When possible, specific in situ diagnostics (pyrometry, optical emission spectroscopy, and so on) are also performed in order to investigate the roles of various parameters (temperature measurements at the crucible surface, along the graphite tube acting as thermal screen,

C_2 radical concentration in the immediate vicinity of the crucible).

Solar Furnace – Results

Some of the results obtained by different groups concerning the influence of the catalyst can be summarized as follows. With Ni/Co, and at low pressure, the sample collected contains mainly MWNTs with bamboo texture, carbon shells, and some bundles of SWNTs [3.59]. At higher pressures, only bundles of SWNTs are obtained, with fewer carbon shells. Relatively long bundles of SWNTs are observed with Ni/Y and at a high pressure. Bundles of SWNTs are obtained in the soot with Co; the diameters of the SWNTs range from 1 to 2 nm. *Laplaze* et al. [3.59] observed very few nanotubes but a large quantity of carbon shells.

In order to proceed to large-scale synthesis of single-wall carbon nanotubes, which is still a challenge for chemical engineers, *Flamant* et al. [3.51] and *Luxembourg* et al. [3.61] recently demonstrated that solar energy-based synthesis is a versatile method for obtaining SWNTs that can be scaled up from $0.1–0.2\,g/h$ to $10\,g/h$ and then to $100\,g/h$ productivity using existing solar furnaces. Experiments performed on a medium scale produced about $10\,g/h$ of SWNT-rich material using various mixtures of catalysts (Ni/Co, Ni/Y, Ni/Ce). A numerical reactor simulation was performed in order to improve the quality of the product, which was subsequently observed to reach 40% SWNT in the soot [3.62].

3.2.2 Gaseous Carbon Source-Based Production Techniques for Carbon Nanotubes

As mentioned in the *Introduction*, the catalysis-enhanced thermal cracking of a gaseous carbon source (hydrocarbons, CO) – commonly referred to as catalytic chemical vapor deposition (CCVD) – has long been known to produce carbon nanofilaments [3.3], so reporting on all of the works published in the field since the beginning of the century is almost impossible. Until the 1990s, however, carbon nanofilaments were mainly produced to act as a core substrate for the subsequent growth of larger (micrometric) carbon fibers – so-called vapor-grown carbon fibers – via thickening in catalyst-free CVD processes [3.63, 64]. We are therefore going to focus instead on more recent attempts to prepare genuine carbon nanotubes.

The synthesis of carbon nanotubes (either single- or multiwalled) by CCVD methods involves the catalytic decomposition of a carbon-containing source on small metallic particles or clusters. This technique involves either an heterogeneous process if a solid substrate is involved or an homogeneous process if everything takes place in the gas phase. The metals generally used for these reactions are transition metals, such as Fe, Co and Ni. It is a rather low-temperature process compared to arc discharge and laser ablation methods, with the formation of carbon nanotubes typically occurring between 600 °C and 1000 °C. Because of the low temperature, the selectivity of the CCVD method is generally better for the production of MWNTs with respect to graphitic particles and amorphous-like carbon, which remain an important part of the raw arc discharge SWNT samples, for example. Both homogeneous and heterogeneous processes appear very sensitive to the nature and the structure of the catalyst used, as well as to the operating conditions. Carbon nanotubes prepared by CCVD methods are generally much longer (a few tens to hundreds of micrometers) than those obtained by arc discharge (a few micrometers). Depending on the experimental conditions, it is possible to grow dense arrays of nanotubes. It is a general statement that MWNTs from CCVD contain more structural defects (exhibit a lower nanotexture) than MWNTs from arc discharge, due to the lower temperature of the reaction, which does not allow any structural rearrangements. These defects can be removed by subsequently applying heat treatments in vacuum or inert atmosphere to the products. Whether such a discrepancy is also true for SWNTs remains questionable. CCVD SWNTs are generally gathered into bundles that are generally of smaller diameter (a few tens of nm) than their arc discharge and laser ablation counterparts (around 100 nm in diameter). CCVD

provides reasonably good perspectives on large-scale and low-cost processes for the mass production of carbon nanotubes, a key point for their application at the industrial scale.

A final word concerns the nomenclature. Because work in the field started more than a century ago, the names of the carbon objects prepared by this method have changed with time with the authors, research areas, and fashions. These same objects have been called vapor-grown carbon fibers, nanofilaments, nanofibers and nanotubes. For multilayered fibrous morphologies (since single-layered fibrous morphologies can only be SWNT anyway), the exact name should be vapor-grown carbon nanofilaments (VGCNF). Whether or not the filaments are tubular is a matter of textural description, which should go with other textural features such as bamboo, herringbone and concentric (see Sect. 3.1.2). In the following, we will therefore use MWNTs for any hollowed nanofilament, whether they contain graphene walls oriented transversally or not. Any other nanofilament will be termed a "nanofiber."

Heterogeneous Processes

Heterogeneous CCVD processes simply involve passing a gaseous flow containing a given proportion of a hydrocarbon (mainly CH_4, C_2H_2, C_2H_4, or C_6H_6, usually as a mixture with either H_2 or an inert gas such as Ar) over small transition metal particles (Fe, Co, Ni) in a furnace. The particles are deposited onto an inert substrate, by spraying a suspension of the metal particles on it or by another method. The reaction is chemically defined as catalysis-enhanced thermal cracking

$$C_xH_y \rightarrow x\,C + y/2\,H_2 \,.$$

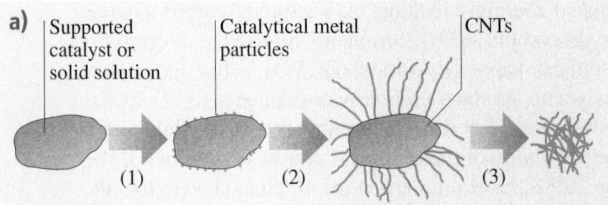

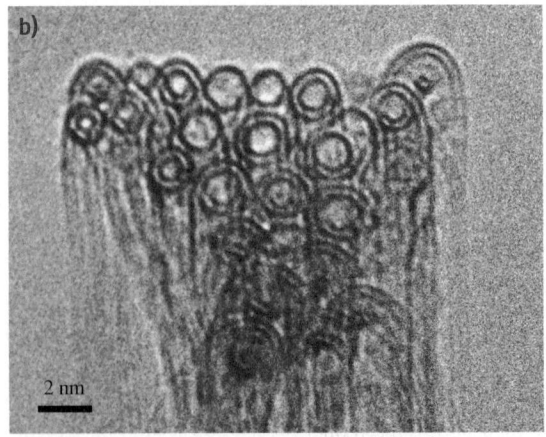

Fig. 3.15 (a) Formation of nanotubes via the CCVD-based impregnation technique. (1) Formation of catalytic metal particles by reduction of a precursor; (2) Catalytic decomposition of a carbon-containing gas, leading to the growth of carbon nanotubes; (3) Removal of the catalyst to recover the nanotubes (from [3.65]). **(b)** Example of a bundle of double-wall nanotubes (DWNTs) prepared this way (from [3.66])

Catalysis-enhanced thermal cracking was used as long ago as the late nineteenth century. Extensive works on this topic published before the 1990s include those by *Baker* et al. [3.5, 67], or *Endo* et al. [3.68, 69]. Several review papers have been published since then, such as [3.70], in addition to many regular papers.

CO can be used instead of hydrocarbons; the reaction is then chemically defined as catalysis-enhanced disproportionation (the so-called the Boudouard equilibrium)

$$2\,CO \rightarrow C + CO_2\,.$$

Heterogeneous Processes – Experimental Devices

The ability of catalysis-enhanced CO disproportionation to make carbon nanofilaments was reported by *Davis* et al. [3.71] as early as 1953, probably for the first time. Extensive follow-up work was performed by *Boehm* [3.72], *Audier* et al. [3.17, 73–75], and *Gadelle* et al. [3.76–79].

Although formation mechanisms for SWNTs and MWNTs can be quite different (see Sect. 3.3, or refer to a review article such as [3.80]), many of the catalytic process parameters play similar and important roles in the type of nanotubes formed: the temperature, the duration of the treatment, the gas composition and flow rate, and of course the catalyst nature and size. At a given temperature, depending mainly on the nature of both the catalyst and the carbon-containing gas, the catalytic decomposition will take place at the surfaces of the metal particles, followed by mass transport of the freshly produced carbon by surface or volume diffusion until the carbon concentration reaches the solubility limit, and the precipitation starts.

It is now agreed that CCVD carbon nanotubes form on very small metal particles, typically in the nanometer range [3.80]. These catalytic metal particles are prepared mainly by reducing transition metal compounds (salts, oxides) by H_2 prior to the nanotube formation step (where the carbon containing gas is required). It is possible, however, to produce these catalytic metal particles in situ in the presence of the carbon source, allowing for a one-step process [3.81]. Because controlling the metal particle size is the key issue (they have to be nanosized), coalescence is generally avoided by placing them on an inert support such as an oxide (Al_2O_3, SiO_2, zeolites, $MgAl_2O_4$, MgO) or more rarely on graphite. A low concentration of the catalytic metal precursor is required to limit the coalescence of the metal particles, which can happen during the reduction step. The supported catalysts can be used as a static phase placed within the gas flow, but can also be used as a fine powder suspended into and by the gas phase, in a so-called fluidised bed process. In the latter, the reactor has to be vertical so that to compensate the effect of gravity by the suspending effect of the gas flow.

There are two main ways to prepare the catalyst: (a) the impregnation of a substrate with a solution of a salt of the desired transition metal catalyst, and (b) the preparation of a solid solution of an oxide of the chosen catalytic metal in a chemically inert and thermally stable host oxide. The catalyst is then reduced to form the metal particles on which the catalytic decomposition of the carbon source will lead to carbon nanotube growth. In most cases, the nanotubes can then be separated from the catalyst (Fig. 3.15).

Heterogeneous Processes – Results with CCVD Involving Impregnated Catalysts

A lot of work had been done in this area even before the discovery of fullerenes and carbon nanotubes, but although the formation of tubular carbon structures by catalytic processes involving small metal particles was clearly identified, the authors did not focus on the preparation of SWNTs or MWNTs with respect to the other carbon species. Some examples will be given here to illustrate the most striking improvements obtained.

With the impregnation method, the process generally involves four different and successive steps: (1) impregnation of the support by a solution of a salt (nitrate, chloride) of the chosen metal catalyst; (2) drying and calcination of the supported catalyst to get the oxide of the catalytic metal; (3) reduction in a H_2-containing atmosphere to make the catalytic metal particles, and lastly (4) the decomposition of a carbon-containing gas over the freshly prepared metal particles that leads to nanotube growth. For example, *Ivanov* et al. [3.82] prepared nanotubes through the decomposition of C_2H_2 (pure or mixed with H_2) on well-dispersed transition metal particles (Fe, Co, Ni, Cu) supported on graphite or SiO_2. Co-SiO_2 was found to be the best catalyst/support combination for the preparation of MWNTs, but most of the other combinations led to carbon filaments, sometimes covered with amorphous-like carbon. The same authors have developed a precipitation-ion-exchange method that provides a better dispersion of metals on silica compared to the classical impregnation technique. The same group then proposed the use of a zeolite-supported Co catalyst [3.83, 84], resulting in very finely dispersed metal particles (from 1 to 50 nm in diameter). They observed MWNTs with a diameter around 4 nm and only two or three walls only on this catalyst. *Dai* et al. [3.85] have prepared SWNTs by CO disproportionation on nanosized Mo particles. The diameters of the nanotubes obtained are closely related to those of the original particles and range from 1 to 5 nm. The nanotubes obtained by this method are free of an amorphous carbon coating. They also found that a synergetic effect occurs for the alloy instead of the components alone, and one of the most striking examples is the addition of Mo to Fe [3.86] or Co [3.87].

Heterogeneous Processes – Results with CCVD Involving Solid Solution–Based Catalysts

A solid solution of two metal oxides is formed when ions of one metalmix with ions of the other metal. For example, Fe_2O_3 can be prepared in solid solution in Al_2O_3 to give a $Al_{2-2x}Fe_{2x}O_3$ solid solution. The use of a solid solution allows a perfectly homogeneous dispersion of one oxide in the other to be obtained. These solid solutions can be prepared in different ways, but coprecipitation of mixed oxalates and combustion synthesis are the most common methods used to prepare nanotubes. The synthesis of nanotubes by the catalytic decomposition of CH_4 over an $Al_{2-2x}Fe_{2x}O_3$ solid solution was originated by *Peigney* et al. [3.81] and then studied extensively by the same group using different oxides such as spinel-based solid solutions ($Mg_{1-x}M_xAl_2O_4$ with M = Fe, Co, Ni, or a binary alloy [3.65, 88]) or magnesia-based solid solutions [3.65, 89] ($Mg_{1-x}M_xO$, with M = Fe, Co or Ni). Because of the very homogeneous dispersion of the catalytic oxide, it is possible to produce very small catalytic metal particles at the high temperature required for the decomposition of CH_4 (which was chosen for its greater thermal stability compared to other hydrocarbons). The method proposed by these authors involves the heating of the solid solution from room temperature to a temperature of between 850 °C and 1050 °C in a mixture of H_2 and CH_4, typically containing 18 mol.% of CH_4. The nanotubes obtained clearly depend upon the nature of both the transition metal (or alloy) used and the inert oxide (matrix); the latter because the Lewis acidity seems to play an important role [3.90]. For example, in the case of solid solutions containing around 10 wt% of Fe, the amount of carbon nanotubes obtained decreases in the following order depending on the matrix oxide: MgO > Al_2O_3 > $MgAl_2O_4$ [3.65]. In the case of MgO-based solid solutions, the nanotubes can be very easily separated from the catalyst by dissolving it (in diluted HCl for example) [3.89]. The nanotubes obtained are typically gathered into small-diameter bundles (less

than 15 nm) with lengths of up to 100 μm. The nanotubes are mainly SWNTs and DWNTs, with diameters of between 1 and 3 nm.

Obtaining pure nanotubes by the CCVD method requires, as for all the other techniques, the removal of the catalyst. When a catalyst supported (impregnated) in a solid solution is used, the supporting – and catalytically inactive – oxide is the main impurity, both in weight and volume. When oxides such as Al_2O_3 or SiO_2 (or even combinations) are used, aggressive treatments involving hot caustic solutions (KOH, NaOH) for Al_2O_3 or the use of HF for SiO_2 are required. These treatments have no effect, however, on other impurities such as other forms of carbon (amorphous-like carbon, graphitized carbon particles and shells, and so on). Oxidizing treatments (air oxidation, use of strong oxidants such as HNO_3, $KMnO_4$, H_2O_2) are thus required and permit the removal of most unwanted forms of carbon, but they result in a low final yield of carbon nanotubes, which are often quite damaged. *Flahaut* et al. [3.89] were the first to use a MgCoO solid solution to prepare SWNTs and DWNTs that could be easily separated without incurring any damage via fast and safe washing with an aqueous HCl solution.

In most cases, only very small quantities of catalyst (typically less than 500 mg) are used, and most claims of "high-yield" productions of nanotubes are based on laboratory experimental data, without taking into account all of the technical problems related to scaling up to a laboratory-scale CCVD reactor. At the present time, although the production ofMWNTs is possible on an industrial scale, the production of affordable SWNTs is still a challenge, and controlling the arrangement of and the number of walls in the nanotubes is also problematic. For example, adding small amounts of molybdenum to the catalyst [3.91] can lead to drastic modifications of the nanotube type (from regular nanotubes to carbon nanofibers–see Sect. 3.1). *Flahaut* et al.have recently shown that the method used to prepare a particular catalyst can play a very important role [3.92]. Double-walled carbon nanotubes (DWNTs) represent a special case: they are at the frontier between single- (SWNTs) and multiwalled nanotubes (MWNTs). Because they are the MWNTs with the lowest possible number of walls, their structures and properties are very similar to those of SWNTs. Any subsequent functionalization, which is often required to improve the compatibility of nanotubes with their external environment (composites) or to give them new properties (solubility, sensors), will partially damage the external wall, resulting in drastic modifications in terms of both electrical and mechanical properties. This is a serious drawback for SWNTs. In the case of DWNTs, the outer wall can be modified (functionalized) while retaining the structure of the inner tube. DWNTs have been recently synthesised on a gram-scale by CCVD [3.66], with a high purity and a high selectivity (around 80% DWNTs) (Fig. 3.15b).

Homogeneous Processes

The homogenous route, also called the "floating catalyst method," differs from the other CCVD-based methods because it uses only gaseous species and does not require the presence of any solid phase in the reactor. The basic principle of this technique, similar to the other CCVD processes, is to decompose a carbon source (ethylene, xylene, benzene, carbon monoxide, and so on) on nanosized transition metal (generally Fe, Co, or Ni) particles in order to obtain carbon nanotubes. The catalytic particles are formed directly in the reactor, however, and are not introduced before the reaction, as occurs in supported CCVD for instance.

Homogeneous Processes –
Experimental Devices

The typical reactor used in this technique is a quartz tube placed in an oven into which the gaseous feedstock, containing the metal precursor, the carbon source, some hydrogen and a vector gas (N_2, Ar, or He), is passed. The first zone of the reactor is kept at a lower temperature, and the second zone, where the formation of tubes occurs, is heated to $700-1200\,°C$. The metal precursor is generally a metal-organic compound, such as a zero-valent carbonyl compound like $[Fe(CO)_5]$ [3.93], or a metallocene [3.94–96] such as ferrocene, nickelocene or cobaltocene. The use of metal salts, such as cobalt nitrate, has also been reported [3.97]. It may be advantageous to make the reactor vertical, so that gravity acts symmetrically on the gaseous volume inside the furnace.

Homogeneous Processes – Results

The metal-organic compound decomposes in the first zone of the reactor, generating nanosized metallic particles that can catalyze nanotube formation. In the second part of the reactor, the carbon source is decomposed to atomic carbon, which is then responsible for the formation of nanotubes.

This technique is quite flexible and SWNTs [3.98], DWNTs [3.99] and MWNTs [3.100] have been obtained, in proportions depending on the carbon feedstock gas. The technique has also been exploited for some time in the production of vapor-grown carbon nanofibers [3.101].

The main drawback of this type of process is again that it is difficult to control the size of the metal nanoparticles, and thus nanotube formation is often accompanied by the production of undesired carbon forms (amorphous carbon or polyaromatic carbon phases found as various phases or as coatings). In particular, encapsulated forms have been often found as the result of the formation of metal particles that are too large to promote nanotube growth (and so they can end up being totally covered with graphene layers instead).

The same kind of parameters have to be controlled as for heterogeneous processes in order to finely tune this process and selectively obtain the desired morphology and structure of the nanotubes formed, such as: the choice of the carbon source; the reaction temperature; the residence time; the composition of the incoming gaseous feedstock, with particular attention paid to the role played by the proportion of hydrogen, which can influence the orientation of the graphene with respect to the nanotube axis, thus switching from c-MWNT to h-MWNT [3.77]; and the ratio of the metallorganic precursor to the carbon source [3.94]. In an independent study [3.102], it was shown that the general tendency is: (i) to synthesize SWNTs when the ferrocene/benzene molar ratio is high, typically $\approx 15\%$; (ii) to produce MWNTs when the ferrocene/benzene molar ratio is between $\approx 4\%$ and $\approx 9\%$, and; (iii) to synthesize carbon nanofibers when the ferrocene/benzene molar ratio is below $\approx 4\%$.

As recently demonstrated, the overall process can be improved by adding other compounds such as ammonia or sulfur-containing species to the reactive gas phase. The former allows aligned nanotubes and mixed C–N nanotubes [3.103] to be obtained, while the latter results in a significant increase in productivity [3.101,104]. An interesting result is the increase in yield and purity brought about by a small input of oxygen, as achieved by using alcohol vapors instead of hydrocarbons as feedstock [3.105]. It is assumed that the oxygen preferably burns the poorly organized carbon out into CO_2, thereby enhancing the purity, and prevents the catalyst particles from being encapsulated in the carbon shells too early, making them inactive, thereby enhancing the nanotube yield. Moreover, it was found to promote the formation of SWNTs over MWNTs, since suppressing carbon shell formation suppresses MWNT formation too.

It should be emphasized that only small amounts have been produced so far, and scale-up to industrial levels seems quite difficult due to the large number of parameters that must be considered. A critical one is to

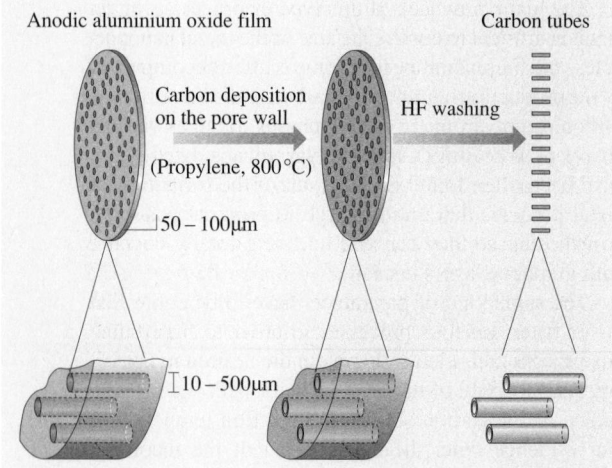

Fig. 3.16 Principle of the templating technique used in the catalyst-free formation of single-walled or concentric-type multiwalled carbon nanotubes (from [3.106])

be able to increase the quantity of metallorganic compound that is used in the reactor, in order to increase production, without obtaining particles that are too big. This problem has not yet been solved. An additional problem inherent in the process is the possibility of clogging the reactor due to the deposition of metallic nanoparticles on the reactor walls followed by carbon deposition.

A significant breakthrough concerning this technique could be the HiPCo process developed at Rice University, which produces SWNTs of very high purity [3.107, 108]. This gas phase catalytic reaction uses carbon monoxide to produce, from [Fe(CO)$_5$], a SWNT material that is claimed to be relatively free of by-products. The temperature and pressure conditions required are applicable to industrial plants. Upon heating, the [Fe(CO)$_5$] decomposes into atoms which condense into larger clusters, and SWNT nucleate and grow on these particles in the gas phase via CO disproportionation (the Boudouard reaction, see *Heterogeneous Processes* in Sect. 3.1.2):

The company Carbon Nanotechnologies Inc. (Houston, TX, USA) currently sells raw SWNT materials prepared in this way, at a market price of 375 \$/g, or 500 \$/g if purified (2005 data). Other companies that specialize in MWNTs include Applied Sciences Inc. (Cedarville, OH, USA), currently has a production facility of ≈ 40 tons/year of ≈ 100 nm large MWNTs (Pyrograf-III), and Hyperion Catalysis (Cambridge, MA, USA), which makes MWNT-based materials.

Though prepared in a similar way by CCVD-related processes, MWNTs remain far less expensive than SWNTs, reaching prices as low as 0.055 \$/g (current ASI fares for Pyrograf-III grade).

Templating

Another interesting technique, although one that is definitely not suitable for mass production (and so we only touch on it briefly here), is the templating technique. It is the only other method aside from the electric arc technique that is able to synthesize carbon nanotubes without any catalyst. Any other work reporting the catalyst-free formation of nanotubes is actually likely to have involved the presence of catalytic metallic impurities in the reactor or some other factors that caused a chemical gradient in the system. Another useful aspect of this approach is that it allows aligned nanotubes to be obtained naturally, without the help of any subsequent alignment procedure. However, the template must be removed (dissolved) to recover the nanotubes, in which case the alignment of the nanotubes is lost.

Templating – Experimental Devices

The principle of this technique is to deposit the solid carbon coating obtained from the CVD method onto the walls of a porous substrate whose pores are arranged in parallel channels. The feedstock is again a hydrocarbon, such as a common source of carbon. The substrate can be alumina or zeolite for instance, which present natural channel pores, while the whole system is heated to a temperature that cracks the hydrocarbon selected as the carbon source (Fig. 3.16).

Templating – Results

Provided the chemical vapor deposition mechanism (which is actually better described as a chemical vapor infiltration mechanism) is well controlled, synthesis results in the channel pore walls being coated with a variable number of graphenes. Both MWNTs (exclusively concentric type) and SWNTs can be obtained. The smallest SWNTs (diameters ≈ 0.4 nm) ever obtained (mentioned in Sect. 3.1) were actually been synthesized using this technique [3.10]. The nanotube lengths are directly determined by the channel lengths; in other words by the thickness of the substrate plate. One main advantage of the technique is the purity of the tubes (no catalyst remnants, and few other carbon phases). On the other hand, the nanotube structure is not closed at both ends, which can be an advantage or a drawback depending on the application. For instance, the porous matrix must be dissolved using one of the chemical treatments previously cited in order to recover the tubes. The fact that the tubes are open makes them even more sensitive to attack from acids.

3.2.3 Miscellaneous Techniques

In addition to the major techniques described in Sects. 3.2.1 and 3.2.2, many attempts to produce nanotubes in various ways, often with a specific goal in mind, such as looking for a low-cost or a catalyst-free production process, can be found in the literature. As yet, none has been convincing enough to be presented as a serious alternative to the major processes described previously. Some examples are provided in the following.

Hsu et al. [3.109] have succeeded in preparing MWNTs (including coiled MWNTs, a peculiar morphology resembling a spring) by a catalyst-free (although Li was present) electrolytic method, by running a 3–5 A current between two graphite electrodes (the anode was a graphite crucible and the cathode a graphite rod). The graphite crucible was filled with lithium chloride, while the whole system was heated in air or argon at $\approx 600\,^{\circ}$C. As with many other techniques, by-products such as encapsulated metal particles, carbon shells, amorphous carbon, and so on, are formed.

Cho et al. [3.110] have proposed a pure chemistry route to nanotubes, using the polyesterification of citric acid onto ethylene glycol at $50\,^{\circ}$C, followed by polymerization at $135\,^{\circ}$C and then carbonization at $300\,^{\circ}$C under argon, followed by oxidation at $400\,^{\circ}$C in air. Despite the latter oxidation step, the solid product contains short MWNTs, although they obviously have poor nanotextures. By-products such as carbon shells and amorphous carbon are also formed.

Li et al. [3.111] have also obtained short MWNTs through a catalyst-free (although Si is present) pyrolytic method which involves heating silicon carbonitride nanograins in a BN crucible to $1200–1900\,^{\circ}$C in nitrogen within a graphite furnace. No details are given about the possible occurrence of by-products, but they are likely considering the complexity of the chemical system (Si-C-B-N) and the high temperatures involved.

Terranova et al. [3.112] have investigated the catalyzed reaction between a solid carbon source and atomic hydrogen. Graphite nanoparticles ($\approx 20\,$nm) are sent with a stream of H_2 onto a Ta filament heated at $2200\,^{\circ}$C. The species produced, whatever they are, then hit a Si polished plate warmed to $900\,^{\circ}$C that supports transition metal particles. The whole chamber is kept in a dynamic

vacuum of 40 torr. SWNTs are supposed to form according to the authors, although their images are not very convincing. One major drawback of the method, besides its complexity compared to the others, is that it is difficult to recover the "nanotubes" from the Si substrates to which they seem to be firmly bonded.

The final example is an attempt to prepare nanotubes by diffusion flame synthesis [3.113]. A regular gaseous hydrocarbon source (ethylene, ...) along with ferrocene vapor is passed into a laminar diffusion flame derived from air and CH_4 of temperature $500–1200\,^{\circ}$C. SWNTs are formed, together with encapsulated metal particles, soot, and so on. In addition to a low yield, the SWNT structure is quite poor.

3.2.4 Synthesis of Carbon Nanotubes with Controlled Orientation

Several applications (such as field emission-based displays, see Sect. 3.6) require that carbon nanotubes grow as highly aligned bunches, in highly ordered arrays, or that they are located at specific positions. In this case, the purpose of the process is not mass production but controlled growth and purity, with subsequent control of nanotube morphology, texture and structure. Generally speaking, the more promising methods for the synthesis of aligned nanotubes are based on CCVD processes, which involve the use of molecular precursors as carbon sources, and the method of thermal cracking assisted by the catalytic activity of transition metal (Co, Ni, Fe) nanoparticles deposited onto solid supports. Although this approach mainly produces MWNT, some attempts have been made to obtain arrays of SWNTs. Generally speaking, SWNTs and DWNTs nucleate at higher temperatures than MWNTs [3.114].

However, the catalyst-free templating methods related to those described in Sect. 3.2.2 are not considered here, due to the lack of support after the template is removed, which means that the previous alignment is not maintained.

During the CCVD growth, nanotubes can self-assemble into nanotube bunches aligned perpendicular to the substrate if the catalyst film on the substrate has a critical thickness [3.115, 116]. The driving forces for this alignment are the van der Waals interactions between the nanotubes, which allow them to grow perpendicularly to the substrates. If the catalyst nanoparticles are deposited onto a mesoporous substrate, the mesoscopic pores may also have an effect on the alignment when the growth starts, thus controlling the growth direction of the nanotubes. Two kinds of substrates have been used so far for this purpose: mesoporous silica [3.117, 118] and anodic alumina [3.119].

Different methods of depositing metal particles onto substrates have been reported in the literature: (i) deposition of a thin film on alumina substrates using metallic salt precursor impregnation followed by oxidation/reduction steps [3.122]; (ii) embedding catalyst particles in mesoporous silica by sol-gel processes [3.117]; (iii) thermal evaporation of Fe, Co, Ni or Co-Ni metal alloys on SiO_2 or quartz substrates under high vacuum [3.123, 124]; (iv) photolithographic patterning of metal-containing photoresist polymer using conventional black and white films as a mask [3.125] or photolithography and the inductive plasma deep etching technique [3.126]; (v) electrochemical deposition into pores in anodic aluminium oxide templates [3.119], (vi) deposition of colloidal suspensions of catalyst particles with tailored diameters on a support [3.127–131], by spin-coating for instance; (vii) stamping a catalyst precursor over a patterned silicon wafer is also possible and has been used to grow networks of nanotubes parallel to the substrate (Fig. 3.17a), or more generally to localize the growth of individual CNTs [3.132]. A technique that combines the advantages of electron beam lithography and template methods has also been reported for the large-scale production of ordered MWNTs [3.133] or AFM tips [3.134].

Depositing the catalyst nanoparticles onto a prepatterned substrate allows one to control the frequency of local occurrence and the arrangement of the nanotube bunches formed. The materials produced mainly consist of arrayed, densely packed, freestanding, aligned MWNTs (Fig. 3.17b), which are quite suitable for field emission-based applications for instance [3.121]. SWNTs have also been produced, and it was reported that the introduction of water vapor during the CVD process allows impurity-free SWNTs to be synthesized [3.135], due to a mechanism related to that previously proposed for the effect of using alcohol instead of hydrocarbon feedstock [3.115].

When a densely packed coating of vertically aligned MWNTs is desired (Fig. 3.18c), another route is the pyrolysis of hydrocarbons in the presence of organometallic precursor molecules like metallocene or iron pentacarbonyl, operating in a dual furnace system (Fig. 3.18a,b). The organo-metallic precursor (such as ferrocene) is first sublimed at low temperature in the first furnace or injected as a solution along with the hydrocarbon feedstock, and then the whole system is pyrolyzed at higher temperature in the second furnace [3.98, 136–139]. The important parameters here are the heating or feeding rate of ferrocene, the flow rates of the vector gas (Ar or N_2) and the gaseous hydrocarbon, and the temperature of pyrolysis (650–1050 °C). Generally speaking, the codeposition process using $[Fe(CO)_5]$ as the catalyst source results in thermal decomposition at elevated temperatures, producing atomic iron that deposits on the substrates in the hot zone of the reactor. Since nanotube growth occurs at the same time as the introduction of $[Fe(CO)_5]$, the temperatures chosen for the growth depend on the carbon feedstock utilized; for example, they can vary from 750 °C for acetylene to 1100 °C for methane. Mixtures of $[FeCp_2]$ and xylene

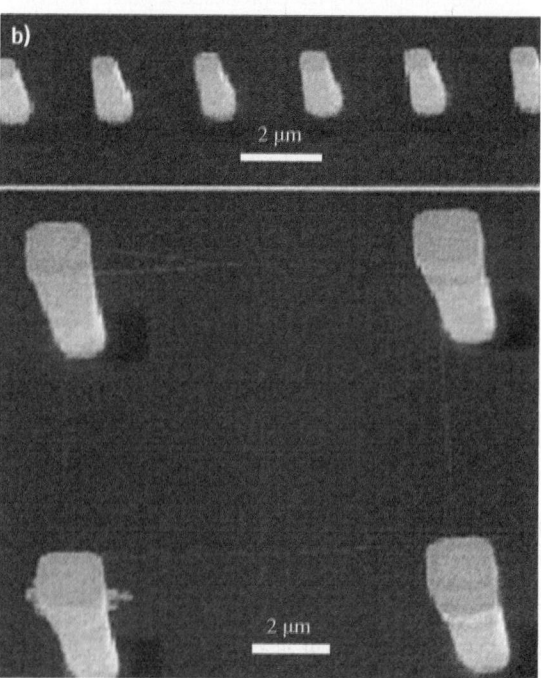

Fig. 3.17 (a) Example of a controlled network of nanotubes grown parallel to the substrate [3.120]; **(b)** Example of a free-standing MWNT array obtained from the pyrolysis of a gaseous carbon source over catalyst nanoparticles previously deposited onto a patterned substrate. Each square-base rod is a bunch of MWNTs aligned perpendicular to the surface (from [3.121])

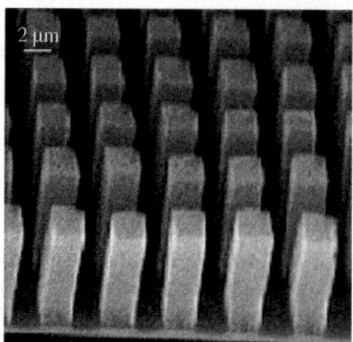

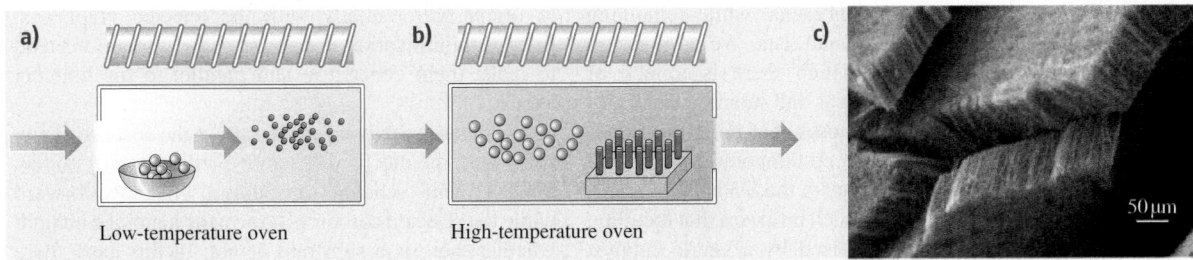

Fig. 3.18a–c Sketch of a double-furnace CCVD device used in the organometallic/hydrocarbon copyrolysis process. (**a**) Sublimation of the precursor. (**b**) Decomposition of the precursor and MWNT growth onto the substrate. (**c**) Example of the densely packed and aligned MWNT material obtained (from [3.137])

or [FeCp$_2$] and acetylene have also been successfully used to produce freestanding MWNTs.

The nanotube yield and quality are directly linked to the amount and size of the catalyst particles, and since the planar substrates used do not exhibit high surface areas, the dispersion of the metal can be a key step in the process. It has been observed that an etching pretreatment of the surface of the deposited catalyst thin film with NH$_3$ may be critical to efficient nanotube growth of nanotubes since it provides the appropriate metal particle size distribution. It may also favor the alignment of

MWNTs and prevent the formation of amorphous carbon due to the thermal cracking of acetylene [3.140]. The application of phthalocyanines of Co, Fe and Ni has also been reported, and in this case the pyrolysis of the organometallic precursors also produces the carbon for the vertically aligned MWNTs [3.141].

Densely packed coatings of vertically aligned MWNTs may also be produced over metal-containing deposits, such as iron oxides on aluminium [3.142], in which case MWNT synthesis takes place on small particles that are formed from the iron oxide deposit.

3.3 Growth Mechanisms of Carbon Nanotubes

The growth mechanisms of carbon nanotubes are still the source of much debate. However, researchers have been impressively imaginative, and have come up with a number of hypotheses. One reason for the debate is that the conditions that allow carbon nanofilaments to grow are very diverse, which means that there are many related growth mechanisms. For a given set of conditions, the true mechanism is probably a combination of or a compromise between some of the proposals. Another reason is that the phenomena that occur during growth are pretty rapid and difficult to observe in situ. It is generally agreed, however, that growth occurs such that the number of dangling bonds is minimized, for energetic reasons.

3.3.1 Catalyst-Free Growth

As already mentioned, in addition to the templating technique, which is merely a chemical vapor infiltration mechanism for pyrolytic carbon, the growth of c-MWNT as a deposit on the cathode in the electric arc method is a rare example of catalyst-free carbon nanofilament growth. The driving force is obviously related to the

electric field; in other words to charge transfer from one electrode to the other via the particles contained in the plasma. It is not clear how the MWNT nucleus is formed, but once it has, it may include the direct incorporation of C$_2$ species into the primary graphene structure, as it was previously proposed for fullerenes [3.143]. This is supported by recent C$_2$ radical concentration measurements that reveal an increasing concentration of C$_2$ from the anode being consumed at the growing cathode (Fig. 3.13). This indicates that C$_2$ are only secondary species and that the C$_2$ species may actually actively participate in the growth of c-MWNTs in the arc method.

3.3.2 Catalytically Activated Growth

Growth mechanisms involving catalysts are more difficult to ascertain, since they are more diverse. Although it involves a more or less extensive contribution from a VLS (vapor-liquid-solid [3.144]) mechanism, it is quite difficult to find comprehensive and plausible explanations that are able to account for both the various conditions used and the various morphologies observed. What follows is an attempt to provide overall explanations of most of the phenomena, while remaining consistent with the experimental data. We do not consider any hypothesis for which there is a lack of experimental evidence, such as the moving nanocatalyst mechanism, which proposes that dangling bonds from a growing SWNT may be temporarily stabilized by a nanosized catalyst located at the SWNT tip [3.25], or the scooter mechanism, which proposes that dangling bonds are temporarily stabilized by a single catalyst atom which moves around the edge of the SWNT , allowing subsequent C atom addition [3.145].

From various results, it appears that the most important parameters are probably the thermodynamic ones (only temperature will be considered here), the catalyst particle size, and the presence of a substrate. Temperature is critical and basically corresponds to the discrepancy between CCVD methods and solid carbon source-based methods.

Low–Temperature Conditions

Low-temperature conditions are typical used in CCVD, where nanotubes are frequently found to grow far below 1000 °C. If the conditions are such that the catalyst is a crystallized solid, the nanofilament is probably formed via a mechanism similar to a VLS mechanism, in which three steps are defined: (i) adsorption then decomposition of C-containing gaseous moieties at the catalyst surface; (ii) dissolution then diffusion of the C species through the catalyst, thus forming a solid solution; (iii) back-precipitation of solid carbon as nanotube walls. The texture is then determined by the orientation of the crystal faces relative to the filament axis (Fig. 3.19), as demonstrated beyond doubt by transmission electron microscopy images such as those in *Rodriguez* et al. [3.16]. This mechanism can therefore provide either c-MWNT, h-MWNT, or platelet nanofibers. The latter, however, are mainly formed in large particle sizes (> 100 nm for example). Platelet nanofibers with low diameters (below 40 nm) have never been observed. The reasons for this are related to graphene energetics, such as the need to reach the optimal ratio between the amount of edge carbon atoms (with dangling bonds) and inner carbon atoms (where all of the σ and π orbitals are satisfied).

If conditions are such that the catalyst is a liquid droplet, due to the use of high temperatures or because a catalyst that melts at a low temperature is employed, a mechanism similar to that described above can still occur, which is really VLS (vapor = gaseous C species, liquid = molten catalyst, S = graphenes), but there are obviously no crystal faces

to orient preferentially with the rejected graphenes. Energy minimization requirements will therefore tend to make them concentric and parallel to the filament axis.

With large catalyst particles (or in the absence of any substrate), the mechanisms above will generally follow a "tip growth" scheme: the catalyst will move forward while the rejected carbon will form the nanotube behind, whether there is a substrate or not. In this case, there is a good chance that one end will be open. On the other hand, when the catalyst particles deposited onto the substrate are small enough (nanoparticles) to be held in place by interaction forces with the substrate, the growth mechanism will follow a "base growth" scheme, where the carbon nanofilament grows away from the substrate, leaving the catalyst nanoparticle attached to the substrate (Fig. 3.20).

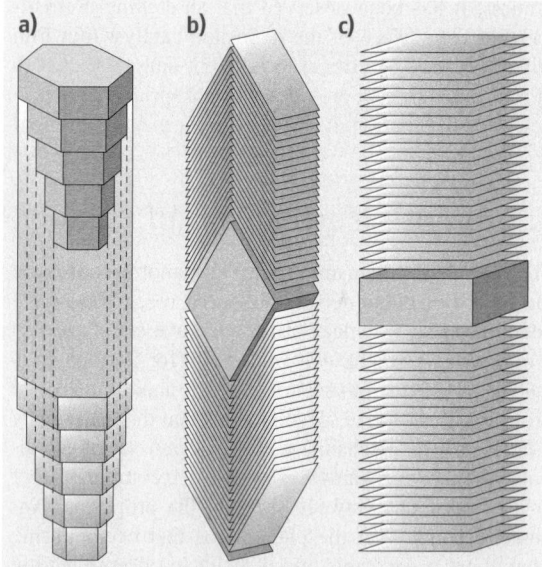

Fig. 3.19a–c Illustration of the possible relationships between the outer morphology of the catalyst crystal and the inner texture of the subsequent carbon nanofilament (adapted from [3.16]). Crystals are drawn with the projected plane perpendicular to the electron beam in a transmission electron microscope; the crystal morphologies and the subsequent graphene arrangements in the out-of-plane dimension are not intended to be accurate representations in these sketches (for example, the graphenes in the herringbone-type nanotubes or nanofibers in (**b**) cannot be arranged like a pile of open books, as sketched here, because it would leave too many dangling bonds)

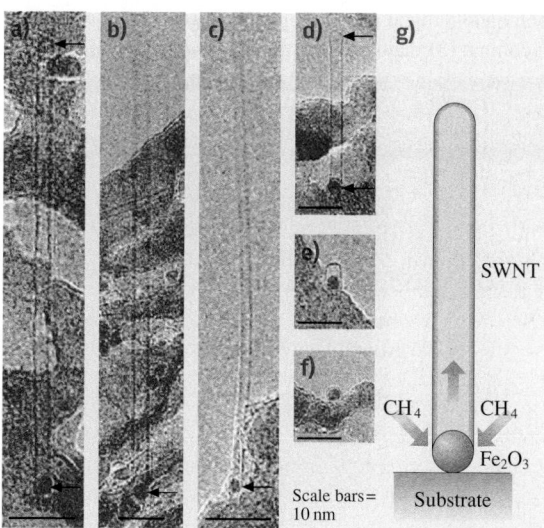

Fig. 3.20 High-resolution transmission electron microscopy images of several SWNTs grown from iron-based nanoparticles using the CCVD method, showing that particle sizes determine SWNT diameters in this case (adapted from [3.147])

The bamboo texture that affects both the herringbone and the concentric texture may reveal a distinguishing aspect of the dissolution-rejection mechanism: the periodic, discontinuous dynamics of the phenomenon. Once the catalyst has reached the saturation threshold in terms of its carbon content, it expels it quite suddenly. Then it becomes able to incorporate a given amount of carbon again without having any catalytic activity for a little while. Then over-saturation is reached again, and so on. An exhaustive study of this phenomenon has been carried out by *Jourdain* et al.[3.146].

Therefore, it is clear that 1 catalyst particle = 1 nanofilament in any of the mechanisms above. This explains why, although it is possible to make SWNTs by CCVD methods, controlling the catalyst particle size is critical, since it influences the nanofilament that grows from it. Achieving a really narrow size distribution in CCVD is quite challenging, particularly when nanosizes are required for the growth of SWNTs. Only particles < 2 nm are useful for this (Fig. 3.20), since larger SWNTs are not favored energetically [3.9]. Another distinguishing aspect of the CCVD method and its related growth mechanisms is that the process can occur all along the isothermal zone of the reactor furnace since it is continuously fed with a carbon-rich feedstock, which is generally in excess, with a constant composition at a given species time of flight. Roughly

speaking, the longer the isothermal zone (in gaseous carbon excess conditions), the longer the nanotubes. This is why the lengths of the nanotubes can be much longer than those obtained using solid carbon source-based methods.

Table 3.2 provides an overview of the relationship between general synthesis conditions and some features of nanotube grown.

High-Temperature Conditions

High-temperature conditions are typical used in solid carbon source- based methods such as the electric arc method, laser vaporization, and the solar furnace method (see Sect. 3.2). The huge temperatures involved (several thousands of °C) atomize both the carbon source and the catalyst. Of course, catalyst-based SWNTs do not form in the areas with the highest temperatures (contrary to c-MWNTs in the electric arc method); medium is a mixture of atoms and radicals, some of which are likely to condense back into the same droplet (since the medium is liquid at the beginning). At some distance from the atomization zone, the medium is therefore made of carbon metal alloy droplets and of secondary carbon species that range from C_2 to higher order molecules such as corannulene, which is made of a central pentagon surrounded by five hexagons. The preferred formation of such a molecule can be explained by the previous association of carbon atoms into a pentagon, because it is the fastest way to limit dangling bonds at low energetic cost, thereby providing a fixation site for other carbon atoms (or C_2) which also will tend to close into a ring, again to limit dangling bonds. Since adjacent pentagons are not energetically favored, these cycles will be hexagons. Such a molecule is thought to be a probable precursor for fullerenes. Fullerenes are actually always produced, even in conditions that produce SWNTs. The same saturation in C described in Sect. 3.3.2 occurs for the carbon-metal alloy droplet as well, resulting in the precipitation of excess C outside the particle due to the effect of the decreasing thermal gradient in the reactor, which decreases the solubility threshold of C in the metal [3.148]. Once the "inner" carbon atoms reach the surface of the catalyst particle, they meet the "outer" carbon species, including corannulene, that will contribute to capping the merging nanotubes. Once formed and capped, nanotubes can grow both from the inner carbon atoms (Fig. 3.21a), according to the VLS mechanism proposed by *Saito* et al. [3.148], and from the outer carbon atoms, according the adatom mechanism proposed by *Bernholc* et al. [3.149]. In the latter, carbon atoms from the surrounding medium in the reactor are

Part A | 3.3

Table 3.2 Guidelines indicating the relationships between possible carbon nanofilament morphologies and some basic synthesis conditions. Columns (1) and (2) mainly relate to CCVD-based methods; column (3) mainly relates to plasma-based methods

		Increasing temperature ... and physical state of catalyst			Substrate		Thermal gradient	
		solid (crystallized) (1)	liquid from melting (2)	liquid from condensing atoms (3)	yes	no	low	high
Catalyst particle size	<~ 3 nm	SWNT	SWNT	?	base-growth			
	>~ 3 nm	MWNT (c,h,b) platelet nanofiber	c-MWNT	SWNT	tip-growth	tip-growth	long length	short length
Nanotube diameter		(heterogeneous related to catalyst particle size)		homogeneous (independent from particle size)				
Nanotube/particle		one nanotube/particle		several SWNTs/particle				

(Substrate column note, vertical: growing from case (3) catalyst) — (except for SWNTs catalyst))

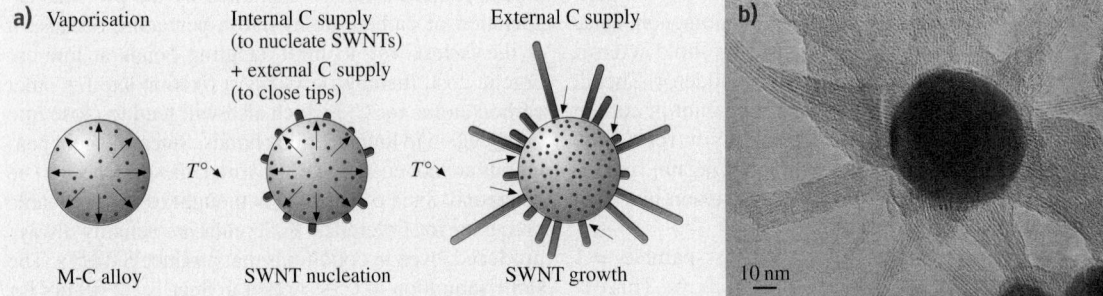

a) Vaporisation   Internal C supply (to nucleate SWNTs) + external C supply (to close tips)   External C supply   **b)**

$T°$ ↘ $T°$ ↘

M-C alloy   SWNT nucleation   SWNT growth   10 nm

Fig. 3.21 (a) Mechanism proposed for SWNT growth (see text). **(b)** Transmission electron microscopy image of SWNT growing radially from the surface of a large Ni catalyst particle in an electric arc experiment. (Modified from [3.18])

attracted then stabilized by the carbon/catalyst interface at the nanotube/catalyst surface contact, promoting their subsequent incorporation at the tube base. The growth mechanism therefore mainly follows the base growth scheme. However, once the nanotubes are capped, any C_2 species that still remains in the medium that meets the growing nanotubes far from the nanotube/catalyst interface may still incorporate the nanotubes from both the side wall or the tip, thereby giving rise to some proportion of Stone–Wales defects [3.42]. The occurrence

of a nanometer-thick surface layer of yttrium carbide (onto the main Ni-containing catalyst core), the lattice distance of which is commensurable with that of the C−C distance in graphene (as recently revealed by *Gavillet* et al. [3.50]), could possibly play a beneficial role in stabilizing the nanotube/catalyst interface, which could explain why the SWNT yield is enhanced by bimetallic alloys (as opposed to single metal catalysts).

A major difference from the low-temperature mechanisms described for CCVD methods is that many nanotubes are formed from a single, relatively large ($\approx$ 10–50 nm) catalyst particle (Fig. 3.21b), whose size distribution is therefore not as critical as it is for the low-temperature mechanisms (particles that are too large, however, induce polyaromatic shells instead of nanotubes). This is why the diameters of SWNTs grown at high temperature are much more homogeneous than those associated with CCVD methods. The reason that the most frequent diameter is $\approx$ 1.4 nm is again a matter of energy balance. Single-wall nanotubes larger than $\approx$ 2.5 nm are not stable [3.9]. On the other hand, the strain on the C−C bond increases as the radius of curvature decreases. The optimal diameter (1.4 nm) should therefore correspond to the best energetic compromise. Another difference from the low-temperature mechanism for CCVD is that temperature gradients in high temperature

methods are huge, and the gas phase composition surrounding the catalysts droplets is also subjected to rapid changes (as opposed to what could happen in a laminar flow of a gaseous feedstock whose carbon source is in excess). This explains why nanotubes from arcs are generally shorter than nanotubes from CCVD, and why mass production by CCVD is favored. In the latter, the metallic particle can act as a catalyst repeatedly as long as the conditions are maintained. In the former, the surrounding conditions change continuously, and the window for efficient catalysis can be very narrow. Decreasing the temperature gradients that occur in solid carbon source-based methods of producing SWNT, such as the electric arc reactor, should therefore increase the SWNT yield and length [3.150]. Amazingly, this is in opposition to what is observed during arc-based fullerene production.

3.4 Properties of Carbon Nanotubes

In previous sections, we noted that the normal planar configuration of graphene can, under certain growth conditions (see Sect. 3.3), be changed into a tubular geometry. In this section, we take a closer look at the properties of these carbon nanotubes, which can depend on whether they are arranged as SWNTs or as MWNTs (see Sect. 3.1).

3.4.1 Overview

The properties of MWNTs are generally similar to those of regular polyaromatic solids (which may exhibit graphitic, turbostratic or intermediate crystallographic structure). Variations are mainly due to different textural types of the MWNTs considered (concentric, herringbone, bamboo) and the quality of the nanotexture (see Sect. 3.1), both of which control the extent of anisotropy. Actually, for polyaromatic solids that consist of stacked graphenes, the bond strength varies significantly depending on whether the in-plane direction is considered (characterized by very strong covalent and therefore very short – 0.142 nm – bonds) or the direction perpendicular to it (characterized by very weak van der Waals and therefore very loose – $\approx$ 0.34 nm – bonds). Such heterogeneity is not found in single (isolated) SWNTs. However, the heterogeneity returns, along with the related consequences, when SWNTs associate into bundles. Therefore, the properties – and applicability – of SWNTs may also change dramatically depending on whether single SWNT or SWNT ropes are involved.

In the following, we will emphasize the properties of SWNTs, since their unique structures often lead to different properties to regular polyaromatic solids. However, we will also sometimes discuss the properties of MWNTs for comparison.

3.4.2 General Properties of SWNTs

The diameters of SWNT-type carbon nanotubes fall in the nanometer regime, but SWNTs can be hundreds of micrometers long. SWNTs are narrower in diameter than the thinnest line that can be obtained in electron beam lithography. SWNTs are stable up to 750 °C in air (but they are usually damaged before this temperature is reached due to oxidation mechanisms, as demonstrated by the fact that they can be filled with molecules (see Sect. 3.5). They are stable up to $\approx$ 1500–1800 °C in inert atmosphere, beyond which they transform into regular, polyaromatic solids (phases built with stacked graphenes instead of single graphenes) [3.151]. They have half the mass density of aluminium. The properties of a SWNT, like any molecule, are heavily influenced by the way that its atoms are arranged. The physical and chemical behavior of a SWNT is therefore related to its unique structural features [3.152].

3.4.3 Adsorption Properties of SWNTs

An interesting feature of a SWNT is that it has the highest surface area of any molecule due to the fact that a graphene sheet is probably the only example of a sheet-like molecule that is energetically stable under normal conditions. If we consider an isolated SWNT with one open end (achieved through oxidation treatment for instance), the surface area is equal to that of a single, flat graphene sheet: $\approx 2700 \, \mathrm{m^2/g}$ (accounting for both sides).

In reality, nanotubes – specifically SWNTs – are usually associated with other nanotubes in bundles, fibers, films, papers, and so on, rather than as a single entity. Each of these associations has a specific range of porosities that determines its adsorption properties (this topic is also covered in Sect. 3.6.2 on applications). It is therefore more appropriate to discuss adsorption onto the outer or the inner surface of a bundle of SWNTs.

Furthermore, theoretical calculations have predicted that the adsorption of molecules onto the surface or inside of a nanotube bundle is stronger than that onto an individual tube. A similar situation exists for MWNTs, where adsorption can occur on or inside the tubes or between aggregated MWNTs. It has also been shown that the curvature of the graphene sheets constituting the nanotube walls results in a lower heat of adsorption compared to planar graphene (see Sect. 3.1.1).

Accessible SWNT Surface Area

Various studies dealing with the adsorption of nitrogen onto MWNTs [3.154–156] and SWNTs [3.157] have highlighted the porous nature of these two materials. The pores in MWNTs can be divided mainly into hollow inner cavities with small diameters (with narrow size distributions, mainly 3–10 nm) and aggregated pores (with wide size distributions, 20–40 nm), formed by interactions between isolated MWNTs. It is also worth noting that the ultra-strong nitrogen capillarity in the aggregated pores dominates the total adsorption, indicating that the aggregated pores are much more important than the inner cavities of the MWNTs during adsorption. Adsorption of N_2 has been studied on as-prepared and acid-treated SWNTs, and the results obtained highlight the microporous nature of SWNT materials, as

opposed to the mesoporous nature of MWNT materials. Also, as opposed to isolated SWNTs (see above), surface areas that are well above $400 \, \mathrm{m^2 \, g^{-1}}$ have been measured for SWNT-bundle-containing materials, with internal surface areas of $300 \, \mathrm{m^2 \, g^{-1}}$ or higher.

The theoretical surface area of a carbon nanotube has a broad range, from 50 to $1315 \, \mathrm{m^2 \, g^{-1}}$ depending on the number of walls, the diameter, and the number of nanotubes in a bundle of SWNTs [3.158]. Experimentally, the surface area of a SWNT is often larger than that of a MWNT. The total surface area of as-grown SWNTs is typically between 400 and $900 \, \mathrm{m^2 \, g^{-1}}$ (micropore volume, $0.15–0.3 \, \mathrm{cm^3 \, g^{-1}}$), whereas values of 200 and $400 \, \mathrm{m^2 \, g^{-1}}$ for as-produced MWNTs are often reported. In the case of SWNTs, the diameters of the tubes and the number of tubes in the bundle will have the most effect on the BET value. It is worth noting that opening or closing the central canal significantly influences the adsorption properties of nanotubes. In the case of MWNTs, chemical treatments such as KOH or NaOH activation are useful for promoting microporosity, and surface areas as high as $1050 \, \mathrm{m^2 g^{-1}}$ have been reported [3.159, 160]. An efficient two-step treatment (acid $+CO_2$ activation) has been reported to open both ends of MWNTs [3.161]. Therefore, it appears that opening or cutting carbon nanotubes, as well as chemically treating them (using purification steps for example) can considerably affect their surface area and pore structure.

Adsorption Sites and Binding Energy of the Adsorbates

An important problem to solve when considering adsorption onto nanotubes is to identify the adsorption sites. The adsorption of gases into a SWNT bundle can occur inside the tubes (pore), in the interstitial triangular channels between the tubes, on the outer surface of the bundle, or in the grooves formed at the contacts between adjacent tubes on the outside of the bundle (Fig. 3.22). Modeling studies have pointed out that the convex surface of the SWNT is more reactive than the concave one and that this difference in reactivity increases as the tube diameter decreases [3.162]. Compared to the highly bent region in fullerenes, SWNTs are only moderately curved and are expected to be much less reactive towards dissociative chemisorption. Models have also predicted enhanced reactivity at the kink sites of bent SWNTs [3.163]. Additionally, it is worth noting that unavoidable imperfections, such as vacancies, Stone–Wales defects, pentagons, heptagons and dopants, are believed to play a role in tailoring the adsorption properties [3.164].

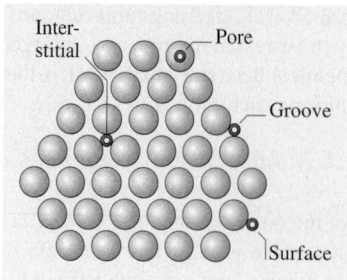

Fig. 3.22 Sketch of a SWNT bundle, illustrating the four different adsorption sites (adapted from [3.153])

Considering closed-end SWNTs first, simple molecules can be adsorbed onto the walls of the outer nanotubes of the bundle and preferably on the external grooves. In the first stages of adsorption (corresponding to the most attractive sites for adsorption), it seems that adsorption or condensation in the interstitial channels of the SWNT bundles depends on the size of the molecule (or on the SWNT diameters) and on their interaction energies [3.165, 166]. Opening the tubes favors gas adsorption (including O_2, N_2 within the inner walls [3.167, 168]). It was found that the adsorption of nitrogen on open-ended SWNT bundles is three times larger than that on closed-ended SWNT bundles [3.169]. The significant influence that the external surface area of the nanotube bundle has on the character of the surface adsorption isotherm of nitrogen (type I, II or even IV of the IUPAC classification) has been demonstrated from theoretical calculations [3.170]. For hydrogen and other small molecules like CO, computational methods have shown that, for open SWNTs, the pore, interstitial and groove sites are energetically more favorable than surface sites [3.171, 172]. In the case of carbon monoxide, aside from physisorbed CO, CO hydrogen bonds to hydroxyl functionalities created on the SWNTs by acid purification have been identified [3.172]. FTIR and temperature-programmed desorption (TPD) experiments have shown that NH_3 or NO_2 adsorb molecularly and that NO_2 is slightly more strongly bound than NH_3 [3.173]. For NO_2, the formation of nitrito (O-bonded) complexes is preferred to nitro (N-bonded) ones. For ozone, a strong oxidizing agent, theoretical calculations have shown that physisorption occurs on ideal, defect-free SWNT, whereas strong chemisorption occurs on Stone–Wales defects, highlighting the key role of defec-

tive sites in adsorption properties [3.174]. Finally, for acetone, TPD experiments have shown that this molecule chemisorbs on SWNT while physisorption occurs on graphite [3.175].

For MWNTs, adsorption can occur in the aggregated pores, inside the tube or on the external walls. In the latter case, the presence of defects, as incomplete graphene layers, must be taken into consideration. Although adsorption between the graphenes (intercalation) has been proposed in the case of hydrogen adsorption in h-MWNTs or platelet nanofibers [3.176], it is unlikely to occur for many molecules due to steric effects and should not prevail for small molecules due to the long diffusion paths involved. In the case of inorganic fluorides (BF_3, TiF_4, NbF_5 and WF_6), accommodation of the fluorinated species into the carbon lattice has been shown to result from intercalation and adsorption/condensation phenomena. In this case, doping-induced charge transfer has been demonstrated [3.177].

Only a few studies deal with adsorption sites in MWNTs, but it has been shown that butane adsorbs more onto MWNTs with smaller outside diameters, which is consistent with another statement that the strain on curved graphene surfaces affects sorption. Most of the butane adsorbs to the external surface of the MWNTs while only a small fraction of the gas condenses in the pores [3.178]. Comparative adsorption of krypton or of ethylene onto MWNTs or onto graphite has allowed scientists to determine the dependence of the adsorption and wetting properties of the nanotubes on their specific morphologies. Nanotubes were found to have higher condensation pressures and lower heats of adsorption than graphite [3.179]. These differences are mainly due

Table 3.3 Adsorption properties and sites of SWNTs and MWNTs. The letters in the *Absorption sites* column refer to Fig. 3.22. The data in the last two columns are from [3.153]

Type of nanotube	Porosity $(cm^3 g^{-1})$	Surface area (m^2/g)	Binding energy of the adsorbate	Adsorption sites	Attractive potential per site (eV)	Surface area per site (m^2/g)
SWNT (bundle)	Microporous V_{micro}: 0.15–0.3	400–900	Low, mainly physisorption 25–75 % > graphite	Surface (A) Groove (B) Pore (C) Interstitial (D)	0.049 0.089 0.062 0.119	483 22 783 45
MWNT	Mesoporous	200–400	Physisorption	Surface Pore Aggregated pores	–	–

to decreased lateral interactions between the adsorbed molecules, related to the curvature of the graphene sheets.

A limited number of theoretical as well as experimental studies on the binding energies of gases onto carbon nanotubes exist. While most of these studies report low binding energies on SWNTs, consistent with physisorption, some experimental results, in particular for hydrogen, are still controversial (see Sect. 3.6.2). For platelet nanofibers, the initial dissociation of hydrogen on graphite edge sites, which constitute most of the nanofiber surface, has been proposed [3.180]. For carbon nanotubes, a mechanism that involves H_2 dissociation on the residual metal catalyst followed by H spillover and adsorption on the most reactive nanotube sites was envisaged [3.181]. Similarly, simply mixing carbon nanotubes with supported palladium catalysts increased the hydrogen uptake of the carbon by a factor of three, due to hydrogen spillover from the supported catalyst [3.182]. Doping nanotubes with alkali may enhance hydrogen adsorption, due to charge transfer from the alkali metal to the nanotube, which polarizes the H_2 molecule and induces dipole interactions [3.183].

Generally speaking, the adsorbates can be either charge donors or acceptors to the nanotubes. Trends in the binding energies of gases with different van der Waals radii suggest that the groove sites of SWNTs are the preferred low coverage adsorption sites due to their higher binding energies. Finally, several studies have shown that, at low coverage, the binding energy of the adsorbate on SWNT is between 25% and 75% higher than the binding energy on a single graphene. This discrepancy can be attributed to an increase of effective coordination at the binding sites, such as the groove sites, in SWNTs bundles [3.184, 185]. Representative results on the adsorption properties of SWNTs and MWNTs are summarized in Table 3.3.

3.4.4 Electronic and Optical Properties

The electronic states in SWNTs are strongly influenced by their one-dimensional cylindrical structures. One-dimensional sub-bands are formed that have strong singularities in the density of states (Van Hove singularities). By rolling the graphene sheet to form a tube, new periodic boundary conditions are imposed on the electronic wavefunctions, which give rise to one-dimensional sub-bands: $Cn K = 2q$ where q is an integer. Cn is the roll-up vector $na1 + ma2$ which defines the

helicity (chirality) and the diameter of the tube (see Sect. 3.1). Much of the electronic band structure of CNTs can be derived from the electronic band structure of graphene by applying the periodic boundary conditions of the tube under consideration. The conduction and the valence bands of the graphene only touch at six corners (K points) of the Brillouin zone [3.186]. If one of these sub-bands passes through the K point, the nanotube is metallic; otherwise it is semiconducting. This is a unique property that is not found in any other one-dimensional system, which means that for certain orientations of the honeycomb lattice with respect to the tube axis (chirality), some nanotubes are semiconducting and others are metallic. The band gap for semiconducting tubes is found to be inversely proportional to the tube diameter. As pointed out in Sect. 3.1, knowing (n, m) allows us, in principle, to predict whether the tube is metallic or not. The energy gap decreases for larger tube diameters and MWNTs with larger diameter are found to have properties similar to other forms of regular, polyaromatic solids. It has been shown that electronic conduction mostly occurs through the external tube for MWNTs; even so, interactions with internal tubes often cannot be neglected and they depend upon the helicity of the neighboring tubes [3.187]. The electronic and optical properties of the tubes are considerably influenced by the environment [3.188]. Under externally applied pressure, the small interaction between the tube walls results in the internal tubes experiencing reduced pressure [3.189]. The electronic transition energies are in the infrared and visible spectral range. The one-dimensional Van Hove singularities have a large influence on the optical properties of CNTs. Visible light is selectively and strongly absorbed. It has been observed that a broad band flash in the visible spectral range can lead to the spontaneous burning of agglomerated SWNTs in air at room temperature [3.190]. The exact energy band positions and the existence and formation of excitons in CNTs are not yet well known and/or understood, and are currently being explored [3.191]. Photoluminescence can be observed in individual SWNT aqueous suspensions stabilized by the addition of surfactants. Detailed photoexcitation maps provide information about the helicity (chirality)-dependent transition energies and the electronic band structures of CNTs [3.192]. Agglomeration of tubes into ropes or bundles influences the electronic states of CNTs. Photoluminescence signals are quenched for agglomerated tubes.

CNTs are model systems for the study of one-dimensional transport in materials. Apart from the singularities in the density of states, electron–electron interactions are expected to show drastic changes at the Fermi edge; the electrons in CNTs are not described by a Fermi liquid, but instead by a Luttinger liquid model [3.193] that describes electronic transport in one-dimensional systems. It is expected that the variation of electronic conductance vs. temperature follows a power law, with zero conductance at low temperatures. Depending on how L_ϕ (the coherence length) on the one hand and L_m (the electronic mean free path) on the other compare to L (the length of the sample), different conduction modes are observed: ballistic if $L \ll L_\phi$, $L \ll L_m$, diffusive if $L_\phi \ll L_m < L$ and localization if $L_m \ll L_\phi \ll L$. Fluctuations in the conductance can be seen when $L \approx L_\phi$. For ballistic conduction (a small number of defects) [3.194–196], the predicted electronic conductance is independent of the tube length. The conductance value is twice the fundamental conductance unit $G0 = 4\,e/h$ due to the existence of two propagating modes. Due to the reduced electron scattering observed for metallic CNTs and their stability at high temperatures, CNTs can support high current densities (max 10^9 A/cm2): about three orders of magnitude higher than Cu. Structural defects can, however, lead to quantum interference of the electronic wave function, which localizes the charge carriers in one-dimensional systems and increases resistivity [3.193, 197, 198]. Localization and quantum interference can be strongly influenced by applying a magnetic field [3.199]. At low temperatures, the discrete energy spectrum leads to a Coulomb blockade resulting in oscillations in the conductance as the gate voltage is increased [3.198]. In order to observe the different conductance regimes, it is important to consider the influence of the electrodes where Schottky barriers are formed. Recently it was shown [3.200] that palladium electrodes form excellent junctions with tubes. The influence of superconducting electrodes or ferromagnetic electrodes on electronic transport in CNTs due to spin polarization has also been explored [3.201, 202].

As a probable consequence of both the small number of defects (at least the kind of defects that oppose phonon transport) and the cylindrical topography, SWNTs exhibit a large phonon mean free path, which results in a high thermal conductivity. The thermal conductivity of SWNTs is comparable to that of a single, isolated graphene layer or high purity diamond [3.203], or possibly higher (≈ 6000 W/mK).

Finally, carbon nanotubes may also exhibit positive or negative magnetoresistance depending on the current, the temperature, and the field, whether they are MWNTs [3.204] or SWNTs [3.198].

3.4.5 Mechanical Properties

While tubular nanomorphology is also observed for many two-dimensional solids, carbon nanotubes are unique due to the particularly strong bonding between the carbons (sp^2 hybridization of the atomic orbitals) of the curved graphene sheet, which is stronger than in diamond (sp^3 hybridization), as revealed by the difference in C−C bond lengths (0.142 vs. 0.154 nm for graphene and diamond respectively). This makes carbon nanotubes – SWNTs or c-MWNTs – particularly stable against deformations. The tensile strength of SWNTs can be 20 times that of steel [3.205] and has actually been measured as ≈ 45 GPa [3.206]. Very high tensile strength values are also expected for ideal (defect-free) c-MWNTs, since combining perfect tubes concentrically is not supposed to be detrimental to the overall tube strength, provided the tube ends are well capped (otherwise, concentric tubes could glide relative to each other, inducing high strain). Tensile strength values as high as ≈ 150 GPa have actually been measured for perfect MWNTs from an electric arc [3.207], although the reason for such a high value compared to that measured for SWNTs is not clear. It probably reveals the difficulties involved in carrying out such measurements in a reliable manner. The flexural modulus of perfect MWNTs should logically be higher than that for SWNTs [3.205], with a flexibility that decreases as the

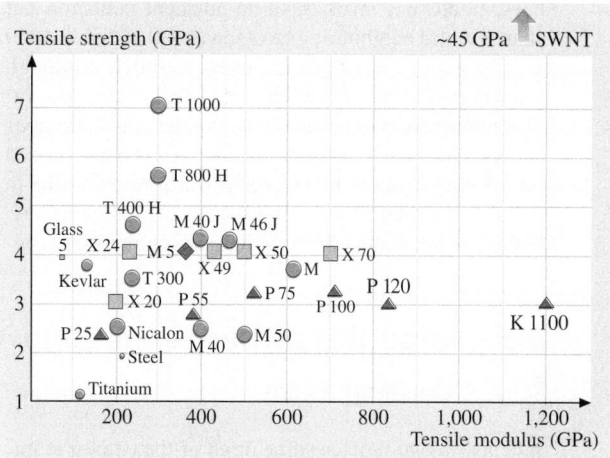

Fig. 3.23 Plot of the tensile strength versus the tensile modulus for various fibrous materials and SWNTs. *Large circles* are PAN-based carbon fibers, which include the fiber with the highest tensile strength available on the market (T1000 from Torayca); *Triangles* are pitch-based carbon fibers, which include the fiber with the highest tensile modulus on the market (K1100 from Amoco)

number of walls increases. On the other hand, measurements performed on defective MWNTs obtained from CCVD exhibit a range of 3–30 GPa [3.208]. Values of tensile modulus are also the highest values known, 1 TPa for MWNTs [3.209], and possibly even higher for SWNTs, up to 1.3 TPa [3.210, 211]. Figure 3.23 illustrates how defect-free carbon nanotubes could spectacularly revolutionize the field of high performance fibrous materials.

3.4.6 Reactivity

The chemical reactivities of graphite, fullerenes, and carbon nanotubes are similar in many ways. Like any small object, carbon nanotubes have a large surface to interact with their environment (see Sect. 3.4.1). It is worth noting, however, that nanotube chemistry differs from that observed for regular polyaromatic carbon materials due the unique shape of the nanotube, its small diameter, and its structural properties. Unlike graphite, perfect SWNTs have no (chemically active) dangling bonds (the reactions of polyaromatic solids is known to occur mainly at graphene edges). Unlike fullerenes, the ratio of "weak" sites (C–C bonds involved in heterocycles) to strong sites (C–C bonds between regular hexagons) is only deviates slightly from 0 for ideal tubes. For C_{60} fullerenes this ratio is $1 - C_{60}$ molecules have 12 pentagons (therefore accounting for $5 \times 12 = 30$ C–C bonds) and 20 hexagons, each of them with three C–C bonds not involved in an adjacent pentagon but shared with a neighboring hexagon (so $20 \times 3 \times 1/2 = 30$ C–C bonds are involved in hexagons only). Although graphene faces are chemically relatively inert, the radius of curvature imposed on the graphene in nanotubes causes the three normally planar C–C bonds caused by sp^2 hybridization to undergo distortions, resulting in bond angles that are closer to three of the four C–C bonds in diamond (characteristic of genuine sp^3 hy-

bridization), as the radius of curvature decreases. Even though it is not enough to make the carbon atoms chemically reactive, one consequence of this is that either nesting sites are created at the concave surface, or strong physisorption sites are created above each carbon atom of the convex surface, both with a bonding efficiency that increases as the nanotube diameter decreases.

As already pointed out in Sect. 3.1, the chemical reactivities of SWNTs (and c-MNWTs) are believed to derive mainly from the caps, since they contain six pentagons each, as opposed to the tube body, which supposedly only contains hexagons. Indeed, applying oxidizing treatments to carbon nanotubes (air oxidation, wet-chemistry oxidation) selectively opens the nanotube tips [3.212]. However, that SWNTs can be opened by oxidation methods and then filled with foreign molecules such as fullerenes (see Sect. 3.5) suggests the occurrence of side defects [3.14], whose identity and occurrence were discussed and then proposed to be an average of one Stone–Wales defect every 5 nm along the tube length, involving about 2% of the carbon atoms in a regular (10,10) SWNT [3.213]. A Stone–Wales defect is formed from four adjacent heterocycles, two pentagons and two heptagons, arranged in pairs opposite each other. Such a defect allows localized double bonds to form between the carbon atoms involved in the defect (instead of these electrons participating in the delocalized electron cloud above the graphene as usual, enhancing the chemical reactivity, for example toward chlorocarbenes [3.213]). This means that the overall chemical reactivity of carbon nanotubes should depend strongly on how they are synthesized. For example, SWNTs prepared by the arc-discharge method are believed to contain fewer structural defects than CCVD-synthesized SWNTs, which are more chemically reactive. Of course, the reactivity of h-MWNT-type nanotubes is intrinsically higher, due to the occurrence of accessible graphene edges at the nanotube surface.

3.5 Carbon Nanotube-Based Nano-Objects

3.5.1 Heteronanotubes

It is possible to replace some or all of the carbon atoms in a nanotube with atoms of other elements without damaging the overall honeycomb lattice-based graphene structure. Nanotubes modified in this way are termed here "heteronanotubes".

The elements used to replace carbon in this case are boron and/or nitrogen. Replacing carbon atoms

in this way can result in new behavior (for example, BN nanotubes are electrical insulators), improved properties (resistance to oxidation for instance), or better control over such properties. For instance, one current challenge in carbon SWNT synthesis is to control the processing so that the desired SWNT structure (metallic or semiconductor) is formed selectively. In this regard, it was demonstrated that replacing some C atoms with N or B atoms leads to

SWNTs with systematically metallic electrical behavior [3.214, 215].

Some examples of heteronanotubes – mainly MWNTs – can be found in the literature. The heteroatom usually involved is nitrogen, due to the ease with which gaseous or solid nitrogen- and/or boron-containing species (such as N_2, NH_3, BN, HfB_2) can be passed into existing equipment for synthesizing MWNTs [3.214, 216] until complete substitution of carbon occurs [3.217, 218]. An amazing result of such attempts to synthesize hetero-MWNTs is the subsequent formation of "multilayered" c-MWNTs: MWNTs made up of coaxial alternate carbon graphene tubes and boron nitride graphene tubes [3.219]. On the other hand, there are only a few examples of hetero-SWNTs. Syntheses of B- or N-containing SWNTs have recently been reported [3.215, 220], while just one successful synthesis of genuine BN-SWNTs has been reported so far [3.221].

3.5.2 Hybrid Carbon Nanotubes

Hybrid carbon nanotubes are defined here as carbon nanotubes, SWNTs or MWNTs that have inner cavities filled (partially or entirely) with foreign atoms, molecules, compounds or crystals. The terminology X@SWNT (or X@MWNT, if appropriate, where X is the atom, molecule and so on involved) is used for such structures [3.222].

Motivation
But why should we want to fill the cavities of carbon nanotubes [3.213]? The very small inner cavity of nanotubes is an amazing tool for preparing and studying the properties of confined nanostructures of any type, such as salts, metals, oxides, gases, or even discrete molecules like C_{60}, for example. Due to the almost one-dimensional structure of carbon nanotubes (particularly for SWNTs), we might expect that encapsulated material might have different physical and/or chemical properties to the unencapsulated material, and that the hybrid nanotube itself may behave differently to a "pure" nanotube. Indeed, if the volume available inside a carbon nanotube is small enough, the foreign material is largely "surface atoms" of reduced coordination. The original motivation to create such hybrids was to obtain metal nanowires that are likely to of interest in electronics (as quantum wires). In this case, the nanotubes were considered to be nanomolds for the metal filler, and it was probably intended that the nanomold was to be removed afterwards. However, it is likely that this removal of the SWNT "container" to liberate the one-dimensional structure inside it

may destroy or at least transform this structure due to the stabilizing effect of interactions with the nanotube wall.

Filling nanotubes while they grow (in situ filling) was one of the pioneering methods of nanotechnology. In most cases, however, the filling step is separate from nanotube synthesis. Three filling methods can then be distinguished: (a) wet chemistry procedures; capillarity-based physical procedures involving (b) a molten material or (c) a sublimated material.

Generally speaking, it is difficult to estimate the filling rate, and this is usually achieved through TEM observation, without obtaining any statistics on the number of tubes observed. Moreover, as far as SWNTs are concerned, the fact that the nanotubes are gathered into bundles makes it difficult to observe the exact number of filled tubes, as well as to estimate the filled length for each tube. It however seems that estimation of filling rates can now be reliably obtained from X-ray studies and Raman spectroscopy.

It is also possible to fill carbon nanotubes with materials that could not have been introduced directly. This is done by first filling the nanotubes with an appropriate precursor (one that is able to sublime, or melt or solubilize) that will later be transformed into the required material by chemical reaction or by a physical interaction, such as electron beam irradiation for example [3.223]. For secondary chemical transformation, reduction by H_2 is often used to obtain nanotubes filled with metals [3.224]. Sulfides can also be obtained if H_2S is used as a reducing agent [3.224].

Because the inner diameters of SWNTs are generally smaller than those of MWNTs, it is more difficult to fill them, and the driving forces involved in this phenomenon are not yet totally understood (see the review paper by *Monthioux* [3.213]). This field is therefore growing fairly rapidly, and so we have chosen to cite the pioneering works and then to focus on more recent works dealing with the more challenging topic of filling SWNTs.

In Situ Filling Method
Initially, most hybrid carbon nanotubes synthesized were based on MWNTs prepared using the electric arc method, and were obtained directly during processing. The filling materials were easily introduced in the system by drilling a central hole in the anode and filling it with the heteroelement. The first hybrid products obtained using this approach were all reported the same year [3.225–229] for heteroelements such as Pb, Bi, YC_2 and TiC. Later on, *Loiseau* et al. [3.230] showed that MWNTs could also be filled to several

μm in length by elements such as Se, Sb, S, and Ge, but only with nanoparticles of elements such as Bi, B, Al and Te. Sulfur was suggested to play an important role during the in situ formation of filled MWNTs using arc discharge [3.231]. This technique is no longer the preferred one because it is difficult to control the filling ratio and yield and to achieve mass production.

Wet Chemistry Filling Method

The wet chemistry method requires that the nanotube tips are opened by chemical oxidation prior to the filling step. This is generally achieved by refluxing the nanotubes in dilute nitric acid [3.233–235], although other oxidizing liquid media may work as well, such as [HCl + CrO$_3$] [3.236] or chlorocarbenes formed from the photolytic dissociation of CHCl$_3$ [3.213], a rare example of a nonacidic liquid route to opening SWNT tips. If a dissolved form (such as a salt or oxide) of the desired metal is introduced during the opening step, some of it will get inside the nanotubes. An annealing treatment (after washing and drying the treated nanotubes) may then lead to the oxide or to the metal, depending on the annealing atmosphere [3.212]. Although the wet chemistry method initially looked promising because a wide variety of materials can be introduced into nanotubes in this way and it operates at temperatures that are not much different from room temperature; however, close attention must be paid to the oxidation method that is used. The damage caused to nanotubes by severe treatments (such by using nitric acid) make them unsuitable

for use with SWNTs. Moreover, the filling yield is not very good, probably due to the solvent molecules that also enter the tube cavity: the filled lengths rarely exceed 100 nm. *Mittal* et al. [3.236] have recently filled SWNTs with CrO$_3$ using wet chemistry with an average yield of ≈ 20%.

Molten State Filling Method

The physical filling method involving a liquid (molten) phase is more restrictive, firstly because some materials can decompose when they melt, and secondly because the melting point must be compatible with the nanotubes, so the thermal treatment temperature should remain below the temperature of transformation or the nanotubes will be damaged. Because the filling occurs due to capillarity, the surface tension threshold of the molten material is 100–200 N/cm^2 [3.237], although this threshold was proposed for MWNTs, whose inner diameters (5–10 nm) are generally larger than those of SWNTs (1–2 nm). In a typical filling experiment, the MWNTs are closely mixed with the desired amount of filler by gentle grinding, and the mixture is then vacuum-sealed in a silica ampoule. The ampoule is then slowly heated to a temperature above the melting point of the filler and slowly cooled. This method does not require that the nanotubes are opened prior to the heat treatment. The mechanism of nanotube opening is yet to be clearly established, but it is certainly related to the chemical reactivities of the molten materials toward carbon, and more precisely toward defects in the tube structure (see Sect. 3.4.4).

Most of the works involving the application of this method to SWNTs come from Oxford University [3.232, 238–241], although other groups have followed the same procedure [3.233, 235, 242]. The precursors used to fill the nanotubes were mainly metal halides. Although little is known about the physical properties of halides crystallized within carbon nanotubes, the crystallization of molten salts within small-diameter SWNTs has been studied in detail, and the one-dimensional crystals have been shown to interact strongly with the surrounding graphene wall. For example, *Sloan* et al. [3.239] described two-layer 4 : 4 coordinated KI crystals that formed within SWNTs that were approximately 1.4 nm in diameter. These two-layer crystals were "all surface" and had no "internal" atoms. Significant lattice distortions occurred compared to the bulk structure of KI, where the normal coordination is 6 : 6 (meaning that each ion is surrounded by six identical close neighbors). Indeed, the distance between two ions across the SWNT capillary is 1.4 times as much as the same distance along

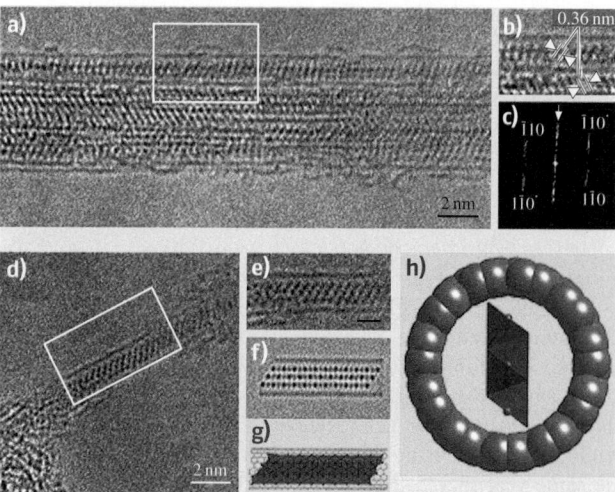

Fig. 3.24 HRTEM images and corresponding structural model for PbI$_2$ filled SWNTs (from [3.232])

the tube axis. This suggests an accommodation of the KI crystal into the confined space provided by the inner nanotube cavity in the constrained crystal direction (across the tube axis). This implies that the interactions between the ions and the surrounding carbon atoms are strong. The volume available within the nanotubes thus somehow controls the crystal structures of inserted materials. For instance, the structures and orientations of encapsulated PbI_2 crystals inside their capillaries were found to differ for SWNTs and DWNTs, depending on the diameter of the confining nanotubes [3.232]. For SWNTs, most of the encapsulated one-dimensional PbI_2 crystals obtained exhibited a strong preferred orientation, with their (110) planes aligning at an angle of around 60° to the SWNT axes, as shown in Fig. 3.24a and 3.24b. Due to the extremely small diameters of the nanotube capillaries, individual crystallites are often only a few polyhedral layers thick, as outlined in Fig. 3.24d to 3.24h. Due to lattice terminations enforced by capillary confinement, the edging polyhedra must be of reduced coordination, as indicated in Fig. 3.24g and 3.24h. Similar crystal growth behavior was generally observed to occur for PbI_2 formed inside DWNTs in narrow nanotubes with diameters comparable to those of SWNTs. As the diameter of the encapsulating capillary increases, however, different preferred orientations are frequently observed (Fig. 3.25). In this example, the PbI_2 crystal is oriented with the [121] direction parallel to the direction of the electron beam (Fig. 3.25a to 3.25d). If the PbI_2@DWNT hybrid is viewed "side-on" (as indicated by the arrow in Fig. 3.25e), polyhedral slabs are seen to arrange along the capillary, oriented at an angle of around 45° with respect to the tubule axis. High-yield filling of CNTs by the capillary method is generally difficult but fillings of more than 60% have been reported for different halides, with filling lengths of up to a couple of hundreds of nm [3.243]. Results from the imaging and characterization of individual molecules and atomically thin, effectively one-dimensional crystals of rock salt and other halides encapsulated within single-walled carbon nanotubes have recently been reviewed by *Sloan* et al.[3.244].

Sublimation Filling Method

This method is even more restrictive than the previous one, since it is only applicable to a very limited number of compounds due to the need for the filling material to sublimate within the temperature range of thermal stability of the nanotubes. Examples are therefore scarce. Actually, except for a few attempts to fill SWNTs with $ZrCl_4$ [3.240] or selenium [3.245], the first and most

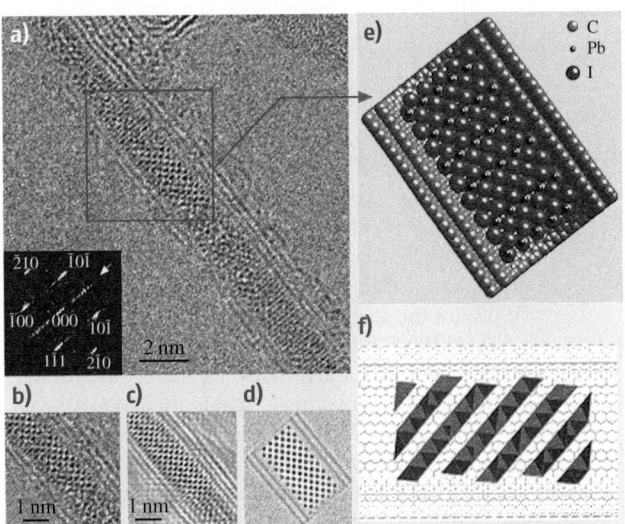

Fig. 3.25a–f HRTEM images (experimental and simulated) and corresponding structural model for a PbI_2-filled double-wall carbon nanotube (from [3.232])

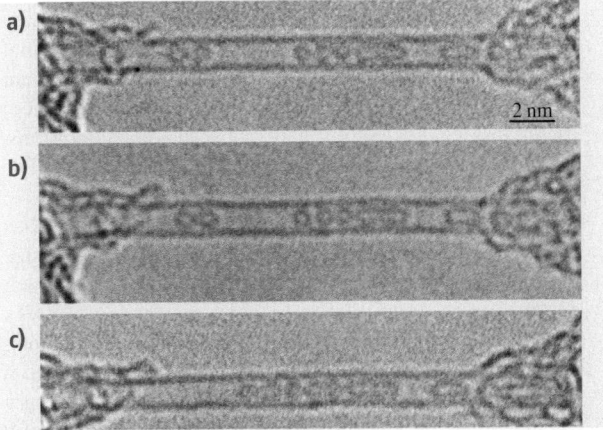

Fig. 3.26a–c HRTEM images of (**a**) an example of five regular C_{60} molecules encapsulated together with two higher fullerenes (C_{120} and C_{180}) as distorted capsules (*on the right*) within a regular 1.4 nm-diameter SWNT. (**a**)–(**c**) Example of the diffusion of the C_{60}-molecules along the SWNT cavity. The time between each image in the sequence is about 10 s. The fact that nothing occurs between (**a**) and (**b**) illustrates the randomness of the ionization events generated by the electron beam that are assumed to be responsible for the molecular displacement

successful example published so far is the formation of C_{60}@SWNT (nicknamed "peapods"), reported for the first time in 1998 [3.246], where regular $\approx$ 1.4 nm-large SWNTs are filled with C_{60} fullerene molecule chains (Fig. 3.26a). Of course, the process requires that the SWNTs are opened by some method, as discussed previously; typically either acid attack [3.247] or heat treatment in air [3.248]. The opened SWNTs are then inserted into a glass tube together with fullerene powder, which is sealed and placed into a furnace heated above the sublimation temperature for fullerite ($> \approx 350\,°C$). Since there are no filling limitations related to Laplace's law or the presence of solvent (only gaseous molecules are involved), filling efficiencies may actually reach $\approx 100\%$ for this technique [3.248].

$C_{60}@SWNT$ has since been shown to possess remarkable behavior traits, such as the ability of the C_{60} molecules to move freely within the SWNT cavity (Fig. 3.26b and 3.26c) upon random ionization effects from electron irradiation [3.249], to coalesce into 0.7 nm-wide elongated capsules upon electron irradiation [3.250], or into a 0.7 nm-wide nanotube upon subsequent thermal treatment above $1200\,°C$ under vacuum [3.249, 251]. Annealing $C_{60}@SWNT$ material could therefore be an efficient way to produce DWNTs with constant inner (≈ 0.7 nm) and outer (≈ 1.4 nm) diameters. Using the coalescence of encapsulated fullerenes through both electron irradiation and thermal treatment, it appears to possible to control subsequent DWNT features (inner tube diameter, intertube distance) by varying the electron energy, flow and dose conditions, the temperature, and the outer tube diameter [3.252]. The smallest MWNTs have been obtained in this way.

By synthesizing "endofullerenes" [3.12], it has been possible to use this process to synthesize more complex nanotube-based hybrid materials such as $La_2@C_{80}@SWNTs$ [3.253], $Gd@C_{82}@SWNTs$ [3.254], and $Er_xSc_{3-x}N@C_{80}@SWNT$ [3.255], among other examples. This suggests even more potential applications for peapods, although they are still speculative since the related properties are still being investigated [3.256–258].

The last example discussed here is the successful attempt to produce peapods by a related method, using accelerated fullerene ions (instead of neutral gaseous molecules) to force the fullerenes to enter the SWNT structure [3.259].

3.5.3 Functionalized Nanotubes

Noting the reactivity of carbon nanotubes (discussed in Sect. 3.4.6), nanotube functionalization reactions can be divided into two main groups. One is based on the chemical oxidation of the nanotubes (tips, structural defects) leading to carboxylic, carbonyl and/or hydroxyl functions. These functions are then used for additional reactions, to attach oligomeric or polymeric functional entities. The second group is based on direct addition to the graphitic-like surface of the nanotubes (without any intermediate step). Examples of the latter reactions include oxidation or fluorination (an important first step for further functionalization with other organic groups). The properties and applications of functionalized nanotubes have been reviewed in [3.260].

Oxidation of Carbon Nanotubes

Carbon nanotubes are often oxidized and therefore opened before chemical functionalization in order to increase their chemical reactivity (to create dangling bonds). The chemical oxidation of nanotubes is mainly performed using either wet chemistry or gaseous oxidants such as oxygen (typically air) or CO_2. Depending on the synthesis used, the oxidation resistance of nanotubes can vary. When oxidation is achieved using a gas phase, thermogravimetric analysis (TGA) is of great use for determining at which temperature the treatment should be applied. It is important to note that TGA accuracy increases as the heating rate diminishes, while the literature often provides TGA analyses obtained in unoptimized conditions, leading to overestimated oxidation temperatures. Differences in the presence of catalyst remnants (metals or, more rarely, oxides), the type of nanotubes used (SWNTs, c-MWNTs, h-MWNTs), the oxidizing agent used (air, O_2 is an inert gas, CO_2, and so on), as well as the flow rate used make it difficult to compare published results. It is generally agreed, however, that amorphous carbon burns first, followed by SWNTs and then multiwall materials (shells, MWNTs), even if TGA is often unable to separate the different oxidation steps clearly. Air oxidation (static or dynamic conditions) can however be used to prepare samples of very high purity – although the yield is generally low – as monitored by in situ Raman spectroscopy [3.261]. Aqueous solutions of oxidizing reagents are often used for nanotube oxidation. The main reagent is nitric acid, either concentrated or diluted (around 3 moles per liter in most cases), but oxidants such as potassium dichromate ($K_2Cr_2O_7$), hydrogen peroxide (H_2O_2) or potassium permanganate ($KMnO_4$) are often used as well. HCl, like HF, does not damage nanotubes because it is not oxidizing.

Functionalization of Oxidized Carbon Nanotubes

The carboxylic groups located at the nanotube tips can be coupled to different chemical groups. Oxidized nanotubes are usually reacted with thionyl chloride ($SOCl_2$) to generate the acyl chloride, even if a direct reaction is theoretically possible with alcohols or amines, for example. The reaction of SWNTs with octadecylamine (ODA) was reported by *Chen* et al. [3.262] after reacting oxidized SWNTs with $SOCl_2$. The functionalized SWNTs are soluble in chloroform ($CHCl_3$), dichloromethane (CH_2Cl_2), aromatic solvents, and carbon bisulfide (CS_2). Many other reactions between functionalized nanotubes (after reaction with $SOCl_2$) and amines have been reported in the literature and will not be reviewed here. Noncovalent reactions between the carboxylic groups of oxidized nanotubes and octadecylammonium ions are possible [3.263], providing solubility in tetrahydrofuran (THF) and CH_2Cl_2. Functionalization by glucosamine using similar procedures [3.264] produced water soluble SWNTs, which is of special interest when considering biological applications of functionalized nanotubes. Functionalization with lipophilic and hydrophilic dendra (with long alkyl chains and oligomeric poly(ethyleneglycol) groups) has been achieved via amination and esterification reactions [3.265], leading to solubility of the functionalized nanotubes in hexane, chloroform, and water. It is interesting to note that, in the latter case, the functional groups could be removed simply by modifying the pH of the solution (base- and acid-catalyzed hydrolysis reaction conditions, [3.266]). One last example is the possible interconnection of nanotubes via chemical functionalization. This has been recently achieved by *Chiu* et al. [3.267] using the acyl chloride method and a bifunctionalized amine to link the nanotubes through the formation of amide bonds.

Sidewall Functionalization of Carbon Nanotubes

Covalent functionalization of nanotube walls is possible through fluorination reactions. It was first reported by *Mickelson* et al. [3.268], based on F_2 gas (the nanotubes can then be defluorinated, if required, with anhydrous hydrazine). As recently reviewed by *Khabashesku* et al. [3.269], it is then possible to use these fluorinated nanotubes to carry out subsequent derivatization reactions. Thus, sidewall-alkylated nanotubes can be prepared by nucleophilic substitution (Grignard synthesis or reaction with alkyllithium precursors [3.270]).

These alkyl sidewall groups can be removed by air oxidation. Electrochemical addition of aryl radicals (from the reduction of aryl diazonium salts) to nanotubes has also been reported by *Bahr* et al. [3.271]. Functionalizations of the external wall of the nanotube by cycloaddition of nitrenes, addition of nuclephilic carbenes or addition of radicals have been described by *Holzinger* et al. [3.272]. Electrophilic addition of dichlorocarbene to SWNTs occurs via a reaction with the deactivated double bonds in the nanotube wall [3.273]. Silanization reactions are another way to functionalize nanotubes, although only tested with MWNTs. *Velasco-Santos* et al. [3.274] have reacted oxidized MWNTs with an organosilane ($RsiR_3$, where R is an organo functional group attached to silicon) and obtained nanotubes with organo functional groups attached via silanol groups.

The noncovalent sidewall functionalization of nanotubes is important because the covalent bonds are associated with changes from sp^2 hybridization to sp^3 carbon hybridization, which corresponds to loss of the graphite-like character. The physical properties of functionalized nanotubes, specifically SWNTs, can therefore be modified. One way to achieve the noncovalent functionalization of nanotubes is to wrap the nanotubes in a polymer [3.275], which permits solubilization (enhancing processing possibilities) while preserving the physical properties of the nanotubes. One reason to functionalize SWNTs is to make them soluble in regular solvents. A promising method to do this was found by *Pénicaud* et al., who made water-soluble by adding charges to SWNTs via the transient and reversible formation of a nanotube salt [3.276].

Finally, it is worth bearing in mind that none of these chemical reactions are specific to nanotubes and so they can affect most of the carbonaceous impurities present in the raw materials as well, making it difficult to characterize the functionalized samples. The experiments must therefore be performed with very pure carbon nanotube samples, which is unfortunately not always the case for the results reported in the literature. On the other hand, purifying the nanotubes to start with may also bias the functionalization experiments, since purification involves chemical treatment. However a demand for such products already exists, and purified then fluorinated SWNTs can be bought for $900/g (Carbon Nanotechnologies Inc., 2005).

3.6 Applications of Carbon Nanotubes

A carbon nanotube is inert, has a high aspect ratio and a high tensile strength, has low mass density, high heat conductivity, a large surface area, and a versatile electronic behavior, including high electron conductivity. However, while these are the main characteristics of individual nanotubes, many of them can form secondary structures such as ropes, fibers, papers and thin films with aligned tubes, all with their own specific properties. These properties make them ideal candidates for a large number of applications provided their cost is sufficiently low. The cost of carbon nanotubes depends strongly on both the quality and the production process. High-quality single-shell carbon nanotubes can cost 50–100 times more than gold. However, carbon nanotube synthesis is constantly improving, and sale prices are falling rapidly. The application of carbon nanotubes is therefore a very fast moving field, with new potential applications found every year, even several times per year. Therefore, creating an exhaustive list of these applications is not the aim of this section. Instead, we will cover the most important applications, and divide them up according to whether they are "current" (Sect. 3.6.1) – they are already on the market, the application is possible in the near future, or because prototypes are currently being developed by profit-based companies – or "expected" applications (Sect. 3.6.2).

3.6.1 Current Applications

Near-Field Microscope Probes

The high mechanical strength of carbon nanotubes makes them almost ideal candidates for use as force sensors in scanning probe microscopy (SPM). They provide higher durability and the ability to image surfaces with a high lateral resolution, the latter being a typical limitation of conventional force sensors (based on ceramic tips). The idea was first proposed and tested by *Dai* et al. [3.85] using c-MWNTs. It was extended to SWNTs by *Hafner* et al. [3.278], since small-diameter SWNTs were believed to give higher resolution than MWNTs due to the extremely short radius of curvature of the tube end. However, commercial nanotube-based tips (such as those made by Piezomax, Middleton, WI, USA) use MWNTs for processing convenience. It is also likely that the flexural modulus of a SWNT is too low, resulting in artifacts that affect the lateral resolution when scanning a rough surface. On the other hand, the flexural modulus of a c-MWNT is believed to increase with the number of walls, although the radius of curvature

of the tip increases at the same time. Whether based on SWNT or MWNT, such SPM tips also offer the potential to be functionalized, leading to the prospect of selective imaging based on chemical discrimination in "chemical force microscopy" (CFM). Chemical function imaging using functionalized nanotubes represents a huge step forward in CFM because the tip can be functionalized very specifically (ideally only at the very tip of the nanotube, where the reactivity is the highest), increasing the spatial resolution. The interaction between the chemical species present at the end of the nanotube tip and the surface containing chemical functions can be recorded with great sensitivity, allowing the chemical mapping of molecules [3.279, 280].

Current nanotube-based SPM tips are quite expensive; typically $\approx \$450/$tip (Nanoscience Co., 2005). This high cost is due to processing difficulties (it is necessary to grow or mount a single MWNT in the appropriate direction at the tip of a regular SPM probe; Fig. 3.27), and the need to individually control the tip quality. The market for nanotube SPM tips has been estimated at $\approx \$20\,$M/year.

Field Emission–Based Devices

In a pioneering work by *de Heer* et al. [3.281], carbon nanotubes were shown to be efficient field emitters and this property is currently being used several applications, including flat panel displays for television sets and computers (the first prototype of such a display was exhibited by Samsung in 1999), and devices

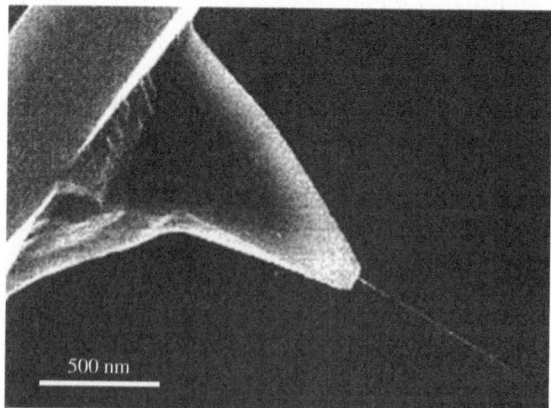

Fig. 3.27 Scanning electron microscopy image of a carbon nanotube (MWNT) mounted onto a regular ceramic tip as a probe for atomic force microscopy (modified from [3.277])

requiring an electron-producing cathode, such as X-ray sources. The principle of a field emission-based screen is demonstrated in Fig. 3.28a. Briefly, a potential difference is set up between the emitting tips and an extraction grid so that electrons are pulled from the tips onto an electron-sensitive screen layer. Replacing the glass support and protecting the screen using a polymer-based material should even permit the development of flexible screens. Unlike regular (metallic) electron-emitting tips, the structural perfection of carbon nanotubes allows higher electron emission stability, higher mechanical resistance, and longer lifetimes. Most importantly, using them saves energy since the tips operate at a lower heating temperature and require much lower threshold voltage than in other set-ups. For example, it is possible to produce a current density of $1 \, \text{mA/cm}^2$ for a threshold voltage of $3 \, \text{V}/\mu\text{m}$ with nanotubes, while it requires $20 \, \text{V}/\mu\text{m}$ for graphite powder and $100 \, \text{V}/\mu\text{m}$ for regular Mo or Si tips. The subsequent reductions in cost and energy consumption are estimated at $1/3$ and $1/10$ respectively. Generally speaking, the maximum current density that can be obtained ranges from $10^6 \, \text{A/cm}^2$ to $10^8 \, \text{A/cm}^2$ depending on the nanotubes involved (SWNT or MWNT, opened or capped, aligned or not, and so on) [3.282–284]. Although the side walls of the nanotubes seem to emit as well as the tips, many works have investigated the growth of nanotubes perpendicular to the substrate surface as regular arrays (Fig. 3.28b). Besides, it does not appear necessary to use SWNTs instead of MWNTs for many of these applications when they are used as bunches. On the other hand, when considering single, isolated nanotubes, SWNTs are generally less preferable since they permit much lower electron doses than MWNTs, although they often provide a more coherent source (an useful feature for devices such as electron microscopes or X-ray generators).

The market associated with this application is huge. With major companies involved, such as Motorola, NEC, NKK, Samsung, Thales and Toshiba, the first flat TV sets and computers using nanotube-based screens should enter the market in 2007 (Samsung data), once a problem with product lifetime (still only about half that required) is fixed. On the other hand, companies such as Oxford Instruments and Medirad are now commercializing miniature X-ray generators for medical applications that use nanotube-based cold cathodes developed by Applied Nanotech Inc.

Chemical Sensors

The electrical conductance of semiconductor SWNTs was recently demonstrated to be highly sensitive to

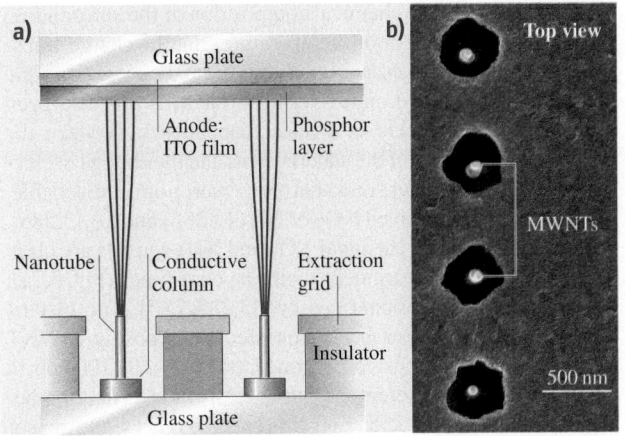

Fig. 3.28 (a) Principle of a field emitter-based screen. (b) Scanning electron microscope image of a nanotube-based emitter system (*top view*). *Round dots* are MWNT tips seen through the holes corresponding to the extraction grid. Courtesy of *P. Legagneux* (Thales Research & Technology, Orsay, France)

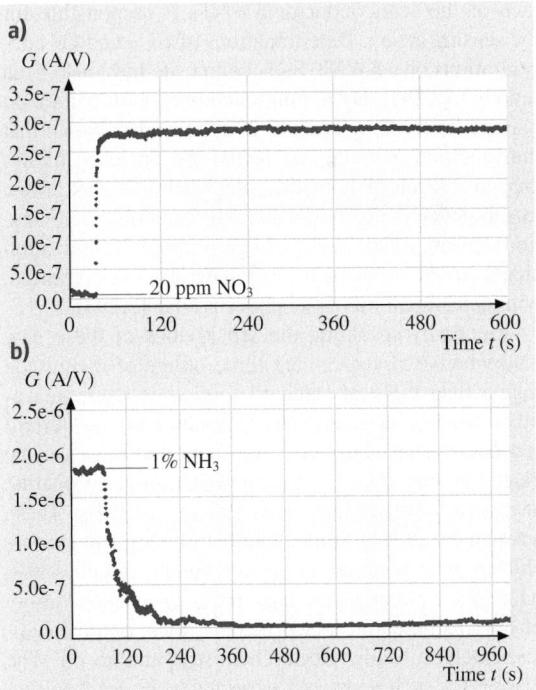

Fig. 3.29a,b Demonstration of the ability of SWNTs to detect trace molecules in inert gases. (a) Increase in the conductance of a single SWNT when 20 ppm of NO_2 are added to an argon gas flow. (b) Same, but with 1% NH_3 added to the argon gas flow (from [3.285])

changes in the chemical composition of the surrounding atmosphere at room temperature, due to charge transfer between the nanotubes and the molecules from the gases adsorbed onto SWNT surfaces. It has also been shown that there is a linear dependence between the concentration of the adsorbed gas and the change in electrical properties, and that the adsorption is reversible. First tries involved NO_2 or NH_3 [3.285] and O_2 [3.286]. SWNT-based chemical NO_2 and NH_3 sensors are characterized by extremely short response times (Fig. 3.29), unlike conventional sensors [3.285, 287]. The electrical response has been measured by exposing MWNT films to sub-ppm NO_2 concentrations (10–100 ppb in dry air) at different operating temperatures ranging between 25 and 215 °C [3.288]. For SWNTs, the sensor responses are linear for similar concentrations, with detection limits of 44 ppb for NO_2 and 262 ppb for nitrotoluene [3.289]. High sensitivity to water or ammonia vapor has been demonstrated on a SWNT-SiO_2 composite [3.290]. This study indicated the presence of p-type SWNTs dispersed among the predominantly metallic SWNTs, and that the chemisorption of gases on the surface of the semiconductor SWNTs is responsible for the sensing action. Determinations of CO_2 and O_2 concentrations on a SWNT-SiO_2 composite have also been reported [3.291]. By doping nanotubes with palladium nanoparticles, *Kong* et al. [3.292] have also shown that the modified material can reveal the presence of hydrogen at levels of up to 400 ppm, whereas the as-grown material was totally ineffective. Miniaturized gas ionization sensors, which work by fingerprinting the ionization characteristics of distinct gases, have also been reported, with detection limits of 25 ppm for NH_3 [3.293].

Generally speaking, the sensitivities of these new nanotube-based sensors are three orders of magnitude higher than those of standard solid state devices. Another reason for using nanotubes instead of current sensors is their simplicity, the facts that they can be placed in very small systems and that they can operate at room temperature, as well as their selectivity. These advantages allow a limited number of sensor device architectures to be built for a variety of industrial purposes, while the current technology requires a large variety of devices based on mixed metal oxides, optomechanics, catalytic beads, electrochemistry, and so on. The market for such devices is expected to be $ 1.6 billion by 2006, including sensing applications in biological fields and the chemical industry. Nanotube-based sensors are currently being developed by large and small companies, such as Nanomix (Emeryville, CA, USA), for example.

Catalyst Support

Carbon-based materials make good supports in heterogeneous catalytic processes due to their ability to be tailored to a specific need: indeed, activated carbons are already currently employed as catalyst supports due to their high surface areas, their stability at high temperatures (under nonoxidizing atmospheres), and the possibility of controlling both their porous structure and the chemical nature of their surfaces [3.294, 295]. Attention has focused on nanosized fibrous morphologies of carbon have appeared over the last decade, that show great potential for use as supports [3.296]. Carbon nanofibers (also incorrectly called graphite nanofibers) and carbon nanotubes have been successfully used in this area, and have been shown to provide, as catalyst-supporting materials, properties superior to those of such other regular catalyst-supports, such as activated carbon, soot or graphite [3.297–299]. Various reactions have been studied [3.297–299]; hydrogenation reactions, hydrocarbon decomposition and use as fuel cell electrocatalysts are among the most popular domains. The application of graphite nanofibers as direct catalysts for oxidative dehydrogenation [3.300, 301] or methane decomposition [3.302] has also been reported.

The morphology and size of the carbon nanotubes (particularly their aspect ratios), can play a significant role in catalytic applications due to their ability to disperse catalytically active metal particles. Their electronic properties are also of primary importance [3.303], since the conductive support may cause electronic perturbations as well as constraining the geometriies of the dispersed metal particles. A recent comparison between the interactions of transition metal atoms with carbon nanotube walls and their interactions with graphite has shown major differences in bonding sites, magnetic moments, and charge transfer direction [3.304]. Thus the possibility of a strong metal–support interaction must be taken into account. Their mechanical strength is also important, and this makes them resistant to attrition when recycled. Their external and internal surfaces are strongly hydrophobic and adsorb organic molecules strongly. For MWNT-based catalyst-supports, the relatively high surface area and the absence of microporosity (pores < 2 nm), associated with a high meso- and macropore volume (see Sect. 3.4.3), result in significant improvements in catalytic activity for liquid phase reactions when compared to catalysts supported on activated carbon. With nanotube supports, the mass transfer of the reactants to the active sites is unlimited, due to the absence of microporosity, and the apparent contact time of the products with the catalyst is dimin-

Table 3.4 Preparation and catalytic performances of some nanotube-supported catalysts

Catalyst	Preparation route	Catalytic reaction	Comments
Ru/MWNT+SWNT [3.296]	Liquid phase impregnation, no pretreatment of the tubes	Liquid phase cinnamalde-hyde hydrogenation	A different kind of metal support interaction compared to activated carbon
Pt/MWNT electrodes [3.305]	Electrodeless plating with prefunctionalization of MWNT	Oxygen reduction for fuel cell applications	High electrocatalytic activity
Rh/MWNT [3.306]	Surface-mediated organo-metallic synthesis, prefunctionalization of MWNT	Liquid phase hydroformyla-tion and hydrogenation	Higher activity of Rh/MWNT compared to Rh/activated carbon
Ru-alkali/MWNT [3.307]	Liquid phase impregnation, no pretreatment of the tubes	Ammonia synthesis, gas phase reaction	Higher activity with MWNT than with graphite
Rh-phosphine/MWNT [3.308]	Liquid phase grafting from [RhH(CO)(PPh$_3$)$_3$]	Liquid phase hydroformyla-tion	Highly active and regiose-lective catalyst

ished, leading to more active and more selective catalytic effects. Finally, as for activated carbon, catalyst-forming is possible and porous granules of carbon nanotubes or electrodes based on carbon nanotubes can be obtained for catalysis or electrocatalysis respectively.

Fig. 3.30 Transmission electron microscopy image showing rhodium nanoparticles supported on the surface of an MWNT (from [3.306])

The technique usually used to prepare carbon nanotube-supported catalysts is incipient wetness impregnation, in which the purified support is impregnated with a solution of the metal precursor and then dried, calcinated and/or reduced in order to obtain metal particles dispersed on the support. Chemical treatment and/or modification of the carbon nanotube surface were found to be useful ways of controlling its hydrophobic or hydrophilic character [3.309]. A strong metal/support interaction can thus be expected from the occurrence of functionalized groups created by the oxidation of the support surface, resulting in smaller particle sizes [3.310]. A more sophisticated technique for achieving the grafting of metal particles onto carbon nanotubes consists of functionalizing the outer surface of the tubes and then performing a chemical reaction with a metal complex, resulting in a good dispersion of the metallic particles (Fig. 3.30) [3.306]. The functionalization of noncovalent carbon nanotubes with polymer multilayers followed by the attachment of gold nanoparticles has also been reported [3.311].

Selected examples of some carbon nanotube-based catalysts together with related preparation routes and catalytic activities are listed in Table 3.4.

The market is important for this application, since it often concerns the heavy chemical industry. It implies and requires mass production of low-cost nanotubes, processed by methods other than those based on solid carbon as the source (see Sect. 3.2.1). Such an application also requires some surface reactivity, making the h-MWNT-type nanotubes, with poor nanotextures (see Sect. 3.1.2), the most suitable starting material for preparing such catalyst supports. Catalysis-enhanced thermal cracking of gaseous carbon precursors is therefore preferred, and pilot plants are already being built by major chemical industrial companies (such as Arkema in France).

3.6.2 Expected Applications Related to Adsorption

Adsorptions of various gases, liquids or metals onto carbon nanotubes, and interactions between them, have attracted much attention recently. The applications resulting from the adsorptive properties of carbon nanotubes can be arbitrarily divided into two groups. The first group is based on the consequences of molecular adsorption on the electronic properties of nanotubes; the main application of this is chemical sensing (see Sect. 3.6.1). The second group includes gas storage, gas separation, the use of carbon nanotubes as adsorbants, and results from morphological investigations of carbon nanotubes (surface areas, aspect ratios, and so forth). Among these latter potential applications, the possibility of storing gases – particularly hydrogen – on carbon nanotubes has received most attention.

Gas Storage – Hydrogen
The development of a lightweight and safe system for hydrogen storage is necessary for the widespead use of highly efficient H_2-air fuel cells in transportation vehicles. The U.S. Department of Energy Hydrogen Plan has provided a commercially significant benchmark for the amount of reversible hydrogen adsorption required. This benchmark requires a system weight efficiency (the ratio of H_2 weight to system weight) of 6.5 wt% hydrogen, and a volumetric density of $63 \, kg \, H_2/m^3$.

The failure to produce a practical storage system for hydrogen has prevented hydrogen from becoming one of the most important transportation fuels. The ideal hydrogen storage system needs to be light, compact, relatively inexpensive, safe, easy to use, and reusable without the need for regeneration. While research and development are continuing into such technologies as liquid hydrogen systems, compressed hydrogen systems, metal hydride systems, and superactivated carbon systems, all have serious disadvantages. For example, liquid hydrogen systems are very expensive, primarily because the hydrogen must be cooled to about $-252\,°C$. This means that a liquid hydrogen system costs about four times as much as an equivalent amount of gasoline. Also, liquid hydrogen must be kept cooled to prevent it from boiling away, even when the vehicle is parked. Compressed hydrogen is

much cheaper than liquid hydrogen, but also much bulkier.

Metal hydride systems store hydrogen as a solid in combination with other materials. For example, metal hydrides are produced by bathing a metal, such as palladium or magnesium, in hydrogen. The metal splits the dihydrogen gas molecules and binds the hydrogen atoms to the metal until released by heating. The disadvantages of a metal hydride system are its weight (typically about eight times more than an equivalent amount of liquid hydrogen or gasoline) and the need to warm it up to release the hydrogen.

Superactivated carbon provides the basis of another system for storing hydrogen that initially showed commercial potential. Superactivated carbon is a material similar to the highly porous activated carbon used in water filters but that can gently adsorb hydrogen molecules through physisorption at subzero temperatures. The colder the carbon, the less heat is needed to disturb the weak forces holding the carbon and hydrogen together. Therefore, again, a major disadvantage of such a system is that it must constantly remain at very low temperatures to stop the hydrogen from escaping, even when the vehicle is parked.

Therefore, there is still a great need for a material that can store hydrogen but is also light, compact, relatively inexpensive, safe, easy to use, and reusable without regeneration. Some recent articles and patents on the very high, reversible adsorption of hydrogen in carbon nanotubes or platelet nanofibers have aroused tremendous interest in the research community, stimulating much experimental and theoretical work. Most of the work done on hydrogen adsorption on carbon nanotubes have been recently reviewed [3.312–317].

A group from Northeastern University [3.176, 318] was the first to report the supposedly successful storage of hydrogen in carbon layered nanostructures possessing some crystallinity. The authors claimed that up to 75 wt % of hydrogen (a C/H ratio of 1/9) can be stored in platelet nanofibers (3–50 nm in width). Complete hydrogen desorption occurs only at very high temperature. However, the same authors recently pointed out the need for the carbon nanofibers to be thermally treated prior to H_2 adsorption [3.318]. This thermal treatment, although simple (1000 °C in an inert gas), is critical for removing chemisorbed gases from the edge and step regions of the carbon structures. Failure to perform this treatment results in a dramatic decrease in the hydrogen adsorption of the material. The authors also performed the experiments at room temperature, and hydrogen uptakes of 20 to 40 wt % were recorded for 100 bar of H_2. The authors also demonstrated (by XRD) that the material was structurally perturbed following hydrogen treatment, as demonstrated by an expansion of the graphitic interplane distance (from 0.34 before H_2 adsorption to 0.347 after H_2 uptake). Another study also underlined the key role played by the structure (tubular or herringbone) and the in situ pretreatment of the material in oxidizing, reducing or neutral environments [3.319]. The presence of residual catalyst in the sample also needs to be taken into account, as it can produce hydrogen spillover [3.320].

Hydrogen adsorption and desorption measurements on nanofibers similar to those used by *Rodriguez* et al. [3.16] were also performed by *Ahn* et al. [3.321]. The authors measured a hydrogen uptake of 2.4 wt %. In fact, in their study, none of the graphite nanofiber material tested gave hydrogen storage abilities above the values already reported for activated carbons.

The adsorption of H_2 onto platelet nanofibers was computed via Monte Carlo simulations [3.322]. The graphene spacing was optimized to maximize the weight fraction of H_2 adsorbed. Given the results of their calculations, the authors concluded that no physically realistic graphite–hydrogen potential can account for the tremendous adsorption ability reported by the group from Penn State.

Rzepka et al. [3.323] have also used theoretical Monte Carlo calculations to calculate the amount of physically adsorbed hydrogen molecules in carbon slit pores with dimensions similar to those used by the Penn State group (an interplanar distance between the graphenes in the nanofibers of 0.34 nm). They reached the conclusion that no hydrogen could be adsorbed at all. Even if the interplanar distance is assumed to expand during the adsorption process, the maximum calculated adsorption is 1 wt % for $d = 0.7$ nm. These authors concluded that the Penn State experimental results could not be attributed to some "abnormal capillary condensation effect" as initially claimed. Since then, nobody has been able to duplicate the results from those first experiments.

Meanwhile, hydrogen storage experiments were also attempted on SWNTs. The first work was reported by *Dillon* et al. [3.324], who used temperature-programmed desorption measurements. A major drawback of their experimental procedure was that they used only 1 mg of *unpurified* soot containing 0.1% of SWNTs. They attributed the reversible hydrogen capacity of 0.01 wt % observed to the SWNTs alone, giving a presumed capacity of $5 - 10$ wt % for a pure sample of nanotubes. TPD experiments indicated that H_2 desorption occurred between -170 and $+100\,°C$. Despite the weakness of the demonstration, the paper, which was not published

not long before that of *Chambers* et al. [3.176], induced worldwide excitement in the field. Hydrogen adsorption studies based on ^1H and ^2H NMR spectroscopy have revealed that adsorption is fast and reversible and must be described as physisorption [3.325].

Hydrogen adsorption measurements were also performed on SWNT samples of high purity [3.326]. At $-193\,°C$, 160 bar of H_2 were admitted and the hydrogen adsorption was found to exceed 8 wt%. In this case, the authors proposed that hydrogen is initially adsorbed on the outer surfaces of the crystalline material. However, no data on H_2 desorption were given.

In other reports, *Dresselhaus* and coworkers [3.312, 327] reported hydrogen storage in SWNTs (average diameter 1.85 nm) at room temperature. A hydrogen storage capacity of 4.2 wt% was achieved reproducibly at 100 bar for about 0.5 g of a SWNT-containing material that had been previously soaked in hydrochloric acid and then heat-treated ($500\,°C$) in vacuum. The purity of the sample was estimated from TGA and TEM observations to be about 50 to 60%. Moreover, 78% of the adsorbed hydrogen could be released under ambient pressure at room temperature, while releasing the residual stored hydrogen required some heating at $200\,°C$.

Valuable input to the topic was provided by *Hirscher* et al. [3.328], who demonstrated that several of the supposedly successful experiments regarding the storage of H_2 in SWNTs were actually misled by the hydrogen storage capacity of Ti nanoparticles originating from the sonoprobe frequently used at one step or another of the procedure, specifically when the SWNT material is purified beforehand. Since then, studies on hydrogen adsorption in SWNTs have reported regularly lower values, ranging between 0.5 and 1.2 wt % at 77 K and atmospheric pressure [3.329–331]. Furthermore, no significant difference has been found by neutron scattering between the hydrogen molecules adsorbed in nanotubes, fibers and coals [3.329]. Hydrogen adsorption has been measured at 0.3 wt % for MWNTs samples as large as 85 g at 300 K and 10 MPa, while 2.27 wt % hydrogen was released at 77 K by nitric acid treated MWNTs [3.332].

Density functional theory (DFT) was used [3.333] to estimate H_2 adsorption in SWNTs. From their calculations, within the regime of operating conditions where adsorptive storage seems attractive, the storage properties of H_2 in a SWNT system appear to fall far short of the DOE target. In their model, rolling graphite into a cylindrical sheet does not significantly alter the nature of the carbon–H_2 interaction, contradicting the latest calculations, which indicate that the physisorption increases as the radius of curvature of the SWNT decreases [3.334].

However, *Lee* et al. [3.335–337] have also used DFT calculations to search for hydrogen adsorption sites and to predict the maximum storage capacity of SWNTs. They found two chemisorption sites: at the exterior and the interior of the tube wall. Thus, they predict that the maximum hydrogen storage capacity could exceed 14 wt % (160 kg H_2/m^3). The authors have also considered H_2 adsorption in multiwall carbon nanotubes and have predicted lower storage capacities for the latter. At 77 K and 4 MPa, DFT calculations predict a total excess gravimetric storage capacity of hydrogen of 7.1 wt % for open 2.719 nm nanotube arrays and 9.5 wt % for open isolated 2.719 nm nanotubes [3.338]. However, at 300 K, the total excess adsorption of hydrogen on both carbon nanotube arrays and isolated nanotubes does not exceed 1 wt %.

These calculations are constrained, however, by the starting hypotheses. While considering the same (10,10) SWNT, calculations based on DFT predict between 14.3% and 1 wt % storage [3.336, 338], calculations based on a geometrical model predict 3.3% [3.312], and calculations based on a quantum mechanical molecular dynamics model predict 0.47% [3.339].

In conclusion, neither experimental results, obviously biased by some problem in the procedures, nor theoretical results are yet able to demonstrate that an efficient storage of H_2 is possible for carbon nanotubes, whatever the type. However, a definitive statement of failure cannot yet be claimed. Further efforts have to be made to enhance the quality of these materials, in particular (i) by adjusting the surface properties, which can be modified by chemical or mechanical treatments, and (ii) by adjusting the structure of the material, such as the pore size [3.340] and possibly the curvature. Whether the best carbon material for H_2 adsorption will then be nanotube-based is another story.

Gas Storage – Gases Other than Hydrogen

Encouraged by the potential applications related to hydrogen adsorption, several research groups have tried to use carbon nanotubes as a means of stocking and transporting other gases such as oxygen, nitrogen, noble gases (argon and xenon) and hydrocarbons (methane, ethane, and ethylene). These studies have shown that carbon nanotubes could become the world's smallest gas cylinders, combining low weight, easy transportability and safe use with acceptable adsorbed quantities. Nanotubes may also be used in medicine, where it would

be extremely useful to physically confine special gases (^{133}Xe for instance) prior to injection.

Kusnetzova et al. [3.341] conducted experiments with xenon and found that the storage capacities of nanotubes can be enhanced by a tremendous amount (a factor of 280, up to a molar ratio of $N_{Xe}/N_C = 0.045$) by opening the SWNT bundles via thermal activation at 800 °C. The gas can be adsorbed inside the nanotubes and the rates of adsorption are also increased using this treatment.

The possibility of storing argon in carbon nanotubes has been studied, with encouraging results, by *Gadd* et al. [3.342]. Their experiments show that large amounts of argon can be trapped in catalytically grown MWNTs (20–150 nm) by hot isostatic pressing (HIPing) for 48 hours at 650 °C under an argon pressure of 1700 bar. Energy-dispersive X-ray spectroscopy was used to determine that the gas was located inside the tubes and not on the tube walls. Further studies determined the argon pressure inside the tubes at room temperature. The authors estimated this to be around 600 bars, indicating that equilibrium pressure was attained in the tubes during the HIP-ing and that MWNTs would be a convenient material for storing the gas.

Gas Separation

As SWNTs or MWNTs have regular geometries that can, to some extent, be controlled, they could be used to develop precise separation tools. If the sorption mechanisms are known, it should be possible to control sorption of various gases through particular combinations of temperature, pressure and nanotube morphology. Since the large-scale production of nanotubes is gradually progressing, and this should ultimately result in low costs, accurate separation methods based on carbon nanotubes are now being investigated.

A theoretical study has aimed to determine the effects of different factors such as tube diameter, density and type of the gas used on the flow of molecules inside nanotubes. An atomistic simulation with methane, ethane and ethylene [3.343] has shown that the molecular mobility decreases with decreasing tube for each of the three gases. Ethane and ethylene have smaller mobilities due to the stronger interactions they seem to have with the nanotube walls. In another theoretical study into the possibility of hydrocarbon mixture separation on SWNT bundles, the authors conclude that carbon nanotubes can be used to separate methane/*n*-butane and methane/isobutene mixtures [3.344] with an efficiency that increases as the average tube diameter decreases. Experimental work was also performed by the same group on the sorption of butane on MWNTs [3.178]. It has been also reported that the Fickian diffusivities of CH_4/H_2 mixtures in SWNT, like their pure component counterparts, are extraordinarily large when compared with adsorbed gases in other nanoporous materials [3.345].

Grand canonical Monte Carlo simulations of the separation of hydrogen and carbon monoxide by adsorption on SWNTs have also been reported [3.346]. In most of the situations studied, SWNTs were found to adsorb more CO than H_2, and excellent separation could again probably be obtained by varying the SWNT average diameter.

Adsorbents

Carbon nanotubes were found to be able to adsorb some toxic gases such as dioxins [3.347], fluoride [3.348], lead [3.349] and alcohols [3.350] better than adsorbent materials in common use, such as activated carbon. These pioneering works opened a new field of applications as cleaning filters for many industrial processes with hazardous by-products. The adsorption of dioxins, which are very common and persistent carcinogenic by-products of many industrial processes, is a good example of the potential of nanotubes in this field. Growing ecological awareness has resulted in the imposition of emission limits on dioxin-generating sources in many countries, but it is difficult to find materials that can act as effective filters, even at extremely low concentrations. *Long* et al. [3.347] found that nanotubes can attract and trap more dioxins than activated carbons or other polyaromatic materials that are currently used as filters. This improvement is probably due to the stronger interaction forces that exist between dioxin molecules and the curved surfaces of nanotubes compared to those for flat graphene sheets.

The capacity of Al_2O_3/MWNT to adsorb fluoride from water has been reported to be 13.5 times that of activated carbon and four times that of Al_2O_3 [3.348]. The same group has also reported a capacity of MWNTs to adsorb lead from water that is higher than that for activated carbon [3.349]. The possibility of using graphite nanofibers to purify water from alcohols has also been explored [3.350]. MWNTs were found to be good adsorbents for the removal of dichlorobenzene from wastewaters over a wide range of pH. Typically, the nanotubes adsorb 30 mg of the organic molecule per gram of MWNTs from a 20 mg/l solution [3.351]. It has also been shown that SWNTs act as molecular sponges for molecules such as CCl_4; the nanotubes were in contact with a support sur-

face which also adsorbs molecules, although more weakly than the nanotubes [3.352]. These experimental results suggest that carbon nanotubes may be promising adsorbents for removing polluting agents from water.

Biosensors

Attaching molecules of biological interest to carbon nanotubes is an excellent way to produce nanometer-sized biosensors. The electrical conductivities of these functionalized nanotubes would depend on the interaction of the probe with the medium being studied, which would be affected by chemical changes or interactions with the target species. The science of attaching biomolecules to nanotubes is rather recent and was inspired by similar research in the fullerene area. Some results have already been patented, and so such systems may become available in the near future. Using the internal cavities of nanotubes to deliver drugs would be another amazing application, but little work has been carried out so far to investigate the toxicity of nanotubes in the human body. Comparison between the effects of nanotubes and asbestos was investigated by *Huczko* et al. [3.353] and they concluded that the tested samples were innocuous. However, a more recent work has shown that contact with nanotubes may lead to dermal toxicity [3.354] or induce lung lesions characterized by the presence of granulomas [3.355]. *Pantarotto* et al. [3.356] reported the translocation of water-soluble SWNT derivatives across cell membranes and have shown that cell death can be induced by functionalised nanotubes (bioactive peptides), depending upon their concentration in the media. Recent results also indicate that nanotubes may lead to an inflammatory response of the immune system by activating the complement system [3.357].

MWNTs have been used by *Mattson* et al. [3.358] as a substrate for neuronal growth. They have compared the activity of untreated MWNTs with that of MWNTs coated with a bioactive molecule (4-hydroxynonenal) and observed that neurons elaborated multiple neurites on these latter functionalized nanotubes. This is an important result that illustrates the feasibility of using nanotubes as a substrate for nerve cell growth.

Davis et al. [3.359] immobilized different proteins (metallothionein, cytochrome c and c_3, β-lactamase I) in MWNTs and checked whether these molecules were still catalytically active compared to the free ones. They have shown that confining a protein within a nanotube provides some protection for the external environment. Protein immobilization via noncovalent sidewall functionalization was proposed by *Chen* et al. [3.360] using a bifunctional molecule (1-pyrenebutanoic acid, succinimidyl ester). This molecule is tied to the nanotube wall by the pyrenyl group, and amine groups or biological molecules can react with the ester function to form amide bonds. This method was also used to immobilize ferritin and streptavidin onto SWNTs. Its main advantages are that it does not modify the SWNT wall and that it does not perturb the sp^2 structure, so the physical properties of the nanotubes are maintained. *Shim* et al. [3.361] have functionalized SWNTs with biotin and observed specific binding with streptavidin, suggesting biomolecular recognition possibilities. *Dwyer* et al. [3.362] have functionalized SWNTs by covalently coupling DNA strands to them using EDC(1-ethyl-3-(3-dimethylaminopropyl) carbodiimide hydrochloride) but did not test biomolecular recognition; other proteins such as bovine serum albumin (BSA) [3.363] have been attached to nanotubes using the same process (diimide-activated amidation with EDC) and most of the attached proteins remained bioactive. Instead of working with individual nanotubes (or more likely nanotube bundles in the case of SWNTs), *Nguyen* et al. [3.364] have functionalized nanotubes arrayed with a nucleic acid, still using EDC as the coupling agent, in order to realize biosensors based on protein-functionalized nanotubes. *Azamian* et al. [3.365] have immobilized a series of biomolecules (cytochrome c, ferritin, and glucose oxidase) on SWNTs, and they observed that the use of EDC was not always necessary, indicating that the binding was predominantly noncovalent. In the case of glucose oxidase, they tested the catalytic activity of functionalized nanotubes immobilized on a glassy carbon electrode and observed a tenfold greater catalytic response compared to that seen in the absence of modified SWNTs.

Functionalization of nanotubes with biomolecules is still in its infancy, and their use as biosensors may lead to practical applications earlier than expected. For example, functionalized nanotubes can be used as AFM tips (see Sect. 3.6.1), allowing single-molecule measurements to be taken using "chemical force microscopy" (CFM). Important improvements in the characterization of biomolecules have even been achieved with unfunctionalized nanotube-based tips (see the review by [3.278]). Nanotube-based biosensors have now been developed. They are based on either field effect transistors [3.366] involving functionalized CNTs (biomolecules) or on electrochemical detection [3.367].

3.6.3 Expected Applications Related to Composite Systems

Because of their exceptional morphological, electrical, thermal, and mechanical characteristics, carbon nanotubes make particularly promising reinforcement materials in composites with metals, ceramics or polymer matrices. Key issues to address include the good dispersion of the nanotubes, the control of the nanotube/matrix bonding, the densification of bulk composites and thin films, and the possibility of aligning the nanotubes. In addition, the nanotube type (SWNT, c-MWNT, h-MWNT, etc.) and origin (arc, laser, CCVD, etc.) are also important variables that help determine the structural perfection, surface reactivity and aspect ratio of the reinforcement.

The application of carbon nanotubes in this field is expected to lead to major advances in composites. The following sections will give overviews of current work on metal-, ceramic- and polymer-matrix composites containing nanotubes. Nanotubes coated with another material are not considered here. Filled nanotubes are discussed in Sect. 3.5.2.

Metal Matrix Composites

Nanotube-metal matrix composites are still rarely studied. Matrices include Al-, Cu-, Mg-, Ni-, Ni-P-, Ti-, WC-Co- and Zr-based bulk metallic glasses. The materials are generally prepared by standard powder metallurgy techniques, but in this case the nanotube dispersion is not optimal. Other techniques such as plasma spray forming [3.368], the so-called nanoscale-dispersion method [3.369] and the rapid solidification technique [3.370] have also been developed. The spark plasma sintering (SPS) technique is sometimes used to densify the composites whilst avoiding matrix-grain growth [3.371, 372]. The room-temperature electrical resistivity of hot-pressed CCVD MWNT-Al composites increases slightly upon increasing the MWNT volume fraction [3.373]. The tensile strengths and elongations of unpurified arc discharge MWNT-Al composites are only slightly affected by annealing at 873 K in contrast to those of pure Al [3.374]. The Young's modulus of nonpurified arc discharge MWNTs-Ti composite is about 1.7 times that of pure Ti [3.375]. The formation of TiC, probably resulting from a reaction between amorphous carbon and the matrix, was observed, but the MWNTs themselves were not damaged. An increase in the Vickers hardness by a factor of 5.5 over that of pure Ti was associated with the suppression of coarsening of the Ti grains, TiC formation, and the addition of MWNTs. Purified nanotube-WC-Co nanocomposites exhibit better hardness-to-toughness relationships than pure nanocrystalline WC-Co [3.372]. MWNT-Mg composites show a higher tensile strength ($+150\%$) and elongation ratio ($+30\%$) than Mg [3.376]. Ni-plated MWNTs give better results than unplated MWNTs in strength tests. Indeed, nanotube coating is a promising way to improve the strength of bonding with the matrix [3.377]. Compressive testing of carbon nanotube-reinforced Zr-based bulk metallic glass composites [3.378] shows that the composites display a high fracture strength. In addition, the composites have strong ultrasonic attenuation characteristics and excellent ability to absorb waves. This implies that such composites may also be useful for shielding acoustic sound or environmental noise. CCVD MWNTs-Cu composites [3.379] also show a higher hardness and a lower friction coefficient and wear loss. Fifty to sixty percent deformation of the composites was observed. Carbon nanotube-Cu composite electrodes have been applied to the amperometric detection of carbohydrates, where they show an enhanced sensitivity compared to detectors based on Cu or nanotubes alone [3.380]. Composite films and coatings on various substrates have also been studied. The addition of up to 15 vol % purified SWNTs to nanocrystalline Al films reduces the coefficient of thermal expansion by as much as 65% and the resulting material could be a promising electronic packaging material [3.381]. Ni-carbon nanotube coatings deposited on carbon steel show significantly increased resistance to corrosion [3.382]. Ni-carbon nanotube coatings with a Zn–Ni interlayer electrodeposited on an Al substrate show an improved wear resistance at lower applied loads. The friction coefficient of the composite coatings decreases with increasing volume fraction of nanotubes due to the reinforcement and self-lubrication of the nanotubes [3.383]. Ni-P-SWNT coatings prepared by electroless plating show not only higher wear resistance but also a lower friction coefficient and a higher corrosion resistance compared to Ni-P coatings [3.384].

Ceramic Matrix Composites

Carbon nanotube-containing ceramic matrix composites have been studied more frequently than their metal matrix equivalents. Much of the effort has focused on obtaining tougher ceramics [3.385]. Composites can be processed using the regular processing route, where the nanotubes are usually mechanically dry- or wet-mixed with the matrix (or a matrix precursor) and then densified using hot-pressing sintering. Zhan et al. [3.386]

ball-milled SWNT bundles (from the HiPCo technique) and nanometric alumina powders, producing a fairly homogeneous dispersion, supposedly without damaging the SWNTs. Like Sun et al. [3.387], they used the SPS technique to prepare fully dense composites without damaging the SWNTs. Ning et al. succeeded in improving the homogeneity of the dispersion of MWNTs in the SiO_2 matrix using surfactants [3.388]. Other original composite processing techniques include the sol-gel route by which bulk and thin film composites were prepared [3.389], and in situ SWNT growth in ceramic foams, using procedures closely related to those described in Sect. 3.2.2 [3.390].

Ma et al. [3.391] prepared MWNT-SiC composites by hot-pressing mixtures of CCVD MWNTs and nano-SiC powders, but the dispersion of the MWNTs looked poor. They claimed that the presence of the MWNTs produces an increase of about 10% in both the bending strength and fracture toughness. Siegel et al. [3.392] claimed that the fracture toughness of 10 vol % MWNT-alumina hot-pressed composites is increased by 24% (to $4.2\,MPa\,m^{1/2}$) over that of pure alumina. However, the density and grain size are not reported, so it is difficult to consider this result to be evidence for the beneficial role of the MWNTs themselves. A confirmation can be found in the work by Sun et al. [3.387], who developed a colloidal route to coating the nanotubes with alumina particles prior to another mixing step with alumina particles and then densification by SPS. A gain in fracture toughness (calculated from Vickers indentation) is reported (from $3.7\,MPa\,m^{1/2}$ for pure Al_2O_3 to $4.9\,MPa\,m^{1/2}$ for an addition of 0.1 wt % of SWNT nanotubes, but it is possible that the beneficial input of the nanotubes is due to the previous coating of the nanotubes before sintering, resulting in improved bonding with the matrix. Zhan et al. [3.386] also used Vickers indentations to determine that the fracture toughness of SWNT-Al_2O_3 composites would be enhanced from $3.3\,MPa\,m^{1/2}$ for Al_2O_3 to $9.7\,MPa\,m^{1/2}$ (for 10 vol % CNTs). However, these results were refuted by Wang et al. [3.393] who reported that CNT-alumina composites are highly contact damage-resistant and showed that the fracture toughness of such composites can be strongly overestimated when measured by the standard indentation method [3.386]. Several detailed studies [3.89, 394–397] have dealt with the preparation of nanotube-Fe-Al_2O_3, nanotube-Co-MgO, and nanotube-Fe/Co-$MgAl_2O_4$ composites by hot-pressing the corresponding composite powders. The nanotubes were grown in situ in the starting powders by a CCVD method [3.65, 81, 398–402] and are therefore very homogeneously dispersed between the metal oxide grains, in a way that may be impossible to achieve by mechanical mixing. The nanotubes (mostly SWNTs and DWNTs) gather in long, branched bundles smaller than 50 nm in diameter, which appear to be very flexible. Depending on the matrix and hot-pressing temperature, a fraction of the nanotubes seems to be destroyed during hot-pressing in a primary vacuum. Increasing the quantity of nanotubes (of approximately $2-12$ wt %) leads to a refinement of the microstructure but also to a strong decrease in relative density (70–93%). *Peigney* et al.[3.403] showed that, during hot-pressing of nanotube-Fe/Co-MgAl$_2$O$_4$ composites, nanotubes do influence the rearrangement step, being favorable to this process at low contents, but detrimental at higher contents because they form a web structure that is too rigid. Moreover, nanotubes are detrimental to the plastic flow controlled by a Nabarro–Herring dislocation creep. Probably because of the insufficient densification, the fracture strength and the toughness of the nanotube-containing composites are generally lower than those of the nanotube-free metal-oxide composites and only marginally higher than those of the corresponding ceramics. SEM observations revealed both the trapping of some of the nanotubes within the matrix grains or at grain boundaries and a relatively good wetting in the case of alumina. Most nanotubes are cut near the fracture surface after some pull-out and could contribute to mechanical reinforcement. *Xia* et al. [3.404] reported microstructural investigations on MWNTs well-aligned in the pores of an alumina membrane. Different possible reinforcement mechanisms induced by the MWNTs have been evidenced, such as crack deflection, crack bridging, MWNT pullout, and MWNT collapse in shear bands. Indeed, although neither the SENB nor the SEVNB results have shown that nanotubes can significantly reinforce alumina ceramics so far, this could be obtained with ceramic-matrix composites in which the nanotubes have been properly organized.

Bulk properties other than mechanical properties are also being investigated. Interestingly, the presence of well-dispersed nanotubes confers an electrical conductivity to the otherwise insulating ceramic-matrix composites. *Rul* et al. [3.405] showed that the percolation threshold of SWNT in an MgAl$_2$O$_4$ matrix was very low ($p_c = 0.64$ vol %) due to the very high aspect ratio of the nanotubes, and that the DC electrical conductivity was well fitted by the scaling law of percolation theory $\sigma = \sigma_0(p - p_c)^t$ with $t = 1.73$, close to the theoretical value ($t = 1.94$) characteristic of a three-dimensional network. This work also showed that the electrical conductivities of such composites can be tailored over a wide scale ($10^{-2}-10$ S.cm^{-1}) using the nanotube content.

Moreover, extrusion at high temperatures in vacuum allows the nanotubes to be aligned in the matrix, thus inducing an anisotropy in the electrical conductivity [3.406]. *Zhan* et al. [3.407] measured DC electrical conductivities of up to 33.45 S.cm^{-1} in Al$_2$O$_3$-based composites containing ropes of SWNTs. *Huang* et al. [3.408] densified MWNT-BaTiO$_3$ composites and showed that their electrical conductivities decreased when the nanotube content was increased. This phenomenon has been attributed to a Schottky barrier constructed at the nanotube–matrix contact, which could be used in the fabrication of new ferroelectric and thermoelectric devices.

One of the early works involved MWNTs in a superconducting Bi$_2$Sr$_2$CaCu$_2$O$_{8+\delta}$ matrix [3.409]. The very high thermal conduction of nanotubes suggests that they could be used to manage the thermal properties of ceramics. *Seeger* et al. [3.410] incorporated MWNTs into SiO$_2$ by partial matrix melting, and they reported that the presence of nanotubes was crucial to the heat absorption and melting of the matrix, showing that the thermal transport of SiO$_2$ was probably greatly enhanced by the nanotubes. *Zhan* et al. [3.411] reported that SWNT-Al$_2$O$_3$ composites exhibit anisotropic thermal diffusivity, with a decrease in the direction transverse to SPS pressing. However, no increase in thermal conductivity due to nanotubes in ceramic-matrix composites has been reported so far and further work is required to investigate the heat transport mechanisms involved. The application of MWNT-lanthanum cobaltate composites in an oxygen electrode used in zinc/air [3.412] was also reported. The friction coefficient was also shown to increase and the wear loss to decrease when the nanotube volume fraction was increased in carbon/carbon composites [3.413].

The latter is one of the few examples where carbon has been considered as the matrix. Another one is the work reported by *Andrews* et al. [3.414], who introduced 5 wt % of SWNTs into a pitch material before spinning it into a SWNT-reinforced pitch-fiber. Subsequent carbonization transformed it into a SWNT-reinforced carbon fiber. Due to the contribution of the SWNTs, which were aligned by the spinning stresses, the resulting gains in tensile strength, modulus, and electrical conductivity with respect to the unreinforced carbon fiber was claimed to be 90, 150 and 340%, respectively.

Polymer Matrix Composites

Nanotube-polymer composites, first reported by *Ajayan* et al. [3.415], have now been intensively studied; especially epoxy- and polymethylmethacrylate (PMMA)-matrix composites. In terms of mechanical characteristics, the three key issues that affect the performance of a fiber-polymer composite are the strength and toughness of the fibrous reinforcement, its orientation, and good interfacial bonding, which is crucial to load transfer [3.416]. The ability of the polymer to form large-diameter helices around individual nanotubes favors the formation of a strong bond with the matrix [3.416]. Isolated SWNTs may be more desirable than MWNTs or bundles for dispersion in a matrix because of the weak frictional interactions between layers of MWNTs and between SWNTs in bundles [3.416]. The main mechanisms of load transfer are micromechanical interlocking, chemical bonding and van der Waals bonding between the nanotubes and the matrix. A high interfacial shear stress between the fiber and the matrix will transfer the applied load to the fiber over a short distance [3.417]. SWNTs longer than $10-100\,\mu$m would be needed for significant load-bearing ability in the case of nonbonded SWNT-matrix interactions, whereas the critical length for SWNTs cross-linked to the matrix is only $1\,\mu$m [3.418]. Defects are likely to limit the working length of SWNTs, however [3.419].

The load transfer to MWNTs dispersed in an epoxy resin was much higher in compression than in tension [3.417]. It was proposed that all of the walls of the MWNTs are stressed in compression, whereas only the outer walls are stressed in tension because all of the inner tubes are sliding within the outer tube. Mechanical tests performed on 5 wt% SWNT-epoxy composites [3.420] showed that SWNT bundles were pulled out of the matrix during the deformation of the material. The influence of the interfacial nanotube/matrix interaction was demonstrated by *Gong* et al. [3.421]. It was also reported that coating regular carbon fiber with MWNTs prior to their dispersion into an epoxy matrix improves the interfacial load transfer, possibly via local stiffening of the matrix near interface [3.422].

As for ceramic matrix composites, the electrical characteristics of SWNT- and MWNT-epoxy composites are described by the percolation theory. Very low percolation thresholds (below 1 wt%) are often reported [3.423–425]. Industrial epoxy loaded with 1 wt% unpurified CCVD-prepared SWNTs showed an increase in thermal conductivity of 70% and 125% at 40 K and at room temperature, respectively [3.426]. Also, the Vickers hardness rose with the SWNT loading up to a factor of 3.5 at 2 wt%. Increasing the amount of MWNTs increased the glass transition temperatures of the MWNT-epoxy composites. The effect is stronger when using samples containing functionalized MWNTs [3.427]. Nanocomposites consisting of double-wall carbon nanotubes in an epoxy matrix were produced by a standard calendering technique with a very good dispersion [3.428]. Increased strength, Young's modulus and strain to failure for a nanotube content of only 0.1 wt% was reported. A significantly improved fracture toughness was also observed for the nanocomposites. *Pecastaings* et al. [3.429] have investigated the role of interfacial effects in carbon nanotube-epoxy nanocomposite behavior. The conductivity of laser-prepared SWNT-PMMA composites increases with the SWNT loading ($1-8$ wt% SWNTs) [3.416]. Thermogravimetric analysis shows that, compared to pure PMMA, the thermal degradation of PMMA films occurs at a slightly higher temperature when 26 wt% of MWNTs are added [3.430]. Improving the wetting between the MWNTs and the PMMA by coating the MWNTs with poly(vinylidene fluoride) prior to melt-blending with PMMA resulted in an increased storage modulus [3.431]. The impact strength in aligned SWNT-PMMA composites increased significantly with only 0.1 wt% of SWNTs, possibly because of weak interfacial adhesion and/or of the high flexibility of the SWNTs and/or the pullout and sliding effects of individual SWNTs within bundles [3.432]. The transport properties of arc discharge SWNT-PMMA composite films ($10\,\mu$m-thick) were studied in great detail [3.433, 434]. The electrical conductivity increases by nine orders of magnitude from 0.1 to 8 wt% SWNTs. The room-temperature conductivity is again well described by the standard percolation theory, confirming the good dispersion of the SWNTs in the matrix. The rheological threshold of SWNT-PMMA composites is about 0.12 wt%, smaller than the percolation threshold of electrical conductivity, about 0.39 wt% [3.435]. This is understood in terms of the smaller nanotube–nanotube distance required for electrical conductivity compared to that required to impede polymer mobility. Furthermore, decreased SWNT alignment, improved SWNT dispersion and/or longer polymer chains increase the elastic response of the nanocomposite. The effects of small quantities of SWNTs (up to 1 wt%) in PMMA on its flammability properties were studied [3.436]. The formation of a continuous SWNTs network layer covering the entire surface without any cracks is critical for obtaining the lowest mass-loss rate of the nanocomposites.

Polymer composites with other matrices include CCVD-prepared MWNT-polyvinyl alcohol [3.437], arc-prepared MWNT-polyhydroxyaminoether [3.438], arc-prepared MWNT-polyurethane acrylate [3.439, 440], SWNT-polyurethane acrylate [3.441], SWNT-polycarbonate [3.442], MWNT-polyaniline [3.443], MWNT-polystyrene [3.444], CCVD double-walled nanotubes-polystyrene-polymethylacrylate [3.445], MWNT-polypropylene [3.446, 447], SWNT-polyethylene [3.448–450], SWNT-poly(vinyl acetate) [3.449, 450], CCVD-prepared MWNT-polyacrylonitrile. [3.451], SWNT-polyacrylonitrile [3.452], MWNT-oxotitanium phthalocyanine [3.453], arc-prepared MWNT-poly(3-octylthiophene) [3.454], SWNT-poly(3-octylthiophene) [3.455] and CCVD MWNT-poly(3-hexylthiophene) [3.456]. These works

deal mainly with films 100–200 micrometer thick, and aim to study the glass transition of the polymer, its mechanical and electrical characteristics, as well as the photoconductivity.

A great deal of work has also been devoted to the applications of nanotube-polymer composites as materials for molecular optoelectronics, using primarily poly(m-phenylenevinylene-co-2,5-dioctoxy-p-phenylenevinylene) (PmPV) as the matrix. This conjugated polymer tends to coil, forming a helical structure. The electrical conductivity of the composite films ([$-w\%t$]436 MWNTs) is increased by eight orders of magnitude compared to that of PmPV [3.458]. Using the MWNT-PmPV composites as the electron transport layer in light-emitting diodes results in a significant increase in bright-

Table 3.5 Applications of nanotube-based multifunctional materials (from [3.457]), courtesy of *B. Maruyama* (WPAFB, Dayton, Ohio)

| Fiber fraction | Applications | Mechanical | | | Electrical | | | Thermal | | Thermo-mechanical | |
		Strength/ stiffness	Specific strength	Through-thickness strength	Static dissipation	Surface conduction[a]	EMI shielding	Service[b] temperature	Conduction/ dissipation[c]	Dimensional stability[d]	CTE reduction[e]
Low volume fraction (fillers)											
Elastomers	Tires	×				×				×	
Thermoplastics	Chip package					×				×	
	Electronics/ Housing	×						×	×	×	
Thermosets	Epoxy products	×	×	×		×				×	×
	Composites					×				×	
High Volume Fraction											
Structural composites	Space/aircraft components	×		×							
High conduction composites	Radiators	×							×	×	
	Heat exchangers	×						×	×		×
	EMI shield	×					×				

[a] For electrostatic painting, to mitigate lightning strikes on aircraft, etc.

[b] To increase service temperature rating of product

[c] To reduce operating temperatures of electronic packages

[d] Reduces warping

[e] Reduces microcracking damage in composites

ness [3.459]. The SWNTs act as a hole-trapping material that blocks the holes in the composites; this is probably induced through long-range interactions within the matrix [3.460]. Similar investigations were carried out on arc discharge SWNT-polyethylene dioxythiophene (PEDOT) composite layers [3.461] and MWNT-polyphenylenevinylene composites [3.462].

To conclude, two critical issues must be considered when using nanotubes as components for advanced composites. One is to chose between SWNTs and MWNTs. The former seem more beneficial to mechanical strengthening, provided that they are isolated or arranged into cohesive yarns so that the load can be conveniently transferred from one SWNT to another. Unfortunately, despite many advances [3.463–468], this is still a technical challenge. The other issue is to tailor the nanotube/matrix interface with respect to the matrix; this is the current focus of many laboratories involved in the field.

Multifunctional Materials

One of the major benefits expected from incorporating carbon nanotubes into other solid or liquid materials is that they endow the material with some electrical conductivity while leaving other properties or behaviors unaffected. As already mentioned in the previous section, the percolation threshold is reached at very low nanotube loadings. Tailoring the electrical conductivity of a bulk material is then achieved by adjusting the nanotube volume fraction in the formerly insulating material while making sure that this fraction is not too large. As demonstrated by *Maruyama* [3.457], there are three areas of interest regarding the electrical conductivity: (i) electrostatic discharge (for example, preventing fire or explosion hazards in combustible environments or perturbations in electronics, which requires an electrical resistivity of less than $10^{12}\,\Omega\,\mathrm{cm}$); (ii) electrostatic painting (which requires the material to be painted to have enough electrical conductivity – an electrical resistivity below $10^6\,\Omega\,\mathrm{cm}$ – to prevent the charged paint droplets from being repelled); (iii) electromagnetic interference shielding (which is achieved for an electrical resistivity of less than $10\,\Omega\,\mathrm{cm}$.

Materials are often required to be multifunctional; for example, to have both high electrical conductivity and high toughness, or high thermal conductivity and high thermal stability. An association of several materials, each of them bringing one of the desired features, generally meets this need. The exceptional features and properties of carbon nanotubes make them likely to be a perfect multifunctional material in many cases. For in-

stance, materials used in satellites are often required to be electrical conductive, mechanically self-supporting, able to transport away excess heat, and often to be robust against electromagnetic interference, while being of minimal weight and volume. All of these properties should be possible with a single nanotube-containing composite material instead of complex multimaterials combining layers of polymers, aluminium, copper, and so on. Table 3.5 provides an overview of various fields in which nanotube-based multifunctional materials should find application.

Nanoelectronics

As reported in Sects. 3.1.1 and 3.4.4, SWNT nanotubes can be either metallic (with an electrical conductivity higher than that of copper), or semiconducting. This has inspired the design of several components for nanoelectronics. First, metallic SWNTs can be used as mere ballistic conductors. Moreover, as early as 1995, realizing a rectifying diode by joining one metallic SWNT to one semiconductor SWNT (hetero-junction) was proposed by *Lambin* et al. [3.469], then later by *Chico* et al. [3.470] and *Yao* et al. [3.471]. Also, field effect transistors (FET) can be built by attaching a semiconductor SWNT across two electrodes (source and drain) deposited on an insulating substrate that serves as a gate electrode [3.472, 473]. The association of two such SWNT-based FETs makes a voltage inverter [3.474].

All of the latter developments are fascinating and provide promising outlets for nanotube-based electronics. However, progress is obviously needed before SWNT-based integrated circuits can be constructed on a routine basis. A key issue is the need to be able to selectively prepare either metallic or semiconductor nanotubes. Although a method of selectively destroying metallic SWNTs in bundles of undifferentiated SWNTs [3.475] has been proposed, the method is not scalable and selective synthesis would be preferable. Also, defect-free nanotubes are required. Generally speaking, this relates to another major challenge, which is to be able to fabricate integrated circuits including nanometer-size components (that only sophisticated imaging methods such as AFM are able to visualize) on an industrial scale. An overview of the issues related to the integration of carbon nanotubes into microelectronics systems has been written by *Graham* et al. [3.476].

Nanotools, Nanodevices and Nanosystems

Due to the ability of graphene to expand slightly when electrically charged, nanotubes have been found to act as actuators. *Kim* et al. [3.477] demonstrated this by designing "nanotweezers", which are able to grab, manipulate and release nano-objects (the "nanobead" that was handled for the demonstration was actually closer to a micrometer in size than a nanometer), as well as to measure their electrical properties. This was made possible by simply depositing two noninterconnected gold coatings onto a pulled glass micropipette (Fig. 3.31), and then attaching two MWNTs (or two SWNT-bundles) $\approx 20-50$ nm in diameter to each of the gold electrodes. Applying a voltage $(0-8.5 V)$ between the two electrodes then makes the tube tips open and close reversibly in a controlled manner.

A similar experiment, again rather simple, was proposed by *Baughman* et al. the same year (1999) [3.478]. This consisted of mounting two SWNT-based paper strips ("bucky-paper") on both sides of insulating double-sided tape. The two bucky-paper strips had been previously loaded with Na^+ and Cl^-, respectively. When 1 V was applied between the two paper strips, both of them expanded, but the strip loaded with Na^+ expanded a bit more, forcing the whole system to bend. Though performed in a liquid environment, this behavior has inspired the authors to predict a future use for their system in "artificial muscles."

Another example of amazing nanotools is the nanothermometer proposed by *Gao* et al. [3.479]. A single MWNT was used, which was partially filled with liquid gallium. Temperature variations in the range $50-500\,^\circ\text{C}$ cause the gallium to reversibly move up and down within the nanotube cavity at reproducible levels with respect to the temperature values applied.

Of course, nanotools such as nanotweezers or nanothermometers are hardly commercial enough to justify

industrial investment. But such experiments are more than just amazing laboratory curiosities. They demonstrate the ability of carbon nanotubes to provide building blocks for future nanodevices, including nanomechanical systems.

Supercapacitors

Supercapacitors consist of two electrodes immersed in an electrolyte (such as 6 M KOH), separated by an insulating ion-permeable membrane. Charging the capacitors is achieved by applying a potential between the two electrodes, which makes the cations and the anions move toward the oppositely charged electrode. Suitable electrodes should exhibit high electrical conductivities and high surface areas, since the capacitance is proportional to these parameters. Actually, the surface area should consist of an appropriate combination of mesopores (to allow the electrolyte components to circulate well, which is related to the charging speed) and micropores (whose walls provide the attractive surfaces and fixation sites for the ions). Based on early work by *Niu* et al. [3.480], such a combination was found to be provided by the specific architecture offered by packed and entangled h-MWNTs with poor nanotextures (see Sect. 3.1.2). However, activation pretreatments were necessary. For instance, a capacitor made from nanotubes with a surface area of $220\,\text{m}^2/\text{g}$ exhibited a capacitance of $20\,\text{F/g}$, which increased to $100\,\text{F/g}$ after an activation treatment was applied to the nanotubes so that their surface area increased to $880\,\text{m}^2/\text{g}$ [3.159]. Alternatively, again due to their remarkable architectures derived from their huge aspect ratios, nanotubes can also be used as supports for conductive polymer coatings, such as polypyrrole or polyaniline [3.481], or additives to regular carbon electrodes [3.482], which make the material more open, allowing easier circulation and penetration of ions. Supercapacitors built from such composites can survive more than 2000 charging cycles, with current densities as high as $350\,\text{mA/g}$ [3.483]. Capacitors including nanotubes have already shown capacitances as high as $180-200\,\text{F/g}$, equivalent to those obtained with electrodes built from regular carbon materials, but they have the advantage of faster charging [3.159]. Current work in this area will certainly lead to further optimization of both the nanotube material architecture and the nanotube-supported conductive polymers, meaning that the outlook for the commercial use of nanotubes as components for supercapacitors is positive, and this is ignoring the potential application of second-generation nanotubes (such as nanotube-based

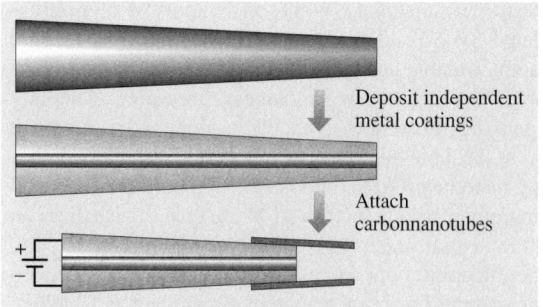

Deposit independent metal coatings

Attach carbonnanotubes

+
−

Fig. 3.31 Sketch explaining how the first nanotweezers were designed. The process involves modifying a glass micropipette (dark cone, *top*). Two Au coatings (in gray, *middle*) are deposited so that they are not in contact. Then a voltage is applied to the electrodes (from [3.477])

nano-objects) in this field. A first attempt to use hybrid nanotubes (see Sect. 3.5.2) has already resulted in improved properties with respect to genuine (undoped) nanotube-based systems [3.484].

3.7 Concluding Remarks

Carbon nanotubes have been the focus of a lot of research work (and therefore a lot of funding) for more than a decade now. Considering this investment of time and money, relatively few nanotube applications have reached the market yet. This may remind some of the disappointments associated with fullerene research, originally believed to be so promising, but which has resulted in no significant application after twenty years.

However, nanotubes exhibit an extraordinary diversity of morphologies, textures, structures and nanotextures, far beyond that provided by fullerenes. Indeed, the properties of nanotubes are yet to be fully identified, and we should not forget the potential of hybrid nanotubes, heteronanotubes and nanotube-containing composites. The history of nanotubes has only just begun.

References

3.1   L. V. Radushkevich, V. M. Lukyanovich: O strukture ugleroda, obrazujucegosja pri termiceskom razlozenii okisi ugleroda na zeleznom kontakte, Zurn. Fisic. Chim. **26**, 88–95 (1952)

3.2   T. V. Hughes, C. R. Chambers: US Patent 405, 480 (1889)

3.3   P. Schützenberger, L. Schützenberger: Sur quelques faits relatifs à l'histoire du carbone, C. R. Acad. Sci. Paris **111**, 774–778 (1890)

3.4   C. Pélabon, H. Pélabon: Sur une variété de carbone filamenteux, C. R. Acad. Sci. (Paris) **137**, 706–708 (1903)

3.5   R. T. K. Baker, P. S. Harris: The formation of filamentous carbon. In: *Chemistry and Physics of Carbon*, Vol. 14, ed. by P. L. Walker Jr., P. A. Thrower (Dekker, New York 1978) pp. 83–165

3.6   S. Iijima: Helical microtubules of graphite carbon, Nature **354**, 56–58 (1991)

3.7   S. Iijima, T. Ichihashi: Single-shell carbon nanotubes of 1-nm diameter, Nature **363**, 603–605 (1993)

3.8   D. S. Bethune, C. H. Kiang, M. S. de Vries, G. Gorman, R. Savoy, J. Vazquez, R. Bayers: Cobalt-catalysed growth of carbon nanotubes with single-atomic-layer walls, Nature **363**, 605–607 (1993)

3.9   J. Tersoff, R. S. Ruoff: Structural properties of a carbon-nanotube crystal, Phys. Rev. Lett. **73**, 676–679 (1994)

3.10 N. Wang, Z. K. Tang, G. D. Li, J. S. Chen: Single-walled 4 Å carbon nanotube arrays, Nature **408**, 50–51 (2000)

3.11 N. Hamada, S. I. Sawada, A. Oshiyama: New one-dimensional conductors, graphite microtubules, Phys. Rev. Lett. **68**, 1579–1581 (1992)

3.12 M. S. Dresselhaus, G. Dresselhaus, P. C. Eklund: *Science of Fullerenes and Carbon Nanotubes* (Academic, San Diego 1995)

3.13 R. C. Haddon: Chemistry of the fullerenes: The manifestation of strain in a class of continuous aromatic molecules, Science **261**, 1545–1550 (1993)

3.14 M. Monthioux, B. W. Smith, B. Burteaux, A. Claye, J. Fisher, D. E. Luzzi: Sensitivity of single-wall nanotubes to chemical processing: An electron microscopy investigation, Carbon **39**, 1261–1272 (2001)

3.15 H. Allouche, M. Monthioux: Chemical vapor deposition of pyrolytic carbon onto carbon nanotubes. Part II – Structure and texture, Carbon **43**, 1265–1278 (2005)

3.16 N. M. Rodriguez, A. Chambers, R. T. Baker: Catalytic engineering of carbon nanostructures, Langmuir **11**, 3862–3866 (1995)

3.17 M. Audier, A. Oberlin, M. Oberlin, M. Coulon, L. Bonnetain: Morphology and crystalline order in catalytic carbons, Carbon **19**, 217–224 (1981)

3.18 Y. Saito: Nanoparticles and filled nanocapsules, Carbon **33**, 979–988 (1995)

3.19 P. J. F. Harris: *Carbon Nanotubes and Related Structures* (Cambridge Univ. Press, Cambridge 1999)

3.20 H. W. Kroto, J. R. Heath, S. C. O'Brien, R. F. Curl, R. E. Smalley: C_{60} Buckminsterfullerene, Nature **318**, 162–163 (1985)

3.21 W. Krätschmer, L. D. Lamb, K. Fostiropoulos, D. R. Huffman: Solid C_{60}: A new form of carbon, Nature **347**, 354–358 (1990)

3.22 L. Fulchieri, Y. Schwob, F. Fabry, G. Flamant, L. F. P. Chibante, D. Laplaze: Fullerene production in a 3-phase AC plasma process, Carbon **38**, 797–803 (2000)

3.23 K. Saidane, M. Razafinimanana, H. Lange, A. Huczko, M. Baltas, A. Gleizes, J. L. Meunier: Fullerene synthesis in the graphite electrode arc process: local plasma characteristics and correlation with yield, J Phys. D: Appl. Phys. **37**, 232–239 (2004)

3.24 T. Guo, P. Nikolaev, A. G. Rinzler, D. Tomanek, D. T. Colbert, R. E. Smalley: Self-assembly of tubular fullerenes, J. Phys. Chem. **99**, 10694–10697 (1995)

3.25 T. Guo, P. Nikolaev, A. Thess, D. T. Colbert, R. E. Smalley: Catalytic growth of single-walled nanotubes by laser vaporisation, Chem. Phys. Lett. **243**, 49–54 (1995)

3.26 A. G. Rinzler, J. Liu, H. Dai, P. Nikolaev, C. B. Huffman, F. J. Rodriguez-Macias, P. J. Boul, A. H. Lu, D. Heymann, D. T. Colbert, R. S. Lee, J. E. Fischer, A. M. Rao, P. C. Eklund, R. E. Smalley: Large scale purification of single wall carbon nanotubes: Process, product and characterization, Appl. Phys. A **67**, 29–37 (1998)

3.27 A. Thess, R. Lee, P. Nikolaev, H. Dai, P. Petit, J. Robert, C. Xu, Y. H. Lee, S. G. Kim, D. T. Colbert, G. Scuseria, D. Tomanek, J. E. Fischer, R. E. Smalley: Crystalline ropes of metallic carbon nanotubes, Science **273**, 487–493 (1996)

3.28 L. M. Chapelle, J. Gavillet, J. L. Cochon, M. Ory, S. Lefrant, A. Loiseau, D. Pigache: A continuous wave CO_2 laser reactor for nanotube synthesis, Proc. Electronic Properties of Novel Materials-XVI Int. Winterschool – AIP Conf. Proc., ed. by H. Kuzmany, J. Fink, M. Mehring, S. Roth (Springer, Berlin Heidelberg 1999) 237–240

3.29 M. Yudasaka, T. Komatsu, T. Ichihashi, S. Iijima: Single wall carbon nanotube formation by laser ablation using double targets of carbon and metal, Chem. Phys. Lett. **278**, 102–106 (1997)

3.30 M. Castignolles, A. Foutel-Richard, A. Mavel, J. L. Cochon, D. Pigache, A. Loiseau, P. Bernier: Combined experimental and numerical study of the parameters controlling the C-SWNT synthesis via laser vaporization, Proc. Electronic Properties of Novel Materials-XVI Int. Winterschool – AIP Conf. Proc., ed. by H. Kuzmany, J. Fink, M. Mehring, S. Roth (Springer, Berlin Heidelberg 2002) 385–389

3.31 T. W. Ebbesen, P. M. Ajayan: Large-scale synthesis of carbon nanotubes, Nature **358**, 220–221 (1992)

3.32 D. Ugarte: Morphology and structure of graphitic soot particles generated in arc-discharge C_{60} production, Chem. Phys. Lett. **198**, 596–602 (1992)

3.33 T. W. Ebbesen: Carbon nanotubes, Ann. Rev. Mater. Sci. **24**, 235–264 (1994)

3.34 T. Beltz, J. Find, D. Herein, N. Pfänder, T. Rühle, H. Werner, M. Wohlers, R. Schlögl: On the production of different carbon forms by electric arc graphite evaporation, Ber. Bunsen. Phys. Chem. **101**, 712–725 (1997)

3.35 C. Journet, W. K. Maser, P. Bernier, A. Loiseau, L. M. de la Chapelle, S. Lefrant, P. Deniard, R. Lee, J. E. Fischer: Large-scale production of single-walled carbon nanotubes by the electric-arc technique, Nature **388**, 756–758 (1997)

3.36 K. Saïdane, M. Razafinimanana, H. Lange, M. Baltas, A. Gleizes, J. J. Gonzalez: Influence of the carbon arc current intensity on fullerene synthesis, Proc. 24th

Int. Conf. on Phenomena in Ionized Gases, ed. by P. Pisarczyk, T. Pisarczyk, J. Wotowski, (Institute of Plasma Physics and Laser Microfusion, Warsaw 1999) 203–204

3.37 H. Allouche, M. Monthioux, M. Pacheco, M. Razafinimanana, H. Lange, A. Huczko, T. P. Teulet, A. Gleizes, T. Sogabe: Physical characteristics of the graphite-electrode electric-arc as parameters for the formation of single-wall carbon nanotubes, Proc. Eurocarbon (Deutsche Keram. Ges.) **2**, 1053–1054 (2000)

3.38 M. Razafinimanana, M. Pacheco, M. Monthioux, H. Allouche, H. Lange, A. Huczko, A. Gleizes: Spectroscopic study of an electric arc with Gd and Fe doped anodes for the carbon nanotube formation, Proc. 25th Int. Conf. on Phenomena in Ionized Gases, ed. by E. Goto (Nagoya Univ., Nagoya 2001) 297–298

3.39 M. Razafinimanana, M. Pacheco, M. Monthioux, H. Allouche, H. Lange, A. Huczko, P. Teulet, A. Gleizes, C. Goze, P. Bernier, T. Sogabe: Influence of doped graphite electrode in electric arc for the formation of single wall carbon nanotubes, Proc. 6th Eur. Conf. on Thermal Plasma Processes – Progress in Plasma Processing of Materials, New York 2000, ed. by P. Fauchais (Begell House, New York 2001) 649–654

3.40 M. Pacheco, H. Allouche, M. Monthioux, A. Razafinimanana, A. Gleizes: Correlation between the plasma characteristics and the morphology and structure of the carbon phases synthesised by electric arc discharge, Proc. 25th Biennial Conf. on Carbon, Lexington (Kentucky) 2001, ed. by F. Derbyshire (American Carbon Society 2001) Extend. Abstr.(CD-Rom), Novel/14.1

3.41 M. Pacheco, M. Monthioux, M. Razafinimanana, L. Donadieu, H. Allouche, N. Caprais, A. Gleizes: New factors controlling the formation of single-wall carbon nanotubes by arc plasma, Proc. Carbon 2002 Int. Conf., Beijing 2002, ed. by H.-M. Cheng (Shanxi Chunqiu Audio-Visual Press, Beijing 2002) (CD-Rom/Oral/I014)

3.42 M. Monthioux, M. Pacheco, H. Allouche, M. Razafinimanana, N. Caprais, L. Donnadieu, A. Gleizes: New data about the formation of SWNTs by the electric arc method, Electronic Properties of Molecular Nanostructures, AIP Conf. Proc., ed. by H. Kuzmany, J. Fink, M. Mehring, S. Roth (Springer, Berlin Heidelberg 2002) 182–185

3.43 H. Lange, A. Huczko, M. Sioda, M. Pacheco, M. Razafinimanana, A. Gleizes: Influence of gadolinium on carbon arc plasma and formation of fullerenes and nanotubes, Plasma Chem. Plasma Process **22**, 523–536 (2002)

3.44 C. Journet: La production de nanotubes de carbone. Ph.D. Thesis (University of Montpellier II, Montpellier 1998)

3.45 T. Sogabe, T. Masuda, K. Kuroda, Y. Hirohaya, T. Hino, T. Ymashina: Preparation of B_4C-mixed graphite by pressureless sintering and its air oxidation behavior, Carbon **33**, 1783–1788 (1995)

3.46 M. Ishigami, J. Cumings, A. Zettl, S. Chen: A simple method for the continuous production of carbon nanotubes, Chem. Phys. Lett. **319**, 457–459 (2000)

3.47 Y. L. Hsin, K. C. Hwang, F. R. Chen, J. J. Kai: Production and in-situ metal filling of carbon nanotube in water, Adv. Mater. **13**, 830–833 (2001)

3.48 H. W. Zhu, X. S. Li, B. Jiang, C. L. Xu, C. L. Zhu, Y. F. Zhu, D. H. Wu, X. H. Chen: Formation of carbon nanotubes in water by the electric arc technique, Chem. Phys. Lett. **366**, 664–669 (2002)

3.49 W. K. Maser, P. Bernier, J. M. Lambert, O. Stephan, P. M. Ajayan, C. Colliex, V. Brotons, J. M. Planeix, B. Coq, P. Molinie, S. Lefrant: Elaboration and characterization of various carbon nanostructures, Synth. Met. **81**, 243–250 (1996)

3.50 J. Gavillet, A. Loiseau, J. Thibault, A. Maigné, O. Stéphan, P. Bernier: TEM study of the influence of the catalyst composition on the formation and growth of SWNT, Proc. Electronic Properties Novel Materials-XVI Int. Winterschool – AIP Conf. Proc., ed. by H. Kuzmany, J. Fink, M. Mehring, S. Roth (Springer, Berlin Heidelberg 2002) 202–206

3.51 G. Flamant, J. F. Robert, S. Marty, J. M. Gineste, J. Giral, B. Rivoire, D. Laplaze: Solar reactor scaling up. The fullerene synthesis case study, Energy **29**, 801–809 (2004)

3.52 T. M. Gruenberger, J. Gonzalez-Aguilar, F. Fabry, L. Fulchieri, E. Grivei, N. Probst, G. Flamant, H. Okuno, J. C. Charlier: Production of carbon nanotubes and other nanostructures via continuous 3-phase AC plasma processing, Fuller. Nanotub. Car. N. **12**, 571–581 (2004)

3.53 H. Okuno, E. Grivel, F. Fabry, T. M. Gruenberger, J. J. Gonzalez-Aguilar, A. Palnichenko, L. Fulchieri, N. Probst, J. C. Chalier: Synthesis of carbon nanotubes and nano-necklaces by thermal plasma process, Carbon **42**, 2543–2549 (2004)

3.54 L. P. F. Chibante, A. Thess, J. M. Alford, M. D. Diener, R. E. Smalley: Solar generation of the fullerenes, J. Phys. Chem. **97**, 8696–8700 (1993)

3.55 C. L. Fields, J. R. Pitts, M. J. Hale, C. Bingham, A. Lewandowski, D. E. King: Formation of fullerenes in highly concentrated solar flux, J. Phys. Chem. **97**, 8701–8702 (1993)

3.56 P. Bernier, D. Laplaze, J. Auriol, L. Barbedette, G. Flamant, M. Lebrun, A. Brunelle, S. Della-Negra: Production of fullerenes from solar energy, Synth. Met. **70**, 1455–1456 (1995)

3.57 M. J. Heben, T. A. Bekkedhal, D. L. Schultz, K. M. Jones, A. C. Dillon, C. J. Curtis, C. Bingham, J. R. Pitts, A. Lewandowski, C. L. Fields: Production of single wall carbon nanotubes using concentrated sunlight, Proc. Symp. Recent Adv. Chem. Phys. Fullerenes Rel. Mater., Pennington 1996, ed. by K. M. Kadish, R. S. Ruoff (Electrochemical Society, Pennington 1996) 803–811

3.58 D. Laplaze, P. Bernier, C. Journet, G. Vié, G. Flamant, E. Philippot, M. Lebrun: Evaporation of graphite using a solar furnace, Proc. 8th Int. Symp. Solar Concentrating Technol., Köln 1996, ed. by M. Becker, M. Balmer (C. F. Müller Verlag, Heidelberg 1997) 1653–1656

3.59 D. Laplaze, P. Bernier, W. K. Maser, G. Flamant, T. Guillard, A. Loiseau: Carbon nanotubes: The solar approach, Carbon **36**, 685–688 (1998)

3.60 T. Guillard, S. Cetout, L. Alvarez, J. L. Sauvajol, E. Anglaret, P. Bernier, G. Flamant, D. Laplaze: Production of carbon nanotubes by the solar route, Eur. Phys. J. **5**, 251–256 (1999)

3.61 D. Luxembourg, G. Flamant, A. Guillot, D. Laplaze: Hydrogen storage in solar produced single-walled carbon nanotubes, Mater. Sci. Eng. B **108**, 114–119 (2004)

3.62 G. Flamant, M. Bijeire, D. Luxembourg: Modelling of a solar reactor for single wall nanotubes synthesis, ASME J. Solar Energy Eng. **128**, 1–124 (2006)

3.63 G. G. Tibbetts, M. Endo, C. P. Beetz: Carbon fibers grown from the vapor phase: A novel material, SAMPE J. **22**, 30 (1989)

3.64 R. T. K. Baker: Catalytic growth of carbon filaments, Carbon **27**, 315–323 (1989)

3.65 E. Flahaut: Synthèse par voir catalytique et caractérisation de composites nanotubes de carbone-metal-oxyde Poudres et matériaux denses. Ph.D. Thesis (Univers. Paul Sabatier, Toulouse 1999)

3.66 E. Flahaut, R. Bacsa, A. Peigney, Ch. Laurent: Gram-scale CCVD synthesis of double-walled carbon nanotubes, Chem. Commun., 1442–1443 (2003)

3.67 R. T. K. Baker, P. S. Harris, R. B. Thomas, R. J. Waite: Formation of filamentous carbon from iron, cobalt, and chromium catalyzed decomposition of acetylene, J. Catal. **30**, 86–95 (1973)

3.68 T. Koyama, M. Endo, Y. Oyuma: Carbon fibers obtained by thermal decomposition of vaporized hydrocarbon, Jap. J. Appl. Phys. **11**, 445–449 (1972)

3.69 M. Endo, A. Oberlin, T. Koyama: High resolution electron microscopy of graphitizable carbon fiber prepared by benzene decomposition, Jap. J. Appl. Phys. **16**, 1519–1523 (1977)

3.70 N. M. Rodriguez: A review of catalytically grown carbon nanofibers, J. Mater. Res. **8**, 3233–3250 (1993)

3.71 W. R. Davis, R. J. Slawson, G. R. Rigby: An unusual form of carbon, Nature **171**, 756 (1953)

3.72 H. P. Boehm: Carbon from carbon monoxide disproportionation on nickel and iron catalysts; morphological studies and possible growth mechanisms, Carbon **11**, 583–590 (1973)

3.73 M. Audier, A. Oberlin, M. Coulon: Crystallographic orientations of catalytic particles in filamentous carbon; case of simple conical particles, J. Cryst. Growth **55**, 546–549 (1981)

3.74 M. Audier, M. Coulon: Kinetic and microscopic aspects of catalytic carbon growth, Carbon **23**, 317–323 (1985)

3.75   M. Audier, A. Oberlin, M. Coulon: Study of biconic microcrystals in the middle of carbon tubes obtained by catalytic disproportionation of CO, J. Cryst. Growth **57**, 524–534 (1981)

3.76   A. Thaib, G. A. Martin, P. Pinheiro, M. C. Schouler, P. Gadelle: Formation of carbon nanotubes from the carbon monoxide disproportionation reaction over CO/Al$_2$O' and Co/SiO' catalysts, Catal. Lett. **63**, 135–141 (1999)

3.77   P. Pinheiro, M. C. Schouler, P. Gadelle, M. Mermoux, E. Dooryhée: Effect of hydrogen on the orientation of carbon layers in deposits from the carbon monoxide disproportionation reaction over Co/Al$_2$O$_3$ catalysts, Carbon **38**, 1469–1479 (2000)

3.78   P. Pinheiro, P. Gadelle: Chemical state of a supported iron-cobalt catalyst during CO disproportionation. I. Thermodynamic study, J. Phys. Chem. Solids **62**, 1015–1021 (2001)

3.79   P. Pinheiro, P. Gadelle, C. Jeandey, J. L. Oddou: Chemical state of a supported iron-cobalt catalyst during CO disproportionation. II. Experimental study, J. Phys. Chem. Solids **62**, 1023–1037 (2001)

3.80   C. Laurent, E. Flahaut, A. Peigney, A. Rousset: Metal nanoparticles for the catalytic synthesis of carbon nanotubes, New J. Chem. **22**, 1229–1237 (1998)

3.81   A. Peigney, C. Laurent, F. Dobigeon, A. Rousset: Carbon nanotubes grown in situ by a novel catalytic method, J. Mater. Res. **12**, 613–615 (1997)

3.82   V. Ivanov, J. B. Nagy, P. Lambin, A. Lucas, X. B. Zhang, X. F. Zhang, D. Bernaerts, G. Van Tendeloo, S. Amelinckx, J. Van Landuyt: The study of nanotubules produced by catalytic method, Chem. Phys. Lett. **223**, 329–335 (1994)

3.83   V. Ivanov, A. Fonseca, J. B. Nagy, A. Lucas, P. Lambin, D. Bernaerts, X. B. Zhang: Catalytic production and purification of nanotubules having fullerene-scale diameters, Carbon **33**, 1727–1738 (1995)

3.84   K. Hernadi, A. Fonseca, J. B. Nagy, D. Bernaerts, A. Fudala, A. Lucas: Catalytic synthesis of carbon nanotubes using zeolite support, Zeolites **17**, 416–423 (1996)

3.85   H. Dai, A. G. Rinzler, P. Nikolaev, A. Thess, D. T. Colbert, R. E. Smalley: Single-wall nanotubes produced by metal-catalysed disproportionation of carbon monoxide, Chem. Phys. Lett. **260**, 471–475 (1996)

3.86   A. M. Cassel, J. A. Raymakers, J. Kong, H. Dai: Large scale CVD synthesis of single-walled carbon nanotubes, J. Phys. Chem. B **109**, 6484–6492 (1999)

3.87   B. Kitiyanan, W. E. Alvarez, J. H. Harwell, D. E. Resasco: Controlled production of single-wall carbon nanotubes by catalytic decomposition of CO on bimetallic Co-Mo catalysts, Chem. Phys. Lett. **317**, 497–503 (2000)

3.88   A. Govindaraj, E. Flahaut, C. Laurent, A. Peigney, A. Rousset, C. N. R. Rao: An investigation of carbon nanotubes obtained from the decomposition of methane over reduced Mg$_{1-x}$M$_x$Al$_2$O$_4$ spinel catalysts, J. Mater. Res. **14**, 2567–2576 (1999)

3.89   E. Flahaut, A. Peigney, C. Laurent, A. Rousset: Synthesis of single-walled carbon nanotube-Co-MgO composite powders and extraction of the nanotubes, J. Mater. Chem. **10**, 249–252 (2000)

3.90   J. Kong, A. M. Cassel, H. Dai: Chemical vapor deposition of methane for single-walled carbon nanotubes, Chem. Phys. Lett. **292**, 567–574 (1998)

3.91   E. Flahaut, A. Peigney, W. S. Bacsa, R. R. Bacsa, Ch. Laurent: CCVD synthesis of carbon nanotubes from (Mg, Co, Mo)O catalysts: Influence of the proportions of cobalt and molybdenum, J. Mater. Chem. **14**, 646–653 (2004)

3.92   E. Flahaut, Ch. Laurent, A. Peigney: Catalytic CVD synthesis of double and triple-walled carbon nanotubes by the control of the catalyst preparation, Carbon **43**, 375–383 (2005)

3.93   R. Marangoni, P. Serp, R. Feurrer, Y. Kihn, P. Kalck, C. Vahlas: Carbon nanotubes produced by substrate free metalorganic chemical vapor deposition of iron catalyst and ethylene, Carbon **39**, 443–449 (2001)

3.94   R. Sen, A. Govindaraj, C. N. R. Rao: Carbon nanotubes by the metallocene route, Chem. Phys. Lett. **267**, 276–280 (1997)

3.95   Y. Y. Fan, H. M. Cheng, Y. L. Wei, G. Su, S. H. Shen: The influence of preparation parameters on the mass production of vapor grown carbon nanofibers, Carbon **38**, 789–795 (2000)

3.96   L. Ci, J. Wei, B. Wei, J. Liang, C. Xu, D. Wu: Carbon nanofibers and single-walled carbon nanotubes prepared by the floating catalyst method, Carbon **39**, 329–335 (2001)

3.97   M. Glerup, H. Kanzow, R. Almairac, M. Castignolles, P. Bernier: Synthesis of multi-walled carbon nanotubes and nano-fibres using aerosol method with metal-ions as the catalyst precursors, Chem. Phys. Lett. **377**, 293–298 (2003)

3.98   O. A. Nerushev, M. Sveningsson, L. K. L. Falk, F. Rohmund: Carbon nanotube films obtained by thermal vapour deposition, J. Mater. Chem. **11**, 1122–1132 (2001)

3.99   Z. Zhou, L. Ci, L. Song, X. Yan, D. Liu, H. Yuan, Y.Gao, J. Wang, L. Liu, W. Zhou, G. Wang, Si Xie: Producing cleaner double-walled carbon nanotubes in a floating catalyst system, Carbon **41**, 2607–2611 (2003)

3.100 F. Rohmund, L. K. L. Falk, F. E. B. Campbell: A simple method for the production of large arrays of aligned carbon nanotubes, Chem. Phys. Lett. **328**, 369–373 (2000)

3.101 G. G. Tibbetts, C. A. Bernardo, D. W. Gorkiewicz, R. L. Alig: Role of sulfur in the production of carbon fibers in the vapor phase, Carbon **32**, 569–576 (1994)

3.102 S. Bai, F. Li, Q. H. Yang, H.-M. Cheng, J. B. Bai: Influence of ferrocene/benzene mole ratio in the synthesis of carbon nanostructures, Chem. Phys. Lett. **376**, 83–89 (2003)

3.103 W. Q. Han, P. Kholer-Riedlich, T. Seeger, F. Ernst, M. Ruhle, N. Grobert, W. K. Hsu, B. H. Chang, Y. Q. Zhu, H. W. Kroto, M. Terrones, H. Terrones: Aligned CN_x nanotubes by pyrolysis of ferrocene under NH_3 atmosphere, Appl. Phys. Lett. **77**, 1807–1809 (2000)

3.104 L. Ci, Z. Rao, Z. Zhou, D. Tang, X. Yan, Y. Liang, D. Liu, H. Yuan, W. Zhou, G. Wang, W. Liu, S. Xie: Double wall carbon nanotubes promoted by sulfur in a floating iron catalyst CVD system, Chem. Phys. Lett. **359**, 63–67 (2002)

3.105 S. Maruyama, R. Kojima, Y. Miyauchi, S. Chiashi, M. Kohno: Low-temperature synthesis of high-purity single-walled carbon nanotubes from alcohol, Chem. Phys. Lett. **360**, 229–234 (2002)

3.106 T. Kyotani, L. F. Tsai, A. Tomita: Preparation of ultrafine carbon tubes in nanochannels of an anodic aluminum oxide film, Chem. Mater. **8**, 2109–2113 (1996)

3.107 R. E. Smalley, J. H. Hafner, D. T. Colbert, K. Smith: (1998) Catalytic growth of single-wall carbon nanotubes from metal particles, US patent US19980601010903

3.108 P. Nikolaev: Gas-phase production of single-walled carbon nanotubes from carbon monoxide: a review of the HiPco process, J. Nanosci. Nanotech. **4**, 307–316 (2004)

3.109 W. K. Hsu, J. P. Hare, M. Terrones, H. W. Kroto, D. R. M. Walton, P. J. F. Harris: Condensed-phase nanotubes, Nature **377**, 687 (1995)

3.110 W. S. Cho, E. Hamada, Y. Kondo, K. Takayanagi: Synthesis of carbon nanotubes from bulk polymer, Appl. Phys. Lett. **69**, 278–279 (1996)

3.111 Y. L. Li, Y. D. Yu, Y. Liang: A novel method for synthesis of carbon nanotubes: Low temperature solid pyrolysis, J. Mater. Res. **12**, 1678–1680 (1997)

3.112 M. L. Terranova, S. Piccirillo, V. Sessa, P. Sbornicchia, M. Rossi, S. Botti, D. Manno: Growth of single-walled carbon nanotubes by a novel technique using nano-sized graphite as carbon source, Chem. Phys. Lett. **327**, 284–290 (2000)

3.113 R. L. Vander Wal, T. Ticich, V. E. Curtis: Diffusion flame synthesis of single-walled carbon nanotubes, Chem. Phys. Lett. **323**, 217–223 (2000)

3.114 H. Cui, G. Eres, J. Y. Howe, A. Puretzki, M. Varela, D. B. Geohegan, D. H. Lowndes: Growth behavior of carbon nanotubes on multilayered metal catalyst film in chemical vapor deposition, Chem. Phys. Lett. **374**, 222–228 (2003)

3.115 Y. Y. Wei, G. Eres, V. I. Merkulov, D. H. Lowdens: Effect of film thickness on carbon nanotube growth by selective area chemical vapor deposition, Appl. Phys. Lett. **78**, 1394–1396 (2001)

3.116 I. T. Han, B. K. Kim, H. J. Kim, M. Yang, Y. W. Jin, S. Jung, N. Lee, S. K. Kim, J. M. Kim: Effect of Al and catalyst thickness on the growth of carbon nanotubes and application to gated field emitter arrays, Chem. Phys. Lett. **400**, 139–144 (2004)

3.117 W. Z. Li, S. S. Xie, L. X. Qian, B. H. Chang, B. S. Zou, W. Y. Zhou, R. A. Zha, G. Wang: Large scale synthesis of aligned carbon nanotubes, Science **274**, 1701–1703 (1996)

3.118 F. Zheng, L. Liang, Y. Gao, J. H. Sukamto, L. Aardahl: Carbon nanotubes synthesis using mesoporous silica templates, Nanolett. **2**, 729–732 (2002)

3.119 S. H. Jeong, O.-K. Lee, K. H. Lee, S. H. Oh, C. G. Park: Preparation of aligned carbon nanotubes with prescribed dimension: Template synthesis and sonication cutting approach, Chem. Mater. **14**, 1859–1862 (2002)

3.120 A. M. Cassel, N. R. Franklin, T. W. Tombler, E. M. Chan, J. Han, H. Dai: Directed growth of free-standing single-walled carbon nanotubes, J. Am. Chem. Soc. **121**, 7975–7976 (1999)

3.121 S. Fan, M. Chapline, N. Franklin, T. Tombler, A. M. Cassel, H. Dai: Self-oriented regular arrays of carbon nanotubes and their field emission properties, Science **283**, 512–514 (1999)

3.122 N. S. Kim, Y. T. Lee, J. Park, H. Ryu, H. J. Lee, S. Y. Choi, J. Choo: Dependence of vertically aligned growth of carbon nanotubes on catalyst, J. Phys. Chem. B **106**, 9286–9290 (2002)

3.123 C. J. Lee, D. W. Kim, T. J. Lee, Y. C. Choi, Y. S. Park, Y. H. Lee, W. B. Choi, N. S. Lee, G.-S. Park, J. M. Kim: Synthesis of aligned carbon nanotubes using thermal chemical vapor deposition, Chem. Phys. Lett. **312**, 461–468 (1999)

3.124 W. D. Zhang, Y. Wen, S. M. Liu, W. C. Tjiu, G. Q. Xu, L. M. Gan: Synthesis of vertically aligned carbon nanotubes on metal deposited quartz plates, Carbon **40**, 1981–1989 (2002)

3.125 S. Huang, L. Dai, A. W. H. Mau: Controlled fabrication of large scale aligned carbon nanofiber/nanotube patterns by photolithography, Adv. Mater. **14**, 1140–1143 (2002)

3.126 T. Sun, G. Wang, H. Liu, L. Feng, D. Zhu: Control over the wettability of an aligned carbon nanotube film, J. Am. Chem. Soc. **125**, 14996–14997 (2003)

3.127 Y. Huh, J. Y. Lee, J. Cheon, Y. K. Hong, J. Y. Koo, T. J. Lee, C. J. Lee: Controlled growth of carbon nanotubes over cobalt nanoparticles by thermal chemical vapor deposition, J. Mater. Chem. **13**, 2297–2300 (2003)

3.128 Y. Kobayashi, H. Nakashima, D. Takagi, Y. Homma: CVD growth of single-walled carbon nanotubes using size-controlled nanoparticle catalyst, Thin Solid Films **464-465**, 286–289 (2004)

3.129 C. L. Cheung, A. Kurtz, H. Park, C. M. Lieber: Diameter-controlled synthesis of carbon nanotubes, J. Phys. Chem B **106**, 2429–2433 (2002)

3.130 Y. Huh, J. Y. Lee, J. Cheon, Y. K. Hong, J. Y. Koo, T. J. Lee, C. J. Lee: Controlled growth of carbon nanotubes over cobalt nanoparticles by thermal chemical vapor deposition, J. Mater. Chem. **13**, 2297–2300 (2003)

Part A | 3

3.131 M. Paillet, V. Jourdain, P. Poncharal, J.-L. Sauvajol,
A. Zahab, J. C. Meyer, S. Roth, N. Cordente, C. Amiens,
B. Chaudret: Versatile synthesis of individual single-
walled carbon nanotubes from nickel nanoparticles
for the study of their physical properties, J. Phys.
Chem. B **108**, 17112–17118 (2004)

3.132 S. Casimirius, E. Flahaut, C. Laurent, C. Vieu,
F. Carcenac, C. Laberty-Robert: Optimized microcontact printing process for the patterned growth of
individual SWNTs, Microelectr. Eng. **73-74**, 564–569
(2004)

3.133 Y. Lei, K. S. Yeong, J. T. L. Thong, W. K. Chim: Large-
scale ordered carbon nanotubes arrays initiated
from highly ordered catalyst arrays on silicon substrates, Chem. Mater. **16**, 2757–2761 (2004)

3.134 Q. Ye, A. M. Cassel, H. Liu, K. J. Chao, J. Han,
M. Meyyappan: Large-scale fabrication of carbon
nanotube probe tips for atomic force microscopy
critical dimension imaging applications, Nano Lett.
4, 1301–1308 (2004)

3.135 K. Hata, D. N. Futaba, K. Mizuno, T. Namai, M. Yu-
mara, S. Iijima: Ware-assisted highly efficient
synthesis of impurity-free single-walled carbon
nanotubes, Science **306**, 1362–1364 (2004)

3.136 C. N. R. Rao, R. Sen, B. C. Satishkumar, A. Govindaraj:
Large aligned carbon nanotubes bundles from ferrocene pyrolysis, Chem. Commun., 1525–1526 (1998)

3.137 R. Andrews, D. Jacques, A. M. Rao, F. Derbyshire,
D. Qian, X. Fan, E. C. Dickey, J. Chen: Continous production of aligned carbon nanotubes: A step closer
to commercial realization, Chem. Phys. Lett. **303**,
467–474 (1999)

3.138 X. Zhang, A. Cao, B. Wei, Y. Li, J. Wei, C. Xu, D. Wu:
Rapid growth of well-aligned carbon nanotube arrays, Chem. Phys. Lett. **362**, 285–290 (2002)

3.139 X. Zhang, A. Cao, Y. Li, C. Xu, J. Liang, D. Wu, B. Wei:
Self-organized arrays of carbon nanotube ropes,
Chem. Phys. Lett. **351**, 183–188 (2002)

3.140 K. S. Choi, Y. S. Cho, S. Y. Hong, J. B. Park, D. J. Kim:
Effects of ammonia on the alignment of carbon
nanotubes in metal-assisted chemical vapor deposition, J. Eur. Ceram. Soc. **21**, 2095–2098 (2001)

3.141 N. S. Kim, Y. T. Lee, J. Park, J. B. Han, Y. S. Choi,
S. Y. Choi, J. Choo, G. H. Lee: Vertically aligned carbon nanotubes grown by pyrolysis of iron, cobalt,
and nickel phtalocyanines, J. Phys. Chem. B **107**,
9249–9255 (2003)

3.142 C. Emmeger, J. M. Bonard, P. Mauron, P. Sudan,
A. Lepora, B. Grobety, A. Züttel, L. Schlapbach:
Synthesis of carbon nanotubes over Fe catalyst on
aluminum and suggested growth mechanism, Carbon **41**, 539–547 (2003)

3.143 M. Endo, H. W. Kroto: Formation of carbon
nanofibers, J. Phys. Chem. **96**, 6941–6944 (1992)

3.144 R. S. Wagner: VLS mechanisms of crystal growth. In:
Whisker Technology, ed. by P. Levit A. (Wiley, New
York 1970) pp. 47–72

3.145 Y. H. Lee, S. G. Kim, D. Tomanek: Catalytic growth of
single-wall carbon nanotubes: An ab initio study,
Phys. Rev. Lett. **78**, 2393–2396 (1997)

3.146 V. Jourdain, H. Kanzow, M. Castignolles, A. Loiseau,
P. Bernier: Sequential catalytic growth of carbon
nanotubes, Chem. Phys. Lett. **364**, 27–33 (2002)

3.147 H. Dai: Carbon Nanotubes: Synthesis, integration,
and properties, Acc. Chem. Res. **35**, 1035–1044 (2002)

3.148 Y. Saito, M. Okuda, N. Fujimoto, T. Yoshikawa,
M. Tomita, T. Hayashi: Single-wall carbon nanotubes
growing radially from Ni fine particles formed by arc
evaporation, Jpn. J. Appl. Phys. **33**, L526–L529 (1994)

3.149 J. Bernholc, C. Brabec, M. Buongiorno Nardelli,
A. Malti, C. Roland, B. J. Yakobson: Theory of growth
and mechanical properties of nanotubes, Appl.
Phys. A **67**, 39–46 (1998)

3.150 M. Pacheco: Synthèse des nanotubes de carbone par
arc electrique. Ph.D. Thesis (Université Toulouse III,
Toulouse 2003)

3.151 K. Méténier, S. Bonnamy, F. Béguin, C. Journet, P. Bernier, L. M. de la Chapelle, O. Chauvet,
S. Lefrant: Coalescence of single walled nanotubes
and formation of multi-walled carbon nanotubes
under high temperature treatments, Carbon **40**,
1765–1773 (2002)

3.152 P. G. Collins, P. Avouris: Nanotubes for electronics,
Sci. Am. **283**, 38–45 (2000)

3.153 K. A. Williams, P. C. Eklund: Monte Carlo simulation
of H_2 physisorption in finite diameter carbon nanotube ropes, Chem. Phys. Lett. **320**, 352–358 (2000)

3.154 Q.-H. Yang, P. X. Hou, S. Bai, M. Z. Wang,
H. M. Cheng: Adsorption and capillarity of nitro-
gen in aggregated multi-walled carbon nanotubes,
Chem. Phys. Lett. **345**, 18–24 (2001)

3.155 S. Inoue, N. Ichikuni, T. Suzuki, T. Uematsu,
K. Kaneko: Capillary condensation of N_2 on multiwall
carbon nanotubes, J. Phys. Chem. **102**, 4689–4692
(1998)

3.156 S. Agnihotri, J. P. Mota, M. Rostam-Abadi, M. J. Rood:
Structural characterization of single-walled carbon
nanotube bundles by experiment and molecular
simulation, Langmuir **21**, 896–904 (2005)

3.157 M. Eswaramoorthy, R. Sen, C. N. R. Rao: A study of
micropores in single-walled carbon nanotubes by
the adsorption of gases and vapors, Chem. Phys.
Lett. **304**, 207–210 (1999)

3.158 A. Peigney, Ch. Laurent, E. Flahaut, R. R. Bacsa,
A. Rousset: Specific surface area of carbon nanotubes and bundles of carbon nanotubes, Carbon **39**,
507–514 (2001)

3.159 E. Frackowiak, S. Delpeux, K. Jurewicz, K. Szostak,
D. Cazorla-Amoros, F. Béguin: Enhanced capacitance
of carbon nanotubes through chemical activation,
Chem. Phys. Lett. **336**, 35–41 (2002)

3.160 E. Raymundo-Piñero, P. Azaïs, T. Cacciaguerra,
D. Cazorla-Amorós, A. Linares-Solano, F. Béguin:
KOH and NaOH activation mechanisms of multi-

walled carbon nanotubes with different structural organisation, Carbon **43**, 786–795 (2005)

3.161 S. Delpeux, K. Szostak, E. Frackowiak, F. Béguin: An efficient two-step process for producing opened multi-walled carbon nanotubes of high purity, Chem. Phys. Lett. **404**, 374–378 (2005)

3.162 Z. Chen, W. Thiel, A. Hirsch: Reactivity of the convex and concave surfaces of single-walled carbon nanotubes (SWCNTs) towards addition reactions: dependence on the carbon-atom pyramidalization, Chem. Phys. Chem. **1**, 93–97 (2003)

3.163 S. Park, D. Srivastava, K. Cho: Generalized reactivity of curved surfaces: carbon nanotubes, Nano Lett. **3**, 1273–1277 (2003)

3.164 X. Lu, Z. Chen, P. Schleyer: Are Stone–Wales defect sites always more reactive than perfect sites in the sidewalls of single-wall carbon nanotubes?, J. Am. Chem. Soc. **127**, 20–21 (2005)

3.165 M. Muris, N. Dupont-Pavlosky, M. Bienfait, P. Zeppenfeld: Where are the molecules adsorbed on single-walled nanotubes?, Surf. Sci. **492**, 67–74 (2001)

3.166 R. B. Hallock, Y. H. Yang: Adsorption of helium and other gases to carbon nanotubes and nanotubes bundles, J. Low Temp. Phys. **134**, 21–30 (2004)

3.167 A. Fujiwara, K. Ishii, H. Suematsu, H. Kataura, Y. Maniwa, S. Suzuki, Y. Achiba: Gas adsorption in the inside and outside of single-walled carbon nanotubes, Chem. Phys. Lett. **336**, 205–211 (2001)

3.168 C. M. Yang, H. Kanoh, K. Kaneko, M. Yudasaka, S. Iijima: Adsorption behaviors of HiPco single-walled carbon nanotubes aggregates for alcohol vapors, J. Phys. Chem. **106**, 8994–8999 (2002)

3.169 D. H. Yoo, G. H. Rue, M. H. W. Chan, Y. W. Hwang, H. K. Kim: Study of nitrogen adsorbed on open-ended nanotube bundles, J. Phys. Chem. B **107**, 1540–1542 (2003)

3.170 J. Jiang, S. I. Sandler: Nitrogen adsorption on carbon nanotubes bundles: role of the external surface, Phys. Rev. B **68**, 245412-1–245412-9 (2003)

3.171 J. Zhao, A. Buldum, J. Han, J. P. Lu: Gas molecule adsorption in carbon nanotubes and nanotube bundles, Nanotechnology **13**, 195–200 (2002)

3.172 C. Matranga, B. Bockrath: Hydrogen-bonded and physisorbed CO in single-walled carbon nanotubes bundles, J. Phys. Chem. B **109**, 4853–4864 (2005)

3.173 M. D. Ellison, M. J. Crotty, D. Koh, R. L. Spray, K. E. Tate: Adsorption of NH_3 and NO_2 on single-walled carbon nanotubes, J. Phys. Chem. B **108**, 7938–7943 (2004)

3.174 S. Picozzi, S. Santucci, L. Lozzi, L. Valentin, B. Delley: Ozone adsorption on carbon nanotubes: the role of Stone–Wales defects, J. Chem. Phys. **120**, 7147–7152 (2004)

3.175 N. Chakrapani, Y. M. Zhang, S. K. Nayak, J. A. Moore, D. L. Carrol, Y. Y. Choi, P. M. Ajayan: Chemisorption of acetone on carbon nanotubes, J. Phys. Chem. B **107**, 9308–9311 (2003)

3.176 A. Chambers, C. Park, R. T. K. Baker, N. Rodriguez: Hydrogen storage in graphite nanofibers, J. Phys. Chem. B **102**, 4253–4256 (1998)

3.177 J. Giraudet, M. Dubois, D. Claves, J. P. Pinheiro, M. C. Schouler, P. Gadelle, A. Hamwi: Modifying the electronic properties of multi-wall carbon nanotubes via charge transfer, by chemical doping with some inorganic fluorides, Chem. Phys. Lett. **381**, 306–314 (2003)

3.178 J. Hilding, E. A. Grulke, S. B. Sinnott, D. Qian, R. Andrews, M. Jagtoyen: Sorption of butane on carbon multiwall nanotubes at room temperature, Langmuir **17**, 7540–7544 (2001)

3.179 K. Masenelli-Varlot, E. McRae, N. Dupont-Pavlosky: Comparative adsorption of simple molecules on carbon nanotubes. Dependence of the adsorption properties on the nanotube morphology, Appl. Surf. Sci. **196**, 209–215 (2002)

3.180 D. J. Browning, M. L. Gerrard, J. B. Lakeman, I. M. Mellor, R. J. Mortimer, M. C. Turpin: Studies into the storage of hydrogen in carbon nanofibers: Proposal of a possible mechanism, Nanolett. **2**, 201–205 (2002)

3.181 F. H. Yang, R. T. Yang: Ab initio molecular orbital study of adsorption of atomic hydrogen on graphite: insight into hydrogen storage in carbon nanotubes, Carbon **40**, 437–444 (2002)

3.182 A. D. Lueking, R. T. Yang: Hydrogen spillover to enhance hydrogen storage – study of the effect of carbon physicochemical properties, Appl. Catal. **A 265**, 259–268 (2004)

3.183 G. E. Froudakis: Why alkali-metal-doped carbon nanotubes possess high hydrogen uptake, Nanolett. **1**, 531–533 (2001)

3.184 H. Ulbricht, G. Moos, T. Hertel: Physisorption of molecular oxygen on single-wall carbon nanotube bundles and graphite, Phys. Rev. B **66**, 075404-1–075404-7 (2002)

3.185 H. Ulbricht, J. Kriebel, G. Moos, T. Hertel: Desorption kinetics and interaction of Xe with single-wall carbon nanotube bundles, Chem. Phys. Lett. **363**, 252–260 (2002)

3.186 R. Saito, G. Dresselhaus, M. S. Dresselhaus: *Physical Properties of Carbon Nanotubes* (Imperial College Press, London 1998)

3.187 A. Charlier, E. McRae, R. Heyd, M. F. Charlier, D. Moretti: Classification for double-walled carbon nanotubes, Carbon **37**, 1779–1783 (1999)

3.188 A. Charlier, E. McRae, R. Heyd, M. F. Charlier: Metal semi-conductor transitions under uniaxial stress for single- and double-walled carbon nanotubes, J. Phys. Chem. Solids **62**, 439–444 (2001)

3.189 P. Puech, H. Hubel, D. Dunstan, R. R. Bacsa, Ch. Laurent, W. S. Bacsa: Discontinuous tangential stress in double wall carbon nanotubes, Phys. Rev. Lett. **93**, 095506 (2004)

3.190 P. M. Ajayan, M. Terrrones, A. de la Guardia, V. Hue, N. Grobert, B. Q. Wei, H. Lezec, G. Ramanath, T. W. Ebbesen: Nanotubes in a flash – Ignition and reconstruction, Science **296**, 705 (2002)

3.191 H. Ajiki, T. Ando: Electronic states of carbon nanotubes, J. Phys. Soc. Jap. **62**, 1255–1266 (1993)

3.192 S. M. Bachilo, M. S. Strano, C. Kittrell, R. H. Hauge, R. E. Smalley, R. B. Weisman: Structure-assigned optical spectra of single-walled carbon nanotubes, Science **298**, 2361 (2002)

3.193 M. Bockrath, D. H. Cobden, J. Lu, A. G. Rinzler, R. E. Smalley, L. Balents, P. L. McEuen: Luttinger liquid behaviour in carbon nanotubes, Nature **397**, 598–601 (1999)

3.194 C. T. White, T. N. Todorov: Carbon nanotubes as long ballistic conductors, Nature **393**, 240–242 (1998)

3.195 S. Frank, P. Poncharal, Z. L. Wang, W. A. de Heer: Carbon nanotube quantum resistors, Science **280**, 1744–1746 (1998)

3.196 W. Liang, M. Bockrath, D. Bozovic, J. H. Hafner, M. Tinkham, H. Park: Fabry–Perot interference in a nanotube electron waveguide, Nature **411**, 665–669 (2001)

3.197 L. Langer, V. Bayot, E. Grivei, J.-P. Issi, J.-P. Heremans, C. H. Olk, L. Stockman, C. van Haesendonck, Y. Buynseraeder: Quantum transport in a multiwalled carbon nanotube, Phys. Rev. Lett. **76**, 479–482 (1996)

3.198 K. Liu, S. Roth, G. S. Duesberg, G. T. Kim, D. Popa, K. Mukhopadhyay, R. Doome, J. B'Nagy: Antilocalization in multiwalled carbon nanotubes, Phys. Rev. B **61**, 2375–2379 (2000)

3.199 G. Fedorov, B. Lassagne, M. Sagnes, B. Raquet, J. M. Broto, F. Triozon, S. Roche, E. Flahaut: Gate-dependent magnetoresistance phenomena in carbon nanotubes, Phys. Rev. Lett. **94**, 66801–66804 (2005)

3.200 A. Javey, J. Guo, Q. Wang, M. Lundstrom, H. Dai: Ballistic carbon nanotube field-effect transistors, Nature **424**, 654–657 (2003)

3.201 Y. A. Kasumov, R. Deblock, M. Kociak, B. Reulet, H. Bouchiat, I. I. Khodos, Y. B. Gorbatov, V. T. Volkov, C. Journet, M. Burghard: Supercurrents through single-walled carbon nanotubes, Science **284**, 1508–1511 (1999)

3.202 B. W. Alphenaar, K. Tsukagoshi, M. Wagner: Magnetoresistance of ferromagnetically contacted carbon nanotubes, Phys. Eng. **10**, 499–504 (2001)

3.203 S. Berber, Y. Kwon, D. Tomanek: Unusually high thermal conductivity of carbon nanotubes, Phys. Rev. Lett. **84**, 4613–4616 (2000)

3.204 S. N. Song, X. K. Wang, R. P. H. Chang, J. B. Ketterson: Electronic properties of graphite nanotubules from galvanomagnetic effects, Phys. Rev. Lett. **72**, 697–700 (1994)

3.205 M.-F. Yu, O. Lourie, M. J. Dyer, K. Moloni, T. F. Kelley, R. S. Ruoff: Strength and breaking mechanism of multiwalled carbon nanotubes under tensile load, Science **287**, 637–640 (2000)

3.206 D. A. Walters, L. M. Ericson, M. J. Casavant, J. Liu, D. T. Colbert, K. A. Smith, R. E. Smalley: Elastic strain of freely suspended single-wall carbon nanotube ropes, Appl. Phys. Lett. **74**, 3803–3805 (1999)

3.207 B. G. Demczyk, Y. M. Wang, J. Cumingd, M. Hetamn, W. Han, A. Zettl, R. O. Ritchie: Direct mechanical measurement of the tensile strength and elastic modulus of multiwalled carbon nanotubes, Mater. Sci. Eng. A **334**, 173–178 (2002)

3.208 R. P. Gao, Z. L. Wang, Z. G. Bai, W. A. De Heer, L. M. Dai, M. Gao: Nanomechanics of individual carbon nanotubes from pyrolytically grown arrays, Phys. Rev. Lett. **85**, 622–625 (2000)

3.209 M. M. J. Treacy, T. W. Ebbesen, J. M. Gibson: Exceptionally high Young's modulus observed for individual carbon nanotubes, Nature **381**, 678–680 (1996)

3.210 N. Yao, V. Lordie: Young's modulus of single-wall carbon nanotubes, J. Appl. Phys. **84**, 1939–1943 (1998)

3.211 O. Lourie, H. D. Wagner: Transmission electron microscopy observations of fracture of single-wall carbon nanotubes under axial tension, Appl. Phys. Lett. **73**, 3527–3529 (1998)

3.212 S. C. Tsang, Y. K. Chen, P. J. F. Harris, M. L. H. Green: A simple chemical method of opening and filling carbon nanotubes, Nature **372**, 159–162 (1994)

3.213 M. Monthioux: Filling single-wall carbon nanotubes, Carbon **40**, 1809–1823 (2002)

3.214 W. K. Hsu, S. Y. Chu, E. Munoz-Picone, J. L. Boldu, S. Firth, P. Franchi, B. P. Roberts, A. Shilder, H. Terrones, N. Grobert, Y. Q. Zhu, M. Terrones, M. E. McHenry, H. W. Kroto, D. R. M. Walton: Metallic behaviour of boron-containing carbon nanotubes, Chem. Phys. Lett. **323**, 572–579 (2000)

3.215 R. Czerw, M. Terrones, J. C. Charlier, X. Blasé, B. Foley, R. Kamalakaran, N. Grobert, H. Terrones, D. Tekleab, P. M. Ajayan, W. Blau, M. Rühle, D. L. Caroll: Identification of electron donor states, in N-doped carbon nanotubes, Nanolett. **1**, 457–460 (2001)

3.216 O. Stephan, P. M. Ajayan, C. Colliex, P. Redlich, J. M. Lambert, P. Bernier, P. Lefin: Doping graphitic and carbon nanotube structures with boron and nitrogen, Science **266**, 1683–1685 (1994)

3.217 A. Loiseau, F. Williaime, N. Demoncy, N. Schramchenko, G. Hug, C. Colliex, H. Pascard: Boron nitride nanotubes, Carbon **36**, 743–752 (1998)

3.218 C. C. Tang, L. M. de la Chapelle, P. Li, Y. M. Liu, H. Y. Dang, S. S. Fan: Catalytic growth of nanotube and nanobamboo structures of boron nitride, Chem. Phys. Lett. **342**, 492–496 (2001)

3.219 K. Suenaga, C. Colliex, N. Demoncy, A. Loiseau, H. Pascard, F. Williaime: Synthesis of nanoparticles and nanotubes with well separated layers of boron-nitride and carbon, Science **278**, 653–655 (1997)

3.220 D. Golberg, Y. Bando, L. Bourgeois, K. Kurashima, T. Sato: Large-scale synthesis and HRTEM analysis of single-walled B- and N-doped carbon nanotube bundles, Carbon **38**, 2017–2027 (2000)

3.221 R. S. Lee, J. Gavillet, M. Lamy de la Chapelle, A. Loiseau, J.-L. Cochon, D. Pigache, J. Thibault, F. Willaime: Catalyst-free synthesis of boron nitride single-wall nanotubes with a preferred zig-zag configuration, Phys. Rev. B **64**, 121405.1–121405.4 (2001)

3.222 B. Burteaux, A. Claye, B. W. Smith, M. Monthioux, D. E. Luzzi, J. E. Fischer: Abundance of encapsulated C$_{60}$ in single-wall carbon nanotubes, Chem. Phys. Lett. **310**, 21–24 (1999)

3.223 D. Ugarte, A. Châtelain, W. A. de Heer: Nanocapillarity and chemistry in carbon nanotubes, Science **274**, 1897–1899 (1996)

3.224 J. Cook, J. Sloan, M. L. H. Green: Opening and filling carbon nanotubes, Fuller. Sci. Technol. **5**, 695–704 (1997)

3.225 P. M. Ajayan, S. Iijima: Capillarity-induced filling of carbon nanotubes, Nature **361**, 333–334 (1993)

3.226 P. M. Ajayan, T. W. Ebbesen, T. Ichihashi, S. Iijima, K. Tanigaki, H. Hiura: Opening carbon nanotubes with oxygen and implications for filling, Nature **362**, 522–525 (1993)

3.227 S. Seraphin, D. Zhou, J. Jiao, J. C. Withers, R. Loufty: Yttrium carbide in nanotubes, Nature **362**, 503 (1993)

3.228 S. Seraphin, D. Zhou, J. Jiao, J. C. Withers, R. Loufty: Selective encapsulation of the carbides of yttrium and titanium into carbon nanoclusters, Appl. Phys. Lett. **63**, 2073–2075 (1993)

3.229 R. S. Ruoff, D. C. Lorents, B. Chan, R. Malhotra, S. Subramoney: Single-crystal metals encapsulated in carbon nanoparticles, Science **259**, 346–348 (1993)

3.230 A. Loiseau, H. Pascard: Synthesis of long carbon nanotubes filled with Se, S, Sb, and Ge by the arc method, Chem. Phys. Lett. **256**, 246–252 (1996)

3.231 N. Demoncy, O. Stephan, N. Brun, C. Colliex, A. Loiseau, H. Pascard: Filling carbon nanotubes with metals by the arc discharge method: The key role of sulfur, Eur. Phys. J. B **4**, 147–157 (1998)

3.232 E. Flahaut, J. Sloan, K. S. Coleman, V. C. Williams, S. Friedrichs, N. Hanson, M. L. H. Green: 1D p-block halide crystals confined into single walled carbon nanotubes, Proc. Mater. Res. Soc. Symp. **633**, A13.15.1–A13.15.6 (2001)

3.233 C. H. Kiang, J. S. Choi, T. T. Tran, A. D. Bacher: Molecular nanowires of 1nm diameter from capillary filling of single-walled carbon nanotubes, J. Phys. Chem. B **103**, 7449–7551 (1999)

3.234 Z. L. Zhang, B. Li, Z. J. Shi, Z. N. Gu, Z. Q. Xue, L. M. Peng: Filling of single-walled carbon nanotubes with silver, J. Mater. Res. **15**, 2658–2661 (2000)

3.235 A. Govindaraj, B. C. Satishkumar, M. Nath, C. N. R. Tao: Metal nanowires and intercalated metal

layers in single-walled carbon nanotubes bundles, Chem. Mater. **12**, 202–205 (2000)

3.236 J. Mittal, M. Monthioux, H. Allouche: Room temperature filling of single-wall carbon nanotubes with chromium oxide in open air, Chem. Phys. Lett. **339**, 311–318 (2001)

3.237 E. Dujardin, T. W. Ebbesen, H. Hiura, K. Tanigaki: Capillarity and wetting of carbon nanotubes, Science **265**, 1850–1852 (1994)

3.238 J. Sloan, A. I. Kirkland, J. L. Hutchison, M. L. H. Green: Integral atomic layer architectures of 1D crystals inserted into single walled carbon nanotubes, Chem. Commun., 1319–1332 (2002)

3.239 J. Sloan, M. C. Novotny, S. R. Bailey, G. Brown, C. Xu, V. C. Williams, S. Friedrichs, E. Flahaut, R. L. Callender, A. P. E. York, K. S. Coleman, M. L. H. Green, R. E. Dunin-Borkowski, J. L. Hutchison: Two layer 4:4 co-ordinated KI crystals grown within single walled carbon nanotubes, Chem. Phys. Lett. **329**, 61–65 (2000)

3.240 G. Brown, S. R. Bailey, J. Sloan, C. Xu, S. Friedrichs, E. Flahaut, K. S. Coleman, J. L. Hutchinson, R. E. Dunin-Borkowski, M. L. H. Green: Electron beam induced in situ clusterisation of 1D ZrCl$_4$ chains within single-walled carbon nanotubes, Chem. Commun., 845–846 (2001)

3.241 J. Sloan, D. M. Wright, H. G. Woo, S. Bailey, G. Brown, A. P. E. York, K. S. Coleman, J. L. Hutchison, M. L. H. Green: Capillarity and silver nanowire formation observed in single walled carbon nanotubes, Chem. Commun., 699–700 (1999)

3.242 X. Fan, E. C. Dickey, P. C. Eklund, K. A. Williams, L. Grigorian, R. Buczko, S. T. Pantelides, S. J. Pennycook: Atomic arrangement of iodine atoms inside single-walled carbon nanotubes, Phys. Rev. Lett. **84**, 4621–4624 (2000)

3.243 G. Brown, S. R. Bailey, M. Novotny, R. Carter, E. Flahaut, K. S. Coleman, J. L. Hutchison, M. L. H. Green, J. Sloan: High yield incorporation and washing properties of halides incorporated into single walled carbon nanotubes, Appl. Phys. A **76**, 457–462 (2003)

3.244 J. Sloan, D. E. Luzzi, A. I. Kirkland, J. L. Hutchison, M. L. H. Green: Imaging and characterization of molecules and one-dimensional crystals formed within carbon nanotubes, Mater. Res. Soc. Bull. **29**, 265–271 (2004)

3.245 J. Chancolon, F. Archaimbault, A. Pineau, S. Bonnamy: Confinement of selenium into carbon nanotubes, Fuller. Nanotub. Car. N. **13**, 189–194 (2005)

3.246 B. W. Smith, M. Monthioux, D. E. Luzzi: Encapsulated C$_{60}$ in carbon nanotubes, Nature **396**, 323–324 (1998)

3.247 B. W. Smith, D. E. Luzzi: Formation mechanism of fullerene peapods and coaxial tubes: A path to large scale synthesis, Chem. Phys. Lett. **321**, 169–174 (2000)

3.248 K. Hirahara, K. Suenaga, S. Bandow, H. Kato, T. Okazaki, H. Shinohara, S. Iijima: One-dimensional metallo-fullerene crystal generated inside single-walled carbon nanotubes, Phys. Rev. Lett. **85**, 5384–5387 (2000)

3.249 B. W. Smith, M. Monthioux, D. E. Luzzi: Carbon nanotube encapsulated fullerenes: A unique class of hybrid material, Chem. Phys. Lett. **315**, 31–36 (1999)

3.250 D. E. Luzzi, B. W. Smith: Carbon cage structures in single wall carbon nanotubes: A new class of materials, Carbon **38**, 1751–1756 (2000)

3.251 S. Bandow, M. Takisawa, K. Hirahara, M. Yudasoka, S. Iijima: Raman scattering study of double-wall carbon nanotubes derived from the chains of fullerenes in single-wall carbon nanotubes, Chem. Phys. Lett. **337**, 48–54 (2001)

3.252 Y. Sakurabayashi, M. Monthioux, K. Kishita, Y. Suzuki, T. Kondo, M. Le Lay: Tayloring double wall carbon nanotubes?. In: *Molecular Nanostructures*, Amer. Inst. Phys. Conf. Proc., Vol. 685, ed. by H. Kuzmany, J. Fink, M. Mehring, S. Roth (Springer, Berlin Heidelberg 2003) pp. 302–305

3.253 B. W. Smith, D. E. Luzzi, Y. Achiba: Tumbling atoms and evidence for charge transfer in $La_2@C_{80}@SWNT$, Chem. Phys. Lett. **331**, 137–142 (2000)

3.254 K. Suenaga, M. Tence, C. Mory, C. Colliex, H. Kato, T. Okazaki, H. Shinohara, K. Hirahara, S. Bandow, S. Iijima: Element-selective single atom imaging, Science **290**, 2280–2282 (2000)

3.255 D. E. Luzzi, B. W. Smith, R. Russo, B. C. Satishkumar, F. Stercel, N. R. C. Nemes: Encapsulation of metallofullerenes and metallocenes in carbon nanotubes, Proc. Electronic Properties of Novel Materials-XVI Int. Winterschool – AIP Conf. Proc., ed. by H. Kuzmany, J. Fink, M. Mehring, S. Roth (Springer, Berlin Heidelberg 2001) 622–626

3.256 D. J. Hornbaker, S.-J. Kahng, S. Misra, B. W. Smith, A. T. Johnson, E. J. Mele, D. E. Luzzi, A. Yazdani: Mapping the one-dimensional electronic states of nanotube peapod structures, Science **295**, 828–831 (2002)

3.257 H. Kondo, H. Kino, T. Ohno: Transport properties of carbon nanotubes encapsulating C_{60} and related materials, Phys. Rev. B **71**, 115413 (2005)

3.258 S. H. Jhang, S. W. Lee, D. S. Lee, Y. W. Park, G. H. Jeong, T. Hirata, R. Hatakeyama, U. Dettlaff, S. Roth, M. S. Kabir, E. E. B. Campbell: Random telegraph noise in carbon nanotube peapod transistors, Fuller. Nanotub. Car. N. **13**, 195–198 (2005)

3.259 G. H. Jeong, R. Hatakeyama, T. Hirata, K. Tohji, K. Motomiya, N. Sato, Y. Kawazoe: Structural deformation of single-walled carbon nanotubes and fullerene encapsulation due to magnetized plasma ion irradiation, Appl. Phys. Lett. **79**, 4213–4215 (2001)

3.260 Y. P. Sun, K. Fu, Y. Lin, W. Huang: Functionalized carbon nanotubes: Properties and applications, Acc. Chem. Res. **35**, 1095–1104 (2002)

3.261 S. Osswald, E. Flahaut, H. Ye, Y. Gogotsi: Elimination of D-band in Raman spectra of double-wall carbon nanotubes by oxidation, Chem. Phys. Lett. **402**, 422–427 (2005)

3.262 J. Chen, M. A. Hamon, M. Hui, C. Yongsheng, A. M. Rao, P. C. Eklund, R. C. Haddon: Solution properties of single-walled carbon nanotubes, Science **282**, 95–98 (1998)

3.263 J. Chen, A. M. Rao, S. Lyuksyutov, M. E. Itkis, M. A. Hamon, H. Hu, R. W. Cohn, P. C. Eklund, D. T. Colbert, R. E. Smalley, R. C. Haddon: Dissolution of full-length single-walled carbon nanotubes, J. Phys. Chem. B **105**, 2525–2528 (2001)

3.264 F. Pompeo, D. E. Resasco: Water solubilization of single-walled carbon nanotubes by functionalization with glucosamine, Nanolett. **2**, 369–373 (2002)

3.265 Y. P. Sun, W. Huang, Y. Lin, K. Fu, A. Kitaygorodskiy, L. A. Riddle, Y. J. Yu, D. L. Carroll: Soluble dendron-functionalized carbon nanotubes: Preparation, characterization, and properties, Chem. Mater. **13**, 2864–2869 (2001)

3.266 K. Fu, W. Huang, Y. Lin, L. A. Riddle, D. L. Carroll, Y. P. Sun: Defunctionalization of functionalized carbon nanotubes, Nanolett. **1**, 439–441 (2001)

3.267 P. W. Chiu, G. S. Duesberg, U. Dettlaff-Weglikowska, S. Roth: Interconnection of carbon nanotubes by chemical functionalization, Appl. Phys. Lett. **80**, 3811–3813 (2002)

3.268 E. T. Mickelson, C. B. Huffman, A. G. Rinzler, R. E. Smalley, R. H. Hauge, J. L. Margrave: Fluorination of single-wall carbon nanotubes, Chem. Phys. Lett. **296**, 188–194 (1998)

3.269 V. N. Khabashesku, W. E. Billups, J. L. Margrave: Fluorination of single-wall carbon nanotubes and subsequent derivatization reactions, Acc. Chem. Res. **35**, 1087–1095 (2002)

3.270 P. J. Boul, J. Liu, E. T. Mickelson, C. B. Huffman, L. M. Ericson, I. W. Chiang, K. A. Smith, D. T. Colbert, R. H. Hauge, J. L. Margrave, R. E. Smalley: Reversible side-wall functionalization of buckytubes, Chem. Phys. Lett. **310**, 367–372 (1999)

3.271 J. L. Bahr, J. Yang, D. V. Kosynkin, M. J. Bronikowski, R. E. Smalley, J. M. Tour: Functionalization of carbon nanotubes by electrochemical reduction of aryl diazonium salts: A bucky paper electrode, J. Am. Chem. Soc. **123**, 6536–6542 (2001)

3.272 M. Holzinger, O. Vostrowsky, A. Hirsch, F. Hennrich, M. Kappes, R. Weiss, F. Jellen: Sidewall functionalization of carbon nanotubes, Angew. Chem. Int. Ed. **40**, 4002–4005 (2001)

3.273 Y. Chen, R. C. Haddon, S. Fang, A. M. Rao, P. C. Eklund, W. H. Lee, E. C. Dickey, E. A. Grulke, J. C. Pendergrass, A. Chavan, B. E. Haley, R. E. Smalley: Chemical attachment of organic functional groups to single-walled carbon nanotube material, J. Mater. Res. **13**, 2423–2431 (1998)

3.274 C. Velasco-Santos, A. L. Martinez-Hernandez, M. Lozada-Cassou, A. Alvarez-Castillo, V. M. Castano: Chemical functionalization of carbon nanotubes through an organosilane, Nanotechnology **13**, 495–498 (2002)

3.275 A. Star, J. F. Stoddart, D. Steuerman, M. Diehl, A. Boukai, E. W. Wong, X. Yang, S. W. Chung, H. Choi, J. R. Heath: Preparation and properties of polymer-wrapped single-walled carbon nanotubes, Angew. Chem. Int. Ed. **41**, 1721–1725 (2002)

3.276 A. Pénicaud, P. Poulin, A. Derré, E. Anglaret, P. Petit: Spontaneous dissolution of a single-wall carbon nanotube salt, J. Amer. Chem. Soc. **127**, 8–9 (2005)

3.277 R. Stevens, C. Nguyen, A. Cassel, L. Delzeit, M. Meyyapan, J. Han: Improved fabrication approach for carbon nanotube probe devices, Appl. Phys. Lett. **77**, 3453–3455 (2000)

3.278 J. H. Hafner, C. L. Cheung, A. T. Wooley, C. M. Lieber: Structural and functional imaging with carbon nanotube AFM probes, Progr. Biophys. Molec. Biol. **77**, 73–110 (2001)

3.279 S. S. Wong, E. Joselevich, A. T. Woodley, C. L. Cheung, C. M. Lieber: Covalently functionalized nanotubes as nanometre-size probes in chemistry and biology, Nature **394**, 52–55 (1998)

3.280 C. L. Cheung, J. H. Hafner, C. M. Lieber: Carbon nanotube atomic force microscopy tips: Direct growth by chemical vapor deposition and application to high-resolution imaging, Proc. Natl. Acad. Sci. USA **97**, 3809–3813 (2000)

3.281 W. A. de Heer, A. Châtelain, D. Ugarte: A carbon nanotube field-emission electron source, Science **270**, 1179–1180 (1995)

3.282 J. M. Bonard, J. P. Salvetat, T. Stockli, W. A. de Heer, L. Forro, A. Chatelâin: Field emission from single-wall carbon nanotube films, Appl. Phys. Lett. **73**, 918–920 (1998)

3.283 W. Zhu, C. Bower, O. Zhou, G. Kochanski, S. Jin: Large curent density from carbon nanotube field emitters, Appl. Phys. Lett. **75**, 873–875 (1999)

3.284 Y. Saito, R. Mizushima, T. Tanaka, K. Tohji, K. Uchida, M. Yumura, S. Uemura: Synthesis, structure, and field emission of carbon nanotubes, Fuller. Sci. Technol. **7**, 653–664 (1999)

3.285 J. Kong, N. R. Franklin, C. Zhou, M. G. Chapline, S. Peng, K. Cho, H. Dai: Nanotube molecular wire as chemical sensors, Science **287**, 622–625 (2000)

3.286 P. G. Collins, K. Bradley, M. Ishigami, A. Zettl: Extreme oxygen sensitivity of electronic properties of carbon nanotubes, Science **287**, 1801–1804 (2000)

3.287 H. Chang, J. D. Lee, S. M. Lee, Y. H. Lee: Adsorption of NH_3 and NO_2 molecules on carbon nanotubes, Appl. Phys. Lett. **79**, 3863–3865 (2001)

3.288 C. Cantalini, L. Valentini, L. Lozzi, I. Armentano, J. M. Kenny, S. Santucci: NO_2 gas sensitivity of carbon nanotubes obtained by plasma enhanced chemical vapor deposition, Sensor. Actuat. B **93**, 333–337 (2003)

3.289 J. Li, Y. Lu, Q. Ye, M. Cinke, J. Han, M. Meyyappan: Carbon nanotubes sensors for gas and organic vapor detection, Nano Lett. **3**, 929–933 (2003)

3.290 O. K. Varghese, P. D. Kichambre, D. Gong, K. G. Ong, E. C. Dickey, C. A. Grimes: Gas sensing characteristics of multi-wall carbon nanotubes, Sensor. Actuat. B **81**, 32–41 (2001)

3.291 K. G. Ong, K. Zeng, C. A. Grimes: A wireless, passive carbon nanotube-based gas sensor, IEEE Sens. J. **2/2**, 82–88 (2002)

3.292 J. Kong, M. G. Chapline, H. Dai: Functionalized carbon nanotubes for molecular hydrogen sensors, Adv. Mater. **13**, 1384–1386 (2001)

3.293 A. Modi, N. Koratkar, E. Lass, B. Wei, P. M. Ajayan: Miniaturized gas ionisation sensors using carbon nanotubes, Nature **424**, 171–174 (2003)

3.294 F. Rodriguez-Reinoso: The role of carbon materials in heterogeneous catalysis, Carbon **36**, 159–175 (1998)

3.295 E. Auer, A. Freund, J. Pietsch, T. Tacke: Carbon as support for industrial precious metal catalysts, Appl. Catal. A **173**, 259–271 (1998)

3.296 J. M. Planeix, N. Coustel, B. Coq, B. Botrons, P. S. Kumbhar, R. Dutartre, P. Geneste, P. Bernier, P. M. Ajayan: Application of carbon nanotubes as supports in heterogeneous catalysis, J. Am. Chem. Soc. **116**, 7935–7936 (1994)

3.297 P. Serp, M. Corrias, P. Kalck: Carbon nanotubes and nanofibers in catalysis, Appl. Catal. A **253**, 337–358 (2003)

3.298 K. P. De Jong, J. W. Geus: Carbon nanofibers: catalytic synthesis and applications, Catal. Rev. **42**, 481–510 (2000)

3.299 N. F. Goldshleger: Fullerene and fullerene-based materials in catalysis, Fuller. Sci. Technol. **9**, 255–280 (2001)

3.300 M. F. R. Pereira, J. L. Figueiredo, J. J. M. Órfão, P. Serp, P. Kalck, Y. Kihn: Catalytic activity of carbon nanotubes in the oxidative dehydrogenation of ethylbenzene, Carbon **42**, 2807–2813 (2004)

3.301 G. Mestl, N. I. Maksimova, N. Keller, V. V. Roddatis, R. Schlögl: Carbon nanofilaments in heterogeneous catalysis: An industrial application for new carbon materials?, Angew. Chem. Int. Ed. Engl. **40**, 2066–2068 (2001)

3.302 N. Muradov: Catalysis of methane decomposition over elemental carbon, Catal. Commun. **2**, 89–94 (2001)

3.303 J. E. Fischer, A. T. Johnson: Electronic properties of carbon nanotubes, Curr. Opin. Solid State Mater. Sci. **4**, 28–33 (1999)

3.304 M. Menon, A. N. Andriotis, G. E. Froudakis: Curvature dependence of the metal catalyst atom interaction with carbon nanotubes walls, Chem. Phys. Lett. **320**, 425–434 (2000)

3.305 Z. Liu, X. Lin, J. Y. Lee, W. Zhang, M. Han, L. M. Gan: Preparation and characterization of platinum-based electrocatalysts on multiwalled carbon nanotubes for proton exchange membrane fuel cells, Langmuir **18**, 4054–4060 (2002)

Part A | 3

3.306 R. Giordano, P. Serp, P. Kalck, Y. Kihn, J. Schreiber, C. Marhic, J.-L. Duvail: Preparation of rhodium supported on carbon canotubes catalysts via surface mediated organometallic reaction, Eur. J. Inorg. Chem., 610–617 (2003)

3.307 H.-B. Chen, J. D. Lin, Y. Cai, X. Y. Wang, J. Yi, J. Wang, G. Wei, Y. Z. Lin, D. W. Liao: Novel multi-walled nanotube-supported and alkali-promoted Ru catalysts for ammonia synthesis under atmospheric pressure, Appl. Surf. Sci. **180**, 328–335 (2001)

3.308 Y. Zhang, H. B. Zhang, G. D. Lin, P. Chen, Y. Z. Yuan, K. R. Tsai: Preparation, characterization and catalytic hydroformylation properties of carbon nanotubes-supported Rh-phosphine catalyst, Appl. Catal. A **187**, 213–224 (1999)

3.309 T. Kyotani, S. Nakazaki, W.-H. Xu, A. Tomita: Chemical modification of the inner walls of carbon nanotubes by HNO_3 oxidation, Carbon **39**, 782–785 (2001)

3.310 Z. J. Liu, Z. Y. Yuan, W. Zhou, L. M. Peng, Z. Xu: Co/carbon nanotubes monometallic system: The effects of oxidation by nitric acid, Phys. Chem. Chem. Phys. **3**, 2518–2521 (2001)

3.311 Carillo, J. A. Swartz, J. M. Gamba, R. S. Kane, N. Chakrapani, B. Wei, P. M. Ajayan: Noncovalent functionalization of graphite and carbon nanotubes with polymer multilayers and gold nanoparticles, Nano Lett. **3**, 1437–1440 (2003)

3.312 M. S. Dresselhaus, K. A. Williams, P. C. Eklund: Hydrogen adsorption in carbon materials, Mater. Res. Soc. Bull. **24**, 45–50 (1999)

3.313 H.-M. Cheng, Q.-H. Yang, C. Liu: Hydrogen storage in carbon nanotubes, Carbon **39**, 1447–1454 (2001)

3.314 G. G. Tibbetts, G. P. Meisner, C. H. Olk: Hydrogen storage capacity of carbon nanotubes, filaments, and vapor-grown fibers, Carbon **39**, 2291–2301 (2001)

3.315 F. L. Darkrim, P. Malbrunot, G. P. Tartaglia: Review of hydrogen storage adsorption in carbon nanotubes, Int. J. Hydrogen Energy **27**, 193–202 (2002)

3.316 G. E. Froudakis: Hydrogen interaction with carbon nanotubes: a review of ab initio studies, J. Phys.: Condens. Matter **14**, R453–R465 (2002)

3.317 M. Hirscher, M. Becher: Hydrogen storage in carbon nanotubes, J. Nanosci. Nanotechnol. **3(1-2)**, 3–17 (2003)

3.318 C. Park, P. E. Anderson, C. D. Tan, R. Hidalgo, N. Rodriguez: Further studies of the interaction of hydrogen with graphite nanofibers, J. Phys. Chem. B **103**, 10572–1058 (1999)

3.319 A. D. Lueking, R. T. Yang, N. M. Rodriguez, R. T. K. Baker: Hydrogen storage in graphite nanofibers: effect of synthesis catalyst and pretreatment conditions, Langmuir **20(3)**, 714–721 (2004)

3.320 A. D. Lueking, R. T. Yang: Hydrogen storage in carbon nanotubes: residual metal content and pretreatment temperature, AIChE J. **49(6)**, 1556–1568 (2003)

3.321 C. C. Ahn, Y. Ye, B. V. Ratnakumar, C. Witham, R. C. Bowman, B. Fultz: Hydrogen adsorption measurements on graphite nanofibers, Appl. Phys. Lett. **73**, 3378–3380 (1998)

3.322 Q. Wang, J. K. Johnson: Computer simulations of hydrogen adsorption on graphite nanofibers, J. Phys. Chem. B **103**, 277–281 (1999)

3.323 M. Rzepka, P. Lamp, M. A. de la Casa-Lillo: Physisorption of hydrogen on microporous carbon and carbon nanotubes, J. Phys. Chem. B **102**, 10894–10898 (1998)

3.324 A. C. Dillon, K. M. Jones, T. A. Bekkedahl, C. H. Kiang, D. S. Bethune, M. J. Heben: Storage of hydrogen in single-walled carbon nanotubes, Nature **386**, 377–379 (1997)

3.325 K. Shen, T. Pietrass: ^1H and ^2H NMR of hydrogen adsorption on carbon nanotubes, J. Phys. Chem. **B 108**, 9937–9942 (2004)

3.326 Y. Ye, C. C. Ahn, C. Witham, R. C. Bowman, B. Fultz, J. Liu, A. G. Rinzler, D. Colbert, K. A. Smith, R. E. Smalley: Hydrogen adsorption and cohesive energy of single-walled carbon nanotubes, Appl. Phys. Lett. **74**, 2307–2309 (1999)

3.327 C. Liu, Y. Y. Fan, M. Liu, H. T. Cong, H. M. Cheng, M. S. Dresselhaus: Hydrogen storage in single-walled carbon nanotubes at room temperature, Science **286**, 1127–1129 (1999)

3.328 M. Hirscher, M. Becher, M. Haluska, U. Dettlaff-Weglikowska, A. Quintel, G. S. Duesberg, Y. M. Choi, P. Dwones, M. Hulman, S. Roth, I. Stepanek, P. Bernier: Hydrogen storage in sonicated carbon materials, Appl. Phys. A **72**, 129–132 (2001)

3.329 H. G. Schimmel, G. J. Kearley, M. G. Nijkamp, C. T. Visser, K. P. de Jong, F. M. Mulder: Hydrogen adsorption in carbon nanostructures: comparison of nanotubes, fibers, and coals, Chem. Eur. J. **9**, 4764–4770 (2003)

3.330 M. R. Smith Jr, E. W. Bittner, W. Shi, J. K. Johnson, B. C. Bockrath: Chemical activation of single-walled nanotubes for hydrogen adsorption, J. Phys. Chem. B **107**, 3752–3760 (2003)

3.331 A. Ansón, M. A. Callejas, A. M. Benito, W. K. Maser, M. T. Izquierdo, B. Rubio, J. Jagiello, M. Tomes, J. B. Parra, M. T. Martinez: Hydrogen adsorption studies on single wall carbon nanotubes, Carbon **42**, 1243–1248 (2004)

3.332 C. Q. Ning, F. Wei, G. H. Luo, Q. X. Wang, Y. L. Wu, H. Yu: Hydrogen storage in multi-wall carbon nanotubes using samples up to 85 g, Appl. Phys. A **78**, 955–959 (2004)

3.333 P. A. Gordon, R. B. Saeger: Molecular modeling of adsorptive energy storage: Hydrogen storage in single-walled carbon nanotubes, Ind. Eng. Chem. Res. **38**, 4647–4655 (1999)

3.334 P. Marinelli, R. Pellenq, J. Conard: *H stocké dans les carbones un site légèrement métastable*, AF-14-020 (National Conference on Materials, Tours, France 2002)

3.335 S. M. Lee, H. Y. Lee, T. Frauenheim, M. Elstner, Y. G. Hwang: Hydrogen storage in single-walled and multi walled carbon nanotubes, Proc. of Material Research Society Symposium **593**, 187–192 (1999)

3.336 S. M. Lee, H. Y. Lee: Hydrogen storage in single-walled carbon nanotubes, Appl. Phys. Lett. **76**, 2877–2879 (2000)

3.337 S. M. Lee, K. S. Park, Y. C. Choi, Y. S. Park, J. M. Bok, D. J. Bae, K. S. Nahm, Y. G. Choi, C. S. Yu, N. Kim, T. Frauenheim, Y. H. Lee: Hydrogen adsorption in carbon nanotubes, Synth. Met. **113**, 209–216 (2000)

3.338 X. Zhang, D. Cao, J. Chen: Hydrogen adsorption storage on single-walled carbon nanotube arrays by a combination of classical potential and density functional theory, J. Phys. Chem. **B 107**, 4942–4950 (2003)

3.339 H. M. Cheng, G. P. Pez, A. C. Cooper: Mechanism of hydrogen sorption in single-walled carbon nanotubes, J. Am. Chem. Soc. **123**, 5845–5846 (2001)

3.340 M. A. de la Casa-Lillo, F. Lamari-Darkrim, D. Cazorla-Amoros, A. Linares-Solano: Hydrogen storage in activated carbons and activated carbon fibers, J. Phys. Chem. B **106**, 10930–10934 (2002)

3.341 A. Kusnetzova, D. B. Mawhinney, V. Naumenko, J. T. Yates, J. Liu, R. E. Smalley: Enhancement of adsorption inside of single-walled nanotubes: Opening the entry ports, Chem. Phys. Lett. **321**, 292–296 (2000)

3.342 G. E. Gadd, M. Blackford, S. Moricca, N. Webb, P. J. Evans, A. M. Smith, G. Jacobsen, S. Leung, A. Day, Q. Hua: The world's smallest gas cylinders?, Science **277**, 933–936 (1997)

3.343 Z. Mao, S. B. Sinnott: A computational study of molecular diffusion and dynamic flow through carbon nanotubes, J. Phys. Chem. B **104**, 4618–4624 (2000)

3.344 Z. Mao, S. B. Sinnott: Separation of organic molecular mixtures in carbon nanotubes and bundles: Molecular dynamics simulations, J. Phys. Chem. B **105**, 6916–6924 (2001)

3.345 H. Chen, D. S. Sholl: Rapid diffusion of CH_4/H_2 mixtures in single-walled carbon nanotubes, J. Am. Chem. Soc. **126**, 7778–7779 (2004)

3.346 C. Gu, G.-H. Gao, Y. X. Yu, T. Nitta: Simulation for separation of hydrogen and carbon monoxide by adsorption on single-walled carbon nanotubes, Fluid Phase Equil. **194/197**, 297–307 (2002)

3.347 R. Q. Long, R. T. Yang: Carbon nanotubes as superior sorbent for dioxine removal, J. Am. Chem. Soc. **123**, 2058–2059 (2001)

3.348 Y. H. Li, S. Wang, A. Cao, D. Zhao, X. Zhang, C. Xu, Z. Luan, D. Ruan, J. Liang, D. Wu, B. Wei: Adsorption of fluoride from water by amorphous alumina supported on carbon nanotubes, Chem. Phys. Lett. **350**, 412–416 (2001)

3.349 Y. H. Li, S. Wang, J. Wei, X. Zhang, C. Xu, Z. Luan, D. Wu, B. Wei: Lead adsorption on carbon nanotubes, Chem. Phys. Lett. **357**, 263–266 (2002)

3.350 C. Park, E. S. Engel, A. Crowe, T. R. Gilbert, N. M. Rodriguez: Use of carbon nanofibers in the removal of organic solvents from water, Langmuir **16**, 8050–8056 (2000)

3.351 X. Peng, Y. Li, Z. Luan, Z. Di, H. Wang, B. Tian, Z. Jia: Adsorption of 1,2-dichlorobenzene from water to carbon nanotubes, Chem. Phys. Lett. **376**, 154–158 (2003)

3.352 P. Kondratyuk, J. T. Yates: Nanotubes as molecular sponges: the adsorption of CCl_4, Chem. Phys. Lett. **383**, 314–316 (2004)

3.353 A. Huczko, H. Lange, E. Calko, H. Grubek-Jaworska, P. Droszcz: Physiological testing of carbon nanotubes: are they asbestos-like?, Full. Sci. Technol. **9**, 251–254 (2001)

3.354 A. A. Shvedova, V. Castranova, E. R. Kisin, D. Schwegler-Berry, A. R. Murray, V. Z. Gandelsman, A. M. Maynard, P. Baron: Exposure to carbon nanotube material: assessment of nanotube cytotoxicity using human keratinocyte cells, Toxical Env. Health A **66**, 1909–1926 (2003)

3.355 C. W. Lam, J. T. James, R. McCluskey, R. L. Hunter: Pulmonary toxicity of single-wall carbon nanotubes in mice 7 and 90 days after intratracheal instillation, Toxicol. Sci. **77**, 126–134 (2004)

3.356 D. Pantarotto, J. P. Briand, M. Prato, A. Bianco: Translocation of bioactive peptides across cell membranes by carbon nanotubes, Chem. Commun., 16–17 (2004)

3.357 C. Salvador-Morales, E. Flahaut, E. Sim, J. Sloan, M. L. H. Green, R. B. Sim: Complement activation and protein adsorption by carbon nanotubes, Molec. Immun. **43**, 193–201 (2006)

3.358 M. P. Mattson, R. C. Haddon, A. M. Rao: Molecular functionalization of carbon nanotubes and use as substrates for neuronal growth, J. Molec. Neurosci. **14**, 175–182 (2000)

3.359 J. J. Davis, M. L. H. Green, H. A. O. Hill, Y. C. Leung, P. J. Sadler, J. Sloan, A. V. Xavier, S. C. Tsang: The immobilization of proteins in carbon nanotubes, Inorg. Chim. Acta **272**, 261–266 (1998)

3.360 R. J. Chen, Y. Zhang, D. Wang, H. Dai: Noncovalent sidewall functionalization of single-walled carbon nanotubes for protein immobilization, J. Am. Chem. Soc. **123**, 3838–3839 (2001)

3.361 M. Shim, N. W. S. Kam, R. J. Chen, Y. Li, H. Dai: Functionalization of carbon nanotubes for biocompatibility and biomolecular recognition, Nanolett. **2**, 285–288 (2002)

3.362 C. Dwyer, M. Guthold, M. Falvo, S. Washburn, R. Superfine, D. Erie: DNA-functionalized single-walled carbon nanotubes, Nanotechnology **13**, 601–604 (2002)

3.363 H. Huang, S. Taylor, K. Fu, Y. Lin, D. Zhang, T. W. Hanks, A. M. Rao, Y. Sun: Attaching proteins to carbon nanotubes via diimide-activated amidation, Nanolett. **2**, 311–314 (2002)

3.364 C. V. Nguyen, L. Delzeit, A. M. Cassell, J. Li, J. Han, M. Meyyappan: Preparation of nucleic acid functionalized carbon nanotube arrays, Nanolett. **2**, 1079–1081 (2002)

3.365 B. R. Azamian, J. J. Davis, K. S. Coleman, C. B. Bagshaw, M. L. H. Green: Bioelectrochemical single-walled carbon nanotubes, J. Am. Chem. Soc. **124**, 12664–12665 (2002)

3.366 E. Katz, I. Willner: Biomolecule-functionalized carbon nanotubes: Applications in nanobioelectronics, Chem. Phys. Chem. **5**, 1084–1104 (2004)

3.367 J. Wang: Carbon-nanotube based electrochemical biosensors: a review, Electroanalysis **17**, 7–14 (2005)

3.368 T. Laha, A. Agarwal, T. McKechnie, S. Seal: Synthesis and characterization of plasma spray formed carbon nanotube reinforced aluminum composite, Mater. Sci. Eng. A **381**, 249–258 (2004)

3.369 T. Noguchi, A. Magario, S. Fukazawa, S. Shimizu, J. Beppu, M. Seki: Carbon nanotube/aluminium composites with uniform dispersion, Mater. Transact. **45**, 602–604 (2004)

3.370 Y. B. Li, Q. Ya, B. Q. Wei, J. Liang, D. H. Wu: Processing of a carbon nanotubes-$Fe_{82}P_{18}$ metallic glass composite, J. Mater. Sci. Lett. **17**, 607–609 (1998)

3.371 K. T. Kim, K. H. Lee, S. I. Cha, C.-B. Mo, S. H. Hong: Characterization of carbon nanotubes/Cu nanocomposites processed by using nano-sized Cu powders, Mater. Res. Soc. Symp. Proc. **821**, 111–116 (2004)

3.372 F. Zhang, J. Shen, J. Sun: Processing and properties of carbon nanotubes-nano-WC-Co composites, Mater. Sci. Eng. A **381**, 86–91 (2004)

3.373 C. L. Xu, B. Q. Wei, R. Z. Ma, J. Liang, X. K. Ma, D. H. Wu: Fabrication of aluminum-carbon nanotube composites and their electrical properties, Carbon **37**, 855–858 (1999)

3.374 T. Kuzumaki, K. Miyazawa, H. Ichinose, K. Ito: Processing of carbon nanotube reinforced aluminum composite, J. Mater. Res. **13**, 2445–2449 (1998)

3.375 T. Kuzumaki, O. Ujiie, H. Ichinose, K. Ito: Mechanical characteristics and preparation of carbon nanotube fiber-reinforced Ti composite, Adv. Eng. Mater. **2**, 416–418 (2000)

3.376 S.-N. Li, S.-Z. Song, T.-Q. Yu, H.-M. Chen, Y.-S. Zhang, J.-L. Shen: Properties of structure of magnesium matrix composite reinforced with CNTs, J. Wuhan Univ. Technol., Mater. Sci. Ed. **19**, 65–68 (2004)

3.377 E. Carreno-Morelli, J. Yang, E. Couteau, K. Hernadi, J. W. Seo, C. Bonjour, L. Forro, R. Schaller: Carbon nanotube/magnesium composites, Phys. Stat. Solidi A **201**, R53–R55 (2004)

3.378 Z. Bian, R. J. Wang, W. H. Wang, T. Zhang, A. Inoue: Carbon-nanotube-reinforced Zr-based bulk metallic glass composites and their properties, Adv. Funct. Mater. **14**, 55–63 (2004)

3.379 S. R. Dong, J. P. Tu, X. B. Zhang: An investigation of the sliding wear behavior of Cu-matrix composite reinforced by carbon nanotubes, Mater. Sci. Eng. A **313**, 83–87 (2001)

3.380 J. Wang, G. Chen, M. Wang, M. P. Chatrathi: Carbon-nanotube/copper composite electrodes for capillary electrophoresis microchip detection of carbohydrates, Analyst (Cambridge) **129**, 512–515 (2004)

3.381 Q. Ngo, B. A. Cruden, A. M. Cassell, M. D. Walker, Q. Ye, J. E. Koehne, M. Meyyappan, J. Li, C. Y. Yang: Thermal conductivity of carbon nanotube composite films, Mater. Res. Soc. Symp. Proc. **812**, 179–184 (2004)

3.382 X. H. Chen, C. S. Chen, H. N. Xiao, F. Q. Cheng, G. Zhang, G. J. Yi: Corrosion behavior of carbon nanotubes-Ni composite coating, Surf. Coat. Technol. **191**, 351–356 (2005)

3.383 J.-P. Tu, T.-Z. Zou, L.-Y. Wang, W.-X. Chen, Z.-D. Xu, F. Liu, X.-B. Zhang: Friction and wear behavior of Ni-based carbon nanotube composite coatings, Zhejiang Daxue Xuebao, Gongxueban **38**, 931–934 (2004)

3.384 Z. Yang, H. Xu, M.-K. Li, Y.-L. Shi, Y. Huang, H.-L. Li: Preparation and properties of Ni-P/single-walled carbon nanotubes composite coatings by means of electroless plating, Thin Solid Films **466**, 86–91 (2004)

3.385 A. Peigney: Tougher ceramics with carbon nanotubes, Nature Mater. **2**, 15–16 (2003)

3.386 G. D. Zhan, J. D. Kuntz, J. Wan, A. K. Mukherjee: Single-wall carbon nanotubes as attractive toughening agents in alumina-based composites, Nature Mater. **2**, 38–42 (2003)

3.387 J. Sun, L. Gao, W. Li: Colloidal processing of carbon nanotube/alumina composites, Chem. Mater. **14**, 5169–5172 (2002)

3.388 J. Ning, J. Zhang, Y. Pan, J. Guo: Surfactants assisted processing of carbon nanotube-reinforced SiO_2 matrix composites, Ceram. Int. **30**, 63–67 (2004)

3.389 V. G. Gavalas, R. Andrews, D. Bhattacharyya, L. G. Bachas: Carbon nanotube sol-gel composite materials, Nanolett. **1**, 719–721 (2001)

3.390 S. Rul, Ch. Laurent, A. Peigney, A. Rousset: Carbon nanotubes prepared in-situ in a cellular ceramic by the gel casting-foam method, J. Eur. Ceram. Soc. **23**, 1233–1241 (2003)

3.391 R. Z. Ma, J. Wu, B. Q. Wei, J. Liang, D. H. Wu: Processing and properties of carbon nanotube/nano-SiC ceramic, J. Mater. Sci. **33**, 5243–5246 (1998)

3.392 R. W. Siegel, S. K. Chang, B. J. Ash, J. Stone, P. M. Ajayan, R. W. Doremus, L. S. Schadler: Mechanical behavior of polymer and ceramic matrix nanocomposites, Scr. Mater. **44**, 2061–2064 (2001)

3.393 X. Wang, N. P. Padture, H. Tanaka: Contact-damage-resistant ceramic/single-wall carbon nanotubes and ceramic/graphite composites, Nature Mater. **3**, 539–544 (2004)

3.394 C. Laurent, A. Peigney, O. Dumortier, A. Rousset: Carbon nanotubes-Fe-alumina nanocomposites.

Part II: Microstructure and mechanical properties of the hot-pressed composites, J. Eur. Ceram. Soc. **18**, 2005–2013 (1998)

3.395   A. Peigney, C. Laurent, A. Rousset: Synthesis and characterization of alumina matrix nanocomposites containing carbon nanotubes, Key Eng. Mater. **132–136**, 743–746 (1997)

3.396   A. Peigney, C. Laurent, E. Flahaut, A. Rousset: Carbon nanotubes in novel ceramic matrix nanocomposites, Ceram. Intern. **26**, 677–683 (2000)

3.397   S. Rul: Synthèse de composites nanotubes de carbone-métal-oxyde. Ph.D. Thesis (Université Toulouse III, Toulouse 2002)

3.398   E. Flahaut, A. Peigney, C. Laurent, C. Marlière, F. Chastel, A. Rousset: Carbon nanotube-metal-oxide nanocomposites: Microstructure, electrical conductivity and mechanical properties, Acta Mater. **48**, 3803–3812 (2000)

3.399   R. R. Bacsa, C. Laurent, A. Peigney, W. S. Bacsa, T. Vaugien, A. Rousset: High specific surface area carbon nanotubes from catalytic chemical vapor deposition process, Chem. Phys. Lett. **323**, 566–571 (2000)

3.400   P. Coquay, A. Peigney, E. De Grave, R. E. Vandenberghe, C. Laurent: Carbon nanotubes by a CVD method. Part II: Formation of nanotubes from (Mg,Fe)O catalysts, J. Phys. Chem. B **106**, 13199–13210 (2002)

3.401   E. Flahaut, Ch. Laurent, A. Peigney: Double-walled carbon nanotubes in composite powders, J. Nanosci. Nanotech. **3**, 151–158 (2003)

3.402   A. Peigney, P. Coquay, E. Flahaut, R. E. Vandenberghe, E. De Grave, C. Laurent: A study of the formation of single- and double-walled carbon nanotubes by a CVD method, J Phys. Chem. B **105**, 9699–9710 (2001)

3.403   A. Peigney, S. Rul, F. Lefevre-Schlick, C. Laurent: Densification during hot-pressing of carbon nanotube metal-ceramic composites, J.Europ. Ceram. Soc., submitted (2006)

3.404   Z. Xia, L. Riester, W. A. Curtin, H. Li, B. W. Sheldon, J. Liang, B. Chang, J. M. Xu: Direct observation of toughening mechanisms in carbon nanotube ceramic matrix composites, Acta Mater. **52**, 931–944 (2004)

3.405   S. Rul, F. Lefevre-Schlick, E. Capria, C. Laurent, A. Peigney: Percolation of single-walled carbon nanotubes in ceramic matrix nanocomposites, Acta Materialia **52**, 1061–1067 (2004)

3.406   A. Peigney, E. Flahaut, C. Laurent, F. Chastel, A. Rousset: Aligned carbon nanotubes in ceramic-matrix nanocomposites prepared by high-temperature extrusion, Chem. Phys. Lett. **352**, 20–25 (2002)

3.407   G.-D. Zhan, J. D. Kuntz, J. E. Garay, A. K. Mukherjee: Electrical properties of nanoceramics reinforced with ropes of single-walled carbon nanotubes, Appl. Phys. Lett. **83**, 1228–1230 (2003)

3.408   Q. Huang, L. Gao: Manufacture and electrical properties of multiwalled carbon nanotube/BaTiO$_3$ nanocomposite ceramics, J. Mater Chem. **14**, 2536–2541 (2004)

3.409   S. L. Huang, M. R. Koblischka, K. Fossheim, T. W. Ebbesen, T. H. Johansen: Microstructure and flux distribution in both pure and carbon-nanotube-embedded Bi$_2$Sr$_2$CaCu$_2$O$_{8+\delta}$ superconductors, Physica C **311**, 172–186 (1999)

3.410   T. Seeger, G. de la Fuente, W. K. Maser, A. M. Benito, M. A. Callejas, M. T. Martinez: Evolution of multi-walled carbon-nanotube/SiO$_2$ composites via laser treatment, Nanotechnology **14**, 184–187 (2003)

3.411   G.-D. Zhan, J. D. Kuntz, H. Wang, C.-M. Wang, A. K. Mukherjee: Anisotropic thermal properties of single-wall-carbon-nanotube-reinforced nanoceramics, Phil. Mag. Lett. **84**, 419–423 (2004)

3.412   A. Weidenkaff, S. G. Ebbinghaus, T. Lippert: Ln$_{1-x}$A$_x$CoO$_3$ (Ln = Er, La; A = Ca, Sr)/carbon nanotube composite materials applied for rechargeable Zn/Air batteries, Chem. Mater. **14**, 1797–1805 (2002)

3.413   D. S. Lim, J. W. An, H. J. Lee: Effect of carbon nanotube addition on the tribological behavior of carbon/carbon composites, Wear **252**, 512–517 (2002)

3.414   R. Andrews, D. Jacques, A. M. Rao, T. Rantell, F. Derbyshire, Y. Chen, J. Chen, R. C. Haddon: Nanotube composite carbon fibers, Appl. Phys. Lett. **75**, 1329–1331 (1999)

3.415   P. M. Ajayan, O. Stephan, C. Colliex, D. Trauth: Aligned carbon nanotube arrays formed by cutting a polymer resin-nanotube composite, Science **265**, 1212–14 (1994)

3.416   R. Haggenmueller, H. H. Gommans, A. G. Rinzler, J. E. Fischer, K. I. Winey: Aligned single-wall carbon nanotubes in composites by melt processing methods, Chem. Phys. Lett. **330**, 219–225 (2000)

3.417   L. S. Schadler, S. C. Giannaris, P. M. Ajayan: Load transfer in carbon nanotube epoxy composites, Appl. Phys. Lett. **73**, 3842–3844 (1998)

3.418   S. J. V. Frankland, A. Caglar, D. W. Brenner, M. Griebel: Molecular simulation of the influence of chemical cross-links on the shear strength of carbon nanotube-polymer interfaces, J. Phys. Chem. B **106**, 3046–3048 (2002)

3.419   H. D. Wagner: Nanotube-polymer adhesion: A mechanics approach, Chem. Phys. Lett. **361**, 57–61 (2002)

3.420   P. M. Ajayan, L. S. Schadler, C. Giannaris, A. Rubio: Single-walled carbon nanotube-polymer composites: Strength and weakness, Adv. Mater. **12**, 750–753 (2000)

3.421   X. Gong, J. Liu, S. Baskaran, R. D. Voise, J. S. Young: Surfactant-assisted processing of carbon nanotube/polymer composites, Chem. Mater. **12**, 1049–1052 (2000)

3.422   E. T. Thostenson, W. Z. Li, D. Z. Wang, Z. F. Ren, T. W. Chou: Carbon nanotube/carbon fiber hybrid multiscale composites, J. Appl. Phys. **91**, 6034–6037 (2002)

3.423 S. Barrau, P. Demont, A. Peigney, C. Laurent, C. Lacabanne: Effect of palmitic acid on the electrical conductivity of carbon nanotubes-polyepoxy composite, Macromolecules **36**, 9678–9680 (2003)

3.424 S. Barrau, P. Demont, A. Peigney, C. Laurent, C. Lacabanne: DC and AC conductivity of carbon nanotubes-polyepoxy composites, Macromolecules **36**, 5187–5194 (2003)

3.425 J. Sandler, M. S. P. Shaffer, T. Prasse, W. Bauhofer, K. Schulte, A. H. Windle: Development of a dispersion process for carbon nanotubes in an epoxy matrix and the resulting electrical properties, Polymer **40**, 5967–5971 (1999)

3.426 M. J. Biercuk, M. C. Llaguno, M. Radosavljevic, J. K. Hyun, A. T. Johnson, J. E. Fischer: Carbon nanotube composites for thermal management, Appl. Phys. Lett. **80**, 2767–2769 (2002)

3.427 F. H. Gojny, K. Schulte: Functionalisation effect on the thermo-mechanical behaviour of multi-wall carbon nanotube/epoxy-composites, Compos. Sci. Technol. **64**, 2303–2308 (2004)

3.428 F. H. Gojny, M. H. G. Wichmann, U. Kopke, B. Fiedler, K. Schulte: Carbon nanotube-reinforced epoxy-composites: enhanced stiffness and fracture toughness at low nanotube content, Compos. Sci. Technol. **64**, 2363–2371 (2004)

3.429 G. Pecastaings, P. Delhaes, A. Derre, H. Saadaoui, F. Carmona, S. Cui: Role of interfacial effects in carbon nanotube/epoxy nanocomposite behavior, J. Nanosci. Nanotech. **4**, 838–843 (2004)

3.430 Z. Jin, K. P. Pramoda, G. Xu, S. H. Goh: Dynamic mechanical behavior of melt-processed multi-walled carbon nanotube/poly(methyl methacrylate) composites, Chem. Phys. Lett. **337**, 43–47 (2001)

3.431 Z. Jin, K. P. Pramoda, S. H. Goh, G. Xu: Poly(vinylidene fluoride)-assisted melt-blending of multi-walled carbon nanotube/poly(methyl methacrylate) composites, Mater. Res. Bull. **37**, 271–278 (2002)

3.432 C. A. Cooper, D. Ravich, D. Lips, J. Mayer, H. D. Wagner: Distribution and alignment of carbon nanotubes and nanofibrils in a polymer matrix, Compos. Sci. Technol. **62**, 1105–1112 (2002)

3.433 J. M. Benoit, B. Corraze, S. Lefrant, W. J. Blau, P. Bernier, O. Chauvet: Transport properties of PMMA-carbon nanotubes composites, Synth. Met. **121**, 1215–1216 (2001)

3.434 J. M. Benoit, B. Corraze, O. Chauvet: Localization, Coulomb interactions, and electrical heating in single-wall carbon nanotubes/polymer composites, Phys. Rev. B **65**, 241405/1–241405/4 (2002)

3.435 F. Du, R. C. Scogna, W. Zhou, S. Brand, J. E. Fischer, K. I. Winey: Nanotube networks in polymer nanocomposites: rheology and electrical conductivity, Macromolecules **37**, 9048–9055 (2004)

3.436 T. Kashiwagi, F. Du, K. I. Winey, K. M. Groth, J. R. Shields, R. H. Harris Jr., J. F. Douglas: Flammability properties of PMMA–single walled carbon

nanotube nanocomposites, Polymer. Mater. Sci. Eng. **91**, 90–91 (2004)

3.437 M. S. P. Shaffer, A. H. Windle: Fabrication and characterization of carbon nanotube/poly(vinyl alcohol) composites, Adv. Mater. **11**, 937–941 (1999)

3.438 L. Jin, C. Bower, O. Zhou: Alignment of carbon nanotubes in a polymer matrix by mechanical stretching, Appl. Phys. Lett. **73**, 1197–1199 (1998)

3.439 H. D. Wagner, O. Lourie, Y. Feldman, R. Tenne: Stress-induced fragmentation of multiwall carbon nanotubes in a polymer matrix, Appl. Phys. Lett. **72**, 188–190 (1998)

3.440 H. D. Wagner, O. Lourie, X. F. Zhou: Macrofragmentation and microfragmentation phenomena in composite materials, Compos. Part A **30A**, 59–66 (1998)

3.441 J. R. Wood, Q. Zhao, H. D. Wagner: Orientation of carbon nanotubes in polymers and its detection by Raman spectroscopy, Compos. Part A **32A**, 391–399 (2001)

3.442 Q. Zhao, J. R. Wood, H. D. Wagner: Using carbon nanotubes to detect polymer transitions, J. Polym. Sci. Part B **39**, 1492–1495 (2001)

3.443 M. Cochet, W. K. Maser, A. M. Benito, M. A. Callejas, M. T. Martinesz, J. M. Benoit, J. Schreiber, O. Chauvet: Synthesis of a new polyaniline/nanotube composite: In-situ polymerisation and charge transfer through site-selective interaction, Chem. Commun., 1450–1451 (2001)

3.444 D. Qian, E. C. Dickey, R. Andrews, T. Rantell: Load transfer and deformation mechanisms in carbon nanotube-polystyrene composites, Appl. Phys. Lett. **76**, 2868–2870 (2000)

3.445 V. Datsyuk, Christelle Guerret-Piecourt, S. Dagreou, L. Billon, J.-C. Dupin, E. Flahaut, A. Peigney, C. Laurent: Double walled carbon nanotube/polymer composites via in-situ nitroxide mediated polymerisation of amphiphilic block copolymers, Carbon **43**, 873–876 (2005)

3.446 R. Blake, Y. K. Gun'ko, J. Coleman, M. Cadek, A. Fonseca, J. B. Nagy, W. J. Blau: A generic organometallic approach toward ultra-strong carbon nanotube polymer composites, J. Am. Chem. Soc. **126**, 10226–10227 (2004)

3.447 T. Kashiwagi, E. Grulke, J. Hilding, K. Groth, R. Harris, K. Butler, J. Shields, S. Kharchenko, J. Douglas: Thermal and flammability properties of polypropylene/carbon nanotube nanocomposites, Polymer **45**, 4227–4239 (2004)

3.448 C. Wei, D. Srivastava, K. Cho: Thermal expansion and diffusion coefficients of carbon nanotube-polymer composites, Los Alamos Nat. Lab., Preprint Archive, Condensed Matter (archiv:cond-mat/0203349), 1–11 (2002)

3.449 J. C. Grunlan, M. V. Bannon, A. R. Mehrabi: Latexbased, single-walled nanotube composites: processing and electrical conductivity, Polym. Prepr. **45**, 154–155 (2004)

3.450 J. C. Grunlan, A. R. Mehrabi, M. V. Bannon, J. L. Bahr: Water-based single-walled-nanotube-filled polymer composite with an exceptionally low percolation threshold, Adv. Mater. (Weinheim) **16**, 150–153 (2004)

3.451 C. Pirlot, I. Willems, A. Fonseca, J. B. Nagy, J. Delhalle: Preparation and characterization of carbon nanotube/polyacrylonitrile composites, Adv. Eng. Mater. **4**, 109–114 (2002)

3.452 H. Lam, H. Ye, Y. Gogotsi, F. Ko: Structure and properties of electrospun single-walled carbon nanotubes reinforced nanocomposite fibrils by co-electrospinning, Polym. Prepr. **45**, 124–125 (2004)

3.453 L. Cao, H. Chen, M. Wang, J. Sun, X. Zhang, F. Kong: Photoconductivity study of modified carbon nanotube/oxotitanium phthalocyanine composites, J. Phys. Chem. B **106**, 8971–8975 (2002)

3.454 I. Musa, M. Baxendale, G. A. J. Amaratunga, W. Eccleston: Properties of regular poly(3-octylthiophene)/multi-wall carbon nanotube composites, Synth. Met. **102**, 1250 (1999)

3.455 E. Kymakis, I. Alexandou, G. A. J. Amaratunga: Single-walled carbon nanotube-polymer composites: Electrical, optical and structural investigation, Synth. Met. **127**, 59–62 (2002)

3.456 K. Yoshino, H. Kajii, H. Araki, T. Sonoda, H. Take, S. Lee: Electrical and optical properties of conducting polymer-fullerene and conducting polymer-carbon nanotube composites, Fuller. Sci. Technol. **7**, 695–711 (1999)

3.457 B. Maruyama, K. Alam: Carbon nanotubes and nanofibers in composite materials, SAMPE J. **38**, 59–70 (2002)

3.458 S. A. Curran, P. M. Ajayan, W. J. Blau, D. L. Carroll, J. N. Coleman, A. B. Dalton, A. P. Davey, A. Drury, B. McCarthy, S. Maier, A. Strevens: A composite from poly(m-phenylenevinylene-co-2,5-dioctoxy-p-phenylenevinylene) and carbon nanotubes. A novel material for molecular optoelectronics, Adv. Mater. **10**, 1091–1093 (1998)

3.459 P. Fournet, D. F. O'Brien, J. N. Coleman, H. H. Horhold, W. J. Blau: A carbon nanotube composite as an electron transport layer for M3EH-PPV based light-emitting diodes, Synth. Met. **121**, 1683–1684 (2001)

3.460 H. S. Woo, R. Czerw, S. Webster, D. L. Carroll, J. Ballato, A. E. Strevens, D. O'Brien, W. J. Blau: Hole blocking in carbon nanotube-polymer composite organic light-emitting diodes based on poly (m-phenylene vinylene-co-2,5-dioctoxy-p-phenylene vinylene), Appl. Phys. Lett. **77**, 1393–1395 (2000)

3.461 H. S. Woo, R. Czerw, S. Webster, D. L. Carroll, J. W. Park, J. H. Lee: Organic light emitting diodes fabricated with single wall carbon nanotubes dispersed in a hole conducting buffer: The role of carbon nanotubes in a hole conducting polymer, Synth. Met. **116**, 369–372 (2001)

3.462 H. Ago, K. Petritsch, M. S. P. Shaffer, A. H. Windle, R. H. Friend: Composites of carbon nanotubes and conjugated polymers for photovoltaic devices, Adv. Mater. **11**, 1281–1285 (1999)

3.463 B. Vigolo, A. Pénicaud, C. Coulon, C. Sauder, R. Pailler, C. Journet, P. Bernier, P. Poulin: Macroscopic fibers and ribbons of oriented carbon nanotubes, Science **290**, 1331–1334 (2000)

3.464 B. Vigolo, P. Poulin, M. Lucas, P. Launois, P. Bernier: Improved structure and properties of single-wall carbon nanotube spun fibers, Appl. Phys. Lett. **11**, 1210–1212 (2002)

3.465 P. Poulin, B. Vigolo, P. Launois: Films and fibers of oriented single wall nanotubes, Carbon **40**, 1741–1749 (2002)

3.466 K. Jiang, Q. Li, S. Fan: Spinning continuous carbon nanotube yarn, Nature **419**, 801 (2002)

3.467 M. Zhang, K. R. Atkinson, R. H. Baughman: Multifunctional carbon nanotube yarns by downsizing an ancient technology, Science **306**, 1356–1361 (2004)

3.468 J. Steinmetz, M. Glerup, M. Paillet, P. Bernier, M. Holzinger: Production of pure nanotube fibers using a modified wet-spinning method, Carbon **43**, 2397–2400 (2005)

3.469 P. Lambin, A. Fonseca, J. P. Vigneron, J. B'Nagy, A. A. Lucas: Structural and electronic properties of bent carbon nanotubes, Chem. Phys. Lett. **245**, 85–89 (1995)

3.470 L. Chico, V. H. Crespi, L. X. Benedict, S. G. Louie, M. L. Cohen: Pure carbon nanoscale devices: Nanotube heterojunctions, Phys. Rev. Lett. **76**, 971–974 (1996)

3.471 Z. Yao, H. W. C. Postma, L. Balents, C. Dekker: Carbon nanotube intramolecular junctions, Nature **402**, 273–276 (1999)

3.472 S. J. Tans, A. R. M. Verschueren, C. Dekker: Room temperature transistor based on single carbon nanotube, Nature **393**, 49–52 (1998)

3.473 R. Martel, T. Schmidt, H. R. Shea, T. Hertel, P. Avouris: Single and multi-wall carbon nanotube field effect transistors, Appl. Phys. Lett. **73**, 2447–2449 (1998)

3.474 V. Derycke, R. Martel, J. Appenzeller, P. Avouris: Carbon nanotube inter- and intramolecular logic gates, Nanolett. **1**, 453–456 (2001)

3.475 P. G. Collins, M. S. Arnold, P. Avouris: Engineering carbon nanotubes using electrical breakdown, Science **292**, 706–709 (2001)

3.476 A. P. Graham, G. S. Duesberg, W. Hoenlein, F. Kreupl, M. Liebau, R. Martin, B. Rajasekharan, W. Pamler, R. Seidel, W. Steinhoegl, E. Unger: How do carbon nanotubes fit into the semiconductor roadmap?, Appl. Phys. A **80**, 1141–1151 (2005)

3.477 P. Kim, C. M. Lieber: Nanotube nanotweezers, Science **286**, 2148–2150 (1999)

3.478 R. H. Baughman, C. Changxing, A. A. Zakhidov, Z. Iqbal, J. N. Barisci, G. M. Spinks, G. G. Wallace,

A. Mazzoldi, D. de Rossi, A. G. Rinzler, O. Jaschinki S. Roth, M. Kertesz: Carbon nanotubes actuators, Science **284**, 1340–1344 (1999)

3.479   Y. Gao, Y. Bando: Carbon nanothermometer containing gallium, Nature **415**, 599 (2002)

3.480   C. Niu, E. K. Sichel, R. Hoch, D. Moy, H. Tennent: High power electro-chemical capacitors based on carbon nanotube electrodes, Appl. Phys. Lett. **70**, 1480–1482 (1997)

3.481   E. Frackowiak, F. Béguin: Electrochemical storage of energy in carbon nanotubes and nanostructured carbons, Carbon **40**, 1775–1787 (2002)

3.482   C. Portet, P. L. Taberna, P. Simon, E. Flahaut: Influence of carbon nanotubes addition on carbon–carbon supercapacitor performances in organic electrolyte, J. Power. Sources **139**, 371–378 (2005)

3.483   E. Frackowiak, K. Jurewicz, K. Szostak, S. Delpeux, F. Béguin: Nanotubular materials as electrodes for supercapacitors, Fuel Process. Technol. **77**, 213–219 (2002)

3.484   J. Mittal, E. Frackowiak, M. Monthioux, G. Lota: High performance supercapacitor from hybrid-nanotube-based electrodes, Nanotechnol. (2006) submitted

4. Nanowires

This chapter provides an overview of recent research on inorganic nanowires, particularly metallic and semiconducting nanowires. Nanowires are one-dimensional, anisotropic structures, small in diameter, and large in surface-to-volume ratio. Thus, their physical properties are different than those of structures of different scale and dimensionality. While the study of nanowires is particularly challenging, scientists have made immense progress in both developing synthetic methodologies for the fabrication of nanowires, and developing instrumentation for their characterization. The chapter is divided into three main sections: Sect. 4.1 the synthesis, Sect. 4.2 the characterization and physical properties, and Sect. 4.3 the applications of nanowires. Yet, the reader will discover many links that make these aspects of nano-science intimately inter-depent.

Part A | 4

Nanowires are attracting much interest from those seeking to apply nanotechnology and (especially) those investigating nanoscience. Nanowires, unlike other low-dimensional systems, have two quantum-confined directions but one unconfined direction available for electrical conduction. This allows nanowires to be used in applications where electrical conduction, rather than tunneling transport, is required. Because of their unique density of electronic states, in the limit of small diameters nanowires are expected to exhibit significantly different optical, electrical and magnetic properties to their bulk 3-D crystalline counterparts. Increased surface area, very high density of electronic states and joint density of states near the energies of their van Hove singularities, enhanced exciton binding energy, diameter-dependent bandgap, and increased surface scattering for electrons and phonons are just some of the ways in which nanowires differ from their corresponding bulk materials. Yet the sizes of nanowires are typically large enough (> 1 nm in the quantum-confined direction) to result in local crystal structures that are closely related to their parent materials, allowing theoretical predictions about their properties to be made based on knowledge of their bulk properties.

Not only do nanowires exhibit many properties that are similar to, and others that are distinctly different from, those of their bulk counterparts, nanowires also have the advantage from an applications standpoint in that some of the materials parameters critical for certain properties can be independently controlled in nanowires but not in their bulk counterparts. Certain properties can also be enhanced nonlinearly in small-diameter nanowires, by exploiting the singular aspects of the 1-D electronic density of states.

Furthermore, nanowires have been shown to provide a promising framework for applying the "bottom-up" approach [4.1] to the design of nanostructures for nanoscience investigations and for potential nanotechnology applications.

Table 4.1 Selected syntheses of nanowires by material

Material	Growth Technique	Reference	Material	Growth Technique	Reference
ABO_4-type	template$^\alpha$	[4.2]	Ge	high-T, high-P liquid-phase, redox	[4.33]
Ag	DNA-template, redox	[4.3]		VLS$^\delta$	[4.34]
	template, pulsed ECD$^\beta$	[4.4]		oxide-assisted	[4.35]
Au	template, ECD$^\beta$	[4.5, 6]	InAs	VLS$^\delta$	[4.36]
Bi	stress-induced	[4.7]	MgO	VLS$^\delta$	[4.37]
	template, vapor-phase	[4.8]	Mo	step decoration, ECD$^\beta$+ redox	[4.38]
	template, ECD$^\beta$	[4.9–11]	Ni	template, ECD$^\beta$	[4.11, 39, 40]
	template, pressure-injection	[4.12–14]	Pb	liquid-phase$^\mu$	[4.41]
BiSb	pulsed ECD$^\beta$	[4.15]	PbSe	liquid phase	[4.42]
Bi_2Te_3	template, dc ECD$^\beta$	[4.16]		self assembly of nanocrystals$^\nu$	[4.43]
CdS	liquid-phase (surfactant), recrystallization	[4.17]	Pd	step decoration, ECD$^\beta$	[4.44]
	template, ac ECD$^\beta$	[4.18, 19]	Se	liquid-phase, recrystallization	[4.45]
CdSe	liquid-phase (surfactant), redox	[4.20]		template, pressure injection	[4.46]
	template, ac ECD$^\beta$	[4.21, 22]	Si	VLS$^\delta$	[4.47]
Cu	vapor deposition	[4.23]		laser-ablation VLS$^\delta$	[4.48]
	template, ECD$^\beta$	[4.24]		oxide-assisted	[4.49]
Fe	template, ECD$^\gamma$	[4.25, 26]		low-T VLS$^\delta$	[4.50]
	shadow deposition	[4.27]	W	vapor transport	[4.51]
GaN	template, CVD$^\gamma$	[4.28, 29]	Zn	template, vapor-phase	[4.52]
	VLS$^\delta$	[4.30, 31]		template, ECD$^\beta$	[4.53]
GaAs	template, liquid/vapor OMCVD$^\epsilon$	[4.32]	ZnO	VLS$^\delta$	[4.54]
				template, ECD$^\beta$	[4.53, 55]

$^\alpha$ Template synthesis
$^\beta$ Electrochemical deposition (ECD)
$^\gamma$ Chemical vapor deposition (CVD)
$^\delta$ Vapor-liquid-solid (VLS) growth
$^\epsilon$ Organometallic chemical vapor deposition (OMCVD)
$^\mu$ Liquid phase synthesis
$^\nu$ Self assembly of nanocrystals (in liquid phase)

Driven by (1) these new research and development opportunities, (2) the smaller and smaller length scales now being used in the semiconductor, optoelectronics and magnetics industries, and (3) the dramatic development of the biotechnology industry where the action is also at the nanoscale, the nanowire research field has developed with exceptional speed in the last few years. Therefore, a review of the current status of nanowire research is of significant broad interest at the present time. It is the aim of this review to focus on nanowire properties that differ from those of their parent crys-

talline bulk materials, with an eye toward possible applications that might emerge from the unique properties of nanowires and from future discoveries in this field.

For quick reference, examples of typical nanowires that have been synthesized and studied are listed in Table 4.1. Also of use to the reader are review articles that focus on a comparison between nanowire and nanotube properties [4.56] and the many reviews that have been written about carbon nanotubes [4.57–59], which can be considered as a model one-dimensional system.

4.1 Synthesis

In this section we survey the most common synthetic approaches that have successfully afforded high-quality nanowires of a large variety of materials (see Table 4.1). In Sect. 4.1.1, we discuss methods which make use of various templates with nanochannels to confine the nanowire growth in two dimensions. In Sect. 4.1.2, we present the synthesis of nanowires by the vapor-liquid-solid mechanism and its many variations. In Sect. 4.1.3, examples of other synthetic methods of general applicability are presented. The last part of this section (Sect. 4.1.4) features several approaches that have been developed to organize nanowires into simple architectures.

4.1.1 Template-Assisted Synthesis

The template-assisted synthesis of nanowires is a conceptually simple and intuitive way to fabricate nanostructures [4.62–64]. These templates contain very small cylindrical pores or voids within the host material, and the empty spaces are filled with the chosen material, which adopts the pore morphology, to form nanowires. In this section, we describe the templates first, and then describe strategies for filling the templates to make nanowires.

Template Synthesis

In template-assisted synthesis of nanostructures, the chemical stability and mechanical properties of the template, as well as the diameter, uniformity and density of the pores are important characteristics to consider. Templates frequently used for nanowire synthesis include anodic alumina (Al_2O_3), nanochannel glass, ion track-etched polymers and mica films.

Porous anodic alumina templates are produced by anodizing pure Al films in selected acids [4.65–67]. Under carefully chosen anodization conditions, the resulting oxide film possesses a regular hexagonal array of parallel and nearly cylindrical channels, as shown in Fig. 4.1a. The self-organization of the pore structure in an anodic alumina template involves two coupled processes: pore formation with uniform diameters and pore ordering. The pores form with uniform diameters because of a delicate balance between electric field-enhanced diffusion which determines the growth rate of the alumina, and dissolution of the alumina into the acidic electrolyte [4.68]. The pores are believed to self-order because of mechanical stress at the aluminium-alumina interface due to expansion during the anodization. This stress produces a repulsive force between the pores, causing them to arrange in a hexagonal lattice [4.69]. Depending on the anodization conditions, the pore diameter can be systematically varied from $\leq 10\,\mathrm{nm}$ up to 200 nm with a pore density in the range of 10^9–10^{11} pores/cm^2 [4.13, 25, 65, 66]. It has been shown by many groups that the pore size distribution and the pore ordering of the anodic alumina templates can be significantly improved by a two-step anodization technique [4.60, 70, 71], where the aluminium oxide layer is dissolved after the first anodization in an acidic solution followed by a second anodization under the same conditions.

Another type of porous template commonly used for nanowire synthesis is the template type fabricated by chemically etching particle tracks originating from ion bombardment [4.72], such as track-etched polycarbonate membranes (Fig. 4.1b) [4.73, 74], and also mica films [4.39].

Other porous materials can be used as host templates for nanowire growth, as discussed by *Ozin* [4.62]. Nanochannel glass (NCG), for example, contains a regular hexagonal array of capillaries similar to the pore structure in anodic alumina with a packing density as high as 3×10^{10} pores/cm^2 [4.63]. Porous Vycor glass that contains an interconnected network of pores less than 10 nm was also employed for the early

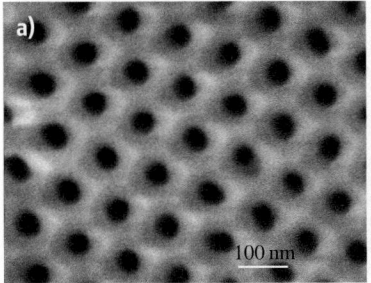
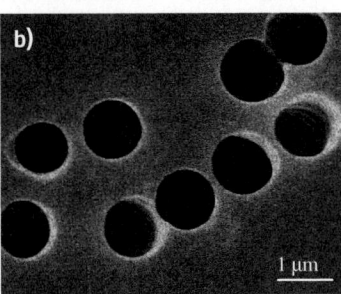

Fig. 4.1 (a) SEM images of the top surfaces of porous anodic alumina templates anodized with an average pore diameter of 44 nm [4.60]. **(b)** SEM image of the particle track-etched polycarbonate membrane, with a pore diameter of 1 μm [4.61]

Part A | 4.1

study of nanostructures [4.75]. Mesoporous molecular sieves [4.76], termed MCM-41, possess hexagonally-packed pores with very small channel diameters which can be varied between 2 nm and 10 nm. Conducting organic filaments have been fabricated in the nanochannels of MCM-41 [4.77]. Recently, the DNA molecule has also been used as a template for growing nanometer-sized wires [4.3].

Diblock copolymers, polymers that consist of two chain segments different properties, have also been utilized as templates for nanowire growth. When two components are immiscible in each other, phase segregation occurs, and depending on their volume ratio, spheres, cylinders and lamellae may self-assemble. To form self-assembled arrays of nanopores, copolymers composed of polystyrene and polymethylmethacrylate [P(S − b − MMA)] [4.79] were used. By applying an electric field while the copolymer was heated above the glass transition temperature of the two constituent polymers, the self-assembled cylinders of PMMA could be aligned with their main axis perpendicular to the film. Selective removal of the PMMA component afforded the preparation of 14 nm-diameter ordered pore arrays with a packing density of 1.9×10^{11} cm^{-3}.

Nanowire Template–Assisted Growth by Pressure Injection

The pressure injection technique is often employed for fabricating highly crystalline nanowires from a low-melting point material and when using porous templates with robust mechanical strength. In the high-pressure injection method, the nanowires are formed by pressure-injecting the desired material in liquid form into the evacuated pores of the template. Due to the heating and pressurization processes, the templates used for the pressure injection method must be chemically stable and be able to maintain their structural integrity at high temperatures and at high pressures. Anodic aluminium oxide films and nanochannel glass are two typical materials used as templates in conjunction with the pressure injection filling technique. Metal nanowires (Bi, In, Sn, and Al) and semiconductor nanowires (Se, Te, GaSb, and Bi_2Te_3) have been fabricated in anodic aluminium oxide templates using this method [4.12, 46, 78].

The pressure P required to overcome the surface tension for the liquid material to fill the pores with a diameter d_W is determined by the Washburn equation [4.80]:

$$d_W = -4\gamma \cos\theta / P \qquad (4.1)$$

where γ is the surface tension of the liquid, and θ is the contact angle between the liquid and the template. To reduce the required pressure and to maximize the filling factor, some surfactants are used to decrease the surface tension and the contact angle. For example, the introduction of Cu into the Bi melt can facilitate filling the pores in the anodic alumina template with liquid Bi and can increase the number of nanowires that are formed [4.13]. However, some of the surfactants might cause contamination problems and should therefore be avoided. Nanowires produced by the pressure injection technique usually possess high crystallinity and a preferred crystal orientation along the wire axis. For example, Fig. 4.2 shows the X-ray diffraction (XRD) patterns of Bi nanowire arrays of three different wire diameters with an injection pressure of ≈ 5000 psi [4.78], showing that the major (> 80%) crystal orientation of the wire axes in the 95 nm and 40 nm diameter Bi nanowire arrays are, respectively, normal to the (202) and (012) lattice planes, which are

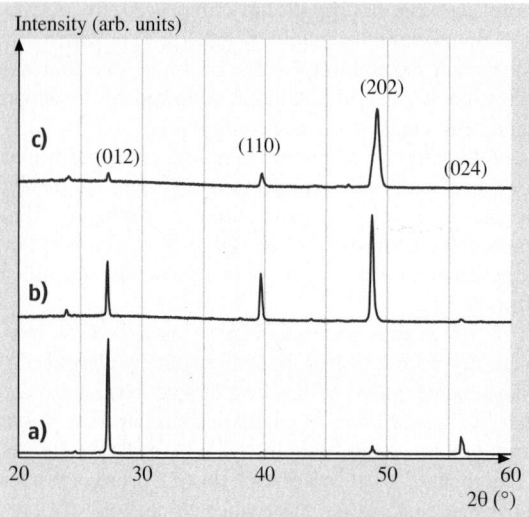

Fig. 4.2a–c XRD patterns of bismuth/anodic alumina nanocomposites with average bismuth wire diameters of (**a**) 40 nm, (**b**) 52 nm, and (**c**) 95 nm [4.78]. The Miller indices corresponding to the lattice planes of bulk Bi are indicated above the individual peaks. The majority of the Bi nanowires are oriented along the $[10\bar{1}1]$ and $[01\bar{1}2]$ directions for $d_W \geq 60$ nm and $d_W \leq 50$ nm, respectively [4.13, 78]. The existence of more than one dominant orientation in the 52 nm Bi nanowires is attributed to the transitional behavior of *intermediate*-diameter nanowires as the preferential growth orientation is shifted from $[10\bar{1}1]$ to $[01\bar{1}2]$ with decreasing d_W

denoted by [10$\bar{1}$1] and [01$\bar{1}$2] when using a hexagonal unit cell, suggesting a wire diameter-dependent crystal growth direction. On the other hand, 30 nm Bi nanowires produced using a much higher pressure of > 20 000 psi show a different crystal orientation of (001) along the wire axis [4.14], indicating that the preferred crystal orientation may also depend on the applied pressure, with the most dense packing direction along the wire axis for the highest applied pressure.

Electrochemical Deposition

The electrochemical deposition technique has attracted increasing attention as a versatile method for fabricating nanowires in templates. Traditionally, electrochemistry has been used to grow thin films on conducting surfaces. Since electrochemical growth is usually controllable in the direction normal to the substrate surface, this method can be readily extended to fabricate 1-D or 0-D nanostructures, if the deposition is confined within the pores of an appropriate template. In the electrochemical methods, a thin conducting metal film is first coated on one side of the porous membrane to serve as the cathode for electroplating. The length of the deposited nanowires can be controlled by varying the duration of the electroplating process. This method has been used to synthesize a wide variety of nanowires, such as metals (Bi [4.9, 74]; Co [4.81, 82]; Fe [4.25, 83]; Cu [4.73, 84]; Ni [4.39, 81]; Ag [4.85]; Au [4.5]); conducting polymers [4.9, 61]; superconductors (Pb [4.86]); semiconductors (CdS [4.19]); and even superlattice nanowires with A/B constituents (such as Cu/Co [4.73, 84]) have been synthesized electrochemically (see Table 4.1).

In the electrochemical deposition process, the chosen template has to be chemically stable in the electrolyte during the electrolysis process. Cracks and defects in the templates are detrimental to the nanowire growth, since the deposition processes primarily occur in the more accessible cracks, leaving most of the nanopores unfilled. Particle track-etched mica films or polymer membranes are typical templates used in simple dc electrolysis. To use anodic aluminium oxide films in the dc electrochemical deposition, the insulating barrier layer which separates the pores from the bottom aluminium substrate has to be removed, and a metal film is then evaporated onto the back of the template membrane [4.87]. Compound nanowire arrays, such as Bi_2Te_3, have been fabricated in alumina templates with a high filling factor using the dc electrochemical deposition [4.16]. Figures 4.3a and b, respectively, show the top view and the axial cross-sectional SEM images of a Bi_2Te_3 nanowire array [4.16]. The light areas are associated with Bi_2Te_3 nanowires, the dark regions denote empty pores, and the surrounding gray matrix is alumina.

Surfactants are also used with electrochemical deposition when necessary. For example, when using templates derived from PMMA/PS diblock copolymers, a methanol surfactant is used to facilitate pore filling [4.79], thereby achieving a $\approx 100\%$ filling factor.

It is also possible to employ an ac electrodeposition method in anodic alumina templates without the removal of the barrier layer, by utilizing the rectifying properties of the oxide barrier. In ac electrochemical deposition, although the applied voltage is sinusoidal and symmetric, the current is greater during the cathodic half-cycles, making deposition dominant over the stripping, which occurs in the subsequent anodic half-cycles. Since no rectification occurs at defect sites, the deposition and stripping rates are equal, and no material is deposited. Hence, the difficulties associated with cracks are avoided. In this fashion, metals, such as Co [4.82] and Fe [4.25, 83], and semiconductors, such as CdS [4.19], have been deposited into the pores of anodic aluminium oxide templates without removing the barrier layer.

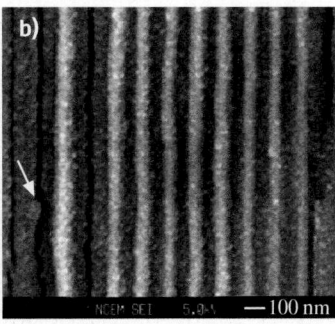

Fig. 4.3 (a) SEM image of a Bi_2Te_3 nanowire array in cross-section showing a relatively high pore filling factor. (b) SEM image of a Bi_2Te_3 nanowire array composite along the wire axis [4.16]

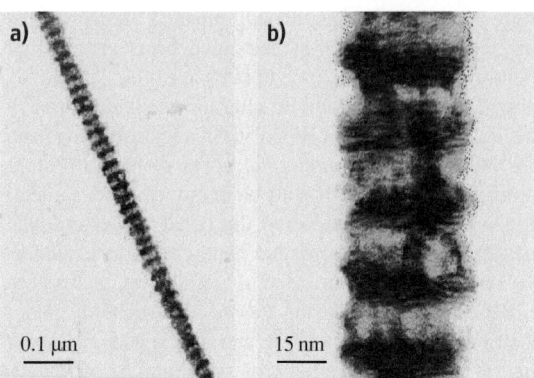

Fig. 4.4 (a) TEM image of a single Co(10 nm)/Cu(10 nm) multilayered nanowire. **(b)** A selected region of the sample at high magnification [4.84]

In contrast to nanowires synthesized by the pressure injection method, nanowires fabricated by the electrochemical process are usually polycrystalline, with no preferred crystal orientations, as observed by XRD studies. However, some exceptions exist. For example, polycrystalline CdS nanowires, fabricated by an ac electrodeposition method in anodic alumina templates [4.19], possibly have a preferred wire growth orientation along the *c*-axis. In addition, *Xu* et al. have prepared a number of single-crystal II–VI semiconductor nanowires, including CdS, CdSe and CdTe, by dc electrochemical deposition in anodic alumina templates with a nonaqueous electrolyte [4.18, 22]. Furthermore, single-crystal Pb nanowires were formed by pulse electrodeposition under overpotential conditions, but no specific crystal orientation along the wire axis was observed [4.86]. The use of pulse currents is believed to be advantageous for the growth of crystalline wires because the metal ions in the solution can be regenerated between the electrical pulses and therefore uniform deposition conditions can be produced for each deposition pulse. Similarly, single-crystal Ag nanowires were fabricated by pulsed electrodeposition [4.4].

One advantage of the electrochemical deposition technique is the possibility of fabricating multilayered structures within nanowires. By varying the cathodic potentials in the electrolyte, which contains two different kinds of ions, different metal layers can be controllably deposited. Co/Cu multilayered nanowires have been synthesized in this way [4.73, 84]. Figure 4.4 shows TEM images of a single Co/Cu nanowire which is about 40 nm in diameter [4.84]. The light bands represent Co-rich regions and the dark bands represent Cu-rich layers. This electrodeposition method provides

a low-cost approach to preparing multilayered 1-D nanostructures.

Vapor Deposition

Vapor deposition of nanowires includes physical vapor deposition (PVD) [4.8], chemical vapor deposition (CVD) [4.29], and metallo-organic chemical vapor deposition (MOCVD) [4.32]. Like electrochemical deposition, vapor deposition is usually capable of preparing smaller-diameter (≤ 20 nm) nanowires than pressure injection methods, since it does not rely on the high pressure and the surface tension involved to insert the material into the pores.

In the physical vapor deposition technique, the material to be filled is first heated to produce a vapor, which is then introduced through the pores of the template and cooled to solidify. Using a specially designed experimental set-up [4.8], nearly single-crystal Bi nanowires in anodic aluminium templates with pore diameters as small as 7 nm have been synthesized, and these Bi nanowires were found to possess a preferred crystal growth orientation along the wire axis, similar to the Bi nanowires prepared by pressure injection [4.8, 13].

Compound materials that result from two reacting gases have also be prepared by the chemical vapor deposition (CVD) technique. For example, single-crystal GaN nanowires have been synthesized in anodic alumina templates through a gas reaction of Ga_2O vapor with a flowing ammonia atmosphere [4.28, 29]. A different liquid/gas phase approach has been used to prepare polycrystalline GaAs and InAs nanowires in a nanochannel glass array [4.32]. In this method, the nanochannels are filled with one liquid precursor (such as Me_3Ga or Et_3In) via a capillary effect and the nanowires are formed within the template by reactions between the liquid precursor and the other gas reactant (such as AsH_3).

4.1.2 VLS Method for Nanowire Synthesis

Some of the recent successful syntheses of semiconductor nanowires are based on the so-called vapor-liquid-solid (VLS) mechanism of anisotropic crystal growth. This mechanism was first proposed for the growth of single crystal silicon whiskers 100 nm to hundreds of microns in diameter [4.88]. The proposed growth mechanism (see Fig. 4.5) involves the absorption of source material from the gas phase into a liquid droplet of catalyst (a molten particle of gold on a silicon substrate in the original work [4.88]). Upon supersaturation of the liquid alloy, a nucleation event generates a solid precipitate of the source material. This seed serves as a preferred site for further deposition of material at the interface of the liquid droplet, promoting the elongation of the seed into a nanowire or a whisker, and suppressing further nucleation events on the same catalyst. Since the liquid droplet catalyzes the incorporation of material from the gas source to the growing crystal, the deposit grows anisotropically as a whisker whose diameter is dictated by the diameter of the liquid alloy droplet. The nanowires thus obtained are of high purity, except for the end containing the solidified catalyst as an alloy particle (see Figs. 4.5 and 4.6a). Real-time observations of the alloying, nucleation, and elongation steps in the growth of germanium nanowires from gold nanoclusters by the VLS method were recorded by in situ TEM [4.89].

Reduction of the average wire diameter to the nanometer scale requires the generation of nanosized catalyst droplets. However, due to the balance between the liquid-vapor surface free energy and the free energy of condensation, the size of a liquid droplet, in equilibrium with its vapor, is usually limited to the micrometer range. This obstacle has been overcome in recent years by several new methodologies. (1) Advances in the synthesis of metal nanoclusters have made monodispersed nanoparticles commercially available. These can be dispersed on a solid substrate in high dilution so that when the temperature is raised above the melting point, the liquid clusters do not aggregate [4.47]. (2) Alternatively, metal islands of nanoscale sizes can self-form when a strained thin layer is grown or heat-treated on a non-epitaxial substrate [4.34]. (3) Laser-assisted catalytic VLS growth is a method used to generate nanowires under non-equilibrium conditions. Using laser ablation of a target containing both the catalyst and the source materials, a plasma is generated from which catalyst nanoclusters nucleate as the plasma cools down. Single

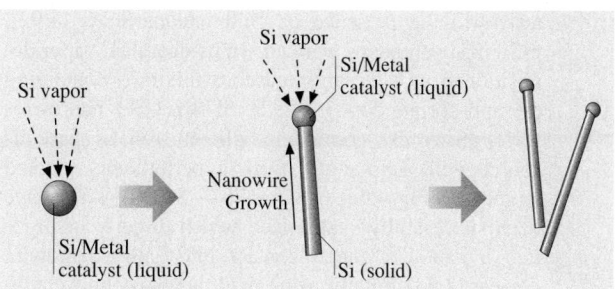

Fig. 4.5 Schematic diagram illustrating the growth of silicon nanowires by the VLS mechanism

crystal nanowires grow as long as the particle remains liquid [4.48]. (4) Interestingly, by optimizing the material properties of the catalyst-nanowire system, conditions can be achieved for which nanocrystals nucleate in a liquid catalyst pool supersaturated with the nanowire material, migrate to the surface due to a large surface tension, and continue growing as nanowires perpendicular to the liquid surface [4.50]. In this case, supersaturated nanodroplets are sustained on the outer end of the nanowire due to the low solubility of the nanowire material in the liquid [4.91].

A wide variety of elemental, binary and compound semiconductor nanowires has been synthesized via the VLS method, and relatively good control over the nanowire diameter and diameter distribution has been achieved. Researchers are currently focusing their attention on the controlled variation of the materials properties along the nanowire axis. In this context, researchers have modified the VLS synthesis apparatus to generate compositionally-modulated nanowires. GaAs/GaP-modulated nanowires have been synthesized by alternately ablating targets of the corresponding ma-

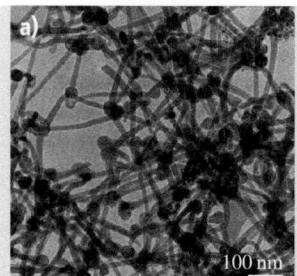 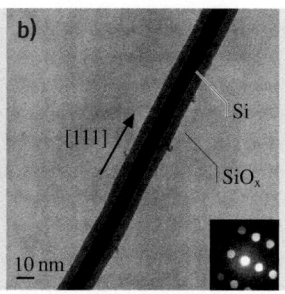 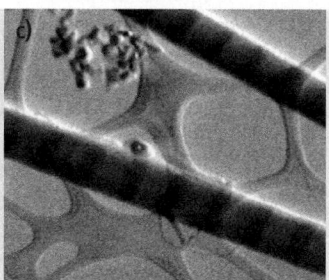

Fig. 4.6 (a) TEM images of Si nanowires produced after laser-ablating a $Si_{0.9}Fe_{0.1}$ target. The dark spheres with a slightly larger diameter than the wires are solidified catalyst clusters [4.48]. **(b)** Diffraction contrast TEM image of a Si nanowire. The crystalline Si core appears darker than the amorphous oxide surface layer. The *inset* shows the convergent beam electron diffraction pattern recorded perpendicular to the wire axis, confirming the nanowire crystallinity [4.48]. **(c)** STEM image of $Si/Si_{1-x}Ge_x$ superlattice nanowires in the bright field mode. The scale bar is 500 nm [4.90]

terials in the presence of gold nanoparticles [4.92]. p-Si/n-Si nanowires were grown by chemical vapor deposition from alternating gaseous mixtures containing the appropriate dopant [4.92]. $Si/Si_{1-x}Ge_x$ nanowires were grown by combining silicon from a gaseous source with germanium from a periodically ablated target (see Fig. 4.6c) [4.90]. $NiSi - Si$ nanowires have been successfully synthesized which directly incorporate a nanowire metal contact into active nanowire devices [4.93]. Finally, using an ultrahigh vacuum chamber and molecular beams, InAs/InP nanowires with atomically sharp interfaces were obtained [4.94]. These compositionally-modulated nanowires are expected to exhibit exciting electronic, photonic, and thermoelectric properties.

Interestingly, silicon and germanium nanowires grown by the VLS method consist of a crystalline core coated with a relatively thick amorphous oxide layer (2–3 nm) (see Fig. 4.6b). These layers are too thick to be the result of ambient oxidation, and it has been shown that these oxides play an important role in the nanowire growth process [4.49, 95]. Silicon oxides were found to serve as a special and highly selective catalyst that significantly enhances the yield of Si nanowires without the need for metal catalyst particles [4.49, 95, 96]. A similar yield enhancement was also found in the synthesis of Ge nanowires from the laser ablation of Ge powder mixed with GeO_2 [4.35]. The Si and Ge nanowires produced from these metal-free targets generally grow along the [112] crystal direction [4.97], and have the benefit that no catalyst clusters are found on either ends of the nanowires. Based on these observations and other TEM studies [4.35, 95, 97], an oxide-enhanced nanowire growth mechanism different from the classical VLS mechanism was proposed, where no metal catalyst is required during the laser ablation-assisted synthesis [4.95]. It is postulated that the nanowire growth is dependent on the presence of SiO (or GeO) vapor, which decomposes in the nanowire tip region into both Si (or Ge), which is incorporated into the crystalline phase, and SiO_2 (or GeO_2), which contributes to the outer coating. The initial nucleation events generate oxide-coated spherical nanocrystals. The [112] crystal faces have the fastest growth rate, and therefore the nanocrystals soon begin elongating along this direction to form one-dimensional structures. The Si_mO or Ge_mO (m > 1) layer on the nanowire tips may be in or at temperatures near their molten states, catalyzing the incorporation of gas molecules in a directional fashion [4.97]. Besides nanowires with smooth walls, a second morphology of chains of unoriented

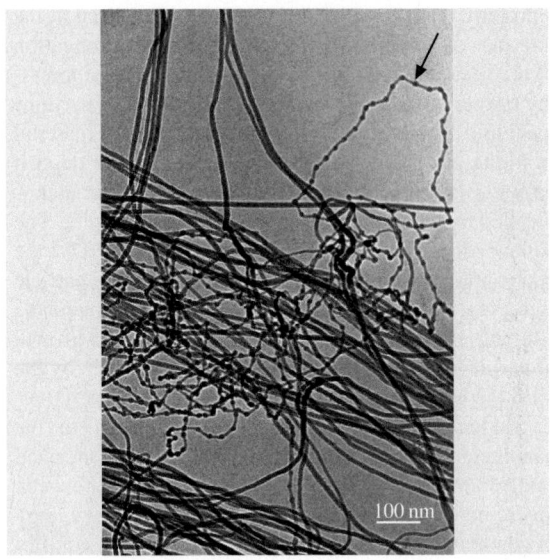

Fig. 4.7 TEM image showing the two major morphologies of Si nanowires prepared by the oxide-assisted growth method [4.95]. Notice the absence of metal particles when compared to Fig. 4.6a. The *arrow* points at an oxide-linked chain of Si nanoparticles

nanocrystals linked by oxide necks is frequently observed (indicated by an arrow in Fig. 4.7). In addition, it was found by STM studies that about 1% of the wires consist of a regular array of two alternating segments, 10 nm and 5 nm in length, respectively [4.98]. The segments, whose junctions form an angle of 30°, are probably a result of alternating growth along different crystallographic orientations [4.98]. Branched and hyperbranched Si nanowire structures have also been synthesized by *Whang* et al. [4.99].

4.1.3 Other Synthesis Methods

In this section we review several other general procedures available for the synthesis of a variety of nanowires. We focus on "bottom-up" approaches, which afford many kinds of nanowires in large numbers, and do not require highly sophisticated equipment (such as scanning microscopy or lithography-based methods), and exclude cases for which the nanowires are not self-sustained (such as in the case of atomic rows on the surface of crystals).

A solution-phase synthesis of nanowires with controllable diameters has been demonstrated [4.45, 100], without the use of templates, catalysts, or surfactants. Instead, *Gates* et al. make use of the anisotropy of the crystal structure of trigonal selenium and tellurium, which can be viewed as rows of 1-D helical atomic chains. Their approach is based on the mass transfer of atoms during an aging step from a high free-energy solid phase (e.g., amorphous selenium) to a seed (e.g., trigonal selenium nanocrystal) which grows preferentially along one crystallographic axis. The lateral dimension of the seed, which dictates the diameter of the nanowire, can be controlled by the temperature of the nucleation step. Furthermore, Se/Te alloy nanowires were synthesized by this method, and Ag_2Se compound nanowires were obtained by treating selenium nanowires with $AgNO_3$ [4.101–103]. In a separate work, tellurium nanowires were transformed into Bi_2Te_3 nanowires by their reaction with $BiPh_3$ [4.104].

More often, however, the use of surfactants is necessary to promote the anisotropic 1-D growth of nanocrystals. Solution phase synthetic routes have been optimized to produce monodispersed quantum dots, (zero-dimensional isotropic nanocrystals) [4.106]. Surfactants are necessary in this case to stabilize the interfaces of the nanoparticles and to retard oxidation and aggregation processes. Detailed studies on the effect of growth conditions revealed that they can be manipulated to induce a directional growth of the nanocrystals, usually generating nanorods (aspect ratio of ≈ 10), and in favorable cases, nanowires with high aspect ratios. Heath and LeGoues synthesized germanium nanowires by reducing a mixture of $GeCl_4$ and phenyl-$GeCl_3$ at high temperature and high pressure. The phenyl ligand was essential for the formation of high aspect ratio nanowires [4.33]. In growing CdSe nanorods [4.20], Alivisatos et al. used a mixture of two surfactants, whose concentration ratio influenced the structure of the nanocrystal. It is believed that different surfactants have different affinities, and different absorption rates, for the different crystal faces of CdSe, thereby regulating the growth rates of these faces. In the liquid phase synthesis of Bi nanowires, the additive $NaN(SiMe_3)_2$ induces the growth of nanowires oriented along the [110] crystal direction from small bismuth seed clusters, while water solely retarded the growth along the [001] direction, inducing the growth of hexagonal-plate particles [4.104]. A coordinating alkyl-diamine solvent was used to grow polycrystalline PbSe nanowires at low temperatures [4.42]. Here, the surfactant-induced directional growth is believed to occur through to the formation of organometallic complexes in which the bidentate ligand assumes the equatorial positions, thus hindering the ions from approaching each other in this plane. Additionally, the alkyl-diamine molecules coat the external

surface of the wire, preventing lateral growth. The aspect ratio of the wires increased as the temperature was lowered in the range $10\,°C < T < 117\,°C$. Ethylenediamine was used to grow CdS nanowires and tetrapods by a solvo-thermal recrystallization process starting with CdS nanocrystals or amorphous particles [4.17]. While the coordinating solvent was crucial for the nanowire growth, its role in the shape and phase control was not clarified.

Stress-induced crystalline bismuth nanowires have been grown from sputtered films of layers of Bi and CrN. The nanowires presumably grow from defects and cleavage fractures in the film, and are up to several millimeters in length with diameters ranging from 30 to 200 nm [4.7]. While the exploration of this technique has only begun, stress-induced unidirectional growth should be applicable to a variety of composite films.

Selective electrodeposition along the step edges in highly oriented pyrolytic graphite (HOPG) was used

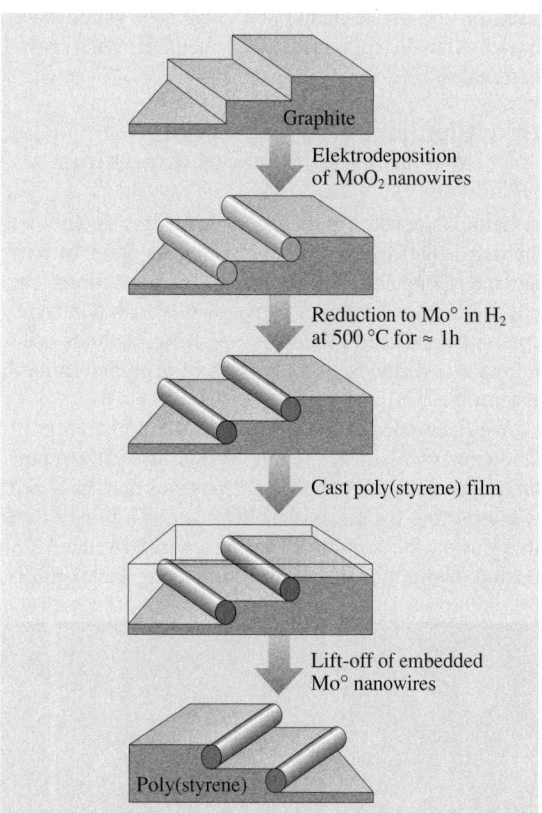

Fig. 4.8 Schematic of the electrodeposition step edge decoration of HOPG (highly oriented pyrolytic graphite) for the synthesis of molybdenum nanowires [4.38, 105]

to obtain MoO$_2$ nanowires as shown in Fig. 4.8. The site-selectivity was achieved by applying a low overpotential to the electrochemical cell in which the HOPG served as cathode, thus minimizing the nucleation events on less favorable sites (plateaux). While these nanowires cannot be removed from the substrate, they can be reduced to metallic molybdenum nanowires, which can then be released as free-standing nanowires. Other metallic nanowires were also obtained by this method [4.38, 105]. In contrast to the template synthesis approaches described above, in this method the substrate only defines the position and orientation of the nanowire, not its diameter. In this context, other surface morphologies, such as self-assembled grooves in etched crystal planes, have been used to generate nanowire arrays via gas-phase shadow deposition (for example: Fe nanowires on (110)NaCl [4.27]). The cross-section of artificially prepared superlattice structures has also been used for site-selective deposition of parallel and closely spaced nanowires [4.107]. Nanowires prepared on the above-mentioned substrates would have semicircular, rectangular, or other unconventional cross-sectional shapes.

4.1.4 Hierarchical Arrangement and Superstructures of Nanowires

Ordering nanowires into useful structures is another challenge that needs to be addressed in order to harness the full potential of nanowires for applications. We will first review examples of nanowires with nontrivial structures, and then proceed to describe methods used to create assemblies of nanowires of a predetermined structure.

We mentioned in Sect. 4.1.2 that the preparation of nanowires with a graded composition or with a superlattice structure along their main axis was demonstrated by controlling the gas phase chemistry as a function of time during the growth of the nanowires by the VLS method. Control of the composition along the axial di-

mension was also demonstrated by a template-assisted method, for example by the consecutive electrochemical deposition of different metals in the pores of an alumina template [4.110]. Alternatively, the composition can be varied along the radial dimension of the nanowire, for example by first growing a nanowire by the VLS method and then switching the synthesis conditions to grow a different material on the surface of the nanowire by CVD. This technique was demonstrated for the synthesis of Si/Ge and Ge/Si coaxial (or core-shell) nanowires [4.111], and it was shown that the outer shell can be formed epitaxially on the inner core by a thermal annealing process. *Han* et al. demonstrated the versatility of MgO nanowire arrays grown by the VLS method as templates for the PLD deposition of oxide coatings to yield MgO/YBCO, MgO/LCMO, MgO/PZT and MgO/Fe$_3$O$_4$ core/shell nanowires, all exhibiting epitaxial growth of the shell on the MgO core [4.37]. A different approach was adopted by *Wang* et al. who generated a mixture of coaxial and biaxial $SiC - SiO_x$ nanowires by the catalyst-free high-temperature reaction of amorphous silica and a carbon/graphite mixture [4.112].

A different category of nontrivial nanowires is that of nanowires with a nonlinear structure, resulting from multiple one-dimensional growth steps. Members of this category are tetrapods, which were mentioned in the context of the liquid phase synthesis (Sect. 4.1.3). In this process, a tetrahedral quantum dot core is first grown, and then the conditions are modified to induce one-dimensional growth of a nanowire from each one of the facets of the tetrahedron. A similar process produced high-symmetry In$_2$O$_3$/ZnO hierarchical nanostructures. From a mixture of heat-treated In$_2$O$_3$, ZnO, and graphite powders, faceted In$_2$O$_3$ nanowires were first obtained, on which oriented shorter ZnO nanowires were crystallized [4.108]. Brush-like structures were obtained as a mixture of 11 structures of different symmetries. For example, two, four, or six rows of ZnO nanorods could be found on different core nanowires, depending on the

Fig. 4.9a–d SEM images of (**a**) sixfold- (**b**) fourfold- and (**c**) twofold-symmetry nanobrushes made of an In$_2$O$_3$ core and ZnO nanowire brushes [4.108], and of (**d**) ZnO nanonails [4.109]

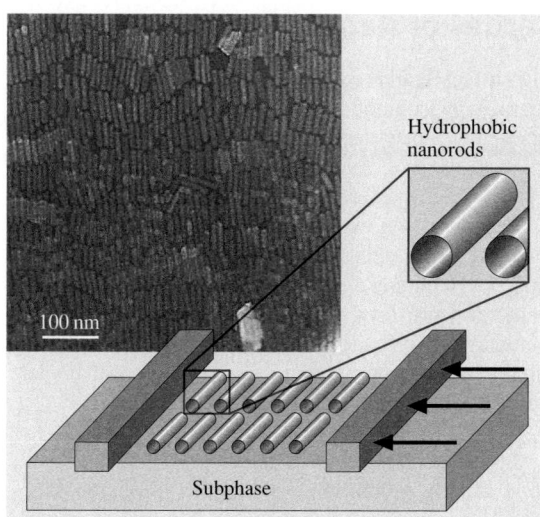

Fig. 4.10 A TEM image of a smectic phase of a BaCrO$_4$ nanorod film (*left inset*) achieved by the Langmuir–Blodgett technique, as depicted by the illustration [4.113]

crystallographic orientation of the main axis of the core nanowire, as shown in Fig. 4.9. Comb-like structures made entirely of ZnO were also reported [4.54].

Controlling the position of a nanowire in the growth process is important for preparing devices or test structures containing nanowires, especially when it involves a large array of nanowires. Post-synthesis methods to align and position nanowires include microfluidic channels [4.114], Langmuir–Blodgett assemblies [4.113], and electric field-assisted assembly [4.115]. The first method involves the orientation of the nanowires by the liquid flow direction when a nanowire solution is injected into a microfluidic channel assembly and by the interaction of the nanowires with the side walls of the channel. The second method involves the alignment of nanowires at a liquid–gas or liquid–liquid interface by the application of compressive forces on the interface (see Fig. 4.10). The aligned nanowire films can then be transferred onto a substrate and lithography methods can be used to define interconnects. This allows the nanowires to be organized with a controlled alignment and spacing over large areas. Using this method, centimeter-scale arrays containing thousands of single silicon nanowire field-effect transistors with high performance could be assembled to make large-scale nanowire circuits and devices [4.99, 116].

The third technique is based on dielectrophoretic forces that pull polarizable nanowires toward regions of high field strength. The nanowires align between two isolated electrodes which are capacitatively coupled to a pair of buried electrodes biased with an AC voltage. Once a nanowire shorts the electrodes, the electric field is eliminated, preventing more nanowires from depositing. The above techniques have been successfully used to prepare electronic circuitry and optical devices out of nanowires (see Sects. 4.3.1 and 4.3.3). Alternatively, alignment and positioning of the nanowires can be specified and controlled during their growth by the proper design of the synthesis method. For example, ZnO nanowires prepared by the VLS method were grown into an array in which both their position on the substrate and their growth direction and orientation were controlled [4.54]. The nanowire growth region was defined by patterning the gold film, which serves as a catalyst for the ZnO nanowire growth, employing soft-lithography, e-beam lithography, or photolithography. The orientation of the nanowires was achieved by selecting a substrate with a lattice structure matching that of the nanowire material to facilitate the epitaxial growth. These conditions result in an array of nanowire posts at predetermined positions, all vertically aligned with the same crystal growth orientation (see Fig. 4.11). Similar rational GaN nanowire arrays have been synthesized epitaxially on $(100) - LiAlO_2$ and (111) MgO single-crystal substrates. In addition, control over the crystallographic growth directions of nanowires was achieved by lattice-matching to different substrates. For example, GaN nanowires on (100) LiAlO$_2$ substrates grow oriented along the [110] direction, whereas (111) MgO substrates result in the growth of GaN nanowires with an [001] orientation, due to the different lattice-matching constraints [4.117]. A similar structure could be obtained by the template-mediated electrochemical synthesis of nanowires (see Sect. 4.1.1), particularly if anodic alumina with its parallel and ordered channels is used. The control over the location of the nucleation of nanowires in the electrochemical deposition is determined by the pore positions and the back-electrode geometry. The pore positions can be precisely controlled by imprint lithography [4.118]. By growing the template on a patterned conductive substrate that serves as a back-electrode [4.119–121] different materials can be deposited in the pores at different regions of the template.

4.2 Characterization and Physical Properties of Nanowires

In this section we review the structure and properties of nanowires and their interrelationship. The discovery and investigation of nanostructures were spurred on by advances in various characterization and microscopy techniques that enabled material characterization to take place at smaller and smaller length scales, reaching length scales down to individual atoms. For applications, characterizing the structural properties of nanowires is especially important, so that a reproducible relationship between their desired functionality and their geometrical and structural characteristics can be established. Due to the enhanced surface-to-volume ratio in nanowires, their properties may depend sensitively on their surface conditions and geometrical configurations. Even nanowires made of the same material may possess dissimilar properties due to differences in their crystal phase, crystalline size, surface conditions, and aspect ratios, which depend on the synthesis methods and conditions used in their preparation.

4.2.1 Structural Characterization

Structural and geometric factors play an important role in determining the various attributes of nanowires, such as their electrical, optical and magnetic properties. Therefore, various novel tools have been developed and employed to obtain this important structural information at the nanoscale. At the micron scale, optical techniques are extensively used for imaging structural features. Since the sizes of nanowires are usually comparable to or, in most cases, much smaller than the wavelength of visible light, traditional optical microscopy techniques are usually limited when characterizing the morphology

and surface features of nanowires. Therefore, electron microscopy techniques play a more dominant role at the nanoscale. Since electrons interact more strongly than photons, electron microscopy is particularly sensitive relative to X-rays for the analysis of tiny samples.

In this section we review and give examples of how scanning electron microscopy, transmission electron microscopy, scanning probe spectroscopies, and diffraction techniques are used to characterize the structures of nanowires. To provide the necessary basis for developing reliable structure–property relations, multiple characterization tools are applied to the same samples.

Scanning Electron Microscopy

SEM usually produces images down to length scales of $\approx 10\,\mathrm{nm}$ and provides valuable information regarding the structural arrangement, spatial distribution, wire density, and geometrical features of the nanowires. The examples of SEM micrographs shown in Figs. 4.1 and 4.3 indicate that structural features at the 10 nm to $10\,\mu\mathrm{m}$ length scales can be probed, providing information on the size, size distribution, shapes, spatial distributions, density, nanowire alignment, filling factors, granularity, etc.. As another example, Fig. 4.11a shows an SEM image of ZnO nanowire arrays grown on a sapphire substrate [4.122], which provides evidence for the nonuniform spatial distribution of the nanowires on the substrate, which was attained by patterning the catalyst film to define high-density growth regions and nanowire-free regions. Figure 4.11b, showing a higher magnification of the same system, indicates that these ZnO nanowires grow perpendicular to the substrate, are well-aligned with approximately equal

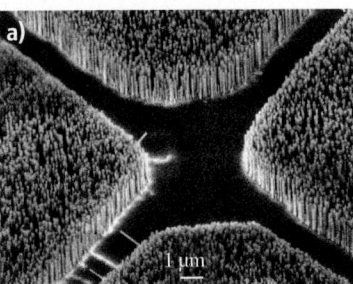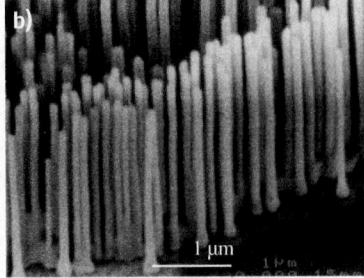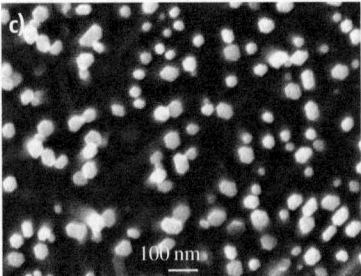

Fig. 4.11a–c SEM images of ZnO nanowire arrays grown on a sapphire substrate, where (**a**) shows patterned growth, (**b**) shows a higher resolution image of the parallel alignment of the nanowires, and (**c**) shows the faceted side-walls and the hexagonal cross-section of the nanowires. For nanowire growth, the sapphire substrates were coated with a 1.0 to 3.5 nm-thick patterned layer of Au as the catalyst, using a TEM grid as the shadow mask. These nanowires have been used for nanowire laser applications [4.122]

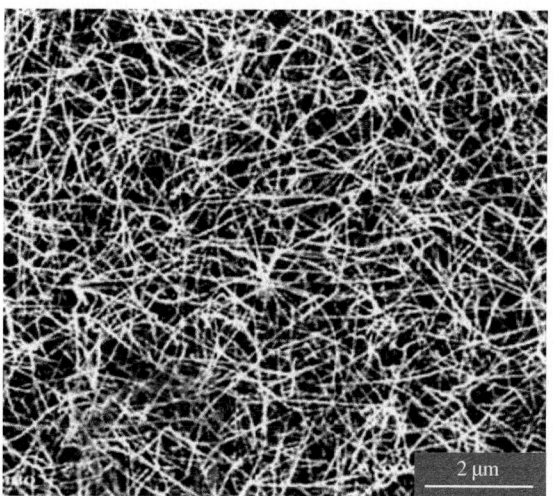

Fig. 4.12 SEM image of GaN nanowires in a mat arrangement synthesized by laser-assisted catalytic growth. The nanowires have diameters and lengths on the order of 10 nm and 10 μm, respectively [4.30]

wire lengths, and have wire diameters in the range $20 \leq d_W \leq 150$ nm. The SEM micrograph in Fig. 4.11c

provides further information about the surface of the nanowires, showing it to be well-faceted, forming a hexagonal cross-section, indicative of nanowire growth along the $\langle 0001 \rangle$ direction. Both the uniformity of the nanowire size, their alignment perpendicular to the substrate, and their uniform growth direction, as suggested by the SEM data, are linked to the good epitaxial interface between the (0001) plane of the ZnO nanowire and the (110) plane of the sapphire substrate. (The crystal structures of ZnO and sapphire are essentially incommensurate, with the exception that the a-axis of ZnO and the c-axis of sapphire are related almost exactly by a factor of 4, with a mismatch of less than 0.08% at room temperature [4.122].) The well-faceted nature of these nanowires has important implications for their lasing action (see Sect. 4.3.2). Figure 4.12 shows an SEM image of GaN nanowires synthesized by a laser-assisted catalytic growth method [4.30], indicating a random spatial orientation of the nanowire axes and a wide diameter distribution for these nanowires, in contrast to the ZnO wires in Fig. 4.11 and to arrays of well-aligned nanowires prepared by template-assisted growth (see Fig. 4.3).

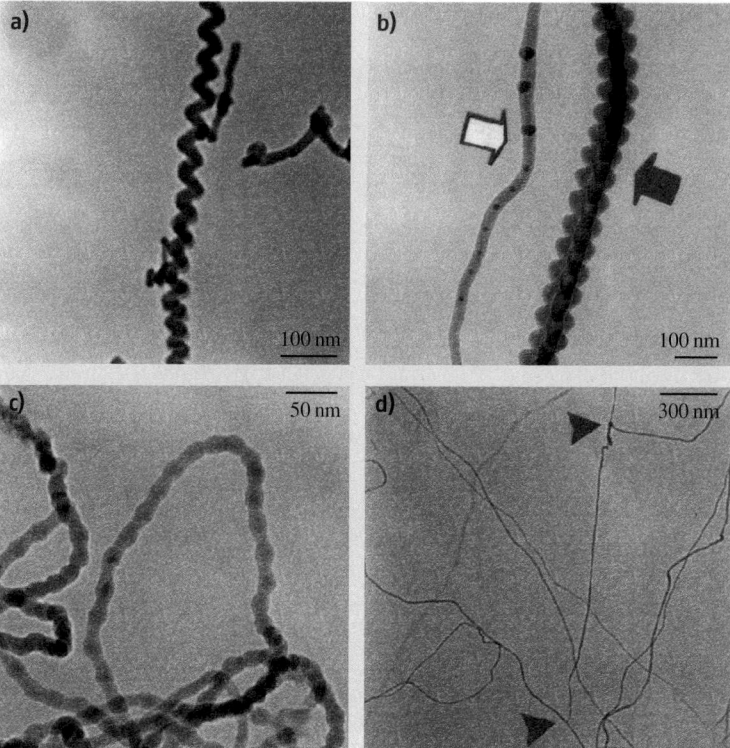

Fig. 4.13a–d TEM morphologies of four special forms of Si nanowires synthesized by the laser ablation of a Si powder target. (**a**) A spring-shaped Si nanowire; (**b**) fishbone-shaped (indicated by a *solid arrow*) and frogs egg-shaped (indicated by a *hollow arrow*) Si nanowires; and (**c**) pearl-shaped nanowires, while (**d**) shows poly-sites for the nucleation of silicon nanowires (indicated by *arrows*) [4.123]

Transmission Electron Microscopy

TEM and high-resolution transmission electron microscopy (HRTEM) are powerful imaging tools for studying nanowires at the atomic scale, and they usually provide more detailed geometrical features than are seen in SEM images. TEM studies also yield information regarding the crystal structure, crystal quality, grain size, and crystal orientation of the nanowire axis. When operating in the diffraction mode, selected area electron diffraction (SAED) patterns can be made to determine the crystal structures of nanowires. As an example, the TEM images in Fig. 4.13 show four different morphologies for Si nanowires prepared by the laser ablation of a Si target [4.123]: (a) spring-shaped; (b) fishbone-shaped (indicated by solid arrow) and frogs egg-shaped (indicated by the hollow arrow), (c) pearl-shaped, while (d) shows the poly-sites of nanowire nucleation. The crystal quality of nanowires is revealed from high-resolution TEM images with atomic resolution, along with selected area electron diffraction (SAED) patterns. For example, Fig. 4.14 shows a TEM image of one of the GaN nanowires from Fig. 4.12, indicating single crystallinity and showing (100) lattice planes, thus indicating

the growth direction of the nanowire. This information is supplemented by the corresponding electron diffraction pattern in the upper right. A more comprehensive review of the application of TEM for growth orientation indexing and crystal defect characterization in nanowires is available elsewhere [4.124].

The high resolution of the TEM also permits the surface structures of the nanowires to be studied. In many cases, the nanowires are sheathed with a native oxide layer, or an amorphous oxide layer that forms during the growth process. This can be seen in Fig. 4.6b for silicon nanowires and in Fig. 4.15 for germanium nanowires [4.35], showing a mass-thickness contrast TEM image and a selected-area electron diffraction pattern of a Ge nanowire. The main TEM image shows that these Ge nanowires possess an amorphous GeO_2 sheath with a crystalline Ge core that is oriented in the [211] direction.

Dynamical processes of the surface layer of nanowires can be studied in-situ using an environmental TEM chamber, which allows TEM observations to be made while different gases are introduced or as the sample is heat-treated at various temperatures, as illustrated in Fig. 4.16. The figure shows high-resolution TEM images of a Bi nanowire with an oxide coating and the effect of a dynamic oxide removal process

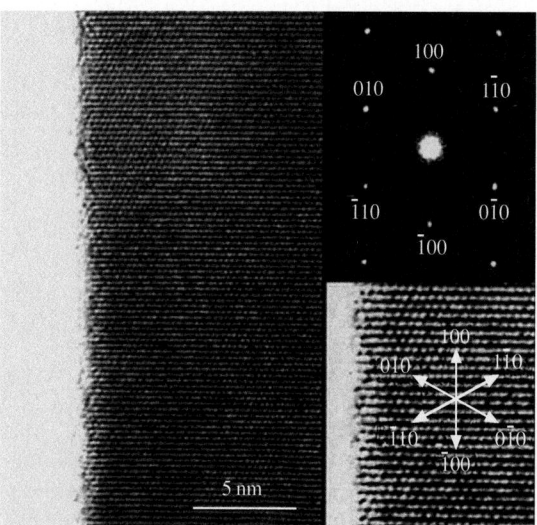

Fig. 4.14 Lattice-resolved high-resolution TEM image of one GaN nanowire (*left*) showing that (100) lattice planes are visible perpendicular to the wire axis. The electron diffraction pattern (*top right*) was recorded along the [001] zone axis. A lattice-resolved TEM image (*lower right*) highlights the continuity of the lattice up to the nanowire edge, where a thin native oxide layer is found. The directions of various crystallographic planes are indicated in the *lower right figure* [4.30]

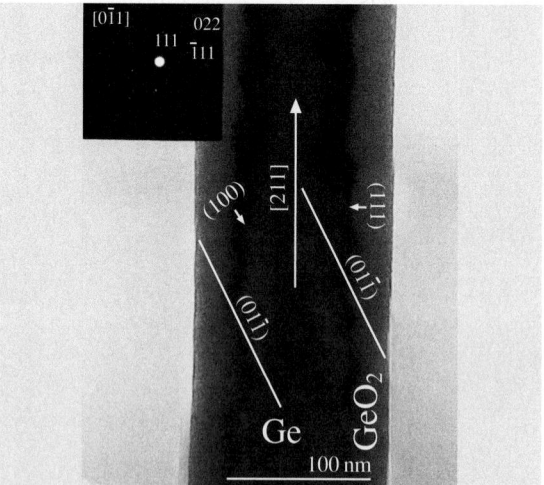

Fig. 4.15 A mass-thickness contrast TEM image of a Ge nanowire taken along the [0$\bar{1}$1] zone axis and a selected-area electron diffraction pattern (*upper left inset*) [4.35]. The Ge nanowires were synthesized by laser ablation of a mixture of Ge and GeO_2 powder. The core of the Ge nanowire is crystalline, while the surface GeO_2 is amorphous

carried out within the environmental chamber of the TEM [4.125]. The amorphous bismuth-oxide layer coating the nanowire (Fig. 4.16a) is removed by exposure to hydrogen gas within the environmental chamber of the TEM, as indicated in Fig. 4.16b.

By coupling the powerful imaging capabilities of TEM with other characterization tools, such as an electron energy loss spectrometer (EELS) or an energy dispersive X-ray spectrometer (EDS) within the TEM instrument, additional properties of the nanowires can be probed with high spatial resolution. With the EELS technique, the energy and momentum of the incident and scattered electrons are measured in an inelastic electron scattering process to provide information on the energy and momentum of the excitations in the nanowire sample. Figure 4.17 shows the dependence on nanowire diameter of the electron energy loss spectra of Bi nanowires. The spectra were taken from the center of the nanowire, and the shift in the energy of the peak position (Fig. 4.17) indicates the effect of the nanowire diameter on the plasmon frequency in the nanowires. The results show that there are changes in the electronic structure of the Bi nanowires as the wire diameter decreases [4.126]. Such changes in electronic structure as a function of nanowire diameter are also observed in their transport (Sect. 4.2.2) and optical (Sect. 4.2.3) properties, and are related to quantum confinement effects.

EDS measures the energy and intensity distribution of X-rays generated by the impact of the electron beam on the surface of the sample. The elemental composition within the probed area can be determined to a high degree of precision. The technique was particularly useful for the compositional characterization of superlattice nanowires [4.90] and core-shell nanowires [4.111] (see Sect. 4.1.2).

Scanning Tunneling Probes

Several scanning probe techniques, such as scanning tunneling microscopy (STM) [4.127], electric field gradient microscopy (EFM) [4.13], magnetic field microscopy (MFM) [4.40], and scanning thermal microscopy (SThM) [4.128], combined with atomic force microscopy (AFM), have been employed to study the structural, electronic, magnetic, and thermal properties of nanowires. A scanning tunneling microscope can be employed to reveal both topographical structural information, such as that illustrated in Fig. 4.18, as well as information on the local electronic density of states of a nanowire, when used in the STS (scanning tunneling spectroscopy) mode. Figure 4.18 shows STM height im-

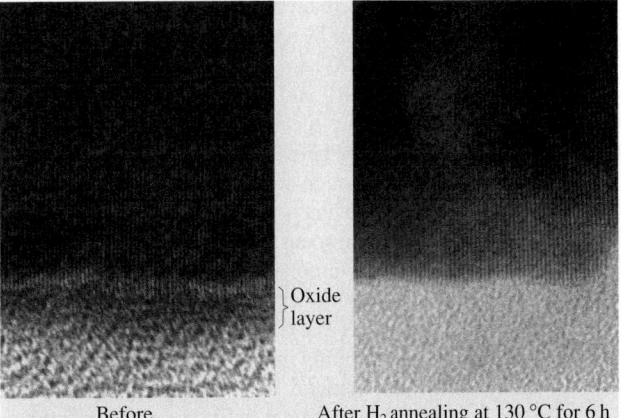

Fig. 4.16 High-resolution transmission electron microscope (HRTEM) image of a Bi nanowire (*left*) before and (*right*) after annealing in hydrogen gas at 130 °C for six hours within the environmental chamber of the HRTEM instrument to remove the oxide surface layer [4.125]

ages (taken in the constant current STM mode) of MoSe molecular wires deposited from a methanol or acetonitrile solution of $Li_2Mo_6Se_6$ onto Au substrates. The STM image of a single MoSe wire (Fig. 4.18a) exhibits a 0.45 nm lattice repeat distance in a MoSe molecular wire. When both STM and STS measurements are made on the same sample, the electronic and structural

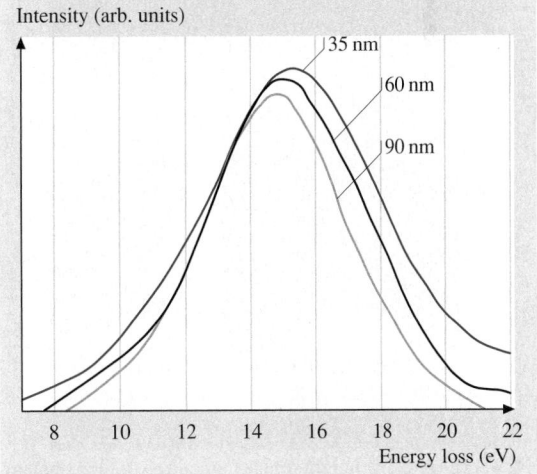

Fig. 4.17 Electron energy loss spectra (EELS) taken from the centers of bismuth nanowires with diameters of 35, 60 and 90 nm. The shift in the volume plasmon peaks is due to the effect of wire diameter on the electronic structure [4.126]

properties can be correlated, as for example in the joint STM/STS studies on Si nanowires [4.98], showing alternating segments of a single nanowire identified with growth along the [110] and [112] directions, and different I–V characteristics measured for the [110] segments as compared with the [112] segments.

Magnetic field microscopy (MFM) has been employed to study magnetic polarization of magnetic nanowires embedded in an insulating template, such as an anodic alumina template. For example, Fig. 4.19a shows the topographic image of an anodic alumina template filled with Ni nanowires, and Fig. 4.19b demonstrates the corresponding magnetic polarization of each nanowire in the template. This micrograph shows that a magnetic field microscopy probe can distinguish between spin-up and spin-down nanowires in the nanowire array, thereby providing a method for measuring interwire magnetic dipolar interactions [4.40].

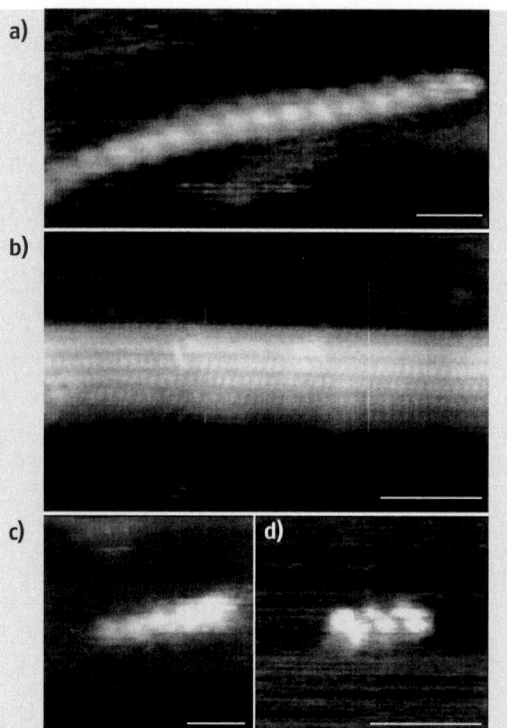

Fig. 4.18a–d STM height images, obtained in the constant current mode, of MoSe chains deposited on an Au(111) substrate. (**a**) A single chain image, and (**b**) a MoSe wire bundle. (**c**) and (**d**) are images of MoSe wire fragments containing five and three unit cells, respectively [4.127]. The scale bars are all 1 nm

X–Ray Analysis

Other characterization techniques that are commonly used to study the crystal structures and chemical compositions of nanowires include X-ray diffraction and X-ray energy dispersion analysis (EDAX). The peak positions in the X-ray diffraction pattern can be used to determined the chemical composition and the crystal phase structure of the nanowires. For example, Fig. 4.2 shows that Bi nanowires have the same crystal structure and lattice constants as bulk bismuth. Both the X-ray diffraction pattern (XRD) for an array of aligned Bi nanowires (Fig. 4.2) and the SAED pattern for individual Bi nanowires [4.13] suggest that the nanowires have a common axis of nanowire alignment.

As another example of an XRD pattern for an array of aligned nanowires, Fig. 4.20 shows the X-ray diffraction pattern of the ZnO nanowires that are displayed in Fig. 4.11. Only (00ℓ) diffraction peaks are observed for these aligned ZnO nanowires, indicating that their preferred growth direction is (001) along the wire axis. Similarly, XRD was used to confirm the different growth directions of GaN nanowire array grown epitaxially on (100) $LiAlO_2$ and (111) MgO substartes [4.117]

EDAX has been used to determine the chemical compositions and stoichiometries of compound nanowires or impurity contents in nanowires. However, the results from EDAX analysis should be interpreted carefully to avoid systematic errors.

4.2.2 Mechanical Properties

Thermal Stability

Due to the large surface area-to-volume ratio in nanowires and other nanoparticles, the thermal stability of nanowires is anticipated to differ significantly from that of the bulk material. Theoretical studies of materials in confined geometries show that the melting point of the material is reduced in nanostructures, as is the latent heat of fusion, and that large hysteresis can be observed in melting–freezing cycles. These phenomena have been studied experimentally in three types of nanowire systems: porous matrices impregnated with a plurality of nanowires, individual nanowires sheathed by a thin coating, and individual nanowires.

The melting freezing of matrix-supported nanowires can be studied by differential scanning calorimetry (DSC), since large volumes of samples can thus be produced. *Huber* et al. investigated the melting of indium in porous silica glasses with mean pore diameters ranging from 6 to 141 nm [4.129]. The melting point of the pore-confined indium shows a linear dependence on inverse

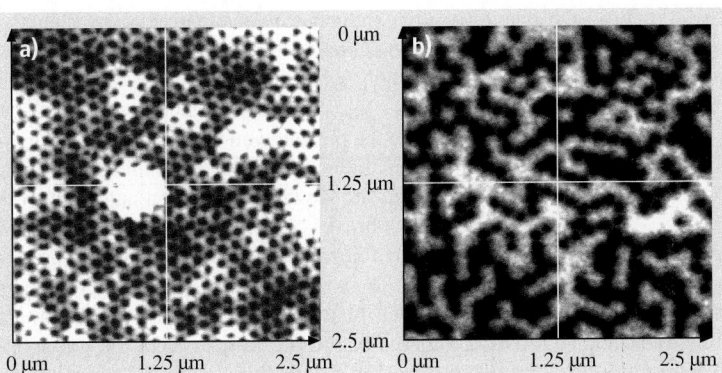

Fig. 4.19 (a) Topographic image of a highly-ordered porous alumina template with a period of 100 nm filled with 35 nm diameter nickel nanowires. (b) The corresponding MFM (magnetic force microscope) image of the nanomagnet array, showing that the pillars are magnetized alternately "up" (*white*) and "down" (*black*) [4.40]

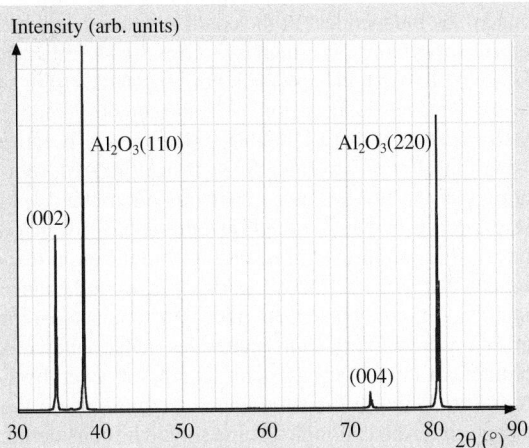

Fig. 4.20 X-ray diffraction pattern of aligned ZnO nanowires (see Fig. 4.11) grown on a sapphire substrate. Only [00ℓ] diffraction peaks are observed for the nanowires, owing to their well-oriented growth orientation. Strong diffraction peaks for the sapphire substrate are found [4.122]

pore diameter, with a maximum melting point depression of 50 °C. They also recorded a 6 °C difference in the melting temperature and the freezing temperature of 12.8 nm diameter indium. The melting profile of the pore-confined indium in these samples is broader in temperature than for bulk indium, as expected for the heterogeneity in the pore diameter and in the indium crystal size aspect ratio within the samples.

Sheathed nanowires provide an opportunity to study the melting and recrystallization of individual nanowires. The shell layer surrounding the nanowire provides confinement to keep the liquid phase within the inner cylindrical volume. However, the shell–nanowire surface interaction should be taken into account when analyzing the phase transition thermodynamics and kinetics. *Yang* et al. produced germanium nanowires coated with a thin (1–5 nm) graphite sheath, by pyrolysis of organic molecules over VLS-grown nanowires, and followed the melting and recrystallization of the germanium by variable temperature TEM imaging [4.130]. The melting of the nanowires was followed by the disappearance of the electronic diffraction pattern. It was found that the nanowires began melting from their ends, with the melting front advancing towards the center of the nanowire as the temperature was increased. During the cool-down part of the cycle, the recrystallization of the nanowire occurred instantaneously following significant supercooling. The authors report both the largest melting point suppression recorded thus far for germanium (≈ 300 °C), and a large melting–recrystallization hysteresis of up to ≈ 300 °C. Similarly, carbon nanotubes have been filled with various low-temperature metals [4.131]. A nanothermometer has been demonstrated using a 10 nm liquid gallium filled-carbon nanotube, showing an expansion coefficient that is linear in temperature and identical to the bulk value [4.132].

A different behavior was observed in free-standing copper nanowires [4.133]. In this system, there is little interaction between the nanowire surface and the surroundings, and the nanowire is not confined in its diameter, as in the case of the sheathed nanowires. Thermal treatment of the free-standing nanowires leads to their fragmentation into a linear array of metal spheres. Thinner nanowires were more vulnerable than thicker nanowires to the thermal treatment, showing constrictions and segmentation at lower temperatures. Analysis of the temperature response of the nanowires indicates that the nanowire segmentation is a result of the Rayleigh instability, starting with oscillatory perturbations of the nanowire diameter, leading to long cylindrical segments, that become more separated and more spherical at higher temperatures. These observations indicate that annealing and melting are dominated by the surface diffusion of atoms on the entire surface of the nanowire (versus tip-initiated melting).

4.2.3 Transport Properties

The study of electrical transport properties of nanowires is important for nanowire characterization, electronic device applications, and the investigation of unusual transport phenomena arising from one-dimensional quantum effects. Important factors that determine the transport properties of nanowires include the wire diameter, (important for both classical and quantum size effects), material composition, surface conditions, crystal quality, and the crystallographic orientation along the wire axis for materials with anisotropic material parameters, such as the effective mass tensor, the Fermi surface, or the carrier mobility.

Electronic transport phenomena in low-dimensional systems can be roughly divided into two categories: ballistic transport and diffusive transport. Ballistic transport phenomena occur when the electrons can travel across the nanowire without any scattering. In this case, the conduction is mainly determined by the contacts between the nanowire and the external circuit, and the conductance is quantized into an integral number of universal conductance units $G_0 = 2e^2/h$ [4.134, 135]. Ballistic transport phenomena are usually observed in very short quantum wires, such as those produced using mechanically controlled break junctions (MCBJ) [4.136, 137] where the electron mean free path is much longer than the wire length and the conduction is a pure quantum phenomenon. To observe ballistic transport, the thermal energy must also obey the relation $k_B T \ll \varepsilon_j - \varepsilon_{j-1}$, where $\varepsilon_j - \varepsilon_{j-1}$ is the energy separation between subband levels j and $j-1$. On the other hand, for nanowires with lengths much larger than the carrier mean free path, the electrons (or holes) undergo numerous scattering events when they travel along the wire. In this case, the transport is in the diffusive regime, and the conduction is dominated by carrier scattering within the wires, due to phonons (lattice vibrations), boundary scattering, lattice and other structural defects, and impurity atoms.

Conductance Quantization in Metallic Nanowires

The ballistic transport of 1-D systems has been extensively studied since the discovery of quantized conductance in 1-D systems in 1988 [4.134, 135]. The phenomena of conductance quantization occur when the diameter of the nanowire is comparable to the electron Fermi wavelength, which is on the order of 0.5 nm for most metals [4.138]. Most conductance quantization experiments up to the present were performed by bringing together and separating two metal electrodes. As the two metal electrodes are slowly separated, a nanocontact is formed before it breaks completely (see Fig. 4.21a), and conductance in integral multiple values of G_0 is observed through these nanocontacts. Figure 4.21b shows the conductance histogram built with 18 000 contact breakage curves between two gold electrodes at room temperature [4.139], with the electrode separation up to ≈ 1.8 nm. The conductance quantization behavior is found to be independent of the contact material, and has been observed in various metals, such as Au [4.139], Ag, Na, Cu [4.140], and Hg [4.141]. For semimetals such as Bi, conductance quantization has also been observed for electrode separations as long as 100 nm at 4 K because of the long Fermi wavelength (≈ 26 nm) [4.138], indicating that the conductance quantization may be due to the existence of well-defined quantum states localized at a constriction instead of resulting from the atom rearrangement as the electrodes separate. Since conductance quantization is only observed in breaking contacts, or for very narrow and very short nanowires, most nanowires of practical interest (possessing lengths of several microns) lie in the diffusive transport regime, where the carrier scattering is significant and should be considered.

I–V Characterization of Semiconducting Nanowires

The electronic transport behavior of nanowires may be categorized based on the relative magnitudes of three length scales: carrier mean free path ℓ_W, the de Broglie wavelength of electrons λ_e, and the wire diameter d_W. For wire diameters much larger than the carrier mean free path ($d_W \gg \ell_W$), the nanowires exhibit transport properties similar to bulk materials, which are independent of the wire diameter, since the scattering due to the wire boundary is negligible compared to other scattering mechanisms. For wire diameters comparable to or smaller than the carrier mean free path ($d_W \approx \ell_W$ or $d_W < \ell_W$), but still much larger than the de Broglie wavelength of the electrons ($d_W \gg \lambda_e$), the transport in nanowires is in the classical finite size regime, where the band structure of the nanowire is still similar to that of bulk, while the scattering events at the wire boundary alter their transport behavior. For wire diameters comparable to the electronic wavelength $d_W \approx \lambda_e$, the electronic density of states is altered dramatically and quantum sub-bands are formed due to the quantum confinement effect at the wire boundary. In this regime, the transport properties are further influenced by the change in the band structure. Therefore, transport properties for nanowires in the classical finite size and quantum size regimes are highly diameter-dependent.

Researchers have investigated the transport properties of various semiconducting nanowires and have demonstrated their potential for diverse electronic devices, such as for p-n diodes [4.142, 143], field effect transistors [4.142], memory cells, and switches [4.144] (see Sect. 4.3.1). So far, the nanowires studied in this context have usually been made from conventional semiconducting materials, such as group IV and III–V compound semiconductors, via the VLS growth method (see Sect. 4.1.2), and their nanowire properties have been compared to their well-established bulk properties. Interestingly, the physical principles for describing bulk semiconductor devices also hold for devices based on these semiconducting nanowires with wire diameters of tens of nanometers. For example, Fig. 4.22 shows the current-voltage (I–V) behavior of a 4-by-1 crossed p-Si/n-GaN junction array at room temperature [4.142]. The long horizontal wire in the figure is a p-Si nanowire (10–25 nm in diameter) and the four short vertical wires are n-GaN nanowires (10–30 nm in diameter). Each of the four nanoscale cross points independently forms a p-n junction with current rectification behavior, as shown by the I–V curves in Fig. 4.22, and the junction behavior (for example the turn-on voltage) can be controlled by varying the oxide coating on these nanowires [4.142].

Huang et al. have demonstrated nanowire junction diodes with a high turn-on voltage ($\approx 5\,\mathrm{V}$) by increasing the oxide thickness at the junctions. The high turn-on voltage enables the use of the junction in a nanoscale FET, as shown in Fig. 4.23 [4.142] where I–V data for a p-Si nanowire are presented, for which the n-GaN nanowire with a thick oxide coating is used as a nanogate. By varying the nanogate voltage, the conductance of the p-Si nanowire can be changed by more than a factor of 10^5 (lower curve in the inset), whereas the conductance changes by only a factor of 10 when a global back-gate is used (top curve in the inset of Fig. 4.23). This behavior may be due to the thin gate dielectric between the crossed nanowires and the better control of the local carrier density through a nanogate. Based on the gate-dependent I–V data from these p-Si nanowires, it is found that the mobility of the holes in the p-Si nanowires may be higher than that for bulk p-Si, although further investigation is required for complete understanding.

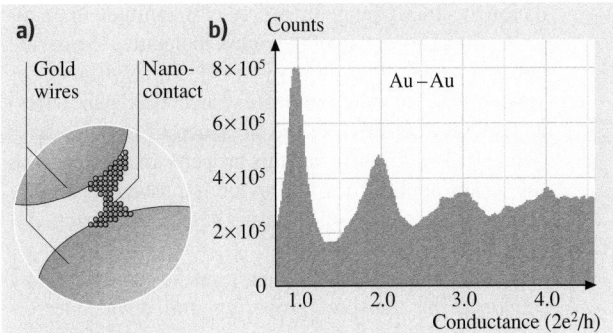

Fig. 4.21 (**a**) Schematic representation of the last stages of the contact breakage process [4.139]. (**b**) Histogram of conductance values built with 18 000 gold contact breakage experiments in air at room temperature, showing conductance peaks at integral values of G_0. In this experiment the gold electrodes approach and separate at 89 000 Å/s [4.139]

Because of the enhanced surface-to-volume ratios of nanowires, their transport behavior may be modified by changing their surface conditions. For example, researchers have found that by coating n-InP nanowires with a layer of redox molecules, such as cobalt phthalocyanine, the conductance of the InP

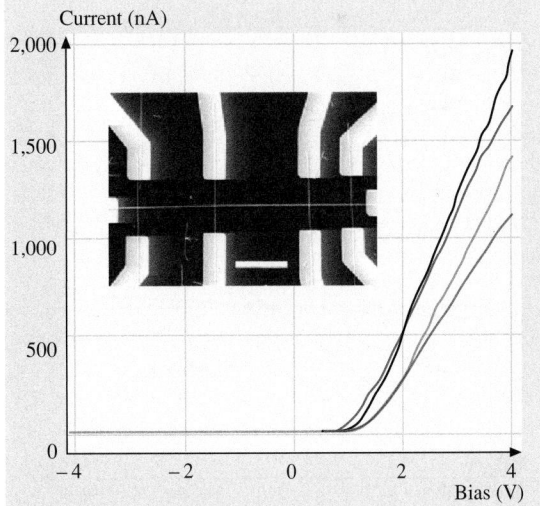

Fig. 4.22 I–V behavior for a 4(p) by 1(n) crossed p-Si/n-GaN junction array shown in the inset. The four curves represent the I–V response for each of the four junctions, showing similar current rectifying characteristics in each case. The length scale bar between the two middle junctions is 2 μm [4.142]. The p-Si and n-GaN nanowires are 10–25 nm and 10–30 nm in diameter, respectively

nanowires may change by orders of magnitude upon altering the charge state of the redox molecules to provide bistable nanoscale switches [4.144]. The resistance (or conductance) of some nanowires (such as Pd nanowires) is also very sensitive to the presence of certain gases (e.g., H_2) [4.145, 146], and this property may be utilized for sensor applications to provide improved sensitivity compared to conventional sensors based on bulk material (see Sect. 4.3.4).

Although it remains unclear how the size effect may influence the transport properties and device performance of semiconducting nanowires, many of the larger diameter semiconducting nanowires are expected to be described by classical physics, since their quantization energies $\hbar^2/(2m_e d_W^2)$ are usually smaller than the thermal energy $k_B T$. By comparing the quantization energy with the thermal energy, the critical wire diameter below which quantum confinement effects become significant is estimated to be 1 nm for Si nanowires at room temperature, which is much smaller than the sizes of many of the semiconducting nanowires that have been investigated so far. By using material systems with much smaller effective carrier masses m_e (such as bismuth), the critical diameter for which such quantum effects can be observed is increased, thereby facilitating the study of quantum confinement effects. It is for this rea-

son that the bismuth nanowire system has been studied so extensively. Furthermore, since the crystal structure and lattice constants of bismuth nanowires are the same as for 3-D crystalline bismuth, it is possible to carry out detailed model calculations to guide and to interpret transport and optical experiments on bismuth nanowires. For these reasons, bismuth can be considered a model system for studying 1-D effects in nanowires.

Temperature–Dependent Resistance Measurements

Although nanowires with electronic properties similar to their bulk counterparts are promising for constructing nanodevices based on well-established knowledge of their bulk counterparts, it is expected that quantum size effects in nanowires will likely be utilized to generate new phenomena absent in bulk materials, and thus provide enhanced performance and novel functionality for certain applications. In this context, the transport properties of bismuth (Bi) nanowires have been extensively studied, both theoretically [4.147] and experimentally [4.8, 10, 78, 148–150] because of their promise for enhanced thermoelectric performance. Transport studies of ferromagnetic nanowire arrays, such as Ni or Fe, have also received much attention because of their potential for high-density magnetic storage applications [4.151].

The very small electron effective mass components and the long carrier mean free paths in Bi facilitate the study of quantum size effects in the transport properties of nanowires. Quantum size effects are expected to become significant in bismuth nanowires with diameters smaller than 50 nm [4.147], and the fabrication of crystalline nanowires with this diameter range is relatively easy.

Figure 4.24a shows the T dependence of the resistance $R(T)$ for Bi nanowires ($7 \leq d_W < 200$ nm) synthesized by vapor deposition and pressure injection [4.8], illustrating the quantum effects in their temperature-dependent resistance. In Fig. 4.24a, the $R(T)$ behavior of Bi nanowires is dramatically different from that of bulk Bi, and is highly sensitive to the wire diameter. Interestingly, the $R(T)$ curves in Fig. 4.24a show a nonmonotonic trend for large-diameter (70 and 200 nm) nanowires, although $R(T)$ becomes monotonic with T for small-diameter (≤ 48 nm) nanowires. This dramatic change in the behavior of $R(T)$ as a function of d_W is attributed to a unique semimetal–semiconductor transition phenomena in Bi [4.78], induced by quantum size effects. Bi is a semimetal in bulk form, in which the T-point valence band overlaps with the L-point con-

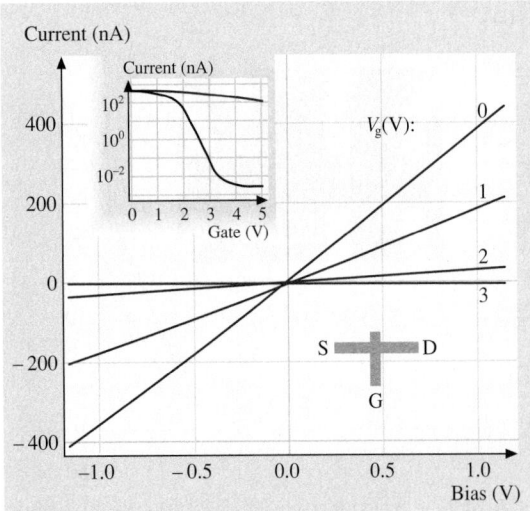

Fig. 4.23 Gate-dependent I–V characteristics of a crossed nanowire field-effect transistor (FET). The n-GaN nanowire is used as the nanogate, with the gate voltage indicated (0, 1, 2, and 3 V). The *inset* shows the current versus V_{gate} for a nanowire gate (*lower curve*) and for a global back-gate (*top curve*) when the bias voltage is set to 1 V [4.142]

duction band by 38 meV at 77 K. As the wire diameter decreases, the lowest conduction sub-band increases in energy and the highest valence sub-band decreases in energy. Model calculations predict that the band overlap should vanish in Bi nanowires (with their wire axes along the trigonal direction) at a wire diameter ≈ 50 nm [4.147].

The resistance of Bi nanowires is determined by two competing factors: the carrier density that increases with T, and the carrier mobility that decreases with T. The nonmonotonic $R(T)$ for large-diameter Bi nanowires is due to a smaller carrier concentration variation at low temperature (≤ 100 K) in semimetals, so that the electrical resistance is dominated by the mobility factor in this temperature range. Based on the semi-classical transport model and the estab-

lished band structure of Bi nanowires, the calculated $R(T)/R(300$ K$)$ for 36 nm and 70 nm Bi nanowires is shown by the solid curves in Fig. 4.24c to illustrate different $R(T)$ trends for semiconducting and semimetallic nanowires, respectively [4.78]. The curves in Fig. 4.24c exhibit trends consistent with experimental results. The condition for the semimetal–semiconductor transition in Bi nanowires can be experimentally determined, as shown by the measured resistance ratio $R(10$ K$)/R(100$ K$)$ of Bi nanowires as a function of wire diameter [4.152] in Fig. 4.25. The maximum in the resistance ratio $R(10$ K$)/R(100$ K$)$ at $d_W \approx 48$ nm indicates the wire diameter for the transition of Bi nanowires from a semimetallic phase to a semiconducting phase. The semimetal–semiconductor transition and the semiconducting phase in Bi nanowires are examples of new

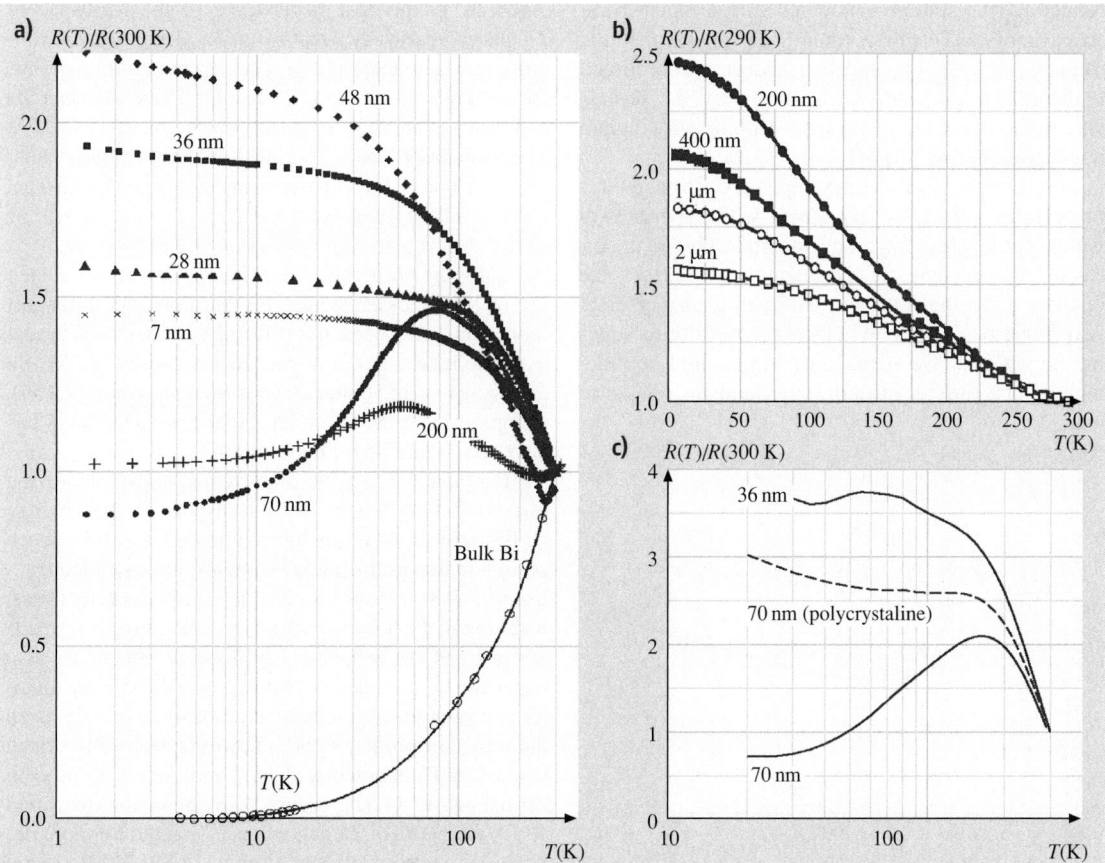

Fig. 4.24 (**a**) Measured temperature dependence of the resistance $R(T)$ normalized to the room temperature (300 K) resistance for bismuth nanowire arrays of various wire diameters d_W [4.8]. (**b**) $R(T)/R(290$ K$)$ for bismuth wires of larger d_W and lower mobility [4.10]. (**c**) Calculated $R(T)/R(300$ K$)$ of 36 nm and 70 nm bismuth nanowires. The *dashed curve* refers to a 70 nm polycrystalline wire with increased boundary scattering [4.78]

transport phenomena resulting from low dimensionality that are absent in the bulk 3-D phase, and these phenomena further increase the possible benefits from the properties of nanowires for desired applications (see Sect. 4.3.2).

It should be noted that good crystal quality is essential for observing the quantum size effect in nanowires, as shown by the $R(T)$ plots in Fig. 4.24a. For example, Fig. 4.24b shows the normalized $R(T)$ measurements of Bi nanowires with larger diameters (200 nm–2 μm) prepared by electrochemical deposition [4.10], and these nanowires possess monotonic $R(T)$ behaviors, quite different from those of the corresponding nanowire diameters shown in Fig. 4.24a. The absence of the resistance maximum in Fig. 4.24b is due to the lower crystalline quality for nanowires prepared by electrochemical deposition, which tends to produce polycrystalline nanowires with a much lower carrier mobility. This monotonic $R(T)$ for semimetallic Bi nanowires with a higher defect level is also confirmed by theoretical calculations, as shown by the dashed curve in Fig. 4.24c for 70 nm wires with increased grain boundary scattering [4.154].

The theoretical model developed for Bi nanowires not only provides good agreement with experimental results, but it also plays an essential role in understanding the influence of the quantum size effect, the boundary scattering, and the crystal quality on their electrical properties. While the electronic density of states may be significantly altered due to quantum confinement effects, various scattering mechanisms related to

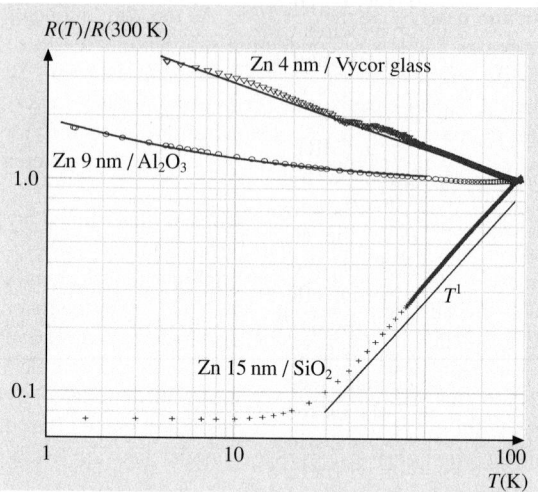

Fig. 4.26 Temperature dependence of the resistance of Zn nanowires synthesized by vapor deposition in various porous templates [4.52]. The data are given as points, the *full lines* are fits to a T^1 law for 15 nm diameter Zn nanowires in an SiO_2 template, denoted by Zn/SiO_2. Fits to a combined T^1 and $T^{-1/2}$ law were made for the smaller nanowire diameter composite samples denoted by Zn 9 nm /Al_2O_3 and Zn 4 nm/Vycor glass

the transport properties of nanowires can be accounted for by Matthiessen's rule. Furthermore, the transport model has also been generalized to predict the transport properties of Te-doped Bi nanowires [4.78], Sb nanowires [4.155], and BiSb alloy nanowires [4.156], and good agreement between experiment and theory has also been obtained for these cases.

For nanowires with diameters comparable to the phase-breaking length, their transport properties may be further influenced by localization effects. It has been predicted that in disordered systems, the extended electronic wavefunctions become localized near defect sites, resulting in the trapping of carriers and giving rise to different transport behavior. Localization effects are also expected to be more pronounced as the dimensionality and sample size are reduced. Localization effects on the transport properties of nanowire systems have been studied on Bi nanowires [4.157] and, more recently, on Zn nanowires [4.52]. Figure 4.26 shows the measured $R(T)/R(300 \text{ K})$ of Zn nanowires fabricated by vapor deposition in porous silica or alumina [4.52]. While 15 nm Zn nanowires exhibit an $R(T)$ behavior with a T^1 dependence as expected for a metallic wire, the $R(T)$ of 9 nm and 4 nm Zn nanowires exhibits a temperature dependence of $T^{-1/2}$ at low temperatures, consistent with

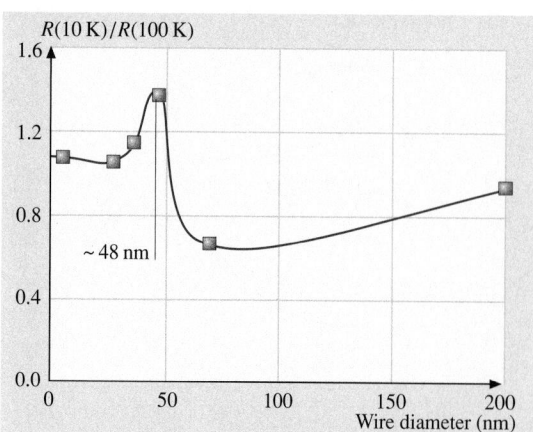

Fig. 4.25 Measured resistance ratio $R(10 \text{ K})/R(100 \text{ K})$ of Bi nanowire array as a function of diameter. The peak indicates the transition from a semimetallic phase to a semiconducting phase as the wire diameter decreases [4.153]

1-D localization theory. Thus, due to this localization effect, the use of nanowires with very small diameters for transport applications may be limited.

Magnetoresistance

Magnetoresistance (MR) measurements provide an informative technique for characterizing nanowires, because these measurements yield a great deal of information about the electron scattering with wire boundaries, the effects of doping and annealing on scattering, and localization effects in the nanowires [4.150]. For example, at low fields the MR data show a quadratic dependence on the B field from which carrier mobility estimates can be made (see Fig. 4.27 at low B field).

Figure 4.27 shows the longitudinal magnetoresistance (B parallel to the wire axis) for 65 nm and 109 nm Bi nanowire samples (before thermal annealing) at 2 K. The MR maxima in Fig. 4.27a are due to the classical size effect, where the wire boundary scattering is reduced as the cyclotron radius becomes smaller than the wire radius in the high field limit, resulting in a decrease in the resistivity. This behavior is typical for the longitudinal MR of Bi nanowires in the diameter range of 45 nm to 200 nm [4.8, 149, 150, 158], and the peak position B_m moves to lower B field values as the wire diameter increases, as shown in Fig. 4.27c [4.158], where B_m varies linearly with $1/d_W$. The condition for the occurrence of B_m is approximately given by $B_m \approx 2c\hbar k_F/ed_W$ where k_F is the wave vector at the Fermi energy. The peak position, B_m, is found to increase linearly with increasing temperature in the range of 2 to 100 K, as shown in Fig. 4.27b [4.158]. As T is increased, phonon scattering becomes increasingly important, and therefore a higher magnetic field is required to reduce the resistivity associated with boundary scattering sufficiently to change the sign of the MR. Likewise, increasing the grain boundary scattering is also expected to increase the value of B_m at a given T and wire diameter.

The presence of the peak in the longitudinal MR of nanowires requires a high crystal quality with long carrier mean free paths along the nanowire axis, so that most scattering events occur at the wire boundary instead of at a grain boundary, at impurity sites, or at defect sites within the nanowire. *Liu* et al. have investigated the MR of 400 nm Bi nanowires synthesized by electrochemical deposition [4.74], and no peak in the longitudinal MR is observed. The absence of a magnetoresistance peak may be attributed to a higher defect level in the nanowires produced electrochemically and to a large wire diameter, much longer than the carrier mean free

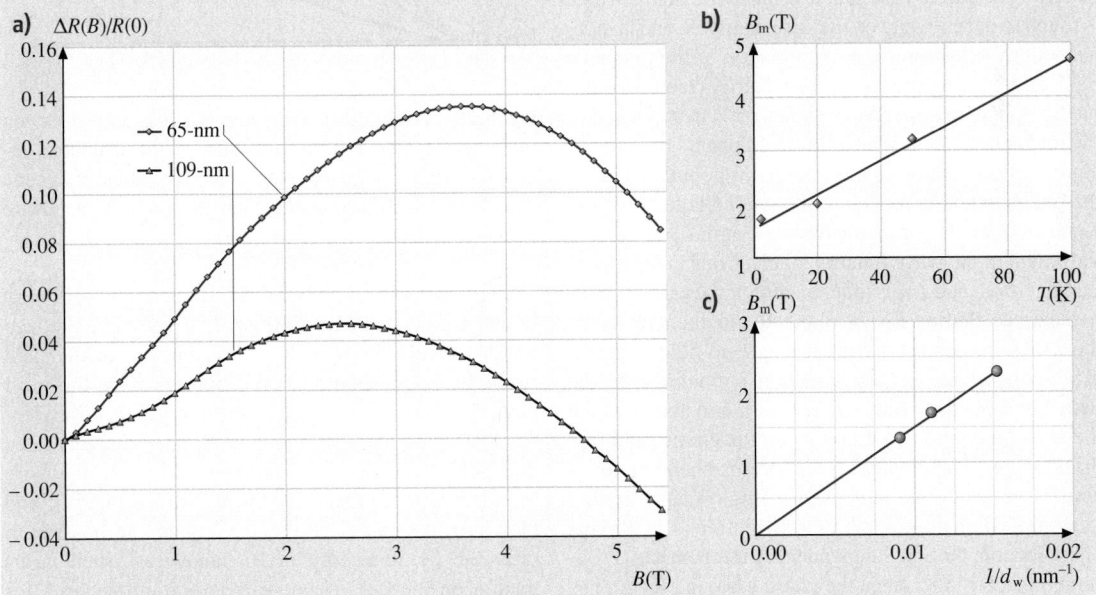

Fig. 4.27 (a) Longitudinal magnetoresistance, $\Delta R(B)/R(0)$, at 2 K as a function of B for Bi nanowire arrays with diameters of 65 and 109 nm before thermal annealing. **(b)** The peak position B_m as a function of temperature for the 109 nm diameter Bi nanowire array after thermal annealing. **(c)** The peak position B_m of the longitudinal MR (after thermal annealing) at 2 K as a function of $1/d_W$, the reciprocal of the nanowire diameter [4.158]

path. The negative MR observed for the Bi nanowire arrays above B_m (see Fig. 4.27) shows that wire boundary scattering is a dominant scattering process for the longitudinal magnetoresistance, thereby establishing that the mean free path is larger than the wire diameter and that a ballistic transport behavior is indeed observed in the high field regime.

In addition to the longitudinal magnetoresistance measurements, transverse magnetoresistance measurements ($\boldsymbol{B}$ perpendicular to the wire axis) have also been performed on Bi nanowire array samples [4.8, 150, 158], where a monotonically increasing B^2 dependence over the entire range $0 \leq B \leq 5.5\,T$ is found for all Bi nanowires studied thus far. This is as expected, since the wire boundary scattering cannot be reduced by a magnetic field perpendicular to the wire axis. The transverse magnetoresistance is also found to be always larger than the longitudinal magnetoresistance in nanowire arrays.

By applying a magnetic field to nanowires at very low temperatures ($\leq 5\,\mathrm{K}$), one can induce a transition from a 1-D confined system at low magnetic fields to a 3-D confined system as the field strength increases, as shown in Fig. 4.28 for the longitudinal MR of Bi nanowire arrays of various nanowire diameters (28–70 nm) for $T < 5\,\mathrm{K}$ [4.150]. In these curves, a subtle step-like feature is seen at low magnetic fields, which is found to depend only on the wire diameter, and is independent of temperature, the orientation of the magnetic field, and even on the nanowire material (see for example Sb nanowires [4.155]). The lack of a dependence of the magnetic field at which the step appears on temperature, field orientation, and material type indicates that the phenomenon is related to the magnetic field length, $L_H = (\hbar/eB)^{1/2}$. The characteristic length L_H is the spatial extent of the wave function of electrons in the lowest Landau level, and L_H is independent of the carrier effective masses. Setting $L_H(B_c)$ equal to the diameter d_W of the nanowire defines a critical magnetic field strength, B_c, below which the wavefunction is confined by the nanowire boundary (the 1-D regime), and above which the wavefunction is confined by the magnetic field (the 3-D regime). The physical basis for this phenomenon is associated with confinement of a single magnetic flux quantum within the nanowire cross-section [4.150]. This phenomenon, though independent of temperature, is observed for $T \leq 5\,\mathrm{K}$, since the phase breaking length has to be larger than the wire diameter. This calculated field strength, B_c, indicated in Fig. 4.28 by vertical lines for the appropriate nanowire diameters, provides a good fit to the step-like features in these MR curves.

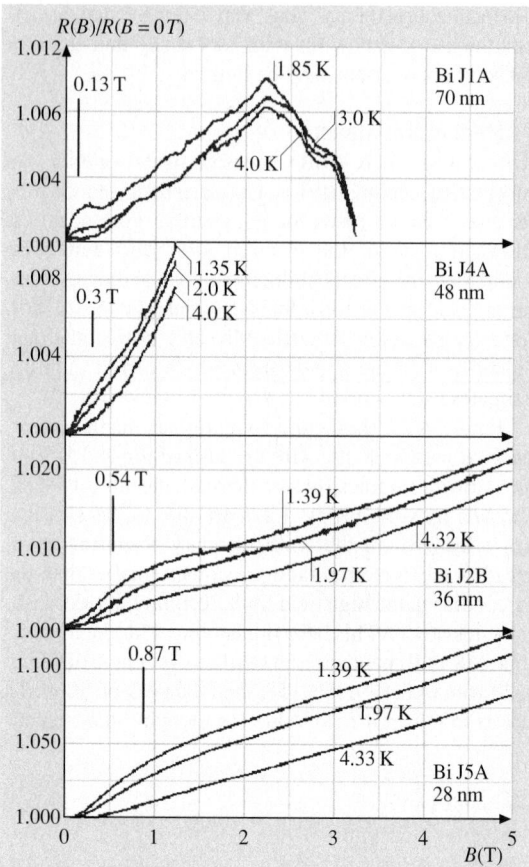

Fig. 4.28 Longitudinal magnetoresistance as a function of magnetic field for Bi nanowires of the diameters indicated. The *vertical bars* indicate the critical magnetic field B_c at which the magnetic length equals the nanowire diameter [4.150]

The Shubnikov–de Haas (SdH) quantum oscillatory effect, which results from the passage of the quantized Landau levels through the Fermi energy as the field strength varies, should, in principle, provide the most direct measurement of the Fermi energy and carrier density. For example, *Heremans* et al. have demonstrated that SdH oscillations can be observed in Bi nanowire samples with diameters down to 200 nm [4.159], and they have demonstrated that Te doping can be used to raise the Fermi energy in Bi nanowires. Such information on the Fermi energy is important because, for certain applications based on nanowires, it is necessary to place the Fermi energy near a sub-band edge where the density of states has a sharp feature. However, due to the unusual 1-D geometry of nanowires, other characterization techniques that are commonly used in bulk materials to determine the Fermi energy and the carrier concentration (such as Hall measurement) cannot be applied to nanowire systems. The observation of the SdH oscillatory effect requires crystal samples of very high quality which allow carriers to execute a complete cyclotron orbit in the nanowire before they are scattered. For small nanowire diameters, large magnetic fields are required to produce cyclotron radii smaller than the wire radius. For some nanowire systems, all Landau levels may have passed through the Fermi level at such a high field strength, and in such a case, no oscillations can be observed. The localization effect may also prevent the observation of SdH oscillations for very small diameter ($\leq 10\,\mathrm{nm}$) nanowires. Observing SdH oscillations in highly doped samples (as may be required for certain applications) may be difficult because impurity scattering reduces the mean free path, requiring high B fields to satisfy the requirement that carriers complete a cyclotron orbit prior to scattering. Therefore, although SdH oscillations provide the most direct method of measuring the Fermi energy and carrier density of nanowire samples, this technique may, however, not work for small-diameter nanowires, nor for nanowires that are heavily doped.

Thermoelectric Properties

Nanowires are predicted to hold great promise for thermoelectric applications [4.147, 161], due to their novel band structure compared to their bulk counterparts and the expected reduction in thermal conductivity associated with enhanced boundary scattering (see below). Due to the sharp density of states at the 1-D subband edges (where the van Hove singularities occur), nanowires are expected to exhibit enhanced Seebeck coefficients compared to their bulk counterparts. Since the Seebeck coefficient measurement is intrinsically independent of the number of nanowires contributing to the signal, the measurements on nanowire arrays of uniform wire diameter are, in principle, as informative as single-wire measurements. The major challenge with measuring the Seebeck coefficients of nanowires lies in the design of tiny temperature probes to accurately determine the temperature difference across the nanowire. Figure 4.29a shows the schematic experimental set-up for the Seebeck coefficient measurement of nanowire arrays [4.160], where two thermocouples are placed on both faces of a nanowire array and a heater is attached to one face of the array to generate a temperature gradient along the nanowire axis. Ideally, the size of the thermocouples should be much smaller than the thickness of

the nanowire array template (i. e. the nanowire length) to minimize error. However, due to the thinness of most templates ($\leq 50\,\mu\mathrm{m}$) and the large size of commercially-available thermocouples ($\approx 12\,\mu\mathrm{m}$), the measured Seebeck coefficient values are usually underestimated.

The thermoelectric properties of Bi nanowire systems have been investigated extensively because of their potential as good thermoelectric materials. Figure 4.29b shows the measured Seebeck coefficients $S(T)$ as a function of temperature for nanowire arrays with diameters of 40 and 65 nm and different isoelectronic Sb alloy concentrations [4.154], and $S(T)$ results for bulk Bi are shown (solid curve) for comparison. Thermopower enhancement is observed in Fig. 4.29b as the wire diameter decreases and as the Sb content increases, which is attributed to the semimetal–semiconductor transition induced by quantum confinement and to Sb alloying effects in $Bi_{1-x}Sb_x$ nanowires. *Heremans* et al. have

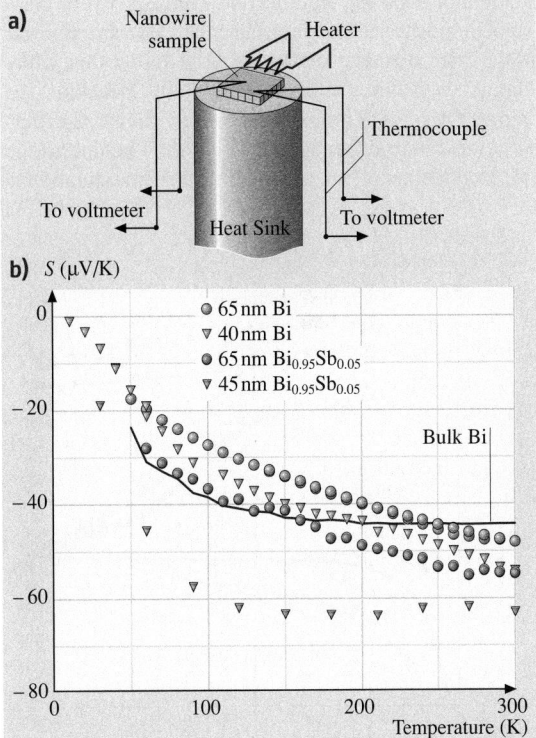

Fig. 4.29 (a) Experimental set-up for the measurement of the Seebeck coefficient in nanowire arrays [4.160]. **(b)** Measured Seebeck coefficient as a function of temperature for Bi ($\circ$, $\triangledown$) and $Bi_{0.95}Sb_{0.05}$ ($\bullet$, $\blacktriangledown$) nanowires with different diameters. The *solid curve* denotes the Seebeck coefficient for bulk Bi [4.154]

observed a substantial increase in the thermopower of Bi nanowires as the wire diameter decreases further, as shown in Fig. 4.30a for 15 nm Bi/silica and 9 nm Bi/alumina nanocomposites [4.52]. The enhancement is due to the sharp density of states near the Fermi energy in a 1-D system. Although the samples in Fig. 4.30a also possess very high electrical resistance ($\sim G\Omega$), the results for the 9 nm Bi/alumina samples show that the Seebeck coefficient can be enhanced by almost 1000 times relative to bulk material. However, for Bi nanowires with very small diameters (≈ 4 nm), the localization effect becomes dominant, which compromises the thermopower enhancement. Therefore, for Bi nanowires, the optimal wire diameter range for the largest thermopower enhancement is found to be between 4 and 15 nm [4.52].

The effect of the nanowire diameter on the thermopower of nanowires has also been observed in Zn nanowires [4.52]. Figure 4.30b shows the Seebeck coefficient of 9 nm Zn/alumina and 4 nm Zn/Vycor glass nanocomposites, also exhibiting enhanced thermopower as the wire diameter decreases. It is found that while 9 nm Zn nanowires still exhibit metallic behavior, the thermopower of 4 nm Zn nanowires shows a different temperature dependence, which may be due to the 1-D localization effect, although further investigation is

required for definitive identification of the conduction mechanism in such small nanowires.

Quantum Wire Superlattices

The studies on superlattice nanowires, which possess a periodic modulation in their materials composition along the wire axis, have attracted much attention recently because of their promise in various applications, such as thermoelectrics (see Sect. 4.3.2) [4.90, 162], nanobarcodes (see Sect. 4.3.3) [4.110], nanolasers (see Sect. 4.3.3) [4.92], one-dimensional waveguides, and resonant tunneling diodes [4.94, 163]. Figure 4.31a shows a schematic structure of a superlattice nanowire consisting of interlaced quantum dots of two different materials, as denoted by A and B. Various techniques have been developed to synthesize superlattice nanowire structures with different interface conditions, as mentioned in Sect. 4.1.1 and Sect. 4.1.2.

In this superlattice (SL) nanowire structure, the electronic transport along the wire axis is made possible by the tunneling between adjacent quantum dots, while the uniqueness of each quantum dot and its 0D characteristic behavior is maintained by the energy difference of the conduction or valence bands between quantum dots of different materials (see Fig. 4.31b), which provides some amount of quantum confinement. Recently,

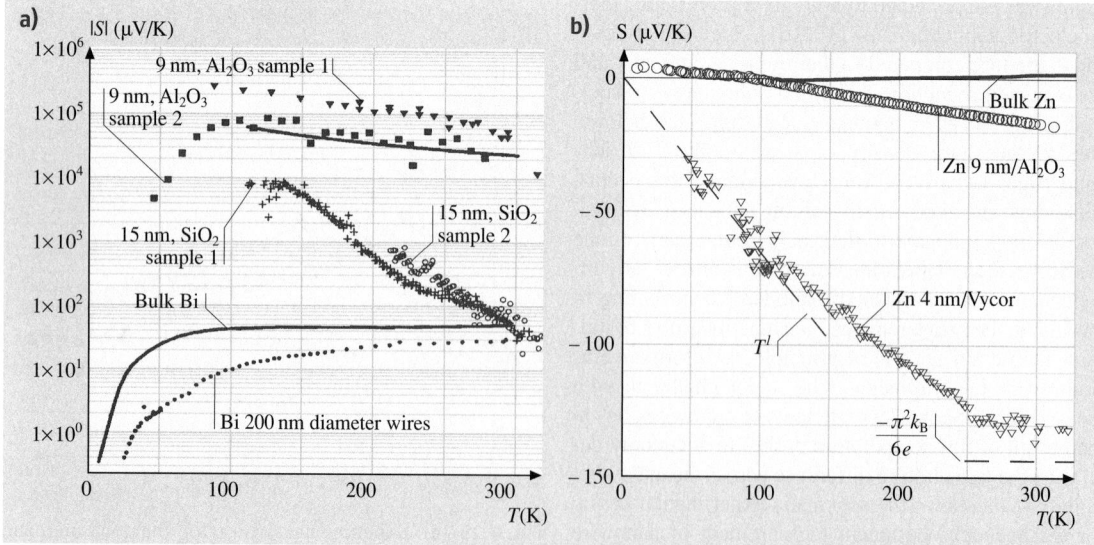

Fig. 4.30 (a) Absolute value of the Seebeck coefficient of two 15 nm Bi/silica and two 9 nm Bi/alumina nanocomposite samples, in comparison to bulk Bi and 200 nm Bi nanowires in the pores of alumina templates [4.52]. The *full line* on top part of the figure is a fit to a T^{-1} law. The Seebeck coefficient of the 9 nm Bi/alumina composite is positive; the rest are negative. (b) The Seebeck coefficient of 9 nm Zn/Al$_2$O$_3$ and 4 nm Zn/Vycor glass nanocomposite samples in comparison to bulk Zn [4.52]

Björk et al. have observed interesting nonlinear I–V characteristics with a negative differential resistance in one-dimensional heterogeneous structures made of InAs and InP, where InP serves as the potential barrier [4.94, 163]. The nonlinear I–V behavior is associated with the double barrier resonant tunneling process in one-dimensional structures, demonstrating that transport phenomena occur in superlattice nanowires via tunneling and the possibility of controlling the electronic band structure of the SL nanowires by carefully selecting the constituent materials. This new kind of structure is especially attractive for thermoelectric applications, because the interfaces between the nanodots can reduce the lattice thermal conductivity by blocking the phonon conduction along the wire axis, while electrical conduction may be sustained and even benefit from the unusual electronic band structures due to the periodic potential perturbation. For example, Fig. 4.32 shows the calculated dimensionless thermoelectric figure of merit $ZT = S^2 \sigma T/\kappa$ (see Sect. 4.3.2) where κ is the total thermal conductivity (including both the lattice and electronic contributions) of 10 nm-diameter PbS/PbSe superlattice nanowires as a function of the segment length. A higher thermoelectric performance than for $PbSe_{0.5}S_{0.5}$ alloy nanowires can be achieved for a 10 nm-diameter superlattice nanowire with segment lengths ≤ 7 nm. However, the localization effect, which may become important for very short segment lengths, may jeopardize this enhancement in the ZT of superlattice nanowires [4.153].

Thermal Conductivity of Nanowires

Experimental measurements of the temperature dependence of the thermal conductivity $\kappa(T)$ of individual

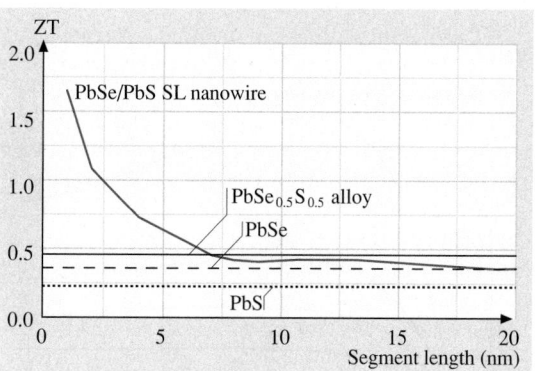

Fig. 4.32 Optimal ZT calculated as a function of segment length for 10 nm diameter PbSe/PbS nanowires at 77 K, where "optimal" refers to the placement of the Fermi level to optimize ZT. The optimal ZT for 10 nm-diameter PbSe, PbS, and $PbSe_{0.5}S_{0.5}$ nanowires are 0.33, 0.22, and 0.48, respectively [4.153]

suspended nanowires have been carried out on study the dependence of $\kappa(T)$ on wire diameter. In this context, measurements have been made on nanowires down to only 22 nm in diameter [4.164]. Such measurements are very challenging and are now possible due to technological development in the micro- and nano-fabrication of miniature thermal sensors, and the use of nanometer-size thermal scanning probes [4.128, 165, 166]. The experiments show that the thermal conductivity of small homogeneous nanowires may be more than one order of magnitude smaller than in the bulk, due mainly to strong boundary scattering effects [4.167]. Phonon confinement effects may eventually become important in nanowires with even smaller diameters. Measurements on mats of nanowires (see, for example, Fig. 4.12) do not generally give reliable results because the contact thermal resistance between adjacent nanowires tends to be high, which is in part due to the thin surface oxide coating which most nanowires have. This surface oxide coating may also be important for thermal conductivity measurements on individual suspended nanowires because of the relative importance of phonon scattering at the lateral walls of the nanowire.

The most extensive experimental thermal conductivity measurements have been done on Si nanowires [4.164], where $\kappa(T)$ measurements have been made on nanowires in the diameter range $22 \leq d_W \leq 115$ nm. The results show a large decrease in the peak of $\kappa(T)$, associated with Umklapp processes as d_W decreases, indicating a growing importance of boundary scattering and a corresponding decreasing im-

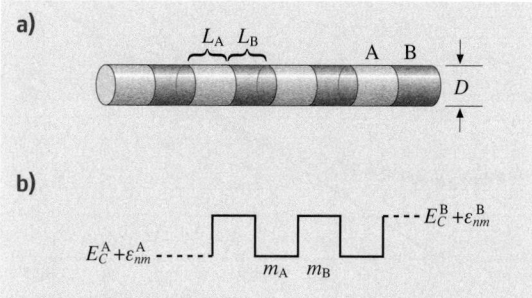

Fig. 4.31 (a) Schematic diagram of superlattice (segmented) nanowires consisting of interlaced nanodots A and B of the indicated length and wire diameter. (b) Schematic potential profile of the sub-bands in the superlattice nanowire [4.162]

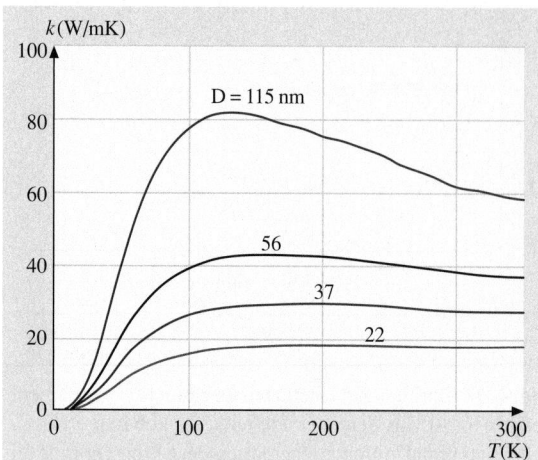

Fig. 4.33 Predicted thermal conductivities of Si nanowires of various diameters [4.169]

portance of phonon-phonon scattering. At the smallest wire diameter of 22 nm, a linear $\kappa(T)$ dependence is found experimentally, consistent with a linear T dependence of the specific heat for a 1-D system, and a temperature-independent mean free path and velocity of sound. Further insights are obtained through studies of the thermal conductivity of Si/SiGe superlattice nanowires [4.168].

Model calculations for $\kappa(T)$ based on a radiative heat transfer model have been carried out for Si nanowires [4.169]. These results show that the predicted $\kappa(T)$ behavior for Si nanowires is similar to that observed experimentally in the range of $37 \leq d_W \leq 115$ nm

regarding both the functional form of $\kappa(T)$ and the magnitude of the relative decrease in the maximum thermal conductivity κ_{max} as a function of d_W. However, the model calculations predict a substantially larger magnitude for $\kappa(T)$ (by 50% or more) than is observed experimentally. Furthermore, the model calculations (see Fig. 4.34) do not reproduce the experimentally observed linear T dependence for the 22 nm nanowires, but rather predict a 3-D behavior for both the density of states and the specific heat in 22 nm nanowires [4.169, 171, 172].

Thermal conductance measurements on GaAs nanowires below 6 K show a power law dependence, but the T dependence becomes somewhat less pronounced below ≈ 2.5 K [4.165]. This deviation from the power law temperature dependence led to a more detailed study of the quantum limit for the thermal conductance. To carry out these more detailed experiments, a mesoscopic phonon resonator and waveguide device were constructed that included four ≈ 200 nm-wide and 85 nm-thick silicon nitride nanowire-like nanoconstrictions (see Fig. 4.33a), and this was used to establish the quantized thermal conductance limit of $g_0 = \pi^2 k_B^2 T/3h$ (see Fig. 4.33b) for ballistic phonon transport [4.170, 173]. For temperatures above 0.8 K, the thermal conductance in Fig. 4.33b follows a T^3 law, but as T is further reduced, a transition to a linear T dependence is observed, consistent with a phonon mean free path of ≈ 1 μm, and a thermal conductance value approaching $16g_0$, corresponding to four massless phonon modes per channel and four channels in their phonon waveguide structure (see Fig. 4.33a). Ballistic phonon

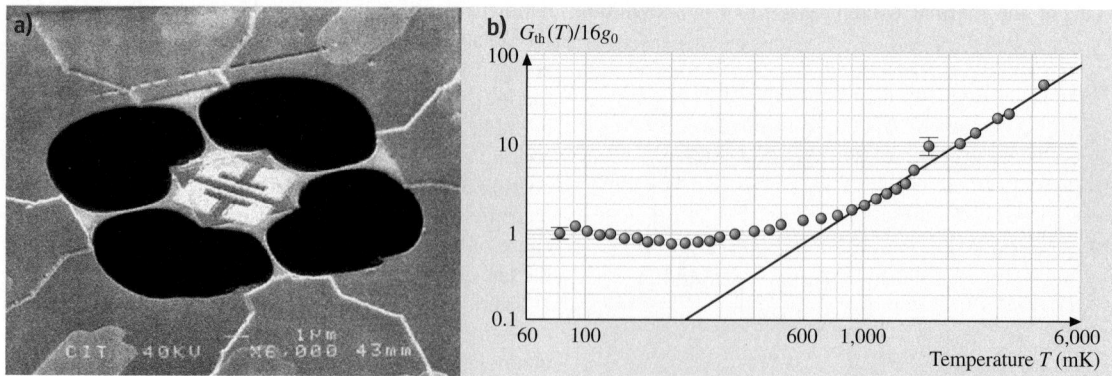

Fig. 4.34 (a) Suspended mesoscopic phonon device used to measure ballistic phonon transport. The device consists of an 4×4 μm "phonon cavity" (*center*) connected to four Si_3N_4 membranes, 60 nm thick and less than 200 nm wide. The two bright "C"-shaped objects on the phonon cavity are thin film heating and sensing Cr/Au resistors, whereas the dark regions are empty space. (b) Log-log plot of the temperature dependence of the thermal conductance G_0 of the structure in (a) normalized to $16G_0$ (see text) [4.170]

transport occurs when the thermal phonon wavelength (380 nm for the experimental structure) is somewhat greater than the width of the phonon waveguide at the waveguide constriction.

4.2.4 Optical Properties

Optical methods provide an easy and sensitive tool for measuring the electronic structures of nanowires, since optical measurements require minimal sample preparation (for example, contacts are not required) and the measurements are sensitive to quantum effects. Optical spectra of 1-D systems, such as carbon nanotubes, often show intense features at specific energies near singularities in the joint density of states that are formed under strong quantum confinement conditions. A variety of optical techniques have shown that the properties of nanowires are different to those of their bulk counterparts, and this section of the review focuses on these differences in the optical properties of nanowires.

Although optical properties have been shown to provide an extremely important tool for characterizing nanowires, the interpretation of these measurements is not always straightforward. The wavelength of light used to probe the sample is usually smaller than the wire length, but larger than the wire diameter. Hence, the probe light used in an optical measurement cannot be focused solely onto the wire, and the wire and the substrate on which the wire rests (or host material, if the wires are embedded in a template) are probed simultaneously. For measurements, such as photoluminescence (PL), if the substrate does not luminescence or absorb in the frequency range of the measurements, PL measures the luminescence of the nanowires directly and the substrate can be ignored. However, in reflection and transmission measurements, even a non-absorbing substrate can modify the measured spectra of nanowires.

In this section we discuss the determination of the dielectric function for nanowires in the context of effective medium theories. We then discuss various optical techniques with appropriate examples that sensitively differentiate nanowire properties from those also found in the parent bulk material, placing particular emphasis on electronic quantum confinement effects. Finally, phonon confinement effects are reviewed.

The Dielectric Function

In this subsection, we review the use of effective medium theory as a method to handle the optical properties of nanowires whose diameters are typically smaller than the wavelength of light, noting that observable optical properties of materials can be related to the complex dielectric function [4.174, 175]. Effective medium theories [4.176, 177] can be applied to model the nanowire and substrate as one continuous composite with a single complex dielectric function ($\epsilon_1 + i\epsilon_2$), where the real and imaginary parts of the dielectric function ϵ_1 and ϵ_2 are related to the index of refraction (n) and the absorption coefficient (K) by the relation $\epsilon_1 + i\epsilon_2 = (n + iK)^2$. Since photons at visible or infrared wavelengths "see" a dielectric function for the composite nanowire array/substrate system that is different from that of the nanowire itself, the optical transmission and reflection are different from what they would be if the light were focused only on the nanowire. One commonly observed consequence of effective medium theory is the shift in the plasma frequency in accordance with the percentage of nanowire material that is contained in the composite [4.178]. The plasma resonance occurs when $\epsilon_1(\omega)$ becomes zero, and the plasma frequency of the nanowire composite will shift to lower (higher) energies when the magnitude of the dielectric function of the host materials is larger (smaller) than that of the nanowire.

Although reflection and transmission measurements probe both the nanowire and the substrate, the optical properties of the nanowires can be determined independently. One technique for separating out the dielectric function of the nanowires from the host is to use an effective medium theory in reverse. Since the dielectric function of the host material is often known, and the dielectric function of the composite material can be measured by the standard method of using reflection and transmission measurements in combination with either the Kramer–Kronig relations or Maxwell's equations, the complex dielectric function of the nanowires can be deduced. An example where this approach has been used successfully is for the determination of the frequency dependence of the real and imaginary parts of the dielectric function $\epsilon_1(\omega)$ and $\epsilon_2(\omega)$ for a parallel array of bismuth nanowires filling the pores of an alumina template [4.179].

Characteristic Optical Properties of Nanowires

A wide range of optical techniques are available for the characterization of nanowires, to distinguish their properties from those of their parent bulk materials. Some differences in properties relate to geometric differences, such as the small diameter size and the large length-to-diameter ratio (also called the aspect ratio), while others focus on quantum confinement issues.

Probably the most basic optical technique is to measure the reflection and/or transmission of a nanowire to determine the frequency- dependent real and imaginary parts of the dielectric function. This technique has been used, for example, to study the band gap and its temperature dependence in gallium nitride nanowires in the 10–50 nm range in comparison to bulk values [4.180]. The plasma frequency, free carrier density, and donor impurity concentration as a function of temperature were also determined from the infrared spectra, which is especially useful for nanowire research, since Hall effect measurements cannot be made on nanowires.

Another common method used to study nanowires is photoluminescence (PL) or fluorescence spectroscopy. Emission techniques probe the nanowires directly and the effect of the host material does not have to be considered. This characterization method has been used to study many properties of nanowires, such as the optical gap behavior, oxygen vacancies in ZnO nanowires [4.55], strain in Si nanowires [4.181], and quantum confinement effects in InP nanowires [4.182]. Figure 4.35 shows the photoluminescence of InP nanowires as a function of wire diameter, thereby providing direct information on the effective bandgap. As the wire diameter of an InP nanowire is decreased so that it becomes smaller than the bulk exciton diameter of 19 nm, quantum confinement effects set in, and the band gap is increased. This results in an increase in the PL peak energy. The smaller the effective mass, the larger the quantum confinement effects. When the shift in the peak energy as a function of nanowire diameter Fig. 4.35 is analyzed using an effective mass model, the reduced effective mass of the exciton is deduced to be $0.052 \, m_0$, which agrees quite well with the literature value of $0.065 \, m_0$ for bulk InP. Although the linewidths of the PL peak for the small-diameter nanowires (10 nm) are smaller at low temperature (7 K), the observation of strong quantum confinement and bandgap tunability effects at room temperature are significant for photonics applications of nanowires (see Sect. 4.3.3).

The resolution of photoluminescence (PL) optical imaging of a nanowire is, in general, limited by the wavelength of light. However, when a sample is placed very close to the detector, the light is not given a chance to diffract, and so samples much smaller than the wavelength of light can be resolved. This technique is known as near-field scanning optical microscopy (NSOM) and has been used to successfully image nanowires [4.183]. For example, Fig. 4.36 shows the topographical (a) and (b) NSOM PL images of a single ZnO nanowire.

Magneto-optics can be used to measure the electronic band structure of nanowires. For example, magneto-optics in conjunction with photoconductance

has been proposed as a tool to determine band parameters for nanowires, such as the Fermi energy, electron effective masses, and the number of sub-bands to be considered [4.184]. Since different nanowire sub-bands have different electrical transmission properties, the electrical conductivity changes when light is used to excite electrons to higher subbands, thereby providing a method for studying the electronic structure of nanowires optically. Magneto-optics can also be used to study the magnetic properties of nanowires in relation to bulk properties [4.27, 185]. For example, the surface magneto-optical Kerr effect has been used to measure the dependence of the magnetic ordering temperature of Fe–Co alloy nanowires on the relative concentration of Fe and Co [4.185], and it was used to find that, unlike in the case of bulk Fe–Co alloys, cobalt in nanowires inhibits magnetic ordering. Nickel nanowires were found to have a strong increase in their magneto-optical activity with respect to bulk nickel. This increase is attributed to the plasmon resonance in the wires [4.186].

Nonlinear optical properties of nanowires have received particular attention since the nonlinear behavior is often enhanced compared to bulk materials and the nonlinear effects can be utilized for many applications. One such study measured the second harmonic generation (SHG) and third harmonic generation (THG) in a single nanowire using near-field optical microscopy [4.187]. ZnO nanowires were shown to have strong SHG and THG effects that are highly polarization-sensitive, and this polarization sensitivity can be explained on the basis of optical and geometrical considerations. Some components of the second harmonic polarization tensor are found to be enhanced in nanowires while others are suppressed as the wire diameter is decreased, and such effects could be of interest for device applications. The authors also showed that the second-order nonlinearities are mostly wavelength-independent for $\lambda < 400$ nm, which is in the transparent regime for ZnO, below the onset of band gap absorption, and this observation is also of interest for device applications.

Reflectivity and transmission measurements have also been used to study the effects of quantum confinement and surface effects on the low-energy indirect transition in bismuth nanowires [4.188]. *Black* et al. investigated an intense and sharp absorption peak in bismuth nanowires, which is not observed in bulk bismuth. The energy position E_p of this strong absorption peak increases with decreasing diameter. However, the rate of increase in energy with decreasing diameter $|\partial E_p / \partial d_W|$ is an order of magnitude less than that predicted for either a direct interband transition or for intersub-band

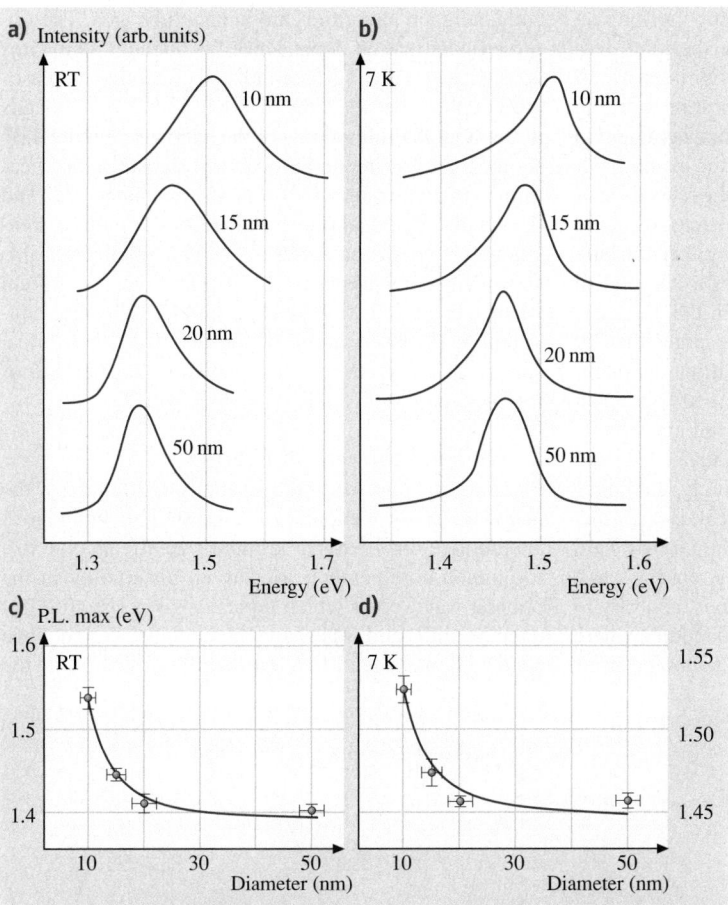

Fig. 4.35a–d Photo-luminescence of InP nanowires of varying diameters at 7 K ((**b**)and (**d**)) and room temperature ((**a**) and (**c**)) showing quantum confinement effects of the exciton for wire diameters of less than 20 nm [4.182]

Part A | 4.2

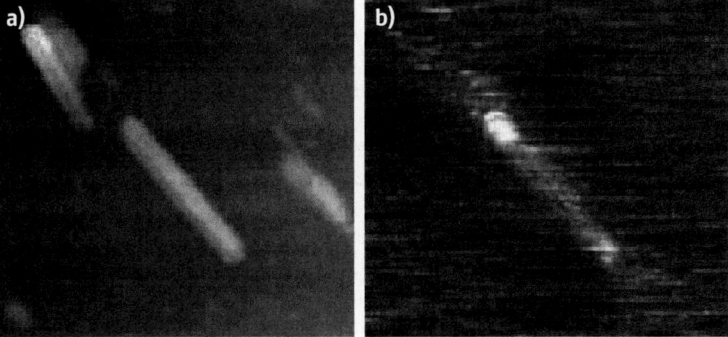

Fig. 4.36 (**a**) Topographical and (**b**) photoluminescence (PL) near-field scanning optical microscopy (NSOM) images of a single ZnO nanowire waveguide [4.183]

transitions in bismuth nanowires. On the other hand, the magnitude of $|\partial E_p/\partial d_W|$ agrees well with that predicted for an indirect L-point valence to T-point valence band transition (see Fig. 4.37). Since both the initial and final states for the indirect L–T point valence band transition downshift in energy as the wire diameter d_W is decreased, the shift in the absorption peak results from a difference between the effective masses and not from the actual value of either of the masses. Hence the diameter dependence of the absorption peak energy is an order of magnitude less for a valence to valence band indirect transition than for a direct interband L-point transition.

Furthermore, the band-tracking effect for the indirect transition gives rise to a large value for the joint density of states, thus accounting for the high intensity of this feature. The enhancement in the absorption resulting from this indirect transition may arise from a gradient in the dielectric function, which is large at the bismuth–air or bismuth–alumina interfaces, or from the relaxation of momentum conservation rules in nanosystems. It should be noted that, in contrast to the surface effect for bulk samples, the whole nanowire contributes to the optical absorption due to the spatial variation in the dielectric function, since the penetration depth is larger than or comparable to the wire diameter. In addition, the intensity can be quite significant because there are abundant initial state electrons, final state holes, and appropriate phonons for making an indirect L–T point valence band transition at room temperature. Interestingly, the polarization dependence of this absorption peak is such that the strong absorption is present when the electric field is perpendicular to the wire axis, but is absent when the electric field is parallel to the wire axis, contrary to a traditional polarizer, such as a carbon nanotube where the optical E field is polarized by the nanotube itself

and is aligned along the carbon nanotube axis. The observed polarization dependence for bismuth nanowires is consistent with a surface-induced effect that increases the coupling between the L-point and T-point bands throughout the full volume of the nanowire. Figure 4.37 shows the experimentally observed transmission spectrum in bismuth nanowires of ≈ 45 nm diameter (a), and the simulated optical transmission from an indirect transition in bismuth nanowires of ≈ 45 nm diameter is also shown for comparison in (b). The indirect L–T point valence band transition mechanism [4.189] is also consistent with observations of the effect on the optical spectra of a decrease in the nanowire diameter and of n-type doping of bismuth nanowires with Te.

Phonon Confinement Effects

Phonons in nanowires are spatially confined by the nanowire cross-sectional area, crystalline boundaries and surface disorder. These finite size effects give rise to phonon confinement, causing an uncertainty in the phonon wavevector which typically gives rise to a frequency shift and lineshape broadening. Since zone center phonons tend to correspond to maxima in the

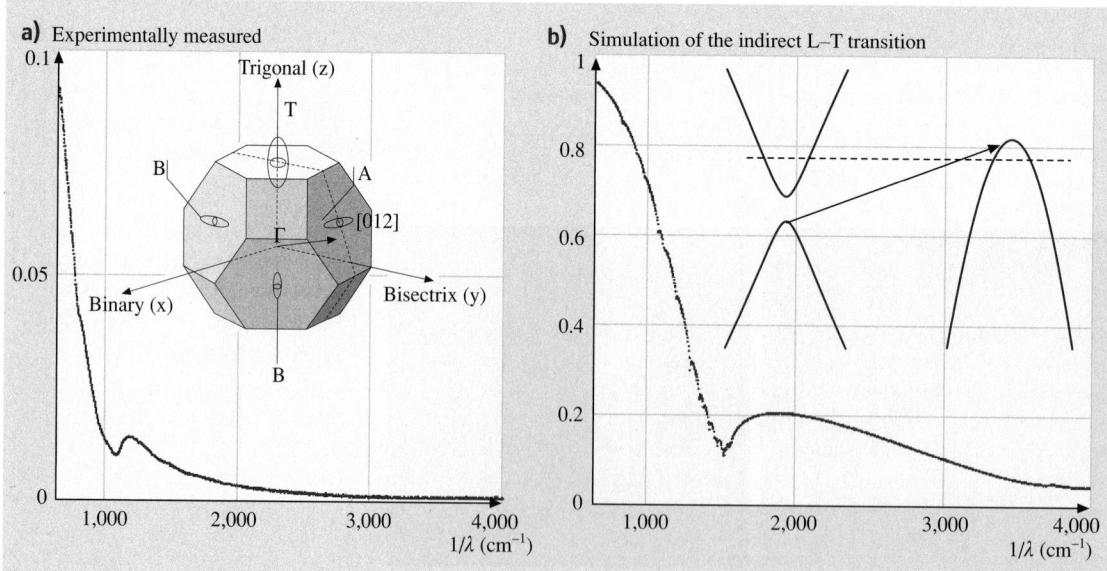

Fig. 4.37 (a) The measured optical transmission spectra as a function of wavenumber ($1/\lambda$) of a ≈ 45 nm-diameter bismuth nanowire array. **(b)** The simulated optical transmission spectrum resulting from an indirect transition of an L point electron to a T point valence sub-band state. The *insert* in **(a)** shows the bismuth Brillouin zone, and the locations of the T-point hole and the three L-point electron pockets, including the nondegenerate A, and the doubly-degenerate B pockets. The *insert* in **(b)** shows the indirect L to T point electronic transition induced by a photon with an energy equal to the energy difference between the initial and final states minus the phonon energy (about 100 cm^{-1}) needed to satisfy conservation of energy in a Stokes process [4.189]

phonon dispersion curves, the inclusion of contributions from a broader range of phonon wave vectors results in both a downshift in frequency and an asymmetric broadening of the Raman line, which develops a low frequency tail. These phonon confinement effects have been theoretically predicted [4.190, 191] and experimentally observed in GaN [4.192], as shown in Fig. 4.38 for GaN nanowires with diameters in the range 10–50 nm. The application of these theoretical models indicates that broadening effects should be noticeable as the wire diameter in GaN nanowires decreases to ≈ 20 nm. When the wire diameter decreases further to ≈ 10 nm, the frequency downshift and asymmetric Raman line broadening effects should become observable in the Raman spectra for the GaN nanowires but are not found in the corresponding spectra for bulk GaN.

The experimental spectra in Fig. 4.38 show the four $A_1 + E_1 + 2E_2$ modes expected from symmetry considerations for bulk GaN crystals. Two types of quantum confinement effects are observed. The first type is the observation of the downshift and the asymmetric broadening effects discussed above. Observations of such downshifts and asymmetric broadening have also been recently reported in 7 nm diameter Si nanowires [4.193]. A second type of confinement effect found in Fig. 4.38 for GaN nanowires is the appearance of additional Raman features not found in the corresponding bulk spectra and associated with combination modes, and a zone boundary mode. Resonant enhancement effects

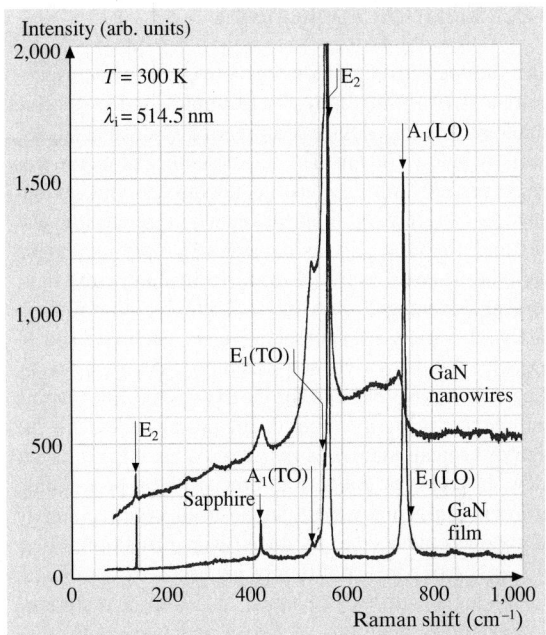

Fig. 4.38 Room-temperature Raman scattering spectra of GaN nanowires and of a 5 μm thick GaN epilayer film with green (514.5 nm) laser excitation. The Raman scattering response was obtained by dividing the measured spectra by the Bose–Einstein thermal factor [4.192]

were also observed for the $A_1(LO)$ phonon at 728 cm^{-1} (see Fig. 4.38) at higher laser excitation energies [4.192].

4.3 Applications

In the preceding sections we have reviewed many of the central characteristics that make nanowires in some cases similar to and in some cases very different from their parent materials. We have also shown that some properties are diameter-dependent, and these properties are therefore tunable during synthesis. Thus, it is of great interest to find applications that could benefit in unprecedented ways from both the unique and tunable properties of nanowires and the small sizes of these nanostructures, especially in the miniaturization of conventional devices. As the synthetic methods for the production of nanowires are maturing (Sect. 4.1) and nanowires can be made in reproducible and cost-effective ways, it is only a matter of time before applications will be seriously explored. This is a timely development, as the semiconductor industry will soon be reaching what seems

to be its limit in feature size reduction, and approaching a classical-to-quantum size transition. At the same time, the field of biotechnology is expanding through the availability of tremendous genome information and innovative screening assays. Since nanowires are similar in size to the shrinking electronic components and to cellular biomolecules, it is only natural for nanowires to be good candidates for applications in these fields. Commercialization of nanowire devices, however, will require reliable mass production, effective assembly techniques and quality control methods.

In this section, applications of nanowires to electronics (Sect. 4.3.1), thermoelectrics (Sect. 4.3.2), optics (Sect. 4.3.3), chemical and biochemical sensing (Sect. 4.3.4), and magnetic media (Sect. 4.3.5) are discussed.

4.3.1 Electrical Applications

The microelectronics industry continues to face technological (in lithography for example) and economic challenges as the device feature size is decreased, especially below 100 nm. The self-assembly of nanowires might present a way to construct unconventional devices that do not rely on improvements in photolithography and, therefore, do not necessarily imply increasing fabrication costs. Devices made from nanowires have several advantages over those made by photolithography. A variety of approaches have been devised to organize nanowires via self-assembly (see Sect. 4.1.4), thus eliminating the need for the expensive lithographic techniques normally required to produce devices the size of typical nanowires that are discussed in this review. In addition, unlike traditional silicon processing, different semiconductors can be used simultaneously in nanowire devices to produce diverse functionalities. Not only can wires of different materials be combined, but a single wire can be made of different materials. For example, junctions of GaAs and GaP show rectifying behavior [4.92], thus demonstrating that good electronic interfaces between two different semiconductors can be achieved in the synthesis of multicomponent nanowires. Transistors made from nanowires could also hold advantages due to their unique morphology. For example, in bulk field effect transistors (FETs), the depletion layer formed below the source and drain region results in a source–drain capacitance which limits the operation speed. However, in nanowires, the conductor is surrounded by an oxide and thus the depletion layer cannot be formed. Thus, depending on the device design, the source–drain capacitance in nanowires could be greatly minimized and possibly eliminated.

Device functionalities common in conventional semiconductor technologies, such as p-n junction diodes [4.142], field-effect transistors [4.144], logic gates [4.142], and light-emitting diodes [4.92, 194], have been recently demonstrated in nanowires, showing their promise as building blocks that could be used to construct complex integrated circuits by employing the "bottom-up" paradigm. Several approaches have been investigated to form nanowire diodes (see Sect. 4.2.2). For example, Schottky diodes can be formed by contacting a GaN nanowire with Al electrodes [4.143]. Furthermore, p-n junction diodes can be formed at the crossing of two nanowires, such as the crossing of n- and p-type InP nanowires doped by Te and Zn, respectively [4.194], or Si nanowires doped by phosphorus (n-type) and boron (p-type) [4.195]. In addition to the

crossing of two distinctive nanowires, heterogeneous junctions have also been constructed inside a single wire, either along the wire axis in the form of a nanowire superlattice [4.92], or perpendicular to the wire axis by forming a core–shell structure of silicon and germanium [4.111]. These various nanowire junctions not only possess the current rectifying properties (see Fig. 4.22) expected of bulk semiconductor devices, but they also exhibit electroluminescence (EL) that may be interesting for optoelectronic applications, as shown in Fig. 4.39 for the electroluminescence of a crossed junction of n- and p-type InP nanowires [4.194] (see Sect. 4.3.3).

In addition to the two-terminal nanowire devices, such as the p-n junctions described above, it is found that the conductance of a semiconductor nanowire can be significantly modified by applying voltage at a third gate terminal, implying the utilization of nanowires in field effect transistors (FETs). This gate terminal can either be the substrate [4.30, 196–199], a separate metal contact located close to the nanowire [4.200], or another nanowire with a thick oxide coating in the crossed nanowire junction configuration [4.142]. The operating principles of these nanowire-based FETs are discussed in Sect. 4.2.2. Various logic devices performing basic logic functions have been demonstrated using nanowire junctions [4.142], as shown in Fig. 4.40 for the OR and AND logic gates constructed from 2-by-1 and 1-by-3 nanowire p-n junctions, respectively. By functionalizing nanowires with redox-active molecules to store charge, nanowire FETs were demonstrated with two-level [4.144] and with eight-level [4.201] memory effects, which may be used for nonvolatile memory or as switches. In another advance, In_2O_3 nanowire FETs with high-k dielectric material were demonstrated, and substantially enhanced performance was obtained due to the highly efficient coupling of the gate [4.202]. A vertical FET with a surrounding gate geometry has also been demonstrated, which has the potential for high-density nanoscale memory and logic devices [4.203].

Nanowires have also been proposed for applications associated with electron field emission [4.204], such as flat panel displays, because of their small diameter and large curvature at the nanowire tip, which may reduce the threshold voltage for electron emission [4.205]. In this regard, the demonstration of very high field emission currents from the sharp tip (≈ 10 nm radius) of a Si cone [4.204], from carbon nanotubes [4.206], from Si nanowires inside a carbon nanotube [4.207], and from Co nanowires [4.208], has stimulated interest in this potential area of application for nanowires.

The concept of constructing electronic devices based on nanowires has already been demonstrated, and the next step for electronic applications would be to devise a feasible method for integration and mass production. We expect that, in order to maintain the growing rate of device density and functionality in the existing electronic industry, new kinds of complementary electronic devices will emerge from this "bottom-up" scheme for nanowire electronics, different from what has been produced by the traditional "top-down" approach pursued by conventional electronics.

4.3.2 Thermoelectric Applications

One proposed application for nanowires is for thermoelectric cooling and for the conversion between thermal and electrical energy [4.171, 209]. The efficiency of a thermoelectric device is measured in terms of a dimensionless figure of merit ZT, where Z is defined as

$$Z = \frac{\sigma S^2}{\kappa}, \qquad (4.2)$$

where σ is the electrical conductivity, S is the Seebeck coefficient, κ is the thermal conductivity, and T is the temperature. In order to achieve a high ZT and therefore efficient thermoelectric performance, a high electrical conductivity, a hugh Seebeck coefficient and

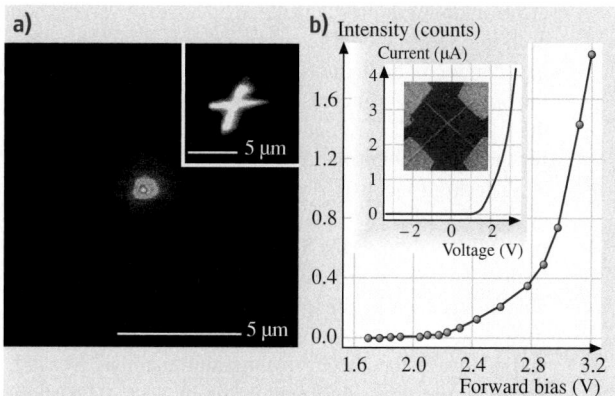

Fig. 4.39a,b Optoelectrical characterization of a crossed nanowire junction formed between 65 nm n-type and 68 nm p-type InP nanowires. (**a**) Electroluminescence (EL) image of the light emitted from a forward-biased nanowire p-n junction at 2.5 V. *Inset*, photoluminescence (PL) image of the junction. (**b**) EL intensity as a function of operation voltage. *Inset*, the SEM image and the I–V characteristics of the junction [4.194]. The scale bar in the inset is 5 μm

a low thermal conductivity are required. In 3-D systems, the electronic contribution to κ is proportional to σ in accordance with the Wiedemann–Franz law, and normally materials with high S have a low σ. Hence an increase in the electrical conductivity (for example by

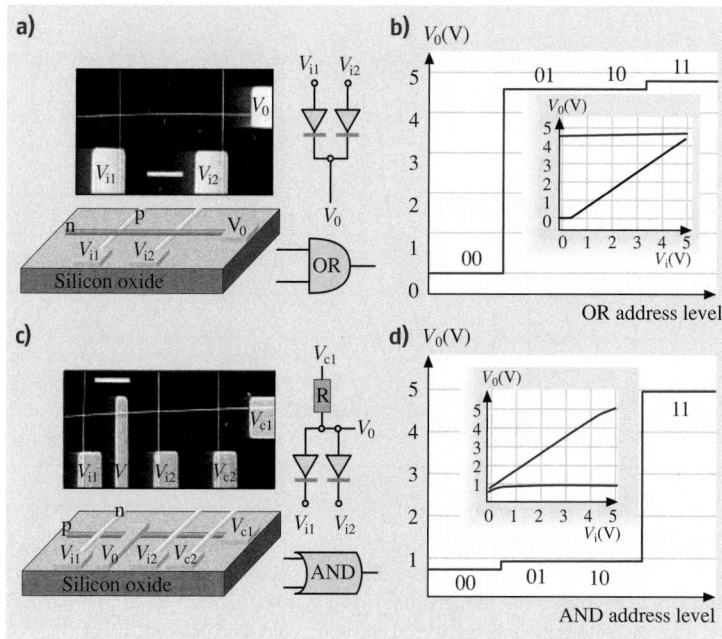

Fig. 4.40a–d Nanowire logic gates: (**a**) Schematic of logic OR gate constructed from a 2(p-Si) by 1(n-GaN) crossed nanowire junction. The *inset* shows the SEM image (bar: 1 μm) of an assembled OR gate and the symbolic electronic circuit. (**b**) The output voltage of the circuit in (**a**) versus the four possible logic address level inputs: (0,0); (0,1); (1,0); (1,1), where logic 0 input is 0 V and logic 1 is 5 V (same for below). (**c**) Schematic of logic AND gate constructed from a 1(p-Si) by 3(n-GaN) crossed nanowire junction. The *inset* shows the SEM image (bar: 1 μm) of an assembled AND gate and the symbolic electronic circuit. (**d**) The output voltage of the circuit in (**c**) versus the four possible logic address level inputs [4.142]

Part A | 4.3

electron donor doping) results in an adverse variation in both the Seebeck coefficient (decreasing) and the thermal conductivity (increasing). These two trade-offs set the upper limit for increasing ZT in bulk materials, with the maximum ZT remaining ≈ 1 at room temperature for the 1960–1995 time frame.

The high electronic density of states in quantum-confined structures is proposed as a promising possibility to bypass the Seebeck/electrical conductivity trade-off and to control each thermoelectric-related variable independently, thereby allowing for increased electrical conductivity, relatively low thermal conductivity, and a large Seebeck coefficient simultaneously [4.211]. For example, Figs. 4.29 and 4.30a in Sect. 4.2.3 show an enhanced in S for bismuth and bismuth-antimony nanowires as the wire diameter decreases. In addition to alleviating the undesired connections between σ, S and the electronic contribution to the thermal conductivity, nanowires also have the advantage that the phonon contribution to the thermal conductivity is greatly reduced because of boundary scattering (see Sect. 4.2), thereby achieving a high ZT. Figure 4.41a shows the theoretical values for ZT versus sample size for both bismuth thin films (2-D) and nanowires (1-D) in the quantum-confined regime, exhibiting a rapidly increasing ZT as the quantum size effect becomes more and more important [4.211]. In addition, the quantum size effect in nanowires can be combined with other parameters to tailor the band structure and electronic transport behav-

ior (for instance, Sb alloying in Bi) to further optimize ZT. For example, Fig. 4.41b shows the predicted ZT for p-type $Bi_{1-x}Sb_x$ alloy nanowires as a function of wire diameter and Sb content x [4.210]. The occurrence of a local ZT maxima in the vicinity of $x \approx 0.13$ and $d_W \approx 45$ nm is due to the coalescence of ten valence bands in the nanowire and the resulting unusual high density of states for holes, which is a phenomenon absent in bulk $Bi_{1-x}Sb_x$ alloys. For nanowires with very small diameters, it is speculated that localization effects will eventually limit the enhancement of ZT. However, in bismuth nanowires, localization effects are not significant for wires with diameters larger than 9 nm [4.52]. In addition to 1-D nanowires, ZT values as high as ≈ 2 have also been experimentally demonstrated in macroscopic samples containing PbSe quantum dots (0D) [4.212] and stacked 2-D films [4.167].

Although the application of nanowires to thermoelectrics appears very promising, these materials are still in the research phase of the development cycle and are far from being commercialized. One challenge for thermoelectric devices based on nanowires lies in finding a suitable host material that will not reduce ZT too much due to the unwanted heat conduction through the host material. Therefore, the host material should have a low thermal conductivity and occupy a volume percentage in the composite material that is as low as possible, while still providing the quantum confinement and the support for the nanowires.

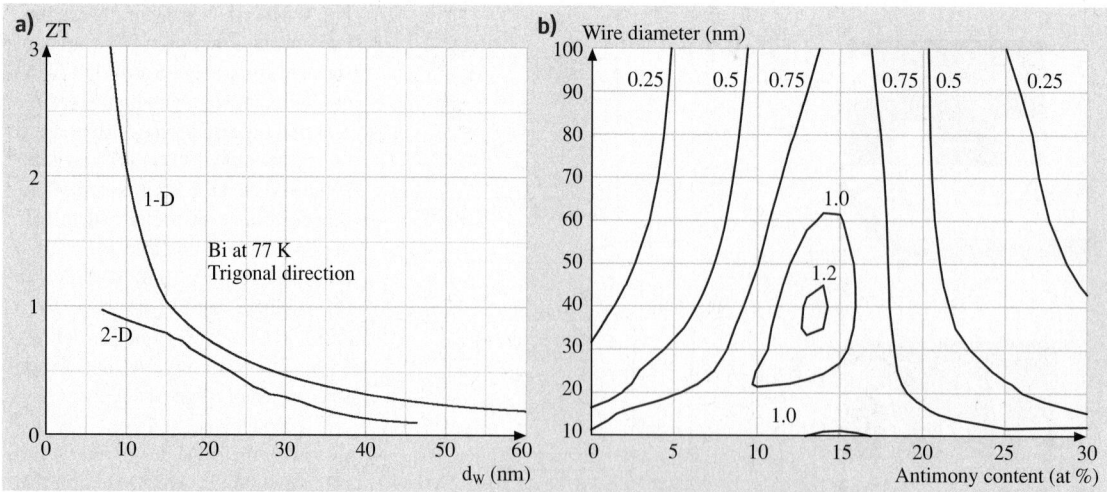

Fig. 4.41 (a) Calculated ZT of 1-D (nanowire) and 2-D (quantum well) bismuth systems at 77 K as a function of d_W, denoting the wire diameter or film thickness. The thermoelectric performance (ZT) is expected to improve greatly when the wire diameter is small enough for the nanowire to become a one-dimensional system. **(b)** Contour plot of optimal ZT values for p-type $Bi_{1-x}Sb_x$ nanowires versus wire diameter and antimony concentration calculated at 77 K [4.210]

4.3.3 Optical Applications

Nanowires also hold promise for optical applications. One-dimensional systems exhibit a singularity in their joint density of states, allowing quantum effects in nanowires to be optically observable, sometimes even at room temperature. Since the density of states of a nanowire in the quantum limit (small wire diameter) is highly localized in energy, the available states quickly fill up with electrons as the intensity of the incident light is increased. This filling up of the sub-bands, as well as other effects that are unique to low-dimensional materials, lead to strong optical non-linearities in quantum wires. Quantum wires may thus yield optical switches with a lower switching energy and increased switching speed compared to currently available optical switches.

Light emission from nanowires can be achieved by photoluminescence (PL) or electroluminescence (EL), distinguished by whether the electronic excitation is achieved by optical illumination or by electrical stimulation across a p-n junction, respectively. PL is often used for optical property characterization, as described in Sect. 4.2.4, but from an applications point of view, EL is a more convenient excitation method. Light-emitting diodes (LEDs) have been achieved in junctions between a p-type and an n-type nanowire (Fig. 4.39) [4.194] and in superlattice nanowires with p-type and n-type segments [4.92]. The light emission was localized to the junction area, and was polarized in the superlattice nanowire. An electrically driven laser was fabricated from CdS nanowires. The wires were assembled by evaporating a metal contact onto an n-type CdS nanowire which resided on a p^+ silicon wafer. The cleaved ends of the wire formed the laser cavity, so that in forward bias, light characteristic of lasing was observed at the end of the wire [4.213]. LEDs have also been achieved with core–shell structured nanowires made of n-GaN/InGaN/p-GaN [4.214].

Light emission from quantum wire p-n junctions is especially interesting for laser applications, because quantum wires can form lasers with lower excitation thresholds than their bulk counterparts and they also exhibit decreased sensitivity of performance to temperature [4.215]. Furthermore, the emission wavelength can be tuned for a given material composition by simply altering the geometry of the wire.

Lasing action has been reported in ZnO nanowires with wire diameters that are much smaller than the wavelength of the light emitted ($\lambda = 385$ nm) [4.122] (see Fig. 4.42). Since the edges and lateral surfaces of

ZnO nanowires are faceted (see Sect. 4.2.1), they form optical cavities that sustain desired cavity modes. Compared to conventional semiconductor lasers, the exciton laser action employed in zinc oxide nanowire lasers exhibits a lower lasing threshold (≈ 40 kW/cm^2) than their 3-D counterparts (≈ 300 kW/cm^2). In order to utilize exciton confinement effects in the lasing action, the exciton binding energy (≈ 60 meV in ZnO) must be greater than the thermal energy (≈ 26 meV at 300 K). Decreasing the wire diameter increases the excitation binding energy and lowers the threshold for lasing. PL NSOM imaging confirmed the waveguiding properties of the anisotropic and the well-faceted structure of ZnO nanowires, limiting the emission to the tips of the ZnO nanowires [4.183]. Time-resolved studies have illuminated the dynamics of the emission process [4.216].

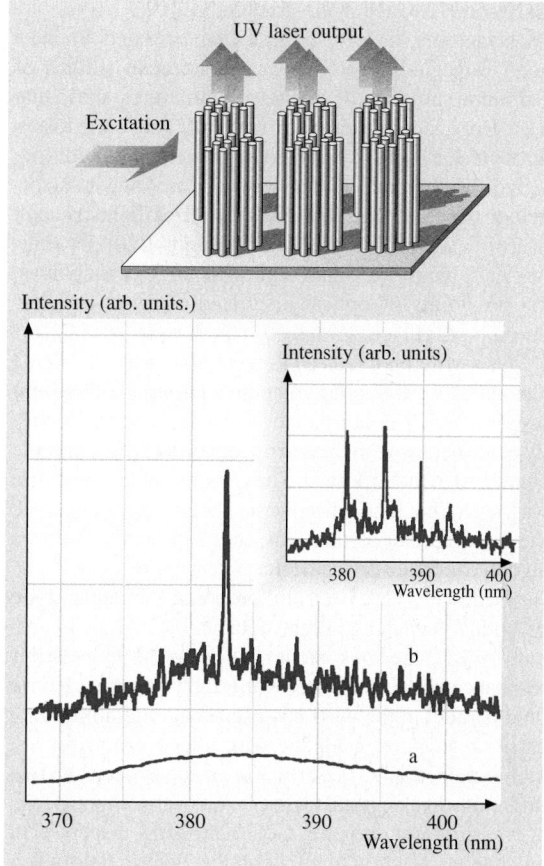

Fig. 4.42 A schematic of lasing in ZnO nanowires and the PL spectra of ZnO nanowires at two excitation intensities. One PL spectrum is taken below the lasing threshold, and the other above it [4.122]

Lasing was also observed in ZnS nanowires in anodic aluminium oxide templates [4.217] and in GaN nanowires [4.218]. Unlike ZnO, GaN has a small exciton binding energy, only ≈ 25 meV. Furthermore, since the wire radii used in this study (15–75 nm) [4.218] are larger than the Bohr radius of excitons in GaN (11 nm), the exciton binding energy is not expected to increase in these GaN wires and quantum confinement effects such as those shown in Fig. 4.35 for InP are not expected. However, some tunability of the center of the spectral intensity was achieved by increasing the intensity of the pump power, causing a redshift in the laser emission, which is explained as a bandgap renormalization as a result of the formation of an electron–hole plasma. Heating effects were excluded as the source of the spectral shift. GaN quantum wire UV lasers with a low threshold for lasing action have been achieved using a self-organized GaN(core)/AlGaN(shell) structure [4.219].

Nanowires have also been demonstrated to have good waveguiding properties. Quantitative studies of cadmium sulfide (CdS) nanowire structures show that light propagation takes place with only moderate losses through sharp and even acute angle bends. In addition, active devices made with nanowires have shown that efficient injection into and modulation of light through nanowire waveguides can be achieved [4.220]. By linking ZnO nanowire light sources to SnO_2 waveguides, the possibility of optical integrated circuitry is introduced [4.221].

Nanowire photodetectors were also proposed. ZnO nanowires were found to display a strong photocurrent response to UV light irradiation [4.222]. The conductivity of the nanowire increased by four orders of magnitude compared to the dark state. The response of the nanowire was reversible, and selective to photon energies above the bandgap, suggesting that ZnO nanowires could be a good candidate for optoelectronic switches.

Nanowires have been also proposed for another type of optical switching. Light with its electric field normal to the wire axis excites a transverse free carrier resonance inside the wire, while light with its electric field parallel to the wire axis excites a longitudinal free carrier resonance inside the wire. Since nanowires are highly anisotropic, these two resonances occur at two different wavelengths and thus result in absorption peaks at two different energies. Gold nanowires dispersed in an aqueous solution align along the electric field when a DC voltage is applied. The energy of the absorption peak can be toggled between the transverse and longitudinal resonance energies by changing the alignment of the nanowires under polarized light illumination us-ing an electric field [4.223, 224]. Thus, electro-optical modulation is achieved.

Nanowires may also be used as barcode tags for optical read-out. Nanowires containing gold, silver, nickel, palladium, and platinum were fabricated [4.110] by electrochemical filling of porous anodic alumina, so that each nanowire consisted of segments of various metal constituents. Thus many types of nanowires can be made from a handful of materials, and identified by the order of the metal segments along their main axis, and the length of each segment. Barcode read-out is possible by reflectance optical microscopy. The segment length is limited by the Rayleigh diffraction limit, and not by synthesis limitations, and thus can be as small as 145 nm. Figure 4.43a shows an optical image of many Au-Ag-Au-Ag barcoded wires, where the silver segments show higher reflectivity. Figure 4.43b is a backscattering mode FE-SEM image of a single nanowire, highlighting the composition and segment length variations along the nanowire.

Both the large surface area and the high conductivity along the length of a nanowire are favorable for its use in inorganic–organic solar cells [4.225], which offer promise from a manufacturing and cost-effectiveness standpoint. In a hybrid nanocrystal–organic solar cell, the incident light forms bound electron–hole pairs (excitons) in both the inorganic nanocrystal and in the surrounding organic medium. These excitons diffuse to the inorganic–organic interface and disassociate to form an electron and a hole. Since conjugated polymers usually have poor electron mobilities, the inorganic phase is chosen to have a higher electron affinity than the organic phase so that the organic phase carries the holes and the semiconductor carries the electrons. The separated electrons and holes drift to the external electrodes through the inorganic and organic materials, respectively. However, only those excitons formed within an exciton diffusion length from an interface can disassociate before recombining, and therefore the distance between the dissociation sites limits the efficiency of a solar cell. A solar cell prepared from a composite of CdSe nanorods inside poly(3-ethylthiophene) [4.225] yielded monochromatic power efficiencies of 6.9% and power conversion efficiencies of 1.7% under A.M. 1.5 illumination (equal to solar irradiance through 1.5 times the air mass of the Earth at direct normal incidence). The nanorods provide a large surface area with good chemical bonding to the polymer for efficient charge transfer and exciton dissociation. Furthermore, they provide a good conduction path for the electrons to reach the electrode. Their enhanced absorption coefficient and

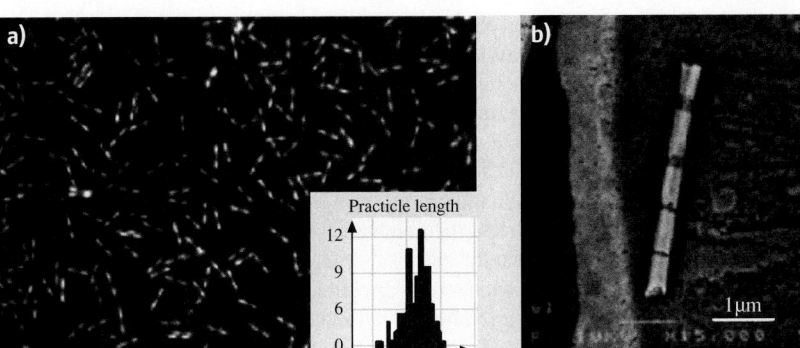

Fig. 4.43 (a) An optical image of many short bar-coded Au-Ag-Au-Au wires and **(b)** an FE-SEM image of an Au/Ag barcoded wire with multiple strips of varying length. The *insert* in **(a)** shows a histogram of the particle lengths for 106 particles in this image [4.110]

their tunable bandgap are also characteristics that can be used to enhance the energy conversion efficiency of solar cells.

4.3.4 Chemical and Biochemical Sensing Devices

Sensors for chemical and biochemical substances with nanowires as the sensing probe are a very attractive application area. Nanowire sensors will potentially be smaller, more sensitive, demand less power, and react faster than their macroscopic counterparts. Arrays of nanowire sensors could, in principle, achieve nanometer-scale spatial resolution and therefore provide accurate real-time information regarding not only the concentration of a specific analyte but also its spatial distribution. Such arrays could be very useful, for example, for dynamic studies on the effects of chemical gradients on biological cells. The operation of sensors made with nanowires, nanotubes, or nanocontacts is based mostly on the reversible change in the conductance of the nanostructure upon absorption of the agent to be detected, but other detection methods, such as mechanical and optical detection, are conceptually plausible. The increased sensitivity and faster response time of nanowires are a result of the large surface-to-volume ratio and the small cross-section available for conduction channels. In the bulk, on the other hand, the abundance of charges can effectively shield external fields, and the abundance of material can afford many alternative conduction channels. Therefore, a stronger chemical stimulus and longer response time are necessary to observe changes in the physical properties of a 3-D sensor in comparison to a nanowire.

It is often necessary to modify the surface of the nanowires to achieve a strong interaction with the analytes that need to be detected. Surface modifications utilize the self-assembly, chemisorption or chemical reactivity of selected organic molecules and polymers towards metal and oxide surfaces. Examples include: thiols on gold, isocyanides on platinum, and siloxanes on silica. These surface coatings regulate the binding and chemical reactivity of other molecules towards the nanowire in a predictable manner [4.226].

Cui et al. placed silicon nanowires made by the VLS method (Sect. 4.1.2) between two metal electrodes and modified the silicon oxide coating of the wire through the addition of molecules that are sensitive to the analyte to be detected [4.227]. For example, a pH sensor was made by covalently linking an amine-containing silane to the surface of the nanowire. Variations in the pH of the solution into which the nanowire was immersed caused protonation and deprotonation of the $-NH_2$ and the $-SiOH$ groups on the surface of the nanowire. The variation in surface charge density regulates the conductance of the nanowire; due to the p-type characteristics of a silicon wire, the conductance increases with the addition of negative surface charge. The combined acid and base behavior of the surface groups results in an approximately linear dependence of the conductance on pH in the pH range 2 to 9, thus leading to a direct readout pH meter. This same type of approach was used for the detection of the binding of biomolecules, such as streptavidin using biotin-modified nanowires (see Fig. 4.44). This nanowire-based device has high sensitivity and could detect streptavidin binding down to a concentration of $10\,pM$ (10^{-12} mole). Subsequent results demonstrated the capabilities of these functionalized Si nanowire sensors as DNA sensors down to the femtomolar range [4.228]. The chemical detection devices were made in a field effect transistor geometry, so that the back-gate potential could be used to regulate the conductance in conjugation with the chemical detection and to provide a real-time direct read-out [4.227]. The extension of this device to detect multiple analytes using multiple nanowires, each sensitized to a differ-

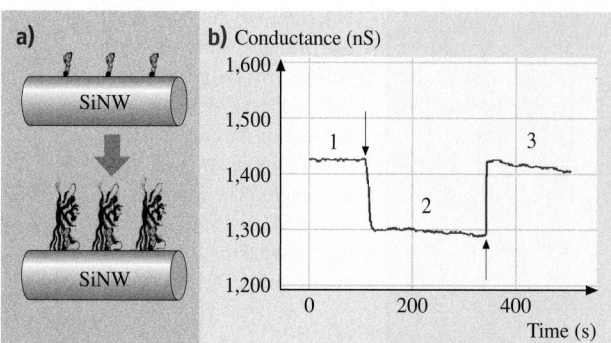

Fig. 4.44 (a) Streptavidin molecules bind to a silicon nanowire functionalized with biotin. The binding of streptavidin to biotin causes the nanowire to change its resistance. **(b)** The conductance of a biotin-modified silicon nanowire exposed to streptavidin in a buffer solution (regions 1 and 3) and with the introduction of a solution of antibiotin monoclonal antibody (region 2) [4.227]

ent analyte, could provide for fast, sensitive, and in situ screening procedures.

A similar approach was used by *Favier* et al., who made a nanosensor for the detection of hydrogen from of an array of palladium nanowires between two metal contacts [4.44]. They demonstrated that nanogaps were present in their nanowire structure, and upon absorption of H_2 and formation of Pd hydride, the nanogap structure would close and improve the electrical contact, thereby increasing the conductance of the nanowire array. The response time of these sensors was 75 msec, and they could operate in the range 0.5–5% H_2 before saturation occurred.

4.3.5 Magnetic Applications

It has been demonstrated that arrays of single-domain magnetic nanowires can be prepared with controlled nanowire diameter and length, aligned along a com-

mon direction and arranged in a close-packed ordered array (see Sect. 4.1), and that the magnetic properties (coercivity, remanence and dipolar magnetic interwire interaction) can be controlled to achieve a variety of magnetic applications [4.40, 79].

The most interesting of these applications is for magnetic storage, where the large nanowire aspect ratio (length/diameter) is advantageous for preventing the onset of the "superparamagnetic" limit at which the magnetization direction in the magnetic grains can be reversed by the thermal energy $k_B T$, thereby resulting in loss of recorded data in the magnetic recording medium. The magnetic energy in a grain can be increased by increasing either the volume or the anisotropy of the grain. If the volume is increased, the particle size increases, so the resolution is decreased. For spherical magnetized grains, the superparamagnetic limit at room temperature is reached at 70 Gbit/in^2. In nanowires, the anisotropy is very large and yet the wire diameters are small, so that the magnetostatic switching energy can easily be above the thermal energy while the spatial resolution is large. For magnetic data storage applications, a large aspect ratio is needed for the nanowires in order to maintain a high coercivity, and a sufficient separation between nanowires is needed to suppress interwire magnetic dipolar coupling. Thus nanowires can form stable and highly dense magnetic memory arrays with packing densities in excess of 10^{11} wires/cm^2.

The onset of superparamagnetism can be prevented in the single-domain magnetic nanowire arrays that have already been fabricated using either porous alumina templates to make Ni nanowires with 35 nm diameters [4.40] or diblock copolymer templates [4.79] to make Co nanowires, with mean diameters of 14 nm and 100% filling of the template pores (see 4.1.1). The ordered magnetic nanowire arrays that have already been demonstrated offer the exciting promise of systems permitting 10^{12} bits/in^2 data storage.

4.4 Concluding Remarks

In this chapter, we reviewed the synthesis, characterization and physical properties of nanowires, placing particular emphasis on nanowire properties that differ from those of the bulk counterparts and potential applications that might result from the special structures and properties of nanowires.

We have shown that the newly emerging field of nanowire research has developed very rapidly over the past few years, driven by the development of a vari-

ety of complementary nanowire synthesis methods and effective tools for measuring nanowire structure and properties (see Sects. 4.1 and 4.2). At present, much of the progress is at the demonstration-of-concept level, with many gaps in knowledge remaining to be elucidated, theoretical models to be developed, and new nanowire systems to be explored. Having demonstrated that many of the most interesting discoveries to date relate to nanowire properties not present in their bulk material counterparts, we can expect future research emphasis to be increasingly focused on smaller diameter nanowires, where new unexplored physical phenomena related to quantum confinement effects are more likely to be found. We can also expect the development of applications to follow, some coming sooner and others later. Many promising applications are now at the early demonstration stage (see Sect. 4.3), but are moving ahead rapidly because of their promise of new functionality, not previously available, in the fields of electronics, optoelectronics, biotechnology, magnetics, and energy conversion and generation, among others. Many exciting challenges remain in advancing both the nanoscience and the nanotechnological promise already demonstrated by the nanowire research described in this review.

References

4.1   R. P. Feynman: There's plenty of room at the bottom, Eng. Sci. (Caltech, February 1960) 22

4.2   Y. Mao, S. S. Wong: General, room-temperature method for the synthesis of isolated as well as arrays of single-crystalline ABO_4-type nanorods, J. Am. Chem. Soc. **126**, 15245–15252 (2004)

4.3   E. Braun, Y. Eichen, U. Sivan, G. Ben-Yoseph: DNA-templated assembly and electrode attachment of a conducting silver wire, Nature **391**, 775–778 (1998)

4.4   G. Sauer, G. Brehm, S. Schneider, K. Nielsch, R. B. Wehrspohn, J. Choi, H. Hofmeister, U. Gösele: Highly ordered monocrystalline silver nanowire arrays, J. Appl. Phys. **91**, 3243–3247 (2002)

4.5   G. L. Hornyak, C. J. Patrissi, C. M. Martin: Fabrication, characterization and optical properties of gold nanoparticle/porous alumina composites: the non-scattering Maxwell-Garnett limit, J. Phys. Chem. B **101**, 1548–1555 (1997)

4.6   X. Y. Zhang, L. D. Zhang, Y. Lei, L. X. Zhao, Y. Q. Mao: Fabrication and characterization of highly ordered Au nanowire arrays, J. Mater. Chem. **11**, 1732–1734 (2001)

4.7   Y.-T. Cheng, A. M. Weiner, C. A. Wong, M. P. Balogh, M. J. Lukitsch: Stress-induced growth of bismuth nanowires, Appl. Phys. Lett. **81**, 3248–3250 (2002)

4.8   J. Heremans, C. M. Thrush, Y.-M. Lin, S. Cronin, Z. Zhang, M. S. Dresselhaus, J. F. Mansfield: Bismuth nanowire arrays: synthesis, galvanomagnetic properties, Phys. Rev. B **61**, 2921–2930 (2000)

4.9   L. Piraux, S. Dubois, J. L. Duvail, A. Radulescu, S. Demoustier-Champagne, E. Ferain, R. Legras: Fabrication and properties of organic, metal nanocylinders in nanoporous membranes, J. Mater. Res. **14**, 3042–3050 (1999)

4.10 K. Hong, F. Y. Yang, K. Liu, D. H. Reich, P. C. Searson, C. L. Chien, F. F. Balakirev, G. S. Boebinger: Giant positive magnetoresistance of Bi nanowire arrays in high magnetic fields, J. Appl. Phys. **85**, 6184–6186 (1999)

4.11 A. J. Yin, J. Li, W. Jian, A. J. Bennett, J. M. Xu: Fabrication of highly ordered metallic nanowire arrays by electrodeposition, Appl. Phys. Lett. **79**, 1039–1041 (2001)

4.12 Z. Zhang, J. Y. Ying, M. S. Dresselhaus: Bismuth quantum-wire arrays fabricated by a vacuum melting and pressure injection process, J. Mater. Res. **13**, 1745–1748 (1998)

4.13 Z. Zhang, D. Gekhtman, M. S. Dresselhaus, J. Y. Ying: Processing and characterization of single-crystalline ultrafine bismuth nanowires, Chem. Mater. **11**, 1659–1665 (1999)

4.14 T. E. Huber, M. J. Graf, P. Constant: Processing and characterization of high-conductance bismuth wire array composites, J. Mater. Res. **15**, 1816–1821 (2000)

4.15 L. Li, G. Li, Y. Zhang, Y. Yang, L. Zhang: Pulsed electrodeposition of large-area, ordered $Bi_{1-x}Sb_x$ nanowire arrays from aqueous solutions, J. Phys. Chem. B **108**, 19380–19383 (2004)

4.16 M. S. Sander, A. L. Prieto, R. Gronsky, T. Sands, A. M. Stacy: Fabrication of high-density, high aspect ratio, large-area bismuth telluride nanowire arrays by electrodeposition into porous anodic alumina templates, Adv. Mater. **14**, 665–667 (2002)

4.17 M. Chen, Y. Xie, J. Lu, Y. J. Xiong, S. Y. Zhang, Y. T. Qian, X. M. Liu: Synthesis of rod-, twinrod-, and tetrapod-shaped CdS nanocrystals using a highly oriented solvothermal recrystallization technique, J. Mater. Chem. **12**, 748–753 (2002)

4.18 D. Xu, Y. Xu, D. Chen, G. Guo, L. Gui, Y. Tang: Preparation of CdS single-crystal nanowires by electrochemically induced deposition, Adv. Mater. **12**, 520–522 (2000)

4.19 D. Routkevitch, T. Bigioni, M. Moskovits, J. M. Xu: Electrochemical fabrication of CdS nanowire arrays in porous anodic aluminum oxide templates, J. Phys. Chem. **100**, 14037–14047 (1996)

4.20 L. Manna, E. C. Scher, A. P. Alivisatos: Synthesis of soluble and processable rod-, arrow-, teardrop-, and tetrapod-shaped CdSe nanocrystals, J. Am. Chem. Soc. **122**, 12700–12706 (2000)

4.21 D. Routkevitch, A. A. Tager, J. Haruyama, D. Al-Mawlawi, M. Moskovits, J. M. Xu: Nonlithographic nano-wire arrays: fabrication, physics, and device applications, IEEE Trans. Electron. Dev. **43**, 1646–1658 (1996)

4.22 D. S. Xu, D. P. Chen, Y. J. Xu, X. S. Shi, G. L. Guo, L. L. Gui, Y. Q. Tang: Preparation of II-VI group semiconductor nanowire arrays by dc electrochemical deposition in porous aluminum oxide templates, Pure Appl. Chem. **72**, 127–135 (2000)

4.23   R. Adelung, F. Ernst, A. Scott, M. Tabib-Azar, L. Kipp, M. Skibowski, S. Hollensteiner, E. Spiecker, W. Jäger, S. Gunst, A. Klein, W. Jägermann, V. Zaporojtchenko, F. Faupel: Self-assembled nanowire networks by deposition of copper onto layered-crystal surfaces, Adv. Mater. **14**, 1056–1061 (2002)

4.24   T. Gao, G. W. Meng, J. Zhang, Y. W. Wang, C. H. Liang, J. C. Fan, L. D. Zhang: Template synthesis of single-crystal Cu nanowire arrays by electrodeposition, Appl. Phys. A **73**, 251–254 (2001)

4.25   D. Al-Mawlawi, N. Coombs, M. Moskovits: Magnetic-properties of Fe deposited into anodic aluminum-oxide pores as a function of particle-size, J. Appl. Phys. **70**, 4421–4425 (1991)

4.26   F. Li, R. M. Metzger: Activation volume of α-Fe particles in alumite films, J. Appl. Phys. **81**, 3806–3808 (1997)

4.27   A. Sugawara, T. Coyle, G. G. Hembree, M. R. Scheinfein: Self-organized Fe nanowire arrays prepared by shadow deposition on NaCl(110) templates, Appl. Phys. Lett. **70**, 1043–1045 (1997)

4.28   G. S. Cheng, L. D. Zhang, Y. Zhu, G. T. Fei, L. Li, C. M. Mo, Y. Q. Mao: Large-scale synthesis of single crystalline gallium nitride nanowires, Appl. Phys. Lett. **75**, 2455–2457 (1999)

4.29   G. S. Cheng, L. D. Zhang, S. H. Chen, Y. Li, L. Li, X. G. Zhu, Y. Zhu, G. T. Fei, Y. Q. Mao: Ordered nanostructure of single-crystalline GaN nanowires in a honeycomb structure of anodic alumina, J. Mater. Res. **15**, 347–350 (2000)

4.30   Y. Huang, X. Duan, Y. Cui, C. M. Lieber: Gallium nitride nanowire nanodevices, Nano Lett. **2**, 101–104 (2002)

4.31   X. Duan, C. M. Lieber: Laser-assisted catalytic growth of single crystal GaN nanowires, J. Am. Chem. Soc. **122**, 188–189 (2000)

4.32   A. D. Berry, R. J. Tonucci, M. Fatemi: Fabrication of GaAs, InAs wires in nanochannel glass, Appl. Phys. Lett. **69**, 2846–2848 (1996)

4.33   J. R. Heath, F. K. LeGoues: A liquid solution synthesis of single-crystal germanium quantum wires, Chem. Phys. Lett. **208**, 263–268 (1993)

4.34   Y. Wu, P. Yang: Germanium nanowire growth via simple vapor transport, Chem. Mater. **12**, 605–607 (2000)

4.35   Y. F. Zhang, Y. H. Tang, N. Wang, C. S. Lee, I. Bello, S. T. Lee: Germanium nanowires sheathed with an oxide layer, Phys. Rev. B **61**, 4518–4521 (2000)

4.36   S. J. May, J.-G. Zheng, B. W. Wessels, L. J. Lauhon: Dendritic nanowire growth mediated by a self-assembled catalyst, Adv. Mater. **17**, 598–602 (2005)

4.37   S. Han, C. Li, Z. Liu, B. Lei, D. Zhang, W. Jin, X. Liu, T. Tang, C. Zhou: Transition metal oxide core-shell

nanowires: Generic synthesis and transport studies, Nano Lett. **4**, 1241–1246 (2004)

4.38   M. P. Zach, K. H. Ng, R. M. Penner: Molybdenum nanowires by electrodeposition, Science **290**, 2120–2123 (2000)

4.39   L. Sun, P. C. Searson, L. Chien: Electrochemical deposition of nickel nanowire arrays in single-crystal mica films, Appl. Phys. Lett. **74**, 2803–2805 (1999)

4.40   K. Nielsch, R. Wehrspohn, S. Fischer, H. Kronmuller, J. Barthel, J. Kirschner, U. Gosele: Magnetic properties of 100 nm nickel nanowire arrays obtained from ordered porous alumina templates, MRS Symp. Proc. **636**, D1.9 1–6 (2001)

4.41   Y. Wang, X. Jiang, T. Herricks, Y. Xia: Single crystalline nanowires of lead: large-scale synthesis, mechanistic studies, and transport measurements, J. Phys. Chem. B **108**, 8631–8640 (2004)

4.42   E. Lifshitz, M. Bashouti, V. Kloper, A. Kigel, M. S. Eisen, S. Berger: Synthesis and characterization of PbSe quantum wires, multipods, quantum rods, cubes, Nano Lett. **3**, 857–862 (2003)

4.43   W. Lu, P. Gao, W. B. Jian, Z. L. Wang, J. Fang: Perfect orientation ordered in-situ one-dimensional self-assembly of Mn-doped PbSe nanocrystals, J. Am. Chem. Soc. **126**, 14816–14821 (2004)

4.44   F. Favier, E. C. Walter, M. P. Zach, T. Benter, R. M. Penner: Hydrogen sensors and switches from electrodeposited palladium mesowire arrays, Science **293**, 2227–2231 (2001)

4.45   B. Gates, B. Mayers, B. Cattle, Y. Xia: Synthesis, characterization of uniform nanowires of trigonal selenium, Adv. Funct. Mater. **12**, 219–227 (2002)

4.46   C. A. Huber, T. E. Huber, M. Sadoqi, J. A. Lubin, S. Manalis, C. B. Prater: Nanowire array composites, Science **263**, 800–802 (1994)

4.47   Y. Cui, L. J. Lauhon, M. S. Gudiksen, J. Wang, C. M. Lieber: Diameter-controlled synthesis of single crystal silicon nanowires, Appl. Phys. Lett. **78**, 2214–2216 (2001)

4.48   A. M. Morales, C. M. Lieber: A laser ablation method for the synthesis of crystalline semiconductor nanowires, Science **279**, 208–211 (1998)

4.49   N. Wang, Y. F. Zhang, Y. H. Tang, C. S. Lee, S. T. Lee: SiO$_2$-enhanced synthesis of Si nanowires by laser ablation, Appl. Phys. Lett. **73**, 3902–3904 (1998)

4.50   M. K. Sunkara, S. Sharma, R. Miranda, G. Lian, E. C. Dickey: Bulk synthesis of silicon nanowires using a low-temperature vapor-liquid-solid method, Appl. Phys. Lett. **79**, 1546–1548 (2001)

4.51   S. Vaddiraju, H. Chandrasekaran, M. K. Sunkara: Vapor phase synthesis of tungsten nanowires, J. Am. Chem. Soc. **125**, 10792–10793 (2003)

4.52   J. P. Heremans, C. M. Thrush, D. T. Morelli, M.-C. Wu: Thermoelectric power of bismuth nanocomposites, Phys. Rev. Lett. **88**, 216801(1–4) (2002)

4.53   Y. Li, G. S. Cheng, L. D. Zhang: Fabrication of highly ordered ZnO nanowire arrays in anodic alumina membranes, J. Mater. Res. **15**, 2305–2308 (2000)

4.54 P. Yang, H. Yan, S. Mao, R. Russo, J. Johnson, R. Saykally, N. Morris, J. Pham, R. He, H.-J. Choi: Controlled growth of ZnO nanowires and their optical properties, Adv. Funct. Mater. **12**, 323–331 (2002)

4.55 M. J Zheng, L. D. Zhang, G. H. Li, W. Z. Shen: Fabrication and optical properties of large-scale uniform zinc oxide nanowire arrays by one-step electrochemical deposition technique, Chem. Phys. Lett. **363**, 123–128 (2002)

4.56 M. S. Dresselhaus, Y.-M. Lin, O. Rabin, A. Jorio, A. G. Souza Filho, M. A. Pimenta, R. Saito, G. G. Samsonidze, G. Dresselhaus: Nanowires and nanotubes, Mater. Sci. Eng. **C 23**, 129–140 (2003) (also in: *Current Trends in Nanotechnologies: From Materials to Systems (Europ. Mater. Res. Soc. Symp. Proc. 140)*, ed. by W. Jantsch, H. Grimmeiss and G. Marietta (Elsevier, Amsterdam 2002)

4.57 R. Saito, G. Dresselhaus, M. S. Dresselhaus: *Physical Properties of Carbon Nanotubes* (Imperial College Press, London 1998)

4.58 M. S. Dresselhaus, G. Dresselhaus, P. Avouris: Carbon nanotubes: synthesis, structure, properties and applications, Springer Ser. Top. Appl. Phys. **80**, 1–447 (2001)

4.59 R. C. Haddon: Special issue on carbon nanotubes, Acc. Chem. Res. **35**, 997–1113 (2002)

4.60 Y.-M. Lin, X. Sun, S. Cronin, Z. Zhang, J. Y. Ying, M. S. Dresselhaus: Fabrication, transport properties of Te-doped bismuth nanowire arrays. In: *Molecular Electronics: MRS Symposium Proceedings*, Vol. 582, ed. by S. T. Pantelides, M. A. Reed, J. Murday, A. Aviran (Materials Research Society Press, Pittsburgh 2000) Chap. H10.3, pp. 1–6

4.61 C. R. Martin: Nanomaterials: A membrane-based synthetic approach, Science **266**, 1961–1966 (1994)

4.62 G. A. Ozin: Nanochemistry: synthesis in diminishing dimensions, Adv. Mater. **4**, 612–649 (1992)

4.63 R. J. Tonucci, B. L. Justus, A. J. Campillo, C. E. Ford: Nanochannel array glass, Science **258**, 783–785 (1992)

4.64 J. Y. Ying: Nanoporous systems and templates, Sci. Spec. **18**, 56–63 (1999)

4.65 J. W. Diggle, T. C. Downie, C. W. Goulding: Anodic oxide films on aluminum, Chem. Rev. **69**, 365–405 (1969)

4.66 J. P. O'Sullivan, G. C. Wood: The morphology and mechanism of formation of porous anodic films on aluminum, Proc. R. Soc. Lond. **A 317**, 511–543 (1970)

4.67 A. P. Li, F. Müller, A. Birner, K. Neilsch, U. Gösele: Hexagonal pore arrays with a 50-420nm interpore distance formed by self-organization in anodic alumina, J. Appl. Phys. **84**, 6023–6026 (1998)

4.68 J. P. Sullivan, G. C. Wood: The morphology, mechanism of formation of porous anodic films on aluminum, Proc. R. Soc. Lond. **A 317**, 511–543 (1970)

4.69 O. Jessensky, F. Müller, U. Gössele: Self-organized formation of hexagonal pore arrays in anodic alumina, Appl. Phys. Lett. **72**, 1173–1175 (1998)

4.70 F. Li, L. Zhang, R. M. Metzger: On the growth of highly ordered pores in anodized aluminum oxide, Chem. Mater. **10**, 2470–2480 (1998)

4.71 H. Masuda, M. Satoh: Fabrication of gold nanodot array using anodic porous alumina as an evaporation mask, Jpn. J. Appl. Phys. **35**, L126–L129 (1996)

4.72 E. Ferain, R. Legras: Track-etched membrane – dynamics of pore formation, Nucl. Instrum. Meth. B **84**, 331–336 (1993)

4.73 A. Blondel, J. P. Meier, B. Doudin, J.-P. Ansermet: Giant magnetoresistance of nanowires of multilayers, Appl. Phys. Lett. **65**, 3019–3021 (1994)

4.74 K. Liu, C. L. Chien, P. C. Searson, Y. Z. Kui: Structural and magneto-transport properties of electrodeposited bismuth nanowires, Appl. Phys. Lett. **73**, 1436–1438 (1998)

4.75 C. A. Huber, T. E. Huber: A novel microstructure: semiconductor-impregnated porous Vycor glass, J. Appl. Phys. **64**, 6588–6590 (1988)

4.76 J. S. Beck, J. C. Vartuli, W. J. Roth, M. E. Leonowicz, C. T. Kresge, K. D. Schmitt, C. T.-W. Chu, D. H. Olson, E. W. Sheppard, S. B. McCullen, J. B. Higgins, J. L. Schlenker: A new family of mesoporous molecular sieves prepared with liquid crystal templates, J. Am. Chem. Soc. **114**, 10834–10843 (1992)

4.77 C.-G. Wu, T. Bein: Conducting polyaniline filaments in a mesoporous channel host, Science **264**, 1757–1759 (1994)

4.78 Y.-M. Lin, S. B. Cronin, J. Y. Ying, M. S. Dresselhaus, J. P. Heremans: Transport properties of Bi nanowire arrays, Appl. Phys. Lett. **76**, 3944–3946 (2000)

4.79 T. Thurn-Albrecht, J. Schotter, G. A. Kästle, N. Emley, T. Shibauchi, L. Krusin-Elbaum, K. Guarini, C. T. Black, M. T. Tuominen, T. P. Russell: Ultrahigh-density nanowire arrays grown in self-assembled diblock copolymer templates, Science **290**, 2126–2129 (2000)

4.80 A. W. Adamson: *Physical Chemistry of Surfaces* (Wiley, New York 1982) p. 338

4.81 R. Ferré, K. Ounadjela, J. M. George, L. Piraux, S. Dubois: Magnetization processes in nickel and cobalt electrodeposited nanowires, Phys. Rev. B **56**, 14066–14075 (1997)

4.82 H. Zeng, M. Zheng, R. Skomski, D. J. Sellmyer, Y. Liu, L. Menon, S. Bandyopadhyay: Magnetic properties of self-assembled Co nanowires of varying length and diameter, J. Appl. Phys **87**, 4718–4720 (2000)

4.83 Y. Peng, H. L. Zhang, S.-L. Pan, H.-L. Li: Magnetic properties and magnetization reversal of α-Fe nanowires deposited in alumina film, J. Appl. Phys. **87**, 7405–7408 (2000)

4.84 L. Piraux, J. M. George, J. F. Despres, C. Leroy, E. Ferain, R. Legras, K. Ounadjela, A. Fert: Giant magnetoresistance in magnetic multilayered nanowires, Appl. Phys. Lett. **65**, 2484–2486 (1994)

4.85   S. Bhattacharrya, S. K. Saha, D. Chakravorty: Nanowire formation in a polymeric film, Appl. Phys. Lett. **76**, 3896–3898 (2000)

4.86   G. Yi, W. Schwarzacher: Single crystal superconductor nanowires by electrodeposition, Appl. Phys. Lett. **74**, 1746–1748 (1999)

4.87   D. Al-Mawlawi, C. Z. Liu, M. Moskovits: Nanowires formed in anodic oxide nanotemplates, J. Mater. Res. **9**, 1014–1018 (1994)

4.88   R. S. Wagner, W. C. Ellis: Vapor-liquid-solid mechanism of single crystal growth, Appl. Phys. Lett. **4**, 89–90 (1964)

4.89   Y. Wu, P. Yang: Direct observation of vapor-liquid-solid nanowire growth, J. Am. Chem. Soc. **123**, 3165–3166 (2001)

4.90   Y. Wu, R. Fan, P. Yang: Block-by-block growth of single-crystalline Si/SiGe superlattice nanowires, Nano Lett. **2**, 83–86 (2002)

4.91   S. Sharma, M. K. Sunkara, R. Miranda, G. Lian, E. C. Dickey: A novel low temperature synthesis method for semiconductor nanowires. In: *Synthesis, Functional Properties and Applications of Nanostructures: Mat. Res. Soc. Symp. Proc., San Francisco, Spring 2001*, Vol. 676, ed. by H. W. Hahn, D. L. Feldheim, C. P. Kubiak, R. Tannenbaum, R. W. Siegel (Materials Research Society Press, Pittsburgh 2001) p. Y1.6

4.92   M. S. Gudiksen, L. J. Lauhon, J. Wang, D. C. Smith, C. M. Lieber: Growth of nanowire superlattice structures for nanoscale photonics and electronics, Nature **415**, 617–620 (2002)

4.93   Y. Wu, J. Xiang, C. Yang, W. Lu, C. M. Lieber: Single-crystal metallic nanowires and metal/semiconductor nanowire heterostructures, Nature **430**, 61–65 (2004)

4.94   M. T. Björk, B. J. Ohlsson, T. Sass, A. I. Persson, C. Thelander, M. H. Magnusson, K. Deppert, L. R. Wallenberg, L. Samuelson: One-dimensional steeplechase for electrons realized, Nano Lett. **2**, 87–89 (2002)

4.95   N. Wang, Y. H. Tang, Y. F. Zhang, C. S. Lee, S. T. Lee: Nucleation and growth of Si nanowires from silicon oxide, Phys. Rev. B **58**, R16024–R16026 (1998)

4.96   Y. F. Zhang, Y. H. Tang, N. Wang, C. S. Lee, I. Bello, S. T. Lee: One-dimensional growth mechanism of crystalline silicon nanowires, J. Cryst. Growth **197**, 136–140 (1999)

4.97   S. T. Lee, Y. F. Zhang, N. Wang, Y. H. Tang, I. Bello, C. S. Lee, Y. W. Chung: Semiconductor nanowires from oxides, J. Mater. Res. **14**, 4503–4507 (1999)

4.98   D. D. D. Ma, C. S. Lee, Y. Lifshitz, S. T. Lee: Periodic array of intramolecular junctions of silicon nanowires, Appl. Phys. Lett. **81**, 3233–3235 (2002)

4.99   D. Whang, S. Jin, C. M. Lieber: Large-scale hierarchical organization of nanowires for functional nanosystems, Jpn. J. Appl. Phys. **43**, 4465–4470 (2004)

4.100   B. Gates, Y. Yin, Y. Xia: A solution-phase approach to the synthesis of uniform nanowires of crystalline selenium with lateral dimensions in the range of 10-30 nm, J. Am. Chem. Soc. **122**, 12582–12583 (2000)

4.101   B. Mayers, B. Gates, Y. Yin, Y. Xia: Large-scale synthesis of monodisperse nanorods of Se/Te alloys through a homogeneous nucleation and solution growth process, Adv. Mater. **13**, 1380–1384 (2001)

4.102   B. Gates, Y. Wu, Y. Yin, P. Yang, Y. Xia: Single-crystalline nanowires of Ag_2Se can be synthesized by templating against nanowires of trigonal Se, J. Am. Chem. Soc. **123**, 11500–11501 (2001)

4.103   B. Gates, B. Mayers, Y. Wu, Y. Sun, B. Cattle, P. Yang, Y. Xia: Synthesis and characterization of crystalline Ag_2Se nanowires through a template-engaged reaction at room temperature, Adv. Funct. Mater. **12**, 679–686 (2002)

4.104   H. Yu, P. C. Gibbons, W. E. Buhro: Bismuth, tellurium and bismuth telluride nanowires, J. Mater. Chem. **14**, 595–602 (2004)

4.105   M. P. Zach, K. Inazu, K. H. Ng, J. C. Hemminger, R. M. Penner: Synthesis of molybdenum nanowires with millimeter-scale lengths using electrochemical step edge decoration, Chem. Mater. **14**, 3206–3216 (2002)

4.106   X. Peng, J. Wickham, A. P. Alivisatos: Kinetics of II-VI, III-V colloidal semiconductor nanocrystal growth: 'Focusing' of size distributions, J. Am. Chem. Soc. **120**, 5343–5344 (1998)

4.107   N. A. Melosh, A. Boukai, F. Diana, B. Gerardot, A. Badolto, P. M. Petroff, J. R. Heath: Ultrahigh-density nanowire lattices and circuits, Science **300**, 112–115 (2003)

4.108   J. Y. Lao, J. G. Wen, Z. F. Ren: Hierarchical ZnO nanostructures, Nano. Lett. **2**, 1287–1291 (2002)

4.109   J. Y. Lao, J. Y. Huang, D. Z. Wang, Z. F. Ren: ZnO nanobridges and nanonails, Nano Lett. **3**, 235–238 (2003)

4.110   S. R. Nicewarner-Peña, R. G. Freeman, B. D. Reiss, L. He, D. J. Peña, I. D. Walton, R. Cromer, C. D. Keating, M. J. Natan: Submicrometer metallic barcodes, Science **294**, 137–141 (2001)

4.111   L. J. Lauhon, M. S. Gudiksen, D. Wang, C. M. Lieber: Epitaxial core-shell and core-multishell nanowire heterostructures, Nature **420**, 57–61 (2002)

4.112   Z. L. Wang, Z. R. Dai, R. P. Gao, Z. G. Bai, J. L. Gole: Side-by-side silicon carbide-silica biaxial nanowires: Synthesis, structure and mechanical properties, Appl. Phys. Lett. **77**, 3349–3351 (2000)

4.113   P. Yang, F. Kim: Langmuir-Blodgett assembly of one-dimensional nanostructures, ChemPhysChem **3**, 503–506 (2002)

4.114   B. Messer, J. H. Song, P. Yang: Microchannel networks for nanowire patterning, J. Am. Chem. Soc. **122**, 10232–10233 (2000)

4.115 P. A. Smith, C. D. Nordquist, T. N. Jackson, T. S. Mayer, B. R. Martin, J. Mbindyo, T. E. Mallouk: Electric-field assisted assembly and alignment of metallic nanowires, Appl. Phys. Lett. **77**, 1399–1401 (2000)

4.116 S. Jin, D. M. Whang, M. C. McAlpine, R. S. Friedman, Y. Wu, C. M. Lieber: Scalable interconnection and integration of nanowire devices without registration, Nano Lett. **4**, 915–919 (2004)

4.117 T. Kuykendall, P. J. Pauzauskie, Y. F. Zhang, J. Goldberger, D. Sirbuly, J. Denlinger, P. D. Yang: Crystallographic alignment of high-density gallium nitride nanowire arrays, Nature Mater. **3**, 524–528 (2004)

4.118 H. Masuda, H. Yamada, M. Satoh, H. Asoh, M. Nakao, T. Tamamura: Highly ordered nanochannel-array architecture in anodic alumina, Appl. Phys. Lett. **71**, 2770–2772 (1997)

4.119 O. Rabin, P. R. Herz, S. B. Cronin, Y.-M. Lin, A. I. Akinwande, M. S. Dresselhaus: Nanofabrication using self-assembled alumina templates. In: *Nonlithographic and Lithographic Methods for Nanofabrication: MRS Symposium Proceedings, Boston, November 2000*, Vol. 636, ed. by J. A. Rogers, A. Karim, L. Merhari, D. Norris, Y. Xia (Materials Research Society Press, Pittsburgh 2001) pp. D4.7(1–6)

4.120 O. Rabin, P. R. Herz, Y.-M. Lin, A. I. Akinwande, S. B. Cronin, M. S. Dresselhaus: Formation of thick porous anodic alumina films and nanowire arrays on silicon wafers and glass., Adv. Funct. Mater. **13**, 631–638 (2003)

4.121 O. Rabin, P. R. Herz, Y.-M. Lin, S. B. Cronin, A. I. Akinwande, M. S. Dresselhaus: Arrays of nanowires on silicon wafers. In: *21st Int. Conf. Thermoelectrics: Proc. ICT '02 Long Beach, CA* (IEEE Inc., Piscataway, NJ 2002) pp. 276–279

4.122 M. H. Huang, S. Mao, H. Feick, H. Yan, Y. Wu, H. Kind, E. Weber, R. Russo, P. Yang: Room-temperature ultraviolet nanowire nanolasers, Science **292**, 1897–1899 (2001)

4.123 Y. H. Tang, Y. F. Zhang, N. Wang, C. S. Lee, X. D. Han, I. Bello, S. T. Lee: Morphology of Si nanowires synthesized by high-temperature laser ablation, J. Appl. Phys. **85**, 7981–7983 (1999)

4.124 Y. Ding, Z. L. Wang: Structure analysis of nanowires and nanobelts by transmission electron microscopy, J. Phys. Chem. B **108**, 12280–12291 (2004)

4.125 S. B. Cronin, Y.-M. Lin, O. Rabin, M. R. Black, G. Dresselhaus, M. S. Dresselhaus, P. L. Gai: Bismuth nanowires for potential applications in nanoscale electronics technology, Microsc. Microanal. **8**, 58–63 (2002)

4.126 M. S. Sander, R. Gronsky, Y.-M. Lin, M. S. Dresselhaus: Plasmon excitation modes in nanowire arrays, J. Appl. Phys. **89**, 2733–2736 (2001)

4.127 L. Venkataraman, C. M. Lieber: Molybdenum selenide molecular wires as one-dimensional conductors, Phys. Rev. Lett. **83**, 5334–5337 (1999)

4.128 A. Majumdar: Scanning thermal microscopy, Annu. Rev. Mater. Sci. **29**, 505–585 (1999)

4.129 K. M. Unruh, T. E. Huber, C. A. Huber: Melting and freezing behavior of indium metal in porous glasses, Phys. Rev. B **48**, 9021–9027 (1993)

4.130 Y. Y. Wu, P. D. Yang: Melting and welding semiconductor nanowires in nanotubes, Adv. Mater. **13**, 520–523 (2001)

4.131 P. M. Ajayan, S. Iijima: Capillarity-induced filling of carbon nanotubes, Nature **361**, 333–334 (1993)

4.132 Y. Gao, Y. Bando: Carbon nanothermometer containing gallium, Nature **415**, 599 (2002)

4.133 M. E. T. Morales, A. G. Balogh, T. W. Cornelius, R. Neumann, C. Trautmann: Fragmentation of nanowires driven by Rayleigh instability, Appl. Phys. Lett. **84**, 5337–5339 (2004)

4.134 D. A. Wharam, T. J. Thornton, R. Newbury, M. Pepper, H. Ahmed, J. E. F. Frost, D. G. Hasko, D. C. Peacock, D. A. Ritchie, G. A. C. Jones: One-dimensional transport and the quantization of the ballistic resistance, J. Phys. C **21**, L209–L214 (1988)

4.135 B. J. van Wees, H. van Houten, C. W. J. Beenakker, J. G. Williamson, L. P. Kouvenhoven, D. van der Marel, C. T. Foxon: Quantized conductance of point contacts in a two-dimensional electron gas, Phys. Rev. Lett. **60**, 848–850 (1988)

4.136 C. J. Muller, J. M. van Ruitenbeek, L. J. deJongh: Conductance and supercurrent discontinuities in atomic-scale metallic constrictions of variable width, Phys. Rev. Lett. **69**, 140–143 (1992)

4.137 C. J. Muller, J. M. Krans, T. N. Todorov, M. A. Reed: Quantization effects in the conductance of metallic contacts at room temperature, Phys. Rev. B **53**, 1022–1025 (1996)

4.138 J. L. Costa-Krämer, N. Garcia, H. Olin: Conductance quantization in bismuth nanowires at 4 K, Phys. Rev. Lett. **78**, 4990–4993 (1997)

4.139 J. L. Costa-Krämer, N. Garcia, H. Olin: Conductance quantization histograms of gold nanowires at 4 K, Phys. Rev. B **55**, 12910–12913 (1997)

4.140 C. Z. Li, H. X. He, A. Bogozi, J. S. Bunch, N. J. Tao: Molecular detection based on conductance quantization of nanowires, Appl. Phys. Lett. **76**, 1333–1335 (2000)

4.141 J. L. Costa-Krämer, N. Garcia, P. Garcia-Mochales, P. A. Serena, M. I. Marques, A. Correia: Conductance quantization in nanowires formed between micro and macroscopic metallic electrodes, Phys. Rev. B **55**, 5416–5424 (1997)

4.142 Y. Huang, X. Duan, Y. Cui, L. J. Lauhon, K.-H. Kim, C. M. Lieber: Logic gates and computation from assembled nanowire building blocks, Science **294**, 1313–1317 (2001)

4.143 J.-R. Kim, H. Oh, H. M. So, J.-J. Kim, J. Kim, C. J. Lee, S. C. Lyu: Schottky diodes based on a single GaN nanowire, Nanotechnology **13**, 701–704 (2002)

4.144 X. Duan, Y. Huang, C. M. Lieber: Nonvolatile memory and programmable logic from molecule-gated nanowires, Nano Lett. **2**, 487–490 (2002)

4.145 E. C. Walter, R. M. Penner, H. Liu, K. H. Ng, M. P. Zach, F. Favier: Sensors from electrodeposited metal nanowires, Surf. Inter. Anal. **34**, 409–412 (2002)

4.146 E. C. Walter, K. H. Ng, M. P. Zach, R. M. Penner, F. Favier: Electronic devices from electrodeposited metal nanowires, Microelectron. Eng. **61-62**, 555–561 (2002)

4.147 Y.-M. Lin, X. Sun, M. S. Dresselhaus: Theoretical investigation of thermoelectric transport properties of cylindrical Bi nanowires, Phys. Rev. B **62**, 4610–4623 (2000)

4.148 K. Liu, C. L. Chien, P. C. Searson: Finite-size effects in bismuth nanowires, Phys. Rev. B **58**, R14681–R14684 (1998)

4.149 Z. Zhang, X. Sun, M. S. Dresselhaus, J. Y. Ying, J. Heremans: Magnetotransport investigations of ultrafine single-crystalline bismuth nanowire arrays, Appl. Phys. Lett. **73**, 1589–1591 (1998)

4.150 J. Heremans, C. M. Thrush, Z. Zhang, X. Sun, M. S. Dresselhaus, J. Y. Ying, D. T. Morelli: Magnetoresistance of bismuth nanowire arrays: A possible transition from one-dimensional to three-dimensional localization, Phys. Rev. B **58**, R10091–R10095 (1998)

4.151 L. Sun, P. C. Searson, C. L. Chien: Finite-size effects in nickel nanowire arrays, Phys. Rev. B **61**, R6463–R6466 (2000)

4.152 Y.-M. Lin, S. B. Cronin, O. Rabin, J. Y. Ying, M. S. Dresselhaus: Transport properties and observation of semimetal-semiconductor transition in Bi-based nanowires. In: *Quantum Confined Semiconductor Nanostructures: MRS Symposium Proceedings, Boston, December 2002*, Vol. 737-C, ed. by J. M. Buriak, D. D. M. Wayner, F. Priolo, B. White, V. Klimov, L. Tsybeskov (Materials Research Society Press, Pittsburgh 2003) p. F3.14

4.153 Y.-M. Lin, M. S. Dresselhaus: Transport properties of superlattice nanowires and their potential for thermoelectric applications. In: *Quantum Confined Semiconductor Nanostructures: MRS Symposium Proceedings, Boston, December 2002*, Vol. 737-C, ed. by J. M. Buriak, D. D. M. Wayner, F. Priolo, B. White, V. Klimov, L. Tsybeskov (Materials Research Society Press, Pittsburgh 2003) p. F8.18

4.154 Y.-M. Lin, O. Rabin, S. B. Cronin, J. Y. Ying, M. S. Dresselhaus: Semimetal-semiconductor transition in $Bi_{1-x}Sb_x$ alloy nanowires and their thermoelectric properties, Appl. Phys. Lett. **81**, 2403–2405 (2002)

4.155 J. Heremans, C. M. Thrush, Y.-M. Lin, S. B. Cronin, M. S. Dresselhaus: Transport properties of antimony nanowires, Phys. Rev. B **63**, 085406(1–8) (2001)

4.156 Y.-M. Lin, S. B. Cronin, O. Rabin, J. Y. Ying, M. S. Dresselhaus: Transport properties of $Bi_{1-x}Sb_x$ alloy nanowires synthesized by pressure injection, Appl. Phys. Lett. **79**, 677–679 (2001)

4.157 D. E. Beutler, N. Giordano: Localization and electron–electron interaction effects in thin Bi wires and films, Phys. Rev. B **38**, 8–19 (1988)

4.158 Z. Zhang, X. Sun, M. S. Dresselhaus, J. Y. Ying, J. Heremans: Electronic transport properties of single crystal bismuth nanowire arrays, Phys. Rev. B **61**, 4850–4861 (2000)

4.159 J. Heremans, C. M. Thrush: Thermoelectric power of bismuth nanowires, Phys. Rev. B **59**, 12579–12583 (1999)

4.160 Y.-M. Lin, S. B. Cronin, O. Rabin, J. Heremans, M. S. Dresselhaus, J. Y. Ying: Transport properties of Bi-related nanowire systems. In: *Anisotropic Nanoparticles: Synthesis, Characterization and Applications: MRS Symposium Proceedings, Boston, December 2000*, Vol. 635, ed. by S. Stranick, P. C. Searson, L. A. Lyon, C. Keating (Materials Research Society Press, Pittsburgh 2001) pp. C4301–C4306

4.161 L. D. Hicks, M. S. Dresselhaus: Thermoelectric figure of merit of a one-dimensional conductor, Phys. Rev. B **47**, 16631–16634 (1993)

4.162 Y.-M. Lin, M. S. Dresselhaus: Thermoelectric properties of superlattice nanowires, Phys. Rev. B **68**, 075304 (2003)

4.163 M. T. Björk, B. J. Ohlsson, C. Thelander, A. I. Persson, K. Deppert, L. R. Wallenberg, L. Samuelson: Nanowire resonant tunneling diodes, Appl. Phys. Lett. **81**, 4458–4460 (2002)

4.164 D. Li, Y. Wu, P. Kim, L. Shi, P. Yang, A. Majumdar: Thermal conductivity of individual silicon nanowires, Appl Phys. Lett. **83**, 2934–2936 (2003)

4.165 T. S. Tighe, J. M. Worlock, M. L. Roukes: Direct thermal conductance measurements on suspended monocrystalline nanostructures, Appl. Phys. Lett. **70**, 2687–2689 (1997)

4.166 S. T. Huxtable, A. R. Abramson, C.-L. Tien, A. Majumdar, C.LaBounty, X. Fan, G. Zeng, J. E. Bowers, A. Shakouri, E. T. Croke: Thermal conductivity of Si/SiGe and SiGe/SiGe superlattices, Appl. Phys. Lett. **80**, 1737–1739 (2002)

4.167 R. Venkatasubramanian, E. Siivola, T. Colpitts, B. O'Quinn: Thin-film thermoelectric devices with high room-temperature figures of merit, Nature **413**, 597–602 (2001)

4.168 D. Li, Y. Wu, R. Fan, P. Yang, A. Majumdar: Thermal conductivity of Si/SiGe superlattice nanowires, Appl. Phys. Lett. **83**, 3186–3188 (2003)

4.169 C. Dames, G. Chen: Modeling the thermal conductivity of a SiGe segmented nanowire. In: *21st Int. Conf. Thermoelectrics: Proc. ICT '02, Long Beach, CA* (IEEE, Piscataway, NJ 2002) pp. 317–320

4.170 K. Schwab, J. L. Arlett, J. M. Worlock, M. L. Roukes: Thermal conductance through discrete quantum channels, Physica E **9**, 60–68 (2001)

4.171 G. Chen, M. S. Dresselhaus, G. Dresselhaus, J.-P. Fleurial, T. Caillat: Recent developments in thermoelectric materials, Int. Mater. Rev. **48**, 45–66 (2003)

4.172 C. Dames, G. Chen: Theoretical phonon thermal conductivity of Si–Ge superlattice nanowires, J. Appl. Phys. **95**, 682–693 (2004)

4.173 K. Schwab, E. A. Henriksen, J. M. Worlock, M. L. Roukes: Measurement of the quantum of thermal conductance, Nature **404**, 974–977 (2000)

4.174 M. Cardona: *Light Scattering in Solids* (Springer, Berlin Heidelberg 1982)

4.175 P. Y. Yu, M. Cardona: *Fundamentals of Semiconductors* (Springer, Berlin Heidelberg 1995) Chap. 7

4.176 J. C. M. Garnett: Colours in metal glasses, in metallic films, and in metallic solutions, Philos. Trans. Roy. Soc. London A **205**, 237–288 (1906)

4.177 D. E. Aspnes: Optical properties of thin films, Thin Solid Films **89**, 249–262 (1982)

4.178 U. Kreibig, L. Genzel: Optical absorption of small metallic particles, Surf. Sci. **156**, 678–700 (1985)

4.179 M. R. Black, Y.-M. Lin, S. B. Cronin, O. Rabin, M. S. Dresselhaus: Infrared absorption in bismuth nanowires resulting from quantum confinement, Phys. Rev. B **65**, 195417(1–9) (2002)

4.180 M. W. Lee, H. Z. Twu, C.-C. Chen, C. H. Chen: Optical characterization of wurtzite gallium nitride nanowires, Appl. Phys. Lett. **79**, 3693–3695 (2001)

4.181 D. M. Lyons, K. M. Ryan, M. A. Morris, J. D. Holmes: Tailoring the optical properties of silicon nanowire arrays through strain, Nano Lett. **2**, 811–816 (2002)

4.182 M. S. Gudiksen, J. Wang, C. M. Lieber: Size-depent photoluminescence from single indium phosphide nanowires, J. Phys. Chem. B **106**, 4036–4039 (2002)

4.183 J. C. Johnson, H. Yan, R. D. Schaller, L. H. Haber, R. J. Saykally, P. Yang: Single nanowire lasers, J. Phys. Chem. B **105**, 11387–11390 (2001)

4.184 S. Blom, L. Y. Gorelik, M. Jonson, R. I. Shekhter, A. G. Scherbakov, E. N. Bogachek, U. Landman: Magneto-optics of electronic transport in nanowires, Phys. Rev. B **58**, 16305–16314 (1998)

4.185 J. P. Pierce, E. W. Plummer, J. Shen: Ferromagnetism in cobalt-iron alloy nanowire arrays on w(110), Appl. Phys. Lett. **81**, 1890–1892 (2002)

4.186 S. Melle, J. L. Menendez, G. Armelles, D. Navas, M. Vazquez, K. Nielsch, R. B. Wehrspohn, U. Gosele: Magneto-optical properties of nickel nanowire arrays, Appl. Phys. Lett. **83**, 4547–4549 (2003)

4.187 J. C. Johnson, H. Yan, R. D. Schaller, P. B. Petersen, P. Yang, R. J. Saykally: Near-field imaging of nonlinear optical mixing in single zinc oxide nanowires, Nano Lett. **2**, 279–283 (2002)

4.188 M. R. Black, P. L. Hagelstein, S. B. Cronin, Y.-M. Lin, M. S. Dresselhaus: Optical absorption from an indirect transition in bismuth nanowires, Phys. Rev. B **68**, 235417 (2003)

4.189 M. R. Black, Y.-M. Lin, S. B. Cronin, M. S. Dresselhaus: Using optical measurements to improve electronic models of bismuth nanowires. In: *21st Int.*

Conf. Thermoelectrics: Proc. ICT '02, Long Beach, CA, Vol. ISSN 1094-2734, ed. by T. Caillat, J. Snyder (IEEE, Piscataway, NJ 2002) pp. 253–256

4.190 H. Richter, Z. P. Wang, L. Ley: The one phonon Raman-spectrum in microcrystalline silicon, Solid State Commun. **39**, 625–629 (1981)

4.191 I. H. Campbell, P. M. Fauchet: The effects of microcrystal size and shape on the one phonon Raman-spectra of crystalline semiconductors, Solid State Commun. **58**, 739–741 (1986)

4.192 H.-L. Liu, C.-C. Chen, C.-T. Chia, C.-C. Yeh, C.-H. Chen, M.-Y. Yu, S. Keller, S. P. DenBaars: Infrared and Raman-scattering studies in single-crystalline GaN nanowires, Chem. Phys. Lett. **345**, 245–251 (2001)

4.193 R. Gupta, Q. Xiong, C. K. Adu, U. J. Kim, P. C. Eklund: Laser-induced Fano resonance scattering in silicon nanowires, Nano Lett. **3**, 627–631 (2003)

4.194 X. Duan, Y. Huang, Y. Cui, J. Wang, C. M. Lieber: Indium phosphide nanowires as building blocks for nanoscale electronic and optoelectronic devices, Nature **409**, 66–69 (2001)

4.195 Y. Cui, C. M. Lieber: Functional nanoscale electronic devices assembled using silicon nanowire building blocks, Science **291**, 851–853 (2001)

4.196 Y. Cui, X. Duan, J. Hu, C. M. Lieber: Doping and electrical transport in silicon nanowires, J. Phys. Chem. B **104**, 101–104 (2000)

4.197 G. F. Zheng, W. Lu, S. Jin, C. M. Lieber: Synthesis and fabrication of high-performance n-type silicon nanowire transistors, Adv. Mater. **16**, 1890–1891 (2004)

4.198 J. Goldberger, D. J. Sirbuly, M. Law, P. Yang: ZnO nanowire transistors, J. Phys. Chem B **109**, 9–14 (2005)

4.199 D. H. Kang, J. H. Ko, E. Bae, J. Hyun, W. J. Park, B. K. Kim, J. J. Kim, C. J. Lee: Ambient air effects on electrical characteristics of gap nanowire transistors, J. Appl. Phys. **96**, 7574–7577 (2004)

4.200 S.-W. Chung, J.-Y. Yu, J. R. Heath: Silicon nanowire devices, Appl. Phys. Lett. **76**, 2068–2070 (2000)

4.201 C. Li, W. Fan, B. Lei, D. Zhang, S. Han, T. Tang, X. Liu, Z. Liu, S. Asano, M. Meyyappan, J. Han, C. Zhou: Multilevel memory based on molecular devices, Appl. Phys. Lett. **84**, 1949–1951 (2004)

4.202 B. Lei, C. Li, D. Q. Zhang, Q. F. Zhou, K. Shung, C. W. Zhou: Nanowire transistors with ferroelectric gate dielectrics: enhanced performance and memory effects, Appl. Phy. Lett. **84**, 4553–4555 (2004)

4.203 H. T. Ng, J. Han, T. Yamada, P. Nguyen, Y. P. Chen, M. Meyyappan: Single crystal nanowire vertical surround-gate field-effect transistor, Nano Lett. **4**, 1247–1252 (2004)

4.204 M. Ding, H. Kim, A. I. Akinwande: Observation of valence band electron emission from n-type silicon field emitter arrays, Appl. Phys. Lett. **75**, 823–825 (1999)

Part A | 4

4.205   F. C. K. Au, K. W. Wong, Y. H. Tang, Y. F. Zhang, I. Bello, S. T. Lee: Electron field emission from silicon nanowires, Appl. Phys. Lett. **75**, 1700–1702 (1999)

4.206   P. M. Ajayan, O. Z. Zhou: Applications of carbon nanotubes. In: *Carbon Nanotubes: Synthesis, Structure, Properties and Applications*, Vol. 80, ed. by M. S. Dresselhaus, G. Dresselhaus, P. Avouris (Springer, Berlin Heidelberg 2001) pp. 391–425 Springer Ser. Top. Appl. Phys.

4.207   M. Lu, M. K. Li, Z. J. Zhang, H. L. Li: Synthesis of carbon nanotubes/si nanowires core-sheath structure arrays and their field emission properties, Appl. Surf. Sci. **218**, 196–202 (2003)

4.208   L. Vila, P. Vincent, L. Dauginet-DePra, G. Pirio, E. Minoux, L. Gangloff, S. Demoustier-Champagne, N. Sarazin, E. Ferain, R. Legras, L. Piraux, P. Legagneux: Growth and field-emission properties of vertically aligned cobalt nanowire arrays, Nano Lett. **4**, 521–524 (2004)

4.209   G. Dresselhaus, M. S. Dresselhaus, Z. Zhang, X. Sun, J. Ying, G. Chen: Modeling thermoelectric behavior in Bi nano-wires. In: *Seventeenth International Conference on Thermoelectrics: Proceedings, ICT'98; Nagoya, Japan*, ed. by K. Koumoto (IEEE, Piscataway 1998) pp. 43–46

4.210   O. Rabin, Y.-M. Lin, M. S. Dresselhaus: Anomalously high thermoelectric figure of merit in $Bi_{1-x}Sb_x$ nanowires by carrier pocket alignment, Appl. Phys. Lett. **79**, 81–83 (2001)

4.211   L. D. Hicks, M. S. Dresselhaus: The effect of quantum well structures on the thermoelectric figure of merit, Phys. Rev. B **47**, 12727–12731 (1993)

4.212   T. C. Harman, P. J. Taylor, M. P. Walsh, B. E. LaForge: Quantum dot superlattice thermoelectric materials and devices, Science **297**, 2229–2232 (2002)

4.213   X. Duan, Y. Huang, R. Agarwal, C. M. Lieber: Single-nanowire electrically driven lasers, Nature **421**, 241 (2003)

4.214   F. Qian, Y. Li, S. Gradecak, D. L. Wang, C. J. Barrelet, C. M. Lieber: Gallium nitride-based nanowire radial heterostructures for nanophotonics, Nano Lett. **4**, 1975–1979 (2004)

4.215   V. Dneprovskii, E. Zhukov, V. Karavanskii, V. Poborchii, I. Salamatini: Nonlinear optical properties of semiconductor quantum wires, Superlattice. Microst. **23(6)**, 1217–1221 (1998)

4.216   J. C. Johnson, K. P. Knutsen, H. Yan, M. Law, Y. Zhang, P. Yang, R. J. Saykally: Ultrafast carrier dynamics in single ZnO nanowire and nanoribbon lasers, Nano Lett. **4**, 197–204 (2004)

4.217   J. X. Ding, J. A. Zapien, W. W. Chen, Y. Lifshitz, S. T. Lee, X. M. Meng: Lasing in ZnS nanowires grown on anodic aluminum oxide templates, App. Phys. Lett. **85**, 2361 (2004)

4.218   J. C. Johnson, H.-J. Choi, K. P. Knutsen, R. D. Schaller, P. Yang, R. J. Saykally: Single gallium nitride nanowire lasers, Nature Mater. **1**, 106–110 (2002)

4.219   H. J. Choi, J. C. Johnson, R. He, S. K. Lee, F. Kim, P. Pauzauskie, J. Goldberger, R. J. Saykally, P. Yang: Self-organized GaN quantum wire uv lasers, J. Phys. Chem. B **107**, 8721–8725 (2003)

4.220   C. J. Barrelet, A. B. Greytak, C. M. Lieber: Nanowire photonic circuit elements, Nano Lett. **4**, 1981–1985 (2004)

4.221   M. Law, D. J. Sirbuly, J. C. Johnson, J. Goldberger, R. J. Saykally, P. Yang: Ultralong nanoribbon waveguides for sub-wavelength photonics integration, Science **305**, 1269–1273 (2004)

4.222   H. Kind, H. Yan, B. Messer, M. Law, P. Yang: Nanowire ultraviolet photodetectors and optical switches, Adv. Mater. **14**, 158–160 (2002)

4.223   B. M. I. van der Zande, M. R. Böhmer, L. G. J. Fokkink, C. Schöneberger: Colloidal dispersions of gold rods: synthesis and optical properties, Langmuir **16**, 451–458 (2000)

4.224   B. M. I. van der Zande, G. J. M. Koper, H. N. W. Lekkerkerker: Alignment of rod-shaped gold particles by electric fields, J. Phys. Chem. B **103**, 5754–5760 (1999)

4.225   W. U. Huynh, J. J. Dittmer, A. P. Alivisatos: Hybrid nanorod-polymer solar cells, Science **295**, 2425–2427 (2002)

4.226   L. A. Bauer, N. S. Birenbaum, G. J. Meyer: Biological applications of high aspect ratio nanoparticles, J. Mater. Chem. **14**, 517–526 (2004)

4.227   Y. Cui, Q. Wei, H. Park, C. Lieber: Nanowire nanosensors for highly sensitive and selective detection of biological and chemical species, Science **293**, 1289–1292 (2001)

4.228   J. Hahm, C. Lieber: Direct ultra-sensitive electrical detection of DNA and DNA sequence variations using nanowire nanosensors, Nano Lett. **4**, 51–54 (2004)

5. Template-Based Synthesis of Nanorod or Nanowire Arrays

This chapter introduces the fundamentals of and various technical approaches developed for template-based synthesis of nanorod arrays. After a brief introduction to various concepts associated with the growth of nanorods, nanowires and nanobelts, the chapter focuses mainly on the most widely used and well established techniques for the template-based growth of nanorod arrays: electrochemical deposition, electrophoretic deposition, template filling via capillary force and centrifugation, and chemical conversion. In each section, the relevant fundamentals are first introduced, and then examples are given to illustrate the specific details of each technique.

Part A | 5

Syntheses, characterizations and applications of nanowires, nanorods, nanotubes and nanobelts (also often referred to as one-dimensional nanostructures) are significant areas of current endeavor in nanotechnology. Many techniques have been developed in these areas, and our understanding of the field has been significantly enhanced [5.1–5]. The field is still evolving rapidly with new synthesis methods and new nanowires or nanorods reported in the literature. Evaporation–condensation growth has been successfully applied to the synthesis of various oxide nanowires and nanorods. Similarly, the dissolution–condensation method has been widely used for the synthesis of various metallic nanowires from solutions. The vapor–liquid–solid (VLS) growth method is a highly versatile approach; various elementary and compound semiconductor nanowires have been synthesized using this method [5.6]. Template-based growth of nanowires or nanorods is an even more versatile method for various materials. Substrate ledge or step-induced growth of nanowires or nanorods has also been investigated intensively [5.7]. Except for VLS and template-based growth, most of the above-mentioned

methods result in randomly oriented nanowires or nanorods (commonly in the form of powder). The VLS method provides the ability to grow well oriented nanorods or nanowires directly attached to substrates, and is therefore often advantageous for characterization and applications; however, catalysts are required to form a liquid capsule at the advancing surface during growth at elevated temperatures. In addition, the possible incorporation of catalyst into nanowires and the difficulty removing such capsules from the tips of nanowires or nanorods are two disadvantages of this technique. Template-based growth often suffers from the polycrystalline nature of the resultant nanowires and nanorods, in addition to the difficulties involved in finding appropriate templates with pore channels of a desired diameter, length and surface chemistry and in removing the template completely without compromising the integrity of grown nanowires or nanorods. The discussion in this chapter will focus on nanorod and nanowire arrays, although nanotube arrays are mentioned briefly in conjunction with nanorod and nanowire fabrication. In addition, the terms of "nanorod" and "nanowire" are used interchangeably without special distinction in this chapter; this is commonplace in the literature.

In comparison with nanostructured materials in other forms, nanorod arrays offer several advantages for studying properties and for practical applications. Significant progress has been made in studies of the physical properties of individual nanowires and nanorods performed by directly measuring the properties of individual nanostructures. However, such studies generally require a lot of experimental preparation. For example, for electrical conductivity measurements, patterned electrodes are first created on a substrate, and then nanowires or nanorods are dispersed in an appropriate solvent or solution. This nanowire colloidal dispersion is then cast on the substrate containing pattern electrodes. Measurements are carried out after identifying individual nanowires or nanorods bridging two electrodes. The options for manipulating nanowires or nanorods are limited, and it is difficult to improve the contact between the sample and the electrodes to ensure the desired ohmic contact. For practical applications, the output or signal generated by single nanowire- or nanorod-based devices is small, and the signal-to-noise ratio is small, which means that highly sensitive instrumentation is required to accommodate such devices.

5.1 Template–Based Approach

The template approach to preparing free-standing, non-oriented and oriented nanowires and nanorods has been investigated extensively. The most commonly used and commercially available templates are anodized alumina membrane (AAM) [5.8] and radiation track-etched polycarbonate (PC) membranes [5.9]. Other membranes have also been used, such as nanochannel array on glass [5.10], radiation track-etched mica [5.11], mesoporous materials [5.12], porous silicon obtained via electrochemical etching of silicon wafer [5.13], zeolites [5.14] and carbon nanotubes [5.15, 16]. Biotemplates have also been explored for the growth of nanowires [5.17] and nanotubes [5.18], such as Cu [5.19], Ni [5.17], Co [5.17], and Au [5.20] nanowires. Commonly used alumina membranes with uniform and parallel pores are produced by the anodic oxidation of aluminium sheet in solutions of sulfuric, oxalic, or phosphoric acids [5.8, 21]. The pores can be arranged in a regular hexagonal array, and densities as high as 10^{11} pores/cm^2 can be achieved [5.22]. Pore size ranging from 10 nm to 100 μm can be achieved [5.22,23]. PC membranes are made by bombarding a nonporous polycarbonate sheet, typically 6 to 20 μm in thickness, with nuclear fission fragments to create damage tracks, and then chemically etching these tracks into pores [5.9]. In these radiation track-etched membranes, the pores are of uniform size (as small as 10 nm), but they are randomly distributed. Pore densities can be as high as 10^9 pores/cm^2.

In addition to the desired pore or channel size, morphology, size distribution and density of pores, template materials must meet certain requirements. First, the template materials must be compatible with the processing conditions. For example, an electrical insulator is required when a template is used in electrochemical deposition. Except in the case of template-directed synthesis, the template materials should be chemically and thermally inert during synthesis and the following processing steps. Secondly, the material or solution being deposited must wet the internal pore walls. Thirdly, for the synthesis of nanorods or nanowires, the deposition should start from the bottom or from one end of the template channel and proceed from one side to the other. However, for the growth of nanotubules, deposition should start from the pore wall and proceed inwardly. Inward growth may result in pore blockage, so this should be avoided during the growth of "solid" nanorods or nanowires. Kinetically, the correct amount of surface relaxation permits maximal packing density, so a diffusion-limited process is preferred. Other considerations include the ease of release of the nanowires or nanorods from the templates and the ease of handling during the experiments.

AAM and PC membranes are most commonly used for the synthesis of nanorod or nanowire arrays. Both templates are very convenient for the growth of nanorods by various growth mechanisms, but each type of template also has its disadvantages. The advantages of using PC as the template are its easy handling and easy removal by means of pyrolysis at elevated temperatures, but the flexibility of PC is more prone to distortion during the heating process, and removal of the template occurs before complete densification of the nanorods. These factors result in broken and deformed nanorods. The advantage of using AAM as the template is its rigidity and resistance to high temperatures, which allows the

nanorods to densify completely before removal. This results in fairly free-standing and unidirectionally-aligned
nanorod arrays with a larger surface area than for PC.

The problem with AAM is the complete removal of
the template after nanorod growth, which is yet to be
achieved when using wet chemical etching.

5.2 Electrochemical Deposition

Electrochemical deposition, also known as electrodeposition, involves the oriented diffusion of charged reactive
species through a solution when an external electric
field is applied, and the reduction of the charged growth
species at the growth or deposition surface (which also
serves as an electrode). In industry, electrochemical deposition is widely used when coating metals in a process
known as electroplating [5.25]. In general, this method
is only applicable to electrically conductive materials
such as metals, alloys, semiconductors, and electrically
conductive polymers. After the initial deposition, the
electrode is separated from the depositing solution by
the deposit and so the deposit must conduct in order
to allow the deposition process to continue. When the
deposition is confined to the pores of template membranes, nanocomposites are produced. If the template
membrane is removed, nanorod or nanowire arrays are
prepared.

When a solid is immersed in a polar solvent or
an electrolyte solution, surface charge will develop.

The electrode potential is described by the Nernst
equation

$$E = E_0 + \frac{RT}{n_i F} \ln(a_i) \,, \tag{5.1}$$

where E_0 is the standard electrode potential (or the
potential difference between the electrode and the solution) when the activity a_i of the ions is unity, F is
Faraday's constant, R is the gas constant, and T is the
temperature. When the electrode potential is higher than
the energy level of a vacant molecular orbital in the
electrolyte, electrons will transfer from the electrode
to the solution and the electrolyte will be reduced, as
shown in Fig. 5.1a [5.24]. On the other hand, if the
electrode potential is lower than the energy level of
an occupied molecular orbital in the electrolyte, the
electrons will transfer from the electrolyte to the electrode, resulting in electrolyte oxidation, as illustrated in
Fig. 5.1b [5.24]. These reactions stop when equilibrium
is achieved.

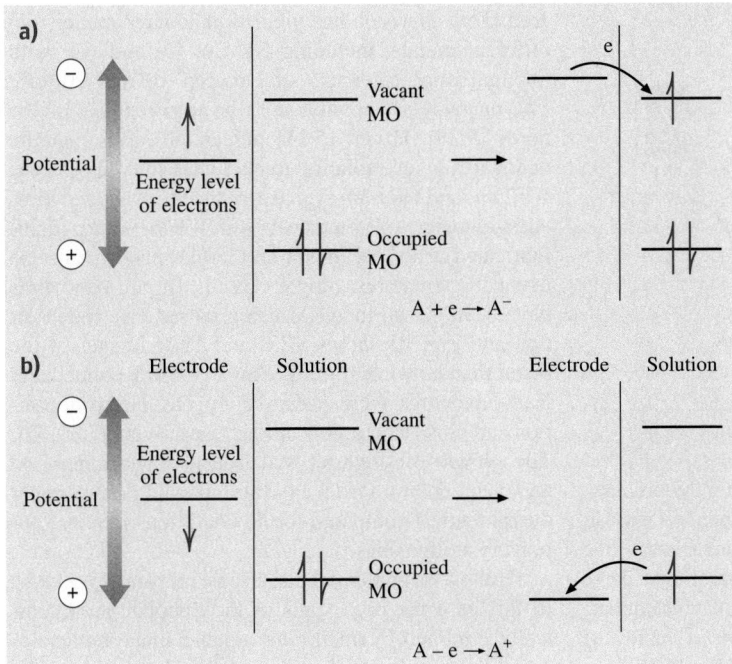

Fig. 5.1a,b Representation of the
(a) reduction and (b) oxidation of
a species A in solution. The molecular orbitals (MO) shown for species
A are the highest occupied MO and
the lowest vacant MO. These approximately correspond to the E^0s
of the A/A$^-$ and A$^+$/A couples,
respectively. (After [5.24])

Part A | 5.2

When an external electric field is applied between two dissimilar electrodes, charged species flow from one electrode to the other, and electrochemical reactions occur at both electrodes. This process, called electrolysis, converts electrical energy to chemical potential. The system used to perform electrolysis is called an electrolytic cell. In this cell, the electrode connected to the positive side of the power supply, termed the "anode", is where an oxidation reaction takes place, whereas the electrode connected to the negative side of the power supply, the "cathode", is where a reduction reaction proceeds, accompanied by deposition. Therefore, electrolytic deposition is also called cathode deposition, but it is most commonly referred to as electrochemical deposition or electrodeposition.

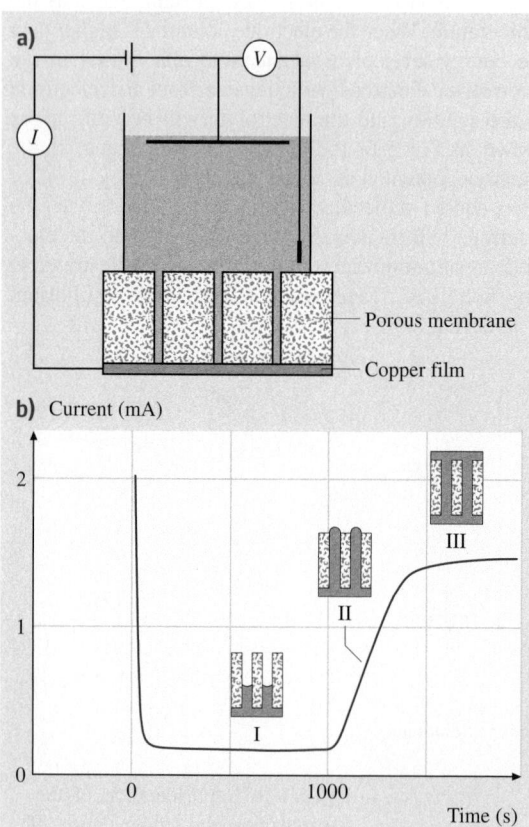

Fig. 5.2a,b Common experimental set-up for the template-based growth of nanowires using electrochemical deposition. (**a**) Schematic illustration of the arrangement of the electrodes for nanowire deposition. (**b**) Current–time curve for electrodeposition of Ni into a polycarbonate membrane with 60 nm diameter pores at − 1.0 V. *Insets* depict the different stages of the electrodeposition. (After [5.26])

5.2.1 Metals

The growth of nanowires of conductive materials in an electric field is a self-propagating process [5.27]. Once the small rods form, the electric field and the density of current lines between the tips of nanowires and the opposing electrode are greater than that between two electrodes, due to the shorter distances between the nanowires and the electrodes. This ensures that the species being deposited is constantly attracted preferentially to the nanowire tips, resulting in continued growth. To better control the morphology and size, templates containing channels in the desired shape are used to guide the growth of nanowires. Figure 5.2 illustrates the a common set-up used for the template-based growth of nanowires [5.26]. The template is attached to the cathode, which is brought into contact with the deposition solution. The anode is placed in the deposition solution, parallel to the cathode. When an electric field is applied, cations diffuse through the channels and deposit on the cathode, resulting in the growth of nanowires inside the template. This figure also shows the current density at different stages of deposition when a constant electric field is applied. The current does not change significantly until the pores are completely filled, at which point the current increases rapidly due to improved contact with the electrolyte solution. The current saturates once the template surface is completely covered. This approach has yielded nanowires made from different metals, including Ni, Co, Cu and Au, with nominal pore diameters of between 10 and 200 nm. The nanowires were found to be true replicas of the pores [5.28]. *Possin* [5.11] prepared various metallic nanowires using radiation track-etched mica. Likewise, *Williams* and *Giordano* [5.29] produced silver nanowires with diameters of less than 10 nm. *Whitney* et al. [5.26] fabricated arrays of nickel and cobalt nanowires, also using PC templates. Single crystal bismuth nanowires have been grown in AAM using pulsed electrodeposition and Fig. 5.3 shows SEM and TEM images of the bismuth nanowires [5.30]. Single crystal copper and lead nanowires were prepared by DC electrodeposition and pulse electrodeposition, respectively [5.31,32]. The growth of single crystal lead nanowires required a greater departure from equilibrium conditions (greater overpotential) compared to the conditions required for polycrystalline ones.

Hollow metal tubules can also be prepared [5.33,34]. In this case the pore walls of the template are chemically modified by anchoring organic silane molecules so that the metal will preferentially deposit onto the

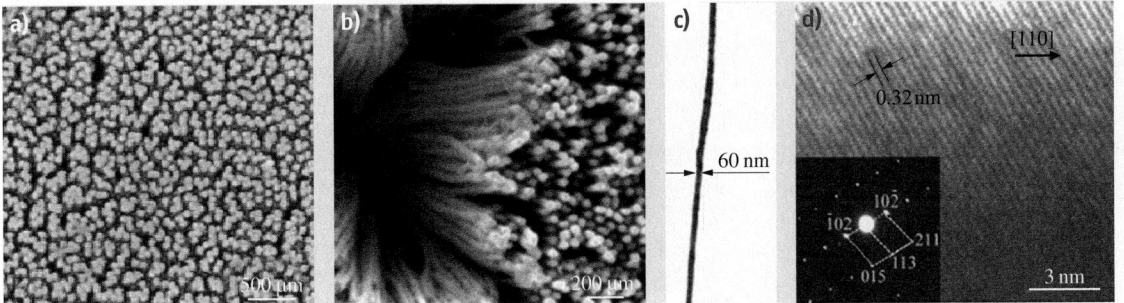

Fig. 5.3a–d SEM images of Bi nanowire arrays: (**a**) top view, (**b**) tilt view. (**c**) TEM image of a typical Bi single nanowire. (**d**) HRTEM image of a typical Bi single nanowire. The *inset* is the corresponding ED pattern. (After [5.30])

pore walls instead of the bottom electrode. For example, the porous surface of an anodic alumina template was first covered with cyanosilanes; subsequent electrochemical deposition resulted in the growth of gold tubules [5.35]. An electroless electrolysis process has also been investigated for the growth of nanowires and nanorods [5.16, 33, 36]. Electroless deposition is actually a chemical deposition process and it involves the use of a chemical agent to coat a material onto the template surface [5.37]. The main differences between electrochemical deposition and electroless deposition are that the deposition begins at the bottom electrode and the deposited materials must be electrically conductive in the former. The electroless method does not require the deposited materials to be electrically conductive, and the deposition starts from the pore wall and proceeds inwardly. Therefore, in general, electrochemical deposition results in the formation of "solid" nanorods or nanowires of conductive materials, whereas electroless deposition often results in hollow fibrils or nanotubules. For electrochemical deposition, the length of nanowires or nanorods can be controlled by the deposition time, whereas in electroless deposition the length of the nanotubules is solely dependent on the length of the deposition channels or pores. Variation of deposition time would result in a different wall thickness of nanotubules. An increase in deposition time leads to a thick wall, but sometimes the hollow tubule morphology persists even after prolonged deposition.

Although many research groups have reported on the growth of uniformly sized nanorods and nanowires on PC template membranes, *Schönenberger* et al. [5.38] reported that the channels of carbonate membranes were not always uniform in diameter. They grew Ni, Co, Cu, and Au nanowires using polycarbonate membranes with nominal pore diameters of between 10 and 200 nm by

an electrolysis method. From both a potentiostatic study of the growth process and a SEM analysis of nanowire morphology, they concluded that the pores were generally not cylindrical with a constant cross-section, but instead were rather cigar-like. For pores with a nominal diameter of 80 nm, the middle section of the pores was wider by up to a factor of 3.

5.2.2 Semiconductors

Semiconductor nanowire and nanorod arrays have been synthesized using AAM templates, such as CdSe and CdTe [5.39]. The synthesis of nanowire arrays of bismuth telluride (Bi_2Te_3) provide a good example of the synthesis of compound nanowire arrays by electrochemical deposition. Bi_2Te_3 is of special interest as a thermoelectric material and Bi_2Te_3 nanowire arrays are believed to offer high figures of merit for thermal-electrical energy conversion [5.40, 41]. Both polycrystalline and single crystal Bi_2Te_3 nanowire arrays have been grown by electrochemical deposition inside anodic alumina membranes [5.42, 43]. *Sander* and coworkers [5.42] fabricated Bi_2Te_3 nanowire arrays with diameters as small as ≈ 25 nm from a solution of 0.075 M Bi and 0.1 M Te in 1 M HNO_3 by electrochemical deposition at -0.46 V versus a Hg/Hg_2SO_4 reference electrode. The resultant Bi_2Te_3 nanowire arrays are polycrystalline in nature, and subsequent melting-recrystallization failed to produce single crystal Bi_2Te_3 nanowires. More recently, single crystal Bi_2Te_3 nanowire arrays have been grown from a solution consisted of 0.035 M Bi $(NO_3)_3$.5H_2O and 0.05 M $HTeO_2^+$; the latter was prepared by dissolving Te powder in 5 M HNO_3 by electrochemical deposition. Figures 5.4 and 5.5 show a SEM image of a cross-section of a Bi_2Te_3 nanowire array and an XRD spectrum showing its crystal orientation, respectively. High-resolution

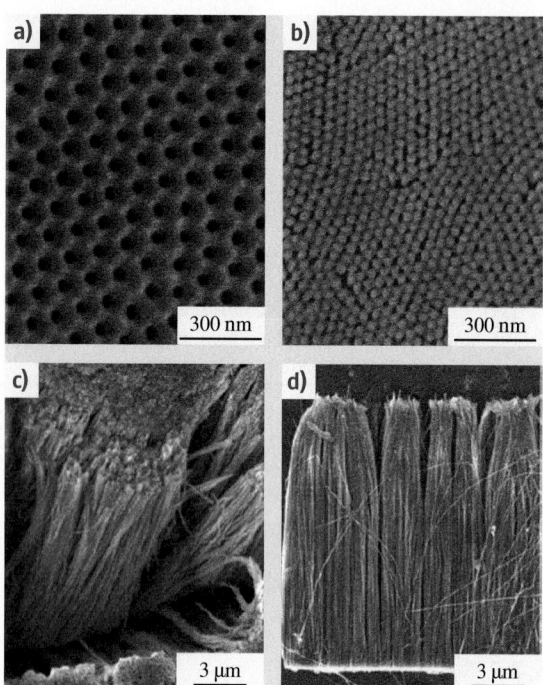

Fig. 5.4a–d SEM photographs of AAM template and Bi$_2$Te$_3$ nanowire arrays. (**a**) A typical SEM photograph of AAM. (**b**) Surface view of Bi$_2$Te$_3$ nanowire array (eroding time: 5 min). (**c**) Surface view of Bi$_2$Te$_3$ nanowire array (eroding time: 15 min). (**d**) Cross-sectional view of Bi$_2$Te$_3$ nanowire array (eroding time: 15 min). (After [5.43])

TEM and electron diffraction, together with XRD, revealed that [110] is the preferred growth direction of Bi$_2$Te$_3$ nanowires. Single crystal nanowire or nanorod arrays can also be made by carefully controlling the initial deposition [5.44]. Similarly, large area Sb$_2$Te$_3$ nanowire arrays have also been successfully grown by template-based electrochemical deposition, but the nanowires grown are polycrystalline and show no clear preferred growth direction [5.45].

5.2.3 Conductive Polymers

Electrochemical deposition has also been explored for the synthesis of conductive polymer nanowire and nanorod arrays [5.46]. Conductive polymers have great potential for plastic electronics and sensor applications [5.47,48]. For example, *Schönenberger* et al. [5.38] have made conductive polyporrole nanowires in PC membranes. Nanotubes are commonly observed for polymer materials, as shown in Fig. 5.6 [5.49], in con-

trast to "solid" metal nanorods or nanowires. It seems that deposition or solidification of polymers inside template pores starts at the surface and proceeds inwardly. *Martin* [5.50] proposed that this phenomenon was caused by the electrostatic attraction between the growing polycationic polymer and the anionic sites along the pore walls of the polycarbonate membrane. In addition, although the monomers are soluble, the polymerized form is insoluble. Hence there is a solvophobic component leading to deposition at the surface of the pores [5.51, 52]. In the final stage, the diffusion of monomers through the inner pores becomes retarded and monomers inside the pores are quickly depleted. The deposition of polymer inside the inner pores stops.

Liang et al. [5.53] reported a direct electrochemical synthesis of oriented nanowires of polyaniline (PANI) – a conducting polymer with a conjugated backbone due to phenyl and amine groups – from solutions using no templates. The experimental design is based on the idea that, in theory, the rate of electropolymerization (or nanowire growth) is related to the current density. Therefore, it is possible to control the nucleation and the polymerization rate simply by adjusting the current density. The synthesis involves electropolymerization of aniline (C$_6$H$_5$NH$_2$) and in situ electrodeposition, resulting in nanowire growth.

5.2.4 Oxides

Similar to metals, semiconductors and conductive polymers, some oxide nanorod arrays can be grown directly from solution by electrochemical deposition. For example, V$_2$O$_5$ nanorod arrays have been grown on ITO substrate from VOSO$_4$ aqueous solution with VO^{++} as the growth species [5.54]. At the interface between the electrode (and therefore the subsequent growth surface) and the electrolyte solution, the ionic cluster (VO^{++}) is oxidized and solid V$_2$O$_5$ is deposited through the following reaction.

$$2VO^{++} + 3H_2O \rightarrow V_2O_5 + 6H^+ + 2e^- \ . \qquad (5.2)$$

A reduction reaction takes place at the counter electrode:

$$2H^+ + 2e^- \rightarrow H_2(g) \ . \qquad (5.3)$$

It is obvious that the pH and the concentration of VO^{++} clusters in the vicinity of the growth surface shift away from that in the bulk solution; both the pH and the VO^{++} concentration decrease.

ZnO nanowire arrays were fabricated by a one-step electrochemical deposition technique based on an

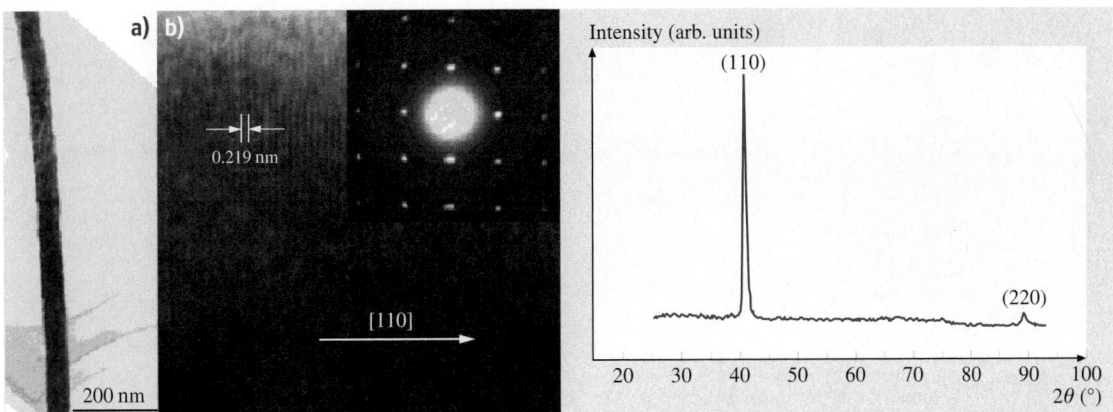

Fig. 5.5a–c TEM images and XRD pattern of a single Bi_2Te_3 nanowire. (**a**) TEM image and (**b**) HRTEM image of the same nanowire. The *inset* is the corresponding ED pattern. (**c**) XRD pattern of Bi_2Te_3 nanowire array (electrodeposition time: 5 min). (After [5.43])

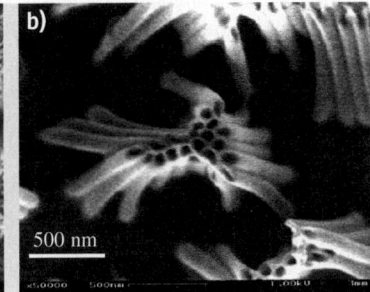

Fig. 5.6 SEM images of polymer nanotubes. (After [5.49])

ordered nanoporous alumina membrane [5.55]. The ZnO nanowire array is uniformly assembled into the nanochannels of an anodic alumina membrane and consists of single crystal particles.

5.3 Electrophoretic Deposition

The electrophoretic deposition technique has been widely explored, particularly for the deposition of ceramic and organoceramic materials onto a cathode from colloidal dispersions [5.56–58]. Electrophoretic deposition differs from electrochemical deposition in several aspects. First, the material deposited in the electrophoretic deposition method does not need to be electrically conductive. Second, nanosized particles in colloidal dispersions are typically stabilized by electrostatic or electrosteric mechanisms. As discussed in the previous section, when dispersed in a polar solvent or an electrolyte solution, the surface of a nanoparticle develops an electrical charge via one or more of the following mechanisms: (1) preferential dissolution, (2) deposition of charges or charged species, (3) preferential reduction or oxidation, and (4) physical adsorption of charged species such as polymers. A combination of electrostatic forces, Brownian motion and osmotic forces results in the formation of a "double layer structure", schematically illustrated in Fig. 5.7. The figure depicts a positively charged particle surface, the concentration profiles of negative ions (counterions) and positive ions (surface charge-determining ions), and the electric potential profile. The concentration of counterions gradually decreases with distance from the particle surface, whereas that of charge-determining ions increases. As a result, the electric potential decreases with distance. Near the particle surface, the electric poten-

Part A | 5.3

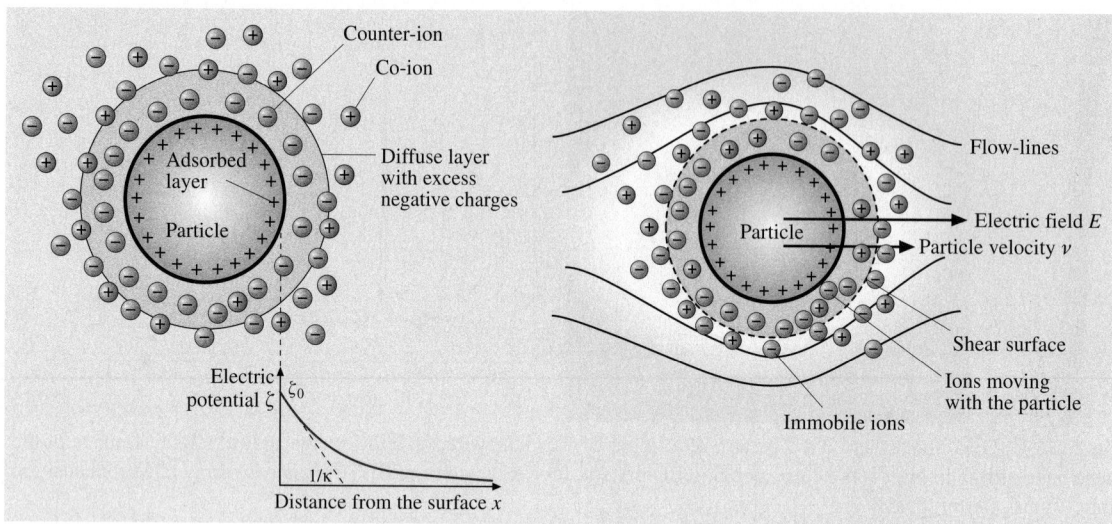

Fig. 5.7 Schematic illustrating electrical double layer structure and the electric potential near the solid surface with both the Stern and Gouy layers indicated. Surface charge is assumed to be positive. (After [5.59])

tial decreases linearly, in the region known as the Stern layer. Outside of the Stern layer, the decrease follows an exponential relationship. The region between the Stern layer and the point where the electric potential equals zero is called the diffusion layer. Taken together, the Stern layer and diffusion layer is known as the double layer structure in the classical theory of electrostatic stabilization.

Upon the application of an external electric field, charged particles are set in motion, as schematically illustrated in Fig. 5.8 [5.59]. This type of motion is referred to as electrophoresis. When a charged particle moves, some of the solvent or solution surrounding the particle will also move with it, since part of the solvent or solution is tightly bound to the particle. The plane that separates the tightly bound liquid layer from the rest of the liquid is called the "slip plane" (Fig. 5.7). The electric potential at the slip plane is known as the "zeta potential", which is an important parameter when determining the stability and transport of a colloidal dispersion or a sol. A zeta potential of more than about 25 mV is typically required to stabilize a system [5.60]. The zeta potential ζ around a spherical particle can be described using the relation [5.61]:

$$\zeta = \frac{Q}{4\pi\varepsilon_r a\,(1+\kappa a)}$$

with

$$\kappa = \left(\frac{e^2 \sum n_i z_i^2}{\varepsilon_r \varepsilon_0 kT}\right)^{1/2}, \tag{5.4}$$

where Q is the charge on the particle, a is the radius of the particle out to the shear plane, ε_r is the relative dielectric constant of the medium, and n_i and z_i are the bulk concentration and valence of the ith ion in the system, respectively.

The mobility of a nanoparticle in a colloidal dispersion or a sol μ, is dependent on the dielectric constant of the liquid medium ε_r, the zeta potential of the nanoparticle ζ, and the viscosity η of the fluid. Several forms for this relationship have been proposed, such as the Hückel equation [5.59, 61–64]:

$$\mu = \frac{2\varepsilon_r \varepsilon_0 \zeta}{3\pi\eta} \tag{5.5}$$

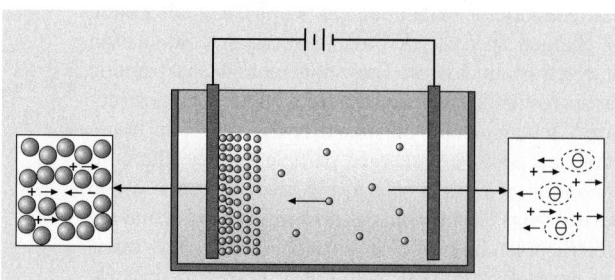

Fig. 5.8 Schematic showing electrophoresis. Upon application of an external electric field to a colloidal system or a sol, the charged nanoparticles or nanoclusters are set in motion. (After [5.1])

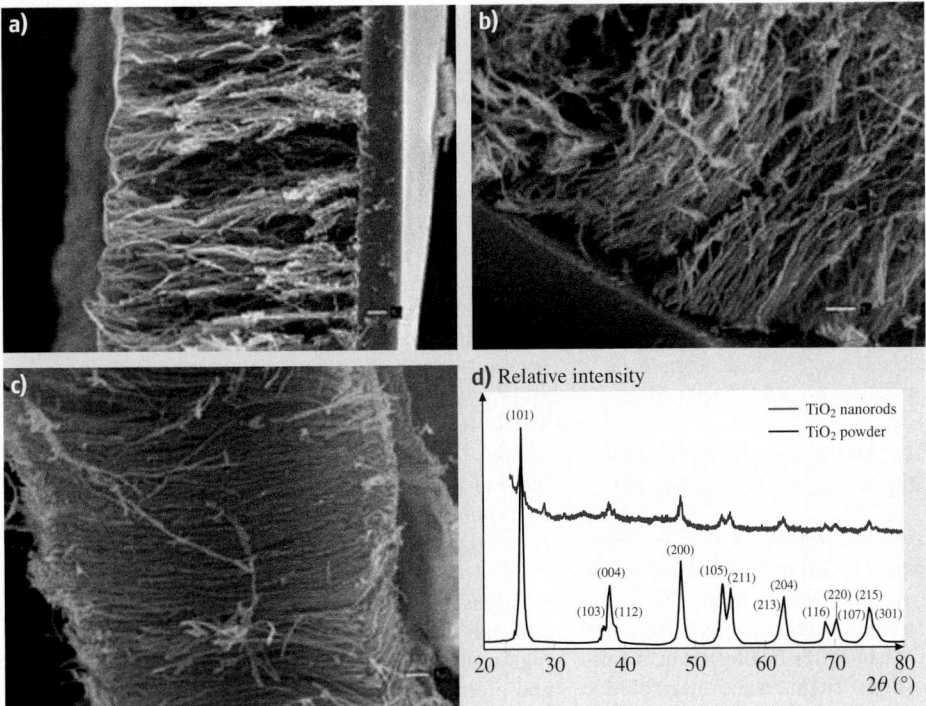

Fig. 5.9a–d SEM micrograph of TiO$_2$ nanorods grown by template-based electrochemically induced sol-gel deposition. The diameters of the nanorods are approximately: (**a**) 180 nm (for the 200 nm polycarbonate membrane); (**b**) 90 nm (for the 100 nm membrane); (**c**) 45 nm (for the 50 nm membrane). (**d**) XRD patterns of both the grown nanorods and a powder derived from the same sol. Both samples consist of the anatase phase only and no peak position shift was observed (after [5.66])

Electrophoretic deposition simply uses the oriented motion of charged particles in an electrical field to grow films or monoliths by transferring the solid particles from a colloidal dispersion or a sol onto the surface of an electrode. If the particles are positively charged (or more precisely, they have a positive zeta potential), deposition of solid particles will occur at the cathode. Otherwise, deposition will be at the anode. The electrostatic double layers collapse at the electrodes and the particles coagulate, producing porous materials made of compacted nanoparticles. Typical packing densities are far less than the theoretical density of 74 vol.% [5.65]. Many theories have been proposed to explain the processes at the deposition surface during electrophoretic deposition. However, the evolution of structure on the deposition surface is not well understood. The electrochemical processes that take place at the deposition surface and at the electrodes are complex and vary from system to system. The final density is dependent upon the concentration of particles in sols or colloidal dispersions, the zeta potential, the external electric field, and the reaction kinetics between the surfaces of the particles. A slow reaction and a slow arrival of nanoparticles onto the surface would allow sufficient particle relaxation on the deposition surface, so a high packing density would be expected.

5.3.1 Polycrystalline Oxides

Limmer et al. [5.66–69] combined sol-gel preparation with electrophoretic deposition to prepare nanorods of various complex oxides. One of the advantages of this technique is the ability to synthesize complex oxides and organic–inorganic hybrids with desired stoichiometric compositions. Another advantage is their applicability to a variety of materials. In their approach, conventional sol-gel processing was applied to the synthesis of various sols. By controlling the sol preparation appropriately, nanometer particles of a desired stoichiometric composition were formed, and electrostatically stabilized by pH adjustment. Using radiation-tracked etched polycarbonate membranes with an electric field of ≈ 1.5 V/cm, they have grown nanowires with diameters ranging from 40 to 175 nm and lengths of 10 μm, corresponding to the thickness of the membrane. The materials include anatase TiO$_2$, amorphous SiO$_2$, perovskite BaTiO$_3$ and Pb(Ti, Zr)O$_3$, and layered perovskite Sr$_2$Nb$_2$O$_7$. Figure 5.9 shows SEM micrographs and XRD patterns of TiO$_2$ nanorod arrays [5.66].

Wang et al. [5.70] used electrophoretic deposition to make nanorods of ZnO from colloidal sols. ZnO colloidal sol was prepared by hydrolyzing an alcoholic solution of zinc acetate with NaOH, with a small amount of zinc nitrate added as a binder. This solution was then introduced into the pores of anodic alumina membranes at voltages of 10–400 V. It was found that lower voltages led to dense, solid nanorods, while higher voltages caused the formation of hollow tubules. They suggested that the higher voltages cause dielectric breakdown of the anodic alumina, causing it to become as positively charged as the cathode. Electrostatic attraction between the ZnO nanoparticles and the pore walls then leads to tubule formation.

5.3.2 Single Crystal Oxide Nanorod Arrays Obtained by Changing the Local pH

A modified version of sol electrophoretic deposition has been used to grow single crystalline titanium oxide and vanadium pentoxide nanorod arrays from TiO^{2+} and VO_2^+ solutions respectively. *Miao* et al. [5.71] prepared single crystalline TiO_2 nanowires by electrochemically induced sol-gel deposition. Titania electrolyte solution was prepared by dissolving Ti powder into a H_2O_2 and NH_4OH aqueous solution to form TiO^{2+} ionic clusters [5.72]. When an electric field was applied, the TiO^{2+} ionic clusters diffused to the cathode and underwent hydrolysis and condensation reactions, resulting in the deposition of nanorods of amorphous TiO_2 gel. After heating at 240 °C for 24 h in air, single crystal anatase nanorods with diameters of 10, 20, and 40 nm and lengths ranging from 2 to 10 μm were synthesized. The formation of single crystal TiO_2 nanorods here is different to that reported by *Martin*'s group [5.73]. It is suggested that the nanoscale crystallites generated during heating assembled epitaxially to form single crystal nanorods.

During typical sol-gel processing, nanoclusters are formed through homogeneous nucleation and subsequent growth through sequential yet parallel hydrolysis and condensation reactions. Sol electrophoretic deposition enriches and deposits these formed nanoclusters at an appropriate electrode surface under an external electric field. The modified process is to limit and induce the condensation reaction at the growth surface by changing local pH value, which is a result of partial water hydrolysis at the electrode or growth surface:

$$2H_2O + 2e^- \rightarrow H_2 + 2OH^- \,, \qquad (5.6)$$

$$2VO_2^+ + 2OH^- \rightarrow V_2O_5 + H_2O \,. \qquad (5.7)$$

Reaction 6, or the electrolysis of water, plays a very important role here. As the reaction proceeds, hydroxyl groups are produced, resulting in increased pH near to the deposition surface. This increase in pH value near to the growth surface initiated and promotes the precipitation of V_2O_5, or reaction 7. The initial pH of the VO_2^+ solution is approximately 1.0, meaning that VO_2^+ is stable. However, when the pH increases to ≈1.8, VO_2^+ is no longer stable and solid V_2O_5 forms. Since the change in pH occurs near to the growth surface, reaction 7 or deposition is likely to occur on the surface of the electrode through heterogeneous nucleation and subsequent growth. It should be noted that the hydrolysis of water has another effect on the deposition of solid V_2O_5. Reaction 6 produces hydrogen on the growth surface. These molecules may poison the growth surface before dissolving into the electrolyte or by forming a gas bubble, which may cause the formation of porous nanorods.

The formation of single crystal nanorods from solutions by pH change-induced surface condensation has been proven by TEM analyses, including high-resolution imaging showing the lattice fringes and electron diffraction. The growth of single crystal nanorods by pH change-induced surface condensation is attributed to evolution selection growth, which is briefly summarized below. The initial heterogeneous nucleation or deposition onto the substrate surface results in the formation of nuclei with random orientations. The subsequent growth of various facets of a nucleus is dependent on the surface energy, and varies significantly from one facet to another [5.74]. For one-dimensional growth, such as film growth, only the highest growth rate with a direction perpendicular to the growth surface will be able to continue to grow. The nuclei with the fastest growth direction perpendicular to the growth surface will grow larger, while nuclei with slower growth rates will eventually cease to grow. Such a growth mechanism results in the formation of columnar structured films where all of the grains have the same crystal orientation (known as textured films) [5.75, 76]. In the case of nanorod growth inside a pore channel, such evolution selection growth is likely to lead to the formation of a single crystal nanorod or a bundle of single crystal nanorods per pore channel. Figure 5.10 shows typical TEM micrographs and selected-area electron diffraction patterns of V_2O_5 nanorods. It is well known that [010] (the *b*-axis) is the fastest growth direction for a V_2O_5 crystal [5.77, 78], which would explain why single crystal vanadium nanorods or a bundle of single crystal nanorods grow along the *b*-axis.

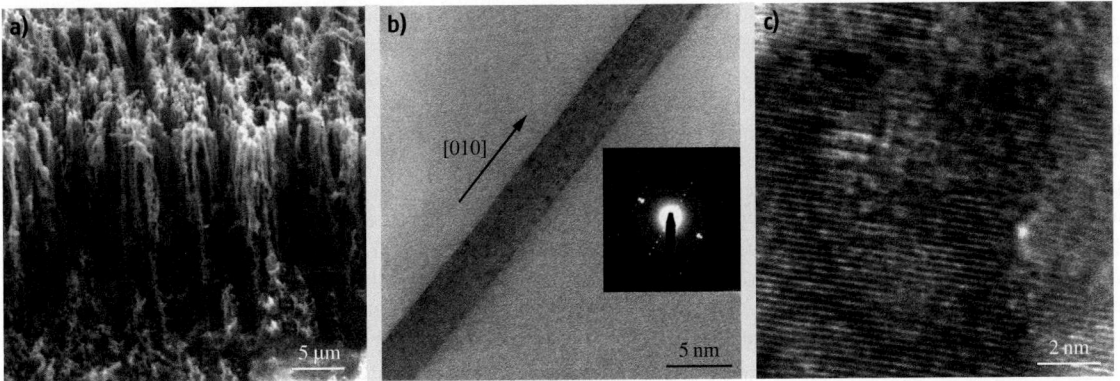

Fig. 5.10 (a) SEM image of V_2O_5 nanorod arrays on an ITO substrate grown in a 200 nm carbonate membrane by sol electrophoretic deposition; **(b)** TEM image of a V_2O_5 nanorod with its electron diffraction pattern; **(c)** high-resolution TEM image of the V_2O_5 nanorod showing the lattice fringes (after [5.54])

5.3.3 Single-Crystal Oxide Nanorod Arrays Grown by Homoepitaxial Aggregation

Single crystal nanorods can also be grown directly by conventional electrophoretic deposition. However, several requirements must be met for such growth. First, the nanoclusters or particles in the sol must have a crystalline structure extended to the surface. Second, the deposition of nanoclusters on the growth surface must have a certain degree of reversibility so that the nanoclusters can rotate or reposition prior to their irreversible incorporation into the growth surface. Thirdly, the deposition rate must be slow enough to permit sufficient time for the nanoclusters to rotate or reposition. Lastly, the surfaces of the nanoclusters must be free of strongly attached alien chemical species. Although precise control of all these parameters remains a challenge, the growth of single crystal nanorods through homoepitaxial aggregation of nanocrystals has been demonstrated [5.79, 80]. The formation of single crystalline vanadium pentoxide nanorods by template-based sol electrophoretic deposition can be attributed to homoepitaxial aggregation of crystalline nanoparticles. Thermodynamically it is favorable for the crystalline nanoparticles to aggregate epitaxially; this growth behavior and mechanism is well documented in the literature [5.81, 82]. In this growth mechanism, an initial weak interaction between two nanoparticles allows rotation and migration relative to each other. Obviously, homoepitaxial aggregation is a competitive process and porous structure is expected to form through this homoepitaxial aggregation (as schematically illustrated in Fig. 5.11). Vanadium oxide particles present in a typ-ical sol are known to easily form ordered crystalline structure [5.83], so it is reasonable to expect that homoepitaxial aggregation of vanadium nanocrystals from sol results in the formation of single crystal nanorods. Single crystal nanorods formed in this way are likely to undergo significant shrinkage when fired at high temperatures due to its original porous nature; 50% lateral shrinkage has been observed in vanadium pentoxide nanorods formed by this method. In addition, it might be possible that the electric field and the internal surfaces of the pore channels play significant roles in the orientation of the nanorods, as suggested in the literature [5.84, 85].

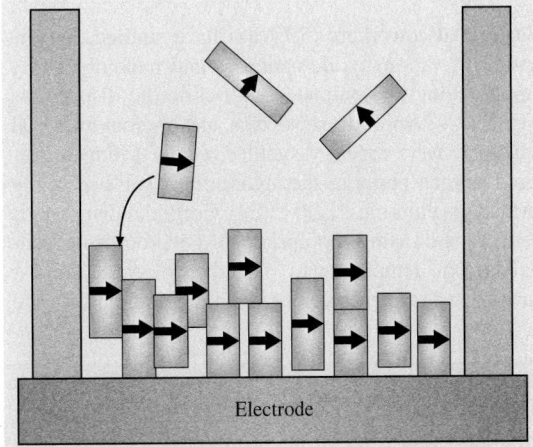

Fig. 5.11 Schematic illustration of the homoepitaxial aggregation growth mechanism of single-crystalline nanorods. (After [5.54])

5.3.4 Nanowires and Nanotubes of Fullerenes and Metallofullerenes

Electrophoretic deposition in combination with template-based growth has also been successfully explored in the formation of nanowires and nanotubes of carbon fullerenes, such as C_{60} [5.86], or metallofullerenes, such as $Sc@C_{82}(I)$ [5.87]. Typical experiments include the purification or isolation of the fullerenes or metallofullerenes required using multiple-step liquid chromatography and dispersion of the fullerenes in a mixed solvent of acetonitrile/toluene

in a ratio of 7:1. The electrolyte solution has a relatively low concentration of fullerenes ($35\,\mu M$) and metallofullerenes ($40\,\mu M$), and the electrophoretic deposition takes place with an externally applied electric field of $100–150\,V$ with a distance of $5\,mm$ between the two electrodes. Both nanorods and nanotubes of fullerenes or metallofullerenes can form and it is believed that initial deposition occurs along the pore surface. A short deposition time results in the formation of nanotubes, whereas extended deposition leads to the formation of solid nanorods. These nanorods possess either crystalline or amorphous structure.

5.4 Template Filling

Directly filling a template with a liquid mixture precursor is the most straightforward and versatile method for preparing nanowire or nanorod arrays. The drawback of this approach is that it is difficult to ensure complete filling of the template pores. Both nanorods and nanotubules can be obtained depending on the interfacial adhesion and the solidification modes. If the adhesion between the pore walls and the filling material is weak, or if solidification starts at the center (or from one end of the pore, or uniformly throughout the rods), solid nanorods are likely to form. If the adhesion is strong, or if the solidification starts at the interfaces and proceeds inwardly, hollow nanotubules are likely to form.

5.4.1 Colloidal Dispersion (Sol) Filling

Martin and coworkers [5.73, 88] have studied the formation of various oxide nanorods and nanotubules by simply filling the templates with colloidal dispersions (Fig. 5.12). Nanorod arrays of a mesoporous material (SBA-15) were recently synthesized by filling an ordered porous alumina membrane with sol containing surfactant (Pluronic P123) [5.89]. Colloidal dispersions were prepared using appropriate sol-gel processing techniques. The template was placed in a stable sol for various periods of time. The capillary force drives the

sol into the pores if the sol has good wettability for the template. After the pores were filled with sol, the template was withdrawn from the sol and dried. The sample was fired at elevated temperatures to remove the template and to densify the sol-gel-derived green nanorods.

A sol typically consists of a large volume fraction of solvent, up to 90% or higher. Although the capillary force may ensure complete filling of the pores with the suspension, the amount of solid occupying the pore space is small. Upon drying and subsequent firing processes, significant shrinkage would be expected. However, the actual shrinkage observed is small when compared with the pore size. These results indicate that an (unknown) mechanism is acting to enrich the concentration of solid inside the pores. One possible mechanism could be the diffusion of solvent through the membrane, leading to the enrichment of solid on the internal surfaces of the template pores, similar to what happens during ceramic slip casting [5.90]. The observed formation of nanotubules (as shown in Fig. 5.12 [5.73]) may imply that this process is indeed present. However, considering the fact that the templates were typically emerged into sol for just a few minutes, diffusion through the membrane and enrichment of the solid inside the pores must be rather rapid processes. It was also noticed that

  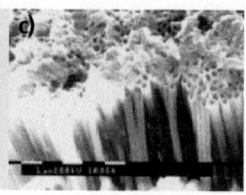

Fig. 5.12a–c SEM micrographs of oxide nanorods created by filling the templates with sol-gels: (**a**) ZnO, (**b**) TiO$_2$ and (**c**) hollow nanotube. (After [5.73])

the nanorods made by template filling are commonly polycrystalline or amorphous, although single crystal TiO_2 nanorods were sometimes observed for nanorods smaller than 20 nm [5.73].

5.4.2 Melt and Solution Filling

Metallic nanowires can also be synthesized by filling a template with molten metals [5.92]. One example is the preparation of bismuth nanowires using pressure injection of molten bismuth into the nanochannels of an anodic alumina template [5.93]. The anodic alumina template was degassed and immersed in the liquid bismuth at 325 °C ($T_m = 271.5$ °C for Bi), and then high pressure Ar gas of ≈ 300 bar was applied in order to inject liquid Bi into the nanochannels of the template for 5 h. Bi nanowires with diameters of 13–110 nm and large aspect ratios (of up to several hundred) have been obtained. Individual nanowires are believed to be single-crystal. When exposed to air, bismuth nanowires are readily oxidized. An amorphous oxide layer ≈ 4 nm in thickness was observed after 48 h. After 4 weeks, the bismuth nanowires were completely oxidized. Nanowires of other metals, such as In, Sn and Al, and the semiconductors Se, Te, GaSb, and Bi_2Te_3, were also prepared by injecting molten liquid into anodic alumina templates [5.94].

Polymeric fibrils have been made by filling the template pores with a monomer solution containing the desired monomer and a polymerization reagent, followed by in situ polymerization [5.14, 95–98]. The polymer preferentially nucleates and grows on the pore walls, resulting in tubules at short deposition times.

Metal, oxide and semiconductor nanowires have recently been synthesized using self-assembled mesoporous silica as the template. For example, *Han* et al. [5.99] have synthesized Au, Ag and Pt nanowires in mesoporous silica templates. The mesoporous templates were first filled with aqueous solutions of the corresponding metal salts (such as $HAuCl_4$). After drying and treatment with CH_2Cl_2, the samples were reduced under H_2 flow to form metallic nanowires. *Liu* et al. [5.100] carefully studied the interface between these nanowires and the matrix using high-resolution electron microscopy and electron energy loss spectroscopy techniques. A sharp interface only exists between noble metal nanowires and the matrix. For magnetic nickel oxide, a core-shell nanorod structure containing a nickel oxide core and a thin nickel silicate shell was observed. The magnetic properties of the templated nickel oxide were found to be significantly different from nickel oxide nanopowders due to the alignment of the nanorods. In another study, *Chen* et al. filled the pores of a mesoporous silica template with an aqueous solution of Cd and Mn salts, dried the sample, and reacted it with H_2S gas to convert it to (Cd,Mn)S [5.101].

5.4.3 Centrifugation

Filling the template with nanoclusters via centrifugation forces is another inexpensive method for mass producing nanorod arrays. Figure 5.13 shows SEM images of lead zirconate titanate (PZT) nanorod arrays with uniform sizes and unidirectional alignment [5.91]. These nanorod arrays were grown in polycarbonate membrane from PZT sol by centrifugation at 1500 rpm for 60 min. The samples were attached to silica glass and fired at 650 °C in air for 60 min. Nanorod arrays of other oxides (silica and titania) were prepared. The advantages of centrifugation include its applicability to any colloidal dispersion system, including those consisting of electrolyte-sensitive nanoclusters or molecules.

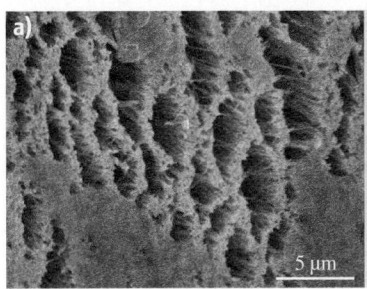 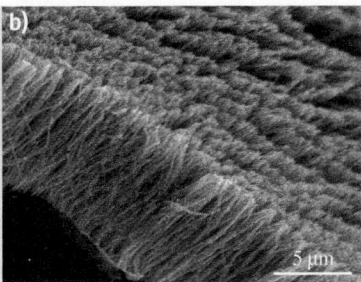

Fig. 5.13 SEM images of the top view (*left*) and side view (*right*) of lead zirconate titanate (PZT) nanorod arrays grown in polycarbonate membrane from PZT sol by centrifugation at 1500 rpm for 60 min. Samples were attached to silica glass and fired at 650 °C in air for 60 min. (After [5.91])

5.5 Converting from Reactive Templates

Nanorods or nanowires can also be synthesized using consumable templates, although the resultant nanowires and nanorods are generally not ordered to form aligned arrays. Nanowires of compounds can be prepared using a template-directed reaction. First nanowires or nanorods of one constituent element are prepared, and then these are reacted with chemicals containing the other element desired in order to form the final product. *Gates* et al. [5.102] converted single crystalline trigonal selenium nanowires into single crystalline nanowires of Ag_2Se by reacting Se nanowires with aqueous $AgNO_3$ solutions at room temperature. Nanorods can also be synthesized by reacting volatile metal halides or oxide species with carbon nanotubes to form solid carbide nanorods with diameters of between 2 and 30 nm and lengths of up to 20 μm [5.103]. ZnO nanowires were prepared by oxidizing metallic zinc nanowires [5.104]. Hollow nanotubules of MoS_2 ≈ 30 μm long and 50 nm in external diameter with wall thicknesses of 10 nm were prepared by filling a solution mixture of the molecular precursors, $(NH_4)_2MoS_4$ and $(NH_4)_2Mo_3S_{13}$, into the pores of alumina membrane templates. Then the template filled with the molecular precursors was heated to an elevated temperature and the molecular precursors were thermally decomposed into MoS_2 [5.105]. Certain polymers and proteins were also used to direct the growth of nanowires of metals or semiconductors. For example, *Braun* et al. [5.106] reported a two-step procedure using DNA as a template for the vectorial growth of a silver nanorods 12 μm in length and 100 nm in diameter. CdS nanowires were prepared by polymer-controlled growth [5.107]. For the synthesis of CdS nanowires, cadmium ions were well distributed in a polyacrylamide matrix. The Cd^{2+}-containing polymer was treated with thiourea (NH_2CSNH_2) solvothermally in ethylenediamine at 170 °C, resulting in degradation of polyacrylamide. Single crystal CdS nanowires 40 nm in diameter and up to 100 μm in length were obtained with preferential [001] orientations.

5.6 Summary and Concluding Remarks

This chapter provides a brief summary of the fundamentals of and techniques used for the template-based synthesis of nanowire or nanorod arrays. Examples were used to illustrate the growth of each nanorod material made with each technique. The literature associated with this field is overwhelming and is expanding very rapidly. This chapter is by no means comprehensive in its coverage of the relevant literature. Four groups of template-based synthesis methods have been reviewed and discussed in detail. Electrochemical deposition or electrodeposition is the method used to grow electrically conductive or semiconductive materials, such as metals, semiconductors, and conductive polymers and oxides. Electrophoretic deposition from colloidal dispersion is the method used to synthesize dielectric nanorods and nanowires. Template filling is conceptually straightforward, although complete filling is often very difficult. Converting reactive templates is a method used to achieve both nanorod arrays and randomly oriented nanowires or nanorods, and it is often combined with other synthetic methods.

This chapter has focused on the growth of solid nanorod and nanowire arrays by template-based synthesis; however, the use of template-based synthesis to synthesize nanotubes, and in particular nanotube arrays, has received increasing attention [5.108]. One of the greatest advantages using template-based synthesis to grow of nanotubes and nanotube arrays is the independent control of the lengths, diameters, and the wall thicknesses of the nanotubes available. While the lengths and the diameters of the resultant nanotubes are dependent on the templates used for the synthesis, the wall thicknesses of the nanotubes can be readily controlled through the duration of growth. Another great advantage of the template-based synthesis of nanotubes is the possibility of multilayered hollow nanotube or solid nanocable structures. For example, $Ni–V_2O_5.nH_2O$ nanocable arrays have been synthesized by a two-step approach [5.109]. First, Ni nanorod arrays were grown in a PC template by electrochemical deposition, and then the PC template was removed by pyrolysis, followed by sol electrophoretic deposition of $V_2O_5.nH_2O$ on the surfaces of the Ni nanorod arrays. It is obvious that there is a lot of scope for more research into template-based syntheses of nanorod, nanotube and nanocable arrays, and their applications.

References

5.1 G. Z. Cao: *Nanostructures and Nanomaterials: Synthesis, Properties and Applications* (Imperial College, London 2004)

5.2 Z. L. Wang: *Nanowires and Nanobelts: Materials, Properties and Devices, Nanowires and Nanobelts of Functional Materials*, Vol. 2 (Kluwer, Boston 2003)

5.3 Y. Xia, P. Yang, Y. Sun, Y. Wu, Y. Yin, F. Kim, H. Yan: One-Dimensional Nanostructures: Synthesis, Characterization and Applications, Adv. Mater. **15**, 353–389 (2003)

5.4 A. Huczko: Template-Based Synthesis of Nanomaterials, Appl. Phys. A **70**, 365–376 (2000)

5.5 C. Burda, X. Chen, R. Narayanan, M. A. El-Sayed: Chemistry and Properties of Nanocrystals of Different Shapes, Chem. Rev. **105**, 1025–1102 (2005)

5.6 X. Duan, C. M. Lieber: General Synthesis of Compound Semiconductor Nanowires, Adv. Mater **12**, 298–302 (2000)

5.7 M. P. Zach, K. H. Ng, R. M. Penner: Molybdenum Nanowires by Electrodeposition, Science **290**, 2120–2123 (2000)

5.8 R. C. Furneaux, W. R. Rigby, A. P. Davidson: The Formation of Controlled-porosity Membranes from Anodically Oxidized Aluminium, Nature **337**, 147–149 (1989)

5.9 R. L. Fleisher, P. B. Price, R. M. Walker: *Nuclear Tracks in Solids* (University of California Press, Berkeley 1975)

5.10 R. J. Tonucci, B. L. Justus, A. J. Campillo, C. E. Ford: Nanochannel Array Glass, Science **258**, 783–787 (1992)

5.11 G. E. Possin: A Method for Forming Very Small Diameter Wires, Rev. Sci. Instrum. **41**, 772–774 (1970)

5.12 C. Wu, T. Bein: Conducting Polyaniline Filaments in a Mesoporous Channel Host, Science **264**, 1757–1759 (1994)

5.13 S. Fan, M. G. Chapline, N. R. Franklin, T. W. Tombler, A. M. Cassell, H. Dai: Self-Oriented Regular Arrays of Carbon Nanotubes and Their Field Emission Properties, Science **283**, 512–514 (1999)

5.14 P. Enzel, J. J. Zoller, T. Bein: Intrazeolite Assembly and Pyrolysis of Polyacrylonitrile, Chem. Commun., 633–635 (1992)

5.15 C. Guerret-Piecourt, Y. Le Bouar, A. Loiseau, H. Pascard: Relation between Metal Electronic Structure and Morphology of Metal Compounds Inside Carbon Nanotubes, Nature **372**, 761–765 (1994)

5.16 P. M. Ajayan, O. Stephan, P. Redlich, C. Colliex: Carbon Nanotubes as Removable Templates for Metal Oxide Nanocomposites, Nanostructures, Nature **375**, 564–567 (1995)

5.17 M. Knez, A. M. Bittner, F. Boes, C. Wege, H. Jeske, E. Maiâ, K. Kern: Biotemplate Synthesis of 3-nm Nickel and Cobalt Nanowires, Nano Lett. **3**, 1079–1082 (2003)

5.18 R. Gasparac, P. Kohli, M. O. M. L. Trofin, C. R. Martin: Template Synthesis of Nano Test Tubes, Nano Lett. **4**, 513–516 (2004)

5.19 C. F. Monson, A. T. Woolley: DNA-Templated Construction of Copper Nanowires, Nano Lett. **3**, 359–363 (2003)

5.20 Y. Weizmann, F. Patolsky, I. Popov, I. Willner: Telomerase-Generated Templates for the Growing of Metal Nanowires, Nano Lett. **4**, 787–792 (2004)

5.21 A. Despic, V. P. Parkhuitik: *Modern Aspects of Electrochemistry*, Vol. 20 (Plenum, New York 1989)

5.22 D. Al Mawiawi, N. Coombs, M. Moskovits: Magnetic Properties of Fe Deposited into Anodic Aluminum Oxide Pores as a Function of Particle Size, J. Appl. Phys. **70**, 4421–4425 (1991)

5.23 C. A. Foss, M. J. Tierney, C. R. Martin: Template-Synthesis of Infrared-Transparent Metal Microcylinders: Comparison of Optical Properties with the Predictions of Effective Medium Theory, J. Phys. Chem. **96**, 9001–9007 (1992)

5.24 A. J. Bard, L. R. Faulkner: *Electrochemical Methods* (Wiley, New York 1980)

5.25 J. B. Mohler, H. J. Sedusky: *Electroplating for the Metallurgist, Engineer and Chemist* (Chemical Publishing, New York 1951)

5.26 T. M. Whitney, J. S. Jiang, P. C. Searson, C. L. Chien: Fabrication and Magnetic Properties of Arrays of Metallic Nanowires, Science **261**, 1316–1319 (1993)

5.27 F. R. N. Nabarro, P. J. Jackson: *Growth and Perfection of Crystals*, ed. by R. H. Doremus, B. W. Roberts, D. Turnbull (Wiley, New York 1958) pp. 11–102

5.28 B. Z. Tang, H. Xu: Preparation, Alignment and Optical Properties of Soluble Poly(phenylacetylene)-Wrapped Carbon Nanotubes, Macromolecules **32**, 2569–2567 (1999)

5.29 W. D. Williams, N. Giordano: Fabrication of 80 Å Metal Wires, Rev. Sci. Instrum. **55**, 410–412 (1984)

5.30 C. G. Jin, G. W. Jiang, W. F. Liu, W. L. Cai, L. Z. Yao, Z. Yao, X. G. Li: Fabrication of Large-Area Single Crystal Bismuth Nanowire Arrays, J. Mater. Chem. **13**, 1743–1746 (2003)

5.31 M. E. T. Molares, V. Buschmann, D. Dobrev, R. Neumann, R. Scholz, I. U. Schuchert, J. Vetter: Single-Crystalline Copper Nanowires Produced by Electrochemical Deposition in Polymeric Ion Track Membranes, Adv. Mater. **13**, 62–65 (2001)

5.32 G. Yi, W. Schwarzacher: Single Crystal Superconductor Nanowires by Electrodeposition, Appl. Phys. Lett. **74**, 1746–1748 (1999)

5.33 C. J. Brumlik, V. P. Menon, C. R. Martin: Synthesis of Metal Microtubule Ensembles Utilizing Chemical,

Part A | 5

Electrochemical and Vacuum Deposition Techniques, J. Mater. Res. **268**, 1174–1183 (1994)

5.34   C. J. Brumlik, C. R. Martin: Template Synthesis of Metal Microtubules, J. Am. Chem. Soc. **113**, 3174–3175 (1991)

5.35   C. J. Miller, C. A. Widrig, D. H. Charych, M. Majda: Microporous Aluminum Oxide Films at Electrodes. 4. Lateral Charge Transport in Self-Organized Bilayer Assemblies, J. Phys. Chem. **92**, 1928–1936 (1988)

5.36   W. Han, S. Fan, Q. Li, Y. Hu: Synthesis of Gallium Nitride Nanorods Through a Carbon Nanotube-Confined Reaction, Science **277**, 1287–1289 (1997)

5.37   G. O. Mallory, J. B. Hajdu (eds.): *Electroless Plating: Fundamentals and Applications* (AESF, Orlando 1990)

5.38   C. Schönenberger, B. M. I. van der Zande, L. G. J. Fokkink, M. Henny, C. Schmid, M. Krüger, A. Bachtold, R. Huber, H. Birk, U. Staufer: Template Synthesis of Nanowires in Porous Polycarbonate Membranes: Electrochemistry and Morphology, J. Phys. Chem. **B101**, 5497–5505 (1997)

5.39   J. D. Klein, R. D. Herrick II, D. Palmer, M. J. Sailor, C. J. Brumlik, C. R. Martin: Electrochemical Fabrication of Cadmium Chalcogenide Microdiode Arrays, Chem. Mater. **5**, 902–904 (1993)

5.40   L. D Hicks, M. S Dresselhaus: Thermoelectric Figure of Merit of a One-Dimensional Conductor, Phys. Rev. B **47**, 679–682 (1993)

5.41   M. S. Dresselhaus, G. Dresselhaus, X. Sun, Z. Zhang, S. B. Cronin, T. Koga: Low-Dimensional Thermoelectric Materials, Phys. Solid State **41**, 679–682 (1999)

5.42   M. S. Sander, R. Gronsky, T. Sands, A. M. Stacy: Structure of Bismuth Telluride Nanowire Arrays Fabricated by Electrodeposition into Porous Anodic Alumina Templates, Chem. Mater. **15**, 335–339 (2003)

5.43   C. Lin, X. Xiang, C. Jia, W. Liu, W. Cai, L. Yao, X. Li: Electrochemical Fabrication of Large-Area, Ordered Bi$_2$Te$_3$ Nanowire Arrays, J. Phys. Chem. B **108**, 1844–1847 (2004)

5.44   D. S. Xu, Y. J. Xu, D. P. Chen, G. L. Guo, L. L. Gui, Y. Q. Tang: Preparation of CdS Single-Crystal Nanowires by Electrochemically Induced Deposition, Adv. Mater. **12**, 520–522 (2000)

5.45   C. Lin, G. Zhang, T. Qian, X. Li, Z. Yao: Large-Area Sb$_2$Te$_3$ Nanowire Arrays, J. Phys. Chem. B **109**, 1430–1432 (2005)

5.46   C. Jérôme, R. Jérôme: Electrochemical Synthesis of Polypyrrole Nanowires, Angew. Chem. Int. Ed. **37**, 2488–2490 (1998)

5.47   A. G. MacDiarmid: Nobel Lecture: "Synthetic Metals": A Novel Role for Organic Polymers, Rev. Mod. Phys. **73**, 701–712 (2001)

5.48   K. Doblhofer, K. Rajeshwar: *Handbook of Conducting Polymers* (Marcel Dekker, New York 1998) Chap. 20

5.49   L. Dauginet, A.-S. Duwez, R. Legras, S. Demoustier-Champagne: Surface Modification of Polycarbonate and Poly(ethylene terephthalate) Films and Membranes by Polyelectrolyte, Langmuir **17**, 3952–3957 (2001)

5.50   C. R. Martin: Membrane-Based Synthesis of Nanomaterials, Chem. Mater **8**, 1739–1746 (1996)

5.51   C. R. Martin: Template Synthesis of Polymeric and Metal Microtubules, Adv. Mater **3**, 457–459 (1991)

5.52   J. C. Hulteen, C. R. Martin: A General Template-Based Method for the Preparation of Nanomaterials, J. Mater. Chem. **7**, 1075–1087 (1997)

5.53   L. Liang, J. Liu, C. F. Windisch Jr., G. J. Exarhos, Y. Lin: Assembly of Large Arrays of Oriented Conducting Polymer Nanowires, Angew. Chem. Int. Ed. **41**, 3665–3668 (2002)

5.54   K. Takahashi, S. J. Limmer, Y. Wang, G. Z. Cao: Growth and Electrochemical Properties of Single-Crystalline V$_2$O$_5$ Nanorod Arrays, Jpn. J. Appl. Phys. B **44**, 662–668 (2005)

5.55   M. J. Zheng, L. D. Zhang, G. H. Li, W. Z. Shen: Fabrication and Optical Properties of Large-Scale Uniform Zinc Oxide Nanowire Arrays by One-Step Electrochemical Deposition Technique, Chem. Phy. Lett. **363**, 123–128 (2002)

5.56   I. Zhitomirsky: Cathodic Electrodeposition of Ceramic and Organoceramic Materials. Fundamental Aspects, Adv. Colloid Interf. Sci. **97**, 279–317 (2002)

5.57   O. O. Van der Biest, L. J. Vandeperre: Electrophoretic Deposition of Materials, Annu. Rev. Mater. Sci. **29**, 327–352 (1999)

5.58   P. Sarkar, P. S. Nicholson: Electrophoretic Deposition (EPD): Mechanism, Kinetics, and Application to Ceramics, J. Am. Ceram. Soc. **79**, 1987–2002 (1996)

5.59   A. C. Pierre: *Introduction to Sol-Gel Processing* (Kluwer, Norwell 1998)

5.60   J. S. Reed: *Introduction to the Principles of Ceramic Processing* (Wiley, New York 1988)

5.61   R. J. Hunter: *Zeta Potential in Colloid Science: Principles and Applications* (Academic, London 1981)

5.62   C. J. Brinker, G. W. Scherer: *Sol-Gel Science: the Physics and Chemistry of Sol-Gel Processing* (Academic, San Diego 1990)

5.63   J. D. Wright, N. A. J. M. Sommerdijk: *Sol-Gel Materials: Chemistry and Applications* (Gordon and Breach, Amsterdam 2001)

5.64   D. H. Everett: *Basic Principles of Colloid Science* (The Royal Society of Chemistry, London 1988)

5.65   W. D. Callister: *Materials Science and Engineering: An Introduction* (Wiley, New York 1997)

5.66   S. J. Limmer, T. P. Chou, G. Z. Cao: A Study on the Growth of TiO$_2$ Using Sol Electrophoresis, J. Mater. Sci. **39**, 895–901 (2004)

5.67   S. J. Limmer, S. Seraji, M. J. Forbess, Y. Wu, T. P. Chou, C. Nguyen, G. Z. Cao: Electrophoretic Growth of Lead Zirconate Titanate Nanorods, Adv. Mater. **13**, 1269–1272 (2001)

5.68   S. J. Limmer, S. Seraji, M. J. Forbess, Y. Wu, T. P. Chou, C. Nguyen, G. Z. Cao: Template-Based Growth of Various Oxide Nanorods by Sol-Gel Electrophoresis, Adv. Func. Mater. **12**, 59–64 (2002)

5.69 S. J. Limmer, G. Z. Cao: Sol-Gel Electrophoretic Deposition for the Growth of Oxide Nanorods, Adv. Mater. **15**, 427–431 (2003)

5.70 Y. C. Wang, I. C. Leu, M. N. Hon: Effect of Colloid Characteristics on the Fabrication of ZnO Nanowire Arrays by Electrophoretic Deposition, J. Mater. Chem. **12**, 2439–2444 (2002)

5.71 Z. Miao, D. Xu, J. Ouyang, G. Guo, Z. Zhao, Y. Tang: Electrochemically Induced Sol-Gel Preparation of Single-Crystalline TiO_2 Nanowires, Nano Lett. **2**, 717–720 (2002)

5.72 C. Natarajan, G. Nogami: Cathodic Electrodeposition of Nanocrystalline Titanium Dioxide Thin Films, J. Electrochem. Soc. **143**, 1547–1550 (1996)

5.73 B. B. Lakshmi, P. K. Dorhout, C. R. Martin: Sol-gel Template Synthesis of Semiconductor Nanostructures, Chem. Mater. **9**, 857–863 (1997)

5.74 A. van der Drift: Evolutionary Selection, a Principle Governing Growth Orientation in Vapor-Deposited Layers, Philips Res. Rep. **22**, 267–288 (1968)

5.75 G. Z. Cao, J. J. Schermer, W. J. P. van Enckevort, W. A. L. M. Elst, L. J. Giling: Growth of {100} Textured Diamond Films by the Addition of Nitrogen, J. Appl. Phys. **79**, 1357–1364 (1996)

5.76 M. Ohring: *Materials Science of Thin Films* (Academic, San Diego 2001)

5.77 D. Pan, Z. Shuyuan, Y. Chen, J. G. Hou: Hydrothermal Preparation of Long Nanowires of Vanadium Oxide, J. Mater. Res. **17**, 1981–1984 (2002)

5.78 V. Petkov, P. N. Trikalitis, E. S. Bozin, S. J. L. Billinge, T. Vogt, M. G. Kanatzidis: Structure of $V_2O_5 \cdot nH_2O$ Xerogel Solved by the Atomic Pair Distribution Function Technique, J. Am. Chem. Soc. **124**, 10157–10162 (2002)

5.79 K. Takahashi, S. J. Limmer, Y. Wang, G. Z. Cao: Synthesis, Electrochemical Properties of Single Crystal V_2O_5 Nanorod Arrays by Template-Based Electrodeposition, J. Phys. Chem. B **108**, 9795–9800 (2004)

5.80 G. Z. Cao: Growth of Oxide Nanorod Arrays through Sol Electrophoretic Deposition, J. Phys. Chem. B **108**, 19921–19931 (2004)

5.81 R. L. Penn, J. F. Banfield: Morphology Development and Crystal Growth in Nanocrystalline Aggregates under Hydrothermal Conditions: Insights from Titania, Geochim. Cosmochim. Acta **63**, 1549–1557 (1999)

5.82 C. M. Chun, A. Navrotsky, I. A. Aksay: Aggregation Growth of Nanometer-sized $BaTiO_3$ Particles, Proc. Microscopy, Microanalysis, 188–189 (1995)

5.83 J. Livage: Synthesis of Polyoxovanadates Via Chimie Douce, Coordination Chem. Rev. **178-180**, 999–1018 (1998)

5.84 K. V. Saban, J. Thomas, P. A. Varughese, G. Varghese: Thermodynamics of Crystal Nucleation in an External Electric Field, Cryst. Res. Technol. **37**, 1188–1199 (2002)

5.85 D. Grier, E. Ben-Jacob, R. Clarke, L. M. Sander: Morphology and Microstructure in Electrochemical Deposition of Zinc, Phys. Rev. Lett. **56**, 1264–1267 (1986)

5.86 C. J. Li, Y. G. Guo, B. S. Li, C. R. Wang, L. J. Wan, C. L. Bai: Template Synthesis of $Sc@C_{82}$ (I) Nanowires and Nanotubes at Room Temperature, Adv. Mater. **17**, 71–73 (2005)

5.87 Y. G. Guo, C. J. Li, L. J. Wan, D. M. Chen, C. R. Wang, C. L. Bai, Y. G. Wang: Well-Defined Fullerene Nanowire Arrays, Adv. Func. Mater. **13**, 626–630 (2003)

5.88 B. B. Lakshmi, C. J. Patrissi, C. R. Martin: Sol-Gel Template Synthesis of Semiconductor Oxide Micro- and Nanostructures, Chem. Mater. **9**, 2544–2550 (1997)

5.89 Q. Lu, F. Gao, S. Komarneni, T. E. Mallouk: Ordered SBA-15 Nanorod Arrays Inside a Porous Alumina Membrane, J. Am. Chem. Soc. **126**, 8650–8651 (2004)

5.90 J. S. Reed: *Introduction to Principles of Ceramic Processing* (Wiley, New York 1988)

5.91 T. Wen, J. Zhang, T. P. Chou, S. J. Limmer, G. Z. Cao: Template-Based Growth of Oxide Nanorod Arrays by Centrifugation, J. Sol-Gel Sci. Tech. **33**, 193–200 (2005)

5.92 C. A. Huber, T. E. Huber, M. Sadoqi, J. A. Lubin, S. Manalis, C. B. Prater: Nanowire Array Composite, Science **263**, 800–802 (1994)

5.93 Z. Zhang, D. Gekhtman, M. S. Dresselhaus, J. Y. Ying: Processing and Characterization of Single-Crystalline Ultrafine Bismuth Nanowires, Chem. Mater. **11**, 1659–1665 (1999)

5.94 E. G. Wolff, T. D. Coskren: Growth, Morphology of Magnesium Oxide Whiskers, J. Am. Ceram. Soc. **48**, 279–285 (1965)

5.95 W. Liang, C. R. Martin: Template-Synthesized Polyacetylene Fibrils Show Enhanced Supermolecular Order, J. Am. Chem. Soc. **112**, 9666–9668 (1990)

5.96 S. M. Marinakos, L. C. Brousseau III, A. Jones, D. L. Feldheim: Template Synthesis of One-Dimensional Au, Au-Poly(pyrrole) and Poly(pyrrole) Nanoparticle Arrays, Chem. Mater. **10**, 1214–1219 (1998)

5.97 H. D. Sun, Z. K. Tang, J. Chen, G. Li: Polarized Raman Spectra of Single-Wall Carbon Nanotubes Mono-Dispersed in Channels of AlPO4-5 Single Crystals, Solid State Commun. **109**, 365–369 (1999)

5.98 Z. Cai, J. Lei, W. Liang, V. Menon, C. R. Martin: Molecular and Supermolecular Origins of Enhanced Electronic Conductivity in Template-Synthesized Polyheterocyclic Fibrils. 1. Supermolecular Effects, Chem. Mater. **3**, 960–967 (1991)

5.99 Y. J. Han, J. M. Kim, G. D. Stucky: Preparation of Noble Metal Nanowires Using Hexagonal Mesoporous Silica SBA-15, Chem. Mater. **12**, 2068–2069 (2000)

5.100 J. Liu, G. E. Fryxell, M. Qian, L.-Q. Wang, Y. Wang: Interfacial Chemistry in Self-Assembled Nanoscale Materials with Structural Ordering, Pure Appl. Chem. **72**, 269–279 (2000)

Part A | 5

5.101 L. Chen, P. J. Klar, W. Heimbrodt, F. Brieler, M. Fröba: Towards Ordered Arrays of Magnetic Semiconductor Quantum Wires, Appl. Phys. Lett. **76**, 3531–3533 (2000)

5.102 B. Gates, Y. Wu, Y. Yin, P. Yang, Y. Xia: Single-crystalline Nanowires of Ag_2Se Can Be Synthesized by Templating Against Nanowires of Trigonal Se, J. Am. Chem. Soc. **123**, 11500–11501 (2001)

5.103 E. W. Wong, B. W. Maynor, L. D. Burns, C. M. Lieber: Growth of Metal Carbide Nanotubes, Nanorods, Chem. Mater. **8**, 2041–2046 (1996)

5.104 Y. Li, G. S. Cheng, L. D. Zhang: Fabrication of Highly Ordered ZnO Nanowire Arrays in Anodic Alumina Membranes, J. Mater. Res. **15**, 2305–2308 (2000)

5.105 C. M. Zelenski, P. K. Dorhout: The Template Synthesis of Monodisperse Microscale Nanofibers, Nan-otubules of MoS_2, J. Am. Chem. Soc. **120**, 734–742 (1998)

5.106 E. Braun, Y. Eichen, U. Sivan, G. Ben-Yoseph: DNA-Templated Assembly and Electrode Attachment of a Conducting Silver Wire, Nature **391**, 775–778 (1998)

5.107 J. Zhan, X. Yang, D. Wang, S. Li, Y. Xie, Y. Xia, Y. Qian: Polymer-Controlled Growth of CdS Nanowires, Adv. Mater. **12**, 1348–1351 (2000)

5.108 Y. Wang, K. Takahashi, H. M. Shang, G. Z. Cao: Synthesis, Electrochemical Properties of Vanadium Pentoxide Nanotube Arrays, J. Phys. Chem. **B109**, 3085–3088 (2005)

5.109 K. Takahashi, Y. Wang, G. Z. Cao: $Ni-V_2O_5 \cdot n$ H_2O Core-Shell Nanocable Arrays for Enhanced Electrochemical Intercalation, J. Phys. Chem. B **109**, 48–51 (2005)

6. Three–Dimensional Nanostructure Fabrication by Focused Ion Beam Chemical Vapor Deposition

In this chapter, we describe three–dimensional nanostructure fabrication using 30 keV Ga$^+$ focused ion beam chemical vapor deposition (FIB–CVD) and a *phenanthrene* (C$_{14}$H$_{10}$) source as a precursor. We also consider microstructure plastic art, which is a new field that has been made possible by microbeam technology, and we present examples of such art, including a "micro wine glass" with an external diameter of 2.75 m and a height of 12 m. The film deposited during such processes is diamond–like amorphous carbon, which has a Young's modulus exceeding 600 GPa, appearing to make it highly desirable for various applications. The production of three–dimensional nanostructures is also discussed. The fabrication of microcoils, nanoelectrostatic actuators, and 0.1 m nanowiring – all potential components of nanomechanical systems – is explained. The chapter ends by describing the realization of nanoinjectors and nanomanipulators, novel nanotools for manipulating and analyzing subcellular organelles.

Part A | 6

Electron beams (EBs) and focused ion beams (FIBs) have been used to fabricate various two-dimensional nanostructure devices such as single electron transistors and MOS transistors with nanometer gate-lengths. Ten nanometer structures can be formed using a commercially available EB or FIB systems with 5–10 nm-diameter beams and high-resolution resists [6.1]. Two-dimensional nanostructure fabrication is therefore already an established process. There are three approaches to three-dimensional fabrication: using a laser, an EB, or a FIB to perform chemical vapor deposition (CVD). FIB- and EB-CVD are superior to laser-CVD [6.2] in terms of spatial resolution and beam-scan control. *Koops* et al. demonstrated some applications such as an AFM tip and a field emitter that were realized using EB-CVD [6.3]. *Blauner* et al. demonstrated pillars and walls with high aspect ratios achieved using FIB-CVD [6.4].

The deposition rate of FIB-CVD is much higher than that of EB-CVD due to factors such as the difference in mass between an electron and an ion. Furthermore, the smaller penetration depth of the ions compared to the electrons makes it easier to create complicated three-dimensional nanostructures. For example, when we attempt to make a coil nanostructure with a linewidth of 100 nm, 10–50 keV electrons pass through the ring of the coil and reach the substrate because of the large range of the electrons (at least a few microns), which makes it difficult to create a coil nanostructure using EB-CVD. On the other hand, since the range of the ions is a few tens of nanometers or less, the ions are deposited inside the ring. Up to now, the realization of complicated nanostructures using FIB-CVD has not been reported. Therefore, this chapter reports on complicated three-dimensional nanostructure fabrication using FIB-CVD.

6.1 Three-Dimensional Nanostructure Fabrication

We used two commercially available FIB systems (SMI9200, SMI2050, SII Nanotechnology Inc., Tokyo, Japan) with a Ga^+ ion beam operating at 30 keV. The FIB-CVD used a precursor of phenanthrene ($C_{14}H_{10}$) as the source material. The beam diameter of SMI9200 was about 7 nm and that of SMI2050 was about 5 nm. The SMI9200 system was equipped with two gas sources in order to increase the gas pressure. The tops of the gas nozzles faced each other and were directed at the beam point. The nozzles were set a distance of 40 μm from each other and positioned about 300 μm above the substrate surface. The inside diameter of a nozzle was 0.3 mm. The phenanthrene gas pressure during pillar growth was typically 5×10^{-5} Pa in the specimen chamber, but the local gas pressure at the beam point was expected to be much higher. The crucible of the source was heated to 85 °C. The SMI2050 system, on the other hand, was equipped with a single gas nozzle. The FIB is scanned in order to be able to write the desired pattern via computer control, and the ion dose is adjusted to deposit a film of the desired thickness. The experiments were carried out at room temperature on a silicon substrate.

The deposited film was characterized by observing it with a transmission electron microscope (TEM) and analyzing its Raman spectra. A thin film of carbon (200 nm thick) was deposited on a silicon substrate by 30 keV Ga^+ FIB using phenanthrene precursor gas. The cross-sections of the structures created and its electron diffraction patterns were observed using a 300 kV TEM. There were no crystal structures in the TEM images and diffraction patterns. It was therefore concluded that the deposited film was amorphous carbon (a-C).

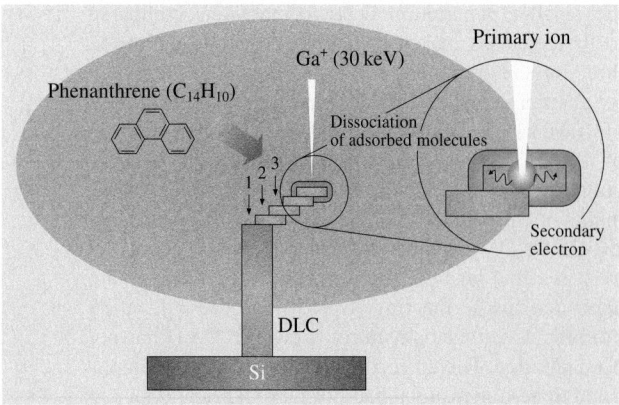

Fig. 6.1 Fabrication process for three-dimensional nanostructure by FIB-CVD

Raman spectra of the a-C films were measured at room temperature with the 514.5 nm line of an argon ion laser. The Raman spectra were recorded using a monochromator equipped with a CCD multichannel detector. Raman spectra were measured at 0.1–1.0 mW to avoid thermal decomposition of the samples. A relatively sharp Raman band at 1550 cm^{-1} and a broad-shouldered band at around 1400 cm^{-1} were observed in the spectra excited by the 514.5 nm line. Two Raman bands were plotted after Gaussian line shape analysis. These Raman bands, located at 1550 cm^{-1} and 1400 cm^{-1}, originate from the trigonal (sp^2) bonding structure of graphite and the tetrahedral (sp^3) bonding structure of diamond. This result suggests that the a-C film deposited by FIB-CVD is diamond-like amorphous carbon (DLC), which has attracted attention due to its hardness, chemical inertness and optical transparency.

6.1.1 Fabrication Process

Beam-induced chemical vapor deposition (CVD) is widely used in the electrical device industry in the repair of chips and masks. This type of deposition is mainly done on two-dimensional (2-D) pattern features, but it can also be used to fabricate a three-dimensional (3-D) object. *Koops* et al. demonstrated nanoscale 3-D structure construction [6.3] by applying electron beam-induced amorphous carbon deposition to a microvacuum tube. However, focused ion beam (FIB)-induced CVD seems to have many advantages for the fabrication of 3-D nanostructures [6.4–6]. The key issue to relaizing such 3-D nanostructures is the short penetration depth of the ions (a few tens of nm) into the target material, where the penetration depth of ions is much shorter than that of electrons (several tens of microns). This short penetration depth reduces the dispersion area of the secondary electrons, and so the deposition area is restricted to roughly several tens of nanometers. A 3-D structure usually contains overhang structures and hollows. Gradual position-scanning of the ion beam during the CVD process causes the position of the growth region around the beam point to shift. When the beam point reaches the edge of the wall, secondary electrons appear at the side of the wall and just below the top surface. The DLC then starts to grow laterally; the width of the vertical growth is about 80 nm. Therefore, by combining the lateral growth mode with rotating beam scanning, it is possible to obtain 3-D structures with rotational symmetry, like a wine glass.

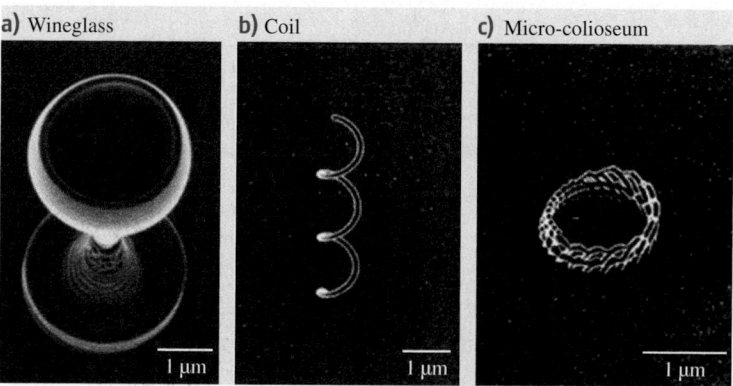

a) Wineglass   **b)** Coil   **c)** Micro-colioseum

1 µm   1 µm   1 µm

Fig. 6.2 (a) Micro wine glass with an external diameter of 2.75 µm and a height of 12 µm, **(b)** Microcoil with a coil diameter of 0.6 µm, a coil pitch of 0.7 µm, and a linewidth of 0.08 µm, **(c)** "MicroColosseum"

The process of fabricating three-dimensional structures by FIB-CVD is illustrated in Fig. 6.1 [6.7]. In FIB-CVD processes, the beam is scanned in digital mode. First, a pillar is formed on the substrate by fixing the beam position (position 1). After that, the beam position is moved to within a diameter of the pillar (position 2) and then fixed until the deposited terrace thickness exceeds the range of the ions (a few tens of nm). This process is repeated to make three-dimensional structures. The key to making three-dimensional structures is

to adjust the beam scan speed so that the ion beam remains within the deposited terrace, which means that the terrace thickness always exceeds the range of the ions. The growth in the x- and y- directions are controlled by both beam deflectors. The growth in the z-direction is determined by the deposition rate; that is, the height of the structure is proportional to the irradiation time when the deposition rate is constant.

We intend to open up a new field of microstructure plastic arts using FIB-CVD. To demonstrate the possibilities of this field, a "micro wine glass" was created on a Si substrate and a human hair as a work of microstructure plastic art, as shown in Figs. 6.2a and 6.3. A micro wine glass with an external diameter of 2.75 µm and a height of 12 µm was formed. The fabrication time was 600 s at a beam current of 16 pA. This beautiful micro wine glass shows the potential of the field of microstructure plastic art. A "micro-Colosseum" and a "micro leaning tower of Pisa" were also fabricated on a Si substrate, as shown in Figs. 6.2c and 6.4.

Various microsystem parts have been fabricated using FIB-CVD. Figure 6.2b shows a microcoil with a coil diameter of 0.6 µm, a coil pitch of 0.7 µm, and a

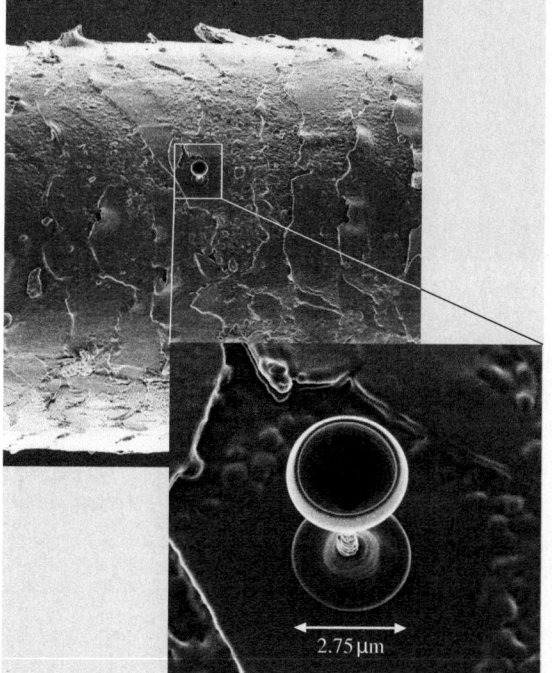

2.75 µm

Fig. 6.3 Micro wine glass with an external diameter of 2.75 µm and a height of 12 µm on a human hair

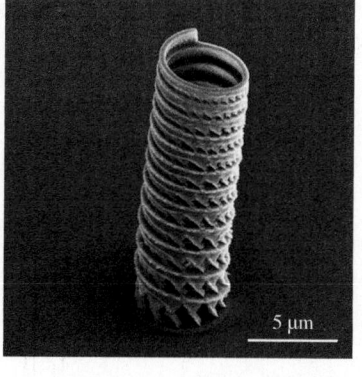

5 µm

Fig. 6.4 "Micro Leaning Tower of Pisa"

linewidth of 0.08 μm. The exposure time was 40 s at a beam current of 0.4 pA. It was easy to change the coil pitch by controlling the speed of growth. A micro drill was formed by reducing the diameter of the microcoil. The diameter, pitch and height of the microcoil were 0.25, 0.20 and 3.8 μ m, respectively. The exposure time was 60 s at a beam current of 0.4 pA. The results show that FIB-CVD is a highly promising technique for realizing parts of a microsystem, although their mechanical performances must be measured.

6.1.2 Three-Dimensional Pattern-Generating Systems

We used ion beam assisted deposition of a source gas to fabricate 3-D structures. The 3-D structure is built-up as a multilayer structure. In the first step of this 3-D pattern-generating system, a 3-D model of the structure, designed using a 3-D CAD system (3-D DXF format), is needed. In this case we realized a structure shaped like a pendulum. The 3-D CAD model, which is a surface model, is cut into several slices, as shown in Fig. 6.5. The thickness of the slices depends upon the resolution in the *z*-direction (the vertical direction). The *x* and *y* coordinates of the slices are then used to create the scan data (voxel data). To fabricate the overhanging structure, the ion beam must irradiate the correct positions in the correct order. If the ion beam irradiates a voxel located in mid-air without a support layer, the ions intended for the voxel will be deposited on the substrate. Therefore, the sequence of irradiation is determined, as shown in Fig. 6.5.

The scan data and the blanking signal therefore include the scan sequence, the dwell time, the interval time, and the irradiation pitch. These parameters are calculated from the beam diameter, the *x–y*

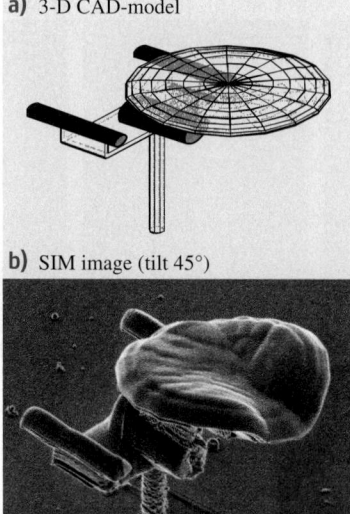

a) 3-D CAD-model

b) SIM image (tilt 45°)

1 μm

Fig. 6.6 "Micro Starship Enterprise NCC-1701D", 8.8 μm long

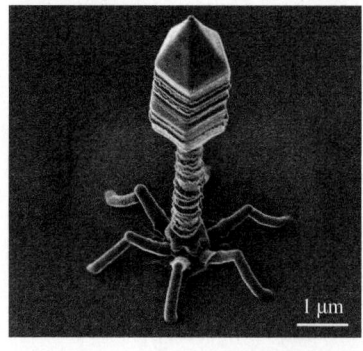

1 μm

Fig. 6.7 T-4 Bacteriophage

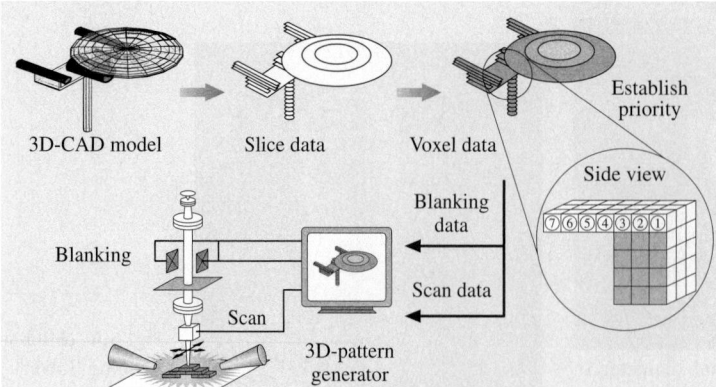

3D-CAD model   Slice data   Voxel data

Establish priority

Side view

⑦⑥⑤④③②①

Blanking data

Blanking

Scan data

Scan

3D-pattern generator

Fig. 6.5 Data flow of 3-D pattern-generating system for FIB-CVD

resolution, and the z resolution of fabrication. The z resolution is proportional to the dwell time and is inversely proportional to the square of the irradiation pitch. The scan data are passed to the beam deflector of the FIB-CVD, as are the blanking data. The blanking signal controls the dwell time and interval time of the ion beam.

Figure 6.6 shows a 3-D CAD model and an SIM image of the starship Enterprise NCC-1701D (from the television series Star Trek), which was fabricated by FIB-CVD at 10~20 pA [6.8]. The nanospaceship is 8.8 μm long and was realized at about 1:100 000 000 scale on silicon substrate. The dwell time (t_d), interval time (t_i), irradiation pitch (p), and total process time (t_p) were 80 μs, 150 μs, 2.4 nm, and 2.5 h, respectively. The horizontal overhang structure was successfully fabricated.

Figure 6.7 shows a "nano T4 bacteriophage", which is an artificial version of the virus, fabricated by FIB-CVD on a silicon surface. The size of the artificial nano "nano T4 bacteriophage" is about ten times as that of the real virus.

6.2 Nanoelectromechanics

6.2.1 Measuring Young's Modulus

An evaluation of the mechanical characteristics of such nanostructures are needed for material physics. *Buks* and *Loukes* reported a simple but useful technique [6.9] for measuring the resonant frequencies of nanoscale objects using a scanning electron microscope (SEM). The secondary electron detector in the SEM can detect frequencies up to around 4 MHz, so the sample vibration is measured as the oscillatory output signal of the detector. Buks and Loukes used this technique to evaluate the Casimir attractive force between the two parallel beams fabricated on a nanoscale. We evaluated the mechanical characteristics of DLC pillars in terms of the Young's modulus, determined using resonant vibration and the SEM monitoring technique [6.10, 11].

The system set-up for monitoring mechanical vibration is shown in Fig. 6.8b. There were two ways of measuring the pillar vibrations. One was active measurement, where the mechanical vibration was induced by a thin piezoelectric device, 300 μm thick and 3 mm square. The piezo device was bonded to the sidewall of the SEM's sample holder with silver paste. The sample holder was designed to observe cross-sections in the SEM (S5000, Hitachi) system. Therefore, the pillar's vibration was observed as a side-view image, as shown in Fig. 6.8a. The range of vibration frequencies involved were 10 kHz up to 2 MHz, which is much faster than the SEM raster scanning speed. Thus, the resonant vibrations of the pillars can be taken as the trace of the pillar's vibration seen in the SEM image. The resonant frequency and amplitude

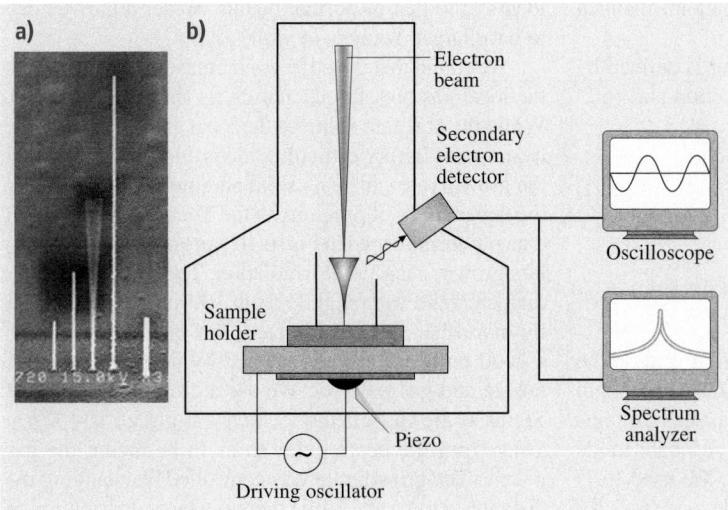

Fig. 6.8 (a) SEM image of the vibration. The resonant frequency was 1.21 MHz. **(b)** Schematic diagram of the vibration monitoring system

were controlled by adjusting the power of the driving oscillator.

The other way to measure pillar vibrations is passive measurement using a spectrum analyzer (Agilent, 4395A), where most of the vibration seemed to derive from environmental noise from rotary pumps and air conditioners. Some parts of the vibration result from the spontaneous vibration associated with thermal excitations [6.9]. Because of this excitation and residual noise, the pillars on the SEM sample holder always vibrate at their fundamental frequency, even if noise isolation is enforced on the SEM system. The amplitude of these spontaneous vibrations was on the order of a few nanometers at the top of the pillar, and high-resolution SEM can easily detect it at a typical magnification of 300 000.

We arranged several pillars that had varying diameters and lengths. The DLC pillars with the smallest diameter of 80 nm were grown using point irradiation. While we used two FIB systems for pillar fabrication, slight differences in the beam diameters of the two systems did not affect the diameters of the pillars. Larger diameter pillars were fabricated using an area-limited raster scan mode. Raster scanning a 160 nm^2 region produced a pillar with a cross-section of about 240 nm^2, and a 400 nm^2 scan resulted in a pillar with a cross-section of 480 nm^2. The typical SEM image taken during resonance is shown in Fig. 6.8a. The FIB-CVD pillars seemed very durable against the mechanical vibration. This kind of measurement usually requires at least 30 min, including a spectrum analysis and photorecording, but the pillars still survived without any change in resonance characteristics. This durability of the DLC pillars should be useful in nanomechanical applications.

The resonant frequency f of the pillar is defined by (6.1) for a pillar with a square cross-section and (6.2) for that with a circular cross-section:

$$f_{\text{square}} = \frac{a\beta^2}{2\pi L^2}\sqrt{\frac{E}{12\rho}}, \tag{6.1}$$

$$f_{\text{circular}} = \frac{a\beta^2}{2\pi L^2}\sqrt{\frac{E}{16\rho}}, \tag{6.2}$$

where a is the width of the square pillar and/or the diameter of the circular-shaped pillar, L is the length of the pillar, ρ is the density, and E is the Young's modulus. The coefficient β defines the resonant mode and β=1.875 for the fundamental mode. We used (6.1) for pillars 240 nm-wide and 480 nm-wide, and (6.2) for

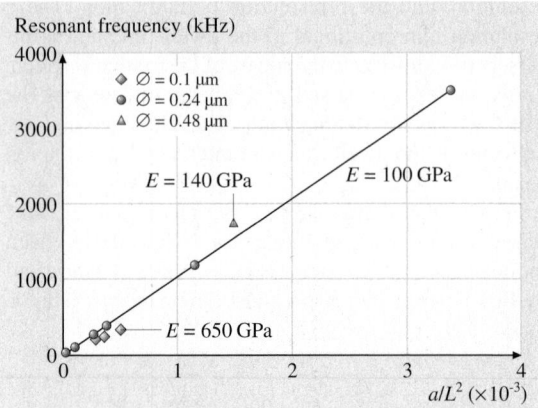

Fig. 6.9 Dependence of resonant frequency on the pillar length

pillars grown by point-beam irradiations. The relationship of the resonant frequency to the Young's modulus, which depends on the ratio of the pillar diameter to the squared length, is summarized in Fig. 6.9. All of the pillars evaluated in this figure were fabricated using the SMI9200 FIB system under rapid growth conditions. Typical growth rates were about 3 μm/min to 5 μm/min for the 100 nm-diameter and 240 μm-wide pillars, and 0.9 μm/min for the 480 nm-wide pillars. When calculating Fig. 6.9, we assumed that the density of the DLC pillars was about 2.3 g/cm^3, which is almost identical to that of graphite and quartz. The inclination of the line in Fig. 6.9 indicates the Young's modulus for each pillar. The Young's modulus of the pillars were distributed over a range from 65 GPa to 140 GPa, which is almost identical to that of normal metals. Wider pillars tended to have larger Young's moduli.

We found that the stiffness increases significantly as the local gas pressure decreases, as shown in Fig. 6.10. While the absolute value of the local gas pressure at the beam point is very difficult to determine, we found that the growth rate can be a useful parameter for describing the dependence of pressure on the Young's modulus. All data points indicated in Fig. 6.10 were obtained from pillars grown using point irradiation. Therefore, the pillar diameters did vary slightly from 100 nm but did deviate by more than 5%. A relatively low gas pressure, with a good uniformity, was obtained by using a single gas nozzle and gas reflector. We use a cleaved side wall of Si tips as the gas reflector, which was placed 10–50 μm away from the beam point so as to be facing the gas nozzle. The growth rate was controlled by changing the distance to the wall. While there is a large distribution of

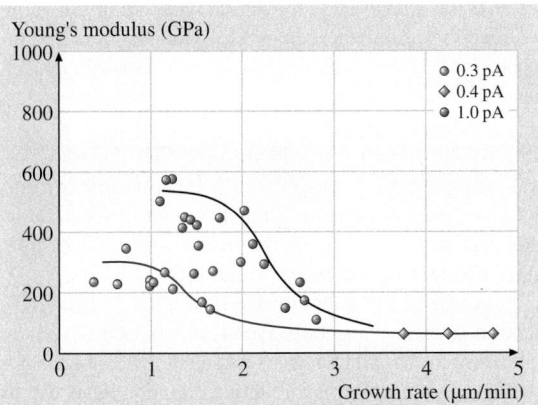

Fig. 6.10 Dependence of Young's modulus on the growth rate

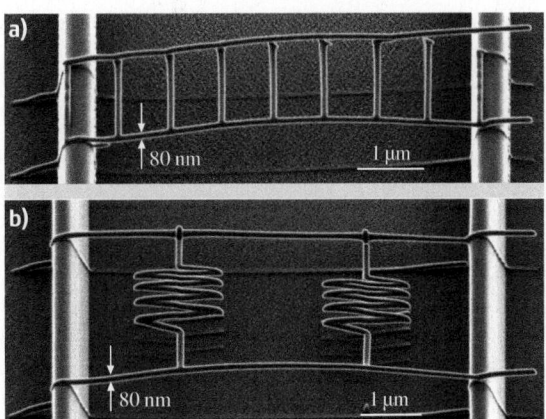

Fig. 6.12 (**a**) DLC free space nanowiring with a bridge shape. (**b**) DLC free space nanowiring with parallel resistances

data points, the stiffness of the pillar tended to become stiffer as the growth rate decreased. The two curves in Fig. 6.10 represent data points obtained for a beam current of 0.3 pA (open circles) and 1 pA (solid circles), respectively. Both curves show the same tendency; the saturated upper levels of the Young's modulus are different for each ion current at low gas pressure (low growth rate). It should be noted that some of the pillars' Young's moduli exceeded 600 GPa, which is of the same order as that of tungsten carbide. In addition, these estimations assume a pillar density of 2.3 g/cm^3, but a finite amount of Ga was incorporated with the pillar growth. If the calculation takes the increase in pillar density due to the Ga concentration into account, the Young's modulus exceeds 800 GPa. Such a high Young's modulus almost reaches that of carbon nanotubes and natural diamond crystals. We think that this high Young's modulus is due to surface modification caused by the direct ion impact.

In contrast, when the gas pressure was high enough to achieve a growth rate of more than 3 μm/min, the pillars became soft but the change in the Young's modulus was small. The uniformity of the Young's modulus (as seen in Fig. 6.9) presumably results from the fact that the growth occurred in this insensitive region where the low levels of source gas limits pillar growth.

6.2.2 Free Space Nanowiring

All experiments were carried out in a commercially available FIB system (SMI9200: SII NanoTechnology Inc.) using a beam of 30 kV Ga$^+$ ions. The beam was focused to a spot size of 7 nm at a beam current of 0.4 pA, and it was incident perpendicular to the surface. The CPG pattern drawing system (CPG-1000: Crestec

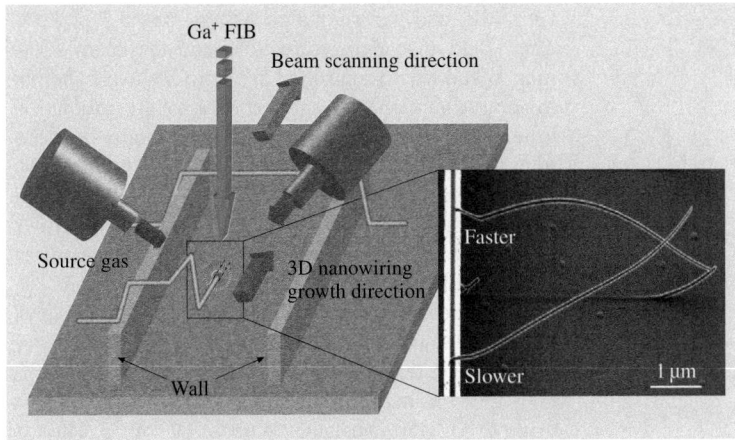

Fig. 6.11 Fabrication of DLC free space nanowiring using both FIB-CVD and CPG

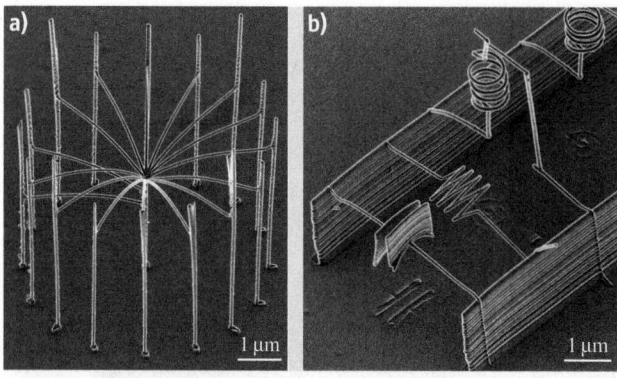

Fig. 6.13 (**a**) Radial DLC free space nanowiring grown in 16 directions from the center. (**b**) Scanning ion microscope (SIM) micrograph of inductor (L), resistor (R) and capacitor (C) structures

Co., Tokyo, Japan) was added to the FIB apparatus to draw any patterns. Using the CPG, it is possible to control beam scan parameters such as scanning speed, x–y direction and blanking of the beam, and so 3-D free space nanowiring can be performed [6.12].

Figure 6.11 illustrates the free space nanowiring fabrication process using both FIB-CVD and CPG. When phenanthrene ($C_{14}H_{10}$) gas or tungsten hexacarbonyl [$W(CO)_6$] gas, which is a reactive organic gas, is evaporated from a heated container and injected into the vacuum chamber by a nozzle located 300 μm above the sample surface at an angle of about 45° with respect to the surface, the gas density of the $C_{14}H_{10}$ or $W(CO)_6$ molecules increase on the substrate near the gas nozzle. The nozzle system creates a local high-pressure region over the surface. The base pressure of the sample chamber was 2×10^{-5} Pa and the chamber pressure upon introducing $C_{14}H_{10}$ and $W(CO)_6$ as source gases were 10^{-4} or 1.5×10^{-3} Pa, respectively. If a Ga$^+$ ion beam is irradiated onto the substrate, $C_{14}H_{10}$ or $W(CO)_6$ molecules adsorbed on the substrate surface are decomposed, and carbon (C) is mainly deposited onto the surface of the substrate. The direction of deposition growth can be controlling through the scanning direction of the beam. The material deposited using $C_{14}H_{10}$ gas was diamond-like carbon, as confirmed by Raman spectra, and it had a very large Young's modulus of 600 GPa [6.7, 10].

After the two walls shown in Fig. 6.11 were formed, free space nanowiring was performed by adjusting the beam scanning speed. The ion beam was used at 30 kV Ga$^+$ FIB, and the irradiation current was 0.8–2.3 pA. The x and y scanning directions and the beam scanning speed were controlled by the CPG. The height in the z-direction was proportional to the irradiation time. Deposition is made to occur horizontally by scanning the beam at a certain fixed speed in the direction of the plane. However, if the beam scanning speed is faster than the nanowiring growth speed, it grows downward or drops, and conversely if the scanning speed is too slow, the deposition grows slanting upward. Therefore, it is very important to carefully control the beam scanning speed when growing a nanowire horizontally. It turns out that the optimal beam scanning speed to realize a nanowire growing horizontally, using two $C_{14}H_{10}$ gas guns, is about 190 nm/s. The ex-

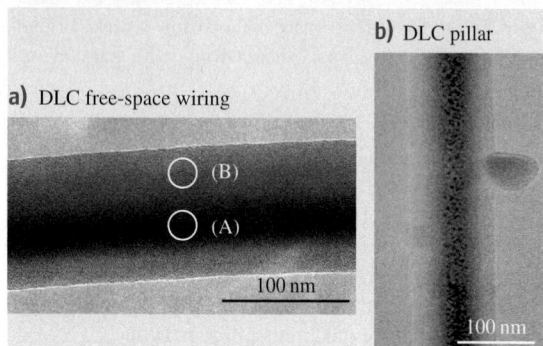

a) DLC free-space wiring

(B)

(A)

100 nm

b) DLC pillar

100 nm

Fig. 6.14 TEM images of (*a*) DLC free space nanowiring and (*b*) DLC pillar

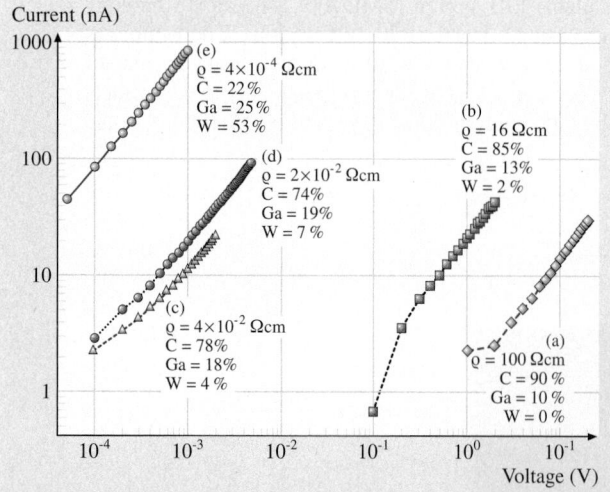

Current (nA)

(e)
$\varrho = 4 \times 10^{-4}$ Ωcm
C = 22%
Ga = 25%
W = 53%

(b)
$\varrho = 16$ Ωcm
C = 85%
Ga = 13%
W = 2%

(d)
$\varrho = 2 \times 10^{-2}$ Ωcm
C = 74%
Ga = 19%
W = 7%

(c)
$\varrho = 4 \times 10^{-2}$ Ωcm
C = 78%
Ga = 18%
W = 4%

(a)
$\varrho = 100$ Ωcm
C = 90%
Ga = 10%
W = 0%

Voltage (V)

Fig. 6.15 Electrical resistivity measurement for nanowiring. The electrical resistivity ρ was calculated using the I–V curve. The elemental contents of C, Ga, W were measured by SEM-EDX

pected pattern resolution achieved using FIB-CVD is around 80 nm, because both the primary Ga^+ ion and the secondary electron scattering occur over distances of around 20 nm [6.10, 13].

Figures 6.12 and 6.13 show examples of free space nanowiring fabricated by FIB-CVD and CPG. All of the structures shown were fabricated using $C_{14}H_{10}$ as the precursor gas.

Figure 6.12a shows nanobridge free space wirings. The growth time was 1.8 min and the wire width was 80 nm. Figure 6.12b shows free space nanowires with parallel resistances. The growth time was 2.8 min, and the wiring width was also 80 nm.

Figure 6.13a shows free space nanowiring grown in sixteen directions from the center. Figure 6.13b shows a scanning ion microscope (SIM) image of an inductor (L), a resistor (R), and a capacitor (C) in a parallel circuit structure with free space nanowiring. A coiled structure was fabricated by circle-scanning Ga^+ FIB. The growth times of the L, R, and C structures were about 6, 2, and 12 min, and the all nanowiring about 110 nm wide. From these structures, it is possible to fabricate nanowiring at an arbitrary position using FIB-CVD and CPG. These results also indicate that various circuit structures can be formed by combining L, C and R.

The free space wiring structures were observed using 200 keV TEM. The analyzed area was 20 nm∅. Figures 6.14a and 6.14b show TEM images of DLC free space wiring and a pillar. It became clear from these EDX measurements that the dark part (A) of Fig. 6.14a corresponds to the Ga core, and the outside part (B) of Fig. 6.14a corresponds to amorphous carbon. This free space wiring therefore consists of amorphous carbon with a Ga core. The center position of the Ga core is actually located below the center of the wiring. However, in the case of the DLC pillar, the Ga core is located at the center of the pillar. To investigate the difference between these core positions, the Ga core distribution in free space wiring was observed in detail by TEM. The center position of the Ga core was about 70 nm from the top, which was 20 nm below the center of the free space wiring. We calculated an ion range of 30 kV Ga ions into amorphous carbon, using TRIM (transport of ions in matter), of 20 nm. This calculation indicates that the displacement of the center of the Ga core in the nanowiring corresponds to the ion range.

The electrical properties of free space nanowiring fabricated by FIB-CVD using a mixture of $C_{14}H_{10}$ and $W(CO)_6$ were measured. Nanowirings were fabricated on an Au electrode. These Au electrodes were formed on 0.2 μm-thick SiO_2 on Si substrate by an EB lithog-

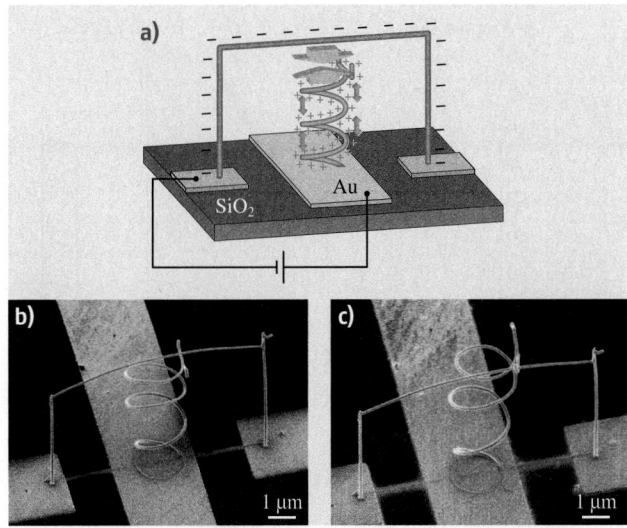

Fig. 6.16 (**a**) Principle of movement for the nanomechanical switch. SIM micrographs of nanomechanical switch (**b**) before applying voltage and (**c**) after applying voltage

raphy and lift-off process. The two-terminal electrode method was used to measure the electrical resistivity of the nanowiring. Figure 6.15a shows the nanowiring fabricated using only $C_{14}H_{10}$ source gas. The growth time here was 65 s and the wiring width was 100 nm. Next, $W(CO)_6$ gas was added to the $C_{14}H_{10}$ gas to create a gas mixture containing a metal in order to obtain a lower electrical resistivity. Figures 6.15b, 6.15c, and 6.15d correspond to increasing $W(CO)_6$ contents in the gas mixture. The $W(CO)_6$ content rate was controlled by the sublimation temperature of the $C_{14}H_{10}$ gas. As the $W(CO)_6$ content was increased, the nanowiring growth time and width become longer: (b) was 195 s and 120 nm, (c) was 237 s and 130 nm, and (d) was 296 s and 140 nm. Finally, we tried to fabricate free space nanowiring using only $W(CO)_6$, but we did not obtain continuous wiring, because the deposition rate for a source gas of just $W(CO)_6$ was very slow.

The electrical resistivity of Fig. 6.15a, fabricated using only $C_{14}H_{10}$ source gas, was $1 \times 10^2 \, \Omega$ cm. The elemental contents were 90% C and 10% Ga, which were measured using a SEM-EDX spot beam. The I–V curves (b), (c) and (d) correspond to increasing $W(CO)_6$ content in the gas mixture. As the $W(CO)_6$ content increases, the electrical resistivity decreases, as shown in the I–V curves (b–d). Moreover, the Ga content also increased because the growth of nanowiring slowed; the irradiation time of Ga^+ FIB became longer. The electrical resistivities of the I–V curves (b), (c) and (d) were

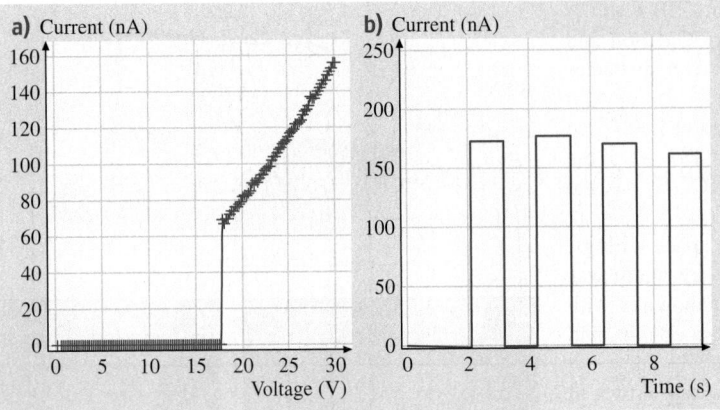

a) Current (nA)

b) Current (nA)

Voltage (V)

Time (s)

Fig. 6.17 (**a**) $I–V$ curve for the nanomechanical switch. (**b**) Pulsed current for the ON/OFF operation for the nanomechanical switch at the applied voltage of 30 V

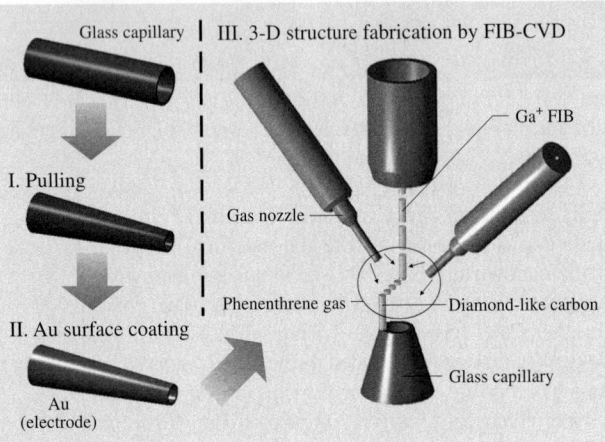

Fig. 6.18 Fabrication of 3-D nanoelectrostatic actuators

Part A | 6.2

16×10^{-2}, 4×10^{-2} and 2×10^{-2} Ωcm, respectively. The electrical resistivity of (e), which was fabricated using only $W(CO)_6$ source gas, was 4×10^{-4} Ω cm. Increasing the Ga and W metallic content decreases the electrical resistivity, as shown by the SEM-EDX measurements seen in Fig. 6.15. These results indicate increasing the metallic content results in lower resistivity.

6.2.3 Nanomechanical Switch

We have also demonstrated a nanomechanical switch fabricated by FIB-CVD [6.14]. Figure 6.16 a shows the principle behind the realization of a nanomechanical switch. First, an Au electrode was formed on a 0.2 μm-thick SiO_2 on Si substrate by an electron beam lithography and lift-off process. After that, a coil and free space nanowiring were fabricated onto the Au electrode to form a switch function using nanowiring fabrication

technology with FIB-CVD and CPG. The coil structure was fabricated by scanning a Ga^+ beam in a circle at fixed speed in $C_{14}H_{10}$ ambient gas. An electric charge (positive or negative) was applied to the coil, and the reverse electric charge was applied to the nanowiring. The coil extended upward when a voltage was applied, because there was now an electrical repulsive force between each loop of the coil. At the same time, the coil and the nanowiring gravitated toward one another, because they had opposite charges. This attraction caused the coil to contact with the nanowiring when a certain threshold voltage was reached.

Next, we evaluated the switch function by measuring the current that flowed when the coil and the nanowiring were in contact. Figures 6.16b and 6.16c show SIM mi-

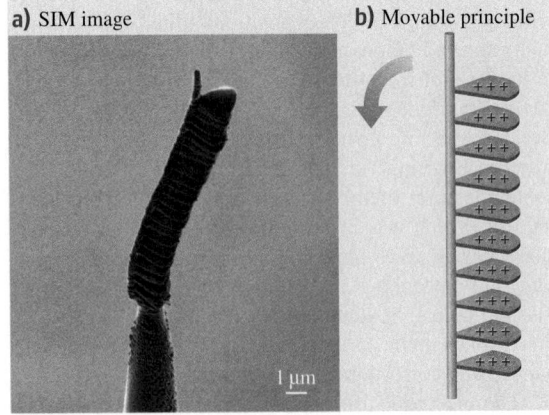

a) SIM image

b) Movable principle

1 μm

Fig. 6.19 Laminated pleats-type electrostatic actuator. (**a**) SIM image of a laminated pleats-type electrostatic actuator fabricated on the tip of a Au-coated glass capillary. (**b**) Illustration of the principle of movement for the actuator

crographs of the nanomechanical switch before and after applying a voltage. These micrographs indicate that the coil and the nanowiring make contact with each other when a voltage is applied to the coil. At the same time, I–V measurements of the nanomechanical switch were carried out as shown in Fig. 6.17a. The current was plotted against the applied voltage at room temperature, and from this graph, it was apparent that the current begins to flow at a threshold voltage of 17.6 V. At this point, the electrical resistance and the resistivity of the nanomechanical switch are about 250 MΩ and 11 Ω cm, respectively. We measured the I–V characteristics for ten nanomechanical switches. The threshold voltage was around 20 V in each case. The switching function was confirmed by performing ON/OFF operations at an applied voltage of 30 V, as shown in Fig. 6.17b. A pulsed current of about 170 nA was detected from this applied voltage.

6.2.4 Nanoelectrostatic Actuator

The fabrication process of 3-D nanoelectrostatic actuators (and manipulators) is very simple [6.15]. Figure 6.18 shows the fabrication process. First, a glass capillary (GD-1: Narishige Co., East Meadow, NY, USA) was pulled using a micropipette puller (PC-10: Narishige Co.). The dimensions of the glass capillary were 90 mm in length and 1 mm in diameter. Using this process, we obtained a glass capillary tip with a diameter of 1 μm diameter. Next, we coated the glass capillary surface with Au by DC sputtering. The Au thickness was approximately 30 nm. This Au coating serves as the electrode that controls the actuator and manipulator. Then the 3-D nanoelectrostatic actuators and manipulators were fabricated by FIB-CVD. This process was carried out in a commercially available FIB system (SIM9200: SII NanoTechnology Inc.) with a Ga$^+$ ion beam operating at 30 keV. FIB-CVD was carried out using a precursor of phenanthrene (C$_{14}$H$_{10}$) as the source material. The beam diameter was about 7 nm. The inner diameter of each nozzle was 0.3 mm. The phenanthrene gas pressure during growth was typically 5×10^{-5} Pa in the specimen chamber. The Ga$^+$ ion beam was controlled by transmitting CAD data on the arbitrary structures to the FIB system.

A laminated pleats-type electrostatic actuator was fabricated by FIB-CVD. Figure 6.19a shows an SIM image of a laminated pleats-type electrostatic actuator fabricated at 7 pA with 60 min of exposure time. Figure 6.19b shows the principle behind the movement of this actuator. The driving force is the repulsive force

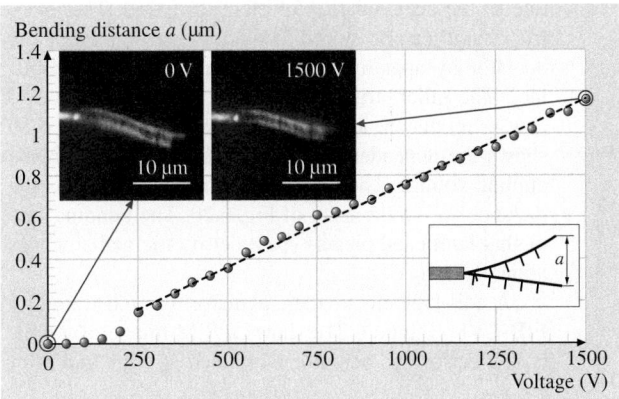

Fig. 6.20 Dependence of bending distance on the applied voltage

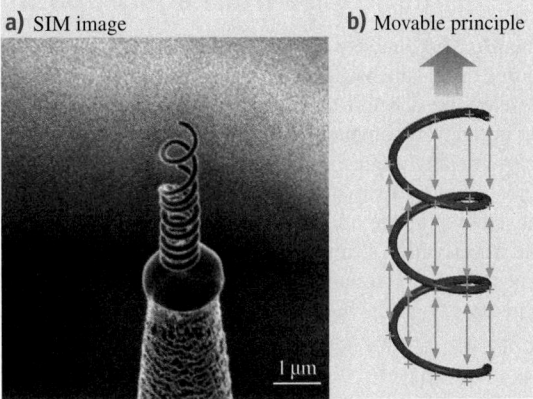

Fig. 6.21 Coil-type electrostatic actuator. (**a**) SIM image of a coil-type electrostatic actuator fabricated on the tip of a Au-coated glass capillary. (**b**) Illustration of the principle of movement for the actuator

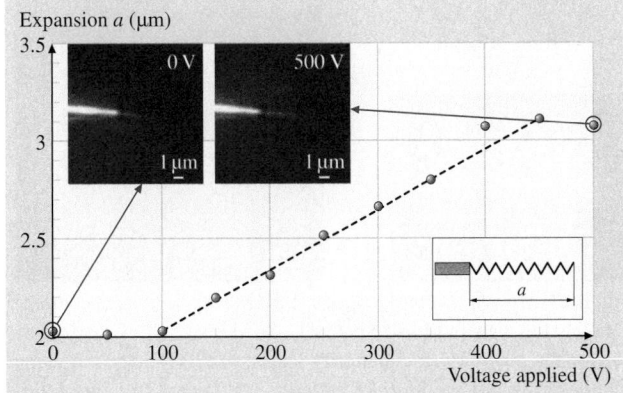

Fig. 6.22 Dependence of coil expansion on the applied voltage

due to the accumulation of electric charge. This electric charge can be stored in the pleat structures of the actuator by applying a voltage across the glass capillary. The pillar structure of this actuator bends due to charge repulsion, as shown in Fig. 6.19b. Figure 6.20 shows the dependence of the bending distance on the applied voltage. The bending distance is defined as the distance "*a*" in the inset of Fig. 6.20. The bending rate of this laminated pleats-type electrostatic actuator was about 0.7 nm/V.

A coil-type electrostatic actuator was fabricated by FIB-CVD. Figure 6.21a shows a SIM image of a coil-type electrostatic actuator fabricated at 7 pA and after

10 min of exposure time. Figure 6.21b shows the principle behind the movement of this actuator, which is very simple. The driving force is the repulsive force induced by electric charge accumulation; the electric charge can be stored in this coil structure by applying a voltage across the glass capillary. This coil structure expands and contracts due to charge repulsion, as shown in Fig. 6.21b. Figure 6.22 shows the dependence of the coil expansion on the applied voltage. The length of the expansion is the distance "*a*" in the inset of Fig. 6.22. The result revealed that the expansion could be controlled in the applied voltage range from 0 to 500 V.

6.3 Nanooptics:
Brilliant Blue from a *Morpho* Butterfly Scale Quasi–Structure

The *Morpho* butterfly has brilliant blue wings, and the source of this intense color has been an interesting topic of scientific debate for a long time. Due to an intriguing optical phenomenon, the scales reflect brilliant blue color for any angle of incidence of white light. This color is called a structural color, meaning that it is not caused by pigment reflection [6.16]. When we observed the scales with a scanning electron microscope (SEM) (Fig. 6.23a), we found three-dimensional (3-D) nanostructures 2 μm in height, 0.7 μm in width, and with a 0.22 μm grating pitch on the scales. These nano-

structures cause a similar optical phenomenon to the iridescence produced by a jewel beetle.

Fig. 6.23 *Morpho* butterfly scales. (**a**) Optical microscope image showing top view of *Morpho* butterfly. SEM image showing a cross-sectional view of *Morpho*-butterfly scales. (**b**) SIM images showing inclined views of the *Morpho* butterfly-scale quasi-structure fabricated by FIB-CVD

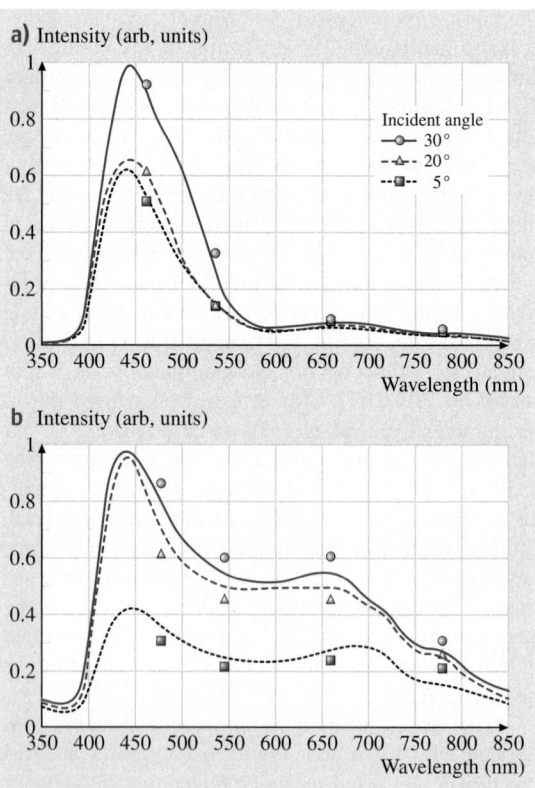

Fig. 6.24a,b Intensity curves of the reflection spectra for (**a**) *Morpho* butterfly scales. (**b**) *Morpho* butterfly scale quasi-structure

We duplicated the *Morpho* butterfly scale quasi-structure with a commercially available FIB system (SMI9200: SII Nanotechnology Inc.) using a Ga$^+$ ion beam operating at 30 kV [6.17]. The beam diameter was about 7 nm at 0.4 pA. The FIB-CVD was performed using a precursor of phenanthrene (C$_{14}$H$_{10}$).

In this experiment, we used a computer-controlled pattern generator, which converted 3-D computer-aided design (CAD) data into a scanning signal, which was passed to an FIB scanning apparatus in order to fabricate a 3-D mold [6.8]. The scattering range of the Ga primary ions is about 20 nm and the range of the secondary electrons induced by the Ga ion beam is about 20 nm, so the expected pattern resolution of the FIB-CVD is about 80 nm.

Figure 6.23b is a scanning ion microscope (SIM) image of the *Morpho* butterfly quasi-structure fabricated by FIB-CVD using 3-D CAD data. This result demonstrates that FIB-CVD can be used to fabricate the quasi-structure.

We measured the reflection intensities from *Morpho* butterfly scales and the *Morpho* butterfly scale quasi-structure optically; white light from a halogen lamp was directed onto a sample with angles of incidence ranging from 5 to 45°. The reflection was concentrated by an optical microscope and analyzed using a commercially available photonic multichannel spectral analyzer system (PMA-11: Hamamatsu Photonics K.K., Hamamatsu City, Japan). The intensity of incident light from the halogen lamp peaked at a wavelength close to 630 nm.

The *Morpho* butterfly scale quasi-structure was made of DLC. The reflectivity and transmittance of a 200 nm-thick DLC film deposited by FIB-CVD, measured by the optical measurement system at a wavelength close to 440 nm (the reflection peak wavelength of the *Morpho* butterfly), were 30% and 60%, respectively. Therefore, the measured data indicated that the DLC film had high reflectivity near 440 nm, which is important for the fabrication of an accurate *Morpho* butterfly scale quasi-structure.

We measured the reflection intensities of the *Morpho* butterfly scales and the quasi-structure with an optical measurement system, and compared their characteristics. Figures 6.24a and 6.24b respectively show the reflection intensities from *Morpho* butterfly scales and the quasi-structure. Both gave a peak intensity near 440 nm and showed very similar reflection intensity spectra for various angles of incidence.

We have thus successfully demonstrated that a *Morpho* butterfly scale quasi-structure fabricated using FIB-CVD can give almost the same optical characteristics as real *Morpho* butterfly scales.

6.4 Nanobiology

6.4.1 Nanoinjector

Three-dimensional nanostructures on a glass capillary have a number of useful applications, such as manipulators and sensors in various microstructures. We have demonstrated the fabrication of a nozzle nanostructure on a glass capillary for a bioinjector using 30 keV Ga$^+$ focused ion beam-assisted deposition with a precursor of phenanthrene vapor and etching [6.18]. It has been demonstrated that nozzle nanostructures of various shapes and sizes can be successfully fabricated. An inner tip diameter of 30 nm for a glass capillary and a tip shape with an inclined angle have been realized. We reported that diamond-like carbon (DLC) pillars grown using FIB-CVD with a precursor of phenanthrene vapor have very large Young' moduli that exceed 600 GPa, which potentially makes them useful for various applications [6.10]. These characteristics are applicable to the fabrication of various biological devices.

In one experiment, nozzle nanostructure fabrication for biological nanoinjector research was studied. The tip diameters of conventional bioinjectors are greater than 100 nm and the tip shapes cannot be controlled. A bionanoinjector with various nanostructures on the top of a glass capillary has the following potential applications (shown in Fig. 6.25): (1) injection of various reagents into a specific organelle in a cell, (2) selective

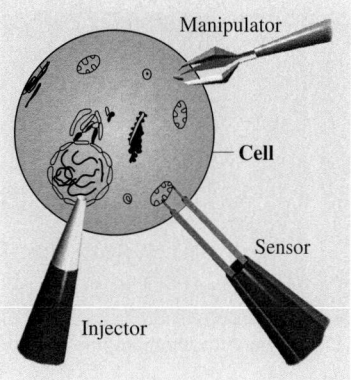

Fig. 6.25
Potential uses for a bionano-injector

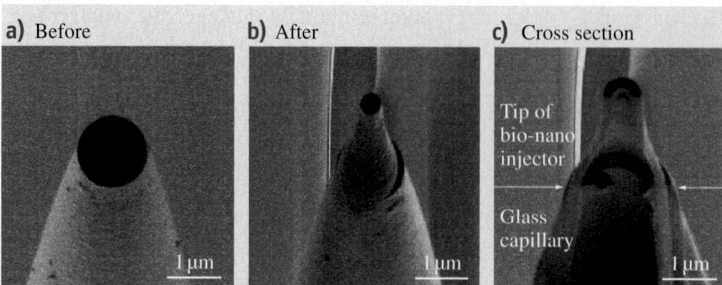

Fig. 6.26a–c SIM images of the bio-nanoinjector fabricated on a glass capillary by FIB-CVD (**a**) before FIB-CVD, (**b**) after FIB-CVD, and (**c**) cross-section of (**b**)

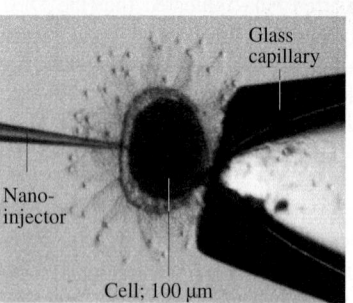

Fig. 6.27 Injection into a egg cell (*Ciona intestinalis*) using a bionano-injector

manipulation of a specific organelle outside of a cell by using the nanoinjector as an aspirator, (3) reducing the mechanical stress produced when operating in the cell by controlling the shape and the size of the bionanoinjector, and (4) measurement of the electric potential of a cell, an organelle, and an ion channel exiting on a membrane, by fabricating an electrode. Thus far, 3-D nanostructure fabrications on a glass capillary have not been reported. We present here nozzle nanostructure fabrication on a glass capillary by FIB-CVD and etching in order to confirm the possibility of bionanoinjector fabrication.

The nozzle structures of the nanoinjector were fabricated using a function generator (wave factory: NF Electronic Instruments, Yokohama, Japan). Conven-

tional microinjectors are fabricated by pulling a glass capillary (GD-1: Narishige Co.) using a micropipette puller (PC-10: Narishige Co.). The glass capillary was 90 mm in length and 1 mm in diameter.

Conventionally, the tip-shape of a microinjector made by pulling a glass capillary, and which is used as an injector into a cell, is controlled by applying mechanical grinding (or not). However, the reliability of this technique for controlling tip-shape is very poor and requires experienced workers.

A bionanoinjector tip was fabricated on a glass capillary by FIB-CVD, as shown in Figs. 6.26a–6.26c. First, FIB etching made the tip surface of the glass capillary smooth. Then a nozzle structure was fabricated at the tip by FIB-CVD. Figure 6.26a shows the surface of a tip smoothed at 120 pA and after 30 s exposure time by FIB-etching, with an inner hole diameter of 870 nm. A nozzle structure fabricated by FIB-CVD with an inner hole diameter of 220 nm is shown in Fig. 6.26b. Figure 6.26c is a cross-section of Fig. 6.26b. These results demonstrate that a bionanoinjector could be successfully fabricated by 3-D nanostructure fabrication using FIB-CVD. The bionanoinjector was used to inject dye into a egg cell (*Ciona intestinalis*), as shown in Fig. 6.27.

6.4.2 Nanomanipulator

An electrostatic 3-D nanomanipulator that can manipulate nanoparts and operate on cells has been developed by FIB-CVD. This 3-D nanomanipulator has four fingers so that it can manipulate a variety of shapes. To move the nanomanipulator, electric charge is accumulated in the structure by applying voltage to the four-fingered structure, and electric charge repulsion causes them to move. Furthermore, we succeeded in catching a microsphere (made from polystyrene latex) with a diameter of 1 μm) using this 3-D nanomanipulator with four fingers [6.19].

The glass capillary (GD-1; Narishige Co.) was pulled using a micropipette puller (PC-10; Narishige Co.). A tip diameter of about 1.0 μm could be obtained

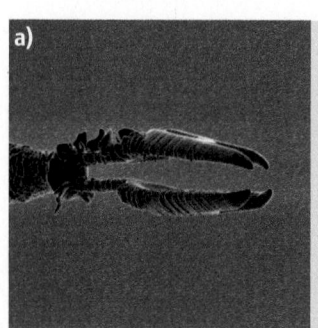
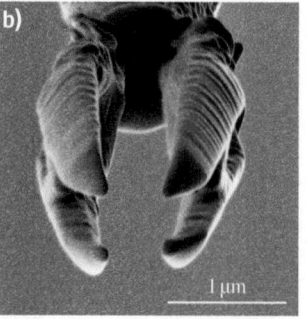

Fig. 6.28 SIM image of the 3-D electrostatic nanomanipulator with four fingers before manipulation

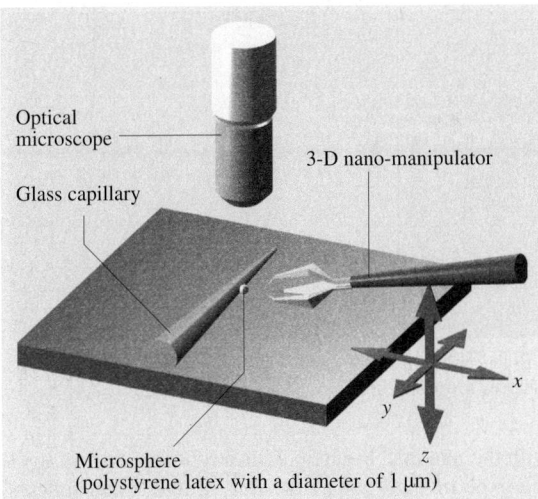

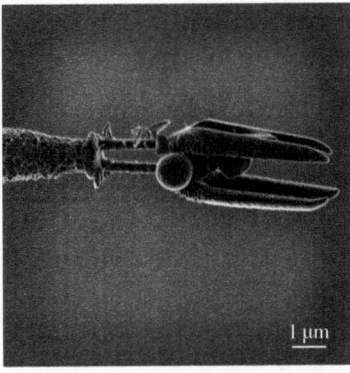

Fig. 6.31 SIM image of the 3-D electrostatic nanomanipulator with four fingers after manipulation

Fig. 6.29 Illustration of 1 μm polystyrene microsphere manipulation by using a 3-D electrostatic nanomanipulator with four fingers

using this process. Then, the glass capillary surface was coated with Au in order to fabricate an electrode for nanomanipulator control. The thickness of the Au coating was about 30 nm. Finally, a 3-D nanomanipulator structure with four fingers (Fig. 6.28) was fabricated by FIB-CVD on the tip of the glass capillary with a single electrode.

Microsphere (a polystyrene latex ball with a diameter of 1 μm) manipulation was carried out using the 3-D nanomanipulator with four fingers. An illustration of this manipulation experiment is shown in Fig. 6.29. By connecting the manipulator fabricated by FIB-CVD to a commercial manipulator (MHW-3; Narishige Co.), the direction of movement along the x-axis, y-axis and z-axis could be controlled. The microsphere target was

fixed to the side of a glass capillary, and the manipulation was observed from the top with an optical microscope.

The optical microscope image in Fig. 6.30 shows the situation during manipulation. First, the 3-D nanomanipulator was made to approach the microsphere; no voltage was applied. Next, the four fingers were opened by applying 600 V in front of the microsphere and the microsphere could be caught by turning off the voltage when the microsphere was in the grasp of the nanomanipulator. The 3-D nanomanipulator was then removed from the side of the glass capillary. Note that the action of catching the microsphere occurs due the elastic force of the manipulator's structure. We succeeded in catching the microsphere, as shown in the SIM image in Fig. 6.31.

6.4.3 Nanonet

Highly functional nanotools are required to perform subcellular operations and analysis in nanospace. For example, nanotweezers have been fabricated on an AFM tip from carbon nanotubes [6.20]. We have produced nanotools with arbitrary structures using FIB-CVD. Currently, we have fabricated a nanonet as a novel nano-tool for the manipulation and analysis of subcellular organelles; subcellular operations like these are easy to perform using a nanonet [6.21].

To realize the nanonet, a glass capillary (GD-1: Narishige Co.) was pulled with a micropipette puller (PC-10: Narishige Co.). The glass capillary was 90 mm in length and 1 mm in diameter. Using this process, we obtained a 1 μm-diameter tip on the glass capillary. Next, the glass capillary's surface was coated with Au for protection during charging with a Ga$^+$ ion beam. In the final process, the nanonet structure was fabricated with FIB-CVD. We used a commercially available FIB system (SIM2050MS2: SII NanoTechnology Inc.) that has

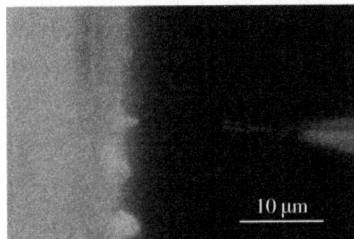

Fig. 6.30 In situ observation of 1 μm polystyrene microsphere manipulation using a 3-D electrostatic nanomanipulator with four fingers

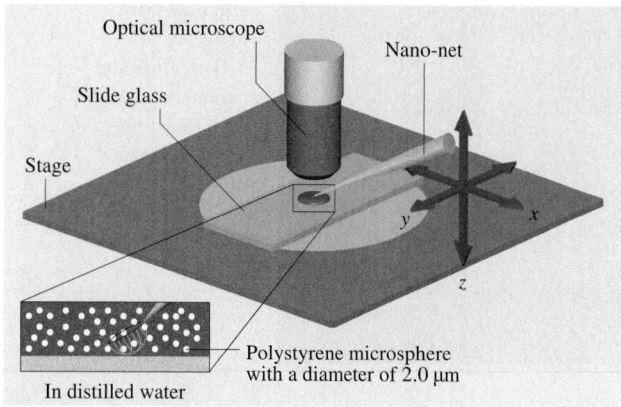

Fig. 6.32 Schematic drawing of the experiment where polystyrene microsphered were captured using a nanonet

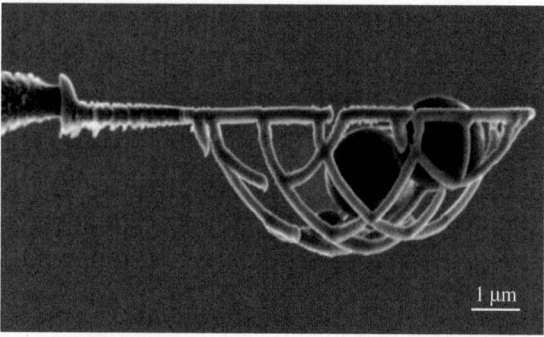

Fig. 6.33 SIM image of the nanonet holding three microspheres after capture

a Ga$^+$ ion beam operating at $30\,kV$. The source material for FIB-CVD was a precursor of phenanthrene ($C_{14}H_{10}$). Diamond-like carbon (DLC) is deposited by using this source material. The minimum beam diameter was about $5\,nm$.

The phenanthrene gas pressure in the specimen chamber during growth was typically $5 \times 10^{-5}\,Pa$. By transmitting the CAD data for the arbitrary structures to the FIB-system, we were able to controlthe Ga$^+$ ion beam, and therefore to fabricate the nanonet structure on the glass capillary.

The flexibility and practicality of the nanonet is enhanced by fabricating it on a glass capillary, since this is used in many fields, including biology and medicine. The FIB-CVD deposition time was about $40\,min$ at a beam current of $7\,pA$. The diameter of the ring used to hang the net was about $7\,\mu m$, and the width of the wire in the net was about $300\,nm$.

We performed an experiment under an optical microscope in which polystyrene microspheres were captured with the nanonet. Figure 6.32 is a schematic drawing of the experimental apparatus. Polystyrene microspheres with a diameter of $2\,\mu m$ were dispersed in distilled water to simulate subcellular organelles. The x-, y- and z-axis movements of the FIB-CVD nanonet were controlled with precision using a commercial manipulator (MHW-3; Narishige Co.).

We performed in situ observations of the capture of $2\,\mu m$-diameter polystyrene microspheres using the nanonet. First, the nanonet was brought to the surface of the water. Next, we placed the nanonet into distilled water by controlling its z-axis movement with a commercial manipulator. Then the microspheres were scooped up by moving the nanonet upward. Finally, the nanonet was removed from the surface of the water. At this point, the nanonet had scooped up three microspheres. After these in situ experiments, we observed that the nanonet contained the $2\,\mu m$-diameter microspheres. Figure 6.33 is a SIM image of the nanonet holding the captured microspheres. This proves that we successfully captured the microspheres.

6.5 Summary

Three-dimensional nanostructure fabrication using $30\,keV$ Ga$^+$ FIB-CVD and a phenanthrene ($C_{14}H_{10}$) source as a precursor has been demonstrated. The film deposited on a silicon substrate was characterized using a transmission microscope and Raman spectra. This characterization indicated that the deposited film is diamond-like amorphous carbon (DLC), which has attracted attention due to its hardness, chemical inertness and optical transparency. Its large Young's modulus, which exceeds $600\,GPa$, makes it highly de-sirable for various applications. A nanoelectrostatic actuator and $0.1\,\mu m$ nanowiring were fabricated and evaluated as parts of a nanomechanical system. Furthermore, a nanoinjector and a nanomanipulator were fabricated as novel nanotools for the manipulation and analysis of subcellular organelles. These results demonstrate that FIB-CVD is one of the key technologies needed to make 3-D nanodevices that can be used in the fields of electronics, mechanics, optics and biology.

References

6.1 S. Matsui: Proc. IEEE **85**, 629 (1997)

6.2 O. Lehmann, F. Foulon, M. Stuke: NATO ASI Series E, Appl. Sci. **265**, 91–102 (1994)

6.3 H. W. Koops: Jpn. J. Appl. Phys. **33**, 7099 (1994)

6.4 A. Wargner, J. P. Levin, J. L. Mauer, P. G. Blauner, S. J. Kirch, P. Long: J. Vacuum. Sci. Technol. **B8**, 1557 (1990)

6.5 I. Utke, P. Hoffmann, B. Dwir, K. Leifer, E. Kapon, P. Doppelt: J. Vacuum. Sci. Technol. **B18**, 3168 (2000)

6.6 A. J. DeMarco, J. Melngailis: J. Vacuum. Sci. Technol. **B17**, 3154 (1999)

6.7 S. Matsui, T. Kaito, J. Fujita, M. Komuro, K. Kanda, Y. Haruyama: J. Vacuum. Sci. Technol. **B18**, 3181 (2000)

6.8 T. Hoshino, K. Watanabe, R. Kometani, T. Morita, K. Kanda, Y. Haruyama, T. Kaito, J. Fujita, M. Ishida, Y. Ochiai, S. Matsui: J. Vacuum. Sci. Technol. **21**, 2732 (2003)

6.9 E. Buks, M. L. Roukes: Phys. Rev. B **63**, 033402 (2001)

6.10 J. Fujita, M. Ishida, T. Sakamoto, Y. Ochiai, T. Kaito, S. Matsui: J. Vacuum. Sci. Technol. **B 19**, 2834 (2001)

6.11 M. Ishida, J. Fujita, Y. Ochiai: J. Vacuum. Sci. Technol. **B20**, 2784 (2002)

6.12 T. Morita, R. Kometani, K. Watanabe, K. Kanda, Y. Haruyama, T. Hoshino, K. Kondo, T. Kaito, T. Ichihashi, J. Fujita, M. Ishida, Y. Ochiai, T. Tajima, S. Matsui: J. Vacuum. Sci. Technol. **B21**, 2737 (2003)

6.13 J. Fujita, M. Ishida, Y. Ochiai, T. Ichihashi, T. Kaito, S. Matsui: J. Vacuum. Sci. Technol. **20**, 2686 (2002)

6.14 T. Morita, K. Nakamatsu, K. Kanda, Y. Haruyama, K. Kondo, T. Hoshino, T. Kaito, J. Fujita, T. Ichihashi, M. Ishida, Y. Ochiai, T. Tajima, S. Matsui: J. Vacuum. Sci. Technol. **B22**, 3137 (2004)

6.15 R. Kometani, T. Hoshino, K. Kondo, K. Kanda, Y. Haruyama, T. Kaito, J. Fujita, M. Ishida, Y. Ochiai, S. Matsui: J. Appl. Phys. **43**(10), 7187 (2004)

6.16 P. Vukusic, J. Roy. Sambles: Nature **424**, 852 (2003)

6.17 K. Watanabe, T. Hoshino, K. Kanda, Y. Haruyama, S. Matsui: Jpn. J. Appl. Phys. **44**, L48 (2005)

6.18 R. Kometani, T. Morita, K. Watanabe, K. Kanda, Y. Haruyama, T. Kaito, J. Fujita, M. Ishida, Y. Ochiai, S. Matsui: Jpn. J. Appl. Phys. **42**, 4107 (2003)

6.19 R. Kometani, T. Hoshino, K. Kondo, K. Kanda, Y. Haruyama, T. Kaito, J. Fujita, M. Ishida, Y. Ochiai, S. Matsui: J. Vacuum. Sci. Technol. **B23**, 298 (2005)

6.20 S. Akita, Y. Nakayama, S. Mizooka, Y. Takano, T. Okawa, K. Y. Miyatake, S. Yamanaka, M. Tsuji, T. Nosaka: Appl. Phys. Lett. **74**, 1691 (2001)

6.21 R. Kometani, T. Hoshino, K. Kanda, Y. Haruyama, T. Kaito, J. Fujita, M. Ishida, Y. Ochiai, S. Matsui: Nuclear Instrum. Meth. Phys. Res. B (2006) in press

Part A | 6

7. Introduction to Micro/Nanofabrication

This chapter outlines and discusses important micro- and nanofabrication techniques. We start with the most basic methods borrowed from the integrated circuit (IC) industry, such as thin film deposition, lithography and etching, and then move on to look at MEMS and nanofabrication technologies. We cover a broad range of dimensions, from the micron to the nanometer scale. Although most of the current research is geared towards the nanodomain, a good understanding of top-down methods for fabricating micron-sized objects can aid our understanding of this research. Due to space constraints, we have focused here on the most important technologies; in the microdomain these include surface, bulk and high aspect ratio micromachining; in the nanodomain we concentrate on e-beam lithography, epitaxial growth, template manufacturing and self-assembly. MEMS technology is maturing rapidly, with some new technologies displacing older ones that have proven to be unsuited to manufacture on a commercial scale. However, the jury is still out on methods used in the nanodomain, although it appears that bottom-up

methods are the most feasible, and these will have a major impact in a variety of application areas such as biology, medicine, environmental monitoring and nanoelectronics.

Recent innovations in the area of micro/nanofabrication have created a unique opportunity to manufacture nanometer- to millimeter-sized structures. This wide (six orders of magnitude) range of sizes is applicable to novel electronic, optical, magnetic, mechanical and chemical/biological devices, with applications ranging from sensors to computation and control. In this chapter, we will introduce major micro/nanofabrication techniques currently used to fabricate structures ranging from nanometers to several hundred microns in size. We will mainly focus on the most important and widely used techniques and will not discuss specialized methods. After a brief introduction to basic microfabrication, we will discuss MEMS-fabrication techniques used to build microstructures down to about $1\,\mu m$ in size. Following this, we will discuss several major top-down and bottom-up nanofabrication methods which have shown great promise in the manufacture of nanostructures (dimensions $<1\,\mu m$).

7.1 Basic Microfabrication Techniques

Most micro/nanofabrication techniques have their roots in standard fabrication methods developed for the semiconductor industry [7.1–3]. Therefore, a clear understanding of these techniques is needed by anyone embarking on a research and development path in the micro/nano area. In this section, we will discuss

Part A | 7

the microfabrication methods most commonly used in the manufacture of micro/nanostructures. Some of these techniques, such as thin-film deposition and etching, are common to the micro/nano and VLSI microchip fabrication disciplines. However, several other techniques that are more specific to the micro/nanofabrication area will also be discussed in this section.

7.1.1 Lithography

Lithography is the technique used to transfer a computer-generated pattern onto a substrate (such as silicon, glass or GaAs). This pattern is then used to etch an underlying thin film (such as an oxide or nitride) for various purposes (including doping and etching). Although photolithography, in other words the lithography using a UV light source, is by far the most common lithography technique in microelectronic fabrication, electron-beam (e-beam) and X-ray lithography are two other alternatives which have attracted considerable attention in the MEMS and nanofabrication areas. We will discuss photolithography in this section and postpone our discussion of e-beam and X-ray techniques to subsequent sections dealing with MEMS and nanofabrication.

The starting point for a specific fabrication sequence (subsequent to the creation of the mask layout on a computer) is the generation of a photomask. This involves a sequence of photographic processes (using optical or e-beam pattern generators) that results in a glass plate that exhibits the desired pattern in the form of a thin (≈ 100 nm) chromium layer. Following the generation of a photomask, the lithography process proceeds as shown in Fig. 7.1. This sequence demonstrates pattern transfer onto a substrate coated with silicon dioxide; however, the same technique is applicable to other materials. After depositing the desired material on the substrate, the process involves spin-coating the substrate with a photoresist. This is a polymeric photosensitive material which can be spun onto the wafer in liquid form (an adhesion promoter such as hexamethyldisilazane, HMDS, is usually used prior to the application of the resist). The spin speed and photoresist viscosity determine the final resist thickness, which is typically between $0.5–2.5\,\mu$m. Two different kinds of photoresist are available: positive and negative. In a positive resist, the UV-exposed areas are dissolved in the subsequent development stage, whereas in a negative photoresist, the exposed areas remain intact after development. Due to the better process control that can be achieved for small geometries, the positive resist is most commonly used in VLSI processes. After spinning the photoresist onto the wafer, the substrate is

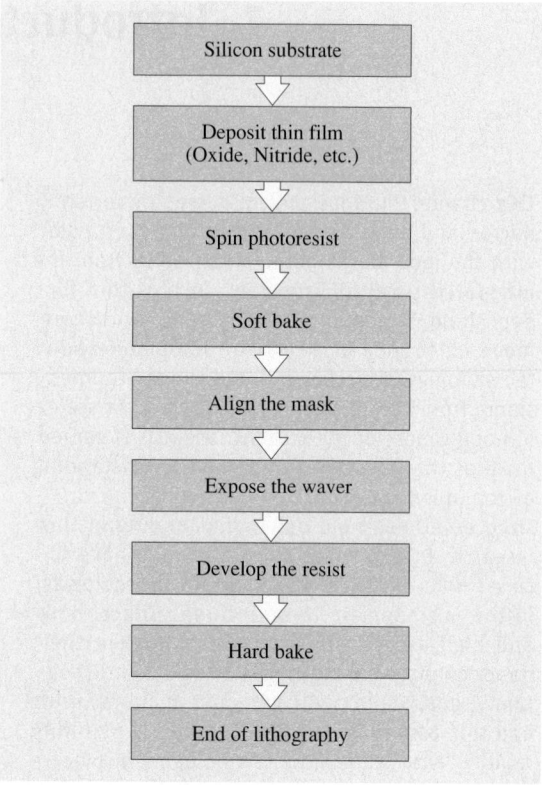

Fig. 7.1 Lithography process flow

soft-baked ($5–30$ min at $60–100\,°$C) in order to remove the solvents from the resist and to improve the adhesion. Subsequently, the mask is aligned to the wafer and the photoresist is exposed to a UV source.

Depending on the separation between the mask and the wafer, three different exposure systems are available: 1) contact, 2) proximity, and 3) projection. Although contact printing gives a better resolution than the proximity technique, the constant contact of the mask with the photoresist reduces the process yield and can damage the mask. Projection printing uses a dual-lens optical system to project the mask image onto the wafer. Since only one die can be exposed at a time, this requires a step and repeat system to completely cover the wafer area. Projection printing is far and away the most popular microfabrication system and it can yield superior resolutions to the contact and proximity methods. The exposure source used for photolithography depends on the resolution. Above $0.25\,\mu$m minimum line width, a high-pressure mercury lamp is adequate (436 nm g-line and 365 nm i-line). However, between 0.25 and $0.13\,\mu$m, deep UV sources such as excimer

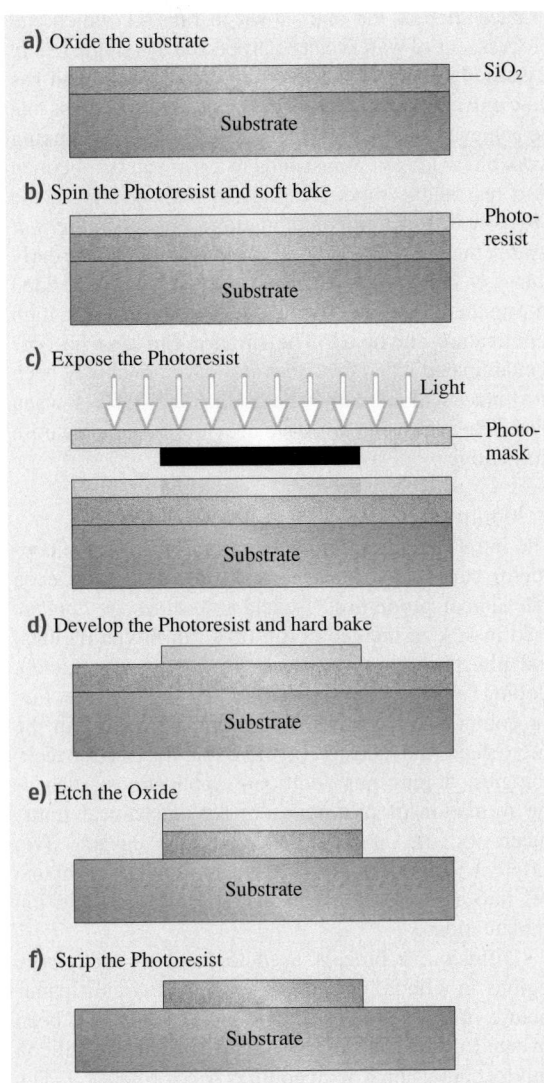

a) Oxide the substrate

SiO$_2$

Substrate

b) Spin the Photoresist and soft bake

Photo-resist

Substrate

c) Expose the Photoresist

Light

Photo-mask

Substrate

d) Develop the Photoresist and hard bake

Substrate

e) Etch the Oxide

Substrate

f) Strip the Photoresist

Substrate

Fig. 7.2 Schematic of photolithography with a positive PR

lasers (248 nm KrF and 193 nm ArF) are required. Although there has been extensive competition between techniques (including e-beam and X-ray) in the below 0.13 μm regime, extreme UV (EUV) (with a wavelength of 10–14 nm) seems to be the preferred technique [7.4].

After exposure, the photoresist is developed in a process similar to the development of photographic films. The resist is subsequently hard baked (20–30 min at 120–180 °C) in order to improve adhesion still further. The hard bake step, which concludes the photolithography process, creates the desired pattern on the wafer. Next, the underlying thin film is etched and the photore-

sist is stripped in acetone or another organic removal solvent. Figure 7.2 shows a schematic of the steps involved in photolithography with a positive photoresist.

7.1.2 Thin Film Deposition and Doping

Thin film deposition and doping are used extensively in micro/nanofabrication technologies. Most of the fabricated micro/nanostructures contain materials other than that of the substrate, which are obtained by various deposition techniques or by modifying the substrate. The following is a list of a few typical applications for the deposited and/or doped materials used in micro/nanofabrication, which gives an idea of the properties required:

- Mechanical structures
- Electrical isolation
- Electrical connection
- Sensing or actuating
- Mask for etching and doping
- Support or mold during the deposition of other materials (sacrificial materials)
- Passivation

Most of the thin films deposited have different properties to those of their corresponding "bulk" forms (for example, metals shows higher resistivities than a thin film). In addition, the techniques utilized to deposit these materials have a huge impact on their final properties. For instance, the internal stress (compressive or tensile) in

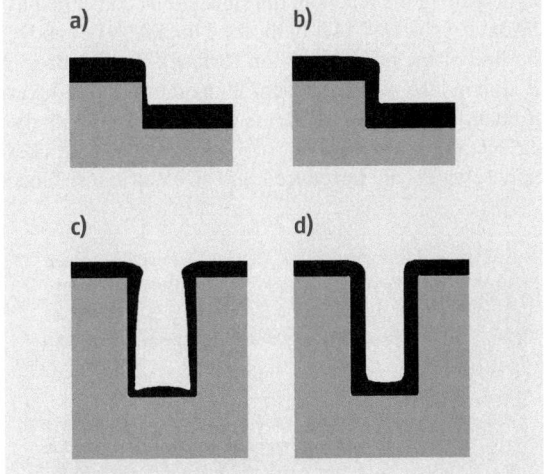

a)   **b)**

c)   **d)**

Fig. 7.3a–d Step coverage and conformality: (**a**) poor step coverage, (**b**) good step coverage, (**c**) nonconformal layer, and (**d**) conformal layer.

a film is strongly process-dependent. Excessive stress may crack or detach the film from the substrate and should therefore be minimized, although this property may also be useful for certain applications. Adhesion is another important issue that needs to be taken into account when depositing thin films. In some cases, such as the deposition of noble metals (including gold), an intermediate layer (chromium or titanium) may be needed to improve the adhesion. Finally, step coverage and conformality are two properties that can also influence the choice of one or another deposition technique. Figure 7.3 illustrates these concepts.

Oxidation

Silicon oxidation is a process used to obtain a thin film of SiO_2 with excellent quality (very low density of defects) and thickness homogeneity. Although it is not strictly a deposition, the result is the same: a thin layer of a new material coats the surface. Oxidation is typically performed at temperatures of 900 °C to 1200 °C in the presence of O_2 (dry oxidation) or H_2O (wet oxidation). The reactions for oxide formation are:

$$Si_{(solid)} + O_{2(gas)} \Rightarrow SiO_{2(solid)}$$

and

$$Si_{(solid)} + 2H_2O_{(steam)} \Rightarrow SiO_{2(solid)} + 2H_{2(gas)}$$

Although the rate of oxide growth is higher for wet oxidation, this is achieved at the expense of lower oxide quality (density). Since silicon atoms from the substrate participate in the reaction, the substrate is consumed as the oxide grows ($\approx 44\%$ of the total thickness lies above the line of the original silicon surface). The oxidation of silicon also occurs at room temperature, but a layer of about 20 Å (native oxide) is enough to passivate the surface and prevent further oxidation. To grow thicker oxides, wafers are introduced into an electric resistance

furnace such as the one shown in Fig. 7.4, which can process tens of wafers can be processed in a single batch. By strictly controlling the timing, temperature and gas flow entering the quartz tube, the desired thickness can be achieved with a high accuracy. Thicknesses ranging from a few tens of Angstroms to 2 μm can be obtained in a reasonable time. Despite the high quality of the SiO_2 obtained by silicon oxidation (also called *thermal oxide*), the use of this process is often limited to the early stages of fabrication, since some of the materials added during the formation of structure may not stand the high temperatures involved. The furnace can also become contaminated when the substrates have previously been in contact with certain etchants such as KOH, or when materials such as metals have been deposited, also pose limitations in most cases.

Doping

The introduction of certain impurities into a semiconductor can change its electrical, chemical, and even mechanical properties. Typical impurities or *dopants* used in silicon include boron (to form p-type regions) and phosphorus or arsenic (to form n-type regions). Doping is the main process used to fabricate major components such us diodes and transistors in the microelectronic industry. In micro/nanofabrication technologies, doping has additional applications, such as the formation of piezoresistors for mechanical transducers or the creation of etch stop layers. Two different techniques are used to introduce the impurities into a semiconductor substrate: diffusion and ion implantation.

Diffusion, a process used to form n- and p-type regions in silicon, dominated integrated circuit manufacture in the period just after such circuits had been invented. The diffusion of impurities into the silicon only occurs at high temperatures (above 800 °C). The furnaces used to carry out this process are similar to the ones used for oxidation. The dopants are introduced into the gaseous atmosphere of the furnace from liquid or solid sources. Figure 7.5 illustrates the process of creating an n-type region by the diffusion of phosphor from the surface into a p-type substrate. A masking material must have been previously deposited and patterned on the surface in order to define the areas to be doped. However, because diffusion is an isotropic process, the doped area will also extend underneath the mask. In microfabrication, diffusion is mainly used in the formation of very highly doped boron regions (p^{++}), which are usually used as etch stops in bulk micromachining.

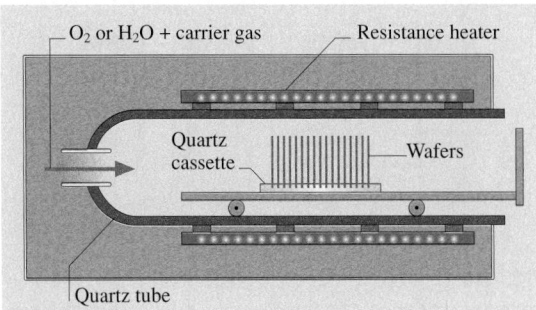

Fig. 7.4 Schematic of a typical oxidation furnace

Ion implantation allows more precise control of the dose (total amount of impurities introduced per area unit) and the impurity profile (concentration versus depth). In ion implantation, the impurities are ionized and accelerated towards the semiconductor surface. The penetration of impurities into the material follows a Gaussian distribution. After implantation, an annealing process is needed to activate the impurities and to repair the damage in the crystal structure produced by the ion collisions. A *drive-in* process to redistribute the impurities, performed in a standard furnace like those used for oxidation or diffusion may also be required.

Chemical Vapor Deposition and Epitaxy

As its name suggests, chemical vapor deposition (CVD) includes all of the deposition techniques that make use of chemical reactions in the gas phase to form the deposited thin film. The energy needed for the chemical reaction to occur is usually supplied by maintaining the substrate at elevated temperatures. Alternative energy sources, such as plasma or optical excitation, are also used because they enable a lower substrate temperature. The most common CVD processes in microfabrication are LPCVD (low pressure CVD) and PECVD (plasma-enhanced CVD).

The LPCVD process is typically carried out in electrically heated tubes, similar to oxidation tubes, equipped with pumps needed to achieve the low pressures (0.1 to 1.0 torr) required. Large numbers of wafers can be processed simultaneously and the material is deposited on both sides of the wafers. The process temperatures depend on the material to be deposited, but are generally in the range of 550 to 900 °C. As in the oxidation, high temperatures and contamination issues can restrict the type of processes used before LPCVD. Typical materials deposited by LPCVD include silicon oxide (for example $SiCl_2H_2 + 2N_2O \Rightarrow SiO_2 +$

$2N_2 + 2HCL$ at 900 °C), silicon nitride (such as $3SiH_4 + 4NH_3 \Rightarrow Si_3N_4 + 12H_2$, at 700–900 °C), and polysilicon (for instance $SiH_4 \Rightarrow Si + 2H_2$ at 600 °C). Due to its faster etch rate in HF, in situ phosphorus-doped LPCVD oxide (phosphosilicate glass or PSG) is used extensively in surface micromachining as the sacrificial layer. The conformality in this process is excellent, even for structures with very high aspect ratios. The mechanical properties of LPCVD materials are good compared to others such as PECVD, and are often used as structural materials in microfabricated devices. The stress in the deposited layers depends on the material, deposition conditions, and subsequent thermal history (for example the post-deposition anneal). Typical values are: 100–300 MPa (compressive) for oxide, ≈ 1 GPa (tensile) for stoichiometric nitride, and ≈ 200–300 MPa (tensile) for polysilicon. The stress present in the nitride layers can be reduced to almost zero using a silicon-rich composition. Since the stress values can vary over a wide range, it is important to measure and characterize the internal stress of deposited thin films for any particular set of equipment and deposition conditions.

The PECVD process is performed in plasma systems such as the one represented in Fig. 7.6. The use of RF energy to create highly reactive species in the plasma allows lower temperatures to be used at the substrate (150 to 350 °C). The parallel-plate plasma reactors normally used in microfabrication can only process a limited number of wafers per batch. The wafers are positioned horizontally on top of the lower electrode so that only one side gets deposited. Typical materials deposited with PECVD include silicon oxide, nitride, and amorphous silicon. Conformality

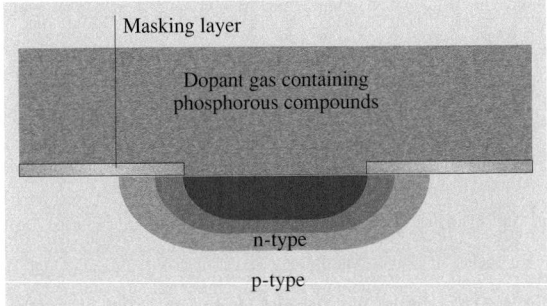

Fig. 7.5 Formation of an n-type region on a p-type silicon substrate by diffusion of phosphorus

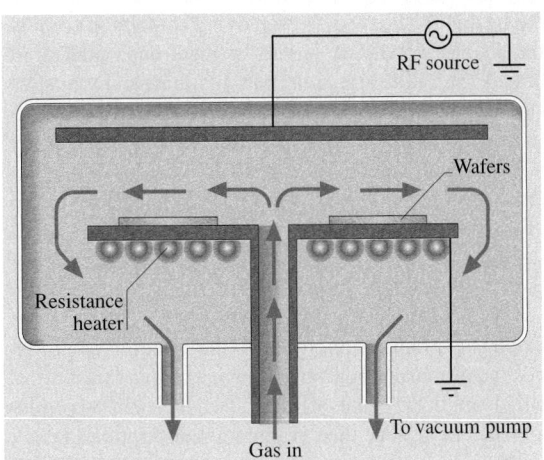

Fig. 7.6 Schematic of a typical PECVD system

is good for structures with low aspect ratios, but becomes very poor for deep trenches (20% of the surface thickness inside through-wafer holes with an aspect ratio of 10). The stress depends on deposition parameters and can be either compressive or tensile. PECVD nitrides are typically nonstoichiometric (Si_xN_y) and are much less resistant to etchants in masking applications.

Another interesting type of CVD is epitaxial growth. In this process, a single-crystalline material is grown as an extension of the crystal structure of the substrate. It is possible to grow dissimilar materials if the crystal structures are somehow similar (lattice-matched). Silicon-on-sapphire (SOS) substrates and some heterostructures are fabricated in this way. However, deposition of silicon on another silicon substrate is the most common CVD technique. Selective epitaxial growth is of particular interest for the formation of microstructures. In this process, the silicon crystal is only allowed to grow in windows patterned on a masking material. Many CVD techniques have been used to produce epitaxial growth. The most common method used for silicon is thermal chemical vapor deposition or vapor-phase epitaxy (VPE). Metallorganic chemical vapor deposition (MOCVD) and molecular beam epitaxy (MBE) commonly used to grow high-quality III–V compound layers with nearly atomic abrupt interfaces. The former uses vapors of organic compounds with group III atoms such as trimethylgallium [$Ga(CH_3)_3$] and group V hydrides such as AsH_3 in a CVD chamber with fast gas switching capabilities. The latter typically uses molecular beams from thermally evaporated elemental sources aimed at the substrate in an ultrahigh vacuum chamber. In this case, rapid on/off control of the beams is achieved by employing shutters in front of the source. Finally, it should be mentioned that many metals (molybdenum, tantalum, titanium and tungsten) can also be deposited using LPCVD. These are attractive due to their low resistivities and their ability to form silicides with silicon. Due to its application in new interconnect technologies, copper CVD is an active area of research.

Physical Vapor Deposition (Evaporation and Sputtering)

In physical deposition systems, the material to be deposited is transported from a source to the wafers, both of which are in the same chamber. Two physical principles are used to achieve this: evaporation and sputtering.

In evaporation, the source is placed in a small container with tapered walls, called a crucible, and is heated

up to a temperature where evaporation occurs. Various techniques are utilized to reach the high temperatures needed, including the induction of high currents with coils wound around the crucible and the bombardment of the material surface with an electron beam (e-beam evaporators). This process is mainly used to deposit metals, although dielectrics can also be evaporated. In a typical system the crucible is located at the bottom of a vacuum chamber, whereas the wafers line the dome-shaped ceiling of the chamber, Fig. 7.7. The main characteristic of this process is very poor step coverage, including shadow effects, as illustrated in Fig. 7.8. As explained in subsequent sections, some microfabrication techniques utilize these effects to pattern the deposited layer. One way to improve the step coverage is to rotate and/or heat the wafers during the deposition.

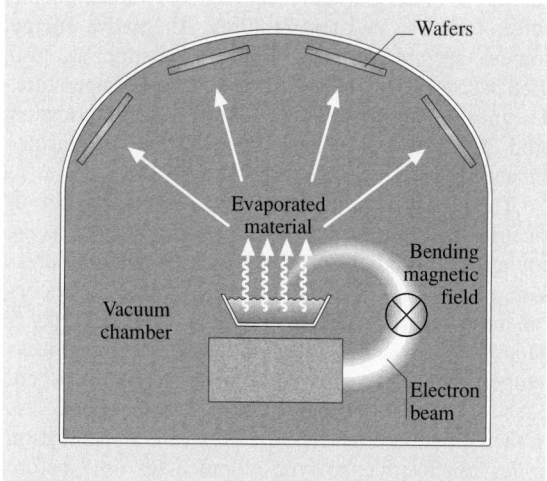

Fig. 7.7 Schematic of an e-beam deposition system

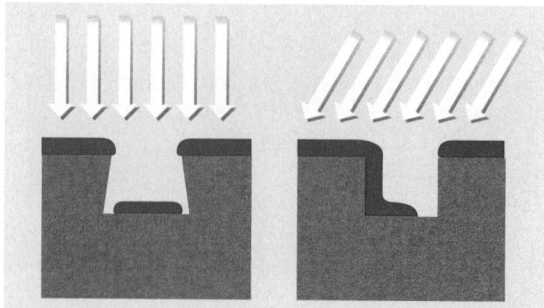

Fig. 7.8 Shadow effects observed in evaporated films. *Arrows* show the trajectory of the material atoms being deposited

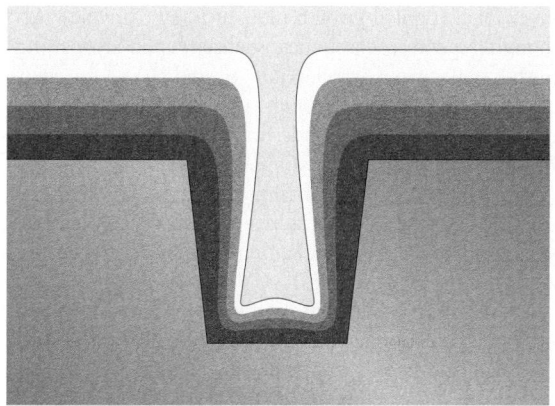

Fig. 7.9 Typical cross-section evolution of a trench being filled via sputter deposition

During sputtering, a target of the material to be deposited is bombarded with high-energy inert ions (usually argon). The outcome of the bombardment is that individual atoms or clusters are removed from the surface and ejected towards the wafer. The physical nature of this process permits its use with virtually any existing material. Examples of interesting materials for microfabrication that are frequently sputtered include metals, dielectrics, alloys (such as shape memory alloys), and all kinds of compounds (for example piezoelectric PZT). The inert ions used to bombard the target are produced in DC or RF plasma. In a simple parallel-plate system the top electrode is the target and the wafers are placed horizontally on top of the bottom electrode. Despite its lower deposition rate, the step coverage achieved via sputtering is much better than that gained from evaporation. However, the films obtained with this deposition process are nonconformal. Figure 7.9 illustrates successive sputtering profiles in a trench.

Both evaporation and sputtering systems are often capable of depositing more than one material simultaneously or sequentially. This ability is very useful for obtaining alloys and multilayers (for instance, multilayer magnetic recording heads are sputtered). For certain low-reactivity metals such as Au and Pt, the previous deposition of a thin layer of another metal is needed to improve the adhesion. Ti and Cr are two adhesion promoters frequently used. The stress in evaporated or sputtered layers is typically tensile. The deposition rates are much higher than most CVD techniques. However, due to stress accumulation and cracking, a thickness of more than 2 μm is rarely achieved with these processes. The technique described in the next section is sometimes used for thicker depositions.

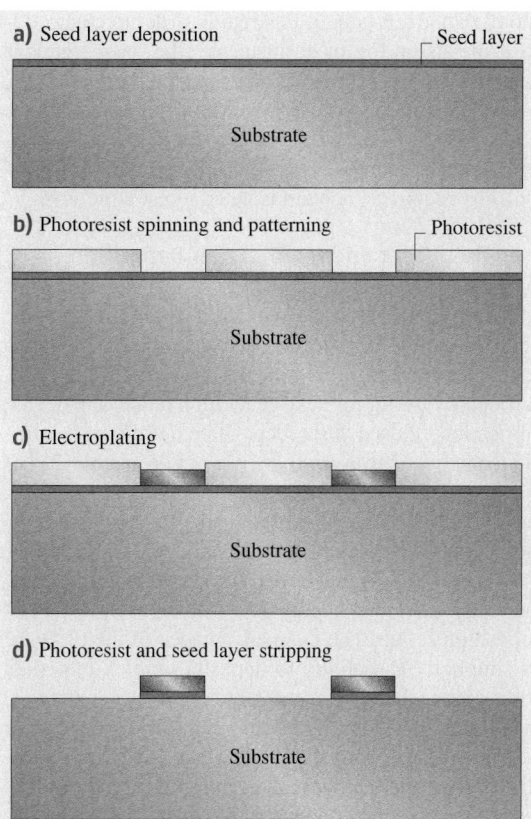

Fig. 7.10a–d Formation of isolated metal structures by electroplating through a mask: (**a**) seed layer deposition, (**b**) photoresist spinning and patterning, (**c**) electroplating, (**d**) photoresist and seed layer stripping

Electroplating

Electroplating (or electrodeposition) is a process typically used to obtain thick (tens of μm) metal structures. The sample to be electroplated is placed into a solution containing a reducible form of the ion of the desired metal, and it is maintained at a negative potential (cathode) relative to a counter electrode (anode). The ions are reduced at the sample surface and the nonsoluble metal atoms are incorporated into the surface. As an example, copper electrodeposition is frequently performed in copper sulfide-based solutions. The reaction that takes place at the surface is $Cu^{2+} + 2e^- \rightarrow Cu_{(s)}$. Recommended current densities for electrodeposition processes are on the order of 5 to 100 mA/cm^2.

As can be deduced from the process mechanism, the surface to be electroplated must be electrically conductive, and preferably of the same material as the deposited one if a good adhesion is desired. In order to electrode-posit metals on top of an insulator (the most frequent case) a thin film of the same metal, called the seed layer, is deposited on the surface first. Masking the seed layer with a resist permits selective electroplating at the patterned areas. Figure 7.10 illustrates the sequence of steps typically required to obtain isolated metal structures.

Pulsed Laser and Atomic Layer Deposition

Pulsed laser and atomic layer deposition techniques have attracted a considerable amount of attention recently. These two techniques offer several unique advantages compared to other thin film deposition methods that are particularly useful for next-generation nanoscale device fabrication. Pulsed laser deposition (PLD) is a simple technique that uses an intense (1 GW within 25 ns) UV laser (such as KrF excimer) to ablate a target material [7.5]. Plasma is subsequently formed from the target and is deposited on the substrate. Multitarget systems with Auger and RHEED spectroscopes are commercially available. Figure 7.11 shows a typical PLD deposition set-up. The main advantages of the PLD are its simplicity and ability to deposit complex materials with preserved stoichiometry ("stoichiometry transfer"). In addition, fine control over the film thickness can also be achieved by controlling the number of pulses. Stoichiometry transfer allows many complex targets, such as ferroelectrics, superconductors and magnetostrictives to be deposited using PLD. Other materials deposited include oxides, carbides, polymers, and metallic systems (such as FeNdB).

Atomic layer deposition (ALD) is a gas-phase self-limiting deposition method capable of depositing atomic layer thin films with excellent large area uniformity and conformality [7.6]. It enables simple and accurate control over film composition and thickness at the atomic

layer level (typical growth rates of a few Å/cycle). Although recent research has focused upon depositing high-k dielectric materials (Al_2O_3, and HfO_2) for next-generation CMOS electronics, other materials can also be deposited. These include transition metals (Cu, Co, Fe, and Ni), metal oxides, sulfides, nitrides and fluorides. Atomic-level control over film thickness and composition are also attractive features for MEMs application, such as conformal 3D packaging and air-gap structures. ALD is a modification of the CVD process and is based on two or more vapor-phase reactants that are introduced into the deposition chamber in a sequential manner. One growth cycle consists of four steps. First, a precursor vapor is introduced into the chamber, resulting in the deposition of a self-limiting monolayer on the surface of the substrate. Then the extra unreacted vapor is pumped out and a vapor dose of a second reactant is introduced. This reacts with the precursor on the surface in a self-limiting fashion. Finally, the extra unreacted vapor is pumped out and the cycle is repeated.

7.1.3 Etching and Substrate Removal

Thin film and bulk substrate etching is another fabrication step that is of fundamental importance to both VLSI processes and micro/nanofabrication. In the VLSI area, various conducting and dielectric thin films deposited for passivation or masking purposes need to be removed at some point. In micro/nanofabrication, in addition to thin film etching, the substrate (silicon, glass, GaAs) usually also needs to be removed in order to create various me-

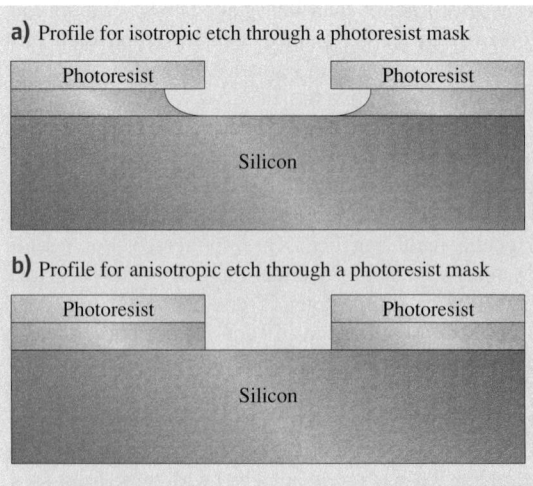

Fig. 7.12a,b Profile for isotropic (**a**) and anisotropic (**b**) etch through a photoresist mask

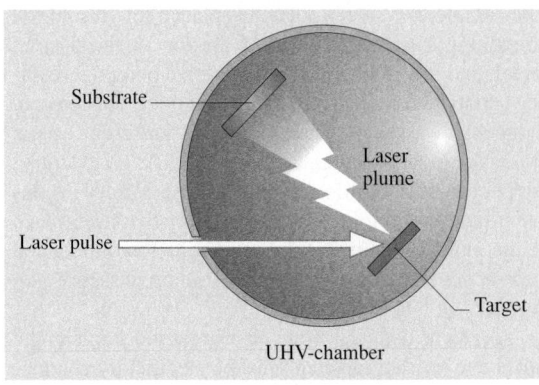

Fig. 7.11 A typical PLD deposition set-up

chanical micro/nanostructures (beams, plates and so on). Two important figures of merit for any etching process are selectivity and directionality. Selectivity is the degree to which the etchant can differentiate between the masking layer and the layer to be etched. Directionality relates to the etch profile under the mask. In an isotropic etch, the etchant attacks the material in all directions at the same rate, hence creating a semicircular profile under the mask, Fig. 7.12a. In an anisotropic etch, the dissolution rate depends on the specific direction and one can obtain straight sidewalls or other noncircular profiles, Fig. 7.12b. One can also divide the various etching techniques into wet and dry categories. We will use this classification in this section, discussing different wet etchants first and then the dry etching techniques used most often in the micro/nanofabrication.

Wet Etching

Historically, wet etching techniques preceded dry techniques. These still constitute an important group of etchants for micro/nanofabrication despite the fact that they are used less frequently in VLSI processes. Wet etchants are, by and large, isotropic, and they show superior selectivity for the masking layer over various dry techniques. In addition, due to the lateral undercut, the minimum feature achievable with wet etchants is limited to $> 3\,\mu m$. Silicon dioxide is commonly etched in a dilute (6:1, 10:1, or 20:1 by volume) or buffered HF (BHF, HF+NH_4F) solution (etch rate of $\approx 1000\,\text{Å/min}$ in BHF). Photoresist and silicon nitride are the two most common masking materials used for the wet oxide etch. The wet etchant used for silicon nitride is hot (140–200 °C) phosphoric acid with silicon oxide used as the masking material. Nitride wet etch is not very common (except for blanket etch) due to the masking difficulties and nonrepeatable etch rates. Metals can be etched using various combinations of acids and base solutions. There are also many commercially available etchant formulations for aluminium, chromium and gold which can be easily used. A comprehensive table of various metal etchants can be found in [7.7].

Anisotropic and isotropic wet etching of crystalline (silicon and gallium arsenide) and noncrystalline (glass) substrates is an important topic in micro/nanofabrication [7.8–11]. In particular, the achievement of anisotropic wet etching of silicon is considered to mark the beginning of the micromachining and MEMS field. Isotropic etching of silicon using $HF/HNO_3/CH_3COOH$ (various different formulations have been used) dates back to the 1950s and is still frequently used to thin down the silicon wafer. The etch

mechanism for this combination has been elucidated and is as follows: HNO_3 is used to oxidize the silicon which is subsequently dissolved away in the HF. The acetic acid is used to prevent the dissociation of HNO_3 (the etch works as well without the acetic acid). For short etch times, silicon dioxide can be used as the masking material; however, one needs to use silicon nitride if a longer etch time is desired. This etch also shows dopant selectivity, with the etch rate dropping at lower doping concentrations ($<10^{17}\,\text{cm}^{-3}$ n- or p-type). Although this effect can potentially be used as an etch stop mechanism in order to fabricate microstructures, the masking problem has prevented widespread application of this approach. Glass can also be isotropically etched using the HF/HNO_3 combination, with the etch surfaces showing considerable roughness. This has been extensively used for fabricating microfluidic components (mainly channels). Although Cr/Au is usually used as the masking layer, long etch times require a more robust mask (bonded silicon has been used for this purpose).

Silicon anisotropic wet etch constitutes an important technique in bulk micromachining. The three most important silicon etchants in this category are potassium hydroxide (KOH), ethylene diamine pyrocatechol (EDP) and tetramethyl ammonium hydroxide (TMAH). These are all anisotropic etchants which attack silicon along preferred crystallographic directions. In addition, they all show a marked reduction in the etch rate in heavily doped ($> 5 \times 10^{19}\,\text{cm}^{-3}$) boron ($p^{++}$) regions. The chemistry behind the action of these etchants is not yet very clear, but it seems that silicon atom oxidation at the surface and the reaction of silicon with hydroxyl ions (OH^-) are responsible for the formation of a soluble silicon complex ($SiO_2(OH)^{2-}$). The etch rate depends on the concentration and temperature and is usually around $1\,\mu m/min$ at temperatures of 85–115 °C. Common masking materials for anisotropic wet etchants are silicon dioxide and nitride, with the latter being superior for longer etch times. The crystallographic plane which shows the slowest etch rate is the (111) plane. Although it has been speculated that the lower atomic concentration along these planes is the reason for this phenomenon, the evidence is inconclusive and other factors must be included to account for this remarkable etch stop property. The anisotropic behavior of these etchants with respect to the (111) plane has been used extensively to create beams, membranes and other mechanical and structural components. Figure 7.13 shows typical cross-sections of (100) and (110) silicon wafers etched with an anisotropic wet etchant. As can be seen, the (111) slow

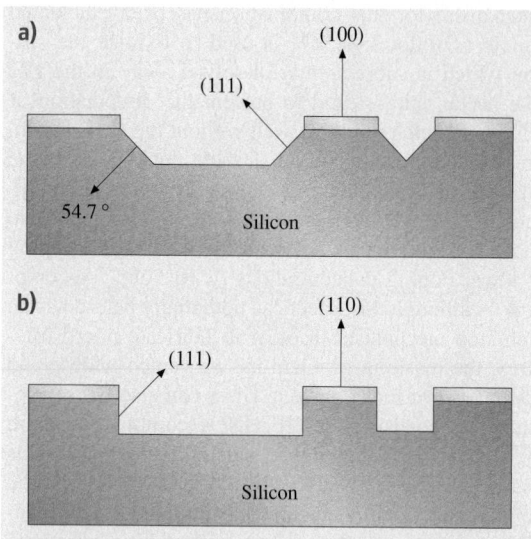

Fig. 7.13a,b Anisotropic etch profiles for: (a) (100), and (b) (110) silicon wafers

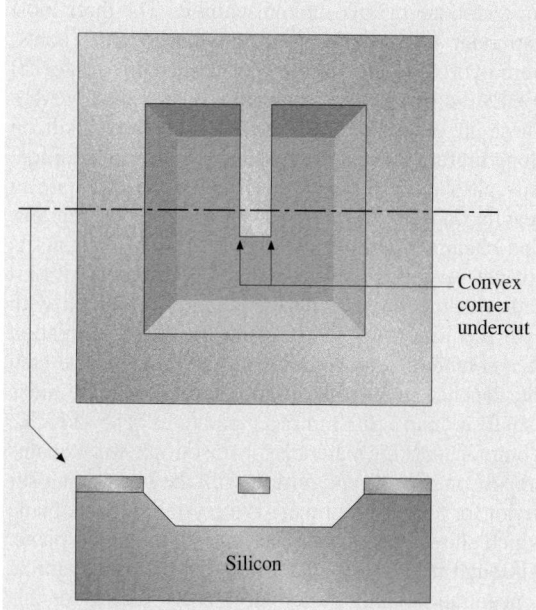

Fig. 7.14 Top view and cross-section of a dielectric cantilever beam fabricated using convex corner undercutting

planes are exposed in both situations, one creating 54.7° sloped sidewalls in the (100) wafer and the other creating vertical sidewalls in the (110) wafer. Depending on the dimensions of the mask opening, a V-groove or

a trapezoidal trench is formed in the (100) wafer. A large enough opening will allow the silicon to be etched all the way through the wafer, thus creating a thin dielectric membrane on the other side. It should be mentioned that the exposed convex corners have a higher etch rate than the concave ones, resulting in an undercut which can be used to create dielectric (such as nitride) cantilever beams. Figure 7.14 shows a cantilever beam fabricated using the convex corner undercut on a (100) wafer.

The three above-mentioned etchants show different directional and dopant selectivities. KOH has the best (111) selectivity (400/1), followed by TMAH and EDP. However, EDP has the highest selectivity with respect to the deep boron diffusion regions. Safety and CMOS compatibility are other important criteria for choosing a particular anisotropic etchant. Among the three mentioned etchants, TMAH is the most benign one, whereas EDP is extremely corrosive and carcinogenic. Silicon can be dissolved in TMAH in order to improve its selectivity with respect to aluminium. This property has made TMAH very appealing for post-CMOS micromachining where aluminium lines have to be protected. Finally, it should be mentioned that one can modulate the etch rate using a reversed biased p-n junction (electrochemical etch stop). Figure 7.15 shows the set-up commonly used to perform electrochemical etching. The silicon wafer under etch consists of an n-epi region on a p-type substrate. Upon the application of a reverse bias voltage to the structure (p substrate is in contact with the solution and n-epi is protected using a watertight fixture), the p substrate is etched away. When the n-epi regions are exposed to the solution an oxide passivation layer is formed and the etch is stopped. This technique can be

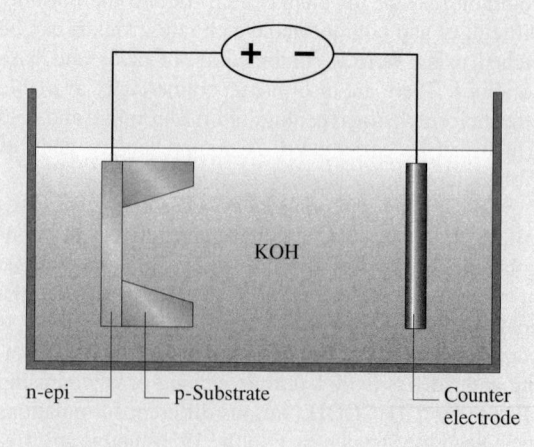

Fig. 7.15 Electrochemical etch set-up

used to fabricate single-crystalline silicon membranes for pressure sensors and other mechanical transducers.

Dry Etching

Most dry etching techniques are plasma-based. They have several advantages when compared with wet etching. These include a smaller undercut (allowing smaller lines to be patterned) and higher anisotropicity (allowing vertical structures with high aspect ratios). However, the selectivities of dry etching techniques are lower than those of wet etchant techniques, and one must take into account the finite etch rate of the masking materials. The three basic dry etching techniques, namely high-pressure plasma etching, reactive ion etching (RIE) and ion milling, utilize different mechanisms to obtain directionality.

Ion milling is a purely physical process which utilizes accelerated inert ions (such as Ar^+) that strike perpendicular to the surface, removing material (pressure $\approx 10^{-4}-10^{-3}$ Torr), Fig. 7.16a. The main characteristics of this technique are very low etch rates

(on the order of a few nm/min) and poor selectivity (close to 1:1 for most materials); hence it is generally used to etch very thin layers. In high-pressure ($10^{-1}-5$ Torr) plasma etchers, highly reactive species are created that react with the material to be etched. The products of the reaction are volatile so they diffuse away and new material is exposed to the reactive species. Directionality can be achieved, if desired, with the sidewall passivation technique (Fig. 7.16b). In this technique, nonvolatile species produced in the chamber deposit and passivate the surfaces. The deposit can only be removed by physical collision with incident ions. Because the movement of the ions has a vertical directionality, the deposit is mainly removed at horizontal surfaces, while vertical walls remain passivated. In this fashion, the vertical etch rate becomes much higher than the lateral one.

RIE etching, also called ion-assisted etching, is a combination of physical and chemical processes. In this technique the reactive species react with the material only when the surfaces are "activated" by the collision of incident ions from the plasma (by breaking bonds

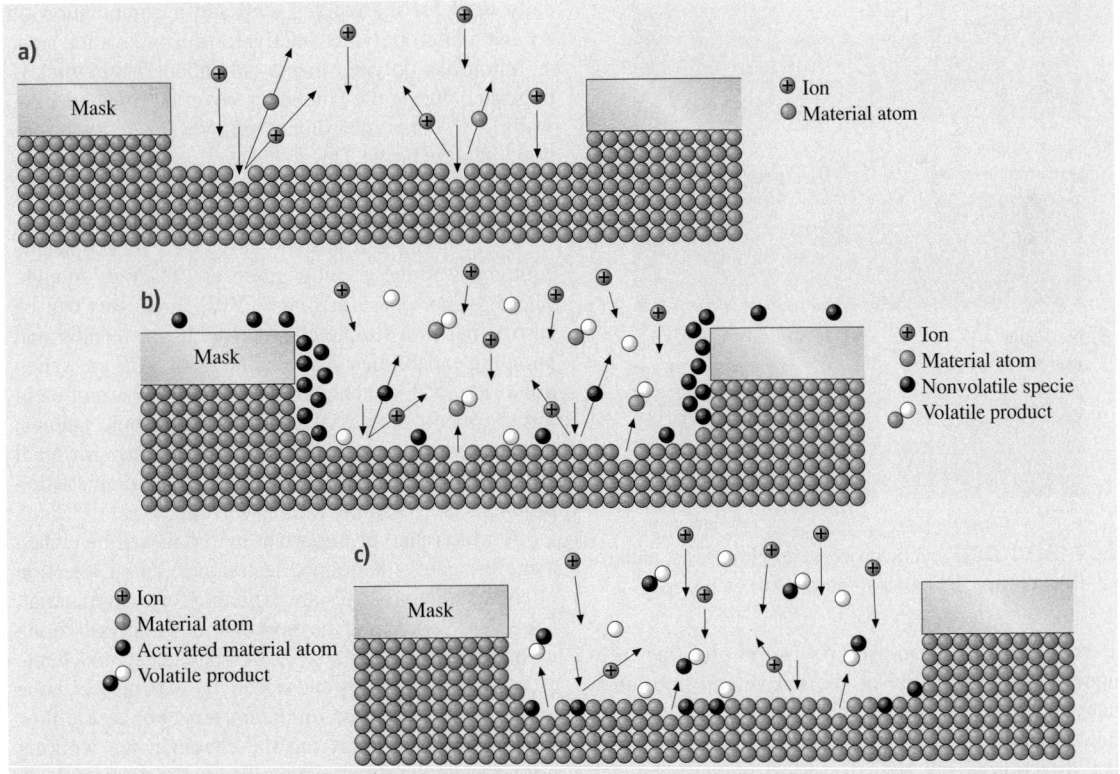

Fig. 7.16a–c Simplified representation of etching mechanisms for (**a**) ion milling, (**b**) high-pressure plasma etching, (**c**) RIE

Table 7.1 Typical dry etch chemistries

Si	CF_4/O_2, CF_2Cl_2, CF_3Cl, $SF_6/O_2/Cl_2$, $Cl_2/H_2/C_2F_6/CCl_4$, C_2ClF_5/O_2, Br_2, SiF_4/O_2, NF_3, ClF_3, CCl_4, CCl_3F_5, C_2ClF_5/SF_6, C_2F_6/CF_3Cl, CF_3Cl/Br_2
SiO2	CF_4/H_2, C_2F_6, C_3F_8, CHF_3/O_2
Si3N4	$CF_4/O_2/H_2$, C_2F_6, C_3F_8, CHF_3
Organics	O_2, CF_4/O_2, SF_6/O_2
Al	BCl_3, BCl_3/Cl_2, $CCl_4/Cl_2/BCl_3$, $SiCl_4/Cl_2$
Silicides	CF_4/O_2, NF_3, SF_6/Cl_2, CF_4/Cl_2
Refractories	CF_4/O_2, NF_3/H_2, SF_6/O_2
GaAs	BCl_3/Ar, $Cl_2/O_2/H_2$, $CCl_2F_2/O_2/Ar/He$, H_2, CH_4/H_2, $CClH_3/H_2$
InP	CH_4/H_2, C_2H_6/H_2, Cl_2/Ar
Au	$C_2Cl_2F_4$, Cl_2, $CClF_3$

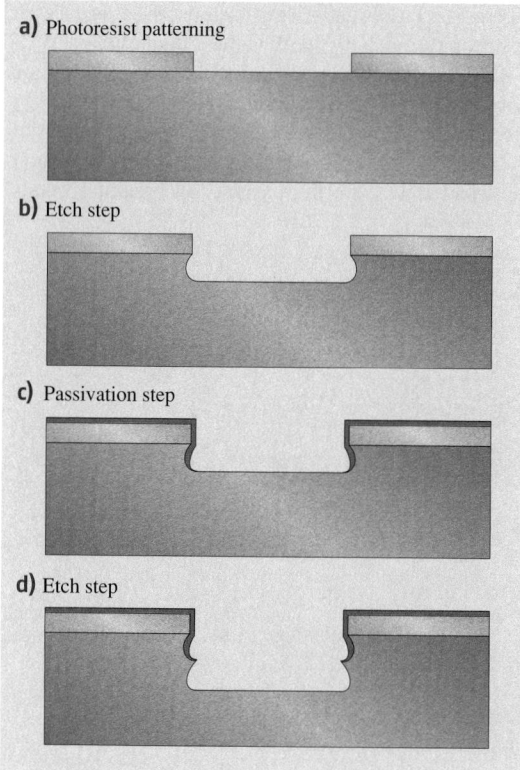

a) Photoresist patterning

b) Etch step

c) Passivation step

d) Etch step

Fig. 7.17a–d DRIE cyclic process: (**a**) photoresist patterning, (**b**) etch step, (**c**) passivation step, (**d**) etch step

at the surface for example). As in the previous technique, the directionality of the ion velocity produces many more collisions in the horizontal surfaces than in the walls, thus generating faster etching rates in the vertical direction (Fig. 7.16c). To further increase the etch anisotropy, sidewall passivation methods are also used in some cases. An interesting case is the deep reactive ion etching (DRIE) technique, capable of achieving aspect ratios of 30:1 and silicon etching rates of 2 to 3 μm/min (through-wafer etch is possible). In this technique, the passivation deposition and etching steps are performed sequentially in a two-step cycle, as shown in Fig. 7.17. In commercial silicon DRIE etchers, SF_6/Ar is typically used for the etching step and a combination of Ar and a fluoropolymer (nCF_2) for the passivation step. A Teflon-like polymer that is only about 50 nm thick is deposited during the latter step, covering only the sidewalls (Ar^+ ion bombardment removes the Teflon on the horizontal surfaces). Due to the cyclic nature of this process, the sidewalls of the etched features show a periodic "wave-shape" roughness of 50 to 400 nm.

Dry etching can also be performed in nonplasma equipment if the etching gases are reactive enough. These "vapor-phase etching" (VPE) processes can be carried out in a simple chamber with gas feeding and pumping capabilities. Two examples of VPE are xenon difluoride (XeF_2) etching of silicon and HF vapor etching of silicon dioxide. Due to their isotropic natures, these processes are typically used for etching sacrificial layers and releasing structures whilst avoiding stiction problems (see sections 7.2.1 and 7.2.2).

A wide range of important materials can be etched using the above-mentioned techniques, and a selection of chemical approaches are available for each material. Table 7.1 lists some of the most common materials along with selected etch recipes [7.12]. For each approach, the etch rate, directionality and selectivity with respect to the mask materials depend on parameters such as the flow rates of the gases entering the chamber, the working pressure and the RF power applied to the plasma.

7.1.4 Substrate Bonding

Substrate (wafer) bonding (silicon–silicon, silicon–glass, and glass–glass) is one of the most important microsystem fabrication techniques [7.13, 14]. It is frequently used to fabricate complex 3-D structures, both as a functional unit and as a part of the final microsystem package and encapsulation. The two most important bonding techniques are silicon–silicon fusion (or silicon direct bonding) and silicon–glass electrostatic (or anodic) bonding. In addition to these techniques, several other alternative methods which utilize an intermediate layer (eutectic, adhesive, and glass frit) have also been investigated. All of these techniques can be used to bond substrates at the wafer level. In this section we will only discuss wafer-level techniques; device-level bonding methods (such as e-beam and laser welding) will not be discussed.

Silicon Direct Bonding

Direct silicon or fusion bonding is used in the fabrication of micromechanical devices and silicon-on-insulator (SOI) substrates. Although it is mostly used to bond two silicon wafers with or without an oxide layer, it has also been used to bond different semiconductors such as GaAs and InP [7.14]. One main requirement for a successful bond is sufficient planarity (<10 Å surface roughness and $< 5\,\mu$m bow across a 4" wafer) and surface cleanliness. Thermal expansion mismatch also needs to be considered if two dissimilar materials are to be bonded. The bonding procedure is as follows: the silicon- or oxide-coated silicon wafers are first thoroughly cleaned. Then the surfaces are hydrated (activated) in HF or boiling nitric acid (an RCA clean also works). This creates an abundance of hydroxyl ions, rendering the surfaces hydrophilic. Then the substrates are brought into close proximity (starting from the center to avoid void formation). The closeness of the bonding surfaces allows short-range attractive van der Waals forces to bring the surfaces into intimate contact on an atomic scale. Following this step, hydrogen bonds between the two hydroxyl-coated silicon wafers bind the substrates together. These steps can be performed at room temperature; however, in order to increase the bond strength, a high-temperature (800–1200 °C) anneal is usually required. A big advantage of silicon fusion bonding is that the substrates are thermally matched.

Anodic Bonding

Silicon-glass anodic (electrostatic) bonding is another major substrate joining technique which has been ex-

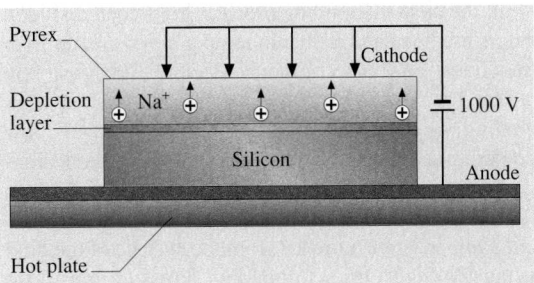

Fig. 7.18 Set-up for anodic bonding of glass to silicon

tensively used for microsensor packaging and device fabrication. The main advantage of this technique is its lower bonding temperature, which is around 300–400 °C. Figure 7.18 shows the bonding set-up. A glass wafer (Pyrex 7740 is often used due to its good thermal expansion match to silicon) is placed on top of a silicon wafer and the sandwich is heated to 300–400 °C. Subsequently, a voltage of ≈ 1000 V is applied to the glass–silicon sandwich with the glass connected to the cathode. The bond starts immediately after the application of the voltage and spreads outward from the cathode contact point. The bond can be visually observed as a dark grayish front which expands throughout the wafer.

The bonding mechanism is as follows. During the heating period, glass sodium ions move toward the cathode and create a depletion layer at the silicon–glass interface. A strong electrostatic force is therefore created at the interface which pulls the substrates into an intimate contact. The exact chemical reaction responsible for anodic bonding is not yet clear but covalent silicon–oxygen bonds at the interface seems to be responsible for the bond. Silicon-to-silicon anodic bonding using sputtered or evaporated glass interlayers is also possible.

Bonding with Intermediate Layers

Various other wafer bonding techniques utilizing an intermediate layer have also been investigated [7.14]. Among the most important ones are adhesive, eutectic, and glass frit bonds. Adhesive bonds, achieved by inserting a polymer (polyimides, epoxies, thermoplast adhesives, and photoresists) between the wafers, have been used to join different wafer substrates [7.15]. Complete curing (in the oven or using dielectric heating) of the polymer before or during the bonding process prevents subsequent solvent outgas and void formation. Although reasonably high bonding strengths can be obtained, these bonds are nonhermetic and unstable over reasonably long periods of time.

In the eutectic bonding process, gold-coated silicon wafers are bonded together at temperatures greater than the silicon–gold eutectic point (363 °C, 2.85% silicon and 97.1% Au) [7.16]. This process can achieve high bonding strength and good stability at relatively low temperatures. For good bond uniformity, silicon dioxide must be removed from the silicon surface prior to the deposition of the gold. In addition, all organic contaminants must be removed from the surface of the gold (using UV light) prior to bonding. Pressure must also be applied in order to achieve a better contact. Although

eutectic bonds can be achieved at low temperatures, attaining uniformity over large areas has proven to be a challenging task.

Glass frit can also be used as an interlayer in substrate bonding. In this technique, a thin layer of glass is first deposited and preglazed. The glass-coated substrates are then brought into contact and the sandwich is heated to above the glass melting temperature (typically < 600 °C). As for the eutectic process, pressure must be applied for an adequate contact to occur [7.17].

7.2 MEMS Fabrication Techniques

In this section, we will discuss various MEMS fabrication techniques commonly used to build various microdevices (microsensors and microactuators) [7.8–11]. The size range of the microstructures that can be fabricated using these techniques spans from 1 mm to 1 μm. As was mentioned in the *Introduction*, we will concentrate on the more important techniques, skimming over specialized methods.

7.2.1 Bulk Micromachining

Bulk micromachining is the oldest MEMS technology and therefore one of the most mature [7.18]. It is currently the most commercially successful MEMS technique by some distance, and is used to manufacture devices such as pressure sensors and ink-jet print heads. Although there are many different variations, the basic concept behind bulk micromachining is selective removal of the substrate (silicon, glass, GaAs, and so on). This allows the creation of various micromechanical components such as beams, plates and membranes which can be used to fabricate a variety of sensors and actuators. The most important microfabrication techniques used in bulk micromachining are wet and dry etch and substrate bonding. Although one can use different criteria to divide bulk micromachining techniques into separate categories, we will use a historical timeline for this purpose. We will start by discussing the more traditional wet etching techniques, and then proceed to discuss the more recent ones using deep RIE and wafer bonding.

Bulk Micromachining
Using Wet Etch and Wafer Bonding
The first use of anisotropic wet etchants to remove silicon is often used to mark the beginning of the micromachining era. Back-side etch was used to create movable structures such as beams, membranes and plates, see Fig. 7.19. Initially, the etching was timed in order to create a specified thickness. However, this technique proved to be inadequate for creating thin structures (<20 μm). Subsequent use of various etch stop techniques allowed the creation of thinner membranes in a more controlled fashion. As was mentioned in Sect. 7.1.3, the use of heavily doped boron regions and electrochemical bias can drastically slow down the etch process and hence create microstructures with controllable thicknesses. Figure 7.20a,b shows the cross-section of two piezoresistive pressure sensors fabricated using electrochemical and P^{++} etch stop techniques. The P^{++} method requires the epitaxial growth of a lightly doped region on top of a P^{++} etch stop layer. This layer is subsequently used for the placement of piezoresistors. However, if no active component is required, one can simply use the P^{++} region to create a thin membrane, Fig. 7.20c.

The P^{++} etch stop technique can also be used to create isolated thin silicon structures via the dissolu-

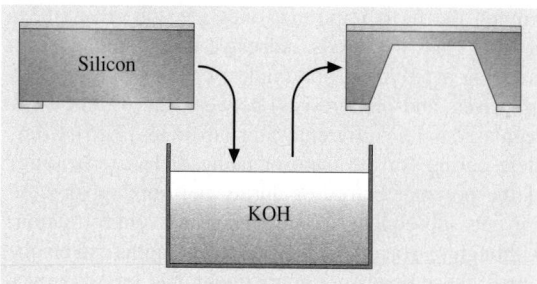

Fig. 7.19 Wet anisotropic silicon back-side etching

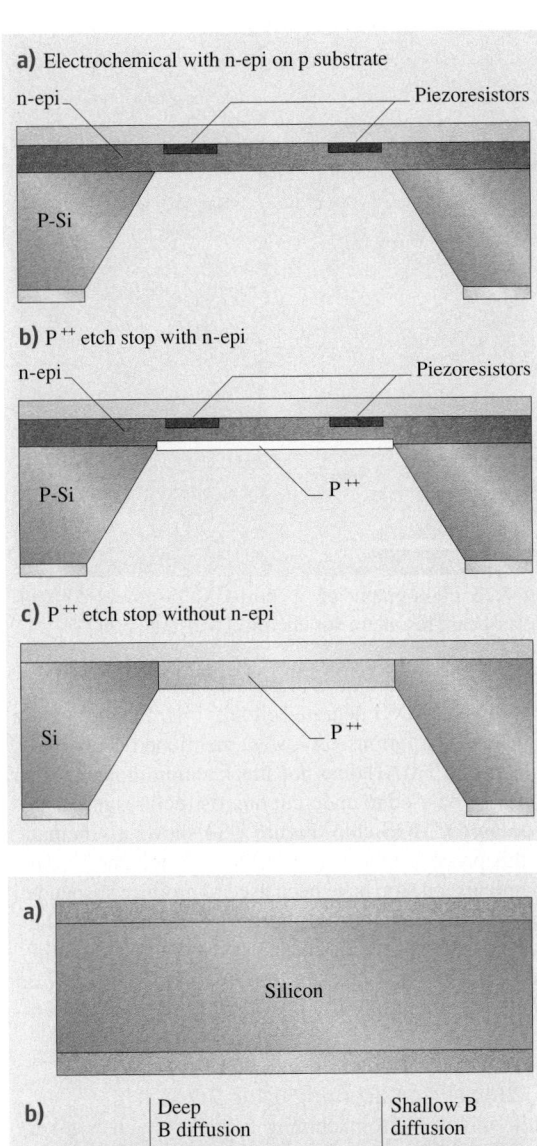

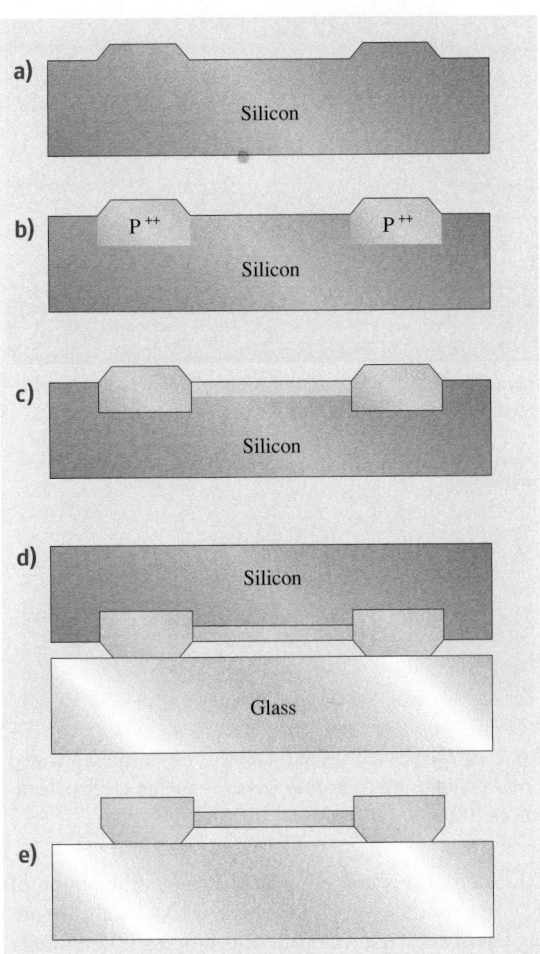

Fig. 7.20a–c Wet micromachining etch stop techniques: (a) electrochemical with n-epi on p substrate, (b) P++ etch stop with n-epi, and (c) P++ etch stop without n-epi◄

Fig. 7.22a–e Dissolved wafer process sequence: (a) KOH etch, (b) deep B diffusion, (c) shallow B diffusion, (d) silicon–glass anodic bond, and (e) release in EDP

tion of the entire lightly doped region [7.19]. This technique was successfully used to fabricate silicon recording and stimulating electrodes for biomedical applications. Figure 7.21 shows the cross-section of one such process which relies on deep (15–20 μm) and shallow boron (2–5 μm) diffusion steps to create microelectrodes with flexible connecting ribbon cables. An

Fig. 7.21 Free-standing microstructure fabrication using deep and shallow boron diffusions and EDP release

Fig. 7.23 SEM photograph of a microaccelerometer fabricated using the dissolved wafer process [7.20]

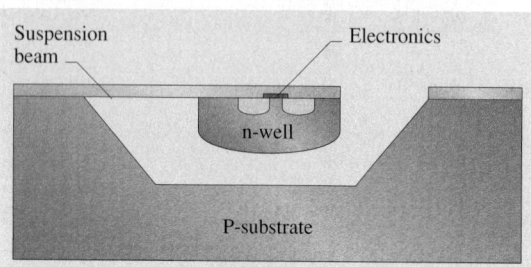

Fig. 7.24 Suspended island created on a prefabricated CMOS chip using front-side wet etch and an electrochemical etch stop

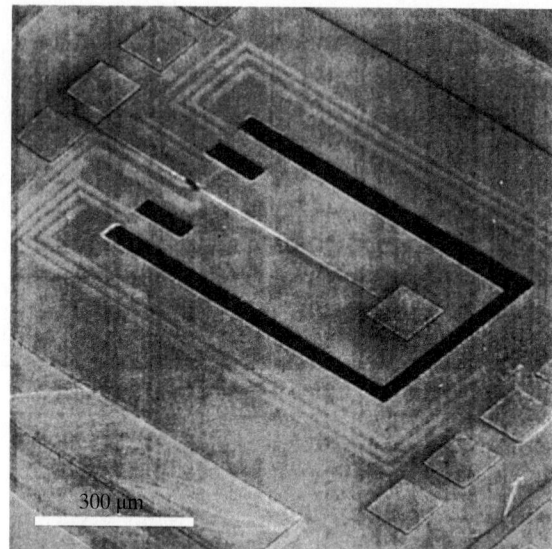

Fig. 7.25 Photograph of a post-CMOS-processed cantilever beam resonator for chemical sensing [7.21]

most popular techniques in this category is the postprocessing of CMOS integrated circuits by frontside etching in TMAH solutions. As was mentioned previously, silicon-rich TMAH does not attack aluminium and it can therefore be used to undercut microstructures into a preprocessed CMOS chip. Figure 7.24 shows a schematic of this process, where a frontside wet etch and electrochemical etch stop have been used to produce suspended beams. This technique has been used extensively to fabricate a variety of microsensors (including humidity, gas, chemical and pressure microsensors). Figure 7.25 shows a photograph of a post-CMOS-processed chemical sensor.

extension of this process which uses a combination of P^{++} etch stop layers and silicon–glass anodic bonding has also been developed. This process is commonly known as the dissolved wafer process and has been used to fabricate a variety of microsensors and microactuators [7.20]. Figure 7.22 shows the cross-section for this process. Figure 7.23 shows an SEM photograph of a microaccelerometer fabricated using the dissolved wafer process.

It is also possible to merge wet bulk micromachining and microelectronics fabrication processes to build micromechanical components on the same substrate as the integrated circuits (CMOS, Bipolar, or BiCMOS) [7.21]. This is very appealing since it allows the integration of interface and signal processing circuitry with MEMS structures on a single chip. However, important fabrication issues such as process compatibility and yield must be considered carefully. One of the

Bulk Micromachining Using Dry Etch

Bulk silicon micromachining using dry etch is a very attractive alternative to the wet techniques described in the previous section. These techniques were developed during the mid-1990s following the successful development of anisotropic dry silicon etch processes. More recent advances in deep silicon RIE and the availability of SOI wafers with a thick top silicon layer have increased the applications of these techniques. They allow the fabrication of vertical structures with high aspect ratios in isolation or along with on-chip electronics. Process compatibility with active microelectronics is less of a concern for dry methods since many of them do not damage the circuit or its interconnect.

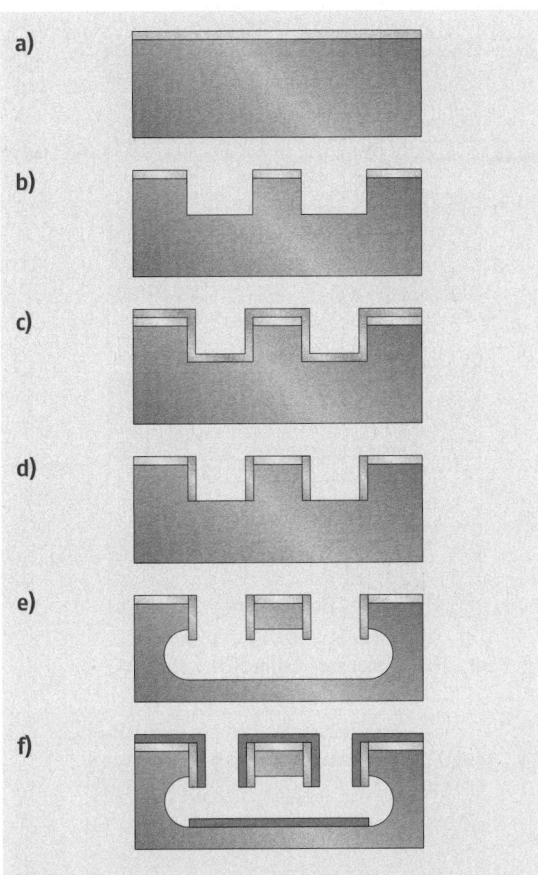

Fig. 7.26 Cross-section of the SCREAM process

Fig. 7.27a–e SEM photographs of structures fabricated using the SCREAM process: (**a**) comb-drive actuator, (**b**) suspended spring, (**c**) spring support, (**d**) moving suspended capacitor plate, and (**e**) fixed capacitor plate [7.25]

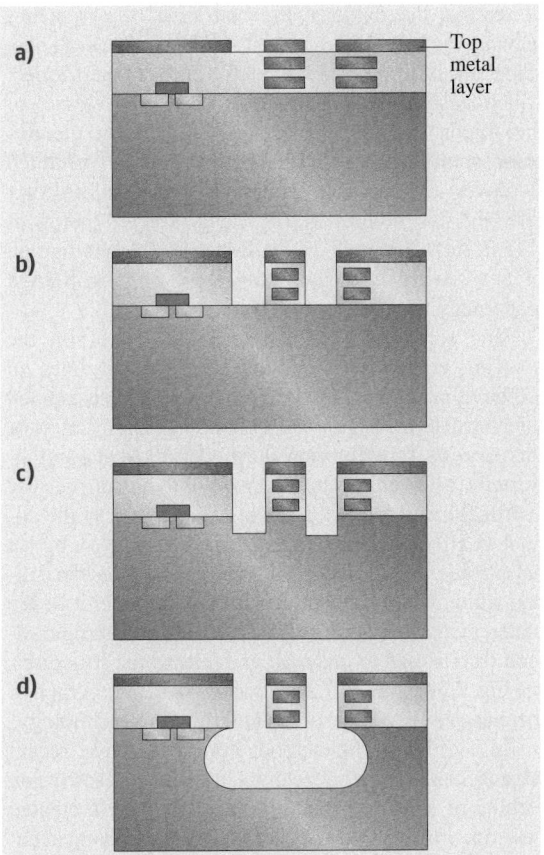

Fig. 7.28 Cross-section of the process flow for post-CMOS dry microstructure fabrication

The most simple dry bulk micromachining technique relies on the frontside undercutting of microstructures using XeF_2 vapor phase etch [7.22]. However, as was mentioned before, this is an isotropic etch and so it has limited applications. A combination of isotropic/anisotropic dry etch is more useful and can be used to create a variety of interesting structures. Two successful techniques using this combination are single-crystal reactive etching and metallization (SCREAM) [7.23] and post-CMOS dry release using aluminium/silicon dioxide laminate [7.24]. The first technique relies on the combination of isotropic/anisotropic dry etch to create single-crystalline suspended structures. Figure 7.26 shows a cross-section of the process. It starts with an anisotropic (Cl_2/BCl_3) silicon etch using an oxide mask (Fig. 7.26b). This is followed by conformal PECVD oxide deposition (Fig. 7.26c). Then an anisotropic oxide etch is used to remove the oxide at the bottom of the trenches, leaving the sidewall oxide intact (Fig. 7.26d). At this stage an isotropic silicon etch (SF_6) is performed, which results in undercutting and the release of the silicon structures (Fig. 7.26e). Finally, if electrostatic actuation is desired, a metal can be sputtered to cover the top and sidewall of the microstructure and the bottom of the cavity formed below it (Fig. 7.26f). Figure 7.27 shows an SEM photograph of a comb-drive actuator fabricated using SCREAM technology.

The second dry release technique relies on the masking ability of aluminium interconnect lines in a CMOS integrated circuit in order to create suspended microstructures. Figure 7.28 shows a cross-section of this process. As can be seen, the third level Al of a prefabricated CMOS chip is used as a mask to anisotropically etch the underlying oxide layers all the way to the silicon (CHF_3/O_2), Fig. 7.28b. This is followed by an anisotropic silicon etch to create a recess in the silicon which will be used in the final step to facilitate the undercut and release Fig. 7.28c. Finally, an isotropic silicon etch is used to undercut and release the structures, see Fig. 7.28d. Figure 7.29 shows an SEM photograph of a comb-drive actuator fabricated using this technology.

In addition to the methods described above, recent advancements in the development of deep reactive ion etching of silicon (DRIE, see Sect. 7.1.3) have created new opportunities for dry bulk micromachining techniques (see Sect. 7.2.3). One of the most important ones uses thick silicon SOI wafers which are commercially available in various top silicon thicknesses [7.27]. Figure 7.30 shows the cross-section of a typical process that uses DRIE and SOI wafers. The top silicon layer is patterned and etched all the way to the buried oxide,

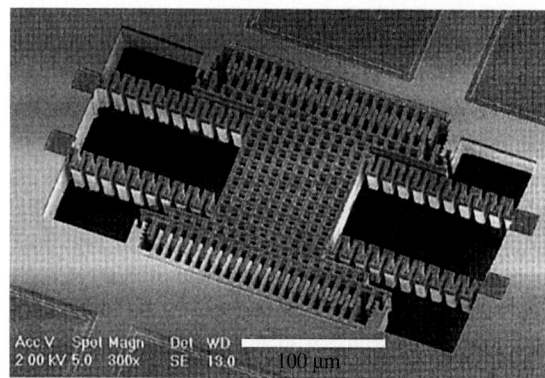

Fig. 7.29 SEM photograph of a comb-drive actuator fabricated using aluminium mask post-CMOS dry release [7.26]

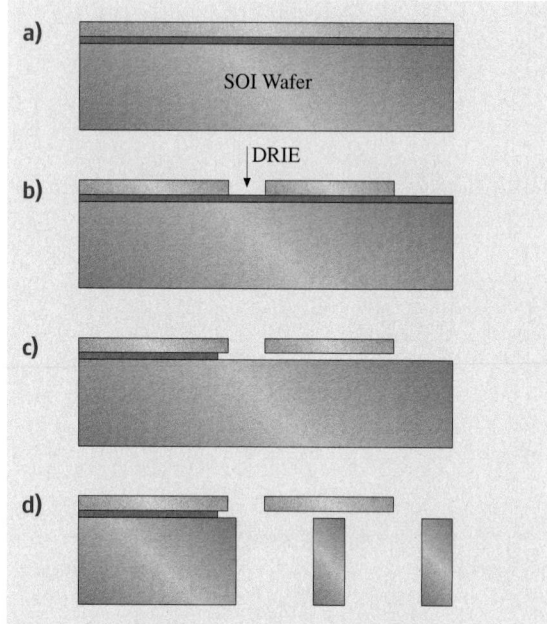

Fig. 7.30 DRIE processes using SOI wafers

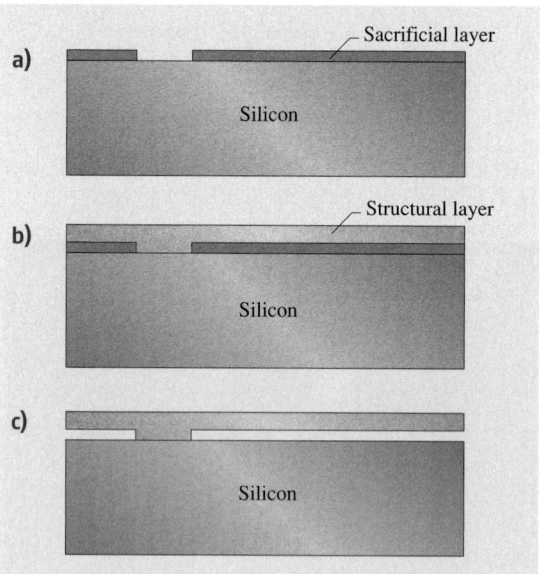

Fig. 7.31 Basic surface micromachining fabrication process

Fig. 7.31b. The oxide is subsequently removed in HF, releasing suspended single-crystalline microstructures, see Fig. 7.31c. In a modified version of this process, the substrate can also be removed from the back-side, allowing easy access from both sides (this makes release easier and prevents stiction), see Fig. 7.31d.

7.2.2 Surface Micromachining

Surface micromachining is another important MEMS microfabrication technique that can be used to create movable microstructures on top of a silicon substrate [7.28]. This technique relies on the deposition of structural thin films on a sacrificial layer which is subsequently etched away resulting in movable micromechanical structures (beams, membranes, plates, and so on). The main advantage of surface micromachining is that extremely small sizes can be obtained. In addition, it is relatively easy to integrate the micromachined structures with on-chip electronics for increased functionality. However, due to the increased surface nonplanarity with any additional layer, there is a limit to the number of layers that can be deposited. Although one of the earliest reported MEMS structures was a surface-micromachined resonant gate transistor [7.29], material-related difficulties resulted in the termination of efforts in this area. In the mid 1980s, improvements in the field of thin film deposition rekindled interest in surface micromachining [7.30]. Polysilicon surface micromachining was introduced later on in the same decade, which opened the door to the fabrication of a variety of microsensors (including accelerometers and gyroscopes) and microactuators (micromirrors, RF switches, and so on). In this section, we will concentrate on the key process steps involved in surface micromachining fabrication and the various materials used in the surface micromachining process. We will also discuss the monolithic integration of CMOS with MEMS structures and 3-D surface micromachining.

Basic Surface Micromachining Processes

The basic surface micromachining process is illustrated in Fig. 7.31. The process begins with a silicon substrate on top of which a sacrificial layer is grown and patterned, Fig. 7.31a. The structural material is then deposited and patterned, Fig. 7.31b. As can be seen, the structural material is anchored to the substrate through the openings created in the sacrificial layer during the previous step. Finally, the sacrificial layer is removed, resulting in the release of the microstructures Fig. 7.32c. In wide structures, it is usually necessary to provide access holes in the structural layer for fast sacrificial layer removal. It is also possible to seal microcavities created by the surface micromachining technique [7.10]. This can be done at the wafer level and is a big advantage in applications such as pressure sensors which require a sealed cavity. Figure 7.32 shows two different techniques that can be used for this purpose. In the first technique, following the etching of the sacrificial layer, a LPCVD dielectric layer (oxide or nitride) is deposited to cover and seal the etch holes in the structural material, Fig. 7.33a. Since the LPCVD deposition is performed at reduced pressures, a subatmospheric pill-box microcavity can be created.

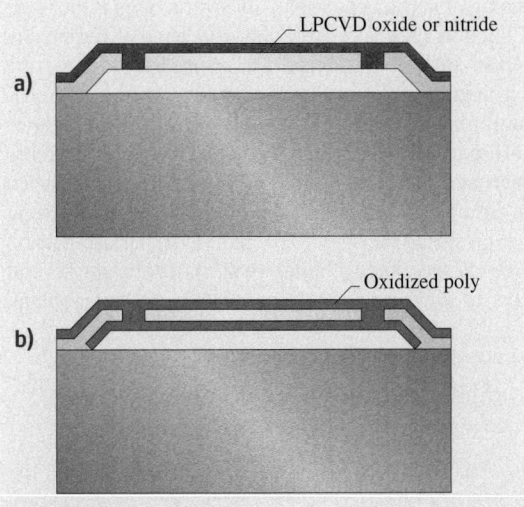

Fig. 7.32 Two sealing techniques for cavities created by surface micromachining

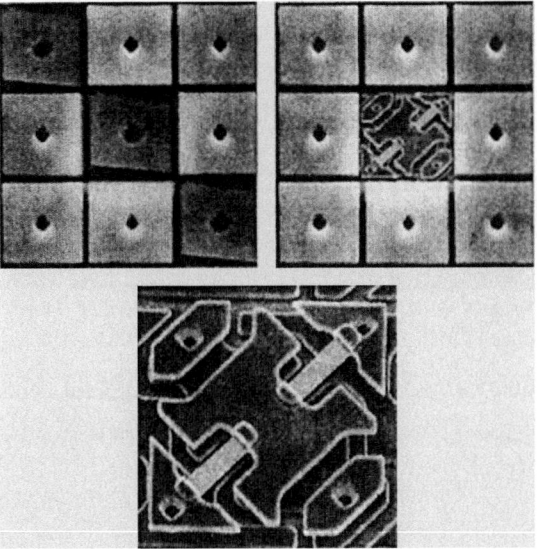

Fig. 7.33 SEM photographs of Texas Instrument's micromirror array [7.28]

In the second technique, also called "reactive sealing", the polysilicon structural material is oxidized following the removal of the sacrificial layer, Fig. 7.33b. If the access holes are small enough, the grown oxide can seal the cavity. Due to the consumption of oxygen during the growth process, the cavity is also subatmospheric in this case.

The most common sacrificial and structural materials are phosphosilicate glass (PSG) and polysilicon, respectively (low temperature oxide or LTO is also frequently used as the sacrificial layer). However, there are several other sacrificial/structural combinations that have been used to create a variety of surface micromachined structures. Important design issues related to the choice of the sacrificial layer are: 1) quality (pinholes and so on), 2) ease of deposition, 3) deposition rate, 4) deposition temperature, and 5) etch difficulty and selectivity (the sacrificial layer etchant should not attack the structural layer). The particular choice of material used for the structural layer depends on the desired properties and specific application. Several important requirements are: 1) ease of deposition, 2) deposition rate, 3) step coverage, 4) mechanical properties (internal stress, stress gradient, Young's moduli, fracture strength and internal damping), 5) etch selectivity, 6) thermal budget and history, 7) electrical conductivity, and 8) optical reflectivity. Two examples from the commercially available surface micromachined devices illustrate various successful sacrificial/structural combinations. Texas Instruments (TI; Dallas, TX) deformable mirror display (DMD) spatial light modulators uses aluminium as the structural material (good optical reflectivity) and photoresist as the sacrificial layer (easy dry etch and low processing temperatures, allowing easy post-IC integration with CMOS) [7.32], Fig. 7.33, whereas Analog Devices' (Cambridge, MA) microgyroscope uses polysilicon structural material and a PSG sacrificial layer, Fig. 7.34. Two recent additions to the collection of available structural layers are polysilicon-germanium and polygermanium [7.33, 34]. These are intended for use as substitutes for polysilicon in applications where the high polysilicon deposition temperature (around $600\,°C$) is prohibitive (during CMOS integration for example). Unlike LPCVD polysilicon, polygermanium (poly-Ge) and polysilicon-germanium (poly-Si$_{1-x}$Ge$_x$) can be deposited at temperatures as low as $350\,°C$ (the poly-Ge deposition temperature is usually lower than poly-SiGe). Table 7.2 summarizes important surface micromachined sacrificial/structural combinations.

An important consideration in the design and processing of surface micromachined structures is the issue of stiction [7.10, 35, 36]. This can happen during the release step if a wet etchant is used to remove the sacrificial layer or during the device lifetime. The reason for stiction during release is the surface tension of the liquid etchant, which can hold the microstructure down and cause stiction. This usually happens when the structure is compliant and does not possess a strong enough spring constant to overcome the surface tension force of the rinsing liquid (water). There are several ways to alleviate the release-related stiction problem.

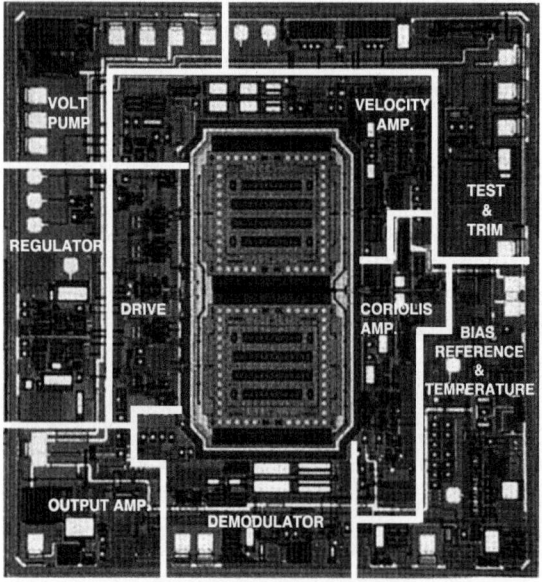

Fig. 7.34 SEM photograph of Analog Devices' gyroscope [7.31]

Table 7.2 Several important surface-micromachined sacrificial/structural combinations

System	Sacrificial layer	Structural layer	Structural layer etchant	Sacrificial layer etchant
1	PSG or LTO	Poly-Si	RIE	Wet or vapor HF
2	Photoresist, polyimide	Metals (Al, Ni, Co, Ni-Fe)	Various metal etchants	Organic solvents, plasma O_2
3	Poly-Si	Nitride	RIE	KOH
4	PSG or LTO	Poly-Ge	H_2O_2 or RCA1	Wet or vapor HF
5	PSG or LTO	Poly-Si-Ge	H_2O_2 or RCA1	Wet or vapor HF

These include: 1) the use of dry or vapor phase etchant, 2) the use of solvents with lower surface tensions, 3) geometrical modifications, 4) CO_2 critical drying, 5) freeze-drying, and 6) self-assembled monolayer (SAM) or organic thin film surface modification. The first technique prevents stiction by circumventing the need for a wet etchant, although in condensation is a possibility in the case of vapor phase release, which can also cause some stiction. The second method uses a rinsing solvent (such as methanol) that has a lower surface tension than water. This is usually followed by rapid evaporation of the solvent using a hot-plate. However, this is not the best technique, because many structures still stick. The third technique is geometrical; dimples are placed in the structural layer in order to reduce the contact surface area and hence reduce the attractive force. The fourth and fifth techniques rely on a phase change (of CO_2 in one case and butyl-alcohol in the other) which avoids the liquid phase altogether by jumping directly to the gas phase. The last technique uses self-assembled monolayers or organic thin films to coat the surfaces with a hydrophobic layer. The stiction that occurs during the operating lifetime of the device (in-use stiction) is due to the condensation of moisture on surfaces, electrostatic charge accumulation, or direct chemical bonding. Surface passivation using self-assembled monolayers or organic thin films can be used to reduce the surface energy and to reduce or eliminate the capillary forces and direct chemical bonding. These organic coatings also reduce electrostatic forces if a thin layer is applied directly to the semiconductor (without the intervening oxide layer). Commonly used organic coatings include fluorinated fatty acids (TI's aluminium micromirrors), silicone polymeric layers (Analog Devices' accelerometers) and siloxane self-assembled monolayers.

Surface Micromachining Integration with Active Electronics

Integration of surface micromachined structures with on-chip circuitry can increase performance and simplify packaging. However, issues related to process compatibility and yield must be considered carefully. The two most common techniques are MEMS-first and MEMS-last techniques. In the MEMS-last technique, the integrated circuit is fabricated and surface micomachined structures are subsequently built on top of the silicon wafer. An aluminium structural layer with a photoresist sacrificial layer is an attractive combination due to the low thermal budget of the process (TI'S

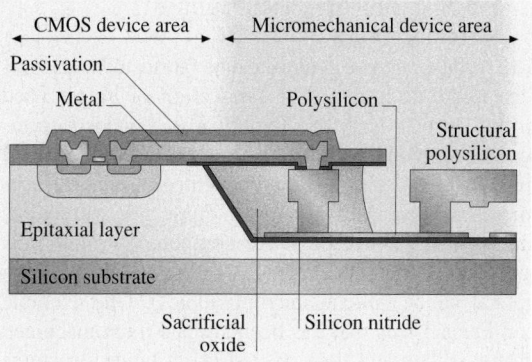

Fig. 7.35 Cross-section of the Sandia MEMS-first integrated fabrication process

micromirror array). However, in applications where the mechanical properties of Al are not adequate, polysilicon structural material must be used with a sacrificial layer of LTO or PSG. Due to the rather high deposition temperature of polysilicon, this combination requires that special attention is paid to the thermal budget. For example, aluminium metallization must be avoided and substituted with refractory metals such as tungsten. This can only be achieved at the cost of greater process complexity and lower transistor performance.

The MEMS-first technique alleviates these difficulties by fabricating the microstructures at the very beginning of the process. However, if the microstructures are processed first, they must be buried in a sealed trench to eliminate interference with microstructures from subsequent CMOS processes. Figure 7.35 shows a cross-section of a MEMS -first fabrication process developed at the Sandia National Laboratory [7.37]. The process starts with a shallow anisotropic etching of trenches in a silicon substrate to accommodate the height of the polysilicon structures fabricated later on. A silicon nitride layer is then deposited to provide isolation at the bottom of the trenches. Next, several layers of polysilicon and sacrificial oxide are deposited and patterned in a standard surface micromachining process. Then the trenches are completely filled with sacrificial oxide and the wafers are planarized with chemical-mechanical polishing (this avoids any complications in the following lithographic steps). After an annealing step, the trenches are sealed with a nitride cap. At this point, a standard CMOS fabrication process is performed. At the end of the CMOS process, the nitride cap is etched and the buried structures are released by etching the sacrificial oxide.

3-D Microstructures From Surface Micromachining

3-D surface microstructures can be fabricated using surface micromachining. The fabrication of hinges used in the vertical MEMS assembly was a major step towards achieving 3-D microstructures [7.39]. Optical microsytems have greatly benefited from surface-micromachined 3-D structures. These microstructures are used as passive or active components (micromirror, Fresnel lens, optical cavity, and so on) on a silicon optical bench (silicon microphotonics). One example is a Fresnel lens that has been surface micromachined in polysilicon and then erected using hinge structures and locked in place using micromachined tabs, thus liberating the structure from the horizontal plane of the wafer [7.38, 40]. Various microactuators (such as comb-drive and vibromotors) have been used to move these structures out of the silicon plane and into position. Figure 7.36 shows an SEM photograph of a bar-code microscanner that uses a silicon optical microbench with 3-D surface micromachined structures.

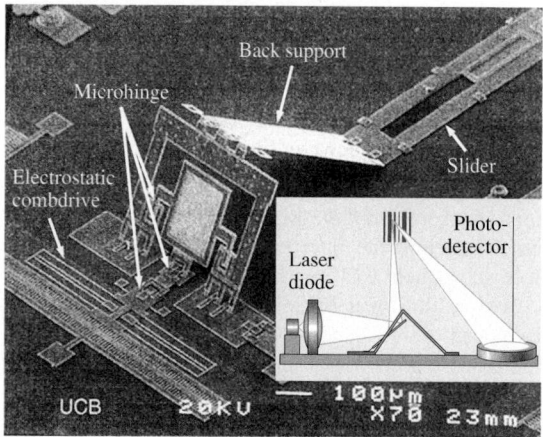

Fig. 7.36 Silicon pin-and-sample hinge scanner with 3-D surface micromachined structures [7.38]

7.2.3 High Aspect Ratio Micromachining

The bulk and surface micromachining technologies presented in the previous sections fulfill the requirements of a large group of applications. Certain applications, however, require the fabrication of high aspect ratio structures that cannot be realized with the aforementioned technologies. In this section we describe three technologies – LIGA, HEXSIL and HARPSS– which are capable of producing structures that have vertical

dimensions that are much larger than their lateral dimensions via X-ray lithography (LIGA) and DRIE etching (HEXSIL and HARPSS).

LIGA

LIGA is a high aspect ratio micromachining process that relies on X-ray lithography and electroplating (in German: Lithographie galvanoformung abformung) [7.41, 42]. We have already introduced the concept of the plating-through-mask technique (in Sect. 7.1.2; see Fig. 7.10). With standard UV photolithography and photoresists, the maximum thickness achievable is on the order of a few tens of microns and the resulting metal structures show tapered walls. LIGA is a technology based on the same plating-through-mask idea but it can be used to fabricate metal structures with thicknesses of a few microns to a few millimeters with almost vertical sidewalls. This is achieved using X-ray lithography and special photoresists. Due to their short wavelengths, X-rays are capable of penetrating through a thick photoresist layer with no scattering and of defining features with lateral dimensions down to $0.2\,\mu m$ (aspect ratio > 100:1).

The photoresists used in LIGA should comply with certain requirements, including sensitivity to X-rays, resistance to electroplating chemicals, and good adhesion to the substrate. Based on such requirements, poly-(methylmethacrylate) (PMMA) is considered to be an optimal choice for the LIGA process. Application of the thick photoresist on top of the substrate can be performed by various techniques, such as multiple spin-coating, precast PMMA sheets, and plasma polymerization coating. The mask structure and materials used for X-ray lithography must also comply with certain requirements. The traditional masks based on glass plates with a patterned chrome thin layer are not suitable because X-rays are not absorbed by the chromium layer and the glass plate is not transparent enough. Instead, X-ray lithography uses a silicon nitride mask with gold as the absorber material (typically formed by electroplating gold to a thickness of $10–20\,\mu m$). The nitride membrane is supported by a silicon frame which can be fabricated using bulk micromachining techniques. Once the photoresist is exposed to the X-rays and developed, the process proceeds with the electroplating of the desired metal. Ni is the most common, although other metals and metallic compounds such us Cu, Au, NiFe, and NiW are also electroplated in LIGA processes. A good agitation of the plating solution is the key to obtaining a uniform and repeatable result during this step. A paddle plating cell, based on a windshield-wiper-like

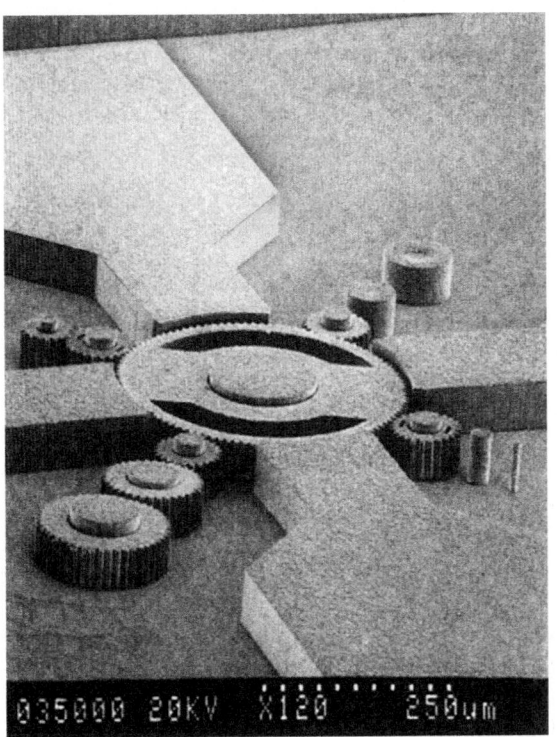

Fig. 7.37 SEM of assembled LIGA-fabricated nickel structures [7.42]

device moving only a millimeter away from the substrate surface, provides extremely reproducible agitation. Figure 7.37 shows an SEM of a LIGA microstructure fabricated by electroplating nickel.

Due to the expensive nature of X-ray sources (related to synchrotron radiation), LIGA technology was initially intended for the fabrication of molds that could be used many times in hot embossing or injection molding processes. However, it has been also used in many applications to form high aspect ratio metal structures directly on top of a substrate. A cheaper alternative to the LIGA process (although somewhat poorer quality) called UV-LIGA or "poor man's LIGA" has been proposed [7.43, 44]. This process uses SU-8 negative photoresists (available for spin-coating at various thickness ranging from 1 to 500 µm) and standard contact lithography equipment. Aspect ratios larger than 20:1 have been demonstrated using this technique. A major problem with this alternative is the removal of the SU-8 photoresist after plating. Various methods have been proposed with different degrees of success. These include: wet etching with special solvents; burning at high tem-

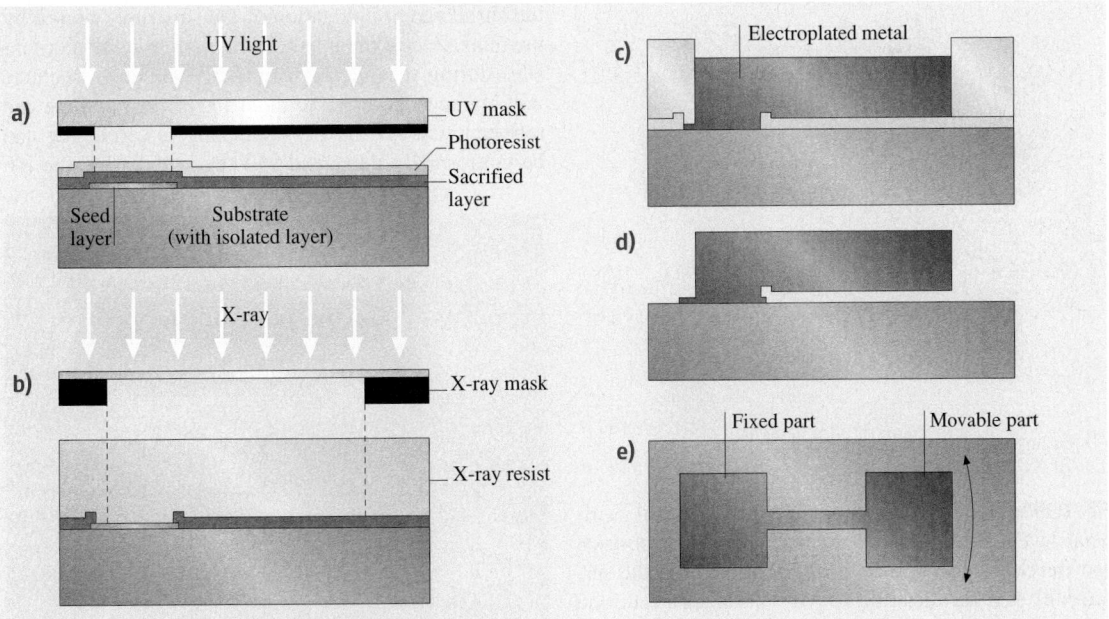

Fig. 7.38a–e Sacrificial LIGA process: (**a**) UV lithography for sacrificial layer patterning, (**b**) X-ray lithography, (**c**) electroplating, (**d**) structure release, (**e**) top view of the movable structure

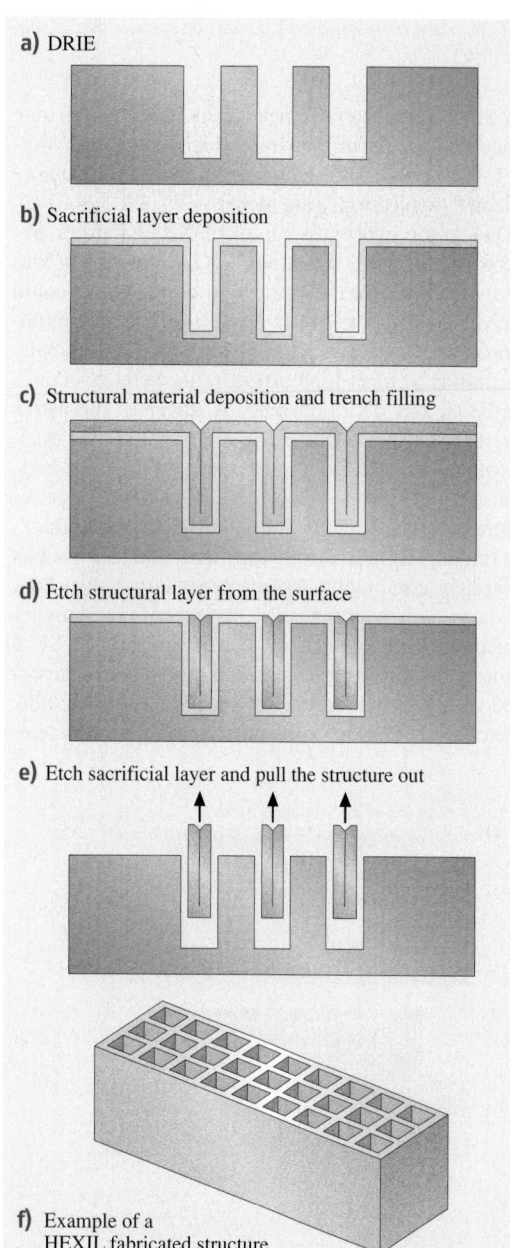

a) DRIE

b) Sacrificial layer deposition

c) Structural material deposition and trench filling

d) Etch structural layer from the surface

e) Etch sacrificial layer and pull the structure out

f) Example of a HEXIL fabricated structure

Fig. 7.39a–f HEXSIL process flow: (**a**) DRIE, (**b**) sacrificial layer deposition, (**c**) structural material deposition and trench filling, (**d**) etch structural layer from the surface, (**e**) etch sacrificial layer and pull the structure out, (**f**) example of a HEXSIL fabricated structure

peratures (600 °C); dry etching; use of a release layer; and high-pressure water jet etching.

A variation on the basic LIGA process, shown in Fig. 7.38, permits the fabrication of electrically isolated movable structures, and thus opens up more possibilities for sensor and actuator design using this technology [7.46]. This so-called "sacrificial LIGA" (SLIGA) starts with the patterning of the seed layer. A sacrificial layer (such as titanium) is then deposited and patterned. The process then proceeds as usual in standard LIGA until the last step, when the sacrificial layer is removed. The electroplated structures that overlap with the sacrificial layer are released in this step.

HEXSIL

The second method for fabricating high aspect ratio structures, which is based on a template replication technology, is HEXSIL (HEXagonal honeycomb polySILicon) [7.47]. Figure 7.39 shows a simplified process flow. A high aspect ratio template is first formed in a silicon substrate using DRIE. Then a sacrificial multilayer is deposited to allow the final release of the structures. The multilayer is composed of one or more PSG nonconformal layers, used for fast etch release ($\approx 20\,\mu m/min$ in 49% HF), alternated with conformal layers of either oxide or nitride to provide enough thickness for proper release of the structures. The total thickness of the sacrificial layer must be larger than the shrinkage or elongation of the structures caused by the relaxation of the internal stress (compressive or tensile) during the release step, otherwise the structures will clamp themselves to the walls of the template and retrieval will not be possible. Any material that can be conformally deposited and released during the HF

Fig. 7.40 SEM micrograph of an angular microactuator fabricated using HEXSIL [7.45]

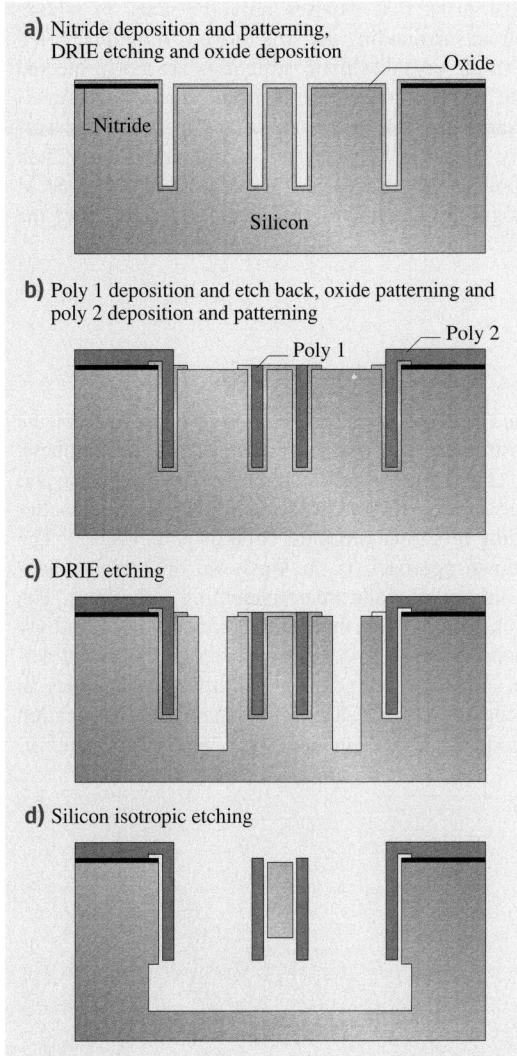

a) Nitride deposition and patterning, DRIE etching and oxide deposition

b) Poly 1 deposition and etch back, oxide patterning and poly 2 deposition and patterning

c) DRIE etching

d) Silicon isotropic etching

Fig. 7.41a–d HARPSS process flow: (**a**) nitride deposition and patterning, DRIE etching and oxide deposition, (**b**) poly 1 deposition and etch back, oxide patterning and poly 2 deposition and patterning, (**c**) DRIE etching, (**d**) silicon isotropic etching

release step without damage is suitable for the structural layer. Structures made of polysilicon, nitride, and electroless nickel [7.48] have been reported. Nickel can only be deposited in combination with polysilicon since a conductive surface is needed for the deposition to occur. After the deposition of the structural materials, a blanket etch (poly or nitride) or a mechanical lapping (nickel) is performed to remove the

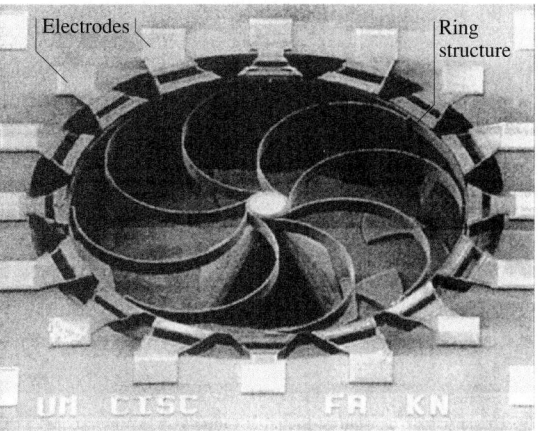

Fig. 7.42 SEM photograph of a microgyroscope fabricated using the HARPSS process [7.49]

excess material from the surface. Finally, 49% HF with surfactant is used to dissolve the sacrificial layers. The process can be repeated many times using the same template, which lowers the fabrication costs considerably. Figure 7.40 shows an SEM photograph of a microactuator fabricated using the HEXSIL process.

HARPSS

The High Aspect Ratio combined Poly- and Single-crystal Silicon (HARPSS) technology is another technique capable of producing high aspect ratio electrically isolated polycrystalline and single-crystal silicon microstructures with capacitive air gaps ranging in size from submicrometer to tens of micrometers [7.50]. The structures, tens to hundreds of micrometers thick, are defined by trenches etched with DRIE and are filled with oxide and poly layers. The release of the microstructures is achieved at the end by means of a directional silicon etch followed by an isotropic etch. The small vertical gaps and thick structures possible with this technology find application during the fabrication of a variety of MEMS devices, particularly inertial sensors [7.51] and RF beam resonators [7.52]. Figure 7.41 shows the process flow at the cross-section of a single-crystal silicon beam resonator. The HARPSS process starts with the deposition and patterning of a silicon nitride layer that will be used to isolate the poly structure's connection pads from the substrate. High aspect ratio trenches ($\approx 5\,\mu m$ wide) are then etched into the substrate using a DRIE etcher. Then a conformal oxide layer (LPCVD) is deposited. This layer has two functions: 1) to protect the structures during the dry etch release, and 2) to define the submicrometer gap between the silicon and polysilicon structures. Following the oxide deposition, the trenches are completely filled with LPCVD polysilicon. The polysilicon is etched back and the oxide beneath is patterned to provide anchor points for the structures. A second layer of polysilicon is then deposited and patterned. Finally, the structures are released using a DRIE step followed by an isotropic silicon etch through a photoresist mask that exposes only the areas of silicon substrate surrounding the structures. It should be noted that single-crystal silicon structures are not protected at the bottom during the isotropic etch. This causes the single-crystal silicon structures to be etched vertically from the bottom, and so they are shorter than the polysilicon structures. Figure 7.42 shows an SEM photograph of a microgyroscope fabricated using the HARPSS process.

7.3 Nanofabrication Techniques

The microfabrication techniques discussed so far are mostly geared towards fabricating devices in the 1 mm to 1 μm size range (although submicron dimensions are possible with certain techniques, such as HARPSS using a dielectric sacrificial layer). Recent years have witnessed a tremendous surge of interest in fabricating submicro- (1 μm-100 nm) and nanostructures (100−1 nm range) [7.53]. This interest arises from both practical and fundamental viewpoints. At the more scientific and fundamental level, nanostructures provide an interesting tool for studying electrical, magnetic, optical, thermal and mechanical properties of matter at the nanometer scale. These include important quantum mechanical phenomena (such as conductance quantization, band-gap modification and coulomb blockade) arising from the confinement of charged carriers in structures such as quantum wells, wires and dots, see Fig. 7.43. On the practical side, employing nanostructures in electronic/optical devices and sensors can lead to significant improvements in performance. In the field of devices, investigators have focused on fabricating nanometer-sized transistors, in anticipation of the predicted technical difficulties with extending Moore's law beyond 100 nm resolution. In addition, optical sources and detectors with nanometer-sized dimensions exhibit enhanced characteristics that are not achievable in larger devices (such as lower threshold currents, improved dynamic behavior, and improved emission line width in quantum dot lasers). These improvements create exciting possibilities for next generation computation and communication devices. In the field of sensing, shrinking dimensions beyond conventional optical lithography can provide major improvements in sensitivity and selectivity.

One can broadly divide various nanofabrication techniques into top-down and bottom-up categories. The first approach starts with a bulk or thin film material and removes selective regions in order to fabricate nanostructures (similar to micromachining techniques). The second method relies on molecular recognition and self-assembly to fabricate nanostructures from smaller building blocks (molecules, colloids and clusters). The top-down approach is obviously an offshoot of standard lithography and micromachining techniques. The bottom-up approach, on the other hand, is more strongly influenced by chemical engineering and material science, and relies on fundamentally different principles. In this chapter, we will discuss five major nanofabrication techniques. These include: i) e-beam and nanoim-

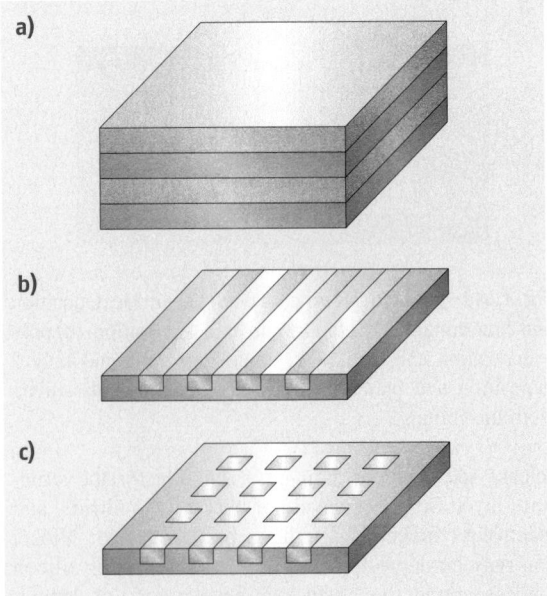

Fig. 7.43a–c Several important quantum confinement structures: (**a**) quantum well, (**b**) quantum wire, and (**c**) quantum dot

print fabrication, ii) epitaxy and strain engineering, iii) scanned probe techniques, iv) self-assembly and template manufacturing, and v) chemical techniques for fabricating nanoparticles and nanowires.

7.3.1 E–Beam and Nanoimprint Fabrication

In previous sections, we discussed several important lithography techniques commonly used in MEMS and microfabrication. These included various forms of UV (regular, deep, and extreme) and X-ray lithographies. However, due to the lack of resolution (in the case of UV) or the difficultlies in manufacturing mask and radiation sources (X-ray), these techniques are not suitable for nanoscale fabrication. E-beam lithography is an alternative and attractive technique for fabricating nanostructures [7.54]. It uses an electron beam to expose an electron-sensitive resist such as polymethyl methacrylate (PMMA) dissolved in trichlorobenzene (positive) or polychloromethylstyrene (negative) [7.55]. The e-beam gun is usually part of a scanning electron microscope (SEM), although transmission electron microscopes (TEM) can also be used. While electron wavelengths of the order of 1 Å are easily achieved, electron scat-

tering in the resist limits the attainable resolutions to >10 nm. The beam control and pattern generation is achieved through a computer interface. E-beam lithography is serial and so it has a low throughput. Although this is not a major concern when fabricating devices used in the study of fundamental microphysics, it severely limits large-scale nanofabrication. E-beam lithography, in conjunction with processes such as lift-off, etching and electro-deposition, can be used to fabricate various nanostructures.

An interesting new technique which circumvents the serial and low throughput limitations of e-beam lithography for fabricating nanostructures is known as nanoimprint technology [7.56]. This technique uses an e-beam-fabricated hard material master (or mold) to stamp and deform a polymeric resist. This is usually followed by a reactive ion etching step to transfer the stamped pattern to the substrate. This technique is economically superior since a single stamp can be used repeatedly to fabricate a large number of nanostructures. Figure 7.44 shows a schematic illustration of nanoimprint fabrication. First, a hard material (such as silicon or SiO_2) stamp is created using e-beam lithography and reactive ion etching. Then a resist-coated substrate is stamped and finally an anisotropic RIE is performed to remove the resist residue in the stamped area. At this stage, the process is complete and one can either etch the substrate, or if metallic nanostructures are desired, evaporate the metal and perform a lift-off.

The resist used in nanoimprint technology can be a thermoplastic, a UV-curable polymer, or a heat-curable

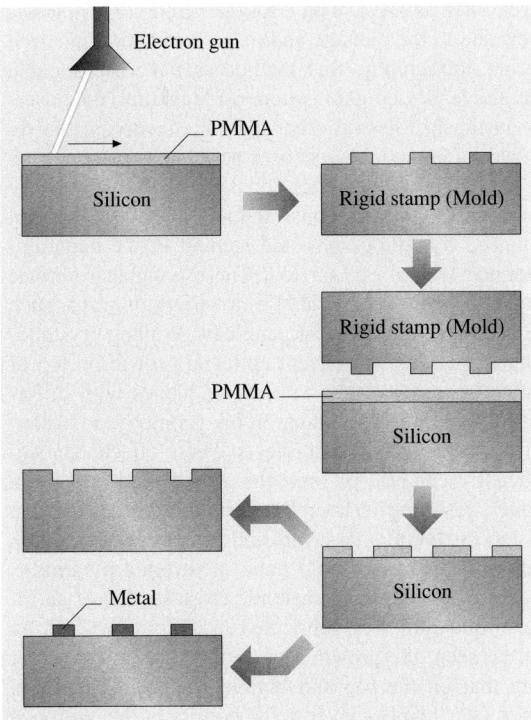

Fig. 7.44 Schematic of nanoimprint fabrication

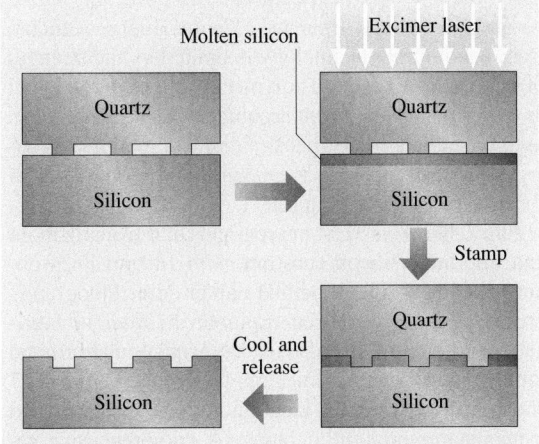

Fig. 7.45 Ultrafast silicon nanoimprinting using an excimer laser

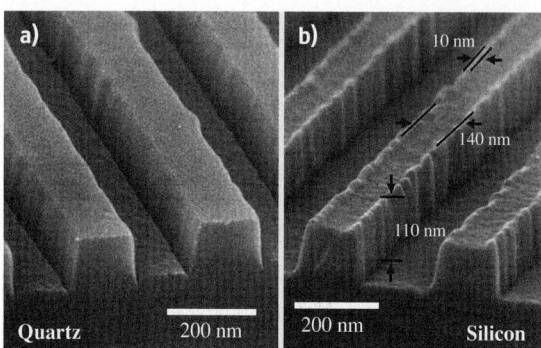

Fig. 7.46a,b SEM micrographs of: (**a**) quartz mold and (**b**) imprinted silicon surface using LADI [7.57]

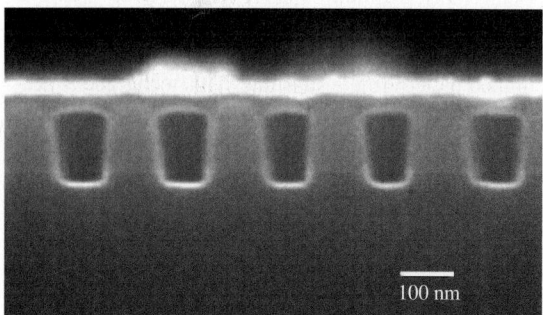

Fig. 7.47 Cross-section of a nanofluidic channel fabricated using nanoimprint lithography [7.60]

polymer. For a thermoplastic resist (such as PMMA), the substrate is heated to above the glass transition temperature (T_g) of the polymer before stamping and is cooled to below T_g before the stamp is removed. Similarly, the UV- and heat-curable resists are fully cured before the stamp is separated. The resolution available from nanoimprint technology is limited by the strength of both the mold and the polymer, and it can be as small as 10 nm. A nanoimprint technique has recently been used to stamp a silicon substrate in less than 250 ns using a XeCl excimer laser (308 nm) and a quartz mask (laser-assisted direct imprint, LADI), Fig. 7.45 [7.57]. Figure 7.46 shows SEM micrographs of the quartz mold and imprinted silicon substrate with 140 nm lines obtained using LADI. Modified nanoimprint lithography processes have also been explored in order to fabricate nanofluidic channels (for DNA manipulation) and multilayer polymeric structures [7.58, 59]. Figure 7.47 shows a cross-section of nanofluidic channels fabricated using nanoimprint lithography. A comprehensive review of this research area by *Guo* [7.60] was recently published.

7.3.2 Epitaxy and Strain Engineering

Atomic precision deposition techniques such as molecular beam epitaxy (MBE) and metallo-organic chemical vapor deposition (MOCVD) have proven to be effective tools for fabricating a variety of quantum confinement structures and devices (including quantum well lasers, photodetectors and resonant tunneling diodes) [7.61–63]. Although quantum wells and superlattices are the structures that lend themselves most easily to these techniques (see Fig. 7.43a), quantum wires and dots have also been fabricated by adding extra steps such as etching and selective growth. The fabrication of quantum well and superlattice structures using epitaxial growth is a mature and well developed field and is therefore not discussed in this section. Instead, we will concentrate on quantum wire and dot nanostructure fabrication using basic epitaxial techniques [7.64, 65].

Quantum Structure Nanofabrication Using Epitaxy on Patterned Substrates

There have been several different approaches to the fabrication of quantum wires and dots using epitaxial layers. The most straightforward technique involves e-beam lithography and etching of an epitaxially grown layer (such as InGaAs on GaAs substrate) [7.66]. However, due to the damage and/or contamination incurred during lithography, this method is not very suitable for active device fabrication (of quantum dot lasers for instance). Several other methods involving the regrowth of epitaxial layers over nonplanar surfaces such as step-edge, cleaved-edge and patterned substrate have been used to fabricate quantum wires and dots without the need for lithography and etching of the quantum-confined structure [7.65, 67]. These nonplanar surface templates can be fabricated in a variety of ways, such as etching through a mask, or cleavage along crystallographic planes. Subsequent epitaxial growth on top of these structures results in a set of planes with different growth rates depending on the geometry or surface diffusion and adsorption effects. These effects can significantly enhance or limit the growth rate on certain planes, resulting in lateral patterning and confinement of deposited epitaxial layers and formation of quantum wires (in V-grooves) and dots (in inverted pyramids). Figure 7.48a shows a schematic cross-section of an InGaAs quantum wire fabricated in a V-groove InP. As can be seen, the growth rate on the sidewalls is lower than that on the top and bottom surfaces. Therefore, the thicker InGaAs layer at the bottom of the V-groove forms a quantum wire confined from the sides by a thin-

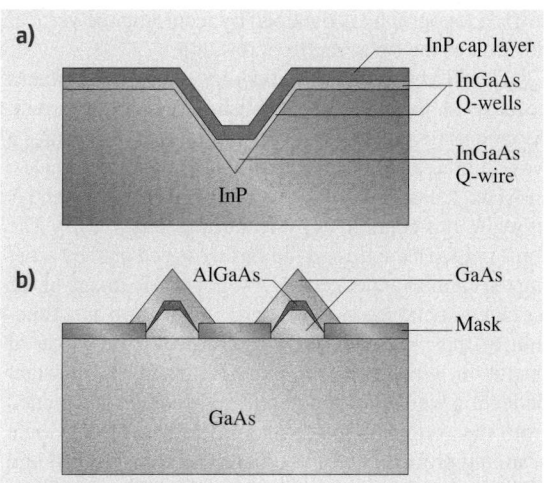

Fig. 7.48 (**a**) InGaAs quantum wire fabricated in a V-groove InP, and (**b**) AlGaAs quantum wire fabricated via epitaxial growth on a masked GaAs substrate

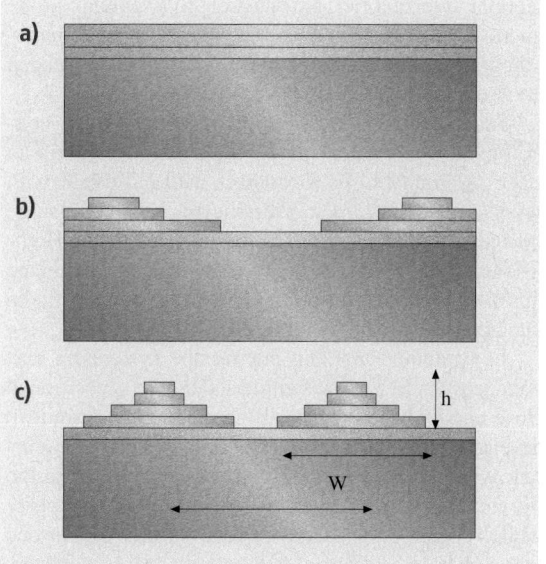

Fig. 7.49a–c Stranski–Krastanow growth mode, (**a**) 2-D wetting layer, (**b**) growth front roughening and break-up, and (**c**) coherent 3-D self-assembly

ner layer with a wider band-gap. Figure 7.48b shows a quantum wire formed using epitaxial growth over a dielectric patterned planar substrate. It is relatively easy to create quantum wells in either of these techniques; however, in order to create quantum wires and dots one still needs e-beam lithography to pattern the grooves and window templates.

Quantum Structure Nanofabrication Using Strain-Induced Self-Assembly

A more recent technique for fabricating quantum wires and dots involves strain-induced self-assembly [7.65, 68]. The term self-assembly represents a process where a strained 2-D system reduces its energy by changing into a 3-D morphology. The material combination most commonly used for this technique is the $In_xGa_{1-x}As/GaAs$ system, which offers a large lattice mismatch (7.2% between InAs and GaAs) [7.69, 70], although Ge dots on Si substrate have recently also attracted considerable attention [7.71]. This method relies on lattice mismatch between an epitaxially grown layer and its substrate, resulting in the formation of an array of quantum dots or wires. Figure 7.49 shows a schematic of the strain-induced self-assembly process. When the lattice constants of the substrate and the epitaxial layer differ markedly, only the first few monolayers deposited crystallize in the form of epitaxially strained layers in which the lattice constants are equal. When a critical thickness is exceeded, significant strain occurs in the layer, leading to the breakdown of this ordered struc-

ture and to the spontaneous formation of randomly distributed islets with regular shapes and similar sizes (usually < 30 nm in diameter). This mode of growth is usually referred to as the Stranski–Krastanow mode. The size, separation and height of the quantum dot depend on the deposition parameters (the total amount of deposited material, the growth rate, and the temperature) and material combinations. As can be seen, this is a very convenient method for growing perfect crystalline nanostructures over a large area without any lithography or etching. One major drawback of this technique is the randomness of the quantum dot distribution. It should be mentioned that this technique can also be used to fabricate quantum wires via strain relaxation bunching at the step edges.

7.3.3 Scanning Probe Techniques

The invention of scanning probe microscopy (SPM) in the 1980s revolutionized atomic-scale imaging and spectroscopy. In particular, scanning tunneling and atomic force microscopes (STM and AFM) have found widespread application in physics, chemistry, material science, and biology. The ability to perform atomic-scale manipulation, lithography and nanomachining using such probes was considered from the beginning and has matured considerably over the past decade. In this section, after a brief introduction to scanning probe microscopes, we will discuss several important nanolithography and machining techniques which have been used to create nanometer-sized structures.

Scanning probe microscopy systems involve controlling the movement of an atomically sharp tip in close proximity to or in contact with a surface with subnanometer accuracy. Piezoelectric positioners are typically used in order to achieve such accuracy. High-resolution images can be acquired by raster scanning the tips over a surface while simultaneously monitoring the interaction of the tip with the surface.

In scanning tunneling microscope systems, a bias voltage is applied to the sample and the tip is positioned close enough to the surface for a tunneling current to develop across the gap (Fig. 7.50a). Because this current is extremely sensitive to the distance between the tip and the surface, scanning the tip in the x-y plane while recording the tunnel current permits the surface topography to be mapped with atomic-scale resolution. In a more common mode of operation, the amplified current signal is connected to the z-axis piezoelectric positioner through a feedback loop, so that the current and therefore the distance is kept constant throughout the scanning. In this configuration, the picture of the surface topography is obtained by recording the vertical position of the tip at each x-y position.

The STM system only works for conductive surfaces because of the need to establish a tunneling current. Atomic force microscopy, on the other hand, provides a way to image conducting and nonconducting surfaces. In AFM, the tip is attached to a flexible cantilever and is brought into contact with the surface (Fig. 7.50b). The force between the tip and the surface is detected by sensing the cantilever deflection. A topographic image of the surface is obtained by plotting the deflection as a function of the x-y position. In a more common mode of operation, a feedback loop is used to maintain a constant deflection while the topographic information is obtained from the vertical displacement of the cantilever. Some scanning probe systems use a combination of AFM and STM modes: the tip is mounted in a cantilever with electrical connection so that both the surface force and the tunneling current are controlled or monitored. STM systems can be operated in ultrahigh vacuum (UHV STM) or in air, whereas AFM systems are typically operated in air. When a scanning probe system is operated in air, water adsorbed onto the sample surface accumulates underneath the tip, forming a meniscus between the tip and the surface. This water meniscus plays an important role in some of the scanning probe techniques described below.

Scanning Probe–Induced Oxidation
The local nanometer-scale oxidation of various materials can be achieved using scanning probes operated in

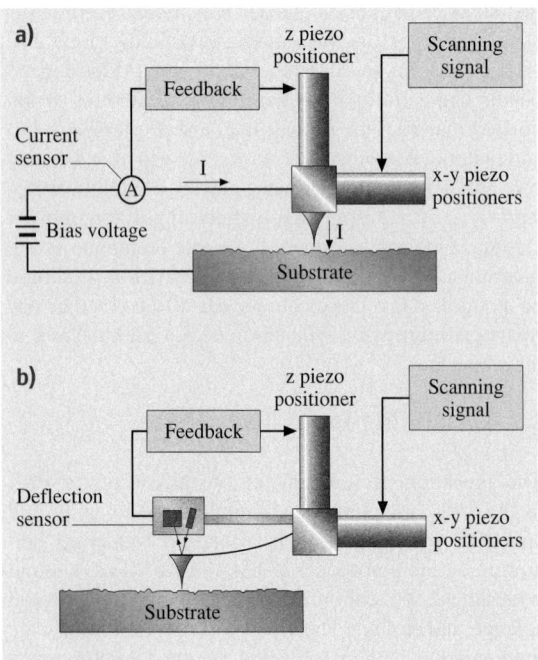

Fig. 7.50a,b Scanning probe systems: (**a**) STM and (**b**) AFM

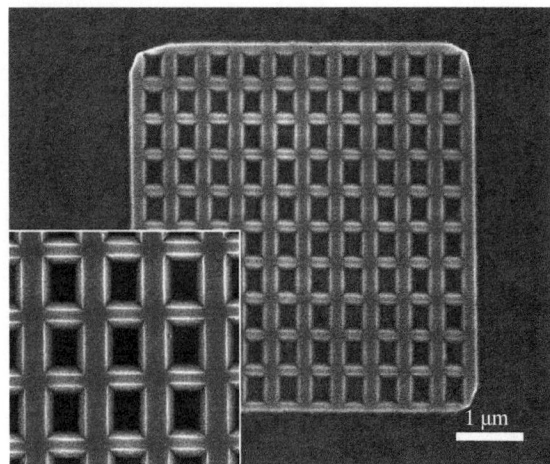

Fig. 7.51 SEM image of an inverted truncated pyramid array fabricated on a silicon SOI wafer by SPM oxidation, and subsequent etch in TMAH (pitch is 500 nm) [7.72]

air and biased at a sufficiently high voltage, Fig. 7.51. Tip biases of -2 to -10 V are normally used, with writing speeds of $0.1-100\,\mu m/s$ in an ambient humidity of 20%–40%. It is believed that the water meniscus formed at the contact point serves as an electrolyte such that the biased tip anodically oxidizes a small region of the surface [7.73]. The most common application of this principle is the oxidation of hydrogen-passivated silicon. To passivate the surface of the silicon with hydrogen atoms, it is often dipped in HF solution. Patterns of oxide "written" on a silicon surface can be used as a mask for wet or dry etching. Patterns with linewidths of 10 nm have been successfully transferred to a silicon substrate in this fashion [7.74]. Various metals have also been locally anodized (with aluminium or titanium for example) using this approach [7.75]. An interesting variant of this process is the anodization of deposited amorphous silicon [7.76]. Amorphous silicon can be deposited at low temperatures onto a variety of materials. The deposited silicon layer can be patterned and used as, for example, the gate of a $0.1\,\mu m$ CMOS transistor [7.77], or it can be used as a mask to pattern an underlying film. The major drawback of this technique is poor reproducibility due to tip wear during the anodization. However, the application of AFM performed in noncontact mode has overcome this problem [7.73].

Scanning Probe Resist Exposure and Lithography

Electrons emitted from a biased SPM tip can be used to expose a resist in a similar way to e-beam lithography (Fig. 7.52) [7.77]. Various systems have been used for this lithographic technique, including constant current STM, noncontact AFM and AFM with constant

tip–resist force and constant current. The systems that use AFM cantilevers have the advantage that imaging and alignment tasks can be performed without exposing the resist. Resists that are well characterized for e-beam lithography (such as PMMA or SAL601) have been used with scanning probe lithography to achieve reliable sub-100 nm lithography. The procedure for this process is as follows. The wafers are cleaned and the native oxide (for silicon or poly) is removed with a HF dip. A $35-100$ nm-thick resist is then spin-coated onto the surface. Exposure is achieved by moving the SPM tip over the surface while applying a bias voltage that is sufficiently high to produce electron emission from the tip (a few tens of volts). The resist is developed in a standard solution following the exposure. Features of less than 50 nm in width have been achieved with this procedure.

Dip-Pen Nanolithography

In dip-pen nanolithography (DPN), the tip of an AFM operated in air is "inked" with a chemical of interest and brought into contact with a surface. The ink molecules flow from the tip onto the surface, as with a fountain pen. The water meniscus that naturally forms between the tip and the surface enables the diffusion and transport of the molecules, as shown in Fig. 7.53. Inking can be done by dipping the tip in a solution containing a low concentration of the molecules followed by a drying step (blowing dry with compressed difluoroethane for instance). Linewidths down to 12 nm, with a spatial resolution of 5 nm, have been demonstrated with this technique [7.78]. Species patterned with DPN include conducting polymers, gold, dendrimers, DNA, organic dyes, antibodies, and alkanethiols. Alkanethiols have been also used as an organic monolayer mask to etch a gold layer and then etch the exposed silicon substrate. One can also use a heated AFM cantilever to control the deposition of a solid organic ink. This tech-

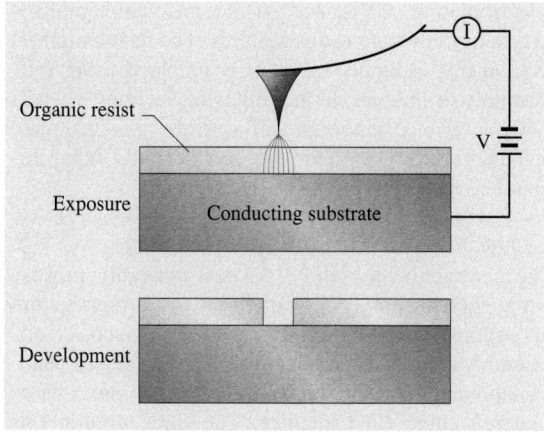

Fig. 7.52 Scanning probe lithography with organic resist

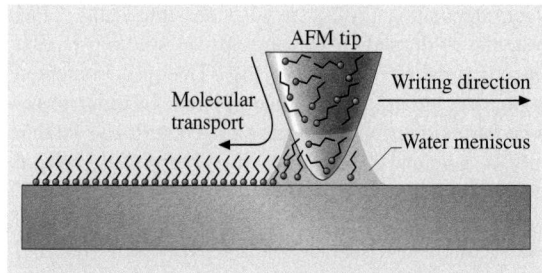

Fig. 7.53 Schematic of the working principles of dip-pen nanolithography

nique was recently reported by *Sheehan* et al. [7.79], where 100 nm lines of octadecylphosphonic acid (melting point: 100 °C) were written using a heated AFM probe [7.79].

Other Scanning Probe Nanofabrication Techniques

A great variety of nanofabrication techniques based on scanning probe systems have been demonstrated. Some of these are proof-of-concept demonstrations and so are yet to be evaluated as viable and repeatable fabrication processes. For example, a substrate can be mechanically machined using STM/AFM tips acting as plows or engraving tools [7.80]. This can be used to directly create structures in the substrate, although it is more commonly used to pattern a resist for a subsequent etch, lift-off or electrodeposition step. Mechanical nanomachining with SPM probes can be facilitated by heating the tip above the glass transition of a polymeric substrate material. This approach has been applied to SPM-based high-density data storage in polycarbonate substrates [7.81].

Electric fields strong enough to induce the emission of atoms from the tip can be easily generated by applying voltage pulses of more than 3 V. This phenomenon has been used to transfer material from the tip to the surface and vice versa. Ten to twenty nanometer mounds of metals such as Au, Ag or Pt have been deposited or removed from a surface in this fashion [7.82]. The same approach has been used to extract single atoms from a semiconductor surface and redeposit them elsewhere [7.83]. The manipulation of nanoparticles, molecules and single atoms on a surface has also been achieved by simply pushing or sliding them with the SPM tip [7.84]. Metals can also be deposited locally by the STM chemical vapor deposition technique [7.85]. In this technique, a precursor organometallic gas is introduced into the STM chamber. A voltage pulse applied between the tip and the surface dissociates the precursor gas into a thin layer of metal. Local electrochemical etching [7.86] and electrodeposition [7.87] are also possible using SPM systems. A droplet of the appropriate solution is first placed on the substrate. Then the STM tip is immersed in the droplet and a voltage is applied. In order to reduce faradaic currents, the tip is coated with wax so that only the very end is exposed to the solution. Sub-100 nm feature sizes have been achieved using this technique.

Using a single tip to serially produce the desired modification in a surface leads to very slow fabrication processes that are impractical for mass production. Many of the scanning probe techniques developed so far, however, could also be performed by an array of tips, which would increase their throughput and make them more competitive with other nanofabrication processes. This approach has been demonstrated for imaging, lithography [7.88] and data storage [7.89] using both one-dimensional and two-dimensional arrays of scanning probes. With the development of larger arrays, with individual advances in force, vertical position and current control, we may see these techniques being used in standard industrial fabrication processes.

7.3.4 Self–Assembly and Template Manufacturing

Self-assembly is a nanofabrication technique that involves aggregation of colloidal nanoparticles into the final desired structure [7.90]. This aggregation can be either spontaneous (entropic) and due to thermodynamic minima (energy minimization) constraints, or chemical and due to the complementary binding of organic molecules and supramolecules (molecular self-assembly) [7.91]. Molecular self-assembly is one of the most important techniques used for the development of complex functional structures in biology. Since these techniques require that the target structures be thermodynamically stable, it tends to produce structures that are relatively defect-free and self-healing. Self-assembly is by no means limited to molecules or the nanodomain, and it can be carried out on just about any scale, making it a powerful bottom-up assembly and manufacturing method (multiscale ordering). Another attractive feature of this technique relates to the possibility of combining self-assembly properties of organic molecules with the electronic, magnetic, and photonic properties of inorganic components. Template manufacturing is another bottom-up technique which utilizes material deposition (electroplating, CVD, and so on) into nanotemplates in order to fabricate nanostructures. The nanotemplates used in this technique are usually prepared using self-assembly techniques. In the following sections, we will discuss various important self-assembly and template manufacturing techniques that are currently being researched extensively.

Physical and Chemical Self–Assembly

The central theme behind the self-assembly process is the spontaneous (physical) or chemical aggregation of colloidal nanoparticles [7.92]. Spontaneous self-assembly exploits the tendency of monodispersed nano- or submicrocolloidal spheres to organize into a face-centered cubic (FCC) lattice. The force driving this process is the desire of the system to achieve a thermodynamically stable state (minimum free energy). In addition to spontaneous thermal self-assembly, gravitational, convective and electrohydrodynamic forces can also be used to induce aggregation into complex 3-D structures. Chemical self-assembly requires the attachment of a single molecular organic layer (self-assembled monolayer or SAM) to the colloidal particles (organic or inorganic) and subsequent self-assembly of these components into a complex structures using molecular recognition and binding.

Physical Self-Assembly. This is an entropy-driven method that relies on the spontaneous organization of colloidal particles into a relatively stable structure through noncovalent interactions. For example, colloidal polystyrene spheres can be assembled into a 3-D structure on a substrate that is held vertically in the colloidal solution, Fig. 7.54 [7.93,94]. Upon the evaporation of the solvent, the spheres aggregate into a hexagonal close-packed (HCP) structure. The interstitial pore size and density are determined by the polymer sphere size. The polymer spheres can be etched into smaller sizes after forming the HCP arrays, thereby altering the template pore separations [7.95]. This technique can fabricate large patterned areas in a quick, simple and cost-effective way. A classic example is the natural assembly of on-chip silicon photonic band-gap crystals [7.93] that are capable of reflecting the light arriving in any direction over a certain wavelength range [7.96]. In this method, a thin layer of silica colloidal spheres are assembled on a silicon substrate. This is achieved by placing a silicon wafer vertically in a vial containing an ethanolic suspension of silica spheres. A temperature gradient across the

Fig. 7.55 Cross-sectional SEM image of a thin planar opal silica template (spheres are 855 nm in diameter) assembled directly onto a Si wafer [7.93]

vial aids the flow of silica spheres. Figure 7.55 shows the cross-sectional SEM image of a thin planar opal template assembled directly on a Si wafer from 855 nm spheres. Once such a template is prepared, LPCVD can used to fill the interstitial spaces with Si, so that the high refractive index of silicon provides the necessary bandgap.

One can also deposit colloidal particles into a patterned substrate (template-assisted self-assembly, TASA) [7.97, 98]. This method is based on the principle that when an aqueous dispersion of colloidal particles is allowed to dewet from a solid surface that is already patterned, the colloidal particles are trapped by the recessed regions and assembled into aggregates of shapes and sizes determined by the geometric confinement provided by the template. The patterned arrays of templates can be fabricated using conventional contact-mode photolithography which provides control over the shapes and dimensions of the templates, thereby allowing the assembly of complex structures from colloidal particles.

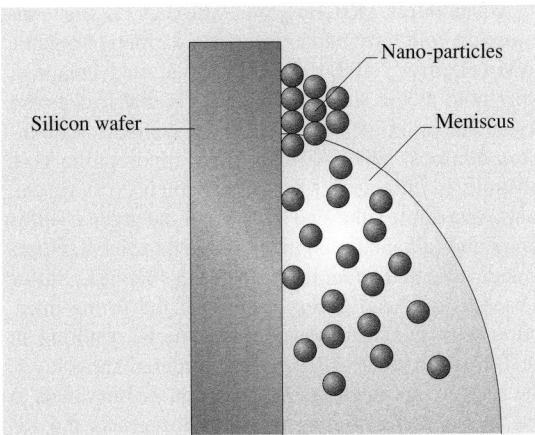

Fig. 7.54 Colloidal particle self-assembly onto solid substrates upon drying in vertical position

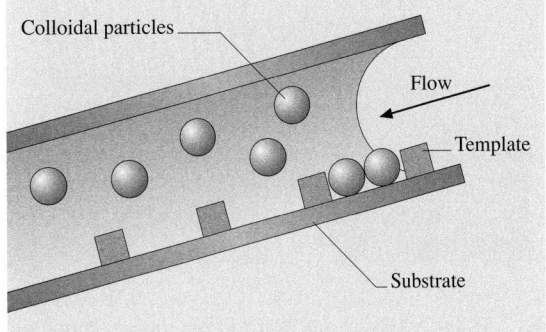

Fig. 7.56 A cross-sectional view of the fluidic cell used for template-assisted self-assembly

Part A | 7.3

A cross-sectional view of a fluidic cell used in TASA is shown in Fig. 7.56. The fluidic cell has two parallel glass substrates to confine the aqueous dispersion of the colloidal particles. The surface of the bottom substrate is patterned with a 2-D array of templates. When the aqueous dispersion is allowed to slowly dewet across the cell, the capillary force exerted on the liquid pushes the colloidal spheres across the surface of the bottom substrate until they are physically trapped by the templates. If the concentration of the colloidal dispersion is high enough, the template will be filled by the maximum number of colloidal particles determined by the geometrical confinement. This method can be used to fabricate a variety of polygonal and polyhedral aggregates that are otherwise difficult to generate [7.99].

Chemical Self-Assembly. Organic and supramolecular SAMs play a critical role in colloidal particle self-assembly. SAMs are robust organic molecules that are chemically adsorbed onto solid substrates [7.100]. They often have a hydrophilic (polar) head which can be bonded to various solid surfaces and a long hydrophobic (nonpolar) tail which extends outward. SAMs are formed by the immersion of a substrate in a dilute solution of the molecule in an organic solvent. The resulting film is a dense organization of molecules arranged to expose the end group. The durability of the SAM is highly dependent on the strength of the anchoring to the surface of the substrate. SAMs have been widely studied because the end group can be functionalized to form precisely arranged molecular arrays for various applications ranging from simple, ultrathin insulators and lubricants to complex biological sensors. Chemical self-assembly uses organic or supramolecular SAMs as the binding and recognition sites for fabricating complex 3-D structures from colloidal nanoparticles. The most commonly used organic monolayers include: 1) organosilicon compounds on glass and native surface oxide layers of silicon, 2) alkanethiols, dialkyl disulfides and dialkyl sulfides on gold, 3) fatty acids on alumina and other metal oxides, and 4) DNA.

Octadecyltrichlorosilane (OTS) is the most common organosilane used in the formation of SAMs mainly due to the fact that it is simple, readily available and forms good, dense layers [7.101, 102]. Alkyltrichlorosilane monolayers can be prepared on clean silicon wafers with SiO_2 on the surface (with almost 5×10^{14} $SiOH$ groups/cm^2). Figure 7.57 shows a schematic representation of the formation of alkylsiloxane monolayers by the adsorption of alkyltrichlorosilanes from solution onto Si/SiO$_2$ substrates. Since the silicon–chlorine bond is susceptible to hydrolysis, the amount of water in the system must be limited in order to obtain good quality monolayers. Monolayers made of methyl- and vinyl-terminated alkylsilanes are autophobic to the hydrocarbon solution and hence emerge uniformly dry from the solution, whereas monolayers made of ester-terminated alkylsilanes emerge wet from the solution used in their formation. The disadvantage of this method is that a cloudy film is deposited on the surface (due to formation of a gel of polymeric siloxane) if the alkyltrichlorosilane in the solvent adhering to the substrate is exposed to water.

Alkanethiols ($X(CH_2)_nSH$, where X is the end-group) on gold form another important group of organic SAM systems [7.100, 103–105]. A major advantage of using gold as the substrate material is that it does not have a stable oxide and so it can be handled in ambient conditions. When a fresh, clean, hydrophilic gold substrate is immersed (for several minutes to several hours) in a dilute solution (10^{-3} M) of the organic sulfur compound (alkanethiols) in an inorganic solvent, close-packed, oriented monolayers can be obtained. Sulfur is used as the head group because of the strong interaction with gold substrate (44 kcal/mol), resulting in the formation of a close-packed, ordered monolayer. The end-group of the alkanethiol can be modified to give hydrophobic or hydrophilic properties to the adsorbed layer. Another method for depositing alkanethiol SAM is soft lithography. This technique is based on ink-

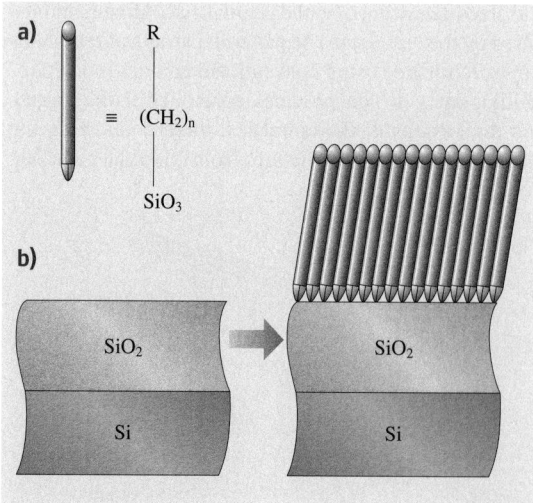

Fig. 7.57 (a) Alkylsiloxane formed from the adsorption of alkyltrichlorosilane onto Si/SiO$_2$ substrates. **(b)** Schematic representation of the process

ing a PDMS stamp with alkanethiol and its subsequent transfer to planar and nonplanar substrates. Alkanethiol-functionalized surfaces (planar, nonplanar, spherical) can also be used to self-assemble a variety of intricate 3-D structures [7.106].

Carboxylic acid derivatives self-assemble on surfaces (such as glass, Al_2O_3 and Ag_2O) through an acid-base reaction, giving rise to monolayers of fatty acids [7.107]. The time required for the formation of a complete monolayer increases with decreased concentration. Higher concentrations of the carboxylic acid are required to form a monolayer on gold than on Al_2O_3. This is due to the different affinities of the COOH group to these substances (more affinity to Al_2O_3 and glass than to gold) and also the surface concentrations of the salt-forming oxides in the two substrates. In the case of amorphous metal oxide surfaces, the chemisorption of alkanoic acids is not unique. For example, on Ag_2O, the two carboxylate oxygen atoms bind to the substrate in an almost symmetrical manner, resulting in ordered monolayers with a chain tilt angle from the surface normal of $15°$ to $25°$. However, on CuO and Al_2O_3, the oxygen atoms bind themselves symmetrically and the chain tilt angle is close to $0°$. The structure of the monolayers is thus the result of a balance between the various interactions taking place in the polymer chains.

Deoxyribonucleic acid (DNA) – the framework on which all life is built – can be used to self-assemble nanomaterials into useful macroscopic aggregates that display a number of desired physical properties [7.108]. DNA consists of two strands that are coiled around each other to form a double helix. Singular strands of nucleotides are left when the two strands are uncoiled . These nucleotides consist of a sugar (a pentose ring), a phosphate (PO_4) and a nitrogenous base. The correct order and architecture of these components is essential to achieving the proper structure of a nucleotide. Four nucleotides are typically found in DNA, adenine (A), guanine (G), cytosine (C), and thymine (T). A key property of the DNA structure is that the nucleotides described bind specifically to another nucleotide when arranged in the two-strand double helix (A to T, and C to G). This specific bonding capability can be used to assemble nanophase material and nanostructures [7.109]. For example, nucleotide-functionalized nanogold particles have been assembled into complex 3-D structures by attaching DNA strands to the gold via an enabler or linker [7.110]. In separate work, DNA was used to assemble nanoparticles into macroscopic materials. This method uses alkane dithiol as the linker molecule to connect the DNA template to the nanoparticle. The thiol groups at each end of the linker molecule covalently attach themselves to the colloidal particles to form aggregate structures [7.111].

Template Manufacturing

Template manufacturing refers to a set of techniques that can be used to fabricate organic or inorganic 3-D structures from a nanotemplate. These templates differ in material, pattern, feature size, overall template size and periodicity. Although nanotemplates can be fabricated using e-beam lithography, the serial nature of this technique prohibits its widespread application. Self-assembly is the preferred technique since it can produce large-area nanotemplates in a massively parallel fashion. Several nanotemplates have been investigated for use in template manufacturing. These include polymer colloidal spheres, alumina membranes, and nuclear track-etched membranes. Colloidal spheres can be deposited in a regular 3-D array using the techniques described in the previous section (see Figs. 7.54–7.56). Porous aluminium oxide membranes can be fabricated by the anodic oxidation of aluminium [7.112]. The oxidized film consists of columnar arrays of hexagonal close-packed pores separated at distances comparable to the pore size. By controlling the electrolyte species, temperature, anodizing voltage and time, different pore sizes, densities and heights can be obtained. The pore size and depth can also be adjusted by etching the oxide in an appropriate acid. Templates of porous polycarbonate or mica membranes can be fabricated by nuclear track-etched membranes [7.113]. This technique is based on the passage of high-energy decay fragments from a radioactive source through a dielectric material. The particles leave behind chemically active damaged tracks which can then be etched to create pores throughout the thickness of the membrane [7.114, 115]. Unlike the other methods, the pore separation and hence the pore density is independent of the pore size. The pore density is only determined by the irradiation process.

Subsequent to template fabrication, the interstitial spaces (in the case of colloidal spheres) or pores (for alumina and polycarbonate membranes) in the template are filled with the desired material [7.95, 116]. This can be achieved using a variety of deposition techniques, such as electroplating and CVD. The final structure can be a composite of nanotemplate and deposited material, or the template can be selectively etched resulting in an air-filled 3-D complex structure. For example, nickel [7.117], iron [7.118] and cobalt [7.119] nanowires have been electrochemically grown in porous template matrices. Three-dimensional photonic crystals

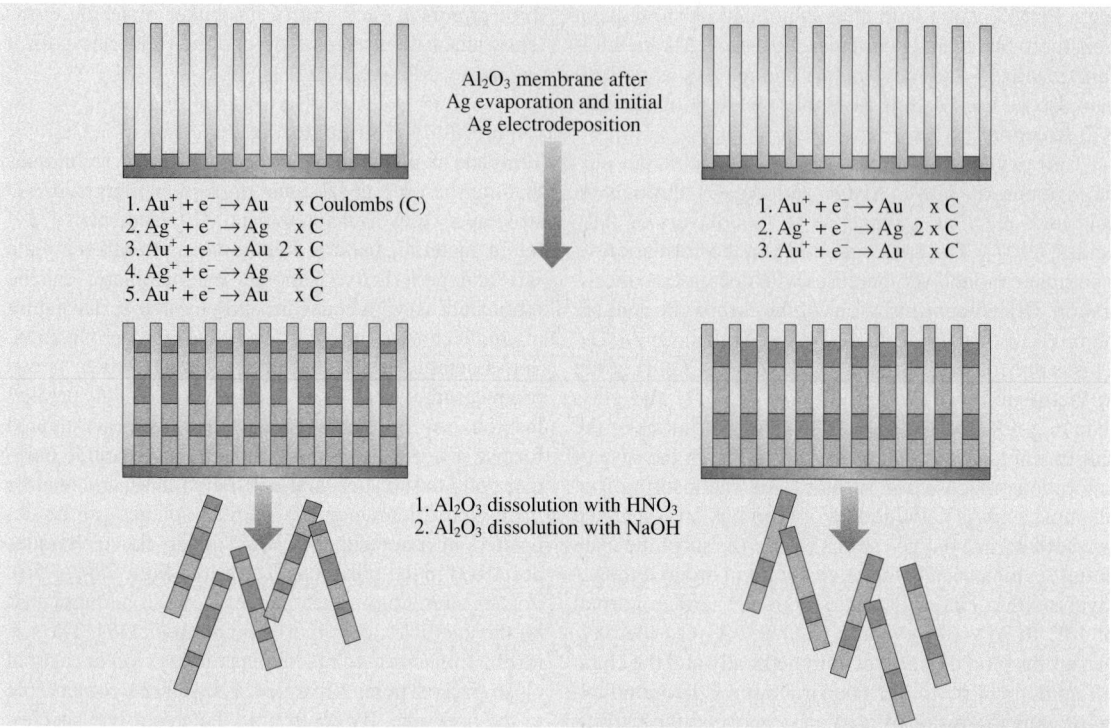

Fig. 7.58 Synthesis of nanobarcode particles

have been fabricated by the electrochemical deposition of CdSe and silicon into polystyrene and silica colloidal assembly templates [7.93, 120]. An interesting example of template-assisted manufacturing is the synthesis of nanometer-sized metallic barcodes [7.121]. These nanobarcodes are prepared by the electrochemical reduction of metallic ions into the pores of an aluminium oxide membrane followed by their release through the etching of the template [7.122–124]. This procedure is schematically illustrated in Fig. 7.58. A back-side silver film is used as the working electrode for the reduction of metallic ions (silver and gold in this case) from solution. Up to seven different metallic segments as short as 10 nm and as long as several micrometers, with 13 distinguishable stripes, have been fabricated using this technique. Optical reflectivity is used to read out the stripe pattern encoded in the metal particles [7.121]. Figure 7.59 shows optical and field emission scanning electron microscope images of a Au-Ag multistripe nanobarcode

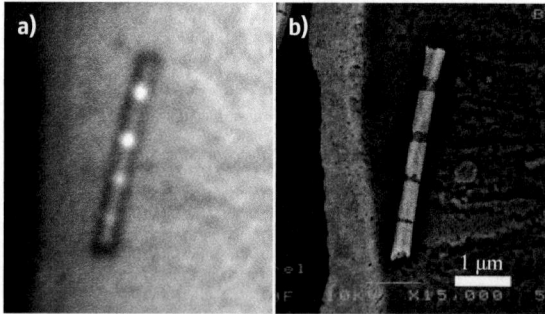

Fig. 7.59 (a) Optical and **(b)** FE-SEM images of Au-Ag multistripe particles [7.121]

(Ag stripes ranging in length from 60 to 240 nm separated by Au segments of 550 nm can be seen). These coded nanoparticles can be used in fluorescence and mass spectrometry-based assays, enabling a wide variety of bioanalytical measurements to be taken.

7.4 Summary and Conclusions

In this chapter, we have discussed various micro/nanofabrication techniques used to manufacture structures covering a wide range of dimensions (mm–nm). Starting with some of the most common microfabrication techniques (lithography, deposition and etching), we presented an array of micromachining and MEMS technologies that can be used to fabricate microstructures down to $\approx 1\,\mu m$. These techniques have attained an adequate level of maturity to allow for a variety of MEMS-based commercial products (pressure sensors, accelerometers, gyroscopes, and so on). More recently, nanometer-sized structures have attracted an enormous amount of interest. This is mainly due to their unique electrical, magnetic, optical, thermal and mechanical properties. These could lead to a variety of electronic, photonic and sensing devices with superior performance compared to their macro counterparts. Subsequent to our discussion on MEMS and micromachining, we presented several important nanofabrication techniques currently under intense investigation. Although e-beam and other high-resolution lithographies can be used to fabricate nanometer-size structures, their serial nature and/or cost preclude their widespread application. This has forced investigators to explore alternative and potentially superior techniques such as strain engineering, self-assembly, and nanoimprint lithography. Among these, self-assembly is the most promising method, due to its low cost and its ability to produce nanostructures at different length scales.

References

7.1   S. A. Campbell: *The Science and Engineering of Microelectronic Fabrication* (Oxford Univ. Press, New York 2001)

7.2   C. J. Jaeger: *Introduction to Microelectronic Fabrication* (Prentice Hall, New Jersey 2002)

7.3   J. D. Plummer, M. D. Deal, P. B. Griffin: *Silicon VLSI Technology* (Prentice-Hall, New Jersey 2000)

7.4   J. E. Bjorkholm: EUV lithography: the successor to optical lithography, Intel Technol. J. **2**, 1–8 (1998)

7.5   H. U. Krebs, M. Störmer, J. Faupel, E. Süske, T. Scharf, C. Fuhse, N. Seibt, H. Kijewski, D. Nelke, E. Panchenko, M. Buback: Pulsed laser deposition (PLD) – a versatile thin film technique, Adv. Solid State Phys. **43**, 505–517 (2003)

7.6   M. Leskela, M. Ritala: Atomic layer deposition chemistry: recent developments and future challenges, Angew. Chem. Int. Ed. **42**, 5548–5554 (2003)

7.7   J. L. Vossen: *Thin Film Processes* (Academic, New York 1976)

7.8   M. Gad-el-Hak: *The MEMS Handbook* (CRC, Boca Raton 2002)

7.9   T-R. Hsu: *MEMS and Microsystems Design and Manufacture* (McGraw-Hill, New York 2002)

7.10 G. T. A. Kovacs: *Micromachined Transducers Sourcebook* (McGraw-Hill, New York 1998)

7.11 P. Rai-Choudhury: *Handbook of Microlithography, Micromachining and Microfabrication. Volume 2: Micromachining and Microfabrication* (SPIE, IEE, Bellingham 1997)

7.12 T. J. Cotler, M. E. Elta: Plasma-etch technology, IEEE Circuits Devices Mag. **6**, 38–43 (1990)

7.13 U. Gosele, Q. Y. Tong: Semiconductor wafer bonding, Annu. Rev. Mater. Sci **28**, 215–241 (1998)

7.14 Q. Y. Tong, U. Gosele: *Semiconductor Wafer Bonding: Science and Technology* (Wiley, New York 1999)

7.15 F. Niklaus, P. Enoksson, E. Kalveston, G. Stemme: Void-free full-wafer adhesive bonding, J. Micromech. Microeng. **11**, 100–107 (2000)

7.16 C. A. Harper: *Electronic Packaging and Interconnection Handbook* (McGraw-Hill, New York 2000)

7.17 W. H. Ko, J. T. Suminto, G. J. Yeh: Bonding techniques for microsensors. In: *Micromachining and Micropackaging for Transducers*, ed. by W. H. Ko (Elsevier, Amsterdam 1985)

7.18 G. T. A. Kovacs, N. I. Maluf, K. A. Petersen: Bulk micromachining of silicon, Proc. IEEE **86**, 1536–1551 (1998)

7.19 K. Najafi, K. D. Wise, T. Mochizuki: A high-yield IC-compatible multichannel recording array, IEEE Trans. Electron Devices **32**, 1206–1211 (1985)

7.20 A. Selvakumar, K. Najafi: A high-sensitivity z-axis capacitive silicon microaccelerometer with a tortional suspension, J. Microelectromech. Sys. **7**, 192–200 (1998)

7.21 H. Baltes, O. Paul, O. Brand: Micromachined thermally based CMOS microsensors, Proc. IEEE **86**, 1660–1678 (1998)

7.22 B. Eyre, K. S. J. Pister, W. Gekelman: Multi-axis microcoil sensors in standard CMOS, Proc. SPIE Conf. Micromachined Devices and Components, Austin, TX, 183–191 (1995)

7.23 K. A. Shaw, Z. L. Zhang, N. C. MacDonald: SCREAM: a single mask, single-crystal silicon process for microelectromechanical structures, Proc. IEEE Workshop Microelectromechanical Systems, Fort Lauderdale, 155–160 (1993)

Part A | 7

7.24   G. K. Fedder, S. Santhanam, M. L. Reed, S. C. Eagle, D. F. Guillo, M. S. C. Lu, L. R. Carley: Laminated high-aspect-ratio microstructures in a conventional CMOS process, Proc. IEEE Workshop Micro Electro Mechanical Systems, San Diego, CA, 13–18 (1996)

7.25   N. C. MacDonald: SCREAM Microelectromechanical systems, Microelectron. Eng. **32**, 51–55 (1996)

7.26   X. Huikai, L. Erdmann, Z. Xu, K. J. Gabriel, G. K. Fedder: Post-CMOS processing for high-aspect-ratio integrated silicon microstructures, J. Microelectromech. Sys. **11**, 93–101 (2002)

7.27   B. P. Van Drieenhuizen, N. I. Maluf, I. E. Opris, G. T. A. Kovacs: Force-balanced accelerometer with mG resolution fabricated using silicon fusion bonding and deep reactive ion etching, Proc. Int. Conf. Solid-State Sensors and Actuators, Chicago, 1229–1230 (1997)

7.28   J. M. Bustillo, R. S. Muller: Surface micromachining for microelectromechanical systems, Proc. IEEE **86**, 1552–1574 (1998)

7.29   H. C. Nathanson, W. E. Newell, R. A. Wickstrom, J. R. Davis: The resonant gate transistor, IEEE Trans. Electron Devices **14**, 117–133 (1967)

7.30   R. T. Howe, R. S. Muller: Polycrystalline silicon micromechanical beams, Proc. Electrochemical Soc., Spring Meeting, Montreal, 184–185 (1982)

7.31   J. A. Geen, S. J. Sherman, J. F. Chang, S. R. Lewis: Single-chip surface-micromachined integrated gyroscope with 50 degrees /hour root Allan variance, IEEE J. Solid-St. Circ. **37**, 1860–1866 (2002)

7.32   P. F. Van Kessel, L. J. Hornbeck, R. E. Meier, M. R. Douglass: A MEMS-based projection display, Proc. IEEE **86**, 1687–1704 (1998)

7.33   A. E. Franke, D. Bilic, D. T. Chang, P. T. Jones, R. T. Howe, G. C. Johnson: Post-CMOS integration of germanium microstructures, Proc. Micro Electro Mechanical Systems, Orlando, FL, 630–637 (1999)

7.34   S. Sedky, P. Fiorini, M. Caymax, S. Loreti, K. Baert, L. Hermans, R. Mertens: Structural and mechanical properties of polycrystalline silicon germanium for micromachining applications, J. Microelectromech. Sys. **7**, 365–372 (1998)

7.35   N. Tas, T. Sonnenberg, H. Jansen, R. Legtenberg, M. Elwenspoek: Stiction in surface micromachining, J. Micromech. Microeng. **6**, 385–397 (1996)

7.36   R. Maboudian, R. T. Howe: Critical review: adhesion in surface micromechanical structures, J. Vacuum Sci. Tech. B **15**, 1–20 (1997)

7.37   J. H. Smith, S. Montague, J. J. Sniegowski, J. R. Murray, P. J. McWhorter: Embedded micromechanical devices for the monolithic integration of MEMS with CMOS, Proc. Int. Electron Devices Meeting, Washington, 609–612 (1995)

7.38   R. S. Muller, K. Y. Lau: Surface-micromachined microoptical elements and systems, Proc. IEEE **86**, 1705–1720 (1998)

7.39   K. S. J. Pister, M. W. Judy, S. R. Burgett, R. S. Fearing: Microfabricated hinges: 1 mm vertical features with surface micromachining, Proc. 6 Int. Conf. Solid-State Sensors and Actuators, San Francisco, 647–650 (1991)

7.40   L. Y. Lin, S. S. Lee, M. C. Wu, K. S. J. Pister: Micromachined integrated optics for free space interconnection, Proc. IEEE Micro-Electromechanical Systems Workshop, Amsterdam, 77–82 (1995)

7.41   E. W. Becker, W. Ehrfeld, P. Hagmann, A. Maner, D. Munchmeyer: Fabrication of microstructures with high aspect ratios and great structural heights by synchrotron radiation lithography galvanoforming, and plastic molding (LIGA process), Microelectron. Eng. **4**, 35–56 (1986)

7.42   H. Guckel Guckel: High-aspect-ratio micromachining via deep X-ray lithography, Proc. IEEE **86**, 1586–1593 (1998)

7.43   K. Y. Lee, N. LaBianca, S. A. Rishton, S. Zolgharnain, J. D. Gelorme, J. Shaw, T. H. P. Chang: Micromachining applications of a high resolution ultra-thick photoresist, J. Vacuum Sci. Tech. B **13**, 3012–3016 (1995)

7.44   K. Roberts, F. Williamson, G. Cibuzar, L. Thomas: The fabrication of an array of microcavities utilizing SU-8 photoresist as an alternative 'LIGA' technology, Proc. 13th Biennial University/Government/Industry Microelectronics Symp. (IEEE), Minneapolis, 139–141 (1999)

7.45   D. A. Horsley, M. B. Cohn, A. Singh, R. Horowitz, A. P. Pisano: Design and fabrication of a angular microactuator for magnetic disk drives, J. Microelectromech. Sys. **7**, 141–148 (1998)

7.46   C. Burbaum, J. Mohr, P. Bley, W. Ehrfeld: Fabrication of capacitive acceleration sensors by the LIGA technique, Sensor. Actuat. A **A27**, 559–563 (1991)

7.47   C. G. Keller, R. T. Howe: Hexsil bimorphs for vertical actuation. In: *Digest of Technical Papers, 8th Int. Conf. on Solid-State Sensors and Actuators and Eurosensors IX* (IEEE, Stockholm 1995) pp. 99–102

7.48   C. G. Keller, R. T. Howe: Nickel-filled hexsil thermally actuated tweezers. In: *Digest of Technical Papers, 8th Int. Conf. on Solid-State Sensors and Actuators and Eurosensors IX* (IEEE, Stockholm 1995) pp. 376–379

7.49   N. Yazdi, F. Ayazi: Micromachined inertial sensor, Proc. IEEE **86**, 1640–1659 (1998)

7.50   F. Ayazi, K. Najafi: High aspect-ratio combined poly and single-crystal silicon (HARPSS) MEMS technology, J. Microelectromech. Sys. **9**, 288–294 (2000)

7.51   F. Ayazi, K. Najafi: A HARPSS polysilicon vibrating ring gyroscope, J. Microelectromech. Sys. **10**, 169–179 (2001)

7.52   Y. S. No, F. Ayazi: The HARPSS process for fabrication of nano-precision silicon electromechanical resonators, Proc. 1st IEEE Conf. on Nanotechnology, Maui, 489–494 (2001)

7.53   G. Timp: *Nanotechnology* (Springer, Berlin Heidelberg 1998)

7.54   P. Rai-Choudhury: *Handbook of Microlithography, Micromachining and Microfabrication. Volume 1: Microlithography* (SPIE, IEE, Bellingham 1997)

7.55   L. Ming, C. Bao-qin, Y. Tian-Chun, Q. He, X. Qiuxia: The sub-micron fabrication technology, Proc. 6th Int. Conf. on Solid-State and Integrated-Circuit Technology, San Francisco, 452–455 (2001)

7.56   S. Y. Chou: Nano-imprint lithography and lithographically induced self-assembly, MRS Bull. **26**, 512–517 (2001)

7.57   S. Y. Chou, C. Keimel, J. Gu: Ultrafast and direct imprint of nanostructures in silicon, Nature **417**, 835–837 (2002)

7.58   H. Cao, Z. Yu, J. Wang, J. O. Tegenfeldt, R. H. Austin, E. Chen, W. Wu, S. Y. Chou: Fabrication of 10 nm enclosed nanofluidic channels, Appl. Phys. Lett. **81**, 174–176 (2002)

7.59   L. R. Bao, X. Cheng, X. D. Huang, L. J. Guo, S. W. Pang, A. F. Yee: Nanoimprinting over topography and multilayer three-dimensional printing, J. Vacuum. Sci. Tech. B **20**, 2881–2886 (2002)

7.60   L. J. Guo: Recent progress in nanoimprint technology and its applications, J. Phys. D **137**, R123–R141 (2004)

7.61   M. A. Herman: *Molecular Beam Epitaxy: Fundamentals and Current Status* (Springer, Berlin Heidelberg 1996)

7.62   J. S. Frood, G. J. Davis, W. T. Tsang: *Chemical Beam Epitaxy and Related Techniques* (Wiley, New York 1997)

7.63   S. Mahajan, K. S. Sree Harsha: *Principles of Growth and Processing of Semiconductors* (McGraw-Hill, New York 1999)

7.64   S. Kim, M. Razegi: Advances in quantum dot structures. In: *Processing and Properties of Compound Semiconductors*, ed. by Willardson, Navawa (Academic, New York 2001)

7.65   D. Bimberg, M. Grundmann, N. N. Ledentsov: *Quantum Dot Heterostructures* (Wiley, New York 1999)

7.66   G. Seebohm, H. G. Craighead: Lithography and patterning for nanostructure fabrication. In: *Quantum Semiconductor Devices and Technologies*, ed. by T. P. Pearsall (Kluwer, Boston 2000)

7.67   E. Kapon: Lateral patterning of quantum well heterostructures by growth on nonplanar substrates. In: *Epitaxial Microstructures*, ed. by A. C. Gossard (Academic, New York 1994)

7.68   F. Guffarth, R. Heitz, A. Schliwa, O. Stier, N. N. Ledentsov, A. R. Kovsh, V. M. Ustinov, D. Bimberg: Strain engineering of self-organized InAs quantum dots, Phys. Rev. B **64**, 085305(1)–085305(7) (2001)

7.69   M. Sugawara: *Self-Assembled InGaAs/GaAs Quantum Dots* (Academic, New York 1999)

7.70   B. C. Lee, S. D. Lin, C. P. Lee, H. M. Lee, J. C. Wu, K. W. Sun: Selective growth of single InAs quantum dots using strain engineering, Appl. Phys. Lett. **80**, 326–328 (2002)

7.71   K. Brunner: Si/Ge nanostructures, Rep. Prog. Phys **65**, 27–72 (2002)

7.72   F. S. S. Chien, W. F. Hsieh, S. Gwo, A. E. Vladar, J. A. Dagata: Silicon nanostructures fabricated by scanning probe oxidation and tetra-methyl ammonium hydroxide etching, J. Appl. Phys. **91**, 10044–10050 (2002)

7.73   M. Calleja, J. Anguita, R. Garcia, K. Birkelund, F. Perez-Murano, J. A. Dagata: Nanometer-scale oxidation of silicon surfaces by dynamic force microscopy: reproducibility, kinetics, nanofabrication, Nanotechnology **10**, 34–38 (1999)

7.74   E. S. Snow, P. M. Campbell, F. K. Perkins: Nanofabrication with proximal probes, Proc. IEEE **85**, 601–611 (1997)

7.75   H. Sugimura, T. Uchida, N. Kitamura, H. Masuhara: Tip-induced anodization of titanium surfaces by scanning tunneling microscopy: a humidity effect on nanolithography, Appl. Phys. Lett. **63**, 1288–1290 (1993)

7.76   N. Kramer, J. Jorritsma, H. Birk, C. Schonenberger: Nanometer lithography on silicon and hydrogenated amorphous silicon with low energy electrons, J. Vacuum Sci. Tech. B **13**, 805–811 (1995)

7.77   H. T. Soh, K. W. Guarini, C. F. Quate: *Scanning Probe Lithography* (Kluwer, Boston 2001)

7.78   C. A. Mirkin: Dip-pen nanolithography: automated fabrication of custom multicomponent, sub-100-nanometer surface architectures, MRS Bull. **26**, 535–538 (2001)

7.79   P. E. Sheehan, L. J. Whitman, W. P. King, B. A. Nelson: Nanoscale deposition of solid inks via thermal dip pen nanolithography, Appl. Phys. Lett. **85**, 1589–1591 (2004)

7.80   L. L. Sohn, R. L. Willett: Fabrication of nanostructures using atomic force microscope-based lithography, Appl. Phys. Lett. **67**, 1552–1554 (1995)

7.81   H. J. Mamin, B. D. Terris, L. S. Fan, S. Hoen, R. C. Barrett, D. Rugar: High-density data storage using proximal probe techniques, IBM J. Res. Dev. **39**, 681–699 (1995)

7.82   K. Bessho, S. Hashimoto: Fabricating nanoscale structures on Au surface with scanning tunneling microscope, Appl. Phys. Lett. **65**, 2142–2144 (1994)

7.83   I. W. Lyo, P. Avouris: Field-induced nanometer- to atomic-scale manipulation of silicon surfaces with the STM, Science **-253**, 173–176 (1991)

7.84   M. F. Crommie, C. P. Lutz, D. M. Eigler: Confinement of electrons to quantum corrals on a metal surface, Science **262**, 218–220 (1993)

7.85   A. de Lozanne: Pattern generation below 0.1 micron by localized chemical vapor deposition with the scanning tunneling microscope, Jpn. J. Appl. Phys. **33**, 7090–7093 (1994)

7.86   L. A. Nagahara, T. Thundat, S. M. Lindsay: Nanolithography on semiconductor surfaces under an etching solution, Appl. Phys. Lett. **57**, 270–272 (1990)

7.87   T. Thundat, L. A. Nagahara, S. M. Lindsay: Scanning tunneling microscopy studies of semiconductor electrochemistry, J. Vacuum Sci. Tech. A **8**, 539–543 (1990)

7.88   S. C. Minne, S. R. Manalis, A. Atalar, C. F. Quate: Independent parallel lithography using the atomic force microscope, J. Vacuum Sci. Tech. B **14**, 2456–2461 (1996)

7.89   M. Lutwyche, C. Andreoli, G. Binnig, J. Brugger, U. Drechsler, W. Haeberle, H. Rohrer, H. Rothuizen, P. Vettiger: Microfabrication and parallel operation of 5*5 2D AFM cantilever arrays for data storage and imaging, Proc. MEMS, 8–11 (1998)

7.90   G. M. Whitesides, B. Grzybowski: Self-assembly at all scales, Science **295**, 2418–2421 (2002)

7.91   P. Kazmaier, N. Chopra: Bridging size scales with self-assembling supramolecular materials, MRS Bull. **25**, 30–35 (2000)

7.92   R. Plass, J. A. Last, N. C. Bartelt, G. L. Kellogg: Self-assembled domain patterns, Nature **412**, 875 (2001)

7.93   Y. A. Vlasov, X–Z. Bo, J. G. Sturm, D. J. Norris: On-chip natural self-assembly of silicon photonic bandgap crystals, Nature **414**, 289–293 (2001)

7.94   C. Gigault, K. Dalnoki-Veress, J. R. Dutcher: Changes in the morphology of self-assembled polystyrene microsphere monolayers produced by annealing, J. Colloid Interf. Sci. **243**, 143–155 (2001)

7.95   J. C. Hulteen, P. Van Duyne: Nanosphere lithography: a materials general fabrication process for periodic particle array surfaces, J. Vacuum Sci. Tech. A **13**, 1553–1558 (1995)

7.96   J. D. Joannopoulos, P. R. Villeneuve, S. Fan: Photonic crystals: putting a new twist on light, Nature **386**, 143–149 (1997)

7.97   T. D. Clark, R. Ferrigno, J. Tien, K. E. Paul, G. M. Whitesides: Template-directed self-assembly of 10-μm-sized hexagonal plates, J. Am. Chem. Soc. **124**, 5419–5426 (2002)

7.98   S. A. Sapp, D. T. Mitchell, C. R. Martin: Using template-synthesized micro- and nanowires as building blocks for self-assembly of supramolecular architectures, Chem. Mater. **11**, 1183–1185 (1999)

7.99   Y. Yin, Y. Lu, B. Gates, Y. Xia: Template assisted self-assembly: a practical route to complex aggregates of monodispersed colloids with well-defined sizes, shapes and structures, J. Am. Chem. Soc. **123**, 8718–8729 (2001)

7.100 J. L. Wilbur, G. M. Whitesides: Self-assembly and self-assembled monolayers in micro and nanofabrication. In: *Nanotechnology*, ed. by G. Timp (Springer, Berlin Heidelberg 1999)

7.101 S. R. Wasserman, Y. T. Tao, G. M. Whitesides: Structure and reactivity of alkylsiloxane monolayers formed by reaction of alkyltrichlorosilanes on silicon substrates, Langmuir **5**, 1074–1087 (1989)

7.102 C. P. Tripp, M. L. Hair: An infrared study of the reaction of octadecyltrichlorosilane with silica, Langmuir **8**, 1120–1126 (1992)

7.103 D. R. Walt: Nanomaterials: top-to-bottom functional design, Nature **1**, 17–18 (2002)

7.104 J. Noh, T. Murase, K. Nakajima, H. Lee, M. Hara: Nanoscopic investigation of the self-assembly processes of dialkyl disulfides and dialkyl sulfides on *Au*(111), J. Phys. Chem. B **104**, 7411–7416 (2000)

7.105 M. Himmelhaus, F. Eisert, M. Buck, M. Grunze: Self-assembly of n-alkanethiol monolayers: a study by IR-visible sum frequency spectroscopy (SFG), J. Phys. Chem. **104**, 576–584 (1999)

7.106 A. K. Boal, F. Ilhan, J. E. DeRouchey, T. Thurn-Albrecht, T. P. Russell, V. M. Rotello: Self-assembly of nanoparticles into structures spherical and network aggregates, Nature **404**, 746–748 (2000)

7.107 A. Ulman: *An Introduction to Ultrathin Organic Films: From Langmuir-Blodgett to Self-Assembly* (Academic, New York 1991)

7.108 E. Winfree, F. Liu, L. A. Wenzler, N. C. Seeman: Design and self-assembly of two-dimensional DNA crystals, Nature **394**, 539–544 (1998)

7.109 J. H. Reif, T. H. LaBean, N. C. Seeman: Programmable assembly at the molecular scale: self-assembly of DNA lattices, Proc. 2001 IEEE Int. Conf. Robotics and Automation, Seoul, 966–971 (2001)

7.110 A. P. Alivisatos, K. P. Johnsson, X. Peng, T. E. Wilson, C. J. Loweth, M. P. Bruchez Jr, P. G. Schultz: Organization of nanocrystal molecules using DNA, Nature **382**, 609–611 (1996)

7.111 C. Y. Cao, R. Jin, C. A. Mirkin: Nanoparticles with Raman spectroscopic fingerprints for DNA and RNA detection, Science **297**, 1536–1540 (2002)

7.112 H. Masuda, H. Yamada, M. Satoh, H. Asoh: Highly ordered nanochannel-array architecture in anodic alumina, Appl. Phys. Lett. **71**, 2770–2772 (1997)

7.113 R. L. Fleischer: *Nuclear Tracks in Solids: Principles and Applications* (Univ. California Press, Berkeley 1976)

7.114 R. E. Packard, J. P. Pekola, P. B. Price, R. N. R. Spohr, K. H. Westmacott, Y. Q. Zhu: Manufacture observation and test of membranes with locatable single pores, Rev. Sci. Instrum. **57**, 1654–1660 (1986)

7.115 L. Sun, P. C. Searson, C. L. Chien: Electrochemical deposition of nickel nanowire arrays in single-crystal mica films, Appl. Phys. Lett. **74**, 2803–2805 (1999)

7.116 Y. Du, W. L. Cai, C. M. Mo, J. Chen, L. D. Zhang, X. G. Zhu: Preparation and photoluminescence of alumina membranes with ordered pore arrays, Appl. Phys. Lett. **74**, 2951–2953 (1999)

7.117 M. Guowen, C. Anyuan, C. Ju-Yin, A. Vijayaraghavan, J. J. Yung, M. Shima, P. M. Ajayan: Ordered Ni nanowire tip arrays sticking out of the anodic aluminum oxide template, J. Appl. Phys. **97**, 64303 (2005)

7.118 S. Yang, H. Zhu, D. Yu, Z. Jin, S. Tang, Y. Du: Preparation and magnetic property of Fe nanowire array, J. Magn. Magn. Mater. **222**, 97–100 (2000)

7.119 M. Sun, G. Zangari, R. M. Metzger: Cobalt island arrays with in-plane anisotropy electrodeposited in highly ordered alumina, IEEE Trans. Magnetics **36**, 3005–3008 (2000)

7.120 P. V. Braun, P. Wiltzius: Electrochemically grown photonic crystals, Nature **402**, 603–604 (1999)

7.121 S. R. Nicewarner-Pena, R. G. Freeman, B. D. Reiss, L. He, D. J. Pena, I. D. Walton, R. Cromer, C. D. Keating, M. J. Natan: Submicrometer metallic barcodes, Science **294**, 137–141 (2001)

7.122 D. Almalawi, C. Z. Ziu, M. Moskovits: Nanowires formed in anodic oxide nanotemplates, J. Mater. Res. **9**, 1014 (1993)

7.123 J. C. Hulteen, C. R. Martin: A general template-based method for the preparation of nanomaterials, J. Mater. Chem. **7**, 1075–1087 (1997)

7.124 B. R. Martin, D. J. Dermody, B. D. Reiss, M. Fang, L. A. Lyon, M. J. Natan, T. E. Mallouk: Orthogonal self-assembly on colloidal gold-platinum nanorods, Adv. Mater. **11**, 1021–1025 (1997)

Part A | 7

8. Nanoimprint Lithography

Nanoimprint lithography is an emerging nanopatterning method, combining nanometer-scale resolution and high throughput. In a top-down approach, a rigid stamp with a surface relief is pressed into a thin film of soft material on a hard substrate. The film is hardened before the stamp is retrieved, and the surface relief is copied into the thin film. A pattern with nano- to micrometer scale features can be replicated in a parallel process, and the stamp may be reused many times. This makes nanoimprint lithography a promising technique for volume manufacturing of nanostructured components. At present, structures with feature sizes down to 5 nm have been realized, and the resolution is limited by the ability to manufacture the stamp relief. For historical reasons, the term nanoimprint lithography (NIL) refers to a hot embossing process, where a thin film of thermoplastic material is softened by heating it, and the embossed film is hardened again when it is cooled down. In ultraviolet (UV)-NIL, a photo-polymerizable resin is used together with a UV-transparent stamp. The resin is liquid at room temperature, allowing easy embossing of the stamp, before the resin is hardened by UV exposure. In this chapter we will give an overview of nanoimprint lithography, with emphasis on NIL. Material aspects of stamps and resists are discussed. Thin-film rheology plays a central role of the understanding of the nanoimprint process, since the resist is patterned by mechanical deformation. We discuss specific applications where imprint methods have significant advantages over other structuring methods. We conclude by

discussing the areas where further development in this field is required.

Take a piece of wax between your fingers and imprint the fingerprints from both sides into it. The pressure is sufficiently high to replicate the soft surface pattern of our skin into the wax by mechanical deformation. The process is facilitated by the heat resulting from our blood circulation, which softens the wax in order to make it deform until it conforms to the three-dimensional (3D)

pattern of our skin. Surely the fidelity of the original pattern is distorted during molding, but even an incomplete molding allows the identification of the person according to the purely two-dimensional (2D) code of its fingerprint. The pattern resolution of below 1 mm is similar to that of the first records fabricated over 100 years ago in celluloid. In 1887 *Emile Berliner* applied for a patent

Fig. 8.1 Printing a seal into viscous wax is a way of replication using hot embossing. The figure shows a seal (stamp), wax tube (candle), and embossed pattern

on a so-called gramophone, which resembled Edison's phonograph with its wax-coated roll [8.1, 2]. The information is inscribed into wax coated onto a zinc disk. The tracks are cut through the wax down to the solid zinc and are etched before using the zinc disk as a mold to press thermoplastic foils. With a playing time of a little more than one minute, these disks had track widths below 1 mm and resolutions in the sub-100-µm range. Over the years the track size was reduced to below 200 µm. The materials changed from shellac to vinyl filled with carbon black. Today's Compact Discs have pit sizes of below 400 nm [in a digital versatile disk (DVD)] and are fabricated in polycarbonate within a few seconds by injection molding. Disc formats with further reduced pit sizes are currently being developed [8.3, 4].

In this introduction some basic concepts of molding polymers are illustrated, ranging from shaping by mechanical pressure, stamps, materials, to pattern transfer. A softened hard material can be deformed by pressure, and even if a soft, flexible stamp is used, the difference in mechanical properties makes it possible to replicate its surface pattern in a parallel, reproducible way. The squeezing of a thin film of wax leads to a lateral flow of material, but because of the high viscosity the process will slow down quickly and a residual layer that

cannot be thinned down to zero will always remain. Furthermore the softness of stamp and the viscosity of the material determine the the completeness of molding and thus the replication fidelity. Similar concepts of molding processes can be observed in daily life, such as imprinting a footprint into snow or clay, or making waffles in a pressure process with subsequent thermo-curing, or by replicating a seal into wax (see Fig. 8.1). However, even these examples show the variety of molding processes. One common important prerequisite of these molding processes is that the mechanical properties of the molded material can be changed by pressure, temperature or chemical processing. The material must be shaped in a viscous state but should keep its form during demolding. The imprint in the snow is a hard molding by local densification, while the clay hardens by squeezing out and evaporation of water. The waffle is cured due to the thermochemical changes in the dough, and the seal can be demolded with high fidelity because the heat of the wax dissipates into the seal and the wax hardens during cooling. The processes described here are very similar to molding of viscous thermoplastic materials in the nanoimprint lithography (NIL) process [8.5, 6], also referred as hot embossing lithography (HEL) [8.7], where a thickness profile in a thin polymer film is generated by pressure, however, with the surprising difference that features below 10 nm can be replicated with unprecedented precision (see Fig. 8.2). In contrast to conventional methods based on exposure and development, limitations imposed by the wavelength of exposure or by chemical reactions can be overcome, simply by inducing a local displacement of material by mechanical force.

The example with the fingerprint may even serve for the illustration of contact printing, see also Chapt. 9. When touching a flat surface, a fingerprint is left. This is because the grease, dirt or ink present on the fingertips is transferred from the elevated areas, and thus covers the surface locally. Even if this layer is not perfectly

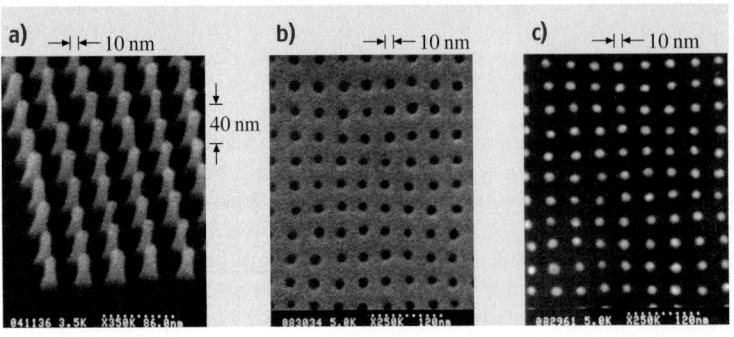

Fig. 8.2a–c Micrographs showing the basic steps of NIL, demonstrated by S. Y. Chou in 1997 [8.5]. (**a**) NIL stamp in silicon with a 40 nm period array of pillars with 40 nm height, (**b**) imprinted 10 nm diameter holes in a thin polymer film (PMMA), (**c**) 10 nm metal dots after pattern transfer (lift-off), using the thin polymer layer as a mask

coated, it can serve for 2D pattern recognition or as a resist for the pattern transfer into the substrate. While for NIL a hard stamp would assure a more complete molding, here the softness of the stamp is essential to assure a conformal contact of any protrusion, but at the expense of a possible reduction of the feature resolution due to deformations in the stamp. These issues are treated in more detail in the Sect. 9.2 on (soft lithography) microcontact printing (SL or μCP) and in [8.8, 9].

In this chapter about NIL we give an overview of the different processes currently called nanoimprint, from hot embossing of thermoplastic materials to imprinting and curing of liquid resins. After this introduction into the basics of molding, Sect. 8.2 places the two main imprint techniques into the context of the emerging nanopatterning methods for lithography. Section 8.2 is the central section where the NIL process is described in detail, beginning with a discussion about polymer properties, giving an insight into squeeze flow of thin films and presenting the major pattern-transfer tech-

niques used in NIL. Sect. 8.3 presents materials and tools for NIL, ranging from materials for stamps and resists to imprint machines. Sect. 8.4 presents typical applications which are currently envisaged, both at an industrial and at laboratory scale. Although for many people the main driving force behind NIL is its use as next-generation lithography for complementary metal-oxide semiconductor (CMOS) chip fabrication, the reader will be introduced to different other applications which profit from molding techniques. We conclude with an outlook in Sect. 8.5, where we discuss the prospects for NIL and aspects of its commercialization. Further information can be found in the references section, in publications dealing with the so-called lithography, electroforming and molding (LIGA from its German abbreviation) technology [8.10] and optical storage fabrication, but not least within this handbook in the chapters about silicon micromachining and soft lithography. In this chapter we restrict ourselves to the lithographic patterning of thin films on hard substrates.

8.1 Emerging Nanopatterning Methods

Nanoimprint lithography (NIL) is a replication technique which has proven to provide a resolution unmatched by many other techniques, while at the same time offering parallel and fast fabrication of micro- and nanostructures [8.11]. On the one hand, this enables a step into fields where large areas are covered by nanostructures, or a number of identical structures for statistical evaluation are needed. This was often impossible due to the low throughput of lithographic research tools. On the other hand the resolution achieved so far by molding is much higher than that used in the industrial fabrication of processors and memory chips with high-end photolithography. This makes NIL a promising technology for next-generation nanolithography [8.12]. Apart from these advantages molding offers more: by creating a three-dimensional resist pattern by mechanical displacement of material the patterning of a range of specific functional materials and polymers becomes possible, without losing their chemical properties during molding. Furthermore this ability can be used to fabricate complex structures, e.g. by building up devices with imbedded channels. These processes are presented in more detail in Sect. 8.3 of this chapter.

In this section we present the basic concepts of NIL and how it can conform to the requirements of state-in-the-art nanofabrication techniques. NIL uses, as do other

lithographic techniques, the concept of resist patterning (which can also be found in different chapters in this book). The resist patterns are generated by molding of a viscous material and fixed by cooling and curing, while in photolithography the resist is patterned by selective chemical modification of a positive or negative resist by exposure and wet development. The two main NIL methods are outlined in Fig. 8.3. For lithographic applications, as required in microelectronics and hard disks, NIL is in competition with other emerging patterning techniques. Its success will mainly depend on the ability to solve processing issues such as resolution, overlay and throughput. It is also important to develop reliable tools with a long lifetime, which are available and can be used in combination with other cleanroom process technologies, and to establish standard processes, which can be scaled up to common wafer sizes.

8.1.1 Next-Generation Lithography

With the integration of nanoimprint lithography into the international technology roadmap (ITRS) on next-generation nanolithography in 2003 for the 32-nm node and beyond, NIL has become more than a simple high-resolution method [8.12] (Table 8.1). It is now considered as a candidate for replacing or complementing

Table 8.1 ITRS roadmap showing the resolution of different lithographic patterning techniques, and practical and actual resolution limits for different lithography methods [8.15]

Lithography type	Practical resolution limit	Ultimate resolution limit
UV / contact / proximity	2500 nm	125 nm
UV projection	150 nm	50 nm
EUV projection (soft X-rays)	90 nm	30 nm
X-rays / proximity / 1:1 mask (with parallel X-rays)	70 nm	10 nm
Ion beam	30–50 nm	resist: 10–20 nm
Electron beam (low-energy beam arrays)	40–50 nm	resist: 7–20 nm
Electron beam projection (SCALPEL)	90 nm	35 nm
Imprinting (embossing)	20–40 nm	5–10 nm
Printing (contact)	30–50 nm	10 nm
Scanning probe microscopy methods	15 nm	0.5 nm (atomic resolution)

advanced optical lithographic methods for the fabrication of processors and solid-state memory chips, which over the years have been developed and pushed to higher resolution with a vast investment of resources. Over more than 40 years, Moore's law has described with an amazing accuracy the reduction of feature sizes, and therefore serves as a roadmap for the developments needed for future microchips [8.13–15]. It is driven by economic considerations, and leads to a competition between different candidate fabrication methods. These do not only have to provide the resolution of the smallest feature sizes (node), but also issues such as alignment, critical dimensions (CD, not identical with CD for Compact Disc), simple mask fabrication, high throughput (mass fabrication), low cost of ownership (i.e., not dependence on large machines such as synchrotrons), which become increasingly difficult to meet if smaller exposure wavelengths have to be used (see Fig. 8.4).

The sheer financial and physical barriers to these techniques are now so high, that alternatives such as NIL are considered as a way out of this spiral of rising investments for the next generation of chips with even smaller feature sizes. This means that all technical issues connected with NIL for integration into chip manufacturing bear the task of full compatibility, similar specifications, yield and throughput. The investments are expected to be lower than for the current frontrunners, extreme ultraviolet (EUV) lithography or parallel electron-beam exposure.

8.1.2 Molding Resists for Lithography

NIL was first reported as thermoplastic molding [8.16–19], and is therefore often referred to as hot embossing lithography (HEL) [8.20, 21]. The unique advantage of a thermoplastic material is that the viscosity can be

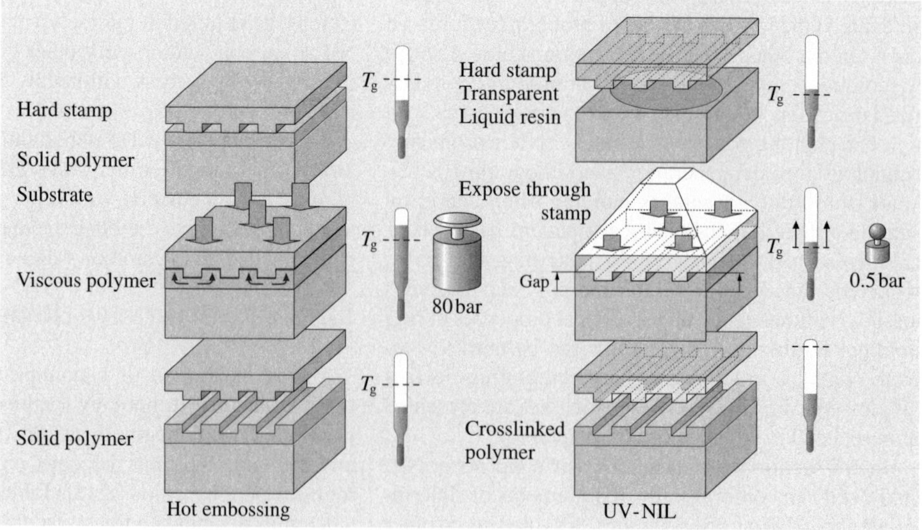

Fig. 8.3a,b Schematic of NIL process: (a) hot embossing, (b) UV-imprint. In both cases a thickness profile is generated in the thin polymer layer. After removing the residual layer, the remaining polymer can serve as a masking resist which can be used for pattern transfer

Table 8.2 Comparison of hot embossing (NIL) and UV imprinting (UV-NIL), with typical parameters of current processes

Type of NIL / properties	NIL (T-NIL) hot embossing	UV-NIL UV-imprint
basic process sequence (see Figs. 8.3 and 8.6)	1) spin-coat thermoplastic film 2) place stamp on film 3) heat until viscous 4) emboss at high pressure 5) cool until solid 6) demold stamp	1) dispense liquid resin 2) parallel alignment of stamp with defined gap 3) imprint at low pressure 4) expose with UV light through stamp and crosslink 5) demold stamp
Pressure p	20–100 bar	0–5 bar
Temperature T_{mold}	100–200 °C	20 °C (ambient)
Temperature T_{demold}	20–80 °C	20 °C (ambient)
Resist	solid, thermoplastic $T_g \approx 60$–100 °C	liquid, UV-curable
Viscosity η	10^3–10^7 Pas	10^{-2}–10^{-3} Pas
Stamp material	Si, SiO_2 opaque	glass, SiO_2 transparent
Stamp area	full wafer, > 200 mm diameter	25×25 cm^2, limited by control of gap
Stamp contact	facilitated by bending	planarization layer
Embossing time	from seconds to minutes	< 1 min (per exposure)
Advantage	low-cost large-area equipment and stamps	low viscosity, low pressure, alignment through stamp
Challenge	process time, thermal expansion due to thermal cycle	step and repeat needed for large areas
Development needed	alignment, residual layer homogeneity	material variety
Hybrid approaches	thermoset resists: embossing and curing before demolding	thermoplastic resists: hot molding and UV-curing before demolding
Advantage	low temperature variation cycle: demolding at high temperature possible	solid resist: full-wafer single imprint possible

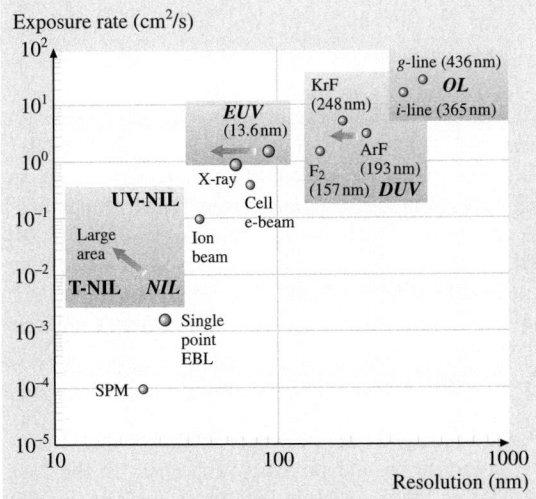

Fig. 8.4 Comparison of exposure rate and resolution of different lithographic techniques reworked from [8.14]. To date, NIL provides a high resolution of below 10 nm, and achieves wafer scale patterning within some minutes

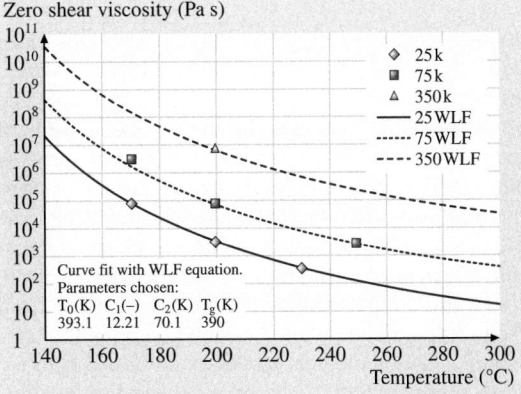

Fig. 8.5 Graph showing zero shear viscosities of PMMA for three molecular weights M_w (rheology measurements carried out at ETH Zürich by T. Schweizer). It shows the potential of rheology and of the large variation of viscosity of thermoplastic polymers with temperature

Part A | 8.1

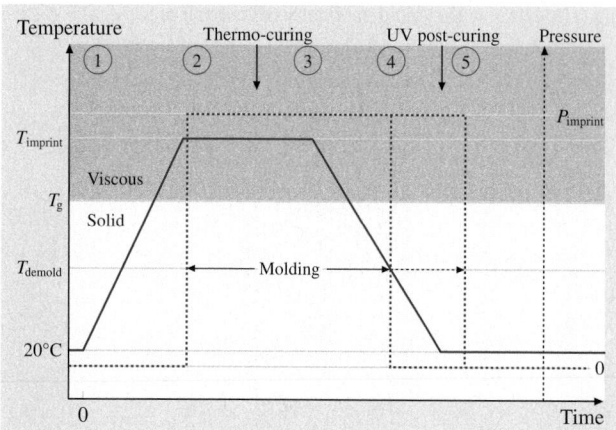

Fig. 8.6 Typical process sequence: schematics of process sequence used for hot embossing (temperature / pressure diagram with time dependence), 1) begin heating, 2) begin embossing, 3) begin cooling, 4) demolding at elevated T, 5) demolding at ambient

changed to a large extent by simply varying the temperature. In Fig. 8.5, the viscosities of different molecular weight poly (methyl methacrylate) (PMMA) are plotted against temperature. Switching between a solid and a highly viscous state is possible within a range of some tens of °C, and can be reversed [8.22]. The first stage of the NIL process is the molding of a thin thermoplastic film using a hard master. During a process cycle the resist material is made viscous by heating, and shaped by

applying pressure (see Fig. 8.6). Here the thermoplastic film is compressed between the stamp and substrate and the viscous polymer is forced to flow into the cavities of the mold, conforming exactly to the surface relief of the stamp. When the cavities of the stamps are filled, the polymer is cooled down, while the pressure is maintained. The molten structure is then frozen. After relieving the pressure, the stamp can be retrieved (demolded) without damage, and reused for the next molding cycle. In a second step, the thickness profile of the polymer film can now be used as a resist for pattern transfer. For this, the residual layer remaining in the thin areas of the resist has to be removed, which is done by homogeneously thinning down the resist in an (ideally) anisotropic etching process. In this way, process windows are opened to the substrate and the polymer can be used as a masking layer for further processing steps.

With the integration of light sources into imprint machines, UV-NIL was developed for curable resists [8.23–26]. The basic difference between UV-NIL and thermal (T-) NIL is that a resin, which is liquid at room temperature, is shaped by a moderate pressure, which is then crosslinked and hardened by curing. All the processes have their specific advantages, e.g. while UV-NIL can be performed at room temperature, hot embossing is low-cost since nontransparent stamps can be used. The major characteristics of the processes, along with those of hybrid approaches, are summarized in Table 8.2.

8.2 Nanoimprint Process

Molding techniques based on imprint processes make use of the differences between the mechanical properties of a structured stamp and a molding material. The viscous molding material is shaped by pressing the hard stamp into it. In order to achieve a reasonable process time and yield, this is normally carried out under pressure and using molding materials with relatively low viscosities. In hot embossing processes we mostly deal with thermoplastic materials whose mechanical properties can be repeatedly and reversibly changed from a solid into a viscous state by simply varying the temperature. The rheological processes described here for thermoplastic materials can be considered to be similar for thermoset or UV-curable materials as long as the thermomechanical properties can be changed without affecting the chemical ones.

In this section we want to take a closer look at the squeeze flow of thin polymer films as used for NIL.

We will give a brief introduction to the theory of polymers [8.27, 28] and discuss the implications for NIL. This will enable the reader to understand rheology in NIL from a practical point of view. More fundamental questions of squeeze flow are discussed in [8.29] and in [8.30]. We conclude this section by presenting pattern-transfer processes used in combination with NIL and show examples of the fabrication of simple devices.

8.2.1 Limits of Molding

Resists used in NIL are polymers defined by their chemical composition and physical properties. In the case of molding these are often long-chain molecules, with a molecular weight of M_w. The polymer molecular weight is important because it determines many physical properties. Some examples include the temperatures for transitions from liquids over viscoelastic rubbers to solids, and mechanical properties such as stiffness, strength, viscoelasticity, toughness, and viscosity. However, if the molecular weight is too low, the transition temperatures will be too low and the mechanical properties of the polymer material will be insufficient to be useful as hard resist for pattern transfer. The examples given in this section are simple and meant to illustrate the specific terms needed for the understanding of polymer behavior in molding.

It has been known for a long time, that polymers can replicate topographies with high fidelity. Up to now 5-nm resolution of polymer ridges with a pitch of 14 nm has been demonstrated [8.31]. In comparison to methods like electron-beam lithography (EBL), where nanoscale polymer structures can be produced by irradiation-induced chain scission, the polymer chains are only moved and deformed during molding, thus keeping their chemical properties such as molecular weight M_w. Molding topographic details down to a few nanometers means that single polymer chains have to deform or flow. This deformation can be illustrated by comparing the polymer with a pot full of cooked spaghetti, and instead of the viscosity change with temperature we simply take the different mobility of the filaments when wet or dry. When a water glass, representing the 10-nm pillar stamp shown in Fig. 8.2a, is pressed into this pot, single spaghetti filaments have to be moved, before the glass can sink into the entangled network. If the polymers can glide along each other, the deformation can be permanent after drying and demolding. If stress is frozen, the matrix around the cylindric hole will relax after demolding. Note that this simple example can also be used to illustrate the difference between totally amorphous and semicrystalline polymers.

A polymer is a large molecule made up of many small, simple chemical units, joined together by chemical reaction. For example, polyethylene [$CH_3-(CH_2)_N-CH_3$] is a long chain-like molecule composed of ethylene molecules ($CH_2=CH_2$). Most artificially produced polymers are a repetitive sequence of particular atomic groups, and take the form ($-A-A-A-$). The basic unit, A, of this sequence is called the monomer unit, and the number of units N in the sequence is called the degree of polymerization. The molecular weight of a polymer is defined by the weight of a molecule expressed in atomic mass units (amu). The molecular weight may be calculated from the molecular formula of the substance; it is the sum of the atomic weights of the atoms making up the molecule. For example, poly (methyl methacrylate) (PMMA), a classical resist material, exhibits very good resolution for both EBL and NIL. A high-molecular-weight PMMA, typically above 500 kg/mol (also 500 k), is normally used for EBL, since the development contrast between exposed and unexposed areas increases with molecular weight [8.32, 33]. A lower molecular weight, of some tens of kg/mol, is patterned in NIL, due to the strong increase in temperature-dependent viscosity with molecular weight [8.34]. Apart from their mobility it is expected that smaller chains, which are normally present as coils, can move more easily into small mold cavities. As an example we take a PMMA macromolecule with a molecular weight of 25 kg/mol. Here the chain contains 250 MMA monomer elements ($C_5H_8O_2$) with a weight of 100 g/mol each and has a total length of about 80 nm. If we anticipate that the polymer occupies the space of a solid sphere, we can estimate its radius R in a simple top-down approach. We take the mass $m_{PMMA} = 4.2 \times 10^{-20}$ g of the PMMA molecule from above, and the density given for PMMA as $\rho_{PMMA} = 1.19$ g/cm^3. Then R_{sphere} is calculated to be 2 nm, which is already close to the resolution achieved so far. However, amorphous polymers are normally present as entangled entities, and the value calculated above can only be used to give a rough estimate of the mean distance between these entities. A convenient way of expressing the size of a macromolecule present as a statistical coil aggregate is the radius of gyration R_g, which is calculated from the statistical mean path of the chain in a random walk model using a self-avoiding walk, and by definition is much larger than R_{sphere}. R_g can be directly measured in experiments by small-angle neutron scattering [8.35], and can also be defined not only for a linear chain but also for polymers with branched structure, etc. It also equals the square of the average distance between the segments and the center of mass of the polymer [8.27]. Since entire coils are both moved and deformed, both R_g and R_{sphere} will only give a rough estimate of the achievable minimum resolution of a pattern in an amorphous polymer film.

A polymeric liquid, whilst retaining the properties of a liquid, displays a rubber-like elasticity. An example is the melted cheese on a pizza. If melted cheese is dripped vertically, it flows slowly, just like a liquid. However, if it is pulled and then the tension removed, melted cheese will contract just like a rubber. In other words, although melted cheese is a liquid, it also has elasticity. Substances like this, which have both viscous and elastic properties, are called viscoelastic substances. In order to calculate the flow of a fluid when an external force is applied, we need an equation relating the stress

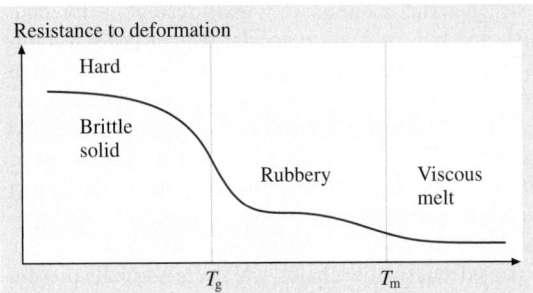

Fig. 8.7 Mechanical resistance to deformation as a function of the temperature for an amorphous polymer. The glass transition is marked by a large change around T_g

in the fluid to its deformation. This type of equation is called a constitutive equation. For example, if a polymeric liquid undergoing a steady flow is stopped, the stress does not immediately become 0, but decays with a relaxation time τ. Here τ depends strongly on the molecular weight of the polymer and the temperature, and can be on the order of several minutes to hours in some cases.

In the case of NIL, the relaxation has the effect that structures can still deform after molding. Considering the fact that that molding is achieved by deformation of a polymer network at a molecular level, the question is how the polymer can be permanently shaped and whether the replicated structure will deform back due to internal reordering and relaxation of polymer chains.

The reduced viscosity of polymers at higher temperatures is a result of the increasing ability of the chains to move freely, while entanglements and Van der Waals interaction of the chains are reduced. The glass transition of a thermoplastic polymer is related to the thermal energy required to allow changes in the conformation of the molecules at a microscopic level, and above the glass-transition temperature T_g there is sufficient thermal energy for these changes to occur. However, the

transition is not sharp, nor is it thermodynamically well defined. It is therefore different from melting, (defined by T_m), which is an equilibrium transition mostly present in polymers with crystalline entities. The glass transition is a thermodynamic transition in the sense that it is marked by discontinuities in thermodynamic quantities (see Fig. 8.7). A distinct change from rubbery (above T_g) to glassy (below T_g) behavior is readily observable in a wide range of polymers over a relatively narrow temperature range. For thin films, however, the T_g can be different from bulk values [8.36, 37].

Most of our considerations here are valid for a range of practical process parameters, as used in current hot embossing processes, where linear behavior can be assumed (Newtonian flow regime). This is in particular the case at molding temperatures well above the T_g (50 °C above T_g).

8.2.2 Squeeze Flow of Thin Films

During embossing the linear movement of a stamp is transformed into a complex squeeze flow of viscous material. In the thin polymer films used in NIL, a small vertical displacement of the stamp results in a large lateral flow. The two surfaces of the stamp and the substrate have to come entirely into contact with each other and maintain this contact until the desired residual layer thickness is reached. Furthermore, new concepts are possible such as roll embossing and soft embossing using flexible stamps. In Fig. 8.8, the embossing of a stamp with line cavities is schematically shown.

Before embossing, the polymer film has an initial thickness h_0 and the depth of the micro-relief is h_r. For a fully inserted stamp, the film thickness under the single stamp protrusions (elevated structures) with width s_i is h_f. Applying the continuity equation with the assumption that the polymer melt is incompressible (conservation of polymer volume), we get a specific residual layer height h_f. This can be directly deduced from the fill factor ν, i.e. the ratio of the area covered

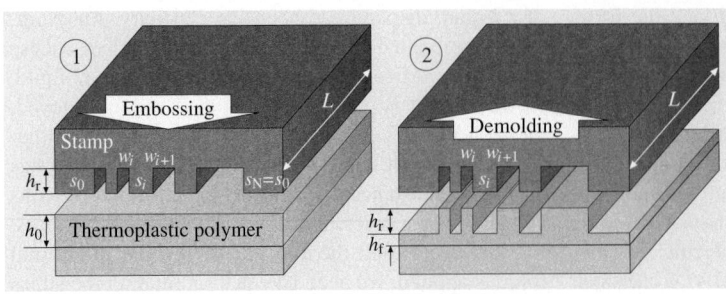

Fig. 8.8 Geometrical definitions used for the description of the flow process for a stamp with line cavities and protrusions. (1) before molding, (2) after demolding

by cavities to the total stamp area.

$$h_f = h_0 - v h_r \text{ with } v = \sum_i w_i \Big/ \sum_i (s_i + w_i) \,. \tag{8.1}$$

This formula only applies for rigid stamps with a constant fill factor.

A simple model for the squeezed polymer flow underneath the stamp protrusion is obtained by treating the polymer as an incompressible liquid of constant viscosity, and solving the Navier–Stokes equation with nonslip boundary conditions at the stamp and substrate surfaces. According to this model, given for line-shaped stamp protrusions and cavities in [8.11, 22, 38, 39], we find the following expression, known as the Stefan equation [8.40], for the film thickness $h(t)$ underneath the stamp protrusion when a constant imprint force F is applied to the stamp protrusion:

$$\frac{1}{h^2(t)} = \frac{1}{h_0^2} + \frac{2F}{\eta_0 L s^3} t \tag{8.2}$$

Inserting the final thickness $h_f \equiv h(t_f)$ in (8.2) gives the embossing time

$$t_f = \frac{\eta_0 L s^3}{2F} \left(\frac{1}{h_f^2} - \frac{1}{h_0^2} \right) \,. \tag{8.3}$$

For many practical cases, where a constant pressure under each stamp protrusion $p = F/(sL)$ is assumed, this formula gives

$$t_f = \frac{\eta_0 s^2}{2p} \left(\frac{1}{h_f^2} - \frac{1}{h_0^2} \right) \,. \tag{8.4}$$

As a direct consequence of the Stefan equation it can be seen that, for the same pressure, small (narrow) stamp protrusions will sink faster than large (wide) ones. The stamp geometry can therefore be optimized by reducing the dimensions of the protrusions. While stamps with nanopillar arrays, as shown in Fig. 8.2, would allow fast embossing of some microseconds, using standard NIL process parameters, protrusions of some hundreds of microns would already increase embossing times to some hours. The strong dependence of the embossing time on the pressing area has the consequence that, for a fully inserted stamp relief (full contact over the total stamp area), the flow practically stops (schematically shown in Fig. 8.9). For this case, s becomes large and the flow continues only towards the stamp borders. It is also evident that the embossing force has only a weak influence ($t_f \approx 1/F$). At first sight there is a similar weak influence for η_0. However, the viscosity can be changed significantly by varying the temperature.

For completeness we now give the expression similar to (8.3), but derived for a cylindrical stamp protrusion with radius R, i. e. with a stamp protrusion area of πR^2:

$$\frac{1}{h^2(t)} = \frac{1}{h_0^2} + \frac{4F}{3\pi \eta_0 R^4} t \tag{8.5}$$

Before we present some rules of thumb which may help to understand the further implications of the rheology involved in squeeze flow, we present an example illustrating the consequences of these equations. In Fig. 8.10a we show a stamp which contains an array of small structures in the center while the large single stamp protrusions surrounding the array dominate the sinking velocity (large s_i). The array in Fig. 8.10a is

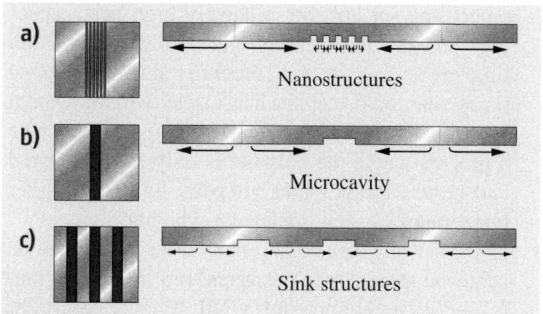

Fig. 8.10a–c Comparison of the squeeze flow for nano- and microcavities (schematics). In the case of an array of nanocavities and (**a**) a single microcavity, (**b**) surrounded by large unstructured stamp areas, the polymer has to flow over large distances, thus leading to long molding time. By the introduction of additional sink microstructures, or a denser arrangement of cavities, (**c**) a faster and more homogeneous molding can be achieved. *Left*: top view. *Right*: side view

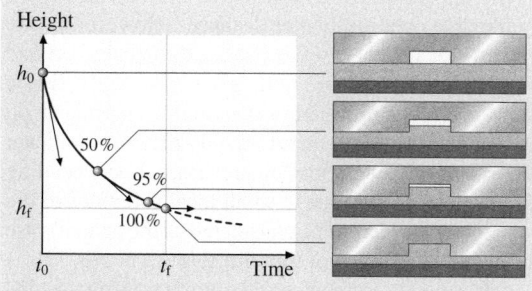

Fig. 8.9 Schematics (*right side*) of the squeeze flow of a compressed polymer film into one cavity. Once the cavity is filled the stamp continues to sink but at a much slower rate (*left side*), as a direct consequence of the Stefan equation

equivalent to the microcavity in Fig. 8.10b, which has the same volume as the total volume of the cavity array. This simplification can be used for the calculation of the embossing times. The fill factor should be kept constant, both locally (at length scales corresponding to the cavity dimensions) and also across the wafer, i. e., for large stamp protrusions, to ensure better flow of the polymer and shorter embossing times. For this purpose, additional protrusions or cavities can be placed in intermediate areas not needed for the device function, or structures repeated several times (Fig. 8.10c). We would also like to draw the reader's attention to the fact that the different sinking rates of protrusions of different sizes means that the stamp, which is normally backed by an elastic silicone mattress, can bend locally. This will result in a residual layer height that is not uniform over the entire embossing area. According to these formulas, there are different areas for optimization:

- Initial film thickness: process times to mold thin polymer layers can be reduced to a few seconds by simply using a high initial layer thickness. The reason for this is easy to understand: for thicker films the squeeze polymer can flow more freely in the central plane of the film unaffected by the friction at the boundaries. Care must be taken in NIL, where a high initial layer thickness usually results in an unacceptable residual thickness, i. e. too high for further pattern transfer.
- Molding temperatures: in NIL, high molding temperatures can lower polymer viscosities and reduce times to fill the mold cavities. However in current embossing set ups, the relatively high heat capacities and the low heating and cooling rates often limit embossing times. In order to reduce embossing times, one could keep the heat capacity of the system low, use efficient heating and cooling devices, take a process where the polymer is only locally heated and reduce the polymer viscosity by using higher temperatures or low-molecular-weight polymers.
- Low-viscosity resists: extremely fast molding can be achieved at moderate pressures, if a liquid polymer is used, as is common in UV-NIL. Precise control of the gap between the stamp and the substrate can be achieved by a stiff adjustable mechanical set up or by spacers on the stamp or substrate. The resin has to be dispensed onto the substrate prior to molding; the density and the size of the droplets has to be adjusted in order to achieve a complete filling of the stamp cavities without local thinning due to the fast lateral displacement.

- Stamp geometry: if resists with higher viscosities are used, the distances of lateral polymer flow should be kept as small as possible. Stamp cavities and protrusions should be distributed homogeneously over the entire stamp area i. e. the fill factor should be kept constant in order to limit the displacement of polymer in the vicinity of each structure. This reduces the tendency for incomplete filling and variation of residual layer thickness over the surface area.

The implications of squeeze flow are discussed in more detail in [8.41–44], including rheological issues [8.45–56], bending of stamps in large-area imprints [8.42, 57–67] and the influence of vacuum and self-assembly [8.68–77]. More information can also be found in Sect. 8.2.6 for pattern transfer and Sect. 8.3.1 for NIL materials.

8.2.3 Residual Layer Thickness Homogeneity

The main difference between NIL and lithography based on exposure and development is that a residual layer is left after demolding. As seen before, this is a result of the molding process slowing down due to the squeeze flow. For many applications, when pattern transfer has to be achieved after the embossing, it is important to know the final residual thickness h_f of the polymer layer (Fig. 8.8). Furthermore it is important to know the thickness variation over the embossed area, otherwise parts of the structure will be lost during pattern transfer. As will be shown in the following, bending of stamps has to be taken into account, as well as effects such as air inclusions and self-assembly of resist.

Bending of stamps

In NIL, the stamp is often considered as a hard tool which is inflexible over millimeter distances. However, this is only true for special cases, e.g. when the density and size of stamp protrusions are homogeneous over the whole stamp surface. Furthermore it is highly dependent on the pressure used, and therefore only plays a significant role in current hot embossing processes. Local bending of a few nanometers due to small local variations of the stamp geometry has to be considered as the general case during hot embossing of thin films. [8.22, 50, 78, 79]

For a typical case where the grating is surrounded by a large unstructured area, stamp bending results in an inhomogeneous residual layer at the border of the grating. Figure 8.10a shows such a case, with a grating area, typically of some square millimeters, is surrounded by a large unstructured area. In the ideal case of a totally rigid stamp, the final thickness would be determined by the fill factor of the grating averaged over the whole stamp area, which could be calculated by the simple argument of conservation of polymer volume. This can only be achieved if the polymer can flow easily over large distances. Otherwise, parts of the grating would not be filled. In the other extreme case of a totally flexible stamp and low lateral transport of polymer, both stamp areas could be calculated independently. While in the center of the grating the stamp would sink to half the depth of the cavities (assuming a fill factor of 50%), in the unstructured area almost no sinking would occur. In between, at the border of the grating, the stamp tries to accommodate this mismatch by bending. Depending on the thickness and elastic behavior of the stamp, as well as its design, characteristic lengths can be calculated.

In Fig. 8.11 we show an example, which demonstrates the implications of a more complex flow within a 2 mm × 2 mm area filled with microstructures. Both fields consist of an imprinted resist pattern resulting from the same initial resist film thickness (300 nm), but imprinted using stamps with inverse polarity (depth of 200 nm). This means that protrusions in the left field (dark) are cavities on the right (bright). The fields are divided into four quarters covered with vertical gratings,

the fill factor varies in the two upper parts, while it is constant over the two lower parts. At first sight, it can be seen that the macroscopic color distribution is opposite in the upper halves of the fields (here in grey tone). This is due to the interference of light within the transparent polymer, and the sensitivity to thickness variations resolves differences between 420-nm (dark) and 380-nm (bright) areas, visible within the protrusions of the grating. In the line structures in the top halves of the fields the fill factor varies from the border (40%) to the center (60%), while the structure-size variation is kept small. The stamp is able to accommodate this variation, and therefore results in an inhomogeneous thickness variation. In the bottom halves the fill factor (50%) stays constant, while the structure-size variation is dominant. However, the distances of the local variations of fill factor are small enough that they do not result in a significant thickness variation. Therefore the figure looks similar for both polarities with a constant thickness in the protrusions of 400 nm, while the residual layer thickness has the 200 nm value calculated by volume conservation. The following conclusions can be drawn from this:

- If the fill-factor variation takes place over large distances, the residual layer thickness can be directly calculated from volume conservation. The transi-

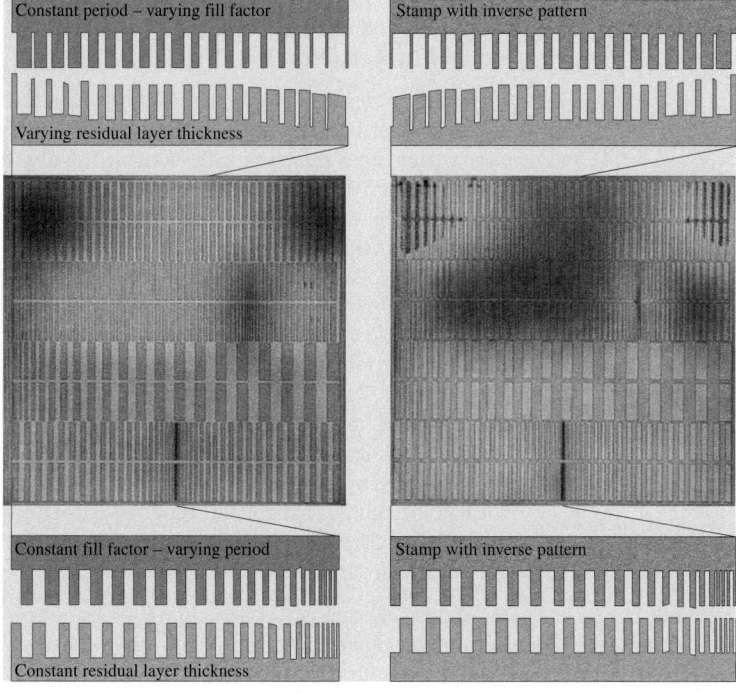

Constant period – varying fill factor

Varying residual layer thickness

Stamp with inverse pattern

Constant fill factor – varying period

Constant residual layer thickness

Stamp with inverse pattern

Fig. 8.11 Effect of structure size and fill factor variation for residual layer thickness for two structured fields of inverse polarity on the left and right side (field size 2 × 2 mm², depth of structures 200 nm, initial film thickness 300 nm). The schematics at the top and the bottom show cross sections of the stamp and the molded resist (in the upper and lower quarter of the field). The color changes show areas of high (dark) and low (bright) protrusion height and show that the fill factor variation leads to a large variation of the residual layer thickness depending on the fill factor

tion region is restricted to a characteristic distance that depends on the stamp depth and the mechanical properties of the stamp. For the 460-μm-thick silicon wafer used here, the bending over 50 nm can be complete over distances of some 100 μm.

- This bending is essential for NIL, because it helps to compensate thickness variations of the substrate and even dust particles of some microns in size. It is not significant at the low pressures used in UV-NIL. Therefore the gap between the stamp and the substrate has to be controlled mechanically before the resist is hardened by UV-curing.
- When fill-factor variation occurs over small distances, but the fill factor is constant over the characteristic length scales, no significant bending will occur. Although bending cannot be eliminated for inhomogeneous structures, it can be greatly reduced by clever design, e.g. in the case of gratings surrounded by a large unstructured protrusion area the bending can be kept out of the grating area by simply placing auxiliary cavities in the vicinity of the grating.

In many cases simulations are needed to predict the filling of both small and large structures in close vicinity. Furthermore the dynamic behavior of filling has to be taken into account. The task becomes even more challenging if embossing over topography has to be considered. In this case, a planarization layer can be used.

Influence of air inclusions and self-assembly

An important question has to be asked: where does the air in the cavities go during hot embossing? From micromolding it is known that air inclusions can lead to defects. This is particularly the case if the air cannot escape from single mold cavities before they are closed by the soft polymer. Although the situation is not totally different from micromolding, in NIL the cavity volume is significantly smaller. Due to the high pressure applied in the hot molding process the air is compressed in each cavity to a fraction of its initial volume [8.22]. Although it appears that a large part of the compressed air diffuses into the polymer, embossing under a vacuum of a few mbar can help to improve the molding [8.68, 69]. In UV-NIL, where a much lower pressure is used, no reports have been published on air inclusions. This might be due to the fact that during the filling of the cavities the liquid resin is able to displace the air entirely.

During hot embossing self-assembled polymer mounds (capillary bridges) can be created. The shape of the mounds depends on local nucleation and compressive effects. The way in which these regular structures may form in squeeze flow is described in [8.70–77]. In any case, no sign has been found that the self-assembly process affects the final quality of the molded structure as long as embossing times are chosen to be long enough to achieve a complete fill of the master cavities. Indeed it may improve the filling of nanocavities.

8.2.4 Demolding

During demolding the rigid stamp is detached from the molded structure. If fully molded, the thickness profile in the resist exhibits the inverse polarity of the relief of the stamp surface. The demolding process, also called de-embossing, is normally performed in the *frozen* state, i.e. when both mold and molded material are considered solid. For thermoplastic materials this happens at a temperature well below T_g, but high enough that frozen stress due to thermal contraction does not lead to damage during demolding. In cases in which the resist is cured before demolding, i.e. crosslinked by exposure or heat, demolding can take place at temperatures similar to the molding temperature. A successful demolding process relies on a controlled balance of forces at the interfaces between stamp, substrate and molded polymer film. Therefore mechanical, physical and chemical mechanisms responsible for adhesion have to be overcome. The following effects have to be avoided or reduced [8.22, 50, 78, 79]:

- Undercuts or negative slopes in the stamp may lead to mechanical interlocking of structures, which in their frozen state are elastically elongated and deformed before ripping. Sidewalls with positive, or at best vertical, inclination are prerequisites for demolding without distortion.
- Friction due to surface roughness may occur during the sliding of molded structures along vertical cavity walls. The effect of this can only be overcome if the surface of the molded material is elastic and enables gliding of the wall without sticking.
- The enlarged surface area of the patterned stamp leads to an increase of hydrogen bridges and Van der Waals forces, or other chemical bonding effects due to ionic, atomic and metallic binding. This effect can only be overcome if the stamp surface can be provided with sufficient anti-adhesive properties.

The most critical point is that demolding forces largely depend on the geometry of the mold, and the overall design of a stamp structure has to be taken into account.

Therefore structures with high aspect ratio may be more prone to ripping, and if many neighboring structures exert high forces to the underlying substrate, whole areas of resist may be detached from the substrate surface. Anti-adhesion layers on the mold can reduce friction forces, but have to be thin and durable. The thermal expansion coefficient of the substrate $\alpha_{substrate}$ and of the stamp α_{stamp} should be similar, to avoid a distorsion due to mechanical stress induced by the cooling. In the case of very thin polymer layers, the lateral thermal expansion of the resist is determined by the substrate. For structures with a higher *aspect ratio* the demolding temperature T_{demold} should be well below T_g, to enable the demolding of a hardened resist without distortion, but as near as possible to T_g, because the stress induced by thermal shrinkage should not be excessive in critical areas where structures tend to break.

8.2.5 Curing of Resists

Curing by UV exposure, by thermal treatment or chemical initiation is a way to crosslink polymers and to make them durable for demolding [8.23,24,26,80–92]. A high reaction speed, as caused by a high exposure dose, high initiator content or curing at high temperatures, leads to fast, but weak crosslinking, whereas a slow reaction leads to highly polymerized, tougher materials because the slow polymerization enables a more complete process. As shown in Sect. 8.1.2, different process strategies have been developed. In most of these the curing step is independent from the molding step, and can be initiated by light or a specific temperature after the molding is complete. Because curing involves a change in the physical conformation of the polymer, it always goes along with a volumetric shrinkage of the polymer, e.g. acrylate polymerization is known to be accompanied by volumetric shrinkage that is the result of chemical bond formation. Consequently, the size, shape, and placement of the replicated features may be affected. In the following the main processes which involve curing are presented in more detail:

- In the UV-NIL process, as used in step-and-flash imprint lithography (SFIL) [8.25,26], the resist is cured after molding, but before demolding of the stamp. The process relies on the photo-polymerization of a low-viscosity, acrylate-based solution. Shrinkage was found to be less than $\approx 10\%$ by volume in most cases. The current liquid is a multicomponent solution. The silylated monomer provides etch resistance in the O_2 transfer etch, and is therefore called an

etch barrier. Crosslinker monomers provide thermal stability to the cured etch barrier and also improve the cohesive strength of the etch barrier. Organic monomers serve as mass-persistent components and lower the viscosity of the etch-barrier formulation. The photoinitiators dissociate to form radicals upon UV irradiation, and these radicals initiate polymerization.

- If a solid curable resist exhibits thermoplastic behavior, it can be molded at an elevated temperature and then crosslinked, either before or after demolding. The advantage of this process is that low-m_w resists with a low T_g are available, which can be processed at moderate temperatures. However, before pattern transfer, hardening is often necessary. They can also be used for mix and match with optical lithography or for polymeric stamp copies.

- Thermoset resists can be crosslinked by heat. Here it is advantageous that the temperature for molding is lower than the curing temperature. Then the structure is first molded and then heated to its crosslinking temperature to induce crosslinking, before the hardened stamp is demolded.

More information about curing and multilayer resists can be found in Sect. 8.3.

8.2.6 Pattern Transfer

In many cases the lithographic process is only complete when the resist pattern is transferred into another material. This process, in which the resist is transformed into a patterned masking layer, allows the substrate to be attacked by plasma, etching solutions, electroplating, deposition of materials and other substrate-altering processes. A unique advantage of molding instead of exposure is that complex stamp profiles, such as staircases, V grooves, pyramids, both convex and concave, can be replicated. They can be used for the generation of 3D structures such as T-gate transistors or contact holes, or serve for the stepwise etching of underlying layers with variation of the opening width. As long as undercuts and 3D patterning is not necessary, in most cases this pattern transfer is therefore similar to EBL. However, in this section we emphasize methods where NIL has some specific process advantages over conventional lithographic methods, or where the use of NIL implies some major changes in the fabrication process or properties of the devices.

- In NIL etching is used both for the removal of the residual layer and for the pattern transfer of the re-

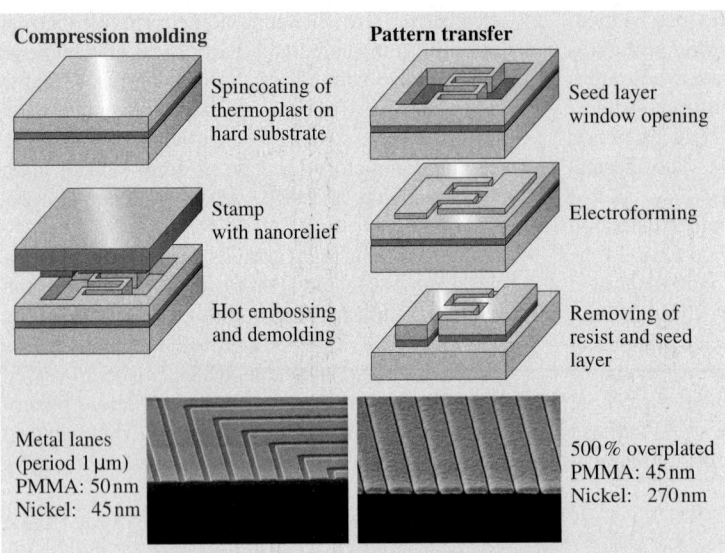

Fig. 8.12 NIL and electroforming: Electrode structures have been fabricated in Ni by using a plating base of Cr and Ge. After plating on top of the Ge, both layers of the plating base can be etched using RIE (Cr: chlorine chemistry; Ge: SF_6). Even with 500% overplating, the thick electrodes stay separated. Reproduced from [8.103]

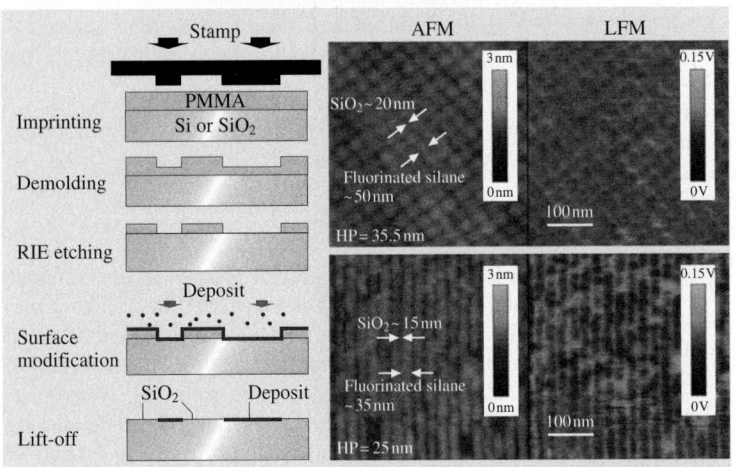

Fig. 8.13a,b NIL and lift-off for the generation of nanopatterns with chemical contrast. (**a**) Process scheme for local silane deposition from gas phase and (**b**) AFM/LFM images for chemically patterned surfaces modified with a fluorinated silane, showing sub-50 nm areas with hydrophobic (silane) and hydrophilic (SiO_2) properties. Reproduced from [8.104]

sist pattern into the underlying substrate [8.93–102]. In the first case the polymer layer has to be homogeneously thinned down until openings to the underlying substrate are generated. This is also called window opening or breakthrough etch. In the second case the thickness contrast of the remaining polymer mask is used to cover the substrate against the etching medium. Both processes have to be highly anisotropic, i.e. during the transfer step the lateral size of the structure has to be preserved, including the slope of the original pattern. Apart from opening windows using reactive-ion etching (RIE), other pattern-transfer strategies have been found which circumvent the residual layer problem.

- Lift-off is a patterning technique adding thin layers of a solid material (e.g. metal) locally to the window openings in the resist [8.104–112]. Undercuts, as can be generated in photolithography and EBL, are a prerequisite for a good lift-off. However, in NIL, where sidewalls are at best vertical, a high thickness contrast (aspect ratio) of the structures is needed. Lift-off resists are a means to generate defined undercuts using a bilayer resist system, by selectively dissolving a sacrificial bottom layer through the structured openings of a top layer.

- Electroforming and electro-plating, like lift-off, are processes that add material in the areas not covered by the resist [8.103, 113, 114]. Electroforming provides a good alternative to the lift-off process because metal structures can be generated with considerable height and good surface quality. If a conductive seed layer is deposited below the resist, during electroplating the metal layer starts to grow from within the window regions and conforms to the outlines of the cavities in the resist. Depending on the extent of electroplating, the structure height can either be preserved or increased.

Some of the examples presented here for pattern transfer already give insight into simple demonstrators, particularly, when the applications is based on simple pattern transfer or NIL is used as the first patterning step of an unstructured surface. Examples of applications are:

- Large-area metal gratings, as needed for polarizers or interdigitated electrode structures, can be fabricated by etching of a metal layer, lift-off or electroplating; in Fig. 8.12, e.g., electrode structures have been fabricated in Ni by using a plating base of Cr and Ge. After plating on top of the Ge, both layers of the plating base can be etched using RIE [8.103].
- Surface patterns with chemical contrast can be generated by locally depositing silanes onto a SiO_2 surface by lift-off (Fig. 8.13). By patterning molecules with biofunctionality, integrated biodevices can be fabricated such as biosensors and biochips. In [8.111] the combination of NIL and molecular assembly patterning by lift-off (MAPL) is demonstrated.
- Using etching the NIL process can be used to draw copies from a stamp original [8.96, 115]. Often the deposition of a metal layer for subsequent etching is used as a hard mask to generate copies with an enhanced aspect ratio.

More specific applications, where one or several of these pattern-transfer processes were used, are shown in more detail in Sect. 8.4.

8.2.7 Mix-and-Match Methods

Mix-and-match approaches are used to combine the advantages of two or more lithographic processes or simply to avoid their mutual disadvantages [8.116–126]. This is also a way to improve throughput and reliability, e.g. since the fabrication of large-area nanostructures is often costly, the definition of microstructures can be done with optical lithography (OL), while the nanopatterning of critical structures in small areas can be done by NIL. In many cases NIL would be used as the first process

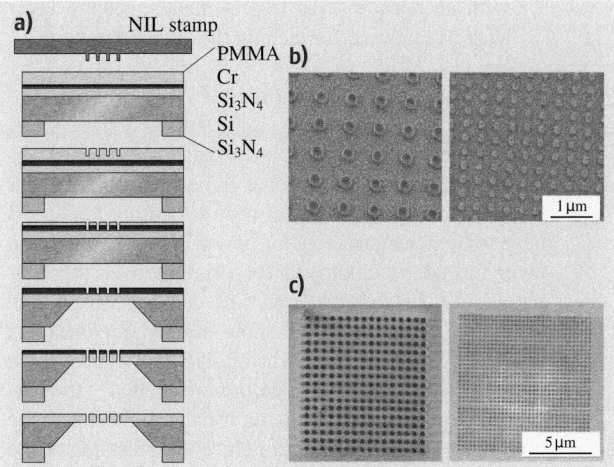

Fig. 8.14a–c Mix and match of NIL and silicon micromachining: **(a)** Process scheme for the fabrication of nanopores in a Si_3N_4 membrane. Scanning electron microscope (SEM) images **(b)** of the NIL stamps (pillars) and **(c)** for the corresponding nanopores. Reproduced from [8.102]

step and, by adding alignment structures along with the nanopatterns, the less-critical structures can be added after the NIL step using OL with an accuracy given by the mask aligner (in the range of 1 μm). NIL allows different variants of mix and match:

- In a sequential approach a second resist pattering process is added onto the first structure after pattern transfer. Specific problems such as overlay, or non-flat surfaces have to be solved. An example of mix-and-match can be seen in Fig. 8.14, where a nanoporous membrane was fabricated by NIL (pore definition) and optical lithography (windows for silicon etching and membrane release) [8.102].
- By using a UV-sensitive thermoplastic resist, the nanopattern can be created by NIL and the micropattern added into the molded resist by OL in subsequent patterning steps. Using this bi-lithographic step, pattern transfer can be done for the whole structure after the resist structuring is complete. The resists used for this purpose are crosslinked during exposure, which makes it possible to dissolve the areas not exposed.
- A specific mix-and-match approach is possible if both molding and UV exposure are done before the stamp is detached from the molded resist. This is possible when parts of the stamp are transparent (e.g. the recessed areas), while the protrusions are coated

with an opaque layer (e.g. a metal masking layer such as that used for etching the stamp structures).

8.2.8 Multilayer and Multilevel Systems

Multilayer resists systems are used if the etching selectivity of a masking layer has to be enhanced, e.g. for the fabrication of high-aspect-ratio structures, undercuts have to be generated, e.g. for lift-off, or a planarization layer has to be employed for printing over topography [8.127–134]. Apart from these applications, which also apply for other lithographic methods, multilayers can also be useful in NIL when polymers with different rheological or chemical properties are needed at the bottom and top of a resist. This can enhance adhesion at the bottom to the substrate, and reduce the adhesion at the top to the stamp. Furthermore the effects of bending of stamps and inhomogeneous residual layer thickness for good pattern transfer can be controlled.

The most important application of double resists is for low pressure processes such as UV-molding (Fig. 8.15). For pre-structured substrates with topography, a planarization layer is needed, because the low pressure of below 1 bar is often not sufficient to achieve a conformal contact of the transparent mask with the non-flat substrate surface. Otherwise parts of the resist stay unmolded. Multilayer resist approaches with a thick polymer planarization layer on top of the substrate require complex processes with multiple steps but also entail deep etching steps to etch through the thick planarization layer, which often degrades the resolution and fidelity of the pattern. Bilayer resists are also used for better lift-off. For this purpose lift-off resists (LOR) have been developed [8.135], which are coated below the top layer, and which can be selectively removed by wet development through the patterned top layer. The developers used are adapted to generate undercuts in LOR layers of some tens of nanometers and some mi-

crons. Then even a curable resist which is crosslinked (equivalent to a negative resist) can be used as a top layer, while the sacrificial bottom layer makes it possible to release the top layer as well as the metal layer used for lift-off.

Top layers can be chosen with high etching resistance, similar to silicon-containing resists. After molding the top layer can be hardened by silanization, and the pattern transferred into the underlying planarization layer. The top layer can be kept thin, while the etching depth can be further increased generated by choosing a thick bottom layer. However, although LORs have shown resolutions of 100 nm, even then the selectivity of double-layer systems is not high enough for high-resolution applications.

Trilayer system use a hard intermediate masking layer (etch stop) between the top and bottom layer to enhance the aspect ratio of a resist structure. Materials such as Ge or metals are often used, since they can be etched by RIE and therefore the pattern is transferred into the thick bottom layer with high fidelity.

The 3D patterning capability of NIL makes it possible to reduce the number of process steps in contact-layer fabrication of microchips by using innovative pattern-transfer processes. The connection of the transistors is done using several levels of lateral wires, each contacted vertically by through-holes. This contact layer of a chip is fabricated using lithography and copper electroplating. For the wiring scheme of a chip, as shown in Fig. 8.16, eight levels of wiring layers are needed, each of which is done in a so-called dual hard damascene process. A process has been proposed which reduces the process steps necessary for one level from 16 to 7 [8.136]. A two-tiered stamp with three height levels makes it possible to pattern the through-holes as well as the wires in one step [8.137]. In this way, several exposure steps can be replaced by a single imprint with patterns of different residual polymer layer thickness.

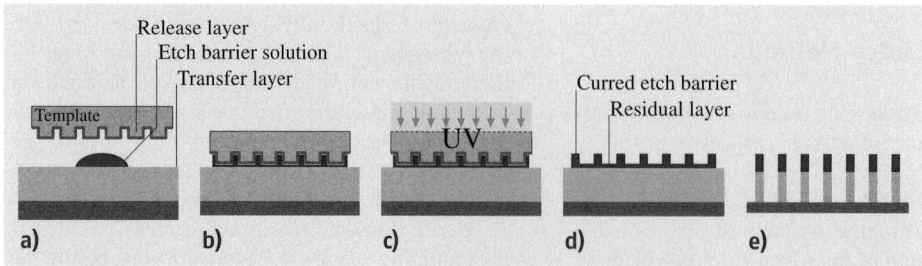

Fig. 8.15 Process scheme of UV-imprint and pattern transfer, using a double layer. The molded top layer, also called etch barrier, is coated on a transfer layer, which serves as a planarization layer. It has also anti-reflective properties for the UV-exposure through the stamp

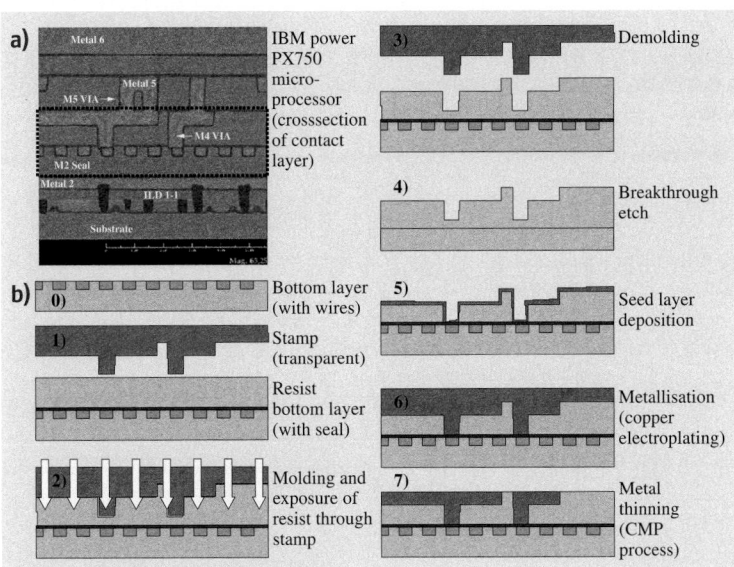

Fig. 8.16a,b Modified S-FIL™ process proposed by Sematech to replace a dual top hard damascene process for copper contact plating by a two-tiered stamp [8.26, 136]. (**a**) left side (*top*): SEM of a contact layer of a microchip (cross section) with interconnecting copper layers, (**b**) process scheme (Source: W. Trybulla, SEMATECH [8.136])

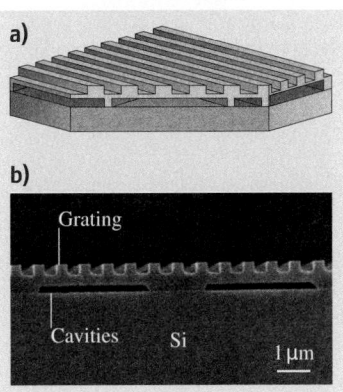

Fig. 8.17a,b Microfluidic channel fabricated by double sided reverse imprint: (**a**) 3D schematic of a resist with a top grating and embedded channels. SEM micrographs of cross sections of imprinted nanofluidic channels: (**b**) 3000 nm (width) × 200 nm (height) channels, with a 700 nm pitch grating on top. Reproduced from [8.138]

In total, the reduction from 128 process steps down to 56 results in cost reduction that justifies the introduc-

tion of a new technology, and this serves as an example that 3D pattern capability can be a decisive argument over resolution for the introduction of NIL into chip manufacture. Figure 8.16 shows the pattering scheme for one level of the contact layer of an IBM Power PC microprocessor.

8.2.9 Reversal Imprint

In contrast to NIL, in reversal imprint the resist is patterned either directly onto the stamp or onto an auxiliary substrate, and then transferred from the mold to a different substrate. Thus, patterned resist structures are obtained, and imbedded channels can be created. The concept is well presented in [8.115, 138–140]. In reversal imprinting it is possible to transfer patterns onto substrates that are not suitable for spin-coating or have surface topographies. However, complete transfer does not only depend on a good balance between the surface energies, but also on the pattern density and roughness of the structures. As an example, embedded channels generated by reversal imprint are shown in Fig. 8.17 [8.138].

8.3 Tools and Materials for Nanoimprint

Mechanical nanofabrication techniques based on molding need tools and materials with matched mechanical properties. The mold has to be made from a material which is sufficiently hard to sustain at least one

processing cycle. In mass-production terms, a mold is considered as a tool which survives the molding process unaltered and uncontaminated, and thus can be reused many times after each molding step. In this way

Table 8.3 Resist materials for thermal nanoimprint

Material (other names)	Solvent	Glass-transition temp. T_g at molecular weight M_w	Viscosity at temp., Young's modulus	Comments
polymethyl-metacrylate (PMMA) [8.135, 141–143]	anisole, ethyllactate (safe solvents)	100 °C at 25–950 k	10^5–10^9 Pas at 170 °C; 380–540 MPas	the classic NIL resist [8.5]
polystyrene (PS) [8.142]	toluene	100 °C at 50 k		integrated optics [8.144]
polycarbonate (PC) [8.145, 146]	cyclohexanone [8.147], 1,1,2,2-tetra-chloroethane	145 °C at 34 k	2350 MPas	integrated optics, $n = 1.6$ high etching resistance [8.144]
cyclo-olefine copolymer (COC) [8.148–150]	toluene [8.150]	60–180 °C [8.149]	2×10^4 Pas at 170 °C [8.150]; 2600 MPas [8.149]	highly transparent, chemically resistant, low water absorption. Lab-on-chip and optical applications [8.150–152]
mr-L 6000 [8.141]	safe solvent	30 °C at 7 k		UV curable, low-T_g NIL resist, used for polymer stamps [8.153] and mix-and-match, multilevel patterning [8.154]
mr-I 7000 [8.141]	safe solvent	60 °C		low-T_g NIL resist
mr-I 8000 [8.141]	safe solvent	115 °C		high-etch-resistance NIL resist with properties as PMMA
mr-I 9000 [8.141]	safe solvent	65 °C at 95 k		thermocurable NIL resist
hybrane [8.155]	toluene [8.156]	0–10 °C		room-temperature imprint, SF_6 RIE etch resistant [8.156]
NEB22 [8.157]		80 °C at 3 k		negative EBL resist. high etch resistance in fluoro- and chloro-based plasmas [8.147]

Table 8.4 Comparison of different materials for stamps

Material	Young's modulus (GPa)	Poisson's ratio	Thermal expansion ($10^{-6}K^{-1}$)	Knoop micro-hardness (kg mm^{-2})	Thermal conductivity (Wm^{-1}K^{-1})	Specific heat (J kg^{-1}K^{-1})
Silicon	131	0.28	2.6	1150	170	705
SiO$_2$ fused silica (bulk)	73	0.17	0.6	500	1–6	700
Quartz (fused)	70–75	0.17	0.6	> 600 (8 GPa)	1.4	670
Silicon nitride (Si$_3$N$_4$)	170–290	0.27	3	1450	15	710
Diamond	1050	0.104	1.5	8000–8500	630	502
Nickel	200	0.31	13.4	700–1000	90	444
TiN	600	0.25	9.4	2000	19	600

many identical replicas can be drawn (copied) from one mold. Due to the conformal molding, the surface of these copies is the negative structure of the original (inverted polarity). Therefore a true replica of the mold is generated when a negative is again molded into a positive structure. Here, we use the terms replica and copy in the more general sense that also negatives are considered as true copies of an original.

As the terms imprint and embossing, molding and replication are often used for the same process, different names for the replication tools exist depending on their origins: mold or mold insert for those coming from polymer processing; master or stamp (stamper) from Compact Disc fabrication; template, mask and die from the lithography community.

In this section we will have a closer look into the concepts for tools, machines and processes used for NIL. We will start with a discussion of resist materials for NIL, and then proceed with materials used for stamps: we describe fabrication methods, both for original stamps and for stamp copies, and the use and application of anti-adhesive coatings. We will then present concepts for imprint machines, and how a homogeneous pressure distribution is achieved for nanoreplication. For thermal imprint as well as UV-imprint, full-wafer processing and step-and-repeat approaches have been developed. The aim is to familiarize the reader with concepts rather than presenting machines and materials sold in the marked (Typical NIL machines can be seen in Fig. 8.20).

8.3.1 Resist Materials for Nanoimprint

Resists used for NIL are either used as an intermediate masking layer for the substrate or as a functional layer for a specific application. Both the processing properties, as well as those for the final application purpose, have to be considered. Many of the resists, such as those used for optical lithography (OL) and electron beam lithography (EBL) [8.158, 159], exhibit thermoplastic behavior. PMMA, a regular linear homopolymer, with a short side-chain, is used as a high-resolution standard material for EBL. PMMA is also used as a bulk material for hot embossing and injection molding. It has been known for a long time that sub-10-nm resolution can be achieved [8.160]. PMMA is a low-cost material, and available in different molecular weights M_w. It is compatible with other cleanroom processes, exhibits good coating properties using safer solvents, and can be coated from solution to a thickness ranging from 20 nm up to several μm. It has well-characterized op-

tical, mechanical and chemical properties, and proven reliability in many different applications. When used as an etching mask, e.g. for Si, it exhibits a sufficient, but not high, etching resistance. The glass-transition temperature T_g of PMMA is low enough to enable molding at temperatures below 200 °C, but high enough to ensure a sufficient thermal stability in etching processes. Acrylate-based polymers can also be used with crosslinking agents.

During the first 10 years of NIL, a number of resists have been developed and characterized. In Table 8.3 we give an overview of NIL resists with references to further information on these materials. Further information can be found in [8.161–172].

8.3.2 Stamp Materials

Not only the mechanical, but also optical and chemical properties are important when choosing a stamp material for NIL. Critical mechanical parameters and their implications for NIL (in brackets) are hardness and thermal stability (lifetime and wear), thermal expansion coefficients and Poisson's ratio (dimension mismatch leading to distortions during demolding), roughness (higher demolding force and damage), Young's modulus (bending), and notch resistance (lifetime and handling). Issues related to fabrication are processability (etching processes, selectivity, cleanroom environment), and surface quality (resolution). The use in a NIL process is determined by additional properties such as transparency, conductivity, anti-sticking properties (with/without anti-adhesive coating, e.g. by covalent coating), availability and cost (standard materials and sizes, tolerances, processing equipment and time), and how easy it is to employ in NIL (e.g. fixing by clamping, thermobonding, gluing). In Table 8.4 we give a brief overview of the mechanical and thermal properties of materials used for stamps. Further information can be found in [8.26, 173–188].

8.3.3 Stamp Fabrication

Any kind of process generating a surface profile in a hard material can be used to fabricate stamps for NIL. The most common lithographic processes are based on resist patterning with subsequent pattern transfer. Therefore the requirements for these processes, such as resolution, aspect ratio, depth homogeneity, sidewall roughness, sidewall inclination, are similar to the processes presented before in this chapter. For the highest resolution, both serial and parallel fabrication methods

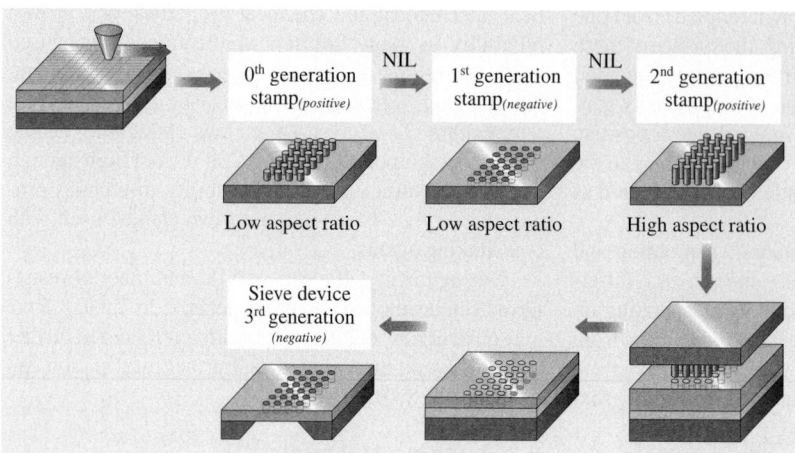

Fig. 8.18 Process chain from stamp origination to application, here as an example the porous membrane chip as shown in Fig. 8.14. The low aspect ratio stamp original fabricated by EBL and RIE is transferred into a high aspect ratio stamp by two consecutive NIL copying steps, providing an increased lifetime of the original and more flexibility

are available, however, with different area, throughput, and freedom of design. The processes are standard processes for nanolithography, which can also be used directly for patterning. When using them for the fabrication of stamps, apart from higher throughput a larger flexibility and reproducibility can be achieved. Using stamp copies instead of the original is a way to enhance the lifetime of a stamp, simply because the original is reserved for the copying process. There are different methods to generate copies from hard masters with proven resolutions below 100 nm:

- Electroplating is a commercially successful method to copy an original into a metal replica. The nickel shims used in compact disk (CD) manufacture support tens of thousands of molding cycles without significant wear. The original, a patterned resist or etched relief on a glass master, is often lost during the transfer to nickel, therefore only after a first-generation nickel copy is drawn can further generations be repeatedly copied from it.
- Using the hard master with an etched surface relief directly as a mold is a straightforward way if the mechanical set up allows or favors the use of silicon or quartz. Stamp copies can be fabricated using NIL and subsequent pattern transfer (see Fig. 8.18). Molds made from silicon wafers are well suited as stamps in NIL, and have even shown their mass fabrication capability in Compact Disc injection molding.
- As a third solution a polymer with an imprinted surface relief can be directly used as a replication tool. This is possible, if the thermomechanical replication process does not exert high forces on the relief structure. Resist hardened by light, heat or by chemical

initiation may support high temperatures and can be repeatedly used in NIL. However, the lifetime of polymeric molds is still low and good solutions for anti-adhesive coatings have to be found.

- Hybrid molds use different materials for the surface relief and the support. They consist of a substrate plate as a mechanical support covered with a thin polymer layer with a nanostructured relief. In the case of NIL they have the advantage that a substrate material can be chosen with thermomechanical properties adapted to the substrate to be patterned. Furthermore it is useful if thin flexible substrates are needed.

These methods differ mostly in the properties of the materials used for the stamps (mechanical robustness, thermal expansion coefficient, transparency, fabrication tolerance) and the surface properties of the patterned relief (possibility of anti-adhesive coating). Although for many applications electroplating of metal molds is favored, because of its high flexibility and robustness compared to that of silicon, the effort to fabricate high-quality mold inserts with defined outlines is often only justified for production tools.

8.3.4 Anti-Adhesive Coatings

One of the most important tasks for NIL is to provide stamps with good anti-sticking surface properties [8.189–197]. The stamp surface should allow the molded surfaces to detach easily from the mold, and once released, provide low friction, resulting in a continuous vertical slipping movement without sticking. Nanoscopic interlocking of structures caused by side-

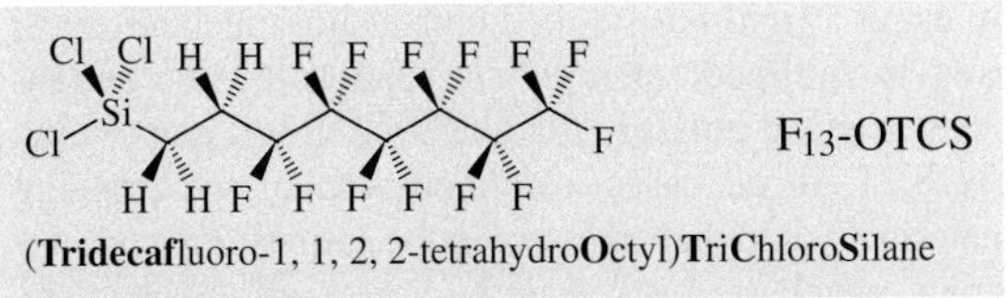

Cl Cl H H F F F F F F
Cl—Si————————————F   F₁₃-OTCS
    H H F F F F F

(Tridecafluoro-1, 1, 2, 2-tetrahydro**O**ctyl)**T**ri**C**hloro**S**ilane

Fig. 8.19 Molecular structures of a fluorinated silane with a reactive trichlorosilane head group and a long alkyl chain with fluorine substituents (length about 2 nm). The silane binds covalently to the silicon oxide of the stamp surface and is used as the standard anti-adhesive coating for silicon stamps in NIL

wall roughness should be elastically absorbed by the molded material, while the surface maintains its anti-sticking properties. Because the molded polymer film is squeezed between the two surfaces of the stamp and substrate, they need to exhibit opposed surface properties. The adhesion at both interfaces must be different to an extent that, while the polymer film adheres perfectly at the substrate surface, the stamp can be separated from the structures without any damage at any location of the stamp. If the stamp material does not exhibit good anti-sticking properties to the molded material, the stamp has to be coated with a thin anti-adhesive layer. A low-surface-energy release layer on stamp surfaces not only helps to improve imprint qualities, but also significantly increases the stamp lifetime by preventing surface contamination. An anti-adhesive coating has to be chemically inert and hydrophobic, but at the same time allow a filling of the mold cavities when the polymer is in its viscous state.

One of the major advantages of using silicon wafers as stamps for NIL is that they can be coated with anti-sticking films using silane chemistry. Damage of the molded structure during demolding is highly dependent on the quality of the anti-adhesive layer. Fluorinated trichlorosilanes with different carbon chain lengths are commonly used due to their low surface energy, high surface reactivity, and high resistance against temperature and pressure. They support multiple, long embossing sequences with repeated temperature cycles higher than $200\,°C$. Currently it seems that, as long as mechanical abrasion can be avoided, the silanes match the normal-use lifetime of a Si stamp, which is some tens of cycles for NIL in a laboratory environment, or thousands if automated step-and-repeat imprint processes or injection-molding processes are used. Apart from silicon wafers, which have the advantage that they are suitable for standard cleanroom processing, other materials to be used as NIL stamps, e.g. Ni shim or duroplastic polymers, can also be coated with silanes if an intermediate SiO_2 layer is deposited onto the materials. The silane coating can be performed by immersion in a solution of iso-octane, or by chemical vapor deposition (CVD), either at ambient pressure by heating the silane on a hot plate or by applying a moderate vacuum of a few mbar. One of the most prominent advantages of the vapor deposition method is that it is not affected by the wetting ability of a surface, so that it is suitable for stamps with extremely small nanostructures.

A commercially available silane used is shown in Fig. 8.19. $F_{13} - OTCS$ =(tridecafluoro-1,1,2,2-tetrahydrooctyl)-trichlorosilane is the standard material

Fig. 8.20a–c Three examples of NIL presses (**a**) Simple hydraulic press, with temperature-controlled pressing plates. (**b**) Semi-automated, hydraulic full-wafer NIL press, based on an anodic bonder. (**c**) Automated step-and-flash UV-NIL production tool

for anti-adhesive coatings on silicon (ABCR SIT 8174) [8.198].

8.3.5 Imprint Machines

An imprint machine needs a precise pressing mechanism with demanding requirements on mechanical stiffness, uniformity and homogeneity over large areas [8.11, 199–206]. At the same time it should adapt to local variations of pressure and temperature, due to imperfections and tolerances in stamps and substrates, and simply because the stamp protrusions are inhomogeneously distributed. In molding of microstructures, where deep channels with lateral and vertical sizes in the range of 50 μm have to be molded, the stamps are made stiff, and precise, reproducible vertical piston movements within some tens of μm have to be realized with good fidelity. NIL would need precisions of a few tens of nm, which does not correspond to the usual tolerances of some μm for the substrates and tools used. Therefore the NIL stamps have to be flexible, and must be made to adapt to the small vertical deviations over a long lateral range, which are on the scale of the tolerances of the common materials used for stamps, substrates, and the density variations of the stamp surface relief.

Embossing machines generate a desired pressure pattern over the total area of the stamp. A high throughput in manufacturing devices on full-wafer scale can be achieved either by parallel patterning of large areas, or by fast repeated patterning using a semi-serial stepping process. The pressure field can also be applied sequentially,

by using a rigid but stepped embossing mechanism, as used in millipede stamps, see Chapt. 46, or continuously scanned pressure field, as used in roll embossing, Fig. 8.21. In all cases a defined area of the molding material is sandwiched between the solid stamp and substrate, which are backed up by a pressing mechanism. The major differences lie in the fact that single-step imprinting processes might not be easily transferable to continuously stepped imprints, where previously structured areas should not be affected by imprints in close vicinity. In photolithography, stepping was needed because of the limitation of the maximum field size to be exposed, and because the continuous reduction of structure sizes and diffraction effects was only possible by optical reduction of the masking structures into the resist by high-resolution optics. Furthermore this enabled to keep the process non-contact, while 1:1 imaging of a mask structure would have lead to an unwanted reduction of the proximity gap.

Full-wafer imprinting

The simplest mechanism for full-wafer imprinting is a parallel-plate embossing system. A linear movement of the piston behind the stamp leads to a local thinning of the polymer under the stamp protrusions, which is possible because the polymer is moved from squeezed areas into voids in the stamp. This movement can be generated using pneumatic, hydraulic, or motor-driven pistons. The pressure must be maintained during the whole molding process, until the voids are filled, and the molded structures are *fixed* during the cooling or curing

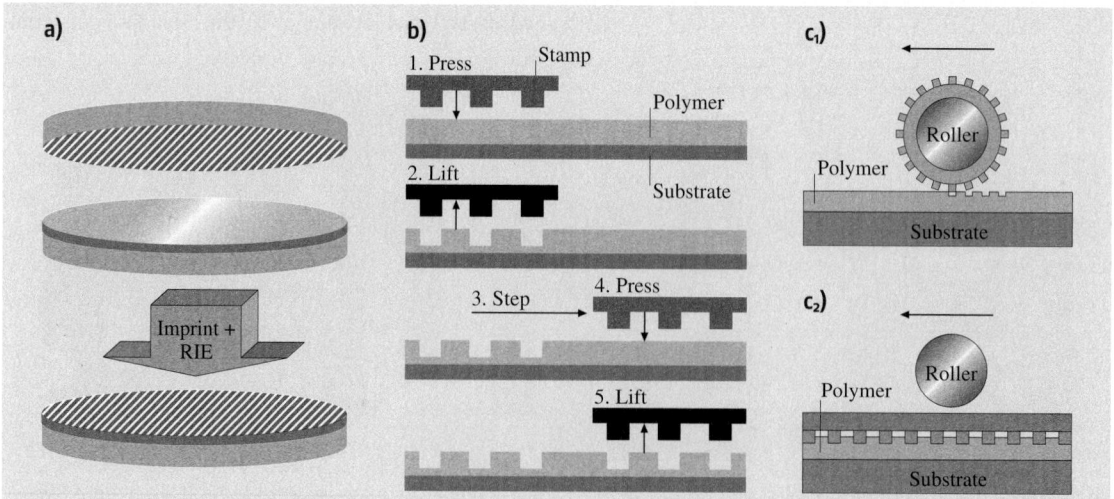

Fig. 8.21a–c Outline of the three most common types of NIL machines (**a**) full-wafer parallel press, (**b**) step-and-repeat press, and (**c**) two roll-embossing set ups

step, depending on the method used. However, under normal process conditions, an embossing with a hard master does not work without a cushioning mechanism. This cushion balances both thickness variations due to tolerances of the set up and the nature of the molding process. The latter is caused by the fact that the size and shape of the stamp surface relief leads to local pressure variations during the squeeze flow and, if the stamp can bend, to a local difference in the sinking velocity. When using thick polymer plates, where the molding leads to a surface modulation of a bulky material, the cushion is formed by the viscous material itself. However, in NIL, a thickness profile has to be generated in a resist with a thickness that is often lower than the thickness tolerances of the substrates and mechanical set up used. Furthermore height defects in the range up to some μm, such as dust particles, have to be equilibrated. Therefore cushioning has to be achieved by the pressing mechanism and its ability to compensate has to be larger than the defects and tolerances of the stamps and substrates. A lateral spreading and dispersion of the applied pressure can be achieved by using a spring mechanism, which can consist of an additional plastic or elastic layer, e.g. a mattress made of rubber [silicone, poly dimethyl siloxane (PDMS), Viton^{TM}], Teflon^{TM} [poly (tetrafluoro-ethylene) or PTFE] or elastic graphite can be used. The thickness has to be chosen in order to achieve an equilibration of a few micrometers, for which some 100 μm are sufficient. Due to the high pressure used in NIL, compensation by a wedge is not needed. The applied pressure of the large backing plate is then spread into infinitesimal small-area elements behind the stamp, and is able to compensate for pressure variations occurring during the lateral flow of the molding material. By using this method the height requirements of the substrate surface and material can be minimized and continuous printing in all areas is enabled. An even better pressure homogeneity can be obtained, when the cushion effect is generated by compressed air or liquid. This can be realized by replacing the upper stamper with a pressure chamber, which uses the stamp as a deformable mambrane. In practice this is realized by placing a metallic or polymeric membrane between the pressure chamber and the stamp, which deforms around the stamp and substrate, and which is sealed with the counterforce of the lower stamper [8.11]. The advantage of this soft stamping method is that very gentle contact between the stamp and substrate can be achieved by adjusting the air pressure, so that the entire stamp surface can assume parallel alignment with respect to the substrate before the molding starts. Dur-

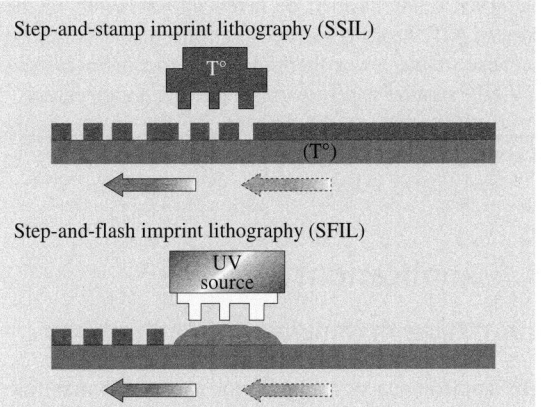

Fig. 8.22a,b Step and repeat processes (**a**) in NIL: step and stamp imprinting lithography (SSIL), and (**b**) in UV-NIL: step and flash imprint lithography (SFIL^{TM}). While in S-FIL the liquid resin is cured locally be exposure through the stamp, in SSIL the resist is locally heated above its glass transition temperature by the hot stamp

ing molding the pressure is equilibrated without delay, which assures a constant press force in all areas of the stamp, only limited by the bending of the stamp.

All press concepts can be realized with heating elements for NIL, or with a UV-exposure tool which enables the exposure of the resist during molding. Furthermore, combinations of thermoplastic molding and UV exposure are possible. The main difference between thermoplastic molding and UV imprinting are the pressures needed for embossing. Pressures of 1–100 bar are used in NIL, while less than 1 bar is sufficient for UV-NIL. Forces of some kN can be achieved using presses with pneumatic systems, however, for large-area imprints, forces of several hundreds of kN can be achieved by using hydraulic systems.

Step-and-repeat imprinting

NIL can be carried out using three different types of machines: single, step-and-repeat, and roller imprint. Single-imprint machines pattern the surface of an entire wafer in one step. Thus the stamp must have the same size as the wafer to be patterned. Step-and-repeat NIL machines pattern a smaller area of a wafer at a time, and then move to an unpatterned area, where the process is repeated. The process is continued until the whole wafer is patterned. While this set up enables the use of smaller and more cost-effective molds, with which higher alignment accuracy can be achieved, higher process times and stitching errors at the boarders of the patterned fields have to be taken into account. In the case of NIL, heating and cooling times can be reduced because of the lower thermal mass, and or in case of UV-NIL smaller exposure fields may be an advantage.

In NIL the thermal mass of the parts being thermally cycled should be minimized, in order to reduce the obtainable process time. This problem is readily addressed in step-and-stamp and roller-embossing approaches, but solutions have also been found in the concepts of heatable stamps [8.206, 207] or by surface heating by means of pulsed laser light [8.208].

8.4 Applications

8.4.1 Types of Nanoimprint Applications

The applications of nanoimprint can be as manifold as those of other lithographic patterning methods. The applications can be divided into two main categories: pattern-transfer applications and polymer devices. In the first category, pattern-transfer applications, the nanoimprinted resist structure is used as a temporary masking layer for a subsequent pattern-transfer step. In the second category, polymer devices, the imprinted pattern adds functionality to the polymer film, which is the end product.

In many pattern-transfer applications, the main issue is high throughput at nanoscale resolution. Disregarding this issue, it is of minor importance whether the resist film is patterned by means of electromagnetic radiation, electrons, or by mechanical deformation. Only a few steps in the process flow are different, for example the dry etch step to remove the 10–100 nm-thick residual polymer layer after the imprint. Both additive and subtractive processes have been demonstrated, as discussed in Sect. 8.2 of this chapter. Sometimes even the resist is the same, for example PMMA, which is a widely used resist for both EBL and NIL. The advantages of NIL come into play if high resolution is needed over a large area. For such applications, NIL is a cost-effective alternative to current cutting-edge lithography techniques such as deep ultraviolet (DUV) lithography [8.209], dedicated to CMOS chip manufacturing. The cost of ownership for next-generation optical lithography technologies, such as extreme ultraviolet (EUV) lithography [8.210], is reaching a level that requires extremely high production volumes to be economically viable. This development has already forced several branches of the electronics industry to explore NIL as an alternative fabrication method. Examples of such products are patterned media for hard-disk drives [8.211, 212], surface-acoustic-wave (SAW) filters for cell phones [8.26, 213] and sub-wavelength wire-grid polarizers for high-definition projection TV [8.214].

Even the semiconductor industry is considering NIL as possible next-generation lithography to deliver the 32 nm node and beyond [8.12]. For chip manufacturing the resolution issue is most important, because NIL simply does not have the restrictions encountered by optical methods and already offers a resolution capability higher than the next technical nodes. Among the major technological challenges to be solved are: overlay accuracy, error detection, fast imprint cycles and critical dimension (CD) control. In addition to its high resolution, the NIL technique also offers the capability for 3D or multilevel imprinting, when the stamp is patterned with structures of different heights (Sect. 8.2.8).

The NIL process offers new possibilities to form polymer devices with microscale to nanoscale features. Nanoscale patterned polymer films find a wide range of applications within optics, electronics and nano-biotechnology. The capability to form 3D polymer structures, with curved surfaces and high aspect ratios, paves the way for new classes of polymer-based passive optical devices, such as lenses and zone plates [8.215], photonic crystals [8.96, 216, 217] and integrated polymer optics [8.144]. The NIL technique allows for choosing a wide range of polymers with optimized optical properties [8.217, 218], and allows for the patterning of thin films of organic light-emitting materials and polymers doped with laser dyes to create organic light-emitting devices (OLED) [8.219, 220] and lasers [8.221–223]. NIL is also suitable for nanoscale patterning of conducting organic films for cost-effective organic electronics [8.224].

Within the rapidly growing field of lab-on-a-chip applications [8.225], NIL offers an attractive, cost-effective method for molding of complex structures, integrating micro- and nanofluidics, optics, mechanics and electronics on a single chip [8.226]. For example, the micro- to nanoscale fabrication capabilities are used to create single-use polymer devices containing nanopillar arrays [8.227] and nanofluidic channels [8.228] for DNA separation and sequencing.

In this section we will give an overview of the different fields of applications. We start with two examples of pattern-transfer applications, which are close to production: patterned media for hard-disk drives, and sub-wavelength metal-wire gratings for high-definition television (HDTV) projectors. We then discuss a few examples of laboratory-scale potentially high-impact applications of NIL. These examples were selected from a large number of NIL applications. The number of laboratory-scale NIL applications is rapidly growing, reflecting the wealth of new possible device architectures becoming feasible by NIL. Some of the applications are directly relevant for industrial production, and others are directed towards research. Even in research the nanostructuring capability of replication processes are needed. Further insight into this field is given in Sect. 8.2.6 about pattern transfer and in Sect. 8.5 about commercialization aspects of NIL.

8.4.2 Patterned Magnetic Media for Hard-Disk Drives

Since the first demonstration of NIL in the mid 1990s, patterned magnetic media for hard-disk drives (HDDs) has been a key application, driving the development of NIL technology [8.229]. Since the invention of the HDD in 1957, the storage capacity, quantified in areal density of bits, has increased to the current (2005) level of $135\,\mathrm{Gb/in^2}$, as described in Chapt. 47. The size and density of the individual bits, defined by the local magnetization of a homogenous, or unpatterned thin magnetic film, was reduced by application of multilayer magnetic films as recording media, the sensitivity of the read head was increased by exploiting the giant-magneto-resistance effect in multilayer thin-film conductors [8.230], and the magnetization was applied perpendicular to the surface of the recording media,

while micro-electromechanical systems (MEMS) technology for the mechanical parts has been developed to an extreme level: In current HDDs the read–write head flies at a height of $2–3\,\mathrm{nm}$ above the surface of the disk plate. An overview of HDD technology is given in [8.231]. This current level of storage density is projected to increase by three orders of magnitude over the next 10 years in order to meet market requirements, see Fig. 8.5.

The possibilities to increase the bit density with the current technology, where bits are written by local magnetization of an unpatterned thin magnetic film, are mainly limited by the read–write width, the positioning of the magnetic head, and by thermal instability, induced by superparamagnetism in the grains of the magnetic film. These challenges are addressed by patterning the magnetic film. Discrete track recording (DTR) media [8.232], where the magnetic film is patterned with a spiral land and groove track have been developed to overcome the problems associated with the read–write width and positioning of the magnetic head, see Fig. 8.23a. In HDDs with unpatterned recording media, the write-to-write track misregistration (TMR) is strongly dependent on the mechanical positioning capability of the magnetic head, and the magnetic write width (MWW) is determined by the write head. Due to the fringing fields from the write head, the magnetic read width (MRW) is only about 60% of the MWW in current unpatterned HDD media. The write-to-write TMR is significantly reduced for a DTR medium, and the MWW is defined by the land width if the write head is equal to or wider than the land width. This allows the increase of the width of the read head beyond the land width or MWW, allowing for an improved signal-to-noise ratio.

The idea of DTR media is more than 40 years old [8.232], but has not been implemented in production due to the lack of a nanolithography process that meets

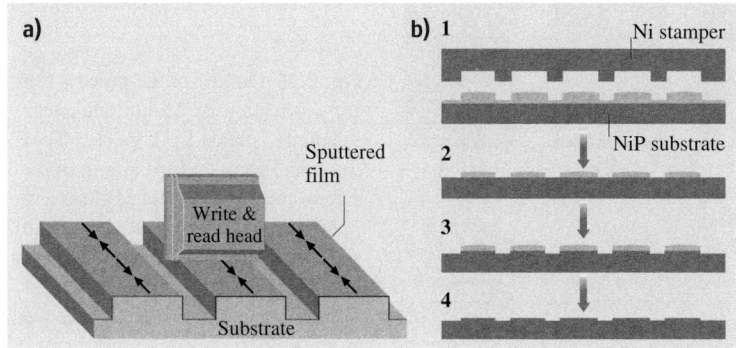

Fig. 8.23 (a) Outline of a DTR medium showing the land-and-groove structure, patterned into a NiP-plated Al:Mg substrate. The magnetic thin film is sputtered onto the patterned substrate. An improved signal-to-noise ratio can be obtained by making the magnetic read and write heads wider than the land width. **(b)** Outline of the NIL-based fabrication process. Reproduced from [8.211]

the demanding requirements for the surface smoothness of the disk surface [8.233], and which is suitable for large-scale low-cost fabrication. Researchers at Komag Inc. have demonstrated a cost-effective process for volume manufacturing of DTR media, based on double-sided thermal NIL with a commercially available resist and wet etching on a 95-nm diameter nickel phosphorous (NiP)-plated Al:Mg disk [8.211]. The process steps are outlined in Fig. 8.23b. The nickel stamps, with track pitches down to 127 nm, corresponding to an areal density of 200 Gb/in^2, were electroformed from a silicon master, which was patterned either by laser-beam or electron-beam writing, equipped with a rotating stage with radial beam positioning. After etching, the polymer was removed by oxygen plasma, and the disk was then sputter coated with a CrX/Co-alloy double layer magnetic thin film. These devices were designed for in-plane, i.e. longitudinal, magnetic polarization, but DTR media for perpendicular polarization have also been realized by EBL and RIE etching of the magnetic film [8.234].

The DTR media technology offers a possibility to regain the loss in electrical signal-to-noise ratio, as the magnetic bit size is reduced. However, with the decreasing bit size, necessary to follow the roadmap in Fig. 48.5, the technology will be limited by thermal instabilities, or superparamagnetism. The magnetic film consists of small, weakly coupled magnetic grains, which behave as single-domain magnetic particles. Each bit consists of the order of 100 grains to obtain a reasonable signal-to-noise ratio. In order to keep this ratio of grains per bit, the grain size must be reduced with bit size. The magnetic energy of a single grain scales with the volume of the grain. This implies that the bit can be erased thermally, when the grain size becomes sufficiently small and weakly coupled to neighboring grains. This is referred to as the superparamagnetic limit.

The superparamagnetic limit can be overcome by lithographically defining each bit, as a magnetic nanoparticle, or *nanomagnet* [8.229, 235, 236]. In such a *quantized magnetic disk* [8.235] each magnetic nanoparticle is a single magnetic domain with a well-defined shape, with a uniaxial magnetic anisotropy, so the magnetization only has two possible stable states, equal in magnitude but opposite in direction, as illustrated in Fig. 8.24. Such defined bits can be thermally stable for sizes down below 10 nm [8.212].

The feasibility of NIL for fabrication of patterns of magnetic nanostructures for quantized magnetic disks has been investigated by several research groups. The imprinted pattern has been transformed to magnetic nanoparticles by electroplating into etched holes [8.237], by lift-off [8.238] and by deposition onto etched pillars [8.212, 239].

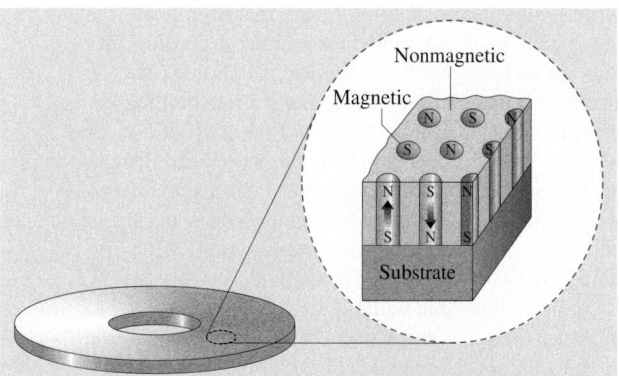

Fig. 8.24 Outline of a patterned magnetic disk for high-density data storage. Each bit is a lithographically defined, single-domain magnetic nanostructure, embedded in a nonmagnetic matrix. Reproduced from [8.229]

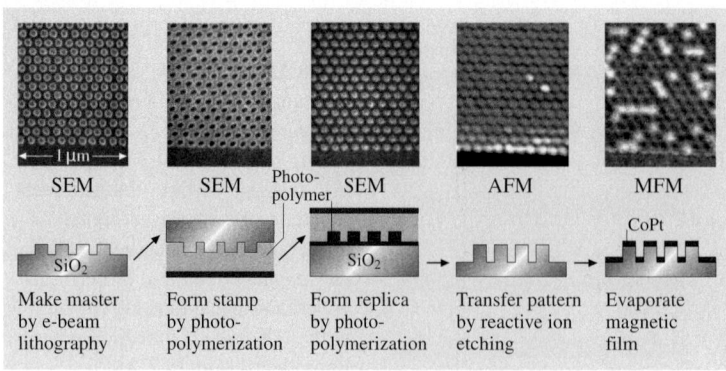

Fig. 8.25 Outline of the process flow for fabrication of 55-nm-diameter magnetic islands by UV-NIL. The top panel shows SEM, atomic force microscope (AFM) and MFM micrographs at the different stages of the process. The MFM micrograph shows quantized up and down magnetization of isolated domains. Reproduced from [8.212]

In Fig. 8.25 we show the outline of the process flow for large-area fabrication of 55-nm-diameter 11-nm-high CoPt magnetic islands [8.212], by means of UV-NIL. A SiO$_2$ master containing three $50 \times 50\,\mu m^2$ areas of hexagonal 100-nm-pitch arrays of 30-nm-high 55-nm-diameter pillars was fabricated by defining the dot pattern by means of EBL in a 160-nm-thick film of PMMA with a molecular weight of 950 kg/mol. The patterned PMMA film was used in a lift-off process to define a Cr etch mask. The pillars were etched by CF RIE, and the metal mask was removed. The master was used to form a stamp in a photopolymer material. This stamp is used to UV-imprint the dot pattern in a photopolymer film on a SiO$_2$ substrate, leaving a replica in the photo-cured polymer, with 28-nm-high pillars on top of a 10-nm-thick residual layer. The pattern was transferred into the SiO$_2$ substrate by CF$_4$ RIE to remove the residual layer, followed by a 7/1 CF$_4$/CH$_4$ RIE. Finally, a CoPt magnetic multilayer structure (Pt$_{1\,nm}$(Co$_{0.3\,nm}$Pt$_{1\,nm}$)$_7$Pt$_{1\,nm}$) was deposited by electron-beam evaporation. The devices were characterized by magnetic force microscopy (MFM), revealing that the film on each pillar is a magnetically isolated single domain that switches independently.

8.4.3 Sub-Wavelength Metal–Strip Gratings

Metallic wire gratings with a pitch below 200 nm can be used to create polarizers, polarization beam splitters and optical isolators in the visible range. Such devices have many applications in compact and integrated optics. One example is the use of sub-wavelength wire-grid polarization beam splitters in liquid crystal on silicon (LCoS) projection displays for high-definition television (HDTV) , yielding higher contrast, uniformity and brightness of the displayed image (Fig. 8.26).

The polarizing functionality of sub-wavelength wire gratings is based on form-birefringence, an optical anisotropy which appears when isotropic material is structured on a length scale much smaller than the wavelength of light, λ. In this limit, the description of light propagation based on the laws of diffraction, refraction and reflection is not valid, and a rigorous solution of Maxwell's equations with the relevant boundary conditions must be applied. For a review of sub-wavelength optics, see [8.240]. The sub-wavelength linear grating of period $d < \lambda/2$, line width a, and height h, as illustrated in Fig. 8.27, will behave as a film of birefringent material with refractive indices n_s and n_p for the s-polarized

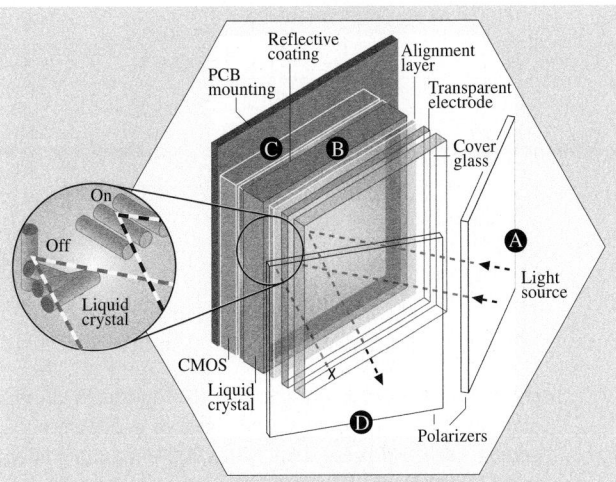

Fig. 8.26 LCoS Display for HDTV projection. A light source shines through an external polarizing layer (A) that blocks all light except waves oriented in one plane. The liquid crystal layer (B) twists some waves and lets others proceed unchanged to the reflective layer (C), depending on each pixel's charge; from there they bounce back to another external polarizing layer (D). Here the untwisted light passes through, and the twisted light is blocked. Reproduced from www.pcmag.com

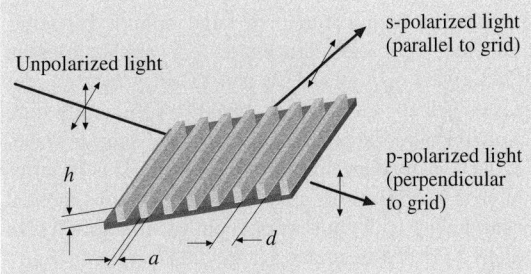

Fig. 8.27 Sub-wavelength wire -rid polarizer. By application of sub-wavelength gratings, with a pitch below 100 nm for visible light, first-order diffraction with a high acceptance angle, and low-dispersion birefringence is obtained

(E-field parallel to the grating) and p-polarized (E-field perpendicular to the grating) light:

$$n_p^2 = \frac{d}{a}n_1^2 + \left(1 - \frac{d}{a}\right)n_2^2 ,$$
$$n_s^2 = \frac{n_1^2 n_2^2}{\frac{d}{a}n_2^2 + \left(1 - \frac{d}{a}\right)n_1^2} . \tag{8.6}$$

Where n_1 and n_2 are the refractive indices of the isotropic grating and fill materials, respectively.

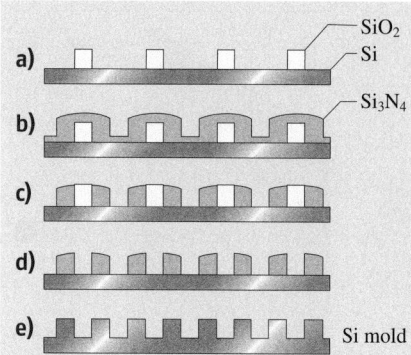

Fig. 8.28a–e Fabrication steps for a 100-nm-pitch grating mold by doubling the number of lines in a 200 nm-pitch grating: (**a**) The process starts with a 200-nm-pitch grating in SiO₂. (**b**) Deposition of CVD Si₃N₄. (**c**) Reactive-ion etching in CHF₃ and O₂. (**d**) Removal of SiO₂ using diluted HF. (**e**) Pattern transfer into the Si substrate. Reproduced from [8.241]

These sub-wavelength wire gratings have several advantages in terms of a large acceptance angle, large extinction (ratio T_p/T_s between the transmittance of the s- and p-polarized components), long-term stability at high light-flux levels, temperature and humidity. They can be manufactured in large volume by semiconductor fabrication processes. For applications in liquid-crystal display (LCD) and LCoS projection devices, it is a key challenge to obtain a sufficiently high extinction ratio, larger than 2000, at the shorter wavelengths, i. e. for blue light ($\lambda \approx 450$ nm). This requires a pitch d of 100 nm or smaller, which is not practical for producing with conventional optical lithography. *Yu* et al. [8.241], demonstrated a large-area (10 cm × 10 cm), $d = 100$ nm pitch by NIL. The stamp gratings were formed by interference lithography using an Ar-ion

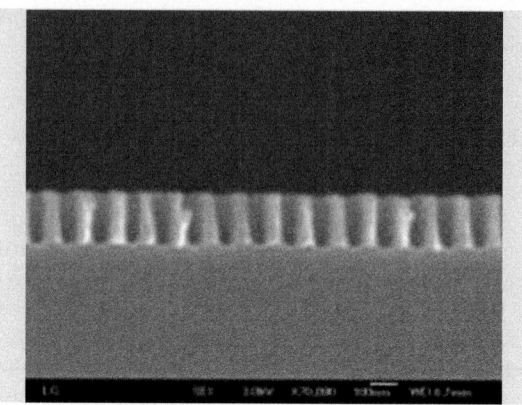

Fig. 8.30 Sub-wavelength wire-grating polarizer with $d = 100$ nm pitch. The aluminium ribs are 100 nm high from K.-D. Lee, LG electronics

laser ($\lambda = 351.1$ nm) to achieve a pitch around 200 nm, which was transferred into a SiO₂ film by RIE. The pitch was subsequently halved by CVD deposition of Si₃N₄, and CHF₃/O₂ RIE, as illustrated in Fig. 8.28. Researchers at LG Electronics have realized $d = 100$ nm Al wire-grating polarizers by thermal NIL. The process is outlined in Fig. 8.29. The large-area grating stamp, see Fig. 8.30, is fabricated by laser interference lithography in photo-resist, and transferred into the underlying 200-nm-thick SiO₂ film using CF₄ and O₂ RIE [8.214]. The 5 cm × 5 cm devices (Fig. 8.31) have an extinction ratio of over 2000 and a transmittance above 85% in the blue, at $\lambda = 450$ nm. In comparison, commercially available $d = 140$ nm wire-grid polarization beam splitters [8.242], fabricated by optical interference lithography, have an extinction ratio around 1000 in the blue. Nanoimprinted sub-wavelength polarizers for the infrared (1.0 μm $< \lambda <$ 1.8 μm) are also commercially available [8.243], with a transmit-

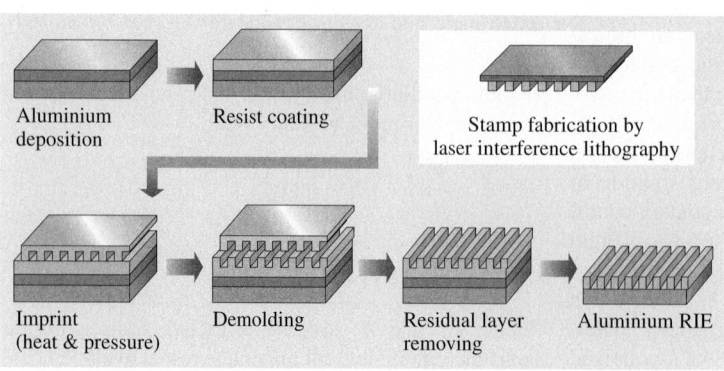

Fig. 8.29 Outline of the NIL process, to fabricate $d = 100$ nm pitch aluminium wire-grating polarizers

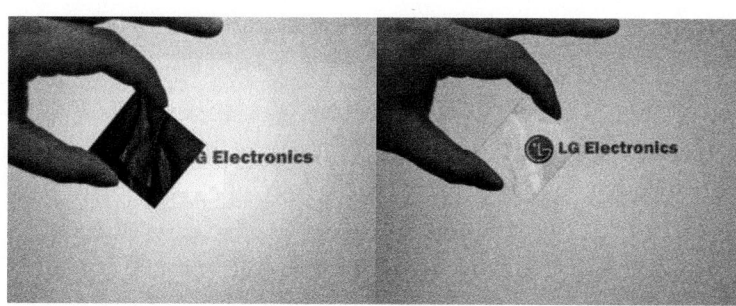

Fig. 8.31 Large-area 100-nm-pitch wire-grid polarizer with 85% transmission and extinction ratio larger than 2000 at wavelength $\lambda = 450$ nm (blue light) From K.-D. Lee, LG electronics

tance above 97%, and a transmission extinction better than 40 dB.

8.4.4 Polymer Optics

NIL is ideally suited for the fabrication of polymer nanophotonics and waveguide devices with submicron critical dimensions, defined over large areas. It is also compatible with many polymer materials, giving large freedom to choose a material with specific optical properties [8.221, 222, 244].

In Fig. 8.32 we show a polymer micro-ring resonator fabricated by NIL [8.144]. This type of device has been realized in PMMA, PC, and PS on SiO_2 substrates. The resonator consists of a planar waveguide and an adjacent micro-ring waveguide. The waveguide and micro-ring are coupled though the evanescent field in the coupling region. Resonant dips in the transmission through the waveguide occur when the phase pick-up in a trip round the micro-ring is equal to $2\pi m$, where m is an integer. The device works as a narrow bandwidth filter, and finds applications within integrated optics and for biosensing [8.45].

The evanescent coupling coefficient between the waveguide and micro-ring depend exponentially on the size of the gap. The devices are realized with 1.5-μm-high waveguides, and a coupling air gap of 100–200 nm. The process flow is outlined in Fig. 8.32c. A thin initial polymer layer is spin-cast onto a SiO_2 substrate layer. The stamp has a very large fill factor and large protrusion areas, implying that a large polymer flow is needed to fill the stamp cavities. A thin residual polymer layer is obtained by combining a high imprint pressure, a high process temperature and a long imprint time. The mode confinement in the polystyrene (PS) waveguides is enhanced by etching the substrate oxide layer isotropically in HF, to create a pedestal structure. The Q-factor of the resonator device depends critically on the surface scattering losses in the waveguides. The surface roughness of the polymer

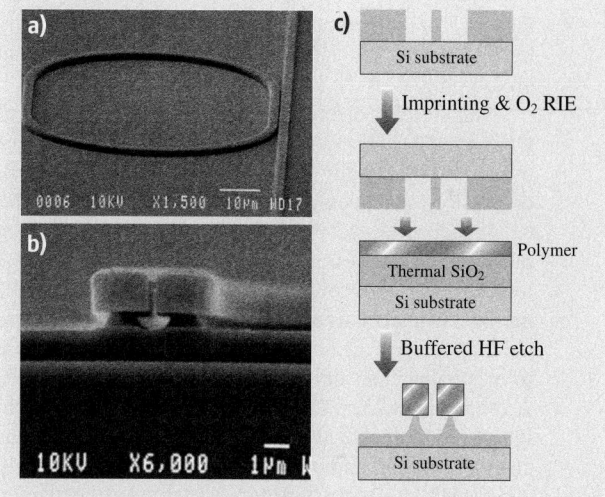

Fig. 8.32a–c Nanoimprinted polymer micro-ring resonator. (**a**) SEM picture of the imprinted device (**b**) Cross-sectional SEM picture of the polymer waveguides in the coupling region of the micro-ring device (**c**) Outline of the process flow. Reproduced from [8.144]

waveguides can be reduced by a controlled thermal reflow. The device is heated to $10–20\,°C$ below the glass-transition temperature, and the surface reflows under the action of surface tension. A loss reduction of more than 70 dB/cm was achieved by this approach [8.245].

8.4.5 Bio Applications

Micro- and nanofabrication technology has enabled methods to manipulate and probe individual molecules and cells on a chip [8.246–250]. This type of applications often requires a large area covered with nanostructures; sometimes a large number of identical devices are needed for statistical evaluation, or to give redundancy, e.g. against clogging of nanofluidic channels. With these requirements, NIL is advan-

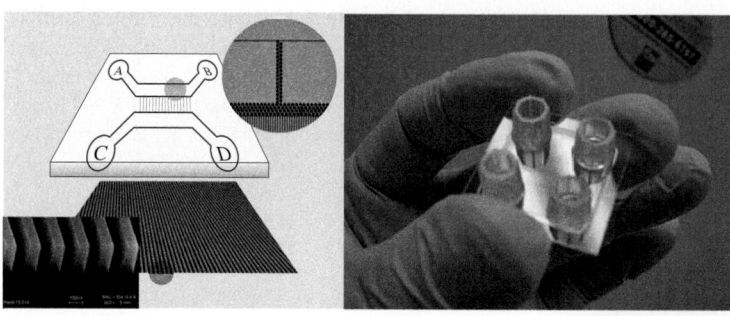

Fig. 8.33 Nanofluidic device for high-throughput linear DNA analysis. Microfluidic channels A–B and C–D are connected via an array of 100-nm-wide nanofluidic channels. The $5 \times 1 \, \text{mm}^2$ nanofluidic channel array is defined by NIL. The picture to the right shows the finished device package. Reproduced from [8.250]

tageous, or sometimes the only viable lithography method, even for laboratory-scale experiments and prototyping.

Nanofluidic channels can be used to stretch DNA [8.250, 251] for high-throughput linear analysis, measuring the length L of individual DNA molecules, or possibly sequencing by detection of fluorescent labels attached to specific DNA sequences [8.228]. The linear analysis relies on uniform stretching of the DNA molecules without coiling as they are driven through the narrow channel. This implies that the nanofluidic channel should have cross-sectional dimensions, D, close to or smaller than the persistence length of the DNA, $L_p \approx 50 \, \text{nm}$ [8.252]. The assumption of uniform stretching of the molecule also places strong demands on the channel sidewall smoothness.

Tegenfeldt et al. [8.250] have investigated the dynamics of genomic-length DNA molecules in 100-nm-wide nanochannels, defined by NIL. The device layout is shown in Fig. 8.33. Two microfluidic channels, A–B and D–E, are connected by a $5 \, \text{mm} \times 1 \, \text{mm}$ array of 100-nm-wide nanofluidic channels. The nanofluidic channel array is defined by NIL, and the pattern is transferred into the silica substrate by metallization, lift-off and $CF_4:H_2$ RIE. The microfluidic channels are defined on a second silica substrate by UV lithography and RIE, and fluidic access ports are sandblasted. The two silica substrates are bonded by cleaning the sur-

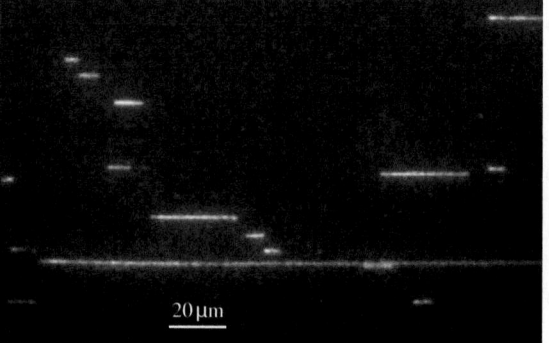

Fig. 8.34 Fluorescent image of stretched DNA in a 100-nm nanochannel array. Reproduced from [8.250]

faces, using the widely used RCA protocol for wafer cleaning, before bonding at room temperature, and annealing at $100\,°\text{C}$. The microfluidic channels allow for fast transport of the DNA from the input port to the nanofluidic channels. External electrodes are fitted in the access ports A–E, in order to apply a driving electric field, pulling the DNA through the nanochannels. The DNA is marked with fluorescent dye molecules, which makes it possible to detect individual DNA molecules optically in the nanochannels by means of an optical microscope. Figure 8.34 shows a fluorescent image of stretched individual DNA in the nanochannel array.

8.5 Conclusion and Outlook

Technological development is much based on so-called enabling techniques. For example, Gutenberg's book printing with movable metal letters was based on a combination of different existing techniques (large wine presses, and metallurgy for letter casting), solving throughput and flexibility problems, and was developed

at a time of globalization when information needed to be spread (1455 A.D., only years before Columbus discovered the sea route to America) [8.253]. In a similar way, a new lithographic technique with micro- and nanopatterning capability, as NIL, is not entirely new, but based on patterning techniques coming from silicon micromachining and Compact Disc molding. In a time of technological dynamics it will lead to advances in different fields:

- In research, as long as machines are affordable and reliable enough that they can replace or complement standard lithographic techniques. Many research institutes and universities now have access to silicon-processing technology, which often comprises tools such as resist process technology, pattern generators, mask aligners, etching and deposition facilities in a cleanroom environment. In the case of NIL it is advantageous that nanostructures can be replicated with simple molding tools, e.g. hot presses without alignment, thus making it possible to integrate NIL into simple device manufacturing. More sophisticated NIL machines are now available, typically for laboratory-type small-scale production. Modified anodic bonders can be used in combination with bond aligners, or micro-embossing presses, which allow alignment and provide increased reproducibility [8.254].

- In industry, if they help to cross technological barriers, reduce cost and enable steps into other fields, which are reserved for high-throughput applications. Success will also depend on whether they fit into the process chain already established in a silicon cleanroom environment. Furthermore substrate sizes, throughput and yield have to correspond to the needs of production. As in research, many of the machines already available can be used for moderate-scale production. They can be scaled up to substrate sizes of 200 mm and higher in combination with batch-mode operation. More sophisticated are machines based on step and repeat, which can help to solve throughput and overlay issues. Further improvements can be expected if new resists and process schemes are developed. In order to achieve a critical mass of technological expertise, the integration of NIL into a consortium of technology providers is of advantage, making it possible for manufacturers to buy standard equipment and materials, along with process knowhow.

Until now NIL was considered a very promising patterning method, because it combines resolution with large area and throughput. As long as it is seen as an alternative to establish high-end optical lithography methods, the strategy will most likely be to replace single lithographic steps by imprinting. The only consequence in the multi-masking process sequence needed in microchip fabrication would then be the modification of pattern transfer, e.g. by adding the residual layer etch (Sect. 8.2.6). The requirements defined by the ITRS roadmap are so high that other more-established lithographic techniques might make faster advances towards the next node, and the introduction of NIL into large-scale fabrication may be further postponed (it is currently scheduled for 2013 [8.12, 255]). However NIL has other capabilities, as demonstrated in Sect. 8.4, even if not all the requirements of the ITRS roadmap are met at once.

- Enterprizes with applications ranging from templates for hard-disk production to SAW filters for mobile phone, polarizers for flat-panel screens and templates for biodevices are now heading into replication techniques based on NIL processes. Most of these processes are based on single layers with nanostructures, mostly regular high-resolution gratings, and need full-wafer replication tools for large areas.

- The 3D patterning capability (Sect. 8.2.8) makes it possible to develop innovative pattern-transfer processes, thus leading to significant cost reduction. Similar advances could be achieved if materials with new properties are patterned. This is mainly due to the fact that NIL uses the concept of displacing material at the nanoscale rather than removing material selectively. Often this goes along with some trade-offs on resolution and alignment, which is justified depending on the application.

NIL has now passed the barrier from the laboratory scale to industrial preproduction. Although it seems that processes based on UV exposure have an advantage over processes based on thermal cycles, to date it is difficult to say which process will become standard and will make it onto the production line. With state-of-the-art UV-NIL equipment [8.256], a throughput of six 200 mm diameter wafers per hour can now be achieved in a step and repeat mode (using a stamp area of $45 \times 60 \, mm^2$). Full-wafer single-step hot embossing has similar capabilities, and can even push throughput further if heatable stamps with low thermal mass are used [8.206]. However, NIL is currently such a fast-moving field that preclusion about the final success of one technique is not possible or advisable. Innovative solutions are still needed to solve process and stamp lifetime issues for many different applications. Probably not only a single NIL process will be successfully implemented, but that many variants of NIL, including hybrid approaches, e.g. in combination with other lithographic processes, will be used.

Part A | 8.5

The aim of this chapter was to give an insight into the concepts used in NIL, along with presenting the advantages and limitations of processes ranging from tool fabrication to pattern transfer. Although referring more to the older thermoplastic molding process, which is the authors' field of expertise, it was intended to be general enough that future developments can be judged. The interested reader, however, will find more-detailed information at technological conferences and in scientific publications, as well as in the patent literature.

References

8.1 E. Berliner: Gramophone, US Patent 372786 (1887) http://www.audioannals.com/berlinere.htm, accessed April 4, 2005

8.2 E. Berliner: Process for producing records of sound, US Patent 382790 (1888) http://www.audioannals.com/berlinere.htm, accessed April 4, 2005.

8.3 K. C. Pohlmann: *The Compact Disc Handbook*, Computer Music and Digital Audio Series, Vol. 5, 2nd edn. (A–R Editions, Middleton, Wisconsin 1992)

8.4 H. Schift, C. David, M. Gabriel, J. Gobrecht, L. J. Heyderman, W. Kaiser, S. Köppel, L. Scandella: Nanoreplication in polymers using hot embossing and injection molding, Microelectron. Eng. **53**, 171–174 (2000)

8.5 S. Y. Chou, P. R. Krauss, W. Zhang, L. Guo, L. Zhuang: Sub-10 nm imprint lithography and applications, J. Vac. Sci. Technol. B **15**, 2897–2904 (1997)

8.6 S. Y. Chou, P. R. Krauss: Imprint lithography with sub-10 nm feature size and high throughput, Microelectron. Eng. **35**, 237–240 (1997)

8.7 R. W. Jaszewski, H. Schift, J. Gobrecht, P. Smith: Hot embossing in polymers as a direct way to pattern resist, Microelectron. Eng. **41–42**, 575–578 (1998)

8.8 Y. Xia, G. M. Whitesides: Soft lithography, Angew. Chem. Int. **37**, 550–575 (1998)

8.9 B. Michel, A. Bernard, A. Bietsch, E. Delamarche, M. Geissler, D. Juncker, H. Kind, J.-P. Renault, H. Rothuizen, H. Schmid, P. Schmidt-Winkel, R. Stutz, H. Wolf: Printing meets lithography: Soft approaches to high-resolution, IBM J. Res. Dev. **45**(5), 697–719 (2001)

8.10 W. Menz, J. Mohr, O. Paul: *Microsystem Technology* (Wiley-VCH, Weinheim 2001)

8.11 C. Sotomayor Torres: *Alternative Lithography— Unleashing the Potential of Nanotechnology*, Nanostructure Science and Technology, ed. by D. J. Lockwood (Kluwer Academic, Plenum Publishers, New York 2003)

8.12 International Technology Roadmap for Semiconductors web site http://public.itrs.net/, accessed April 4, 2005

8.13 H. Moore: Cramming more components onto integrated circuits with unit cost falling as the number of components per circuit rises, Electronics **38**(8), 114–117 (1965)

8.14 S. Okazaki: Resolution limits of optical lithography, J. Vac. Sci. Technol. B **9**(6), 2829–2833 (1991)

8.15 R. Compaño (ed): *European Commission IST programme, Future and Emerging Technologies, Technology Roadmap for Nanoelectronics*, 2nd edn. (Office for Official Publications of the European Commission, Luxembourg 2001)

8.16 S. Y. Chou, P. R. Krauss, P. J. Renstrom: Imprint of sub-25 nm vias and trenches in polymers, Appl. Phys. Lett. **67**(21), 3114–3116 (1995)

8.17 S. Y. Chou, P. R. Krauss, P. J. Renstrom: Imprint lithography with 25-nanometer resolution, Science **272**, 85–87 (1996)

8.18 S. Y. Chou, P. R. Krauss, P. J. Renstrom: Nanoimprint lithography, J. Vac. Sci. Technol. B **14**(6), 4129–4133 (1996)

8.19 S. Y. Chou: Nanoimprint lithography, US Patent 5772905 (1995)

8.20 L. Baraldi, R. Kunz, J. Meissner: High-precision molding of integrated optical structures, Proc. SPIE **1992**, 21–29 (1993)

8.21 R. W. Jaszewski, H. Schift, J. Gobrecht, P. Smith: Hot embossing in polymers as a direct way to pattern resist, Microelectron. Eng. **41–42**, 575–578 (1998)

8.22 L. J. Heyderman, H. Schift, C. David, J. Gobrecht, T. Schweizer: Flow behaviour of thin polymer films used for hot embossing lithography, Microelectron. Eng. **54**, 229–245 (2000)

8.23 J. Haisma, M. Verheijen, K. van den Heuvel, J. van den Berg: Mold-assisted lithography: A process for reliable pattern replication, J. Vac. Sci. Technol. B **14**, 4124–4128 (1996)

8.24 M. Colburn, S. Johnson, M. Stewart, S. Damle, T. Bailey, B. Choi, M. Wedlake, T. Michealson, S. V. Sreenivasan, J. Ekerdt, C. G. Willson: Step and flash imprint lithography: A new approach to high-resolution patterning, Proc. SPIE **3676**, 379–389 (1999)

8.25 D. J. Resnick, W. J. Dauksher, D. Mancini, K. J. Nordquist, T. C. Bailey, S. Johnson, N. Stacey, J. G. Ekerdt, C. G. Willson, S. V. Sreenivasan, N. Schumaker: Imprint lithography: Lab curiosity or the real NGL?, Proc. SPIE **5037**, 12–23 (2003)

8.26 D. J. Resnick, S. V. Sreenivasan, C. G. Willson: Step & flash imprint lithography, Mater. Today **8**, 34–42 (2005)

8.27   M. Doi: *Introduction to Polymer Physics* (Clarendon, Oxford 1996)

8.28   D. W. van Krevelen: *Properties of Polymers* (Elsevier, Amsterdam 1990)

8.29   H. Schift, L. J. Heyderman: Nanorheology–squeezed flow in hot embossing of thin films. In: *Alternative Lithography*, Nanostructure science and technology, ed. by C. Sotomayor Torres (Kluwer Plenum, New York 2003) pp. 46–76

8.30   H.-C. Scheer, H. Schulz, T. Hoffmann, C. M. Sotomayor Torres: Nanoimprint techniques. In: *Handbook of Thin Film Materials*, Vol. 5, ed. by H. S. Nalva (Academic, New York 2002) Chap. 1, pp. 1–60

8.31   M. D. Austin, H. Ge, W. Wu, M. Li, Z. Yu, D. Wasserman, S. A. Lyon, S. Y. Chou: Fabrication of 5 nm linewidth and 14 nm pitch features by nanoimprint lithography, Appl. Phys. Lett. **84**(26), 5299–5301 (2004)

8.32   E. A. Dobisz, S. L. Brandow, R. Bass, J. Mitterender: Effects of molecular properties on nanolithography in polymethyl methacrylate, J. Vac. Sci. Technol. B **18**, 107–111 (2000)

8.33   A. Olzierski, I. Raptis: Development and molecular-weight issues on the lithographic performance of poly-(methyl methacrylate), Microelectron. Eng. **73–74**, 244–251 (2004)

8.34   M. Khoury, D. K. Ferry: Effect of molecular weight on poly(methyl-methacrylate) resolution, J. Vac. Sci. Technol. B **14**, 75–79 (1996)

8.35   L. J. Fetters, D. J. Lohse, D. Richter, T. A. Witten, A. Zirkel: Connection between polymer molecular weight, density, chain dimensions, and melt viscoelastic properties, Macromolecules **27**, 4639–4647 (1994)

8.36   C. B. Roth, J. R. Dutcher: Mobility on different length scales in thin polymer films. In: *Soft Materials: Structure and Dynamics*, ed. by J. R. Dutcher, A. G. Marangoni (Dekker, New York 2004)

8.37   J. N. D'Amour, U. Okoroanyanwu, C. W. Frank: Influence of substrate chemistry on the properties of ultrathin polymer films, Microelectron. Eng. **73–74**, 209–217 (2004)

8.38   R. B. Bird, C. F. Curtis, R. C. Armstrong, O. Hassager: *Dynamics of Polymeric Liquids*, Vol. 1, Fluid Mechanics (Wiley, New York 1987)

8.39   L. G. Baraldi: Heißprägen in Polymeren für die Herstellung integriert-optischer Systemkomponenten. Ph.D. Thesis (ETH Zürich, Zürich 1994)

8.40   M. J. Stefan: Parallel Platten Rheometer, Akad Wiss. Math.-Natur., Wien **2**(69), 713–735 (1874)

8.41   J.-H. Jeong, Y.-S. Choi, Y.-J. Shin, J.-J. Lee, K.-T. Park, E.-S. Lee, S.-R. Lee: Flow behavior at the embossing stage of nanoimprint lithography, Fibers Polym. **3**(3), 113–119 (2002)

8.42   H. Schift, S. Park, J. Gobrecht: Nano-Imprint–molding resists for lithography, J. Photopolym. Sci. Technol. (Japan) **16**(3), 435–438 (2003)

8.43   H.-C. Scheer, H. Schulz, T. Hoffmann, C. M. Sotomayor Torres: Problems of the nanoimprinting technique for nanometer scale pattern definition, J. Vac. Sci. Technol. B **16**, 3917–3921 (1998)

8.44   H.-C. Scheer, H. Schulz: A contribution to the flow behaviour of thin polymer films during hot embossing lithography, Microelectron. Eng. **56**, 311–332 (2001)

8.45   L. J. Guo: Recent progress in nanoimprint technology and its applications, J. Phys. D: Appl. Phys. **37**, R123–R141 (2004)

8.46   C. Gourgon, C. Perret, G. Micouin, F. Lazzarino, J. H. Tortai, O. Joubert, J.-P. E. Grolier: Influence of pattern density in nanoimprint lithography, J. Vac. Sci. Technol. B **21**(1), 98–105 (2003)

8.47   A. Lebib, Y. Chen, J. Bourneix, F. Carcenac, E. Cambril, L. Couraud, H. Launois: Nanoimprint lithography for a large area pattern replication, Microelectron. Eng. **46**, 319–322 (1999)

8.48   C. Gourgon, J. H. Tortai, F. Lazzarino, C. Perret, G. Micouin, O. Joubert, S. Landis: Influence of residual solvent in polymers patterned by nanoimprint lithography, J. Vac. Sci. Technol. B **22**(6), 602–606 (2004)

8.49   Y. Hirai, M. Fujiwara, T. Okuno, Y. Tanaka, M. Endo, S. Irie, K. Nakagawa, M. Sasago: Study of the resist deformation in nanoimprint lithography, J. Vac. Sci. Technol. B **19**(6), 2811–2815 (2001)

8.50   Y. Hirai, T. Konishi, T. Yoshikawa, S. Yoshida: Simulation and experimental study of polymer deformation in nanoimprint lithography, J. Vac. Sci. Technol. B **22**(6), 3288–3293 (2002)

8.51   H. D. Rowland, W. P. King: Polymer deformation and filling modes during microembossing, J. Micromech. Microeng. **14**, 1625 (2004)

8.52   S. Zankovych, T. Hoffmann, J. Seekamp, J.-U. Bruch, C. M. Sotomayor Torres: Nanoimprint lithography: Challenges and prospects, Nanotechnology **12**(2), 91–95 (2001)

8.53   M. Beck, M. Graczyk, I. Maximov, E.-L. Sarwe, T. G. I. Ling, M. Keil, L. Montelius: Improving stamps for 10 nm level wafer scale nanoimprint, lithography, Microelectron. Eng. **61–62**, 441–448 (2002)

8.54   D.-Y. Khang, H. H. Lee: Room-temperature imprint lithography by solvent vapor treatment, Appl. Phys. Lett. **76**(7), 870–872 (2000)

8.55   D.-Y. Khang, H. Yoon, H. H. Lee: Room-temperature imprint lithography, Adv. Mater. **13**(10), 749–751 (2001)

8.56   D.-Y. Khang, H. Kang, T.-I. Kim, H. H. Lee: Low-pressure nanoimprint lithography, Nanoletters **4**(4), 633–637 (2004)

8.57   L. J. Guo: Recent progress in nanoimprint technology and its applications, J. Phys. D: Appl. Phys. **37**, R123–R141 (2004)

8.58   H. Lee, G. Y. Jung: Full wafer scale near zero residual nano-imprinting lithography using UV curable

Part A | 8

monomer solution, Microelectron. Eng. **77**(1), 42–47 (2005)

8.59 L. Tan, Y. P. Kong, S. W. Pang, A. F. Yee: Imprinting of polymer at low temperature and pressure, J. Vac. Sci. Technol. B **22**(5), 2486–2492 (2004)

8.60 C. Finder, C. Mayer, H. Schulz, H.-C. Scheer, M. Fink, K. Pfeiffer: Non-contact fluorescence measurements for inspection and imprint depth control in nanoimprint lithography, Proc. SPIE **4764**, 218–223 (2002)

8.61 D. Jucius, V. Grigaliunas, A. Guobiene: Rapid evaluation of imprint quality using optical scatterometry, Microelectron. Eng. **71**, 190–196 (2004)

8.62 A. Fuchs, B. Vratzov, T. Wahlbrink, Y. Georgiev, H. Kurz: Interferometric in situ alignment for UV-based nanoimprint, J. Vac. Sci. Technol. B **22**(6), 3242–3245 (2002)

8.63 Z. Yu, H. Gao, S. Y. Chou: In situ real time process characterisation in nanoimprint lithography using time-resolved diffractive scatterometry, Appl. Phys. Lett. **85**(18), 4166–4168 (2004)

8.64 F. Lazzarino, C. Gourgon, P. Schiavone, C. Perret: Mold deformation in nanoimprint lithography, J. Vac. Sci. Technol. B **22**(6), 3318–3322 (2002)

8.65 C. Perret, C. Gourgon, F. Lazzarino, J. Tallal, S. Landis, R. Pelzer: Characterization of 8-in. wafers printed by nanoimprint lithography, Microelectron. Eng. **73–74**, 172–177 (2004)

8.66 C. Gourgon, C. Perret, J. Tallal, F. Lazzarino, S. Landis, O. Joubert, R. Pelzer: Uniformity across 200 mm silicon wafers printed by nanoimprint lithography, J. Phys. D: Appl. Phys. **38**, 70–73 (2005)

8.67 U. Plachetka, M. Bender, A. Fuchs, B. Vratzov, T. Glinsner, F. Lindner, H. Kurz: Wafer scale patterning by soft UV-nanoimprint lithography, Microelectron. Eng. **73–74**, 167–171 (2004)

8.68 N. Roos, M. Wissen, T. Glinsner, H.-C. Scheer: Impact of vacuum environment on the hot embossing process, Proc. SPIE **5037**, 211–218 (2003)

8.69 D. Pisignano, A. Melcarne, D. Mangiullo, R. Cingolani, G. Gigli: Nanoimprint lithography of chromophore molecules under high-vacuum conditions, J. Vac. Sci. Technol. B **22**(1), 185–188 (2004)

8.70 H. Schift, L. J. Heyderman, M. auf der Maur, J. Gobrecht: Pattern formation in hot embossing of thin polymer films, Nanotechnology **12**, 173–177 (2001)

8.71 S. Y. Chou, L. Zhuang: Lithographically induced self-assembly of periodic polymer micropillar arrays, J. Vac. Sci. Technol. B **17**, 3197–3202 (1999)

8.72 S. Y. Chou, L. Zhuang, L. J. Guo: Lithographically induced self-construction of polymer microstructures for resistless patterning, Appl. Phys. Lett. **75**, 1004–1006 (1999)

8.73 L. Wu, S. Y. Chou: Electrohydrodynamic instability of a thin film of viscoelastic polymer underneath a lithographically manufactured mask, J. Non-Newtonian Fluid Mech. **125**, 91–99 (2005)

8.74 E. Schäffer, T. Thurn-Albrecht, T. P. Russell, U. Steiner: Electrically induced structure formation and pattern transfer, Nature **403**, 874–877 (2000)

8.75 E. Schäffer, T. Thurn-Albrecht, T. P. Russell, U. Steiner: Method and apparatus for forming submicron patterns on films. US Patent Registration 07880075001 (1999)

8.76 E. Schäffer, U. Steiner: Methods and apparatus for the formation of patterns in films using temperature gradients. European Patent Application PCT 124205.6 (2000)

8.77 K. Y. Suh, H. H. Lee: Capillary force lithography: large-area patterning, self-organization, and anisotropic dewetting, Adv. Funct. Mater. **12**(6,7), 405–413 (2002)

8.78 Y. Hirai, S. Yoshida, N. Takagi: Defect analysis in thermal nanoimprint lithography, J. Vac. Sci. Technol. B **21**(6), 2765–2770 (2003)

8.79 Y. Hirai, T. Yoshikawa, N. Takagi, S. Yoshida: Mechanical properties of poly-methyl methacrylate (PMMA) for nanoimprint lithography, J. Photopolym. Sci. Technol. (Japan) **16**(4), 615–620 (2003)

8.80 M. Colburn, I. Suez, B. J. Choi, M. Meissl, T. Bailey, S. V. Sreenivasan, J. G. Ekerdt, C. G. Willson: Characterization and modelling of volumetric and mechanical properties for step and flash imprint lithography photopolymers, J. Vac. Sci. Technol. B **19**(6), 2685–2689 (2001)

8.81 D. J. Resnick, W. J. Dauksher, D. Mancini, K. J. Nordquist, T. C. Bailey, S. Johnson, N. Stacey, J. G. Ekerdt, C. G. Willson, S. V. Sreenivasan, N. Schumaker: Imprint lithography for integrated circuit fabrication, J. Vac. Sci. Technol. B **21**(6), 2624–2631 (2003)

8.82 M. Otto, M. Bender, B. Hadam, B. Spangenberg, H. Kurz: Characterization and application of a UV-based imprint technique, Microelectron. Eng. **57–58**, 361–366 (2001)

8.83 B. Vratzov, A. Fuchs, M. Lemme, W. Henschel, H. Kurz: Large scale ultraviolet-based nanoimprint lithography, J. Vac. Sci. Technol. B **21**(6), 2760–2764 (2003)

8.84 M. Komuro, J. Taniguchi, S. Inoue, N. Kimura, Y. Tokano, H. Hiroshima, S. Matsui: Imprint characteristics by photo-induced solidification of liquid polymer, Jpn. J. Appl. Phys. **39**, 7075–7079 (2000)

8.85 H. Schulz, H.-C. Scheer, T. Hoffmann, C. M. Sotomayor Torres, K. Pfeiffer, G. Bleidiessel, G. Grützner, C. Cardinaud, F. Gaboriau, M.-C. Peignon, J. Ahopelto, B. Heidari: New polymer materials for nanoimprinting, J. Vac. Sci. Technol. B **18**(4), 1861–1865 (2000)

8.86 H. Schulz, D. Lyebyedyev, H.-C. Scheer, K. Pfeiffer, G. Bleidiessel, G. Grützner, J. Ahopelto: Master replication into thermosetting polymers for nanoimprinting, J. Vac. Sci. Technol. B **18**(6), 3582–3585 (2000)

8.87 K. Pfeiffer, M. Fink, G. Bleidiessel, G. Gruetzner, H. Schulz, H.-C. Scheer, T. Hoffmann, C. M. Sotomayor Torres, F. Gaboriau, C. Cardinaud: Novel

Part A | 8

linear and crosslinking polymers for nanoimprinting with high etch resistance, Microelectron. Eng. **53**, 411–414 (2000)

8.88   S. Rudschuck, D. Hirsch, K. Zimmer, K. Otte, A. Braun, R. Mehnert, F. Bigl: Replication of 3-D-micro- and nanostrucutures using different UV-curable polymers, Microelectron. Eng. **53**, 557–560 (2000)

8.89   M. Sagnes, L. Malaquin, F. Carcenac, C. Vieu, C. Fournier: Imprint lithography using thermo-polymerisation of MMA, Microelectron. Eng. **61–62**, 429–433 (2002)

8.90   A. Abdo, S. Schuetter, G. Nellis, A. Wei, R. Engelstad, V. Truskett: Predicting the fluid behavior during the dispensing process for step-and-flash imprint lithography, J. Vac. Sci. Technol. **22**(6), 3279–3282 (2002)

8.91   Y. Hirai, H. Kikuta, T. Sanou: Study on optical intensity distribution in photocuring nanoimprint lithography, J. Vac. Sci. Technol. B **21**(6), 2777–2782 (2003)

8.92   C.-H. Chang, R. K. Heilmann, R. C. Fleming, J. Carter, E. Murphy, M. L. Schattenburg, T. C. Bailey, J. G. Ekerdt, R. D. Frankel, R. Voisin: Fabrication of sawtooth diffraction gratings using nanoimprint lithography, J. Vac. Sci. Technol. B **21**(6), 2755–2759 (2003)

8.93   P. R. Krauss, S. Y. Chou: Nano-compact disks with 400 Gbit/in^2 storage density fabricated using nanoimprint lithography and read with proximal probe, Appl. Phys. Lett. **71**(21), 3174–3176 (1997)

8.94   W. Wu, B. Cui, X. Sun, W. Zhang, L. Zhuang, L. Kong, S. Y. Chou: Large area high density quantized magnetic disks fabricated using nanoimprint lithography, J. Vac. Sci. Technol. B **16**(6), 3825 (1998)

8.95   R. W. Jaszewski, H. Schift, J. Gobrecht, P. Smith: Hot embossing in polymers as a direct way to pattern resist, Microelectron. Eng. **41–42**, 575–578 (1998)

8.96   H. Schift, S. Park, C.-G. Choi, C.-S. Kee, S.-P. Han, K.-B. Yoon, J. Gobrecht: Fabrication process for polymer photonic crystals using nanoimprint lithography, Nanotechnology **16**, S261–S265 (2005)

8.97   M. Hartney, D. Hess, D. Soane: Oxygen plasma etching for resist stripping and multilayer lithography, J. Vac. Sci. Technol. B **7**, 1–13 (1989)

8.98   W. Pilz, J. Janes, K. P. M. Müller, J. Pelka: Oxygen reactive ion etching of polymers—profile evolution and process mechanisms, Proc. SPIE **1392**, 84–94 (1990)

8.99   B. Heidari, I. Maximov, E.-L. Sarwe, L. Montelius: Large scale nanolithography using imprint lithography, J. Vac. Sci. Technol. B **17**, 2961–2964 (1999)

8.100 D. Lyebyedyev, H.-C. Scheer: Mask definition by nanoimprint lithography, Proc. SPIE **4349**, 82–85 (2001)

8.101 X.-M. Yan, S. Kwon, A. M. Contreras, J. Bokor, G. A. Somorjai: Fabrication of large number density platinum nanowire arrays by size reduction lithography and nanoimprint lithography, Nanoletters **web-publication**, (web access April 2005) (2005)

8.102 L. J. Heyderman, B. Ketterer, D. Bächle, F. Glaus, B. Haas, H. Schift, K. Vogelsang, J. Gobrecht, L. Tiefenauer, O. Dubochet, P. Surbled, T. Hessler: High volume fabrication of customised nanopore membrane chips, Microelectron. Eng. **67–68**, 208–213 (2003)

8.103 L. J. Heyderman, H. Schift, C. David, B. Ketterer, M. auf der Maur, J. Gobrecht: Nanofabrication using hot embossing lithography and electroforming, Microelectron. Eng. **57–58**, 375–380 (2001)

8.104 H. Schift, L. J. Heyderman, C. Padeste, J. Gobrecht: Chemical nano-patterning using hot embossing lithography, Microelectron. Eng. **61–62**, 423–428 (2002)

8.105 H. Schift, R. W. Jaszewski, C. David, J. Gobrecht: Nanostructuring of polymers and fabrication of interdigitated electrodes by hot embossing lithography, Microelectron. Eng. **46**, 121–124 (1999)

8.106 L. Montelius, B. Heidari, M. Graczik, T. Ling, I. Maximov, E.-L. Sarwe: Large area nanoimprint fabrication of sub-100 nm interdigitated metal arrays, Proc. SPIE **3997**, 442–452 (2000)

8.107 L. Montelius, B. Heidari, M. Graczyk, E.-L. Sarwe, T. G. I. Ling: Nanoimprint- and UV-lithography: mix & match process for fabrication of interdigitated nanobiosensors, Microelectron. Eng. **53**, 521–524 (2000)

8.108 M. Beck, F. Persson, P. Carlberg, M. Graczyk, I. Maximov, T. G. I. Ling, L. Montelius: Nanoelectrochemical transducers for (bio-)chemical sensor applications fabricated by nanoimprint lithography, Microelectron. Eng. **73–74**, 837 (2004)

8.109 H. Schift, C. Park, C. Padeste, J. Gobrecht: Nanostructuring of anti-adhesive layer by hot embossing lithography, Microelectron. Eng. **67–68**, 252–258 (2003)

8.110 S. Park, S. Saxer, C. Padeste, H. H. Solak, J. Gobrecht, H. Schift: Chemical patterning of sub 50 nm half pitches via nanoimprint lithography, Microelectron. Eng. **78–79**, 682–688 (2005)

8.111 D. Falconnet, D. Pasqui, S. Park, R. Eckert, H. Schift, J. Gobrecht, R. Barbucci, M. Textor: A novel approach to produce protein nanopatterns by combining nanoimprint, lithography and molecular self-assembly, Nanoletters **4**(10), 1909–1914 (2004)

8.112 J. D. Hoff, L.-J. Cheng, E. Meyhofer, L. J. Guo, A. J. Hunt: Nanoscale protein patterning by imprint lithography, Nanoletters **4**(5), 853 (2004)

8.113 T. Schliebe, G. Schneider, H. Aschoff: Nanostructuring high resolution phase zone plates in nickel and germanium using cross-linked polymers, Microelectron. Eng. **30**, 513–516 (1996)

8.114 G. Simon, A. M. Haghiri-Gosnet, F. Carcenac, H. Launois: Electroplating: An alternative transfer technology in the 20 nm range, Microelectron. Eng. **35**, 51–54 (1997)

Part A | 8

8.115 D. Suh, J. Rhee, H. H. Lee: Bilayer reversal imprint lithography: Direct metal–polymer transfer, Nanotechnology **15**, 1103–1107 (2004)

8.116 F. Reuther, K. Pfeiffer, M. Fink, G. Gruetzner, H. Schulz, H.-C. Scheer, F. Gaboriau, C. Cardinaud: Mix and match of nanoimprint and UV lithography, Proc. SPIE **4343**, 802–809 (2001)

8.117 K. Pfeiffer, M. Fink, G. Gruetzner, G. Bleidiessel, H. Schulz, H.-C. Scheer: Multistep profiles by mix and match of nanoimprint and UV-lithography, Microelectron. Eng. **57–58**, 381–387 (2001)

8.118 X. Cheng, L. J. Guo: A combined nanoimprint and photolithography patterning technique, Microelectron. Eng. **3–4**, 277–282 (2004)

8.119 X. Cheng, L. J. Guo: One-step lithography for various size patterns with a hybrid mask-mold, Microelectron. Eng. **3–4**, 288–293 (2004)

8.120 N. Kehagias, S. Zankovych, A. Goldschmidt, R. Kian, M. Zelsmann, C. M. Sotomayor Torres, K. Pfeiffer, G. Ahrens, G. Gruetzner: Embedded polymer waveguides: design and fabrication approaches, Superlattices Microstruct. **36**(1–3), 201 (2004)

8.121 W. Zhang, S. Y. Chou: Multilevel nanoimprint lithography with submicron alignment over 4 in. Si wafers, Appl. Phys. Lett. **79**(6), 845 (2001)

8.122 H. Schulz, M. Wissen, N. Roos, H.-C. Scheer, K. Pfeiffer, G. Gruetzner: Low-temperature wafer-scale 'warm' embossing for mix & match with UV-lithography, SPIE Proc. **4688**, 223–231 (2002)

8.123 I. Martini, J. Dechow, M. Kamp, A. Forchel, J. Koeth: GaAs field effect transistors fabricated by imprint lithography, Microelectron. Eng. **60**(3-4), 451–455 (2002)

8.124 A. P. Kam, J. Seekamp, V. Solovyev, A. Goldschmidt, C. M. Sotomayor Torres: Nanoimprinted organic field-effect transistors: Fabrication, transfer mechanism and solvent effects on device characteristics, Microelectron. Eng. **73–74**, 809–813 (2004)

8.125 H. Schulz, A. S. Körbes, H.-C. Scheer, L. J. Balk: Combination of nanoimprint and scanning force lithography for local tailoring of sidewalls of nanometer devices, Microelectron. Eng. **53**, 221–224 (2000)

8.126 M. Tormen, L. Businaro, M. Altissimo, F. Romanato, S. Cabrini, F. Perennes, R. Proietti, Hong-Bo Sun, S. Kawata, E. Di Fabrizio: 3-D patterning by means of nanoimprinting, X-ray and two-photon lithography, Microelectron. Eng. **73–74**, 535–541 (2004)

8.127 X. Sun, L. Zhuang, W. Zhang, S. Y. Chou: Multilayer resist methods for nanoimprint lithography on non-flat surfaces, J. Vac. Sci. Technol. B **16**(6), 3922–3925 (1998)

8.128 F. van Delft: Bilayer resist used in e-beam lithography for deep narrow structures, Microelectron. Eng. **46**, 369–373 (1999)

8.129 L. Tan, Y. P. Kong, L.-L. Bao, X. D. Huang, L. J. Guo, S. W. Pang, A. F. Yee: Imprinting polymer film on patterned substrates, J. Vac. Sci. Technol. B **21**(6), 2742–2748 (2003)

8.130 B. Faircloth, H. Rohrs, R. Tiberio, R. Ruoff, R. R. Krchnavek: Bilayer nanoimprint lithography, J. Vac. Sci. Technol. B **18**(4), 1866–1873 (2000)

8.131 A. Lebib, M. Natali, S. P. Li, E. Cambril, L. Manin, Y. Chen, H. M. Janssen, R. P. Sijbesma: Control of the critical dimension with a trilayer nanoimprint lithography procedure, Microelectron. Eng. **57–58**, 411–416 (2001)

8.132 Y. Chen, K. Peng, Z. Cui: A lift-off process for high resolution patterns using PMMA/LOR resist stack, Microelectron. Eng. **73–74**, 278–281 (2004)

8.133 P. Carlberg, M. Graczyk, E.-L. Sawe, I. Maximov, M. Beck, L. Montelius: Lift-off process for nanoimprint lithography, Microelectron. Eng. **67–68**, 203–207 (2003)

8.134 W. Li, J. O. Tegenfeldt, L. Chen, R. H. Austin, S. Y. Chou, P. A. Kohl, J. Krotine, J. C. Sturm: Sacrificial polymers for nanofluidic channels in biological applications, Nanotechnology **14**, 578–583 (2003)

8.135 MicroChem Corp., 1254 Chestnut Street, Newton, MA 02464, USA: http://www.microchem.com/

8.136 W. Trybula: Sematech, AMRC, and Nano. Nanoprint and Nanoimprint Technology (NNT) conference, Vienna, Austria, December 2, 2004. Oral presentation

8.137 S. Johnson, D. J. Resnick, D. Mancini, K. J. Nordquist, W. J. Dauksher, K. Gehoski, J. H. Baker, L. Dues, A. Hooper, T. C. Bailey, S. V. Sreenivasan, J. G. Ekerdt, C. G. Willson: Fabrication of multi-tiered structures on step and flash imprint lithography templates, Microelectron. Eng. **67–68**, 221–228 (2003)

8.138 Y. P. Kong, H. Y. Lowa, S. W. Pang, A. F. Yee: Duo-mold imprinting of three-dimensional polymeric structures, J. Vac. Sci. Technol. B **22**(6), 3251–3265 (2004)

8.139 T. Borzenko, M. Tormen, G. Schmidt, L. W. Molenkamp: Polymer bonding process for nanolithography, Appl. Phys. Lett. **79**(14), 2246–2248 (2001)

8.140 X. D. Huang, L.-R. Bao, X. Cheng, L. J. Guo, S. W. Panga, A. F. Yee: Reversal imprinting by transferring polymer from mold to substrate, J. Vac. Sci. Technol. B **20**(6), 2872–2876 (2002)

8.141 micro resist technology GmbH, Köpenicker Str. 325, D-12555 Berlin, Germany: http://www. microresist.de/

8.142 Polysciences, Inc., 400 Valley Road, Warrington, PA 18976, USA: http://www.polysciences.com

8.143 ALLRESIST GmbH, Am Biotop 14, D-15344 Strausberg, Germany: http://www.allresist.de

8.144 C.-Y. Chao, L. J. Guo: Polymer microring resonators fabricated by nanoimprint technique, J. Vac. Sci. Technol. B **20**, 2862–2866 (2002)

8.145 Bayer AG, Bayer Material Science, Building B207, D-51368 Leverkusen, Germany: http://plastics .bayer.com

8.146 LG Dow Polycarbonate Ltd., 762-1, Jungheung-Dong, Yeosu, Chunnam, Republic of Korea: http://www.lg-dow.com

8.147 J. Tallal, D. Peyrade, F. Lazzarino, K. Berton, C. Perret, M. Gordon, C. Gourgon, P. Schiavone: Replication of sub-40 nm gap nanoelectrodes over an 8-in. substrate by nanoimprint lithography, Microelectron. Eng. **78–79**, 676–681 (2005)

8.148 Zeon Chemicals L. P., 4111 Bells Lane, Louisville, KY 40211, USA: http://www.zeonchemicals.com

8.149 Ticona GmbH, Professor-Staudinger-Straße, 65451 Kelsterbach, Germany: http://www.ticona.com/

8.150 T. Nielsen, D. Nilsson, F. Bundgaard, P. Shi, P. Szabo, O. Geschke, A. Kristensen: Nanoimprint lithography in the cyclic olefin copolymer, Topas, a highly UV-transparent and chemically resistant thermoplast, J. Vac. Sci. Technol. B **22**, 1770–1775 (2004)

8.151 B. Simmons, B. Lapizco-Encinas, R. Shediac, J. Hachman, J. Chames, J. Brazzle, J. Ceremuga, G. Fiechtner, E. Cummings, Y. Fintschenko: Polymeric insulator-based (electrodeless) dielectrophoresis (iDEP) for the monitoring of water-borne pathogens, Proceedings of MicroTAS 2004, the Eight International Conference on Miniaturised Systems for Chemistry and Life Sciences, Malmö, Sweden 2004, ed. by T. Laurell, J. Nilsson, K. Jensen, D. J. Harrison, J. P. Kutter (Royal Society of Chemistry, Cambridge, United Kingdom 2004) 171–173

8.152 D. Nilsson, S. Balslev, A. Kristensen: A microfluidic dye laser fabricated by nanoimprint lithography in a highly transparent and chemically resistant cyclo-olefin copolymer (COC), J. Micromech. Microeng. **15**, 296–300 (2005)

8.153 K. Pfeiffer, M. Fink, G. Ahrens, G. Gruetzner, F. Reuther, J. Seekamp, S. Zankovych, C. M. Sotomayor Torres, I. Maximov, M. Beck, M. Graczyk, L. Montelius, H. Schulz, H.-C. Scheer, F. Steingrueber: Polymer stamps for nanoimprinting, Microelectron. Eng. **61–62**, 393–398 (2002)

8.154 M. Wissen, H. Schulz, N. Bogdanski, H.-C. Scheer, Y. Hirai, H. Kikuta, G. Ahrens, F. Reuther, K. Pfeiffer: UV curing of resists for warm embossing, Microelectron. Eng. **73–74**, 184–189 (2004)

8.155 http://www.dsm.com/en_US/html/hybrane/hybhome.htm

8.156 A. Lebib, Y. Chen, E. Cambril, P. Youinou, V. Studer, M. Natali, A. Pepin, H. M. Janssen, R. P. Sijbesma: Room-temperature and low-pressure nanoimprint lithography, Microelectron. Eng. **61–62**, 371–377 (2002)

8.157 Sumitomo Chemical Corp., 27-1, Shinkawa 2-chome, Chuo-ku,Tokyo 104-8260, Japan: http://www.sumitomo-chem.co.jp/

8.158 C. G. Willson, R. A. Dammel, A. Reiser: Photoresist materials: A historical perspective, Proc. SPIE **3049**, 28–41 (1997)

8.159 M. D. Stewart, C. G. Willson: Photoresists. In: *Encyclopedia of Materials: Science and Technology*, ed.

by K. H. J. Buschow, R. W. Cahn, M. C. Flemings, B. Ilschner, E. J. Kramer, S. Mahajan, P. Veyssière (Elsevier Science, Amsterdam 2001) pp. 6973–6978

8.160 M. Khoury, D. K. Ferry: Effect of molecular weight on poly(methyl methacrylate) resolution, J. Vac. Sci. Technol. B **14**, 75–79 (1996)

8.161 K. Pfeiffer, G. Bleidiessel, G. Gruetzner, H. Schulz, T. Hoffmann, H.-C. Scheer, C. M. Sotomayor Torres, J. Ahopelto: Suitability of new polymer materials with adjustable glass temperature for nanoimprinting, Microelectron. Eng. **46**, 431–434 (1999)

8.162 K. Pfeiffer, M. Fink, G. Bleidiessel, G. Gruetzner, H. Schulz, H.-C. Scheer, T. Hoffmann, C. M. Sotomayor Torres, F. Gaboriau, C. Cardinaud: Novel linear and crosslinking polymers for nanoimprinting with high etch resistance, Microelectron. Eng. **53**, 411–414 (2000)

8.163 F. Gaboriau, M. C. Peignon, G. Turban, C. Cardinaud, K. Pfeiffer, G. Bleidiessel, G. Grutzner: Etch behaviour of resists suitable for new patterning processes in nanotechnologies, Proc. CIP, Antibes, France 1999

8.164 F. Gaboriau, M.-C. Peignon, A. Barreau, G. Turban, C. Cardinaud, K. Pfeiffer, G. Bleidiessel, G. Grutzner: High density fluorocarbon plasma etching of new resists suitable for nanoimprint lithography, Microelectron. Eng. **53**, 501–505 (2000)

8.165 F. Gottschalch, T. Hoffmann, C. M. Sotomayor Torres, H. Schulz, H.-C. Scheer: Polymer issues in nanoprinting technique, Solid State Elec. **43**, 1079–1083 (1999)

8.166 H. Schulz, H.-C. Scheer, T. Hoffmann, C. M. Sotomayor Torres, K. Pfeiffer, G. Bleidiessel, G. Grützner, C. Cardinaud, F. Gaboriau, M.-C. Peignon, J. Ahopelto, B. Heidari: New polymer materials for nanoimprinting, J. Vac. Sci. Technol. B **18**(4), 1861–1865 (2000)

8.167 H. Schulz, D. Lyebyedyev, H.-C. Scheer, K. Pfeiffer, G. Bleidiessel, G. Grützner, J. Ahopelto: Master replication into thermosetting polymers for nanoimprinting, J. Vac. Sci. Technol. B **18**(6), 3582–3585 (2000)

8.168 D. Lyebyedyev, H. Schulz, H.-C. Scheer: Characterisation of new thermosetting polymer materials for nanoimprint lithography, Mater. Sci. Eng. **15**(1-2), 241–243 (2001)

8.169 K. Pfeiffer, F. Reuther, M. Fink, G. Gruetzner, P. Carlberg, I. Maximov, L. Montelius, J. Seekamp, S. Zankovych, C. M. Sotomayor-Torres, H. Schulz, H.-C. Scheer: A comparison of thermally and photochemically cross-linked polymers for nanoimprinting, Microelectron. Eng. **67–68**, 266–273 (2003)

8.170 M. Colburn, I. Suez, B.J. Choi, M. Meissl, T. Bailey, S. V. Sreenivasan, J. G. Ekerdt, C. G. Willson: Characterization and modelling of volumetric and mechanical properties for step and flash imprint lithography photopolymers, J. Vac. Sci. Tech. B **19**(6), 2685–2689 (2001)

8.171 C. D. Schaper, A. Miahnahri: Polyvinyl alcohol templates for low cost, high resolution, complex printing, J. Vac. Sci. Technol. B **22**(6), 3323–3326 (2002)

8.172 R. M. Reano, Y. P. Kong, H. Y. Low, L. Tan, F. Wang, S. W. Pang, A. F. Yee: Stability of functional polymers after plasticizer-assisted imprint lithography, J. Vac. Sci. Technol. B **22**(6), 3294–3299 (2002)

8.173 M. Köhler: *Etching in Microsystem Technology* (Wiley-VCH, Weinheim 1999)

8.174 H. Schift, J. Gobrecht, B. Satilmis, J. Söchtig, F. Meier, W. Raupach: Nanoreplikation im Verbund: Ein Schweizer Netzwerk, Kunststoffe **94**, 22–26 (2004) English version: Nanoreplication in a Network, Kunststoffe Plast Europe **94** (2004) 1–4

8.175 H. Schift, S. Park, C.-G. Choi, C.-S. Kee, S.-P. Han, K.-B. Yoon, J. Gobrecht: Fabrication process for polymer photonic crystals using nanoimprint lithography, Nanotechnology **16**, S261–S265 (2005)

8.176 S. Park, H. Schift, H. H. Solak, J. Gobrecht: Stamps for nanoimprint lithography by extreme ultraviolet interference lithography, J. Vac. Sci. Technol. B **22**(6), 3246–3250 (2004)

8.177 K. A. Lister, B. G. Casey, P. S. Dobson, S. Thoms, D. S. Macintyre, C. D. W. Wilkinson, J. M. R. Weaver: Pattern transfer of a 23 nm-period grating and sub-15 nm dots into CVD diamond, Microelectron. Eng. **73–74**, 319–322 (2004)

8.178 J. Taniguchi, Y. Tokano, I. Miyamoto, M. Komuro, H. Hiroshima: Diamond nanoimprint lithography, Nanotechnology **13**, 592–596 (2002)

8.179 Y. Hirai, S. Yoshida, N. Takagi, Y. Tanaka, H. Yabe, K. Sasaki, H. Sumitani, K. Yamamoto: High aspect pattern fabrication by nano imprint lithography using fine diamond mold, Jpn. J. Appl. Phys. **42**(6B), 3863–3866 (2003)

8.180 S. W. Pang, T. Tamamura, M. Nakao, A. Ozawa, H. Masuda: Direct nano-printing on Al substrate using SiC mold, J. Vac. Sci. Technol. B **16**, 1145 (1998)

8.181 J. Gao, M. B. Chan-Park, D. Xie, Y. Yan, W. Zhou, B. K. A. Ngoi, C. Y. Yue: UV embossing of submicron patterns on biocompatible polymeric films using a focused ion beam fabricated mold, Chem. Mater. **16**(6), 956–958 (2004)

8.182 M. M. Alkaisi, R. J. Blaikie, S. J. McNab: Low temperature nanoimprint lithography using silicon nitride molds, Microelectron. Eng. **57–58**, 367–373 (2001)

8.183 Y. Hirai, S. Harada, S. Isaka, M. Kobayashi, Y. Tanaka: Nano-imprint lithography using replicated mold by Ni electroforming, Jpn. J. Appl. Phys. **41**(6B), 4186–4189 (2002)

8.184 Z. Yu, L. Chen, W. Wu, H. Ge, S. Y. Chou: Fabrication of nanoscale gratings with reduced line edge roughness using nanoimprint lithography, J. Vac. Sci. Technol. B **21**(5), 2089–2092 (2003)

8.185 N. Roos, H. Schulz, L. Bendfeldt, M. Fink, K. Pfeiffer, H.-C. Scheer: First and second generation purely thermoset stamps for hot embossing, Microelectron. Eng. **61–62**, 399–405 (2002)

8.186 N. Roos, H. Schulz, M. Fink, K. Pfeiffer, F. Osenberg, H.-C. Scheer: Performance of 4″ wafer-scale thermoset working stamps in hot embossing lithography, Proc. SPIE **4688**, 232–239 (2002)

8.187 M. Fink. Pfeiffer, G. Ahrens, G. Grützner, F. Reuther, J. Seekamp, S. Zankovych, C. M. Sotomayor Torres, I. Maximov, M. Beck, M. Graczyk, L. Montelius, H. Schulz, H.-C. Scheer, F. Steingrueber: Polymer stamps for nanoimprinting, Microelectron. Eng. **61–62**, 393–398 (2002)

8.188 H. Schift, S. Park, J. Gobrecht, S. Saxer, F. Meier, W. Raupach, K. Vogelsang: Hybrid bendable stamp copies for molding fabricated by nanoimprint, Microelectron. Eng. **78–79**, 605–611 (2005)

8.189 R. W. Jaszewski, H. Schift, B. Schnyder, A. Schneuwly, P. Gröning: The deposition on anti-adhesive ultra-thin teflon-like films and their interaction with polymers during hot embossing, Appl. Surf. Sci. **143**, 301–308 (1999)

8.190 R. W. Jaszewski, H. Schift, P. Gröning, G. Margaritondo: Properties of thin anti-adhesive films used for the replication of microstructures in polymers, Microelectron. Eng. **35**, 381–384 (1997)

8.191 U. Srinivasan, M. R. Houston, R. T. Howe, R. Maboudian: Alkyltrichlorosilane-based self-assembled monolayer films for stiction reduction in silicon micromachines, J. Microelectromech. Syst. **7**, 252–260 (1998)

8.192 H. Schulz, F. Osenberg, J. Engemann, H.-C. Scheer: Mask fabrication by nanoimprint lithography using antisticking layers, Proc. SPIE **3996**, 244–249 (2000)

8.193 M. Beck, M. Graczyk, I. Maximov, E.-L. Sarwe, T. G. I. Ling, M. Keil, L. Montelius: Improving stamps for 10 nm level wafer scale nanoimprint lithography, Microelectron. Eng. **61–62**, 441–448 (2002)

8.194 S. Park, H. Schift, C. Padeste, A. Scheybal, T. Jung, B. Schnyder, R. Kötz, J. Gobrecht: Improved anti-adhesive coating for nanoimprint lithography by co-evaporation of tri- and monochlorosilanes, Mater. Res. Soc. Proc. **EXS-2**, 37–39 (2004)

8.195 H. Schift, S. Saxer, S. Park, C. Padeste, U. Pieles, J. Gobrecht: Controlled co-evaporation of silanes for nanoimprint stamps, Nanotechnology **16**, S171–S175 (2005)

8.196 M. Keil, M. Beck, G. Frennesson, E. Theander, E. Bolmsjö, L. Montelius, B. Heidari: Process development and characterization of antisticking layers on nickel-based stamps designed for nanoimprint lithography, J. Vac. Sci. Technol. B **22**(6), 3283–3287 (2002)

8.197 S. Park, H. Schift, C. Padeste, B. Schnyder, R. Kötz, J. Gobrecht: Anti-adhesive layers on nickel stamps for nanoimprint lithography, Microelectron. Eng. **73–74**, 196–201 (2004)

8.198 ABCR GmbH & Co. KG, Im Schlehert 10, D-76187 Karlsruhe, Germany: http://www.abcr.de/

8.199 B. Heidari, I. Maximov, E.-L. Sarwe, L. Montelius: Large scale nanolithography using imprint lithography, J. Vac. Sci. Technol. B **17**, 2961–2964 (1999)

8.200 B. Heidari, I. Maximov, L. Montelius: Nanoimprint lithography at the 6 in. wafer scale, J. Vac. Sci. Technol. B **18**(6), 3557–3560 (2000)

8.201 N. Roos, T. Luxbacher, T. Glinsner, K. Pfeiffer, H. Schulz, H.-C. Scheer: Nanoimprint lithography with a commercial 4 inch bond system for hot embossing, SPIE **4343**, 427–436 (2001)

8.202 C. Gourgon, C. Perret, J. Tallal, F. Lazzarino, S. Landis, O. Joubert, R. Pelzer: Uniformity across 200 mm silicon wafers printed by nanoimprint lithography, J. Phys. D: Appl. Phys. **38**, 70–73 (2005)

8.203 L. Bendfeldt, H. Schulz, N. Roos, H.-C. Scheer: Groove design of vacuum chucks for hot embossing lithography, Microelectron. Eng. **61–62**, 455–459 (2002)

8.204 T. Haatainen, J. Ahopelto, G. Grützner, M. Fink, K. Pfeiffer: Step & stamp imprint lithography using a commercial flip chip bonder, Proc. SPIE **3997**, 874–879 (2000)

8.205 H. Tana, A. Gilbertson, S. Y. Chou: Roller nanoimprint lithography, J. Vac. Sci. Technol. B **16**(6), 3926–3928 (1998)

8.206 L. Olsson: Method and device for transferring a pattern. European patent PCT/SE2003/001003 (2002)

8.207 M. Tormen: A nano impression lithographic process which involves the use of a die having a region able to generate heat. European patent PCT/IB2004/002120 (2004)

8.208 S. Y. Chou, C. Keimel, J. Gu: Ultrafast and direct imprint of nanostructures in silicon, Nature **417**, 835–837 (2002)

8.209 J. J. Shamaly, V. F. Bunze: I-line to DUV transition for critical levels, Microelectron. Eng. **30**, 87–93 (1996)

8.210 J. Bjorkholm: EUV lithography—the successor to optical lithography? Intel Technology Journal, Q3/98. http://www.intel.com/technology/itj/q31998/articles/art_4.htm

8.211 D. Wachenschwanz, W. Jiang, E. Roddick, A. Homola, P. Dorsey, B. Harper, D. Treves, C. Bajorek: Design of a manufacturable discrete track recording medium, IEEE Trans. Mag. **41**, 670–675 (2005)

8.212 G. M. McClelland, M. W. Hart, C. T. Rettner, M. E. Best, K. R. Carter, B. D. Terris: Nanoscale patterning of magnetic islands by imprint lithography using a flexible mold, Appl. Phys. Lett. **81**, 1483–1485 (2002)

8.213 G. F. Cardinale, J. L. Skinner, A. A. Talin, R. W. Brocato, D. W. Palmer, D. P. Mancini, W. J. Dauksher, K. Gehoski, N. Le, K. J. Nordquist, D. J. Resnick: Fabrication of a surface acoustic wave-based correlator using step-and-flash imprint lithography, J. Vac. Sci. Technol. B **22**, 3265–3270 (2004)

8.214 S.-W. Ahn, K.-D. Lee, J.-S. Kim, S. H. Kim, S. H. Lee, J.-D. Park, P.-W. Yoon: Fabrication of subwavelength aluminum wire grating using nanoimprint lithography and reactive ion etching, Microelectron. Eng. **78–79**, 314–318 (2005)

8.215 M. Tormen, L. Businaro, M. Altissimo, F. Romanato, S. Cabrini, F. Perennes, R. Proietti, Hong-Bo Sun, S. Kawata, E. Di Fabrizio: 3-D patterning by means of nanoimprinting, X-ray and two-photon lithography, Microelectron. Eng. **73–74**, 535–541 (2004)

8.216 J. Seekamp, S. Zankovych, A. H. Helfer, P. Maury, C. M. Sotomayor Torres, G. Böttger, C. Liguda, M. Eich, B. Heidari, L. Montelius, J. Ahopelto: Nanoimprinted passive optical devices, Nanotechnology **13**, 581–586 (2002)

8.217 C. M. Sotomayor Torres, S. Zankovych, J. Seekamp, A. P. Kam, C. Clavijo Cedeño, T. Hoffmann, J. Ahopelto, F. Reuther, K. Pfeiffer, G. Bleidiessel, G. Gruetzner, M. V. Maximov, B. Heidari: Nanoimprint lithography: An alternative nanofabrication approach, Mater. Sci. Eng. C **23**, 23–31 (2003)

8.218 T. Nielsen, D. Nilsson, F. Bundgaard, P. Shi, P. Szabo, O. Geschke, A. Kristensen: Nanoimprint lithography in the cyclic olefin copolymer, Topas, a highly UV-transparent and chemically resistant thermoplast, J. Vac. Sci. Technol. B **22**, 1770–1775 (2004)

8.219 J. Wang, X. Sun, L. Chen, S. Y. Chou: Direct nanoimprint of submicron organic light-emitting structures, Appl. Phys. Lett. **75**, 2767–2769 (1999)

8.220 X. Cheng, Y. Hong, J. Kanicki, L. J. Guo: High-resolution organic polymer light-emitting pixels fabricated by imprinting technique, J. Vac. Sci. Technol. B **20**, 2877–2880 (2002)

8.221 D. Pisignano, L. Persano, E. Mele, P. Visconti, R. Cingolani, G. Gigli, G. Barbarella, L. Favaretto: Emission properties of printed organic semiconductor lasers, Opt. Lett. **30**, 260–262 (1995)

8.222 D. Nilsson, T. Nielsen, A. Kristensen: Solid state micro-cavity dye lasers fabricated by nanoimprint lithography, Rev. Sci. Instr. **75**, 4481–4486 (2004)

8.223 D. Nilsson, S. Balslev, A. Kristensen: A microfluidic dye laser fabricated by nanoimprint lithography in a highly transparent and chemically resistant cyclo-olefin copolymer (COC), J. Micromech. Microeng. **15**, 296–300 (2005)

8.224 C. Clavijo Cedeno, J. Seekamp, A. P. Kam, T. Hoffmann, S. Zankovych, C. M. Sotomayor Torres, C. Menozzi, M. Cavallini, M. Murgia, G. Ruani, F. Biscarini, M. Behl, R. Zentel, J. Ahopelto: Nanoimprint lithography for organic electronics, Microelectron. Eng. **61–62**, 25–31 (2002)

8.225 A. Manz, N. Graber, H. M. Widmer: Miniaturized total chemical analysis systems: A novel concept for chemical sensing, Sens. Actuators B**1**, 244–248 (1990)

8.226 E. Verpoorte, N. F. De Rooij: Microfluidics meets MEMS, Proc. IEEE **91**, 930–953 (2003)

8.227 A. Pepin, P. Youinou, V. Studer, A. Lebib, Y. Chen: Nanoimprint lithography for the fabrication of DNA electrophoresis chips, Microelectron. Eng. **61–62**, 927–932 (2002)

8.228 J.O. Tegenfeldt, C. Prinz, H. Cao, R.L. Huang, R.H. Austin, S.Y. Chou, E.C. Cox, J.C. Sturm: Micro- and nanofluidics for DNA analysis, Anal. Bioanal. Chem. **378**, 1678–1692 (2004)

8.229 S.Y. Chou: Patterned Magnetic Nanostructures and Quantized Magnetic Disks, Proc. IEEE **85**, 652–671 (1997)

8.230 M.N. Baibich, J.M. Broto, A. Fert, F. Nguyen Van Dau, F. Petroff, P. Eitenne, G. Creuzet, A. Friederich, J. Chazelas: Giant magnetoresistance of (001)Fe/(001)Cr magnetic superlattices, Phys. Rev. Lett. **61**, 2472–2475 (1988)

8.231 Y. Li, A.K. Menon: Magnetic recording technologies: Overview. In: *Encyclopedia of Materials: Science and Technology*, ed. by K.H.J. Buschow, R.W. Cahn, M.C. Flemings, B. Ilschner, E.J. Kramer, S. Mahajan, P. Veyssière (Elsevier, Amsterdam 2001) pp. 4948–4957

8.232 L.F. Shew: Discrete tracks for saturation magnetic recording, IEEE Trans. Broadcast Television Recievers **BTR-9**, 56–62 (1963)

8.233 A.K. Menon: Interface tribology for 100 Gb/in², Tribology Int. **33**, 299–308 (2000)

8.234 Y. Soeno, M. Moriya, K. Ito, K. Hattori, A. Kaizu, T. Aoyama, M. Matsuzaki, H. Sakai: Feasibility of discrete track perpendicular media for high track density recording, IEEE Trans. Magn. **39**, 1967–1971 (2003)

8.235 S.Y. Chou, M. Wei, P.R. Krauss, P.B. Fisher: Study of nanoscale magnetic structures fabricated using electron beam lithography and quantum magnetic disk, J. Vac. Sci. Technol. B **12**, 3695–3698 (1994)

8.236 R.L. White, R.M.H. Newt, R.F.W. Pease: Patterned media: A viable route to 50 Gbit/in² and up for magnetic recording?, IEEE Trans. Magn. **33**, 990–995 (1997)

8.237 W. Wu, B. Cui, X.-Y. Sun, W. Zhang, L. Zhuang, L. Kong, S.Y. Chou: Large area high density quantized magnetic disks fabricated using nanoimprint lithography, J. Vac. Sci. Technol. B **16**, 3825–3829 (1998)

8.238 M. Natali, A. Lebib, E. Cambril, Y. Chen, I.L. Prejbeanu, K. Ounadjela: Nanoimprint lithography of high-density cobalt dot patterns for fine tuning of dipole interactions, J. Vac. Sci. Technol. B **19**, 2779–2783 (2001)

8.239 J. Moritz, B. Dieny, J.P. Nozieres, S. Landis, A. Lebib, Y. Chen: Domain structure in magnetic dots prepared by nanoimprint and e-beam lithography, J. Appl. Phys. **91**, 7314–7316 (2002)

8.240 P. Lalanne, M. Hutley: Artificial media optical properties – subwavelength scale. In: *Enclopedia of Optical Engineering*, ed. by R. Driggers (Dekker, New York 2003) pp. 62–71

8.241 Z. Yu, W. Wu, L. Chen, S. Chou: Fabrication of large area 100 nm pitch grating by spatial frequency doubling and nanoimprint lithography for sub-

wavelength optical applications, J. Vac. Sci. Technol. B **19**, 2816–2819 (2001)

8.242 MOXTEK, Inc., 452 West 1260 North, Orem, UT 84057, USA: http://www.moxtek.com/

8.243 NanoOpto Corporation, 1600 Cottontail Lane, Somerset, NJ 08873-5117, USA: http://www.nanoopto.com/

8.244 L.J. Guo, X. Cheng, C.Y. Chao: Fabrication of photonic nanostructures in nonlinear optical polymers, J. Mod. Opt. **49**, 663–673 (2002)

8.245 C.-Y. Chao, L.J. Guo: Reduction of surface scattering loss in polymer microrings using thermal-reflow technique, IEEE Photonics Technol. Lett. **16**, 1498–1500 (2004)

8.246 H.C. Hoch, L.W. Jelinski, H.C. Craighead (eds): *Nanofabrication and Biosystems: Integrating Materials Science, Engineering, and Biology* (Cambridge Univ. Press, Cambridge 1996)

8.247 H.G. Craighead: Nanoelectromechanical systems, Science **290**, 1532–1535 (2000)

8.248 L.R. Huang, J.O. Tegenfeldt, J.J. Kraeft, J.C. Sturm, R.H. Austin, E.C. Cox: A DNA prism for high-speed continous frationation of large DNA molecules, Nature Biotechnol. **20**, 1048–1051 (2002)

8.249 H.G. Craighead: Nanostructure science and technology: impact and prospects for biology, J. Vac. Sci. Technol. A **21**, S216–S221 (2003)

8.250 J.O. Tegenfeldt, C. Prinz, H. Cao, S. Chou, W.W. Reisner, R. Riehn, Y.M. Wang, E.C. Cox, J.C. Sturm, P. Silberzan, R.H. Austin: The dynamics of genomic-length DNA molecules in 100-nm channels, Proc. Nat. Acad. Sci. USA **101**, 10979–10983 (2004)

8.251 L.J. Guo, X. Cheng, C.-F. Chou: Fabrication of size-controllable nanofluidic channels by nanoimprinting and its application for DNA stretching, Nano Lett. **4**, 69–73 (2004)

8.252 C. Bustamante, J.F. Marko, E.D. Siggia, S. Smith: Entropic elasticity of λ-phage DNA, Science **265**, 1599–1600 (1994)

8.253 A. Kapr: *Johann Gutenberg: The Man and His Invention* (Scolar Press, London 1996) English translation of J. Gutenberg: http://www.gutenberg.de/publ.htm

8.254 The enterprises offering equipment and materials for NIL:
EVG (http://www.evgroup.com/),
SÜSS (http://www.suss.com/),
Obducat (http://www.obducat.com/),
Jenoptik (http://www.jenoptik.com/),
Molecular Imprints (http://www.molecularimprints.com/),
Microresist Technology GmbH (http://www.microresist.de/),
Nanonex (http://www.nanonex.com/)

8.255 Sematech: http://www.sematech.org/, accessed April 27, 2005

8.256 Molecular Imprints: http://www.molecularimprints.com/

9. Stamping Techniques for Micro- and Nanofabrication

Soft-lithographic techniques that use rubber stamps and molds provide simple means to generate patterns with lateral dimensions that can be much smaller than one micron and can even extend into the single nanometer regime. These methods rely on the use of soft elastomeric elements typically made out of the polymer poly(dimethylsiloxane). The first section of this chapter presents the fabrication techniques for these elements together with data and experiments that provide insights into the fundamental resolution limits. Next, several representative soft-lithography techniques based on the use of these elements are presented: (i) microcontact printing, which uses molecular 'inks' that form self-assembled monolayers, (ii) near- and proximity-field photolithography for producing two- and three-dimensional structures with subwavelength resolution features, and (iii) nano-

transfer printing, where soft or hard stamps print single or multiple layers of solid inks with feature sizes down to 100 nm. The chapter concludes with descriptions of some device-level applications that highlight the patterning capabilities and potential commercial uses of these techniques.

There is considerable interest in methods that can be used to build structures that have micron or nanometer dimensions. Historically, research and development in this area has been driven mainly by the needs of the microelectronics industry. The spectacularly successful techniques that have emerged from those efforts – such as photolithography and electron beam lithography – are extremely well suited to the tasks for which they were principally designed: forming structures of radiation-sensitive materials (including photoresists or electron beam resists) on ultraflat glass or semiconductor surfaces. Significant challenges exist in adapting these methods for new emerging applications and areas of research that require patterning of unusual systems and materials, (including those in biotechnology and plastic electronics), structures with nanometer dimensions (below 50–100 nm), large areas in a single step (larger than a few square centimeters), or nonplanar (rough or curved) surfaces. These established techniques also have the disadvantage of high capital and operational costs. As a result, some of the oldest and conceptually simplest forms of lithography – embossing, molding, stamping,

writing, and so on – are now being re-examined for their potential to serve as the basis for nanofabrication techniques that can avoid these limitations [9.1]. Considerable progress has been made in the last few years, mainly by combining these approaches or variants of them with new materials, chemistries, and processing techniques. This chapter highlights some recent advances in high-resolution printing methods, in which a "stamp" forms a pattern of "ink" on a surface that it contacts. It focuses on approaches whose capabilities, level of development, and demonstrated applications indicate a strong potential for widespread use, especially in areas where conventional methods are unsuitable.

Contact printing involves the use of an element with surface relief (the "stamp") to transfer material applied to its surface (the "ink") to locations on a substrate that it contacts. The printing press, one of the earliest manufacturable implementations of this approach, was introduced by Gutenberg in the fifteenth century. Since then, this general approach has been used almost exclusively for producing printed text or images with features that are one hundred microns or larger in their smallest dimension. The resolution is determined by the nature of the ink and its interaction with the stamp and/or substrate, the resolution of the stamp, and the processing conditions that are used for printing or to convert the pattern of ink into a pattern of functional material. This chapter focuses on (1) printing techniques that are capable of micron and nanometer resolution, and (2) their use for fabricating key elements of active electronic or optical devices and subsystems. It begins with an overview of some methods for fabricating high-resolution stamps and then illustrates two different ways that these stamps can be used to print patterns of functional materials. Applications that highlight the capabilities of these techniques and the performances of systems that are constructed with them are also presented.

9.1 High-Resolution Stamps

The printing process can be separated into two parts: fabrication of the stamp and the use of this stamp to pattern features defined by the relief on its surface. These two processes are typically quite different, although it is possible in some cases to use patterns generated by a stamp to produce a replica of that stamp. The structure from which the stamp is derived, which is known as the "master", can be fabricated with any technique that is capable of producing well-defined structures of relief on a surface. This master can then be used directly as the stamp, or to produce stamps via molding or printing procedures. It is important to note that the technique for producing the master does not need to be fast or low in cost. It also does not need to possess many other characteristics that might be desirable for a given patterning task: it is used just once to produce a master, which is directly or indirectly used to fabricate stamps. Each one of these stamps can then be used many times for printing.

In a common approach for the high-resolution techniques that are the focus of this chapter, an established lithographic technique, such as one of those developed for the microelectronics industry, defines the master. Figure 9.1 schematically illustrates typical processes. Here, photolithography patterns a thin layer of resist onto a silicon wafer. Stamps are generated from this structure in one of two ways: by casting against this master, or by etching the substrate with the patterned resist as a mask. In the first approach, the master itself can be used multiple times to produce many stamps, typically using a light or heat-curable prepolymer. In the second, the etched substrate serves as the stamp. Additional stamps can be generated either by repeating the lithography and etching, or by using the original stamp to print replica stamps. For minimum lateral feature sizes that are greater than ≈ 1−2 microns, contact- or proximity-mode photolithography with a mask produced by direct write photolithography represents a convenient method of fabricating the master. For features smaller than ≈ 2 microns, several different techniques can be used [9.2], including: (1) projection mode photolithography [9.3], (2) direct write electron beam (or focused

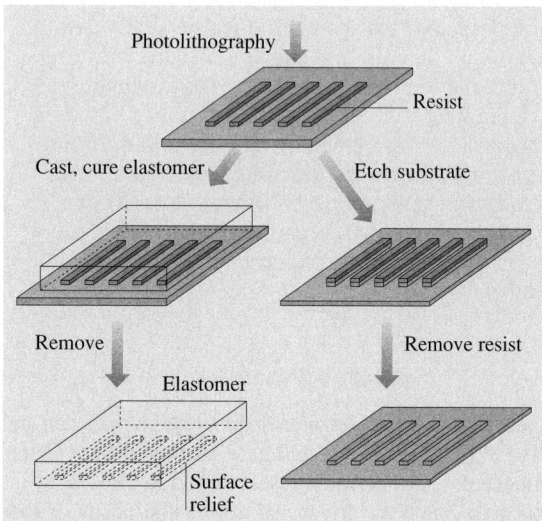

Fig. 9.1 Schematic illustration of two methods for producing high-resolution stamps. The first step involves patterning a thin layer of some radiation-sensitive material, known as the resist, on a flat substrate, such as a silicon wafer. It is convenient to use an established technique, such as photolithography or electron beam lithography, for this purpose. This structure, known as the 'master', is converted to a stamp either by etching or by molding. In the first case, the resist acts as a mask for etching the underlying substrate. Removing the resist yields a stamp. This structure can be used directly as a stamp to print patterns or to produce additional stamps. In the molding approach, a prepolymer is cast against the relief structure formed by the patterned resist on the substrate. Curing (thermally or optically) and then peeling the resulting polymer away from the substrate yields a stamp. In this approach, many stamps can be made with a single 'master' and each stamp can be used many times

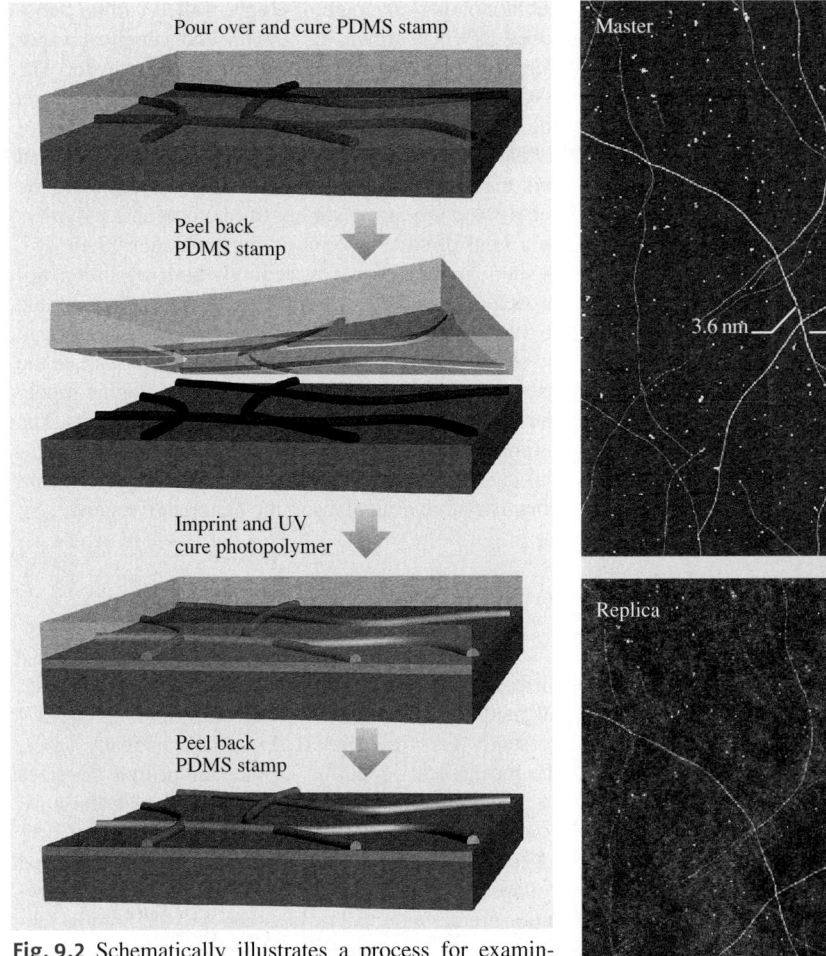

Pour over and cure PDMS stamp

Peel back
PDMS stamp

Imprint and UV
cure photopolymer

Peel back
PDMS stamp

Fig. 9.2 Schematically illustrates a process for examining the ultimate limits in resolution of soft lithographic methods. The approach uses a SWNT master to create a PDMS mold with nanoscale relief features. Soft nanoimprint lithography transfers the relief on the PDMS to that on the surface of an ultraviolet curable photopolymer film

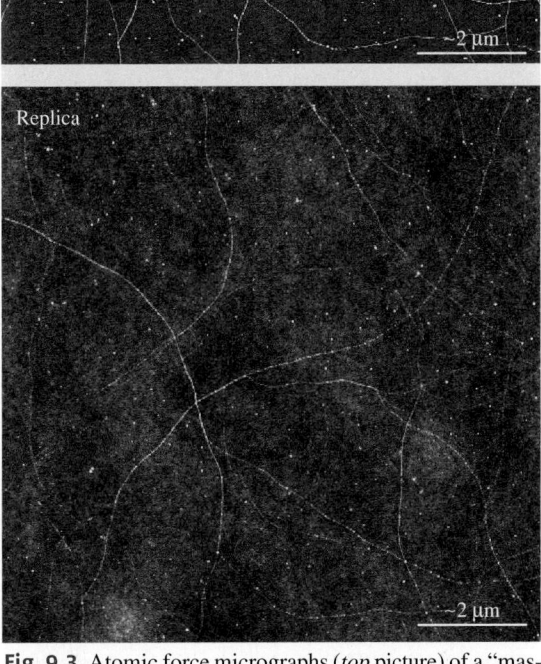

Master

~1 nm

3.6 nm

~2 μm

Replica

~2 μm

Fig. 9.3 Atomic force micrographs (*top* picture) of a "master" that consists of a submonolayer of single-walled carbon nanotubes (SWNTs; diameter between 0.5 and 5 nm) grown on a SiO_2/Si wafer. The *bottom* atomic force micrograph shows a replica of the relief structures in poly(urethane). These results indicate effective operation of a PDMS stamp for soft imprint lithography at the single nanometer scale

ion beam) lithography [9.4, 5], (3) scanning probe lithography [9.6–9] or (4) laser interference lithography [9.10]. The first approach requires a photomask generated by some other method, such as direct write photolithography or electron beam lithography. The reduction (typically 4×) provided by the projection optics relaxes the resolution requirements on the mask and enables features as small as ≈ 90 nm when deep ultraviolet radiation and phase shifting masks are used. The costs for these systems are, however, very high and their availability for general research purposes is limited. The second method is flexible in the geometry of pat-

terns that can be produced, and the writing systems are highly developed: 30–50 nm features can be achieved with commercial systems [9.11], and < 10 nm features are possible with research tools, as first demonstrated more than 25 years ago by *Broers* [9.12]. The main draw-backs of this method are that it is relatively slow and it is difficult to pattern large areas. Like projection-mode photolithography, it can be expensive. The third method, scanning probe lithography, is quite powerful in principle, but the tools are not as well established as those for other approaches. This technique has atomic resolution, but its writing speed can be lower and the areas that can be patterned are smaller than electron beam systems. Interference lithography provides a powerful, low-cost tool for generating periodic arrays of features with dimensions down to 100–200 nm; smaller sizes demand ultraviolet lasers, and patterns with aperiodic or nonregular features are difficult to produce.

In order to evaluate the ultimate resolution limit of the soft lithography methods, masters with relief structures in the single nanometer range must be fabricated. A simple method, presented in Fig. 9.2, uses submonolayer coverage of single-walled carbon nanotubes (SWNT grown, by established chemical vapor deposition techniques, on an ultraflat silicon wafer. The SWNT, which have diameters (heights and widths) in the 0.5–5 nm range, are molded on the bottom surface of a PDMS stamp generated by casting and curing against this master. Such a mold can be used to replicate the relief structure into a variety of photocurable polymers in a kind of soft nanoimprinting technique [9.13–15]. A single mold can give exceedingly high resolution, approaching the single nanometer scale range (comparable to a few bond lengths in the polymer backbone), as can be seen in Fig. 9.3 [9.16]. These results demonstrate the extreme efficiency of the basic soft lithographic procedure for generating and using elastomeric elements. The ultimate limits are difficult to predict, due to substantial uncertainties surrounding the polymer physics and chemistry that dominates in the nanometer regime.

9.2 Microcontact Printing

Microcontact printing (μCP) [9.17] is one of several soft lithographic techniques – replica molding, micromolding in capillaries, microtransfer molding, near-field conformal photolithography using an elastomeric phase-shifting mask, and so on – that have been developed as alternatives to established methods for micro- and nanofabrication [9.18–22]. μCP uses an elastomeric element (usually polydimethylsiloxane – PDMS) with high-resolution features of relief as a stamp to print patterns of chemical inks. It was mainly developed for use with inks that form self-assembled monolayers (SAMs) of alkanethiolates on gold and silver. The procedure for carrying out μCP in these systems is remarkably simple: a stamp, inked with a solution of alkanethiol, is brought into contact with the surface of a substrate in order to transfer ink molecules to regions where the stamp and substrate contact. The resolution and effectiveness of μCP rely on conformal contact between the stamp and the surface of the substrate, rapid formation of highly ordered monolayers [9.23], and the autophobicity of the SAM, which effectively blocks the reactive spreading of the ink across the surface [9.24]. It can pattern SAMs over relatively large areas ($\approx$ up to 0.25 ft^2 has been demonstrated in prototype electronic devices) in a single impression [9.25]. The edge resolution of SAMs printed onto thermally evaporated gold films is on the order of 50 nm, as determined by lateral force microscopy [9.26]. Microcontact printing has been used with a range of different SAMs on various substrates [9.18]. Of these, alkanethiolates on gold, silver, and palladium [9.27] presently give the highest resolution. In many cases, the mechanical properties of the stamp limit the sizes of the smallest features that can be achieved: the most commonly used elastomer (Sylgard 184, Dow Corning) has a low modulus, which can lead to mechanical collapse or sagging for features of relief with aspect ratios greater than ≈ 2 or less than ≈ 0.05. Stamps fabricated with high modulus elastomers avoid some of these problems [9.28, 29]. Conventional stamps are also susceptible to in-plane mechanical strains that can cause distortions in the printed patterns. Composite stamps that use thin elastomer layers on stiff supports are effective at minimizing this source of distortion [9.30]. Methods for printing that avoid direct mechanical manipulation of the stamp can reduce distortions with conventional and composite stamps [9.25]. This approach has proven effective in large-area flexible circuit applications that require accurate multilevel registration.

The patterned SAM can be used either as a resist in selective wet etching or as a template in selective deposition to form structures of a variety of materials: metals, silicon, liquids, organic polymers and even biological species. Figure 9.4 schematically illustrates the use of μCP and wet etching to pattern a thin film of Au. Figure 9.5 shows SEM images of nanostructures of gold (20 nm thick, thermally evaporated with

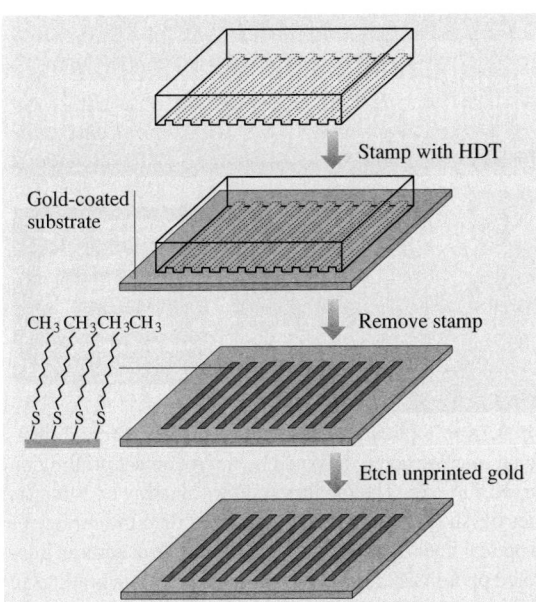

Fig. 9.4 Schematic illustration of microcontact printing. The first step involves "inking" a "stamp" with a solution of a material that is capable of forming a self-assembled monolayer (SAM) on a substrate that will be printed. In the case illustrated here, the ink is a millimolar concentration of hexadecanethiol (HDT) in ethanol. Directly applying the ink to the surface of the stamp with a pipette prepares the stamp for printing. Blowing the surface of the stamp dry and contacting it to a substrate delivers the ink to areas where the stamp contacts the substrate. The substrate consists of a thin layer of Au on a flat support. Removing the stamp after a few seconds of contact leaves a patterned SAM of HDT on the surface of the Au film. The printed SAM can act as a resist for the aqueous-based wet etching of the exposed regions of the Au. The resulting pattern of conducting gold can be used to build devices of various types

a 2.5 nm layer of Ti as an adhesion promoter) and silver ($\approx 100\,\text{nm}$ thick formed by electroless deposition using commercially available plating baths) [9.31] that were fabricated using this approach. In the first and second examples, the masters for the stamps consisted of photoresist patterned on silicon wafers with projection and contact mode photolithography, respectively. Placing these masters in a desiccator for $\approx 1\,\text{h}$ with a few drops of tridecafluoro-1,1,2,2-tetrahydrooctyl-1-trichlorosilane forms a silane monolayer on the exposed native oxide of the silicon. This monolayer prevents adhesion of the master to PDMS (Sylgard 184), which is cast and cured from a 10:1 mixture of prepolymer

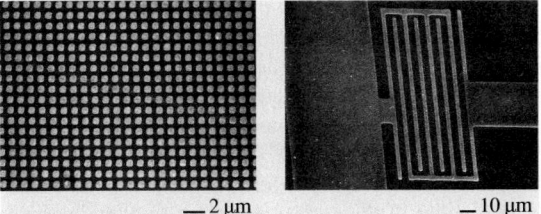

Fig. 9.5 Scanning electron micrographs of typical structures formed by microcontact printing a self-assembled monolayer ink of hexadecanethiol onto a thin metal film followed by etching of the unprinted areas of the film. The *left* frame shows an array of Au (20 nm thick) dots with $\approx 500\,\text{nm}$ diameters. The *right* frame shows a printed structure of Ag (100 nm thick) in the geometry of interdigitated source/drain electrodes for a transistor in a simple inverter circuit. The edge resolution of patterns that can be easily achieved with microcontact printing is 50–100 nm

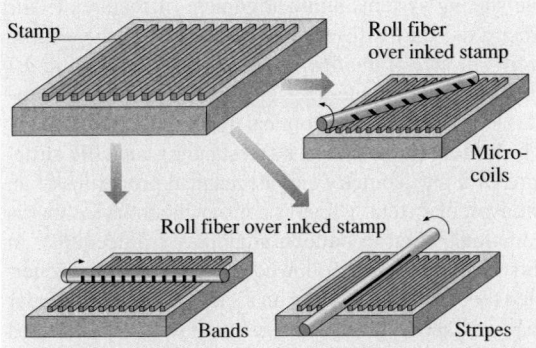

Fig. 9.6 Schematic illustration of a simple method to print lines on the surfaces of optical fibers. Rolling a fiber over the inked stamp prints a pattern onto the fiber surface. Depending on the orientation of the fiber axis with the line stamp illustrated here, it is possible, in a single rotation of the fiber, to produce a continuous microcoil, or arrays of bands or stripes

and curing agent. Placing a few drops of a $\approx 1\,\text{mM}$ solution of hexadecanethiol (HDT) in ethanol on the surface of the stamps and then blowing them dry with a stream of nitrogen prepares them for printing. Contacting the metal film for a few seconds with the stamp produces a patterned self-assembled monolayer (SAM) of HDT. An aqueous etchant (1 mM $K_4Fe(CN)_6$, 10 mM $K_3Fe(CN)_6$, and 0.1 M $Na_2S_2O_3$) removes the unprinted regions of the silver [9.32]. A similar solution (1 mM $K_4Fe(CN)_6$, 10 mM $K_3Fe(CN)_6$, 1.0 M KOH, and 0.1 M $Na_2S_2O_3$) can be used to etch the bare gold [9.33]. The results in Fig. 9.5 show that the roughness on the edges of the patterns is ≈ 50–100 nm. The resolution is determined by the grain size of the metal films, the isotropic etching process, slight reactive spreading of the inks, and edge disorder in the patterned SAMs.

The structures of Fig. 9.5 were formed on the flat surfaces of silicon wafers (left image) and glass slides (right image). An attractive feature of μCP and certain other contact printing techniques is their ability to pattern features with high resolution on highly curved or rough surfaces [9.22, 34, 35]. This type of patterning task is difficult or impossible to accomplish with photolithography due to its limited depth of focus and the difficulty involved with casting uniform films of photoresist on nonflat surfaces. Figure 9.6 shows, as an example, a straightforward approach for high-resolution printing on the highly curved surfaces of optical fibers. Here, simply rolling the fiber over an inked stamp prints a pattern on the entire outer surface of the fiber. Simple staging systems allow alignment of features to the fiber axis; they also ensure registration of the pattern from one side of the fiber to the other [9.20]. Figure 9.7 shows 3 micron-wide lines and spaces printed onto the surface of a single mode optical fiber (diameter 125 μm). The bottom frame shows a freestanding metallic structure with the geometry and mechanical properties of an intravascular stent, which is a biomedical device that is commonly used in balloon angioplasty procedures. In this latter case μCP followed by electroplating generated the Ag microstructure on a sacrificial glass cylinder that was subsequently etched away with concentrated hydrofluoric acid [9.36]. Other examples of microcontact printing on nonflat surfaces (low cost plastic sheets

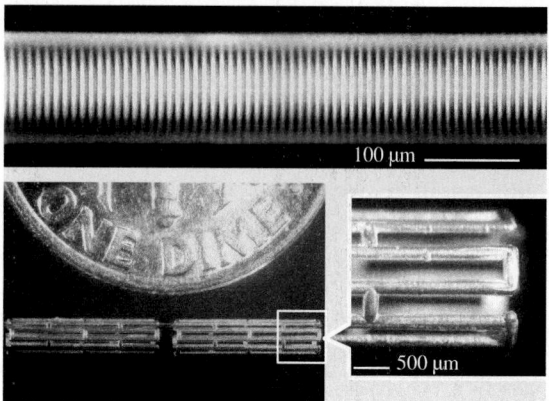

Fig. 9.7a–c Optical micrographs of some three-dimensional microstructures formed by microcontact printing on curved surfaces. The *top* frame shows an array of 3 micron lines of Au (20 nm)/Ti (1.5 nm) printed onto the surface of an optical fiber. This type of structure can be used as an integrated photomask for producing mode-coupling gratings in the core of the fiber. The *bottom* frames show a free-standing metallic microstructure formed by (**a**) microcontact printing and etching a thin (100 nm thick) film of Ag on the surface of a glass microcapillary tube, (**b**) electroplating the Ag to increase its thickness (to tens of microns) and (**c**) etching away the glass microcapillary with concentrated hydrofluoric acid. The structure shown here has the geometry and mechanical properties of an intravascular stent, which is a biomedical device commonly used in balloon angioplasty

and optical ridge waveguides) appear in the *Applications* section of this chapter.

9.3 Nanotransfer Printing

Nanotransfer printing (nTP) is a more recent high-resolution printing technique, which uses surface chemistries as interfacial "glues" and "release" layers (rather than "inks", as in μCP) to control the transfer of solid material layers from relief features on a stamp to a substrate [9.37–39]. This approach is purely additive (material is only deposited in locations where it is needed), and it can generate complex patterns of single or multiple layers of materials with nanometer resolution over large areas in a single process step. It does not suffer from surface diffusion or edge disorder in the patterned inks of μCP, nor does it require postprinting etching or deposition steps to produce structures

of functional materials. The method involves four components: (1) a stamp (rigid, flexible, or elastomeric) with relief features in the geometry of the desired pattern, (2) a method for depositing a thin layer of solid material onto the raised features of this stamp, (3) a means of bringing the stamp into intimate physical contact with a substrate, and (4) surface chemistries that prevent adhesion of the deposited material to the stamp and promote its strong adhesion to the substrate. nTP has been demonstrated with SAMs and other surface chemistries for printing onto flexible and rigid substrates with hard inorganic and soft polymer stamps. Figure 9.8 presents a set of procedures for using nTP to pattern a thin metal bilayer of Au/Ti with a surface transfer chemistry that relies on a dehydration reaction [9.37]. The process begins with fabrication of a suitable stamp. Elastomeric stamps can be built using the same casting and curing procedures described for μCP. Rigid stamps can be fabricated by (1) patterning resist (such as electron beam resist or photoresist) on a substrate (such as Si or GaAs), (2) etching the exposed regions of the substrate with an anisotropic reactive ion etch, and (3) removing the resist, as illustrated in Fig. 9.1. For both types of stamps, careful control of the lithography and the etching steps yields features of relief with nearly vertical or slightly re-entrant sidewalls. The stamps typically have depths of relief $> 0.2\,\mu$m for patterning metal films with thicknesses < 50 nm.

Electron beam evaporation of Au (20 nm; 1 nm/s) and Ti (5 nm; 0.3 nm/s) generates uniform metal bilayers on the surfaces of the stamp. A vertical, collimated flux of metal from the source ensures uniform deposition only on the raised and recessed regions of relief. The gold adheres poorly to the surfaces of stamps made of GaAs, PDMS, glass, or Si. In the process of Fig. 9.8, a fluorinated silane monolayer acts to reduce the adhesion further when a Si stamp (with native oxide) is used. The Ti layer serves two purposes: (1) it promotes adhesion between the Au layer and the substrate after pattern transfer, and (2) it readily forms a ≈ 3 nm oxide layer at ambient conditions, which provides a surface where the dehydration reaction can take place. Exposing the titanium oxide (TiO_x) surface to an oxygen plasma breaks bridging oxygen bonds, thus creating defect sites where water molecules can adsorb. The result is a titanium oxide surface with some fractional coverage of hydroxyl ($-$OH) groups (titanol).

In the case of Fig. 9.8, the substrate is a thin film of PDMS ($10-50\,\mu$m thick) cast onto a sheet of poly(ethylene terephthalate) (PET; 175 μm thick). Exposing the PDMS to an oxygen plasma produces surface ($-$OH) groups (silanol). Placing the plasma-oxidized, Au/Ti-coated stamp on top of these substrates leads to intimate, conformal contact between the raised regions of the stamp and the substrate, without the application of any external pressure. (The soft, conformable PDMS is important in this regard.) It is likely that a dehydration reaction takes place at the ($-$OH)-bearing interfaces during contact; this reaction results in permanent Ti$-$O$-$Si bonds that produce strong adhesion between the two surfaces. Peeling the substrate and stamp apart transfers the Au/Ti bilayer from the raised regions of the stamp (to which the metal has extremely poor adhesion) to the substrate. Complete pattern transfer from an elastomeric

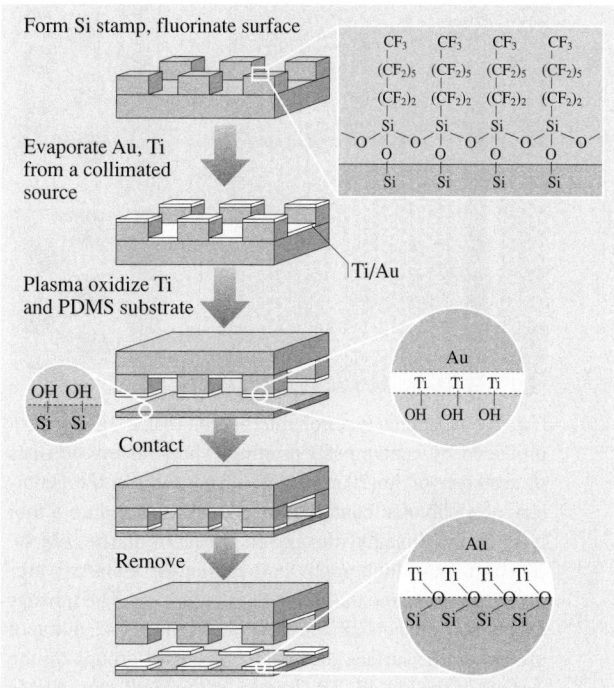

Fig. 9.8 Schematic illustration of nanotransfer printing procedure. Here, interfacial dehydration chemistries control the transfer of a thin metal film from a hard inorganic stamp to a conformable elastomeric substrate (thin film of polydimethylsiloxane (PDMS) on a plastic sheet). The process begins with fabrication of a silicon stamp (by conventional lithography and etching) followed by surface functionalization of the native oxide with a fluorinated silane monolayer. This layer ensures poor adhesion between the stamp and a bilayer metal film (Au and Ti) deposited by electron beam evaporation. A collimated flux of metal oriented perpendicular to the surface of the stamp avoids deposition on the sidewalls of the relief. Exposing the surface Ti layer to an oxygen plasma produces titanol groups. A similar exposure for the PDMS produces silanol groups. Contacting the metal-coated stamp to the PDMS results in a dehydration reaction that links the metal to the PDMS. Removing the stamp leaves a pattern of metal in the geometry of the relief features

stamp to a thin elastomeric substrate occurs readily at room temperature in open air with contact times of less than 15 seconds. When a rigid stamp is employed, slight heating is needed to induce transfer. While the origin of this difference is unclear, it may reflect the comparatively poor contact when rigid stamps are used; similar differences are also observed in cold welding of gold films [9.40].

Figure 9.9 shows scanning electron micrographs of a pattern produced using a GaAs stamp generated by

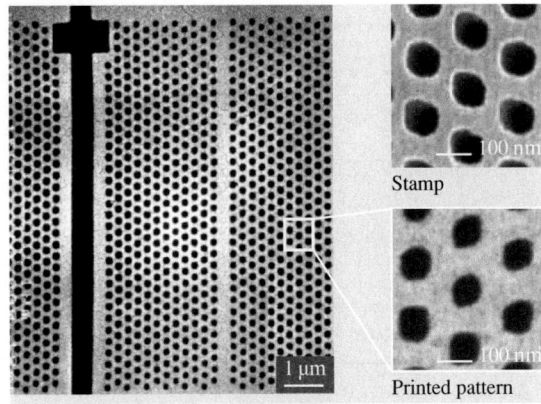

Fig. 9.9 Scanning electron micrograph (SEM) of a pattern produced by nanotransfer printing. The structure consists of a bilayer of Au(20 nm)/Ti (1 nm) (*white*) in the geometry of a photonic bandgap waveguide printed onto a thin layer of polydimethylsiloxane on a sheet of plastic (*black*). Electron beam lithography and etching of a GaAs wafer produced the stamp that was used in this case. The transfer chemistry relied on condensation reactions between titanol groups on the surface of the Ti and silanol groups on the surface of the PDMS. The frames on the *right* show SEMs of the Au/Ti-coated stamp (*top*) before printing and on the substrate (*bottom*) after printing. The electron beam lithography and etching used to fabricate the stamp limit the minimum feature size (≈ 70 nm) and the edge resolution (≈ 5–10 nm) of this pattern

electron beam lithography and etching. The frames on the right show images of the metal-coated stamp before printing (top) and the transferred pattern (bottom). The resolution appears to be limited only by the resolution of the stamp itself, and perhaps by the grain size of the metal films. Although the accuracy in multilevel registration that is possible with nTP has not yet been quantified, its performance is likely similar to that of embossing techniques when rigid stamps are used [9.41].

A wide range of surface chemistries can be used for the transfer. SAMs are particularly attractive due to their chemical flexibility. Figure 9.10 illustrates the use of a thiol-terminated SAM and nTP for forming patterns of Au on a silicon wafer [9.38]. Here, the vapor phase cocondensation of the methoxy groups of molecules of 3-mercaptopropyltrimethoxysilane (MPTMS) with the $-$OH-terminated surface of the wafer produces a SAM of MPTMS with exposed thiol ($-$SH) groups. PDMS stamps can be prepared for printing on this surface by coating them with a thin film (≈ 15 nm) of Au using conditions (thermal evaporation 1.0 nm/s; $\approx 10^{-7}$ torr

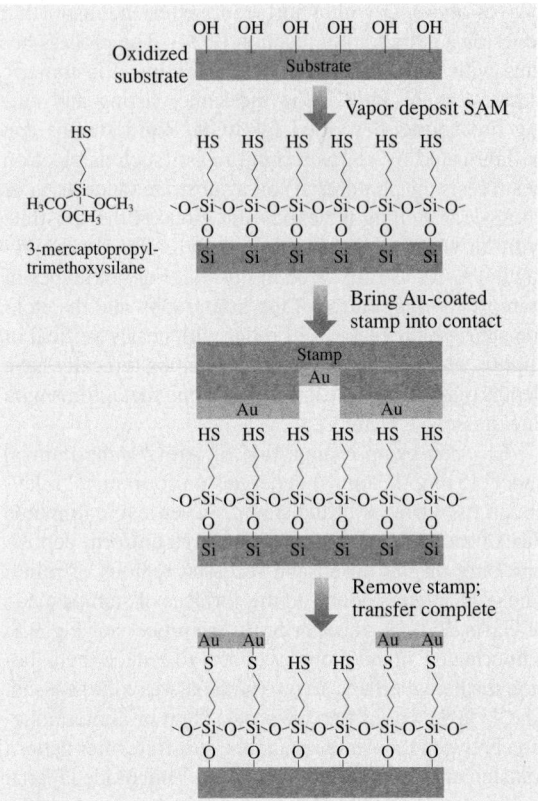

Fig. 9.10 Schematic illustration of steps involved in nanotransfer printing a pattern of a thin layer of Au onto a silicon wafer using a self-assembled monolayer (SAM) surface chemistry. Plasma oxidizing the surface of the wafer generates OH groups. Solution or vapor phase exposure of the wafer to 3-mercaptopropyltrimethoxysilane yields a SAM with exposed thiol groups. Contacting an Au-coated stamp to this surface produces thiol linkages that bond the gold to the substrate. Removing the stamp completes the transfer printing process

base pressure) that yield optically smooth, uniform films without the buckling that has been observed in the past with similar systems [9.42]. Nanocracking that sometimes occurs in the films deposited in this way can be reduced or eliminated by evaporating a small amount of Ti onto the PDMS before Au deposition and/or by exposing the PDMS surface briefly to an oxygen plasma. Bringing this coated stamp into contact with the MPTMS SAM leads to the formation of sulfur–gold bonds in the regions of contact. Removing the stamp after a few seconds efficiently transfers the gold from the raised regions of the stamp (Au does not adhere to the PDMS)

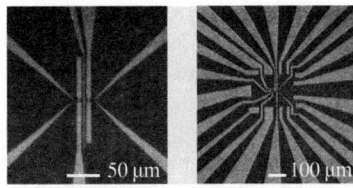

Fig. 9.11 Optical micrographs of patterns of Au (15 nm thick) formed on plastic (*left* frame) and silicon (*right* frame) substrates with nanotransfer printing. The transfer chemistries in both cases rely on self-assembled monolayers with exposed thiol groups. The minimum feature sizes and the edge resolution are both limited by the photolithography used to fabricate the stamps

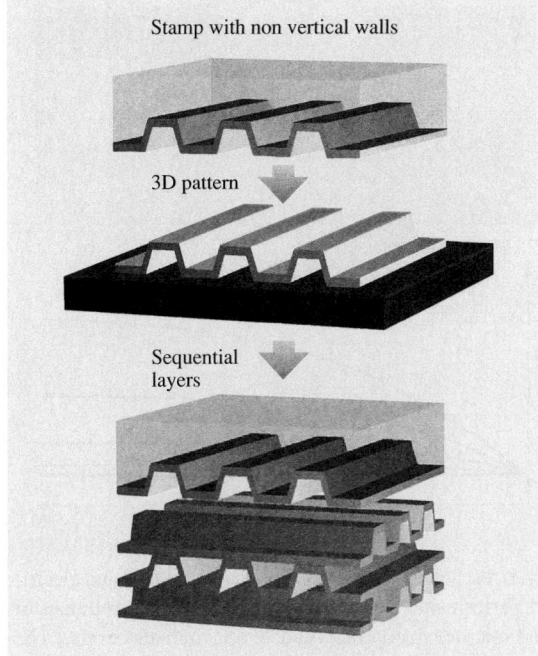

Fig. 9.12 Schematic illustration of the nanotransfer printing (nTP) process for generating continuous 3D structures when the stamp relief side walls are not vertical. Successive transfer by cold welding the gold films on top of each other yields complex multilayer structures

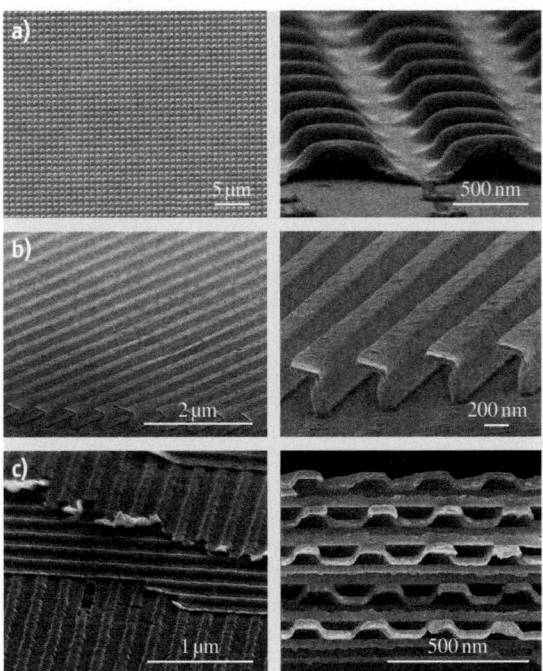

Fig. 9.13a–c Scanning electron micrographs of three-dimensional metal structures obtained by nanotransfer printing gold metal films. Part (**a**) shows closed gold nanocapsules. Part (**b**) shows free-standing *L* structures obtained using a stamp coated with a steeply angled flux of metal. Part (**c**) shows a multilayer 3-D structure obtained by the successive transfer and cold welding of continuous gold nanocorrugated films

to the substrate. Covalent bonding of the SAM glue to both the substrate and the gold leads to good adhesion of the printed patterns: They easily pass Scotch tape adhesion tests. Similar results can be obtained with other substrates containing surface −OH groups. For example, Au patterns can be printed onto ≈ 250 μm-thick sheets of poly(ethylene terephthalate) (PET) by

first spin-casting and curing (130 °C for 24 h) a thin film of an organosilsesquioxane on the PET. Exposing the cured film (≈ 1 μm thick) to an oxygen plasma and then to air produces the necessary surface (−OH) groups. Figure 9.11 shows some optical micrographs of typical printed patterns in this case [9.38].

Similar surface chemistries can guide transfer to other substrates. Alkanedithiols, for example, are useful for printing Au onto GaAs wafers [9.39]. Immersing these substrates (freshly etched with 37% HCl for ≈ 2 min to remove the surface oxide) in a 0.05 M solution of 1,8-octanedithiol in ethanol for 3 h produces a monolayer of dithiol on the surface. Although the chemistry of this system is not completely clear, it is generally believed that the thiol end groups bond chemically to the surface. Surface spectroscopy suggests the formation of Ga−S and As−S bonds. Contacting an Au-coated PDMS stamp with the treated substrate causes the exposed thiol endgroups to react with Au in the regions of contact. This reaction produces permanent Au−S bonds at the stamp/substrate interface (see insets in Fig. 9.4 for idealized chemical reaction schemes). Figure 9.12 schematically illustrates how this procedure can generate continuous 3D metal patterns using stamps with nonvertical side walls. Several layers can be transferred on top of each other by successively cold-welding the different gold metal layers. Figure 9.13 shows high- and low-magnification scanning electron microscope images of nanotransfer printed single- and multi-layer 3D metal structures [9.43,44]. The integrity of these free-standing 3D metal structures is remarkable but depends critically on the careful optimization of the metal evaporation conditions and stamp & substrate surface chemistries.

9.4 Applications

Although conventional patterning techniques, such as photolithography or electron beam lithography, have the required resolution, they are not appropriate because they are expensive and generally require multiple processing steps with resists, solvents and developers that can be difficult to use with organic active materials and plastic substrates. Microcontact and nanotransfer printing are both particularly well suited for this application. They can be combined and matched with other techniques, such as ink-jet or screen printing, to form a complete system for patterning all layers in practical plastic electronic devices [9.45]. We have focused our efforts partly on unusual electronic systems such as flexible plastic circuits and devices that rely on electrodes patterned on curved objects such as microcapillaries and optical fibers. We have also explored photonic systems such as distributed feedback structures for lasers and other integrated optical elements that demand submicron features. The sections below highlight several examples in each of these areas.

9.4.1 Unconventional Electronic Systems

A relatively new direction in electronics research seeks to establish low-cost plastic materials, substrates and printing techniques for large-area flexible electronic devices, such as paper-like displays. These types of novel devices can complement those (including high-density memories and high-speed microprocessors) that are well suited to existing inorganic (such as silicon) electronics technologies. High-resolution patterning methods for defining the separation between the source and drain electrodes (the channel length) of transistors in these plastic circuits are particularly important because this dimension determines current output and other important characteristics [9.46].

Figure 9.14 illustrates schematically a cross-sectional view of a typical organic transistor. The frame on the right shows the electrical switching charac-

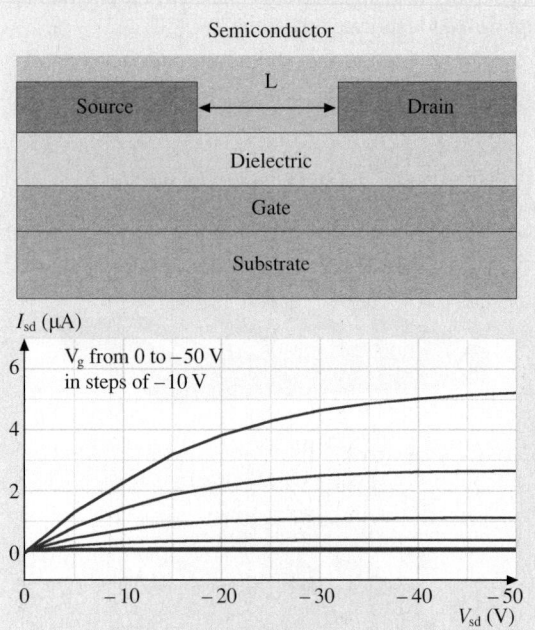

Fig. 9.14 Schematic cross-sectional view (*left*) and electrical performance (*right*) of an organic thin film transistor with microcontact printed source and drain electrodes. The structure consists of a substrate (PET), a gate electrode (indium tin oxide), a gate dielectric (spin-cast layer of organosilsesquioxane), source and drain electrodes (20 nm Au and 1.5 nm Ti), and a layer of the organic semiconductor pentacene. The electrical properties of this device are comparable to or better than those that use pentacene with photolithographically defined source/drain electrodes and inorganic dielectrics, gates and substrates

teristics of a device that uses source/drain electrodes of Au patterned by μCP, a dielectric layer of an organosilsesquioxane, a gate of indium tin oxide (ITO), and a PET substrate. The effective semiconductor mobility extracted from these data is comparable to those measured in devices that use the same semiconductor (pentacene in this case) with inorganic substrates and dielectrics, and gold source/drain electrodes defined by photolithography. Our recent work [9.1,31,47] with μCP in the area of plastic electronics demonstrates: (1) methods for using cylindrical "roller" stamps mounted on fixed axles for printing in a continuous reel-to-reel fashion, high-resolution source/drain electrodes in ultrathin gold and silver deposited from solution at room temperature using electroless deposition, (2) techniques for performing registration and alignment of the printed features with other elements of a circuit over large areas, (3) strategies for achieving densities of defects that are as good as those observed with photolithography when the patterning is performed outside of clean room facilities, (4) methods for removing the printed SAMs to allow good electrical contact of the electrodes with organic semiconductors deposited on top of them, and (5) materials and fabrication sequences that can efficiently exploit these printed electrodes for working organic TFTs in large-scale circuits.

Figure 9.15 provides an image of a large-area plastic circuit with critical features defined by μCP. This circuit is a flexible active matrix backplane for a display. It consists of a square array of interconnected transistors, each of which serves as a switching element that

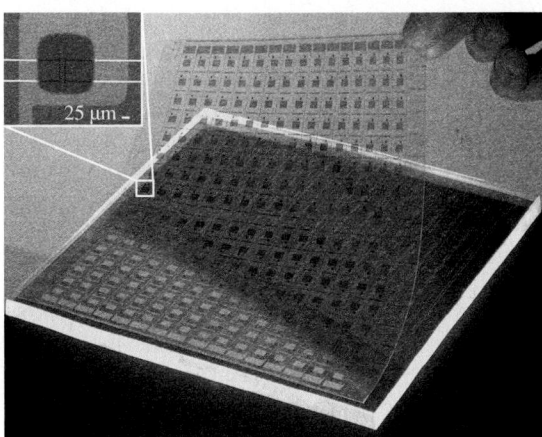

Fig. 9.15 Image of a flexible plastic active matrix backplane circuit whose finest features (transistor source/drain electrodes and related interconnects) are patterned by microcontact printing. The circuit rests partly on the elastomeric stamp that was used for printing. The circuit consists of a square array of interconnected organic transistors, each of which acts locally as a voltage-controlled switch to control the color of an element in the display. The *inset* shows an optical micrograph of one of the transistors

controls the color of a display pixel [9.25,48]. The transistors themselves have the layout illustrated in Fig. 9.11, and they use similar materials. The semiconductor in this image is blue (pentacene), the source/drain level is Au, the ITO appears green in the optical micrograph

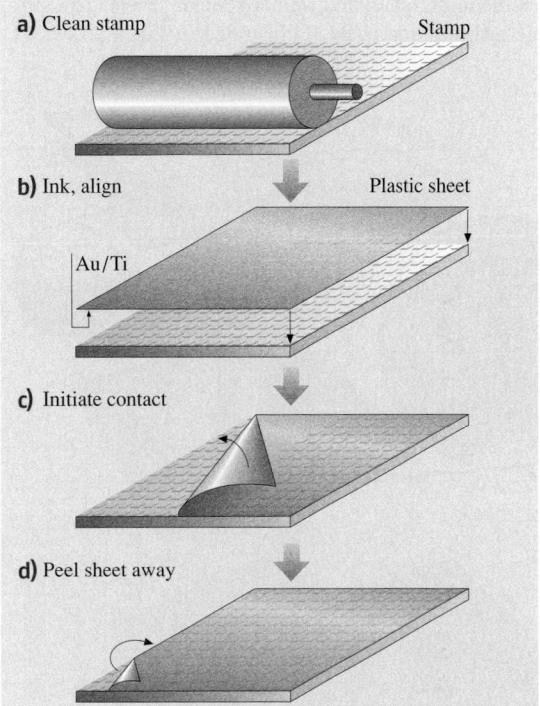

Fig. 9.16 Schematic illustration of fabrication steps for microcontact printing over large areas onto plastic sheets. The process begins with cleaning the stamp using a conventional adhesive roller lint remover. This procedure effectively removes dust particles. To minimize distortions, the stamp rests face-up on a flat surface and it is not manipulated directly during the printing. Alignment and registration are achieved with alignment marks on one side of the substrate and the stamp. By bending the plastic sheet, contact is initiated on one side of the stamp; the contact line is then allowed to progress gradually across the stamp. This approach avoids formation of air bubbles that can frustrate good contact. After the substrate is in contact with the stamp for a few seconds, the plastic substrate is separated from the stamp by peeling it away beginning in one corner. Good registration (maximum cumulative distortions of less than 50 microns over an area of 0.25 square feet) and low defect density can be achieved with this simple approach. It is also well suited for use with rigid composite stamps designed to reduce the level of distortions even further

in the inset. Part of the circuit rests on the stamp that was used for μCP. The smallest features are the source and drain electrodes ($\approx 15\,\mu$m lines), the interconnecting lines ($\approx 15\,\mu$m lines), and the channel length of the transistor ($\approx 15\,\mu$m). This circuit incorporates five layers of material patterned with good registration of the source/drain, gate, and semiconductor levels. The simple printing approach is illustrated in Fig. 9.16 [9.25].

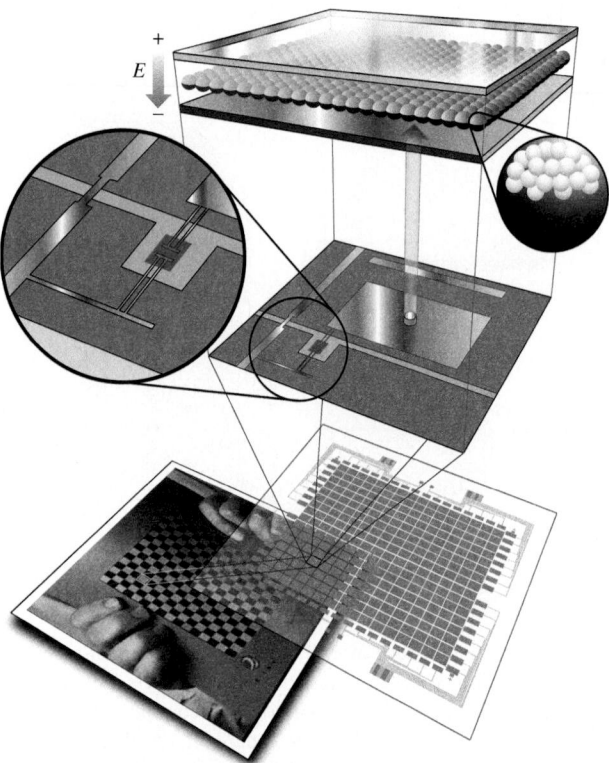

Fig. 9.17 Schematic exploded view of the components of a pixel in an electronic paperlike display (*bottom frame*) that uses a microcontact printed flexible active matrix backplane circuit (illustration near the *bottom frame*). The circuit is laminated against an unpatterned thin sheet of electronic ink (*top frame*) that consists of a monolayer of transparent polymer microcapsules (diameter ≈ 100 microns). These capsules contain a heavily dyed black fluid and a suspension of charged white pigment particles (see *right inset*). When one of the transistors turns on, electric fields develop between an unpatterned transparent frontplane electrode (indium tin oxide) and a backplane electrode that connects to the transistor. Electrophoretic flow drives the pigment particles to the front or the back of the display, depending on the polarity of the field. This flow changes the color of the pixel, as viewed from the front of the display, from black to white or vica versa

Just before use, the surface of the stamp is cleaned using a conventional adhesive roller lint remover; this procedure removes dust from the stamp in such a way that does not contaminate or damage its surface. Inking the stamp and placing it face-up on a flat surface prepares it for printing. Matching the cross-hair alignment marks on the corners of one edge of the stamp with those patterned in the ITO brings the substrate into registration with the stamp. During this alignment, features on the stamp are viewed directly through the semitransparent substrate. By bending the PET sheet, contact with the stamp is initiated on the edge of the substrate that contains the cross-hair marks. Gradually unbending the sheet allows contact to progress across the rest of the surface. This printing procedure is attractive because it avoids distortions that can arise when directly manipulating the flexible rubber stamp. It also minimizes the number and size of trapped air pockets that can form between the stamp and substrate. Careful measurements performed after etching the unprinted areas of the gold show that over the entire $6''\times 6''$ area of the circuit, (1) the overall alignment accuracy for positioning the stamp relative to the substrate (the offset of the center of the distribution of registration errors) is ≈ 50–$100\,\mu$m, even with the simple approach used here, and (2) the distortion in the positions of features in the source/drain level, when referenced to the gate level, can be as small as $\approx 50\,\mu$m (the full width at half maximum of the distribution of registration errors). These distortions represent the cumulative effects of deformations in the stamp and distortions in the gate and column electrodes that may arise during the patterning and processing of the flexible PET sheet. The density of defects in the printed patterns is comparable to (or smaller than) that in resist patterned by contact-mode photolithography when both procedures are performed outside of a clean-room facility (when dust is the dominant source of defects).

Figure 9.17 shows an "exploded" view of a paperlike display that consists of a printed flexible plastic backplane circuit, like the one illustrated in Fig. 9.16, laminated against a thin layer of "electronic ink" [9.25, 49]. The electronic ink is composed of a monolayer of transparent polymer microcapsules that contain a suspension of charged white pigment particles suspended in a black liquid. The printed transistors in the backplane circuit act as local switches, which control electric fields that drive the pigments to the front or back of the display. When the particles flow to the front of a microcapsule, it appears white; when they flow to the back, it appears black. Figure 9.18 shows a working sheet of active matrix electronic paper that uses this design.

Fig. 9.18 Electronic paperlike display showing two different images. The device consists of several hundred pixels controlled by a flexible active matrix backplane circuit formed by microcontact printing. The relatively coarse resolution of the display is not limited by material properties or by the printing techniques. Instead, it is set by practical considerations for achieving high pixel yields in the relatively uncontrolled environment of the chemistry laboratory in which the circuits were fabricated

This prototype display has several hundred pixels and an optical contrast that is both independent of the viewing angle and significantly better than newsprint. The device is ≈ 1 mm thick, is mechanically flexible, and weighs $\approx 80\%$ less than a conventional liquid crystal display of similar size. Although these displays have only a relatively coarse resolution, all of the processing techniques, the μCP method, the materials, and the electronic inks, are suitable for the large numbers of pixels required for high-information content electronic newspapers and other systems.

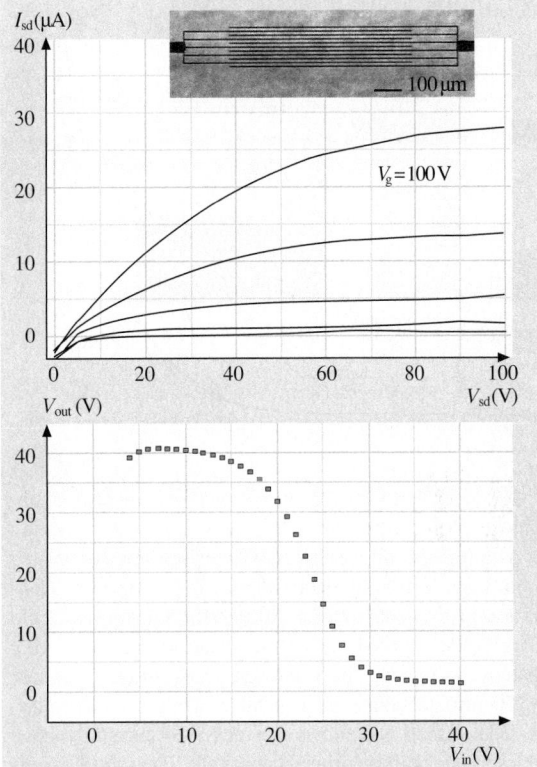

Fig. 9.19 The *upper* frame shows current–voltage characteristics of an n-channel transistor formed with electrodes patterned by nanotransfer printing that are laminated against a substrate that supports an organic semiconductor, a gate dielectric and a gate. The *inset* shows an optical micrograph of the interdigitated electrodes. The *lower* frame shows the transfer characteristics of a simple CMOS inverter circuit that uses this device and a similar one for the p-channel transistor

Like μCP, nTP is well suited to forming high-resolution source/drain electrodes for plastic electronics. nTP of Au/Ti features in the geometry of the drain and source level of organic transistors, and with appropriate interconnects on a thin layer of PDMS on PET it yields a substrate that can be used in an unusual but powerful way for building circuits: soft, room temperature lamination of such a structure against a plastic substrate that supports the semiconductor, gate dielectric, and gate levels yields a high performance circuit embedded between two plastic sheets [9.37, 50]. (Details of this lamination procedure are presented elsewhere.) The left frame of Fig. 9.19 shows the current–voltage characteristics of a laminated n-channel transistor that uses the

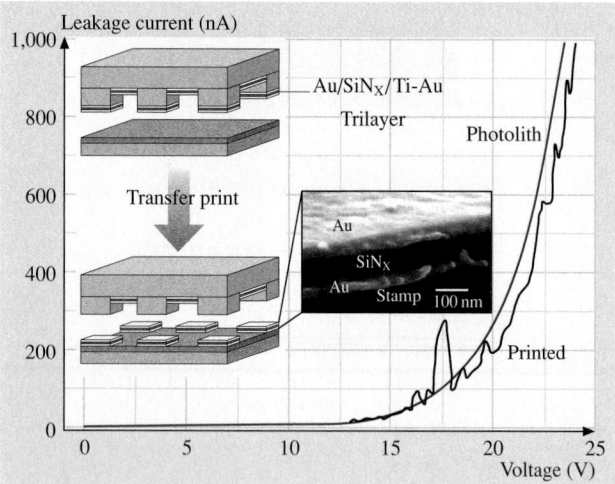

Fig. 9.20 Multilayer thin film capacitor structure printed in a single step onto a plastic substrate using the nanotransfer printing technique. A multilayer of Au/SiN − x/Ti/Au was first deposited onto a silicon stamp formed by photolithography and etching. Contacting this stamp to a substrate of Au/PDMS/PET forms a cold weld that bonds the exposed Au on the stamp to the Au-coating on the substrate. Removing the stamp produces arrays of square (250 mm × 250 mm) metal/insulator/metal capacitors on the plastic support. The *dashed line* shows the measured current–voltage characteristics of one of these printed capacitors. The *solid line* corresponds to a similar structure formed on a rigid glass substrate using conventional photolithographic procedures. The characteristics are the same for these two cases. The slightly higher level of noise in the printed devices results, at least partly, from the difficulties involved with making good electrical contacts to structures on the flexible plastic substrate

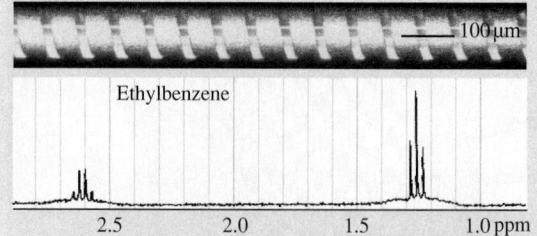

Fig. 9.21 The *top frame* shows an optical micrograph of a continuous conducting microcoil formed by microcontact printing onto a microcapillary tube. This type of printed microcoil is well suited for excitation and detection of nuclear magnetic resonance spectra from nanoliter volumes of fluid housed in the bore of the microcapillary. The *bottom frame* shows a spectra trace collected from an ≈ 8 nL volume of ethyl benzene using a structure similar to the one shown in the *top frame*

organic semiconductor copper hexadecafluorophthalocyanine (n-type) and source/drain electrodes patterned with nTP. The inset shows an optical micrograph of the printed interdigitated source/drain electrodes of this device. The bottom frame of Fig. 9.16 shows the transfer characteristics of a laminated complementary organic inverter circuit whose electrodes and connecting lines are defined by nTP. The p-channel transistor in this circuit used pentacene for the semiconductor [9.37].

In addition to high-resolution source/drain electrodes, it is possible to use nTP to form complex multilayer devices with electrical functionality on plastic substrates [9.38]. Figure 9.20 shows a metal/insulator/metal (MIM) structure of Au (50 nm), SiN_x (100 nm; by plasma enhanced vapor deposition, PECVD), Ti (5 nm) and Au (50 nm) formed by transfer printing with a silicon stamp that is coated

sequentially with these layers. In this case, a short reactive ion etch (with CF_4) after the second Au deposition removes the SiN_x from the sidewalls of the stamp. nTP transfers these layers in a patterned geometry to a substrate of Au(15 nm)/Ti(1 nm)-coated PDMS(50 μm)/PET(250 μm). Interfacial cold-welding between the Au on the surfaces of the stamp and substrate bonds the multilayers to the substrate. Figure 9.8 illustrates the procedures, the structures (lateral dimensions of 250 μm × 250 μm, for ease of electrical probing), and their electrical characteristics. These MIM capacitors have performances similar to devices fabricated on silicon wafers by photolithography and lift-off. This example illustrates the ability of nTP to print patterns of materials whose growth conditions (high-temperature SiN_x by PECVD, in this case) prevent their direct deposition or processing on the substrate of interest (PET, in this case). The cold-welding transfer approach has also been exploited in other ways for patterning components for plastic electronics [9.51, 52].

Another class of unusual electronic/optoelectronic devices relies on circuits or circuit elements on curved surfaces. This emerging area of research was stimulated primarily by the ability of μCP to print high-resolution features on fibers and cylinders. Figure 9.21 shows a conducting microcoil printed with μCP on a microcapillary tube using the approach illustrated in Fig. 9.4. The coil serves as the excitation and detection element for high-resolution proton nuclear magnetic resonance of nanoliter volumes of fluid that are housed in the bore of the microcapillary [9.53]. The high fill factor and other considerations lead to extremely high sensitiv-

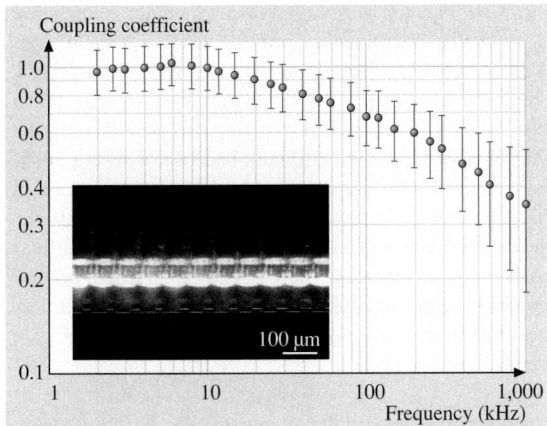

Fig. 9.22 The *inset* shows a concentric microtransformer formed using microcoils printed onto two different microcapillary tubes. The smaller of the tubes (outer diameter 135 microns) has a ferromagnetic wire threaded through its core. The larger one (outer diameter 350 microns) has the smaller tube threaded through its core. The resulting structure is a microtransformer that shows good coupling coefficients at frequencies up to $\approx$ 1 MHz. The graph shows its performance

ity with such printed coils. The bottom frame of Fig. 9.21 shows the spectrum of an $\approx$ 8 nL volume of ethylbenzene. The narrow lines demonstrate the high resolution that is possible with this approach. Similar coils can be used as magnets [9.54], springs [9.36], and electrical transformers [9.55]. Figure 9.22 shows an optical micrograph and the electrical measurements from a concentric cylindrical microtransformer that uses a microcoil printed on a microcapillary tube with a ferromagnetic wire threaded through its core. Inserting this structure into the core of a larger microcapillary that also supports a printed microcoil completes the transformer [9.55]. This type of device shows good coupling coefficients up to relatively high frequencies. Examples of other optoelectronic components appear in fiber optics where microfabricated on-fiber structures serve as integrated photomasks [9.20] and distributed thermal actuators [9.22].

9.4.2 Lasers and Waveguide Structures

In addition to integral components of unconventional electronic systems, useful structures for integrated optics can be built by using µCP and nTP to print sacrificial resist layers for etching glass waveguides. These printing techniques offer the most significant potential

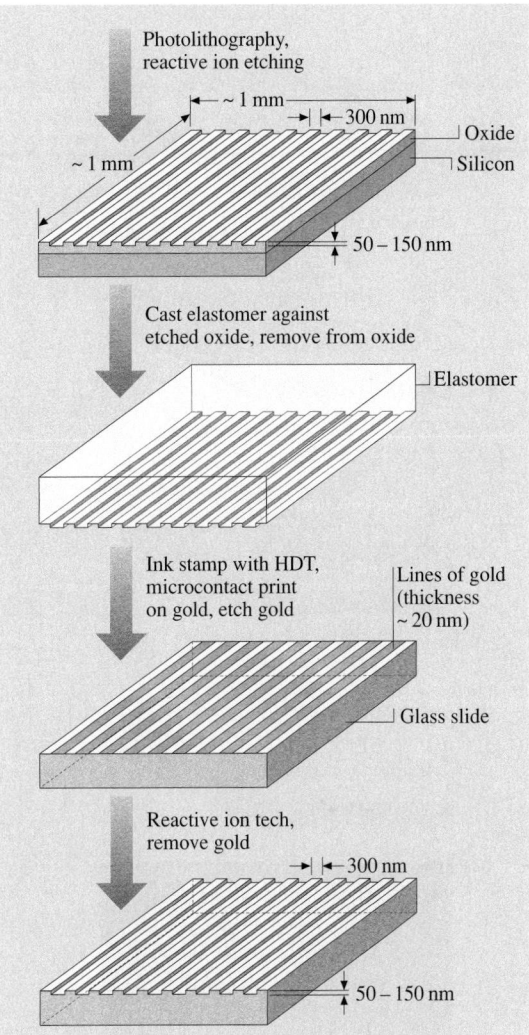

Fig. 9.23 Schematic illustration of the use of microcontact printing (µCP) for fabricating high-resolution gratings that can be incorporated into distributed planar laser structures or other components for integrated optics. The geometries illustrated here are suitable for third-order distributed feedback (DFB) lasers that operate in the red

value for this area when they are used to pattern features that are smaller than those that can be achieved with contact mode photolithography. Mode coupling gratings and distributed laser resonators are two such classes of structures. We have demonstrated µCP for forming distributed feedback (DFB) and distributed Bragg reflector (DBR) lasers that have narrow emission line widths [9.56]. This challenging fabrication

Part A | 9.4

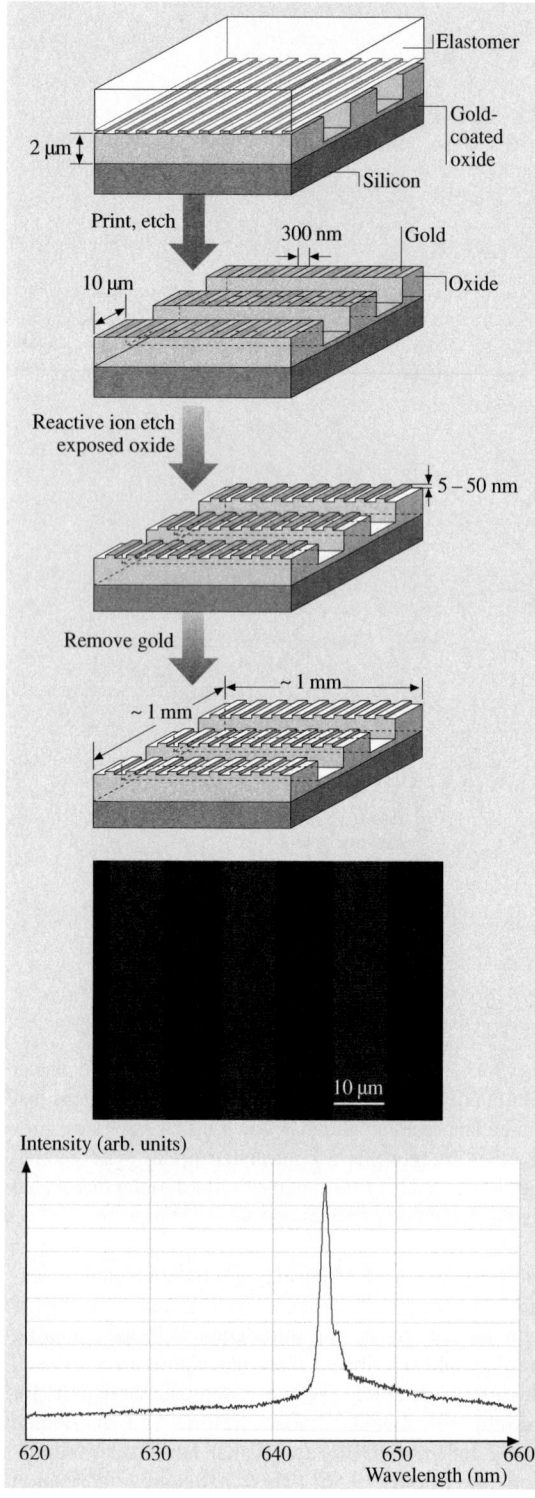

Fig. 9.24 The *top frames* gives a schematic illustration of steps for microcontact printing high-resolution gratings directly onto the top surfaces of ridge waveguides. The printing defines a sacrificial etch mask of gold which is subsequently removed. Producing this type of structure with photolithography is difficult because of severe thickness nonuniformities that appear in photoresist spin-cast on this type of nonplanar substrate. The *upper bottom* frame shows a top view optical micrograph of printed gold lines on the ridge waveguides. The *lower bottom* frame shows the emission output of a plastic photopumped laser that uses the printed structure and a thin evaporated layer of gain media

demonstrates the suitability of μCP for building structures that have (1) feature sizes of significantly less than one micron ($\approx 300\,\text{nm}$), *and* (2) long-range spatial coherence ($\approx 1\,\text{mm}$). The lasers employ optically pumped gain material deposited onto DFB or DBR resonators formed from periodic relief on a transparent substrate. The gain media confines light to the surface of the structure; its thickness is chosen to support a single transverse mode. To generate the required relief, lines of gold formed by μCP on a glass slide act as resists for reactive ion etching of the glass. Removing the gold leaves a periodic pattern of relief (600 nm period, 50 nm depth) on the surface of the glass (see Fig. 9.23). Figure 9.24 shows the performance of plastic lasers that use printed DFB and DBR resonators with gain media consisting of thin films of PBD doped with 1% by weight of coumarin 490 and DCMII, photopumped with 2 ns pulses from a nitrogen laser with intensities $> 5\,\text{kW/cm}^2$ [9.56]. Multimode lasing at resolution-limited line widths was observed at wavelengths corresponding to the third harmonic of the gratings. These characteristics are similar to those observed in lasers that use resonators generated with photolithography and are better than those that use imprinted polymers [9.57].

Contact printing not only provides a route to low-cost equivalents of gratings fabricated with other approaches, but also allows the fabrication of structures that would be difficult or impossible to generate with photolithography. For example, μCP can be used to form DFB resonators directly on the top surfaces of ridge waveguides [9.58]. Figure 9.25 illustrates the procedures. The bottom left frame shows an optical micrograph of the printed gold lines. Sublimation of a $\approx 200\,\text{nm}$ film of tris(8-hydroxyquinoline) aluminium (Alq) doped with 0.5–5.0 weight percent of the laser dye DCMII onto the resonators produces waveguide DFB lasers. The layer of

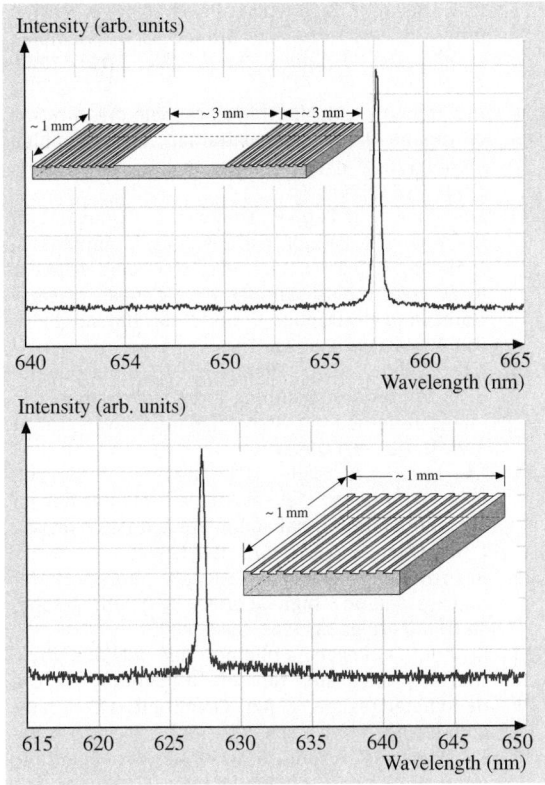

Fig. 9.25 Schematic illustrations and lasing spectra of plastic lasers that use microcontact printed resonators based on surface relief distributed Bragg reflectors (DBRs) and distributed feedback gratings (DFBs) on glass substrates. The grating periods are $\approx 600\,\text{nm}$ in both cases. The lasers use thin film plastic gain media deposited onto the printed gratings. This layer forms a planar waveguide that confines the light to the surface of the substrate. The laser shows emission over a narrow wavelength range, with a width that is limited by the resolution of the spectrometer used to characterize the output. In both cases, the emission profiles, the lasing thresholds and other characteristics of the devices are comparable to similar lasers that use resonators formed by high-resolution projection-mode photolithography

gain material itself provides a planar waveguide with air and polymer as the cladding layers. The relief waveguide provides lateral confinement of the light. Photopumping these devices with the output of a pulsed nitrogen laser ($\approx 2\,\text{ns}$, $337\,\text{nm}$) causes lasing due to Bragg reflections induced by the DFB structures on the top surfaces of the ridge waveguides. Some of the laser emission scatters out of the plane of the waveguide at an angle allowed by phase matching conditions. In this way, the grating also functions as an output coupler and offers a convenient way to characterize the laser emission. The bottom right frame of Fig. 9.22 shows the emission profile.

9.5 Conclusions

This chapter provides an overview of two contact printing techniques that are capable of micron and submicron resolution. It also illustrates some applications of these methods that may provide attractive alternatives to more established lithographic methods. The growing interest in nanoscience and technology makes it crucial to develop new methods for fabricating the relevant test structures and devices. The simplicity of these techniques together with the interesting and subtle materials science, chemistry, and physics associated with them make this a promising area for basic and applied study.

References

9.1   C. A. Mirkin, J. A. Rogers: Emerging methods for micro- and nanofabrication, MRS Bull. **26**, 506–507 (2001)

9.2   H. I. Smith, H. G. Craighead: Nanofabrication, Phys. Today **43**, 24–43 (February 1990)

9.3   W. M. Moreau (Ed.): *Semiconductor Lithography: Principles and Materials* (Plenum, New York 1988)

9.4   S. Matsui, Y. Ochiai: Focused ion beam applications to solid state devices, Nanotechnology **7**, 247–258 (1996)

9.5   J. M. Gibson: Reading and writing with electron beams, Phys. Today **50**, 56–61 (1997)

9.6   L. L. Sohn, R. L. Willett: Fabrication of nanostructures using atomic-force microscope-based lithography, Appl. Phys. Lett. **67**, 1552–1554 (1995)

9.7   E. Betzig, K. Trautman: Near-field optics – microscopy, spectroscopy, and surface modification beyond the diffraction limit, Science **257**, 189–195 (1992)

9.8   A. J. Bard, G. Denault, C. Lee, D. Mandler, D. O. Wipf: Scanning electrochemical microscopy: a new technique for the characterization and modification of surfaces, Acc. Chem. Res. **23**, 357 (1990)

9.9   J. A. Stroscio, D. M. Eigler: Atomic and molecular manipulation with the scanning tunneling microscope, Science **254**, 1319–1326 (1991)

9.10   J. Nole: Holographic lithography needs no mask, Laser Focus World **33**, 209–212 (1997)

9.11   A. N. Broers, A. C. F. Hoole, J. M. Ryan: Electron beam lithography – resolution limits, Microelectron. Eng. **32**, 131–142 (1996)

9.12   A. N. Broers, W. Molzen, J. Cuomo, N. Wittels: Electron-beam fabrication of 80 Å metal structures, Appl. Phys. Lett. **29**, 596 (1976)

9.13   G. D. Aumiller, E. A. Chandross, W. J. Tomlinson, H. P. Weber: Submicrometer resolution replication of relief patterns for integrated optics, J. Appl. Phys. **45**, 4557–4562 (1974)

9.14   Y. Xia, J. J. McClelland, R. Gupta, D. Qin, X.-M. Zhao, L. L. Sohn, R. J. Celotta, G. M. Whiteside: Replica molding using polymeric materials: a practical step toward nanomanufacturing, Adv. Mater. **9**, 147–149 (1997)

9.15   T. Borzenko, M. Tormen, G. Schmidt, L. W. Molenkamp, H. Janssen: Polymer bonding process for nanolithography, Appl. Phys. Lett. **79**, 2246–2248 (2001)

9.16   H. Hua, Y. Sun, A. Gaur, M. A. Meitl, L. Bilhaut, L. Rotinka, J. Wang, P. Geil, M. Shim, J. A. Rogers: Polymer imprint lithography with molecular-scale resolution, Nano Lett. **4(12)**, 2467–2471 (2004)

9.17   A. Kumar, G. M. Whitesides: Features of gold having micrometer to centimeter dimensions can be formed through a combination of stamping with an elastomeric stamp and an alkanethiol ink followed by chemical etching, Appl. Phys. Lett. **63**, 2002–2004 (1993)

9.18   Y. Xia, G. M. Whitesides: Soft lithography, Angew. Chem. Int. Ed. **37**, 550–575 (1998)

9.19   Y. Xia, J. A. Rogers, K. E. Paul, G. M. Whitesides: Unconventional methods for fabricating and patterning nanostructures, Chem. Rev. **99**, 1823–1848 (1999)

9.20   J. A. Rogers, R. J. Jackman, J. L. Wagener, A. M. Vengsarkar, G. M. Whitesides: Using microcontact printing to generate photomasks on the surface of optical fibers: a new method for producing in-fiber gratings, Appl. Phys. Lett. **70**, 7–9 (1997)

9.21   B. Michel, A. Bernard, A. Bietsch, E. Delamarche, M. Geissler, D. Juncker, H. Kind, J. P. Renault, H. Rothuizen, H. Schmid, P. Schmidt-Winkel, R. Stutz, H. Wolf: Printing meets lithography: soft approaches to high-resolution printing, IBM J. Res. Dev. **45**, 697–719 (2001)

9.22   J. A. Rogers: Rubber stamping for plastic electronics and fiber optics, MRS Bull. **26**, 530–534 (2001)

9.23   N. B. Larsen, H. Biebuyck, E. Delamarche, B. Michel: Order in microcontact printed self-assembled monolayers, J. Am. Chem. Soc. **119**, 3017–3026 (1997)

9.24   H. A. Biebuyck, G. M. Whitesides: Self-organization of organic liquids on patterned self-assembled monolayers of alkanethiolates on gold, Langmuir **10**, 2790–2793 (1994)

9.25   J. A. Rogers, Z. Bao, K. Baldwin, A. Dodabalapur, B. Crone, V. R. Raju, V. Kuck, H. Katz, K. Amundson, J. Ewing, P. Drzaic: Paper-like electronic displays: Large area, rubber stamped plastic sheets of electronics and electrophoretic inks, Proc. Nat. Acad. Sci. USA **98**, 4835–4840 (2001)

9.26   J. L. Wilbur, H. A. Biebuyck, J. C. MacDonald, G. M. Whitesides: Scanning force microscopies can image patterned self-assembled monolayers, Langmuir **11**, 825–831 (1995)

9.27   J. C. Love, D. B. Wolfe, M. L. Chabinyc, K. E. Paul, G. M. Whitesides: Self-assembled monolayers of alkanethiolates on palladium are good etch resists, J. Am. Chem. Soc. **124**, 1576–1577 (2002)

9.28   H. Schmid, B. Michel: Siloxane polymers for high-resolution, high-accuracy soft lithography, Macromolecules **33**, 3042–3049 (2000)

9.29   K. Choi, J. A. Rogers: A photocurable poly(dimethylsiloxane) chemistry for soft lithography in the nanometer regime, J. Am. Chem. Soc. **125**, 4060–4061 (2003)

9.30   J. A. Rogers, K. E. Paul, G. M. Whitesides: Quantifying distortions in soft lithography, J. Vacuum. Sci. Tech. B **16**, 88–97 (1998)

9.31   J. Tate, J. A. Rogers, C. D. W. Jones, W. Li, Z. Bao, D. W. Murphy, R. E. Slusher, A. Dodabalapur, H. E. Katz, A. J. Lovinger: Anodization and microcontact printing on electroless silver: solution-based fabrication procedures for low voltage organic electronic systems, Langmuir **16**, 6054–6060 (2000)

9.32   Y. Xia, E. Kim, G. M. Whitesides: Microcontact printing of alkanethiols on silver and its application to microfabrication, J. Electrochem. Soc. **143**, 1070–1079 (1996)

9.33   Y. N. Xia, X. M. Zhao, E. Kim, G. M. Whitesides: A selective etching solution for use with patterned self-assembled monolayers of alkanethiolates on gold, Chem. Mater. **7**, 2332–2337 (1995)

9.34   R. J. Jackman, J. Wilbur, G. M. Whitesides: Fabrication of submicrometer features on curved substrates by microcontact printing, Science **269**, 664–666 (1995)

9.35   R. J. Jackman, S. T. Brittain, A. Adams, M. G. Prentiss, G. M. Whitesides: Design and fabrication of topologically complex, three-dimensional microstructures, Science **280**, 2089–2091 (1998)

9.36   J. A. Rogers, R. J. Jackman, G. M. Whitesides: Microcontact printing and electroplating on curved substrates: a new means for producing free-standing three-dimensional microstructures with possible applications ranging from micro-coil springs to coronary stents, Adv. Mater. **9**, 475–477 (1997)

9.37   Y.-L. Loo, R. W. Willett, K. Baldwin, J. A. Rogers: Additive, nanoscale patterning of metal films with a stamp and a surface chemistry mediated transfer process: applications in plastic electronics, Appl. Phys. Lett. **81**, 562–564 (2002)

9.38   Y.-L. Loo, R. W. Willett, K. Baldwin, J. A. Rogers: Interfacial chemistries for nanoscale transfer printing, J. Am. Chem. Soc. **124**, 7654–7655 (2002)

9.39   Y.-L. Loo, J. W. P. Hsu, R. L. Willett, K. W. Baldwin, K. W. West, J. A. Rogers: High-resolution transfer printing on GaAs surfaces using alkane dithiol self-assembled monolayers, J. Vacuum. Sci. Tech. B. **20**, 2853–2856 (2002)

9.40   G. S. Ferguson, M. K. Chaudhury, G. B. Sigal, G. M. Whitesides: Contact adhesion of thin gold-films on elastomeric supports – cold welding under ambient conditions, Science **253**, 776–778 (1991)

9.41   W. Zhang, S. Y. Chou: Multilevel nanoimprint lithography with submicron alignment over 4 in Si wafers, Appl. Phys. Lett. **79**, 845–847 (2001)

9.42   N. Bowden, S. Brittain, A. G. Evans, J. W. Hutchinson, G. M. Whitesides: Spontaneous formation of ordered structures in thin films of metals supported on an elastomeric polymer, Nature **393**, 146–149 (1998)

9.43   E. Menard, L. Bilhaut, J. Zaumseil, J. A. Rogers: Improved surface chemistries, thin film deposition techniques, and stamp designs for nanotransfer printing, Langmuir **20**, 6871–6878 (2004)

9.44   J. Zaumseil, M. A. Meitl, J. W. P. Hsu, B. Acharya, K. W. Baldwin, Y.-L. Loo, J. A. Rogers: Three-dimensional and multilayer nanostructures formed by nanotransfer printing, Nano. Lett. **3**, 1223–1227 (2003)

9.45   Z. Bao, J. A. Rogers, H. E. Katz: Printable organic and polymeric semiconducting materials and devices, J. Mater. Chem. **9**, 1895–1904 (1999)

9.46   J. A. Rogers, Z. Bao, A. Dodabalapur, A. Makhija: Organic smart pixels and complementary inverter circuits formed on plastic substrates by casting, printing and molding, IEEE Electron Dev. Lett. **21**, 100–103 (2000)

9.47   J. A. Rogers, Z. Bao, A. Makhija: Non-photolithographic fabrication sequence suitable for reel-to-reel production of high performance organic transistors and circuits that incorporate them, Adv. Mater. **11**, 741–745 (1999)

9.48   P. Mach, S. Rodriguez, R. Nortrup, P. Wiltzius, J. A. Rogers: Active matrix displays that use printed organic transistors and polymer dispersed liquid crystals on flexible substrates, Appl. Phys. Lett. **78**, 3592–3594 (2001)

9.49   J. A. Rogers: Toward paperlike displays, Science **291**, 1502–1503 (2001)

9.50   Y.-L. Loo, T. Someya, K. W. Baldwin, P. Ho, Z. Bao, A. Dodabalapur, H. E. Katz, J. A. Rogers: Soft, conformable electrical contacts for organic transistors: high resolution circuits by lamination, Proc. Nat. Acad. Sci. USA **99**, 10252–10256 (2002)

9.51   C. Kim, P. E. Burrows, S. R. Forrest: Micropatterning of organic electronic devices by cold-welding, Science **288**, 831–833 (2000)

9.52   C. Kim, M. Shtein, S. R. Forrest: Nanolithography based on patterned metal transfer and its application to organic electronic devices, Appl. Phys. Lett. **80**, 4051–4053 (2002)

9.53   J. A. Rogers, R. J. Jackman, G. M. Whitesides, D. L. Olson, J. V. Sweedler: Using microcontact printing to fabricate microcoils on capillaries for high resolution ^{1}H–NMR on nanoliter volumes, Appl. Phys. Lett. **70**, 2464–2466 (1997)

9.54   J. A. Rogers, R. J. Jackman, G. M. Whitesides: Constructing single and multiple helical microcoils and characterizing their performance as components of microinductors and microelectromagnets, J. Microelectromech. Sys. (JMEMS) **6**, 184–192 (1997)

9.55   R. J. Jackman, J. A. Rogers, G. M. Whitesides: Fabrication and characterization of a concentric, cylindrical microtransformer, IEEE Trans. Magn. **33**, 2501–2503 (1997)

9.56   J. A. Rogers, M. Meier, A. Dodabalapur: Using stamping and molding techniques to produce distributed feedback and Bragg reflector resonators for plastic lasers, Appl. Phys. Lett. **73**, 1766–1768 (1998)

9.57   M. Berggren, A. Dodabalapur, R. E. Slusher, A. Timko, O. Nalamasu: Organic solid-state lasers with imprinted gratings on plastic substrates, Appl. Phys. Lett. **72**, 410–411 (1998)

9.58   J. A. Rogers, M. Meier, A. Dodabalapur: Distributed feedback ridge waveguide lasers fabricated by nanoscale printing and molding on non-planar substrates, Appl. Phys. Lett. **74**, 3257–3259 (1999)

10. Material Aspects of Micro- and Nanoelectromechanical Systems

One of the more significant technological achievements during the last 20 years has been the development of MEMS and its new offshoot, NEMS. These developments were made possible by significant advancements in the materials and processing technologies used in the fabrication of MEMS and NEMS devices. While initial developments capitalized on a mature Si infrastructure built for the integrated circuit (IC) industry, recent advances have come about using materials and processes not associated with IC fabrication, a trend that is likely to continue as new application areas emerge.

A well-rounded understanding of MEMS and NEMS technology requires a basic knowledge of the materials used to construct the devices, since material properties often govern device performance and dictate fabrication approaches. An understanding of the materials used in MEMS and NEMS involves an understanding of material systems, since such devices are rarely constructed of a single material but rather a collection of materials working in conjunction with each other to provide critical functions. It is from this perspective that the following chapter is constructed. A preview of the materials selected for inclusion in this chapter is presented in Table 10.1. It should be clear from this table that this chapter is not a summary of all materials used in MEMS and NEMS, as such a work would itself constitute a text of

significant size. It does, however, present a selection of some of the more important material systems, and especially those that illustrate the importance of viewing MEMS and NEMS in terms of material systems.

10.1 Silicon

10.1.1 Single-Crystal Silicon

Use of silicon (Si) as a material for microfabricated sensors dates back to the middle of the 20th century when the piezoresistive effect in germanium (Ge) and Si was first identified [10.1]. It was discovered that the piezoresistive coefficients of Si were sig-

nificantly higher than those associated with metals used in conventional strain gauges; and this finding initiated the development of Si-based strain gauge devices, and along with Si bulk micromachining techniques, piezoresistive Si pressure sensors during the 1960s and 1970s. The subsequent development of Si surface micromachining techniques along with the

Table 10.1 Distinguishing characteristics and application examples of selected materials for MEMS and NEMS

Material	Distinguishing characteristics	Application examples
Single-crystal silicon (Si)	High-quality electronic material, selective anisotropic etching	Bulk micromachining, piezoresistive sensing
Polycrystalline Si (polysilicon)	Doped Si films on sacrificial layers	Surface micromachining, electrostatic actuation
Silicon dioxide (SiO$_2$)	Insulating, etched by HF, compatible with polysilicon	Sacrificial layer in polysilicon surface micromachining, passivation layer for devices
Silicon nitride (Si$_3$N$_4$, Si$_x$N$_y$)	Insulating, chemically resistant, mechanically durable	Isolation layer for electrostatic devices, membrane and bridge material
Polycrystalline germanium (polyGe), Polycrystalline silicon-germanium (poly Si $-$ Ge)	Deposited at low temperatures	Integrated surface micromachined MEMS
Gold (Au), aluminum (Al)	Conductive thin films, flexible deposition techniques	Innerconnect layers, masking layers, electromechanical switches
Bulk ti	High strength, corrosion resistant	Optical MEMS
Nickel-iron (NiFe)	Magnetic alloy	Magnetic actuation
Titanium-nickel (TiNi)	Shape-memory alloy	Thermal actuation
Silicon carbide (SiC) diamond	Electrically and mechanically stable at high temperatures, chemically inert, high Young's modulus to density ratio	Harsh-environment MEMS, high-frequency MEMS/NEMS
Gallium arsenide (GaAs), indium phosphide (InP), indium arsenide (InAs) and related materials	Wide bandgap, epitaxial growth on related ternary compounds	RF MEMS, optoelectronic devices, single-crystal bulk and surface micromachining
Lead zirconate titanate (PZT)	Piezoelectric material	Mechanical sensors and actuators
Polyimide	Chemically resistant, high-temperature polymer	Mechanically flexible MEMS, bioMEMS
SU-8	Thick, photodefinable resist	Micromolding, High-aspect-ratio structures
Parylene	Biocompatible polymer, deposited at room temperature by CVD	Protective coatings, molded polymer structures
Liquid crystal polymer	Chemically resistant, low moisture permeability, insulating	bioMEMS, RF MEMS

recognition that micromachined Si structures could be integrated with Si IC devices marked the advent of MEMS with Si firmly positioned as the primary MEMS material.

For MEMS applications, single-crystal Si serves several key functions. Single-crystal Si is one of the most versatile materials for bulk micromachining due to the availability of anisotropic etching processes in conjunction with good mechanical properties. Single-crystal Si has favorable mechanical properties (i.e., a Young's modulus of about 190 GPa), enabling its use as a material for membranes, resonant beams, and other such structures. For surface micromachining applications, single-crystal Si substrates are used primarily as mechanical platforms on which device structures are fabricated, although the advent of silicon-on-insulator (SOI) substrates enables the fabrication of single-crystal Si surface micromachined structures by using the buried oxide as a sacrificial layer. Use of high-quality single-crystal wafers enables the fabrication of integrated MEMS devices, at least for materials and processes that are compatible with Si ICs.

From the materials perspective, single-crystal Si is a relatively easy material to bulk micromachine due to the availability of anisotropic etchants such as potassium hydroxide (KOH) and tetramethyl-aluminum hydroxide (TMAH) that attack the (100) and (110) Si crystal planes significantly faster than the (111) crystal planes. For example, the etching rate ratio of (100) to (111) planes in Si is about 400:1 for a typical KOH/water etching solution. Silicon dioxide (SiO_2), silicon nitride (Si_3N_4), and some metallic thin films (e.g., Cr, Au, etc.) provide good etch masks for most Si anisotropic etchants. Heavily boron-doped Si is an effective etch stop for some liquid reagents. Boron-doped etch stops are often less than $10\,\mu m$ thick, since the boron concentration in Si must exceed $7 \times 10^{19}\,cm^3$ for the etch stop to be effective and the doping is done by thermal diffusion. Ion implantation can be used to create a subsurface etch stop layer; however, the practical limit is a few microns.

In contrast to anisotropic etching, isotropic etching exhibits no selectivity to the various crystal planes. Commonly used isotropic Si etchants consist of hydrofluoric (HF) and nitric (HNO_3) acid mixtures in water or acetic acid (CH_3COOH), with the etch rate dependent on the ratio of HF to HNO_3. From a processing perspective, isotropic etching of Si is commonly used for removal of work-damaged surfaces, creation of structures in single-crystal slices, and patterning of single-crystal or polycrystalline films.

Well-established dry etching processes are routinely used to pattern single-crystal Si. The process spectrum ranges from physical techniques such as sputtering and ion milling to chemical techniques such as plasma etching. Reactive ion etching (RIE) is the most commonly used dry etching technique for Si patterning. By combining both physical and chemical processes, RIE is a highly effective anisotropic Si etching technique that can be used to generate patterns that are independent of crystalline orientation. Fluorinated compounds such as CF_4, SF_6, and NF_3, or chlorinated compounds such as

Fig. 10.1 A collection of Si nanoelectromechanical beam resonators fabricated from a single-crystal Si substrate (courtesy M. Roukes, Caltech.)

CCl_4 or Cl_2, sometimes mixed with He, O_2, or H_2, are commonly used in Si RIE. The RIE process is highly directional, which enables direct lateral pattern transfer from an overlying masking material to the etched Si surface. SiO_2 thin films are often used as masking and sacrificial layers owing to its chemical durability under these plasma conditions. Process limitations (i. e., etch rates) restrict the etch depths of conventional Si RIE to less than 10 microns; however, a process called deep reactive ion etching (DRIE) has extended the use of anisotropic dry etching to depths well beyond several hundred microns.

Using the aforementioned processes and techniques, a wide variety of microfabricated devices have been made from single-crystal Si, such as piezoresistive pressure sensors, accelerometers, and mechanical resonators, to name a few. Using nearly the same approaches but on a smaller scale, "top-down" nanomachining techniques have been used to fabricate nanoelectromechanical devices from single-crystal Si. Single-crystal Si is particularly well suited for nanofabrication because high crystal quality substrates with very smooth surfaces are readily available. By coupling electron-beam (e-beam) lithographic techniques with conventional Si etching, device structures with submicron dimensions have been fabricated. Submicron, single-crystal Si nanomechanical structures have been successfully micromachined from bulk Si wafers [10.2] and silicon-on-insulator (SOI) wafers [10.3]. In the former case, an isotropic Si etch was performed to release the device structures, whereas in the latter case, the 50 nm to 200 nm structures were released by dissolving the underlying oxide layer in HF. An example of nanoelectromechanical beam structures fabricated from a single-crystal Si substrate is shown in Fig. 10.1.

10.1.2 Polycrystalline and Amorphous Silicon

Surface micromachining is a process where a sequence of thin films, often of different materials, is deposited and selectively etched to form the desired micromechanical (or microelectromechanical) structure. In contrast to bulk micromachining, the substrate serves primarily as a device-supporting platform. For Si-based surface micromachined MEMS, polycrystalline Si (polysilicon) is most often used as the structural material, SiO_2 as the sacrificial material, silicon nitride (Si_3N_4) for electrical isolation of device structures, and single-crystal Si as the substrate. Like single-crystal Si, polysilicon can be doped during or after film deposition. SiO_2 can be thermally grown or deposited on polysilicon over a broad temperature range (e.g., 200 °C to 1150 °C) to meet various process and material requirements. SiO_2 is readily dissolvable in hydrofluoric acid (HF), which does not etch polysilicon and thus can be used to dissolve SiO_2 sacrificial layers. Si_3N_4 is an insulating film that is highly resistant to oxide etchants. The polysilicon micromotor shown in Fig. 10.2 was surface micromachined using a process that included these materials.

For MEMS and IC applications, polysilicon films are commonly deposited using a process known as low-pressure chemical vapor deposition (LPCVD). The typical polysilicon LPCVD reactor is based on a hot-wall, resistance-heated furnace. Typical processes are performed at temperatures ranging from 580 °C to 650 °C and pressures from 100 to 400 mtorr. The most commonly used source gas is silane (SiH_4). The microstructure of polysilicon thin films consist of a collection of small grains whose microstructure and orientation is a function of the deposition conditions [10.4]. For typical LPCVD processes (e.g., 200 mtorr), the amorphous-to-polycrystalline transition temperature is about 570 °C, with polycrystalline films deposited above the transition temperature. At 600 °C, the grains are small and equiaxed, while at 625 °C, the grains are large and columnar [10.4]. The crystal orientation is predominantly (110) Si for temperatures between 600 °C and 650 °C, while the (100) orientation is dominant for temperatures between 650 °C and 700 °C.

The resistivity of polysilicon can be modified using the doping methods developed for single-crystal Si. Diffusion is an effective method for doping polysilicon films, especially for heavy doping of thick films. Phosphorus, which is the most commonly used dopant in polysilicon MEMS, diffuses significantly faster in polysilicon than in single-crystal Si due primarily to enhanced diffusion rates along grain boundaries. The diffusivity of phosphorus in polysilicon thin films with small equiaxed grains is about 1×10^{12} cm^2/s. Ion implantation is also used to dope polysilicon films. A high-temperature annealing step is usually required to electrically activate the implanted dopants as well as to repair implant-related damage in the polysilicon films. In general, the conductivity of implanted polysilicon films is not as high as films doped by diffusion.

In situ doping of polysilicon is performed by simply including a dopant gas, usually diborane (B_2H_6) or phosphine (PH_3), in the CVD process. The addition of dopants during the deposition process not only modifies the conductivity but also affects the deposition rate of the polysilicon films. As shown in Fig. 10.3, the inclusion of boron generally increases the deposition rate of polysilicon relative to undoped films (McMahon et al.), while phosphorus (not shown) reduces the rate. In situ doping can be used to produce conductive films with uniform doping profiles without requiring the high-temperature steps commonly associated with diffusion or ion implantation. Although commonly used to produce doped polysilicon for electrostatic devices, *Cao* et al. [10.5]

Fig. 10.2 SEM micrograph of a surface micromachined polysilicon micromotor fabricated using a SiO_2 sacrificial layer

have used in situ phosphorus-doped polysilicon films in piezoresistive strain gauges, achieving gauge factors as high as 15 for a single strip sensor.

The thermal conductivity of polysilicon is a strong function of its microstructure, and therefore the conditions used during deposition [10.4]. For fine-grained films, the thermal conductivity is about 25% of the value of single-crystal Si. For thick films with large grains, the thermal conductivity ranges between 50% and 85% of the single-crystal value.

Like the electrical and thermal properties of polysilicon, the as-deposited residual stress in polysilicon films depends on microstructure. For films deposited under typical conditions (200 mtorr, 625 °C), the as-deposited polysilicon films have compressive residual stresses. The highest compressive stresses are found in amorphous Si films and polysilicon films with a strong, columnar (110) texture. For films with fine-grained microstructures, the stress tends to be tensile. Annealing can be used to reduce the compressive stress in as-deposited polysilicon films. For instance, compressive residual stresses on the order of 500 MPa can be reduced to less than 10 MPa by annealing the as-deposited films at 1000 °C in a N_2 ambient [10.6, 7]. Rapid thermal annealing (RTA) provides an effective method of stress reduction in polysilicon films on temperature-sensitive substrates. *Zhang* et al. [10.8] reported that a 10-s anneal at 1100 °C was sufficient to completely relieve the stress in films that originally had a compressive stress of about 340 MPa. RTA is particularly attractive in situations where the process parameters require a low thermal budget.

As an alternative to high-temperature annealing, *Yang* et al. [10.9] have developed an approach that actually utilizes the residual stress characteristics of polysilicon deposited under various conditions to construct polysilicon multilayers that have the desired thickness and stress values. The multilayers are comprised of alternating tensile and compressive polysilicon layers that are deposited in a sequential manner. The tensile layers consist of fine-grained polysilicon grown at a temperature of 570 °C, while the compressive layers are made up of columnar polysilicon deposited at 615 °C. The overall stress in the composite film depends on the number of alternating layers and the thickness of each layer. With the proper set of parameters, a composite polysilicon multilayer can be deposited with near zero residual stress and no stress gradient. The process achieves stress reduction without high-temperature annealing, a considerable advantage for integrated MEMS processes.

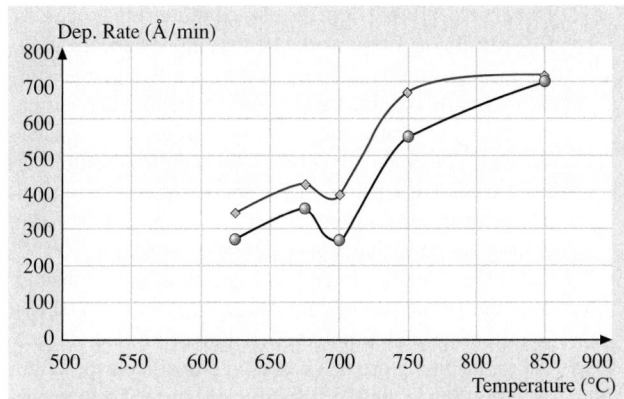

Fig. 10.3 Deposition rate versus substrate temperature for in situ boron-doped (◇) and undoped (•) polysilicon films grown by atmospheric pressure chemical vapor deposition (McMahon et al. 2001)

Many device designs require polysilicon thicknesses that are not readily achievable using conventional LPCVD polysilicon due to the low deposition rates associated with such systems. For these applications, epitaxial Si reactors can be used to grow polysilicon films. Unlike conventional LPCVD processes with deposition rates of less than 100 Å/min, epitaxial processes have deposition rates on the order of 1 micron/min [10.10]. The high deposition rates result from the much higher substrate temperatures (> 1000 °C) and deposition pressures (> 50 torr) used in these processes. The polysilicon films are usually deposited on SiO_2 sacrificial layers to enable surface micromachining. An LPCVD polysilicon seed layer is sometimes used in order to control nucleation, grain size, and surface roughness. As with conventional polysilicon, the microstructure and residual stress of the epi-poly films, as they are known, are related to deposition conditions. Compressive films generally have a mixture of [110] and [311] grains [10.11, 12], while tensile films have a random mix of [110], [100], [111], and [311] grains [10.11]. The Young's modulus of epi-poly measured from micromachined test structures is comparable with LPCVD polysilicon [10.12]. Mechanical properties test structures [10.10–12], thermal actuators [10.10], electrostatically actuated accelerometers [10.10], and gryoscopes [10.13] have been fabricated from these films.

As a low-temperature alternative to LPCVD polysilicon, physical vapor deposition (PVD) techniques have been developed to produce Si thin films on temperature-sensitive substrates. *Abe* et al. [10.14] and

Honer et al. [10.15] have developed sputtering processes for polysilicon. Early work [10.14] emphasized the ability to deposit very smooth (2.5 nm) polysilicon films on thermally oxidized wafers at reasonable deposition rates (19.1 nm/min) and with low residual compressive stresses. The process involved DC magnitron sputtering from a Si target using an Ar sputtering gas, a chamber pressure of 5 mtorr, and a power of 100 W. The authors reported that a postdeposition anneal at 700 °C in N_2 for 2 h was needed to crystallize the deposited film and perhaps lower the stress. *Honer* et al. [10.15] sought to develop a polymer-friendly, Si-based surface micromachining process based on polysilicon sputtered onto polyimide and PSG sacrificial layers. To improve the conductivity of the micromachined Si structures, the sputtered Si films were sandwiched between two TiW cladding layers. The device structures on polyimide were released using oxygen plasma etching. The processing step with the highest temperature was, in fact, the polyimide cure at 350 °C. To test the robustness of the process, sputter-deposited Si microstructures were fabricated on substrates containing CMOS devices. As expected from thermal budget considerations, the authors reported no measurable degradation of device performance.

PECVD has emerged as an alternative to LPCVD for the production of Si-based surface micromachined structures on temperature-sensitive substrates. *Gaspar* et al. [10.16] recently reported on the development of surface micromachined microresonators fabricated from hydrogenated amorphous Si (a-Si:H) thin films deposited by PECVD. The vertically actuated resonators consisted of doubly-clamped microbridges suspended over fixed Al electrodes. The a-Si:H films were deposited using SiH_4 and H_2 precursors and PH_3 as a doping gas. The substrate temperature was held to around 100 °C, which enabled the use of photoresist as a sacrificial layer. The microbridges consisted of a large paddle suspended by two thin paddle supports, with the paddle providing a large reflective surface for optical detection of resonant frequency. The megahertz-frequency resonators exhibited quality factors in the 1×10^5 range when tested in vacuum.

10.1.3 Porous Silicon

Porous Si is produced by room temperature electrochemical etching of Si in HF. If configured as an electrode in an HF-based electrochemical circuit, positive charge carriers (holes) at the Si surface facilitate the exchange of F atoms with H atoms terminating the Si surface.

The exchange continues in the subsurface region, leading to the eventual removal of the fluorinated Si. The quality of the etched surface is related to the density of holes at the surface, which is controlled by the applied current density. For high current densities, the density of holes is high and the etched surface is smooth. For low current densities, the hole density is low and clustered in highly localized regions associated with surface defects. Surface defects become enlarged by etching, which leads to the formation of pores. Pore size and density are related to the type of Si used and the conditions of the electrochemical cell. Both single-crystal and polycrystalline Si can be converted to porous Si.

The large surface-to-volume ratios make porous Si attractive for gaseous and liquid applications, including filter membranes and absorbing layers for chemical and mass sensing [10.17]. When single-crystal substrates are used, the unetched porous layer remains single crystalline and is suitable for epitaxial Si growth. It has been shown that CVD coatings do not generally penetrate the porous regions, but rather overcoat the pores at the surface of the substrate [10.18]. The formation of localized Si-on-insulator structures is therefore possible by simply combining pore formation with epitaxial growth, followed by dry etching to create access holes to the porous region and thermal oxidation of the underlying porous region. A third application uses porous Si as a sacrificial layer for polysilicon and single-crystalline Si surface micromachining. As shown by *Lang* et al. [10.18], the process involves the electrical isolation of the solid structural Si layer by either pn-junction formation through selective doping or use of electrically insulating thin films since the formation of pores only occurs on electrically charged surfaces. A weak Si etchant will aggressively attack the porous regions with little damage to the structural Si layers and can be used to release the devices.

Porous polysilicon is currently being developed as a structural material for chip-level vacuum packaging [10.19]. In this example, a 1.5-μm-thick polysilicon is deposited onto a supporting PSG sacrificial layer, electrochemically etched in an HF solution to render it porous, and then annealed by RTA to reduce stress in the porous layer. When fabricated locally over a prefabricated device structure (prior to release), the porous Si forms a localized shell that will serve as a mechanical support for the main packaging structure. The porous structure enables an HF etch to remove the supporting PSG layer as well as any sacrificial oxide layers associated with the prefabricated MEMS device. After the sacrificial etch, the packaging sequence is completed by depositing a polysilicon film by LPCVD at 179 mtorr on the porous shell, thus fully encapsulating the device under vacuum conditions. This technique was used to package a microfabricated Pirani vacuum gauge, which enabled an in situ measurement of pressure versus time. The authors found no detectable change in pressure over a 3-month period.

10.1.4 Silicon Dioxide

Silicon dioxide (SiO_2) is one of the most widely used materials in the fabrication of MEMS. In polysilicon surface micromachining, SiO_2 is used as a sacrificial material since it can be easily dissolved using etchants that do not attack polysilicon. SiO_2 is widely used as an etch mask for dry etching of thick polysilicon films since it is chemically resistant to dry etching processes for polysilicon. SiO_2 films are also used as passivation layers on the surfaces of environmentally sensitive devices.

The most common processes used to produce SiO_2 films for polysilicon surface micromachining are thermal oxidation and LPCVD. Thermal oxidation of Si is performed at temperatures of 900 °C to 1200 °C in the presence of oxygen or steam. Since thermal oxidation is a self-limiting process, the maximum practical film thickness that can be obtained is about 2 μm, which is sufficient for many sacrificial applications. As noted by its name, thermal oxidation of Si can only be performed on Si surfaces.

SiO_2 films can be deposited on a wide variety of substrate materials by LPCVD. In general, LPCVD provides a means for depositing thick (> 2 μm) SiO_2 films at temperatures much lower than thermal oxidation. Known as low-temperature oxides, or LTO for short, these films have a higher etch rate in HF than thermal oxides, which translates to significantly faster release times when LTO films are used as sacrificial layers. Phosphosilicate glass (PSG) can be formed using nearly the same deposition process as LTO by adding a phosphorus-containing gas to the precursor flows. PSG films are useful as sacrificial layers since they generally have higher etching rates in HF than LTO films.

PSG and LTO films are deposited in hot-wall, low-pressure, fused-silica furnaces in systems similar to those described previously for polysilicon. Precursor gases include SiH_4 as a Si source, O_2 as an oxygen source, and, in the case of PSG, PH_3 as a source of phosphorus. LTO and PSG films are typically deposited at temperatures of 425 °C to 450 °C and pressures ranging from 200 mtorr to 400 mtorr. The low deposition temperatures result in LTO and PSG films that are slightly less dense than thermal oxides due to the incorporation of hydrogen in the films. LTO films can, however, be densified by an annealing step at high temperature (1000 °C). The low density of LTO and PSG films is partially responsible for the increased etch rate in HF.

Thermal SiO_2 and LTO are electrical insulators used in numerous MEMS applications. The dielectric constants of thermal oxide and LTO are 3.9 and 4.3, respectively. The dielectric strength of thermal SiO_2 is 1.1×10^6 V/cm, and for LTO it is about 80% of that value [10.20]. The stress in thermal SiO_2 is compressive with a magnitude of about 300 MPa [10.20]. For LTO, however, the typical as-deposited residual stress is tensile, with a magnitude of about 100 MPa to 400 MPa [10.20]. The addition of phosphorus to LTO decreases the tensile residual stress to about 10 MPa for phosphorus concentrations of 8% [10.21]. As with polysilicon, the properties of LTO and PSG are dependent on processing conditions.

Plasma enhanced chemical vapor deposition (PECVD) is another common method to produce oxides of silicon. Using a plasma to dissociate the gaseous precursors, the deposition temperatures needed to deposit PECVD oxide films is lower than for LPCVD films. For this reason, PECVD oxides are quite commonly used as masking, passivation, and protective layers, especially on devices that have been coated with metals.

Quartz is the crystalline form of SiO_2 and has interesting properties for MEMS. Quartz is optically transparent, piezoelectric, and electrically insulating. Like single-crystal Si, quartz substrates are available as high-quality, large-area wafers that can be bulk micromachined using anisotropic etchants. A short review of the basics of quartz etching was written by *Danel* et al. [10.22] and is recommended for those interested in the subject. Quartz has recently become a popular substrate material for microfluidic devices due to its optical, electronic, and chemical properties.

Another SiO_2-related material that has recently found uses in MEMS is spin-on-glass (SOG). SOG is a polymeric material with a viscosity suitable for spin coating. Two recent publications illustrate the potential for SOG in MEMS fabrication. In the first example, *Yasseen* et al. [10.23] detailed the development of SOG as a thick-film sacrificial molding material for thick polysilicon films. The authors reported a process to deposit, polish, and etch SOG films that were 20 μm thick. The thick SOG films were patterned into molds and filled with $10\,\mu m$ thick LPCVD polysilicon films, planarized by selective CMP, and subsequently dissolved in a wet etchant containing HCl, HF, and H_2O to reveal the patterned polysilicon structures. The cured SOG films were completely compatible with the polysilicon deposition process. In the second example, *Liu* et al. [10.24] fabricated high aspect ratio channel plate microstructures from SOG. Electroplated nickel (Ni) was used as a molding material, with Ni channel plate molds fabricated using a conventional LIGA process. The Ni molds were then filled with SOG, and the sacrificial Ni molds were removed in a reverse electroplating process. In this case, the fabricated SOG structures (over $100\,\mu m$ tall) were micromachined glass structures fabricated using a molding material more commonly used for structural components.

10.1.5 Silicon Nitride

Silicon nitride (Si_3N_4) is widely used in MEMS for electrical isolation, surface passivation, etch masking, and as a mechanical material typically for membranes and other suspended structures. Two deposition methods are commonly used to deposit Si_3N_4 thin films, LPCVD, and PECVD. PECVD silicon nitride is generally nonstoichiometric (sometimes denoted as Si_xN_y:H) and may contain significant concentrations of hydrogen. Use of PECVD silicon nitride in micromachining applications is somewhat limited because it has a high etch rate in HF (e.g., often higher than that of thermally grown SiO_2). However, PECVD offers the ability to deposit nearly stress-free silicon nitride films, an attractive property for encapsulation and packaging.

Unlike its PECVD counterpart, LPCVD Si_3N_4 is extremely resistant to chemical attack, thereby making it the material of choice for many Si bulk and surface micromachining applications. LPCVD Si_3N_4 is commonly used as an insulating layer because it has a resistivity

of $10^{16}\,\Omega\,cm$ and field breakdown limit of $10^7\,V/cm$. LPCVD Si_3N_4 films are deposited in horizontal furnaces similar to those used for polysilicon deposition. Typical deposition temperatures and pressures range between $700\,°C$ and $900\,°C$ and $200\,mtorr$ and $500\,mtorr$, respectively. The standard source gases are dichlorosilane (SiH_2Cl_2) and ammonia (NH_3). To produce stoichiometric Si_3N_4 a NH_3 to SiH_2Cl_2 ratio 10:1 is commonly used. The microstructure of films deposited under these conditions is amorphous.

The residual stress in stoichiometric Si_3N_4 is large and tensile, with a magnitude of about $1\,GPa$. Such a large residual stress causes films thicker than a few thousand angstroms to crack. Nonetheless thin stoichiometric Si_3N_4 films have been used as mechanical support structures and electrical insulating layers in piezoresistive pressure sensors [10.25]. To enable the use of Si_3N_4 films for applications that require micron-thick, durable, and chemically resistant membranes, Si_xN_y films can be deposited by LPCVD. These films, often referred to as Si-rich or low-stress nitride, are intentionally deposited with an excess of Si by simply decreasing the ratio of NH_3 to SiH_2Cl_2 during deposition. Nearly stress-free films can be deposited using a NH_3-to-SiH_2Cl_2 ratio of 1/6, a deposition temperature of $850\,°C$, and a pressure of $500\,mtorr$ [10.26]. The increase in Si content not only leads to a reduction in tensile stress, but also a decrease in the etch rate in HF. Such properties have enabled the development of fabrication techniques that would otherwise not be feasible with stoichiometric Si_3N_4. For example, low-stress silicon nitride has been surface micromachined using polysilicon as the sacrificial material [10.27]. In this case, Si anisotropic etchants such as KOH and EDP were used for dissolving the sacrificial polysilicon. *French* et al. [10.28] used PSG as a sacrificial layer to surface micromachine low-stress nitride, capitalizing on the HF resistance of the nitride films.

10.2 Germanium–Based Materials

10.2.1 Polycrystalline Ge

Like Si, Ge has a long history as a semiconductor device material, dating back to the development of the earliest transistors and semiconductor strain gauges. Issues related to germanium oxide, however, stymied the development of Ge for microelectronic devices. Nonetheless, there is a renewed interest in using Ge in surface micromachined devices due to the relatively low

processing temperatures required to deposit the material and its compatibility with Si.

Thin polycrystalline Ge (poly-Ge) films can be deposited by LPCVD at temperatures as low as $325\,°C$ on Si, Ge, and SiGe substrates [10.29]. Ge does not nucleate on SiO_2 surfaces, which prohibits the use of thermal oxides and LTO films as sacrificial layers but enables the use of these films as sacrificial molds. Residual stress in poly-Ge films deposited on Si substrates can be reduced to nearly zero after short anneals at modest temperatures (30 s at 600 °C). Poly-Ge is essentially impervious to KOH, TMAH, and BOE, enabling the fabrication of Ge membranes on Si substrates [10.29]. The mechanical properties of poly-Ge are comparable to those of polysilicon, having a Young's modulus of 132 GPa and a fracture stress ranging between 1.5 GPa and 3.0 GPa [10.30]. Mixtures of HNO_3, H_2O, and HCl and H_2O, H_2O_2, and HCl, as well as the RCA SC-1 cleaning solution, isotropically etch Ge. Since these mixtures do not etch Si, SiO_2, Si_3N_4, and SiN, poly-Ge can be used as a sacrificial substrate layer in polysilicon surface micromachining. Using these techniques, devices such as poly-Ge-based thermistors and Si_3N_4 membrane-based pressure sensors made using poly-Ge sacrificial layers have been fabricated [10.29]. *Franke* et al. [10.30] found no performance degradation in Si CMOS devices following the fabrication of surface micromachined poly-Ge structures, thus demonstrating the potential for on-chip integration of Ge electromechanical devices with Si circuitry.

10.2.2 Polycrystalline SiGe

Like poly-Ge, polycrystalline SiGe (poly-SiGe) is a material that can be deposited at temperatures lower than polysilicon. Deposition processes include LPCVD, APCVD, and RTCVD (rapid thermal CVD) using SiH_4 and GeH_4 as precursor gases. Deposition temperatures range between 450 °C for LPCVD [10.31] and 625 °C by rapid thermal CVD (RTCVD) [10.32]. In general, the deposition temperature is related to the concentration of Ge in the films, with higher Ge concentrations resulting in lower deposition temperatures. Like polysilicon, poly-SiGe can be doped with boron and phosphorus to modify its conductivity. In situ boron doping can be performed at temperatures as low as 450 °C [10.31]. *Sedky* et al. [10.32] showed that the deposition temperature of conductive films doped with boron could be further reduced to 400 °C if the Ge content was kept at or above 70%.

Unlike poly-Ge, poly-SiGe can be deposited on a number of sacrificial substrates, including

SiO_2 [10.32], PSG [10.30], and poly-Ge [10.30]. For Ge-rich films, a thin polysilicon seed layer is sometimes used on SiO_2 surfaces since Ge does not readily nucleate on oxide surfaces. Like many compound materials, variations in film composition can change the physical properties of the material. For instance, etching of poly-SiGe by H_2O_2 becomes significant for Ge concentrations over 70%. *Sedky* et al. [10.32] has shown that the microstructure, film conductivity, residual stress, and residual stress gradient are related to the concentration of Ge in the material. With respect to residual stress, *Franke* et al. [10.31] produced in situ boron-doped films with residual compressive stresses as low as 10 MPa.

The poly-SiGe, poly-Ge material system is particularly attractive for surface micromachining since H_2O_2 can be used as a release agent. It has been reported that poly-Ge etches at a rate of 0.4 microns/min in H_2O_2, while poly-SiGe with Ge concentrations below 80% have no observable etch rate after 40 h [10.33]. The ability to use H_2O_2 as a sacrificial etchant makes the combination of poly-SiGe and poly-Ge extremely attractive for surface micromachining from processing, safety, and materials compatibility points of view. Due to the conformal nature of LPCVD processing, poly-SiGe structural elements, such as gimbal-based microactuator structures have been made by high-aspect-ratio micromolding [10.33]. Capitalizing on the low deposition temperatures, poly-SiGe MEMS integrated with Si ICs has been demonstrated [10.31]. In this process, CMOS structures are first fabricated on Si wafers. Poly-SiGe mechanical structures are then surface micromachined using a poly-Ge sacrificial layer. A significant advantage of this design lies in the fact that the MEMS structure is positioned directly above the CMOS structure, thus reducing the parasitic capacitance and contact resistance characteristic of interconnects associated with side-by-side integration schemes. Use of H_2O_2 as the sacrificial etchant eliminates the need for layers to protect the underlying CMOS structure during release. In addition to its utility as a material for integrated MEMS devices, poly-SiGe has been identified as a material well suited for micromachined thermopiles [10.34] to its lower thermal conductivity relative to Si.

10.3 Metals

It can be argued that of all the material categories associated with MEMS, metals may be among the most enabling, since metallic thin films are used in many different capacities, from etch masks used in device

fabrication to interconnects and structural elements in microsensors and microactuators. Metallic thin films can be deposited using a wide range of techniques, including evaporation, sputtering, CVD, and electroplating. Since a complete review of the metals used in MEMS is far beyond the scope of this chapter, the examples presented in this section were selected to represent a broad cross section where metals have found uses in MEMS.

Aluminum (Al) and gold (Au) are among the most widely employed metals in microfabricated electronic and electromechanical devices as a result of their use as innerconnect and packaging materials. In addition to these critical electrical functions, Al and Au are also desirable as electromechanical materials. One such example is the use of Au micromechanical switches for RF MEMS. For conventional RF applications, chip level switching is currently performed using FET and PIN diode-based solid state devices fabricated from gallium arsenide (GaAs) substrates. Unfortunately, these devices suffer from insertion losses and poor electrical isolation. In an effort to develop replacements for GaAs-based solid state switches, *Hyman* et al. [10.35] reported the development of an electrostatically actuated, cantilever-based micromechanical switch fabricated on GaAs substrates. The device consisted of a silicon-nitride-encased Au cantilever constructed on a sacrificial silicon dioxide layer. The silicon nitride and silicon dioxide layers were deposited by PECVD, and the Au beam was electroplated from a sodium sulfite solution inside a photoresist mold. A thin multilayer of Ti and Au was sputter deposited in the mold prior to electroplating. The trilayer cantilever structure was chosen to minimize the deleterious effects of thermal- and process-related stress gradients in order to produce unbent and thermally stable beams. After deposition and pattering, the cantilevers were released in HF. The processing steps proved to be completely compatible with GaAs substrates. The released cantilevers demonstrated switching speeds of better than 50 µs at 25 V with contact lifetimes exceeding 10^9 cycles.

In a second example from RF MEMS, *Chang* et al. [10.36] reported the fabrication of an Al-based micromachined switch as an alternative to GaAs FETs and PIN diodes. In contrast to the work by *Hyman* et al. [10.35], this switch utilizes the differences in the residual stresses in Al and Cr thin films to create bent cantilever switches that capitalize on the stress differences in the materials. Each switch is comprised of a series of linked bimorph cantilevers designed in such a way that the resulting structure bends significantly out of the plane of the wafer due to the stress differences in the bimorph. The switch is drawn closed by electrostatic attraction. The bimorph consists of metals that can easily be processed with GaAs wafers, thus making integration with GaAs devices possible. The released switches were relatively slow, at 10 ms, but an actuation voltage of only 26 V was needed to close the switch.

Direct bulk micromachining of metal substrates is being developed for MEMS applications requiring structures with the dimensional complexity associated with Si DRIE and the physical properties of metals. One such example is Ti, which has a higher fracture toughness, a greater biocompatibility, and a more stable passivating oxide than Si. A process to fabricate high-aspect-ratio, three-dimensional structures from bulk Ti substrates has recently been developed [10.37]. This process involves inductively coupled plasma etching of a TiO_2-capped Ti substrate. The TiO_2 capping layer is deposited by DC reactive sputtering and photolithographically patterned using a CHF_3-based dry etch. The deep Ti etch is then performed using a Cl/Ar-based plasma that exhibits a selectivity of 40:1 with the masking TiO_2 layer. The etch process consists of a series of two-step sequences, where the first step involves Ti removal by the Cl/Ar plasma while the second step involves sidewall passivation using an oxygen plasma. After the prescribed etch period, the masking thin film can be removed by HF etching. High-aspect-ratio comb-drive actuators and other beam-based structures have been fabricated directly from bulk Ti using this method.

Thin-film metallic alloys that exhibit the shape-memory effect are of particular interest to the MEMS community for their potential in microactuators. The shape-memory effect relies on the reversible transformation from a ductile martensite phase to a stiff austenite phase in the material with the application of heat. The reversible phase change allows the shape-memory effect to be used as an actuation mechanism since the material changes shape during the transition. It has been found that high forces and strains can be generated from shape-memory thin films at reasonable power inputs, thus enabling shape memory actuation to be used in MEMS-based microfluidic devices such as microvalves and micropumps. Titanium-nickel (TiNi) is among the most popular of the shape-memory alloys owing to its high actuation work density, $(50 \, MJ/m^3)$, and large bandwidth (up to 0.1 kHz) [10.38]. TiNi is also attractive because conventional sputtering techniques can be employed to deposit thin films, as detailed in a recent report by *Shih* et al. [10.38]. In this study, TiNi films were deposited by cosputtering elemental Ti and Ni targets and cosputtering TiNi alloy and elemental Ti targets. It was reported that cosputtering from TiNi and Ti targets produced better films due to process variations related to roughening of the Ni target in the case of Ti and Ni cosputtering. The TiNi/Ti cosputtering process has been used to produce shape-memory material for a silicon spring-based microvalve [10.39].

Use of thin-film metal alloys in magnetic actuator systems is another example of the versatility of metallic materials in MEMS. Magnetic actuation in microdevices generally requires the magnetic layers to be relatively thick (tens to hundreds of microns) to generate magnetic fields of sufficient strength to generate the desired actuation. To this end, magnetic materials are often deposited by thick-film methods such as electroplating. The thicknesses of these layers exceeds what can feasibly be patterned by etching, so plating is often performed in microfabricated molds made from materials such as polymethylmethacrylate (PMMA). The PMMA mold thickness can exceed several hundred microns, so x-rays are used as the exposure source during the patterning steps. When necessary a metallic thin-film seed layer is deposited prior to plating. After plating, the mold is dissolved, which frees the metallic component. Known as LIGA (short for lithography, galvanoforming, and abformung), this process has been used to produce a wide variety of high-aspect-ratio structures from plateable materials, such as nickel-iron (NiFe) magnetic alloys [10.40] and Ni [10.41].

In addition to elemental metals and simple compound alloys, more complex metallic alloys commonly used in commerical macroscopic applications are finding their way into MEMS applications. One such example is an alloy of titanium known as Ti-6Al-4V. Composed of 88% titanium, 6% aluminum, and 4% vanadium, this alloy is widely used in commercial avation due to its weight, strength, and temperature tolerance. *Pornsin-Sirirak* et al. [10.42] have explored the use of this alloy in the manufacture of MEMS-based winged structures for micro aerial vehicles. The authors considered this alloy not only because of its weight and strength, but also because of its ductility and its etching rate at room temperature. The designs for the wing prototype were modeled after the wings of bats and various flying insects. For this application, Ti-alloy structures patterned from bulk ($250\,\mu$m-thick) material by an $HF/HO_3/H_2O$ etching solution were used rather than thin films. Parylene-C (detailed in a later section) was deposited on the patterned alloy to serve as the wing membrane. The miniature micromachined wings were integrated into a test setup, and several prototypes actually demonstrated short duration flight.

10.4 Harsh-Environment Semiconductors

10.4.1 Silicon Carbide

Silicon carbide (SiC) has long been recognized as the leading semiconductor for use in high-temperature and high-power electronics and is currently being developed as a material for harsh-environment MEMS. SiC is a polymorphic material that exists in cubic, hexagonal, and rhombehedral polytypes. The cubic polytype, called 3C-SiC, has an electronic bandgap of 2.3 eV, which is over twice that of Si. Numerous hexagonal and rhombehedral polytypes have been identified, with the two most common being 4H-SiC and 6H-SiC. The electronic bandgaps of 4H- and 6H-SiC are even higher than 3C-SiC, being 2.9 and 3.2 eV, respectively. SiC films can be doped to create n-type and p-type materials. The Young's modulus of SiC is still the subject of research, but most reported values range from 300 GPa to 450 GPa, depending on the microstructure and measurement technique. SiC is not etched in any wet Si etchants and is not attacked by XeF_2, a popular dry Si etchant used for releasing device structures [10.43]. SiC is a material that does not melt, but rather sublimes at temperatures

in excess of 1800 °C. Single-crystal 4H- and 6H-SiC wafers are commercially available, but they are smaller in diameter (3 inch) and much more expensive than Si wafers.

SiC thin films can be grown or deposited using a number of different techniques. For high-quality single-crystal films, APCVD and LPCVD processes are most commonly employed. Homoepitaxial growth of 4H- and 6H-SiC yields high-quality films suitable for microelectronic applications but typically only on substrates of the same polytype. These processes usually employ dual precursors, such as SiH_4 and C_3H_8, and are performed at temperatures ranging from 1500 °C to 1700 °C. Epitaxial films with p-type or n-type conductivity can be grown using Al and B for p-type films and N and P for n-type films. Nitrogen is so effective at modifying the conductivity of SiC that growth of undoped SiC films is extremely challenging because the concentrations of residual nitrogen in typical deposition systems are sufficient for n-type doping.

APCVD and LPCVD can also be used to deposit 3C-SiC on Si substrates. Heteroepitaxy is possible despite a 20% lattice mismatch because 3C-SiC and Si have the same lattice structure. The growth process involves two key steps. The first step, called carbonization, converts the near surface region of the Si substrate to 3C-SiC by simply exposing it to a hydrocarbon/hydrogen mixture at high substrate temperatures ($> 1200\,°C$). The carbonized layer forms a crystalline template on which a 3C-SiC film can be grown by adding a silicon-containing gas to the hydrogen/hydrocarbon mix. The lattice mismatch between Si and 3C-SiC results in the formation of crystalline defects in the 3C-SiC film, with the density being highest in the carbonization layer and decreasing with increasing thickness. The crystal quality of 3C-SiC films is nowhere near that of epitaxially grown 4H- and 6H-SiC films; however, the fact that 3C-SiC can be grown on Si substrates enables the use of Si bulk micromachining techniques for fabrication of a host of 3C-SiC-based mechanical devices. These include microfabricated pressure sensors [10.44] and nanoelectromechanical resonant structures [10.45]. For designs that require electrical isolation from the substrate, 3C-SiC devices can be made directly on SOI substrates [10.44] or by wafer bonding and etchback, such as the capacitive pressure sensor developed by *Young* et al. [10.46].

Polycrystalline SiC (poly-SiC) is a more versatile material for SiC MEMS than its single-crystal counterparts. Unlike single-crystal versions of SiC, poly-SiC can be deposited on a variety of substrate types, including common surface micromachining materials such as polysilicon, SiO_2, and Si_3N_4. Commonly used deposition techniques include LPCVD [10.43, 47, 48] and APCVD [10.49, 50]. The deposition of poly-SiC requires much lower substrate temperatures than epitaxial films, ranging from roughly $700\,°C$ to $1200\,°C$. Amorphous SiC can be deposited at even lower temperatures ($25\,°C$ to $400\,°C$) by PECVD [10.51] and sputtering [10.52]. The microstructure of poly-SiC films is temperature, substrate, and process dependent. For amorphous substrates such as SiO_2 and Si_3N_4, APCVD poly-SiC films deposited from SiH_4 and C_3H_8 are randomly oriented with equiaxed grains [10.50], whereas for oriented substrates such as polysilicon, the texture of the poly-SiC film matches that of the substrate itself [10.49]. By comparison, poly-SiC films deposited by LPCVD from SiH_2Cl_2 and C_2H_2 are highly textured (111) films with a columnar microstructure [10.47], while films deposited from disilabutane have a distribution of orientations [10.43]. This variation suggests that device performance can be tailored by selecting the proper substrate and deposition conditions.

SiC films deposited by AP- and LPCVD generally suffer from large tensile stresses on the order of several hundred MPa. Moreover, the residual stress gradients in these films tend to be large, leading to significant out-of-plane bending of structures that are anchored at a single location. The thermal stability of SiC makes a postdeposition annealing step impractical for films deposited on Si substrates, since the temperatures needed to significantly modify the film are likely to exceed the melting temperature of the wafer. For LPCVD processes using SiH_2Cl_2 and C_2H_2 precursors, *Fu* et al. [10.53] has described a relationship between deposition pressure and residual stress that enables the deposition of undoped poly-SiC films with nearly zero residual stresses and negligible stress gradients. This work has recently been extended to include films doped with nitrogen [10.54].

Direct bulk micromachining of SiC is very difficult, due to its chemical inertness. Although conventional wet chemical techniques are not effective, several electrochemical etch processes have been demonstrated and used in the fabrication of 6H-SiC pressure sensors [10.55]. The etching processes are selective to the conductivity of the material, so dimensional control of the etched structures depends on the ability to form doped layers, which can only be formed by in situ or ion-implantation processes since solid source diffusion is not possible at reasonable processing temperatures. This constraint somewhat limits the geometrical complexity of the patterned structures as compared with conventional plasma-based etching. To fabricate thick (hundreds of microns), 3D, high-aspect-ratio SiC structures, a molding technique has been developed [10.41]. The molds are fabricated from Si substrates using deep reactive ion etching and then filled with SiC using a combination of thin epitaxial and thick polycrystalline film CVD processes. The thin-film process is used to protect the mold from pitting during the more aggressive mold-filling SiC growth step. The mold-filling process coats all surfaces of the mold with a SiC film as thick as the mold is deep. To release the SiC structure, the substrate is first mechanically polished to expose sections of the Si mold; then the substrate is immersed in a Si etchant to completely dissolve the mold. This process has been used to fabricate solid SiC fuel atomizers [10.41], and a variant has been used to fabricate SiC structures for micropower systems [10.56]. Recently, *Min* et al. [10.57] reported a process to fabricate reusable glass press molds made from SiC structures that were patterned using Si molding masters. SiC was selected as the material for the glass press mold because the application requires a hard, mechanically strong, and chemically stable material that can withstand and maintain its properties at temperatures between 600 °C and 1400 °C.

In addition to CVD processes, bulk micromachined SiC structures can be fabricated using sintered SiC powders. *Tanaka* et al. [10.59] describe a process where SiC components, such as micro gas turbine engine rotors, can be fabricated from SiC powders using a microreaction-sintering process. The molds are microfabricated from Si using DRIE and filled with SiC and graphite powders mixed with a phenol resin. The molds are then reaction-sintered using a hot isostatic pressing technique. The SiC components are then released from the Si mold by wet chemical etching. The authors reported that the component shrinkage was less than 3%. The bending strength and Vickers hardness of the microreaction-sintered material was roughly 70 to 80% of commercially available reaction-sintered SiC, the difference being attributed to the presence of unreacted Si in the microscale components.

In a related process, *Liew* et al. [10.60] detail a technique to create silicon carbon nitride (SiCN) MEMS structures by molding injectable polymer precursors. Unlike the aforementioned processes, this technique uses SU-8 photoresists for the molds. To be detailed later in this chapter, SU-8 is a versatile photodefinable polymer in which thick films (hundreds of microns) can be patterned using conventional UV photolithographic techniques. After patterning, the molds are filled with the SiCN-containing polymer precursor, lightly polished, and then subjected to a multistep heat-treating process. During the thermal processing steps, the SU-8 mold decomposes and the SiCN structure is released. The resulting SiCN structures retain many of the same properties of stoichiometric SiC.

Although SiC cannot be etched using conventional wet etch techniques, SiC can be patterned using conventional dry etching techniques. RIE processes using fluorinated compounds such as CHF_3 and SF_6 combined with O_2 and sometimes with an inert gas or H_2 are used to pattern thin films. The high oxygen content in these plasmas generally prohibits the use of photoresist as a masking material; therefore, hard masks made of Al, Ni, and ITO are often used. RIE-based SiC surface micromachining processes with polysilicon and SiO_2 sacrificial layers have been developed for single-layer devices [10.61, 62]. ICP RIE of SiC using SF_6 plasmas and Ni or ITO etch masks has been developed for bulk micromachining SiC substrates, with structural depths in excess of 100 μm reported [10.63].

Until recently, multilayer thin-film structures were very difficult to fabricate by direct RIE because the etch rates of the sacrificial layers were much higher than the SiC structural layers, making dimensional control very difficult. To address this issue, a micromolding process for patterning SiC films on sacrificial-layer substrates was developed [10.64]. In essence, the micromolding technique is the thin-film analog to the molding-based, bulk micromachining technique presented earlier. The micromolding process utilizes polysilicon and SiO_2 films as both molds and sacrificial substrate layers, with SiO_2 molds used with polysilicon sacrificial layers and vice versa. These films are deposited and patterned using conventional methods, thus leveraging the well-characterized and highly selective processes developed for polysilicon MEMS. Poly-SiC films are simply deposited into the micromolds and mechanical polishing is used to remove poly-SiC from atop the molds. Appropriate etchants are then used to dissolve the molds and sacrificial layers. The micromolding method utilizes the differences in chemical properties of the three materials in this system in a way that bypasses the difficulties associated with chemical etching of SiC. This technique has been developed specifically for multilayer processing and has been used successfully to fabricate SiC micromotors [10.64] and the lateral resonant structure shown in Fig. 10.4 [10.58].

Recent advancements in the area of SiC RIE show that significant progress has been made in developing etch recipes with selectivities to nonmetal mask and sac-

Fig. 10.4 SEM micrograph of a poly-SiC lateral resonant structure fabricated using a multilayer, micromolding-based micromachining process [10.58]

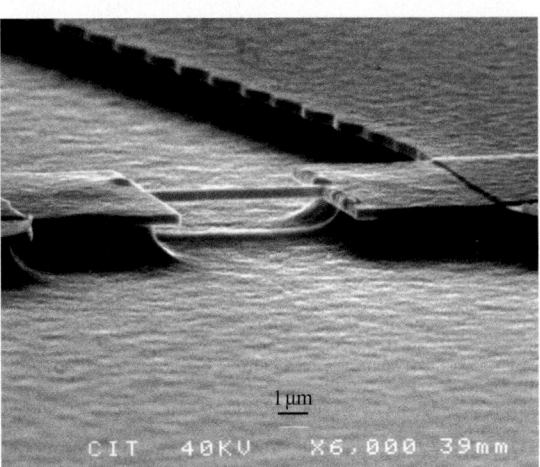

Fig. 10.5 SEM micrograph of a 3C − SiC nanomechanical beam resonator fabricated by electron-beam lithography and dry etching processes (courtesy M. Roukes, Caltech)

rificial layers that are suitable for multilayer SiC surface micromachining. For instance, *Gao* et al. [10.65] have developed a transformer-coupled RIE process using a HBr-based chemistry for thin-film poly-SiC etching. The recipe exhibits a SiC-to-SiO_2 selectivity of 20:1 and a SiC-to-Si_3N_4 selectivity of 22:1, which are the highest reported thus far. In addition, the anisotropy of the etch was quite high, and micromasking, a common problem when metal masks are used, was not an issue. This process has since been used to fabricate multilayered lateral resonant structures that utilize poly-SiC as the main structural material and polysilicon as a conducting plane that underlies the resonating shuttle [10.65].

Yang et al. [10.45] have recently shown that the chemical inertness of SiC facilitates the fabrication of NEMS devices. In this work, the authors present a fabrication method to realize SiC mechanical resonators with submicron thickness and width dimensions. The resonators were fabricated from ≈ 260 − nm-thick 3C-SiC films epitaxially grown on (100) Si wafers. The films were patterned into 150-nm-wide beams ranging in length from 2 to 8 µm. The beams were etched in a $NF_3/O_2/Ar$ plasma using an evaporated Cr etch mask. After patterning, the beams were released by etching the underlying Si isotropically using a NF_3/Ar plasma. The inertness of the SiC film to the Si etchant enables the dry release of the nanomechanical beams. An example of a 3C-SiC nanomechanical beam is shown in Fig. 10.5.

10.4.2 Diamond

Diamond is commonly known as nature's hardest material, making it ideal for high wear environments. Diamond has a very large electronic bandgap (5.5 eV), which makes it attractive for high temperature electronics. Undoped diamond is a high-quality insulator with a dielectric constant of 5.5; however, it can be relatively easily doped with boron to create p-type conductivity. Diamond has a very high Young's modulus (1035 GPa), making it suitable for high-frequency micromachined resonators, and it is among nature's most chemically inert materials, making it well suited for harsh chemical environments.

Unlike SiC, fabrication of diamond MEMS is currently restricted to polycrystalline and amorphous material, since single-crystal diamond wafers are not yet commercially available. Polycrystalline diamond films can be deposited on Si and SiO_2 substrates by CVD methods, but the surfaces must often be seeded by diamond powders or biased with a negative charge to initiate growth. In general, diamond nucleates much more readily on Si surfaces than on SiO_2 surfaces, an effect that has been used to selectively pattern diamond films into micromachined AFM cantilever probes using SiO_2 molding masks [10.66].

Bulk micromachining of diamond using wet and dry etching is extremely difficult given its extreme chemical inertness. Diamond structures have nevertheless been fabricated using bulk micromachined Si molds to pattern the structures [10.67]. The Si molds were fabricated using conventional micromachining techniques and filled with polycrystalline diamond deposited by hot filament chemical vapor deposition (HFCVD). The HFCVD process uses H_2 and CH_4 precursors. The process was performed at a substrate temperature of 850 °C to 900 °C and a pressure of 50 mtorr. The Si substrate was seeded prior to deposition using a diamond particle/ethanol solution. After deposition, the top surface of the structure was polished using a hot iron plate. After polishing, the Si mold was removed in a Si etchant, leaving behind the micromachined diamond structure. This process was used to produce high-aspect-ratio capillary channels for microfluidic applications [10.68] and components for diffractive optics, laser-to-fiber alignment, and power device cooling structures [10.69].

Due to the nucleation processes associated with diamond film growth, surface micromachining of polycrystalline diamond thin films requires modifications to conventional micromachining to facilitate film growth on sacrificial substrates. Initially, conventional RIE methods were generally ineffective, so work was focused on developing selective deposition techniques. One early method used selective seeding to form patterned templates for diamond nucleation. The selective seeding process employed the lithographic patterning of photoresist that contained diamond powders [10.70]. The diamond-loaded photoresist was deposited and patterned onto a Cr-coated Si wafer. During the onset of diamond growth, the patterned photoresist rapidly evaporates, leaving behind the diamond seed particles in the desired locations. A patterned diamond film is then selectively grown on these locations.

A second process utilized selective deposition directly on sacrificial substrate layers. This process combined conventional diamond seeding with photolithographic patterning and etching to fabricate micromachined diamond structures on SiO_2 sacrificial layers [10.71]. The process was performed in one of two ways. The first approach begins with the seeding of an oxidized Si wafer. The wafer is coated with a photoresist and photolithographically patterned. Unmasked regions of the seeded SiO_2 film are then partially etched, forming a surface unfavorable for diamond growth. The photoresist is then removed and a diamond film is deposited on the seeded regions. The second approach also begins with an oxidized Si wafer. The wafer is coated with a photoresist, photolithographically patterned, and then seeded with diamond particles. The photoresist is removed, leaving behind a patterned seed layer suitable for selective growth. These techniques have been successfully used to fabricate cantilever beams and bridge structures.

A third method to surface micromachine polycrystalline diamond films follows the conventional approach of film deposition, dry etching, and release. The chemical inertness of diamond renders most conventional plasma chemistries useless; however, oxygen-based ion-beam plasmas can be used to etch diamond thin films [10.72]. A simple surface micromachining process begins with the deposition of a polysilicon sacrificial layer on a Si_3N_4-coated Si wafer. The polysilicon layer is seeded using diamond slurry, and a diamond film is deposited by HFCVD. Since photoresists are not resistant to O_2 plasmas, an Al masking film is deposited and patterned. The diamond films are then etched in the O_2 ion-beam plasma, and the structures are released by etching the polysilicon with KOH. This process has been used to create lateral resonant structures, but a significant stress gradient in the films rendered the devices inoperable.

In general, conventional HFCVD requires that the substrate be pretreated with a seeding layer prior to diamond film growth. However, a method called biased enhanced nucleation (BEN) has been developed that enables the growth of diamond on unseeded Si surfaces. *Wang* et al. [10.73] have shown that if Si substrates are masked with patterned SiO_2 films, selective diamond growth will occur primarily on the exposed Si surfaces, and a slight HF etch is sufficient to remove the adventitious diamond from the SiO_2 mask. This group was able to use this method to fabricate diamond micromotor rotors and stators on Si surfaces.

Diamond is a difficult, but not impossible, material to etch using conventional RIE techniques. It is well known that diamond can be etched in oxygen plasmas, but these plasmas can be problematic for device fabrication because the etching tends to be isotropic. A recent development, however, suggests that RIE processes for diamond are close at hand. *Wang* et al. [10.73] describe a process to fabricate a vertically actuated, doubly clamped micromechanical diamond beam resonator using RIE. The process outlined in this paper addresses two key issues related to diamond surface micromachining, namely, residual stress gradients in the diamond films and diamond patterning techniques. A microwave plasma CVD (MPCVD) reactor was used to grow the diamond films on sacrificial SiO_2 layers pretreated with a nanocrystalline diamond powder, resulting in a uniform nucleation density at the diamond/SiO_2 interface. The diamond films were etched in a CF_4/O_2 plasma using Al as a hard mask. Reasonably straight sidewalls were created, with roughness attributable to the surface roughness of the faceted diamond film. An Au/Cr drive electrode beneath the sacrificial oxide remained covered throughout the diamond-patterning steps and thus was undamaged during the diamond-etching process. This work has since been extended to develop a 1.51-GHz diamond micromechanical disk resonator [10.73]. In this instance, the nanocrystalline diamond film was deposited my MPCVD, coated with an oxide film that had been patterned into an etch mask, and then etched in a O_2/CF_4 RIE plasma under conditions that yielded a fairly anisotropic etch with a diamond-to-oxide selectivity of 15:1. The disk was suspended over the substrate on a polysilicon stem using an oxide sacrificial layer. Polysilicon was also used as the drive and sense electrodes. The material mismatch between the step and the resonating disk substantially reduced anchor losses, thus allowing for very high-quality factors (11, 500) for 1.5-GHz resonators tested in a vacuum.

In conjunction with recent advances in RIE and micromachining techniques, work is being performed to develop diamond-deposition processes specifically for

Fig. 10.6 SEM micrograph of folded beam truss of diamond lateral resonator. The diamond film was deposited using a seeding-based hot filament CVD process. The micrograph illustrates the challenges currently facing diamond MEMS, namely, roughened surfaces and residual stress gradients

MEMS applications. Diamond films grown using conventional techniques, especially processes that require pregrowth seeding, tend to have high residual stress gradients and roughened surface morphologies as a result of the highly faceted, large-grain polycrystalline films that are produced by these methods (Fig. 10.6). The rough surface morphology degrades the patterning process, resulting in roughened sidewalls in etched structures and roughened surfaces of films deposited over these layers. Unlike polysilicon and SiC, a postdeposition polishing process is not technically feasible for diamond due to its extreme hardness. For the fabrication of multilayer diamond devices, methods to reduce the surface roughness of the as-deposited films are highly desirable. Along these lines, *Krauss* et al. [10.74] have reported on the development of an ultrananocrystalline diamond (UCND) film that exhibits a much smoother surface morphology than comparable diamond films grown using conventional methods. Unlike conventional CVD diamond films that are grown using a mixture of H_2 and

CH_4, the ultrananocrystalline diamond films are grown from mixtures of Ar, H_2, and C_{60} or Ar, H_2, and CH_4. Films produced by this method have proven to be effective as conformal coatings on Si surfaces and have been used successfully in several surface micromachining processes. Recently, this group has extended the UCND deposition technology to low deposition temperatures, with high-quality nanocrystalline diamond films being deposited at rates of $0.2 \mu m/h$ at substrate temperatures of $400\,°C$, making these films compatible from a thermal budget perspective with Si IC technology [10.75].

Another alternative deposition method that is proving to be well suited for diamond MEMS is based on pulsed laser deposition [10.76]. The process is performed in a high vacuum chamber and uses a pulsed eximer laser to ablate a pyrolytic graphite target. Material from the ejection plume deposits on a substrate, which is kept at room temperature. Background gases composed of N_2, H_2, and Ar can be introduced to adjust the deposition pressure and film properties. The as-deposited films consist of tetrahedrally bonded carbon that is amorphous in microstructure, hence the name amorphous diamond. Nominally stress-free films can be deposited by proper selection of deposition parameters [10.77] or by a short postdeposition annealing step [10.76]. The amorphous diamond films exhibit many of the properties of single-crystal diamond, such as a high hardness (88 GPa), a high Young's modulus (1100 GPa), and chemical inertness. Many single-layer surface micromachined structures have been fabricated using these films, in part because the films can be readily deposited on oxide sacrificial layers and etched in an oxygen plasma. Recently, amorphous diamond films have been used as a dielectric isolation layer in vertically actuated microbridges in micromachined RF capacitive switches [10.78]. The diamond films sit atop fixed tungsten electrodes to provide dielectric isolation from an Au microbridge that spans the fixed electrode structure. The diamond films are particularly attractive for such applications since the surfaces are hydrophobic and thus do not suffer from stiction and are highly resistant to wear over repeated use.

10.5 GaAs, InP, and Related III–V Materials

Gallium arsenide (GaAs), indium phosphide (InP), and related III–V compounds have favorable piezoelectric and optoelectric properties, high piezoresistive constants, and wide electronic bandgaps relative to Si, making them attractive for various sensor and optoelectronic applications. Like Si, significant research in bulk crystal growth has led to the development of GaAs and InP substrates that are commercially available as high-quality, single-crystal wafers. Unlike compound semiconductors such as SiC, III–V materials can be deposited as ternary and quaternary alloys with lattice constants that closely match the binary compounds from which they are derived (i. e., $Al_xGa_{1-x}As$ and GaAs), thus permitting the fabrication of a wide variety of heterostructures that facilitate device performance.

Crystalline GaAs has a zinc blend crystal structure with an electronic bandgap of 1.4 eV, enabling GaAs electronic devices to function at temperatures as high as 350 °C [10.79]. High-quality, single-crystal wafers are commercially available, as are well-developed metalorganic chemical vapor deposition (MOCVD) and molecular beam epitaxy (MBE) growth processes for epitaxial layers of GaAs and its alloys. GaAs does not outperform Si in terms of mechanical properties; however, its stiffness and fracture toughness are still suitable for micromechanical devices.

Micromachining of GaAs is relatively straightforward, since many of its lattice-matched ternary and quaternary alloys have sufficiently different chemical properties to allow their use as sacrificial layers [10.80]. For example, the most common ternary alloy for GaAs is $Al_xGa_{1-x}As$. For values of x less than or equal to 0.5, etchants containing mixtures of HF and H_2O will etch $Al_xGa_{1-x}As$ without attacking GaAs, while etchants containing NH_4OH and H_2O_2 attack GaAs isotropically but do not etch $Al_xGa_{1-x}As$. Such selectivity enables the micromachining of GaAs wafers using lattice-matched etch stops and sacrificial layers. Devices fabricated using these methods include comb drive lateral resonant structures [10.80], pressure sensors [10.81, 82], thermopile sensors [10.82], Fabry–Perot detectors [10.83], and cantilever-based sensors and actuators [10.84, 85]. In addition, nanoelectromechanical devices, such as suspended micromechanical resonators [10.86] and tethered membranes [10.87], have been fabricated using these techniques. An example of a nanoelectromechanical beam structure fabricated from GaAs is shown in Fig. 10.7.

In addition to using epitaxial layers as etch stops, ion-implantation methods can also be used to produce etch stops in GaAs layers. *Miao* et al. [10.88] describe a process that uses electrochemical etching to selectively remove n-type GaAs layers. The process relies on the creation of a highly resistive near-surface GaAs layer on an n-type GaAs substrate by low-dose nitrogen implantation in the MeV energy range. A pulsed electrochemical etch method using an H_2PtCl_6, H_3PO_4, H_2SO_4

Fig. 10.7 SEM micrograph of a GaAs nanomechanical beam resonator fabricated by epitaxial growth, electron-beam lithography, and selective etching (courtesy of M. Roukes, Caltech)

platinum electrolytic solution at 40 °C with 17-V, 100-ms pulses is sufficient to selectively remove n-type GaAs at about 3 μm/min. Using this method, stress-free, tethered membranes could readily be fabricated from the highly resistive GaAs layer. The high implant energies enable the fabrication of membranes several microns thick. Moreover, the authors demonstrated that if the GaAs wafer were etched in such a way as to create an undulating surface prior to ion implantation, corrugated membranes could be fabricated. These structures can sustain much higher deflection amplitudes than flat structures.

Micromachining of InP closely resembles the techniques used for GaAs. Many of the properties of InP are similar to GaAs in terms of crystal structure, mechanical stiffness, and hardness; however, the optical properties of InP make it particularly attractive for microoptomechanical devices to be used in the 1.3- to 1.55-μm wavelength range [10.89]. Like GaAs, single-crystal wafers of InP are readily available, and ternary and quaternary lattice-matched alloys, such as InGaAs, InAlAs, InGaAsP, and InGaAlAs, can be used as either etch stop and/or sacrificial layers depending on the etch chemistry [10.80]. For instance, InP structural layers deposited on $In_{0.53}Al_{0.47}As$ sacrificial layers can be released using etchants containing $C_6H_8O_7$, H_2O_2, and H_2O. In addition, InP films and substrates can be etched in solutions containing HCl and H_2O using $In_{0.53}Ga_{0.47}As$ films as etch stops. Using InP-based micromachining techniques, multiair gap filters [10.90] bridge structures [10.89], and torsional membranes [10.83] have been fabricated from InP and its related alloys.

In addition to GaAs and InP, materials such as indium arsenide (InAs) can be micromachined into device structures. Despite a 7% lattice mismatch between InAs and (111) GaAs, high-quality epitaxial layers can be grown on GaAs substrates. As described by *Yamaguchi* et al. [10.91], the surface Fermi level

of InAs/GaAs structures is pinned in the conduction band, enabling the fabrication of very thin conductive membranes. In fact, the authors have successfully fabricated free-standing InAs structures that range in thickness from 30 to 300 nm. The thin InAs films were grown directly on GaAs substrates by MBE and etched using a solution containing H_2O, H_2O_2, and H_2SO_4. The structures, mainly doubly clamped cantilevers, were released by etching the GaAs substrate using an $H_2O/H_2O_2/NH_4OH$ solution.

10.6 Ferroelectric Materials

Piezoelectric materials play an important role in MEMS technology for sensing and mechanical actuation applications. In a piezoelectric material, mechanical stress produces a polarization, and conversely a voltage-induced polarization produces a mechanical stress. Many asymmetric materials, such as quartz, GaAs, and zinc oxide (ZnO), exhibit some piezoelectric behavior. Recent work in MEMS has focused on the development of ferroelectric compounds such as lead zirconate titanate, $Pb(Zr_xTi_{1-x})O_3$, or PZT for short, because such compounds have high piezoelectric constants that result in high mechanical transduction. It is relatively straightforward to fabricate a PZT structure on top of a thin free-standing structural layer (i. e., cantilever, diaphragm). Such a capability enables the piezoelectric material to be used in sensor applications or actuator applications where piezoelectric materials are particularly well suited. Like Si, PZT films can be patterned using dry etch techniques based on chlorine chemistries, such as Cl_2/CCl_4, as well as ion-beam milling using inert gases like Ar.

PZT has been successfully deposited in thin-film form using cosputtering, CVD, and sol-gel processing. So-gel processing is particularly attractive because the composition and homogeneity of the deposited material over large surface areas can be readily controlled. The sol gel process outlined by *Lee* et al. [10.92] uses PZT solutions made from liquid precursors containing Pb, Ti, Zr, and O. The solution is deposited by spin coating on a Si wafer that has been coated with a $Pt/Ti/SiO_2$ thin-film multilayer. The process is executed to produce a PZT film in layers, with each layer consisting of a spin-

coated layer that is dried at 110 °C for 5 min and then heat-treated at 600 °C for 20 min. After building up the PZT layer to the desired thickness, the multilayer was heated at 600 °C for up to 6 h. Prior to this anneal, a PbO top layer was deposited on the PZT surface. An Au/Cr electrode was then sputter-deposited on the surface of the piezoelectric stack. This process was used to fabricate a PZT-based force sensor. *Xu* et al. [10.93] describe a similar sol-gel process to produce 12-μm-thick, crack-free PZT films on Pt-coated Si wafers and 5-μm-thick films on insulating ZrO_2 layers to produce micromachined MHz-range two-dimensional transducer arrays for acoustic imaging.

Thick-film printing techniques for PZT have been developed to produce thick films in excess of 100 μm. Such thicknesses are desired for applications that require actuation forces that cannot be achieved with the much thinner sol-gel films. *Beeby* et al. [10.94] describe a thick-film printing process whereby a PZT paste is made from a mixture of 95% PZT powder, 5% lead borosilicate powder, and an organic carrier. The paste was then printed through a stainless steel screen using a thick-film printer. Printing was performed on an oxidized Si substrate that is capped with a Pt electrode. After printing, the paste was dried and then fired at 850 °C to 950 °C. Printing could be repeated to achieve the desired thickness. The top electrode consisted of an evaporated Al film. The authors found that it was possible to perform plasma-based processing on the printed substrates but that the porous nature of the printed PZT films made them unsuitable for wet chemical processing.

10.7 Polymer Materials

10.7.1 Polyimide

Polyimides comprise an important class of durable polymers that are well suited for many of the techniques used in conventional MEMS processing. In general, polyimides can be acquired in bulk or deposited as thin films by spin coating, and they can be patterned using conventional dry etching techniques and processed at relatively high temperatures. These attributes make polyimides an attractive group of polymers for MEMS that require polymer structural and/or substrate layers, such as microfabricated biomedical devices where inertness and flexibility are important parameters.

Shearwood et al. [10.95] explored the use of polyimides as a robust mechanical material for microfabricated audio membranes. The authors fabricated 7-μm-thick, 8-mm-diameter membranes on GaAs substrates by bulk micromachining the GaAs substrate using a NH_3/H_2O_2 solution. They realized 100% yield and, despite a low Young's modulus (≈ 3 GPa), observed flat membranes to within 1 nm after fabrication.

Jiang et al.[10.96] capitalized on the strength and flexibility of polyimides to fabricate a flexible sheer-stress sensor array based on Si sensors. The sensor array consisted of a collection of Si islands linked by two polyimide layers. Each Si sensor island was $250 \times 250\,\mu m^2$ in area and 80 μm in thickness. Al was used as an electrical innerconnect layer. The two polyimide layers served as highly flexible hinges, making it possible to mount the sensor array on curved surfaces. The sensor array was successful in profiling the shear-stress distribution along the leading edge of a rounded delta wing.

The chemical and temperature durability of polyimides enables their use as a sacrificial layer for a number of commonly used materials, such as evaporated or sputter-deposited metals. *Memmi* et al. [10.97] developed a fabrication process for capacitive micromechanical ultrasonic transducers using a polyimide as a sacrificial layer. The authors showed that the polyimide could withstand the conditions used to deposit silicon monoxide by evaporation and silicon nitride by PECVD at 400 °C. Recent work by *Bagolini* et al. [10.98] has shown that polyimides can even be used as sacrificial layers for PECVD SiC.

In the area of microfabricated biomedical devices, polyimides are receiving attention as a substrate material for implantable devices, owing to their potential biocompatiblity and mechanical flexibility. *Stieglitz* [10.99]

reported on the fabrication of multichannel microelectrodes on polyimide substrates. Instead of using polyimide sheets as starting substrates, Si carrier wafers coated with a 5-μm-thick polyimide film were used. Pt microelectrodes were then fabricated on these substrates using conventional techniques. Thin polyimide layers were deposited between various metal layers to serve as insulating layers. A capping polyimide layer was then deposited on the top of the substrates, and then the entire polyimide/metal structure was peeled off the Si carrier wafers. Backside processing was then performed on the free-standing polyimide structures to create devices that have exposed electrodes on both surfaces. In a later paper, *Stieglitz* et al. [10.100] describe a variation of this process for neural prostheses.

10.7.2 SU-8

SU-8 is a negative-tone epoxylike photoresist that is receiving much attention for its versatility in MEMS processing. It is a high-aspect-ratio, UV-sensitive resist designed for applications requiring single-coat resists with thicknesses on the order of 500 μm [10.101]. SU-8 has favorable chemical properties that enable it to be used as a molding material for high-aspect-ratio electroplated structures (as an alternative to LIGA) and as a structural material for microfluidics [10.101]. In terms of mechanical properties, *Lorenz* et al. [10.102] reported that SU-8 has a modulus of elasticity of 4.02 GPa, which compares favorably with a commonly-used polyamid (3.4 GPa).

In addition to the above-mentioned conventional uses for SU-8, several interesting alternative uses are beginning to appear in the literature. *Conradie* et al. [10.103] have used SU-8 to trim the mass of silicon paddle oscillators as a means to adjust the resonant frequency of the beams. The trimming process involves the patterning of SU-8 posts on Si paddles. The process capitalizes on the relative chemical stability of the SU-8 resin in conjunction with the relatively large masses that can be patterned using standard UV exposure processes.

SU-8 is also of interest as a bonding layer material for wafer bonding processes using patterned bonding layers. *Pan* et al. [10.104] compared several UV photodefinable polymeric materials and found that SU-8 exhibited the highest bonding strength (20.6 MPa) for layer thicknesses up to 100 μm.

10.7.3 Parylene

Parylene (poly-paraxylylene) is another emerging polymeric MEMS material due in large part to its biocompatibility. It is particularly attractive from the fabrication point of view because it can be deposited by CVD at room temperature. Moreover, the deposition process is conformal, which enables parylene coatings to be applied to prefabricated structures, such as Si microneedles [10.105], low-stress silicon nitride membrane particle filters [10.72], and micromachined polyimide/Au optical scanners [10.106]. In the former cases, the parylene coating served to strengthen the microfabricated structures, while in the latter case it served to protect the structure from condensing water vapor.

In addition to its function as a protective coating, parylene can actually be micromachined into free-standing components. *Noh* et al. [10.107] demonstrated a method to create bulk micromachined parylene microcolumns for miniature gas chromatographs. The structure is fabricated using a micromolding technique where Si molds are fabricated by DRIE and coated with parylene to form three sides of the microcolumn. A second wafer is coated with parylene, and the two are bonded together via a fusion bonding process. After bonding, the structure is released from the Si mold by KOH etching. In a second example, *Yao* et al. [10.108] describe a dry release process for parylene surface micromachining. In this process, sputtered Si is used as a sacrificial layer onto which a thick sacrificial photoresist is deposited. Parylene is then deposited on the photoresist and patterned into the desired structural shape. The release procedure is a two-step process. First the photoresist is dissolved in acetone. This results in the parylene structure sticking to the sputtered Si. Next, a dry BrF_3 etch is performed that dissolves the Si and releases the parylene structures. Parylene beams that were 1 mm long and 4.5 μm thick were successfully fabricated using this technique.

10.7.4 Liquid Crystal Polymer

Liquid crystal polymer (LCP) is a high-performance thermoplastic currently being used in printed circuit board and electronics packaging applications and has recently been investigated for use in MEMS applications requiring a material that is mechanically flexible, electrically insulating, chemically durable, and impermeable to moisture. LCP can be bonded to itself and other substrate materials such as glass and Si by thermal lamination. It can be micromachined using an oxygen plasma and yet is highly resistant to HF and many metal etchants [10.109]. The moisture absorption is less than 0.02% as compared with about 1% for polyimide [10.110], making it well suited as a packaging material.

Applications where LCPs are used as a key component in a MEMS device are beginning to emerge. *Faheem* et al. [10.111] reported on the use of LCP for encapsulation of variable RF MEMS capacitors. In this example, LCP, dispensed in liquid form, was used to join and seal a glass microcap to a prefabricated, microbridge capacitor. LCP was chosen in part because in addition to the aforementioned properties, it has very low RF loss characteristics, making it very well suited as an RF MEMS packaging material. *Wang* et al. [10.109] showed that LCP is a very versatile material that is highly compatible with many standard Si-based processing techniques. They also showed that micromachining techniques can be used to make LCP cantilever flow sensors that incorporate metal strain gauges and LCP membrane tactile sensors using NiCr strain gauges. *Lee* et al. [10.110] has developed a LCP-based, mechanically flexible, multichannel microelectrode array structure for neural stimulation and recording.

10.8 Future Trends

The rapid expansion of MEMS in recent years is due in large part to the inclusion of new materials that have expanded the functionality of microfabricated devices beyond what is achievable in silicon. This trend will certainly continue as new application areas for micro- and nanofabricated devices are identified. Many of these applications will likely require both new materials and new processes to fabricate the micro- and nanomachined devices for these yet-to-be-identified applications. Currently, conventional micromachining techniques employ a "top-down" approach that begins with either bulk substrates or thin films. Future MEMS and NEMS will likely incorporate materials that are created using a "bottom-up" approach. A significant challenge facing device design and fabrication engineers alike will be how to marry top-down and bottom-up approaches to create devices and systems that cannot be made using either process alone.

References

10.1   C. S. Smith: Piezoresistive effect in germanium and silicon, Phys. Rev. **94**, 1–10 (1954)

10.2   A. N. Cleland, M. L. Roukes: Fabrication of high frequency nanometer scale mechanical resonators from bulk Si crystals, Appl. Phys. Lett. **69**, 2653–2655 (1996)

10.3   D. W. Carr, H. G. Craighead: Fabrication of nano-electromechanical systems in single crystal silicon using silicon on insulator substrates and electron beam lithography, J. Vac. Sci. Technol. B **15**, 2760–2763 (1997)

10.4   T. Kamins: *Polycrystalline Silicon for Integrated Circuits and Displays*, 2 edn. (Kluwer, Boston 1988)

10.5   L. Cao, T. S. Kin, S. C. Mantell, D. Polla: Simulation and fabrication of piezoresistive membrane type MEMS strain sensors, Sens. Actuators **80**, 273–279 (2000)

10.6   H. Guckel, T. Randazzo, D. W. Burns: A simple technique for the determination of mechanical strain in thin films with application to polysilicon, J. Appl. Phys. **57**, 1671–1675 (1983)

10.7   R. T. Howe, R. S. Muller: Stress in polysilicon and amorphous silicon thin films, J. Appl. Phys. **54**, 4674–4675 (1983)

10.8   X. Zhang, T. Y. Zhang, M. Wong, Y. Zohar: Rapid thermal annealing of polysilicon thin films, J. Microelectromech. Syst. **7**, 356–364 (1998)

10.9   J. Yang, H. Kahn, A.-Q. He, S. M. Phillips, A. H. Heuer: A new technique for producing large-area as-deposited zero-stress LPCVD polysilicon films: The Multipoly Process, J. Microelectromech. Syst. **9**, 485–494 (2000)

10.10   P. Gennissen, M. Bartek, P. J. French, P. M. Sarro: Bipolar-compatible epitaxial poly for smart sensors: stress minimization and applications, Sens. Actuators **A62**, 636–645 (1997)

10.11   P. Lange, M. Kirsten, W. Riethmuller, B. Wenk, G. Zwicker, J. R. Morante, F. Ericson, J. A. Schweitz: Thick polycrystalline silicon for surface-micromechanical applications: deposition, structuring, and mechanical characterization, Sens. Actuators **A54**, 674–678 (1996)

10.12   S. Greek, F. Ericson, S. Johansson, M. Furtsch, A. Rump: Mechanical characterization of thick polysilicon films: Young's modulus and fracture strength evaluated with microstructures, J. Micromech. Microeng. **9**, 245–251 (1999)

10.13   K. Funk, H. Emmerich, A. Schilp, M. Offenberg, R. Neul, F. Larmer: *A surface micromachined silicon gyroscope using a thick polysilicon layer*, Proceedings of the 12th International Conference on Microelectromechanical Systems (IEEE, Piscataway NJ 1999) pp. 57–60

10.14   T. Abe, M. L. Reed: *Low Strain Sputtered Polysilicon for Micromechanical Structures*, Proceedings of the 9th International Workshop on Microelectromechanical Systems (IEEE, Piscataway NJ 1996) pp. 258–262

10.15   K. Honer, G. T. A. Kovacs: Integration of sputtered silicon microstructures with pre-fabricated CMOS circuitry, Sens. Actuators A **91**, 392–403 (2001)

10.16   J. Gaspar, T. Adrega, V. Chu, J. P. Conde: *Thin-Film Paddle Microresonators with High Quality Factors Fabricated at Temperatures Below* 110 °C, Proceedings of the 18th International Conference on Microelectromechanical Systems (IEEE, Piscataway NJ 2005) pp. 125–128

10.17   R. Anderson, R. S. Muller, C. W. Tobias: Porous polycrystalline silicon: a new material for MEMS, J. Microelectromech. Syst. **3**, 10–18 (1994)

10.18   W. Lang, P. Steiner, H. Sandmaier: Porous silicon: a novel material for microsystems, Sens. Actuators **A51**, 31–36 (1995)

10.19   R. He, C. J. Kim: *On-Chip Hermetic Packaging Enabled by Post-Deposition Electrochemical Etching of Polysilicon*, Proceedings of the 18th International Conference on Microelectromechanical Systems (IEEE, Piscataway NJ 2005) pp. 544–547

10.20   S. K. Ghandhi: *VLSI Fabrication Principles – Silicon and Gallium Arsenide* (Wiley, New York 1983)

10.21   W. A. Pilskin: Comparison of properties of dielectric films deposited by various methods, J. Vac. Sci. Technol. **21**, 1064–1081 (1977)

10.22   J. S. Danel, F. Michel, G. Delapierre: Micromachining of quartz and its application to an acceleration sensor, Sens. Actuators **A21–A23**, 971–977 (1990)

10.23   A. Yasseen, J. D. Cawley, M. Mehregany: Thick glass film technology for polysilicon surface micromachining, J. Microelectromech. Syst. **8**, 172–179 (1999)

10.24   R. Liu, M. J. Vasile, D. J. Beebe: The fabrication of nonplanar spin-on glass microstructures, J. Microelectromech. Syst. **8**, 146–151 (1999)

10.25   B. Folkmer, P. Steiner, W. Lang: Silicon nitride membrane sensors with monocrystalline transducers, Sens. Actuators **A51**, 71–75 (1995)

10.26   M. Sekimoto, H. Yoshihara, T. Ohkubo: Silicon nitride single-layer x-ray mask, J. Vacuum Sci. Technol. **21**, 1017–1021 (1982)

10.27   D. J. Monk, D. S. Soane, R. T. Howe: *Enhanced Removal of Sacrificial Layers for Silicon Surface Micromachining*, Technical Digest – The 7th International Conference on Solid State Sensors and Actuators (Institute of Electrical Engineers of Japan, Tokyo 1993) pp. 280–283

10.28   P. J. French, P. M. Sarro, R. Mallee, E. J. M. Fakkeldij, R. F. Wolffenbuttel: Optimization of a low-stress silicon nitride process for surface micromachining applications, Sens. Actuators **A58**, 149–157 (1997)

10.29   B. Li, B. Xiong, L. Jiang, Y. Zohar, M. Wong: Germanium as a versatile material for low-temperature micromachining, J. Microelectromech. Syst. **8**, 366–372 (1999)

10.30   A. Franke, D. Bilic, D. T. Chang, P. T. Jones, T. J. King, R. T. Howe, C. G. Johnson: *Post-CMOS Integration Of Germanium Microstructures*, Proceedings of the 12th International Conference on Microelectromechanical Systems (IEEE, Piscataway NJ 1999) pp. 630–637

10.31   A. E. Franke, Y. Jiao, M. T. Wu, T. J. King, R. T. Howe: *Post-CMOS Modular Integration of Poly-SiGe Microstructures Using Poly-Ge Sacrificial Layers*, Technical Digest – Solid State Sensor and Actuator Workshop (Transducers Research Foundation, Hilton Head 2000) pp. 18–21

10.32   S. Sedky, P. Fiorini, M. Caymax, S. Loreti, K. Baert, L. Hermans, R. Mertens: Structural and mechanical properties of polycrystalline silicon germanium for micromachining applications, J. Microelectromech. Syst. **7**, 365–372 (1998)

10.33   J. M. Heck, C. G. Keller, A. E. Franke, L. Muller, T.-J. King, R. T. Howe: *High Aspect Ratio Polysilicon-Germanium Microstructures*, Proceedings of the 10th International Conference on Solid State Sensors and Actuators (Institute of Electrical Engineers of Japan, Tokyo 1999) pp. 328–334

10.34   P. Van Gerwen, T. Slater, J. B. Chevrier, K. Baert, R. Mertens: Thin-film boron-doped polycrystalline silicon$_{70\%}$-germanium$_{30\%}$ for thermopiles, Sens. Actuators A **53**, 325–329 (1996)

10.35   D. Hyman, J. Lam, B. Warneke, A. Schmitz, T. Y. Hsu, J. Brown, J. Schaffner, A. Walson, R. Y. Loo, M. Mehregany, J. Lee: Surface micromachined RF MEMS switches on GaAs substrates, Int. J. Radio Frequency Microwave Commun. Eng. **9**, 348–361 (1999)

10.36   C. Chang, P. Chang: Innovative micromachined microwave switch with very low insertion loss, Sens. Actuators **79**, 71–75 (2000)

10.37   M. F. Aimi, M. P. Rao, N. C. MacDonald, A. S. Zuruzi, D. P. Bothman: High-aspect-ratio bulk micromachining of Ti, Nat. Mater. **3**, 103–105 (2004)

10.38   C. L. Shih, B. K. Lai, H. Kahn, S. M. Phillips, A. H. Heuer: A robust co-sputtering fabrication procedure for TiNi shape memory alloys for MEMS, J. Microelectromech. Syst. **10**, 69–79 (2001)

10.39   G. Hahm, H. Kahn, S. M. Phillips, A. H. Heuer: *Fully Microfabricated Silicon Spring Biased Shape Memory Actuated Microvalve*, Technical Digest- Solid State Sensor and Actuator Workshop (Transducers Research Foundation, Hilton Head Island 2000) pp. 230–233

10.40   S. D. Leith, D. T. Schwartz: High-rate through-mold electrodeposition of thick (> 200 micron) NiFe MEMS components with uniform composition, J. Microelectromech. Syst. **8**, 384–392 (1999)

10.41   N. Rajan, M. Mehregany, C. A. Zorman, S. Stefanescu, T. Kicher: Fabrication and testing of micromachined silicon carbide and nickel fuel atomizers for gas turbine engines, J. Microelectromech. Syst. **8**, 251–257 (1999)

10.42   T. Pornsin-Sirirak, Y. C. Tai, H. Nassef, C. M. Ho: Titanium-alloy MEMS wing technology for a microaerial vehicle application, Sens. Actuators A **89**, 95–103 (2001)

10.43   C. R. Stoldt, C. Carraro, W. R. Ashurst, D. Gao, R. T. Howe, R. Maboudian: A low temperature CVD process for silicon carbide MEMS, Sens. Actuators A **97-98**, 410–415 (2002)

10.44   M. Eickhoff, H. Moller, G. Kroetz, J. von Berg, R. Ziermann: A high temperature pressure sensor prepared by selective deposition of cubic silicon carbide on SOI substrates, Sens. Actuators **74**, 56–59 (1999)

10.45   Y. T. Yang, K. L. Ekinci, X. M. H. Huang, L. M. Schiavone, M. L. Roukes, C. A. Zorman, M. Mehregany: Monocrystalline silicon carbide nanoelectromechanical systems, Appl. Phys. Lett. **78**, 162–164 (2001)

10.46   D. Young, J. Du, C. A. Zorman, W. H. Ko: High-temperature single crystal 3C-SiC capacitive pressure sensor, IEEE Sens. J. **4**, 464–470 (2004)

10.47   C. A. Zorman, S. Rajgopal, X. A. Fu, R. Jezeski, J. Melzak, M. Mehregany: Deposition of polycrystalline 3C-SiC films on 100 mm-diameter (100) Si wafers in a large-volume LPCVD furnace, Electrochem. Solid State Lett. **5**, G99–G101 (2002)

10.48   I. Behrens, E. Peiner, A. S. Bakin, A. Schlachetzski: Micromachining of silicon carbide on silicon fabricated by low-pressure chemical vapor deposition, J. Micromech. Microeng. **12**, 380–384 (2002)

10.49   C. A. Zorman, S. Roy, C. H. Wu, A. J. Fleischman, M. Mehregany: Characterization of polycrystalline silicon carbide films grown by atmospheric pressure chemical vapor deposition on polycrystalline silicon, J. Mater. Res. **13**, 406–412 (1996)

10.50   C. H. Wu, C. H. Zorman, M. Mehregany: Growth of polycrystalline SiC films on SiO_2 and Si_3N_4 by APCVD, Thin Solid Films **355-356**, 179–183 (1999)

10.51   P. Sarro: Silicon carbide as a new MEMS technologytuators, Sens. Actuators A **82**, 210–218 (2000)

10.52   N. Ledermann, J. Baborowski, P. Muralt, N. Xantopoulos, J. M. Tellenbach: Sputtered silicon carbide thin films as protective coatings for MEMS applications, Surface Coatings Technol. **125**, 246–250 (2000)

10.53   X. A. Fu, R. Jezeski, C. A. Zorman, M. Mehregany: Use of deposition pressure to control the residual stress in polycrystalline SiC films, Appl. Phys. Lett. **84**, 341–343 (2004)

10.54   J. Trevino, X. A. Fu, M. Mehregany, C. Zorman: *Low-Stress, Heavily-Doped Polycrystalline Silicon Carbide for MEMS Applications*, Proceedings of the 18th International Conference on Microelectromechanical Systems (IEEE, Piscataway NJ 2005) pp. 451–454

10.55 R. S. Okojie, A. A. Ned, A. D. Kurtz: Operation of a 6H–SiC pressure sensor at 500 °C, Sens. Actuators A **66**, 200–204 (1998)

10.56 K. Lohner, K. S. Chen, A. A. Ayon, M. S. Spearing: Microfabricated silicon carbide microengine structures, Mater. Res. Soci. Symp. Proc. **546**, 85–90 (1999)

10.57 K. O. Min, S. Tanaka, M. Esashi: *Micro/Nano Glass Press Molding Using Silicon Carbide Molds Fabricated by Silicon Lost Molding*, Proceedings of the 18th International Conference on Microelectromechanical Systems (IEEE, Miami 2005) pp. 475–478

10.58 X. Song, S. Rajgolpal, J. M. Melzak, C. A. Zorman, M. Mehregany: Development of a multilayer SiC surface micromachining process with capabilities and design rules comparable with conventional polysilicon surface micromachining, Mater. Sci. Forum **389–393**, 755–758 (2001)

10.59 S. Tanaka, S. Sugimoto, J.-F. Li, R. Watanabe, M. Esashi: Silicon carbide micro-reaction-sintering using micromachined silicon molds, J. Microelectromech. Syst. **10**, 55–61 (2001)

10.60 L. A. Liew, W. Zhang, V. M. Bright, A. Linan, M. L. Dunn, R. Raj: Fabrication of SiCN ceramic MEMS using injectable polymer-precursor technique, Sens. Actuators A **89**, 64–70 (2001)

10.61 A. J. Fleischman, S. Roy, C. A. Zorman, M. Mehregany: *Polycrystalline Silicon Carbide for Surface Micromachining*, Proceedings of the 9th International Workshop on Microelectromechanical Systems (IEEE, San Diego 1996) pp. 234–238

10.62 A. J. Fleischman, X. Wei, C. A. Zorman, M. Mehregany: Surface micromachining of polycrystalline SiC deposited on SiO$_2$ by APCVD, Mater. Sci. Forum **264–268**, 885–888 (1998)

10.63 G. Beheim, C. S. Salupo: Deep RIE process for silicon carbide power electronics and MEMS, Mater. Res. Soc. Symp. Proc. **622**, T8.8.1–T8.8.6. (2000)

10.64 A. Yasseen, C. H. Wu, C. A. Zorman, M. Mehregany: Fabrication and testing of surface micromachined polycrystalline SiC micromotors, Electron. Device Lett. **21**, 164–166 (2000)

10.65 D. Gao, M. B. Wijesundara, C. Carraro, R. T. Howe, R. Maboudian: Recent progress toward and manufacturable polycrystalline SiC surface micromachining technology, IEEE Sens. J. **4**, 441–448 (2004)

10.66 T. Shibata, Y. Kitamoto, K. Unno, E. Makino: Micromachining of diamond film for MEMS applications, J. Microelectromech. Syst. **9**, 47–51 (2000)

10.67 H. Bjorkman, P. Rangsten, P. Hollman, K. Hjort: Diamond replicas from microstructured silicon masters, Sens. Actuators **73**, 24–29 (1999)

10.68 P. Rangsten, H. Bjorkman, K. Hjort: *Microfluidic Components in Diamond*, Proceedings of the 10th International Conference on Solid State Sensors and Actuators (IEEE, Sendai 1999) pp. 190–193

10.69 H. Bjorkman, P. Rangsten, K. Hjort: Diamond microstructures for optical microelectromechanical systems, Sens. Actuators **78**, 41–47 (1999)

10.70 M. Aslam, D. Schulz: *Technology of Diamond Microelectromechanical Systems*, Proceedings of the 8th International Conference on Solid State Sensors and Actuators (IEEE, Stockholm 1995) pp. 222–224

10.71 R. Ramesham: Fabrication of diamond microstructures for microelectromechanical systems (MEMS) by a surface micromachining process, Thin Solid Films **340**, 1–6 (1999)

10.72 X. Yang, J. M. Yang, Y. C. Tai, C. M. Ho: Micromachined membrane particle filters, Sens. Actuators **73**, 184–191 (1999)

10.73 X. D. Wang, G. D. Hong, J. Zhang, B. L. Lin, H. Q. Gong, W. Y. Wang: Precise patterning of diamond films for MEMS application, J. Mater. Process. Technol. **127**, 230–233 (2002)

10.74 A. R. Krauss, O. Auciello, D. M. Gruen, A. Jayatissa, A. Sumant, J. Tucek, D. C. Mancini, N. Moldovan, A. Erdemire, D. Ersoy, M. N. Gardos, H. G. Busmann, E. M. Meyer, M. Q. Ding: Ultrananocrystalline diamond thin films for MEMS and moving mechanical assembly devices, Diamond Related Mater. **10**, 1952–1961 (2001)

10.75 X. Xiao, J. Birrell, J. E. Gerbi, O. Auciello, J. A. Carlisle: Low temperature growth of ultrananocrystalline diamond, J. Appl. Phys. **96**, 2232–2239 (2004)

10.76 T. A. Friedmann, J. P. Sullivan, J. A. Knapp, D. R. Tallant, D. M. Follstaedt, D. L. Medlin, P. B. Mirkarimi: Thick stress-free amorphous-tetrahedral carbon films with hardness near that of diamond, Appl. Phys. Lett. **71**, 3820–3822 (1997)

10.77 J. P. Sullivan, T. A. Friedmann, K. Hjort: Diamond and amorphous carbon MEMS, MRS Bull. **26**, 309–311 (2001)

10.78 J. R. Webster, C. W. Dyck, J. P. Sullivan, T. A. Friedmann, A. J. Carton: Performance of amorphous diamond RF MEMS capacitive switch, Electron. Lett. **40**, 43–44 (2004)

10.79 K. Hjort, J. Soderkvist, J.-A. Schweitz: Galium arsenide as a mechanical material, J. Micromech. Microeng. **4**, 1–13 (1994)

10.80 K. Hjort: Sacrificial etching Of III–V compounds for micromechanical devices, J. Micromech. Microeng. **6**, 370–365 (1996)

10.81 K. Fobelets, R. Vounckx, G. Borghs: A GaAs pressure sensor based on resonant tunnelling diodes, J. Micromech. Microeng. **4**, 123–128 (1994)

10.82 A. Dehe, K. Fricke, H. L. Hartnagel: Infrared thermopile sensor based on AlGaAs-GaAs micromachining, Sens. Actuators **A46–A47**, 432–436 (1995)

10.83 A. Dehe, J. Peerlings, J. Pfeiffer, R. Riemenschneider, A. Vogt, K. Streubel, H. Kunzel, P. Meissner, H. L. Hartnagel: III–V compound semiconductor micromachined actuators for long resonator tunable Fabry-Perot detectors, Sens. Actuators **A68**, 365–371 (1998)

10.84   T. Lalinsky, S. Hascik, Mozolova, E. Burian, M. Drzik: The improved performance of GaAs micromachined power sensor microsystem, Sens. Actuators **76**, 241–246 (1999)

10.85   T. Lalinsky, E. Burian, M. Drzik, S. Hascik, Z. Mozolova, J. Kuzmik, Z. Hatzopoulos: Performance of GaAs micromachined microactuator, Sens. Actuators **85**, 365–370 (2000)

10.86   H. X. Tang, X. M. H. Huang, M. L. Roukes, M. Bichler, W. Wegscheider: Two-dimensional electron-gas actuation and transduction for GaAs nanoelectromechanical systems, Appl. Phys. Lett. **81**, 3879–3881 (2002)

10.87   T. S. Tighe, J. M. Worlock, M. L. Roukes: Direct thermal conductance measurements on suspended monocrystalline nanostructure, Appl. Phys. Lett. **70**, 2687–2689 (1997)

10.88   J. Miao, B. L. Weiss, H. L. Hartnagel: Micromachining of three-dimensional GaAs membrane structures using high-energy nitrogen implantation, J. Micromech. Microeng. **13**, 35–39 (2003)

10.89   C. Seassal, J. L. Leclercq, P. Viktorovitch: Fabrication of inp-based freestanding microstructures by selective surface micromachining, J. Micromech. Microeng. **6**, 261–265 (1996)

10.90   J. Leclerq, R. P. Ribas, J. M. Karam, P. Viktorovitch: III–V micromachined devices for microsystems, Microelectron. J. **29**, 613–619 (1998)

10.91   H. Yamaguchi, R. Dreyfus, S. Miyashita, Y. Hirayama: Fabrication and elastic properties of InAs freestanding structures based on InAs/GaAs(111) a heteroepitaxial systems, Phys. E **13**, 1163–1167 (2002)

10.92   C. Lee, T. Itoh, T. Suga: Micromachined piezoelectric force sensors based on PZT thin films, IEEE Trans. on Ultrasonics, Ferroelectrics, and Frequency Control **43**, 553–559 (1996)

10.93   B. Xu, L. E. Cross, J. J. Bernstein: Ferroelectric and antiferroelectric films for microelectromechanical systems applications, Thin Solid Films **377-378**, 712–718 (2000)

10.94   S. P. Beeby, A. Blackburn, N. M. White: Processing of PZT piezoelectric thick films on silicon for microelectromechanical systems, J. Micromech. Microeng. **9**, 218–229 (1999)

10.95   C. Shearwood, M. A. Harradine, T. S. Birch, J. C. Stevens: Applications of polyimide membranes to MEMS technology, Microelectron. Eng. **30**, 547–550 (1996)

10.96   F. Jiang, G. B. Lee, Y. C. Tai, C. M. Ho: A flexible micromachine-based shear-stress sensor array and its application to separation-point detection, Sens. Actuators **79**, 194–203 (2000)

10.97   D. Memmi, V. Foglietti, E. Cianci, G. Caliano, M. Pappalardo: Fabrication of capacitive micromechanical ultrasonic transducers by low-temperature process, Sens. Actuators A **99**, 85–91 (2002)

10.98   A. Bagolini, L. Pakula, T. L. M. Scholtes, H. T. M. Pham, P. J. French, P. M. Sarro: Polyimide sacrificial layer and novel materials for post-processing surface micromachining, J. Micromech. Microeng. **12**, 385–389 (2002)

10.99   T. Stieglitz: Flexible biomedical microdevices with double-sided electrode arrangements for neural applications, Sens. Actuators A **90**, 203–211 (2001)

10.100   T. Stieglitz, G. Matthias: Flexible BioMEMS with electrode arrangements on front and back side as key component in neural prostheses and biohybrid systems, Sens. Actuators B **83**, 8–14 (2002)

10.101   H. Lorenz, M. Despont, N. Fahrni, J. Brugger, P. Vettiger, P. Renaud: High-aspect-ratio, ultrathick, negative-tone-near–UV photoresist and its applications in MEMS, Sens. Actuators A **64**, 33–39 (1998)

10.102   H. Lorenz, M. Despont, N. Fahrni, N. LaBianca, P. Renaud, P. Vettiger: SU-8: A low-cost negative resist for MEMS, J. Micromech. Microeng. **7**, 121–124 (1997)

10.103   E. H. Conradie, D. F. Moore: SU-8 thick photoresist processing as a functional material for MEMS applications, J. Micromech. Microeng. **12**, 368–374 (2002)

10.104   C. T. Pan, H. Yang, S. C. Shen, M. C. Chou, H. P. Chou: A low-temperature wafer bonding technique using patternable materials, J. Micromech. Microeng. **12**, 611–615 (2002)

10.105   P. A. Stupar, A. P. Pisano: *Silicon, Parylene, and Silicon/Parylene Micro-Needles for Strength and Toughness*, Technical Digest of the 11th International Conference on Solid State Sensors and Actuators (Springer, Berlin 2001) pp. 1368–1389

10.106   J. M. Zara, S. W. Smith: Optical scanner using a MEMS actuator, Sens. Actuators A **102**, 176–184 (2002)

10.107   H. S. Noh, P. J. Hesketh, G. C. Frye-Mason: Parylene gas chromatographic column for rapid thermal cycling, J. Microelectromech. Syst. **11**, 718–725 (2002)

10.108   T. J. Yao, X. Yang, Y. C. Tai: BrF_3 dry release technology for large freestanding parylene microstructures and electrostatic actuators, Sens. Actuators A **97-98**, 771–775 (2002)

10.109   X. Wang, J. Engel, C. Liu: Liquid crystal polymer (LCP) for MEMS: processing and applications, J. Micromech. Microeng. **13**, 628–633 (2003)

10.110   C. J. Lee, S. J. Oh, J. K. Song, S. J. Kim: Neural signal recording using microelectrode arrays fabricated on liquid crystal polymer material, Mater. Sci. Eng. C **4**, 265–268 (2004)

10.111   F. F. Faheem, K. C. Gupta, Y. C. Lee: Flip-chip assembly and liquid crystal polymer encapsulation for variable MEMS capacitors, IEEE Trans. Microwave Theory Tech. **51**, 2562–2567 (2003)

11. Complexity and Emergence as Design Principles for Engineering Decentralized Nanoscale Systems

In this chapter, we will investigate the principles of complex adaptive and emergent nanoscale systems, the rules that govern them, some functional examples, and, ultimately, how these rules can produce systems with novel functionality. First, we will define appropriate terminology and common characteristics among complex systems and describe the principles by which they operate. Next, we will see how these principles manifest in biological nanoscale systems. We will also examine emergent systems computationally to glean applicable problem-solving strategies for engineering decentralized complexity. Using this information, a framework for nanoscale self-assembly and adaptive system construction will become apparent. We will begin with the language relevant to nanoscale emergent systems.

The 17th-century discourse of Rene Descartes introduced the world to the ghost in the machine [11.1]. He presented a revolutionary mechanistic paradigm, the human body as a biological device. Today we can expand this idea to the realm of nanoscale molecular interactions. The tissues, organs, neural pathways, and complex regulatory systems of the human body are a product of a collection of many biological structures – nucleic acids, proteins, lipids, cells – all interacting in a cooperative and adaptive manner. But we are not alone; adaptation and complexity are fundamental to evolution and, as a result, biological systems on many scales.

One of the first people to apply evolutionary principles to machines was the mathematician John von Neumann. As early as the 1940s, von Neumann proposed nonbiological, self-reproducing machines, an idea which Arthur Burks later published on his behalf in 1966 [11.2]. Complex systems have many similarities, regardless of scale, as well as definitive characteristics. To approach engineering nanoscale complex adaptive systems, we must examine several categories of emergent behavior.

11.1 Definitions

Complexity

Complex dynamical systems can be both intelligent and adaptive. Also, a system can be said to be more complex, but absent additional functionality. Functional complex systems can be further divided into organized and disorganized categories, the primary difference being the stochastic nature of the system. The characteristic complexity of an object, as John Casti explains " ... is directly proportional to the length of the shortest possible description of that object ... . Another way of expressing this is to say that something is random if it is incompressible." [11.3]. Randomness is sometimes viewed as simply an inadequate knowledge of the causality of events. However, in terms of engineering nanoscale systems, it may be more prudent to view it as a driving force for the incompressibility of information within the system. That is, if it is neither noise nor redundant, it is complex information.

For our purposes it should suffice that a complex nanoscale system offers emergent behavior in the form of greater functionality. This need not be smart or adaptive. However, there are obvious benefits to this adaptation.

But to begin, what do we mean when we say a system is complex? *Depew* and *Weber* [11.4] offer a concise definition:

"Complex systems are not just complicated systems. A snowflake is complicated, but the rules for generating it are simple. The structure of a snowflake, moreover, persists unchanged and crystalline, from the first moment of its existence until it melts, while complex systems change over time. It is true that a turbulent river rushing through the narrow channel of rapids changes over time, too, but it changes chaotically. The kind of change characteristic of complex systems lies somewhere between the pure order of crystalline snowflakes and the disorder of chaotic or turbulent flow. So identified, complex systems are systems that have a large number of components that can interact simultaneously in a sufficiently rich number of parallel ways so that the system shows spontaneous self-organization and produces global, emergent structures."

Emergence

Emergence can be defined as a decentralized complexity that materializes from a simple but abundant number of constituents. *Johnson* [11.5] explains:

"In the simplest terms emergent systems solve problems by drawing on masses of relatively stupid elements, rather than a single, intelligent "executive branch." They are bottom-up systems, not top-down ... In these systems, agents residing on one scale start producing behavior that lies one scale above them: ants create colonies; urbanites create neighborhoods; simple pattern-recognition software learns how to recommend new books. The movement from low-level rules to higher-level sophistication is what we call emergence."

So a sufficient number of components working in a spontaneously coordinated fashion are required for global behavior to arise or "emerge."

Intelligence

It is often misleading to assume that collective behavior implies central authority. Swarm behavior is a result of the simple rules of interaction between many parts, much like central governance and emergent nanosystems, which are no different. As we shall see, amoeboid movement exhibits such intelligence but is merely a cascade of events linked to actin-based motility.

Eventually, intelligence implies that a system can change and adapt its responses to user input or environmental cues. Also, self-replication or patterns of self-organization control can be seen, as in Wolfram's type 4 cellular automata, an example of a generative engine that is simultaneously a producer and a product [11.6]. A nanobiological example can be found in the form of DNA transposons. These small segments of DNA are capable of self-splicing and relocation, removing themselves from one location within the host genome and reinserting in another. Transposons represent a product in the form of the nucleotides that are replicated by the host but also a producer, since the nucleotides alone are sufficient for cutting and rearrangement.

Adaptation

For a system to intelligently respond to its environment, it must be able to adapt and move beyond purely reactive mechanisms. The system needs to be able to produce output that is patently novel. Evolution requires reproduction, inheritance, variation, and selection. Additionally, artificial evolution requires two other important features: initialization and selective termination. Initialization usually refers to the starting values assigned to variables with a computational model. Selective termination is the criterion for choosing numerical products that are good enough to persist in the next round or end an optimization search.

Because it is not the actual processing but how the processing evolves, learns, and adapts which will ultimately reinforce the cooperative participation, nanoscale complex emergent systems must be understood in a time-dependent manner.

As mentioned above, the metric of complexity is redundancy; the globally complex behavior that emerges cannot be reduced to its parts. Nevertheless, it is evident that there are general principles that complex emergent systems have in common. Most complex adaptive systems are generated by simple rules. From cellular automata (CA) to social insects, many of these systems can be examined for their congruencies.

It should be noted that there is some disagreement on the precise use of the terms: swarm, agent, intelligence, adaptation, and complexity. What is clear is that many natural phenomena, like schooling fish, foraging insects, flock formation, amoeba movement, and social interactions, are all products of complex adaptive systems, and these systems have commonality on many levels.

11.1.1 Rules

John Holland addresses the complimentary traits of complex emergent systems [11.7], which include:

1. *A large number of interacting agents.* Certainly it is correct that a minimum number of agents are required, but it is their connectivity that contributes the most to complexity. Weaver elaborates on this point when he notes the transition between modeling two-variable equations and multibillion-variable equations. He says, "Much more important than the mere number of variables is the fact that these variables are all interrelated ... These problems, as contrasted with the disorganized situations with which statistics can cope, show the essential feature of organization. We will therefore refer to this group of problems as those of organized complexity" [11.3].
    Kauffman explains this feature with his model of a Boolean network of light bulbs [11.8, 9]: "We might, for example, study the pool of networks with 1000 bulbs (we'll call this variable N) and 20 inputs per bulb (the variable K). Given $N = 1000$ and $K = 20$, a vast ensemble of networks can be built." Sampled randomly, he finds that for a variable N, where $K = 4$ or 5, the system becomes chaotic. While for the same number of bulbs, if only

2 connections are allowed per node, the system approaches an organization he terms "combinatorial optimization" [11.9].

2. *The aggregate behavior of the agents.* A global purpose or result from the individual agents.

3. *Agent interaction.* A nonlinear process, which makes aggregate behavior unequal to the summation of individual agents. For many nanoscale systems this involves self-assembly. Furthermore, an internal system of rules enables these interactions to produce assumptions about the system environment.

4. *Agent diversity.* This leads to adaptation, robustness, and perpetual novelty in global behavior. Aggregate behavior is a very important variable for nanoscale systems since it is most likely the collective behavior that will be harnessed. From afar we see that agent interaction forms a complex whole that cannot be understood in terms of the elements. This irreducibility is described in more detail by *Johnson* in his analysis of several conventions of emergent behavior [11.5]. These include (1) more is different, (2) ignorance is useful, (3) encourage random encounters, (4) look for patterns in signals, and (5) pay attention to your neighbors. It is obvious that such rules are useful in explaining social insects. Consider the ant example he uses, which echoes Holland's first rule: "This old slogan of complexity theory actually has two meanings that are relevant to our ant colonies. First the statistical nature of ant interactions demands that there be a critical mass of ants for the colony to make intelligent assessments of its global state" [11.5]. Furthermore, he states: " *'More is different' also applies to the distinction between micromotives and macrobehavior: individual ants don't 'know' that they're prioritizing pathways between different food sources when they lay down a pheromone gradient ... It's only by observing the entire system at work that the global behavior becomes apparent.*"

This highlights the autonomous but uninformed nature of the system components. Obviously we must expect our nanosized devices to be autonomous and uninformed as well.

Ignorance is useful to a system with many interacting parts because simple rules allow for a densely structured and reliable system, since individuals with global understanding would serve to destabilize the group. Random encounters are essential for a decentralized system of elements because it allows the system to explore previously undefined boundaries and adapt to new environments.

The fourth rule, "the patterns in signs," amounts to the language of the system. Just as slime mold navigates toward cAMP, signal distribution throughout the network is essential for cooperative behavior. Additionally, it is the timing, position, and strength of these "signs" that dictate the resulting global response.

11.1.2 Engineering Principles

Emergent behavior frequently refers to the observed function of the system rather than the underlying structure or form. But we have seen examples where form is the observed function, as well as where function takes on form. These behaviors exist because of the nature of the interaction between the parts and their environment, a relationship that cannot be reduced to its constituents.

When designing an emergent nanoscale system, one must consider its purpose. Is the system designed for calculation or for a more physical purpose such as nanoscale assembly, and will the system require adaptation and/or growth? Optimization of emergent systems is often considered in terms of the best solutions for interdependent networks. These networked problems can range from the visual cortex of the brain, to the traveling salesman, to the game of life [11.10]. These mathematical models provide a language to discuss paradigms of emergence, which can be applied to building or designing these types of systems. We will examine several of these models to glean applicable information about emergence. These models include several classic examples from cellular automata (CA) [11.6], such as the game of life [11.10]; swarm optimization [11.3] and the iterative prisoner's dilemma model [11.7]. We will also consider Latane's dynamic social impact theory [11.11], Hutchins' parallel constraint satisfaction networks [11.12], fractal interpolation functions [11.13], and biological examples of nanoscale emergent systems.

When we think of emergent nanoscale systems, we must consider these information terms. Not only can elementary electronic components represent logic gates, but an ensemble of nanoscale devices might represent a computationally equivalent structure if only the information could be extracted.

The principle of computational equivalence is given support by numerous CA examples [11.6]. Wolfram defines four classes of CA (see Fig. 11.2): (1) Simple behavior in the sense that all initial conditions lead to exactly the same uniform final state. (2) Many different final states, but the system remains a simple set of structures, which repeat every few steps or toward in-

finity. (3) More random in nature but discrete, repeated structures can be seen. (4) This class is a mixture of order and randomness, but structures reorganize and interact in a complicated fashion [11.6]. Interestingly, type 4 CA exhibits characteristics of living, learning systems. The programs appear to evolve, with areas of defined structure and repetition, alongside neighborhoods of perpetual randomness. Also, elements within the graphic can be seen to "react" to other structures when coming into contact. This type of action is indicative of systems that can both store and process information (Fig. 11.1). The game of life is a classic example of cellular automata producing rich and complex results [11.6]. In it, patterns of organization can be seen to manifest from initially chaotic seed values. This tells us that very simple interactions, such as the game's rules of binary (live or dead) output derived from neighboring states, can eventually establish organized patterns involving many elements. Similarly, Dynamic Social Impact theory demonstrates that a population with sufficient numbers and randomness will still create clustering and local areas of "polarizations." Based on what we've seen from type 4 CA, clustering and local structures should be expected design considerations for the system construction.

The algorithm for particle swarm optimization is based on human social behavior. Agents interact, learn, and apply this knowledge by moving toward a space of better optimization [11.3]. To utilize this, a nanoscale system would have to be capable of storing information and allowing this information to be shared among the group members. Sharing is an important factor that is also demonstrated in the prisoner's dilemma.

The prisoners' dilemma is a problem involving two agents, each individual having two choices, to cooperate or defect. The rules of the game state that an individual's payoff is dependent on the cooperation state of the other, while this state is not known at the time of the decision. Like Kauffman's light bulb example, this game becomes enormously complex very quickly, with a three-step horizon yielding 2^{64} possible outcomes. Without access to the choices of other individuals, the system converges on the minimax solution of both members defecting. This result occurs because defecting has the fewest possible side effects. On the other hand, if past experience can be shared between members, such as in the iterative prisoner's dilemma, the solution tends to produce individuals who cooperate for the highest payoff. Like swarm optimization, this example illustrates that, given the ability to exchange and store information, cooperative optimization is facilitated [11.7].

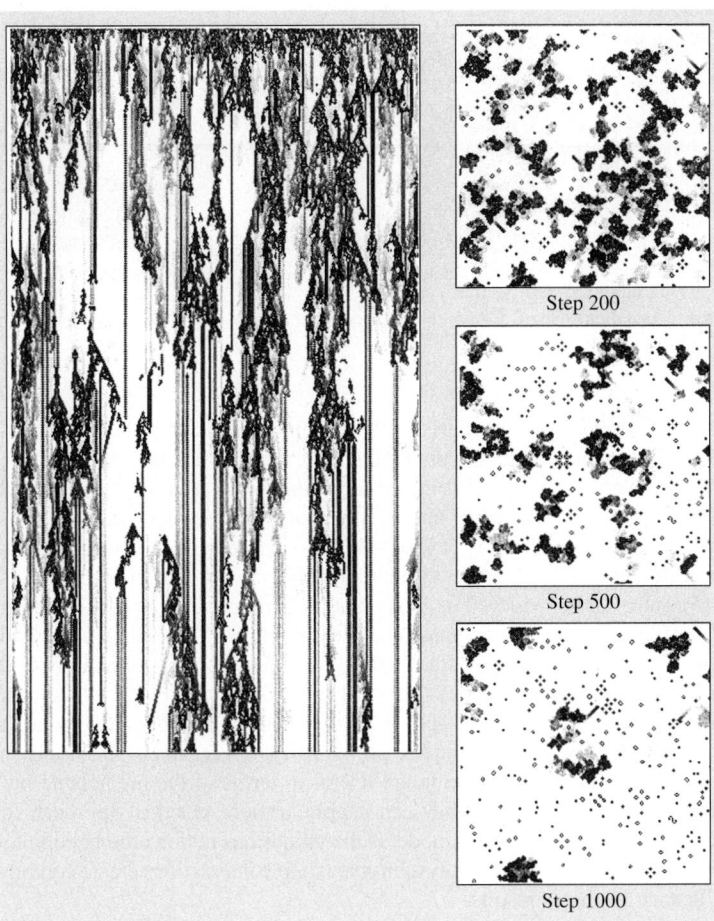

Step 200

Step 500

Step 1000

Fig. 11.1 Two-dimensional type 4 CA. This program shows an evolving mixture of order and randomness, producing structures that appear to interact in a complicated manner

Hutchins' parallel constraints network offers one last insight into an engineering consideration. He found that a system solution tended to gravitate toward nodes of greater connectivity, even if these nodes represented unsatisfactory solutions [11.12]. This implies that if some agents can be networked with greater efficiency, then these individuals will have the greatest influence on global behavior. Thus, when purposefully designing more influential elements, emphasis should be placed on connectivity.

Interconnected systems coevolve, a correlative change that is subject to a fitness landscape, which chooses between adaptations that persist and those that do not. Hofstadter describes the careful balance that is created as the interconnecting nodes of a system grow " ... to a natural state between order and chaos, a grand compromise between structure and surprise" [11.14].

In order for systems to adapt, they must learn from the past and be able to anticipate future events. This requires memory or experience from the previous generation. "To be able to adapt effectively, systems must move toward conditions far from equilibrium where they have enough flexibility to change yet enough stability to survive modifications and transformations" [11.14].

So it is clear that flexibility, connectivity, exchanging/storing information, and randomness are all design criteria for complex emergent systems on the nanoscale.

To have explicit calculations in the end, they must be manifested in a physical process, which is therefore subject to the limitations of that process. This has two meanings. First, the actual act of calculating requires a finite system. Much of the CA viewed as emergent is still the result of physical calculations. Nanoscale systems present the prospect that these calculations can be both a product of the system as well as a prod-

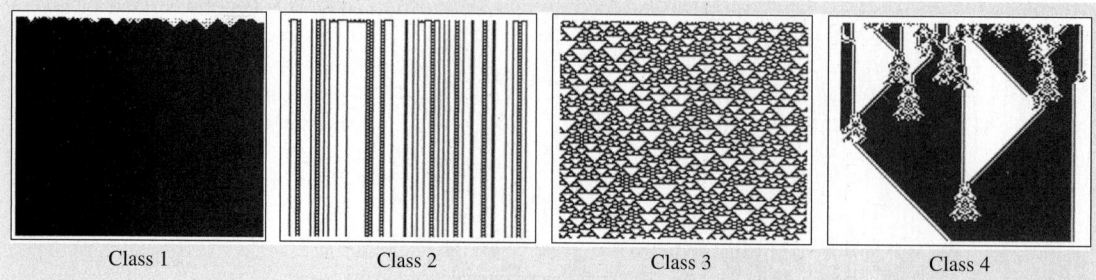

Fig. 11.2 Four classes of cellular automata. Classification of CA into four general types by visual inspection by *S. Wolfram* [11.6]

uct within the system but are, in the end, still limited in reduction to the actual building blocks. The second meaning is that the physically discrete events must give rise to the calculation. We have already seen that does not necessarily imply that continuous results cannot be obtained but merely that emergent nanoscale systems may be better at approximating or performing tasks that are discrete. Wolfram writes that the "principle of computational equivalence" is concerned with " ... the computational sophistication of complete systems, but also with the computational sophistication of specific processes within systems." And that even highly complex computational sophistication can have a fundamental equivalence in other kinds of processes. To expand this, even systems with very different underlying structures, rules can be found that will produce computation of equal sophistication. The significance of this notion is not so much that computations of immense complexity can evolve from simple initial conditions, but rather, identical solutions can be found by very different means. For example, consider CA rule 132 from Wolfram's cellular automata (Fig. 11.3) [11.6]. If the system starts with an odd number of black cells, the system will evolve to have one surviving black cell. Conversely, if an even number of black cells starts, then 0 are left after a few steps of evolution. Clearly this rule demonstrates a binary output of whether the initial

conditions for the system were odd or even. While computationally elementary, the rule demonstrates the very powerful notion of computational equivalence, which holds an interesting analogy for our nanoscale systems. That is, it should be possible to build structures that are both very similar but perform vastly different processes, as well as those that are structurally very different that can compute complex problems. This dualism need not be separated either. Such a system could contain both properties (physical tasks performed by growth and assembly) and, in so doing, form the qualitative solution of a complex problem. However, since most rules in engineering are useful in terms of the predictable outcomes they can supply, a more practical approach for CA as a model is the ubiquitous notion among complex adaptive systems that simple interactions create complex results.

Ultimately, the analogies of computation are limited by continuity. As mentioned previously, nanoscale systems, while extremely small, still represent discrete elements on the atomic scale, whereas equations tend to involve continuous quantities. For example, consider the kneading process, where material is stretched to twice its length, cut in two, stacked, and combined (Fig. 11.4). An algorithm of this type will take the distance between neighboring points and double it. In mathematical terms this presents a system very sensitive to initial condi-

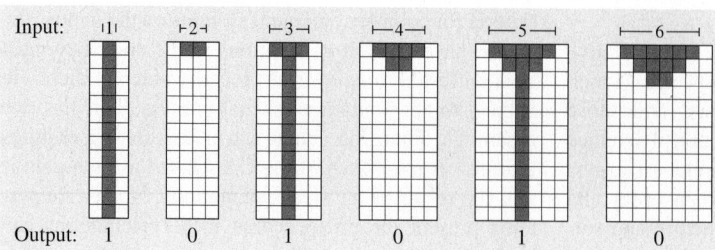

Fig. 11.3 CA rule 132. A CA program showing the initial system number parity in the form of a binary output

tions, with associated chaotic dependencies. However, in a physical system this process amplifies distance and after only a few dozen cycles would have dramatic scale change on the order of nanometers to meters. Thus, when analyzing analogous forms of computation or information in physical structure, engineers must consider the limits in terms of practical constraints on the system.

That is not to say that a nanoscale system of discrete components can't exhibit continuous behavior. One need only look at water to observe discrete water molecules in a continuous fluid. The essential factor in creating an emergent continuity in these systems is randomness.

Randomness is necessary, but not sufficient, for complex emergent behavior. *Wolfram* notes how initial conditions are considered in the process [11.6]: " ... the amount of time over which the details of initial conditions can ever be considered the dominant source of randomness will inevitably be limited by the level of separation that exists between the large-scale features that one observes and the small-scale features one cannot control." This implies that systems that contain an inherent internal randomness will evolve in a manner indistinguishable from chaotic initial conditions [11.6]. While this is useful in terms of random number generation, it is difficult to extract information from such a system because, on the whole, it will appear as noise.

Wolfram discusses four general types of cellular automata, classified first on general appearance. Moreover, another distinguishing factor is if the system is built by simple rules or quarantined by constraints. This is a fundamental difference, as engineered nanoscale systems are more likely to be the former, while systems in nature the latter. Constraint systems are not given simple rules to follow but rather limitations that must not be exceeded. Constraint systems build complexity from the lawless random nature of the system, while one built on explicit rules may evolve into complex behavior from three types of randomness: one, where the system is constantly subjected to randomness or noise; two, where only the initial conditions are randomized, and this initial variety propagates along with system evolution; or three, a system that is intrinsically random.

Furthermore, continuous behavior can lead to discrete complex behavior. For example, thermal protein unfolding (a process that yields a stark, and usually irreversible, shape change in a protein) results from the continuous application of heat. What this tells us is that a continuous operation can produce discrete results. Usually we think of nanoscale system construction as bottom-up, but we see that it should also be possible to plan a system that yields continuous performance.

Logically, complex behavior in nature can be considered a product of constraints. Random mutations evolve into specialized genetic and phenotypic traits, but these traits should provide an increased level of fitness for the organism in order for the mutations to survive. Essentially, maximizing fitness is the same as satisfying

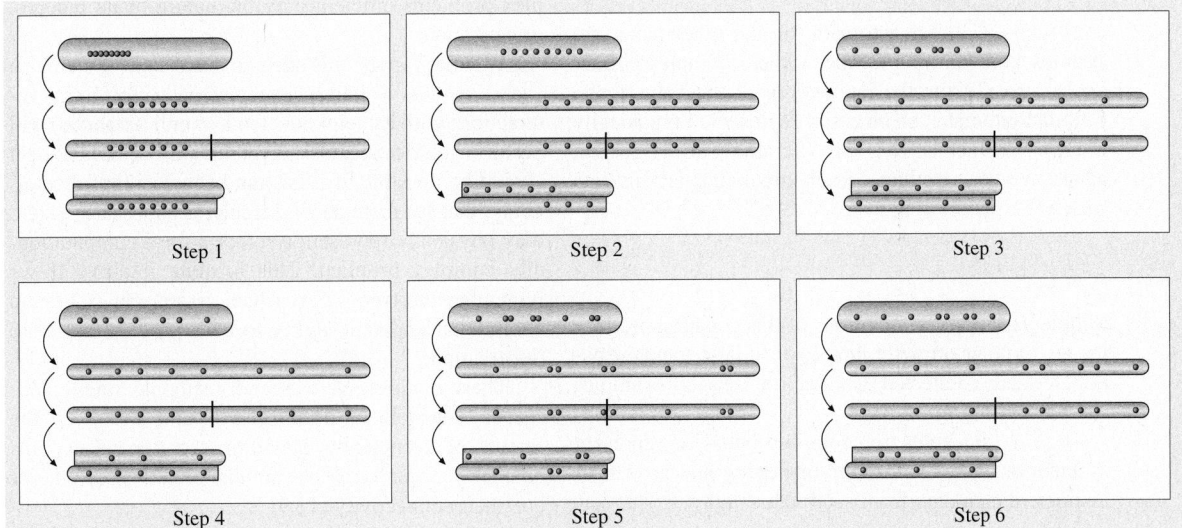

Fig. 11.4 The simple kneading process quickly randomizes system elements

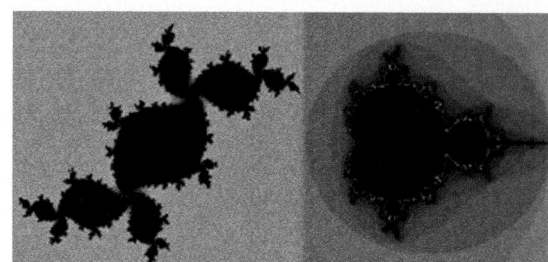

Fig. 11.5 Julia set (*left*) and Mandelbrot set (*right*). (Rendered with Mandelbrot and Julia set explorer http://aleph0. clarku.edu/~djoyce/julia/explorer.html)

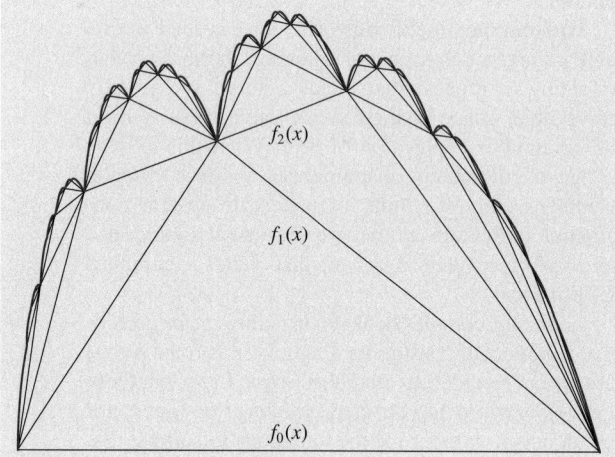

Fig. 11.6 Fractal interpolation (after [11.13])

prising; however, in fractal terms, it is quite expected. How can this similarity be applied?

Fractal interpolation seeks to obtain a converging nonempty compact subset that can describe the attractor for an iterated function system (IFS). As mentioned previously, many physical processes are discrete, and thus approximating them using continuous functions such as partial differential equations can yield results that smooth out and remove descriptive information. The advantage fractal interpolation has over traditional trigonometric functions can be thought of in graphic form. While traditional Euclidean approximation is valuable for polynomial curve fitting (like least-squares), fractal interpolation can describe assemblages of objects as a single iterative system [11.13] (Fig. 11.6). Since fractal functions describe sets rather than points, the power of this method lies within the contraction mapping of an entire system to within the Hausdorff metric. That is, if the distance between sets can be made small enough, the system, which in least-squares terms might not readily conform to a polynomial fit, can be computed rapidly by merely a sequence of iterative steps of the fractal function. But this type of analysis need not be constrained to one dimension; multidimensional fractal interpolation has been investigated and the utility of fractal interpolation is commonly used by stock analysts for price predictions [11.15]. This echoes what we found in CA, clearly akin to fractal geometry, which tells us that a system itself can represent a computational equivalent entity, an emergent system that can be applied to complex problems efficiently by the nature of its inherent organization.

We may apply this idea to nanoscale systems in such a way as to build a component, which by its interactions with its neighbors (and several iterations) will perform the desired task or computation. Meaning, it should be possible, if it has not been accomplished already, that by mere self-assembly, a nanoscale system may physically represent a solution for a computationally complex problem. This is quite intuitive if we remember the type 4 CA, which as a complex system exhibits an apparent ability to both process and store information.

These models give us search methods, fitness landscapes, multiobject optimization, and many of the factors of complexity including the dependence (as Kauffman observed) of the number of network elements and their connectivity [11.9].

constraints. Cellular automata, on the other hand, are systems built on simple rules, where complex behavior is derived from the interactions of these automata. Cellular automata can be easily represented graphically in terms of fractals. Recursive iterations of the general equation below produce complexity on a nearly infinite scale [11.13]:

$$f_\lambda(z) = z^2 - \lambda \ .$$

Simple variations on iterations of this equation produce the well-known attractor images of Julia sets and the parameterized, connected Mandelbrot subsets (examples can be seen in Fig. 11.5).

Just as our rules for complex behavior represent systems on many scales, the branching and growth of distinctive elements from seed values may appear sur-

11.2 Examples and Experimental Analysis of Decentralized Systems in Nature

Examples of emergent and decentralized systems that span from the macro- to the nanoscale are prevalent in nature, where independent components with no central governing unit function in a coordinated fashion to produce a higher-order behavior that is absent in the component alone. As a result, these cohesive entities possess an internal intelligence or adaptability to their surroundings that results in a dynamic behavior to a certain set of stimuli. Many of these aspects can be seen in the natural phenomenon of chemotaxis, where embedded sensors and actuators within a cell effect a cascade of downstream reactions that enable the organism to sense their surroundings and move. The underlying concept of emergence has been of much interest to researchers in the field of nanotechnology. Evolving and advancing capabilities in interrogating biological processes that contribute to emergent behavior may enable the eventual mimicry of such phenomena in vitro.

11.2.1 Termite Mounds as Macroscale Decentralized Systems

Macroscale observations of emergent properties found in nature are prevalent and serve as representations of identical phenomena that transcend the length scales down to the nanoscale regime [11.17]. A fundamental macroscopic example is based upon coordinated activity in insect colonies. More specifically, mound formation by termites serves as an example of self-organization, where termites, acting on their own, are capable of collectively forming extremely complex structures. These individually large mounds that possess extremely elaborate architectures are essentially small pellets of dirt that are "glued" together. In a similar fashion to slime mold stalk formation, the formation of termite mounds occurs without a central organizer. Instead, the concept of "stigmergy" is introduced. It is characterized by stimulus-response pairs. For example, as shown in Fig. 11.7, a condition of A elicits a reaction by the termite R. This reaction converts A to A_1. Condition A_1 elicits a response R_1 from *another* termite, which in turn converts A_1 to A_2. This condition then continues to produce a new response from another termite. This process explains how a mound can undergo morphological changes to produce arcs or other unique properties based upon the multitude of stimuli that can be received. This example of mound formation was simulated using the StarLogo program [11.16]. The conditions stipulated by this model

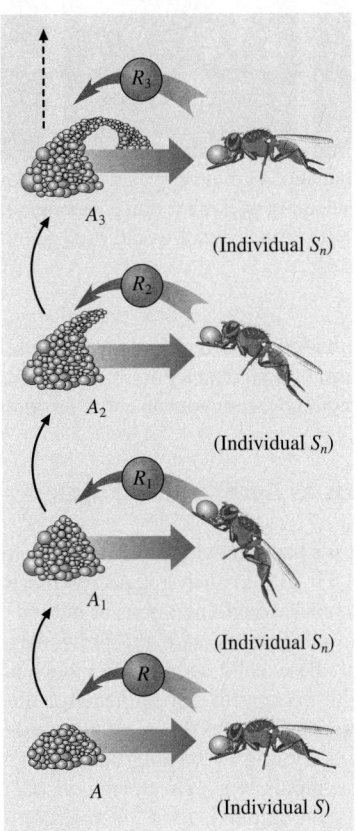

Fig. 11.7 Process of termite mound formation. Condition of A elicits a reaction by the termite R. This reaction converts A to A_1. Condition A_1 elicits a response R_1 from *another* termite, which in turn converts A_1 to A_2. This condition then continues to produce a new response from another termite. This process explains how a mound can undergo morphological changes to produce arcs or other unique properties based upon the multitude of stimuli that can be received (after [11.16])

utilized randomly walking termites among a set of randomly scattered wood chips. The rules declared that if a termite bumped into a wood chip, it would carry the woodship around and unload it at the location of other stacked woodchips. Figure 11.8 shows the StarLogo depiction of the transition from a random state toward a more organized state that was created without the need for a central computer or organizer. In this example, a set of simple rules resulted in the formation of com-

Fig. 11.8 StarLogo simulation of mound formation. To demonstrate a decentralized behavior, the conditions stipulated by this model utilized randomnly walking termites among a set of randomnly scattered wood chips. The rules defined that if a termite bumped into a wood chip, it would carry the woodchip around and unload it at the location of other stacked woodchips (after [11.16])

plex macrostructures based upon termite–environment interactions in the form of a feedback circuit that could be propagated and amplified throughout an entire termite colony.

11.2.2 Slime Molds as Decentralized Systems

Slime mold behavior is a key example of a decentralized system in nature [11.19–21]. In such a capacity, these molds can be considered as conglomerations of individually acting cells. In congruence with the previously mentioned notion of these cells serving as independent components with no central governing unit, the slime molds are emergent structures that react to assemble and disassemble according to existing parameters. In the case of slime molds, e.g., *Dictyostelium discoideum*, one of the prime stimuli is food. In conditions where food is readily available, the slime mold cells exist independently as microscale amoebas. They are capable of migrating through the environment to feed on bacteria and can reproduce. However, under conditions when food is scarce, the slime mold behavior changes and is characterized by an aggregative process. At this point, the individual cells stop reproducing and begin a synchronization process through the emission of cyclic AMP chemical signals, or "waves," and move toward one another in a spiral fashion. More specifically, the individual cells secrete the chemicals, and the cells aggregate in regions of higher chemical concentrations. A resultant clustered formation containing tens of thousands of cells, often called a "pseudoplasmodium," is created. At this point, the cell aggregate behaves as a single, multicellular unit instead of a mere collection of unicellular creatures. The slime mold aggregate is now capable of undergoing morphological changes as well as crawling to seek out a more favorable environment that may contain food. Once this

area is located, the aggregate then differentiates into a stalk supporting a round mass of spores, which can give rise to another group of slime mold cells. This phenomenon was modeled using NetLogo, a program that was developed to simulate cluster formation with no coordination from a central control complex. The central theme of this simulation was very similar to the Belousov–Zhabotinsky (BZ) reaction based on chem-

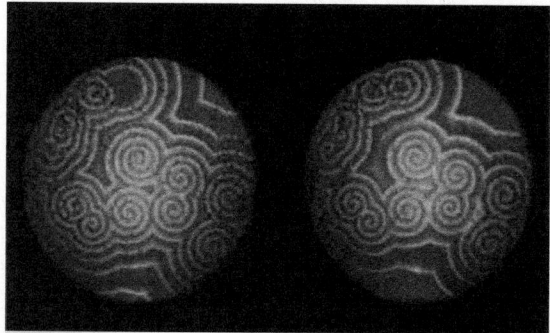

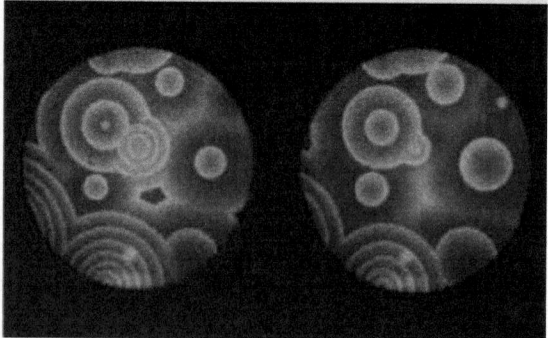

Fig. 11.9 Chemical waves and synchronization. Chemicals released from cells elicit pattern formations that propagate the signal to surrounding cells, which eventually undergo the same behavior, thus resulting in a self-organization of the system (after [11.18])

ical wave oscillation. Under this condition, chemicals released from cells elicit pattern formations that propagate the signal to surrounding cells that eventually undergo the same behavior, resulting in self-organization of the system. Figure 11.9 shows this process as performed by *Winfree* and colleagues [11.18]. A similar approach was simulated by *Wilensky* and colleagues based upon a sick/healthy/infected wave propagation study [11.22]. Using this approach, the components cells were given one of three states represented by different colors, where black represented a "healthy" state, red represented an "infected" state, and white represented a "sick" state. Each cell was programmed to follow three specific rules: (1) If the cell was sick, it became healthy; (2) if the cell was healthy, it became infected as a function of its neighboring cells; and (3) an infected cell would increase infection levels as a function of its neighbors. This phenomenon of chemical oscillations used to induce pattern formation and self-organization is shown in Fig. 11.10. Figure 11.11 shows the resultant NetLogo correlation with the slime-mold-aggregation process based upon the BZ reaction. The cells (red circles) emit a chemical represented by green shades. The cells then move toward regions of concentrated chemical to form the aggregation. In summary, the slime-mold-formation process represents

a decentralized system, where the spiral formation is not orchestrated by specialized or designated cells. The combination of chemical wave propagation and local cell communication is essentially a collection of simple reactions that drive the development of a complex adaptive system. This phenomenon has been well characterized in nature and may serve as a central theme for design principles that are applied to engineering a cell mimetic complexity and intelligence into future nanosystems.

11.2.3 The Complex Adaptation of Cellular Behavior

The previous examples of emergent systems were based upon the analyses of amoebas or termites as independently functioning units that could coalesce without the guidance of a central organizer to produce a higher-order functionality. In nature, another core example of a similar relationship is found in the neutrophil, or white blood cell. Taken individually, cellular components may possess fairly simple functions or no function at all. For example, receptor-mediated chemical detectors on the membrane surface are not capable of performing complex activity on their own, for any kind of introduced stimulus (e.g. cAMP) would have no sig-

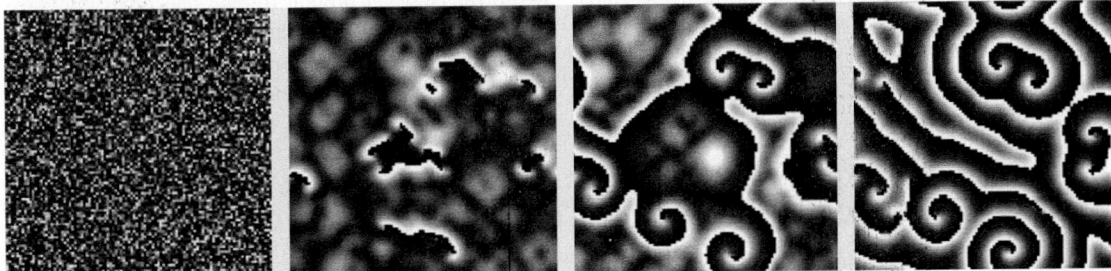

Fig. 11.10 NetLogo modeling of chemical waves. This phenomenon of chemical oscillations is used to induce pattern formation and self-organization. This phenomenon was modeled using NetLogo, a program that was developed to simulate cluster formation without any coordination from a central control complex (after [11.18])

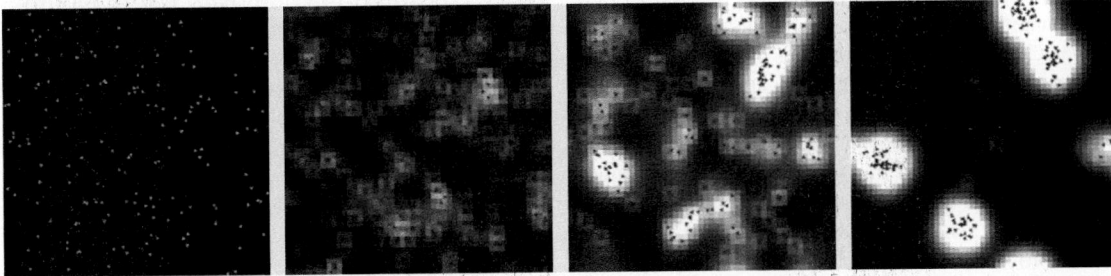

Fig. 11.11 Computer simulation of slime mold behavior. The cells (*red circles*) emit a chemical (*green shades*). The cells then move toward regions of concentrated chemical to form the aggregation (after [11.22])

nal transduction pathway on which to act. However, when the coordinated activity of the embedded sensors (e.g., chemoattractant-responsive membrane-bound elements), actuators (cytoskeleton), and internal circuitry (signal transduction pathways) coordinate their functionalities, an extremely complex process can be realized. In this case, a foreign agent such as a bacterium releases a chemoattractant that is sensed by the cell membrane. This elicits a response to form membrane-based projections around the foreign particle, which is supported by the cytoskeleton. The resultant engulfing of the bacterium represents a vital immunological process known as phagocytosis. Achieving this type of coordinative behavior is one of the goals of synthetic biology-based nanotechnology. The decentralized aggregation of information presented by each component of the phagocytic process is a key example of a design principle with which these engineered systems may be realized.

11.2.4 Protein Folding as an Emergent Process

Proteins serve as the machinery and building blocks that drive the processes of life. There are on the order of 100 000 different proteins that are synthesized within the human body, and the range of the different morphologies they possess is correlated to the diversity of their activity. Proteins usually self-assemble to attain the conformational structure that they utilize to perform

their respective jobs. For example, pore proteins fold into three-dimensional cylinderlike configurations. Furthermore, these pores usually have the added complexity in design that lines the pores with specific amino acid residues to enable the protein to be selective for specific ions. Other proteins, such as hemoglobin, exist to transport oxygen within the blood. ATP synthase possesses moving parts that mimic an engine rotor while providing the bioenergetic fuel for a cell's survival. The amino acids that comprise the proteins represent the individually acting molecules that have been described for the slime mold and termite mound. There is no central organizer that resides within a newly synthesized protein molecule. The resultant amino acids autonomously interact to find the folding conformation that possesses the lowest energy, or most stable state. An example of this would be the hydrophobic (water fearing) amino acid residues forming clusters that reside within the core of the folded protein and away from water while the hydrophilic amino acids line the surface of the protein and pore. Interactive bonds between amino acids in the form of hydrogen bonding also result to preserve the folded structure of the protein. The final shape of the protein correlates with its emergent functionality as a result of a synchronization of the individual amino acid components. A high degree of correlation can be drawn between protein folding and other naturally occurring decentralized processes. Through a combination of localized interactions and molecular clustering, a higher-order complexity is derived.

11.3 Engineering Emergent Behavior into Nanoscale Systems: Thematic Examples of Synthetic Decentralized Nanostructures

The ability to engineer complex behavior into artificial systems will be key to the progressive course of nanotechnological development. With a fundamental understanding of complexity, emergence, intelligence, and adaptability, the ability to derive a higher-order functionality from a collection of interfaced components will be closer to fruition. Examples of the experimental observation of emergent behavior can span several areas. However, one of the central themes underlying the engineered adaptability is the ability to derive a higher-order functionality or output (e.g., motility, energy production, self-organization) based upon a stimulating input (e.g., chemical additive, light, altered solvent condition). Evaluated individually, the components that make up these engineered systems would not exhibit the properties observed from the collective. As such,

without a central governing body, the observation of these outputs represents an engineered realization of complexity.

11.3.1 Engineering Composite Nanosystems for Biomotility Applications

To illustrate the rules of engineering complex emergent systems consider the major locomotion mechanism on the cellular level, actin-based motility.

Actin-based motility is the most important cytoskeleton-mediated systems for cell movement and cell shape change. Actin is one of the most prevalent and tightly conserved proteins found in eukaryotes. It is responsible for the rigidity of cell walls, as well as their flexibility and is found in abundance in all cells of the human body. Given the appropriate factors, such as actin binding proteins, ATP, and magnesium, actin will self-assemble to form long crosslinked polymer chains. This network provides the scaffold of the cytoskeleton as well as the driving force behind amoeba lamellipodium. Additionally, the ActA transmembrane protein of *Listeria monocytogenes* can induce actin assembly on the bacterial surface in their host cell cytoplasm [11.23]. The ActA protein interacts with vasodilator-stimulated rofilin-protein (VASP), a eukaryotic cytoplasm protein responsible for localizing rofiling–actin protein complexes at the barbed end of the actin filament [11.24]. In addition, ActA can interact with Arp 2/3, the actin polymerization nucleation complex responsible for filament branching and network formation [11.25]. Consequently, ActA can initiate actin polymerization as a core actin nucleation factor in conjunction with several other contributing actin binding proteins. These include cofilin for actin depolymerization, α-actinin for stabilizing filamentous actin, and capping protein for blocking the barbed end sites of growing filaments to provide stability [11.26]. Amoeboid movement requires all of these essential proteins, with the exception of ActA, for movement (WASP performs the identical function). A system of at least seven proteins interacting in an independent but self-assembled manner yields a network capable of up to 4 nN of polymerization force. We have shown that this force can be used to do work on hybrid nanoscale devices in the form of physical displacement [11.27]. Actin motility allows single-celled organisms, like slime mold, to perform chemotaxis by moving up a cAMP gradient toward their food. Signal transduction occurs via cAMP membrane receptors (G-proteins) that start a cascade of activated motile machinery, such as the actin nucleation factor WASP [11.28]. Eventually the proteins begin to assemble more rapidly in the direction of the greater cAMP concentration, moving the amoeba in its desired direction (Fig. 11.12).

We can see that this coordinated protein assembly has all the makings of a complex emergent and adaptive system on the nanoscale. It contains a large pool of agents in the form of actin protein molecules, as well as a diverse number of other actin binding proteins. This interaction creates a global behavior that is manifested in locomotion, and this motion is perceived as intelligent because the G-protein cascades permit directed responses to the organism's environment.

11.3.2 Self-Directed Supramolecular Assembly from Synthetic Materials as a Model for Emergent Behavior

Amphiphilic block polymeric chains retain the ability to self-assemble into various shapes and morphologies depending on various factors including solvent composition or modifications to the copolymer block properties. This fact demonstrates an important concept that the input of external stimuli (e.g., varying solvent ratios) can result in differing outcomes such as torus formation, polymer membrane nanotube formation, or vesicle formation, to name a few. Factors that govern the formation of these systems include charge properties of the block residues and how these properties interact with the solvents in which they are immersed. In addition, folding mechanics and thermodynamics also play a role in the attained conformation. In fact, the self-aggregative nature of these synthetic molecules can in some respects be viewed as a mimetic demonstration of a mechanism found in protein folding. This intrinsic response and complexity within these polymers for self-directed activity as well as tunable properties offers a compelling argument for the observation of adaptability and higher-order behavior in engineered systems.

A fundamental piece of work that first demonstrated the diverse morphologies that could be attained from identically composed linear block copolymer chains that varied only by block lengths was published by *Zhang* et al., who reported an unprecedented number of morphologies that could be obtained using a polystyrene-polyacrylic acid (PS-PAA) diblock copolymer [11.29]. Specifically, six different shapes were obtained including spheres, rods, lamellae, vesicles, micellelike aggregates, and microscale spheres with hydrophilic surfaces and enclosed micellelike aggregates. In addition, when aqueous solutions of the spherical micelles were dried, a needlelike structure could be obtained as well. By manipulating the asymmetry between the PS-PAA blocks, shape changes could be induced (Fig. 11.13). For example, the most symmetrical structure defined as $200 - b - 21$ (PS-b-PAA) formed spherical micelles. Upon decreasing the PAA block

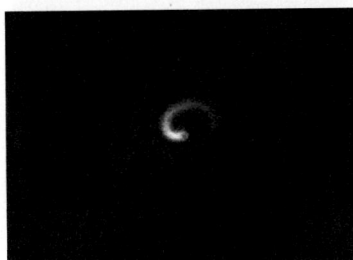

Fig. 11.12 Actin polymerization on polystyrene bead (after [11.27])

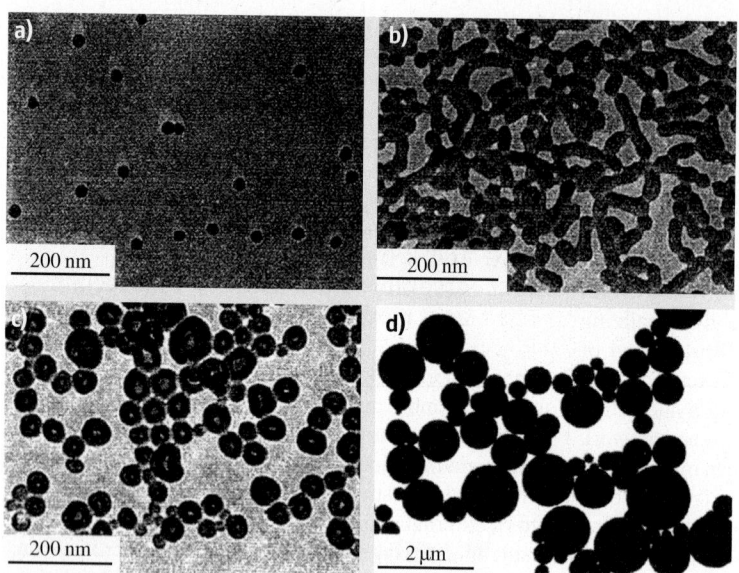

Fig. 11.13a–d Morphologies of PS-PAA block copolymer. Various morphologies that were observed based upon varying the length of the PAA block. (**a**) Spherical micelles. (**b**) Transition from a spherical to rodlike configuration. (**c**) Rodlike to vesicular transition. (**d**) Transition from vesicular to spherical, large, and highly polydisperse group of particles (after [11.29])

$(200 - b - 15)$, the spherical morphology changed from spherical to a rod-shaped structure. Using a $200 - b - 8$ complex, vesicular objects resembling hollow spheres were observed instead of rods. A feasible explanation from a thermodynamic standpoint for how the shape changes are effected was offered. Three major components served as contributing factors to the thermodynamics of aggregation including the PAA-based corona interaction with the solvent, PS core, and the core-solvent interaction. It was observed that the core underwent a large degree of stretching depending on which morphology was created by the polymer. This degree of stretching may have accounted for the shape transitions of the polymer complexes. It was shown that the greatest degree of core stretching was encountered during the spherical state, as the ratio of the core radius to the chain end-to-end distance was 1.4. This ratio decreased to 1.2 as the morphology changed from that of a sphere to a rodlike structure. This ratio decreased further to 1.0 as vesicles or lamellae were formed. Previous experimental and theoretical work [11.30, 31] has shown that the radius of the spherical core increases with a decrease in the PAA component while the PS block length is held constant (which could be correlated to an increased stretching of the PS block). As this stretching decreases the PS chain entropy in the core, a breaking point is eventually reached. Beyond a certain magnitude of stretching, the thermodynamic stretching constraint drives the transition from shape to shape. This elasticity concept was also explored

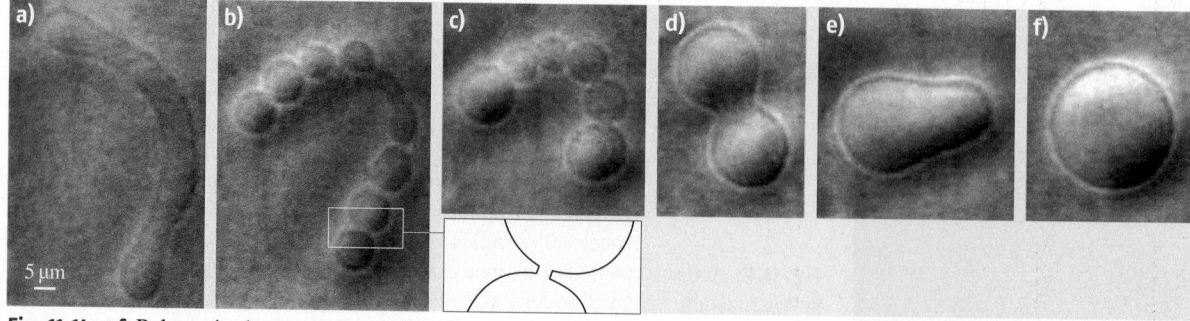

Fig. 11.14a–f Polymeric shape transformation. The changes in shape of the polymer were driven by osmotic swelling, which was manipulated by progressively diluting the external sucrose concentration to prompt polymer swelling (**a**)–(**f**). The tube-shaped polymer transitioned from a series of interconnected, smaller-diameter vesicles to a large, single vesicle (after [11.32])

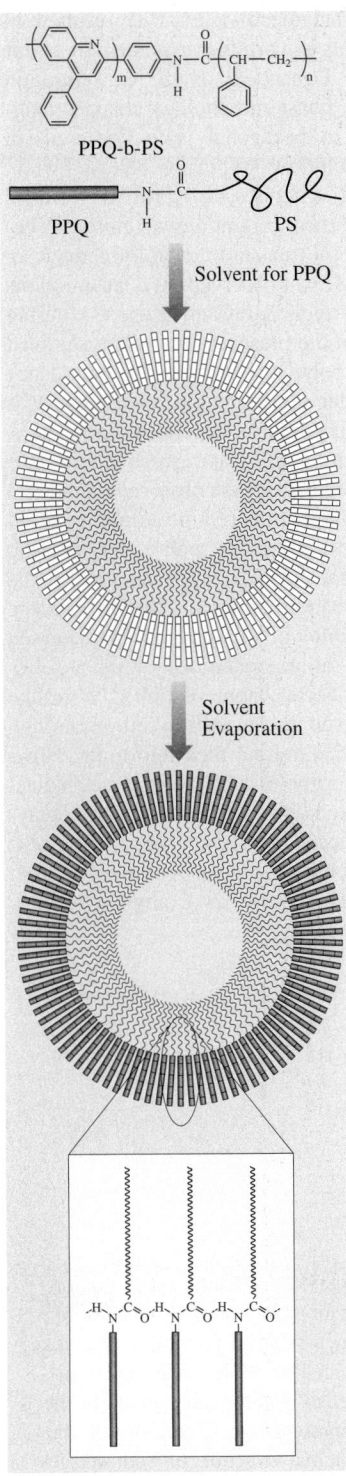

Fig. 11.15 Schematic of polymer structure. The structure of the PPQ-PS copolymer and how it self-assembled into a hollow core aggregate (after [11.33])

by *Discher* and colleagues who tuned the morphology of diblock copolymer vesicles (polyethyleneoxide-polyethyletheylene; $EO_{40}EE_{37}$) [11.32]. By manipulating the external osmolarity of a deflated tubular vesicle, water permeated and swelled the tubule to induce a transitional cascade from multiple small vesicular pearls to pears to buds and finally the large vesicle/tensed state. This metamorphosis was attributed to the energetic limitations found within the vesicle area, the number of membrane molecules comprising the area, intravesicular volume, and the membrane elasticity (Fig. 11.14).

Continued work has also demonstrated that amphiphilic linear polymers can self-assemble to form various morphologies depending on the *solubilization parameters* as well as block lengths and compositions. *Jenekhe* and colleagues utilized a poly(phenylquinoline)-*block*-polystyrene rod-coil diblock copolymer (PPQ-b-PS) to observe induced formation of spheres, vesicles, cylinders, and lamellar accumulations and have also observed photoactive polymeric properties (Figs. 11.15 and 11.16) [11.33, 34]. By modifying the lengths of the rigid rod region of the polymer (PPQ), size scales of the resultant structures could be altered. Typically, the reduction of the PPQ

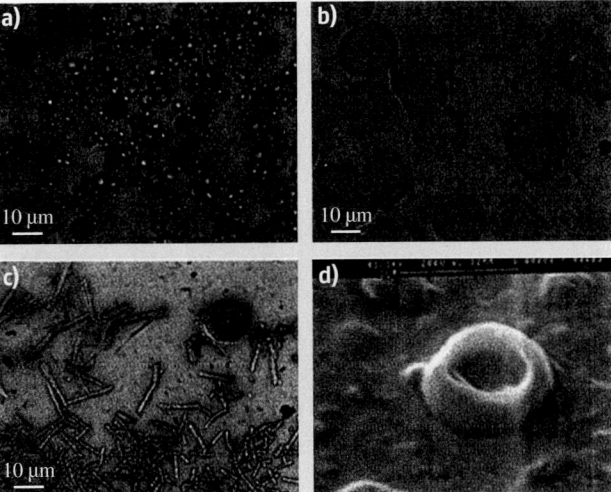

Fig. 11.16a–d Morphology change in polymer based on change in solvent concentrations. Optical and scanning electron micrographs were used to acquire images of changes in polymeric morphologies attributed to altering the concentration of solvents [Trifluoroacetic acid (TFA):Dichloromethane (DCM)] in the system. In (**a**), spherical aggregates were observed under the conditions of 1:1 TFA:DCM v/v at 95 °C. (**b**) Lamellar structures using 1:1 TFA:DCM v/v at 25 °C. (**c**) Cylindrical structures using 9:! TFA/DCM, at 25 °C. (**d**) Vesicles using 1:1-1:4 TFA:DCM at 25 °C (after [11.33])

blocks resulted in a decrease in shape dimensions. Ultimately, this work demonstrated the tuning properties of the amphiphilic polymer according to PPQ length and ionization. As an example, it was shown in [11.35] that protonation of the imine nitrogen the PPQ block could be converted to a polyelectrolyte. In addition, it was shown that the π-conjugated nature of the PPQ block presented various electro- or photoactive properties to the polymers. As such, the ability to reconcile polymer morphologies due to a range of experimental conditions made these diblock copolymers an ideal example of stimulus-induced, self-directed molecular activity.

The use of PPQ-selective solvents in varying ratios such as trifluoroacetic acid (TFA) or dichloromethane (DCM) resulted in spherical, vesicular, cylindrical, and lamellar aggregate formation. On the other hand, the use of nonselective solvents including nitromethane with GaCl$_3$ did not result in any self-assembly. Fluorescence measurements of a labeled PPQ block of the packing conformations of the polymer resulted in varying emission spectra that were dependent upon the shape being studied. This in turn indicated that the morphologies exhibited altered packing conformations of the PPQ block. Other environmental parameters including solvent drying conditions produced predominantly nonspherical structures (lamellar, cylindrical). The tuning of morphological outcomes of these polymers demonstrates a set of properties (protonation, block length) that can determine the emergent attributes of linear block copolymers.

Other work by *Won* and colleagues using the low-molecular-weight diblock copolymer poly-(ethylenoxide)-poly(butadiene) (PEO-PB) resulted in the formation of large wormlike micelles in water (Fig. 11.17) [11.36]. Depending on the concentration of PEO-PB in wt %, phase morphology changes ranging from isotropic to hexagonal were observed. In low-concentration solutions, typically below 5 wt %, nonlinear cylindrical structures, or wormlike micelles were observed. As the concentration was raised to between 5 and 10 wt %, a transition toward a nematic or one-dimensional phase was observed. At concentrations above 10 wt %, small-angle X-ray and neutron scattering were used to confirm the presence of a self-assembled hexagonally packed polymer structure. The emergence of domain or molecular aggregation in this instance was brought about by concentration-dependent parameters. In addition, through the addition of a crosslinker in solution, these wormlike micelles were observed to undergo a crosslinking reaction that altered the solubility properties of the micelle network as well as viscoelastic attributes. Free radical polymerization resulted in the inability to solubilize in chloroform. Furthermore, the dynamic elastic modulus of the material G' increased by over two orders of magnitude to attain a solid-like property. To elaborate, noncrosslinked wormlike micelles were observed to be able to move through other micelles by breaking and then reforming. However, the crosslinked samples underwent a mechanical stiffening to attain fixed rubberlike properties. Displaying both concentration and crosslink reaction-dependent morphological responses, diblock copolymer-based wormlike micelles also serve as examples of how

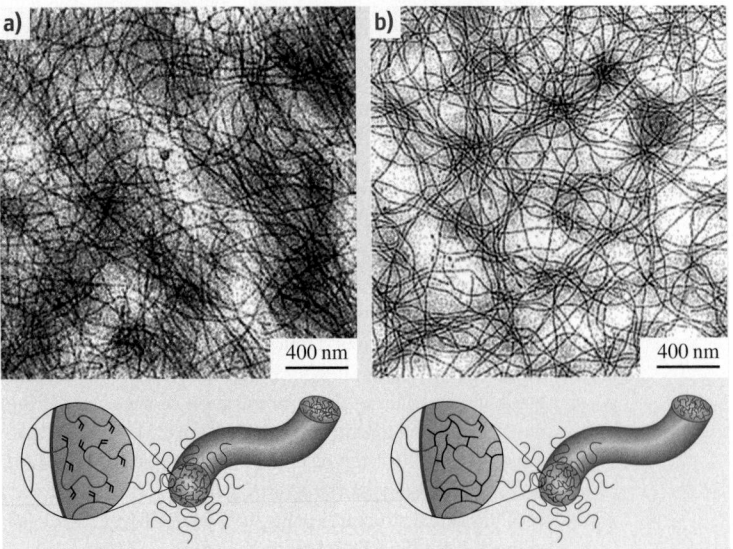

Fig. 11.17 Wormlike micelle formation in water. Formation of uncrosslinked polymeric wormlike micelles, while crosslinked micelles. Below each image is the representative illustration of the internal structure of both species (after [11.36])

F(CF₂)₈(CH₂)₄O

1

F(CF₂)₈(CH₂)₄O

F(CF₂)₈(CH₂)₄O

F(CF₂)₈(CH₂)₄O

2

3   **R:C₁₂H₂₅**

4   **R:C₁₂H₂₅**

copolymers can exhibit intrinsic higher-order behaviors induced by modifications to their surroundings.

Fig. 11.18 Monodendron structure. The structure of the tapered monodendrons is represented by 1 and 2, while the conical monodendron is represented by 3 and 4 (after [11.37])

Supramolecular construction of synthetic materials through molecular recognition events and self-assembly guided by thermodynamic control has been applied to liquid crystalline materials as shown by *Hudson* and colleagues [11.37]. They utilized a liquid crystalline monodendron (Fig. 11.18) to produce self-organized cylindrical and spherical structures. The initial structure of the monodendron was a determining factor in the resultant supramolecular geometry that was observed as shown in Fig. 11.19. The flat and tapered molecular structure contributed to the resultant circular aggregation, which could be stacked to form a cylindrical dendrimer. These dendrimers in turn were able to assemble into thermotropic hexagonal columnar structures classified as a p6mm Φ_m liquid crystal supramolecular assembly. The conical monodendrons were capable of assembling into spherical dendromers that in turn aggregated to form a Pm3̄n cubic liquid crystalline supramolecular configuration (Fig. 11.20). Figure 11.21 shows the phase contrast micrographs of these configu-

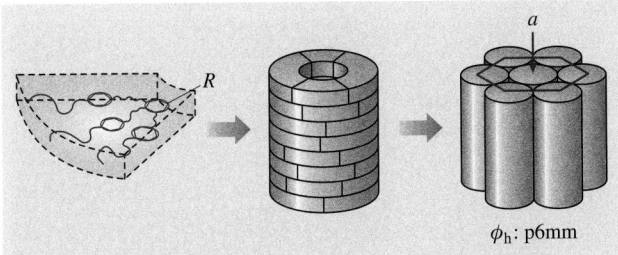

Fig. 11.19 Cylindrical dendrimer formation. Formation of the cylindrical dendrimer from the planar, tapered monodendron and the subsequent formation of the LC supramolecular assembly are represented in this figure (after [11.37])

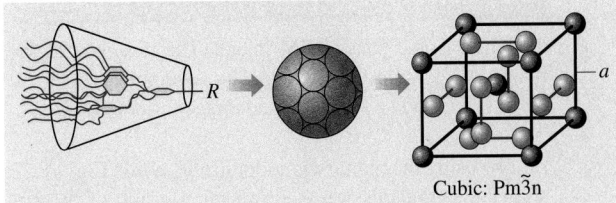

Fig. 11.20 Formation of spherical dendrimer from conical monodendrons. The formation of the spherical dendrimer from the conical monodendrimer is shown here followed by the subsequent formation of the cubic supramolecular structure (after [11.37])

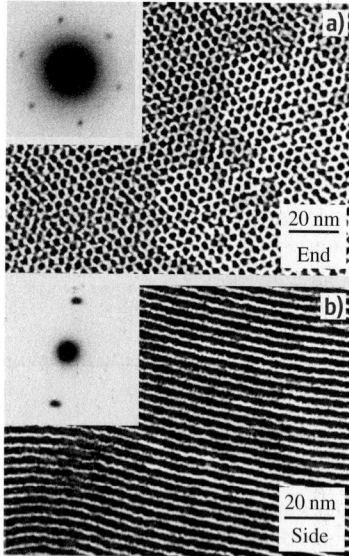

Fig. 11.21a,b Phase contrast imagery of cylindrical structures. (**a**) Homeotropic view of LC supramolecular assemblies self-organized by planar tapered molecule (structure 1). (**b**) Images of planar-aligned LC assembly. In both instances the *inset* represents the electron diffraction (ED) pattern. (after [11.37])

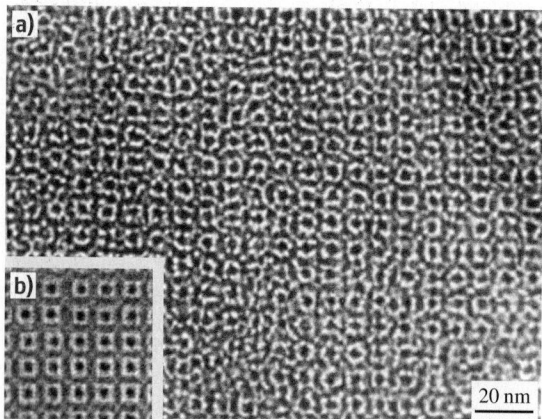

Fig. 11.22 Cubic phase TEM image. The raw data of the image of the cubic supramolecular assembly are shown here. The *inset* is a reconstructed image (after [11.37])

rations using the planar monodendrons, while Fig. 11.22 shows the transmission electron micrograph of the cubic liquid crystalline phase. This work elucidated the fundamental concept that utilizing basic design principles that can be governed by thermodynamic control and molecular recognition can serve as an efficient and

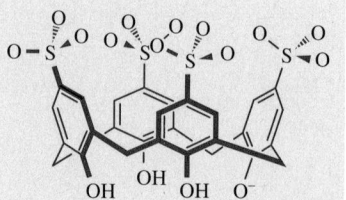

Fig. 11.23 Molecular structure. The structural illustration of the *p*-sulfanatocalix[4]arene anion is represented here with a pyramidlike shape and hydrophobic cavity connected by four aromatic rings (after [11.38])

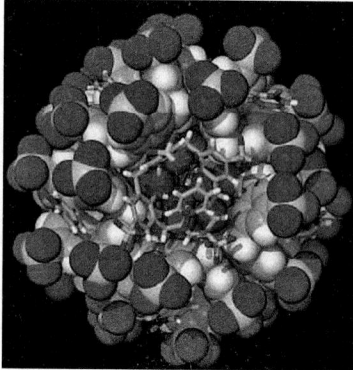

Fig. 11.24 Spherical conformation. Spherical suprastructures consisting of 12 macrocyclic anions can be formed in the presence of pyridine and the Ln^{3+} in a 2:2:1 ratio (after [11.38])

simple mechanism for the assembly of higher-order systems. This process is not unlike those found in nature that govern the assembly of biological systems.

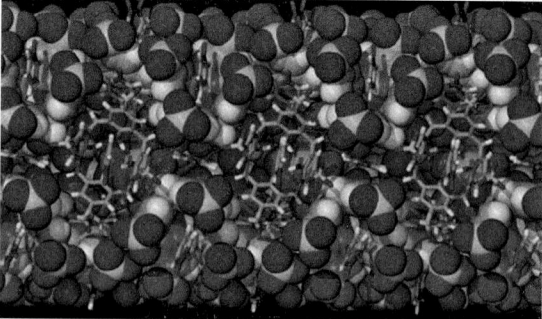

Fig. 11.25 Tubular assembly. The steric conformations created by the lanthanide and pyridine ions resulted in the formation of cylindrical or tubular structures characterized by a sulfonate-group-lined polar exterior and a hydroxyl-group-based polar interior (after [11.38])

Developing another self-assembly strategy using amphiphilic macrocyclic anions, *Orr* et al. utilized a *p*-sulfanatocalix [11.4] arene anion to create nanoscale spheres and tubules based upon the addition of pyridine *N*-oxide and lanthanide ions [11.38]. In its native form, the anion retained a pyramidlike shape with a hydrophobic cavity that were connected by four aromatic rings (Fig. 11.23). The basal and apical locations of the molecule contain sulfonate and phenolic hydroxyl groups, respectively. These residues serve as the reactive elements of the molecule. Initiators of the aggregation process (pyridine, lanthanide) coordinated with the sulfonic and phenolic hydroxyl groups of the *p*-sulfanatocalix[4]arene to form either spherical or cylindrical morphologies. The determination of which shape was achieved was dependent upon the ratio of *p*-sulfanatocalix[4]arene : pyridine *N*-oxide : lanthanide ions. As an example, when these components were mixed in a 2:2:1 molar ratio, it was observed that two anions had reacted with a lanthanide coordinating ion via their sulfonate groups to form a dimeric assembly. In addition, the pyridine-*N*-oxide molecules interacted with the lanthanide ions and were bound within the calixarene cavities of the other dimeric assembly. As such, it was noted that the lanthanide ion acted as a "hinge" and the steric properties of the pyridine molecules resulted in an autonomous formation of a 60° dihedral angle between the *p*-sulfanatocalix[4]arene molecules. This arrangement in turn enabled the formation of morphologies possessing curved surfaces such as spheres (Fig. 11.24). A similar general process explains how the tubules were formed as the ratio of the coordinating ions was altered. The steric conformations induced by the lanthanide and pyridine ions resulted in the formation of cylindrical structures characterized by a sulfonate-group-lined polar exterior and a hydroxyl-group-based polar interior (Fig. 11.25). In this arrangement, the pyridine molecules intercalated between the aromatic rings of adjacent *p*-sulfanatocalix[4]arene molecules via π-stacking interactions (Fig. 11.26).

This work was important in further elaborating upon the mimicry of assembly processes found in nature through the cohesive interaction of three diverse molecules. The proposed approach offers the opportunity to utilize properties such as metal-ligand coordination, hydrogen bonding, and van der Waals forces to drive the nanoscale assembly of diverse structures with well-defined morphologies with extreme molecular precision. The ability of the macrocyclic anions to self-coordinate based upon the input of the pyridine and lanthanide ions demonstrates that the diversity of

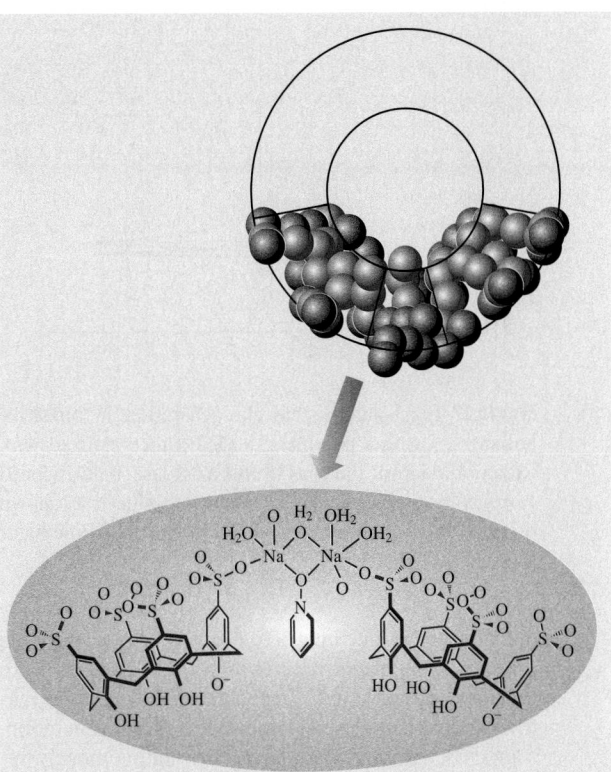

Fig. 11.26 π-stacking configuration. The pyridine molecules intercalated between the aromatic rings of adjacent *p*-sulfanatocalix[4]arene molecules via π-stacking interactions are depicted here using a partial cross-sectional view of a tubular structure (after [11.38])

reactive materials and exploitable properties of the component molecules is shared among the type of synthetic materials exhibiting higher-order behaviors that can be developed. Other studies involving supramolecular self-assembly have also translated the same stimulus-output relationship to block-copolymer-based systems. These experiments have shown that artificial materials possess the dynamic capabilities of adjusting to their environments that are required for potential applications in the areas of cosmetics, drug delivery, chemical sequestering, and beyond.

11.3.3 Engineered Bioenergy Transduction as a Decentralized Process

Coupling energy-transducing behavior of proteins in a polymeric membrane, *Ho* and colleagues produced an independent source of electron generation [11.39,40]. In

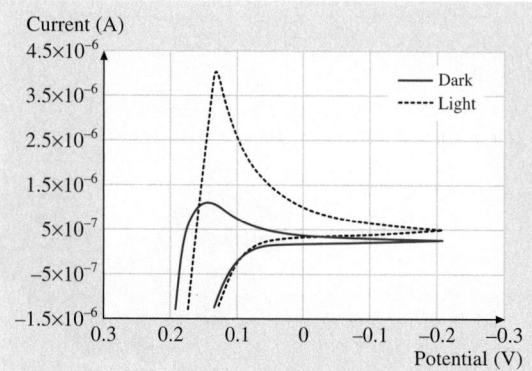

Fig. 11.27 Light-induced current generation in bioactive polymers. Using a polymeric vesicle functionalized with bacteriorhodopsin and cytochrome c oxidase, light-induced currents in excess of $10\,\mu A$ could be produced as shown using cyclic voltammetry and a cytochrome c mediator (after [11.39])

Fig. 11.28 Self-assembling "smart dust." Injection of a dichloromethane solvent droplet in a solution of asymmetric microfabricated particles resulted in the hydrophobic side of the particles interacting with the droplet and the hydrophilic side being exposed (after [11.41])

this experiment, the input of light into a system where the primary components were two bacterial proteins from independent hosts and a reconstitution polymeric matrix resulted in the measurement of electron generation, a process that could not have been autonomously realized had the proteins been evaluated individually. As such, this coupled bioenergy transduction process could then be viewed as a derivation of higher-order behavior.

Bacteriorhodopsin was selected as a light-dependent transmembrane proton gradient generator. Cytochrome c oxidase was selected as the electron-generating species. In its native forward reaction, cytochrome c oxidase serves as the last step of the respiratory chain, whereby it receives an electron from the cytochrome c mediator. This electron transport mechanism is coupled with a proton-pumping function of the protein. However, it was previously established by *Wikstrom* and colleagues that by introducing a high electron acceptor potential from the cytochrome c mediator, as well as a high electrochemical proton gradient (in this case, produced by bacteriorhodopsin), the activity of cytochrome c oxidase protein could be partially reversed [11.42]. This would entail that the protons from the external gradient would travel back through the protein, and this process would be coupled with the transfer of electrons back to the cytochrome c mediator. This work intended to harvest those electrons for current measurement. Through the use of composite polymeric vesicles containing both proteins that resided in a solution with a high concentration of cytochrome c, it was shown that light-induced

current increases in excess of $10\,\mu A$ could be generated (Fig. 11.27). While one of the chief potential impacts behind bioenergy sources is based on the generation of high-power density systems, several key issues would need to be addressed to reach that point in the roadmap. This would include the capability of prolonging protein functionality in the ambient environment as well as demonstrating that current output of the composite system could be enhanced such that practical everyday electronics could be powered by protein-based activity. While there is still work to be done, this study did succeed in serving as an example of how, using a synthetic biology approach, the resultant structures, in this case the composite polymeric vesicles, could process an input (light) to produce outputs (current) that would not normally be present if the system were deconstructed and each component analyzed individually. As such, the composite "nanomachine" comprised of three components (bacteriorhodopsin, cytochrome c oxidase, and the polymeric membrane) exhibited a complex behavior to produce useful work.

11.3.4 Emergent Behavior from "Smart Dust"

Work performed by *Link* and colleagues showed that chemically modified microfabricated structures could respond (self-directed aggregation) to an external stimulus (dichloromethane droplet) in a decentralized fashion [11.41]. To demonstrate this concept, an asymmetric silicon plate was constructed whereby one side of the wafer was made porous (green) through a chemical etch and filled with hydrosilylate to induce a hydrophobic property. The other side of the water was chemically etched and oxidized to produce a hydrophilic property (red). The wafer was then sonicated to produce "smart dust" amphiphilic particles in solution. The addition of a droplet of dichloromethane solvent caused the particles to cluster around the droplet. The hydrophobic side of the particles interacted with the droplet to expose the red component of the wafer Fig. 11.28. Again, through the use of simple rules, an emergent assembly and synchronization of individual particles was achieved through the introduction of external stimuli. The capacity to enhance the resultant complexity of functionalities that are achieved through these engineered systems will push the boundaries of synthetic emergence.

11.4 Conclusion

This work has outlined the underlying concept of how emergent and complex behavior resides within a range of different systems, both living and nonliving. An in-depth view of current studies involving various materials, from polymers to liquid crystalline structures, has revealed that inherent intelligent and emergent behavior can be extracted based upon an equally wide-ranging set of stimuli, from manipulating molecular charge properties to introducing secondary reactive chemical species. To add another level of complexity to these materials, the added functionality from proteins and other biocomponents has yielded higher-order outputs based on mimicry of natural phenomena. Before the fruits of engineered emergence can be harvested, however, a deeper understanding of each systemic component must be attained. Current nanoscale interrogative modalities such as atomic force microscopy and nanoscale scanning optical microscopy, to name a few, will help the scientific community to probe biological structures and processes with unprecedented capabilities. The culmination of this set of parameters will ultimately establish the field of engineered emergence through synthetic biology as a driving force behind the advent of nanotechnology as an accepted and profitable industry.

References

11.1   R. Descartes: *Meditations on First Philosophy*, J. Cottingham translator (Cambridge University press, UK 1996)

11.2   J. von Neumann: *The Theory of Self-Reproducing Automata*, ed. by A. Burks (University of Illinois, Urbana 1966)

11.3   J. Kennedy, R. C. Eberhart, S. Yuhui: *Swarm Intelligence*, ed. by P. Bentley (Academic, San Diego 2001)

11.4   D. D. a. B. Weber: *Darwinism Evolving: Systems Dynamics and the Genealogy of Natural Selection* (MIT Press, Cambridge, MA 1996)

11.5   S. Johnson: *Emergence: The connected lives of ants, brains, cities and software* (Simon and Schuster, New York 2001)

11.6   S. Wolfram: *The New Science* (Wolfram Media, Champaign, IL 2002)

11.7   J. Holland: *Hidden Order* (Perseus, Cambridge, MA 1995)

11.8   S. A. Kauffman: *The origins of order: self-orginization and selection in evalution* (Oxford University Press, New York 1993)

11.9   S. A. Kauffman: *At Home in the universe: The Search for the Laws of Self-Organization and Complexity* (Oxford University Press, New York 1995)

11.10   M. Gardner: Mathmatical Games, Sci. Am. **223**, 120–123 (1970)

11.11   B. Latane: The psychology of social impact, Am. Psychol. **36**, 343–356 (1981)

11.12   E. Hutchins: *Cognition in the Wild* (MIT Press, Cambridge, MA 1995)

11.13   M. F. Barnsley: *Fractals Everywhere* (Academic, Cambridge, MA 1993)

11.14   M. C. Taylor: *The moment of complexity emerging network culture* (University of Chicago Press, Chicago 2001)

11.15   C. M. Wittenbrink: *IFS fractal interpolation for 2D and 3D visualization* (Proceedings of the IEEE Visualization Conference, Atlanta, GA 1995) pp. 77–84

11.16   This model was developed at the MIT Media Lab. See M. Resnick, "Turtles, Termites and Traffic Jams: Explorations in Massively Parallel Microworlds" (1994), Cambridge, MA: MIT Press. Adapted to StarLogoT, 1997, as part of the Connected Mathematics Project. Adapted to NetLogo, 2000, as part of the Participatory Simulations Project

11.17   R. Doursat: *Computational Models of Complex Systems* (Seminar University of Nevada, Reno 2005)

11.18   A. T. Winfree: Spiral waves of chemical activity, Science **175**, 634–636 (1972)

11.19   E. F. Keller, L. A. Segel: Initiation of Slime Mold. Aggregation Viewed as an Instability, J. Theor. Biol. **26**, 399–415 (1970)

11.20   T. Hofer, P. K. Maini: Streaming instability of slime mold amoebae: an analytical model, Phys. Rev. E. **56**, 2074–2080 (1997)

11.21   T. Hofer, J. A. Sherratt, P. Maini: Cellular pattern formation during dictyostelium aggregation, Physica D. **85**, 425–444 (1995)

11.22   U. Wilensky: *NetLogo Slime model* (Center for Connected Learning and Computer-Based Modeling, Northwestern University, Evanston, IL 1998) http://ccl.northwestern.edu/netlogo/models/Slime

11.23   C. Kocks, E. G. M. Tabouret, P. Berche, J. Ohayon, P. Cossart: L. monocytogenes-induced actin assembly requires the actA gene product, a surface protein., Cell **68**, 521–531 (1992)

11.24   V. Laurent, T. P. L. B. Harbeck, A. Wehman, L. Grobe, B. M. Jockusch, J. Wehland, F. B. Gertler, M-F. Carlier: Role of proteins of the Ena/VASP family in actin-based motility of Listera monocytogenes, J. Cell Biol. **144**, 1245–1258 (1999)

11.25   M. D. Welch, J. R. J. Skoble, D. A. Portnoy, T. J. Mitchison: Interaction of human Arp2/3 complex and the Listeria monocytogenes ActA protein in actin filament nucleation, Science **281**, 105–108 (1998)

11.26   D. Pantaloni, C. C. M.-F. Carlier: Mechanism of actin–based motility, Science **292**, 1502–1506 (2001)

11.27   D. Wendell, J. Yi, S. Schmidt. Freirez, J. H. Nerves, Carlo Montemagno+: Microsphere Dynamics for Actin Based Nanorobotic Motility, IEEE Proceedings on nanotechnology **2**, 725–728 (2003)

11.28   D. Bray: *Cell Movements: From Molecules to Motility* (Garland, New York 2001)

11.29   L. Zhang, A. Eisenberg: Multiple Morphologies of "Crew-Cut" Aggregate of Polystyrene-b-Poly(acrylic acid) Block Copolymers, Science **268**, 727–731 (1995)

11.30   M. D. Whitmore, J. Noolandi: Theory of micelle formation in block copolymer-homopolymer blend, Macromolecules **18**, 657–665 (1985)

11.31   R. Nagarajan, K. Ganesh: Block copolymer self-assembly in selective solvents: Spherical micelles

11.32   B. M. Discher, Y. Won, D. Ege, J. Lee, F. Bates, D. Discher, D. Hammer: Polymersomes: vesicles made from diblock copolymers, Science **284**, 1143–1146 (1999)

11.33   S. A. Jenekhe, X. L. Chen: Self-assembled aggregates of rod-coil block copolymers, their solubilization, encapsulation of fullerenes, Science **279**, 1903–1907 (1998)

11.34   S. A. Jenekhe, K. J. Wynne (Eds.): *Photonic and Optoelectronic Polymers* (American Chemical Society, Washington, DC 1997)

11.35   P. D. Sybert, W. H. Beever, J. K. Stille: Synthesis and properties of rigid-rod polyquinolines, Macromolecules **14**, 493–502 (1981)

11.36   Y. Won, H. T. Davis, F. S. Bates: Giant wormlike rubber micelles, Science **283**, 960–963 (1999)

11.37   S. D. Hudson, H.-T. Jung, V. Percec, W.-D. Cho, G. Johansson, G. Ungar, V. S. K. Balagurusamy: Direct visualization of individual cylindrical and spherical supramolecular dendrimers, Science **278**, 449–452 (1997)

11.38   G. W. Orr, L. J. Barbour, J. L. Atwood: Controlling molecular self-organization: formation of nanometer-scale spheres and tubules, Science **285**, 1049–1052 (1999)

11.39   J. Xi, D. Ho, B. Chu, C. D. Montemagno: Learned From Engineering Biologically-Active Hybrid Nano/Micro-devices, Adv. Funct. Mater. **15**(8), 1233–1240 (2005)

11.40   D. Ho, B. Chu, H. Lee, E. Brooks, K. Kuo, C. D. Montemagno: Light-dependent current generation based on a coupled protein functionality, Nanotechnol. **16**(12), 3120–3132 (2005)

11.41   J. R. Link, M. J. Sailor: Smart dust: self-assembling, self-orienting photonic crystals of porous Si, Proc. Natl. Acad. Sci. USA **100**, 10607–10610 (2003)

11.42   M. Wikstrom: Energy-dependent reversal of the cytochrome oxidase reaction, Proc. Natl. Acad. Sci. USA **78**, 4051–4054 (1981)

12. Nanometer-Scale Thermoelectric Materials

The thermal and electrical transport properties of solids are strongly affected by a system's dimensionality. While many other reviews deal with the electrical conductivity of nanowires, nanotubes, and solids that include quantum dots, we concentrate here on their thermal and thermoelectric properties. Not only is this of fundamental scientific interest, but nanometer-scaled solids also hold the promise of leading to a breakthrough in thermoelectric technology.

The theory of transport properties essentially derives from Boltzmann's theory: the amount of charge or heat transported in a solid is given by the amount carried by each individual electron or phonon, times the flux of those particles. This latter quantity depends on the density of particles—and thus on the density of available states—and on their group velocity, which is limited by their mass and the scattering mechanisms. The dimensionality affects both the density of states and the scattering mechanisms.

As far as practical applications are concerned, it has recently been experimentally demonstrated that the thermoelectric efficiency of superlattices and of superlattices containing quantum dots can be double that of conventional solids. In three-dimensional solids, the three quantities that determine the efficiency, the electrical and thermal conductivities and the Seebeck coefficient are related and must be optimized as a group. The dimensionality of the system adds an independent

variable that favorably modifies this relation and thus enables a more refined optimization.

The first transport property investigated in low-dimensional systems is typically the electrical resistivity. Indeed, it is usually quite difficult to contact nanoscale solids, and electrical contacts are simpler to make than thermal contacts. The thermal and thermoelectric properties are at least as fundamental; in particular, the product of the charge and the thermoelectric power is the entropy of the charge carrier [12.1]. The thermoelectric power, or Seebeck coefficient, is also more sensitive than the electrical conductivity to the details of the energy-band structure and the scattering mechanisms because it is related via the *Mott* [12.2] relation to the energy derivative of the electrical conductivity of the charge carrier system. On top of this fundamental interest in low-dimensional thermoelectric phenomena is a practical aspect in energy conversion. Thermoelectric materials can generate electrical power from heat and use electricity to function as heat pumps providing active cooling or heating. The "working fluid" in these energy converters consists of the conduction electrons.

The field of thermoelectric and thermomagnetic phenomena has grown in spurts. After the early experimental discoveries by *Seebeck* et al. in 1823–1836 came the contributions of thermodynamics to the understanding of thermoelectricity, around 1851–1887, with work by Lord Kelvin, Boltzmann, and Nernst. Next came the development of the theory of thermoelectric transport phenomena in semiconductors in the period 1947–1960 with contributions by *Telkes* [12.3] and *Ioffe* [12.4, 5]. The conventional thermoelectric materials developed then [12.6] still dominate the field today (except for the recent developments in skutterudite materials [12.7, 8]). The classical monographs [12.9, 10] on the subject can be traced to that period as well. It is quite possible that the recent progress made in nanometer-scale thermoelectric materials will be seen as the next such advances: it was suggested [12.11, 12] in the 1990s that low-dimensional systems should result in materials with much better efficiencies than bulk materials, through low-dimensional effects on both charge carriers and lattice waves. Such improvements have been experimentally demonstrated on Bi_2Te_3/Sb_2Te_3 superlattices [12.13] and on PbTe/PbSeTe and PbSnTe/PbSnSeTe quantum-dot superlattices [12.14] in the early 2000s. We describe these advances in this chapter.

The coefficient of performance of a thermoelectric heat pump and the efficiency of a thermoelectric generator are both described by one fundamental material property, the thermoelectric figure of merit [12.10]:

$$Z = \frac{S^2\sigma}{\kappa}, \qquad (12.1)$$

where S is the thermoelectric power or Seebeck coefficient, κ is the thermal conductivity, and σ is the electrical conductivity. The Seebeck coefficient relates the temperature gradient across the element, ∇T, to the electric field E via $E = S\nabla T$. The product qS, where q is the charge of the carrier, is the entropy of the carrier [12.1, this is actually only rigorously true in high magnetic field [12.15]]; the Seebeck coefficient is thus an equation of state and, unlike the electrical conductivity, does not depend on the path taken by the carrier through the material. The thermoelectric figure of merit Z has the dimensions of inverse temperature; it is usual to multiply it by the average temperature T to yield a dimensionless number ZT. Conventional thermoelectric materials have been limited to $ZT = 1$ at $T = 300$ K; the new nanoscale materials have reached $ZT = 2$ at 300 K [12.14], and

even $ZT = 3$ at 450 K [12.16]. In the next paragraph, we will show the technological significance of this progress.

In a classical bulk thermoelectric material, the three ingredients in the figure of merit, S, σ, and κ, are dependent on each other, and this impedes the optimization of Z. The numerator of Z is the *power factor* $S^2\sigma$, which is dominated by charge carriers. The denominator of Z is the thermal conductivity; as most thermoelectric materials are semiconductors, κ is dominated by conduction of heat via acoustic phonons. A higher density of electrons results in a higher electrical conductivity σ but in a lower S. Longer electron mean free paths improve σ without decreasing S, but usually this is achieved in crystals with a lower density of defects. This in turn implies longer phonon mean free paths, and thus an increase in κ. Before moving on to nanostructured materials, we must point out that it has been formally shown [12.17] that a macroscopic composite of thermoelectric materials cannot have a ZT value larger than one of its components. One cannot optimize ZT by mixing compounds with a high value of S with others that have a high σ or a low κ in the absence of interactions between the compounds, such as size-related effects or interactions between the charge and heat-carrying mechanisms. The power factor $S^2\sigma$, though, can [12.18] be enhanced in a mixture.

The main advantage of using nanotechnologies in the design of thermoelectric materials lies in the fact that the addition of the dimensionality of the system as a new parameter can influence the interdependence of the transport coefficients. There are two basic mechanisms through which ZT is improved in low-dimensional systems:

1. An increase in the power factor $S^2\sigma$, or at least the avoidance of a decrease in $S^2\sigma$ with the decrease in κ, through size quantization [12.11, 12] or through energy filtering [12.19] of the electrons.
2. A reduction in the lattice thermal conductivity due to the scattering [12.13, 20] or to the refraction [12.21] of phonons on the physical boundaries of the nanoscale structure.

The structure of this review is as follows. Firstly, we introduce the technological benefits that can result from the development of new nanoscale thermoelectric materials with $ZT = 2$. Then, we outline a few of the basic transport equations in 3, 2, and 1 dimensions and show how low dimensionality can improve ZT. In the subsequent sections, we review the thermoelectric properties in a few prototype systems of lower dimensionality. These are 2D systems (quantum wells), 1D systems (quantum wires), and quasi-zero-dimensional

systems (quantum-dot superlattices and solids containing nanoscale inclusions).

12.1 The Promise of Thermoelectricity

As outlined in the introduction, conventional thermoelectric materials have a value of ZT around unity at 300 K, while nanometer-scale materials have been shown to have $ZT = 2$. Even though there are, at present, no commercial materials that achieve this value, the fact that one has the proof of principle that nanotechnology can result in such performance is significant, as the value of $ZT = 2$ is the threshold [12.22] at which one can envision large-scale applications of thermoelectricity with performances comparable to those of conventional mechanical machines.

12.1.1 Cooling Applications

Peltier coolers are solid-state heat pumps that have, as a basic building block, two semiconductor elements,

called "legs," one doped p-type and one doped n-type connected as shown in Fig. 12.1. The p-type material has a positive Seebeck coefficient, and the n-type material has a negative one. The inverse of the Seebeck effect is the Peltier effect: when a current density j passes through a material, it generates a heat flux q_f related to j by $q_f = \Pi j$. The Peltier coefficient, Π, is related to the Seebeck coefficient via the Onsager relation $\Pi = ST$. The two legs in a practical Peltier couple are connected electrically in series: the current I is passed through each leg, giving rise to two heat flows Q_N and Q_P that both cool down the "cold" plate and pump the heat $Q = Q_N + Q_P$, plus the amount of Joule heating that is generated by the current in the legs, into the "hot" plate. Heat exchangers must be used to extract the relevant heat flows on both the hot and cold plates.

Every cooling technology has an efficiency characterized by a "coefficient of performance", or COP. The

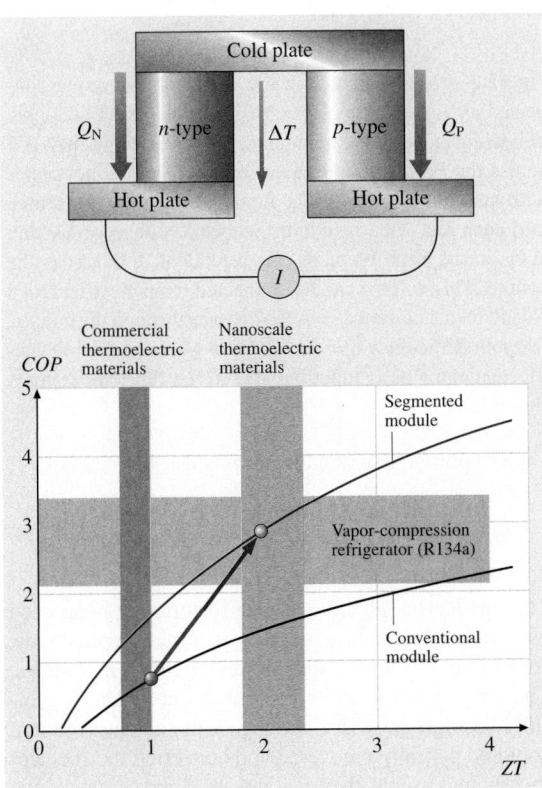

Fig. 12.1 *(Top)* Schematic view of a Peltier couple, consisting of a leg of p-type and a leg of n-type semiconductor, connected for a cooling application. Passing a current I (which takes a voltage V and a power VI) through both p- and n-type legs results in heat fluxes Q_P and Q_N that pump heat out of cold plate into hot plates; heat exchangers are expected to extract heat from both sides. *(Bottom)* Comparison of coefficient of performance, defined as $(Q_P + Q_N)/VI$, of thermoelectric refrigerators, as a function of figure of merit ZT, with that of a conventional vapor-compression cycle using a conventional refrigerant (R134a). The comparison is made at $T = 300$ K, for a ΔT of 40 °C, and for two types of modules, depending on how multiple Peltier couples are connected. In conventional modules, multiple Peltier couples are connected thermally in parallel, with one common hot plate and one common cold plate. The "segmented" modules are a more advanced system designed to extract heat from moving fluids in applications such as air conditioning. Commercial thermoelectric materials have $ZT = 1$, and the COP achieved with them in a conventional isothermal module is not competitive with that of vapor-compression refrigerators. The situation changes when we consider using segmented modules in conjunction with the recently developed nanoscale thermoelectric materials that have $ZT = 2$

COP is, by definition, the amount of heat extracted (here $Q_N + Q_P$) divided by the electrical or mechanical work used (here VI, where V is the total voltage drop across both the p- and the n-type legs). The *COP* is a function of the temperature difference ΔT over which the cooler operates, as well as the hot and cold end temperatures, T_{HOT} and T_{COLD}. The best possible *COP* is that of a Carnot cycle, which has $COP = T_{COLD}/\Delta T$; actual devices have a lower *COP*. For a thermoelectric cooler, the *COP* depends on the current I through the Peltier elements as well as on the average temperature and ΔT. The current can be optimized to either maximize the *COP* or the cooling power. The dependence of the optimum *COP*, COP_{max}, on the figure of merit ZT of the Peltier elements (assumed equal for the n and p leg and taken at the average temperature T) is given in classical textbooks as [12.10]:

$$COP_{max} = \frac{T_{COLD}}{\Delta T} \frac{\sqrt{1+ZT} - \frac{T_{HOT}}{T_{COLD}}}{\sqrt{1+ZT} + 1}. \quad (12.2)$$

Figure 12.1 shows (12.2) as the curve labeled "conventional module," which holds for a single Peltier couple. When a much larger cooling power is desired, there are several schemes that can be used to connect Peltier couples. Classically, multiple Peltier couples are connected thermally in parallel, between two isothermal plates, and connected electrically in series. Since each couple in the entire module has the same thermal environment, the module has the same performance as an individual couple. However, in air conditioning applications in which a fairly large amount of heat must be exchanged between fluids, Peltier couples, or groups of couples, can be connected [12.23–25] so as to optimize the *COP* of each couple locally, resulting in the curve labeled "segmented module"; this is the configuration that should be compared to a classical vapor-compression cooling system operating under the same gradient. The range of *COP* obtained by conventional vapor systems is also shown in Fig. 12.1. It is clear that if a thermoelectric material were available with a $ZT = 2$ at 300 K, thermoelectric cooling would become competitive with vapor-compression cooling.

12.1.2 Power Generation

Thermoelectric couples are also used to generate electrical power from heat. For that purpose, they are connected as shown in Fig. 12.2. The efficiency of such a generator, η, is again calculated in classical handbooks and is a function of the electrical load applied to the generator.

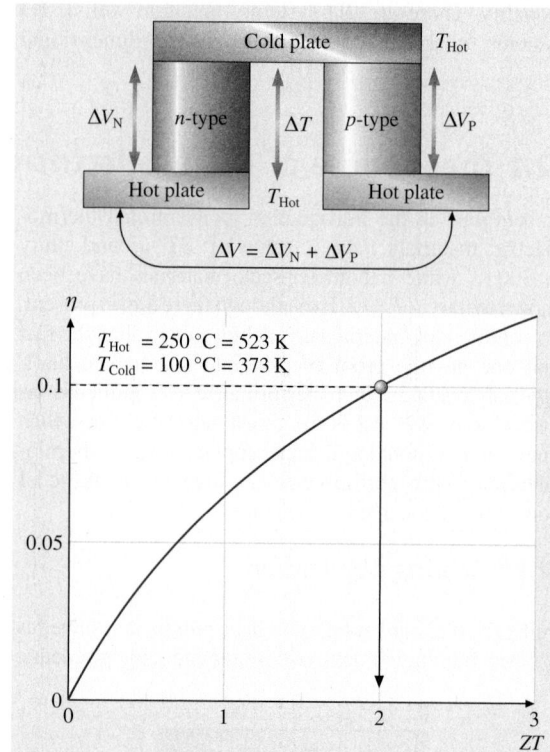

Fig. 12.2 *(Top)* Schematic view of a Peltier couple connected as a generator of electrical power. External source of heating and cooling maintain a temperature gradient ΔT across the Peltier couple, between the cold and hot plates with heat fluxes Q_P and Q_N flowing out of the hot source into each leg. This results in Seebeck voltages ΔV_N and ΔV_P across each element, for a total of ΔV across the couple. This voltage can be connected to an electrical load R, and then a current $I = \Delta V/R$ passes through the couple. *(Bottom)* Efficiency $\eta = I\Delta V/(Q_P + Q_N)$ of a thermoelectric generator as a function of ZT for the temperatures shown

Under optimal load conditions, η is [12.10]:

$$\eta_{max} = \frac{\Delta T}{T_{HOT}} \frac{\sqrt{1+ZT} - 1}{\sqrt{1+ZT} + \frac{T_{COLD}}{T_{HOT}}}. \quad (12.3)$$

Thermoelectric generators have historically been used with radioisotope heat sources in space applications. An example of a new application would be as follows. The heat lost by internal combustion engines, typically through the water-cooling or the exhaust system, could be partially recovered and converted thermoelectrically into usable electrical power. Approximately one third of the heat of combustion of the fuel in an automotive engine is used to do mechanical work; one third is wasted as heat in the cooling system and one third in the exhaust. Considerable savings in fossil-fuel consumption, perhaps comparable to the conversion of the vehicle fleet to diesel engines, could be achieved if 10% of the waste heat could be transformed into usable electrical energy. The exhaust gases could be used for this purpose at temperatures around 400 °C

to 500 °C, but the heat exchangers necessary to extract the heat are expected to lower the hot-side temperature of a thermoelectric generator to 250 °C. Similarly, to avoid condensation of water, the cold-side temperatures are limited to 100 °C. The efficiency of a thermoelectric generator is plotted as a function of the ZT of the materials in Fig. 12.2 under these conditions: once again, a threshold value of $ZT = 2$ makes this approach practical.

12.2 Theory of Thermoelectric Transport in Low–Dimensional Solids

It is convenient to view the problem of low-dimensional thermoelectricity in terms of the different length scales that enter the transport equations and to compare these with the characteristic sample dimension d. For 2D structures, or quantum wells, d is the thickness of the well. For quantum wires and dots, d is the wire or dot diameter. Control of this dimension d on the nanometer scale can be put to good use through two mechanisms. Firstly, one tries to decrease the lattice thermal conductivity while minimizing the reduction of electrical conductivity. Secondly, size-quantization effects can result in an increase in the product of the Seebeck coefficient (squared) and the electrical conductivity, $S^2\sigma$. Unfortunately, there is a third consideration: the effects of disorder and defects in nanostructures are much more pronounced than in conventional bulk solids. We outline now how these considerations translate into length scales.

Firstly, in the regime of diffusive transport, the electrical conductivity and the lattice thermal conductivities

are typically determined by the electron and the phonon mean free paths, ℓ_e and ℓ_ϕ, respectively. Both mean free paths are on the order of tens to hundreds of nanometers. Since it is beneficial to increase the electrical conductivity and to decrease the thermal conductivity, one wants to work in samples with critical dimensions d so that ℓ_ϕ is limited by d but not ℓ_e. This sets a condition on the system: $\ell_e < \ell_\phi$. It was realized early on [12.5] in the example of the PbSe$_x$Te$_{1-x}$ alloy system that alloy scattering induces a strong decrease in ℓ_ϕ, accompanied by only a moderate decreases in ℓ_e, and thus in electrical conductivity.

Secondly, the origin of the field of low-dimensional thermoelectricity lies in the realization [12.11, 12] that size-quantization effects can increase the Seebeck coefficient while not affecting the density of charge carriers and thus the electrical conductivity. The density of electric states is then a strong function of the dimensionality of the system, as can be seen in Fig. 12.3. As we will show later, S is a function of the energy derivative of the

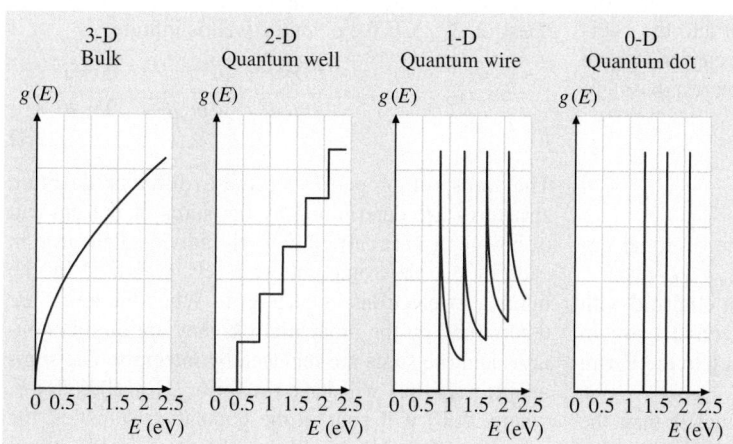

Fig. 12.3 Electronic density of states $g(E)$ as function of carrier energy E, for (**a**) a bulk 3D crystal, (**b**) a 2D crystal, or quantum well, (**c**) a 1D crystal or quantum wire, (**d**) a 0D crystal or quantum dot

density of states and is thus enhanced when the latter is a sharply peaked function of energy, as is the case for lower-dimensional crystals in Fig. 12.3. This effect can compensate for an eventual decrease in ℓ_e. However, it sets the condition that the characteristic sample dimension d (the diameter of a quantum dot or wire or the thickness of a quantum well) must be on the order of the electron wavelength λ_e.

Thirdly, for this to work, the system must be sufficiently free of disorder so that the band structure model holds. The smaller the diameter of a quantum wire, the higher the likelihood that one defect will localize the electron wavefunction to one section of the wire, impeding transport. The band structure picture then no longer holds and the conduction becomes dominated by localization effects, characterized by the phase coherence length L_ϕ of the electrons in the system. Localization effects are typically identified by magnetoresistance measurements, in which L_ϕ is compared to the sample dimension d but also to the magnetic length $\ell_H = \sqrt{\hbar/2eH}$, which represents the spatial extent of the wavefunction on a Landau level. As L_ϕ becomes of the order of d only at low temperatures, localization effects are not encountered at room temperature in thermoelectric materials with $d > 5$ nm.

12.2.1 Density of States and Energy Bands in Three, Two, One, and Zero Dimensions

When $\lambda \ll \ell_e$, the classical band model for solids holds, and the energy states of the electrons are described by Bloch functions. In any direction in which the electron motion is free, it is characterized by the momentum $\mathbf{k}$. The dispersion relation between the electron energy and its momentum is parabolic $E = \hbar^2 k_I^2/(2m_I^*)$, where the index I denotes the direction of motion and the coefficient m_I^* is the effective mass of the carrier (for an electron in vacuum, $m_I^* = m_e$, the electron rest mass). In 3D bulk solids, we have thus:

$$E_{3-\mathrm{Dim}} = \frac{\hbar^2 k_x^2}{2m_x^*} + \frac{\hbar^2 k_y^2}{2m_y^*} + \frac{\hbar^2 k_z^2}{2m_z^*} \,. \tag{12.4}$$

When (12.4) is plotted as a 3D surface of equal energy as a function of momentum in (k_x, k_y, k_z)-space, this surface, which we label $\Sigma(E)$, is an ellipsoid with axes inversely proportional to the effective mass in each direction. The surface corresponding to the Fermi energy, $\Sigma(E_F)$, is the Fermi surface.

In quantum wells of thickness d smaller than the wavelength of the electron, the motion of the electron in the direction normal to the well is impossible. In that direction, say z, k_z is not a good quantum number. The electron energy levels are quantized following the classical problem of the potential well in quantum mechanics into sublevels E_i, where $i(= 1, 2, 3 \ldots)$ is the new quantum number. Assuming further that the material on both sides of the slab represents an infinitely high potential well, the sublevels are $E_i = \frac{\hbar^2 \pi^2 i^2}{2m_z d^2}$ and the energy dispersion relation is then:

$$\begin{aligned} E_{2-\mathrm{Dim}} &= \frac{\hbar^2 k_x^2}{2m_x^*} + \frac{\hbar^2 k_y^2}{2m_y^*} + E_i \\ &= \frac{\hbar^2 k_x^2}{2m_x^*} + \frac{\hbar^2 k_y^2}{2m_y^*} + \frac{\hbar^2 \pi^2 i^2}{2m_z^* d^2} \,. \end{aligned} \tag{12.5}$$

The equi-energetic surfaces $\Sigma(E)$ and the Fermi surface in particular are now flat ellipses in the (k_x, k_y) plane.

In quantum wires of diameter d, the only direction along which the motion is free and the momentum is a good quantum number is the longitudinal direction, say x. The confinement in both the y and z directions results in quantized energy levels $E_{i,j}$ using two quantum numbers i and $j(= 1, 2, 3 \ldots)$. Assuming again that the potential wells are infinite, we have:

$$\begin{aligned} E_{1-\mathrm{Dim}} &= \frac{\hbar^2 k_x^2}{2m_x^*} + E_{i,j} \\ &= \frac{\hbar^2 k_x^2}{2m_x^*} + \frac{\hbar^2 \pi^2 i^2}{2m_y^* d^2} + \frac{\hbar^2 \pi^2 j^2}{2m_z^* d^2} \,. \end{aligned} \tag{12.6}$$

The equi-energetic surfaces $\Sigma(E)$ such as the Fermi surface $\Sigma(E_F)$ are tubes aligned with the k_x direction.

Finally, in quantum dots shaped as spheres of diameter d, there is no free motion, and the electrons are confined to a series of quantized energy levels $E_{i,j,l}$ characterized by three quantum numbers i, j, and $l(= 1, 2, 3 \ldots)$. If the potential well is infinite:

$$E_{0-\mathrm{Dim}} = E_{i,j,l} = \frac{\hbar^2 \pi^2 i^2}{2m_x d^2} + \frac{\hbar^2 \pi^2 j^2}{2m_y d^2} + \frac{\hbar^2 \pi^2 l^2}{2m_z d^2} \,. \tag{12.7}$$

The transport properties of these different quantum structures are determined by the sums of the amount of charge or heat carried by each individual electron in each of the electronic energy levels, weighted by the number of electrons in each level. When the levels are determined by the momentum $\mathbf{k}$, they are quasicontinuous and the sums are replaced by integrals. The sums or integrals thus will have the form $\int_0^\infty g(E)X(E)\,\mathrm{d}E$, where $X(E)$ will contain the population statistics, the scattering probability, and the quantity carried (charge or heat). The factor $g(E)$ is the *density of states*, defined by the fact that $g(E)dE$ is the number, per unit volume of material, of allowed wave vectors in a given energy band in the energy range between E and $E + dE$.

The exact derivation of the density of states is given by Eq. (8.63) in *Ashcroft* and *Mermin* [12.26]:

$$g(E) = \int_{\Sigma(E)} \frac{d\Sigma}{4\pi^3} \frac{1}{|\nabla E(\mathbf{k})|} , \qquad (12.8)$$

where $\Sigma(E)$ is the equi-energetic surface as described in each case above and $\nabla E(\mathbf{k})$ is the gradient of E with respect to k. The dependence of the density of states on the inverse of the gradient of E can result in divergences in the function $g(E)$ in lower dimensions, and these are particularly favorable to thermoelectric properties, as we will see in the following analysis. More explicitly, using the dispersion relations equations (12.4–12.7), the expressions for the density of states in the different dimensions are:

$$g(E)_{3-\text{Dim}} = \frac{\sqrt{m_x^* m_y^* m_z^*}}{\hbar^3 \pi^2} \sqrt{2E} ; \qquad (12.9a)$$

$$g(E)_{2-\text{Dim}} = \sum_i g_i(E) ,$$

$$g_i(E) = \frac{\sqrt{m_x^* m_y^*}}{\pi \hbar^2} \frac{1}{d} \text{ if } E > E_i ,$$

$$g_i(E) = 0 \qquad \text{if } E \le E_i ; \qquad (12.9b)$$

$$g(E)_{1-\text{Dim}} = \sum_{i,j} g_{i,j}(E) ,$$

$$g_{i,j}(E) = \frac{\sqrt{2m_x}}{\pi \hbar} \frac{1}{d^2} \frac{1}{\sqrt{E - E_{ij}}} \text{ if } E > E_{ij} ,$$

$$g_i(E) = 0 \qquad \text{if } E \le E_{ij} ; \qquad (12.9c)$$

$$g(E)_{0-\text{Dim}} = \sum_{i,j,l} g_{i,j,l}(E) ,$$

$$g_{i,j,l} = 1/d^3 \qquad \text{if } E = E_{i,j,l} ,$$

$$g_{i,j,l} = 0 \qquad \text{otherwise} . \qquad (12.9d)$$

The factors $1/d^D$, where D is the dimensionality, permit the use of the same dimension (inverse volume in m^{-3}) for quantum wells, wires, and dots, rather than having to use units of m^{-2} for quantum wells, m^{-1} for quantum wires, and a dimensionless number for quantum dots. This sequence of densities of states for different dimensions is shown graphically in Fig. 12.3. In real systems, disorder will broaden the spikes in the predicted density

of states for quantum wires and quantum dots, but the general features are expected to remain.

The density of charge carriers, n, is the first quantity that depends on the density of states, as well as on the population statistics of these states, the Fermi distribution function f_0:

$$f_0(E) \equiv \frac{1}{1 + e^{\frac{E - E_F}{k_B T}}} . \qquad (12.10)$$

The density (in m^{-3}) can be expressed as an integral that is a function of f_0, and the integral can be integrated by parts and expressed as a function of the energy derivative of f_0:

$$n = \int_0^\infty g(E) f_0(E) dE$$

$$= \int_0^\infty \left[\int_0^E g(E') dE' \right] \left(-\frac{\partial f_0(E)}{\partial E} \right) dE$$

$$= A \int_0^\infty (E - E_A) g(E) \left(-\frac{\partial f_0(E)}{\partial E} \right) dE , \qquad (12.11)$$

where, in 3-D: $A = 2/3$, $E_A = 0$,

in 2-D: $A = 1$, $E_A = E_i$,

in 1-D: $A = 2$, $E_A = E_{ij}$.

In general, thermoelectric materials are narrow-gap semiconductors. The equations given here (12.4–12.7) are only approximately valid for electrons and holes in the conduction and valence bands of degenerately doped narrow-gap semiconductors at energy levels above the band edge comparable to the energy gap (i. e., when $E_F \approx E_g$, where E_g is the gap energy). In that case, the energy E in the left-hand side of (12.4–12.7) is to be replaced by $\gamma(E) = E(1 + E/E_g)$, the *nonparabolic band model*. We ignore this in this review article, as it does not affect the qualitative conclusions illustrated here, but the nonparabolic model should be used for quantitative calculations.

12.2.2 Electronic Transport in the Relaxation–Time Approximation

The equations governing the transport of heat and electricity in a solid derive from those developed for the kinetic theory of gases, but with different expression for the density of states and carrier statistics. The electrical current density involves integrals in which each carrier transports an electron charge q, and the heat flux involves integrals in which each carrier transports a thermal energy ($E - E_F$). In solids of good crystalline quality, the Boltzmann equation can be solved in the relaxation time approximation. Considering only elastic collisions, the relaxation time is assumed to have an energy dependence given by:

$$\tau(E) = \tau_0 E^{\lambda - \frac{1}{2}} , \qquad (12.12)$$

where λ is the *scattering exponent*, approximated [12.27] for various scattering mechanisms as follows:

- for scattering of electrons on acoustic phonons, $\lambda = 0$,
- for scattering of electrons on neutral impurities, $\lambda = 1/2$,
- for scattering of electrons on ionized impurities, $\lambda = 2$.

We take the further five simplifying assumptions: (1) carriers are in one single band; (2) the dispersion relation is of the form (12.4–12.7), with the mass tensor independent of energy; (3) the phonon distribution is at equilibrium; (4) the magnetic field has no quantizing effect on the dispersion relation; (5) the fields and temperature gradient are sufficiently weak that the devi-

ations of the distribution function are linear functions of them. The transport integrals are now given for solids of dimensionality $D = 3$ to 1 in Table 12.1 in terms of the Fermi integrals $F_n = \frac{1}{n!} \int\limits_0^\infty \frac{x^n \, dx}{e^{x-\eta}+1}$ and $\eta = \frac{E_F}{k_B T}$ where E_F is the Fermi energy. We now discuss the quantities given in Table 12.1 in some detail.

The electrical conductivity σ is given as the product of the carrier density n, charge q, and mobility μ:

$$\sigma = nq\mu . \qquad (12.13)$$

The mobility is, by definition, the group velocity of the carrier divided by the electric field. For transport in the x direction, it is related to the relaxation time by:

$$\mu = |q|\frac{\tau}{m_x} . \qquad (12.14)$$

Another concept already used is that of carrier mean free path, ℓ_e, which is related to the relaxation time and the electron group velocity v by:

$$\ell_e = v\tau . \qquad (12.15)$$

The thermal conductivity κ of solids is the sum of two contributions, the electronic thermal conductivity κ_E and the lattice thermal conductivity κ_L:

$$\kappa = \kappa_E + \kappa_L . \qquad (12.16)$$

Table 12.1 Transport integrals in 3D, 2D, and 1D systems for transport along x direction; electrical conductivity is $\sigma = nq\mu$

	3D system: bulk	2D system: quantum well	1D system: quantum wire
$g(E)$	$\dfrac{\sqrt{m_x^* m_y^* m_z^*}}{\hbar^3 \pi^2}\sqrt{2E}$	$\begin{array}{l} E < E_i : 0 \\[4pt] E > E_i : \dfrac{\sqrt{m_x^* m_y^*}}{\hbar^2 \pi}\dfrac{1}{d} \end{array}$	$\begin{array}{l} E < E_{i,j} : 0 \\[4pt] E > E_{i,j} : \dfrac{\sqrt{2m_x^*}}{\hbar \pi d^2}\dfrac{1}{\sqrt{E - E_{i,j}}} \end{array}$
n	$\dfrac{(2k_B T)^{3/2}\sqrt{m_x^* m_y^* m_z^*}}{2\pi^2 \hbar^3} F_{1/2}$	$\dfrac{k_B T \sqrt{m_x^* m_y^*}}{\pi \hbar^2 d} F_0$	$\dfrac{\sqrt{2 k_B T m_x^*}}{\pi \hbar d^2} F_{-1/2}$
$\mu \dfrac{1}{\tau_0}\dfrac{m_x}{q^2}$	$\dfrac{2}{3}\left(\dfrac{3}{2}+\lambda\right)\dfrac{F_{1/2+\lambda}}{F_{1/2}}(k_B T)^{\frac{2}{3}\left(\frac{3}{2}+\lambda\right)}$	$(1+\lambda)\dfrac{F_\lambda}{F_0}(k_B T)^{(1+\lambda)}$	$2\left(\dfrac{1}{2}+\lambda\right)\dfrac{F_{-1/2+\lambda}}{F_{-1/2}}(k_B T)^{2\left(\frac{1}{2}+\lambda\right)}$
$R_H nq$	$\pm\dfrac{3}{2}\dfrac{\frac{3}{2}+2\lambda}{\left(\frac{3}{2}+\lambda\right)^2}\dfrac{F_{1/2+2\lambda}F_{1/2}}{F_{1/2+\lambda}}$	$\pm\dfrac{1+2\lambda}{(1+\lambda)^2}\dfrac{F_{2\lambda}F_0}{F_\lambda}$	$\pm\dfrac{1}{2}\dfrac{\frac{1}{2}+2\lambda}{\left(\frac{1}{2}+\lambda\right)^2}\dfrac{F_{-1/2+2\lambda}F_{-1/2}}{F_{-1/2+\lambda}}$
$S\dfrac{q}{k_B}$	$\pm\left(\dfrac{\frac{5}{2}+\lambda}{\frac{3}{2}+\lambda}\dfrac{F_{3/2+\lambda}}{F_{1/2+\lambda}}-\dfrac{E_F}{k_B T}\right)$	$\pm\left(\dfrac{2+\lambda}{1+\lambda}\dfrac{F_{1+\lambda}}{F_\lambda}-\dfrac{E_F}{k_B T}\right)$	$\pm\left(\dfrac{\frac{3}{2}+\lambda}{\frac{1}{2}+\lambda}\dfrac{F_{1/2+\lambda}}{F_{-1/2+\lambda}}-\dfrac{E_F}{k_B T}\right)$
$L_0\left(\dfrac{q}{k_B}\right)^2$	$\dfrac{\frac{7}{2}+\lambda}{\frac{3}{2}+\lambda}\dfrac{F_{5/2+\lambda}}{F_{1/2+\lambda}}-\left(\dfrac{\frac{5}{2}+\lambda}{\frac{3}{2}+\lambda}\dfrac{F_{3/2+\lambda}}{F_{1/2+\lambda}}\right)^2$	$\dfrac{3+\lambda}{1+\lambda}\dfrac{F_{2+\lambda}}{F_\lambda}-\left(\dfrac{2+\lambda}{1+\lambda}\dfrac{F_{1+\lambda}}{F_\lambda}\right)^2$	$\dfrac{\frac{5}{2}+\lambda}{\frac{1}{2}+\lambda}\dfrac{F_{3/2+\lambda}}{F_{-1/2+\lambda}}-\left(\dfrac{\frac{3}{2}+\lambda}{\frac{1}{2}+\lambda}\dfrac{F_{3/2+\lambda}}{F_{-1/2+\lambda}}\right)^2$
$\dfrac{N}{R_H \sigma}\dfrac{q}{k_B}$	$\dfrac{\frac{5}{2}+2\lambda}{\frac{3}{2}+2\lambda}\dfrac{F_{3/2+2\lambda}}{F_{1/2+2\lambda}}-\dfrac{\frac{5}{2}+\lambda}{\frac{3}{2}+\lambda}\dfrac{F_{3/2+\lambda}}{F_{1/2+\lambda}}$	$\dfrac{2+2\lambda}{1+2\lambda}\dfrac{F_{1+2\lambda}}{F_{2\lambda}}-\dfrac{2+\lambda}{1+\lambda}\dfrac{F_{1+\lambda}}{F_\lambda}$	$\dfrac{\frac{3}{2}+2\lambda}{\frac{1}{2}+2\lambda}\dfrac{F_{1/2+2\lambda}}{F_{-1/2+2\lambda}}-\dfrac{\frac{3}{2}+\lambda}{\frac{1}{2}+\lambda}\dfrac{F_{1/2+\lambda}}{F_{-1/2+\lambda}}$

The lattice thermal conductivity will be addressed separately in Sect. 12.2.4. The electronic thermal conductivity is related to the electrical conductivity σ via the Lorenz number L_o by the Wiedemann–Franz law:

$$\kappa_E = L_0 \sigma T . \tag{12.17}$$

For a free electron in vacuum, $L_0 = (\pi^2/3)(k_B/q)^2$, and in solids in which electrons undergo only elastic collisions, this holds as well. More generally, the Lorenz number L_0 is one of the classical electronic transport properties and is given in Table 12.1.

The two properties shown in Table 12.1 that are not directly involved in thermoelectric transport are the Hall coefficient R_H and the isothermal transverse

Nernst–Ettingshausen coefficient N. These quantities are transverse transport properties observed in the presence of a magnetic field H_z oriented along the z-axis. The electric field measured in Hall and Nernst measurements is along the y direction, while the applied fluxes are along x. The Hall effect is observed in the presence of a longitudinal current density j_x: $R_H = E_y/(j_x H_z)$. The Nernst effect is measured in the presence of a temperature gradient along x: $N = E_y/[(\mathrm{d}T/\mathrm{d}x)H_z]$. Their usefulness is in determining important materials parameters, R_H being mostly sensitive to the carrier density n and N to the scattering parameter λ.

Finally, the Seebeck coefficient S is the most sensitive of all to the details of the band structure, and in particular to the effects of low dimensionality. The detailed expressions in the relaxation time approximation are given in Table 12.1. However, there is another way to express the Seebeck coefficient that is independent of the concept of band structure. This expression, the *Mott relation,* only uses the concept of density of states (even when (12.9a–12.9d) are not valid) and the Fermi distribution function. It is therefore valid irrespective of the dominant transport mechanism. Just as we associated a density, $g(E)$, to electrons that fill the energy levels between E and $E + \mathrm{d}E$, we can associate an electrical conductivity $\sigma(E)$ to the same electrons, irrespective of the mechanism that limits $\sigma(E)$. The total electrical conductivity is then the integral of this over the entire energy range, moderated by the Fermi distribution function. Integrating by parts as for (12.11), the total conductivity is:

$$\sigma = \int_0^\infty \sigma(E)\left(-\frac{\partial f_0(E)}{\partial E}\right)\mathrm{d}E . \tag{12.18}$$

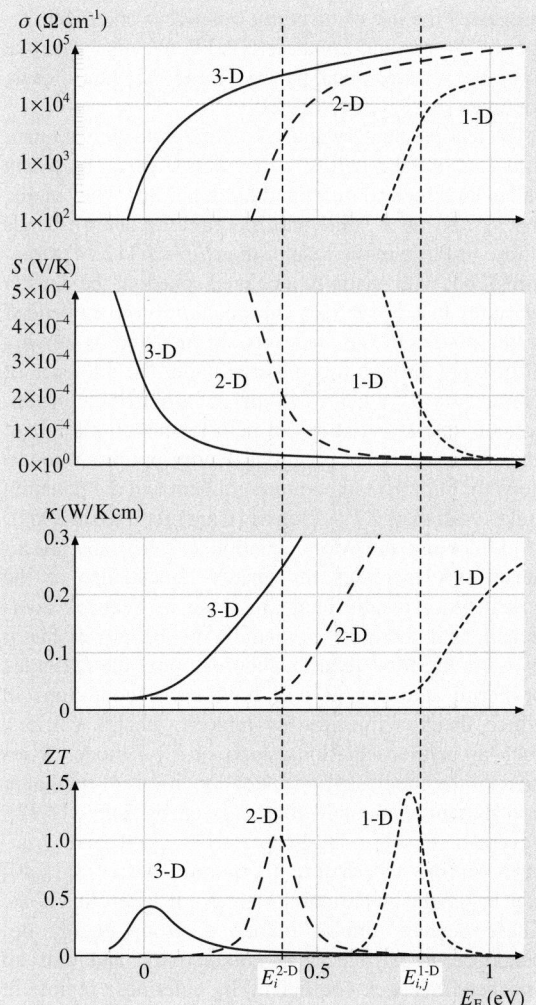

Fig. 12.4 Transport properties of 3D *(full line)*, 2D *(long dashes)*, and 1D *(short dashes)* solids as function of Fermi energy. Calculations are made at 300 K and assume acoustic phonon scattering ($\lambda = 0$) and an anisotropic Fermi surface with $m^* = 0.15\, m_e$. The thickness of the quantum well and the diameter of the quantum wire are $d = 2.5$ nm. The value of the first size-quantized subband edge, E_i for 2D systems and $E_{i,j}$ for 1D systems, are shown. The Fermi energy in each system has to be chosen by doping to maximize the value of ZT, and the location is different for the different dimensionalities. For the calculations of ZT, a constant $\kappa_L = 2$ W/m·K was used. The maximum achievable ZT is improved as the dimensionality of the system is decreased

Cutler and *Mott* [12.28] derive the Seebeck coefficient in this formalism to be the differential form of the Mott relation:

$$S = \frac{k_B}{q}\frac{1}{\sigma}\int\limits_0^\infty \sigma(E)\left(\frac{E-E_F}{k_B T}\right)\left(\frac{\partial f_0(E)}{\partial E}\right) dE .$$ (12.19)

In systems in which the Fermi statistics are degenerate, such as metals and degenerately doped semiconductors, (12.19) simplifies to the better-known form of the Mott relation:

$$S = \frac{\pi^2}{3}\frac{k_B}{q}k_B T\left(\frac{d\{\ln[\sigma(E)]\}}{dE}\right)_{E=E_F} ,$$ (12.20)

which is generally valid, whether conduction is through band states, localized states, hopping, or other mechanisms.

Returning now to the case of band conduction, we first show, in Fig. 12.4, how the doping level of a thermoelectric material is optimized. Figure 12.4 illustrates the room-temperature behavior of thermoelectric materials of different dimensionality for quantum structures in which the energy scale of the features shown in the density of states in Fig. 12.3 is larger than the thermal smearing energy $k_B T$, by about a factor 10. In Fig. 12.4, we plot the equations given for the electrical conductivity and the Seebeck coefficient, assuming acoustic phonon scattering ($\lambda = 0$), as a function of the Fermi energy of the carriers. The carrier density increases as E_F increases following (12.11). A constant value κ_L of $2\,W/(Km)$ is also assumed, and (12.16) and (12.17) are used for the thermal conductivity. Figure 12.4 shows how, at low carrier density, the Seebeck coefficient is large, but the conductivity low, and vice versa at high carrier density. Therefore, there exists an optimum doping level at which the value of ZT is maximized.

The most striking result shown by Fig. 12.4 is that a decrease in dimensionality of the system results in a considerable increase in the value of the maximum ZT. This observation, first made by the *Dresselhaus* group [12.11, 12], marked the birth of the field of low-dimensional thermoelectricity. It can be understood from the shape of the density of states curves shown in Fig. 12.3, coupled with the Mott formula, (12.20). The Seebeck coefficient is sensitive to the energy derivative of the conductivity. The sharp features in the density of states that are the hallmark of low-dimensional systems therefore increase the Seebeck coefficient for a given carrier density and conductivity. However, the range of Fermi energies over which ZT is optimized is narrower in low-dimensional systems than in bulk: finding the optimum doping level is more critical.

A brief historical note about the evolution of this thinking is now added. There is an abundant older literature on transport, and on the Seebeck coefficient in particular, in low-dimensional systems. The earliest paper this author found dates from 1973 [12.29]: it suggests enhancements of the Seebeck coefficient in 2D systems. The early work by *Friedman* and his coworker [12.30, 31] shows how the energy (or doping-level) dependence of the Seebeck coefficient can in fact outline the features of the 2D density of states, and in particular that, at low temperature, the Seebeck coefficient vanishes when the Fermi energy lies on a plateau of $g(E)$, consistently with the Mott equation. *Hicks* et al. [12.32] generalized the calculation to solids with two bands, and suggested the use of bismuth; this was worked out further by *Broido* and *Reinecke* [12.33]. We also note that there are several other papers [12.33–35] that closely followed [12.11] and [12.12].

A complementary analysis of the influence of quantum dots in thermoelectric materials is given by recent calculation by *Humphrey* and *Linke* [12.36]. They model the system as a solid with continuous energy bands (those of PbTe in the case of the *Harman* [12.14] structures), but with spatially localized spikes in the density of states that arise from the quantum dots embedded in the material. The electrons are modeled as moving diffusively through this material under the influence of a temperature gradient. The charge carriers therefore encounter spatial variations in both temperature gradient and electrochemical potential. Under optimum conditions for both the temperature gradient and the potential spikes, values of ZT as high as 10 are predicted at 300 K.

Following the Mott equation (12.20), any mechanism that increases the energy dependence of the conductivity results in an increase in Seebeck coefficient, and therefore potentially in the power factor $S^2\sigma$. As the Mott relation supposes only the existence of Fermi statistics, it will hold under conditions in which there is injection of minority charge carriers, such as in the depletion region of a *pn* diode. A review of thermoelectricity under conditions of nonlinear transport in semiconductors is given by *Tauc* [12.37]. Several authors [12.38,39] recently suggested that structures similar to vacuum thermionic devices [12.40] can be made in the solid state from heterojunctions. These devices aim to induce a strong energy dependence of the electrical conductivity, and thus an enhanced Seebeck coefficient, by entering a regime in which the current/voltage curves deviate from linear.

Device-optimization calculations [12.41] show that conceptually values of ZT up to 5 could be achieved in very specific heterostructures, which have been labeled heterojunction integrated thermionic (HIT) coolers.

12.2.3 Beyond Band Conduction: Weak Localization

The equations in Sect. 12.2.2 treat electrons as classical particles, which assumes that their mean free path ℓ_e is longer than their wavelength, i. e., $\ell_e \gg \lambda_e = 1/|\mathbf{k}|$, where $\mathbf{k}$ is the wavevector in (12.4–12.6). Quantum corrections must be applied to the conductivity when this condition does not hold, which brings us in the realm of *Anderson localization* [12.42]. In the limit for $T = 0$, each electron will become localized at one point in the conductor, and conduction will cease. At finite but low temperatures, the conductance decreases, and the conduction regime is called *weak localization*. The propagation of the waves that represent the electrons is not rigidly blocked by defects: the waves can circumvent defects in the solid by propagating around them through quantum mechanical interference, provided that these waves interfere constructively on the other side of the defect. This can only happen as long as the waves do retain the memory of their relative phase. For that reason, the critical length scale for weak localization is the distance over which this phase information is retained, the *phase coherence length L_ϕ* that is on the order of a few nanometers in most solids near liquid helium temperatures. The phase coherence length is related to a *phase coherence time*, τ_{Phase}, by the electron velocity, as in (12.15).

Corrections to the conductance due to this effect are much more important in low-dimensional solids than in classical bulk solids, because transport in low-dimensional solids is more sensitive to defects. The reason for this can be understood from the following simple analogy, given by *Abrikosov* [12.42]. Picture the electron as an airplane flying over mountains, which represent the defects in the solid. The airplane's altitude corresponds to the electron's energy or temperature; the mountain's height corresponds to the defect's potential to inelastically scatter the electron. As long as the plane is above the highest mountaintop, the relief simply creates turbulence, in an airplane because of thermal convection in the atmosphere, in the solid because of quantum-mechanical reflections of the electron wavefunction on the defects. Consider now the case where the plane flies lower than the mountaintops. If the motion is two-dimensional, the airplane can veer around the highest mountains, until its altitude is lower than that of the mountain passes between the tops, at which altitude it must stop. If the motion is 1D, with the airplane on a fixed course, its motion will be stopped at the first mountaintop it encounters that is higher than its altitude, at a higher altitude thus than in the 2D case.

The quantum corrections to the classical Boltzmann transport equation in the electrical conductivity are given in textbooks [12.42], and their derivation is not repeated here. To summarize, the temperature dependence of the weak-localization correction to the conductivity comes from the temperature dependence of the phase coherence time or length, which is a power law:

$$\tau_{Phase} = \tau_{Phase,0} T^{-p} \; ; \; L_\phi = L_{\phi,0} T^{-p} \,, \qquad (12.21)$$

where the exponent $p = 2$ when the phase-breaking mechanism in electron–electron interactions. For electron–phonon interactions, $p = 3$. Mostly, p is an adjustable parameter.

The dimensionality dependence comes from the dimensions of the space that electrons can diffuse into. For solids made from light elements, one can neglect spin-orbit scattering, and the quantum corrections to the conductivity can be summarized as:

$$
\begin{aligned}
\left. \frac{\Delta\sigma}{\sigma} \right|_{3-\mathrm{Dim}} &\approx \frac{\rho_e}{L_{\phi,0}} \frac{q^2}{\hbar\pi^3} T^{p/2} \,, \\
\left. \frac{\Delta\sigma}{\sigma} \right|_{2-\mathrm{Dim}} &\approx \rho_e \frac{q^2}{\hbar\pi^2} \frac{p}{2} \ln\left(\frac{T}{T_0}\right) \,, \\
\left. \frac{\Delta\sigma}{\sigma} \right|_{1-\mathrm{Dim}} &\approx -\rho_e L_{\phi,0} \frac{q^2}{\hbar\pi} T^{-p/2} \,,
\end{aligned}
\qquad (12.22)
$$

where ρ_e is the resistivity in the absence of weak localization, the "elastic" resistivity. Equation (12.22) gives a logarithmic temperature dependence of 2D localization and power law in one dimension. The case of solids, like Bi, in which spin-orbit interactions are dominant, is somewhat more complicated: the signature logarithmic or power-law dependences are maintained, but the sign of the temperature dependence can be reversed (an increase instead of a decrease with increasing temperature). This situation is labeled *antilocalization*.

The temperature dependence is a function of both the nature of the phase-breaking mechanism and the dimensionality of the system. Therefore, the system is underdetermined if we only have the temperature dependence as an experimental parameter. It is usual to add the magnetoresistance as a second experimental quantity in order to identify the influence of weak localization, and the reader is again referred to *Abrikosov* [12.42].

The qualitative features of the magnetoresistance in 1D localization are shown in Fig. 12.5. In essence, the magnetoresistance saturates at a magnetic field (the vertical line) such that one quantum of magnetic flux (the *fluxon*) fits inside one wire diameter. Once again, the sign of the magnetoresistance is inverted for localization and anti-

localization: when spin-orbit scattering is small, as in Zn, the magnetoresistance is negative [12.43], and the reverse is true when spin-orbit scattering dominates, as it does in Bi [12.22].

The Seebeck coefficient in systems subjected to weak localization has been the subject of numerous research papers [12.44], but it is most simply described using the Mott relation. We suggest a very simple quasiempirical approach [12.43], illustrated in Figs. 12.6 and 12.7. In Fig. 12.6, we show the experimental temperature dependence of the electrical resistivity of Zn nanowires; in the narrowest wires (4 nm), it follows a power law $R(T) \propto T^{-1/2}$ consistent with (12.22) for one dimension and $p = 1$. We associate an energy dependence to the conductivity that mirrors the experimental temperature dependence: $\sigma = \sigma_0 E^p$, for energies in excess of a critical energy $E_c = k_B T_c$, which is necessary in order to divide the temperature range into two distinct regimes. At low temperatures ($T < T_c$), the Mott formula gives a Seebeck coefficient that is linear in T, with a slope that is a function of p. At high temperatures ($T > T_c$), the Seebeck coefficient saturates at a value of $S = \left(-\pi^2/3\right)(k_B/q)(p/2)$. The experimental values observed for the Seebeck coefficients are shown in Fig. 12.7 and, for the 4 nm wires, follow this argument quite well. These values being on the order of a multiple of the free-electron Seebeck coefficient (k_B/q), no large

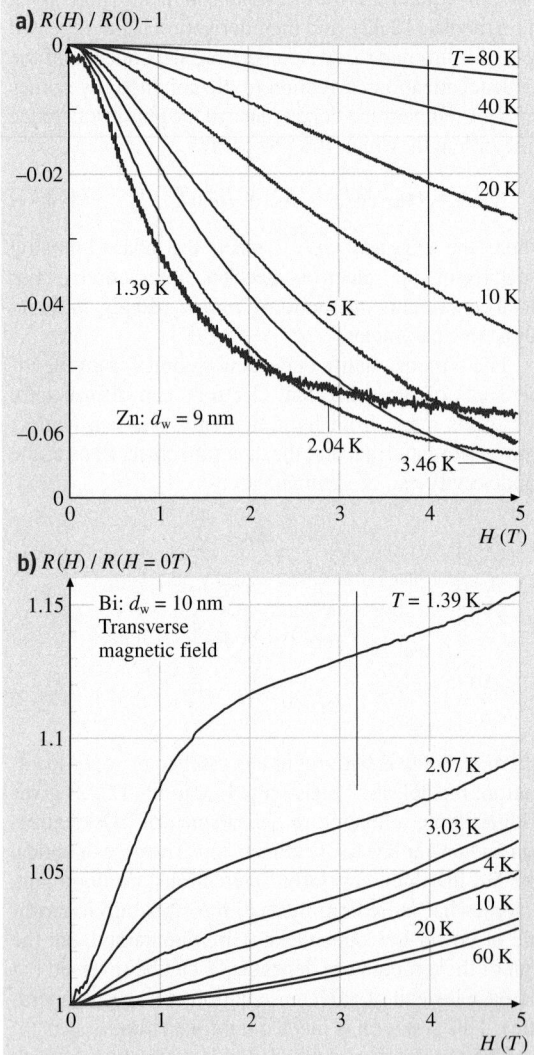

a) $R(H) / R(0) - 1$

Zn: $d_w = 9$ nm

$T = 80$ K
40 K
20 K
1.39 K
5 K
10 K
2.04 K
3.46 K

b) $R(H) / R(H = 0T)$

Bi: $d_w = 10$ nm
Transverse magnetic field

$T = 1.39$ K
2.07 K
3.03 K
4 K
10 K
20 K
60 K

Fig. 12.5 Magnetic-field dependence of magnetoresistance of Zn (**a**) and Bi (**b**) nanowires of diameters shown, in regime of weak localization. In Zn, spin-orbit scattering is weak, and weak-localization effects result in a negative magnetoresistance; in Bi, where spin-orbit scattering is strong, the magnetoresistance is positive, and the effect is called antilocalization

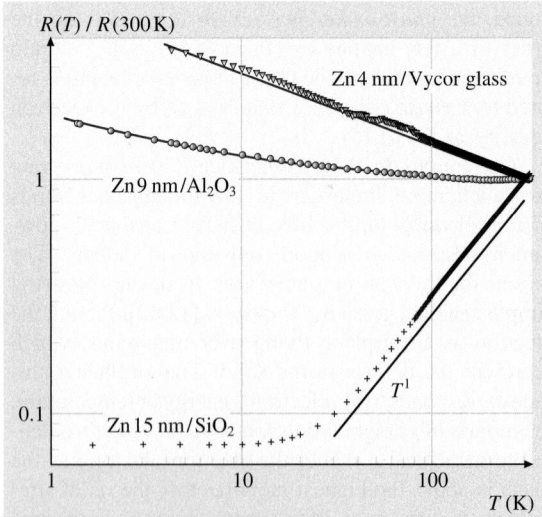

$R(T) / R(300$ K$)$

Zn 4 nm / Vycor glass
Zn 9 nm / Al$_2$O$_3$
Zn 15 nm / SiO$_2$
T^1

Fig. 12.6 Temperature dependence on resistance of Zn nanowires (normalized to 300 K). The 15-nm wires have a metallic behavior. The full lines through the data for the 9- and 4-nm wires are $T^{-1/2}$ power laws, consistent with 1D localization

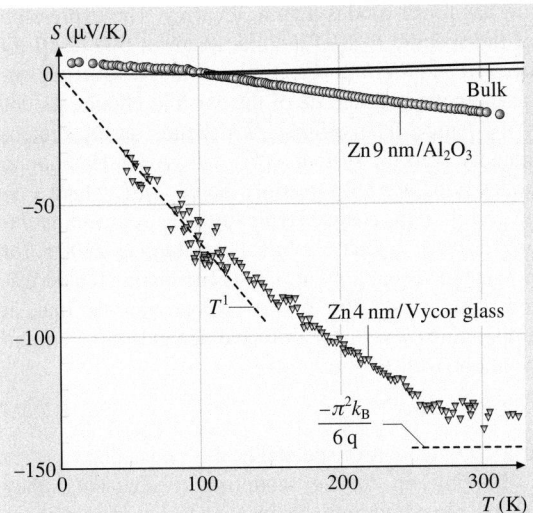

Fig. 12.7 Temperature dependence of Seebeck coefficient of Zn nanowires

enhancement of the thermoelectric performance is expected in systems in which weak localization dominates.

12.2.4 Phonon Transport in Low Dimensions

As most thermoelectric materials are degenerately doped semiconductors, the dominant term in the thermal conductivity is the lattice term κ_L, though the electronic part cannot be neglected in the doping-optimization process, as we saw above. The problem of phonon heat transport in low-dimensional systems has, once again, a long and distinguished history and was the subject of an excellent recent review [12.45]. We only touch upon the subject cursorily here.

Firstly, there is a question about the very applicability of the concept of conductivity in the usual sense to heat transport in low-dimensional solids. In short, this arises because the very concept of temperature is an equilibrium concept and therefore requires a system of sufficient size [12.45] to be defined. Obviously, the concept of temperature gradient, which is essential as the driving potential for the transport of a flux of thermal energy, is then also sensitive to dimensionality. The early work of *Fermi* et al. [12.46] pointed out that classical lattice recurrences of the initial state prevented thermalization in strongly nonlinear 1D classical lattices. The theory was followed by the work of *Prosen* and *Campbell* [12.47] in 1D systems, later [12.48] generalized to 1 and 2 dimensions. To summarize their conclusions,

which are based on a consideration of momentum conservation: (1) the thermal conductivity is expected to decrease with a decrease in the system dimensionality; (2) the thermal conductivity *diverges* with the finite size of the system along its free direction, e.g., the length L of a 1D system. The divergence of the conductivity with the length L of the system follows a power law $L^{1/3}$ in a 1D system and is logarithmic for a 2D system [12.48]. There is no divergence in conventional 3D systems. Numerical simulations [12.49] confirm this. Some experimental attempt has been made [12.50] at checking the length dependence of the thermal conductivity of carbon nanotubes, but no effect was found; the issue is still very much open, however.

Returning now, in spite of the caveat above, to the classical Boltzmann model for transport properties, one can write the lattice thermal conductivity as the product of the heat capacity C, the phonon velocity v_ϕ, and the phonon mean free path ℓ_ϕ by analogy again with the kinetic theory of gases:

$$\kappa_L = \frac{1}{2}Cv_\phi\ell_\phi = \frac{1}{2}Cv_\phi^2\tau_\phi . \tag{12.23}$$

In this equation, τ_ϕ is the phonon relaxation time. This simple-looking Boltzmann transport equation needs to be solved for each of the various phonon modes as well as the different scattering mechanisms. In nanostructures of physical sizes similar to the phonon wavelength, these modes, being normal modes of vibration, depend on the geometry of the structure. The problem quickly becomes intractable, and various theoretical approaches have been suggested [12.51–54], including the use of equations derived for radiative heat transfer [12.55].

Based on the latter approach [12.55], the heat transfer along a 1D system can be treated using the *Landauer* [12.56] formalism. In this representation, a 1D system behaves as a phonon waveguide, coupled into two heat reservoirs [12.57]. The waveguide sustains a finite number of phonon modes. The system can strictly be considered 1D for heat conduction only at temperatures sufficiently low that the thermal wavevector, $k_B T/\hbar\omega$ (where ω is the phonon's angular frequency), is smaller than the spacing between the allowed normal modes of vibration sustained by the phonon waveguide. When these conditions are met, there are only four modes to the 1D chain of atoms, one dilatational, one torsional, and two flexural. A particular property of the 1D system is that there is a cancellation between the density of states and the group velocity for the modes. The transmission coefficients between the reservoirs and the chain also become unity, and the thermal conductance G_{th} becomes quantized in terms of a materials-independent phonon conductance g_0:

$$G_{th} = g_0 = \frac{\pi^2}{3}\frac{k_B^2 T}{h} = 9.456 \times 10^{-13} T \left[\frac{W}{K}\right].$$
(12.24)

Experimental evidence for the existence of this quantum thermal conductance is given in the same paper [12.57] on a 60 nm-thick microfabricated silicon nitride membrane narrowed down to a 200 nm-wide channel. The temperature dependence of the thermal conductance of a strictly 1D system is shown in Fig. 12.8; it stands in sharp contrast to the usual T^3 behavior observed in bulk solids at comparable temperatures. For reasons explained in the original publication, there are 16 occupied modes in the measured structure, and the experimental conductance is quantized at $16g_0$.

We return now to (12.23) for systems of arbitrary dimension. We start by neglecting the coupling between the phonon modes and the scattering mechanisms. This will enable us to derive very qualitative conclusions, from which, however, we can at least predict a temperature dependence of the thermal conductivity. We separate the effects of dimensionality on the specific heat and phonon density of states from those on scattering.

The lattice-specific heat of a D-dimensional system depends on D. A 1D chain of atoms can sustain a smaller density of phonon modes than a 2D sheet, which in

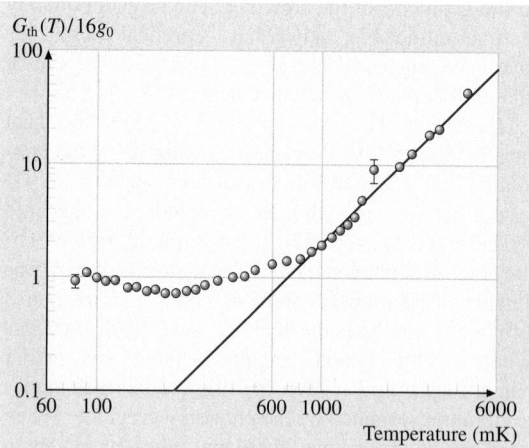

Fig. 12.8 Temperature dependence of 1D phonon thermal conductance of 60-nm-thick silicon nitride "phonon waveguide", which sustains, below 80 mK, only 16 modes of vibration. The thermal conductance is quantized in terms of the quantity g_0 below 80 mK and displays the characteristic T^1 above

turn has fewer modes than a 3D array. The expression for the specific heat of a solid, as given by (12.16) in *Ashcroft* and *Mermin* [12.26], is an integral over the entire momentum space $d\boldsymbol{k}$ of the existing phonon modes of the lattice. The volume $d\boldsymbol{k}$ depends on the dimensionality D of the system as $k^{D-1}dk$. Available phonon modes populate only energies below $k_B\theta$, where θ is the Debye temperature (typically near room temperature for thermoelectric solids, but as high as 2400 K for in-plane modes in graphite). By generalizing the derivation made in [12.26] for $D = 3$, one can see that, at temperatures $T < \theta$, specific heat depends on T and on the phonon velocity as:

$$C \sim T^D v_\phi^{-D}.$$
(12.25)

At $T \gg \theta$, the lattice specific heat is constant.

Moving on to the scattering mechanisms, they can be separated into intrinsic mechanisms, such as Umklapp processes, and extrinsic ones due to the scattering of phonons on the boundaries of the samples. *Peierls* [12.58] pointed out, in 1929, that anharmonic scattering processes are not present in 1D phonon systems. The reason is that, if there is any dispersion at all in the phonon spectrum (if the energy is not rigorously linear with momentum), energy and momentum cannot be simultaneously conserved unless the momentum vector can change angle, which is impossible in one dimension. This applies in particular to Umklapp processes, which dominate the scattering at temperatures near or above about one tenth of the Debye temperature θ of a bulk 3D solid. As a result, one expects extrinsic mechanisms to dominate at any temperature in 1D systems, unlike the situation in bulk solids. If scattering on the boundaries dominates, then the temperature dependence of the lattice thermal conductivity in 1D systems should follow that of the specific heat, being essentially temperature independent above θ. Experimental confirmation of such predictions exists, for instance, in single-crystal polymers [12.59].

Finally, extrinsic scattering mechanisms of the phonons are expected to be more prevalent in any nanostructure than in bulk crystals, based on the fact that a nanometer-scale device size d is of the order of magnitude of the mean free path ℓ_ϕ. We neglect here the effect that d has on the phonon dispersion, which strictly holds when the phonon wavelength is smaller than d. Additional scattering reduces κ_L even in the absence of any changes in heat capacity or sound velocity. Following *Dames* and *Chen* [12.60], let us investigate the case of a segment of length L of a quantum wire of diameter d. We can define an effective phonon mean free path, $\ell_{A,eff}$ for the segment, limited by three scattering terms, $3L/4$ for interface scattering and d for wire boundary scattering:

$$\ell_{A,eff}^{-1} = \ell_{\phi}^{-1} + \frac{4}{3}L^{-1} + \frac{1}{\beta}d^{-1}, \tag{12.26}$$

where β is a dimensionless geometrical factor that includes the specularity of the wire boundary. The thermal conductivity of the segment is reduced by a factor $\ell_{A,eff}/\ell_{\phi}$.

Combining the temperature and velocity dependence of the specific heat (12.25) into (12.23), and also the fact that we expect the scattering mechanisms to be mostly temperature independent in low-dimensional solids, we expect the thermal conductivity to be proportional to:

$$\kappa_{L,D-dim}(T < \theta) \propto T^D v_\phi^{2-D}. \tag{12.27}$$

From this we can draw two conclusions. Firstly, the exponent of the power law of the low-temperature temperature dependence of the thermal conductivity gives the dimensionality of the system. Secondly, while it is the slower phonon modes that dominate conduction in bulk solids because of their high density of modes, that does not hold in lower dimensions.

Considering specifically 2D systems, *Daly* et al. [12.61] report molecular dynamics calculations of the thermal conductivity of superlattices in the direction normal to the layers, including the effect of interface roughness and the Kapitza resistance. A minimum thermal conductivity is observed for one specific period: in the particular case of GaAs/AlAs superlattices, the conductivity at 300 K can be reduced to about 30% of that in the bulk in superlattices of 10 monolayer thicknesses. When the superlattice period is much larger than the phonon wavelength and mean free path, a formula analogous to (12.26) holds, with an additional resistance in series: the Kapitza resistance [12.62], which is the resistance an interface presents to the propagation of an acoustic wave between two materials with different acoustic properties.

In summary, the theory of low-dimensional heat transfer has many unresolved fundamental issues. The experiments are quite challenging. Nevertheless, the general theoretical predictions and early observations point to the fact that the lattice thermal conductivity of low-dimensional systems is expected to be reduced compared to that of bulk solids.

12.3 Two–Dimensional Thermoelectric Transport in Quantum Wells

Increases in the thermoelectric figures of merit of multilayer systems over that of the equivalent bulk materials have been experimentally observed in both of the classical thermoelectric materials systems, the $(Bi_{1-x}Sb_x)_2(Se_{1-y}Te_y)_3$ system [12.13], the Group-IV chalcogenides, and the Si/Ge superlattices [12.63]. We briefly review the observations on both the Group-V chalcogenides and the lead-chalcogenide systems.

Venkatasubramaniam and coworkers [12.13] reported in 2001 that a figure of merit ZT up to 2.4 could be achieved on multiple-quantum-well systems built from Bi_2Te_3/Sb_2Te_3 superlattices. The films were rather thin, so that the measurement of ZT were taken using a series of data as a function of thickness, and the results are indirect. In these superlattices, the claimed increase in ZT comes from a strong decrease in κ_L. Random $Bi_{2-x}Sb_xTe$ ($x = 0.5$) alloys have a lattice thermal conductivity of $\kappa_L = 0.49$ W/m·K in the direction parallel to the c-axis. The authors observe a reduction of κ_L by a factor of 2.2, as they compare the value above to the value $\kappa_L = 0.22$ W/m·K measured on a Bi_2Te_3/Sb_2Te_3 superlattice grown along the same direction with alternating layer thicknesses of 1 nm and 5 nm. The reduction in thermal conductivity has been confirmed on $Bi_2Te_3/Bi_2(Se_xTe_{1-x})$ superlattices [12.64–66].

Lead salt films have been extensively studied [12.67, 68], and, once again, the primary effect is a reduction in the thermal conductivity, for instance on $PbTe/PbSe_{0.2}Te_{0.8}$ superlattices [12.64]. Both this effect and an enhancement of the power factor were demonstrated in the $PbTe/Pb_{0.93}Eu_{0.07}Te$ quantum-well system [12.69, 70]. The maximum $S^2\sigma$ value achieved for the quantum wells is four times higher than that for bulk and is shifted to higher doping levels; the measured values of the power factor are in good agreement with model calculations [12.71]. Using a value for the lattice thermal conductivity of 2 W/m·K for the wells alone, and ignoring the influence of the PbEuTe barrier layers, this corresponds to a room-temperature ZT value of 1.2 for the n-type material and 1.5 for the p-type material. Slightly different results were obtained in PbTe/PbSrTe quantum-well structures [12.64], in which the Seebeck coefficient and the power factor are increased when the quantum well width decreases below 3 to 5 nm. Unfortunately, the power factor of the wells does not exceed that of bulk because the narrower wells have a reduced electron mobility. The ZT values of the PbTe/PbSrTe quantum wells they report is, in one case, in excess of 1, but this is ascribed to a reduction of the lattice thermal conductivity to about 0.30 W/m·K.

12.4 One-Dimensional Thermoelectric Transport in Quantum Wires

12.4.1 Bismuth Nanowires

Bismuth (Bi) is particularly suitable for the study of transport properties at the nanometer scale. Dealing with a single element rather than with a compound simplifies the sample preparation. The small effective mass of electrons in that material results in a long electron de Broglie wavelength; their high mobility gives a long mean free path. When the material is in the form of nanowires, size quantization of the band structure of Bi leads to singularities in the density of states, as illustrated in Fig. 12.3, and this leads to a strong dependence of the density of states (DOS) on energy. Thus the energy derivative of the electrical conductivity σ is large, and, as we saw in Fig. 12.4, this leads to an enhanced Seebeck coefficient S via the Mott formula (12.19) and (12.20). A full calculation is presented by *Lin* et al. [12.72]. In the case of Bi, there is an additional twist. Bulk Bi is a semimetal, with a conduction band and a valence band that overlap by about 38 meV at 4 K. Size-quantization effects raise the energies of both electron and hole bands. As a result, in nanowires with progressively smaller diameters, first the overlap decreases, then crosses zero in nanowires with diameters on the order of 50 nm, and finally turns into an energy gap. Thus we have a size-quantization-driven metal-to-semiconductor transition. The energy-band gap of semiconducting Bi increases as the wire diameter d decreases following roughly a d^{-2} law. This semiconductor is predicted [12.72] to have values of ZT at 77 K that, if nothing interferes with the size-quantization effects, reach 2 at 10 nm diameter and 6 at 5 nm diameter. The optimum ZT is reached when the material contains 10^{18} cm^{-3} charge carriers. As was mentioned in paragraph 2, this optimum is quite sensitive to the doping level because the enhancement holds when the Fermi energy is quite exactly located near one of the discontinuities in the DOS. Disorder will decrease ZT. Firstly, disorder will smear the discontinuities in DOS, thus reducing S. Secondly, localization effects will compete with size-quantization effects, reduce σ, and influence S. These effects will become more pronounced as the wire diameter decreases.

The preparation techniques used on Bi nanowires are already reviewed in the section on nanowires of this handbook. The nanowires are imbedded in porous, electrically insulating host materials, of four types:

1. Anodic Al_2O_3, grown on aluminium, with pores ranging from 200 to 7 nm, oriented parallel to each other and perpendicularly to the anodic film plane;
2. Porous Al_2O_3 in small grains, with randomly oriented 9 nm pores;
3. Porous SiO_2, silica gel, also in small grains, with 15 nm randomly oriented pores;
4. Porous Vycor glass, > 95%SiO_2, with 4 nm randomly oriented pores.

Three different techniques have been used to grow Bi nanowires in these pores: electrochemical deposition [12.73, 74], high-pressure liquid injection [12.75], and vapor-phase growth [12.76]. All three techniques produce highly crystalline wires, and, in the oriented porous anodic alumina host, even single-crystal wires along a preferred orientation (this is the [0112] direction for vapor-grown wires [12.76]). Here, we only review the vapor-phase-preparation technique that we

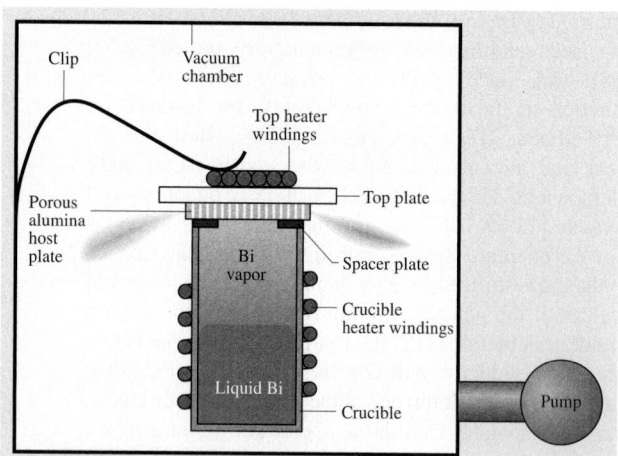

Fig. 12.9 Vapor-phase preparation of nanowires embedded in a porous host material

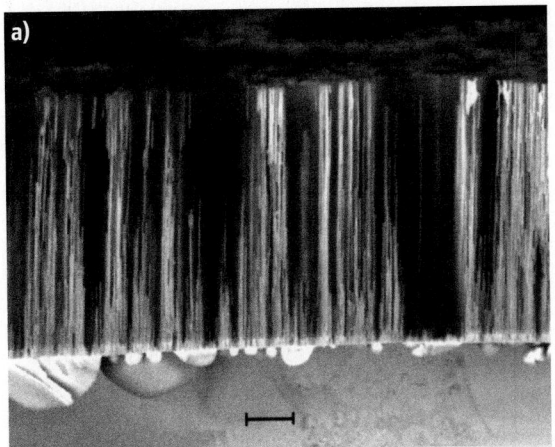

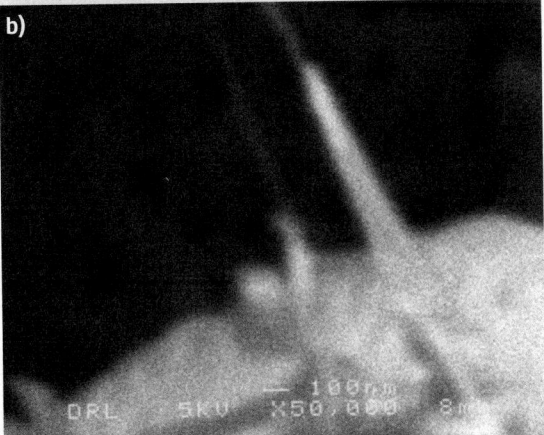

Fig. 12.10 Scanning electron micrographs of (**a**) aligned Bi nanowires in anodic Al_2O_3 with wire diameter $d = 200$ nm and (**b**) SiO_2 with $d = 15 -$ nm pores containing a random network of Bi nanowires. Some wires can be seen protruding out of a grain. The scale marker in (**a**) is $10\,\mu m$, in (**b**) 100 nm (after [12.76, 77])

introduced [12.76] and that is applicable in general to elemental nanowires made from materials that have a sufficient vapor pressure such as Bi, Sb, or Zn. This technique results in wires with a very low defect density.

The vapor-phase technique is shown in Fig. 12.9. One can think of this approach in the following way: many of the porous host materials are commercially available as desiccators. The host material is placed in a vacuum chamber, thoroughly dried out, and then put in the presence of a metal vapor, which it absorbs just like it would water vapor out of the atmosphere. The material is then cooled and the metal vapor solidifies inside the

pores. In detail, referring to Fig. 12.9, the process proceeds as follows: in a cryopumped vacuum chamber, the porous host material is placed on top of a crucible, which is resistively heated. A top plate with a separate heater caps the assembly, which is held in place by a clip. In a first step, after the assembly has been pumped down to a base pressure of about 10^{-8} Torr, the heaters are used to outgas the host material at about 650 °C. In the second step, both heaters are powered up to an operating temperature of about 480 °C for Zn and 590 °C for both Bi and Sb. The charge of the guest metal (Bi) is heated so that metal vapor fills the crucible, passes through the porous host, and escapes into the vacuum. This cleans the pores. After a few minutes, the temperature of the top heater is decreased first, and then the whole assembly is slowly cooled, while a temperature gradient across the porous host is maintained so that nanowires of the guest metal condense in the pores, from top to bottom. The process is stopped once the pores are filled with metal.

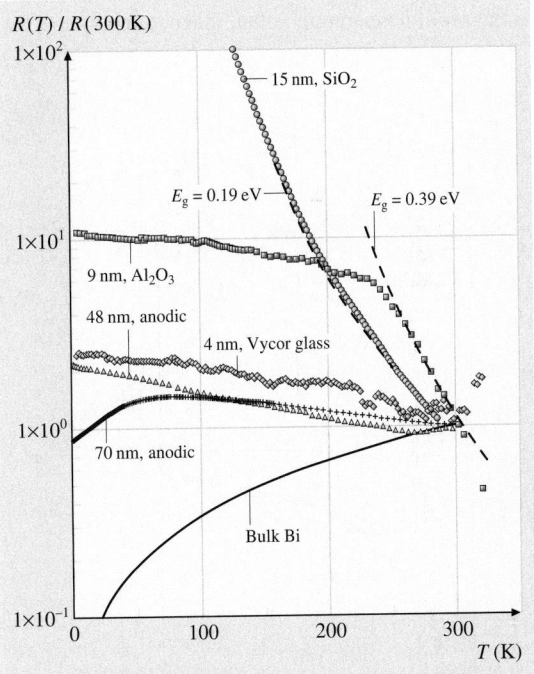

Fig. 12.11 Temperature dependence of resistance ratio $R(T)/R(300\,K)$ of Bi nanowires of various diameters d. The wires cover the transition from the semimetallic ($d > 50$ nm) to the semiconducting ($d < 50$ nm) regime. The semiconducting wires show an activated temperature dependence of $R(T)$, with the activation energy corresponding to the energy gap shown (after [12.76, 77])

Part A | 12.4

Samples prepared as described above are macroscopic nanocomposites and can be handled and measured as bulk samples. Scanning electron micrographs of Bi-impregnated anodic Al_2O_3 (nanowire diameter $d_w = 200$ nm) and SiO_2 ($d_w = 15$ nm) are shown in Fig. 12.10. The anodic alumina is a plate, $50\,\mu$m thick, in which the pores are aligned and traverse the sample from end to end, as can be seen in Fig. 12.10a. The porous SiO_2 grains are visible in Fig. 12.10b, with the pores visible as dark spots. Some pores are filled with Bi, and a few Bi nanowires are seen extending out of the foreground grain in Fig. 12.10(b).

The samples are mounted in conventional cryostats, and their electrical resistance and magnetoresistance [12.76], Seebeck coefficient [12.77, 79], and thermal conductance [12.79] have been reported. As it is impossible to assess how many nanowires are electrically connected through the nanocomposite, it is not possible to estimate the effective cross-section of the nanowires and, thus, the electrical resistivity. The authors therefore report only resistance values normalized to the room temperature value, magnetoresistance values normalized to the resistance at zero magnetic field at the same temperature, and thermal conductance values.

Figure 12.11 reports a compilation of the temperature dependence of the normalized resistance of various nanowires in anodic Al_2O_3 and SiO_2 for the various diameters indicated. Three distinct behaviors are observed.

1. Bulk Bi has a positive dR/dT slope, because, even though the electron and hole densities increase by about one order of magnitude between 70 and 300 K in that semimetal [12.80], the phonon-limited mobility decreases with temperature almost as a T^{-4} law [12.80]. The 200 nm- and 70 nm-diameter wires are also semimetals: they have $dR/dT > 0$ below 100 K, where the carrier density is rather temperature independent and phonon scattering contributes to a negative temperature coefficient of mobility, even though scattering on the wire boundaries dominates. Above 100 K, the temperature dependence of the carrier density dominates, while that of the mobility is much weaker than in bulk.

2. Nanowires with diameters $d_w < 50$ nm have a $dR/dT < 0$ over the entire temperature range, implying that the carrier density is now temperature dependent at all temperatures. This is evidence for the metal-to-semiconductor transition that occurs at a diameter of about 50 nm [12.76]. In this semiconductor regime (50 nm $< d <$ 9 nm), the samples, including the 9 nm Bi/Al_2O_3 samples at high temperature, display an activated behavior following a law:

$$R(T) = R_0 \exp(-E_g/kT). \tag{12.28}$$

The fitted values of the activation energy E_g compare [12.22] very well to the calculated values [12.72] for the energy gap that opens up in Bi nanowires as the wire diameter is reduced.

3. At narrower diameters yet, or at lower temperatures for the 9 nm wires, the resistance follows roughly a $T^{-1/2}$ law, indicative of localization effects. These localization effects have been confirmed by magnetoresistance studies [12.76, 81] and are also present in Sb [12.82] and Zn [12.43] nanowires, as shown in Figs. 12.5 and 12.6.

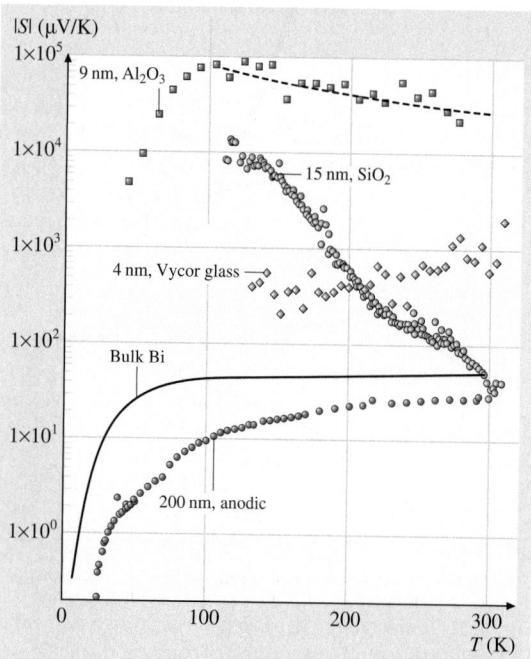

Fig. 12.12 Temperature dependence of absolute value of Seebeck coefficient of Bi nanowires of various diameters d. *Solid curve:* $S(T)$ for bulk Bi [12.78]. The Seebeck coefficient of 9-nm wires follows a $1/T$ law (*dashed line*) at high temperature (after [12.77, 79])

The absolute value of the thermoelectric power S of various Bi nanowires is compiled in Fig. 12.12 from references [12.79] and [12.77]; data for bulk Bi are those of *Gallo* et al. [12.78]; $S > 0$ for the $d_w = 9$ nm wires, and $S < 0$ for the others. Again, we identify three regimes.

The metallic 200 nm-diameter Bi wires were single crystals in oriented anodic alumina [12.79]. Their magnetoresistance at 4.2 K shows Schubnikov–de Haas oscillations, which were used to deduce the electron and hole densities and the Fermi energies in the conduction and valence bands. From this, the authors calculated the partial electron and hole thermopowers and thus the total thermoelectric power. This calculation resulted in an excellent agreement with the measured values.

The absolute value of the thermoelectric power of the semiconducting Bi nanowires of the diameters indicated is very much enhanced over that of the metallic wires. The temperature dependence of the 9 nm wires follows a T^{-1} law at high temperatures, as expected for nearly intrinsic semiconductors, but the thermopower decreases at low temperatures. This is again consistent with the theory [12.12, 72] that attributes the thermopower enhancement to quantum size effects, the same mechanism that leads to the opening of an energy gap in Bi nanowires. The 15 nm wires are probably compensated, with contributions of both electrons and holes, which leads to a very pronounced temperature dependence.

The nanowires in which the resistivity follows a $T^{-1/2}$ law, and in which localization effects are important, such as the 4 nm wires and the 9 nm wires at low temperatures, have a strongly decreased thermopower compared to the semiconducting ones.

The thermal conductance measured in [12.79] on samples with Bi and Zn nanowires is shown in Fig. 12.13 as a function of temperature. The experimentally observed dependence shows a decrease from room temperature to a plateau, followed by a T^n law upon further cooling, with $n \approx 1.2$ to 1.4. In contrast, the thermal conductivity of bulk Bi, which is phonon dominated, is quite different and follows a conventional temperature dependence. The Zn nanocomposite has exactly the same behavior as the Bi nanocomposite. These two observations suggest that the alumina matrix conducts most of the heat, even though the thermal conductivity of Bi is two orders of magnitude larger than that of glasses at room temperature.

In summary, there is a remarkable agreement between the experimental data and the theoretical work [12.12, 72] on the bismuth nanowire system. There is an optimum nanowire diameter, around 9 nm, where size-quantization effects are important but below which the thermopower decreases, along with the temperature dependence of the resistance, possibly because of localization effects. The evidence for

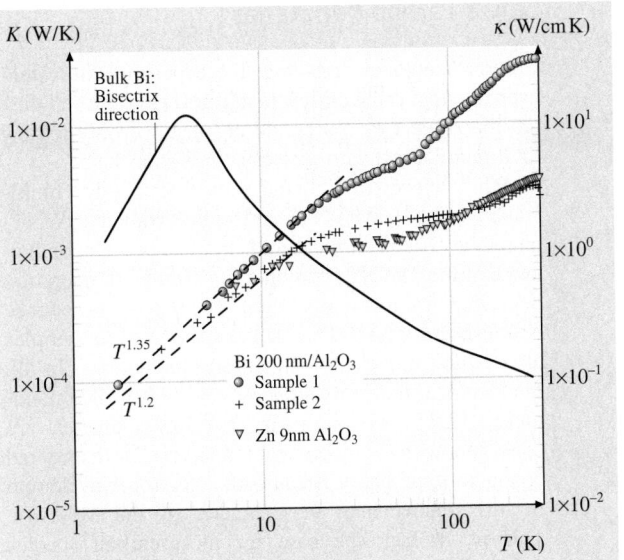

Fig. 12.13 Temperature dependence of thermal conductance of two samples of 200-nm Bi nanowires imbedded in oriented anodic alumina and of one sample of 9-nm Zn nanowires in randomly oriented porous alumina (*left ordinate axis*). The thermal conductivity of bulk Bi is also shown (*right ordinate*) (after [12.83])

localization comes from magnetoresistance measurements. Unfortunately, the heat conduction in the Bi and Zn nanowire systems is dominated by that through the host material. Combined with the fact that the electrical conductance is low, because few wires connect electrically all the way through the insulating host, this severely limits the figure of merit of the composite.

$Bi_{1-x}Sb_x$ bulk solid solutions are semiconductors [12.84] in the concentration range $5\% < x < 22\%$. $Bi_{1-x}Sb_x$ nanowires (for $x < 0.2$) are predicted to have a better thermoelectric performance than pure Bi nanowires of the same diameter. The semimetal-to-semiconductor transition is moved to higher diameters in the $Bi_{1-x}Sb_x$ system. $Bi_{1-x}Sb_x$ nanowires can be prepared [12.85] using the same basic electrochemical and pressure-injection synthesis methods as described above for preparing bismuth nanowires, but the vapor-phase method was not successful, because the very vapor pressures Bi and Sb are very different at any given temperature. A phase diagram giving the semiconductor-to-semimetal transition, and the optimum ZT calculated [12.86] as a function of nanowire diameter and x are reviewed in the Chapt. 4 on nanowires in this handbook.

12.4.2 Carbon Nanotubes

Carbon nanotubes constitute the experimental system available that is the closest to a true 1D system. As they are the subject of a separate chapter, we only review their thermal and thermoelectric properties here.

The electrical conductivity of both multi-walled [12.87] (MWNT) and single-walled (SWNT) nanotubes has been measured. In essence, the conductance of the MWNTs exhibit a logarithmic temperature dependence at low temperature. The magnetoresistance is consistent with 2D weak localization. The samples also exhibit universal conductance fluctuations. In all, the electrical transport properties of multiwalled nanotubes [12.87], despite their very low diameter, are consistent with the theoretical models for 2D disordered conductors, probably because the electron wavelength is still smaller than the diameter of the MWNTs. This is, in fact, the case for most carbon species with an appreciable degree of disorder [12.88, 89]. The situation of SWNTs can be different. It has been suggested by *White* and *Todorov* [12.90] that armchair single-walled nanotubes should be insensitive to localization and have exceptional ballistic behavior over long length scales. Similarly, metallic nanotubes might be insensitive [12.91] to certain types of long-range disorder.

The one dimensionality of carbon nanotubes is more evident in the specific heat and lattice thermal conductivity, probably because the phonon wavelength (unlike the electron wavelength) is larger than typical tube diameters. The specific heat of MWNTs was shown [12.92] to be linear with T over the entire measurement range (10 to 300 K), which is below the in-chain Debye temperature, consistent with (12.25).

The thermal conductivity has been measured by many groups [12.50,92], but we concentrate on three sets of results. *Kim* et al. have measured individual MWNTs in an elegant experiment involving a microfabricated silicon device [12.93]. The low-temperature thermal conductivity follows a $\kappa \propto T^{2.5}$ law for $8\,\text{K} < T < 50\,\text{K}$. The temperature exponent slows to a $\kappa \propto T^2$ law for $(50\,\text{K} < T < 150\,\text{K})$ and then flattens, reaching a flat maximum of 3000 W/m·K near 300 K. Above 320 K, the thermal conductivity decreases. The low-temperature power laws in excess of 1, as well as the high-temperature maximum followed by a decrease in κ reminiscent of Umklapp processes, leads to the suggestion that this system is not one-dimensional. In-plane graphitelike phonon modes that result in a $\kappa \propto T^2$ law, and even interplane phonon modes that lead to an even higher exponent, probably exist. Also, the very high value of the thermal conductivity, much in excess of what was measured before on samples that consisted of mats on nanotubes, is not consistent with the paucity of available modes in a purely 1D system. In fact, the results are somewhat reminiscent of those measured [12.94] on conventional vapor-grown graphite fibers heat-treated to 3000 C, where a classical graphitelike $\kappa \propto T^{2.28}$ law was observed, and a maximum around 2300 W/m·K.

As far as measurements on SWNT are concerned, there are data on ropes [12.95] and films [12.96]. The temperature behavior is now clearly different from that of graphite: a monotonous increase in κ with T is observed, and no clear power law can be attributed over the entire measurement range (8 to 350 K). The curves are best described as "linear by part," with a linear $\kappa \propto T$ dependence below 25 K, and another one above 150 K. A linear dependence is consistent with a true 1D system, (12.27), as is the absence of Umklapp processes near room temperature. The absolute magnitude of the conductivity is not reported, as it is indeed very difficult to determine from the intricate geometry of the wire bundle.

The thermopower of carbon nanotubes is much easier to measure quantitatively than the conductivities, as it does not depend on the path of the heat flux. The Seebeck coefficient of graphite is small due to the delicate balance between the electron and hole thermopowers that partially compensate each other. It is therefore not surprising that in nanotubes, in which the electronic structure is greatly affected by the finite size, the thermopower is a very sensitive probe of small variations in both the charge density and the scattering mechanisms.

The same MWNTs whose thermal conductivity was reminiscent of graphite fibers [12.94] show a strikingly different thermopower [12.93]. The Seebeck coefficient of MWNTs is linear in temperature, positive, and can reach 75 µV/K at 300 K, while the Seebeck coefficient of graphite, including that of heat-treated fibers, is negative, has a phonon-drag peak around 20 K and only reaches −10 µV/K at 300 K.

Films of SWNTs [12.96] have a Seebeck coefficient that is a strong function of annealing conditions and alignment. This behavior is the result of the sensitivity of the Seebeck coefficient of carbon nanotubes to external perturbations. The density of states of carbon nanotubes has extrema such as those seen in Fig. 12.3 for 1D systems (the difference between semiconducting and metallic tubes depends on the existence of a gap, i. e., the absence of states, in the trough between the peaks). The Seebeck coefficient being, via the Mott formula, sensitive to the energy dependence of the conductivity, it is a function of the energy dependence of either the density of states or of the scattering mechanisms. The density of states at the Fermi level can thus be expected to vary dramatically with any shift in it, brought about, for instance, by applying an external electric field to the side of the nanotubes. As a result, there are reports in the literature about the modulation of the Seebeck coefficient of carbon nanotubes with gate electrodes [12.97]. The presence of an external gas around the nanotubes is also known to strongly affect the Seebeck coefficient [12.98, 99], and both mechanisms (changes of the Fermi level in the density of states, or changes in scattering) are potentially responsible. A more detailed examination of this case follows.

With their large exposed surface area and the small density of charge carriers per atom, SWNTs offer a unique opportunity for studying [12.98] the interaction of molecules with the electric properties of the substrate, of which, as pointed out above, the Seebeck coefficient is the most sensitive. Several groups [12.98, 100, 101] have observed the sensitivity of the Seebeck coefficient of carbon nanotubes to environmental gases, in particular oxygen and hydrocarbons. The material studied usually consists of films or bundles of SWNTs; it is likely that the Seebeck coefficient measurements are dominated by the metallic nanotubes in the mixture, since they have the highest conductivity. Indeed, composite systems in which several materials with different Seebeck coefficients are connected in parallel behave like voltage sources connected in parallel: the average Seebeck coefficient of the composite equals the average of the Seebeck coefficients of each component weighted by its electrical conductivity. To summarize the early observations [12.98, 100, 101], bundles of SWNTs exposed to O_2 have a large positive thermopower at room temperature. Upon outgassing, the room-temperature Seebeck coefficient can become small and negative (-6 to $-10 \mu V/K$), i.e., graphite-like. Exposure to cyclic hydrocarbons C_6H_{2n} increases the room-temperature Seebeck coefficient, to perhaps $-4 \mu V/K$ for $n = 5$, $-2 \mu V/K$ for $n = 4$ and $+2 \mu V/K$ for $n = 3$. The precise mechanism through which the tubes and the gases interact was the subject of further investigation, as the most intuitive mechanism seemed to be a charge transfer between the carbon and the molecules of adsorbed gas. Recently, however, *Romero* and coworkers [12.99] established that the Seebeck coefficient of nanotube felts are primarily sensitive to two parameters, the molecular weight and the pressure of the gas they are exposed to, but not the chemical nature of the gas. The dependence on gas pressure saturates at a pressure on the order of 0.3 to 0.5 atm. The changes in Seebeck coefficient after saturation depend on the molecular weight M of the gas as $M^{1/3}$, for He, Ar, Ne, Kr, Xe, N_2, and CH_4. The simultaneously observed changes in resistivity are consistent with the Mott relation. The temporal rate of change of the Seebeck coefficient is consistent with the dif-

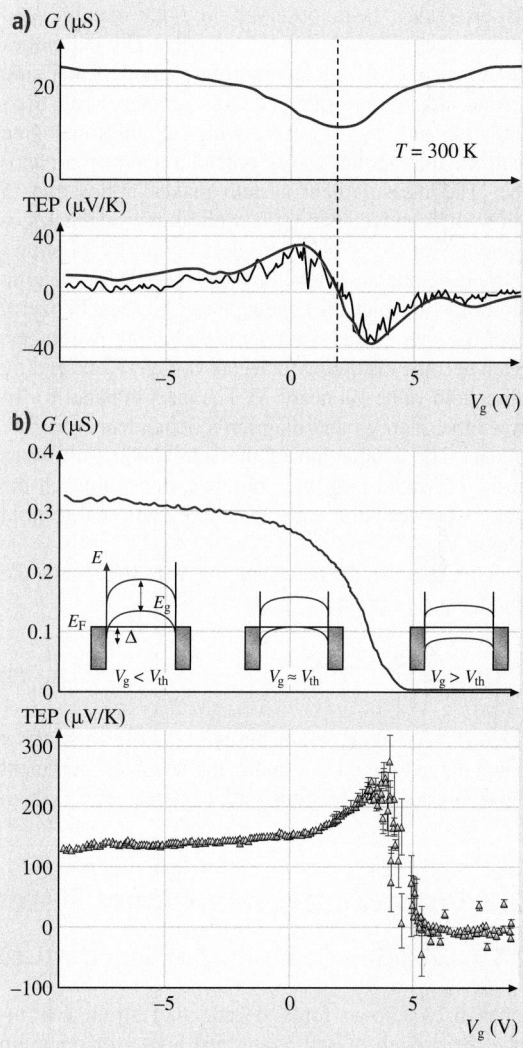

Fig. 12.14a,b Gate voltage dependence at 300 K of conductance and Seebeck coefficient of (**a**) metallic and (**b**) semiconducting single-walled carbon nanotubes (after [12.97])

fusion of gas into the pores of the nanotube bundles. All this leads the authors to conclude that the primary effect of gases on the Seebeck coefficient is through a kinetic energy transfer between the gas molecules and the nanotubes: the impact of the gas molecules excites a particular "squash" phonon mode of the tube walls, and this phonon changes the scattering of the electrons, to which the Seebeck coefficient is very sensitive.

Returning now to the dependence of the Seebeck coefficient on an applied gate electrode potential [12.97], this effect has been observed in both metallic and semiconducting single SWNT samples. The nanotubes are attached to a silicon microfabricated heater and thermal sink system equipped with the appropriate thermometry, and, in this case, with an additional gate electrode that applies a bias potential across the nanotube. The measurement system makes it possible to follow the conductance G as well as the Seebeck coefficient S of the SWNT samples. Figure 12.14 shows the room-temperature results: panel a holds for metallic nanotubes, in which the conductance never reaches zero; panel b shows G and S for semiconducting nanotubes, which become insulating for a gate voltage V_g exceeding a threshold value V_{th} near 5 V. The inset in panel b also shows the energy-band diagram's dependence on gate voltage. The conduction of the semiconducting wires can be classified into three regimes, depending on the relation between the gate voltage V_g and a threshold voltage V_{th}.

At $V_g < V_{th}$, the device is in the "on state, and the conductance is given by:

$$G(E) = \frac{4e^2}{h} e^{-\alpha\sqrt{\Delta+E}} , \qquad (12.29)$$

where α is a scaling parameter and Δ is the band offset shown in the figure. Once again, the Seebeck coefficient is given by the Mott relation (with G instead of σ), which gives:

$$S = \frac{\pi^2}{6} \frac{k_B}{q} \frac{k_B T}{\Delta} \ln \left(\frac{G_{measured}}{\frac{4q^2}{h}} \right) . \qquad (12.30)$$

At $V_g \approx V_{th}$, the region where the Seebeck coefficient goes through a maximum, the conductance is given by:

$$G(E) = e^{-\frac{E-\Delta}{k_B T}} , \qquad (12.31)$$

and the Mott relation gives:

$$S = \frac{\pi^2}{3} \frac{k_B}{q} = 270 \frac{\mu V}{K} , \qquad (12.32)$$

much as is observed. In the third region, $V_g > V_{th}$, both G and S vanish.

Two essential conclusions can be drawn from the observations in the paper. Firstly, the relation between G and S follows the Mott equation. Secondly, while the absolute value of the Seebeck coefficient of the metallic tubes is on the order of $|S| \approx 10$ to $40\,\mu V/K$, with a sign depending on the gate voltage and therefore presumably on the type of dominant carrier, the semiconducting nanotubes have Seebeck coefficients that can reach $280\,\mu V/K$. The authors suggest that such material might be suitable for thermoelectric applications, as it is clear that the power factor can be quite favorable. Of course, (12.1) makes this suggestion seem incompatible with the claim that carbon nanotubes may have one of the highest thermal conductivities [12.93] (3000 W/Km) of any solid, but, given the experimental difficulties with the geometrical factors involved in each experiment, there is plenty of opportunity for further investigations. The geometrical factors cancel in the expressions of ZT because both the electrical and thermal conductivity have the same factor and the Seebeck coefficient is path independent. A simultaneous measurement of S, σ, and κ on the same sample is the next necessary step, however technically challenging.

This work was recently extended [12.102] to silicon nanowires, with much the same conclusions.

12.5 Quasi-Zero-Dimensional Systems, Solids Containing Quantum Dots

12.5.1 Lead–Salt Quantum–Dot Superlattices

It has been known for a decade that strain can induce the growth of self-assembled arrays of quantum dots at the interfaces between two slightly mismatched films. This technique has been applied successfully in lead chalcogenides [12.103, 104], in which PbSe quantum dots can be grown [12.105] with a uni-

formity in size distribution of better than 2% on a $4\,\mu m$-thin PbTe layer on Si (111). This knowledge was applied to create optimal thermoelectric materials [12.14, 16], grown by molecular beam epitaxy (MBE). The structure of a PbTe/PbSe$_{1-x}$Te$_x$ quantum-dot superlattice (QDSL) is shown in Fig. 12.15, along with a plan-view scanning electron micrograph showing the quantum dots [12.14, 106]. When viewed from

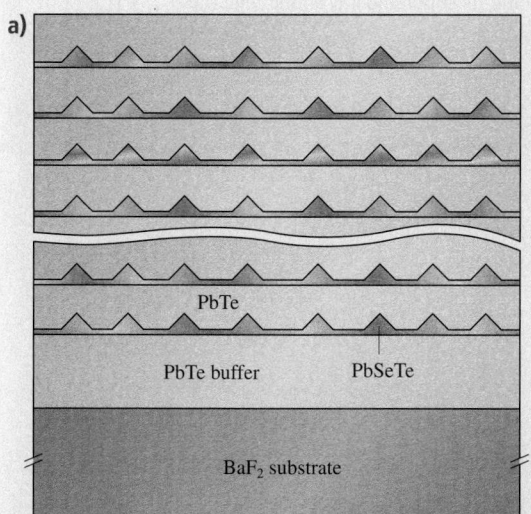

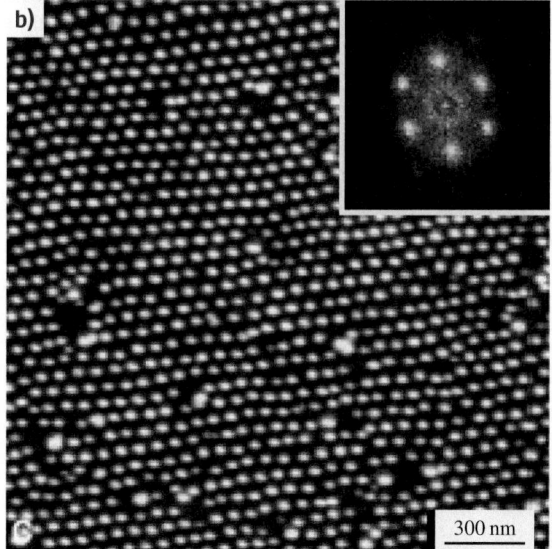

Fig. 12.15 Schematic structure (**a**) and scanning electron micrograph (**b**) of a PbTe quantum-dot superlattice (after [12.103, 104, 106])

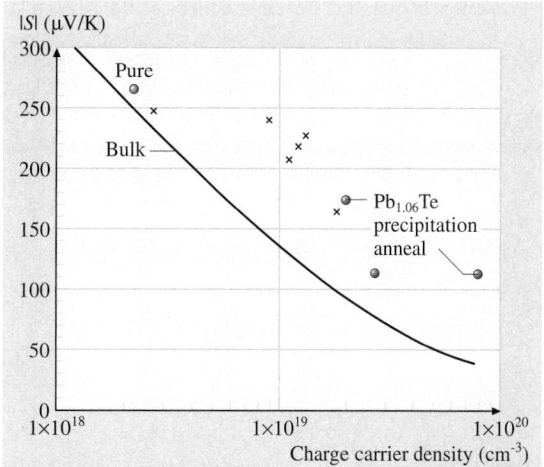

Fig. 12.16 Dependence of Seebeck coefficient of PbTe on carrier density, for typical bulk PbTe (*full line*). The data obtained by Harman et al. [12.14] on PbTe quantum dot superlattices (*crosses*) and the data measured on bulk PbTe containing Pb-metal nanoscale inclusions exceed the values for bulk (after [12.107])

a 3D standpoint, these QDSLs can be interpreted as a PbTe host into which quantum dots are introduced in a periodic way in all three directions. Systematic improvements of the growth procedures have led to dramatic enhancements in ZT, and most recently values of ZT exceeding 3 at elevated temperatures were reported [12.16]. This can be considered as one of the great successes of thermoelectric nanotechnology.

In Table 12.2 we summarize the properties of both n- and p-type QDSLs, taken from [12.14] and its supplement. The measurements are taken with the current in the plane of the film on freestanding films and are thus very reliable. There are two mechanisms through which the QDSL structure improves ZT.

The thermal conductivity for the n-type PbSeTe/PbTe QDSL is measured to be 0.58 to 0.62 W/m·K. Subtracting an estimate for κ_E from the Wiedemann–Franz law (12.8), this leaves $\kappa_L \approx 0.33$ W/m·K, with possibly lower values yet for the quaternary Sn-containing QDSLs. The lattice thermal conductivity of an equivalent random alloy is [12.5] 1.25 W/m·K. The authors conclude that the primary effect of the quantum dots is reduce κ_L by a factor of almost four. At 300 K the electron mobility of a similar alloy at a very low doping level is [12.5] 1000 cm^2/Vs, and the mobility in lead salts is not [12.108] a strong function of doping up to 2×10^{19} cm^{-3}. Therefore, the QDSL structure reduces the electron mobility by less than a factor of three, so that the QDSL structure results in a net gain in the ratio μ/κ_L. This is the first mechanism through which quantum dots improve ZT.

The second mechanism is an improvement in the power factor. Figure 12.16 shows the absolute value of the Seebeck coefficient of the samples described in Table 12.2 (full diamonds) as a function of the carrier

Table 12.2 Properties of quantum-dot superlattices at 300 K from [12.14] and its supplemental online material

Sample	ZT	S	n, p	μ	ρ	E_g
		(μV/K)	(10^{19} cm^{-3})	$\frac{cm^2}{Vs}$	(10^{-3} Ωcm)	(meV)
n-type (Bi-doped) PbSe$_{0.98}$Te$_{0.02}$/PbTe	1.6	−219	1.2	370	1.4	
n-type (Bi-doped) PbSe$_{0.98}$Te$_{0.02}$/PbTe	1.3	−208	1.1	300	1.9	
n-type (Bi-doped) Pb$_{0.84}$Sn$_{0.16}$Se$_{0.98}$Te$_{0.02}$/PbTe	1.6	−228	1.3	320	1.5	120
n-type (Bi-doped) Pb$_{0.84}$Sn$_{0.16}$Se$_{0.98}$Te$_{0.02}$/PbTe	2.0	−241	0.89	540	1.3	170
p-type (Tl-doped) PbSe$_{0.98}$Te$_{0.02}$/PbTe	0.8	248	0.27	440	5.3	
p-type (Tl-doped) PbSe$_{0.98}$Te$_{0.02}$/PbTe	0.9	165	1.8	260	1.3	

density, alongside data for bulk samples (open circles) and other samples that will be described later. The thermopower is considerably enhanced in the QDSLs compared to bulk samples with the same doping level. The exact physical mechanism that leads to this enhancement is not clearly identified, though theories are emerging [12.36] as discussed in Sect. 12.2. Since the mobilities reported in Table 12.2 are only a factor of 2 to 4 (on average 3) lower than those in PbTe alloys, the power factor is unaffected by the mobility decrease. It is, of course, the combination of both effects that leads to $ZT = 2$.

12.5.2 AgPb$_m$SbTe$_{2+m}$

For the field of low-dimensional thermoelectricity to make economic sense, one must be able to produce kilograms or even tons of thermoelectric nanomaterials, which therefore must self-assemble. The recently reported AgPb$_m$SbTe$_{2+m}$ bulk alloys [12.109] have $ZT = 2.1$ at 800 K and also contain nanoscale inclusions. From a strictly metallurgical point, these bulk alloys can be viewed [12.110] as solutions of PbTe, much like TAGS alloys [12.6] are alloys of GeTe and AgSbTe$_2$. The latter compound is also well known to have nanometer inclusions [12.111]. The role of the nanoinclusions in AgPb$_m$SbTe$_{2+m}$ is discussed in a subsequent paper [12.112, 113]. The alloys are not really solid solutions of PbTe and the compound AgSbTe$_2$ but should be regarded as a PbTe lattice with Ag and Sb inclusions. Ag atoms introduce new states near the top of the valence band of PbTe; isolated Sb atoms introduce resonant states near the bottom of the PbTe conduc-

tion band. The Ag–Sb pairs result in an increase in the density of states, compared to that of pure PbTe, right around the band gap. As a result of the Mott equation once again, the Seebeck coefficient, and thus the power factor $S^2\sigma$, is increased. The ubiquitous second mechanism, the reduction of the lattice thermal conductivity, is also at work.

12.5.3 Nanometer-Scale Inclusions in Bulk Solids

Several other bulk materials containing nanometer-scale inclusions have already yielded encouraging preliminary results: for instance, the increase in the Seebeck coefficient is observed in CoSb$_3$/fullerene compos-

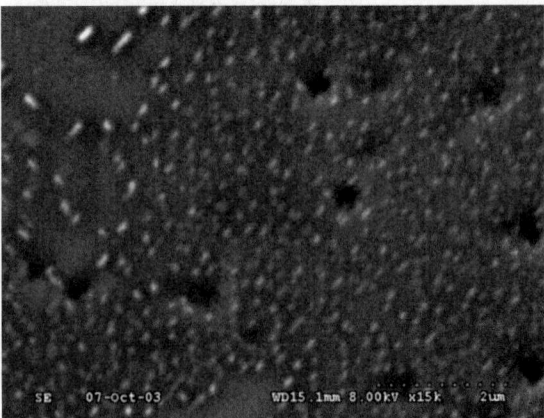

Fig. 12.17 Scanning electron micrograph of bulk PbTe containing nanoscale inclusions of PbTe (after [12.107])

ites [12.114] as well as sintered PbTe samples made of nanometer-sized grains [12.115].

According to the older literature [12.116], lead-rich PbTe can contain small inclusions of metallic lead. Inspired by this, we have shown recently that nanometer-diameter metallic Pb inclusions can be made to precipitate [12.107] in PbTe with 6% excess Pb by conventional metallurgical heat treatments engendered by the precipitation hardening of aluminium [12.117, 118]. The evidence for the presence of metallic Pb is threefold: X-ray diffraction (XRD), scanning electron micrography, and a superconductive phase transition at the same transition temperature as metallic Pb. The SEM image showing the inclusions is presented in Fig. 12.17. XRD patterns show small peaks that correspond to Pb metal, and a Scherer analysis [12.119] established the metallic inclusions to have sizes to be on the order of 30 nm. The temperature dependence of the resistivity and the magnetoresistance show a superconductive phase transition with a critical temperature $T_c = 7$ K, and a critical field ($H_c = 0.06$ to 0.11 T) similar to those of metallic Pb [12.120] ($T_c = 7.2$ K, $H_c = 0.08$ T). Metallic Pb in porous glass [12.121], in which small grains of superconductor are separated spatially but connected by electron tunneling, gives much the same T_c but enhanced values of H_{c2} inversely proportional to the grain size. Samples of PbTe doped with Ag in the high-10^{19} cm^{-3} range are n-type and display XRD spectra, SEM images, and a partial superconductive transition similar to the Pb-rich samples. As was the case of MBE-prepared QDSLs, the Seebeck coefficient of the Pb-rich Pb$_{1.06}$Te samples is much enhanced over that of bulk PbTe samples, in both n- and p-type-doped PbTe: the measured Seebeck coefficients have been added to Fig. 12.16. A priori, one can imagine two possible mechanisms for this enhancement: (1) an increase in the density of states such as could be induced by size-quantization effects or by the mechanism reported [12.112, 113] for the AgPb$_m$SbTe$_{2+m}$ alloys or (2) a change in scattering mechanisms.

To answer this question, we applied [12.115] to the "method of the four coefficients." At each temperature, there are four unknown parameters: the density of carriers, their mobility, the density of states effective mass (which can be increased by size-quantization effects or by the abovementioned effects [12.112, 113] of the nanoinclusions on the density of states in the host material), and the energy dependence of the scattering (which can be affected by electron-energy-filtering ef-

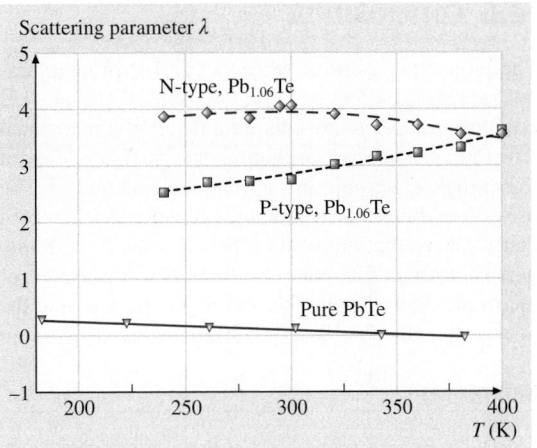

Fig. 12.18 Temperature dependence of scattering parameter λ of bulk PbTe with and without (o) nanoinclusions (after [12.107])

fects [12.19, 39] or by scattering on small inclusions). This latter dependence can conveniently, if not entirely rigorously, be represented by the scattering exponent λ defined by (12.12). To uniquely determine these four unknowns at each temperature, four experimental data points are needed, and these are the resistivity, the Seebeck coefficient, and the Hall and transverse Nernst–Ettingshausen coefficients. In the case of PbTe with 30 nm metallic inclusions, the analysis showed that the density of states mass was not enhanced over that of PbTe, but the scattering exponent was: the values for λ are shown in Fig. 12.18 for homogeneous material ($\lambda \approx 0.7$) and for material with nanoprecipitates ($\lambda \approx 3.5$). The complete model used includes nonparabolic band effects [12.108], which explains why the value for the homogeneous material is 0.7 and higher than 0, as expected for acoustic phonon scattering when the energy bands have parabolic dispersion relations. The value observed in the material with nanoprecipitates is much higher than even ionized-impurity scattering would give, and it is clear that this increased energy dependence of the scattering is the origin of the increased Seebeck coefficient. Since the mobility of the n-type samples remains on the order of 350 cm^2/Ws (compared to 1000 in bulk uniform polycrystalline PbTe), the presence of nanoprecipitates improves the power factor, just as in bulk AgPb$_m$SbTe$_{2+m}$ alloys. On the other hand, no decrease in κ_L was observed in the samples with nanoprecipitates.

12.6 Conclusions

The pioneering work of *Harman* [12.14] in quantum-dot superlattices and of *Ventakasubramaniam* [12.13] establishes the experimental proof that low-dimensional structures can improve the thermoelectric figure of merit over what is possible in conventional 3D solids. The use of low-dimensional structures constitutes a new tool that weakens the interrelation between the three transport parameters that enter the thermoelectric figure of merit, the electrical and thermal conductivities, and the Seebeck coefficient. Two mechanisms are at work in low-dimensional structures: the lattice thermal conductivity is reduced, and the power factor is increased, through an increase in the density of states, or through an increase in the energy dependence of the scattering. We conclude by suggesting that the technological future for thermoelectric energy conversion devices lies in "bulk" low-dimensional systems, in which the quantum dots of wires self-assemble inside a bulk host material, and we reviewed several metallurgical techniques that can lead to this effect.

References

12.1 H. Oji: Thermopower and thermal conductivity in two-dimensional systems in a quantizing magnetic field, Phys. Rev. B **29**, 3148–3152 (1984)

12.2 N. H. Mott, E. A. Davis: *Electronic Processes in Non-Crystalline Materials* (Clarendon, Oxford 1979)

12.3 M. Telkes: The efficiency of thermoelectric generators, Int. J. Appl. Phys. **18**, 1116 (1947)

12.4 A. F. Ioffe: *Semiconductor Thermoelements and Thermoelectric Cooling* (Infosearch, London 1957)

12.5 A. F. Ioffe: *Physics of Semiconductors* (Academic, New York 1960)

12.6 F. D. Rosi, J. P. Dismukes, E. F. Hockings: Semiconductor materialsfor thermoelectric power generation up to 700 C, Electrical Engineering **79**, 450–459 (1960)

12.7 J–P. Fleurial, A. Borshchevsky, T. Caillat, D. T. Morelli, G. P. Meisner: High figure of merit in Ce-filled skutterudites, Proc. 15th Int. Conf. on Thermoelectrics, Piscataway, NJ 1996, ed. by T. Caillat, J.-P. Fleurial, A. Borshchevsky (IEEE, Piscataway, NJ 1996) IEEE catalog number 96TH8169

12.8 D. T. Morelli, T. Caillat, J.-P. Fleurial, A. Borshchevsky, J. Vandersande, B. Chen, C. Uher: Low temperature transport properties of *p*-type CoSb$_3$, Phys. Rev. B **51**, 9622–9628 (1995)

12.9 H. J. Goldsmid: *Applications of Thermoelectricity* (Methuen, London 1960)

12.10 G. S. Nolas, J. Sharp, H. J. Goldsmid: *Thermoelectricity* (Springer, Berlin Heidelberg 2000)

12.11 L. D. Hicks, M. S. Dresselhaus: Effect of quantum-well structures on the thermoelectric figure of merit, Phys. Rev. B **47**, 12727–31 (1993)

12.12 L. D. Hicks, M. S. Dresselhaus: Thermoelectric figure of merit of a one-dimensional conductor, Phys. Rev. B **47**, 16631–4 (1993)

12.13 R. Venkatasubramanian, E. Siivola, T. Colpitts, B. O'Quinn: Thin–film thermoelectric devices with high room-temperature figures of merit, Nature **413**, 597 (2001)

12.14 T. C. Harman, P. J. Taylor, M. P. Walsh, B. E. LaForge: Quantum dot superlattice thermoelectric materials and devices, Science **297**, 2229–32 (2002)

12.15 W. Zawadzki, R. Lassnig: Magnetization, specific heat, magneto-thermal effect and thermoelectric power of two-dimensional electron gas in a quantizing magnetic field, Surface Science **142**, 225–235 (1984)

12.16 T. C. Harman: *Oral Report and Invited Paper* (Materials Research Society Meeting, Boston 2003)

12.17 D. J. Bergman, O. Levy: Thermoelectric properties of a composite medium, J. Appl. Phys. **70**, 6821 (1991)

12.18 D. J. Bergman, L. G. Fel: Enhancement of thermoelectric power factor in composite thermoelectrics, J Appl. Phys. **85**, 8205–16 (1999)

12.19 D. Vashaee, A. Shakouri: Improved thermoelectric power factor in metal-based superlattices, Phys. Rev. Lett. **92**, 106103 (2004)

12.20 N. V. Simkin, G. D. Mahan: Minimum thermal conductivity of superlattices, Phys. Rev. Lett. **84**, 927–930 (2000)

12.21 C. Dames, M. S. Dresselhaus, G. Chen: Phonon thermal conductivity of superlattice nanowires for thermoelectric applications, Mat. Res. Soc. Symp. Proc. **793**, S1.2.1 (2004)

12.22 J. P. Heremans: Thermoelectric power, electrical and thermal resistance, and magnetoresistance of nanowire composites, Mat. Res. Soc. Symp. Proc. **793**, S1.1 (2003)

12.23 J. W. Fenton, J. S. Lee, R. J. Buist: Counter flow Thermoelectric Heat Pump with Discrete Sections, (1978) USPN 4,065,936

12.24 J. Stockholm: Large-scale Cooling: Integrated Thermoelectric Element Technology. In: *CRC Handbook of Thermoelectrics*, ed. by D. M. Rowe (CRC Press, Boca Raton 1995) p. 661

12.25 L. E. Bell: *Use of Thermal Isolation to Improve Thermoelectric System Operating Efficiency* (Proc 20th Int. Conf. Thermoelectricity, IEEE, Long Beach 2002)

12.26 N. W. Ashcroft, N. D. Mermin: *Solid State Physics* (Holt, Rinehart and Winston, New York 1976)

12.27 W. F. Leonard, T. J. Martin Jr.: *Electronic Structure and Transport Properties of Crystals* (Krieger, Malabar 1979)

12.28 M. Cutler, N. F. Mott: Observation of Anderson localization in an electron gas, Phys. Rev. **181**, 1336 (1969)

12.29 A. Ya. Shik: Transport in one-dimensional superlattices, Sov. Phys. – Semiconductors **7**, 187–92 (1973) [translated from Fiz. Technol. Poluprov. **7** (1973) 261]

12.30 L. Friedman: Thermopower of superlattices as a probe of the density of states distribution, J. Phys. C: Solid State Phys. **17**, 3999–4008 (1984)

12.31 Z. Tao, L. Friedman: Thermoelectric power of superlattices II, J. Phys. C: Solid State Phys. **18**, L455–71 (1985)

12.32 L. D. Hicks, T. C. Harman, M. S. Dresselhaus: Use of quantum-well superlattices to obtain a high figure of merit from nonconventional thermoelectric materials, Appl. Phys. Lett. **63**, 3230–1 (1993)

12.33 D. A. Broido, T. L. Reinecke: Use of quantum-well superlattices to obtain a high figure of merit from nonconventional thermoelectric materials, Appl. Phys. Lett. **67**, 1170–1 (1995)

12.34 J. O. Soho, G. D. Mahan: Thermoelectric figure of merit of superlattices, Appl. Phys. Lett. **65**, 2690–2 (1994)

12.35 P. J. Lin-Chung, T. L. Reinecke: Thermoelectric figure of merit of composite superlattice systems, Phys. Rev. B **51**, 13244–8 (1995)

12.36 T. E. Humphrey, H. Linke: Reversible thermoelectric nanomaterials, Phys. Rev. Lett. **94**, 096601 (2005)

12.37 J. Tauc: *Photo and Thermoelectric Effects in Semiconductors* (Pergamon, New York 1962)

12.38 G. D. Mahan, L. M. Woods: Multilayer thermionic refrigeration, Phys. Rev. Lett. **80**, 4061 (1998)

12.39 A. Shakouri, J. D. Bowers: Heterostructure integrated thermionic coolers, Appl. Phys. Lett. **71**, 1234 (1997)

12.40 G. N. Hatsopoulos, J. A. Welsh: Thermionic Emission Thermoelectricity. In: *Thermoelectric Materials and Devices*, ed. by I. B. Cadoff, E. Miller (Reinhold, New York 1960) pp. 47–54

12.41 D. Vashaee, A. Shakouri: Improved thermoelectric power factor in metal-based superlattices, Phys. Rev. Lett. **92**(1-4), 106103 (2004)

12.42 A. A. Abrikosov: *Fundamentals of the Theory of Metals* (North-Holland, Amsterdam 1988)

12.43 J. P. Heremans, C. M. Thrush, D. T. Morelli, M–C. Wu: Resistance, magnetoresistance, and thermopower of zinc nanowire composites, Phys. Rev. Lett. **91**, 106103 (2003)

12.44 C. S. Ting, A. Houghton, T. R. Senna: Thermoelectric power in a disordered two-dimensional electron system, Phys. Rev. B **25**, 1439 (1982)

12.45 D. G. Cahill, W. K. Ford, K. E. Goodson, G. D. Mahan, A. Majumdar, H. J. Maris, R. Merlin, S. R. Phillpot: Nanoscale thermal transport, J. Appl. Phys. **93**, 793–818 (2003)

12.46 E. Fermi, J. Pasta, S. Ulam: *E. Fermi Collected Papers*, Vol. 2 (University of Chicago, Chicago 1965) p. 978

12.47 T. Prozen, D. K. Campbell: Momentum conservation implies anomalous energy transport in 1D classical lattices, Phys. Rev. Lett. **84**, 2857 (2000)

12.48 O. Narayan, S. Ramaswamy: Anomalous heat conduction in one-dimensional momentum-conserving systems, Phys. Rev. Lett. **89**, 200601 (2002)

12.49 S. Lepri, R. Livi, A. Politi: Thermal conduction in classical low-dimensional lattices, Phys. Rep. **377**, 1–80 (2003)

12.50 D. J. Yang, Q. Zhang, G. Chen, S. F. Yoon, J. Ahn, Q. Zhou, Q. Wang, Q. J. Li: Thermal conductivity of multiwalled carbon nanotubes, Phys. Rev. B **66**, 165440 (2002)

12.51 A. Majumdar: Microscale heat conduction in dielectric thin films, J. Heat Transfer **115**, 7 (1993)

12.52 K. E. Goodson: Thermal conduction in nonhomogeneous CVD diamond layers in electronic microstructures, J. Heat Transfer **118**, 279 (1996)

12.53 G. Chen, C. L. Tien: Thermal conductivity of quantum well structures, J. Thermophys. Heat Transfer **7**, 311 (1993)

12.54 G. Chen, S. Q. Zhou, D.-Y. Yao, C. J. Kim, X. Y. Zheng, Z. L. Liu, K. L. Wang: *Proc. 17th Int. Conf. on Thermoelectrics* (IEEE, Piscataway, NJ 1998) pp. 202–205

12.55 T. Klitsner, J. E. VanCleve, J. E. Fischer, R. O. Pohl: Phonon radiative heat transfer and surface scattering, Phys. Rev. B **38**, 7576 (1988)

12.56 L. G. C. Rego, G. Kirczenow: Quantized thermal conductance of dielectric quantum wires, Phys. Rev. Lett. **81**, 232 (1998)

12.57 K. Schwab, E. A. Henrickson, J. M. Worlock, M. L. Roukes: Measurement of the quantum of thermal conductance, Nature (London) **404**, 974 (2000)

12.58 R. Peierls: Zur Theorie der galvanomagnetischen Effekte, Ann. Physik (Leipzig) **3**, 1055 (1929)

12.59 D. T. Morelli, J. P. Heremans, M. Sakamoto, C. Uher: Anisotropic heat conduction in diacetylenes, Phys. Rev. Lett. **57**, 869 (1986)

12.60 C. Dames, G. Chen: Theoretical phonon thermal conductivity of Si/Ge superlattice nanowires, J. Appl. Phys. **95**, 683 (2003)

12.61 B. C. Daly, H. J. Maris, K. Imamura, S. Tamura: Molecular dynamics calculation of the thermal conductivity of superlattices, Phys. Rev. B **66**, 024301 (2002)

12.62 E. T. Schwartz, R. O. Pohl: Thermal boundary resistance, Rev. Mod. Phys. **61**, 605 (1989)

12.63 T. Koga, S. B. Cronin, M. S. Dresselhaus, J. L. Liu, K. L. Wang: Experimental proof-of-principle investigation of enhanced $Z_{3D}T$ in (001) oriented Si/Ge superlattices, Appl. Phys. Lett. **77**, 1490–1492 (2000)

12.64   H. Beyer, J. Nurnus, H. Boettner, A. Lambrecht, E. Wagner, G. Bauer: High Thermoelectric Figure of Merit ZT in PbTe and Bi_2Te_3-based Superlattices by a Reduction of the Thermal Conductivity, Physica E **13**, 965–968 (2002)

12.65   J. Nurnus, C. Künzel, H. Beyer, A. Lambrecht, H. Böttner, A. Meier, M. Blumers, F. Völklein, N. Herres: *Int. Conf. on Thermoelectricity* (IEEE, Long Beach 2000)

12.66   A. Lambrecht, H. Beyer, J. Nurnus, C. Künzel, H. Böttner: *Int. Conf. on Thermoelectricity* (IEEE, Long Beach 2001)

12.67   A. Dauscher, B. Lenoir, A. Jacquot, C. Bellouard, M. Dinescu: Microstructure and electrical properties of PbTe based films prepared by pulsed laser deposition, Mat. Res. Soc. Symp. Proc. **691**, G8.3.1 (2001)

12.68   H. Beyer, J. Nurnus, H. Boettner, A. Lambrecht, L. Schmitt, F. Voelklein: Thermoelectric properties of PbSr(Se,Te)-based low dimensional structures, Mat. Res. Soc. Symp. Proc. **626**, Z2.5.1 (2000)

12.69   L. D. Hicks, T. C. Harman, X. Sun, M. S. Dresselhaus: Experimental study of the effect of quantum–well structures on the thermoelectric figure of merit, Phys. Rev. B **53**, R10493–R10496 (1996)

12.70   T. C. Harman, M. S. Dresselhaus, D. L. Spears, M. P. Walsh, S. B. Cronin, X. Sun, T. Koga: Superlattice structures for use in thermoelectric devices, US Patent 6452206 B1 (2002)

12.71   A. Casian, I. Sur, H. Scherrer, Z. Dashevsky: Thermoelectric properties of n-type PbTe/$Pb_{1-x}Eu_x$Te quantum wells, Phys. Rev. B **61**, 15965 (2000)

12.72   Y.-M. Lin, X. X. Sun, M. S. Dresselhaus: Theoretical investigation of thermoelectric transport properties of cylindrical Bi nanowires, Phys. Rev. B **62**, 4610 (2000)

12.73   O. Rabin, P. R. Herz, Y-M. Lin, S. B. Cronin, A. I. Akinwande, M. S. Dresselhaus: *Arrays of nanowires on silicon wafers, 21st Int. Conf. Thermoelectrics Symp. Proc.* (IEEE, Long Beach 2002) pp. 276–279

12.74   O. Rabin, Y-M. Lin, S. B. Cronin, M. S. Dresselhaus: Thermoelectric Nanowires by Electrochemical Deposition. In: *Thermoelectric Materials 2001 – Research and Applications: MRS Symp. Proc.*, ed. by G. Nolas, D. C. Johnson, D. G. Mandus (Materials Research Society, Pittsburgh 2002) p. G8.20

12.75   M. S. Dresselhaus, Y.-M. Lin, S. B. Cronin, O. Rabin, M. R. Black, G. Dresselhaus, T. Koga: Quantum Wells and Quantum Wires for Potential Thermoelectric Applications. In: *Semiconductors and Semimetals: Recent Trends in Thermoelectric Materials Research III*, ed. by T. M. Tritt (Academic, San Diego 2001) pp. 1–121

12.76   J. P. Heremans, C. M. Thrush, Y-M. Lin, S. B. Cronin, Z. Zhang, M. S. Dresselhaus, J. F. Mansfield: Bismuth nanowire arrays: Synthesis and galvano-

magnetic properties, Phys. Rev. B **61**, 2921–2930 (2000)

12.77   J. P. Heremans, C. M. Thrush, D. T. Morelli, M-C. Wu: Thermoelectric power of bismuth nanocomposites, Phys. Rev. Lett. **88**(1-4), 216801 (2002)

12.78   C. F. Gallo, B. S. Chandrasekhar, P. H. Sutter: Transport properties of bismuth single crystals, J. Appl. Phys. **34**, 144–152 (1963)

12.79   J. Heremans, C. M. Thrush: Thermoelectric power of bismuth nanowires, Phys. Rev. B **59**, 12579 (1999)

12.80   J-P. Michenaud, J-P. Issi: Electron and hole transport in bismuth, J. Phys. C: Solid State Phys. **5**, 3061 (1972)

12.81   J. Heremans, C. M. Thrush, J. Zhang, X. Sun, M. S. Dresselhaus, J. Ying, D. T. Morelli: Magnetoresistance of bismuth nanowire arrays: A possible transition from one-dimensional to three-dimensional localization, Phys. Rev. B **58**, R10091–5 (1998)

12.82   J. Heremans, C. M. Thrush, Y.-M. Lin, S. B. Cronin, M. S. Dresselhaus: Transport properties of antimony nanowires, Phys. Rev. B **63**, 085406 (2001)

12.83   J. Heremans: Thermal Transport in Bismuth Nanowires. In: *Thermal Conductivity 25*, ed. by D. T. Morelli, C. Uher (CRC Pr Llc, Lancaster 1999) p. 114

12.84   B. Lenoir, A. Dauscher, X. Devaux, R. Martin-Lopez, Yu. I. Ravich, H. Scherrer, S. Scherrer: Bi–Sb *Alloys: An Update* (Proc. 15th Int. Conf. Thermoelectrics, Pasadena, CA, IEEE, Piscataway 1996) pp. 1–13

12.85   Yu-Ming Lin: *Thermoelectric Properties of Bi1–xSbx and Superlattice Nanowires* (Ph. D. thesis, Massachusetts Institute of Technology, Department of Electrical Engineering and Computer Science, Cambridge 2003)

12.86   O. Rabin, Yu-Ming Lin, M. S. Dresselhaus: Anomalously high thermoelectric figure of merit in $Bi_{1-x}Sb_x$ nanowires by carrier pocket alignment, Appl. Phys. Lett. **79**, 81–83 (2001)

12.87   L. Langer, V. Bayot, E. Grivey, J.-P. Issi, J. P. Heremans, C. H. Olk, L. Stockman, C. Van Haesendonck, Y. Bruynseraede: Quantum transport in a multi-walled carbon nanotube, Phys. Rev. Lett. **76**, 479 (1996)

12.88   V. Bayot, L. Piraux, J.-P. Michenaud, J.-P. Issi, M. Lelaurain, A. Moore: Two-dimensional weak localization in partially graphitic carbons, Phys. Rev. B **41**, 11770 (1990)

12.89   V. Bayot, L. Piraux, J.-P. Michenaud, J.-P. Issi: Weak localization in pregraphitic carbon fibers, Phys. Rev. B **40**, 3514 (1989)

12.90   C. T. White, T. N. Tedorov: Carbon nanotubes as long ballistic conductors, Nature (London) **393**, 240 (1998)

12.91   P. L. McEuen, M. Bockrath, D. H. Cobden, Y.-G. Yoon, S. G. Louie: Disorder, pseudospins, and backscattering in carbon nanotubes, Phys. Rev. Lett. **83**, 5098 (1999)

12.92 W. Yi, L. Lu, Z. Dang-lin, Z. W. Pan, S. S. Xie: Linear specific heat of carbon nanotubes, Phys. Rev. B **59**, 9015 (1999)

12.93 P. Kim, L. Shi, A. Majumdar, P. L. McEuen: Thermal transport measurements of individual multiwalled nanotubes, Phys. Rev. Lett. **87**, 215502 (2001)

12.94 J. Heremans, C. P. Beetz Jr.: Thermal conductivity and thermopower of vapor-grown graphite fibers, Phys. Rev B **32**, 1981 (1985)

12.95 J. Hone, M. Whitney, C. Piskoty, A. Zettl: Thermal conductivity of single-walled carbon nanotubes, Phys. Rev. B **59**, 2514 (1999)

12.96 J. Hone, M. C. Liaguno, N. M. Nemes, A. T. Johnson, J. E. Fischer, D. A. Walters, M. J. Casavant, J. Schmidt, R. E. Smalley: Electrical and thermal transport properties of magnetically aligned single wall carbon nanotube films, Appl. Phys. Lett. **77**, 666 (2000)

12.97 J. P. Small, K. M. Perez, P. Kim: Modulation of thermoelectric power of individual carbon nanotubes, Phys. Rev. Lett. **91**, 256801 (2003)

12.98 G. U. Sumanasekera, B. K. Pradan, H. E. Romero, K. W. Adu, P. C. Ecklund: Giant thermopower effects from molecular physisorption on carbon nanotubes, Phys. Rev. Lett. **89**, 166801 (2002)

12.99 H. E. Romero, K. Bolton, A. Rosén, P. C. Eklund: Atom collision-induced resistivity of carbon nanotubes, Science **307**, 89–93 (2005)

12.100 P. G. Collins, K. Bradley, M. Ishigami, A. Zettl: Extreme oxygen sensitivity of electronic properties of carbon nanotubes, Science **287**, 1801 (2000)

12.101 G. U. Sumanasekera, C. A. K. Adu, S. Fang, P. C. Ecklund: Effects of gas adsorption and collisions on electrical transport in single-walled carbon nanotubes, Phys. Rev. Lett. **85**, 1096 (2002)

12.102 J. Small, M. Purewall, P. Kim: Measurement of gate-modulated thermoelectric power in Si nanowires, Bull. Am. Phys. Soc. **50**, 1266 (2005)

12.103 G. Springholz, V. Holy, M. Pinczolits, G. Bauer: Self-organized growth of three-dimensional quantum-dot crystals with fcc-like stacking and a tunable lattice constant, Science **282**, 734–737 (1998)

12.104 G. Springholz, M. Pinczolits, P. Mayer, V. Holy, G. Bauer, H. H. Kang, L. Salamanca-Riba: Tuning of vertical and lateral correlations in self-organized PbSe/Pb$_{1-x}$Eu$_x$Te Quantum Dot Superlattices, Phys. Rev. Lett. **84**, 4669–4672 (2000)

12.105 K. Alchalabi, D. Zimin, G. Kostorz, H. Zogg: Self-assembled semiconductor quantum dots with nearly uniform sizes, Phys. Rev. Lett. **90**(1-4), 026104 (2003)

12.106 T. C. Harman, P. J. Taylor, M. P. Walsh: Nanostructured thermoelectric materials and devices, US Patent 6605772 B2 (2003)

12.107 J. P. Heremans, C. M. Thrush, D. T. Morelli: Phys. Rev. Lett. **86**(10), 2098–2101 (2001)

12.108 Yu. I. Ravich, B. A. Efimova, V. I. Tamarchenko: Scattering of current carriers and transport phenomena in lead chalcogenides, Phys. Stat. Sol. (b) **43**, 453–469 (1971)

12.109 K. F. Hsu, S. Loo, F. Guo, W. Chen, J. S. Dyck, C. Uher, T. Hogan, E. K. Polychroniadis, M. G. Kanaztidis: Cubic AgPb$_m$SbTe$_{2+m}$: Bulk thermoelectric materials with high figure of merit, Science **303**, 818 (2004)

12.110 T. Irie, T. Takahama, T. Ono: The thermoelectric properties of AgSbTe$_2$-AgBiTe$_2$, AgSbTe$_2$-PbTe and -SnTe systems, Jpn. J. Appl. Phys. **2**, 72–82 (1963)

12.111 R. W. Armstrong, J. W. Faust, W. A. Tiller: A structural study of the compound AgSbTe$_2$, J. Appl. Phys. **31**, 1954–1959 (1960)

12.112 D. Bilc, S. D. Mahanti, E. Quarez, K.-F. Hsu, R. Pcionek, M. G. Kanatzidis: Resonant states in the electronic structure of the high performance thermoelectrics AgPb$_m$SbTe$_{2+m}$: The role of Ag-Sb microstructures, Phys. Rev. Lett. **93**, 146403 (2004)

12.113 D. Bilc, S. D. Mahanti, E. Quarez, K.-F. Hsu, R. Pcionek, M. G. Kanatzidis: Bull. Am. Phys. Soc. **50**, 1407 (2005)

12.114 X. Shi, L. Chen, J. Yang, G. P. Meisner: Enhanced thermoelectric figure of merit of CoSb$_3$ via large-defect scattering, Appl. Phys. Lett. **84**, 2301–2304 (2004)

12.115 J. P. Heremans, C. M. Thrush, D. T. Morelli: Thermopower enhancement in lead telluride nanostructures, Phys. Rev. B **70**, 115334 (2004)

12.116 A. Lasbley, R. Granger, S. Rolland: High-temperature superconducting behaviour in PbTe-Pb, Solid State Commun. **13**, 1045–8 (1973)

12.117 A. Guinier: Structure of age-hardened aluminium-copper alloys, Nature **142**, 569 (1938)

12.118 G. D. Preston: Structure of age-hardened aluminium-copper alloys, Nature **142**, 570 (1938)

12.119 Cullity: *Elements of X-Ray Diffraction* (Addison-Wesley, Reading, MA 1956)

12.120 B. W. Roberts: Survey of superconductive materials and critical evaluation of selected properties, J. Phys. Chem. Ref. Data **5**, 581–821 (1976)

12.121 N. K. Hindley, J. H. P. Watson: Superconducting metals in porous glass as granular superconductors, Phys. Rev. **183**, 525 (1969)

Part A | 12

13. Nano- and Microstructured Semiconductor Materials for Macroelectronics

This chapter reviews the potential applications of micro- and nanostructured materials in the emerging area of electronic systems that involve flexible thin film transistors (TFTs) on large area plastic substrates, often referred to as 'macro-electronics'. Approaches for fabricating various one-dimensional or two-dimensional semi-conductor nanostructures (including nanowires, nanoribbons, and single-walled carbon nano-tubes) are discussed. Techniques such as dry transfer printing and solution phase assembly are compared for the purpose of generating high-quality thin films of semiconductor nano-structures on plastic substrates. High-performance TFTs and complementary logic gates are demon-strated. The combined use of these materials and techniques may advance the development of practical technologies for emerging applications of macroelectronics in consumer, space and military systems.

Macroelectronic systems represent a new form of electronics that consists of integrated circuits on substrates with sizes much larger than those of conventional semiconductor wafers. This class of device is very different to established microelectronics and nanoelectronics, where progress is driven primarily by shrinking the size of the functional elements (for example, standard transistors on bulk semiconductor wafers) in order to increase the speed and memory capacity. Macroelectronics, instead, uses thin-film transistors (TFTs) distributed over large areas where, in existing applications that use glass substrates, they provide switching elements in liquid crystal displays or medical x-ray imaging devices [13.1–4]. This direction in electronics, in which the overall size of the systems represents the primary scaling metric, instead of minimum feature size of an individual circuit component, is of rapidly increasing

importance. Amorphous silicon, for example, is now the second most economically significant semiconductor behind single crystal silicon. A single generation VII fabrication line at Samsung will, at full operation, generate 10% of the entire worldwide output of conventional silicon integrated circuits, as measured by system area. There is an emerging effort in research to develop the means to form such large area TFT-based circuits using printing-type techniques on thin, lightweight substrates that are mechanically flexible [13.5]. Potential application drivers are for active sensory skins and electronic textiles with locally-embedded logic, foldable and expandable antennas for military and space applications, and even sensing and display systems that can be integrated into the structures of buildings used for living environments. These systems require superflexible, lightweight, rugged and low-cost circuits that offer high performance and reliable behavior. Rolled plastics made of electrically inert polymers represent ideal classes of materials for these substrates. Although it is possible to integrate low temperature semiconductors based on small molecule organics, polymers or amorphous silicon onto these substrates, the performance that can be obtained from them and (for the organics) their uncertain reliability represent drawbacks. New types of laser annealing approaches may be useful for converting, on the plastic substrate, amorphous silicon to large-grained polycrystalline silicon for improved performance. Fluidic self-assembly and massively parallel 'pick-and-place' approaches for bonding arrays of preformed circuits or circuit elements onto plastic substrates also have some promise. An ideal approach involves direct printing of high-quality semiconductors onto the plastic followed by depositing and patterning other materials at relatively low temperatures ($< 200\,°C$) in order to build up circuits in a way that is compatible with economically efficient, continuous roll-to-roll processing. Separately growing the semiconductors, processing them into micro/nanostructured form, and then printing them onto the plastic substrate represents one way to achieve this approach. This sequence also allows certain high-temperature steps that might be required (such as forming ohmic contacts) to be performed on platforms (e.g. semiconductor wafers) that are compatible with this type of processing. The semiconductor can be derived from conventional inorganic materials or single-walled carbon nanotubes, both of which have carrier mobilities that are much higher than those of conventional organics or amorphous silicon. In order to maintain good mechanical flexibility, ultrathin films of inorganic semiconductors and thin plastic substrates must be used [13.6]. Figure 13.1 outlines the major steps and strategies for fabricating high-performance circuits on flexible plastics over large area. One- or two-dimensional (1-D or 2-D) semiconductor nanostructures with single crystallinity, including nanowires, nanoribbons and nanotubes, represent promising building blocks for the semiconductor, in which the transport of carriers from source to drain can occur without any interruption caused by grain boundaries [13.7]. We present here micro- and nanomaterials with different morphologies and constituent materials, methods for patterning them onto plastic substrates, and strategies for building high-performance devices. We evaluate the potential use of these technologies for macroelectronics.

This chapter briefly reviews the preparation, assembly and printing of inorganic semiconductor nanomaterials (including single-walled carbon nano-

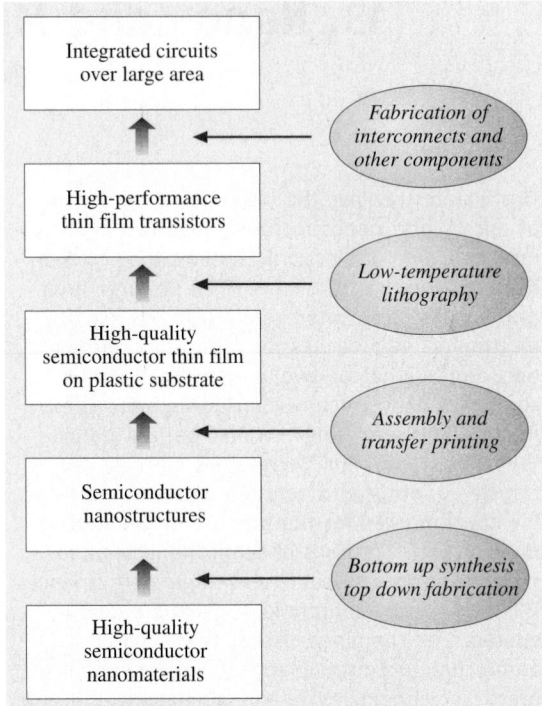

Fig. 13.1 Schematic outline of key challenges (*open rectangles*) and applicable techniques (*ellipses*) required to fabricate low-cost, large-area, flexible macroelectronics using nanomaterials

tubes). It is organized into four main sections. The first summarizes general approaches for preparing semiconductor nanomaterials with various morphologies, such as nanoparticles, nanowires/nanoribbons, and carbon nanotubes. Two principle concepts – synthetic, "bottom-up" methods (growth of nanostructures through assembly and crystallization of atoms of precursors) and complementary "top-down" processes (preparation of small objects on the nanometer scale by carving, slicing, or etching macroscale material sources) – are compared. The second section outlines strategies for the formation of high-quality thin films consisting of ensembles of aligned 1-D and 2-D nanostructures on plastic substrates. The applicable approaches include aligning solution-based nanostructures into monolayer or submonolayer arrays with various external forces and then transferring them onto plastic substrates, and dry transfer printing of nanowire/nanoribbon arrays generated from "top-down" processes and networks of single-walled carbon nanotubes. The third section describes the use of effective thin films of this type on plastic substrates for high-performance transistors and circuits. Several examples of devices and their electronic and mechanical characteristics demonstrate the current state of the technology. The last section concludes with personal perspectives on the trends for future work related to flexible macroelectronics.

13.1 Classes of Semiconductor Nanomaterials and their Preparation

Achieving high performance with a semiconductor that can be deposited onto plastic is one of the most challenging technical aspects of flexible macroelectronics. The most commonly used semiconductors are the elemental semiconductors (such as silicon, germanium, carbon, boron, and so on), and the compound semiconductors including III–V compounds (such as GaAs, InP, GaN, chalcogenides (CdSe, CdS, ...) and oxides of many metals. In the following sections we provide an overview of micro- and nanostructured semiconductor elements that show promise for high-performance transistors.

13.1.1 Nanoparticles

Semiconductor nanoparticles, also known as quantum dots, are a unique class of material because their sizes, which range from several nanometers to tens of nanometers, strongly influence their electronic properties [13.8]. Size-reducing processes, such as ball milling and ultrasonic fracturing, provide a straightforward route to prepare nanoparticles [13.9]. The resulting particles, however, usually have a broad distribution in size and shape, which might make it challenging to use them in reliable, high-performance devices. On the other hand, high-temperature pyrolysis of precursors and/or chemical reactions between various precursors in solvents with high boiling points enables the synthesis of nanoparticles with uniform sizes in the presence of capping reagents. *Alivisatos* and coworkers and *Peng* and coworkers systematically studied the growth mechanisms and developed several methods for growing various quantum dots (such as CdSe, CdS, CdTe, InP, InAs) in large quantities [13.10–13]. The as-obtained nanoparticles in solution can generate thin films, which serve as carrier layers for TFTs, through sedimentation, spin casting and layer-by-layer self-assembly [13.14, 15]. For example, *Cui* et al. demonstrated that thin films of 15 nm SnO_2 nanoparticles could be assembled in a layer-by-layer fashion with the assistance of positively charged polyions (such as poly(dimethyldially ammonium chloride)) and negatively charged polyions (such as sodium poly(styrenesulfonate)). These films serve as n-type semiconductors for metal-oxide-semiconductor field-effect transistors (MOSFET) with device mobil-

ities of $2.1 \times 10^{-2}\,cm^2/V \cdot s$ and on/off current ratios of 10^4 [13.15]. Nanomaterials of this type are compatible with low-cost, high-throughput, non-vacuum casting processes and plastic substrates. The grain boundaries formed between individual particles can, however, lead to low device mobilities and a broad distribution of device performance. In addition, formation of gaps between adjacent particles can strongly influence transport behavior of devices on plastics when the substrates are bent. It is for this reason that 1-D or 2-D single crystalline micro/nanostructures of semiconductors that are long enough to span the gaps between electrodes have advantages for flexible devices.

13.1.2 Nanowires/Nanoribbons

Single crystalline nanowires/nanoribbons made of inorganic semiconducting materials are of great interest for fabricating TFTs because the transport behavior of carriers through them is not disturbed by grain boundaries. For instance, single crystalline Si nanowires show transport properties of electrons and holes that are similar to those of bulk materials [13.7, 16]. A wide range of approaches have been developed to generate high-quality nanowires of various materials (including Si, Ge, III–V compounds, II–VI compounds and oxides) [13.17]. The techniques can be classified into two general categories: "bottom-up" and "top-down" approaches.

"Bottom-Up" Approaches
Assembled atoms of semiconductor materials can condense and grow into nanowires/nanoribbons by breaking the symmetry of their crystalline lattices. Because a large number of publications have reviewed these synthetic approaches [13.18–21], this chapter only summarizes the concepts. The most successful methods rely on the catalysis of metal nanoparticles in gaseous and liquid media by controlling reaction conditions. Figure 13.2a presents a typical phase diagram of a binary system consisting of metal (M) and semiconductor (Sc), which guides the selection of appropriate compositions (semiconductor percentage of the system higher than that of the eutectic binary alloy), catalysts and reaction temperatures (temperature between the eutectic point

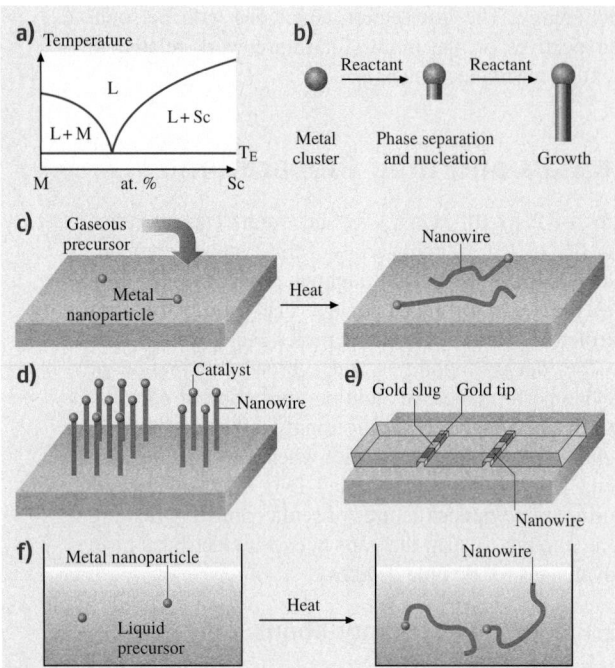

Fig. 13.2a–f Synthesis of single crystalline nanowires of semiconductors by use of metal nanoparticle catalysts. (**a**) Binary metal (M)-semiconductor (Sc) phase diagram used to select a catalyst for growth of semiconductor nanowires. (**b**) Schematic illustration of the steps of catalytic growth of semiconductor nanowires. (**c**) Growth of semiconductor nanowires on substrates, which can withstand high temperatures (higher than the eutectic point of alloy of metal and semiconductor), by feeding the system with gaseous precursors. This synthesis is referred to as the vapor-liquid-solid (VLS) process. (**d**) Preferential growth of a vertically oriented nanowire array on a single crystal substrate through the VLS process. (**e**) Growth of horizontal nanowires in template channels through the VLS process. (**f**) Growth of nanowires in a solvent with high boiling point by delivering liquid precursors to catalyst

solid semiconductor phase separates out from the alloy droplet. This process is referred to as nucleation. The nucleus guides the growth of semiconductor nanowire with the continuous delivery of semiconductor atoms to the liquid alloy droplet. The diameter of the droplet, determined by the size of the metal nanoparticle, confines the lateral dimension of the resultant semiconductor nanowire.

This type of growth, when performed in the gas phase, is called the vapor-liquid-solid (VLS) process, and was originally developed by *Wagner* and coworkers [13.22]. The results demonstrated by *Lieber*, *Yang* and many other research groups confirm that the VLS process has the capability to grow single crystalline nanowires of various semiconductor materials in large scale. The vapor of semiconductor atoms, which is delivered to the catalyst droplets to determine the composition of semiconductor nanowires, can be generated by laser ablation [13.23] or thermal decomposition of organometallic compounds [13.24]. In a typical synthesis (Fig. 13.2c), target materials (such as InP, CdS and Si) and silicon substrates covered with gold nanoparticles (serving as catalysts) lie at the upstream end and central part of a tube furnace, respectively. The substrate is heated to temperatures of 600–1200 °C while the temperature of target material remains at room temperature. A pulsed laser ablates the target to generate a vapor which flows, with the aid of a stream of argon, to the surface of the substrate to grow into nanowires from the gold nanoparticles [13.25]. This method has, in principle, the ability to synthesize nanowires of most semiconductor materials that can be evaporated by tuning the parameters (such as power, wavelength and pulse width) of the laser. The growth time and ablation conditions control the lengths of the nanowires [13.26]. The as-grown nanowires are usually in the form of random assemblies.

By choosing a substrate with a specific crystalline orientation and lattice constant, nanowires can grow normal to the surface epitaxially. As a result, aligned nanowire arrays can be grown in situ via the VLS process by patterning metal nanoparticle catalysts on an appropriate substrate (Fig. 13.2d) [13.27–30]. For example, an array of Si nanowires with the longitudinal axis in the <111> direction can grow perpendicular to the surface of a (111) Si wafer through catalysis of Au nanoparticles (>20 nm) [13.27]. In this case, the growing plane of the wires is the same as that of the surface of the substrate. On the other hand, ZnO nanowire arrays can grow vertically on an *a*-plane (110) sapphire substrate, even though ZnO nanowires tend to grow along the <001>

of alloy and the melting point of semiconductor) for growing semiconductor nanowires. Figure 13.2b depicts the concept of wire growth with the assistance of metal clusters and semiconductor precursors. At a proper temperature, atoms of semiconductor condense on the surface of a metal nanoparticle, driven by a decrease in the total surface energy. The particle melts when the concentration of semiconductor atoms around the metal particle reaches a critical value, resulting in the formation of alloyed droplets. As the concentration of semiconductor atoms increases, the particle system shifts to the right side of the phase diagram, and the pure

direction. The epitaxial growth of ZnO nanowires along the (110) plane of sapphire is attributed to the fact that the a-axis of wurzite ZnO and the c-axis of sapphire are related by a factor of four (a_{ZnO}= 3.24 Å versus $c_{sapphire}$ = 12.99 Å) with a mismatch of less than 0.08% at room temperature [13.28]. Locations, patterns and densities of nanowire arrays grown in this way can be easily controlled by manipulating the Au nanoparticles through various lithographic techniques including soft lithography, e-beam lithography and photolithography.

Nanowire arrays can be grown in the plane of the substrate surface if nanochannels are patterned to confine the growth. *Fonash* and coworkers demonstrated this concept through a so-called "grow-in-place" approach that has the ability to produce self-assembled, crystalline Si nanowires via the VLS process [13.31]. As shown in Fig. 13.2e, Au lines that have heights and widths on the nanometer scale are first generated on an insulating substrate through e-beam lithography, deposition of metal and lift-off. A capping layer (such as silicon nitride) is then deposited over the substrate, followed by photolithographic patterning and selective dry etching. The Au lines under the capping layer are shortened to slugs with lengths of several microns by wet etching, resulting in the formation of nanochannels for wire growth. In the VLS process, the remaining Au in the center of the channels serves as the catalyst for the growth of semiconductor nanowires (Si), while the nanochannels act as templates to confine the dimensions of nanowires. When the Au slug is longer than 10 μm, the silicon atoms that diffuse to the surface of the Au cannot saturate the whole slug, but instead locally saturate the ends. The nanodroplet of Au–Si alloy at each end induces the growth of one wire using $SiH_4 - H_2$ gas. As a result, two Si nanowires separated by the remaining Au slug form in each nanochannel. These two wires join together to form a continuous wire when the Au slug is sufficiently short (e.g., < 2 μm). The resultant wires have well-controlled size, shape, orientation and position. On the other hand, the use of e-beam lithography and the excess Au make the process expensive for generating wire arrays over large areas.

An analog of the VLS process can be operated in the liquid phase when nanoparticles of metals (such as In, Ga or Bi) with low melting points and solvents with high boiling points are chosen as catalysts and reaction media, respectively. In this case, organometallic compounds serve as the source of semiconductor atoms through a thermal decomposition reaction. This process (Fig. 13.2f) is referred to as the solution-liquid-solid process, or SLS. *Buhro* and

coworkers have used it to synthesize nanowires made of III–V compounds [13.32, 33]. For example, refluxing 1,3-diisopropylbenzene solutions of specific precursors $[(^tBu)_3Ga$ and $As(SiMe_3)_3]$ in the presence of In nanoparticles generates nanowires of GaAs with small diameters (for example, 6–17 nm) and relatively narrow diameter distributions. However, the eutectic temperatures of most metal-semiconductor alloys are higher than the boiling temperatures of traditional solvents that can dissolve organometallic precursors and metal nanoparticles. This problem limits the composition of nanowires that can be prepared. Most recently, *Korgel* and coworkers employed supercritical fluids as the reaction medium for wire growth [13.34, 35]. Because supercritical fluid can remain in a liquid state at very high temperatures, metal nanoparticles with high melting points can be used to catalyze the growth of semiconductor nanowires. For example, GaAs nanowires have been synthesized in supercritical hexane at 500 °C and 37 MPa by reacting $(^tBu)_3Ga$ and $As(SiMe_3)_3$ in the presence of dodecanethiol-stabilized 7 nm Au nanoparticles. The SLS process is mainly limited by the availability of suitable organometallic precursors.

Nanowires synthesized via "bottom-up" approaches usually have broad distributions in both lateral and longitudinal dimensions, although the "grow-in-place" approach can generate wires with uniform lateral dimensions. Lengths greater than 100 μm are difficult to achieve with any of these approaches. The surface properties of the wires, their compositional purity, doping uniformity and concentration, which all strongly influence the electrical properties, are also less well defined than those of wafers used in the semiconductor industry.

"Top-Down" Approaches

Nanowires/nanoribbons can be fabricated from high-quality, single-crystal, bulk wafers via "top-down" approaches in which lithographic patterning and etching creates the necessary structures from the top surfaces of the wafers. Nanostructures generated in this manner can have well controlled doping type, carrier concentration, dimensions and crystallinity, which, in turn, give rise excellent electrical properties for applications in high-performance electronics. The critical step in "top-down" approaches is the generation of patterned mask lines on the nanometer or micrometer scale using established photolithographic techniques or, more recently, low-cost methods based on phase-shift photolithography with soft phase masks made of poly(dimethylsiloxane) (PDMS) or nanoimprinting [13.36, 37]. When applied to layered wafers such as silicon-on-insulator (SOI) wafers, resist lines can be defined such that the etching of underlying layers can produce free-standing beams, wires and other structures. Figure 13.3a shows the steps of such a process. In this case, photoresist exposed through an elastomeric phase mask generates patterned lines with widths ranging from 50 to 300 nm to serve as the mask pattern for etching the underlying SOI wafer [13.38]. Reactive-ion etching (RIE) or wet-etching of the top Si

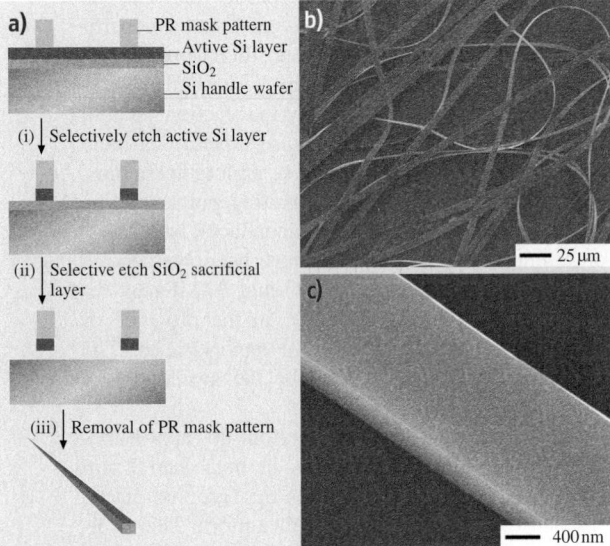

Fig. 13.3 (a) Process for fabricating rectangular nanowires (or nanoribbons) from high-quality wafers with multiple layers: active semiconductor layer (Si), sacrificial layer (SiO$_2$) and handle wafer (Si). **(b),(c)** SEM images of silicon ribbons with thicknesses of 100 nm and widths of 1 μm. These ribbons were fabricated from commercial SOI wafers through the processes in **(a)**. PR: photoresist

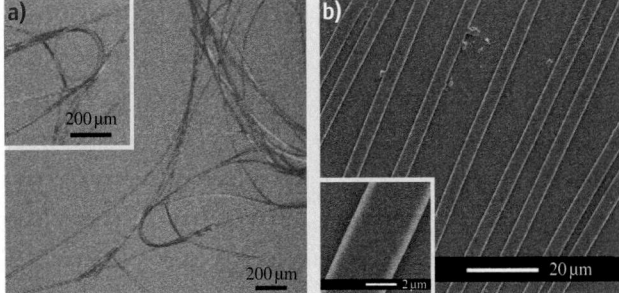

Fig. 13.4a,b GaN (with epitaxial AlN layer) ribbons fabricated from multiple-layered wafers (GaN/AlN/Si), through shallow dry etching combined with isotropic wet etching of Si. **(a)** Optical image of collected free-standing ribbons on a new Si wafer, showing their excellent bendability. **(b)** SEM image of some ribbons printed on a PET substrate

layer (step i) can be accomplished. After the buried SiO$_2$ layer is selectively dissolved with HF solution (step ii) and photoresist mask stripes are removed (step iii), pristine Si nanowires are released from the mother wafer. The resulting wires can be further thinned through oxidation and selective etching of the outer SiO$_2$ layer. These types of structures as well as analogous ones in the form of suspended beams or integrated Si waveguides have been used in a variety of applications in microelectromechanical systems, optics and in specialized electronic devices in which the undercut etching reduces the capacitive coupling to the substrate. Applications in flexible macroelectronic devices are more recent [13.39]. In these and other systems, it is not necessary to reduce all of the lateral dimensions of resultant structures down to the nanometer scale. Figure 13.3b and 13.3c show scanning electron microscopy (SEM) images of Si ribbons with thicknesses of 100 nm and widths of 1 μm fabricated using SOI wafer and traditional photolithography, illustrating that these ribbons have extremely smooth surfaces and excellent mechanical flexibility. Ribbons of compound semiconductor materials, such as GaN, can also be fabricated through a similar process. Lithographically defined photoresist lines on the GaN/AlN/Si wafer serve as the etch mask to etch through thin layers of GaN and AlN via RIE. Further selectively isotropic etching underneath the Si wafer can release GaN/AlN composite ribbons [13.40]. Figure 13.4 shows typical images of as-prepared GaN ribbons, which can be used as printable elements for device fabrication. This approach is, however, unsuitable for some applications due to the relatively high cost of these kinds of multilayered wafers (such as ≈$500 for a 6 inch SOI wafer versus ≈$5 for a 6 inch Si wafer). Developing convenient methods for generating nanowires from inexpensive wafers is therefore crucial to commercializing electronics with large areas (macroelectronics).

Anisotropic etching of single crystalline wafers over specific crystalline planes provides a route to fabricating nanowires/nanoribbons from inexpensive wafers with uniform compositions. For example, anisotropic etching of a Si wafer with patterned mask lines on its (110) plane in a hot aqueous KOH solution can generate vertical profiles [13.41]. Ribbons can be formed in this manner by etching completely through an ultrathin (110) wafer, or by combining the anisotropic etching process with an isotropic etch to free ribbons defined at the top of a bulk wafer. Alternatively, isotropic etching of trenches on the surface of a bulk (111) Si wafer followed by passivation of the sidewalls [terminated by (110) planes] and anisotropic etching can produce thin ribbons from the top of the wafer. Many other related approaches can generate similar structures.

Anisotropic etching of a semiconductor wafer in a manner that leads to the formation of reverse mesas (structures with newly formed sidewalls with an acute angle relative to the original surface of the wafer) seems to be a more straightforward way to generate nanowires since it only requires a single etching step, thereby decreasing the fabrication time and cost. Compound semiconductor materials made of group III and V elements are usually crystallized in the form of zinc blende face-centered cubic (f.c.c.) lattices, whose {111} planes are the most stable [13.42]. The {111} surface terminated by the Group-V atoms, also called {111}B, is high in electron density compared to the {111} surface terminated by the Group-III atoms, or {111}A. This difference causes the {111}A surface to be more stable toward chemical oxidation and dissolution. As a result, sidewalls with an acute angle (rather than obtuse angle) relative to the top surface can be generated by controlling the orientation of the stripes of the etching mask. Figure 13.5a depicts the major steps for preparing GaAs wires through anisotropic etching of a (100) GaAs wafer patterned with mask lines in an aqueous solution of H_2O_2 and H_3PO_4 [13.43, 44]. In the process, photolithography defines parallel photoresist lines with controlled widths and oriented along the $(0\,\overline{1}\,\overline{1})$ crystallographic direction on the surface of GaAs wafer (step i). Anisotropic etching of GaAs wafer through the photoresist lines generates reverse mesas with side surfaces terminated by $(\overline{1}\,\overline{1}\,1)$A and $(\overline{1}\,1\,\overline{1})$A planes, respectively (step ii). With long etching times, the two side walls of each reverse mesa connect to release a wire with a triangular cross-section (step iii). Removing the etching mask yields high-quality wires with single crystallinity, uniform doping concentration and well-defined crystalline surfaces (step iv). They can be collected into a solution suspension, or they can be transferred onto other substrates using approaches that preserve the positions and orientations defined by the lithography (step v).

Figure 13.5b–e show SEM images of a GaAs wafer patterned with $2\,\mu m$-wide photoresist lines before and after etching for different times. The photoresist lines, as shown in Fig. 13.5b, have straight and uniform widths along their longitudinal axis. The structure shown in panel c exhibits a mushroom-like morphology, clearly showing the formation of reverse mesa and lateral undercutting along with the vertical etching. Figure 13.5d shows an SEM image of a released wire with triangular cross-section, which sits on the mother wafer against one edge of the wire and one edge of the photoresist stripe.

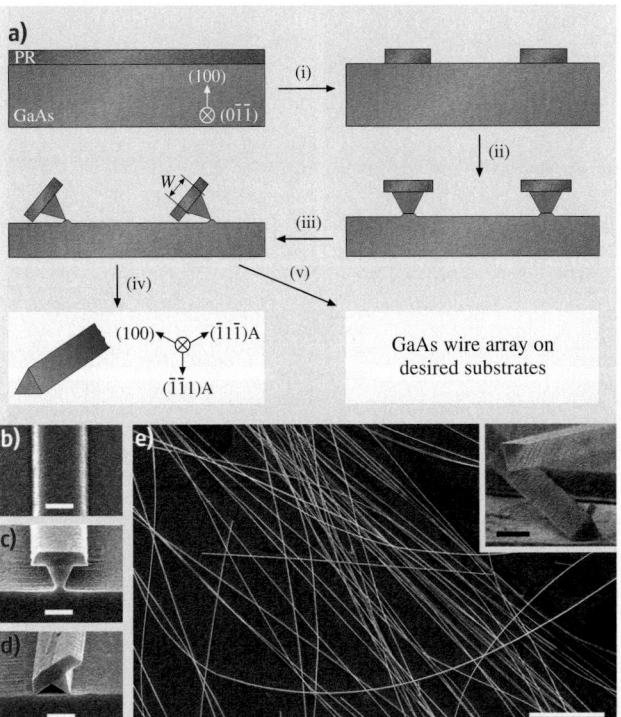

Fig. 13.5a–e Fabrication of nanowires of III–V compound semiconductor materials (such as GaAs) through one-step anisotropic chemical etching. (**a**) Schematic illustration of the main steps involved in the fabrication process: (i) defining patterned photoresist (PR) lines along $(0\,\overline{1}\,\overline{1})$ direction on the surface of (100) wafer; (ii) anisotropic etching of GaAs wafer in $H_3PO_4-H_2O_2-H_2O$ using PR pattern as etch mask; (iii) continuous etching of GaAs; (iv) removal of PR; (v) printing or casting of GaAs wires onto a desired substrate. (**b**)–(**e**) represent SEM images of etching results at different stages: (**b**) a part of one PR stripe; (**c**) GaAs reverse mesa capped with PR stripes; (**d**) GaAs wire with PR stripes released from mother wafer; (**e**) a random assemble of free-standing GaAs wires after removing the PR stripes. The *scale bars* in (**b**), (**c**), (**d**) and inset of (**e**) represent $1\,\mu m$; and the *scale bar* in (**e**) represents $50\,\mu m$. (After [44] with permission, 2005 Wiley-VCH)

Figure 13.2e shows an SEM image of free-standing GaAs wires randomly assembled on a GaAs substrate. The wires exhibit uniform widths (W) of $840\,nm$ along their longitudinal axis. The SEM image shown in the inset reveals the roughness of the side walls and the flatness of the top surface. The roughness on the side walls originates mainly from the edge roughness of the mask lines. The curved configurations that many of the free-standing wires exhibit clearly indicate that they are flexible. The bendability increases with decreasing lat-

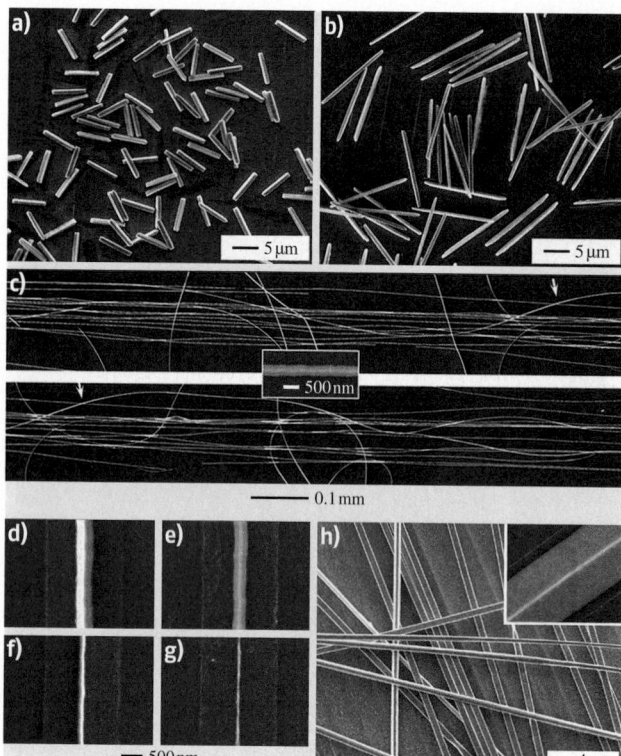

Fig. 13.6 (a),(b) SEM image of GaAs wires with different lengths fabricated by controlling the length of the mask stripes. (c) SEM images of GaAs wires with lengths of several millimeters. The arrows indicate that same position of the sample. (d)-(g) SEM images of GaAs wires with various widths obtained by adjusting the etching time. (h) SEM image of InP wires fabricated by anisotropic etching of a (100) InP wafer masked with SiO_2 stripes in a Br_2–methanol solution

eral dimensions. Wires with widths of ≈ 400 nm can form arc shapes with radii as small as $\approx 20 \, \mu$m [13.43]. The mechanical flexibility of the nanowires allows them to serve as active components in flexible macroelectronic devices.

The lengths of GaAs wires can be easily controlled by tuning the lengths of photoresist mask stripes. Figures 13.6a and 13.6b show SEM images of GaAs rods with average lengths of 6.4 and 15.5 μm, respectively. In principle, we can prepare wires with lengths as long as tens of centimeters, limited only by the diameter of the original wafers. Figure 13.6c shows a series of SEM images of long GaAs wires, indicating that they can be continuous even when their lengths are as long as several millimeters. It is also worth noting that the etching generates undercutting and maintains high anisotropy even

after the wires lift off the substrate [13.43]. As a result, the lateral dimensions of the wires can be decreased to tens of nanometers by controlling the etching time even when the original mask lines have widths on the micrometer scale. Figures 13.6d–g show SEM images of individual wires obtained by etching a GaAs wafer patterned with 2 μm-wide mask lines. The widths (W) of the wires decrease to 50 nm for the longest etching times, while the triangular cross-sections are preserved. This general type of anisotropic etching can be used to fabricate wires consisting of other compositions, such as InP, with suitable etchants and etch masks. Figure 13.6h shows a typical image of InP wires obtained by etching a (100) InP wafer patterned with 2 μm SiO_2 mask stripes in 1% (v/v) methanol solution of Br_2. The inset indicates that the cross-sections of individual InP wires are also triangular.

In general, the physical parameters of nanowires/nanoribbons fabricated via "top-down" approaches exhibit good uniformity (comparable to their high-quality wafer precursors) and enable easy control over the processing. The cost of the wires can be greatly decreased if the mother wafers are polished and reused to generate more wires [13.44]. The major disadvantage of the "top-down" approach is that the composition of the wires is limited to materials (such as Si, Ge, GaAs, InP) that are already available in wafer or thin film forms. In addition, wires of this type always have some degree of roughness on their etched surfaces and their diameters are difficult to narrow to dimensions of less than $\approx$ 20 nm.

13.1.3 Single-Walled Carbon Nanotubes

Single-walled carbon nanotubes (SWNTs) represent another class of nanowire that has potential applications in macroelectronics. SWNTs can be considered as being generated from a sheet of graphene that is rolled up along a direction given by the circumference vector $c = na_1 + ma_2$ (where a_1 and a_2 are the lattice vectors in graphene). SWNTs are of great interest for many applications because of their unique electronic and mechanical properties [13.45, 46]. For example, SWNTs can exhibit either metallic or semiconducting transport behaviors by tuning their symmetry (the value of n and m). If the difference between n and m is divisible by three, then the nanotubes are metallic; otherwise they are semiconducting. Previous studies reveal that the intrinsic hole mobilities of SWNTs can exceed $100\,000 \, \text{cm}^2/\text{Vs}$ at room temperature, which is the highest known value for a material [13.47]. In addition, the very small lateral dimensions ($\approx$ 1 nm) and superhigh Young's moduli enable the SWNTs to withstand, without breaking, very small bend radii, and very high strains (such as elongation strains at break of $\approx$ 30% and threshold strains for intra-tube electronic scattering of 5–10%) [13.48]. As a result, SWNTs represent an interesting type of building block for fabricating flexible macroelectronics with high performance. One of the challenges associated with this strategy is that SWNTs formed with known growth techniques contain mixtures of semiconducting and metallic tubes, the latter of which can form electrical shorts in transistor channels. There are, however, several electrical-, chemical- and optical-based approaches that appear to have the potential to overcome this challenge [13.49, 50].

Single-walled carbon nanotubes can be grown through two types of approaches. The first one, classified as a "bulk method", produces large amounts of nanotubes, up to gram scale, in the form of nanotube-containing soot. The soot can be purified and the resultant nanotubes are then deposited onto appropriate substrates for device fabrication. Typical processes include arc-discharge [13.51] and laser ablation [13.52] of a carbon target doped with a catalyst consisting of powdered Ni and Co, and high-pressure decomposition of carbon monoxide (HiPCO), in which CO gas with a small amount of $Fe(CO)_5$ is sprayed through a nozzle into a reactor at high temperature and pressure [13.53]. The second class of approach directly grows SWNTs on a suitable substrate by predepositing catalyst on the substrates and then exposing the substrate to carbon-containing feedstock gas at high temperatures. The chemical vapor deposition (CVD) process represents a typical approach of this class [13.54]. Nanotubes grown from "bulk methods" usually exhibit highly entangled bundles, which are difficult to purify and deposit uniformly onto substrates. Carbon nanotubes synthesized through the CVD process can have uniform dimensions and spatial distributions because the uniformity and coverage of nanosized catalyst particles can be controlled by optimizing the related processes [13.55]. For a variety of possible reasons, they often show better performance in devices than SWNTs derived from "bulk methods".

Figure 13.7 shows atomic force microscopy (AFM) images of SWNTs with different densities and orientations prepared through a CVD process catalyzed with iron oxide nanoparticles generated from the oxidation of ferritin [13.46, 56]. The images indicate that the as-grown nanotubes have a narrow distribution of diameters and that they exist mainly in the form of individual nano-

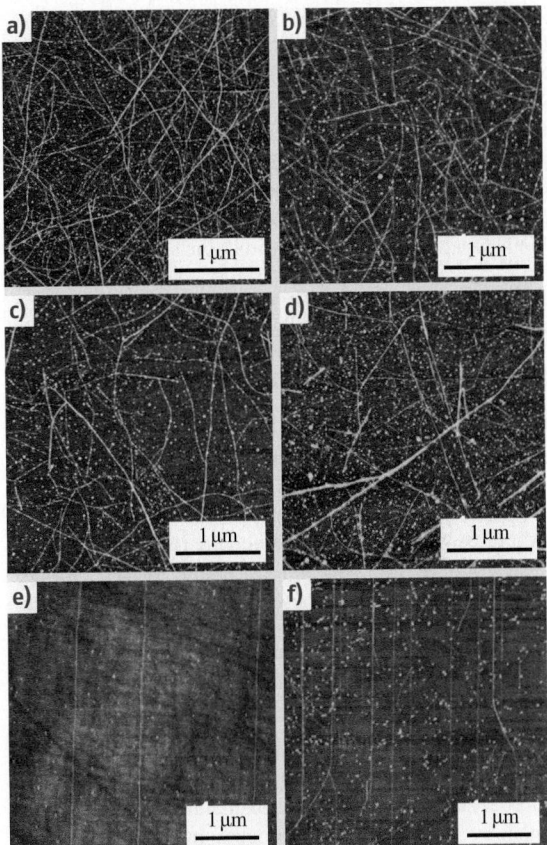

Fig. 13.7a–f AFM image of random assembly of single-walled carbon nanotubes on SiO_2/Si substrates grown from (**a**) 20× diluted ferritin catalyst for 1 hr (with density of $\approx$ 28 tubes/μm^2); (**b**) 20× diluted ferritin catalyst for 10 min (with density of $\approx$ 20 tubes/μm^2); (**c**) 100× diluted ferritin catalyst for 1 h (with density of $\approx$ 14 tubes/μm^2); (**d**) 100× diluted ferritin catalyst for 10 min (with density of $\approx$ 6 tubes/μm^2). Aligned single-walled carbon nanotubes on Y-cut single crystalline quartz grown from (**e**) 1000× diluted ferritin catalyst for 10 min; (**f**) 200× diluted ferritin catalyst for 10 min. Reaction conditions for tube growth: temperature - 900 °C; flow rate of CH_4 - 500 sccm; and flow rate of H_2 - 75 sccm

tubes, rather than bundles of tubes. The small particles on the surface are the catalyst. Tube density – the number of nanotubes per unit surface area, which strongly influences the performance of transistors fabricated with them (such as device mobility, on/off ratio) – can be easily controlled by tuning the concentration of catalyst and growth time. As shown in Fig. 13.7a–d, the tube density can be increased from several tubes to tens of tubes per μm^2 (on SiO_2/Si wafer) by increasing the concentration of ferritin and extending the growth time. Arrays of SWNTs with uniaxial alignment, rather than disordered random networks, can be generated through a similar process by carefully choosing growth substrates [13.56]. For instance, ferritin deposited on a quartz crystal (trigonal symmetry) with a Y-cut leads to the growth of SWNTs along a particular direction. The densities of

nanotube arrays generated in this manner are influenced by the growth conditions, such as the concentration of catalyst and growth time. Figures 13.7e and 13.7f show AFM images of arrayed nanotubes formed on quartz by using low and high concentrations of ferritin to generate catalyst, respectively. All of the nanotubes have good alignment in the plane of the substrate surface, which makes them easy to integrate into electronic devices.

13.2 Generation of Thin Films of Ordered Nanostructures on Plastic Substrates

Effective thin films formed with well-defined architectures of nanostructures on plastic substrates are the cornerstone for fabrication of high-performance devices. Fully dense monolayers of aligned 1-D or 2-D nanostructures with uniform end-to-end registry represent an ideal configuration. In addition to the quartz templating approach described above, a number of methods can organize nanowires dispersed in solutions into parallel arrays by applying various external forces (such as microfludic shear force, surface compression, a electric field or magnetic field) to overcome Brownian motion and to introduce well defined order [13.57–62]. Nanowires/nanoribbons fabricated using "top-down" approaches can be designed to maintain the order defined by the lithographic process by including local bonding sites to the substrate from which they are defined. The resultant arrays of nanostructures can be transfer-printed onto a wide range of desired sub-

strates including plastic sheets, without disrupting the order.

13.2.1 Assembly of Solution–Based Nanowires/Nanoribbons Using External Forces

Nanowires/nanoribbons synthesized through "bottom-up" approaches are usually dispersed in solvents and then deposited onto substrates for further applications. Generating organized arrays, which have lower entropy than their random assembly, requires external forces to manipulate the nanowires. For instance, flow of a nanowire suspension through microfludic channels formed between a PDMS stamp and substrate has been demonstrated by *Lieber* and coworkers to align nanowires of various materials [13.57]. In this case, a shear flow formed near the surface of the substrate can orient the wires parallel to the flow direction on the substrate (Fig. 13.8a). The degree of alignment, which is represented by the angle (θ) of nanowires with respect to the flow direction, can be controlled by the flow rate. Meanwhile, the spacing between individual aligned nanowires can be narrowed by increasing the time for deposition and concentration of the nanowire solution. A disadvantage of this approach is that it is difficult to produce high coverage in the form of a monolayer in a reasonable amount of time.

Nanowires at the water–air interface can form nematic liquid crystal type structures through Langmuir–Blodgett (LB) type techniques [13.58, 59]. In this case, the surfaces of the nanowires are first modified with surfactant molecules that have hydrophobic heads facing the water. The repulsive interaction between these hydrophobic groups and the water molecules lifts the nanowires to the water–air interface. A force is then applied to squeeze the nanowire film floating on the wa-

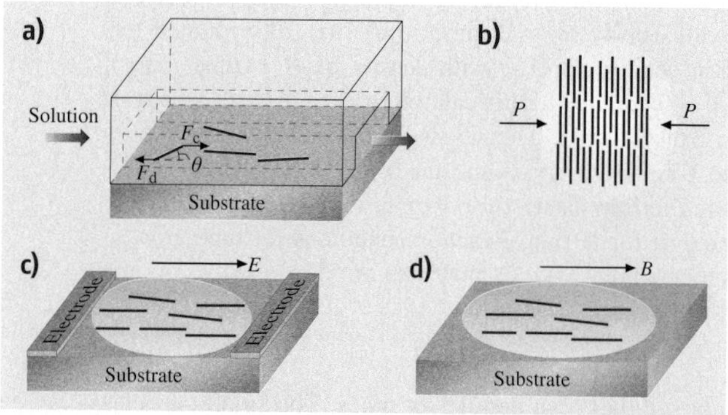

Fig. 13.8a–d Strategies for assembly of as-synthesized 1-D and 2-D nanostructures (such as nanowires, nanoribbons, nanotubes) dispersed in solvents through manipulation of different external forces: (**a**) flow-directed force in microfludic channels; (**b**) surface pressure at solution/air interface generated through the Langmuir-Blodgett technique; (**c**) electric field; and (**d**) magnetic field

ter surface, resulting in an increase of surface pressure (P). Once the surface pressure reaches a critical value, nanowires rotate perpendicular to the pressure direction (Fig. 13.8b) in the LB trough. In this manner, an array of aligned nanowires forms on the surface of the water. These wire arrays can be transferred onto other substrates for further applications. The advantage of this approach is that it can achieve high density and large area coverage, both of which are critical requirements for fabrication of high performance macroelectronics. Surfactant modification on the surfaces of nanowires can, however, affect their electrical properties and thus their performance in applications.

When the nanowires have suitable electrical and magnetic properties, they can be aligned on a substrate by applying electric and/or magnetic fields and drying the dispersion at relatively low speed. For example, some nanowires can be induced to generate polarization when they are exposed to an electric field. The interaction between the induced wire polarization and the electric field aligns the wires parallel to the field direction (Fig. 13.8c) [13.60, 61]. Magnetic fields can also easily concentrate and align ribbons or wires that are coated with a ferromagnetic material (such as Ni) (Fig. 13.8d) [13.62]. In addition, the Ni stripes can serve as electrodes in the further TFT fabrication. If Si is used as channel material, Ni can form a silicide with Si to generate ohmic contacts at high temperature and thus to increase the performance of devices fabricated using these elements [13.63].

These types of alignment processes must be developed further, however, for generating large-area thin films of nanowires/nanoribbons with high throughput. The use of solvent (for dispersing nanostructures) and surfactant molecules (for modifying the interaction between nanostructures and solvent or substrate) has the potential to cause adverse effects on the electrical properties of the nanostructures. In addition, with existing techniques, the resultant nanowire arrays have relatively poor end-to-end registry and uniformity of spacing between adjacent wires.

13.2.2 "Dry Transfer Printing" of Wire/Ribbon Arrays Derived from Wafers

Nanowire arrays grown on substrates via the VLS process (see Sect. 13.1.2 and Fig. 13.2d and 2e) have good alignment, uniformity and position registry. The nanowires perpendicular to the surface of the substrate are, however, difficult to transfer, without disrupting their order, onto plastics in the type of horizontal configuration that would be most effective for transistor applications. The "grow-in-place" approach for generating nanowire arrays in the plane of the substrate has some promise, but its ability to scale up to high-density arrays over large areas has not been demonstrated. In contrast, nanowires fabricated via "top-down" approaches can inherit the order of the lithographic pattern used to produce them and they can be defined over large areas. Ordered nanowires produced in this manner can be transfer-printed onto nearly any desired substrate, including plastic sheets.

Figure 13.9 presents steps for transferring nanowire arrays of GaAs (generated from GaAs wafer via anisotropic chemical etching) to a plastic substrate, poly(ethylene terephthalate) (PET) sheet [13.43]. When etch masks of patterned SiO_2 lines are surrounded by a uniform SiO_2 film on the surface of a (100) wafer, chemical etching generates GaAs nanowires with their ends connected to the mother wafer. This connection confines the wires and preserves the spatial orientation defined by the pattern of SiO_2 even after the wires are completely undercut. In this manner, GaAs wire arrays can be transfer-printed to plastic sheets with retention of their translation and orientational order using a PDMS stamp. In the transfer process, the PDMS stamp is slightly oxidized with an oxygen plasma or

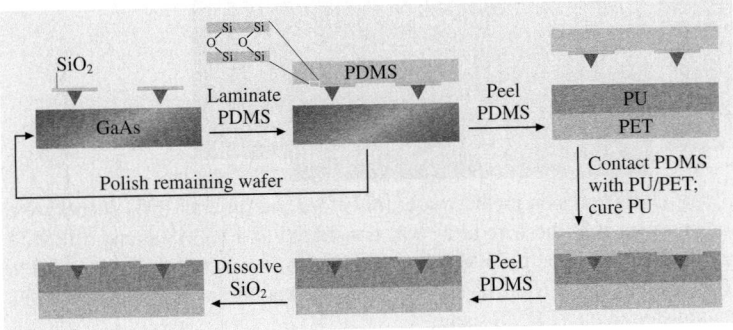

Fig. 13.9 Schematic illustration of transfer printing of GaAs wire arrays from mother wafer onto a plastic substrate (PET sheet covered with thin layer of PU) using a flat PDMS stamp as a transfer element

ozone to generate a hydrophilic surface by transforming Si-H groups on the surface of the stamp to $Si-O$ or $Si-OH$ groups. Contacting the oxidized PDMS stamp with SiO_2-coated GaAs wires on the wafer generates strong chemical bonding due to the formation of covalent siloxane $(Si-O-Si)$ linkages between PDMS and SiO_2 via a condensation reaction. This bonding is strong enough to break the crystalline connections at the ends of wires. Peeling the PDMS stamp away from the GaAs substrate transfers all of the wires to the stamp. The remaining GaAs wafer can be polished using standard processes, thereby preparing it for another round of wire fabrication. In this manner, it is possible to generate a huge number of GaAs wires from a single wafer, thus significantly decreasing the cost per wire.

GaAs wire arrays on the PDMS stamp can be transferred onto a plastic substrate using a photocurable polymer such as polyurethane (PU). Placing the PDMS stamp (with GaAs wires) against a PET sheet coated with a thin layer of PU (thickness between one and tens of microns, determined by spin speed) forms conformal contact by pushing the surfaces of the wires into the liquid prepolymer of PU. Illuminating the sample with an ultraviolet (UV) lamp crosslinks and solidifies the PU layer, resulting in the formation of a strong bond between the cured PU, the GaAs wires and the underlying PET substrate. Peeling off the PDMS stamp and dissolving the SiO_2 stripes leaves the GaAs wire arrays embedded in the matrix of cured PU with order and crystallographic orientation similar to those of the wires prior to transfer. The top flat surfaces of the wires face upward, in an orientation that can easily be exploited for subsequent device fabrication steps. Dissolution of the SiO_2 stripes leads to the formation of trenches on the surface of the resultant sample, which might affect the processes of device fabrication. The step depth can be reduced by decreasing the thickness of SiO_2 mask stripes and/or stripping a layer of cured PU with thickness similar to that of the SiO_2 stripes before removing the SiO_2, which can serve as etch mask for etching PU.

Figure 13.10a and 13.10b give SEM images of arrays of GaAs wires with widths of $\approx 400\,nm$ on the mother wafer before and after the SiO_2 etch mask was removed. The images clearly show that the nanowires

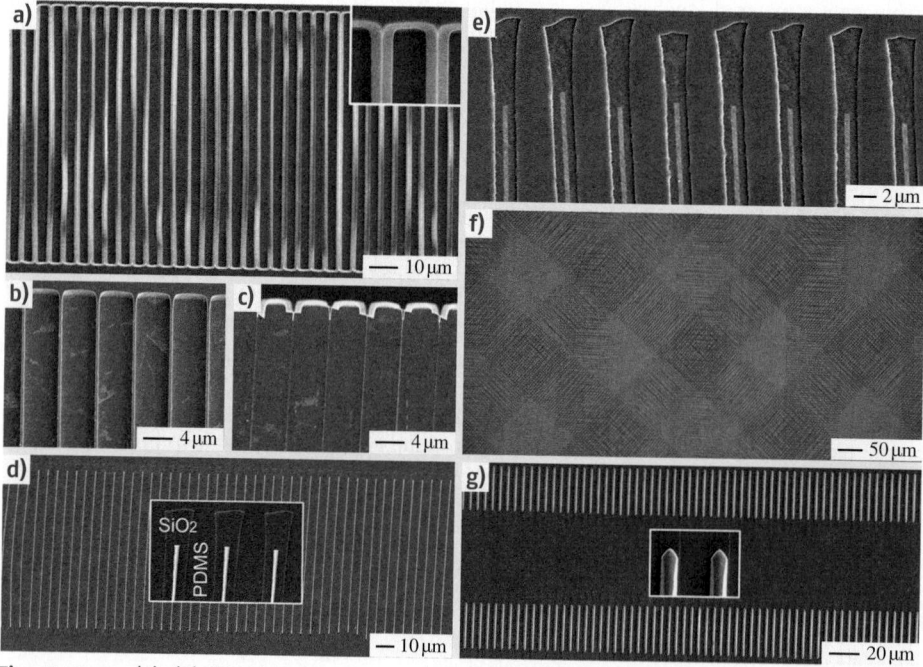

Fig. 13.10a–g (a), (b) SEM images of arrays of GaAs wires on mother wafer (a) before and (b) after SiO_2 stripes were removed. (c) SEM image of remaining mother wafer after the wire array was transferred to a PDMS stamp. (d) SEM image of a wire array bonded to the surface of a PDMS stamp. (e) SEM image of the ends of arrayed wires embedded in a cured PU layer cast on a PET substrate. (f) Optical image of triple layers of GaAs wire arrays transfer-printed on a PET substrate. (g) SEM image of InP wire arrays transferred on a PDMS stamp

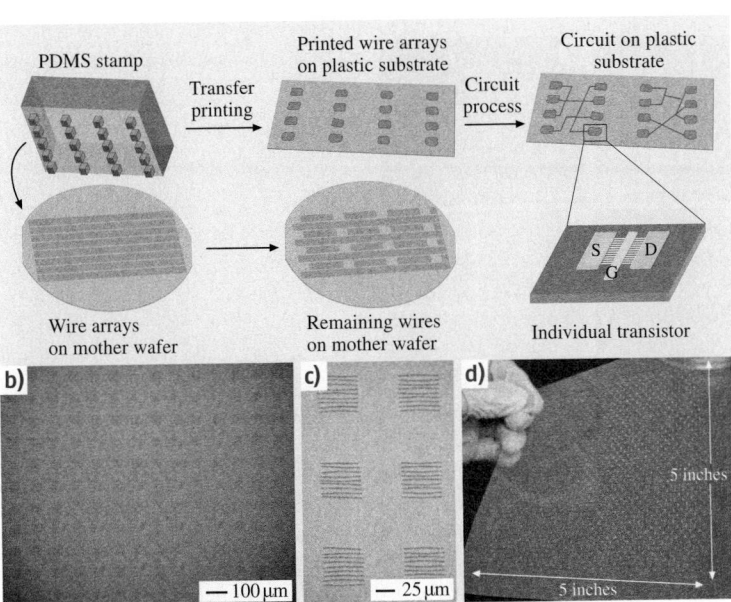

Fig. 13.11a–d Transfer printing patches of wire arrays on plastic substrates using PDMS stamps with patterned posts. (**a**) Schematic illustration of the printing process and circuit integration. The interconnections in the drawing are used to demonstrate the concept, not for any particular circuit. (**b**),(**c**) Optical images of patches of arrays of 50 μm long GaAs wires on PET substrates. (**d**) A large-area sample with transfer-printed Si ribbons on a PET substrate

have excellent alignment and that all of the ends of the nanowires connect to the mother wafer. Figure 13.10c presents an SEM image of the remaining wafer after transferring the wire arrays to a PDMS stamp; all of the wires were broken at points close to their ends. The wire array on the PDMS stamp retains the order present on the mother wafer (Fig. 13.10d). The inset shows that the SiO_2 mask stripes serve as an adhesive layer to bond the GaAs nanowires to the PDMS stamp. Figure 13.10e gives an SEM image of a PU/PET substrate with a GaAs wire array obtained by transfer-printing the sample of Fig. 13.10d onto the PU/PET substrate. This image indicates that all of the GaAs wires preserve the order and spatial orientation defined by lithography and that they are well embedded in the cured PU. The flat top surface of each wire is exposed for further applications. The transfer printing process can be repeated to print multiple layers of GaAs wire arrays on the same PET substrate. Figure 13.10f presents an optical image of a sample with three layers of wire arrays formed by rotating the second layer and third layer with $\approx 45\,°C$ and $\approx 90\,°C$ relative to the first, respectively. The cured PU between the levels of GaAs wire arrays can electrically insulate circuits or devices built on different layers. The same procedure can transfer-print wire arrays of other materials (InP, GaN, Si, ...) generated via "top-down" approaches. Figure 13.10g shows an SEM image of InP wire arrays transferred to a PDMS stamp. The wires can be further transfer-printed onto a PU/PET substrate for fabricating flexible devices.

This "dry transfer printing" process combined with the "top-down" fabrication approach is promising for applications related to high-performance macroelectronics. The approach provides a means to control the crystallographic orientation of transferred wires and it leaves the top (100) surfaces facing up to provide an extremely flat surface (similar to that of the original wafer) for device fabrication. The use of SiO_2 stripes as etch masks and adhesive layers for the transfer prevents the (100) surfaces of the GaAs wires from being contaminated by the organics used in the processing. Conformal contact between the PDMS stamp and the liquid prepolymer of PU generates GaAs wires conformally bound to and embedded in cured PU, which prevents the wires from moving, especially when the plastic substrates are bent or twisted. In addition, the "top-down" approach can easily fabricate wire arrays compatible with the layouts of complex circuits.

In most cases, circuits for macroelectronics are fabricated on substrates with sizes much larger than semiconductor wafers. Furthermore, active semiconductor films (for TFTs) usually occupy only 10–20% area of the whole substrate, and the remaining regions are used to build interconnections between individual device units (such as transistors) and other elements (such as resistors or capacitors). It is therefore useful to be able to selectively transfer-print wire arrays from cer-

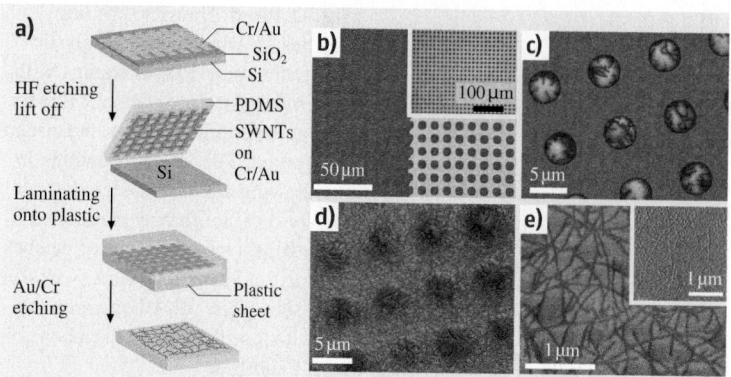

Fig. 13.12 (a) Schematic illustration of the dry transfer of CVD-grown single-walled carbon nanotubes onto plastic substrates. The first step involves evaporating a layer of Cr (2 nm)/Au (20 nm) onto a film of single-walled carbon nanotubes and then patterning the metal by photolithography or microcontact printing. Exposing the resulting substrate to HF removes the SiO_2 layer underneath the nanotubes. A piece of flat PDMS can remove the metal/nanotube layer and transfer it to a plastic substrate. Etching away the metal leaves a layer of nanotubes on the plastic. SEM images of **(b)** metal patterns on Si after HF etching and lifting off with PDMS, *inset* shows an optical image of a metal pattern on PDMS, **(c)** nanotube/metal patterns printed on a layer of epoxy, **(d)** nanotube patterns on epoxy after removing the Cr/Au by wet etching, **(e)** nanotubes on epoxy after Cr/Au etching, *inset* is the AFM image of the same sample

tain regions of a mother wafer in order to minimize the cost of fabrication. Figure 13.11a shows a diagram for printing nanowire arrays generated from a single wafer over large areas on a plastic substrate. This process uses a PDMS stamp with patterned posts. Placing the stamp against the wafer and peeling it off picks up only the wires that contact with the surface of the posts. The printing process can transfer patches of wire arrays onto plastic substrates. Repeating the steps over the remaining wires on the mother wafer can yield organized arrays of wires over large areas on a plastic substrate. Figure 13.11b gives an optical image of patches consisting of GaAs wire arrays that were transfer-printed on a PU/PET substrate. The image with relatively high magnification (as shown in Fig. 13.11c) shows that the wires of each patch are well aligned and have good end-to-end registry. This transfer printing process can be extended to wire arrays of other materials (including InP, Si, and so on) fabricated via "top-down" approaches. For instance, Fig. 13.11d shows an image of a $5'' \times 5''$ PET sheet with patterned patches of Si ribbons generated from a small piece of SOI wafer [13.64].

13.2.3 Transfer Printing of Thin Films of Single–Walled Carbon Nanotubes

Single-walled carbon nanotubes generated from "bulk methods" can be easily dispersed in various solvents

with the assistance of surfactants. Such a solution is compatible with the deposition of nanotubes onto plastic substrates for device fabrication at low temperatures. A controlled flocculation approach and printing scheme provides considerable flexibility in the coverage, orientation and form of the deposited tubes, and it can be used with a wide range of substrates, including curved surfaces [13.65]. Usually, devices formed with solution-deposited tubes show poor performance compared to similar devices made with tubes synthesized by CVD [13.65, 66]. High-performance TFTs on plastic can be achieved by the transfer of films of CVD-grown SWNTs from their high-temperature growth substrates to low temperature plastic substrates.

Figure 13.12a schematically illustrates the steps for transfer-printing CVD nanotubes from their growth substrate (SiO_2/Si in this case) to a plastic substrate [13.67]. A submonolayer of SWNTs is first grown on a SiO_2 (100 nm)/Si wafer using well-established CVD methods. By controlling the growth conditions, the coverage can be varied from ≈ 1 tube/μm^2 to ≈ 30 tubes/μm^2. Blanket electron beam evaporation over this wafer forms thin layers of Cr (2 nm) and then Au (20 nm) to cover the nanotubes. Photolithography combined with lift-off or microcontact printing (μCP) and chemical etching pattern this metal layer to expose the nanotubes in certain regions. The regions where metal is removed provide chemical access of a concentrated hydrofluoric acid etchant to attack the SiO_2 layer underneath the nanotubes. Undercutting by the etchant enables the SiO_2 to be completely removed, resulting in the release of carbon nanotubes from the growth substrate. The substrate is then blown dry with nitrogen gas. Contacting a flat PDMS stamp against the substrate and then quickly peeling the stamp away transfers the metal pattern, with the embedded nanotubes, to the stamp. Contacting this "inked" stamp to a plastic substrate (epoxy, PET, polyimide) for 30 min at 80 °C and then slowly peeling it back transfers the metal/nanotube structure to plastic. Removing the metals with suitable wet etchants (TFA, Transene Co. for Au and CR-7, Cyantek Co. for Cr) leaves the network of CVD-grown SWNTs on the plastic substrate for further use in device fabrication.

Figure 13.12b shows a metal pattern on nanotube/Si after HF etching; the left side of this image corresponds to parts of the metal that have been removed with the PDMS stamp. Metal patterns with sufficiently high density of openings could be undercut-etched over large areas and successfully transferred to the PDMS; areas of up to 1 cm × 1 cm were possible. Figure 13.12c shows metal patterns after transfer-printing onto a layer of cured epoxy (1.7 μm-thick SU8) spin-cast on a piece of indium tin oxide (ITO, 100 nm)/PET (175 μm) sheet. After etching away the metal, networks of nanotubes can be clearly observed, as shown in Fig. 13.12d and 13.12e. High-resolution SEM and AFM images show that the tube density on the plastic substrate is ≈ 15 tubes/μm², which is comparable to the density measured on the growth substrates. It is notable that the metal etchants do not tend to wash away the tubes or destroy the tubes. The van der Waals forces between the tubes and the substrate provide sufficient adhesion for them to remain on the substrate.

13.3 Applications for Macroelectronics

High-quality monolayer films of nanostructures with desired architectures on plastic substrates can be used to produce flexible TFTs via traditional lithographic processing steps carried out at relatively low temperatures (less than 150 °C). Depending on the properties of the semiconductor nanostructures, the geometries of the TFTs (such as thin film type MOSFETs or metal-semiconductor field-effect transistors, MESFETs) and their corresponding fabrication processes are different, as described below.

13.3.1 Bottom Gate TFTs

Figure 13.13 summarizes the steps for fabricating bottom gate TFTs using arrays of nanowires/nanoribbons on plastic substrates. A thin conductive layer of metal (such as Au) or oxide (such as ITO) is first deposited to serve as gate electrode, on a thin plastic sheet which can withstand the temperature of the lithographic process. A thin dielectric layer (such as aluminium oxide, SiO_2, resin epoxy) is then deposited on the surface of conductive film to isolate the gate electrode from the device channel. Aligned nanowire/nanoribbon films can be formed by directly assembling solution-based objects on the surface of the dielectric layer or by transferring wire/ribbon arrays from their mother substrates to the surface of the dielectric layer using the approaches described in the previous section. Finally, large-area lithographic techniques define the pattern of

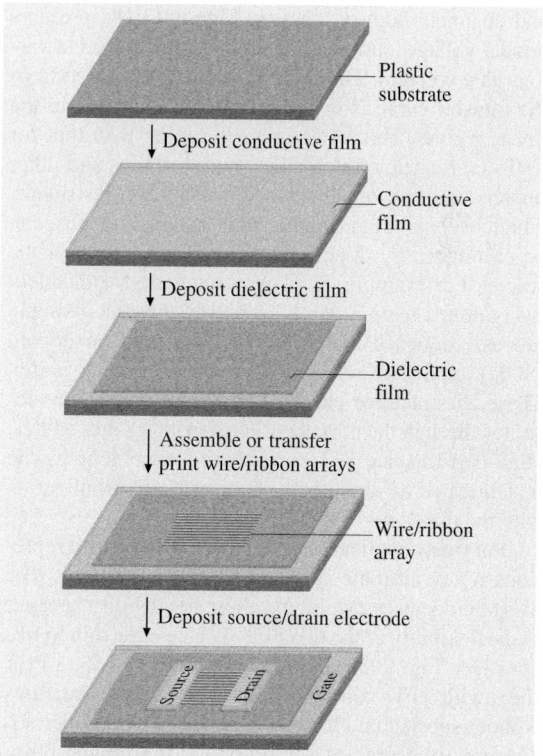

Fig. 13.13 Scheme for fabricating thin-film transistors with bottom-gate geometry using well-aligned arrays of nanowires/nanoribbons on plastic substrates

source-drain electrodes over the thin film of oriented nanowires/nanoribbons [13.68].

TFTs Formed with Si Nanowires/Nanoribbons

Rogers and co-workers have recently developed an approach to fabricating TFTs with high performance using "dry transfer-printed" Si ribbon arrays on plastic substrates [13.39, 69, 70]. In their demonstration, Si ribbons with thicknesses of $20-100$ nm and widths of $1-100$ μm were fabricated from high-quality SOI wafers via the "top-down" approach. These Si structures, with good order and orientation, were then transferred onto an ITO (100 nm)-coated PET (175 μm) substrate through the dry transfer-printing technique depicted in Figs. 13.9 and 13.11a. Before the Si ribbons were transferred, a thin layer of polymeric resin of SU8 photoresist (with a thickness of 600 nm) was spin-cast on the PET sheet to serve not only as the glue to bind the Si film and PET substrate together but also as a gate dielectric. Patterns of 100 nm-thick Ti films formed the source and drain electrodes. Figure 13.14a shows an optical image of an individual transistor with a channel length of 20 μm and channel width of 250 μm. Figure 13.14c presents typical voltage–current characteristics measured at various gate voltages. The electron mobility extracted from the transfer curve is $\approx 230 \, \text{cm}^2/\text{V} \cdot \text{s}$, evaluated in the linear regime. This value is much higher than that for TFTs fabricated with organics, nanoparticles and other polycrystalline inorganic thin films. Contact resistance, which becomes significant at short channel lengths, can be eliminated by doping the source/drain areas of the silicon. For example, Si ribbons integrated with selectively doped regions can be achieved through a simple process sequence that uses layers of spin-on dopant (SOD) applied to selected areas of the mother wafer. These Si structures can be transferred to plastic substrates through the aforementioned printing procedures. High-performance TFTs with short channel lengths can be fabricated by aligning the source/drain metallization with the doped regions [13.70].

For many applications, mechanical flexibility represents a key attribute of macroelectronic systems. The Si-ribbon devices exhibit excellent bendability because of the flexibility of the thin PET sheet and the thin Si ribbons (see Fig. 13.3b). As shown in Fig. 13.14b, a PET sheet with TFTs can be bent to form curved structures without substantial changes in the electrical properties. The precise degree of flexibility and the fatigue properties have been evaluated using the set-up shown in Fig. 13.14d. The strain generated in the Si ribbons can

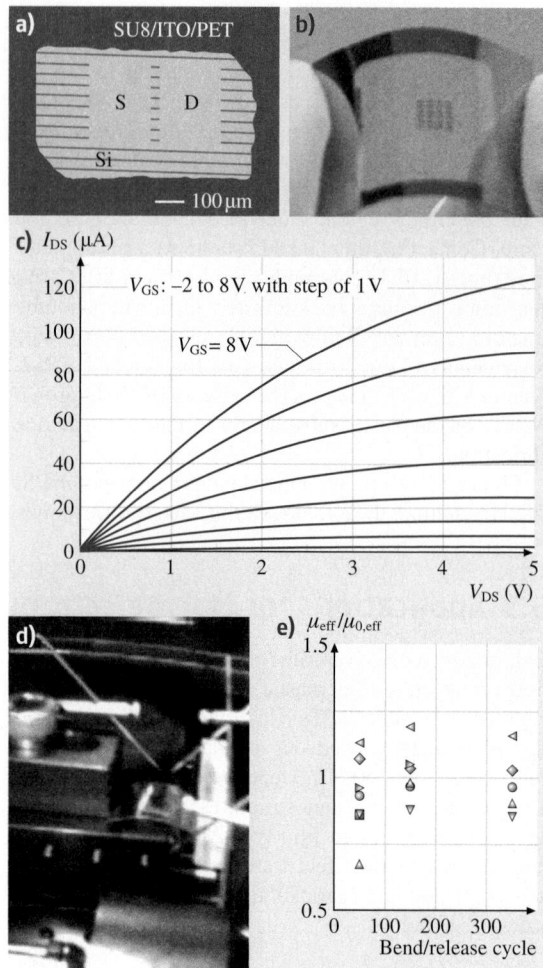

Fig. 13.14 (**a**) Optical image of an individual TFT built using aligned Si ribbons on a 600 nm-thick cured SU8 epoxy layer spin-cast on an ITO-coated PET sheet. The two rectangular pads consist of Ti film of thickness 100 nm, which serve as source and drain electrodes. The ITO layer underneath the epoxy layer serves as a gate electrode. (**b**) Photograph of a piece of a PET sheet with an array of TFTs, indicating their mechanical flexibility. (**c**) Typical $I_{DS} - V_{DS}$ characteristics of a Si ribbon transistor on a PET substrate with modulation of different gate voltages, V_{GS}. From bottom to top, V_{GS} varies from -2 V to 8 V with steps of 1 V. (**d**) Set-up for fatigue testing of Si ribbon transistors on a PET substrate. (**e**) Variation of electron mobility of Si ribbon transistors with different channel lengths (represented with different symbols) after repeated bend/release cycles

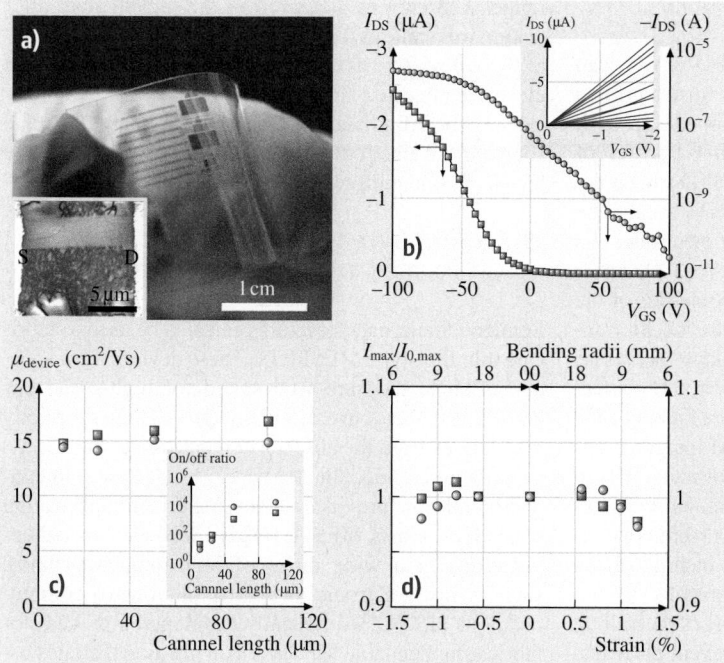

Fig. 13.15 (a) An optical photograph of an array of single-walled carbon nanotube TFTs on a PET substrate. The *inset* shows an SEM image of the channel of an individual transistor. (b) Transfer characteristics of a nanotube TFT with a channel length of $100\,\mu m$ and a channel width of $250\,\mu m$. The source-drain driving voltage, V_{DS}, was $-0.5\,V$. The *inset* gives the current–voltage characteristics of the same transistor with modulation of different gate voltages, V_{GS}; V_{GS} varies from $-100\,V$ to $0\,V$ from the top to bottom with steps of $-10\,V$, (c) Comparison of device mobilities on plastic (*squares*) and on the SiO_2/Si wafer (the substrate for nanotube growth) (*circles*). The *inset* presents the on/off ratio of this series of devices. A $100\,nm$-thick layer of SiO_2 formed the gate dielectric in the latter case while a $1.7\,\mu m$-thick layer of epoxy formed the gate dielectric in the former case. (d) Variation of current output of individual nanotube TFTs at different bending radii. The channel lengths of the two measured transistors were $5\,\mu m$ (*squares*) and $100\,\mu m$ (*circles*), respectively. The thickness of the gate dielectric (the epoxy layer) was $1\,\mu m$ for both transistors

be determined by measuring the geometric parameters of the substrate. The data indicate that the electrical properties do not change significantly with 350 bend/release cycles to strains of 0.98% (bend radii of $\approx 9\,mm$ for this case). Figure 13.14e shows the variation in the mobilities of transistors with different channel lengths after hundreds of bend/release cycles, indicating good fatigue stability of Si transistors with mobility changes of less than 20%.

An alternative route to TFTs on plastic substrates is to use monolayer (or submonolayer) films of nanowires/nanoribbons in the form of aligned arrays that are generated by assembling liquid dispersions of nanowires/nanoribbons. For example, *Duan* and coworkers assembled Si nanowires, which were synthesized via a "bottom-up" approach (via the laser-assisted VLS process), on a polyetheretherke-

tone (PEEK) sheet (with a thickness of $125\,\mu m$) by passing a nanowire suspension through microfluidic channels (detailed description referred to in a previous section) [13.71]. Before assembling the nanowires, thin strips of Cr/Au ($10/30\,nm$) and a $30\,nm$ layer of aluminium oxide were deposited on the PEEK sheet to serve as gate electrodes and dielectric. Ti/Au ($60/80\,nm$) patches served as source and drain electrodes. The resultant devices show a threshold voltage of $\approx 3.0\,V$, an on-off ratio higher than 10^5, and a subthreshold swing of $500–800\,mV$ per decade. The performance did not significantly change even when the bend radius was $\approx 55\,mm$. In spite of the relatively high per wire mobility ($\approx 150\,cm^2/V \cdot s$), the device mobilities are low ($\approx 3\,cm^2/V \cdot s$) due to the low densities of these arrays and their random end-to-end registries.

TFTs Formed with Single-Walled Carbon Nanotubes

Flexible TFTs based on networks of SWNTs can be fabricated with CVD nanotubes transfer-printed on plastic substrates via the aforementioned approach (Sect. 13.2.3) [13.67]. For example, after transfer of nanotubes onto PET coated with 100 nm ITO and epoxy (SU8, 1.7 μm), bottom gate nanotube TFTs can be fabricated by defining source and drain electrodes via lithography and deposition of Ti (2 nm)/Au (20 nm) or by nanotransfer printing [13.72]. The geometries of the nanotube transistors are similar to those of Si transistors shown in Fig. 13.14a. The ITO layer serves as a gate electrode and the epoxy layer as a gate dielectric. Figure 13.15a presents a photograph of a bent PET sheet with an array of devices. The SEM image (inset) clearly shows the network organization of nanotubes in the channel region. These types of nanotube networks have been demonstrated to serve as good semiconductors for TFTs [13.73, 74]. Reactive ion etching of the carbon nanotubes through photolithographically defined patterns of resists created network stripes oriented along the transistor channels. These stripes prevent electrical crosstalk between devices, and they can facilitate breakdown procedures that can eliminate metallic pathways through the networks [13.74]. Figure 13.15b plots the transfer characteristics of a device whose channel length and channel width are 100 μm and 250 μm, respectively; the inset shows current–voltage measurements at various gate voltages. These devices showed unipolar p-channel behavior, similar to those made at original growth SiO_2/Si substrates with SWNTs. The carbon nanotube-based devices on plastic substrates showed mobilities ($\approx 15\,cm^2/V \cdot s$) and on-off current ratios similar to those of devices fabricated on the growth substrates for the entire range of channel lengths (L between 10 μm and 100 μm). Figure 13.15c gives the results. The slightly lower device mobility, μ_{device}, of the transferred devices compared to that of the original substrates could be due to incomplete transfer of tubes, particularly near the open areas in the metal films.

These types of devices exhibit very good mechanical flexibility because of the excellent mechanical properties associated with nanotubes. The experimental set-up used for the bending tests is the same as that shown in Fig. 13.14d. Figure 13.15d shows the change in the on current, normalized by the value in the unbent state $I_{0,max}$ as a function of strain (or bending radius). Negative and positive strains correspond to tension and compression, respectively. For both compression and tension, the I_{max} varied over only a narrow ($\pm 5\%$) range. The current measured in the unbent state after testing was same as that before testing. The bending limit of these devices was defined by fracture of the ITO gate electrodes because the tubes themselves are extremely robust in bending. The addition of SWNT-based gate and source/drain electrodes could yield devices with extreme levels of bendability [13.75].

13.3.2 Top Gate MESFETs Formed with GaAs Wire Arrays

Semiconductor nanostructures can also be used to fabricate thin film-type MESFETs. These devices have some appeal since they are well suited for high-speed operation and they have a level of processing simplicity that derives from the absence of a gate dielectric. GaAs, a compound semiconductor widely used in the semiconductor industry, provides an ideal material for fabricating MESFETs. GaAs MESFETs have well established applications in a wide range of high-frequency fields, such as fast electronic switching microwave communications [13.76]. We recently developed a process for fabricating bendable MESFETs on plastic substrates using GaAs wires that have integrated ohmic source/drain contacts [13.77]. In this process, a (100) semi-insulating GaAs (SI-GaAs) wafer with an epitaxial Si-doped n-type GaAs layer (carrier concentration of $4.0 \times 10^{17}\,/cm^3$) is used as the source material for generating GaAs wires. Photolithography and metallization via electron-beam (and/or thermal) evaporation generate arrays of narrow metal stripes (with a width of 2 μm and a spacing of 13 μm) composed of conventional multilayer stacks (AuGe (120 nm)/Ni (20 nm)/Au (120 nm)) for ohmic contacts. Annealing the wafer at elevated temperatures (450 °C for 1 min) in a quartz tube with flowing N_2 forms ohmic contacts to the n-GaAs. The "top down" approach as depicted in Fig. 13.5 generates GaAs wires with ohmic stripes; the resultant composite wire arrays can then be transferred onto a PU/PET substrate via the "dry printing" technique. Each wire has two ohmic stripes separated by a gap that defines the channel length of the resultant MESFET. Removing the adhesive layer (for example the photoresist or SiO_2) used in the transfer printing process exposes the surfaces of ohmic stripes and GaAs wires for device fabrication. As shown in Fig. 13.16a, the wire arrays on plastic substrate can be used to fabricate flexible MESFETs using the traditional lithographic process performed at temperatures less than 150 °C, which is lower than the glass transition temperatures (T_g) of either the cured PU or PET. Deposition of 250 nm Au pads connects the ohmic stripes to form

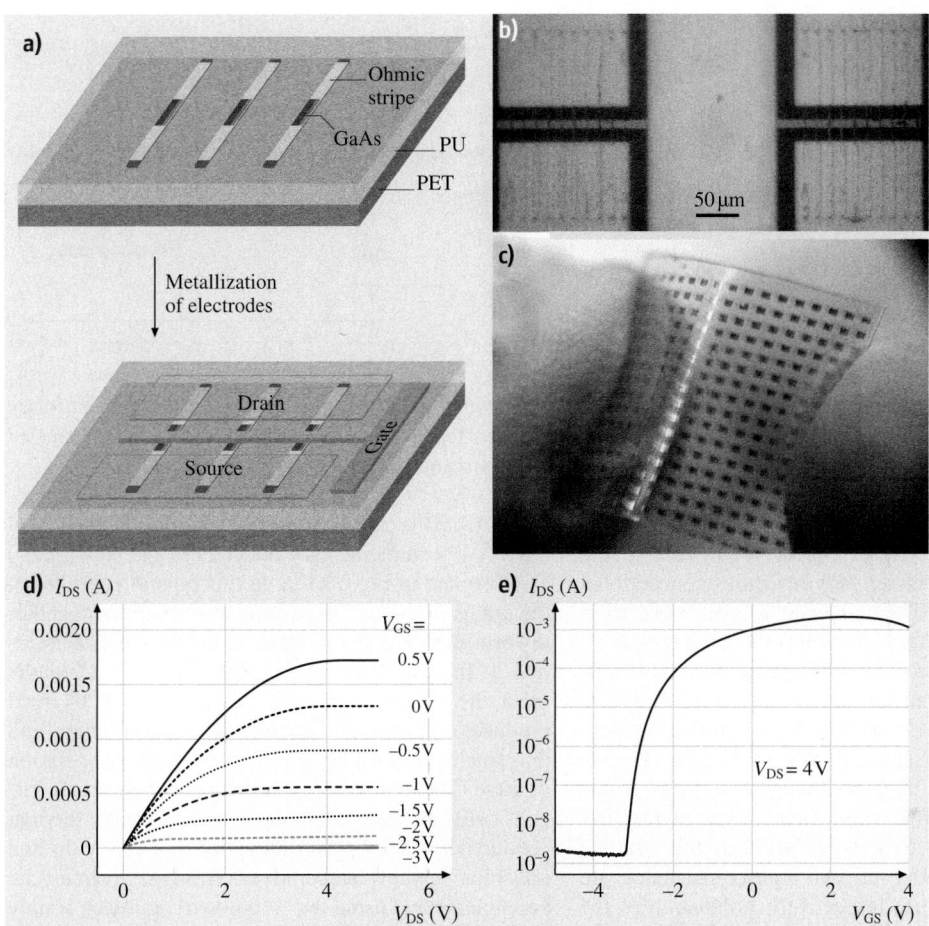

Fig. 13.16a–e Fabrication and characterization of MESFETs on plastic substrates formed with arrays of GaAs wires integrated with ohmic stripes. (**a**) Layout of a MESFET on a PET substrate. (**b**) Optical image of two MESFETs with identical parameters: channel length 50 μm; channel width 150 μm; gate length 10 μm. These two transistors share the same large pad (in the *center of image*). (**c**) Photograph of a bent PET sheet with a number of MESFETs. (**d**) Current–voltage characteristics of a GaAs MESFET shown in (b). From *top* to *bottom*, V_{GS} varies from 0.5 V to -3.0 V with steps of -0.5 V. (**e**) Transfer curve for the same transistor with $V_{DS} = 4$ V

the source and drain. Narrow stripes of Ti (150 nm)/Au (150 nm) deposited between the source and drain serve as gate electrodes. The resultant MESFETs are mechanically flexible due to the bendability of the PU/PET sheet (thickness of ≈ 200 μm) and the GaAs wires (widths and thicknesses less than 5 μm).

Figure 13.16b shows a representative image of two GaAs wire-based MESFETs on plastic. The wires have well-aligned orientations and uniform widths of ≈ 1.8 μm. Au pads with widths of 150 μm and lengths of 250 μm connect the ohmic stripes on ten GaAs wires to form source and drain electrodes for each individual

MESFET. A Ti/Au stripe with a width of 10 μm (gate length) deposited in the 50 μm gap (transistor channel) between source and drain electrodes provides the gate electrode. These stripes connect to a larger metal pad for probing. Figure 13.16c presents a photograph of a 2 cm × 2 cm PET sheet with a number of transistors, clearly showing its flexibility. Various parameters, such as the width of the GaAs wires, the channel width and length, the gate length and gate metal can be easily adjusted to yield MESFETs with desired output characteristics. The DC performance of an individual MESFET is good. Typical $I-V$ curves of a transistor with a channel length

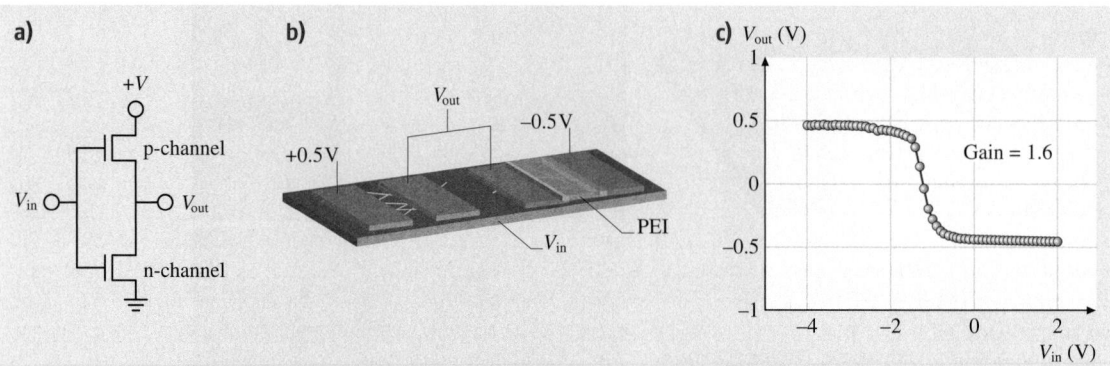

Fig. 13.17 (**a**) Typical circuit of a complementary invert. Illustration (**b**) and characterization (**c**) of a simple complementary inverter on a plastic substrate with TFTs based on carbon nanotubes. Input-output curves from an inverter constructed with single-walled carbon nanotube TFTs with channel lengths of 50 μm and channel widths of 250 μm

of 50 μm and a gate length of 10 μm (similar to the one shown in Fig. 13.16b) are presented in Fig. 13.16d. The $I_{DS} - V_{DS}$ characteristics resemble conventional wafer-based MESFETs built with n-type GaAs layer and standard techniques (I_{DS} saturates in the regions of high V_{DS} and I_{DS} decreases with decrease of gate voltage) [13.78]. The transfer curve (I_{DS} versus V_{GS}) was measured at $V_{DS} = 4$ V, where the current flow through source and drain is saturated, as shown in Fig. 13.16e. The drop in I_{DS} at high positive gate voltages is due to the leakage current from gate to source that develops through the Schottky contact in this regime. The on/off (current) ratio and transconductance are 1.15×10^6 and 670 μS, respectively. Evaluation of the mechanical flexibility of these GaAs wire MESFETs on PET substrates reveals that these transistors can survive multiple bend/release cycles with strain of ≈1.2% with a change in current of less than 20%. In addition, multiple printing steps and/or wire fabrication runs (as shown in Figs. 13.9 and 13.11a) can generate large numbers of wires patterned over large areas on plastic substrates to fabricate low-cost, large area, flexible TFTs.

13.3.3 Complementary Metal–Oxide–Semiconductor (CMOS) Circuits

Complete macroelectronic systems require sophisticated circuitry formed from the integration of many transistors as well as other passive circuit elements such as capacitors, inductors, diodes, resistors, and so on. A particularly useful building block for digital applications

results from the combination of n-channel and p-channel devices in complementary inverters (logic NOT gates) as illustrated in Fig. 13.17a. In this type of component, the drains of the two transistors are connected together to form the output; the input is the common connection to the transistor gates. With a binary "1" at the input, the output is in the "0" state, whereas a "0" input produces a "1" output. The low power consumption, and thus small heat generation, associated with the operation of these CMOS inverters is beneficial for electronic circuits built on plastic substrates, which have poor thermal conductivity for thermal dissipation and can withstand only relatively low temperatures. This type of circuit can be implemented using the "top-down" approach simply by adjusting the doping species to obtain either n- or p-type contacts and/or channel regions. Similar components can be achieved in SWNTs by room-temperature chemical modification, which can be directly processed on plastic substrate. As shown in Fig. 13.17b, soaking the nanotubes in a methanol solution of polyethylenimine (PEI) can change the channel from p-type to n-type [13.46, 74]. The dependence of V_{out} on V_{in} in a SWNT-based TFT inverter is shown in Fig. 13.17c. The voltage gain of this inverter is 1.6, which is similar to inverters made with an individual nanotube by potassium doping [13.79]. The tuning ability of channel-type transistors on plastic substrates enables us to generate circuits with more complex functions for intelligent macroelectronic systems. It can also be exploited to build tunable pn-diodes and other more complex device structures [13.74].

13.4 Outlook

Nanomaterials with various morphologies and compositions have been briefly reviewed in terms of preparation approaches and routes to generating high-quality thin films of them on plastic substrates. Their potential applications in macroelectronics and some preliminary examples were also discussed. For inorganic nanowires/nanoribbons, two different paradigms can be used to form TFTs on plastic substrates. In the first, nanowires synthesized via "bottom-up" approaches form aligned arrays through post-grown assembly and/or transfer onto plastic substrates. These "bottom-up" approaches can produce nanowires with a wide range of materials, providing a flexibility to choose suitable nanowires/nanoribbons to meet specific applications. The uncertain quality of the nanowires/nanoribbons (in terms of dimensional uniformity, purity, dopant concentration, doping uniformity and surface crystallography) and of the films assembled from them (in terms of density, orientation and end-to-end registry) represent issues that will demand further attention in order to build reliable, large arrays of high-performance devices. The current lack of effective approaches for assembling high-quality nanowire/nanoribbon thin films represents the main challenge for successful technology applications of these systems. In the second strategy, the nanostructures are fabricated from high-quality, single-crystal, bulk wafers via "top-down" approaches. The structures in this case adopt the organization, orientation and spatial layout defined by the pattern of the mask, which is easily controlled via the initial lithography steps. The resultant nanowire/nanoribbon arrays can be transferred onto plastic substrates through "dry transfer printing" techniques. A drawback of this approach is that it is limited to materials that currently exist in bulk wafer or thin film form. Single-walled carbon nanotubes, which continue to attract significant research attention, are an interesting class of materials because the tubes have been shown to have extremely high carrier mobilities, high mechanical strength and flexibility, and their electronic properties can be tuned chemically. They are also electrically and chemically stable. Their use in the form of large-scale networks and arrays for applications such as macroelectronics represents a recently emerging trend in research. One of the most significant challenges, which is the control of electronic purity of carbon nanotubes, is alleviated to some extent by the architecture of the devices themselves, which average over the ensemble of tubes involved in the transport. Elimination of metallic tubes, while not a solved problem, appears to be moving toward working solutions that involve chemical, electrical and/or optical approaches.

Although high performance is important, reliable, robust operation of macroelectronic devices is crucial for many of their envisioned applications. In particular, the nature of the interfaces between the plastic sheet (or plastic sheet covered with cured polymer) and the semiconductor, and between the metal and semiconductor are critical from both electronic and mechanical points of view. Appropriate adhesive layers between a semiconductor nanowire film and plastic sheet are important to achieve reliable device structures, especially when the substrate is bent and relaxed. The long term reliability of low-temperature dielectrics is also a topic that must be investigated. Development of low-cost, large-area lithographic techniques and reliable processes for fabricating passive electronic elements, including resistors, capacitors, metal interconnects, and so on, on plastic substrates also represent important issue for building robust integrated circuits of macroelectronic systems. In spite of these challenges, we believe that printable inorganic semiconductor nanostructures represent one of the most promising approaches to building these and other classes of large-area electronic devices.

References

13.1   S. Ucjikoga: Low-temperature polycrystalline silicon thin-film transistor technologies for system-on-glass displays, MRS Bull. **27**, 881–886 (2002)

13.2   J. A. Rogers, Z. Bao, K. Baldwin, A. Dodabalapur, B. Crone, V. R. Raju, V. Kuck, H. Katz, K. Amundson, J. Ewing, P. Drzaic: Paper-like electronic displays: large-area rubber-stamped plastic sheets of electronics, microencapsulated electrophoretic inks, Proc. Natl. Acad. Sci. U.S.A. **98**, 4835–4840 (2001)

13.3   Y. Izumi, Y. Yamane: Solid-state x-ray imagers, MRS Bull. **27**, 889–893 (2002)

13.4   P. T. Kazlas, M. D. McCreary: Paperlike microencapsulated electrophoretic materials, displays, MRS Bull. **27**, 894–897 (2002)

13.5   J. A. Rogers, R. G. Nuzzo: Recent progress in soft lithography, Materials Today **8(2)**, 50–56 (2005)

13.6   Y. Isono, T. Namazu, T. Tanaka: AFM bending testing of nanometric single crystal silicon wire at intermediate temperatures for MEMS. In: *The 14th*

IEEE International Conference on MEMS (IEEE, Interlaken, Switzerland 2001) pp. 135–138

13.7   Y. Cui, Z. Zhong, D. Wang, W. Wang, C. M. Lieber: High performance silicon nanowire field effect transistors, Nano Lett. **3**, 149–152 (2003)

13.8   P. Alivisatos, P. F. Barbara, A. W. Castleman, J. Chang, D. A. Dixon, M. L. Klein, G. L. McLendon, J. S. Miller, M. A. Ratner, P. J. Rossky, S. I. Stupp, M. E. Thompson: From molecules to materials: current trends, future directions, Adv. Mater. **10**, 1297–1336 (1998)

13.9   R. A. Bley, S. M. Kauzlarich, J. E. Davis, H. W. H. Lee: Characterization of silicon nanoparticles prepared from porous silicon, Chem. Mater. **8**, 1881–1888 (1996)

13.10   X. Peng, L. Manna, W. Yang, J. Wickham, E. Scher, A. Kadavanich, A. P. Alivisatos: Shape control of CdSe nanocrystals, Nature **404**, 59–61 (2000)

13.11   L. Manna, E. C. Scher, A. P. Alivisatos: Synthesis of soluble, processable rod-, arrow-, teardrop-, tetrapod-shaped CdSe nanocrystals, J. Am. Chem. Soc. **122**, 12700–12706 (2000)

13.12   W. W. Yu, X. Peng: Formation of high-quality CdS, other II–VI semiconductor nanocrystals in noncoordinating solvents: tunable reactivity of monomers, Angew. Chem. Int. Ed. **41**, 2368–2371 (2002)

13.13   D. Battaglia, X. Peng: Formation of high quality InP, InAs nanocrystals in noncoordinating solvent, Nano Lett. **2**, 1027–1030 (2002)

13.14   B. A. Ridley, B. Nivi, J. M. Jacobson: All-inorganic field effect transistors fabricated by printing, Science **286**, 746–749 (1999)

13.15   T. Cui, F. Hua, Y. Lvov: FET fabricated by layer-by-layer nanoassembly, IEEE Trans. Electron Devices **51**, 503–506 (2004)

13.16   G. Zheng, W. Lu, S. Jin, C. M. Lieber: Synthesis, fabrication of high-performance n-type silicon nanowire transistors, Adv. Mater. **16**, 1890–1893 (2004)

13.17   Z. L. Wang (Eds.): *Nanowires and Nanobelts: Materials, Properties and Devices*, Vol. I and II (Kluwer, Boston 2003)

13.18   Y. Xia, P. Yang, Y. Sun, Y. Wu, B. Mayers, B. Gates, Y. Yin, F. Kim, H. Yan: One-dimensional nanostructures: Synthesis, characterization, applications, Adv. Mater. **15**, 353–389 (2003)

13.19   C. M. Lieber: One-dimensional nanostructures: Chemistry, physics and applications, Solid State Commun. **107**, 607–616 (1998)

13.20   M. Law, J. Goldberger, P. Yang: Semiconductor nanowires and nanotubes, Ann. Rev. Mater. Res. **34**, 83–122 (2004)

13.21   Z. R. Dai, Z. W. Pan, Z. L. Wang: Novel nanostructures of functional oxides synthesized by thermal evaporation, Adv. Funct. Mater. **13**, 9–24 (2003)

13.22   R. S. Wagner, W. C. Ellis: Vapor-liquid-solid mechanism of single crystal growth, Appl. Phys. Lett. **4**, 89–90 (1964)

13.23   A. M. Morales, C. M. Lieber: A laser ablation method for the synthesis of crystalline semiconductor nanowires, Science **279**, 208–211 (1998)

13.24   K. Hiruma, M. Yazawa, T. Katsuyama, K. Ogawa, K. Haraguchi, M. Koguchi, H. Kaklbayashi: Growth, optical properties of nanometer-scale GaAs, InAs whiskers, J. Appl. Phys. **77**, 447–462 (1995)

13.25   X. Duan, C. M. Lieber: General synthesis of compound semiconductor nanowires, Adv. Mater. **12**, 298–302 (2000)

13.26   M. S. Gudiksen, J. Wang, C. M. Lieber: Synthetic control of the diameter, length of single crystal semiconductor nanowires, J. Phys. Chem. B **105**, 4062–4064 (2001)

13.27   Y. Wu, H. Yan, M. Huang, B. Messer, J. H. Song, P. Yang: Inorganic semiconductor nanowires: Rational growth, assembly, novel properties, Chem. Eur. J. **8**, 1261–1268 (2002)

13.28   M. H. Huang, S. Mao, H. Feick, H. Yan, Y. Wu, H. Kind, E. Weber, R. Russo, P. Yang: Room-temperature ultraviolet nanowire nanolasers, Science **292**, 1897–1899 (2001)

13.29   S. Bhunia, T. Kawamura, S. Fujikawa, K. Tokushima, Y. Watanabe: Free-standing, vertically aligned InP nanowires grown by metalorganic vapor phase epitaxy, Physica E **21**, 583–587 (2004)

13.30   Z. H. Wu, X. Y. Mei, D. Kim, M. Blumin, H. E. Ruda: Growth of Au-catalyzed ordered GaAs nanowire arrays by molecular-beam epitaxy, Appl. Phys. Lett. **81**, 5177–5179 (2002)

13.31   Y. Shan, K. Kalkan, C.-Y. Peng, S. J. Fonash: From Si source gas directly to positioned, electrically contacted Si nanowires: the self-assembling "grow-in-place" approach, Nano Lett. **4**, 2085–2089 (2004)

13.32   T. J. Trentler, K. M. Hickman, S. C. Goel, A. M. Viano, P. C. Gibbons, W. E. Buhro: Solution-liquid-solid growth of crystalline III–V semiconductors: An analogy to vapor-liquid-solid growth, Science **270**, 1791–1794 (1995)

13.33   H. Yu, W. E. Buhro: Solution-liquid-solid growth of soluble GaAs nanowires, Adv. Mater. **15**, 416–419 (2003)

13.34   F. M. II. I. Davidson, A. D. Schricker, R. J. Wiacek, B. A. Korgel: Supercritical fluid-liquid-solid synthesis of gallium arsenide nanowires seeded by alkanethiol-stabilized gold nanocrystals, Adv. Mater. **16**, 646–649 (2004)

13.35   H.-Y. Tuan, D. C. Lee, T. Hanrath, B. A. Korgel: Catalytic solid-phase seeding of silicon nanowires by nickel nanocrystals in organic solvents, Nano Lett. **5**, 681–684 (2005)

13.36   J. A. Rogers, K. E. Paul, R. J. Jackman, G. M. Whitesides: Using an elastomeric phase mask for sub-100 nm photolithography in the optical near field, Appl. Phys. Lett. **70**, 2658–2660 (1997)

13.37   F. Hua, Y. Sun, A. Gaur, M. A. Meitl, L. Bilhaut, L. Rotkina, J. Wang, P. Geil, M. Shim, J. A. Rogers: Polymer imprint lithography with molecular-scale resolution, Nano Lett. **4**, 2467–2471 (2004)

13.38   Y. Yin, B. Gates, Y. Xia: A soft lithography approach to the fabrication of nanostructures of single crystalline silicon with well-defined dimensions, shapes, Adv. Mater. **12**, 1426–1430 (2000)

13.39   E. Menard, K. J. Lee, D.-Y. Khang, R. Z. Nuzzo, J. A. Rogers: A printable form of silicon for high performance thin film transistors on plastic substrates, Appl. Phys. Lett. **84**, 5398–5400 (2004)

13.40   K. J. Lee, J. Lee, H. Hwang, Z. J. Reitmeier, R. F. Davis, J. A. Rogers, R. G. Nuzzo: A printable form of single-crystalline gallium nitride for flexible optoelectronic systems, Small **1**, 1164–1168 (2005)

13.41   S. C. Lee, S. R. J. Brueck: Nanoscale two-dimensional patterning on Si (001) by large-area interferometric lithography, anisotropic wet etching, J. Vac. Sci. Technol. B **22**, 1949–1952 (2004)

13.42   Z. L. Wang: Transmission electron microscopy of shape-controlled nanocrystals, their assemblies, J. Phys. Chem. B **104**, 1153–1175 (2000)

13.43   Y. Sun, J. A. Rogers: Fabricating semiconductor nano/microwires, transfer printing ordered arrays of them onto plastic substrates, Nano Lett. **4**, 1953–1959 (2004)

13.44   Y. Sun, D.-Y. Khang, F. Hua, K. Hurley, R. G. Nuzzo, J. A. Rogers: Photolithographic route to the fabrication of micro/nanowires of III–V semiconductors, Adv. Funct. Mater. **15**, 30–40 (2005)

13.45   T. Dürkop, B. M. Kim, M. S. Fuhrer: Properties, applications of high-mobility semiconducting nanotubes, J. Phys. Condens. Matter **16**, R553–R580 (2004)

13.46   S. H. Hur, C. Kocabas, A. Gaur, O. O. Park, J. A. Rogers: Printed thin film transistors and complementary logic gates that use polymer coated single-walled carbon nanotube networks, J. Appl. Phys. **98**, 114302 (2005)

13.47   T. Durkop, S. A. Getty, E. Cobas, M. S. Fuhrer: Extraordinary mobility in semiconducting carbon nanotubes, Nano Lett. **4**, 35–39 (2004)

13.48   D. Bozovic, M. Bockrath, J. H. Hafner, C. M. Lieber, H. Park, M. Tinkham: Plastic deformations in mechanically strained single-walled carbon nanotubes, Phys. Rev. B **67**, 033407 (2003)

13.49   C. Kocabas, M. A. Meitl, A. Gaur, M. Shim, J. A. Rogers: Aligned arrays of single-walled carbon nanotubes generated from random networks by orientationally selective laser ablation, Nano Lett. **4**, 2421–2426 (2004)

13.50   C. Wang, Q. Cao, T. Ozel, A. Gaur, J. A. Rogers, M. Shim: Electronically selective chemical functionalization of carbon nanotubes: correlation

between Raman spectral and electrical responses, J. Am. Chem. Soc. **127**, 2005 (11460-11468)

13.51   M. S. Dresselhaus, G. Dresselhaus, P. C. Eklund: *Science of Fullerenes and Carbon Nanotubes* (Academic, San Diego 1996)

13.52   A. Thess, R. Lee, P. Nikolaev, H. Dai, P. Petit, J. Robert, C. Xu, Y. H. Lee, S. G. Kim, A. G. Rinzler, D. T. Colbert, G. E. Scuseria, D. Tom'anek, J. E. Fischer, R. E. Smalley: Crystalline ropes of metallic carbon nanotubes, Science **273**, 483–487 (1996)

13.53   M. J. Bronikowski, P. A. Willis, D. T. Colbert, K. A. Smith, R. E. Smalley: Gas-phase production of carbon single-walled nanotubes from carbon monoxide via the HiPco process: a parametric study, J. Vac. Sci. Technol. A **19**, 1800–1805 (2001)

13.54   H. Dai, A. G. Rinzler, P. Nikolaev, A. Thess, D. T. Colbert, R. E. Smalley: Single-wall nanotubes produced by metal-catalyzed disproportionation of carbon monoxide, Chem. Phys. Lett. **260**, 471–475 (1996)

13.55   Y. Li, W. Kim, Y. Zhang, M. Rolandi, D. Wang, H. Dai: Growth of single-walled carbon nanotubes from discrete catalytic nanoparticles of various sizes, J. Phys. Chem. **105**, 11424–11431 (2001)

13.56   C. Kocabas, S. H. Hur, A. Gaur, M. A. Meitl, M. Shim, J. A. Rogers: Guided growth of large scale, horizontally aligned arrays of single walled carbon nanotubes and their use in thin film transistors, Small **1**, 1110–1116 (2005)

13.57   Y. Huang, X. Duan, Q. Wei, C. M. Lieber: Directed assembly of one-dimensional nanostructures into functional networks, Science **291**, 630–633 (2001)

13.58   D. Whang, S. Jin, Y. Wu, C. M. Lieber: Large-scale hierarchical organization of nanowire arrays for integrated nanosystems, Nano Lett. **3**, 1255–1259 (2003)

13.59   A. Tao, F. Kim, C. Hess, J. Goldberger, R. He, Y. Sun, Y. Xia, P. Yang: Langmuir-Blodgett silver nanowire monolayers for molecular sensing using surface-enhanced Raman spectroscopy, Nano Lett. **3**, 1229–1233 (2003)

13.60   P. A. Smith, C. D. Nordquist, T. N. Jackson, T. S. Mayer, B. R. Martin, J. Mbindyo, T. E. Mallouk: Electric-field assisted assembly, alignment of metallic nanowires, Appl. Phys. Lett. **77**, 1399–1401 (2000)

13.61   X. Duan, Y. Huang, Y. Cui, J. Wang, C. M. Lieber: Indium phosphide nanowires as building blocks for nanoscale electronic, optoelectronic devices, Nature **409**, 66–69 (2001)

13.62   A. K. Bentley, J. S. Trethewey, A. B. Ellis, W. C. Crone: Magnetic manipulation of copper-tin nanowires capped with nickel ends, Nano Lett. **4**, 487–490 (2004)

13.63   Y. Wu, J. Xinag, C. Yang, W. Lu, C. M. Lieber: Single-crystal metallic nanowires, metal/semiconductor nanowires heterostructures, Nature **430**, 61–65 (2004)

13.64   K. J. Lee, M. J. Motala, M. A. Meitl, W. R. Childs, E. Menard, A. Shim, J. A. Rogers, R. G. Nuzzo: Large area, selective transfer of microstructured silicon (μs-Si): a printing-based approach to high performance thin film transistors supported on flexible substrates, submitted./CEnotePlease update this reference, or remove if paper has not been accepted yet

13.65   M. A. Meitl, Y. Zhou, A. Gaur, S. Jeon, M. L. Usrey, M. S. Strano, J. A. Rogers: Solution casting, transfer printing single-walled carbon nanotube films, Nano Lett. **4**, 1643–1647 (2004)

13.66   E. Artukovic, M. Kaempgen, D. S. Hecht, S. Roth, G. Grüner: Transparent, flexible carbon nanotube transistors, Nano Lett. **5**, 757–760 (2005)

13.67   S. H. Hur, O. O. Park, J. A. Rogers: Extreme bendability of single-walled carbon nanotube networks transferred from high-temperature growth substrates to plastic and their use in thin-film transistors, Appl. Phys. Lett. **86**, 243502 (2005)

13.68   C. R. Kagan, P. Andry (Eds.): *Thin Film Transistors* (Dekker, New York 2003)

13.69   E. Menard, R. G. Nuzzo, J. A. Rogers: Bendable single crystal silicon thin film transistors formed by printing on plastic substrates, Appl. Phys. Lett. **86**, 093507 (2005)

13.70   Z. Zhu, E. Menard, K. Hurley, R. G. Nuzzo, J. A. Rogers: Spin on dopants for high performance single crystalline silicon transistor on flexible plastic substrates, Appl. Phys. Lett. **86**, 133507 (2005)

13.71   X. Duan, C. Niu, V. Sahi, J. Chen, J. W. Parce, S. Empedocles, J. L. Goldman: High-performance thin-film transistors using semiconductor nanowires, nanoribbons, Nature **425**, 274–278 (2003)

13.72   S. H. Hur, D. Y. Khang, C. Kocabas, J. A. Rogers: Nanotransfer printing by use of noncovalent surface forces: Applications to thin-film transistors that use single-walled carbon nanotube networks and semiconducting polymers, Appl. Phys. Lett. **85**, 5730–5732 (2004)

13.73   E. S. Snow, J. P. Novak, P. M. Campbell, D. Park: Random networks of carbon nanotubes as an electronic material, Appl. Phys. Lett. **82**, 2145–2147 (2003)

13.74   Y. Zhou, A. Gaur, S.-H. Hur, C. Kocabas, M. A. Meitl, A. Shim, J. A. Rogers: p-Channel, n-channel thin film transistors, p-n diodes based on single wall carbon nanotube networks, Nano Lett. **4**, 2031–2035 (2004)

13.75   Z. Wu, Z. Chen, X. Du, J. M. Logan, J. Sippel, M. Nikolou, K. Kamaras, J. R. Reynolds, D. B. Tanner, A. F. Hebanrd, A. G. Rinzler: Transparent, conductive carbon nanotube film, Science **305**, 1273–1276 (2004)

13.76   C. Y. Chang, F. Kai: *GaAs High-Speed Devices: Physics, Technology, and Circuit Applications* (Wiley, New York 1994)

13.77   Y. Sun, S. Kim, I. Adesida, J. A. Rogers: Bendable GaAs metal-semiconductor field effect transistors formed with printed GaAs wire arrays on plastic substrates, Appl. Phys. Lett. **87**, 083501 (2005)

13.78   S. M. Sze: *Semiconductor Devices, Physics and Technology* (Wiley, New York 1985)

13.79   V. Derycke, R. Martel, J. Appenzeller, Ph. Avouris: Carbon nanotube inter-, intramolecular logic gates, Nano Lett. **1**, 453–456 (2001)

Part B
MEMS/NEM

Part B MEMS/NEMS and BioMEMS/NEMS

14. Next-Generation DNA Hybridization and Self-Assembly Nanofabrication Devices

The new era of nanotechnology presents many challenges and opportunities. One area of considerable challenge is nanofabrication, in particular the development of fabrication technologies that can evolve into viable manufacturing processes. Considerable efforts are being expended to refine classical top-down approaches, such as photolithography, to produce silicon-based electronics with nanometer-scale features. So-called bottom-up or self-assembly processes are also being researched and developed as new ways of producing heterogeneous nanostructures, nanomaterials and nanodevices. It is also hoped that there are novel ways to combine the best aspects of both top-down and bottom-up processes to create a totally unique paradigm change for the integration of heterogeneous molecules and nanocomponents into higher order structures. Over the past decade, sophisticated microelectrode array devices produced by the top-down process (photolithography) have been developed and commercialized for DNA diagnostic genotyping applications. These devices have the ability to produce electric field geometries on their surfaces that allow DNA molecules to be transported to or from any site on the surface of the array. Such devices are also able to assist in the self-assembly (via hybridization) of DNA molecules at specific locations on the array surface. Now a new generation of these microarray devices are available that contain integrated CMOS components within their underlying silicon structure. The integrated CMOS allows more precise control over the voltages and currents sourced to the individual microelectrode sites. While such microelectronic array devices

have been used primarily for DNA diagnostic applications, they do have the intrinsic ability to transport almost any type of charged molecule or other entity to or from any site on the surface of the array. These include other molecules with self-assembling properties such as peptides and proteins, as well as nanoparticles, cells and even micron-scale semiconductor components. Microelectronic arrays thus have the potential to be used in a highly parallel electric field "pick and place" fabrication process allowing a variety of molecules and nanostructures to be organized into higher order two- and three-dimensional structures. This truly represents a synergy of combining the best aspects of "top-down" and "bottom-up" technologies into a novel nanomanufacturing process.

Nanotechnology and nanoscience are producing a wide range of new ideas and concepts, and are likely to enable novel nanoelectronics, nanophotonics, nanomaterials, energy conversion processes and a new generation of biomaterials, biosensors and other biomedical devices. Many of the challenges and opportunities in nanotechnology have been identified through the efforts of the National Nanotechnology Initiative [14.1]. While many opportunities exist, there are also considerable challenges that must be met and overcome in order to obtain the benefits (see Table 14.1). Most challenging will be those areas that relate to nanofabrication, in particular the development of viable fabrication technologies which will lead to cost-effective nanomanufacturing

Table 14.1 Some challenges for nanotechnology and nano-fabrication (National Nanotechnology Initiative & NSF)

(1)   Better understanding of scaling problems and phenomena

(2)   Better synthetic methods for nano building blocks

(3)   Control of nanoscale building blocks (such as size and shape)

(4)   Enormous complexity, heterogeneous materials and sizes

(5)   Surfaces for nanostructure assembly

(6)   Directed hierarchical self-assembly (mimic biological)

(7)   Need for highly parallel processes

(8)   Equipment for parallel directed self-assembly

(9)   Integrate bottom-up and top-down approaches

(10)   Need better analytical capabilities

(11)   Tools for modeling and simulations

(12)   Scale-up issues for manufacturing

processes. Enormous efforts are now being carried out to refine classical top-down or photolithography processes to produce silicon (CMOS) integrated electronic devices with nanometer-scale features. While this goal is being achieved, this type of process requires billion-dollar fabrication facilities and it appears to be reaching some fundamental limits. So-called "bottom-up" self-assembly processes are also being studied and developed as possible new ways of producing nanoelectronics as well as new nanomaterials and nanodevices. Generally, self-assembly-based nanoelectronics are envisioned as one of the more revolutionary outcomes of nanotechnology. There are now numerous examples of promising nanocomponents such as organic electron transfer molecules, quantum dots, carbon nanotubes and nanowires, and also some limited success in first-level assembly of such nanocomponents into simple structures with higher order electronic properties [14.2–4]. Nevertheless, the issue of developing a viable cost-effective self-assembly nanofabrication process that allows billions of nanocomponents to be assembled into useful logic and memory devices still remains a considerable challenge. In addition to the nanoelectronic applications, other new nanomaterials and nanodevices with higher order photonic, mechanical, mechanistic,

sensory, chemical, catalytic and therapeutic properties are also envisioned as an outcome of nanotechnology efforts [14.1–4]. Again, a key problem in enabling such new materials and devices will most likely be in developing effective nanomanufacturing technologies for organizing and integrating heterogeneous components of different sizes and compositions into these higher level structures and devices.

Living systems provide some of the best examples of self-assembly or self-organization processes that should be considered very closely when developing strategies for "bottom-up" nanofabrication. The molecular biology of living systems includes many molecules which have high fidelity recognition properties such as DNA, RNA, and many types of protein macromolecules. Proteins can serve as structural elements, as binding recognition moieties (antibodies), and as highly efficient chemomechanical catalytic macromolecules (enzymes). Such biomolecules are able to interact and organize into second-order macromolecules and nanostructures which store and translate genetic information (involving DNA and RNA structural proteins as well as enzymes), and perform biomolecular syntheses and energy conversion metabolic processes (involving enzymes and structural proteins). All of these biomolecules, macromolecules, nanostructures and nanoscale processes are integrated and contained within higher order membrane-encased structures called cells. Cells in turn can then replicate and differentiate (via these nanoscale processes) to form and maintain living organisms. Thus, biology has developed the ultimate "bottom-up" nanofabrication processes that allow component biomolecules and nanostructures with intrinsic self-assembly and catalytic properties to be organized into highly intricate living systems.

Of all the different biomolecules that could be useful for nanofabrication, nucleic acids, with their high fidelity recognition and intrinsic self-assembly properties, represent a most promising material that can be used to create nanoelectronic, nanophotonic and many other types of organized nanostructures [14.5–8]. The nucleic acids, which include deoxyribonucleic acid (DNA), ribonucleic acid (RNA) and other synthetic DNA analogs (peptide nucleic acids and so on) are "programmable" molecules, which have intrinsic molecular recognition and self-assembly properties via their nucleotide base (A,T,G,C) sequence. Short DNA sequences called oligonucleotides are readily synthesized by automated techniques. They can additionally be modified with a variety of functional groups such as amines, biotin moieties, fluorescent or chromophore groups, and charge transfer molecules. Additionally, synthetic DNA molecules can be attached to quantum dots, metallic nanoparticles and carbon nanotubes, as well as surfaces like glass, silicon, gold and semiconductor materials. Synthetic DNA molecules (oligonucleotides) represent an ideal type of "molecular Lego" for the self-assembly of nanocomponents into more complex two- and three-dimensional higher order structures. Initially, DNA sequences can be used as a kind of template for assembly on solid surfaces. The technique involves taking complementary DNA sequences and using them as a kind of selective glue to bind other DNA-modified macromolecules or nanostructures together. The base pairing property of DNA allows one single strand of DNA with a unique base sequence to recognize and bind together with its complementary DNA strand to form a stable double-stranded DNA structures. While high-fidelity recognition molecules like DNA allow one to "self-assemble" higher order structures, the process has some significant limitations. First, for in vitro applications (in a test tube), DNA and other high-fidelity recognition molecules like antibodies, streptavidins, and lectins work most efficiency when the complexity of the system is relatively low. In other words, as the complexity of the system increases (more unrelated DNA sequences, proteins, and other biomolecules), the high-fidelity recognition properties of DNA molecules are overcome by nonspecific binding and other entropy-related factors, and the specificity and efficiency of DNA hybridization is considerably reduced. Under in vivo conditions (inside living cells), the binding interactions of high-fidelity recognition molecules like DNA are much more controlled and compartmentalized, and the DNA hybridization process is assisted by structural protein elements and active dynamic enzyme molecules. Thus, new nanofabrication processes based on self-assembly or self-organization using high-fidelity recognition molecules like DNA should also incorporate strategies for assisting and controlling the overall process.

Active microelectronic arrays have been developed for a number of applications in bioresearch and DNA clinical diagnostics [14.8–17]. These active microarrays are able to produce electric fields on the array surface that allow charged reagent molecules (DNA, RNA, proteins, enzymes), nanostructures, cells and micron-scale structures to be transported to any of the microscopic sites on the device surface. When DNA hybridization is carried out on the microarray, the device allows electric fields to direct the self-assembly of the DNA hybrid at the test site. In principle, these active microarray devices can serve as "motherboards or hostboards" to assist in the self-assembly of DNA molecules, as well as other moieties such as nanostructures or even microscale components [14.18–25]. Active microarray electric field assembly is thus a type of "pick and place" process that has the potential to be used for heterogeneous integration and nanofabrication of molecular and nanoscale components into higher order materials, structures and devices [14.24].

14.1 Electronic Microarray Technology

In the last decade, the development of microarray technologies has greatly expanded our analytical capabilities of carrying out both DNA and protein analysis [14.26]. Many of these novel microarray technologies now allow us to analyze thousands of DNA sequences with very high specificity and sensitivity. Examples include Affymetrix's (Santa Clara, CA, USA) GeneChip [14.27–29] Nanosphere's (Northbrook, IL, USA) [14.29] technology, and Nanogen's (San Diego, CA, USA) electronically active Nanochip [14.8–24, 30–43] technologies. Many assay techniques have been developed to carry out genotyping, gene expression analysis, forensics analysis and for a variety of other assay procedures.

Nanogen, Inc. has developed electronic microarray technology that utilizes electric fields to accelerate and manipulate biomolecules such as DNA, RNA, and proteins on a microarray surface. Each test site or microlocation on the microarray has an underlying platinum microelectrode which can be activated independently. The original 100 test site microarray with $80\,\mu m$-diameter platinum microelectrodes is fabricated on a silicon substrate. In this device, each of the 100 test site microelectrodes has a separate wire contact with only the microelectrode surface exposed to the sample solution. The newer 400 test site microarray with $50\,\mu m$-diameter platinum microelectrodes has CMOS control elements fabricated into the underlying silicon. This integrated CMOS is used to independently regulate the currents and voltages to each of the 400 test sites on the microarray surface. Both the 100 test site and 400 test site CMOS microarray chips are embedded within a disposable plastic fluidic cartridge that provides for automated control of sample or reagents injection onto the electronic microarray.

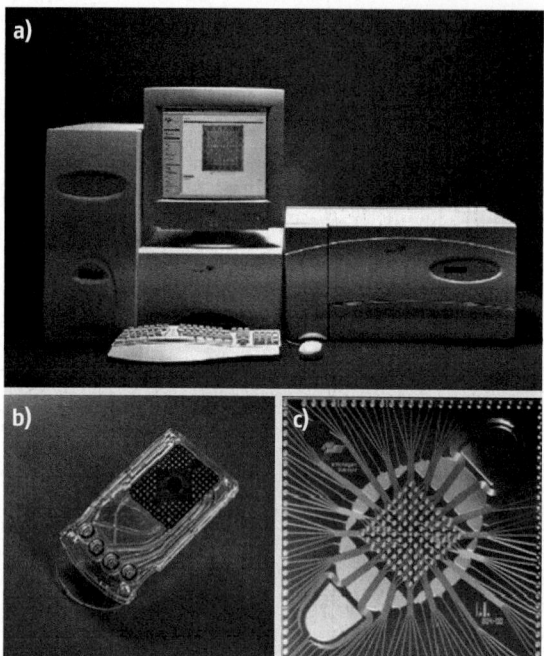

Fig. 14.1a–c First commercialized version of the Nanogen Molecular Biology Workstation. This includes the controller and fluorescent detection component ((**a**) *upper left*) and the loader system ((**b**) *upper right*) which can be used to address four 100-test site cartridges with DNA samples or DNA probes. The cartridge component containing the a 100-test site chip is shown in the *lower left* (**b**), and the 100-test site chip is show in the *lower right* (**c**)

The first generation platform, the Molecular Biology Workstation, was developed for a 100 test site microarray cartridge. Figure 14.1a shows the Molecular Biology Workstation, Fig. 14.1b shows the 100 test site microarray cartridge, and Fig. 14.1c the 100 test site microarray itself. The Workstation platform consists of two separate instruments, a loader that is capable of processing (addressing) up to four cartridges, and a single cartridge fluorescent reader. The newer Nanochip 400 System integrates both the loader and the reader into a single instrument, and the unique 400 test site CMOS microarray has been embedded within a new cartridge design (see Fig. 14.2). The Nanochip 400 System and 400 CMOS electronic microarray provide a tremendous amount of flexibility and control; each of the 400 test sites on the microarray can be easily configured and modified for a range of electronic assay formats.

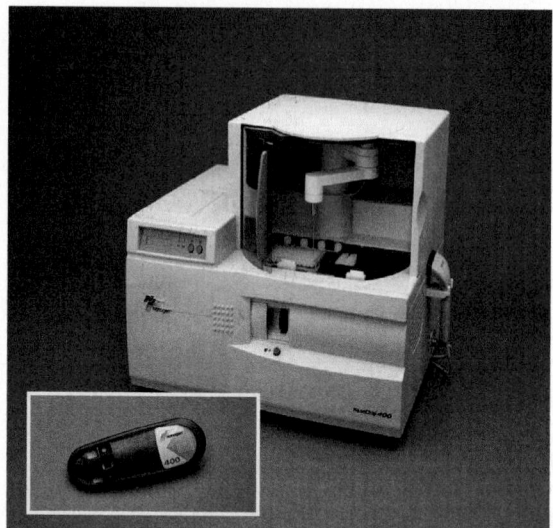

Fig. 14.2 The NanoChip 400 (NC400) is a fully integrated system capable of electronically loading samples onto a microarray and interrogating samples using the built-in fluorescent reader. The system is fully automated and has the capacity to analyze up to 364 samples in a single run. The 400 site electronic microarray, which is contained within a plastic cartridge, can be used up to ten times, allowing for greater user flexibility. A permlayer, into which streptavidin has been embedded, is molded on top of the microarray. The *insert* shows a NC 400 cartridge with 400 array sites or electrodes

14.1.1 400 Test Site CMOS Microarray

The external control of different voltages and currents via individual wires to a large number of microelectrodes can be a cumbersome process. Thus, for higher density microarrays (> 100 test sites) it is advantageous to integrate the microelectrode bias control circuitry directly into the microchip silicon structure itself. In the new 400 test site microarray, standard CMOS circuitry has been used to integrate digital communication, memory, temperature sensing, voltage/current sourcing and measuring circuits "on-chip". After standard CMOS fabrication within the underlying silicon, thin film deposition and patterning techniques were used to fabricate the platinum microelectrodes on the surface of the standard CMOS chip [14.30]. The CMOS microelectrode array chip developed for the Nanochip 400 system consists of a 16 by 25 array of 50 micron-diameter microelectrodes spaced 150 microns center to center. Figure 14.3 shows the 400 test site CMOS microelectronic array device which is only 5 mm by 7 mm in size. Figure 14.4 shows a close up of the 400 site microarray with several of the 50 micron diameter platinum microelectrodes. Underneath each of the 400 microelectrodes is an analog sample and hold circuit which maintains a predefined voltage on the electrode. A digital to analog converter sequentially interrogates the digitally stored bias value for each of the microelectrodes and refreshes each sample and hold circuit accordingly. A separate loop sequentially measures the voltage and current at each of the microelectrodes. Microelectrodes can be operated at a fixed voltage, or by means of a feed back loop, at constant current or at a fixed voltage offset from a reference electrode. The microchip also contains a p-n junction temperature sensor and EEPROM memory to store thermal calibration coefficients, serial numbers, and assay-related data. Figure 14.5 shows the Nanochip 400 CMOS circuitry block diagram. The chip has only 12 external electrical connections, +5 V and ground for the digital circuits, +5 V and ground for the analog circuits, digital signal in, digital signal out, clock signal, reset, and two terminals for an external current sam-

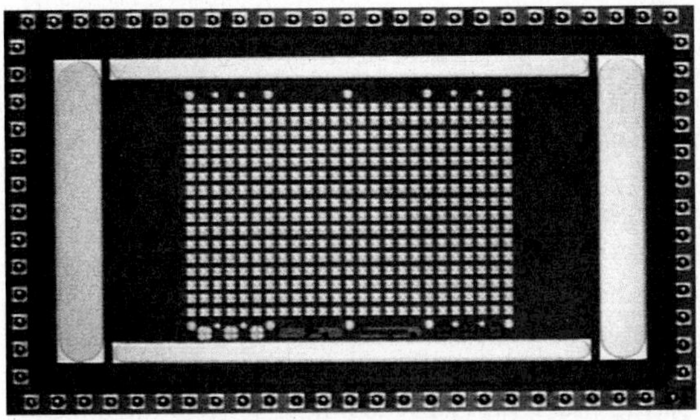

Fig. 14.3 Photograph of the 400-site CMOS ACV400-chip array. Four counter-electrodes, two positioned longitudinally and two horizontally, surround the active working electrode array

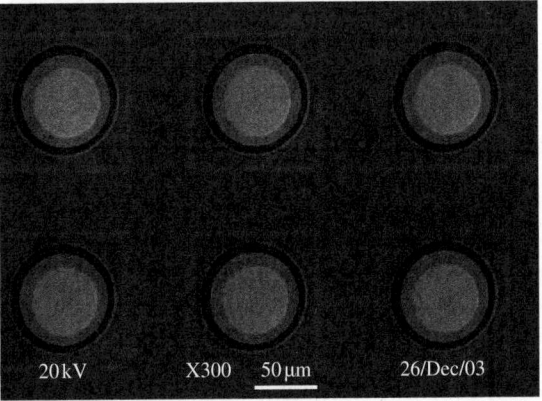

Fig. 14.4 Close-up of the current 400-site chip array. The CMOS chip has an array of 16×25 (400) sites; each electrode is 50 μm in diameter with a 150 μm center-to-center distance

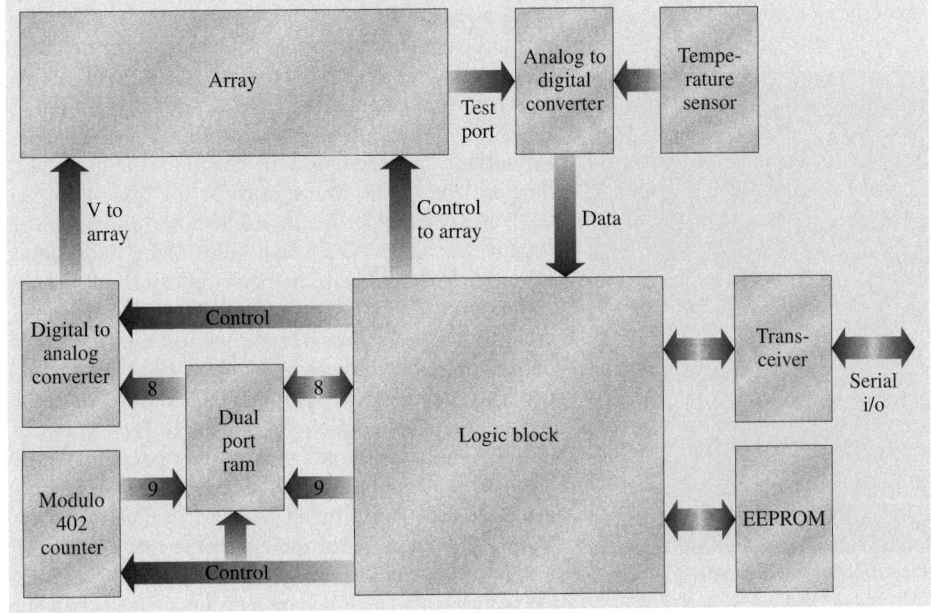

Fig. 14.5 Block diagram of the 400 CMOS chip control circuitry

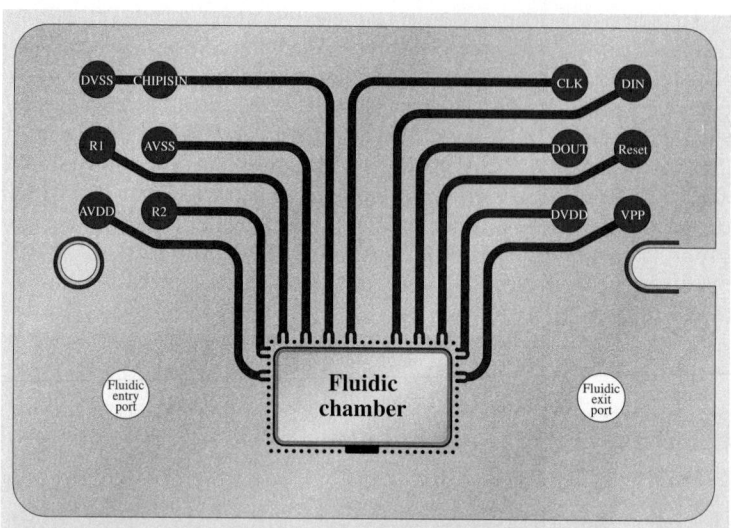

Fig. 14.6 Ceramic substrate and electrical connections to the chip

pling resistor. For structural support, 76 flip-chip solder bonds are used between the chip and a ceramic substrate which doubles as both a fluidic chamber and as an electrical contact interface (see Fig. 14.6). In order prevent any compromise in the quality of either the circuitry or the microelectrode array, the CMOS is fabricated at one foundry while the platinum microelectrode array is fabricated at a second foundry. The CMOS process requires 16 masking step followed by an additional three masking steps for the platinum electrodes. Figure 14.7 shows a high-magnification cross-section of an individual microelectrode and the underlying CMOS circuitry. After

the finished wafers have been inspected for defects, the wafers are diced into individual chips and then flip-chip bonded onto the ceramic substrates. The flip-chip on substrates (FCOS) are thermally calibrated and then finally assembled into the NanoChip 400 plastic cartridge housing which provides the fluidic delivery system and outside electrical connections (Fig. 14.2, inset).

14.1.2 Electric Field Technology Description

Nanogen's microarray technology is unique among DNA microarrays due to the use of electrophoretically driven or active transport of the DNA target or DNA probe molecules on the microarray surface. This active transport over the microarray is electronically controlled by biasing different microelectrodes on the microarray surface. Depending on the charge of the molecule, they will be rapidly transported to the oppositely biased microelectrode. Figure 14.8 shows a diagram of negatively charged DNA molecules being transported to a positively charged microelectrode. The electronic addressing of DNA and other biomolecules onto the microarray test sites can accelerate hybridization and other molecular binding processes by up to 1000 times compared to traditional passive methods (such as those of Affymetrix and GeneChip) (Table 14.2). By way of example, hybridization on a passive microarray may take up to several hours for the low concentrations of target DNA frequently found in many clinical samples. During the operation of the electronic microarray, the bias potential is sufficiently high (> 1.2 V) to cause the electrolysis of water. Oxidation occurs on the positively

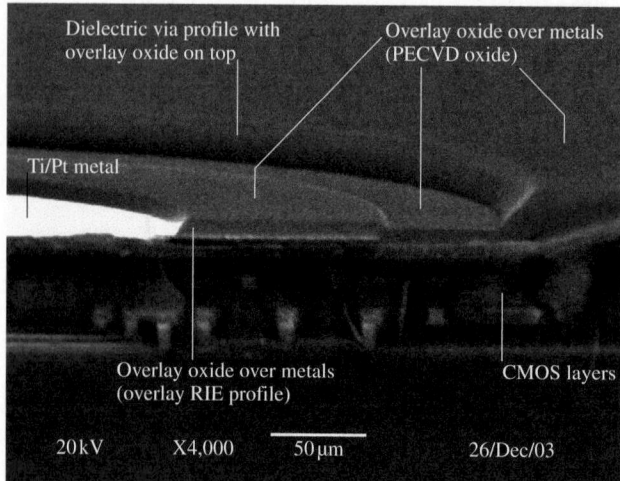

Fig. 14.7 Cross-section of an electrode and accompanying CMOS circuitry

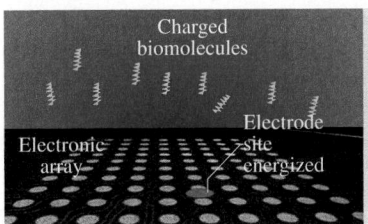 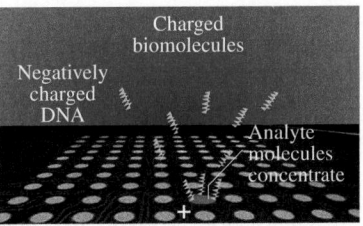

Fig. 14.8 Diagram of negatively charged DNA molecules being transported to a positively charged microelectrode

biased microelectrode and reduction occurs on the negatively biased microelectrode surface (see 14.1 and 14.2 below).

$$\text{Oxidation} \quad H_2O \rightarrow 2H^+ + 1/2O_2 + 2e^- \,, \quad (14.1)$$

$$\text{Reduction} \quad 2e^- + 2H_2O \rightarrow 2OH^- + H_2 \,. \quad (14.2)$$

Important to the operation of the device is a thin $1-2\,\mu m$ hydrogel permeation layer which covers the platinum microelectrode surface. This hydrogel permeation layer is designed to protect the more sensitive DNA and other biomolecules from the electrolysis reactions and products that occur on the platinum microarray surface. The permeation layer is usually made of either agarose or polyacrylamide. The layer also contains streptavidin which facilitates the capture and binding of biotinylated DNA probes or DNA target molecules onto the hydrogel surface.

The electric field microarray technology also takes advantage of the H^+ ions (low pH) generated at the positive electrode to carry out a unique "electronic hybridization" process. Electronic hybridization is carried out under low salt and low conductance conditions and it allows denatured DNA molecules to be hybridized only in the microscopic area around the activated microelectrode test site. A second electronic process called "electronic stringency" is achieved by reversing the

polarity at the microelectrode (negative activation) causing nonspecifically bound (negatively charged) DNA molecules to be driven away from the test site. Electronic stringency can also aid in the differentiation of DNA binding strengths, allowing better single base discrimination for the determination of single nucleotide polymorphisms (SNPs).

14.1.3 Electronic DNA Hybridization and Assay Design

The electronic hybridization technology and the Nanochip 400 System provide the researcher with a completely flexible and open platform for performing DNA hybridizations. It offers users the option to either generate or create user-defined microarrays specific for their own targets using electronic addressing of the biotinylated probes on the array, or to perform target- or single-base-specific analysis of DNA or RNA molecules. Both processes are performed very rapidly because the electronic addressing enables hybridization of the target DNA to occur within 30 to 60 s, as compared to competitive microarray technologies where hybridization is performed within one to several hours. Electronic microarrays provide an open platform and offer flexibility in the assay design [14.31–36]. This includes either the *PCR amplicon down format* to screen

Table 14.2 Comparison of electronic hybridization speed with that of conventional passive hybridization on microarrays

	Hybridization time	Concentration of targets	Concentration factor at a site	Stringency control
NanoChip active hybridization	10–100 s	Directed and localized at the array sites; ability to control individual sites	> 1000 times	Electronic thermal chemical
Passive hybridization technologies	1–2 h	Undirected; sites cannot be controlled independently	Low, diffusion-	Thermal chemical dependent

one or more patients for one or more SNPs, or the *capture probe down format* to screen many patients for one or more SNPs. In addition, *sandwich-type* assays are easy to perform where oligonucleotide discriminators and/or fluorescent probe labeled oligonucleotides are hybridized, electronically or passively, to captured PCR amplicons or probes. These assay design tools enable the users to increase the discrimination at single- base resolution as well as to minimize the nonspecific binding. Multiple probes or discriminators can be attached at a single site which facilitates the detection of multiple targets on a single electrode site. This enables the worker to use over 1000 characteristic genes or single nucleotide polymorphisms (SNPs) on a single array. In multiple DNA target detection the user can use blocker nucleotides that block SNPs which are not reported at a particular location. This allows the same universal reporter and specific discriminators to be used to report one or more SNPs on another site. Two lasers are used to recognize fluorescent signal from two reporter dyes (green and red). All of the steps are extremely fast, so the users can design and generate their own assays and incorporate the preparation of the array with oligonucleotides specific to multiple targets and samples as a part of the assay.

Other flexible electronic hybridization format designs allow several types of *multiplexed DNA analyses* to be carried out, such as the determination of multiple genes in one sample, multiple samples with one gene, or multiple samples with multiple genes. The ability to control individual test sites permits genetically unrelated DNA molecules to be used simultaneously on the same microchip. In contrast, sites on a conventional DNA array cannot be controlled separately, and all process steps must be performed on an entire array. Types of multiplexed analysis include: (1) determination of multiple genes in one sample addressed on the chip; (2) determination of multiple samples with one gene of interest addressed on the chip; (3) determination of multiple samples with multiple genes of interest addressed on the chip; and (4) single site multiplexing where several targets are discriminated on the same site using different fluorescent probes.

14.1.4 DNA Genotyping Applications

Electronic hybridization technology has been developed and commercialized and is now used for a number of practical applications including clinical diagnostic DNA genotyping [14.31–40]. In all applications, the DNA probe detection step is accomplished with fluorophore reporters and laser-based fluorescence detection. In addition, due to its open platform character, Nanogen's users have developed about 200 additional assays using the electronic microarray system and technology. The DNA assay areas developed by the users include diagnostics related to the following diseases and applications: coronary artery disease, cardiovascular disease, hypertension, cardiac function, venous thrombotic disease, metabolism, drug metabolism/cancer, cancer, cytokine, transcription factor, bacterial ID, Rett syndrome, thrombophilia, thalassemia and deafness.

14.1.5 On–Chip Strand Displacement Amplification

It has also been demonstrated that a complex DNA amplification technique can be performed on separate test sites of the microarray chip. This approach significantly reduces the time for DNA analysis because it incorporates DNA amplification and detection into a single platform. The assays were demonstrated using an isothermal strand displacement amplification (SDA is licensed technology from Becton, Dickinson and Co., Franklin Lakes, NJ, USA). In the SDA amplification, DNA polymerase recognizes the nicked strand of DNA and initiates resynthesis of that strand, displacing the original strand. The released amplicons then travel in solution to primers of the complementary strand which are either in solution or anchored on the test site. Oligonucleotide primers without nicking sites, called "bumper primers", are synthesized in the regions flanking the amplicons that were just produced, and assist in strand displacement and initial template replication [14.41].

14.1.6 Cell Separation on Microelectronic Arrays

Microelectronic arrays have also used been for cell separation applications. Disease diagnostics frequently involve identifying a small number of specific bacteria or viruses in a blood sample (infectious disease), fetal cells in maternal blood (genetic diseases) or tumor cells among a background of normal cells (early cancer detection). One powerful electric field technique used for cell separation is called dielectrophoresis (DEP). The DEP process involves the application of an asymmetric alternating current (ac) electric field to the cell population. Active microelectronic arrays have been used to achieve the separation of bacteria from whole blood [14.42], for the separation of cervical carcinoma cells from blood [14.43], and for gene expression analysis [14.44]. Microelectronic array devices utilizing high frequency ac fields have been used to carry out the DEP separation of Listeria bacterial cells ($\approx 1\,\mu m$) from whole blood cells ($\approx 10\,\mu m$) in a highly parallel manner. At an ac frequency of about 10 kHz the Listeria bacterial cells can be positioned on specific microlocations at high-field regions and the blood cells can be positioned in the low-field regions between the microelectrodes. The relative positioning of the cells between the high- and low-field regions is based on dielectric differences between the cell types. While maintaining the ac field, the microarray can be washed with a buffer solution that removes the blood cells (low-field regions) from the more firmly bound bacteria (high-field regions) near the microelectrodes. The bacteria can then be released and collected or electronically lysed to release the genomic DNA or RNA for further manipulation and analysis [14.42]. DEP represents a particularly useful process that allows difficult cell separation applications to be carried out rapidly and with high selectivity. The DEP process may also be useful for nanofabrication purposes [14.23, 24].

14.2 Electric Field–Assisted Nanofabrication Processes

Many examples of individual molecular and nanoscale components with basic electronic and photonic properties exist, including such entities as metallic nanoparticles, quantum dots, carbon nanotubes, nanowires and various organic molecules with electronic switching capabilities. However, the larger issue with enabling self-assembly-based nanoelectronics and nanophotonics is more likely to be the development of "viable" processes that will allow billions of molecular and/or nanoscale components to be assembled and interconnected into useful materials and devices. In addition to the electronic and photonic applications, nanostructures, nanomaterials and nanodevices with higher order mechanical/mechanistic, chemical/catalytic, biosensory and therapeutic properties are also envisioned [14.1–4]. The biggest challenges in enabling such devices and systems will most likely come from the stage of organizing components for higher level functioning, rather than the availability of the molecular components. Thus, a key problem with mimicking this type of nanotechnology is the lack of a viable "bottom-up" nanofabrication process to carry out the precision integration of diverse molecular and nanoscale components into viable higher order structures.

14.2.1 Electric Field–Assisted Self-Assembly Nanofabrication

As was described earlier, microelectronic array devices have been developed for applications in DNA genotyping diagnostics. These active microarray devices are able to produce reconfigurable electric field geometries on the surface of the device. The resulting electric fields are able to transport any type of charged molecule or structure, including DNA, RNA, proteins, antibodies, enzymes, nanostructures, cells or micron-

scale devices to or from any of the sites on the array surface. When DNA hybridization reactions are carried out using the device, the electric fields are actually assisting in the self-assembly of DNA molecules at the specified test site. In principle, these active devices serve as a "motherboard or hostboard" for the assisted assembly of DNA molecules into higher order or more complex structures. Since DNA molecules have intrinsic programmable self-assembly properties and can be derivatized with electronic or photonic groups or attached to larger nanostructures (quantum dots, metallic nanoparticles and nanotubes), we have the basis for a unique bottom-up nanofabrication process. Active microelectronic arrays serving as motherboards allow one to carry out a highly parallel electric field "pick and place" process for the heterogeneous integration of molecular, nanoscale and micron-scale components into complex three-dimensional structures. If desired, this process can be used to assemble molecules and/or nanocomponents within the defined perimeters of larger silicon or other semiconductor structures. Electric field-assisted self-assembly technology is based on three key physical principles: (1) the use of functionalized DNA or other high-fidelity recognition components as "molecular Lego" blocks for nanofabrication; (2) the use of DNA or other high-fidelity recognition components as a "selective glue" that provides intrinsic self-assembly properties to other molecular, nanoscale or micron-scale components (metallic nanoparticles, quantum dots, carbon nanotubes, organic molecular electronic switches, micron and submicron silicon lift-off devices and components); and (3) the use of active microelectronic array devices to provide electric field assistance or control of the intrinsic self-assembly of any modified electronic/photonic components and structures [14.18–24].

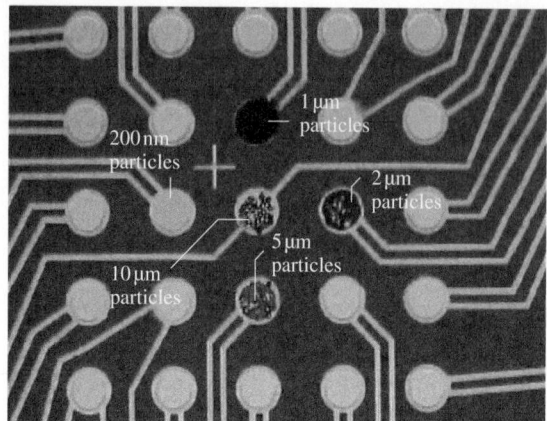

Fig. 14.9 Electronic addressing of five different types of microspheres and nanospheres to the microelectronic array test sites

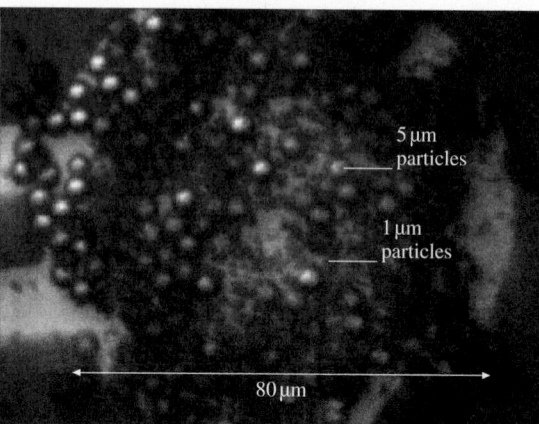

Fig. 14.11 Electronic addressing of two different layers of microspheres to the microelectronic array test sites

Microelectronic arrays have been used to direct the binding of derivatized nanospheres and microspheres onto selected locations on the microarray surface. In this case, fluorescent and nonfluorescent polystyrene nanospheres and microspheres derivatized with specific DNA oligonucleotides are transported and bound to selected test sites or microlocations derivatized with the specific complementary oligonucleotide sequences [14.8, 23, 24]. Microelectronic array devices have also been used for selective transport and addressing of larger nanoparticles and microspheres, and even objects as large as 20-micron light emitting diode structures [14.21–24]. In this context, Fig. 14.9 shows the electric field addressing of five differently

sized negatively charged polystyrene microspheres and nanospheres (100 nm) to selectively activated microlocations on a 25-test site microelectronic array. The rate of transport is related to the strength of the electric field and the charge/mass ratio of the molecule or structure. Figure 14.10 shows the results for the parallel transport and positioning of two different types of microspheres onto the microelectronic array surface. Finally, Fig. 14.11 now shows the results for the initial addressing of

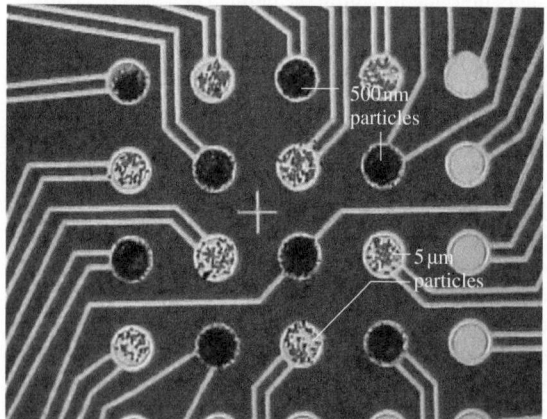

Fig. 14.10 Parallel electronic addressing of two different types of microspheres to the microelectronic array test sites

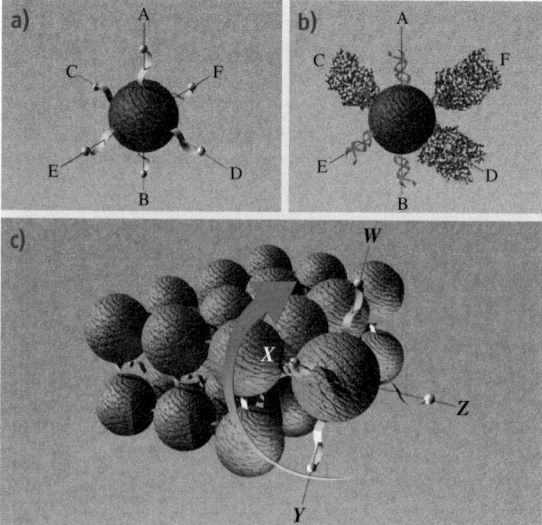

Fig. 14.12 (a),(b) Precision nanosphere functionalization scheme. **(c)** Type of heterogeneous 3-D higher order structure that can only be obtained using precision nanostructures

derivatized negatively charged one-micron polystyrene microspheres to a selectively activated microlocation on a 25-site microelectronic array, and the subsequent covering of the layer of one-micron polystyrene microspheres by larger five-micron microspheres. Thus, it is possible to use electric field transport and addressing to form multiple layers of particles and other materials, allowing fabrication in the third dimension.

Present nanofabrication methods do not allow most nanostructures to be modified in a controlled or precise manner. For example, it would be extremely difficult to attach different DNA sequences or different kinds of protein molecules in precise locations around quantum dots or other nanoparticles (Fig. 14.12a,b). Unfortunately, without this first-order property it becomes even more difficult to then assemble these nanostructures into higher order heterogeneous 3-D structures, even though the core structure is derivatized with

high-fidelity recognition components (see Fig. 14.12c). Microelectronic array devices may offer the opportunity to develop processes that will allow core nanostructures to be selectively modified in a precise fashion [14.24]. The proposed electric field microarray techniques may provide the ability to carry out the precision functionalization of nanostructures by processes which involve transporting and orienting the nanostructures onto surfaces containing the selected ligand molecules which are then reacted only with a selected portion of the nanostructure. By repeating the process and reorienting the nanostructures, it will be possible to functionalize the core structure selectively with most biological and/or chemical groups. Such devices and processes allow one to design and create functionalized nanostructures with binding groups arranged in tetrahedral, hexagonal or other coordinate positions around the core nanostructure.

14.3 Conclusions

Active microelectronic array technology provides a number of advantages for carrying out DNA hybridization diagnostics and other affinity-based assays for molecular biology research and clinical diagnostic applications. The technology has also demonstrated the potential for assisted self-assembly and other nanofabrication applications. Microelectronic arrays have been designed and fabricated with 25 to 10 000 microscopic test sites, and devices with 100 and 400 test sites have been commercialized. The newer 400-test site devices have CMOS elements incorporated into the underlying silicon structure that provide on-board control of current and voltage to each of the test sites on the device. Microelectronic chips are incorporated into a cartridge-type device so that they can be conveniently used with a probe loading station and fluorescent detection system. Active microelectronic arrays are fundamentally different from other DNA chip or microarray devices, which are essentially passive. Active microelectronic

arrays allow DNA molecules, RNA, oligonucleotide probes, PCR amplicons, proteins, nanostructures, cells and even microscale devices to be rapidly transported and selectively addressed to any of the test sites on the microelectronic array surface. Active microarray devices have considerable potential for nanofabrication by directed self-assembly of molecular, nanoscale and microscale components into higher order mechanisms, structures, and devices. This electric field technology makes possible a type of "pick and place" process for the heterogeneous integration of diverse molecular and nanoscale components into higher order structures within defined perimeters of larger silicon or semi-conductor structures. The technology provides the best aspects of a top-down and bottom-up process, and has the inherent hierarchical logic of allowing one to control the organization and assembly of components from the molecular level to the nanoscale level to microscale three-dimensional integrated structures and devices.

References

14.1 National Research Council: *Small Wonders, Endless Frontiers: Review of the Nanotional Nanotechnology Initiative* (National Research Council, Washington D.C. 2002)

14.2 M. P. Hughes (Ed.): *Nanoelectromechanics in Engineering and Biology* (CRC Press, Boca Raton 2003)

14.3 W. A. Goddard, D. Brenner, S. Lyshevski, G. Lafrate (Eds.): *Handbook of Nanoscience, Engineering and Technology* (CRC Press, Boca Raton 2003)

14.4   V. Balzani, M. Venturi, A. Credi (Eds.): *Molecular Devices and Mechanics – Journey into the Nanoworld* (Wiley-VCH, Weinheim 2003)

14.5   R. Bashir: Biologically mediated assembly of artificial nanostructures and microstructures. In: *Handbook of Nanoscience, Engineering and Technology*, ed. by W.A. Goddard, D. Brenner, S. Lyshevski, G. Lafrate (CRC Press, Boca Raton 2003) Chap. 15, pp. 15–1 – 15–31

14.6   M.J. Heller, R.H. Tullis: Self-organizing molecular photonic structures based on functionalized synthetic DNA polymers, Nanotechnology **2**, 165–171 (1991)

14.7   D.M. Hartmann, D. Schwartz, G. Tu, M. Hellerand S.C. Esener: Selective DNA attachment of particles to substrates, J. Mater. Res. **17**, 473–478 (2002)

14.8   M.J. Heller: An active microelectronics device for multiplex DNA analysis, IEEE Eng. Med. Biol. **15**, 100–103 (1996)

14.9   R.G. Sosnowski, E. Tu, W.F. Butler, J.P. O'Connell, M.J. Heller: Rapid determination of single base mismatch in DNA hybrids by direct electric field control, Proc. Natl. Acad. Sci. USA **94**, 1119–1123 (1997)

14.10   C.F. Edman, D.E. Raymond, D.J. Wu, E. Tu, R.G. Sosnowski, W.F. Butler, M. Nerenberg, M.J. Heller: Electric field directed nucleic acid hybridization on microchips, Nucl. Acids Res. **25**, 4907–4914 (1997)

14.11   M.J. Heller: An integrated microelectronic hybridization system for genomic research and diagnostic applications. In: *Micro Total Analysis Systems*, ed. by D.J. Harrison, A. van den Berg (Kluwer Academic, Dordrecht 1998) pp. 221–224

14.12   M.J. Heller, E. Tu, A. Holmsen, R.G. Sosnowski, J.P. O'Connell: Active microelectronic arrays for DNA hybridization analysis. In: *DNA Microarrays: A Practical Approach*, ed. by M. Schena (Univ. Press, Oxford 1999) pp. 167–185

14.13   M.J. Heller, A.H. Forster, E. Tu: Active microelectronic chip devices which utilize controlled electrophoretic fields for multiplex DNA hybridization and genomic applications, Electrophoresis **21**, 157–164 (2000)

14.14   C. Gurtner, E. Tu, N. Jamshidi, R. Haigis, T. Onofrey, C.F. Edman, R. Sosnowski, B. Wallace, M.J. Heller: Microelectronic array devices and techniques for electric field enhanced DNA hybridization in low-conductance buffers, Electrophoresis **23**, 1543–1550 (2002)

14.15   M.J. Heller: DNA microarray technology: devices, systems and applications, Ann. Rev. Biomed. Eng. **4**, 129–153 (2002)

14.16   M.J. Heller, E. Tu, R. Martinsons, R.R. Anderson, C. Gurtner, A. Forster, R. Sosnowski: Active microelectronic array systems for DNA hybridization, genotyping, pharmacogenomics and nanofabrication applications. In: *Integrated Microfabricated Devices*, ed. by M.J. Heller, A. Guttman (Marcel Dekker, New York 2002) Chap. 10, pp. 223–270

14.17   S.K. Kassengne, H. Reese, D. Hodko, J.M. Yang, K. Sarkar, P. Swanson, D.E. Raymond, M.J. Heller, M.J. Madou: Numerical modeling of transport and accumulation of DNA on electronically active biochips, Sensors Actuat. B **94**, 81–98 (2003)

14.18   S.C. Esener, D. Hartmann, M.J. Heller, J.M. Cable: *DNA Assisted Micro-Assembly: A Heterogeneous Integration Technology for Optoelectronics*, Proc. SPIE, Vol. CR 70, ed. by A. Hussain (SPIE, Bellingham 1998) Chap. 7

14.19   C. Gurtner, C.F. Edman, R.E. Formosa, M.J. Heller: Photoelectrophoretic transport and hybridization of DNA on unpatterned silicon substrates, J. Am. Chem. Soc. **122**, (36) 8589–8594 (2000)

14.20   Y. Huang, K.L. Ewalt, M. Tirado, R. Haigis, A. Forster, D. Ackley, M.J. Heller, J.P. O'Connell, M. Krihak: Electric manipulation of bioparticles and macromolecules on microfabricated electrodes, Anal. Chem. **73**, 1549–1559 (2001)

14.21   C.F. Edman, C. Gurtner, R.E. Formosa, J.J. Coleman, M.J. Heller: Electric-field-directed pick-and-place assembly, HDI **3**, 10: 30–35 (2000)

14.22   C.F. Edman, R.B. Swint, C. Gurthner, R.E. Formosa, S.D. Roh, K.E. Lee, P.D. Swanson, D.E. Ackley, J.J. Colman, M.J. Heller: Electric field directed assembly of an InGaAs LED onto silicon circuitry, IEEE Photon. Tech. Lett. **12**, (9) 1198–1200 (2000)

14.23   C.F. Edman, M.J. Heller, R. Formosa, C. Gurtner: Methods and apparatus for the electronic homogeneous assembly and fabrication of devices, US Patent 6,569,382 (2003)

14.24   M.J. Heller, J.M. Cable, S.C. Esener: Methods for the electronic assembly and fabrication of devices, US Patent 6,652,808 (2003)

14.25   C.F. Edman, M.J. Heller, C. Gurtner, R. Formosa: Systems and devices for the photoelectrophoretic transport and hybridization of oligonucleotides, US Patent 6,706,473 (2004)

14.26   A. Taton, C. Mirkin, R. Letsinger: Scanometric DNA array detection with nanoparticle probes, Science **289**, 1757–1760 (2000)

14.27   M. Chee, R. Yang, E. Hubbell, A. Berno, X. Huang, D. Stern, J. Winkler, D. Lockhart, M. Morris, S. Fodor: Accessing genetic information with high-density DNA arrays, Science **274**, 610–614 (1996)

14.28   A. Pease, D. Solas, E. Sullivan, M. Cronin, C. Holmes, S. Fodor: Light-generated oligonucleotide arrays for rapid DNA sequence analysis, Proc. Natl. Acad. Sci. USA **99**, 5022–5026 (1994)

14.29   R.J. Lipshutz, D. Morris, M. Chee, E. Hubbell, M.J. Kozal, N. Shah, N. Shen, R. Yang, S.P. Fodor: Using oligonucleotide probe arrays to access genetic diversity, Biotechniques **19**, (3) 442–447 (1995)

14.30   P. Swanson, R. Gelbart, E. Atlas, L. Yang, T. Grogan, W.F. Butler, D.E. Ackley, E. Sheldon: A fully multiplexed CMOS biochip for DNA analysis, Sensors Actuat. B **64**, 22–30 (2000)

14.31 P.N. Gilles, D.J. Wu, C.B. Foster, P.J. Dillion, S.J. Channock: Single nucleotide polymorphic discrimination by an electronic dot blot assay on semiconductor microchips, Nature Biotechnol. **17**, (4) 365–370 (1999)

14.32 N. Narasimhan, D. O'Kane: Validation of SNP genotyping for human serum paraoxonase gene, Clin. Chem. **34**, (7) 589–592 (2001)

14.33 R. Sosnowski, M.J. Heller, E. Tu, A. Forster, R. Radtkey: Active microelectronic array system for DNA hybridization, genotyping and pharmacogenomic applications, Psychiatr. Genet. **12**, 181–192 (2002)

14.34 Y.R. Sohni, J.R. Cerhan, D.J. O'Kane: Microarray and microfluidic methodology for genotyping cytokine gene polymorphisms, Hum. Immunol. **64**, 990–997 (2003)

14.35 E.S. Pollak, L. Feng, H. Ahadian, P. Fortina: Microarray-based genetic analysis for studying susceptibility to arterial and venous thrombotic disorders, Ital. Heart J. **2**, 569–572 (2001)

14.36 W.A. Thistlethwaite, L.M. Moses, K.C. Hoffbuhr, J.M. Devaney, E.P. Hoffman: Rapid genotyping of common MeCP2 mutations with an electronic DNA microchip using serial differential hybridization, J. Mol. Diagnos. **5**, (2) 121–126 (2003)

14.37 V.R. Mas, R.A. Fisher, D.G. Maluf, D.S. Wilkinson, T.G. Carleton, A. Ferreira-Gonzalez: Hepatic artery thrombosis after liver transplantation and genetic factors: prothrombin G20210A polymorphism, Transplantation **76**, (1) 247–249 (2003)

14.38 R. Santacroce, A. Ratti, F. Caroli, B. Foglieni, A. Ferraris, L. Cremonesi, M. Margaglione, M. Seri, R. Ravazzolo, G. Restagno, B. Dallapiccola, E. Rappaport, E.S. Pollak, S. Surrey, M. Ferrari, P. Fortina: Analysis of clinically relevant single-nucleotide polymorphisms by use of microelectric array technology, Clin. Chem. **48**, (12) 2124–2130 (2002)

14.39 A. Åsberg, K. Thorstensen, K. Hveem, K. Bjerve: Hereditary hemochromatosis: the clinical significance of the S64C mutation, Genet. Test. **6**, (1) 59–62 (2002)

14.40 J.G. Evans, C. Lee-Tataseo: Determination of the factor V Leiden single-nucleotide polymorphism in a commercial clinical laboratory by use of NanoChip microelectric array technology, Clin. Chem. **48**, (9) 1406–1411 (2002)

14.41 T. Walker, J. Nadeau, P. Spears, J. Schram, C. Nycz, D. Shank: Multiplex strand displacement amplification (SDA) and detection of DNA sequences from mycobacterium tuberculosis and other mycobacteria, Nucl. Acids Res. **22**, (13) 2670–2677 (1994)

14.42 J. Cheng, E.L. Sheldon, L. Wu, A. Uribe, L.O. Gerrue, J. Carrino, M.J. Heller, J.P. O'Connell: Electric field controlled preparation and hybridization analysis of DNA/RNA from *E. coli* on microfabricated bioelectronic chips, Nat. Biotechnol. **16**, 541–546 (1998)

14.43 J. Cheng, E.L. Sheldon, L. Wu, M.J. Heller, J. O'Connell: Isolation of cultured cervical carcinoma cells mixed with peripheral blood cells on a bioelectronic chip, Anal. Chem. **70**, 2321–2326 (1998)

14.44 Y. Huang, J. Sunghae, M. Duhon, M.J. Heller, B. Wallace, X. Xu: Dielectrophoretic separation and gene expression profiling on microelectronic chip arrays, Anal. Chem. **74**, 3362–3371 (2002)

Part B | 14

15. MEMS/NEMS Devices and Applications

Microelectromechanical systems (MEMS) have played key roles in many important areas, for example transportation, communication, automated manufacturing, environmental monitoring, health care, defense systems, and a wide range of consumer products. MEMS are inherently small, thus offering attractive characteristics such as reduced size, weight, and power dissipation and improved speed and precision compared to their macroscopic counterparts. Integrated circuit (IC) fabrication technology has been the primary enabling technology for MEMS besides a few special etching, bonding and assembly techniques. Microfabrication provides a powerful tool for batch processing and miniaturizing electromechanical devices and systems to a dimensional scale that is not accessible by conventional machining techniques. As IC fabrication technology continues to scale toward deep submicron and nanometer feature sizes, a variety of nanoelectromechanical systems (NEMS) can be envisioned in the foreseeable future. Nanoscale mechanical devices and systems integrated with nanoelectronics will open a vast number of new exploratory research areas in science and engineering. NEMS will most likely serve as an enabling technology, merging engineering with the life sciences in ways that are not currently feasible with microscale tools and technologies.

MEMS has been applied to a wide range of fields. Hundreds of microdevices have been developed for specific applications. It is thus difficult to provide an overview covering every aspect of the

topic. In this chapter, key aspects of MEMS technology and applications are illustrated by selecting a few demonstrative device examples, such as pressure sensors, inertial sensors, optical and wireless communication devices. Microstructure examples with dimensions on the order of submicron are presented with fabrication technologies for future NEMS applications.

Although MEMS has experienced significant growth over the past decade, many challenges still remain. In broad terms, these challenges can be grouped into three general categories: (1) fabrication challenges; (2) packaging challenges; and (3) application challenges. Challenges in these areas will, in large measure, determine the commercial success of a particular MEMS device in both technical and economic terms. This chapter presents a brief discussion of some of these challenges as well as possible approaches to addressing them.

Microelectromechanical Systems, generally referred to as MEMS, has had a history of research and development over a few decades. Besides the traditional microfabricated sensors and actuators, the field covers micromechanical components and systems integrated or micro-assembled with electronics on the same substrate or package, achieving high-performance functional systems. These devices and systems have played key roles in many important areas such as transportation, communication, automated manufacturing, environmental monitoring, health care, defense systems, and a wide range of consumer products. MEMS are inherently small, thus offering attractive characteristics such as reduced size, weight, and power dissipation and improved speed and precision compared to their macroscopic counterparts. The development of MEMS requires

Fig. 15.1 SEM micrograph of a polysilicon microelectromechanical motor [15.1]

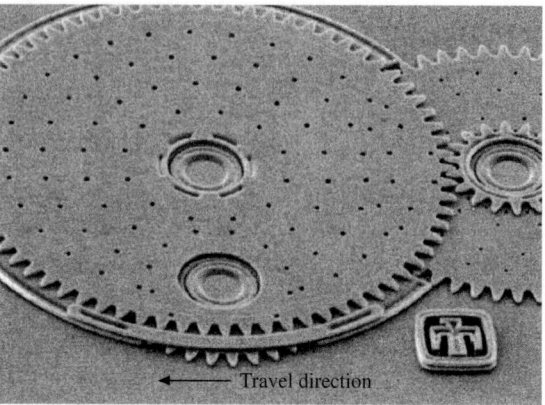

Fig. 15.2 SEM micrograph of polysilicon microgears [15.3]

appropriate fabrication technologies that enable the definition of small geometries, precise dimension control, design flexibility, interfacing with microelectronics, repeatability, reliability, high yield, and low cost. Integrated circuits (IC) fabrication technology meets all of the above criteria and has been the primary enabling fabrication technology for MEMS besides a few special etching, bonding and assembly techniques. Microfabrication provides a powerful tool for batch processing and miniaturization of electromechanical devices and systems into a dimensional scale, which is not accessible by conventional machining techniques. Most MEMS devices exhibit a length or width ranging from micrometers to several hundreds of micrometers with a thickness from sub-micrometer up to tens of micrometers depending upon fabrication technique employed. A physical displacement of a sensor or an actuator is typically on the same order of magnitude. Figure 15.1 shows an SEM micrograph of a microelectromechanical motor developed in late 1980s [15.1]. Polycrystalline silicon (polysilicon) surface micromachining technology was used to fabricate the micromotor achieving a diameter of 150 μm and a minimum vertical feature size on the order of a micrometer. A probe tip is also shown in the micrograph for a size comparison. This device example and similar others [15.2] demonstrated at that time what MEMS

technology could accomplish in microscale machining and served as a strong technology indicator for continued MEMS development. The field has expanded greatly in recent years along with rapid technology advances. Figure 15.2, for example, shows a photo of micro-gears fabricated in mid-1990s using a five-level polysilicon surface micromachining technology [15.3]. This device represents one of the most advanced surface micromachining fabrication process developed to date. One can imagine that a wide range of sophisticated microelectromechanical devices and systems can be realized through applying such technology in the future. As IC fabrication technology continues to scale toward deep sub-micron and nano-meter feature sizes, a variety of nanoelectromechanical systems (NEMS) can be envisioned in the foreseeable future. Nano-scale mechanical devices and systems integrated with nanoelectronics will open a vast number of new exploratory research areas in science and engineering. NEMS will most likely serve as an enabling technology merging engineering with the life sciences in ways that are not currently feasible with the microscale tools and technologies.

This chapter will provide a general overview on MEMS and NEMS devices along with their applications. MEMS technology has been applied to a wide range of fields. Over hundreds of micro-devices have been developed for specific applications. Thus, it is difficult to provide an overview covering every aspect of the topic. It is the authors' intent to illustrate key aspects of MEMS technology and its impact to specific applications by selecting a few demonstrative device examples in this chapter. For a wide-ranging discussion of nearly all types of micromachined sensors and actuators, books by Kovacs [15.4] and Senturia [15.5] are recommended.

15.1 MEMS Devices and Applications

MEMS devices have played key roles in many areas of development. Microfabricated sensors, actuators, and electronics are the most critical components required to implement a complete system for a specific function. Microsensors and actuators can be fabricated by various micromachining processing technologies. In this section, a number of selected MEMS devices are presented to illustrate the basic device operating principles as well as to demonstrate key aspects of the microfabrication technology and application impact.

15.1.1 Pressure Sensor

Pressure sensors are one of the early devices realized by silicon micromachining technologies and have become successful commercial products. The devices have been widely used in various industrial and biomedical applications. The sensors can be based on piezoelectric, piezoresistive, capacitive, and resonant sensing mechanisms. Silicon bulk and surface micromachining techniques have been used for sensor batch fabrication, thus achieving size miniaturization and low cost. Two types of pressure sensors, piezoresistive and capacitive, are described here for an illustration purpose.

Piezoresistive Sensor

The piezoresisitve effect in silicon has been widely used for implementing pressure sensors. A pressure-induced strain deforms the silicon band structure, thus changing the resistivity of the material. The piezoresistive effect is typically crystal orientation dependent and is also affected by doping and temperature. A practical piezoresistive pressure sensor can be implemented by fabricating four sensing resistors along the edges of a thin silicon diaphragm, which acts as a mechanical

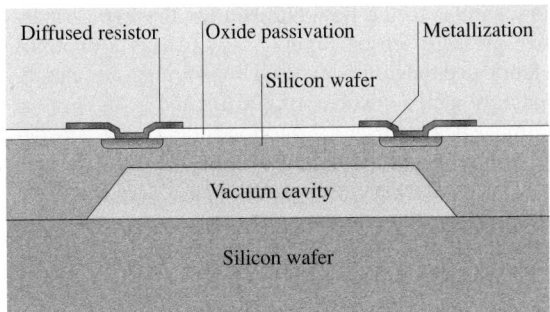

Fig. 15.3 Cross-sectional schematic of a piezoresistive pressure sensor

amplifier to increase the stress and strain at the sensor site. The four sensing elements are connected in a bridge configuration with push-pull signals to increase the sensitivity. The measurable pressure range for such a sensor can be from 10^{-3} to 10^6 Torr depending upon the design. An example of a piezoresistive pressure sensor is shown in Fig. 15.3. The device consists of a silicon diaphragm suspended over a reference vacuum cavity to form an absolute pressure sensor. An external pressure applied over the diaphragm introduces a stress on the sensing resistors, thus resulting in a resistance value change corresponding to the pressure. The fabrication sequence is outlined as follows. The piezoresistors are typically first formed through a boron diffusion process followed by a high temperature annealing step in order to achieve a resistance value on the order of a few kilo-ohms. The wafer is then passivated with a silicon dioxide layer and contact windows are opened for metallization. At this point, the wafer is patterned on the backside, followed by a timed silicon wet etch to form the diaphragm, typically having a thickness around a few tens of micrometers. The diaphragm can have a length of several hundreds of micrometers. A second silicon wafer is then bonded to the device wafer in vacuum to form a reference vacuum cavity, thus completing the fabrication process. The second wafer can also be further etched through to form an inlet port, implementing a gauge pressure sensor [15.6]. The piezoresistive sensors are simple to fabricate and can be readily interfaced with electronic systems. However, the resistors are temperature dependent and consume DC power. Long-term characteristic drift and resistor thermal noise ultimately limit the sensor resolution.

Capacitive Sensor

Capacitive pressure sensors are attractive because they are virtually temperature independent and consume zero DC power. The devices do not exhibit initial turn-on drift and are stable over time. Furthermore, CMOS microelectronic circuits can be readily interfaced with the sensors to provide advanced signal conditioning and processing, thus improving overall system performance. An example of a capacitive pressure sensor is shown in Fig. 15.4. The device consists of an edge clamped silicon diaphragm suspended over a vacuum cavity. The diaphragm can be square or circular with a typical thickness of a few micrometers and a length or radius of a few hundreds micrometers, respectively. The vacuum cavity typically has a depth of a few micrometers. The

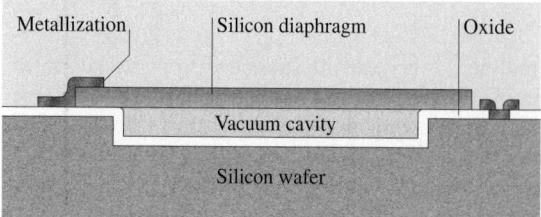

Fig. 15.4 Cross-sectional schematic of a capacitive pressure sensor

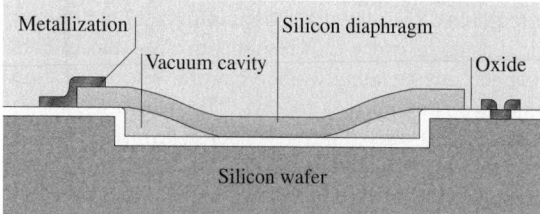

Fig. 15.5 Cross-sectional schematic of a touch-mode capacitive pressure sensor

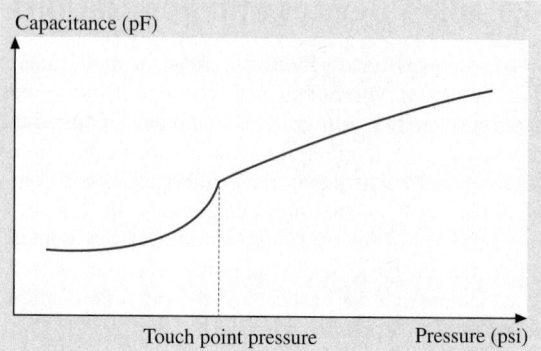

Fig. 15.6 Touch-mode capacitive pressure sensor characteristic response

diaphragm and substrate form a pressure dependent air-gap variable capacitor. An increased external pressure causes the diaphragm to deflect towards the substrate, thus resulting in an increase in the capacitance value. A simplified fabrication process can be outlined as follows. A silicon wafer is first patterned and etched to form the cavity. The wafer is then oxidized followed by bonding to a second silicon wafer with a heavily-doped boron layer, which defines the diaphragm thickness, at the surface. The bonding process can be performed in vacuum to realize the vacuum cavity. If the vacuum bonding is not performed at this stage, a low pressure sealing process can be used to form the vacuum cavity after patterning the sensor diaphragm, provided that sealing channels are available. The silicon substrate above the boron layer is then removed through a wet etching process, followed by patterning to form the sensor diaphragm, which serves as the device top electrode. Contact pads are formed by metallization and patterning. This type of pressure sensor exhibits a nonlinear characteristic and a limited dynamic range. These phenomena, however, can be alleviated through applying an electrostatic force-balanced feedback architecture. A common practice is to introduce another electrode above the sensing diaphragm through wafer bonding [15.7], thus forming two capacitors in series with the diaphragm being the middle electrode. The capacitors are interfaced with electronic circuits, which convert the sensor capacitance value to an output voltage corresponding to the diaphragm position. This voltage is further processed to generate a feedback signal to the top electrode, thus introducing an electrostatic pull up force to maintain the deflectable diaphragm at its nominal position. This negative feedback loop would substantially minimize the device nonlinearity and also extend the sensor dynamic range.

A capacitive pressure sensor achieving an inherent linear characteristic response and a wide dynamic range can be implemented by employing a touch-mode architecture [15.8]. Figure 15.5 shows the cross-sectional view of a touch-mode pressure sensor. The device consists of an edge-clamped silicon diaphragm suspended over a vacuum cavity. The diaphragm deflects under an increasing external pressure and touches the substrate, causing a linear increase in the sensor capacitance value beyond the touch point pressure. Figure 15.6 shows a typical device characteristic curve. The touch point pressure can be designed through engineering the sensor geometric parameters such as the diaphragm size, thickness, cavity depth, etc. for various application requirements. The device can be fabricated using a process flow similar to the flow outlined for the basic capacitive pressure sensor. Figure 15.7 presents a photo of a fabricated touch-mode sensor employing a circular diaphragm with a diameter of 800 μm and a thickness of 5 μm suspended over a 2.5 μm vacuum cavity. The device achieves a touch point pressure of 8 psi and exhibits a linear capacitance range from 33 pF at 10 psi to 40 pF at 32 psi (absolute pressures). Similar sensor structures have been demonstrated by using single-crystal 3C-SiC diaphragm achieving a high-temperature pressure sensing capability up to 400 °C [15.9].

The above processes use bulk silicon materials for machining and are usually referred to as bulk microma-

Suspended diaphragm (0.8 mm diameter)

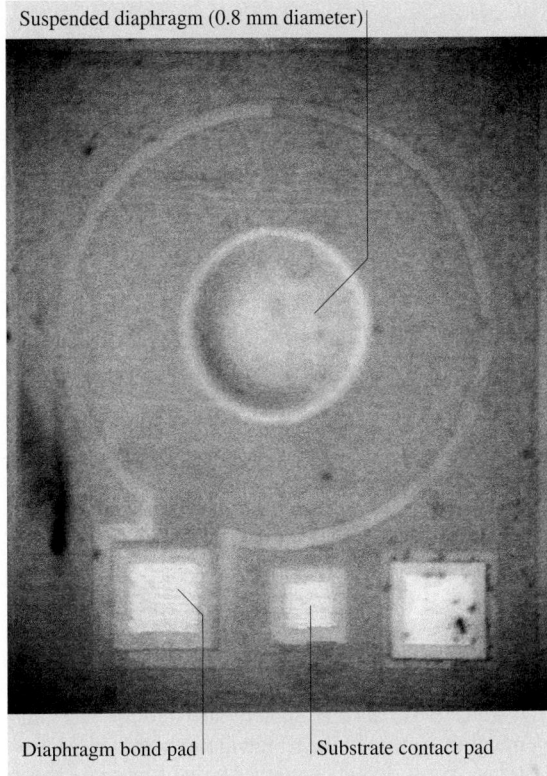

Diaphragm bond pad   Substrate contact pad

Fig. 15.7 Photo of a touch-mode capacitive pressure sensor [15.8]

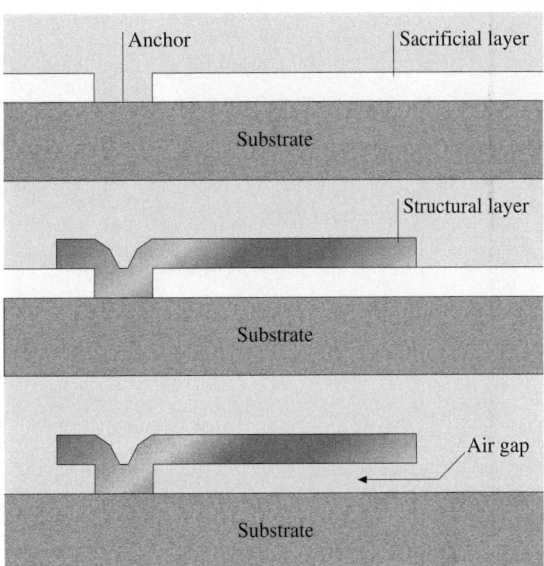

Fig. 15.8 Simplified fabrication sequence of surface micromachining technology

chining. The same devices can also be fabricated using so called surface micromachining. Surface micromachining technology is attractive for integrating MEMS sensors with on-chip electronic circuits. As a result, advanced signal processing capabilities such as data conversion, offset and noise cancellation, digital calibration, temperature compensation, etc. can be implemented adjacent to micro-sensors on a same substrate, providing a complete high performance microsystem solution. The single chip approach also eliminates external wiring, which is critical for minimizing noise pick up and enhancing system performance. Surface micromachining, simply stated, is a method of fabricating MEMS through depositing, patterning, and etching a sequence of thin films with thickness on the order of a micrometer. Figure 15.8 illustrates a typical surface micromachining process flow [15.10]. The process starts by depositing a layer of sacrificial material such as silicon dioxide over a wafer followed by anchor formation. A structural layer, typically a polysilicon film, is deposited and patterned. The underlying sacrificial layer is then removed to freely

release the suspended microstructure and to complete the fabrication sequence. The processing materials and steps are compatible with standard integrated circuit process, thus can be readily incorporated as an add-on module to an IC process [15.10–12]. A similar surface micromachining technology has been developed to produce monolithic pressure sensor systems [15.13]. Figure 15.9 shows an SEM micrograph of an array of MEMS capacitive pressure sensors fabricated with BiCMOS electronics on the same substrate. Each sensor consists of a $0.8\,\mu m$ thick circular polysilicon membrane with a diameter on the order of $20\,\mu m$ suspended over a $0.3\,\mu m$ deep vacuum cavity. The devices operate using the same principle as the sensor shown in Fig. 15.4. A close view of the sensor cross-section is shown in Fig. 15.10, which shows the suspended membrane and underneath air gap. These sensors have demonstrated operations in pressure ranges up to 400 bar with an accuracy of 1.5%.

15.1.2 Inertial Sensor

Micromachined inertial sensors consist of accelerometers and gyroscopes. These devices are one of the important types of silicon-based MEMS sensors that have been successfully commercialized. MEMS accelerometers alone have the second largest sales volume after pressure sensors. Gyroscopes are expected to reach a comparable sales volume in a foreseeable fu-

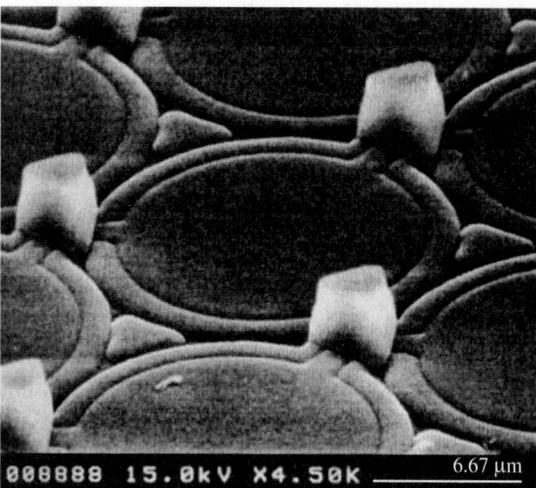

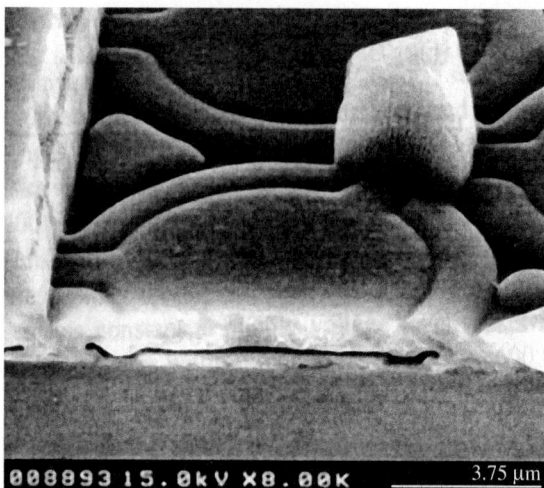

Fig. 15.9 SEM micrograph of polysilicon surface-micromachined capacitive pressure sensors [15.13]

Fig. 15.10 SEM micrograph of a close-up view of a polysilicon surface-micromachined capacitive pressure sensor [15.13]

ture. Accelerometers have been used in a wide range of applications including automotive application for safety systems, active suspension and stability con-

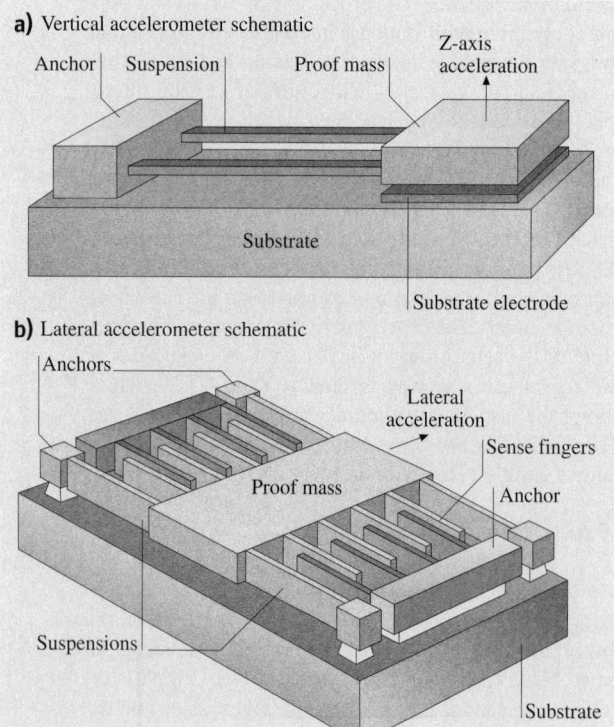

Fig. 15.11 Schematics of vertical and lateral accelerometers

trol, biomedical application for activity monitoring, and numerous consumer products such as head-mount displays, camcorders, three-dimensional mouse, etc. High-sensitivity accelerometers are crucial for implementing self-contained navigation and guidance systems. A gyroscope is another type of inertial sensor that measures rate or angle of rotation. The devices can be used along with accelerometers to provide heading information in an inertial navigation system. Gyroscopes also are useful in applications such as automotive ride stabilization and rollover detection, camcorder stabilization, virtual reality, etc. Inertial sensors fabricated by micromachining technology can achieve reduced size, weight, and cost, all which are critical for consumer applications. More importantly, these sensors can be integrated with microelectronic circuits to achieve a functional microsystem with high performance.

Accelerometer

An accelerometer generally consists of a proof mass suspended by compliant mechanical suspensions anchored to a fixed frame. An external acceleration displaces the support frame relative to the proof mass. The displacement can result in an internal stress change in the suspension, which can be detected by piezoresistive sensors as a measure of the external acceleration. The displacement can also be detected as a capacitance change in capacitive accelerometers. Capacitive sensors are attractive for various applications because they exhibit high sensitivity and low temperature dependence,

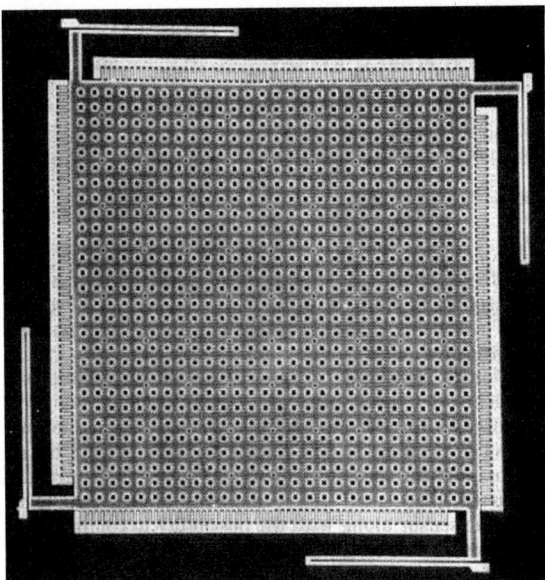

Fig. 15.12 SEM micrograph of a polysilicon surface-micromachined z-axis accelerometer [15.14]

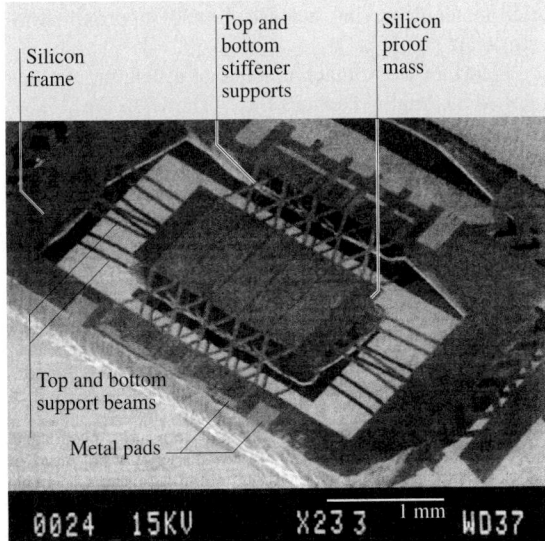

Fig. 15.13 SEM micrograph of a MEMS z-axis accelerometer fabricated by using a combined surface and bulk micromachining technology [15.15]

turn-on drift, power dissipation, and noise. The sensors can also be readily integrated with CMOS electronics to perform advanced signal processing for high system performance. Capacitive accelerometers may be divided into two categories as vertical and lateral type sensors.

Figure 15.11 shows sensor structures for the two versions. In a vertical device, the proof mass is suspended above the substrate electrode by a small gap typically on the order of a micrometer, forming a parallel-plate sense capacitance. The proof mass moves in the direction perpendicular to the substrate (z-axis) upon a vertical input acceleration, thus changing the gap and hence the capacitance value. The lateral accelerometer consists of a number of movable fingers attached to the proof mass, forming a sense capacitance with an array of fixed parallel fingers. The sensor proof mass moves in a plane parallel to the substrate when subjected to a lateral input acceleration, thus changing the overlap area of these fingers; hence the capacitance value. Figure 15.12 shows an SEM top view of a surface-micromachined polysilion z-axis accelerometer [15.14]. The device consists of a 400 μm × 400 μm proof mass with a thickness of 2 μm suspended above the substrate electrode by four folded beam suspensions with an air gap around 2 μm, thus achieving a sense capacitance of approximately 500 fF. The visible holes are used to ensure complete removal of the sacrificial oxide underneath the proof mass at the end of the fabrication process. The sensor can be interfaced with a microelectronic charge amplifier converting the capacitance value to an output voltage for further signal processing and analysis. Force feedback architecture can be applied to stabilize the proof mass position. The combs around the periphery of the proof mass can exert an electrostatic levitation force on the proof mass to achieve the position control, thus improving the system frequency response and linearity performance [15.14].

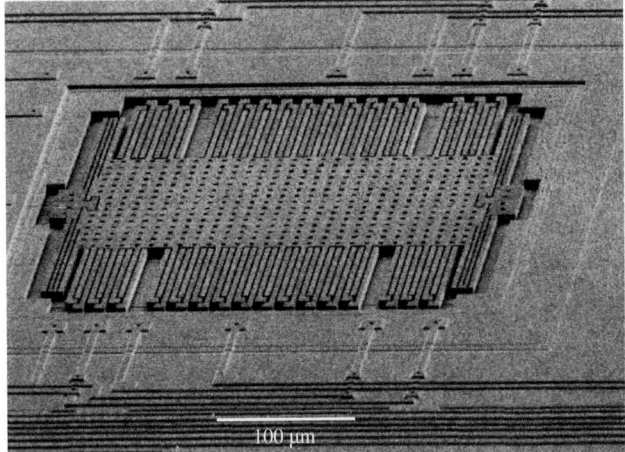

Fig. 15.14 SEM micrograph of a polysilicon surface-micromachined lateral accelerometer (courtesy of Analog Devices Inc.)

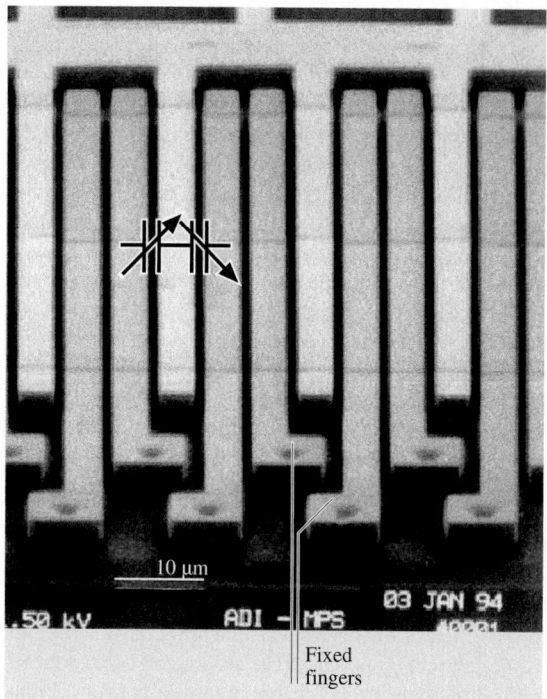

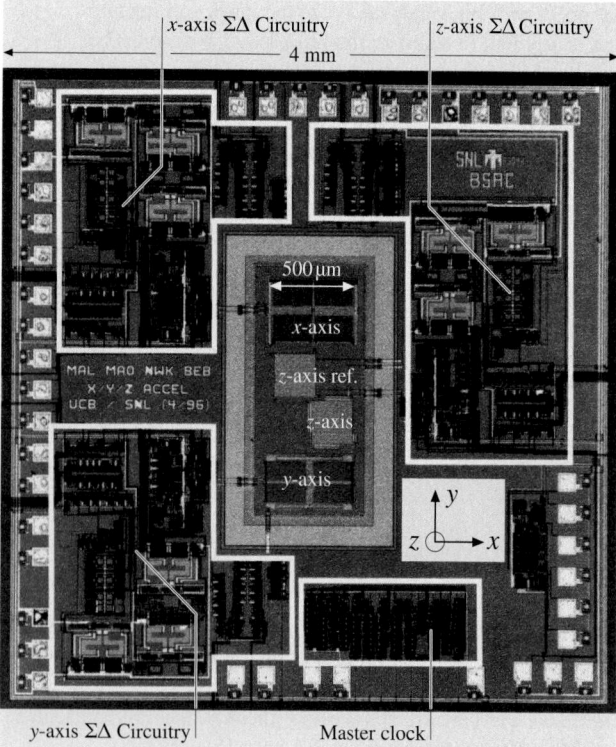

Fig. 15.15 SEM micrograph of a capacitive sensing finger structure

Surface micromachined accelerometers typically suffer from severe mechanical thermal vibration, commonly referred to as Brownian motion [15.16], due to the small proof mass, thus resulting in a high mechanical noise floor which ultimately limits the sensor resolution. Vacuum packaging can be employed to minimize this adverse effect but with a penalty of increasing system complexity and cost. Accelerometers using large proof masses fabricated by bulk micromachining or a combination of surface and bulk micromachining techniques are attractive for circumventing this problem. Figure 15.13 shows an SEM micrograph of an all-silicon z-axis accelerometer fabricated through a single silicon wafer by using combined surface and bulk micromachining process to obtain a large proof mass with dimensions of approximately $2\,\text{mm} \times 1\,\text{mm} \times 450\,\mu\text{m}$ [15.15]. The large mass suppresses the Brownian motion effect, achieving a high performance with a resolution on the order of several μg. Similar fabrication techniques have been used to demonstrate a three-axis capacitive accelerometer achieving a noise floor of approximately $1\,\mu\text{g}/\sqrt{\text{Hz}}$ [15.17].

A surface-micromachined lateral accelerometer developed by Analog Devices Inc. is shown in Fig. 15.14. The sensor consists of a center proof mass supported by folded beam suspensions with arrays of attached movable fingers, forming a sense capacitance with the fixed parallel fingers. The device is fabricated using a 6 um thick polysilicon structural layer with a small air gap on the order of a micrometer to increase the sensor capacitance value, thus improving the device resolution. Figure 15.15 shows a close-up view of the finger structure for a typical lateral accelerometer. Each movable finger forms differential capacitances with two adjacent fixed fingers. This sensing capacitance configuration is attractive for interfacing with differential electronic detection circuits to suppress common-mode noise and other undesirable signal coupling. Monolithic accelerometers with a three-axis sensing capability integrated with on-chip electronic detection circuits have been realized using surfacing micromachining and CMOS microelectronics fabrication technologies [15.18]. Figure 15.16 shows a photo of one of these microsystem chips, which has an area

Fig. 15.16 Photo of a monolithic three-axis polysilicon surface-micromachined accelerometer with integrated interface and control electronics [15.18]

of 4 mm × 4 mm. One vertical accelerometer and two lateral accelerometers are placed at the chip center with corresponding detection electronics along the periphery. A z-axis reference device, which is not movable, is used with the vertical sensor for electronic interfacing. The prototype system achieves a sensing resolution on the order of 1 mG with a 100 Hz bandwidth along each axis. The level of performance is adequate for automobile safety activation systems, vehicle stability and active suspension control, and various consumer products. Recently, monolithic MEMS accelerometers fabricated by using post-CMOS surface micromachining fabrication technology have been developed to achieve an acceleration noise floor of $50 \, \mu g/\sqrt{Hz}$ [15.19]. This technology can enable MEMS capacitive inertial sensors to be integrated with interface electronics in a commercial CMOS process, thus minimizing prototyping cost.

Gyroscope

Most of micromachined gyroscopes employ vibrating mechanical elements to sense rotations. The sensors rely on energy transfer between two vibration modes of a structure caused by Coriolis acceleration. Figure 15.17 presents a schematic of a z-axis vibratory rate gyroscope. The device consists of an oscillating mass electrostatically driven into resonance along the drive-mode axis using comb fingers. An angular rotation along the vertical axis (z-axis) introduces a Coriolis acceleration, which results in a structure deflection along the sense-mode axis, shown in the figure. The deflection changes the differential sense capacitance value, which can be detected as a measure of input angular rotation. A z-axis vibratory rate gyroscope operating upon this principle is fabricated using surface micromachining technology and integrated together with electronic detection circuits, as illustrated in Fig. 15.18 [15.20]. The micromachined sensor is fabricated using polysilicon structural material with a thickness around 2 μm and occupies an area of 1 mm × 1 mm. The sensor achieves a resolution of approximately $1°/s/\sqrt{Hz}$ under a vacuum pressure around 50 mTorr. Other MEMS single-axis gyroscopes integrated in commercial IC processes were demonstrated recently achieving an enhance performance [15.21, 22].

A dual-axis gyroscope based on a rotational disk at its resonance can be used to sense angular rotation along two lateral axes (x-axis and y-axis). Figure 15.19 shows a device schematic demonstrating the operating principle. A rotor disk supported by four mechani-

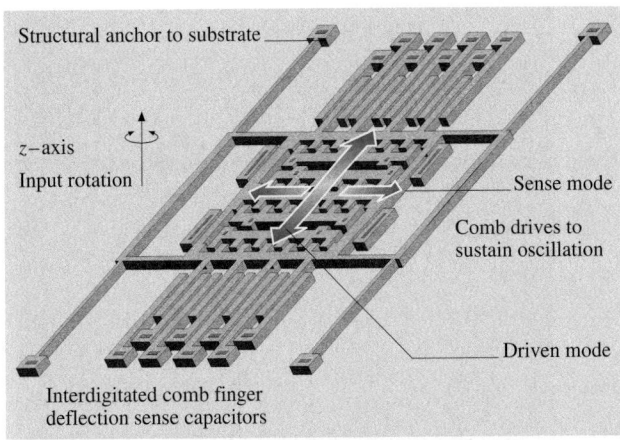

Fig. 15.17 Schematic of a vibratory rate gyroscope

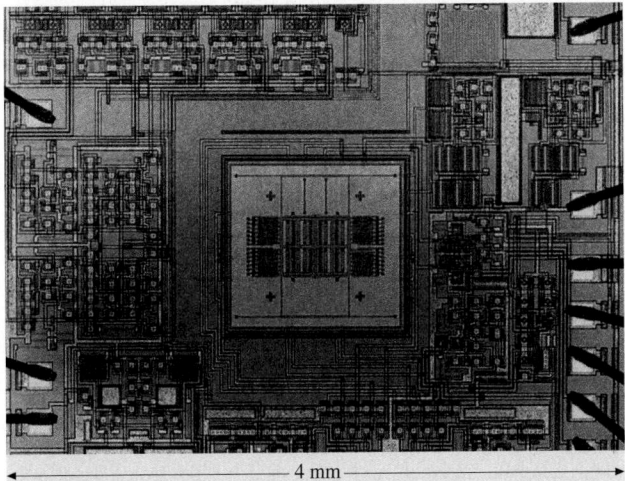

Fig. 15.18 Photo of a monolithic polysilicon surface-micromachined z-axis vibratory gyroscope with integrated interface and control electronics [15.20]

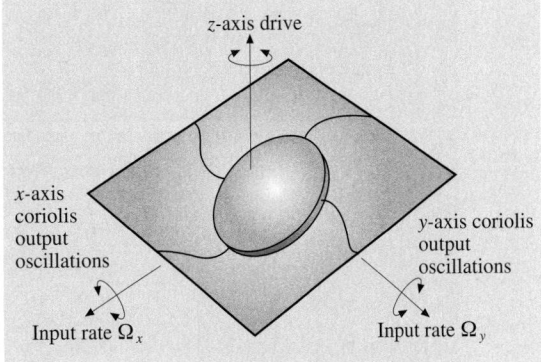

Fig. 15.19 Schematic of a dual-axis gyroscope

Fig. 15.20 Photo of a polysilicon surface-micromachined dual-axis gyroscope [15.23]

cal suspensions can be driven into angular resonance along the z-axis. An input angular rotation along the x-axis will generate a Coriolis acceleration causing the disk to rotate along the y-axis; and vice versa. This Coriolis-acceleration-induced rotation will change the sensor capacitance values between the disk and different sensing electrodes underneath. The capacitance change can be detected and processed by electronic interface circuits. Angular rotations along the two lateral axes can be measured simultaneously using this device architecture. Figure 15.20 shows a photo of a dual-axis gyroscope fabricated using a $2\,\mu m$ thick polysilicon surface micromachining technology [15.23]. As shown in the figure, curved electrostatic drive combs are positioned along the circumference of the rotor dick to drive it into resonance along the vertical axis. The gyroscope exhibits a low random walk of $1°/\sqrt{h}$ under a vacuum pressure around 60 mTorr. With accelerometers and gyroscopes each capable of three-axis sensing, a micormachine-based inertial measurement system providing a six-degree-of-freedom sensing capability can be realized. Figure 15.21 presents a photo of such a system containing a dual-axis gyroscope, a z-axis gyroscope, and a three-axis accelerometer chip integrated with microelectronic circuitry. Due to the precision in device layout and fabrication, the system can measure angular rotation and acceleration without the need to align individual sensors.

15.1.3 Optical MEMS

Surface micromachining has served as a key enabling technology to realize microeletromechancal optical devices for various applications ranging from sophisticated visual information displays and fiber-optic telecommunication to bar-code reading. Most of the existing optical systems are implemented using conventional optical components, which suffer from bulky size, high cost, large power consumption, poor efficiency and reliability issues. MEMS technology is promising for producing miniaturized, reliable, inexpensive optical components

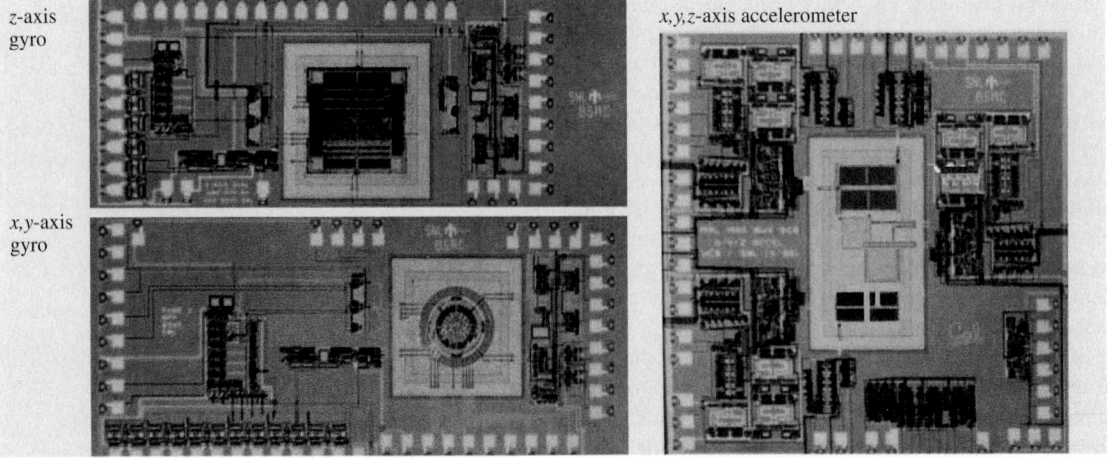

Fig. 15.21 Photo of a surface-micromachined inertial measurement system with a six-degree sensing capability

to revolutionize conventional optical systems [15.24]. In this section of the chapter, a few selected MEMS optical devices will be presented to illustrate their impact in the fields of visual display, precision optical platform, and data switching for optical communication.

Visual Display

An early MEMS device successfully used for various display applications is the Texas Instruments Digital Micromirror Device (DMD). The DMD technology can achieve higher performance in terms of resolution, brightness, contrast ratio, and convergence than the conventional cathode ray tube and is critical for digital high-definition television applications. A DMD consists of a large array of small mirrors with a typical area of $16\,\mu m \times 16\,\mu m$ as illustrated in Fig. 15.22. A probe tip is shown in the figure for a size comparison. Figure 15.23 shows an SEM micrograph of a close-up view of a DMD pixel array [15.25]. Each mirror is capable of rotating by $+/-10$ degrees corresponding to either the "on" or "off" position due to an electrostatic actuation force. Light reflected from any on-mirrors passes through a projection lens and creates images on a large screen. Light from the remaining off-mirrors is reflected away from the projection lens to an absorber. The proportion of time during each video frame that a mirror remains in the on-state determines shades of gray, from black for zero on-time to white for a hundred percent on-time. Color can be added by a color wheel or a three-DMD chip setup. The three DMD chips are used for projecting red, green and blue colors. Each DMD pixel consists of a mirror connected by a mirror support post to an underlying yoke. The yoke, in turn is connected by torsion hinges to hinge support posts, as shown in Fig. 15.24 [15.26].

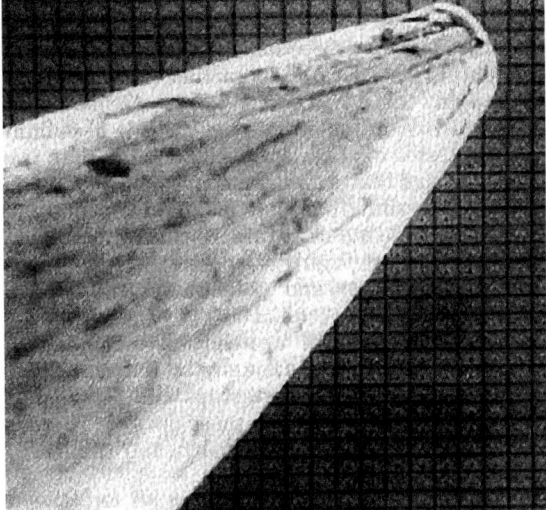

Fig. 15.22 Photo of a digital micromirror device (DMD) array (courtesy of Texas Instruments)

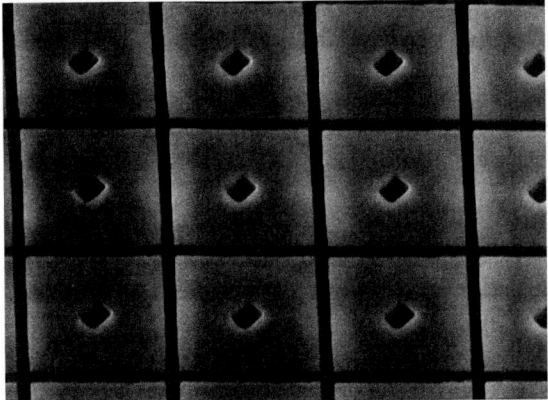

Fig. 15.23 SEM micrograph of a close-up view of a DMD pixel array [15.25]

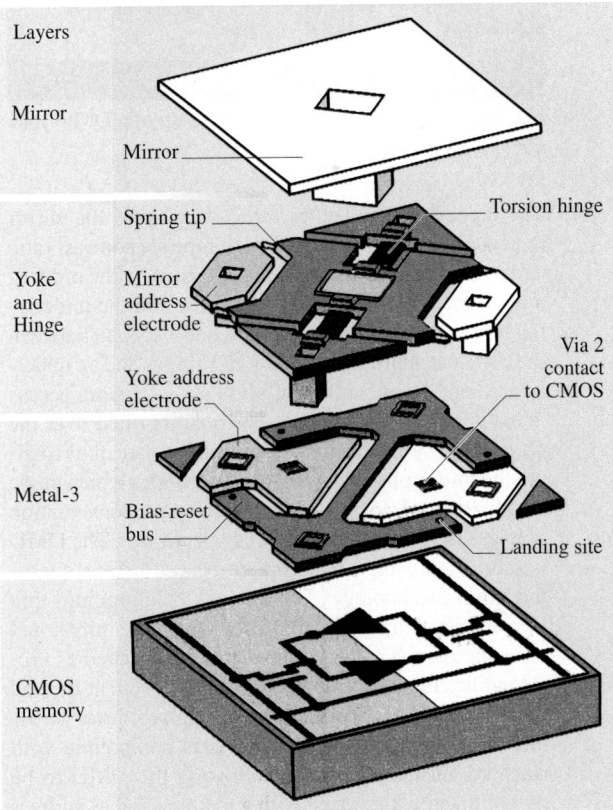

Fig. 15.24 Detailed structure layout of a DMD pixel [15.26]

Fig. 15.25 SEM micrograph of a DMD pixel after removing half of the mirror plate using ion milling (courtesy of Texas Instruments)

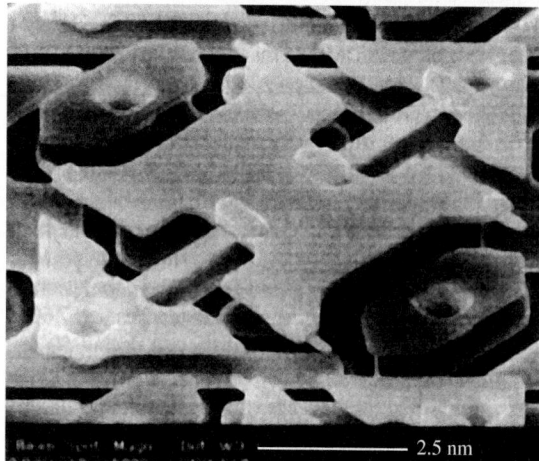

Fig. 15.26 SEM micrograph of a close-up of a DMD yoke and hinges [15.26]

The support post and hinges are hidden under the mirror to avoid light diffraction and thus improve contrast ratio and optical efficiency. There are two gaps on the order of a micrometer, one between the mirror and the underlying hinges and address electrodes, and a second between the coplanar address electrodes and hinges and an underlying metal layer from the CMOS static random access memory (SRAM) structure. The yoke is tilted over the second gap by an electrostatic actuation force, thus rotating the mirror plate. The SRAM determines which angle the mirror needs to be tilted by applying proper actuation voltages to the mirror and address electrodes. The DMD is fabricated using an aluminum-based surface micromachining technology. Three layers of aluminum thin film are deposited and patterned to form the mirror and its suspension system. Polymer material is used as sacrificial layer and is removed by a plasma etch at the end of the process to freely release the micromirror structure. The micromachining process is compatible with standard CMOS fabrication, allowing the DMD to be monolithically integrated with a mature CMOS address circuit technology, thus achieving high yield and low

cost. Figure 15.25 shows an SEM micrograph of a fabricated DMD pixel revealing its cross section after an ion milling. A close-up view on the yoke and hinge support under the mirror is shown in Fig. 15.26.

Precision Optical Platform

The growing optical communication and measurement industry require low cost, high performance optoelectronic modules such as laser-to-fiber couplers, scanners, interferometers, etc. A precision alignment and the ability to actuate optical components such as mirrors, gratings, and lenses with sufficient accuracy are critical for high performance optical applications. Conventional hybrid optical integration approaches, such as the silicon-optical-bench, suffers from a limited alignment tolerance of $+/- 1 \mu m$ and also lacks of component actuation capability [15.27, 28]. As a result, only simple optical systems can be constructed with no more than a few components, thus severely limiting the performance. Micromachining, however, provides a critical enabling technology, allowing movable optical components to be fabricated on a silicon substrate. Component movement with high precision can be achieved through electrostatic actuation. By combining micromachined movable optical components with lasers, lenses, and fibers on the same substrate, an on-chip complex self-aligning optical system can be realized. Figure 15.27a shows an SEM micrograph of a surface-micromachined, electrostatically actuated microreflector for laser-to-fiber coupling and external-cavity-laser applications [15.29]. The device consists of a polysilicon mirror plate hinged to a support beam. The mirror and the support, in turn, are hinged to a vibromotor-actuator slider. The microhinge technology [15.30] allows the joints to rotate out of the substrate plane to achieve large aspect ratios. Common-mode actuation of the sliders results in a translational motion, while differential slider motion produces an out-of-plane mirror rotation. These motions permit the microreflector to redirect an optical beam in a desirable location. Each of the two slides is actuated with an integrated microvibromotor shown in Fig. 15.27b. The vibromotor consists of four electrostatic comb resonators with attached impact arms driving a slider through oblique impact. The two opposing impacters are used for each travel direction to balance the forces. The resonator is a capacitively driven mass anchored to the substrate through a folded beam flexure. The flexure compliance determines the resonant frequency and travel range of the resonator. When the comb structures are driven at their resonant frequency (around 8 KHz), the slider exhibits a maximum velocity of over 1 mm/s. Characterization of the vibromotor also shows that a slider step resolution of less than 0.3 μm can be achieved [15.31], making it attractive for precision alignment of various optical components. The prototype microreflector can obtain an angular travel range over 90 degrees and a translational travel range of 60 μm. By using this device, beam steering, fiber coupling, and optical scanning have been demonstrated.

Optical Data Switching

High-speed communication infrastructures are highly desirable for transferring and processing real-time voice and video information. Optical fiber communication technology has been identified as the critical backbone to support such systems. A high-performance optical data switching network, which routes various optical signals from sources to destinations, is one of the key building blocks for system implementation. At present, optical signal switching is performed by using hybrid optical-electronic-optical (O-E-O) switches. These devices first convert incoming light from input fibers to electrical signals first and then route the electrical signals to the proper output ports after signal analyses. At the output ports, the electrical signals are converted back to streams of photons or optical signals for further transmission over the fibers. The O-E-O switches are expensive to build, integrate, and maintain. Furthermore, they consume substantial amount of power and introduce additional latency. It is therefore highly desirable to develop an all-optical switching network in which optical signals can be routed without intermediate conversion into electrical form, thus minimizing power dissipation and system delay. While a number of approaches are being considered for building all-optical switches, MEMS technology is attractive because it can provide arrays of tiny movable mirrors which can redirect incoming beams from input fibers to corresponding output fibers. As described in the previous sections, these micromirrors can be batch fabricated

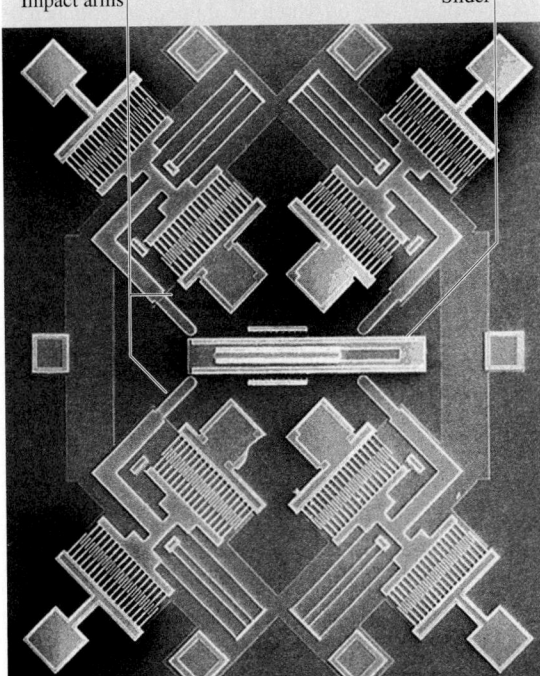

Fig. 15.27 (a) SEM micrograph of a surface-micromachined, electrostatically actuated microreflector; (b) SEM micrograph of a surface-micromachined vibromotor [15.29]

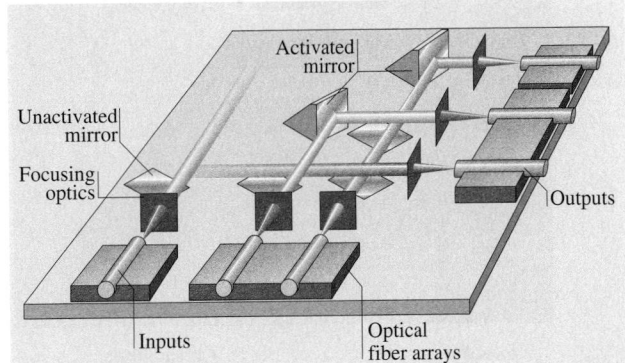

Fig. 15.28 Schematic of a two-dimensional micromirror-based fiber optic switching matrix

using silicon micromachining technologies, thus achieving an integrated solution with the potential for low cost. A significant reduction in power dissipation is also expected.

Figure 15.28 shows an architecture of a two-dimensional micromirror array forming a switching matrix with rows of input fibers and columns of output fibers (or vice versa). An optical beam from an input fiber can be directed to an output fiber through activating the corresponding reflecting micromirror. Switches with eight inputs and eight outputs can be readily implemented using this technique, which can be further extended to a 64×64 matrix. The micromirrors are moved between two fixed stops by digital control, thus eliminating the need for precision motion control. Figure 15.29 presents an SEM micrograph of a simple 2×2 MEMS fiber optic switching network prototype for an illustration purpose [15.32]. The network includes a mirror chip passively integrated with a silicon submount, which contains optical fibers and ball lenses. The mirror chip consists of four surface-micromachined vertical torsion mirrors. The four mirrors are arranged such that in the "reflection" mode, the input beams are reflected by two 45-degree vertical torsion mirrors and coupled into the output fibers located on the same side of the chip. In the "transmission" mode, the vertical torsion mirrors are rotated out of the optical paths, thus allowing the input beams to be coupled into the opposing output fibers. Figure 15.30 shows an SEM micrograph of a polysilicon vertical torsion

mirror. The device consists of a mirror plate attached to a vertical supporting frame by torsion beams and a vertical back electrode plate. The mirror plate is approximately $200\,\mu m$ wide, $160\,\mu m$ long, and $1.5\,\mu m$ thick. The mirror surface is coated with a thin layer of gold to improve the optical reflectivity. The back plate is used to electrostatically actuate the mirror plate so that the mirror can be rotated out of the optical path in the "transmission" mode. Surface micromachining with micro-hinge technology is used to realize the overall structure. The back electrode plate is integrated with a scratch drive actuator array [15.33] for self-assembly. The self-assembly approach is critical when multiple vertical torsion mirrors are used to implement more advanced functions.

A more sophisticated optical switching network with a large scaling potential can be implemented by using a three-dimensional (3-D) switching architecture as shown in Fig. 15.31. The network consists of arrays of two-axis mirrors to steer optical beams from input fibers to output fibers. A precision analog closed-loop mirror position control is required to accurately direct a beam along two angles so that one input fiber can be optically connected to any output fiber. The optical length depends little on which set of fibers are connected, thus achieving a more uniform switching characteristic, which is critical for implementing large scale network. Two-axis mirrors are the crucial components for implementing the 3-D architecture. Figure 15.32 shows an SEM micrograph of a surface-micromachined two-axis beam-steering mirror positioned by using self-assembly technique [15.34]. The self-assembly is accomplished

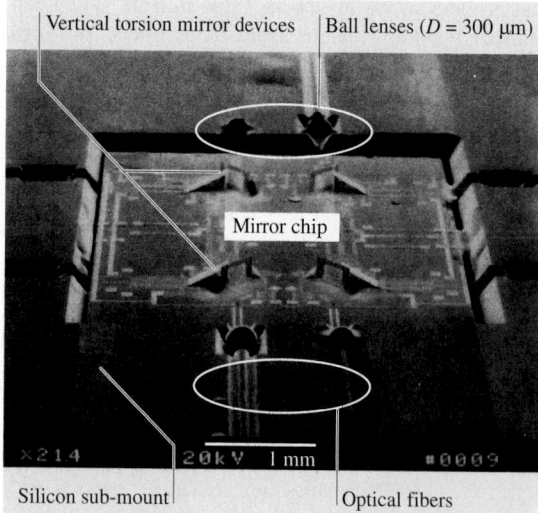

Vertical torsion mirror devices   Ball lenses ($D = 300\,\mu m$)

Mirror chip

Silicon sub-mount   Optical fibers

Fig. 15.29 SEM micrograph of a 2×2 MEMS fiber optic switching network [15.32]

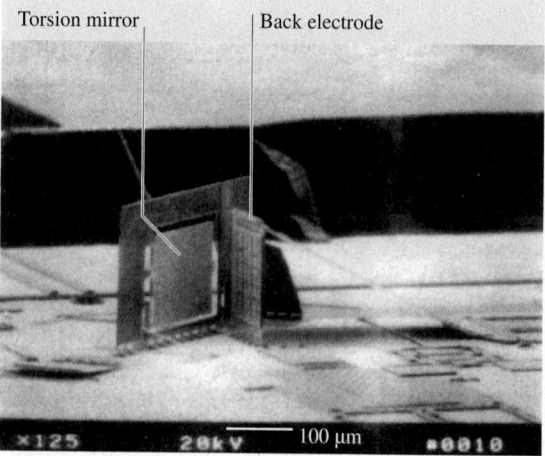

Torsion mirror   Back electrode

Fig. 15.30 SEM micrograph of a polysilicon surface-micromachined vertical torsion mirror [15.32]

during the final release step of the mirror processing sequence. Mechanical energy is stored in a special high-stress layer during the deposition, which is put on top of the four assembly arms. Immediately after the assembly arms are released, the tensile stress in this layer causes the arms to bend up, pushing the mirror frame and lifting it above the silicon substrate. All mirrors used in the switching network can be fabricated simultaneously without any human intervention or external power supply.

15.1.4 RF MEMS

The increasing demand for wireless communication applications, such as cellular and cordless telephony, wireless data networks, two-way paging, global positioning system, etc., motivates a growing interest in building miniaturized wireless transceivers with multistandard capabilities. Such transceivers will greatly enhance the convenience and accessibility of various wireless services independent of geographic location. Miniaturizing current single-standard transceivers, through a high-level of integration, is a critical step towards building transceivers that are compatible with multiple standards. Highly integrated transceivers will also result in reduced package complexity, power consumption, and cost. At present, most radio transceivers rely on a large number of discrete frequency-selection components, such as radio-frequency (RF) and intermediate-frequency (IF) band-pass filters, RF voltage-controlled oscillators (VCOs), quartz crystal oscillators, solid-state switches, etc. to perform the necessary analog signal processing. Figure 15.33 shows a schematic of a super-heterodyne radio architecture, in which discrete components are shaded in dark color. Theses off-chip devices occupy the majority of the system area, thus severely hindering transceiver miniaturization. MEMS technology, however, offers a potential solution to integrate these discrete components onto silicon substrates with microelectronics, achieving a size reduction of a few orders of magnitude. It is therefore expected to become an enabling technology to ultimately miniaturize radio transceivers for future wireless communications.

MEMS Variable Capacitors
Integrated high-performance variable capacitors are critical for low noise VCOs, antenna tuning, tunable matching networks, etc. Capacitors with high quality factors (Q), large tuning range and linear

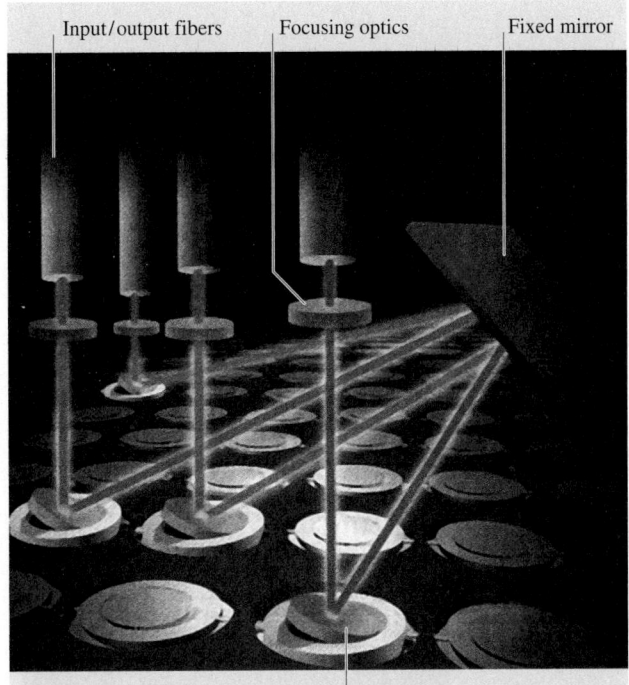

Fig. 15.31 Schematic of a three-dimensional micromirror-based fiber optic switching matrix

Fig. 15.32 SEM micrograph of a surface-micromachined two-axis beam-steering micromirror positioned using a self-assembly technique [15.34]

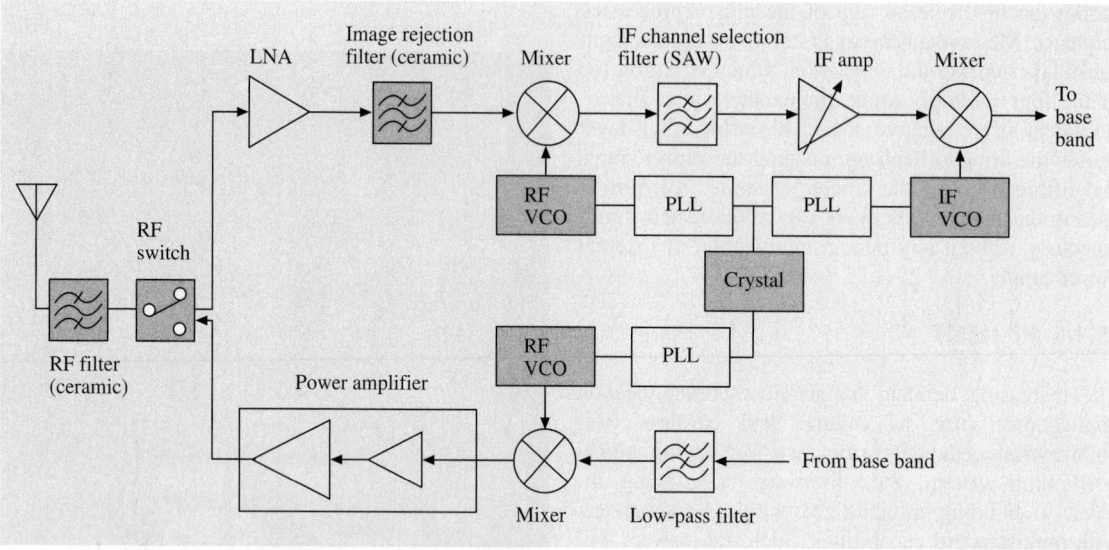

Fig. 15.33 Schematic of a super-heterodyne radio architecture

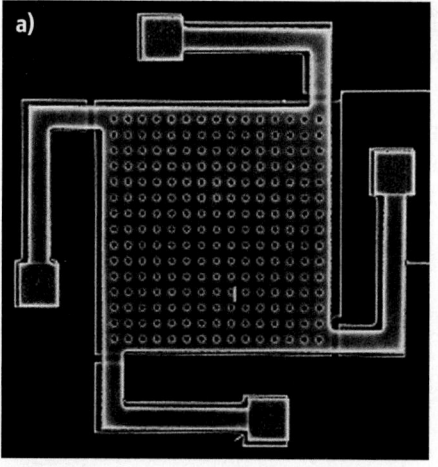

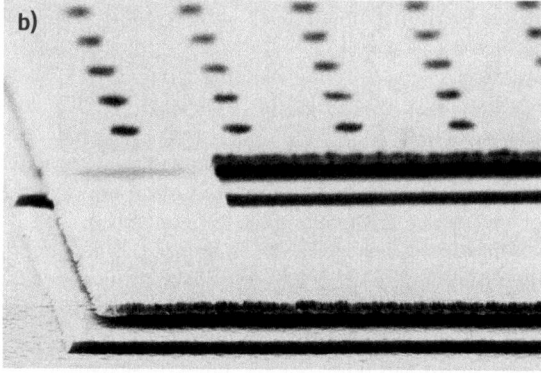

Fig. 15.34 (a) SEM micrograph of a top view of an aluminium surface-micromachined variable capacitor; **(b)** SEM micrograph of a cross-sectional view of the variable capacitor [15.35]

characteristics are crucial for achieving system performance requirements. On-chip silicon PN junction and MOS based variable capacitors suffer from low quality factors (below 10 at 1 GHz), limited tuning range and poor linearity, thus are inadequate for building high-performance transceivers. MEMS technology has demonstrated monolithic variable capacitors achieving stringent performance requirements. These devices typically reply on an electrostatic actuation method to vary

Fig. 15.35 SEM micrograph of four MEMS aluminium variable capacitors connected in parallel [15.35]

Fig. 15.36 SEM micrograph of a silicon tunable capacitor using a comb drive actuator [15.39]

Fig. 15.37 SEM micrograph of a close view of a tunable capacitor comb fingers [15.39]

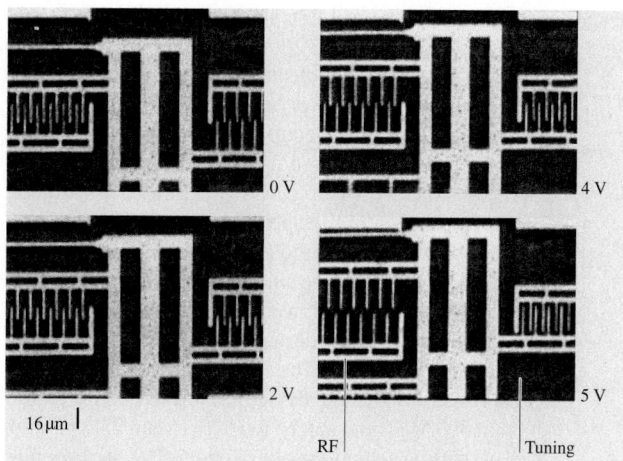

Fig. 15.38 Photos of comb fingers at different actuation voltages [15.39]

the air gap between a set of parallel plates [15.35–38] or vary the capacitance area between a set of conductors [15.39] or mechanically displace a dielectric layer in an air-gap capacitor [15.40]. Improved tuning

ranges have been achieved with various device configurations. Figure 15.34 shows SEM micrographs of an aluminum micromachined variable capacitor fabricated on a silicon substrate [15.35]. The device consists of a 200 μm × 200 μm aluminum plate with a thickness of 1 μm suspended above the bottom electrode by an air gap of 1.5 μm. Aluminum is selected as the structural material due to its low resistivity, critical for achieving a high quality factor at high frequencies. A DC voltage applied across the top and bottom electrodes introduces an electrostatic pull-down force, which pulls the top plate towards the bottom electrode, thus changing the device capacitance value. The capacitors are fabricated using aluminum-based surface micromachining technology. Sputtered aluminum is used for building the capacitor top and bottom electrodes. Photoresist is served as the sacrificial layer, which is then removed through an oxygen-based plasma dry etch to release the microstructure. The processing technology requires a low thermal budget, thus allowing the variable capacitors to be fabricated on top of wafers with completed electronic circuits without degrading the performance of active devices. Figure 15.35 presents an SEM micrograph of four MEMS tunable capacitors connected in parallel. This device achieves a nominal capacitance value of 2 pF and a tuning range of 15% with 3 V. A quality factor of 62 has been demonstrated at 1 GHz, which matches or exceeds that of discrete varactor diodes and is at least an order of magnitude larger than that of a typical junction capacitor implemented in a standard IC process.

MEMS tunable capacitors based upon varying capacitance area between a set of conductors have been demonstrated. Figure 15.36 shows an SEM micrograph of a such device [15.39]. The capacitor comprises arrays of interdigitated electrodes, which can be electrostatically actuated to vary the electrode overlap area. A close-up view of the electrodes is shown in Fig. 15.37. The capacitor is fabricated using a silicon-on-insulator (SOI) substrate with a top silicon layer thickness around 20 μm to obtain a high aspect ratio for the electrodes, critical for achieving a large capacitance density and reduced tuning voltage. The silicon layer is etched to form the device structure followed by removing the underneath oxide to release the capacitor. A thin aluminum layer is then sputtered over the capacitor to reduce the series resistive loss. The device exhibits a quality factor of 34 at 500 MHz and can be tuned between 2.48 pF and 5.19 pF with an actuation voltage under 5 V, corresponding to a tuning range over 100%. Figure 15.38 shows the variation of electrode overlap area under different tuning voltages.

Tunable capacitors relying on a movable dielectric layer have been fabricated using MEMS technology. Figure 15.39 presents an SEM micrograph of a copper-based micromachined tunable capacitor [15.40]. The device consists of an array of copper top electrodes suspended above a bottom copper plate with an air gap of approximately 1 μm. A thin nitride layer is deposited, patterned, and suspended between the two copper layers by lateral mechanical spring suspensions after sacrificial release. A DC voltage applied across the copper layers introduces a lateral electrostatic pull-in force on the nitride, thus resulting in a movement which changes the overlapping area between each copper electrode and the bottom plate, and hence the device capacitance. The tunable capacitor achieves a quality factor over 200 at 1 GHz with 1 pF capacitance due to the highly conductive copper layers and a tuning range around 8% with 10 V.

Micromachined Inductors

Integrated inductors with high quality factors are as critical as the tunable capacitors for high performance RF system implementation. They are the key components for building low noise oscillators, low loss matching networks, etc. Conventional on-chip spiral inductors suffer from limited quality factors of around 5 at 1 GHz, an order of magnitude lower than the required values from discrete counterparts. The poor performance is mainly caused by substrate loss and metal resistive loss at high frequencies. Micromachining technology provides an attractive solution to minimize these loss contributions; hence enhancing the device quality factors. Figure 15.40 shows an SEM micrograph of a 3-D coil inductor fabricated on a silicon substrate [15.41]. The device consists of 4-turn 5 μm thick copper traces electroplated around an insulting core with a 650 μm by 500 μm cross section. Compared to spiral inductors, this geometry minimizes the coil area which is in close proximity to the substrate and hence the eddy-current loss, resulting in a maximized Q-factor and device self-resonant frequency.

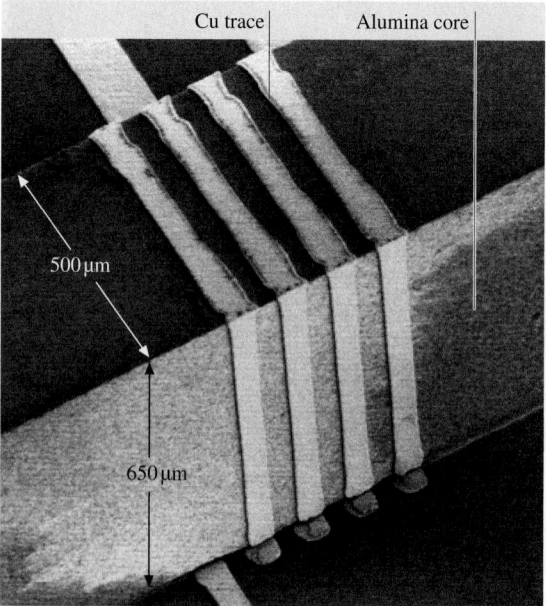

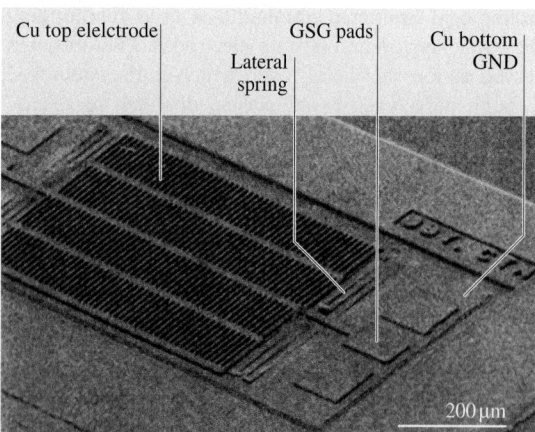

Fig. 15.39 SEM micrograph of a copper surface-micromachined tunable capacitor with a movable dielectric layer [15.40]

Fig. 15.40 SEM micrograph of a 3-D coil inductor fabricated on a silicon substrate [15.41]

Copper is selected as the interconnect metal because of its low sheet resistance, critical for achieving a high Q-factor. The inductor achieves a 14 nH inductance value with a quality factor of 16 at 1 GHz. A single-turn 3-D device exhibits a Q-factor of 30 at 1 GHz, which matches the performance of discrete counterparts. The high-Q 3-D inductor and MEMS tunable capacitors, shown in Fig. 15.35, have been employed to implement a RF CMOS VCO achieving a low phase noise performance suitable for typical wireless communication applications such as GMS cellular telephony [15.42].

Other 3-D inductor structures such as the levitated spiral inductors have been demonstrated using micromachining fabrication technology. Figure 15.41 shows an

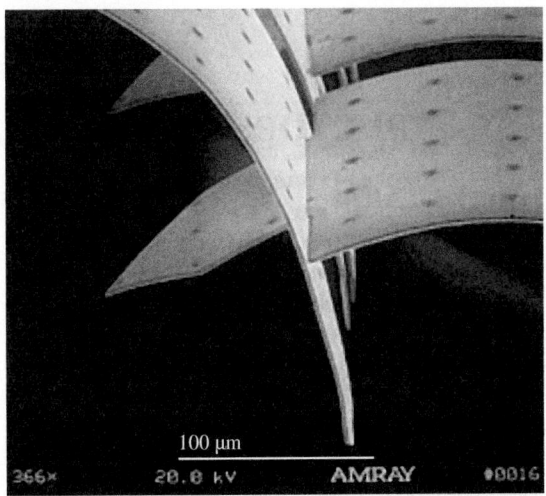

Fig. 15.43 SEM micrograph of an interlocking trace from a self-assembled out-of-plane oil inductor [15.43]

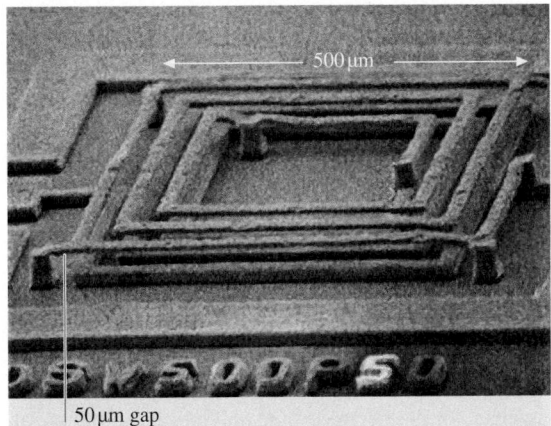

Fig. 15.41 SEM micrograph of a levitated spiral inductor fabricated on a glass substrate [15.40]

SEM micrograph of a levitated copper inductor, which is suspended above the substrate through supporting posts [15.44]. The levitated geometry can minimize the substrate loss, thus achieving an improved quality factor. The inductor shown in the figure achieves a 1.4 nH inductance value with a Q-factor of 38 at 1.8 GHz using a glass substrate. Similar inductor structures have been demonstrated on standard silicon substrates achieving a nominal inductance value of approximately 1.4 nH with a Q-factor of 70 measured at 6 GHz [15.45].

A self-assembled out-of-plane coil has been fabricated using micromachining technology. The inductor winding traces are made of refractory metals with controlled built-in stress such that the traces can curl out of the substrate surface upon release and interlock into each other to form coil windings. Figure 15.42 shows an SEM micrograph of a self-assembled out-of-plane coil inductor [15.43]. A close-up view of an interlocking trace is shown in Fig. 15.43. Copper is plated on the interlocked traces to form highly conductive windings at the end of processing sequence. The inductor shown in Fig. 15.42 achieves a quality factor around 40 at 1 GHz.

MEMS Switches

The microelectromechanical switch is another potentially attractive miniaturized component enabled by micromachining technologies. These switches offer superior electrical performance in terms of insertion loss, isolation, linearity, etc. and are intended to replace off-chip solid state counterparts, which provide switching between the receiver and transmitter signal paths.

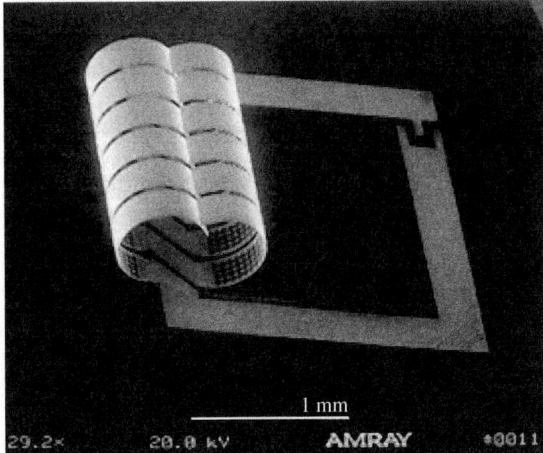

Fig. 15.42 SEM micrograph of a self-assembled out-of-plane coil inductor [15.43]

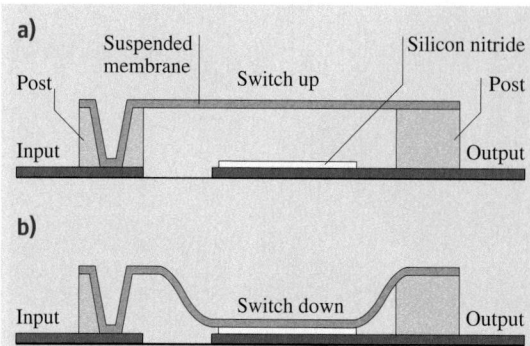

Fig. 15.44 Cross-sectional schematics of an RF MEMS capacitive switch

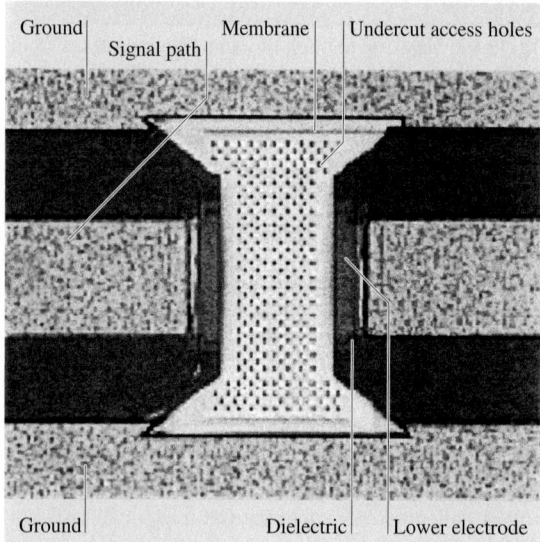

Fig. 15.45 Top view photo of a fabricated RF MEMS capacitive switch [15.46]

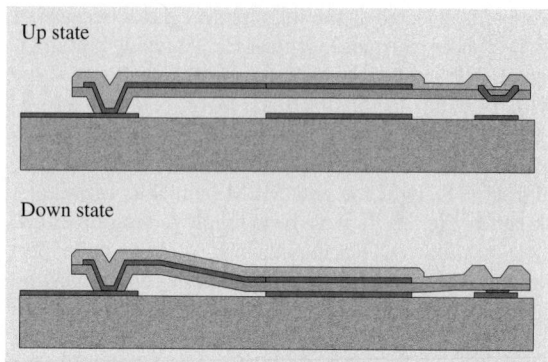

Fig. 15.46 Cross-sectional schematic of a metal-to-metal contact switch [15.47]

They are also critical for building phase shifters, tunable antennas, and filters. The MEMS switches can be characterized into two categories: capacitive and metal-to-metal contact types. Figure 15.44 presents a cross-sectional schematic of an RF MEMS capacitive switch. The device consists of a conductive membrane, typically made of aluminum or gold alloy suspended above a coplanar electrode by an air gap of a few micrometers. For RF or microwave applications, actual metal-to-metal contact is not necessary; rather, a step change in the plate-to-plate capacitance realizes the switching function. A thin silicon nitride layer with a thickness on the order of 1000 Å is typically deposited

above the bottom electrode. When the switch is in the on-state, the membrane is high resulting in a small plate-to-plate capacitance; hence, a minimum high-frequency signal coupling (high isolation) between the two electrodes. In the off-state (with a large enough applied DC voltage), the switch provides a large capacitance due to the thin dielectric layer, thus causing a strong signal coupling (low insertion loss). The capacitive switch consumes near-zero power, which is attractive for low power portable applications. Switching cycles over millions for this type of device have been demonstrated. Figure 15.45 shows a top view photo of a fabricated MEMS capacitive switch [15.46]. Surface micromachining technology, using metal for the electrodes and polymer as the sacrificial layer, is used to fabricate the device. The switch can be actuated with a DC voltage on the order of 50 V and exhibits a low insertion loss of approximately −0.28 dB at 35 GHz and a high isolation of −35 dB at the same frequency.

Metal-to-metal contact switches are important for interfacing large bandwidth signals including DC. This type of device typically consists of a cantilever beam or clamped-clamped bridge with a metallic contact pad positioned at the beam tip or underneath bridge center. Through an electrostatic actuation, a contact can be formed between the suspended contact pad and an underlying electrode on the substrate [15.47–49]. Figure 15.46 shows a cross-sectional schematic of a metal-to-metal contact switch [15.47]. The top view of the fabricated device is presented in Fig. 15.47. The switch exhibits an actuation voltage of 30 V, a response time of 20 μs, and mechanical strength to withstand 10^9 actuations. An isolation greater than 50 dB below 2 GHz and insertion loss less than 0.2 dB from DC through 40 GHz has been demonstrated. Metal-to-metal contact switches

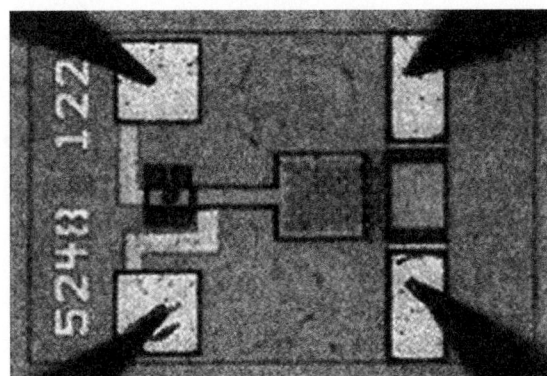

Fig. 15.47 Top view photo of a fabricated metal-to-metal contact switch [15.47]

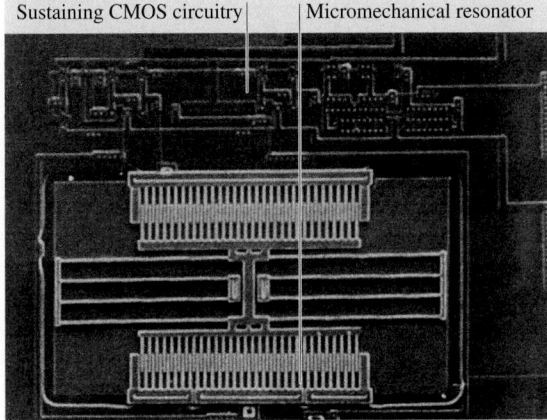

Fig. 15.48 SEM micrograph of a surface-micromachined comb drive resonator integrated with CMOS sustaining electronics [15.52]

relying on electro-thermal actuation method have also been developed to demonstrate a low actuation voltage around 3 V, however, at an expense of reduced switching speed of 300 μs and increased power dissipation in the range of 60–100 mW [15.50, 51]. The fabricated switches achieve an off-state isolation of −20 dB at 40 GHz and an insertion loss of − 0.1 dB up to 50 GHz.

MEMS Resonators

Microelectromechanical resonators based upon polysilicon comb-drive structures, suspended beams, and center-pivoted disk configurations have been demonstrated for performing analog signal processing [15.50, 52–56]. These microresonators can be excited into mechanical resonance through an electrostatic drive. The mechanical motion causes a change of device capac-

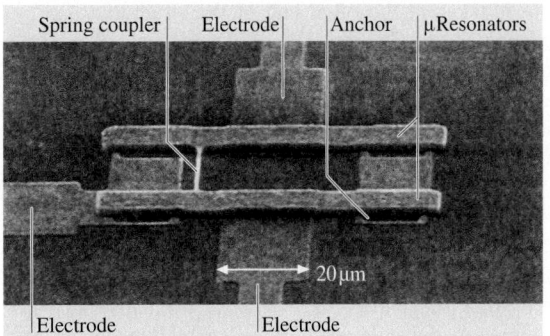

Fig. 15.49 SEM micrograph of a polysilicon surface-micromachined two-resonator spring-coupled bandpass micromechanical filter [15.53]

itance resulting in an output electrical current when a proper DC bias voltage is used. The output current exhibits the same frequency as the mechanical resonance, thus achieving an electrical filtering function through the electromechanical coupling. Micromachined polysilicon flexural-mode mechanical resonators have demonstrated a quality factor greater than 80 000 in a 50 μTorr vacuum [15.57]. This level of performance is comparable to a typical quartz crystal and is thus attractive for implementing monolithic low noise and low drift reference signal sources. Figure 15.48 shows an SEM micrograph of a surface-micromachined comb drive resonator integrated with CMOS sustaining electronics on a same substrate to form a monolithic high-Q MEMS resonator-based oscillator [15.52]. The oscillator achieves an operating frequency of 16.5 KHz with a clean spectral purity. A chip area of approximately 420 μm × 230 μm is consumed for fabricating the overall system, representing a size reduction by orders of magnitude compared to conventional quartz crystal oscillators.

Micromachined high-Q resonators can be coupled to implement low-loss frequency selection filters. Figure 15.49 shows an SEM micrograph of a surface-micromachined polysilicon two-resonator, spring-coupled bandpass micromechanical filter [15.53]. The filter consists of two silicon micromechanical clamped-clamped beam resonators, coupled mechanically by a soft spring, all suspended 0.1 μm above the substrate. Polysilicon strip lines underlie the central regions of each resonator and serve as capacitive transducer electrodes positioned to induce resonator vibration in a direction perpendicular to the substrate. Under a normal operation, the device is excited capacitively by a signal voltage applied to the input electrode. The output is taken at the other end of the structure, also by capacitive transduction. The filter achieves a center frequency of 7.81 MHz, a bandwidth of 0.23%, and an insertion loss less than 2 dB. The achieved performance is attractive for implementing filters in the low MHz range.

To obtain a higher mechanical resonant frequency with low losses, a surface-micromachined contour-mode disk resonator has been proposed, as shown in Fig. 15.50 [15.56]. The resonator consists of a polysilicon disk suspended 5000 Å above the substrate with a single anchor at its center. Plated metal input electrodes surround the perimeter of the disk with a narrow separation of around 1000 Å, which defines the capacitive, electromechanical transducer of the device. To operate the device, a DC bias voltage is applied to the structure with an AC input signal applied to the electrodes, resulting in a time varying electrostatic force acting radially on the disk. When the input signal matches the device resonant frequency, the resulting electrostatic force is amplified by the Q factor of the resonator, producing expansion and contraction of the disk along its radius. This motion, in turn, produces a time varying output current at the same frequency, thus achieving the desirable filtering. The prototype resonator demonstrates an operating frequency of 156 MHz with a Q factor of 9400 in vacuum. The increased resonant frequency is comparable to the first intermediate frequency used in a typical wireless transceiver design and is thus suitable for implementing

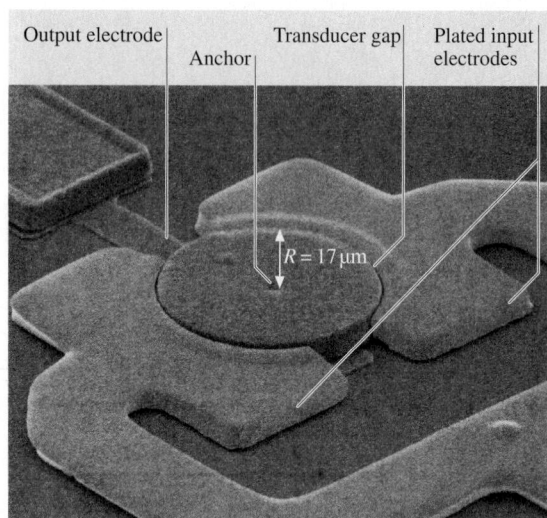

Fig. 15.50 SEM micrograph of a polysilicon surface-micromachined contour-mode disk resonator [15.56]

IF bandpass filters. Recently, self-aligned MEMS fabrication technique was developed to demonstrate vibrating radial-contour mode polysilicon micromechanical disk resonators with resonant frequencies up to 1.156 GHz and measured Q's close to 3000 in both vacuum and air [15.51]. The achieved performance is attractive for potentially replacing RF frequency selection filters in current wireless transceivers with MEMS versions.

15.2 Nanoelectromechanical Systems (NEMS)

Unlike their microscale counterparts, nanoelectromechanical systems (NEMS) are made of electromechanical devices that have critical structural dimensions at or below 100 nm. These devices are attractive for applications where structures of very small mass and/or very large surface area-to-volume ratios provide essential functionality, such as force sensors, chemical sensors, biological sensors, and ultrahigh frequency resonators to name a few. NEMS fabrication processes can be classified into two general categories based on the approach used to create the structures. "Top-down" approaches utilize submicron lithographic techniques to fabricate device structures from "bulk" material, either thin films or thick substrates. "Bottom-up" approaches involve the fabrication of nanoscale devices in much the same way that nature constructs objects, by sequential assembly using atomic and/or molecular building

blocks. While advancements in bottom-up approaches are developing at a very rapid pace, most advanced NEMS devices are currently created utilizing top-down techniques that combine existing process technologies, such as electron-beam lithography, conventional film growth and chemical etching. Top-down approaches make integration with microscale packaging relatively straightforward since the only significant difference between the nano-scale and microscale processing steps is the method used to pattern the various features.

In large measure, NEMS has followed a developmental path similar to the route taken in the development of MEMS in that both have leveraged existing processing techniques from the IC industry. For instance, the electron-beam lithographic techniques used in top-down NEMS fabrication are the same techniques that have become standard in the fabrication of submicron transistors. Furthermore, the materials used in many of the first generation, top-down NEMS devices, (Si, GaAs, Si_3N_4, SiC) were first used in IC's and then in MEMS. Like the first MEMS devices, the first generation NEMS structures consisted of free-standing nanomechanical beams, paddle oscillators, and tethered plates made using simple bulk and single layer surface "nano" machining processes. Recent advancements have focused on incorporating "nano" materials such as nanotubes and nanowires synthesized using bottom-up approaches into NEMS devices by integrating these materials into top-down nano- and micromachining processes. The following text serves only at a brief introduction to the technology, highlighting the key materials, fabrication approaches, and emerging application areas. For additional details and perspectives, readers are encouraged to consult an excellent review on the subject [15.58].

15.2.1 Materials and Fabrication Techniques

Like Si MEMS, Si NEMS capitalizes on well-developed processing techniques for Si and the availability of high quality substrates. Cleland [15.59] reported a relatively simple process to fabricate nanomechanical clamped-clamped beams directly from single crystal (100) Si substrates. As illustrated in Fig. 15.1, the process begins with the thermal oxidation of a Si substrate (Fig. 15.1a). Large Ni contact pads were then fabricated using optical lithography and lift-off. A polymethyl methacrylate (PMMA) lift-off mold was then deposited and patterned using electron-beam lithography into the shape of nanomechanical beams (Fig. 15.1b). Ni was then deposited and patterned by lifting off the PMMA (Fig. 15.1c). Next, the underlying oxide film was patterned by RIE using the Ni film as an etch mask. After oxide etching, nanomechanical beams were patterned by etching the Si substrate using RIE, as shown in Fig. 15.1d. Following Si RIE, the Ni etch mask was removed and the sidewalls of the Si nanomechanical beams were lightly oxidized in order to protect them during the release step (Fig. 15.1e). After performing an anisotropic SiO_2 etch to clear any oxide from the field areas, the Si beams were released using an isotropic Si RIE step, as shown in Fig. 15.1f. After release, the protective SiO_2 film was removed by wet etching in HF (Fig. 15.1g). Using this process, the authors reported the successful fabrication of nanomechanical Si beams with micron-scale lengths ($\approx 8\,\mu$m) and submicron widths (330 nm) and heights (800 nm).

The advent of silicon-on-insulator (SOI) substrates with high quality, submicron-thick silicon top layers en-

ables the fabrication of nanomechanical Si beams with fewer processing steps than the aforementioned technique, since the buried oxide layer makes these device structures relatively easy to pattern and release. Additionally, the buried SiO_2 layer electrically isolates the beams from the substrate. *D.W. Carr* [15.60] detail a process that uses SOI substrates to fabricate submicron clamped-clamped mechanical beams and suspended plates with submicron tethers. The process, presented in Fig. 15.2, begins with the deposition of PMMA on an SOI substrate. The SOI substrate has a top Si layer that was either 50 nm or 200 nm in thickness. The PMMA is patterned into a metal lift-off mask by electron beam lithography (Fig. 15.2b). An Al film is then deposited and patterned by lift-off into a Si etch mask, as shown in Fig. 15.2c. The nanomechanical beams are then patterned by Si RIE and released by etching the underlying SiO_2 in a buffered hydrofluoric acid solution as shown in Fig. 15.2d and Fig. 15.2e, respectively. Using this process, nanomechanical beams that were 7 to 16 μm in length, 120 to 200 nm in width and 50 or 200 nm in thickness were successfully fabricated.

Fabrication of NEMS structures is not limited to Si. In fact, III–V compounds, such as gallium arsenide (GaAs), make particularly good NEMS materials from a fabrication perspective because thin epitaxial GaAs films can be grown on lattice-matched materials that can be used as sacrificial release layers. A collection of clamped-clamped nanomechanical GaAs beams fabricated on lattice-matched sacrificial layers having micron-scale lengths and submicron widths and thicknesses is shown in Fig. 15.3. *Tighe* [15.61] reported on the fabrication of GaAs plates suspended with nanomechanical tethers. The structures were made from single crystal GaAs films that were epitaxially grown on aluminum arsenide (AlAs) sacrificial layers. Ni etch masks were fabricated using electron beam lithography and lift-off as described previously. The GaAs films were patterned into beams using a chemically assisted ion beam etching process and released using a highly selective AlAs etchant. In a second example, *Tang* [15.62] has capitalized on the ability to grow high quality GaAs layers on ternary compounds such as $Al_xGa_{1-x}As$ to fabricate complex GaAs-based structures, such as submicron clamped-clamped beams from GaAs/AlGaAs quantum well heterostructures. As with the process described by *Tighe* [15.61], this process exploits a lattice matched sacrificial layer, in this case $Al_{0.8}Ga_{0.2}As$, which can be selectively etched to release the heterostructure layers.

Silicon carbide and diamond NEMS structures have been developed for applications requiring a material with a higher acoustic velocity and/or a higher degree of chemical inertness than Si. Silicon carbide nanomechanical resonators have been successfully fabricated from both epitaxial 3C-SiC films grown on Si substrates [15.63] and bulk 6H-SiC substrates [15.64]. In the case of the 3C-SiC devices, the ultra-thin epitaxial films were grown by atmospheric pressure chemical vapor deposition (APCVD) on (100) Si substrates. Nanomechanical beams were patterned using a metal RIE mask that was itself patterned by e-beam lithography. Reactive ion etching was performed using two NH_3-based plasma chemistries, with the first recipe performing an anisotropic SiC etch down to the Si substrate and the second performing an isotropic Si etch used to release the SiC beams. The two etches were performed sequentially, thereby eliminating a separate wet or dry release step. For the 6H-SiC structures, a suitable sacrificial layer was not available since the structures were fabricated directly on commercially available bulk wafers. To fabricate the structures, a metal etch mask was lithographically patterned by e-beam techniques on the 6H-SiC surface. The anisotropic SiC etch mentioned above was then performed, but with the substrate tilted roughly 45° with respect to the direction of the plasma using a special fixture to hold the wafer. A second such etch was performed on the substrate tilted back 90° with respect to the first etch, resulting in released beams with triangular cross sections.

Nanomechanical resonators have also been fashioned out of thin nanocrystalline diamond thin films [15.65]. In this case, the diamond films were deposited on SiO_2-coated Si substrates by microwave plasma chemical vapor deposition using CH_4 and H_2 as feedstock. The diamond films were patterned by RIE using metal masks patterned by e-beam lithography. The plasma chemistry in this case was based on CF_4 and O_2. The devices were then released in a buffered HF solution. It is noteworthy that the structures did not require a critical-point drying step after the wet chemical release, owing to the chemical inertness of the diamond surface.

NEMS structures are not restricted to those that can be made from patterned thin films using top-down techniques. In fact, carbon nanotubes (CNT) have been incorporated into NEMS devices using an approach that combines both bottom-up and top-down processing techniques. An example illustrating the promise and challenges of merging bottom-up with top-down

techniques is the CNT-based electrostatic rotational actuator developed by *Fennimore* et al. [15.66]. In this example, multiwalled CNT's (MWCNT) are grown using a conventional arc discharge process, which typically produces an assortment of CNTs. The CNT's are then transferred to a suitable SiO_2 coated Si in a 1,2 dichlorobenzene suspension. An AFM or SEM is then used to select a properly positioned CNT as determined by prefabricated alignment marks on the substrate. Conventional electron-beam lithography and lift-off techniques are then used to pattern an Au film into contact/anchor pads on the two ends of the CNT, a rotor pad at its center and two counter electrodes at 90° to the anchor pads. Anchoring is accomplished by sandwiching the CNT between the Au contact and the underlying SiO_2 film. The rotor is released by simply etching the sacrificial SiO_2 layer, taking care not to completely undercut the anchors yet allowing for adequate clearance for the rotor. Under proper conditions, the outer wall of the MWCNT's could be detached from the inner walls in order to allow for free rotation of the rotor plate.

15.2.2 Transduction Techniques

Several unique approaches have been developed to actuate and sense the motion of NEMS devices. Electrostatic actuation can be used to actuate beams [15.67], tethered meshes [15.68], and paddle oscillators [15.69]. *Sekaric* [15.70] has shown that low power lasers can be used to drive paddle oscillators into self-oscillation by induced thermal effects on the structures. In these examples, an optical detection scheme based on the modulation of incident laser light by a vibrating beam is used to detect the motion of the beams. *Cleland* [15.59] describe a magnetomotive transduction technique that capitalizes on a time-varying Lorentz force created by an alternating current in the presence of a strong magnetic field. In this case, the nanomechanical beam is positioned in the magnetic field so that an AC current passing through the beam is transverse to the field lines. The resulting Lorentz force causes the beam to oscillate, which creates an electromotive force along the beam that can be detected as a voltage. Thus, in this method, the excitation and detection are performed electrically. In all of the above-mentioned cases, the measurements were performed in vacuum, presumably to minimize the effects of squeeze-film damping as well as mass loading due to adsorbates from the environment.

15.2.3 Application Areas

For the most part, NEMS technology is still in the initial stage of development. Technological challenges related to fabrication and packaging will require innovative solutions before such devices make a significant commercial impact. Nevertheless, NEMS devices have already been used for precision measurements [15.71] enabling researchers to probe the properties of matter on a nanoscopic level [15.72, 73]. Sensor technologies based on NEMS structures, most notably for atto-gram scale mass detection [15.74, 75], atto-newton force detection [15.76], virus detection [15.77], and gaseous chemical detection [15.78] have emerged and will continue to mature. Without question, NEMS structures will prove to be useful platforms for a host of experiments and scientific discoveries in fields ranging from physics to biology, and with advancements in process integration and packaging, there is little doubt that NEMS technology will find its way into commercial micro/nanosystems as well.

15.3 Current Challenges and Future Trends

Although the field of MEMS has experienced significant growth over the past decade, many challenges still remain. In broad terms, these challenges can be grouped into three general categories; (1) fabrication challenges, (2) packaging challenges, and (3) application challenges. Challenges in these areas will, in large measure, determine the commercial success of a particular MEMS device both in technical and economic terms. The following presents a brief discussion of some of these challenges as well as possible approaches to address them.

In terms of fabrication, MEMS is currently dominated by planar processing techniques which find their roots in silicon IC fabrication. The planar approach and the strong dependence on silicon worked well in the early years, since many of the processing tools and methodologies commonplace in IC fabrication could be directly utilized in the fabrication of MEMS devices. This approach lends itself to the integration of MEMS with silicon ICs. Therefore, it still is popular for various applications. However, modular process integration of micromachining with standard IC fabrication is not straightforward and represents a great challenge in terms of processing material compatibility, thermal budget requirements, etc. Furthermore, planar processing places significant geometric restrictions on device designs, especially for complex mechanical components requiring high aspect ratio three-dimensional geometries, which are certain to increase as the application areas for MEMS continue to grow. Along the same lines, new applications will likely demand materials other than silicon, which may not be compatible with the conventional microfabrication approach, posing a significant challenge if integration with silicon microelectronics is required. Micro-assembly technique can become an attractive solution to alleviate these issues. Multifunctional microsystems can be implemented by assembling various MEMS devices and electronic building blocks fabricated through disparate processing technologies. Microsystems on a common substrate will likely become the ultimate solution. Development of sophisticated modeling programs for device design and performance will become increasingly important as fabrication processes and device designs become more complex. In terms of NEMS, the most significant challenge is likely the integration of nano- and microfabrication techniques into a unified process, since NEMS devices are likely to consist of both nano-scale and microscale structures. Integration will be particularly challenging for nano-scale devices fabricated using a "bottom-up" approach, since no analog is found in microfabrication. Nevertheless, hybrid systems consisting of nano-scale and microscale components will become increasingly common as the field continues to expand.

Fabrication issues notwithstanding, packaging is and will continue to be a significant challenge to the implementation of MEMS. MEMS is unlike IC packaging which benefits from a high degree of standardization. MEMS devices inherently require interaction with the environment, and since each application has in some way a unique environment, standardization of packaging becomes extremely difficult. This lack of standardization tends to drive up the costs associated with packaging, making MEMS less competitive with alternative approaches. In addition, packaging tends to negate the effects of miniaturization based upon microfabrication, especially for MEMS devices requiring protection from certain environmental conditions. Moreover, packaging can cause performance degradation of MEMS devices, especially in situations where the environment exerts mechanical stresses on the package, which in turn results in a long-term device performance drift. To address many of these issues, wafer level packaging schemes that are customized to the device of interest will likely become more common. In essence, packaging of MEMS will move away from the conventional IC methods that utilize independently manufactured packages toward custom packages, which are created specifically for the device as a part of the batch fabrication process.

Without question, the increasing advancement of MEMS will open many new potential application areas to the technology. In most cases, MEMS will be one of several alternatives available for implementation. For cost sensitive applications, the trade off between technical capabilities and cost will challenge those who desire to commercialize the technology. The biggest challenge to the field will be to identify application areas that are well suited for MEMS/NEMS technology and have no serious challengers. As MEMS technology moves away from component level and more towards microsystems solutions, it is likely that such application areas will come to the fore.

References

15.1   M. Mehregany, S. F. Bart, L. S. Tavrow, J. H. Lang, S. D. Senturia: Principles in design and microfabrication of variable-capacitance side-drive motors, J. Vacuum. Sci. Technol. A. **8**, 3614–3624 (1990)

15.2   Y.-C. Tai, R. S. Muller: IC-processed electrostatic synchronous micromotors, Sensors Actuat. **20**, 49–55 (1989)

15.3   J. J. Sniegowski, S. L. Miller, G. F. LaVigne, M. S. Roders, P. J. McWhorter: Monolithic geared-mechanisms driven by a polysilicon surface-micromachined on-chip electrostatic microengine, IEEE Solid-St. Sens. Actuat. Workshop, 178–182 (1996)

15.4   G. T. A. Kovacs: *Micromachined Transducer Sourcebook* (McGraw Hill, Boston 1998)

15.5   S. D. Senturia: *Microsystem Design* (Kluwer, Dordrecht 1998)

15.6   J. E. Gragg, W. E. McCulley, W. B. Newton, C. E. Derrington: Compensation and calibration of a monolithic four terminal silicon pressure transducer, IEEE Solid-St. Sens. Actuat. Workshop, 21–27 (1984)

15.7   Y. Wang, M. Esashi: A novel electrostatic servo capacitive vacuum sensor, IEEE Int. Conf. Solid-St. Sens. Actuat., 1457–1460 (1997)

15.8   W. H. Ko, Q. Wang: Touch mode capacitive pressure densors, Sensors Actuat. **75**, 242–251 (1999)

15.9   D. J. Young, J. Du, C. A. Zorman, W. H. Ko: High-temperature single-crystal 3C-SiC capacitive pressure sensor, IEEE Sensor J. **4**, 464–470 (2004)

15.10   J. M. Bustillo, R. T. Howe, R. S. Muller: Surface micromachining for microelectromechanical systems, Proc. IEEE **86**(8), 1552–1574 (1998)

15.11   J. H. Smith, S. Montague, J. J. Sniegowski, J. R. Murray, P. J. McWhorter: Embedded micromechanical devices for the monolithic integration of MEMS with CMOS, IEEE Int. Electron Dev. Meeting, 609–612 (1993)

15.12   T. A. Core, W. K. Tsang, S. J. Sherman: Fabrication technology for an integrated surface-micromachined sensor, Solid State Technol. **36**(10), 39–40, 42, 46–47 (1993)

15.13   H. Kapels, R. Aigner, C. Kolle: Monolithic surface-micromachined sensor system for high pressure applications, Int. Conf. Solid-St. Sens. Actuat., 56–59 (2001)

15.14   C. Lu, M. Lemkin, B. E. Boser: A monolithic surface micromachined accelerometer with digital output, ISSCC, 160–161 (1995)

15.15   N. Yazdi, K. Najafi: An all-silicon single-wafer fabrication technology for precision microaccelerometers, IEEE Int. Conf. Solid-St. Sens. Actuat., 1181–1184 (1997)

15.16   T. B. Gabrielson: Mechanical-thermal noise in micromachined acoustic and vibration sensors, IEEE Trans. Electron Dev. **40**(5), 903–909 (1993)

15.17   J. Chae, H. Kulah, K. Najafi: A monolithic three-axis micro-g micromachined silicon capacitive accelerometer, IEEE J. Solid-St. Circ. **14**, 235–242 (2005)

15.18   M. Lemkin, M. A. Ortiz, N. Wongkomet, B. E. Boser, J. H. Smith: A 3-axis surface micromachined $\Sigma\Delta$ accelerometer, ISSCC, 202–203 (1997)

15.19   J. Wu, G. K. Fedder, L. R. Carley: A low-noise low-offset capacitive sensing amplifier for a $50\mu g/\sqrt{Hz}$ monolithic CMOS MEMS accelerometer, IEEE J. Solid-St. Circ. **39**, 722–730 (2004)

15.20   W. A. Clark, R. T. Howe: Surface micromachined Z-axis vibratory rate gyroscope, IEEE Solid-St. Sens. Actuat. Workshop, 283–287 (1996)

15.21   J. A. Geen, S. J. Sherman, J. F. Chang, S. R. Lewis: Single-chip surface micromachined integrated gyroscope with 50°/h Allan deviation, IEEE J. Solid-St. Circ. **37**, 1860–1866 (2002)

15.22 H. Xie, G. K. Fedder: Fabrication, characterization, and analysis of a DRIE CMOS-MEMS gyroscope, IEEE Sensors J. **3**, 622–631 (2003)

15.23 T. Juneau, A. P. Pisano: Micromachined dual input axis angular rate sensor, IEEE Solid-St. Sens. Actuat. Workshop, 299–302 (1996)

15.24 R. S. Muller, K. Y. Lau: Surface-micromachined microoptical elements and systems, Proc. IEEE **86**(8), 1705–1720 (1998)

15.25 L. J. Hornbeck: Current status of the digital micromirror device (DMD) for projection television applications, IEEE Int. Electron Dev. Meeting, 381–384 (1993)

15.26 P. F. Van Kessel, L. J. Hornbeck, R. E. Meier, M. R. Douglass: A MEMS-based projection display, Proc. IEEE **86**(8), 1687–1704 (1998)

15.27 M. S. Cohen, M. F. Cina, E. Bassous, M. M. Opyrsko, J. L. Speidell, F. J. Canora, M. J. DeFranza: Packaging of high density fiber/laser modules using passive alignment techniques, IEEE Trans. Comp. Hybrids Manufact. Technol **15**, 944–954 (1992)

15.28 M. J. Wale, C. Edge: Self-aligned flip-chip assembly of photonic devices with electrical and optical connections, IEEE Trans. Comp. Hybrids Manufact. Technol **13**, 780–786 (1990)

15.29 M. J. Daneman, N. C. Tien, O. Solgaard, K. Y. Lau, R. S. Muller: Linear vibromotor-actuated micromachined microreflector for integrated optical systems, IEEE Solid-St. Sens. Actuat. Workshop, 109–112 (1996)

15.30 K. S. J. Pister, M. W. Judy, S. R. Burgett, R. S. Fearing: Microfabricated hinges, Sensors Actuat. **33**(3), 249–256 (1992)

15.31 O. Solgaard, M. Daneman, N. C. Tien, A. Friedberger, R. S. Muller, K. Y. Lau: Optoelectronic packaging using silicon surface-micromachined alignment mirrors, IEEE Photon. Technol. Lett. **7**(1), 41–43 (1995)

15.32 S. S. Lee, L. S. Huang, C. J. Kim, M. C. Wu: 2 × 2 MEMS fiber optic switches with silicon sub-mount for low-cost packaging, IEEE Solid-St. Sens. Actuat. Workshop, 281–284 (1998)

15.33 T. Akiyama, H. Fujita: A quantitative analysis of scratch drive actuator using Buckling motion, Technical Digest, 8th IEEE International MEMS Workshop, 310–315 (1995)

15.34 V. A. Aksyuk, F. Pardo, D. J. Bishop: Stress-induced curvature engineering in surface-micromachined devices, SPIE **3680**, 984 (1999)

15.35 D. J. Young, B. E. Boser: A micromachined variable capacitor for monolithic low-noise VCOs, IEEE Solid-St. Sens. Actuat. Workshop, 86–89 (1996)

15.36 A. Dec, K. Suyama: Micromachined electromechanically tunable capacitors and their applications to RF IC's, IEEE Trans. Microw. Theory. Techniques **46**, 2587–2596 (1998)

15.37 Z. Li, N. C. Tien: A high tuning-ratio silicon-micromachined variable capacitor with low driving voltage, IEEE Solid-St. Sens. Actuat. Workshop, 239–242 (2002)

15.38 Z. Xiao, W. Peng, R. F. Wolffenbuttel, K. R. Farmer: Micromachined variable capacitor with wide tuning range, IEEE Solid-St. Sens.d Actuat. Workshop, 346–349 (2002)

15.39 J. J. Yao, S. T. Park, J. DeNatale: High tuning-ratio MEMS-based tunable capacitors for RF communications applications, IEEE Solid-St. Sens. Actuat. Workshop, 124–127 (1998)

15.40 J. B. Yoon, C. T.-C. Nguyen: A high-Q tunable micromechanical capacitor with movable dielectric for RF applications, IEEE Int. Electron Dev. Meeting, 489–492 (2000)

15.41 D. J. Young, V. Malba, J. J. Ou, A. F. Bernhardt, B. E. Boser: Monolithic high-performance three-dimensional coil inductors for wireless communication applications, IEEE Int. Electron Dev. Meeting, 67–70 (1997)

15.42 D. J. Young, B. E. Boser, V. Malba, A. F. Bernhardt: A micromachined RF low phase noise voltage-controlled oscillator for wireless communication, Int. J. RF Microw. Comput.-Aid. Eng. **11**(5), 285–300 (2001)

15.43 C. L. Chua, D. K. Fork, K. V. Schuylenbergh, J. P. Lu: Self-Assembled Out-Of-Plane High Q Inductors, IEEE Solid-St. Sens. Actuat. Workshop, 372–373 (2002)

15.44 J. B. Yoon, C. H. Han, E. Yoon, K. Lee, C. K. Kim: Monolithic high-Q overhang inductors fabricated on silicon and glass substrates, IEEE Int. Electron Dev. Meeting, 753–756 (1999)

15.45 J. B. Yoon, Y. Choi, B. Kim, Y. Eo, E. Yoon: CMOS-compatible surface-micromachined suspended-spiral inductors for multi-GHz silicon RF ICs, IEEE Electron Dev. Lett. **23**, 591–593 (2002)

15.46 C. L. Goldsmith, Z. Yao, S. Eshelman, D. Denniston: Performance of low-loss RF MEMS capacitive switches, IEEE Microw. Guided Wave Lett. **8**(8), 269–271 (1998)

15.47 D. Hyman, J. Lam, B. Warneke, A. Schmitz, T. Y. Hsu, J. Brown, J. Schaffner, A. Walston, R. Y. Loo, M. Mehregany, J. Lee: Surface-micromachined RF MEMs switches on GaAs substrates, Int. J. RF Microw. Comput.-Aid. Eng. **9**(4), 348–361 (1999)

15.48 J. J. Yao, M. F. Chang: A surface micromachined miniature switch for telecommunication applications with signal frequencies from DC up to 40 GHz, 8th Int. Conf. Solid-St. Sens. Actuat., 384–387 (1995)

15.49 P. M. Zavracky, N. E. McGruer, R. H. Morriosn, D. Potter: Microswitches and microrelays with a view toward microwave applications, Int. J. RF Microw. Comput.-Aid. Eng. **9**(4), 338–347 (1999)

15.50 J. Wang, Z. Ren, C. T. C. Nguyen: 1.156-GHz self-aligned vibrating micromechanical disk resonator, IEEE Trans. Ultrason. Ferr. Freq. Control **51**, 1607–1628 (2004)

Part B | 15

15.51 Y. Wang, Z. Li, D.T. McCormick, N.C. Tien: A low-voltage lateral MEMS switch with high RF performance, J. Microelectromech. Syst. **13**, 902–911 (2004)

15.52 C.T.C. Nguyen, R.T. Howe: CMOS microelectromechanical resonator oscillator, IEEE Int. Electron Dev. Meeting, 199–202 (1993)

15.53 F.D. Bannon III, J.R. Clark, C.T.C. Nguyen: High frequency micromechanical filter, IEEE J. Solid-St. Circ. **35**(4), 512–526 (2000)

15.54 L. Lin, R.T. Howe, A.P. Pisano: Microelectromechanical Filters for Signal Processing, IEEE J. Microelectromech. Syst. **7**(3), 286–294 (1998)

15.55 K. Wang, Y. Yu, A.C. Wong, C.T.C. Nguyen: VHF free-free beam high-Q micromechanical resonators, The 12th IEEE Int. Conf. Micro Electro Mechanical Systems, 453–458 (1999)

15.56 J.R. Clark, W.T. Hsu, C.T.C. Nguyen: High-Q VHF micromechanical contour-mode disk resonators, IEEE Int. Electron Dev. Meeting, 493–496 (2000)

15.57 C.T.C. Nguyen, R.T. Howe: Quality factor control for micromechanical resonator, IEEE Int. Electron Dev. Meeting, 505–508 (1992)

15.58 M.L. Roukes: Plenty of room, indeed, Sci. Am. **285**, 48–57 (2001)

15.59 A.N. Cleland, M.L. Roukes: Fabrication of high frequency nanometer scale mechanical resonators from bulk Si crystals, Appl. Phys. Lett. **69**, 2653–2655 (1996)

15.60 D.W. Carr, H.G. Craighead: Fabrication of nanoelectromechanical systems in single crystal silicon using silicon on insulator substrates and electron beam lithography, J. Vacuum. Sci. Technol. B **15**, 2760–2763 (1997)

15.61 T.S. Tighe, J.M. Worlock, M.L. Roukes: Direct thermal conductance measurements on suspended monocrystalline nanostructures, Appl. Phys. Lett. **70**, 2687–2689 (1997)

15.62 H.X. Tang, X.M.H. Huang, M.L. Roukes, M. Bichler, W. Wegsheider: Two-dimensional electron-gas actuation and transduction for GaAs nanoelectromechanical systems, Appl. Phys. Lett. **81**, 3879–3881 (2002)

15.63 Y.T. Yang, K.L. Ekinci, X.M.H. Huang, L.M. Schiavone, M.L. Roukes, C.A. Zorman, M. Mehregany: Monocrystalline silicon carbide nanoelectromechanical systems, Appl. Phys. Lett. **78**, 162–164 (2001)

15.64 X.M.H. Huang, X.L. Feng, M.K. Prakash, S. Kumar, C.A. Zorman, M. Mehregany, M.L. Roukes: Fabrication of suspended nanomechanical structures from bulk 6H-SiC substrates, Mater. Sci. Forum **457–460**, 1531–1534 (2004)

15.65 L. Sekaric, M. Zalalutdinov, S.W. Turner, A.T. Zehnder, J.M. Parpia, H.G. Craighead: Nanomechanical resonant structures as tunable passive modulators, Appl. Phys. Lett. **80**, 3617–3619 (2002)

15.66 A.M. Fennimore, T.D. Yuzvinsky, W.Q. Han, M.S. Fuhrer, J. Cummings, A. Zettl: Rotational actuators based on carbon nanotubes, Nature **424**, 408–410 (2003)

15.67 D.W. Carr, S. Evoy, L. Sekaric, H.G. Craighead, J.M. Parpia: Measurement of mechanical resonance and losses in nanometer scale silicon wires, Appl. Phys. Lett. **75**, 920–922 (1999)

15.68 D.W. Carr, L. Sekaric, H.G. Craighead: Measurement of nanomechanical resonant structures in single-crystal silicon, J. Vacuum. Sci. Technol. B **16**, 3821–3824 (1998)

15.69 S. Evoy, D.W. Carr, L. Sekaric, A. Olkhovets, J.M. Parpia, H.G. Craighead: Nanofabrication and electrostatic operation of single-crystal silicon paddle oscillators, J. Appl. Phys. **86**, 6072–6077 (1999)

15.70 L. Sekaric, J.M. Parpia, H.G. Craighead, T. Feygelson, B.H. Houston, J.E. Butler: Nanomechanical resonant structures in nanocrystalline diamond, Appl. Phys. Lett. **81**, 4455–4457 (2002)

15.71 A.N. Cleland, M.L. Roukes: A nanometre-scale mechanical electrometer, Nature **392**, 160–162 (1998)

15.72 K. Schwab, E.A. Henriksen, J.M. Worlock, M.L. Roukes: Measurement of the quantum of thermal conductance, Nature **404**, 974–977 (2000)

15.73 S. Evoy, A. Olkhovets, L. Sekaric, J.M. Parpia, H.G. Craighead, D.W. Carr: Temperature-dependent internal friction in silicon nanoelectromechanical systems, Appl. Phys. Lett. **77**, 2397–2399 (2000)

15.74 K.L. Ekinci, X.M.H. Huang, M.L. Roukes: Ultrasensitive nanoelectromechanical mass detection, Appl. Phys. Lett. **84**, 4469–4471 (2004)

15.75 B. Illic, H.G. Craighead, S. Krylov, W. Senaratne, C. Ober, P. Neuzil: Attogram detection using nanoelectromechanical oscillators, J. Appl. Phys. **95**, 3694–3703 (2004)

15.76 T.D. Stowe, K. Yasumura, T.W. Kenny, D. Botkin, K. Wago, D. Rugar: Attonewton force detection using ultrathin silicon cantilevers, Appl. Phys. Lett. **71**, 288–290 (1997)

15.77 B. Illic, Y. Yang, H.G. Craighead: Virus detection using nanoelectromechanical devices, Appl. Phys. Lett. **85**, 2604–2606 (2004)

15.78 H. Liu, J. Kameoka, D.A. Czaplewski, H.G. Craighead: Polymeric nanowire chemical sensor, Nano Lett. **4**, 617–675 (2004)

16. Nanomechanical Cantilever Array Sensors

Microfabricated cantilever sensors have attracted much interest in recent years as devices for the fast and reliable detection of small concentrations of molecules in air and solution. In addition to application of such sensors for gas and chemical-vapor sensing, for example as an artificial nose, they have also been employed to measure physical properties of tiny amounts of materials in miniaturized versions of conventional standard techniques such as calorimetry, thermogravimetry, weighing, photothermal spectroscopy, as well as for monitoring chemical reactions such as catalysis on small surfaces. In the past few years, the cantilever-sensor concept has been extended to biochemical applications and as an analytical device for measurements of biomaterials. Because of the label-free detection principle of cantilever sensors, their small size and scalability, this kind of device is advantageous for diagnostic applications and disease monitoring, as well as for genomics or proteomics purposes. The use of microcantilever arrays enables detection of several analytes simultaneously and solves the inherent problem of thermal drift often present when using single microcantilever sensors, as some of the cantilevers can be used as sensor cantilevers for detection, and other cantilevers serve as passivated reference cantilevers that do not exhibit affinity to the molecules to be detected.

Part B | 16

16.1 Technique

Sensors are devices that detect, or sense, a signal. Moreover, a sensor is also a transducer, i.e. it transforms one form of energy into another or responds to a physical parameter. Most people will associate sensors with electrical or electronic devices that produce a change in response when an external physical parameter is changed. However, many more types of transducers exist, such as electrochemical (pH probe), electromechanical (piezoelectric actuator, quartz, strain gauge), electroacoustic (gramophone pick-up, microphone), photoelectric (photodiode, solar cell), electromagnetic (antenna), magnetic (Hall-effect sensor, tape or hard-disk head for storage applications), electrostatic (electrometer), thermoelectric (thermocouple, thermoresistors), and electrical (capacitor, resistor). Here we want to concentrate on a further type of sensor not yet mentioned: the mechanical sensor. It responds to changes of an external parameter, such as temperature changes or molecule adsorption, by a mechanical response, e.g. by bending or deflection.

16.1.1 Cantilevers

Mechanical sensors consist of a fixed and a movable part. The movable part can be a thin membrane, a plate or a beam, fixed at one or both ends. The structures described here are called cantilevers. A cantilever is regarded here as a microfabricated rectangular bar-shaped structure that is longer than it is wide and has a thickness that is much smaller than its length or width. It is a horizontal structural element supported only at one end on a chip body; the other end is free (Fig. 16.1). Most often it is used as a mechanical probe to image the topography of a sample using a technique called atomic force microscopy (AFM) or scanning force microscopy (SFM) [16.1], invented by *Binnig, Quate* and *Gerber* in the mid 1980s [16.1]. For AFM a microfabricated sharp tip is attached to the apex of the cantilever and serves as a local probe to scan the sample surface. The distance between tip and surface is controlled via sensitive measurement of interatomic forces in the piconewton range.

By scanning the tip across a conductive or non-conductive surface using an *x-y-z* actuator system (e.g. a piezoelectric scanner), an image of the topography is obtained by recording the correction signal that has to be applied to the *z*-actuation drive to keep the interaction between tip and sample surface constant. SFM methods are nowadays well established in scientific research, education and, to a certain extent, also in industry. Beyond imaging of surfaces, cantilevers have been used for many other purposes. However, here we focus on their application as sensor devices.

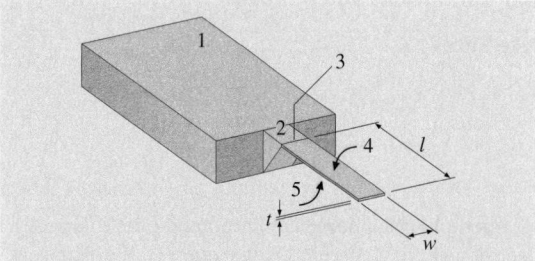

Fig. 16.1 Schematic of a cantilever: (1) rigid chip body, (2) solid cantilever-support structure, (3) hinge of cantilever, (4) upper surface of the cantilever, which is usually functionalized with a sensor layer for detection of molecules, (5) lower surface of the cantilever, usually passivated in order not to show affinity to the molecules to be detected. The geometrical dimensions, length *l*, width *w* and thickness *t*, are indicated

16.1.2 History of Cantilever Sensors

The idea of using beams of silicon as sensors to measure deflections or changes in resonance frequency is actually quite old. First reports go back to 1968, when *Wilfinger* et al. [16.2] investigated silicon cantilever structures of $50\,\text{mm} \times 30\,\text{mm} \times 8\,\text{mm}$, i. e. quite large structures, for detecting resonances. On the one hand, they used localized thermal expansion in diffused resistors (piezoresistors) located near the cantilever support to create a temperature gradient for actuating the cantilever at its resonance frequency. On the other hand, the piezoresistors could also be used to sense mechanical deflection of the cantilever. This early report already contains concepts for sensing and actuation of cantilevers. In the following years only a few reports are available on the use of cantilevers as sensors, e.g. *Heng* [16.3], who fabricated gold cantilevers capacitively coupled to microstrip lines in 1971 to mechanically trim high-frequency oscillator circuits. In 1979, *Petersen* [16.4] constructed cantilever-type micromechanical membrane switches in silicon that should have filled the gap between silicon transistors and mechanical electromagnetic relays. *Kolesar* [16.5] suggested the use of cantilever structures as electronic nerve-agent detectors in 1985.

Only with the availability of microfabricated cantilevers for AFM [16.1] did reports on the use of cantilevers as sensors become more frequent. In 1994, *Itoh* et al. [16.6] presented a cantilever coated with a thin film of zinc oxide and proposed piezoresistive deflection readout as an alternative to optical beam-deflection readout. *Cleveland* et al. [16.7] reported the tracking of cantilever resonance frequency to detect nanogram changes in mass loading when small particles are deposited onto AFM probe tips. *Thundat* et al. [16.8] showed that the resonance frequency as well as static bending of microcantilevers are influenced by ambient conditions, such as moisture adsorption, and that deflection of metal-coated cantilevers can be further influenced by thermal effects (bimetallic effect). The first chemical sensing applications were presented by *Gimzewski* et al. [16.9], who used static cantilever bending to detect chemical reactions with very high sensitivity. Later *Thundat* et al. [16.10] observed changes in the resonance frequency of microcantilevers due to adsorption of analyte vapor on exposed surfaces. Frequency changes have been found to be caused by mass loading or adsorption-induced changes in the cantilever spring constant. By coating cantilever surfaces with hygroscopic materials, such as phosphoric acid or gelatin, the cantilever can sense water vapor with picogram mass resolution.

The deflection of individual cantilevers can easily be determined using AFM-like optical beam-deflection electronics. However, single cantilever responses can be prone to artifacts such as thermal drift or unspecific ad-sorption. For this reason the use of passivated reference cantilevers is desirable. The first use of cantilever arrays with sensor and reference cantilevers was reported in 1998 [16.11], and represented significant progress for the understanding of true (difference) cantilever responses.

16.2 Cantilever Array Sensors

16.2.1 Concept

For the use of a cantilever as a sensor, neither a sharp tip at the cantilever apex nor a sample surface is required. The cantilever surfaces serve as sensor surfaces and al-low the processes taking place on the surface of the beam to be monitored with unprecedented accuracy, in particular the adsorption of molecules. The formation of molecule layers on the cantilever surface will generate surface stress, eventually resulting in a bending of the cantilever, provided the adsorption preferentially occurs on one surface of the cantilever. Adsorption is controlled by coating one surface (typically the upper surface) of a cantilever with a thin layer of a material that exhibits affinity to molecules in the environment (sensor surface). This surface of the cantilever is referred to as the func-tionalized surface. The other surface of the cantilever (typically the lower surface) may be left uncoated or be coated with a passivation layer, i. e. a chemical surface that does not exhibit significant affinity to the molecules in the environment to be detected. To enable function-alized surfaces to be established, often a metal layer is evaporated onto the surface designed as sensor surface. Metal surfaces, e.g. gold, may be used to covalently bind a monolayer that represents the chemical surface sensi-tive to the molecules to be detected from environment. Frequently, a monolayer of thiol molecules covalently bound to a gold surface is used. The gold layer is also favorable for use as a reflection layer if the bending of the cantilever is read out via an optical beam-deflection method.

16.2.2 Compressive and Tensile Stress

Given a cantilever coated with gold on its upper surface for adsorption of alkanethiol molecules and left uncoated on its lower surface (consisting of silicon and silicon ox-ide), the adsorption of thiol molecules will take place on the upper surface of the cantilever, resulting in a down-ward bending of the cantilever due to the formation of surface stress. We will call this process development of compressive surface stress, because the forming self-assembled monolayer produces a downward bending of the cantilever (away from the gold coating). In the op-posite situation, i. e. when the cantilever bends upwards, we would speak of tensile stress. If both the upper and lower surfaces of the cantilevers are involved in the re-action, then the situation will be much more complex, as a predominant compressive stress formation on the lower cantilever surface might appear like tensile stress on the upper surface. For this reason, it is of utmost im-portance that the lower cantilever surface is passivated in order that ideally no processes take place on the lower surface of the cantilever.

16.2.3 Disadvantages of Single Microcantilevers

Single microcantilevers are susceptible to parasitic de-flections that may be caused by thermal drift or chemical interaction of a cantilever with its environment, in particular if the cantilever is operated in a liquid. Of-ten, a baseline drift is observed during static-mode measurements. Moreover, nonspecific physisorption of molecules on the cantilever surface or nonspecific bind-ing to receptor molecules during measurements may contribute to the drift.

16.2.4 Reference and Sensor Cantilevers in an Array

To exclude such influences, simultaneous measurement of reference cantilevers aligned in the same array as the sensing cantilevers is crucial [16.11]. As the differ-ence in signals from the reference and sensor cantilevers shows the net cantilever response, even small sensor responses can be extracted from large cantilever de-flections without being dominated by undesired effects. When only single microcantilevers are used, no thermal-drift compensation is possible. To obtain useful data under these circumstances, both microcantilever sur-faces have to be chemically well defined. One of the

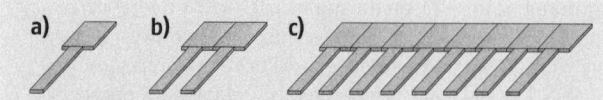

Fig. 16.2 (a) Single cantilever; **(b)** a pair of cantilevers, one to be used as a sensor cantilever, the other as a reference cantilever, and **(c)** an array of cantilevers with several sensor and reference cantilevers

surfaces, typically the lower one, has to be passivated; otherwise the cantilever response will be convoluted with undesired effects originating from uncontrolled reactions taking place on the lower surface (Fig. 16.2a). With a pair of cantilevers, reliable measurements are obtained. One cantilever is used as the sensor cantilever (typically coated on the upper side with a molecule layer exhibiting affinity to the molecules to be detected), whereas the other cantilever serves as the reference cantilever. It should be coated with a passivation layer on the upper surface so as not to exhibit affinity to the molecules to be detected. Thermal drifts are canceled out if difference responses, i.e. difference in deflections of sensor and reference cantilevers, are taken. Alternatively, both cantilevers are used as sensor cantilevers (sensor layer on the upper surfaces), and the lower surface has to be passivated (Fig. 16.2b). It is best to use a cantilever array (Fig. 16.2c), in which several cantilevers are used either as sensor or as reference cantilevers so that multiple difference signals can be evaluated simultaneously. Thermal drift is canceled out as one surface of all cantilevers, typically the lower one, is left uncoated or coated with the same passivation layer.

16.3 Modes of Operation

In analogy to AFM, various operating modes for cantilevers are described in the literature. The measurement of static deflection upon the formation of surface stress during adsorption of a molecular layer is termed the *static mode*. *Ibach* used cantilever-like structures to study adsorbate-induced surface stress [16.12] in 1994. Surface-stress-induced bending of cantilevers during the adsorption of alkanethiols on gold was reported by *Berger* et al. in 1997 [16.13]. The mode corresponding to noncontact AFM, termed the *dynamic mode*, in which a cantilever is oscillated at its resonance frequency, was described by *Cleveland* et al. [16.7]. They calculated mass changes from shifts in the cantilever resonance frequency upon the mounting of tiny tungsten particle spheres at the apex of the cantilever. The so-called *heat mode* was pioneered by *Gimzewski* et al. [16.9], who took advantage of the bimetallic effect that produces a bending of a metal-coated cantilever when heat is produced on its surface. Therewith they constructed a miniaturized calorimeter with picojoule sensitivity. Further operating modes exploit other physical effects such as the production of heat from the absorption of light by materials deposited on the cantilever (photothermal spectroscopy) [16.14], or cantilever bending caused by electric or magnetic forces.

16.3.1 Static Mode

The continuous bending of a cantilever with increasing coverage by molecules is referred to as operation in the static mode, see Fig. 16.3a. Adsorption of molecules onto the functional layer produces stress at the interface between the functional layer and the molecular layer forming. Because the forces within the functional layer try to keep the distance between molecules constant, the cantilever beam responds by bending because of its extreme flexibility. This property is described by the spring constant k of the cantilever, which for a rectangular microcantilever of length l, thickness t and width w is calculated as

$$k = \frac{Ewt^3}{4l^3} \, , \tag{16.1}$$

where E is the Young's modulus [$E_{Si} = 1.3 \times 10^{11} \, \text{N/m}^2$ for Si(100)].

As a response to surface stress, e.g. owing to adsorption of a molecular layer, the microcantilever bends, and its shape can be approximated as part of a circle with radius R. This radius of curvature is given by [16.15, 16]

$$\frac{1}{R} = \frac{6(1-\nu)}{Et^2} \, . \tag{16.2}$$

The resulting surface stress change is described using Stoney's formula [16.15]

$$\Delta\sigma = \frac{Et^2}{6R(1-\nu)} \, , \tag{16.3}$$

where E is Young's modulus, t the thickness of the cantilever, ν the Poisson's ratio ($\nu_{Si} = 0.24$), and R the bending radius of the cantilever.

Static-mode operation has been reported in various environments. In its simplest configuration, molecules

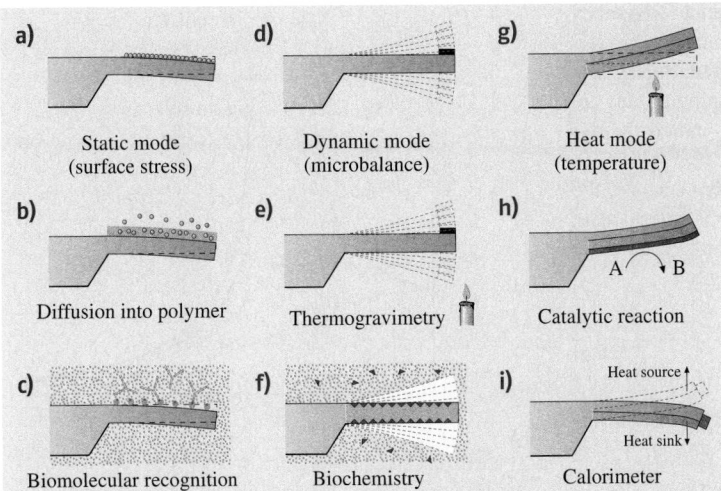

Fig. 16.3a–i Basic cantilever operation modes: (**a**) static bending of a cantilever on adsorption of a molecular layer. (**b**) Diffusion of molecules into a polymer layer leads to swelling of the polymer and eventually to a bending of the cantilever. (**c**) Highly specific molecular recognition of biomolecules by receptors changes the surface stress on the upper surface of the cantilever and results in bending. (**d**) Oscillation of a cantilever at its resonance frequency (dynamic mode) allows information on mass changes taking place on the cantilever surface to be obtained (application as a microbalance). (**e**) Changing the temperature while a sample is attached to the apex of the cantilever allows information to be gathered on decomposition or oxidation process. (**f**) Dynamic-mode measurements in liquids yield details on mass changes during biochemical processes. (**g**) In the heat mode, a bimetallic cantilever is employed. Here bending is due to the difference in the thermal expansion coefficients of the two materials. (**h**) A bimetallic cantilever with a catalytically active surface bends due to heat production during a catalytic reaction. (**i**) A tiny sample attached to the apex of the cantilever is investigated, taking advantage of the bimetallic effect. Tracking the deflection as a function of temperature allows the observation of phase transitions in the sample in a calorimeter mode

from the gaseous environment adsorb on the functionalized sensing surface and form a molecular layer (Fig. 16.3a), provided the molecules exhibit some affinity to the surface. In the case of alkanethiol covalently binding to gold, the affinity is very high, resulting in a fast bending response within minutes [16.13]. Polymer sensing layers only exhibit a partial sensitivity, i. e. polymer-coated cantilevers always respond to the presence of volatile molecules, but the magnitude and temporal behavior are specific to the chemistry of the polymer. Molecules from the environment diffuse into the polymer layer at different rates, mainly depending on the size and solubility of the molecules in the polymer layer (Fig. 16.3b). A wide range of hydrophilic/hydrophobic polymers can be selected, differing in their affinity to polar/unpolar molecules. Thus, the polymers can be chosen according to what an application requires.

Static-mode operation in liquids, however, usually requires rather specific sensing layers, based on molecular recognition, such as DNA hybridization [16.17] or antigen–antibody recognition (Fig. 16.3c). Cantilevers functionalized by coating with biochemical sensing layers respond very specifically using biomolecular key–lock principles of molecular recognition. However, whether molecular recognition will actually lead to a bending of the cantilever depends on the efficiency of transduction, because the surface stress has to be generated very close to the cantilever surface to produce bending. By just scaling down standard gene-chip strategies to cantilever geometry utilizing long spacer molecules so that DNA molecules become more accessible for hybridization, the hybridization takes place at a distance of several nanometers from the cantilever surface. In such experiments, no cantilever bending was observed [16.18].

16.3.2 Dynamic Mode

Mass changes can be determined accurately by using a cantilever actuated at its eigenfrequency. The eigenfrequency is equal to the resonance frequency of an oscillating cantilever if the elastic properties of the cantilever remain unchanged during the molecule-adsorption process and if damping effects are insignificant. This mode of operation is called the dynamic mode (e.g., the use as a microbalance, Fig. 16.3d). Owing to mass addition on the cantilever surface, the cantilever's eigenfrequency will shift to a lower value. The frequency change per mass change on a rectangular cantilever is calculated [16.19] according to

$$\Delta f / \Delta m = \frac{1}{4\pi n_l l^3 w} \times \sqrt{\frac{E}{\rho^3}} \,, \tag{16.4}$$

where $\rho = m/lwt$ is the mass density of the microcantilever and the deposited mass, and $n_l \approx 1$ is a geometrical factor.

The mass change is calculated [16.8] from the frequency shift using

$$\Delta m = \frac{k}{4\pi^2} \times \left(\frac{1}{f_1^2} - \frac{1}{f_0^2} \right) \,, \tag{16.5}$$

where f_0 is the eigenfrequency before the mass change occurs, and f_1 the eigenfrequency after the mass change.

Mass-change determination can be combined with varying environment temperature conditions (Fig. 16.3e) to obtain a method introduced in the literature as *micromechanical thermogravimetry* [16.20]. A tiny piece of sample to be investigated has to be mounted at the apex of the cantilever. Its mass should not exceed several hundred nanograms. Adsorption, desorption and decomposition processes, occurring while changing the temperature, produce mass changes in the picogram range that can be observed in real time by tracking the resonance-frequency shift.

Dynamic-mode operation in a liquid environment is more difficult than in air, because of the large damping of the cantilever oscillation due to the high viscosity of the surrounding media (Fig. 16.3f). This results in a low quality factor Q of the oscillation, and thus the resonance frequency shift is difficult to track with high resolution. The quality factor is defined as

$$Q = 2\Delta f / f_0 \,. \tag{16.6}$$

Whereas in air the resonance frequency can easily be determined with a resolution of below 1 Hz, only a frequency resolution of about 20 Hz is expected for measurements in a liquid environment.

The damping or altered elastic properties of the cantilever during the experiment, e.g. by a stiffening

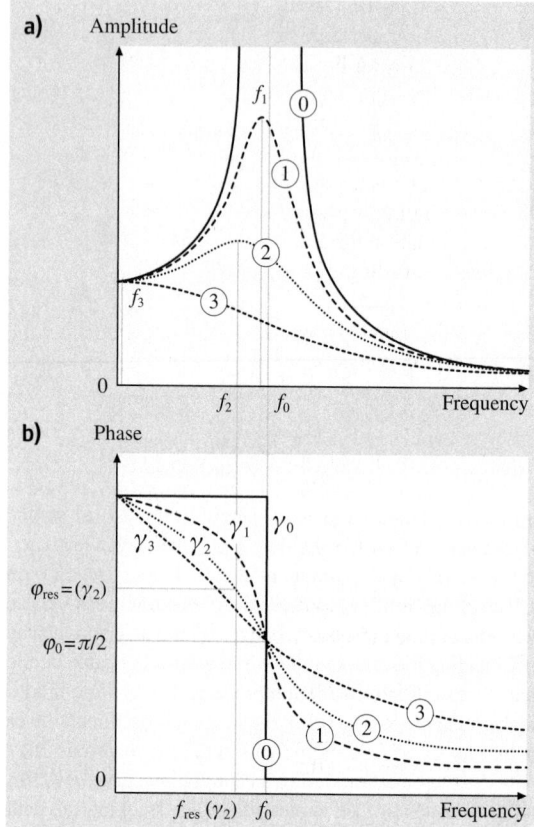

Fig. 16.4 (a) Resonance curve with no damping (0), and increasing damping (1)–(3). The undamped curve with resonance frequency f_0 exhibits a very high amplitude, whereas the resonance peak amplitude decreases with damping. This also involves a shift in resonance frequencies from f_1 to f_3 to lower values. (b) Corresponding phase curves showing no damping (0), and increasing damping (1)–(3). The step-like phase jump at resonance of the undamped resonance gradually broadens with increasing damping

or softening of the spring constant caused by the adsorption of a molecule layer, result in the fact that the measured resonance frequency will not be exactly equal to the eigenfrequency of the cantilever, and therefore the mass derived from the frequency shift will be inaccurate. In a medium, the vibration of a cantilever is described by the model of a driven damped harmonic oscillator:

$$m^* \frac{\mathrm{d}^2 x}{\mathrm{d}t^2} + \gamma \frac{\mathrm{d}x}{\mathrm{d}t} + kx = F \cos(2\pi f t) \,, \tag{16.7}$$

where $m^* = \text{const}(m_c + m_l)$ is the effective mass of the cantilever (for a rectangular cantilever the constant is 0.25). Especially in liquids, the mass of the co-moved liquid m_l adds significantly to the mass of the cantilever m_c. The term $\gamma \frac{dx}{dt}$ is the drag force due to damping, $F \cos(2\pi ft)$ is the driving force executed by the piezo-oscillator, and k is the spring constant of the cantilever.

If no damping is present, the eigenfrequencies of the various oscillation modes of a bar-shaped cantilever are calculated according to

$$f_n = \frac{\alpha_n^2}{2\pi} \sqrt{\frac{k}{2(m_c + m_l)}} \,, \qquad (16.8)$$

where f_n are the eigenfrequencies of the n-th mode, α_n are constants depending on the mode: $\alpha_1 = 1.8751$, $\alpha_2 = 4.6941$, $\alpha_n = \pi(n - 0.5)$; k is the spring constant of the cantilever, m_c the mass of the cantilever, and m_l the mass of the medium surrounding the cantilever, e.g. liquid [16.21].

Addition of mass to the cantilever due to adsorption will change the effective mass as follows:

$$m^* = \text{const}(m_c + m_l + \Delta m) \,, \qquad (16.9)$$

where Δm is the additional mass adsorbed. Typically, the co-moved mass of the liquid is much larger than the adsorbed mass.

Figure 16.4 clearly shows that the resonance frequency is only equal to the eigenfrequency if no damping is present. With damping, the frequency at which the peak of the resonance curve occurs is no longer identical to that at which the turning point of the phase curve occurs. For example, resonance curve 2 with damping γ_2 has its maximum amplitude at frequency f_2. The corresponding phase would be $\varphi_{\text{res}}(\gamma_2)$, which is not equal to $\pi/2$, as would be expected in the undamped case. If direct resonance-frequency tracking or a phase-locked loop is used to determine the frequency of the oscillating cantilever, then only its resonance frequency is detected, but not its eigenfrequency. Remember that the eigenfrequency, and not the resonance frequency, is required to determine mass changes.

16.3.3 Heat Mode

If a cantilever is coated with metal layers, thermal expansion differences in the cantilever and the coating layer will further influence cantilever bending as a function of temperature. This mode of operation is referred to as the *heat mode* and causes cantilever

bending because of differing thermal expansion coefficients in the sensor layer and cantilever materials [16.9] (Fig. 16.3g):

$$\Delta z = \frac{5}{4}(\alpha_1 - \alpha_2)\frac{t_1 + t_2}{t_2^2 \kappa}\frac{l^3}{(\lambda_1 t_1 + \lambda_2 t_2)w}P \,. \qquad (16.10)$$

Here α_1, α_2 are the thermal expansion coefficients of the cantilever and coating materials, respectively, λ_1, λ_2 their thermal conductivities, t_1, t_2 the material thicknesses, P is the total power generated on the cantilever, and κ is a geometry parameter of the cantilever device.

Heat changes are either caused by external influences (change in temperature, Fig. 16.3g), occur directly on the surface by exothermal, e.g. catalytic, reactions (Fig. 16.3h), or are due to material properties of a sample attached to the apex of the cantilever (micromechanical calorimetry, Fig. 16.3i). The sensitivity of the cantilever heat mode is orders of magnitude higher than that of traditional calorimetric methods performed on milligram samples, as it only requires nanogram amounts of sample and achieves nanojoule [16.20], picojoule [16.22] and femtojoule [16.23] sensitivity.

These three measurement modes have established cantilevers as versatile tools to perform experiments in nanoscale science with very small amounts of material.

16.3.4 Further Operation Modes

Photothermal Spectroscopy

When a material adsorbs photons, a fraction of the energy is converted into heat. This photothermal heating can be measured as a function of the light wavelength to provide optical absorption data of the material. The interaction of light with a bimetallic microcantilever creates heat on the cantilever surface, resulting in a bending of the cantilever [16.14]. Such bimetallic-cantilever devices are capable of detecting heat flows due to an optical heating power of 100 pW, which is two orders of magnitude better than in conventional photothermal spectroscopy.

Electrochemistry

A cantilever coated with a metallic layer (measurement electrode) on one side is placed in an electrolytic medium, e.g. a salt solution, together with a metallic reference electrode, usually made of a noble metal. If the voltage between the measurement and the reference electrode is changed, electrochemical processes on the measurement electrode (cantilever) are induced, such as adsorption or desorption of ions from the electrolyte solution onto the measurement electrode. These processes lead to a bending of the cantilever due to changes in surface stress and in the electrostatic forces [16.24].

Detection of Electrostatic and Magnetic Forces
The detection of electrostatic and magnetic forces is possible if charged or magnetic particles are deposited on the cantilever [16.25, 26]. If the cantilever is placed in the vicinity of electrostatic charges or magnetic particles, attractive or repulsion forces occur according to the polarity of the charges or magnetic particles present on the cantilever. These forces will result in an upward or a downward bending of the cantilever. The magnitude of the bending depends on the distribution of charged or magnetic particles on both the cantilever and in the surrounding environment according to the laws of electrostatics and magnetism.

16.4 Microfabrication

Silicon cantilever sensor arrays have been microfabricated using a dry-etching silicon-on-insulator (SOI) fabrication technique developed in the micro-/nano-mechanics department at the IBM Zurich Research Laboratory. One chip comprises eight cantilevers, having a length of $500\,\mu m$, a width of $100\,\mu m$, and a thickness of $0.5\,\mu m$, and arranged on a pitch of $250\,\mu m$. For dynamic-mode operation, the cantilever thickness may be up to $7\,\mu m$. The resonance frequencies of the cantilevers vary by 0.5% only, demonstrating the high reproducibility and precision of cantilever fabrication. A scanning electron microscopy image of a cantilever sensor-array chip is shown in Fig. 16.5.

Fig. 16.5 Scanning electron micrograph of a cantilever-sensor array. Image courtesy of Viola Barwich, University of Basel, Switzerland

16.5 Measurement Set-Up

16.5.1 Measurements in Gaseous or Liquid Environments

A measurement set-up for cantilever arrays consists of four major parts: (1) the measurement chamber containing the cantilever array, (2) an optical or electrical system to detect the cantilever deflection (e.g. laser sources, collimation lenses and a position-sensitive detector (PSD), or piezoresistors and Wheatstone-bridge detection electronics), (3) electronics to amplify, process and acquire the signals from the detector, and (4) a gas- or liquid-handling system to inject samples reproducibly into the measurement chamber and purge the chamber.

Figure 16.6 shows the schematic set-up for experiments performed in a gaseous (Fig. 16.6(a)) and a liquid, biochemical (Fig. 16.6(b)) environment for the optical beam-deflection embodiment of the measurement set-up. The cantilever sensor array is located in an analysis chamber with a volume of $3-90\,\mu l$, which has inlet and outlet ports for gases or liquids. The cantilever deflection is determined by means of an array of eight vertical-cavity surface-emitting lasers (VCSELs) arranged at a linear pitch of $250\,\mu m$ that emit at a wavelength of $760\,nm$ into a narrow cone of 5 to $10°$.

The light of each VCSEL is collimated and focused onto the apex of the corresponding cantilever by a pair of achromatic doublet lenses, $12.5\,mm$ in diameter. This size has to be selected in such a way that all eight laser beams pass through the lens close to its center to minimize scattering, chromatic and spherical aberration artifacts. The light is then reflected off the gold-coated surface of the cantilever and hits the surface of a position-sensing detector (PSD). PSDs are light-sensitive photo-potentiometer-like devices that produce photocurrents at two opposing electrodes. The magnitude of the photocurrents depends linearly on the distance of the

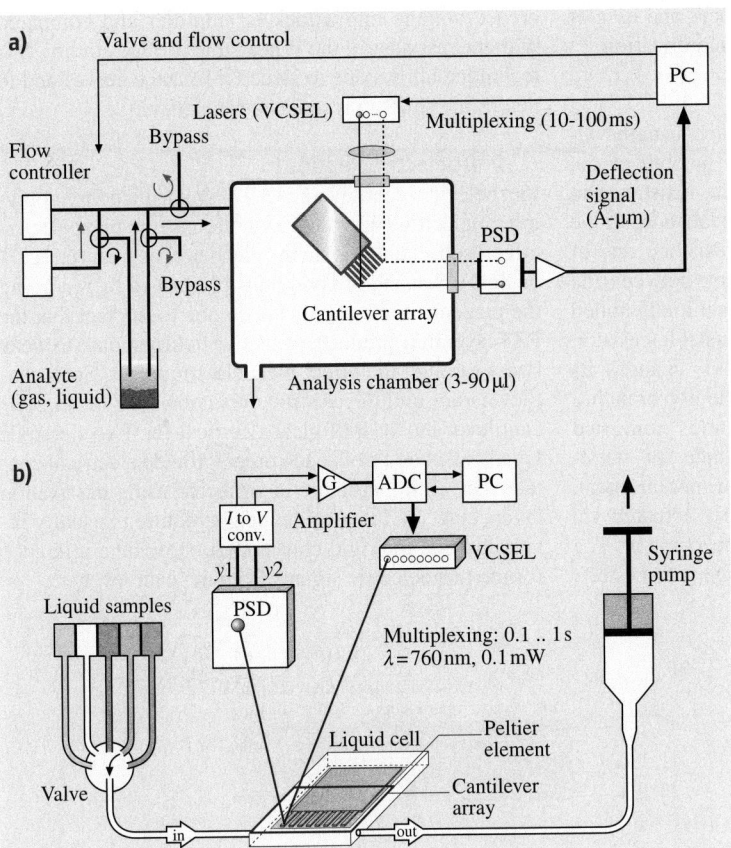

Fig. 16.6 Schematic of measurement set-ups for (**a**) a gaseous (artificial nose) and (**b**) a liquid environment (biochemical sensor)

impinging light spot from the electrodes. Thus the position of an incident light beam can easily be determined with micrometer precision. The photocurrents are transformed into voltages and amplified in a preamplifier. As only one PSD is used, the eight lasers cannot be switched on simultaneously. Therefore, a time-multiplexing procedure is used to switch the lasers on and off sequentially at typical intervals of 10–100 ms. The resulting deflection signal is digitized and stored together with time information on a personal computer (PC), which also controls the multiplexing of the VCSELs as well as the switching of the valves and mass flow controllers used for setting the composition ratio of the analyte mixture.

The measurement set-up for liquids (Fig. 16.6b) consists of a poly-etheretherketone (PEEK) liquid cell, which contains the cantilever array and is sealed by a viton O-ring and a glass plate. The VCSELs and the PSD are mounted on a metal frame around the liquid cell. After preprocessing the position of the deflected light beam in a current-to-voltage converter and ampli-

fier stage, the signal is digitized in an analog-to-digital converter and stored on a PC. The liquid cell is equipped with inlet and outlet ports for liquids. They are connected via 0.18-mm-inner-diameter Teflon tubing to individual thermally equilibrated glass containers, in which the biochemical liquids are stored. A six-position valve allows the inlet to the liquid chamber to be connected to each of the liquid-sample containers separately. The liquids are pulled (or pushed) through the liquid chamber by means of a syringe pump connected to the outlet of the chamber. A Peltier element is situated very close to the lumen of the chamber to allow temperature regulation within the chamber. The entire experimental set-up is housed in a temperature-controlled box regulated with an accuracy of 0.01 K to the target temperature.

16.5.2 Readout Principles

This section describes various ways to determine the deflection of cantilever sensors. They differ in sensitivity, effort for alignment and set-up, robustness and ease of readout as well as their potential for miniaturization.

Piezoresistive readout

Piezoresistive cantilevers [16.6, 20] are usually U-shaped, having diffused piezoresistors in both of the legs close to the hinge (Fig. 16.7a). The resistance in the piezoresistors is measured by a Wheatstone-bridge technique employing three reference resistors, one of which is adjustable. The current flowing between the two branches of the Wheatstone bridge is initially nulled by changing the resistance of the adjustable resistor. If the cantilever bends, the piezoresistor changes its value and a current will flow between the two branches of the Wheatstone bridge. This current is converted via a differential amplifier into a voltage for static-mode measurement. For dynamic-mode measurement, the piezoresistive cantilever is externally actuated via a frequency generator connected to a piezocrystal. The alternating current (AC) actuation voltage is fed as ref-

erence voltage into a lock-in amplifier and compared with the response of the Wheatstone-bridge circuit. This technique allows one to sweep resonance curves and to determine shifts in resonance frequency.

Piezoelectric Readout

Piezoelectric cantilevers [16.27] are actuated by applying an electric AC voltage via the inverse piezoelectric effect (self-excitation) to the piezoelectric material (PZT or ZnO). Sensing of bending is performed by recording the piezoelectric current change due to the fact that the PZT layer may produce a sensitive field response to weak stress through the direct piezoelectric effect. Such cantilevers are multilayer structures consisting of an SiO_2 cantilever and the PZT piezoelectric layer. Two electrode layers, insulated from each other, provide electrical contact. The entire structure is protected using passivation layers (Fig. 16.7b). An identical structure is usually integrated into the rigid chip body to provide a reference for the piezoelectric signals from the cantilever.

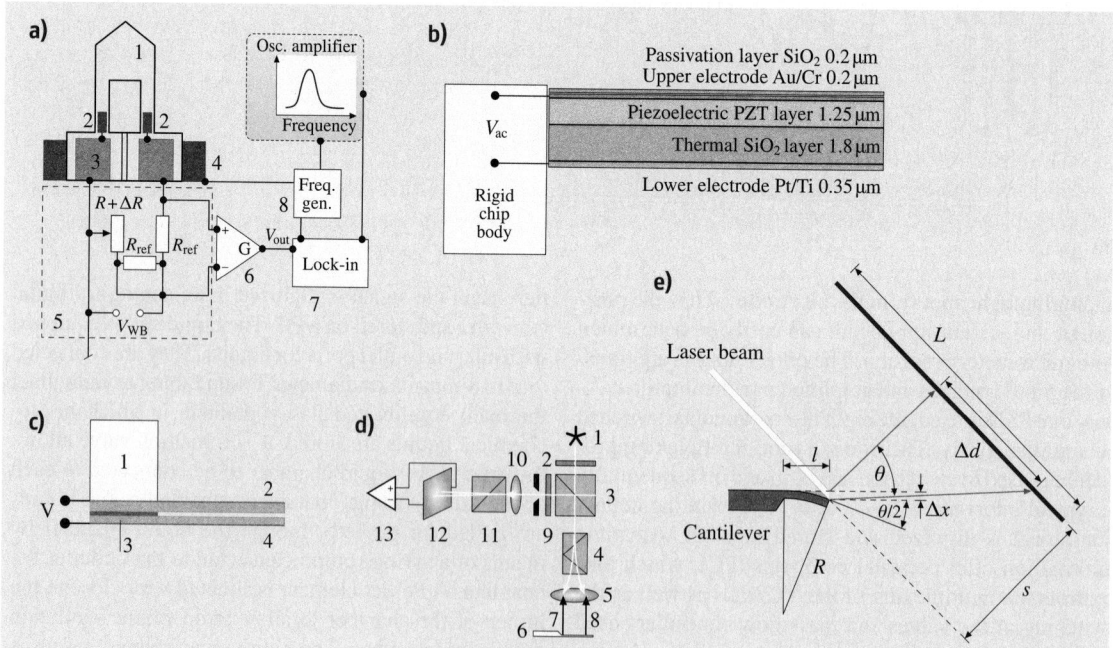

Fig. 16.7 (a) Piezoresistive readout: (1) cantilever, (2) piezoresistors, (3) Au contact pads, (4) external piezocrystal for actuation, (5) Wheatstone-bridge circuit, (6) differential amplifier, (7) lock-in amplifier, (8) function generator. **(b)** Piezoelectric readout. **(c)** Capacitive readout: (1) solid support, (2) rigid beam with counter-electrode, (3) insulation layer (SiO_2), (4) flexible cantilever with electrode. **(d)** Interferometric readout: (1) laser diode, (2) polarizer, (3) nonpolarizing beam splitter, (4) Wollaston prism, (5) focusing lens, (6) cantilever, (7) reference beam (near cantilever hinge), (8) object beam (near cantilever apex), (9) diaphragm and $\lambda/4$ plate, (10) focusing lens, (11) Wollaston prism, (12) quadrant photodiode, (13) differential amplifier. **(e)** Beam-deflection readout

Capacitive Readout

For capacitive readout (Fig. 16.7c), a rigid beam with an electrode mounted on the solid support and a flexible cantilever with another electrode layer are used [16.28, 29]. Both electrodes are insulated from each other. Upon bending of the flexible cantilever the capacitance between the two electrodes changes and allows the deflection of the flexible cantilever to be determined. Both static- and dynamic-mode measurements are possible.

Optical (Interferometric) Readout

Interferometric methods [16.30, 31] are most accurate for the determination of small movements. A laser beam passes through a polarizer plate (polarization 45°) and is partially transmitted by a nonpolarized beam splitter (Fig. 16.7d). The transmitted beam is divided in a Wollaston prism into a reference and an object beam. These mutually orthogonally polarized beams are then focused onto the cantilever. Both beams (the reference beam from the hinge region and the object beam from the apex region of the cantilever) are reflected back to the objective lens, pass the Wollaston prism, where they are recombined into one beam, which is then reflected into the other arm of the interferometer, where after the $\lambda/4$ plate a phase shift of a quarter wavelength between object and reference beam is established. Another Wollaston prism separates the reference and object beams again for analysis with a four-quadrant photodiode. A differential amplifier is used to obtain the cantilever deflection with high accuracy. However, the interferometric set-up is quite bulky and difficult to handle.

Optical (Beam–Deflection) Readout

The most frequently used approach to read out cantilever deflections is optical beam deflection [16.32], because it is a comparatively simple method with an excellent lateral resolution. A schematic of this method is shown in Fig. 16.7e.

The actual cantilever deflection Δx scales with the cantilever dimensions; therefore the surface stress $\Delta\sigma$

in N/m is a convenient quantity to measure and compare cantilever responses. It takes into account the cantilever material properties, such as Poisson's ratio v, Young's modulus E and the cantilever thickness t. The radius of curvature R of the cantilever is a measure of bending, (16.2). As shown in the drawing in Fig. 16.7e, the actual cantilever displacement Δx is transformed into a displacement Δd on the PSD. The position of a light spot on a PSD is determined by measuring the photocurrents from the two facing electrodes. The movement of the light spot on the linear PSD is calculated from the two currents I_1 and I_2 and the size L of the PSD by

$$\Delta d = \frac{I_1 - I_2}{I_1 + I_2} \cdot \frac{L}{2} \, . \tag{16.11}$$

As all angles are very small, it can be assumed that the bending angle of the cantilever is equal to half of the angle θ of the deflected laser beam, i.e. $\theta/2$. Therefore, the bending angle of the cantilever can be calculated to be

$$\frac{\theta}{2} = \frac{\Delta d}{2s} \, , \tag{16.12}$$

where s is the distance between the PSD and the cantilever. The actual cantilever deflection Δx is calculated from the cantilever length l and the bending angle $\theta/2$ by

$$\Delta x = \frac{\theta/2}{2} \cdot l \, . \tag{16.13}$$

Combination of (16.12) and (16.13) relates the actual cantilever deflection Δx to the PSD signal:

$$\Delta x = \frac{l \Delta d}{4s} \, . \tag{16.14}$$

The relation between the radius of curvature and the deflection angle is

$$\frac{\theta}{2} = \frac{l}{R} \, , \tag{16.15}$$

and after substitution becomes

$$R = \frac{2ls}{\Delta d} \, , \tag{16.16}$$

or $R = \frac{2\Delta x}{l^2}$.

16.6 Functionalization Techniques

16.6.1 General Strategy

To serve as sensors, cantilevers have to be coated with a sensor layer that is either highly specific, i.e. is able

to recognize target molecules in a key–lock process, or partially specific, so that the sensor information from several cantilevers yields a pattern that is characteristic of the target molecules.

To provide a platform for specific functionalization, the upper surface of these cantilevers is typically coated with 2 nm of titanium and 20 nm of gold, which yields a reflective surface and an interface for attaching functional groups of probe molecules, e.g. for anchoring molecules with a thiol group to the gold surface of the cantilever. Such thin metal layers are believed not to contribute significantly to bimetallic bending, because the temperature is kept constant.

16.6.2 Functionalization Methods

There are numerous ways to coat a cantilever with material, both simple and more advanced ones. The method of choice should be fast, reproducible, reliable and allow one or both of the surfaces of a cantilever to be coated separately.

Simple Methods
Obvious methods to coat a cantilever are thermal or electron-beam-assisted evaporation of material, electrospray or other standard deposition methods. The disadvantage of these methods is that they only are suitable for coating large areas, but not individual cantilevers in an array, unless shadow masks are used. Such masks need to be accurately aligned to the cantilever structures, which is a time-consuming process.

Other methods to coat cantilevers use manual placement of particles onto the cantilever [16.9, 20, 33–35], which requires skillful handling of tiny samples. Cantilevers can also be coated by directly pipetting solutions of the probe molecules onto the cantilevers [16.36] or by employing air-brush spraying and shadow masks to coat the cantilevers separately [16.37].

All these methods have only limited reproducibility and are very time-consuming if a larger number of cantilever arrays has to be coated.

Microfluidics
Microfluidic networks (μFN) [16.38] are structures of channels and wells, etched several ten to hundred micrometer deep into silicon wafers. The wells can be filled easily using a laboratory pipette, so that the fluid with the probe molecules for coating the cantilever is guided through the channels towards openings at a pitch matched to the distance between individual cantilevers in the array (Fig. 16.8a).

The cantilever array is then introduced into the open channels of the μFN that are filled with a solution of the probe molecules. The incubation of the cantilever array in the channels of the μFN takes from a few seconds (self-assembly of alkanethiol monolayers) to several tens of minutes (coating with protein solutions). To prevent evaporation of the solutions, the channels are covered by a slice of poly(dimethylsiloxane) (PDMS). In addition, the microfluidic network may be placed in an environment filled with saturated vapor of the solvent used for the probe molecules.

Array of Dimension-matched Capillaries
A similar approach is insertion of the cantilever array into an array of dimension-matched disposable glass capillaries. The outer diameter of the glass capillaries is 240 μm so that they can be placed neatly next to each other to accommodate the pitch of the cantilevers in the array (250 μm). Their inner diameter is 150 μm, providing sufficient room to insert the cantilevers (width: 100 μm) safely (Fig. 16.8b). This method has been successfully applied for the deposition of a variety of materials onto cantilevers, such as polymer solutions [16.37], self-assembled monolayers [16.39], thiol-functionalized single-stranded DNA oligonucleotides [16.40], and protein solutions [16.41].

Inkjet Spotting
All of the above techniques require manual alignment of the cantilever array and functionalization tool, and are therefore not ideal for coating a large number of cantilever arrays. The inkjet-spotting technique, however, allows rapid and reliable coating of cantilever arrays [16.42, 43]. An x-y-z positioning system allows a fine nozzle (capillary diameter: 70 μm) to be positioned with an accuracy of approximately 10 μm over a cantilever. Individual droplets (diameter: 60–80 μm, volume 0.1–0.3 nl) can be dispensed individually by means of a piezo-driven ejection system in the inkjet nozzle. When the droplets are spotted with a pitch smaller than 0.1 mm, they merge and form continuous films. By adjusting the number of

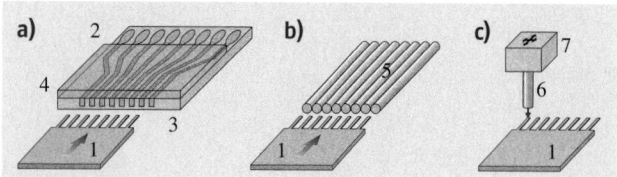

Fig. 16.8 (a) Cantilever functionalization in microfluidic networks. (b) Incubation in dimension-matched microcapillaries. (c) Coating with an inkjet spotter: (1) cantilever array, (2) reservoir wells, (3) microfluidic network with channels, (4) PDMS cover to avoid evaporation, (5) microcapillaries, (6) inkjet nozzle, (7) inkjet x-y-z positioning unit

droplets deposited on the cantilevers, the resulting film thickness can be controlled precisely. The inkjet-spotting technique allows a cantilever to be coated within seconds and yields very homogeneous, reproducibly deposited layers of well-controlled thickness. Successful coating of self-assembled alkanethiol monolayers, polymer solutions, self-assembled DNA single-stranded oligonucleotides [16.43], and protein layers has been demonstrated. In conclusion, inkjet spotting has turned out to be a very efficient and versatile method for functionalization, which can even be used to coat arbitrarily shaped sensors reproducibly and reliably [16.44, 45].

16.7 Applications

In recent years the field of cantilever sensors has been very active, as the bar chart in Fig. 16.9 of the number of publications between 1993 and 2004 on microcantilevers, cantilever sensors and cantilever arrays demonstrates. This section gives a short overview of the research topics in the field of microcantilever sensors in the literature. Early reports involve the adsorption of alkyl thiols on gold [16.13, 46], detection of mercury vapor and relative humidity [16.47], dye molecules [16.34], monoclonal antibodies [16.48], sugar and proteins [16.49], solvent vapors [16.36, 37, 50, 51], fragrance vapors [16.52] as well as the pH-dependent response of carboxy-terminated alkyl thiols [16.39], label-free DNA hybridization detection [16.17, 40], and biomolecular recognition of proteins relevant in cardiovascular diseases [16.41]. The more recent literature is reviewed in [16.53–56]. Major topics published in 2003 and 2004 include the following studies: fabrication of silicon, piezoresistive [16.57, 58] or polymer [16.59] cantilevers, detection of vapors and volatile compounds, e.g. mercury vapor [16.60], HF vapor [16.61, 62], chemical vapors [16.63], as well as the development of gas sensors utilizing the piezoresistive method [16.64]. Pd-based sensors for hydrogen [16.65], deuterium and tritium [16.66] are reported, as well as sensors based on hydrogels [16.67] or zeolites [16.68]. A humidity sensor is suggested in [16.69]. Further topics include the detection of explosives [16.70], pathogens [16.71], nerve agents [16.72], viruses [16.73], bacteria, e.g. *E. coli* [16.74], and pesticides such as dichlorodiphenyltrichloroethane(DDT) [16.75]. The

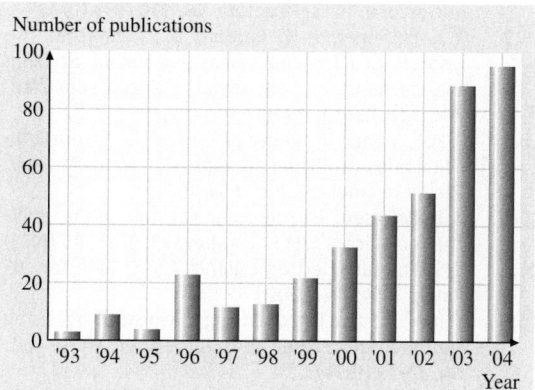

Fig. 16.9 Number of publications from 1993 to 2004 in the field of microcantilevers, cantilever sensors and cantilever arrays

issues of detection of environmental pollutants are discussed in [16.76]. A chemical-vapor sensor based on the bimetal technique is described in [16.77]. Also electrochemical redox reactions have been measured with cantilevers [16.78]. In biochemical applications, detection of DNA [16.18, 79], proteins [16.80], prostate-specific antigen (PSA) [16.81], peptides using antibodies [16.82] and living cells [16.83] has been reported. Medical applications include diagnostics [16.84], drug discovery [16.85], and detection of glucose [16.86]. To increase the complexity of microcantilever applications, two-dimensional microcantilever arrays have been proposed for multiplexed biomolecular analysis [16.87, 88].

16.8 Conclusions and Outlook

Cantilever-sensor array techniques have turned out to be a very powerful and highly sensitive tool to study physisorption and chemisorption processes, as well as to determine material-specific properties such as heat transfer during phase transitions. Experiments in liquids have provided new insights into such complex biochemical reactions as the hybridization of DNA or molecular recognition in antibody–antigen systems or proteomics. Future developments must go towards technological applications, in particular to find new ways to characterize real-world samples such as clinical samples. The development of medical diagnosis tools requires an improvement of the sensitivity of a large number of genetic tests to be performed with small amounts of single donor-blood or body-fluid samples at low cost. From a scientific point of view, the challenge lies in optimizing cantilever sensors to improve their sensitivity to the ultimate limit: the detection of individual molecules.

References

16.1   G. Binnig, C. F. Quate, Ch. Gerber: Atomic force microscope, Phys. Rev. Lett. **56**, 930–933 (1986)

16.2   R. J. Wilfinger, P. H. Bardell, D. S. Chhabra: The resonistor, a frequency sensitive device utilizing the mechanical resonance of a silicon substrate, IBM J. **12**, 113–118 (1968)

16.3   T. M. S. Heng: Trimming of microstrip circuits utilizing microcantilever air gaps, IEEE Trans. Microw. Theory Technol. **19**, 652–654 (1971)

16.4   K. E. Petersen: Micromechanical membrane switches on silicon, IBM J. Res. Dev. **23**, 376–385 (1979)

16.5   E. S. Kolesar: Electronic Nerve Agent Detector, US Patent 4,549,427 (1983)

16.6   T. Itoh, T. Suga: Force sensing microcantilever using sputtered zinc-oxide thin-film, Appl. Phys. Lett. **64**, 37–39 (1994)

16.7   J. P. Cleveland, S. Manne, D. Bocek, P. K. Hansma: A nondestructive method for determining the spring constant of cantilevers for scanning force microscopy, Rev. Sci. Instrum. **64**, 403–405 (1993)

16.8   T. Thundat, R. J. Warmack, G. Y. Chen, D. P. Allison: Thermal, ambient-induced deflections of scanning force microscope cantilevers, Appl. Phys. Lett. **64**, 2894–2896 (1994)

16.9   J. K. Gimzewski, Ch. Gerber, E. Meyer, R. R. Schlittler: Observation of a chemical reaction using a micromechanical sensor, Chem. Phys. Lett. **217**, 589–594 (1994)

16.10   T. Thundat, G. Y. Chen, R. J. Warmack, D. P. Allison, E. A. Wachter: Vapor detection using resonating microcantilevers, Anal. Chem. **67**, 519–521 (1995)

16.11   H. P. Lang, R. Berger, C. Andreoli, J. Brugger, M. Despont, P. Vettiger, Ch. Gerber, J. K. Gimzewski, J.-P. Ramseyer, E. Meyer, H.-J. Güntherodt: Sequential position readout from arrays of micromechanical cantilever sensors, Appl. Phys. Lett. **72**, 383–385 (1998)

16.12   H. Ibach: Adsorbate-induced surface stress, J. Vac. Sci. Technol. A **12**, 2240–2245 (1994)

16.13   R. Berger, E. Delamarche, H. P. Lang, Ch. Gerber, J. K. Gimzewski, E. Meyer, H.-J. Güntherodt: Surface stress in the self-assembly of alkanethiols on gold, Science **276**, 2021–2024 (1997)

16.14   J. R. Barnes, R. J. Stephenson, M. E. Welland, Ch. Gerber, J. K. Gimzewski: Photothermal spectroscopy with femtojoule sensitivity based on micromechanics, Nature **372**, 79–81 (1994)

16.15   G. G. Stoney: The tension of metallic films deposited by electrolysis, Proc. R. Soc. London **82**, 172–177 (1909)

16.16   F. J. von Preissig: Applicability of the classical curvature-stress relation for thin films on plate substrates, J. Appl. Phys. **66**, 4262–4268 (1989)

16.17   J. Fritz, M. K. Baller, H. P. Lang, H. Rothuizen, P. Vettiger, E. Meyer, H.-J. Güntherodt, Ch. Gerber, J. K. Gimzewski: Translating biomolecular recognition into nanomechanics, Science **288**, 316–318 (2000)

16.18   M. Alvarez, L. G. Carrascosa, M. Moreno, A. Calle, A. Zaballos, L. M. Lechuga, C.-A. Martinez, J. Tamayo: Nanomechanics of the formation of DNA self-assembled monolayers, hybridization on microcantilevers, Langmuir **20**, 9663–9668 (2004)

16.19   D. Sarid: *Scanning Force Microscopy with Applications to Electric, Magnetic, and Atomic Forces* (Oxford Univ. Press, New York 1991)

16.20   R. Berger, H. P. Lang, Ch. Gerber, J. K. Gimzewski, J. H. Fabian, L. Scandella, E. Meyer, H.-J. Güntherodt: Micromechanical thermogravimetry, Chem. Phys. Lett. **294**, 363–369 (1998)

16.21   J. E. Sader: Frequency response of cantilever beams immersed in viscous fluids with applications to the atomic force microscope, J. Appl. Phys. **84**, 64–76 (1998)

16.22   T. Bachels, F. Tiefenbacher, R. Schafer: Condensation of isolated metal clusters studied with a calorimeter, J. Chem. Phys. **110**, 10008–10015 (1999)

16.23   J. R. Barnes, R. J. Stephenson, C. N. Woodburn, S. J. O'Shea, M. E. Welland, T. Rayment, J. K. Gimzewski, Ch. Gerber: A femtojoule calorimeter using micromechanical sensors, Rev. Sci. Instrum. **65**, 3793–3798 (1994)

16.24   T. A. Brunt, T. Rayment, S. J. O'Shea, M. E. Welland: Measuring the surface stresses in an electrochemically deposited monolayer: Pb on Au(111), Langmuir **12**, 5942–5946 (1996)

16.25   R. Puers, D. Lapadatu: Electrostatic forces and their effects on capacitive mechanical sensors, Sens. Actuators A **56**, 203–210 (1996)

Part B | 16

16.26 J. Fricke, C. Obermaier: Cantilever beam accelerometer based on surface micromachining technology, J. Micromech. Microeng. **3**, 190–192 (1993)

16.27 C. Lee, T. Itoh, T. Ohashi, R. Maeda, T. Suga: Development of a piezoelectric self-excitation, self-detection mechanism in PZT microcantilevers for dynamic scanning force microscopy in liquid, J. Vac. Sci. Technol. B **15**, 1559–1563 (1997)

16.28 T. Göddenhenrich, H. Lemke, U. Hartmann, C. Heiden: Force microscope with capacitive displacement detection, J. Vac. Sci. Technol. A **8**, 383–387 (1990)

16.29 J. Brugger, R.A. Buser, N.F. de Rooij: Micromachined atomic force microprobe with integrated capacitive read-out, J. Micromech. Microeng. **2**, 218–220 (1992)

16.30 C. Schönenberger, S.F. Alvarado: A differential interferometer for force microscopy, Rev. Sci. Instrum. **60**, 3131–3134 (1989)

16.31 M.J. Cunningham, S.T. Cheng, W.W. Clegg: A differential interferometer for scanning force microscopy, Meas. Sci. Technol. **5**, 1350–1254 (1994)

16.32 G. Meyer, N.M. Amer: Novel optical approach to atomic force microscopy, Appl. Phys. Lett. **53**, 2400–2402 (1988)

16.33 R. Berger, Ch. Gerber, J.K. Gimzewski, E. Meyer, H.-J. Güntherodt: Thermal analysis using a micromechanical calorimeter, Appl. Phys. Lett. **69**, 40–42 (1996)

16.34 L. Scandella, G. Binder, T. Mezzacasa, J. Gobrecht, R. Berger, H.P. Lang, Ch. Gerber, J.K. Gimzewski, J.H. Koegler, J.C. Jansen: Combination of single crystal zeolites and microfabrication: Two applications towards zeolite nanodevices, Microporous Mesoporous Mater. **21**, 403–409 (1998)

16.35 R. Berger, Ch. Gerber, H.P. Lang, J.K. Gimzewski: Micromechanics: a toolbox for femtoscale science: towards a laboratory on a tip, Microelectron. Eng. **35**, 373–379 (1997)

16.36 H.P. Lang, R. Berger, F.M. Battiston, J.-P. Ramseyer, E. Meyer, C. Andreoli, J. Brugger, P. Vettiger, M. Despont, T. Mezzacasa, L. Scandella, H.-J. Güntherodt, Ch. Gerber, J.K. Gimzewski: A chemical sensor based on a micromechanical cantilever array for the identification of gases and vapors, Appl. Phys. A **66**, 61–64 (1998)

16.37 M.K. Baller, H.P. Lang, J. Fritz, Ch. Gerber, J.K. Gimzewski, U. Drechsler, H. Rothuizen, M. Despont, P. Vettiger, F.M. Battiston, J.-P. Ramseyer, P. Fornaro, E. Meyer, H.-J. Güntherodt: A cantilever array based artificial nose, Ultramicroscopy **82**, 1–9 (2000)

16.38 S. Cesaro-Tadic, G. Dernick, D. Juncker, G. Buurman, H. Kropshofer, B. Michel, C. Fattinger, E. Delamarche: High-sensitivity miniaturized immunoassays for tumor necrosis factor α using microfluidic systems, Lab on a Chip **4**, 563–569 (2004)

16.39 J. Fritz, M.K. Baller, H.P. Lang, T. Strunz, E. Meyer, H.-J. Güntherodt, E. Delamarche, Ch. Gerber, J.K. Gimzewski: Stress at the solid–liquid interface of self-assembled monolayers on gold investigated with a nanomechanical sensor, Langmuir **16**, 9694–9696 (2000)

16.40 R. McKendry, J. Zhang, Y. Arntz, T. Strunz, M. Hegner, H.P. Lang, M.K. Baller, U. Certa, E. Meyer, H.-J. Güntherodt, Ch. Gerber: Multiple label-free biodetection and quantitative DNA-binding assays on a nanomechanical cantilever array, Proc. Nat. Acad. Sci. USA **99**, 9783–9787 (2002)

16.41 Y. Arntz, J.D. Seelig, H.P. Lang, J. Zhang, P. Hunziker, J.-P. Ramseyer, E. Meyer, M. Hegner, Ch. Gerber: Label-free protein assay based on a nanomechanical cantilever array, Nanotechnology **14**, 86–90 (2003)

16.42 A. Bietsch, M. Hegner, H.P. Lang, Ch. Gerber: Inkjet deposition of alkanethiolate monolayers and DNA oligonucleotides on gold: evaluation of spot uniformity by wet etching, Langmuir **20**, 5119–5122 (2004)

16.43 A. Bietsch, J. Zhang, M. Hegner, H.P. Lang, Ch. Gerber: Rapid functionalization of cantilever array sensors by inkjet printing, Nanotechnology **15**, 873–880 (2004)

16.44 D. Lange, C. Hagleitner, A. Hierlemann, O. Brand, H. Baltes: Complementary metal oxide semiconductor cantilever arrays on a single chip: mass-sensitive detection of volatile organic compounds, Anal. Chem. **74**, 3084–3085 (2002)

16.45 C.A. Savran, T.P. Burg, J. Fritz, S.R. Manalis: Microfabricated mechanical biosensor with inherently differential readout, Appl. Phys. Lett. **83**, 1659–1661 (2003)

16.46 R. Berger, E. Delamarche, H.P. Lang, Ch. Gerber, J.K. Gimzewski, E. Meyer, H.-J. Güntherodt: Surface stress in the self-assembly of alkanethiols on gold probed by a force microscopy technique, Appl. Phys. A **66**, 55 (1998)

16.47 E.A. Wachter, T. Thundat: Micromechanical sensors for chemical and physical measurements, Rev. Sci. Instrum. **66**, 3662–3667 (1995)

16.48 R. Raiteri, G. Nelles, H.J. Butt, W. Knoll, P. Skladal: Sensing of biological substances based on the bending of microfabricated cantilevers, Sens. Actuators B **61**, 213–217 (1999)

16.49 A.M. Moulin, S.J. O'Shea, M.E. Welland: Microcantilever-based biosensors, Ultramicroscopy **82**, 23–31 (2000)

16.50 G.G. Bumbu, G. Kircher, M. Wolkenhauer, R. Berger, J.S. Gutmann: Synthesis and characterization of polymer brushes on micromechanical cantilevers, Macromol. Chem. Phys. **205**, 1713–1720 (2004)

16.51 H.P. Lang, M.K. Baller, R. Berger, Ch. Gerber, J.K. Gimzewski, F.M. Battiston, P. Fornaro, J.P. Ramseyer, E. Meyer, H.-J. Güntherodt: An artificial nose based on a micromechanical cantilever array, Anal. Chim. Acta **393**, 59–65 (1999)

16.52   F. M. Battiston, J.-P. Ramseyer, H. P. Lang, M. K. Baller, Ch. Gerber, J. K. Gimzewski, E. Meyer, H.-J. Güntherodt: A chemical sensor based on a microfabricated cantilever array with simultaneous resonance-frequency and bending readout, Sens. Actuators B **77**, 122–131 (2001)

16.53   C. Ziegler: Cantilever-based biosensors, Anal. Bioanal. Chem. **379**, 946–959 (2004)

16.54   N. V. Lavrik, M. J. Sepaniak, P. G. Datskos: Cantilever transducers as a platform for chemical and biological sensors, Rev. Sci. Instrum. **75**, 2229–2253 (2004)

16.55   A. Majumdar: Bioassays based on molecular nanomechanics, Dis. Markers **18**, 167–174 (2002)

16.56   H. P. Lang, M. Hegner, Ch. Gerber: Cantilever array sensors, Mater. Today **8**, 30–36 (2005)

16.57   Y. J. Tang, J. Fang, X. D. Yan, H. F. Ji: Fabrication, characterization of SiO_2 microcantilever for microsensor applicatio, Sens. Actuators B **97**, 109–113 (2004)

16.58   E. Forsen, S. G. Nilsson, P. Carlberg, G. Abadal, F. Perez-Murano, J. Esteve, J. Montserrat, E. Figueras, F. Campabadal, J. Verd, L. Montelius, N. Barniol, A. Boisen: Fabrication of cantilever based mass sensors integrated with CMOS using direct write laser lithography on resist, Nanotechnology **15**, 628 (2004)

16.59   A. W. McFarland, M. A. Poggi, L. A. Bottomley, J. S. Colton: Production and characterization of polymer microcantilevers, Rev. Sci. Instrum. **75**, 2756–2758 (2004)

16.60   B. Rogers, L. Manning, M. Jones, T. Sulchek, K. Murray, B. Beneschott, J. D. Adams, Z. Hu, T. Thundat, H. Cavazos, S. C. Minne: Mercury vapor detection with a self-sensing, resonating piezoelectric cantilever, Rev. Sci. Instrum. **74**, 4899–4901 (2003)

16.61   J. Mertens, E. Finot, M. H. Nadal, V. Eyraud, O. Heintz, E. Bourillot: Detection of gas trace of hydrofluoric acid using microcantileve, Sens. Actuators B. **99**, 58–65 (2004)

16.62   Y. J. Tang, J. Fang, X. H. Xu, H. F. Ji, G. M. Brown, T. Thundat: Detection of femtomolar concentrations of HF using an SiO_2 microcantilever, Anal. Chem. **76**, 2478–2481 (2004)

16.63   N. Abedinov, C. Popov, Z. Yordanov, T. Ivanov, T. Gotszalk, P. Grabiec, W. Kulisch, I. W. Rangelow, D. Filenko, Y. J. Shirshov: Chemical recognition based on micromachined silicon cantilever array, Vac. Sci. Technol. B **21**, 2931–2936 (2003)

16.64   J. Zhou, P. Li, S. Zhang, Y. P. Huang, P. Y. Yang, M. H. Bao, G. Ruan: Self-excited piezoelectric microcantilever for gas detection, Microelectron. Eng. **69**, 37–46 (2003)

16.65   D. R. Baselt, B. Fruhberger, E. Klaassen, S. Cemalovic, C. L. Britton Jr., S. V. Patel, T. E. Mlsna, D. McCorkle, B. Warmack: Design and performance of a microcantilever-based hydrogen sensor, Sens. Actuators B **88**, 120–131 (2003)

16.66   A. Fabre, E. Finot, J. Demoment, S. Contreras: Monitoring the chemical changes in Pd induced by hydrogen absorption using microcantilevers, Ultramicroscopy **97**, 425–432 (2003)

16.67   Y. F. Zhang, H. F. Ji, G. M. Brown, T. Thundat: Detection of CrO_4^{2-} using a hydrogel swelling microcantilever sensor, Anal. Chem. **75**, 4773–4777 (2003)

16.68   J. Zhou, P. Li, S. Zhang, Y. C. Long, F. Zhou, Y. P. Huang, P. Y. Yang, M. H. Bao: Zeolite-modified microcantilever gas sensor for indoor air quality control, Sens. Actuators B **94**, 337–342 (2003)

16.69   C. Y. Lee, G. B. Lee: Micromachine-based humidity sensors with integrated temperature sensors for signal drift compensation, J. Micromech. Microeng. **13**, 620–627 (2003)

16.70   L. A. Pinnaduwage, A. Wig, D. L. Hedden, A. Gehl, D. Yi, T. Thundat, R. T. Lareau: Detection of trinitrotoluene via deflagration on a microcantilever, J. Appl. Phys. **95**, 5871–5875 (2004)

16.71   B. L. Weeks, J. Camarero, A. Noy, A. E. Miller, L. Stanker, J. J. De Yoreo: A microcantilever-based pathogen detector, Scanning **25**, 297–299 (2003)

16.72   Y. M. Yang, H. F. Ji, T. Thundat: Nerve agents detection using a Cu^{2+}/L-cysteine bilayer-coated microcantilever, J. Am. Chem. Soc. **125**, 1124–1125 (2003)

16.73   R. L. Gunter, W. G. Delinger, K. Manygoats, A. Kooser, T. L. Porter: Viral detection using an embedded piezoresistive microcantilever sensor, Sens. Actuators A **107**, 219–224 (2003)

16.74   K. Y. Gfeller, N. Nugaeva, M. Hegner: Micromechanical oscillators as rapid biosensor for the detection of active growth of Escherichia coli, Biosens. Bioelectron. **21**, 528–533 (2005)

16.75   M. Alvarez, A. Calle, J. Tamayo, L. M. Lechuga, A. Abad, A. Montoya: Development of nanomechanical biosensors for detection of the pesticide DD, Biosens. Bioelectron. **18**, 649–653 (2003)

16.76   S. Cherian, R. K. Gupta, B. C. Mullin, T. Thundat: Detection of heavy metal ions using protein-functionalized microcantilever sensors, Biosens. Bioelectron. **19**, 411–416 (2003)

16.77   J. D. Adams, G. Parrott, C. Bauer, T. Sant, L. Manning, M. Jones, B. Rogers, D. McCorkle, T. L. Ferrell: Nanowatt chemical vapor detection with a self-sensing, piezoelectric microcantilever array, Appl. Phys. Lett. **83**, 3428–3430 (2003)

16.78   F. Quist, V. Tabard-Cossa, A. Badia: Nanomechanical cantilever motion generated by a surface-confined redox reaction, J. Phys. Chem. B **107**, 10691–10695 (2003)

16.79   R. L. Gunter, R. Zhine, W. G. Delinger, K. Manygoats, A. Kooser, T. L. Porter: Investigation of DNA sensing using piezoresistive microcantilever probes, IEEE Sens. J. **4**, 430–433 (2004)

16.80 J. H. Lee, T. S. Kim, K. H. Yoon: Effect of mass and stress on resonant frequency shift of functionalized $Pb(Zr_{0.52}Ti_{0.48})O_3$ thin film microcantilever for the detection of C-reactive protein, Appl. Phys. Lett. **84**, 3187–3189 (2004)

16.81 G. Wu, R. H. Datar, K. M. Hansen, T. Thundat, R. J. Cote, A. Majumdar: Bioassay of prostate-specific antigen (PSA) using microcantilevers, Nature Biotechnol. **19**, 856–860 (2001)

16.82 B. H. Kim, O. Mader, U. Weimar, R. Brock, D. P. Kern: Detection of antibody peptide interaction using microcantilevers as surface stress sensors, J. Vac. Sci. Technol. B. **21**, 1472–1475 (2003)

16.83 M. T. A. Saif, C. R. Sager, S. Coyer: Functionalized biomicroelectromechanical systems sensors for force response study at local adhesion sites of single living cells on substrates, Annals Biomed. Eng. **31**, 950–961 (2003)

16.84 S. Kumar, R. P. Bajpai, L. M. Bharadwaj: Microcantilever based diagnostic chip for multiple analytes, IETE Techn. Rev. **20**, 361–368 (2003)

16.85 Y. F. Zhang, S. P. Venkatachalan, H. Xu, X. H. Xu, P. Joshi, H. F. Ji, M. Schulte: Micromechanical measurement of membrane receptor binding for label-free drug discovery, Biosens. Bioelectron. **19**, 1473–1478 (2004)

16.86 J. H. Pei, F. Tian, T. Thundat: Glucose biosensor based on the microcantilever, Anal. Chem. **76**, 292–297 (2004)

16.87 K. Khanafer, A. R. A. Khaled, K. Vafai: Spatial optimization of an array of aligned microcantilever based sensors, J. Micromech. Microeng. **14**, 1328–1336 (2004)

16.88 M. Yue, H. Lin, D. E. Dedrick, S. Satyanarayana, A. Majumdar, A. S. Bedekar, J. W. Jenkins, S. Sundaram: A 2-D microcantilever array for multiplexed biomolecular analysis, J. Microelectromech. Syst. **13**, 290–299 (2004)

Part B | 16

17. Therapeutic Nanodevices

Therapeutic nanotechnology offers minimally invasive therapies with high densities of function concentrated in small volumes, features that may reduce patient morbidity and mortality. Unlike other areas of nanotechnology, novel physical properties associated with nanoscale dimensionality are not the raison d'être of therapeutic nanotechnology, whereas the aggregation of multiple biochemical (or comparably precise) functions into controlled nanoarchitectures is. Multifunctionality is a hallmark of emerging nanotherapeutic devices, and multifunctionality can allow nanotherapeutic devices to perform multistep work processes, with each functional component contributing to one or more nanodevice subroutine such that, in aggregate, subroutines sum to a cogent work process. Cannonical nanotherapeutic subroutines include tethering (targeting) to sites of disease, dispensing measured doses of drug (or bioactive compound), detection of residual disease after therapy and communication with an external clinician/operator. Emerging nanotherapeutics thus blur the boundaries between medical devices and traditional pharmaceuticals. Assembly of therapeutic nanodevices generally exploits either (bio)material self-assembly properties or chemoselective bioconjugation techniques, or both. Given the complexity, composition, and the necessity for their tight chemical and structural definition inherent in the nature of nanotherapeutics, their cost of goods (COGs) might exceed that of (already expensive) biologics. Early therapeutic nanodevices will likely be applied to disease states which exhibit significant unmet patient need (cancer and cardiovascular disease), while application to other disease states well-served by conventional therapy may await perfection of nanotherapeutic design and assembly protocols.

Part B | 17

17.1 Definitions and Scope of Discussion

Nanotechnology is a field in rapid flux and development, as cursory examination of this volume shows, and definition of its meets and bounds, as well as identification of subdisciplines embraced by it, can be elusive. The word means many things to many people, and aspects of multiple disciplines, from physics to information technology to biotechnology, legitimately fall into the intersection of the Venn diagram of disciplines that defines nanotechnology. The breadth of the field allows almost any interested party to contribute to it, but the same ambiguity can render the field diffuse and amorphous. If nanotechnology embraces everything, then what is it? The ambiguity fuels cognitive dissonance that can result in frustrating interactions between investigators and funders, authors and editors, and entrepreneurs and in-

vestors. Some consideration of the scope of the field is therefore useful.

To frame the discussion, we will define nanotechnology as the discipline that aims to satisfy desired objectives using materials and devices whose valuable properties are based on a specific nanometer-scale element of their structures. The field is unabashedly application-oriented, so its raison d'être is fulfillment of tasks of interest: technical information is important primarily to the extent that it bears on device design, function, or application.

The meaning of "therapeutic" is largely self-explanatory and refers here to intervention in *human* disease processes (although many of the approaches discussed are equally applicable to veterinary medicine). Our discussion will be confined primarily to therapeutics used in vivo, because such applications clearly benefit from the low invasiveness that ultrasmall, but multipotent, nanotherapeutics potentially offer. It is debatable whether imaging, diagnostic, or sensing devices can be considered therapeutic in this context, though, as we will see, sensing/diagnostic functionalities are often inextricable elements of therapeutic nanodevices, and it is difficult to consider so-called smart nanotherapeutics without discussion of their sensing capabilities.

Our definition of nanotechnology projects several corollaries. First, it embraces macroscale structures whose useful properties derive from their nanoscale aspects. Second, the modifier "specific" (as in, "specific nanometer-scale elements") is intended to exclude materials whose utility derives solely from properties inherent in being finely divided (high surface-to-volume ratios, for instance), or other bulk chemical and physical properties. We made the exclusion based on our assessment that therapeutic nanodevices are more intriguing than nanomaterials per se (see below), though we will engage these attributes where they are germane to specific devices or therapeutic applications. Third, our definition implies that limited nanotechnology has been available since the 1970s in the form of biotechnology. Based on their nanoscale structures, individual biological macromolecules (such as proteins) often exhibit the coordinated, modular multifunctionality that is characteristic of purpose-built devices (Fig. 17.1). An analogous, but perhaps less persuasive, argument can be made that organic chemistry is an early form of nanotechnology. Compared to small organic molecules, protein functional capabilities and properties are generally more complex and extremely dependent on their

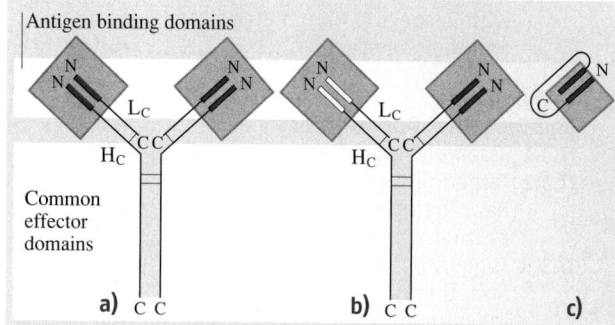

Fig. 17.1a–c Antibodies resemble purpose-built devices with distinct functional domains [17.1]. Native antibodies are composed of four polypeptide chains: two heavy chains (H_C) and two light chains (L_C), joined by interchain disulfide linkages (lines between H_C and L_C moieties). Amino and carboxy termini of individual polypeptide chains are indicated (by N and C). Antigen binding domains are responsible for specific antigen recognition, vary from antibody to antibody, and are indicated by the thicker lines. Common effector functions (F_C receptor binding, complement fixation, and so on) are delimited to domains of the antibodies that are constant from molecule to molecule. (**a**) A native IgG antibody is monospecific but bivalent in its antigen binding capacity. (**b**) An engineered, bispecific, bivalent antibody capable of recognizing two distinct antigens. (**c**) An engineered antibody fragment (single chain Fv or SCFv) that is monospecific and monovalent can recognize only one antigenic determinant and is engineered to lack common effector functions. This construct is translated as a single, continuous polypeptide chain (hence the name SCFv) because a peptide linker (indicated by the connecting line in the figure) is incorporated to connect the carboxy end of the H_C fragment and the amino end of the L_C fragment

conformation in three-dimensional space at nanometer scale. The nanotechnology sobriquet, therefore, may be more appropriate to biotechnology than organic chemistry.

Biological macromolecules rely on the deployment of specific chemical functionalities to specific relative distributions in space with nanometer (and greater) resolution for their function, so the inclusion of molecular engineering aspects of biotechnology practice under the nanotechnology rubric is legitimate, despite the discomfort it may cause traditionally trained engineers. As we will see, intervention in human disease often requires inclusion of biomolecules in therapeutic devices; frequently no functional synthetic analog of active proteins and nucleic acids is available.

As described above, this chapter focuses primarily on nanoscale therapeutic devices as opposed to therapeutic nanomaterials. Devices are integrated functional structures and not mixtures of materials. Devices exhibit desirable emergent properties inherent in their design: the properties emerge as the result of the spatial and/or temporal organization, and coordination and regulation of action of individual components. The organization of components in devices allows them to perform multistep, cogent work processes that can't be mimicked by simple admixtures of individual components. In fact, if device functions can be mimicked well by simple mixtures of components, the labor involved in configuring and constructing a nanoscale device is not warranted. Our device definition thus excludes nanomaterials used as drug formulation excipients (pharmacologically inert materials included in formulations that improve pharmacophore uptake, biodistribution, pharmacokinetic, handling, storage, or other properties), but embraces those same materials as integral components of drug delivery or other clinical devices.

17.1.1 Design Issues

The biotechnology industry has historically focused on the production of individual soluble protein and nucleic acid molecules for pharmaceutical use, with only limited attention paid to functional supramolecular structures [17.2–7]. This bias toward free molecules flies in the face of the obvious importance of integrated supramolecular structures in biology and, to the casual observer, may seem an odd gap in attention and emphasis on the part of practicing biotechnologists. The bias toward single molecule, protein therapeutics, however, follows from the fact that biotechnology is an industrial activity, governed by market considerations. Of the

myriad potential therapeutics that might be realized from biotechnology, single protein therapeutics are among the easiest to realize from both technical and regulatory perspectives and so warrant extensive industrial attention. This is changing, however, and more complex entities (actual supramolecular therapeutic devices) have and will appear with increasing frequency in the twenty-first century.

New top-down and bottom-up materials derived from micro/nanotechnology provide the opportunity to complement the traditional limits of biotechnology by providing scaffolds that can support higher level organization of multiple biomolecules to perform work activities they could not perform as free, soluble molecules. Such supramolecular structures have been called nanobiotechnological devices [17.8], nanobiological devices [17.2–7], or semi-synthetic nanodevices, and figure prominently in therapeutic nanotechnology.

Incorporation of / Interaction with Biomolecules

In general, designing nanodevices is a similar process to designing other engineered structures, provided that the special properties of the materials (relating to their nanoscale aspects such as quantum, electrical, mechanical and biological properties), as well as their impact in therapy, are considered. Therapeutics can interact with patients on multiple levels, ranging from organismal to molecular, but it is reasonable to expect that most nanotherapeutics will interface with patients at the nanoscale, at least to some extent [17.2–5, 9–14]. Typically, this means interaction between therapeutics and biological macromolecules, supramolecular structures and organelles, which, in turn, often dictates the incorporation of biological macromolecules (and other biostructures) into nanodevices [17.2, 5, 13–15]. Incorporating biological structures into (nanobiological) devices presents special challenges that do not occur in other aspects of engineering practice.

Unlike fully synthetic devices, semibiological nanodevices must incorporate prefabricated biological components (or derivatives thereof), and therefore the intact nanodevices are seldom made entirely de novo. As a corollary, knowledge of properties of biological device components is often incomplete (as they were not made by human design), and therefore the range of activities inherent in any nanobiological device design may be much less obvious and less well-defined than it is for fully synthetic devices. Further complicating the issue, the activities of biological molecules are often multifaceted (many genes and proteins exhibit pleiotropic activities), and the full range of functionality of individual biological molecules in interactions with other biological systems (as in nanotherapeutics) is often not known. This makes design and prototyping of biological nanodevices an empirically intensive, iterative process [17.3–5, 14].

Biological macromolecules have properties, particularly those relating to their stability, that can limit their use in device contexts. In general, proteins, nucleic acids, lipids, and other biomolecules are more labile to physical insult than are synthetic materials. With the possible exceptions of topical agents or oral delivery and endosomal uptake of nanotherapeutics (both involving exposure to low pH), patients can tolerate conditions encountered by nanobiological therapeutics in vivo, and device lability in the face of *physical* insult is generally a major consideration only in ex vivo settings (relating to storage, sterilization, ex vivo cell culture, and so on). Living organisms remodel themselves constantly in response to stress, development, pathology, and external stimuli. For instance, epithelial tissues and blood components are constantly eliminated and regenerated, and bone and vasculature are continuously remodeled. The metabolic facilities responsible (such as circulating and tissue-bound proteases and other enzymes, various clearance organs and the immune system) can potentially process biological components of nanobiological therapeutic devices as well as endogenous materials, leading to partial or complete degradation of nanotherapeutic structure, function, or both. Furthermore, the host immune and wound responses protect the host against pathogenic organism incursions by mechanisms that involve sequestering and degrading the pathogens. Nanobiological therapeutics are subject to the actions of these host defense systems and to normal remodeling processes. As we will discuss, various strategies to stabilize biomolecules and structures in heterologous in vivo environments are applicable to nanobiological therapeutics [17.3, 16–20]. Conversely, instability of active biocomponents can offer a valuable and simple way to delimit the activity of nanotherapeutics containing biomolecules.

Nanotherapeutic Design Paradigms

Several early attempts to codify the canonical properties of ideal nanobiological devices, and therapeutic nanodevices in particular, have been made [17.5, 13, 15, 21] and are summarized in Table 17.1. In general, nanobiological devices contain biological components that retain their function in new (device) contexts. In other words, one must abstract enough of a functional biological unit from its native context to allow it to perform the function for which it was selected. If one wishes, for example, to appropriate the specific antigen recognition property of an antibody for a device function (say, in targeting, discussed later), it is not necessary to incorporate the entire 150 000 atomic mass unit (AMU) antibody, the bulk of which is devoted to functions other than antigen recognition (see Fig. 17.1 [17.1]), but it *is* critical to incorporate the approximately 20 000 AMU of the antibody essential for specific antibody–antigen binding. Device function is the result of the summed

Table 17.1 Some ideal characteristics of nanodevices. (**A**) Characteristics of all nanobiological devices [17.2, 5, 13–15]. (**B**) Desirable characteristics of therapeutic platforms [17.2, 5, 13–15]

A
Biological molecules must retain function.
Device function is the result of the summed activities of device components.
The relative organization of device components drives device function.
Device functions can be unprecedented in the biological world.

B
Therapeutics should be minimally invasive.
Therapeutics should have the capacity to target sites of disease.
Therapeutics should be able to sense disease states in order to:
– report conditions at the disease site to clinicians.
– administer metered therapeutic interventions.
Therapeutic functions should be segregated into standardized modules.
Modules should be interchangeable to tune therapeutic function.

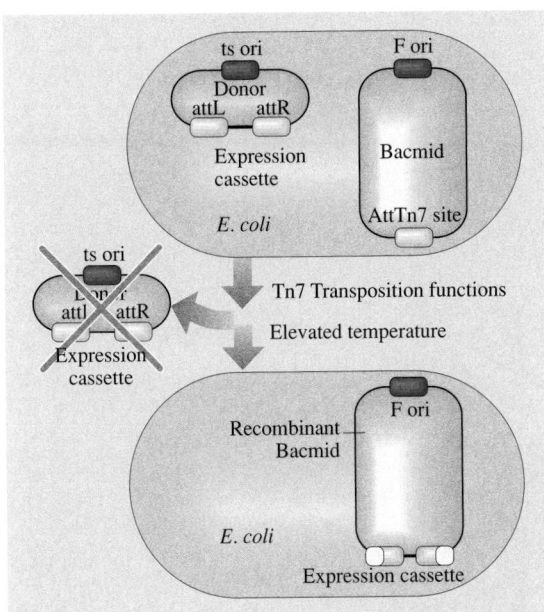

Part B | 17.1

Fig. 17.2 The bacmid molecular cloning system is a molecular device designed to allow efficient production of recombinant insect viruses (baculovirus) in *Escherichia coli* [17.22–24]. Baculovirus is replicated in *E. coli* by the F plasmid origin of replication (F ori), and as such, is called a bacmid. The bacmid also includes an engineered transposable DNA element 7 (Tn7) attachment site isolated from the chromosome of an enteric bacteria (AttTn7). AttTn7 can receive Tn7 elements transposed from other cellular locations. A donor plasmid (donor) is replicated by a temperature-sensitive plasmid pSC101 origin of replication (ts ori). The donor also incorporates an expression cassette containing both the gene of interest for ultimate expression in insect cells and a selectable genetic marker operable in *E. coli*. The expression cassette is flanked by DNA sequences (attL and attR) that are recognized by the Tn7 transposition machinery. Tn7 transposition machinery resides elsewhere in the same *E. coli* cell. When donor plasmid is introduced into *E. coli* containing bacmid, Tn7 transposition machinery causes the physical relocation of expression cassettes from donor plasmid to bacmid. Unreacted donor plasmid is conveniently removed by elevating the incubation temperature, causing the ts pSC101 replicon to cease to function, in turn causing the donor to be lost. If selection for the genetic markers within the expression cassette is applied at this point, the only *E. coli* that survive are those containing recombinant bacmid (those that have received the gene for insect cell expression by transposition from the donor). Recombinant bacmid are conveniently isolated from *E. coli* and introduced into insect cell culture, where expression of the gene of interest occurs

and various activities of biological and synthetic device components, as well, though functional biological components generally exist in the context of higher order systems that support the organisms of which they are a part. The control the nanobiological device designer can exert on the relative organization of biological device components allows biomolecules abstracted from their native context and incorporated into nanobiotechnological devices to contribute to functions entirely different from those they performed in their organismal contexts. All of these features are illustrated in the bacmid, or Bac-to-Bac, system, a commercially available molecular cloning device ([17.22–24] and Fig. 17.2). This system configures prokaryotic genetic elements from multiple sources into a device for producing recombinant eukaryotic viruses, a function that is unprecedented in nature. The system is feasible because of the modularity of the genetic elements involved and because of the strict control of the relative arrangement of genetic elements allowed by recombinant DNA technology. Analogous devices based on bacterial and eukaryotic regulatory elements used to preprogram the micro- and nanoscale architectural properties and physiological behavior of living things are now being realized [17.25, 26].

Bacmid provides an example of a nontherapeutic nanobiological device and illustrates some specific design approaches for building functional devices with biocomponents. Hypothetical properties of nanoscale

devices specifically intended for therapeutic purposes have been codified (Table 17.1) and bear examination [17.13, 15].

In nanotherapeutic applications, devices should be noninvasive and target therapeutic payloads to sites of disease to maximize therapeutic benefit while minimizing undesired side-effects. This of course implies the existence of therapeutic effector functions in these nanodevices, to give devices the ability to remediate a physiologically undesirable condition. Beyond that, several attributes relate to sensing of biomolecules, cells, or physical conditions (sensing disease itself, identification of residual disease, and, potentially, targeting capacity, responding to intrinsic or externally supplied triggers for payload release). Other properties relate to communication between device subunits (for instance, between sensor and effector domains of the device) or between the device and an external operator (external triggering and data

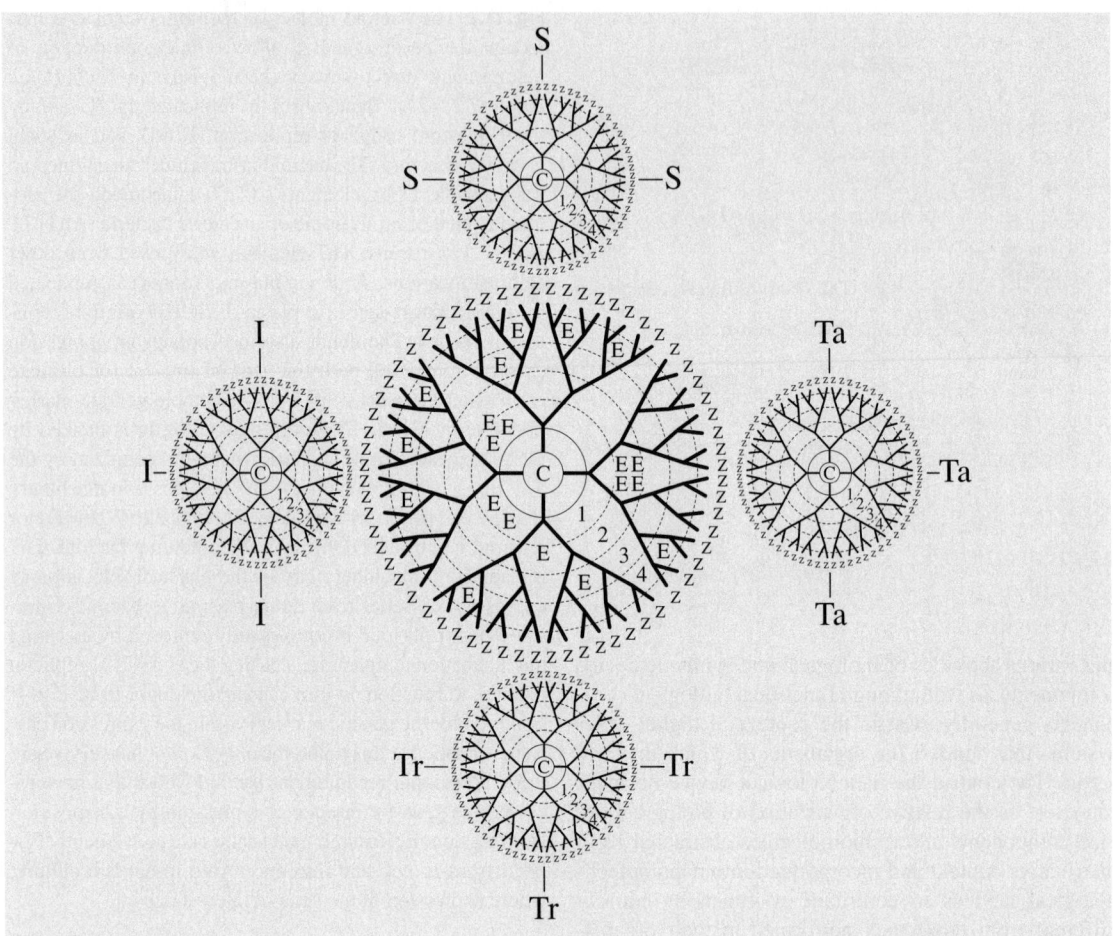

Fig. 17.3 A hypothetical, modular nanotherapeutic patterned after the dendrimer-based cluster agent for oncology of Baker [17.15]. As described in the text, each dendrimer subunit is grown from an initiator core (C), and the tunable surface groups of the dendrimers are represented by Z. Each dendrimer subunit has a specific, dedicated function in the device: the central dendrimer encapsulates small molecule therapeutics (E), whereas other functional components are segregated to other dendrimer components. These include biochemical targeting/tethering functions (Ta), therapeutic triggering functions to allow activation of prodrug portions of the device by an external operator (Tr), metal or other constituents for imaging (I), and sensing functions (S) to mediate intrinsically controlled activation/release of therapeutic. This design constitutes a therapeutic platform [17.13, 15] because of its modular design. The depicted device is only one possible configuration of an almost infinite number of analogous therapeutics that can be tuned to fit particular therapeutic needs by interchanging functional modules

documentation capability). With appropriate design, device functions can be modular [17.13, 15]. As discussed below, this approach allows construction of nanotherapeutic platforms (such as the dendrimer-based therapeutic of Fig. 17.3, dendrimer synthesis and assembly are discussed later), as opposed to single one-off devices that are capable of only one therapeutic task.

The vision of nanoscale therapeutic platforms arose from collaboration between the National Cancer Institute (NCI) and the National Aeronautics and Space Agency (NASA). NASA is concerned with minimal mass therapeutics: therapeutics, along with almost everything else used by astronauts, must be launched from Earth. NCI is interested in early detection of disease to improve prognosis. Since this requires screening a population predominantly of healthy patients, the screening mode must be minimally invasive. The considerations of both agencies might be met with ultrasmall (micro- or nanoscale) multipotent therapeutic devices. Furthermore, the proposed therapeutic platforms should not only remediate undesired physiological conditions but also have the capacity to recognize them and report them. Extensive capability for molecular recognition and communication with external clinicians/operators is integral to the requirements of NASA and NCI. These capacities would allow drugs or other therapeutic interventions to be provided in a controlled fashion, to maximize benefit and minimize side-effects, and is the essence of "smart" therapeutics (see the discussion of targeting and triggering below).

Though substantial progress has been made in device design and realization, no fully realized multipotent nanoscale therapeutic platform has yet been commercialized, but the therapeutic platform paradigm, in which devices are modular, with functional tasks segregated into individual modules, has the potential to be extremely powerful. Classes of broadly similar devices could be tailored to specific disease states by interchang-

ing modules (including targeting modules and drug dispensers) as appropriate to the disease state or therapeutic course. Thus, the hypothetical device represents a possible therapeutic platform composed of functional modules that could be of use in particular indications or in specific individuals.

17.1.2 Utility and Scope of Therapeutic Nanodevices

Therapeutic nanotechnology will be useful, of course, when the underlying biology of the disease states involved is amenable to intervention at the nanoscale. As we will discuss, while several disease states and physiological conditions (such as cancer, vaccination, cardiovascular disease) are particularly accessible to nanoscale interventions, some nanotechnological approaches may be applicable more broadly. Much as was the case with the introduction of recombinant protein therapeutics over the last 20 years, nanotherapeutics may present regulatory and pharmacoeconomic challenges related to their novelty and their cost of goods (COGs).

17.2 Synthetic Approaches: "Top-Down" Versus "Bottom-Up" Approaches for Nanotherapeutic Device Components

Synthesis of nanomaterials is commonly thought of in terms of "top-down" or "bottom-up" processes. Top-down approaches begin with larger starting materials and, in a more or less controlled fashion (depending on the technique), remove material until the desired structure is achieved. Most microfabrication techniques for inorganic materials (such as lithography and milling techniques) fit this description. In contrast, bottom-up approaches begin with smaller subunits that are assembled, again with varying levels of control, depending on technique, into the final product. Key examples of materials made by bottom-up approaches include some inorganic structures. This includes "handmade" structures created by using direct atomic or molecular placement by force microscopy (discussed below), structures built using various deposition or growth methods, as well as all polymerization methods of synthesis. Thus, almost all biologic macromolecules and most biogenic structures, including mineralized biomaterials, are made by bottom-up methods. To make the distinction between top-down and bottom-up approaches more concrete, we will consider an example of each.

17.2.1 Production of Nanoporous Membranes by Microfabrication Methods: A Top-Down Approach

Lithography can be summarized by three basic steps: 1) pattern design (generation of masks), 2) pattern definition (exposure), and 3) pattern transfer (etching/liftoff). Optical lithography uses masks to form patterns on resist/substrate surfaces to produce features. The technique's power lies in its reproducibility and its capacity to manufacture via highly parallel processes. The key limitation of photolithography lies in the fact that resolution of features is diffraction-limited by the wavelength of light used. To address this limitation, short-wave radiation (X-rays with wavelengths of about 1 nm) can be generated by synchrotron or other sources (from X-ray tubes, discharge plasma, or laser plasma), controlled, and focused for use in X-ray lithographic techniques [17.27]. The process is identical conceptually to optical lithography but requires special masks and resists amenable to the high-frequency radiation used. Combinations of filters and mirrors can produce resolutions in feature size of less than 100 nm, with fabrication throughputs congruent with those of other optical lithography processes. Limited access to synchrotron sources in turn limits the wide application of the method, however. The relatively long wavelengths used in conventional photolithography are generally unsuitable for formation of nanoscale features unless some clever technical expedient, like the use of a sacrificial layer (described below for generating nanoporous membranes) is employed.

As we will discuss when considering applications of nanotherapeutic devices, there are numerous potential applications for tunable nanopore membranes [17.16–18]. For certain applications, such as immunoisolation (see below), the distribution of pore sizes must be tight and nearly perfect. Until the late 1990s, this was an unattainable objective.

The key technical innovation facilitating the microfabrication of highly defined nanoporous membranes was an approach featuring a sacrificial oxide layer sandwiched between two structural layers that ultimately is etched away to define the pore pathway and diameter [17.16–18]. The sacrificial layer (SiO_2) is sandwiched between a silicon wafer and a polysilicon layer: pore channel diameter is determined by

the SiO_2 thickness. In the process (Fig. 17.4), the top of a silicon wafer is doped with boron to increase its mechanical robustness, and p+ silicon is overlaid. Pore exit holes are plasma-etched through the p+ and doped material. In the pore dimension-determining step, SiO_2 is grown on the wafer by dry thermal oxidation, a process allowing control of oxide layer thickness to within 1 nm. This oxide constitutes the critical sacrificial layer. A thick polysilicon layer (again, boron-doped for mechanical strength) is then deposited over the oxide. Pore entry holes are etched by plasma etch, offset from the exit pores. Ultimately, the offset will require the diffusion pathway of the finished structure to pass through a "bottleneck" whose diameter is determined by the thickness of the sacrificial layer. The backside of the wafer is anisotropically etched to expose the doped layer of the wafer (now the bottom of the membrane structure). Pores are opened by removing exposed sacrificial layer using concentrated hydrofluoric acid.

17.2.2 Synthesis of Poly(amido) Amine (PAMAM) Dendrimers: A Bottom-Up Approach

PAMAM dendrimers are remarkably defined synthetic molecules made using polymer chemistry (Fig. 17.5). Their unique structural attribute is their fractal geometry, and their unique physical property is their high monodispersity. Clever use of orthogonal conjugation strategies in their synthesis drives their monodispersity [17.28–30].

Orthogonal conjugation involves the mutually exclusive reactivity of the chemical specificities present in the reaction: when the orthogonal reactants are present in a vessel, and the reaction is carried at appropriate stoichiometry, only a monodisperse single product is formed. PAMAM dendrimers are among the most monodisperse synthetic materials available and, within limits, their sizes, surface chemistries, and shapes can be controlled at the synthetic level. To a greater or lesser extent, these desirable material properties of dendrimers are caused by an orthogonal synthetic strategy.

PAMAM dendrimers are usually built from an initiator core (an amine) in sequential shells, called generations ($G_0, G_1, G_2, ...$). Each generation comprises two synthetic half-steps, each of which is self-limiting in that the reagents performing each addition step are reactive only with the distinct chemical functionalities added to the growing dendrimer in the *previous* addition reaction. Reagent is added until the chemical functionalities

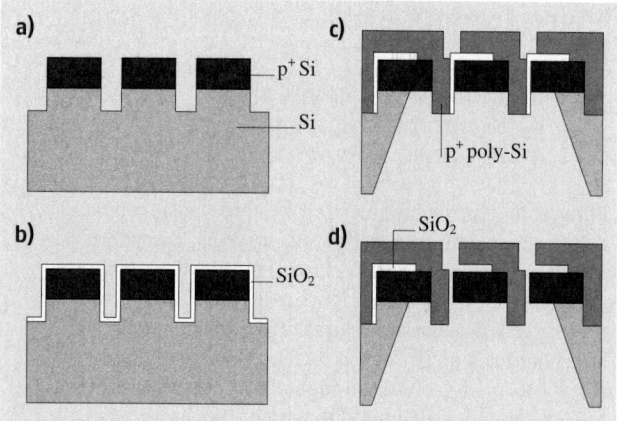

Fig. 17.4a–d Nanoporous silicon membranes are fabricated by a top-down approach [17.16–18]. (**a**) Pore exit holes are plasma-etched through p+ silicon (*dark brown*) and boronated silicon (*gray*). (**b**) A controlled depth silicon oxide (*white*) layer is grown on the structure. (**c**) A thick polysilicon layer (*light brown*) is deposited over the oxide, and pore exits are established by back-side plasma etch. (**d**) Pores are opened by removal of the sacrificial layer with hydrofluoric acid. The membranes themselves are macroscale objects that fit under the nanotechnology rubric because they derive their useful properties from controlled nanoscale architecture

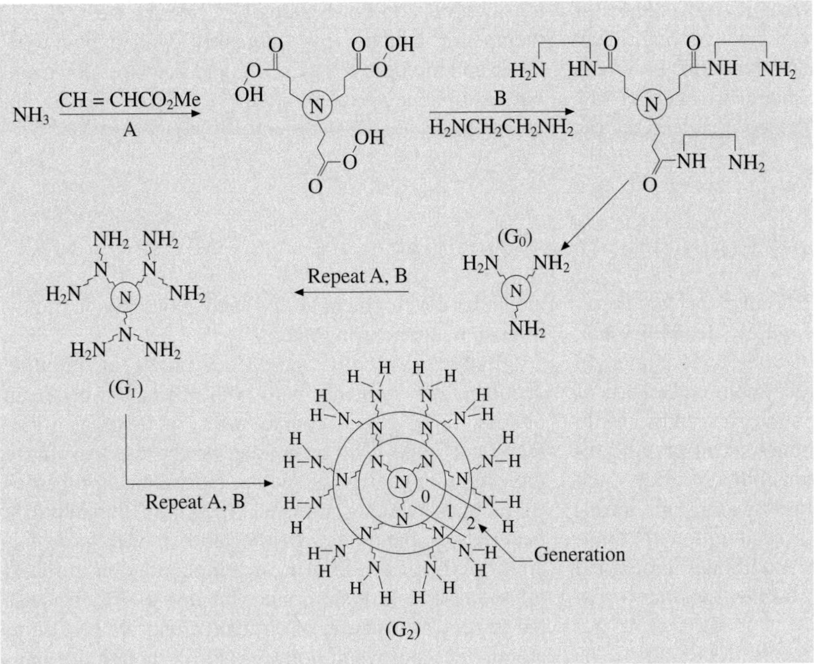

Fig. 17.5 PAMAM dendrimers are grown bottom-up using sequential orthogonal synthetic steps [17.28–30]. Dendrimers are grown from amine cores, in this case ammonia, by sequential addition of methyl acrylate, followed by an independent reaction addition reaction involving ethylene diamine. Together, these two half-steps constitute a full generation of dendrimer growth. PAMAM dendrimers are classified by their generation of growth as being G_0, G_1, G_2, ..., as indicated in the figure

from the previous step are fully consumed (Fig. 17.5). In the first synthetic half-step, methyl acrylate is added to the amine precursor, adding a carboxy-terminated functionality to all amine branch points. Polymer intermediates terminating in such carboxy functionalities are called "half-generations." In the second half-step, the generation is completed. Ethylene diamine is added, resulting in a branched amine-terminated adduct for each carboxylate group of the half-generation precursor. In general, before proceeding to downstream synthetic operations, products of individual reactions are purified so that only functionalities incorporated into growing dendrimer are available for subsequent reactions. Under ideal conditions, only a single chemical structure can result from each generation of growth, and with adequate purification between addition reactions; in the absence of limitations on the completeness of each reaction (such as steric limitations which become manifest in dendrimers of generation six and above), dendrimers of each generation are perfectly identical to each other.

17.2.3 The Limits of Top-Down and Bottom-Up Distinctions with Respect to Nanomaterials and Nanodevices

Some materials can be produced by alternate means, some of which are bottom-up, some of which are top-down approaches. For instance, carbon nanotubes can be synthesized in an arguably top-down approach from graphite sheets in an arc oven or can be grown bottom-up, by a metal-catalyzed polymerization method [17.31]. Additionally, not all finished materials can be classified as either top-down or bottom-up: synthetic protocols can contain both steps. For instance, while proteins are synthesized from lower molecular weight amino acid precursors by chemically or biologically mediated polymerization (bottom-up), they are often made as precursor molecules that are processed to a final product by chemical or enzymatic cleavage (top-down, see [17.32] for an overview). While the question of whether any given material is made top-down or bottom-up can be ambiguous, the synthetic provenance of multicomponent nanodevices can be even more so. In analogy to biogenic materials, synthetic polymers are bottom-up materials per se, but fabrication of raw polymeric materials into final device architectures of-ten involves top-down steps [17.33]. As we will see, therapeutic nanodevices frequently contain both synthetic and biologic components, and strict top-down and bottom-up categorization of these devices is often not applicable.

17.3 Technological and Biological Opportunities

This section considers selected enablers for therapeutic nanodevices. Some are purely technological: nanomaterial self-assembly properties, bioconjugation methods, engineered polymers for conditional release of therapeutics, external triggering strategies, and so forth. Others relate to disease-state tissue or cell-specific biology that can be exploited by nanotherapeutics, such as the emerging vascular address system and intrinsic triggering approaches. In association with those applications, we will consider additional biological opportunities for nanoscale approaches specific to particular disease states.

17.3.1 Assembly Approaches

Assembly of components into devices is amenable to multiple approaches. In the case of devices comprising a single molecule or processed from a single crystal (some microfabricated structures, single polymers, or grafted polymeric structures), assembly may not be an issue. Integration of multiple, separately microfabricated components may sometimes be necessary (as in the immunoisolation capsule discussed below) and may sometimes drive the need for assembly, even for silicon devices. Furthermore, many therapeutic nanodevices contain multiple, chemically diverse components that must be assembled precisely to support their harmonious contribution to device function.

"One-Off" Nanostructures and Low-Throughput Construction Methods

Direct-write technologies can obtain high (nanometer scale) resolution. For instance, electron beam (e-beam) lithography is a technique requiring no mask, and that can yield resolutions on the order of tens of nanometers, depending on the resist materials used [17.34]. Resolution in e-beam lithography is ultimately limited by electron scattering in the resist and electron optics, and like most direct-write approaches, e-beam lithography is limited in its throughput. Parallel approaches involving simultaneous writing with up to 1000 shaped e-beams are under development [17.34] and may mitigate limitations in manufacturing rate.

Force microscopy approaches utilize an ultrafine cantilever tip (typically with point diameters of 50 nm or less, Fig. 17.6) in contact with, or tapping, a surface or a stage. The technique can be used to image molecules, to analyze molecular biochemical properties (like ligand-receptor affinity [17.35]), or to manipulate materials at nanoscale. In the latter mode, force microscopy has been used to manipulate atoms to build individual nanostructures since the mid-1980s. This has led to the construction of structures that are precise to atomic levels of resolution (Fig. 17.6), though the manufacturing throughput of "manual" placement of atoms by force microscopy is limited.

Dip-pen nanolithography (DPN) is a force microscopy methodology that can achieve high-resolution features (features of 100 nm or less) in a single step. In DPN, the AFM tip is coated with molecules to be deployed on a surface, and the molecules are transferred from the AFM tip to the surface as the coated tip contacts it. DPN also can be used to functionalize surfaces with two or more constituents and is well suited for deployment of functional biomolecules on synthetic surfaces with nanoscale precision [17.36, 37]. DPN suffers the limitations of synthetic throughput typical of AFM construction strategies.

Much as multibeam strategies might improve throughput in e-beam lithography [17.34], multiple tandem probes may increase assembly throughput for construction methods that depend on force microscopy significantly, but probably not enough to allow manufacture of bulk quantities of nanostructures, as will likely be needed for consumer nanotherapeutic devices. However, as standard of care evolves increasingly toward tailored courses of therapy [17.38] and individual therapeutics become increasingly multicapable and powerful, relatively low-throughput synthesis/assembly methods may become more desirable. For the moment, though, ideal manufacturing approaches for nanotherapeutic devices resemble either industrial polymer chemistry, occurring

a)

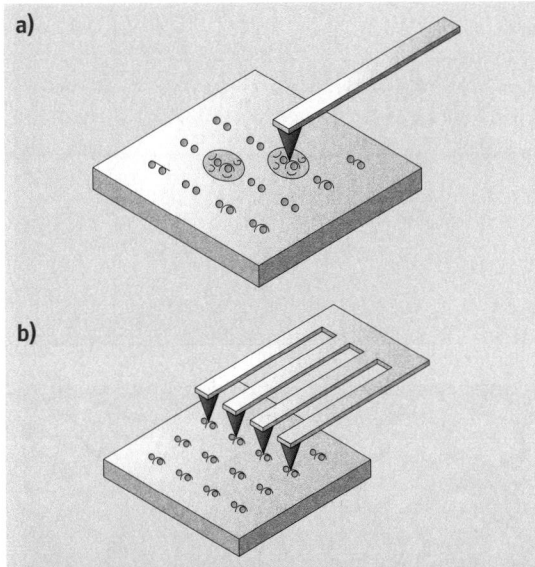

b)

c)

Fig. 17.6 (a) A schematic depiction of an atomic force microscope cantilever and tip interacting with materials on a surface. Tips typically have points of 50 nm or less in diameter [17.31, 39]. **(b)** Schematic of multiplexed AFM tips performing multiple operations in parallel [17.31, 39]. **(c)** AFM image of a quantum corral, a structure built using AFM manipulation of individual atoms (from the IBM Image Gallery)

in bulk, in convenient buffer systems, or in massively parallel industrial microfabrication approaches. In any case, therapy for a single patient may involve billions and billions of individual nanotherapeutic units, so each individual nanotherapeutic structure must require only minimal input from a human synthesis/manufacturing technician.

Self-Assembly of Nanostructures

Self-assembly has been long recognized as a potentially critical labor-saving approach to construction of nanostructures [17.40], and many organic and inorganic materials have self-assembly properties that can be exploited to build structures with controlled configurations. Self-assembly processes are driven by thermodynamic forces and generally result in structures that are not covalently linked. The intra/intermolecular forces driving assembly can be electrostatic or hydrophobic interactions, hydrogen bonds, and van der Waals interactions between and within subunits of the self-assembling structures and the assembly environment. Thus, final configurations are limited by the ability to "tune" the properties of the subunits and control the assembly environment to generate particular structures.

Self-Assembly of Carbon Nanostructures

Carbon nanotubes (Fig. 17.7) spontaneously assemble into higher order [17.31] structures (nanoropes) as the result of hydrophobic interactions between individual tubes. Multiwall carbon nanotubes (MWCNTs) are well-known structures that can be viewed as self-assembled, nested structures of nanotubes with tube diameters decreasing serially from the outermost to innermost tubes. The striking resemblance that MWCNTs have to macroscale bearings has been noted and exploited [17.41]. MWCNT linear bearings can be actuated by applying mechanical force to the inner nanotubes of the MWCNT assembly. Actuation causes the assembly to undergo a reversible telescoping motion. Interestingly, these linear MWCNT bearings exhibit essentially no wear as the result of friction between bearing components.

C_{60} fullerenes and single-wall carbon nanotubes (SWCNT) also spontaneously assemble (Fig. 17.7) into higher order nanostructures called "peapods" [17.42] in which fullerene molecules are encapsulated in nanotubes. The fullerenes of peapods modulate the local electronic properties of the SWCNTs in which they are encapsulated and may allow tuning of carbon nanotube electrical properties. The potentially fine control of nanotube properties may prove useful in nanotube-containing electrical devices, particularly in cases where nanotubes are serving as molecular wires. In this capacity, carbon nanotubes have been incorporated into FETs (field-effect transistors, discussed below under sensing architectures [17.43]), and other molecular electronic structures. Ultimately, these architectures may result in powerful, ultrasmall computers to provide the

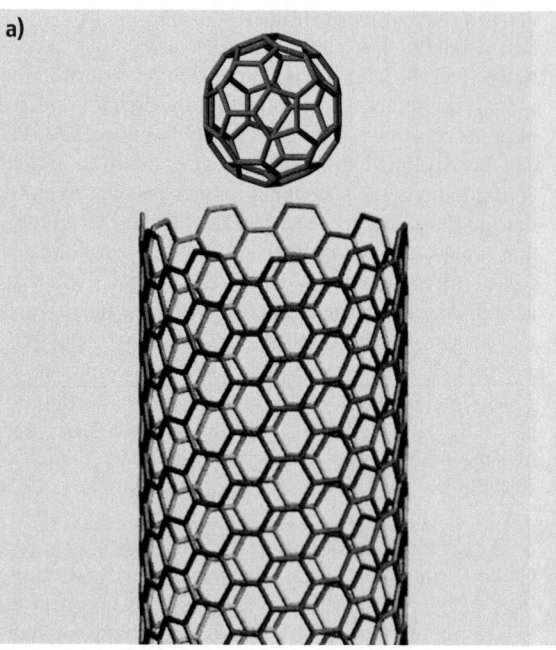

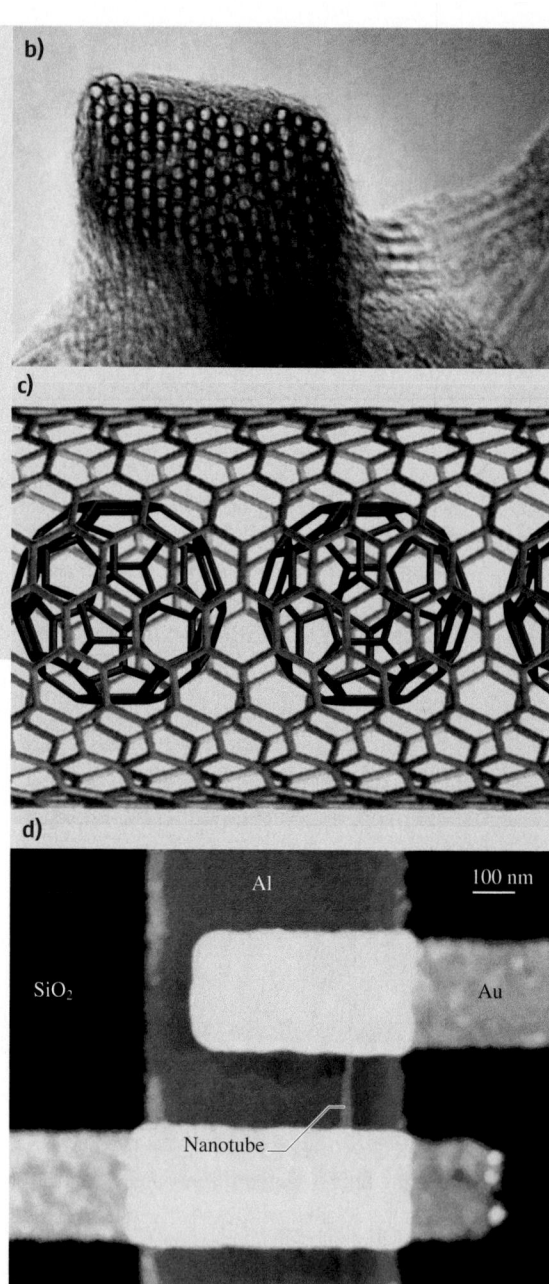

Fig. 17.7 (a) Shown to scale are two highly defined carbon nanostructures: a C_{60} fullerene and (10, 10) single-wall carbon nanotube (SWCNT). **(b)** A self-assembled nanorope composed of carbon nanotubes that assemble by virtue of hydrophobic interactions [17.31, 39]. **(c)** Schematic depiction of another self-assembled carbon nanostructure (a peapod) consisting of the fullerenes and SWCNT of **(a)**, and wherein the fullerenes are encapsulated in the SWCNT [17.42]. Fullerene encapsulation in the peapods modulates local electronic properties of the SWCNT. **(d)** A nanotube field effect transistor (FET) consisting of gold source and drain electrodes on an aluminium stage with a carbon nanotube serving as the FET channel [17.43]

intelligence of "smart," indwelling nanotherapeutic devices. In general, though, fullerenes, nanotubes carbon nanoropes, nanotube bearings, and peapods have yet to find extensive biological application, in part because of their extreme hydrophobicity, presumably poor biocompatibility, and high chemical stability [17.31, 39]. However, controlled derivatization of nanotubes may be possible through a number of approaches [17.31,39], including controlled introduction of bond strain to render individual carbons of the tubes selectively chemically reactive (so-called mechanosynthesis). Carbon nanotubes and fullerenes, however, owe many of their remarkable properties (chemical stability, mechanical robustness, some electrical properties) to the fact that all of the valences of the constituent carbon atoms (except those at the ends of nanotubes that are "open") are satisfied. Thus, to derivatize carbon nanostructures is to degrade them, a fact one must consider when the design purposes that drove incorporation of carbon structures into a therapeutic depends on the chemical perfection of the material.

Self-Assembly of Materials
Made by Traditional Polymer Chemistry

In the realms of drug delivery and biomedical micro and nanodevices, the most familiar self-assembled structures are micelles [17.44–46]. These structures are formed from the association of block copolymer subunits (Fig. 17.8), each individual subunit containing hydrophobic and hydrophilic domains. Micelles spontaneously form when the concentration of their subunits exceeds the critical micelle concentration (cmc) in a solvent in which one of the polymeric domains is immiscible (Fig. 17.8). The cmc is determined by the immiscible polymeric domain and can be adjusted by controlling the chemistry and length of the immiscible domain, as well as by controlling solvent conditions. Micelles formed at low concentrations from low-cmc polymers are stable at high dilution. Micelles formed from polymer monomers with high cmcs can dissociate upon dilution, a phenomenon that might be exploited to control release of therapeutic cargos. If desired, micelles can be stabilized by covalent cross-linking to generate shell-stabilized structures [17.44–47].

The size dispersity and other properties of micelles can be manipulated by controlling solvent conditions, incorporating excipients (to modulate polymer packing properties), temperature and agitation. From the standpoint of size, reasonably monodisperse preparations (polydispersity of 1–5%) of nanoscale micellar structures can be prepared [17.44–47]. The immense versatility of industrial polymer chemistry allows micellar structures to be tuned chemically to suit the task at hand. They can be modified for targeting or to support higher order assembly properties. They can be made to imbibe therapeutic or other molecules for delivery and caused to dissociate or disgorge themselves of payloads at desired times or bodily sites under the influence of local physical/chemical conditions. The tunability of these and other properties at the level of monomeric polymer subunits as well as the level of assembled higher order structures make micelles potentially powerful nanoscale drug delivery and imaging vehicles.

Fractal materials, such as dendritic polymers, whose synthesis and structure were discussed previously, exhibit packing properties that can be exploited to assemble higher order aggregate structures called "tecto(dendrimers)" [17.48]. In fact, these self-assembly properties are being exploited in oncological nanotherapeutics, as Fig. 17.3 shows [17.15, 49]. In principle, these self-assembling therapeutic complexes need not be preformed prior to administration. Individual functional modules of the therapeutic assembly might be administered sequentially, potentially to tailor therapies more precisely to individual patient responses.

Stoichiometric Control and Self-Assembly

As the preceding examples demonstrate, self-assembly approaches sometimes do not feature precise control of subunit identity and stoichiometry in the assembled complexes. This can be a limitation when the stoichiometry and relative arrangement of differentiable individual subunits is critical to device function. Stoichiometry is less an issue when the self-assembling components are identical and functionally fungible, as in the synthetic, peptidyl anti-infective illustrated in Fig. 17.9 [17.50,51].

In the anti-infective architecture, individual peptide components are flat, circular molecules. The planar character of the toroidal subunits is a consequence of the alternating chirality of alternating D-L amino acids (aas) in the primary sequence of the peptide rings. Alternating D and L aas is not possible in proteins made by ribosomal synthesis [17.32]. Ribosomes recognize and incorporate into nascent polypeptides only L amino acids, and so, as the result of aa chirality and bond strain, peptides made by ribosomes cannot be made flat, closed toroids like those of the peptidyl anti-infectives. Much as in α-helical domains of ribosomally synthesized proteins, however, the aa R-groups (which are of varying hydrophobic or hydrophilic chemical specificities [17.32]) are arranged in the plane of the closed D, L rings extending out from the centers of the rings. Hydrogen bonds between individual rings govern self-assembly of the toroids into rod-like stacks, while the R-groups dominate interactions between multiple stacks of toroids and other macromolecules and structures (see below).

The planar toroidal subunits can be administered as monomers and self-assemble into multitoroid rods at the desired site of action (in biological membranes). But the peptide toroids' R-groups are chemically tuned so that the rod structures into which they spontaneously assemble intercalate preferentially in specific lipid bilayers (in pathogen vs. host membranes). Moreover, the assembled rods may undergo an additional level of self-assembly into multirod structures, spanning pathogen membranes [17.50, 51]. Whether as single-rod or multirod assemblies, membrane intercalation by stacked toroids reduces the integrity of pathogen membranes selectively, and therefore particular toroid species exhibit selective toxicity to specific pathogens.

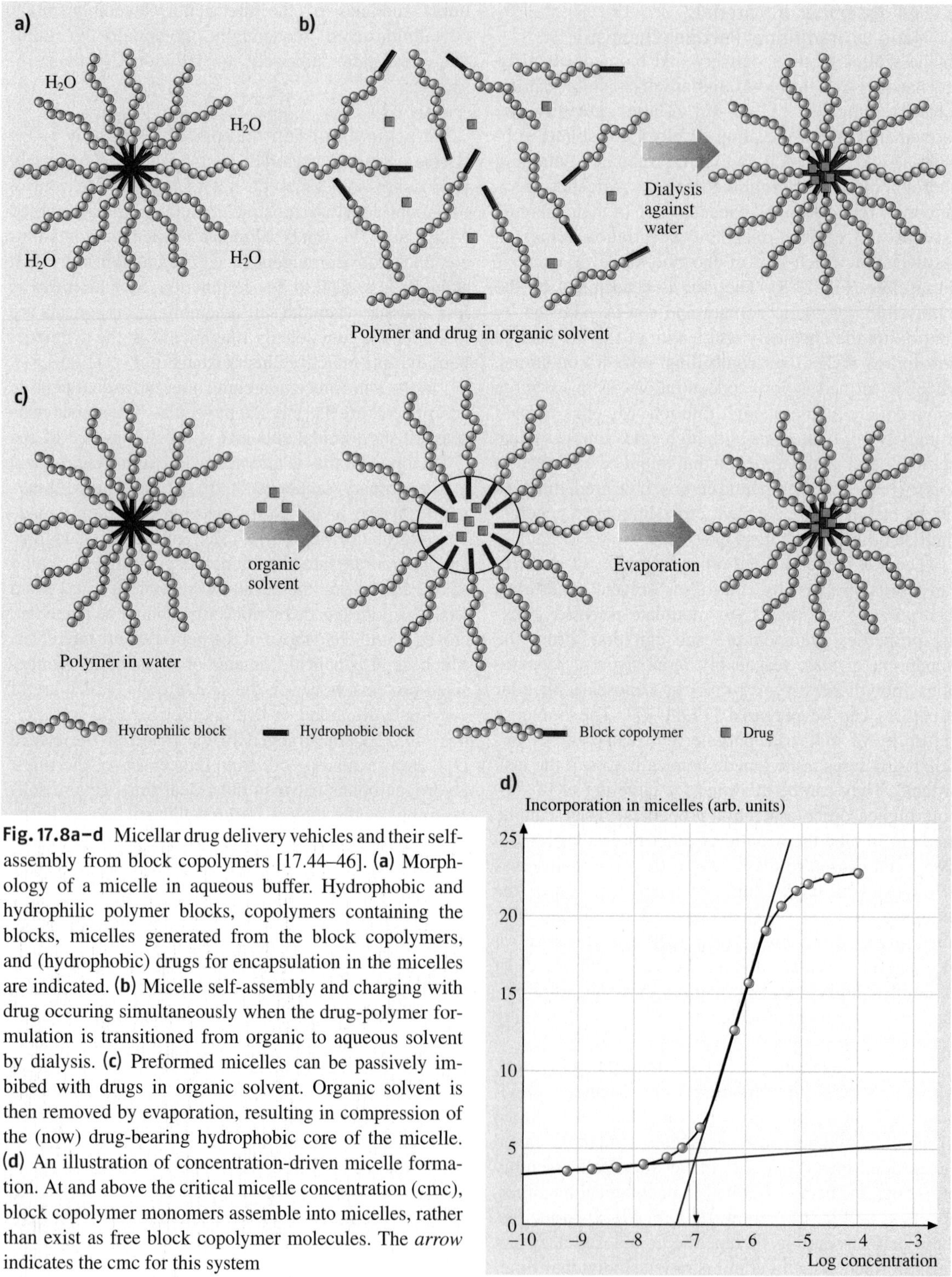

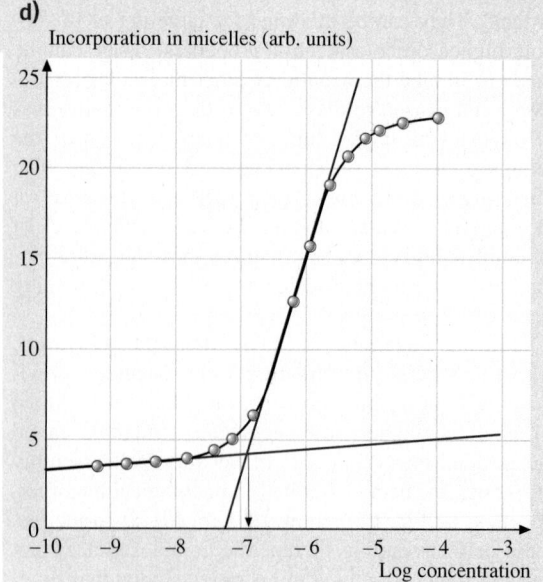

Fig. 17.8a–d Micellar drug delivery vehicles and their self-assembly from block copolymers [17.44–46]. (**a**) Morphology of a micelle in aqueous buffer. Hydrophobic and hydrophilic polymer blocks, copolymers containing the blocks, micelles generated from the block copolymers, and (hydrophobic) drugs for encapsulation in the micelles are indicated. (**b**) Micelle self-assembly and charging with drug occuring simultaneously when the drug-polymer formulation is transitioned from organic to aqueous solvent by dialysis. (**c**) Preformed micelles can be passively imbibed with drugs in organic solvent. Organic solvent is then removed by evaporation, resulting in compression of the (now) drug-bearing hydrophobic core of the micelle. (**d**) An illustration of concentration-driven micelle formation. At and above the critical micelle concentration (cmc), block copolymer monomers assemble into micelles, rather than exist as free block copolymer molecules. The *arrow* indicates the cmc for this system

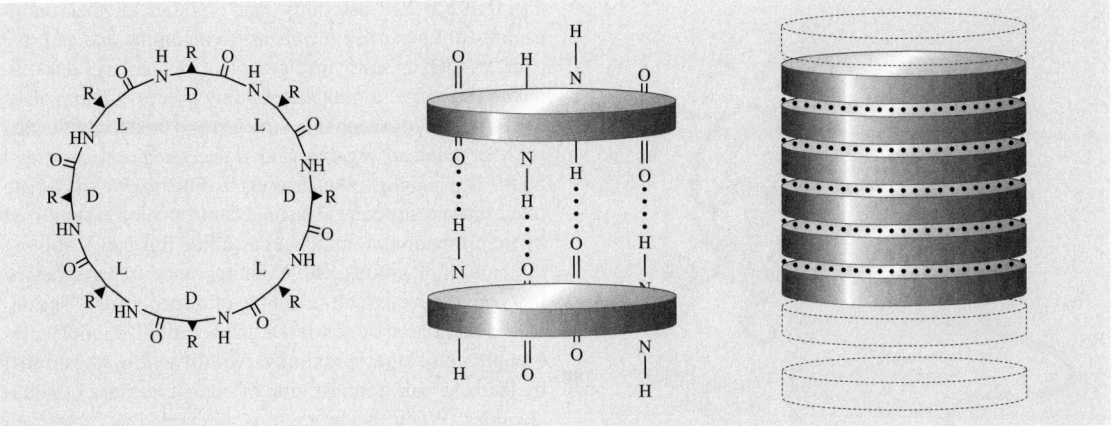

Fig. 17.9 A self-assembling peptide antibiotic nanostructure [17.50, 51]. Peptide linkages and the α-carbons and their pendant R groups are indicated. The synthetic peptide rings are planar as the result of the alternating chirality (D or L) of their amino acid (aa) constituents. R groups of aas radiate out from the center of the toroid structure. Individual toroids self-assemble (stack) as the result of hydrogen bonding interactions between amine and carboxy groups of the peptide backbones of adjacent toroids. The surface chemistry of multitoroid stacks is tuned at the level of the aa sequence and therefore the R group content of the synthetic peptide rings. The chemical properties of the stacked toroid surfaces allow them to intercalate into the membranes of pathogenic organisms with lethal consequences. The specific membrane preferences for intercalation of the compound are tuned by controlling the R group contents of toroids

These toroidal, synthetic antibiotics, and other nanoscale antimicrobials represent critically needed, novel antibacterial agents. Resistance to traditional, microbially derived antibiotics is often tied to detoxifying functions associated with secondary metabolite synthesis; these detoxifying functions are essential for the viability of many antibiotic-producing organisms (for instance, see [17.52]). The genes encoding such detoxifying functions are rapidly disseminated to other microorganisms, accounting for the rapid evolution of drug resistant organisms that has bedeviled antimicrobial chemotherapy for the last 25 years. Synthetic nanoscale antibiotics, like the peptide toroids [17.50, 51] and the N8N antimicrobial nanoemulsion [17.53], act by mechanisms entirely distinct from those of traditional secondary metabolite antibiotics, and no native detoxifying gene exists. Therefore, novel nanoscale antimicrobials may not be subject to the unfortunately rapid rise in resistant organisms associated with most secondary metabolite antibiotics, although this remains to be seen. As bacterial infection continues to re-emerge as a major cause of morbidity and mortality in the developed world, a consequence of increasing antibiotic-resistant pathogens, novel nanoscale antibiotics will become more important.

Biomolecules in Therapeutic Nanodevices: Self-Assembly and Orthogonal Conjugation

Biological macromolecules undergo self-assembly at multiple levels, and like all instances of such construction, biological self-assembly processes are driven by thermodynamic forces. Some biomolecules undergo intramolecular self-assembly (as in protein folding from linear peptide sequences, Fig. 17.10). Higher order structures are, in turn, built by self-assembly of smaller self-assembled subunits (for instance, structures assembled by hybridization of multiple oligonucleotides, enzyme complexes, fluid mosaic membranes, ribosomes, organelles, cells and tissues).

Proteins are nonrandom copolymers of 20 chemically distinct amino acid (aa) subunits [17.32]. The precise order of aas (i. e., via interactions between aa side chains) drives the linear polypeptide chains to form specific secondary structures (the α helices and β sheet structures seen in Fig. 17.10). The secondary structures have their own preferences for association, which, in turn, leads to the formation of the tertiary and quarternary structures that constitute the folded protein structures. In its entirety, this process produces consistent structures that derive their biological functions from strict control of the deployment of chemical specificities (the aa side chains) in three-dimensional space.

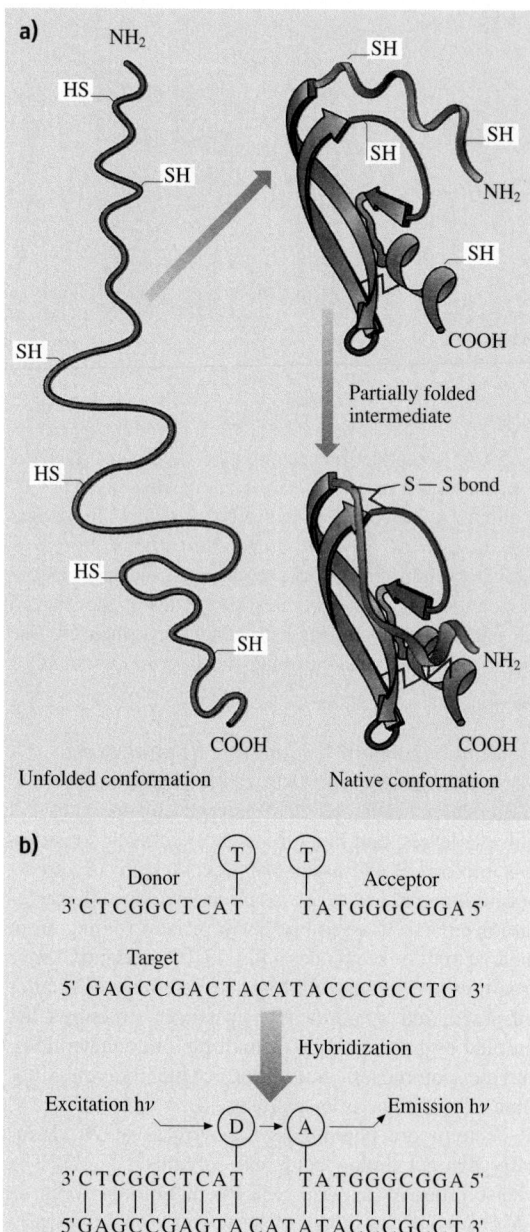

a)

Unfolded conformation

Partially folded intermediate

Native conformation

S—S bond

b)

Donor   Acceptor

3' CTCGGCTCAT   TATGGGCGGA 5'

Target

5' GAGCCGACTACATACCCGCCTG 3'

Hybridization

Excitation hν Emission hν

D → A

3' CTCGGCTCAT   TATGGGCGGA 5'
   | | | | | | | | | | | | | | | | | | | |
5' GAGCCGAGTA CATATACCCGCCT 3'

Fig. 17.10a,b Self-assembly and biological macromolecules. **(a)** Linear peptide chains (with amino and carboxy ends, as well as sulfhydral groups of cysteine residues indicated) undergo a multistep folding process that involves the formation of secondary structures (α-helices, indicated by heavy helical regions, and β-sheet regions, indicated by the heavy antiparallel arrows) that themselves associate into a tertiary structure. The final conformation is stabilized by the formation of intrachain disulfide linkages involving cysteine thiol groups. **(b)** A fluorescence transfer device that depends on the self-assembly of biomolecules. The device is composed of donor (D) and acceptor (A) molecules brought into close proximity (within a few angstroms) by the base pair hybridization of complementary oligonucleotides. When the structure is assembled, acceptor and donor are energetically coupled, and fluorescence transfer can occur [17.54–57]

Biomolecules can be used to drive assembly of nanostructures, either as free molecules or conjugated to heterologous nanomaterials. For instance, three-dimensional nanostructures can be made by DNA hybridization [17.54–57]. Such DNA nanostructures can exhibit tightly controlled topographies but limited integrity in terms of geometry [17.57], due to the flexibility of DNA strands. Oligonucleotides, antibodies and other specific biological affinity reagents can also be used to assemble nanostructures (Fig. 17.10, see also the targeting and triggering discussions below). Often the domains of biomolecules responsible for assembly and recognition are small, continuous, and discrete enough that they can be abstracted from their native context as modules and appended to other nanomaterials of interest to direct formation of controlled nanoscale architectures (for instance, see the SCFv antibody fragment of Fig. 17.1).

Several orthogonal bioconjugate approaches have arisen from the field of protein semisynthesis [17.20, 58, 59]. These protein synthetic chemistries allow site-specific conjugation of polypeptides to heterologous materials in bulk, as the result of conjugation between exclusively, mutually reactive electrophile–nucleophile pairs (analogous to dendrimer synthesis discussed above, see Table 17.2). They have been applied to the synthesis of multiple therapeutic nanodevices [17.2–5, 7, 14, 58, 59].

As described above, proteins are profoundly dependent on their three-dimensional shapes: chemical derivatization at critical aa sites can profoundly impact protein bioactivity. Because conjugation can be directed to preselected sites via orthogonal approaches, and since the sites of conjugation in the protein can be chosen because the proteins involved tolerate adducts at those positions, proteins coupled to nanomaterials by such orthogonal methodologies often retain their biological activity. In contrast, protein bioactivity in conjugates is generally lost or profoundly impaired when proteins are coupled to nanomaterials using promiscuous chemistries. For instance, proteins such

Electrophile	Nucleophile	Product
1. (ketone/aldehyde) R, R' with C=O	H_2N-O-R'' Aminooxy	O–R'' oxime Oxime
2. (ketone/aldehyde) R, R' with C=O	H_2N–N(H)–C(=O)–R'' Hydrazide	Hydrazone
3. (ketone/aldehyde) R, R' with C=O	H_2N–N(H)–C(=S)–N(H)–R'' Thiosemicarbazide	Thiosemicarbazone
4. R–CHO (R, H)	H_2N–CH(R')–CH$_2$–SH β-Aminothiol	Thiazolidine
5. R–C(=O)–S^- Thiocarboxylate	Br–CH$_2$–C(=O)–R' α-Halocarbonyl	Thioester
6. R–C(=O)–S–R Thioester	H_2N–CH(C(=O)R / S^-) Cysteine	$R-C(=O)-NH-CH(SH)-C(=O)-R'$ Amide

Table 17.2 Orthogonal conjugation chemistries originally derived for protein semisynthesis but applicable to conjugation of proteins to nanomaterials [17.20, 58, 59] (R' = H: aldehyde, R = alkyl: ketone)

as cytokines (and other protein hormones) elicit their effects by interacting with a receptor, and a large fraction of their surface (20% or more) is involved in receptor binding, directly or indirectly. Promiscuous chemistries (1-ethyl-3-(3-diaminopropyl) carbodiimide or EDC conjugation, see [17.60]) used to conjugate cytokines to nanoparticles tend to inactivate hIL-3 and other cytokines whereas the same protein/particle bioconjugates retain bioactivity if judiciously chosen orthogonal conjugation strategies are used (Lee and Parthasarathy, unpublished). Proteins for which only a small portion of their surfaces contribute to the interesting portions of their bioactivities (from the standpoint of the nanodevice designer), such as some enzymes or intact antibodies (Fig. 17.1), may be somewhat less sensitive to promiscuity of the bioconjugate strategy used [17.20, 58, 59], but the benefits of orthogonal conjugation strategies can also apply to these protein bioconjugates [17.20, 58, 59]. The potential utility of orthogonal conjugation for incorporation of active biological structures into semisynthetic nanodevices is becoming more fully recognized and cannot be overestimated [17.58, 59].

17.3.2 Targeting: Delimiting Nanotherapeutic Action in Three-Dimensional Space

Delivery of therapeutics to sites of action is a key strategy for enhancing clinical benefit, particularly for drugs useful within only narrow windows of concentration due to their toxicity (drugs with narrow therapeutic windows). Diverse targeting approaches are available, ranging from methods exploiting differential extravasation limits of vasculature of different tissues (see the discussion of oncology below), sizes, and surface chemistry preferences for cellular uptake (see discussion of vaccines below); preferential partition of molecules and particles into specific tissues by virtue of their charges, sizes, surface chemistry, or extent of opsonization (see below); or the affinity of biological molecules decorating the nanodevice for counter-receptors on the cells or tissues of interest.

The Reticuloendothelial System and Clearance of Foreign Materials

Physical properties such as surface chemistry and particle size can drive targeting of nanomaterials (and presumably nanodevices containing them) to some tissues. For instance, the pharmacokinetic (P_k) and biodistribution (B_d) properties [17.32] of many drugs and nanomaterials are driven by their clearance in urine, which is in turn governed by the filtration preferences of the kidney. Most molecules making transit into urine have masses of less than 25 to 50 kilodaltons (kDa; 25–50 kDa particles corresponding *loosely* to effective diameters of about 5 nm or less) and are preferably positively charged; these parameters are routinely modulated to control clearance rates of administered drugs. Clearance of low molecular weight (nano)materials in urine

can be suppressed by tuning their molecular weights and effective diameters, which is typically accomplished by chemical conjugation to polymers such as poly(ethylene glycol) [17.19]. Polymer conjugation (pegylation) has been applied to many different materials and may provide some degree of charge shielding. Pegylation also increases the effective molecular weights of small materials above the kidney exclusion limit, diverting them from rapid clearance in urine.

Coating foreign particles with serum proteins (opsonization [17.61]) is the first step in the clearance of foreign materials. Opsonized particles are recognized and taken up by tissue dendritic cells (DCs) and specific clearance organs. These tissues (thymus, liver and spleen, constituting the organs of the reticuloendothelial system or RES) extract materials from circulation by both passive diffusion and active processes (receptor-mediated endocytosis). Charge-driven, receptor-mediated uptake of synthetic nanomaterials occurs in the RES and can result in partition of positively charged nanoparticles into the RES. For instance, PAMAM dendritic polymers exhibit high positive charge densities related to the large number of primary amines on their surfaces [17.28–30]. In experimental animals, biodistribution of unmodified PAMAM dendrimers is limited nearly exclusively to RES organs [17.62]. This unfavorable biodistribution can be modulated by "capping" the dendrimers (derivatizing the dendrimer to another chemical specificity, such as carboxy or hydroxyl functionalities [17.15, 62]).

Despite legitimate applications of targeting to the kidney and the RES (for instance in glomerular disease [17.63]), intrinsic targeting to clearance sites is of interest primarily as a technical problem that impedes therapeutic delivery to other sites. In such cases, numerous targeting strategies are available, some of which depend on synthetic nanomaterial properties (for instance, see the discussion of the enhanced permeability and retention or EPR effect in the context of oncology, below) to minimize uptake of nanotherapeutic devices by clearance systems and maximize delivery to desired sites. Targeting via biological affinity reagents decorating the surfaces of therapeutic nanodevices may be the most direct approach.

Nanotherapeutic Targeting Exploiting Biological Affinity Properties

Tissue-specific delivery by biological affinity requires the presence of tissue-specific surface features, most commonly proteins or glycoproteins (tissue-specific antigens). Historically, the search for tissue-specific

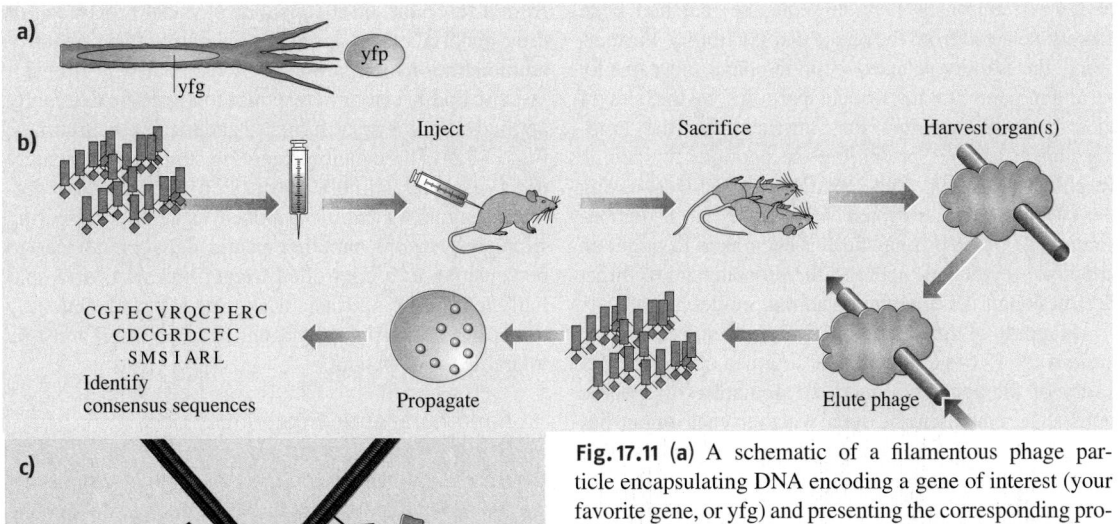

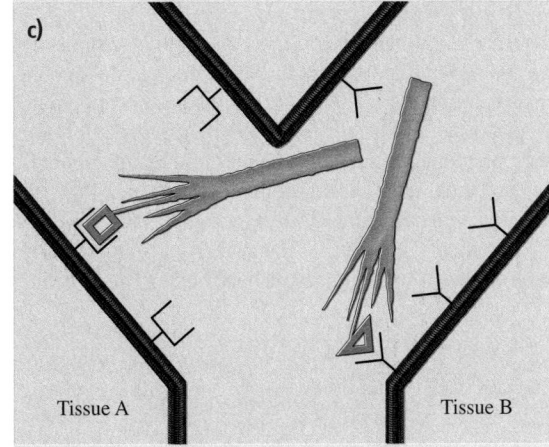

Fig. 17.11 (a) A schematic of a filamentous phage particle encapsulating DNA encoding a gene of interest (your favorite gene, or yfg) and presenting the corresponding protein (your favorite protein or yfp) on its surface. The linkage between yfp and yfg provided by the phage allows simultaneous affinity isolation and recovery of proteins of interest along with replicable genetic elements encoding them. When variant protein libraries are built by this method, they can be sorted for their ability to bind receptors by affinity. **(b)** A library of filamentous phage presenting random peptides is shown schematically. The library is injected into the vasculature of an animal (in this case a mouse), where individual phage bind specific receptors present in the vasculature of different tissues. The animals are sacrificed, and organs of interest are harvested. Phage particles bound to receptors of vasculature of the organ at hand can be eluted by any of several methods (low pH elution is schematically shown here). *E. coli* is infected with the eluted phage particles and clonally propagated, allowing identification of the peptides encoded by the eluted phage which mediated binding to the vascular receptors. Ultimately, this allows the identification of consensus peptide sequences that recognize specific receptors on the vasculature of the organ or tissue involved. These receptors constitute the molecular addresses for these organs, and the peptides isolated from phage eluted from them are biological affinity reagents that can be used to direct nanotherapeutic devices to vasculature of particular organs. **(c)** Specific recognition of vascular beds of two distinct tissues (A and B) via recognition of organ-specific vascular addresses (the Y and box-shaped receptors shown on vascular endothelium) by specific peptides (the *triangle* and the *square*) presented on the phage surface is depicted [17.64–66]

antigens for drug targeting, whether associated with tumors or other cells, organs or tissues, has been arduous and not entirely gratifying. The primary problems are specificity (few antigens are uniquely present in any single tissue), availability (some tissues may not have their own unique antigenic signature or marker), and therapeutic extravasation or directed migration in tissue spaces (markers in tissue may not be accessible from vasculature). An exciting recent development in biochemical targeting is the discovery of a vascular address system [17.64–66].

The vascular address system has been characterized by administering a peptide phage display to library animals, resecting individual organs, and extracting phage from the vasculature of the isolated organs (Fig. 17.11 [17.64–66]; see [17.67] for a discussion of display technology). Amazingly, phage isolated from different organs exhibited distinct consensus presented peptide sequences, indicating that the vasculature

of individual organs presented unique cognate receptors, each bound by a different short (ten amino acids or fewer) consensus peptide sequence that had been affinity-selected from the phage display library. Furthermore, the affinity-selected peptides have the capacity to tether nano- to microscale particles to the site of their cognate receptors (as illustrated by the binding phage particles presenting the peptides to specific vascular locations). Site-specific drug delivery using the vascular address system has already been demonstrated [17.68, 69]: it has further been used to target an apoptotic (cytocidal) agent to the prostate and to direct destruction of the organ in an animal model [17.68].

Mapping of the vascular address system is currently underway [17.64] and holds the promise of specific delivery of therapeutic agents to vasculature of specific organs. It remains to be seen whether each organ has a single molecular marker constituting its address that is amenable to binding a single peptide sequence; organs may instead have unique constellations of antigenic markers. If so, specific targeting may be possible using multiple peptides, each peptide binding its cognate receptor on target organ vasculature very weakly. Peptides used in such a multivalent targeting strategy would be chosen to reflect the unique constellation of address markers present in the target tissue. Affinities of cognates for linear peptides are often very low [17.67], though their aggregate affinity may be substantially higher than that of any peptide-vascular address cognate alone. Under ideal conditions, the affinity of such a multipeptide, multicognate complex should be the equivalent of the products of the affinities of each constituent peptide for its individual constituent cognate. Such multivalent interaction avidities can be extremely high (and the corresponding effective affinity constants are also high) but seldom fully realize their theoretical maximums (Lee and Parthasarathy, unpublished; see [17.32] for a review of receptor biochemistry).

It should be noted that most of the vascular addresses identified to date deliver materials to the organ vasculature: extravasation and access of organ tissue spaces by nanotherapeutics remains a separate issue.

17.3.3 Triggering: Delimiting Nanotherapeutic Action in Space and Time

Controlled triggering of therapeutic action is the other side of the targeting coin. If the site and time of nanotherapeutic delivery cannot be adequately controlled, the site of therapeutic action can be delimited by spatially- or temporally-specific triggering. The triggering event might drive release of the active therapeutic

from a reservoir, or chemical or physical processing of drug materials from an inert to an active form (inactive administrations that are converted to active forms at a specific bodily site or in response to a specific externally applied stimulus are referred to generically as prodrugs, Fig. 17.12). Three major triggering strategies are widely used: external stimuli, intrinsic triggering, and secondary signaling (multicomponent systems). Triggering strategies require nanotherapeutic delivery devices to be sensitive to a controlled triggering event, or a spatially/temporally intrinsic triggering event mediated by the host. Obviously, the triggering event itself must be tolerable to the patient.

Nanotherapeutic Triggering Using External Stimuli

External stimuli are provided by an external nanodevice operator/clinician, usually in the form of a site-specific energy input, typically light, ultrasound, or magnetic or electrical fields. Organic polymeric structures are very amenable to interaction with these energy sources. For instance, micellar structures can be reversibly dissociated with ultrasound, in which case they disgorge their contents or expose their internal spaces to the environment during ultrasound pulses (20 to 90 kilo-

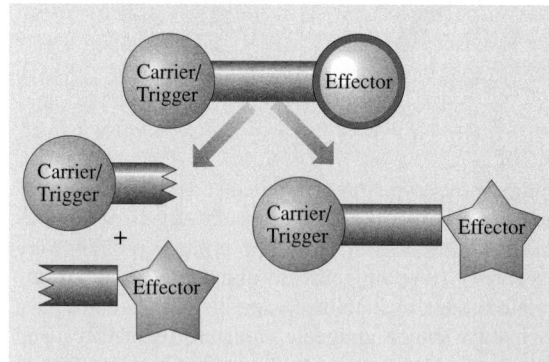

Fig. 17.12 One possible configuration of a prodrug, in which carrier/trigger and effector functions are separate functional domains of the therapeutic. The key feature of prodrugs is that they are therapeutically inert (as indicated by the colorless effector domain) until an activation event occurs (mediated here through the carrier/trigger domain, with effector activation indicated by its change to a star shape). Activation events can involve cleavage of inhibitory carrier/trigger domains from effectors. Other activation strategies involve a chemical change or shift in conformation of the effector, mediated through the carrier/trigger domain and in response to an environmental condition

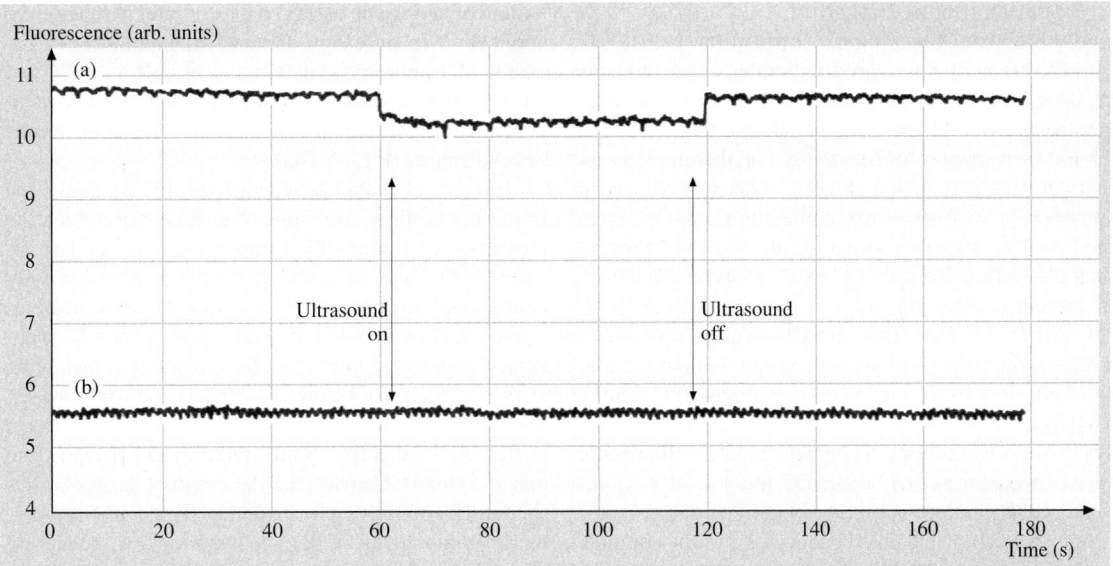

Fig. 17.13 Externally applied stimuli can trigger drug release: hydrophobic drug molecules are reversibly deployed from micelles in response to acoustic stimuli [17.70]. Intact pluoronic micelles maintain the hydrophobic cytotoxin doxorubicin in a hydrophobic environment (the core of the micelle, as in Fig. 17.4). Doxorubicin fluorescence is quenched in aqueous environments, and hence changes in the integrity of pluoronic micelles carrying doxorubicin can be monitored by doxorubicin fluorescence. Here, an ultrasound pulse is applied to suspensions of such micelles or to doxorubicin in solution. Application of ultrasound triggers exposure of micelle-encapsulated doxorubicin to the aqueous solvent (drug release), as demonstrated by the reduction in fluorescence. After cessation of the acoustic pulse, doxorubicin is repackaged into the micelles, as evidenced by the increase in fluorescence after the ultrasound administration (*line* a). Ultrasound has no impact on fluorescence of doxorubicin in solution, as expected (*line* b)

hertz range). The process has been used to control the release of cytotoxin (doxorubicin) from micelles (Fig. 17.13), and short ultrasound transients might be used for pulsatile or intermittent exposure of patients to therapeutics [17.70].

Light is another popular external triggering modality. Bioactive materials can be covalently associated with a nanoscale-delivery vehicle by photolabile linkages [17.15, 71, 72], or micelles can be constructed so that their permeability is altered as the result of exposure to light [17.72]. In the latter case, light input can cause photopolymerization resulting in micelle compaction that drives release of therapeutic cargo, or photo-oxidation, which causes loss of micellar integrity to release encapsulated materials. The ability of light to penetrate dense tissues is a clear limitation on this approach. This concern can be accommodated by polymer systems responsive to wavelengths that penetrate tissue efficiently (usually in the red-infrared region of the spectrum), or by using systems such as fiber optics to deliver light to deep tissues [17.15, 71, 72].

Externally applied magnetic fields can also be used to control nanotherapeutic activity. For instance, eddy currents induced by alternating magnetic fields can heat nanometallic particles and their immediate vicinity, an approach that has been successfully applied to control the bioactivity of individual biological macromolecules [17.73]. Colloidal gold particles are covalently conjugated to biomolecules (nucleic acid, or NA, duplexes or proteins), and alternating magnetic fields are used to induce heating sufficient to cause dissociation of hybridized NA structures or denaturation of three-dimensional protein structures. Within certain parameters, the process is reversible and so allows the construction of semibiological assembly or release switches. The modality has clear application to temporally specific triggered release, but the current inability to direct magnetic fields to preselected tissue locations may limit its use for spatially specific triggering.

Nanotherapeutic Triggering by Intrinsic Physiological Conditions

Intrinsic triggering is a prodrug strategy that depends on the conditions at the desired site of action to trigger the activity of the nanotherapeutic. Much as the three-dimensional conformation of, and therefore the activity of, proteins can be controlled in external triggering strategies [17.73], proteins can be engineered to make their conformations sensitive to intrinsic conditions at their desired sites of activity (see [17.3] for a discussion of protein engineering for use in nanobiotechnological devices). For instance, noncovalent complexes of a diphtheria toxin (DT) protein variant and a bispecific antibody (Fig. 17.1) have been engineered for specific toxicity to cells that can take up the complexes [17.75]. The bispecific antibody recognizes both a cell-specific surface receptor as well as the DT in an inactive (nontoxic) configuration. Antibody binding to the receptor results in receptor-mediated endocytosis of the complex in target cells. As a result of the lower pH of the endosomal compartment (reaching pH 4–5 in endosomes, as opposed to the constant pH 7–8 in circulation) into which the bound complex is taken up, the DT variant protein undergoes a three-dimensional conformational shift to its toxic form. This conformation of DT is not recognized by the bispecific antibody, so the toxin dissociates from the complex and kills the cell. Hypoxia-triggered prodrugs have also been developed that are activated either by chemical reduction or by enzymatic activities induced in hypoxic tissues. Extreme hypoxia [17.76] is a unique

feature of neoplastic tissue, so these strategies have clear application in oncology. Clinical manipulation of the extent of tumor oxygenation is possible [17.77] and may also be possible in other tissues. If so, redox state-dependent triggering approaches may have applications beyond cancer (Fig. 17.14a).

Enzyme-activated delivery (EAD [17.74], Fig.17.14) is another prodrug-like triggering strategy in which the properties of a nanoscale drug-delivery vehicle are altered at the site of action by an enzymatic activity endogenous to that site. Most typically, this involves a liposomal or micellar nanostructure from which designed pendant groups can be cleaved by metabolic enzymes (such as alkaline phosphatase, phospholipases, proteases, glycosidic enzymes) that are highly expressed at the site of therapy. Nanostructures are designed so that enzymatic cleavage of the pendant group causes a conformational or electrostatic change in the polymeric components of the delivery device, rendering the micellar structure fusogenic, leaky, or causing partial or complete dissociation of the structure [17.12] (Fig. 17.14c). This has the result of delivering or releasing therapeutic payloads at preselected sites.

Nanotherapeutic Triggering Using Secondary Signaling

Secondary signaling or multicomponent delivery systems are more complex delivery strategies (Fig. 17.15). These systems can feature site-specific and systemic delivery of one or more components. One class of such systems is the so-called ADEPTS devices (antibody directed enzyme-prodrug therapy, Fig. 17.15 [17.78]). ADEPTS systems feature affinity-based targeting of an enzymatic prodrug activator to a desired site, followed by systemic administration of a prodrug. ADEPTS is used primarily for oncology, and so the prodrug involved

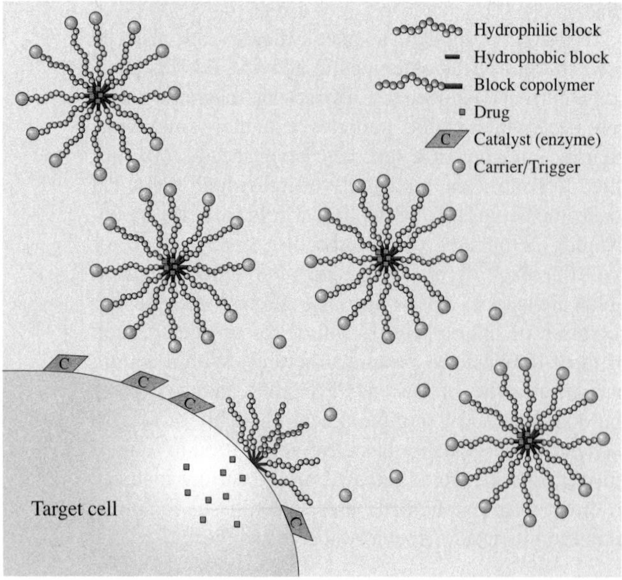

Fig. 17.14 Intrinsic conditions at desired sites of therapeutic action can trigger release of active drug molecules. Enzyme-activated delivery (EAD) is an intrinsic triggering strategy that depends on chemical cleavage of the carrier/trigger to induce changes in the properties of the delivery vehicle to allow delivery of the drug [17.74]. Here a surface functionality is cleaved from a micellar drug delivery vehicle, rendering the micelle fusogenic and allowing it to fuse with the cell membrane. As a result, the therapeutic payload of the micelle is delivered to the cytosol of target cells. Tuning of micellar surface functionalities and their linkages to the block copolymers constituting micelles can theoretically be used as target delivery to a variety of intracellular compartments

Legend (figure):
- Hydrophilic block
- Hydrophobic block
- Block copolymer
- Drug
- Catalyst (enzyme)
- Carrier/Trigger

Target cell

Fig. 17.15a,b Secondary biochemical signals can trigger release of active therapeutics. (**a**) Antibody directed enzyme-prodrug therapy (ADEPT) exploits enzyme-monoclonal antibody (mAB) bioconjugates to activate prodrugs [17.78]. Here, a mAB (like the native IgG of Fig. 17.1) is covalently linked to a catalyst (*C in a rhombus*, usually an enzyme). The binding specificity of the monoclonal antibody causes the bioconjugate to bind to a desired target site (possibly by recognition of a vascular address, as shown in Fig. 17.11). The catalyst (enzyme) specificity of the bioconjugate is capable of triggering activation of a prodrug, much as in Fig. 17.12. The corresponding prodrug is then administered (represented as in Fig. 17.12). The enzyme localized at the desired target site of therapeutic action then activates the prodrug (represented here by cleavage of carrier/trigger moiety and change in shape of the effector to a *star*, as in Fig. 17.12). (**b**) Another secondary signaling triggering strategy [17.55]. Here, single-stranded nucleic acid (SSNA) drives assembly of an activating complex for a prodrug. A catalyst (indicated again by *C*) is conjugated to an oligonucleotide that hybridizes to a single-stranded nucleic acid (SSNA). A prodrug, containing an effector domain (*E*) and a second oligonucleotide, in this case homologous to a region of the same SSNA, and proximal to the site of hybridization of the first oligonucleotide, is also administered. Catalyst specificity is tuned to correspond to linkage between the prodrug effector and oligonucleotide. The catalyst in the assembled structure is sufficiently close to the prodrug to cleave the linkage between effector and oligonucleotide, releasing active drug (*star*). Potentially, prodrug and catalyst oligonucleotides might dissociate from the SSNA and the cycle could be repeated. Note the similarity to Fig. 17.10

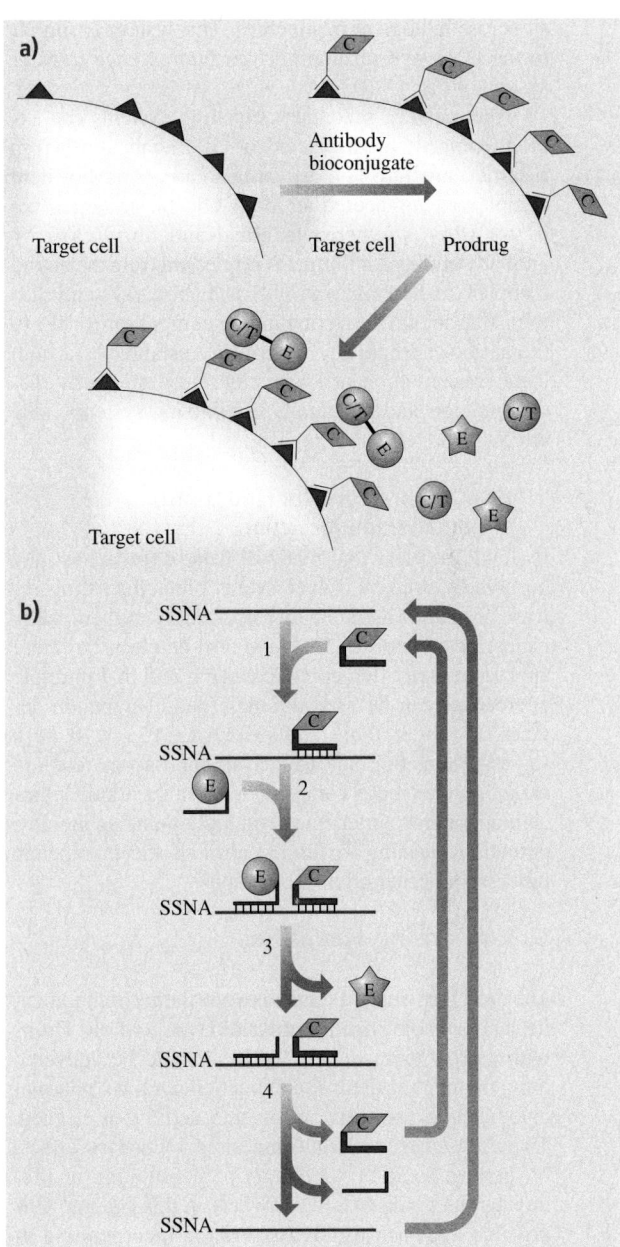

is generally one that can be activated to cytotoxicity. Since prodrug dosing is systemic, prodrug toxicity must be de minimus in the inactive form, and the activating catalyst must not be present at locations where cytotoxicity would be deleterious. To address this issue, the nanotherapeutic designer can choose to engineer an activation mechanism that has no physiological analog in the host. In addition, tuned, high lability of the active drug can potentially delimit the site and extent of cytotoxic effect mediated by ADEPTS systems.

Other secondary signaling systems rely on DNA hybridization to assemble therapeutic components into catalytically active complexes at their sites of action [17.55]: in this system, the individual components lack therapeutic activity until they are brought within a few nanometers of each other by hybridization to a target single-stranded nucleic acid molecule (SSNA

in Fig. 17.15, potentially a RNA species expressed in a tissue or cell type of interest). At this point, the catalytic moiety and the prodrug moieties are sufficiently close in space to allow catalytic cleavage of the active component from the prodrug. As depicted, this strategy makes no provision for delivery of drug and catalyst bioconjugates to the site of interest, nor for their transit across biological membranes. The system is similar to the DNA hybridization-driven fluorescence transfer system of Fig. 17.10.

Alternatively, secondary signaling systems can exploit competitive displacement of therapeutics from a carrier structure [17.79]. In this case, a noncovalent complex of engineered antibody and plasminogen activator (PA) is tethered to blood and fibrin clots by antibody affinity for fibrin. PA is released from the bound complex (to dissociate the clot) using bolus systemic administration of a nontoxic binding competitor for PA to the antibody complex. The strategy establishes a high local concentration of PA at clot sites, efficiently dissolving clots and potentially minimizing systemic side effects.

Layering Strategies for Fine Control of Nanotherapeutic Action

It should be clear that many of these approaches may be broadly used to trigger events other than drug release or to drive assembly or disassembly of therapeutic nanostructures in situ. It should also be clear that these approaches are often complementary, and that multiple approaches can be used in single nanotherapeutic devices (Fig. 17.3, Table 17.1 [17.11–13, 15, 71, 78, 80]). Layering targeting and triggering approaches tends to make devices more complex, but it also allows clinician/operators to intervene at multiple points in therapy, potentially leading to finer control of the therapeutic process and better clinical outcomes.

17.3.4 Sensing Modalities

The need for "smart" therapy is a key theme of therapeutic nanotechnology and pharmacology as a whole. Drugs with narrow therapeutic windows should be delivered only to their desired site of action and be pharmacologically active only when that activity is needed. These strategies can limit undesired secondary effects of therapy, some of which can be debilitating or life-threatening, as discussed elsewhere in this chapter. One possible approach to this issue is the incorporation of sensing capability (specifically, the capacity to recognize appropriate contexts for therapeutic activity) into nanotherapeutic devices. Sensing capability may allow self-regulation of a therapeutic device, reporting to an external clinician/device operator (though this begins to touch on imaging applications, see below), or both. In the context of our discussion of therapeutic nano- and microscale devices, we will consider primarily electrical and electrochemical sensor systems, particularly micro-

fabricated (field-effect transistor or FET, and cantilever) and conducting polymer sensors.

Sensor Systems

Sensing is predominantly a higher order device functionality, depending on multiple device components, though one could argue that some targeting/triggering strategies, particularly targeting by bioaffinity and intrinsic triggering strategies, must, a priori, incorporate at least limited sensing capability. However, biosensors, as they are typically considered, are multifunctional, multicomponent devices [17.81]. Usually a biosensor system is composed of a signal transducer, a sensor interface, a biological detection (bioaffinity) agent, and an associated assay methodology, with each system component governed by its inherent operational considerations.

The transducer component determines the physical size and portability of the biosensor system. Signal transducers are moieties that are sensitive to a physicochemical change in their environment and that undergo some detectable change in chemistry, structure, or state as the result of analyte (the thing to be sensed) recognition. Analytes for nanotherapeutic application could be biomolecules, like proteins, small molecules (organic or inorganic), ions (salts or hydrogen ions), or physical conditions (such as redox state, temperature). Interfaces are the sensor components that interact directly with the analyte. For sensor use in nanotherapeutic devices, immobilized or otherwise captured biological molecules (proteins, nucleic acids) often constitute the sensor interface. Whatever the chemical nature of the interface, it determines the selectivity, sensitivity, and stability of the sensing system and is also a dominant determinant of sensor operational limits. Assay methodology determines the need for (or lack thereof) analyte tracers, the number of analytical reagents, and the complexity and rapidity of the sensing process. Nanotherapeutics are of interest at least in part because they can be minimally invasive, low complexity, yet robust and accurate; convenient assay methods are therefore highly desirable.

Cantilever Biosensors

Micromechanical cantilevers (discussed above in conjunction with force microscopy, Fig. 17.6) transduce sensed events by mechanical means [17.82]. Both changes in the resonant frequency and deflection of cantilevers resulting from analyte binding or dissociation can be conveniently and sensitively detected. These changes in cantilever state can be conveniently detected by optical, capacitive, interferometry, or piezoresistive/piezoelectric methods, among others. Microfabricated cantilever dimensions range from micron to submicron range, with potential for further dimensional optimization (by carbon nanotubes appended to them, for instance [17.82]). They are operationally versatile and can be used in air, vacuum, or liquid, although they suffer some degradation in performance in liquid media. Like most micromachined structures, they can be batch fabricated and conveniently multiplexed.

Cantilevers used in atomic force microscopy (AFM, see also discussion of one-off nanostructures above) approaches can be used to study individual biomolecular interactions [17.35, 81]. In this approach, cantilever tips are derivatized with biomolecules (effectively, one member of a receptor/counter-receptor pair), and the tip-bound biomolecule is allowed to bind its counter-receptor (itself bound on a surface). Under nonequilibrium conditions (conditions that result in thermodynamically irreversible changes in analyte molecular structures), the force required to disrupt single molecular interactions can be measured and related to classical biochemical parameters of receptor binding. The method has been applied to interactions between hormones and their receptors, sugars and lectins, as well as hybridizing DNA strands.

Cantilever systems sensitively detect changes in mass at their surfaces: changes as small as a mass density of $0.67\,\text{ng/cm}^{-3}$ are theoretically detectable. This can allow detection of binding of extremely small objects to the cantilever and has been applied to the detection and enumeration of prokaryotic and eukaryotic cells, as well as small numbers of macromolecules [17.82]. The incorporation of biological receptors or affinity reagents on the cantilever surface can drive specific binding events for particular sensing tasks. Microcantilevers are also highly sensitive to temperature, detecting changes as low as $10^{-5}\,\text{K}$; they can also detect small changes in pH.

Field–Effect Transistor Biosensors

Field-effect transistor (FET) architectures are another sensing architecture that can be conveniently produced by micro/nanofabrication. FETs consist of a current source, a current drain, a conductive path (sensing channel) between them, and a sensing gate to which a bias can be applied. Analyte binding to the sensing channel induces a charge transfer resulting in a dipole between the surface and the underlying depletion region of the semiconductor: current that passes between the source and drain of a semiconductor FET is quite sensitive to the charge state and potential of the surface in the connecting channel region. Moving a standard silicon FET from depletion to strong inversion (shifting the surface potential by $\gtrsim 0.5\,\text{eV}$) requires less than $\approx 10^{-7}\,\text{C/cm}^2$ or $\approx 6 \times 10^{12}$ charges/cm^2, corresponding to the transfer of $6.25 \times 10^{11}\,\text{e/cm}^2$. With FETs of 2000 square micrometers, detection of biological analytes in subnanomolar concentrations is easily feasible. Specificity for binding of macromolecular analytes of interest can be provided by deployment of biological affinity reagents in the FET sensing channel. Submicron FETs are routinely manufactured; use of carbon nanotubes in FETS will offer still greater miniaturization [17.43, 83].

Carbon nanotubes also have excellent mechanical properties and chemical stability in addition to potentially tunable electrical properties, making them highly desirable electrode/nanoelectrical materials for any number of nanoelectrical applications [17.84]. Biomolecules can be bound to carbon nanotubes, particularly in FET and nanoelectrode applications. Most biomolecules bound to carbon nanotubes are not covalently bound (as discussed above) and do not exhibit direct electrical communication with the nanotube, though redox enzymes bound to nanotubes and other conductive nanomaterials may [17.84, 85]. Flavin adenine dinucleotide (FAD) and flavoenyzme glucose oxidase (Gox) both display quasi-reversible one-electron transfer when absorbed onto unannealed carbon nanotubes in glassy carbon electrodes. Gox immobilized in this way retains its substrate-specific (glucose) oxidative activity, leading to applications in sensing circulating glucose for diabetes and, perhaps, to a strategy of harvesting electrical power from metabolic energy.

Conducting Polymers and Sensor Biocompatibility

The biocompatibility of most metallic structures (as might be used in the bioelectrical sensors described) is limited at best; metal structures rapidly foul with serum proteins (become opsonized), undergo electrochemical degradation, or have other problematic properties. Polymer chemistry, however, has the capacity to tune composition to enhance biocompatibility properties and is commonly used to make synthetic surfaces more biologically tolerable. Electrically conductive polymers are potentially attractive in this context.

Polymeric materials are available with intrinsic conductivities comparable to that of metals (up to $1.5 \times 10^7\,(\Omega\,\text{m})^{-1}$, which, by weight, is about twice that of copper [17.86]). Significant conductivity has been documented for a dozen or so polymers, including polyacetylene, polyparaphenylene, polypyrrole, and polyaniline, doped with various impurities. Careful control of doping can tune electrical conductivity properties over several orders of magnitude [17.87].

Conductive polymeric materials may be extremely well suited to biosensing applications. The polymeric materials themselves are often compatible with proteins and other biomolecules in solution. Furthermore, polymers are amenable to very simple assembly of sensor transducer-interface components by deposition of the polymer and trapped protein (sensor interface) directly on a metallic micro- or nanoelectrode surface [17.87,88]. Combined with a facile nanoelectrode array microfabrication method [17.89], simultaneous conducting polymer-interface protein deposition may offer an extremely simple way to fabricate multiplexed sensor arrays.

These electrochemical and electrophysical sensing modalities are of little use in and of themselves: they must communicate with either other device components (such as drug-dispensing effector components) or with external observers or operators. Sensor coupling in autonomously operating devices produced by microfabrication can be done directly, and the coupling linkages incorporated into the fabrication protocol. Coupling with biological device components can also be direct, as when conductive materials are conjugated to biomolecules, and can directly modulate their activity [17.73].

Sensor-device coupling can also be indirect, through electrochemically produced mediator molecules [17.85]. Sensors might be independent of nanotherapeutics and report conditions at the site of therapeutic action to an external operator, who would use any one of the external triggering strategies discussed above to engage therapy when and where appropriate. Communication/sensor interrogation might be accomplished most crudely by direct electrical wiring of sensors to an external observation station. Alternatively, if the event to be detected is transduced optically, as in colorimetric smart polymers that change optical properties in the presence of analyte [17.90], sensors might be interrogated by fiber optics. These modalities are most applicable to sensor arrays delivered to the site of interest by a catheter. The use of a catheter may be justified in some therapeutic applications, but it is an invasive procedure and is not optimal.

We have already seen multiple examples of prodrugs that are activated by cleavage of an inhibitory domain from the complex. One noninvasive approach to communicating with external operators could exploit this phenomenon by detecting the cleaved fragment in bodily fluids. If the cleaved moiety cleared through urine, the extent of drug activation could be monitored non-invasively via urinalysis. There is no a priori need to connect the cleavage event to drug activation. For instance, an operator might administer a catalyst that cleaves a detectable material from a nanotherapeutic. If this secondary signaling moeity was independently targeted to the desired site of therapy, presence of the detectable cleaved product would provide information regarding the bodily location of the therapeutic nano-device. The cleaved product would not be detectable unless the therapeutic and secondary signaling moiety colocalized at a single site. Other sophisticated communication involves ultrasound or electromagnetic radiation to carry information. These sorts of modalities are currently applied to in vivo imaging approaches.

17.3.5 Imaging Using Nanotherapeutic Contrast Agents

Imaging is a minimally invasive procedure that allows visualization of organs and tissues following the administration of a detectable moiety (contrast agent). The contrast agent is then exposed to some condition that interacts with the contrast agent so as to produce an emission or response detectable to an external monitoring device. Nanosized particles (5–100 nm in diameter) have found application as contrast-enhancing agents for medical imaging modalities such as magnetic resonance imaging (MRI, our primary topic, reviewed in [17.91, 92]). MRI currently provides cross-sectional and volumetric images with high spatial resolution (< 1 mm) and is potentially applicable to many clinical purposes, though enhanced imaging capabilities are desirable. For instance, blood flow measurements of healthy and diseased arteries can be quantified better with improved contrast agents that highlight blood flow at the vessel wall. Similarly, some tissues (certain tumors) do not exhibit strong contrast within the MRI field and cannot be readily identified or characterized at present. It is of great interest, therefore, to develop nanoscale particles that provide enhanced contrast for many applications.

Magnetic Resonance Imaging (MRI): the Basics

Objects to be imaged are exposed to a strong magnetic field and a well-defined radio frequency pulse. The external magnetic field (B_0) serves to loosely align protons either with (low energy level) or against (high energy level) the field, the difference between the two energy levels being proportional to B_0. Once the protons are separated into these two populations, a short multi-wavelength burst (or pulse) of radio frequency energy is applied. Any particular proton will absorb only the frequency that matches its particular energy (the Larmor frequency). This *resonance* absorption is followed by the excitation of protons from the low to the high energy level and of equivalent protons moving from high to low energy levels. After the radio frequency pulse, protons rapidly return to their original equilibrium energy levels. This process is called relaxation and involves the release of absorbed energy. Once equilibrium is again established, another pulse can be applied.

Data is collected by positioning a receiver perpendicular to the transmitter: relaxation energy release induces a detectable, quantifiable signal (in amplitude, phase, and frequency) at the receiver coil. Since multiple protons in multiple chemical environments are involved, the signal at the receiver includes many frequencies. The received signal consists of multiple, superimposed signals (called the free induction decay or FID) signal, resulting from the relaxation of multiple, chemically distinct protons. FID is converted from the time domain to the frequency domain (by Fourier transformation), within which individual proton types can be identified. Iterations of this procedure in two or three dimensions can create a high-resolution image of anatomical cross-section or volume.

Intrinsic factors affecting image quality include the proton density of the tissues, local blood flow, and two relaxation time constants [17.93]: the longitudinal relaxation time (T_1) and the transverse relaxation time (T_2). Control of T_1 and T_2 relaxation effects are most critical for high-resolution MR images. T_1 relaxation measures energy transfer from an excited proton to its environment. In tissues, protons of fats and cholesterol molecules (relatively movement-constrained macromolecules) relax efficiently after a pulse and exhibit a short T_1 time. Water in solution (a small molecule tumbling relatively freely in solution) has a much longer T_1 time. T_2 relaxation measures the duration of coherency between resonating protons after a pulse, prior to their return to equilibrium. Tightly packed, solid tissues with closely interacting hydrogen nuclei relax more quickly than loosely structured liquids, so tissues such as skeletal muscle have a short T_2s, while cerebrospinal fluid has a very long T_2. Intrinsic factors are manipulated through extrinsic factors, such as the external magnetic field strength and the specific pulse sequence, allowing the collection of meaningful MR images.

Nanoparticle Contrast Agents

Signal intensity in tissue is influenced linearly by proton density, while changes in T_1 or T_2 result in exponential changes in signal intensity. T_1 and T_2, therefore, are manipulated to enhance imaging by administration of exogenous contrast-enhancing agents. MRI contrast agents are divided into paramagnetic, ferromagnetic, or superparamagnetic materials. Metal ion toxicity is an unfortunate consequence of physiologic administration of contrast agents but can be mitigated somewhat by complexation of the metals with organic molecules.

Paramagnetic metals used for enhanced MRI contrast (gadolinium, Gd, iron, Fe, chromium, Cr, and manganese, Mn) have permanent magnetic fields, though the magnetic moments of individual domains are unaligned [17.94]. Upon exposure to an external magnetic field, individual domain moments become aligned, generating a strong local field (up to 10^4 gauss [17.95]). Paramagnetic metal ions interact with water molecules, causing an enhanced relaxation of the water molecules via tumbling of the water-metal complex, dramatically decreasing the T_1 value for the water molecules and enhancing the proton signal [17.94]. Contrast enhancement by paramagnetics is thus due to the indirect effect the contrast agent has on water and its magnetic resonance properties.

Ferromagnetic and superparamagnetic materials both contain iron (Fe) clusters, which generate magnetic moments 10 to 1000 times greater than those of individual iron ions. Clusters greater than 30 nm in diameter are ferromagnetic, whereas smaller particles are superparamagnetic [17.93]. Ferromagnetic materials maintain their magnetic moment after the external field is removed, but superparamagnetic materials lose their magnetic field after the field is removed, as do paramagnetics. Both ferromagnetic and superparamagnetic substances minimize the proton signal by shortening T_2 [17.96], resulting in negative contrast (darkening of the image [17.97]).

Nanobiotechnological Contrast Agent Design

First-generation contrast agents often contained a single metal ion/complex, whereas emerging agents incorporate nanoscale metal clusters, crystals, or aggregates, sometimes encapsulated within a synthetic or biopolymer matrix or shell [17.98–100]. These metal cluster agents improve contrast effects and, hence, output MR images profoundly. Furthermore, surface chemical groups (from the matrix or shell) can be derivatized to improve biocompatibility or allow targeting to a tissue or site of interest.

Particles are typcially prepared from colloidal suspension, where metallic cores are thoroughly mixed with the matrix material before being aggregated out of solution with a nonsolvent. Dextran (a polymer of 1,6-β-D-glucose) is a typical matrix material used in commercial imaging reagents: Combidex is coated with 10 000 molecular weight dextran [17.99, 101, 102], Feridex has an incomplete, variable dextran coating [17.102], and Resovist is coated with carboxy-dextran [17.97]. Other polymeric materials are also used (oxidized starch [17.98, 100]), and matrixless particles are also produced [17.103].

Nanoparticle contrast agents must be purified under tightly controlled conditions (generally by centrifugation or high-pressure liquid chromatography [17.104]) and accurately characterized (for size dispersity by light scattering, chromatography, photon correlation, or electron microscopy [17.100, 104]) to assure the reproducibility of imaging agent production lots. Elemental analysis, X-ray powder diffraction, and Mössbauer spectroscopy have also been used to characterize metallic cores [17.103, 105]. Tight definition of the finished particles ensures accurate correlation between structural properties of imaged materials prior in vitro and in vivo studies, allowing collection and interpretation of meaningful images. Metallic cores generally range from 4 to 20 nm in diameter [17.106], while coated particles can be up to 100 nm or more in diameter [17.105]. Compared to larger cores, however, nanoparticles less than 20 nm in diameter exhibit considerably longer blood half-lives and improved T_1 and T_2 relaxivity effects [17.107].

As previously discussed, clearance of nanoparticles via the RES is a critical problem usually approached by surface modifications to mitigate nonspecific adsorption of proteins to the biomaterial surface (the opsonization of synthetic materials). Neutral, hydrophilic surfaces tend to adsorb less serum protein than hydrophobic or charged surfaces. Bisphosphonate and phosphorylcholine-derived thin film coatings have been applied to nanoparticles to stabilize iron oxide particles against pH, opsonization, and aggregation [17.108]. Such thin films do not fully eliminate protein adsorption, and dense layers or thick brushes of polysaccharides or hydrophilic polymers may avoid opsonization more effectively [17.109].

Other, as yet incompletely understood, biological factors influence the use of nanoparticles in vivo. For instance, a direct correlation between the circulating half-life of nanoparticles (their $T_{1/2}$s) and the age of the animals used has been reported [17.106]. The observed increase in $T_{1/2}$ may be correlated to age-related changes in phagocytic activity [17.106]. Local environments also influence particle stability: iron oxide particles are degraded at a pH of 4.5 or less [17.100], a condition sometimes attained in some intracellular vesicles.

First-generation contrast agents were primarily blood-pool agents that moved freely through the entire vasculature. Targeting contrast-enhancing nanoparticles to sites of interest can reduce heavy metal toxicity associated with the commonly used agents by diminishing the dose required to obtain an acceptable image. Contrast agent targeting can also provide enhanced diagnostic information. For instance, nanoparticles that bind to molecular fibrin at a clot site on a vessel wall have been developed [17.110], potentially allowing differentiation between vulnerable and stable atherosclerotic plaques. Similarly, prognostically valuable data regarding disruptions of the blood–brain barrier (BBB) have been visualized with MRI as the result of delivery of contrast-enhancing nanoparticles to the affected site [17.102].

17.4 Applications of Nanotherapeutic Devices

As discussed above, nanotherapeutic devices are novel, emerging therapeutics with properties that are not fully understood or predictable. Nanotherapeutics, therefore, must be justifiable on at least two levels. As we have seen, the nature of the therapeutic task and the state of current nanoscale-materials technology make the incorporation of biological macromolecules unavoidable for many nanotherapeutics. Proteins, for example, are typically substantially more expensive than small molecule therapeutics, and precise nanostructures containing proteins will be more costly still. Nanotherapeutics must justify their high COGs. Secondly, as new therapeutic modalities, nanotherapeutics may carry significantly larger risks than those associated with more conventional therapies. Expensive, novel moieties, such as nanobiotechnological therapeutic devices, are therefore most likely to be accepted for treatment of conditions that are not only accessible to intervention at the nanoscale but also for which existing therapeutic modalities have acknowledged shortcomings in patient morbidity or mortality. We have selected two disease states sufficiently grave and sufficiently unserved to warrant nanotherapy: cancer and cardiovascular disease. Modulation of immune responses and vaccination is our third application area.

17.4.1 Nanotherapeutic Devices in Oncology

The economic burden imposed by cancer is immense, measured in the billions of dollars annually in the United States alone. Existing therapies such as surgical resection, radiotherapy, and chemotherapy have profoundly limited efficacy and frequently provide unfavorable outcomes as the result of catastrophic therapeutic side-effects. Additionally, the biology, chemistry, and physics of cancer, in general, and solid tumors in particular, provide therapeutic avenues accessible only by nanoscale therapeutics. Oncology is thus an ideal arena for emerging nanotechnological therapies.

Tumor Architecture and Properties
Tumors as tissues are relatively chaotic structures exhibiting vast structural heterogeneity as a function of both time and space [17.9, 77, 111, 112]. In healthy tissues, vasculature resembles a regular mesh in which the mean distance of tissue spaces to the nearest vessel is tightly controlled and highly uniform. On the other hand, the vasculature of tumors resembles a percolation network containing regions experiencing vastly different levels of perfusion. Tumors of 1 mm^3 or larger typically contain measurably hypoxic domains, with pO_2 values as much as two- to threefold lower than in normal tissue. High levels of hypoxia are characteristic of enhanced metastatic potential and tumor progression. Due to insufficiency of perfusion, tumors also frequently contain necrotic domains.

The average tortuosity of vascular flow paths in tumors is also much greater than in healthy tissue, and transient thrombotic events lead to enhanced resistance to flow and ongoing vascular remodeling. Aside from its plasticity, tumor vasculature itself is highly irregular, may be incompletely lined with endothelial cells, and often exhibits significantly higher extravasation limits (the highest molecular weight of materials that can leave the vasculature and diffuse into the interstitial spaces) than normal vasculature. Tumor tissue is also poorly drained by lymphatics, so extravasated materials tend to remain in situ in tumors and are not cleared efficiently.

These biological phenomena all differentiate tumor tissue from normal tissue and can be exploited in therapy. For instance, the special vascular integrity and lymphatic drainage properties of tumors constitute the enhanced permeability and retention effect (EPR) [17.11, 12, 80, 113]. EPR presents an obvious opportunity for intervention with nanoscale therapeutics (Fig. 17.3). Extravasation limits for normal tissues are variable, but structures larger than a few nanometers in diameter do not leave circulation efficiently in most tissues, whereas tumor vasculature frequently allows egress of materials in the tens to hundreds of nanometers range. Further, nanomaterials, once extravasated, are not cleared by lymphatic drainage. EPR provides tumor targeting that does not depend on biological affinity reagents: tumor tissue provides preferential depot sites for extravasated drug delivery devices. Targeting to a desired site of action is highly desirable for cytotoxic therapeutics, and EPR provides the basis for a growing class of polymer therapeutics [17.11, 12, 80, 113]. That said, EPR and targeting by biological affinity are not necessarily mutually exclusive. A number of antigens more or less specifically related to tumors are known (tumor associated antigens or TAAs), and as discussed above, tumor-/organ-specific vascular addresses might be exploited for delivery of therapeutics [17.65, 66]. For instance, various biological reagents that recognize TAAs (such as antibody to carcinoembryonic antigen, transferrin) have been conjugated to nanoscale contrast agents to enhance contrast agent localization to tumoral sites [17.104].

Tumor (EPR) Properties and Nanotherapeutics
PK1 (Fig. 17.16) is an early example of a growing and increasingly sophisticated class of polymer therapeutics [17.11, 12, 80, 113]. It is a nanoscale molecular therapeutic device wherein the therapeutic moiety (doxorubicin) is covalently linked to the polymeric backbone. As we will see, each component of the polymer therapeutic fulfills a discrete function, so the device rubric is warranted [17.6].

PK1 consists of a polymeric backbone (HPMA, hydroxyl polymethacrylamide) which, as discussed below, targets desired sites by virtue of its specific nanoscale size (by EPR effect) and a cytotoxin (doxorubicin), covalently linked together by a peptide. PK1 exhibits multiple design features that allow it to preferentially deliver doxorubicin to tumors. The size of the complex is tuned to be too large to extravasate in healthy tissue but still small enough for renal clearance. Effectively, tumors and the excretory system compete for PK1. This aspect of the relationship between therapeutic size and systemic tolerability is illustrated in Fig. 17.16. The therapeutic device has to be large enough to have limited access to healthy tissues but must have relatively unimpaired access to tumors and the excretory system via the kidneys.

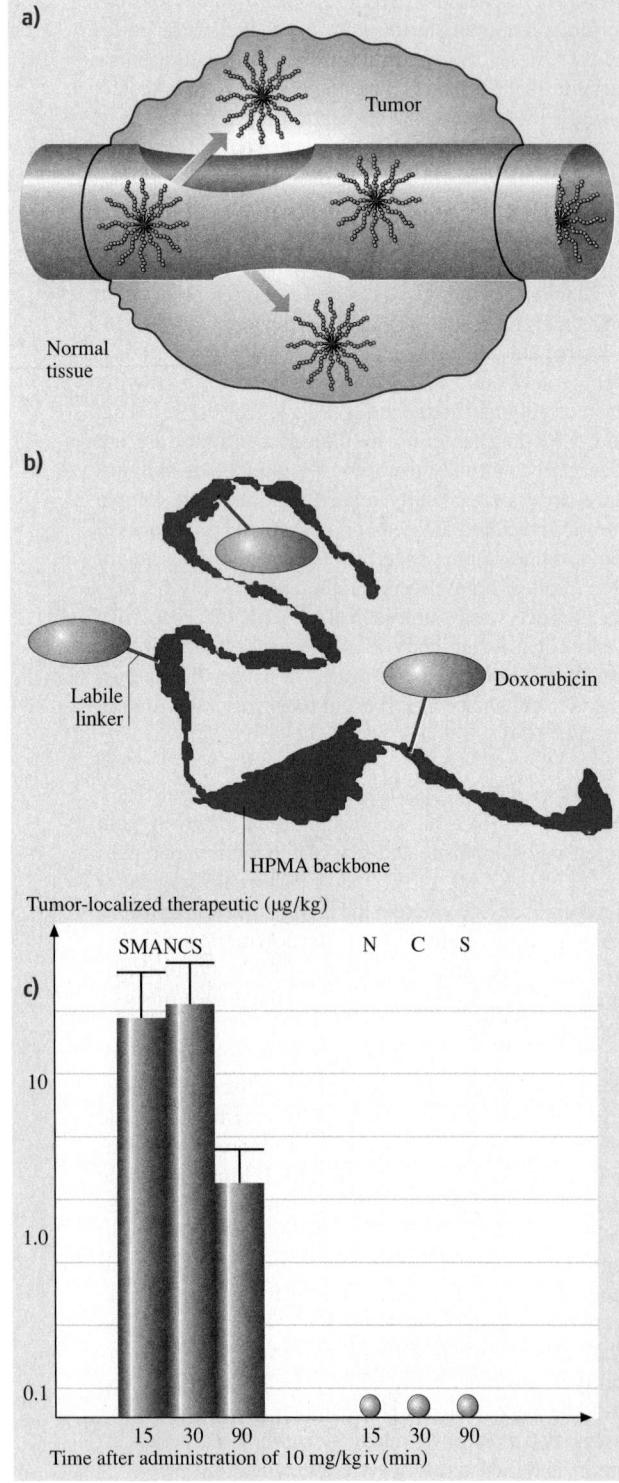

a)

Tumor

Normal
tissue

b)

Labile
linker

Doxorubicin

HPMA backbone

Tumor-localized therapeutic (μg/kg)

c)

SMANCS   N   C   S

10

1.0

0.1

15 30 90   15 30 90
Time after administration of 10 mg/kg iv (min)

Fig. 17.16a–c Therapeutic nanodevices can be specifically delivered to tumors by virtue of their size [17.11, 12, 80, 113]. (**a**) An illustration of the relatively low integrity of tumor vasculature. Therapeutic nanodevices (in this case, micelles carrying cytotoxins) cannot leave (extravasate from) the vasculature of normal tissue, but the vasculature of tumors is sufficiently leaky to allow specific delivery to tumoral interstitial spaces. This is the basis of size-dependent targeting to tumors by the enhanced permeability and retention (EPR) effect. (**b**) A polymer nanotherapeutic (PK1, [17.80]) that exploits the EPR effect to deliver a cytotoxin (doxorubicin) to tumors. The overall mass of the complex is tuned by controlling the mass of the N-2-(hydroxypropyl) methacrylimide backbone, so as to be above extravasation limits for normal tissues but below both tumor extravasation limits and the size limit for excretion through the kidney. A final feature of the device that favors delimiting doxorubicin cytotoxicity to tumoral sites is the fact that the device is a prodrug. The cytotoxicity of doxorubicin is severely curtailed when the doxorubicin is covalently linked to the polymer therapeutic, though it manifests its full toxicity when it is released from the complex by cleavage of the labile (peptidic) linker. The labile linker contains a cleavage site recognized by a protease known to be overexpressed in the target tumor type. Thus, the device produces a depot of inactive doxorubicin in tumors that is activated preferentially to full toxicity by conditions prevalent in the tumor. (**c**) EPR can drive localization of appropriately sized therapeutics to tumoral sites [17.113]. The polymer therapeutic SMANCS accumulates rapidly in murine tumors, whereas the NCS, the parent therapeutic (not engineered for EPR delivery), does not

A second engineered feature of PK1 relates to tumor-specific release of cytotoxin. Doxorubicin is toxic as a free molecule but is nontoxic as a part of a polymer complex. In PK1, doxorubicin is linked covalently to the polymeric backbone by a peptide chosen specifically because it is the substrate for a highly expressed protease in the target tumor. Thus PK1 deposits cytotoxin to tumors in an inactive form that is processed to an active form preferentially by tumoral metabolic activity. In all, the design features of PK1 act to minimize systemic exposure to cytotoxin and to augment tumoral exposure, thereby enhancing therapeutic benefit and diminishing undesired side effects.

Therapeutics need not be covalently linked complexes to be delivered by EPR. In some cases, bioactives can be engineered to interact with blood constituents noncovalently so as to bring the apparent size of the complex into a range that allows preferential tumor deposition. This is the mechanism of delivery for the oldest EPR-exploiting therapeutic in clinical use, SMANCS (*S*-methacryl-neocarzinostatin) [17.113]. Neocarzinostatin (NCS) is a small (12 kDa, a few nm in diameter) cytotoxic protein that elicits its toxicity by generating intracellular superoxide radicals enzymatically, which in turn damage DNA. Unmodified NCS is cleared rapidly and exhibits dose-limiting bone-marrow toxicity. Both problems are mitigated by the site-specific conjugation of two small (about 2 kDa and less than 1 nm in length) styrene-comaleic acid polymer chains, resulting in SMANCS. The $T_{1/2}$ of SMANCS is about tenfold higher than that of NCS, though SMANCS itself is too small to accumulate preferentially in tumors by EPR. This is mitigated by the fact that SMANCS (but not NCS) is efficiently bound by serum albumin. The resulting noncovalent complex is sufficiently large not to extravasate into normal tissue, though it still partitions efficiently into many tumors. The EPR-driven partitioning of the noncovalent SMANCS-albumin complexes into tumors suggests that other noncovalent complexes (such as the multidendrimer device of Fig. 17.3, were it to be assembled prior to administration) might also target to tumors via EPR.

Delivery of antitumor agents using EPR can be highly efficacious. As shown in Fig. 17.17, conjugation of therapeutics to produce controlled-size delivery complexes can enhance tumoral accumulation by multiple orders of magnitude relative to free therapeutics [17.11, 12, 80, 113]. In effect, EPR targeting broadens the therapeutic window (the range of concentration between the lowest effective and highest tolerable dose), allowing more intensive therapy. The increased therapeutic intensity allowed by EPR strategies also produces significantly enhanced clinical outcomes [17.11, 12, 80, 113]. The benefits of EPR strategies seem to be general in that a broad range of materials can be targeted to tumors and almost any particulate material falling within the appropriate size range will partition into tumor tissue via EPR. This includes not only linear and dendritic polymers but also higher order, self-assembled structures such as micelles or shell cross-linked structures and inorganic materials such as ferromagnetic particles. In addition to EPR tumor targeting, each of these structures has intrinsic properties that might be exploited in therapeutic nanodevices (see the discussion of imaging above). Targeted contrast agents may be particularly important for early tumor detection.

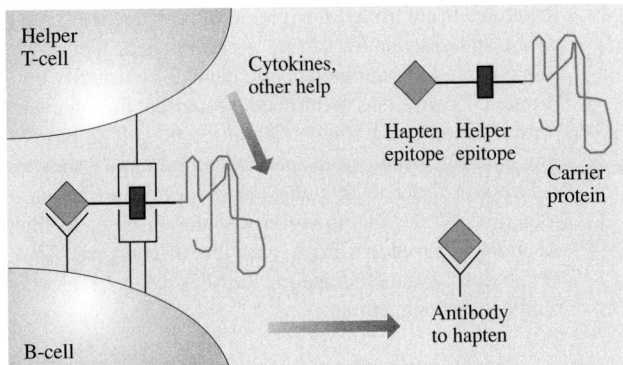

Fig. 17.17 Antibody responses [17.1] are efficiently triggered by antigens that contain both B-cell epitopes (*diamond*) and helper T-cell epitopes (*rectangle*). The B-cell epitope binds a receptor on the B-cell surface (the *Y shape* shown interacting with the B-cell epitope) that is essentially a membrane-bound form of the antibody species that the particular B-cell can produce. The peptidic T-helper epitope engages another receptor complex on the B-cell surface (class II MHC, the two-part receptor schematically depicted as interacting with the T-helper epitope, see text). When presented in the context of the class II MHC complex of the B-cell, the T-helper epitope can be recognized by a cognate T-cell receptor on the helper T-cell surface (shown as a receptor on the T-cell surface interacting with the T-helper epitope/class II MHC complex). In response to recognition of its appropriately presented, cognate T-helper antigen, the helper T-cell provides multiple stimulatory signals that cause the B-cell to proliferate, secrete antibody, and differentiate. This is a classical hapten-carrier configuration of an immunogen [17.1], in which a synthetic B-cell epitope is the hapten (in our cases, a PAMAM dendrimer or SWCNT epitope), and the protein covalently linked to the hapten is the carrier, providing the T-helper epitope. Note that the antibody produced recognizes *only* the B-cell epitope and *not* the T-helper epitope. The T-helper epitope also need not be present for antibody recognition of carrier epitopes, nor is its presence necessary to maintain the ongoing antibody response once antibody production is triggered. Note that whether the T-helper epitope remains covalently linked to the B-cell epitope during antigen presentation (here, the intact carrier protein is shown linked to the B-cell epitope) is debatable. Proximity of the B-cell epitope/receptor complex and T-helper epitope/class II MHC complex on the B-cell surface may be sufficient to trigger a robust antibody response

17.4.2 Cardiovascular Applications of Nanotherapeutics

The cardiovascular system is one of the most dynamic and vital systems of the body. In principle, every cell in the body is accessible to the bloodstream via the vasculature. If controlled navigation of, and extravasation from, the vasculature can be accomplished, therefore, the potential of nanotherapeutic devices is virtually unlimited. However, as mentioned throughout this chapter, opsonization and immune clearance, as well as targeting and triggering of therapeutics, remain key issues to address in order to facilitate wide application of nanotherapeutic devices to cardiovascular disease. In this section we provide a brief overview of categories and examples of nanotherapeutic devices used in blood-contacting applications.

Cardiovascular Tissue Engineering

A usable, immunologically compatible artery for coronary bypass is not always available, so synthetic or semisynthetic artery substitutes are highly valuable. Tissue engineering approaches might mitigate the lack of acceptable homologous artery, as long as an acceptable artery substitute can be produced. Current tissue engineering approaches involve synthesis of three-dimensional porous scaffolds that allow adhesion, growth, and proliferation of seeded cells to generate a functional vessel. Cellular organization and growth in synthetic scaffolds are often chaotic and random, and sufficient blood supply is seldom achieved within the scaffold. These conditions often result in death of the seeded cells and compromise the integrity of the vascular prosthesis.

Microfabrication techniques have allowed miniaturization of the tissue engineering scaffolds. These techniques allow the molecular-level control of cellular adhesion and propagation by controlling surface topography, surface chemistry, and the arrangement of morphogenic and proliferative signaling ligands and cytokines. Microfabrication schemes are being developed for the elucidation of parameters necessary for cellular attachment and orientation on well-defined silicon- and polymer-based surfaces. One particularly powerful application of controlled cellular growth is for the development of artificial capillaries to remediate vascular disease [17.114] (Moldovan, unpublished). Micromachined silicon or polymer channels of the dimensions of capillaries have shown promise as effective conduits to direct endothelial cells to form tubes through which a fluid (blood) could eventually flow.

MEMS technology and nanoscale control of molecular events and interactions has also been applied to the development of cardiovascular sensors. For instance, blood cell counters have successfully detected red and white blood cells and show promise for quick, reliable blood cell analysis [17.115]. Additionally, force trans-

ducers have been developed to measure the contractile force of a single cardiac myocyte [17.116]. Basic research into cellular function and response to stimuli such as these will eventually support more efficacious and robust cardiovascular tissue engineering approaches.

Nanoparticulate Carriers for Therapy and Imaging

Two major cardiovascular applications of nanoparticles are targeted contrast agents for imaging (MRI, CT, X-ray, and so on) and targeted drug delivery vehicles to address vascular lesions (atherosclerotic plaques) and other vascular disease states. In either application, nanoparticle surfaces must be appropriately modified to avoid RES clearance and to drive interaction with the desired effector cells. Decades of research in the area of molecular and cell biology has provided researchers with many tools to make the appropriate protein component selection for nanotherapeutic contrast agents. For instance, knowing that the procoagulant protein tissue factor is expressed only at the surface of injured endothelial cells provides a means to differentiate healthy from diseased vasculature. It also provides a method to target disease sites. Similar sorts of information regarding disease and normal vasculature at specific tissue sites can be critically important for targeting specialized vascular beds, such as that which occurs at the BBB. Several groups have begun efforts to develop nanoparticles that selectively pass across the BBB to deliver chemotherapeutic or other drugs to the brain [17.117, 118], and this may improve outcomes from stroke.

With regard to MRI approaches, particles used for targeted contrast enhancement in specific tissues or vascular beds may provide a powerful way to detect unstable atherosclerotic plaques and to quantitatively analyze blood flow. For instance, nonspecific accumulation of nanoparticles within human atherosclerotic plaques has been reported [17.119, 120] and may ultimately provide a way to distinguish vulnerable and stable plaques. Such data might be used to direct therapy, with the intent, as always, to improve therapeutic outcome. Furthermore, nanotherapeutic devices have been designed to include both contrast and drug delivery functions (Fig. 17.3). Single-platform, multifunctional particles for cardiovascular therapy represent the cutting edge in the cardiovascular field. These delivery vehicles are continuously being redesigned and improved to improve circulation times, targeting [17.66, 121], degradability and biocompatibility [17.122].

17.4.3 Nanotherapeutics and Specific Host Immune Responses

Immune responses directed to immunogens (molecules or structures capable of eliciting specific immune responses, also called antigens) are a pervasive aspect of the responses of vertebrates to exposure to foreign macromolecules, proteins, particles, and organisms (for overviews, see [17.1, 123, 124]). Undesired antibody responses, in particular, are the single most important hurdle for clinical use of recombinant proteins. On the other hand, undesired cellular immune responses are key mediators of frequently devastating disorders for patients, whether the responses are directed to self (as in autoimmune disease) or to foreign antigens (as in transplant rejection). As we shall see, host immune responses to nanotherapeutic devices are a critical determinant of their efficacy. Any strategy (including the biotechnological strategies) that allows controlled manipulation of immune responses either to augment desirable responses, as in vaccines, or to mitigate deleterious responses holds substantial potential benefit.

Basic Immunology

Specific immune responses are directed to individual macromolecules or assemblages of them and can be categorized into cellular responses, mediated by cytotoxic T-lymphocytes (CTLs), and humeral responses, mediated by soluble proteins (found in blood serum and other biological fluids) secreted by B-cells (antibodies). Regardless of the effector (CTLs or antibodies), specific responses are directed to individual molecular features of antigens called epitopes. The three types of epitopes are: cytotoxic T-cell (CTL) epitopes, helper T-cell epitopes, and B-cell epitopes. The first two types are peptides that must be presented on cell surfaces in the context of the class I (in the case of CTL epitopes for CTL responses) or class II (in the case of T-helper epitomes, involved in both antibody and CTL responses) major histocompatibility complex (MHC) molecules. B-cell epitopes are a chemically and structurally heterogeneous group of epitopes and include synthetic molecules as well as peptides. In addition to CTL and B-cell epitopes, which are recognized by immune effectors, T-helper epitopes are recognized by regulatory components of the immune system (helper T-cells). Vibrant CTL and antibody responses require immunogens containing T-helper epitopes along with either B-cell or CTL epitopes.

While epitope-presenting MHC class I and class II complexes must be present on cell surfaces to trigger immune responses, the machinery for antigen processing and loading of MHC class I or class II proteins for antigen presentation reside in distinct membrane-bound compartments in the cell [17.1]. For instance, processing for class I presentation (to trigger CTL responses) occurs through the cytosol, but charging of MHC class I complexes is accomplished in the endoplasmic reticulum. On the other hand, processing for class II presentation (to augment antibody and CTL responses) as well as the loading of class II occur in endosomal compartments. Although these are membrane-bounded compartments in the cytosol, topographically the interiors of endosomal compartments are part of the extracellular environment and are not inside the cell at all.

In the case of both class I and class II, the antigen-presenting complexes are moved to the cell surface after charging. The moving of molecules between different cellular locations (called trafficking) is tightly regulated by and driven by specific cellular proteins, systems, and structures. Proper trafficking is essential to assure that all cellular components assume their correct position in the cellular organization, which is to assure that antigens are directed to the compartments that will (most often) produce most protective immune responses in this case. These complex antigen processing and presentation systems can be perturbed and exploited to offer novel opportunities in vaccine design.

Nanotherapeutic Vaccines: Eliciting Desired Host Immune Responses

Both therapeutic and prophylactic vaccines are naturals for nanobiological approaches. Vaccines are therapeutics that modulate individual, specific host immune responses (antibody or cytotoxic lymphocyte responses, or both) in order to produce the desired immunity to pathogens in infectious disease or modulate host responses to self antigens in autoimmunity. Ideally they are chemically well defined, safe, and induce long-term (preferably, multiyear) protective immune responses [17.125]. Recent developments in immunobiology have opened up opportunities for nanobiological vaccine design [17.126, 127].

Particulate Vaccines

The immune system, through dendritic cells (DCs), other antigen-presenting cells (APCs), and immune effectors, performs ongoing surveillance for non-self-antigens using the MHC antigen presentation system described above [17.1]. It has long been known that APCs sample serum for antigens that they process and present using the MHC class II pathway for antibody production [17.1]. Research over the last few years has shown that DCs, macrophages, and other APCs also take up particulate materials and process and present the constituent peptide epitopes via the MHC class I pathway in order to trigger specific CTL responses. Previous dogma held that only endogenous antigens (such as those produced in virally infected cells) could be presented through the class I MHC pathway. Both soluble and particulate antigens can be presented via the class I pathway, but particulate antigens are orders of magnitude more potent than the same immunogens formulated in soluble form [17.125–128].

APCs are not particularly fastidious regarding the chemical compositions of the particles they take up: latex beads [17.126, 128] as well as iron beads [17.126] have been used. Self-assembled nanoparticles made of lipidated epitope peptides also deliver orders of magnitude of better immunization than do soluble formulations of the same peptides [17.129]. The specific size for optimal CTL responses is not fully defined but is clearly in the nanometer to micrometer size range [17.125, 126, 128, 129]. One could imagine a systematic exploration of optimum size and compositional preferences for uptake by APCs using such chemically and morphologically tunable polymeric nanomaterials as dendrimers and tectodendrimers.

Nanobiological Design of Nanotherapeutic Vaccines

As mentioned earlier, antigen trafficking determines which specific MHC complex (class I or class II) epitopes are presented on and, therefore, which immune effectors are produced in the subsequent response. The process, while tightly regulated, is amenable to intervention using nanotherapeutic vaccines. Nanobiological design strategies (discussed above) can take advantage of the numerous proteins and peptides now known to mediate the trafficking of carried materials to specific sites. In the device context, these trafficking moieties might assure delivery of antigens to the appropriate cellular compartments to trigger desired class I- or class II-mediated responses.

Vaccine polypeptides, for instance, might be fused or formulated with pathogen polypeptides that will deliver them across epithelium from outside the host for use in mucosal vaccination [17.130]. Still other peptides are known to carry injected materials across the cell envelope into the cytosol [17.131]. Alternatively, vaccine entities taken up by receptor-mediated endocytosis (into endosomal vesicles, the traditional site of uptake for antigens destined for MHC class II presentation) could be shuttled into the cytosol (and hence to the

class I MHC pathway) by the incorporation of peptide domains of bacterial toxins that mediate endosomal escape to cytoplasm [17.132]. Potentially, nanobiological therapeutic design strategies [17.5, 13, 15] can combine knowledge of polymer chemistry, nanomaterials, chemistry, and cell, molecular and immunobiology to generate highly effective nanotherapeutic vaccines.

Nanotechnology and Modulation of Immune Responses

On the other hand, protein immunogenicity represents a critical limitation on the therapeutic use of recombinant proteins, particularly when specific antibody responses to the therapeutics are engendered [17.32]. The consequences of antibody responses to proteins are variable, ranging from no effect to decidedly negative consequences to the host. Antibodies can neutralize therapeutic bioactivity, rendering drug entities inactive in repeated rounds of therapy, distort drug P_k and B_d properties, and, in the worst case, trigger autoimmune responses with cross-reactive host antigens (host antigens that share some identity with the immunizing epitopes).

Antibody Responses to Nanotherapeutic Devices

Antibody responses to protein components are a major challenge for nanobiological therapeutic devices in vivo. We, and others [17.6, 14, 21, 133–136], have demonstrated that protein components of bioconjugates to diverse nanomaterials can not only retain their intrinsic immunogenicity but, worse, can render the otherwise nonimmunogenic synthetic nanomaterials to which they are linked capable of inducing antibody responses [17.6, 7, 14, 21, 126, 133–135]. The mechanism of immunogenicity seems to be haptenization, a well-known phenomenon in immunology. Haptenization occurs when antigens containing only B-cell epitopes are covalently linked to proteins containing T-helper epitopes, thereby creating hybrid antigens that are fully immunogenic with respect to antibody responses. The process is illustrated in Fig. 17.17.

As a result of the chemically repetitive structure of many synthetic nanomaterials, the antibodies directed to them have interesting properties. Most notably, antibodies generated to nanostructures are often cross-reactive to other chemically similar nanostructures. For example, antibodies raised using haptenized fullerenes also recognize carbon nanotubes [17.133], and antibodies raised to haptenized generation 0 (G_0) PAMAM dendrimers also recognize higher generation PAMAM dendrimers (G_1, G_2, G_3, ... [17.14, 21, 133–136]. Presumably, cross-reactivity results from the presence of common B-cell epitopes in the immunogen used and in structurally similar cross-reactive antigens. Thus, patients immunized with any therapeutic containing one haptenized nanostructure could potentially raise antibodies that would recognize any nanotherapeutic containing a material structurally similar to the haptenized nanomaterial.

This is a potentially serious problem, in light of the documented ability of antibody responses to interfere with therapy by neutralizing drug or distorting P_k and B_d properties. As discussed before, molecular weights of serum antibodies are around 150 000 AMU, with an effective diameter in excess of 10 nm [17.1, 123, 124]. Many of the desirable properties of nanotherapeutics derive from their precisely controlled sizes, and size-dependent properties like EPR targeting could be jeopardized if the therapeutic to be delivered were noncovalently complexed with antibodies directed to its synthetic components. Due to long-term persistence of memory immune responses, exposure to haptenized nanotherapeutics could endanger patient responsiveness to structurally similar materials for the duration of their lives. Obviously, immunogenicity is a critical challenge to the nanobiological therapeutic paradigm that has yet to be satisfactorily addressed.

Polymer Conjugation to Mitigate Immune Responses

There are multiple approaches to mitigating immunogenicity, some of which are applicable to nanobiotechnological therapeutic devices. Polymer conjugation, particularly conjugation of proteins to poly(ethylene glycol) (PEG, a flexible, biologically well-tolerated hydrophilic polymer) has been used to control the immunogenicity of a broad class of proteins and synthetic molecules since the mid-1970s [17.19, 137]. Typically, pegylation involves covalent linkage (often by promiscuous conjugation chemistry) of variable numbers (usually two to six polymer molecules) of size-polydisperse (usually 3000 to 10 000 AMU) linear or branched PEG chains to the potential immunogen. This approach results in polydisperse bioconjugates, which flies in the face of the molecular level structural precision often considered essential for nanotechnology. Still, pegylation may be useful in some limited nanobiological applications.

The mechanism of molecule-specific immunosuppression remains obscure, despite almost 30 years of research. The flexibility of the PEG strands allows them to wrap around the molecules they are conjugated to, perhaps delimiting access of immune effectors and immune processing machinery to the potential immunogens. Pegylated molecules exhibit dramatically enhanced apparent diameters and vastly enhanced solubility in aqueous buffer. Both of these phenomena result in substantially diminished clearance of pegylated molecules by the RES and may account for the retention of in vivo bioactivity despite the apparent steric hindrance that may result from wrapping PEG moieties around their protein bioconjugate partners. Pegylation-related, bioconjugate-specific immunosuppressive phenomena cannot be fully accounted for by sterics, though. Immunological experiments in syngeneic mice show that the protein-specific immunosuppression associated with pegylated proteins is transmissible from one animal to another by surgically harvested splenocytes. This finding is more consistent with a mechanism of suppression involving immunologically induced tolerance of foreign antigens as opposed to one caused by simple steric considerations [17.137].

Synthetic Nanoporous Membranes for Immunoisolation

A second approach to avoiding immune responses to nanodevices or biocomponents is immunoisolation by encapsulation [17.16, 17]. In this strategy, potentially immunogenic structures (nanodevices, cells, macromolecules) are sequestered from high molecular weight, large host immune effectors, and immunosurveillance, while small molecules (such as glucose and small proteins) diffuse freely (Fig. 17.18). While conceptually simple, immunoisolation is technically challenging in that the number of pores must be high to maximize flux, and the pores must be highly uniform in dimension.

Because of the exquisite sensitivity of the immune system, immunoisolation must be nearly perfect if it is to work at all. Particularly in the case of immunoisolation of cells for xenografts, if only a tiny fraction (as few as 1%) of the pores of an immunoisolation capsule are large enough to admit immune effectors (for instance, antibodies and complement components, exclusion of which requires pores of 50 nm or less), the viability of the graft will be compromised [17.16, 17, 138]. Thus, the strategy requires fabrication of membranes with pores that have very tightly controlled dispersities in terms of their maximum diameters. Until recently, immunoisolation for xenografts has failed for lack of such membranes.

The technological innovations that made the synthesis of these high-quality nanoporous membranes possible were described above and in Fig. 17.4. The nanoporous membranes can be used to form immunoisolation capsules either when two such membranes are joined so that they bind an enclosed space or when

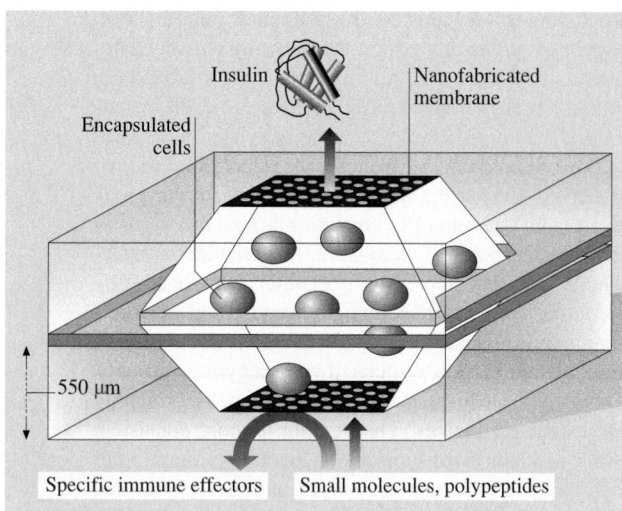

Fig. 17.18 Biological components can be protected from unfavorable environments by encapsulation [17.16–18]. Shown is an immunoisolation capsule containing heterologous (immunogenic) cells producing insulin (a four helical bundle protein, see text). The cells are protected from immune responses by membranes with 10 nm pores that allow free diffusion of small molecules (oxygen, nutrients) into the capsule and small proteins (insulin, shown here not to scale) out of it. However, immune effectors with diameters greater than pore size (immune cells like CTLs and APCs, antibodies, other effectors) don't access the capsule interior, protecting the xenograft from immune rejection. The structure integrates both micro- and nanoscale components (including the nanoporous membranes of Fig. 17.4) into a macroscale construct

a nanoporous membrane is used to enclose a compartment of some sort in a therapeutic structure. The enclosed spaces can be charges with functional nanodevices or, for that matter, living cells.

Such immunoisolation capsules charged with heterologous pancreatic islet cells have demonstrated immense promise for regulating normoglycemia in diabetic animal models [17.16–18].

Biogenic Nanoporous Immunoisolation Membranes

The novel microfabrication strategy has made it possible to make these highly defined membranes in an industrial process, and they are applicable to numerous biomedical applications beyond immunoisolation [17.16, 17]. Still, the fabrication process is rigorous, and an alternative biogenic method to derive selectively porous silicon capsules has been proposed [17.139, 140]. Diatoms are encapsulated in biogenic amorphous silicon shells (called frustules), which the living cells secrete. Frustules must necessarily be porous to allow the cells to take on nutrients and excrete wastes, and the porosity, pore dimensions, and other micro- and nanoscale morphological frustule features are tightly controlled at genetic and physiological levels. Using techniques well known to microbiologists (manipulation of culture conditions), geneticists (mutagenesis), and cell biologists (optically driven cell sorting in flow systems), it is theoretically possible to select and propagate diatoms whose frustule morphology, size, and pore characteristics fall within preselected limits [17.139, 140]. Frustules can be isolated intact from diatom cultures, so biogenic production of silicon membranes with porosity characteristics similar to the microfabricated membranes discussed above may be feasible. Though they have yet to be realized, immunoisolation capsules made from genetically/physiologically tuned frustules have been proposed [17.140]: one wonders which other biomineralized structures eventually might also be put to nanobiotechnological use.

17.5 Concluding Remarks: Barriers to Practice and Prospects

The effort to produce nanoscale therapeutic devices is clearly highly interdisciplinary. As we have seen, it touches on numerous established disciplines, encompassing elements of physiology, biotechnology, bioconjugate chemistry, electrical engineering, and materials science, to name just a few of the fields involved. Obviously, this broad sweep of knowledge is difficult for any one investigator to master fully. The breadth of the effort constitutes just one of the major barriers to entry in the field. Other challenges include the raw complex-

ity of biology, the fashion in which biologists hold and distribute information, and cultural differences between engineers and biological scientists.

17.5.1 Complexity in Biology

Biology is characterized by particularity: nuances of biological systems are often unique to the system at hand and highly idiosyncratic. This follows from the fact that biological systems are not purpose-built, unlike designed devices (although they can give that appearance, as seen in Fig. 17.1), but rather arose as the consequence of evolutionary processes. Evolution is a highly chaotic business, with the outcome of any given evolutionary process highly sensitive to initial conditions (including populations of organisms subjected to selection, other organisms in the environment, the biologies of all organisms involved, resource availability and other environmental factors). Moreover, many of the variables affecting natural selection processes are dynamic and change, as a function of time or space or both, *while* selection is exerted against a population of organisms. Conditions leading to one evolutionary adaptation or another are therefore seldom duplicated exactly, so that individual adaptive features are idiosyncratic, with elements relating not only to their biological functions but also to their evolutionary history. Individual systems and adaptive features thereof can be almost baroquely complex because the structures themselves arose under unique conditions, under unique selective pressures, and from unique initial biological systems. The extent of the complexity of biological systems is apparent even in individual macromolecular constituents of biological systems and can be made clear by comparing synthetic nanostructures to biological nanostructures with similar dimensional aspects.

SWCNTs (single-wall carbon nanotubes) have a minimal diameter of about 1.3 nm, and many proteins likewise have diameters of a few nanometers. SWCNTs are regular and homogeneous polymeric structures composed entirely of carbon atoms (Fig. 17.6). Proteins, on the other hand, are not polymers of one repeating subunit, but rather are nonrandom copolymers of 20 chemically distinct amino acids (aas). The order of the aas drives specific folding events that produce three-dimensional structures that, in turn, present specific aas and their side chains at specific positions in space (Fig. 17.10). This control of aa position in space drives protein activity: small changes in aa sequence can perturb structure and function profoundly. Therefore, though they fall within the same broad size regime, proteins are structurally and functionally much more complex than SWCNTs. The extent of biological complexity can be glimpsed when one considers that individual living things are ordered aggregates of multiple macromolecules and supramolecular structures (potentially, billions and billions of them), belonging to distinct chemical classes (such as lipids, proteins, nucleic acids), with each individual macromolecule being at least as complex in structure and function as the protein of Fig. 17.10. The nearly irreducible complexity of

biological systems is a central fact of practice in the biological sciences and drives the way biologists gather and disseminate information. It is the key informant of the culture of biologists.

17.5.2 Dissemination of Biological Information

Biologists typically consider themselves primarily as scientists, as opposed to being engineers or technology developers. The product biologists generate, then, is information rather than devices or structures. Their interest in technological applications of the knowledge they produce is typically secondary to their desire to develop the information itself. Additionally, as we have just seen, biology is a ferociously complex discipline.

As a practical matter, biological data itself is often much more rich (and often more ambiguous) than data from harder scientific endeavors. The data can be so very rich that biologists must make choices as to what data is relevant to a given phenomena and, therefore, what data they will publish. With absolutely no intent to conceal or mislead, biologists are often driven to publish only a small fraction of the data they gather.

Since they see themselves primarily as scientists, biologists tend to publish data they believe to be of broad scientific significance. Typically, the chosen information does not include data that might be critical to technology development, frequently making the biological literature an inadequate resource for engineers. The omitted information thus becomes lore. That is to say, the information critical for the practice or development of a given technology but is not usually accessible to persons outside the field (as many nanotherapeutic device producers may be). This can be illustrated by recent experiences around phage display-mediated affinity maturation of four helical bundle proteins [17.141–143].

Four helical bundle proteins are loosely related small proteins consisting of four α helices arranged in a specific configuration (see the dynamite stick representation of insulin that occurs in Fig. 17.18). They include insulin, growth hormone, most cytokines, and various other molecules involved in cell-to-cell signaling. They have a wide range of therapeutically valuable properties that might be made even more desirable if their potencies could be enhanced. There is thus significant interest in variant four-helical bundle proteins with improved bioactivities.

Phage display [17.67] is a method that can be used to sort protein variant libraries for variants with enhanced affinity for a target receptor. This is usually accomplished by iterative rounds of affinity selection on the receptor followed by propagation of the selected phage. As the rounds of selection and propagation proceed, the mean affinity of the selected variants for the receptor increases (so-called affinity maturation). A phage display method is depicted in Fig. 17.11.

Affinity maturation by phage display can be used to identify variant proteins from libraries that have enhanced biological activities, providing that the affinity for receptor is limiting in the overall activity of the parent protein [17.134]. Phage display affinity maturation has been successfully used to select some enhanced activity variants of four helical bundle proteins [17.144, 145] but is *not* applicable to engineering increased activity for *all* four helical bundle proteins [17.134]. The reason for the inapplicability of display methods to engineer high-activity human growth hormone (hGH) variants became apparent in a recent publication from Genentech [17.142].

In the case of hGH, the affinity of the parent molecule for receptor is more than that required to drive maximal biological responses [17.142]. This fact was inferred after a multiyear effort (beginning in the late 1980s and culminating in 1999) to derive more active hGH proteins, involving literally thousands of hGH variants. Many of the variants did indeed exhibit enhanced affinity for the hGH binding protein (receptor), but none exhibited significantly enhanced biological activity. Each data point (the biological activity of an individual variant) essentially constituted a failure of a technology to provide a desired result and was therefore seen, not unreasonably, as negative (and therefore uninteresting) data by the investigators. Consequently, the information was not broadly disseminated until a significant conclusion could be presented (that biological responses to hGH are not limited by ligand–receptor affinity). Thus, between the late 1980s and the 1999 publication, this information was known primarily and unambiguously only to the investigators involved, even though it would have been critical to any technologist planning to affinity-mature hGH by phage display for enhanced activity [17.134].

17.5.3 Cultural Differences Between Technologists and Biologists

Biology is the realm of particularity, whereas more physical sciences are realms within which general rules can be derived and broadly applied. The vast complexity of biology usually drives biologists to be highly spe-

cialized, perhaps making a career from the study of particular macromolecules (specific enzymes, growth factors, genetic elements, and so on) or biological systems. Biologists are, therefore, neither trained nor encouraged by funding agencies to see themselves as fungible, and frequently they are very reluctant to step away from their primary research foci into new areas. Doing so is a major undertaking (and a substantial risk) for most biologists.

Engineers, on the other hand, often see themselves as operating from first principles and are more (if not entirely) willing to step into new areas. After all, crossing disciplines is feasible for engineers working in the physical sciences, whereas it is much more difficult, not only for reasons of training but also because of the biological realities discussed above, for biologists. Also, by reason of their training and experience, engineers often vastly underestimate the complexity of biological systems and sometimes exhibit what biologists perceive as naïveté in their approach to biological systems. Unfortunately (or fortunately, depending on your perspective), therapeutic nanodevices require both biological and engineering expertise. An often bruising debate as to how best to prepare students for practice in this extremely challenging field is currently playing out in biomedical engineering and other related academic departments across the country.

The traditional approach for training biomedical engineers has been to train conventional engineers with a smattering of biological knowledge. This "primarily engineers" model of biomedical engineering has been reasonably successful in multiple endeavors (such as orthopedics and imaging technologies), but it seems unlikely in the extreme that it will suffice for robust practice in nanotherapeutic devices. As this chapter shows, the role of biology is pivotal to success in the field, and sophisticated nanotherapeutics are virtually impossible to design without deep knowledge of the biology involved. A reasonable argument can be made that successful engineering of therapeutic nanostructures will demand intense focus on biology and a smattering of engineering knowledge (a "primarily biologists" approach, as opposed to the historically prevalent "primarily engineers" training model). This view is often strongly and passionately resisted in colleges of engineering, which are the usual homes of biomedical engineering and related efforts. The outcome of the debate is not finalized, but it is critically important: it will determine whether or not a ready cadre of researchers is prepared for a career in the area. If not, the field will progress much as it has, very slowly bringing the benefits of the research to patients. It seems apparent to us, and experience to date also seems to show, that a substantial biological component of training for the discipline is simply unavoidable if we are committed to realizing the vast potential of nanotechnology in human therapy.

References

17.1 C. A. Janeway, P. Travers, M. Walport, J. D. Capra: *Immunobiology* (Elsevier, London 1999)

17.2 S. C. Lee: Biotechnology for nanotechnology, Trends Biotechnol. **16**, 239–240 (1998)

17.3 S. C. Lee: Engineering the protein components of nanobiological devices. In: *Biological Molecules in Nanotechnology: The Convergence of Biotechnology, Polymer Chemistry and Materials Science*, ed. by S. C. Lee, L. Savage (IBC, Southborough 1998) pp. 67–74

17.4 S. C. Lee: How a molecular biologist can wind up organizing nanotechnology meetings. In: *Biological Molecules in Nanotechnology: The Convergence of Biotechnology, Polymer Chemistry and Materials Science*, ed. by S. C. Lee, L. Savage (IBC, Southborough 1998)

17.5 S. C. Lee: The nanobiological strategy for construction of nanodevices. In: *Biological Molecules in Nanotechnology: The Convergence of Biotechnology, Polymer Chemistry and Materials Science*, ed. by S. C. Lee, L. Savage (IBC, Southborough 1998) pp. 3–14

17.6 S. C. Lee: A biological nanodevice for drug delivery. In: *National Science and Technology Council. IWGN Workshop Report: Nanotechnology Research Directions. International Technology Research Institute, World Technology Division* (Kluwer, Baltimore 1999) pp. 91–92

17.7 S. C. Lee, R. Parthasarathy, K. Botwin: Protein-polymer conjugates: Synthesis of simple nanobiotechnological devices, Polymer Preprints **40**, 449–450 (1999)

17.8 L. Jelinski: Biologically related aspects of nanoparticles, nanostructured materials and nanodevices. In: *Nanostructure Science and Technology*, ed. by R. W. Siegel, E. Hu, M. C. Roco (Kluwer, Dordrecht 1999) pp. 113–130

17.9 J. Baish, Y. Gazit, D. Berk, M. Nozue, L. T. Baxter: Role of tumor vascular architecture in nutrient and drug delivery: An invasion percolation-based network model, Microvasc. Res. **51**, 327–346 (1996)

17.10 J. R. Baker Jr.: Therapeutic nanodevices. In: *Biological Molecules in Nanotechnology: The Convergence of Biotechnology, Polymer Chemistry and Materials Science*, ed. by S. C. Lee, L. Savage (IBC, Southborough 1998) pp. 173–183

17.11 R. Duncan: Drug targeting: Where are we now and where are we heading?, J. Drug Targeting **5**, 1–4 (1997)

17.12 R. Duncan, S. Gac-Breton, R. Keane, Y. N. Sat, R. Satchi, F. Searle: Polymer-drug conjugates, PDEPT and PELT: Basic principles for design and transfer from the laboratory to clinic, J. Cont. Release **74**, 135–146 (2001)

17.13 D. S. Goldin, C. A. Dahl, K. L. Olsen, L. H. Ostrach, R. D. Klausner: Biomedicine. The NASA-NCI collaboration on biomolecular sensors, Science **292**, 443–444 (2001)

17.14 S. C. Lee: Dendrimers in nanobiological devices. In: *Dendrimers and other Dendritic Polymers*, ed. by D. Tomalia, J. Frechet (Wiley, London 2001) pp. 548–557

17.15 J. R. Baker Jr., A. Quintana, L. Piehler, M. Banazak-Holl, D. Tomalia, E. Racka: The synthesis and testing of anti-cancer therapeutic nanodevices, Biomed. Microdevices **3**, 61–69 (2001)

17.16 T. Desai, D. Hansford, L. Kulinsky, A. Nashat, G. Rasi, J. Tu, Y. Wang, M. Zhang, M. Ferrari: Nanopore technology for biomedical applications, Biomed. Microdevices **21**, 11–40 (1999)

17.17 T. A. Desai, W. H. Chu, J. K. Tu, G. M. Beattie, A. Hayek, M. Ferrari: Microfabricated immunoisolating biocapsules, Biotechnol. Bioeng. **57**, 118–120 (1998)

17.18 M. Ferrari, W. H. Chu, T. A, Desai, J. Tu: *Microfabricated silicon biocapsule for immunoisolation of pancreatic islets*, Advanced Manufacturing Systems, CISM Courses and Lectures, Vol. 372, ed. by E. Kuljanic (Springer, Wien 1996) pp. 559–568

17.19 J. M. Harris, N. E. Martin, M. Modi: Pegylation: A novel process for modifying pharmacokinetics, Clin. Pharmacokin. **40**, 539–551 (2001)

17.20 S. B. H. Kent: Building proteins through chemistry: Total chemical synthesis of protein molecules by chemical ligation of unprotected protein segments. In: *Biological Molecules in Nanotechnology: The Convergence of Biotechnology, Polymer Chemistry and Materials Science*, ed. by S. C. Lee, L. Savage (IBC, Southborough 1998) pp. 75–92

17.21 S. C. Lee, R. Parthasarathy, T. Duffin, K. Botwin, T. Beck, G. Lange, J. Zobel, D. Jansson, D. Kunneman, E. Rowold, C. F. Voliva: Antibodies to PAMAM dendrimers: Reagents for immune detection assembly and patterning of dendrimers. In: *Dendrimers and other Dendritic Polymers*, ed. by D. Tomalia, J. Frechet (Wiley, London 2001) pp. 559–566

17.22 S. C. Lee, M. S. Leusch, V. A. Luckow, P. Olins: Method of producing recombinant viruses in bacteria, (1993)US Patent 5,348,886

17.23 M. S. Leusch, S. C. Lee, P. O. Olins: A novel host-vector system for direct selection of recombinant baculoviruses (bacmids) in E. coli, Gene. **160**, 191–194 (1995)

17.24   V. A. Luckow, S. C. Lee, G. F. Barry, P. O. Olins: Efficient generation of infectious recombinant baculoviruses by site-specific, transposon-mediated insertion of foreign DNA into a baculovirus genome propagated in E. coli, J. Virol. **67**, 4566–4579 (1993)

17.25   T. Gardner, C. R. Cantor, J. J. Collins: Construction of a genetic toggle switch in E. coli, Nature **403**, 339–342 (2000)

17.26   J. Hasty, F. Isaacs, M. Dolnik, D. McMillen, J. J. Collins: Designer gene networks: Towards fundamental cellular control, Chaos **11**, 107–220 (2001)

17.27   E. Di Fabrisio, A. Nucara, M. Gentili, R. Cingolani: Design of a beamline for soft and deep lithography on third generation synchrotron radiation source, Rev. Sci. Instrum. **70**, 1605–1613 (1999)

17.28   G. M. Dykes: Dendrimers: A review of their appeal and applications, J. Chem. Tech. Biotech. **76**, 903–918 (2001)

17.29   R. Spindler: PAMAM starburst dendrimers: Designed nanoscopic reagents for biological applications. In: *Biological Molecules in Nanotechnology: The Convergence of Biotechnology, Polymer Chemistry and Materials Science*, ed. by S. C. Lee, L. Savage (IBC, Southborough 1998) pp. 15–32

17.30   D. Tomalia, H. M. Brothers II: Regiospecific conjugation to dendritic polymers to produce nanodevices. In: *Biological Molecules in Nanotechnology: The Convergence of Biotechnology, Polymer Chemistry and Materials Science*, ed. by S. C. Lee, L. Savage (IBC, Southborough 1998) pp. 107–120

17.31   B. I. Yacobson, R. E. Smalley: Fullerene nanotubes: C1000000 and beyond, A. Scient. **85**, 324–337 (1997)

17.32   D. J. A. Crommelin, R. D. Sindelar: *Pharmaceutical Biotechnology* (Harwood Academic Publishers, Amsterdam 1997)

17.33   L. J. Lee: BioMEMS and micro-/nano-processing of polymers – An overview, Chin. J. Chem. Eng. **34**, 25–46 (2003)

17.34   T. R. Groves, D. Pickard, B. Rafferty, N. Crosland, D. Adam, G. Schubert: Maskless electron beam lithography: Propects, progress and challenges, Microelectron. Eng. **61**, 285–293 (2002)

17.35   M. Guthold, R. Superfine, R. Taylor: The rules are changing: Force measurements on single molecules and how they relate to bulk reaction kinetics and energies, Biomed. Microdevices **3**, 9–18 (2001)

17.36   L. M. Demers, D. S. Ginger, S.-J. Park, Z. Li, S.-W. Chung, C. A. Mirkin: Direct patterning of modified oligonucleotides on metals and insulatos by dip-pen nanolithography, Science **296**, 1836–1838 (2002)

17.37   K.-B. Lee, S.-J. Park, C. A. Mirkin, J. C. Smith, M. Mrksich: Protein nanoarrays generated by dip-pen nanolithography, Science **295**, 1702–1705 (2002)

17.38   M. Ferrari, J. Liu: The engineered course of treatment, Mech. Eng. **123**, 44–47 (2001)

17.39   H. W. Rohrs, R. S. Ruoff: The use of carbon nanotubes in hybrid nanometer-scale devices. In: *Biological Molecules in Nanotechnology: The Convergence of Biotechnology, Polymer Chemistry and Materials Science*, ed. by S. C. Lee, L. Savage (IBC, Southborough 1998) pp. 33–38

17.40   K. E. Drexler: *Engines of Creation: The Coming Era of Nanotechnology* (Anchor Books, New York 1986)

17.41   J. Cumings, A. Zetti: Low-friction nanoscale linear bearing realized from multiwall carbon nanotubes, Science **289**, 602–604 (2000)

17.42   D. J. Hornbaker, S.-J. Kahng, S. Mirsa, B. W. Smith, A. T. Johnson, E. J. Mele, D. E. Luzzi, A. Yazdoni: Mapping the one-dimensional electronic states of nanotube peapod structures, Science **295**, 828–831 (2002)

17.43   C. Dekker: Carbon nanotubes as molecular quantum wires, Phys. Today **28**, 22–28 (1999)

17.44   M.-C. Jones, J.-C. Leroux: Polymeric micelles-a new generation of colloidal drug carriers, Eur. J. Pharma. Biopharma. **48**, 101–111 (1999)

17.45   I. Uchegbu: Parenteral drug delivery: 1, Pharma. J. **263**, 309–318 (1999)

17.46   I. Uchegbu: Parenteral drug delivery: 2, Pharma. J. **263**, 355–359 (1999)

17.47   K. B. Thurmond II, H. Huang, K. L. Wooley: Stabilized micellar structures in nanodevices. In: *Biological Molecules in Nanotechnology: The Convergence of Biotechnology, Polymer Chemistry and Materials Science*, ed. by S. C. Lee, L. Savage (IBC, Southborough 1998) pp. 39–43

17.48   S. Uppuluri, D. R. Swanson, L. T. Peihler, J. Li, G. Hagnauer, D. A. Tomalia: Core shell tecto(dendrimers). I. Synthesis and characterization of saturated shell models, Adv. Mater. **12**, 796–800 (2000)

17.49   A. K. Patri, I. J. Majoros, J. R. Baker Jr.: Dendritic polymer macromolecular carriers for drug delivery, Curr. Opin. Chem. Biol. **6**, 466–471 (2002)

17.50   S. Fernandez-Lopez, H.-S. Kim, E. C. Choi, M. Delgado, J. R. Granja, A. Khasanov, K. Kraehenbuehl, G. Long, D. A. Weinberger, K. M. Wilcoxen, M. Ghardiri: Antibacterial agents based on the cyclic D,L-alpha-peptide architecture, Nature **412**, 452–455 (2001)

17.51   A. Saghatelian, Y. Yokobayashi, K. Soltani, M. R. Ghadiri: A chiroselective peptide replicator, Nature **409**, 777–778 (2001)

17.52   J. Davies: Aminoglycoside-aminocyclitol antibiotics and their modifying enzymes. In: *Antibiotics in Laboratory Medicine*, ed. by V. Lorian (Williams and Wilkins, Baltimore 1984) pp. 474–489

17.53   T. Hamouda, A. Myc, B. Donovan, A. Y. Shih, J. D. Reuter, J. R. Baker Jr.: A novel surfactant nanoemulsion with a unique non-irritant topical antimicrobial activity against bacteria, enveloped viruses and fungi, Microbiol. Res. **156**, 1–7 (2001)

17.54 M. J. Heller: Utilization of synthetic DNA for molecular electronic and photonic-based device applications. In: *Biological Molecules in Nanotechnology: The Convergence of Biotechnology, Polymer Chemistry and Materials Science*, ed. by S. C. Lee, L. Savage (IBC, Southborough 1998) pp. 59–66

17.55 Z. Ma, S. Taylor: Nucleic acid triggered catalytic drug release, Proc. Nat. Acad. Sci. USA **97**, 11159–11163 (2000)

17.56 R. C. Merkle: Biotechnology as a route to nanotechnology, Trends Biotechnol. **17**, 271–274 (1999)

17.57 N. C. Seeman, J. Chen, Z. Zhang, B. Lu, H. Qiu, T.-J. Fu, Y. Wang, X. Li, J. Qi, F. Liu, L. A. Wenzler, S. Du, J. E. Mueller, H. Wang, C. Mao, W. Sun, Z. Shen, M. H. Wong, R. Sha: A bottom-up approach to nanotechnology using DNA. In: *Biological Molecules in Nanotechnology: The Convergence of Biotechnology, Polymer Chemistry and Materials Science*, ed. by S. C. Lee, L. Savage (IBC, Southborough 1998) pp. 45–58

17.58 G. Lemieux, C. Bertozzi: Chemoselective ligation reactions with proteins, oligosaccharides and cells, Trends Biotechnol. **16**, 506–512 (1998)

17.59 R. Offord, K. Rose: Multicomponent synthetic constructs. In: *Biological Molecules in Nanotechnology: The Convergence of Biotechnology, Polymer Chemistry and Materials Science*, ed. by S. C. Lee, L. Savage (IBC, Southborough 1998) pp. 93–105

17.60 G. T. Hermanson: *Bioconjugate Chemistry* (Academic, San Diego 1996)

17.61 S. S. Davis: Biomedical applications of nanotechnology-implications for drug targeting and gene therapy, Trends Biotechnol. **15**, 217–224 (1997)

17.62 J. C. Roberts, M. K. Bhalgat, R. T. Zera: Preliminary biological evaluation of polyamidoamine (PAMAM) starburst dendrimers, J. Biomed. Mater. Res. **30**, 53–65 (1996)

17.63 N. S. Nahman, T. Drost, U. Bhatt, T. Sferra, A. Johnson, P. Gamboa, G. Hinkle, A. Haynam, V. Bergdall, C. Hickey, J. D. Bonagura, L. Brannon-Pappas, J. Ellison, A. Mansfield, S. Shiwe, N. Shen: Biodegradable microparticles for in vivo glomerular targeting: Implications for gene therapy of glomerular disease, Biomed. Microdevices **4**, 189–196 (2002)

17.64 W. Arap, M. Kolonin, M. Trepel, J. Lahdenranta, M. Cardo-Vila, R. Giordano, P. J. Mintz, P. Ardelt, V. Yao, C. Vidal, L. Chen, A. Flamm, H. Valtanen, L. M. Weavind, M. E. Hicks, R. Pollock, G. H. Botz, C. D. Bucana, E. Koivunen, D. Cahil, P. Troncosco, K. A. Baggerly, R. D. Pentz, K.-A. Do, C. Logothetis, R. Pasqualini: Steps towards mapping the human vasculature by phage display, Nature Med. **8**, 121–127 (2002)

17.65 M. Kolonin, R. Pasqualini, W. Arap: Molecular addresses in blood vessels as targets for therapy, Curr. Opin. Chem. Biol. **5**, 308–313 (2001)

17.66 E. Ruoslahti: Special delivery of drugs by targeting to tissue-specific receptors in vasculature, Pharmaceutical News **7**, 35–40 (2000)

17.67 B. K. Kay, J. Winter, J. McCofferty: *Phage Display of Peptides and Proteins* (Academic, San Diego 1996)

17.68 W. Arap, W. Haedicke, M. Bernasconi, R. Kain, D. Rajotte, S. Krajewski, M. Ellerby, R. Pasqualini, E. Ruoslahti: Targeting the prostate for destruction through a vascular address, Proc. Nat. Acad. Sci. USA **99**, 1527–1531 (2002)

17.69 M. Essier, E. Ruoslahti: Molecular specialization of breast vasculature: A breast homing phage displayed peptide binds to aminopeptidase P in breast vasculature, Proc. Nat. Acad. Sci. USA **99**, 2252–2257 (2002)

17.70 G. A. Husseini, G. D. Myrup, W. G. Pitt, D. Christensen, N. Y. Rapoport: Factors affecting acoustically triggered release of drugs from polymeric micelles, J. Cont. Release **69**, 43–52 (2000)

17.71 A. Quintana, E. Raczka, L. Piehler, I. Lee, A. Myc, I. Majoros, A. K. Patri, T. Thomas, J. Mule, J. R. Baker Jr.: Design and function of a dendrimer-based therapeutic nanodevice targeted to tumor cells through the folate receptor, Pharma. Res. **19**, 1310–1316 (2002)

17.72 P. Shum, J.-M. Kim, D. H. Thompson: Phototriggering of liposomal delivery systems, Adv. Drug Deliv. Rev. **53**, 273–284 (2001)

17.73 K. Hamad-Schifferli, J. J. Schwartz, A. T. Santos, S. Zhang, J. M. Jacobson: Remote electronic control of DNA hybridization through inductive coupling to an attached metal nanocrystal antenna, Nature **415**, 152–155 (2002)

17.74 P. Meers: Enzyme-activated targeting of liposomes, Adv. Drug Deliv. Rev. **53**, 265–272 (2001)

17.75 V. Raso, M. Brown, J. McGrath: Intracellular triggering with low pH-triggered bispecific antibodies, J. Biol. Chem. **272**, 27623–27628 (1997)

17.76 W. A. Denny: The role of hypoxia activated prodrugs in cancer therapy, The Lancet Oncol. **1**, 25–29 (2000)

17.77 P. Vaupel, D. K. Kelleher, O. Thews: Modulation of tumor oxygenation, Int. J. Radiation Oncology Biol. Phys. **42**, 843–848 (1998)

17.78 P. D. Senter, C. J. Springer: Selective activation of anticancer prodrugs by monoclonal antibody-enzyme conjugates, Adv. Drug Deliv. Rev. **53**, 247–264 (2001)

17.79 H. Wang, H. Song, V. C. Yang: A recombinant prodrug type approach for triggered delivery of streptokinase, J. Cont. Release **59**, 119–122 (1999)

17.80 R. Duncan: Polymer therapeutics for tumor specific delivery, Chem. Industry **7**, 262–264 (1997)

17.81 K. Rogers: Principles of affinity-based biosensors, Mol. Biotechnol. **14**, 109–129 (2000)

17.82   R. Raiteri, M. Grattarola, H.-J. Butt, P. Skladl: Micromechanical cantilever-based biosensors, Sensors Actuat. B **79**, 115–126 (2001)

17.83   R. Martel, T. Schmidt, H.R. Shea, T. Hertel, P. Avouris: Single and multiwall carbon nanotube field-effect transistors, Appl. Phys. Lett. **73**, 2447–2449 (1998)

17.84   A. Guiseppi-Elie, C. Lei, R.H. Baughman: Direct electron transfer of glucose oxidase on carbon nanotubes, Nanotechnology **13**, 559–564 (2002)

17.85   C.N. Campbell: How far are we from detecting single bioconjugation events?. In: *Biological Molecules in Nanotechnology: The Convergence of Biotechnology Polymer Chemistry and Materials Science*, ed. by S.C. Lee, L. Savage (IBC, Southborough 1998) pp. 163–171

17.86   W.D. Callister Jr.: *Material Science and Engineering: an Introduction* (Wiley, New York 1997)

17.87   M. Gerard, A. Chaubey, B.D. Malhotra: Applications of conducting polymers to biosensors, Biosens. Bioelectron. **17**, 345–349 (2002)

17.88   P.N. Bartlett, Y. Astier: Microelectrochemical enzyme transistors, Chem. Commun. **2**, 105–112 (2000)

17.89   I. Kleps, A. Angelscu, R. Valisco, D. Dascalu: New micro and nanoelectrode arrays for biomedical applications, Biomed. Microdevices **3**, 29–33 (2001)

17.90   J. Song, Q. Cheng, S. Zhu, R.C. Stevens: "smart" materials for biosensing devices: cell-mimicking supramolecular assemblies and colorometric detection of pathogenic agents, Biomed. Microdevices **4**, 213–222 (2002)

17.91   M. Brown, R. Semelka: *MRI: Basic Principles and Applications* (Wiley, New York 1999)

17.92   A. Elster, S. Handel, A. Goldman: *Magnetic Resonance Imaging: A Reference Guide and Atlas* (Lippincott, Philadelphia 1997)

17.93   H. Paajanen, M. Kormano: Contrast agents in magnetic resonance imaging. In: *Radiographic Contrast Agents*, ed. by J. Skucas (Aspen, Rockville 1989) pp. 377–406

17.94   L. Thunus, R. Lejeune: Overview of transition metal and lanthanide complexes as diagnostic tools, Coord. Chem. Rev. **184**, 125–155 (1999)

17.95   M. Mendoca-Dias, E. Gaggelli, P. Lauterbur: Paramagnetic contrast agents in nuclear magnetic resonance medical imaging, Sem. Nuclear Med. **13**, 364–376 (1983)

17.96   M. Ollsen, B. Persson, L. Salford: Ferromagnetic particles as contrast agent in T2 NMR imaging, Magn. Reson. Imag. **4**, 437–440 (1986)

17.97   D. Kehagias, A. Gouliamos, V. Smyrniotis, L. Vlahos: Diagnostic efficacy and safety of MRI of the liver with superparamagnetic iron oxide particles (SH U 555 A), J. Magn. Reson. Im. **14**, 595–601 (2001)

17.98   C. Nolte-Ernsting, G. Adam, A. Bucker, S. Berges, A. Bjornerud, R. Gunther: Abdominal MR angiography performed using blood pool contrast agents, Am. J. Roentgenol. **171**, 107–113 (1998)

17.99   F. Rety, O. Clement, N. Siauve, C.-A. Cuenod, F. Carnot, M. Sich, A. Buisine, G. Frija: MR lymphography using iron oxide nanoparticles in rats: pharmacokinetics in the lymphatic system after intravenous injection, J. Magn. Reson. Im. **12**, 734–739 (2000)

17.100 T. Skotland, P. Sontum, I. Oulie: In vitro stability analyses as a model for metabolism of ferromagnetic particles (Clariscan™), a contrast agent for magnetic resonance imaging, J. Pharm. Biomed. Anal. **28**, 323–329 (2002)

17.101 D. Hogemann, L. Josephson, R. Weissleder, J. Basilion: Improvement of MRI probes to allow efficient detection of gene expression, Bioconjugate Chem. **11**, 941–946 (2000)

17.102 L.L. Muldoon, M.A. Pagel, R.A. Kroll, S. Roman-Goldstein, R.S. Jones, E.A. Neuwelt: A physiological barrier distal to the anatomic blood-brain barrier in a model of transvascular delivery, AJNR Am. J. Neuroradial. **20**, 217–222 (1999)

17.103 G. Biddlecombe, Y. Gun'ko, J. Kelly, S. Pillai, J. Coey, M. Ventatesan, A. Douvalis: Preparation of magnetic nanoparticles and their assemblies using a new Fe(II) alkoxide precursor, J. Mater. Chem. **11**, 2937–2939 (2001)

17.104 L. Tiefenauer, G. Kuhne, R. Andres: Antibody-magnetite nanoparticles: in vitro characterization of a potential tumor-specific contrast agent for magnetic resonance imaging, Bioconjugate Chem. **4**, 347–352 (1993)

17.105 C. Liu, Z. Zhang: Size-dependent superparamagnetic properties of Mn spinel ferrite nanoparticles synthesized from reverse micelles, Chem. Mater. **13**, 2092–2096 (2001)

17.106 J. Schnorr, M. Taupitz, S. Wagner, H. Pilgrimm, J. Hansel, B. Hamm: Age-related blood half-life of particulate contrast materials: Experimental results with a USPIO in rats, J. Magn. Reson. Im. **12**, 740–744 (2000)

17.107 R. Weissleder, G. Elizondo, J. Wittenberg, C. Rabito, H. Bengele, L. Josephson: Ultrasmall superparamagnetic iron oxide: Characterization of a new class of contrast agents for MR imaging, Radiology **175**, 489–493 (1990)

17.108 D. Portet, B. Denizot, E. Rump, J.-J. Lejeune, P. Jallet: Nonpolymeric coatings of iron oxide colloids for biological use as magnetic resonance imaging contrast agents, J. Coll. Inter. Sci. **238**, 37–42 (2001)

17.109 M. Ruegsegger, R. Marchant: Reduced protein adsorption and platelet adhesion by controlled variation of oligomaltose surfactant polymer coatings, J. Biomed. Mater. Res. **56**, 159–167 (2001)

17.110 S. Flack, S. Fischer, M. Scott, R. Fuhrhop, J. Allen, M. McLean, P. Winter, G. Sicard, P. Gaffney, S. Wickline, G. Lanza: Novel MRI contrast agent for molecular fibrin, Circulation **104**, 1280–1285 (2001)

17.111 D. F. Baban, L. W. Seymour: Control of tumor vascular permeability, Adv. Drug Deliv. Rev. **34**, 109–119 (1998)

17.112 R. K. Jain: Delivery of molecular and cellular medicine to tumors, J. Cont. Release **53**, 49–67 (1998)

17.113 H. Maeda, T. Sawa, T. Konno: Mechanism of tumor-targeted delivery of macromolecular drugs, including the EPR effect in solid tumor and clinical overview of the prototype polymeric drug SMANCS, J. Cont. Release **74**, 47–61 (2001)

17.114 J. Borenstein, H. Terai, K. King, E. Weinberg, M. Kaazempour-Mofrad, J. Vacanti: Microfabrication technology for vascularized tissue engineering, Biomed. Microdevices **4**, 167–175 (2002)

17.115 D. Satake, H. Ebi, N. Oku, K. Matsuda, H. Takao, M. Ashiki, M. Ishida: A sensor for blood cell counters using MEMS technology, Sensors Actuat. B – Chem. **83**, 77–81 (2002)

17.116 G. Lin, R. Palmer, K. Pister, K. Roos: Miniature heart cell force transducer system implemented in MEMS technology, IEEE Trans. Biomed. Engin. **48**, 996–1006 (2001)

17.117 J. Kreuter, D. Shamenkov, V. Petrov, P. Ramge, K. Cychutek, C. Koch-Brandt, R. Alyautdin: Apolipoprotein-mediated transport of nanoparticle-bound drugs across the blood-brain barrier, J. Drug Target **10**, 317–325 (2002)

17.118 P. Lockman, R. Mumper, M. Khan, D. Allen: Nanoparticle technology for drug delivery across the blood-brain barrier, Drug Devel. Indust. Pharm. **28**, 1–13 (2002)

17.119 H. Quick, J. Debatin, M. Ladd: MR imaging of the vessel wall, Eur. Radiol. **12**, 889–900 (2002)

17.120 S. Schmitz, M. Taupitz, S. Wagner, K.-J. Wolf, D. Beyersdorff, D. Hamm: Magnetic resonance imaging of atherosclerotic plaques using supermagnetic iron oxide particles, J. Magn. Reson. Im. **14**, 355–361 (2001)

17.121 D. Ranney: Biomimetic transport and rational drug delivery, Biochem. Pharm. **59**, 105–114 (2000)

17.122 K. Soppimath, T. Aminabhavi, A. Kulkarni, W. Rudzinski: Biodegradable polymeric nanoparticles as drug delivery devices, J. Cont. Release **70**, 1–20 (2001)

17.123 F. Breitling, S. Dubel: *Recombinant Antibodies* (Wiley, London 1998)

17.124 E. Harlow, D. Lane: *Antibodies: A Laboratory Manual* (Cold Spring Harbor Press, Cold Spring Harbor 1988)

17.125 S. Raychanduri, K. L. Rock: Fully mobilizing host defense: Building better vaccines, Nature Biotech. **16**, 1025–1031 (1998)

17.126 M. Kovasovics-Bankowski, K. Clark, B. Benacerraf, K. L. Rock: Efficient major histocompatibility complex class I presentation of exogenous antigen upon phagocytosis by macrophages, Proc. Nat. Acad. Sci. USA **90**, 4942–4946 (1993)

17.127 K. Rock, S. Gamble, L. Rothstein: Presentation of exogenous antigen with class I major histocompatibility complex molecules, Science **249**, 918–921 (1990)

17.128 C. V. Harding, R. Song: Phagocytic processing of exogenous particulate antigens by macrophages for presentation by class I MHC molecules, J. Immunol. **152**, 4925–4933 (1994)

17.129 C. Oseroff, A. Sette, P. Wentworth, E. Celis, A. Maewal, C. Dahlberg, J. Fikes, R. T. Kubo, R. W. Chestnut, H. M. Grey, J. Alexander: Pools of lapidated HTL-CTL constructs prime for multiple HBV and HCV CTL epitope responses, Vaccine **16**, 823–833 (1998)

17.130 R. Mrsny, A. L. Daughtery, M. Mckee, D. Fitzgerald: Bacterial toxins as tools for mucosal vaccination, Drug Disc. Today **7**, 247–257 (2000)

17.131 S. Fawell, J. Seery, Y. Daikh, C. Moore, L. L. Chen, B. Pepinsky, J. Barsoum: Tat-mediated delivery of heterologous proteins to cells, Proc. Nat. Acad. Sci. USA **91**, 664–668 (1994)

17.132 T. J. Golentz, K. Klimpel, S. Leppla, J. M. Keith, J. A. Berzofsky: Delivery of antigens to the MHC class I pathway using bacterial toxins, Hum. Immunol. **54**, 129–136 (1997)

17.133 B.-X. Chen, S. R. Wilson, M. Das, D. J. Coughlin, B. F. Erlanger: Antigenicity of fullerenes: antibodies specific for fullerenes and their characteristics, Proc. Nat. Acad. Sci. USA **95**, 10809–10813 (1998)

17.134 S. C. Lee, R. Ibdah, C. D. van Valkenburgh, E. Rowold, A. Donelly, A. Abegg, J. Klover, S. Merlin, J. McKearn: Phage display mutagenesis of the chimeric dual cytokine receptor agonist myelopoietin, Leukemia **15**, 1277–1285 (2001)

17.135 S. C. Lee, R. Parthasarathy, K. Botwin, D. Kunneman, E. Rowold, G. Lange, J. Zobel, T. Beck, T. Miller, C. F. Voliva: Humeral immune responses to polymeric nanomaterials. In: *Functional Condensation Polymers*, ed. by C. Carraher, G. Swift (Kluwer, New York 2002) pp. 31–41

17.136 S. C. Lee, R. Parthasarathy, T. Duffin, K. Botwin, T. Beck, G. Lange, J. Zobel, D. Kunneman, E. Rowold, C. F. Voliva: Recognition properties of antibodies to PAMAM dendrimers and their use in immune detection of dendrimers, Biomed. Microdevices **3**, 51–57 (2001)

17.137 A. H. Sehon: Suppression of antibody responses by conjugates of antigens and monomethoxypoly(ethylene glycol). In: *Poly(ethylene glycol) Chemistry*, ed. by J. M. Harris (Plenum, New York 1992) pp. 139–151

17.138 R. P. Lanza, S. J. Sullivan, W. L. Chick: Perspectives in diabetes. Islet transplantation with immunoisolation, Diabetes **41**, 1503–1510 (1992)

17.139 R. Gordon: Computer controlled evolution of diatoms: Design for a compustat, Nova Hedwigia **112**, 215–219 (1996)

17.140 J. Parkinson, R. Gordon: Beyond micromachining: The potential of diatoms, Trends Biotechnol. **17**, 190–196 (1999)

17.141 S. C. Lee: Antibody responses to nanomaterials. In: *Nanospace 2001: Exploring Interdisciplinary Frontiers*, ed. by T. Nicodemus (Institute for Advanced Interdisciplinary Research, Houston 2002)

17.142 K. Pearce, B. Cunningham, G. Fuh, T. Teeri, J. A. Wells: Growth hormone affinity for its receptor surpasses the requirements for cellular activity, Biochem. **38**, 81–89 (1999)

17.143 T. L. Ciardelli: Reengineering growth factors through the looking glass, Nat. Biotechnol. **14**, 1652 (1996)

17.144 P. J. Buchli, Z. Wu, T. L. Ciardelli: Functional display of interleukin-2 on filamentous phage, Arch. Biochem. Biophys. **339**, 79–84 (1997)

17.145 I. Saggio, I. Gloaguen, R. Laufer: Functional phage display of cilliary neurotropic factor, Gene **152**, 35–39 (1995)

18. G-Protein Coupled Receptors: Surface Display and Biosensor Technology

Signal transduction by G-protein coupled receptors (GPCRs) underpins a multitude of physiological processes. Ligand recognition by the receptor leads to the activation of a generic molecular switch involving heterotrimeric G-proteins and guanine nucleotides. With growing interest and commercial investment in GPCRs in areas such as drug targets, orphan receptors, high-throughput screening of drugs and biosensors, greater attention will focus on assay development to allow for miniaturization, ultrahigh-throughput and, eventually, microarray/biochip assay formats that will require nanotechnology-based approaches. Stable, robust, cell-free signaling assemblies comprising receptor and appropriate molecular switching components will form the basis of future GPCR/G-protein platforms, which should be able to be adapted to such applications as microarrays and biosensors. This chapter focuses on cell-free GPCR assay nanotechnologies and describes some molecular biological approaches for the construction of more sophisticated, surface-immobilized, homogeneous, functional GPCR sensors. The latter points should greatly extend the range of applications to which technologies based on GPCRs could be applied.

Part B | 18

The Superfamily of GPCRs

G-protein coupled receptors (GPCRs) represent a superfamily of intra-membrane proteins (polypeptides) which initiate many signal transduction pathways in virtually all eukaryotic cells. GPCRs are structurally characterized by their seven transmembrane ("serpentine") spanning domains (Fig. 18.1). GPCR activation can be initiated by a wide variety of extracellular stimuli such as light, odorants, neurotransmitters and hormones. In most cases the GPCR uses a transmembrane signaling system which involves three separate components (systems). Firstly, the extracellular ligand and is specifically detected by a cell surface GPCR. Once recognition takes place, the GPCR in turn, triggers the activation of a heterotrimeric G-protein complex located on the peripheral intracellular (cytoplasmic) surface of the cell membrane (the term "G-protein" is used since these proteins bind guanine nucleotides such as guanosine di- and triphosphate present in cells, as will be discussed in detail later). Finally, the "signal transduction" cascade involves the activated G-protein altering the activity of some downstream "effector" protein(s), which can be enzymes or ion channels located in the cell membrane. This

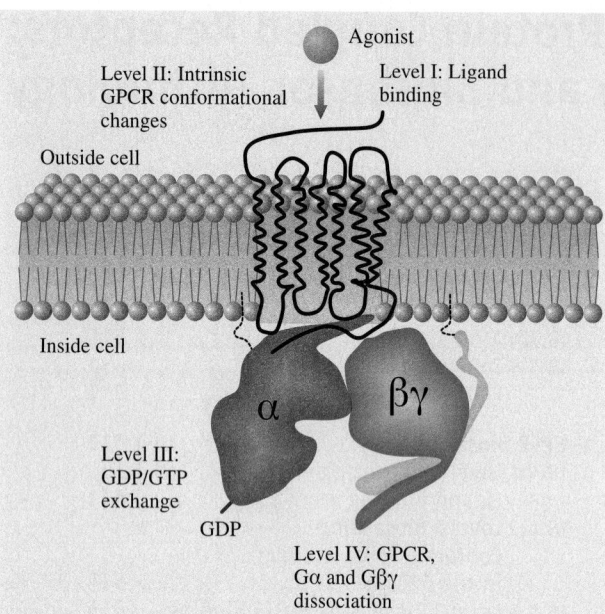

Fig. 18.1 Transmembrane topology of a typical "serpentine" G-protein coupled receptor (GPCR) and representation of the levels of biosensing. The receptor's amino terminal (N-terminal) is extracellular (outside of the cell) and its carboxy (C) terminal is within the cytoplasm (intracellular). The receptor polypeptide chain traverses the plane of the membrane phospholipid bilayer seven times. The hydrophobic transmembrane segments of the GPCR are indicated by spirals. The agonist approaches the receptor from the extracellular surface and binds, depending on the receptor type, to a site near the N-terminal or to a site deep within the receptor, surrounded by the transmembrane regions of the receptor protein. The G-proteins (Gα and G$\beta\gamma$) interact with cytoplasmic regions of the receptor, including the 3rd intracellular loop between transmembrane regions *V* and *VI*. The Gα and Gγ subunits contain fatty acid modifications (myristate and palmitate on the Gα and isoprenylate on the Gγ) to help anchor the proteins at the lipid bilayer (shown as dotted lines). The "levels" of signaling that may be exploited for detection in a cell-free mode are shown (GDP, guanosine diphosphate)

then leads to a change in the cellular concentration of cAMP, calcium ions, or metabolites such as phosphoinositides within the cell, resulting in a physiological response such as stronger and faster contraction of the heart.

Since many disease processes involve aberrant or altered GPCR signaling dynamics, GPCRs represent a significant target for medicinal pharmaceuticals. Furthermore, approximately 70% of all drugs currently marketed worldwide are directed against GPCRs [18.1] (see Table 18.1 for some commonly prescribed drugs acting on GPCRs). GPCRs are associated with almost every major therapeutic category or disease class, including pain, asthma, inflammation, obesity, cancer, as well as cardiovascular, metabolic, gastrointestinal and central nervous system diseases [18.2]. It is this vitally important function (transduction of exogenous signals into an intracellular response) of these cell-surface receptors that makes the GPCRs so physiologically significant. Indeed, there are reported to be approximately 747 different human GPCRs (as predicted from gene sequencing analyses), 380 of which are thought to be chemosensory receptors, whereas the remaining 367 GPCRs are predicted to bind endogenous ligands such as neurotransmitters, hormones, fatty acids and peptides [18.3]. About 230 of these GPCRs having been identified already (they have known ligands), which currently leaves about 140 "orphan" GPCRs with as-yet undiscovered ligands. For a summary of some of the known GPCR ligands see Table 18.2.

This chapter focuses firstly on possible cell-free approaches which could be used in biosensor applications, diagnostic platforms and for the high-throughput screening (HTS) of GPCR ligands, with particular emphasis on GPCR signaling complexes and associated enabling nanotechnologies (Fig. 18.2). Additionally, we have incorporated molecular biology approaches involving G-proteins and GPCRs with reference to biosensor and HTS applications, since one of the most important breakthroughs permitting these developments for GPCR and GPCR signaling is the ability to produce these GPCRs and G-proteins in relatively high amounts and in a purified form using recombinant DNA techniques. Also, it is becoming increasingly routine to produce recombinant modifications of such proteins using basic molecular biological approaches. These modifications include biotin tags, hexahistidine tags and FLAG tags, which allow site-specific interaction of the recombinant protein(s) with appropriately derivatized biosensor surfaces such as glass or gold.

18.1 The GPCR:G-Protein Activation Cycle

In order to understand how we measure the activation of GPCRs and their associated G-proteins, a first step is to review the GPCR:G-protein activation cycle in more detail. At the cellular level, the GPCRs are integral membrane proteins which reside within the cell membrane lipid bilayer and are closely associated with the peripheral G-protein heterotrimeric complex consisting of the subunits $G\alpha$ and the $G\beta\gamma$ dimer (also see Fig. 18.1). Owing to the very high affinity between $G\beta$ and $G\gamma$, these two subunits are almost exclusively considered as the $G\beta\gamma$ dimer. The $G\alpha$ subunits are approximately 41 kDa and have a theoretical diameter of approximately 4.7 nm, whilst β subunits are approximately 37 kDa and γ subunits 8 kDa, giving $G\beta\gamma$ dimers an approximate diameter of 4.6 nm. Figure 18.3 depicts the cycle of activation/inactivation of the heterotrimeric G-protein complex. In the resting inactive state (when there is no agonist bound to the receptor), the G-proteins $G\alpha$ and $G\beta\gamma$ have high affinity for each other and remain tightly bound, forming the heterotrimeric G-protein complex. In this state, guanosine diphosphate (GDP) is tightly bound to the $G\alpha$-subunit associated with the $G\beta\gamma$ dimer. Both $G\alpha$ and $G\beta\gamma$ subunits can bind to the GPCR. When the agonist (a GPCR ligand which activates the GPCR signaling pathway) approaches the

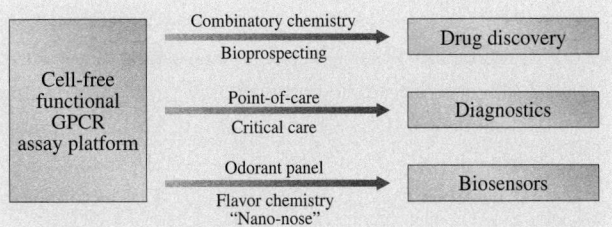

Fig. 18.2 Future applications of GPCR platforms. The challenge of being able to develop functional assay platforms for integral membrane proteins, particularly those of the GPCR class, which are compatible with future high-throughput, microarray formats, will offer significant opportunities in a number of areas. It would be expected that such advances in assay technology will likely impact on drug discovery, diagnostics and biosensors. There is a strong requirement for technologies that enable screening of multiple GPCR targets simultaneously (multiplexing). Therefore, it would be advantageous in the future to design new biosensor platforms using miniaturized nanotechnology approaches. Furthermore, to achieve this aim successfully, it will be an absolute requirement that cross-disciplinary fields of research (including biology, physics and chemistry, as well as mathematics for molecular modeling and bioinformatics) will need to be highly integrated to achieve such goals

Table 18.1 Some examples of prescription drugs which target GPCRs for the indicated disease state

Brand Name	Generic Name	G-protein coupled receptor(s)	Indication
Zyprexa	Olanzapine	Serotonin 5-HT$_2$ and dopamine	Schizophrenia, Antipsychotic
Risperdal	Risperidone	Serotonin 5-HT$_2$	Schizophrenia
Claritin	Loratidine	Histamine H$_1$	Rhinitis, Allergies
Imigran	Sumatriptan	Serotonin 5-HT$_{1B/1D}$	Migraine
Cardura	Doxazosin	$\alpha-$adrenoceptor	Prostate hypertrophy
Tenormin	Atenolol	β_1-adrenoceptor	Coronary heart disease
Serevent	Salmeterol	β_2-adrenoceptor	Asthma
Duragesic	Fentanyl	Opioid	Pain
Imodium	Loperamide	Opioid	Diarrhea
Cozaar	Losartan	Angiotensin II	Hypertension
Zantac	Ranitidine	Histamine H$_2$	Peptic ulcer
Cytotec	Misoprostol	Prostaglandin PGE$_1$	Ulcer
Zoladex	Goserelin	Gonadotrophin-releasing factor	Prostate cancer
Requip	Ropinirole	Dopamine	Parkinson's disease
Atrovent	Ipratropium	Muscarinic	Chronic obstructive pulmonary disease (COPD)

Table 18.2 A partial list of some of the known endogenous and exogenous GPCR ligands

Acetylcholine	Glucagon	Opioids
Adenosine	Glutamate	Orexin
Adrenaline	Gonadotropin-releasing hormone	Oxytocin
Adrenocorticotropic hormone	Growth hormone-releasing factor	Parathyroid hormone
Angiotensin II	Growth-hormone secretagogue	Photons (light)
Bradykinin	Histamine	Platelet activating factor
Calcitonin	Luteinising hormone	Prolactin releasing peptide
Chemokines	Lymphotactin	Prostaglandins
Cholecystokinin	Lysophospholipids	Secretin
Corticotropin releasing factor	Melanocortin	Serotonin
Dopamine	Melanocyte-stimulating hormone	Somatostatin
Endorphins	Melatonin	Substances P, K
Endothelin	Neuromedin-K	Thrombin
Enkephalins	Neuromedin-U	Thromboxanes
Fatty acids	Neuropeptide-FF	Thyrotropin
Follitropin	Neuropeptide-Y	Thyrotropin releasing hormone
GABA	Neurotensin	Tyramine
Galanin	Noradrenaline	Urotensin
Gastric inhibitory peptide	Odorants	Vasoactive intestinal peptide
Gastrin		Vasopressin
Ghrelin		

GPCR from the extracellular fluid and binds to the active site on the GPCR, the GPCR is in turn "activated", possibly leading to a change in its conformation. The GDP-liganded Gα-subunit responds with a conformational change which results in a decreased affinity, so that GDP is no longer bound to the Gα-subunit. At this point guanosine triphosphate (GTP), which is at a higher concentration in the cell than GDP, can rapidly bind to the Gα-subunit, thus replacing the GDP. This replace-

ment of GDP with GTP activates the Gα-subunit causing it to dissociate from the Gβγ-subunit as well as from the receptor. This, in effect, results in exposure of new surfaces on the Gα and Gβγ subunits which can interact with cellular effectors such as the enzyme adenylate cy-

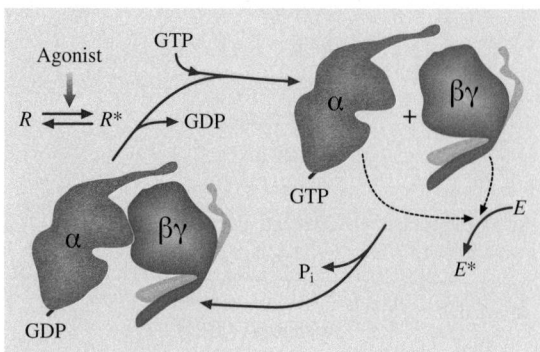

Fig. 18.3 Molecular switching: the regulatory cycle of agonist-induced (receptor-activated) heterotrimeric G-proteins. The binding of the agonist to the unoccupied receptor (R) causes a change in conformation, thus activating the receptor (R^*) which promotes the release of GDP from the heterotrimeric G-protein complex and rapid exchange with GTP into the nucleotide binding site on the Gα subunit. In its GTP-bound state, the G-protein heterotrimer dissociates into the Gα and Gβγ subunits exposing new surfaces, allowing interaction with specific downstream effectors (E). The signal is terminated by hydrolysis of GTP to GDP (and P_i) by the intrinsic GTPase activity of the Gα subunit followed by return of the system to the basal unstimulated state (* indicates activated state of receptor (R) or effector (E); P_i, inorganic phosphate, *GDP*, guanosine diphosphate; *GTP*, guanosine triphosphate)

clase, which converts adenosine triphosphate (ATP) to cAMP. The activated state of the Gα-subunit lasts until the GTP is hydrolyzed to GDP by the intrinsic GTPase activity of the Gα-subunit. The various families of Gα subunits (Gα_s, G$\alpha_{i/o}$, G$\alpha_{q/11}$, G$\alpha_{12/13}$) are all GTPases although the intrinsic rate of GTP hydrolysis varies greatly from one type of Gα-subunit to another. Following the hydrolysis of GTP to GDP on the Gα-subunit, the Gα and G$\beta\gamma$ subunits reassociate, and return to the receptor-associated state.

18.2 Preparation of GPCRs and G-proteins

GPCR function can be assayed using cell-based or cell-free methods. Cell-based methods usually use a mammalian cell-based expression system to display the required receptor on the cell surface coupled with an intrinsic or genetically-constructed reporter system incorporated into the cellular machinery. For example, the reporter system may be a fluorescently measured change in the intracellular level of Ca^{2+} or the enhanced expression of a fluorescent protein. Cell-free assay systems for GPCR function make use of fragments of cells that contain various components of the receptor and signaling machinery.

In cell-free assays, host cells are transfected with DNA that allows high levels of expression of the GPCR of interest (in a similar manner to that of whole-cell assays); however, partial purification of the GPCRs is usually carried out in order to obtain a supply of the GPCRs. This results in small (nanometer-scale) crude membrane fragments being produced. The GPCR membrane fragments, which will usually contain hydrophobic membrane lipids (which are required for functionality) as well as other native "contaminating" proteins can then be manipulated and immobilized by various means (discussed later) to appropriate surfaces for use as "biosensors".

To date GPCRs have proven to be extremely difficult to purify due primarily to the lipophilic (hydrophobic) nature of these receptors and the fact that they are usually irreversibly denatured (inactivated) when they are removed from their native lipid environment using detergent treatment. On the other hand, the G-proteins, which are classified as peripheral as opposed to integral membrane proteins do not require an absolute lipid environment for activity, can be routinely purified in relatively large amounts (mg quantities) when expressed using recombinant DNA-based technologies.

18.3 Measurement of GPCR Signaling

The basic requirement of a biosensor is the use of a biological element, such as an immobilized protein, to act as a sensor for a specific binding analyte. This is coupled with a reporter system which amplifies the initial signal to produce some form of output. Depending on the type of output required for a given screening process (such as ligand binding to a GPCR or a functional assay such as G-protein activation), a number of protocols are available to target the site of interest. In this chapter we will refer to these as "levels" of GPCR activation (Fig. 18.1). Examples of each of these levels will be discussed below. Additionally, in this section, the levels of biosensing referred to represent those "cell-free" samples or biological preparations which are derived from cells and are used in the cell-free mode; in other words the GPCRs and G-proteins have either been partially or fully purified from cells expressing the GPCRs or G-proteins, and then subsequently "reconstituted" at known concentrations usually within the nanomolar range.

Before discussing cell-free biosensor GPCR signaling complexes at differing levels of measurement, we briefly review some of the technologies used to perform such measurements, and to display GPCRs in a format suitable for cell-free biosensors.

18.3.1 Flow Cytometry

Flow cytometry is a technique used to analyze the fluorescence of individual cells or particles (for example dextran beads). Fluorescence can arise from intrinsic properties of the cell but the molecules/particles of interest are generally fluorescently labeled. Hydrodynamic focusing is used to force the cells or particles into a single file, where they are then passed through a laser beam,

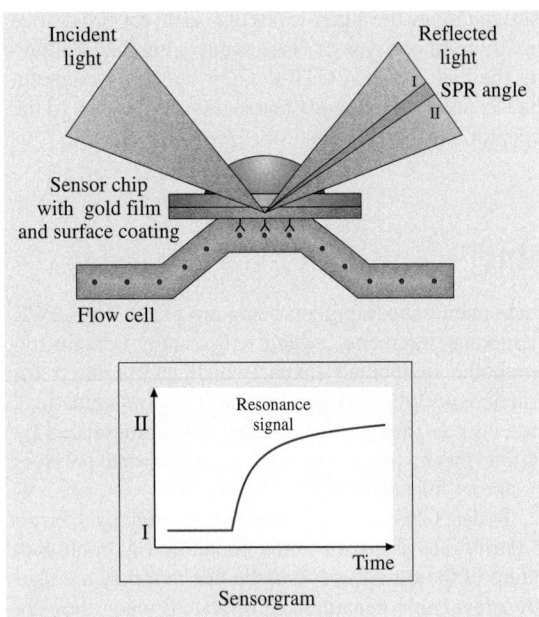

Resonance
signal

II

I

Time

Sensorgram

Fig. 18.4 Surface plasmon resonance (SPR) provides a mass detector. Most importantly, this technique does not require labeling of the interacting components. Since it is the evanescent field wave and not the incident light which penetrates the sample, measurements can be made on turbid or even opaque samples. The detection principle of SPR relies on electron charge density wave phenomena arising at the surface of a metallic film when light is reflected at the film under specific conditions (surface plasmon resonance). The resonance is a result of energy and momentum being transformed from incident photons into surface plasmons, and is sensitive to the refractive index of the medium on the opposite side of the film from the reflected light. Quantitative measurements of the binding interaction between one or more molecules are dependent on the immobilization of a target molecule to the sensor chip surface. Binding partners to the target can be captured from a complex mixture as they pass over the chip. Interactions between proteins, nucleic acids, lipids, carbohydrates and even whole cells can be studied. The sensor chip consists of a glass surface coated with a thin layer of gold. This forms the basis for a range of specialized surfaces designed to optimize the binding of a variety of molecules. The gold layer in the sensor chip creates the physical conditions required for SPR. The *upper figure* shows a detector with sensor chip. When molecules in the test solution bind to a target molecule the mass increases; when they dissociate the mass falls. This simple principle forms the basis of the sensorgram – a continuous, real-time monitoring of the association and dissociation of the interacting molecules (see *lower figure*). The sensorgram provides quantitative information in real time on specificity of binding, active concentration of molecules in a sample, kinetics and affinity. Molecules as small as 100 Da can be studied

and the emitted light is then measured. A benefit of this method is that simultaneous measurements can be performed on individual particles.

18.3.2 Surface Plasmon Resonance

One of the most versatile techniques for measuring biospecific interactions in real time are biosensors based on the optical phenomenon of surface plasmon resonance (SPR, Fig. 18.4). Surface plasmon resonance occurs when light interacts with a conducting surface (plasmon interaction) which is positioned between two materials with different refractive indices. At a specific angle the intensity of the reflected light decreases with this angle; this is dependent on (among other things) the refractive index of the material on the opposite side to which the light is applied. The binding and dissociation of molecules to this material (such as receptor to surface or ligand to receptor) will change the refractive index of the material and be detected by measuring the reflected light. The instrument detects the change in angle of the reflected light minimum. The technique can be used to study interactions between ligands, GPCR, and G-proteins. SPR experiments do not require a large amount of sample and detection does not require fluorescent or radioisotopic-labeling. A variety of available surface chemistries permits the immobilization of many types of proteins using a range of strategies.

18.3.3 Plasmon Waveguide Resonance Spectroscopy

A recently developed method, plasmon-waveguide resonance (PWR) spectroscopy, measures real-time binding of free molecules to immobilized molecules such as GPCRs without the application of specific labels [18.4]. PWR has several significant advantages compared with conventional surface plasmon resonance, including enhanced sensitivity and spectral resolution, as well as the ability to distinguish between mass and conformational changes. This latter property is a consequence of the use of both p- and s-polarized excitation to produce resonances. This allows for measurement of refractive index anisotropy, which reflects changes in mass distribution

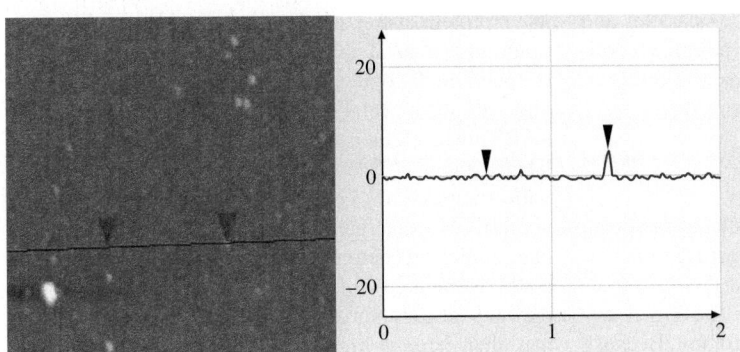

Fig. 18.5 AFM height image of a $G\beta_1\gamma_2$ dimer on mica. $\beta_1\gamma_2$ was applied to the mica surface. The height image was taken in air using tapping mode AFM with a scan rate of 1606 Hz, a scan size of 2 µm (data scale of 20 nm). The mean height of the protein was 4.30±0.14 nm (mean±SEM, $n = 15$) (this image was contributed by Ms Amanda Aloia, Flinders University, South Australia, Australia)

and, therefore, changes in molecular orientation and conformation.

18.3.4 Atomic Force Microscopy

Atomic force microscopy (AFM), as discussed elsewhere in this book, uses an atomically sharp tip positioned on a flexible cantilever to produce images of a surface over which the tip scans. As the tip moves across the surface, a laser positioned on the head of the tip records fluctuations in surface topography by monitoring the reflected beam through a photodiode. AFM can be used to image surface-immobilized proteins, record protein–protein interactions (through the use of a fluid cell) and determine force interactions between molecules. AFM has been used, among other applications, for analysis of interactions between many biomolecular partners (one example being streptavidin–biotin) and to gain structural information (for example on the $GABA_A$ and nicotinic acetylcholine receptors) [18.5,6]. The GPCR rhodopsin has been characterized using AFM [18.7]. Here the authors were able to image the GPCR in native disc membranes and show that mouse rhodopsin exists in dimers and oligomers. Additionally, it is possible to image purified G-proteins bound to a mica surface, as shown in Fig. 18.5.

18.3.5 Total Internal Reflection Fluorescence (TIRF)

Total Internal Reflection Fluorescence takes advantage of refractive index differences at a solid/liquid interface, with the solid surface being either glass or plastic, such as cell culture containers. At a critical angle, when total internal reflection occurs, an evanescent wave is produced in the liquid medium. This electromagnetic field decays exponentially with increasing distance from the surface. The range of this field limits background fluorescence, as only fluorophores in close proximity to the surface are excited. The technique is therefore used to examine interactions between the molecule of interest and the surface, such as receptors binding to a surface.

18.4 GPCR Biosensing

18.4.1 Level 1 Biosensing – Ligand Binding

We have defined level 1 biosensing as ligand binding to the receptor. It includes such techniques as radioligand binding (not discussed here) and fluorescent (and fluorescent polarization) ligand binding assays. Ligand binding can also be detected by such techniques as fluorescence-activated cell sorting (FACS) and surface plasmon resonance (SPR). This level of biosensing does not discriminate between compounds that can be pharmacologically defined as agonists, antagonists, partial agonists or inverse agonists. Therefore, when used in biosensing, the "activation" of a signaling pathway is somewhat limited. However, it is still useful for some specific purposes, such as screening for compounds which "interact" with a particular GPCR.

Fluorescence Polarization
Polarization is a general property of most fluorescent molecules. Polarization-based experiments are less dye-dependent and less susceptible to environmental interferences (such as pH changes) than assays based on fluorescence intensity measurements. The degree of polarization (or anisotropy) can be determined from

measurements of fluorescence intensities parallel and perpendicular to the plane of linearly polarized excitation light [18.8]. The data are expressed in terms of fluorescence polarization (P) or anisotropy (r):

$$P = \frac{(F \parallel - F \perp)}{(F \parallel + F \perp)} \quad r = \frac{(F \parallel - F \perp)}{(F \parallel + 2F \perp)} , \quad (18.1)$$

where $F \parallel$ = fluorescence intensity parallel to the plane of excitation $F \perp$ = fluorescence intensity perpendicular to the plane of excitation.

Both P and r are ratio quantities and are therefore independent of fluorophore concentration. Because of the ratio formulation, fluorescence intensity variations due to the presence of samples which may be colored (as can occur during drug screenings of compound libraries) tend to cancel and produce relatively minor interferences. Measured values of P in bioanalytical applications typically range from 0.01 to 0.3. Although this measurement window appears narrow, the data obtained are usually very precise measurements (for example, $P \pm 0.002$) and are readily obtained with modern instrumentation.

Fluorescence and fluorescence polarization (FP) assays, which are based on specific binding of the ligand to a GPCR, can offer an alternative to the traditional radioligand binding assays, which utilize radionuclides (radioisotopes) [18.9]. FP assays usually take the form of a homogeneous or "mix and read" type of assay (and an example of a level 1 assay), which indicates that they are readily transferable from assay development to high-throughput screening (HTS). FP allows for the development of protocols which are both real-time measurements (kinetic assays), and insensitive to variations in concentrations. Furthermore, FP is compatible with the development of homogeneous assay formats in which no separation steps are required to separate free from bound ligand. One of the disadvantages of this assay format is the lack of adaptability to all GPCR ligands due to the fact that only a small number of ligands can be chemically tagged with an appropriate fluorophore and still retain their intrinsic binding qualities. Finally, the choice of fluorophore is important; as well as obviously needing to have good polarizing properties, the intensity of the fluorescent compound must be of a sufficient magnitude [18.10].

Total Internal Reflection Fluorescence (TIRF)

Martinez et al. [18.11] used TIRF to demonstrate ligand binding to the neurokinin-1 GPCR by surface immobilization of membrane fragments containing this receptor protein. In this study, the GPCR expressed as a biotiny-

lated protein using mammalian cells was selectively immobilized on a quartz sensor surface coated with streptavidin (streptavidin binds biotin with extremely high affinity). Total internal reflection fluorescence (TIRF) measurements were made using a fluorescent-labeled agonist (the cognate agonist substance-P labeled with fluorescein). Using this approach, it was not necessary to detergent-solubilize and reconstitute the neurokinin-1 receptors, thus avoiding the deleterious effect(s) associated with such processes. This receptor, in the form of a mammalian cell membrane homogenate, was then surface-immobilized without further purification. The selective high-affinity interaction between biotin and streptavidin allowed a template-directed and uniform orientation of the neurokinin-1 receptor on the support matrix. Additionally, the highly selective TIRF fluorescence detection methodology was able to resolve the binding of fluorescently-tagged agonist to as little as 1 attomol of receptor molecules.

Microbead Approaches

Waller et al. [18.12] have conjugated dextran beads with the cognate ligand dihydroalprenolol, which allowed for the capture of solubilized β_2 adrenergic receptors (β-AR) to this immobilized surface ligand. To measure the specific binding of the receptor to the bead in this flow cytometry-based assay system, the receptor was expressed as a fusion protein with green fluorescent protein (GFP). It was then possible to screen for ligands (either agonists or antagonists) to the β-AR using a competition assay. Another successful bead-based approach used paramagnetic beads [18.13]. In that study the authors built up a surface containing the captured CCR5 receptor from a cell lysate held within a lipid bilayer. In this instance the CCR5 receptor was not able to freely move laterally in the bilayer as it was tethered via an antibody (directed at the CCR5 receptor) conjugated to the paramagnetic beads ("paramagnetic proteoliposome").

Microspotting of GPCRs on Glass

The intrinsic difficulties in producing, purifying and manipulating membrane proteins have delayed their introduction into microarray platforms. Hence there are no reports to date describing purified membrane protein (GPCR) microarrays and their use in functional screening or biosensor applications. However, as a first step towards such display technologies, researchers at Corning Inc. (New York, USA) have recently described the fabrication of GPCR membrane arrays for the screening of GPCR ligands [18.14–16]. The arraying of membrane GPCRs required appropriate surface chemistry for the

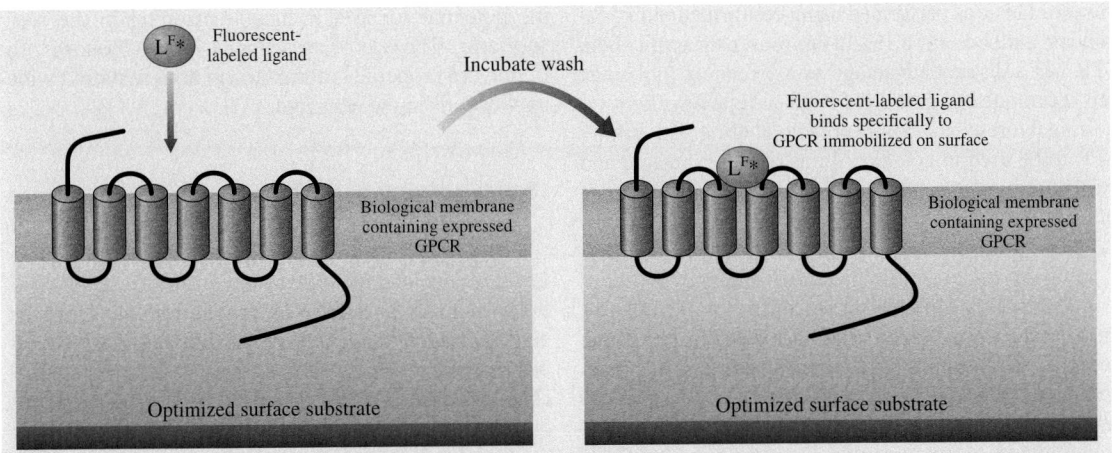

Fig. 18.6 Idealized schematic of an immobilized GPCR with associated G-proteins. The fabricated surface array is printed on a γ-aminopropylsilane (GAPS)-presenting surface. The height of the supported lipid bilayer is approximately 5 nm. Fluorescently labeled (L^{F*}) ligands (such as BODIPY-TMR-neurotensin) will bind specifically to the GPCR (for example, a neurotensin receptor) at nanomolar concentrations. The fluorescence is measured following an incubation/washing step to remove unbound fluorescent ligands. When compounds of unknown activity are added to the incubation step, as in drug screening programs, fluorescent-labeled ligand binding is blocked by agents that bind to the GPCR (for example GPCR antagonists). Adapted from [18.14]

immobilization of the lipid phase containing the GPCR of interest (Fig. 18.6). They reported surface modification with γ-aminopropylsilane (an amine-presenting surface) provided the best combination of properties to allow surface capture of the GPCR-G-protein complex from crude membrane preparations, resulting in microspots of approximately $100\,\mu m$ diameter. AFM demonstrated that the height of the supported lipid bilayer was approximately 5 nm, corresponding to GPCRs confined in a single, supported lipid layer scaffold. Using these chemically-derivatized surfaces, it was possible to demonstrate capture of the β_1, β_2 and α_{2A} subtypes of the adrenergic receptor, as well as neurotensin-1 receptors and D1-dopamine receptors. This was achieved using ligands with covalently attached fluorescent labels by detecting fluorescence binding to the GPCRs with a fluorescence-based microarray scanner. Dose-response curves using the fluorescently-labeled ligands gave IC_{50} values in the nM range, suggesting that the GPCR-G-protein complex was largely preserved and biologically intact in the microspot. There was no change in the performance of the arrays over a 60 day time period, indicating good long-term stability. Although the use of glass slides for the printing of the GPCR arrays was promoted by this research group, in some instances gold surfaces were required due to nonspecific binding of fluorescent ligands. A current limitation of this technology is the inability to carry out a functional (signaling) assay, which would allow test compounds to be classified as agonists or antagonists. Furthermore, although there are increasing numbers of commercially available fluorescently-labeled ligands, the need to always structurally modify the ligand to accommodate some reporter moiety may limit the implementation of the technology. Nevertheless, these GPCR microarrays may find application as "functional" GPCR assays when complexed with G-proteins and integrated with appropriate signal generation and detection methods.

Antibody Capture of GPCRs

Ligand binding to a GPCR attached to a surface has been reported for the chemokine CCR5 receptor using SPR methodology [18.17]. For such displays, purification of the GPCR has not always been necessary and crude membrane preparations have either been fused with an alkylthiol monolayer (approximately 3 nm thickness) formed on a gold-coated glass surface, or onto a carboxymethyl-modified dextran sensor surface [18.18]. One problem with surface-based assays is the orientation of the receptor once attached to the surface. One means of overcoming this problem was to specifically select only those proteoliposomes (≈ 300 nm-diameter vesicles) in which the carboxy terminus of the receptor was aligned to the outside of the vesicle. This was performed using conformationally dependent antibodies [18.19]. In this biosensor application, SPR has a distinct advantage as a screening tool since this technique can detect the cognate ligand without requiring fluorescent labeling or radiolabeling. This allows SPR to be used in complex fluids of natural origin and simplifies (and potentially speeds up) the development of assay technologies.

Plasmon Waveguide Resonance Spectroscopy and the β_2-Adrenergic Receptor

In a recent study, ligand binding to the β_2-adrenergic receptor was demonstrated using plasmon waveguide resonance (PWR) [18.20]. Using this technique, changes in the refractive index upon ligand binding to surface-immobilized receptor results in a shift in the PWR spectra. The authors used ligands with similar molecular weights in order to study structural changes in the receptor caused by agonist, inverse agonist and partial agonist binding. The technique was used to produce binding curves for five ligands using shifts in the PWR spectra (with both s- and p-polarized light) with increasing ligand concentration, with the results from PWR being compared to those obtained by traditional radioligand binding assays. Differences in s- and p-polarized light measurements demonstrated changes in receptor structure which varied depending on whether the ligand was a full, partial or inverse agonist. Previous work using PWR technology has been reported for the detection of conformational changes in a proteolipid membrane containing the human δ-opioid receptor following binding of nonpeptide agonists, partial agonists, antagonists and inverse agonists [18.21]. Although the ligands in the above study were similar molecular weights, there were distinctly different refractive index changes induced by ligand binding and these were too large to be accounted for by differences in the mass alone. The inference from this finding was that a ligand-specific conformation change in the receptor protein may have been detected. Therefore, this methodology may find use in a future biosensor, particularly with regard to the GPCRs.

Piezoelectric Crystal Sensing

Piezoelectric crystal sensing measures the change in mass upon molecules binding to a surface, due to a change in the resonance frequency of the crystal. The technique has been used in an "electronic nose" with olfactory receptors which are typically GPCRs [18.22], where an array of six sensor elements could be used to characterize each of six test compounds, emphasizing

the potential for GPCR ligand screening in the sensory area. The use of an artificial nose ("bionose") to mimic the properties of the human nose may find wide applications in the near future.

18.4.2 Level 2 Biosensing – Conformational Changes in the GPCR

Level 2 involves the detection of intrinsic conformational changes in the GPCR protein following agonist activation, and may involve the use of fluorescence-based techniques. Measurements of conformational changes in the GPCR following ligand (usually agonist or partial agonist) binding have been limited to date.

One study demonstrated the immobilization of β_2-adrenergic receptors onto glass and gold surfaces [18.23]. This provides a good example of a "level 2" cell-free assay. In that study, the receptors were site-specifically labeled with the fluorophore tetramethyl-rhodamine-maleimide at cysteine 265 (Cys265) using a series of molecular biology approaches. It was then possible to show agonist (isoproterenol)-induced conformational changes within the vicinity of the fluorescent moiety (tetramethyl-rhodamine) at position Cys265 of the recombinant β_2-adrenergic receptors. Moreover, the agonist-induced signal was large enough to detect using a simple ICCD camera image. Thus, it was suggested that the technique may be useful for drug screening with GPCR arrays. Indeed, this method did not require the formation of lipid bilayers and did not require the use of purified G-proteins or fluorescent ligands to detect receptor activation.

18.4.3 Level 3 Biosensing – GTP Binding

Measurement of GPCR activation further downstream from level 2 are considered, for the purposes of this chapter, to be truly "functional" assays since the transducer G-proteins are the first differentiated site of signaling initiated from the GPCR. This means that the GPCR must be in a "functional" form that enables it to interact and activate a G-protein signaling pathway. Level 3 biosensing involves the use of nonhydrolyzable GTP-analogs such as the radiolabeled ^{35}S-GTP or fluorescent-tagged Europium-GTP which bind to the receptor-activated form of the Gα-subunit targeting the site of guanine nucleotide exchange (GDP for GTP on the Gα subunit of the G$\alpha\beta\gamma$ heterotrimer). The guanine nucleotide exchange process is generally considered to be the first major point of G-protein activation following GPCR stimulation (refer to Figs. 18.1 and 18.3).

Guanine nucleotide exchange is a very early generic event in the signal transduction process of GPCR activation and is therefore an attractive event to monitor as it is less subject to regulation by further downstream

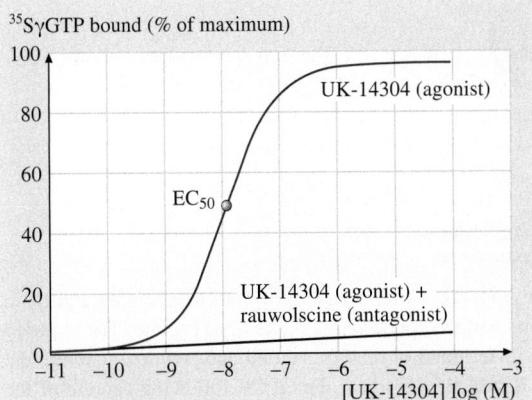

Fig. 18.7 Activation of GPCR-induced GTP binding. The data show results from an experiment which was conducted by incubating 20 nM purified G-proteins ($G\alpha_{i1}$ and $G\beta_1\gamma_2$) reconstituted with 0.4 nM recombinant α_{2A}-adrenergic receptor-expressed membranes (these receptors normally bind adrenaline with high affinity). The assay also contained 0.2 nM $^{35}S\gamma$GTP (a radioactive nonhydrolyzable analog of GTP). An adrenaline analog (UK-14304) was then added to the reconstituted α_{2A}-adrenergic receptor membrane, at the concentrations indicated on the x-axis (0.01 nM to 100 μM) in the presence or absence of the α_{2A}-adrenergic receptor antagonist, rauwolscine (10 μM). Following a filtration step to remove the bound $^{35}S\gamma$GTP:$G\alpha_{i1}$ complex from unbound $^{35}S\gamma$GTP, the filters were subsequently counted in a scintillation counter to measure the level of radioactivity. As the concentration of the agonist (UK-14304) was increased above 1 nM (10^{-9}M), the characteristic sigmoidal dose-response effect was seen. This result shows an increase in receptor-activated binding of $^{35}S\gamma$GTP to the $G\alpha_{i1}$ subunits as the UK-14304 is increased in concentration, indicating functional signaling of the receptor through the G-protein complex. The concentration at which 50% (also the point of inflexion) of the signaling response (effective concentration) was observed (EC_{50}) was approximately 12 nM. In the presence of an excess of the α_{2A}-adrenergic receptor antagonist (rauwolscine), the signal was completely blocked at the receptor. Therefore, this type of biosensing application demonstrates sensitivity as well as specificity

cellular processes (we have depicted this as "level 3" biosensing, Fig. 18.1). The radiolabeled $^{35}S\gamma$GTP or fluorescent Europium-GTP binding assays measure the level of G-protein activation following agonist activation of a GPCR by determining the binding of these nonhydrolyzable analogs of GTP to the $G\alpha$ subunit. Therefore, they are defined as "functional" assays of GPCR activation. Ligand regulation of the binding of $^{35}S\gamma$GTP is one of the assay methods most widely used to measure receptor activation of heterotrimeric G-proteins, as discussed elsewhere in detail [18.24, 25]. This methodology also provides the basis for measurement of such pharmacological characteristics as potency, efficacy and the antagonist affinity of compounds [18.25] in cell-free assays and artificial expression systems for GPCRs (an example of typical data is shown in Fig. 18.7). However, despite the highly desirable attributes of this methodology and its widespread use to date, ligand regulation of ^{35}S-GTP binding has been largely restricted to those receptors which signal through the $G\alpha_{i/o}$ proteins (pertussis toxin-sensitive) and, to a lesser extent, the $G\alpha_s$ and $G\alpha_q$ families of G-proteins. The use of these assay platforms can therefore be problematic for high-throughput screening as they are not homogeneous (they require a separation step to remove bound from free ^{35}S-GTP). Additionally, the use of radioactive-based assays (including ligand binding assays) has led to safety, handling, waste disposal and cost concerns. Therefore, the newly developed and fluorescent-based Europium-GTP assay partly overcomes some of the above limitations and has already been successfully used with the following GPCRs: motilin, neurotensin, muscarinic-M_1 and α_{2A}-adrenergic receptors.

18.4.4 Level 4 Biosensing – GPCR, G-Protein Dissociation

Procedures which utilize only ligand binding (level 1) do not distinguish between agonist (activates receptor), antagonist (blocks the action of the agonist at the receptor binding site) or inverse agonist (inhibits the intrinsic – non-agonist stimulated – activity of the receptor signaling, often observed in over-expressed receptors). However, if a functional GPCR assay is constructed in which G-protein "activation" is an end-point – in other words level 4 biosensing – then it is possible to distinguish between these functionally distinct ligands. For cell-free assays, both methodologies (levels 1 and 4) are important in HTS programs for example, and may have differing levels of applicability. Indeed, novel nanotechnology approaches will be required to achieve level 4

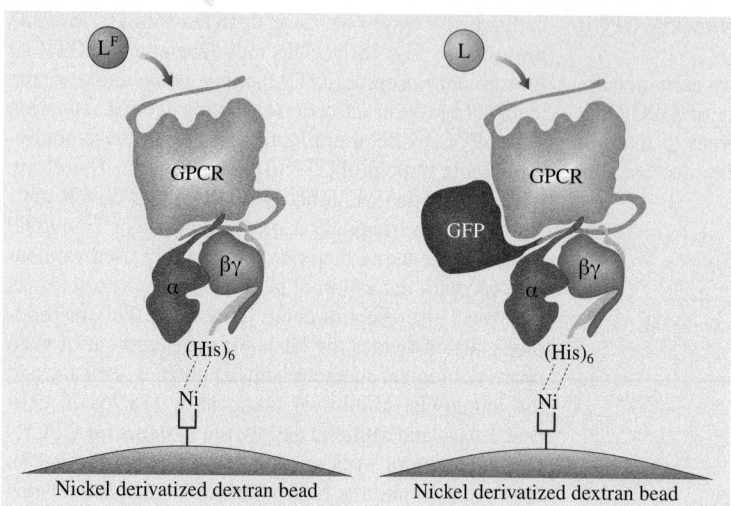

Fig. 18.8 Schematic diagram of two flow cytometry modes for detection of the ligand:receptor:G-protein assembly on nickel-coated beads. The G-proteins are immobilized on the bead surface containing a nickel (Ni) chelate. The exposed Ni binds to an engineered hexahistidine $(His)_6$ sequence on the N-terminal of the $G\gamma$ subunit and is able to capture the heterotrimeric G-protein complex. (*left figure*) The fluorescent ligand (L^F) binds to the GPCR following capture of the GPCR with appropriate G-proteins complexed on the surface of dextran beads. This technique is useful for biosensing the interaction of specific GPCR ligands (agonists and antagonists) and may be useful for demonstrating receptor:G-protein specificity and screening of ligands. In (*right figure*), the assembly uses a GPCR fusion protein containing enhanced green fluorescent protein (GFP). This technique demonstrates the requirement of the heterotrimeric complex for ligand activation and allows quantification of the receptor without the use of fluorescent ligands. Adapted from [18.26]

biosensing, including suitable surface derivatization for immobilization of GPCR and G-protein complexes.

Level 4 biosensing encompasses those assays which measure the final stage of activation of the G-protein heterotrimeric complex, that being the putative dissociation or rearrangement of the subunits following GPCR-induced G-protein activation [18.27]. This level of GPCR activation has currently not been investigated in great detail and may prove to be extremely valuable in future functional biosensor applications. Assay methodologies which are examples of level 4 biosensing have been reported that use surface plasmon resonance and FACS technologies to demonstrate receptor dissociation from the G-protein complex.

Bieri et al. [18.28] used carbohydrate-specific biotinylation chemistry to achieve appropriate orientation and functional immobilization of the solubilized bovine rhodopsin receptor with high-contrast micropatterns of the receptor being used to spatially separate protein regions. This reconstituted GPCR:G-protein system provided relatively stable results (over hours) with the added advantage of obtaining repeated activation/deactivation cycles of the GPCR:G-protein system. Measurements

were made using SPR detection of G-protein dissociation from the receptor surface following the positioning of the biotinylated form of the rhodopsin receptor onto a self-assembled monolayer containing streptavidin. Using this approach, G-protein activation could be directly monitored, giving a functional output, as opposed to ligand:receptor binding interactions which yield little information on the receptor-activated pathway when screening agonists and antagonists. Although SPR is useful for the study of G-protein interactions, it may not be well suited to detecting the binding of small ligand molecules directly due to its reliance on changes in mass concentration. An advantage of repeated activation/deactivation cycles of GPCRs is that different compounds can be tested serially with the same receptor preparation. The above approach appears promising for future applications of chip-based technologies in the area of GPCR biosensor applications.

Modifying the surface of epoxy-activated dextran beads by forming a Ni^{2+}-NTA conjugate was shown to produce beads with a surface capable of binding hexahistidine (his)-tagged $\beta_1\gamma_2$ subunits (Fig. 18.8). Tethered $\beta_1\gamma_2$ subunits were then used to capture

$G\alpha_s$ subunits which in turn were capable of binding membrane preparations with expressed β_2 adrenergic receptor containing a GFP fusion protein (see later section for a detailed description of fusion proteins); alternatively, a fluorescent-labeled ligand could be detected binding to the tethered β_2 adrenergic receptor, with the whole complex being measured using flow cytometry. Additionally, quantitative solubilization and reassembly of the (hexahistidine-tagged) N-formyl peptide receptor (FPR) has been demonstrated on Ni^{2+}-silica particles using flow cytometry with dodecyl maltoside as the detergent [18.29]. Using such approaches, it may be possible to screen ligands for a known solubilized GPCR or, alternatively, to test which G-proteins preferentially couple with a particular solubilized, reconstituted GPCR. The flow cytometry system used above had a sampling rate of approximately 50–100 samples per minute; however, flow cytometry's greatest advantage is its ability to be multiplexed, where different molecular assemblies can be made with one sample and yet be discriminated by their unique spectral characteristics [18.12, 26, 30]. In more detailed studies, the assembly and disassembly of the FPR and the his-tagged G-proteins complexed on Ni^{2+}-silica particles provided insight into the activation kinetics of the ternary complex (receptors and heterotrimeric G-proteins) [18.26, 31]. The study by *Simons* et al. [18.26] extended the knowledge of ligand:GPCR interactions to involve G-protein:GPCR:ligand interactions assayed in a homogeneous format with a bead-based approach amenable to high-throughput flow cytometry. Indeed, HTS and proteomic applications could easily be based on such bead arrays with the potential for color-coded particles and multiplexing (using quantum-dot technology for instance [18.32]). Particle-based screening constitutes an enabling technology for the identification of agonists promoting the assembly of G-protein:GPCR interactions as well as antagonists that inhibit such assemblies.

18.5 Protein Engineering in GPCR Signaling

18.5.1 Concept

GPCR engineering has been widely used in cell-based assays, most often to determine receptor function. Molecular engineering of proteins is likely to be of great importance for the production of receptors which have modified structures or functions amenable for use in cell-free biosensing applications. Modifications may include tagging of receptors with specific peptide sequences or the alteration of wild-type sequences to modify functional or structural characteristics of receptors to meet a particular requirement, such as to confer fluorescent properties in order to elucidate molecular interactions. Currently, many receptor or G-protein modifications are aimed at enhancing the purification of proteins or facilitating attachment to a specific surface. At present, much effort is being directed towards high-throughput approaches to determine the optimum conditions for solubilization and refolding of membrane-bound proteins, in particular GPCRs [18.33].

18.5.2 GPCR:Gα Fusion Proteins

A common approach to elucidating the molecular mechanisms regulating GPCR pharmacology is to use PCR-based mutagenic approaches to alter specific receptor amino acids and to subsequently determine how such changes affect receptor activity. Another common technique for receptor modification is the production of fusion proteins, whereby DNA sequences coding the receptor and another protein are joined such that a single protein (composed of two normally separate proteins) can be expressed (Fig. 18.9). An early example of this approach was a β_2-adrenergic receptor-$G\alpha_s$ fusion protein which was more potent than that expressed in wild-type cells where the alpha subunit and the GPCR were not fused, indicating that the covalent linkage between the proteins provided an advantage in terms of activation of cellular effectors [18.34].

Various GPCR-$G\alpha$ fusion proteins constructed by fusing the C-terminus of the GPCR with the N-terminus of the alpha subunit have proved useful for the elucidation of various aspects of GPCR pharmacology (see reviews [18.35, 36]). The main advantage of GPCR-$G\alpha$ fusions is the 1 : 1 stoichiometry between the fusion partners, allowing parameters associated with agonist binding to be directly comparable between expression systems (although such constructs may not be a correct representation of the physiological situation). It should also be noted that some fusion partners have been reported to be more effective than others in terms of $G\alpha$ affinity for GDP upon ligand binding [18.37].

Part B | 18.5

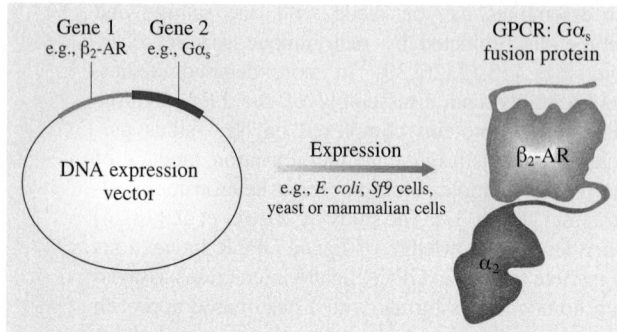

Fig. 18.9 Generation of a fusion protein. Two separate genes of interest are cloned and subsequently ligated into a DNA expression vector "in frame". In this example, the DNA sequence encoding the GPCR (β_2-adrenergic receptor; β_2-AR) is incorporated into the expression vector within the multiple cloning site. The DNA sequence encoding the Gα_s protein is also cloned into this vector. The resultant recombinant expression vector contains the (carboxy) C-terminus of the β_2-AR fused in frame to the (amino) N-terminus of the Gα_s protein. The recombinant DNA expression vector is then transfected into an appropriate cell line and the fusion protein is expressed

18.5.3 Engineering of Promiscuous Gα Proteins

A major impediment to the production of homogeneous, cell-free, GPCR-based screening systems is the coupling between a given GPCR and a subset of Gα subunits. For example, the muscarinic receptor subtypes M$_1$, M$_3$ and M$_5$ typically couple to G$\alpha_{q/11}$ whilst M$_2$ and M$_4$ subtypes couple to G$_i$ or G$_o$ [18.38]. Biologically, this discrimination is the basis for correct cellular signaling but needs to be modified from the in vivo situation to allow production of a generic GPCR biosensing system. In this regard, recent attempts have been made to produce 'promiscuous' Gα subunits capable of transducing signals resulting from extracellular interactions involving any GPCR [18.39, 40]. Many of the current promiscuous subunits constructed thus far are based on variants of the human Gα_{16} (a member of the Gα_q subfamily). This protein was first isolated from hematopoietic cells [18.41] and was shown to couple to a wider range of receptors than other known alpha subunits, and to transduce ligand-mediated signaling through phospholipase C (PLC) resulting in the modification of intracellular calcium concentrations [18.42–46]. Molecular biology approaches have also been utilized to increase the promiscuity of various Gα subunits by altering the sequence of amino acids within the protein [18.40, 47–50]. Although cell-free applications have

not been routinely used to date, it is expected that promiscuous G-proteins will be used in the near future in a similar manner to those used in whole-cell applications. In this regard the following points are noted:

1. Replacing the five C-terminal amino acids of Gα_q with the corresponding residues from Gα_i resulted in a chimeric subunit which signals through PLC-β on (some) receptors that normally couple exclusively to Gα_i [18.47, 48]).
2. Similarly, replacement of the five C-terminal amino acids of Gα_q with those from Gα_s allowed some Gα_s-coupled receptors to signal through PLC-β [18.48]. Also the Gα_q-coupled receptors bombesin and V1a vasopressin could stimulate adenylate cyclase by coupling to a modified Gα_q subunit containing the five N-terminal amino acids from Gα_z in place of the wild-type residues.
3. The N-terminus of Gα_q has also been modified to enhance promiscuity. Deletion of the six N-terminal amino acids produced a subunit allowing transduction of agonist-induced signaling through PLC-β and inositol (1,4,5)-trisphosphate by G$_i$- and G$_s$-linked receptors, an approach that was successful for the β_2-adrenergic and sst$_1$ receptors [18.49].

More recently, the promiscuity of human Gα_{16} has been increased by creating chimeric Gα_{16} subunits containing various lengths of C-terminal amino acids from rat Gα_s or Gα_z, in place of the wild-type human sequences [18.40, 50]. Therefore, it is conceivable that one or more modified Gα subunits may be developed which are capable of coupling to a large range of GPCRs, allowing for the development of truly generic GPCR biosensors.

18.5.4 Expression Systems for Recombinant GPCRs/G-proteins

A prerequisite for molecular approaches to the design of cell-free GPCR assays is an expression system which produces large amounts of recombinant proteins with the required activity and level of expression. Expression systems utilizing either bacteria, yeast, mammalian or insect cells are detailed in Table 18.3. These systems are generally well characterized and show the greatest promise in terms of their ability to produce large amounts of functional proteins which can be relatively easily purified for GPCR biosensing assay formats.

Table 18.3 Comparison of the main advantages and disadvantages of various expression systems commonly used to obtain GPCRs and/or G-proteins

Expression System	Advantage	Disadvantage
Bacteria e.g., *Eschericia coli spp.*	• Many host species to chose from • Many DNA expression vectors available • Relatively cheap • Fast process and easy to scale-up • Yield can be very high	• Prokaryotic, not eukaryotic • Truncated proteins can be produced • The expressed proteins often do not fold properly and so are biologically inactive • Insufficient post-translational modifications made e.g., GPCR glycosylation, G-protein palmitoylation • Overexpression can be toxic to the host cells
Yeast e.g., *Saccharomyces cerevisiae*	• Eukaryotic • Fast process and relatively easy to scale-up • Yield can be very high • Relatively cheap • Performs many of the post-translational modifications made to human proteins	• Cell wall may hinder recovery of expressed proteins • Presence of active proteases that degrade foreign (expressed) proteins, therefore may reduce yield
Insect e.g., *Spodoptera frugiperda Sf*9, Hi-5	• High levels of expression • Correct folding • Post-translational modifications similar to those in mammalian cells	• Expensive to scale up • Slow generation time • Difficult to work with
Mammalian e.g., CHO, HEK, COS	• Good levels of expression • Correct folding and post-translational modifications	• Relatively low yields • Very expensive to scale up • Slow generation time • Difficult to work with • Health and safety implications involved

Part B | 18.5

18.5.5 Fluorescent Proteins

Green fluorescent protein (GFP) was first isolated from jellyfish (*Aequorea victoria*) in the early 1960s [18.51]. GFPs have been used as fusion proteins (see Fig. 18.8) and have been widely exploited in molecular/cell biology research applications due to their efficient fluorescence emission properties. GFPs are particularly useful because they do not require unusual substrates, external catalysis or accessory cofactors for fluorescence unlike many other natural pigments [18.52].

Whilst fluorescent proteins provide many advantages, they are limited in their use as protein labels due to being large multimeric proteins. The huge array of untapped, naturally-occurring fluorescent proteins, combined with the protein engineering techniques used to produce novel fluorescent proteins or quenchers, suggests that it is likely that fluorescence-based assay development involving these compounds will increase in efficiency and flexibility, allowing such methods to be at the forefront of future technologies for determining molecular interactions using cell-free systems.

18.6 The Future of GPCRs in Nanobiotechnologies

Although this chapter has focused on the GPCR signaling system for biosensing applications, many other potential biological systems could be equally exploited for biosensing applications, including those involving antibodies, ion channels and enzymes. We have emphasized that molecular biology, combined with nanobiotechnologies, are important tools by which every facet of designing and investigating cell-free biosensing approaches can be improved. GPCR and G-protein engineering is a technique which has been employed not only to study GPCR interactions, but to enhance the measurement of GPCR activation, which will interface with future biosensing applications. Fusion proteins, promiscuous and chimeric Gα proteins

and molecular tagging are some of the molecular attributes that have been described here. Structural enhancements to GPCRs and G-protein subunits or effectors are only limited by the creativity of the researcher, and these enhancements will be imperative in the design of novel cell-free assay technologies. Microarray and chip-based technologies, recombinant protein design and production, assay automation and new assay methodologies for studying GPCR signaling are developing rapidly. The involvement of GPCR signaling in such a multitude of cellular processes indicates that it is unlikely that the current interest in GPCRs will diminish in the foreseeable future.

References

18.1   A. Wise, K. Gearing, S. Rees: Target validation of G-protein coupled receptors, Drug Discov. Today. **7**, 235–246 (2002)

18.2   K. L. Pierce, R. T. Premont, R. J. Lefkowitz: Seven-transmembrane receptors, Nat. Rev. Mol. Cell Biol. **3**, 639–650 (2002)

18.3   D. K. Vassilatis, J. G. Hohmann, H. Zeng, F. Li, J. E. Ranchalis, M. T. Mortrud, A. Brown, S. S. Rodriguez, J. R. Weller, A. C. Wright, J. E. Bergmann, G. A. Gaitanaris: The G protein-coupled receptor repertoires of human and mouse, Proc. Natl. Acad. Sci. USA **100**, 4903–4908 (2003)

18.4   G. Tollin, Z. Salamon, V. J. Hruby: Techniques: plasmon-waveguide resonance (PWR) spectroscopy as a tool to study ligand–GPCR interactions, Trends Pharmacol. Sci. **24**, 655–659 (2003)

18.5   J. M. Edwardson, R. M. Henderson: Atomic force microscopy and drug discovery, Drug Discov. Today. **9**, 64–71 (2004)

18.6   N. C. Santos, M. A. Castanho: An overview of the biophysical applications of atomic force microscopy, Biophys. Chem. **107**, 133–149 (2004)

18.7   D. Fotiadis, Y. Liang, S. Filipek, D. A. Saperstein, A. Engel, K. Palczewski: The G protein-coupled receptor rhodopsin in the native membrane, FEBS Lett. **564**, 281–288 (2004)

18.8   J. C. Owicki: Fluorescence polarization and anisotropy in high throughput screening: perspectives and primer, J. Biomol. Screen. **5**, 297–306 (2000)

18.9   C. J. Daly, J. C. McGrath: Fluorescent ligands, antibodies, and proteins for the study of receptors, Pharmacol. Ther. **100**, 101–118 (2003)

18.10 P. Banks, M. Harvey: Considerations for using fluorescence polarization in the screening of G protein-coupled receptors, J. Biomol. Screen. **7**, 111–117 (2002)

18.11 K. L. Martinez, B. H. Meyer, R. Hovius, K. Lundstrom, H. Vogel: Ligand binding to G protein-coupled receptors in tethered cell membranes, Langmuir **19**, 10925–10929 (2003)

18.12 A. Waller, P. Simons, E. R. Prossnitz, B. S. Edwards, L. A. Sklar: High throughput screening of G-protein coupled receptors via flow cytometry, Comb. Chem. High-T. Screen. **6**, 389–397 (2003)

18.13 T. Mirzabekov, H. Kontos, M. Farzan, W. Marasco, J. Sodroski: Paramagnetic proteoliposomes containing a pure, native, and oriented seven-transmembrane segment protein, CCR5, Nat. Biotechnol. **18**, 649–654 (2000)

18.14 Y. Fang, A. G. Frutos, J. Lahiri: Membrane protein microarrays, J. Am. Chem. Soc. **124**, 2394–2395 (2002)

18.15   Y. Fang, A. G. Frutos, B. Webb, Y. Hong, A. Ferrie, F. Lai, J. Lahiri: Membrane biochips, BioTechniques **Suppl**, 62–65 (2002)

18.16   Y. Fang, A. G. Frutos, J. Lahiri: G-protein-coupled receptor microarrays, Chembiochem **3**, 987–991 (2002)

18.17   N. M. Rao, V. Silin, K. D. Ridge, J. T. Woodward, A. L. Plant: Cell membrane hybrid bilayers containing the G-protein-coupled receptor CCR5, Anal. Biochem. **307**, 117–130 (2002)

18.18   O. Karlsson, L. Stefan: Flow-mediated on-surface reconsititution of G-protein coupled receptors for applications in surface plasmon resonance biosensors, Anal. Biochem. **300**, 132–138 (2002)

18.19   P. Stenlund, G. J. Babcock, J. Sodroski, D. G. Myszka: Capture and reconstitution of G protein-coupled receptors on a biosensor surface, Anal. Biochem. **316**, 243–250 (2003)

18.20   S. Devanathan, Z. Yao, Z. Salamon, B. Kobilka, G. Tollin: Plasmon-waveguide resonance studies of ligand binding to the human beta 2-adrenergic receptor, Biochemistry **43**, 3280–3288 (2004)

18.21   I. D. Alves, S. M. Cowell, Z. Salamon, S. Devanathan, G. Tollin, V. J. Hruby: Different structural states of the proteolipid membrane are produced by ligand binding to the human delta-opioid receptor as shown by plasmon-waveguide resonance spectroscopy, Mol. Pharmacol. **65**, 1248–1257 (2004)

18.22   T. Z. Wu: A piezoelectric biosensor as an olfactory receptor for odour detection: electronic nose, Biosens. Bioelectron. **14**, 9–18 (1999)

18.23   L. Neumann, T. Wohland, R. J. Whelan, R. N. Zare, B. K. Kobilka: Functional immobilization of a ligand-activated G-protein-coupled receptor, Chembiochem **3**, 993–998 (2002)

18.24   G. Milligan: Principles: extending the utility of [35S]GTP gamma S binding assays, Trends Pharmacol. Sci. **24**, 87–90 (2003)

18.25   C. Harrison, J. R. Traynor: The [^{35}S]GTPgammaS binding assay: approaches and applications in pharmacology, Life Sci. **74**, 489–508 (2003)

18.26   P. C. Simons, M. Shi, T. Foutz, D. F. Cimino, J. Lewis, T. Buranda, W. K. Lim, R. R. Neubig, W. E. McIntire, J. Garrison, E. Prossnitz, L. A. Sklar: Ligand-receptor-G-protein molecular assemblies on beads for mechanistic studies and screening by flow cytometry, Mol. Pharmacol. **64**, 1227–1238 (2003)

18.27   M. Bunemann, M. Frank, M. J. Lohse: Gi protein activation in intact cells involves subunit rearrangement rather than dissociation, Proc. Natl. Acad. Sci. USA **100**, 16077–16082 (2003)

18.28   C. Bieri, O. P. Ernst, S. Heyse, K. P. Hofmann, H. Vogel: Micropatterned immobilization of a G protein-coupled receptor and direct detection of G protein activation, Nat. Biotechnol. **17**, 1105–1108 (1999)

18.29   L. A. Sklar, J. Vilven, E. Lynam, D. Neldon, T. A. Bennett, E. Prossnitz: Solubilization and display of

18.30   G protein-coupled receptors on beads for real-time fluorescence and flow cytometric analysis, BioTechniques **28**, 976–5 (2000)

18.30   A. Waller, P. C. Simons, S. M. Biggs, B. S. Edwards, E. R. Prossnitz, L. A. Sklar: Techniques: GPCR assembly, pharmacology and screening by flow cytometry, Trends Pharmacol. Sci. **25**, 663–669 (2004)

18.31   T. A. Bennett, T. A. Key, V. V. Gurevich, R. Neubig, E. R. Prossnitz, L. A. Sklar: Real-time analysis of G protein-coupled receptor reconstitution in a solubilized system, J. Biol. Chem. **276**, 22453–22460 (2001)

18.32   X. Michalet, F. F. Pinaud, L. A. Bentolila, J. M. Tsay, S. Doose, J. J. Li, G. Sundaresan, A. M. Wu, S. S. Gambhir, S. Weiss: Quantum dots for live cells and in vivo imaging and diagnostics, Science **307**, 538–544 (2005)

18.33   K. Lundstrom: Structural genomics on membrane proteins: mini review, Comb. Chem. High-T. Screen. **7**, 431–439 (2004)

18.34   B. Bertin, M. Freissmuth, R. Jockers, A. D. Strosberg, S. Marullo: Cellular signaling by an agonist-activated receptor/Gs alpha fusion protein, Proc. Natl. Acad. Sci. USA **91**, 8827–8831 (1994)

18.35   R. Seifert, K. Wenzel-Seifert, B. K. Kobilka: GPCR-Galpha fusion proteins: molecular analysis of receptor-G-protein coupling, Trends Pharmacol. Sci. **20**, 383–389 (1999)

18.36   G. Milligan: Insights into ligand pharmacology using receptor-G-protein fusion proteins, Trends Pharmacol. Sci. **21**, 24–28 (2000)

18.37   Z. D. Guo, H. Suga, M. Okamura, S. Takeda, T. Haga: Receptor-Galpha fusion proteins as a tool for ligand screening, Life Sci. **68**, 2319–2327 (2001)

18.38   M. P. Caulfield, N. J. Birdsall: International Union of Pharmacology. XVII. Classification of muscarinic acetylcholine receptors, Pharmacol. Rev. **50**, 279–290 (1998)

18.39   A. M. Liu, M. K. Ho, C. S. Wong, J. H. Chan, A. H. Pau, Y. H. Wong: Galpha(16/z) chimeras efficiently link a wide range of G protein-coupled receptors to calcium mobilization, J. Biomol. Screen. **8**, 39–49 (2003)

18.40   A. Hazari, V. Lowes, J. H. Chan, C. S. Wong, M. K. Ho, Y. H. Wong: Replacement of the alpha5 helix of Galpha16 with Galphas-specific sequences enhances promiscuity of Galpha16 toward Gs-coupled receptors, Cell. Signal. **16**, 51–62 (2004)

18.41   D. A. Steele, V. Z. Slepak, M. I. Simon: G alpha 16, a G protein alpha subunit specifically expressed in hematopoietic cells, Proc. Natl. Acad. Sci. USA **88**, 5587–5591 (1991)

18.42   S. Offermanns, M. I. Simon: G alpha 15 and G alpha 16 couple a wide variety of receptors to phospholipase C, J. Biol. Chem. **270**, 15175–15180 (1995)

18.43   X. Zhu, L. Birnbaumer: G protein subunits and the stimulation of phospholipase C by Gs-and Gi-coupled receptors: Lack of receptor selectivity of

Part B | 18

Galpha(16) and evidence for asynergic interaction between Gbeta gamma and the alpha subunit of a receptor activated G protein, Proc. Natl. Acad. Sci. USA **93**, 2827–2831 (1996)

18.44   G. Milligan, F. Marshall, S. Rees: G16 as a universal G protein adapter: implications for agonist screening strategies, Trends Pharmacol. Sci. **17**, 235–237 (1996)

18.45   J. W. Lee, S. Joshi, J. S. Chan, Y. H. Wong: Differential coupling of mu-, delta-, and kappa-opioid receptors to G alpha16-mediated stimulation of phospholipase C, J. Neurochem. **70**, 2203–2211 (1998)

18.46   E. Kostenis: Is Galpha16 the optimal tool for fishing ligands of orphan G-protein-coupled receptors?, Trends Pharmacol. Sci. **22**, 560–564 (2001)

18.47   B. R. Conklin, Z. Farfel, K. D. Lustig, D. Julius, H. R. Bourne: Substitution of three amino acids switches receptor specificity of Gq alpha to that of Gi alpha, Nature **363**, 274–276 (1993)

18.48   B. R. Conklin, P. Herzmark, S. Ishida, T. A. Voyno-Yasenetskaya, Y. Sun, Z. Farfel, H. R. Bourne:

Carboxyl-terminal mutations of Gq alpha and Gs alpha that alter the fidelity of receptor activation, Mol. Pharmacol. **50**, 885–890 (1996)

18.49   E. Kostenis, F. Y. Zeng, J. Wess: Functional characterization of a series of mutant G protein alphaq subunits displaying promiscuous receptor coupling properties, J. Biol. Chem. **273**, 17886–17892 (1998)

18.50   S. M. Mody, M. K. Ho, S. A. Joshi, Y. H. Wong: Incorporation of Galpha(z)-specific sequence at the carboxyl terminus increases the promiscuity of galpha(16) toward G(i)-coupled receptors, Mol. Pharmacol. **57**, 13–23 (2000)

18.51   O. Shimomura, F. H. Johnson, Y. Saiga: Extraction, purification and properties of aequorin, a bioluminescent protein from the luminous hydromedusan, Aequorea, J. Cell. Comp. Physiol. **59**, 223–239 (1962)

18.52   V. V. Verkhusha, K. A. Lukyanov: The molecular properties and applications of Anthozoa fluorescent proteins and chromoproteins, Nat. Biotechnol. **22**, 289–296 (2004)

Part B | 18

19. Microfluidics and Their Applications to Lab-on-a-Chip

Various microfluidic components and their charac-teristics, along with the demonstration of two recent achievements of lab-on-chip systems are reviewed and discussed. Many microfluidic devices and components have been developed during the past few decades, as introduced earlier for various applications. The design and de-velopment of microfluidic devices still depend on the specific purposes of the devices (ac-tuation and sensing) due to a wide variety of application areas, which encourages researchers to develop novel, purpose-specific microfluidic devices and systems. Microfluidics is the mul-tidisciplinary research field that requires basic knowledge in fluidics, micromachining, electro-magnetics, materials, and chemistry for various applications.

Among the various application areas of mi-crofluidics, one of the most important is the lab-on-a-chip system. Lab-on-a-chip is be-coming a revolutionary tool for many different applications in chemical and biological analyses due to its fascinating advantages (fast speed and low cost) over conventional chemical or biolog-ical laboratories. Furthermore, the simplicity of lab-on-a-chip systems will enable self-testing

capability for patients or health consumers by overcoming space limitations.

Part B | 19

Microfluidics covers the science of fluidic behaviors on the micro/nanoscales and the design engineering, simulation, and fabrication of fluidic devices for the transport, delivery, and handling of fluids on the order of microliters or smaller volumes. It is the backbone of biological or biomedical microelectromechanical sys-tems (BioMEMS) and lab-on-a-chip concept, as most biological analyses involve fluid transport and reaction. Biological or chemical reactions on the micro/nanoscale are usually rapid since small amounts of samples and reagents are used, which offers quick and low-cost analysis.

A fluidic volume of 1 nanoliter (1 nl) can be under-stood as the volume in a cube surrounded by $100\,\mu m$ in each direction. It is much smaller than the size of a grain

of table salt. Microfluidic devices and systems handle sample fluids in this range for various applications, including inkjet printing, blood analysis, biochemical detection, chemical synthesis, drug screening/delivery, protein analysis, DNA sequencing, and so on.

Microfluidic systems consist of microfluidic plat-forms or devices for fluidic sampling, control, monitoring, transport, mixing, reaction, incubation, and analysis. To construct microfluidic systems, or labs-on-a-chip, microfluidic devices must be functionally integrated on a microfluidic platform using proper micro/nanofabrication techniques. In this chapter, the basics of microfluidic devices and their applications to lab-on-a-chip are briefly reviewed and summarized. Basic materials and fabrication techniques for microfluidic devices will be introduced first and various active and passive microfluidic components will be described.

Then, their applications to lab-on-a-chip, or biochemical analysis, will be discussed.

19.1 Materials for Microfluidic Devices and Micro/Nanofabrication Techniques

Various materials are being used for the fabrication of microfluidic devices and systems. Silicon is one of the most popular materials in micro/nanofabrication because its micromachining has been well established over a period of decades. In general, the advantages of using silicon as a substrate or structural material include good mechanical properties, excellent chemical resistance, well-characterized processing techniques, and the capability for integration of control/sensing circuitry in the semiconductor. Other materials such as glass, quartz, ceramics, metals, and polymers are also being used for substrates and structures in micro/nanofabrication, depending on the application. Among these materials, polymers or plastics have recently become one of the more promising materials for lab-on-a-chip applications, due to their excellent material properties for biochemical fluids and their low-cost manufacturability. The main issues in the fabrication techniques of microfluidic devices and systems usually lie in forming microfluidic channels, which are key micro/nanostructures of lab-on-a-chip. In this section, the basic micro/nanofabrication techniques for silicon, glass, and polymers are described.

19.1.1 Silicon

Microfluidic channels on silicon substrates are usually formed either by wet (chemical) etching or by dry (plasma) etching. Crystalline silicon has a preferential etch direction, depending on which crystalline plane is exposed to an etchant. Etch rate is slowest in the $\langle 111 \rangle$ crystalline direction—approximately 100:1 etch rate anisotropy compared with $\langle 100 \rangle$:$\langle 111 \rangle$ or $\langle 110 \rangle$:$\langle 111 \rangle$.

Potassium hydroxide (KOH), tetra-methyl ammonium hydroxide (TMAH), and ethylene diamine pyrocatechol (EDP) are commonly used silicon anisotropic etchants. In most cases, silicon dioxide (SiO_2) or silicon nitride (Si_3N_4) is used as a masking material during the etching process. Anisotropic etchants and basic etching mechanisms are summarized by *Ristic* et al. [19.1]. There is also an isotropic wet-etching process available using a mixture of hydrofluoric acid (HF), nitric acid (HNO_3), and acetic acid (CH_3COOH): the so-called hydrofluoric–nitric–acetic ("HNA") etch. HNA etches in all directions with almost the same etch rate regardless of crystalline directions. Figure 19.1 illustrates wet anisotropic and isotropic etching profiles.

Reactive ion etching (RIE) is also one of the most commonly used dry-etching processes to generate microfluidic channels or deep trench structures on silicon substrate. In this dry-etching technique, radio-frequency (RF) energy is used to excite ions in a gas to an energetic state. The energized ions supply the necessary energy to generate physical and chemical reactions on the exposed area of the substrate, which starts the etching process. RIE can generate strong anisotropic, as well as isotropic profiles, depending on the gases used, the condition of plasma, and the applied power. Further information on reactive ion etching process, including deep reactive ion etching (DRIE), on silicon substrates can be found in the literature [19.2–5].

Many microfluidic devices have been realized using silicon as a substrate material, including microvalves and micropumps, which are covered in the next section.

19.1.2 Glass

Glass substrate has been widely used for the fabrication of microfluidic systems and lab-on-a-chip due to its excellent optical transparency and ease of electroosmotic flow (EOF). Chemical wet etching and thermal fusion bonding are the common fabrication techniques for glass substrate. Chemical wet etching and the bonding technique have also been widely reported [19.6, 7]. The most commonly used etchants are hydrofluoric acid

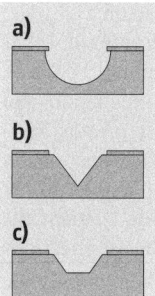

Fig. 19.1a–c Wet etching of silicon substrate for anisotropic and isotropic etching: (**a**) isotropic etching profile, (**b**) anisotropic etching profile (long term), and (**c**) anisotropic etching profile (short term)

(HF), buffered hydrofluoric acid, and a mixture of hydrofluoric acid, nitric acid, and deionized water (HF, HNO_3, H_2O). Gold with an adhesion layer of chrome is most often used as an etch mask for wet etching of a glass substrate. Since glass has no crystalline structure, only isotropic etch profiles are obtained, such as forming a hemispherical-shaped channel. Often the problem is that stresses within the surface layers of the glass cause preferential etching, and scratches created by polishing or handling errors cause spikes to be etched in the channels. Pre-etching is one method to release stress, which causes defects in channels after etching. Another way to improve the channel-etching process is to anneal the glass wafers before etching. Annealing can be done close to the glass-transition temperature for at least a couple of hours. Figure 19.2 shows two examples of poorly etched channels without the pre-etching and annealing steps compared with an etched channel without defects. Other fabrication techniques for glass substrate include photoimageable glass, as reported by *Dietrich* et al. [19.8]. Anisotropy is introduced into glass by making the glass photosensitive using lithium/aluminium/silicates in its composition.

One of the most successful examples of using glass as a substrate material in lab-on-a-chip applications is the capillary electrophoresis (CE) chip, which is fabricated using the glass etching and fusion bonding techniques, since the most advantageous property of glass is its excellent optical transparency, which is required for most lab-on-a-chip applications that use optical detection, including capillary electrophoresis microchips [19.6, 7, 9].

19.1.3 Polymer

Among the various substrates available for lab-on-a-chip, polymers, or plastics, have recently become one of the most popular and promising substrates due to their low cost, ease of fabrication, and favorable biochemical reliability and compatibility. Plastic substrates, such as polyimide, poly(methyl methacrylate) (PMMA), poly(methyl siloxane) (PDMS), polyethylene or polycarbonate, offer a wide range of physical and chemical material parameters for the applications of lab-on-a-chip, generally at low cost using replication approaches. Polymers and plastics are promising materials in microfluidic and lab-on-a-chip applications because they can be used for mass production using casting, hot embossing, and injection-molding techniques. This mass-production capability allows the successful commercialization of lab-on-a-chip technology, including disposable lab-on-a-chip. While several fabrication methods have recently been developed, three fabrication techniques—casting, hot embossing, and injection molding—are major techniques of great interest. Figure 19.3 shows schematic illustrations of these polymer microfabrication techniques.

For polymer or plastic micro/nanofabrication, a mold master is essential for replication. Mold masters are fabricated using photolithography, silicon/glass bulk etching, and metal electroplating, depending on the application. Figure 19.4 summarizes the mold masters from different fabrication techniques.

Photolithography, including LIGA (from the German words Lithografie, Galvanoformung, Abformung, meaning lithography, electroforming and molding) [19.10] and UV-LIGA (ultraviolet LIGA) [19.11, 12], is used to fabricate mold masters for casting or soft lithography replication, while silicon-based and electroplated mold masters are used for hot-embossing replication. For injection molding, electroplated metallic mold masters are preferable.

Casting or soft lithography [19.13, 14] usually offers flexible access to microfluidic structures using mostly

Fig. 19.2a–c Isotropically etched microfluidic channels on glass substrate: (**a**) poorly etched channel with large underetching, (**b**) poorly etched channel with spikes, and (**c**) well-etched channel without any defects

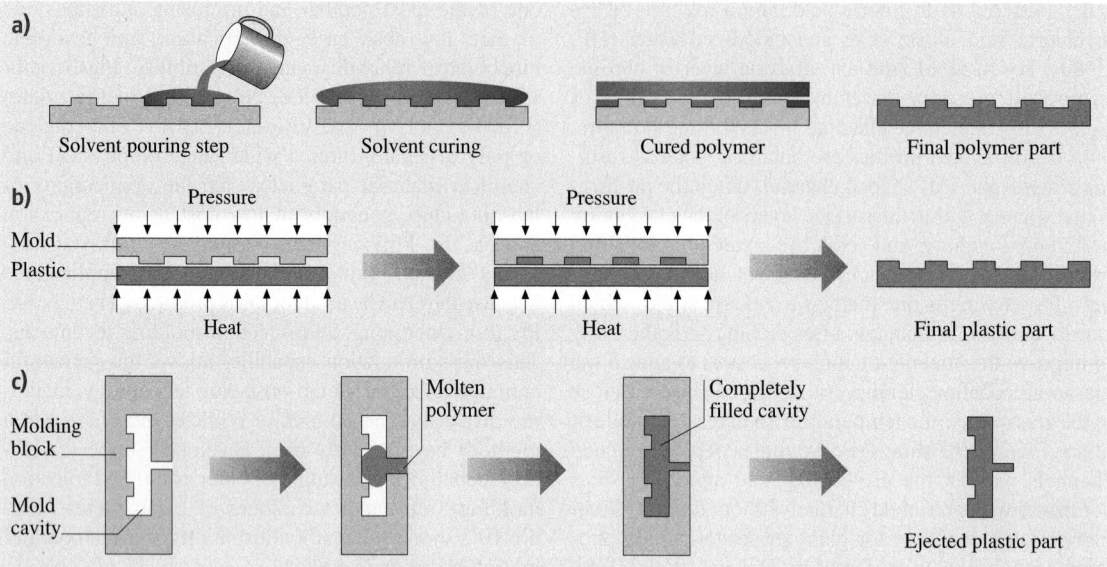

Fig. 19.3a–c Concept of polymer micro/nanofabrication techniques: (**a**) casting, (**b**) hot embossing, and (**c**) injection molding

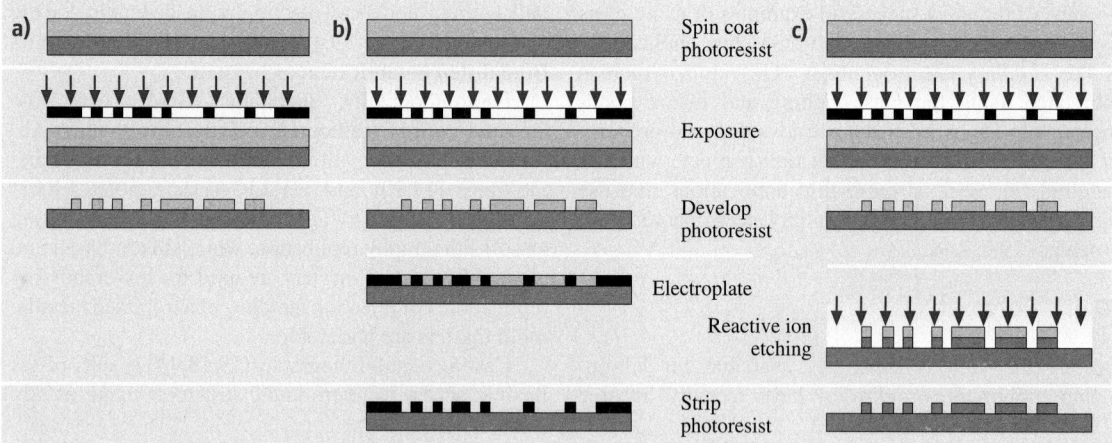

Fig. 19.4a–c Mold masters in polymer/plastic fabrication: (**a**) photolithography-based mold master, (**b**) silicon-based mold master, and (**c**) mold master by electroplating

poly(methyl siloxane) (PDMS) as a casting material. A mixture of the elastomer precursor and curing agent is poured over a master mold structure. After curing, the replicated elastomer is released from the mold master, having transferred a reverse structure of the mold master. Patterns of a few nm can be achieved using this technique.

While casting can be carried out at room temperature, hot embossing requires a slightly higher temperature—up to the glass-transition temperature of the plastic substrate to be replicated. The hot-embossing technique has been developed by several research groups [19.15–17]. A mold master is placed in the chamber of a hot-embossing system with the plastic substrate, then heated plates press both the plastic substrate and the mold master, as illustrated in Fig. 19.3b. After a certain amount of time (typically 5–20 mins, depending on the plastic substrate), the plates are cooled down to

Table 19.1 Overview of the different polymer micro/nanofabrication techniques

Fabrication type	Casting	Hot embossing	Injection molding
Investment	Low	Moderate	High
Manufacturability	Low	Moderate	High
Cycle time	8–10 hrs	1 hr	1 min
Polymer choices	Low	Moderate	Moderate
Mold replication	Good	Good	Good
Reusability of mold	No (photolithography- based molds)	Yes	Yes

release the replicated plastic substrate. Hot embossing offers mass production of polymer microstructures, as its cycle time is less than an hour.

Injection molding is a technique to fabricate polymer microstructures at low cost and high volumes [19.18]. A micromachined mold master is placed in the molding block of the injection molding machine, as illustrated in Fig. 19.3c. The plastic in granular form is melted and then injected into the cavity of a closed mold block, where the mold master is located. The molten plastic continues to flow into and fill the mold cavity until the plastic cools down to a highly viscous melt, and a cooled plastic part is ejected. In order to ensure good flow properties during theinjection molding process, thermoplastics with low or medium viscosity are desirable. The filling of the mold cavity, and subsequently the microstructures, de-

pend on the viscosity of the plastic melt, injection speed, molding-block temperature, and nozzle temperature of the injection unit. This technique allows very rapid replication and high-volume mass production. The typical cycle time is several seconds for most applications. However, due to the high shear force on the mold master inside the mold cavity during injection molding, metallic mold masters are highly recommended. Poly(methyl methacrylate) (PMMA), polyethylene (PE), polystyrene (PS), polycarbonate (PC), and cyclic olefin copolymers (COC) are common polymer/plastic materials for both hot embossing and injection molding replication.

All of these polymer/plastic replication techniques are summarized in Table 19.1. Since each technique differs from the others, fabrication techniques and materials have to be selected according to the application.

19.2 Active Microfluidic Devices

Microfluidic devices are essential for the development of lab-on-a-chip or micro-total analysis systems (μTAS). A number of different microfluidic devices have been developed with basic structures analogous to macroscale fluidic devices. Such devices include microfluidic valves, microfluidic pumps, microfluidic mixers, etc. The devices listed above have been developed both as active and passive devices. While passive microfluidic devices are generally easy to fabricate, they do not offer the same functional diversity that the active microfluidic devices provide. For example, passive microvalves based on the surface tension effect [19.19] can operate a few times to hold the liquid, which is acceptable for a disposable format. Once the air–liquid interface passes over the valve mechanism, the operation

characteristics of the valve will differ, due to the change of surface energy over the channel. Similarly, passive check valves [19.20–22] are dependent on the pressure of the fluid for operation. Since the active microvalves can be triggered on/off depending on an external signal regardless of the status of the fluid system, there has been considerable research effort to develop active microfluidic devices. However, active devices are usually more expensive due to their desired functional and fabrication complexity.

This section reviews some of the active microfluidic devices, such as microvalves, micropumps, and active microfluidic mixers. Passive counterparts of these microfluidic devices will be discussed in the next section.

19.2.1 Microvalves

Active microvalves have been an area of intense research over the past decade, and a number of novel design and actuation schemes have been developed. This makes the categorization of active microvalves a confusing enterprize. Classification schemes for active microvalves include:

1. Fluidics handled: liquid/gas/liquid and gas
2. Materials used for the structure: silicon/polysilicon/glass/polymer
3. Actuation mechanisms: electrostatic/pneumatic/thermpneumatic, etc.
4. Physical actuating microstructures: membrane-type, flap-type, ball valve, etc.

All of the classification schemes listed above are valid, but the most commonly used method [19.23, 24] is the classification based on the actuation mechanisms.

In this section, the various microvalves are discussed in terms of their actuation mechanisms and their relevance to the valve mechanism, as well as fluid handled and special design criteria.

Pneumatically/Thermopneumatically Actuated Microvalves

Pneumatic actuation uses an external air line (or pneumatic source) to actuate a flexible diaphragm. Pneumatic actuation offers such attractive features as high force, high displacement, and rapid response time. Figure 19.5 illustrates a schematic concept of pneumatically actuated microvalves. *Schomburg* et al. [19.25] demonstrated pneumatically actuated microvalves.

Pneumatically actuated microvalves have also been demonstrated using polymeric substrate. *Hosokawa* et al. [19.26] demonstrated a pneumatically actuated three-way microvalve system using a PDMS platform. The microfluidic lines and the pneumatic lines are fabricated on separate layers.

Thermopneumatic actuation is typically performed by heating a fluid (usually a gas) in a confined cavity, as illustrated in Fig. 19.6. The increase in temperature leads to a rise in the pressure of the gas, and this pressure is used to deflect a membrane for valve operation. Thermopneumatic actuation is an inherently slow technique but offers very high forces when compared to other techniques [19.23]. Thermopneumatically actuated microvalves have been realized by many researchers using various substrates and diaphragm materials [19.27–29].

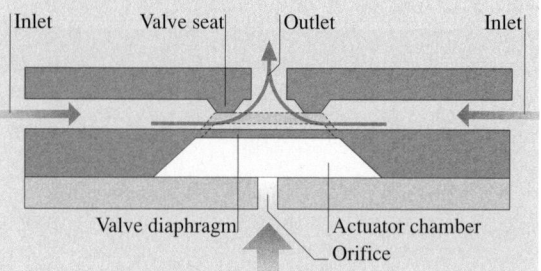

Fig. 19.5 Schematic concept of pneumatically actuated microvalve

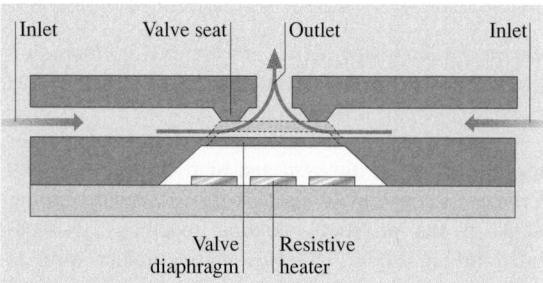

Fig. 19.6 Schematic concept of thermopneumatically actuated microvalve

Electrostatically Actuated Microvalves

Electrostatic actuation has been widely explored for a number of applications, including pressure sensors, comb drives, active mirror arrays, etc. Electrostatically actuated devices typically have a fairly simple structure and are easy to fabricate. A number of fabrication issues, such as stiction and release problems of membranes and valve flaps, need to be addressed to realize practical electrostatic microvalves. *Sato* et al. [19.30] have developed a novel membrane design in which the deflection *propagates* through the membrane, rather than deforming it entirely. Figure 19.7 shows a schematic sketch of the actuation mechanism and valve design. The use of this S-shaped design allows them to have relatively large gaps across the two surfaces, as the electrostatic force need only be concentrated at the edges of the S-shape where the membrane is deflected. *Robertson* et al. [19.31] have developed an array of electrostatically actuated valves using a flap design (rather than a membrane) to seal the fluid flow. The demonstrated system is suitable for very-low-pressure gas control systems such as those needed in a clean-room environment.

Wijngaart et al. [19.32] have developed a high-stroke high-pressure electrostatic actuator for valve applications. This reference provides a good overview of the

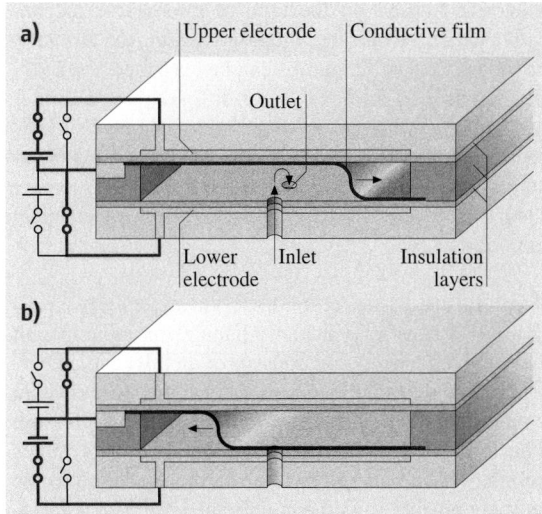

Fig. 19.7a,b Electrostatically actuated microvalve with an S-shaped film element: propagation of bend in the film as (**a**) open and (**b**) closed (adapted from [19.30]). Reproduced with permission from IOP Publishing Limited

theoretical design parameters used to design and analyze an electrostatically actuated microvalve.

Piezoelectrically Actuated Microvalve

Piezoelectric actuation schemes offer a significant advantage in terms of operating speed; they are typically the fastest actuation scheme at the expense of a reduced actuator stroke. Also, piezoelectric materials are more challenging to incorporate into fully integrated MEMS devices. *Watanabe* et al. [19.34] and *Stehr* et al. [19.35] demonstrated piezoelectric actuators for valve applications. A film of piezoelectric material is deposited on the movable membrane, and upon application of an electric potential, a small deformation occurs in the piezo film that is transmitted to the valve membrane.

Electromagnetically Actuated Microvalves

Electromagnetic actuators are typically capable of delivering high force and range of motion. A significant advantage of electromagnetic microvalves is that they are relatively insensitive to external interference. However, electromagnetic actuators involve a fairly complex fabrication process. Usually, a soft electromagnetic material such as NiFe (nickel iron, also known as permalloy) is used as a membrane layer, and an external electromagnet is used to actuate this layer. *Sadler* et al. [19.33] have developed a microvalve using the electromagnetic actuation scheme shown in Fig. 19.8.

They have demonstrated a fully integrated magnetic actuator with magnetic interconnection vias to guide the magnetic flux. The valve seat design, also shown in Fig. 19.8, allows for very intimate contact between the NiFe valve membrane and the valve seat, hence, achieving an ultra-low leak rate when the valve is closed.

Jackson et al. [19.36] demonstrate an electromagnetic microvalve using *magnetic* PDMS as the membrane material. For this application, the PDMS pre-polymer is loaded with soft magnetic particles and then cured to form the valve membrane. The PDMS membrane is then assembled over the valve body, and miniature electromagnets are used to actuate the membrane.

Other Microvalve Actuation Schemes

Microvalves have most commonly been implemented with one of the actuation schemes listed above. However, these are not the only actuation schemes that are used for microvalves. Some others include the use of (shape memory alloys) (SMA) [19.37], electrochemi-

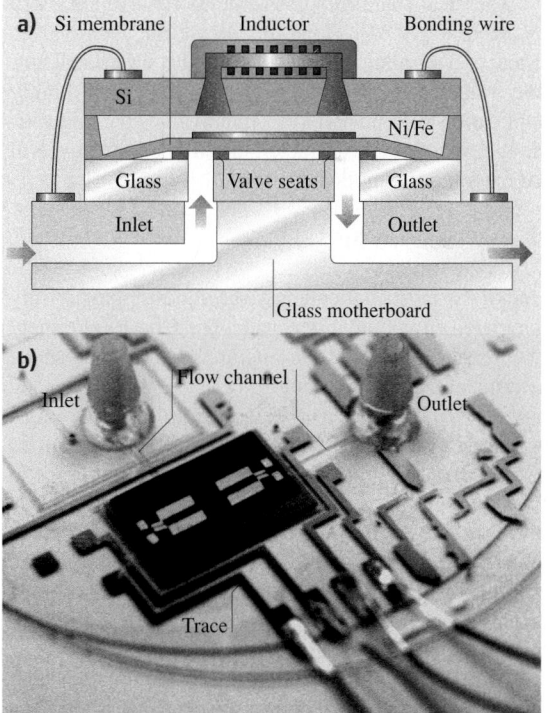

Fig. 19.8a,b Electromagnetically actuated microvalves: (**a**) schematic illustration and (**b**) photograph of the electromagnetically actuated microvalve as a part of lab-on-a-chip (after [19.33])

cal actuation [19.38], etc. SMA actuation schemes offer the advantage of generating very large forces when the SMA material is heated to its original state. *Neagu* et al. [19.38] presented an electrochemically actuated microvalve. In their device, an electrolysis reaction is used to generate oxygen in a confined chamber. This chamber is sealed by a deformable membrane that is deflected due to increased pressure. The reported microvalve has a relatively fast actuation time and can generate very high pressures. *Yoshida* et al. [19.39] present a novel approach to the microvalve design: a micro-electrorheological valve (ER valve). An electrorheological fluid is loaded into the microchannel, and, depending on the strength of an applied electric field, the viscosity of the ER fluid changes considerably. A higher viscosity is achieved when an electric field is applied perpendicular to the flow direction. This increased viscosity leads to a drop in the flow rate, allowing the ER fluid to act as a valve. Of course, this technique is limited to fluids that can exhibit such properties; nevertheless, it provides a novel idea to generate on-chip microvalves.

The ideal characteristics for a microvalve are listed by Kovacs [19.24]. However, of all the microvalves listed above, none can satisfy all the criteria. Thus, microvalve design, fabrication, and utility are highly application-specific and most microvalves try to generate the performance characteristics that are most useful for the intended application.

19.2.2 Micropumps

One of the most challenging tasks in developing a fully integrated microfluidic system has been the development of efficient and reliable micropumps. On the macroscale, a number of pumping techniques exist, such as peristaltic pumps, vacuum-driven pumps, Venturi-effect pumps, etc. However, in microscale, most mechanical pumps rely on pressurizing the working fluid and forcing it to flow through the system. Practical vacuum pumps are not available on the microscale. There are also some effects such as electro-osmotic pumping, that are only possible on the microscale. Consequently, electrokinetic driving mechanisms have also been widely studied for microfluidic pumping applications.

In the previous section, various microvalves that are used for microfluidic control were discussed. They were presented based on the actuation schemes that they employ. A similar classification can also be adopted for micropumps. However, rather than using the same classification scheme, the micropumps are

categorized based on the type of microvalve mechanism used as part of the pumping mechanism. Broadly, the mechanical micropumps can be classified as check-valve-controlled microvalves or diffuser pumps. Either mechanism can use various actuation schemes such as electrostatic, electromagnetic, piezoelectric, etc. Micropumps driven by direct electrical control form a separate category and are discussed following the mechanical micropumps.

Micropumps Using a Check-Valve Design
Figure 19.9 shows a typical mechanical micropump with check valves. This pump consists of an inlet and an outlet check valve with a pumping chamber in between. A membrane is deflected upwards, and a low-pressure zone is created in the pumping chamber. This forces the inlet check valve open, and fluid is sucked into the pumping chamber. As the membrane returns to its original state and continues to travel downwards, a positive pressure builds up, which seals the inlet valve while simultaneously opening the outlet valve. The fluid is then ejected and the pump is ready for another cycle.

A number of other techniques have been used to realize mechanical micropumps with check valves and a pumping chamber. *Jeon* et al. [19.40] present a micropump that uses PDMS flap valves to control the pumping mechanism, *Koch* et al. [19.41], *Cao* et al. [19.42], *Park* et al. [19.43], and *Koch* et al. [19.44] present piezoelectrically driven micropumps. *Xu* et al. [19.45] and *Makino* et al. [19.46] present SMA-driven micropumps. *Chou* et al. [19.47] present a novel rotary pump using a soft-lithography approach. As explained in the active valve section, valves can be created on a PDMS layer using separate liquid and air layers. When a pressure is applied to the air lines, the membranes deflect to seal the fluidic path. *Chou* et al. [19.47] have implemented a series of such valves in a loop. When they are deflected in

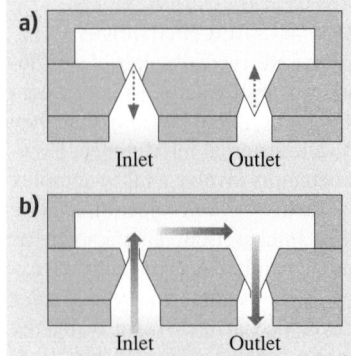

Fig. 19.9a,b Operation of a micropump using a check-valve design: (**a**) check valves are closed and (**b**) check valves are open

a set sequence, the liquid within the ring is pumped by peristaltic motion.

An interesting approach toward the development of a bidirectional micropump has been used by *Zengerle* et al. [19.48]. Their design has two flap valves at the inlet and outlet of the micropump, which work in the forward mode at low actuation frequencies and in the reverse direction at higher frequencies. *Zengerle* et al. [19.48] attribute the change in pumping direction to the phase shift between the response of the valves and the pressure difference that drives the fluid. *Carrozza* et al. [19.49] use a different approach to generate the check valves. Rather than using the conventional membrane or flap-type valves, they use ball valves by employing a stereolithographic approach. The developed pump is actuated using a piezoelectric actuation scheme.

Diffuser Micropumps

The use of nozzle-diffuser sections, or pumps with fixed valves, or pumps with dynamic valves has been extensively researched. The basic principle of these pumps is based on the idea that the geometrical structure used as an inlet valve has a preferential flow direction toward the pumping chamber, and the outlet valve structure has a preferential flow direction away from the pumping chamber. An illustrative example of this concept is shown in Fig. 19.10. As can be readily seen, these pumps are designed to work with liquids only. When the pump is in suction mode, the flow in the inlet diffuser structure is primarily directed toward the pump, and a slight back-flow occurs from the outlet diffuser (acting as a nozzle) section. When the pump is in pressure mode, the outlet diffuser section allows most of the flow out of the pump, whereas the inlet diffuser section allows a slight back-flow. The net effect is that liquid pumping occurs from left to right.

Diffuser micropumps are simple and valveless structures that improve pumping reliability [19.50], but they cannot eliminate back-flow problems.

These micropumps can also be described as flow rectifiers, analogously to diodes in electrical systems. *Forster* et al. [19.51] have used the Tesla-valve geometry, instead of the diffuser section, and the reference presents a detailed discussion of the design parameters and operational characteristics of their fixed-valve micropumps.

Electric/Magnetic-Field-Driven Micropumps

An electric field can be used to directly pump liquids in microchannels using such techniques as electro-osmosis

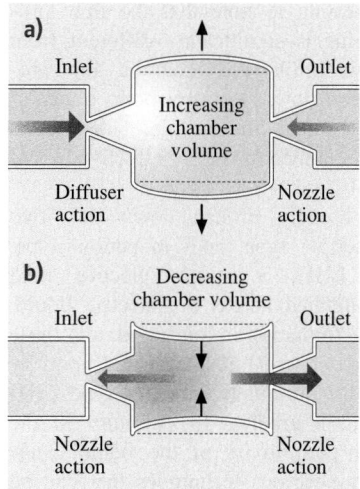

Fig. 19.10a,b Operating principle of a diffuser micropump: (a) suction mode and (b) pumping mode

(EO), electrohydrodynamic (EHD) pumping, magneto-hydrodynamic (MHD) pumping, etc. These pumping techniques rely on creating an attractive force for some of the ions in the liquid, and the remaining liquid is dragged along to form a bulk flow.

When a liquid is introduced into a microchannel, a double-layer charge exists at the interface of the liquid and the microchannel wall. The magnitude of this charge is governed by the zeta potential of the channel–liquid pair. Figure 19.11 shows a schematic view of the electro-osmotic transport phenomenon.

As shown in Fig. 19.11, the channel wall is negatively charged, which attracts the positive ions in the solution. When a strong electric field is applied along the length of the microchannel, the ions at the interface experience an attractive force toward the cathode. As the positive ions move toward the cathode, they exert a drag force on the bulk fluid, and net fluid transport occurs from the anode to the

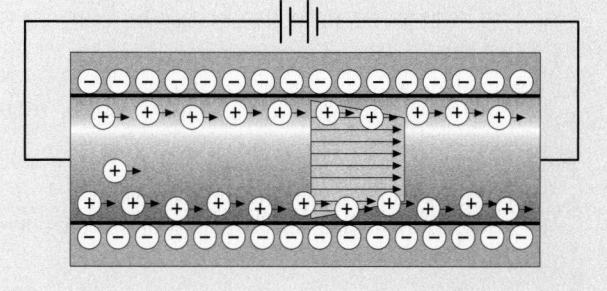

Fig. 19.11 Schematic sketch explaining the principle of electro-osmotic fluid transport

cathode. It is interesting to note that the flow profile of the liquid plug is significantly different from pressure-driven flow. Unlike the parabolic flow profile of pressure-driven flow, electro-osmotic transport leads to an almost vertical flow profile. The electro-osmotic transport phenomenon is only effective across very narrow channels.

EHD can be broadly broken down into two subcategories: injection type and induction type. In injection-type EHD, a strong electric field ($\approx 100\,\text{kV/cm}$) is applied across a dielectric liquid. This induces charge formation in the liquid, and these induced (or injected) charges are then acted upon by the electrical field for pumping. Induction-type EHD relies on generating a gradient/discontinuity in the conductivity and/or permittivity of the liquid. *Fuhr* et al. [19.52] explain various techniques that can be used to generate gradients for non-injection-type EHD pumping.

MHD pumps rely on creating a Lorentz force on the liquid particles in the presence of an externally applied electric field. MHD has been demonstrated using both direct-current (DC) [19.53] and alternating-current (AC) [19.54] excitations. MHD, EHD, and electro-osmotic transport share one feature in common that makes them very appealing for microfluidic systems, namely that none of them requires microvalves to regulate the flow. This makes these pumping techniques very reliable, as there is no concern about wear and tear on the microvalves, or any other moving parts of the micropump. However, it has been difficult to implement these actuation schemes fully on the microscale, owing to the high voltages, electromagnets, etc. required for these actuation schemes.

19.3 Smart Passive Microfluidic Devices

Passive microfluidics is a powerful technique for the rapidly evolving discipline of BioMEMS. It is a fluid control topology in which the physical configuration of the microfabricated system primarily determines the functional characteristics of the device/system. Typically, passive microfluidic devices do not require an external power source, and the control exerted by the devices is based, in part, on energy drawn from the working fluid, or based purely on surface effects, such as surface tension, selective hydrophobic/hydrophilic control, etc. Most passive microfluidic devices exploit various physical properties such as shape, contact angle, and flow characteristics to achieve the desired function. Passive microfluidic systems are usually easier to implement and allow for a simple microfluidic system with little or no control circuitry. A further list of advantages and disadvantages of passive microfluidic systems (or devices) is considered toward the end of this section.

Passive microfluidic devices can be categorized based on:

- Function: microvalves, micromixers, filters, reactors, etc.
- Fluidic medium: gas or liquid
- Application: biological, chemical, or other
- Substrate material: silicon, glass, polysilicon, polymer, or others

In this section, we will study various passive microfluidic devices that are categorized based on their function. Passive microfluidic devices include, but are not limited to, microvalves, micromixers, filters, dispensers, etc. [19.24].

19.3.1 Passive Microvalves

Passive microvalves have been a subject of great interest ever since the inception of the lab-on-a-chip concept. Microvalves are a key component of any microfluidic system and are essential for fluidic sequencing operations. Since most chemical and biochemical reactions require about five to six reaction steps, passive microvalves with limited functionality are ideally suited for such simple tasks. Passive microvalves can be broadly categorized as follows:

- Silicon/polysilicon or polymer-based check valves
- Passive valves based on surface tension effects
- Hydrogel-based biomimetic valves

Passive Check Valves
Shoji et al. [19.23] provide an excellent review of check-type passive microvalves. Some of the valves, shown in Fig. 19.12, illustrate the various techniques that can be used to fabricate check valves.

Figure 19.12a shows a microvalve fabricated using silicon bulk-etching techniques. A through-hole (pyramidal cavity) is etched through a silicon wafer that is sandwiched between two glass wafers. The normally closed valve is held in position by the spring effect of the silicon membrane. Upon applying pressure to the lower fluidic port, the membrane deflects upwards, allowing fluid flow through the check valve. The same working principle is employed by the microvalve shown in Fig. 19.12b. However, instead of using a membrane supported on all sides, a cantilever structure is used for the flap. This reduces the burst pressure, i. e., the minimum pressure required to open the microvalve. Figure 19.12c shows that the membrane structure can also be realized using a polysilicon layer deposited on a bulk-etched silicon wafer. Polysilicon processes typically allow tighter control over dimensions and, consequently, offer more repeatable operating characteristics. Figure 19.12d shows the simplest type of check valve where the V-groove etched in a bulk silicon substrate acts as a check valve. However, the low contact area between the flaps of the microvalve leads to nontrivial leakage rates in the forward direction. Figures 19.12e and f show check-valve designs that are realized using polymer/metal films, in addition to the traditional glass/silicon platform. This technique offers a significant advantage in terms of biocompatibility characteristics and controllable operating characteristics. The surface properties of polymers can be easily tailored using a wide variety of techniques such as plasma treatment and surface adsorption [19.55]. Thus, in applications for which the biocompatibility requirements are very stringent, it is preferable to have polymers as the fluid-contacting material. Furthermore, polymer properties such as stiffness can also be controlled in some cases based on the composition and/or processing conditions. Thus, it may be possible to fabricate microvalves with different burst pressures by using different processing conditions for the same polymer.

In addition to the passive check valves reviewed by *Shoji* et al. [19.23], other designs include check valves using composite titanium/polyimide membranes [19.20, 21], polymeric membranes such as Mylar or KAPTON [19.56], or PDMS [19.40], and metallic membranes such as [19.22].

Terray et al. [19.57] present an interesting approach to fabricating ultra-small passive valve structures. They have demonstrated a technique to polymerize colloidal particles into linear structures using an optical trap to form microscale particulate valves.

Passive Valves Based on Surface Tension Effects

The passive valves listed in the previous section use the forced motion of the membrane or flap to control the flow of fluids. These valves are prone to such prob-

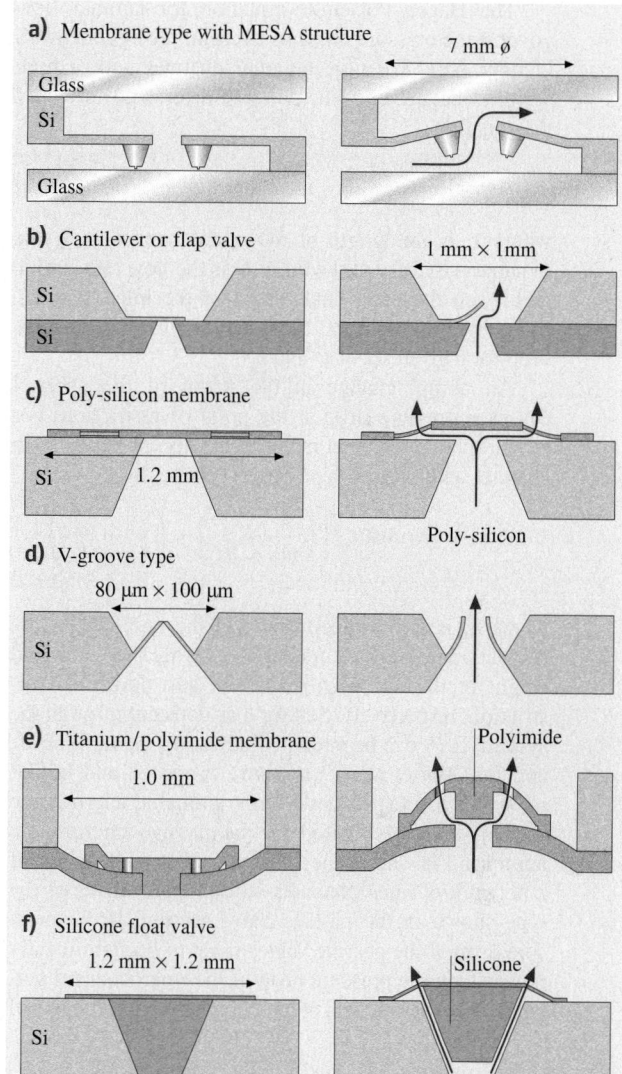

Fig. 19.12a–f Various types of passive microvalve designs: (**a**) membrane type with a mesa structure; (**b**) cantilever or flap valve; (**c**) polysilicon membrane; (**d**) V-groove type; (**e**) titanium/polyimide membrane; and (**f**) silicone float valve. Adapted from [19.23]. Reproduced with permission from IOP Publishing Limited

lems as clogging and mechanical wear and tear. Passive valves based on surface tension effects, on the other hand, have no moving parts and control the fluid motion based on their physical structure and the surface property of the substrate. Figure 19.13 shows a schematic sketch of a passive microvalve on a hydrophobic substrate [19.19].

Part B | 19.3

The Hagen–Poiseuille equation for laminar flow governs the pressure drops in microfluidic systems with laminar flow. For a rectangular channel with a high width-to-height ratio, the pressure drop is governed by the equation

$$\Delta P = \frac{12 L \mu Q}{w h^3} \, , \qquad (19.1)$$

where L is the length of the microchannel, μ is the dynamic viscosity of the fluid, Q is the flow rate, and w and h are the width and height of the microchannel, respectively. Varying L or Q can control the pressure drop for a given set of w and h.

An abrupt change in the width of the channel causes a pressure drop at the point of restriction. For a hydrophobic channel material, an abrupt decrease in channel width causes a positive pressure drop:

$$\Delta P_2 = 2\sigma_1 \cos(\theta_c) \left[\left(\frac{1}{w_1} + \frac{1}{h_1} \right) - \left(\frac{1}{w_2} + \frac{1}{h_2} \right) \right] , \qquad (19.2)$$

where σ_1 is the surface tension of the liquid, θ_c is the contact angle, and w_1, h_1 and w_2, h_2 are the width and height of the two sections before and during the restriction, respectively. Setting h as constant through the system, ΔP_2 can be varied by adjusting the ratio of w_1 and w_2. *Ahn* et al. [19.19] have proposed and implemented a novel structurally programmable microfluidic system (sPROMs) based on the passive microfluidic approach. In short, the sPROMs system consists of a network of microchannels with passive valves of the type shown in the passive valve section. If the pressure drop of the passive valves is set to be significantly higher than the pressure drop of the microchannel network, then the position of the liquid in the microchannel

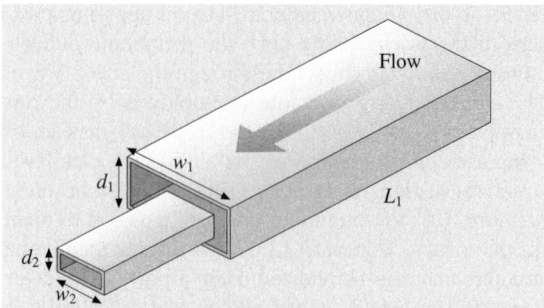

Fig. 19.13 Structure of a passive microvalve based on surface tension effects (adapted from [19.19])

network can be controlled accurately. By applying sequentially higher pressure pulses, the liquid is forced to move from one passive valve to another. Thus, the movement of the fluid within the microfluidic channels is *programmed* using the physical structure of the microfluidic system, and this forms the basic idea behind sPROMs.

The abrupt transition from a wide channel to a narrow channel can also be affected along the height of the microchannel. Furthermore, the passive valve geometry shown in Fig. 19.13 is not exclusive. *Puntambekar* et al. [19.58] have demonstrated different geometries of passive valves, as shown in Fig. 19.14a. As shown in Fig. 19.14b, the various geometries of the passive valves in Fig. 19.14a can act as effective passive valves without having an abrupt transition. This is important in order to avoid the dead volume that is commonly encountered across an abrupt step junction.

The use of surface tension to control the operation of passive valves is not limited to hydrophobic substrates. *Madou* et al. [19.59] demonstrate a capillarity-driven

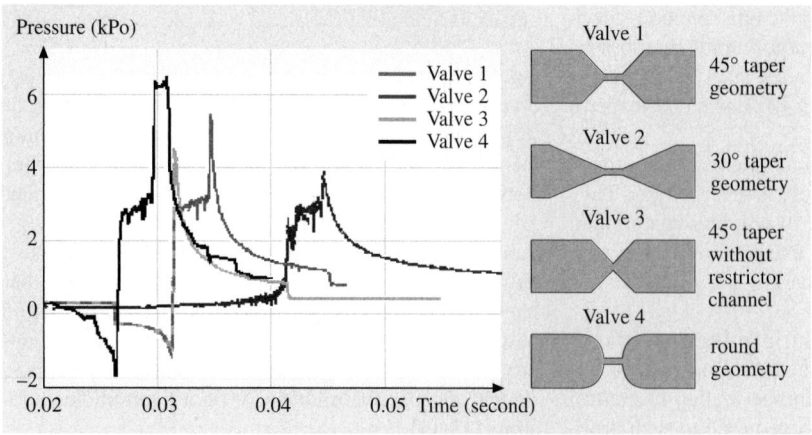

Fig. 19.14 Analysis of different geometries of passive valves. After [19.58], reproduced with permission from The Royal Society of Chemistry

stop valve on a hydrophilic substrate. On a hydrophilic substrate, the fluid can easily *wick through* in the narrow region. However, at the abrupt transition to a larger channel section, the surface tension effects will not allow the fluid to leave the narrow channel. Thus, in this case, the fluid is held at the transition from the narrow capillary to the wide outlet channel.

Another mechanism to implement passive valves is the use of hydrophobic patches on a normally hydrophilic channel. *Handique* et al. [19.60] demonstrate the use of this technique to implement passive valves for a DNA analysis system. The fluid is sucked into the microfluidic channels via capillary suction force. The hydrophobic patch exerts a negative capillary pressure that stops further flow of the fluid. The use of hydrophobic patches as passive valves is reported by *Andersson* et al. [19.61].

Other Passive Microvalves

A novel approach to realizing passive valves that are responsive to their surrounding environment is demonstrated by *Low* et al. [19.62] and *Yu* et al. [19.63]. *Yu* et al. [19.63] have developed customized polymer *cocktails* that are polymerized in situ around prefabricated posts. The specialized polymer is selectively responsive to stimuli such as pH, temperature, electric fields, light, carbohydrates, and antigens.

Another interesting approach in developing passive valves has been adopted by *Foster* et al. [19.51]. They have developed the so-called no-moving-part (NMP) valves, which are based on a physical configuration that allows a higher flow rate along one direction compared to the reverse direction.

19.3.2 Passive Micromixers

The successful implementation of microfluidic systems for many lab-on-a-chip systems is partly owing to the significant reduction in volumes handled by such systems. This reduction in volume is made possible by the use of microfabricated features and channel dimensions ranging from a few μm to several hundred μm. Despite the advantage offered by the micron-sized channels, one of the significant challenges has been the implementation of effective microfluidic mixers on the microscale. Mixing on the macroscale is a turbulent-flow-regime process. However, on the microscale, because of the low Reynolds numbers as a result of the small channel dimensions, most flow streams are laminar in nature, which does not allow for efficient mixing. On the other hand, diffusion is an im-

portant factor in mixing because of the short diffusion lengths.

There have been numerous attempts to realize micromixers using both active and passive techniques. Active micromixers rely on creating localized turbulence to enhance the mixing process, whereas passive micromixers usually enhance the diffusion process. The diffusion process can be modeled by the following equation:

$$\tau = \frac{d^2}{D} \,, \tag{19.3}$$

where τ is the mixing time, d is the distance, and D is the diffusion coefficient. Equation (19.3) illustrates the diffusion-dominated mixing at the microscale. Because of the small diffusion lengths (d), the mixing times can be made very short. The simplest category of micromixers is illustrated in Fig. 19.15. In these mixers, creating a convoluted path increases the path length that the two fluids share, leading to higher diffusion and more complete mixing. However, these mixers only exhibit good mixing performance at low flow rates, in the range of a few μl/min.

Mitchell et al. [19.64] have demonstrated three-dimensional micromixers that can achieve better performance by alternately laminating the two fluid streams to be mixed. *Beebe* et al. [19.65] have created a chaotic mixer that has the convoluted channel along three dimensions. This micromixer works on the principle of forced advection resulting from repeated turns in three dimensions. Furthermore, at each turn, eddies are generated because of the difference in flow veloci-

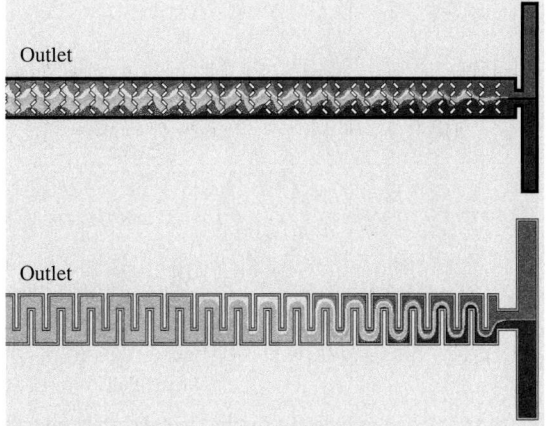

Fig. 19.15 Diffusion-enhanced mixers based on a long, convoluted flow path

ties along the inner and outer radii, which enhances the mixing.

Stroock et al. [19.68] demonstrate a passive micromixer that uses chaotic mixing by superimposing a transverse flow component on the axial flow. Ridges are fabricated at the bottom of microchannels. The flow resistance is lower along the ridges (peak/valley) and higher in the axial direction. This generates a helical flow pattern that is superimposed on the laminar flow. The demonstrated mixer shows good mixing performance over a wide range of flow velocities.

Hong et al. [19.66] have demonstrated a passive micromixer based on the Coanda effect. Their design uses the effects of diffusion mixing at low flow velocities; at high flow velocities, a convective component is added perpendicular to the flow direction, allowing for rapid mixing. This mixer shows excellent mixing performances across a wide range of flow rates because of the dual mixing effects. Figure 19.16 shows a schematic sketch of the mixer structure.

The mixer works on the principle of superimposing a parabolic flow profile in a direction perpendicular to the flow direction. The parabolic profile creates a Taylor dispersion pattern across the cross section of the flow path. The dispersion is directly proportional to the flow velocity, and higher flow rates generate more dispersion mixing.

Brody et al. [19.67] have used the laminar flow characteristics in a microchannel to develop a diffusion-based extractor. When two fluid streams, where fluid 1 is

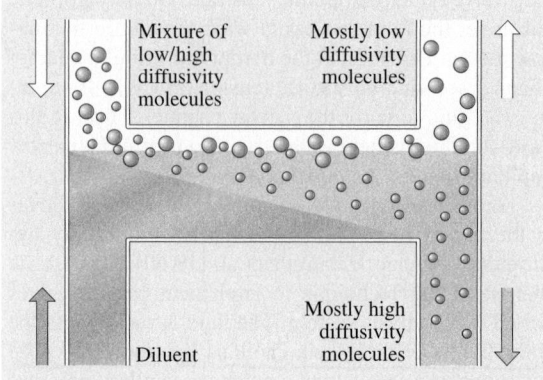

Fig. 19.17 Conceptual illustration of the H-sensor. Adapted from [19.67]

loaded with particles of different diffusivity and fluid 2 is a diluent, are forced to flow together in a microchannel, they form two laminar streams with little mixing. If the length that the two streams are in contact is carefully adjusted, only particles with high diffusivity (usually small molecules) can diffuse across into the diluent stream, as shown in Fig. 19.17. The same idea can be extended to a T-filter. *Weigl* et al. [19.69] have demonstrated a rapid diffusion immunoassay using the T-filter.

19.3.3 Passive Microdispensers

The principle of the structurally programmable microfluidic systems (sPROMs) was introduced earlier in the passive valve section. One of the key components of the sPROMs system is the microdispenser, which is designed to accurately and repeatedly dispense fluidic volumes in the micro- to nanoliter range. This would allow the dispensing of a controlled amount of the analyte into the system that could be used for further biochemical analysis. Figure 19.18 shows a schematic sketch of the microdispenser design [19.58].

The microdispenser works on the principle of graduated volume measurement. The fluid fills up the exact fixed volume of the reservoir, and the second passive valve at the other end of the microdispenser stops further motion. When the reservoir is filled with fluid, the fluidic actuation is stopped, and, simultaneously, pneumatic actuation from the air line causes a split in the fluid column at point A (Fig. 19.18). Thus, the accuracy of the reservoir decides the accuracy of the dispensed volume. Since the device is manufactured using UV-LIGA lithography techniques, highly accurate and reproducible volumes can be defined. The

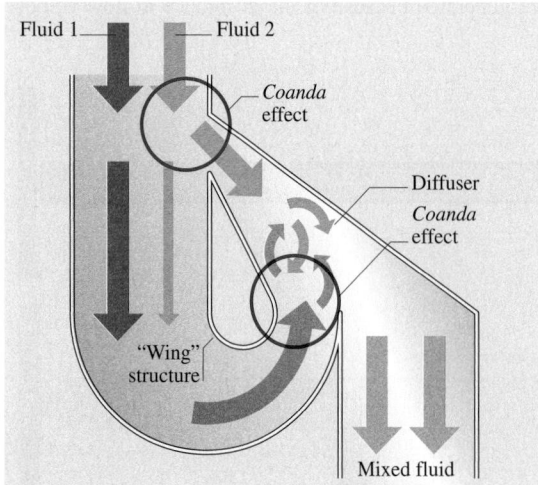

Fig. 19.16 Mixing unit design for the Coanda effect mixer. The actual mixer has mixing unit pairs in series (after [19.66])

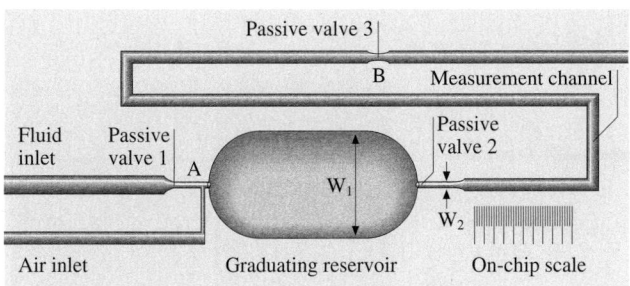

Fig. 19.18 Schematic sketch of the microdispenser. Adapted from [19.58], reproduced with permission from The Royal Society of Chemistry

precisely measured volume of fluid is expelled to the right from the reservoir. The expelled fluid then starts to fill up the measuring channel. When the fluid reaches point B (Fig. 19.18), the third microvalve holds the fluid column.

Figure 19.19 shows an actual operation sequence of the microdispenser. Figure 19.19f shows that the dispensed volume is held by passive valve 3, and at this stage, the length (and hence volume) can be calculated using the on-chip scale. In experiments only, the region in the immediate vicinity of the scale was viewed using a stereomicroscope to measure the length of the fluid column. The microdispenser demonstrated above is reported to have dispensing variation of less than 1% between multiple dispensing cycles.

19.3.4 Microfluidic Multiplexer Integrated with Passive Microdispenser

Ahn et al. [19.19] have demonstrated the sPROMs technology to be an innovative method of controlling liquid movement in a programmed fashion in a microfluidic network. By integrating this technique with the microdispensers, a more functionally useful microfluidic system can be realized. Figure 19.20a shows a schematic illustration of the microfluidic multiplexer with the integrated dispenser, and Fig. 19.20b shows an actual device fabricated using rapid prototyping techniques [19.70].

The operation of the microdispenser has been explained earlier in this section. Briefly, the fluid is loaded in the fixed-volume metering reservoir via a syringe pump. The fluid is locked in the reservoir by the pas-

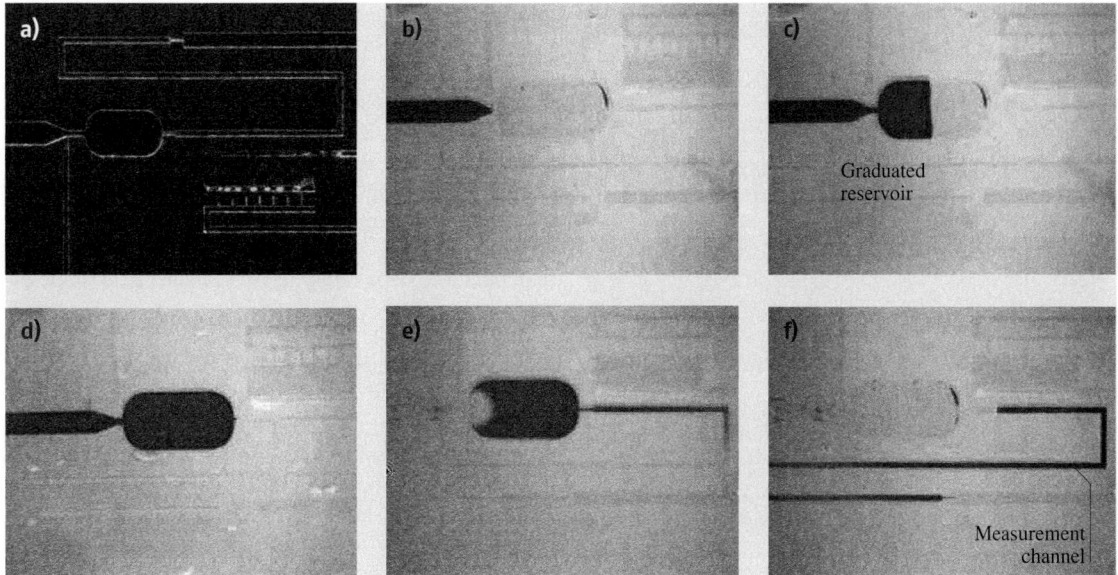

Fig. 19.19a–f Microphotographs of the microdispenser sequence: (**a**) fabricated device; (**b**) fluid at reservoir inlet; (**c**) reservoir filling; (**d**) reservoir filled; (**e**) split in liquid column due to pneumatic actuation; and (**f**) fluid ejected to measurement channel and locked in by passive valve. After [19.58], reproduced with permission from The Royal Society of Chemistry

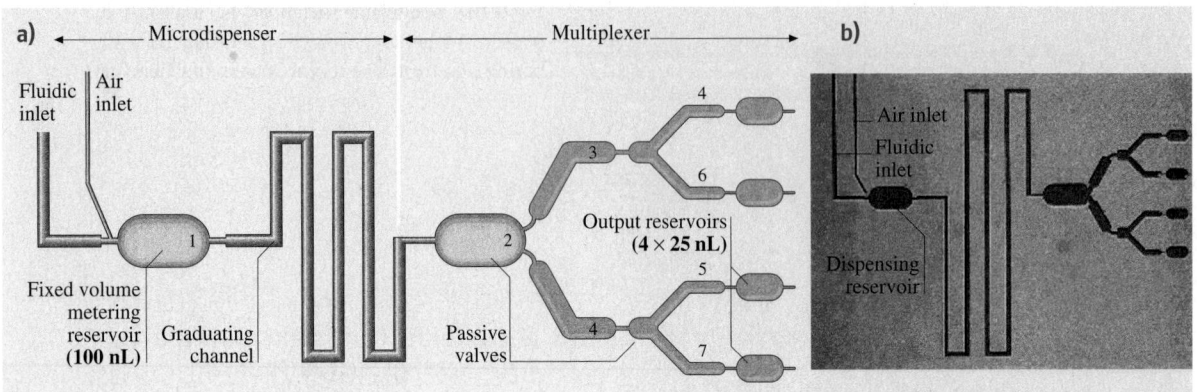

Fig. 19.20a,b Microfluidic multiplexer with integrated dispenser: (**a**) schematic sketch and (**b**) fabricated device filled with dye (after [19.58])

sive valve at the outlet of the reservoir. When a higher pressure is applied via the air inlet line, the liquid column is split and the fluid is dispensed into the graduating channel.

The microfluidic multiplexer is designed to have a programmed delivery sequence, as shown in Fig. 19.20a, where the numbers on each branch of the multiplexer indicates the filling sequence. This sequential filling is achieved by using different ratios of passive valves along the multiplexer section. For instance, at the first branching point, the passive valve at the upper branch offers less resistance than the passive valve at the beginning of the lower branch. Thus, the dispensed fluid will first fill the top branch. After filling the top branch, the liquid encounters another passive valve at the end of the top branch. The pressure needed to push beyond this valve is higher than the pressure needed to push liquid into the lower branch of the first split. The liquid will then fill the lower branch. This sequence of nonsymmetrical passive valves continues along all the branches of the multiplexer, as shown in Fig. 19.21.

The ability to sequentially divide and deliver liquid volumes was demonstrated for the first time using a passive microfluidic system. This approach has the potential to deliver very simple microfluidic control systems that are capable of a number of sequential microfluidic manipulation steps required in a biochemical analysis system.

19.3.5 Passive Micropumps

A passive system is defined inherently as one that does not require an external energy source. Thus, the term passive micropump might seem a misnomer upon initial inspection. However, there have been some efforts dedicated to realizing a passive micropump that essentially does not draw energy from an external source, but stores the required actuation energy in some form and converts it to mechanical energy on demand.

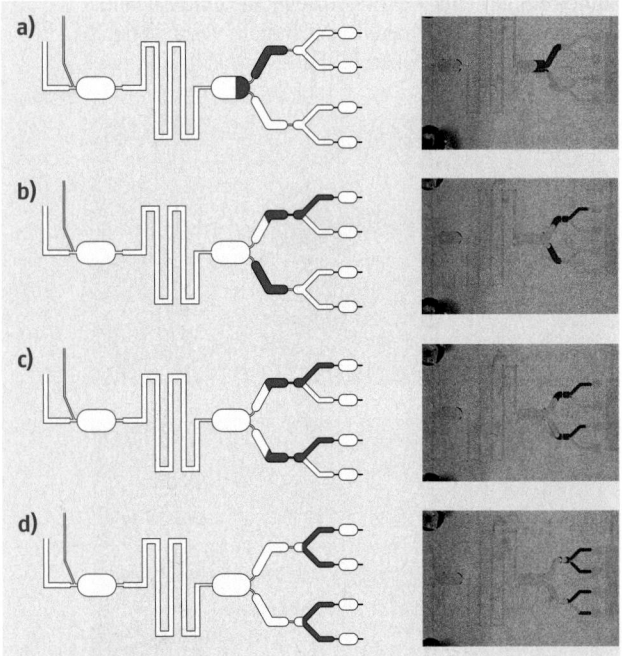

Fig. 19.21a–d Microphotographs showing operation of sequential multiplexer: (**a**) first-level division; (**b**) second-level division; (**c**) continued multiplexing; and (**d**) end of sequential multiplexing sequence (after [19.58])

Passive Micropumps Based on Osmotic Pressure

Nagakura et al. [19.71] have demonstrated a mesoscale osmotic actuator that converts chemical energy to mechanical displacement. Osmosis is a well-known phenomenon by which liquid is transported across a semipermeable membrane to achieve a uniform concentration distribution across the membrane. If the membrane is flexible, such as the one used by *Nagakura* et al. [19.71], then the transfer of liquid would cause the membrane to deform and act as an actuator. The inherent drawback of using osmosis as an actuation mechanism is that it is a very slow process: typical response times (on a macroscale) are on the order of several hours. However, osmotic transport scales favorably to the microscale, and it is expected that these devices will have response times on the order of several minutes, rather than hours. Based on this idea, *Nagakura* et al. [19.71] are developing a miniature insulin pump. *Su* et al. [19.72] have demonstrated a microscale osmotic actuator that is capable of developing pressures as high as 35 MPa. This is still a relatively unexplored realm in BioMEMS actuation, and it has good potential for applications such as sustained drug delivery.

Passive Pumping Based on Surface Tension

Walker and *Beebe* [19.73] have demonstrated pumping action using the difference between the surface tension pressure at the inlet and outlet of a microfluidic channel. In the simplest case, a small drop of a fluid is placed at one end of a straight microchannel, and a much larger drop of fluid is placed at the opposite end of the microchannel. The pressure within the small drop is significantly higher than the pressure within the large drop, due to the difference in the surface tension effects across the two drops. Consequently, the liquid will flow from the small drop and add to the larger drop. The flow rate can be varied by changing various parameters such as the volume of the pumping drop, the surface free energy of the liquid, or the resistance of the microchannel, etc. This pumping scheme is very easy to realize and can be used for a wide variety of fluids.

Evaporation-Based Continuous Micropumps

Effenhauser et al. [19.74] have demonstrated a continuous-flow micropump based on a controlled evaporation approach. Their concept is based on the controlled evaporation of a liquid through a membrane into a gas reservoir. The reservoir contains a suitable adsorption agent that draws out the liquid vapors and maintains a low vapor pressure conducive to further evaporation. If the liquid being pumped is replenished from a reservoir, capillary forces will ensure that the fluid is continuously pumped through the microchannels as it evaporates at the other end into the adsorption reservoir. Though the pump suffers from inherent disadvantages such as strong temperature dependence and operation only in suction mode, it offers a very simple technique for fluidic transport.

19.3.6 Advantages and Disadvantages of the Passive Microfluidic Approach

This chapter has covered a number of different passive microfluidic devices and systems. Passive microfluidic devices have only recently been a subject of considerable research effort. One of the reasons for this interest is the long list of advantages that passive microfluidic devices offer. However, since most microfluidic devices are very application-specific (and even more so for passive microfluidic devices), the advantages are not to be considered universally applicable for all the devices/systems. Some of the advantages that are commonly found are:

● avoiding the need for an *active* control system;
● they are usually very easy to fabricate;
● passive microfluidic systems with no moving parts are inherently more reliable because of the lack of mechanical wear and tear;
● they offer very repeatable performance once the underlying phenomena are well understood and characterized;
● they are highly suited for BioMEMS applications; they can easily handle a limited number of microfluidic manipulation sequences;
● well suited for low-cost mass production;
● their low cost offers the possibility of having disposable microfluidic systems for specific applications, such as working with blood;
● they can offer other interesting possibilities, such as biomimetic responses.

However, like all MEMS devices, passive microfluidic devices or systems are not the solution to the microfluidic handling problem. Usually they are very application-specific; they cannot be reconfigured for another application easily. Other disadvantages are listed below:

● they are suited for well-understood, niche applications for which the fluidic sequencing steps are well decided,

- they are strongly dependent on variations in the fabrication process,

- they are usually not very suitable for a wide range of fluidic mediums.

19.4 Lab-on-a-Chip for Biochemical Analysis

Recent development in MEMS (microelectromechanical systems) has brought a new and revolutionary tool in biological or chemical applications: *lab-on-a-chip*. New terminology, such as micro-total analysis systems and lab-on-a-chip, was introduced in the last decade, and several prototype systems have been reported.

The idea of lab-on-a-chip is basically to reduce biological or chemical laboratories to a microscale system, hand-held size or smaller. Lab-on-a-chip systems can be made out of silicon, glass, and polymeric materials, and the typical microfluidic channel dimensions are in the range of several tens to hundreds of μm.Liquid samples or reagents can be transported through the microchannels from reservoirs to reactors using electrokinetic, magnetic, or hydrodynamic pumping methods. Fluidic motion or biochemical reactions can also be monitored using various sensors, which are often used for biochemical detection of products.

There are many advantages to using lab-on-a-chip over conventional chemical or biological laboratories. One of the important advantages lies in its low cost. Many reagents and chemicals used in biological and chemical reactions are expensive, so the prospect of using very small amounts (in the micro- to nanoliter ranges) of reagents and chemicals for an application is very appealing. Another advantage is that lab-on-a-chip requires very small amounts of reagents/chemicals (which enables rapid mixing and reaction) because biochemical reaction is mainly involved in the diffusion of two chemical or biological reagents, and microscale fluidics reduces diffusion time as it increases reaction probabilities. In practical terms, reaction products can be produced in a matter of seconds/minutes, whereas laboratory-scale reactions can take hours, or even days. In addition, lab-on-a-chip systems minimize harmful by-products since their volume is so small. Complex reactions with many reagents could happen on a lab-on-a-chip, ultimately with potential in DNA analysis, biochemical warfare-agent detection, biological cell/molecule sorting, blood analysis, drug screening/development, combinatorial chemistry, and protein analysis. In this section, two recent developments of microfluidic systems for lab-on-a-chip applications will be introduced: (a) a magnetic micro/nano-bead-based bio-

chemical detection system and (b) a disposable smart lab-on-a-chip for blood analysis.

19.4.1 Magnetic Micro/Nano-Bead-Based Biochemical Detection System

In the past few years, a large number of microfluidic prototype devices and systems have been developed, specifically for biochemical warfare detection systems and portable diagnostic applications. The BioMEMS team at the University of Cincinnati has been working on the development of a remotely accessible generic microfluidic system for biochemical detection and biomedical analysis, based on the concepts of surface-mountable microfluidic motherboards, sandwich immunoassays, and electrochemical detection techniques [19.75, 76]. The limited goal of this work is to develop a generic MEMS-based microfluidic system and to apply the fluidic system to detect biomolecules, such as specific proteins and/or antigens, in liquid samples. Figure 19.22 illustrates the schematic diagram of a generic microfluidic system for biochemical detection using a magnetic-bead approach for both sampling and manipulating the target biomolecules [19.77, 78].

The analytical concept is based on sandwich immunoassay and electrochemical detection [19.79], as illustrated in Fig. 19.23. Magnetic beads are used as both substrates for the antibodies and carriers for the target antigens. A simple concept of magnetic-bead-based biosampling with an electromagnet for the case of sandwich immunoassay is shown in Fig. 19.24.

Antibody-coated beads are introduced on the electromagnet and separated by applying magnetic fields. While holding the antibody-coated beads, antigens are injected into the channel. Only target antigens are immobilized and, thus, separated on the magnetic bead surface due to antibody/antigen reaction. Other antigens get washed out with the flow. Next, enzyme-labeled secondary antibodies are introduced and incubated, along with the immobilized antigens. The chamber is then rinsed to remove all unbound secondary antibodies. A substrate solution, which will react with the enzyme, is injected into the channel, and the electrochemical detection is performed. Finally, the magnetic beads are

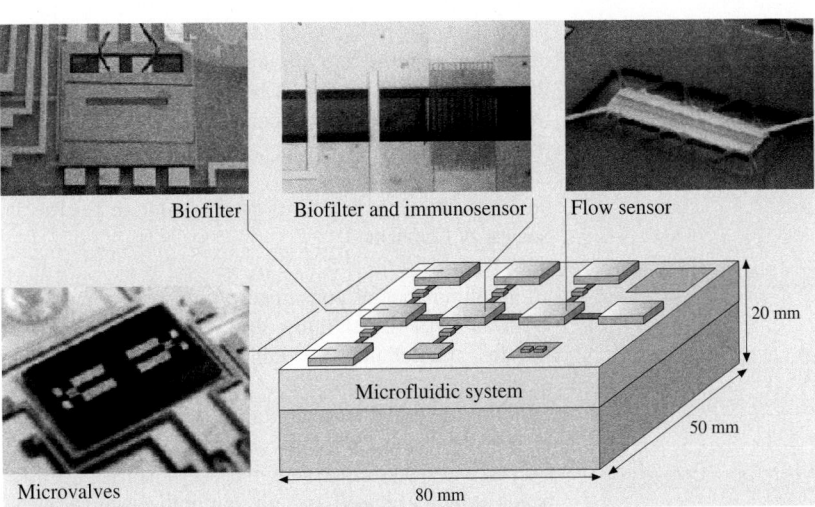

Fig. 19.22 Schematic diagram of a generic microfluidic system for biochemical detection (after [19.76])

released to the waste chamber, and the bio-separator is ready for another immunoassay. Alkaline phosphatase (AP) and p-aminophenyl phosphate (PAPP) were chosen as the enzyme and electrochemical substrate, respectively. Alkaline phosphatase makes PAPP turn into its electrochemical product, p-aminophenol (PAP). By applying a potential, PAP gives up electrons and turns into 4-quinoneimine (4QI), which is the oxidant form of PAP.

For a successful immunoassay, the biofilter [19.77] and the immunosensor were fabricated separately and integrated together. The integrated biofilter and immunosensor were surface-mounted using a fluoropolymer bonding technique [19.80] on a microfluidic motherboard, which contains microchannels fabricated using the glass-etching and glass-to-glass direct-bonding technique. Each the inlet and outlet were connected to sample reservoirs through custom-designed microvalves. Figure 19.25 shows the integrated microfluidic biochemical detection system for the magnetic-bead-based immunoassay.

After a fluidic sequencing test, full immunoassays were performed in the integrated microfluidic system to prove magnetic-bead-based biochemical detection and sampling function. Magnetic beads (Dynabeads® M-280, Dynal Biotech Inc.) coated with biotinylated sheep anti-mouse immunoglobulin G (IgG) were injected into the reaction chamber and separated on the surface of the biofilter by applying magnetic fields. While holding the magnetic beads, antigen (mouse IgG) was injected into the chamber and incubated. Then secondary antibody with label (rat anti-mouse IgG conjugated alkaline phosphatase) and electrochemical substrate (PAPP) to alkaline phosphatase were sequentially injected and incubated to ensure production of PAP. Electrochemical detection using an amperometric time-based detection method was performed during incubation. After detection, magnetic beads with all the reagents were washed away, and the system was ready for another immunoassay. This sequence was repeated for every new immunoassay. The flow rate was set to 20 μl/min in every step. After calibration of the electrochemical immunosensor, full immunoassays were performed following the sequence stated above for different antigen concentrations: 50, 75, 100, 250, and 500 ng/ml. Concentration of the primary antibody-coated magnetic beads and conjugated secondary antibody was 1.02×10^7 beads/ml and 0.7 μg/ml, respectively. Immunoassay results for different antigen concentrations are shown in Fig. 19.26.

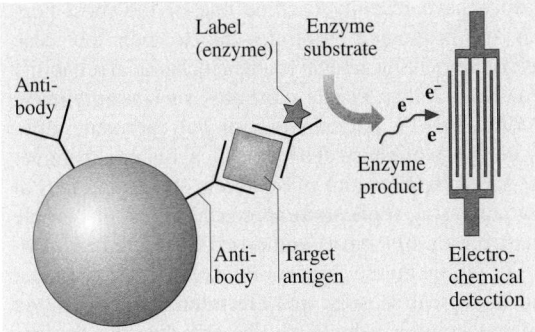

Fig. 19.23 Analytical concept based on sandwich immunoassay and electrochemical detection. (After [19.78], reproduced with permission from The Royal Society of Chemistry)

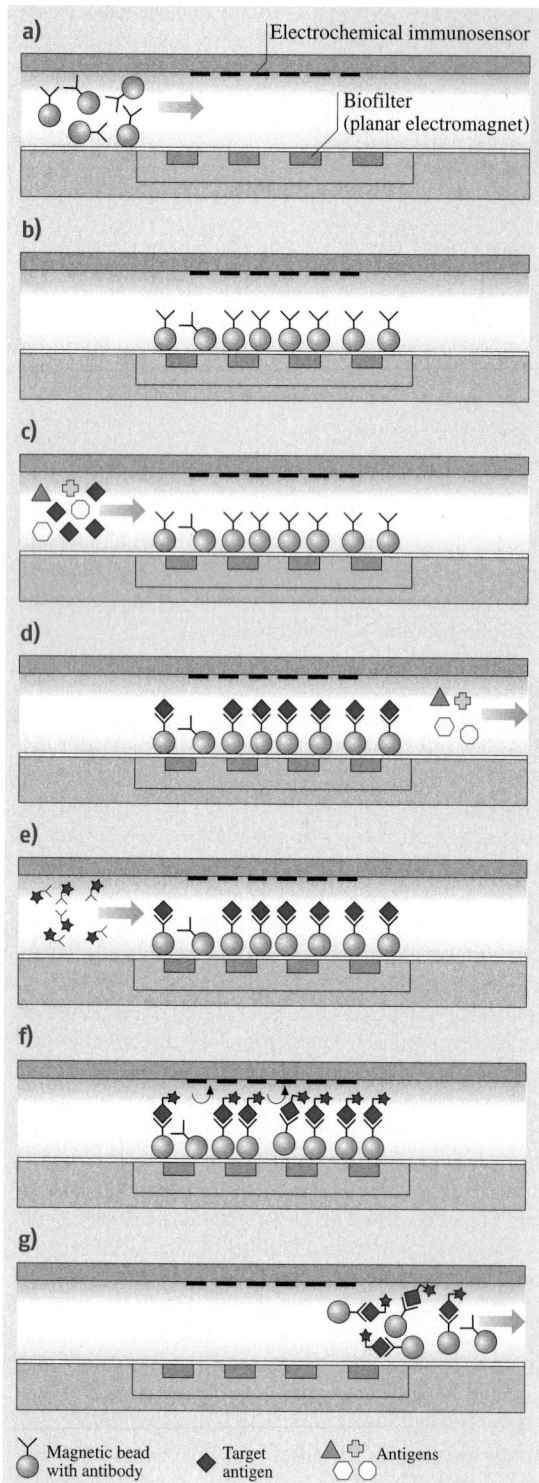

a)
Electrochemical immunosensor
Biofilter (planar electromagnet)

b)

c)

d)

e)

f)

g)

Y⬢ Magnetic bead with antibody   ◆ Target antigen   △✛⬡◯ Antigens

Fig. 19.24a–g Conceptual illustration of biosampling and immunoassay procedure: (**a**) injection of magnetic beads; (**b**) separation and holding of beads; (**c**) flowing samples; (**d**) immobilization of target antigen; (**e**) flowing labeled antibody; (**f**) electrochemical detection; and (**g**) washing out magnetic beads and ready for another immunoassay (after [19.78], reproduced with permission from The Royal Society of Chemistry)

Immunoreactant consumed during one immunoassay was $10\,\mu l$ ($20\,\mu l/min \times 30\,s$), and total assay time was less than 20 min, including all incubation and detection steps.

The integrated microfluidic biochemical detection system has been successfully developed and fully tested for fast and low-volume immunoassays using magnetic beads, which are used as both immobilization surfaces and biomolecule carriers. Magnetic-bead-based immunoassay, as a typical example of biochemical detection and analysis, has been performed on the integrated microfluidic biochemical analysis system that includes a surface-mounted biofilter and immunosensor on a glass microfluidic motherboard. Protein-sampling capability has been demonstrated by capturing target antigens.

The methodology and system can also be applied to generic biomolecule detection and analysis systems by replacing the antibody/antigen with appropriate bioreceptors/reagents, such as DNA fragments or oligonucleotides, for application to DNA analysis and/or high-throughput protein analysis.

19.4.2 Disposable Smart Lab-on-a-Chip for Blood Analysis

One of several substrates available for biofluidic chips, plastics have recently become one of the most popular and promising substrates due to their low cost, ease of fabrication, and favorable biochemical reliability and compatibility. Plastic substrates, such as polyimide, PMMA, PDMS, polyethylene, or polycarbonate, offer a wide range of physical and chemical material parameters for the applications of biofluidic chips, generally at low cost using replication approaches. The disposable smart plastic biochip is composed of integrated modules of plastic fluidic chips for fluid regulation, chemical and biological sensors, and electronic controllers. As a demonstration vehicle, the biochip has the specific goal of detecting and identifying three metabolic parameters such as PO_2 (partial pressure of oxygen), lactate, and glucose from a blood sample. The schematic concept of the cartridge-type disposable lab-on-a-chip for

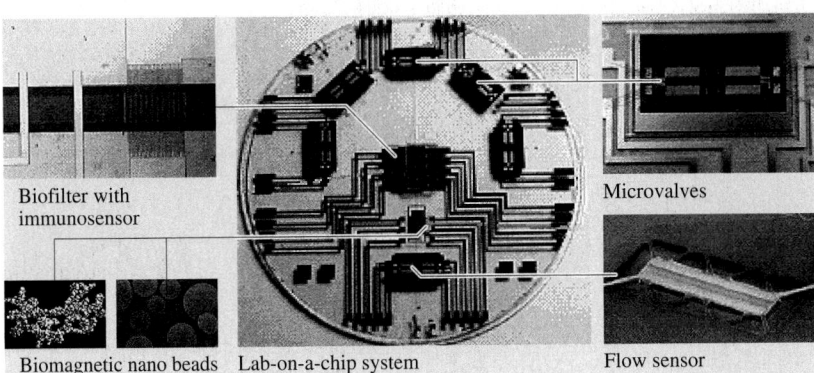

Fig. 19.25 Photograph of the fabricated lab-on-a-chip for magnetic-bead-based immunoassay

Biofilter with immunosensor

Microvalves

Biomagnetic nano beads Lab-on-a-chip system

Flow sensor

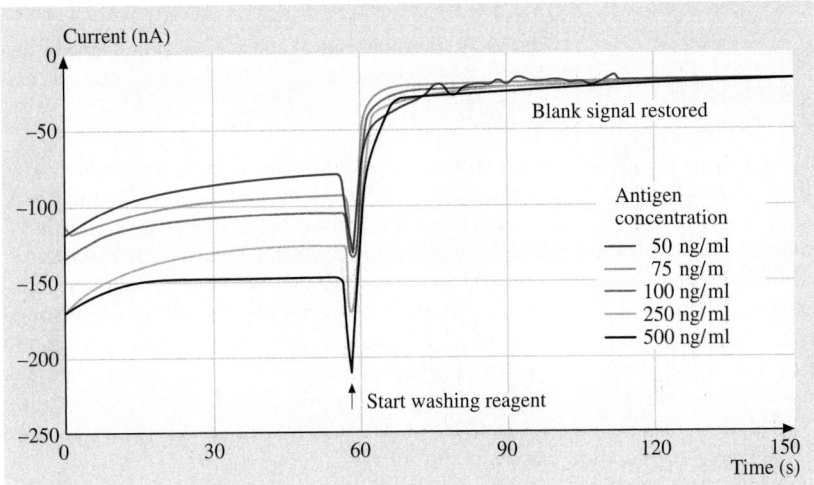

Fig. 19.26 Immunoassay results measured by amperometric time-based detection method (after [19.78], reproduced with permission from The Royal Society of Chemistry)

blood analysis is illustrated in Fig. 19.27. The disposable lab-on-a-chip cartridge has been fabricated using plastic micro-injection molding and plastic-to-plastic direct-bonding techniques. The biochip cartridge consists of a fixed-volume microdispenser based on the structurally programmable microfluidic system (sPROMs) technique [19.70], an air-bursting on-chip pressure source [19.81], and electrochemical biosensors [19.82].

A passive microfluidic dispenser measures exact amounts of sample to be analyzed, and then the air-bursting on-chip power source is detonated to push the graduated sample fluid from the dispenser reservoir. Upon air-bursting, the graduated sample fluid travels through the microfluidic channel into sensing reservoirs, under which the biosensor array is located, as shown in Fig. 19.28.

An array of disposable biosensors consisting of an oxygen sensor, a glucose sensor and a lactate sensor has been fabricated using screen-printing technology [19.82]. Measurements from the developed biosensor array can be done based on tiny amounts of sample (as low as 100 nl). One of the most fundamental sensor designs is the oxygen sensor, which is the basic sensing structure for many other metabolic products such as glucose and lactate. The principle of the oxygen sensor is based on amperometric detection. Figure 19.29 shows a schematic representation of an oxygen sensor. When the diffusion profile for oxygen from the sample to the electrode surface is saturated, a constant oxygen gradient profile is generated. Under these circumstances the detection current is proportional only to the oxygen concentration in the sample.

The gel-based electrolyte is essential for the ion-exchange reactions at the anode of the electrochemical pair. The oxygen semipermeable membrane ensures that mainly oxygen molecules permeate through this layer and that the electrochemical cell is not exposed to other ions. A silicone layer was spin-coated and utilized as

Part B | 19.4

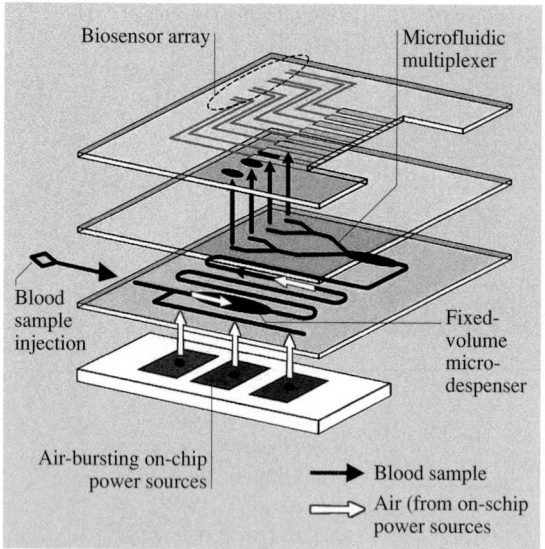

Fig. 19.27 Schematic illustration of smart and disposable plastic lab-on-a-chip by *Ahn* et al. (after © 2004 IEEE [19.83])

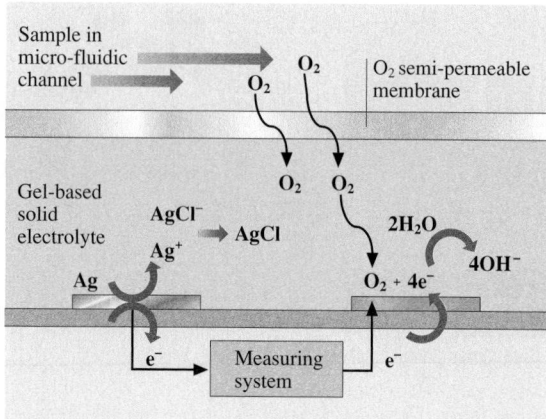

Fig. 19.29 Electrochemical and analytical principle of the developed disposable biosensor for partial oxygen concentration sensing (after, © 2002 IEEE [19.82])

an oxygen semipermeable membrane because of its high permeability and low signal-to-noise ratio. Water molecules pass through the silicone membrane and reconstitute the gel-based electrolyte so the Cl⁻ ions can move close to the anode to coalesce with Ag⁺ ions. The number of electrons in this reaction is counted by the measuring system.

For the glucose sensor, additional layers—a glucose semipermeable membrane (polyurethane) and immobilized glucose oxidase (GOD) in a polyacrylamide gel for the glucose sensor—allow direct conversion of the oxygen sensor into a glucose sensor. A similar modification is made for the lactate sensor by replacing the immobilized glucose oxidase with lactate oxidase. The glucose molecules will pass through the semipermeable layer and be oxidized immediately. The oxygen sensor will measure hydrogen peroxide, which is a by-product of glucose oxidation. The level of hydrogen peroxide

is proportional to the glucose level in the sample. The fabricated disposable plastic lab-on-a-chip cartridge was inserted into a hand-held biochip analyzer for analysis of human blood samples, as shown in Fig. 19.30. The prototype biochip analyzer consists of biosensor detection circuitry, timing/sequence circuitry for the air-bursting, on-chip power source, and a display unit. The hand-held biochip analyzer initiated the sensing sequence and displayed readings in one minute. The measured glucose and lactate levels in human blood samples are also shown in Fig. 19.31.

The development of disposable smart microfluidic-based biochips is of immediate relevance to several patient-monitoring systems, specifically for point-of-care health monitors. Since the developed biochip is a low-cost plastic-based system, we envision a disposable application for monitoring clinically significant parameters such as PO_2, glucose, lactate, hematocrit, and pH. These health indicators provide an early warning system for the detection of patient status and can also serve as markers for disease and toxicity monitoring.

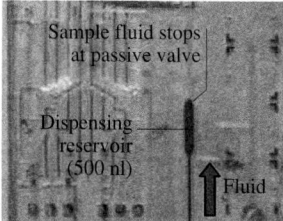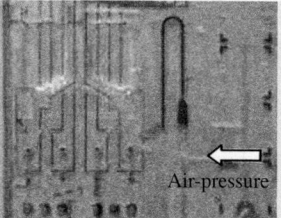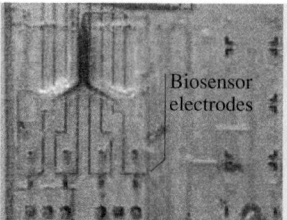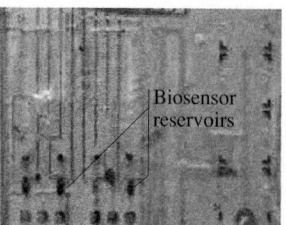

Fig. 19.28 Upon air-bursting, sample fluid travels through the microfluidic channel into the biosensor detection chamber. In sequence: loading the dispenser; dispensing; multiplexing, and delivered volume to biosensor array

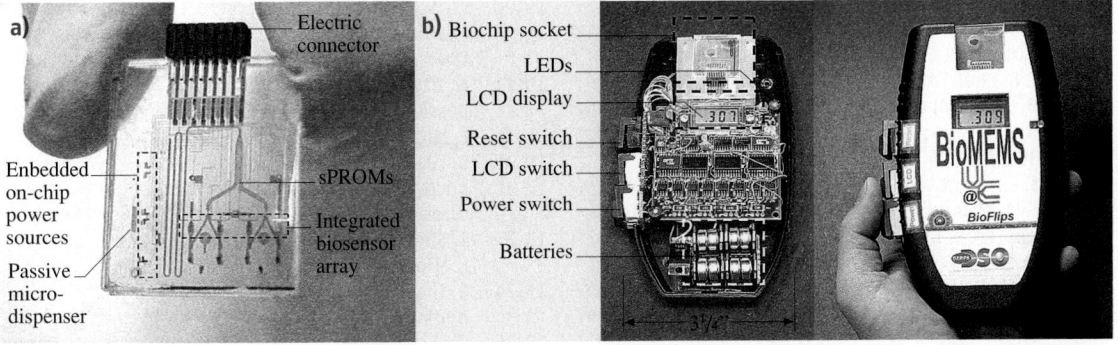

Fig. 19.30a,b Disposable biochip and hand-held analyzer: (**a**) developed smart and disposable biochip cartridge and (**b**) hand-held analyzer developed at the University of Cincinnati

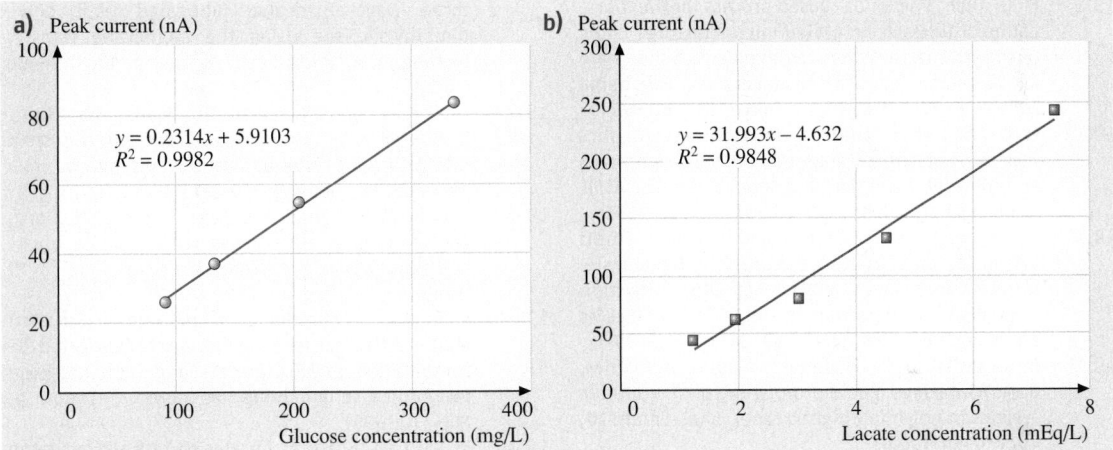

Fig. 19.31a,b Measurement results from the biochip cartridge and analyzer: (**a**) glucose level and (**b**) lactate level (after © 2004 IEEE [19.83])

References

19.1   L. Ristic, H. Hughes, F. Shemansky: Bulk micro-machining technology. In: *Sensor Technology and Devices*, ed. by L. Ristic (Artech House, Norwood 1994) pp. 49–93

19.2   M. Esashi, M. Takinami, Y. Wakabayashi, K. Minami: High-rate directional deep dry etching for bulk silicon micromachining, J. Micromech. Microeng. **5**, 5–10 (1995)

19.3   H. Jansen, M. de Boer, R. Legtenberg, M. Elwenspoek: The black silicon method: A universal method for determining the parameter setting of a fluorine-based reactive ion etcher in deep silicon trench etching with profile control, J. Micromech. Microeng. **5**, 115–120 (1995)

19.4   Z. L. Zhang, N. C. MacDonald: A RIE process for submicron silicon electromechanical structures, J. Micromech. Microeng. **2**, 31–38 (1992)

19.5   C. Linder, T. Tschan, N. F. de Rooij: Deep dry etching techniques as a new IC compatible tool for silicon micromachining, Technical Digest of the 6th International Conference on Solid-State Sensors and Actuators (Transducers '91), San Francisco June 24–27, 1991 (IEEE, Piscataway 1991) 524–527

19.6   D. J. Harrison, A. Manz, Z. Fan, H. Ludi, H. M. Widmer: Capillary electrophoresis and sample injection systems integrated on a planar glass chip, Anal. Chem. **64**, 1926–1932 (1992)

19.7   A. Manz, D. J. Harrison, E. M. J. Verpoorte, J. C. Fettinger, A. Paulus, H. Ludi, H. M. Widmer: Planar chips technology for miniaturization and integration of separation technique into monitoring systems, J. Chromatogr. **593**, 253–258 (1992)

19.8   T. R. Dietrich, W. Ehrfeld, M. Lacher, M. Krämer, B. Speit: Fabrication technologies for microsystems utilizing photoetchable glass, Microelectron. Eng. **30**, 497–504 (1996)

19.9   S. C. Jacobson, R. Hergenroder, L. B. Koutny, J. M. Ramsey: Open channel electrochromatography on a microchip, Anal. Chem. **66**, 2369–2373 (1994)

19.10   E. W. Becker, W. Ehrfeld, P. Hagmann, A. Maner, D. Munchmeyer: Fabrication of microstructures with high aspect ratios and great structural heights by synchrotron radiation lithography, galvanofoming, and plastic moulding (LIGA process), Microelectron. Eng. **4**, 35–36 (1986)

19.11   M. G. Allen: Polyimide-based process for the fabrication of thick electroplated microstructures, Proc. of the 7th International Conference on Solid-State Sensors and Actuators (Transducers '93), Yokohama June 7–10, 1993 (IEE of Jpn., Tokyo 1993) 60–65

19.12   C. H. Ahn, M. G. Allen: A fully integrated surface micromachined magnetic microactuator with a multilevel meander magnetic core, IEEE/ASME J. Microelectromech. Syst. (MEMS) **2**, 15–22 (1993)

19.13   C. S. Effenhauser, G. J. M. Bruin, A. Paulus, M. Ehrat: Integrated capillary electrophoresis on flexible silicone microdevices: Analysis of DNA restriction fragments and detection of single DNA molecules on microchips, Anal. Chem. **69**, 3451–3457 (1997)

19.14   D. C. Duffy, J. C. McDonald, O. J. A. Schueller, G. M. Whitesides: Rapid prototyping of microfluidic systems in poly(dimethylsiloxane), Anal. Chem. **70**, 4974–4984 (1998)

19.15   L. Martynova, L. Locascio, M. Gaitan, G. W. Kramer, R. G. Christensen, W. A. MacCrehan: Fabrication of plastic microfluid channels by imprinting methods, Anal. Chem. **69**, 4783–4789 (1997)

19.16   H. Becker, W. Dietz, P. Dannberg: Microfluidic manifolds by polymer hot embossing for micro-TAS applications, Proc. of Micro-Total Analysis Systems '98, Banff, Canada 1998, ed. by D. J. Harrison, A. van den Berg (Kluwer, Dordrecht 1998) 253–256

19.17   H. Becker, U. Heim: Hot embossing as a method for the fabrication of polymer high aspect ratio structures, Sens. Actuators A **83**, 130–135 (2000)

19.18   R. M. McCormick, R. J. Nelson, M. G. Alonso-Amigo, J. Benvegnu, H. H. Hooper: Microchannel electrophoretic separations of DNA in injection-molded plastic substrates, Anal. Chem. **69**, 2626–2630 (1997)

19.19   C. H. Ahn, A. Puntambekar, S. M. Lee, H. J. Cho, C.-C. Hong: Structurally programmable microfluidic systems, Proc. of Micro-Total Analysis Systems 2000, Enschede 2000, ed. by A. van den Berg, W. Olthius, P. Bergveld (Kluwer, Dordrecht 2000) 205–208

19.20   W. Schomburg, B. Scherrer: 3.5 mm thin valves in titanium membranes, J. Micromech. Microeng. **2**, 184–186 (1992)

19.21   A. Ilzofer, B. Ritter, C. Tsakmasis: Development of passive microvalves by the finite element method, J. Micromech. Microeng. **5**, 226–230 (1995)

19.22   B. Paul, T. Terharr: Comparison of two passive microvalve designs for microlamination architectures, J. Micromech. Microeng. **10**, 15–20 (2000)

19.23   S. Shoji, M. Esashi: Microflow device and systems, J. Micromech. Microeng. **4**, 157–171 (1994)

19.24   G. T. A. Kovacs: *Micromachined Transducers Sourcebook* (McGraw–Hill, New York 1998)

19.25   W. Schomburg, J. Fahrenberg, D. Maas, R. Rapp: Active valves and pumps for microfluidics, J. Micromech. Microeng. **3**, 216–218 (1993)

19.26   K. Hosokawa, R. Maeda: A pneumatically-actuated three way microvalve fabricated with polydimethylsiloxane using the membrane transfer technique, J. Micromech. Microeng. **10**, 415–420 (2000)

19.27   J. Fahrenberg, W. Bier, D. Maas, W. Menz, R. Ruprecht, W. Schomburg: A microvalve system fabricated by thermoplastic molding, J. Micromech. Microeng. **5**, 169–171 (1995)

19.28   C. Goll, W. Bacher, B. Bustgens, D. Maas, W. Menz, W. Schomburg: Microvalves with bistable buckled diaphragms, J. Micromech. Microeng **6**, 77–79 (1996)

19.29   A. Ruzzu, J. Fahrenberg, M. Heckele, T. Schaller: Multi-functional valve components fabricated by combination of LIGA process and high precision mechanical engineering, Microsyst. Technol. **4**, 128–131 (1998)

19.30   K. Sato, M. Shikida: An electrostatically actuated gas valve with S-shaped film element, J. Micromech. Microeng. **4**, 205–209 (1994)

19.31   J. Robertson, K. D. Wise: A low pressure micromachined flow modulator, Sens. Actuators A **71**, 98–106 (1998)

19.32   W. Wijngaart, H. Ask, P. Enoksson, G. Stemme: A high-stroke, high-pressure electrostatic actuator for valve applications, Sens. Actuators A **100**, 264–271 (2002)

19.33   D. J. Sadler, K. W. Oh, C. H. Ahn, S. Bhansali, H. T. Henderson: A new magnetically actuated microvalve for liquid and gas control applications, Proc. of the 6th International Conference on Solid-State Sensors and Actuators (Transducers '99), Sendai 1999 (IEE of Jpn., Tokyo 1999) 1812–1815

19.34   T. Watanabe, H. Kuwano: A microvalve matrix using piezoelectric actuators, Microsyst. Technol. **3**, 107–111 (1997)

19.35   M. Stehr, S. Messner, H. Sandmaier, R. Zangerle: The VAMP—A new device for handling liquids and gases, Sens. Actuators A **57**, 153–157 (1996)

19.36   W. Jackson, H. Tran, M. O'Brien, E. Rabinovich, G. Lopez: Rapid prototyping of active microfluidic components based on magnetically modified elastomeric materials, J. Vac. Sci. Technol. B **19**, 596 (2001)

19.37 M. Kohl, K. Skrobanek, S. Miyazaki: Development of stress-optimized shape memory microvalves, Sens. Actuators A **72**, 243–250 (1999)

19.38 C. Neagu, J. Gardeniers, M. Elwenspoek, J. Kelly: An electrochemical active valve, Electrochemica Acta **42**, 3367–3373 (1997)

19.39 K. Yoshida, M. Kikuchi, J. Park, S. Yokota: Fabrication of micro-electro-rheological valves (ER valves) by micromachining and experiments, Sens. Actuators A **95**, 227–233 (2002)

19.40 N. Jeon, D. Chiu, C. Wargo, H. Wu, I. Choi, J. Anderson, G. M. Whitesides: Design and fabrication of integrated passive valves and pumps for flexible polymer 3-dimensional microfluidic systems, Biomed. Microdevices **4**, 117–121 (2002)

19.41 M. Koch, N. Harris, A. Evans, N. White, A. Brunnschweiler: A novel micropump design with thick-film piezoelectric actuation, Meas. Sci. Technol. **8**, 49–57 (1997)

19.42 L. Cao, S. Mantell, D. Polla: Design and simulation of an implantable medical drug delivery system using microelectromechanical systems technology, Sens. Actuators A **94**, 117–125 (2001)

19.43 J. Park, K. Yoshida, S. Yokota: Resonantly driven piezoelectric micropump fabrication of a micropump having high power density, Mechatronics **9**, 687–702 (1999)

19.44 M. Koch, A. Evans, A. Brunnschweiler: The dynamic micropump driven with a screen printed PZT actuator, J. Micromech. Microeng. **8**, 119–122 (1998)

19.45 D. Xu, L. Wang, G. Ding, Y. Zhou, A. Yu, B. Cai: Characteristics and fabrication of NiTi/Si diaphragm micropump, Sens. Actuators A **93**, 87–92 (2001)

19.46 E. Makino, T. Mitsuya, T. Shibata: Fabrication of TiNi shape memory micropump, Sens. Actuators A **88**, 256–262 (2001)

19.47 H. Chou, M. Unger, S. Quake: A microfabricated rotary pump, Biomed. Microdevices **3**, 323–330 (2001)

19.48 R. Zengerle, J. Ulrich, S. Kulge, M. Richter, A. Richter: A bidirectional silicon micropump, Sens. Actuators A **50**, 81–86 (1995)

19.49 M. Carrozza, N. Croce, B. Magnani, P. Dario: A piezoelectric driven stereolithography-fabricated micropump, J. Micromech. Microeng. **5**, 177–179 (1995)

19.50 H. Andersson, W. Wijngaart, P. Nilsson, P. Enoksson, G. Stemme: A valveless diffuser micropump for microfluidic analytical systems, Sens. Actuators B **72**, 259–265 (2001)

19.51 F. Forster, R. Bardell, M. Afromowitz, N. Sharma: Design, fabrication and testing of fixed-valve micropumps, Proc. ASME Fluids Engineering Division, ASME Int. Mechanical Engineering Congress and Exposition **234**, 39–44 (1995)

19.52 G. Fuhr, T. Schnelle, B. Wagner: Travelling wave driven microfabricated electrohydrodynamic pumps for liquids, J. Micromech. Microeng. **4**, 217–226 (1994)

19.53 J. Jang, S. Lee: Theoretical and experimental study of MHD micropump, Sens. Actuators A **80**, 84–89 (2000)

19.54 A. Lemoff, A. P. Lee: An AC magnetohydrodynamic micropump, Sens. Actuators B **63**, 178–185 (2000)

19.55 J. DeSimone, G. York, J. McGrath: Synthesis and bulk, surface and microlithographic characterization of poly(1-butene sulfone)-g-poly(dimethylsiloxane), Macromolecules **24**, 5330–5339 (1994)

19.56 A. Wego, L. Pagel: A self-filling micropump based on PCB technology, Sens. Actuators A **88**, 220–226 (2001)

19.57 A. Terray, J. Oakey, D. Marr: Fabrication of colloidal structures for microfluidic applications, Appl. Phys. Lett. **81**, 1555–1557 (2002)

19.58 A. Puntambekar, J.-W. Choi, C. H. Ahn, S. Kim, V. B. Makhijani: Fixed-volume metering microdispenser module, Lab Chip **2**, 213–218 (2002)

19.59 M. Madou, Y. Lu, S. Lai, C. Koh, L. Lee, B. Wenner: A novel design on a CD disc for 2-point calibration measurement, Sens. Actuators A **91**, 301–306 (2001)

19.60 K. Handique, D. Burke, C. Mastrangelo, C. Burns: On-chip thermopneumatic pressure for discrete drop pumping, Anal. Chem. **73**, 1831–1838 (2001)

19.61 H. Andersson, W. Wijngaart, P. Griss, F. Niklaus, G. Stemme: Hydrophobic valves of plasma deposited octafluorocyclobutane in DRIE channels, Sens. Actuators B **75**, 136–141 (2001)

19.62 L. Low, S. Seetharaman, K. He, M. Madou: Microactuators toward microvalves for responsive controlled drug delivery, Sens. Actuators B **67**, 149–160 (2000)

19.63 Q. Yu, J. Bauer, J. Moore, D. Beebe: Responsive biomimetic hydrogel valve for microfluidics, Appl. Phys. Lett. **78**, 2589–2591 (2001)

19.64 M. Mitchell, V. Spikmans, F. Bessoth, A. Manz, A. de Mello: Towards organic synthesis in microfluidic devices: Multicomponent reactions for the construction of compound libraries, Proc. of Micro-Total Analysis Systems 2000, Enschede 2000, ed. by A. van den Berg, W. Olthius, P. Bergveld (Kluwer, Dordrecht 2000) 463–465

19.65 D. Beebe, R. Adrian, M. Olsen, M. Stremler, H. Aref, B. Jo: Passive mixing in microchannels: Fabrication and flow experiments, Mecanique & Industries **2**, 343–348 (2001)

19.66 C.-C. Hong, J.-W. Choi, C. H. Ahn: A novel in-plane passive micromixer using Coanda effect, Proc. of the μTAS 2001 Symposium, Monterey 2001, ed. by J. M. Ramsey, A. van den Berg (Kluwer, Dordrecht 2001) 31–33

19.67   J. Brody, P. Yager: Diffusion-based extraction in a microfabricated device, Sens. Actuators A **54**, 704–708 (1996)

19.68   A. Stroock, S. Dertinger, A. Ajdari, I. Mezic, H. Stone, G. M. Whitesides: Chaotic mixer for microchannels, Science **295**, 647–651 (2002)

19.69   B. Weigl, P. Yager: Microfluidic diffusion-based separation and detection, Science **283**, 346–347 (1999)

19.70   A. Puntambekar, J.-W. Choi, C. H. Ahn, S. Kim, S. Bayyuk, V. B. Makhijani: An air-driven fluidic multiplexer integrated with microdispensers, Proc. of the μTAS 2001 Symposium, Monterey 2001, ed. by J. M. Ramsey, A. van den Berg (Kluwer, Dordrecht 2001) 78–80

19.71   T. Nagakura, K. Ishihara, T. Furukawa, K. Masuda, T. Tsuda: Auto-regulated osmotic pump for insulin therapy by sensing glucose concentration without energy supply, Sens. Actuators B **34**, 229–233 (1996)

19.72   Y. Su, L. Lin, A. Pisano: Water-powered osmotic microactuator, Proc. of the 14th IEEE MEMS Workshop (MEMS 2001), Interlaken 2001 (IEEE, Piscataway 393–396

19.73   G. M. Walker, D. J. Beebe: A passive pumping method for microfluidic devices, Lab Chip **2**, 131–134 (2002)

19.74   C. Effenhauser, H. Harttig, P. Kramer: An evaporation-based disposable micropump concept for continuous monitoring applications, Biomed. Devices **4**, 27–32 (2002)

19.75   C. H. Ahn, H. T. Henderson, W. R. Heineman, H. B. Halsall: Development of a generic microfluidic system for electrochemical immunoassay-based remote bio/chemical sensors, Proc. of Micro-Total Analysis Systems '98, Banff, Canada 1998, ed. by D. J. Harrison, A. van den Berg (Kluwer, Dordrecht 1998) 225–230

19.76   J.-W. Choi, K. W. Oh, A. Han, N. Okulan, C. A. Wijayawardhana, C. Lannes, S. Bhansali, K. T. Schlueter, W. R. Heineman, H. B. Halsall, J. H. Nevin, A. J. Helmicki, H. T. Henderson, C. H. Ahn: Development and characterization of microfluidic devices and systems for magnetic bead-based biochemical detection, Biomed. Microdevices **3**, 191–200 (2001)

19.77   J.-W. Choi, C. H. Ahn, S. Bhansali, H. T. Henderson: A new magnetic bead-based, filterless bio-separator with planar electromagnet surfaces for integrated bio-detection systems, Sens. Actuators B **68**, 34–39 (2000)

19.78   J.-W. Choi, K. W. Oh, J. H. Thomas, W. R. Heineman, H. B. Halsall, J. H. Nevin, A. J. Helmicki, H. T. Henderson, C. H. Ahn: An integrated microfluidic biochemical detection system for protein analysis with magnetic bead-based sampling capabilities, Lab Chip **2**, 27–30 (2002)

19.79   O. Niwa, Y. Xu, H. B. Halsall, W. R. Heineman: Small-volume voltammetric detection of 4-aminophenol with interdigitated array electrodes and its application to electrochemical enzyme immunoassay, Anal. Chem. **65**, 1559–1563 (1993)

19.80   K. W. Oh, A. Han, S. Bhansali, H. T. Henderson, C. H. Ahn: A low-temperature bonding technique using spin-on fluorocarbon polymers to assemble microsystems, J. Micromech. Microeng. **12**, 187–191 (2002)

19.81   C.-C. Hong, J.-W. Choi, C. H. Ahn: A disposable on-chip air detonator for driving fluids on point-of-care systems, Proc. of the 6th International Conference on Micro Total Analysis Systems (μ-TAS), Nara 2002, ed. by Y. Baba, S. Shoji, A. van den Berg (Kluwer, Dordrecht 2002)

19.82   C. Gao, J.-W. Choi, M. Dutta, S. Chilukuru, J. H. Nevin, J. Y. Lee, M. G. Bissell, C. H. Ahn: A fully integrated biosensor array for measurement of metabolic parameters in human blood, Proc. of the 2nd Second Annual International IEEE–EMBS Special Topic Conference on Microtechnologies in Medicine and Biology, Madison May 2–4, 2002, ed. by A. Dittmar, D. Beebe (IEEE, New York 2002) 223–226

19.83   C. H. Ahn, J.-W. Choi, G. Beaucage, J. H. Nevin, J.-B. Lee, A. Puntambekar, J. Y. Lee: Disposable Smart Lab on a Chip for Point-of-Care Clinical Diagnostics, Proc. IEEE 2004) 154–173

20. Centrifuge–Based Fluidic Platforms

In this chapter centrifuge–based microfluidic platforms are reviewed and compared with other popular microfluidic propulsion methods. The underlying physical principles of centrifugal pumping in microfluidic systems are presented and the various centrifuge fluidic functions such as valving, decanting, calibration, mixing, metering, heating, sample splitting, and separation are introduced. Those fluidic functions have been combined with analytical measurements techniques such as optical imaging, absorbance and fluorescence spectroscopy and mass spectrometry to make the centrifugal platform a powerful solution for medical and clinical diagnostics and high–throughput screening (HTS) in drug discovery. Applications of a compact disc (CD)–based centrifuge platform analyzed in this review include: two–point calibration of an optode–based ion sensor, an automated immunoassay platform, multiple parallel screening assays and cellular–based assays. The use of modified commercial CD drives for high-resolution optical imaging is discussed as well. From a broader perspective, we compare the technical barriers involved in applying microfluidics for sensing and diagnostic as opposed to applying such techniques to HTS. The latter poses less challenges and explains why HTS products based on a CD fluidic platform are already commercially available, while we might have to wait longer to see commercial CD–based diagnostics.

Once it became apparent that individual chemical or biological sensors used in complex samples would not attain the hoped for sensitivity or selectivity, wide commercial use became severely hampered and sensor arrays and sensor instrumentation were proposed instead. It was projected that by using orthogonal sensor array elements (e.g., in electronic noses and tongues) selectivity would be improved dramatically [20.1]. It was envisioned that instrumentation would reduce matrix complexities through filtration, separation, and concentration of the target compound while at the same time ameliorating selectivity and sensitivity of the overall system by frequent recalibration and washing of the sensors. With microfluidics, the miniaturization of analytical equipment may potentially alleviate the shortcomings associated with large and expensive instrumentation through the reduction in reagent volumes, favorable scaling properties of several important instrument processes (basic theory of hydrodynamics and diffusion predicts faster heating and cooling and more efficient chromatographic and electrophoretic separations in miniaturized equipment) and batch fabrication which may enable low-cost disposable instruments to be used once and then thrown away to prevent sample contamination [20.2]. Micromachining [micro-electromechanical systems (MEMS)] might also allow co-fabrication of many integrated functional instrument blocks. Tasks that are now performed in a series of conventional bench top instruments could then be combined into one unit, reducing labor and minimizing the risk of sample contamination.

Today it appears that sensor-array development in electronic noses and tongues has slowed down because of the lack of highly stable chemical and biological sensors: too frequent recalibration of the sensors and relearning of the pattern recognition software is putting a damper on the original enthusiasm for this sensor approach. In the case of miniaturization of instrumentation through the application of microfluidics,

progress was made in the development of platforms for high-throughput screening (HTS) as evidenced by new products introduced by, for example, Caliper and Tecan Boston [20.3, 4]. In contrast, progress with miniaturized analytical equipment remains limited; platforms have been developed for a limited amount of human and veterinary diagnostic tests that do not require complex fluidic design, see for example Abaxis [20.5]. In this review paper we are, in a narrow sense, summarizing the state of the art of compact disc (CD)-based microfluidics and in a broader sense we are comparing the technical barriers involved in applying microfluidics to sensing and diagnostic as opposed to applying such techniques to HTS. It will quickly become apparent that the former poses the more severe technical challenges and as a result the promise of lab-on-a-chip has not been fulfilled yet.

20.1 Why Centripetal Force for Fluid Propulsion?

There are various technologies for moving small quantities of fluids or suspended particles from reservoirs to mixing and reaction sites, to detectors, and eventually to waste or to a next instrument. Methods to accomplish this include syringe and peristaltic pumps,

electrochemical bubble generation, acoustics, magnetics, direct-current (DC) and alternating-current (AC) electrokinetics, centrifuge, etc. In Table 20.1 we compare four of the more important and promising fluid propulsion means [20.6]. The pressure that mechani-

Table 20.1 Comparison of microfluidics propulsion techniques

| Comparison | Fluid propulsion mechanism | | | |
	Centrifuge	Pressure	Acoustic	Electrokinetic
Valving solved?	Yes for liquids, no for vapor	Yes for liquids and vapor	No solution shown yet for liquid or vapor	Yes for liquids, no for vapor
Maturity	Products available	Products available	Research	Products available
Propulsion force influenced by	Density and viscosity	Generic	Generic	pH, ionic strength
Power source	Rotary motor	Pump, mechanical roller	5 to 40 V	10 kV
Materials	Plastics	Plastics	Piezoelectrics	Glass, plastics
Scaling	L^3	L^3	L^2	L^2
Flow rate	From less than 1 nL/s to greater than 100 μl/s	Very wide range (less than nL/s to L/s)	20 μl/s	0.001–1 μl/s
General remarks	Inexpensive CD drive, mixing is easy, most samples possible (including cells). Better for diagnostics.	Standard technique. Difficult to miniaturize and multiplex.	Least mature of the four techniques. Might be too expensive. Better for smallest samples.	Mixing difficult. High-voltage source is dangerous and many parameters influence propulsion, better for smallest samples (HTS)

cal pumps have to generate to propel fluids through capillaries is higher the narrower the conduit. Pressure and centripetal force are both volume-dependent forces, which scale as L^3 (in this case L is the characteristic length corresponding to the capillary diameter). Piezoelectric, electro-osmotic, electrowetting and electrohydrodynamic (EHD) pumping (the latter two are not shown in Table 20.1) all scale as surface forces (L^2), which represent more favorable scaling behavior in the micro-domain (propulsion forces scaling with a lower power of the critical dimension become more attractive in the micro-domain) and lend themselves better to pumping in smaller and longer channels. In principle, this should make pressure- and centrifuge-based systems less favorable but other factors turn out to be more decisive; despite better scaling of the non-mechanical pumping approaches in Table 20.1, almost all biotechnology equipment today remain based on traditional external syringe or peristaltic pumps. The advantages of these approach are that they rely on well-developed, commercially available components and that a very wide range of flow rates is attainable. Although integrated micromachined pumps based on two one-way valves may achieve precise flow control on the order of $1\,\mu l/min$ with fast response, high sensitivity, and negligible dead volume, these pumps generate only modest flow rates and low pressures, and consume a large amount of chip area and considerable power.

Acoustic streaming is a constant (DC) fluid motion induced by an oscillating sound field at a solid/fluid boundary. A disposable fluidic manifold with capillary flow channels can simply be laid on top of the acoustic pump network in the reader instrument. The method is considerably more complex to implement than electro-osmosis (see next paragraph) but the in-

sensitivity of acoustic streaming to the chemical nature of the fluids inside the fluidic channels and its ability to mix fluids make it a potentially viable approach. A typical flow rate measured for water in a small metal pipe lying on a piezoelectric plate is $0.02\,cm^3/s$ at 40 V, peak to peak [20.7]. Today acoustic streaming as a propulsion mechanism remains in the research stage.

Electro-osmotic pumping (DC electrokinetics) in a capillary does not involve any moving parts and is easily implemented. All that is needed is a metal electrode in some type of a reservoir at each end of a small flow channel. Typical electro-osmotic flow velocities are on the order of 1 mm/s with a 1200 V/cm applied electric field. For example, in free-flow capillary electrophoresis work by *Jorgenson*, electro-osmotic flow of 1.7 mm/s was reported [20.8]. This is fast enough for most analytical purposes. *Harrison* et al. achieved electro-osmotic pumping with flow rates up to 1 cm/s in $20-\mu m$ capillaries that were micromachined in glass [20.9]. They also demonstrated the injection, mixing and reaction of fluids in a manifold of micromachined flow channels without the use of valves. The key aspect for tight valving of liquids at intersecting capillaries in such a manifold is the suppression of convective and diffusion effects. The authors demonstrated that these effects can be controlled by the appropriate application of voltages to the intersecting channels simultaneously. Some disadvantages of electro-osmosis are the high voltage required (1–30 kV power supply) and direct electrical–fluid contact with resulting sensitivity of flow rate to the charge of the capillary wall and the ionic strength and pH of the solution. It is consequently more difficult to make it into a generic propulsion method. For example, liquids with high ionic strength cause excessive Joule heating;

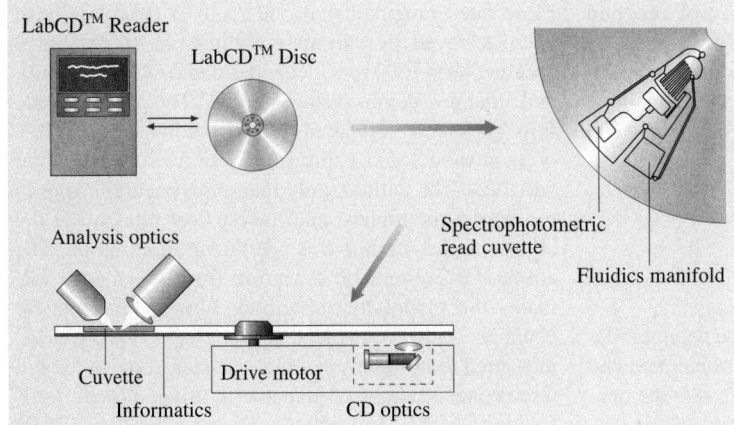

Fig. 20.1 LabCD™ instrument and disposable disc. Here, the analytical result is obtained through reflection spectrophotometry

it is therefore difficult or impossible to pump biological fluids such as blood and urine.

Using a rotating disc, centrifugal pumping provides flow rates ranging from less than 10 nL/s to greater than 100 μL/s depending on disc geometry, rotational rate [revolutions per minute (RPM)], and fluid properties (Fig. 20.1) [20.10]. Pumping is relatively insensitive to physicochemical properties such as pH, ionic strength, or chemical composition (in contrast to AC and DC electrokinetic means of pumping). Aqueous solutions, solvents [e.g., dimethyl sulfoxide (DMSO)], surfactants, and biological fluids (blood, milk, and urine) have all been pumped successfully. Fluid gating, as we will describe in more detail further below, is accomplished using capillary valves in which capillary forces pin fluids at an enlargement in a channel until rotationally induced pressure is sufficient to overcome the capillary pressure (at the so-called burst frequency) or by hydrophobic methods. Since the types and the amounts of fluids one can pump on a centrifugal platform spans a greater dynamic range than for electrokinetic and acoustic pumps, this approach seems more amenable to sample preparation tasks than electrokinetic and acoustic approaches. Moreover miniaturization and multiplexing are quite easily implemented. A whole range of fluidic functions including valving, decanting, calibration, mixing, metering, sample splitting, and separation can be implemented on this platform and analytical measurements may be electrochemical, fluorescent or absorption based, and informatics embedded on the same disc could provide test-specific information.

An important deciding factor in choosing a fluidic system is the ease of implementing valves; the method that solves the valving issue most elegantly (traditional pumps) is already commercially accepted, even if it is not the most easily scaled method. In traditional pumps two one-way valves form a barrier for both liquids and vapors. In the case of the micro-centrifuges, valving is accomplished by varying the rotation speed and capillary diameter. Thus, no real physical valve is required to stop water flow, but as in the case of acoustic and electrokinetic pumping, there is no simple means to stop vapors from spreading over the whole fluidic platform. If the liquids need to be stored for a long time, the valves, which are often disposable in sensing and diagnostics applications, must be barriers to both liquid and vapor. Some initial attempts at implementing vapor barriers on CDs will be reported in this review.

From the preceding comparison of fluidic propulsion methods for sensing and diagnostic applications, centrifugation in fluidic channels and reservoirs crafted in a CD-like plastic substrate as shown in Fig. 20.1 constitute an attractive fluidic platform.

20.2 Compact Disc or Micro-Centrifuge Fluidics

20.2.1 How it Works

CD fluid propulsion is achieved through centrifugally induced pressure and depends on rotation rate, geometry and the location of channels and reservoirs, and fluid properties. *Madou* et al. [20.11] and *Duffy* et al. [20.10] characterized the flow rate of aqueous solutions in fluidic CD structures and compared the results to simple centrifuge theory. The average velocity of the liquid (U) from centrifugal theory is given as:

$$U = D_h^2 \rho \omega^2 \bar{r} \Delta r / 32 \mu L \,, \tag{20.1}$$

and the volumetric flow rate (Q) as:

$$Q = UA \,, \tag{20.2}$$

where D_h is the hydraulic diameter of the channel (defined as $4A/P$, where A is the cross-sectional area and P is the wetted perimeter of the channel), ρ is the density of the liquid, ω is the angular velocity of the CD, $\bar{r}$ is the average distance of the liquid in the channels to the center of the disc, Δr is the radial extent of the fluid, μ is the viscosity of the solution, and L is the length of the liquid in the capillary channel (Fig. 20.2). Flow rates ranging from 5 nL/s to > 0.1 mL/s have been achieved by various combinations of rotational speeds (400–1600 rpm), channel widths (20–500 μm), and channel depths (16–340 μm). The experimental flow rates were compared to rates predicted by the theoretical model and exhibited an 18.5% coefficient of variation. The authors note that experimental errors in measuring the highest and lowest flow rates made the largest contribution to this coefficient of variation. The absence of systematic deviation from the theory validates the model for describing flow in microfluidic channels under centripetal force. *Duffy* et al. [20.10] measured flow rates of water, plasma, bovine blood, three concentrations of hematocrit, urine, dimethyl sulfoxide (DMSO), and polymerase chain reaction (PCR)

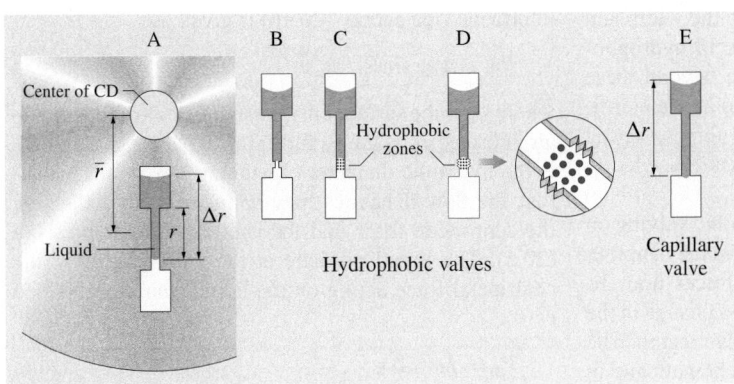

Fig. 20.2 A–E Schematic illustrations for the description of CD microfluidics. **(A)** Two reservoirs connected by a microfluidic chamber. **(B)** Hydrophobic valve made by a constriction in a chamber made of hydrophobic material. **(C)** Hydrophobic valve made by the application of hydrophobic material to a zone in the channel. **(D)** Hydrophobic channel made by the application of hydrophobic material to a zone in a channel made with structured vertical walls (*inset*). **(E)** Capillary valve made by a sudden expansion in channel diameter such as when a channel meets a reservoir

products and report that centrifugal pumping is relatively insensitive to such physicochemical properties as ionic strength, pH, conductivity, and the presence of various analytes, noting good agreement between experiment and theory for all the liquids.

20.2.2 Some Simple Fluidic Functions Demonstrated on a CD

Fluid Mixing
In the work by *Madou* et al. [20.11] and *Duffy* et al. [20.10], different means to mix liquids were designed, implemented, and tested. Observations of flow velocities in narrow channels on the CD enabled Reynolds numbers (R_e) calculations that established that the flow remained laminar in all cases. Even in the largest fluidic channels tested R_e was smaller than 100, well below the transition regime from laminar to turbulent flow ($R_e \approx 2300$) [20.12]. The laminar flow condition necessitates mixing by simple diffusion or by creating special features on the CD that enable advection or turbulence. In one scenario, fluidic diffusional mixing was implemented by emptying two microfluidic channels together into a single long meandering fluidic channel. Proper design of channel length and reagent reservoirs allowed for stoichiometric mixing in the meandering channel by maintaining equal flow rates of the two streams joining in the mixing channel. Concentration profiles may be calculated from the diffusion rates of the reagents and the time required for the liquids to flow through the tortuous path. Mixing can also be achieved by chaotic advection [20.6]. Chaotic advection is a result of the rapid distortion and elongation of the fluid/fluid interface, increasing the interfacial area where diffusion occurs, which increases the mean values of the diffusion gradients that drive the diffusion process; one may call this process an enhanced diffusional process. In addition to the simple

and enhanced diffusional processes, one can create turbulence on the CD by emptying two narrow streams to be mixed into a common chamber. The streams violently splash against a common chamber wall, causing their effective mixing (no continuity of the liquid columns is required on the CD-based system, in contrast to the case of electrokinetics platforms where a broken liquid column would cause a voltage overload).

Valving
Valving is an important function in any type of fluidic platform. Both hydrophobic and capillary valves have been integrated into the CD platform [20.10,11,13–23]. Hydrophobic valves feature an abrupt decrease in the hydrophobic channel cross section, i.e., a hydrophobic surface prevents further fluid flow (Fig. 20.2b–d). In contrast, in capillary valves (Fig. 20.2e), liquid flow is stopped by a capillary pressure barrier at junctions where the channel diameter suddenly expands.

Hydrophobic Valving. The pressure drop in a channel with laminar flow is given by the Hagen–Poiseuille equation [20.12]

$$\Delta P = \frac{12L\mu Q}{wh^3} \,, \tag{20.3}$$

where L is the microchannel length, μ is the dynamic viscosity, Q is the flow rate, and w and h are the channel width and height, respectively. The pressure required to overcome a sudden narrowing in a rectangular channel is given by [20.6]:

$$\Delta P = 2\sigma_l \cos\theta_c \left(\frac{1}{w_1} + \frac{1}{h_1} \right) - \left(\frac{1}{w_2} + \frac{1}{h_2} \right) \,, \tag{20.4}$$

where σ_l is the liquid's surface tension, θ_c is the contact angle, w_1 and h_1 are the width and height of the channel before the restriction, and w_2 and h_2 are the width and height after the restriction, respectively. In hydrophobic valving, in order for liquid to move beyond these pressure barriers, the CD must be rotated above a critical speed, at which point the centripetal forces exerted on the liquid column overcome the pressure needed to move past the valve.

Ekstrand et al. [20.13] used hydrophobic valving on a CD to control discrete sample volumes in the nanoliter range with centripetal force. Capillary forces draw liquid into the fluidic channel until there is a change in the surface properties at the hydrophobic valve region. The valving was implemented as described schematically in Fig. 20.2c. *Tiensuu* et al. [20.14] introduced localized hydrophobic areas in CD microfluidic channels by ink-jet printing of hydrophobic polymers onto hydrophilic channels. In this work, hydrophobic lines were printed onto the bottom wall of channels with both unstructured (Fig. 20.2c) and structured (Fig. 20.2d) vertical channel walls. Several channel width-to-depth ratios were investigated. The CDs were made by injection molding of polycarbonate and were subsequently rendered hydrophilic by oxygen plasma treatment. Ink-jet printing was used for the introduction of the hydrophobic polymeric material at the valve position. The parts were capped with polydimethylsiloxane (PDMS) to form the fourth wall of the channel. In testing of unstructured channels (without the sawtooth pattern) there were no valve failures for 300- and 500 − µm-wide channels but some failures for the 100 − µm channels, however, in structured vertical walls (with sawtooth patterns), there were no valve failures. The authors attribute the better results of the structured vertical walls to both the favorable distribution of hydrophobic polymer within the channel and the sharper sidewall geometry to be wetted (the side walls are hydrophilic since the printed hydrophobic material is only on the bottom of the channel) compared to the unstructured vertical channel walls.

Capillary Valving. Capillary valves have been implemented frequently on CD fluidic platforms [20.10, 11, 15–18, 21, 22]. The physical principle involved is based on the surface tension, which develops when the cross section of a hydrophilic capillary expands abruptly as illustrated in Fig. 20.2e. As shown in this figure, a capillary channel connects two reservoirs, and the top reservoir (the one closest to the center of the CD) and the connecting capillary is filled with liquid. For capillaries with axisymmetric cross sections, the maximum pressure at the capillary barrier expressed in terms of the

interfacial free energy [20.16] is given by

$$P_{cb} = 4\gamma_{al} \sin\theta_c / D_h , \tag{20.5}$$

where γ_{al} is the surface energy per unit area of the liquid–air interface, θ_c is the equilibrium contact angle, and D_h is the hydraulic diameter. Assuming low liquid velocities, the flow dynamics may be modeled by balancing the centripetal force and the capillary barrier pressure (20.5). The liquid pressure at the meniscus, from the centripetal force acting on the liquid, can be described as

$$P_m = \rho\omega^2 \bar{r}\Delta r , \tag{20.6}$$

where, ρ is the density of the liquid, ω is the angular velocity, $\bar{r}$ is the average distance from the liquid element to the center of the CD, and Δr is the radial length of the liquid sample (Fig. 20.2a,e). Liquid will not pass a capillary valve as long as the pressure at the meniscus P_m is less than or equal to the capillary barrier pressure P_{cb}. *Kellogg* and coworkers [20.16] named the point at which P_m equals P_{cb}, the critical burst condition and the rotational frequency at which it occurs they called the burst frequency. Experimental values of critical burst frequencies versus channel geometry, for rectangular cross sections over a range of channel sizes, show good agreement with simulation over the entire range of diameters studied. Since these simulations did not assume an axisymmetric capillary with a circular contact line and a diameter D_h, the meniscus contact line may be a complex shape. Burst frequencies were shown to be cross section dependent for equal hydraulic diameters. The theoretical burst-frequency equation was modified as follows to account for variation of the channel cross section:

$$\rho\omega^2 \bar{r}\Delta r < 4\gamma_{al} \sin\theta_c / (D_h)^n , \tag{20.7}$$

where $n = 1.08$ for an equilateral triangular cross section and $n = 1.14$ for a rectangular cross section. For *pipe flow* (circular cross section) an additional term is used in the burst-frequency expression:

$$\rho\omega^2 \bar{r}\Delta r < 4\gamma_{al} \sin\theta_c / D_h \\ + \gamma_{al} \sin\theta_c (1/D_h - 1/D_0) , \tag{20.8}$$

where the empirically determined constant $D_0 = 40\,\mu m$. The physical reason for the additional pipe-flow term, used to get a fit to the simulation results, is not well understood at this time.

Duffy et al. [20.10] modeled capillary valving by balancing the pressure induced by the centripetal force $(\rho\omega^2 \bar{r}\Delta r)$ at the exit of the capillary with the pressure inside the liquid droplet being formed at the capillary outlet and the pressure required to wet the chamber beyond the valve. The pressure inside a droplet is given by the Young–Laplace equation [20.24];

$$\Delta P = \gamma (1/R_1 + 1/R_2) \, , \tag{20.9}$$

where γ is the surface tension of the liquid and R_1 and R_2 are the meniscus radii of curvature in the x- and y-dimensions of the capillary cross section. In the case of small circular capillary cross sections with spherical droplet shapes, $R_1 = R_2 \cong$ channel cross-section radius and (20.9) can be rewritten as:

$$\Delta P = 4\gamma / D_h \, . \tag{20.10}$$

On this basis Duffy et al. [20.10] derived a simplified expression for the critical burst frequency (ω_c) as

$$\rho \omega_c^2 \bar{r} \Delta r = a(4\gamma / D_h) + b \tag{20.11}$$

with the first term on the right representing the pressure inside the liquid droplet being formed at the capillary outlet scaled by a factor a (for non-spherical droplet shapes) and the second term on the right, b, representing the pressure required to wet the chamber beyond the valve. The b term depends on the geometry of the chamber to be filled and the wettability of its walls.

A plot of the centripetal pressure ($\rho \omega_c^2 \bar{r} \Delta r$) at which the burst occurs verses $1/D_h$ was linear, as expected from (20.11), with a 4.3% coefficient of variation. The authors note a potential limitation with capillary valves due to the fact that liquids with low surface tension tend to wet the walls of the chamber at the capillary valve opening, resulting in the inability to gate the flow. The b term in (20.11) is beneficial in gating flow unless the surface walls at the abrupt enlargement of the capillary valve are so hydrophilic that the liquid is drawn past the valve and into the reservoir.

Madou et al. [20.17, 18] have designed a CD to sequentially valve fluids through a monotonic increase of rotational rate with progressively higher burst frequencies. The CD, shown in Fig. 20.3, was designed to carry out an assay for ions based on an optode-based detection scheme. The CD design employed five serial capillary valves opening at different times as actuated by rotational speed. Results showed good agreement between the observed and the calculated burst frequencies (see later).

It is very important to realize that the valves we mentioned thus far constitute liquid barriers and that they are not barriers for vapors. Vapor barriers must be implemented in any fluidic platform where reagents need to be stored for long periods of time. This is especially important for a disposable diagnostic assay platform. A multi-month, perhaps multi-year, shelf life would require vapor locks in order to prevent reagent solutions from drying or liquid evaporation and condensation in undesirable areas of the fluidic pathway. Tecan, Boston have investigated vapor-resistant valves made of wax that was melted to actuate valve opening [20.25].

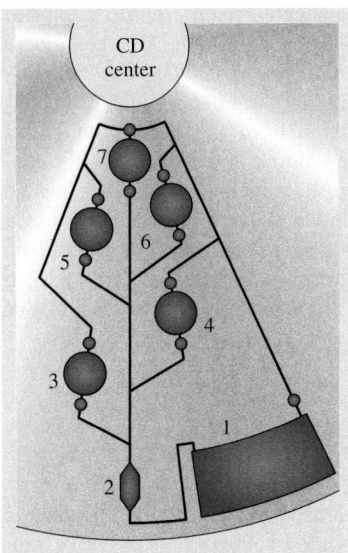

Fig. 20.3 Schematic illustration of the microfluidic structure employed for the ion-selective optode CD platform. The fluidic structure contains five solution reservoirs (numbered **1** – **5**), a detection chamber (**6**), and a waste reservoir (**7**). Reservoir (**1**) and (**3**) contain the first and second calibrant, respectively, reservoirs (**2**) and (**4**) contain wash solutions, and reservoir (**5**) contains the sample. Upon increasing rotation rates, calibrant 1, wash 1, calibrant 2, rinse 2, and then sample were serially gated into the optical detection chamber. Absorption of the calibrants and sample was measured

Volume Definition (Metering) and Common Distribution Channels

The CD centrifugal microfluidic platform enables very fine volume control (or metering) of liquids. Precise volume definition is one of the important functions, necessary in many analytical sample-processing protocols, which has been added, for example, to the fluidic design in the Gyrolab MALDI SP1 CD [20.20]. In this CD, developed for matrix assisted laser desorption ionization (MALDI) sample preparation, a common distribution channel feeds several parallel individual sample-preparation fluidic structures (Fig. 20.4).

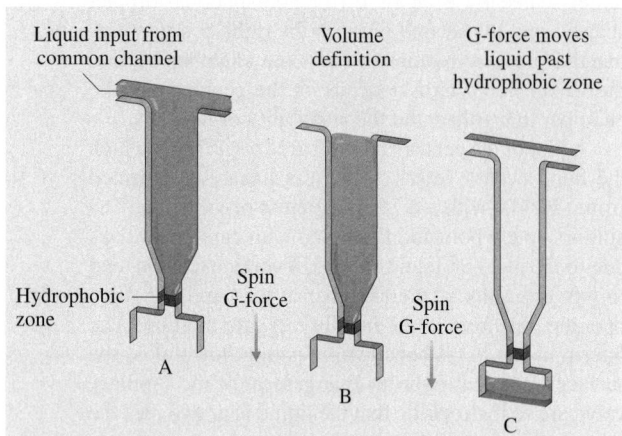

Fig. 20.4a–c Schematic illustration of liquid metering. (**a**) The common distribution channel and liquid metering reservoirs are filled (by capillary forces) with a reagent to be metered. Liquid entering the reservoir does not pass the hydrophobic zone (valve) because of surface tension forces. (**b**) The CD is rotated at a rate that supplies enough centripetal force to empty the common distribution but not enough to force the liquid through the hydrophobic zone. The volume of the fluid metered is determined by the volume of the reservoir. (**c**) A further increase in the rotational speed provides enough force to move the well-defined volume of solution past the hydrophobic valve. (After [20.20])

Reagents are introduced by the capillary force exerted by the hydrophilic surfaces into the common channel and defined volume (200 − nl) chambers until a hydrophobic valve stops the flow. When all of the defined-volume chambers are filled, the CD is spun at a velocity large enough to move the excess liquid from the common channel into the waste. Although there is sufficient centripetal force to empty the common channel, the velocity is not high enough to allow liquid to move past the hydrophobic valve and the well-defined-volume chambers remain filled. These precisely defined volumes can be introduced into the subsequent fluidic structures by increasing the CD angular momentum until the centripetal force allows the liquid to move past the hydrophobic barriers.

Packed Columns

Many commercial products are now available that use conventional centrifuges to move liquid, in a controlled manner, through a chromatographic column. One example is the Quick Spin™ protein desalting column (Roche Diagnostics Corp, Indianapolis, IN), based on the size-exclusion principle. There is an obvious fit for this same type of separation experiment to be carried out on a CD fluidic device (we sometimes refer to the CD platform as a smart, miniaturized centrifuge). Affinity chromatography has been implemented in the fluidic design of the Gyrolab MALDI SP1 CD [20.20]. A reverse phase chromatography column material (SOURCE™ 15 RPC) is packed into a microfluidic channel and protein is adsorbed on the column from an aqueous sample as it passes through the column under centrifugally controlled flow rates. A rinse solution is subsequently passed through the column and finally an elution buffer is flown through to remove the protein and carry it into the fluidic system for further processing. The complete Gyrolab MALDI SP1 CD is discussed in a later section of this review.

20.3 CD Applications

20.3.1 Two-Point Calibration of an Optode-Based Detection System

A CD based system with ion-selective optode detection and a two-point-calibration structure for the accurate detection of a wide variety of ions has been developed [20.15, 17, 18]. The microfluidic architecture, depicted in Fig. 20.3, is comprised of channels, five solution reservoirs, a chamber for colorimetric measurement of the optode membrane, and a waste reservoir, all manufactured onto a poly(methyl methacrylate) disc. Ion-selective optode membranes, composed of plasticized poly(vinyl chloride) impregnated with an ionophore, a chromoionophore, and a lipophilic anionic additive, were cast, with a spin-on device, onto a support layer and then immobilized on the disc. With this system, it is possible to deliver calibrant solutions, washing buffers, and unknown solutions (e.g., saliva, blood, urine, etc.) to the measuring chamber where the optode membrane is located. Absorbance measurements on a potassium optode indicate that optodes immobilized on the platform exhibit the theoretical absorbance response. Samples of unknown concentration can be quantified to within 3% error by fitting the response curve for a given optode membrane using an acid (for measuring the signal for a fully protonated chromoionophore), a base (for fully deprotonated chromoionophore), and two standard solutions. Further, the ability to measure ion concentrations employing one standard solution in conjunction with an acid and base, and with two standards alone were studied to delineate whether the current architecture could be simplified. Finally, the efficacy of incorporating washing steps into the calibration protocol was investigated.

This work was further extended to include anion-selective optodes and fluorescence rather than absorbance detection [20.17]. Furthermore, in addition to employing a standard excitation source where a fiber optic probe is coupled to a lamp, laser diodes were evaluated as excitation sources to enhance the fluorescence signal.

20.3.2 CD Platform for Enzyme–Linked Immunosorbant Assays (ELISA)

The automation of immunoassays on microfluidic platforms presents multiple challenges because of the high number of fluidic processes and the many different liquid reagents involved. Often there is also the need for highly accurate quantitative results at extremely low concentration and care must be taken to prevent nonspecific binding of reporter enzymes and to deliver well-defined volumes of reagents consistently. An enzyme-linked immunosorbant assay (i.e., ELISA) is one of the most common immunoassay methods and is often carried out in microtiter plates using labor-intensive manual pipetting techniques. Recently, *Lee* et al. [20.21] have implemented an automated enzyme-linked immunosorbant assay on the CD platform. This group used a five-step flow sequence in the same CD design illustrated in Fig. 20.3. A capture antibody (anti-rat IgG) was applied to the detection reservoir (reservoir 2 in Fig. 20.3) by adsorption to the PMMA CD surface, then the surface was blocked to prevent nonspecific binding. Antigen/sample (rat IgG), wash solution, second antibody, and substrate solutions were loaded into reservoirs 3–7 (Fig. 20.3) respectively. Using capillary valving techniques, the sample and reagents were pumped, one at a time, through the detection chamber. First, the sample was introduced for antibody antigen binding (reservoir 3), then a wash solution (reservoir 4), then an enzyme-labeled secondary antibody (reservoir 5), then another wash solution (reservoir 6), and finally the substrate was added (reservoir 7). The U-shaped bend in the fluidic path allows the solutions to incubate in the capture zone/detection chamber until the next solution is released into the chamber. Detection of the fluorescence

is performed after the substrate is introduced into the detection reservoir (reservoir 2). Endpoint measurements (completion of enzyme–substrate reaction) were made and compared to conventional microtiter plate methods using similar protocols. The CD ELISA platform was shown to have advantages such as lower reagent consumption, and shorter assay times, explained in terms of larger surface-to-volume ratios, which favor diffusion-limited processes. Since the reagents were all loaded into the CD at the same time, there was no need for manual operator interventions in between fluidic assay steps. The consistent control and repeatability of liquid propulsion removes experimental errors associated with inconsistent manual pipetting methods, for example, rinsing/washing can be carried out not only with equal volumes, but equal flow conditions.

20.3.3 Multiple Parallel Assays

The ability to obtain simultaneous and identical flow rates, incubation times, mixing dynamics, and detection makes the CD an attractive platform for multiple parallel assays. *Kellogg* et al. [20.22] have reported on a CD system that performs multiple (48) enzymatic assays simultaneously by combining centrifugal pumping in microfluidic channels with capillary valving and colorimetric detection. The investigation of multiplexed parallel enzyme inhibitor assays are needed for high-throughput screening in diagnostics and in screening of drug libraries. For example, enzymatic dephosphorylation of colorless *p*-nitrophenol phosphate by alkaline phosphatase results in the formation of the yellow-colored *p*-nitrophenol and inhibition of this reaction may be quantified by light absorption measurement. Theophylline, a known inhibitor of the reaction, was used as the model inhibitory compound in *Kellogg* et al.'s [20.22] feasibility study. A single assay element on the CD contains three reservoirs: one for the enzyme, one for the inhibitor, and one for the substrate. Rotation of the CD allows the enzyme and inhibitor to pass capillary valves, mix in a meandering $100-\mu$m-wide channel, and then move to a point where flow is stopped by another capillary valve. A further increase in the rotational speed allows the enzyme/inhibitor mixture and substrate to pass through the next set of capillary valves where they are mixed in a second meandering channel and emptied into an on-disc planar cuvette. The CD is slowed and absorption through each of the 48 parallel assay cuvettes is measured by reflectance, all in a period of 60 s, the entire fluidic process including measurement took about 3 min. The CDs were fabricated using

PDMS replication techniques [20.26], with the addition of a white pigment to the PDMS polymerization for enhanced reflectivity in the colorimetric measurements. The flow rates and meandering channel widths were selected such that the diffusion rate would allow 90% mixing of the solutions.

The variation in performance between the individual fluidic CD structures was quantified by carrying out the same assay 45 times simultaneously on a CD. The background-corrected absorbance was measured and the coefficient of variation in the assay was ≈ 3.2%. When the experiment was repeated on different discs the coefficient of variation was 3–3.5%. Furthermore, variation of absorption across a single cuvette was less than 1%, confirming complete mixing. In experiments to show enzyme inhibition, 45 simultaneous reactions were carried out on the CD using fixed concentrations of enzyme and substrate and 15 concentrations of theophylline in triplicate and a complete isotherm was generated for the inhibition of alkaline phosphatase. The three remaining structures were used for calibration with known concentrations of p-nitrophenol. A dose response was seen over three logs of theophylline concentration in the range of 0.1–100 mM. The authors concluded that a large number of identical assays, with applications in rapid, high-throughput screening, can be carried out on the CD platform simultaneously because of the symmetric force acting on the fluids in high-quality identical microfluidic structures and that detection was simplified by rotating all the reaction mixtures under a fixed detector. In later work [20.22], the same group has extended the number of assays to 96 per CD and has investigated fluorescent enzymatic assays.

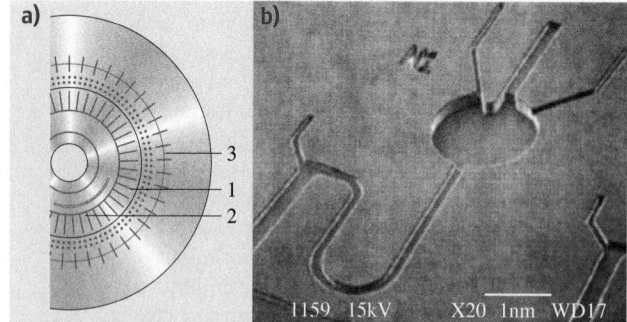

Fig. 20.5a,b Microfabricated cell-culture CD. (**a**) The CD caries a number of cell growth chambers (**1**) radially arranged around a common distribution channel (**2**) and is sealed with a silicone cover (**3**). (**b**) SEM close-up of an individual cell growth chamber and microfluidic connections. (After [20.23])

20.3.4 Cellular–Based Assays on CD Platform

Cell-based assays are often used in drug screening [20.27] and rely on labor-intensive microtiter plate technologies. Microtiter plate methods may be difficult to automate without the use of large and expensive liquid-handling systems and they present problems with evaporation when scaled down to small volumes. *Thomas* et al. [20.23] have reported on a CD platform-based automated adherent cell system. This adherent cell assay involved introducing the compounds to be screened to a cell culture, then determining if the cells were killed (a cell viability assay).

Reagents for cell growth, rinsing and viability staining were serially loaded into an annular, common distribution chamber and centripetal force was used for reagent loading, exchange, and rinsing of the cell growth chamber (Fig. 20.5). Individual inlets were used for the addition of compounds to be screened. The plastic channels, Fig. 20.5b, were capped with a polydimethyl-siloxane (PDMS) sheet capable of fast gas transport in and out of the culture reservoirs.

HeLA, L929, CHO-M1, and MRC-5 cell lines were cultivated on the CD device. Cell viability assays were performed, on the CD, by removing the growth medium from the cells, washing the cells with PBS, and introducing a solution of the fluorescence assay reagents into the growth chamber. The LIVE/DEAD® Viability Assay (Molecular Probes, Inc., Eugene, OR) uses a mixture of calcein green-fluorescent nucleic-acid stain and the red-fluorescent nucleic-acid stain, ethidium. The assay performance is based on the differing abilities of the stains to penetrate healthy bacterial cells. The calcein green-fluorescent dye will label all cells, live or dead. The red-fluorescent ethidium stain will only label cells with damaged membranes. The red stain causes a reduction in the green stain fluorescence when both dyes are present. When the appropriate mixture of green and red stains is used, cells with intact membranes will have a green fluorescence and cells with damaged membranes will have a red fluorescence. The background remains almost completely non-fluorescent (Fig. 20.6). All liquid transfers were carried out using centripetal force from CD rotation with angular frequencies of 200–600 rpm. Quantitative detection of multiple cell viability assays, within 30 s, was carried out by measurement of calcien fluorescence with a charge-coupled device (CCD)-based fluorescence imaging system. These experimental results show linear fluorescence intensity across the range of 200–4000 cells and give an indication of the potential

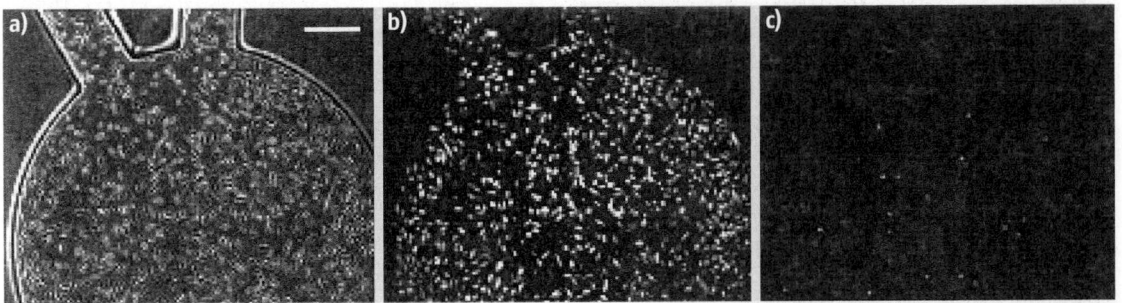

Fig. 20.6a–c L929 fibroblasts cultured for 48 hours in CD growth chambers. (**a**) phase contrast (scale bar 100 μm), (**b**) epifluorescence image of calcein-stained viable cells, (**c**) epifluorescence image of ethidium-stained nonviable cells. (After [20.23])

of this platform for miniaturized quantitative cell-based assays.

In the same work, the authors reported the results of experiments designed to investigate the effect on cells of using centripetal force to move liquids. The cells tested were shown to be compatible with centripetal forges of at least $600 \times g$, much larger than the $50-100 \times g$ needed for filling and emptying cell chambers. Furthermore, it was reported that cells grown in such devices appear to show the same cell morphology as cells grown under standard conditions.

In separate work done by our group in collaboration with NASA Ames [20.28] the LIVE/DEAD®BacLight™ Bacterial Viability Kit (Molecular Probes, Inc., Eugene, OR) has been integrated to a completely automated process on CD. Disposable and reusable CD structures, hardware, and software were developed for the LIVE/DEAD assay.

The CD design for assay automation must have the following functions or properties: contain separate reservoirs for each dye and the sample, retain those solutions in the reservoir until the disc is rotated at a certain velocity, evenly and completely mix the two dyes, evenly and completely mix the dye mixture with the sample containing the cells, collect this final mixture in a reservoir with good optical properties. Two methods for quick fabrication of prototype CDs were used. One method used molded PDMS structures. In a second method, a dry film photoresist (DF 8130, Think & Tinker, Palmer Lake, CO) was laminated onto a $1-$mm-thick polycarbonate disc with pre-drilled holes for sample introduction. The microfluidic pattern was made using a photolithographic pattern on the negative photoresist. The fluidic system was capped with a polycarbonate disc that had been laminated with an optical-quality pressure-sensitive adhesive (3M 8142, 3M, Minneapolis, MN). Figure 20.7

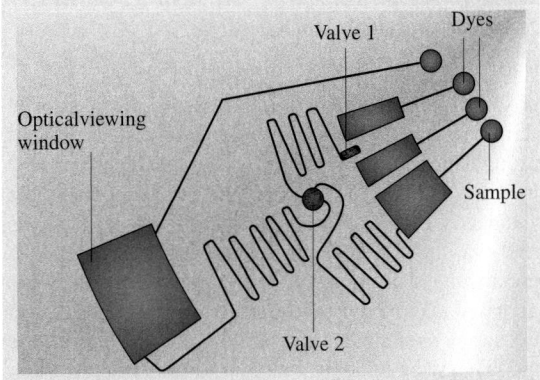

Fig. 20.7 Microfluidic pattern for LIVE/DEAD®BacLight™ bacterial viability assay. The dyes and sample are introduced into the reservoir chambers using a pipette. The dyes fill the chamber stopping at a capillary valve (valve 1). Similarly, the sample containing cells is introduced into the sample reservoir. The disc is rotated to a velocity of 800 rpm, the dyes are forced through the capillary valves and they are mixed as they flow through the switchback turns of the microfluidic channels. Simultaneously, the sample passes from the reservoir into a fluid channel where it meets the dye mixture at valve 2. The velocity of the disc is increased to 1600 rpm and the dye mixture and sample combine and mix in the switchback microfluidic path leading to the optical viewing window

shows the fluidic pattern for this assay. This pattern is based on the structure developed in a similar approach used to demonstrate multiple enzymatic assays on CD [20.10].

The dyes and sample were introduced into reservoir chambers using a pipette. The dyes fill the chamber stopping at a capillary valve (valve 1 in Fig. 20.7). Simi-

Fig. 20.8 Fluorescent microscopy overlaid images of *red*- and *green*-stained *E. coli* on CD from LIVE/DEAD® BacLight™ bacterial viability assay

larly, the sample containing cells was introduced into the sample reservoir. Upon rotation, the dyes were forced through the capillary valves and were mixed as they flowed through the switchback turns of the microfluidic channels. Simultaneously, the sample passed from its reservoir into a fluid channel where it met the dye mixture at valve 2 of Fig. 20.7. The velocity of the disc was increased and the dye mixture and sample combine and mix in the switchback microfluidic path leading to the optical viewing window. The dye–sample mixture is allowed to incubate in the dark at room temperature for 5 min. The optical viewing chamber was imaged twice, once with optics for the green signal and then with optics for the red signal. A typical fluorescence microscopy image of an overlay of the red and green images of stained *E. coli* is shown in Fig. 20.8.

The instrument for disc rotation and fluorescence imaging (Fig. 20.9) used a programmable rotational motor for various velocities and acceleration/deceleration rates. The use of standard microscope objectives enabled magnification selection. An automatic focusing system was used. The light source was a mercury lamp, which used standard low-pass excitation filters for fluorescent excitation. A CCD camera was combined with standard emission filter cubes for imaging.

20.3.5 Integrated Nucleic–Acid Sample Preparation and PCR Amplification

Nucleic-acid analysis is often facilitated by the polymerase chain reaction (PCR) and requires substantial sample preparation that, unless automated, is labor extensive. After the initial sample preparation step of cell lyses to release the deoxyribonucleic acid (DNA)/ribonucleic acid (RNA), a step must be taken to prevent PCR inhibitors, usually proteins such as hemoglobin, from entering into the PCR thermocycle reaction. This can be done by further purification methods such as precipitation and centrifugation, solid-phase extraction, or by denaturing the inhibitory proteins. Finally, the sample must be mixed with the PCR reagents followed by thermocycling, a process that presents difficulty in a microfluidic environment because of the relatively high temperatures (up to 95 °C) required. In a small-volume microfluidic reaction chamber, the liquid will easily evaporate unless care is taken to prevent vapor from escaping.

Kellogg et al. [20.22] combine sample preparation with PCR on the CD. The protocol involves the following steps: (1) mixing raw sample (5 μL of dilute whole bovine blood or *E. coli* suspension) with 5 μL of 10 mM NaOH; (2) heating to 95 °C for 1−2 min (cell lyses and inhibitory protein denaturization); (3) neutralization of basic lysate by mixing with 5 μL of 16 mM tris-HCl ($pH = 7.5$); (4) neutralized lysate is mixed with 8−10 μL of liquid PCR reagents and primers of interest; and (5) thermal cycling. The CD fluidic design is shown schematically in Fig. 20.10. Three mixing channels are used in series to mix small volumes. A spinning platen allows control of the temperature by positioning thermoelectric devices against the appropriate fluidic

Fig. 20.9 *Left:* Optical disc drive/imager with cover removed. Size of unit is made to fit in specific cargo bay of Space-Lab. *Right:* zoom of microscope objectives and a disc loaded in the drive

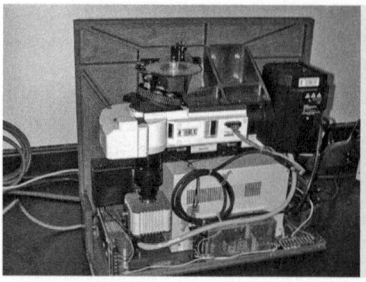

Fig. 20.10 Schematic illustration of the CD microfluidic PCR structure. The center of the disc is above the figure. The elements are (a) sample, (b) NaOH, (c) tris-HCl, (d) capillary valves, (e) mixing channels, (f) lysis chamber, (g) tris-HCl holding chamber, (h) neutralization lysate holding chamber, (i) PCR reagents, (j) thermal cycling chamber, (k) air gap. Fluids loaded in (a), (b), and (c) are driven at a first revolutions per minute (RPM) into reservoirs (g) and (f), at which time (g) is heated to 95 °C. The RPM is increased and the fluids are driven into (h). The RPM is increased and fluids in (h) and (i) flow into (j). On the *right*, the cross section shows the disc body (m), air gap (k), sealing layers (n), heat sink (l), thermoelectric (p), PC-board (q) and thermistor (o). (After [20.13])

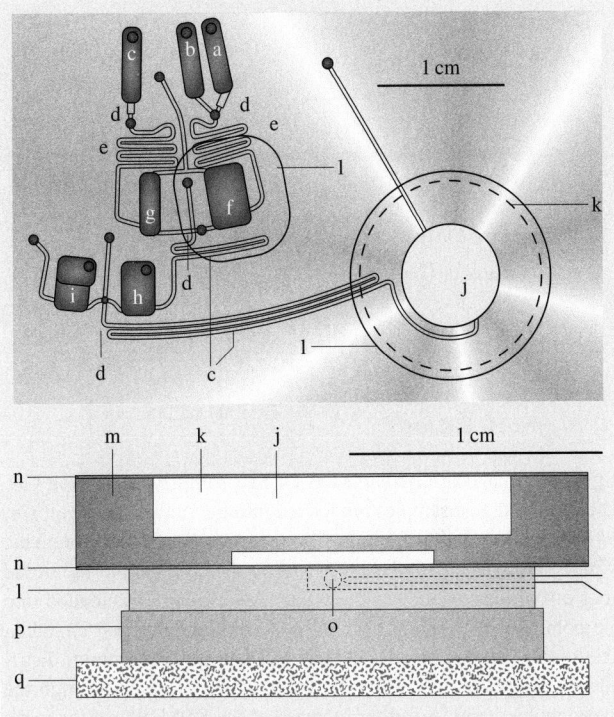

chambers. The CD contacts the PC board platen on the spindle of a rotary motor, with the correct angular alignment, which is connected by a slip ring to stationary power supplies and a temperature controller. Thermocouples are used for closed-loop temperature control and air sockets are used as insulators to isolate heating to reservoirs of interest. The thermoelectric at the PCR chamber both heats and cools and since the PCR reaction chamber is thin, 0.5 mm, fast thermocycling is achieved. Slew rates of $\pm 2\,°C/s$ with fluid volumes of $25\,\mu L$ and thermal gradients across the liquid of $0.5\,°C$ are reported. It is important to note here that the PCR chambers were not sealed; vapor generated inside the PCR chamber condensed on the cooler surfaces of the connecting microfluidic chamber and, since the CD is rotating, the condensed drops are centrifuged back into the hot PCR chamber. This micro-condensation apparatus is unique for the centrifugal CD platform. Details of the experimental parameters used can be found in the original reference [20.22], but to summarize, sample preparation and PCR amplification for two types of samples, whole blood and *E. coli*, were demonstrated on the CD platform and shown to be comparable to conventional methods.

20.3.6 Sample Preparation for MALDI MS Analysis

MALDI MS peptide mapping is a commonly used method for protein identification. Correct identification and highly sensitive MS analysis require careful sample preparation. Manual sample preparation is quite tedious, time-consuming, and can introduce errors common to multi-step pipetting. MALDI MS sample preparation protocols employ a protein digest followed by sample concentration, purification, and recrystalliza-

tion with minimal loss of protein. Automation of the sample-preparation process, without sample loss or contamination, has been enabled on the CD platform by the Gyrolab MALDI SP1 CD and the Gyrolab Workstation (Gyros AB, Sweden) [20.20].

The Gyrolab MALDI SP1 sample-preparation CD will process up to 96 samples simultaneously using separate microfluidic structures. Protein digest from gels or solutions are concentrated, desalted, and eluted with matrix onto a MALDI target area. The CD is then transferred to a MALDI instrument for analysis without the need for further transfer to a separate target plate. The CD fluidic structure contains functions for common reagent distribution, volume definition (metering), valving, reverse phase column (RPC) for concentration and desalting, washing, and target areas for external calibrants. Figure 20.11 shows the Gyrolab MALDI SP1 sample preparation CD. The CDs are loaded with reagents and processed in a completely automated, custom workstation capable of holding up to five microtiter plates containing samples and reagents and up to five CD micro-laboratories. The reagents are taken from the microplates to the CD inlets using a precision robotic arm fitted with multiple needles, the liquid is drawn into specific inlets by capillary forces, and then the needles

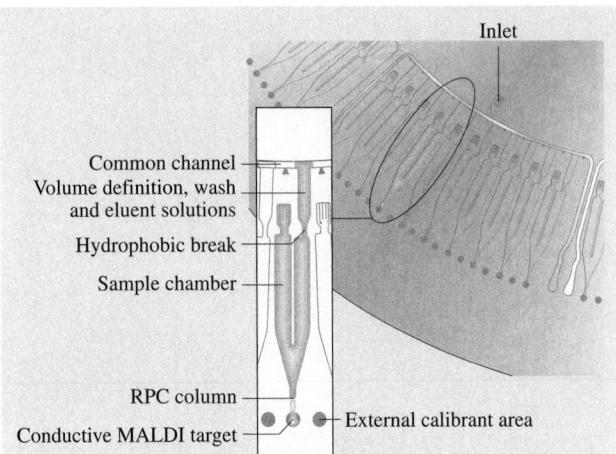

Fig. 20.11 Image of Gyrolab MALDI SP1 sample-preparation CD. The protein digest samples are loaded into the sample reservoir (*inset*) by capillary action. Upon rotation, the sample passes through the RPC column. The peptides are bound to the column and the liquid goes out of the system into the waste. A wash buffer is loaded into the common distribution channel and volume-definition chamber. The disc is rotated at a RPM that will empty the common distribution channel but not allow the wash solution to pass through the hydrophobic zone. A further increase of the RPM allows the well-defined volume of wash solution to pass the hydrophobic break and wash the RPC column then be discarded as waste. Next, a well-defined volume of the elution/matrix solution is loaded and passed through the column, taking the peptides to the MALDI target zone. The flow rate is controlled to optimize the evaporation of the solvent crystallization of the protein and matrix at the target zone. (After [20.20])

are cleaned by rinsing at a wash station. Samples are applied in aliquots from 200 nl up to 5 μl sequentially to each channel where it is contained using hydrophobic surface valves. The CD is then rotated, at an optimized rate, causing the sample to flow through an imbedded reverse phase chromatography column and liquid that passes through the column is collected in a waste container. Controlling the angular-velocity-dependent liquid flow rate maximizes protein binding to the column. A wash solution is introduced by capillary action into common distribution channels connected to groups of microstructures. The wash solution fills a volume definition chamber (200 nl) until it reaches a hydrophobic valve and the CD is rotated to clear the excess liquid in the distribution channel. Not until the rotational velocity is further increased is the defined wash volume able to pass through the hydrophobic valve and into the RPC column (SOURCE™15 RPC). The peptides are

eluted from the column and directly onto the MALDI target area using a solution that contains α-cyano-4-hydroxycinnamic acid and acetonitrile using the same common distribution channel and defined volume as the previous wash step. Optimization of rotational velocity during elution enables maximum recovery and balances the rate of elution with the rate of solvent evaporation from the target surface. Areas in and around the targets are gold-plated to prevent charging of the surface that would cause spectral mass shift and ensures uniform field strength. Well-defined matrix/peptide crystals form in the CD MALDI target area. Gyros reports high reproducibility, high sensitivity, and improved performances when compared to conventional pipette tip technologies. Data was shown that includes: comparison of 23 identical samples, processed in parallel on the same CD, from a bovine serum albumin (BSA) tryptic digest and analysis of identical samples processed on different CDs, run on different days. Sensitivities were shown in the attomole to femtomole range, indicating the ability to identify low-abundance proteins. The report attributed the superior performance of this platform to the pretreatment of the CD surface to minimized nonspecific adsorption of peptides, reproducible wash volume and flow, and reproducible elution (volume, flow, and evaporation) and crystallization.

20.3.7 Modified Commercial CD/DVD Drives in Analytical Measurements

The commercial CD/digital versatile disc (DVD) drive, commonly used for data storage and retrieval, can be thought of as a laser scanning imager. The CD drive retrieves optically generated electrical signals from the reflection of a highly focused laser light (spot size: full width at half maximum ≈ 1 μm), from a 1.2 − mm-thick polycarbonate disc that contains a spiral optical track feature. The track is fabricated by injection molding and is composed of a series of pits that are 1−4 μm long, 0.15 μm deep, and about 0.5 μm wide. The upper surface of a CD is made reflective by gold or aluminium metallization and protected with a thin plastic coating. Information is generated as the focused laser follows the spiral track by converting the reflected light signal into digital information. A flat surface gives a value of zero, an edge of a pit gives a value of one. The data is retrieved at a constant acquisition rate and the serial values (0/1) are converted to data of different kinds for various applications (music, data, etc.). In addition to the code generated by the spacing of the pits, optical signals necessary for focusing, laser tracking of the spiral track, and radial position determination of the read head are monitored and used in feedback loops for proper CD operation. The laser is scanned in a radial direction toward the outer diameter of the disc with an elaborate servo that maintains both lateral tracking and vertical focusing.

Researchers [20.29, 30] have taken advantage of this low-cost high-resolution optical platform in analytical DNA array applications. *Barathur* et al. [20.29] from Burstein Technologies (Irvine, CA), for example, have modified the normal CD drive for use as a sophisticated laser-scanning microscope for analysis of a Bio Compact Disk^{TM} assay, where all analysis is carried out in microfluidic chambers on the CD. The assay is carried out concurrently with the normal optical scanning capabilities of a regular CD drive. The authors report on the application of this device for DNA micro-spot-array hybridization assays and comment on its use in other diagnostic and clinical research applications. For the DNA spot-array application, arrays of captures probes for specific DNA sequences are immobilized on the surface of the CD in microfluidic chambers. Sample preparation and multiplexed PCR, using biotinylated primers, are carried out off-disc, then the biotinylated amplicons are introduced into the array chamber and hybridization occurs if amplicons with the correct sequence are present. Hybridization detection is achieved by monitoring the optical signal from the CD photo detector, while the CD is rotating. To generate an optical signal when hybridization has occurred a reporter is used, for the Bio Compact Disk^{TM} assay the reporter is a streptavidin-labeled microsphere that will bind only to the array spots which have successfully captured biotinylated amplicons. The unbound microsphere reporters are removed from the array using simple centrifugation and no further rinsing is needed. As the laser is scanned across the CD surface, the microparticle scatters light that would have normally been reflected to the photodetector resulting in less light on the detector (bright-field microscopy) and a distinctive electronic signal is generated. The electronic signal-intensity data can be stored in memory then deconvoluted into an image. A 1 cm^2 microarray can be scanned in 20–30 s with a data-reduction time of 5 min and custom algorithms that perform the interpretations in real time. Data was shown for identification of three different species of the *Brucella* coccobacilli on the CD platform. Human infection occurs by transmission from animals by ingestion of infected food products, contact with an infected animal, or inhalation of aerosols. Multiplex PCR-amplified DNA from all three species (common forward primers and specific reverse primers resulting in amplicons of different length for various

species, were used for verification of PCR on external gels) were incubated on arrays with species-specific capture probes. Removal of one of the species in the sample resulted in no probes present on that specific array spot, verifying the specificity of the assay.

Alexandre et al. [20.30] at Advanced Array Technology (Namur, Belgium), utilize the inner diameter area of a CD and standard servo optics for numerical information and operational control and employ a second scanning laser system to image DNA arrays on transparent surfaces at the outer perimeter of a CD. The second laser system, consisting of a laser-diode module that illuminates a 50 – μm spot on the CD surface, is scanned radially at a constant linear velocity of 20 mm/min while the CD is rotating. Each CD contains 15 arrays arranged in a single ring on the CD perimeter that extends in the radial direction for 15 mm. The arrays are rectangular and consist of four rows and 11 columns of 300 – μm spots. The normal CD servo optics are located below the disc and the added imaging optics are above the disc. A photodiode head follows the imaging laser and the refracted light intensity is stored digitally at a high sampling rate. An image of each array on the disc is reconstructed by deconvolution of the light-intensity data. The entire CD can be scanned in less than one minute, producing a total of 6 MB of information. Sample preparation and PCR amplification was carried out off-disc. Specific DNA capture probes were spotted on the surface of the CD using a custom arrayer that transfers the probes from a multi-well plate onto the surface of up to 12 discs using a robotic arm. Biotinylated amplicons are introduced onto the array chambers (one chamber for each array) and hybridization occurs if amplicons with the correct sequence are present. In order to get an optical signal that can be detected, after a rinse step, a solution of streptavidin-labeled colloidal gold particles is applied to the array followed by a Silver Blue solution (AAT, Namur, Belgium). The silver solution causes silver metal to grow on the gold particles, thereby making the hybridization-positive micro-array spots refractive to the incident laser light. Results were shown for the detection of the five most common species of *Staphylococci* and an antibiotic-resistant strain. The *fem A* and *mec A* genes of the various species of *Staphylococci* were amplified by primers common to all *Staphylococci* species then hybridized to a micro-array containing spots with probes specific for the different *Staphylococci* species. The array also included a capture probe for the genus *Staphylococci* and a probe for the *mec A* gene that is associated with methicillin resistance of the *Staphylococci* species. The results were digitized and quantified with software that is part of the custom Bio-CD™ workstation. Signal-to-noise ratios were above 50 for all positive signals.

20.3.8 Microarray Hybridization for Molecular Diagnosis of Infectious Diseases

In recent years, microarrays have become important tools for nucleic-acid analysis and gene-expression profiling. The expression of thousands of genes can be monitored in a single experiment using this technology. A number of investigators have attempted to adapt this technology to rapidly detect infectious agents in clinical specimens for diagnostic purposes [20.31–35]. However, such systems are still in their infancy and most of them require technologically complex biochips with integrated heating/cooling systems [20.31, 32, 36]. The Madou group at UCI together with the Bergeron group at Laval University have reported [20.37] a CD-based microfluidic platform for DNA microarray analysis of infectious disease, presenting an elegant solution to automate and speed up microarray hybridization. Staphiloccocal-specific oligonucleotides were used as capture probes immobilized in 4×5 arrays of $125 - \mu m$ spots on a standard 3×1 in. glass slide. The layout of the array is shown in Fig. 20.12a. A flow cell is designed to realize the self-contained hybridization process in the CD platform. As shown in Fig. 20.12b, the flow cell consists of a hybridization column 1, aligned with the DNA micro-array on the glass slide, a sample chamber 2, and a rinsing chambers 3 and 4. The reagent chambers are connected to the hybridization column with a microchannel which is $50 \, \mu m$ in width and $25 \, \mu m$ in depth. The flow cell is aligned with and adhered to the glass slide to form a DNA hybridization detection unit, up to five of which can be mounted into the CD platform fabricated from acrylic plastic using computer numerical control (CNC) machining (Fig. 20.12c). The reagents are positioned to be pumped through the hybridization column by centrifugal force in a sequence beginning with chamber 2 up to chamber 4 and this flow sequence is achieved by manipulating the balance between the capillary force and centrifugal pressure. The sample (chamber 2) is released first and flows over the $140 - nl$ hybridization chamber (chamber 1) where the oligonucleotide capture probes are spotted onto the glass support. The rinsing buffers (chamber 3 and 4) are then released sequentially at a higher angular velocity and are

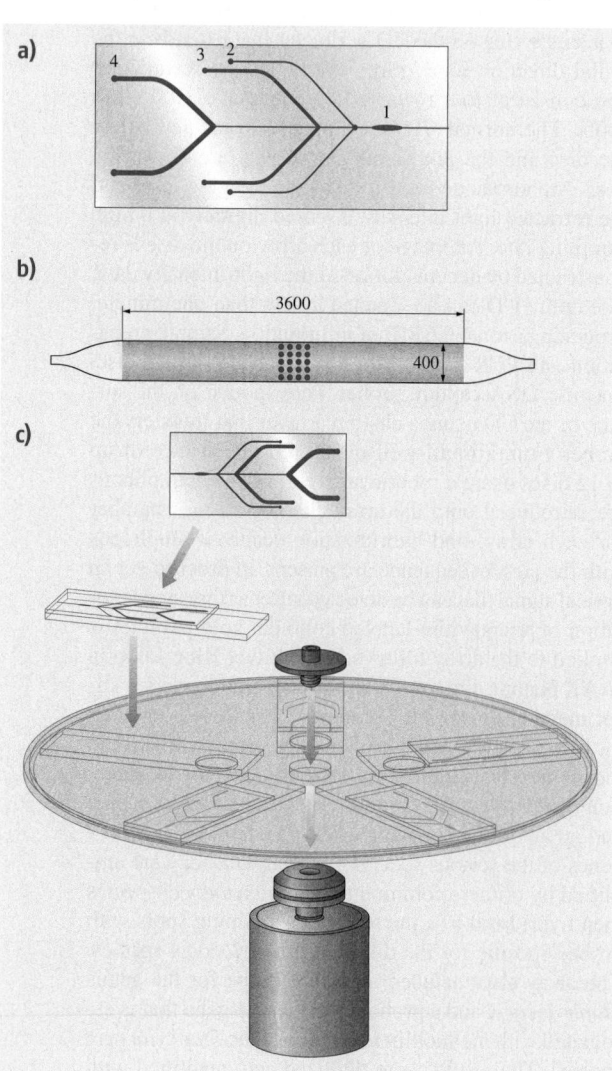

Fig. 20.12a–c Schematic representation of the microfluidic system. (**a**) PDMS microfluidic unit: The test sample (chamber 2) is released first and flows over the hybridization chamber (chamber 1) where the oligonucleotide capture probes are spotted onto the glass support. The wash buffer in chamber 3 and the rinsing buffer in chamber 4 then start to flow at a higher angular velocity. (**b**) Schematic view of the hybridization chamber showing the dimension in μm and the area of the chamber (*shaded section*) that can accommodate up to 150 microarray spots. Layout of the staphylococcal microarray used in the present study is also showed (five capture probes for each species). (**c**) Engraved PDMS is applied to a glass slide on which are arrayed nucleic-acid capture probes. The glass slide is placed on a compact disc support that can hold up to five slides

used to wash the nonspecifically bound targets following the hybridization process.

This custom microarray hybridization microfluidic platform is easy to use, automated, and rapid. It uses standard glass slides which are compatible with commercial arrayers and standard commercial scanners found in most academic departments. In this removable microfluidic system, the hybridization chamber is composed of a low-cost elastomeric material, PDMS using standard moulding methods [20.26], engrafted with a microfluidic network. This elastomeric material reversibly sticks to the glass slide without any adhesives or chemical reactions, forming the microfluidic unit. Placed onto a plastic compact disc-like support, the microfluidic units are spun at different speeds to control fluid movements. To simplify hybridization experiments using this device, buffer compositions and capture probe sequences were optimized to be compatible with room-temperature hybridizations to prevent the need for a heating device. Furthermore, this microfluidic system allows one to drastically reduce the volume of reagents needed for microarray hybridizations and does not require a PCR amplicon purification step, which may be time-consuming.

In a passive hybridization system, a hybridization event requiring collision between a capture probe and the analyte relies solely on diffusion. In such systems, sensitivity is increased by using longer hybridization periods [20.38, 39]. One advantage of flow through hybridization is that the probability of collision between the probe and the analyte is increased by the much shorter diffusion distance allowed by the shallow hybridization chamber, thereby accelerating the hybridization kinetics [20.39–41]. In the study it was shown that for the same concentration of 15-mer oligonucleotides or 368 − bp amplicons, a five-minute flow through hybridization increased the kinetics of hybridization respectively by a factor of 2.5 and 7.5, respectively, in comparison with the passive hybridization. These results are in line with a previous study. Using a microfluidic system, *Chung* et al. have shown a sixfold rate increase between flow-through hybridization versus passive hybridization. However, this system required a 30 − min hybridization step [20.42]. Interestingly, the difference between passive and flow-through hybridization was about three times more important for the amplicons compared to the shorter 15-mer oligonucleotides [20.38]. This could be explained by the higher diffusion coefficient of the smaller oligonucleotide molecules.

To be used for clinical applications, in addition to being rapid and inexpensive, a molecular test should be sensitive and specific. In five minutes of hybridization, the CD system showed a detection limit of 500 atmol of amplified target. This result is comparable with results obtained with more complex microfluidic devices [20.31, 43]. One system using chemiluminescence shows a detection limit of 250 atmol, but requires a three-hour hybridization time [20.44]. In order to detect a significant fluorescent signal, an amplification step is required with microarray technology. The CD microfluidic system reported allows detection of amplicons amplified from 10 bacterial genome copies, which is at least 1000 times more sensitive than results obtained by other groups showing microarray hybridization using microfluidic devices [20.45].

In terms of specificity, the CD system was able to discriminate four different *Staphylococcus* species using a post-PCR hybridization protocol of only 15 min. The *S. aureus* probe designed with only one mismatch in the *S. epidermidis* amplicon sequence, did not show any significant cross-hybridization. This clearly demonstrates the possibility to discriminate one SNP using the CD system at room temperature and with only 10 μl of washing and rinsing buffer. This SNP discrimination capacity will allow rapid identification of bacteria and their antibiotic-resistance genes.

20.3.9 Cell Lysis on CD

There are many types of cell-lysis methods used today that are based on mechanical [20.46], physicochemical [20.47], chemical [20.48] and enzymatic [20.48] principles. The most commonly used methods in biology research labs rely on chemical and enzymatic principles. The main drawbacks of those procedures include intensive labor, adulteration of cell lysate, and the need for additional purification steps. In order to minimize the required steps for cell lysis, a rapid and reagentless cell-lysis method would be very useful. Recently, cell lysis has been demonstrated by *Kim* et al. [20.49] on a microfluidic CD platform. In this purely mechanical lysis method, spherical particles (beads) in a lysis chamber microfabricated in a CD caused disruption of mammalian (CHO-K1), bacterial *Escherichia coli*, and yeast (*Saccharomyces cerevisiae*) cells. Investigators took advantage of interactions between beads and cells generated in rimming flow [20.50, 51] established inside a partially filled annular chamber in the CD rotating around a horizontal axis (Fig. 20.13). To maximize bead–cell interactions in the lysis chamber, the CD was spun forward and backward around this axis, using high accelerations for 5–7 min. Cell disruption efficiency

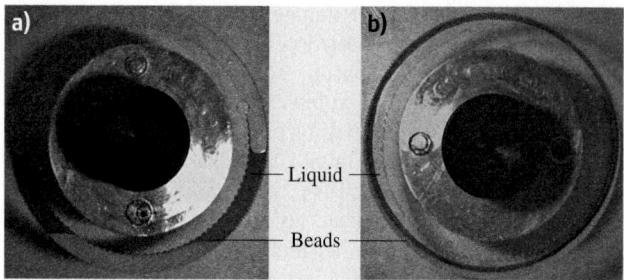

Fig. 20.13a,b Flow patterns for two rotational states of the CD (**a**) At rest: beads sediment at the bottom of the annular chamber (**b**) While spinning: two circumferential bands of beads (*lighter*) and liquid (*darker*) are observed

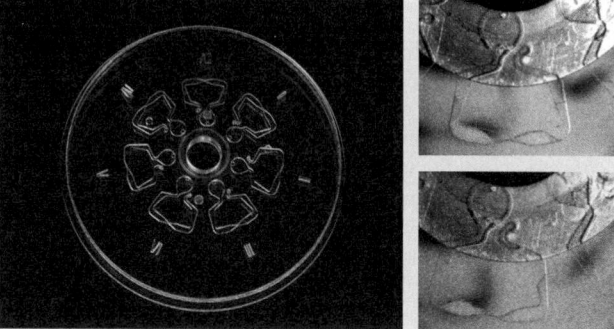

Fig. 20.14 *Left:* photograph of CD. *Right:* still images of a rotating CD (zirconia–silica beads and water-loaded). *Upper right:* more beads are observed on the left because of a rapid stop from a clockwise rotation. *Lower right:* more beads are on the right because of a rapid stop from a counterclockwise rotation

was verified either through direct microscopic viewing or measurement of the DNA concentration after cell lysing. Lysis efficiency relative to a conventional lysis protocol was approximately 65%. Experiments identified the relative contribution of control parameters such as bead density, angular velocity, acceleration rate, and solid-volume fraction.

More recent work [20.52] by the same investigators used the multiplexed lysis design shown in Fig. 20.14. Bead–cell interactions for lysing arise while the beads and cells are pushed back and forth (by switching the CD rotational direction) through a continuously narrowing chamber wall. This phenomenon is called the *keystone effect*. There are two interaction forces associated with the keystone effect: collision induced by the geometry and friction due to a velocity gradient set up along the chamber (i.e., fast in the core and slow around the wall). The investigators used real-time PCR to characterize the performance of this CD de-

sign and achieved 95% lysis efficiency of *B. globigii* spores.

All prototype CDs in this work were fabricated using photolithography and PDMS molding. For the purpose of mechanical cell disruption, an ultra-thick SU-8 process was developed to fabricate a mold featuring high structures (≈ 1 mm) so that sufficiently high lysing chambers could be formed in the PDMS.

In the long term, this work is geared toward CD-based sample-to-answer nucleic-acid analysis which will include cell lysis, DNA purification, DNA amplification, and DNA hybridization detection.

20.3.10 CD Automated Culture of *C. Elegans* for Gene Expression Studies

Kim et al. [20.53] are developing a CD platform for automated cultivation and gene-expression studies of *C. elegans* nematodes. In this research, funded by NASA, the ultimate goal was to understand how a space environment, such as microgravity, hypergravity and radiation, affect various living creatures. The space environment can cause various physiological changes in organisms that have evolved in unit gravity ($1 \times g$) [20.54]. The CD platform is of particular interest in space studies because of its ability to provide a $1 \times g$ control using centripetal force, however its use is not particularly limited to these space study applications. A CD capable of the automated culture of *C. elegans* has been developed and is discussed in this section.

The culture system for *C. elegans* contains cultivation chambers, waste chambers, microchannels, and venting holes. Feeding and waste-removal processes are achieved automatically using centrifugal-force-driven fluidics. In this microfluidic system, the nutrient, Escherichia coli (*E. coli*) and the liquid media are automatically managed for the feeding and waste-removal processes. *C. elegans* was selected as a model organism for the gene-expression experiment in space due to its short lifespan (2–3 weeks), availability of green fluorescent protein (GFP) mutants, ease of laboratory cultivation and completely sequenced genome. Moreover, one can observe its transparent body with a microscope.

The main fabrication material of microfluidic platform is polydimethylsiloxane (PDMS), which is highly permeability to gases (a requirement for any aerobic culture), a chemically inert surface and optically transparent down to 300 nm such that it can be used to observe the behavior of *C. elegans*.

Fig. 20.15 (a) Schematic illustration of the microfluidic structure employed for the CD cultivation system. The fluidic structure contains a nutrient reservoir (**1**), a cultivation chamber (**2**), and a waste reservoir (**3**). A liquid nutrient is loaded in a nutrient reservoir (**1**). Upon increasing the rotation rate of the system, the nutrient solution is gated into the cultivation chamber and some of the waste from the cultivation chamber can drain through the microchannels (50 μm × 40 μm). **(b)** A cross section of the waste chamber (**4**) waste-removal channels (**5**) in **(b)**. Note that **(a)** **3** is the same as **(b)** **4**

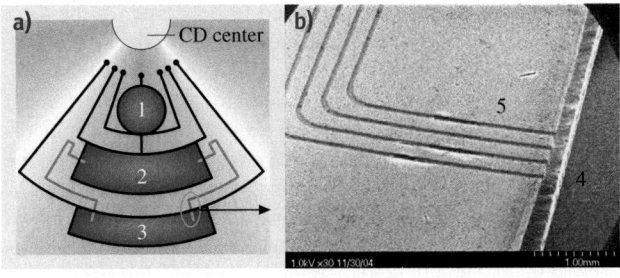

Fig. 20.16 The number of *C. elegans* nematodes cultivated on *E. coli* in S-medium in a CD-based culture under unit gravity over a 14 − d period. Values are mean ± standard error of the mean (SEM); $n = 9$ (cultivation discs)

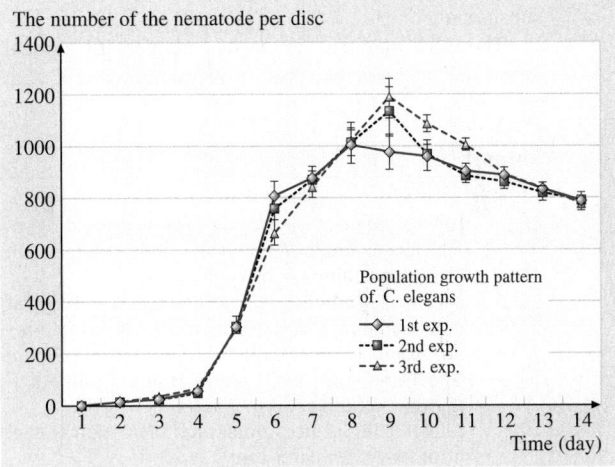

The CD assembly has a two-layer PDMS structure (Fig. 20.15) [20.55]. One layer contains the low-height channels (40 μm, see Figs. 20.2, 20.5) for draining waste from the cultivation chamber, and the other layer contains a cultivation chamber, a loading chamber and microfluidic connections all with a height of 1 mm. This design allows only wastes such as ammonia, not adult worms (80 μm diameter), to be moved from the cultivation chamber to the waste chamber. Cultivation of *C. elegans* was successfully carried out in the CD cultivation system for a period of up to two weeks. *C. elegans* show a specific population growth pattern (Fig. 20.15). Based on these results, the Madou Group and NASA have begun further development of the CD platform for gene-expression experiments to evaluate gene-expression changes in *C. elegans* upon exposure to altered gravity conditions and other factors.

20.4 Conclusion

In comparing miniaturized centrifugal fluidic platforms to other available microfluidic propulsion methods we have demonstrated how CD-based centrifugal methods are advantageous in many analytical situations because of their versatility in handling a wide variety of sample types, ability to gate the flow of liquids (valving), simple rotational motor requirements, ease and economic fabrication methods, and the large range of flow rates attainable. Most analytical functions required for a lab-on-a-disc, including metering, dilution, mixing, calibration, separation, etc., have all been successfully demonstrated in the laboratory. Moreover, the possibility of maintaining simultaneous and identical flow rates, to perform identical volume additions, to establish identical incubation times, mixing dynamics, and detection in a multitude of parallel CD assay elements makes the CD an attractive platform for multiple parallel assays. The platform has been commercialized by Tecan Boston for high-throughput screening (HTS) [20.4], by Gyros AB for sample-preparation techniques for MALDI [20.20] and by *Abaxis* (in a somewhat larger and less-integrated rotor format compared to the CD format) for human and veterinary diagnostic blood analysis [20.5]. The Abaxis system for human and veterinary medicine uses only dry reagents, but for many diagnostic assays, requiring more fluidic steps, there are severe limitations in progressing toward the lab-on-a-disc goal, as liquid storage on the disc becomes necessary. In high-throughput screening (HTS) applications, the CD platform is being coupled to automated liquid-reagent loading systems and no liquids/reagents need to be stored on the disc. The latter makes the commercial introduction of the CD platform for HTS somewhat simpler [20.4, 20, 30]. There is an urgent need though for the development of methods for long-term reagent storage that incorporates both liquid and vapor barriers to enable the introduction of lab-on-a-disc platforms for a wide variety of fast diagnostic tests. One possible solution to this problem involves the use of lyophilized reagents with common hydration reservoir feeds, but the issue in this situation becomes the speed of the test as the time required to redissolve the lyophilized reagents is often substantial.

The CD platform is easily adapted to optical detection methods because it is manufactured with high-optical-quality plastics, enabling absorption, fluorescence, and microscopy techniques. Additionally, the technology developed by the optical disc industry is being used to image the CD at micron resolution and move to DVD and high-density (HD) DVD will allow submicron resolution. The latter evolution will continue to open up new applications for the CD-based fluid platform. Whereas today the CD fluidic platform may be considered a smart micro-centrifuge, we believe that in the future the integration of fluidics and informatics on DVDs and HD DVDs may lead to a merging of informatics and fluidics on the same disc. One can then envision making very sharp images of the bacteria under test and correlate both test and images with library data on the disc.

References

20.1   For commercial electronic noses and tongues see for example Alpha-MOS, Hillsborough, NJ: http://www.alpha-mos.com

20.2   A. Manz, E. Verpoorte, C. S. Effenhauser, N. Burggraf, D. E. Raymond, D. J. Harrison, H. M. Widmer: Miniaturization of Separation Techniques Using Planar Chip Technology, HRC J High Resolut. Chromatogr. **16**, 433–36 (1993)

20.3   Caliper Life Science, Hopkinton, MA: Homepage: http://www.caliperLS.com

20.4   Tecan, Boston Home Page: Look for LabCD-ADMET system: http://www.tecan-us.com/us-index.htm

20.5   Abaxis, Union City, CA. Home page: http://www.abaxis.com

20.6   M. J. Madou: *Fundamentals of Microfabrication*, 2nd edn. (CRC, Boca Raton 2002)

20.7   S. Miyazaki, T. Kawai, M. Araragi: A piezo-electric pump driven by a flexural progressive wave. In: *Proc. IEEE Micro Electro Mechanical Systems (MEMS '91)* (Nara, Japan 1991) pp. 283–288

20.8   J. W. Jorgenson, E. J. Guthrie: Liquid chromatography in open-tubular columns, J. Chromatogr. **255**, 335–48 (1983)

20.9   D. J. Harrison, Z. Fan, K. Fluri, K. Seiler: Integrated electrophoresis systems for biochemical analyses. In: *Solid State Sensor and Actuator Workshop* (Tech. Dig., Hilton Head Island, S.C. 1994) pp. 21–24

20.10   D. C. Duffy, H. L. Gills, J. Lin, N. F. Sheppard, G. J. Kellogg: Microfabicated centrifugal microfluidic systems: Characterization and multiple enzymatic assays, Anal. Chem. **71**(20), 4669–4678 (1999)

20.11   M. J. Madou, G. J. Kellogg: A centrifuge-based microfluidic platform for diagnostics, The LabCD™ **3259**, 80–93 (1998)

20.12   G. T. A. Kovacs: *Micromachined Transducers Sourcebook* (Dordrecht, WCB/McGraw-Hill, Boston 1998) pp. 787–793

20.13   G. Ekstrand, C. Holmquist, A. Edman Örlefors, B. Hellman, A. Larsson, P. Anderson: Microfluidics in a rotating CD. In: *Micro Total Analysis Systems 2000*, ed. by A. van den Berg, W. Olthuis, P. Bergveld (Kluwer, Dordrecht 2000) pp. 311–314

20.14   A.-L. Tiensuu, O. Öhman, L. Lundbladh, O. Larsson: Hydrophobic valves by ink-jet printing on plastic CDs with integrated microfluidics. In: *Micro Total Analysis Systems 2000*, ed. by A. van den Berg, W. Olthuis, P. Bergveld (Kluwer, Dordrecht 2000) pp. 575–578

20.15   M. J. Madou, Y. Lu, S. Lai, J. Lee, S. Daunert: A centrifugal microfluidic platform – A comparison. In: *Micro Total Analysis Systems 2000*, ed. by A. van den Berg, W. Olthuis, P. Bergveld (Kluwer, Dordrecht 2000) pp. 565–570

20.16   J. Zeng, D. Banerjee, M. Deshpande, J. R. Gilbert, D. C. Duffy, G. J. Kellogg: Design analysis of capillary burst valves in centrifugal microfluidics. In: *Micro Total Analysis Systems 2000*, ed. by A. van den Berg, W. Olthuis, P. Bergveld (Kluwer, Dordrecht 2000) pp. 579–582

20.17   I. H. A. Badr, R. D. Johnson, M. J. Madou, L. G. Bachas: Fluorescent ion-selective optode membranes incorporated onto a centrifugal microfluidics platform, Anal. Chem. **74**(21), 5569–5575 (2002)

20.18   R. D. Johnson, I. H. A. Badr, G. Barrett, S. Lai, Y. Lu, M. J. Madou, L. G. Bachas: Development of a fully integrated analysis system for ions based on ion-selective optodes and centrifugal microfluidics, Anal. Chem. **73**(16), 3940–3946 (2001)

20.19   M. McNeely, M. Spute, N. Tusneem, A. Oliphant: Hydrophobic microfluidics, Proc. Microfluidic Dev. Syst. **3877**, 210–220 (1999)

20.20   Gyros AB: Gyrolab MALDA SP1, Application Report 101, Gyros AB, Uppsala

20.21   S. Lai, S. Wang, J. Luo, J. Lee, S. Yang, M. J. Madou: Design of a compact disk-like microfluidic platform for enzyme-linked immunosorbent assay, Anal. Chem. **76**(7), 1832–1837 (2004)

20.22   G. J. Kellogg, T. E. Arnold, B. L. Carvalho, D. C. Duffy, N. F. Sheppard: Centrifugal microfluidics: Applications. In: *Micro Total Analysis Systems 2000*, ed. by A. van den Berg, W. Olthuis, P. Bergveld (Kluwer, Dordrecht 2000) pp. 239–242

20.23   N. Thomas, A. Ocklind, I. Blikstad, S. Griffiths, M. Kenrick, H. Derand, G. Ekstrand, C. Ellström, A. Larsson, P. Anderson: Integrated cell based assays in microfabricated disposable CD devices. In: *Micro Total Analysis Systems 2000*, ed. by A. van den Berg, W. Olthuis, P. Bergveld (Kluwer, Dordrecht 2000) pp. 249–252

20.24   A. W. Anderson: *Physical Chemistry of Surfaces* (Wiley, New York 1960) pp. 5–6

20.25   Private communication with Gregory J. Kellogg of Tecan, Boston

20.26   D. C. Duffy, J. C. McDonald, O. J. A. Schueller, G. M. Whitesides: Rapid prototyping of microfluidic systems in poly(dimethylsiloxane), Anal. Chem. **70**, 4974–4984 (1998)

20.27   J. Burbaum: Miniaturization technologies in HTS: how fast, how small, how soon?, Drug Discov. Today **3**(7), 313–322 (1998)

20.28   J. V. Zoval, R. Boulanger, C. Blackwell, B. Borchers, M. Flynn, D. Smernoff, R. Landheim, R. Mancinelli, M. J. Madou: Work done by this group, SETI Institute, Orbital Sciences, Dysyscon Inc., Stanford University and NASA Ames.

20.29   R. Barathur, J. Bookout, S. Sreevatsan, J. Gordon, M. Werner, G. Thor, M. Worthington: New disc-based technologies for diagnostic and research applications, Psychiatr. Genet. **12**(4), 193–206 (2002)

20.30   I. Alexandre, Y. Houbion, J. Collet, S. Hamels, J. Demarteau, J.-L. Gala, J. Remacle: Compact disc with both numeric and genomic information as DNA microarray platform, BioTechniques **33**(2), 435–439 (2002)

20.31   Y. Wang, B. Vaidya, H. D. Farquar, W. Stryjewski, R. P. Hammer, R. L. McCarley, S. A. Soper, Y. W. Cheng, F. Barany: Microarrays assembled in microfluidic chips fabricated from poly(methyl methacrylate) for the detection of low-abundant DNA mutations, Anal. Chem. **75**, 1130–1140 (2003)

20.32   V. Mikhailovich, S. Lapa, D. Gryadunov, A. Sobolev, B. Strizhkov, N. Chernyh, O. Skotnikova, O. Irtuganova, A. Moroz, V. Litvinov, M. Vladimirskii, M. Perelman, L. Chernousova, V. Erokhin, A. Zasedatelev, A. Mirzabekov: Identification of rifampin-resistant mycobacterium tuberculosis strains by hybridization, PCR, and ligase detection reaction on oligonucleotide microchips, J. Clin. Microbiol. **39**, 2531–2540 (2001)

20.33   S. Bekal, R. Brousseau, L. Masson, G. Prefontaine, J. Fairbrother, J. Harel: Rapid identification of Escherichia coli pathotypes by virulence gene detection with DNA microarrays, J. Clin. Microbiol. **41**, 2113–2125 (2003)

20.34   S. G. Bavykin, J. P. Akowski, V. M. Zakhariev, V. E. Barsky, A. N. Perov, A. D. Mirzabekov: Portable system for microbial sample preparation and oligonucleotide microarray analysis, Appl. Environ. Microbiol. **67**, 922–928 (2001)

20.35   L. Westin, C. Miller, D. Vollmer, D. Canter, R. Radtkey, M. Nerenberg, J. P. O'Connell: Antimicrobial resistance and bacterial identification utilizing a microelectronic chip array, J. Clin. Microbiol. **39**, 1097–104 (2001)

20.36   H. Z. Fan, S. Mangru, R. Granzow, P. Heaney, W. Ho, Q. Dong, R. Kumar: Dynamic DNA hybridization on a chip using paramagnetic beads, Anal. Chem. **71**, 4851–4859 (1999)

20.37   R. Peytavi, F. R. Raymond, K. Boissinot, F. J. Picard, M. Boissinot, L. Bissonnette, M. Ouellette, M. G. Bergeron, G. Jia, J. Zoval, M. Madou: Anal. Chem. (2005) in press

20.38   V. Chan, D. J. Graves, S. E. McKenzie: The biophysics of DNA hybridization with immobilized oligonucleotide probes, Biophys. J. **69**, 2243–2255 (1995)

20.39   M. K. McQuain, K. Seale, J. Peek, T. S. Fisher, S. Levy, M. A. Stremler, F. R. Haselton: Chaotic mixer improves microarray hybridization, Anal. Biochem. **325**, 215–226 (2004)

20.40   E. Bringuier, A. Bourdon: Colloid transport in nonuniform temperature, Phys. Rev. E **67**, 011404 (2003)

20.41   D. Axelrod, M. D. Wang: Reduction-of-dimensionality kinetics at reaction-limited cell-surface receptors, Biophys. J. **66**, 588–600 (1994)

20.42   Y. C. Chung, W. N. Chang, Y. C. Lin, M. Z. Shiu: Microfluidic chip for fast nucleic acid hybridization, Lab on Chip **3**, 228–233 (2003)

20.43   R. H. Liu, R. Lenigk, R. L. Druyor-Sanchez, J. Yang, P. Grodzinski: Hybridization enhancement using cavitation microstreaming, Anal. Chem. **75**, 1911–1917 (2003)

20.44   B. J. Cheek, A. B. Steel, M. P. Torres, Y. Y. Yu, H. Yang: Chemiluminescence detection for hybridization assays on the flow-thru chip, a three-dimensional microchannel biochip, Anal. Chem. **73**, 5777–5783 (2001)

20.45   R. Lenigk, R. H. Liu, M. Athavale, Z. Chen, D. Ganser, J. Yang, C. Rauch, Y. Liu, B. Chan, H. Yu, M. Ray, R. Marrero, P. Grodzinski: Plastic biochannel hybridization devices: A new concept for microfluidic DNA arrays, Anal. Biochem. **311**, 40–49 (2002)

20.46   D. Di Carlo, K.-H. Jeong, L. P. Lee: Reagentless mechanical cell lysis by nanoscale barbs in microchannels for sample preparation, Lab Chip **3**, 287–291 (2003)

20.47   S. W. Lee, Y.-C. Tai: A micro cell lysis device, Sens. Actuators A **73**, 74–79 (1999)

20.48   J. Sambrook, D. W. Russell: *Molecular Cloning* (CSHL Press, Cold Spring Harbor 2001)

20.49   J. Kim, S. H. Jang, G. Jia, J. V. Zoval, N. A. Da Silva, M. J. Madou: Cell lysis on a microfluidic CD (compact disc), Lab Chip **4**, 516–522 (2004)

20.50   K. K. J. Ruschak, L. E. Scriven: Rimming flow of liquid in a rotating horizontal cylinder, J. Fluid Mech. **76**, 113–125 (1976)

20.51   S. T. Thoroddsen, L. Mahadevan: Experimental study of coating flows in a partially-filled horizontally rotating cylinder, Exp. Fluids **23**, 1–13 (1997)

20.52   J. Kim, H. Kido, J. Zoval, R. Peytavi, F. J. Picard, D. Gagné, M. G. Bergeron

20.53   N. Kim, J. Kim, G. Jia, C. Deng, J. Zoval, C. Dempsey, J. Sze, M. Madou

20.54   E. Le Bourg: A review of the effects of microgravity and of hypergravity on aging and longevity, Exp. Gerontl. **34**, 319–336 (1999)

20.55   H. Wu, T. W. Odom, D. T. Chiu, G. M. Whitesides: Fabrication of complex three-dimensional microchannel systems in PDMS, J. Am. Chem. Soc. **125**, 554–559 (2003)

21. Micro/Nanodroplets in Microfluidic Devices

Fluid is often transported in the form of droplets in nature. From the formation of clouds to the condensation of dew on leaves, droplets are formed spontaneously in air, on solids and in immiscible fluids. In biological systems, droplets with lipid bilayer membranes are used to transport subnanoliter amounts of reagents between organelles, between cells, and between organs, in processes that control our day-to-day metabolic activities. The precision of such systems is self-evident and proves that droplet-based systems provide intrinsically efficient ways to perform controlled transport, reactions and signaling.

This precision and efficiency can be utilized in many lab-on-a-chip applications by manipulating individual droplets using microfabricated force gradients. Complex segmented flow processes involving generating, fusing, splitting and sorting droplets have been developed to digitally control fluid volumes and concentrations to nanoliter levels. In this chapter, microfluidic techniques for manipulating droplets are reviewed and analyzed.

Droplet microfluidics, also termed "digital microfluidics", has been the focus of much interest due to its potential for manipulating small quantities of reagent volumes and controlling complex reaction processes. Conventional microfluidics, or analog microfluidics, is governed by diffusion and low Reynold number laminar flow that prevent rapid mixing of miscible liquids. When a liquid is dispersed into another immiscible liquid as droplets, mixing is on the millisecond scale due to the decreased striation area between miscible fluid interfaces. The ability to split and fuse individual droplets further improves the simplicity with which the reagent volume and concentrations can be controlled. Multiphase interactions, such as polymerization at the monomer–initiator fluid interface, lipid self-assembling at the oil–water interface and ion exchange at pH-sensitive interfaces, all add to the benefits of droplet-based systems. These advantages have trans-

formed the application of droplet technology, making it useful not just for making bulk quantities of particles, but also as an integrated multidisciplinary tool that can be applied to fields ranging from materials synthesis to molecular biology.

The versatility of droplet microfluidic systems lies essentially in the ability to selectively "dial-in" the fluidic volume and concentration of a single droplet. This can be achieved either actively or passively. Active devices use dynamic modulation of forces around the droplet to achieve reconfigurable droplet flows. Passive devices use the passive gradients to manipulate droplets through a set of operations governed by channel geometries and predefined wetting properties. In the next two sections, mechanisms for the actuation and control of droplets will be introduced, and important parameters and design criteria will be reviewed.

Part B | 21

21.1 Active or Programmable Droplet System

Active droplet microfluidic systems dynamically modulate droplet motions and offer real-time control of changes in droplet chemistry. Individual droplets may be transported on-demand in opposite directions and then fused to mix reagents in order to induce a chemical reaction. The system can also be programmed to split droplets that would breakup the reaction products into desired volumes.

While a vast literature exists on different droplet actuation mechanisms, ranging from using Marangoni flow [21.2], active chemical gradient [21.3], acoustic energy, pneumatic pressure [21.4], continuous electrowetting [21.1,5–18], and thermocapillary flows [21.19–23], given the span of this chapter it is not possible to review all of these methods of droplet actuation. Instead, this section will focus on actuation mechanisms that have demonstrated the ability to accurately dispense liquid volumes and to consistently split and mix droplet volumes, as well as rapid actuation schemes.

21.1.1 Electrowetting on Dielectric (EWOD)-Based Droplet Microfluidic Devices

Electrowetting is a method that modulates the surface tension of material by applying an electric field. Since most forces diminish at the microscale, surface tension becomes a dominant force, and thus controlling the interfacial tension is an attractive way to actuate fluids at the micro level. While there are several different types of electrowetting configurations, electrowetting on dielectric (EWOD) is the method most widely used to control droplets. EWOD can be applied to virtually any aqueous liquids [21.10]. Furthermore, since the interfacial energy is varied on the dielectric layer in EWOD, the problem of electrode electrolysis is avoided [21.11], and

it can be used to drive both droplets in oil and droplets in air. While applications of electrowetting (EW) and continuous electrowetting (CEW) have been mostly limited to droplet dispensing and actuation of metallic liquids, EWOD has found application in droplet generation, fusion and fission operations for a wide range of chemicals and biological fluids.

21.1.2 Operational principle of EWOD

EW, CEW, and EWOD are all based upon Lippman's principle [21.24],

$$\gamma_{SL} = \gamma_{SL}^0 - \frac{\varepsilon V^2}{2d} , \tag{21.1}$$

where γ_{SL}^0 is the interfacial tension in the absence of the applied potential V, ε is the dielectric constant and d is the thickness of the insulating layer [21.12]. When the droplet is on a dielectric surface covering an electrode source, the contact angle of the droplet may be lowered by applying an electric field. The change in contact angle is predicted by Lippman and Young's equation [21.6,10],

$$\cos\theta = \cos\theta_0 + \frac{\varepsilon V^2}{2d\gamma_{LG}} , \tag{21.2}$$

where θ_0 is the contact angle when the electric field across the interfacial layer is zero [21.6]. While many have suggested that this change in contact angle is required for droplet transport, it has recently been suggested by *Zeng* et al. [21.25] that the change in contact angle merely reflects the difference in interfacial energy and is not required for droplet transport.

For droplet-based EWOD, the device consists of a top ground electrode layer on glass and a bottom layer of control electrodes underneath an insulating layer of dielectric material. A thin hydrophobic layer is coated

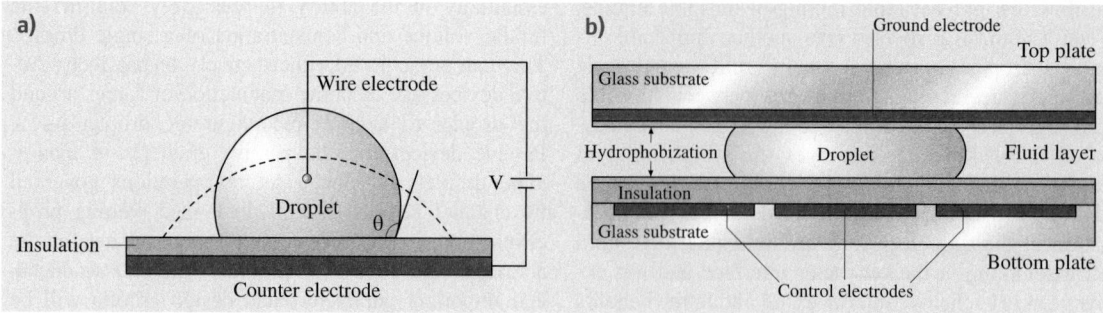

Fig. 21.1 (a) The conductive path of an EWOD device. (b) A schematic of an EWOD device. After [21.1]

on the surface of the electrodes and the insulating material. The hydrophobic coating acts to prevent droplets from spreading into the channel and is not considered to be insulative. The typical EWOD set-up is shown in Fig. 21.1 [21.1].

During operation, the activation of one electrode induces a local interfacial energy difference between adjacent electrodes. When the droplet experiences the energy difference, it moves toward the surface of lower energy, and so through the sequential activation and deactivation of electrodes the droplet can be transported, as shown in Fig. 21.2 [21.1]. Due to the localized activated interfacial energy difference, *Cho* et al. [21.17] have shown that a droplet volume large enough to cover the edge between two electrode surfaces is required to actuate the droplet [21.12]. In addition, *Pollack* et al. [21.12] reported that a threshold voltage is required to actuate the droplet, and it was found that the threshold voltage for a water droplet dispersed in silicon oil is much lower than for a water droplet in air. It was also shown by *Moon* et al. [21.6] that the threshold voltage decreases with the thickness and the capacitance of the dielectric layer. Using barium strontium titanate (BST) as the dielectric material, an the actuation voltage as low as 15 V can be used to transport droplets [21.6]. Once the applied voltage exceeds the threshold voltage, the droplet transport speed increases with the driving voltage. A transport speed of 250 mm/s with an ac potential of > 150 V has been reported by *Cho* et al. [21.17], and the transport of droplet volumes as small as 5 nl have been reported by *Lee* et al. [21.10].

Similarly, more complex droplet manipulations can be achieved through the simultaneous operation of multiple droplet control sequences. Droplets can be fused by

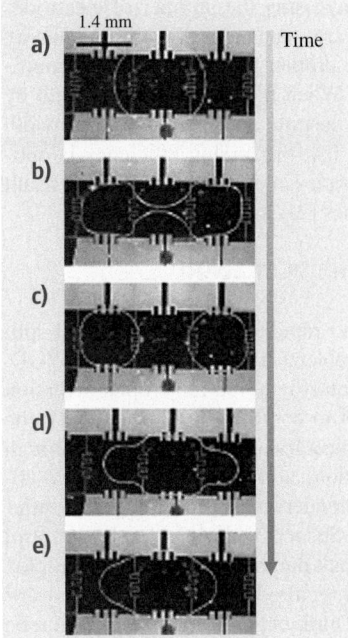

Fig. 21.3 Split and fuse operation for a droplet in EWOD. A single droplet is split by activating the two side electrodes, and then merged through the simultaneous activation of the middle electrode and the subsequent deactivation of the side electrodes. After [21.17] (©2003 IEEE)

transporting two droplets toward the same electrode, or a single droplet can be split into smaller droplets by simultaneously transporting the two halves of the droplet in different directions. Droplet operations can also be combined to form sequential fuse and split operations, as shown in Fig. 21.3. The efficiency of the droplet splitting process, due to the energy required to overcome the capillary pressure, is dependent upon the geometry of the electrodes. *Cho* et al. [21.17] have shown that smaller channel gaps and larger electrode sizes favor droplet fission. The geometric constraint for a square electrode is $d/R_2 < 0.22$, where d is the gap size, and R_2 is roughly half the width of the electrode [21.17].

21.1.3 Reagent Mixing in EWOD

Reagent mixing in droplets is efficient in EWOD; *Paik* et al. demonstrated that the complete mixing of two 800 nl droplets can be achieved as rapidly as 1.7 s [21.26]. It was shown that the droplet mixing rate is dependent on the subsequent motion of the coalesced droplet. In a linear array like that shown in Fig. 21.4, movement of the droplet in one direction promotes mixing while movement in the reverse direction undoes the mixing, due to flow reversibility at low Reynold's number [21.13]. The mixing rate can be further improved by increasing the rate of oscillation of the droplet be-

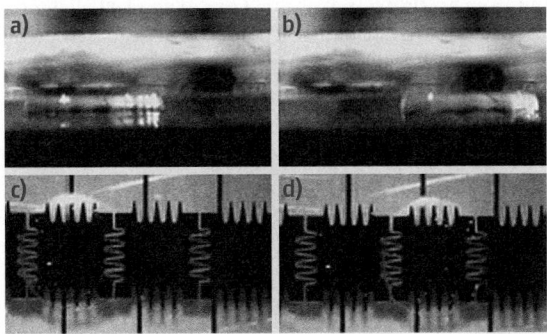

Fig. 21.2a,b Time-lapse images of the droplet transport process. The electrodes are sequentially activated to transfer the droplet from the left (**a**) to the right (**b**). After [21.1], © R. Soc. Chem. by perm.

tween electrodes, increasing the number of electrodes for a larger transport area, and increasing the complexity of movement of the droplet through the use of multidimensional arrays. When the aspect ratio, defined by *Paik* et al. [21.26] as the ratio of the gap size to the width of the electrodes, is 0.4, the mixing efficiency is optimized; lower aspect ratios inhibit vertical flow and result in longer mixing times [21.26].

21.1.4 Improvements in EWOD

Despite the ability to rapidly transport, mix and split droplets, there are problems that limit the use of EWOD. First, the surface contact required for droplet actuation limits its application to complex biological fluids; the absorbance of biomolecules onto the electrodes through electrostatic interactions and passive hydrophobic absorptions eventually renders the electrode unusable after repeated use [21.11]. Second, while it is simple to control the actuation of a small number of electrodes, complex logarithms would be required to control large arrays of electrodes [21.14]. Third, only aqueous droplets can be actuated due to the need for conductivity between the droplet and the ground electrode. Lastly, a droplet volume larger than the size of the electrode is required for EWOD to work, which limits the fluidic volume that can be used. Nevertheless, improvements in EWOD have addressed the first two limitations. *Yoon* et al. [21.11] reported that reducing the duration of the square wave electric field applied improved the durability of the electrode, and *Chiou* et al. [21.14] demonstrated that the addition of a photoconductive layer to EWOD permits the use of light signals to control up to 20 000 electrodes.

21.1.5 Droplet Manipulation via Dielectrophoresis (DEP)

Droplet actuation by dielectrophoresis originates from the polarization of droplets under a nonuniform electric field. The electric field induced on the droplet interacts with the imposed spatially varying electric field to produce controlled droplet motion. The DEP force (F_{DEP}) for a droplet of volume (V) that is suspended in a medium of dielectric constant ε_S and is under the effect of an inhomogeneous electric field (E) can be mathematically described as

$$F_{DEP} = \frac{3}{2} V \varepsilon_S f_{CM} \nabla E^2 , \tag{21.3}$$

where f_{CM} is the real part of the Clausius–Mossotti factor [21.27]. For typical droplet processing applications,

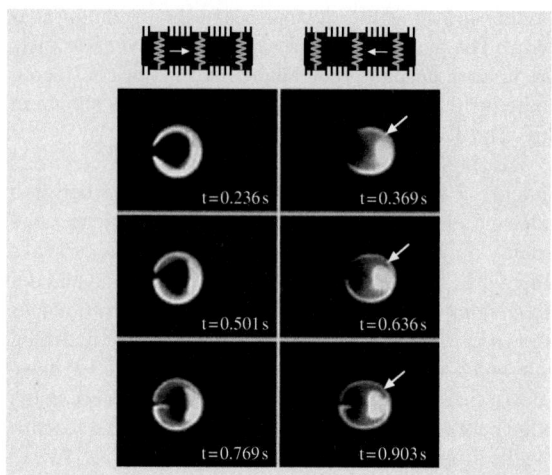

Fig. 21.4 Time-lapse images of flow reversibility. The pattern created was undone by moving the mixed droplet in the reverse direction. After [21.13], © R. Soc. Chem. by perm.

the difference in permittivity between the droplet and the medium is large, which allows f_{CM} to be expressed as

$$f_{CM} = Re \left(\frac{\varepsilon_d - \varepsilon_s}{\varepsilon_d + 2\varepsilon_s} \right) , \tag{21.4}$$

where ε_d is the dielectric constant of the droplet [21.27].

When $\varepsilon_d > \varepsilon_s$, positive dielectrophoresis causes the droplet to move toward the region of high field strength. In contrast, when $\varepsilon_d < \varepsilon_s$, the droplet is displaced toward the high field. Since DEP is conducted through bulk liquid, no physical contact between the droplet and the substrate is required to actuate the droplet. This allows droplet DEP to work with polar, nonpolar, aqueous and organic droplets [21.25]. *Gascoyne* et al. [21.27] has recently demonstrated a complete DEP-based platform that is capable of transporting, dispensing, fusing, and splitting droplets. Shown in Fig. 21.5, the device consists of microfabricated electrodes, all of which

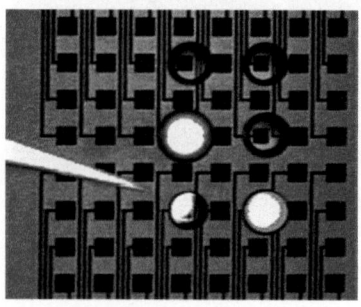

Fig. 21.5 Device used for DEP droplet actuation [21.27], © R. Soc. Chem. by perm.

are independently addressable through a user interface. The surface of the electrode is coated with an electrically insulating material to prevent current leakage and a droplet-repelling layer to minimize droplet surface contact. When $\varepsilon_d > \varepsilon_s$ the droplets move towards the highest field region and when $\varepsilon_d < \varepsilon_s$ droplets can be trapped inside energy "cages", as indicated in Fig. 21.6 [21.27]. Thus, through the controlled excitation of the electrodes, the droplet can be shifted from position to position.

To dispense a droplet, the liquid is initially pushed out from a small orifice; Laplace pressure on the dispensing medium causes this liquid to bulge. Then an electric field is applied in order to induce a DEP force on the liquid droplet so that the injection can be modulated. For a liquid with positive DEP characteristics, the DEP force acts to pull the liquid volume from the tube and the size of the droplet is controlled by the duration that the electric field is applied [21.27]. For liquids with negative DEP, DEP force pulls the liquid into the tube and an increase in the threshold pressure is required to maintain the curvature of the droplet. In this case, the net threshold pressure difference causes fluid injection once the electric field is removed.

Droplet fusion is achieved by moving two droplets into close proximity in order to cause spontaneous coalescence due to the reduction of surface energy. Droplet fission is achieved by using small dielectric beads to remove a portion of the droplet under DEP [21.27].

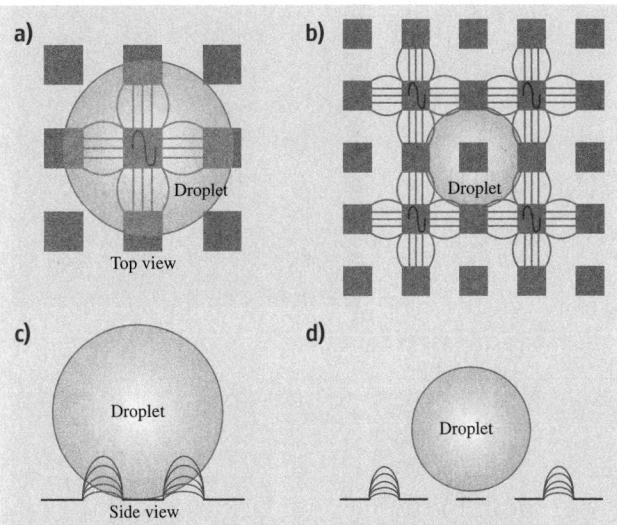

Fig. 21.6a–d Two types of droplet DEP control are possible. (**a**) and (**c**) show top and side views of positive droplet DEP. (**b**) and (**d**) show top and side viewd of negative droplet DEP, where the droplet is "caged" by the electrodes. After [21.27], © R. Soc. Chem. by perm.

Using this system, a sample volume as small as 4 pL can be dispensed, and a droplet 0.29 nL in volume can be moved at up to 670 μm/s across a 60 μm-pitch electrode array actuated by 130 V [21.27].

21.2 Passive Droplet Control Techniques

Unlike the active microfluidic system, where one actuation mechanism and thus a single component set-up is responsible for multipurpose processing, the passive microfluidic device is controlled by immiscible flows. Likewise, almost all passive microfluidic devices are driven by fluidic pumps that provide constant output flow rates or deliver fluids at constant pressure. By manipulating the properties of the immiscible fluid flow within different channel geometries, the operations of droplet generation, fusing, splitting and sorting can be achieved. Droplet transport is naturally achieved within any fluid flow carrying droplets. Since these different operations are driven by the same fluid actuation mechanism, they can be combined to form complex processors for sequential droplet operations in one integrated device. In addition, the device does not have any moving components, which makes fabrication simple and more reliable.

21.2.1 Generation of Monodispersed Droplets

The key features of passive droplet generation system are that the generated droplet sizes can be much smaller than the features of the device, the generated droplets have narrow size distributions, and the generation of droplets is a continuous process. Droplet size distributions of $< 2\%$ have been reported [21.28], and droplet sizes as small as 100 nm have been demonstrated [21.28, 29]. Most passive microfluidic droplet-generating devices utilize either the shear stress created at the immiscible flow interface or the pressure gradient created at a junction of narrow pores to initiate the continuous break-up of droplets. The shear break-up device uses either an asymmetric or a symmetric channel junction to introduce the immiscible fluids, and the pressure gradient break-up device uses either straight-through holes

Fig. 21.7
T-junction
devices from
Nisiako et al.
[21.32],
© R. Soc. Chem.
by perm.

of the droplet, η is the viscosity of the continuous phase, ε is the shear rate in the channel and σ is the interfacial tension at the immiscible fluid interface [21.34]. Assuming that Ca= 1 is the critical condition when the shear stress is large enough to break-up the liquid thread, $r \sim \sigma/\eta\varepsilon$, which is the scaling of droplet sizes observed by *Thorsen* et al. [21.35]. This corroborates with the observations of *Nisisako* et al. [21.32]. As the diameter of the droplet increases to beyond the width of the channel, the effect of the wall becomes dominant over the effect of shear stress [21.36] and hence the size is weakly dependent on the flow rate, as reported by *Tan* et al. [21.30] and *Tice* et al. [21.37, 38]. This is especially true for Ca< 1, where Ca can also be defined by Ca $= \eta U/\sigma$, and the observed droplets become confined to elongated plugs [21.37, 38].

Where Ca is less useful for predicting droplet size, the conservation of mass flow can be used to determine the droplet volume. Tice et al. demonstrated that the length of the plug can be predicted from $l = p[V_d/(V_d + V_c)]$, where l is the length of the plug, p is the period of break-up, V_d is the volume flow of the dispersed phase, and V_c is the volume of the continuous phase [21.37]. In addition, *Tice* et al. [21.37] observed different regimes of droplet formation. Stable droplet formation occurs within a limited regime. This change in behavior for different regimes was also observed by *Anna* et al. [21.29] and *Dreyfus* et al. [21.39]. The flow regime behavior observed by Dreyfus et al. indicates that when the flow rate of the dispersed phase is much lower than the flow rate of the continuous phase, isolated drop formation is observed, but as the flow rate of the dispersed phase increases the stratified regime is observed.

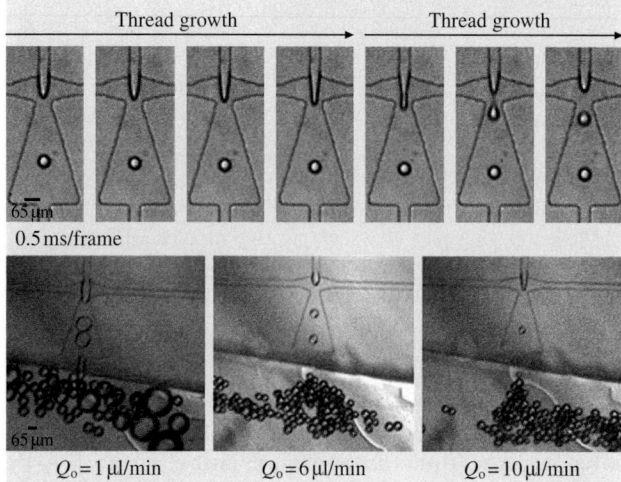

Fig. 21.8 The shear-focusing design demonstrated by *Tan* et al. [21.30, 33], © R. Soc. Chem. by perm.

or microcapillary channels with flat terraces to pressurize the dispersed phase. In the microfluidic flow, the Reynold's number Re (inertial force/viscous force) is much less than 1, and so viscous forces dominate over inertial forces. The viscous force is also weaker in magnitude when compared to the surface tension force, and thus, regardless of the channel geometries, in order to get steady generation of droplets, it is critical that the dispersed phase does not wet the droplet generating surface [21.30, 31].

Shear-Induced Droplet Generation

The generation of droplets in a microfluidic device is governed by the interaction of shear stress with surface tension. The shear stress exerted by the continuous phase acts to deform the liquid surface, while the interfacial force at the immiscible fluid interface acts to restore the deformation. This can be described by the dimensionless capillary number Ca $= \eta\varepsilon r/\sigma$, where r is the radius

Asymmetric and Symmetric Shearing Designs

The asymmetric shearing of immiscible fluids is achieved using T-type intersections, as shown in Fig. 21.7 [21.32]. In both of these devices, the dispersed phase is sheared at the junction by the continuous phase, and the droplet size and frequency of generation are controlled by controlling the continuous phase and disperse phase flows. Due to the asymmetric nature of the droplet break-up process, the reagents injected for mixing are partially exchanged when the liquid droplets break off. This effect is known as "twirling" and will be discussed more detail later [21.37]. Twirling can be an advantage if quick mixing is desired but a disadvantage if the reagents need to be aligned to create bi-color polymeric beads. Furthermore, the simplicity of the design allows two or more generator units to be aligned adjacent or opposite to each other in the same channel for the synchronized generation of droplets. This geometry has been exploited in order to index mixing conditions [21.28,40].

Droplet generation from symmetric fluidic junction generates monodispersed droplets at controlled frequency [21.29,30,33,42]. Figure 21.8 shows an example of the droplet generation process [21.30,33]. As the flow passes through the junction, the narrow width creates maximum velocity and after the flow has passed the junction, fluid velocity decreases due to the increase in channel width. This creates a shear gradient that is maximized at the orifice, producing "shear focusing" break-up. Droplets generated from the "shear focusing" break-up precisely at the orifice; the droplet size and the generation frequency are controlled by the relative flow rates.

21.2.2 Devices Based on Microcapillary Arrays

Straight silicon through-holes and microcapillary arrays (MCs) are attractive designs for forming emulsions. Both of these methods can be performed under conditions of no external flow to produce narrow size distributions. Variations of $< 2\%$ have been reported for straight silicon through-holes [21.43] and $\approx 5\%$ for MC designs [21.41].

In these devices, the production rate is rather low; *Kobayashi* et al. [21.43] reported a production rate of $3-11$ drops per second for a device based on silicon through-holes. Furthermore, the size of the generated droplet is limited by the size of the pores [21.44].

In the silicon through-hole method, the dispersed phase is pushed through small holes etched in the silicon wafer. The droplet detaches when the interfacial tension pinches off the dispersed phase. The generated droplets are monodispersed in terms of droplet size [21.43].

In the microcapillary array design (MC), a glass plate is bonded to the etched silicon channel. The channel connects to a flat terrace that leads to an indented well. The droplet generation process is characterized by the two-step process of filling the terrace with the dispersed phase and the detachment of the liquid as it deforms at the terrace. Upon increasing the pressure of the dispersed phase, the dispersed liquid is entrained on the terrace surface. When liquid reaches the indented well, the edge of the liquid deforms, causing the surface tension to pinch off liquid near the edge of the terrace.

When designing the channel terrace, *Sugiura* et al. [21.41] reported that the diameters of the droplets

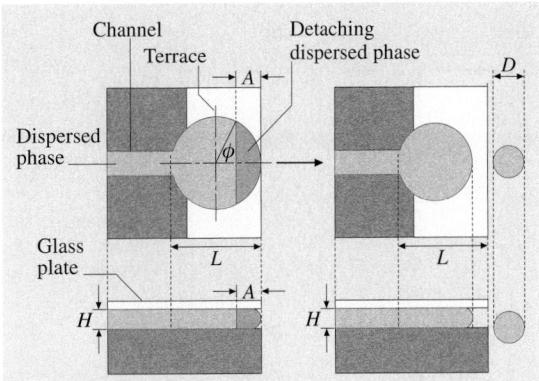

Fig. 21.9 Model used to predict droplet diameter. After [21.41]

generated from MC can be predicted using the terrace length L and the depth H, according to the geometric parameters presented in Fig. 21.9.

$$D = \left(\frac{6(H+0.626)}{\pi} \left\{ \frac{L^2}{4} \cos^{-1}\left(\frac{L-13.76H-8.61}{L} \right) \right. \right.$$
$$- \frac{L(L-13.76H-8.61)}{4}$$
$$\left. \left. \times \sin\left[\cos^{-1}\left(\frac{L-13.76H-8.61}{L} \right) \right] \right\} \right)^{\frac{1}{3}}.$$
$$(21.5)$$

The effect of external flow can also reduce the droplet size, as reported by *Kawakatsu* et al. [21.44]. The droplet size decreases from 47.6 to 37.3 μm as the flow is increased from 1.4×10^{-2} to 2.4 ml/min. However, for 16 and 20 μm droplets generated with a smaller MC channel, varying the flow has no effect on the droplet size.

21.2.3 Double Emulsions

Double emulsions are formed when a liquid is dispersed in an immiscible fluid that is further dispersed in another immiscible fluid. Since the generated double emulsion contains both organic and aqueous phases, it is a versatile way of encapsulating and delivering polar or nonpolar substances. *Kawakatsu* et al. [21.45] was able to generate monodispersed double emulsions from MC devices where the core droplets were generated by the homogenization of water in oil. Controlled formation of both the inner phase and the middle phase has recently been demonstrated by *Okushima* et al. [21.46] using sequential break-up through the two T-junctions

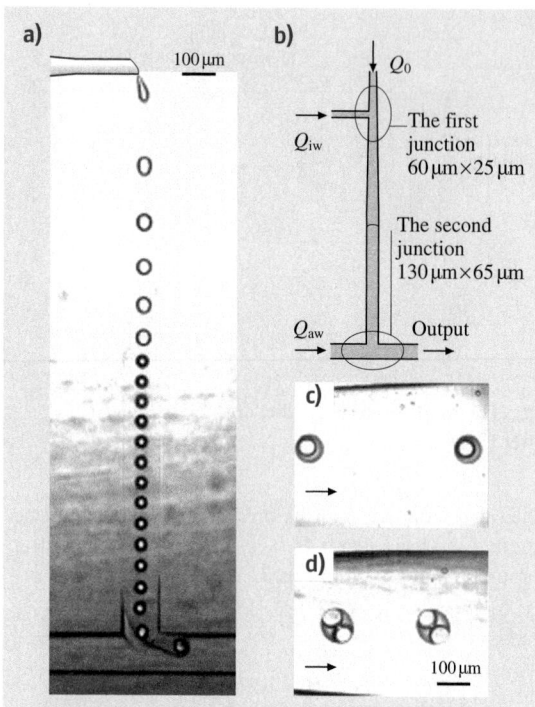

Fig. 21.10 Generation of compound drops through a sequential droplet break-up process. After [21.46], © Am. Soc. Chem. by perm.

methods provide the ability to control the size of the inner droplet, the thickness of the middle or shell layer, and the number of encapsulated inner droplets. Ratios of shell thickness to outer drop radius as low as 3% and as high as 40% have been reported [21.47].

Reagent Mixing

There are two ways to mix reagents in a passive microfluidic system. In the first technique, the reagents to be mixed are introduced as adjacent laminar streams that are broken into single droplets using either an asymmetric or a symmetric shearing system. The reagents are then mixed by diffusion or convection induced either by the surrounding flow or the walls of the channel. The other mixing mechanism is based on fusing two droplets in the microfluidic channel, and this is detailed in the next section.

Mixing in moving plugs is facilitated by recirculation flow, which distributes reagents from the center of the droplet to the edge of the droplet [21.48]. When the reagent gradient is perpendicular to the direction of transport, the recirculation is not as effective at accelerating the mixing, as illustrated in Fig. 21.12 [21.37]. As shown in Fig. 21.13, for droplets generated from a symmetric shearing system, the reagent gradient from the laminar stream is directly transferred into the droplet [21.49]. However, for droplets generated from an asymmetric shearing system, mixing is facilitated by an effect called "twirling", which is an eddy that transports reagents to different parts of the droplet. The effect of twirling is finite and so it increases the mixing rate for short plugs but does not significantly increase the mixing rate for long plugs, as shown in Fig. 21.14 [21.37].

shown in Fig. 21.10. *Utada* et al. [21.47] also demonstrated the precise generation of double emulsions using coaxial flow in order to simultaneously break up an immiscible fluid interface, as shown in Fig. 21.11. Both

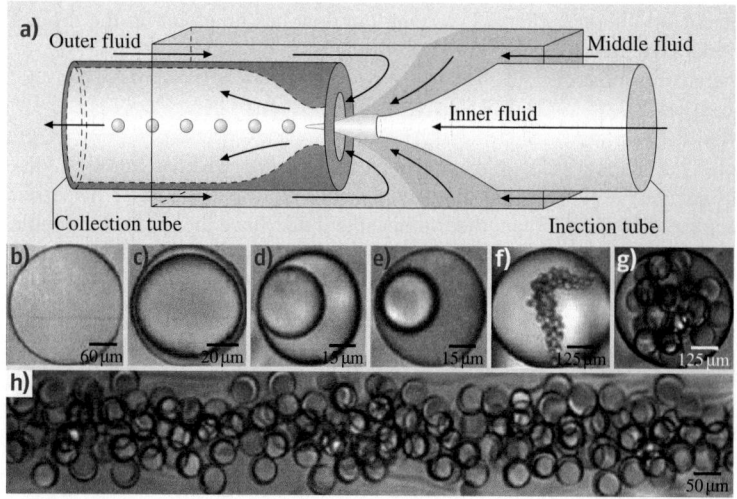

Fig. 21.11 Generation of compound drops through coaxial flow generated in a microfluidic device. Parameters such as the shell thickness, the internal droplet number and the sizes of the internal droplets could be individually controlled. After [21.47], ©AAAS

Mixing can be further improved through the use of winding channels. As the droplet passes through these winding channels, it is stretched, folded and reoriented to induce chaotic mixing inside the droplets [21.50,51]. The time of mixing is verified experimentally to be

$$t_{\text{mix,ca}} \sim (aw/U)\log(Pe)\ , \qquad (21.6)$$

$$Pe = \frac{wU}{D}\ , \qquad (21.7)$$

where w is the cross-section of the microchannel, a is the dimensionless length of the plug measured relative to w, U is the flow velocity, and Pe is the Peclet number [21.50]. Submillisecond mixing times have been reported for mixing in winding channels [21.50].

Droplet Fusion

Similar to the principle of pipetting volumes in and out of a single mixing well, the fission (splitting) and fusion of droplets in a microfluidic channel control both the concentration of reagents and the volume of the mixed samples. While in the channel, droplet manipulations are spatially dependent such that the droplets must travel through fixed channel geometries to be split or fused, the rate operation is controlled by the velocity of the continuous phase, and millisecond-scale operations are possible.

Droplet fusion or coalescence is due to film drainage, which has been reviewed elsewhere [21.36]. Film drainage occurs when drops are close to each other. In the microfluidic channel, however, droplets are separated by plugs of immiscible fluids, meaning that film drainage is unlikely to occur between droplets. The challenge is to control the flow of the liquid separating the droplets. There are several ways to achieve this. *Song* et al. [21.51] utilized the difference in traveling velocity between droplet sizes in straight channels to fuse large and small drops. *Köhler* et al. [21.52] and *Tan* et al. [21.28] used passive channel geometries to temporarily trap and fuse droplets at fluidic junctions. Alternatively, by designing the channel geometry appropriately, the fluid separating the droplets can be continuously drained at bifurcating junctions to achieve the coalescence of a series of droplets, as demonstrated by *Tan* et al. [21.28] with the "flow-rectifying junction". This design, shown in Fig. 21.15, allows droplets to coalesce in various numbers at various generation frequencies and traveling velocities.

Droplet fusion mixing is analogous to the digital mixing of droplets in active devices. Since the mixing process can be made to be weakly dependent on the generation process, it allows reagent volumes and con-

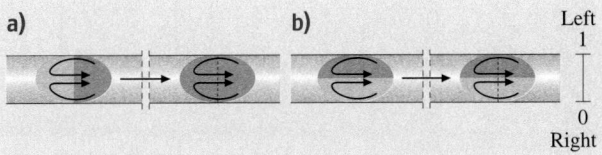

Fig. 21.12 (a) When the concentration gradient is parallel to the direction of transport, recirculation flow mixes the reagents efficiently. **(b)** When the concentration gradient is perpendicular to the direction of transport, mixing by recirculation flow is not efficient. (After [21.37,47], ©AAAS)

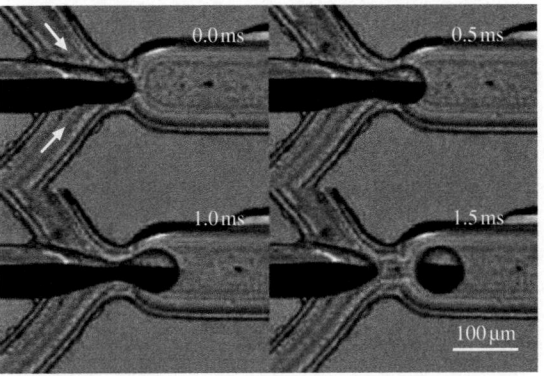

Fig. 21.13 Distinct patterns of laminar flow distribution are transferred into the droplets in a symmetric generation system. After [21.49], © Elsevier, by perm.

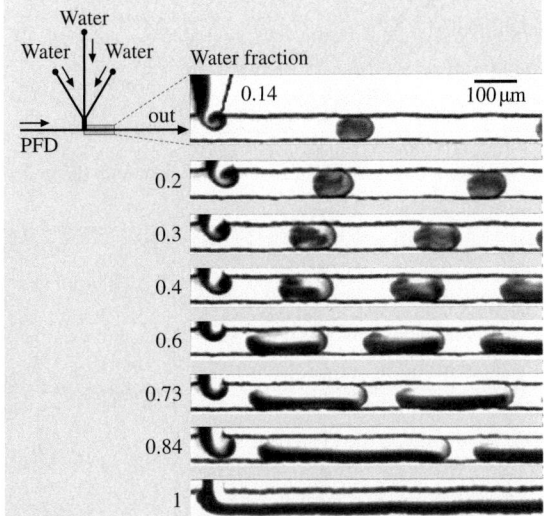

Fig. 21.14 The "twirling effect" transports small amounts of the reagents across the interface immediately after breakup. After [21.37], © Am. Chem. Soc. by perm.

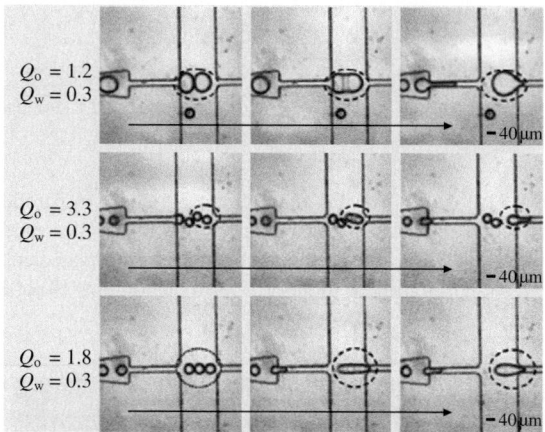

Fig. 21.15 Controlled fusion of droplets in a microfluidic channel using the flow rectifying design. After [21.28], © R. Soc. Chem. by perm.

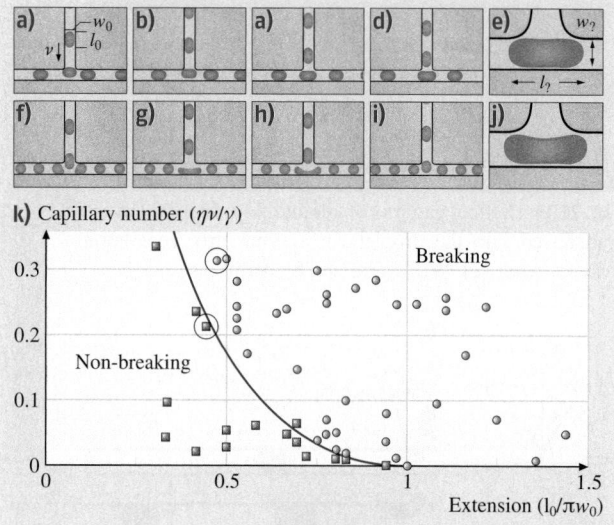

centrations to be controlled independently. Controlled mixing of two different reagents by fusion is shown in Fig. 21.17 [21.28, 53].

Droplet Fission

Droplet fission occurs at the bifurcating junction of the channel. Similar to the splitting of the droplet from the thread, droplets at the bifurcating junction are continuously elongated by the extensional shear stress exerted by the flow, eventually reaching a critical length that can no longer be sustained by the interfacial tension of the droplet surface, which results in droplet break-up. Droplet break-up can be symmetric, where a single droplet is broken into two equal halves, or asymmetric, where a single droplet is broken into multiple parts. Symmetric fission is achieved at a junction of equal bifurcating flow. For a channel with a square cross-sectional geometry, the critical break-up conditions can be expressed according to the initial droplet length and the width of the channel, as indicated in Fig. 21.16. Link et al. [21.54] derived the critical capillary number as

$$C_{cr} = \alpha \varepsilon_0 \left(1/\varepsilon_0^{2/3} - 1 \right)^2 , \qquad (21.8)$$

$$\varepsilon_0 = \frac{\ell_0}{\pi w_0} \qquad (21.9)$$

which has been shown to be in good agreement with experimental results, as shown in Fig. 21.16 [21.28, 54]. ε_0 is the initial extension ratio of the length ℓ_0 to the cir-

Fig. 21.16a–f (*Top*) The parameters used to predict droplet break-up conditions are shown in (**a**). (**a**)–(**f**) show time-lapse images of the break-up process. (*Bottom*) The predicted capillary number $C_{cr} = \alpha \varepsilon_0 (1/\varepsilon_0^{2/3} - 1)^2$ agrees with experimental results. After [21.54] ◄

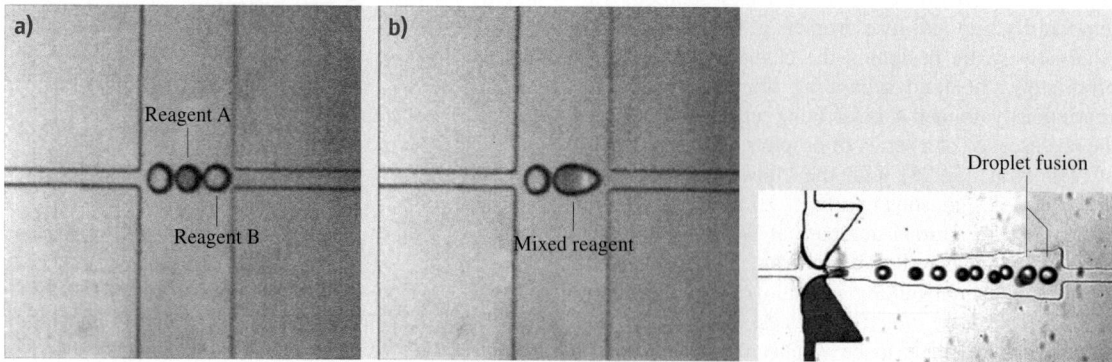

Fig. 21.17 Fusion mixing via a flow-rectifying design (*left*) [21.28] and a channel expansion design (*right*) [21.53]

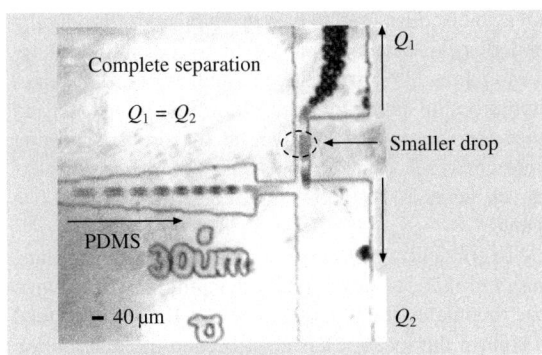

Fig. 21.18 When the flow rates exiting the bifurcating junctions are balanced, droplets are transported toward regions of higher shear stress created by the narrow inlet channel. After [21.28], © R. Soc. Chem. by perm.

cumference (πw_0), and α is the fitting parameter, which is equal to 1 for square channels [21.54].

Droplets of various sizes can be created through asymmetric break-up, such that even submicron-sized droplets can be split from large droplets tens of microns in diameter [21.28]. It was shown that the sizes of the split droplets depend on the size of the original droplet [21.28] and the bifurcating flow distribution at the junction [21.28, 50, 54]. Using droplet break-off from unmixed reagents, by controlling the location and the time of fission, the reagent concentration inside the droplet can be redistributed according to the mixed gradient at the time of break-up to produce arrays of split droplets with different reagent concentrations [21.28].

Droplet Sorting

In an active droplet control system, the complex droplet transport process is modulated by logarithms that control the electrode switches. In a passive system, the transport is guided by the flow distribution in the channel, and controlled multi-pathway transport is difficult to achieve because individual droplets are simply distributed according to the flow rates. The ability to switch droplets between different continuous phases is useful because it makes it possible to filter contaminants from the droplet stream, it allows the concentrations of reagents present in the continuous phase to be changed, it means that we can organize unknown particulates by size, and it allows us to set up a passive monitoring system for variations in droplet size.

Tan et al. [21.28, 56] demonstrated droplet sorting in a microfluidic channel by controlling the shear stress gradient at bifurcating junctions. As shown in Fig. 21.18, a droplet at the junction will be transported toward a region of higher shear stress. Since the shear force experienced by the droplet depends on the surface area of the droplet, passive sorting by droplet size can also be achieved, as shown in the figure, where larger droplets and smaller droplets are separated at the bifurcating junction [21.56].

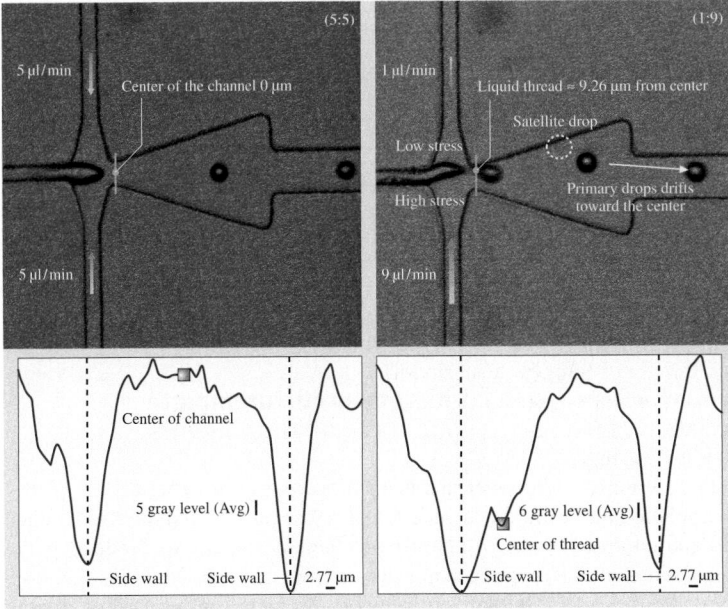

Fig. 21.19 The location of satellite break-up is controlled by the balance of stresses on the liquid thread. (*Left*) Satellite droplets are generated at the center of the channel when the flow is balanced between the two oil inlets located at the top and bottom. (*Right*) Adjusting the flow balance shifts the position of the satellite droplets to the upper part of the channel. After [21.55], © R. Soc. Chem. by perm.

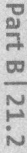

21.2.4 Satellite Droplets

During the droplet generation process, the continuous break-up of the neck connecting the droplet to the liquid thread leads to the formation of small satellite droplets. The presence of satellite droplets increases the size distribution and decreases the mixing accuracy due to the fusion of satellite droplets with primary droplets.

New microfluidic techniques recently demonstrated by *Tan* et al. [21.28, 55, 56], and *Nisisako* et al. [21.57] have shown success in filtering out these satellite droplets using bifurcating channel geometries. *Tan* et al. [21.28, 56] used a combination of a shear stress gradient at the bifurcating junction and a balance of bifurcating flows to passively sort the primary droplets from the satellite droplets. *Nisisako* et al. [21.57] used the difference in positions between primary droplets and satellite droplets generated from the T-junction to filter satellite droplets.

While the generation of satellite droplets is considered to be a problem in most microfluidic devices, it can also be considered to be a mechanism for producing monodispersed nanodroplets. Due to the similar nature of each thread break-up, the distribution of droplet sizes should be the same during each droplet break-off, and since the size of the interface at the droplet splitting point could reach the atomic scale, submicron droplets can be generated without the need for surface active agents.

To efficiently collect the individual satellite droplets, *Tan* et al. [21.55] developed a dynamic flow technique that modulates the stress exerted on the liquid thread to control the location of break-up and the distribution of different satellite droplets in the channel, as shown in Fig. 21.19 and Fig. 21.20. With this design, satellite droplets could be individually selected according to their size during one single break-up event and continuously collected during the generation process. Satellite droplets < 100 nm in diameter were generated using this process, monodispersed satellite droplets with submicron diameters were collected, and a size distribution of ≈ 6% was reported [21.55].

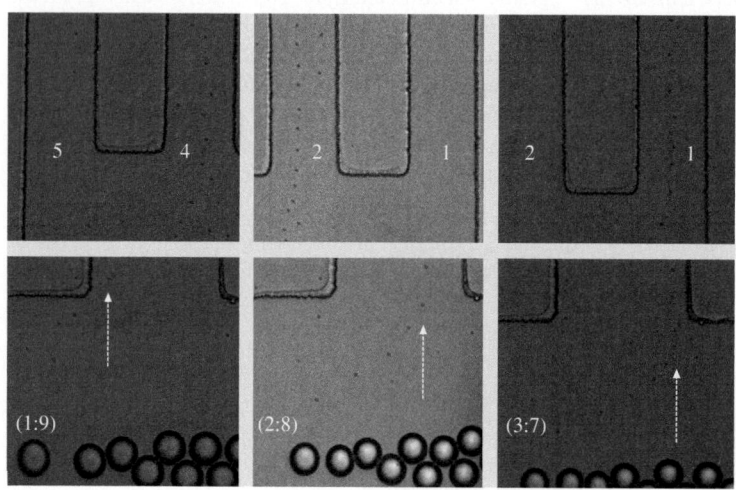

Fig. 21.20 Satellite droplets of the desired size can be selectively generated in the desired collection zone. After [21.55], © R. Soc. Chem. by perm.

21.3 Applications

Droplet-based microfluidic systems possess tremendous potential to improve current emulsion technologies that are widely used in industry to produce sol gels, drugs, synthetic materials and food products. In addition, a wide variety of new applications can be developed due to its precise metering capabilities, rapid and controllable mixing response, and automated combinatorial capabilities.

21.3.1 Droplets as Microtemplate and Encapsulation Agents

Polymer precursor droplets can be generated in microfluidic devices, and subsequently polymerized either by UV [21.49] or chemical agents to produce polymeric beads with narrow size distributions [21.58]. Alternatively, aqueous droplets can be used as a tem-

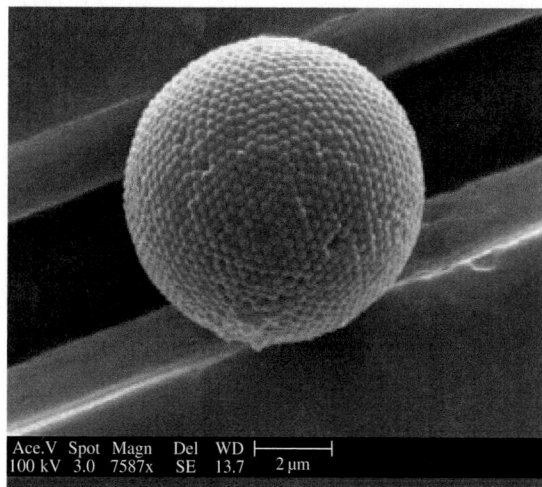

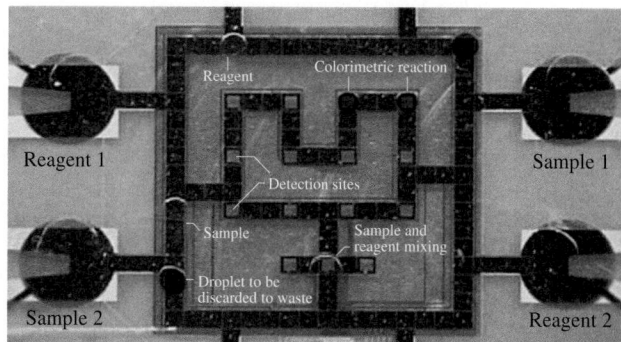

Fig. 21.22 Electrowetting-based device used for detecting glucose in human physiological fluids. After [21.62], © R. Soc. Chem. by perm.

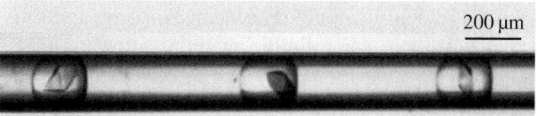

Fig. 21.21 SEM image of latex photonic ball[21.59], © Elsevier, by perm.

Fig. 21.23 Polarized-light micrograph of crystals generated in droplets. After [21.63], © Wiley

plate for synthesizing uniform colloid structures. *Yi* et al. [21.59–61] developed a technique to generate monodisperse aqueous droplets containing latex beads. Evaporation of the droplets results in the formation of photonic balls, as shown in Fig. 21.21 [21.59], and other colloid structures that show unique responses to flows and magnetic fields [21.59–61].

By transferring the laminar patterns to droplets containing two reagents using the symmetric shearing device, *Nisisako* et al. [21.49] demonstrated the production of polymeric beads with bichromal and oriented charge polarities. Similarly, as demonstrated by *Millman* et al. [21.64] using dielectrophoretic-based digital fusion of different polymeric droplets, anisotropic particles with tailored properties can be synthesized. For biological applications, single and multiple cells and organelles can be trapped inside droplets [21.65, 66] for analysis or to provide scaffolds for cell [21.52] and tissue growth.

21.3.2 Droplets as Real-Time Chemical Processors and Combinatorial Synthesizers

Mixing assays that result in photodetectable changes can be rapidly carried out by mixing reagents in microfluidic droplets. Utilizing an EWOD-based droplet microfluidic system, *Srinivasan* et al. [21.62, 67] have demonstrated a rapid on-chip glucose essay involving three steps – dispensing, mixing, and detection – such that glucose concentrations in the range of 25–300 mg/dl could

be detected in less than 60 s [21.67]. Subsequently, a similar programmable lab-on-a-chip device (shown in Fig. 21.22) was applied to the detection of glucose in human physiological fluids, including human whole blood, serum, plasma, urine, saliva, sweat and tears. This latter device has shown great sustainability, lasting for > 25 000 continuous cycles of reagent transport performed at a frequency of 20 Hz and actuated by less than 65 V [21.62].

Droplet-based systems have also been applied to millisecond-scale nanoparticle synthesis [21.68], DNA polymerization chain reactions (PCR) [21.69], DNA analysis [21.70] and to screen protein crystallization conditions [21.40, 63, 71, 72]. In these systems, the variation in the flow rate automatically produces a time-dependent concentration gradient for the reagents inside the droplets. These properties can be used to screen protein crystallization conditions, as demonstrated by *Zheng* et al. [21.71]. < 4 nL of reagent could be used for each trial inside a 7.5 nL aqueous droplet, and hundreds of trials performed at rates of several trials per second can be achieved with computer control [21.71]. Subsequent crystal growth by vapor diffusion from droplets generated X-ray analyzable crystals, as shown in Fig. 21.23 [21.63]. To mitigate the complexity of resolving the time-dependent screening concentrations, an indexing stream of fluorescent droplets generated at the same time as the screening droplets can be used to indicate the concentration of the protein [21.71].

21.3.3 Droplets as Micromechanical Components

Since droplets can be deformed by the flow, they can be fixed into a variety of shapes and sizes that could be used as as microcomponents; furthermore, these structures can be made into permanent micro building blocks for device assembly if they are polymerized [21.73]. Microcomponents such as colloidal microspheres can be made into pumps and valves in microfluidic channels [21.74]. Droplets driven by electrowetting have also been used as the main driving components for liquid micromotors [21.9], and to regulate the frequencies of optical fiber devices [21.8, 75, 76].

21.4 Conclusion

The field of droplet-based microfluidic technology is diverse in terms of the actuation mechanisms and methods of operation used. Essentially, all droplet-based systems are used to control liquid dispensing, mixing, splitting and localization. Microfluidic methods allow droplets to be processed individually, provide accuracy dispensing of fluid volume, and improve the speed of reagent mixing. These factors have made droplet technology into a valuable new tool for controlling micro- and nanointeractions.

References

21.1   M. G. Pollack, A. D. Shenderov, R. B. Fair: Electrowetting-based actuation of droplets for integrated microfluidics, Lab on a Chip **2**, 96–101 (2002)

21.2   R. H. Farahi, A. Passian, T. L. Ferrell, T. Thundat: Microfluidic manipulation via Marangoni forces, Appl. Phys. Lett. **85**(18), 4237–4239 (2004)

21.3   B. S. Gallardo, V. K. Gupta, F. D. Eagerton, L. I. Jong, V. S. Craig, R. R. Shah, N. L. Abbott: Electrochemical principles for active control of liquids on submillimeter scales, Science **283**, 57–60 (1999)

21.4   K. Hosokawa, T. Fujii, I. Endo: Handling of picoliter liquid samples in a poly(dimethysiloxane)-based microfluidic device, Anal. Chem. **71**, 4781–4785 (1999)

21.5   C. Quillet, B. Berge: Electrowetting: A recent outbreak, Curr. Opin. Colloid Sci. **6**, 34–39 (2001)

21.6   H. Moon, S. K. Cho, R. L. Garrell, C. Kim: Low voltage electrowetting-on-dielectric, J. Appl. Phys. **92**(7), 4080–4087 (2002)

21.7   J. Ding, K. Chakrabarty, R. B. Fair: Scheduling of microfluidic operations for reconfigurable two-dimensional electrowetting arrays, **20**(12), 1463–1468 (2001)

21.8   J. Hsieh, P. Mach, F. Cattaneo, S. Yang, T. Krupenkine, K. Baldwin, J. A. Rogers: Tunable microfluidic optical-fiber devices based on electrowetting pumps and plastic microchannels, IEEE Photon. Technol. Lett. **15**(1), 81–83 (2003)

21.9   J. Lee, C. Kim: Surface-tension-driven microactuation based on continuous electrowetting, J. Microelectromech. Syst. **9**(2), 171–180 (2000)

21.10   J. Lee, H. Moon andJ. Fowler, T. Schoellhammer, C. Kim: Electrowetting and electrowetting-on-dielectric for microscale liquid handling, Sens. Actuat. A **95**, 259–268 (2002)

21.11   J. Yoon, R. L Garrell: Preventing biomolecular adsorption in electrowetting-based biofluidic chips, Anal. Chem. **75**(19), 5097–5102 (2003)

21.12   M. G. Pollack, R. B. Fair: Electrowetting-based actuation of liquid droplets for microfluidic applications, Appl. Phys. Lett. **77**(11), 1725–1726 (2000)

21.13   P. Paik, V. K. Pamula, M. G. Pollack, R. B. Fair: Electrowetting-based droplet mixers for microfluidic systems, Lab on a Chip **3**, 28–33 (2003)

21.14   P. Y. Chiou, H. Moon, H. Toshiyoshi, C. Kim, M. C. Wu: Light actuation of liquid by optoelectrowetting, Sens. Actuat. A **104**, 222–228 (2003)

21.15   R. A. Hayes, B. J. Feenstra: Video-speed electronic paper based on electrowetting, Nature **425**, 383–385 (2003)

21.16   S. K. Cho, H. Moon, J. Fowler, C. Kim: *Splitting a liquid droplet for electrowetting-based microfluidics*, ASME International Mechanical Engineering Congress and Exposition (ASME International, New York 2001)

21.17   S. K. Cho, H. Moon, C. Kim: Creating, transporting, cutting, and merging liquid droplets by electrowetting-based actuation for digital microfluidic circuits, J. Microelectromech. Syst. **12**(1), 70–80 (2003)

21.18   T. B. Jones, J. D. Fowler, Y. S. Chang, C. Kim: Frequency-based relationship of electrowetting and dielectrophoretic liquid microactuation, Langmuir **19**, 7646–7651 (2003)

21.19   A. A. Darhuber, J. M. Davis, S. M. Troian, W. W. Reisner: Thermocapillary actuation of liquid flow on chemically patterned surfaces, Phys. Fluids **15**(5), 1295–1304 (2003)

21.20 A. A. Darhuber, J. P. Valentino, S. M. Troian, S. Wagner: Thermocapillary actuation of droplets on chemically patterned surfaces by programmable microheater arrays, J. Microelectromech. Syst. **12**(6), 873–879 (2003)

21.21 A. A. Darhuber, J. P. Valentino, J. M. Davis, S. M. Troian: Microfluidic actuation by modulation of surface stresses, Appl. Phys. Lett. **82**(4), 657–659 (2003)

21.22 A. A. Darhuber, J. Z. Chen, J. M. Davis, S. M. Troian: A study of mixing in thermocapillary flows on micropatterned surfaces, Phil. Trans. Roy. Soc. Lond. A **362**, 1037–1058 (2003)

21.23 A. A. Darhuber, S. M. Troian: Dynamics of capillary spreading along hydrophilic microstripes, Phys. Rev. E. **64**, 031603 (2001)

21.24 M. G. Lippmann: Relations entre les phénomènes électriques et capillaires, Anal. Chim. Phys. **5**(11), 494–549 (1875)

21.25 J. Zeng, T. Korsmeyer: Principles of droplet electro-hydrodynamics for lab-on-a-chip, Lab on a Chip **4**(4), 265–277 (2004)

21.26 P. Paik, V. K. Pamula, R. B. Fair: Rapid droplet mixers for digital microfluidic systems, Lab on a Chip **3**, 253–259 (2003)

21.27 P. R. C. Gascoyne, J. V. Vykoukal, J. A. Schwartz, T. J. Anderson, D. M. Vykoukal, K. W. Current, C. McConaghy, F. F. Becker, C. Andrews: Dielectrophoresis-based programmable fluidic processors, Lab on a Chip **4**(4), 299–309 (2004)

21.28 Y. C. Tan, J. S. Fisher, A. I. Lee, V. Cristini, A. P. Lee: Design of microfluidic channel geometries for the control of droplet volume, chemical concentration, and sorting, Lab on a Chip **4**(4), 292–298 (2004)

21.29 S. L. Anna, N. Bontoux, H. A. Stone: Formation of dispersions using "flow focusing" in microchannels, Appl. Phys. Lett. **82**(3), 364–366 (2003)

21.30 Y. C. Tan, V. Cristini, A. P. Lee: Monodispersed microfluidic droplet generation by shear focusing microfluidic device, Sensors Actuat. B **114**, 350–356 (2006)

21.31 T. Kawakatsu, G. Trägårdh, Ch. Trägårdh, M. Nakajima, N. Oda, T. Yonemoto: The effect of the hydrophobicity of microchannels and components in water and oil phases on droplet formation in microchannel water-in-oil emulsification, Colloids Surfaces **179**, 29–37 (2001)

21.32 T. Nisisako, T. Tori, T. Higuchi: Droplet formation in a microchannel network, Lab on a Chip **2**, 24–26 (2002)

21.33 Y. C. Tan, A. P. Lee: *Nanojet controlled droplet emulsion in microfluidic channels*, 7th International Conference on Micro Total Analysis Systems (Transducers Research Foundation, Lake Tahoe 2003)

21.34 B. J. Briscoe, C. J. Lawrence, W. G. P. Mietus: A review of immiscible fluid mixing, Adv. Colloid Interf. Sci. **81**(1), 1–17 (1999)

21.35 T. Thorsen, R. W. Roberts, F. H. Arnold, S. R. Quake: Dynamic pattern formation in a vesicle-generating microfluidic device, Phys. Rev. Lett. **86**(18), 4163–4166 (2001)

21.36 V. Cristini, Y. C. Tan: Theory and numerical simulation of droplet dynamics in complex flows – a review, Lab on a Chip **4**(4), 257–264 (2004)

21.37 J. D. Tice, H. Song, A. D. Lyon, R. F. Ismagilov: Formation of droplets and mixing in multiphase microfluidics at low values of the Reynolds and the capillary numbers, Langmuir **19**(22), 9127–9133 (2003)

21.38 J. D. Tice, A. D. Lyon, R. F. Ismagilov: Effects of viscosity on droplet formation and mixing in microfluidic channels, Anal. Chim. Acta **507**, 73–77 (2003)

21.39 R. Dreyfus, P. Tabeling, H. Willaime: Ordered and disordered patterns in two-phase flows in microchannels, Phys. Rev. Lett. **90**(14), 144505 (2003)

21.40 B. Zheng, J. D. Tice, R. F. Ismagilov: Formation of droplets of alternating composition in microfludic channels and applications to indexing of concentration in droplet-based assays, Anal. Chem. **76**(17), 4977–4982 (2004)

21.41 S. Sugiura, M. Nakajima, M. Seki: Prediction of droplet diameter for microchannel emulsification, Langmuir **18**, 3854–3859 (2002)

21.42 Q. Xu, M. Nakajima: The generation of highly monodisperse droplets through the breakup of hydrodynamically focused microthread in a microfluidic device, Appl. Phys. Lett. **85**(17), 3726–3728 (2004)

21.43 I. Kobayashi, M. Nakajima, K. Chun, Y. Kikuchi, H. Fujita: Silicon array of elongated through-holes for monodisperse emulsion droplets, AIChE J. **48**(8), 1639–1644 (2002)

21.44 T. Kawakatsu, H. Komori, M. Nakajima, Y. Kikuchi, T. Yonemoto: Production of monodispersed oil-in-water emulsion using crossflow-type silicon microchannel plate, J. Chem. Eng. Jpn. **32**(2), 241–244 (1999)

21.45 T. Kawakatsu, G. Trägårdh, C. Trägårdh: Production of W/O/W emulsions and S/O/W pectin microcapsules by microchannel emulsification, Colloids Surfaces **189**, 257–264 (2001)

21.46 S. Okushima, T. Nisisako, T. Torii, T. Higuchi: Controlled production of monodisperse double emulsions by two-step droplet breakup in microfluidic devices, Langmuir **20**, 9905–9908 (2004)

21.47 A. S. Utada, E. Lorenceau, D. R. Link, P. D. Kaplan, H. A. Stone, D. A. Weitz: Monodisperse double emulsions generated from a microcapillary device, Science **308**, 537–541 (2005)

21.48 K. Handique, M.A. Burns: Mathematical modeling of drop mixing in a slit-type microchannel, J. Micromech. Microeng. **11**(5), 548–554 (2001)

21.49 T. Nisisako, T. Torii, T. Higuchi: Novel microreactors for functional polymer beads, Chem. Eng. J. **101**, 23–29 (2004)

21.50 H. Song, M.R. Bringer, J.D. Tice, C.J. Gerdts, R.F. Ismagilov: Experimental test of scaling of mixing by chaotic advection in droplets moving through microfluidic channels, Appl. Phys. Lett. **83**(22), 4664–4666 (2003)

21.51 H. Song, J.D. Tice, R.F. Ismagilov: A microfluidic system for controlling reaction networks in time, Angew. Chem. Int. Ed. **42**(7), 768–772 (2003)

21.52 J.M. Kohler, Th. Henkel, A. Grodrian, Th. Kirner, M. Roth, K. Martin, J. Metze: Digital reaction technology by micro segmented flow – components, concepts and applications, Chem. Eng. J. **101**(1–3), 201–216 (2004)

21.53 L.H. Hung, W.Y. Tseng, K. Choi, Y.C. Tan, K.J. Shea, A.P. Lee: *Controlled droplet fusion in microfluidic devices*, 8th International Conference on Micro Total Analysis Systems (R. Soc. Chem., Cambridge 2004)

21.54 D.R. Link, S.L. Anna, D.A. Weitz, H.A. Stone: Geometrically mediated breakup of drops in microfluidic devices, Phys. Rev. Lett. **92**(5), 054503 (2004)

21.55 Y.C. Tan, A.P. Lee: Microfluidic filtering and sorting of satellite droplets as the basis of a monodispersed micron and submicron emulsification system, Lab on a Chip **10**, 1178–1183 (2005)

21.56 Y.C. Tan, A.P. Lee: *Droplet sorting by size in microfluidic channels*, 8th International Conference on Micro Total Analysis Systems (R. Soc. Chem., Cambridge 2004)

21.57 T. Nisisako, T. Torii, T. Higuchi: *Separation of satellite droplets using branch microchannel configuration*, 8th International Conference on Micro Total Analysis Systems (R. Soc. Chem., Cambridge 2004)

21.58 S. Sugiura, M. Nakajima, H. Itou, M. Seki: Synthesis of polymeric microspheres with narrow size distributions employing microchannel emulsification, Macromol. Rapid Commun. **22**(10), 773–778 (2001)

21.59 G. Yi, S. Jeon, T. Thorsen, V.N. Manoharan, S.R. Quake, D.J. Pine, S. Yang: Generation of uniform photonic balls by template-assisted colloidal crystallization, Synthetic Met. **139**, 803–806 (2003)

21.60 G. Yi, T. Thorsen, V.N. Manoharan, M. Hwang, S. Jeon, D.J. Pine, S.R. Quake, S. Yang: Generation of uniform colloidal assemblies in soft microfluidic devices, Adv. Mater. **15**(15), 1300–1304 (2003)

21.61 G. Yi, V.N. Manoharan, S. Klein, K.R. Brzezinska, D.J. Pine, F.F. Lange, S. Yang: Monodisperse micrometer-scale spherical assemblies of polymer particles, Adv. Mater. **14**(16), 1137–1140 (2002)

21.62 V. Srinivasan, V.K. Pamula, R.B. Fair: An integrated digital microfluidic lab-on-a-chip for clinical diagnostics on human physiological fluids, Lab on a Chip **4**(4), 310–315 (2004)

21.63 B. Zheng, J.D. Tice, R.F. Ismagilov: Formation of arrayed droplets by soft lithography and two-phase fluid flow, and application in protein crystallization, Adv. Mater. **16**(15), 1365–1368 (2004)

21.64 J.R. Millman, K.H. Bhatt, B.G. Prevo, O.D. Velev: Anisotropic particle synthesis in dielectrophoretically controlled microdroplet reactors, Nature **4**, 98–102 (2005)

21.65 M. He, J.S. Edgar, G.D. Jeffries, R.M. Lorenz, J.P. Shelby, D.T. Chiu: Selective encapsulation of single cells and subcellular organelles into picoliter- and femtoliter-volume droplets, Anal. Chem. **77**, 1539–1544 (2005)

21.66 J.S. Fisher, A.E. Lee: *Cell encapsulation on a microfluidic platform*, 8th International Conference on Micro Total Analysis Systems (R. Soc. Chem., Cambridge 2004) p. 67

21.67 V. Srinivasan, V.K. Pamula, R.B. Fair: Droplet-based microfluidic lab-on-a-chip for glucose detection, Anal. Chim. Acta **507**, 145–150 (2004)

21.68 I. Shestopalov, J.D. Tice, R.F. Ismagilov: Multi-step synthesis of nanoparticles performed on millisecond time scale in a microfluidic droplet-based system, Lab on a Chip **4**, 316–321 (2004)

21.69 Z. Guttenberg, H. Muller, H. Habermuller, A. Geisbauer, J. Pipper, J. Felbel, M. Kielpinski, J. Scriba, A. Wixforth: Planar chip device for PCR and hybridization with surface acoustic wave pump, Lab on a Chip **5**, 308–317 (2004)

21.70 M.A. Burns, C.H. Mastrangelo, T.S. Sammarco, F.P. Man, J.R. Webster, B.N. Johnson, B. Foerster, D. Jones, Y. Fields, A.R. Kaiser, D.T. Burke: Microfabricated structures for integrated DNA analysis, Proc. Natl. Acad. Sci. USA **93**, 5556–5561 (1996)

21.71 B. Zheng, L.S. Roach, R.F. Ismagilov: Screening of protein crystallization conditions on a microfluidic chip using nanoliter-size droplets, J. Am. Chem. Soc. **125**, 11170–11171 (2003)

21.72 B. Zheng, J.D. Tice, L.S. Roach, R.F. Ismagilov: A droplet-based, composite PDMS/glass capillary microfluidic system for evaluating protein crystallization conditions by microbatch and vapor-diffusion methods with on-chip X-ray diffraction, Angew. Chem. Int. Ed. **43**, 2508–2511 (2004)

21.73 D. Dendukuri, K. Tsoi, T.A. Hatton, P.S. Doyle: Controlled synthesis of nonspherical microparticles using microfluidics, Langmuir **21**, 2113–2116 (2005)

21.74 A. Terray, J. Oakey, D.W.M. Marr: Microfluidic control using colloidal devices, Science **296**(7), 1841–1844 (2005)

21.75 B. R. Acharya, T. Krupenkin, S. Ramachandran, Z. Wang, C. C. Huang, J. A. Rogers: Tunable optical fiber devices based on broadband long-period gratings and pumped microfluidics, Appl. Phys. Lett. **83**(24), 4912–4914 (2003)

21.76 F. Cattaneo, K. Baldwin, S. Yang, T. Krupenkine, S. Ramachandran, J. A. Rogers: Digitally tunable microfluidic optical fiber devices, J. Microelectromech. Syst. **12**(6), 907–912 (2003)

Part C
Scanning

Part C Scanning Probe Microscopy

22. Scanning Probe Microscopy – Principle of Operation, Instrumentation, and Probes

Since the introduction of the STM in 1981 and the AFM in 1985, many variations of probe-based microscopies, referred to as SPMs, have been developed. While the pure imaging capabilities of SPM techniques initially dominated applications of these methods, the physics of probe–sample interactions and quantitative analyses of tribological, electronic, magnetic, biological, and chemical surfaces using SPMs have become of increasing interest in recent years. SPMs are often associated with nanoscale science and technology, since they allow investigation and manipulation of surfaces down to the atomic scale. As our understanding of the underlying interaction mechanisms has grown, SPMs have increasingly found application in many fields beyond basic research fields. In addition, various derivatives of all these methods have been developed for special applications, some of them intended for areas other than microscopy.

This chapter presents an overview of STM and AFM and various probes (tips) used in these instruments, followed by details on AFM instrumentation and analyses.

The scanning tunneling microscope (STM), developed by *Dr. Gerd Binnig* and his colleagues in 1981 at the IBM Zurich Research Laboratory in Rueschlikon (Switzerland), was the first instrument capable of directly obtaining three-dimensional (3-D) images of solid surfaces with atomic resolution [22.1]. *Binnig* and *Rohrer* received a Nobel Prize in Physics in 1986 for their discovery. STMs can only be used to study surfaces which are electrically conductive to some degree. Based on their design of the STM, in 1985, *Binnig* et al. developed an atomic force microscope (AFM) to measure ultrasmall forces (less than 1 μN) between the AFM tip surface and the sample surface [22.2] (also see [22.3]). AFMs can be used to measure any engineering surface, whether it is electrically conductive or insulating. The AFM has become a popular surface profiler for topographic and normal force measurements on the micro- to nanoscale [22.4]. AFMs

modified in order to measure both normal and lateral forces are called lateral force microscopes (LFMs) or friction force microscopes (FFMs) [22.5–11]. FFMs have been further modified to measure lateral forces in two orthogonal directions [22.12–16]. A number of researchers have modified and improved the original AFM and FFM designs, and have used these improved systems to measure the adhesion and friction of solid and liquid surfaces on micro- and nanoscales [22.4, 17–30]. AFMs have been used to study scratching and wear, and to measure elastic/plastic mechanical properties (such as indentation hardness and the modulus of elasticity) [22.4, 10, 11, 21, 23, 26–29, 31–36]. AFMs have been used to manipulate individual atoms of xenon [22.37], molecules [22.38], silicon surfaces [22.39] and polymer surfaces [22.40]. STMs have been used to create nanofeatures via localized heating or by inducing chemical reactions under the STM tip [22.41–43]

Table 22.1 Comparison of various conventional microscopes with SPMs

	Optical	SEM/TEM	Confocal	SPM
Magnification	10^3	10^7	10^4	10^9
Instrument Price (U.S. $)	$10k	$250k	$30k	$100k
Technology Age	200 yrs	40 yrs	20 yrs	20 yrs
Applications	Ubiquitous	Science and technology	New and unfolding	Cutting edge
Market 1993	$800 M	$400 M	$80 M	$100 M
Growth Rate	10%	10%	30%	70%

and through nanomachining [22.44]. AFMs have also been used for nanofabrication [22.4, 10, 45–47] and nanomachining [22.48].

STMs and AFMs are used at extreme magnifications ranging from 10^3 to 10^9 in the x-, y- and z-directions in order to image macro to atomic dimensions with high resolution and for spectroscopy. These instruments can be used in any environment, such as ambient air [22.2, 49], various gases [22.17], liquids [22.50–52], vacuum [22.1, 53], at low temperatures (lower than about 100 K) [22.54–58] and at high temperatures [22.59, 60]. Imaging in liquid allows the study of live biological samples and it also eliminates the capillary forces that are present at the tip–sample interface when imaging aqueous samples in ambient air. Low-temperature (liquid helium temperatures) imaging is useful when studying biological and organic materials and low-temperature phenomena such as superconductivity or charge-density waves. Low-temperature operation is also advantageous for high-sensitivity force mapping due to the reduced thermal vibration. They also have been used to image liquids such as liquid crystals and lubricant molecules on graphite surfaces [22.61–64]. While applications of SPM techniques initially focused on their pure imaging capabilities, research into the physics and chemistry of probe–sample interactions and SPM-based quantitative analyses of tribological, electronic, magnetic, biological, and chemical surfaces have become increasingly popular in recent years. Nanoscale science and technology is often tied to the use of SPMs since they allow investigation and manipulation of surfaces down to the atomic scale. As our understanding of the underlying interaction mechanisms has grown, SPMs and their derivatives have found applications in many fields beyond basic research fields and microscopy.

Families of instruments based on STMs and AFMs, called scanning probe microscopes (SPMs), have been developed for various applications of scientific and industrial interest. These include STM, AFM, FFM (or LFM), scanning electrostatic force microscopy (SEFM) [22.65, 66], scanning force acoustic microscopy (SFAM) (or atomic force acoustic microscopy (AFAM)) [22.21, 22, 36, 67–69], scanning magnetic microscopy (SMM) (or magnetic force microscopy (MFM)) [22.70–73], scanning near-field optical microscopy (SNOM) [22.74–77], scanning thermal microscopy (SThM) [22.78–80], scanning electrochemical microscopy (SEcM) [22.81], scanning Kelvin probe microscopy (SKPM) [22.82–86], scanning chemical potential microscopy (SCPM) [22.79], scanning ion conductance microscopy (SICM) [22.87, 88] and scanning capacitance microscopy (SCM) [22.82, 89–91]. When the technique is used to measure forces (as in AFM, FFM, SEFM, SFAM and SMM) it is also referred to as scanning force microscopy (SFM). Although these instruments offer atomic resolution and are ideal for basic research, they are also used for cutting-edge industrial applications which do not require atomic resolution. The commercial production of SPMs started with the STM in 1987 and the AFM in 1989 by Digital Instruments Inc (Santa Barbara, CA, USA). For comparisons of SPMs with other microscopes, see Table 22.1 (Veeco Instruments, Inc., Santa Barbara, CA, USA). Numbers of these instruments are equally divided between the U.S., Japan and Europe, with the following split between industry/university and government laboratories: 50/50, 70/30, and 30/70, respectively. It is clear that research and industrial applications of SPMs are expanding rapidly.

Scanning Probe Microscopy – Principle of Operation, Instrumentation, and Probes | 22.1 Scanning Tunneling Microscope   593

Part C | 22.1

22.1 Scanning Tunneling Microscope

The principle of electron tunneling was first proposed by *Giaever* [22.93]. He envisioned that if a potential difference is applied to two metals separated by a thin insulating film, a current will flow because of the ability of electrons to penetrate a potential barrier. To be able to measure a tunneling current, the two metals must be spaced no more than 10 nm apart. *Binnig* et al. [22.1] introduced vacuum tunneling combined with lateral scanning. The vacuum provides the ideal barrier for tunneling. The lateral scanning allows one to image surfaces with exquisite resolution – laterally to less than 1 nm and vertically to less than 0.1 nm – sufficient to define the position of single atoms. The very high vertical resolution of the STM is obtained because the tunnel current varies exponentially with the distance between the two electrodes; that is, the metal tip and the scanned surface. Typically, the tunneling current decreases by a factor of 2 as the separation is increased by 0.2 nm. Very high lateral resolution depends upon sharp tips. *Binnig* et al. overcame two key obstacles by damping external vibrations and moving the tunneling probe in close proximity to the sample. Their instrument is called the scanning tunneling microscope (STM). Today's STMs can be used in ambient environments for atomic-scale imaging of surfaces. Excellent reviews on this subject have been presented by *Hansma* and *Tersoff* [22.92], *Sarid* and *Elings* [22.94], *Durig* et al. [22.95]; *Frommer* [22.96], *Güntherodt* and *Wiesendanger* [22.97], *Wiesendanger* and *Güntherodt* [22.98], *Bonnell* [22.99], *Marti* and *Amrein* [22.100], *Stroscio* and *Kaiser* [22.101], and *Güntherodt* et al. [22.102].

The principle of the STM is straightforward. A sharp metal tip (one electrode of the tunnel junction) is brought close enough (0.3–1 nm) to the surface to be investigated (the second electrode) to make the tunneling current measurable at a convenient operating voltage (10 mV–1 V). The tunneling current in this case varies from 0.2 to 10 nA. The tip is scanned over the surface at a distance of 0.3–1 nm, while the tunneling current between it and the surface is measured. The STM can be operated in either the constant current mode or the constant height mode, Fig. 22.1. The left-hand column of Fig. 22.1 shows the basic constant current mode of operation. A feedback network changes the height of the tip z to keep the current constant. The displacement of the tip, given by the voltage applied to the piezoelectric drive, then yields a topographic map of the surface. Alternatively, in the constant height mode, a metal tip

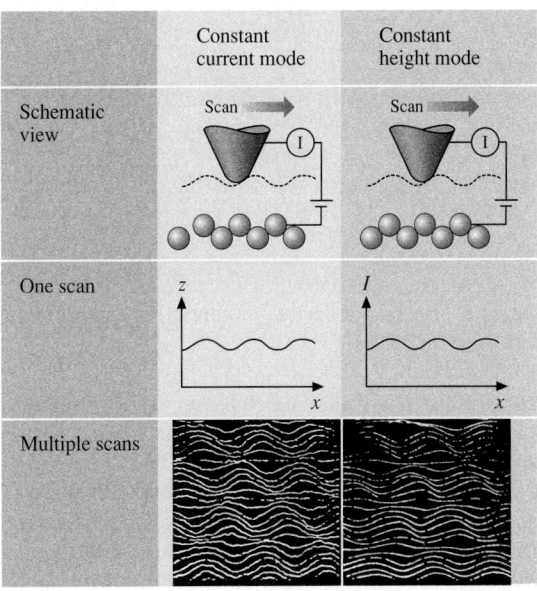

Fig. 22.1 An STM can be operated in either the constant-current or the constant-height mode. The images are of graphite in air [22.92]

can be scanned across a surface at nearly constant height and constant voltage while the current is monitored, as shown in the right-hand column of Fig. 22.1. In this case, the feedback network responds just rapidly enough to keep the average current constant. The current mode is generally used for atomic-scale images; this mode is not practical for rough surfaces. A three-dimensional picture $[z(x, y)]$ of a surface consists of multiple scans $[z(x)]$ displayed laterally to each other in the y direction. It should be noted that if different atomic species are present in a sample, the different atomic species within a sample may produce different tunneling currents for a given bias voltage. Thus the height data may not be a direct representation of the topography of the surface of the sample.

22.1.1 The STM Design of Binnig et al.

Figure 22.2 shows a schematic of an AFM designed by Binnig and Rohrer and intended for operation in ultrahigh vacuum [22.1, 103]. The metal tip was fixed to rectangular piezodrives P_x, P_y, and P_z made out of commercial piezoceramic material for scanning. The sample is mounted via either superconducting magnetic levitation or a two-stage spring system to achieve

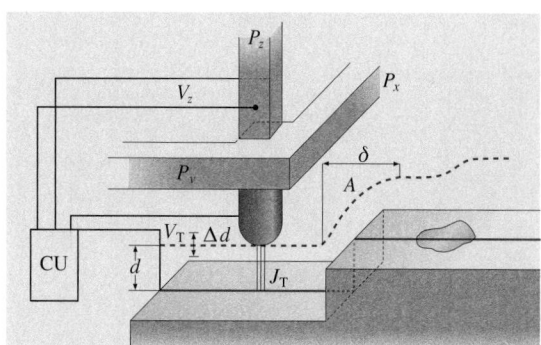

Fig. 22.2 Principle of operation of the STM, from *Binnig* and *Rohrer* [22.103]

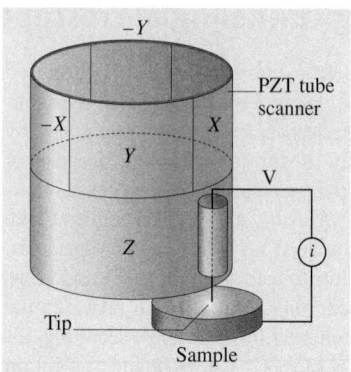

Fig. 22.3
Principle of operation of a commercial STM. A sharp tip attached to a piezoelectric tube scanner is scanned on a sample

a stable gap width of about 0.02 nm. The tunnel current J_T is a sensitive function of the gap width d where $J_T \propto V_T \exp(-A\phi^{1/2}d)$. Here V_T is the bias voltage, ϕ is the average barrier height (work function) and the constant $A = 1.025\,\text{eV}^{-1/2}\text{Å}^{-1}$. With a work function of a few eV, J_T changes by an order of magnitude for an angstrom change in d. If the current is kept constant to within, for example, 2%, then the gap d remains constant to within 1 pm. For operation in the constant current mode, the control unit CU applies a voltage V_z to the piezo P_z such that J_T remains constant when scanning the tip with P_y and P_x over the surface. At a constant work function ϕ, $V_z(V_x, V_y)$ yields the roughness of the surface $z(x, y)$ directly, as illustrated by a surface step at A. Smearing the step, δ (lateral resolution) is on the order of $(R)^{1/2}$, where R is the radius of the curvature of the tip. Thus, a lateral resolution of about 2 nm requires tip radii on the order of 10 nm. A 1-mm-diameter solid rod ground at one end at roughly 90° yields overall tip radii of only a few hundred nanometers, the presence of rather sharp microtips on the relatively dull end yields a lateral resolution of about 2 nm. In situ sharpening of the tips, achieved by gently touching the surface, brings the resolution down to the 1-nm range; by applying high fields (on the order of 10^8 V/cm) for, say, half an hour, resolutions considerably below 1 nm can be reached. Most experiments have been performed with tungsten wires either ground or etched to a typical radius of 0.1–10 μm. In some cases, in situ processing of the tips has been performed to further reduce tip radii.

22.1.2 Commercial STMs

There are a number of commercial STMs available on the market. Digital Instruments, Inc. introduced the first

commercial STM, the Nanoscope I, in 1987. In the recent Nanoscope IV STM, intended for operation in ambient air, the sample is held in position while a piezoelectric crystal in the form of a cylindrical tube (referred to as a PZT tube scanner) scans the sharp metallic probe over the surface in a raster pattern while sensing and relaying the tunneling current to the control station (Fig. 22.3). The digital signal processor (DSP) calculates the tip–sample separation required by sensing the tunneling current flowing between the sample and the tip. The bias voltage applied between the sample and the tip encourages the tunneling current to flow. The DSP completes the digital feedback loop by relaying the desired voltage to the piezoelectric tube. The STM can operate in either the "constant height" or the "constant current" mode, and this can be selected using the control panel. In the constant current mode, the feedback gains are set high, the tunneling tip closely tracks the sample surface, and the variation in the tip height required to maintain constant tunneling current is measured by the change in the voltage applied to the piezo tube. In the constant height mode, the feedback gains are set low, the tip remains at a nearly constant height as it sweeps over the sample surface, and the tunneling current is imaged.

Physically, the Nanoscope STM consists of three main parts: the head, which houses the piezoelectric tube scanner which provides three-dimensional tip motion and the preamplifier circuit for the tunneling current (FET input amplifier) mounted on the top of the head; the base on which the sample is mounted; and the base support, which supports the base and head [22.4]. The base accommodates samples which are up to 10 mm by 20 mm and 10 mm thick. Scan sizes available for the STM are 0.7 μm (for atomic resolution), 12 μm, 75 μm and 125 μm square.

The scanning head controls the three-dimensional motion of the tip. The removable head consists of a piezo

Scanning Probe Microscopy – Principle of Operation, Instrumentation, and Probes | 22.1 Scanning Tunneling Microscope   595

Part C | 22.1

tube scanner, about 12.7 mm in diameter, mounted into an Invar shell, which minimizes vertical thermal drift because of the good thermal match between the piezo tube and the Invar. The piezo tube has separate electrodes for x, y and z motion, which are driven by separate drive circuits. The electrode configuration (Fig. 22.3) provides x and y motions which are perpendicular to each other, it minimizes horizontal and vertical coupling, and it provides good sensitivity. The vertical motion of the tube is controlled by the Z electrode, which is driven by the feedback loop. The x and y scanning motions are each controlled by two electrodes which are driven by voltages of the same magnitude but opposite signs. These electrodes are called $-y$, $-x$, $+y$, and $+x$. Applying complimentary voltages allows a short, stiff tube to provide a good scan range without the need for a large voltage. The motion of the tip that arises due to external vibrations is proportional to the square of the ratio of vibration frequency to the resonant frequency of the tube. Therefore, to minimize the tip vibrations, the resonant frequencies of the tube are high: about 60 kHz in the vertical direction and about 40 kHz in the horizontal direction. The tip holder is a stainless steel tube with an inner diameter of 300 μm when 250 μm-diameter tips are used, which is mounted in ceramic in order to minimize the mass at the end of the tube. The tip is mounted either on the front edge of the tube (to keep the mounting mass low and the resonant frequency high) (Fig. 22.3) or the center of the tube for large-range scanners, namely 75 and 125 μm (to preserve the symmetry of the scanning). This commercial STM accepts any tip with a 250 μm-diameter shaft. The piezotube requires x–y calibration, which is carried out by imaging an appropriate calibration standard. Cleaved graphite is used for heads with small scan lengths while two-dimensional grids (a gold-plated rule) can be used for long-range heads.

The Invar base holds the sample in position, supports the head, and provides coarse x–y motion for the sample. A sprung-steel sample clip with two thumb screws holds the sample in place. An x–y translation stage built into the base allows the sample to be repositioned under the tip. Three precision screws arranged in a triangular pattern support the head and provide coarse and fine adjustment of the tip height. The base support consists of the base support ring and the motor housing. The stepper motor enclosed in the motor housing allows the tip to be engaged and withdrawn from the surface automatically.

Samples to be imaged with the STM must be conductive enough to allow a few nanoamperes of current to flow from the bias voltage source to the area to be scanned. In many cases, nonconductive samples can be coated with a thin layer of a conductive material to facilitate imaging. The bias voltage and the tunneling current depend on the sample. Usually they are set to a standard value for engagement and fine tuned to enhance the quality of the image. The scan size depends on the sample and the features of interest. A maximum scan rate of 122 Hz can be used. The maximum scan rate is usually related to the scan size. Scan rates above 10 Hz are used for small scans (typically 60 Hz for atomic-scale imaging with a 0.7 μm scanner). The scan rate should be lowered for large scans, especially if the sample surfaces are rough or contain large steps. Moving the tip quickly along the sample surface at high scan rates with large scan sizes will usually lead to a tip crash. Essentially, the scan rate should be inversely proportional to the scan size (typically 2–4 Hz for a scan size of 1 μm, 0.5–1 Hz for 12 μm, and 0.2 Hz for 125 μm). The scan

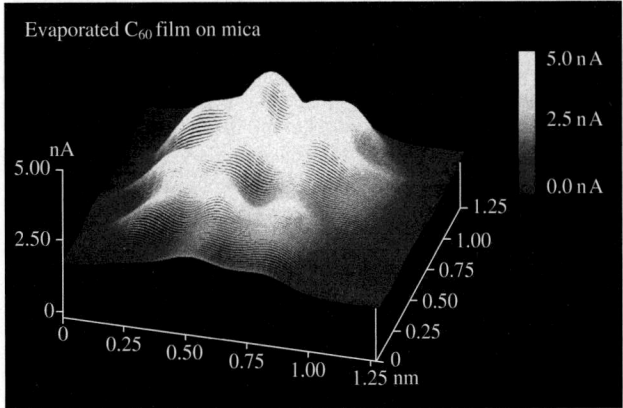

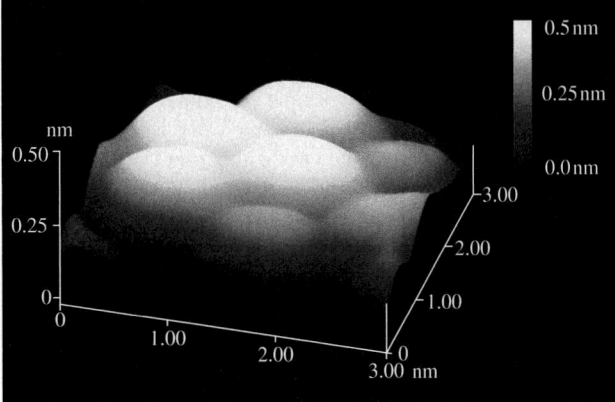

Fig. 22.4 STM images of evaporated C_{60} film on gold-coated freshly cleaved mica obtained using a mechanically sheared Pt–Ir (80-20) tip in constant height mode [22.104]

rate (in length/time) is equal to the scan length divided by the scan rate in Hz. For example, for a scan size of $10\,\mu m \times 10\,\mu m$ scanned at $0.5\,Hz$, the scan rate is $10\,\mu m/s$. 256×256 data formats are the most common. The lateral resolution at larger scans is approximately equal to scan length divided by 256.

Figure 22.4 shows sample STM images of an evaporated C_{60} film on gold-coated freshly-cleaved mica taken at room temperature and ambient pressure [22.104]. Images were obtained with atomic resolution at two scan sizes. Next we describe some STM designs which are available for special applications.

Electrochemical STM

The electrochemical STM is used to perform and monitor the electrochemical reactions inside the STM. It includes a microscope base with an integral potentiostat, a short head with a $0.7\,\mu m$ scan range and a differential preamp as well as the software required to operate the potentiostat and display the result of the electrochemical reaction.

Standalone STM

Standalone STMs are available to scan large samples. In this case, the STM rests directly on the sample. It is available from Digital Instruments in scan ranges of 12 and $75\,\mu m$. It is similar to the standard STM design except the sample base has been eliminated.

22.1.3 STM Probe Construction

The STM probe has a cantilever integrated with a sharp metal tip with a low aspect ratio (tip length/tip shank) to minimize flexural vibrations. Ideally, the tip should be atomically sharp, but in practice most tip preparation methods produce a tip with a rather ragged profile that consists of several asperities where the one closest to the surface is responsible for tunneling. STM cantilevers with sharp tips are typically fabricated from metal wires (the metal can be tungsten (W), platinum-iridium (Pt-Ir), or gold (Au)) and are sharpened by grinding, cutting with a wire cutter or razor blade, field emission/evaporation, ion milling, fracture, or electrochemical polishing/etching [22.105, 106]. The two most commonly used tips are made from either Pt-Ir (80/20) alloy or tungsten wire. Iridium is used to provide stiffness. The Pt-Ir tips are generally formed mechanically and are readily available. The tungsten tips are etched from tungsten wire by an electrochemical process, for example by using 1 M KOH solution with a platinum electrode in a electrochemical cell at about 30 V. In

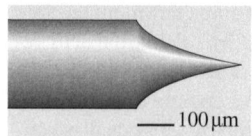

Fig. 22.5 Schematic of a typical tungsten cantilever with a sharp tip produced by electrochemical etching

—— $100\,\mu m$

general, Pt-Ir tips provide better atomic resolution than tungsten tips, probably due to the lower reactivity of Pt. However, tungsten tips are more uniformly shaped and may perform better on samples with steeply sloped features. The tungsten wire diameter used for the cantilever is typically $250\,\mu m$, with the radius of curvature ranging from 20 to $100\,nm$ and a cone angle ranging from 10 to $60°$ (Fig. 22.5). The wire can be bent in an L shape, if so required, for use in the instrument. For calculations of the normal spring constant and the natural frequency of round cantilevers, see *Sarid and Elings* [22.94].

High aspect ratio, controlled geometry (CG) Pt-Ir probes are commercially available to image deep trenches (Fig. 22.6). These probes are electrochemically etched from Pt-Ir (80/20) wire and are polished to a specific shape which is consistent from tip to tip. The probes have a full cone angle of approximately $15°$, and a tip radius of less than $50\,nm$. To image very deep trenches ($> 0.25\,\mu m$) and nanofeatures, focused ion beam (FIB)-milled CG probes with extremely sharp tips (radii $< 5\,nm$) are used. The Pt/Ir probes are coated with a nonconducting film (not shown in the figure) for electrochemistry. These probes are available from Materials Analytical Services (Raleigh, NC, USA).

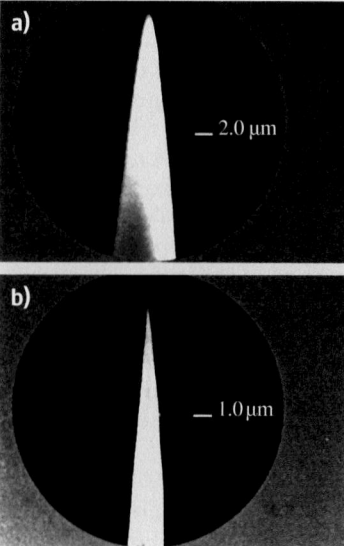

a)

—— $2.0\,\mu m$

b)

—— $1.0\,\mu m$

Fig. 22.6a,b Schematics of (**a**) CG Pt–Ir probe, and (**b**) CG Pt–Ir FIB milled probe

22.2 Atomic Force Microscope

Like the STM, the AFM relies on a scanning technique to produce very high resolution, 3-D images of sample surfaces. The AFM measures ultrasmall forces (less than 1 nN) present between the AFM tip surface and a sample surface. These small forces are measured by measuring the motion of a very flexible cantilever beam with an ultrasmall mass. While STMs require the surface being measured be electrically conductive, AFMs are capable of investigating the surfaces of both conductors and insulators on an atomic scale if suitable techniques for measuring the cantilever motion are used. During the operation of a high-resolution AFM, the sample is generally scanned instead of the tip (unlike for STM) because the AFM measures the relative displacement between the cantilever surface and the reference surface and any cantilever movement from scanning would add unwanted vibrations. However, for measurements of large samples, AFMs are available where the tip is scanned and the sample is stationary. As long as the AFM is operated in the so-called contact mode, little if any vibration is introduced.

The AFM combines the principles of the STM and the stylus profiler (Fig. 22.7). In an AFM, the force between the sample and tip is used (rather than the tunneling current) to sense the proximity of the tip to the sample. The AFM can be used either in the static or the dynamic mode. In the static mode, also referred to as the repulsive or contact mode [22.2], a sharp tip at the end of the cantilever is brought into contact with the surface of the sample. During initial contact, the atoms at the end of the tip experience a very weak repulsive force due to electronic orbital overlap with the atoms in the surface of the sample. The force acting on the tip causes the cantilever to deflect, which is measured by tunneling, capacitive, or optical detectors. The deflection can be measured to within 0.02 nm, so a force as low as 0.2 nN (corresponding to a normal pressure of ∼ 200 MPa for a Si_3N_4 tip with a radius of about 50 nm against single-crystal silicon) can be detected for typical cantilever spring constant of 10 N/m. (To put these number in perspective, individual atoms and human hair are typically a fraction of a nanometer and about 75 μm in diameter, respectively, and a drop of water and an eyelash have masses of about 10 μN and 100 nN, respectively.) In the dynamic mode of operation, also referred to as attractive force imaging or noncontact imaging mode, the tip is brought into close proximity to (within a few nanometers of), but not in contact with, the sample. The cantilever is deliberately vibrated in either amplitude modulation (AM) mode [22.65] or frequency modulation (FM) mode [22.65, 94, 108, 109]. Very weak van der Waals attractive forces are present at the tip–sample interface. Although the normal pressure exerted at the interface is zero in this technique (in order to avoid any surface deformation), it is slow and difficult to use, and is rarely used outside of research environments. The surface topography is measured by laterally scanning the sample under the tip while simultaneously measuring the separation-dependent force or force gradient (derivative) between the tip and the surface (Fig. 22.7). In the contact (static) mode, the interaction force between tip and sample is measured by monitoring the cantilever deflection. In the noncontact (or dynamic) mode, the force gradient is obtained by vibrating the cantilever and measuring the shift in the resonant frequency of the cantilever. To obtain topographic information, the interaction force is either recorded directly, or used as a control parameter for a feedback circuit that maintains the force or force derivative at a constant value. Using an AFM operated in the contact mode, topographic images with a vertical resolution of less than 0.1 nm (as low as 0.01 nm) and a lateral resolution of about 0.2 nm have been obtained [22.3, 50, 110–114]. Forces of 10 nN to 1 pN are measurable with a displacement sensitivity of 0.01 nm. These forces are comparable to the forces associated with chemical bonding, for example 0.1 μN for an ionic bond and 10 pN for a hydrogen bond [22.2]. For further reading, see [22.94–96, 100, 102, 115–119].

Lateral forces applied at the tip during scanning in the contact mode affect roughness measurements [22.120]. To minimize the effects of friction and

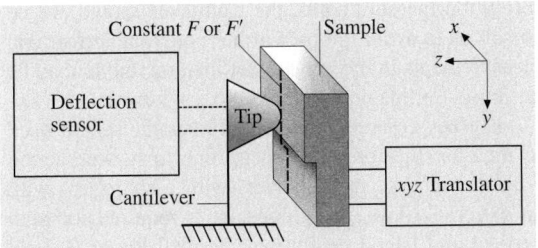

Fig. 22.7 Principle of operation of the AFM. Sample mounted on a piezoelectric scanner is scanned against a short tip and the cantilever deflection is usually measured using a laser deflection technique. The force (in contact mode) or the force gradient (in noncontact mode) is measured during scanning

other lateral forces on topography measurements in the contact mode, and to measure the topographies of soft surfaces, AFMs can be operated in the so-called tapping or force modulation mode [22.32, 121].

The STM is ideal for atomic-scale imaging. To obtain atomic resolution with the AFM, the spring constant of the cantilever should be weaker than the equivalent spring between atoms. For example, the vibration frequencies ω of atoms bound in a molecule or in a crystalline solid are typically 10^{13} Hz or higher. Combining this with an atomic mass m of approximately 10^{-25} kg gives an interatomic spring constant k, given by $\omega^2 m$, of around 10 N/m [22.115]. (For comparison, the spring constant of a piece of household aluminium foil that is 4 mm long and 1 mm wide is about 1 N/m.) Therefore, a cantilever beam with a spring constant of about 1 N/m or lower is desirable. Tips must be as sharp as possible, and tip radii of 5 to 50 nm are commonly available.

Atomic resolution cannot be achieved with these tips at normal loads in the nN range. Atomic structures at these loads have been obtained from lattice imaging or by imaging the crystal's periodicity. Reported data show either perfectly ordered periodic atomic structures or defects on a larger lateral scale, but no well-defined, laterally resolved atomic-scale defects like those seen in images routinely obtained with a STM. Interatomic forces with one or several atoms in contact are 20–40 or 50–100 pN, respectively. Thus, atomic resolution with an AFM is only possible with a sharp tip on a flexible cantilever at a net repulsive force of 100 pN or lower [22.122]. Upon increasing the force from 10 pN, *Ohnesorge* and *Binnig* [22.122] observed that monoatomic steplines were slowly wiped away and a perfectly ordered structure was left. This observation explains why mostly defect-free atomic resolution has been observed with AFM. Note that for atomic-resolution measurements, the cantilever should not be so soft as to avoid jumps. Further note that performing measurements in the noncontact imaging mode may be desirable for imaging with atomic resolution.

The key component in an AFM is the sensor used to measure the force on the tip due to its interaction with the sample. A cantilever (with a sharp tip) with an extremely low spring constant is required for high vertical and lateral resolutions at small forces (0.1 nN or lower), but a high resonant frequency is desirable (about 10 to 100 kHz) at the same time in order to minimize the sensitivity to building vibrations, which occur at around 100 Hz. This requires a spring with an extremely low vertical spring constant (typically 0.05 to 1 N/m) as well as a low mass (on the order of

1 ng). Today, the most advanced AFM cantilevers are microfabricated from silicon or silicon nitride using photolithographic techniques. Typical lateral dimensions are on the order of 100 μm, with thicknesses on the order of 1 μm. The force on the tip due to its interaction with the sample is sensed by detecting the deflection of the compliant lever with a known spring constant. This cantilever deflection (displacement smaller than 0.1 nm) has been measured by detecting a tunneling current similar to that used in the STM in the pioneering work of *Binnig* et al. [22.2] and later used by *Giessibl* et al. [22.56], by capacitance detection [22.123, 124], piezoresistive detection [22.125, 126], and by four optical techniques, namely (1) optical interferometry [22.5, 6, 127, 128] using optical fibers [22.57, 129] (2) optical polarization detection [22.72, 130], (3) laser diode feedback [22.131] and (4) optical (laser) beam deflection [22.7, 8, 53, 111, 112]. Schematics of the four more commonly used detection systems are shown in Fig. 22.8. The tunneling method originally used by *Binnig* et al. [22.2] in the first version of the AFM uses a second tip to monitor the deflection of the cantilever with its force sensing tip. Tunneling is rather sensitive to contaminants and the interaction between the tunneling tip and the rear side of the cantilever can become comparable to the interaction between the tip and sample. Tunneling is rarely used and is mentioned mainly for historical reasons. *Giessibl* et al. [22.56] have used it for a low-temperature AFM/STM design. In contrast to tunneling, other deflection sensors are placed far from the cantilever, at distances of microns to tens of millimeters. The optical techniques are believed

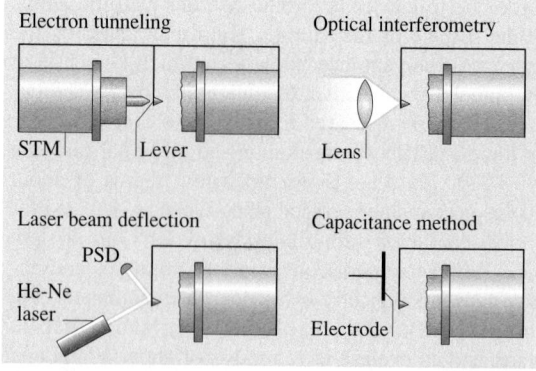

Fig. 22.8 Schematics of the four detection systems to measure cantilever deflection. In each set-up, the sample mounted on piezoelectric body is shown on the right, the cantilever in the middle, and the corresponding deflection sensor on the left [22.118]

to be more sensitive, reliable and easily implemented detection methods than the others [22.94, 118]. The optical beam deflection method has the largest working distance, is insensitive to distance changes and is capable of measuring angular changes (friction forces); therefore, it is the most commonly used in commercial SPMs.

Almost all SPMs use piezo translators to scan the sample, or alternatively to scan the tip. An electric field applied across a piezoelectric material causes a change in the crystal structure, with expansion in some directions and contraction in others. A net change in volume also occurs [22.132]. The first STM used a piezo tripod for scanning [22.1]. The piezo tripod is one way to generate three-dimensional movement of a tip attached at its center. However, the tripod needs to be fairly large ($\sim 50\,\text{mm}$) to get a suitable range. Its size and asymmetric shape makes it susceptible to thermal drift. Tube scanners are widely used in AFMs [22.133]. These provide ample scanning range with a small size. Electronic control systems for AFMs are based on either analog or digital feedback. Digital feedback circuits are better suited for ultralow noise operation.

Images from the AFMs need to be processed. An ideal AFM is a noise-free device that images a sample with perfect tips of known shape and has a perfectly linear scanning piezo. In reality, scanning devices are affected by distortions and these distortions must be corrected for. The distortions can be linear and nonlinear. Linear distortions mainly result from imperfections in the machining of the piezo translators, causing crosstalk between the Z-piezo to the x- and y-piezos, and vice versa. Nonlinear distortions mainly result from the presence of a hysteresis loop in piezoelectric ceramics. They may also occur if the scan frequency approaches the upper frequency limit of the x- and y-drive amplifiers or the upper frequency limit of the feedback loop (z-component). In addition, electronic noise may be present in the system. The noise is removed by digital filtering in real space [22.134] or in the spatial frequency domain (Fourier space) [22.135].

Processed data consists of many tens of thousand of points per plane (or data set). The outputs from the first STM and AFM images were recorded on an x-y chart recorder, with the z-value plotted against the tip position in the fast scan direction. Chart recorders have slow responses, so computers are used to display the data these days. The data are displayed as wire mesh displays or grayscale displays (with at least 64 shades of gray).

22.2.1 The AFM Design of Binnig et al.

In the first AFM design developed by *Binnig* et al. [22.2], AFM images were obtained by measuring the force exerted on a sharp tip created by its proximity to the surface of a sample mounted on a 3-D piezoelectric scanner. The tunneling current between the STM tip and the backside of the cantilever beam to which the tip was attached was measured to obtain the normal force. This force was kept at a constant level with a feedback mechanism. The STM tip was also mounted on a piezoelectric element to maintain the tunneling current at a constant level.

22.2.2 Commercial AFMs

A review of early designs of AFMs has been presented by *Bhushan* [22.4]. There are a number of commercial AFMs available on the market. Major manufacturers of AFMs for use in ambient environments are: Digital Instruments Inc., Topometrix Corp. and other subsidiaries of Veeco Instruments, Inc., Molecular Imaging Corp. (Phoenix, AZ, USA), Quesant Instrument Corp. (Agoura Hills, CA, USA), Nanoscience Instruments Inc. (Phoenix, AZ, USA), Seiko Instruments (Chiba, Japan); and Olympus (Tokyo, Japan). AFM/STMs for use in UHV environments are manufactured by Omicron Vakuumphysik GMBH (Taunusstein, Germany).

We describe here two commercial AFMs – small-sample and large-sample AFMs – for operation in the contact mode, produced by Digital Instruments, Inc., with scanning lengths ranging from about $0.7\,\mu\text{m}$ (for atomic resolution) to about $125\,\mu\text{m}$ [22.9, 111, 114, 136]. The original design of these AFMs comes from *Meyer* and *Amer* [22.53]. Basically, the AFM scans the sample in a raster pattern while outputting the cantilever deflection error signal to the control station. The cantilever deflection (or the force) is measured using a laser deflection technique (Fig. 22.9). The DSP in the workstation controls the z position of the piezo based on the cantilever deflection error signal. The AFM operates in both "constant height" and "constant force" modes. The DSP always adjusts the distance between the sample and the tip according to the cantilever deflection error signal, but if the feedback gains are low the piezo remains at an almost "constant height" and the cantilever deflection data is collected. With high gains, the piezo height changes to keep the cantilever deflection nearly constant (so the force is constant), and the change in piezo height is collected by the system.

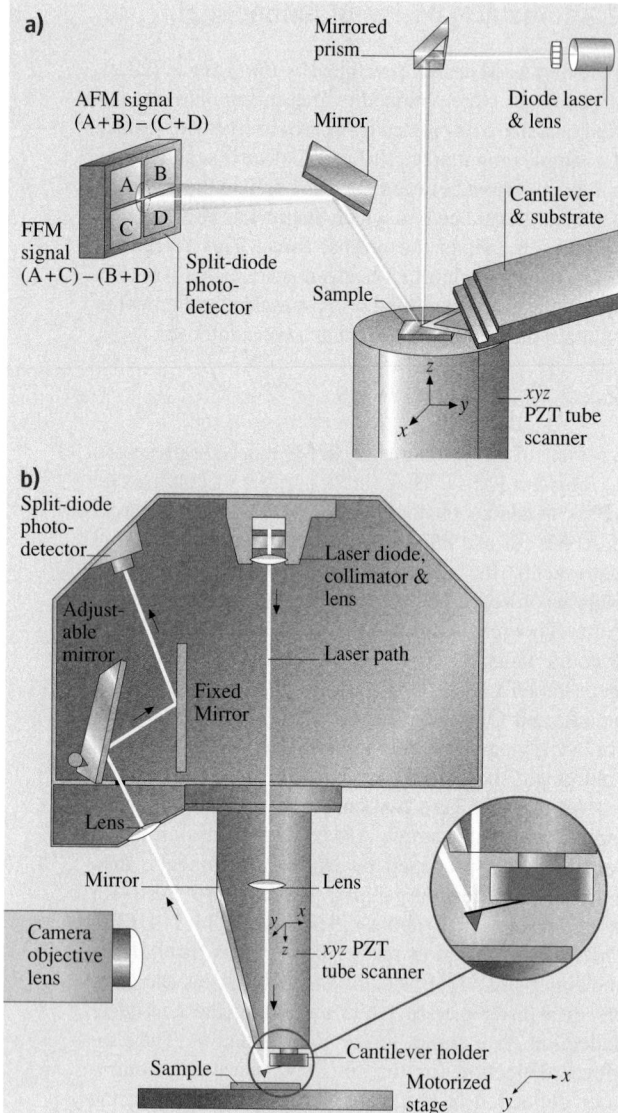

Fig. 22.9a,b Principles of operation of (**a**) a commercial small-sample AFM/FFM, and (**b**) a large-sample AFM/FFM

tical and lateral directions as the sample moves under the tip. A laser beam from a diode laser (5 mW max peak output at 670 nm) is directed by a prism onto the back of a cantilever near its free end, tilted downward at about 10° with respect to the horizontal plane. The reflected beam from the vertex of the cantilever is directed through a mirror onto a quad photodetector (split photodetector with four quadrants) (commonly called a position-sensitive detector or PSD, produced by Silicon Detector Corp., Camarillo, CA, USA). The difference in signal between the top and bottom photodiodes provides the AFM signal, which is a sensitive measure of the cantilever vertical deflection. The topographic features of the sample cause the tip to deflect in the vertical direction as the sample is scanned under the tip. This tip deflection will change the direction of the reflected laser beam, changing the intensity difference between the top and bottom sets of photodetectors (AFM signal). In a mode of operation called the height mode, used for topographic imaging or for any other operation in which the normal forceapplied is to be kept constant, a feedback circuit is used to modulate the voltage applied to the PZT scanner in order to adjust the height of the PZT, so that the cantilever vertical deflection (given by the intensity difference between the top and bottom detector) will remain constant during scanning. The PZT height variation is thus a direct measure of the surface roughness of the sample.

In a large-sample AFM, force sensors based on optical deflection methods or scanning units are mounted on the microscope head (Fig. 22.9b). Because of the unwanted vibrations caused by cantilever movement, the lateral resolution of this design is somewhat poorer than

In the operation of a commercial small-sample AFM (as shown in Fig. 22.9a), the sample (which is generally no larger than 10 mm × 10 mm) is mounted on a PZT tube scanner, which consists of separate electrodes used to precisely scan the sample in the x–y plane in a raster pattern and to move the sample in the vertical (z) direction. A sharp tip at the free end of a flexible cantilever is brought into contact with the sample. Features on the sample surface cause the cantilever to deflect in the ver-

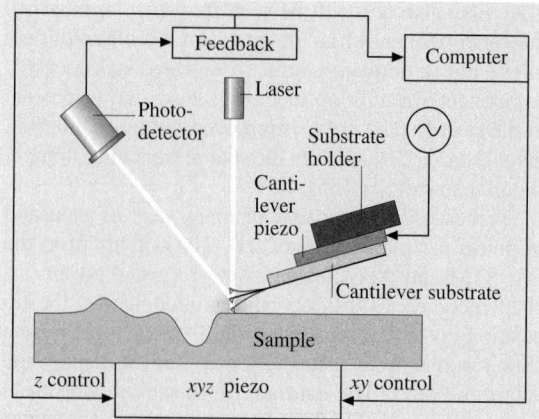

Fig. 22.10 Schematic of tapping mode used for surface roughness measurements

the design in Fig. 22.9a in which the sample is scanned instead of the cantilever beam. The advantage of the large-sample AFM is that large samples can be easily measured.

Most AFMs can be used for topography measurements in the so-called tapping mode (intermittent contact mode), in what is also referred to as dynamic force microscopy. In the tapping mode, during the surface scan, the cantilever/tip assembly is sinusoidally vibrated by a piezo mounted above it, and the oscillating tip slightly taps the surface at the resonant frequency of the cantilever (70–400 kHz) with a constant (20–100 nm) amplitude of vertical oscillation, and a feedback loop keeps the average normal force constant (Fig. 22.10). The oscillating amplitude is kept large enough that the tip does not get stuck to the sample due to adhesive attraction. The tapping mode is used in topography measurements to minimize the effects of friction and other lateral forces to measure the topography of soft surfaces.

Topographic measurements can be made at any scanning angle. At first glance, the scanning angle may not appear to be an important parameter. However, the friction force between the tip and the sample will affect the topographic measurements in a parallel scan (scanning along the long axis of the cantilever). This means that a perpendicular scan may be more desirable. Generally, one picks a scanning angle which gives the same topographic data in both directions; this angle may be slightly different to that for the perpendicular scan.

The left-hand and right-hand quadrants of the photodetector are used to measure the friction force applied at the tip surface during sliding. In the so-called friction mode, the sample is scanned back and forth in a direction orthogonal to the long axis of the cantilever beam. Friction force between the sample and the tip will twist the cantilever. As a result, the laser beam will be deflected out of the plane defined by the incident beam and the beam is reflected vertically from an untwisted cantilever. This produces a difference in laser beam intensity between the beams received by the left-hand and right-hand sets of quadrants of the photodetector. The intensity difference between the two sets of detectors (FFM signal) is directly related to the degree of twisting and hence to the magnitude of the friction force. This method provides three-dimensional maps of the friction force. One problem associated with this method is that any misalignment between the laser beam and the photodetector axis introduces errors into the measurement. However, by following the procedures developed by *Ruan* and *Bhushan* [22.136], in which the average FFM signal for the sample scanned in two opposite directions is subtracted from the friction profiles of each of the two scans, the misalignment effect can be eliminated. By following the friction force calibration procedures developed by *Ruan* and *Bhushan* [22.136], voltages corresponding to friction forces can be converted to force units. The coefficient of friction is obtained from the slope of the friction force data measured as a function of the normal load, which typically ranges from 10 to 150 nN. This approach eliminates any contributions from adhesive forces [22.10]. To calculate the coefficient of friction based on a single point measurement, the friction force should be divided by the sum of the normal load applied and the intrinsic adhesive force. Furthermore, it should be pointed out that the coefficient of friction is not independent of load for single-asperity contact,. This is discussed in more detail later.

The tip is scanned in such a way that its trajectory on the sample forms a triangular pattern (Fig. 22.11). Scanning speeds in the fast and slow scan directions depend on the scan area and scan frequency. Scan sizes ranging from less than 1 nm × 1 nm to 125 μm × 125 μm and scan rates of less than 0.5 to 122 Hz are typically used. Higher scan rates are used for smaller scan lengths. For example, the scan rates in the fast and slow scan directions for an area of 10 μm × 10 μm scanned at 0.5 Hz are 10 μm/s and 20 nm/s, respectively.

We now describe the construction of a small-sample AFM in more detail. It consists of three main parts: the optical head which senses the cantilever deflection; a PZT tube scanner which controls the scanning motion of the sample mounted on one of its ends; and the base, which supports the scanner and head and includes circuits for the deflection signal (Fig. 22.12a). The AFM connects directly to a control system. The optical head consists of a laser diode stage, a photodiode stage preamp board, the cantilever mount and its

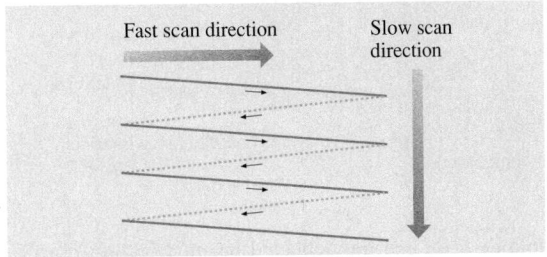

Fig. 22.11 Schematic of triangular pattern trajectory of the AFM tip as the sample is scanned in two dimensions. During imaging, data are only recorded during scans along the solid scan lines

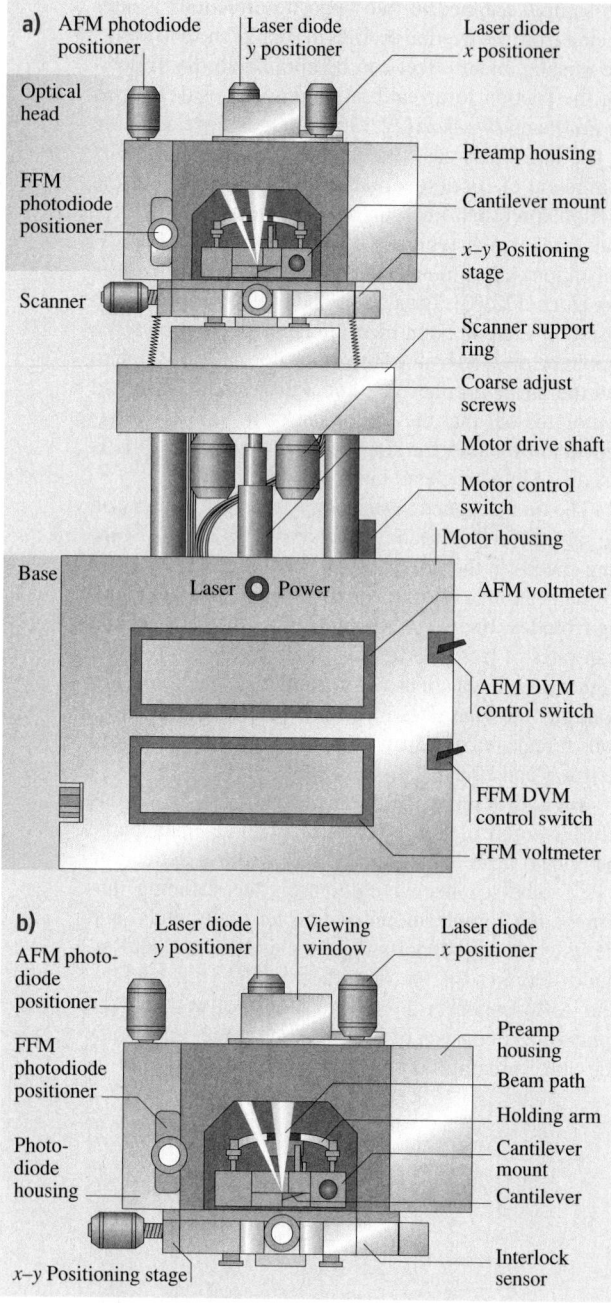

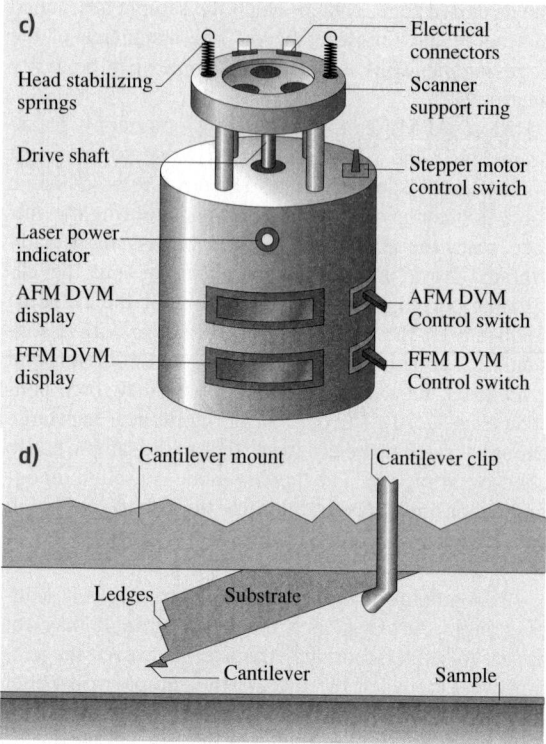

Fig. 22.12a–d Schematics of a commercial AFM/FFM made by Digital Instruments Inc. (**a**) Front view, (**b**) optical head, (**c**) base, and (**d**) cantilever substrate mounted on cantilever mount (not to scale)

lens, baseplate, and the x and y laser diode positioners. The positioners are used to place the laser spot on the end of the cantilever. The photodiode stage is an adjustable stage used to position the photodiode elements relative to the reflected laser beam. It consists of the split photodiode, the base plate, and the photodiode positioners. The deflected beam reflecting mirror is mounted on the upper left in the interior of the head. The cantilever mount is a metal (for operation in air) or glass (for operation in water) block which holds the cantilever firmly at the proper angle (Fig. 22.12d). Next, the tube scanner consists of an Invar cylinder holding a single tube made of piezoelectric crystal which imparts the necessary three-dimensional motion to the sample. Mounted on top of the tube is a magnetic cap on which the steel sample puck is placed. The tube is rigidly held at one end with the sample mounted on the other end of the tube. The scanner also contains three fine-pitched screws which form the mount for the optical head. The optical head rests on the tips of the screws, which are used to ad-

holding arm, and the deflected beam reflecting mirror, which reflects the deflected beam toward the photodiode (Fig. 22.12b). The laser diode stage is a tilt stage used to adjust the position of the laser beam relative to the cantilever. It consists of the laser diode, collimator, focusing

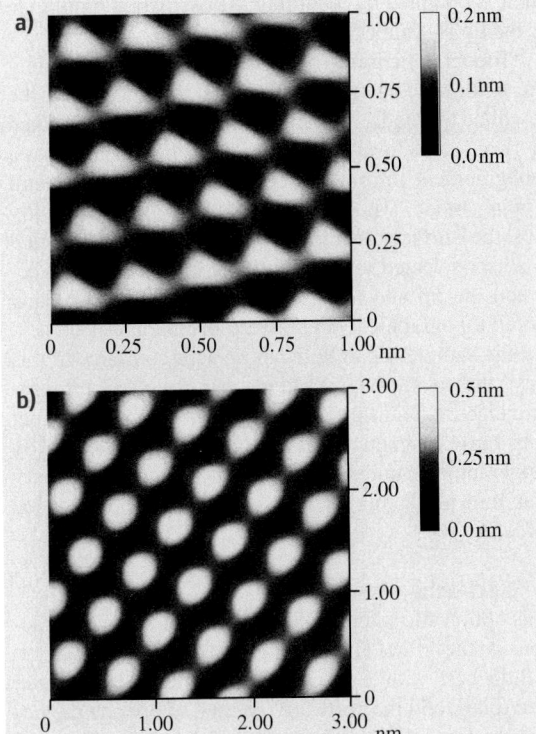

Fig. 22.13a,b Typical AFM images of freshly-cleaved (**a**) highly oriented pyrolytic graphite and (**b**) mica surfaces taken using a square pyramidal Si_3N_4 tip

fast and slow scan directions are $10\,\mu m/s$ and $20\,nm/s$, respectively. Normally 256×256 data points are taken for each image. The lateral resolution at larger scans is approximately equal to the scan length divided by 256. The piezo tube requires x–y calibration, which is carried out by imaging an appropriate calibration standard. Cleaved graphite is used for small scan heads, while two-dimensional grids (a gold-plated rule) can be used for long-range heads.

Examples of AFM images of freshly cleaved highly oriented pyrolytic (HOP) graphite and mica surfaces are shown in Fig. 22.13 [22.50, 110, 114]. Images with near-atomic resolution are obtained.

The force calibration mode is used to study interactions between the cantilever and the sample surface. In the force calibration mode, the x and y voltages applied to the piezo tube are held at zero and a sawtooth voltage is applied to the z electrode of the piezo tube, Fig. 22.14a. At the start of the force measurement the cantilever is in its rest position. By changing the applied voltage, the sample can be moved up and down relative to the stationary cantilever tip. As the piezo moves the sample up and down, the cantilever deflection signal from the photodiode is monitored. The force–distance curve, a plot of the cantilever tip deflection signal as a function of the voltage applied to the piezo tube, is obtained. Figure 22.14b shows the typical features of a force–distance curve. The arrowheads indicate the direction of piezo travel. As the piezo extends, it approaches

just the position of the head relative to the sample. The scanner fits into the scanner support ring mounted on the base of the microscope (Fig. 22.12c). The stepper motor is controlled manually with the switch on the upper surface of the base and automatically by the computer during the tip-engage and tip-withdraw processes.

The scan sizes available for these instruments are $0.7\,\mu m$, $12\,\mu m$ and $125\,\mu m$. The scan rate must be decreased as the scan size is increased. A maximum scan rate of $122\,Hz$ can be used. Scan rates of about $60\,Hz$ should be used for small scan lengths ($0.7\,\mu m$). Scan rates of 0.5 to $2.5\,Hz$ should be used for large scans on samples with tall features. High scan rates help reduce drift, but they can only be used on flat samples with small scan sizes. The scan rate or the scanning speed (length/time) in the fast scan direction is equal to twice the scan length multiplied by the scan rate in Hz, and in the slow direction it is equal to the scan length multiplied by the scan rate in Hz divided by number of data points in the transverse direction. For example, for a scan size of $10\,\mu m \times 10\,\mu m$ scanned at $0.5\,Hz$, the scan rates in the

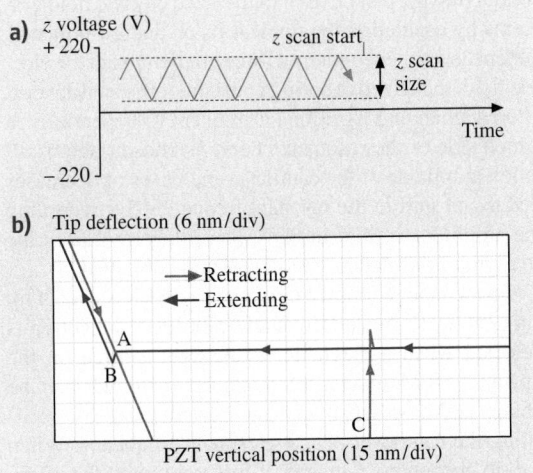

Fig. 22.14 (**a**) Force calibration Z waveform, and (**b**) a typical force–distance curve for a tip in contact with a sample. Contact occurs at point B; tip breaks free of adhesive forces at point C as the sample moves away from the tip

the tip, which is in mid-air at this point and hence shows no deflection. This is indicated by the flat portion of the curve. As the tip approaches the sample to within a few nanometers (point A), an attractive force kicks in between the atoms of the tip surface and the atoms of the surface of the sample. The tip is pulled towards the sample and contact occurs at point B on the graph. From this point on, the tip is in contact with the surface, and as the piezo extends further, the tip gets deflected further. This is represented by the sloped portion of the curve. As the piezo retracts, the tip moves beyond the zero deflection (flat) line due to attractive forces (van der Waals forces and long-range meniscus forces), into the adhesive regime. At point C in the graph, the tip snaps free of the adhesive forces, and is again in free air. The horizontal distance between points B and C along the retrace line gives the distance moved by the tip in the adhesive regime. Multiplying this distance by the stiffness of the cantilever gives the adhesive force. Incidentally, the horizontal shift between the loading and unloading curves results from the hysteresis in the PZT tube [22.4].

Multimode Capabilities

The multimode AFM can be used for topography measurements in the contact mode and tapping mode, described earlier, and for measurements of lateral (friction) force, electric force gradients and magnetic force gradients.

The multimode AFM, when used with a grounded conducting tip, can be used to measure electric field gradients by oscillating the tip near its resonant frequency. When the lever encounters a force gradient from the electric field, the effective spring constant of the cantilever is altered, changing its resonant frequency. Depending on which side of the resonance curve is chosen, the oscillation amplitude of the cantilever increases or decreases due to the shift in the resonant frequency. By recording the amplitude of the cantilever, an image revealing the strength of the electric field gradient is obtained.

In the magnetic force microscope (MFM), used with a magnetically coated tip, static cantilever deflection is detected when a magnetic field exerts a force on the tip, and MFM images of magnetic materials can be obtained. MFM sensitivity can be enhanced by oscillating the cantilever near its resonant frequency. When the tip encounters a magnetic force gradient, the effective spring constant (and hence the resonant frequency) is shifted. By driving the cantilever above or below the resonant frequency, the oscillation amplitude varies as the resonance shifts. An image of the magnetic field gradient is obtained by recording the oscillation amplitude as the tip is scanned over the sample.

Topographic information is separated from the electric field gradient and magnetic field images using the so-called lift mode. In lift mode, measurements are taken in two passes over each scan line. In the first pass, topographical information is recorded in the standard tapping mode, where the oscillating cantilever lightly taps the surface. In the second pass, the tip is lifted to a user-selected separation (typically 20–200 nm) between the tip and local surface topography. By using stored topographical data instead of standard feedback, the tip–sample separation can be kept constant. In this way, the cantilever amplitude can be used to measure electric field force gradients or relatively weak but long-range magnetic forces without being influenced by topographic features. Two passes are made for every scan line, producing separate topographic and magnetic force images.

Electrochemical AFM

This option allows one to perform electrochemical reactions on the AFM. The technique involves a potentiostat, a fluid cell with a transparent cantilever holder and electrodes, and the software required to operate the potentiostat and display the results of the electrochemical reaction.

22.2.3 AFM Probe Construction

Various probes (cantilevers and tips) are used for AFM studies. The cantilever stylus used in the AFM should meet the following criteria: (1) low normal spring constant (stiffness); (2) high resonant frequency; (3) high cantilever quality factor Q; (4) high lateral spring constant (stiffness); (5) short cantilever length; (6) incorporation of components (such as mirror) for deflection sensing, and; (7) a sharp protruding tip [22.137]. In order to register a measurable deflection with small forces, the cantilever must flex with a relatively low force (on the order of few nN), requiring vertical spring constants of 10^{-2} to 10^2 N/m for atomic resolution in the contact profiling mode. The data rate or imaging rate in the AFM is limited by the mechanical resonant frequency of the cantilever. To achieve a large imaging bandwidth, the AFM cantilever should have a resonant frequency of more than about 10 kHz (30–100 kHz is preferable), which makes the cantilever the least sensitive part of the system. Fast imaging rates are not just a matter of convenience, since the effects of thermal drifts are more pronounced with slow scanning speeds. The combined

Table 22.2 Relevant properties of materials used for cantilevers

Property	Young's Modulus (E) (GPa)	Density (ρg) (kg/m^3)	Microhardness (GPa)	Speed of sound ($\sqrt{E/\rho}$) (m/s)
Diamond	900–1050	3515	78.4–102	17 000
Si$_3$N$_4$	310	3180	19.6	9900
Si	130–188	2330	9–10	8200
W	350	19 310	3.2	4250
Ir	530	–	≈ 3	5300

requirements of a low spring constant and a high resonant frequency are met by reducing the mass of the cantilever. The quality factor Q ($= \omega_R/(c/m)$, where ω_R is the resonant frequency of the damped oscillator, c is the damping constant and m is the mass of the oscillator) should have a high value for some applications. For example, resonance curve detection is a sensitive modulation technique for measuring small force gradients in noncontact imaging. Increasing the Q increases the sensitivity of the measurements. Mechanical Q values of 100–1000 are typical. In contact modes, the Q value is of less importance. A high lateral cantilever spring constant is desirable in order to reduce the effect of lateral forces in the AFM, as frictional forces can cause appreciable lateral bending of the cantilever. Lateral bending results in erroneous topography measurements. For friction measurements, cantilevers with reduced lateral rigidity are preferred. A sharp protruding tip must be present at the end of the cantilever to provide a well-defined interaction with the sample over a small area. The tip radius should be much smaller than the radii of the corrugations in the sample in order for these to be measured accurately. The lateral spring constant depends critically on the tip length. Additionally, the tip should be centered at the free end.

In the past, cantilevers have been cut by hand from thin metal foils or formed from fine wires. Tips for these cantilevers were prepared by attaching diamond fragments to the ends of the cantilevers by hand, or in the case of wire cantilevers, electrochemically etching the wire to a sharp point. Several cantilever geometries for wire cantilevers have been used. The simplest geometry is the L-shaped cantilever, which is usually made by bending a wire at a 90° angle. Other geometries include single-V and double-V geometries, with a sharp tip attached at the apex of the V, and double-X configuration with a sharp tip attached at the intersection [22.31, 138]. These cantilevers can be constructed with high vertical spring constants. For example, a double-cross cantilever with an effective spring constant of 250 N/m was used

by *Burnham* and *Colton* [22.31]. The small size and low mass needed in the AFM make hand fabrication of the cantilever a difficult process with poor reproducibility. Conventional microfabrication techniques are ideal for constructing planar thin-film structures which have submicron lateral dimensions. The triangular (V-shaped) cantilevers have improved (higher) lateral spring constants in comparison to rectangular cantilevers. In terms of spring constants, the triangular cantilevers are approximately equivalent to two rectangular cantilevers placed in parallel [22.137]. Although the macroscopic radius of a photolithographically patterned corner is seldom much less than about 50 nm, microscopic asperities on the etched surface provide tips with near-atomic dimensions.

Cantilevers have been used from a whole range of materials. Cantilevers made of Si$_3$N$_4$, Si, and diamond are the most commonl. The Young's modulus and the density are the material parameters that determine the resonant frequency, aside from the geometry. Table 22.2 shows the relevant properties and the speed of sound, indicative of the resonant frequency for a given shape. Hardness is an important indicator of the durability of the cantilever, and is also listed in the table. Materials used for STM cantilevers are also included.

Silicon nitride cantilevers are less expensive than those made of other materials. They are very rugged and well suited to imaging in almost all environments. They are especially compatible with organic and biological materials. Microfabricated triangular silicon nitride beams with integrated square pyramidal tips made using plasma-enhanced chemical vapor deposition (PECVD) are the most common [22.137]. Four cantilevers, marketed by Digital Instruments, with different sizes and spring constants located on cantilever substrate made of boron silicate glass (Pyrex), are shown in Figs. 22.15a and 22.16. The two pairs of cantilevers on each substrate measure about 115 and 193 μm from the substrate to the apex of the triangular cantilever, with base widths of 122 and 205 μm, respectively. The cantilever legs,

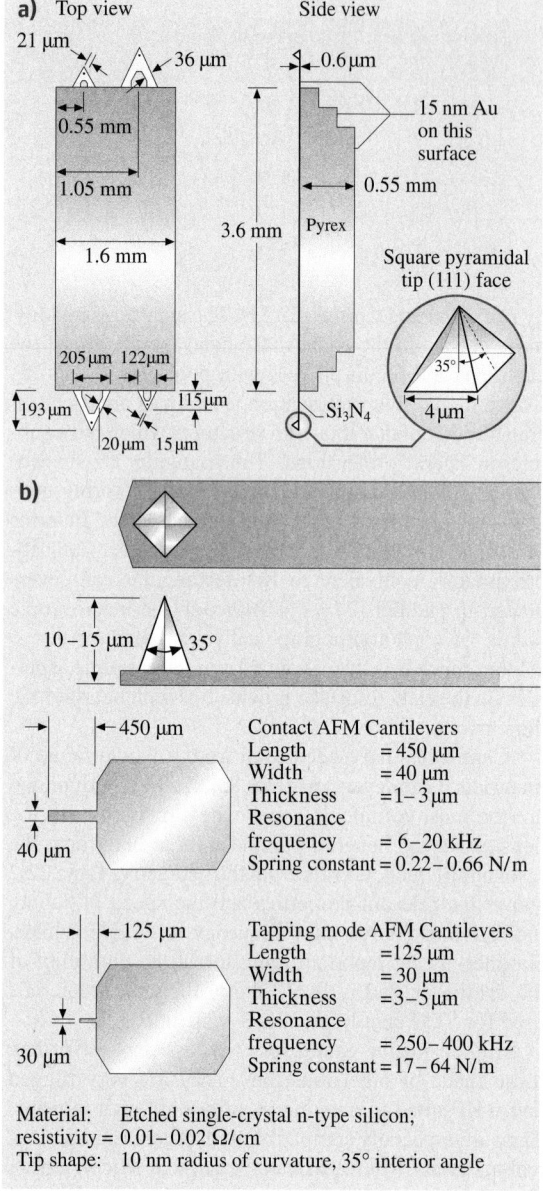

a) Top view   Side view

21 μm
36 μm
0.6 μm
15 nm Au on this surface
0.55 mm
1.05 mm
0.55 mm
3.6 mm   Pyrex
1.6 mm
Square pyramidal tip (111) face
205 μm 122 μm
115 μm
35°
193 μm
Si₃N₄
4 μm
20 μm 15 μm

b)

10–15 μm   35°

450 μm   Contact AFM Cantilevers
Length   = 450 μm
Width   = 40 μm
Thickness   = 1–3 μm
40 μm   Resonance
                frequency   = 6–20 kHz
                Spring constant = 0.22–0.66 N/m

125 μm   Tapping mode AFM Cantilevers
Length   = 125 μm
Width   = 30 μm
Thickness   = 3–5 μm
30 μm   Resonance
                frequency   = 250–400 kHz
                Spring constant = 17–64 N/m

Material: Etched single-crystal n-type silicon;
resistivity = 0.01–0.02 Ω/cm
Tip shape: 10 nm radius of curvature, 35° interior angle

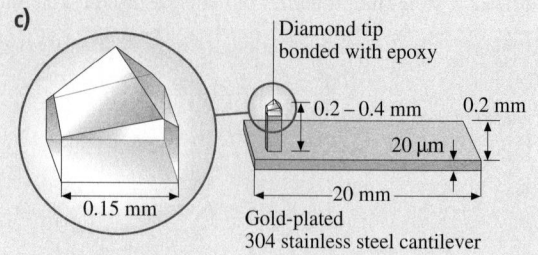

c)

Diamond tip bonded with epoxy
0.2–0.4 mm   0.2 mm
20 μm
0.15 mm   20 mm
Gold-plated 304 stainless steel cantilever

Fig. 22.15a–c Schematics of (**a**) triangular cantilever beam with square-pyramidal tips made of PECVD Si₃N₄, (**b**) rectangular cantilever beams with square-pyramidal tips made of etched single-crystal silicon, and (**c**) rectangular cantilever stainless steel beam with three-sided pyramidal natural diamond tip

constants should be used on softer samples. The pyramidal tip is highly symmetric, and the end has a radius of about 20–50 nm. The side walls of the tip have a slope of 35 deg and the lengths of the edges of the tip at the cantilever base are about 4 μm.

An alternative to silicon nitride cantilevers with integrated tips are microfabricated single-crystal silicon cantilevers with integrated tips. Si tips are sharper than Si₃N₄ tips because they are formed directly by anisotropic etching of single-crystal Si, rather than through the use of an etch pit as a mask for the deposited material [22.139]. Etched single-crystal n-type

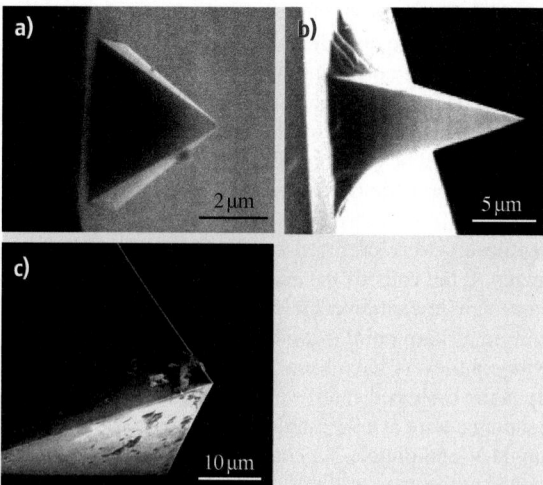

Fig. 22.16a–c SEM micrographs of a square-pyramidal PECVD Si₃N₄ tip (**a**), a square-pyramidal etched single-crystal silicon tip (**b**), and a three-sided pyramidal natural diamond tip (**c**)

which are of the same thickness (0.6 μm) in all the cantilevers, are available in wide and narrow forms. Only one cantilever is selected and used from each substrate. The calculated spring constants and measured natural frequencies for each of the configurations are listed in Table 22.3. The most commonly used cantilever beam is the 115 μm-long, wide-legged cantilever (vertical spring constant = 0.58 N/m). Cantilevers with smaller spring

Table 22.3 Measured vertical spring constants and natural frequencies of triangular (V-shaped) cantilevers made of PECVD Si_3N_4 (data provided by Digital Instruments, Inc.)

Cantilever dimension	Spring constant (k_z) (N/m)	Natural frequency (ω_0) (kHz)
115-μm long, narrow leg	0.38	40
115-μm long, wide leg	0.58	40
193-μm long, narrow leg	0.06	13–22
193-μm long, wide leg	0.12	13–22

Table 22.4 Vertical (k_z), lateral (k_y), and torsional (k_{yT}) spring constants of rectangular cantilevers made of Si (IBM) and PECVD Si_3N_4 (source: Veeco Instruments, Inc.)

Dimensions/stiffness	Si cantilever	Si_3N_4 cantilever
Length (L) (μm)	100	100
Width (b) (μm)	10	20
Thickness (h) (μm)	1	0.6
Tip length (ℓ) (μm)	5	3
k_z (N/m)	0.4	0.15
k_y (N/m)	40	175
k_{yT} (N/m)	120	116
ω_0 (kHz)	~ 90	~ 65

Note: $k_z = Ebh^3/4L^3$, $k_y = Eb^3h/4\ell^3$, $k_{yT} = Gbh^3/3L\ell^2$, and $\omega_0 = [k_z/(m_c + 0.24bhL\rho)]^{1/2}$, where E is Young's modulus, G is the modulus of rigidity [$= E/2(1+\nu)$], ν is Poisson's ratio], ρ is the mass density of the cantilever, and m_c is the concentrated mass of the tip (~ 4 ng) [22.94]. For Si, $E = 130$ GPa, $\rho g = 2300$ kg/m^3, and $\nu = 0.3$. For Si_3N_4, $E = 150$ GPa, $\rho g = 3100$ kg/m^3, and $\nu = 0.3$

silicon rectangular cantilevers with square pyramidal tips of radii <10 nm for contact and tapping mode (tapping-mode etched silicon probe or TESP) AFMs are commercially available from Digital Instruments and Nanosensors GmbH, Aidlingen, Germany, Figs. 22.15b and 22.16. Spring constants and resonant frequencies are also presented in the Fig. 22.15b.

Commercial triangular Si_3N_4 cantilevers have a typical width:thickness ratio of 10 to 30, which results in spring constants that are 100 to 1000 times stiffer in the lateral direction than in the normal direction. Therefore, these cantilevers are not well suited for torsion. For friction measurements, the torsional spring constant should be minimized in order to be sensitive to the lateral force. Rather long cantilevers with small thicknesses and large tip lengths are most suitable. Rectangular beams have smaller torsional spring constants than the triangular (V-shaped) cantilevers. Table 22.4 lists the spring constants (with the full length of the beam used) in three directions for typical rectangular beams. We note that the lateral and torsional spring constants are about two orders of magnitude larger than the normal spring constants. A cantilever beam required for the tapping mode is quite stiff and may not be sensitive enough for friction measurements. *Meyer* et al. [22.140] used a specially designed rectangular silicon cantilever with length $= 200\,\mu$m, width $= 21\,\mu$m, thickness $= 0.4\,\mu$m, tip length $= 12.5\,\mu$m and shear modulus $= 50$ GPa, giving a normal spring constant of 0.007 N/m and a torsional spring constant of 0.72 N/m, which gives a lateral force sensitivity of 10 pN and an angle of resolution of 10^{-7} rad. Using this particular geometry, the sensitivity to lateral forces can be improved by about a factor of 100 compared with commercial V-

shaped Si_3N_4 or the rectangular Si or Si_3N_4 cantilevers used by *Meyer* and *Amer* [22.8], with torsional spring constants of ~ 100 N/m. *Ruan* and *Bhushan* [22.136] and *Bhushan* and *Ruan* [22.9] used 115 μm-long, wide-legged V-shaped cantilevers made of Si_3N_4 for friction measurements.

For scratching, wear and indentation studies, single-crystal natural diamond tips ground to the shape of a three-sided pyramid with an apex angle of either 60° or 80° and a point sharpened to a radius of about 100 nm are commonly used [22.4, 10] (Figs. 22.15c and 22.16). The tips are bonded with conductive epoxy to a gold-plated 304 stainless steel spring sheet (length $= 20$ mm, width $= 0.2$ mm, thickness $= 20$ to $60\,\mu$m) which acts as a cantilever. The free length of the spring is varied in order to change the beam stiffness. The normal spring constant of the beam ranges from about 5 to 600 N/m for a 20 μm-thick beam. The tips are produced by R-DEC Co., Tsukuba, Japan.

High aspect ratio tips are used to image within trenches. Examples of two probes used are shown in Fig. 22.17. These high aspect ratio tip (HART) probes are produced from conventional Si_3N_4 pyramidal probes. Through a combination of focused ion beam (FIB) and high-resolution scanning electron microscopy (SEM) techniques, a thin filament is grown at the apex of the pyramid. The probe filament is approximately

a)

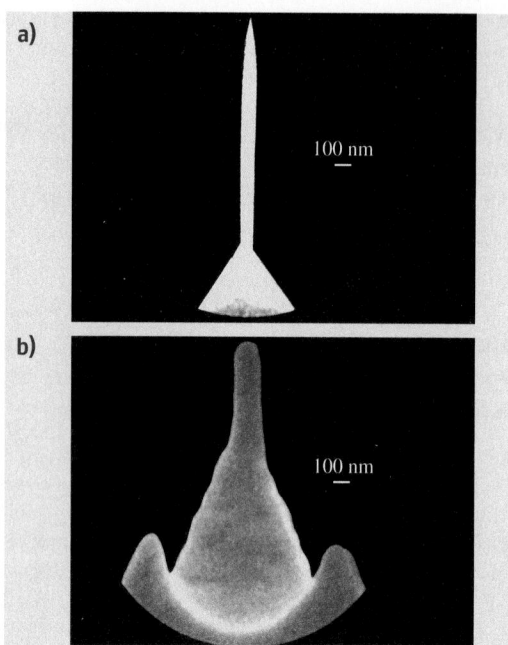

100 nm

b)

100 nm

Fig. 22.17a,b Schematics of (**a**) HART Si₃N₄ probe, and (**b**) an FIB-milled Si₃N₄ probe

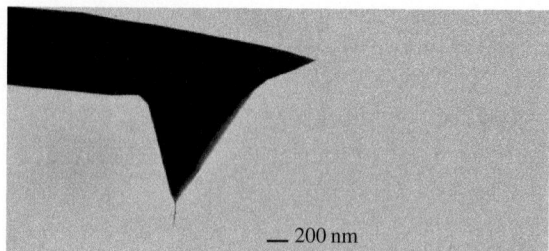

— 200 nm

Fig. 22.18 SEM micrograph of a multiwall carbon nanotube (MWNT) tip physically attached to a single-crystal silicon, square-pyramidal tip (courtesy of Piezomax Technologies, Inc.)

1 μm long and 0.1 μm in diameter. It tapers to an extremely sharp point (with a radius that is better than the resolutions of most SEMs). The long thin shape and sharp radius make it ideal for imaging within "vias" of microstructures and trenches (> 0.25 μm). This is, however, unsuitable for imaging structures at the atomic level, since probe flexing can create image artefacts. A FIB-milled probe is used for atomic-scale imaging, which is relatively stiff yet allows for closely spaced topography. These probes start out as conventional Si₃N₄ pyramidal probes, but the pyramid is FIB-milled until a small cone shape is formed which has a high aspect ratio and is 0.2–0.3 μm in length. The milled probes permit nanostructure resolution without sacrificing rigidity. These types of probes are manufactured by various manufacturers including Materials Analytical Services.

Carbon nanotube tips with small diameters and high aspect ratios are used for high-resolution imaging of surfaces and of deep trenches, in the tapping mode or the noncontact mode. Single-wall carbon nanotubes (SWNTs) are microscopic graphitic cylinders that are 0.7 to 3 nm in diameter and up to many microns in length. Larger structures called multiwall carbon nanotubes (MWNTs) consist of nested, concentrically arranged

SWNTs and have diameters of 3 to 50 nm. MWNT carbon nanotube AFM tips are produced by manual assembly [22.141], chemical vapor deposition (CVD) synthesis, and a hybrid fabrication process [22.142]. Figure 22.18 shows a TEM micrograph of a carbon nanotube tip, ProbeMax, commercially produced by mechanical assembly by Piezomax Technologies, Inc. (Middleton, WI, USA). To fabricate these tips, MWNTs are produced using a carbon arc and they are physically attached to the single-crystal silicon, square-pyramidal tips in the SEM, using a manipulator and the SEM stage to independently control the nanotubes and the tip. When the nanotube is first attached to the tip, it is usually too long to image with. It is shortened by placing it in an AFM and applying voltage between the tip and the sample. Nanotube tips are also commercially produced by CVD synthesis by NanoDevices (Santa Barbara, CA, USA).

22.2.4 Friction Measurement Methods

The two methods for performing friction measurements that are based on the work by *Ruan* and *Bhushan* [22.136] are now described in more detail (also see [22.8]). The scanning angle is defined as the angle relative to the y-axis in Fig. 22.19a. This is also the long axis of the cantilever. The zero-degree scanning angle corresponds to the sample scan in the y-direction, and the 90-degree scanning angle corresponds to the sample scan perpendicular to this axis in the xy-plane (along x-axis). If both the y- and −y-directions are scanned, we call this a "parallel scan". Similarly, a "perpendicular scan" means that both the x- and −x-directions are scanned. The direction of sample travel for each of these two methods is illustrated in Fig. 22.19b.

Using method 1 ("height" mode with parallel scans) in addition to topographic imaging, it is also possible to measure friction force when the sample scanning direction is parallel to the y-direction (parallel scan). If there was no friction force between the tip and the moving sample, the topographic feature would be the only factor that would cause the cantilever to be deflected vertically. However, friction force does exist on all surfaces that are in contact where one of the surfaces is moving relative to the other. The friction force between the sample and the tip will also cause the cantilever to

be deflected. We assume that the normal force between the sample and the tip is W_0 when the sample is stationary (W_0 is typically $10\,\text{nN}$ to $200\,\text{nN}$), and the friction force between the sample and the tip is W_f as the sample is scanned by the tip. The direction of the friction force (W_f) is reversed as the scanning direction of the sample is reversed from the positive (y) to the negative ($-y$) direction ($W_{f(y)} = -W_{f(-y)}$).

When the vertical cantilever deflection is set at a constant level, it is the total force (normal force and friction force) applied to the cantilever that keeps the cantilever deflection at this level. Since the friction force is directed in the opposite direction to the direction of travel of the sample, the normal force will have to be adjusted accordingly when the sample reverses its traveling direction, so that the total deflection of the cantilever will remain the same. We can calculate the difference in the normal force between the two directions of travel for a given friction force W_f. First, since the deflection is constant, the total moment applied to the cantilever is constant. If we take the reference point to be the point where the cantilever joins the cantilever holder (substrate), point P in Fig. 22.20, we have the following relationship:

$$(W_0 - \Delta W_1)L + W_f\ell = (W_0 + \Delta W_2)L - W_f\ell \tag{22.1}$$

or

$$(\Delta W_1 + \Delta W_2)L = 2W_f\ell . \tag{22.2}$$

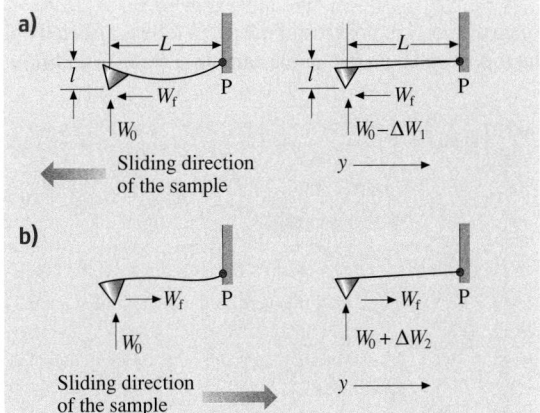

Fig. 22.20 (a) Schematic showing an additional bending of the cantilever due to friction force when the sample is scanned in the y- or $-y$-directions (*left*). **(b)** This effect can be canceled out by adjusting the piezo height using a feedback circuit (*right*) [22.136]

Fig. 22.19 (a) Schematic defining the x- and y-directions relative to the cantilever, and showing the direction of sample travel in two different measurement methods discussed in the text. **(b)** Schematic of deformation of the tip and cantilever shown as a result of sliding in the x- and y-directions. A twist is introduced to the cantilever if the scanning is performed in the x-direction (**(b)**, *lower part*) [22.136]

Thus

$$W_f = (\Delta W_1 + \Delta W_2)L/(2\ell), \qquad (22.3)$$

where ΔW_1 and ΔW_2 are the absolute values of the changes in normal force when the sample is traveling in the $-y$- and y-directions, respectively, as shown in Fig. 22.20; L is the length of the cantilever; ℓ is the vertical distance between the end of the tip and point P. The coefficient of friction (μ) between the tip and the sample is then given as

$$\mu = \frac{W_f}{W_0} = \left(\frac{(\Delta W_1 + \Delta W_2)}{W_0}\right)\left(\frac{L}{2\ell}\right). \qquad (22.4)$$

There are adhesive and interatomic attractive forces between the cantilever tip and the sample at all times. The adhesive force can be due to water from the capillary condensation and other contaminants present at the surface, which form meniscus bridges [22.4, 143, 144] and the interatomic attractive force includes van der Waals attractions [22.18]. If these forces (and the effect of indentation too, which is usually small for rigid samples) can be neglected, the normal force W_0 is then equal to the initial cantilever deflection H_0 multiplied by the spring constant of the cantilever. $(\Delta W_1 + \Delta W_2)$ can be derived by multiplying the same spring constant by the change in height of the piezo tube between the two traveling directions (y- and $-y$-directions) of the sample. This height difference is denoted as $(\Delta H_1 + \Delta H_2)$, shown schematically in Fig. 22.21. Thus, (22.4) can be rewritten as

$$\mu = \frac{W_f}{W_0} = \left(\frac{(\Delta H_1 + \Delta H_2)}{H_0}\right)\left(\frac{L}{2\ell}\right). \qquad (22.5)$$

Since the vertical position of the piezo tube is affected by the topographic profile of the sample surface in addition

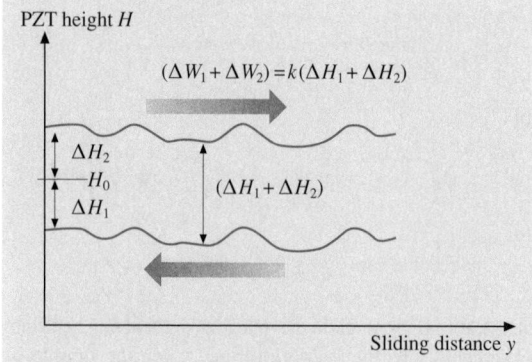

Fig. 22.21 Schematic illustration of the height difference for the piezoelectric tube scanner as the sample is scanned in the y- and $-y$-directions

to the friction force being applied at the tip, this difference must be found point-by-point at the same location on the sample surface, as shown in Fig. 22.21. Subtraction of point-by-point measurements may introduce errors, particularly for rough samples. We will come back to this point later. In addition, precise measurements of L and ℓ (which should include the cantilever angle) are also required.

If the adhesive force between the tip and the sample is large enough that it cannot be neglected, it should be included in the calculation. However, determinations of this force can involve large uncertainties, which is introduced into (22.5). An alternative approach is to make the measurements at different normal loads and to use $\Delta(H_0)$ and $\Delta(\Delta H_1 + \Delta H_2)$ in (22.5). Another comment on (22.5) is that, since only the ratio between $(\Delta H_1 + \Delta H_2)$ and H_0 enters this equation, the vertical position of the piezo tube H_0 and the difference in position $(\Delta H_1 + \Delta H_2)$ can be in volts as long as the vertical travel of the piezo tube and the voltage applied to have a linear relationship. However, if there is a large nonlinearity between the piezo tube traveling distance and the applied voltage, this nonlinearity must be included in the calculation.

It should also be pointed out that (22.4) and (22.5) are derived under the assumption that the friction force W_f is the same for the two scanning directions of the sample. This is an approximation, since the normal force is slightly different for the two scans and the friction may be direction-dependent. However, this difference is much smaller than W_0 itself. We can ignore the second-order correction.

Method 2 ("aux" mode with perpendicular scan) of measuring friction was suggested by *Meyer* and *Amer* [22.8]. The sample is scanned perpendicular to the long axis of the cantilever beam (along the x- or $-x$-direction in Fig. 22.19a) and the outputs from the two horizontal quadrants of the photodiode detector are measured. In this arrangement, as the sample moves under the tip, the friction force will cause the cantilever to twist. Therefore, the light intensity between the left and right (L and R in Fig. 22.19b, right) detectors will be different. The differential signal between the left and right detectors is denoted the FFM signal [(L − R)/(L + R)]. This signal can be related to the degree of twisting, and hence to the magnitude of friction force. Again, because possible errors in measurements of the normal force due to the presence of adhesive force at the tip–sample interface, the slope of the friction data (FFM signal vs. normal load) needs to be measured for an accurate value of the coefficient of friction.

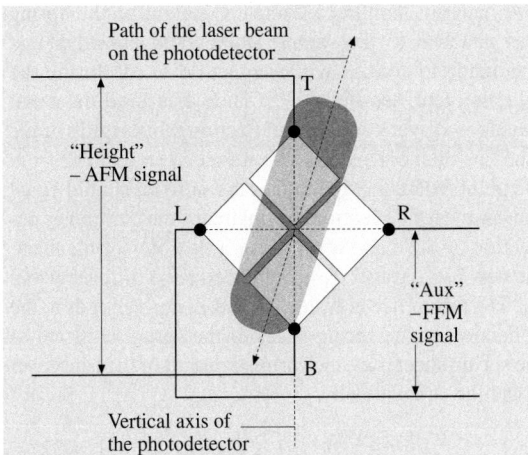

Fig. 22.22 The trajectory of the laser beam on the photodetectors as the cantilever is vertically deflected (with no torsional motion) with respect to the laser beam for a misaligned photodetector. For a change of normal force (vertical deflection of the cantilever), the laser beam is projected to a different position on the detector. Due to a misalignment, the projected trajectory of the laser beam on the detector is not parallel with the detector vertical axis (the line T–B) [22.136]

While friction force contributes to the FFM signal, friction force may not be the only contributing factor in commercial FFM instruments (for example, NanoScope IV). One can see this if we simply engange the cantilever tip with the sample. The left and right detectors can be balanced beforehand by adjusting the positions of the detectors so that the intensity difference between these two detectors is zero (FFM signal is zero). Once the tip is engaged with the sample, this signal is no longer zero, even if the sample is not moving in the xy-plane with no friction force applied. This would be a detrimental effect. It has to be understood and eliminated from the data acquisition before any quantitative measurement of friction force is made.

One of the reasons for this observation is as follows. The detectors may not have been properly aligned with respect to the laser beam. To be precise, the vertical axis of the detector assembly (the line joining T–B in Fig. 22.22) is not in the plane defined by the incident laser beam and the beam reflected from the untwisted cantilever (we call this plane the "beam plane"). When the cantilever vertical deflection changes due to a change in the normal force applied (without the sample being scanned in the xy-plane), the laser beam will be reflected up and down and form a projected trajectory on the

detector. (Note that this trajectory is in the defined beam plane.) If this trajectory is not coincident with the vertical axis of the detector, the laser beam will not evenly bisect the left and right quadrants of the detectors, even under the condition of no torsional motion of the cantilever, see Fig. 22.22. Thus, when the laser beam is reflected up and down due a change in the normal force, the intensity difference between the left and right detectors will also change. In other words, the FFM signal will change as the normal force applied to the tip is changed, even if the tip is not experiencing any friction force. This (FFM) signal is unrelated to friction force or to the actual twisting of the cantilever. We will call this part of the FFM signal "FFM_F", and the part which is truly related to friction force "FFM_T".

The FFM_F signal can be eliminated. One way of doing this is as follows. First the sample is scanned in both the x- and the $-x$-directions and the FFM signals for scans in each direction are recorded. Since the friction force reverses its direction of action when the scanning direction is reversed from the x- to the $-x$-direction, the FFM_T signal will change signs as the scanning direction of the sample is reversed ($FFM_T(x) = -FFM_T(-x)$). Hence the FFM_T signal will be canceled out if we take the sum of the FFM signals for the two scans. The average value of the two scans will be related to FFM_F due to the misalignment,

$$FFM(x) + FFM(-x) = 2FFM_F . \qquad (22.6)$$

This value can therefore be subtracted from the original FFM signals of each of these two scans to obtain the true FFM signal (FFM_T). Or, alternately, by taking the difference of the two FFM signals, one gets the FFM_T value directly:

$$FFM(x) - FFM(-x) = FFM_T(x) - FFM_T(-x)$$
$$= 2FFM_T(x) . \qquad (22.7)$$

Ruan and *Bhushan* [22.136] have shown that the error signal (FFM_F) can be very large compared to the friction signal FFM_T, so correction is required.

Now we compare the two methods. The method of using the "height" mode and parallel scanning (method 1) is very simple to use. Technically, this method can provide 3-D friction profiles and the corresponding topographic profiles. However, there are some problems with this method. Under most circumstances, the piezo scanner displays hysteresis when the traveling direction of the sample is reversed. Therefore, the measured surface topographic profiles will be shifted relative to each other along the y-axis for the two opposite (y and $-y$)

scans. This would make it difficult to measure the local difference in height of the piezo tube for the two scans. However, the average difference in height between the two scans and hence the average friction can still be measured. The measurement of average friction can serve as an internal means of friction force calibration. Method 2 is a more desirable approach. The subtraction of the FFM$_F$ signal from FFM for the two scans does not introduce any error into local friction force data. An ideal approach when using this method would be to add the average values of the two profiles in order to get the error component (FFM$_F$) and then subtract this component from either profile to get true friction profiles in either directions. By performing measurements at various loads, we can get the average value of the coefficient of friction which then can be used to convert the friction profile to the coefficient of friction profile. Thus, any directionality and local variations in friction can be easily measured. In this method, since topography data are not affected by friction, accurate topography data can be measured simultaneously with friction data and a better localized relationship between the two can be established.

22.2.5 Normal Force and Friction Force Calibrations of Cantilever Beams

Based on *Ruan* and *Bhushan* [22.136], we now discuss normal force and friction force calibrations. In order to calculate the absolute values of normal and friction forces in Newtons using the measured AFM and FFM$_T$ voltage signals, it is necessary to first have an accurate value of the spring constant of the cantilever (k_c). The spring constant can be calculated using the geometry and the physical properties of the cantilever material [22.8, 94, 137]. However, the properties of the PECVD Si$_3$N$_4$ (used to fabricate cantilevers) can be different from those of the bulk material. For example, using ultrasonics, we found the Young's modulus of the cantilever beam to be about 238 ± 18 GPa, which is less than that of bulk Si$_3$N$_4$ (310 GPa). Furthermore, the thickness of the beam is nonuniform and difficult to measure precisely. Since the stiffness of a beam goes as the cube of thickness, minor errors in precise measurements of thickness can introduce substantial stiffness errors. Thus one should measure the spring constant of the cantilever experimentally. *Cleveland* et al. [22.145] measured normal spring constants by measuring resonant frequencies of beams.

For normal spring constant measurement, *Ruan* and *Bhushan* [22.136] used a stainless steel spring sheet of known stiffness (width $= 1.35$ mm, thickness $= 15$ μm,

free hanging length$= 5.2$ mm). One end of the spring was attached to the sample holder and the other end was made to contact with the cantilever tip during the measurement, see Fig. 22.23. They measured the piezo travel for a given cantilever deflection. For a rigid sample (such as diamond), the piezo travel Z_t (measured from the point where the tip touches the sample) should equal the cantilever deflection. To maintain the cantilever deflection at the same level using a flexible spring sheet, the new piezo travel $Z_{t'}$ would need to be different from Z_t. The difference between $Z_{t'}$ and Z_t corresponds to the deflection of the spring sheet. If the spring constant of the spring sheet is k_s, the spring constant of the cantilever k_c can be calculated by

$$(Z_{t'} - Z_t)k_s = Z_t k_c$$

or

$$k_c = k_s(Z_{t'} - Z_t)/Z_t . \tag{22.8}$$

The spring constant of the spring sheet (k_s) used in this study is calculated to be 1.54 N/m. For the wide-legged cantilever used in our study (length $= 115$ μm, base width $= 122$ μm, leg width $= 21$ μm and thickness $= 0.6$ μm), k_c was measured to be 0.40 N/m instead of the 0.58 N/m reported by its manufacturer – Digital Instruments Inc. To relate the photodiode detector output to the cantilever deflection in nanometers, they used the same rigid sample to push against the AFM tip. Since the cantilever vertical deflection equals the sample traveling distance measured from the point where the tip touches

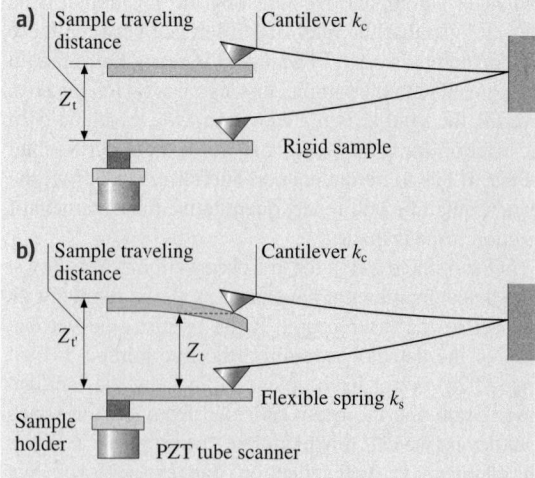

Fig. 22.23a,b Illustration showing the deflection of the cantilever as it is pushed by (**a**) a rigid sample, (**b**) a flexible spring sheet [22.136]

the sample for a rigid sample, the photodiode output observed as the tip is pushed by the sample can be converted directly to the cantilever deflection. For these measurements, they found the conversion factor to be 20 nm/V.

The normal force applied to the tip can be calculated by multiplying the cantilever vertical deflection by the cantilever spring constant for samples that have very small adhesion with the tip. If the adhesive force between the sample and the tip is large, it should be included in the normal force calculation. This is particularly important in atomic-scale force measurements, because the typical normal force that is measured in this region is in the range of a few hundreds of nN to a few mN. The adhesive force could be comparable to the applied force.

The conversion of friction signal (from FFM$_T$) to friction force is not as straightforward. For example, one can calculate the degree of twisting for a given friction force using the geometry and the physical properties of the cantilever [22.53, 144]. One would need information about the detector such as its quantum efficiency, laser power, gain and so on in order to be able convert the signal into the degree of twisting. Generally speaking, this procedure can not be accomplished without having some detailed information about the instrument. This information is not usually provided by the manufacturer. Even if this information is readily available, errors may still occur when using this approach because there will always be variations as a result of the instrumental set-up. For example, it has been noticed that the measured FFM$_T$ signal varies for the same sample when different AFM

microscopes from the same manufacturer are used. This means that one can not calibrate the instrument experimentally using this calculation. *O'Shea* et al. [22.144] did perform a calibration procedure in which the torsional signal was measured as the sample was displaced a known distance laterally while ensuring that the tip did not slide over the surface. However, it is difficult to verify that tip sliding does not occur.

A new method of calibration is therefore required. There is a simpler, more direct way of doing this. The first method described above (method 1) of measuring friction can provide an absolute value of the coefficient of friction directly. It can therefore be used as an internal calibration technique for data obtained using method 2. Or, for a polished sample, which introduces the least error into friction measurements taken using method 1, method 1 can be used to calibrate the friction force for method 2. Then this calibration can be used for measurements taken using method 2. In method 1, the length of the cantilever required can be measured using an optical microscope; the length of the tip can be measured using a scanning electron microscope. The relative angle between the cantilever and the horizontal sample surface can be measured directly. This enables the coefficient of friction to be measured with few unknown parameters. The friction force can then be calculated by multiplying the coefficient of friction by the normal load. The FFM$_T$ signal obtained using method 2 is then converted into the friction force. For their instrument, they found the conversion to be 8.6 nN/V.

22.3 AFM Instrumentation and Analyses

The performance of AFMs and the quality of AFM images greatly depend on the instrument available and the probes (cantilever and tips) in use. This section describes the mechanics of cantilevers, instrumentation and analysis of force detection systems for cantilever deflections, and scanning and control systems.

22.3.1 The Mechanics of Cantilevers

Stiffness and Resonances
of Lumped Mass Systems
All of the building blocks of an AFM, including the body of the microscope itself and the force-measuring cantilevers, are mechanical resonators. These resonances can be excited either by the surroundings or by the rapid movement of the tip or the sample. To avoid problems due to building- or air-induced oscillations,

it is of paramount importance to optimize the design of the AFM for high resonant frequencies. This usually means decreasing the size of the microscope [22.146]. By using cube-like or sphere-like structures for the microscope, one can considerably increase the lowest eigenfrequency. The fundamental natural frequency, ω_0, of any spring is given by

$$\omega_0 = \frac{1}{2\pi}\sqrt{\frac{k}{m_{eff}}}, \tag{22.9}$$

where k is the spring constant (stiffness) in the normal direction and m_{eff} is the effective mass. The spring constant k of a cantilever beam with uniform cross-section (Fig. 22.24) is given by [22.147]

$$k = \frac{3EI}{L^3}, \tag{22.10}$$

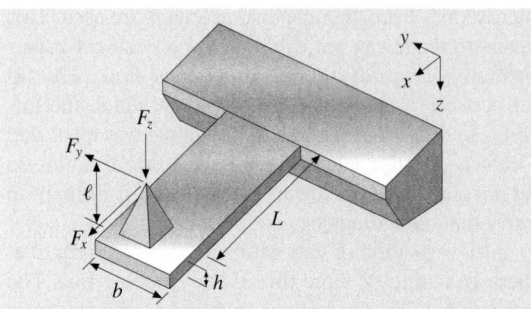

Fig. 22.24 A typical AFM cantilever with length L, width b, and height h. The height of the tip is ℓ. The material is characterized by the Young's modulus E, the shear modulus G and the mass density ρ. Normal (F_z), axial (F_x) and lateral (F_y) forces exist at the end of the tip

where E is the Young's modulus of the material, L is the length of the beam and I is the moment of inertia of the cross-section. For a rectangular cross-section with a width b (perpendicular to the deflection) and a height h one obtains the following expression for I

$$I = \frac{bh^3}{12} . \tag{22.11}$$

Combining (22.9), (22.10) and (22.11), we get an expression for ω_0

$$\omega_0 = \sqrt{\frac{Ebh^3}{4L^3 m_{\text{eff}}}} . \tag{22.12}$$

The effective mass can be calculated using Raleigh's method. The general formula using Raleigh's method for the kinetic energy T of a bar is

$$T = \frac{1}{2} \int_0^L \frac{m}{L} \left(\frac{\partial z(x)}{\partial t} \right)^2 dx . \tag{22.13}$$

For the case of a uniform beam with a constant cross-section and length L, one obtains for the deflection $z(x) = z_{\text{max}} \left[1 - (3x/2L) + (x^3/2L^3) \right]$. Inserting z_{max} into (22.13) and solving the integral gives

$$T = \frac{1}{2} \int_0^L \frac{m}{L} \left[\frac{\partial z_{\text{max}}(x)}{\partial t} \left(1 - \frac{3x}{2L} \right) + \left(\frac{x^3}{L^3} \right) \right]^2 dx$$

$$= \frac{1}{2} m_{\text{eff}} (z_{\text{max}} t)^2 ,$$

which gives

$$m_{\text{eff}} = \frac{9}{20} m . \tag{22.14}$$

Substituting (22.14) into (22.12) and noting that $m = \rho L b h$, where ρ is the mass density, one obtains the following expression:

$$\omega_0 = \left(\frac{\sqrt{5}}{3} \sqrt{\frac{E}{\rho}} \right) \frac{h}{L^2} . \tag{22.15}$$

It is evident from (22.15) that one way to increase the natural frequency is to choose a material with a high ratio E/ρ; see Table 22.2 for typical values of $\sqrt{E/\rho}$ for various commonly used materials. Another way to increase the lowest eigenfrequency is also evident in (22.15). By optimizing the ratio h/L^2, one can increase the resonant frequency. However, it does not help to make the length of the structure smaller than the width or height. Their roles will just be interchanged. Hence the optimum structure is a cube. This leads to the design rule that long, thin structures like sheet metal should be avoided. For a given resonant frequency, the quality factor Q should be as low as possible. This means that an inelastic medium such as rubber should be in contact with the structure in order to convert kinetic energy into heat.

Stiffness and Resonances of Cantilevers

Cantilevers are mechanical devices specially shaped to measure tiny forces. The analysis given in the previous section is applicable. However, to better understand the intricacies of force detection systems, we will discuss the example of a cantilever beam with uniform cross-section, Fig. 22.24. The bending of a beam due to a normal load on the beam is governed by the Euler equation [22.147]

$$M = EI(x) \frac{d^2 z}{dx^2} , \tag{22.16}$$

where M is the bending moment acting on the beam cross-section. $I(x)$ is the moment of inertia of the cross-section with respect to the neutral axis, defined by

$$I(x) = \int_z \int_y z^2 \, dy \, dz . \tag{22.17}$$

For a normal force F_z acting at the tip,

$$M(x) = (L - x) F_z \tag{22.18}$$

since the moment must vanish at the endpoint of the cantilever. Integrating (22.16) for a normal force F_z acting at the tip and observing that EI is a constant for beams with a uniform cross-section, one gets

$$z(x) = \frac{L^3}{6EI} \left(\frac{x}{L} \right)^2 \left(3 - \frac{x}{L} \right) F_z . \tag{22.19}$$

The slope of the beam is

$$z'(x) = \frac{Lx}{2EI}\left(2 - \frac{x}{L}\right)F_z .$$ (22.20)

From (22.19) and (22.20), at the end of the cantilever (for $x = L$), for a rectangular beam, and by using an expression for I in (22.11), one gets,

$$z(L) = \frac{4}{Eb}\left(\frac{L}{h}\right)^3 F_z ,$$ (22.21)

$$z'(L) = \frac{3}{2}\left(\frac{z}{L}\right) .$$ (22.22)

Now, the stiffness in the normal (z) direction, k_z, is

$$k_z = \frac{F_z}{z(L)} = \frac{Eb}{4}\left(\frac{h}{L}\right)^3 .$$ (22.23)

and the change in angular orientation of the end of cantilever beam is

$$\Delta\alpha = \frac{3}{2}\frac{z}{L} = \frac{6}{Ebh}\left(\frac{L}{h}\right)^2 F_z .$$ (22.24)

Now we ask what will, to a first-order approximation, happen if we apply a lateral force F_y to the end of the tip (Fig. 22.24). The cantilever will bend sideways and it will twist. The stiffness in the lateral (y) direction, k_y, can be calculated with (22.23) by exchanging b and h

$$k_y = \frac{Eh}{4}\left(\frac{b}{L}\right)^3 .$$ (22.25)

Therefore, the bending stiffness in the lateral direction is larger than the stiffness for bending in the normal direction by $(b/h)^2$. The twisting or torsion on the other hand is more complicated to handle. For a wide, thin cantilever ($b \gg h$) we obtain torsional stiffness along y-axis, k_{yT}

$$k_{yT} = \frac{Gbh^3}{3L\ell^2} ,$$ (22.26)

where G is the modulus of rigidity ($= E/2(1+\nu)$; ν is Poisson's ratio). The ratio of the torsional stiffness to the lateral bending stiffness is

$$\frac{k_{yT}}{k_y} = \frac{1}{2}\left(\frac{\ell b}{hL}\right)^2 ,$$ (22.27)

where we assume $\nu = 0.333$. We see that thin, wide cantilevers with long tips favor torsion while cantilevers with square cross-sections and short tips favor bending. Finally, we calculate the ratio between the torsional stiffness and the normal bending stiffness,

$$\frac{k_{yT}}{k_z} = 2\left(\frac{L}{\ell}\right)^2 .$$ (22.28)

Equations (22.26) to (22.28) hold in the case where the cantilever tip is exactly in the middle axis of the cantilever. Triangular cantilevers and cantilevers with tips which are not on the middle axis can be dealt with by finite element methods.

The third possible deflection mode is the one from the force on the end of the tip along the cantilever axis, F_x (Fig. 22.24). The bending moment at the free end of the cantilever is equal to $F_x\ell$. This leads to the following modification of (22.18) for forces F_z and F_x

$$M(x) = (L - x)F_z + F_x\ell .$$ (22.29)

Integration of (22.16) now leads to

$$z(x) = \frac{1}{2EI}\left[Lx^2\left(1 - \frac{x}{3L}\right)F_z + \ell x^2 F_x\right]$$ (22.30)

and

$$z'(x) = \frac{1}{EI}\left[\frac{Lx}{2}\left(2 - \frac{x}{L}\right)F_z + \ell x F_x\right] .$$ (22.31)

Evaluating (22.30) and (22.31) at the end of the cantilever, we get the deflection and the tilt

$$z(L) = \frac{L^2}{EI}\left(\frac{L}{3}F_z - \frac{\ell}{2}F_x\right) ,$$

$$z'(L) = \frac{L}{EI}\left(\frac{L}{2}F_z + \ell F_x\right) .$$ (22.32)

From these equations, one gets

$$F_z = \frac{12EI}{L^3}\left[z(L) - \frac{Lz'(L)}{2}\right] ,$$

$$F_x = \frac{2EI}{\ell L^2}\left[2Lz'(L) - 3z(L)\right] .$$ (22.33)

A second class of interesting properties of cantilevers is their resonance behavior. For cantilever beams, one can calculate the resonant frequencies [22.147, 148]

$$\omega_n^{\text{free}} = \frac{\lambda_n^2}{2\sqrt{3}}\frac{h}{L^2}\sqrt{\frac{E}{\rho}}$$ (22.34)

with $\lambda_0 = (0.596864\ldots)\pi$, $\lambda_1 = (1.494175\ldots)\pi$, $\lambda_n \to (n+1/2)\pi$. The subscript n represents the order of the frequency, such as the fundamental, the second mode, and the nth mode.

A similar equation to (22.34) holds for cantilevers in rigid contact with the surface. Since there is an additional restriction on the movement of the cantilever, namely the location of its endpoint, the resonant frequency increases. Only the terms of λ_n change to [22.148]

$$\lambda_0' = (1.2498763\ldots)\pi, \quad \lambda_1' = (2.2499997\ldots)\pi,$$
$$\lambda_n' \to (n+1/4)\pi .$$ (22.35)

The ratio of the fundamental resonant frequency during contact to the fundamental resonant frequency when not in contact is 4.3851.

For the torsional mode we can calculate the resonant frequencies as

$$\omega_0^{\text{tors}} = 2\pi \frac{h}{Lb} \sqrt{\frac{G}{\rho}} . \qquad (22.36)$$

For cantilevers in rigid contact with the surface, we obtain the following expression for the fundamental resonant frequency: [22.148]

$$\omega_0^{\text{tors, contact}} = \frac{\omega_0^{\text{tors}}}{\sqrt{1+3(2L/b)^2}} . \qquad (22.37)$$

The amplitude of the thermally induced vibration can be calculated from the resonant frequency using

$$\Delta z_{\text{therm}} = \sqrt{\frac{k_{\text{B}} T}{k}} , \qquad (22.38)$$

where k_{B} is Boltzmann's constant and T is the absolute temperature. Since AFM cantilevers are resonant structures, sometimes with rather high Q values, the thermal noise is not as evenly distributed as (22.38) suggests. The spectral noise density below the peak of the response curve is [22.148]

$$z_0 = \sqrt{\frac{4 k_{\text{B}} T}{k \omega_0 Q}} \quad (\text{in m}/\sqrt{\text{Hz}}) , \qquad (22.39)$$

where Q is the quality factor of the cantilever, described earlier.

22.3.2 Instrumentation and Analyses of Detection Systems for Cantilever Deflections

A summary of selected detection systems was provided in Fig. 22.8. Here we discuss the pros and cons of various systems in detail.

Optical Interferometer Detection Systems

Soon after the first papers on the AFM [22.2] appeared, which used a tunneling sensor, an instrument based on an interferometer was published [22.149]. The sensitivity of the interferometer depends on the wavelength of the light employed in the apparatus. Figure 22.25 shows the principle of such an interferometeric design. The light incident from the left is focused by a lens onto the cantilever. The reflected light is collimated by the same lens and interferes with the light reflected at the

flat. To separate the reflected light from the incident light, a $\lambda/4$ plate converts the linearly polarized incident light into circularly polarized light. The reflected light is made linearly polarized again by the $\lambda/4$-plate, but with a polarization orthogonal to that of the incident light. The polarizing beam splitter then deflects the reflected light to the photodiode.

Homodyne Interferometer. To improve the signal-to-noise ratio of the interferometer, the cantilever is driven by a piezo near its resonant frequency. The amplitude Δz of the cantilever as a function of driving frequency Ω is

$$\Delta z (\Omega) = \Delta z_0 \frac{\Omega_0^2}{\sqrt{\left(\Omega^2 - \Omega_0^2\right)^2 + \frac{\Omega^2 \Omega_0^2}{Q^2}}} , \qquad (22.40)$$

where Δz_0 is the constant drive amplitude and Ω_0 the resonant frequency of the cantilever. The resonant frequency of the cantilever is given by the effective potential

$$\Omega_0 = \sqrt{\left(k + \frac{\partial^2 U}{\partial z^2}\right) \frac{1}{m_{\text{eff}}}} , \qquad (22.41)$$

where U is the interaction potential between the tip and the sample. Equation (22.41) shows that an attractive potential decreases Ω_0. The change in Ω_0 in turn results in a change in Δz (22.40). The movement of the cantilever changes the path difference in the interferometer. The light reflected from the cantilever with amplitude $A_{\ell,0}$ and the reference light with amplitude $A_{r,0}$ interfere on the detector. The detected intensity

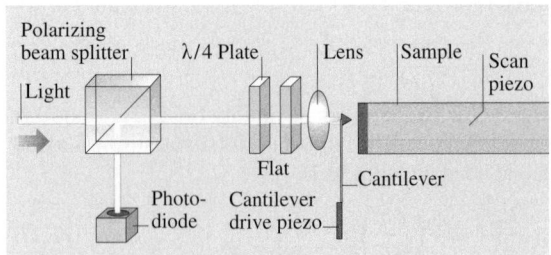

Fig. 22.25 Principle of an interferometric AFM. The light from the laser light source is polarized by the polarizing beam splitter and focused onto the back of the cantilever. The light passes twice through a quarter-wave plate and is hence orthogonally polarized to the incident light. The second arm of the interferometer is formed by the flat. The interference pattern is modulated by the oscillating cantilever

$I(t) = [A_\ell(t) + A_r(t)]^2$ consists of two constant terms and a fluctuating term

$$2A_\ell(t) A_r(t)$$

$$= A_{\ell,0} A_{r,0} \sin\left[\omega t + \frac{4\pi\delta}{\lambda} + \frac{4\pi\Delta z}{\lambda} \sin(\Omega t)\right] \sin(\omega t) .$$

(22.42)

Here ω is the frequency of the light, λ is the wavelength of the light, δ is the path difference in the interferometer, and Δz is the instantaneous amplitude of the cantilever, given according to (22.40) and (22.41) as a function of Ω, k, and U. The time average of (22.42) then becomes

$$\langle 2A_\ell(t) A_r(t)\rangle_T \propto \cos\left[\frac{4\pi\delta}{\lambda} + \frac{4\pi\Delta z}{\lambda} \sin(\Omega t)\right]$$

$$\approx \cos\left(\frac{4\pi\delta}{\lambda}\right) - \sin\left[\frac{4\pi\Delta z}{\lambda} \sin(\Omega t)\right]$$

$$\approx \cos\left(\frac{4\pi\delta}{\lambda}\right) - \frac{4\pi\Delta z}{\lambda} \sin(\Omega t) .$$

(22.43)

Here all small quantities have been omitted and functions with small arguments have been linearized. The amplitude of Δz can be recovered with a lock-in technique. However, (22.43) shows that the measured amplitude is also a function of the path difference δ in the interferometer. Hence, this path difference δ must be very stable. The best sensitivity is obtained when $\sin(4\delta/\lambda) \approx 0$.

Heterodyne Interferometer. This influence is not present in the heterodyne detection scheme shown in Fig. 22.26. Light incident from the left with a frequency ω is split into a reference path (upper path in Fig. 22.26) and a measurement path. Light in the measurement path is shifted in frequency to $\omega_1 = \omega + \Delta\omega$ and focused onto the cantilever. The cantilever oscillates at the frequency Ω, as in the homodyne detection scheme. The reflected light $A_\ell(t)$ is collimated by the same lens and interferes on the photodiode with the reference light $A_r(t)$. The fluctuating term of the intensity is given by

$$2A_\ell(t) A_r(t)$$

$$= A_{\ell,0} A_{r,0} \sin\left[(\omega + \Delta\omega) t + \frac{4\pi\delta}{\lambda}\right.$$

$$\left. + \frac{4\pi\Delta z}{\lambda} \sin(\Omega t)\right] \sin(\omega t) ,$$

(22.44)

where the variables are defined as in (22.42). Setting the path difference $\sin(4\pi\delta/\lambda) \approx 0$ and taking the time average, omitting small quantities and linearizing functions with small arguments, we get

$$\langle 2A_\ell(t) A_r(t)\rangle_T$$

$$\propto \cos\left[\Delta\omega t + \frac{4\pi\delta}{\lambda} + \frac{4\pi\Delta z}{\lambda} \sin(\Omega t)\right]$$

$$= \cos\left(\Delta\omega t + \frac{4\pi\delta}{\lambda}\right) \cos\left[\frac{4\pi\Delta z}{\lambda} \sin(\Omega t)\right]$$

$$- \sin\left(\Delta\omega t + \frac{4\pi\delta}{\lambda}\right) \sin\left[\frac{4\pi\Delta z}{\lambda} \sin(\Omega t)\right]$$

$$\approx \cos\left(\frac{4\pi\delta}{\lambda}\right) - \sin\left[\frac{4\pi\Delta z}{\lambda} \sin(\Omega t)\right]$$

$$\approx \cos\left(\Delta\omega t + \frac{4\pi\delta}{\lambda}\right) \left[1 - \frac{8\pi^2\Delta z^2}{\lambda^2} \sin(\Omega t)\right]$$

$$- \frac{4\pi\Delta z}{\lambda} \sin\left(\Delta\omega t + \frac{4\pi\delta}{\lambda}\right) \sin(\Omega t)$$

$$= \cos\left(\Delta\omega t + \frac{4\pi\delta}{\lambda}\right) - \frac{8\pi^2\Delta z^2}{\lambda^2} \cos\left(\Delta\omega t + \frac{4\pi\delta}{\lambda}\right)$$

$$\times \sin(\Omega t) - \frac{4\pi\Delta z}{\lambda} \sin\left(\Delta\omega t + \frac{4\pi\delta}{\lambda}\right) \sin(\Omega t)$$

$$= \cos\left(\Delta\omega t + \frac{4\pi\delta}{\lambda}\right) - \frac{4\pi^2\Delta z^2}{\lambda^2} \cos\left(\Delta\omega t + \frac{4\pi\delta}{\lambda}\right)$$

$$+ \frac{4\pi^2\Delta z^2}{\lambda^2} \cos\left(\Delta\omega t + \frac{4\pi\delta}{\lambda}\right) \cos(2\Omega t)$$

$$- \frac{4\pi\Delta z}{\lambda} \sin\left(\Delta\omega t + \frac{4\pi\delta}{\lambda}\right) \sin(\Omega t)$$

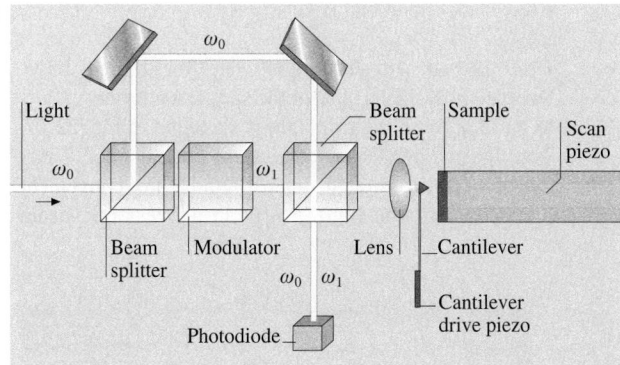

Fig. 22.26 Principle of a heterodyne interferometric AFM. Light with frequency ω_0 is split into a reference path (upper path) and a measurement path. The light in the measurement path is frequency shifted to ω_1 by an acousto-optical modulator (or an electro-optical modulator). The light reflected from the oscillating cantilever interferes with the reference beam on the detector

$$= \cos\left(\Delta\omega t + \frac{4\pi\delta}{\lambda}\right)\left(1 - \frac{4\pi^2\Delta z^2}{\lambda^2}\right)$$
$$+ \frac{2\pi^2\Delta z^2}{\lambda^2}\left\{\cos\left[(\Delta\omega + 2\Omega)\, t + \frac{4\pi\delta}{\lambda}\right]\right.$$
$$\left. + \cos\left[(\Delta\omega - 2\Omega)\, t + \frac{4\pi\delta}{\lambda}\right]\right\}$$
$$+ \frac{2\pi\Delta z}{\lambda}\left\{\cos\left[(\Delta\omega + \Omega)\, t + \frac{4\pi\delta}{\lambda}\right]\right.$$
$$\left. + \cos\left[(\Delta\omega - \Omega)\, t + \frac{4\pi\delta}{\lambda}\right]\right\}. \tag{22.45}$$

Multiplying electronically the components oscillating at $\Delta\omega$ and $\Delta\omega + \Omega$ and rejecting any product except the one oscillating at Ω we obtain

$$A = \frac{2\Delta z}{\lambda}\left(1 - \frac{4\pi^2\Delta z^2}{\lambda^2}\right)\cos\left[(\Delta\omega + 2\Omega)\, t + \frac{4\pi\delta}{\lambda}\right]$$
$$\times \cos\left(\Delta\omega t + \frac{4\pi\delta}{\lambda}\right)$$
$$= \frac{\Delta z}{\lambda}\left(1 - \frac{4\pi^2\Delta z^2}{\lambda^2}\right)\left\{\cos\left[(2\Delta\omega + \Omega)\, t + \frac{8\pi\delta}{\lambda}\right]\right.$$
$$\left. + \cos\left(\Omega t\right)\right\}$$
$$\approx \frac{\pi\Delta z}{\lambda}\cos\left(\Omega t\right). \tag{22.46}$$

Unlike in the homodyne detection scheme, the recovered signal is independent from the path difference δ of the interferometer. Furthermore, a lock-in amplifier with the reference set $\sin(\Delta\omega t)$ can measure the path difference δ independent of the cantilever oscillation. If necessary, a feedback circuit can keep $\delta = 0$.

Fiber–Optical Interferometer. The fiber-optical interferometer [22.129] is one of the simplest interferometers to build and use. Its principle is sketched in Fig. 22.27. The light of a laser is fed into an optical fiber. Laser diodes with integrated fiber pigtails are convenient light sources. The light is split in a fiber-optic beam

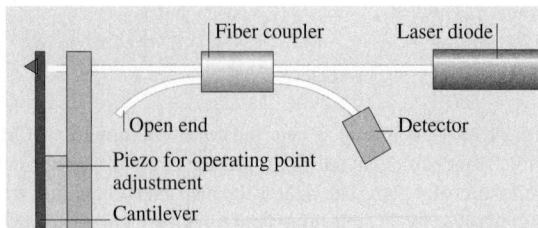

Fig. 22.27 A typical set-up for a fiber-optic interferometer readout

splitter into two fibers. One fiber is terminated by index-matching oil to avoid any reflections back into the fiber. The end of the other fiber is brought close to the cantilever in the AFM. The emerging light is partially reflected back into the fiber by the cantilever. Most of the light, however, is lost. This is not a big problem since only 4% of the light is reflected at the end of the fiber, at the glass–air interface. The two reflected light waves interfere with each other. The product is guided back into the fiber coupler and again split into two parts. One half is analyzed by the photodiode. The other half is fed back into the laser. Communications grade laser diodes are sufficiently resistant to feedback to be operated in this environment. They have, however, a bad coherence length, which in this case does not matter, since the optical path difference is in any case no larger than $5\,\mu\mathrm{m}$. Again the end of the fiber has to be positioned on a piezo drive to set the distance between the fiber and the cantilever to $\lambda(n + 1/4)$.

Nomarski–Interferometer. Another way to minimize the optical path difference is to use the Nomarski interferometer [22.130]. Figure 22.28 shows a schematic of the microscope. The light from a laser is focused on the cantilever by lens. A birefringent crystal (for instance calcite) between the cantilever and the lens, which has its optical axis 45° off the polarization direction of the light, splits the light beam into two paths, offset by a distance given by the length of the crystal. Birefringent crystals have varying indices of refraction. In calcite, one

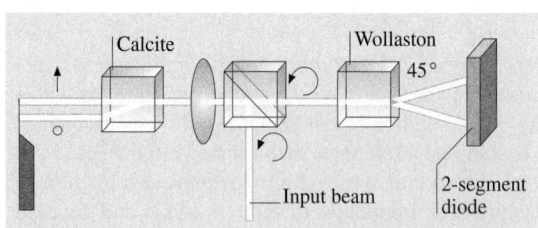

Fig. 22.28 Principle of Nomarski AFM. The circularly polarized input beam is deflected to the left by a nonpolarizing beam splitter. The light is focused onto a cantilever. The calcite crystal between the lens and the cantilever splits the circular polarized light into two spatially separated beams with orthogonal polarizations. The two light beams reflected from the lever are superimposed by the calcite crystal and collected by the lens. The resulting beam is again circularly polarized. A Wollaston prism produces two interfering beams with a $\pi/2$ phase shift between them. The minimal path difference accounts for the excellent stability of this microscope

crystal axis has a lower index than the other two. This means that certain light rays will propagate at different speeds through the crystal than others. By choosing the correct polarization, one can select the ordinary ray or the extraordinary ray or one can get any mixture of the two rays. A detailed description of birefringence can be found in textbooks (e.g., [22.150]). A calcite crystal deflects the extraordinary ray at an angle of 6° within the crystal. Any separation can be set by choosing a suitable length for the calcite crystal.

The focus of one light ray is positioned near the free end of the cantilever while the other is placed close to the clamped end. Both arms of the interferometer pass through the same space, except for the distance between the calcite crystal and the lever. The closer the calcite crystal is placed to the lever, the less influence disturbances like air currents have.

Sarid [22.116] has given values for the sensitivities of different interferometeric detection systems. Table 22.5 presents a summary of his results.

Optical Lever

The most common cantilever deflection detection system is the optical lever [22.53, 111]. This method, depicted in Fig. 22.29, employs the same technique as light beam deflection galvanometers. A fairly well collimated light beam is reflected off a mirror and projected to a receiving target. Any change in the angular position of the mirror will change the position where the light ray hits the target. Galvanometers use optical path lengths of several meters and scales projected onto the target wall are also used to monitor changes in position.

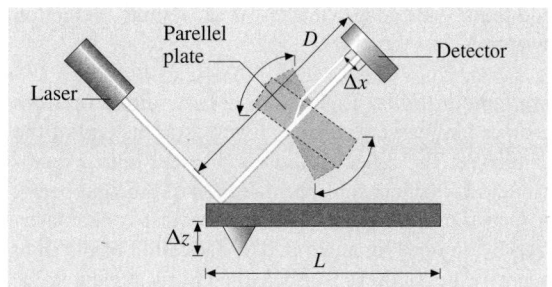

Fig. 22.29 Set-up for an optical lever detection microscope

In an AFM using the optical lever method, a photodiode segmented into two (or four) closely spaced devices detects the orientation of the end of the cantilever. Initially, the light ray is set to hit the photodiodes in the middle of the two subdiodes. Any deflection of the cantilever will cause an imbalance of the number of photons reaching the two halves. Hence the electrical currents in the photodiodes will be unbalanced too. The difference signal is further amplified and is the input signal to the feedback loop. Unlike the interferometeric AFMs, where a modulation technique is often necessary to get a sufficient signal-to-noise ratio, most AFMs employing the optical lever method are operated in a static mode. AFMs based on the optical lever method are universally used. It is the simplest method for constructing an optical readout and it can be confined in volumes that are smaller than 5 cm in side length.

The optical lever detection system is a simple yet elegant way to detect normal and lateral force signals simultaneously [22.7, 8, 53, 111]. It has the

Table 22.5 Noise in interferometers. F is the finesse of the cavity in the homodyne interferometer, P_i the incident power, P_d is the power on the detector, η is the sensitivity of the photodetector and RIN is the relative intensity noise of the laser. P_R and P_S are the power in the reference and sample beam in the heterodyne interferometer. P is the power in the Nomarski interferometer, $\delta\theta$ is the phase difference between the reference and the probe beam in the Nomarski interferometer. B is the bandwidth, e is the electron charge, λ is the wavelength of the laser, k the cantilever stiffness, ω_0 is the resonant frequency of the cantilever, Q is the quality factor of the cantilever, T is the temperature, and δi is the variation in current i

	Homodyne interferometer, fiber-optic interferometer	Heterodyne interferometer	Nomarski interferometer
Laser noise $\left\langle \delta i^2 \right\rangle_L$	$\frac{1}{4}\eta^2 F^2 P_i^2$ RIN	$\eta^2 \left(P_R^2 + P_S^2 \right)$ RIN	$\frac{1}{16}\eta^2 P^2 \delta\theta$
Thermal noise $\left\langle \delta i^2 \right\rangle_T$	$\frac{16\pi^2}{\lambda^2}\eta^2 F^2 P_i^2 \frac{4k_B T B Q}{\omega_0 k}$	$\frac{4\pi^2}{\lambda^2}\eta^2 P_d^2 \frac{4k_B T B Q}{\omega_0 k}$	$\frac{\pi^2}{\lambda^2}\eta^2 P^2 \frac{4k_B T B Q}{\omega_0 k}$
Shot noise $\left\langle \delta i^2 \right\rangle_S$	$4e\eta P_d B$	$2e\eta \left(P_R + P_S \right) B$	$\frac{1}{2}e\eta PB$

additional advantage that it is a remote detection system.

Implementations. Light from a laser diode or from a super luminescent diode is focused on the end of the cantilever. The reflected light is directed onto a quadrant diode that measures the direction of the light beam. A Gaussian light beam far from its waist is characterized by an opening angle β. The deflection of the light beam by the cantilever surface tilted by an angle α is 2α. The intensity on the detector then shifts to the side by the product of 2α and the separation between the detector and the cantilever. The readout electronics calculates the difference in the photocurrents. The photocurrents, in turn, are proportional to the intensity incident on the diode.

The output signal is hence proportional to the change in intensity on the segments

$$I_{\text{sig}} \propto 4\frac{\alpha}{\beta} I_{\text{tot}} .\tag{22.47}$$

For the sake of simplicity, we assume that the light beam is of uniform intensity with its cross-section increasing in proportion to the distance between the cantilever and the quadrant detector. The movement of the center of the light beam is then given by

$$\Delta x_{\text{Det}} = \Delta z \frac{D}{L} .\tag{22.48}$$

The photocurrent generated in a photodiode is proportional to the number of incoming photons hitting it. If the light beam contains a total number of N_0 photons, then the change in difference current becomes

$$\Delta (I_{\text{R}} - I_{\text{L}}) = \Delta I = \text{const} \, \Delta z \, D \, N_0 .\tag{22.49}$$

Combining (22.48) and (22.49), one obtains that the difference current ΔI is independent of the separation of the quadrant detector and the cantilever. This relation is true if the light spot is smaller than the quadrant detector. If it is greater, the difference current ΔI becomes smaller with increasing distance. In reality, the light beam has a Gaussian intensity profile. For small movements Δx (compared to the diameter of the light spot at the quadrant detector), (22.49) still holds. Larger movements Δx, however, will introduce a nonlinear response. If the AFM is operated in a constant force mode, only small movements Δx of the light spot will occur. The feedback loop will cancel out all other movements.

The scanning of a sample with an AFM can twist the microfabricated cantilevers because of lateral forces [22.5,7,8] and affect the images [22.120]. When the tip is subjected to lateral forces, it will twist the cantilever and the light beam reflected from the end of the cantilever will be deflected perpendicular to the ordinary deflection direction. For many investigations this influence of lateral forces is unwanted. The design of the triangular cantilevers stems from the desire to minimize the torsion effects. However, lateral forces open up a new dimension in force measurements. They allow, for instance, two materials to be distinguished because of their different friction coefficients, or adhesion energies to be determined. To measure lateral forces, the original optical lever AFM must be modified. The only modification compared with Fig. 22.29 is the use of a quadrant detector photodiode instead of a two-segment photodiode and the necessary readout electronics, see Fig. 22.9a. The electronics calculates the following signals:

$$U_{\text{Normal Force}} = \alpha \left[\left(I_{\text{Upper Left}} + I_{\text{Upper Right}} \right) \right.$$
$$\left. - \left(I_{\text{Lower Left}} + I_{\text{Lower Right}} \right) \right] ,$$
$$U_{\text{Lateral Force}} = \beta \left[\left(I_{\text{Upper Left}} + I_{\text{Lower Left}} \right) \right.$$
$$\left. - \left(I_{\text{Upper Right}} + I_{\text{Lower Right}} \right) \right] .\tag{22.50}$$

The calculation of the lateral force as a function of the deflection angle does not have a simple solution for cross-sections other than circles. An approximate formula for the angle of twist for rectangular beams is [22.151]

$$\theta = \frac{M_t L}{\beta G b^3 h} ,\tag{22.51}$$

where $M_t = F_y \ell$ is the external twisting moment due to lateral force, F_y, and β, a constant determined by the value of h/b. For the equation to hold, h has to be larger than b.

Inserting the values for a typical microfabricated cantilever with integrated tips

$$b = 6 \times 10^{-7} \, \text{m} ,$$
$$h = 10^{-5} \, \text{m} ,$$
$$L = 10^{-4} \, \text{m} ,$$
$$\ell = 3.3 \times 10^{-6} \, \text{m} ,$$
$$G = 5 \times 10^{10} \, \text{Pa} ,$$
$$\beta = 0.333 \tag{22.52}$$

into (22.51) we obtain the relation

$$F_y = 1.1 \times 10^{-4} \, \text{N} \times \theta .\tag{22.53}$$

Typical lateral forces are of the order of $10^{-10} \, \text{N}$.

Sensitivity. The sensitivity of this set-up has been calculated in various papers [22.116, 148, 152]. Assuming a Gaussian beam, the resulting output signal as a function of the deflection angle is dispersion-like. Equation (22.47) shows that the sensitivity can be increased by increasing the intensity of the light beam I_{tot} or by decreasing the divergence of the laser beam. The upper bound of the intensity of the light I_{tot} is given by saturation effects on the photodiode. If we decrease the divergence of a laser beam we automatically increase the beam waist. If the beam waist becomes larger than the width of the cantilever we start to get diffraction. Diffraction sets a lower bound on the divergence angle. Hence one can calculate the optimal beam waist w_{opt} and the optimal divergence angle β [22.148, 152]

$$w_{opt} \approx 0.36b \,,$$

$$\theta_{opt} \approx 0.89\frac{\lambda}{b} \,. \tag{22.54}$$

The optimal sensitivity of the optical lever then becomes

$$\varepsilon \,[\text{mW/rad}] = 1.8\frac{b}{\lambda}I_{tot}\,[\text{mW}] \,. \tag{22.55}$$

The angular sensitivity of the optical lever can be measured by introducing a parallel plate into the beam. Tilting the parallel plate results in a displacement of the beam, mimicking an angular deflection.

Additional noise sources can be considered. Of little importance is the quantum mechanical uncertainty of the position [22.148, 152], which is, for typical cantilevers at room temperature

$$\Delta z = \sqrt{\frac{\hbar}{2m\omega_0}} = 0.05\,\text{fm} \,, \tag{22.56}$$

where $\hbar$ is the Planck constant ($= 6.626 \times 10^{-34}$ J s). At very low temperatures and for high-frequency cantilevers this could become the dominant noise source. A second noise source is the shot noise of the light. The shot noise is related to the particle number. We can calculate the number of photons incident on the detector using

$$n = \frac{I\tau}{\hbar\omega} = \frac{I\lambda}{2\pi B\hbar c} = 1.8 \times 10^9\,\frac{I[\text{W}]}{B[\text{Hz}]} \,, \tag{22.57}$$

where I is the intensity of the light, τ the measurement time, $B = 1/\tau$ the bandwidth, and c the speed of light. The shot noise is proportional to the square root of the number of particles. Equating the shot noise signal with the signal resulting from the deflection of the cantilever

one obtains

$$\Delta z_{shot} = 68\frac{L}{w}\sqrt{\frac{B\,[\text{kHz}]}{I\,[\text{mW}]}}\,\,[\text{fm}] \,, \tag{22.58}$$

where w is the diameter of the focal spot. Typical AFM set-ups have a shot noise of 2 pm. The thermal noise can be calculated from the equipartition principle. The amplitude at the resonant frequency is

$$\Delta z_{therm} = 129\sqrt{\frac{B}{k\,[\text{N/m}]\,\omega_0 Q}}\,\,[\text{pm}] \,. \tag{22.59}$$

A typical value is 16 pm. Upon touching the surface, the cantilever increases its resonant frequency by a factor of 4.39. This results in a new thermal noise amplitude of 3.2 pm for the cantilever in contact with the sample.

Piezoresistive Detection

Implementation. A piezoresistive cantilever is an alternative detection system which is not as widely used as the optical detection schemes [22.125, 126, 132]. This cantilever is based on the fact that the resistivities of certain materials, in particular Si, change with the applied stress. Figure 22.30 shows a typical implementation of a piezo-resistive cantilever. Four resistances are integrated on the chip, forming a Wheatstone bridge. Two of the resistors are in unstrained parts of the cantilever, and the other two measure the bending at the point of the maximal deflection. For instance, when an AC voltage is applied between terminals a and c, one can measure the detuning of the bridge between terminals b and d. With such a connection the output signal only varies due to bending, not due to changes in the ambient temperature and thus the coefficient of the piezoresistance.

Sensitivity. The resistance change is [22.126]

$$\frac{\Delta R}{R_0} = \Pi\delta \,, \tag{22.60}$$

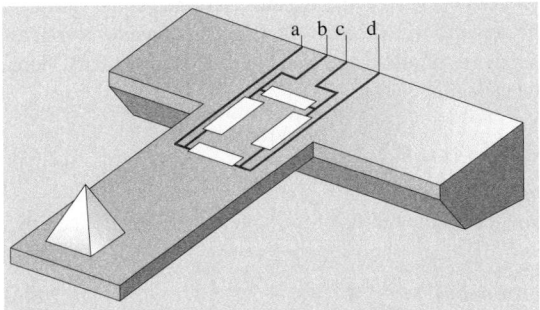

Fig. 22.30 A typical set-up for a piezoresistive readout

where Π is the tensor element of the piezo-resistive coefficients, δ the mechanical stress tensor element and R_0 the equilibrium resistance. For a single resistor, they separate the mechanical stress and the tensor element into longitudinal and transverse components:

$$\frac{\Delta R}{R_0} = \Pi_t \delta_t + \Pi_l \delta_l . \tag{22.61}$$

The maximum values of the stress components are $\Pi_t = -64.0 \times 10^{-11} \, \text{m}^2/\text{N}$ and $\Pi_l = -71.4 \times 10^{-11} \, \text{m}^2/\text{N}$ for a resistor oriented along the (110) direction in silicon [22.126]. In the resistor arrangement of Fig. 22.30, two of the resistors are subject to the longitudinal piezo-resistive effect and two of them are subject to the transversal piezo-resistive effect. The sensitivity of that set-up is about four times that of a single resistor, with the advantage that temperature effects cancel to first order. The resistance change is then calculated as

$$\frac{\Delta R}{R_0} = \Pi \frac{3Eh}{2L^2} \Delta z = \Pi \frac{6L}{bh^2} F_z , \tag{22.62}$$

where $\Pi = 67.7 \times 10^{-11} \, \text{m}^2/\text{N}$ is the averaged piezo-resistive coefficient. Plugging in typical values for the dimensions (Fig. 22.24) ($L = 100 \, \mu\text{m}$, $b = 10 \, \mu\text{m}$, $h = 1 \, \mu\text{m}$), one obtains

$$\frac{\Delta R}{R_0} = \frac{4 \times 10^{-5}}{\text{nN}} F_z . \tag{22.63}$$

The sensitivity can be tailored by optimizing the dimensions of the cantilever.

Capacitance Detection

The capacitance of an arrangement of conductors depends on the geometry. Generally speaking, the capacitance increases for decreasing separations. Two parallel plates form a simple capacitor (see Fig. 22.31, upper left), with capacitance

$$C = \frac{\varepsilon \varepsilon_0 A}{x} , \tag{22.64}$$

where A is the area of the plates, assumed equal, and x is the separation. Alternatively one can consider a sphere versus an infinite plane (see Fig. 22.31, lower left). Here the capacitance is [22.116]

$$C = 4\pi\varepsilon_0 R \sum_{n=2}^{\infty} \frac{\sinh(\alpha)}{\sinh(n\alpha)} \tag{22.65}$$

where R is the radius of the sphere, and α is defined by

$$\alpha = \ln\left(1 + \frac{z}{R} + \sqrt{\frac{z^2}{R^2} + 2\frac{z}{R}}\right) . \tag{22.66}$$

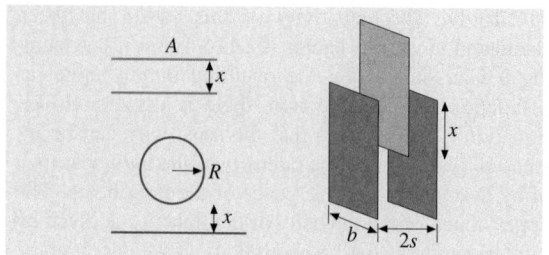

Fig. 22.31 Three possible arrangements of a capacitive readout. The *upper left* diagram shows a cross-section through a parallel plate capacitor. The *lower left* diagram shows the geometry of a sphere versus a plane. The *right-hand* diagram shows the linear (but more complicated) capacitive readout

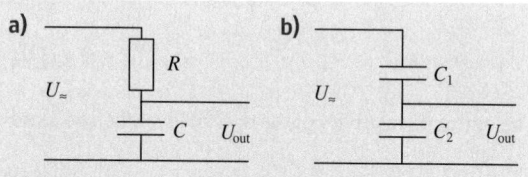

Fig. 22.32a,b Measuring the capacitance. (**a**) Low pass filter, (**b**) capacitive divider. C (*left*) and C_2 (*right*) are the capacitances under test

One has to bear in mind that the capacitance of a parallel plate capacitor is a nonlinear function of the separation. One can circumvent this problem using a voltage divider. Figure 22.32a shows a low-pass filter. The output voltage is given by

$$U_{\text{out}} = U_{\approx} \frac{\frac{1}{j\omega C}}{R + \frac{1}{j\omega C}} = U_{\approx} \frac{1}{j\omega CR + 1}$$

$$\cong \frac{U_{\approx}}{j\omega CR} . \tag{22.67}$$

Here C is given by (22.64), ω is the excitation frequency and j is the imaginary unit. The approximate relation at the end is true when $\omega CR \gg 1$. This is equivalent to the statement that C is fed by a current source, since R must be large in this set-up. Plugging (22.64) into (22.67) and neglecting the phase information, one obtains

$$U_{\text{out}} = \frac{U_{\approx} x}{\omega R \varepsilon \varepsilon_0 A} , \tag{22.68}$$

which is linear in the displacement x.

Scanning Probe Microscopy – Principle of Operation, Instrumentation, and Probes | 22.3 AFM Instrumentation and Analyses 623

Part C | 22.3

Figure 22.32b shows a capacitive divider. Again the output voltage U_{out} is given by

$$U_{out} = U_{\approx} \frac{C_1}{C_2 + C_1} = U_{\approx} \frac{C_1}{\frac{\varepsilon\varepsilon_0 A}{x} + C_1} . \quad (22.69)$$

If there is a stray capacitance C_s then (22.69) is modified as

$$U_{out} = U_{\approx} \frac{C_1}{\frac{\varepsilon\varepsilon_0 A}{x} + C_s + C_1} . \quad (22.70)$$

Provided $C_s + C_1 \ll C_2$, one has a system which is linear in x. The driving voltage $U_{\approx}$ must be large (more than 100 V) to gave an output voltage in the range of 1 V. The linearity of the readout depends on the capacitance C_1 (Fig. 22.33).

Another idea is to keep the distance constant and to change the relative overlap of the plates (see Fig. 22.31, right side). The capacitance of the moving center plate versus the stationary outer plates becomes

$$C = C_s + 2 \frac{\varepsilon\varepsilon_0 bx}{s} , \quad (22.71)$$

where the variables are defined in Fig. 22.31. The stray capacitance comprises all effects, including the capacitance of the fringe fields. When the length x is comparable to the width b of the plates, one can safely assume that the stray capacitance is constant and independent of x. The main disadvantage of this set-up is that it is not as easily incorporated into a microfabricated device as the others.

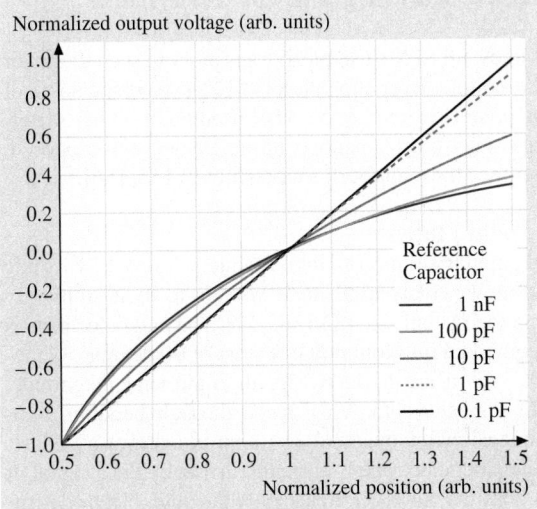

Fig. 22.33 Linearity of the capacitance readout as a function of the reference capacitor

Sensitivity. The capacitance itself is not a measure of the sensitivity, but its derivative is indicative of the signals one can expect. Using the situation described in Fig. 22.31 (upper left) and in (22.64), one obtains for the parallel plate capacitor

$$\frac{dC}{dx} = -\frac{\varepsilon\varepsilon_0 A}{x^2} . \quad (22.72)$$

Assuming a plate area A of $20\,\mu m$ by $40\,\mu m$ and a separation of $1\,\mu m$, one obtains a capacitance of $31\,fF$ (neglecting stray capacitance and the capacitance of the connection leads) and a dC/dx of $3.1 \times 10^{-8}\,F/m = 31\,fF/\mu m$. Hence it is of paramount importance to maximize the area between the two contacts and to minimize the distance x. The latter however is far from being trivial. One has to go to the limits of microfabrication to achieve a decent sensitivity.

If the capacitance is measured by the circuit shown in Fig. 22.32, one obtains for the sensitivity

$$\frac{dU_{out}}{U_{\approx}} = \frac{dx}{\omega R \varepsilon\varepsilon_0 A} . \quad (22.73)$$

Using the same value for A as above, setting the reference frequency to $100\,kHz$, and selecting $R = 1\,G\Omega$, we get the relative change in the output voltage U_{out} as

$$\frac{dU_{out}}{U_{\approx}} = \frac{22.5 \times 10^{-6}}{\text{Å}} \times dx . \quad (22.74)$$

A driving voltage of $45\,V$ then translates to a sensitivity of $1\,mV/Å$. A problem in this set-up is the stray capacitances. They are in parallel to the original capacitance and decrease the sensitivity considerably.

Alternatively, one could build an oscillator with this capacitance and measure the frequency. RC-oscillators typically have an oscillation frequency of

$$f_{res} \propto \frac{1}{RC} = \frac{x}{R \varepsilon\varepsilon_0 A} . \quad (22.75)$$

Again the resistance R must be of the order of $1\,G\Omega$ when stray capacitances C_s are neglected. However C_s is of the order of $1\,pF$. Therefore one gets $R = 10\,M\Omega$. Using these values, the sensitivity becomes

$$df_{res} = \frac{C\,dx}{R(C + C_s)^2 x} \approx \frac{0.1\,Hz}{\text{Å}}\,dx . \quad (22.76)$$

The bad thing is that the stray capacitances have made the signal nonlinear again. The linearized set-up in Fig. 22.31 has a sensitivity of

$$\frac{dC}{dx} = 2 \frac{\varepsilon\varepsilon_0 b}{s} . \quad (22.77)$$

Substituting typical values ($b = 10\,\mu m$, $s = 1\,\mu m$), one gets $dC/dx = 1.8 \times 10^{-10}\,F/m$. It is noteworthy that the sensitivity remains constant for scaled devices.

Implementations. Capacitance readout can be achieved in different ways [22.123, 124]. All include an alternating current or voltage with frequencies in the 100 kHz to 100 MHz range. One possibility is to build a tuned circuit with the capacitance of the cantilever determining the frequency. The resonance frequency of a high-quality Q tuned circuit is

$$\omega_0 = (LC)^{-1/2} .\qquad(22.78)$$

where L is the inductance of the circuit. The capacitance C includes not only the sensor capacitance but also the capacitance of the leads. The precision of a frequency measurement is mainly determined by the ratio of L and C

$$Q = \left(\frac{L}{C}\right)^{1/2} \frac{1}{R} .\qquad(22.79)$$

Here R symbolizes the losses in the circuit. The higher the quality, the more precise the frequency measurement. For instance, a frequency of 100 MHz and a capacitance of 1 pF gives an inductance of $250\,\mu H$. The quality then becomes 2.5×10^8. This value is an upper limit, since losses are usually too high.

Using a value of $dC/dx = 31\,fF/\mu m$, one gets $\Delta C/\text{Å} = 3.1\,aF/\text{Å}$. With a capacitance of 1 pF, one gets

$$\frac{\Delta\omega}{\omega} = \frac{1}{2}\frac{\Delta C}{C} ,$$

$$\Delta\omega = 100\,MHz \times \frac{1}{2}\frac{3.1\,aF}{1\,pF} = 155\,Hz .\qquad(22.80)$$

This is the frequency shift for a deflection of 1 Å. The calculation shows that this is a measurable quantity. The quality also indicates that there is no physical reason why this scheme should not work.

22.3.3 Combinations for 3−D Force Measurements

Three-dimensional force measurements are essential if one wants to know all of the details of the interaction between the tip and the cantilever. The straightforward attempt to measure three forces is complicated, since force sensors such as interferometers or capacitive sensors need a minimal detection volume, which is often too large. The second problem is that the force-sensing tip has to be held in some way. This implies that one of the three Cartesian axes is stiffer than the others.

However, by combining different sensors it is possible to achieve this goal. Straight cantilevers are employed for these measurements, because they can be handled analytically. The key observation is that the optical lever method does not determine the position of the end of the cantilever. It measures the orientation. In the previous sections, one has always made use of the fact that, for a force along one of the orthogonal symmetry directions at the end of the cantilever (normal force, lateral force, force along the cantilever beam axis), there is a one-to-one correspondence of the tilt angle and the deflection. The problem is that the force along the cantilever beam axis and the normal force create a deflection in the same direction. Hence, what is called the normal force component is actually a mixture of two forces. The deflection of the cantilever is the third quantity, which is not considered in most of the AFMs. A fiber-optic interferometer in parallel with the optical lever measures the deflection. Three measured quantities then allow the separation of the three orthonormal force directions, as is evident from (22.27) and (22.33) [22.12–16].

Alternatively, one can put the fast scanning direction along the axis of the cantilever. Forward and backward scans then exert opposite forces F_x. If the piezo movement is linearized, both force components in AFM based on optical lever detection can be determined. In this case, the normal force is simply the average of the forces in the forward and backward direction. The force, F_x, is the difference in the forces measured in the forward and backward directions.

22.3.4 Scanning and Control Systems

Almost all SPMs use piezo translators to scan the tip or the sample. Even the first STM [22.1, 103] and some of its predecessors [22.153, 154] used them. Other materials or set-ups for nanopositioning have been proposed, but they have not been successful [22.155, 156].

Piezo Tubes

A popular solution is tube scanners (Fig. 22.34). They are now widely used in SPMs due to their simplicity and their small size [22.133, 157]. The outer electrode is segmented into four equal sectors of 90 degrees. Opposite sectors are driven by signals of the same magnitude, but opposite sign. This gives, through bending, two-dimensional movement on (approximately) a sphere. The inner electrode is normally driven by the z signal. It is possible, however, to use only the outer electrodes for scanning and for the z-movement. The main drawback of applying the z-signal to the outer electrodes is that the

Scanning Probe Microscopy – Principle of Operation, Instrumentation, and Probes | 22.3 AFM Instrumentation and Analyses   625

Part C | 22.3

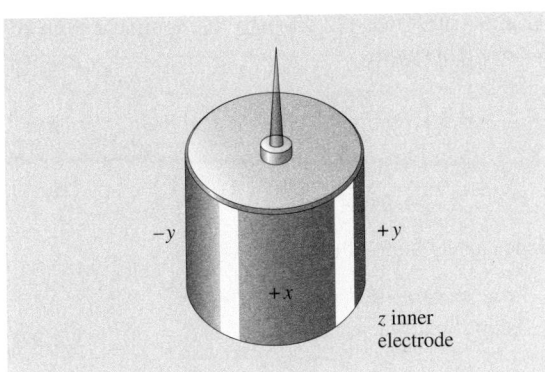

Fig. 22.34 Schematic drawing of a piezoelectric tube scanner. The piezo ceramic is molded into a tube form. The outer electrode is separated into four segments and connected to the scanning voltage. The z-voltage is applied to the inner electrode

applied voltage is the sum of both the x- or y-movements and the z-movement. Hence a larger scan size effectively reduces the available range for the z-control.

Piezo Effect

An electric field applied across a piezoelectric material causes a change in the crystal structure, with expansion in some directions and contraction in others. Also, a net volume change occurs [22.132]. Many SPMs use the transverse piezo electric effect, where the applied electric field E is perpendicular to the expansion/contraction direction.

$$\Delta L = L\,(E \cdot n)\,d_{31} = L\frac{V}{t}d_{31}\,, \tag{22.81}$$

where d_{31} is the transverse piezoelectric constant, V is the applied voltage, t is the thickness of the piezo slab or the distance between the electrodes where the voltage is applied, L is the free length of the piezo slab, and n is the direction of polarization. Piezo translators based on the transverse piezoelectric effect have a wide range of sensitivities, limited mainly by mechanical stability and breakdown voltage.

Scan Range

The the scanning range of a piezotube is difficult to calculate [22.157–159]. The bending of the tube depends on the electric fields and the nonuniform strain induced. A finite element calculation where the piezo tube was divided into 218 identical elements was used [22.158] to calculate the deflection. On each node, the mechanical stress, the stiffness, the strain and the piezoelectric

stress were calculated when a voltage was applied on one electrode. The results were found to be linear on the first iteration and higher order corrections were very small even for large electrode voltages. It was found that, to first order, the x- and z-movement of the tube could be reasonably well approximated by assuming that the piezo tube is a segment of a torus. Using this model, one obtains

$$dx = (V_+ - V_-)\,|d_{31}|\,\frac{L^2}{2td}\,, \tag{22.82}$$

$$dz = (V_+ + V_- - 2V_z)\,|d_{31}|\,\frac{L}{2t}\,, \tag{22.83}$$

where $|d_{31}|$ is the coefficient of the transversal piezoelectric effect, L is the tube's free length, t is the tube's wall thickness, d is the tube's diameter, V_+ is the voltage on the positive outer electrode, while V_- is the voltage of the opposite quadrant negative electrode and V_z is the voltage of the inner electrode.

The cantilever or sample mounted on the piezotube has an additional lateral movement because the point of measurement is not in the endplane of the piezotube. The additional lateral displacement of the end of the tip is $\ell \sin \varphi \approx \ell \varphi$, where ℓ is the tip length and φ is the deflection angle of the end surface. Assuming that the sample or cantilever is always perpendicular to the end of the walls of the tube, and calculating with the torus model, one gets for the angle

$$\varphi = \frac{L}{R} = \frac{2dx}{L}\,, \tag{22.84}$$

where R is the radius of curvature of the piezo tube. Using the result of (22.84), one obtains for the additional x-movement

$$\begin{aligned} dx_{add} = \ell\varphi &= \frac{2\,dx\ell}{L} \\ &= (V_+ - V_-)\,|d_{31}|\,\frac{\ell L}{td} \end{aligned} \tag{22.85}$$

and for the additional z-movement due to the x-movement

$$\begin{aligned} dz_{add} = \ell - \ell \cos\varphi &= \frac{\ell\varphi^2}{2} = \frac{2\ell\,(dx)^2}{L^2} \\ &= (V_+ - V_-)^2\,|d_{31}|^2\,\frac{\ell L^2}{2t^2d^2}\,. \end{aligned} \tag{22.86}$$

Carr [22.158] assumed for his finite element calculations that the top of the tube was completely free to move and, as a consequence, the top surface was distorted, leading to a deflection angle that was about half that of the geometrical model. Depending on the attachment of the sample or the cantilever, this distortion may be smaller, leading to a deflection angle in-between that of the geometrical model and the one from the finite element calculation.

Nonlinearities and Creep

Piezo materials with a high conversion ratio (a large d_{31} or small electrode separations with large scanning ranges) are hampered by substantial hysteresis resulting in a deviation from linearity by more than 10%. The sensitivity of the piezo ceramic material (mechanical displacement divided by driving voltage) decreases with reduced scanning range, whereas the hysteresis is reduced. Careful selection of the material used for the piezo scanners, the design of the scanners, and of the operating conditions is necessary to obtain optimum performance.

Passive Linearization: Calculation. The analysis of images affected by piezo nonlinearities [22.160–163] shows that the dominant term is

$$x = AV + BV^2 , \qquad (22.87)$$

where x is the excursion of the piezo, V is the applied voltage and A and B are two coefficients describing the sensitivity of the material. Equation (22.87) holds for scanning from $V = 0$ to large V. For the reverse direction, the equation becomes

$$x = \tilde{A}V - \tilde{B}(V - V_{max})^2 , \qquad (22.88)$$

where $\tilde{A}$ and $\tilde{B}$ are the coefficients for the back scan and V_{max} is the applied voltage at the turning point. Both equations demonstrate that the true x-travel is small at the beginning of the scan and becomes larger towards the end. Therefore, images are stretched at the beginning and compressed at the end.

Similar equations hold for the slow scan direction. The coefficients, however, are different. The combined action causes a greatly distorted image. This distortion can be calculated. The data acquisition systems record the signal as a function of V. However the data is measured as a function of x. Therefore we have to distribute the x-values evenly across the image. This can be done by inverting an approximation of (22.87). First we write

$$x = AV\left(1 - \frac{B}{A}V\right). \qquad (22.89)$$

For $B \ll A$ we can approximate

$$V = \frac{x}{A}. \qquad (22.90)$$

We now substitute (22.90) into the nonlinear term of (22.89). This gives

$$x = AV\left(1 + \frac{Bx}{A^2}\right),$$

$$V = \frac{x}{A}\frac{1}{(1 + Bx/A^2)} \approx \frac{x}{A}\left(1 - \frac{Bx}{A^2}\right). \qquad (22.91)$$

Hence an equation of the type

$$x_{true} = x\left(\alpha - \beta x/x_{max}\right)$$
$$\text{with} \quad 1 = \alpha - \beta \qquad (22.92)$$

takes out the distortion of an image. α and β are dependent on the scan range, the scan speed and on the scan history, and have to be determined with exactly the same settings as for the measurement. x_{max} is the maximal scanning range. The condition for α and β guarantees that the image is transformed onto itself.

Similar equations to the empirical one shown above (22.92) can be derived by analyzing the movements of domain walls in piezo ceramics.

Passive Linearization: Measuring the Position. An alternative strategy is to measure the positions of the piezo translators. Several possibilities exist.

1. The interferometers described above can be used to measure the elongation of the piezo elongation. The fiber-optic interferometer is especially easy to implement. The coherence length of the laser only limits the measurement range. However, the signal is of a periodic nature. Hence direct use of the signal in a feedback circuit for the position is not possible. However, as a measurement tool and, especially, as a calibration tool, the interferometer is without competition. The wavelength of the light, for instance that in a HeNe laser, is so well defined that the precision of the other components determines the error of the calibration or measurement.

2. The movement of the light spot on the quadrant detector can be used to measure the position of a piezo [22.164]. The output current changes by $0.5\,\text{A/cm} \times P(\text{W})/R(\text{cm})$. Typical values ($P = 1\,\text{mW}$, $R = 0.001\,\text{cm}$) give $0.5\,\text{A/cm}$. The noise limit is typically $0.15\,\text{nm} \times \sqrt{\Delta f(\text{Hz})/H(\text{W/cm}^2)}$. Again this means that the laser beam above would have a $0.1\,\text{nm}$ noise limitation for a bandwidth of $21\,\text{Hz}$. The advantage of this method is that, in principle, one can linearize two axes with only one detector.

3. A knife-edge blocking part of a light beam incident on a photodiode can be used to measure the position of the piezo. This technique, commonly used in optical shear force detection [22.75, 165], has a sensitivity of better than 0.1 nm.
4. The capacitive detection [22.166, 167] of the cantilever deflection can be applied to the measurement of the piezo elongation. Equations (22.64) to (22.79) apply to the problem. This technique is used in some commercial instruments. The difficulties lie in the avoidance of fringe effects at the borders of the two plates. While conceptually simple, one needs the latest technology in surface preparation to get a decent linearity. The electronic circuits used for the readout are often proprietary.
5. Linear variable differential transformers (LVDT) are a convenient way to measure positions down to 1 nm. They can be used together with a solid state joint set-up, as often used for large scan range stages. Unlike capacitive detection, there are few difficulties in implementation. The sensors and the detection circuits LVDTs are available commercially.
6. A popular measurement technique is the use of strain gauges. They are especially sensitive when mounted on a solid state joint where the curvature is maximal. The resolution depends mainly on the induced curvature. A precision of 1 nm is attainable. The signals are low – a Wheatstone bridge is needed for the readout.

Active Linearization. Active linearization is done with feedback systems. Sensors need to be monotonic. Hence all of the systems described above, with the exception of the interferometers, are suitable. The most common solutions include the strain gauge approach, capacitance measurement or the LVDT, which are all electronic solutions. Optical detection systems have the disadvantage that the intensity enters into the calibration.

Alternative Scanning Systems

The first STMs were based on piezo tripods [22.1]. The piezo tripod (Fig. 22.35) is an intuitive way to generate the three-dimensional movement of a tip attached to its center. However, to get a suitable stability and scanning range, the tripod needs to be fairly large (about 50 mm). Some instruments use piezo stacks instead of monolithic piezoactuators. They are arranged in a tripod. Piezo stacks are thin layers of piezoactive materials glued together to form a device with up to 200 μm of

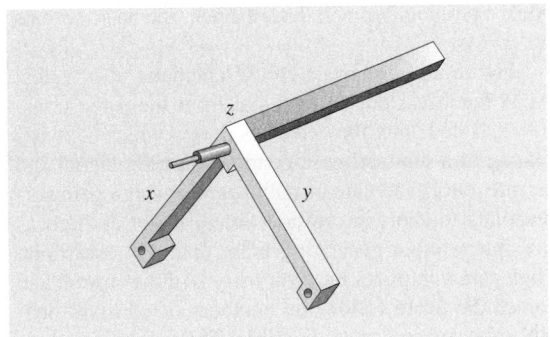

Fig. 22.35 An alternative type of piezo scanner: the tripod

actuation range. Preloading with a suitable metal casing reduces the nonlinearity.

If one tries to construct a homebuilt scanning system, the use of linearized scanning tables is recommended. They are built around solid state joints and actuated by piezo stacks. The joints guarantee that the movement is parallel with little deviation from the predefined scanning plane. Due to the construction it is easy to add measurement devices such as capacitive sensors, LVDTs or strain gauges, which are essential for a closed loop linearization. Two-dimensional tables can be bought from several manufacturers. They have linearities of better than 0.1% and a noise level of 10^{-4} to 10^{-5} for the maximal scanning range.

Control Systems

Basics. The electronics and software play an important role in the optimal performance of an SPM. Control electronics and software are supplied with commercial SPMs. Electronic control systems can use either analog or digital feedback. While digital feedback offers greater flexibility and ease of configuration, analog feedback circuits might be better suited for ultralow

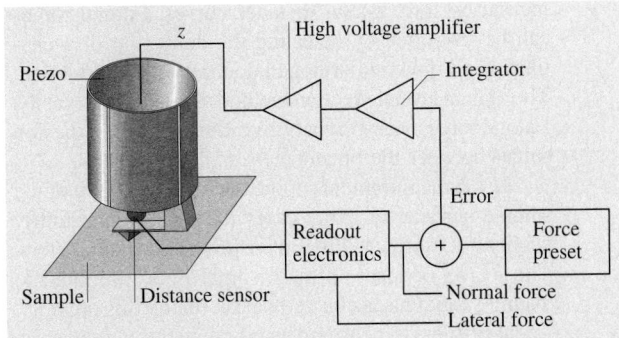

Fig. 22.36 Block schematic of the feedback control loop of an AFM

noise operation. We will describe here the basic set-ups for AFMs.

Figure 22.36 shows a block schematic of a typical AFM feedback loop. The signal from the force transducer is fed into the feedback loop, which consists mainly of a subtraction stage to get an error signal and an integrator. The gain of the integrator (high gain corresponds to short integration times) is set as high as possible without generating more than 1% overshoot. High gain minimizes the error margin of the current and forces the tip to follow the contours of constant density of states as well as possible. This operating mode is known as constant force mode. A high-voltage amplifier amplifies the outputs of the integrator. As AFMs using piezotubes usually require ± 150 V at the output, the output of the integrator needs to be amplified by a high-voltage amplifier.

In order to scan the sample, additional voltages at high tension are required to drive the piezo. For example, with a tube scanner, four scanning voltages are required, namely $+V_x$, $-V_x$, $+V_y$ and $-V_y$. The x- and y-scanning voltages are generated in a scan generator (analog or computer-controlled). Both voltages are input to the two respective power amplifiers. Two inverting amplifiers generate the input voltages for the other two power amplifiers. The topography of the sample surface is determined by recording the input voltage to the high-voltage amplifier for the z-channel as a function of x and y (constant force mode).

Another operating mode is the variable force mode. The gain in the feedback loop is lowered and the scanning speed increased such that the force on the cantilever is no longer constant. Here the force is recorded as a function of x and y.

Force Spectroscopy. Four modes of spectroscopic imaging are in common use with force microscopes: measuring lateral forces, $\partial F/\partial z$, $\partial F/\partial x$ spatially resolved, and measuring force versus distance curves. Lateral forces can be measured by detecting the deflection of a cantilever in a direction orthogonal to the normal direction. The optical lever deflection method does this most easily. Lateral force measurements give indications of adhesion forces between the tip and the sample.

$\partial F/\partial z$ measurements probe the local elasticity of the sample surface. In many cases the measured quantity originates from a volume of a few cubic nanometers. The $\partial F/\partial z$ or local stiffness signal is proportional to Young's modulus, as far as one can define this quantity. Local stiffness is measured by vibrating the cantilever by a small amount in the z-direction. The expected signal

for very stiff samples is zero: for very soft samples one also gets, independent of the stiffness, a constant signal. This signal is again zero for the optical lever deflection and equal to the driving amplitude for interferometric measurements. The best sensitivity is obtained when the compliance of the cantilever matches the stiffness of the sample.

A third spectroscopic quantity is the lateral stiffness. It is measured by applying a small modulation in the x-direction on the cantilever. The signal is again optimal when the lateral compliance of the cantilever matches the lateral stiffness of the sample. The lateral stiffness is, in turn, related to the shear modulus of the sample.

Detailed information on the interaction of the tip and the sample can be gained by measuring force versus distance curves. The cantilevers need to have enough compliance to avoid instabilities due to the attractive forces on the sample.

Using the Control Electronics as a Two-Dimensional Measurement Tool. Usually the control electronics of an AFM is used to control the x- and y-piezo signals while several data acquisition channels record the position-dependent signals. The control electronics can be used in another way: they can be viewed as a two-dimensional function generator. What is normally the x- and y-signal can be used to control two independent variables of an experiment. The control logic of the AFM then ensures that the available parameter space is systematically probed at equally spaced points. An example is friction force curves measured along a line across a step on graphite.

Figure 22.37 shows the connections. The z-piezo is connected as usual, like the x-piezo. However, the y-output is used to command the desired input parameter. The offset of the y-channel determines the position of the tip on the sample surface, together with the x-channel.

Some Imaging Processing Methods

The visualization and interpretation of images from AFMs is intimately connected to the processing of these images. An ideal AFM is a noise-free device that images a sample with perfect tips of known shape and has perfect linear scanning piezos. In reality, AFMs are not that ideal. The scanning device in an AFM is affected by distortions. The distortions are both linear and nonlinear. Linear distortions mainly result from imperfections in the machining of the piezotranslators causing crosstalk from the z-piezo to the x- and y-piezos, and vice versa. Among the linear distortions, there are two kinds which are very important. First, scanning piezos

Scanning Probe Microscopy – Principle of Operation, Instrumentation, and Probes | 22.3 AFM Instrumentation and Analyses   629

Part C | 22.3

invariably have different sensitivities along the different scan axes due to variations in the piezo material and uneven electrode areas. Second, the same reasons might cause the scanning axes to be non-orthogonal. Furthermore, the plane in which the piezoscanner moves for constant height z is hardly ever coincident with the sample plane. Hence, a linear ramp is added to the sample data. This ramp is especially bothersome when the height z is displayed as an intensity map.

The nonlinear distortions are harder to deal with. They can affect AFM data for a variety of reasons. First, piezoelectric ceramics do have a hysteresis loop, much like ferromagnetic materials. The deviations of piezoceramic materials from linearity increase with increasing amplitude of the driving voltage. The mechanical position for one voltage depends on the previously applied voltages to the piezo. Hence, to get the best positional accuracy, one should always approach a point on the sample from the same direction. Another type of nonlinear distortion of images occurs when the scan frequency approaches the upper frequency limits of the x- and y-drive amplifiers or the upper frequency limit of the feedback loop (z-component). This distortion, due to the feedback loop, can only be minimized by reducing the scan frequency. On the other hand, there is a simple way to reduce distortions due to the x- and y-piezo drive amplifiers. To keep the system as simple as possible, one normally uses a triangular waveform to drive the scanning piezos. However, triangular waves contain frequency components as multiples of the scan frequency. If the cut-off frequencies of the x- and y-drive electronics or of the feedback loop are too close to the scanning frequency (two or three times the scanning frequency), the triangular drive voltage is rounded off at the turning points. This rounding error causes, first, a distortion of the scan linearity and, second, through phase lags, the projection of part of the backward scan onto the forward scan. This type of distortion can be minimized by carefully selecting the scanning frequency and by using driving voltages for the x- and y-piezos with waveforms like trapezoidal waves, which are closer to a sine wave. The values measured for x, y or z piezos are affected by noise. The origin of this noise can be either electronic, disturbances, or a property of the sample surface due to adsorbates. In addition to this incoherent noise, interference with main and other equipment nearby might be present. Depending on the type of noise, one can filter it in real space or in Fourier space. The most important part of image processing is to visualize the measured data. Typical AFM data sets can consist of many thousands to over a million points per

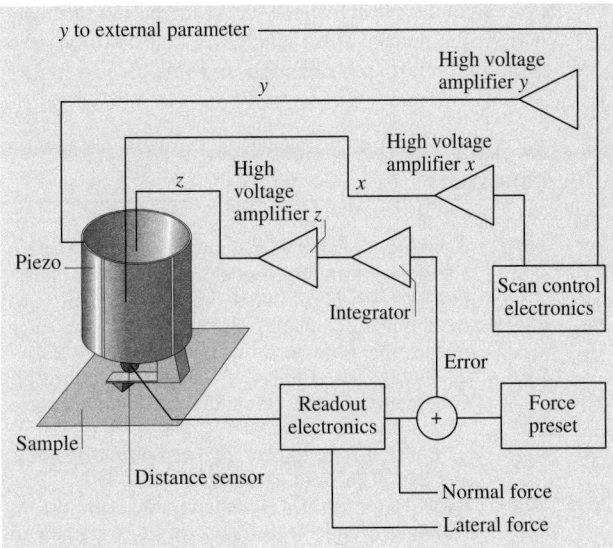

Fig. 22.37 Wiring of an AFM to measure friction force curves along a line

plane. There may be more than one image plane present. The AFM data represents a topography in various data spaces.

Most commercial data acquisition systems implicitly use some kind of data processing. Since the original data is commonly subject to slopes on the surface, most programs use some kind of slope correction. The least disturbing way is to subtract a plane $z(x, y) = Ax + By + C$ from the data. The coefficients are determined by fitting $z(x, y)$ to the data. Another operation is to subtract a second-order function such as $z(x, y) = Ax^2 + By^2 + Cxy + Dx + Ey + F$. Again, the parameters are determined with a fit. This function is appropriate for almost planar data, where the nonlinearity of the piezos caused the distortion.

In the image processing software from Digital Instruments, up to three operations are performed on the raw data. First, a zero-order flatten is applied. The flatten operation is used to eliminate image bow in the slow scan direction (caused by a physical bow in the instrument itself), slope in the slow scan direction, and bands in the image (caused by differences in the scan height from one scan line to the next). The flattening operation takes each scan line and subtracts the average value of the height along each scan line from each point in that scan line. This brings each scan line to the same height. Next, a first-order plane fit is applied in the fast scan direction. The plane-fit operation is used to eliminate bow and slope in the fast scan direction. The plane fit operation calculates a best fit plane for the image and subtracts it from the image. This plane has a constant non-zero slope in the fast scan direction. In some cases a higher order polynomial "plane" may be required. Depending upon the quality of the raw data, the flattening operation and/or the plane fit operation may not be required at all.

References

22.1   G. Binnig, H. Rohrer, Ch. Gerber, E. Weibel: Surface studies by scanning tunneling microscopy, Phys. Rev. Lett. **49**, 57–61 (1982)

22.2   G. Binnig, C. F. Quate, Ch. Gerber: Atomic force microscope, Phys. Rev. Lett. **56**, 930–933 (1986)

22.3   G. Binnig, Ch. Gerber, E. Stoll, T. R. Albrecht, C. F. Quate: Atomic resolution with atomic force microscope, Europhys. Lett. **3**, 1281–1286 (1987)

22.4   B. Bhushan: *Handbook of Micro/Nanotribology*, 2nd edn. (CRC, Boca Raton 1999)

22.5   C. M. Mate, G. M. McClelland, R. Erlandsson, S. Chiang: Atomic-scale friction of a tungsten tip on a graphite surface, Phys. Rev. Lett. **59**, 1942–1945 (1987)

22.6   R. Erlandsson, G. M. McClelland, C. M. Mate, S. Chiang: Atomic force microscopy using optical interferometry, J. Vacuum Sci. Technol. A **6**, 266–270 (1988)

22.7   O. Marti, J. Colchero, J. Mlynek: Combined scanning force and friction microscopy of mica, Nanotechnology **1**, 141–144 (1990)

22.8   G. Meyer, N. M. Amer: Simultaneous measurement of lateral and normal forces with an optical-beam-deflection atomic force microscope, Appl. Phys. Lett. **57**, 2089–2091 (1990)

22.9   B. Bhushan, J. Ruan: Atomic-scale friction measurements using friction force microscopy: Part II – Application to magnetic media, ASME J. Tribol. **116**, 389–396 (1994)

22.10   B. Bhushan, V. N. Koinkar, J. Ruan: Microtribology of magnetic media, Proc. Inst. Mech. Eng., Part J: J. Eng. Tribol. **208**, 17–29 (1994)

22.11   B. Bhushan, J. N. Israelachvili, U. Landman: Nanotribology: Friction, wear, and lubrication at the atomic scale, Nature **374**, 607–616 (1995)

22.12   S. Fujisawa, M. Ohta, T. Konishi, Y. Sugawara, S. Morita: Difference between the forces measured by an optical lever deflection and by an optical interferometer in an atomic force microscope, Rev. Sci. Instrum. **65**, 644–647 (1994)

22.13   S. Fujisawa, E. Kishi, Y. Sugawara, S. Morita: Fluctuation in 2-dimensional stick-slip phenomenon observed with 2-dimensional frictional force microscope, Jpn. J. Appl. Phys. **33**, 3752–3755 (1994)

22.14   S. Grafstrom, J. Ackermann, T. Hagen, R. Neumann, O. Probst: Analysis of lateral force effects on the to-

pography in scanning force microscopy, J. Vacuum Sci. Technol. B **12**, 1559–1564 (1994)

22.15   R. M. Overney, H. Takano, M. Fujihira, W. Paulus, H. Ringsdorf: Anisotropy in friction and molecular stick-slip motion, Phys. Rev. Lett. **72**, 3546–3549 (1994)

22.16   R. J. Warmack, X. Y. Zheng, T. Thundat, D. P. Allison: Friction effects in the deflection of atomic force microscope cantilevers, Rev. Sci. Instrum. **65**, 394–399 (1994)

22.17   N. A. Burnham, D. D. Domiguez, R. L. Mowery, R. J. Colton: Probing the surface forces of monolayer films with an atomic force microscope, Phys. Rev. Lett. **64**, 1931–1934 (1990)

22.18   N. A. Burham, R. J. Colton, H. M. Pollock: Interpretation issues in force microscopy, J. Vacuum Sci. Technol. A **9**, 2548–2556 (1991)

22.19   C. D. Frisbie, L. F. Rozsnyai, A. Noy, M. S. Wrighton, C. M. Lieber: Functional group imaging by chemical force microscopy, Science **265**, 2071–2074 (1994)

22.20   V. N. Koinkar, B. Bhushan: Microtribological studies of unlubricated and lubricated surfaces using atomic force/friction force microscopy, J. Vacuum Sci. Technol. A **14**, 2378–2391 (1996)

22.21   V. Scherer, B. Bhushan, U. Rabe, W. Arnold: Local elasticity and lubrication measurements using atomic force and friction force microscopy at ultrasonic frequencies, IEEE Trans. Magn. **33**, 4077–4079 (1997)

22.22   V. Scherer, W. Arnold, B. Bhushan: Lateral force microscopy using acoustic friction force microscopy, Surf. Interf. Anal. **27**, 578–587 (1999)

22.23   B. Bhushan, S. Sundararajan: Micro/Nanoscale friction and wear mechanisms of thin films using atomic force and friction force microscopy, Acta Mater. **46**, 3793–3804 (1998)

22.24   U. Krotil, T. Stifter, H. Waschipky, K. Weishaupt, S. Hild, O. Marti: Pulse force mode: A new method for the investigation of surface properties, Surf. Interf. Anal. **27**, 336–340 (1999)

22.25   B. Bhushan, C. Dandavate: Thin-film friction and adhesion studies using atomic force microscopy, J. Appl. Phys. **87**, 1201–1210 (2000)

22.26   B. Bhushan: *Micro/Nanotribology and its Applications* (Kluwer, Dordrecht 1997)

22.27   B. Bhushan: *Principles and Applications of Tribology* (Wiley, New York 1999)

22.28   B. Bhushan: *Modern Tribology Handbook Vol. 1: Principles of Tribology* (CRC, Boca Raton 2001)

22.29   B. Bhushan: *Introduction to Tribology* (Wiley, New York 2002)

22.30   M. Reinstaedtler, U. Rabe, V. Scherer, U. Hartmann, A. Goldade, B. Bhushan, W. Arnold: On the nanoscale measurement of friction using atomic force microscope cantilever torsional resonances, Appl. Phys. Lett. **82**, 2604–2606 (2003)

22.31   N. A. Burnham, R. J. Colton: Measuring the nanomechanical properties and surface forces of materials using an atomic force microscope, J. Vacuum Sci. Technol. A **7**, 2906–2913 (1989)

22.32   P. Maivald, H. J. Butt, S. A. C. Gould, C. B. Prater, B. Drake, J. A. Gurley, V. B. Elings, P. K. Hansma: Using force modulation to image surface elasticities with the atomic force microscope, Nanotechnology **2**, 103–106 (1991)

22.33   B. Bhushan, A. V. Kulkarni, W. Bonin, J. T. Wyrobek: Nano/Picoindentation measurements using capacitive transducer in atomic force microscopy, Philos. Mag. A **74**, 1117–1128 (1996)

22.34   B. Bhushan, V. N. Koinkar: Nanoindentation hardness measurements using atomic force microscopy, Appl. Phys. Lett. **75**, 5741–5746 (1994)

22.35   D. DeVecchio, B. Bhushan: Localized surface elasticity measurements using an atomic force microscope, Rev. Sci. Instrum. **68**, 4498–4505 (1997)

22.36   S. Amelio, A. V. Goldade, U. Rabe, V. Scherer, B. Bhushan, W. Arnold: Measurements of mechanical properties of ultra-thin diamond-like carbon coatings using atomic force acoustic microscopy, Thin Solid Films **392**, 75–84 (2001)

22.37   D. M. Eigler, E. K. Schweizer: Positioning single atoms with a scanning tunnelling microscope, Nature **344**, 524–528 (1990)

22.38   A. L. Weisenhorn, J. E. MacDougall, J. A. C. Gould, S. D. Cox, W. S. Wise, J. Massie, P. Maivald, V. B. Elings, G. D. Stucky, P. K. Hansma: Imaging and manipulating of molecules on a zeolite surface with an atomic force microscope, Science **247**, 1330–1333 (1990)

22.39   I. W. Lyo, Ph. Avouris: Field-induced nanometer-to-atomic-scale manipulation of silicon surfaces with the STM, Science **253**, 173–176 (1991)

22.40   O. M. Leung, M. C. Goh: Orientation ordering of polymers by atomic force microscope tip-surface interactions, Science **225**, 64–66 (1992)

22.41   D. W. Abraham, H. J. Mamin, E. Ganz, J. Clark: Surface modification with the scanning tunneling microscope, IBM J. Res. Dev. **30**, 492–499 (1986)

22.42   R. M. Silver, E. E. Ehrichs, A. L. de Lozanne: Direct writing of submicron metallic features with a scanning tunnelling microscope, Appl. Phys. Lett. **51**, 247–249 (1987)

22.43   A. Kobayashi, F. Grey, R. S. Williams, M. Ano: Formation of nanometer-scale grooves in silicon with

22.44   B. Parkinson: Layer-by-layer nanometer scale etching of two-dimensional substrates using the scanning tunneling microscopy, J. Am. Chem. Soc. **112**, 7498–7502 (1990)

22.45   A. Majumdar, P. I. Oden, J. P. Carrejo, L. A. Nagahara, J. J. Graham, J. Alexander: Nanometer-scale lithography using the atomic force microscope, Appl. Phys. Lett. **61**, 2293–2295 (1992)

22.46   B. Bhushan: Micro/Nanotribology and its applications to magnetic storage devices and MEMS, Tribol. Int. **28**, 85–96 (1995)

22.47   L. Tsau, D. Wang, K. L. Wang: Nanometer scale patterning of silicon(100) surface by an atomic force microscope operating in air, Appl. Phys. Lett. **64**, 2133–2135 (1994)

22.48   E. Delawski, B. A. Parkinson: Layer-by-layer etching of two-dimensional metal chalcogenides with the atomic force microscope, J. Am. Chem. Soc. **114**, 1661–1667 (1992)

22.49   B. Bhushan, G. S. Blackman: Atomic force microscopy of magnetic rigid disks and sliders and its applications to tribology, ASME J. Tribol. **113**, 452–458 (1991)

22.50   O. Marti, B. Drake, P. K. Hansma: Atomic force microscopy of liquid-covered surfaces: atomic resolution images, Appl. Phys. Lett. **51**, 484–486 (1987)

22.51   B. Drake, C. B. Prater, A. L. Weisenhorn, S. A. C. Gould, T. R. Albrecht, C. F. Quate, D. S. Cannell, H. G. Hansma, P. K. Hansma: Imaging crystals, polymers and processes in water with the atomic force microscope, Science **243**, 1586–1589 (1989)

22.52   M. Binggeli, R. Christoph, H. E. Hintermann, J. Colchero, O. Marti: Friction force measurements on potential controlled graphite in an electrolytic environment, Nanotechnology **4**, 59–63 (1993)

22.53   G. Meyer, N. M. Amer: Novel optical approach to atomic force microscopy, Appl. Phys. Lett. **53**, 1045–1047 (1988)

22.54   J. H. Coombs, J. B. Pethica: Properties of vacuum tunneling currents: Anomalous barrier heights, IBM J. Res. Dev. **30**, 455–459 (1986)

22.55   M. D. Kirk, T. Albrecht, C. F. Quate: Low-temperature atomic force microscopy, Rev. Sci. Instrum. **59**, 833–835 (1988)

22.56   F. J. Giessibl, Ch. Gerber, G. Binnig: A low-temperature atomic force/scanning tunneling microscope for ultrahigh vacuum, J. Vacuum Sci. Technol. B **9**, 984–988 (1991)

22.57   T. R. Albrecht, P. Grutter, D. Rugar, D. P. E. Smith: Low temperature force microscope with all-fiber interferometer, Ultramicroscopy **42–44**, 1638–1646 (1992)

22.58   H. J. Hug, A. Moser, Th. Jung, O. Fritz, A. Wadas, I. Parashikor, H. J. Güntherodt: Low temperature magnetic force microscopy, Rev. Sci. Instrum. **64**, 2920–2925 (1993)

22.59 C. Basire, D. A. Ivanov: Evolution of the lamellar structure during crystallization of a semicrystalline-amorphous polymer blend: Time-resolved hot-stage SPM study, Phys. Rev. Lett. **85**, 5587–5590 (2000)

22.60 H. Liu, B. Bhushan: Investigation of nanotribological properties of self-assembled monolayers with alkyl and biphenyl spacer chains, Ultramicroscopy **91**, 185–202 (2002)

22.61 J. Foster, J. Frommer: Imaging of liquid crystal using a tunneling microscope, Nature **333**, 542–547 (1988)

22.62 D. Smith, H. Horber, C. Gerber, G. Binnig: Smectic liquid crystal monolayers on graphite observed by scanning tunneling microscopy, Science **245**, 43–45 (1989)

22.63 D. Smith, J. Horber, G. Binnig, H. Nejoh: Structure, registry and imaging mechanism of alkylcyanobiphenyl molecules by tunnelling microscopy, Nature **344**, 641–644 (1990)

22.64 Y. Andoh, S. Oguchi, R. Kaneko, T. Miyamoto: Evaluation of very thin lubricant films, J. Phys. D **25**, A71–A75 (1992)

22.65 Y. Martin, C. C. Williams, H. K. Wickramasinghe: Atomic force microscope-force mapping and profiling on a sub 100-A scale, J. Appl. Phys. **61**, 4723–4729 (1987)

22.66 J. E. Stern, B. D. Terris, H. J. Mamin, D. Rugar: Deposition and imaging of localized charge on insulator surfaces using a force microscope, Appl. Phys. Lett. **53**, 2717–2719 (1988)

22.67 K. Yamanaka, H. Ogisco, O. Kolosov: Ultrasonic force microscopy for nanometer resolution subsurface imaging, Appl. Phys. Lett. **64**, 178–180 (1994)

22.68 K. Yamanaka, E. Tomita: Lateral force modulation atomic force microscope for selective imaging of friction forces, Jpn. J. Appl. Phys. **34**, 2879–2882 (1995)

22.69 U. Rabe, K. Janser, W. Arnold: Vibrations of free and surface-coupled atomic force microscope: Theory and experiment, Rev. Sci. Instrum. **67**, 3281–3293 (1996)

22.70 Y. Martin, H. K. Wickramasinghe: Magnetic imaging by force microscopy with 1000 Å resolution, Appl. Phys. Lett. **50**, 1455–1457 (1987)

22.71 D. Rugar, H. J. Mamin, P. Guethner, S. E. Lambert, J. E. Stern, I. McFadyen, T. Yogi: Magnetic force microscopy – General principles and application to longitudinal recording media, J. Appl. Phys. **63**, 1169–1183 (1990)

22.72 C. Schoenenberger, S. F. Alvarado: Understanding magnetic force microscopy, Z. Phys. B **80**, 373–383 (1990)

22.73 U. Hartmann: Magnetic force microscopy, Annu. Rev. Mater. Sci. **29**, 53–87 (1999)

22.74 D. W. Pohl, W. Denk, M. Lanz: Optical stethoscopy-image recording with resolution lambda/20, Appl. Phys. Lett. **44**, 651–653 (1984)

22.75 E. Betzig, J. K. Troutman, T. D. Harris, J. S. Weiner, R. L. Kostelak: Breaking the diffraction barrier – optical microscopy on a nanometric scale, Science **251**, 1468–1470 (1991)

22.76 E. Betzig, P. L. Finn, J. S. Weiner: Combined shear force and near-field scanning optical microscopy, Appl. Phys. Lett. **60**, 2484 (1992)

22.77 P. F. Barbara, D. M. Adams, D. B. O'Connor: Characterization of organic thin film materials with near-field scanning optical microscopy (NSOM), Annu. Rev. Mater. Sci. **29**, 433–469 (1999)

22.78 C. C. Williams, H. K. Wickramasinghe: Scanning thermal profiler, Appl. Phys. Lett. **49**, 1587–1589 (1986)

22.79 C. C. Williams, H. K. Wickramasinghe: Microscopy of chemical-potential variations on an atomic scale, Nature **344**, 317–319 (1990)

22.80 A. Majumdar: Scanning thermal microscopy, Annu. Rev. Mater. Sci. **29**, 505–585 (1999)

22.81 O. E. Husser, D. H. Craston, A. J. Bard: Scanning electrochemical microscopy – high resolution deposition and etching of materials, J. Electrochem. Soc. **136**, 3222–3229 (1989)

22.82 Y. Martin, D. W. Abraham, H. K. Wickramasinghe: High-resolution capacitance measurement and potentiometry by force microscopy, Appl. Phys. Lett. **52**, 1103–1105 (1988)

22.83 M. Nonnenmacher, M. P. O'Boyle, H. K. Wickramasinghe: Kelvin probe force microscopy, Appl. Phys. Lett. **58**, 2921–2923 (1991)

22.84 J. M. R. Weaver, D. W. Abraham: High resolution atomic force microscopy potentiometry, J. Vacuum Sci. Technol. B **9**, 1559–1561 (1991)

22.85 D. DeVecchio, B. Bhushan: Use of a nanoscale Kelvin probe for detecting wear precursors, Rev. Sci. Instrum. **69**, 3618–3624 (1998)

22.86 B. Bhushan, A. V. Goldade: Measurements and analysis of surface potential change during wear of single-crystal silicon (100) at ultralow loads using Kelvin probe microscopy, Appl. Surf. Sci. **157**, 373–381 (2000)

22.87 P. K. Hansma, B. Drake, O. Marti, S. A. C. Gould, C. B. Prater: The scanning ion-conductance microscope, Science **243**, 641–643 (1989)

22.88 C. B. Prater, P. K. Hansma, M. Tortonese, C. F. Quate: Improved scanning ion-conductance microscope using microfabricated probes, Rev. Sci. Instrum. **62**, 2634–2638 (1991)

22.89 J. Matey, J. Blanc: Scanning capacitance microscopy, J. Appl. Phys. **57**, 1437–1444 (1985)

22.90 C. C. Williams: Two-dimensional dopant profiling by scanning capacitance microscopy, Annu. Rev. Mater. Sci. **29**, 471–504 (1999)

22.91 D. T. Lee, J. P. Pelz, B. Bhushan: Instrumentation for direct, low frequency scanning capacitance microscopy, and analysis of position dependent stray capacitance, Rev. Sci. Instrum. **73**, 3523–3533 (2002)

22.92 P. K. Hansma, J. Tersoff: Scanning tunneling microscopy, J. Appl. Phys. **61**, R1–R23 (1987)

22.93 I. Giaever: Energy gap in superconductors measured by electron tunneling, Phys. Rev. Lett. **5**, 147–148 (1960)

22.94 D. Sarid, V. Elings: Review of scanning force microscopy, J. Vacuum Sci. Technol. B **9**, 431–437 (1991)

22.95 U. Durig, O. Zuger, A. Stalder: Interaction force detection in scanning probe microscopy: Methods and applications, J. Appl. Phys. **72**, 1778–1797 (1992)

22.96 J. Frommer: Scanning tunneling microscopy and atomic force microscopy in organic chemistry, Angew. Chem. Int. Ed. **31**, 1298–1328 (1992)

22.97 H. J. Güntherodt, R. Wiesendanger (eds): *Scanning Tunneling Microscopy I: General Principles and Applications to Clean and Adsorbate-Covered Surfaces* (Springer, Berlin, Heidelberg 1992)

22.98 R. Wiesendanger, H. J. Güntherodt (eds): *Scanning Tunneling Microscopy, II: Further Applications and Related Scanning Techniques* (Springer, Berlin, Heidelberg 1992)

22.99 D. A. Bonnell (ed): *Scanning Tunneling Microscopy and Spectroscopy – Theory, Techniques, and Applications* (VCH, New York 1993)

22.100 O. Marti, M. Amrein (eds): *STM and SFM in Biology* (Academic, San Diego 1993)

22.101 J. A. Stroscio, W. J. Kaiser (eds): *Scanning Tunneling Microscopy* (Academic, Boston 1993)

22.102 H. J. Güntherodt, D. Anselmetti, E. Meyer (eds): *Forces in Scanning Probe Methods* (Kluwer, Dordrecht 1995)

22.103 G. Binnig, H. Rohrer: Scanning tunnelling microscopy, Surf. Sci. **126**, 236–244 (1983)

22.104 B. Bhushan, J. Ruan, B. K. Gupta: A scanning tunnelling microscopy study of fullerene films, J. Phys. D **26**, 1319–1322 (1993)

22.105 R. L. Nicolaides, W. E. Yong, W. F. Packard, H. A. Zhou: Scanning tunneling microscope tip structures, J. Vacuum Sci. Technol. A **6**, 445–447 (1988)

22.106 J. P. Ibe, P. P. Bey, S. L. Brandon, R. A. Brizzolara, N. A. Burnham, D. P. DiLella, K. P. Lee, C. R. K. Marrian, R. J. Colton: On the electrochemical etching of tips for scanning tunneling microscopy, J. Vacuum Sci. Technol. A **8**, 3570–3575 (1990)

22.107 R. Kaneko, S. Oguchi: Ion-implanted diamond tip for a scanning tunneling microscope, Jpn. J. Appl. Phys. **28**, 1854–1855 (1990)

22.108 F. J. Giessibl: Atomic resolution of the silicon(111)–(7×7) surface by atomic force microscopy, Science **267**, 68–71 (1995)

22.109 B. Anczykowski, D. Krueger, K. L. Babcock, H. Fuchs: Basic properties of dynamic force spectroscopy with

the scanning force microscope in experiment and simulation, Ultramicroscopy **66**, 251–259 (1996)

22.110 T. R. Albrecht and C. F. Quate: Atomic resolution imaging of a nonconductor by atomic force microscopy, J. Appl. Phys. **62**, 2599–2602 (1987)

22.111 S. Alexander, L. Hellemans, O. Marti, J. Schneir, V. Elings, P. K. Hansma: An atomic-resolution atomic-force microscope implemented using an optical lever, J. Appl. Phys. **65**, 164–167 (1989)

22.112 G. Meyer, N. M. Amer: Optical-beam-deflection atomic force microscopy: The NaCl(001) surface, Appl. Phys. Lett. **56**, 2100–2101 (1990)

22.113 A. L. Weisenhorn, M. Egger, F. Ohnesorge, S. A. C. Gould, S. P. Heyn, H. G. Hansma, R. L. Sinsheimer, H. E. Gaub, P. K. Hansma: Molecular resolution images of Langmuir–Blodgett films and DNA by atomic force microscopy, Langmuir **7**, 8–12 (1991)

22.114 J. Ruan, B. Bhushan: Atomic-scale and microscale friction of graphite and diamond using friction force microscopy, J. Appl. Phys. **76**, 5022–5035 (1994)

22.115 D. Rugar, P. K. Hansma: Atomic force microscopy, Phys. Today **43**, 23–30 (1990)

22.116 D. Sarid: *Scanning Force Microscopy* (Oxford Univ. Press, Oxford 1991)

22.117 G. Binnig: Force microscopy, Ultramicroscopy **42–44**, 7–15 (1992)

22.118 E. Meyer: Atomic force microscopy, Surf. Sci. **41**, 3–49 (1992)

22.119 H. K. Wickramasinghe: Progress in scanning probe microscopy, Acta Mater. **48**, 347–358 (2000)

22.120 A. J. den Boef: The influence of lateral forces in scanning force microscopy, Rev. Sci. Instrum. **62**, 88–92 (1991)

22.121 M. Radmacher, R. W. Tillman, M. Fritz, H. E. Gaub: From molecules to cells: Imaging soft samples with the atomic force microscope, Science **257**, 1900–1905 (1992)

22.122 F. Ohnesorge, G. Binnig: True atomic resolution by atomic force microscopy through repulsive and attractive forces, Science **260**, 1451–1456 (1993)

22.123 G. Neubauer, S. R. Coben, G. M. McClelland, D. Horne, C. M. Mate: Force microscopy with a bidirectional capacitance sensor, Rev. Sci. Instrum. **61**, 2296–2308 (1990)

22.124 T. Goddenhenrich, H. Lemke, U. Hartmann, C. Heiden: Force microscope with capacitive displacement detection, J. Vacuum Sci. Technol. A **8**, 383–387 (1990)

22.125 U. Stahl, C. W. Yuan, A. L. Delozanne, M. Tortonese: Atomic force microscope using piezoresistive cantilevers and combined with a scanning electron microscope, Appl. Phys. Lett. **65**, 2878–2880 (1994)

22.126 R. Kassing, E. Oesterschulze: Sensors for scanning probe microscopy. In: *Micro/Nanotribology and Its*

Applications, ed. by B. Bhushan (Kluwer, Dordrecht 1997) pp. 35–54

22.127 C. M. Mate: Atomic-force-microscope study of polymer lubricants on silicon surfaces, Phys. Rev. Lett. **68**, 3323–3326 (1992)

22.128 S. P. Jarvis, A. Oral, T. P. Weihs, J. B. Pethica: A novel force microscope and point contact probe, Rev. Sci. Instrum. **64**, 3515–3520 (1993)

22.129 D. Rugar, H. J. Mamin, P. Guethner: Improved fiber-optical interferometer for atomic force microscopy, Appl. Phys. Lett. **55**, 2588–2590 (1989)

22.130 C. Schoenenberger, S. F. Alvarado: A differential interferometer for force microscopy, Rev. Sci. Instrum. **60**, 3131–3135 (1989)

22.131 D. Sarid, D. Iams, V. Weissenberger, L. S. Bell: Compact scanning-force microscope using laser diode, Opt. Lett. **13**, 1057–1059 (1988)

22.132 N. W. Ashcroft, N. D. Mermin: *Solid State Physics* (Holt Reinhart and Winston, New York 1976)

22.133 G. Binnig, D. P. E. Smith: Single-tube three-dimensional scanner for scanning tunneling microscopy, Rev. Sci. Instrum. **57**, 1688 (1986)

22.134 S. I. Park, C. F. Quate: Digital filtering of STM images, J. Appl. Phys. **62**, 312 (1987)

22.135 J. W. Cooley, J. W. Tukey: An algorithm for machine calculation of complex Fourier series, Math. Comput. **19**, 297 (1965)

22.136 J. Ruan, B. Bhushan: Atomic-scale friction measurements using friction force microscopy: Part I – General principles and new measurement techniques, ASME J. Tribol. **116**, 378–388 (1994)

22.137 T. R. Albrecht, S. Akamine, T. E. Carver, C. F. Quate: Microfabrication of cantilever styli for the atomic force microscope, J. Vacuum Sci. Technol. A **8**, 3386–3396 (1990)

22.138 O. Marti, S. Gould, P. K. Hansma: Control electronics for atomic force microscopy, Rev. Sci. Instrum. **59**, 836–839 (1988)

22.139 O. Wolter, T. Bayer, J. Greschner: Micromachined silicon sensors for scanning force microscopy, J. Vacuum Sci. Technol. B **9**, 1353–1357 (1991)

22.140 E. Meyer, R. Overney, R. Luthi, D. Brodbeck: Friction force microscopy of mixed Langmuir–Blodgett films, Thin Solid Films **220**, 132–137 (1992)

22.141 H. J. Dai, J. H. Hafner, A. G. Rinzler, D. T. Colbert, R. E. Smalley: Nanotubes as nanoprobes in scanning probe microscopy, Nature **384**, 147–150 (1996)

22.142 J. H. Hafner, C. L. Cheung, A. T. Woolley, C. M. Lieber: Structural and functional imaging with carbon nanotube AFM probes, Prog. Biophys. Mol. Biol. **77**, 73–110 (2001)

22.143 G. S. Blackman, C. M. Mate, M. R. Philpott: Interaction forces of a sharp tungsten tip with molecular films on silicon surface, Phys. Rev. Lett. **65**, 2270–2273 (1990)

22.144 S. J. O'Shea, M. E. Welland, T. Rayment: Atomic force microscope study of boundary layer lubrication, Appl. Phys. Lett. **61**, 2240–2242 (1992)

22.145 J. P. Cleveland, S. Manne, D. Bocek, P. K. Hansma: A nondestructive method for determining the spring constant of cantilevers for scanning force microscopy, Rev. Sci. Instrum. **64**, 403–405 (1993)

22.146 D. W. Pohl: Some design criteria in STM, IBM J. Res. Dev. **30**, 417 (1986)

22.147 W. T. Thomson, M. D. Dahleh: *Theory of Vibration with Applications*, 5th edn. (Prentice Hall, Upper Saddle River 1998)

22.148 J. Colchero: Reibungskraftmikroskopie. Ph.D. Thesis (University of Konstanz, Konstanz 1993)

22.149 G. M. McClelland, R. Erlandsson, S. Chiang: Atomic force microscopy: General principles and a new implementation. In: *Review of Progress in Quantitative Nondestructive Evaluation*, Vol. 6B, ed. by D. O. Thompson, D. E. Chimenti (Plenum, New York 1987) pp. 1307–1314

22.150 Y. R. Shen: *The Principles of Nonlinear Optics* (Wiley, New York 1984)

22.151 T. Baumeister, S. L. Marks: *Standard Handbook for Mechanical Engineers*, 7th edn. (McGraw-Hill, New York 1967)

22.152 J. Colchero, O. Marti, H. Bielefeldt, J. Mlynek: Scanning force and friction microscopy, Phys. Stat. Sol. **131**, 73–75 (1991)

22.153 R. Young, J. Ward, F. Scire: Observation of metal-vacuum-metal tunneling, field emission, and the transition region, Phys. Rev. Lett. **27**, 922 (1971)

22.154 R. Young, J. Ward, F. Scire: The topographiner: An instrument for measuring surface microtopography, Rev. Sci. Instrum. **43**, 999 (1972)

22.155 C. Gerber, O. Marti: Magnetostrictive positioner, IBM Tech. Discl. Bull. **27**, 6373 (1985)

22.156 R. Garcìa Cantù, M. A. Huerta Garnica: Long-scan imaging by STM, J. Vacuum Sci. Technol. A **8**, 354 (1990)

22.157 C. J. Chen: In situ testing and calibration of tube piezoelectric scanners, Ultramicroscopy **42–44**, 1653–1658 (1992)

22.158 R. G. Carr: Finite element analysis of PZT tube scanner motion for scanning tunnelling microscopy, J. Microsc. **152**, 379–385 (1988)

22.159 C. J. Chen: Electromechanical deflections of piezoelectric tubes with quartered electrodes, Appl. Phys. Lett. **60**, 132 (1992)

22.160 N. Libioulle, A. Ronda, M. Taborelli, J. M. Gilles: Deformations and nonlinearity in scanning tunneling microscope images, J. Vacuum Sci. Technol. B **9**, 655–658 (1991)

22.161 E. P. Stoll: Restoration of STM images distorted by time-dependent piezo driver aftereffects, Ultramicroscopy **42–44**, 1585–1589 (1991)

22.162 R. Durselen, U. Grunewald, W. Preuss: Calibration and applications of a high precision piezo scanner for nanometrology, Scanning **17**, 91–96 (1995)

22.163 J. Fu: In situ testing and calibrating of Z-piezo of an atomic force microscope, Rev. Sci. Instrum. **66**, 3785–3788 (1995)

22.164 R. C. Barrett, C. F. Quate: Optical scan-correction system applied to atomic force microscopy, Rev. Sci. Instrum. **62**, 1393 (1991)

22.165 R. Toledo-Crow, P. C. Yang, Y. Chen, M. Vaez-Iravani: Near-field differential scanning optical microscope with atomic force regulation, Appl. Phys. Lett. **60**, 2957–2959 (1992)

22.166 J. E. Griffith, G. L. Miller, C. A. Green: A scanning tunneling microscope with a capacitance-based position monitor, J. Vacuum Sci. Technol. B **8**, 2023–2027 (1990)

22.167 A. E. Holman, C. D. Laman, P. M. L. O. Scholte, W. C. Heerens, F. Tuinstra: A calibrated scanning tunneling microscope equipped with capacitive sensors, Rev. Sci. Instrum. **67**, 2274–2280 (1996)

23. Probes in Scanning Microscopies

Scanning probe microscopy (SPM) provides nanometer-scale mapping of numerous sample properties in essentially any environment. This unique combination of high resolution and broad applicability has lead to the application of SPM to many areas of science and technology, especially those interested in the structure and properties of materials at the nanometer scale. SPM images are generated through measurements of a tip-sample interaction. A well-characterized tip is the key element to data interpretation and is typically the limiting factor.

Commercially available atomic force microscopy (AFM) tips, integrated with force sensing cantilevers, are microfabricated from silicon and silicon nitride by lithographic and anisotropic etching techniques. The performance of these tips can be characterized by imaging nanometer-scale standards of known dimension, and the resolution is found to roughly correspond to the tip radius of curvature, the tip aspect ratio, and the sample height. Although silicon and silicon nitride tips have a somewhat large radius of curvature, low aspect ratio, and limited lifetime due to wear, the widespread use of AFM today is due in large part to the broad availability of these tips. In some special cases, small asperities on the tip can provide resolution much higher than the tip radius of curvature for low-Z samples such as crystal surfaces and ordered protein arrays.

Several strategies have been developed to improve AFM tip performance. Oxide sharpening improves tip sharpness and enhances tip asperities. For high-aspect-ratio samples such as integrated circuits, silicon AFM tips can be modified by focused ion beam (FIB) milling. FIB tips reach 3 degree cone angles over lengths of several microns and can be fabricated at arbitrary angles.

Other high resolution and high-aspect-ratio tips are produced by electron beam deposition (EBD) in which a carbon spike is deposited onto the tip apex from the background gases in an electron microscope. Finally, carbon nanotubes have been employed as AFM tips. Their nanometer-scale diameter, long length, high stiffness, and elastic buckling properties make carbon nanotubes possibly the ultimate tip material for AFM. Nanotubes can be manually attached to silicon or silicon nitride AFM tips or "grown" onto tips by chemical vapor deposition (CVD), which should soon make them widely available. In scanning tunneling microscopy (STM), the electron tunneling signal decays exponentially with tip-sample separation, so that in principle only the last few atoms contribute to the signal. STM tips are, therefore, not as sensitive to the nanoscale tip geometry and can be made by simple mechanical cutting or electrochemical etching of metal wires. In choosing tip materials, one prefers hard, stiff metals that will not oxidize or corrode in the imaging environment.

In scanning probe microscopy (SPM), an image is created by raster scanning a sharp probe tip over a sample and measuring some highly localized tip-sample interaction as a function of position. SPMs are based on several interactions, the major types including scanning tunneling microscopy (STM), which measures an electronic tunneling current; atomic force microscopy (AFM), which measures force interactions; and near-field scanning optical microscopy (NSOM), which measures local optical properties by exploiting near-field effects (Fig. 23.1). These methods allow the characterization of many properties (structural, mechanical, electronic, optical) on essentially any material (metals, semiconductors, insulators, biomolecules) and in essentially any environment (vacuum, liquid, or ambient air conditions). The unique combination of nanoscale resolution, previously the domain of electron microscopy, *and broad applicability* has led to the proliferation of SPM into virtually all areas of nanometer-scale science and technology.

Several enabling technologies have been developed for SPM, or borrowed from other techniques. Piezoelectric tube scanners allow accurate, sub-angstrom positioning of the tip or sample in three dimensions. Optical deflection systems and microfabricated cantilevers can detect forces in AFM down to the picoNewton range. Sensitive electronics can measure STM currents less than 1 picoamp. High transmission fiber optics and sensitive photodetectors can manipulate and detect small optical signals of NSOM. Environmental control has been developed to allow SPM imaging in UHV, cryogenic temperatures, at elevated temperatures, and in fluids. Vibration and drift have been controlled such that a probe tip can be held over a single molecule for hours of observation. Microfabrication techniques have been developed for the mass production of probe tips, making SPMs commercially available and allowing the development of many new SPM modes and combi-

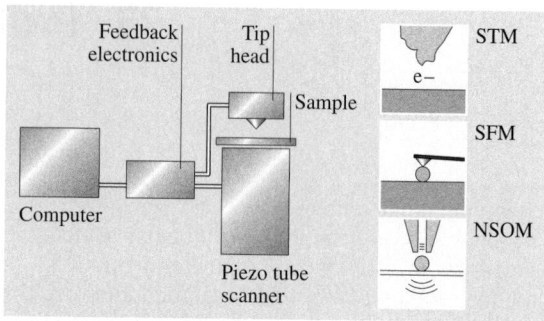

Fig. 23.1 A schematic of the components of a scanning probe microscope and the three types of signals observed: STM senses electron tunneling currents, AFM measures forces, and NSOM measures near-field optical properties via a sub-wavelength aperture

nations with other characterization methods. However, of all this SPM development over the past 20 years, what has received the least attention is perhaps the most important aspect: the probe tip.

Interactions measured in SPMs occur at the tip-sample interface, which can range in size from a single atom to tens of nanometers. The size, shape, surface chemistry, electronic and mechanical properties of the tip apex will directly influence the data signal and the interpretation of the image. Clearly, the better characterized the tip the more useful the image information. In this chapter, the fabrication and performance of AFM and STM probes will be described.

23.1 Atomic Force Microscopy

AFM is the most widely used form of SPM, since it requires neither an electrically conductive sample, as in STM, nor an optically transparent sample or substrate, as in most NSOMs. Basic AFM modes measure the topography of a sample with the only requirement being that the sample is deposited on a flat surface and rigid enough to withstand imaging. Since AFM can measure a variety of forces, including van der Waals forces, electrostatic forces, magnetic forces, adhesion forces and friction forces, specialized modes of AFM can characterize the electrical, mechanical, and chemical properties of a sample in addition to its topography.

23.1.1 Principles of Operation

In AFM, a probe tip is integrated with a microfabricated force-sensing cantilever. A variety of silicon and

silicon nitride cantilevers are commercially available with micron-scale dimensions, spring constants ranging from 0.01 to 100 N/m, and resonant frequencies ranging from 5 kHz to over 300 kHz. The cantilever deflection is detected by optical beam deflection, as illustrated in Fig. 23.2. A laser beam bounces off the back of the cantilever and is centered on a split photodiode. Cantilever deflections are proportional to the difference signal $V_A - V_B$. Sub-angstrom deflections can be detected and, therefore, forces down to tens of picoNewtons can be measured. A more recently developed method of cantilever deflection measurement is through a piezoelectric layer on the cantilever that registers a voltage upon deflection [23.1].

A piezoelectric scanner rasters the sample under the tip while the forces are measured through deflections of the cantilever. To achieve more controlled imaging con-

Fig. 23.2 An illustration of the optical beam deflection system that detects cantilever motion in the AFM. The voltage signal $V_A - V_B$ is proportional to the deflection

ditions, a feedback loop monitors the tip-sample force and adjusts the sample Z-position to hold the force constant. The topographic image of the sample is then taken from the sample Z-position data. The mode described is called contact mode, in which the tip is deflected by the sample due to repulsive forces, or "contact". It is generally only used for flat samples that can withstand lateral forces during scanning. To minimize lateral forces and sample damage, two AC modes have been developed. In these, the cantilever is driven into AC oscillation near its resonant frequency (tens to hundreds of kHz) with amplitudes of 5 to tens of s. When the tip approaches the sample, the oscillation is damped, and the reduced amplitude is the feedback signal, rather than the DC deflection. Again, topography is taken from the varying Z-position of the sample required to keep the tip oscillation amplitude constant. The two AC modes differ only in the nature of the interaction. In intermittent contact mode, also called tapping mode, the tip contacts the sample on each cycle, so the amplitude is reduced by ionic repulsion as in contact mode. In non-contact mode, long-range van der Waals forces reduce the amplitude by effectively shifting the spring constant experienced by the tip and changing its resonant frequency.

23.1.2 Standard Probe Tips

In early AFM work, cantilevers were made by hand from thin metal foils or small metal wires. Tips were created by gluing diamond fragments to the foil cantilevers or electrochemically etching the wires to a sharp point. Since these methods were labor intensive and not highly reproducible, they were not amenable to large-

scale production. To address this problem, and the need for smaller cantilevers with higher resonant frequencies, batch fabrication techniques were developed (see Fig. 23.3). Building on existing methods to batch fabricate Si_3N_4 cantilevers, *Albrecht* et al. [23.2] etched an array of small square openings in an SiO_2 mask layer over a (100) silicon surface. The exposed square (100) regions were etched with KOH, an anisotropic etchant that terminates at the (111) planes, thus creating pyramidal etch pits in the silicon surface. The etch pit mask was then removed and another was applied to define the cantilever shapes with the pyramidal etch pits at the end. The Si wafer was then coated with a low stress Si_3N_4 layer by LPCVD. The Si_3N_4 fills the etch pit, using it as a mold to create a pyramidal tip. The silicon was later removed by etching to free the cantilevers and tips. Further steps resulting in the attachment of the cantilever to a macroscopic piece of glass are not described here. The resulting pyramidal tips were highly symmetric and had a tip radius of less than 30 nm, as determined by scanning electron microscopy (SEM). This procedure has likely not changed significantly, since commercially available Si_3N_4 tips are still specified to have a curvature radius of 30 nm.

Wolter et al. [23.3] developed methods to batch fabricate single-crystal Si cantilevers with integrated tips. Microfabricated Si cantilevers were first prepared using previously described methods, and a small mask was formed at the end of the cantilever. The Si around the mask was etched by KOH, so that the mask was under cut. This resulted in a pyramidal silicon tip under the mask, which was then removed. Again, this partial description of the full procedure only describes tip fabrication. With some refinements the silicon tips were made in high yield with curvature radii of less than 10 nm. Si tips are sharper than Si_3N_4 tips, because they are directly

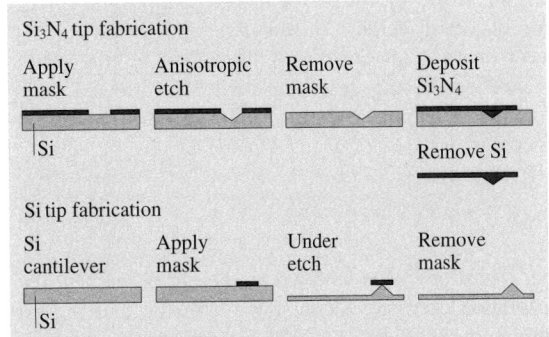

Fig. 23.3 A schematic overview of the fabrication of Si and Si_3N_4 tip fabrication as described in the text

formed by the anisotropic etch in single-crystal Si, rather than using an etch pit as a mask for deposited material. Commercially available silicon probes are made by similar refined techniques and provide a curvature typical radius of < 10 nm.

23.1.3 Probe Tip Performance

In atomic force microscopy the question of resolution can be a rather complicated issue. As an initial approximation, resolution is often considered strictly in geometrical terms that assume rigid tip-sample contact. The topographical image of a feature is broadened or narrowed by the size of the probe tip, so the resolution is approximately the width of the tip. Therefore, the resolution of AFM with standard commercially available tips is on the order of 5 to 10 nm. *Bustamante* and *Keller* [23.4] carried the geometrical model further by drawing an analogy to resolution in optical systems. Consider two sharp spikes separated by a distance d to be point objects imaged by AFM (see Fig. 23.4). Assume the tip has a parabolic shape with an end radius R. The tip-broadened image of these spikes will appear as inverted parabolas. There will be a small depression between the images of depth Δz. The two spikes are considered "resolved" if Δz is larger than the instrumental noise in the z direction. Defined in this manner, the resolution d, the minimum separation at which the spikes are resolved, is

$$d = 2\sqrt{2R(\Delta z)}, \tag{23.1}$$

where one must enter a minimal detectable depression for the instrument (Δz) to determine the resolution. So for a silicon tip with radius 5 nm and a minimum detectable Δz of 0.5 nm, the resolution is about 4.5 nm. However, the above model assumes the spikes are of equal height. *Bustamante* and *Keller* [23.4] went on to point out that if the height of the spikes is not equal, the resolution will be affected. Assuming a height difference of Δh, the resolution becomes:

$$d = \sqrt{2R}\left(\sqrt{\Delta z} + \sqrt{\Delta z + \Delta h}\right). \tag{23.2}$$

For a pair of spikes with a 2 nm height difference, the resolution drops to 7.2 nm for a 5 nm tip and 0.5 nm minimum detectable Δz. While geometrical considerations are a good starting point for defining resolution, they ignore factors such as the possible compres-

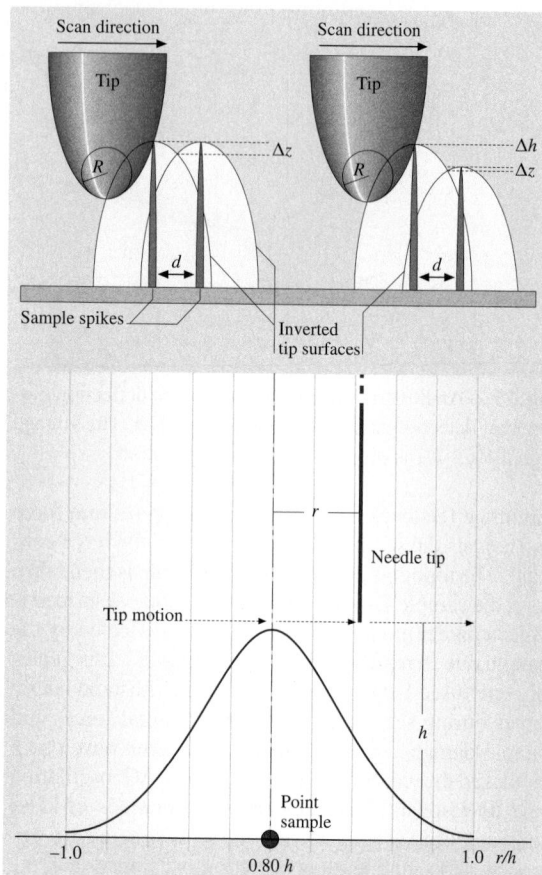

Fig. 23.4 The factors that determine AFM imaging resolution in contact mode (*top*) and noncontact mode (*bottom*), adapted from [23.4]

sion and deformation of the tip and sample. *Vesenka* et al. [23.5] confirmed a similar geometrical resolution model by imaging monodisperse gold nanoparticles with tips characterized by transmission electron microscopy (TEM).

Noncontact AFM contrast is generated by long-range interactions such as van der Waals forces, so resolution will not simply be determined by geometry because the tip and sample are not in rigid contact. *Bustamante* and *Keller* [23.4] have derived an expression for the resolution in noncontact AFM for an idealized, infinitely thin "line" tip and a point particle as the sample (Fig. 23.4). Noncontact AFM is sensitive to the gradient of long-range forces, so the van der Waals force gradient was calculated as a function of position for the tip at height h above the surface. If the resolution d is defined as the full width at half maximum of this curve, the resolution is:

$$d = 0.8h \, . \tag{23.3}$$

This shows that even for an ideal geometry, the resolution is fundamentally limited in noncontact mode by the tip-sample separation. Under UHV conditions, the tip-sample separation can be made very small, so atomic resolution is possible on flat, crystalline surfaces. Under ambient conditions, however, the separation must be larger to keep the tip from being trapped in the ambient water layer on the surface. This larger separation can lead to a point where further improvements in tip sharpness do not improve resolution. It has been found that imaging 5 nm gold nanoparticles in noncontact mode with carbon nanotube tips of 2 nm diameter leads to particle widths of 12 nm, larger than the 7 nm width one would expect assuming rigid contact [23.8]. However, in tapping mode operation, the geometrical definition of resolution is relevant, since the tip and sample come into rigid contact. When imaging 5 nm gold particles with 2 nm carbon nanotube tips in tapping mode, the expected 7 nm particle width is obtained [23.9].

The above descriptions of AFM resolution cannot explain the sub-nanometer resolution achieved on crystal surfaces [23.10] and ordered arrays of biomolecules [23.11] in contact mode with commercially available probe tips. Such tips have nominal radii of curvature ranging from 5 nm to 30 nm, an order of magnitude larger than the resolution achieved. A detailed model to explain the high resolution on ordered membrane proteins has been put forth by [23.6]. In this model, the larger part of the silicon nitride tip apex balances the tip-sample interaction through electrostatic forces, while a very small tip asperity interacts with the sample to provide contrast (see Fig. 23.5). This model is supported by measurements at varying salt concentrations to vary the electrostatic interaction strength and the observation of defects in the ordered samples. However, the existence of such asperities has never been confirmed by independent electron microscopy images of the tip. Another model, considered especially applicable to atomic resolution on crystal surfaces, assumes the tip is in contact with a region of the sample much larger than the resolution observed, and that force components matching the periodicity of the sample are transmitted to the tip, resulting in an "averaged" image of the periodic lattice. Regardless of the mechanism, the structures determined are accurate and make this a highly valuable method for membrane proteins. However, this level of

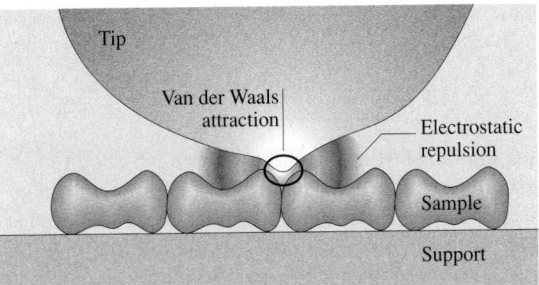

Fig. 23.5 A tip model to explain the high resolution obtained on ordered samples in contact mode, from [23.6]

resolution should not be expected for most biological systems.

23.1.4 Oxide–Sharpened Tips

Both Si and Si$_3$N$_4$ tips with increased aspect ratio and reduced tip radius can be fabricated through oxide sharpening of the tip. If a pyramidal or cone-shaped silicon tip is thermally oxidized to SiO$_2$ at low temperature ($< 1050\,^\circ$C), Si-SiO$_2$ stress formation reduces the oxidation rate at regions of high curvature. The result is a sharper, higher-aspect-ratio cone of silicon at the high curvature tip apex inside the outer pyramidal layer of SiO$_2$ (see Fig. 23.6). Etching the SiO$_2$ layer with HF then leaves tips with aspect ratios up to 10:1 and radii down to 1 nm [23.7], although 5–10 nm is the nominal specification for most commercially available tips. This oxide sharpening technique can also be applied to Si$_3$N$_4$ tips by oxidizing the silicon etch pits that are used as molds. As with tip fabrication, oxide sharpening is not quite as effective for Si$_3$N$_4$. Si$_3$N$_4$ tips were reported to have an 11 nm radius of curvature [23.12], while commercially available

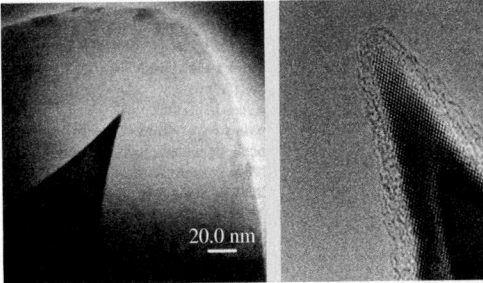

Fig. 23.6 Oxide sharpening of silicon tips. The *left image* shows a sharpened core of silicon in an outer layer of SiO$_2$. The *right image* is a higher magnification view of such a tip after the SiO$_2$ is removed. Adapted from [23.7]

oxide-sharpened Si_3N_4 tips have a nominal radius of < 20 nm.

23.1.5 FIB tips

A common AFM application in integrated circuit manufacture and MEMs is to image structures with very steep sidewalls such as trenches. To accurately image these features, one must consider the micron-scale tip structure, rather than the nanometer-scale structure of the tip apex. Since tip fabrication processes rely on anisotropic etchants, the cone half-angles of pyramidal tips are approximately 20 degrees. Images of deep trenches taken with such tips display slanted sidewalls and may not reach the bottom of the trench due to the tip broadening effects. To image such samples more faithfully, high-aspect-ratio tips are fabricated by focused ion beam (FIB) machining a Si tip to produce a sharp spike at the tip apex. Commercially available FIB tips have half cone angles of < 3 degrees over lengths of several microns, yielding aspect ratios of approximately 10:1. The radius of curvature at the tip end is similar to that of the tip before the FIB machining. Another consideration for high-aspect-ratio tips is the tip tilt. To ensure that the pyramidal tip is the lowest part of the tip-cantilever assembly, most AFM designs tilt the cantilever about 15 degrees from parallel. Therefore, even an ideal "line tip" will not give an accurate image of high steep sidewalls, but will produce an image that depends on the scan angle. Due to the versatility of the FIB machining, tips are available with the spikes at an angle to compensate for this effect.

23.1.6 EBD tips

Another method of producing high-aspect-ratio tips for AFM is called electron beam deposition (EBD). First developed for STM tips [23.13, 14], EBD tips were introduced for AFM by focusing an SEM onto the apex of a pyramidal tip arranged so that it pointed along the electron beam axis (see Fig. 23.7). Carbon material was deposited by the dissociation of background gases in the SEM vacuum chamber. *Schiffmann* [23.15] systematically studied the following parameters and how they affected EBD tip geometry:

Deposition time :	0.5 to 8 min
Beam current :	3–300 pA
Beam energy :	1–30 keV
Working distance :	8–48 mm .

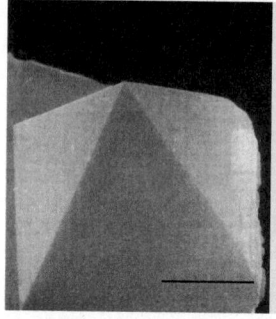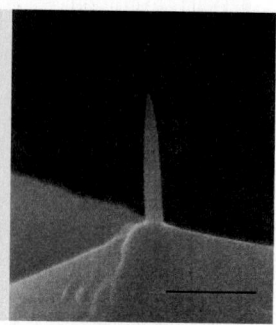

Fig. 23.7 A pyramidal tip before (*left*, 2-μm-scale bar) and after (*right*, 1-μm-scale bar) electron beam deposition, adapted from [23.13]

EBD tips were cylindrical with end radii of 20–40 nm, lengths of 1 to 5 μm, and diameters of 100 to 200 nm. Like FIB tips, EBD tips were found to achieve improved imaging of steep features. By controlling the position of the focused beam, the tip geometry can be further controlled. Tips were fabricated with lengths over 5 μm and aspect ratios greater than 100:1, yet these were too fragile to use as a tip in AFM [23.13].

23.1.7 Carbon Nanotube Tips

Carbon nanotubes are microscopic graphitic cylinders that are nanometers in diameter, yet many microns in length. Single-walled carbon nanotubes (SWNT) consist of single sp^2 hybridized carbon sheets rolled into seamless tubes and have diameters ranging from 0.7 to 3 nm.

Carbon Nanotube Structure
Larger structures called multiwalled carbon nanotubes (MWNT) consist of nested, concentrically arranged SWNT and have diameters ranging from 3 to 50 nm. Figure 23.8 shows a model of nanotube structure, as well as TEM images of a SWNT and a MWNT. The small diameter and high aspect ratio of carbon nanotubes suggests their application as high resolution, high-aspect-ratio AFM probes.

Carbon Nanotube Mechanical Properties
Carbon nanotubes possess exceptional mechanical properties that impact their use as probes. Their lateral stiffness can be approximated from that of a solid elastic rod:

$$k_{\text{lat}} = \frac{3\pi Y r^4}{4l^3} \, , \tag{23.4}$$

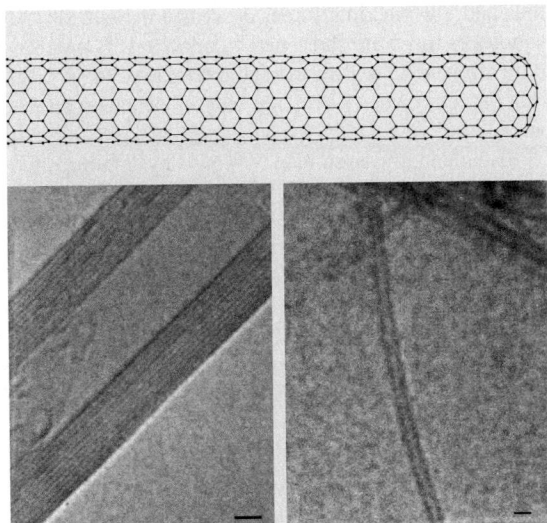

Fig. 23.8 The structure of carbon nanotubes, including TEM images of a MWNT (*left*) and a SWNT (*right*), from [23.16]

where the spring constant k_{lat} represents the restoring force per unit lateral displacement, r is the radius, l is the length, and Y is the Young's modulus (also called the elastic modulus) of the material. For the small diameters and extreme aspect ratios of carbon nanotube tips, the thermal vibrations of the probe tip at room temperature can become sufficient to degrade image resolution. These thermal vibrations can be approximated by equating $\frac{1}{2}k_B T$ of thermal energy to the energy of an oscillating nanotube:

$$\frac{1}{2}k_B T = \frac{1}{2}k_{lat}a^2 \, , \tag{23.5}$$

where k_B is Boltzmann's constant, T is the temperature, and a is the vibration amplitude. Substituting for k_{lat} from (23.4) yields:

$$a = \sqrt{\frac{4k_B T l^3}{3\pi Y r^4}} \, . \tag{23.6}$$

The strong dependence on radius and length reveals that one must carefully control the tip geometry at this size scale. Equation (23.6) implies that the stiffer the material, i.e., the higher its Young's modulus, the smaller the thermal vibrations and the longer and thinner a tip can be. The Young's moduli of carbon nanotubes have been determined by measurements of the thermal vibration amplitude by TEM [23.18, 19] and by directly measuring

the forces required to deflect a pinned carbon nanotube in an AFM [23.20]. These experiments revealed that the Young's modulus of carbon nanotubes is $1-2$ TPa, in agreement with theoretical predictions [23.21]. This makes carbon nanotubes the stiffest known material and, therefore, the best for fabricating thin, high-aspect-ratio tips. A more detailed and accurate derivation of the thermal vibration amplitudes was derived for the Young's modulus measurements [23.18, 19].

Carbon nanotubes elastically buckle under large loads, rather than fracture or plastically deform like most materials. Nanotubes were first observed in the buckled state by transmission electron microscopy [23.17], as shown in Fig. 23.9. The first experimental evidence that nanotube buckling is elastic came from the application of nanotubes as probe tips [23.22], described in detail below. A more direct experimental observation of elastic buckling was obtained by deflecting nanotubes pinned to a low friction surface with an AFM tip [23.20]. Both reports found that the buckling force could be approximated with the macroscopic Euler buckling formula for an elastic column:

$$F_{Euler} = \frac{\pi^3 Y r^4}{4l^2} \, . \tag{23.7}$$

The buckling force puts another constraint on the tip length: If the nanotube is too long the buckling force will be too low for stable imaging. The elastic buckling

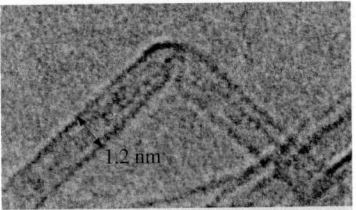

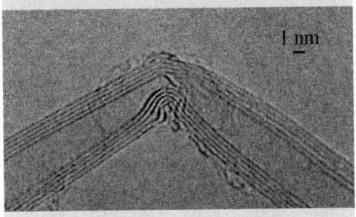

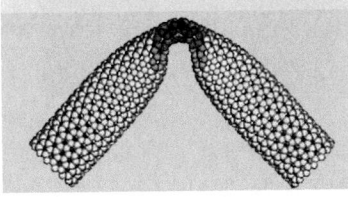

Fig. 23.9 TEM images and a model of a buckled nanotube, adapted from [23.17]

property of carbon nanotubes has significant implications for their use as AFM probes. If a large force is applied to the tip inadvertently, or if the tip encounters a large step in sample height, the nanotube can buckle to the side, then snap back without degraded imaging resolution when the force is removed, making these tips highly robust. No other tip material displays this buckling characteristic.

Manually Assembled Nanotube Probes

The first carbon nanotube AFM probes [23.22] were fabricated by techniques developed for assembling single-nanotube field emission tips [23.23]. This process, illustrated in Fig. 23.10, used purified MWNT material synthesized by the carbon arc procedure. The raw material, which must contain at least a few percent of long nanotubes ($> 10\,\mu$m) by weight, purified by oxidation to approximately 1% of its original mass. A torn edge of the purified material was attached to a micromanipulator by carbon tape and viewed under a high power optical microscope. Individual nanotubes and nanotube bundles were visible as filaments under dark field illumination. A commercially available AFM tip was attached to another micromanipulator opposing the nanotube material. Glue was applied to the tip apex from high vacuum carbon tape supporting the nanotube material. Nanotubes were then manually attached to the tip apex by micromanipulation. As assembled, MWNT tips were often too long for imaging due to thermal vibra-

tions and low buckling forces described in Sect. 23.1.7. Nanotubes tips were shortened by applying 10 V pulses to the tip while it was near a sputtered niobium surface. This process etched ~ 100 nm lengths of nanotube per pulse.

The manually assembled MWNT tips demonstrated several important nanotube tip properties [23.22]. First, the high aspect ratio of the MWNT tips allowed the accurate imaging of trenches in silicon with steep sidewalls, similar to FIB and EBD tips. Second, elastic buckling was observed indirectly through force curves (see Fig. 23.11). Note that as the tip taps the sample, the amplitude drops to zero and a DC deflection is observed, because the nanotube is unbuckled and is essentially rigid. As the tip moves closer, the force on the nanotube eventually exceeds the buckling force. The nanotube buckles, allowing the vibration amplitude to partially recover, and the deflection remains constant. Numeric tip trajectory simulations could only reproduce these force curves if elastic buckling was included in the nanotube response. Finally, the nanotube tips were highly robust. Even after "tip crashes" or hundreds of controlled buckling cycles, the tip retained its resolution and high aspect ratio.

Manual assembly of carbon nanotube probe tips is straightforward, but has several limitations. It is labor intensive and not amenable to mass production. Although

Fig. 23.10 A schematic drawing of the setup for manual assembly of carbon nanotube tips (*top*) and optical microscopy images of the assembly process (the cantilever was drawn in for clarity)

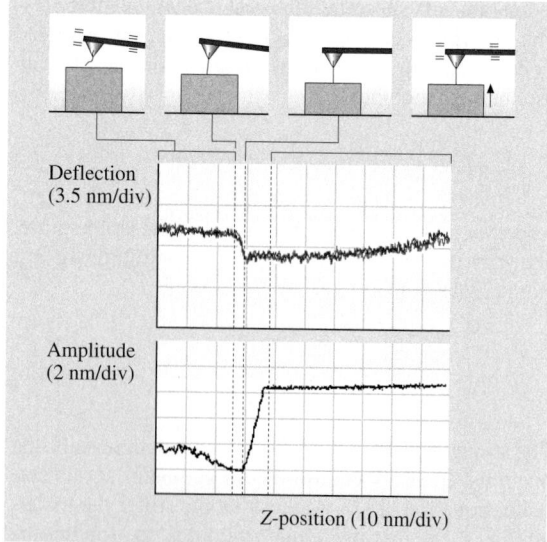

Fig. 23.11 Nanotube tip buckling. *Top* diagrams correspond to labeled regions of the force curves. As the nanotube tip buckles, the deflection remains constant and the amplitude increases, from [23.16]

MWNT tips have been made commercially available by this method, they are about ten times more expensive than silicon probes. The manual assembly method has also been carried out in an SEM, rather than an optical microscope [23.24]. This eliminates the need for pulse-etching, since short nanotubes can be attached to the tip, and the "glue" can be applied by EBD. But this is still not the key to mass production, since nanotube tips are made individually. MWNT tips provided a modest improvement in resolution on biological samples, but typical MWNT radii are similar to that of silicon tips, so they cannot provide the ultimate resolution possible with a SWNT tip. SWNT bundles can be attached to silicon probes by manual assembly. Pulse etching at times produces very high resolution tips that likely result from exposing a small number of nanotubes from the bundle, but this is not reproducible [23.25]. Even if a sample could be prepared that consisted of individual SWNT for manual assembly, such nanotubes would not be easily visible by optical microscopy or SEM.

CVD Nanotube Probe Synthesis

The problems of manual assembly of nanotube probes discussed above can largely be solved by directly growing nanotubes onto AFM tips by metal-catalyzed chemical vapor deposition (CVD). The key features of the nanotube CVD process are illustrated in Fig. 23.12. Nanometer-scale metal catalyst particles are heated in a gas mixture containing hydrocarbon or CO. The gas molecules dissociate on the metal surface, and carbon is adsorbed into the catalyst particle. When this carbon precipitates, it nucleates a nanotube of similar diameter to the catalyst particle. Therefore, CVD allows control over nanotube size and structure, including the production of SWNTs [23.26] with radii as low as 3.5 Angstrom [23.27].

Several key issues must be addressed to grow nanotube AFM tips by CVD: (1) the alignment of the nanotubes at the tip, (2) the number of nanotubes that grow at the tip, and (3) the length of the nanotube tip. *Li* et al. [23.28] found that nanotubes grow perpendicular to a porous surface containing embedded catalyst. This approach was exploited to fabricate nanotube tips by CVD [23.29] with the proper alignment, as illustrated in Fig. 23.13. A flattened area of approximately $1-5\,\mu m^2$ was created on Si tips by scanning in contact mode at high load ($1\,\mu N$) on a hard, synthetic diamond surface. The tip was then anodized in HF to create 100 nm-diameter pores in this flat surface [23.30]. It is important to only anodize the last $20-40\,\mu m$ of

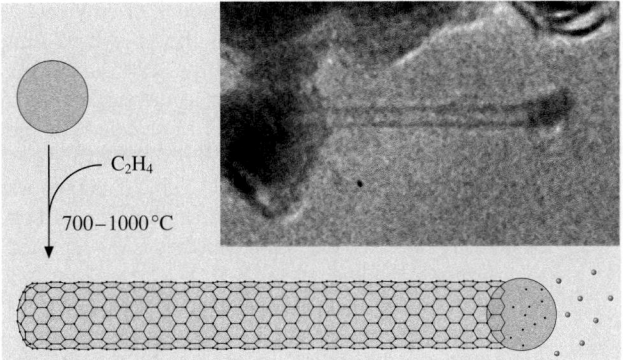

Fig. 23.12 CVD nanotube synthesis. Ethylene reacts with a nanometer-scale iron catalyst particle at high temperature to form a carbon nanotube. The *inset* in the upper right is a TEM image showing a catalyst particle at the end of a nanotube, from [23.16]

the cantilever, which includes the tip, so that the rest of the cantilever is still reflective for use in the AFM. This was achieved by anodizing the tip in a small drop of HF under the view of an optical microscope. Next, iron was electrochemically deposited into the pores to form catalyst particles [23.31]. Tips prepared in this

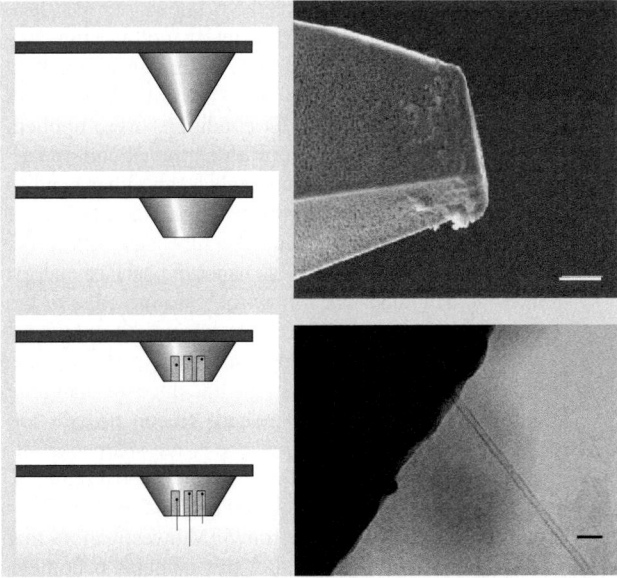

Fig. 23.13 Pore-growth CVD nanotube tip fabrication. The *left panel*, from top to bottom, shows the steps described in the text. The *upper right* is an SEM image of such a tip with a small nanotube protruding from the pores (scale bar is 1 μm). The *lower right* is a TEM of a nanotube protruding from the pores (scale bar is 20 nm), from [23.16]

way were heated in low concentrations of ethylene at 800 °C, which is known to favor the growth of thin nanotubes [23.26]. When imaged by SEM, nanotubes were found to grow perpendicular to the surface from the pores as desired (Fig. 23.13). TEM revealed that the nanotubes were thin, individual, multiwalled nanotubes with typical radii ranging from 3–5 nm. If nanotubes did not grow in an acceptable orientation, the carbon could be removed by oxidation, and then CVD repeated to grow new nanotube tips.

These "pore-growth" CVD nanotube tips were typically several microns in length – too long for imaging – and were pulse-etched to a usable length of < 500 nm. The tips exhibited elastic buckling behavior and were very robust in imaging. In addition, the thin, individual nanotube tips enabled improved resolution [23.29] on isolated proteins. The pore-growth method demonstrated the potential of CVD to simplify the fabrication of nanotube tips, although there were still limitations. In particular, the porous layer was difficult to prepare and rather fragile.

An alternative approach for CVD fabrication of nanotube tips involves direct growth of SWNTs on the surface of a pyramidal AFM tip [23.32, 33]. In this "surface-growth" approach, an alumina/iron/molybdenum-powdered catalyst known to produce SWNT [23.26] was dispersed in ethanol at 1 mg/mL. Silicon tips were dipped in this solution and allowed to dry, leaving a sparse layer of ~ 100 nm catalyst clusters on the tip. When CVD conditions were applied, single-walled nanotubes grew along the silicon tip surface. At a pyramid edge, nanotubes can either bend to align with the edge, or protrude from the surface. If the energy required to bend the tube and follow the edge is less than the attractive nanotube-surface energy, then the nanotube will follow the pyramid edge to the apex. Therefore, nanotubes were effectively steered toward the tip apex by the pyramid edges. At the apex, the nanotube protruded from the tip, since the energetic cost of bending around the sharp silicon tip was too high. The high aspect ratio at the oxide-sharpened silicon tip apex was critical for good nanotube alignment. A schematic of this approach is shown in Fig. 23.14. Evidence for this model came from SEM investigations that show that a very high yield of tips contains nanotubes only at the apex, with very few protruding elsewhere from the pyramid. TEM analysis demonstrated that the tips typically consist of small SWNT bundles that are formed by nanotubes coming together from different edges of the pyramid to join at the apex, supporting the surface growth model described above (Fig. 23.14).

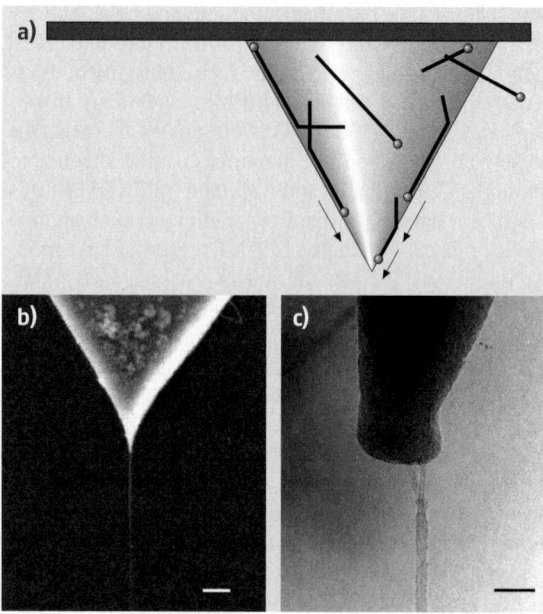

Fig. 23.14a–c Surface-growth nanotube tip fabrication. (**a**) Schematic represents the surface growth process in which nanotubes growing on the pyramidal tip are guided to the tip apex. (**b**),(**c**) Images show (**b**) SEM (200-nm-scale bar) and (**c**) TEM (20-nm-scale bar) images of a surface growth tip, from [23.16]

The "surface growth" nanotube tips exhibit a high aspect ratio and high resolution imaging, as well as elastic buckling.

The surface growth method has been expanded to include wafer-scale production of nanotube tips with yields of over 90% [23.34], yet one obstacle remains to the mass production of nanotube probe tips. Nanotubes protruding from the tip are several microns long, and since they are so thin, they must be etched to less than 100 nm. While the pulse-etching step is fairly reproducible, it must be carried out on nanotube tips in a serial fashion, so surface growth does not yet represent a true method of batch nanotube tip fabrication.

Hybrid Nanotube Tip Fabrication: Pick-up Tips

Another method of creating nanotube tips is something of a hybrid between assembly and CVD. The motivation was to create AFM probes that have an *individual* SWNT at the tip to achieve the ultimate imaging resolution. In order to synthesize isolated SWNT, they must be nucleated at sites separated farther than their typical length. The alumina-supported catalyst contains a high density of catalyst particles per 100 nm cluster, so nanotube bundles cannot be avoided. To fabricate completely isolated nanotubes, isolated catalyst particles were formed by dipping a silicon wafer in an isopropyl alcohol solution of $Fe(NO_3)_3$. This effectively left a submonolayer of iron on the wafer, so that when it was heated in a CVD furnace, the iron became mobile and aggregated to form isolated iron particles. During CVD conditions, these particles nucleated and grew SWNTs. By controlling the reaction time, the SWNT lengths were kept shorter than their typical separation, so that the nanotubes never had a chance to form bundles. AFM analysis of these samples revealed 1–3 nm-diameter SWNT and un-nucleated particles on the surface (Fig. 23.15). However, there were tall objects that were difficult to image at a density of about 1 per $50\,\mu m^2$. SEM analysis at an oblique angle demonstrated that these were SWNTs that had grown perpendicular to the surface (Fig. 23.15).

In the "pick-up tip" method, these isolated SWNT substrates were imaged by AFM with silicon tips in air [23.9]. When the tip encountered a vertical SWNT, the oscillation amplitude was damped, so the AFM pulled the sample away from the tip. This pulled the SWNT into contact with the tip along its length, so that it became attached to the tip. This assembly process happened automatically when imaging in tapping mode – no special tip manipulation was required. When imaging a wafer with the density shown in Fig. 23.15, one nanotube was attached per $8\,\mu m \times 8\,\mu m$ scan at 512×512 and 2 Hz, so a nanotube tip could be made in about 5 min. Since the as-formed SWNT tip continued to image, there was usually no evidence in the topographic image that a nanotube had been attached. However, the pick-up event was identified when the Z-voltage suddenly stepped to larger tip-

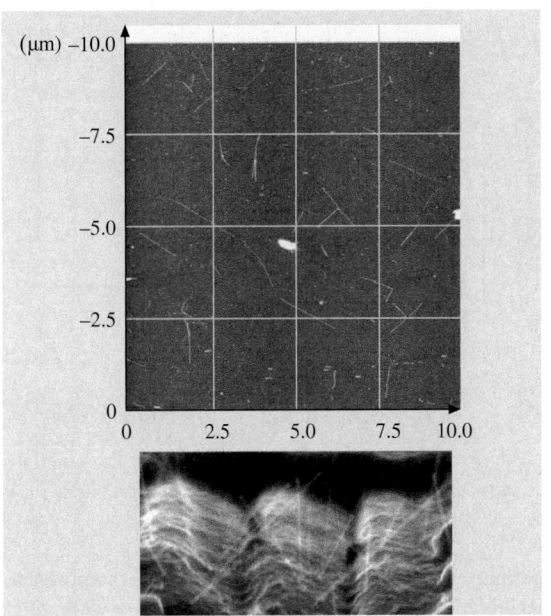

Fig. 23.15 Atomic force microscopy image (*top*) of a wafer with isolated nanotubes synthesized by CVD. The SEM view provides evidence that some of these nanotubes are arranged vertically

sample separation due to the effective increase in tip length.

Individual SWNT tips must be quite short, typically less than 50 nm in length, for reasons outlined above. Pulse-etching, which removes 50–100 nm of nanotube length at a time, lacked the necessary precision for shortening pick-up tips, so the tips were shortened through force curves. A pick-up tip force curve is shown in Fig. 23.17. When the tip first interacted with the sample,

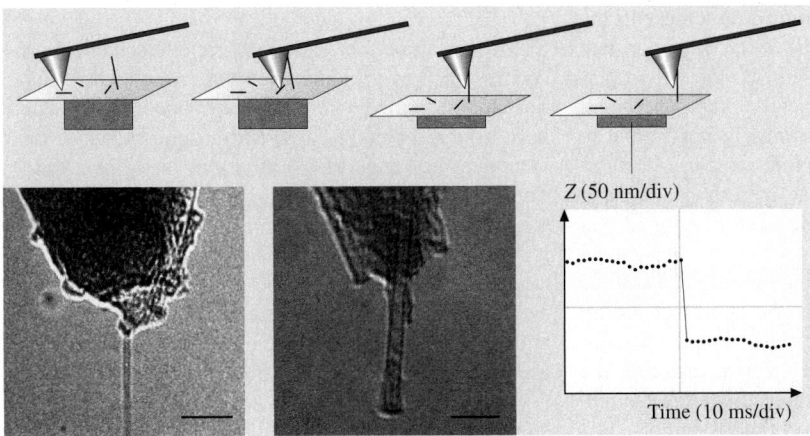

Fig. 23.16 Pick-up tip assembly of nanotube probes. The *top* illustrates the nanotube pick-up process that occurs while imaging vertical nanotubes in an AFM, including a trace of the Z-position during a pick-up event. The *lower left* TEM images show single nanotubes (diameters 0.9 nm and 2.8 nm) on an AFM tip fabricated by this method, adapted from [23.9]

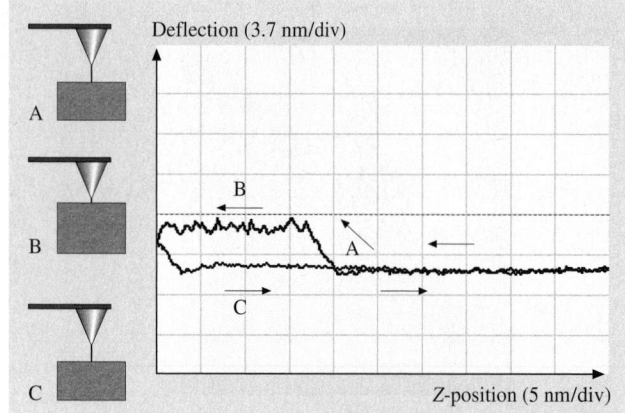

Fig. 23.17 The process by which nanotube tips can be shortened in AFM force curves. The hysteresis in the deflection trace (*bottom*) reveals that ~ 17 nm were removed

the amplitude decreased to zero, and further approach generated a small deflection that ultimately saturated.

However, this saturation was not due to buckling. Note that the amplitude did not recover, as in the buckling curves. This leveling was due to the nanotube sliding on the pyramidal tip, which was confirmed by the hysteresis in the amplitude and deflection curves. If the force curve was repeated, the tip showed no deflection or amplitude drop until further down, because the tip was essentially shorter.

Pick-up SWNT tips achieve the highest resolution of all nanotube tips, since they are always individuals rather than bundles. In tapping mode, they produce images of 5 nm gold particles that have a *full width* of ~ 7 nm, the expected geometrical resolution for a 2 nm cylindrical probe [23.9]. Although the pick-up method is serial in nature, it may still be the key to the mass production of nanotube tips. Note that the original nanotube tip length can be measured electronically from the size of the Z-piezo step. The shortening can be electronically controlled through the hysteresis in the force curves. Therefore, the entire procedure (including tip exchange) can be automated by computer.

23.2 Scanning Tunneling Microscopy

Scanning tunneling microscopy (STM) was the original scanning probe microscopy and generally produces the highest resolution images, routinely achieving atomic resolution on flat, conductive surfaces. In STM, the probe tip consists of a sharpened metal wire that is held 0.3 to 1 nm from the sample. A potential difference of 0.1 V to 1 V between the tip and sample leads to tunneling currents on the order of 0.1 to 1 nA. As in AFM, a piezo-scanner rasters the sample under the tip, and the Z-position is adjusted to hold the tunneling current constant. The Z-position data represents the "topography", or in this case the surface, of constant electron density. As with other SPMs, the tip properties and performance greatly depend on the experiment being carried out. Although it is nearly impossible to prepare a tip with a known atomic structure, a number of factors are known to affect tip performance, and several preparation methods have been developed that produce good tips.

The nature of the sample being investigated and the scanning environment will affect the choice of the tip material and how the tip is fabricated. Factors to consider are mechanical properties – a hard material that will resist damage during tip-sample contact is desired. Chemical properties should also be considered – for-

mation of oxides, or other insulating contaminants will affect tip performance. Tungsten is a common tip material because it is very hard and will resist damage, but its use is limited to ultrahigh vacuum (UHV) conditions, since it readily oxidizes. For imaging under ambient conditions an inert tip material such as platinum or gold is preferred. Platinum is typically alloyed with iridium to increase its stiffness.

23.2.1 Mechanically Cut STM Tips

STM tips can be fabricated by simple mechanical procedures such as grinding or cutting metal wires. Such tips are not formed with highly reproducible shapes and have a large opening angle and a large radius of curvature in the range of 0.1 to 1 μm (see Fig. 23.18a). They are not useful for imaging samples with surface roughness above a few nanometers. However, on atomically flat samples, mechanically cut tips can achieve atomic resolution due to the nature of the tunneling signal, which drops exponentially with tip-sample separation. Since mechanically cut tips contain many small asperities on the larger tip structure, atomic resolution is easily achieved as long as one atom of the tip is just a few angstroms lower than all of the others.

23.2.2 Electrochemically Etched STM Tips

For samples with more than a few nanometers of surface roughness, the tip structure in the nanometer-size range becomes an issue. Electrochemical etching can provide tips with reproducible and desirable shapes and sizes (Fig. 23.18), although the exact atomic structure of the tip apex is still not well controlled. The parameters of electrochemical etching depend greatly on the tip material and the desired tip shape. The following is an entirely general description. A fine metal wire (0.1–1 mm diameter) of the tip material is immersed in an appropriate electrochemical etchant solution. A voltage bias of 1–10 V is applied between the tip and a counterelectrode such that the tip is etched. Due to the enhanced etch rate at the electrolyte-air interface, a neck is formed in the wire. This neck is eventually etched thin enough so that it cannot support the weight

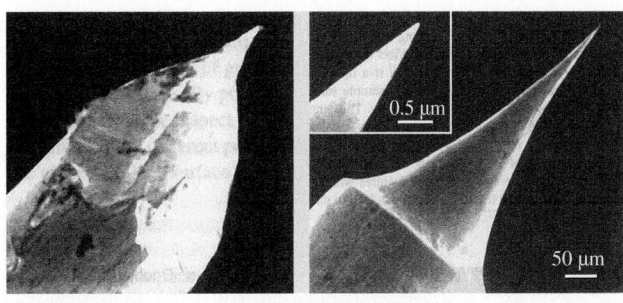

Fig. 23.18 A mechanically cut STM tip (*left*) and an electrochemically etched STM tip (*right*), from [23.35]

of the part of the wire suspended in the solution, and it breaks to form a sharp tip. The widely varying parameters and methods will be not be covered in detail here, but many recipes are found in the literature for common tip materials [23.36–39].

Part C | 23

References

23.1 R. Linnemann, T. Gotzalk, I. W. Rangelow, P. Dumania, E. Oesterschulze: Atomic force microscopy and lateral force microscopy using piezoresistive cantilevers, J. Vac. Sci. Technol. B **14**(2), 856–860 (1996)

23.2 T. R. Albrecht, S. Akamine, T. E. Carver, C. F. Quate: Microfabrication of cantilever styli for the atomic force microscope, J. Vac. Sci. Technol. A **8**(4), 3386–3396 (1990)

23.3 O. Wolter, T. Bayer, J. Greschner: Micromachined silicon sensors for scanning force microscopy, J. Vac. Sci. Technol. B **9**(2), 1353–1357 (1991)

23.4 C. Bustamante, D. Keller: Scanning force microscopy in biology, Phys. Today **48**(12), 32–38 (1995)

23.5 J. Vesenka, S. Manne, R. Giberson, T. Marsh, E. Henderson: Colloidal gold particles as an incompressible atomic force microscope imaging standard for assessing the compressibility of biomolecules, Biophys. J. **65**, 992–997 (1993)

23.6 D. J. Muller, D. Fotiadis, S. Scheuring, S. A. Muller, A. Engel: Electrostatically balanced subnanometer imaging of biological specimens by atomic force microscope, Biophys. J. **76**(2), 1101–1111 (1999)

23.7 R. B. Marcus, T. S. Ravi, T. Gmitter, K. Chin, D. Liu, W. J. Orvis, D. R. Ciarlo, C. E. Hunt, J. Trujillo: Formation of silicon tips with <1 nm radius, Appl. Phys. Lett. **56**(3), 236–238 (1990)

23.8 J. H. Hafner, C. L. Cheung, C. M. Lieber: unpublished results (2001)

23.9 J. H. Hafner, C. L. Cheung, T. H. Oosterkamp, C. M. Lieber: High-yield assembly of individual single-walled carbon nanotube tips for scanning probe microscopies, J. Phys. Chem. B **105**(4), 743–746 (2001)

23.10 F. Ohnesorge, G. Binnig: True atomic resolution by atomic force microscopy through repulsive and attractive forces, Science **260**, 1451–1456 (1993)

23.11 D. J. Muller, D. Fotiadis, A. Engel: Mapping flexible protein domains at subnanometer resolution with the atomic force microscope, FEBS Lett. **430**(1–2 Special Issue SI), 105–111 (1998)

23.12 S. Akamine, R. C. Barrett, C. F. Quate: Improved atomic force microscope images using microcantilevers with sharp tips, Appl. Phys. Lett. **57**(3), 316–318 (1990)

23.13 D. J. Keller, C. Chih-Chung: Imaging steep, high structures by scanning force microscopy with electron beam deposited tips, Surf. Sci. **268**, 333–339 (1992)

23.14 T. Ichihashi, S. Matsui: In situ observation on electron beam induced chemical vapor deposition by transmission electron microscopy, J. Vac. Sci. Technol. B **6**(6), 1869–1872 (1988)

23.15 K. I. Schiffmann: Investigation of fabrication parameters for the electron-beam-induced deposition of contamination tips used in atomic force microscopy, Nanotechnology **4**, 163–169 (1993)

23.16   J. H. Hafner, C. L. Cheung, A. T. Woolley, C. M. Lieber: Structural and functional imaging with carbon nanotube AFM probes, Prog. Biophys. Mol. Biol. **77**(1), 73–110 (2001)

23.17   S. Iijima, C. Brabec, A. Maiti, J. Bernholc: Structural flexibility of carbon nanotubes, J. Chem. Phys. **104**(5), 2089–2092 (1996)

23.18   M. M. J. Treacy, T. W. Ebbesen, J. M. Gibson: Exceptionally high Young's modulus observed for individual carbon nanotubes, Nature **381**, 678–680 (1996)

23.19   A. Krishnan, E. Dujardin, T. W. Ebbesen, P. N. Yianilos, M. M. J. Treacy: Young's modulus of single-walled nanotubes, Phys. Rev. B **58**(20), 14013–14019 (1998)

23.20   E. W. Wong, P. E. Sheehan, C. M. Lieber: Nanobeam mechanics – elasticity, strength, and toughness of nanorods and nanotubes, Science **277**(5334), 1971–1975 (1997)

23.21   J. P. Lu: Elastic properties of carbon nanotubes and nanoropes, Phys. Rev. Lett. **79**(7), 1297–1300 (1997)

23.22   H. J. Dai, J. H. Hafner, A. G. Rinzler, D. T. Colbert, R. E. Smalley: Nanotubes as nanoprobes in scanning probe microscopy, Nature **384**(6605), 147–150 (1996)

23.23   A. G. Rinzler, Y. H. Hafner, P. Nikolaev, L. Lou, S. G. Kim, D. Tomanek, D. T. Colbert, R. E. Smalley: Unraveling nanotubes: Field emission from atomic wire, Science **269**, 1550 (1995)

23.24   H. Nishijima, S. Kamo, S. Akita, Y. Nakayama, K. I. Hohmura, S. H. Yoshimura, K. Takeyasu: Carbon-nanotube tips for scanning probe microscopy: Preparation by a controlled process and observation of deoxyribonucleic acid, Appl. Phys. Lett. **74**(26), 4061–4063 (1999)

23.25   S. S. Wong, A. T. Woolley, T. W. Odom, J. L. Huang, P. Kim, D. V. Vezenov, C. M. Lieber: Single-walled carbon nanotube probes for high-resolution nanostructure imaging, Appl. Phys. Lett. **73**(23), 3465–3467 (1998)

23.26   J. H. Hafner, M. J. Bronikowski, B. R. Azamian, P. Nikolaev, A. G. Rinzler, D. T. Colbert, K. A. Smith, R. E. Smalley: Catalytic growth of single-wall carbon nanotubes from metal particles, Chem. Phys. Lett. **296**(1–2), 195–202 (1998)

23.27   P. Nikolaev, M. J. Bronikowski, R. K. Bradley, F. Rohmund, D. T. Colbert, K. A. Smith, R. E. Smalley: Gas-phase catalytic growth of single-walled

carbon nanotubes from carbon monoxide, Chem. Phys. Lett. **313**(1–2), 91–97 (1999)

23.28   W. Z. Li, S. S. Xie, L. X. Qian, B. H. Chang, B. S. Zou, W. Y. Zhou, R. A. Zhao, G. Wang: Large-scale synthesis of aligned carbon nanotubes, Science **274**(5293), 1701–1703 (1996)

23.29   J. H. Hafner, C. L. Cheung, C. M. Lieber: Growth of nanotubes for probe microscopy tips, Nature **398**(6730), 761–762 (1999)

23.30   V. Lehmann: The physics of macroporous silicon formation, Thin Solid Films **255**, 1–4 (1995)

23.31   F. Ronkel, J. W. Schultze, R. Arensfischer: Electrical contact to porous silicon by electrodeposition of iron, Thin Solid Films **276**(1–2), 40–43 (1996)

23.32   J. H. Hafner, C. L. Cheung, C. M. Lieber: Direct growth of single-walled carbon nanotube scanning probe microscopy tips, J. Am. Chem. Soc. **121**(41), 9750–9751 (1999)

23.33   E. B. Cooper, S. R. Manalis, H. Fang, H. Dai, K. Matsumoto, S. C. Minne, T. Hunt, C. F. Quate: Terabit-per-square-inch data storage with the atomic force microscope, Appl. Phys. Lett. **75**(22), 3566–3568 (1999)

23.34   E. Yenilmez, Q. Wang, R. J. Chen, D. Wang, H. Dai: Wafer scale production of carbon nanotube scanning probe tips for atomic force microscopy, Appl. Phys. Lett. **80**(12), 2225–2227 (2002)

23.35   A. Stemmer, A. Hefti, U. Aebi, A. Engel: Scanning tunneling and transmission electron microscopy on identical areas of biological specimens, Ultramicroscopy **30**(3), 263 (1989)

23.36   R. Nicolaides, L. Yong, W. E. Packard, W. F. Zhou, H. A. Blackstead, K. K. Chin, J. D. Dow, J. K. Furdyna, M. H. Wei, R. C. Jaklevic, W. J. Kaiser, A. R. Pelton, M. V. Zeller, J. J. Bellina: Scanning tunneling microscope tip structures, J. Vac. Sci. Technol. A **6**(2), 445–447 (1988)

23.37   J. P. Ibe, P. P. Bey, S. L. Brandow, R. A. Brizzolara, N. A. Burnham, D. P. DiLella, K. P. Lee, C. R. K. Marrian, R. J. Colton: On the electrochemical etching of tips for scanning tunneling microscopy, J. Vac. Sci. Technol. A **8**, 3570–3575 (1990)

23.38   L. Libioulle, Y. Houbion, J.-M. Gilles: Very sharp platinum tips for scanning tunneling microscopy, Rev. Sci. Instrum. **66**(1), 97–100 (1995)

23.39   A. J. Nam, A. Teren, T. A. Lusby, A. J. Melmed: Benign making of sharp tips for STM and FIM: Pt, Ir, Au, Pd, and Rh, J. Vac. Sci. Technol. B **13**(4), 1556–1559 (1995)

24. Noncontact Atomic Force Microscopy and Related Topics

Scanning probe microscopy (SPM) methods such as scanning tunneling microscopy (STM) and non-contact atomic force microscopy (NC-AFM) are the basic technologies for nanotechnology and also for future bottom-up processes. In Sect. 24.1, the principles of AFM such as its operating modes and the NC-AFM frequency-modulation method are fully explained. Then, in Sect. 24.2, applications of NC-AFM to semiconductors, which make clear its potential in terms of spatial resolution and function, are introduced. Next, in Sect. 24.3, applications of NC-AFM to insulators such as alkali halides, fluorides and transition-metal oxides are introduced. Lastly, in Sect. 24.4, applications of NC-AFM to molecules such as carboxylate (RCOO$^-$) with R=H, CH$_3$, C(CH$_3$)$_3$ and CF$_3$ are introduced. Thus, NC-AFM can observe atoms and molecules on various kinds of surfaces such as semiconductors, insulators and metal oxides with atomic or molecular resolution. These sections are essential to understand the state of the art and future possibilities for NC-AFM, which is the second generation of atom/molecule technology.

The scanning tunneling microscope (STM) is an atomic tool based on an electric method that measures the tunneling current between a conductive tip and a conductive surface. It can electrically observe individual atoms/molecules. It can characterize or analyze the electronic nature around surface atoms/molecules. In addition, it can manipulate individual atoms/molecules. Hence, the STM is the first generation of atom/molecule technology. On the other hand, the atomic force microscopy (AFM) is a unique atomic tool based on a mechanical method that can even deal with insulator surfaces. Since the invention of noncontact AFM (NC-AFM) in 1995, the NC-AFM and NC-AFM-based

methods have rapidly developed into powerful surface tools on the atomic/molecular scales, because NC-AFM has the following characteristics: (1) it has true atomic resolution, (2) it can measure atomic force (so-called atomic force spectroscopy), (3) it can observe even insulators, and (4) it can measure mechanical responses such as elastic deformation. Thus, NC-AFM is the second generation of atom/molecule technology. Scanning probe microscopy (SPM) such as STM and NC-AFM is the basic technology for nanotechnology and also for future bottom-up processes.

In Sect. 24.1, the principles of NC-AFM will be fully introduced. Then, in Sect. 24.2, applications to semiconductors will be presented. Next, in Sect. 24.3, applications to insulators will be described. And, in Sect. 24.4, applications to molecules will be

introduced. These sections are essential to understanding the state of the art and future possibilities for NC-AFM.

24.1 Atomic Force Microscopy (AFM)

The atomic force microscope (AFM), invented by *Binnig* [24.1] and introduced in 1986 by *Binnig, Quate* and *Gerber* [24.2] is an offspring of the scanning tunneling microscope (STM) [24.3]. The STM is covered in several books and review articles, e.g. [24.4–9]. Early in the development of STM it became evident that relatively strong forces act between a tip in close proximity to a sample. It was found that these forces could be put to good use in the atomic force microscope (AFM). Detailed information about the noncontact AFM can be found in [24.10–12].

24.1.1 Imaging Signal in AFM

Figure 24.1 shows a sharp tip close to a sample. The potential energy between the tip and the sample V_{ts} causes a z component of the tip–sample force $F_{ts} = -\partial V_{ts}/\partial z$. Depending on the mode of operation, the AFM uses F_{ts}, or some entity derived from F_{ts}, as the imaging signal.

Unlike the tunneling current, which has a very strong distance dependence, F_{ts} has long- and short-range contributions. We can classify the contributions by their range and strength. In vacuum, there are van-der-Waals, electrostatic and magnetic forces with a long range (up to 100 nm) and short-range chemical forces (fractions of nm).

The van-der-Waals interaction is caused by fluctuations in the electric dipole moment of atoms and their mutual polarization. For a spherical tip with radius R next to a flat surface (z is the distance between the plane connecting the centers of the surface atoms and the center of the closest tip atom) the van-der-Waals potential

is given by [24.13]:

$$V_{vdW} = -\frac{A_H}{6z} \ . \tag{24.1}$$

The Hamaker constant A_H depends on the type of materials (atomic polarizability and density) of the tip and sample and is of the order of 1 eV for most solids [24.13].

When the tip and sample are both conductive and have an electrostatic potential difference $U \neq 0$, electrostatic forces are important. For a spherical tip with radius R, the force is given by [24.14]:

$$F_{electrostatic} = -\frac{\pi \varepsilon_0 R U^2}{z} \ . \tag{24.2}$$

Chemical forces are more complicated. Empirical model potentials for chemical bonds are the Morse potential (see e.g. [24.13]).

$$V_{Morse} = -E_{bond}\left(2\,e^{-\kappa(z-\sigma)} - e^{-2\kappa(z-\sigma)}\right) \tag{24.3}$$

and the Lennard–Jones potential [24.13]:

$$V_{Lennard-Jones} = -E_{bond}\left(2\frac{\sigma^6}{z^6} - \frac{\sigma^{12}}{z^{12}}\right) \tag{24.4}$$

These potentials describe a chemical bond with bonding energy E_{bond} and equilibrium distance σ. The Morse potential has an additional parameter: a decay length κ.

24.1.2 Experimental Measurement and Noise

Forces between the tip and sample are typically measured by recording the deflection of a cantilever beam

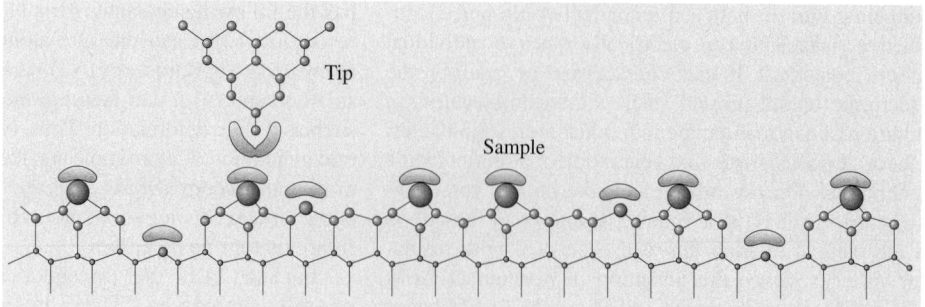

Fig. 24.1 Schematic view of an AFM tip close to a sample

that has a tip mounted on its end (see Fig. 24.2). Today's microfabricated silicon cantilevers were first created in the group of *Quate* [24.15–17] and at IBM [24.18].

The cantilever is characterized by its spring constant k, eigenfrequency f_0 and quality factor Q.

For a rectangular cantilever with dimensions w, t and L (see Fig. 24.2), the spring constant k is given by [24.6]:

$$k = \frac{E_Y w t^3}{4L^3} \tag{24.5}$$

where E_Y is the Young's modulus. The eigenfrequency f_0 is given by [24.6]:

$$f_0 = 0.162 \frac{t}{L^2} \sqrt{\frac{E}{\rho}} \tag{24.6}$$

where ρ is the mass density of the cantilever material. The Q-factor depends on the damping mechanisms present in the cantilever. For micromachined cantilevers operated in air, Q is typically a few hundred, while Q can reach hundreds of thousands in vacuum.

In the first AFM, the deflection of the cantilever was measured with an STM; the back side of the cantilever was metalized, and a tunneling tip was brought close to it to measure the deflection [24.2]. Today's designs use optical (interferometer, beam-bounce) or electrical methods (piezoresistive, piezoelectric) to measure the cantilever deflection. A discussion of the various techniques can be found in [24.19], descriptions of piezoresistive detection schemes are found in [24.17,20] and piezoelectric methods are explained in [24.21–24].

The quality of the cantilever deflection measurement can be expressed in a schematic plot of the deflection noise density versus frequency as in Fig. 24.3.

The noise density has a $1/f$ dependence for low frequency and merges into a constant noise density (white noise) above the $1/f$ corner frequency.

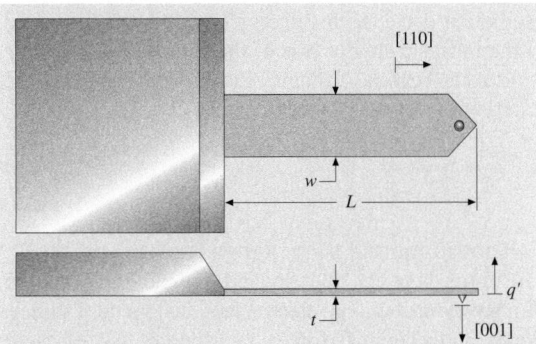

Fig. 24.2 Top view and side view of a microfabricated silicon cantilever (schematic)

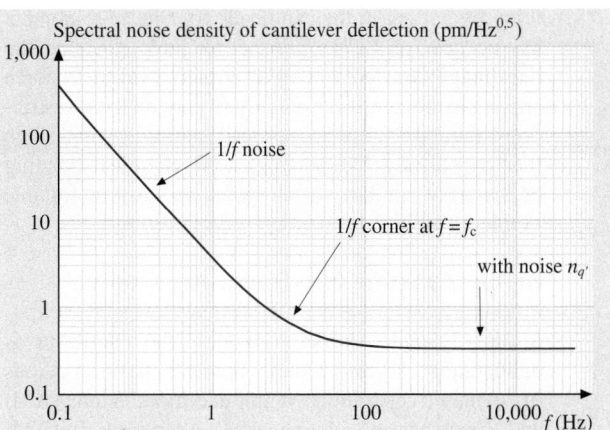

Fig. 24.3 Schematic view of $1/f$ noise apparent in force detectors. Static AFMs operate in a frequency range from 0.01 Hz to a few hundred Hz, while dynamic AFMs operate at frequencies around 10 kHz to a few hundred kHz. The noise of the cantilever deflection sensor is characterized by the $1/f$ corner frequency f_c and the constant deflection noise density $n_{q'}$ for the frequency range where white noise dominates

24.1.3 Static AFM Operating Mode

In the static mode of operation, the force translates into a deflection $q' = F_{ts}/k$ of the cantilever, yielding images as maps of $z(x, y, F_{ts} = \text{const.})$. The noise level of the force measurement is then given by the cantilever's spring constant k times the noise level of the deflection measurement. In this respect, a small value for k increases force sensitivity. On the other hand, instabilities are more likely to occur with soft cantilevers (see Sect. 24.1.1). Because the deflection of the cantilever should be significantly larger than the deformation of the tip and sample, the cantilever should be much softer than the bonds between the bulk atoms in the tip and sample. Interatomic force constants in solids are in the range 10–100 N/m; in biological samples, they can be as small as 0.1 N/m. Thus, typical values for k in the static mode are 0.01–5 N/m.

Even though it has been demonstrated that atomic resolution is possible with static AFM, the method can only be applied in certain cases. The detrimental effects of $1/f$-noise can be limited by working at low temperatures [24.25], where the coefficients of thermal expansion are very small or by building the AFM using a material with a low thermal-expansion coefficient [24.26]. The long-range attractive forces have to be canceled by immersing the tip and sample in a liquid [24.26] or by partly compensating the attractive force by pulling at the cantilever after jump-to-contact has occurred [24.27]. *Jarvis* et al. have canceled the long-range attractive force with an electromagnetic force applied to the cantilever [24.28]. Even with these restrictions, static AFM does not produce atomic resolution on reactive surfaces like silicon, as the chemical bonding of the AFM tip and sample poses an unsurmountable problem [24.29, 30].

24.1.4 Dynamic AFM Operating Mode

In the dynamic operation modes, the cantilever is deliberately vibrated. There are two basic methods of dynamic operation: amplitude-modulation (AM) and frequency-modulation (FM) operation. In AM-AFM [24.31], the actuator is driven by a fixed amplitude A_{drive} at a fixed frequency f_{drive} where f_{drive} is close to f_0. When the tip approaches the sample, elastic and inelastic interactions cause a change in both the amplitude and the phase (relative to the driving signal) of the cantilever. These changes are used as the feedback signal. While the AM mode was initially used in a noncontant mode, it was later implemented very successfully at a closer distance range in ambient conditions involving repulsive tip–sample interactions.

The change in amplitude in AM mode does not occur instantaneously with a change in the tip–sample interaction, but on a timescale of $\tau_{AM} \approx 2Q/f_0$ and the AM mode is slow with high-Q cantilevers. However, the use of high Q-factors reduces noise. *Albrecht* et al. found a way to combine the benefits of high Q and high

speed by introducing the frequency-modulation (FM) mode [24.32], where the change in the eigenfrequency settles on a timescale of $\tau_{FM} \approx 1/f_0$.

Using the FM mode, the resolution was improved dramatically and finally atomic resolution [24.33, 34] was obtained by reducing the tip–sample distance and working in vacuum. For atomic studies in vacuum, the FM mode (see Sect. 24.1.6) is now the preferred AFM technique. However, atomic resolution in vacuum can also be obtained with the AM mode, as demonstrated by *Erlandsson* et al. [24.35].

24.1.5 The Four Additional Challenges Faced by AFM

Some of the inherent AFM challenges are apparent by comparing the tunneling current and tip–sample force as a function of distance (Fig. 24.4).

The tunneling current is a monotonic function of the tip–sample distance and has a very sharp distance dependence. In contrast, the tip–sample force has long- and short-range components and is not monotonic.

Jump-to-Contact and Other Instabilities

If the tip is mounted on a soft cantilever, the initially attractive tip–sample forces can cause a sudden jump-to-contact when approaching the tip to the sample. This instability occurs in the quasistatic mode if [24.36, 37]

$$k < \max\left(-\frac{\partial^2 V_{ts}}{\partial z^2}\right) = k_{ts}^{max} . \tag{24.7}$$

Jump-to-contact can be avoided even for soft cantilevers by oscillating at a large enough amplitude A [24.38]:

$$kA > \max\left(-F_{ts}\right) . \tag{24.8}$$

If hysteresis occurs in the $F_{ts}(z)$-relation, energy ΔE_{ts} needs to be supplied to the cantilever for each oscillation cycle. If this energy loss is large compared to the intrinsic energy loss of the cantilever, amplitude control can become difficult. An additional approximate criterion for k and A is then

$$\frac{kA^2}{2} \geq \frac{\Delta E_{ts} Q}{2\pi} . \tag{24.9}$$

Contribution of Long-Range Forces

The force between the tip and sample is composed of many contributions: electrostatic, magnetic, van-der-Waals and chemical forces in vacuum. All of these force types except for the chemical forces have strong long-range components which conceal the atomic force

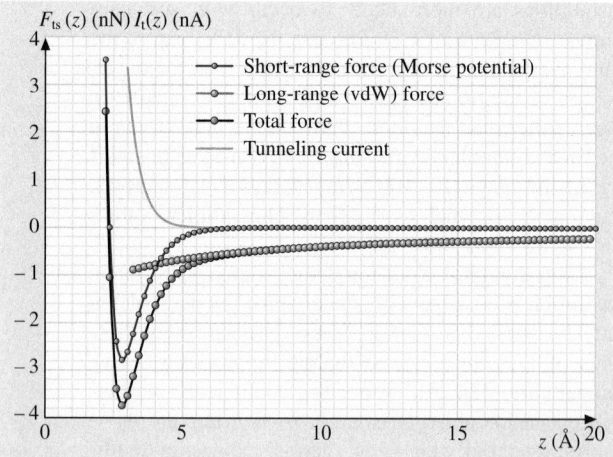

Fig. 24.4 Plot of the tunneling current I_t and force F_{ts} (typical values) as a function of the distance z between the front atom and surface atom layer

components. For imaging by AFM with atomic resolution, it is desirable to filter out the long-range force contributions and only measure the force components which vary on the atomic scale. While there is no way to discriminate between long- and short-range forces in static AFM, it is possible to enhance the short-range contributions in dynamic AFM by proper choice of the oscillation amplitude A of the cantilever.

Noise in the Imaging Signal

Measuring the cantilever deflection is subject to noise, especially at low frequencies ($1/f$ noise). In static AFM, this noise is particularly problematic because of the approximate $1/f$ dependence. In dynamic AFM, the low-frequency noise is easily discriminated when using a bandpass filter with a center frequency around f_0.

Non-monotonic Imaging Signal

The tip–sample force is not monotonic. In general, the force is attractive for large distances and, upon decreasing the distance between tip and sample, the force turns repulsive (see Fig. 24.4). Stable feedback is only possible on a monotonic subbranch of the force curve.

Frequency-modulation AFM helps to overcome challenges. The non-monotonic imaging signal in AFM is a remaining complication for FM-AFM.

24.1.6 Frequency-Modulation AFM (FM–AFM)

In FM-AFM, a cantilever with eigenfrequency f_0 and spring constant k is subject to controlled positive feedback such that it oscillates with a constant amplitude A [24.32], as shown in Fig. 24.5.

Experimental Set-Up

The deflection signal is phase-shifted, routed through an automatic gain control circuit and fed back to the actuator. The frequency f is a function of f_0, its quality factor Q, and the phase shift ϕ between the mechanical excitation generated at the actuator and the deflection of the cantilever. If $\phi = \pi/2$, the loop oscillates at $f = f_0$. Three physical observables can be recorded: (1) a change in the resonance frequency Δf, (2) the control signal of the automatic gain control unit as a measure of the tip–sample energy dissipation, and (3) an average tunneling current (for conducting cantilevers and tips).

Applications

FM-AFM was introduced by *Albrecht* and coworkers in magnetic force microscopy [24.32]. The noise level

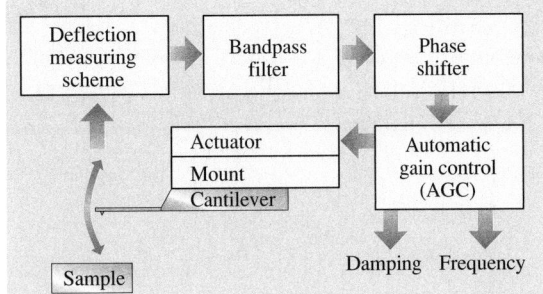

Fig. 24.5 Block diagram of a frequency-modulation force sensor

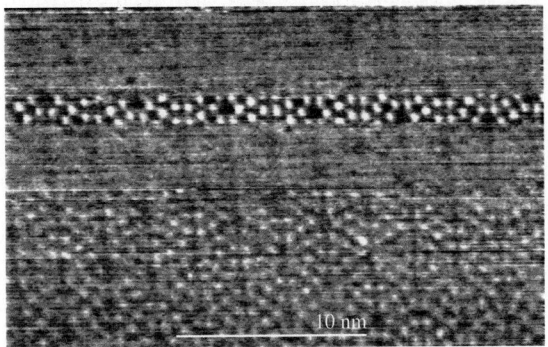

Fig. 24.6 First AFM image of the Si(111)-(7×7) surface. Parameters: $k = 17\,\mathrm{Nm}$, $f_0 = 114\,\mathrm{kHz}$, $Q = 28\,000$, $A = 34\,\mathrm{nm}$, $\Delta f = -70\,\mathrm{Hz}$, $V_\mathrm{t} = 0\,\mathrm{V}$

and imaging speed was enhanced significantly compared to amplitude-modulation techniques. Achieving atomic resolution on the Si(111)-(7×7) surface has been an important step in the development of the STM [24.39] and, in 1994, this surface was imaged by AFM with true atomic resolution for the first time [24.33] (see Fig. 24.6).

The initial parameters which provided true atomic resolution (see caption of Fig. 24.6) were found empirically. Surprisingly, the amplitude necessary to obtain good results was very large compared to atomic dimensions. It turned out later that the amplitudes had to be so large to fulfill the stability criteria listed in Sect. 24.1.5. Cantilevers with $k \approx 2000\,\mathrm{N/m}$ can be operated with amplitudes in the Å range [24.24].

24.1.7 Relation Between Frequency Shift and Forces

The cantilever (spring constant k, effective mass m^*) is a macroscopic object and its motion can be described by

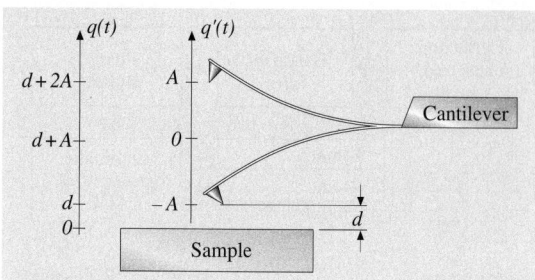

Fig. 24.7 Schematic view of an oscillating cantilever and definition of geometric terms

classical mechanics. Figure 24.7 shows the deflection $q'(t)$ of the tip of the cantilever: it oscillates with an amplitude A at a distance $q(t)$ from a sample.

Generic Calculation

The Hamiltonian of the cantilever is:

$$H = \frac{p^2}{2m^*} + \frac{kq'^2}{2} + V_{ts}(q) \qquad (24.10)$$

where $p = m^* \, dq'/dt$. The unperturbed motion is given by:

$$q'(t) = A\cos(2\pi f_0 t) \qquad (24.11)$$

and the frequency is:

$$f_0 = \frac{1}{2\pi}\sqrt{\frac{k}{m^*}} . \qquad (24.12)$$

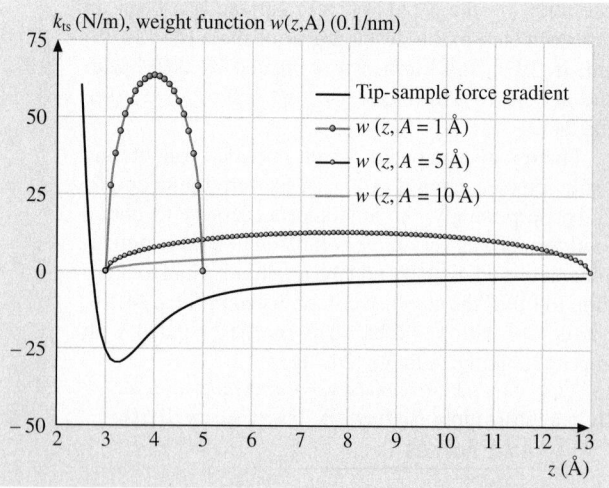

Fig. 24.8 The tip–sample force gradient k_{ts} and weight function for the calculation of the frequency shift

If the force gradient $k_{ts} = -\partial F_{ts}/\partial z = \partial^2 V_{ts}/\partial z^2$ is constant during the oscillation cycle, the calculation of the frequency shift is trivial:

$$\Delta f = \frac{f_0}{2k} k_{ts} . \qquad (24.13)$$

However, in classic FM-AFM k_{ts} varies over orders of magnitude during one oscillation cycle and a perturbation approach, as shown below, has to be employed for the calculation of the frequency shift.

Hamilton–Jacobi Method

The first derivation of the frequency shift in FM-AFM was achieved in 1997 [24.38] using canonical perturbation theory [24.40]. The result of this calculation is:

$$\begin{aligned}
\Delta f &= -\frac{f_0}{kA^2}\left\langle F_{ts} q' \right\rangle \\
&= -\frac{f_0}{kA^2} \int_0^{1/f_0} F_{ts}(d + A + q'(t))q'(t)\,dt .
\end{aligned}$$
$$(24.14)$$

The applicability of first-order perturbation theory is justified because, in FM-AFM, E is typically in the range of several keV, while V_{ts} is of the order of a few eV. Dürig [24.41] found a generalized algorithm that even allows one to reconstruct the tip–sample potential if not only the frequency shift, but the higher harmonics of the cantilever oscillation are known.

A Descriptive Expression for Frequency Shifts as a Function of the Tip–Sample Forces

With integration by parts, the complicated expression (24.14) is transformed into a very simple expression that resembles (24.13) [24.42].

$$\Delta f = \frac{f_0}{2k} \int_{-A}^{A} k_{ts}(z - q') \frac{\sqrt{A^2 - q'^2}}{\frac{\pi}{2}kA^2}\,dq' . \qquad (24.15)$$

This expression is closely related to (24.13): the constant k_{ts} is replaced by a weighted average, where the weight function $w(q', A)$ is a semicircle with radius A divided by the area of the semicircle $\pi A^2/2$ (see Fig. 24.8). For $A \to 0$, $w(q', A)$ is a representation of Dirac's delta function and the trivial zero-amplitude result of (24.13) is immediately recovered. The frequency shift results from a convolution between the tip–sample force gradient and weight function. This convolution can easily be reversed with a linear transformation and the tip–sample force can be recovered from the curve of frequency shift versus distance [24.42].

The dependence of the frequency shift on amplitude confirms an empirical conjecture: small amplitudes increase the sensitivity to short-range forces. Adjusting the amplitude in FM-AFM is comparable to tuning an optical spectrometer to a passing wavelength. When short-range interactions are to be probed, the amplitude should be in the range of the short-range forces. While using amplitudes in the Å range has been elusive with conventional cantilevers because of the instability problems described in Sect. 24.1.5, cantilevers with a stiffness of the order of 1000 N/m like those introduced in [24.23] are well suited for small-amplitude operation.

24.1.8 Noise in Frequency Modulation AFM: Generic Calculation

The vertical noise in FM-AFM is given by the ratio between the noise in the imaging signal and the slope of the imaging signal with respect to z:

$$\delta z = \frac{\delta \Delta f}{\left| \frac{\partial \Delta f}{\partial z} \right|} \, . \tag{24.16}$$

Figure 24.9 shows a typical curve of frequency shift versus distance. Because the distance between the tip and sample is measured indirectly through the frequency shift, it is clearly evident from Fig. 24.9 that the noise in the frequency measurement $\delta \Delta f$ translates into vertical noise δz and is given by the ratio between $\delta \Delta f$ and the slope of the frequency shift curve $\Delta f(z)$ (24.16). Low vertical noise is obtained for a low-noise frequency measurement and a steep slope of the frequency-shift curve.

The frequency noise $\delta \Delta f$ is typically inversely proportional to the cantilever amplitude A [24.32, 43]. The derivative of the frequency shift with distance is constant for $A \ll \lambda$ where λ is the range of the tip–sample

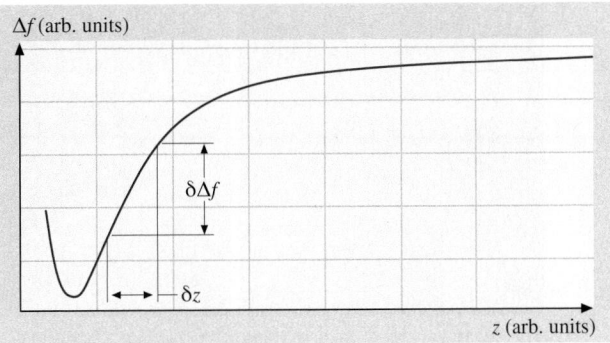

Fig. 24.9 Plot of the frequency shift Δf as a function of the tip–sample distance z. The noise in the tip–sample distance measurement is given by the noise of the frequency measurement $\delta \Delta f$ divided by the slope of the frequency shift curve

interaction and proportional to $A^{-1.5}$ for $A \gg \lambda$ [24.38]. Thus, minimal noise occurs if [24.44]:

$$A_{\text{optimal}} \approx \lambda \tag{24.17}$$

for chemical forces, $\lambda \approx 1$ Å. However, for stability reasons, (Sect. 24.1.5) extremely stiff cantilevers are needed for small-amplitude operation. The excellent noise performance of the stiff cantilever and the small-amplitude technique has been verified experimentally [24.24].

24.1.9 Conclusion

Dynamic force microscopy, and in particular frequency-modulation atomic force microscopy has matured into a viable technique that allows true atomic resolution of conducting and insulating surfaces and spectroscopic measurements on individual atoms [24.10, 45]. Even true atomic resolution in lateral force microscopy is now possible [24.46]. Challenges remain in the chemical composition and structural arrangement of the AFM tip.

24.2 Applications to Semiconductors

For the first time, corner holes and adatoms on the Si(111)-(7×7) surface have been observed in very local areas by *Giessible* using pure noncontact AFM in ultrahigh vacuum (UHV) [24.33]. This was the breakthrough of true atomic-resolution imaging on a well-defined clean surface using the noncontact AFM. Since then, Si(111)-(7×7) [24.34, 35, 45, 47], InP(110) [24.48] and

Si(100)-2×1 [24.34] surfaces have been successively resolved with true atomic resolution. Furthermore, thermally induced motion of atoms or atomic-scale point defects on a InP(110) surface have been observed at room temperature [24.48]. In this section we will describe typical results of atomically resolved noncontact AFM imaging of semiconductor surfaces.

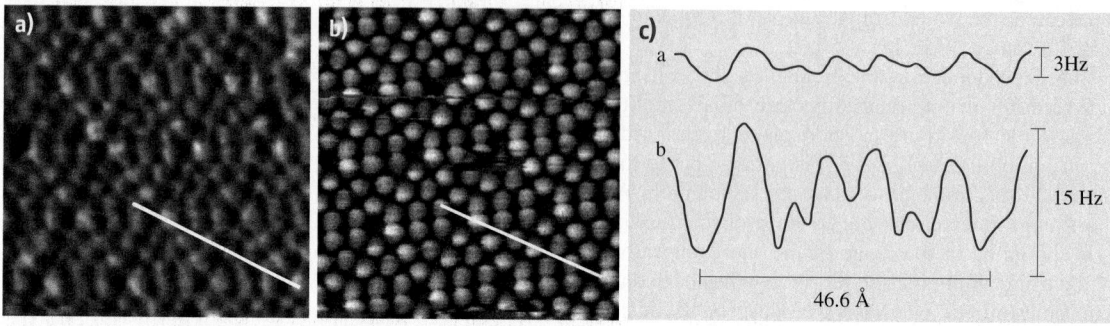

Fig. 24.10a–c Noncontact-mode AFM images of a Si(111)-(7×7) reconstructed surface obtained using the Si tips (**a**) without and (**b**) with a dangling bond. The scan area is 99 Å × 99 Å. (**c**) The cross-sectional profiles along the long diagonal of the 7×7 unit cell indicated by the *white lines* in (**a**) and (**b**)

24.2.1 Si(111)–(7×7) Surface

Figure 24.10 shows the atomic-resolution images of the Si(111)-(7×7) surface[24.49]. Here, Fig. 24.10a (type I) was obtained using the Si tip without dangling, which is covered with an inert oxide layer. Figure 24.10b (type II) was obtained using the Si tip with a dangling bond, on which the Si atoms were deposited due the mechanical soft contact between the tip and the Si surface. The variable frequency shift mode was used. We can see not only adatoms and corner holes but also missing adatoms described by the dimer–adatom–stacking (DAS) fault model. We can see that the image contrast in Fig. 24.10b is clearly stronger than that in Fig. 24.10a.

Interestingly, by using the Si tip with a dangling bond, we observed contrast between inequivalent halves and between inequivalent adatoms of the 7×7 unit cell. Namely, as shown in Fig. 24.11a, the faulted halves

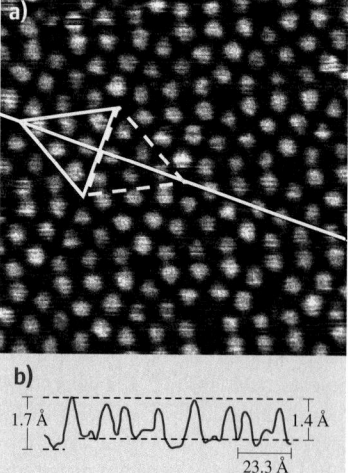

Fig. 24.11 (**a**) Noncontact mode AFM image with contrast of inequivalent adatoms and (**b**) a cross-sectional profile indicated by the *white line*. The halves of the 7×7 unit cell surrounded by the *solid line* and *broken line* correspond to the faulted and unfaulted halves, respectively. The scan area is 89 Å × 89 Å

(surrounded with a solid line) are brighter than the unfaulted halves (surrounded with a broken line). Here, the positions of the faulted and unfaulted halves were determined from the step direction. From the cross-sectional profile along the long diagonal of the 7×7 unit cell in Fig. 24.11b, the heights of the corner adatoms are slightly higher than those of the adjacent center adatoms in the faulted and unfaulted halves of the unit cell. The measured corrugation are in the following decreasing order: $Co-F > Ce-F > Co-U > Ce-U$, where $Co-F$ and $Ce-F$ indicate the corner and center adatoms in faulted halves, and $Co-U$ and $Ce-U$ indicate the corner and center adatoms in unfaulted halves, respectively. Averaging over several units, the corrugation height differences are estimated to be 0.25 Å, 0.15 Å and 0.05 Å for $Co-F$, $Ce-F$ and $Co-U$, respectively, with respect to to $Ce-U$. This tendency, that the heights of the corner adatoms are higher than those of the center adatoms, is consistent with the experimental results using a silicon tip [24.47], although they could not determine the faulted and unfaulted halves of the unit cell in the measured AFM images. However, this tendency is completely contrary to the experimental results using a tungsten tip [24.35]. This difference may originate from the difference between the tip materials, which seems to affect the interaction between the tip and the reactive sample surface. Another possibility is that the tip is in contact with the surface during the small fraction of the oscillating cycle in their experiments [24.35].

We consider that the contrast between inequivalent adatoms is not caused by tip artifacts for the following reasons: (1) each adatom, corner hole and defect was clearly observed, (2) the apparent heights of the adatoms are the same whether they are located adjacent to defects or not, and (3) the same contrast in several images for the different tips has been observed.

It should be noted that the corrugation amplitude of adatoms ≈ 1.4 Å in Fig. 24.11b is higher than that of $0.8-1.0$ Å obtained with the STM, although the depth of the corner holes obtained with noncontact AFM is almost the same as that observed with STM. Moreover, in noncontact-mode AFM images, the corrugation amplitude of adatoms was frequently larger than the depth of the corner holes. The origin of such large corrugation of adatoms may be due to the effect of the chemical interaction, but is not yet clear.

The atom positions, surface energies, dynamic properties and chemical reactivities on the Si(111)-(7×7) reconstructed surface have been extensively investigated theoretically and experimentally. From these investigations, the possible origins of the contrast between inequivalent adatoms in AFM images are the followings: the true atomic heights that correspond to the adatom core positions, the stiffness (spring constant) of interatomic bonding with the adatoms corresponding to the frequencies of the surface mode, the charge on the adatom, and the chemical reactivity of the adatoms. Table 24.1 summarizes the decreasing orders of the inequivalent adatoms for individual property. From Table 24.1, we can see that the calculated adatom heights and the stiffness of interatomic bonding cannot explain the AFM data, while the amount of charge of adatom and the chemical reactivity of adatoms can explain the our data. The contrast due to the amount of charge of adatom means that the AFM image is originated from the difference of the vdW or electrostatic physical interactions between the tip and the valence electrons at the adatoms. The contrast due to the chemical reactivity of adatoms means that the AFM image is originated from the difference of covalent bonding chemical interaction between the atoms at the tip apex and dangling bond of adatoms. Thus, we can see there are two possible interactions which explain the strong contrast between inequivalent adatoms of 7×7 unit cell observed using the Si tip with dangling bond.

The weak-contrast image in Fig. 24.10a is due to vdW and/or electrostatic force interactions. On the other hand, the strong-contrast images in Figs. 24.10b and 24.11a are due to a covalent bonding formation between the AFM tip with Si atoms and Si adatoms. These results indicate the capability of the noncontact-mode AFM to image the variation in chemical reactivity of Si adatoms. In the future, by controlling an atomic species at the tip apex, the study of chemical reactivity on an atomic scale will be possible using noncontact AFM.

24.2.2 Si(100)–(2×1) and Si(100)–(2×1):H Monohydride Surfaces

In order to investigate the imaging mechanism of the noncontact AFM, a comparative study between a reactive surface and an insensitive surface using the same tip is very useful. Si(100)-(2×1):H monohydride surface is a Si(100)-(2×1) reconstructed surface that is terminated by a hydrogen atom. It does not reconstruct as metal is deposited on the semiconductor surface. The surface structure hardly changes. Thus, the Si(100)-(2×1):H monohydride surface is one of most useful surface for a model system to investigate the imaging mechanism, experimentally and theoretically. Furthermore, whether the interaction between a very small atom such as hydrogen and a tip apex is observable with noncontact AFM is interested. Here, we show noncontact AFM images measured on a Si(100)-(2×1) reconstructed surface with a dangling bond and on a Si(100)-(2×1):H monohydride surface on which the dangling bond is terminated by a hydrogen atom [24.50].

Figure 24.12a shows the atomic-resolution image of the Si(100)-(2×1) reconstructed surface. Pairs of bright spots arranged in rows with a 2×1 symmetry were observed with clear contrast. Missing pairs of bright spots were also observed, as indicated by arrows. Furthermore, the pairs of bright spots are shown by the white dashed arc and appear to be the stabilize-buckled asymmetric

Part C | 24.2

Table 24.1 Comparison between the adatom heights observed in an AFM image and the variety of properties for inequivalent adatoms

	Decreasing order	Agreement
AFM image	Co−F > Ce−F > Co−U > Ce−U	−
Calculated height	Co−F > Co−U > Ce−F > Ce−U	×
Stiffness of inter-atomic bonding	Ce−U > Co−U > Ce−F > Co−F	×
Amount of charge of adatom	Co−F > Ce−F > Co−U > Ce−U	○
Calculated chemical reactivity	Faulted > Unfaulted	○
Experimental chemical reactivity	Co−F > Ce−F > Co−U > Ce−U	○

a)

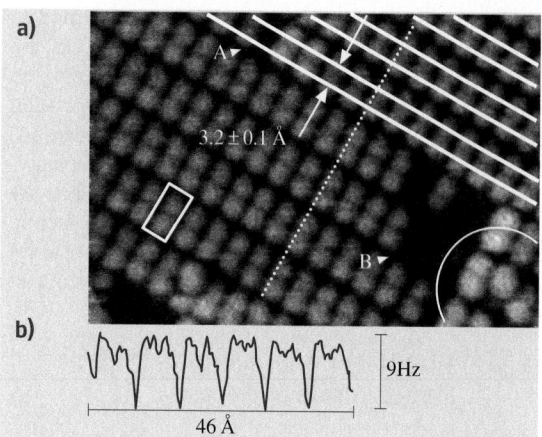

b)

Fig. 24.12 (a) Noncontact AFM image of a Si(001)(2 × 1) reconstructed surface. The scan area was 69 × 46 Å. One 2 × 1 unit cell is outlined with a *box*. *White rows* are super-imposed to show the bright spots arrangement. The distance between the bright spots on the dimer row is 3.2 ± 0.1 Å. On the *white arc*, the alternative bright spots are shown. (b) Cross-sectional profile indicated by the *white dotted line*

a)

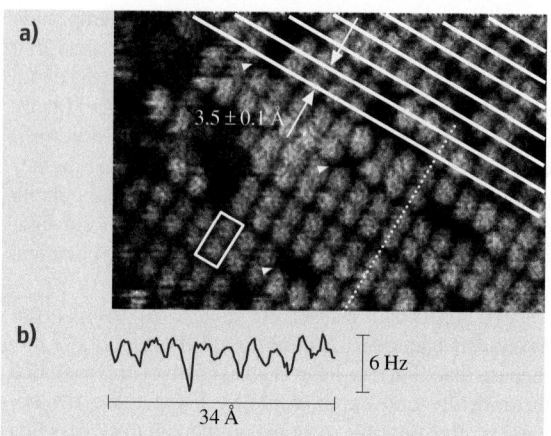

b)

Fig. 24.13 (a) Noncontact AFM image of Si(001)-(2 × 1):H surface. The scan area was 69 × 46 Å. One 2 × 1 unit cell is outlined with a *box*. *White rows* are superimposed to show the bright spots arrangement. The distance between the bright spots on the dimer row is 3.5 ± 0.1 Å. (b) Cross-sectional profile indicated by the *white dotted line*

dimer structure. Furthermore, the distance between the pairs of bright spots is 3.2 ± 0.1 Å.

Figure 24.13a shows the atomic-resolution image of the Si(100)-(2 × 1):H monohydride surface. Pairs of bright spots arranged in rows were observed. Missing paired bright spots as well as those paired in rows and single bright spots were observed, as indicated by arrows. Furthermore, the distance between paired bright spots is 3.5 ± 0.1 Å. This distance of 3.5 ± 0.1 Å is 0.2 Å larger than that of the Si(100)-(2 × 1) reconstructed surface. Namely, it is found that the distance between bright spots increases in size due to the hydrogen termination.

The bright spots in Fig. 24.12 do not merely image the silicon-atom site, because the distance between the bright spots forming the dimer structure of Fig. 24.12a, 3.2 ± 0.1 Å, is lager than the distance between silicon atoms of every dimer structure model. (The maximum is the distance between the upper silicones in an asymmetric dimer structure 2.9 Å.) This seems to be due to the contribution to the imaging of the chemical bonding interaction between the dangling bond from the apex of the silicon tip and the dangling bond on the Si(100)-(2 × 1) reconstructed surface. Namely, the chemical bonding interaction operates strongly, with strong direction dependence, between the dangling bond pointing out of the silicon dimer structure on the Si(100)-(2 × 1) recon-

structed surface and the dangling bond pointing out of the apex of the silicon tip; a dimer structure is obtained with a larger separation than between silicones on the surface.

The bright spots in Fig. 24.13 seem to be located at hydrogen atom sites on the Si(100)-(2 × 1):H monohydride surface, because the distance between the bright spots forming the dimer structure (3.5 ± 0.1 Å) approximately agrees with the distance between the hydrogens, i. e., 3.52 Å. Thus, the noncontact AFM atomically resolved the individual hydrogen atoms on the topmost layer. On this surface, the dangling bond is terminated by a hydrogen atom, and the hydrogen atom on the topmost layer does not have chemical reactivity. Therefore, the interaction between the hydrogen atom on the topmost layer and the apex of the silicon tip does not contribute to the chemical bonding interaction with strong direction dependence as on the silicon surface, and the bright spots in the noncontact AFM image correspond to the hydrogen atom sites on the topmost layer.

24.2.3 Metal Deposited Si Surface

In this section, we will introduce the comparative study of force interactions between a Si tip and a metal-deposited Si surface, and between a metal adsorbed Si tip and a metal-deposited Si surface [24.51, 52]. As for the metal-deposited Si surface, Si(111)-($\sqrt{3} \times \sqrt{3}$)-

Ag (hereafter referred to as $\sqrt{3}$-Ag) surface was used.

For the $\sqrt{3}$-Ag surface, the honeycomb-chained trimer (HCT) model has been accepted as the appropriate model. As shown in Fig. 24.5, this structure contains a Si trimer in the second layer, 0.75 Å below the Ag trimer in the topmost layer. The topmost Ag atoms and lower Si atoms form covalent bonds. The interatomic distances between the nearest-neighbor Ag atoms forming the Ag trimer and between the lower Si atoms forming the Si trimer are 3.43 Å and 2.31 Å, respectively. The apexes of the Si trimers and Ag trimers face the [11$\bar{2}$] direction and the direction tilted a little to the [$\bar{1}$1$\bar{2}$] direction, respectively.

In Fig. 24.15, we show the noncontact AFM images measured using a normal Si tip at a frequency shift of (a) −37 Hz, (b) −43 Hz and (c) −51 Hz, respectively. These frequency shifts correspond to tip–sample distances of about 0–3 Å. We defined the zero position of the tip–sample distance, i.e., the contact point, as the point at which the vibration amplitude began to decrease. The rhombus indicates the $\sqrt{3} \times \sqrt{3}$ unit cell. When the tip approached the surface, the contrast of the noncontact AFM images become strong and the pattern changed remarkably. That is, by approaching the tip toward the sample surface, the hexagonal pattern, the trefoil-like pattern composed of three dark lines, and the triangle pattern can be observed sequentially. In Fig. 24.15a, the distance between the bright spots is 3.9 ± 0.2 Å. In Fig. 24.15c, the distance between the bright spots is 3.0 ± 0.2 Å, and the direction of the apex of all the triangles composed of three bright spots is [11$\bar{2}$].

In Fig. 24.16, we show the noncontact AFM images measured by using Ag-absorbed tip at a frequency shift of (a) −4.4 Hz, (b) −6.9 Hz and (c) −9.4 Hz, respectively. The tip–sample distances Z are roughly estimated to be $Z = 1.9$ Å, 0.6 Å and ≈ 0 Å (in the noncontact region), respectively. When the tip approached the surface, the pattern of the noncontact AFM images did not

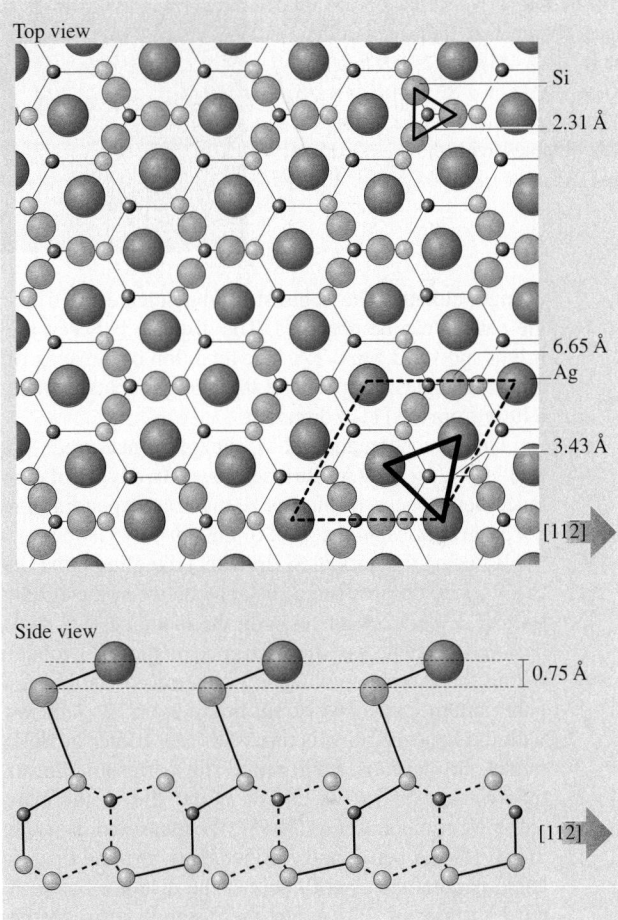

Fig. 24.14 HCT model for the structure of the Si(111)-($\sqrt{3} \times \sqrt{3}$)–Ag surface. *Black closed circle, gray closed circle, open circle,* and *closed circle with horizontal line* indicate Ag atom at the topmost layer, Si atom at the second layer, Si atom at the third layer, and Si atom at the fourth layer, respectively. The *rhombus* indicates the $\sqrt{3} \times \sqrt{3}$ unit cell. The *thick, large, solid triangle* indicates an Ag trimer. The *thin, small, solid triangle* indicates a Si trimer

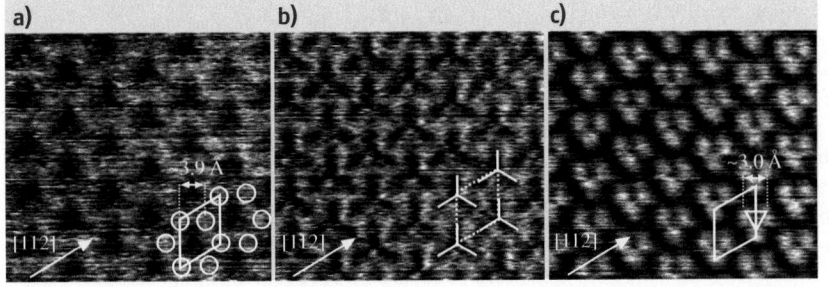

Fig. 24.15a–c Noncontact AFM images obtained at frequency shifts of (**a**) −37 Hz, (**b**) −43 Hz, and (**c**) −51 Hz on a Si(111)-($\sqrt{3} \times \sqrt{3}$)–Ag surface. This distance dependence was obtained with a Si tip. The scan area is 38 Å × 34 Å. A *rhombus* indicates the $\sqrt{3} \times \sqrt{3}$ unit cell

Fig. 24.16a–c Noncontact AFM images obtained at frequency shifts of (**a**) $-4.4\,\mathrm{Hz}$, (**b**) $-6.9\,\mathrm{Hz}$, and (**c**) $-9.4\,\mathrm{Hz}$ on a Si(111)-($\sqrt{3}\times\sqrt{3}$)-Ag surface. This distance dependence was obtained with the Ag-adsorbed tip. The scan area is $38\,\text{Å}\times34\,\text{Å}$

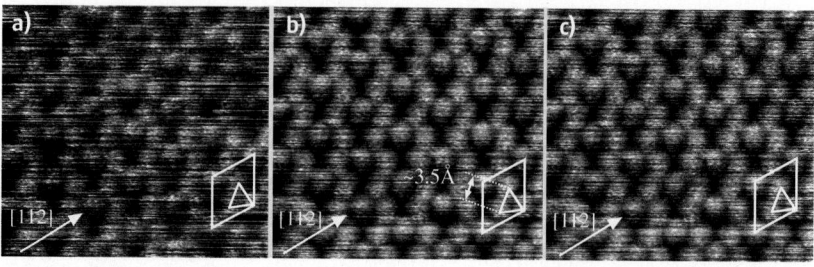

change, although the contrast become clearer. A triangle pattern can be observed. The distance between the bright spots is $3.5\pm0.2\,\text{Å}$. The direction of the apex of all the triangles composed of three bright spots is tilted a little from the $[\bar{1}\bar{1}2]$ direction.

Thus, noncontact AFM images measured on Si(111)-($\sqrt{3}\times\sqrt{3}$)-Ag surface showed two types of distance dependence in the image patterns depending on the atom species on the apex of the tip.

By using the normal Si tip with a dangling bond, in Fig. 24.15a, the measured distance between the bright spot of $3.9\pm0.2\,\text{Å}$ agrees with the distance of $3.84\,\text{Å}$ between the centers of the Ag trimers in the HCT model within the experimental error. Furthermore, the hexagonal pattern composed of six bright spots also agrees with the honeycomb structure of the Ag trimer in HCT model. So the most appropriate site corresponding to the bright spots in Fig. 24.15a is the site of the center of Ag trimers. In Fig. 24.15c, the measured distance of $3.0\pm0.2\,\text{Å}$ between the bright spots forming the triangle pattern agrees with neither the distance between the Si trimer of $2.31\,\text{Å}$ nor the distance between the Ag trimer of $3.43\,\text{Å}$ in the HCT model, while the direction of the apex of the triangles composed of three bright spots agrees with the $[11\bar{2}]$ direction of the apex of the Si trimer in the HCT model. So the most appropriate site corresponding to the bright spots in Fig. 24.15c is the intermediate site between the Si atoms and Ag atoms. On the other hand, by using the Ag-adsorbed tip, the measured distance between the bright spots of $3.5\pm0.2\,\text{Å}$ in Fig. 24.16 agrees with the distance of $3.43\,\text{Å}$ between the nearest-neighbor Ag atoms forming the Ag trimer in the topmost layer in the HCT model within the experimental error. Furthermore, the direction of the apex of the triangles composed of three bright spots also agrees with the direction of the apex of the Ag trimer, i.e., tilted $[\bar{1}\bar{1}2]$, in the HCT model. So, the most appropriate site corresponding to the bright spots in Fig. 24.16 is the site of individual Ag atoms forming the Ag trimer in the topmost layer.

It should be noted that, by using the noncontact AFM with a Ag-adsorbed tip, for the first time, the individual Ag atom on the $\sqrt{3}$-Ag surface could be resolved in real space, although by using the noncontact AFM with an Si tip, it could not be resolved. So far, the $\sqrt{3}$-Ag surface has been observed by a scanning tunneling microscope (STM) with atomic resolution. However, the STM can also measure the local charge density of states near the Fermi level on the surface. From first-principle calculations, it was proven that unoccupied surface states are densely distributed around the center of the Ag trimer. As a result, bright contrast is obtained at the center of the Ag trimer with the STM.

Finally, we consider the origin of the atomic-resolution imaging of the individual Ag atoms on the $\sqrt{3}$-Ag surface. Here, we discuss the difference between the force interactions when using the Si tip and the Ag-adsorbed tip. As shown in Fig. 24.17a, when using the Si

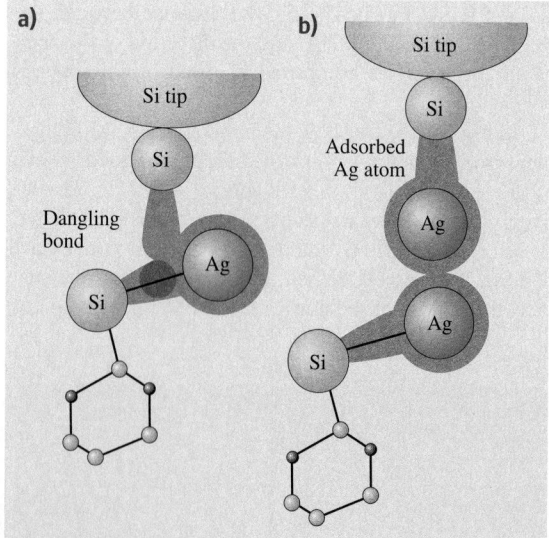

Fig. 24.17a,b Schematic illustration of (**a**) the Si atom with dangling bond and (**b**) the Ag-adsorbed tip above the Si–Ag covalent bond on a Si(111)-($\sqrt{3}\times\sqrt{3}$)-Ag surface

tip, there is a dangling bond pointing out of the topmost Si atom on the apex of the Si tip. As a result, the force interaction is dominated by physical bonding interactions, such as the Coulomb force, far from the surface and by chemical bonding interaction very close to the surface. Namely, if a reactive Si tip with a dangling bond approaches a surface, at distances far from the surface the Coulomb force acts between the electron localized on the dangling bond pointing out of the topmost Si atom on the apex of the tip, and the positive charge distributed around the center of the Ag trimer. At distances very close to the surface, the chemical bonding interaction will occur due to the onset of orbital hybridization between the dangling bond pointing out of the topmost Si atom on the apex of the Si tip and a Si–Ag covalent bond on the surface. Hence, the individual Ag atoms will not be resolved and the image pattern will change depending on the tip–sample distance. On the other hand,

as shown in Fig. 24.17b, by using the Ag-adsorbed tip, the dangling bond localized out of topmost Si atom on the apex of the Si tip is terminated by the adsorbed Ag atom. As a result, even at very close tip–sample distances, the force interaction is dominated by physical bonding interactions such as the vdW force. Namely, if the Ag-adsorbed tip approaches the surface, the vdW force acts between the Ag atom on the apex of the tip and the Ag or Si atom on the surface. Ag atoms in the topmost layer of the $\sqrt{3}$-Ag surface are located higher than the Si atoms in the lower layer. Hence, the individual Ag atoms (or their nearly true topography) will be resolved, and the image pattern will not change even at very small tip–sample distances. It should be emphasized that there is a possibility to identify or recognize atomic species on a sample surface using noncontact AFM if we can control the atomic species at the tip apex.

24.3 Applications to Insulators

Insulators such as alkali halides, fluorides, and metal oxides are key materials in many applications, including optics, microelectronics, catalysis, and so on. Surface properties are important in these technologies, but they are usually poorly understood. This is due to their low conductivity, which makes it difficult to investigate them using electron- and ion-based measurement techniques such as low-energy electron diffraction, ion-scattering spectroscopy, and scanning tunneling microscopy (STM). Surface imaging by noncontact atomic force microscopy (NC-AFM) does not require a sample surface with a high conductivity because NC-AFM detects a force between the tip on the cantilever and the surface of the sample. Since the first report of atomically resolved NC-AFM on a Si(111)-(7×7) surface [24.33], several groups have succeeded in obtaining "true" atomic resolution images of insulators, including defects, and it has been shown that NC-AFM is a powerful new tool for atomic-scale surface investigation of insulators.

In this section we will describe typical results of atomically resolved NC-AFM imaging of insulators such as alkali halides, fluorides and metal oxides. For the alkali halides and fluorides, we will focus on contrast formation, which is the most important issue for interpreting atomically resolved images of binary compounds on the basis of experimental and theoretical results. For the metal oxides, typical examples of

atomically resolved imaging will be exhibited and the difference between the STM and NC-AFM images will be demonstrated. Also, theoretical studies on the interaction between realistic Si tips and representative oxide surfaces will be shown. Finally, we will describe an antiferromagnetic NiO(001) surface imaged with a ferromagentic tip to explore the possibility of detecting short-range magnetic interactions using the NC-AFM.

24.3.1 Alkali Halides, Fluorides and Metal Oxides

The surfaces of alkali halides were the first insulating materials to be imaged by NC-AFM with "true" atomic resolution [24.53]. To date, there have been reports on atomically resolved images of (001) cleaved surfaces for single-crystal NaF, RbBr, LiF, KI, NaCl, [24.54], KBr [24.55] and thin films of NaCl(001) on Cu(111) [24.56]. In this section we describe the contrast formation of alkali halides surfaces on the basis of experimental and theoretical results.

Alkali Halides
In experiments on alkali halides, the symmetry of the observed topographic images indicates that the protrusions exhibit only one type of ions, either the positive or negatively charged ions. This leads to the conclusion that the atomic contrast is dominantly caused by electro-

static interactions between a charged atom at the apex of the tip and the surface ions, i.e. long-range forces between the macroscopic tip and the sample, such as the van der Waals force, are modulated by an alternating short-range electrostatic interaction with the surface ions. Theoretical work employing the atomistic simulation technique has revealed the mechanism for contrast formation on an ionic surface [24.57]. A significant part of the contrast is due to the displacement of ions in the force field, not only enhancing the atomic corrugations, but also contributing to the electrostatic potential by forming dipoles at the surface. The experimentally observed atomic corrugation height is determined by the interplay of the long- and short-range forces. In the case of NaCl, it has been experimentally demonstrated that a blunter tip produces a lager corrugation when the tip–sample distance is shorter [24.54]. This result shows that the increased long-range forces induced by a blunter tip allow for more stable imaging closer to the surface. The stronger electrostatic short-range interaction and lager ion displacement produce a more pronounced atomic corrugation. At steps and kinks on an NaCl thin film on Cu(111), the corrugation amplitude of atoms with low coordination number has been observed to increase by a factor of up to two more than that of atomically flat terraces [24.56]. The low coordination number of the ions results in an enhancement of the electrostatic potential over the site and an increase in the displacement induced by the interaction with the tip.

Theoretical study predicts that the image contrast depends on the chemical species at the apex of the tip. *Bennewitz* et al. [24.56] have performed the calculations using an MgO tip terminated by oxygen and an Mg ion. The magnitude of the atomic contrast for the Mg-terminated tip shows a slight increase in comparison with an oxygen-terminated tip. The atomic contrast with the oxygen-terminated tip is dominated by the attractive electrostatic interaction between the oxygen on the tip apex and the Na ion, but the Mg-terminated tip attractively interacts with the Cl ion. In other words, these results demonstrated that the species of the ion imaged as the bright protrusions depends on the polarity of the tip apex.

These theoretical results emphasized the importance of the atomic species at the tip apex for the alkali halide (001) surface, while it is not straightforward to define the nature of the tip apex experimentally because of the high symmetry of the surface structure. However, there are a few experiments exploring the possibilities to determine the polarity of the tip apex. *Bennewitz*

et al. [24.58] studied imaging of surfaces of a mixed alkali halide crystal, which was designed to observe the chemically inhomogeneous surface. The mixed crystal is composed of 60% KCl and 40% KBr, with the Cl and Br ions interfused randomly in the crystal. The image of the cleaved $KCl_{0.6}Br_{0.4}(001)$ surface indicates that only one type of ion is imaged as protrusions, as if it were a pure alkali halide crystal. However, the amplitude of the atomic corrugation varies strongly between the positions of the ions imaged as depressions. This variation in the corrugations corresponds to the constituents of the crystal, i.e. the Cl and Br ions, and it is concluded that the tip apex is negatively charged. Moreover, the deep depressions can be assigned to Br ions by comparing the number with the relative density of anions. The difference between Cl and Br anions with different masses is enhanced in the damping signal measured simultaneously with the topographic image [24.59]. The damping is recorded as an increase in the excitation amplitude necessary to maintain the oscillation amplitude of the cantilever in the constant-amplitude mode [24.56]. Although the dissipation phenomena on an atomic scale are a subject under discussion, any dissipative interaction must generally induce energy losses in the cantilever oscillation [24.60, 61]. The measurement of energy dissipation has the potential to enable chemical discrimination on an atomic scale. Recently, a new procedure for species recognition on a alkali halide surface was proposed [24.62]. This method is based on a comparison between theoretical results and the site-specific measurement of frequency versus distance. The differences in the force curves measured at the typical sites, such as protrusion, depression, and their bridge position, are compared to the corresponding differences obtained from atomistic simulation. The polarity of the tip apex can be determined, leading to the identification of the surface species. This method is applicable to highly symmetric surfaces and is useful for determining the sign of the tip polarity.

Fluorides

Fluorides are important materials for the progress of an atomic-scale-resolution NC-AFM imaging of insulators. There are reports in the literature of surface images for single-crystal BaF_2, SrF_2 [24.63], CaF_2 [24.64–66] and a CaF bilayer on Si(111) [24.67]. Surfaces of fluorite-type crystals are prepared by cleaving along the (111) planes. Their structure is more complex than the structure of alkali halides, which have a rock-salt structure. The complexity is of great interest for atomic-resolution imaging using NC-AFM and also for theoretical predictions of the interpretation of the atomic-scale contrast information.

The first atomically resolved images of a CaF$_2$(111) surface were obtained in topographic mode [24.65], and the surface ions mostly appear as spherical caps. *Barth* et al. [24.68] have found that the CaF$_2$(111) surface images obtained by using the constant-height mode, in which the frequency shift is recorded with a very low loop gain, can be categorized into two contrast patterns. In the first of these the ions appear as triangles and in the second they have the appearance of circles, similar to the contrast obtained in a topographic image. Theoretical studies demonstrated that these two different contrast patterns could be explained as a result of imaging with tips of different polarity [24.68–70]. When imaging with a positively charged (cation-terminated) tip, the triangular pattern appears. In this case, the contrast is dominated by the strong short-range electrostatic attraction between the positive tip and the negative F ions. The cross section along the [121] direction of the triangular image shows two maxima: one is a larger peak over the F(I) ions located in the topmost layer and the other is a smaller peak at the position of the F(III) ions in the third layer. The minima appear at the position of the Ca ions in the second layer. When imaging with a negatively charged (anion-terminated) tip, the spherical image contrast appears and the main periodicity is created by the Ca ions between the topmost and the third F ion layers. In the cross section along the [121] direction, the large maxima correspond to the Ca sites because of the strong attraction of the negative tip and the minima appear at the sites of maximum repulsion over the F(I) ions. At a position between two F ions, there are smaller maxima. This reflects the weaker repulsion over the F(III) ion sites compared to the protruding F(I) ion sites and a slower decay in the contrast on one side of the Ca ions.

The triangular pattern obtained with a positively charged tip appears at relatively large tip–sample distance, as shown in Fig. 24.18a. The cross section along the [121] direction, experiment results and theoretical studies both demonstrate the large-peak and small-shoulder characteristic for the triangular pattern image (Fig. 24.18d). When the tip approaches the surface more closely, the triangular pattern of the experimental images is more vivid (Fig. 24.18b), as predicted in the theoretical works. As the tip approaches, the amplitude of the shoulder increases until it is equal to that of the main peak, and this feature gives rise to the honeycomb pattern image, as shown in Fig. 24.18c. Moreover, theoretical results predict that the image contrast changes again

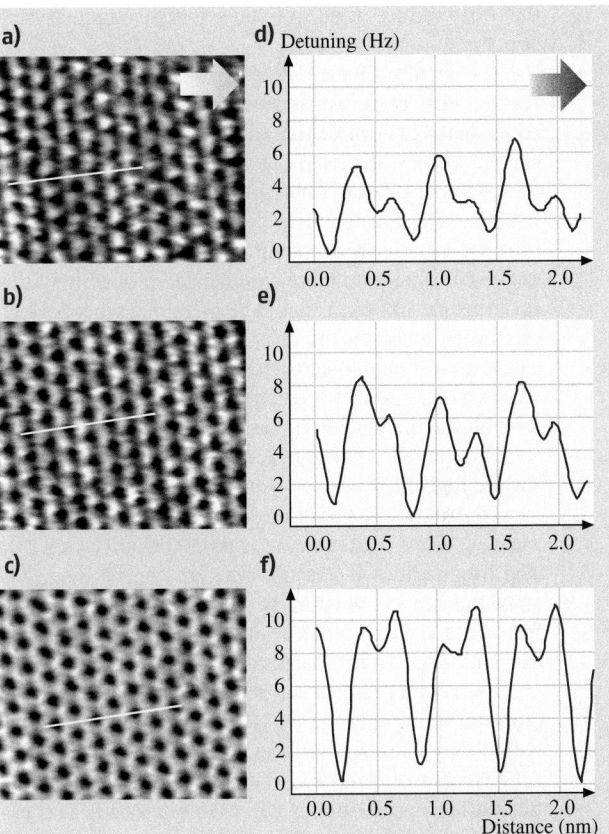

Fig. 24.18 (a)–(c) CaF$_2$(111) surface images obtained by using the constant-height mode. From (a) to (c) the frequency shift was lowered. The *white lines* represent the positions of the cross section. (d)–(f) The cross section extracted from the Fourier-filtered images of (a)–(c). The *white* and *black arrows* represent the scanning direction. The images and the cross sections are from [24.68]

when the tip apex is in close proximity to surface. Recently, *Giessibl* et al. [24.71] achieved atomic imaging in the repulsive region and proved experimentally the predicted change of the image contrast. As described here, there is good correspondence in the distance dependency of the image obtained by experimental and theoretical investigations.

From detailed theoretical analysis of the electrostatic potential [24.72], it was suggested that the change in displacement of the ions due to the proximity of the tip plays an important role in the formation of the image contrast. Such a drastic change in image contrast, depending on both the polarity of the terminated tip atom and on the tip–sample distance, is inherent to the fluoride (111) surface, and this image-contrast feature cannot be seen on the (001) surface of alkali halides with a simple crystal structure.

The results of careful experiments show another feature: that the cross sections taken along the three equivalent [121] directions do not yield identical results [24.68]. It is thought that this can be attributed to the asymmetry of the nanocluster at the tip apex, which leads to different interactions in the equivalent directions. A better understanding of the asymmetric image contrast may require more complicated modeling of the tip structure. In fact, it should be mentioned that perfect tips on an atomic scale can occasionally be obtained. These tips do yield identical results in forward and backward scanning, and cross sections in the three equivalent directions taken with this tip are almost identical [24.74].

The fluoride (111) surface is an excellent standard surface for calibrating tips on an atomic scale. The polarity of the tip-terminated atom can be determined from the image contrast pattern (spherical or triangular pattern). The irregularities in the tip structure can be detected, since the surface structure is highly symmetric. Therefore, once such a tip has been prepared, it can be used as a calibrated tip for imaging unknown surfaces.

The polarity and shape of the tip apex play an important role in interpreting NC-AFM images of alkali halide and fluorides surfaces. It is expected that the achievement of good correlation between experimental and theoretical studies will help to advance surface imaging of insulators by NC-AFM.

Metal Oxides

Most of the metal oxides that have attracted strong interest for their technological importance are insulating. Therefore, in the case of atomically resolved imaging

of metal oxide surfaces by STM, efforts to increase the conductivity of the sample are needed, such as, the introduction of anions or cations defects, doping with other atoms and surface observations during heating of the sample. However, in principle, NC-AFM provides the possibility of observing nonconductive metal oxides without these efforts. In cases where the conductivity of the metal oxides is high enough for a tunneling current to flow, it should be noted that most surface images obtained by NC-AFM and STM are not identical.

Since the first report of atomically resolved images on a $TiO_2(110)$ surface with oxygen point defects [24.75], they have also been reported on rutile $TiO_2(100)$ [24.76–78], anatase $TiO_2(001)$ thin film on $SrTiO_3(100)$ [24.79] and on $LaAO_3(001)$ [24.80], $SnO_2(110)$ [24.81], $NiO(001)$ [24.82, 83], $SrTiO_3(100)$ [24.84], $CeO_2(111)$ [24.85] and $MoO_3(010)$ [24.86] surfaces. Also, *Barth* et al. have succeeded in obtaining atomically resolved NC-AFM images of a clean $\alpha-Al_2O_3(0001)$ surface [24.73] and of a UHV cleaved $MgO(001)$ [24.87] surface, which are impossible to investigate using STM. In this section we describe typical results of the imaging of metal oxides by NC-AFM.

The $\alpha-Al_2O_3(0001)$ surface exists in several ordered phases that can reversibly be transformed into each other by thermal treatments and oxygen exposure. It is known that the high-temperature phase has a large $(\sqrt{31} \times \sqrt{31})R \pm 9°$ unit cell. However, the details of the atomic structure of this surface have not been revealed, and two models have been proposed. *Barth* et al. [24.73] have directly observed this reconstructed $\alpha-Al_2O_3(0001)$ surface by NC-AFM. They confirmed that the dominant contrast of the low-

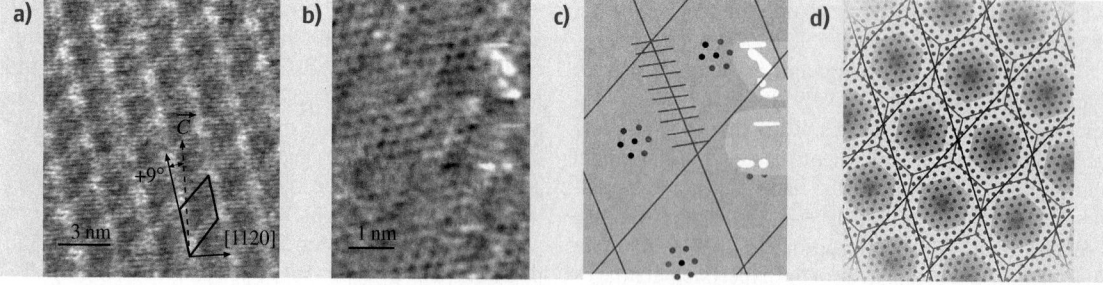

Fig. 24.19 (a) Image of the high-temperature, reconstructed clean $\alpha-Al_2O_3$ surface obtained by using the constant-height mode. The *rhombus* represents the unit cell of the $(\sqrt{31} \times \sqrt{31})R + 9°$ reconstructed surface. **(b)** Higher-magnification image of **(a)**. Imaging was performed at a reduced tip–sample distance. **(c)** Schematic representation of the indicating regions of hexagonal order in the center of reconstructed rhombi. **(d)** Superposition of the hexagonal domain with reconstruction rhombi found by NC-AFM imaging. Atoms in the *gray shaded regions* are well ordered. The images and the schematic representations are from [24.73]

magnification image corresponds to a rhombic grid representing a unit cell of $(\sqrt{31} \times \sqrt{31})R + 9°$, as shown in Fig. 24.19a. Also, more details of the atomic structures were determined from the higher-magnification image (Fig. 24.19b), which was taken at a reduced tip–sample distance. In this atomically resolved image, it was revealed that each side of the rhombus is intersected by ten atomic rows, and that a hexagonal arrangement of atoms exists in the center of the rhombi (Fig. 24.19c). This feature agrees with the proposed surface structure that predicts order in the center of the hexagonal surface domains and disorder at the domain boundaries. Their result is an excellent demonstration of the capabilities of the NC-AFM for the atomic-scale surface investigation of insulators.

The atomic structure of the $SrTiO_3(100)$-$(\sqrt{5}x \times \sqrt{5})R26.6°$ surface, as well as that of $Al_2O_3(0001)$ can be determined on the basis of the results of NC-AFM imaging [24.84]. $SrTiO_3$ is one of the perovskite oxides, and its (100) surface exhibits the many different kinds of reconstructed structures. In the case of the $(\sqrt{5} \times \sqrt{5})R26.6°$ reconstruction, the oxygen vacancy–Ti^{3+}–oxygen model (where the terminated surface is TiO_2 and the observed spots are related to oxygen vacancies) was proposed from the results of STM imaging. As shown in Fig. 24.20, *Kubo* et al. [24.84] have performed measurements using both STM and NC-AFM, and have found that the size of the bright spots as observed by NC-AFM is always smaller than that for STM measurement, and that the dark spots, which are not observed by STM, are arranged along the [001] and [010] directions in the NC-AFM image. A theoretical simulation of the NC-AFM image using first-principles calculations shows that the bright and dark spots correspond to Sr and oxygen atoms, respectively. It has been proposed that the structural model of the reconstructed surface consists of an ordered Sr adatom located at the oxygen fourfold site on the TiO_2-terminated layer (Fig. 24.20c).

Because STM images are related to the spatial distribution of the wave functions near the Fermi level, atoms without a local density of states near the Fermi level are generally invisible even on conductive materials. On the other hand, the NC-AFM image reflects the strength of the tip–sample interaction force originating from chemical, electrostatic and other interactions. Therefore, even STM and NC-AFM images obtained using an identical tip and sample may not be identical generally. The simultaneous imaging of a metal oxide surface enables the investigation of a more detailed surface structure. The images of a $TiO_2(110)$ surface simultaneously obtained with STM and NC-AFM [24.78] are a typical example. The STM image shows that the dangling-bond states at the tip apex overlap with the dangling bonds of the $3d$ states protruding from the Ti atom, while the NC-AFM primarily imaged the uppermost oxygen atom.

Recently, calculations of the interaction of a Si tip with metal oxides surfaces, such as $Al_2O_3(0001)$, $TiO_2(110)$, and $MgO(001)$, were reported [24.88, 89]. Previous simulations of AFM imaging of alkali halides and fluorides assume that the tip would be oxides or contaminated and hence have been performed with a model of ionic oxide tips. In the case of imaging a metal oxide surface, pure Si tips are appropriate for a more realistic tip model because the tip is sputtered for cleaning in many experiments. The results of ab initio calculations for a Si tip with a dangling bond demonstrate that the balance between polarization of the tip and covalent bonding between the tip and the surface should determine the tip–surface force. The interaction force can be related to the nature of the surface electronic structure. For wide-gap insulators with a large valence-band offset that prevents significant electron-density transfer between the tip and the sample, the force is dominated by polarization of the tip. When the gap is narrow, the

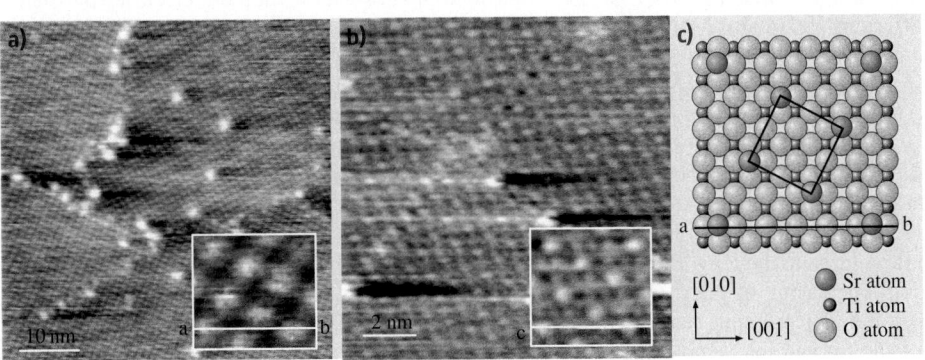

Fig. 24.20 (a) STM and (b) NC-AFM images of a $SrTiO_3(100)$ surface. (c) A proposed model of the $SrTiO_3(100)$-$(\sqrt{5} \times \sqrt{5})R26.6°$ surface reconstruction. The images and the schematic representations are from [24.84]

charge transfer increase and covalent bonding dominates the tip–sample interaction. The forces over anions (oxygen ions) in the surface are larger than over cations (metal ions), as they play a more significant role in charge transfer. This implies that a pure Si tip would always show the brightest contrast over the highest anions in the surface. In addition, *Foster* et al. [24.88] suggested the method of using applied voltage, which controls the charge transfer, during an AFM measurement to define the nature of tip apex.

The collaboration between experimental and theoretical studies has made great progress in interpreting the imaging mechanism for binary insulators surface and reveals that a well-defined tip with atomic resolution is preferable for imaging a surface. As described previously, a method for the evaluation of the nature of the tip has been developed. However, the most desirable solution would be the development of suitable techniques for well-defined tip preparation and a few attempts at controlled production of Si tips have been reported [24.24, 90, 91].

24.3.2 Atomically Resolved Imaging of a NiO(001) Surface

The transition metal oxides, such as NiO, CoO, and FeO, feature the simultaneous existence of an energy gap and unpaired electrons, which gives rise to a variety of magnetic property. Such magnetic insulators are widely used for the exchange biasing for magnetic and spintronic devices. NC-AFM enables direct surface imaging of magnetic insulators on an atomic scale. The forces detected by NC-AFM originate from several kinds of interaction between the surface and the tip, including magnetic interactions in some cases. Theoretical studies predict that short-range magnetic interactions such as the exchange interaction should enable the NC-AFM to image magnetic moments on an atomic scale. In this section, we will describe imaging of the antiferromagnetic NiO(001) surface using a ferromagnetic tip. Also, theoretical studies of the exchange force interaction between a magnetic tip and a sample will be described.

Theoretical Studies of the Exchange Force
In the system of a magnetic tip and sample, the interaction detected by NC-AFM includes the short-range magnetic interaction in addition to the long-range magnetic dipole interaction. The energy of the short-range interaction depends on the electron spin states of the atoms on the apex of the tip and the sample surface,

and the energy difference between spin alignments (parallel or anti-parallel) is referred to as the exchange interaction energy. Therefore, the short-range magnetic interaction leads to the atomic-scale magnetic contrast, depending on the local energy difference between spin alignments.

In the past, extensive theoretical studies on the short-range magnetic interaction between a ferromagnetic tip and a ferromagnetic sample have been performed by a simple calculation [24.92], a tight-binding approximation [24.93] and first-principles calculations [24.94]. In the calculations performed by *Nakamura* et al. [24.94], three-atomic-layer Fe(001) films are used as a model for the tip and sample. The exchange force is defined as the difference between the forces in each spin configuration of the tip and sample (parallel and anti-parallel). The result of this calculation demonstrates that the amplitude of the exchange force is measurable for AFM (about 0.1 nN). Also, they forecasted that the discrimination of the exchange force would enable direct imaging of the magnetic moments on an atomic scale. Foster et al. [24.95] have theoretically investigated the interaction between a spin-polarized H atom and a Ni atom on a NiO(001) surface. They demonstrated that the difference in magnitude in the exchange interaction between opposite-spin Ni ions in a NiO surface could be sufficient to be measured in a low-temperature NC-AFM experiment. Recently, first-principles calculation of the interaction of a ferromagnetic Fe tip with an NiO surface has demonstrated that it should be feasible to measure the difference in exchange force between opposite-spin Ni ions [24.96].

Atomically Resolved Imaging Using Non-coated and Fe-coated Si Tips
The detection of the exchange interaction is a challenging task for NC-AFM applications. An antiferromagnetic insulator NiO single crystal that has regularly aligned atom sites with alternating electron spin states is one of the best candidates to prove the feasibility of detecting the exchange force for the following reason. NiO has an antiferromagnetic AF_2 structure as the most stable below the Néel temperature of 525 K. This well-defined magnetic structure, in which Ni atoms on the (001) surface are arranged in a checkerboard pattern, leads to the simple interpretation of an image containing the atomic-scale contrast originating in the exchange force. In addition, a clean surface can easily be prepared by cleaving.

Figure 24.21a shows an atomically resolved image of a NiO(001) surface with a ferromagnetic Fe-coated

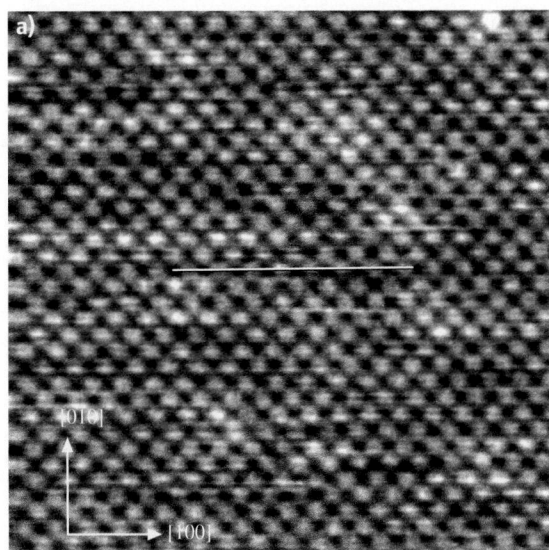

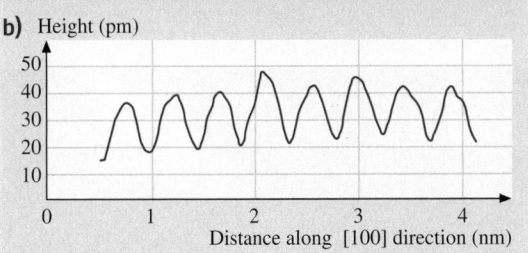

Fig. 24.21 (a) Atomically resolved image obtained with an Fe-coated tip. (b) Shows the cross sections of the middle part in (a). Their corrugations are about 30 pm

tip [24.97]. The bright protrusions correspond to atoms spaced about 0.42 nm apart, consistent with the expected periodic arrangement of the NiO(001) surface. The corrugation amplitude is typically 30 pm, which is comparable to the value previously reported [24.82, 83, 98–100], as shown in Fig. 24.21b. The atomic-resolution image (Fig. 24.21b), in which there is one maximum and one minimum within the unit cell, resembles that of the alkali halide (001) surface. The symmetry of the image reveals that only one type of atom appears to be at the maximum. From this image, it seems difficult to distinguish which of the atoms are observed as protrusions. The theoretical works indicate that a metal tip interacts strongly with the oxygen atoms on the MgO(001) surface [24.95]. From this result, it is presumed that the bright protrusions correspond to the oxygen atoms. However, it is still questionable which of the atoms are visible with a Fe-coated tip.

If the short-range magnetic interaction is included in the atomic image, the corrugation amplitude of the atoms should depend on the direction of the spin over the atom site. From the results of first-principles calculations [24.94], the contribution of the short-range magnetic interaction to the measured corrugation amplitude is expected to be about a few percent of the total interaction. Discrimination of such small perturbations is therefore needed. In order to reduce the noise, the corrugation amplitude was added on the basis of the periodicity of the NC-AFM image. In addition, the topographical asymmetry, which is the index characterizing the difference in atomic corrugation amplitude, has been defined [24.101]. The result shows that the value of the topographical asymmetry calculated from the image obtained with an Fe-coated Si tip depends on the direction of summing of the corrugation amplitude, and that the dependency corresponds to the antiferromagnetic spin ordering of the NiO(001) surface [24.101, 102]. Therefore, this result implies that the dependency of the topographical asymmetry originates in the short-range magnetic interaction. However, in some cases the topographic asymmetry with uncoated Si tips has a finite value [24.103]. The possibility that the asymmetry includes the influence of the structure of tip apex and of the relative orientation between the surface and tip cannot be excluded. In addition, it is suggested that the absence of unambiguous exchange contrast is due to the fact that surface ion instabilities occur at tip–sample distances that are small enough for a magnetic interact [24.100]. Another possibility is that the magnetic properties of the tips are not yet fully controlled because the topographic asymmetries obtained by Fe- and Ni-coated tips show no significant difference [24.103]. In any cases, a careful comparison is needed to evaluate the exchange interaction included in an atomic image.

From the aforementioned theoretical works, it is presumed that a metallic tip has the capability to image an oxygen atom as a bright protrusion. Recently, the magnetic properties of the NiO(001) surface were investigated by first-principles electronic-structure calculations [24.104]. It was shown that the surface oxygen has finite spin magnetic moment, which originates from symmetry breaking. We must take into account the possibility that a metal atom at the ferromagnetic tip apex may interact with a Ni atom on the second layer through a magnetic interaction mediated by the electrons in an oxygen atom on the surface.

The measurements presented here demonstrate the feasibility of imaging magnetic structures on an atomic scale by NC-AFM. In order to realize explicit detection of exchange force, further experiments and a theoretical study are required. In particular, the development of a tip with well-defined atomic structure and magnetic properties is essential for *exchange force microscopy*.

24.4 Applications to Molecules

In the future, it is expected that electronic, chemical, and medical devices will be downsized to the nanometer scale. To achieve this, visualizing and assembling individual molecular components is of fundamental importance. Topographic imaging of nonconductive materials, which is beyond the range of scanning tunneling microscopes, is a challenge for atomic force microscopy (AFM). Nanometer-sized domains of surfactants terminated with different functional groups have been identified by lateral force microscopy (LFM) [24.107] and by chemical force microscopy (CFM) [24.108] as extensions of AFM. At a higher resolution, a periodic array of molecules, Langmuir–Blodgett films [24.109] for example, was recognized by AFM. However, it remains difficult to visualize an isolated molecule, molecule vacancy, or the boundary of different periodic domains, with a microscope with the tip in contact.

24.4.1 Why Molecules and Which Molecules?

Access to individual molecules has not been a trivial task even for noncontact atomic force microscopy (NC-AFM). The force pulling the tip into the surface is less sensitive to the gap width (r), especially when chemically stable molecules cover the surface. The attractive potential between two stable molecules is shallow and exhibits r^{-6} decay [24.13].

High-resolution topography of formate ($HCOO^-$) [24.110] was first reported in 1997 as a molecular adsorbate. The number of imaged molecules is now increasing because of the technological importance of molecular interfaces. To date, the following studies on molecular topography have been published: C_{60} [24.105, 111], DNAs [24.106, 112], adenine and thymine [24.113], alkanethiols [24.113, 114], a perylene derivative (PTCDA) [24.115], a metal porphyrin (Cu–TBPP) [24.116], glycine sulfate [24.117], polypropylene [24.118], vinylidene fluoride [24.119], and a series of carboxylates ($RCOO^-$) [24.120–126].

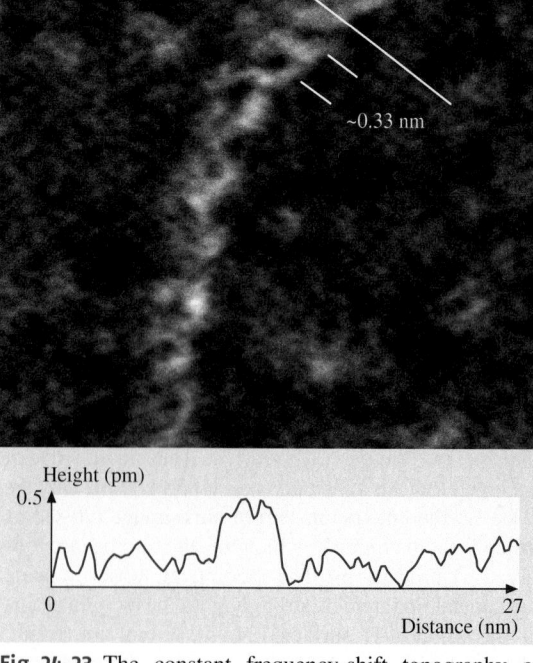

~0.33 nm

Height (pm)

0.5

0

0   27
                              Distance (nm)

Fig. 24.23 The constant frequency-shift topography of a DNA helix on a mica surface based on [24.106]. Image size: 43×43 nm^2. The image revealed features with a spacing of 3.3 nm, consistent with the helix turn of B-DNA

Fig. 24.22 The constant frequency-shift topography of domain boundaries on a C_{60} multilayered film deposited on a Si(111) surface based on [24.105]. Image size: 35×35 nm^2

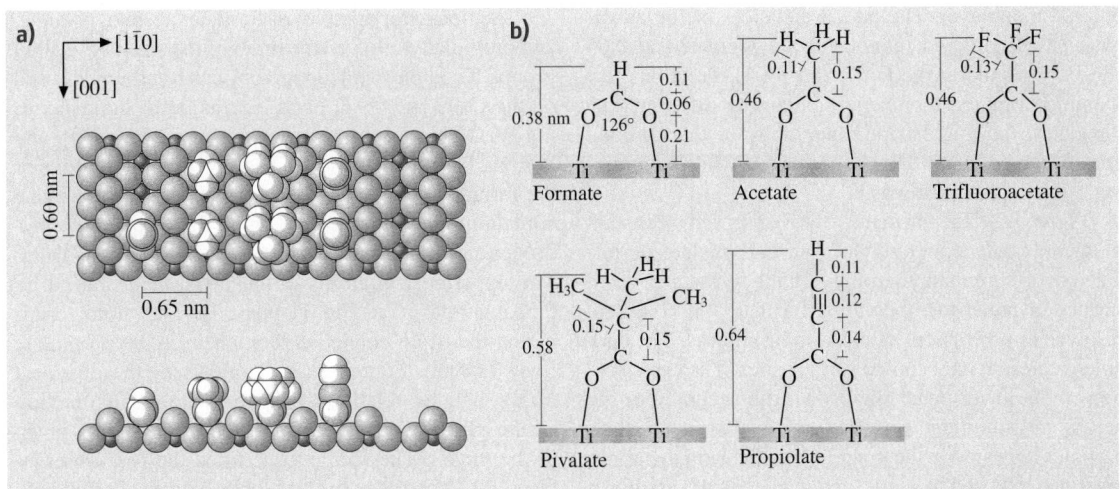

Fig. 24.24a,b The carboxylates and TiO$_2$ substrate. (**a**) Top and side view of the ball model. *Small shaded* and *large shaded balls* represent Ti and O atoms in the substrate. Protons yielded in the dissociation reaction are not shown. (**b**) Atomic geometry of formate, acetate, pivalate, propiolate, and trifluoroacetate adsorbed on the TiO$_2$(110) surface. The O—Ti distance and O—C—O angle of the formate were determined in the quantitative analysis using photoelectron diffraction [24.127]

Two of these are presented in Figs. 24.22 and 24.23 to demonstrate the current stage of achievement. The proceedings of the annual NC-AFM conference represent a convenient opportunity for us to update the list of molecules imaged.

24.4.2 Mechanism of Molecular Imaging

A systematic study of carboxylates (RCOO$^-$) with R = H, CH$_3$, C(CH$_3$)$_3$, C≡CH, and CF$_3$ revealed that the van der Waals force is responsible for the molecule-dependent microscope topography despite its long-range (r^{-6}) nature. Carboxylates adsorbed on the (110) surface of rutile TiO$_2$ have been extensively studied as a prototype for organic materials interfaced with an inorganic metal oxide [24.128]. A carboxylic acid molecule (RCOOH) dissociates on this surface to a carboxylate (RCOO$^-$) and a proton (H$^+$) at room temperature, as illustrated in Fig. 24.24. The pair of negatively charged oxygen atoms in the RCOO$^-$ coordinate two positively charged Ti atoms on the surface. The adsorbed carboxylates create a long-range

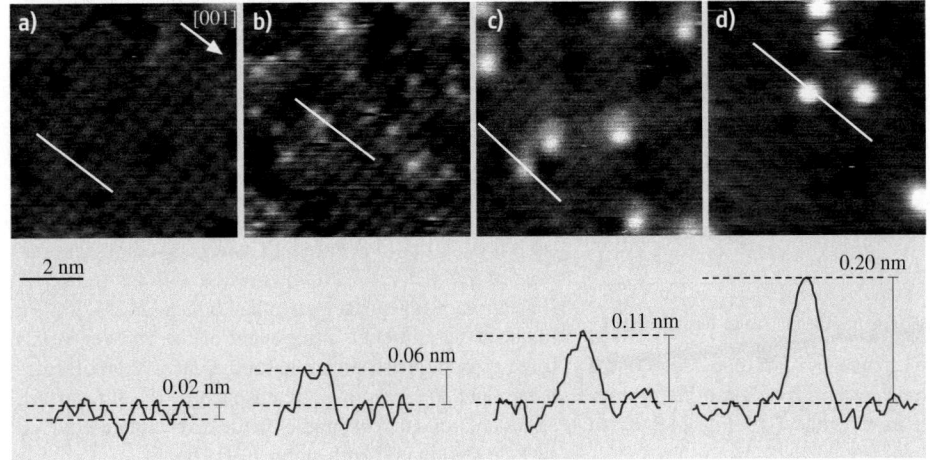

Fig. 24.25a–d The constant frequency-shift topography of carboxylate monolayers prepared on the TiO$_2$(110) surface based on [24.121, 123, 125]. Image size: 10 × 10 nm^2. (**a**) Pure formate monolayer; (**b**) formate–acetate mixed layer; (**c**) formate–pivalate mixed layer; (**d**) formate–propiolate mixed layer. Cross sections determined on the *lines* are shown in the lower panel

ordered monolayer. The lateral distances of the adsor-bates in the ordered monolayer are regulated at 0.65 and 0.59 nm along the [110] and [001] directions. By scanning a mixed monolayer containing different car-boxylates, the microscope topography of the terminal groups can be quantitatively compared while minimiz-ing tip-dependent artifacts.

Figure 24.25 presents the observed constant frequency-shift topography of four carboxylates termi-nated by different alkyl groups. On the formate-covered surface of panel (a), individual formates (R=H) were resolved as protrusions of uniform brightness. The dark holes represent unoccupied surface sites. The cross sec-tion in the lower panel shows that the accuracy of the height measurement was 0.01 nm or better. Brighter particles appeared in the image when the formate mono-layer was exposed to acetic acid (CH$_3$COOH) as shown in panel (b). Some formates were exchanged with ac-etates (R=CH$_3$) impinging from the gas phase [24.129]. Because the number of brighter spots increased with exposure time to acetic acid, the brighter particle was assigned to the acetate [24.121]. Twenty-nine acetates and 188 formates were identified in the topography. An isolated acetate and its surrounding formates exhibited an image height difference of 0.06 nm. Pivalate is termi-nated by bulky R=(CH$_3$)$_3$. Nine bright pivalates were surrounded by formates of ordinary brightness in the im-age of panel (c) [24.123]. The image height difference of an isolated pivalate over the formates was 0.11 nm. Propiolate with C≡CH is a needle-like adsorbate of single-atom diameter. That molecule exhibited in panel (d) a microscope topography 0.20 nm higher than that of the formate [24.125].

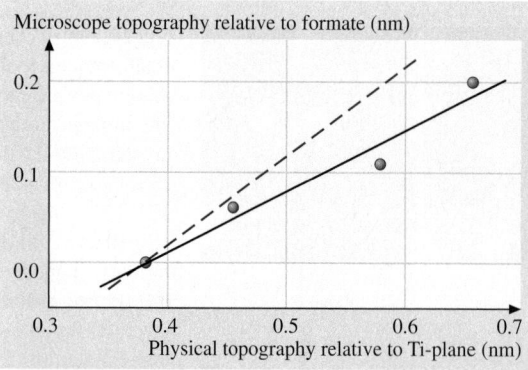

Fig. 24.26 The constant frequency-shift topography of the alkyl-substituted carboxylates as a function of their phys-ical topography given in the model of Fig. 24.3 based on [24.123]

The image topography of formate, acetate, pivalate, and propiolate followed the order of the size of the alkyl groups. Their physical topography can be assumed based on the C−C and C−H bond lengths in the correspond-ing RCOOH molecules in the gas phase [24.130], and is illustrated in Fig. 24.24. The top hydrogen atom of the formate is located 0.38 nm above the surface plane containing the Ti atom pair, while three equivalent hy-drogen atoms of the acetate are more elevated at 0.46 nm. The uppermost H atoms in the pivalate are raised by 0.58 nm relative to the Ti plane. The H atom termi-nating the triple-bonded carbon chain in the propiolate is at 0.64 nm. Figure 24.26 summarizes the observed image heights relative to the formate, as a function of the physical height of the topmost H atoms given in the model. The straight line fitted the four observa-tions [24.122]. When the horizontal axis was scaled with other properties (molecular weight, the number of atoms in a molecule, or the number of electrons in valence states), the correlation became poor.

On the other hand, if the tip apex traced the con-tour of a molecule composed of hard-sphere atoms, the image topography would reproduce the physical topog-raphy in a one-to-one ratio, as shown by the broken line in Fig. 24.26. However, the slope of the fitted line was 0.7. A slope of less than unity is interpreted as the long-range nature of the tip–molecule force. The observable frequency shift reflects the sum of the forces between the tip apex and individual molecules. When the tip passes above a tall molecule embedded in short molecules, it is pulled up to compensate for the increased force originat-ing from the tall molecule. Forces between the lifted tip and the short molecules are reduced due to the increased tip–surface distance. Feedback regulation pushes down the probe to restore the lost forces.

This picture predicts that microscope topography is sensitive to the lateral distribution of the molecules, and that was in fact the case. Two-dimensionally clustered acetates exhibited enhanced image height over an iso-lated acetate [24.121]. The tip–molecule force therefore remained nonzero at distances over the lateral separa-tion of the carboxylates on this surface (0.59−0.65 nm). Chemical bond interactions cannot be important across such a wide tip–molecule gap, whereas atom-scale im-ages of Si(111)(7×7) are interpreted with the fractional formation of tip–surface chemical bonds [24.24, 45, 49]. Instead, the attractive component of the van der Waals force is probable responsible for the observed molecule-dependent topography. The absence of the tip–surface chemical bond is reasonable on the carboxylate-covered surface terminated with stable C−H bonds.

The attractive component of the van der Waals force contains electrostatic terms caused by permanent-dipole/permanent-dipole coupling, permanent-dipole/induced-dipole coupling, and induced-dipole/induced-dipole coupling (dispersion force). The four carboxylates examined are equivalent in terms of their permanent electric dipole, because the alkyl groups are non-polar. The image contrast of one carboxylate relative to another is thus ascribed to the dispersion force and/or the force created by the coupling between the permanent dipole on the tip and the induced dipole on the molecule. If we further assume that the Si tip used exhibits the smallest permanent dipole, the dispersion force remains dominant to create the NC-AFM topography dependent on the non-polar groups of atoms. A numerical simulation based on this assumption [24.125] successfully reproduced the propiolate topography of Fig. 24.25d. A calculation that does not include quantum chemical treatment is expected to work, unless the tip approaches the surface too closely, or the molecule possesses a dangling bond.

In addition to the contribution of the dispersion force, the permanent dipole moment of molecules may perturb the microscope topography through electrostatic coupling with the tip. Its possible role was demonstrated by imaging a fluorine-substituted acetate. The strongly polarized $C-F$ bonds were expected to perturb the electrostatic field over the molecule. The constant frequency-shift topography of acetate ($R=CH_3$) and trifluoroacetate ($R=CF_3$) was indeed sensitive to the fluorine substitution. The acetate was observed to be 0.05 nm higher than the trifluoroacetate [24.122], although the F atoms in the trifluoroacetate as well as the H atoms in the acetate were lifted by 0.46 nm from the surface plane, as illustrated in Fig. 24.24.

24.4.3 Perspectives

The experimental results summarized in this section prove the feasibility of using NC-AFM to identify individual molecules. A systematic study on the constant frequency-shift topography of carboxylates with $R=CH_3$, $C(CH_3)_3$, $C \equiv CH$, and CF_3 has revealed the mechanism behind the high-resolution imaging of the chemically stable molecules. The dispersion force is primarily responsible for the molecule-dependent topography. The permanent dipole moment of the imaged molecule, if it exists, perturbs the topography through the electrostatic coupling with the tip. A tiny calculation containing empirical force fields works when simulating the microscope topography.

These results make us optimistic about analyzing physical and chemical properties of nanoscale supramolecular assemblies constructed on a solid surface. If the accuracy of topographic measurement is developed by one more order of magnitude, which is not an unrealistic target, it may be possible to identify structural isomers, chiral isomers, and conformational isomers of a molecule. Kelvin probe force microscopy (KPFM), an extension of NC-AFM, provides a nanoscale analysis of molecular electronic properties [24.118, 119]. Force spectroscopy with chemically modified tips seems promising for the detection of a selected chemical force. Operation in a liquid atmosphere [24.131] is required for the observation of biochemical materials in their natural environment.

Part C | 24

References

24.1   G. Binnig: Atomic force microscope, method for imaging surfaces with atomic resolution, US Patent 4,724,318 (1986)

24.2   G. Binnig, C.F. Quate, C. Gerber: Atomic force microscope, Phys. Rev. Lett. **56**, 930–933 (1986)

24.3   G. Binnig, H. Rohrer, C. Gerber, E. Weibel: Surface studies by scanning tunneling microscopy, Phys. Rev. Lett. **49**, 57–61 (1982)

24.4   G. Binnig, H. Rohrer: The scanning tunneling microscope, Sci. Am. **253**, 50–56 (1985)

24.5   G. Binnig, H. Rohrer: In touch with atoms, Rev. Mod. Phys. **71**, S320–S330 (1999)

24.6   C.J. Chen: *Introduction to Scanning Tunneling Microscopy* (Oxford Univ. Press, Oxford 1993)

24.7   H.-J. Güntherodt, R. Wiesendanger (Eds.): *Scanning Tunneling Microscopy I–III* (Springer, Berlin, Heidelberg 1991)

24.8   J.A. Stroscio, W.J. Kaiser (Eds.): *Scanning Tunneling Microscopy* (Academic, Boston 1993)

24.9   R. Wiesendanger: *Scanning Probe Microscopy and Spectroscopy: Methods and Applications* (Cambridge Univ. Press, Cambridge 1994)

24.10 S. Morita, R. Wiesendanger, E. Meyer (Eds.): *Noncontact Atomic Force Microscopy* (Springer, Berlin, Heidelberg 2002)

24.11 R. Garcia, R. Perez: Dynamic atomic force microscopy methods, Surf. Sci. Rep. **47**, 197–301 (2002)

24.12 F.J. Giessibl: Advances in atomic force microscopy, Rev. Mod. Phys. **75**, 949–983 (2003)

24.13   J. Israelachvili: *Intermolecular and Surface Forces*, 2nd edn. (Academic, London 1991)

24.14   L. Olsson, N. Lin, V. Yakimov, R. Erlandsson: A method for in situ characterization of tip shape in AC-mode atomic force microscopy using electrostatic interaction, J. Appl. Phys. **84**, 4060–4064 (1998)

24.15   S. Akamine, R. C. Barrett, C. F. Quate: Improved atomic force microscopy images using cantilevers with sharp tips, Appl. Phys. Lett. **57**, 316–318 (1990)

24.16   T. R. Albrecht, S. Akamine, T. E. Carver, C. F. Quate: Microfabrication of cantilever styli for the atomic force microscope, J. Vac. Sci. Technol. A **8**, 3386–3396 (1990)

24.17   M. Tortonese, R. C. Barrett, C. Quate: Atomic resolution with an atomic force microscope using piezoresistive detection, Appl. Phys. Lett. **62**, 834–836 (1993)

24.18   O. Wolter, T. Bayer, J. Greschner: Micromachined silicon sensors for scanning force microscopy, J. Vac. Sci. Technol. **9**, 1353–1357 (1991)

24.19   D. Sarid: *Scanning Force Microscopy*, 2nd edn. (Oxford Univ. Press, New York 1994)

24.20   F. J. Giessibl, B. M. Trafas: Piezoresistive cantilevers utilized for scanning tunneling and scanning force microscope in ultrahigh vacuum, Rev. Sci. Instrum. **65**, 1923–1929 (1994)

24.21   P. Güthner, U. C. Fischer, K. Dransfeld: Scanning near-field acoustic microscopy, Appl. Phys. B **48**, 89–92 (1989)

24.22   K. Karrai, R. D. Grober: Piezoelectric tip–sample distance control for near field optical microscopes, Appl. Phys. Lett. **66**, 1842–1844 (1995)

24.23   F. J. Giessibl: High-speed force sensor for force microscopy and profilometry utilizing a quartz tuning fork, Appl. Phys. Lett. **73**, 3956–3958 (1998)

24.24   F. J. Giessibl, S. Hembacher, H. Bielefeldt, J. Mannhart: Subatomic features on the silicon (111)-(7×7) surface observed by atomic force microscopy, Science **289**, 422–425 (2000)

24.25   F. Giessibl, C. Gerber, G. Binnig: A low-temperature atomic force/scanning tunneling microscope for ultrahigh vacuum, J. Vac. Sci. Technol. B **9**, 984–988 (1991)

24.26   F. Ohnesorge, G. Binnig: True atomic resolution by atomic force microscopy through repulsive and attractive forces, Science **260**, 1451–1456 (1993)

24.27   F. J. Giessibl, G. Binnig: True atomic resolution on KBr with a low-temperature atomic force microscope in ultrahigh vacuum, Ultramicroscopy **42–44**, 281–286 (1992)

24.28   S. P. Jarvis, H. Yamada, H. Tokumoto, J. B. Pethica: Direct mechanical measurement of interatomic potentials, Nature **384**, 247–249 (1996)

24.29   L. Howald, R. Lüthi, E. Meyer, P. Guthner, H.-J. Güntherodt: Scanning force microscopy on the Si(111)7×7 surface reconstruction, Z. Phys. B **93**, 267–268 (1994)

24.30   L. Howald, R. Lüthi, E. Meyer, H.-J. Güntherodt: Atomic-force microscopy on the Si(111)7×7 surface, Phys. Rev. B **51**, 5484–5487 (1995)

24.31   Y. Martin, C. C. Williams, H. K. Wickramasinghe: Atomic force microscope–force mapping and profiling on a sub 100 scale, J. Appl. Phys. **61**, 4723–4729 (1987)

24.32   T. R. Albrecht, P. Grutter, H. K. Horne, D. Rugar: Frequency modulation detection using high-Q cantilevers for enhanced force microscope sensitivity, J. Appl. Phys. **69**, 668–673 (1991)

24.33   F. J. Giessibl: Atomic resolution of the silicon (111)-(7×7) surface by atomic force microscopy, Science **267**, 68–71 (1995)

24.34   S. Kitamura, M. Iwatsuki: Observation of silicon surfaces using ultrahigh-vacuum noncontact atomic force microscopy, Jpn. J. Appl. Phys. **35**, 668–L671 (1995)

24.35   R. Erlandsson, L. Olsson, P. Martensson: Inequivalent atoms and imaging mechanisms in AC-mode atomic-force microscopy of Si(111)7×7, Phys. Rev. B **54**, R8309–R8312 (1996)

24.36   N. Burnham, R. J. Colton: Measuring the nanomechanical and surface forces of materials using an atomic force microscope, J. Vac. Sci. Technol. A **7**, 2906–2913 (1989)

24.37   D. Tabor, R. H. S. Winterton: Direct measurement of normal and related van der Waals forces, Proc. R. Soc. London A **312**, 435 (1969)

24.38   F. J. Giessibl: Forces and frequency shifts in atomic resolution dynamic force microscopy, Phys. Rev. B **56**, 16011–16015 (1997)

24.39   G. Binnig, H. Rohrer, C. Gerber, E. Weibel: 7×7 reconstruction on Si(111) resolved in real space, Phys. Rev. Lett. **50**, 120–123 (1983)

24.40   H. Goldstein: *Classical Mechanics* (Addison Wesley, Reading 1980)

24.41   U. Dürig: Interaction sensing in dynamic force microscopy, New J. Phys. **2**, 5.1–5.12 (2000)

24.42   F. J. Giessibl: A direct method to calculate tip-sample forces from frequency shifts in frequency-modulation atomic force microscopy, Appl. Phys. Lett. **78**, 123–125 (2001)

24.43   U. Dürig, H. P. Steinauer, N. Blanc: Dynamic force microscopy by means of the phase-controlled oscillator method, J. Appl. Phys. **82**, 3641–3651 (1997)

24.44   F. J. Giessibl, H. Bielefeldt, S. Hembacher, J. Mannhart: Calculation of the optimal imaging parameters for frequency modulation atomic force microscopy, Appl. Surf. Sci. **140**, 352–357 (1999)

24.45   M. A. Lantz, H. J. Hug, R. Hoffmann, P. J. A. van Schendel, P. Kappenberger, S. Martin, A. Baratoff, H.-J. Güntherodt: Quantitative measurement of short-range chemical bonding forces, Science **291**, 2580–2583 (2001)

24.46   F. J. Giessibl, M. Herz, J. Mannhart: Friction traced to the single atom, Proc. Nat. Acad. Sci. USA **99**, 12006–12010 (2002)

24.47   N. Nakagiri, M. Suzuki, K. Oguchi, H. Sugimura: Site discrimination of adatoms in Si(111)-7×7 by noncontact atomic force microscopy, Surf. Sci. Lett. **373**, L329–L332 (1997)

24.48   Y. Sugawara, M. Ohta, H. Ueyama, S. Morita: Defect motion on an InP(110) surface observed with noncontact atomic force microscopy, Science **270**, 1646–1648 (1995)

24.49   T. Uchihashi, Y. Sugawara, T. Tsukamoto, M. Ohta, S. Morita: Role of a covalent bonding interaction in noncontact-mode atomic-force microscopy on Si(111)7×7, Phys. Rev. B **56**, 9834–9840 (1997)

24.50   K. Yokoyama, T. Ochi, A. Yoshimoto, Y. Sugawara, S. Morita: Atomic resolution imaging on Si(100)2× 1 and Si(100)2×1-H surfaces using a non-contact atomic force microscope TS11, Jpn. J. Appl. Phys. **39**, L113–L115 (2000)

24.51   Y. Sugawara, T. Minobe, S. Orisaka, T. Uchihashi, T. Tsukamoto, S. Morita: Non-contact AFM images measured on Si(111)$\sqrt{3} \times \sqrt{3}$-Ag and Ag(111) surfaces, Surf. Interface Anal. **27**, 456–461 (1999)

24.52   K. Yokoyama, T. Ochi, Y. Sugawara, S. Morita: Atomically resolved Ag imaging on Si(111)$\sqrt{3} \times \sqrt{3}$-Ag surface with noncontact atomic force microscope, Phys. Rev. Lett. **83**, 5023–5026 (1999)

24.53   M. Bammerlin, R. Lüthi, E. Meyer, A. Baratoff, J. Lü, M. Guggisberg, Ch. Gerber, L. Howald, H.-J. Güntherodt: True atomic resolution on the surface of an insulator via ultrahigh vacuum dynamic force microscopy, Probe Microscopy **1**, 3–7 (1997)

24.54   M. Bammerlin, R. Lüthi, E. Meyer, A. Baratoff, J. Lü, M. Guggisberg, C. Loppacher, Ch. Gerber, H.-J. Güntherodt: Dynamic SFM with true atomic resolution on alkali halide surfaces, Appl. Phys. A **66**, S293–S294 (1998)

24.55   R. Hoffmann, M.A. Lantz, H.J. Hug, P.J.A. van Schendel, P. Kappenberger, S. Martin, A. Baratoff, H.-J. Güntherodt: Atomic resolution imaging and force versus distance measurements on KBr(001) using low temperature scanning force microscopy, Appl. Surf. Sci. **188**, 238–244 (2002)

24.56   R. Bennewitz, A.S. Foster, L.N. Kantotovich, M. Bammerlin, Ch. Loppacher, S. Schär, M. Guggisberg, E. Meyer, A.L. Shluger: Atomically resolved edges and kinks of NaCl islands on Cu(111): Experiment and theory, Phys. Rev. B **62**, 2074–2084 (2000)

24.57   A.I. Livshits, A.L. Shluger, A.L. Rohl, A.S. Foster: Model of noncontact scanning force microscopy on ionic surfaces, Phys. Rev. **59**, 2436–2448 (1999)

24.58   R. Bennewitz, O. Pfeiffer, S. Schär, V. Barwich, E. Meyer, L.N. Kantorovich: Atomic corrugation in nc-AFM of alkali halides, Appl. Surf. Sci. **188**, 232–237 (2002)

24.59   R. Bennewitz, S. Schär, E. Gnecco, O. Pfeiffer, M. Bammerlin, E. Meyer: Atomic structure of alkali halide surfaces, Appl. Phys. A **78**, 837–841 (2004)

24.60   M. Gauthier, L. Kantrovich, M. Tsukada: Theory of energy dissipation into surface viblationsed. In: *Noncontact Atomic Force Microscopy*, ed. by S. Morita, R. Wiesendanger, E. Meyer (Springer, Berlin, Heidelberg 2002) pp. 371–394

24.61   H.J. Hug, A. Baratoff: Measurement of dissipation induced by tip–sample interactions. In: *Noncontact Atomic Force Microscopy*, ed. by S. Morita, R. Wiesendanger, E. Meyer (Springer, Berlin, Heidelberg 2002) pp. 395–431

24.62   R. Hoffmann, L.N. Kantorovich, A. Baratoff, H.J. Hug, H.-J. Güntherodt: Sublattice identification in scanning force microscopy on alkali halide surfaces, Phys. Rev. B **92**, 146103/1–4 (2004)

24.63   C. Barth, M. Reichling: Resolving ions and vacancies at step edges on insulating surfaces, Surf. Sci. **470**, L99–L103 (2000)

24.64   R. Bennewitz, M. Reichling, E. Matthias: Force microscopy of cleaved and electron-irradiated CaF$_2$(111) surfaces in ultra-high vacuum, Surf. Sci. **387**, 69–77 (1997)

24.65   M. Reichling, C. Barth: Scanning force imaging of atomic size defects on the CaF$_2$(111) surface, Phys. Rev. Lett. **83**, 768–771 (1999)

24.66   M. Reichling, M. Huisinga, S. Gogoll, C. Barth: Degradation of the CaF$_2$(111) surface by air exposure, Surf. Sci. **439**, 181–190 (1999)

24.67   A. Klust, T. Ohta, A.A. Bostwick, Q. Yu, F.S. Ohuchi, M.A. Olmstead: Atomically resolved imaging of a CaF bilayer on Si(111): Subsurface atoms and the image contrast in scanning force microscopy, Phys. Rev. B **69**, 035405/1–5 (2004)

24.68   C. Barth, A.S. Foster, M. Reichling, A.L. Shluger: Contrast formation in atomic resolution scanning force microscopy of CaF$_2$(111): experiment and theory, J. Phys. Condens. Matter **13**, 2061–2079 (2001)

24.69   A.S. Foster, C. Barth, A.L. Shulger, M. Reichling: Unambiguous interpretation of atomically resolved force microscopy images of an insulator, Phys. Rev. Lett. **86**, 2373–2376 (2001)

24.70   A.S. Foster, A.L. Rohl, A.L. Shluger: Imaging problems on insulators: What can be learnt from NC-AFM modeling on CaF$_2$?, Appl. Phys. A **72**, S31–S34 (2001)

24.71   F.J. Giessibl, M. Reichling: Investigating atomic details of the CaF$_2$(111) surface with a qPlus sensor, Nanotechnology **16**, S118–S124 (2005)

24.72   A.S. Foster, C. Barth, A.L. Shluger, R.M. Nieminen, M. Reichling: Role of tip structure and surface relaxation in atomic resolution dynamic force microscopy: CaF$_2$(111) as a reference surface, Phys. Rev. B **66**, 235417/1–10 (2002)

24.73   C. Barth, M. Reichling: Imaging the atomic arrangements on the high-temperature reconstructed α-Al$_2$O$_3$ surface, Nature **414**, 54–57 (2001)

24.74   M. Reichling, C. Barth: Atomically resolution imaging on Fluorides. In: *Noncontact Atomic Force Microscopy*, ed. by S. Morita, R. Wiesendanger, E. Meyer (Springer, Berlin, Heidelberg 2002) pp. 109–123

24.75 K. Fukui, H. Ohnishi, Y. Iwasawa: Atom-resolved image of the TiO$_2$(110) surface by noncontact atomic force microscopy, Phys. Rev. Lett. **79**, 4202–4205 (1997)

24.76 H. Raza, C. L. Pang, S. A. Haycock, G. Thornton: Non-contact atomic force microscopy imaging of TiO$_2$(100) surfaces, Appl. Surf. Sci. **140**, 271–275 (1999)

24.77 C. L. Pang, H. Raza, S. A. Haycock, G. Thornton: Imaging reconstructed TiO$_2$(100) surfaces with non-contact atomic force microscopy, Appl. Surf. Sci. **157**, 223–238 (2000)

24.78 M. Ashino, T. Uchihashi, K. Yokoyama, Y. Sugawara, S. Morita, M. Ishikawa: STM and atomic-resolution noncontact AFM of an oxygen-deficient TiO$_2$(110) surface, Phys. Rev. B **61**, 13955–13959 (2000)

24.79 R. E. Tanner, A. Sasahara, Y. Liang, E. I. Altmann, H. Onishi: Formic acid adsorption on anatase TiO$_2$(001)-(1×4) thin films studied by NC-AFM and STM, J. Phys. Chem. B **106**, 8211–8222 (2002)

24.80 A. Sasahara, T. C. Droubay, S. A. Chambers, H. Uetsuka, H. Onishi: Topography of anatase TiO$_2$ film synthesized on LaAlO$_3$(001), Nanotechnology **16**, S18–S21 (2005)

24.81 C. L. Pang, S. A. Haycock, H. Raza, P. J. Møller, G. Thornton: Structures of the 4×1 and 1×2 reconstructions of SnO$_2$(110), Phys. Rev. B **62**, R7775–R7778 (2000)

24.82 H. Hosoi, K. Sueoka, K. Hayakawa, K. Mukasa: Atomic resolved imaging of cleaved NiO(100) surfaces by NC-AFM, Appl. Surf. Sci. **157**, 218–221 (2000)

24.83 W. Allers, S. Langkat, R. Wiesendanger: Dynamic low-temperature scanning force microscopy on nickel oxide (001), Appl. Phys. A **72**, S27–S30 (2001)

24.84 T. Kubo, H. Nozoye: Surface Structure of SrTiO3(100)-($\sqrt{5}\times\sqrt{5}$) − R26.6°, Phys. Rev. Lett. **86**, 1801–1804 (2001)

24.85 K. Fukui, Y. Namai, Y. Iwasawa: Imaging of surface oxygen atoms and their defect structures on CeO$_2$(111) by noncontact atomic force microscopy, Appl. Surf. Sci. **188**, 252–256 (2002)

24.86 S. Suzuki, Y. Ohminami, T. Tsutsumi, M. M. Shoaib, M. Ichikawa, K. Asakura: The first observation of an atomic scale noncontact AFM image of MoO$_3$(010), Chem. Lett. **32**, 1098–1099 (2003)

24.87 C. Barth, C. R. Henry: Atomic resolution imaging of the (001) surface of UHV cleaved MgO by dynamic scanning force microscopy, Phys. Rev. Lett. **91**, 196102/1–4 (2003)

24.88 A. S. Foster, A. Y. Gal, J. M. Airaksinen, O. H. Pakarinen, Y. J. Lee, J. D. Gale, A. L. Shluger, R. M. Nieminen: Towards chemical identification in atomic-resolution noncontact AFM imaging with silicon tips, Phys. Rev. B **68**, 195420/1–8 (2003)

24.89 A. S. Foster, A. Y. Gal, J. D. Gale, Y. J. Lee, R. M. Nieminen, A. L. Shluger: Interaction of silicon dangling bonds with insulating surfaces, Phys. Rev. Lett. **92**, 036101/1–4 (2004)

24.90 T. Eguchi, Y. Hasegawa: High resolution atomic force microscopic imaging of the Si(111)-(7×7) surface: Contribution of short-range force to the images, Phys. Rev. Lett. **89**, 266105/1–4 (2002)

24.91 T. Arai, M. Tomitori: A Si nanopillar grown on a Si tip by atomic force microscopy in ultrahigh vacuum for a high-quality scanning probe, Appl. Phys. Lett. **86**, 073110/1–3 (2005)

24.92 K. Mukasa, H. Hasegawa, Y. Tazuke, K. Sueoka, M. Sasaki, K. Hayakawa: Exchange interaction between magnetic moments of ferromagnetic sample and tip: Possibility of atomic-resolution images of exchange interactions using exchange force microscopy, Jpn. J. Appl. Phys. **33**, 2692–2695 (1994)

24.93 H. Ness, F. Gautier: Theoretical study of the interaction between a magnetic nanotip and a magnetic surface, Phys. Rev. B **52**, 7352–7362 (1995)

24.94 K. Nakamura, H. Hasegawa, T. Oguchi, K. Sueoka, K. Hayakawa, K. Mukasa: First-principles calculation of the exchange interaction and the exchange force between magnetic Fe films, Phys. Rev. B **56**, 3218–3221 (1997)

24.95 A. S. Foster, A. L. Shluger: Spin-contrast in non-contact SFM on oxide surfaces: Theoretical modeling of NiO(001) surface, Surf. Sci. **490**, 211–219 (2001)

24.96 T. Oguchi, H. Momida: Electronic structure and magnetism of antiferromagnetic oxide surface– First-principles calculations, J. Surf. Sci. Soc. Jpn. **26**, 138–143 (2005)

24.97 H. Hosoi, M. Kimura, K. Sueoka, K. Hayakawa, K. Mukasa: Non-contact atomic force microscopy of an antiferromagnetic NiO(100) surface using a ferromagnetic tip, Appl. Phys. A **72**, S23–S26 (2001)

24.98 H. Hölscher, S. M. Langkat, A. Schwarz, R. Wiesendanger: Measurement of three-dimensional force fields with atomic resolution using dynamic force spectroscopy, Appl. Phys. Lett. **81**, 4428–4430 (2002)

24.99 S. M. Langkat, H. Hölscher, A. Schwarz, R. Wiesendanger: Determination of site specific interaction forces between an iron coated tip and the NiO(001) surface by force field spectroscopy, Surf. Sci. **527**, 12–20 (2003)

24.100 R. Hoffmann, M. A. Lantz, H. J. Hug, P. J. A. van Schendel, P. Kappenberger, S. Martin, A. Baratoff, H.-J. Güntherodt: Atomic resolution imaging and frequency versus distance measurement on NiO(001) using low-temperature scanning force microscopy, Phys. Rev. B **67**, 085402/1–6 (2003)

24.101 H. Hosoi, K. Sueoka, K. Hayakawa, K. Mukasa: Atomically resolved imaging of a NiO(001) surface. In: *Noncontact Atomic Force Microscopy*, ed. by

S. Morita, R. Wiesendanger, E. Meyer (Springer, Berlin, Heidelberg 2002) pp. 125–134

24.102 K. Sueoka, A. Subagyo, H. Hosoi, K. Mukasa: Magnetic imaging with scanning force microscopy, Nanotechnology **15**, S691–S698 (2004)

24.103 H. Hosoi, K. Sueoka, K. Mukasa: Investigations on the topographic asymmetry of non-contact atomic force microscopy images of NiO(001) surface observed with a ferromagnetic tip, Nanotechnology **15**, 505–509 (2004)

24.104 H. Momida, T. Oguchi: First-principles studies of antiferromagnetic MnO and NiO surfaces, J. Phys. Soc. Jpn. **72**, 588–593 (2003)

24.105 K. Kobayashi, H. Yamada, T. Horiuchi, K. Matsushige: Structures and electrical properties of fullerene thin films on Si(111)-7×7 surface investigated by noncontact atomic force microscopy, Jpn. J. Appl. Phys. **39**, 3821–3829 (2000)

24.106 T. Uchihashi, M. Tanigawa, M. Ashino, Y. Sugawara, K. Yokoyama, S. Morita, M. Ishikawa: Identification of B-form DNA in an ultrahigh vacuum by noncontact-mode atomic force microscopy, Langmuir **16**, 1349–1353 (2000)

24.107 R. M. Overney, E. Meyer, J. Frommer, D. Brodbeck, R. Lüthi, L. Howald, H.-J. Güntherodt, M. Fujihira, H. Takano, Y. Gotoh: Friction measurements on phase-separated thin films with amodified atomic force microscope, Nature **359**, 133–135 (1992)

24.108 D. Frisbie, L. F. Rozsnyai, A. Noy, M. S. Wrighton, C. M. Lieber: Functional group imaging by chemical force microscopy, Science **265**, 2071–2074 (1994)

24.109 E. Meyer, L. Howald, R. M. Overney, H. Heinzelmann, J. Frommer, H.-J. Guntherodt, T. Wagner, H. Schier, S. Roth: Molecular-resolution images of Langmuir–Blodgett films using atomic force microscopy, Nature **349**, 398–400 (1992)

24.110 K. Fukui, H. Onishi, Y. Iwasawa: Imaging of individual formate ions adsorbed on TiO$_2$(110) surface by non-contact atomic force microscopy, Chem. Phys. Lett. **280**, 296–301 (1997)

24.111 K. Kobayashi, H. Yamada, T. Horiuchi, K. Matsushige: Investigations of C$_{60}$ molecules deposited on Si(111) by noncontact atomic force microscopy, Appl. Surf. Sci. **140**, 281–286 (1999)

24.112 Y. Maeda, T. Matsumoto, T. Kawai: Observation of single- and double-strand DNA using non-contact atomic force microscopy, Appl. Surf. Sci. **140**, 400–405 (1999)

24.113 T. Uchihashi, T. Ishida, M. Komiyama, M. Ashino, Y. Sugawara, W. Mizutani, K. Yokoyama, S. Morita, H. Tokumoto, M. Ishikawa: High-resolution imaging of organic monolayers using noncontact AFM, Appl. Surf. Sci **157**, 244–250 (2000)

24.114 T. Fukuma, K. Kobayashi, T. Horiuchi, H. Yamada, K. Matsushige: Alkanethiol self-assembled monolayers on Au(111) surfaces investigated by non-contact AFM, Appl. Phys. A **72**, S109–S112 (2001)

24.115 B. Gotsmann, C. Schmidt, C. Seidel, H. Fuchs: Molecular resolution of an organic monolayer by dynamic AFM, Euro. Phys. J. B **4**, 267–268 (1998)

24.116 Ch. Loppacher, M. Bammerlin, M. Guggisberg, E. Meyer, H.-J. Güntherodt, R. Lüthi, R. Schlittler, J. K. Gimzewski: Forces with submolecular resolution between the probing tip and Cu-TBPP molecules on Cu(100) observed with a combined AFM/STM, Appl. Phys. A **72**, S105–S108 (2001)

24.117 L. M. Eng, M. Bammerlin, Ch. Loppacher, M. Guggisberg, R. Bennewitz, R. Lüthi, E. Meyer, H.-J. Güntherodt: Surface morphology, chemical contrast, and ferroelectric domains in TGS bulk single crystals differentiated with UHV non-contact force microscopy, Appl. Surf. Sci. **140**, 253–258 (1999)

24.118 S. Kitamura, K. Suzuki, M. Iwatsuki: High resolution imaging of contact potential difference using a novel ultrahigh vacuum non-contact atomic force microscope technique, Appl. Surf. Sci. **140**, 265–270 (1999)

24.119 H. Yamada, T. Fukuma, K. Umeda, K. Kobayashi, K. Matsushige: Local structures and electrical properties of organic molecular films investigated by non-contact atomic force microscopy, Appl. Surf. Sci **188**, 391–398 (2000)

24.120 K. Fukui, Y. Iwasawa: Fluctuation of acetate ions in the (2×1)-acetate overlayer on TiO$_2$(110)-(1×1) observed by noncontact atomic force microscopy, Surf. Sci. **464**, L719–L726 (2000)

24.121 A. Sasahara, H. Uetsuka, H. Onishi: Singlemolecule analysis by non-contact atomic force microscopy, J. Phys. Chem. B **105**, 1–4 (2001)

24.122 A. Sasahara, H. Uetsuka, H. Onishi: NC-AFM topography of HCOO and CH$_3$COO molecules co-adsorbed on TiO$_2$(110), Appl. Phys. A **72**, S101–S103 (2001)

24.123 A. Sasahara, H. Uetsuka, H. Onishi: Image topography of alkyl-substituted carboxylates observed by noncontact atomic force microscopy, Surf. Sci. **481**, L437–L442 (2001)

24.124 A. Sasahara, H. Uetsuka, H. Onishi: Noncontact atomic force microscope topography dependent on permanent dipole of individual molecules, Phys. Rev. B **64**, 121406(R) (2001)

24.125 A. Sasahara, H. Uetsuka, T. Ishibashi, H. Onishi: A needle-like organic molecule imaged by noncontact atomic force microscopy, Appl. Surf. Sci. **188**, 265–271 (2002)

24.126 H. Onishi, A. Sasahara, H. Uetsuka, T. Ishibashi: Molecule-dependent topography determined by noncontact atomic force microscopy: Carboxylates on TiO$_2$(110), Appl. Surf. Sci. **188**, 257–264 (2002)

24.127 S. Thevuthasan, G. S. Herman, Y. J. Kim, S. A. Chambers, C. H. F. Peden, Z. Wang, R. X. Ynzunza, E. D. Tober, J. Morais, C. S. Fadley: The structure of formate on TiO$_2$(110) by scanned-energy and scanned-angle photoelectron diffraction, Surf. Sci. **401**, 261–268 (1998)

24.128 H. Onishi: Carboxylates adsorbed on TiO_2(110). In: *Chemistry of Nano-molecular Systems*, ed. by T. Nakamura (Springer, Berlin, Heidelberg 2002) pp. 75–89

24.129 H. Uetsuka, A. Sasahara, A. Yamakata, H. Onishi: Microscopic identification of a bimolecular reaction

intermediate, J. Phys. Chem. B **106**, 11549–11552 (2002)

24.130 D. R. Lide: *Handbook of Chemistry and Physics*, 81st edn. (CRC, Boca Raton 2000)

24.131 K. Kobayashi, H. Yamada, K. Matsushige: Dynamic force microscopy using FM detection in various environments, Appl. Surf. Sci. **188**, 430–434 (2002)

25. Low-Temperature Scanning Probe Microscopy

This chapter is dedicated to scanning probe microscopy (SPM) operated at cryogenic temperatures, where the more fundamental aspects of phenomena important in the field of nanotechnology can be investigated with high sensitivity under well-defined conditions. In general, scanning probe techniques allow the measurement of physical properties down to the nanometer scale. Some techniques, such as the scanning tunneling microscope and the scanning force microscope even go down to the atomic scale. Various properties are accessible. Most importantly, one can image the arrangement of atoms on conducting surfaces by scanning tunneling microscopy and on insulating substrates by scanning force microscopy. But the arrangement of electrons (scanning tunneling spectroscopy), the force interaction between different atoms (scanning force spectroscopy), magnetic domains (magnetic force microscopy), the local capacitance (scanning capacitance microscopy), the local temperature (scanning thermo microscopy), and local light-induced excitations (scanning near-field microscopy) can also be measured with high spatial resolution. In addition, some techniques even allow the manipulation of atomic configurations.

Probably the most important advantage of the low-temperature operation of scanning probe techniques is that they lead to a significantly better signal-to-noise ratio than measuring at room temperature. This is why many researchers work below 100 K. However, there are also physical reasons to use low-temperature equipment. For example, the manipulation of atoms or scanning tunneling spectroscopy with high energy resolution can only be realized at low temperatures. Moreover, some physical effects such as superconductivity or the Kondo effect are restricted to low temperatures. Here, we describe the design criteria of low-temperature scanning probe equipment and summarize some of the most spectacular results achieved since the invention of the method about 20 years ago. We first focus on the scanning tunneling microscope, giving examples of atomic manipulation and the analysis of electronic

Part C | 25

properties in different material arrangements. Afterwards, we describe results obtained by scanning force microscopy, showing atomic-scale imaging on insulators, as well as force spectroscopy analysis. Finally, the magnetic force microscope, which images domain patterns in ferromagnets and vortex patterns in superconductors, is discussed. Although this list is far from complete, we feel that it gives an adequate impression of the fascinating possibilities of low-temperature scanning probe instruments.

In this chapter low temperatures are defined as lower than about 100 K and are normally achieved by cooling with liquid nitrogen or liquid helium. Applications in which SPMs are operated close to 0 °C are not covered in this chapter.

More than two decades ago, the first design of an experimental setup was presented where a sharp tip was systematically scanned over a sample surface in order to obtain local information on the tip-sample interaction down to the atomic scale. This original instrument used the tunneling current between a conducting tip and a conducting sample as a feedback signal and was named *scanning tunneling microscope* accordingly [25.1]. Soon after this historic breakthrough, it became widely recognized that virtually any type of tip-sample interaction can be used to obtain local information on the sample by applying the same general principle, provided that the selected interaction was reasonably short-ranged. Thus, a whole variety of new methods has been introduced, which are denoted collectively as *scanning probe methods*. An overview is given by *Wiesendanger* [25.2].

The various methods, especially the above mentioned scanning tunneling microscopy (STM) and scanning force microscopy (SFM) – which is often further classified into subdisciplines such as the topography-reflecting atomic force microscopy (AFM), the magnetic force microscopy (MFM), or the electrostatic force microscopy (EFM) – have been established as standard methods for surface characterization on the nanometer scale. The reason is that they feature extremely high resolution (often down to the atomic scale for STM and AFM), despite a principally simple, compact, and comparatively inexpensive design.

A side effect of the simple working principle and the compact design of many scanning probe microscopes (SPMs) is that they can be adapted to different environments such as air, all kinds of gaseous atmospheres, liquids, or vacuum with reasonable effort. Another advantage is their ability to work within a wide temperature range. A microscope operation at higher temperatures is chosen to study surface diffusion, surface reactivity, surface reconstructions that only manifest at elevated temperatures, high-temperature phase transitions, or to simulate conditions as they occur, e.g., in engines, catalytic converters, or reactors. Ultimately, the upper limit for the operation of an SPM is determined by the stability of the sample, but thermal drift, which limits the ability to move the tip in a controlled manner over the sample, as well as the depolarization temperature of the piezoelectric positioning elements might further restrict successful measurements.

On the other hand, low-temperature (LT) application of SPMs is much more widespread than operation at high temperatures. Essentially five reasons make researchers adapt their experimental setups to low-temperature compatibility. These are: (1) the reduced thermal drift, (2) lower noise levels, (3) enhanced stability of tip and sample, (4) the reduction in piezo hysteresis/creep, and (5) probably the most obvious, the fact that many physical effects are restricted to low temperature. Reasons (1) to (4) only apply unconditionally if the whole microscope body is kept at low temperature (typically in or attached to a bath cryostat, see Sect. 25.2). Setups in which only the sample is cooled may show considerably less favorable operating characteristics. As a result of (1) to (4), ultrahigh resolution and long-term stability can be achieved on a level that significantly exceeds what can be accomplished at room temperature even under the most favorable circumstances. Typical examples for (5) are superconductivity [25.3] and the Kondo effect [25.4].

25.1 Microscope Operation at Low Temperatures

Nevertheless, before we devote ourselves to a short overview of experimental LT-SPM work, we will take a closer look at the specifics of microscope operation at low temperatures, including a discussion of the corresponding instrumentation.

25.1.1 Drift

Thermal drift originates from thermally activated movements of the individual atoms, which are reflected by the thermal expansion coefficient. At room temperature, typical values for solids are on the order of $(1–50) \times 10^{-6} \, \text{K}^{-1}$. If the temperature could be kept precisely constant, any thermal drift would vanish, regardless of the absolute temperature of the system. The close coupling of the microscope to a large temperature bath that maintains a constant temperature ensures a significant reduction in thermal drift and allows for distortion-free long-term measurements. Microscopes that are efficiently attached to sufficiently large bath cryostats, therefore, show a one-to two-order-of-magnitude increase in thermal stability compared with non-stabilized set-ups operated at room temperature.

A second effect also helps suppress thermally induced drift of the probing tip relative to a specific location on the sample surface. The thermal expansion coefficients at liquid-helium temperatures are two or more orders of magnitude smaller than at room temperature. Consequently, the thermal drift during low-temperature operation decreases accordingly.

For some specific scanning probe methods, there may be additional ways in which a change in temperature can affect the quality of the data. In *frequency-modulation SFM* (FM-SFM), for example, the measurement principle relies on the accurate determination of the eigenfrequency of the cantilever, which is determined by its spring constant and its effective mass. However, the spring constant changes with temperature due to both thermal expansion (i. e., the resulting change in the cantilever dimensions) and the variation of the Young's modulus with temperature. Assuming drift rates of about $2\,mK/min$, as is typical for room-temperature measurements, this effect might have a significant influence on the obtained data.

25.1.2 Noise

The theoretically achievable resolution in SPM often increases with decreasing temperature due to a decrease in thermally induced noise. An example is the thermal noise in SFM, which is proportional to the square root of the temperature [25.5, 6]. Lowering the temperature from $T = 300\,K$ to $T = 10\,K$ thus results in a reduction of the thermal frequency noise by more than a factor of five. Graphite, e.g., has been imaged with atomic resolution only at low temperatures due to its extremely low corrugation, which was below the room-temperature noise level [25.7, 8].

Another, even more striking, example is the spectroscopic resolution in *scanning tunneling spectroscopy* (STS). This depends linearly on the temperature [25.2] and is consequently reduced even more at LT than the thermal noise in AFM. This provides the opportunity to study structures or physical effects not accessible at room temperature such as spin and Landau levels in semiconductors [25.9].

Finally, it might be worth mentioning that the enhanced stiffness of most materials at low temperatures (increased Young's modulus) leads to a reduced coupling to external noise. Even though this effect is considered small [25.6], it should not be ignored.

25.1.3 Stability

There are two major stability issues that considerably improve at low temperature. First, low temperatures close to the temperature of liquid helium inhibit most of the thermally activated diffusion processes. As a consequence, the sample surfaces show a significantly increased long-term stability, since defect motion or adatom diffusion is massively suppressed. Most strikingly, even single xenon atoms deposited on suitable substrates can be successfully imaged [25.10, 11], or even manipulated [25.12]. In the same way, the low temperatures also stabilize the atomic configuration at the tip end by preventing sudden jumps of the most loosely bound, foremost tip atom(s). Secondly, the large cryostat that usually surrounds the microscope acts as an effective cryo-pump. Thus samples can be kept clean for several weeks, which is a multiple of the corresponding time at room temperature (about $3–4\,h$).

25.1.4 Piezo Relaxation and Hysteresis

The last important benefit from low-temperature operation of SPMs is that artifacts from the response of the piezoelectric scanners are substantially reduced. After applying a voltage ramp to one electrode of a piezoelectric scanner, its immediate initial deflection, l_0, is followed by a much slower relaxation, Δl, with a logarithmic time dependence. This effect, known as piezo relaxation or *creep*, diminishes substantially at low temperatures, typically by a factor of ten or more. As a consequence, piezo nonlinearities and piezo hysteresis decrease accordingly. Additional information is given by *Hug* et al.[25.13].

25.2 Instrumentation

The two main design criteria for all vacuum-based scanning probe microscope systems are: (1) to provide an efficient decoupling of the microscope from the vacuum system and other sources of external vibrations, and (2) to avoid most internal noise sources through the high mechanical rigidity of the microscope body itself. In vacuum systems designed for low-temperature applications, a significant degree of complexity is added, since, on the one hand, close thermal contact of the SPM and cryogen is necessary to ensure the (approximately) drift-free conditions described above, while, on the other hand, good vibration isolation (both from the outside world, as well as from the boiling or flowing cryogen) has to be maintained.

Plenty of microscope designs have been presented in the last 10–15 years, predominantly in the field of STM. Due to the variety of the different approaches, we will, somewhat arbitrarily, give two examples at different levels of complexity that might serve as illustrative model designs.

25.2.1 A Simple Design for a Variable-Temperature STM

A simple design for a variable-temperature STM system is presented in Fig. 25.1 (similar systems are also offered by Omicron (Germany) or Jeol (Japan)). It should

give an impression of what the minimum requirements are, if samples are to be investigated successfully at low temperatures. It features a single ultrahigh vacuum (UHV) chamber that houses the microscope in its center. The general idea to keep the set-up simple is that only the sample is cooled, by means of a flow cryostat that ends in the small liquid-nitrogen (LN) reservoir. This reservoir is connected to the sample holder with copper braids. The role of the copper braids is to attach the LN reservoir thermally to the sample located on the sample holder in an effective manner, while vibrations due to the flow of the cryogen should be blocked as much as possible. In this way, a sample temperature of about 100 K is reached. Alternatively, with liquid-helium operation, a base temperature of below 30 K can be achieved, while a heater that is integrated into the sample stage enables high-temperature operation up to 1000 K.

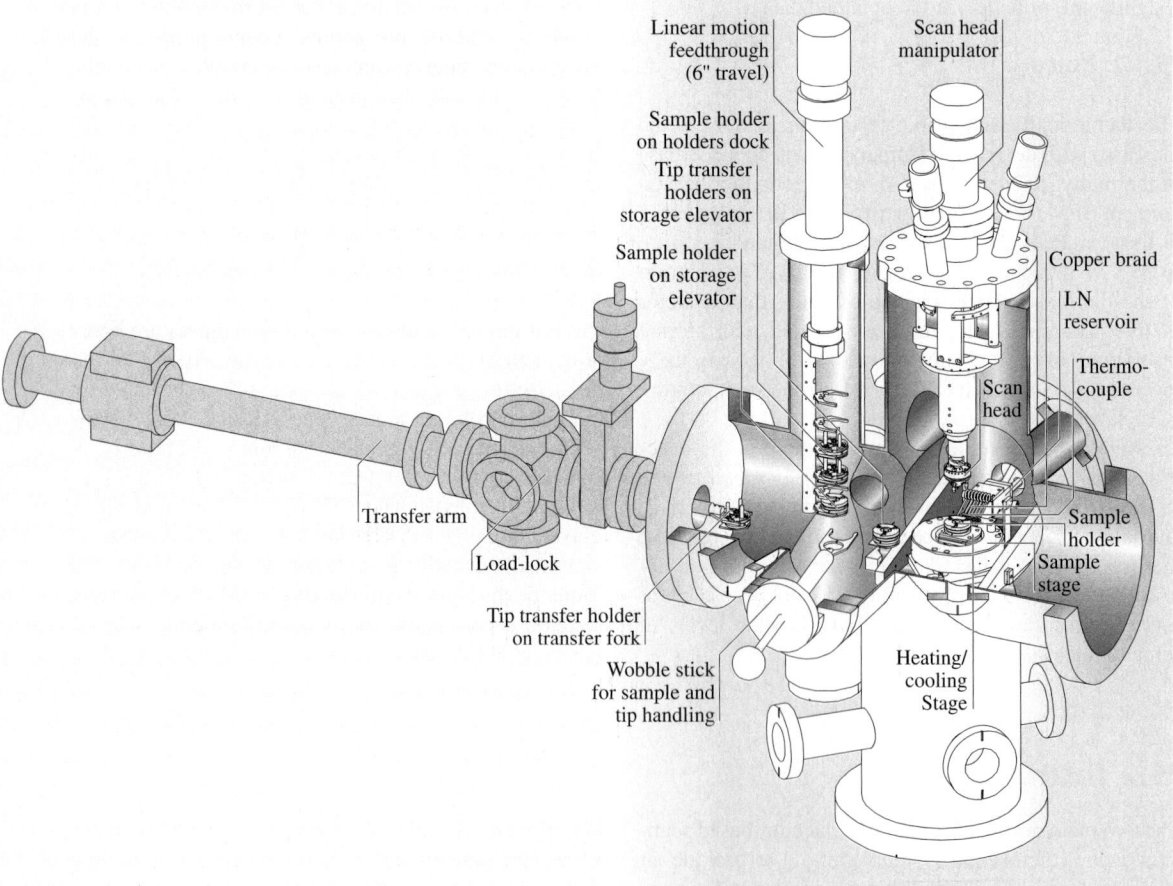

Fig. 25.1 One-chamber UHV system with variable-temperature STM based on a flow cryostat design. (Courtesy of RHK Technology, USA)

A typical experiment would run as follows. First, the sample is brought into the system by placing it in the so-called *load-lock*. This small part of the chamber can be separated from the rest of the system by a valve, so that the main part of the system can remain under vacuum at all times (i. e., even if the load-lock is opened to introduce the sample). After vacuum is reestablished, the sample is transferred to the main chamber using the transfer arm. A linear-motion feedthrough enables the storage of sample holders or, alternatively, specialized holders that carry replacement tips for the STM. Extending the transfer arm further, the sample can be placed on the sample stage and subsequently cooled down to the desired temperature. The scan head, which carries the STM tip, is then lowered with the scan-head manipulator onto the sample holder (see Fig. 25.2). The special design of the scan head (see [25.14] for details) allows not only a flexible positioning of the tip on any desired location on the sample surface, but also compensates to a certain degree for the thermal drift that inevitably occurs in such a design due to temperature gradients.

In fact, thermal drift is often much more prominent in LT-SPM designs, where only the sample is cooled, than in room-temperature designs. Therefore, to benefit fully from the high stability conditions described in the introduction, it is mandatory to keep the whole microscope at the exact same temperature. This is mostly

realized by using bath cryostats, which add a certain degree of complexity.

25.2.2 A Low Temperature SFM Based on a Bath Cryostat

As an example of an LT-SPM set-up based on a bath cryostat, let us take a closer look at the LT-SFM system sketched in Fig. 25.3, which has been used to acquire the images on graphite, xenon, NiO, and InAs presented in Sect. 25.4. The force microscope is built into a UHV system that comprises three vacuum chambers: one for cantilever and sample preparation, which also serves as a transfer chamber, one for analysis purposes, and a main chamber that houses the microscope. A specially designed vertical trans-

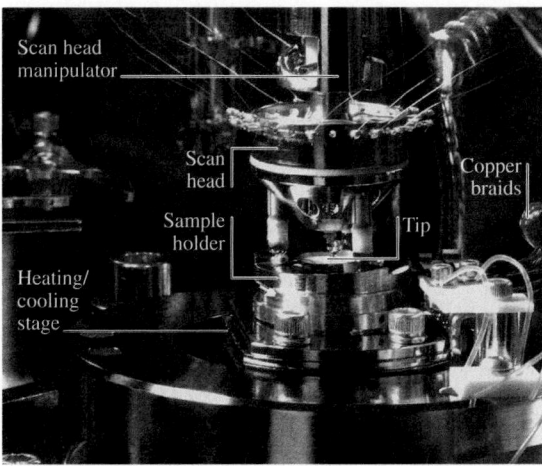

Fig. 25.2 Photograph of the STM located inside the system sketched in Fig. 25.1. After the scan head has been lowered onto the sample holder, it is fully decoupled from the scan head manipulator and can be moved laterally using the three piezo legs on which it stands. (Courtesy of RHK Technology, USA)

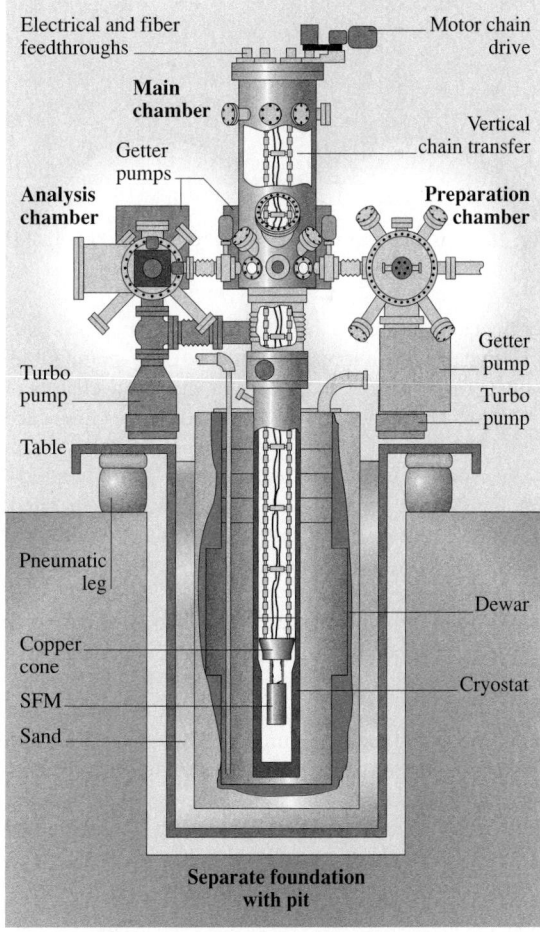

Fig. 25.3 Three-chamber UHV and bath cryostat system for scanning force microscopy, front view

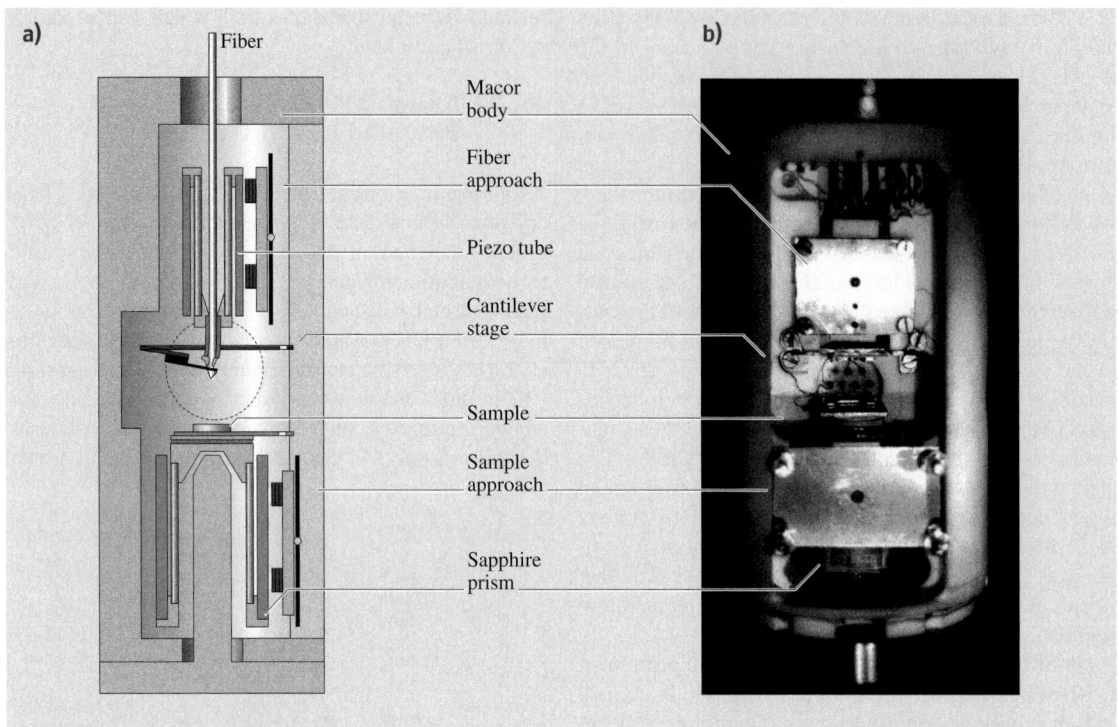

Fig. 25.4a,b The scanning force microscope incorporated into the system presented in Fig. 25.3. (**a**) Section along plane of symmetry. (**b**) Photo from the front

fer mechanism based on a double chain allows the lowering of the microscope into a UHV-compatible bath cryostat attached underneath the main chamber. To damp the system, it is mounted on a table carried by pneumatic damping legs, which, in turn, stand on a separate foundation to decouple it from building vibrations. The cryostat and dewar are separated from the rest of the UHV system by a bellow. In addition, the dewar is surrounded by sand for acoustic isolation.

In this design, tip and sample are exchanged at room temperature in the main chamber. After the transfer into the cryostat, the SFM can be cooled by either liquid nitrogen or liquid helium, reaching temperatures down to 10 K. An all-fiber interferometer as the detection mechanism for the cantilever deflection ensures high resolution, while simultaneously allowing the construction of a comparatively small, rigid, and symmetric microscope.

Figure 25.4 highlights the layout of the SFM body itself. Along with the careful choice of materials, the symmetric design eliminates most of the problems with drift inside the microscope encountered when cooling or warming it up. The microscope body has an overall cylindrical shape with a height of 13 cm and a diameter of 6 cm and exact mirror symmetry along the cantilever axis. The main body is made of a single block of macor, a machinable glass ceramic, which ensures a rigid and stable design. For most of the metallic parts titanium was used, which has a temperature coefficient similar to macor. The controlled but stable accomplishment of movements, such as coarse approach and lateral positioning in other microscope designs, is a difficult task at low temperatures. The present design uses a special type of piezo motor that moves a sapphire prism (see the *fiber approach* and the *sample approach* labels in Fig. 25.3); it is described in detail in [25.15]. More information regarding this design is given in [25.16].

25.3 Scanning Tunneling Microscopy and Spectroscopy

In this section, we review some of the most important results achieved by LT-STM. After summarizing the results, placing emphasis on the necessity for LT equipment, we turn to the details of the different experiments and the physical meaning of the results obtained.

As described in Sect. 25.1, the LT equipment has basically three advantages for scanning tunneling microscopy (STM) and spectroscopy (STS). First, the instruments are much more stable with respect to thermal drift and coupling to external noise, allowing the establishment of new functionalities of the instrument. In particular, the LT-STM has been used to move atoms on a surface [25.12], cut molecules into pieces [25.17], reform bonds [25.18], and, consequently, establish new structures on the nanometer scale. Also, the detection of light resulting from tunneling into a particular atom [25.19] and the visualization of thermally induced atomic movements [25.20] partly require LT instrumentation.

Second, the spectroscopic resolution in STS depends linearly on temperature and is, therefore, considerably reduced at LT. This provides the opportunity to study physical effects unaccessible at room temperature. Obvious examples are the resolution of spin and Landau levels in semiconductors [25.9], or the investigation of lifetime broadening effects of particular electronic states on the nanometer scale [25.21]. More spectacularly, electronic wave functions have been imaged for the first time in real space using an LT-STM [25.22], and vibrational levels giving rise to additional inelastic tunneling have been detected [25.23] and localized within particular molecules [25.24].

Third, and most obviously, many physical effects, in particular, effects guided by electronic correlations, are restricted to low temperature. Typical examples are superconductivity [25.3], the Kondo effect [25.4], and many of the electron phases found in semiconductors [25.25]. Here, LT-STM provides the possibility to study electronic effects on a local scale, and intensive work has been done in this field, the most elaborate with respect to high-temperature superconductivity [25.26].

25.3.1 Atomic Manipulation

Although manipulation of surfaces on the atomic scale can be achieved at room temperature [25.27], only the use of LT-STM allows the placement of individual atoms at desired atomic positions [25.28].

The usual technique to manipulate atoms is to increase the current above a certain atom, which reduces the tip–atom distance, then to move the tip with the atom to a desired position, and finally to reduce the current again in order to decouple the atom and tip. The first demonstration of this technique was performed by *Eigler* and *Schweizer* [25.12], who used Xe atoms on a Ni(110) surface to write the three letters "IBM" (their employer) on the atomic scale (Fig. 25.5a). Nowadays, many laboratories are able to move different kinds of atoms and molecules on different surfaces with high precision. An example featuring CO molecules on Cu(110) is shown in Fig. 25.5b–g. Basic modes of controlled motion, pushing, pulling, and sliding of the molecules, have been established that depend on the tunneling current, i. e., the distance and the particular molecule–substrate combination [25.29]. It is believed that the electric field between the tip and molecule is the strongest force moving the molecules, but other mechanisms such as electromigration caused by the high current density [25.28] or modifications of the surface potential due to the presence of the tip [25.30] have been put forth as important for some of the manipulation modes.

Meanwhile, other types of manipulation on the atomic scale have been developed. Some of them require inelastic tunneling into vibrational or rotational modes of the molecules or atoms. They lead to controlled desorption [25.31], diffusion [25.32], pick-up of molecules by the tip [25.18], or rotation of individual entities [25.33, 34]. Also, dissociation of molecules by voltage pulses [25.17], conformational changes induced by dramatic change of the tip–molecule distance [25.35], and association of pieces into larger molecules by reducing their lateral distance [25.18] have been shown. Figure 25.5h–m shows the production of biphenyl from two iodobenzene molecules [25.36]. The iodine is abstracted by voltage pulses (Fig. 25.5i,j), then the iodine is moved to the terrace by the pulling mode (Fig. 25.5k,l), and finally the two phenyl parts are slid along the step edge until they are close enough to react (Fig. 25.5m). The chemical identification of the components is not deduced straightforwardly from STM images and partly requires detailed calculations of their apparent shape.

Low temperatures are not always required in these experiments, but they increase reproducibility because of the higher stability of the instrument, as discussed in Sect. 25.1. Moreover, rotation or diffusion of entities could be excited at higher temperatures, making the intentionally produced configurations unstable.

Part C | 25.3

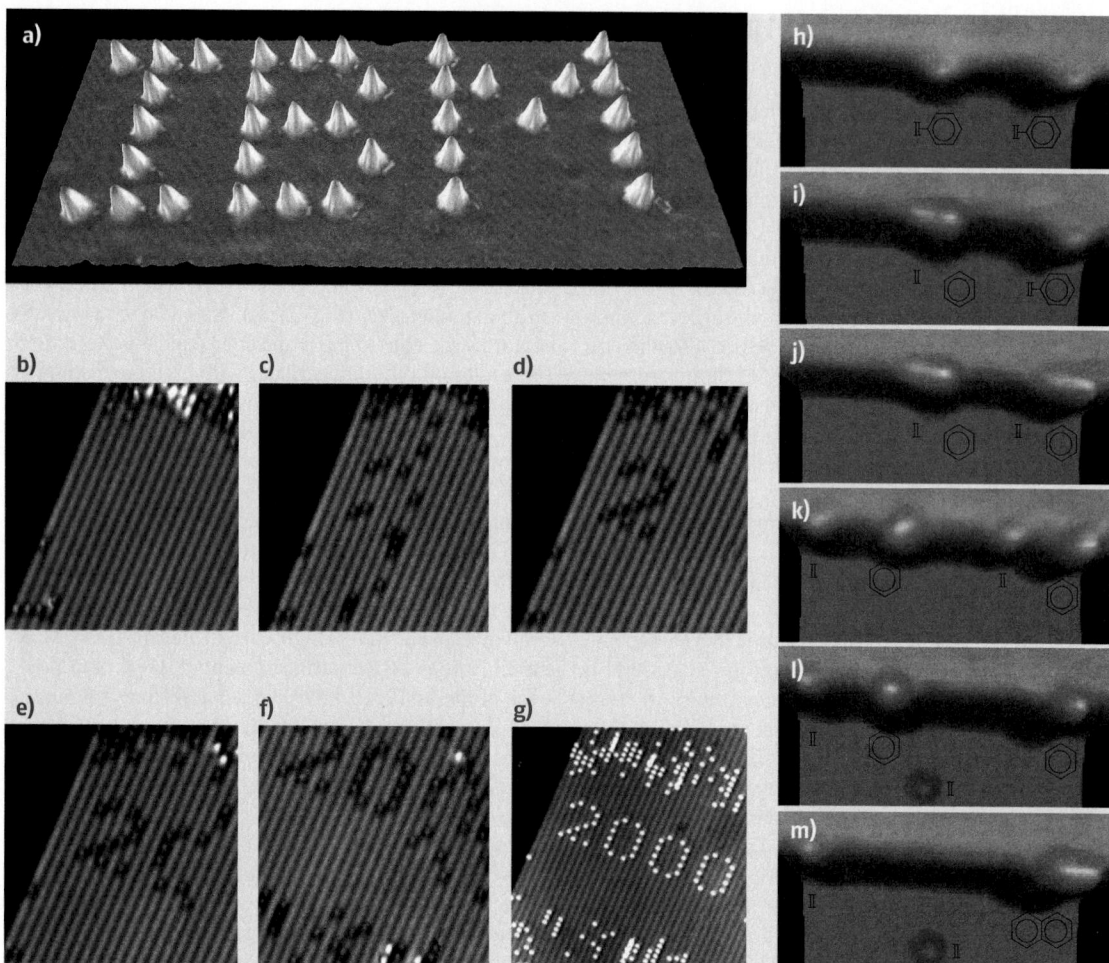

Fig. 25.5 (a) STM image of single Xe atoms positioned on a Ni(110) surface in order to realize the letters IBM on the atomic scale (courtesy of Eigler, IBM); (b)–(f) STM images recorded after different positioning processes of CO molecules on a Cu(110) surface; (g) final artwork greeting the new millennium on the atomic scale ((b)–(g) courtesy of Meyer, Berlin). (h)–(m) Synthesis of biphenyl from two iodobenzene molecules on Cu(111): First, iodine is abstracted from both molecules (i),(j), then the iodine between the two phenyl groups is removed from the step (k), and finally one of the phenyls is slid along the Cu-step (l) until it reacts with the other phenyl (m); the line drawings symbolize the actual status of the molecules ((h)–(m) courtesy of Hla and Rieder, Ohio)

25.3.2 Imaging Atomic Motion

Since individual manipulation processes last seconds to minutes, they probably cannot be used to manufacture large and repetitive structures. A possibility to construct such structures is self-assembled growth. This partly relies on the temperature dependence of different diffusion processes on the surface. A detailed knowledge of the diffusion parameters is required, which can be deduced from sequences of STM images measured at temperatures close to the onset of the process of interest [25.37]. Since many diffusion processes have their onset at LT, LT are partly required [25.20]. Consecutive images of so-called hexa-tert-butyl-decacyclene (HtBDC) molecules on Cu(110) recorded at $T = 194$ K are shown in Fig. 25.6a–c [25.38]. As indicated by the arrows, the position of the molecules changes with time, implying diffusion. Diffusion parameters are ob-

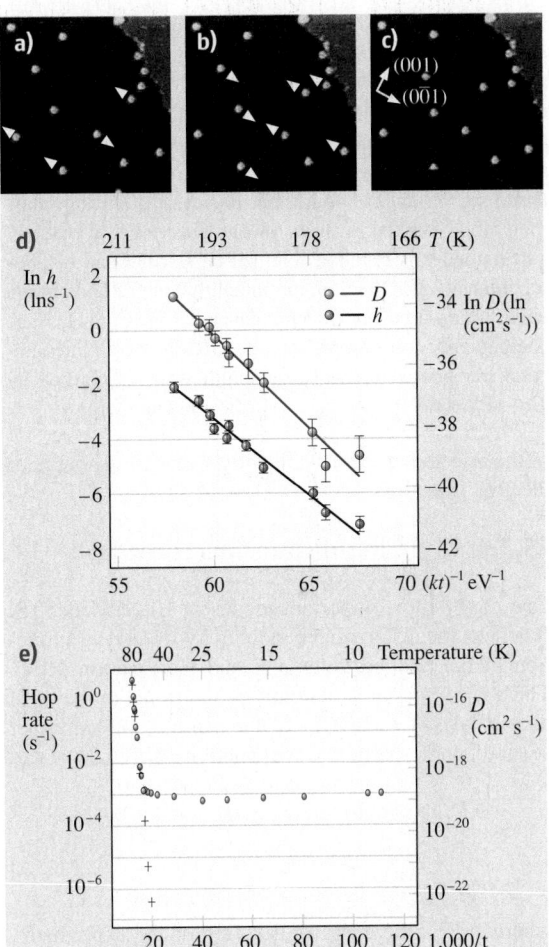

Fig. 25.6 (a)–(c) Consecutive STM images of hexa-tert-butyl decacyclene molecules on Cu(110) imaged at $T = 194$ K; *arrows* indicate the direction of motion of the molecules between two images. (d) Arrhenius plot of the hopping rate h determined from images like (a)–(c) as a function of inverse temperature (*grey symbols*); the *brown symbols* show the corresponding diffusion constant D; *lines* are fit results revealing an energy barrier of 570 meV for molecular diffusion ((a)–(d) courtesy of M. Schuhnack and F. Besenbacher, Aarhus). (e) Arrhenius plot for D (*crosses*) and H (*circles*) on Cu(001). The constant hopping rate of H below 65 K indicates a nonthermal diffusion process, probably tunneling (courtesy of Ho, Irvine)

tained from Arrhenius plots of the determined hopping rate h, as shown in Fig. 25.6d. Of course, one must make sure that the diffusion process is not influenced by the presence of the tip, since it is known

from manipulation experiments that the presence of the tip can move a molecule. However, particularly at low tunneling voltages, these conditions can be fulfilled.

Besides the determination of diffusion parameters, studies of the diffusion of individual molecules showed the importance of mutual interactions in diffusion, which can lead to concerted motion of several molecules [25.20], or, very interestingly, the influence of quantum tunneling [25.39]. The latter is deduced from the Arrhenius plot of hopping rates of H and D on Cu(001), as shown in Fig. 25.6e. The hopping rate of H levels off at about 65 K, while the hopping rate of the heavier D atom goes down to nearly zero, as expected from thermally induced hopping.

Other diffusion processes such as the movement of surface vacancies [25.40] or of bulk interstitials close to the surface [25.41], and the Brownian motion of vacancy islands [25.42] have also been displayed.

25.3.3 Detecting Light from Single Atoms and Molecules

It had already been realized in 1988 that STM experiments are accompanied by light emission [25.43]. The fact that molecular resolution in the light intensity was achieved at LT (Fig. 25.7a,b) [25.19] raised the hope of performing quasi-optical experiments on the molecular scale. Meanwhile, it is clear that the basic emission process observed on metals is the decay of a local plasmon induced in the area around the tip by inelastic tunneling processes [25.44, 45]. Thus, the molecular resolution is basically a change in the plasmon environment, largely given by the increased height of the tip with respect to the surface above the molecule [25.46]. However, the electron can, in principle, also decay via single-particle excitations. Indeed, signatures of single-particle levels are observed. Figure 25.7c shows light spectra measured at different tunneling voltages V above a nearly complete Na monolayer on Cu(111) [25.47]. The plasmon-mode peak energy (arrow) is found, as usual, to be proportional to V, but an additional peak that does not move with V appears at 1.6 eV (p). Plotting the light intensity as a function of photon energy and V (Fig. 25.7d) clearly shows that this additional peak is fixed in photon energy and corresponds to the separation of quantum-well levels of the Na ($E_n - E_m$).

Light has also been detected from semiconductors [25.48], including heterostructures [25.49]. There, the light is mostly caused by single-particle relaxation

Part C | 25.3

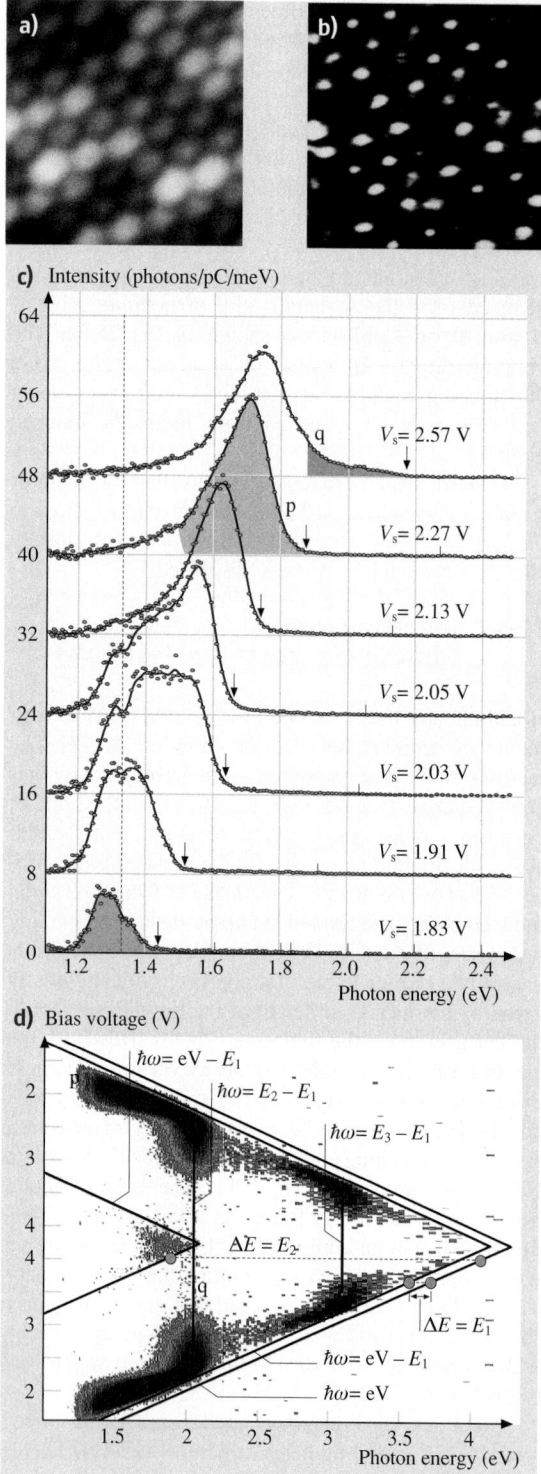

Fig. 25.7 (a) STM image of C_{60} molecules on Au(110) imaged at $T = 50$ K. **(b)** STM-induced photon intensity map of the same area; all photons from 1.5 eV to 2.8 eV contribute to the image, tunneling voltage $V = -2.8$ V ((**a**),(**b**) courtesy of Berndt, Kiel). **(c)** STM-induced photon spectrum measured on 0.6 monolayer of Na on Cu(111) at different tunneling voltages as indicated. Besides the shifting plasmon mode marked by an *arrow*, an energetically constant part named p is recognizable. **(d)** Greyscale map of photon intensity as a function of tunnelling voltage and photon energy measured on 2.0 monolayer Na on Cu(111). The energetically constant photons are identified with intersubband transitions of the Na quantum well, as marked by $E_n - E_m$ ((**c**),(**d**) courtesy of Hoffmann, Hamburg)

of the injected electrons, allowing a very local source of electron injection.

25.3.4 High–Resolution Spectroscopy

One of the most important modes of LT-STM is STS. It detects the differential conductivity dI/dV as a function of the applied voltage V and the position (x, y). The dI/dV signal is basically proportional to the local density of states (LDOS) of the sample, the sum over squared single-particle wave functions Ψ_i [25.2]

$$\frac{dI}{dV}(V, x, y) \propto \text{LDOS}(E, x, y)$$
$$= \sum_{\Delta E} |\Psi_i(E, x, y)|^2 , \quad (25.1)$$

where ΔE is the energy resolution of the experiment. In simple terms, each state corresponds to a tunneling channel if it is located between the Fermi levels (E_F) of the tip and the sample. Thus, all states located in this energy interval contribute to I, while $dI/dV(V)$ detects only the states at the energy E corresponding to V. The local intensity of each channel depends further on the LDOS of the state at the corresponding surface position and its decay length into vacuum. For s-like tip states, *Tersoff* and *Hamann* have shown that it is simply proportional to the LDOS at the position of the tip [25.50]. Therefore, as long as the decay length is spatially constant, one measures the LDOS at the surface (25.1). Note that the contributing states are not only surface states, but also bulk states. However, surface states usually dominate if present. *Chen* has shown that higher orbital tip states lead to the so-called derivation rule [25.51]: p_z-type tip states detect $d(\text{LDOS})/dz$, d_{z^2}-states detect $d^2(\text{LDOS})/dz^2$, and so on. As long as the decay into vacuum is exponential and spatially constant, this leads only to an additional factor in dI/dV. Thus, it is still the LDOS that is measured (25.1). The requirement of a spatially constant decay is usually fulfilled on larger length scales, but not on the atomic scale [25.51]. There, states located close to the atoms show a stronger decay into vacuum than the less localized states in the interstitial region. This effect can lead to corrugations that are larger than the real LDOS corrugations [25.52].

The voltage dependence of dI/dV is sensitive to a changing decay length with V, which increases with V. Additionally, $dI/dV(V)$-curves might be influenced by possible structures in the DOS of the tip, which also contributes to the number of tunneling channels [25.53]. However, these structures can usually be identified, and only tips free of characteristic DOS structures are used for quantitative experiments.

Importantly, the energy resolution ΔE is largely determined by temperature. It is defined as the smallest energy distance of two δ-peaks in the LDOS that can still be resolved as two individual peaks in $dI/dV(V)$-curves and is $\Delta E = 3.3\,kT$ [25.2]. The temperature dependence is nicely demonstrated in Fig. 25.8, where the tunneling gap of the superconductor Nb is measured at different

temperatures [25.54]. The peaks at the rim of the gap get wider at temperatures well below the critical temperature of the superconductor ($T_c = 9.2\,\text{K}$).

Lifetime Broadening

Besides ΔE, intrinsic properties of the sample lead to a broadening of spectroscopic features. Basically, the finite lifetime of the electron or hole in the corresponding state broadens its energetic width. Any kind of interaction such as electron–electron interaction can be responsible. Lifetime broadening has usually been measured by photoemission spectroscopy (PES), but it turned out that lifetimes of surface states on noble metal surfaces determined by STS (Fig. 25.9a,b) are up to a factor of three larger than those measured by PES [25.55]. The reason is probably that defects broaden the PES spectrum. Defects are unavoidable in a spatially integrating technique such as PES, thus, STS has the advantage of choosing a particularly clean area for lifetime measurements. The STS results can be successfully compared to theory, highlighting the dominating influence of intraband transitions for the surface-state lifetime on Au(111) and Cu(111), at least close to the onset of the surface band [25.21].

With respect to band electrons, the analysis of the width of the band onset in $dI/dV(V)$-curves has the disadvantage of being restricted to the onset energy. Another method circumvents this problem by measuring the decay of standing electron waves scattered from a step edge as a function of energy [25.56]. Figure 25.9c,d shows the resulting oscillating dI/dV-signal measured for two different energies. To deduce the coherence length L_Φ, which is inversely proportional to the lifetime τ_Φ, one has to consider that the finite ΔE in the experiment also leads to a decay of the standing wave away from the step edge. The dotted fit line using $L_\Phi = \infty$ indicates this effect and, more importantly, shows a discrepancy from the measured curve. Only including a finite coherence length of 6.2 nm results in good agreement, which, in turn, determines L_Φ and thus τ_Φ, as displayed in Fig. 25.9c. The found $1/E^2$-dependence of τ_Φ points to a dominating influence of electron–electron interactions at higher energies in the surface band.

Landau and Spin Levels

Moreover, the increased energy resolution at LT allows the resolution of electronic states that are not resolvable at room temperature (RT). For example, Landau and spin quantization appearing in a magnetic field B have been probed on InAs(110) [25.9,57]. The corresponding quantization energies are given by $E_{\text{Landau}} = \hbar eB/m_{\text{eff}}$

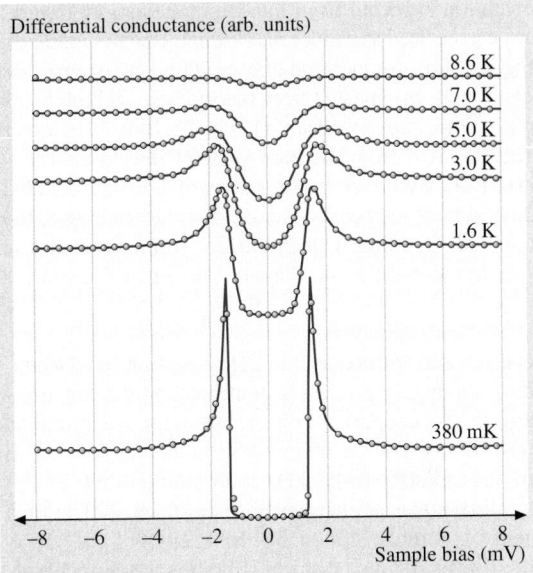

Fig. 25.8 Differential conductivity curve $dI/dV(V)$ measured on a Au surface by a Nb tip (*circles*). Different temperatures are indicated; the lines are fits according to the superconducting gap of Nb folded with the temperature-broadened Fermi distribution of the Au (courtesy of Pan, Houston)

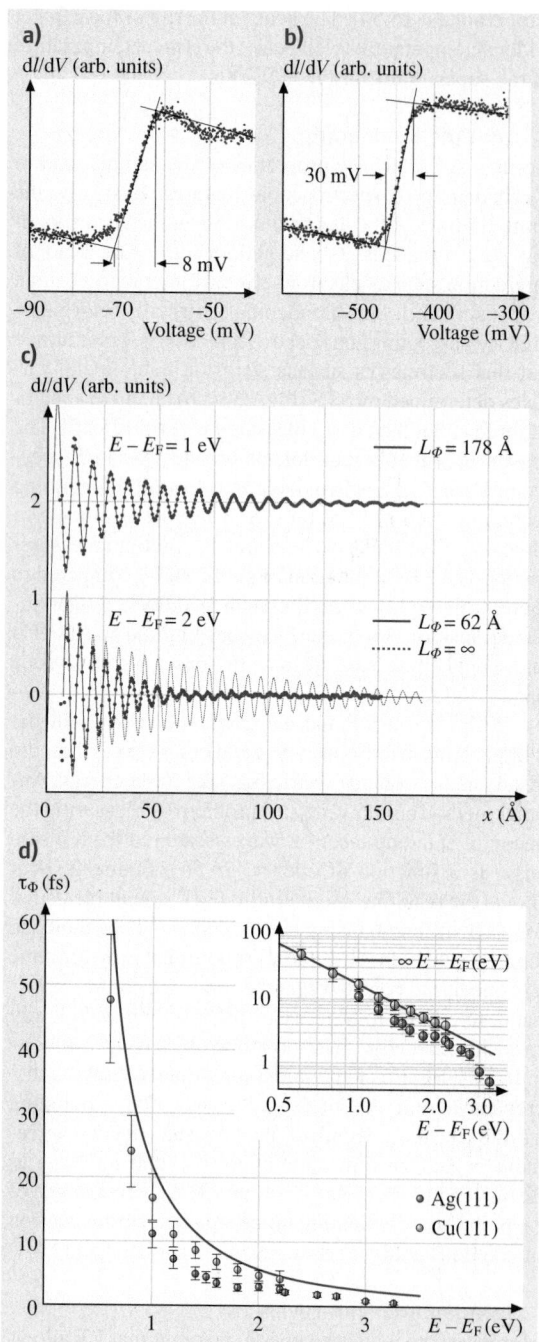

Fig. 25.9 (a),(b) spatially averaged $dI/dV(V)$-curves of Ag(111) and Cu(111); both surfaces exhibit a surface state with parabolic dispersion starting at -65 meV and -430 meV, respectively. The lines are drawn to determine the energetic width of the onset of these surface bands ((**a),(b)** courtesy of Berndt, Kiel); (**c**) dI/dV-intensity as a function of position away from a step edge of Cu(111) measured at the voltages ($E - E_F$), as indicated (*points*); the lines are fits assuming standing electron waves with a phase coherence length L_Φ as marked; (**d**) resulting phase coherence time as a function of energy for Ag(111) and Cu(111). *Inset* shows the same data on a double logarithmic scale evidencing the E^{-2}-dependence (*line*) ((**c),(d)** courtesy of Brune, Lausanne)

energy splittings of only 1.25 meV and 1.2 meV at $B = 10$ T. This is obviously lower than the typical lifetime broadenings discussed in the previous section and also close to $\Delta E = 1.1$ meV achievable at $T = 4$ K.

Fortunately, the electron density in doped semiconductors is much lower, and thus the lifetime increases significantly. Figure 25.10a shows a set of spectroscopy curves obtained on InAs(110) in different magnetic fields [25.9]. Above E_F, oscillations with increasing intensity and energy distance are observed. They show the separation expected from Landau quantization. In turn, they can be used to deduce m_{eff} from the peak separation (Fig. 25.10b). An increase of m_{eff} with increasing E has been found, as expected from theory. Also, at high fields spin quantization is observed (Fig. 25.10c). It is larger than expected from the bare g-factor due to contributions from exchange enhancement [25.58]. A detailed discussion of the peaks revealed that they belong to the so-called tip-induced quantum dot resulting from the work function difference between the tip and sample.

Vibrational Levels

As discussed with respect to light emission in STM, inelastic tunneling processes contribute to the tunneling current. The coupling of electronic states to vibrational levels is one source of inelastic tunneling [25.23]. It provides additional channels contributing to $dI/dV(V)$ with final states at energies different from V. The final energy is simply shifted by the energy of the vibrational level. If only discrete vibrational energy levels couple to a smooth electronic DOS, one expects a peak in d^2I/dV^2 at the vibrational energy. This situation appears for molecules on noble-metal surfaces. As usual, the isotope effect on the vibrational energy can be used to verify the vibrational origin of the peak. First indications of vibrational levels have been found for H_2O

and $E_{spin} = g\mu B$. Thus InAs is a good choice, since it exhibits a low effective mass $m_{eff}/m_e = 0.023$ and a high g-factor of 14 in the bulk conduction band. The values in metals are $m_{eff}/m_e \approx 1$ and $g \approx 2$, resulting in

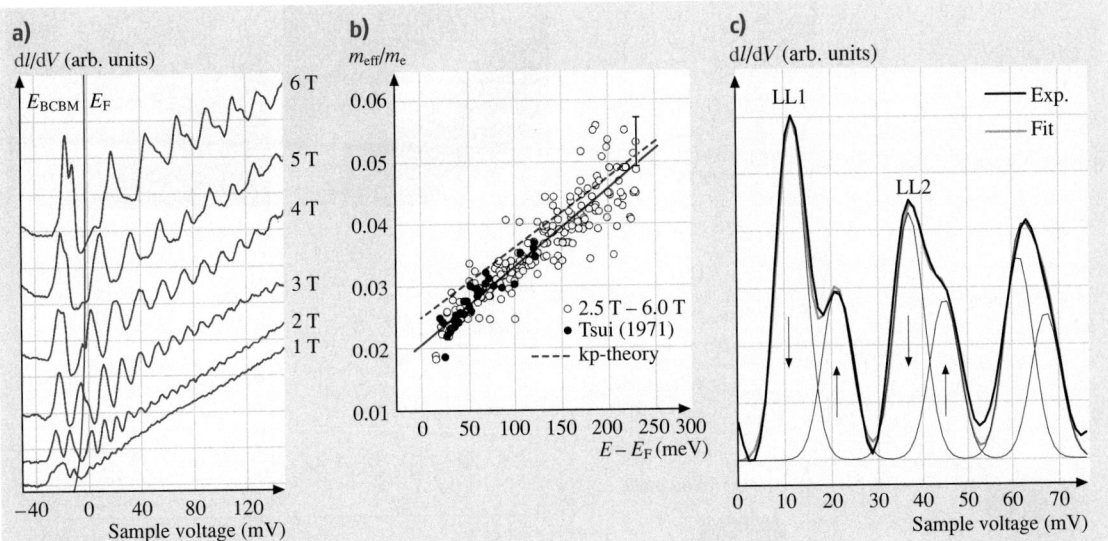

Fig. 25.10 (a) dI/dV-curves of n-InAs(110) at different magnetic fields as indicated; E_{BCBM} marks the bulk conduction band minimum; oscillations above E_{BCBM} are caused by Landau quantization; the double peaks at $B = 6\,T$ are caused by spin quantization. (b) Effective-mass data deduced from the distance of adjacent Landau peaks ΔE according to $\Delta E = h\,eB/m_{eff}$ (*open symbols*); *filled symbols* are data from planar tunnel junctions (Tsui), the *solid line* is a mean-sqare fit of the data and the *dashed line* is the expected effective mass of InAs according to kp-theory. (c) Magnification of a dI/dV-curve at $B = 6\,T$ exhibiting spin splitting; the Gaussian curves marked by *arrows* are the fitted spin levels

and D_2O on TiO_2 [25.59], and completely convincing work has been performed for C_2H_2 and C_2D_2 on Cu(001) [25.23] (Fig. 25.11a). The technique can be used to identify individual molecules on the surface by their characteristic vibrational levels. In particular, surface reactions, as described in Fig. 25.5h–m, can be directly verified. Moreover, the orientation of complexes with respect to the surface can be determined to a certain extent, since the vibrational excitation depends on the position of the tunneling current within the molecule. Finally, the excitation of certain molecular levels can induce such corresponding motions as hopping [25.32], rotation [25.34] (Fig. 25.11b–e), or desorption [25.31], leading to additional possibilities for manipulation on the atomic scale.

Kondo Resonance
A rather intricate interaction effect is the Kondo effect. It results from a second-order scattering process between itinerate states and a localized state [25.60]. The two states exchange some degree of freedom back and forth, leading to a divergence of the scattering probability at the Fermi level of the itinerate states. Due to the divergence, the effect strongly modifies sample properties. For example, it leads to an unexpected increase in resis-

tance with decreasing temperature for metals containing magnetic impurities [25.4]. Here, the exchanged degree of freedom is the spin. A spectroscopic signature of the Kondo effect is a narrow peak in the DOS at the Fermi level, disappearing above a characteristic temperature (the Kondo temperature). STS provides the opportunity to study this effect on the local scale [25.61, 62].

Figure 25.12a–d shows an example of Co clusters deposited on a carbon nanotube [25.63]. While a small dip at the Fermi level, which is probably caused by curvature influences on the π-orbitals, is observed without Co (Fig. 25.12b) [25.64], a strong peak is found around a Co cluster deposited on top of the tube (Fig. 25.12a, arrow). The peak is slightly shifted with respect to $V = 0\,mV$ due to the so-called Fano resonance [25.65], which results from interference of the tunneling processes into the localized Co-level and the itinerant nanotube levels. The resonance disappears within several nanometers of the cluster, as shown in Fig. 25.12d.

The Kondo effect has also been detected for different magnetic atoms deposited on noble-metal surfaces [25.61, 62]. There, it disappears at about 1 nm from the magnetic impurity, and the effect of the Fano resonance is more pronounced, contributing to dips in $dI/dV(V)$-curves instead of peaks.

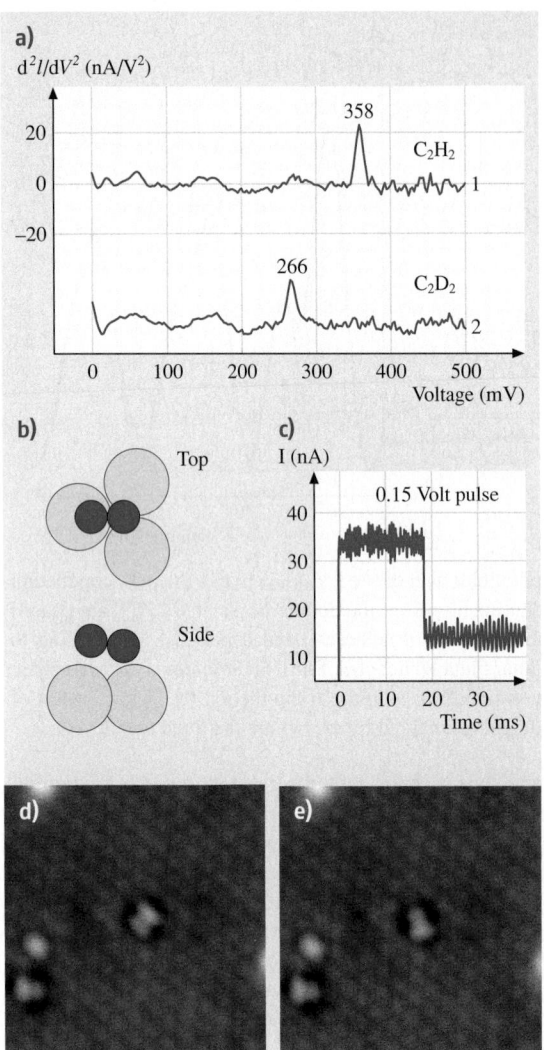

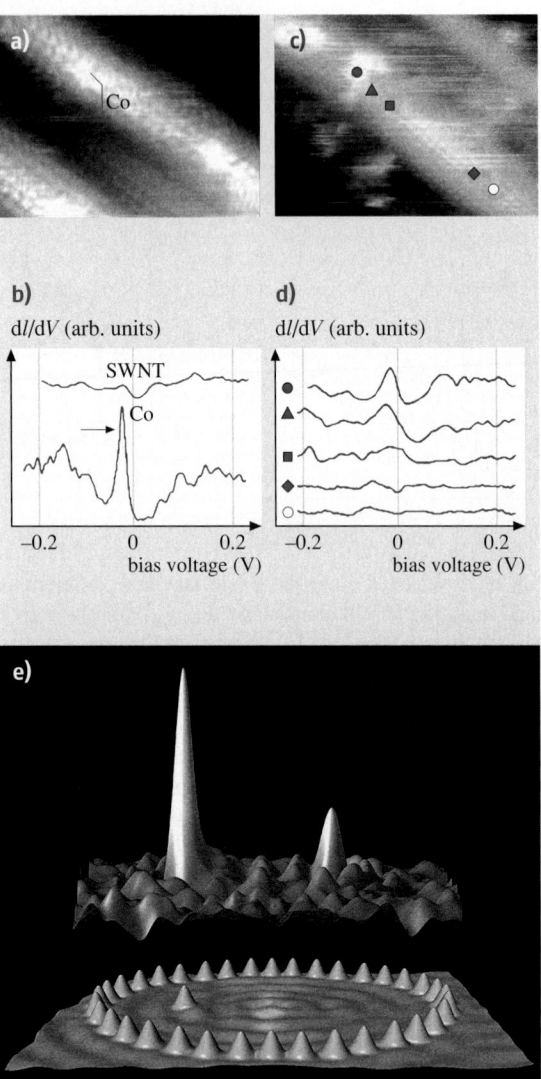

Fig. 25.11 (**a**) d^2I/dV^2-curves taken above a C_2H_2 and a C_2D_2 molecule on Cu(100); the peaks correspond to the C–H, respectively, C–D stretch-mode energy of the molecule. (**b**) Sketch of O_2 molecule on Pt(111). (**c**) Tunnelling current above an O_2 molecule on Pt(111) during a voltage pulse of 0.15 V; the jump in current indicates rotation of the molecule. (**d**), (**e**) STM image of an O_2 molecule on Pt(111) ($V = 0.05$ V) prior and after rotation induced by a voltage pulse to 0.15 V ((**a**)–(**e**) courtesy of Ho, Irvine)

Fig. 25.12 (**a**) STM image of a Co cluster on a single-wall carbon nanotube (SWNT). (**b**) dI/dV-curves taken directly above the Co cluster (Co) and far away from the Co cluster (SWNT); the *arrow* marks the Kondo peak. (**c**) STM image of another Co cluster on a SWNT with *symbols* marking the positions where the dI/dV-curves displayed in (**d**) are taken. (**d**) dI/dV-curves taken at the positions marked in (**c**) ((**a**)–(**d**) courtesy of Lieber, Cambridge). (**e**) *Lower part:* STM image of a quantum corral of elliptic shape made from Co atoms on Cu(111); one Co atom is placed in one of the foci of the ellipse. *Upper part:* map of the strength of the Kondo signal in the corral; note that there is also a Kondo signal in the focus, which is not covered by a Co atom ((**e**) courtesy of Eigler, Almaden)

A fascinating experiment has been performed by *Manoharan* et al. [25.66], who used manipulation to form an elliptic cage for the surface states of Cu(111) (Fig. 25.12e, bottom). This cage was constructed to have a quantized level at E_F. Then, a cobalt atom was placed

Fig. 25.13 (**a**) Low-voltage STM image of Cu(111), including two defect atoms; the waves are electronic Bloch waves scattered at the defects. (**b**) Low-voltage STM image of a rectangular quantum corral made from single atoms on Cu(111); the pattern inside the corral is the confined state of the corral close to E_F; ((**a**),(**b**) courtesy of Eigler, Almaden). (**c**) STM image of GaAs(110) around a Si donor, $V = -2.5\,\mathrm{V}$; the line scan along A shown in (**d**) exhibits an additional oscillation around the donor caused by a standing Bloch wave; the grid like pattern corresponds to the atomic corrugation of the Bloch wave ((**c**),(**d**) courtesy of van Kempen, Nijmegen). (**e**)–(**g**) STM images of an InAs/ZnSe-core/shell-nanocluster at different V. The image is measured in the so-called constant-height mode, i. e., the images display the tunneling current at constant height above the surface; the hill in (**e**) corresponds to the s-state of the cluster, the ring in (**f**) to the degenerate p_x- and p_y-state and the hill in (**g**) to the p_z-state ((**e**)–(**g**) courtesy of Millo, Jerusalem). (**h**) STM-image of a short-cut carbon nanotube. (**i**) Colour plot of the dI/dV intensity inside the short-cut nanotube as a function of position and tunneling voltage; four wavy patterns of different wavelength are visible in the voltage range from -0.1 to $0.15\,\mathrm{V}$ ((**h**),(**i**) courtesy of Dekker, Delft)

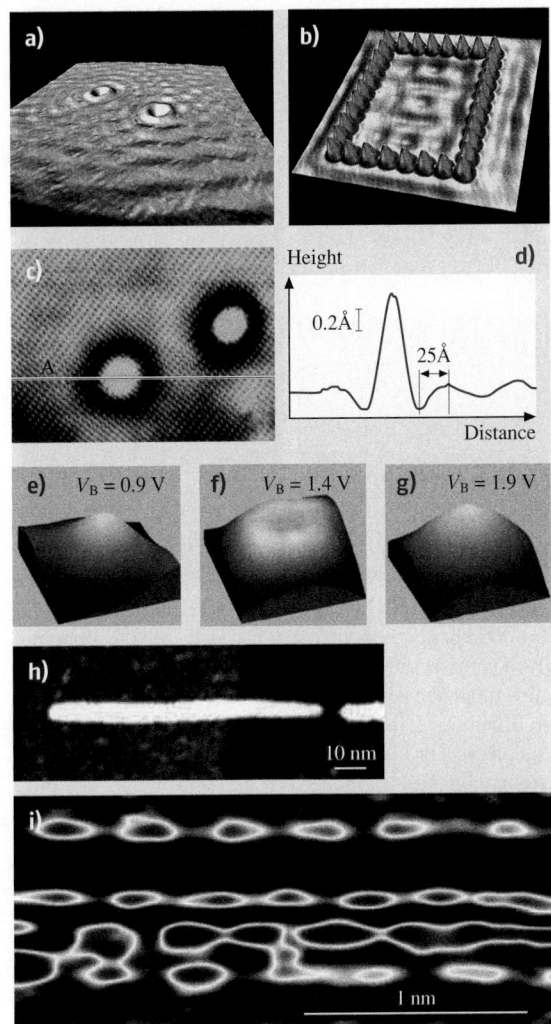

in one focus of the elliptic cage, producing a Kondo resonance. Surprisingly, the same resonance reappeared in the opposite focus, but not away from the focus (Fig. 25.12e, top). This shows amazingly that complex local effects such as the Kondo resonance can be guided to remote points.

Orbital scattering as a source of the Kondo resonance has also been found around a defect on Cr(001) [25.67]. Here, it is believed that itinerate sp-levels scatter at a localized d-level to produce the Kondo peak.

25.3.5 Imaging Electronic Wave Functions

Since STS measures the sum of squared wave functions (25.1), it is an obvious task to measure the local appearance of the most simple wave functions in solids, namely, Bloch waves.

Bloch Waves

The atomically periodic part of the Bloch wave is always measured if atomic resolution is achieved (inset of Fig. 25.14a). However, the long-range wavy part requires the presence of scatterers. The electron wave impinges on the scatterer and is reflected, leading to self-interference. In other words, the phase of the Bloch wave becomes fixed by the scatterer.

Such self-interference patterns were first found on Graphite(0001) [25.68] and later on noble-metal surfaces, where adsorbates or step edges scatter the surface states (Fig. 25.13a) [25.22]. Fourier transforms of the real-space images reveal the k-space distribution of the corresponding states [25.69], which may include additional contributions besides the surface state [25.70]. Using particular geometries as the so-called quantum corrals to form a cage for the electron wave, the scattering state can be rather complex (Fig. 25.13b). Anyway, it can usually be reproduced by simple calculations involving single-particle states [25.71].

Bloch waves in semiconductors scattered at charged dopants (Fig. 25.13c,d) [25.72], Bloch states confined

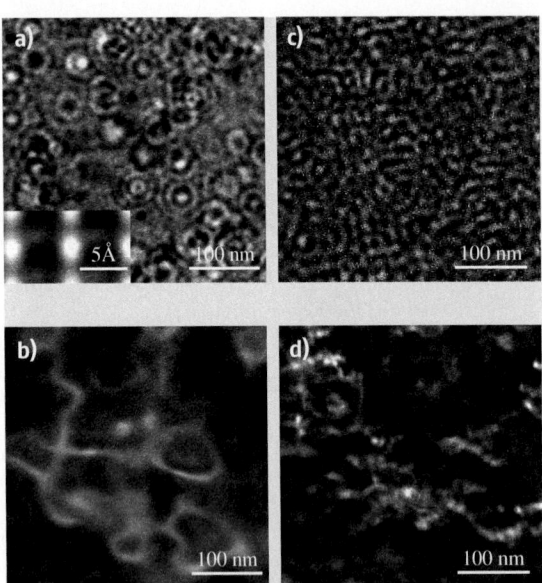

Fig. 25.14 (a) dI/dV-image of InAs(110) at $V = 50$ mV, $B = 0$ T; circular wave patterns corresponding to standing Bloch waves around each S donor are visible; inset shows a magnification revealing the atomically periodic part of the Bloch wave. (b) Same as (a), but at $B = 6$ T; the stripe structures are drift states. (c) dI/dV-image of a 2-D electron system on InAs(110) induced by the deposition of Fe, $B = 0$ T. (d) Same as (c) but at $B = 6$ T; note that the contrast in (a) is increased by a factor of ten with respect to (b)–(d)

in semiconductor quantum dots (Fig. 25.13e–g) [25.73], and Bloch waves confined in short-cut carbon nanotubes (Fig. 25.13h,i) [25.74, 75] have been visualized.

Drift States

More-complex wave functions result from interactions. A nice playground to study such interactions is doped semiconductors. The reduced electron density with respect to metals increases the importance of electron interactions with potential disorder and other electrons. Applying a magnetic field quenches the kinetic energy, further enhancing the importance of interactions. A dramatic effect can be observed on InAs(110), where three-dimensional (3-D) bulk states are displayed. While the usual scattering states around individual dopants are observed at $B = 0$ T (Fig. 25.14a) [25.76], stripe structures are found in high magnetic field (Fig. 25.14b) [25.77]. They run along equipotential lines of the disorder potential. This can be understood by recalling that the electron tries to move in a cyclotron orbit, which is accelerated and decelerated in electrostatic po-

tential, leading to a drift motion along an equipotential line [25.78].

The same effect has been found in two-dimensional (2-D) electron systems (2-DES) of the same substrate, where the scattering states at $B = 0$ T are, however, found to be more complex (Fig. 25.14c) [25.79]. The reason is the tendency of a 2-DES to exhibit closed scattering paths [25.80]. Consequently, the self-interference does not result from scattering at individual scatterers, but from complicated self-interference paths involving many scatterers. However, drift states are also observed in the 2-DES at high magnetic fields (Fig. 25.14d) [25.81].

Charge Density Waves

Another interaction modifying the LDOS is the electron–phonon interaction. Phonons scatter electrons between different Fermi points. If the wave vectors connecting Fermi points exhibit a preferential orientation, a so-called Peierls instability occurs [25.82]. The corresponding phonon energy goes to zero, the atoms are slightly displaced with the periodicity of the corresponding wave vector, and a charge density wave (CDW) with the same periodicity appears. Essentially, the CDW increases the overlap of the electronic states with the phonon by phase-fixing with respect to the atomic lattice. The Peierls transition naturally occurs in one-dimensional (1-D) systems, where only two Fermi points are present, and, hence, preferential orientation is pathological. It can also occur in 2-D systems if large areas of the Fermi line run in parallel.

STS studies of CDWs are numerous (e.g., [25.83, 84]). Examples of a 1-D CDW on a quasi-1-D bulk material and of a 2-D CDW are shown in Fig. 25.15a–d and Fig. 25.15e, respectively [25.85, 86]. In contrast to usual scattering states, where LDOS corrugations are only found close to the scatterer, the corrugations of CDWs are continuous across the surface. Heating the substrate toward the transition temperature leads to a melting of the CDW lattice, as shown in Fig. 25.15f–h.

CDWs have also been found on monolayers of adsorbates such as a monolayer of Pb on Ge(111) [25.87]. These authors performed a nice temperature-dependent study revealing that the CDW is nucleated by scattering states around defects, as one might expect [25.88]. 1-D systems have also been prepared on surfaces showing Peierls transitions [25.89, 90]. Finally, the energy gap occurring at the transition has been studied by measuring $dI/dV(V)$-curves [25.91].

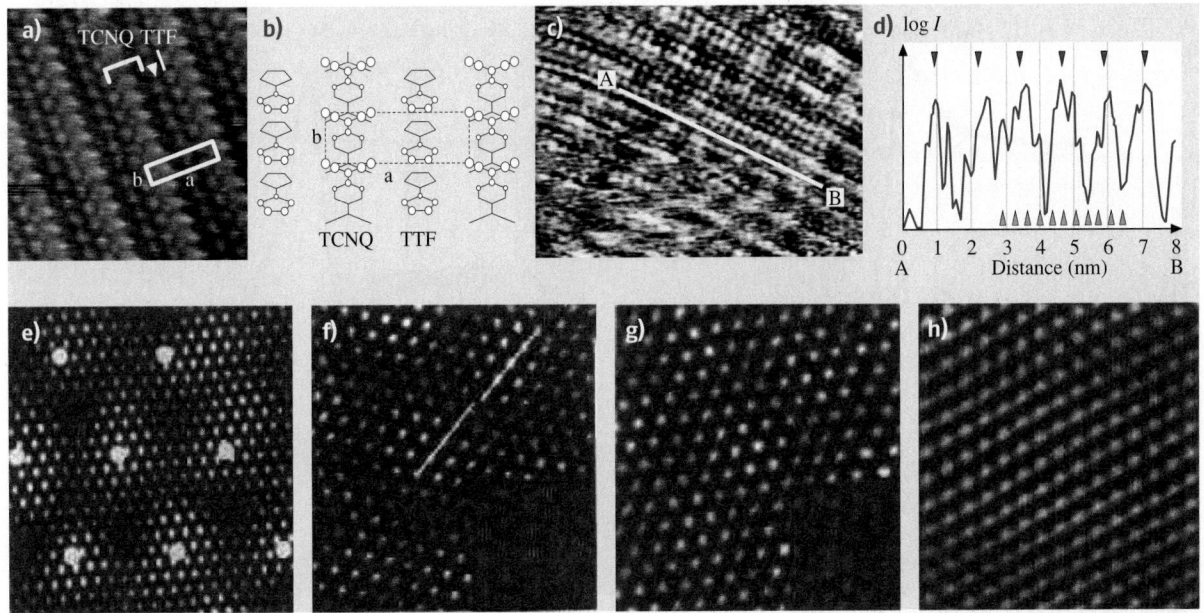

Fig. 25.15 (a) STM image of the ab-plane of the organic quasi-1-D conductor TTF-TCNQ, $T = 300$ K; while the TCNQ chains are conducting, the TTF chains are insulating. (b) Stick-and-ball model of the ab-plane of TTF-TCNQ. (c) STM image taken at $T = 61$ K, the additional modulation due to the Peierls transition is visible in the profile along line A shown in (d); the *brown triangles* mark the atomic periodicity and the *black triangles* the expected CDW periodicity ((a)–(d) courtesy of Kageshima, Kanagawa). (e)–(h) Low-voltage STM images of the two-dimensional CDW-system 1 T-TaS₂ at $T = 242$ K (e), 298 K (f), 349 K (g), 357 K (h). A long-range, hexagonal modulation is visible besides the atomic spots; its periodicity is highlighted by *large white dots* in (e); the additional modulation obviously weakens with increasing T, but is still apparent in (f) and (g), as evidenced in the lower magnification images in the insets ((e)–(h) courtesy of Lieber, Cambridge)

Superconductors

An intriguing effect resulting from electron–phonon interaction is superconductivity. Here, the attractive part of the electron–phonon interaction leads to the coupling of electronic states with opposite wave vector and mostly opposite spin [25.92]. Since the resulting Cooper pairs are bosons, they can condense at LT, forming a coherent many-particle phase, which can carry current without resistance. Interestingly, defect scattering does not influence the condensate if the coupling along the Fermi surface is homogeneous (s-wave superconductor). The reason is that the symmetry of the scattering of the two components of a Cooper pair effectively leads to a scattering from one Cooper pair state to another without affecting the condensate. This is different if the scatterer is magnetic, since the different spin components of the pair are scattered differently, leading to an effective pair breaking, which is visible as a single-particle excitation within the superconducting gap. On a local scale, this effect was first demonstrated by putting Mn, Gd, and Ag atoms on a Nb(110) surface [25.93]. While the nonmag-netic Ag does not modify the gap shown in Fig. 25.16a, it is modified in an asymmetric fashion close to Mn or Gd adsorbates, as shown in Fig. 25.16b. The asymmetry of the additional intensity is caused by the breaking of the particle–hole symmetry due to the exchange interaction between the localized Mn state and the itinerate Nb states.

Another important local effect is caused by the relatively large coherence length of the condensate. At a material interface, the condensate wave function cannot stop abruptly, but overlaps into the surrounding material (proximity effect). Consequently, a superconducting gap can be measured in areas of non-superconducting material. Several studies have shown this effect on the local scale using metals and doped semiconductors as surrounding materials [25.94, 95].

While the classical type-I superconductors are ideal diamagnets, the so-called type-II superconductors can contain magnetic flux. The flux forms vortices, each containing one flux quantum. These vortices are accompanied by the disappearance of the superconducting

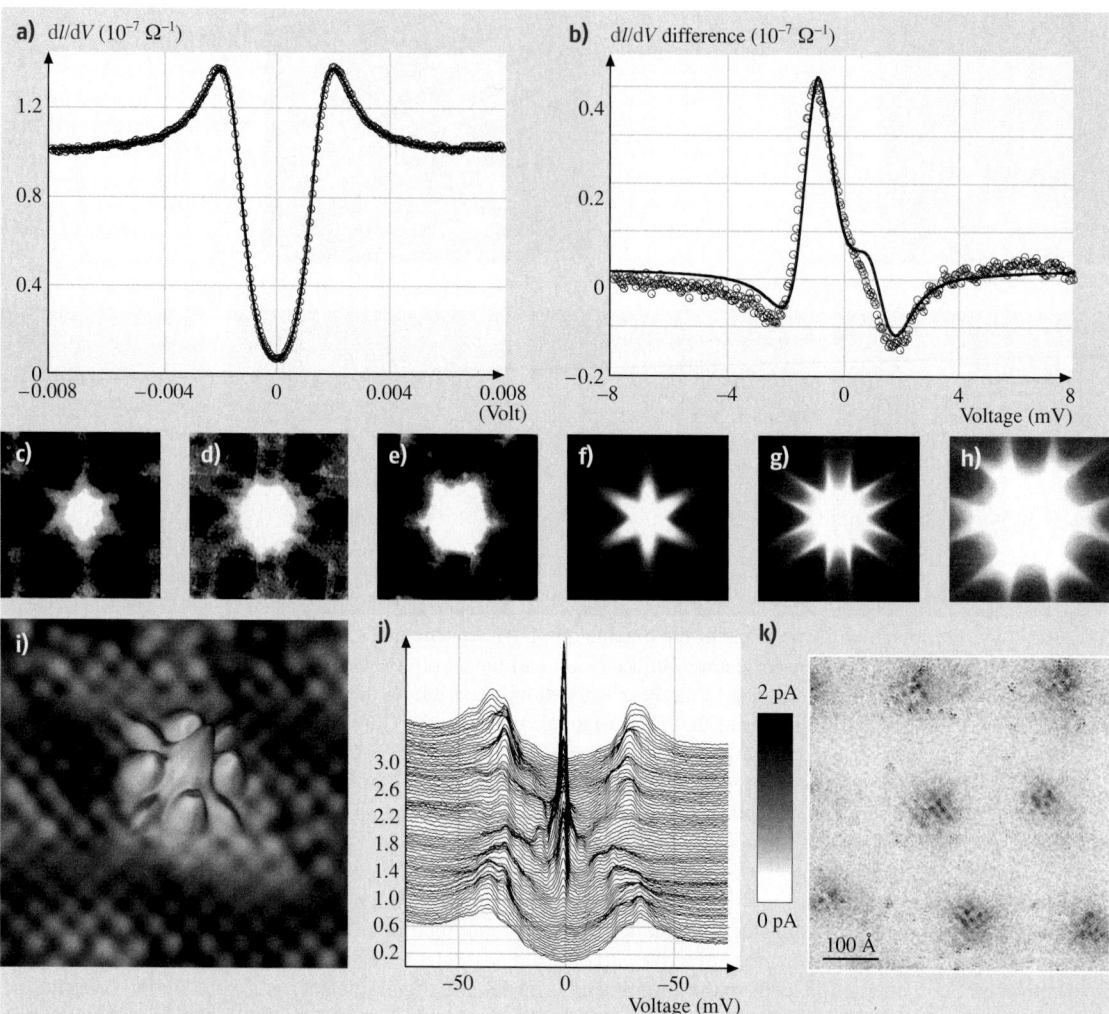

Fig. 25.16 (a) dI/dV-curve of Nb(110) at $T = 3.8$ K (*symbols*) in comparison with a BCS fit of the superconducting gap of Nb (*line*). (b) Difference between the dI/dV-curve taken directly above a Mn-atom on Nb(110) and the dI/dV-curve taken above the clean Nb(110) (*symbols*) in comparison with a fit using the Bogulubov-de Gennes equations (*line*) ((a),(b) courtesy of Eigler, Almaden). (c)–(e) dI/dV-images of a vortex core in the type-II superconductor 2H-NbSe$_2$ at 0 mV (c), 0.24 mV (d), and 0.48 mV (e) ((c)–(e) courtesy of H. F. Hess). (f)–(h) Corresponding calculated LDOS images within the Eilenberger framework ((f)–(h) courtesy of Machida, Okayama). (i) Overlap of an STM image at $V = -100$ mV (background 2-D image) and a dI/dV-image at $V = 0$ mV (overlapped 3-D image) of optimally doped Bi$_2$Sr$_2$CaCu$_2$O$_{8+\delta}$ containing 0.6% Zn impurities. The STM image shows the atomic structure of the cleavage plane, while the dI/dV-image shows a bound state within the superconducting gap, which is located around a single Zn impurity. The fourfold symmetry of the bound state reflects the d-like symmetry of the superconducting pairing function; (j) dI/dV-curves taken at different positions across the Zn impurity; the bound state close to 0 mV is visible close to the Zn atom; (k) LDOS in the vortex core of slightly overdoped Bi$_2$Sr$_2$CaCu$_2$O$_{8+\delta}$, $B = 5$ T; the dI/dV-image taken at $B = 5$ T is integrated over $V = 1–12$ mV, and the corresponding dI/dV-image at $B = 0$ T is subtracted to highlight the LDOS induced by the magnetic field. The checkerboard pattern within the seven vortex cores exhibits a periodicity, which is fourfold with respect to the atomic lattice shown in (i) and is thus assumed to be a CDW ((i)–(k) courtesy of S. Davis, Cornell and S. Uchida, Tokyo)

Fig. 25.17 (a)–(d) Spin-polarized STM images of 1.65 monolayer of Fe deposited on a stepped W(110) surface measured at different *B*-fields, as indicated. Double-layer and monolayer Fe stripes are formed on the W substrate; only the double-layer stripes exhibit magnetic contrast with an out-of-plane sensitive tip, as used here. *White* and *grey areas* correspond to different domains. Note that more white areas appear with increasing field. (e) STM image of an antiferromagnetic Mn monolayer on W(110). (f) Spin-polarized STM-image of the same surface (in-plane tip). The insets in (e) and (f) show the calculated STM and spin-polarized STM images, respectively, and the stick-and-ball models symbolize the atomic and the magnetic unit cell ((a)–(f) courtesy of M. Bode, Hamburg). (g) Spin-polarized STM image of a 6-nm-high Fe island on W(110) (in-plane tip). Four different areas are identified as four different domains with domain orientations, as indicated by the *arrows*. (h) Spin-polarized STM image of the central area of an island; the size of the area is indicated by the rectangle in (g); the measurement is performed with an out-of-plane sensitive tip showing that the magnetization turns out-of-plane in the center of the island

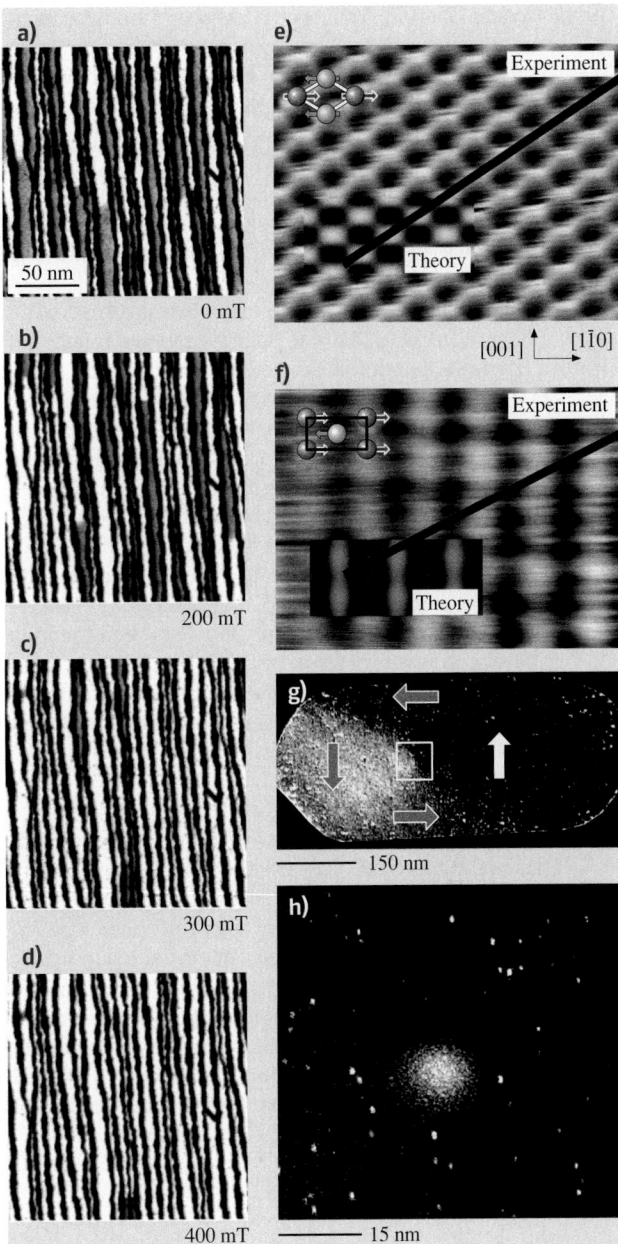

gap and, therefore, can be probed by STS [25.96]. LDOS maps measured inside the gap lead to bright features in the area of the vortex core. Importantly, the length scale of these features is different from the length scale of the magnetic flux due to the difference between London's penetration depth and the coherence length. Thus, STS probes another property of the vortex than the usual magnetic imaging techniques (see Sect. 25.4.4). Surprisingly, first measurements of the vortices on NbSe$_2$ revealed vortices shaped as a sixfold star [25.97] (Fig. 25.16c). With increasing voltage inside the gap, the orientation of the star rotates by 30° (Fig. 25.16d,e). The shape of these stars could finally be reproduced by theory, assuming an anisotropic pairing of electrons in the superconductor (Fig. 25.16f–h) [25.98]. Additionally, bound states inside the vortex core, which result from confinement by the surrounding superconducting material, are found [25.97]. Further experiments investigated the arrangement of the vortex lattice, including transitions between hexagonal and quadratic lattices [25.99], the influence of pinning centers [25.100], and the vortex motion induced by current [25.101].

The understanding of high-temperature superconductivity (HTCS) is still an important topic. An almost accepted property of HTCS is its d-wave pairing symmetry. In contrast to s-wave superconductors, scattering can lead to pair breaking, since the Cooper-pair density

vanishes in certain directions. Indeed, scattering states (bound states in the gap) around nonmagnetic Zn impurities have been observed in Bi$_2$Sr$_2$CaCu$_2$O$_{8+\delta}$ (BSCCO) (Fig. 25.16i,j) [25.26]. They reveal a *d*-like symmetry, but not the one expected from simple Cooper-pair scattering. Other effects such as magnetic polarization in the environment probably have to be taken into account [25.102]. Moreover, it has been found that magnetic Ni impurities exhibit a weaker scattering structure than Zn impurities [25.103]. Thus, BSCCO shows exactly the opposite behavior to that of Nb discussed above (Fig. 25.16a,b). An interesting topic is the importance of inhomogeneities in HTCS materials. Evidence for inhomogeneities has indeed been found in underdoped materials, where puddles of the superconducting phase are shown to be embedded in non-superconducting areas [25.104].

Of course, vortices have also been investigated in HTCS materials [25.105]. Bound states are found, but at energies that are in disagreement with simple models, assuming a Bardeen–Cooper–Schrieffer (BCS)-like d-wave superconductor [25.106, 107]. Theory predicts, instead, that the bound states are magnetic-field-induced spin density waves, stressing the competition between antiferromagnetic order and superconductivity in HTCS materials [25.108]. Since the spin density wave is accompanied by a charge density wave of half wavelength, it can be probed by STS [25.109]. Indeed, a checkerboard pattern of the right periodicity has been found in and around vortex cores in BSCCO (Fig. 25.16k). It exceeds the width of an individual vortex core, implying that the superconducting coherence length is different from the antiferromagnetic one.

Complex Systems (Manganites)

Complex phase diagrams are not restricted to HTCS materials (cuprates). They exist with similar complexity for other doped oxides such as manganites. Only a few studies of these materials have been performed by STS, mainly showing the inhomogeneous evolution of metallic and insulating phases [25.110, 111]. Similarities to the granular case of an underdoped HTCS material are obvious. Since inhomogeneities seem to be crucial in many of these materials, a local method such as STS might continue to be important for the understanding of their complex properties.

25.3.6 Imaging Spin Polarization: Nanomagnetism

Conventional STS couples to the LDOS, i. e., the charge distribution of the electronic states. Since electrons also have spin, it is desirable to also probe the spin distribution of the states. This can be achieved if the tunneling tip is covered by a ferromagnetic material [25.112]. The coating acts as a spin filter or, more precisely, the tunneling current depends on the relative angle α_{ij} between the spins of the tip and the sample according to $\cos(\alpha_{ij})$. In ferromagnets, the spins mostly have one preferential orientation along the so-called easy axis, i. e., a particular tip is not sensitive to spin orientations of the sample that are perpendicular to the spin orientation of the tip. Different tips have to be prepared to detect different spin orientations of the sample. Moreover, the magnetic stray field of the tip can perturb the spin orientation of the sample. To avoidthis, a technique using antiferromagnetic Cr as a tip coating material has been developed [25.113]. This avoids stray fields, but still provides a preferential spin orientation of the few atoms at the tip apex that dominate the tunneling current. Depending on the thickness of the Cr coating, spin orientations perpendicular or parallel to the sample surface are prepared.

So far, the described technique has been used to image the evolution of magnetic domains with increasing B-field (Fig. 25.17a–d) [25.114], the antiferromagnetic order of a Mn monolayer on W(110) (Fig. 25.17e,f) [25.115], and the out-of-plane orientation predicted for a magnetic vortex core as it exists in the center of a Fe island exhibiting four domains in the flux closure configuration (Fig. 25.17g,h) [25.116].

Besides the obvious strong impact on nanomagnetism, the technique might also be used to investigate other electronic phases such as the proposed spin density wave around a HTCS vortex core.

25.4 Scanning Force Microscopy and Spectroscopy

The examples discussed in the previous section show the wide variety of physical questions that have been tackled with the help of LT-STM. Here, we turn to the other prominent scanning probe method that is applied at low temperatures, namely, SFM, which gives complementary information on sample properties on the atomic scale.

The ability to detect *forces* sensitively with spatial resolution down to the atomic scale is of great interest, since force is one of the most fundamental quantities in physics. Mechanical force probes usually consist of a cantilever with a tip at its free end that is brought close to the sample surface. The cantilever can be mounted parallel or perpendicular to the surface (general aspects of force probe designs are described in Chapt. 23). Basically, two methods exist to detect forces with cantilever-based probes: the *static* and the *dynamic* mode (see Chapt. 22). They can be used to generate a laterally resolved image (*microscopy* mode) or determine its distance dependence (*spectroscopy* mode). One can argue about the terminology, since spectroscopy is usually related to energies and not to distance dependencies. Nevertheless, we will use it throughout the text, because it avoids lengthy paraphrases and is established in this sense throughout the literature.

In the static mode, a force that acts on the tip bends the cantilever. By measuring its deflection Δz the tip–sample force F_{ts} can be directly calculated with Hooke's law: $F_{ts} = c_z \cdot \Delta z$, where c_z denotes the spring constant of the cantilever. In the various dynamic modes, the cantilever is oscillated with amplitude A at or near its eigenfrequency f_0, but in some applications also off-resonance. At ambient pressures or in liquids, amplitude modulation (AM-SFM) is used to detect amplitude changes or the phase shift between the driving force and cantilever oscillation. In vacuum, the frequency shift Δf of the cantilever due to a tip–sample interaction is measured by the frequency-modulation technique (FM-SFM). The nomenclature is not standardized. Terms like tapping mode or intermittent contact mode are used instead of AM-SFM, and NC-AFM (noncontact atomic force microscopy) or DFM (dynamic force microscopy) instead of FM-SFM or FM-AFM. However, all these modes are *dynamic*, i.e., they involve an oscillating cantilever and can be used in the noncontact, as well as in the contact, regime. Therefore, we believe that the best and most consistent way is to distinguish them by their different detection schemes. Converting the measured quantity (amplitude, phase, or frequency shift) into a physically meaningful quantity, e.g., the tip–sample interaction force F_{ts} or the force gradient $\partial F_{ts}/\partial z$, is not always straightforward and requires an analysis of the equation of motion of the oscillating tip (see Chaps. 24 and 27).

Whatever method is used, the resolution of a cantilever-based force detection is fundamentally limited by its intrinsic *thermomechanical* noise. If the cantilever is in thermal equilibrium at a temperature T, the equipartition theorem predicts a thermally induced *root mean square* (rms) motion of the cantilever in the z direction of $z_{rms} = (k_B T/c_{eff})^{1/2}$, where k_B is the Boltzmann constant and $c_{eff} = c_z + \partial F_{ts}/\partial z$. Note that usually $dF_{ts}/dz \gg c_z$ in the contact mode and $dF_{ts}/dz < c_z$ in the noncontact mode. Evidently, this fundamentally limits the force resolution in the static mode, particularly if

operated in the noncontact mode. Of course, the same is true for the different dynamic modes, because the thermal energy $k_B T$ excites the eigenfrequency f_0 of the cantilever. Thermal noise is *white* noise, i. e., its spectral density is flat. However, if the cantilever transfer function is taken into account, one can see that the thermal energy mainly excites f_0. This explains the term *thermo* in thermomechanical noise, but what is the *mechanical* part?

A more detailed analysis reveals that the thermally induced cantilever motion is given by

$$z_{rms} = \sqrt{\frac{2k_B TB}{\pi c_z f_0 Q}}, \tag{25.2}$$

where B is the measurement bandwidth and Q is the quality factor of the cantilever. Analogous expressions can be obtained for all quantities measured in dynamic modes, because the deflection noise translates, e.g., into frequency noise [25.5]. Note that f_0 and c_z are correlated with each other via $2\pi f_0 = (c_z/m_{eff})^{1/2}$, where the effective mass m_{eff} depends on the geometry, density, and elasticity of the material. The Q-factor of the cantilever is related to the external damping of the cantilever motion in a medium and on the intrinsic damping within the material. This is the *mechanical* part of the fundamental cantilever noise.

It is possible to operate a low-temperature force microscope directly immersed in the cryogen [25.117, 118] or in the cooling gas [25.119], whereby the cooling is simple and very effective. However, it is evident from (25.2) that the smallest fundamental noise is achievable in vacuum, where the Q-factors are more than 100 times larger than in air, and at low temperatures.

The best force resolution up to now, which is better than 1×10^{-18} N/Hz$^{1/2}$, has been achieved by *Mamin* et al. [25.120] in vacuum at a temperature below 300 mK. Due to the reduced thermal noise and the lower thermal drift, which results in a higher stability of the tip–sample gap and a better signal-to-noise ratio, the highest resolution is possible at low temperatures in ultrahigh vacuum with FM-SFM. A vertical rms noise below 2 pm [25.121, 122] and a force resolution below 1 aN [25.120] have been reported.

Besides the reduced noise, the application of force detection at low temperatures is motivated by the increased stability and the possibility to observe phenomena that appear below a certain critical temperature T_c, as outlined on page 680. The experiments, which have been performed at low temperatures until now, were motivated by at least one of these reasons and can be roughly divided into four groups: (i) atomic-scale imaging, (ii) force spectroscopy, (iii) investigation of quantum phenomena by measuring electrostatic forces, and (iv) utilizing magnetic probes to study ferromagnets, superconductors, and single spins. In the following, we describe some exemplary results.

25.4.1 Atomic-Scale Imaging

In a simplified picture, the dimensions of the tip end and its distance to the surface limit the lateral resolution of force microscopy, since it is a near-field technique. Consequently, atomic resolution requires a stable single atom at the tip apex that has to be brought within a distance of some tenths of a nanometer to an atomically flat surface. The latter condition can only be fulfilled in the dynamic mode, where the additional restoring force $c_z A$ at the lower turnaround point prevents the jump-to-contact. As described in Chapt. 24, by preventing the so-called jump-to-contact *true* atomic resolution is nowadays routinely obtained in vacuum by FM-AFM. The nature of the short-range tip–sample interaction during imaging with atomic resolution has been studied experimentally as well as theoretically. Si(111)-(7×7) was the first surface on which true atomic resolution was achieved [25.123], and several studies have been performed at low temperatures on this well-known material [25.124–126]. First-principles simulations performed on semiconductors with a silicon tip revealed that *chemical* interactions, i.e., a significant charge redistribution between the dangling bonds of the tip and sample, dominate the atomic-scale contrast [25.127–129]. On V–III semiconductors, it was found that only one atomic species, the group V atoms, is imaged as protrusions with a sili-

con tip [25.128, 129]. Furthermore, these simulations revealed that the sample, as well as the tip atoms are noticeably displaced from their equilibrium position due to the interaction forces. At low temperatures, both aspects could be observed with silicon tips on indium arsenide [25.121, 130]. On weakly interacting surfaces the short-range interatomic van der Waals force has been believed responsible for the atomic-scale contrast [25.131–133].

Chemical Sensitivity of Force Microscopy
The (110) surface of the III–V semiconductor indium arsenide exhibits both atomic species in the top layer (see Fig. 25.18a). Therefore, this sample is well suited to study the chemical sensitivity of force microscopy [25.121]. In Fig. 25.18b, the usually observed atomic-scale contrast on InAs(110) is displayed. As predicted, the arsenic atoms, which are shifted by 80 pm above the indium layer due to the (1×1) relaxation, are imaged as protrusions. While this general appearance was similar for most tips, two other distinctively different contrasts were also observed: a second protrusion (c) and a sharp depression (d). The arrangement of these two features corresponds well to the zigzag configuration of the indium and arsenic atoms along the [1$\bar{1}$0]-direction. A sound explanation would be as follows: the contrast usually obtained with one feature per surface unit cell corresponds to a silicon-terminated tip, as predicted by simulations. A different atomic species at the tip apex, however, can result in a very different charge redistribution. Since the atomic-scale contrast is due to a chemical interaction, the two other contrasts would then correspond to a tip that has been accidentally contaminated with sample material (an arsenic or indium-terminated tip apex). Nevertheless, this expla-

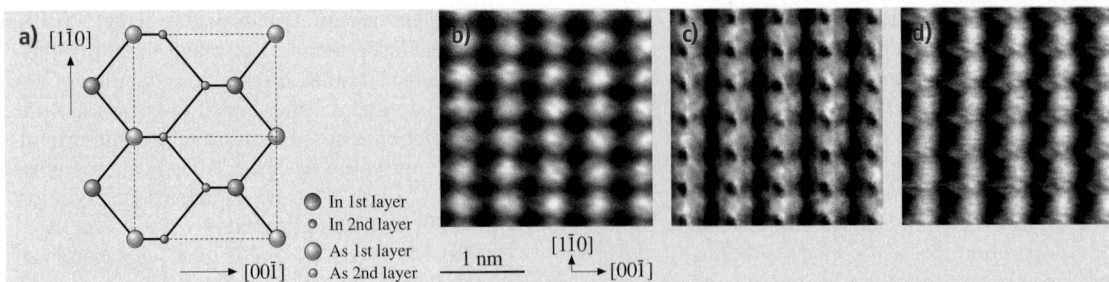

Fig. 25.18a–d The structure of InAs(110) as seen from above (**a**) and three FM-AFM images of this surface obtained with different tips at 14 K (**b**)–(**d**). In (**b**), only the arsenic atoms are imaged as protrusions, as predicted for a silicon tip. The two features in (**c**) and (**d**) corresponds to the zigzag arrangement of the indium and arsenic atoms. Since force microscopy is sensitive to short-range chemical forces, the appearance of the indium atoms can be associated with a chemically different tip apex

nation has not yet been verified by simulations for this material.

Tip-Induced Atomic Relaxation

Schwarz et al. [25.121] were able to visualize directly the predicted tip-induced relaxation during atomic-scale imaging near a point defect. Figure 25.19 shows two FM-AFM images of the same point defect recorded with different constant frequency shifts on InAs(110), i.e., the tip was closer to the surface in (b) compared to (a). The arsenic atoms are imaged as protrusions with the silicon tip used. From the symmetry of the defect, an indium-site defect can be inferred, since the

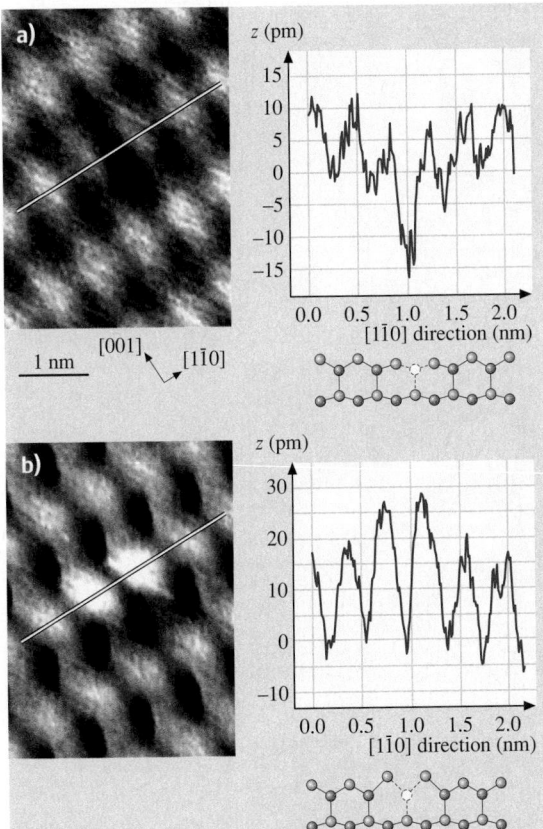

distance-dependent contrast is consistent with what is expected for an indium vacancy. This expectation is based on calculations performed for the similar III–V semiconductor GaP(110), where the two surface gallium atoms around a P-vacancy were found to relax downward [25.134]. This corresponds to the situation in Fig. 25.19a, where the tip is relatively far away and an inward relaxation of the two arsenic atoms is observed. The considerably larger attractive force in Fig. 25.19b, however, pulls the two arsenic atoms toward the tip. All other arsenic atoms are also pulled, but they are less displaced, because they have three bonds to the bulk, while the two arsenic atoms in the neighborhood of an indium vacancy have only two bonds. This direct experimental proof of the presence of tip-induced relaxations is also relevant for STM measurements, because the tip–sample distances are similar during atomic-resolution imaging. Moreover, the result demonstrates that FM-AFM can probe elastic properties on an atomic level.

Imaging of Weakly Interacting van der Waals Surfaces

For weakly interacting van der Waals surfaces, much smaller atomic corrugation amplitudes are expected compared to strongly interacting surfaces of semiconductors. A typical example is graphite, a layered material, where the carbon atoms are covalently bonded and arranged in a honeycomb structure within the (0001) plane. Individual graphene layers stick together by van der Waals forces. Due to the *ABA* stacking, three distinctive sites exist on the (0001) surface: carbon atoms with (*A*-type) and without (*B*-type) neighbor in the next graphite layer and the *hollow site* (H-site) in the hexagon center. In static contact force microscopy as well as in STM the contrast exhibits usually a trigonal symmetry with a periodicity of 246 pm, where *A*- and *B*-site carbon atoms could not be distinguished. However, in high-resolution FM-AFM images acquired at low temperatures, a large maximum and two different minima have been resolved, as demonstrated by the profiles along the three equivalent [1-100] directions in Fig. 25.20a. A simulation using the Lennard–Jones potential, given by the short-range interatomic van der Waals force, reproduced these three features very well (dotted line). Therefore, the large maximum could be assigned to the H-site, while the two different minima represent *A*- and *B*-type carbon atoms [25.132].

Compared to graphite, the carbon atoms in a single-walled carbon nanotube (SWNT), which consists of

Fig. 25.19a,b Two FM-AFM images of the identical indium-site point defect (presumably an indium vacancy) recorded at 14 K. If the tip is relatively far away, the theoretically predicted inward relaxation of two arsenic atoms adjacent to an indium vacancy is visible (**a**). At a closer tip–sample distance (**b**), the two arsenic atoms are pulled farther toward the tip compared to the other arsenic atoms, since they have only two instead of three bonds

a single rolled up graphene layer, are indistinguishable. For the first time *Ashino* et al. [25.133] successfully imaged the curved surface of a SWNT with atomic resolution. Note that for geometric reasons, atomic resolution is only achieved on the top (see Fig. 25.20b). Indeed, as shown in Fig. 25.20b, all profiles between two hollow sites across two neighboring carbon atoms are symmetric [25.135]. Particularly, curve 1 and 2 exhibit two minima of equal depth, as predicted by theory (cf., dotted line). The assumption used in the simulation (dotted lines in the profiles of Fig. 25.20) that interatomic van der Waals forces are responsible for the atomic-scale contrast has been supported by a quantitative evaluation of force spectroscopy data obtained on SWNTs [25.133].

Interestingly, the image contrast on graphite and SWNTs is inverted with respect to the arrangement of the atoms, i.e., the minima correspond to the position of the carbon atoms. This can be related to the small carbon–carbon distance of only 142 pm, which is in fact the smallest interatomic distance that has been resolved with FM-AFM so far. The van der Waals radius of the front tip atom, (e.g., 210 pm for silicon) has a radius that is significantly larger than the intercarbon distance. Therefore, next-nearest-neighbor interactions become important and result in a contrast inversion [25.135].

While experiments on graphite and SWNTs basically take advantage of the increased stability and signal-to-noise ratio at low temperatures, solid xenon (melting temperature $T_m = 161$ K) can only be observed at sufficient low temperatures [25.8]. In addition, xenon is a pure van der Waals crystal and, since it is an insulator, FM-AFM is the only real-space method available today that allows the study of solid xenon on the atomic scale.

Allers et al. [25.8] adsorbed a well-ordered xenon film on cold graphite(0001) ($T < 55$ K) and studied it subsequently at 22 K by FM-AFM (see Fig. 25.20c). The sixfold symmetry and the distance between the protrusions corresponds well with the nearest-neighbor

distance in the close-packed (111) plane of bulk xenon, which crystallizes in a face-centered cubic structure. A comparison between experiment and simulation confirmed that the protrusions correspond to the position of the xenon atoms [25.132]. However, the simulated corrugation amplitudes do not fit as well as for graphite (see sections in Fig. 25.20c). A possible reason is that tip-induced relaxations, which were not considered in the simulations, are more important for this pure van der Waals crystal xenon than they are for graphite, because in-plane graphite exhibits strong covalent bonds.

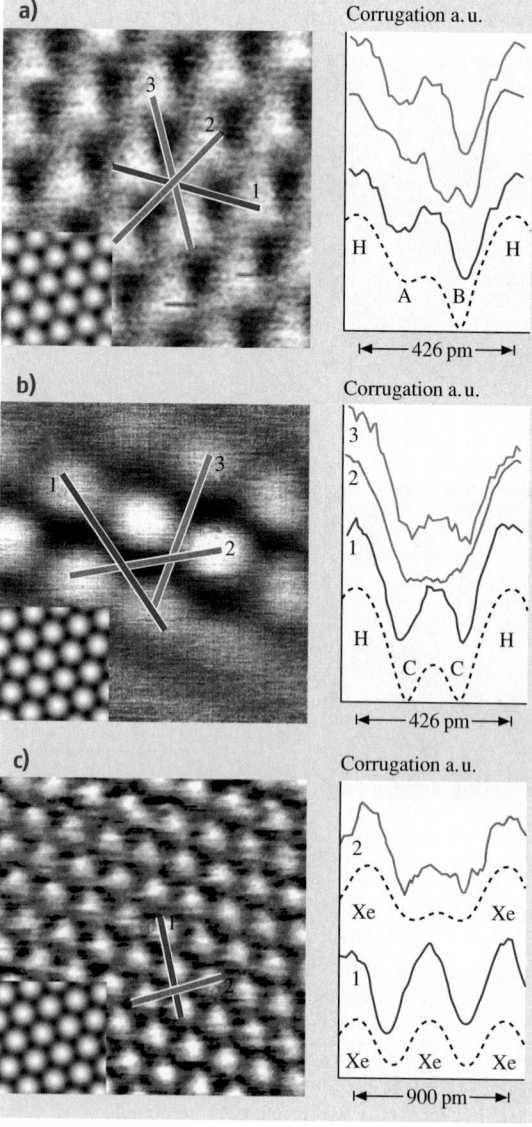

Fig. 25.20a–c FM-AFM images of graphite(0001) (**a**) a single-walled carbon nanotube (SWNT) (**b**) and Xe(111) (**c**) recorded at 22 K. On the right side, line sections taken from the experimental data (*solid lines*) are compared to simulations (*dotted lines*). A- and B-type carbon atoms, as well as the hollow site (H-site) on graphite can be distinguished, but are imaged with inverted contrast, i.e., the carbon sites are displayed as minima. Such an inversion does not occur on Xe(111)

Nevertheless, the results demonstrated for the first time that a weakly bonded van der Waals crystal could be imaged nondestructively on the atomic scale. Note that on Xe(111) no contrast inversion exists, presumably because the separation between Xe sites is about 450 pm, i. e., twice as large as the van der Waals radius of a silicon atom at the tip end.

Atomic Resolution
Using Small Oscillation Amplitudes

All the examples above described used spring constants and amplitudes on the order of $40\,\text{N/m}$ and 10 nm, respectively, to obtain atomic resolution. However, *Giessibl* et al. [25.137] pointed out that the optimal amplitude should be on the order of the characteristic decay length λ of the relevant tip–sample interaction. For short-range interactions, which are responsible for the atomic-scale contrast, λ is on the order of 0.1 nm. On the other hand, stable imaging without a jump-to-contact is only possible as long as the restoring force $c_z A$ at the lower turnaround point of each cycle is larger than the maximal attractive tip–sample force. Therefore, reducing the desired amplitude by a factor of 100 requires a 100 times larger spring constant. Indeed, *Hembacher* et al. [25.136] could demonstrate atomic resolution with small amplitudes (about 0.25 nm) and large spring constants (about 1800 N/m) utilizing a qPlus sensor [25.138]. Figure 25.21 shows a constant-height image of graphite recorded at 4.9 K within the repulsive regime. Note that compared to Fig. 25.20a,b the contrast is inverted, i. e., the carbon atoms appear as maxima. This is expected, because the imaging interaction is switched from attractive to repulsive regime [25.131, 135].

25.4.2 Force Spectroscopy

A wealth of information about the nature of the tip–sample interaction can be obtained by measuring its distance dependence. This is usually done by recording the measured quantity (deflection, frequency shift, amplitude change, phase shift) and applying an appropriate voltage ramp to the z-electrode of the scanner piezo, while the z-feedback is switched off. According to (25.2), low temperatures and high Q-factors (vacuum) considerably increase the force resolution. In the static mode, long-range forces and contact forces can be examined. Force measurements at small tip–sample distances are inhibited by the *jump-to-contact* phenomenon: If the force gradient $\partial F_{ts}/\partial z$ becomes larger than the spring constant c_z, the cantilever can-

not resist the attractive tip–sample forces and the tip snaps onto the surface. Sufficiently large spring constants prevent this effect, but reduce the force resolution. In the dynamic modes, the jump-to-contact can be avoided due to the additional restoring force $(c_z A)$ at the lower turnaround point. The highest sensitivity can be achieved in vacuum by using the FM technique, i. e., by recording $\Delta f(z)$-curves. An alternative FM spectroscopy method, the recording of $\Delta f(A)$-curves, has been suggested by *Hölscher* et al. [25.139]. Note that, if the amplitude is much larger than the characteristic decay length of the tip–sample force, the frequency shift cannot simply be converted into force gradients by using $\partial F_{ts}/\partial z = 2c_z \cdot \Delta f/f_0$ [25.140]. Several methods have been published to convert $\Delta f(z)$ data into the tip–sample potential $V_{ts}(z)$ and tip–sample force $F_{ts}(z)$ (see, e.g., [25.141–144]).

Measurement of Interatomic Forces
at Specific Atomic Sites

FM force spectroscopy has been successfully used to measure and determine quantitatively the short-range chemical force between the foremost tip atom and specific surface atoms [25.109, 145, 146]. Figure 25.22

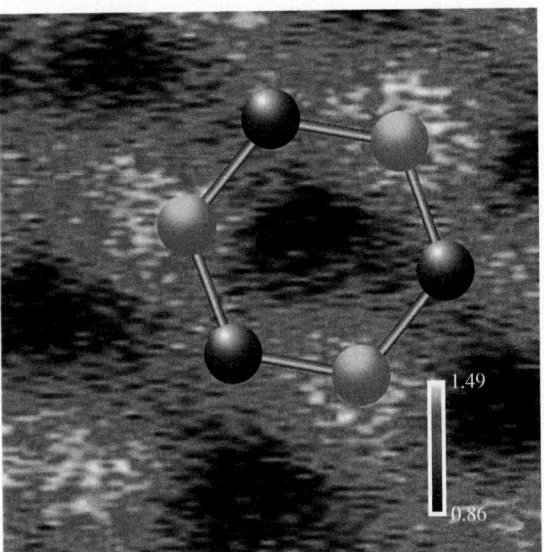

1.49

0.86

Fig. 25.21 Constant-height FM-AFM image of graphite(0001) recorded at 4.9 K using a small amplitude ($A = 0.25$ nm) and a large spring constant ($c_z = 1800$ N/m). As in Fig. 25.20a, *A*- and *B*-site carbon atoms can be distinguished. However, they appear as maxima, because imaging has been performed in the repulsive regime (courtesy of F. J. Giessibl; cf. [25.136])

displays an example for the quantitative determination of the short-range force. Figure 25.22a shows two $\Delta f(z)$-curves measured with a silicon tip above a corner hole and above an adatom. Their position is indicated by arrows in the inset, which displays the atomically resolved Si(111)-(7×7) surface. The two curves differ from each other only for small tip0-sample distances, because the long-range forces do not contribute to the atomic-scale contrast. The low, thermally induced lateral drift and the high stability at low temperatures were required to precisely address the two specific sites. To extract the short-range force, the long-range van der Waals and/or electrostatic forces can be subtracted from the total force. The grey curve in Fig. 25.22b has been reconstructed from the $\Delta f(z)$-curve recorded above an adatom and represents the total force. After removing the long-range contribution from the data, the much steeper black line is obtained, which corresponds to the short-range force between the adatom and the atom at the tip apex. The measured maximum attractive force $(-2.1\,\text{nN})$ agrees well with first-principles calculations $(-2.25\,\text{nN})$.

Three-Dimensional Force Field Spectroscopy

Further progress with the FM technique has been made by *Hölscher* et al. [25.147]. They acquired a complete 3-D force field on NiO(001) with atomic resolution (*3-D force field spectroscopy*). In Fig. 25.23a, the atomically resolved FM-AFM image of NiO(001) is shown together with the coordinate system used and the tip to illustrate the measurement principle. NiO(001) crystallizes in the rock-salt structure. The distance between the protrusions corresponds to the lattice constant of 417 pm, i. e., only one type of atom (most likely the oxygen) is imaged as a protrusion. In an area of 1 nm × 1 nm, 32 × 32 individual $\Delta f(z)$-curves have been recorded at every (x, y) image point and converted into $F_{ts}(z)$-curves. The $\Delta f(x, y, z)$ data set is thereby converted into the 3-D force field $F_{ts}(x, y, z)$. Figure 25.23b, where a specific x–z-plane is displayed, demonstrates that atomic resolution is achieved. It represents a 2-D cut $F_{ts}(x, y = \text{const}, z)$ along the [100]-direction (corresponding to the shaded slice marked in Fig. 25.23a). Since a large number of curves have been recorded, *Langkat* et al. [25.146] could evaluate the whole data set by standard statistical means to extract the long- and short-range forces. A possible future application of 3-D force field spectroscopy could be to map the short-range forces of complex molecules with functionalized tips in order to resolve locally their chemical reactivity. A first step in this direction has been accomplished on SWNTs. Its structural unit, a hexagonal carbon ring, is common to all aromatic molecules. Like the constant frequency-shift image of an SWNT shown in Fig. 25.20b the force map shows clear differences between hollow sites and carbon sites [25.133]. Analyzing site-specific individual force curves extracted from the 3-D data revealed a maximum attractive force of about $-0.106\,\text{nN}$ above H-sites and about $-0.075\,\text{nN}$ above carbon sites. Since the attraction is one order of magnitude weaker than on Si(111)-(7×7) (cf., Fig. 25.22b), it has been inferred that the short-range interatomic van der Waals force and not a chemical force is responsible

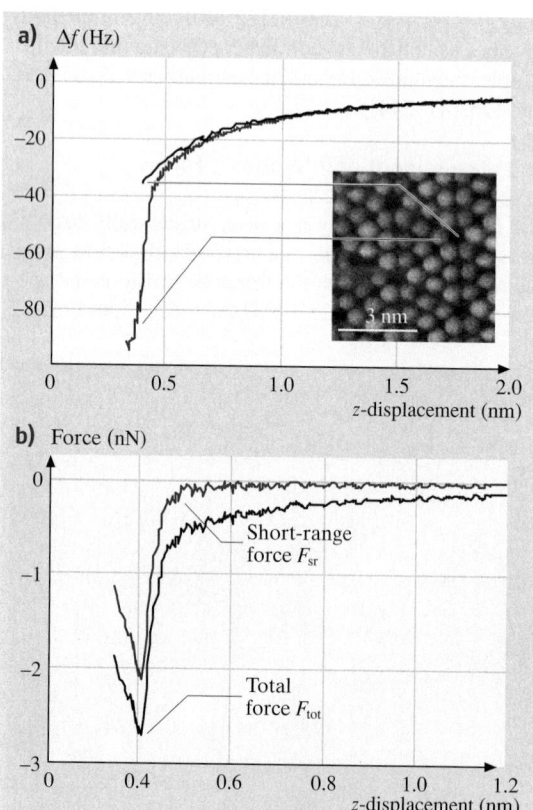

Fig. 25.22a,b FM force spectroscopy on specific atomic sites at 7.2 K. In (**a**), an FM-SFM image of the Si(111)-(7×7) surface is displayed together with two $\Delta f(z)$-curves, which have been recorded at the positions indicated by the *arrows*, i. e., above the corner hole (*brown*) and above an adatom (*black*). In (**b**), the total force above an adatom (*brown line*) has been recovered from the $\Delta f(z)$-curve. After subtraction of the long-range part, the short-range force can be determined (*black line*) (courtesy of H. J. Hug; cf. [25.145])

for atomic-scale contrast formation on such nonreactive surfaces.

Noncontact Friction

Another approach to achieve small tip–sample distances in combination with high force sensitivity is to use soft springs in a perpendicular configuration. The much higher cantilever stiffness along the cantilever axis prevents the jump-to-contact, but the lateral resolution is limited by the magnitude of the oscillation amplitude. However, with such a set-up at low temperatures, *Stipe* et al. [25.148] measured the distance dependence of the very small force due to noncontact friction between the tip and sample in vacuum. The effect was attributed to electric charges, which are moved parallel to the surface by the oscillating tip. Since the topography was not recorded in situ, the influence of contaminants or surface steps remained unknown.

25.4.3 Atomic Manipulation

Nowadays, atomic-scale manipulation is routinely performed using an STM tip (see Sect. 25.3.1). In most of these experiments an adsorbate is dragged with the tip due to an attractive force between the foremost tip apex atoms and the adsorbate. By adjusting a large or a small tip–surface distance via the tunneling resistance, it is possible to switch between imaging and manipulation. Recently, it has been demonstrated that controlled manipulation of individual atoms is also possible in the dynamic mode of atomic force microscopy, i. e., FM-AFM. Vertical manipulation was demonstrated by pressing the tip in a controlled manner into the Si(111)-(7×7) surface [25.149]. The strong repulsion leads to the removal of the selected silicon atom. The process could be traced by recording the frequency shift and the damping signal during the approach. For lateral manipulation a *rubbing* technique has been utilized [25.150], where the slow scan axis is halted above a selected atom, while the tip–surface distance is gradually reduced until the selected atom hops to a new stable position. Figure 25.24 shows a Ge adatom on Ge(111)-c(2×8) that was moved during scanning in two steps from its original position (a) to its final position (c). In fact, manipulation by FM-AFM is reproducible and fast enough to write nanostructures in a bottom-up process with single atoms [25.151].

25.4.4 Electrostatic Force Microscopy

Electrostatic forces are readily detectable by a force microscope, because the tip and sample can be regarded

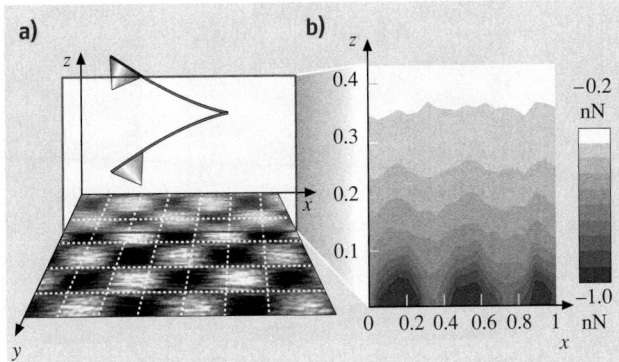

Fig. 25.23a,b Principle of the 3-D force field spectroscopy method (**a**) and a 2-D cut through the 3-D force field $F_{ts}(x,y,z)$ recorded at 14 K (**b**). At all 32×32 image points of the 1 nm × 1 nm scan area on NiO(001), a $\Delta f(z)$-curve has been recorded. The $\Delta f(x, y, z)$ data set obtained is then converted into the 3-D tip–sample force field $F_{ts}(x, y, z)$. The *shaded slice* $F_{ts}(x, y = \text{const}, z)$ in (**a**) corresponds to a cut along the [100]-direction and demonstrates that atomic resolution has been obtained, because the distance between the protrusions corresponds well to the lattice constant of nickel oxide

as two electrodes of a capacitor. If they are electrically connected via their back sides and have different work functions, electrons will flow between the tip and sample until their Fermi levels are equalized. As a result, an electric field and, consequently, an attractive electrostatic force exists between them at zero bias. This *contact potential difference* can be balanced by applying an appropriate bias voltage. It has been demonstrated that individual doping atoms in semiconducting materials can be detected by electrostatic interactions due to the local variation of the surface potential around them [25.152, 153].

Detection of Edge Channels in the Quantum Hall Regime

At low temperatures, electrostatic force microscopy has been used to measure the electrostatic potential in the quantum Hall regime of a *two-dimensional electron gas* (2-DEG) buried in epitaxially grown GaAs/AlGaAs heterostructures [25.154–157]. In the 2-DEG, electrons can move freely in the x–y-plane, but they cannot move in z-direction. Electrical transport properties of a 2-DEG are very different compared to normal metallic conduction. Particularly, the Hall resistance $R_H = h/ne^2$ (where h represents Planck's constant, e is the electron charge, and $n = 1, 2, \ldots$) is quantized in the quantum Hall regime, i. e., at sufficiently low temperatures

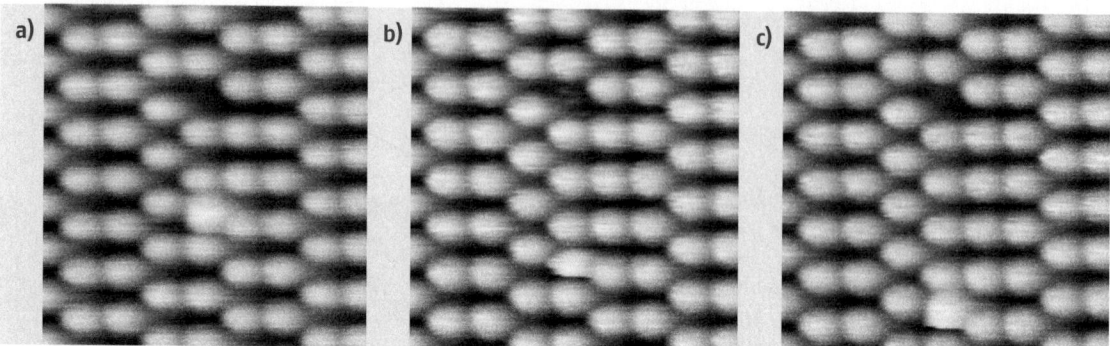

Fig. 25.24 Consecutively recorded FM-AFM images showing the tip-induced manipulation of a Ge adatom on Ge(111)-c(2×8) at 80 K. Scanning was performed from bottom to top (courtesy of N. Oyabu; cf. [25.150])

($T < 4$ K) and high magnetic fields (up to 20 T). Under these conditions, theoretical calculations predict the existence of *edge channels* in a Hall bar. A Hall bar is a strip conductor that is contacted in a specific way to allow longitudinal and transversal transport measurements in a perpendicular magnetic field. The current is not evenly distributed over the cross section of the bar, but passes mainly along rather thin paths close to the edges. This prediction has been verified by measuring profiles of the electrostatic potential across a Hall bar in different perpendicular external magnetic fields [25.154–156].

Figure 25.25a shows the experimental set-up used to observe these edge channels on top of a Hall bar with a force microscope. The tip is positioned above the surface of a Hall bar under which the 2-DEG is buried.

The direction of the magnetic field is oriented perpendicular to the 2-DEG. Note that, although the 2-DEG is located several tens of nanometers below the surface, its influence on the electrostatic surface potential can be detected. In Fig. 25.25b, the results of scans perpendicular to the Hall bar are plotted against the magnitude of the external magnetic field. The value of the electrostatic potential is grey-coded in arbitrary units. In certain field ranges, the potential changes linearly across the Hall bar, while in other field ranges the potential drop is confined to the edges of the Hall bar. The predicted edge channels can explain this behavior. The periodicity of the phenomenon is related to the filling factor v, i.e., the number of Landau levels that are filled with electrons (see also Sect. 25.3.4). Its value depends on $1/B$

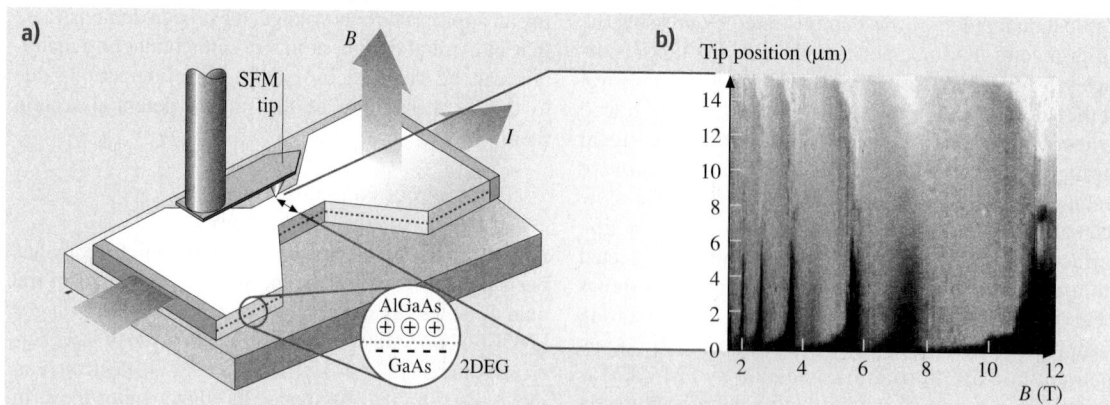

Fig. 25.25a,b Configuration of the Hall bar within a low temperature ($T < 1$ K) force microscope (**a**) and profiles (*y*-axis) at different magnetic field (*x*-axis) of the electrostatic potential across a 14-μm-wide Hall bar in the quantum Hall regime (**b**). The external magnetic field is oriented perpendicular to the 2-DEG, which is buried below the surface. *Bright* and *dark regions* reflect the characteristic changes of the electrostatic potential across the Hall bar at different magnetic fields and can be explained by the existence of the theoretically predicted edge channels (courtesy of E. Ahlswede; cf. [25.156])

and is proportional to the electron concentration n_e in the 2-DEG ($\nu = n_e h / eB$, where h represents Planck's constant and e the electron charge).

25.4.5 Magnetic Force Microscopy

To detect magnetostatic tip–sample interactions with magnetic force microscopy (MFM), a ferromagnetic probe has to be used. Such probes are readily prepared by evaporating a thin magnetic layer, e.g., 10 nm iron, onto the tip. Due to the in-plane shape anisotropy of thin films, the magnetization of such tips lies predominantly along the tip axis, i.e., perpendicular to the surface. Since magnetostatic interactions are long-range, they can be separated from the topography by scanning in a certain constant height (typically around 20 nm) above the surface, where the z-component of the sample stray field is probed (see Fig. 25.26a). Therefore, MFM is always operated in noncontact mode. The signal from the cantilever is directly recorded while the z-feedback is switched off. MFM can be operated in the static mode or in the dynamic modes (AM-MFM at ambient pressures and FM-MFM in vacuum). A lateral resolution below 50 nm can be routinely obtained.

Observation of Domain Patterns

MFM is widely used to visualize domain patterns of ferromagnetic materials. At low temperatures, *Moloni* et al. [25.158] observed the domain structure of magnetite below its Verwey transition temperature ($T_V = 122$ K), but most of the work concentrated on thin films of $La_{1-x}Ca_xMnO_3$ [25.159–161]. Below T_V, the conductivity decreases by two orders of magnitude and a small structural distortion is observed. The domain structure of this mixed-valence manganite is of great interest, because its resistivity strongly depends on the external magnetic field, i.e., it exhibits a large colossal-magnetoresistive effect. To investigate the field dependence of the domain patterns under ambient conditions, electromagnets have to be used. They can cause severe thermal drift problems due to Joule heating of the coils by large currents. Fields on the order of 100 mT can be achieved. In contrast, much larger fields (more than 10 T) can be rather easily produced by implementing a superconducting magnet in low-temperature set-ups. With such a design, *Liebmann* et al. [25.161] recorded the domain structure along the major hysteresis loop of $La_{0.7}Ca_{0.3}MnO_3$ epitaxially grown on $LaAlO_3$ (see Fig. 25.26b–f). The film geometry (the thickness is 100 nm) favors an in-plane magnetization, but the lattice mismatch with the substrate induces an out-of-plane anisotropy. Thereby, an irregular pattern of strip domains appears at zero field. If the external magnetic field is increased, the domains with antiparallel orientation shrink and finally disappear in saturation (see Fig. 25.26b,c). The residual contrast in saturation (d) reflects topographic features. If the field is decreased after saturation (see Fig. 25.26e,f), cylindrical domains first nucleate and then start to grow. At zero field, the maze-type domain pattern has evolved again. Such data sets can be used to analyze domain nucleation and the domain growth mode. Moreover, due to the negligible drift, domain structure and surface morphology can be directly compared, because every MFM can be used as a regular topography-imaging force microscope.

Detection of Individual Vortices in Superconductors

Numerous low-temperature MFM experiments have been performed on superconductors [25.162–169]. Some basic features of superconductors have been mentioned already in Sect. 25.3.5. The main difference of STM/STS compared to MFM is its high sensitivity to the electronic properties of the surface. Therefore, careful sample preparation is a prerequisite. This is not so important for MFM experiments, since the tip is scanned at a certain distance above the surface.

Superconductors can be divided into two classes with respect to their behavior in an external magnetic field. For type-I superconductors, any magnetic flux is entirely excluded below their critical temperature T_c (Meissner effect), while for type-II superconductors, cylindrical inclusions (*vortices*) of normal material exist in a superconducting matrix (*vortex* state). The radius of the vortex *core*, where the Cooper-pair density decreases to zero, is on the order of the coherence length ξ. Since the superconducting gap vanishes in the core, they can be detected by STS (see Sect. 25.3.5). Additionally, each vortex contains one magnetic quantum flux $\Phi = h/2e$ (where h represents Planck's constant and e the electron charge). Circular supercurrents around the core screen the magnetic field associated with a vortex; their radius is given by the London penetration depth λ of the material. This magnetic field of the vortices can be detected by MFM. Investigations have been performed on the two most popular copper oxide high-T_c superconductors, $YBa_2Cu_3O_7$ [25.162, 163, 165] and $Bi_2Sr_2CaCu_2O_8$ [25.163, 169], on the only elemental conventional type-II superconductor Nb [25.166, 167], and on the layered compound crystal $NbSe_2$ [25.164, 166].

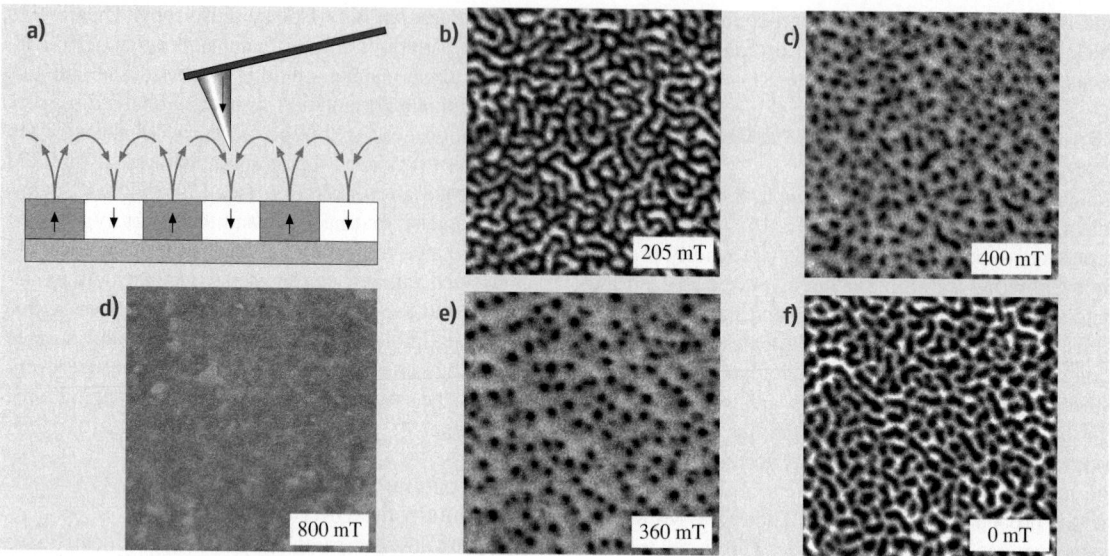

Fig. 25.26a–f Principle of MFM operation (**a**) and field-dependent domain structure of a ferromagnetic thin film (**b**)–(**f**) recorded at 5.2 K with FM-MFM. All images were recorded on the same $4\,\mu m \times 4\,\mu m$ scan area. The $La_{0.7}Ca_{0.3}MnO_3/LaAlO_3$ system exhibits a substrate-induced out-of-plane anisotropy. *Bright* and *dark areas* are visible and correspond to attractive and repulsive magnetostatic interactions, respectively. The series shows how the domain pattern evolves along the major hysteresis loop from, i. e., zero field to saturation at 600 mT and back to zero field

Most often, vortices have been generated by cooling the sample from the normal state to below T_c in an external magnetic field. After such a *field-cooling* procedure, the most energetically favorable vortex arrangement is a regular triangular Abrikosov lattice. *Volodin* et al. [25.164] were able to observe such an Abrikosov lattice on $NbSe_2$. The intervortex distance d is related to the external field during B cool down via $d = (4/3)^{1/4}(\Phi/B)^{1/2}$. Another way to introduce vortices into a type-II superconductor is vortex penetration from the edge by applying a magnetic field at temperatures below T_c. According to the Bean model, a vortex density gradient exists under such conditions within the superconducting material. *Pi* et al. [25.169] slowly increased the external magnetic field until the vortex front approaching from the edge reached the scanning area.

If the vortex configuration is dominated by the *pinning* of vortices at randomly distributed structural defects, no Abrikosov lattice emerges. The influence of pinning centers can be studied easily by MFM, because every MFM can be used to scan the topography in its AFM mode. This has been done for natural growth defects by *Moser* et al. [25.165] on $YBa_2Cu_3O_7$ and for $YBa_2Cu_3O_7$ and niobium thin films, respectively, by *Volodin* et al. [25.168].

Roseman et al. [25.170] investigated the formation of vortices in the presence of an artificial structure on niobium films, while *Pi* et al. [25.169] produced columnar defects by heavy-ion bombardment in a $Bi_2Sr_2CaCu_2O_8$ single crystal to study the strong pinning at these defects.

Figure 25.27 demonstrates that MFM is sensitive to the polarity of vortices. In Fig. 25.27a, six vortices have been produced in a niobium film by field cooling in $+0.5\,mT$. The external magnetic field and tip magnetization are parallel, and, therefore, the tip–vortex interaction is attractive (bright contrast). To remove the vortices, the niobium was heated above T_c ($\approx 9\,K$). Thereafter, vortices of opposite polarity were produced by field-cooling in $-0.5\,mT$, which appear dark in Fig. 25.27b. The vortices are probably bound to strong pinning sites, because the vortex positions are identical in both images of Fig. 25.27. By imaging the vortices at different scanning heights, *Roseman* et al. [25.167] tried to extract values for the London penetration depth from the scan height dependence of their profiles. While good qualitative agreement with theoretical predictions has been found, the absolute values do not agree with published literature values. The disagreement was attributed to the convolution between the tip and vortex stray fields. Better values might be obtained with calibrated tips.

Single Spin Detection

So far, only collective magnetic phenomena like ferromagnetic domains have been observed via magnetostatic tip–sample interactions detected by MFM. However, magnetic ordering exists due to the magnetic exchange interaction between the electron spins of neighboring atoms in a solid. The most energetically favorable situation can be either ferromagnetic (parallel orientation) or antiferromagnetic (antiparallel orientation) ordering. It has been predicted that the magnetic exchange force between an individual spin of a magnetically ordered sample and the spin of the foremost atom of a magnetic tip can be detected at sufficiently small tip–sample distances [25.172, 173].

The experimental realization, however, is very difficult, because the magnetic exchange force is about a factor of ten weaker and of even shorter range than the chemical interactions that are responsible for the atomic-scale contrast. FM-AFM experiments with a ferromagnetic tip have been performed on the antiferromagnetic NiO(001) surface at room temperature [25.174] and with a considerable better signal-to-noise ratio at low temperatures [25.122]. Although it was possible to achieve atomic resolution, a periodic contrast that could be attributed to the antiferromagnetically ordered spins of the nickel atoms could not be observed.

Even more ambitious is the proposed detection of individual nuclear spins by magnetic resonance force

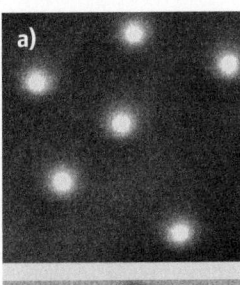

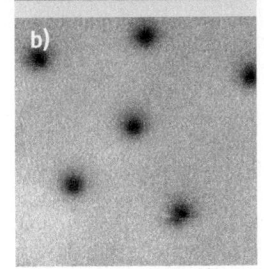

Fig. 25.27a,b Two 5 μm × 5 μm FM-MFM images of vortices in a niobium thin film after field-cooling at 0.5 mT (**a**) and − 0.5 mT (**b**), respectively. Since the external magnetic field was parallel in (**a**) and antiparallel in (**b**) with respect to the tip magnetization, the vortices exhibit reversed contrast. Strong pinning dominates the position of the vortices, since they appear at identical locations in (**a**) and (**b**) and are not arranged in a regular Abrikosov lattice (courtesy of P. Grütter; cf. [25.167])

microscopy (MRFM) using a magnetic tip [25.175, 176]. Conventionally, nuclear spins are investigated by nuclear magnetic resonance (NMR), a spectroscopic technique to obtain microscopic chemical and physical information about molecules. An important application of NMR for medical diagnostics of the inside of humans is magnetic resonance imaging (MRI). This tomographic imaging technique uses the NMR signal from thin slices through the body to reconstruct its three-dimensional structure. Currently, at least 10^{12} nuclear spins must be present in a given volume to obtain a significant MRI signal. The ultimate goal of MRFM is to combine aspects of force microscopy with MRI to achieve true 3-D imaging with atomic resolution and elemental selectivity.

The experimental set-up is sketched in Fig. 25.28. An oscillating cantilever with a magnetic tip at its end points toward the surface. The spherical resonant slice within the sample represents those points where the stray field from the tip and the external field match the condition for magnetic resonance. The cyclic spin flip causes a slight shift of the cantilever frequency due to the magnetic force exerted by the spin on the tip. Since the forces are extremely small, very low temperatures are required.

To date, no individual nuclear spins have been detected by MRFM. However, the design of ultrasensitive cantilevers has made considerable progress, and the detection of forces below 1×10^{-18} N has been achieved [25.120]. Therefore, it has become possible to perform nuclear magnetic resonance [25.177], and ferromagnetic resonance [25.178] experiments of spin ensembles with micrometer resolution. Moreover, in

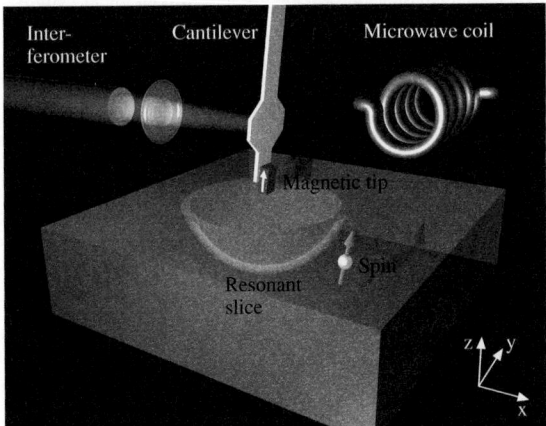

Fig. 25.28 MRFM set-up. The cantilever with the magnetic tip oscillates parallel to the surface. Only electron spins within a hemispherical slice, where the stray field of the tip plus the external field matches the condition for magnetic resonance, can contribute to the MRFM signal due to cyclic spin inversion (courtesy of D. Rugar; cf. [25.171])

SiO$_2$ the magnetic moment of a single electron, which is three orders of magnitude larger than the nuclear magnetic moment, could be detected [25.171] using the set-up shown in Fig. 25.28 at 1.6 K. This major breakthrough demonstrates the capability of force microscopy to detect single spins.

References

25.1   G. Binnig, H. Rohrer, Ch. Gerber, E. Weibel: Surface studies by scanning tunneling microscopy, Phys. Rev. Lett. **49**, 57–61 (1982)

25.2   R. Wiesendanger: *Scanning Probe Microscopy and Spectroscopy* (Cambridge Univ. Press, Cambridge 1994)

25.3   M. Tinkham: *Introduction to Superconductivity* (McGraw–Hill, New York 1996)

25.4   J. Kondo: Theory of dilute magnetic alloys, Solid State Phys. **23**, 183–281 (1969)

25.5   T. R. Albrecht, P. Grütter, H. K. Horne, D. Rugar: Frequency modulation detection using high-Q cantilevers for enhanced force microscope sensitivity, J. Appl. Phys. **69**, 668–673 (1991)

25.6   F. J. Giessibl, H. Bielefeld, S. Hembacher, J. Mannhart: Calculation of the optimal imaging parameters for frequency modulation atomic force microscopy, Appl. Surf. Sci. **140**, 352–357 (1999)

25.7   W. Allers, A. Schwarz, U. D. Schwarz, R. Wiesendanger: Dynamic scanning force microscopy at low temperatures on a van der Waals surface: graphite(0001), Appl. Surf. Sci. **140**, 247–252 (1999)

25.8   W. Allers, A. Schwarz, U. D. Schwarz, R. Wiesendanger: Dynamic scanning force microscopy at low temperatures on a noble-gas crystal: atomic resolution on the xenon(111) surface, Europhys. Lett. **48**, 276–279 (1999)

25.9   M. Morgenstern, D. Haude, V. Gudmundsson, C. Wittneven, R. Dombrowski, R. Wiesendanger: Origin of Landau oscillations observed in scanning tunneling spectroscopy on n-InAs(110), Phys. Rev. B **62**, 7257–7263 (2000)

25.10   D. M. Eigler, P. S. Weiss, E. K. Schweizer, N. D. Lang: Imaging Xe with a low-temperature scanning tunneling microscope, Phys. Rev. Lett. **66**, 1189–1192 (1991)

25.11   P. S. Weiss, D. M. Eigler: Site dependence of the apparent shape of a molecule in scanning tunneling microscope images: Benzene on Pt{111}, Phys. Rev. Lett. **71**, 3139–3142 (1992)

25.12   D. M. Eigler, E. K. Schweizer: Positioning single atoms with a scanning tunneling microscope, Nature **344**, 524–526 (1990)

25.13   H. Hug, B. Stiefel, P. J. A. van Schendel, A. Moser, S. Martin, H.-J. Güntherodt: A low temperature ultrahigh vacuum scanning force microscope, Rev. Sci. Instrum. **70**, 3627–3640 (1999)

25.14   S. Behler, M. K. Rose, D. F. Ogletree, F. Salmeron: Method to characterize the vibrational response of a beetle type scanning tunneling microscope, Rev. Sci. Instrum. **68**, 124–128 (1997)

25.15   C. Wittneven, R. Dombrowski, S. H. Pan, R. Wiesendanger: A low-temperature ultrahigh-vacuum scanning tunneling microscope with rotatable magnetic field, Rev. Sci. Instrum. **68**, 3806–3810 (1997)

25.16   W. Allers, A. Schwarz, U. D. Schwarz, R. Wiesendanger: A scanning force microscope with atomic resolution in ultrahigh vacuum and at low temperatures, Rev. Sci. Instrum. **69**, 221–225 (1998)

25.17   G. Dujardin, R. E. Walkup, Ph. Avouris: Dissociation of individual molecules with electrons from the tip of a scanning tunneling microscope, Science **255**, 1232–1235 (1992)

25.18   H. J. Lee, W. Ho: Single-bond formation and characterization with a scanning tunneling microscope, Science **286**, 1719–1722 (1999)

25.19   R. Berndt, R. Gaisch, J. K. Gimzewski, B. Reihl, R. R. Schlittler, W. D. Schneider, M. Tschudy: Photon emission at molecular resolution induced by a scanning tunneling microscope, Science **262**, 1425–1427 (1993)

25.20   B. G. Briner, M. Doering, H. P. Rust, A. M. Bradshaw: Microscopic diffusion enhanced by adsorbate interaction, Science **278**, 257–260 (1997)

25.21   J. Kliewer, R. Berndt, E. V. Chulkov, V. M. Silkin, P. M. Echenique, S. Crampin: Dimensionality effects in the lifetime of surface states, Science **288**, 1399–1401 (2000)

25.22   M. F. Crommie, C. P. Lutz, D. M. Eigler: Imaging standing waves in a two-dimensional electron gas, Nature **363**, 524–527 (1993)

25.23   B. C. Stipe, M. A. Rezaei, W. Ho: Single-molecule vibrational spectroscopy and microscopy, Science **280**, 1732–1735 (1998)

25.24   H. J. Lee, W. Ho: Structural determination by single-molecule vibrational spectroscopy and microscopy: Contrast between copper and iron carbonyls, Phys. Rev. B **61**, R16347–R16350 (2000)

25.25   C. W. J. Beenakker, H. van Houten: Quantum transport in semiconductor nanostructures, Solid State Phys. **44**, 1–228 (1991)

25.26   S. H. Pan, E. W. Hudson, K. M. Lang, H. Eisaki, S. Uchida, J. C. Davis: Imaging the effects of individual zinc impurity atoms on superconductivity in Bi$_2$Sr$_2$CaCu$_2$O$_{8+\bar{z}}$, Nature **403**, 746–750 (2000)

25.27   R. S. Becker, J. A. Golovchenko, B. S. Swartzentruber: Atomic-scale surface modifications using a tunneling microscope, Nature **325**, 419–42 (1987)

25.28 J. A. Stroscio, D. M. Eigler: Atomic and molecular manipulation with the scanning tunneling microscope, Science **254**, 1319–1326 (1991)

25.29 L. Bartels, G. Meyer, K. H. Rieder: Basic steps of lateral manipulation of single atoms and diatomic clusters with a scanning tunneling microscope, Phys. Rev. Lett. **79**, 697–700 (1997)

25.30 J. J. Schulz, R. Koch, K. H. Rieder: New mechanism for single atom manipulation, Phys. Rev. Lett. **84**, 4597–4600 (2000)

25.31 T. C. Shen, C. Wang, G. C. Abeln, J. R. Tucker, J. W. Lyding, Ph. Avouris, R. E. Walkup: Atomic-scale desorption through electronic and vibrational excitation mechanisms, Science **268**, 1590–1592 (1995)

25.32 T. Komeda, Y. Kim, M. Kawai, B. N. J. Persson, H. Ueba: Lateral hopping of molecules induced by excitations of internal vibration mode, Science **295**, 2055–2058 (2002)

25.33 Y. W. Mo: Reversible rotation of antimony dimers on the silicon(001) surface with a scanning tunneling microscope, Science **261**, 886–888 (1993)

25.34 B. C. Stipe, M. A. Rezaei, W. Ho: Inducing and viewing the rotational motion of a single molecule, Science **279**, 1907–1909 (1998)

25.35 F. Moresco, G. Meyer, K. H. Rieder, H. Tang, A. Gourdon, C. Joachim: Conformational changes of single molecules by scanning tunneling microscopy manipulation: a route to molecular switching, Phys. Rev. Lett. **86**, 672–675 (2001)

25.36 S. W. Hla, L. Bartels, G. Meyer, K. H. Rieder: Inducing all steps of a chemical reaction with the scanning tunneling microscope tip: Towards single molecule engineering, Phys. Rev. Lett. **85**, 2777–2780 (2000)

25.37 E. Ganz, S. K. Theiss, I. S. Hwang, J. Golovchenko: Direct measurement of diffusion by hot tunneling microscopy: Activations energy, anisotropy, and long jumps, Phys. Rev. Lett. **68**, 1567–1570 (1992)

25.38 M. Schuhnack, T. R. Linderoth, F. Rosei, E. Laegsgaard, I. Stensgaard, F. Besenbacher: Long jumps in the surface diffusion of large molecules, Phys. Rev. Lett. **88**, 156102, 1–4 (2002)

25.39 L. J. Lauhon, W. Ho: Direct observation of the quantum tunneling of single hydrogen atoms with a scanning tunneling microscope, Phys. Rev. Lett. **85**, 4566–4569 (2000)

25.40 N. Kitamura, M. Lagally, M. B. Webb: Real-time observation of vacancy diffusion on Si(001)-(2×1) by scanning tunneling microscopy, Phys. Rev. Lett. **71**, 2082–2085 (1993)

25.41 M. Morgenstern, T. Michely, G. Comsa: Onset of interstitial diffusion determined by scanning tunneling microscopy, Phys. Rev. Lett. **79**, 1305–1308 (1997)

25.42 K. Morgenstern, G. Rosenfeld, B. Poelsema, G. Comsa: Brownian motion of vacancy islands on Ag(111), Phys. Rev. Lett. **74**, 2058–2061 (1995)

25.43 B. Reihl, J. H. Coombs, J. K. Gimzewski: Local inverse photoemission with the scanning tunneling microscope, Surf. Sci. **211–212**, 156–164 (1989)

25.44 R. Berndt, J. K. Gimzewski, P. Johansson: Inelastic tunneling excitation of tip-induced plasmon modes on noble-metal surfaces, Phys. Rev. Lett. **67**, 3796–3799 (1991)

25.45 P. Johansson, R. Monreal, P. Apell: Theory for light emission from a scanning tunneling microscope, Phys. Rev. B **42**, 9210–9213 (1990)

25.46 J. Aizpurua, G. Hoffmann, S. P. Apell, R. Berndt: Electromagnetic coupling on an atomic scale, Phys. Rev. Lett. **89**, 156803, 1–4 (2002)

25.47 G. Hoffmann, J. Kliewer, R. Berndt: Luminescence from metallic quantum wells in a scanning tunneling microscope, Phys. Rev. Lett. **78**, 176803, 1–4 (2001)

25.48 A. Downes, M. E. Welland: Photon emission from Si(111)-(7×7) induced by scanning tunneling microscopy: atomic scale and material contrast, Phys. Rev. Lett. **81**, 1857–1860 (1998)

25.49 M. Kemerink, K. Sauthoff, P. M. Koenraad, J. W. Geritsen, H. van Kempen, J. H. Wolter: Optical detection of ballistic electrons injected by a scanning-tunneling microscope, Phys. Rev. Lett. **86**, 2404–2407 (2001)

25.50 J. Tersoff, D. R. Hamann: Theory and application for the scanning tunneling microscope, Phys. Rev. Lett. **50**, 1998–2001 (1983)

25.51 C. J. Chen: *Introduction to Scanning Tunneling Microscopy* (Oxford Univ. Press, Oxford 1993)

25.52 J. Winterlin, J. Wiechers, H. Brune, T. Gritsch, H. Hofer, R. J. Behm: Atomic-resolution imaging of close-packed metal surfaces by scanning tunneling microscopy, Phys. Rev. Lett. **62**, 59–62 (1989)

25.53 A. L. Vazquez de Parga, O. S. Hernan, R. Miranda, A. Levy Yeyati, N. Mingo, A. Martin-Rodero, F. Flores: Electron resonances in sharp tips and their role in tunneling spectroscopy, Phys. Rev. Lett. **80**, 357–360 (1998)

25.54 S. H. Pan, E. W. Hudson, J. C. Davis: Vacuum tunneling of superconducting quasiparticles from atomically sharp scanning tunneling microscope tips, Appl. Phys. Lett. **73**, 2992–2994 (1998)

25.55 J. T. Li, W. D. Schneider, R. Berndt, O. R. Bryant, S. Crampin: Surface-state lifetime measured by scanning tunneling spectroscopy, Phys. Rev. Lett. **81**, 4464–4467 (1998)

25.56 L. Bürgi, O. Jeandupeux, H. Brune, K. Kern: Probing hot-electron dynamics with a cold scanning tunneling microscope, Phys. Rev. Lett. **82**, 4516–4519 (1999)

25.57 J. W. G. Wildoer, C. J. P. M. Harmans, H. van Kempen: Observation of Landau levels at the InAs(110) surface by scanning tunneling spectroscopy, Phys. Rev. B **55**, R16013–R16016 (1997)

25.58 M. Morgenstern, V. Gudmundsson, C. Wittneven, R. Dombrowski, R. Wiesendanger: Nonlocality of the exchange interaction probed by scanning tunneling spectroscopy, Phys. Rev. B **63**, 201301(R), 1–4 (2001)

25.59 M. V. Grishin, F. I. Dalidchik, S. A. Kovalevskii, N. N. Kolchenko, B. R. Shub: Isotope effect in the vibrational spectra of water measured in experiments with a scanning tunneling microscope, JETP Lett. **66**, 37–40 (1997)

25.60 A. Hewson: *From the Kondo Effect to Heavy Fermions* (Cambridge Univ. Press, Cambridge 1993)

25.61 V. Madhavan, W. Chen, T. Jamneala, M. F. Crommie, N. S. Wingreen: Tunneling into a single magnetic atom: Spectroscopic evidence of the Kondo resonance, Science **280**, 567–569 (1998)

25.62 J. Li, W. D. Schneider, R. Berndt, B. Delley: Kondo scattering observed at a single magnetic impurity, Phys. Rev. Lett. **80**, 2893–2896 (1998)

25.63 T. W. Odom, J. L. Huang, C. L. Cheung, C. M. Lieber: Magnetic clusters on single-walled carbon nanotubes: the Kondo effect in a one-dimensional host, Science **290**, 1549–1552 (2000)

25.64 M. Ouyang, J. L. Huang, C. L. Cheung, C. M. Lieber: Energy gaps in metallic single-walled carbon nanotubes, Science **292**, 702–705 (2001)

25.65 U. Fano: Effects of configuration interaction on intensities and phase shifts, Phys. Rev. **124**, 1866–1878 (1961)

25.66 H. C. Manoharan, C. P. Lutz, D. M. Eigler: Quantum mirages formed by coherent projection of electronic structure, Nature **403**, 512–515 (2000)

25.67 O. Y. Kolesnychenko, R. de Kort, M. I. Katsnelson, A. I. Lichtenstein, H. van Kempen: Real-space observation of an orbital Kondo resonance on the Cr(001) surface, Nature **415**, 507–509 (2002)

25.68 H. A. Mizes, J. S. Foster: Long-range electronic perturbations caused by defects using scanning tunneling microscopy, Science **244**, 559–562 (1989)

25.69 P. T. Sprunger, L. Petersen, E. W. Plummer, E. Laegsgaard, F. Besenbacher: Giant Friedel oscillations on beryllium(0001) surface, Science **275**, 1764–1767 (1997)

25.70 P. Hofmann, B. G. Briner, M. Doering, H. P. Rust, E. W. Plummer, A. M. Bradshaw: Anisotropic two-dimensional Friedel oscillations, Phys. Rev. Lett. **79**, 265–268 (1997)

25.71 E. J. Heller, M. F. Crommie, C. P. Lutz, D. M. Eigler: Scattering and adsorption of surface electron waves in quantum corrals, Nature **369**, 464–466 (1994)

25.72 M. C. M. M. van der Wielen, A. J. A. van Roij, H. van Kempen: Direct observation of Friedel oscillations around incorporated Si_{Ga} dopants in GaAs by low-temperature scanning tunneling microscopy, Phys. Rev. Lett. **76**, 1075–1078 (1996)

25.73 O. Millo, D. Katz, Y. W. Cao, U. Banin: Imaging and spectroscopy of artificial-atom states in core/shell nanocrystal quantum dots, Phys. Rev. Lett. **86**, 5751–5754 (2001)

25.74 L. C. Venema, J. W. G. Wildoer, J. W. Janssen, S. J. Tans, L. J. T. Tuinstra, L. P. Kouwenhoven, C. Dekker: Imaging electron wave functions of quantized energy levels in carbon nanotubes, Nature **283**, 52–55 (1999)

25.75 S. G. Lemay, J. W. Jannsen, M. van den Hout, M. Mooij, M. J. Bronikowski, P. A. Willis, R. E. Smalley, L. P. Kouwenhoven, C. Dekker: Two-dimensional imaging of electronic wavefunctions in carbon nanotubes, Nature **412**, 617–620 (2001)

25.76 C. Wittneven, R. Dombrowski, M. Morgenstern, R. Wiesendanger: Scattering states of ionized dopants probed by low temperature scanning tunneling spectroscopy, Phys. Rev. Lett. **81**, 5616–5619 (1998)

25.77 D. Haude, M. Morgenstern, I. Meinel, R. Wiesendanger: Local density of states of a three-dimensional conductor in the extreme quantum limit, Phys. Rev. Lett. **86**, 1582–1585 (2001)

25.78 R. Joynt, R. E. Prange: Conditions for the quantum Hall effect, Phys. Rev. B **29**, 3303–3317 (1984)

25.79 M. Morgenstern, J. Klijn, C. Meyer, M. Getzlaff, R. Adelung, R. A. Römer, K. Rossnagel, L. Kipp, M. Skibowski, R. Wiesendanger: Direct comparison between potential landscape and local density of states in a disordered two-dimensional electron system, Phys. Rev. Lett. **89**, 136806, 1–4 (2002)

25.80 E. Abrahams, P. W. Anderson, D. C. Licciardello, T. V. Ramakrishnan: Scaling theory of localization: absence of quantum diffusion in two dimensions, Phys. Rev. Lett. **42**, 673–676 (1979)

25.81 M. Morgenstern, J. Klijn, R. Wiesendanger: Real space observation of drift states in a two-dimensional electron system at high magnetic fields, Phys. Rev. Lett. **90**, 056804, 1–4 (2003)

25.82 R. E. Peierls: *Quantum Theory of Solids* (Clarendon, Oxford 1955)

25.83 C. G. Slough, W. W. McNairy, R. V. Coleman, B. Drake, P. K. Hansma: Charge-density waves studied with the use of a scanning tunneling microscope, Phys. Rev. B **34**, 994–1005 (1986)

25.84 X. L. Wu, C. M. Lieber: Hexagonal domain-like charge-density wave of TaS_2 determined by scanning tunneling microscopy, Science **243**, 1703–1705 (1989)

25.85 T. Nishiguchi, M. Kageshima, N. Ara-Kato, A. Kawazu: Behaviour of charge density waves in a one-dimensional organic conductor visualized by scanning tunneling microscopy, Phys. Rev. Lett. **81**, 3187–3190 (1998)

25.86 X. L. Wu, C. M. Lieber: Direct observation of growth and melting of the hexagonal-domain charge-density-wave phase in 1T-TaS_2 by scanning tunneling microscopy, Phys. Rev. Lett. **64**, 1150–1153 (1990)

25.87   J.M. Carpinelli, H.H. Weitering, E.W. Plummer, R. Stumpf: Direct observation of a surface charge density wave, Nature **381**, 398–400 (1996)

25.88   H.H. Weitering, J.M. Carpinelli, A.V. Melechenko, J. Zhang, M. Bartkowiak, E.W. Plummer: Defect-mediated condensation of a charge density wave, Science **285**, 2107–2110 (1999)

25.89   H.W. Yeom, S. Takeda, E. Rotenberg, I. Matsuda, K. Horikoshi, J. Schäfer, C.M. Lee, S.D. Kevan, T. Ohta, T. Nagao, S. Hasegawa: Instability and charge density wave of metallic quantum chains on a silicon surface, Phys. Rev. Lett. **82**, 4898–4901 (1999)

25.90   K. Swamy, A. Menzel, R. Beer, E. Bertel: Charge-density waves in self-assembled halogen-bridged metal chains, Phys. Rev. Lett. **86**, 1299–1302 (2001)

25.91   J.J. Kim, W. Yamaguchi, T. Hasegawa, K. Kitazawa: Observation of Mott localization gap using low temperature scanning tunneling spectroscopy in commensurate $1T-TaSe_2$, Phys. Rev. Lett. **73**, 2103–2106 (1994)

25.92   J. Bardeen, L.N. Cooper, J.R. Schrieffer: Theory of superconductivity, Phys. Rev. **108**, 1175–1204 (1957)

25.93   A. Yazdani, B.A. Jones, C.P. Lutz, M.F. Crommie, D.M. Eigler: Probing the local effects of magnetic impurities on superconductivity, Science **275**, 1767–1770 (1997)

25.94   S.H. Tessmer, M.B. Tarlie, D.J. van Harlingen, D.L. Maslov, P.M. Goldbart: Probing the superconducting proximity effect in $NbSe_2$ by scanning tunneling micrsocopy, Phys. Rev. Lett **77**, 924–927 (1996)

25.95   K. Inoue, H. Takayanagi: Local tunneling spectroscopy of Nb/InAs/Nb superconducting proximity system with a scanning tunneling microscope, Phys. Rev. B **43**, 6214–6215 (1991)

25.96   H.F. Hess, R.B. Robinson, R.C. Dynes, J.M. Valles, J.V. Waszczak: Scanning-tunneling-microscope observation of the Abrikosov flux lattice and the density of states near and inside a fluxoid, Phys. Rev. Lett. **62**, 214–217 (1989)

25.97   H.F. Hess, R.B. Robinson, J.V. Waszczak: Vortex-core structure observed with a scanning tunneling microscope, Phys. Rev. Lett. **64**, 2711–2714 (1990)

25.98   N. Hayashi, M. Ichioka, K. Machida: Star-shaped local density of states around vortices in a type-II superconductor, Phys. Rev. Lett. **77**, 4074–4077 (1996)

25.99   H. Sakata, M. Oosawa, K. Matsuba, N. Nishida: Imaging of vortex lattice transition in YNi_2B_2C by scanning tunneling spectroscopy, Phys. Rev. Lett. **84**, 1583–1586 (2000)

25.100 S. Behler, S.H. Pan, P. Jess, A. Baratoff, H.-J. Güntherodt, F. Levy, G. Wirth, J. Wiesner: Vortex pinning in ion-irradiated $NbSe_2$ studied by scanning tunneling microscopy, Phys. Rev. Lett. **72**, 1750–1753 (1994)

25.101 R. Berthe, U. Hartmann, C. Heiden: Influence of a transport current on the Abrikosov flux lattice observed with a low-temperature scanning tunneling microscope, Ultramicroscopy **42–44**, 696–698 (1992)

25.102 A. Polkovnikov, S. Sachdev, M. Vojta: Impurity in a d-wave superconductor: Kondo effect and STM spectra, Phys. Rev. Lett. **86**, 296–299 (2001)

25.103 E.W. Hudson, K.M. Lang, V. Madhavan, S.H. Pan, S. Uchida, J.C. Davis: Interplay of magnetism and high-T_c superconductivity at individual Ni impurity atoms in $Bi_2Sr_2CaCu_2O_{8+z}$, Nature **411**, 920–924 (2001)

25.104 K.M. Lang, V. Madhavan, J.E. Hoffman, E.W. Hudson, H. Eisaki, S. Uchida, J.C. Davis: Imaging the granular structure of high-T_c superconductivity in underdoped $Bi_2Sr_2CaCu_2O_{8+z}$, Nature **415**, 412–416 (2002)

25.105 I. Maggio-Aprile, C. Renner, E. Erb, E. Walker, Ø. Fischer: Direct vortex lattice imaging and tunneling spectroscopy of flux lines on $YBa_2Cu_3O_{7-z}$, Phys. Rev. Lett. **75**, 2754–2757 (1995)

25.106 C. Renner, B. Revaz, K. Kadowaki, I. Maggio-Aprile, Ø. Fischer: Observation of the low temperature pseudogap in the vortex cores of $Bi_2Sr_2CaCu_2O_{8+z}$, Phys. Rev. Lett. **80**, 3606–3609 (1998)

25.107 S.H. Pan, E.W. Hudson, A.K. Gupta, K.W. Ng, H. Eisaki, S. Uchida, J.C. Davis: STM studies of the electronic structure of vortex cores in $Bi_2Sr_2CaCu_2O_{8+z}$, Phys. Rev. Lett. **85**, 1536–1539 (2000)

25.108 D.P. Arovas, A.J. Berlinsky, C. Kallin, S.C. Zhang: Superconducting vortex with antiferromagnetic core, Phys. Rev. Lett. **79**, 2871–2874 (1997)

25.109 J.E. Hoffmann, E.W. Hudson, K.M. Lang, V. Madhavan, H. Eisaki, S. Uchida, J.C. Davis: A four unit cell periodic pattern of quasi-particle states surrounding vortex cores in $Bi_2Sr_2CaCu_2O_{8+z}$, Science **295**, 466–469 (2002)

25.110 M. Fäth, S. Freisem, A.A. Menovsky, Y. Tomioka, J. Aarts, J.A. Mydosh: Spatially inhomogeneous metal–insulator transition in doped manganites, Science **285**, 1540–1542 (1999)

25.111 C. Renner, G. Aeppli, B.G. Kim, Y.A. Soh, S.W. Cheong: Atomic-scale images of charge ordering in a mixed-valence manganite, Nature **416**, 518–521 (2000)

25.112 M. Bode, M. Getzlaff, R. Wiesendanger: Spin-polarized vacuum tunneling into the exchange-split surface state of Gd(0001), Phys. Rev. Lett. **81**, 4256–4259 (1998)

25.113 A. Kubetzka, M. Bode, O. Pietzsch, R. Wiesendanger: Spin-polarized scanning tunneling microscopy with antiferromagnetic probe tips, Phys. Rev. Lett. **88**, 057201, 1–4 (2002)

25.114 O. Pietzsch, A. Kubetzka, M. Bode, R. Wiesendanger: Observation of magnetic hysteresis at the nanometer scale by spin–polarized scanning tunneling spectroscopy, Science **292**, 2053–2056 (2001)

25.115 S. Heinze, M. Bode, A. Kubetzka, O. Pietzsch, X. Xie, S. Blügel, R. Wiesendanger: Real-space imaging of two-dimensional antiferromagnetism on the atomic scale, Science **288**, 1805–1808 (2000)

25.116 A. Wachowiak, J. Wiebe, M. Bode, O. Pietzsch, M. Morgenstern, R. Wiesendanger: Internal spin-structure of magnetic vortex cores observed by spin-polarized scanning tunneling microscopy, Science **298**, 577–580 (2002)

25.117 M. D. Kirk, T. R. Albrecht, C. F. Quate: Low-temperature atomic force microscopy, Rev. Sci. Instrum. **59**, 833–835 (1988)

25.118 D. Pelekhov, J. Becker, J. G. Nunes: Atomic force microscope for operation in high magnetic fields at milliKelvin temperatures, Rev. Sci. Instrum. **70**, 114–120 (1999)

25.119 J. Mou, Y. Jie, Z. Shao: An optical detection low temperature atomic force microscope at ambient pressure for biological research, Rev. Sci. Instrum. **64**, 1483–1488 (1993)

25.120 H. J. Mamin, D. Rugar: Sub-attoNewton force detection at milliKelvin temperatures, Appl. Phys. Lett. **79**, 3358–3360 (2001)

25.121 A. Schwarz, W. Allers, U. D. Schwarz, R. Wiesendanger: Dynamic mode scanning force microscopy of n-InAs(110)-(1×1) at low temperatures, Phys. Rev. B **61**, 2837–2845 (2000)

25.122 W. Allers, S. Langkat, R. Wiesendanger: Dynamic low-temperature scanning force microscopy on nickel oxide(001), Appl. Phys. A **72**, S27–S30 (2001)

25.123 F. J. Giessibl: Atomic resolution of the silicon(111)-(7×7) surface by atomic force microscopy, Science **267**, 68–71 (1995)

25.124 M. A. Lantz, H. J. Hug, P. J. A. van Schendel, R. Hoffmann, S. Martin, A. Baratoff, A. Abdurixit, H.-J. Güntherodt: Low temperature scanning force microscopy of the Si(111)-(7×7) surface, Phys. Rev. Lett. **84**, 2642–2465 (2000)

25.125 K. Suzuki, H. Iwatsuki, S. Kitamura, C. B. Mooney: Development of low temperature ultrahigh vacuum force microscope/scanning tunneling microscope, Jpn. J. Appl. Phys. **39**, 3750–3752 (2000)

25.126 N. Suehira, Y. Sugawara, S. Morita: Artifact and fact of Si(111)-(7×7) surface images observed with a low temperature noncontact atomic force microscope (LT-NC-AFM), Jpn. J. Appl. Phys. **40**, 292–294 (2001)

25.127 R. Peréz, M. C. Payne, I. Štich, K. Terakura: Role of covalent tip–surface interactions in noncontact atomic force microscopy on reactive surfaces, Phys. Rev. Lett. **78**, 678–681 (1997)

25.128 S. H. Ke, T. Uda, R. Pérez, I. Štich, K. Terakura: First principles investigation of tip–surface interaction on GaAs(110): Implication for atomic force and tunneling microscopies, Phys. Rev. B **60**, 11631–11638 (1999)

25.129 J. Tobik, I. Štich, R. Peréz, K. Terakura: Simulation of tip–surface interactions in atomic force microscopy of an InP(110) surface with a Si tip, Phys. Rev. B **60**, 11639–11644 (1999)

25.130 A. Schwarz, W. Allers, U. D. Schwarz, R. Wiesendanger: Simultaneous imaging of the In and As sublattice on InAs(110)-(1×1) with dynamic scanning force microscopy, Appl. Surf. Sci. **140**, 293–297 (1999)

25.131 H. Hölscher, W. Allers, U. D. Schwarz, A. Schwarz, R. Wiesendanger: Interpretation of 'true atomic resolution' images of graphite (0001) in noncontact atomic force microscopy, Phys. Rev. B **62**, 6967–6970 (2000)

25.132 H. Hölscher, W. Allers, U. D. Schwarz, A. Schwarz, R. Wiesendanger: Simulation of NC-AFM images of xenon(111), Appl. Phys. A **72**, S35–S38 (2001)

25.133 M. Ashino, A. Schwarz, T. Behnke, R. Wiesendanger: Atomic-resolution dynamic force microscopy and spectroscopy of a single-walled carbon nanotube: characterization of interatomic van der Waals forces, Phys. Rev. Lett. **93**, 136101, 1–4. (2004)

25.134 G. Schwarz, A. Kley, J. Neugebauer, M. Scheffler: Electronic and structural properties of vacancies on and below the GaP(110) surface, Phys. Rev. B **58**, 1392–1499 (1998)

25.135 M. Ashino, A. Schwarz, H. Hölscher, U. D. Schwarz, R. Wiesendanger: Interpretation of the atomic scale contrast obtained on graphite and single-walled carbon nanotubes in the dynamic mode of atomic force microscopy, Nanotechnology **16**, 134–137 (2005)

25.136 S. Hembacher, F. J. Giessibl, J. Mannhart, C. F. Quate: Local spectroscopy and atomic imaging of tunneling current, forces, and dissipation on graphite, Phys. Rev. Lett. **94**, 056101, 1–4 (2005)

25.137 F. J. Giessibl, H. Bielefeldt, S. Hembacher, J. Mannhart: Calculation of the optimal imaging parameters for frequency modulation atomic force microscopy, Appl. Surf. Sci. **140**, 352–357 (1999)

25.138 F. J. Giessibl: High-speed force sensor for force microscopy and profilometry utilizing a quartz tuning fork, Appl. Phys. Lett. **73**, 3956–3958 (1998)

25.139 H. Hölscher, W. Allers, U. D. Schwarz, A. Schwarz, R. Wiesendanger: Determination of tip–sample interaction potentials by dynamic force spectroscopy, Phys. Rev. Lett. **83**, 4780–4783 (1999)

25.140 H. Hölscher, U. D. Schwarz, R. Wiesendanger: Calculation of the frequency shift in dynamic force microscopy, Appl. Surf. Sci. **140**, 344–351 (1999)

25.141 B. Gotsman, B. Anczykowski, C. Seidel, H. Fuchs: Determination of tip–sample interaction forces from measured dynamic force spectroscopy curves, Appl. Surf. Sci. **140**, 314–319 (1999)

25.142 U. Dürig: Extracting interaction forces and complementary observables in dynamic probe microscopy, Appl. Phys. Lett. **76**, 1203–1205 (2000)

25.143 F. J. Giessibl: A direct method to calculate tip-sample forces from frequency shifts in frequency-modulation atomic force microscopy, Appl. Phys. Lett. **78**, 123–125 (2001)

25.144 J. E. Sader, S. P. Jarvis: Accurate formulas for interaction force and energy in frequency modulation force spectroscopy, Appl. Phys. Lett. **84**, 1801–1803 (2004)

25.145 M. A. Lantz, H. J. Hug, R. Hoffmann, P. J. A. van Schendel, P. Kappenberger, S. Martin, A. Baratoff, H.-J. Güntherodt: Quantitative measurement of short-range chemical bonding forces, Science **291**, 2580–2583 (2001)

25.146 S. M. Langkat, H. Hölscher, A. Schwarz, R. Wiesendanger: Determination of site specific forces between an iron coated tip and the NiO(001) surface by force field spectroscopy, Surf. Sci. (2002)in press

25.147 H. Hölscher, S. M. Langkat, A. Schwarz, R. Wiesendanger: Measurement of three-dimensional force fields with atomic resolution using dynamic force spectroscopy, Appl. Phys. Lett. (2002)in press

25.148 B. C. Stipe, H. J. Mamin, T. D. Stowe, T. W. Kenny, D. Rugar: Noncontact friction and force fluctuations between closely spaced bodies, Phys. Rev. Lett. **87** (2001)

25.149 N. Oyabu, O. Custance, I. Yi, Y. Sugawara, S. Morita: Mechanical vertical manipulation of selected single atoms by soft nanoindentation using near contact atomic force microscopy, Phys. Rev. Lett. **90**, 176102, 1–4 (2004)

25.150 N. Oyabu, Y. Sugimoto, M. Abe, O. Custance, S. Morita: Lateral manipulation of single atoms at semiconductor surfaces using atomic force microscopy, Nanotechnology **16**, 112–117 (2005)

25.151 Y. Sugimoto, M. Abe, S. Hirayama, N. Oyabu, O. Custance, S. Morita: Atom inlays performed at room temperature using atomic force microscopy, Nature Mater. **4**, 156–160 (2005)

25.152 C. Sommerhalter, T. W. Matthes, T. Glatzel, A. Jäger-Waldau, M. C. Lux-Steiner: High-sensitivity quantitative Kelvin probe microscopy by noncontact ultra-high-vacuum atomic force microscopy, Appl. Phys. Lett. **75**, 286–288 (1999)

25.153 A. Schwarz, W. Allers, U. D. Schwarz, R. Wiesendanger: Dynamic mode scanning force microscopy of n-InAs(110)-(1×1) at low temperatures, Phys. Rev. B **62**, 13617–13622 (2000)

25.154 K. L. McCormick, M. T. Woodside, M. Huang, M. Wu, P. L. McEuen, C. Duruoz, J. S. Harris: Scanned potential microscopy of edge and bulk currents in the quantum Hall regime, Phys. Rev. B **59**, 4656–4657 (1999)

25.155 P. Weitz, E. Ahlswede, J. Weis, K. v. Klitzing, K. Eberl: Hall-potential investigations under quantum Hall conditions using scanning force microscopy, Physica E **6**, 247–250 (2000)

25.156 E. Ahlswede, P. Weitz, J. Weis, K. v. Klitzing, K. Eberl: Hall potential profiles in the quantum Hall regime measured by a scanning force microscope, Physica B **298**, 562–566 (2001)

25.157 M. T. Woodside, C. Vale, P. L. McEuen, C. Kadow, K. D. Maranowski, A. C. Gossard: Imaging interedge-state scattering centers in the quantum Hall regime, Phys. Rev. B **64** (2001)041310-1–041310-4

25.158 K. Moloni, B. M. Moskowitz, E. D. Dahlberg: Domain structures in single crystal magnetite below the Verwey transition as observed with a low-temperature magnetic force microscope, Geophys. Res. Lett. **23**, 2851–2854 (1996)

25.159 Q. Lu, C. C. Chen, A. de Lozanne: Observation of magnetic domain behavior in colossal magnetoresistive materials with a magnetic force microscope, Science **276**, 2006–2008 (1997)

25.160 G. Xiao, J. H. Ross, A. Parasiris, K. D. D. Rathnayaka, D. G. Naugle: Low-temperature MFM studies of CMR manganites, Physica C **341–348**, 769–770 (2000)

25.161 M. Liebmann, U. Kaiser, A. Schwarz, R. Wiesendanger, U. H. Pi, T. W. Noh, Z. G. Khim, D. W. Kim: Domain nucleation and growth of $La_{07}Ca_{0.3}MnO_{3-\tilde{z}}/LaAlO_3$ films studied by low temperature MFM, J. Appl. Phys. **93**, 8319–8321 (2003)

25.162 A. Moser, H. J. Hug, I. Parashikov, B. Stiefel, O. Fritz, H. Thomas, A. Baratoff, H. J. Güntherodt, P. Chaudhari: Observation of single vortices condensed into a vortex-glass phase by magnetic force microscopy, Phys. Rev. Lett. **74**, 1847–1850 (1995)

25.163 C. W. Yuan, Z. Zheng, A. L. de Lozanne, M. Tortonese, D. A. Rudman, J. N. Eckstein: Vortex images in thin films of $YBa_2Cu_3O_{7-x}$ and $Bi_2Sr_2Ca_1Cu_2O_{8-x}$ obtained by low-temperature magnetic force microscopy, J. Vac. Sci. Technol. B **14**, 1210–1213 (1996)

25.164 A. Volodin, K. Temst, C. van Haesendonck, Y. Bruynseraede: Observation of the Abrikosov vortex lattice in $NbSe_2$ with magnetic force microscopy, Appl. Phys. Lett. **73**, 1134–1136 (1998)

25.165 A. Moser, H. J. Hug, B. Stiefel, H. J. Güntherodt: Low temperature magnetic force microscopy on $YBa_2Cu_3O_{7-\tilde{z}}$ thin films, J. Magn. Magn. Mater. **190**, 114–123 (1998)

25.166 A. Volodin, K. Temst, C. van Haesendonck, Y. Bruynseraede: Imaging of vortices in conventional superconductors by magnetic force microscopy images, Physica C **332**, 156–159 (2000)

25.167 M. Roseman, P. Grütter: Estimating the magnetic penetration depth using constant-height magnetic force microscopy images of vortices, New J. Phys. **3**, 24.1–24.8 (2001)

25.168 A. Volodin, K. Temst, C. van Haesendonck, Y. Bruynseraede, M. I. Montero, I. K. Schuller: Magnetic force microscopy of vortices in thin niobium films: Correlation between the vortex distribution and the thickness-dependent film morphology, Europhys. Lett. **58**, 582–588 (2002)

Part C | 25

25.169 U. H. Pi, T. W. Noh, Z. G. Khim, U. Kaiser, M. Lieb-mann, A. Schwarz, R. Wiesendanger: Vortex dynamics in $Bi_2Sr_2CaCu_2O_8$ single crystal with low density columnar defects studied by magnetic force microscopy, J. Low Temp. Phys. **131**, 993–1002 (2003)

25.170 M. Roseman, P. Grütter, A. Badia, V. Metlushko: Flux lattice imaging of a patterned niobium thin film, J. Appl. Phys. **89**, 6787–6789 (2001)

25.171 D. Rugar, R. Budakian, H. J. Mamin, B. W. Chui: Single spin detection by magnetic resonance force microscopy, Nature **430**, 329–332 (2004)

25.172 K. Nakamura, H. Hasegawa, T. Oguchi, K. Sueoka, K. Hayakawa, K. Mukasa: First-principles calcula-tion of the exchange interaction and the exchange force between magnetic Fe films, Phys. Rev. B **56**, 3218–3221 (1997)

25.173 A. S. Foster, A. L. Shluger: Spin-contrast in non-contact AFM on oxide surfaces: Theoretical mod-eling of NiO(001) surface, Surf. Sci. **490**, 211–219 (2001)

25.174 H. Hoisoi, M. Kimura, K. Hayakawa, K. Sueoka, K. Mukasa: Non-contact atomic force microscopy of an antiferromagnetic NiO(100) surface using a ferromagnetic tip, Appl. Phys. A **72**, S23–S26 (2001)

25.175 J. A. Sidles, J. L. Garbini, G. P. Drobny: The theory of oscillator-coupled magnetic resonance with po-tential applications to molecular imaging, Rev. Sci. Instrum. **63**, 3881–3899 (1992)

25.176 J. A. Sidles, J. L. Garbini, K. J. Bruland, D. Ru-gar, O. Züger, S. Hoen, C. S. Yannoni: Magnetic resonance force microscopy, Rev. Mod. Phys. **67**, 249–265 (1995)

25.177 D. Rugar, O. Züger, S. Hoen, C. S. Yannoni, H. M. Vieth, R. D. Kendrick: Force detection of nu-clear magnetic resonance, Science **264**, 1560–1563 (1994)

25.178 Z. Zhang, P. C. Hammel, P. E. Wigen: Observation of ferromagnetic resonance in a microscopic sample using magnetic resonance force microscopy, Appl. Phys. Lett. **68**, 2005–2007 (1996)

26. Higher-Harmonic Force Detection in Dynamic Force Microscopy

In atomic force microscopy, a force-sensing cantilever probes a sample and thereby creates a topographic image of its surface. The simplest implementation uses the static deflection of the cantilever to probe the forces. More recently, dynamic operation modes have been introduced. The dynamic modes either work at a constant oscillation frequency and sense the amplitude variations caused by tip–sample forces (amplitude-modulation or tapping mode) or operate at a constant amplitude and varying frequency (frequency-modulation mode). Here, we report on new operational concepts that capture the higher harmonics in either amplitude-modulation or frequency-modulation mode. Higher-harmonic detection in AM force microscopy allows the measurement of time-resolved tip–sample forces that contain detailed information about the material characteristics of the sample, while higher-harmonic detection in small-amplitude frequency-modulation mode allows a significant improvement in spatial resolution, in particular when operating in vacuum at low temperatures.

The most widely used mode of operation of atomic force microscopy (AFM) is the tapping mode, because in this mode the lateral tip–sample interaction forces are minimized. The gentle interaction between the AFM tip and the sample under test reduces wear on the sample and localizes the deformations to give nanometer, or even molecular, resolution [26.1, 2]. In tapping mode, the AFM cantilever is vibrated at resonance in the vicinity of the sample so that the tip makes contact with the sample once during each cycle. The tip–sample forces reduce the vibration amplitude of the cantilever. The vibrating cantilever is scanned across the surface while a feedback mechanism adjusts the height of the cantilever base to maintain the vibration amplitude at a constant set-point value. The topography of the surface is then obtained by recording the feedback signal.

Tapping-mode AFM has the potential to measure much more than just the topography of a surface, however. As can be seen from Fig. 26.1, the tip–sample interaction forces as the AFM tip approaches, interacts and retracts from the surface has complex time dependence. This time dependence reflects the attractive and repulsive forces that act between the tip and the sam-

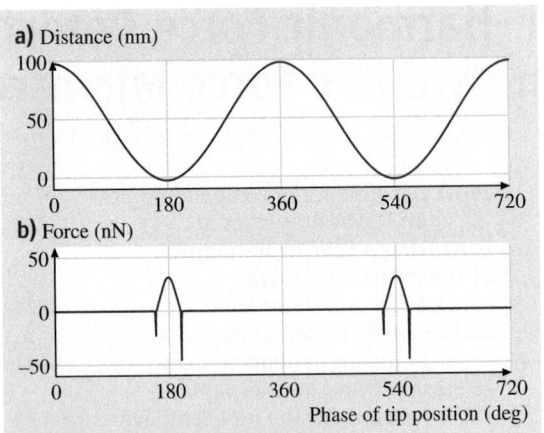

Fig. 26.1a,b Calculated tip–sample distance (**a**) and tip–sample interaction forces (**b**) over two cycles of cantilever oscillation. The *dashed line* in (a) is the surface rest level. Negative displacements correspond to sample indentation. Attractive (negative) and repulsive (positive) forces appear during the tip–sample interaction. The magnitude and duration of these forces depend on the physical properties of the sample

ple, and contains information about the chemical and physical properties of the sample.

In the remaining sections of this chapter we describe methods that enable measurement of the time-resolved

tip–sample forces in tapping-mode AFM. We first present a simple model for the time-resolved tip–sample forces and show how higher harmonics of the force can be used to distinguish between samples with different physical properties. Then we will show how to measure specific higher harmonics that contain information about physical properties, by using higher-order flexural vibration modes of the AFM cantilever. We then describe how the torsional vibration modes of the AFM cantilever can be used to measure not only one, but a large number of harmonics of the force, so that the tip–sample force of Fig. 26.1b can be measured with good time resolution. As application examples we present: (1) time-resolved force measurements that allow quantitative comparisons of material stiffness, and (2) observation of the glass transition of polymer blends with nanometer-scale lateral resolution.

After the discussion of time-resolved force measurements in standard AFM tapping mode, we introduce higher-harmonic imaging in AFM with small vibration amplitudes. In small-amplitude AFM imaging, the tip is in the force field of the sample during most of its vibration cycle. Relatively low-order harmonics of the tip–sample force then contain information about the higher-order gradients of the tip–sample interaction force field. These low-order harmonics in small-amplitude dynamic AFM imaging can be measured directly, yielding excellent spatial resolution.

26.1 Modeling of Tip–Sample Interaction Forces in Tapping-Mode AFM

26.1.1 Tip-Sample Forces as a Periodic Waveform

In the tapping mode the cantilever is periodically driven at its fundamental resonant frequency. Under typical operating conditions, the periodic driving force results in a periodic motion of the cantilever and a periodic tip–sample force waveform [26.3]. This allows us to use frequency-domain techniques to understand the motion of the cantilever and the tip–sample forces [26.4].

Because the tip–sample force F_{ts} is a periodic waveform we can expand it as a Fourier series as follows

$$F_{ts}(t) = \sum_{n=0}^{\infty} a_n \cos(n\omega t) + b_n \sin(n\omega t)$$

$$(n = 0, 1, 2, \dots).$$

(26.1)

Here the frequency, ω, is the driving frequency, which is chosen to be close to the fundamental resonance frequency of the cantilever. The coefficients a_n and b_n are given as

$$a_n = \frac{\omega}{\pi} \int_0^{2\pi/\omega} F_{ts} \cos(n\omega t)\,dt$$

(26.2a)

$$b_n = \frac{\omega}{\pi} \int_0^{2\pi/\omega} F_{ts} \sin(n\omega t)\,dt.$$

(26.2b)

The k-th harmonic force can be written as

$$F_{tsn} \cos(n\omega t + \theta_n) = a_n \cos(n\omega t) + b_n \sin(n\omega t)$$

(26.3)

Here $F_{tsn} = \sqrt{a_n^2 + b_n^2}$ and θ_n are the magnitude and phase of the n-th harmonic force, respectively. The

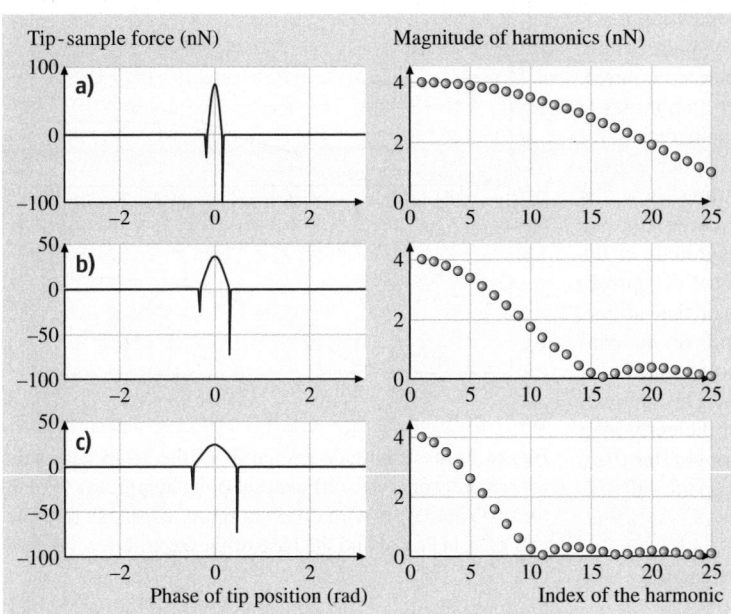

Tip-sample force (nN)

Magnitude of harmonics (nN)

Fig. 26.2a–c Interaction forces between the tip and the sample for three different samples; (a) stiff, (b) medium and (c) compliant. The amplitudes of the harmonics for the three tip–sample forces are shown on the *right*

phase, θ_n, of a higher harmonic is defined relative to a reference harmonic at the same frequency that is in phase with the fundamental displacement, i.e. if we represent the tip displacement by $A_s \cos(\omega t)$, the reference signal is $\cos(n\omega t)$.

The tip–sample forces can be seen as a superposition of harmonic forces, each at an integer multiple of the driving frequency. Later we will show how we can measure these higher harmonics by tuning a higher-order resonance of the cantilever to an integer multiple of its fundamental resonance frequency (i.e. $n\omega = \omega_k$, where ω_k is the resonance frequency of the k-th flexural mode of the cantilever). The higher-harmonic force will then drive the higher-order resonance, and the deflection of the cantilever in the higher-order resonance is a measure of the higher-harmonic force.

26.1.2 Frequency Spectrum of the Tip–Sample Force

Figure 26.2 shows the calculated periodic tip–sample forces and their respective harmonic force components, F_{tsn}, for stiff, medium and compliant samples. Details of the tip–sample force calculations can be found in [26.4]. The harmonic force components are calculated for a cantilever with spring constant $K_1 = 10$, quality factor $Q_1 = 100$, free amplitude $A_0 = 100$ nm, and setpoint amplitude $A_s = 80$ nm. We model the tip–sample

interaction forces by the equations below

$$f_{ts}(r) = \frac{HR}{6\sigma^2}\left[-\left(\frac{\sigma}{r}\right)^2 + \frac{1}{30}\left(\frac{\sigma}{r}\right)^8 \right], \quad (26.4a)$$

$$f_{ts}(d) = \frac{4}{3}E\sqrt{R}d^{3/2}. \quad (26.4b)$$

Here r is the tip–sample separation, and d is the sample indentation. H, R, σ, and E are the Hamaker constant, tip radius, typical atomic distance for the tip and the sample, and the reduced Young's modulus for the tip and the sample, respectively. Tip sample forces are governed by (26.4a) when the tip is away from the sample and by (26.4b) when the tip is indenting the sample. The reduced Young's modulus E of the samples is the main factor that determines the tip–sample contact duration. In Fig. 26.2, E values for the three samples are chosen to give contact durations of 5%, 10% and 15% of the period on the stiff, medium and compliant samples, respectively. Attractive forces on all the samples are assumed to be equal, giving equal amounts of energy dissipation at contact. For the Hamaker constants we used $H_a = 10 \times 10^{-20}$ J for approach and $H_r = 30 \times 10^{-20}$ J for retraction. The parameters R and σ are chosen to be 10 nm and 0.1 nm, respectively. These values result in an energy dissipation of approximately 30 eV per contact. The parameters other than E are chosen to be the same for all three materials in order to simplify the analysis.

For the specific example shown in Fig. 26.2, harmonics above the tenth are strongly dependent on the stiffness of the sample. The tip–sample force is approximately a periodic clipped sine wave, so the pulse width, or contact duration, determines the harmonic content. Shorter-duration contacts generate larger amplitudes at higher harmonics. The contact duration increases for more compliant samples, resulting in smaller magnitudes at higher harmonics. These calculations show that the first harmonics for each of the three cases have the same magnitude. This is because the magnitude of the first harmonic [$n = 1$ in Eqs. (26.1–26.2b)] is approximately equal to the average tapping force on the surface. The average tapping force mainly depends on the cantilever, the drive amplitude and the set-point amplitude, which are all equal for the three samples. This analysis shows that only the harmonics that are sufficiently high to have periods that are comparable to, or shorter than, the contact duration, contain information on the stiffness of the samples.

26.1.3 Dependence of Force Harmonics on Elastic Properties

In order to interpret the measurements of higher-harmonics force components, we must consider the dependency of higher-harmonic force components on the physical properties of the sample. To understand the amplitude response of higher harmonics on samples with varying E, we calculated the response under varying experimental conditions. In particular, we altered the spring constant and set-point amplitude, which, according to our calculations, influence the higher-harmonic response.

We use two cantilevers with spring constants $K_1 = 1\,\text{N/m}$ and $10\,\text{N/m}$, both with quality factors $Q_1 = 100$. The indices of K and Q represent the first flexural mode. The free amplitude and set-point amplitude are chosen to be $A_0 = 100\,\text{nm}$ and $A_0 = 80\,\text{nm}$. The calculations for $K_1 = 1\,\text{N/m}$ are repeated for a set-point amplitude of $60\,\text{nm}$. The Hamaker constants are $H_a = 10 \times \times 10^{-20}\,\text{J}$ for the approach and $H_r = 30 \times \times 10^{-20}\,\text{J}$ for the retraction. We use $\sigma = 0.1\,\text{nm}$ and $R = 10\,\text{nm}$ for the typical interatomic distance and tip radius.

Using (26.3), we calculate the 16-th harmonic and we assume that this frequency coincides with the third flexural resonance frequency of the cantilever. This implies that the cantilever beam has been altered to tune the third flexural resonance frequency, ω_3, to 16 times the fundamental (i. e. $\omega_3 = 16\omega$). In a later section we will describe how such cantilevers are made. Note that the choice of the 16-th harmonic is arbitrary and that

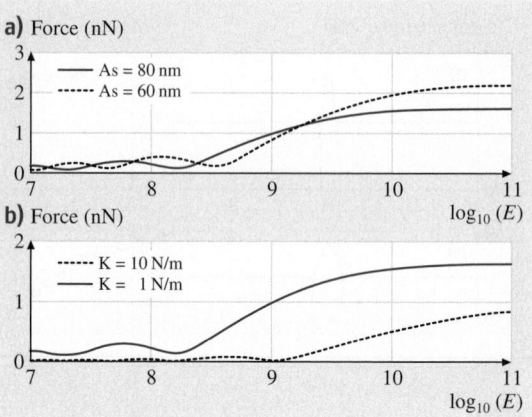

Fig. 26.3a,b Amplitude responses of the 16-th harmonic for a cantilever at two different set-point amplitudes (**a**) and for two cantilevers with different spring constants (**b**). The unit of E is Pascal and the base of the logarithm is 10

the main results of this analysis are valid for any higher harmonic. Details can be found in [26.4].

Figure 26.3 shows plots of the amplitudes of the resonant harmonics with respect to the reduced Young's modulus for two different set-point amplitudes (a) and two different cantilever spring constants (b). In all these graphs, the amplitude of the higher harmonic reaches a maximum at very high E values. From this maximum the amplitude drops monotonously to a minimum, as the E value gets smaller. After this first minimum, the harmonic response is oscillatory with respect to E. In this range of E it is more difficult to interpret the measurements. If the materials under investigation are known to have E values within the monotonous range of the harmonic response, the results can be easily interpreted.

Figures 26.3a and 26.2b show that, for stiffer cantilevers and smaller set-point amplitudes (with free amplitude held constant), the graphs shift toward higher E. These dependencies give additional flexibility for matching the E values of the samples under investigation to the monotonous region of the harmonic amplitude response. Both of these experimental parameters (spring constant and set-point amplitude), will alter the tip–sample force and contact duration. These results indicate that, while higher harmonics are sensitive to variations in the stiffness of materials, knowledge of the cantilever spring constant, the set-point vibration amplitudes, and the nominal stiffness of the samples is essential for quantitative measurements.

For imaging, we can simply record the amplitude of a particular harmonic while scanning the surface

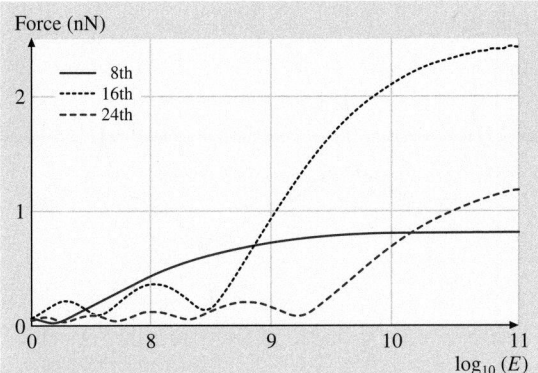

Fig. 26.4 Amplitude responses at the 8-th, 16-th and 24-th harmonics of the driving frequency. The amplitude response at the 8-th harmonic is much higher than the others; therefore, we divided it by 10 in order to see all the responses clearly within one graph. Units are as in Fig. 26.3

in tapping-mode [26.5, 6]. According to the results in Fig. 26.3, the amplitude signals are correlated with the stiffness of the surface. Therefore, an image generated by monitoring the resonant harmonic will map the elastic properties of the sample. Preferably, the local stiffness values should fall in the monotonously increasing region of the harmonic amplitude responses in Fig. 26.3. Knowledge of the nominal stiffness of the sample allows us to design a cantilever with the correct spring constant. For example, according to Fig. 26.3b, a cantilever with $K_1 = 10\,\mathrm{N/m}$ is most sensitive to stiffness variations around 10 GPa, while a cantilever with $K_1 = 1\,\mathrm{N/m}$ is less sensitive in that range. On the other hand, a cantilever with $K_1 = 1\,\mathrm{N/m}$ is sensitive to variations around 1 GPa. A proper value for the spring constant is therefore crucial for operating in the monotonically increasing and highly sensitive region. This is, however, not a very limiting constraint, because the monotonically increasing region extends over almost two orders of magnitude (Fig. 26.3). It is uncommon that the variations in a given sample are this large. Although we need to use soft cantilevers for compliant samples and stiff cantilevers for

hard samples, adjusting the set-point amplitude to tune the sensitivity provides some flexibility in the measurements (Fig. 26.3a).

So far the calculations were carried out for the 16-th harmonic of the driving frequency. Now we would like to discuss the case where the frequency of the harmonic is equal to other integer multiples of the fundamental resonance frequency. We have calculated the harmonic response for cantilevers with their higher-order resonant frequencies at the 8-th, 16-th and 24-th harmonic. It is assumed that the cantilevers all have the same spring constant for the fundamental vibration mode, $K_1 = 2\,\mathrm{N/m}$. We are only interested in the general behavior of the resonant-harmonic response at different integer multiples of the fundamental. The spring constant and quality factors of the higher-order resonances affect the amplitude values, but they do not change the general dependence of the amplitude and phase variations on the hardness of the surface. For these calculations, the set-point and free amplitudes are chosen as 80 nm and 100 nm, respectively. In Fig. 26.4 we summarize the results for the calculated harmonic response at different higher harmonics. The amplitude responses of all three harmonics converge to their maximum as the stiffness of the surface increases. As the surface stiffness is reduced, the harmonic amplitude first drops to a minimum and then shows oscillatory behavior as the stiffness is further reduced. While the general trends of these curves are the same, the sensitive and monotonously increasing regions appear at different effective Young's modulus values. The 24-th harmonic has its first amplitude minimum at a higher stiffness than the 16-th, and the 16-th harmonic has its first minimum at a stiffer surface than the 8-th harmonic. This result indicates that the 24-th harmonic gives better contrast for harder samples, and the 8-th harmonic gives better contrast for softer samples. This feature guides us in the choice of harmonics. With a proper choice of the higher harmonic, the cantilever spring constant, and the tapping parameters, one can study the physical properties of a wide range of materials with high resolution.

26.2 Enhancing a Specific Harmonic of the Interaction Force Using a Flexural Resonance

In this section we will present a theoretical model for the response of the cantilever to higher-harmonic forces and then we will describe a method to tune the frequency response of the cantilever to enhance

the vibration signals at a particular harmonic frequency. Finally we will present the implementation of a cantilever designed according to the described method.

26.2.1 Cantilever Response to Higher Harmonic Forces

We have discussed the origin of higher-harmonic forces and how they depend on the physical properties of the samples. In a practical experimental setting we cannot directly measure the higher-harmonic forces. We can only measure the vibrations of the cantilever, so it is necessary to understand how the cantilever responds to higher-harmonic force components. To obtain good signal-to-noise ratios for the measurements, the cantilever must have a good response to higher harmonics.

In order to calculate the cantilever response to the harmonic forces of the tip–sample interaction, we need to go beyond the simple harmonic oscillator models [26.7] and model the cantilever as a continuum mechanical system [26.8]. The AFM cantilever is fixed at the base and free to move at the tip end. The external forces acting on the cantilever are the drive force at the base and the tip–sample forces at the tip. The motion of the cantilever is governed by the Euler–Bernoulli equation. The solution of this equation for a rectangular cantilever can be found in [26.9]. The Euler–Bernoulli equation is linear in time, so we can describe the cantilever response in terms of the eigenmodes of the cantilever. These eigenmodes are the flexural vi-

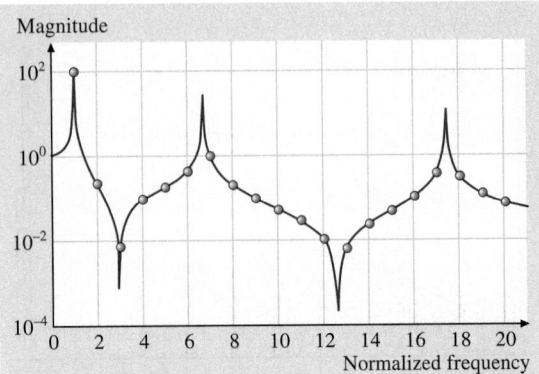

Fig. 26.6 Calculated frequency response of a rectangular cantilever. The frequency axis is normalized to the first resonance frequency. The magnitudes represent the optical signal at the position-sensitive detector. This optical signal is proportional to the slope of the cantilever at the laser spot. The *crosses* are located at the integer multiples of the first resonance frequency

bration modes, each of which has a specific resonance frequency and mode shape.

Simulated mode shapes of a rectangular beam fixed at one end and free at the other end are given in Fig. 26.5. In a typical tapping-mode experiment, the cantilever is driven at the resonance frequency of the first flexural mode shown in Fig. 26.5a. The other flexural modes are excited when the tip interacts with the surface. With tip–sample interaction as the driving force, the motion of the cantilever can be expressed as a superposition of the responses of the eigenmodes. The response of the cantilever $y(t)$ to an external harmonic force applied to the tip at the free end can be approximated as

$$y(t) = e^{i\omega t} \frac{F}{M} \sum_{k=1}^{\infty} \frac{4}{\omega_k^2 - \omega^2 + i\omega\omega_k/Q_k} . \quad (26.5)$$

Here $y(t)$ is the displacement of the tip (free end) of the cantilever at time t, and F and ω are the magnitude and frequency of the harmonic force acting on the tip, respectively. The parameters ω_k and Q_k are the resonance frequency and quality factor of the k-th eigenmode.

Figure 26.6 shows the frequency response of the optical-lever signal for a rectangular cantilever. These calculations are based on (26.4a) while taking into account that the optical-lever signals in the position-sensitive detector are proportional to the slopes of the free end of the cantilever. The peaks in the response

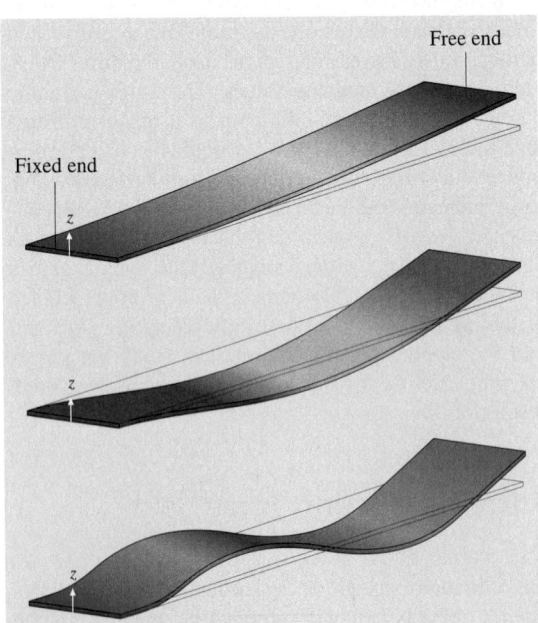

Fig. 26.5 Mode shapes of a rectangular cantilever fixed at one end and free at the other end

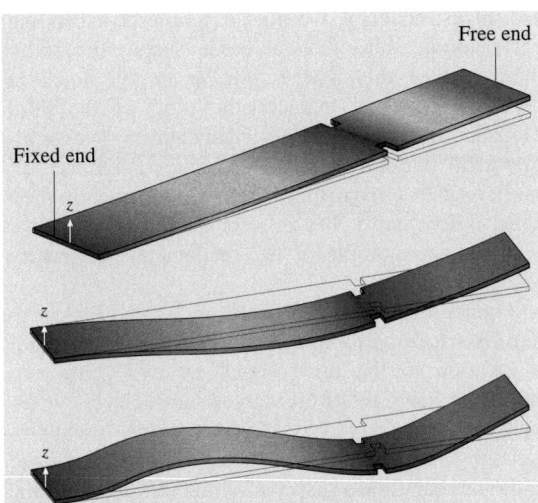

Fig. 26.7 Mode shapes of a rectangular cantilever with a notch. The notch is located at the high-bending region in the third flexural mode shape and is approximately one third of the total length away from the free end

curves are the resonances of each flexural vibration mode. The frequency axis is normalized to the first resonance frequency, which equals the driving frequency in tapping-mode AFM. The higher harmonics, marked with crosses on the frequency response curve, are therefore located at integer multiples of the first resonance frequency. The cantilever responds to each force harmonic with a displacement given by this frequency

Fig. 26.8 Scanning electron microscopy (SEM) image of a harmonic cantilever. Width, length and thickness of the cantilever are 50, 300, and 2.2 μm, respectively. The rectangular opening is 22 by 18 μm and is centered 190 μm away from the cantilever base

response. The harmonics close to a resonance frequency will have larger deflections. We see that the higher harmonics, which contain information on the stiffness of the sample, yield relatively low cantilever responses, limiting our ability to measure the higher harmonics.

26.2.2 Improving the Mechanical Response of the AFM Cantilever

To improve the response of a given force harmonic, the resonance frequency of the third-order mode is tuned to an integer multiple of the fundamental resonance frequency. This is done by appropriately removing mass from regions where the cantilever has high mechanical stress in that particular mode. When this cantilever is driven at the fundamental resonance frequency in tapping mode, the chosen higher harmonic will be exactly at the third resonance frequency, leading to resonant enhancement of this harmonic [26.10].

Removing mass from regions of high mechanical stress reduces the elastic energy and the resonance frequency of that mode without strongly affecting the other resonant modes of the system. Figure 26.7 illustrates the first three flexural mode shapes of a rectangular cantilever with a notch. The position of the notch corresponds to a highly curved region of the third mode, but not to highly curved regions of the first two modes. Therefore, the effect of the notch in reducing the elastic energy is more prominent in the third mode. Highly curved regions of a mode are also highly displaced, so the removal of mass from these regions will also reduce the kinetic energy of that mode and increase the resonance frequency. This effect will however affect both the first and third mode relatively equally, because the displacements at the notch are similar for the two modes, as can be seen from Fig. 26.7. The net effect of the notch is therefore to lower the resonance frequency of the third flexural mode relative to the first flexural mode. The size of the notch can therefore be chosen to obtain an integer ratio between the resonance frequencies.

26.2.3 Implementation of Harmonic Cantilevers

A rectangular cantilever fixed at the base and free at the tip end has its second resonance frequency at 6.7 times the first resonance frequency and third resonance frequency at 17.4 times the first resonance frequency. In

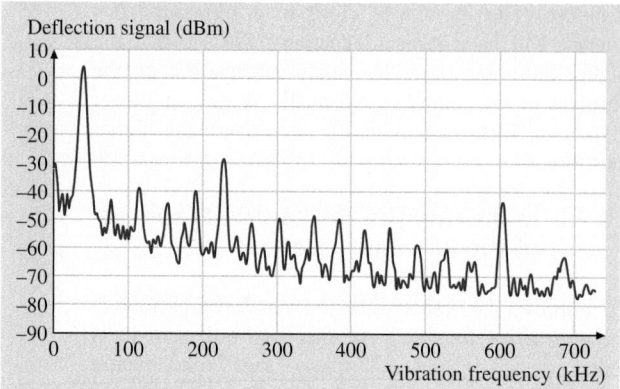

Fig. 26.9 Vibration spectrum of a harmonic cantilever in tapping-mode AFM. The cantilever is driven at its fundamental resonance frequency (37.4 kHz), and higher-harmonic generation is observed. The second (240 kHz) and third (598 kHz) harmonics coincide with higher resonances and have relatively large signal power

order to obtain a higher-order resonance frequency that is an integer multiple of the first, one has to modify the geometry of the cantilever.

In one design intended to obtain a flexural resonance at the 16-th integer multiple of the fundamental

resonance frequency, we placed a hole at a bending region in the third flexural mode shape (Fig. 26.8). The effect of this hole is similar to the notch in Fig. 26.7. The hole reduces the ratio of the third resonance frequency to the fundamental to give an integer ratio. We use the name *harmonic cantilever* for cantilevers that have this property that one of their higher-order modes has a resonance frequency that is an integer multiple of the fundamental resonance frequency.

Figure 26.9 shows the measured vibration spectrum of a harmonic cantilever in tapping-mode AFM. In addition to the drive signal at, two peaks (#6 and #16) have relatively large signal levels compared to their neighbors. These are the harmonics that are closest to the resonance frequencies of the harmonic cantilever. Especially the 16-th harmonic has a much higher signal level relative to its neighbors. This is because the frequency of that particular harmonic matches the third resonance frequency of the harmonic cantilever. Such cantilevers can be fabricated with conventional silicon-based microfabrication techniques. A more detailed discussion of the fabrication of the cantilever in Fig. 26.8 is given in [26.10].

26.3 Recovering the Time–Resolved Tip–Sample Forces with Torsional Vibrations

AFM cantilevers have a second type of vibration mode, called the torsional modes, in addition to the flexural modes discussed above. Vibrations in these modes result in angular deflections of the cantilever. These modes are excited as a result of torque acting on the cantilever. The tip of a typical cantilever is located on the longitudinal axis, preventing tip–sample forces from creating torque on the cantilever when tapping on a sample. In this section we will describe a class of cantilevers, called coupled torsional cantilevers, that enable the excitation of torsional vibration modes [26.11]. Torsional vibration modes are very sensitive to tip–sample forces and allow simultaneous measurement of a large number of higher harmonics, so that the tip–sample forces can be recreated with high temporal resolution. We will begin with a theoretical discussion of the torsional response of an AFM cantilever with an offset tip and then show experimental results from the vibration measurements of a coupled torsional harmonic cantilever.

26.3.1 Torsional Response of Coupled Torsional Cantilevers

The coupled torsional harmonic cantilever has a tip that is offset from the long axis of the cantilever. An example of such a cantilever is shown in Fig. 26.10a. When a coupled torsional harmonic cantilever is vibrated in tapping mode, tip–sample interaction forces generate torque around the long axis of the cantilever and excite the torsional modes (Fig. 26.10b). The overall motion of the cantilever is a combination of flexural and torsional vibrations. The vibration at the fundamental flexural resonance frequency is still the dominant component. The motion of the cantilever is detected with a laser beam reflected from the backside of the cantilever falling onto a four-quadrant position-sensitive diode (Fig. 26.10c). The difference in optical powers in the upper and lower halves is proportional to longitudinal (flexural mode) deflection and the difference in the left and right halves is proportional to torsional deflection.

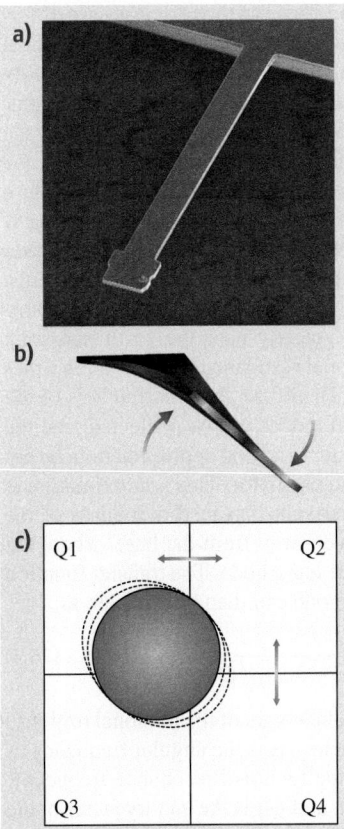

Fig. 26.10 (a) Scanning electron micrograph image of a torsionally coupled harmonic cantilever. The cantilever is nominally 300 μm long, 2 μm thick and 30 μm wide. The tip is offset 15 μm from the centerline of the cantilever. **(b)** Simulated first-torsional-mode shape of a rectangular cantilever fixed at the base. **(c)** Illustration of the laser spot on the four-quadrant position-sensitive photo-detector. The optical power difference $(Q1+Q2)-(Q3+Q4)$ is proportional to vertical cantilever deflection, and the optical power difference $(Q1+Q3)-(Q2+Q4)$ is proportional to torsional angle

When the tip interacts with the surface as it approaches and retracts, the torsional vibration mode acts as a force sensor that measures the force acting on the tip. The torsional resonance frequency is much higher than the first flexural resonance frequency, so the torsional mode responds to the variations in the tip–sample force over a wide frequency range. Figure 26.11 shows the calculated frequency response of the torsional and flexural modes of the coupled torsional harmonic cantilever. These curves correspond to the lateral

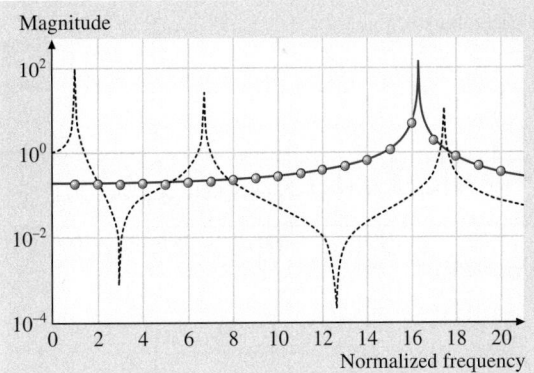

Fig. 26.11 Calculated frequency response of a rectangular cantilever in the torsional and flexural modes. The frequency axis is normalized to the first flexural resonance frequency. The magnitudes represent the optical signal at the position-sensitive detector. These optical signals are proportional to the slope of the cantilever at the laser spot. The *crosses* are located at the integer multiples of the first resonance frequency. Note that the torsional response is much higher than flexural response at higher harmonics

and vertical deflection signals at the position-sensitive detector.

At frequencies below the first flexural resonance frequency, the flexural response is much higher than the torsional response. This is because the effective spring constant of the first flexural mode is much smaller than the first torsional mode. On the other hand, at higher frequencies where the higher harmonics of the tip–sample forces are located, the torsional response is larger than the flexural response.

Figures 26.12a and 26.12b shows the flexural and torsional vibration spectra of a coupled torsional harmonic cantilever while tapping on a polystyrene sample. The cantilever is driven near the first flexural resonance frequency (52.5 kHz) by a piezo element from the base. The free vibration amplitude and set-point amplitude are 100 nm and 60 nm, respectively. While tapping on the surface the cantilever simultaneously moves in flexural and torsional modes as shown by the vibration spectra. The dominant component of the cantilever motion is the first peak in the flexural vibration spectrum. This is the motion at the drive frequency. The other harmonics are generated by the tip–sample interaction.

The higher harmonics in the flexural vibration spectrum have signal-to-noise ratios that are too low for practical measurements. In the torsional vibration spectrum in Fig. 26.12b, on the other hand, the signal-to-

Part C | 26.3

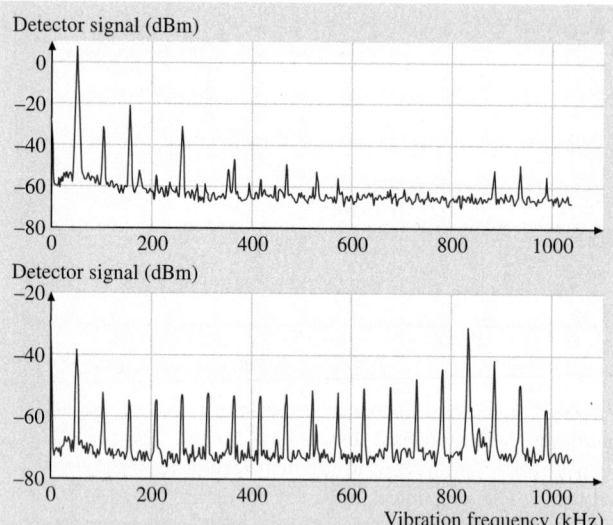

Fig. 26.12 (a) Flexural and **(b)** torsional vibration spectra of a torsionally coupled harmonic cantilever while tapping on a polystyrene sample. The first peak of the flexural spectrum in (a) is at the driving frequency. It is the largest component of the cantilever motion. The other flexural and torsional peaks are the higher harmonics generated by the tip–sample interaction forces. The torsional peaks have much higher signal levels at higher harmonics

noise ratios are sufficient for practical measurements for the first 19 harmonics. The signal levels around the 16-th harmonic increase due to first torsional resonance of the cantilever located at 16.2 times the drive frequency. The vibration spectrum in Fig. 26.12b show that the torsional vibrations provide good signal levels up to the 19-th harmonic of the first flexural resonance frequency. This means that this particular torsional-mode force sensor can resolve tip–sample forces with a temporal resolution roughly 20 times shorter than the fundamental flexural oscillation period.

26.3.2 Time–Resolved Force Measurements

The mechanical bandwidth of the torsional mode determines the response to variations in the tip–sample forces as the tip vibrates. This bandwidth is determined by the first torsional resonance frequency, which is 16.2 times the drive frequency for this cantilever. In general, it is possible to measure the first few harmonics beyond the first torsional resonance frequency without significant attenuation. This high bandwidth allows the torsional mode to respond to high-frequency tip–sample forces. While the cantilever responds to harmonic forces up to

19, the magnitude and phase of the responses are different for each harmonic. This can be seen in the torsional frequency response of the cantilever given in Fig. 26.12b. Therefore, forces at different harmonics cannot be compared directly. Instead, it is necessary to measure the frequency response and adjust the measurements by the mechanical gain introduced by the resonant response of the cantilever. (*Stark* et al. performed a similar experiment with the flexural vibrations of the cantilever and demonstrated time-resolved force measurements, albeit with lower signal levels [26.12].) The first torsional resonance is typically near the 15-th harmonic and the second torsional resonance is about three times higher in frequency. Therefore, the contributions of the higher-order torsional modes can be neglected, and the first torsional mode can, to a good approximation, be described as a harmonic oscillator. This approximation is even better if the laser spot is placed two thirds of the length of the cantilever away from the base, where the second torsional mode has a node. The transfer function of the first torsional mode can then be modeled as:

$$H_T(\omega) = \frac{1}{\omega_T^2 - \omega^2 + i\omega\omega_T/Q} \tag{26.6}$$

Here H_T is the mechanical gain of the torsional response as a function of frequency, ω is the angular frequency of the vibration, ω_T is the torsional resonance frequency, i is the imaginary unit, and Q_T is the quality factor of the torsional resonance. The two parameters that determine the frequency response, ω_T and Q_T, are easily measured for a given AFM system. The photo-detector gain, the location of the laser spot on the cantilever and the offset distance of the tip will all multiply H_T in a scalar fashion, but they will not affect the relative enhancement of the different harmonics.

Once the torsional frequency response is determined, it is possible to recover the time-resolved tip–sample forces by measurement of torsional vibrations of the cantilever and digitally correcting for the mechanical gain introduced by the torsional frequency response.

An example of this procedure is shown in Fig. 26.13. The torsional deflection signals at the position sensitive detector are recorded with a digital oscilloscope. The data is averaged over 128 samples to achieve an approximate noise bandwidth of 500 Hz. The measured vibration signals coming from both flexural and torsional modes over one period is given in Fig. 26.13a. The effect of nonlinear mechanical gain due to the torsional frequency response is removed digitally with the aid of (26.6). The resulting waveform is given in Fig. 26.13b. In this waveform the tip–sample forces are not zero even

Fig. 26.13a–c Vibration signals from flexural and torsional motions, and tip–sample forces. (**a**) The signals at the four-quadrant photo-detector for vertical and torsional displacements. The *solid curve* is the torsional signal. We multiplied the torsional signal by a factor of 10 to view the two curves clearly in one graph. (**b**) The torsional vibration signal after being divided with the torsional frequency response. Except for the pulse located between 300-th and 400-th time steps, the tip–sample forces should have been close to zero, because the tip is far away from the surface at those times. The measured signals when not in contact come from cross talk from the flexural deflection signal. The *dashed curve* estimates the error introduced by these sources. When it is subtracted from the *solid curve* we get the time-resolved forces plotted in (**c**) ▶

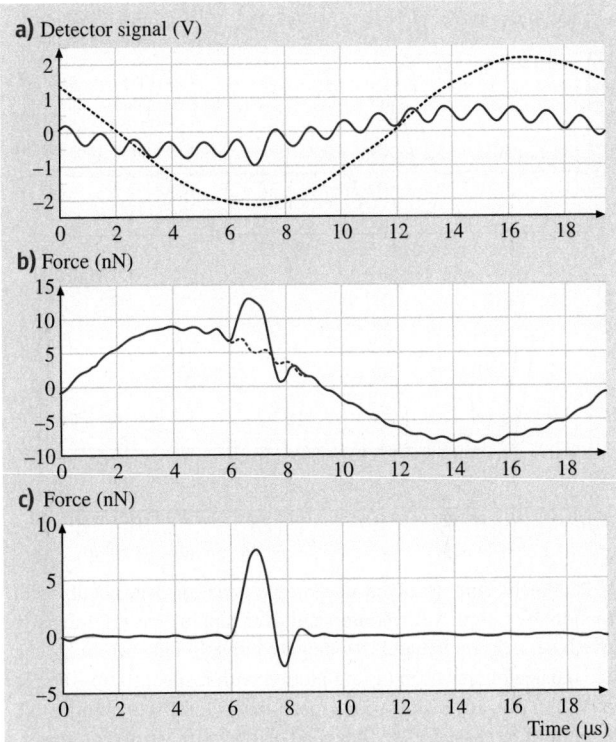

when the tip is away from the surface. This error is due to the nonlinearity of the detection circuit and crosstalk from the large flexural signal, which produces additional signals at the first few harmonics. Those components are removed with a signal-processing procedure that assumes that the tip–sample interaction forces are zero when the cantilever is away from the surface, and subtracts any additional signal at the first few harmonics that results in a nonzero tip–sample force. In Fig. 26.13b the computed correction signal that arises from crosstalk and nonlinearities is also given. Notice that two curves overlap during the times when the tip is not in contact. Once this additional component is subtracted we get the corrected tip–sample force waveform in Fig. 26.13c. It is important to note that, while the measurements of torsional vibrations are in units of Volts, the tip–sample forces are indicated in Newtons. Conversions of meas-

ured voltages into force units are done by using the relation between time-average tip–sample forces and the cantilever spring constant, quality factor, and free and tapping amplitudes and phase [(7) in [26.13]]. Also note that forces are linearly proportional to measured voltages.

26.4 Application Examples

In this section we will present application examples of the use of harmonic forces to study various material systems. In a first experiment we show how time-resolved force measurements can measure stiffness variations and reveal the origin of the hysteresis in tip–sample interaction forces that results in energy dissipation. Analysis of the measured data shows that quantitative comparisons of the stiffness of materials can be made without the knowledge of cantilever spring constant, tip geometry, vibration amplitude, and the gain of the photo-detector. In a second experiment we present imaging of the glass transition of a component in a binary polymer blend by mapping the magnitude of a higher harmonic across

the surface at different temperatures. Then we analyze the changes in the surface with the aid of time-resolved force measurements.

26.4.1 Time–Resolved Force Measurements on Different Materials

The time-resolved tip–sample forces obtained with a coupled torsional harmonic cantilever tapping on a high-density polyethylene ($0.92\,g/cm^3$), highly oriented pyrolytic graphite and low-density polyethylene ($0.86\,g/cm^3$) are given in Fig. 26.14a. The cantilever has a nominal spring constant of $1\,N/m$ and the free vi-

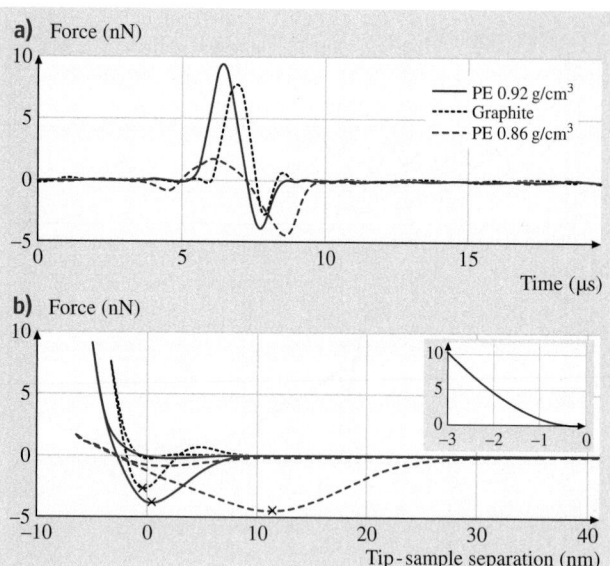

a) Force (nN)

— PE 0.92 g/cm³
------- Graphite
----- PE 0.86 g/cm³

Time (µs)

b) Force (nN)

Tip-sample separation (nm)

Fig. 26.14a,b Time-resolved tip–sample force measurements with a torsionally coupled harmonic cantilever tapping on high-density polyethylene, graphite, and low-density polyethylene (**a**). In (**b**) the same measurements as in (a) are plotted with respect to tip–sample separation. Negative separations mean that the sample is indented. The rates of increase in forces depend on the stiffness of the sample. Larger negative forces arise during the retraction of the tip. The *inset* shows the tip–sample forces during retraction (first 3 nm to the left of the *crosses*) with the forces on high-density polyethylene multiplied by 3.0 and low-density polyethylene multiplied by 27.0. The good correspondence between the shapes of these curves indicates that the contact mechanisms are the same on these samples. This allows quantitative measurements of the ratios of the elastic parameters for these materials

bration amplitude and set-point amplitude are 100 nm and 60 nm, respectively. These measurements are obtained with the signal-processing procedures described in Fig. 26.13. Here the positive forces are the repulsive tip–sample interactions that arise from the indentation of the sample. Negative forces are due to capillary forces and van der Waals forces [26.14].

These time waveforms of tip–sample forces point to many differences in these materials. First of all, the contact durations and peak forces differ from one material to the other. The magnitude and duration of the attractive interactions are also different for each material. These differences are due to variations in stiffness and van der Waals parameters and wettability of these materials.

By plotting the tip–sample forces with respect to tip–sample separation, we get a better understanding of the

differences in these samples. If the tapping amplitude, or the set-point amplitude, is known we can determine the vertical position of the tip as a function of time. We can then determine the tip–sample forces with respect to the tip–sample distance, as shown in Fig. 26.14b. Two important features of the curves plotted in this graph are the rate of increase of the forces during the indentation (negative separation) and the hysteresis in the forces.

The rate of increase in the forces during the indentation (negative tip–sample separation) is determined by the stiffness of the sample. Larger loading forces are required to produce a given depth of indentation on stiffer materials. Therefore, the forces increase faster on graphite than the polyethylene samples in Fig. 26.14b. The indentation forces do not follow a linear relationship. An approximate model predicts that the repulsive forces vary with a power-law relation [26.15],

$$F_{\text{rep}} = \gamma d^n . \tag{26.7}$$

Here F_{rep} is the repulsive tip–sample force, γ is a constant that depends on the reduced Young's modulus and tip diameter, d is the indentation depth, and n is a number that depends on the tip and surface geometry. For a spherical tip and flat surface, n is 1.5 [26.16]. If the same tip is used, the value of n is expected to be the same for experiments on different materials. The differences will arise only in the multiplicative term, γ, because the reduced Young's modulus values are different.

The curves in Fig. 26.14b show double values for forces mainly at positive separations. This is because the tip–sample forces are different during the approach and retraction of the tip. The upper branches of the curves show the approach of the tip, whereas the lower branches show retraction. Capillary forces are larger in retraction due to attractive van der Waals forces, which pull the surface and raise it above its equilibrium level during retraction. This results in the observation of attractive forces at larger tip–sample separations. There is also the possibility of forming a liquid neck between the tip and the sample during retraction, but the samples used for the measurements are hydrophobic, so this is not likely to be the origin of the hysteresis observed in the force curves of Fig. 26.14.

26.4.2 Quantitative Comparison of Material Properties

The ratio of the Young's modulus values of different materials can be quantitatively measured by comparing the tip–sample force curves for those materials, because for a given depth of indentation, the ratio of forces is equal the ratio of the respective γ coefficients. For these quantitative measurements we do not need to assume a power-law functional dependency as in (26.7). The approximation that the ratio of the Young's modulus values equals the ratio of the measured forces at a given depth of indentation is valid as long as the functional dependency is the same for the two materials.

Absolute measurements require a reference. To demonstrate this, we have chosen graphite as our reference material for measurements of the indentation forces for high-density and low-density polyethylene. To compare the forces curves obtained on the three samples, we need to eliminate the contribution of attractive forces because they are not governed by the elastic properties of the materials. On the graphs of Fig. 26.14b, the points with the highest attractive forces are marked with crosses. Tip–sample mechanical contact is broken at approximately these points as the tip is retracted. There are no repulsive forces at these points, so the net negative force is due to the attractive forces. To the left of these crosses, the tip–sample forces increase as the tip indents the samples. In the insert in Fig. 26.14b, we have plotted the indentation forces for the first three nanometers to the left of the crosses. We choose the points of maximum attractive force as the origin of the graphs because this is where the indentation forces and indentation depths are all zero. By scaling the force on high-density polyethylene by a factor of 3.0 and the forces on low-density polyethylene by a factor of 27.0, they both match the reference forces on graphite. The matching of these curves supports the assumption that the functional dependency of forces with respect to indentation is similar. Graphite has an approximate elastic modulus of 5 GPa, so our measured values for the elastic moduli of high-density polyethylene and low-density polyethylene are 1.7 GPa and 180 MPa, respectively. High-density polyethylene is more ordered and it is expected to be significantly stiffer than the low-density polyethylene. Unfortunately, there is a relatively large spread in published values for the elastic modulus for graphite, so a well-calibrated reference material is still needed for accurate measurements.

These quantitative comparisons and absolute measurements are made without knowledge of many parameters of the AFM experiment, such as cantilever spring constant, tip geometry, drive force, set-point amplitude, photo-detector gain, position of the laser spot, and tip offset distance. This is possible because the calculations for quantitative measurements use only the voltages at the output of the photo-detector, together with the flexural and torsional resonance frequencies. Because we take the ratio of force values in the force–distance curves, the ratio of measured voltages is all that is required, while knowledge of the relationship between the measured voltages and actual forces is not needed. This technique eliminates all the instrument variables, which are difficult to measure and control in AFM experiments.

26.4.3 Imaging the Glass Transition of a Binary Polymer Blend with a Single Harmonic Force

According to the theoretical analysis carried out in the first section, the magnitudes of the higher harmonics depend on the stiffness of the samples. If carefully chosen, the amplitude of a particular harmonic will monotonically increase with increasing stiffness of the samples.

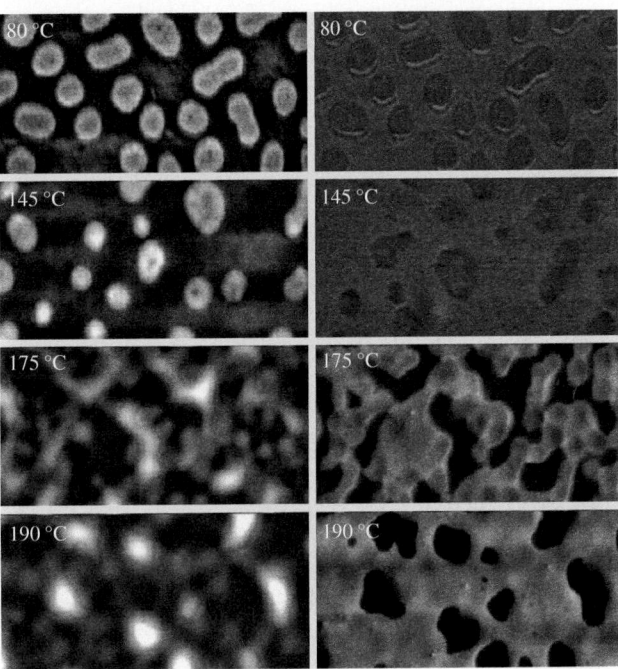

Fig. 26.15 Topography (*left column*) and 12-th harmonic (*right column*) images of a thin polymer film composed of polystyrene and PMMA. The harmonic images are generated by recording the magnitude of the 12-th harmonic as the cantilever is scanned across the sample at 80 °C, 145 °C, 175 °C, and 190 °C. *Darker color* indicates a lower signal. The increased contrast in the harmonic image is due to softening of one material component in the film, indicating a phase transition. The image at 80 and 145 °C is 2.5 by 5 μm, and the images at 175 °C and 190 °C are 5 by 10 μm

Part C | 26.4

We are interested in mapping mechanical property variations, so imaging the amplitude of a single higher harmonic is the simplest solution. The coupled torsional harmonic cantilever provides the first 20 torsional harmonics with good signal levels, so we can measure any one of these harmonics with a lock-in amplifier to produce the corresponding harmonic image of the surface.

By recording the torsional harmonic amplitudes while scanning the surface in tapping-mode we have generated the harmonic force images of an ultra-thin (about 50 nm) binary polymer film on a silicon substrate and studied the glass transition of the two components, polystyrene and poly-methyl-methacrylate (PMMA). These components with different glass-transition temperatures (around 100 °C and 130 °C, respectively) form submicron domains within the film. As the temperature is elevated, polystyrene goes through the glass transition before PMMA. The regions of rubbery phase will be softer than the glassy regions, so that we can observe the glass transitions of individual components of the composite polymer film. For the measurements

the same coupled torsional harmonic cantilever used in Sect. 26.4.1 is employed with different vibration amplitudes. The free amplitude and set-point amplitude are chosen as 150 nm and 100 nm, respectively.

Images of the topography and 12-th harmonic at 80 °C, 145 °C, 175 °C, and 190 °C are presented in Fig. 26.15. The two components of the polymer blend are easily distinguishable in the topography image at 80 °C and 145 °C because of the height differences. The higher regions are PMMA domains and lower regions are composed of polystyrene. In the harmonic image different components are distinguishable at 80 °C and 145 °C, however the contrast is small. At 175 °C, one component has softened and a large contrast in the harmonic image is observed. The softening results in lower harmonic amplitudes. This indicates the glass transition. As the temperature is raised to 190 °C, the contrast increases further. While topography images show some changes at the surface at different temperatures, they do not reveal the nature of the changes. The topographical images become blurry at elevated temperatures because the glass transition is accompanied by morphological changes due to increased molecular mobility. The harmonic images, on the other hand, allow tracking of the different materials on the surface and illuminate the origin of the changes on the surface. A more detailed understanding of the changes on the surface can be gained by using the full potential of torsionally coupled harmonic cantilevers to measure all the harmonics and generate time-resolved tip–sample forces.

26.4.4 Detailed Analysis with Time–Resolved Forces

The changes in the polymer components were studied in greater detail by measuring the time-resolved forces and generating force–distance curves, as previously described in Fig. 26.14b. In Fig. 26.16 we plot the curves of force versus distance as obtained from time-resolved forces on polystyrene and PMMA regions at each temperature. On polystyrene, the rate of increase in the repulsive forces decreases with increasing temperatures. Between 145 °C and 175 °C, the material softened almost an order of magnitude, identifying the glass transition. On PMMA, however, the curves remain unchanged below 190 °C. At higher temperatures there is a partial dewetting of the surface. This, together with material pile up, prevented us from observing the glass transition of PMMA.

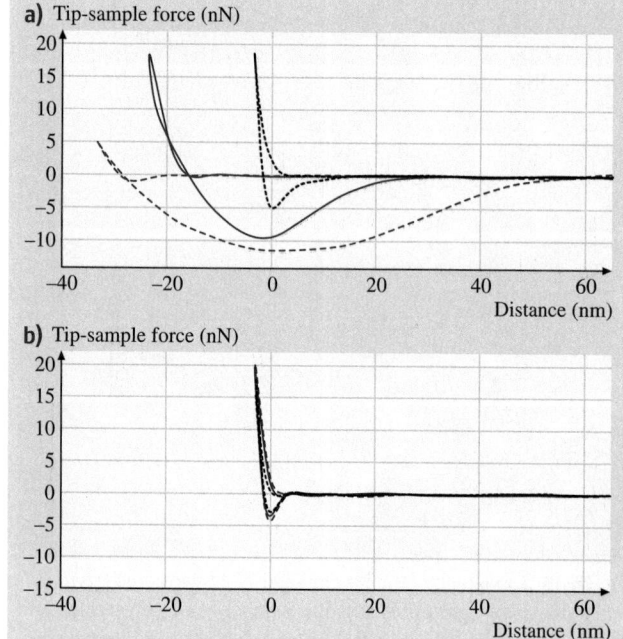

Fig. 26.16a,b Time-resolved tip–sample force measurements plotted against tip–sample separation (**a**) on polystyrene regions, and (**b**) on PMMA regions. While measurements on polystyrene resulted in drastic changes in interaction forces, forces on PMMA showed almost no changes. The reduced slope and increased hysteresis of the force on polystyrene indicate softening of the sample

Another notable phenomenon observed in the force distance curves on polystyrene is the increase in the hysteresis. As the material gets softer, attractive forces will pull the surface above the equilibrium level while the tip is retracting from the surface. Therefore we observe attractive forces at larger tip–sample separations. The same behavior in hysteresis was also observed in Fig. 26.13b where graphite and polyethylene samples were compared.

The relatively high glass-transition temperatures measured in these experiments are expected consequences of the frequency dependence of the glass transition. Our technique measures the mechanical properties at a drive frequency of 50 kHz, and the interaction time is hundreds of nanoseconds. Glass transitions of bulk materials are determined either through thermal measurements or dynamic mechanical measurements, with frequencies below 100 Hz [26.17].

26.5 Higher Harmonic/Atomic Force Microscopy with Small Amplitudes

26.5.1 Principle

AFM cantilevers that oscillate freely show a sinusoidal motion where the deflection of the end of the cantilever q' is described by $q'(t) = A \cos(2\pi f t)$. When the tip of the oscillating cantilever is in the force field of the sample, the potential is generally no longer harmonic, giving rise to anharmonic components where the deflection is described by a Fourier series:

$$q'(t) = \sum_{n=0}^{\infty} a_n \cos(2\pi f t) \qquad (26.8)$$

with $a_1 = A$. When the oscillation amplitude of the cantilever is large compared to the range of the tip–sample potential, the lower orders of these anharmonic components are proportional to the frequency shift [(11) in [26.18]], so in principle, no new information is available over frequency modulation AFM (FM-AFM) Chapt. 24. Only if higher-order components with periods comparable to, or shorter than, the contact duration are measured, as outlined in the first part of this chapter, does higher-harmonic AFM with large amplitudes yield advantages over FM-AFM. For small oscillation amplitudes the tip is in the force field during all or most of the vibration cycle, so the lower-order higher harmonics are no longer proportional to the frequency shift and offer physical content on their own right. *Dürig* [26.19] has found that, in small-amplitude AFM, the full tip–sample potential can be immediately recovered over the z-range covered by the oscillating cantilever if the amplitudes and phases of all higher harmonics are available. *Dürig* has expressed the higher harmonics as a Tchebychev expansion of the tip–sample force [26.19]. *Sahin* [26.4] and *de Lozanne* [26.20] expand the temporal dependence of the tip–sample force in a Fourier series with base frequency f and express the higher harmonics as

the response of the cantilever to an excitation at frequency $n \times f$. Mathematically, these two notions are of course equivalent. We start with Dürig's formula for the amplitude of the n-th harmonic:

$$a_n = \frac{2}{\pi k} \frac{1}{(1-n^2)} \int_{-1}^{1} F_{ts}(z + Au) \frac{T_n(u)}{\sqrt{1-u^2}} \, du \, , \qquad (26.9)$$

where $T_n(u)$ is the n-th Chebychev polynomial of the first kind. We can show, by applying integration by parts n-times ($\int gh' = -\int g'h + gh$), that

$$a_n = \frac{2}{\pi k} \frac{1}{(1-n^2)} \frac{1}{(2n+1) \dots 3 \cdot 1} A^n$$
$$\times \int_{-1}^{1} \frac{d F_{ts}^n(z + Au)}{dz^n} (1-u^2)^{n-1/2} \, du \, , \qquad (26.10)$$

because the n-th integral of $T_n(u)/(1-u^2)^{1/2}$ is $(1 - u^2)^{n-1/2}/[(2n+1) \dots 3 \cdot 1]$. Thus, the n-th harmonic can be expressed by a convolution of the n-th force gradient $d F_{ts}^n/dz^n$ with a bell-shaped weight function $(1 - u^2)^{n-1/2}$. Figure 26.17 shows three snapshots of the oscillating cantilever in the upper turnaround point, the neutral position and the closest sample approach. The graph on the left part of the figure shows various weight functions for deriving the frequency shift and higher harmonics from the force gradient and higher-order gradients. The semicircular weight function is convoluted with the force gradient, yielding the frequency shift (Chapt. 24). The bell-shaped weight functions are convoluted with higher-order force gradients to derive the higher harmonics (26.8). The weight functions have their maxima at $u = 0$, i. e. a distance A further away from the surface than the minimal tip–sample distance z_{min}.

Part C | 26.5

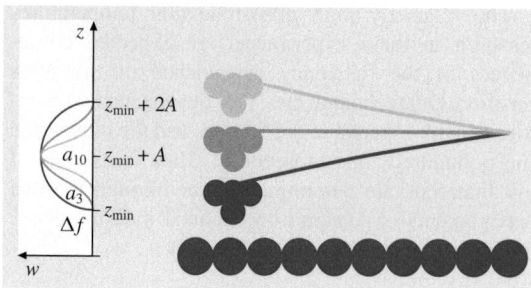

Fig. 26.17 Cantilever in three phases of its oscillation cycle: minimal distance z_{min}, average distance $z_{min} + A$ and maximal distance $z_{min} + 2A$. The frequency shift is calculated by the convolution of the force gradient with a semi-spherical weight function indicated by Δf, the second-harmonic amplitude a_2 is calculated by convoluting the second-order force gradient by a weight function $(1 - u^2)^{3/2}$, and the n-th harmonic a_n is computed by convoluting the n-th order force gradient by $(1 - u^2)^{n-1/2}$. The *left part* shows the weight functions for Δf (*dark brown*), a_3 (*brown*) and a_{10} (*light brown*)

If the amplitude A is small enough that higher-order force gradients still have a reasonable magnitude at distance $z_{min} + A$, the higher harmonics are not just proportional to the frequency shift Δf, but contain information about higher-order force gradients.

The benefit of higher-order force-gradient maps is expressed in an elementary but instructive example illuminating the contrast achievable by tip–sample inter-

action potential, force, force gradient and higher-order gradients shown in Fig. 26.18. Figure 26.18a shows a model of the charge distribution where the AFM tip is replaced by a single electron with a charge $-e$. The sample ion probed by the tip is modeled by a central ion with charge $+9e$, surrounded by eight electrons with charge $-e$ that point towards the corners of a cubic lattice.

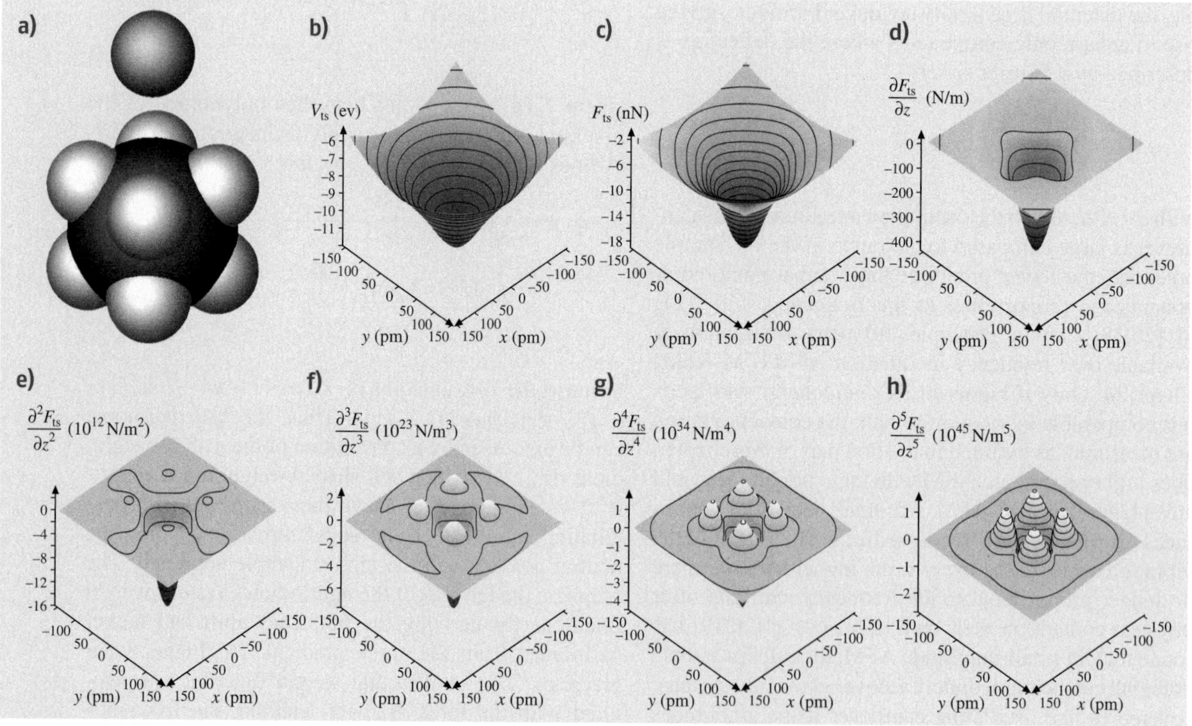

Fig. 26.18 (a) Simple electrostatic model for the tip–sample interaction in AFM: the tip is a test charge of $-e$, the sample atom that is probed consists of a central ion with charge $+9e$ and the valence charge distribution is modeled by eight surrounding charges of $-e$ oriented towards the corners of a cube. The tip "atom" is located 132 pm above the center of the sample ion, and the point charges are located at a distance of 55 pm from the center of the sample ion. (b) Tip–sample interaction potential for a distance of 132 pm as a function of the lateral positions x and y. (c) Tip–sample force, (d) force gradient, (e) second derivative of force versus z, (f) third derivative, (g) fourth derivative, and (h) fifth derivative of force versus z

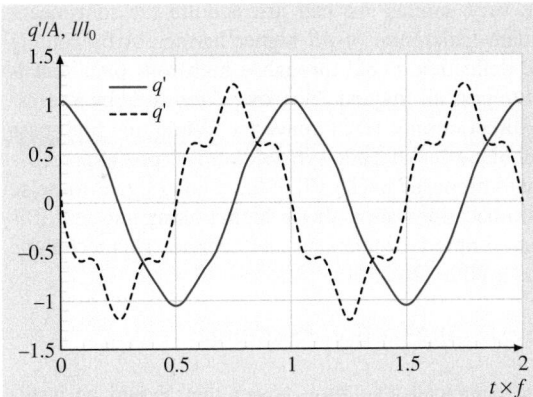

Fig. 26.19 Deflection of a cantilever versus time (*black*) for a cantilever according to $q'(t) = A[\cos(2\pi ft) + 0.04\cos(2\pi 5 ft)]$ and current (*dashed*) $I(t) = -\text{const} \cdot A[\sin(2\pi ft) + 0.2\sin(2\pi 5 ft)]$ produced by a piezoelectric force detector such as the qPlus sensor [26.8]. Because piezoelectric detection is proportional to frequency, higher harmonics produce more current than the oscillation at the base frequency f

Fig. 26.20 Simultaneous images of tunneling current (*left column*) and higher harmonics (*center column*) in a constant-height measurement. A graphite sample was imaged by a tungsten tip in a 4 K scanneling tunneling microscopy (STM)/AFM in ultrahigh vacuum ([26.23] for details). The highly resolved details in the higher-harmonic images are caused by the electronic structure of the tungsten tip atom. Because tungsten crystallizes in a bcc symmetry, the high-symmetry configuration are the two-fold [110] symmetry shown in the *first line*, the three-fold symmetry [111] in the *second line* and the four-fold [001] symmetry in the third line. The *right column* shows the Wigner–Seitz unit cell of bcc materials such as tungsten. Typical acquisition speed for each set of images is 0.5 lines/s at 256×256 pixels, i.e. a typical time of 10 min per image

This charge distribution for the sample ion is motivated by the charge-density calculations by *Posternak* [26.21] and *Mattheis* [26.22] for the (001) surface of tungsten. While tungsten (001) is a very special surface, it is reasonable to assume that most surface atoms do not have a spherically symmetric charge distribution as free atoms with filled shells, but local charge maxima that reflect the chemical bonding symmetry. Figure 26.18b shows the potential energy as a function of lateral displacements in x and y for a constant height z. It is interesting to note that both the energy and the force shown in Fig. 26.18c are almost symmetric with respect to rotations around the z-axis. However, the symmetry of the underlying charge distribution becomes more apparent with increasing order of the force gradient, as shown in Fig. 26.18d–h. In theory, it is possible to record (x, y) maps of the tip–sample force for various z values and calculate gradients and higher-order gradients from these maps later. However, experimental noise would render these maps totally useless for obtaining reliable data for higher-order gradients. Instead, a direct method that couples higher-order gradients to experimental observables is needed. The higher harmonics described in (26.8) are perfect for this purpose. However, the magnitude of higher harmonics is typically very small; therefore, they have to be enhanced either by resonant enhancement as described in the previous sections or by using small amplitudes and

enhancing the magnitude of higher harmonics by other means, as outlined below.

The mechanical bending of AFM cantilevers is transformed into an electrical signal by a deflection sensor. Optical and piezoresistive deflection sensors generate an output signal that is proportional to the cantilever deflection. For example, if the mechanical deflection is composed of a sinusoidal oscillation with frequency f and amplitude A plus an oscillation at $5 \times f$ with an amplitude of 4% of the amplitude A at the base frequency, both the deflection signal and the electrical signal look

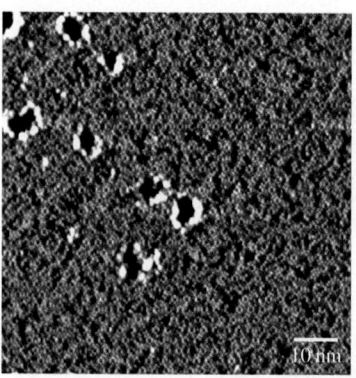

Fig. 26.21 Topographic data of higher-harmonic AFM image of damaged Si(111)-(7×7), recorded at room temperature. A qPlus sensor with a stiffness of 1800 N/m and an amplitude of 780 pm and $f_0 = 16\,740$ Hz was used in this image. The z-control feedback was set such that the second-harmonic amplitude was constant at $a_2 = 4.4$ pm. The frequency shift was also recorded and had an average value of -25 Hz (Δf data not shown here). The acquisition speed was very low: 0.1 lines/s at 512×512 pixels, i.e. it took 85 min to acquire this image [26.11]

like the dark brown curve in Fig. 26.19. However, in piezoelectric detectors such as the qPlus sensor [26.24], a charge accumulates at the electrodes of the "cantilever" that is proportional to the deflection [(2) in [26.24]]. When the sensor oscillates, an alternative current (AC) results that is proportional to deflection times frequency. Typically, the AC current is transformed to a voltage by a transimpedance amplifier [26.24]. Because of the proportionality between current and deflection times frequency, higher harmonics generate greater signal strength in piezoelectric sensors. The grey curve in Fig. 26.19 shows the current that is generated by a qPlus sensor oscillating at a base frequency and the fifth harmonic with a relative amplitude of 4%. Due to the enhancement of the sensitivity for higher harmonics, the current at frequency $5 \times f$ amounts to 20% of the base frequency.

In principle, one can pick an individual higher harmonic by analyzing the deflection of the cantilever with a lock-in amplifier that is triggered at the base frequency f and set to the n-th harmonic of f. Of course, one could also apply a battery of lock-in amplifiers and record as many higher harmonics as practical for a full potential recovery, as proposed by *Dürig* [26.19]. Because the higher force gradient images in Fig. 26.18e–h

are very similar, we can just acquire the root-mean-square (rms) sum of *all* higher harmonics by routing the deflection signal through a high-pass filter that is set to pass all frequencies above f followed by an rms-to-direct current (DC) converter. While this high-pass technique does not allow immediate full-potential recovery as proposed in [26.19], it has a good signal-to-noise ratio because it uses all the higher harmonics and it is very simple to implement and operate. This technique was used in [26.23].

26.5.2 Application Examples

The enhanced resolution power that should be available to higher-harmonic AFM with small amplitude is best demonstrated by a direct comparison. Figure 26.20 shows simultaneous constant-height data where a graphite sample was imaged by a tungsten tip mounted on an oscillating qPlus sensor. The left column maps the tunneling current, and the center column shows the intensity of the higher harmonics data, clearly showing the enhanced spatial resolution. The intensity of the higher harmonics can not only be monitored in constant-height mode or when the feedback uses a tunneling current or frequency shift for distance regulation, but can also be used for z-feedback. Figure 26.21 shows an image of a Si(111)-(7×7) surface that has been acquired by higher-harmonic AFM. The deflection signal of the cantilever was fed into a lock-in amplifier, and the second harmonic was used as the signal.

26.5.3 Conclusion

Higher-harmonic AFM with small amplitudes is an interesting AFM mode because it allows greatly increased spatial resolution. The signal levels of the higher harmonics are small, therefore bandwidth reduction can become necessary to increase the signal-to-noise ratio. A small signal bandwidth requires slow scanning, therefore low-temperature environments are helpful because the thermal drift effects that can plague images acquired at low scanning speeds at room temperature are usually much less severe at low temperatures. It is possible that a combination of small-amplitude higher-harmonic AFM with the resonance-enhancement technique described in the previous section might allow one to retain the high-resolution capability of small-amplitude higher-harmonic AFM while increasing the possible scanning speed.

References

26.1   Q. Zhong, D. Inniss, K. Kjoller, V. B. Elings: Fractured polymer/silica fiber surface studied by tapping mode atomic force microscopy, Surf. Sci. **280**, L688–L692 (1993)

26.2   D. Klinov, S. Magonov: True molecular resolution in tapping mode atomic force microscopy, Appl. Phys. Lett. **84**, 2697–2699 (2004)

26.3   M. V. Salapaka, D. J. Chen, J. P. Cleveland: Phys. Rev. B **61**, 1106–1115 (2000)

26.4   O. Sahin, A. Atalar, C. F. Quate, O. Solgaard: Resonant harmonic response in tapping-mode atomic force microscopy, Phys. Rev. B **69**, 16 5416–16 5424 (2004)

26.5   R. Hillenbrand, M. Stark, R. Guckenberger: Higher-harmonics generation in tapping-mode atomic force microscopy: Insights into tip–sample interaction, Appl. Phys. Lett. **76**, 3478–3480 (2000)

26.6   R. W. Stark, W. M. Heckl: Higher harmonics imaging in tapping-mode atomic-force microscopy, Rev. Sci. Instrum. **74**, 5111–5114 (2003)

26.7   A. S. Paulo, R. Garcia: Unifying theory of tapping-mode atomic force microscope, Phys. Rev. B **66**, 041406–041409(R) (2002)

26.8   R. W. Stark, W. M. Heckl: Fourier transformed atomic force microscopy: Tapping mode atomic force microscopy beyond the Hookian approximation, Surf. Sci. **457**, 219–228 (2000)

26.9   U. Rabe, K. Janser, W. Arnold: Rev. Sci. Instrum. **67**, 3281–3293 (1996)

26.10   O. Sahin, G. Yaralioglu, R. Grow, S. F. Zappe, A. Atalar, C. Quate, O. Solgaard: High resolution imaging of elastic properties using harmonic cantilevers, Sensors Actuators A **114**, 183–190 (2004)

26.11   F. J. Giessibl: unpublished result

26.12   M. Stark, R. W. Stark, W. M. Heckl, R. Guckenberger: Inverting dynamic force microscopy: From signals to time resolved forces, Proc. Nat. Acad. Sci. **99**, 8473–8478 (2002)

26.13   A. S. Paulo, R. Garcia: Tip–surface forces, amplitude, energy dissipation in amplitude modulation (tapping mode) force microscopy, Phys. Rev. B **64**, 193 411–193 414 (2001)

26.14   L. Zitzler, S. Herminghaus, F. Mugele: Capillary forces in tapping-mode atomic force microscopy, Phys. Rev. B **66**, 155 436–155 443 (2002)

26.15   I. N. Sneddon: The relation between load and penetration in the axisymmetric Boussinesq problem for a punch of arbitrary profile, Int. J. Eng. Sci. **3**, 47–57 (1965)

26.16   J. Isrelachvili: *Intermolecular and Surface Forces* (Academic, London 2003)

26.17   I. M. Ward, J. Sweeney: *An Introduction to the Mechanical Properties of Solid Polymers* (Wiley, Chichester 2004)

26.18   U. Dürig: Relations between interaction force, frequency shift in large-amplitude dynamic force microscopy, Appl. Phys. Lett. **75**, 433–435 (2004)

26.19   U. Dürig: Interaction sensing in dynamic force microscopy, New J. Phys. **2**, 5.1–5.12 (2000)

26.20   A. de Lozanne: Music of the spheres at the atomic scale, Science **305**, 348 (2004)

26.21   M. Posternak, H. Krakauer, A. J. Freeman, D. D. Koelling: Self-consistent electronic structure of surfaces: Surface states, surface resonances on W(001), Phys. Rev. B **21**, 5601–5612 (1980)

26.22   F. Mattheiss, D. R. Hamann: Electronic structure of the tungsten (001) surface, Phys. Rev. B **20**, 5372–5381 (1984)

26.23   S. Hembacher, F. J. Giessibl, J. Mannhart: Force microscopy with light-atom probes, Science **305**, 380 (2004)

26.24   F. J. Giessibl: Atomic resolution on Si(111)–(7×7) by noncontact atomic force microscopy with a force sensor based on a quartz tuning fork, Appl. Phys. Lett. **76**, 1470–1472 (2000)

27. Dynamic Modes of Atomic Force Microscopy

This chapter presents an introduction to the concept of the dynamic operational modes of the atomic force microscope (AFM). While the static (or contact-mode) AFM is a widespread technique to obtain nanometer-resolution images on a wide variety of surfaces, true atomic-resolution imaging is routinely observed only in the dynamic mode. We will explain the jump-to-contact phenomenon encountered in static AFM and present the dynamic operational mode as a solution to avoid this effect. The dynamic force microscope is modeled as a harmonic oscillator to gain a basic understanding of the underlying physics in this mode.

Under closer inspection dynamic AFM comprises a whole family of operational modes. A systematic overview of the different modes typically found in force microscopy is presented, and special attention is paid to the distinct features of each mode. Two modes of operation dominate the application of dynamic AFM. First, the amplitude-modulation mode (also called the tapping mode) is shown to exhibit an instability, which separates the purely attractive force interaction regime from the attractive–repulsive regime. Second, the self-excitation mode is derived and its experimental realization is outlined. While the tapping mode is primarily used for imaging in air and liquid, the self-excitation mode is typically used under ultrahigh vacuum (UHV) conditions for atomic-resolution imaging. In particular, we explain the influence of different forces on spectroscopy curves obtained in dynamic force microscopy. A quantitative link between the experimental spectroscopy curves and the interaction forces is established.

Force microscopy in air suffers from the small quality factors of the force sensor (i. e., the

cantilever beam), which are shown to limit the resolution. Also, the aforementioned instability in the amplitude-modulation mode often hinders imaging of soft and fragile samples. A combination of the amplitude modulation with the self-excitation mode is shown to increase the quality, or Q-factor, and extend the regime of stable operation. This so-called Q-control module allows one to increase as well as decrease the Q-factor. Apart from the advantages of dynamic force microscopy as a non-destructive, high-resolution imaging method, it can also be used to obtain information about energy-dissipation phenomena at the nanometer scale. This measurement channel can provide crucial information on electric and magnetic surface properties. Even atomic-resolution imaging has been obtained in the dissipation mode. Therefore, in the last section, the quantitative relation between the experimental measurement channels and the dissipated power is derived.

Part C | 27

27.1 Motivation: Measurement of a Single Atomic Bond

The direct measurement of the force interaction between two distinct molecules has been a challenge for scientists for many years now. The fundamental forces responsible for the solid state of matter can be directly investigated ultimately between defined single molecules. However, it has not been until very recently that the chemical forces could be quantitatively measured for a single atomic bond [27.1]. How can we reliably measure forces that may be as small as one billionth of 1 N? How can we identify one single pair of atoms as the source of the force interaction?

The same mechanical principle that is used to measure the gravitational force exerted by your body weight (e.g., with the scale in your bathroom) can be employed to measure the forces between single atoms. A spring with a defined elasticity is compressed by an arbitrary force (e.g., your weight). The compression Δz of the spring (with spring constant k) is a direct measure of the force F exerted, which in the regime of elastic deformation obeys Hooke's law:

$$F = k \cdot \Delta z . \tag{27.1}$$

The only difference with regard to your bathroom scale is the sensitivity of the measurement. Typically springs with a stiffness of $0.1–10\,\text{N/m}$ are used, which will be deflected by $0.1–100\,\text{nm}$ upon application of an interatomic force of some nN. Experimentally, a laser-deflection technique is used to measure the movement of the spring. The spring is a bendable cantilever microfabricated from silicon wafers. If a sufficiently sharp tip, usually directly attached to the cantilever, approaches within some nanometers of a surface, we can measure the interacting forces through changes in the deflected laser beam. This is a static measurement; hence it is called *static AFM*. Alternatively, the cantilever can be excited to vibrate at its resonant frequency. Under the influence of tip–sample forces the resonant frequency (and consequently also the amplitude and phase) of the cantilever will change and serve as measurement parameters. This is called *dynamic AFM*. Due to the multitude of possible operational modes, expressions such as non-contact mode, intermittent-contact mode, tapping-mode,

frequency-modulation (FM)-mode, amplitude modulation (AM)-mode, self-excitation, constant-excitation, or constant-amplitude-mode AFM are found in the literature, which will be systematically categorized in the following paragraphs.

In fact, the first AFMs were operated in the dynamic mode. In 1986, *Binnig*, *Quate* and *Gerber* presented the concept of the atomic force microscope [27.2]. The deflection of the cantilever with the tip was measured with sub-angstrom precision by an additional scanning tunneling microscope (STM). While the cantilever was externally oscillated close to its resonant frequency, the amplitude and phase of the oscillation were measured. If the tip approaches the surface, the oscillation parameters, amplitude and phase, are influenced by the tip–surface interaction, and can, therefore, be used as feedback channels. Typically, a certain set point for the amplitude is defined, and the feedback loop will adjust the tip–sample distance such that the amplitude remains constant. The controller parameter is recorded as a function of the lateral position of the tip with respect to the sample, and the scanned image essentially represents the surface topography.

What then is the difference between the static and the dynamic mode of operation for the AFM? The static-deflection AFM directly gives the interaction force between tip and sample using (27.1). In the dynamic mode, we find that the resonant frequency, amplitude, and phase of the oscillation change as a consequence of the interaction forces (and also dissipative processes, as discussed in the last section).

In order to get a basic understanding of the underlying physics, it is instructive to consider a very simplified case. Assume that the vibration amplitude is small compared to the range of force interaction. Since van der Waals forces range over typical distances of 10 nm, the vibration amplitude should be less than 1 nm. Furthermore, we require that the force gradient $\partial F_{ts}/\partial z$ does not vary significantly over one oscillation cycle. We can view the AFM set-up as a coupling of two springs (see Fig. 27.1). Whereas the cantilever is represented by a spring with spring constant k, the force interaction between tip and surface can be modeled by a second spring. The derivative of the force with respect to the tip–sample distance is the force gradient and represents the spring constant k_{ts} of the interaction spring. This spring constant k_{ts} is constant only with respect to one oscillation cycle, but varies with the average tip–sample distance as the probe approaches the sample. The two springs are effectively coupled in parallel, since the sample and tip supports are rigidly connected for a given value of z_0.

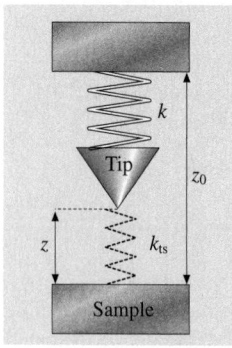

Fig. 27.1 Model of the AFM tip while experiencing tip–sample forces. The tip is attached to a cantilever with spring constant k, and the force interaction is modeled by a spring with a stiffness equal to the force gradient. Note that the force interaction spring is not constant, but depends on the tip–sample distance z

Therefore, we can write for the total spring constant of the AFM system:

$$k_{total} = k + k_{ts} = k - \frac{\partial F_{ts}}{\partial z} . \tag{27.2}$$

From the simple harmonic oscillator (neglecting any damping effects) we find that the resonant frequency ω of the system is shifted by $\Delta\omega$ from the free resonant frequency ω_0 due to the force interaction:

$$\omega^2 = (\omega_0 + \Delta\omega)^2 = k_{total}/m^* = \left(k - \frac{\partial F_{ts}}{\partial z}\right)/m^* . \tag{27.3}$$

Here m^* represents the effective mass of the cantilever. A detailed analysis of how m^* is related to the geometry and total mass of the cantilever can be found in the literature [27.3]. In the approximation that $\Delta\omega$ is much smaller than ω_0, we can write:

$$\frac{\Delta\omega}{\omega_0} \cong -\frac{1}{2k} \cdot \frac{\partial F_{ts}}{\partial z} . \tag{27.4}$$

Therefore, we find that the frequency shift of the cantilever resonance is proportional to the force gradient of the tip–sample interaction.

Although this consideration is based on a very simplified model, it shows qualitatively that in dynamic force microscopy we will find that the oscillation frequency depends on the force gradient, while static force microscopy measures the force itself. In principle, we can calculate the force curve from the force gradient and vice versa (neglecting a constant offset). It seems, therefore, that the two methods are equivalent, and our choice will depend on whether we can measure the beam deflection or the frequency shift with better precision at the cost of technical effort.

However, we have neglected one important issue for the operation of the AFM so far: the mechanical stability of the measurement. In static AFM, the tip approaches the surface slowly. The force between the tip and the surface will always be counteracted by the restoring force of the cantilever. In Fig. 27.2, you can see a typical force–distance curve. As the tip approaches the sample, the negative attractive forces, representing van der Waals or chemical interaction forces, increase until a maximum is reached. This turnaround point is due to the onset of repulsive forces caused by Coulomb repulsion, which will start to dominate upon further approach. The spring constant of the cantilever is represented by the slope of the straight line. The position of the z-transducer (typically a piezo element), which moves the probe, is at the intersection of the line with the horizontal axis. The position of the tip, shifted from the probe's base due to

the lever bending, can be found at the intersection of the cantilever line with the force curve. Hence, the total force is zero, i. e., the cantilever is in its equilibrium position (note that the spring constant line here shows attractive forces, although in reality the forces are repulsive, i. e., pulling the tip back from the surface). As soon as the position A in Fig. 27.2 is reached, we find two possible intersection points, and upon further approach there are even three force equilibrium points. However, between points A and B the tip is at a local energy minimum and, therefore, will still follow the force curve. But at point B, when the adhesion force upon further approach would become larger than the spring restoring force, the tip will suddenly jump to point C. We can then probe the predominantly repulsive force interaction by further reducing the tip–sample distance. When retracting the tip, we will pass point C, because the tip is still in a local energy minimum. Only at position D will the tip suddenly jump to point A again, since the restoring force now exceeds the adhesion. From Fig. 27.2 we

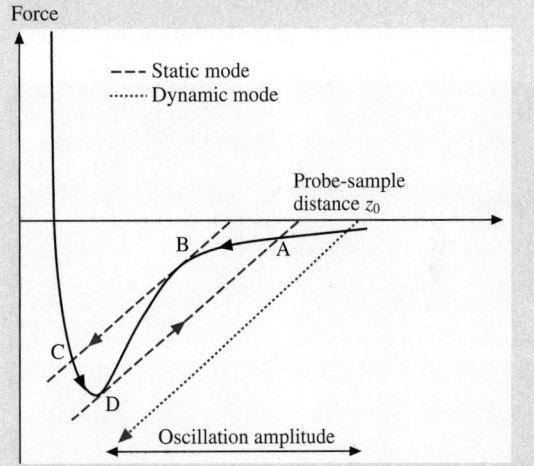

Fig. 27.2 Force–distance curve of a typical tip–sample interaction. In static-mode AFM the tip would follow the force curve until point B is reached. If the slope of the force curve becomes larger than the spring constant of the cantilever (*dashed line*) the tip will suddenly jump to position C. Upon retraction a different path will be followed along D and A again. In dynamic AFM the cantilever oscillates with a preset amplitude. Although the equilibrium position of the oscillation is far from the surface, the tip will experience the maximum attractive force at point D during some parts of the oscillation cycle. However, the total force is always pointing away from the surface, therefore avoiding an instability

can see that the sudden instability will happen at exactly the point where the slope of the adhesion force exceeds the slope of the spring constant. Therefore, if the negative force gradient of the tip–sample interaction at any point exceeds the spring constant, a mechanical instability occurs. Mathematically speaking, we demand that for a stable measurement:

$$\frac{\partial F_{ts}}{\partial z}\big|_z > k \text{ for all points } z \qquad (27.5)$$

The phenomenon of mechanical instability is often referred to as the *jump-to-contact*.

Looking at Fig. 27.2, we realize that large parts of the force curve cannot be measured if the jump-to-contact phenomenon occurs. We will not be able to measure the point at which the attractive forces reach their maximum, representing the temporary chemical bonding of the tip and the surface atoms. Secondly, the sudden instability, the jump-to-contact, will often cause the tip to change the very last tip or surface atoms. A smooth, careful approach needed to measure the full force curve does not seem feasible. Our goal of measuring the chemical

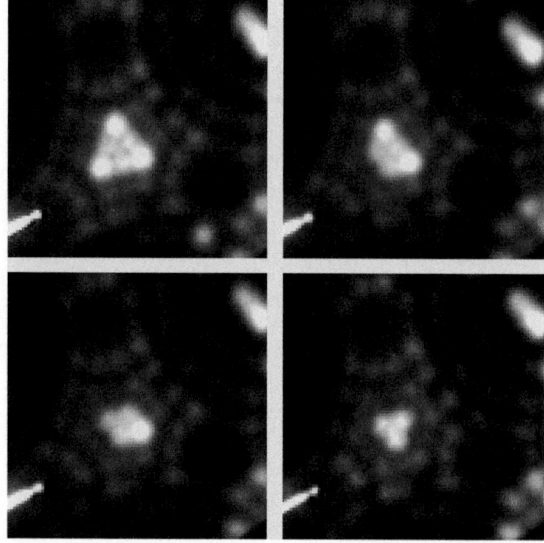

Fig. 27.3 Manipulation of the apex atoms of an AFM tip using field ion microscopy (FIM). Images are acquired at a tip bias of 4.5 kV. The last six atoms of the tip can be inspected in this example. Field evaporation to remove single atoms is performed by increasing the bias voltage for a short time to 5.2 kV. Each of the outer three atoms can be consecutively removed, eventually leaving a trimer tip apex

interaction forces of two single molecules may become impossible.

There are several solutions to the jump-to-contact problem. On one hand, we can simply choose a sufficiently stiff spring, so that (27.5) is fulfilled at all points of the force curve. On the other hand, we can resort to a trick to enhance the counteracting force of the cantilever: we can oscillate the cantilever with large amplitude, thereby making it virtually stiffer at the point of strong force interaction.

Consider the first solution, which seems simpler at first glance. Chemical bonding forces extend over a distance range of about 0.1 nm. Typical binding energies of a couple of eV will lead to adhesion forces on the order of some nN. Force gradients will, therefore, reach values of some 10 N/m. A spring for stable force measurements will have to be as stiff as 100 N/m to ensure that no instability occurs (a safety factor of ten seems to be a minimum requirement, since usually one cannot be sure a priori that only one atom will dominate the interaction). In order to measure the nN interaction force, a static cantilever deflection of 0.01 nm has to be detected. With standard beam deflection AFM setups this becomes a challenging task.

This problem was solved [27.4, 5] using an *in-situ* optical interferometer measuring the beam deflection at liquid-nitrogen temperature in a UHV environment. In order to ensure that the force gradients are smaller than the lever spring constant (50 N/m), the tips were fabricated to terminate in only three atoms, therefore, minimizing the total force interaction. The field ion microscope (FIM) is a tool that allows one to engineer scanning probe microscopy (SPM) tips down to atomic dimensions. This technique not only allows imaging of the tip apex with atomic precision, but can also be used to manipulate the tip atoms by field evaporation [27.6], as shown in Fig. 27.3. Atomic interaction forces were measured with sub-nanonewton precision, revealing force curves of only a few atoms interacting without mechanical hysteresis. However, the technical effort to achieve this type of measurement is considerable, and most researchers today have resorted to the second solution.

The alternative solution can be visualized in Fig. 27.2. The straight, dashed line now represents the force values of the oscillating cantilever, with amplitude A assuming Hooke's law is valid. This is the tensile force of the cantilever spring pulling the tip away from the sample. The restoring force of the cantilever is at all points stronger than the adhesion force. For example, the total force at point D is still pointing away from the sample, although the spring has the same stiffness as before. Mathematically speaking, the measurement is stable as long as the cantilever spring force $F_{cb} = kA$ is larger than the attractive tip–sample force F_{ts} [27.7]. In the static mode we would already experience an instability at that point. However, in the dynamic mode, the spring is preloaded with a force stronger than the attractive tip–sample force. The equilibrium point of the oscillation is still far away from the point of closest contact of the tip and surface atoms. The total force curve can now be probed by varying the equilibrium point of the oscillation, i. e., by adjusting the z-piezo.

The diagram also shows that the oscillation amplitude has to be quite large if fairly soft cantilevers are to be used. With lever spring constants of $10\,N/m$, the amplitude must be at least 1 nm to ensure that forces of 1 nN can be reliably measured. In practical applications, amplitudes of $10–100\,nm$ are used to incorporate a safety margin. This means that the oscillation amplitude is much larger than the force interaction range. The above simplification, that the force gradient remains constant within one oscillation cycle, does not hold anymore. Measurement stability is gained at the cost of a simple quantitative analysis of the experiments. In fact, dynamic AFM was first used to obtain atomic resolution images of clean surfaces [27.8], and it took another six years [27.1] before quantitative measurements of single bond forces were obtained.

The technical realization of dynamic mode AFMs is based on the same key components as a static AFM set-up. The most common principle is the method of laser deflection sensing (see, e.g., Fig. 27.4). A laser beam is focused on the back side of a microfabricated cantilever. The reflected laser spot is detected with a positional-sensitive diode (PSD). This photodiode is sectioned into two parts that are read out separately (usually even a four-quadrant diode is used to detect torsional movements of the cantilever for lateral friction measurements). With the cantilever in equilibrium, the spot is adjusted such that the two sections show the same intensity. If the cantilever bends up or down, the spot moves,

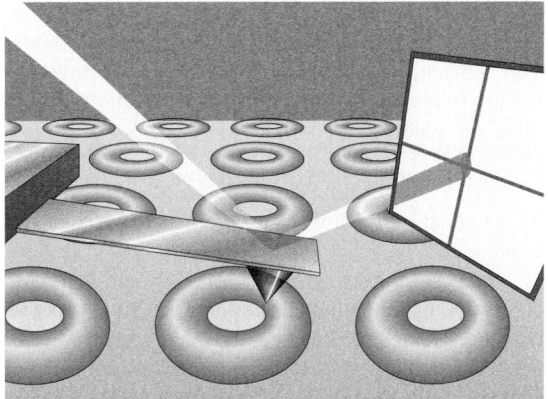

Fig. 27.4 Representation of an AFM set-up with the laser-beam deflection method. Cantilever and tip are microfabricated from silicon wafers. A laser beam is deflected from the backside of the cantilever and again focused onto a photosensitive diode. The diode is segmented into four quadrants, which allows measurement of vertical and torsional bending of the cantilever (artwork from J. Heimel rendered with POV-Ray 3.0)

and the difference signal between the upper and lower sections is a measure of the bending.

In order to enhance sensitivity, several groups have adopted an interferometer system to measure the cantilever deflection. A thorough comparison of different measurement methods with analysis of sensitivity and noise level is given in reference [27.3].

The cantilever is mounted on a device that allows oscillation of the beam. Typically a piezo-element directly underneath the cantilever beam serves this purpose. The reflected laser beam is analyzed for oscillation amplitude, frequency and phase difference. Depending on the mode of operation, a feedback mechanism will adjust the oscillation parameters and/or tip–sample distance during the scanning. The set-up can be operated in air, UHV, and even in fluids. This allows measurement of a wide range of surface properties from atomic-resolution imaging [27.8] up to studying biological processes in liquids [27.9, 10].

27.2 Harmonic Oscillator: A Model System for Dynamic AFM

The oscillating cantilever has three degrees of freedom: the amplitude, the frequency, and the phase difference between excitation and oscillation. Let us consider the damped driven harmonic oscillator. The cantilever is mounted on a piezoelectric element that is oscillating

with amplitude A_d at frequency ω:

$$z_d(t) = A_d \cos(\omega t). \qquad (27.6)$$

We assume that the cantilever spring obeys Hooke's law. Secondly, we introduce a friction force that is proportional to the speed of the cantilever motion, with α denoting the damping coefficient (Amontons's law). With Newton's first law we find for the oscillating system the following equation of motion for the position $z(t)$ of the cantilever tip (see also Fig. 27.1):

$$m\ddot{z}(t) = -\alpha\dot{z}(t) - kz(t) - kz_{\mathrm{d}}(t); \qquad (27.7)$$

We define $\omega_0^2 = k/m^*$, which turns out to be the resonant frequency of the free (undamped, i.e., $\alpha = 0$) oscillating beam. We further define the dimensionless quality factor $Q = m^*\omega_0/\alpha$, which is antiproportional to the damping coefficient. The quality factor describes the number of oscillation cycles after which the damped oscillation amplitude decays to $1/e$ of the initial amplitude with no external excitation ($A_{\mathrm{d}} = 0$). After some basic manipulation, this results in the following differential equation:

$$\ddot{z}(t) + \frac{\omega_0}{Q}\dot{z}(t) + \omega_0^2 z(t) = A_{\mathrm{d}}\omega_0^2\cos(\omega t). \qquad (27.8)$$

The solution is a linear combination of two regimes [27.11]. Starting from rest and switching on the piezo-excitation at $t = 0$, the amplitude will increase from zero to the final magnitude and reach a steady state, where the amplitude, phase, and frequency of the oscillation stay constant over time. The steady-state solution $z_1(t)$ is reached after $2Q$ oscillation cycles and follows the external excitation with amplitude A_0 and phase difference φ:

$$z_1(t) = A_0\cos(\omega t + \varphi) \qquad (27.9)$$

The oscillation amplitude in the transient regime during the first $2Q$ cycles follows:

$$z_2(t) = A_t \cdot \mathrm{e}^{-\omega_0 t/2Q} \cdot \sin(\omega_0 t + \varphi_t). \qquad (27.10)$$

We emphasize the important fact that the exponential term causes $z_2(t)$ to diminish exponentially with time constant τ:

$$\tau = 2Q/\omega_0 \qquad (27.11)$$

In vacuum conditions, only the internal dissipation due to bending of the cantilever is present, and Q reaches values of $10\,000$ at typical resonant frequencies of $100\,000\,\mathrm{Hz}$. This results in a relatively long transient regime of $\tau \cong 30\,\mathrm{ms}$, which limits the possible operational modes for dynamic AFM (detailed analysis by Albrecht et al., 1991). Changes in the measured amplitude, which reflect a change of atomic forces, will have a time lag of 30 ms, which is very slow considering one wants to scan a 200×200 point image within a few minutes. In air, however, viscous damping due to air friction

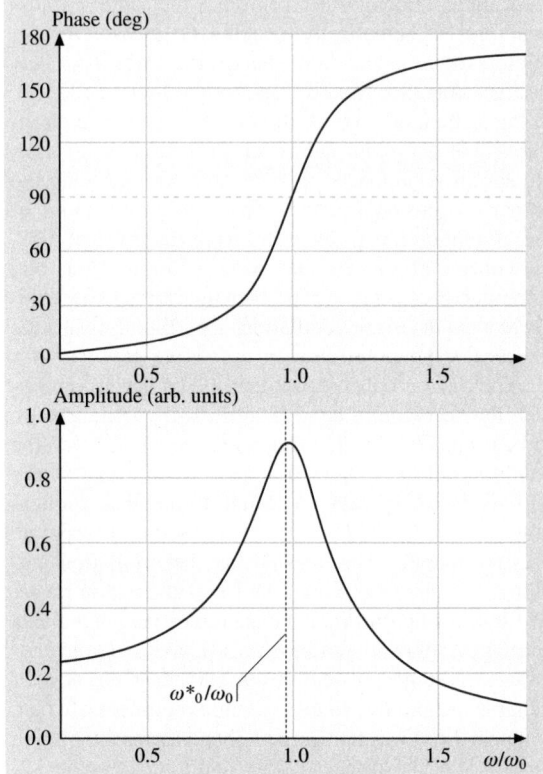

Fig. 27.5 Curves of amplitude and phase versus excitation frequency for the damped harmonic oscillator, with a quality factor of $Q = 4$

dominates and Q goes down to less than 1000, resulting in a time constant below the millisecond level. This response time is fast enough to use the amplitude as a measurement parameter.

If we evaluate the steady-state solution $z_1(t)$ of the differential equation, we find the following well-known solution for amplitude and phase of the oscillation as a function of the excitation frequency ω:

$$A_0 = \frac{A_{\mathrm{d}} \cdot Q \cdot \omega_0^2}{\sqrt{\omega^2\omega_0^2 + Q^2(\omega_0^2 - \omega^2)^2}} \qquad (27.12)$$

$$\varphi = \arctan\left(\frac{\omega \cdot \omega_0}{Q \cdot (\omega_0^2 - \omega^2)}\right). \qquad (27.13)$$

Amplitude and phase diagrams are depicted in Fig. 27.5. As can be seen from (27.12) the amplitude will reach its maximum at a frequency different from ω_0, if Q has a finite value. The damping term of the harmonic oscillator causes the resonant frequency to shift from ω_0

to ω_0^*:

$$\omega_0^* = \omega_0 \sqrt{1 - \frac{1}{2Q^2}} \qquad (27.14)$$

The shift is negligible for Q-factors of 100 and above, which is the case for most applications in vacuum or air. However, for measurements in liquids, Q can be smaller than 10 and ω_0 differs significantly from ω_0^*. As we will discuss later, it is also possible to enhance Q by using a special excitation method called Q-control.

In the case that the excitation frequency is equal to the resonant frequency of the undamped cantilever $\omega = \omega_0$, we find the useful relation:

$$A_0 = Q \cdot A_d \quad \text{for} \quad \omega = \omega_0 . \qquad (27.15)$$

Since $\omega_0^* \approx \omega_0$ for most cases, we find that (27.15) holds true when exciting the cantilever at its resonance. From a similar argument, the phase becomes approximately 90 deg for the resonance case. We also see that, in order to reach vibration amplitudes of some 10 nm, the excitation can be as small as 1 pm for typical cantilevers operated in vacuum.

So far, we have not considered an additional force term that describes the interaction between the probing tip and the sample. For typical large-vibration amplitudes of 10–100 nm no general solution for this analytical problem has been found yet. The cantilever tip will experience a whole range of force interactions during one single oscillation cycle, rather than one defined tip–sample force. Only in the special case of a self-excited cantilever oscillation can the problem be solved analytically, as we will see later.

27.3 Dynamic AFM Operational Modes

While the quantitative interpretation of force curves in contact AFM is straightforward using (27.1), we explained in the previous paragraphs that its application to assess short-range attractive interatomic forces is rather limited. The dynamic mode of operation seems to open a viable direction towards achieving this task. However interpretation of the measurements generally appears to be more difficult. Different operational modes are employed in dynamic AFM, and the following paragraphs are intended to distinguish these modes and categorize them in a systematic way.

The oscillation trajectory of a dynamically driven cantilever is determined by three parameters: the amplitude, the phase, and the frequency. Tip–sample interactions can influence all three parameters, which are termed the internal parameters in the following. The oscillation is driven externally, with excitation amplitude A_d and excitation frequency ω. These variables will be referred to as the external parameters. The external parameters are set by the experimentalist, whereas the internal parameters are measured and contain the crucial information about the force interaction. In scanning probe applications, it is common to control the probe–surface distance z_0 in order to keep an internal parameter constant (i. e., the tunneling current in STM or the beam deflection in contact AFM), which represents a certain tip–sample interaction. In z-spectroscopy mode, the distance is varied in a certain range, and the change of the internal parameters is measured as a fingerprint of the tip–sample interactions.

In dynamic AFM the situation is rather complex. Any of the internal parameters can be used for feedback of the tip–sample distance z_0. However, we already realized that, in general, the tip–sample forces can only be fully assessed by measuring all three parameters. Therefore, dynamic AFM images are difficult to interpret. A solution to this problem is to establish additional feedback loops, which keep the internal parameters constant by adjusting the external variables. In the simplest set-up, the excitation frequency is set to a predefined value, and the excitation amplitude remains constant by a feedback loop. This is called the amplitude-modulation (AM) or tapping mode. As stated before, in principle, any of the internal parameters can be used for feedback to the tip–sample distance; in AM mode the amplitude signal is used. A certain amplitude (smaller than the free oscillation amplitude) at a frequency close to the resonance of the cantilever is chosen; the tip is approached towards the surface under investigation, and the approach is stopped as soon as the set-point amplitude is reached. The oscillation phase is usually recorded during the scan, however, the shift of the resonant frequency of the cantilever cannot be directly accessed, since this degree of freedom is blocked by the external excitation at a fixed frequency. It turns out that this mode is simple to operate from a technical perspective, but quantitative information about the tip–sample interaction forces has so far not been reliably extracted from AM-mode AFM. Despite this, it is one of the most commonly used modes in dynamic AFM operated in air, and even in liquid. The strength of this mode is the qualitative imaging of a large variety of surfaces.

It is interesting to discuss the AM mode in the situation that the external excitation frequency is much lower than the resonant frequency [27.12, 13]. This results in a quasistatic measurement, although a dynamic oscillation force is applied, and, therefore, this mode can be viewed as a hybrid between static and dynamic AFM. Unfortunately, it has the drawbacks of the static mode, namely, that stiff spring constants must be used and, therefore, the sensitivity of the deflection measurement must be very good, typically employing a high-resolution interferometer. However, it has the advantage of the static measurement in quantitative interpretation, since in the regime of small amplitudes (< 0.1 nm) a direct interpretation of the experiments is possible. In particular, the force gradient at the tip–sample distance z_0 is given by the change of the amplitude A and the phase angle φ

$$\frac{\partial F_{\text{ts}}}{\partial z}\Big|_{z_0} = k\left(1 - \frac{A_0}{A}\cos\varphi\right) . \tag{27.16}$$

In effect, the modulated AFM technique can profit from enhanced sensitivity due to the use of lock-in techniques, which allow the measurement of the amplitude and phase of the oscillation signal with high precision.

As stated before, the internal parameters can be fed back to the external excitation variables. One of the most useful applications in this direction is the self-excitation system. Here the resonant frequency of the cantilever is detected and selected again as the excitation frequency. In a typical set-up, this is done with a phase shift of 90 deg by feeding back the detector signal to the excitation piezo, i.e., the cantilever is always excited in resonance. Tip–sample interaction forces then only influence the resonant frequency, but do not change the two other parameters of the oscillation: amplitude and phase. Therefore, it is sufficient to measure the frequency shift induced by the tip–sample interaction. Since the phase remains at a fixed value, the oscillating system is much better defined than before, and the degrees of freedom for the oscillation are reduced. To reduce the last degree of freedom even further an additional feedback loop can be incorporated to keep the oscillation amplitude A constant by varying the excitation amplitude A_d. Now, all internal parameters have a fixed relation to the external excitation variables, the system is well defined, and all parameters can be assessed during the measurement. As it turns out, this mode is the only dynamic mode in which a quantitative relation between the tip–sample

forces and the change of the resonant frequency can be established.

In the following section we want to discuss the two most popular operational modes, the tapping mode and the self-excitation mode, in more detail.

27.3.1 Amplitude-Modulation/Tapping-Mode AFM

In the tapping mode, or AM-AFM, the cantilever is excited externally at a constant frequency close to its resonance. The oscillation amplitude and phase during the approach of the tip and the sample serve as the experimental observation channels. Figure 27.6 shows a diagram of a typical tapping-mode AFM set-up. The oscillation of the cantilever is detected with the photodiode, whose output signal is analyzed with a lock-in amplifier to obtain amplitude and phase information. The amplitude is then compared to the set point, and the resulting difference or error signal is fed into the proportional–integral–derivative (PID) controller, which adjusts the z-piezo, i.e., the probe–sample distance, accordingly. The external modulation unit supplies the

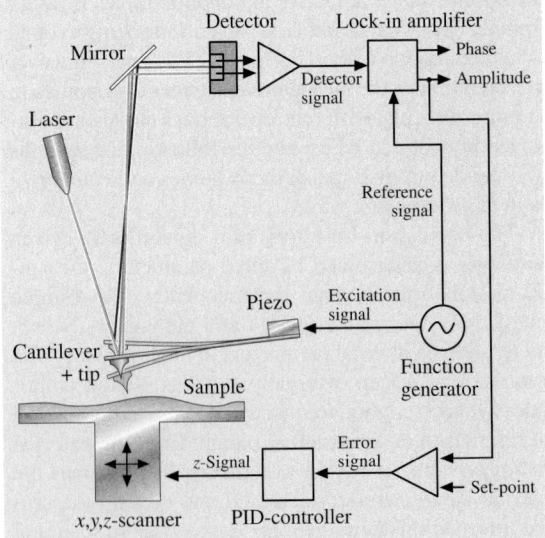

Fig. 27.6 Set-up of a dynamic force microscope operated in the AM or tapping mode. A laser beam is deflected by the backside of the cantilever and the deflection is detected by a split photodiode. The excitation frequency is chosen externally with a modulation unit, which drives the excitation piezo. A lock-in amplifier analyses the phase and amplitude of the cantilever oscillation. The amplitude is used as the feedback signal for the probe–sample distance control

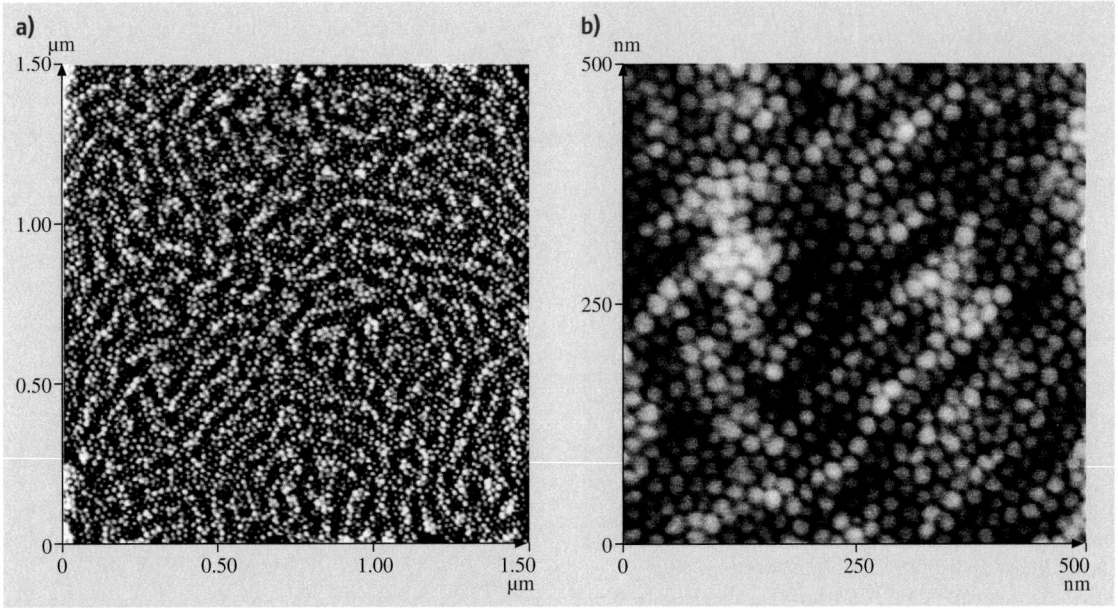

Fig. 27.7a,b Tapping-mode images of $BC_{0.26}$-$3A_{0.53}$F8H10 at (**a**) low resolution and (**b**) high resolution. The height scale is 10 nm. (Reprinted in part with permission from [27.14]. Copyright (2001) American Chemical Society)

signal for the excitation piezo, and, at the same time, the oscillation signal serves as the reference for the lock-in amplifier. As shown by the following applications the tapping mode is typically used to measure surface topography and other material parameters on the nanometer scale. The tapping mode is mostly used in ambient conditions and in liquids.

High-resolution imaging has been extensively performed in the area of material science. Due to its technical relevance the investigation of polymers has been the focus of many studies (see e.g. a recent review about AFM imaging of polymers in [27.15]. In Fig. 27.7 the topography of a diblock copolymer ($BC_{0.26}$-$3A_{0.53}$F8H10) at different magnifications is shown [27.14]. On the large scan (a) the large-scale structure of the microphase-separated polystyrene (PS) cylinders [within a polyisoprene (PI) matrix] lying parallel to the substrate can be seen. In the high-resolution image (b) a surface substructure of regular domes can be seen, which were found to be related to the cooling process during the polymer preparation.

Imaging in liquids opens up the avenue for the investigation of biological samples in their natural environment. For example Möller et al. [27.16] have obtained high-resolution images of the topography of hexagonally packed intermediate (HPI) layer of Deinococcus Radiodurans with tapping-mode AFM. Another inter-

esting example is the imaging of deoxyribonucleic acid (DNA) in liquid, as shown in Fig. 27.8. Jiao et al. [27.10] measured the time evolution of a single DNA strand interacting with a molecule, as shown by a sequence of images acquired in liquid over a time period of several minutes.

For a quantitative interpretation of tip–sample forces one has to consider that during one oscillation cycle with amplitudes of 10–100 nm the tip–sample interaction will range over a wide distribution of forces, including attractive as well as repulsive forces. We will, therefore, measure a convolution of the force–distance curve with the oscillation trajectory. This complicates the interpretation of AM-AFM measurements appreciably.

At the same time, the resonant frequency of the cantilever will change due to the appearing force gradients, as could already be seen in the simplified model from (27.4). If the cantilever is excited exactly at its resonant frequency before interaction forces are encountered, it will be excited off-resonance after they are encountered. This, in turn, changes the amplitude and phase (see (27.12) and (27.13)), which serve as the measurement signals. Consequently, a different amplitude will cause a change in the encountered effective force. We can see already from this simple *gedanken*-experiment that the interpretation of force curves will be highly complicated. In fact, no quantitative theory for AM-AFM that

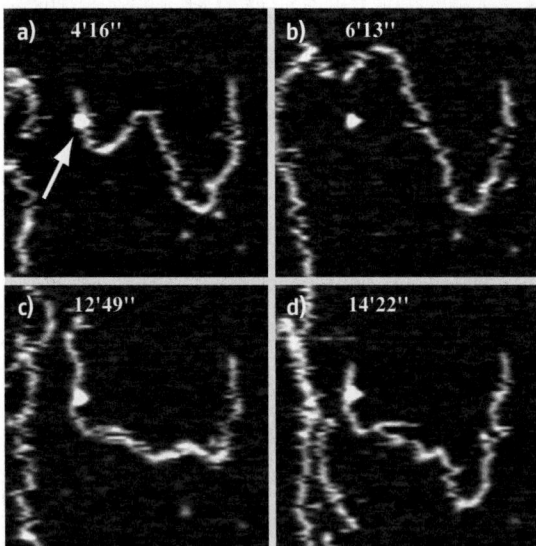

Fig. 27.8a–d Dynamic p53-DNA interactions observed by time-lapse tapping-mode AFM imaging in solution. Both p53 protein and DNA were weakly adsorbed to a mica surface by balancing the buffer conditions. (**a**) A p53 protein molecule (arrow) was bound to a DNA fragment. The protein (**b**) dissociated from and then (**c**) re-associated with the DNA fragment. (**d**) A downward movement of the DNA with respect the protein occurred, constituting a *sliding* event whereby the protein changes its position on the DNA. Image size: 620 nm. Colorscale (height) range: 4 nm. Time units: minutes, seconds. (Image courtesy of Tilman Schäffer, University of Münster)

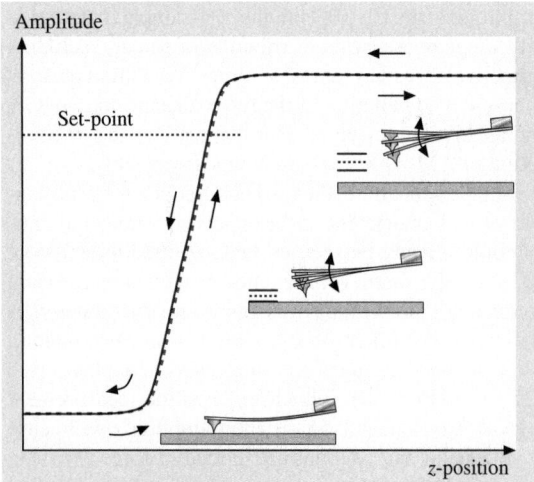

Fig. 27.9 Simplified model showing the oscillation amplitude in tapping-mode AFM for various probe–sample distances

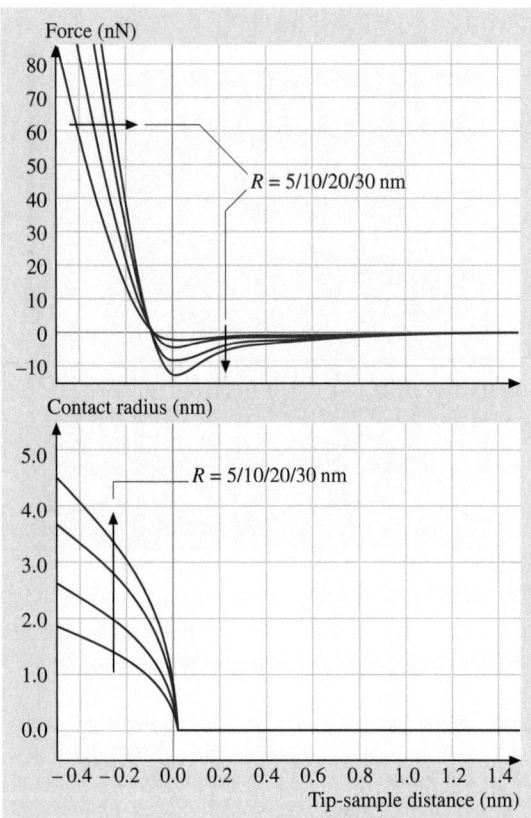

Fig. 27.10 Force curves and corresponding contact radius calculated with the MYD/BHW model as a function of tip radius for a Si–Si contact. These force curves are used for the tapping-mode AFM simulations

allows the experimentalist to convert the experimental data to a force–distance relationship unambiguously is available.

The qualitative behavior for amplitude versus z_0-position curves is depicted in Fig. 27.9. At large distances, where the forces between the tip and the sample are negligible, the cantilever oscillates with its free oscillation amplitude. As the probe approaches the surface the interaction forces cause the amplitude to change, typically resulting in an amplitude that decreases with continuously decreasing tip–sample distance. This is expected, since the force–distance curve will eventually reach the repulsive part and the tip is hindered from indenting further into the sample, resulting in smaller oscillation amplitudes.

However, in order to gain some qualitative insight into the complex relationship between forces and oscillation parameters, we resort to numerical simulations.

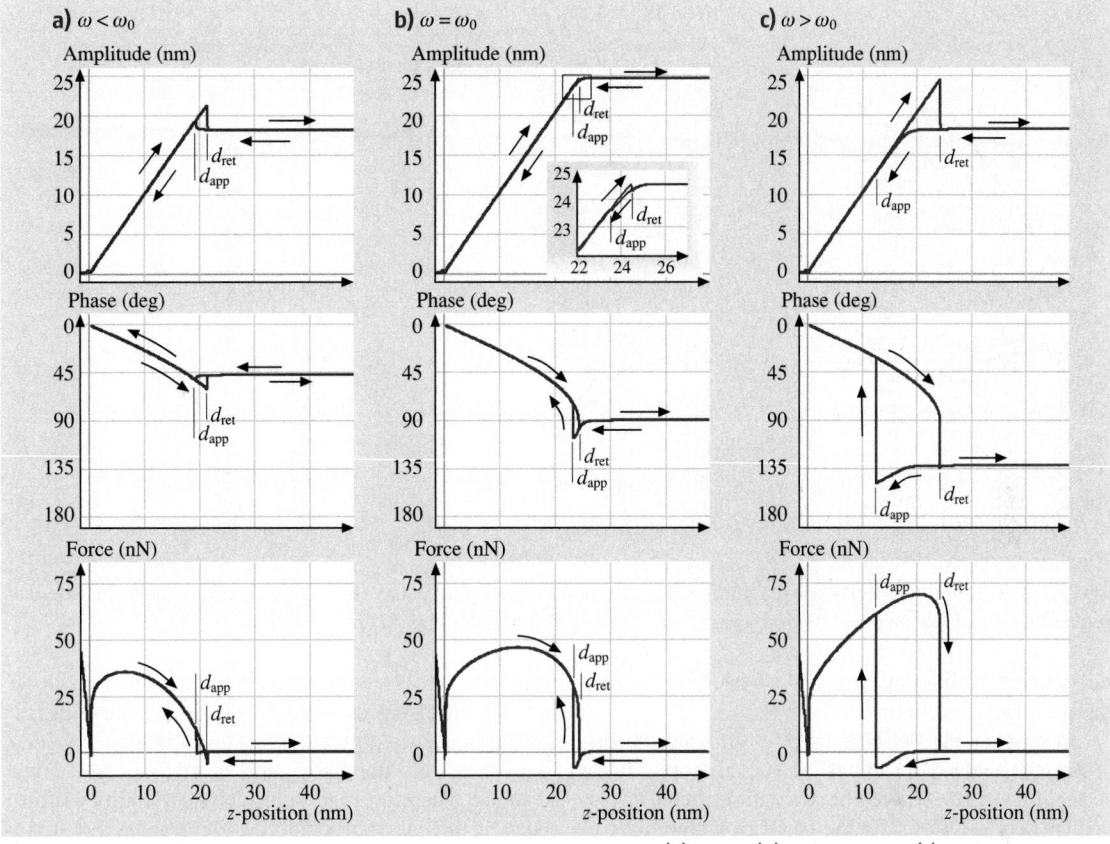

Fig. 27.11a–c Amplitude and phase diagrams with excitation frequency (**a**) below (**b**) exactly at and (**c**) above the resonant frequency for tapping-mode AFM from the numerical simulations. Additionally, the lower diagrams show the interaction forces at the point of closest tip–sample distance, i.e. the lower turnaround point of the oscillation

Anczykowski et al. [27.17, 18] have calculated the oscillation trajectory of the cantilever under the influence of a given force model. Van der Waals interactions were considered the only effective, attractive forces, and the total interaction resembled a Lennard–Jones-type potential. Mechanical relaxations of the tip and sample surface were treated in the limits of continuum theory with the numerical Muller-Yushchenko-Derjaguin/Burgess-Hughes-White (MYD/BHW) [27.19, 20] approach, which allows the simulations to be compared to experiments. Figure 27.10 shows the corresponding force–distance curves used in the simulations for different tip radii.

The cantilever trajectory was analyzed by solving the differential equation (27.7) extended by the force–distance relations from Fig. 27.10 using the numerical Verlet algorithm [27.21, 22]. The results of the simulation for the amplitude and phase of the tip oscillation as a function of z-position of the probe are presented in Fig. 27.11. One has to keep in mind that the z-position of the probe is not equivalent to the real tip–sample distance at the equilibrium position, since the cantilever might bend statically due to the interaction forces. The behavior of the cantilever can be subdivided into three different regimes. We distinguish the cases in which the beam is oscillated below its resonant frequency ω_0, exactly at ω_0, and above ω_0. In the following, we will refer to ω_0 as the resonant frequency, although the correct resonant frequency is ω_0^* if one takes into account the finite Q-value.

Clearly, Fig. 27.12 exhibits more features than were anticipated from the initial, simple arguments. Amplitude and phase seem to change rather abruptly at certain points when the z_0-position is decreased. Besides, the amplitude or phase–distance curves do not resemble the force–distance curves from Fig. 27.11 in a simple, di-

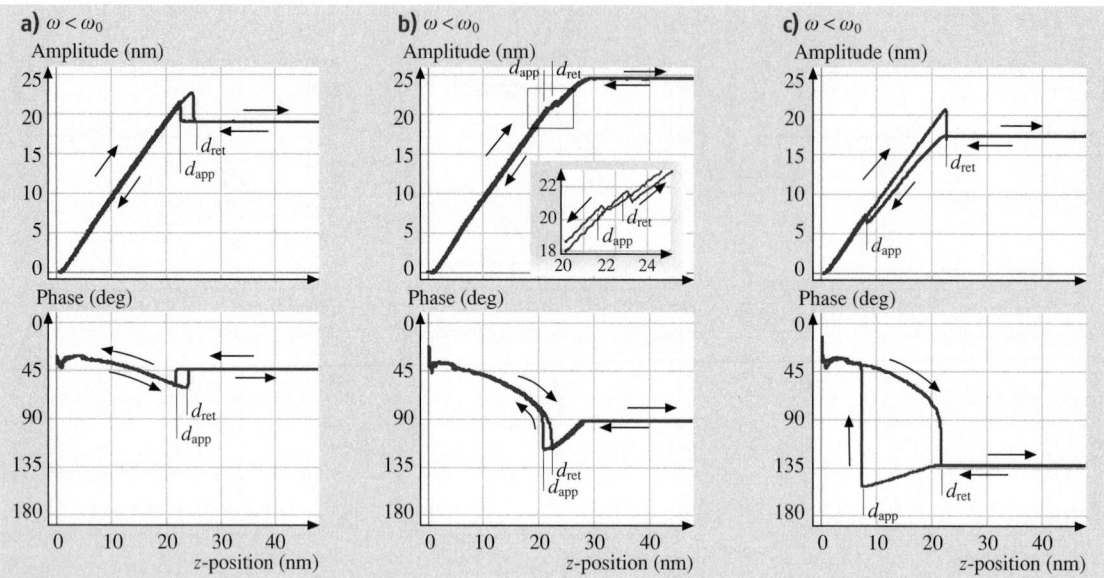

Fig. 27.12a–c Amplitude and phase diagrams with excitation frequency (**a**) below (**b**) exactly at and (**c**) above the resonant frequency for tapping-mode AFM from experiments with a Si cantilever on a Si wafer in air

rect manner. Additionally, we find a hysteresis between approach and retraction.

As an example, let us start by discussing the discontinuous features in the AFM spectroscopy curves of the first case, where the excitation frequency is smaller than ω_0. Consider the oscillation amplitude as a function of the excitation frequency in Fig. 27.5 in conjunction with a typical force curve, as depicted in Fig. 27.10. Upon approach of probe and sample, attractive forces will lower the effective resonant frequency of the oscillator. Therefore, the excitation frequency will now be closer to the resonant frequency, causing the vibration amplitude to increase. This, in turn, reduces the tip–sample distance, which again gives rise to a stronger attractive force. The system becomes unstable until the point $z_0 = d_{app}$ is reached, where repulsive forces stop the self-enhancing instability. This can be clearly observed in Fig. 27.11a. Large parts of the force–distance curve cannot be measured due to this instability.

In the second case, where the excitation equals the free resonant frequency, only a small discontinuity is observed upon reduction of the z-position. Here, a shift of the resonant frequency towards smaller values, induced by the attractive force interaction, will reduce the oscillation amplitude. The distance between the tip and sample is, therefore, reduced as well, and the self-amplifying effect with the sudden instability does not occur as long as

repulsive forces are not encountered. However, at closer tip–sample distances, repulsive forces will cause the resonant frequency to shift again towards higher values, increasing the amplitude with decreasing tip–sample distance. Therefore, a self-enhancing instability will also occur in this case, but at the crossover from purely attractive forces to the regime where repulsive forces occur. Correspondingly, a small kink in the amplitude curve can be observed in Fig. 27.11b. An even clearer indication of this effect is manifested by the sudden change in the phase signal at d_{app}.

In the last case, with $\omega > \omega_0$, the effect of amplitude reduction due to the resonant frequency shift is even larger. Again, we find no instability in the amplitude signal during approach in the attractive force regime. However, as soon as the repulsive force regime is reached, the instability occurs due to the induced positive frequency shift. Consequently, a large jump in the phase curve from values smaller than 90 deg to values larger than 90 deg is observed. The small change in the amplitude curve is not resolved in the simulated curves in Fig. 27.11c, however, it can be clearly seen in the experimental curves.

Figure 27.12 depicts the corresponding experimental amplitude and phase curves. The measurements were performed in air with a Si cantilever approaching a Si wafer, with a cantilever resonant frequency of 299.95 kHz. Qualitatively, all prominent features of the simulated curves can also be found in the experimental data sets. Hence, this model seems to capture the important factors necessary for an appropriate description of the experimental situation.

But what is the reason for this unexpected behavior? We have to turn to the numerical simulations again, where we have access to all physical parameters, in order to understand the underlying processes. The lower part of Fig. 27.11 also shows the interaction force between the tip and the sample at the point of closest approach, i. e., the sample-side turnaround point of the oscillation. We see that exactly at the points of the discontinuities the total interaction force changes from the net-attractive regime to the attractive–repulsive regime, also termed the intermittent contact regime. The term net-attractive is used to emphasize that the total force is attractive, despite the fact that some minor contributions might still originate from repulsive forces. As soon as a minimum distance is reached, the tip also starts to experience repulsive forces, which completely changes the oscillation behavior. In other words, the dynamic system switches between two oscillatory states.

Directly related to this fact is the second phenomenon: the hysteresis effect. We find separate curves for the approach of the probe towards the surface and the retraction. This seems to be somewhat counterintuitive, since the tip is constantly approaching and retreating from the surface, and the average values of amplitude and phase should be independent of the direction of the average tip–sample distance movement. A hysteresis between approach and retraction within one oscillation due to dissipative processes should directly influence amplitude and phase. However, no dissipation models were included in the simulation. In this case, the hysteresis in Fig. 27.11 is due to the fact that the oscillation jumps into different modes; the system exhibits bistability. This effect is often observed in oscillators under the influence of nonlinear forces [27.23].

For the interpretation of these effects it is helpful to look at Fig. 27.13, which shows the behavior of the simulated tip trajectory and the force during one oscillation cycle over time. The data is shown for the z-positions where hysteresis is observed, while (a) was taken during the approach and (b) during the retraction. Excitation was in resonance, where the amplitude shows a small hysteresis. Also note that the amplitude is almost exactly the same in (a) and (b). We see that the oscillation at the same z-position exhibits two different modes: while in (a) the experienced force is net-attractive, in (b) the tip is exposed to attractive and repulsive interactions. Experimental and simulated data show that the change

between the net-attractive and intermittent contact mode takes place at different z-positions (d_{app} and d_{ret}) for approach and retraction. Between d_{app} and d_{ret} the system is in a bistable mode. Depending on the history of the measurement, e.g., whether the position d_{app} during the approach (or d_{ret} during retraction) has been reached, the system flips to the other oscillation mode. While the amplitude might not be influenced strongly, the phase is a clear indicator of the mode switch. On the other hand, if the point d_{app} is never reached during the approach, the system will stay in the net-attractive regime and no hysteresis is observed, i. e., the system remains stable.

In conclusion, we find that, although a qualitative interpretation of the interaction forces is possible, the AM-AFM is not suitable to gain direct quantitative knowledge of tip–sample force interactions. However, it is a very useful tool for imaging nanometer-sized structures in a wide variety of setups, in air or even in liquid. We find that two distinct modes exist for the externally excited oscillation – the net-attractive and the intermittent contact mode – which describe what kind of forces govern the tip–sample interaction. The phase can be used as an indicator of the current mode of the system.

In particular, it can be easily seen that, if the free resonant frequency of the cantilever is higher than the excitation frequency, the system cannot stay in the net-attractive regime due to a self-enhancing instability. Since in many applications involving soft and deli-

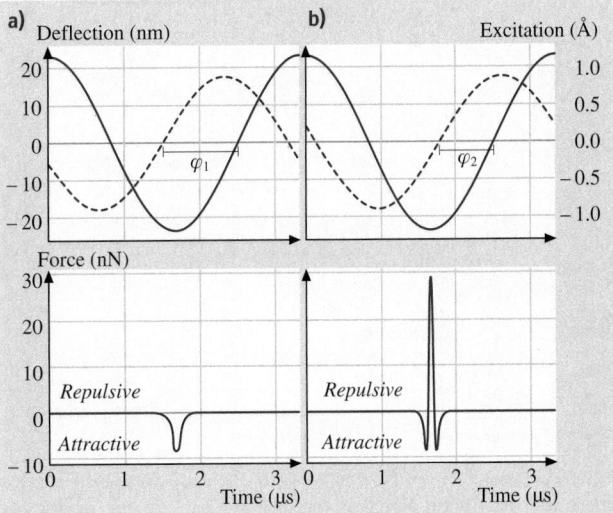

Fig. 27.13a,b Simulation of the tapping-mode cantilever oscillation in the (**a**) net-attractive and (**b**) the intermittent contact regime. The *dashed line* represents the excitation amplitude and the *solid line* is the oscillation amplitude

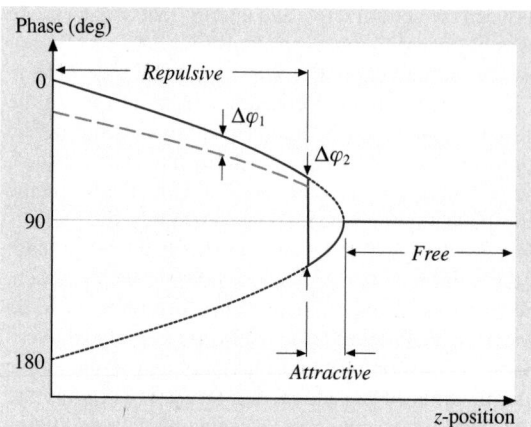

Fig. 27.14 Phase shift in tapping mode as a function of tip–sample distance

cate biological samples strong repulsive forces should be avoided, the tapping-mode AFM should be operated at frequencies equal to or above the free resonant frequency [27.24]. Even then, statistical changes of tip–sample forces during the scan might induce a sudden jump into the intermittent contact mode, and the previously explained hysteresis will tend to keep the system

in this mode. It is therefore of high importance to tune the oscillation parameters in such a way that the AFM stays in the net-attractive regime [27.25]. A concept that achieves this task is the Q-control system, which will be discussed in some detail in the forthcoming paragraphs.

A last word concerning the overlap of simulation and experimental data: while the qualitative agreement down to the detailed shape of hysteresis and instabilities is rather striking, we still find some quantitative discrepancies between the positions of the instabilities d_{app} and d_{ret}. This is probably due to the simplified force model, which only takes into account van der Waals and repulsive forces. Especially at ambient conditions, an omnipresent water meniscus between the tip and sample will give rise to much stronger attractive and also dissipative forces than considered in the model. A very interesting feature is that the simulated phase curves in the intermittent contact regime tend to have a steeper slope in the simulation than in the experiments (see also Fig. 27.14). We will later show that this effect is a fingerprint of an effect that was not included in this simulation at all: dissipative processes during the oscillation, giving rise to an additional loss of oscillation energy.

27.3.2 Self-Excitation Modes

Despite the wide range of technical applications of the AM mode of dynamic AFM, it has been found unsuitable for measurements in an environment extremely useful for scientific research: vacuum or ultrahigh vacuum (UHV) with pressures reaching 1×10^{-10} mbar. The STM has already shown how much insight can be gained from some highly defined experiments under those conditions.

Consider (27.11) from the previous section. The time constant τ for the amplitude to adjust to a different tip–sample force scales with $1/Q$. In vacuum applications, Q of the cantilever is on the order of 10 000, which means that τ is in the range of some 10 ms. This is clearly too long for a scan of at least (100×100) data points. On the other hand, the resonant frequency of the system will react instantaneously to tip–sample forces. This has led *Albrecht* et al. [27.11] to use a modified excitation scheme.

The system is always oscillated at its resonant frequency. This is achieved by feeding back the oscillation signal from the cantilever into the excitation piezoelement. Figure 27.15 illustrates the method in a block diagram. The signal from the PSD is phase-shifted by 90 deg (and, therefore, always exciting in resonance) and used as the excitation signal of the cantilever. An addi-

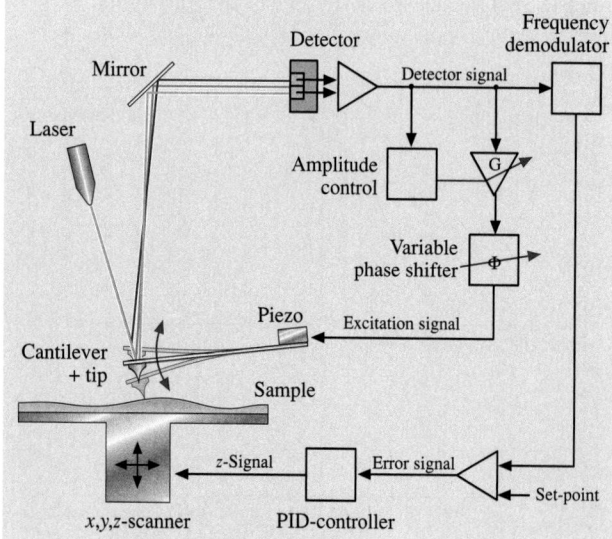

Fig. 27.15 Dynamic AFM operated in the self-excitation mode, where the oscillation signal is directly fed back to the excitation piezo. The detector signal is amplified with the variable gain G and phase shifted by phase φ. The frequency demodulator detects the frequency shift due to tip–sample interactions, which serves as the control signal for the probe–sample distance

tional feedback loop adjusts the excitation amplitude in such a way that the oscillation amplitude remains constant. This ensures that the tip–sample distance is not influenced by changes in the oscillation amplitude. The only degree of freedom that the oscillation system still has, which can react to the tip–sample forces, is the change of the resonant frequency. This shift of the frequency is detected and used as the set-point signal for surface scans. Therefore, this mode is also called the frequency-modulated (FM) mode.

Let us take a look at the sensitivity of the dynamic AFM. If electronic noise, laser noise, and thermal drift can be neglected, the main noise contribution will come from thermal excitations of the cantilever. A detailed analysis of a dynamic system yields, for the minimum detectable force gradient, the following relation [27.11]:

$$\frac{\partial F}{\partial z}\Big|_{MIN} = \sqrt{\frac{4k \cdot k_B T \cdot B}{\omega_0 Q \langle z_{osc}^2 \rangle}} \qquad (27.17)$$

Here B is the bandwidth of the measurement, T the temperature, and $\langle z_{osc}^2 \rangle$ is the mean-square amplitude of the oscillation. Please note that this sensitivity limit was deliberately calculated for the FM mode. A similar analysis of the AM mode, however, yields virtually the same result [27.26]. We find that the minimum detectable force gradient, i.e., the measurement sensitivity, is inversely proportional to the square root of the Q-factor of the cantilever. This means that it should be possible to achieve very high-resolution imaging under vacuum conditions where the Q-factor is very high.

A breakthrough in high-resolution AFM imaging was the atomic-resolution imaging of the Si(111)–(7×7) surface reconstruction by Giessibl [27.8] under UHV conditions. Moreover, Sugawara et al. [27.27] observed the motion of single atomic defects on InP with true atomic resolution. However, imaging on conducting or semiconducting surfaces is also possible with the scanning tunneling microscope (STM) and these first non-contact AFM (NC-AFM) images provided little new information on surface properties. The true potential of NC-AFM lies in the imaging of nonconducting surface with atomic precision, which was first demonstrated by Bammerlin et al. [27.28] on NaCl. A long-standing question about the surface reconstruction of the technologically relevant material aluminium oxide could be answered by Barth et al. [27.29], who imaged the atomic structure of the high-temperature phase of α-Al$_2$O$_3$ (0001).

The high-resolution capabilities of non-contact atomic force microscopy are nicely demonstrated by

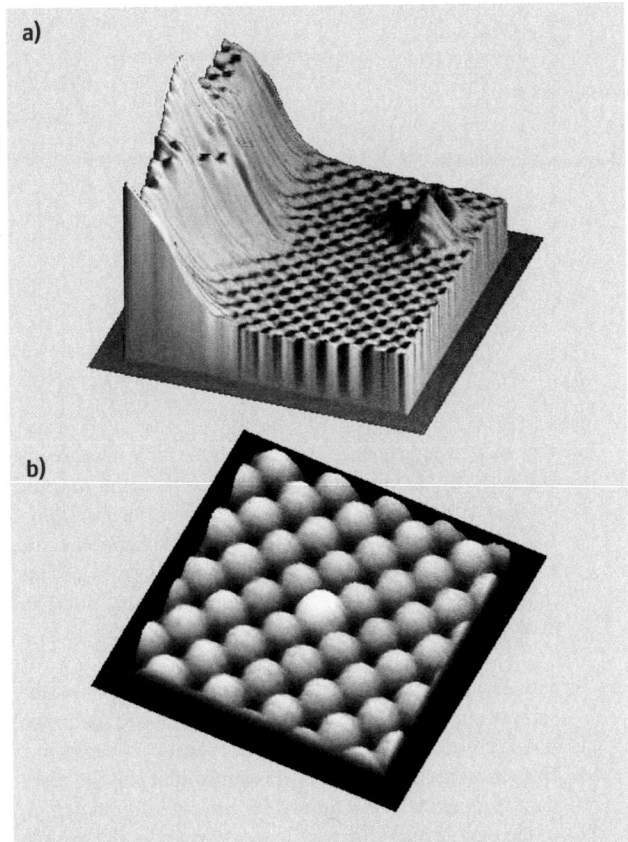

Fig. 27.16a,b Imaging of a NiO(001) sample surface with non-contact AFM. (**a**) Surface step and an atomic defect. The lateral distance between two atoms is 4.17 Å. (**b**) A dopant atom is imaged as a light protrusion about 0.1 Å higher than the other atoms. (Images courtesy of W. Allers and S. Langkat, University of Hamburg)

the images shown in Fig. 27.16. *Allers* et al. [27.30] imaged steps and defects on the insulator nickel oxide with atomic resolution. Such a resolution is routinely obtained today by different research groups (for an overview see, e.g., [27.31–33]).

However, we are concerned with measuring atomic force potentials of a single pair of molecules. Clearly, FM-mode AFM will allow us to identify single atoms, and with sufficient care we will be able to ensure that only one atom from the tip contributes to the total force interaction. Can we, therefore, fill in the last bit of information and find a quantitative relation between the oscillation parameters and the force?

A good insight into the cantilever dynamics can be drawn from the tip potential displayed in

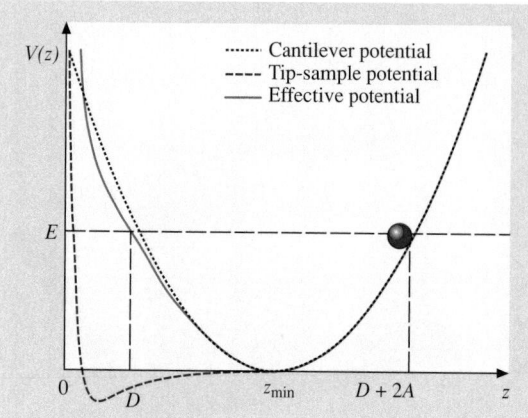

Fig. 27.17 The frequency shift in dynamic force microscopy is caused by the tip–sample interaction potential (*dashed line*), which alters the harmonic cantilever potential (*dotted line*). Therefore, the tip moves in an anharmonic and asymmetric effective potential (*solid line*). Here, z_{min} is the minimum position of the effective potential (from Hölscher et al. [27.34])

Fig. 27.17 [27.34]. If the cantilever is far away from the sample surface, the tip moves in a symmetric parabolic potential (dotted line), and the oscillation is harmonic. In such a case, the tip motion is sinusoidal and the resonant frequency is determined by the eigenfrequency f_0 of the cantilever. If the cantilever approaches the sample surface the potential is changed, given by an effective potential V_{eff} (solid line) which is the sum of the parabolic potential and the tip–sample interaction potential V_{ts} (dashed line). This effective potential differs from the original parabolic potential and shows an asymmetric shape. As a result the oscillation becomes anharmonic, and the resonant frequency of the cantilever depends on the oscillation amplitude.

Gotsmann et al. [27.35] investigated this relation with a numerical simulation. During each oscillation cycle the tip experiences a whole range of forces. For each step during the approach the differential equation for the whole oscillation loop (also including the feedback system) was evaluated and finally the quantitative relation between the force and frequency shift was revealed.

However, there is also an analytical relationship, if some approximations are accepted [27.7, 36, 37]. Here, we will follow the route indicated by [27.37], although alternative ways have also been proven successful. Consider the tip oscillation trajectory reaches over a large part of the force gradient curve in Fig. 27.2. We model the tip–sample interaction as a spring con-

stant of stiffness $k_{ts}(z) = \partial F / \partial z \big|_{z_0}$ as in Fig. 27.1. For small oscillation amplitudes we already found that the frequency shift is proportional to the force gradient in (27.4). For large amplitudes, we can calculate an effective force gradient k_{eff} as a convolution of the force and the fraction of time, the tip spends between the positions x and $x + dx$:

$$k_{eff}(z) = \frac{2}{\pi A^2} \int_z^{z+2A} F(x) \cdot g\left(\frac{x-z}{A} - 1\right) dx$$

(27.18)

with $\quad g(u) = -\dfrac{u}{\sqrt{1-u^2}}$.

In the approximation that the vibration amplitude is much larger than the range of the tip sample forces the above equation can be simplified to:

$$k_{eff}(z) = \frac{\sqrt{2}}{\pi} A^{3/2} \cdot \int_z^{\infty} \frac{F(x)}{\sqrt{x-z}} dx .$$

(27.19)

This effective force gradient can now be used in (27.4), the relation between the frequency shift and force gradient. We find:

$$\Delta f = \frac{f_0}{\sqrt{2\pi} \cdot kA^{3/2}} \cdot \int_z^{\infty} \frac{F(x)}{\sqrt{x-z}} dx .$$

(27.20)

If we separate the integral from other parameters, we can define:

$$\Delta f = \frac{f_0}{kA^{3/2} \cdot \gamma(z)}$$

(27.21)

with $\quad \gamma(z) = \dfrac{1}{\sqrt{2\pi}} \cdot \displaystyle\int_z^{\infty} \frac{F(x)}{\sqrt{x-z}} dx$.

This means we can define $\gamma(z)$, which is only dependent on the shape of the force curve $F(z)$ but independent of the external parameters of the oscillation. The function $\gamma(z)$ is also referred to as the *normalized frequency shift* [27.7], a very useful parameter, which allows us to compare measurements independent of resonant frequency, amplitude and spring constant of the cantilever.

The dependence of the frequency shift on the vibration amplitude is an especially useful relation, since this parameter can be easily varied during one experiment. A nice example is depicted in Fig. 27.18, where frequency shift curves for different amplitudes were found to coincide very well in the $\gamma(z)$ diagrams [27.38].

This relationship has been nicely exploited for the calibration of the vibration amplitude by *Guggisberg* [27.39], which is a problem often encountered

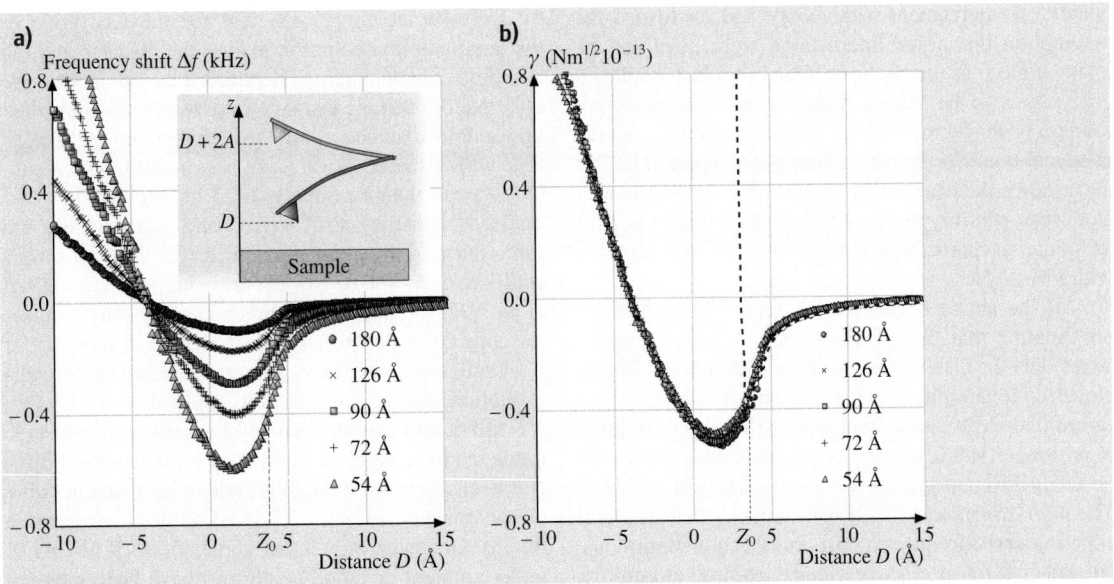

Fig. 27.18 (a) Frequency shift curves for different oscillation amplitudes for a silicon tip on a graphite surface in UHV, **(b)** γ-curves calculated from the Δf-curves in (a) (reprinted from Hölscher et al. [27.38] with permission, Copyright (2000) by The American Physical Society)

in dynamic AFM operation and worthwhile discussing. One approaches tip and sample and records frequency shift versus distance curves, which show a reproducible shape. Then, with the z-feedback disabled, several curves with different amplitudes are acquired. The amplitudes are typically chosen by adjusting the amplitude set point in volts. One has to take care that drift in the z-direction is negligible. An analysis of the corresponding $\gamma(z)$-curves will show the same curves (as in Fig. 27.18), but the curves will be shifted in the horizontal axis. These shifts correspond to the change in amplitude, allowing one to correlate the voltage values with the z-distances.

For the often-encountered force contributions from electrostatic, van der Waals, and chemical binding forces the frequency shift has been calculated from the force laws. In the approximation that the tip radius R is larger than the tip–sample distance z, an electrostatic potential V will yield a normalized frequency shift of (adapted from Guggisberg [27.40]):

$$\gamma(z) = \frac{\pi \varepsilon_0 R \cdot V^2}{\sqrt{2}} \cdot z^{-1/2} \,. \tag{27.22}$$

For van der Waals forces with Hamaker constant H and also with R larger than z we find accordingly:

$$\gamma(z) = \frac{H \cdot R}{12\sqrt{2}} \cdot z^{-3/2} \tag{27.23}$$

Finally, short-range chemical forces represented by the Morse potential (with the parameters binding energy U_0, decay length λ and equilibrium distance z_{equ} yield:

$$\gamma(z) = \frac{U_0 \sqrt{2}}{\sqrt{\pi \lambda}} \cdot \exp\left(-\frac{(z - z_{\text{equ}})}{\lambda}\right) \,. \tag{27.24}$$

These equations allow the experimentalist to interpret the spectroscopic measurements directly. For example, the contributions of the electrostatic and van der Waals forces can be easily distinguished by their slope in a double logarithmic plot (for an example, see Guggisberg et al. [27.40]).

Alternatively, if the force law is not known beforehand, the experimentalist wants to analyze the experimental frequency-shift data curves and extract the force or energy potential curves. We, therefore, have to invert the integral in (27.21) to find the tip–sample interaction potential V_{ts} from the $\gamma(z)$-curves [27.37]:

$$V_{\text{ts}}(z) = \sqrt{2} \cdot \int_{z}^{\infty} \frac{\gamma(x)}{\sqrt{x-z}} \, \mathrm{d}x \,. \tag{27.25}$$

Using this method, quantitative force curves were extracted from Δf-spectroscopy measurements on different, atomically resolved sites of the Si(111)-(7×7) reconstruction [27.1]. Comparison to theoretical molecular dynamics (MD) simulations showed good quantitative agreement with theory and confirmed the assumption that force interactions were governed by a single atom at the tip apex. Our initially formulated goal seems to be achieved: with FM-AFM we have found a powerful method that allows us to measure the chemical bond formation of single molecules. The last uncertainty, the exact shape and identity of the tip apex atom, can possibly be resolved by employing the FIM technique to characterize the tip surface in combination with FM-AFM.

All the above equations are only valid in the approximation that the oscillation amplitudes are much larger than the distance range of the encountered forces. However, for amplitudes of, e.g., 10 nm and long-ranged forces like electrostatic interactions this approximation is no longer valid. Several approaches have been proposed by different authors to solve this issue [27.41–43]. The *matrix method* [27.42] uses the fact that in a real experiment the frequency shift curve is not continuous, but rather a set of discrete values acquired at equidistant points. Therefore the integral in (27.18) can be substituted by a sum and the equation can be rewritten as a linear equation system, which in turn can be easily inverted by appropriate matrix operations. This *matrix method* is a very simple and general method for the AFM user to extract force curves from experimental frequency-shift curves without the restrictions of the large-amplitude approximation.

In this context it is worthwhile to point out a slightly different dynamic AFM method. While in the typical FM-AFM set-up the oscillation amplitude is controlled to stay constant by a dedicated feedback circuit, one could simply keep the excitation amplitude constant (this has been termed the constant-excitation (CE) mode as opposed to the constant-amplitude (CA) mode. It is expected that this mode is more gentle to the surface, because any dissipative interaction will reduce the amplitude and therefore prevent a further reduction of the effective tip–sample distance. This mode has been employed to image soft biological molecules like DNA or thiols in UHV [27.44].

At first glance, quantitative interpretation of the obtained frequency spectra seems more complicated, since the amplitude as well as the tip–sample distance is altered during the measurement. However, it was found

by Hölscher et al. [27.45] that for the CE mode in the large-amplitude approximation the distance and the amplitude channel can be decoupled by calculating the effective tip–sample distance from the piezo-controlled tip–sample distance z and the change in the amplitude with distance $A(z): z_{eff}(z) = z - A(z)$. As a result, (27.21) can then be directly used to calculate the normalized frequency shift $\gamma(z_{eff})$ and consequently the force curve can be obtained from (27.25). This concept has been verified in experiments by Schirmeisen et al. [27.46] through a direct comparison of spectroscopy curves acquired in the CE and CA modes.

Until now, we have always associated the self-excitation scheme with vacuum applications. Although it is difficult to operate FM-AFM in constant-amplitude mode in air, since large dissipative effects make it difficult to ensure a constant amplitude, it is indeed possible to use constant-excitation FM-AFM in air or even in liquids. However, only a few applications of FM-AFM under ambient or liquid conditions have been reported so far. Interestingly, a low-budget construction set (employing a tuning-fork force sensor) for a CE-mode dynamic AFM set-up has been published on the internet (*http://www.sxm4.uni-muenster.de*).

If it is possible to measure atomic scale forces with the NC-AFM, it should vice versa also be possible to exert forces with a similar precision. In fact, the new and exciting field of nanomanipulation could be driven to a whole new dimension, if defined forces can be reliably applied to single atoms or molecules. In this respect, Loppacher et al. [27.47] were able to push on different parts of an isolated Cu-tetrabromobisphenol (TBBP) molecule, which is known to possess four rotatable legs. They measured the force–distance curves while pushing one of the legs with the AFM tip. From the force curves they were able to determine the energy which was dissipated during the *switching* process of the molecule. The manipulation of single silicon atoms with NC-AFM was demonstrated by Oyabu et al. [27.48], who removed single atoms from a Si(111)-7×7 surface with an AFM tip and could subsequently deposit atoms from the tip onto the surface again. The possibility to exert and measure forces simultaneously during single-atom or molecule manipulation is an exciting new application of high-resolution NC-AFM experiments.

27.4 Q-Control

We have already discussed the virtues of a high Q value for high-sensitivity measurements: the minimum detectable force gradient was inversely proportional to the square root of Q. In vacuum, Q mainly represents the internal dissipation of the cantilever during oscillation, an internal damping factor. Low damping is obtained by using high-quality cantilevers, which are cut (or etched) from defect-free single-crystal silicon wafers. Under ambient or liquid conditions, the quality factor is dominated by dissipative interactions between the cantilever and the surrounding medium, and *Q* values can be as low as 100 for air or even 5 in liquid. Still, we ask if it is somehow possible to compensate for the damping effect by exciting the cantilever in a sophisticated way?

It turns out that the shape of the resonance curves in Fig. 27.5 can be influenced towards higher (or lower) *Q* values by an amplitude feedback loop. In principle, there are several mechanisms to couple the amplitude signal back to the cantilever, e.g., by the photothermal effect [27.49] or capacitive forces [27.50]. Figure 27.19 shows a method in which the amplitude feedback is mediated directly by the excitation piezo [27.51]. This has the advantage that no additional mechanical setups are necessary.

The working principle of the feedback loop can be understood by analyzing the equation of motion of the modified dynamic system:

$$m^* \ddot{z}(t) + \alpha \dot{z}(t) + kz(t) - F_{ts}(z_0 + z(t))$$
$$= F_{ext} \cos(\omega t) + Ge^{i\phi} z(t). \qquad (27.26)$$

This ansatz takes into account the feedback of the detector signal through a phase shifter, amplifier and adder as an additional force, which is linked to the cantilever deflection $z(t)$ through the gain G and the phase shift $e^{i\phi}$. We assume that the oscillation can be described by a harmonic oscillation trajectory. With a phase shift of $\phi = \pm\pi/2$ we find:

$$e^{\pm i\pi/2} z(t) = \pm \frac{1}{\omega} \dot{z}(t). \qquad (27.27)$$

This means, that the additional feedback force signal $Ge^{i\phi} z(t)$ is proportional to the velocity of the cantilever, just like the damping term in the equation of motion. We can define an effective damping constant α_{eff}, which combines the two terms:

$$m^* \ddot{z}(t) + \alpha_{eff} \dot{z}(t) + kz(t) - F_{ts}(z_0 + z(t))$$
$$= F_{ext} \cos(\omega t), \qquad (27.28)$$

$$\text{with} \quad \alpha_{eff} = \alpha \pm \frac{1}{\omega} G \quad \text{for} \quad \phi = \pm \frac{\pi}{2}.$$

Equation (27.28) shows that the damping of the oscillator can be enhanced or weakened by choosing $\phi = +\pi/2$ or $\phi = -\pi/2$, respectively. The feedback loop therefore allows us to vary the effective quality factor $Q_{eff} = m\omega_0/\alpha_{eff}$ of the complete dynamic

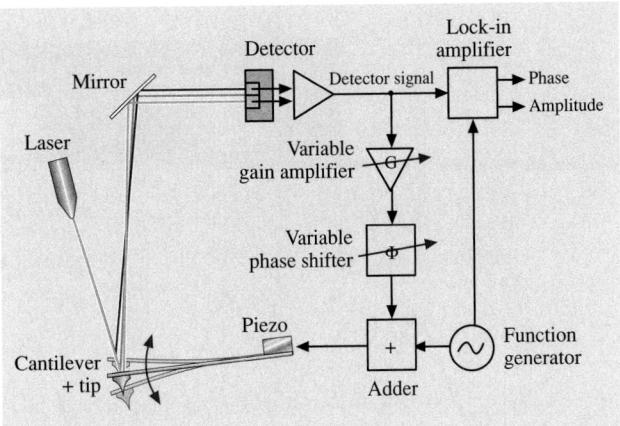

Fig. 27.19 Schematic diagram of operating a *Q-control* feedback circuit with an externally driven dynamic AFM. The tapping-mode set-up is in effect extended by an additional feedback loop

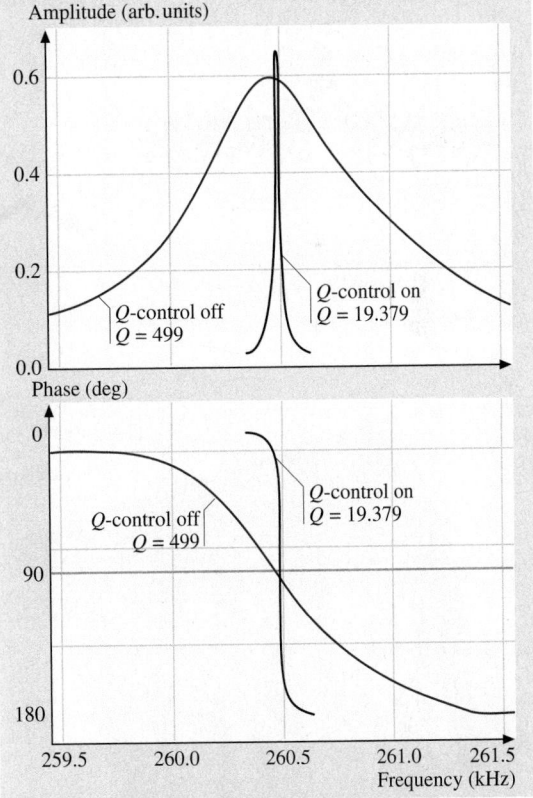

Fig. 27.20 Amplitude and phase diagrams measured in air with a Si cantilever far away from the sample. The quality factor can be increased from 450 to 20 000 by using the *Q-control* feedback method

Part C | 27.4

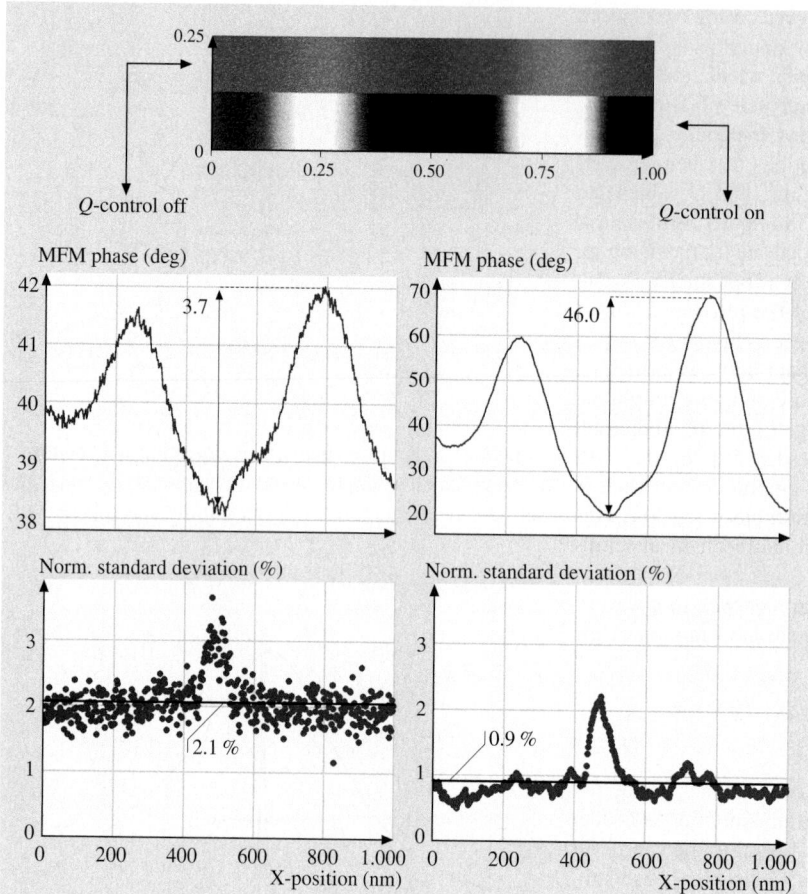

Fig. 27.21 Signal-to-noise analysis with a magnetic tip in tapping mode AFM on a magnetic tape sample with *Q-control*

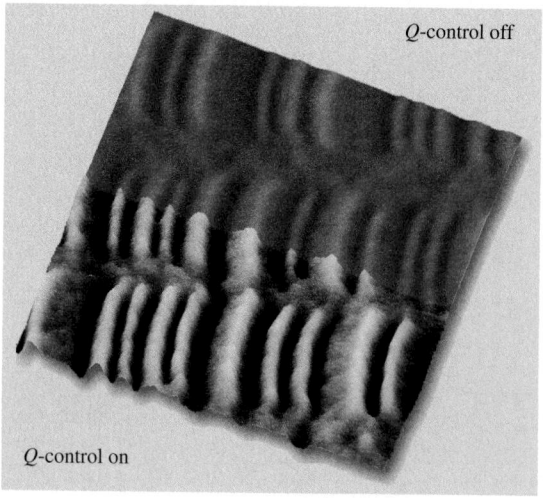

system. Hence, this system was termed *c-Control*. Figure 27.20 shows experimental data regarding the effect of *Q-control* on the amplitude and phase as a function of the external excitation frequency [27.51]. In this example, *Q-control* was able to increase the *Q*-value by a factor of over 40.

The effect of improved image contrast is demonstrated in Fig. 27.22. Here, a computer hard disk was analyzed with a magnetic tip in tapping mode, where the magnetic contrast is observed in the phase image. The upper part shows the recorded magnetic data structures in standard mode, whereas in the lower part of the image *Q-control* feedback was activated, giving rise to an im-

Fig. 27.22 Enhancement of the contrast in the phase channel due to *Q-control* on a magnetic hard disk measured with a magnetic tip in tapping-mode AFM in air. Scan size 5×5 μm, phase range 10 deg (www.nanoanalytics.com)

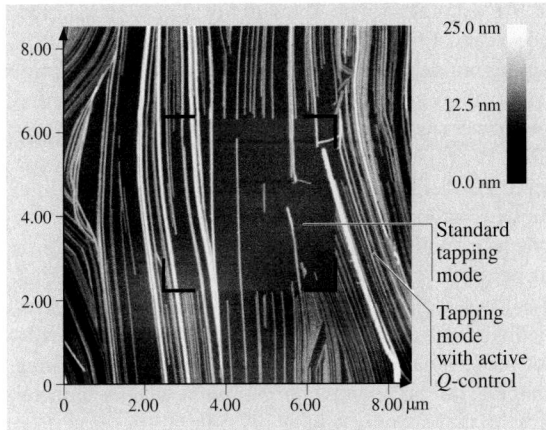

Fig. 27.23 Imaging of a delicate organic surface with Q-Control. Sample was a Langmuir–Blodgett film (ethyl-2,3-dihydroxyoctadecanoate) on a mica substrate. The topographical image clearly shows that the highly sensitive sample surface can only be imaged non-destructively with active *Q-control*, whereas the periodic repulsive contact with the probe in standard operation without *Q-control* leads to a significant modification or destruction of the surface structure (data courtesy of Lifeng Chi and coworkers, University of Münster, Germany)

increased by a factor of 12.4 by the *Q-control* feedback. The lower image shows a noise analysis of the signal, indicating an improvement of the signal-to-noise ratio by a factor of 2.3.

Note that the diagrams represent measurements in air with an AFM operated in AM mode. Only then can we make a distinction between excitation and vibration frequency, since in the FM mode these two frequencies are

proved signal, i. e., magnetic contrast. A more detailed analysis of measurements on a magnetic tape shows that the signal amplitude (upper diagrams in Fig. 27.21) was

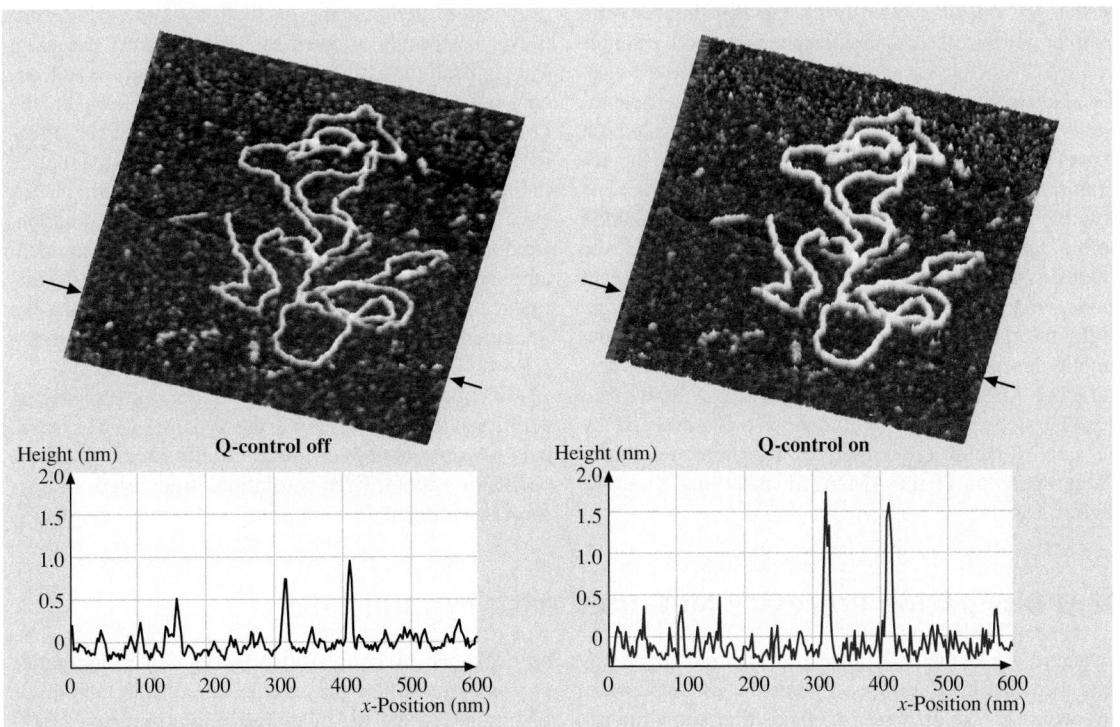

Fig. 27.24 AFM images of DNA on mica scanned in buffer solution (600 nm × 600 nm). Each scan line was scanned twice: in standard tapping mode during the first scan of the line (*left data*) and with *Q-control* activated by a trigger signal during the subsequent scan of the same line (*right data*). This interleave technique allows a direct comparison of the results of the two modes obtained on the same surface area while minimizing drift effects. Cross sections of the topographic data reveal that the observed DNA height is significantly higher in the case of imaging under *Q-control* (data courtesy of Daniel Ebeling and Hendrik Hölscher, University of Münster, Germany)

equal by definition. Although the relation between sensitivity and Q-factor in (27.17) is the same for AM and FM mode, it must be critically investigated whether the enhanced quality factor by Q-control can be inserted in the equation for FM-mode AFM. In vacuum applications, Q is already very high, which makes it unnecessary to operate an additional Q-control module.

As stated before, we can also use Q-control to enhance the damping in the oscillating system. This would decrease the sensitivity of the system. But on the other hand, the response time of the amplitude change is decreased as well. For tapping-mode applications, where high-speed scanning is the goal, Q-control was able to reduce the scan speed that limits relaxation time [27.52].

A large quality factor Q does not only have the virtue of increasing the force sensitivity of the instrument. It also has the advantage of increasing the parameter space of stable AFM operation in AM-mode AFM. Consider the resonance curve of Fig. 27.5. When approaching the tip towards the surface there are two competing mechanisms: on one hand, we bring the tip closer to the sample, which results in an increase in attractive forces (see Fig. 27.2). On the other hand, for the case $\omega > \omega_0$, the resonant frequency of the cantilever is shifted towards smaller values due to the attractive forces, which causes the amplitude to become smaller. This is the desirable regime, where stable operation of the AFM is possible in the net-attractive regime. But as explained before, below a certain tip–sample separation d_{app}, the system switches suddenly into the intermittent contact mode, where surface modifications are more likely due to the onset of strong repulsive forces. The steeper the amplitude curve the larger the regime of stable, net-attractive AFM operation. Looking at Fig. 27.20 we find that the slope of the amplitude curve is governed by the quality factor Q. A high Q, therefore, facilitates stable operation of the AM-AFM in the net-attractive regime.

An example can be found in Fig. 27.23. Here, a surface scan of an ultrathin organic film is acquired in tapping mode under ambient conditions. First, the inner square is scanned without the Q enhancement, and then a wider surface area was scanned with Q-control applied. The high quality factor provides a larger parameter space for operating the AFM in the net-attractive regime, allowing good resolution of the delicate organic surface structure. Without the Q-control the surface structures are deformed and even destroyed due to the strong repulsive tip–sample interactions [27.53–55]. The Q-control feedback also allowed imaging of DNA structures without predominantly depressing the soft material during imaging. It was then possible to observe a DNA diameter close to the theoretical value [27.56].

The same technique has been successfully employed to minimize the interaction forces during scanning in liquids. This is of special relevance for imaging delicate biological samples in environments such as water or buffer solution. When the AFM probe is submerged in a liquid medium, the oscillation of the AFM cantilever is strongly affected by hydrodynamic damping. This typically leads to quality factors below 10 and, accordingly, to a loss in force sensitivity. However, the Q-control technique allows one to increase the effective quality factor by about three orders of magnitude in liquids. Figure 27.24 shows results of scanning DNA structures on a mica substrate under a buffer solution. Comparison of the topographic data obtained in standard tapping mode and under Q-Control, in particular the difference in the observed DNA height, indicates that the imaging forces were successfully reduced by employing Q-control.

In conclusion, we have shown that, by applying an additional feedback circuit to the dynamic AFM system, it is possible to influence the quality factor Q of the oscillator system. High-resolution, high-speed, or low-force scanning is then possible.

27.5 Dissipation Processes Measured with Dynamic AFM

Dynamic AFM methods have proven their great potential for imaging surface structures at the nanoscale, and we have also discussed methods that allow the assessment of forces between distinct single molecules. However, there is another physical mechanism that can be analyzed with the dynamic mode and has been mentioned in some previous paragraphs: energy dissipation.

In Fig. 27.12, we have already shown an example, where the phase signal in tapping mode cannot be explained by conservative forces alone; dissipative processes must also play a role. In constant-amplitude FM mode, where the quantitative interpretation of experiments has proven to be less difficult, an intuitive distinction between conservative and dissipative tip–sample interaction is possible. We have shown the correlation between forces and frequency shifts of the oscillating system, but we have neglected one experimental input channel; the excitation amplitude, which is necessary to keep the oscillation amplitude constant, is a direct indication of the energy dissipated during one oscillation cycle. *Dürig* [27.57] has shown that in self-excitation mode (with an excitation–oscillation phase difference of 90 deg), conservative and dissipative interactions can be strictly separated. Part of this energy is dissipated in the cantilever itself, another part is due to external viscous forces in the surrounding medium. But more interestingly, some energy is dissipated at the tip–sample junction. This is the focus of the following paragraphs.

In contrast to conservative forces acting at the tip–sample junction, which at least in vacuum can be understood in terms of van der Waals, electrostatic, and chemical interactions, the dissipative processes are poorly understood. *Stowe* et al. [27.58] have shown that, if a voltage potential is applied between tip and sample, charges are induced in the sample surface, which will follow the tip motion (in their set-up the oscillation was parallel to the surface). Due to the finite resistance of the sample material, energy will be dissipated during the charge movement. This effect has been exploited to image the doping level of semiconductors. Energy dissipation has also been observed in imaging magnetic materials. *Liu* et al. [27.59] found that energy dissipation due to magnetic interactions was enhanced at the boundaries of magnetic domains, which was attributed to domain-wall oscillations, and even a simple system such as two clean metal surfaces which are moved in close proximity can give rise to frictional forces. *Stipe* et al. [27.60] have measured the energy dissipation due to fluctuating electromagnetic fields between two closely spaced gold surfaces, which was later interpreted by Volokitin and Persson [27.61] in terms of van der Waals friction.

Energy dissipation was also observed, in the absence of external electromagnetic fields, when the tip and sample were in close proximity, within 1 nm. Clearly, mechanical surface relaxations must give rise to energy losses. One could model the AFM tip as a small hammer, hitting the surface at high frequency, possibly resulting in phonon excitations. From a continuum mechanics point of view, we assume that the mechanical relaxation of the surface is not only governed by elastic responses. Viscoelastic effects of soft surfaces will also render a significant contribution to energy dissipation. The whole area of phase imaging in tapping mode is concerned with those effects [27.62–64].

In the atomistic view, the last tip atom can be envisaged as changing position while experiencing the tip–sample force field. A strictly reversible change of position would not result in a loss of energy. Still, it has been pointed out by *Sasaki* et al. [27.65] that a change in atom position would result in a change in the force interaction itself. Therefore, it is possible that the tip atom changes position at different tip–surface distances during approach and retraction, effectively causing an atomic-scale hysteresis to develop. *Hoffmann* et al. [27.13] and *Hembacher* et al. [27.66] have measured the short-range energy dissipation for different combinations of tip and surface materials in UHV. For atomic-resolution experiments at low temperatures on graphite [27.66] it was found that the energy dissipation is a step-like function. A similar shape of dissipation curves was found in a theoretical analysis by *Kantorovich* and *Trevethan* [27.67], where the energy dissipation was directly associated with atomic instabilities at the sample surface.

The dissipation channel has also been used to image surfaces with atomic resolution [27.68]. Instead of feeding back the distance on the frequency shift, the excitation amplitude in FM mode has been used as the control signal. The Si(111)-(7×7) reconstruction was successfully imaged in this mode. The step edges of mono-atomic NaCl islands on single-crystalline copper have also rendered atomic-resolution contrast in the dissipation channel [27.69]. The dissipation processes discussed so far are mostly in the configuration in which the tip is oscillated perpendicular to the surface. Friction is usually referred to as the energy loss due to lateral movement of solid bodies in contact. It is interesting to note in this context that *Israelachivili* [27.70] has pointed out a quantitative relationship between lateral and vertical (with respect to the surface) dissipation. He states that the hysteresis in vertical force–distance curves should equal the energy loss in lateral friction. An experimental confirmation of this conjecture at the molecular level is still missing.

Physical interpretation of energy dissipation processes at the atomic scale seems to be a daunting task at this point. Nevertheless, we can find a quantitative relation between the energy loss per oscillation cycle and the experimental parameters in dynamic AFM, as will be shown in the following section.

In static AFM it was found that permanent changes of the sample surface by indentations can cause a hysteresis between approach and retraction. The area between the approach and retraction curves in a force–distance diagram represents the lost or dissipated energy caused by the irreversible change of the surface structure. In dynamic-mode AFM, the oscillation parameters such as amplitude, frequency, and phase must contain

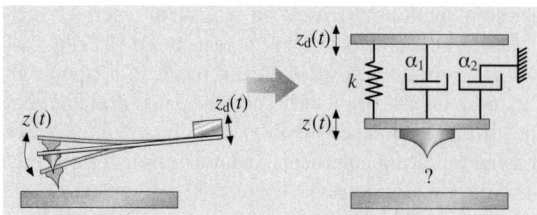

Fig. 27.25 Rheological models applied to describe the dynamic AFM system, comprising the oscillating cantilever and tip interacting with the sample surface. The movement of the cantilever base and the tip is denoted by $z_d(t)$ and $z(t)$, respectively. The cantilever is characterized by the spring constant k and the damping constant α. In a first approach damping is broken into two pieces α_1 and α_2: firstly, intrinsic damping caused by the movement of the cantilever's tip relative to its base, and secondly, damping related to the movement of the cantilever body in a surrounding medium, e.g. air damping

the information about the dissipated energy per cycle. So far, we have resorted to a treatment of the equation of motion of the cantilever vibration in order to find a quantitative correlation between forces and the experimental parameters. For the dissipation it is useful to treat the system from the point of view of energy conservation.

Assuming that a dynamic system is in equilibrium, the average energy input must equal the average energy output or dissipation. Applying this rule to an AFM running in a dynamic mode means that the average power fed into the cantilever oscillation by an external driver, denoted by $\overline{P}_{in}$, must equal the average power dissipated by the motion of the cantilever beam $\overline{P}_0$ and by the tip–sample interaction $\overline{P}_{tip}$:

$$\overline{P}_{in} = \overline{P}_0 + \overline{P}_{tip} . \tag{27.29}$$

The term $\overline{P}_{tip}$ is what we are interested in, since it gives us a direct physical quantity to characterize the tip–sample interaction. Therefore, we first have to calculate and then measure the two other terms in (27.29) in order to determine the power dissipated when the tip periodically probes the sample surface. This requires an appropriate rheological model to describe the dynamic system. Although there are investigations in which the complete flexural motion of the cantilever beam has been considered [27.71], a simplified model, comprising a spring and two dashpots (Fig. 27.25), represents a good approximation in this case [27.72].

The spring, characterized by the constant k according to Hooke's law, represents the only channel through which power P_{in} can be delivered to the oscillating tip

$z(t)$ by the external driver $z_d(t)$. Therefore, the instantaneous power fed into the dynamic system is equal to the force exerted by the driver times the velocity of the driver (the force that is necessary to move the base side of the dashpot can be neglected, since this power is directly dissipated and, therefore, does not contribute to the power delivered to the oscillating tip):

$$P_{in}(t) = F_d(t)\dot{z}_d(t) = k[z(t) - z_d(t)]\dot{z}_d(t) . \tag{27.30}$$

Assuming a sinusoidal steady-state response and that the base of the cantilever is driven sinusoidally (see (27.6)) with amplitude A_d and frequency ω, the deflection from equilibrium of the end of the cantilever follows (27.9), where A and $0 \le \varphi \le \pi$ are the oscillation amplitude and phase shift, respectively. This allows us to calculate the average power input per oscillation cycle by integrating (27.30) over one period $T = 2\pi/\omega$:

$$\overline{P}_{in} = \frac{1}{T}\int_0^T P_{in}(t)\,\mathrm{d}t = \frac{1}{2}k\omega A_d A \sin\varphi . \tag{27.31}$$

This contains the familiar result that the maximum power is delivered to an oscillator when the response is 90 deg out of phase with the drive.

The simplified rheological model, as depicted in Fig. 27.25, exhibits two major contributions to the damping term $\overline{P}_0$. Both are related to the motion of the cantilever body and assumed to be well modeled by viscous damping with coefficients α_1 and α_2. The dominant damping mechanism in UHV conditions is intrinsic damping, caused by the deflection of the cantilever beam, i. e., the motion of the tip relative to the cantilever base. Therefore the instantaneous power dissipated by such a mechanism is given by

$$P_{01}(t) = |F_{01}(t)\dot{z}(t)| = |\alpha_1[\dot{z}(t) - \dot{z}_d(t)]\dot{z}(t)| . \tag{27.32}$$

Note that the absolute value has to be calculated, since all dissipated power is *lost* and, therefore, cannot be returned to the dynamic system.

However, when running an AFM in ambient conditions an additional damping mechanism has to be considered; damping due to the motion of the cantilever body in the surrounding medium, e.g., air damping, is in most cases the dominant effect. The corresponding instantaneous power dissipation is given by

$$P_{02}(t) = |F_{02}(t)\dot{z}(t)| = \alpha_2\dot{z}^2(t) . \tag{27.33}$$

In order to calculate the average power dissipation, (27.32) and (27.33) have to be integrated over one

complete oscillation cycle. This yields

$$\overline{P}_{01} = \frac{1}{T} \int_0^T P_{01}(t)\,\mathrm{d}t$$

$$= \frac{1}{\pi} \alpha_1 \omega^2 A \left[(A - A_\mathrm{d} \cos\varphi) \right.$$

$$\times \arcsin\left(\frac{A - A_\mathrm{d} \cos\varphi}{\sqrt{A^2 + A_\mathrm{d}^2 - 2 A A_\mathrm{d} \cos\varphi}} \right)$$

$$\left. + A_\mathrm{d} \sin\varphi \right] \qquad (27.34)$$

and

$$\overline{P}_{02} = \frac{1}{T} \int_0^T P_{02}(t)\,\mathrm{d}t = \frac{1}{2} \alpha_2 \omega^2 A^2 . \qquad (27.35)$$

Considering the fact that commonly used cantilevers exhibit a quality factor of at least several hundreds (in UHV even several ten thousands), we can assume that the oscillation amplitude is significantly larger than the drive amplitude when the dynamic system is driven at or near its resonance frequency: $A \gg A_\mathrm{d}$. Therefore (27.34) can be simplified in a first-order approximation to an expression similar to (27.35). Combining the two equations yields the total average power dissipated by the oscillating cantilever

$$\overline{P}_0 = \frac{1}{2} \alpha \omega^2 A^2 \quad \text{with} \quad \alpha = \alpha_1 + \alpha_2 , \qquad (27.36)$$

where α denotes the overall effective damping constant.

We can now solve (27.29) for the power dissipation localized to the small interaction volume of the probing tip with the sample surface, represented by the question mark in Fig. 27.25. Furthermore by expressing the damping constant α in terms of experimentally accessible quantities such as the spring constant k, the quality factor Q and the natural resonant frequency ω_0 of the free oscillating cantilever, $\alpha = \frac{k}{Q \cdot \omega_0}$, we obtain:

$$\overline{P}_\mathrm{tip} = \overline{P}_\mathrm{in} - \overline{P}_0$$

$$= \frac{1}{2} \frac{k\omega}{Q} \left[Q_\mathrm{cant} A_\mathrm{d} A \sin\varphi - A^2 \frac{\omega}{\omega_0} \right] . \qquad (27.37)$$

Note that so far no assumptions have been made on how the AFM is operated, except that the motion of the oscillating cantilever has to remain sinusoidal to

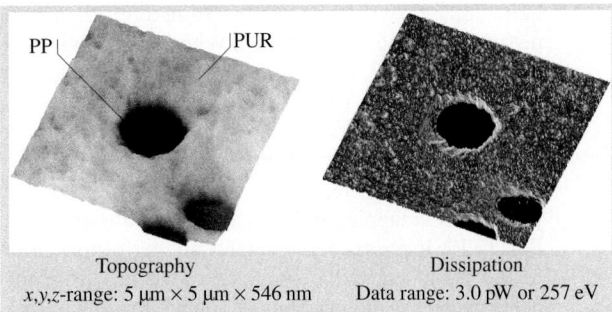

PP | PUR

Topography
x,y,z-range: 5 μm × 5 μm × 546 nm

Dissipation
Data range: 3.0 pW or 257 eV

Fig. 27.26 Topography and phase image in tapping-mode AFM of a polymer blend composed of polypropylene (PP) particles embedded in a polyurethane (PUR) matrix. The dissipation image shows a strong contrast between the harder PP (*little dissipation, dark*) to the softer PUR (*large dissipation, bright*) surface

a good approximation. Therefore (27.37) is applicable to a variety of different dynamic AFM modes.

For example, in FM-mode AFM the oscillation frequency ω changes due to tip–sample interaction while at the same time the oscillation amplitude A is kept constant by adjusting the drive amplitude A_d. By measuring these quantities, one can apply (27.37) to determine the average power dissipation related to the tip–sample interaction. In spectroscopy applications $A_\mathrm{d}(z)$ is usually not measured directly, but a signal $G(z)$ proportional to $A_\mathrm{d}(z)$ is acquired that represents the gain factor applied to the excitation piezo. With the help of (27.15) we can write:

$$A_\mathrm{d}(z) = \frac{A_0 \cdot G(z)}{Q \cdot G_0} , \qquad (27.38)$$

while A_0 and G_0 are the amplitude and gain at large tip–sample distances where the tip–sample interactions are negligible.

Now let us consider the tapping-mode AFM. In this case the cantilever is driven at a fixed frequency and with constant drive amplitude, while the oscillation amplitude and phase shift may change when the probing tip interacts with the sample surface. Assuming that the oscillation frequency is chosen to be ω_0, (27.37) can be further simplified by employing (27.15) for the free oscillation amplitude A_0. This yields

$$\overline{P}_\mathrm{tip} = \frac{1}{2} \frac{k\omega_0}{Q_\mathrm{cant}} \left[A_0 A \sin\varphi - A^2 \right] . \qquad (27.39)$$

Equation (27.39) implies that, if the oscillation amplitude A is kept constant by a feedback loop, as is commonly done in tapping-mode, simultaneously ac-

Part C | 27.5

quired phase data can be interpreted in terms of energy dissipation [27.63, 64, 73, 74]. When analyzing such phase images [27.75–77] one also has to consider the fact that the phase may also change due to the transition from the net-attractive ($\varphi > 90$ deg) to intermittent contact ($\varphi < 90$ deg) interaction between the tip and the sample [27.18, 51, 78, 79]. For example, consider the phase shift in tapping mode as a function of z-position, Fig. 27.12. If phase measurements are performed close to the point where the oscillation switches from the net-attractive to the intermittent contact regime, a large

contrast in the phase channel is observed. However, this contrast is not due to dissipative processes. Only a variation of the phase signal within the intermittent contact regime will give information about the tip–sample dissipative processes.

An example of a dissipation measurement is depicted in Fig. 27.26. The surface of a polymer blend was imaged in air, simultaneously acquiring the topography and dissipation. The dissipation on the softer polyurethane matrix is significantly larger than on the embedded, mechanically stiffer polypropylene particles.

27.6 Conclusion

Dynamic force microscopy is a powerful tool, which is capable of imaging surfaces with atomic precision. It also allows us to look at surface dynamics and it can operate in vacuum, air or even in liquid. However, the oscillating cantilever system introduces a level of complexity, which prevents a straightforward interpretation of acquired images. An exception is the self-excitation mode, where tip–sample forces can be successfully extracted from spectroscopic experiments. However, not only conservative forces can be investigated with dy-

namic AFM, energy dissipation also influences the cantilever oscillation and can therefore serve as a new information channel.

Open questions remain, concerning the exact geometric and chemical identity of the probing tip, which significantly influences the imaging and spectroscopic results. Using predefined tips like single-walled nanotubes or using atomic-resolution techniques like field ion microscopy to image the tip itself are possible approaches to address this issue.

References

27.1   M. A. Lantz, H. J. Hug, R. Hoffmann, P. J. A. van Schendel, P. Kappenberger, S. Martin, A. Baratoff, H.-J. Güntherodt: Quantitative measurement of short-range chemical bonding forces, Science **291**, 2580–2583 (2001)

27.2   G. Binnig, C. F. Quate, Ch. Gerber: Atomic force microscope, Phys. Rev. Lett. **56**, 930–933 (1986)

27.3   O. Marti: AFM Instrumentation and Tips. In: *Handbook of Micro/Nanotribology*, 2nd edn., ed. by B. Bushan (CRC, Boca Raton 1999) pp. 81–144

27.4   G. Cross, A. Schirmeisen, A. Stalder, P. Grütter, M. Tschudy, U. Dürig: Adhesion interaction between atoically defined tip and sample, Phys. Rev. Lett. **80**, 4685–4688 (1998)

27.5   A. Schirmeisen, G. Cross, A. Stalder, P. Grütter, U. Dürig: Metallic adhesion and tunneling at the atomic scale, New J. Phys. **2**, 1–29 (2000)

27.6   PhD thesis, Metallic Adhesion and Tunneling at the Atomic Scale, McGill University, Montréal, Canada, 29–38

27.7   F. J. Giessibl: Forces and frequency shifts in atomic-resolution dynamic-force microscopy, Phys. Rev. B **56**, 16010–16015 (1997)

27.8   F. J. Giessibl: Atomic resolution of the silicon (111)-(7x7) surface by atomic force microscopy, Science **267**, 68–71 (1995)

27.9   M. Bezanilaa, B. Drake, E. Nudler, M. Kashlev, P. K. Hansma, H. G. Hansma: Motion and enzymatic degradation of DNA in the atomic force microscope, Biophys. J. **67**, 2454–2459 (1994)

27.10 Y. Jiao, D. I. Cherny, G. Heim, T. M. Jovin, T. E. Schäffer: Dynamic interactions of p53 with DNA in solution by time-lapse atomic force microscopy, J. Mol. Biol. **314**, 233–243 (2001)

27.11 T. R. Albrecht, P. Grütter, D. Horne, D. Rugar: Frequency modulation detection using high-Q cantilevers for enhanced force microscopy sensitivity, J. Appl. Phys. **69**, 668–673 (1991)

27.12 S. P. Jarvis, M. A. Lantz, U. Dürig, H. Tokumoto: Off resonance AC mode force spectroscopy and imaging with an atomic force microscope, Appl. Surf. Sci. **140**, 309–313 (1999)

27.13 P. M. Hoffmann, S. Jeffery, J. B. Pethica, H. Ö. Özer, A. Oral: Energy dissipation in atomic force microscopy and atomic loss processes, Phys. Rev. Lett. **87**, 265502–265505 (2001)

27.14 E. Sivaniah, J. Genzer, G. H. Fredrickson, E. J. Kramer, M. Xiang, X. Li, C. Ober, S. Magonov: Periodic surface topology of three-arm semifluorinated alkane monodendron diblock copolymers, Langmuir **17**, 4342–4346 (2001)

27.15 S. N. Magonov: Visualization of Polymer Structures with Atomic Force Microscopy. In: *Applied Scanning Probe Methods*, ed. by H. Fuchs, M. Hosaka, B. Bhushan (Springer, Berlin 2004) pp. 207–250

27.16 C. Möller, M. Allen, V. Elings, A. Engel, D. J. Müller: Tapping-mode atomic force microscopy produces faithful high-resolution images of protein surfaces, Biophys. J. **77**, 1150–1158 (1999)

27.17 B. Anczykowski, D. Krüger, H. Fuchs: Cantilever dynamics in quasinoncontact force microscopy: spectroscopic aspects, Phys. Rev. B **53**, 15485–15488 (1996)

27.18 B. Anczykowski, D. Krüger, K. L. Babcock, H. Fuchs: Basic properties of dynamic force spectroscopy with the scanning force microscope in experiment and simulation, Ultramicroscopy **66**, 251–259 (1996)

27.19 V. M. Muller, V. S. Yushchenko, B. V. Derjaguin: On the influence of molecular forces on the deformation of an elastic sphere and its sticking to a rigid plane, J. Coll. Interf. Sci. **77**, 91–101 (1980)

27.20 B. D. Hughes, L. R. White: 'Soft' contact problems in linear elasticity, Quart. J. Mech. Appl. Math. **32**, 445–471 (1979)

27.21 L. Verlet: Computer experiments on classical fluids. I. Thermodynamical properties of Lennard–Jones Molecules, Phys. Rev. **159**, 98–103 (1967)

27.22 L. Verlet: Computer experiments on classical fluids. II. Equilibrium correlation functions, Phys. Rev. **165**, 201–214 (1968)

27.23 P. Gleyzes, P. K. Kuo, A. C. Boccara: Bistable behavior of a vibrating tip near a solid surface, Appl. Phys. Lett. **58**, 2989–2991 (1991)

27.24 A. San Paulo, R. Garcia: High-resolution imaging of antibodies by tapping-mode atomic force microscopy: Attractive and repulsive tip–sample interaction regimes, Biophys. J. **78**, 1599–1605 (2000)

27.25 D. Krüger, B. Anczykowski, H. Fuchs: Physical properties of dynamic force microscopies in contact and noncontact operation, Ann. Phys. **6**, 341–363 (1997)

27.26 Y. Martin, C. C. Williams, H. K. Wickramasinghe: Atomic force microscope–force mapping and profiling on a sub 100-A scale, J. Appl. Phys. **61**, 4723–4729 (1987)

27.27 Y. Sugawara, M. Otha, H. Ueyama, S. Morita: Defect motion on an InP(110) surface observed with noncontact atomic force microscopy, Science **270**, 1646–1648 (1995)

27.28 M. Bammerlin, R. Lüthi, E. Meyer, A. Baratoff, J. Lue, M. Guggisberg, Ch. Gerber, L. Howald, H.-J. Güntherodt: True atomic resolution on the surface of an insulator via ultrahigh vacuum dynamic force microscopy, Probe Microsc. **1**, 3–9 (1996)

27.29 C. Barth, M. Reichling: Imaging the atomic arrangement on the high-temperature reconstructed α-Al$_2$O$_3$(0001) Surface, Nature **414**, 54–57 (2001)

27.30 W. Allers, S. Langkat, R. Wiesendanger: Dynamic low-temperature scanning force microscopy on nickel oxide(001), Appl. Phys. A [Suppl.] **72**, S27–S30 (2001)

27.31 S. Morita, R. Wiesendanger, E. Meyer: *Noncontact Atomic Force Microscopy* (Springer, Berlin, Heidelberg 2002)

27.32 R. Garcia, R. Pérez: Dynamic atomic force microscopy methods, Surf. Sci. Rep. **47**, 197–301 (2002)

27.33 F. J. Giessibl: Advances in atomic force microscopy, Rev. Mod. Phys. **75**, 949–983 (2003)

27.34 H. Hölscher, U. D. Schwarz, R. Wiesendanger: Calculation of the frequency shift in dynamic force microscopy, Appl. Surf. Sci. **140**, 344–351 (1999)

27.35 B. Gotsmann, H. Fuchs: Dynamic force spectroscopy of conservative and dissipative forces in an Al-Au(111) tip–sample system, Phys. Rev. Lett. **86**, 2597–2600 (2001)

27.36 H. Hölscher, W. Allers, U. D. Schwarz, A. Schwarz, R. Wiesendanger: Determination of tip–sample interaction potentials by dynamic force spectroscopy, Phys. Rev. Lett. **83**, 4780–4783 (1999)

27.37 U. Dürig: Relations between interaction force and frequency shift in large-amplitude dynamic force microscopy, Appl. Phys. Lett. **75**, 433–435 (1999)

27.38 H. Hölscher, A. Schwarz, W. Allers, U. D. Schwarz, R. Wiesendanger: Quantitative analysis of dynamic-force-spectroscopy data on graphite (0001) in the contact and noncontact regime, Phys. Rev. B **61**, 12678–12681 (2000)

27.39 Guggisberg, M. (2000): PhD thesis, Lokale Messung von atomaren Kräften, University of Basel, Switzerland, 9-11

27.40 M. Guggisberg, M. Bammerlin, E. Meyer, H.-J. Güntherodt: Separation of interactions by noncontact force microscopy, Phys. Rev. B **61**, 11151–11155 (2000)

27.41 U. Dürig: Extracting interaction forces and complementary observables in dynamic probe microscopy, Appl. Phys. Lett. **76**, 1203–1205 (2000)

27.42 F. J. Giessibl: A direct method to calculate tip–sample forces from frequency shifts in frequency-modulation atomic force microscopy, Appl. Phys. Lett. **78**, 123–125 (2001)

27.43 J. E. Sader, S. P. Jarvis: Accurate formulas for interaction force and energy in frequency modulation force spectroscopy, Appl. Phys. Lett. **84**, 1801–1803 (2004)

27.44 T. Uchihasi, T. Ishida, M. Komiyama, M. Ashino, Y. Sugawara, W. Mizutani, K. Yokoyama, S. Morita, H. Tokumoto, M. Ishikawa: High-resolution imaging of organic monolayers using noncontact AFM, Appl. Surf. Sci. **157**, 244–250 (2000)

27.45   H. Hölscher, B. Gotsmann, A. Schirmeisen: Dynamic force spectroscopy using the frequency modulation technique with constant excitation, Phys. Rev. B **68**, 153401/1–4 (2003)

27.46   A. Schirmeisen, H. Hölscher, B. Anczykowski, D. Weiner, M. M. Schäfer, H. Fuchs: Dynamic force spectroscopy using the constant-excitation and constant-amplitude modes, Nanotechnology **16**, 13–17 (2005)

27.47   Ch. Loppacher, M. Guggisberg, O. Pfeiffer, E. Meyer, M. Bammerlin, R. Lüthi, R. Schlittler, J. K. Gimzewski, H. Tang, C. Joachim: Direct determination of the energy required to operate a single molecule switch, Phys. Rev. Lett. **90**, 066107/1–4 (2003)

27.48   N. Oyabu, O. Custance, I. Yi, Y. Sugawara, S. Morita: Mechanical vertical manipulation of selected single atoms by soft nanoindentation using near contact atomic force microscopy, Phys. Rev. Lett. **90**, 176102 (2003)

27.49   J. Mertz, O. Marti, J. Mlynek: Regulation of a microcantilever response by force feedback, Appl. Phys. Lett. **62**, 2344–2346 (1993)

27.50   D. Rugar, P. Grütter: Mechanical parametric amplification and thermomechanical noise squeezing, Phys. Rev. Lett. **67**, 699–702 (1991)

27.51   B. Anczykowski, J. P. Cleveland, D. Krüger, V. B. Elings, H. Fuchs: Analysis of the interaction mechanisms in dynamic mode SFM by means of experimental data and computer simulation, Appl. Phys. A **66**, 885 (1998)

27.52   T. Sulchek, G. G. Yaralioglu, C. F. Quate, S. C. Minne: Characterization and optimisation of scan speed for tapping-mode atomic force microscopy, Rev. Sci. Instr. **73**, 2928–2936 (2002)

27.53   L. F. Chi, S. Jacobi, B. Anczykowski, M. Overs, H.-J. Schäfer, H. Fuchs: Supermolecular periodic structures in monolayers, Adv. Mater. **12**, 25–30 (2000)

27.54   S. Gao, L. F. Chi, S. Lenhert, B. Anczykowski, C. Niemeyer, M. Adler, H. Fuchs: High-quality mapping of DNA–protein complexes by dynamic scanning force microscopy, ChemPhysChem **6**, 384–388 (2001)

27.55   B. Zou, M. Wang, D. Qiu, X. Zhang, L. F. Chi, H. Fuchs: Confined supramolecular nanostructures of mesogen-bearing amphiphiles, Chem. Commun. **9**, 1008–1009 (2002)

27.56   B. Pignataro, L. F. Chi, S. Gao, B. Anczykowski, C. Niemeyer, M. Adler, H. Fuchs: Dynamic scanning force microscopy study of self-assembled DNA-protein nanostructures, Appl. Phys. A **74**, 447–452 (2002)

27.57   U. Dürig: Interaction sensing in dynamic force microscopy, New J. Phys. **2**, 1–5 (2000)

27.58   T. D. Stowe, T. W. Kenny, D. J. Thomson, D. Rugar: Silicon dopant imaging by dissipation force microscopy, Appl. Phys. Lett. **75**, 2785–2787 (1999)

27.59   Y. Liu, P. Grütter: Magnetic dissipation force microscopy studies of magnetic materials, J. Appl. Phys. **83**, 7333–7338 (1998)

27.60   B. C. Stipe, H. J. Mamin, T. D. Stowe, T. W. Kenny, D. Rugar: Noncontact friction and force fluctuations between closely spaced bodies, Phys. Rev. Lett. **87**, 96801/1–4 (2001)

27.61   A. I. Volokitin, B. N. J. Persson: Resonant photon tunneling enhancement of the van der Waals friction, Phys. Rev. Lett. **91**, 106101/1–4 (2003)

27.62   J. Tamayo, R. Garcia: Effects of elastic and inelastic interactions on phase contrast images in tapping-mode scanning force microscopy, Appl. Phys. Lett. **71**, 2394–2396 (1997)

27.63   J. P. Cleveland, B. Anczykowski, A. E. Schmid, V. B. Elings: Energy dissipation in tapping-mode atomic force microscopy, Appl. Phys. Lett. **72**, 2613–2615 (1998)

27.64   B. Anczykowski, B. Gotsmann, H. Fuchs, J. P. Cleveland, V. B. Elings: How to measure energy dissipation in dynamic mode atomic force microscopy, Appl. Surf. Sci. **140**, 376–382 (1999)

27.65   N. Sasaki, M. Tsukada: Effect of microscopic nonconservative process on noncontact atomic force microscopy, Jpn. J. Appl. Phys. **39**, 1334 (2000)

27.66   S. Hembacher, F. J. Giessibl, J. Mannhart, C. F. Quate: Local spectroscopy and atomic imaging of tunneling current, forces, and dissipation on graphite, Phys. Rev. Lett. **94**, 056101/1–4 (2005)

27.67   L. N. Kantorovich, T. Trevethan: General theory of microscopic dynamical response in surface probe microscopy: From imaging to dissipation, Phys. Rev. Lett. **93**, 236102/1–4 (2004)

27.68   R. Lüthi, E. Meyer, M. Bammerlin, A. Baratoff, L. Howald, C. Gerber, H.-J. Güntherodt: Ultrahigh vacuum atomic force microscopy: True atomic resolution, Surf. Rev. Lett. **4**, 1025–1029 (1997)

27.69   R. Bennewitz, A. S. Foster, L. N. Kantorovich, M. Bammerlin, Ch. Loppacher, S. Schär, M. Guggisberg, E. Meyer, A. L. Shluger: Atomically resolved edges and kinks of NaCl islands on Cu(111): Experiment and theory, Phys. Rev. B **62**, 2074–2084 (2000)

27.70   J. Israelachvili: *Intermolecular and Surface Forces* (Academic, London 1992)

27.71   U. Rabe, J. Turner, W. Arnold: Analysis of the high-frequency response of atomic force microscope cantilevers, Appl. Phys. A **66**, 277 (1998)

27.72   T. R. Rodriguez, R. Garcia: Tip motion in amplitude modulation (tapping-mode) atomic-force microscopy: Comparison between continuous and point-mass models, Appl. Phys. Lett **80**, 1646–1648 (2002)

27.73   J. Tamayo, R. Garcia: Relationship between phase shift and energy dissipation in tapping-mode scanning force microscopy, Appl. Phys. Lett. **73**, 2926–2928 (1998)

27.74 R. Garcia, J. Tamayo, A. San Paulo: Tapping mode scanning force microsopy, Surf. Interface Anal. **27**, 312–316 (1999)

27.75 S. N. Magonov, V. B. Elings, M. H. Whangbo: Phase imaging and stiffness in tapping-mode atomic force microscopy, Surf. Sci. **375**, 385–391 (1997)

27.76 J. P. Pickering, G. J. Vancso: Apparent contrast reversal in tapping mode atomic force microscope images on films of polystyrene-b-polyisoprene-b-polystyrene, Polymer Bulletin **40**, 549–554 (1998)

27.77 X. Chen, S. L. McGurk, M. C. Davies, C. J. Roberts, K. M. Shakesheff, S. J. B. Tendler, P. M. Williams, J. Davies, A. C. Dwakes, A. Domb: Chemical and morphological analysis of surface enrichment in a biodegradable polymer blend by phase-detection imaging atomic force microscopy, Macromolecules **31**, 2278–2283 (1998)

27.78 A. Kühle, A. H. Sørensen, J. Bohr: Role of attractive forces in tapping tip force microscopy, J. Appl. Phys. **81**, 6562–6569 (1997)

27.79 A. Kühle, A. H. Sørensen, J. B. Zandbergen, J. Bohr: Contrast artifacts in tapping tip atomic force microscopy, Appl. Phys. A **66**, 329–332 (1998)

Part C | 27

28. Molecular Recognition Force Microscopy: From Simple Bonds to Complex Energy Landscapes

Atomic force microscopy (AFM), developed in the late eighties to explore atomic details on hard material surfaces, has evolved to an imaging method capable of achieving fine structural details on biological samples. Its particular advantage in biology is that the measurements can be carried out in aqueous and physiological environment, which opens the possibility to study the dynamics of biological processes in vivo. The additional potential of the AFM to measure ultra-low forces at high lateral resolution has paved the way for measuring inter- and intra-molecular forces of bio-molecules on the single molecule level. Molecular recognition studies using AFM open the possibility to detect specific ligand–receptor interaction forces and to observe molecular recognition of a single ligand–receptor pair. Applications include biotin–avidin, antibody-antigen, NTA nitrilotriacetate–hexahistidine 6, and cellular proteins, either isolated or in cell membranes.

The general strategy is to bind ligands to AFM tips and receptors to probe surfaces (or vice versa), respectively. In a force–distance cycle, the tip is first approached towards the surface whereupon a single receptor–ligand complex is formed, due to the specific ligand receptor recognition. During subsequent tip–surface retraction a temporarily increasing force is exerted to the ligand–receptor connection thus reducing its lifetime until the interaction bond breaks at a critical force (unbinding force). Such experiments allow for estimation of affinity, rate constants, and structural data of the binding pocket. Comparing them with values ob-

tained from ensemble-average techniques and binding energies is of particular interest. The dependences of unbinding force on the rate of load increase exerted to the receptor–ligand bond reveal details of the molecular dynamics of the recognition process and energy landscapes. Similar experimental strategies were also used for studying intra-molecular force properties of polymers and unfolding–refolding kinetics of filamentous proteins. Recognition imaging, developed by combing dynamic force microscopy with force spectroscopy, allows for localization of receptor sites on surfaces with nanometer positional accuracy.

Part C | 28

Molecular recognition plays a pivotal role in nature. Signaling cascades, enzymatic activity, genome replication and transcription, cohesion of cellular structures, interaction of antigens and antibodies and metabolic pathways all rely critically on specific recognition. In fact, every process in which molecules interact with each other in a specific manner depends on this trait.

Molecular recognition studies emphasize specific interactions between receptors and their cognitive ligands. Despite a growing body of literature on the structure and function of receptor ligand complexes, it is still not possible to predict reaction kinetics or energetics for any given complex formation, even when the structures are known. Additional insights, in particular about the molecular dynamics (MD) within the complex during the association and dissociation processes, are needed. The high-end strategy is to probe the energy landscape that underlies the interactions between molecules, whose structure is known at atomic resolution.

Receptor ligand complexes are usually formed by a few, non-covalent weak interactions between contacting chemical groups in complementary determining regions, supported by framework residues providing structurally conserved scaffolding. Both the complementary determining regions and the framework have a considerable amount of plasticity and flexibility, allowing for conformational movements during association and dissociation. In addition to knowledge about structure, energies, and kinetic constants, information about these movements is required for the understanding of the recognition process. Deeper insight into the nature of these movements as well as the spatiotemporal action of the many weak interactions, in particular the cooperativity of bond formation, is the key for the understanding of receptor ligand recognition.

For this, experiments at the single-molecule level, on time scales typical for receptor ligand complex formation and dissociation appear to be required. The potential of the atomic force microscope (AFM) [28.1] to measure ultra-low forces at high lateral resolution has paved the way for single-molecule recognition force microscopy studies. The particular advantage of AFM in biology is that the measurements can be carried out in an aqueous and physiological environment, which opens the possibility for studying biological processes in vivo. The methodology described in this paper for investigating the molecular dynamics of receptor ligand interactions, molecular recognition force microscopy (MRFM) [28.2–4], is based on scanning probe microscopy (SPM) technology [28.1]. A force is exerted on a receptor ligand complex and the dissociation process is followed over time. Dynamic aspects of recognition are addressed in force spectroscopy (FS) experiments, where distinct force–time profiles are applied to give insight into changes of conformations and states during receptor ligand dissociation. It will be shown that MRFM is a versatile tool to explore kinetic and structural details of receptor ligand recognition.

28.1 Ligand Tip Chemistry

In MRFM experiments, the binding of ligands immobilized on AFM tips to surface-bound receptors (or vice versa) is studied by applying a force to the receptor ligand complex that reduces its lifetime until the bond breaks at a measurable unbinding force. This requires a careful AFM tip sensor design, including tight attachment of the ligands to the tip surface. In the first pioneering demonstrations of single-molecule recognition force measurements [28.2,3], strong physical adsorption of bovine serum albumin (BSA) was used to directly coat the tip [28.3] or a glass bead glued to it [28.2]. This physisorbed protein layer may then serve as a matrix for biochemical modifications with chemically active ligands (Fig. 28.1). In spite of the large number of probe molecules on the tip ($10^3–10^4 \, nm^{-2}$) the low fraction of properly oriented molecules, or internal blocks of most reactive sites (see Fig. 28.1), allowed the measurement of single receptor ligand unbinding forces. Nevertheless, parallel breakage of multiple bonds was predominately observed with this configuration.

To measure interactions between isolated receptor ligand pairs, strictly defined conditions need to be fulfilled. Covalently coupling ligands to gold-coated tip surfaces via freely accessible SH groups guarantees a sufficiently stable attachment because these bonds are about ten times stronger than typi-

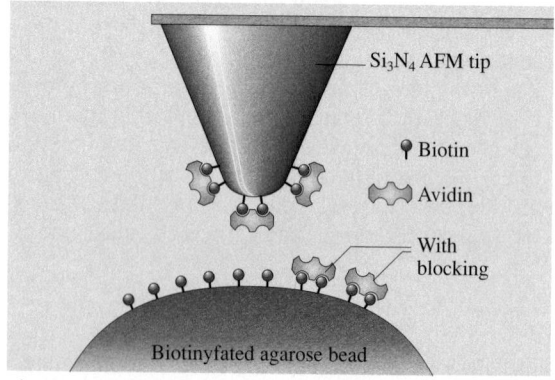

Fig. 28.1 Avidin-functionalized AFM tip. A dense layer of biotinylated BSA was adsorbed to the tip and subsequently saturated with avidin. The biotinylated agarose bead opposing the tip also contained a high surface density of reactive sites. These were partly blocked with avidin to achieve single-molecule binding events. After [28.3]

cal ligand-receptor interactions [28.5]. This chemistry has been used to detect the forces between complementary DNA strands [28.6] as well as between isolated nucleotides [28.7]. Self-assembled monolayers of dithio-bis(succinimidylundecanoate) were formed to enable covalent coupling of biomolecules via amines [28.8] and were used to study the binding strength between cell adhesion proteoglycans [28.9] and between biotin-directed IgG antibodies and biotin [28.10]. A vectorial orientation of Fab molecules on gold tips was achieved by site-directed chemical binding via their SH groups [28.11], without the need for additional linkers. To this end, antibodies were digested with papain and subsequently purified to generate Fab fragments with freely accessible SH groups in the hinge region.

Gold surfaces provide a unique and selective affinity for thiols, although the adhesion strength of the resulting bonds is comparatively weak [28.5]. Since all commercially available AFM tips are etched from silicon nitride or silicon oxide material, the deposition of a gold layer onto the tip surface is required prior to using this chemistry. Therefore, designing a sensor with covalent attachments of biomolecules to the silicon surface may be more straightforward. Amine functionalization procedures, a strategy widely used in surface biochemistry, were applied using ethanolamine [28.4, 12] and various silanization methods [28.13–15] as a first step in thoroughly developed surface-anchoring protocols suitable for single-molecule experiments. Since the amine surface density determines, to a large extent, the number of ligands on the tip that can specifically bind to the receptors on the surface, it has to be sufficiently low to guarantee single-molecular recognition events [28.4, 12]. Typically, these densities are kept between 200 and 500 molecules per square micron, which, for AFM tips with radii of $\approx 5-20$ nm, amounts to about one molecule per effective tip area. A striking example of a minimally ligated tip was given by *Wong* et al. [28.16] who derivatized a few carboxyl groups present at the open end of carbon nanotubes attached to the tips of gold-coated Si cantilevers.

In a number of laboratories, a distensible and flexible linker was used to distance the ligand molecule from the tip surface (e.g. [28.4, 14]) (Fig. 28.2). At a given low number of spacer molecules per tip, the ligand can then freely orient and diffuse within a certain volume, provided by the length of the tether, to achieve unconstrained binding to its receptor. The unbinding process occurs with little torque and the ligand molecule escapes the danger of being squeezed between the tip and the surface. It also opens the possibility of site-directed cou-

pling for a defined orientation of the ligand relative to the receptor at receptor ligand unbinding. As a crosslinking element, polyethylene glycol (PEG), a water-soluble, nontoxic polymer with a wide range of applications in surface technology and clinical research, was often used [28.17]. PEG is known to prevent surface adsorption of proteins and lipid structures and appears therefore ideally suited for this purpose. Glutaraldehyde [28.13] and DNA [28.6] were also successfully applied in recognition force studies as molecular spacers. Crosslinker lengths, ideally arriving at a good compromise between high tip-molecule mobility and narrow lateral resolution of the target recognition site, varied from 2 to 100 nm.

For coupling to the tip surface and to the ligand, the crosslinker typically carries two different functional ends, e.g. an amine reactive N-hydroxysuccinimidyl (NHS) group on one end, and 2-pyridyldithiopropionyl (PDP) [28.18] or vinyl sulfone [28.19] groups, which can be covalently bound to thiols, on the other (Fig. 28.2). This sulfur chemistry is highly advantageous since it is very reactive and readily enables site-directed coupling. However, free thiols are hardly available on native ligands and must, therefore, be added.

Different strategies have been used to achieve this goal. Lysine residues were derivatized with the short heterobifunctional linker N-succinimidyl-3-

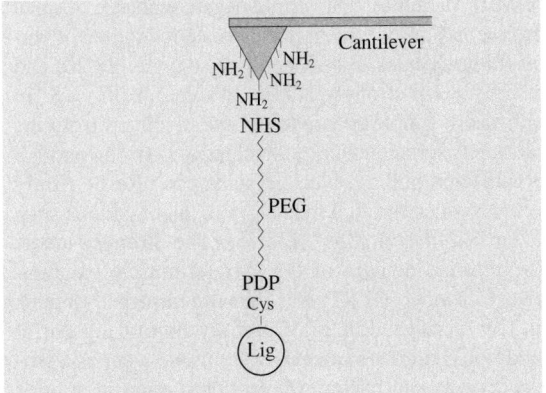

Fig. 28.2 Linkage of ligands to AFM tips. Ligands were covalently coupled to AFM tips via a heterobifunctional polyethylene glycol (PEG) derivative of 8-nm length. Silicon tips were first functionalized with ethanolamine $(NH_2-C_2H_4OH\cdot HCl)$. Then, the N-hydroxy-succinimide (NHS)-end of the PEG linker was covalently bound to amines on the tip surface before ligands were attached to the pyridoyldithiopropionate (PDP)-end via a free thiol or cysteine

(S-acetylthio)propionate (SATP) [28.18]. Subsequent deprotection with NH_2OH led to reactive SH groups. Alternatively, lysins can be directly coupled via aldehyde groups (manuscript in preparation). A problem with the latter two methods is that they do not allow for site-specific coupling of the crosslinker, since lysine residues are quite abundant. Several protocols are commercially available (Pierce, Rockford, IL) to generate active antibody fragments with free cysteines. Half-antibodies are produced by cleaving the two disulfide bonds in the central region of the heavy chain using 2-mercaptoethylamine HCl [28.20] and Fab fragments are generated from papain digestion [28.11]. The most elegant methods are to introduce a cysteine into the primary sequence of proteins or to append a thiol group to the end of a DNA strand [28.6], allowing for a well-defined sequence-specific coupling of the ligand to the crosslinker.

An attractive alternative for covalent coupling is provided by the widely used nitrilotriacetate (NTA)-His$_6$ system. The strength of binding in this system, which is routinely used in chromatographic and biosensor matrices, is significantly larger than that between most ligand-receptor pairs [28.21–23]. Since a His$_6$ tag can be readily appended to proteins, a crosslinker containing an NTA residue is ideally suited for coupling proteins to the AFM tip. This generic, site-specific coupling strategy also allows rigorous and ready control of binding specificity by using Ni^{++} as a molecular switch of the NTA-His$_6$ bond.

28.2 Immobilization of Receptors onto Probe Surfaces

To enable force detection, the receptors recognized by the ligand-functionalized tip need to be firmly attached to the probed surface. Loose association will unavoidably lead to a pull-off of the receptors from the surface by the tip-immobilized ligands, precluding detection of the interaction force.

Freshly cleaved muscovite mica is a perfectly pure and atomically flat surface and, therefore, ideally suited for MRFM studies. The strong negative charge of mica also accomplishes very tight electrostatic binding of various biomolecules. For example, lysozyme [28.20] and avidin [28.24] strongly adhere to mica at pH < 8. In such cases, simple adsorption of the receptors from the solution is sufficient, since attachment is strong enough to withstand pulling. Nucleic acids can also be firmly bound to mica through mediatory divalent cations such as Zn^{2+}, Ni^{2+} or Mg^{2+} [28.25]. The strongly acidic sarcoplasmic domain of the skeletal-muscle calcium-release channel (RYR1) was likewise absorbed to mica via Ca^{2+} bridges [28.26]. Carefully optimizing buffer conditions, similar strategies were used to deposit protein crystals and bacterial layers onto mica in defined orientations [28.27, 28].

The use of nonspecific electrostatic-mediated binding is however quite limited and generally offers no means to orient the molecules over the surface in a desirable direction. Immobilization through covalent attachment must therefore be frequently explored. When glass, silicon or mica are used as probe surfaces, immobilization is essentially the same as described above for tip functionalization. The number of reactive SiOH groups of the chemically relatively inert mica can optionally be increased by water plasma treatment [28.29]. As with tips, crosslinkers are also often used to provide receptors with motional freedom and to prevent surface-induced protein denaturation [28.4]. Immobilization can be controlled, to some extent, by using photoactivatable crosslinkers, such as N-5-azido-2-nitrobenzoyloxysuccinimide [28.30].

A major limitation of silicon chemistry is that it does not allow for high surface densities, i. e., $> 1000/\mu m^2$. By comparison, the surface density of a monolayer of streptavidin is about 60 000 molecules per μm^2 and that of a phospholipid monolayer may exceed 10^6 molecules per μm^2. The latter high density is also achievable by chemisorption of alkanethiols to gold. Tightly bound functionalized alkanethiol monolayers formed on ultraflat gold surfaces provide excellent probes for AFM [28.10] and readily allow for covalent and non-covalent attachment of biomolecules [28.10,31] (Fig. 28.3).

Recently, *Kada* et al. [28.32] reported on a new strategy to immobilize proteins on gold surfaces using phosphatidyl choline or phosphatidyl ethanolamine analogues containing dithio-phospholipids at their hydrophobic tail. Phosphatidyl ethanolamine, which is chemically reactive, was derivatized with a long-chain biotin for the molecular recognition of streptavidin molecules in an initial study [28.32]. These self-assembled phospholipid monolayers closely mimic the cell surface and minimize nonspecific adsorption. Additionally, they can be spread as insoluble monolayers at an

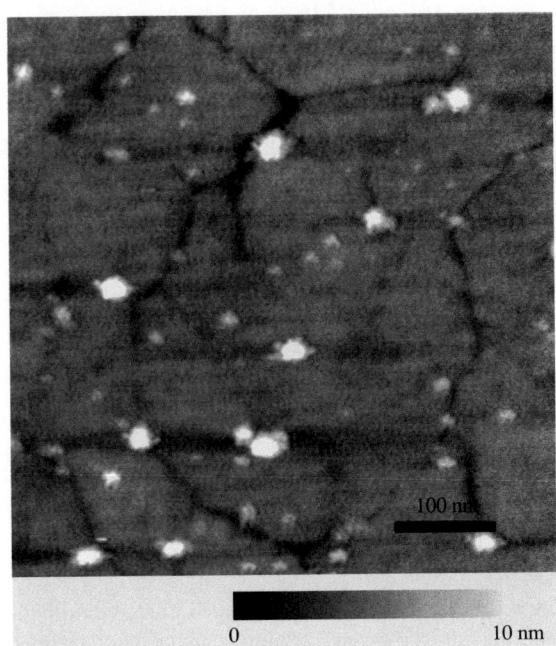

Fig. 28.3 AFM image of hisRNAP molecules specifically bound to nickel-NTA domains on a functionalized gold surface. Alkanethiols terminated with ethylene glycol groups to resist unspecific protein adsorption served as a host matrix and were doped with 10% nickel-NTA alkenthiols. The sample was prepared to achieve full monolayer coverage. Ten individual hisRNAP molecules can be clearly visualized bound to the surface. The more abundant, smaller, lower features are NTA islands with no bound molecules. The underlying morphology of the gold can also be distinguished. (After [28.31].)

air/water interface. Thereby, the ratio of functionalized thio-lipids to host lipids accurately defines the surface density of bioreactive sites in the monolayer. Subsequent transfer onto gold substrates leads to covalent and, hence, tight attachment of the monolayer.

MRFM has also been used to study the interactions between ligands and cell surface receptors in situ, on fixed or unfixed cells. In these studies it was found that the immobilization of cells strongly depends on cell type. Adherent cells are readily usable for MRFM whereas cells that grow in suspension need to be adsorbed onto the probe surface. Various protocols for tight immobilization of cells over a surface are available. For adherent cells, the easiest way is to grow the cells directly on glass or other surfaces suitable for MRFM [28.33]. Firm immobilization of non- and weakly adhering cells can be achieved by various adhesive coatings such as Cell-Tak [28.34], gelatin, or polylysine. Hydrophic surfaces like gold or carbon are also very useful to immobilize non-adherent cells or membranes [28.35]. Covalent attachment of cells to surfaces can be accomplished by crosslinkers that carry reactive groups, such as those used for immobilization of molecules [28.34]. Alternatively, one can use crosslinkers carrying a fatty-acid moiety that can penetrate into the lipid bilayer of the cell membrane. Such linkers provide sufficiently strong fixation without interference with membrane proteins [28.34].

28.3 Single-Molecule Recognition Force Detection

Measurements of interaction forces traditionally rely on ensemble techniques such as shear flow detachment (SFD) [28.36] and the surface force apparatus (SFA) [28.37]. In SFD, receptors are fixed to a surface to which ligands carried by beads or presented on the cell surface bind specifically. The surface-bound particles are then subjected to a fluid shear stress that disrupts the ligand-receptor bonds. However, the force acting between single molecular pairs can only be estimated because the net force applied to the particles can only be approximated and the number of bonds per particle is unknown.

SFA measures the forces between two surfaces to which different interacting molecules are attached using a cantilever spring as a force probe and interferometry for detection. The technique, which has a distance resolution of $\approx 1\,\text{Å}$, allows one to measure adhesive and compressive forces and to follow rapid transient effects in real time. However, the force sensitivity of the technique ($\approx 10\,\text{nN}$) does not allow for single-molecule measurements of non-covalent interaction forces.

The biomembrane force probe (BFP) technique uses pressurized membrane capsules rather than mechanical springs as the force transducer (Fig. 28.4; for a recent paper see [28.38]). To form the transducer, a red blood cell or a lipid bilayer vesicle is pressurized into the tip of a glass micropipette. The spring constant of

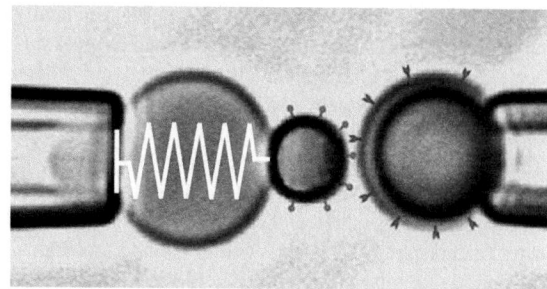

Fig. 28.4 Experimental set-up of the biomembrane force probe (BFP). The spring in the BFP is a pressurized membrane capsule. Its spring constant is set by membrane tension, which is controlled by micropipette suction. The BFP tip is formed by a glass microbead with a diameter of $1-2\,\mu m$ chemically glued to the membrane. The BFP (*on the left*) was kept stationary and the test surface, formed by another microbead (*on the right*), was translated to or from contact with the BFP tip by precision piezo control. (After [28.38].)

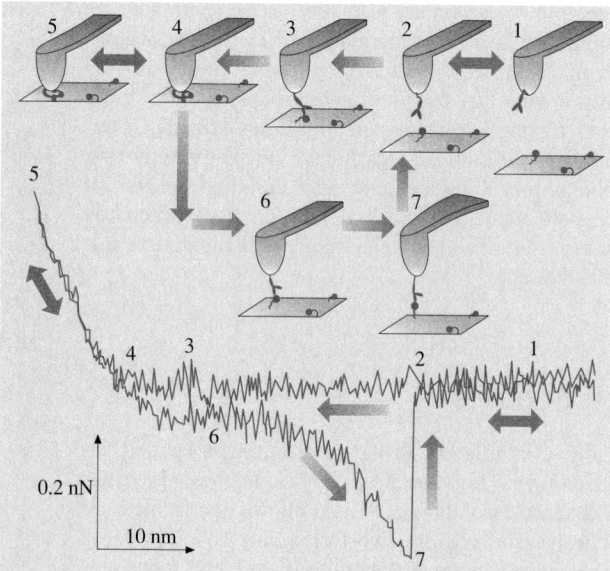

Fig. 28.5 Single-molecule recognition event detected with AFM. A force–distance cycle, measured with and amplitude of 100 nm at a sweep frequency of 1 Hz, for an antibody antigen pair in PBS. Binding of the antibody immobilized on the tip to the antigen on the surface, which occurs during the approach (*trace points 1 to 5*), results in a parabolic retract force curve (*points 6 to 7*) reflecting the extension of the distensible crosslinker antibody antigen connection. The force increases until unbinding occurs at a force of 268 pN (*points 7 to 2*). (After [28.4].)

the capsule can then be varied over several orders of magnitude by suction. This simple, but highly effective configuration, enables the measurement of forces ranging from 0.1 pN to 1000 pN with a force resolution of about 1 pN, allowing probing of single molecular bonds.

Optical tweezers (OT) belong to a family of techniques that rely on external fields to manipulate single particles. Thus, unlike mechanical transducers, force is exerted from a distance. In optical tweezers (OT), small particles (beads) are manipulated by optical traps [28.39]. Three-dimensional light-intensity gradients of a focused laser beam are used to pull or push particles with positional accuracy of a few nanometers. Using this technique, forces in the range of $10^{-13}-10^{-10}\,N$ can accurately be measured. Optical tweezers have been used extensively to measure the force-generating properties of various molecular motors at the single-molecule level [28.40] and to obtain force-extension profiles of single DNA [28.41] or protein [28.42] molecules. Defined, force-controlled twisting of DNA using rotating magnetically manipulated particles gave even further insights into DNA viscoelastic properties [28.43].

The atomic force microscope (AFM; [28.1]) is the force-measuring method with the smallest force sensor and therefore provides the highest lateral resolution. Radii of commercially available AFM tips vary between 2 and 50 nm. In contrast, the particles used for force sensing in SFD, BFP, and OT are in the $1-10\,\mu m$ range, and the surfaces used in SFA exceed millimeter extensions. The small apex of the AFM tip allows the visualization of single biomolecules with molecular to submolecular resolution [28.25, 27, 28].

In addition to imaging, AFM has successfully been used to measure interaction forces between various single molecular pairs [28.2–4]. In these measurements, one of the binding partners is immobilized onto a tip mounted at the end of a flexible cantilever that functions as a force transducer and the other is immobilized over a hard surface such as mica or glass. The tip is initially brought to, and subsequently retracted from the surface, and the interaction (unbinding) force is measured by following the cantilever deflection, which is monitored by measuring the reflection of a laser beam focused on the back of the cantilever using a split photodiode. Approach and retract traces obtained from the unbinding of a single molecular pair are shown in Fig. 28.5 [28.4]. In this experiment, the binding partners were immobilized onto their respective surfaces through a distensible PEG tether.

Cantilever deflection, Δx, relates directly to the force F, acting on it directly through Hook's law $F = k\Delta x$, where k is the spring constant of the cantilever. During most of the approach phase (trace, and lines 1 to 5), when the tip and the surface are sufficiently far away from each other (1 to 4), cantilever deflection remains zero because the molecules are as yet unbound to each other. Upon contact (4) the cantilever bends upwards (4 to 5) due to a repulsive force that increases linearly as the tip is pushed further into the surface. If the cycle was futile, and no binding had occurred, retraction of the tip from the surface (retrace, 5 to 7) will lead to a gradual relaxation of the cantilever to its rest position (5 to 4). In such cases, the retract curve will look very much like the approach curve. On the other hand, if binding had occurred, the cantilever will bend downwards as the cantilever is retracted from the surface (retrace, 4 to 7). Since the receptor and ligand were tethered to the surfaces through flexible crosslinkers, the shape of the attractive force–distance profile is nonlinear, in contrast to the profile obtained during contact (4 to 7). The exact shape of the retract curve depends on the elastic properties of the crosslinker used for immobilization [28.17, 44] and exhibits parabolic-like characteristics, reflecting an increase of the spring constant of the crosslinker during extension. The downward bending of the retracting cantilever continues until the ramping force reaches a critical value that dissociates the ligand-receptor complex (unbinding force, 7). Un-binding of the complex is indicated by a sharp spike in the retract curve that reflects an abrupt recoil of the cantilever to its rest position. Specificity of binding is usually demonstrated by block experiments in which free ligands are added to mask receptor sites over the surface.

The force resolution of the AFM, $\Delta F = (k_B T k)^{1/2}$, is limited by the thermal noise of the cantilever which, in turn, is determined by its spring constant. A way to reduce thermal fluctuations of cantilevers without changing their stiffness or lowering the temperature is to increase the apparent damping constant. Applying an actively controlled external dissipative force to cantilevers to achieve such an increase, *Liang* et al. [28.45] reported a 3.4-fold decrease in thermal noise amplitude. The smallest forces that can be detected with commercially available cantilevers are in the few piconewton range. Decreasing cantilever dimensions enables one to push the range of detectable forces to smaller forces since small cantilevers have lower coefficients of viscous damping [28.46]. Such miniaturized cantilevers also have much higher resonance frequencies than conventional cantilevers and, therefore, allow for faster measurements.

Besides the detection of intermolecular forces, the AFM also shows great potential in measuring forces acting within molecules. In these experiments, the molecule is clamped between the tip and the surface and its viscoelastic properties are studied by force–distance cycles.

28.4 Principles of Molecular Recognition Force Spectroscopy

Molecular recognition is mediated by a multitude of non-covalent interactions the energy of which is only slightly higher than thermal energy. Due to the power law dependence of these interactions on the distance, the attractive forces between non-covalently interacting molecules are extremely short-ranged. A close geometrical and chemical fit within the binding interface is therefore a prerequisite for productive association. The weak bonds that govern molecular cohesion are believed to be formed in a spatially and temporarily correlated fashion. Protein binding often involves structural rearrangements that can be either localized or global. These rearrangements often bear functional significance by modulating the activity of the interactants. Signaling pathways, enzyme activity, and the activation and inactivation of genes all depend on conformational changes induced in proteins by ligand binding.

The strength of binding is usually given by the binding energy E_B, which amounts to the free-energy difference between the bound and the free state, and which can readily be determined by ensemble measurements. E_B determines the ratio of bound complexes [RL] to the product of free reactants [R][L] at equilibrium and is related to the equilibrium dissociation constant K_D through $E_B = -RT \ln(K_D)$, where R is the gas constant. K_D itself is related to the empirical association (k_{on}) and dissociation (k_{off}) rate constants through $K_D = k_{off}/k_{on}$. In order to get an estimate for the interaction forces, f, from the binding energies E_B, the depth of the binding pocket may be used as a characteristic length scale l. Using typical values of $E_B = 20k_B T$ and $l = 0.5$ nm, an order-of-magnitude estimate of $f(= E_B/l) \approx 170$ pN is obtained for the binding strength of a single molecular pair. Classical mechanics describes bond strength as the gradient in energy along the direction of separation. Unbinding therefore occurs when the applied force exceeds the steepest gradient in energy. This purely mechanical description of molecular bonds, however, does not provide insights into the microscopic determinants of bond formation and rupture.

Non-covalent bonds have limited lifetimes and will therefore break even in the absence of external force on characteristic time scales needed for spontaneous dissociation ($\tau(0) = k_{off}^{-1}$). Pulled faster than $\tau(0)$, however, bonds will resist detachment. Notably, the unbinding force may approach and even exceed the adiabatic limit given by the steepest energy gradient of the interaction potential, if rupture occurs in less time than needed for diffusive relaxation ($10^{-10}–10^{-9}$ s for biomolecules in viscous aqueous medium) and friction effects become dominant [28.48]. Therefore, unbinding forces do not resemble unitary values and the dynamics of the experiment critically affects the measured bond strengths. At the time scale of AFM experiments (milliseconds to seconds), thermal impulses govern the unbinding process. In the thermal activation model, the lifetime of a molecular complex in solution is described by a Boltzmann ansatz, $\tau(0) = \tau_{osc} \exp(E_b/k_B T)$ [28.49], where τ_{osc} is the inverse of the natural oscillation frequency and E_b is the height of the energy barrier for dissociation. This gives a simple Arrhenius dependency of dissociation rate on barrier height.

A force acting on a complex deforms the interaction free-energy landscape and lowers barriers for dissociation (Fig. 28.6). As a result of the latter, bond lifetime is shortened. The lifetime $\tau(f)$ of a bond loaded with a constant force f is given by: $\tau(f) = \tau_{osc} \exp[(E_b - x_\beta f)/k_B T]$ [28.49], where x_β marks the thermally averaged projection of the energy barrier along the direction of the force. A detailed analysis of the relation between bond strength and lifetime was performed by *Evans* et al. [28.50], using Kramers' theory for overdamped kinetics. For a sharp barrier, the lifetime $\tau(f)$ of a bond subjected to a constant force f relates to its characteristic lifetime, $\tau(0)$, according to: $\tau(f) = \tau(0) \exp(-x_\beta f/k_B T)$ [28.4]. However, in most pulling experiments the applied force is not constant. Rather, it increases in a complex, non-linear manner, which depends on the pulling velocity, the spring constant of the cantilever, and the force–distance profile of the molecular complex. Nevertheless, contributions arising from thermal activation manifest themselves mostly near the point of detachment. Therefore, the change of force with time or the loading rate $r(= df/dt)$ can be derived from the product of

the pulling velocity and the effective spring constant at the end of the force curve, just before unbinding occurs.

The dependence of the rupture force on the loading rate (force spectrum), in the thermally activated regime was first derived by *Evans* and *Ritchie* [28.50] and described further by *Strunz* et al. [28.47]. Forced dissociation of receptor ligand complexes using AFM or BFP can often be regarded as an irreversible process because the molecules are kept moving away from each other after unbinding had occurred (rebinding can be safely neglected when measurements are made with soft springs). Rupture itself is a stochastic process and the likelihood of bond survival is expressed in the master equation as a time-dependent probability $N(t)$ to be in the bound state under a steady ramp of force, namely $dN(t)/dt = -k_{off}(rt)N(t)$ [28.47]. This results in a distribution of unbinding forces $P(F)$ parameterized by the loading rate [28.47, 50, 51]. The most probable force for unbinding f^*, given by the maximum of the distribution, relates to the loading rate through $f^* = f_\beta \ln(rk_{off}^{-1}/f_\beta)$, where the force scale f_β is set by the ratio of thermal energy to x_β [28.47, 50]. Thus, the unbinding force scales linearly with the logarithm of the loading rate. For a single barrier, this would give rise to a simple, linear force spectrum f^* versus $\log(r)$. In cases where the escape path is traversed by several barriers, the curve will follow a sequence of linear regimes, each marking a particular barrier [28.38, 50, 51]. Transition from one

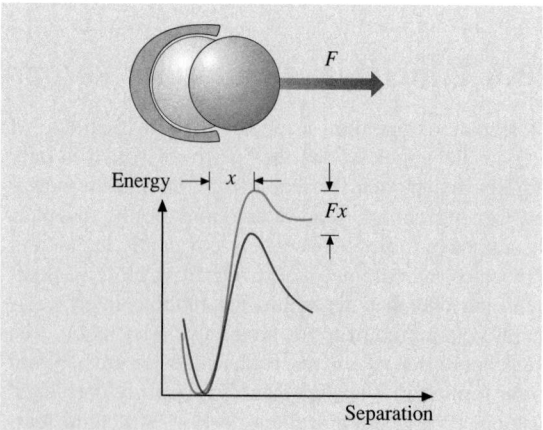

Fig. 28.6 Dissociation over a single sharp energy barrier. Under a constant force, the barrier is lowered by the applied force F. This gives rise to a characteristic length scale x_β that is interpreted as the distance of the energy barrier from the energy minimum along the projection of the force. (After [28.47].)

regime to the other is associated with an abrupt change of slope determined by the characteristic barrier length scale and signifying that a crossover between barriers has occurred.

Dynamic force spectroscopy (DFS) exploits the dependence of bond strength on the loading rate to obtain detailed insights into intra- and intermolecular interactions. By measuring bond strength over a broad range of loading rates, length scales and relative heights of energy barriers traversing the free-energy surface can be readily obtained. The lifetime of a bond at any given force is likewise contained in the complete force

distribution [28.4]. Finally, one may attempt to extract dissociation rate constants by extrapolation to zero force [28.52]. However, the application of force acts to select the dissociation path. Since the kinetics of reactions is pathway-dependent, such a selection implies that kinetic parameters extracted from force-probe experiments may differ from those obtained from assays conducted in the absence of external force. Under extremely fast complexation/decomplexation kinetics the forces can be independent of the loading rate, indicating that the experiments were carried out under thermodynamic equilibrium [28.53].

28.5 Recognition Force Spectroscopy: From Isolated Molecules to Biological Membranes

28.5.1 Forces, Energies, and Kinetic Rates

Conducted at fixed loading rates, pioneering measurements of interaction forces represent single points in continuous spectra of bond strengths [28.38]. Not unexpectedly, the first interaction studied was that between biotin and its extremely high-affinity receptors avidin [28.3] and streptavidin [28.2]. The unbinding forces measured for these interactions were 250–300 pN and 160 pN for streptavidin and avidin, respectively. During this initial phase it was also revealed that differ-

ent unbinding forces can be obtained for the same pulling velocity if the spring constant of the cantilever is varied [28.2], which is consistent with the aforementioned dependency of bond strength on the loading rate. The interaction force between several biotin analogues and avidin or streptavidin [28.54] and between biotin and a set of streptavidin mutants [28.55] was investigated and found generally to correlate with the equilibrium binding enthalpy and the enthalpic activation barrier. No correlation with the equilibrium free energy of binding or the activation free-energy barrier to dissociation

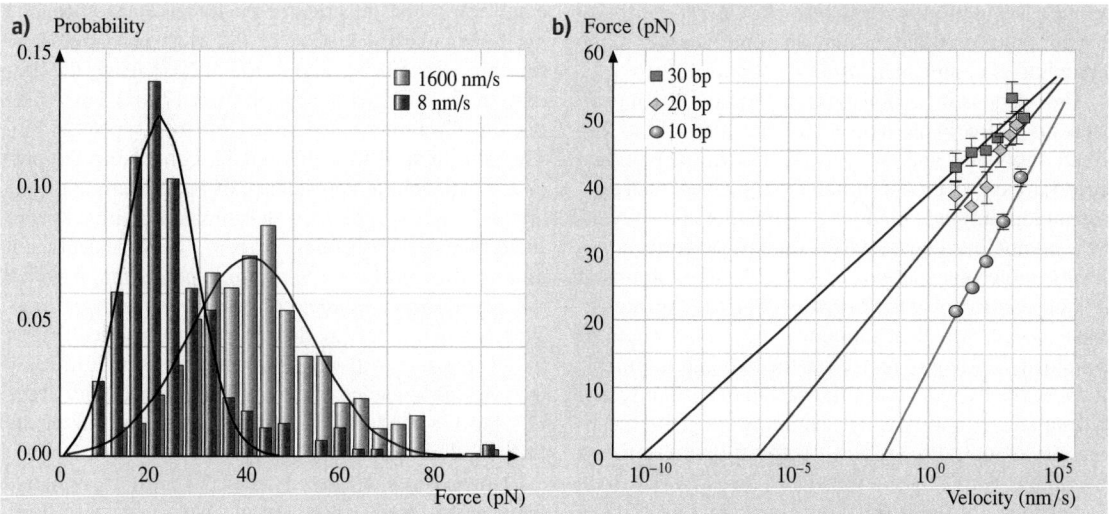

Fig. 28.7 Dependence of the unbinding force between DNA single-strand duplexes on the retract velocity. In addition to the expected logarithmic behavior on the loading rate, the unbinding force scales with the length of the strands, increasing from the 10- to 20- to 30-base-pair duplexes. (After [28.15].)

was observed, suggesting that internal energies rather than entropic contributions were probed by the force measurements [28.55].

In another pioneering study, *Lee* et al. [28.6] measured the forces between complementary 20-base DNA strands covalently attached to a spherical probe and surface. The interaction forces fell into three different distributions amounting to the rupture of duplexes consisting of 12, 16, and 20 base pairs. The average rupture force per base pair was $\approx 70\,\text{pN}$. When a long, single-stranded DNA was analyzed, both intra- and interchain forces were observed, the former probing the elastic properties of the molecule. Hydrogen bonds between nucleotides have been probed for all 16 combinations of the four DNA bases [28.7]. Directional hydrogen-bonding interactions were measured only when complementary bases were present on the tip and probe surfaces, indicating that AFM can be used to follow specific pairing of DNA strands.

Strunz et al. [28.15] measured the forces required to separate individual double-stranded DNA molecules of 10, 20, and 30 base pairs (Fig. 28.7). The parameters describing the energy landscape, i.e., the distance of the energy barrier from the minimum energy along the separation path and the logarithm of the thermal dissociation rate, were found to be proportional to the number of base pairs of the DNA duplex. Such scaling suggests that unbinding proceeds in a highly cooperative manner characterized by one length scale and one time scale. Studying the dependence of rupture forces on the temperature, it was proposed by *Schumakovitch* et al. [28.56] that entropic contributions play an important role in the unbinding of complementary DNA strands [28.56].

Prevalent as it is, molecular recognition has mostly been discussed in the context of the interactions between antibodies and antigens. To maximize motional freedom and to overcome problems associated with misorientation and steric hindrance, antibodies and antigens were immobilized onto AFM tip and probe surfaces via flexible molecular spacers [28.4, 10, 13, 14]. Optimizing antibody density over the AFM tip [28.4, 12], the interaction between individual antibody antigen pairs could be examined. Binding of antigen to the two Fab fragments of the antibody was shown to occur independently and with equal probability. Single antibody antigen recognition events were also recorded with tip-bound antigens interacting with intact antibodies [28.10, 13] or with single-chain wild-type (Fv) fragments [28.14]. The latter study also showed that an Fv mutant whose affinity to the antigen was attenuated by about 10 folds dissociated from the antigen under applied forces that were 20% lower than those required to unbind the wild-type (Fv) antibody.

Besides measurements of interaction forces, single-molecule force spectroscopy also allows estimation of association and dissociation rate constants with the concern stated above withstanding [28.4, 12, 22, 52, 57, 58], and measurement of structural parameters of the binding pocket [28.4, 12, 15, 57, 58]. Quantification of the association rate constant k_{on} requires determination of the interaction time needed for half-maximal probability of binding $(t_{1/2})$. This can be obtained from experiments where the encounter time between receptor and ligand is varied over a broad range [28.57]. Given that the concentration of ligand molecules on the tip available for interaction with the surface-bound receptors c_{eff} is known, the association rate constant can be derived from $k_{on} = t_{0.5}^{-1} c_{eff}^{-1}$. Determination of the effective ligand concentration requires knowledge of the effective volume V_{eff} explored by the tip-tethered ligand which, in turn, depends on the tether length. Therefore, only order-of-magnitude estimates of k_{on} can be gained from such measurements [28.57].

Additional information about the unbinding process is contained in the distributions of the unbinding forces. Concomitant with the shift of maxima to higher unbinding forces, increasing the loading rate also leads to an increase in the width σ of the distributions [28.22, 38], indicating that, at lower loading rates, the system adjusts closer to equilibrium. The lifetime $\tau(f)$ of a bond under an applied force was estimated by the time the cantilever spends in a regime the force window spanned by the standard deviation of the most probable force for unbinding [28.4]. In the case of Ni^{2+}-His_6, the lifetime of the complex decreased from 17 to 2.5 ms when the force was increased from 150 to 194 pN [28.22]. The data fit well to Bell's model, confirming the predicted exponential dependence of bond lifetime on the applied force, and yielded an estimated lifetime at zero force of about 15 seconds. A more direct measurement of τ is afforded by force-clamp experiments in which the applied force is kept constant by a feedback loop. This configuration was first adapted for use with AFM by *Oberhauser* et al. [28.59], who employed it to study the force dependence of the unfolding probability of the I27 and I28 modules of cardiac titin as well as of the full-length protein [28.59].

However, as discussed above, in most experiments the applied force is not constant but varies with time, and the measured bond strength depends on the loading rate [28.48, 50, 60]. In accordance with this, experimentally measured unbinding forces do not assume

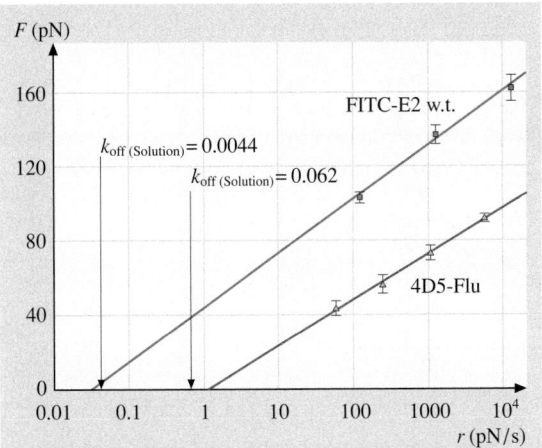

Fig. 28.8 The dependence of the unbinding force on the loading rate for two anti-fluorescein antibodies. For both FITC-E2 w.t. and 4D5-Flu a strictly single-exponential dependence was found in the range accessed, indicating that only a single energy barrier was probed. The same energy barrier dominates dissociation without forces applied because extrapolation to zero force matches kinetic off-rates determined in solution (indicated by the arrow). (After [28.58].)

unitary values but rather vary with both pulling velocity [28.52, 57] and cantilever spring constant [28.2]. The predicted logarithmic dependence of the unbinding force on the loading rate in the thermally activated regime was likewise confirmed by a large number of unbinding and unfolding experiments [28.15, 22, 38, 52, 57, 58, 61]. The slopes of the force–loading rate curves contain information about the length scale x_β of prominent energy barriers along the force-driven dissociation pathway, which may be related to the depth of the binding pocket of the interaction [28.57].

The force spectra may also be used to derive the dissociation rate constant k_{off} by extrapolation to zero force [28.52, 57, 58]. As mentioned above, values derived in this manner may differ from those obtained from bulk measurements because only a subset of dissociation pathways defined by the force is sampled. Nevertheless, a simple correlation between unbinding forces and thermal dissociation rates was obtained for a set consisting of nine different Fv fragments constructed from point mutations of three unrelated anti-fluorescein antibodies [28.58]. This correlation, which implies a close similarity between the force- and thermally driven pathways explored during dissociation, was probably due to the highly rigid nature of the interaction, which proceeds

in a lock-and-key fashion. The force spectra obtained for the different constructs exhibited a single linear regime, indicating that in all cases unbinding was governed by a single prominent energy barrier (Fig. 28.8). Interestingly, the position of the energy barrier along the forced-dissociation pathway was found to be proportional to the height of the barrier and, thus, most likely includes contributions arising from elastic stretching of the antibodies during the unbinding process.

28.5.2 Complex Bonds and Energy Landscapes

The energy landscapes that describe proteins are generally not smooth. Rather, they are traversed by multiple energy barriers of various heights that render them highly corrugated or rugged. All these barriers affect the kinetics and conformational dynamics of proteins and any one of them may govern interaction lifetime and strength on certain time scales. Dynamic force spectroscopy provides an excellent tool to detect energy barriers which are difficult or impossible to detect by conventional, near-equilibrium assays and to probe the free-energy surface of proteins and protein complexes. It also provides a natural means to study interactions which are normally subjected to varying mechanical loads [28.52, 57, 62–64].

A beautiful demonstration of the ability of dynamic force spectroscopy to reveal hidden barriers was provided by *Merkel* et al. [28.38] who used BFP to probe bond formation between biotin and streptavidin or avidin over a broad range of loading rates. In contrast to early studies, which reported fixed values of bond strength [28.54, 55], a continuous spectrum of unbinding forces ranging from 5 to 170 pN was obtained (Fig. 28.9). Concomitantly, interaction lifetime decreased from about 1 min to 0.001 s, revealing the reciprocal relation between bond strength and lifetime expected for thermally activated kinetics under a rising force. Most notably, depending on the loading rate, unbinding kinetics was dominated by different activation energy barrier positioned along the force-driven unbinding pathway. Barriers emerged sequentially, with the outermost barrier appearing first, each giving rise to a distinct linear regime in the force spectrum. Going from one linear regime to the next was associated with an abrupt change in slope, indicating that a crossover between an outer to (more) inner barrier had occurred. The position of two of the three barriers identified in the force spectra was consistent with the location of prominent transition states revealed by molecular dy-

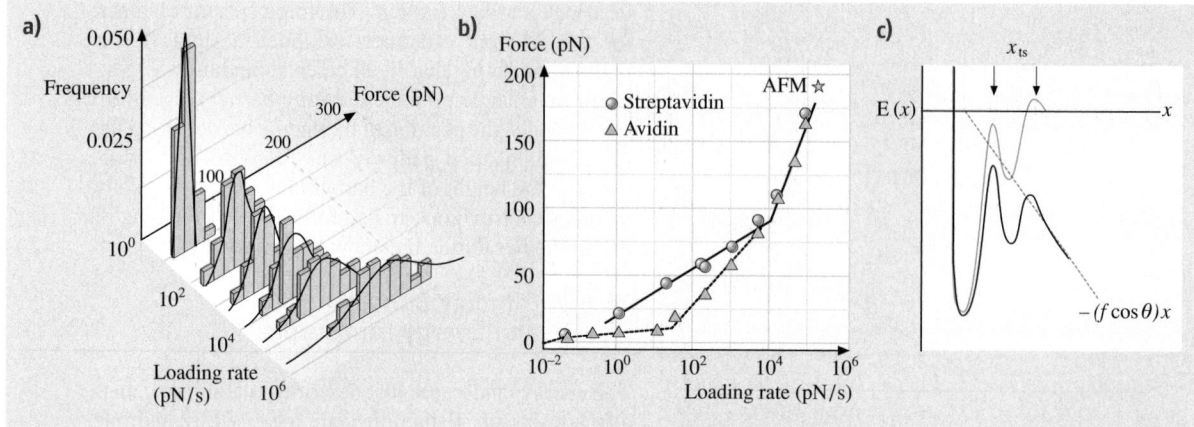

Fig. 28.9a–c Unbinding force distributions and energy landscape of a complex molecular bond. (**a**) Force histograms of single biotin streptavidin bonds recorded at different loading rates. The shift in peak location and the increase in width with increasing loading rate is clearly demonstrated. (**b**) Dynamic force spectra for biotin streptavidin (*open circles*) and biotin avidin (*closed triangles*). The slopes of the linear regimes mark distinct activation barriers along the direction of force. (**c**) Conceptual energy landscape traversed along a reaction coordinate under force. The external force f adds a mechanical potential that tilts the energy landscape and lowers the barriers. The inner barrier starts to dominate when the outer has fallen below it due to the applied force. (After [28.38].)

namics simulations [28.48, 60]. However, as mentioned earlier, unbinding is not necessarily confined to a single, well-defined path, and may take different routes even when directed by an external force. Molecular dynamics simulations of force-driven unbinding of an antibody antigen complex characterized by a highly flexible binding pocket revealed a large heterogeneity of enforced dissociation pathways [28.65].

The rolling of leukocytes on activated endothelium is a first step in the emergence of leukocytes out of the blood stream into sites of inflammation. This rolling, which occurs under hydrodynamic shear forces, is mediated by selectins, a family of extended, calcium-dependent lectin receptors present on the surface of endothelial cells. To fulfill their function, selectins and their ligands exhibit a unique combination of mechanical properties: they associate rapidly and avidly and can tether cells over very long distances by their long, extensible structure. In addition, complexes made between selectins and their ligands can withstand high tensile forces and dissociate in a controllable manner, which allows them to maintain rolling without being pulled out of the cell membrane.

Fritz et al. [28.52] used dynamic force spectroscopy to study the interaction between P-selectin and its leukocyte-expressed surface ligand P-selectin glycoprotein ligand-1 (PSGL-1). Modeling both intermolecular and intramolecular forces, as well as adhesion prob-

ability, they were able to obtain detailed information about rupture forces, elasticity, and the kinetics of the interaction. Complexes were able to withstand forces up to 165 pN and exhibited a chain-like elasticity with a molecular spring constant of 5.3 pN nm^{-1} and a persistence length of 0.35 nm. Rupture forces and the lifetime of the complexes exhibited the predicted logarithmic dependence on the loading rate.

An important characteristics of the interaction between P-selectin and PSGL-1, which is highly relevant to the biological function of the complex, was found by investigating the dependence of the adhesion probability between the two molecules on the velocity of the AFM probe. Counterintuitively and in contrast to experiments with avidin–biotin [28.54], antibody antigen [28.4], or cell adhesion proteoglycans [28.9], the adhesion probability between P-selectin and PSGL-1 was found to *increase* with increasing velocities [28.52]. This unexpected dependency explains the increase in the leukocyte tethering probability with increased shear flow observed in rolling experiments. Since the adhesion probability approached 1 at fast pulling velocities, it was concluded that binding occurs instantaneously as the tip reaches the surface and, thus, proceeds with a very fast on-rate. The complex also exhibited a fast forced off-rate. Such fast-on fast-off kinetics is probably important for the ability of leukocytes to rapidly bind and detach from the endothelial cell surface. Likewise, the long contour length of the complex together with its high elasticity reduces the mechanical loading on the complex upon binding and allows leukocyte rolling even at high shear rates.

Evans et al. [28.62] used BFP to study the interaction between PSGL-1 and another member of the selectin family, L-selectin. The force spectra, obtained over a range of loading rates extending from 10 to $100\,000\,\mathrm{pN\,s^{-1}}$, revealed two prominent energy barriers along the unbinding pathway: an outer barrier, probably constituted by an array of hydrogen bonds, that impeded dissociation under slow detachment and an inner, Ca^{2+}-dependent barrier that dominated dissociation under rapid detachment. The observed hierarchy of inner and outer activation barriers was proposed to be important for multibond recruitment during selectin-mediated function.

Using force clamp AFM [28.59], bond lifetimes were directly measured as a function of a constantly applied force. For this, lifetime force relations of P-selectin complexed to two forms of P-selectin glycoprotein ligand-1 (PSGL-1) and to G1, a blocking monoclonal antibody against P-selectin, respectively, were determined [28.64]. Both monomeric (sPSGL-1) and dimeric PSGL-1 exhibited a biphasic relationship between lifetime and force in their interaction to P-selectin (Fig. 28.10a,b). The bond lifetimes initially increased, indicating the presence of catch bonds. After reaching a maximum, the lifetimes decreased with force, indicating a catch bond. In contrast, the P-selectin/G1 bond lifetimes decreased exponentially with force (Fig. 28.10c), displaying typical slip-bond characteristics that are well described by the single-energy-barrier Bell model. The curves of lifetime against force for the two forms of PSGL1-1 had similar biphasic shapes (Fig. 28.10a,b), but the PSGL-1 curve (Fig. 28.10b) was shifted relative to the sPSGL-1 curve (Fig. 28.10a), approximately doubling the force and the lifetime. These data suggest that that sPSGL-1 forms monomeric bonds with P-selectin, whereas PSGL-1 forms dimeric bonds with P-selectin. In agreement with the studies described above, it was concluded that the use of force-induced switching from catch to slip bonds might be physiologically relevant for the tethering and rolling process of leukocytes on selectins [28.64].

Baumgartner et al. [28.57]) used AFM to probe specific trans-interaction forces and conformational changes of recombinant vascular/enothelial (VE)-cadherin strand dimers. Vascular endothelial-cadherins (VE-cadherin) are cell surface proteins that mediate the adhesion of cells in the vascular endothelium through Ca^{2+}-dependent homophilic interactions of their N-terminal extracellular domains. Acting as such they play an important role in the regulation of intercellular adhesion and communication in the inner surface of blood vessels. Unlike selectin-mediated adhesion, association between trans-interacting VE dimers was slow and independent of probe velocity, and complexes were ruptured at relatively low forces. These differences were attributed to the fact that, as opposed to selectins, cadherins mediate adhesion between resting cells. Mechanical stress on the junctions is thus less intense and high-affinity binding is not required to establish and maintain intercellular adhesion. Determination of the Ca^{2+}-dependency of recognition events between tip- and surface-bound VE-cadherins revealed a surprisingly high K_D (1.15 mM), which is very close to the free ex-

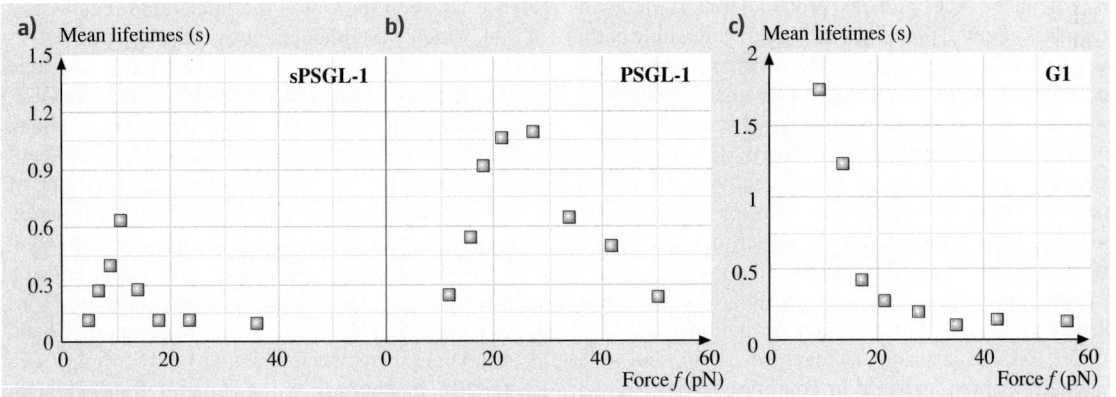

Fig. 28.10a–c Lifetimes of bonds of single-molecular complexes, depending on a constantly applied force. (a) sPSGL-1 / P-selectin: catch bond and slip bond. (b) PSGL-1 / P-selectin: catch bond and slip bond. (c) G1 / P-selectin: slip bond only. (After [28.64].)

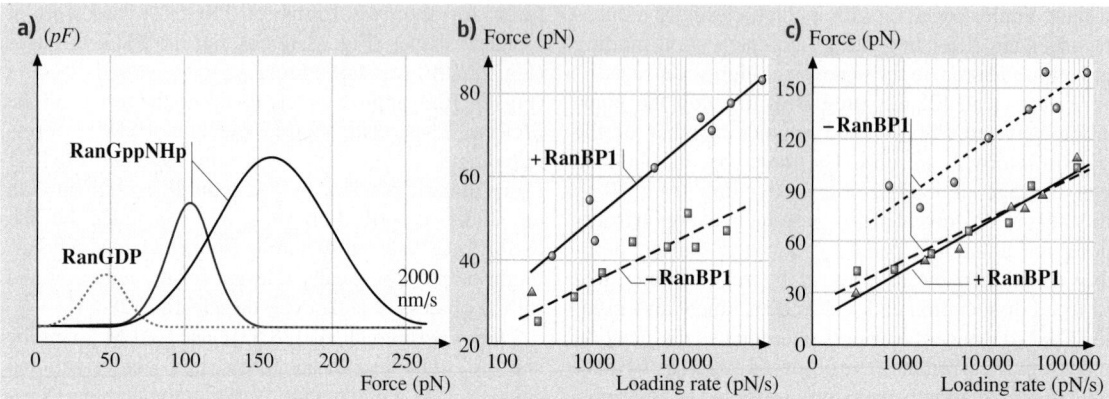

Fig. 28.11a,b Protein activation revealed by force spectroscopy. Ran and importinβ (impβ) were immobilized onto the AFM cantilevered tip and mica, respectively, and the interaction force was measured at different loading rates in the absence or presence of RanBP1, which was added as a mobile substrate to the solution in the AFM liquid cell. Unbinding force distributions obtained for impβ-Ran complexes at pulling velocity of 2000 nm/s. Association of impβ with Ran loaded with GDP (**a**) or with non-hydrolyzable GTP analogue (GppNHp) (**b**) gives rise to uni- or bimodal force distributions, respectively, reflecting the presence of one and two bound states. (b–c) Force spectra obtained for complexes of impβ with RanGDP or with RanGppNHp, in the absence (*dashed lines*) or presence (*solid lines*) of RanBP1. The results indicate that activation of impβ-RanGDP and imp-RanGTP complexes by RanBP1 proceeds through induced-fit and dynamic population-shift mechanisms, respectively (*see text for details*). (After [28.66, 67].)

tracellular Ca^{2+} concentration in the body. Binding also revealed a strong dependence on calcium concentrations, giving rise to an unusually high Hill coefficient of ≈ 5. This steep dependency suggests that local changes of free extracellular Ca^{2+} in the narrow intercellular space may facilitate rapid remodeling of intercellular adhesion and permeability.

Nevo et al. [28.66, 67] used single-molecule force spectroscopy to discriminate between alternative mechanisms of protein activation (Fig. 28.11). The activation of proteins by other proteins, protein domains, or small ligands is a central process in biology for signalling pathways, and enzyme activity. Moreover, activation and inactivation of genes all depend on the switching of proteins between alternative functional states. Two general mechanisms have been proposed. The induced-fit model assigns changes in protein activity to conformational changes triggered by effector binding. The population-shift model, on the other hand, ascribes these changes to a redistribution of *pre-existing* conformational isomers. According to this model, known also as the pre-equilibrium or conformational selection model, protein structure is regarded as an ensemble of conformations existing in equilibrium. The ligand binds to one of these conformations, i.e., the one to which it is most complementary, thus shifting the equilibrium in favor of this conformation. Discrimination

between the two models of activation requires that the distribution of conformational isomers in the ensemble is known. Such information, however, is very hard to obtain from conventional bulk methods because of ensemble averaging.

Using AFM, Nevo and coworkers measured the unbinding forces of two related protein complexes in the absence or presence of a common effector. The complexes consisted of the nuclear transport receptor importin β (imp β) and the small GTPase Ran. The difference between them was the nucleotide bound state of Ran, which was either guanosine diphosphate (GDP) or guanosine triphosphate (GTP). The effector molecule was the Ran-binding protein RanBP1. Loaded with GDP, Ran associated weakly with impβ to form a single bound state characterized by unimodal distributions of small unbinding forces (Fig. 28.11a, dotted line). Addition of RanBP1 resulted in a marked shift of the distribution to higher unbinding forces (Fig. 28.11b, dotted to solid line). These results were interpreted to be consistent with an induced-fit mechanism where binding of RanBP1 induces a conformational change in the complex, which, in turn, strengthens the interaction between impβ and Ran(GDP). In contrast, association of RanGTP with impβ was found to lead to alternative bound states of relatively low and high adhesion strength represented by partially overlapping force distributions (Fig. 28.11a, solid line). When RanBP1 was added to the solution, the higher-strength population, which predominated the ensemble in the absence of the effector (Fig. 28.11c, dotted lines), was diminished, and the lower-strength conformation became correspondingly more populated (Fig. 28.11c, solid line). The means of the distributions, however, remain unchanged, indicating that the strength of the interaction in the two states of the complex has not been altered by the effector. These data fit a dynamic population-shift mechanism in which RanBP1 binds selectively to the lower-strength conformation of RanGTP-impβ, changing the properties and function of the complex by shifting the equilibrium between its two states.

The complex between impβ and RanGTP was also used in studies aimed to measure the energy landscape roughness of proteins. The roughness of the energy landscapes that describe proteins has numerous effects on their folding and binding as well as on their behavior at equilibrium, since undulations in the free-energy surface can attenuate diffusion dramatically. Thus, to understand how proteins fold, bind and function, one needs to know not only the energy of their initial and final states, but also the roughness of the energy surface that connects them. However, for a long time, knowledge of protein energy-landscape roughness came solely from theory and simulations of small model proteins.

Adopting *Zwanzig*'s theory of diffusion in rough potentials [28.68], *Hyeon* and *Thirumalai* [28.69] proposed that the energy-landscape roughness of proteins can be measured from single-molecule mechanical unfolding experiments conducted at different temperatures. In particular, their simulations showed that, at constant loading rate, the most probable force for unfolding increases because of roughness that acts to attenuate diffusion. Because this effect is temperature dependent, an overall energy scale of roughness, ε, can be derived from plots of force versus loading rate acquired at two arbitrary temperatures. Extending this theory to the case of unbinding, and performing single-molecule force spectroscopy measurements, *Nevo* et al. [28.70] extracted the overall energy scale of roughness ε for RanGTP-impβ. The results yielded $\varepsilon > 5k_B T$, indicating a bumpy energy surface, which is consistent with the unusually high structural flexibility of impβ and its ability to interact with different, structurally distinct ligands in a highly specific manner. This mechanistic principle may also be applicable to other proteins whose function demands highly specific and regulated interactions with multiple ligands.

28.5.3 Live Cells and Membranes

Thus far, there have been only a few attempts to apply recognition force spectroscopy to cells. In one of the early studies, *Lehenkari* et al. [28.71] measured the unbinding forces between integrin receptors present on the surface of intact cells and several RGD-containing (Arg-Gly-Asp) ligands. The unbinding forces measured were found to be cell- and amino acid sequence-specific and sensitive to the pH and divalent cation composition of the cellular culture medium. In contrast to short linear RGD hexapeptides, larger peptides and proteins containing the RGD sequence showed different binding affinities, demonstrating that the context of the RGD motif within a protein has considerable influence upon its interaction with the receptor. In another study, *Chen* et al. [28.72] used AFM to measure the adhesive strength between concanavalin A (Con A) coupled to an AFM tip and Con A receptors on the surface of NIH3T3 fibroblasts. Crosslinking of receptors with either glutaraldehyde or 3,3'-dithiobis(sulfosuccinimidylproprionate) (DTSSP) led to an increase in adhesion that was attributed to enhanced cooperativity among adhesion complexes. The results support the notion that receptor crosslinking can increase adhesion strength by creating a shift towards cooperative binding of receptors. *Pfister* et al. [28.73] investigated the surface localization of HSP60 on stressed and unstressed human umbilical venous endothelial cells (HUVECs). By detecting specific single-molecule binding events between the monoclonal antibody AbII-13 tethered to AFM tips and HSP60 molecules on cells, clear evidence was found for the occurrence of HSP60 on the surface of stressed HUVECs, but not on unstressed HUVECs.

The sidedness and accessibility of protein epitopes of the Na^{2+} D-glucose co-transporter 1 (SGLT1) was probed in intact brush border membranes by a tip-bound antibody directed against an amino acid sequence close to the glucose binding site [28.35]. Binding of glucose and transmembrane transport altered both the binding probability and the most probable unbinding force, suggesting changes in the orientation and conformation of the transporter. These studies were extended to live SGLT1-transfected CHO cells *Puntheeranurak* et al. [28.74] Using AFM tips carrying the substrate 1-β-thio D-glucose, direct evidence could be obtained that, in the presence of sodium, a sugar binding site appears on the SGLT1 surface. It was shown that this binding site accepts the sugar residue of the glucoside phlorizin, free d-Glucose and D-galactose, but not free L-glucose.

The data indicate the importance of stereo-selectivity for sugar binding and transport.

Zhang et al. [28.63] studied the interaction between leukocyte function-associated antigen-1 (LFA-1) and its cognate ligand, intercellular adhesion molecule-1 (ICAM-1), which play a crucial role in leukocyte adhesion. The experimental system consisted of an LFA-1-expressing T cell hybridoma attached to the end of the AFM cantilever and an apposing surface expressing ICAM-1. The force spectra revealed fast and slow loading regimes, amounting to a sharp, inner energy barrier and a shallow, outer barrier, respectively. Addition of Mg^{2+} led to an increase of the unbinding force in the slow loading regime whereas ethylenediaminetetraacidic acid (EDTA) suppressed the inner barrier. These results suggest that the dissociation of LFA-1/ICAM-1 is governed by the outer activation barrier of the complex, while the ability of the complex to resist a pulling force is determined by the divalent cation-dependent inner barrier.

28.6 Recognition Imaging

Besides measuring interaction strengths, locating binding sites over biological surfaces such as cells or membranes is of great interest. To achieve this goal, force detection must be combined with high-resolution imaging.

Ludwig et al. [28.75] used chemical force microscopy to image a streptavidin pattern with a biotinylated tip. An approach–retract cycle was performed at each point of a raster and topography, adhesion, and sample elasticity were extracted from the local force ramps. This strategy was also used to map binding sites on cells [28.76, 77] and to differentiate between red blood cells of different blood groups (A and 0) using AFM tips functionalized with a group-A-specific lectin [28.78].

Identification and localization of single antigenic sites was achieved by recording force signals during the scanning of an AFM tip coated with antibodies along a single line across a surface immobilized with a low density of antigens [28.4, 12]. Using this method, antigens could be localized over the surface with positional accuracy of 1.5 nm. A similar configuration used by *Willemsen* et al. [28.79] enabled the simultaneous acquisition of height and adhesion-force images with near-molecular resolution.

The aforementioned strategies of force mapping either lack high lateral resolution [28.75] and/or are much slower [28.4, 12, 79] than conventional topographic imaging since the frequency of the force-sensing retract–approach cycles is limited by hydrodynamic damping. In addition, the ligand needs to be detached from the receptor in each retract approach cycle, necessitating large working amplitudes (50 nm). Therefore, the surface-bound receptor is inaccessible to the tip-immobilized ligand on the tip during most of the time of the experiment. This problem however, should be overcome with the use of small cantilevers [28.46] which should increase the speed for force mapping because the hydrodynamic forces are significantly reduced and the resonance frequency is higher than that of commercially available cantilevers. Short cantilevers were recently applied to follow the association and dissociation of individual chaperonin proteins, GroES to GroEL, in real time using dynamic force microscopy topography imaging [28.80].

An imaging method for mapping antigenic sites on surfaces was developed [28.20] by combining molecular recognition force spectroscopy [28.4] with dynamic force microscopy (DFM) [28.25, 81]. In DFM, the AFM tip is oscillated across a surface and the amplitude reduction arising from tip–surface interactions is held constant by a feedback loop that lifts or lowers the tip according to the detected amplitude signal. Since the tip contacts the surface only intermittently, this technique provides very gentle tip–surface interactions and the specific interaction of the antibody on the tip with the antigen on the surface can be used to localize antigenic sites for recording recognition images. The AFM tip is magnetically coated and oscillated by an alternating magnetic field at very small amplitudes while being scanned along the surface. Since the oscillation frequency is more than a hundred times faster than typical frequencies in conventional force mapping, the data acquisition rate is much higher. This method was recently extended to yield fast, simultaneous acquisition of two independent maps, i. e. a topography image and a lateral map of recognition sites, recorded with nm resolution at experimental times equivalent to normal AFM imaging [28.82–84].

Topography and recognition images were simultaneously obtained (TREC imaging) using a special electronic circuit (PicoTrec, Molecular Imaging, Tempe, AZ) (Fig. 28.12a). Maxima (U_{up}) and minima (U_{down}) of each sinusoidal cantilever deflection period were depicted in a peak detector, filtered, and amplified.

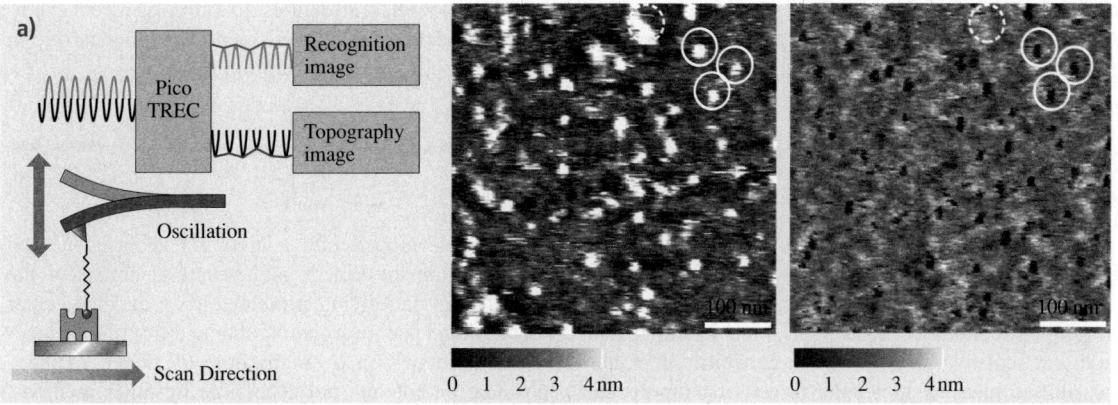

Fig. 28.12a,b Simultaneous topography and recognition (TREC) imaging (**a**) Principle. The cantilever oscillation is split into lower and upper parts, resulting in simultaneously acquired topography and recognition images. (**b**) Avidin was electrostatically adsorbed to mica and imaged with a biotin-tethered tip. A good correlation between topography (*left image, bright spots*) and recognition (*right image, dark spots*) was found (*solid circles*). Topographical spots without recognition denote structures lacking specific interaction (*dashed circle*). Scan size was 500 nm. (After [28.84].)

Direct-current (DC) offset signals were used to compensate for the thermal drifts of the cantilever. U_{up} and U_{down} were fed into the AFM controller, with U_{down} driving the feedback loop to record the height (i. e. topography) image and U_{up} providing the data for constructing the recognition image (Fig. 28.12a). Since we used cantilevers with low Q-factor (≈ 1 in liquid) driven at frequencies below resonance the two types of information were independent. In this way, topography and recognition image were recorded simultaneously and independently.

The circuit was applied to mica containing singly distributed avidin molecules using a biotinylated AFM tip [28.84]. The sample was imaged with an antibody-containing tip, yielding the topography (Fig. 28.12b, left image) and the recognition image (Fig. 28.12b, right image) at the same time. The tip oscillation amplitude (5 nm) was chosen to be slightly smaller than the extended crosslinker length (8 nm), so that both the antibody remained bound while passing a binding site and the reduction of the upwards deflection was significant compared to the thermal noise. Since the spring constant of the polymeric crosslinker increases non-linearly with the tip–surface distance (Fig. 28.5), the binding force is only sensed close to full extension of the crosslinker (given at the maxima of the oscillation period). Therefore, the recognition signals were well separated from the topographic signals arising from the surface, both in space ($\Delta z \approx 5$ nm) and time (half oscillation period ≈ 0.1 ms).

The bright dots, with a height of 2–3 nm and diameter of 15–20 nm, that are visible in the topography image (Fig. 28.12b, left image) represent single avidin molecules stably adsorbed onto the flat mica surface. The recognition image shows black dots at positions of avidin molecules (Fig. 28.12b, right image) because the oscillation maxima are lowered due to the physical avidin–biotin connection established during recognition. Spatial correlation between the lateral positions of the avidin molecules obtained in the topography image and the recognition signals of the recognition image is indicated by solid circles in the images (Fig. 28.12). Recognition between the antibody on the tip and the avidin on the surface took place for almost all avidin molecules, most likely because avidin contains four biotin binding sites, two on either side. Thus, one would assume that there will always be binding epitopes oriented away from the mica surface and accessible to the biotinylated tip, resulting in a high binding efficiency. Structures observed in the topography image and not detected in the recognition image were very rare (dotted circle in Fig. 28.12b).

It is important to note that topography and recognition images were recorded at speeds typical for standard AFM imaging and were therefore considerably faster than conventional force mapping. With this methodology, topography and recognition images can be obtained at the same time and distinct receptor sites in the recognition image can be assigned to structures from the topography image. It is applicable with any ligand and,

Part C | 28.6

therefore, it should prove possible to recognize many types of proteins or protein layers and carry out epitope mapping on the nm scale on membranes and cells, and complex biological structures. In a striking recent example, histone proteins H3 were identified and localized in a complex chromatin preparation [28.83].

28.7 Concluding Remarks

Atomic force microscopy has evolved into an imaging method that yields the greatest structural details on live, biological samples in their native, aqueous environment at ambient conditions. Due to its high lateral resolution and sensitive force detection capability, it is now possible to measure molecular forces of biomolecules on the single-molecule level. Well beyond the proof-of-principle stage of the pioneering experiments, AFM has now developed into a high-end analysis method for exploring the kinetic and structural details of the interactions underlying protein folding and molecular recognition. The information obtained from force spectroscopy, being on a single-molecule level, includes physical parameters not accessible by other methods. In particular, it opens up new perspectives to explore the dynamics of biological processes and interactions.

References

28.1   G. Binnig, C. F. Quate, Ch. Gerber: Atomic force microscope, Phys. Rev. Lett. **56**, 930–933 (1986)

28.2   G. U. Lee, D. A. Kidwell, R. J. Colton: Sensing discrete streptavidin–biotin interactions with atomic force microscopy, Langmuir **10**, 354–357 (1994)

28.3   E. L. Florin, V. T. Moy, H. E. Gaub: Adhesion forces between individual ligand receptor pairs, Science **264**, 415–417 (1994)

28.4   P. Hinterdorfer, W. Baumgartner, H. J. Gruber, K. Schilcher, H. Schindler: Detection and localization of individual antibody–antigen recognition events by atomic force microscopy, Proc. Natl. Acad. Sci. U.S.A. **93**, 3477–3481 (1996)

28.5   M. Grandbois, M. Dettmann, M. Benoit, H. E. Gaub: How strong is a covalent bond, Science **283**, 1727–1730 (1999)

28.6   G. U. Lee, A. C. Chrisey, J. C. Colton: Direct measurement of the forces between complementary strands of DNA, Science **266**, 771–773 (1994)

28.7   T. Boland, B. D. Ratner: Direct measurement of hydrogen bonding in DNA nucleotide bases by atomic force microscopy, Proc. Natl. Acad. Sci. USA **92**, 5297–5301 (1995)

28.8   P. Wagner, M. Hegner, P. Kernen, F. Zaugg, G. Semenza: Covalent immobilization of native biomolecules onto Au(111) via N-hydroxysuccinimide ester functionalized self assembled monolayers for mscanning probe microscopy, Biophys. J **70**, 2052–2066 (1996)

28.9   U. Dammer, O. Popescu, P. Wagner, D. Anselmetti, H.-J. Güntherodt, G. M. Misevic: Binding strength between cell adhesion proteoglycans measured by atomic force microscopy, Science **267**, 1173–1175 (1995)

28.10   U. Dammer, M. Hegner, D. Anselmetti, P. Wagner, M. Dreier, H. J. Güntherodt, W. Huber: Specific antigen/antibody interactions measured by force microscopy, Biophys. J. **70**, 2437–2441 (1996)

28.11   Y. Harada, M. Kuroda, A. Ishida: Specific and quantized antibody–antigen interaction by atomic force microscopy, Langmuir **16**, 708–715 (2000)

28.12   P. Hinterdorfer, K. Schilcher, W. Baumgartner, H. J. Gruber, H. Schindler: A mechanistic study of the dissociation of individual antibody–antigen pairs by atomic force microscopy, Nanobiology **4**, 39–50 (1998)

28.13   S. Allen, X. Chen, J. Davies, M. C. Davies, A. C. Dawkes, J. C. Edwards, C. J. Roberts, J. Sefton, S. J. B. Tendler, P. M. Williams: Spatial mapping of specific molecular recognition sites by atomic force microscopy, Biochem. **36**, 7457–7463 (1997)

28.14   R. Ros, F. Schwesinger, D. Anselmetti, M. Kubon, R. Schäfer, A. Plückthun, L. Tiefenauer: Antigen binding forces of individually addressed single-chain Fv antibody molecules, Proc. Natl. Acad. Sci. USA **95**, 7402–7405 (1998)

28.15   T. Strunz, K. Oroszlan, R. Schäfer, H.-G. Güntherodt: Dynamic force spectroscopy of single DNA molecules, Proc. Natl. Acad. Sci. USA **96**, 11277–11282 (1999)

28.16   S. S. Wong, E. Joselevich, A. T. Woolley, C. L. Cheung, C. M. Lieber: Covalently functionalyzed nanotubes as nanometre-sized probes in chemistry and biology, Nature **394**, 52–55 (1998)

28.17   P. Hinterdorfer, F. Kienberger, A. Raab, H. J. Gruber, W. Baumgartner, G. Kada, C. Riener, S. Wielert-Badt, C. Borken, H. Schindler: Poly(ethylene glycol): An ideal spacer for molecular recognition force microscopy/spectroscopy, Single Mol. **1**, 99–103 (2000)

28.18   Th. Haselgrübler, A. Amerstorfer, H. Schindler, H. J. Gruber: Synthesis and applications of a new poly(ethylene glycol) derivative for the crosslink-ing of amines with thiols, Bioconjugate Chem. **6**, 242–248 (1995)

28.19   C. K. Riener, G. Kada, C. Borken, F. Kien-berger, P. Hinterdorfer, H. Schindler, G. J. Schütz, T. Schmidt, C. D. Hahn, H. J. Gruber: Bioconjuga-tion for biospecific detection of single molecules in atomic force microscopy (AFM) and in single dye tracing (SDT), Recent Res. Devel. Bioconj. Chem. **1**, 133–149 (2002)

28.20   A. Raab, W. Han, D. Badt, S. J. Smith-Gill, S. M. Lindsay, H. Schindler, P. Hinterdorfer: An-tibody recognition imaging by force microscopy, Nature Biotech. **17**, 902–905 (1999)

28.21   M. Conti, G. Falini, B. Samori: How strong is the coordination bond between a histidine tag and Ni-Nitriloacetate? An experiment of mechanochemistry on single molecules, Angew. Chem. **112**, 221–224 (2000)

28.22   F. Kienberger, G. Kada, H. J. Gruber, V. Ph. Pas-tushenko, C. Riener, M. Trieb, H.-G. Knaus, H. Schindler, P. Hinterdorfer: Recognition force spectroscopy studies of the NTA–His6 bond, Single Mol. **1**, 59–65 (2000)

28.23   L. Schmitt, M. Ludwig, H. E. Gaub, R. Tampé: A metal-chelating microscopy tip as a new tool-box for single-molecule experiments by atomic force microscopy, Biophys. J. **78**, 3275–3285 (2000)

28.24   C. Yuan, A. Chen, P. Kolb, V. T. Moy: Energy land-scape of avidin–biotin complexes measured by atomic force microscopy, Biochemistry **39**, 10219–10223 (2000)

28.25   W. Han, S. M. Lindsay, M. Dlakic, R. E. Harrington: Kinked DNA, Nature **386**, 563 (1997)

28.26   G. Kada, L. Blaney, L. H. Jeyakumar, F. Kienberger, V. Ph. Pastushenko, S. Fleischer, H. Schindler, F. A. Lai, P. Hinterdorfer: Recognition force microscopy/spectroscopy of ion channels: Ap-plications to the skeletal muscle Ca^{2+} re-lease channel (RYR1), Ultramicroscopy **86**, 129–137 (2001)

28.27   D. J. Müller, W. Baumeister, A. Engel: Controlled unzipping of a bacterial surface layer atomic force microscopy, Proc. Natl. Acad. Sci. USA **96**, 13170–13174 (1999)

28.28   F. Oesterhelt, D. Oesterhelt, M. Pfeiffer, A. Engel, H. E. Gaub, D. J. Müller: Unfolding pathways of in-dividual bacteriorhodopsins, Science **288**, 143–146 (2000)

28.29   E. Kiss, C.-G. Gölander: Chemical derivatization of muscovite mica surfaces, Coll. Surf. **49**, 335–342 (1990)

28.30   S. Karrasch, M. Dolder, F. Schabert, J. Ramsden, A. Engel: Covalent binding of biological samples to solid supports for scanning probe microscopy in buffer solution, Biophys. J. **65**, 2437–2446 (1993)

28.31   N. H. Thomson, B. L. Smith, N. Almquist, L. Schmitt, M. Kashlev, E. T. Kool, P. K. Hansma: Oriented, ac-tive *escherichia coli* RNA polymerase: An atomic force microscopy study, Biophys. J. **76**, 1024–1033 (1999)

28.32   G. Kada, C. K. Riener, P. Hinterdorfer, F. Kienberger, C. M. Stroh, H. J. Gruber: Dithio-phospholipids for biospecific immobilization of proteins on gold sur-faces, Single Mol. **3**, 119–125 (2002)

28.33   C. LeGrimellec, E. Lesniewska, M. C. Giocondi, E. Finot, V. Vie, J. P. Goudonnet: Imaging of the surface of living cells by low-force contact-mode atomic force microscopy, Biophys. J. **75**(2), 695–703 (1998)

28.34   K. Schilcher, P. Hinterdorfer, H. J. Gruber, H. Schindler: A non-invasive method for the tight anchoring of cells for scanning force microscopy, Cell. Biol. Int. **21**, 769–778 (1997)

28.35   S. Wielert-Badt, P. Hinterdorfer, H. J. Gruber, J.-T. Lin, D. Badt, H. Schindler, R. K.-H. Kinne: Single molecule recognition of protein binding epitopes in brush border membranes by force microscopy, Biophys. J. **82**, 2767–2774 (2002)

28.36   P. Bongrand, C. Capo, J.-L. Mege, A.-M. Benoliel: *Use of Hydrodynamic Flows to Study Cell Adhesion*, ed. by P. Bongrand (CRC, Boca Raton, Florida 1988) pp. 125–156

28.37   J. N. Israelachvili: *Intermolecular and Surface Forces*, 2nd edn. (Academic Press, London & New York 1991) p. 2

28.38   R. Merkel, P. Nassoy, A. Leung, K. Ritchie, E. Evans: Energy landscapes of receptor-ligand bonds ex-plored by dynamic force spectroscopy, Nature **397**, 50–53 (1999)

28.39   A. Askin: Optical trapping and manipulation of neutral particles using lasers, Proc. Natl. Acad. Sci. USA **94**, 4853–4860 (1997)

28.40   K. Svoboda, C. F. Schmidt, B. J. Schnapp, S. M. Block: Direct observation of kinesin stepping by opti-cal trapping interferometry, Nature **365**, 721–727 (1993)

28.41   S. Smith, Y. Cui, C. Bustamante: Overstretching B–DNA: The elastic response of individual double-stranded and single-stranded DNA molecules, Science **271**, 795–799 (1996)

28.42   M. S. Z. Kellermayer, S. B. Smith, H. L. Granzier, C. Bustamante: Folding–unfolding transitions in single titin molecules characterized with laser tweezwers, Sience **276**, 1112–1216 (1997)

28.43   T. R. Strick, J. F. Allemend, D. Bensimon, A. Ben-simon, V. Croquette: The elasticity of a single supercoiled DNA molecule, Biophys. J. **271**, 1835–1837 (1996)

28.44   F. Kienberger, V. Ph. Pastushenko, G. Kada, H. J. Gruber, C. Riener, H. Schindler, P. Hinter-dorfer: Static and dynamical properties of single poly(ethylene glycol) molecules investigated by force spectroscopy, Single Mol. **1**, 123–128 (2000)

28.45   S. Liang, D. Medich, D. M. Czajkowsky, S. Sheng, J.-Y. Yuan, Z. Shao: Thermal noise reduction of mechanical oscillators by actively controlled external dissipative forces, Ultramicroscopy **84**, 119–125 (2000)

28.46   M. B. Viani, T. E. Schäffer, A. Chand, M. Rief, H. E. Gaub, P. K. Hansma: Small cantilevers for force spectroscopy of single molecules, J. Appl. Phys. **86**, 2258–2262 (1999)

28.47   T. Strunz, K. Oroszlan, I. Schumakovitch, H.-G. Güntherodt, M. Hegner: Model energy landscapes and the force-induced dissociation of ligand-receptor bonds, Biophys. J. **79**, 1206–1212 (2000)

28.48   H. Grubmüller, B. Heymann, P. Tavan: Ligand binding: Molecular mechanics calculation of the streptavidin-biotin rupture force, Science **271**, 997–999 (1996)

28.49   G. I. Bell: Models for the specific adhesion of cells to cells, Science **200**, 618–627 (1978)

28.50   E. Evans, K. Ritchie: Dynamic strength of molecular adhesion bonds, Biophys. J. **72**, 1541–1555 (1997)

28.51   E. Evans, K. Ritchie: Strength of a weak bond-connecting flexible polymer chains, Biophys. J. **76**, 2439–2447 (1999)

28.52   J. Fritz, A. G. Katopidis, F. Kolbinger, D. Anselmetti: Force-mediated kinetics of single P-selectin/ligand complexes observed by atomic force microscopy, Proc. Natl. Acad. Sci. USA **95**, 12283–12288 (1998)

28.53   T. Auletta, M. R. de Jong, A. Mulder, F. C. J. M. van Veggel, J. Huskens, D. N. Reinhoudt, S. Zou, S. Zapotocny, H. Schönherr, G. J. Vancso, L. Kuipers: β-cyclodextrin host-guest complexes probed under thermodynamic equilibrium: Thermodynamics and force spectroscopy, J. Am. Chem. Soci. **126**, 1577–1584 (2004)

28.54   V. T. Moy, E.-L. Florin, H. E. Gaub: Adhesive forces between ligand and receptor measured by AFM, Science **266**, 257–259 (1994)

28.55   A. Chilkoti, T. Boland, B. Ratner, P. S. Stayton: The relationship between ligand-binding thermodynamics and protein-ligand interaction forces measured by atomic force microscopy, Biophys. J. **69**, 2125–2130 (1995)

28.56   I. Schumakovitch, W. Grange, T. Strunz, P. Bertoncini, H.-J. Güntherodt, M. Hegner: Temperature dependence of unbinding forces between complementary DNA strands, Biophys. J. **82**, 517–521 (2002)

28.57   W. Baumgartner, P. Hinterdorfer, W. Ness, A. Raab, D. Vestweber, H. Schindler, D. Drenckhahn: Cadherin interaction probed by atomic force microscopy, Proc. Natl. Acad. Sci. USA **8**, 4005–4010 (2000)

28.58   F. Schwesinger, R. Ros, T. Strunz, D. Anselmetti, H.-J. Güntherodt, A. Honegger, L. Jermutus, L. Tiefenauer, A. Plückthun: Unbinding forces of single antibody–antigen complexes correlate with their thermal dissociation rates, Proc. Natl. Acad. Sci. USA **29**, 9972–9977 (2000)

28.59   A. F. Oberhauser, P. K. Hansma, M. Carrion-Vazquez, J. M. Fernandez: Stepwise unfolding of titin under force-clamp atomic force microscopy, Proc. Natl. Acad. Sci. USA **16**, 468–472 (2000)

28.60   S. Izraelev, S. Stepaniants, M. Balsera, Y. Oono, K. Schulten: Molecular dynamics study of unbinding of the avidin-biotin complex, Biophys. J. **72**, 1568–1581 (1997)

28.61   M. Rief, F. Oesterhelt, B. Heyman, H. E. Gaub: Single molecule force spectroscopy on polysaccharides by atomic force microscopy, Science **275**, 1295–1297 (1997)

28.62   E. Evans, E. Leung, D. Hammer, S. Simon: Chemically distinct transition states govern rapid dissociation of single L-selectin bonds under force, Proc. Natl. Acad. Sci. USA **98**, 3784–3789 (2001)

28.63   X. Zhang, E. Woijcikiewicz, V. T. Moy: Force spectroscopy of the leukocyte function-associated antigen-1/intercellular adhesion molecule-1 interaction, Biophys. J. **83**, 2270–2279 (2002)

28.64   B. T. Marshall, M. Long, J. W. Piper, T. Yago, R. P. McEver, Z. Zhu: Direct observation of catch bonds involving cell adhesion molecules, Nature **423**, 190–193 (2003)

28.65   B. Heymann, H. Grubmüller: Molecular dynamics force probe simulations of antibody/antigen unbinding: Entropic control and non additivity of unbinding forces, Biophys. J. **81**, 1295–1313 (2001)

28.66   R. Nevo, C. Stroh, F. Kienberger, D. Kaftan, V. Brumfeld, M. Elbaum, Z. Reich, P. Hinterdorfer: A molecular switch between two bound states in the RanGTP-importinβ1 interaction, Nat. Struct. & Mol. Biol. **10**, 553–557 (2003)

28.67   R. Nevo, V. Brumfeld, M. Elbaum, P. Hinterdorfer, Z. Reich: Direct discrimination between models of protein activation by single-molecule force measurements, Biophys. J. **87**, 2630–2634 (2004)

28.68   R. Zwanzig: Diffusion in a rough potential, Proc. Natl. Acad. Sci. USA **85:**, 2029–2030 (1988)

28.69   C. B. Hyeon, D. Thirumalai: Can energy landscape roughness of proteins and RNA be measured by using mechanical unfolding experiments?, Proc. Natl. Acad. Sci. USA **100**, 10249–10253 (2003)

28.70   R. Nevo, V. Brumfeld, P. Hinterdorfer, Z. Reich: Direct measurement of protein energy landscape roughness, EMBO Rep. **6**, 482–486 (2005)

28.71   P. P. Lehenkari, M. A. Horton: Single integrin molecule adhesion forces in intact cells measured by atomic force microscopy, Biochem. Biophys. Res. Com. **259**, 645–650 (1999)

28.72   A. Chen, V. T. Moy: Cross-linking of cell surface receptors enhances cooperativity of molecular adhesion, Biophys. J. **78**, 2814–2820 (2000)

28.73   G. Pfister, C. M. Stroh, H. Perschinka, M. Kind, M. Knoflach, P. Hinterdorfer, G. Wick: Detection of HSP60 on the membrane surface of stressed human endothelial cells by atomic force and confocal microscopy, J. Cell. Sci. **118**, 1587–1594 (2005)

28.74 T. Puntheeranurak, L. Wildling, H.J. Gruber, R.K.H. Kinne, P. Hinterdorfer: Ligands on the string: Single molecule studies on the interaction of antibodies and substrates with the surface of the Na$^+$-glucose cotransporter SGLT1 *in vivo*, J. Cell Sci., in press

28.75 M. Ludwig, W. Dettmann, H.E. Gaub: Atomic force microscopy imaging contrast based on molecuar recognition, Biophys. J. **72**, 445–448 (1997)

28.76 P.P. Lehenkari, G.T. Charras, G.T. Nykänen, M.A. Horton: Adapting force microscopy for cell biology, Ultramicroscopy **82**, 289–295 (2000)

28.77 N. Almqvist, R. Bhatia, G. Primbs, N. Desai, S. Banerjee, R. Lal: Elasticity and adhesion force mapping reveals real-time clustering of growth factor receptors and associated changes in local cellular rheological properties, Biophys. J. **86**, 1753–1762 (2004)

28.78 M. Grandbois, M. Beyer, M. Rief, H. Clausen-Schaumann, H.E. Gaub: Affinity imaging of red blood cells using an atomic force microscope, J. Histochem. Cytochem. **48**, 719–724 (2000)

28.79 O.H. Willemsen, M.M.E. Snel, K.O. van der Werf, B.G. de Grooth, J. Greve, P. Hinterdorfer, H.J. Gruber, H. Schindler, Y. van Kyook, C.G. Figdor: Simultaneous height and adhe-

sion imaging of antibody antigen interactions by atomic force microscopy, Biophys. J. **57**, 2220–2228 (1998)

28.80 B.V. Viani, L.I. Pietrasanta, J.B. Thompson, A. Chand, I.C. Gebeshuber, J.H. Kindt, M. Richter, H.G. Hansma, P.K. Hansma: Probing protein-protein interactions in real time, Nature Struct. Biol. **7**, 644–647 (2000)

28.81 W. Han, S.M. Lindsay, T. Jing: A magnetically driven oscillating probe microscope for operation in liquid, Appl. Phys. Lett. **69**, 1–3 (1996)

28.82 C.M. Stroh, A. Ebner, M. Geretschläger, G. Freudenthaler, F. Kienberger, A.S.M. Kamruzzahan, S.J. Smith-Gill, H.J. Gruber, P. Hinterdorfer: Simultaneous topography and recognition imaging using force microscopy, Biophys. J. **87**, 1981–1990 (2004)

28.83 C. Stroh, H. Wang, R. Bash, B. Ashcroft, J. Nelson, H.J. Gruber, D. Lohr, S.M. Lindsay, P. Hinterdorfer: Single-molecule recognition imaging microscope, Proc. Natl. Acad. Sci. **101**, 12503–12507 (2004)

28.84 A. Ebner, F. Kienberger, G. Kada, C.M. Stroh, M. Geretschläger, A.S.M. Kamruzzahan, L. Wildling, W.T. Johnson, B. Ashcroft, J. Nelson, S.M. Lindsay, H.J. Gruber, P. Hinterdorfer: Localization of single avidin biotin interactions using simultaneous topography and molecular recognition imaging, Chem. Phys. Chem. **6**, 897–900 (2005)

Part D Nanotribo

Part D Nanotribology and Nanomechanics

29. Nanotribology, Nanomechanics and Materials Characterization

Nanotribology and nanomechanics studies are needed to develop fundamental understanding of interfacial phenomena on a small scale and to study interfacial phenomena in micro/nanoelectromechanical systems (MEMS/NEMS), magnetic storage devices, and other applications. Friction and wear of lightly loaded micro/nanocomponents are highly dependent on the surface interactions (few atomic layers). These structures are generally coated with molecularly thin films. Nanotribology and nanomechanics studies are also valuable in fundamental understanding of interfacial phenomena in macrostructures, and provide a bridge between science and engineering. An atomic force microscope (AFM) tip is used to simulate a single asperity contact with a solid or lubricated surface. AFMs are used to study the various tribological phenomena, which include surface roughness, adhesion, friction, scratching, wear, detection of material transfer, and boundary lubrication. In situ surface characterization of local deformation of materials and thin coatings can be carried out using a tensile stage inside an AFM. Mechanical properties such as hardness, Young's modulus of elasticity and creep/relaxation behavior can be determined on the micro- to picoscales using a depth-sensing indentation system in an AFM. Localized surface elasticity and viscoelastic mapping can be obtained of near-surface regions with nanoscale lateral resolution. Finally, an AFM can be used for nanofabrication/nanomachining.

The mechanisms and dynamics of the interactions of two contacting solids during relative motion, ranging from atomic- to microscale, need to be understood in order to develop fundamental understanding of adhesion, friction, wear, indentation, and lubrication processes. At most solid–solid interfaces of technological relevance, contact occurs at many asperities. Consequently the importance of investigating single asperity contacts in studies of the fundamental micro/nanomechanical and micro/nanotribological properties of surfaces and interfaces has long been recognized. The recent emergence and proliferation of proximal probes, in particular scanning probe microscopies (the scanning tunneling microscope and the atomic force microscope), the surface force apparatus, and of computational techniques for simulating tip–surface interactions and interfacial properties, has allowed systematic investigations of interfacial problems with high resolution as well as means for modifying and manipulating nanoscale structures. These advances have led to the appearance of the new field of nanotribology, which pertains to experimental and theoretical investigations of interfacial processes on scales ranging from the atomic- and molecular- to the microscale, occurring during adhesion, friction, scratching, wear, indentation, and thin-film lubrication at sliding surfaces [29.1–14]. Proximal probes have also been used for mechanical characterization, in situ characterization of local deformation, and other nanomechanics studies.

Nanotribological and nanomechanics studies are needed to develop fundamental understanding of interfacial phenomena on a small scale and to study interfacial phenomena in micro/nanostructures used in magnetic storage devices, micro/nanoelectromechanical systems (MEMS/NEMS), and other applications [29.3–12, 15–17]. Friction and wear of lightly loaded micro/nanocomponents are highly dependent on the surface interactions (a few atomic layers). These structures are generally coated with molecularly thin films. Nanotribological and nanomechanics studies are also valuable in fundamental understanding of interfacial phenomena in macrostructures, and provide a bridge between science and engineering.

The surface force apparatus (SFA), the scanning tunneling microscope (STM), atomic force and friction force microscopes (AFM and FFM) are widely used in nanotribological and nanomechanics studies. Typical operating parameters are compared in Table 29.1. The SFA was developed in 1968 and is commonly employed to study both static and dynamic properties of molecularly thin films sandwiched between two molecularly smooth surfaces. The STM, developed in 1981, allows imaging of electrically conducting surfaces with atomic resolution, and has been used for imaging of clean surfaces as well as of lubricant molecules. The introduction of the AFM in 1985 provided a method for measuring ultra-small forces between a probe tip and an engineering (electrically conducting or insulating) surface, and has been used for morphological and surface roughness measurements of surfaces on the nanoscale, as well as for adhe-

Table 29.1 Comparison of typical operating parameters in SFA, STM and AFM/FFM used for micro/nanotribological studies

Operating parameter	SFA	STM[a]	AFM/FFM
Radius of mating surface/tip	≈ 10 mm	$5-100$ nm	$5-100$ nm
Radius of contact area	$10-40\,\mu$m	N/A	$0.05-0.5$ nm
Normal load	$10-100$ mN	N/A	< 0.1 nN $- 500$ nN
Sliding velocity	$0.001-100\,\mu$m/s	$0.02-200\,\mu$m/s (scan size ≈ 1 nm $\times 1$ nm to $125\,\mu$m $\times 125\,\mu$m; scan rate $< 1-122$ Hz)	$0.02-200\,\mu$m/s (scan size ≈ 1 nm $\times 1$ nm to $125\,\mu$m $\times 125\,\mu$m; scan rate $< 1-122$ Hz)
Sample limitations	Typically atomically smooth, optically transparent mica; opaque ceramic, smooth surfaces can also be used	Electrically conducting samples	None

[a] Can only be used for atomic-scale imaging

sion measurements. Subsequent modifications of the AFM led to the development of the FFM, designed for atomic-scale and microscale studies of friction. This instrument measures forces in the scanning direction. The AFM is also being used for various investigations including scratching, wear, indentation, detection of transfer of material, boundary lubrication, and fabrication and machining [29.13, 18, 19]. Meanwhile, significant progress in understanding the fundamental nature of bonding and interactions in materials, combined with advances in computer-based modeling and simulation methods, has allowed theoretical studies of complex interfacial phenomena with high resolution in space and time. Such simulations provide insights into atomic-scale energetics, structure, dynamics, thermodynamics, transport and rheological aspects of tribological processes.

The nature of interactions between two surfaces brought close together, and those between two surfaces in contact as they are separated, have been studied experimentally with the surface force apparatus. This has led to a basic understanding of the normal forces between surfaces, and the way in which these are modified by the presence of a thin liquid or a polymer film. The frictional properties of such systems have been studied by

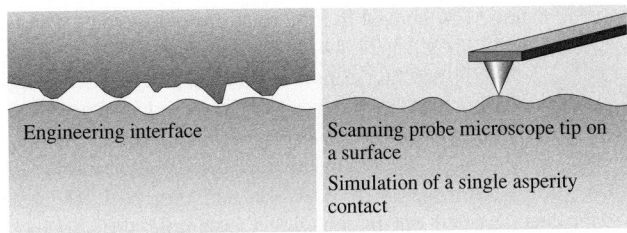

Fig. 29.1 Schematics of an engineering interface and scanning probe microscope tip in contact with an engineering interface

moving the surfaces laterally, and such experiments have provided insights into the molecular-scale operation of lubricants such as thin liquid or polymer films. Complementary to these studies are those in which the AFM tip is used to simulate a single asperity contact with a solid or lubricated surface, Fig. 29.1. These experiments have demonstrated that the relationship between friction and surface roughness is not always simple or obvious. AFM studies have also revealed much about the nanoscale nature of intimate contact during wear, indentation, and lubrication.

In this chapter, we present a review of significant aspects of nanotribological, nanomechanical and materials characterization studies conducted using AFM/FFM.

29.1 Description of AFM/FFM and Various Measurement Techniques

An AFM was developed by *Binnig* and his colleagues in 1985. It is capable of investigating surfaces of scientific and engineering interest on an atomic scale [29.20, 21]. The AFM relies on a scanning technique to produce very-high-resolution, three-dimensional images of sample surfaces. It measures ultra-small forces (less than 1 nN) present between the AFM tip surface mounted on a flexible cantilever beam, and a sample surface. These small forces are obtained by measuring the motion of a very flexible cantilever beam having an ultra-small mass, by a variety of measurement techniques including optical deflection, optical interference, capacitance, and tunneling current. The deflection can be measured to within 0.02 nm, so for a typical cantilever spring constant of 10 N/m, a force as low as 0.2 nN can be detected. To put these numbers in perspective, individual atoms and human hair are typically a fraction of a nanometer and about 75 μm in diameter, respectively, and a drop of water and an eyelash have a mass of about 10 μN and 100 nN, respectively. In the operation of high-resolution AFM, the sample is generally scanned rather than the

tip because any cantilever movement would add vibrations. AFMs are available for measurement of large samples, where the tip is scanned and the sample is stationary. To obtain atomic resolution with the AFM, the spring constant of the cantilever should be weaker than the equivalent spring between atoms. A cantilever beam with a spring constant of about 1 N/m or lower is desirable. For high lateral resolution, tips should be as sharp as possible. Tips with a radius in the range 10–100 nm are commonly available. Interfacial forces, adhesion, and surface roughness, including atomic-scale imaging, are routinely measured using the AFM.

A modification to AFM, providing a sensor to measure the lateral force, led to the development of the friction force microscope (FFM) or the lateral force microscope (LFM), designed for atomic-scale and microscale studies of friction [29.3–5, 7, 8, 13, 22–37] and lubrication [29.38–43]. This instrument measures lateral or friction forces (in the plane of the sample surface and in the scanning direction). By using a standard or a sharp diamond tip mounted on a stiff cantilever beam, AFM is used in investigations of scratching and wear [29.6, 9, 13, 27, 44–47], indentation [29.9, 13, 17, 27, 48–51], and fabrication/machining [29.4, 13, 27]. An oscillating cantilever is used for localized surface elasticity and viscoelastic mapping, referred to as dynamic AFM [29.35, 52–60]. In situ surface characterization of local deformation of materials and thin coatings has been carried out by imaging the sample surfaces us-

ing an AFM, during tensile deformation using a tensile stage [29.61–63].

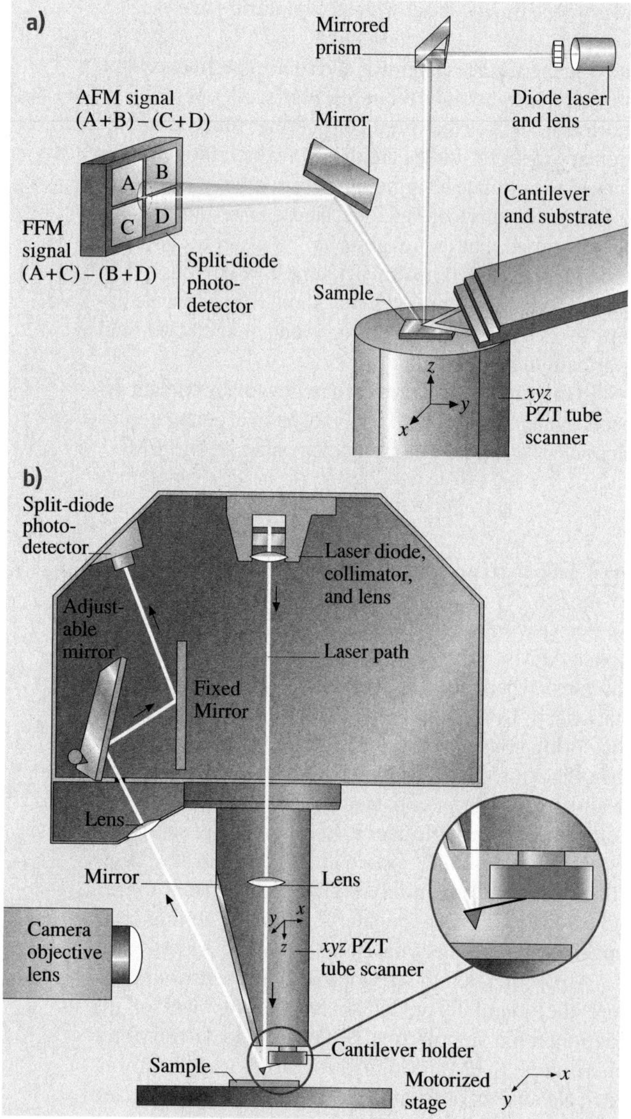

29.1.1 Surface Roughness and Friction Force Measurements

Surface height imaging down to atomic resolution of electrically conducting surfaces is carried out using an STM. An AFM is also used for surface height imaging and roughness characterization down to the nanoscale. Commercial AFM/FFM are routinely used for simultaneous measurements of surface roughness and friction force [29.4, 11]. These instruments are available for measurement of small and large samples. In a small-sample AFM, shown in Fig. 29.2a, the sample, generally no larger than 10 mm × 10 mm, is mounted on a piezoelectric crystal in the form of a cylindrical tube [referred to as a lead zirconate titanate (PZT) tube scanner] which consists of separate electrodes to scan the sample precisely in the x–y plane in a raster pattern and to move the sample in the vertical (z) direction. A sharp tip at the free end of a flexible cantilever is brought in contact with the sample. Normal and frictional forces being applied at the tip–sample interface are measured using a laser beam deflection technique. A laser beam from a diode laser is directed by a prism onto the back of a cantilever near its free end, tilted downward at about 10° with respect to the horizontal plane. The reflected beam from the vertex of the cantilever is directed through a mirror onto a quad photodetector (split photodetector with four quadrants). The differential signal from the top and bottom photodiodes provides the AFM signal, which is a sensitive measure of the cantilever vertical deflection. Topographic features of the sample cause the tip to deflect in the vertical direction as the sample is scanned under the tip. This tip deflection will change the direction of the reflected laser beam, changing the intensity difference between the top and bottom sets of photodetectors (AFM signal). In the AFM operating mode called the height mode, for topographic imaging or for any other operation in which the applied normal force is to be kept constant, a feedback circuit is used to modulate the voltage applied to the PZT scanner to adjust the height of the PZT, so that the cantilever vertical deflection (given by the intensity difference between the top and bottom detector) will remain constant during scanning. The PZT height variation is thus a direct measure of the surface roughness of the sample.

In a large-sample AFM, both force sensors using optical deflection method and scanning unit are mounted on the microscope head, Fig. 29.2b. Because of vibra-

Fig. 29.2a,b Schematics (a) of a commercial small-sample atomic force microscope/friction force microscope (AFM/FFM), and (b) of a large-sample AFM/FFM

tions added by cantilever movement, lateral resolution of this design is somewhat poorer than the design in Fig. 29.2a in which the sample is scanned instead of cantilever beam. The advantage of the large-sample AFM is that large samples can be measured readily.

Most AFMs can be used for surface roughness measurements in the so-called tapping mode (intermittent contact mode), also referred to as dynamic (atomic) force microscopy. In the tapping mode, during scanning over the surface, the cantilever/tip assembly, with a normal stiffness of 20–100 N/m (tapping mode etched Si probe or DI TESP) is sinusoidally vibrated at its resonance frequency (350–400 kHz) by a piezo mounted above it, and the oscillating tip slightly taps the surface. The piezo is adjusted using feedback control in the Z direction to maintain a constant (20–100 nm) oscillating amplitude (set point) and constant average normal force (Fig. 29.3) [29.4, 11]. The feedback signal to the Z direction sample piezo (to keep the set point constant) is a measure of surface roughness. The cantilever/tip assembly is vibrated at some amplitude, here referred to as the free amplitude, before the tip engages the sample. The tip engages the sample at some set point, which may be thought of as the amplitude of the cantilever as influenced by contact with the sample. The set point is defined as a ratio of the vibration amplitude after engagement to the vibration amplitude in free air before engagement. A lower set point gives a reduced amplitude and closer mean tip-to-sample distance. The amplitude should be kept large enough so that the tip does not get stuck to the sample because of adhesive attractions. Also the oscillating amplitude applies a lower average (normal) load compared to the contact mode and reduces sample damage. The tapping mode is used in topography measurements to minimize the effects of friction and other lateral forces and to measure the topography of soft surfaces.

For the measurement of the friction force at the tip surface during sliding, left-hand and right-hand sets of quadrants of the photodetector are used. In the so-called friction mode, the sample is scanned back and forth in a direction orthogonal to the long axis of the cantilever beam. A friction force between the sample and the tip will produce a twisting of the cantilever. As a result, the laser beam will be reflected out of the plane defined by the incident beam and the beam reflected vertically from an untwisted cantilever. This produces an intensity difference of the laser beam received in the left-hand and right-hand sets of quadrants of the photodetector. The intensity difference between the two sets of detectors (FFM signal) is directly related to the degree of twisting

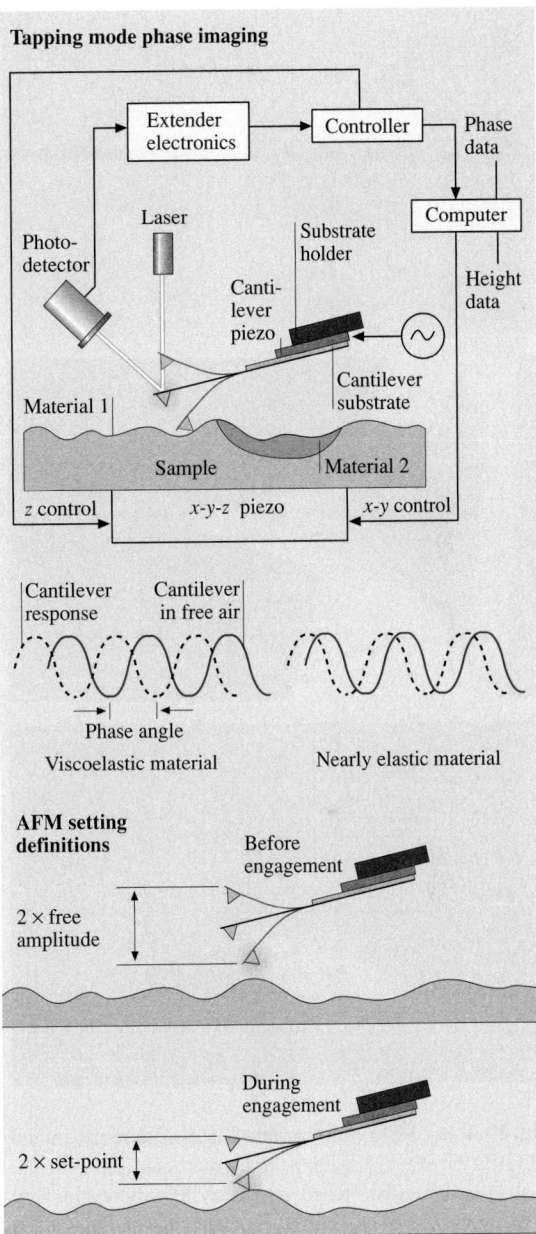

Fig. 29.3 Schematic of tapping mode used to obtain height and phase data and definitions of free amplitude and set point. During scanning, the cantilever is vibrated at its resonance frequency and the sample x–y–z piezo is adjusted by feedback control in the z-direction to maintain a constant set point. The computer records height (which is a measure of surface roughness) and phase angle (which is a function of the viscoelastic properties of the sample) data

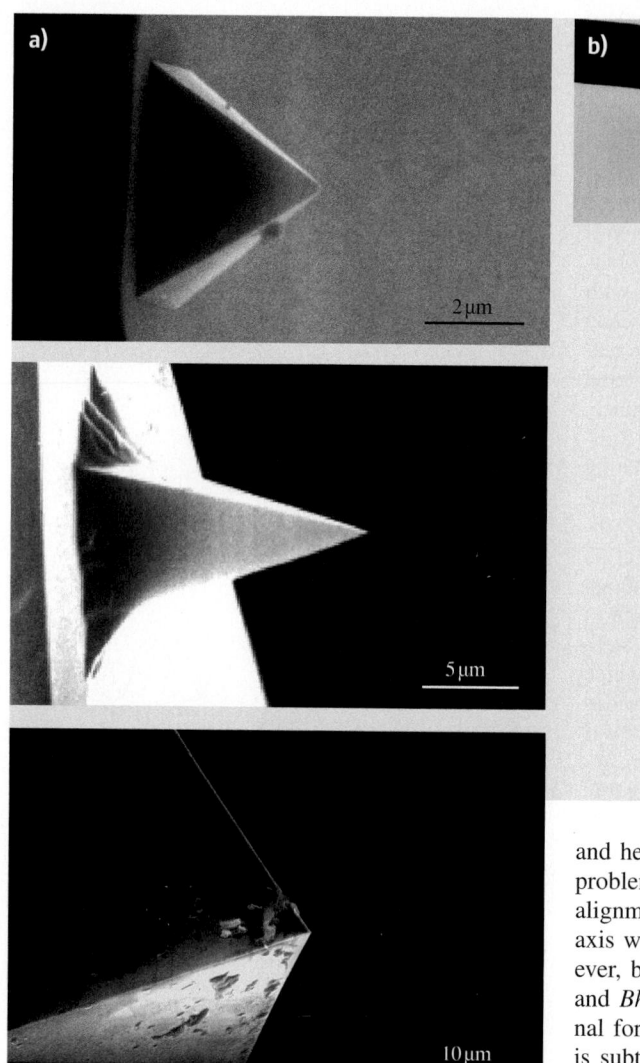

Fig. 29.4 (a) SEM micrographs of a square-pyramidal PECVD Si_3N_4 tip with a triangular cantilever beam, a square-pyramidal etched single-crystal silicon tip with a rectangular silicon cantilever beam, and a three-sided pyramidal natural diamond tip with a square stainless-steel cantilever beam, **(b)** SEM micrograph of a multiwalled carbon nanotube (MWNT) physically attached to the single-crystal silicon, square-pyramidal tip, and **(c)** optical micrographs of commercial Si_3N_4 tip and two modified tips showing SiO_2 spheres mounted over the sharp tip, at the end of the triangular Si_3N_4 cantilever beams (radii of the tips are given in the figure)

and hence to the magnitude of the friction force. One problem associated with this method is that any misalignment between the laser beam and the photodetector axis would introduce error in the measurement. However, by following the procedures developed by *Ruan* and *Bhushan* [29.24], in which the average FFM signal for the sample scanned in two opposite directions is subtracted from the friction profiles of each of the two scans, the misalignment effect is eliminated. This method provides three-dimensional maps of friction force. By following the friction force calibration procedures developed by *Ruan* and *Bhushan* [29.24], voltages corresponding to friction forces can be converted to force units. The coefficient of friction is obtained from the slope of friction force data measured as a function of normal loads typically ranging from 10 to 150 nN. This approach eliminates any contributions due to the adhesive forces [29.27]. For calculation of the coefficient of friction based on a single-point measurement, friction force should be divided by the sum of applied normal load and intrinsic adhesive force. Furthermore it should be pointed out that, for a single asperity contact, the

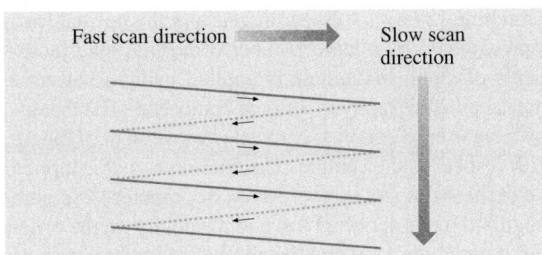

Fig. 29.5 Schematic of triangular pattern trajectory of the tip as the sample (or the tip) is scanned in two dimensions. During scanning, data are recorded only during scans along the *solid scan lines*

coefficient of friction is not independent of load (see discussion later).

Surface roughness measurements in the contact mode are typically made using a sharp, microfabricated square-pyramidal Si_3N_4 tip with a radius of 30–50 nm on a triangular cantilever beam (Fig. 29.4a) with normal stiffness on the order of 0.06–0.58 N/m with a normal natural frequency of 13–40 kHz (silicium nitride probe or DI NP) at a normal load of about 10 nN, and friction measurements are carried out in the load range of 1–100 nN. Surface roughness measurements in the tapping mode utilize a stiff cantilever with high resonance frequency; typically a square-pyramidal etched single-crystal silicon tip, with a tip radius of 5–10 nm, integrated with a stiff rectangular silicon cantilever beam (Fig. 29.4a) with a normal stiffness on the order of 17–60 N/m and a normal resonance frequency of 250–400 kHz (DI TESP), is used. Multiwalled carbon nanotube tips having small diameter (a few nm) and a length of about 1 μm (high aspect ratio), attached to the single-crystal silicon square-pyramidal tips are used for high-resolution imaging of surfaces and of deep trenches in the tapping mode (noncontact mode) (Fig. 29.4b) [29.64]. To study the effect of the radius of a single asperity (tip) on adhesion and friction, microspheres of silica with radii ranging from about 4 to 15 μm are attached at the end of cantilever beams. Optical micrographs of two of the microspheres at the ends of triangular cantilever beams are shown in Fig. 29.4c.

The tip is scanned in such a way that its trajectory on the sample forms a triangular pattern, Fig. 29.5. Scanning speeds in the fast and slow scan directions depend on the scan area and scan frequency. Scan sizes ranging from less than 1 nm × 1 nm to 125 μm × 125 μm and scan rates from less than 0.5 to 122 Hz can typically be used. Higher scan rates are used for smaller scan lengths. For example, scan rates in the

fast and slow scan directions for an area of 10 μm × 10 μm scanned at 0.5 Hz are 10 μm/s and 20 nm/s, respectively.

29.1.2 Adhesion Measurements

Adhesive force measurements are performed in the so-called force calibration mode. In this mode, force–distance curves are obtained; for an example see Fig. 29.6. The horizontal axis gives the distance the piezo (and hence the sample) travels and the vertical axis gives the tip deflection. As the piezo extends, it approaches the tip, which is at this point in free air and hence shows no deflection. This is indicated by the flat portion of the curve. As the tip approaches the sample within a few nanometers (point A), an attractive force exists between the atoms of the tip surface and the atoms of the sample surface. The tip is pulled towards the sample and contact occurs at point B on the graph. From this point on, the tip is in contact with the surface and, as the piezo further extends, the tip is further deflected. This is represented by the sloped portion of the curve. As the piezo retracts, the tip goes beyond the zero-deflection (flat) line because of attractive forces (van der Waals forces and long-range meniscus forces), into the adhesive regime. At point C on the graph, the tip snaps free of the adhesive forces, and is again in free air. The horizontal distance between points B and C along the retrace line gives the distance moved by the tip in the adhesive regime. This distance multiplied by the stiffness of the cantilever gives the adhesive force. Incidentally, the horizontal shift between the loading and unloading curves results from the hysteresis in the PZT tube [29.4, 11].

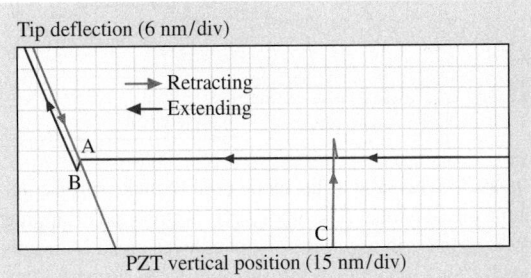

Fig. 29.6 Typical force–distance curve for a contact between Si_3N_4 tip and single-crystal silicon surface in measurements made in the ambient environment. Contact between the tip and silicon occurs at point B; the tip breaks free of adhesive forces at point C as the sample moves away from the tip

29.1.3 Scratching, Wear and Fabrication/Machining

For microscale scratching, microscale wear, nanofabrication/nanomachining and nanoindentation hardness measurements, an extremely hard tip is required. A three-sided pyramidal single-crystal natural-diamond tip with an apex angle of 80° and a radius of about 100 nm mounted on a stainless-steel cantilever beam with normal stiffness of about 25 N/m is used at relatively higher loads (1–150 μN), Fig. 29.4a. For scratching and wear studies, the sample is generally scanned in a direction orthogonal to the long axis of the cantilever beam (typically at a rate of 0.5 Hz) so that friction can be measured during scratching and wear. The tip is mounted on the cantilever such that one of its edges is orthogonal to the long axis of the beam; therefore, wear during scanning along the beam axis is higher (about 2 to 3 times) than that during scanning orthogonal to the beam axis. For wear studies, an area on the order of 2 μm × 2 μm is scanned at various normal loads (ranging from 1 to 100 μN) for a selected number of cycles [29.4, 11, 45].

Scratching can also be performed at ramped loads and the coefficient of friction can be measured during

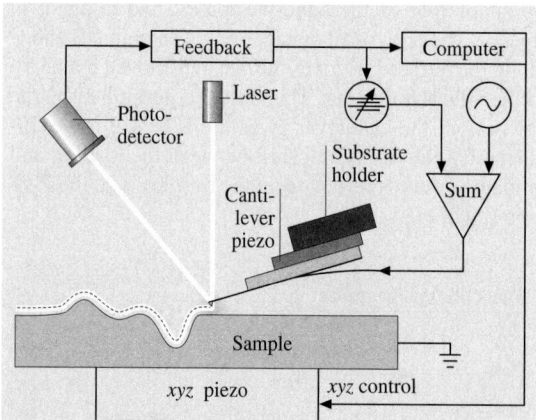

Fig. 29.7 Schematic of lift mode used to make surface potential measurement. The topography is collected in tapping mode in the primary scan. The cantilever piezo is deactivated. Using topography information of the primary scan, the cantilever is scanned across the surface at a constant height above the sample. An oscillating voltage at the resonant frequency is applied to the tip and a feedback loop adjusts the DC bias of the tip to maintain the cantilever amplitude at zero. The output of the feedback loop is recorded by the computer and becomes the surface potential map

scratching [29.47]. A linear increase in the normal load approximated by a large number of normal load increments of small magnitude is applied using a software interface (lithography module in Nanoscope III) that allows the user to generate controlled movement of the tip with respect to the sample. The friction signal is tapped out of the AFM and is recorded on a computer. A scratch length on the order of 25 μm and a velocity on the order of 0.5 μm/s are used and the number of loading steps is usually taken to be 50.

Nanofabrication/nanomachining is conducted by scratching the sample surface with a diamond tip at specified locations and scratching angles. The normal load used for scratching (writing) is on the order of 1–100 μN with a writing speed on the order of 0.1–200 μm/s [29.4, 6, 11–13, 27, 65].

29.1.4 Surface Potential Measurements

To detect wear precursors and to study the early stages of localized wear, the multimode AFM can be used to measure the potential difference between the tip and the sample by applying a DC bias potential and an oscillating (AC) potential to a conducting tip over a grounded substrate in a Kelvin probe microscopy or so-called *nano-Kelvin probe* technique [29.66–68].

Mapping of the surface potential is made in the so-called *lift mode* (Fig. 29.7). These measurements are made simultaneously with the topography scan in the tapping mode, using an electrically conducting (nickel-coated single-crystal silicon) tip. After each line of the topography scan is completed, the feedback loop controlling the vertical piezo is turned off, and the tip is lifted from the surface and traced over the same topography at a constant distance of 100 nm. During the lift mode, a DC bias potential and an oscillating potential (3–7 Volts) is applied to the tip. The frequency of oscillation is chosen to be equal to the resonance frequency of the cantilever (≈80 kHz). When a DC bias potential equal to the negative value of surface potential of the sample (on the order of ±2 Volts) is applied to the tip, it does not vibrate. During scanning, a difference between the DC bias potential applied to the tip and the potential of the surface will create DC electric fields that interact with the oscillating charges (as a result of the AC potential), causing the cantilever to oscillate at its resonance frequency, as in the tapping mode. However, a feedback loop is used to adjust the DC bias on the tip to exactly nullify the electric field, and thus the vibrations of the cantilever. The required bias voltage follows the localized potential of the surface. The surface potential

Nanotribology, Nanomechanics and Materials Characterization | 29.1 Description of AFM/FFM and Various Measurement Techniques 799

Part D | 29.1

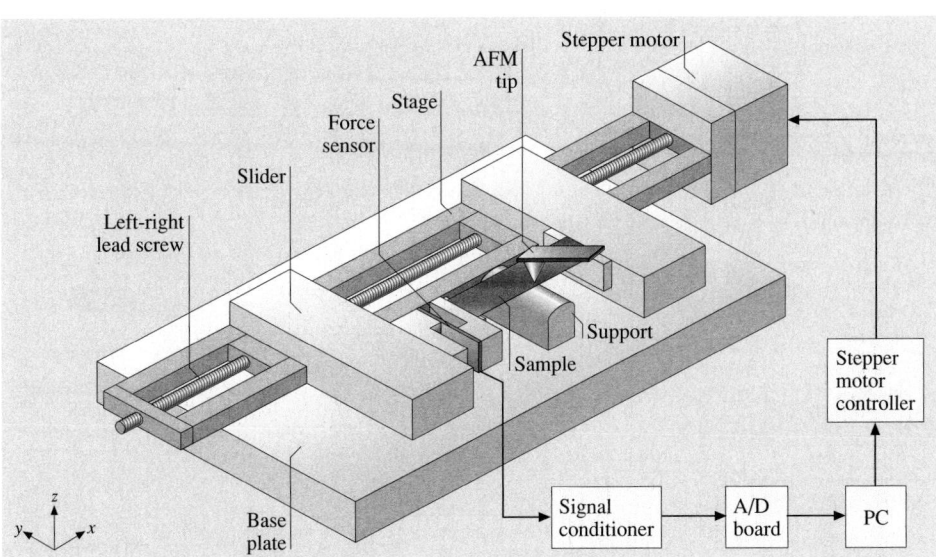

was obtained by reversing the sign of the bias potential provided by the electronics [29.67, 68]. Surface and subsurface changes of structure and/or chemistry can cause changes in the measured potential of a surface. Thus, mapping of the surface potential after sliding can be used to detect wear precursors and study the early stages of localized wear.

29.1.5 In Situ Characterization of Local Deformation Studies

In situ characterization of local deformation of materials can be carried out by performing tensile, bending, or compression experiments inside an AFM and by observing nanoscale changes during the deformation experiment [29.17]. In these experiments, small defor-

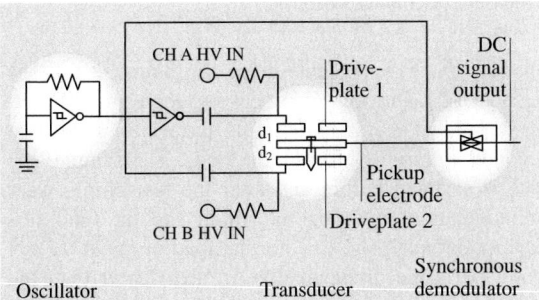

Fig. 29.9 Schematic of a nano/picoindentation system with three-plate transducer with electrostatic actuation hardware and capacitance sensor [29.49]

mation stages are used to deform the samples inside an AFM. In tensile testing of the polymeric films carried out by *Bobji* and *Bhushan* [29.61, 62] and *Tambe* and *Bhushan* [29.63] a tensile stage was used (Fig. 29.8). The stage with a left–right combination lead screw (which helps to move the slider in the opposite direction) was used to stretch the sample to minimize the movement of the scanning area, which was kept close to the center of the tensile specimen. One end of the sample was mounted on the slider via a force sensor to monitor the tensile load. The samples were stretched for various strains using a stepper motor and the same control area at different strains was imaged. In order to better locate the control area for imaging, a set of four markers was created at the corners of a $30\,\mu m \times 30\,\mu m$ square at the center of the sample by scratching the sample with a sharp silicon tip. The scratching depth was controlled such that it did not affect the cracking behavior of the coating. A minimum displacement of $1.6\,\mu m$ could be obtained. This corresponded to a strain increment of $8 \times 10^{-3}\%$ for a sample length of 38 mm. The maximum travel was about 100 mm. The resolution of the force sensor was 10 mN with a capacity of 45 N. During stretching, a stress–strain curve was obtained during the experiment to study any correlation between the degree of plastic strain and propensity of cracks.

29.1.6 Nanoindentation Measurements

For nanoindentation hardness measurements the scan size is set to zero and then a normal load is ap-

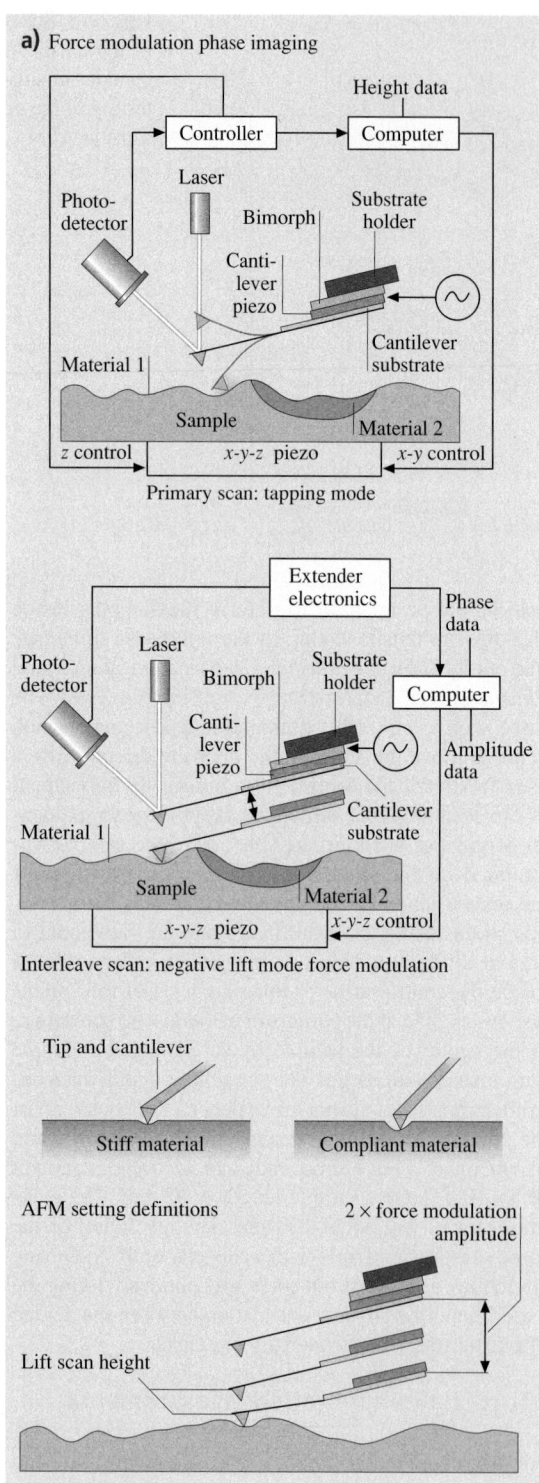

a) Force modulation phase imaging

Primary scan: tapping mode

Interleave scan: negative lift mode force modulation

AFM setting definitions

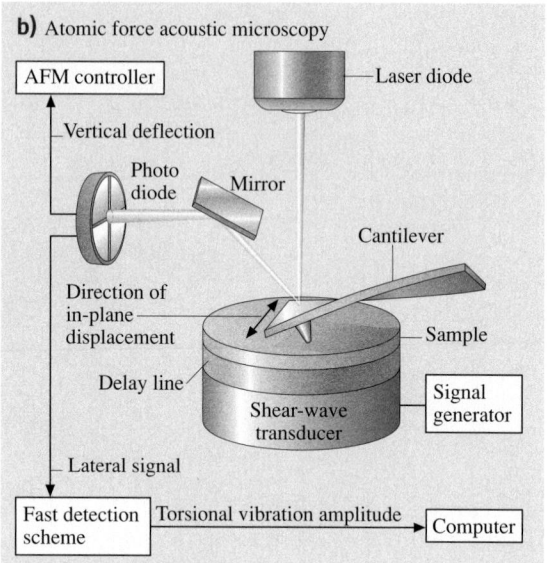

b) Atomic force acoustic microscopy

Fig. 29.10 (a) Schematic of force modulation mode used to obtain amplitude (stiffness) and definitions of force modulation amplitude and lift scan height. During primary scanning, height data is recorded in tapping mode. During interleave scanning, the entire cantilever/tip assembly is vibrated at the bimorph's resonance frequency and the z-direction feedback control for the sample x–y–z piezo is deactivated. During this scanning, height information from the primary scan is used to maintain a constant lift scan height. The computer records amplitude (which is a function of material stiffness) during the interleave scan, and **(b)** Schematic of an AFM incorporating shear wave transducer which generates in-plane lateral sample surface vibrations. Because of the forces between the tip and the surface, torsional vibrations of the cantilever are excited [29.33]. The shift in contact resonance frequency is a measure of contact stiffness

plied to make the indents using the diamond tip (see Sect. 29.1.5). During this procedure, the tip is continuously pressed against the sample surface for about two seconds at various indentation loads. The sample surface is scanned before and after the scratching, wear or indentation to obtain the initial and the final surface topography, at a low normal load of about $0.3\,\mu\mathrm{N}$ using the same diamond tip. An area larger than the indentation region is scanned to observe the indentation marks. Nanohardness is calculated by dividing the indentation load by the projected residual area of the indents [29.50].

Direct imaging of the indent allows one to quantify piling up of ductile material around the indenter. However, it becomes difficult to identify the boundary of the indentation mark with great accuracy. This makes the direct measurement of contact area somewhat inaccurate. A technique with the dual capability of depth-sensing as well as in situ imaging, which is most appropriate in nanomechanical property studies, is used for accurate measurement of hardness with shallow depths [29.4, 11, 49]. This nano/picoindentation system is used to make load–displacement measurements and subsequently carry out in situ imaging of the indent, if required. The indentation system, shown in Fig. 29.9, consists of a three-plate transducer with electrostatic actuation hardware used for direct application of normal load and a capacitive sensor used for measurement of vertical displacement. The AFM head is replaced with this transducer assembly while the specimen is mounted on the PZT scanner, which remains stationary during indentation experiments. The transducer consists of a three ($Be - Cu$) plate capacitive structure and the tip is mounted on the center plate. The upper and lower plates serve as drive electrodes and the load is applied by applying appropriate voltage to the drive electrodes. Vertical displacement of the tip (indentation depth) is measured by measuring the displacement of the center plate relative to the two outer electrodes using capacitance technique. The indent area and consequently hardness value can be obtained from the load–displacement data. The Young's modulus of elasticity is obtained from the slope of the unloading curve.

29.1.7 Localized Surface Elasticity and Viscoelasticity Mapping

Localized Surface Elasticity

Indentation experiments provide a single-point measurement of the Young's modulus of elasticity calculated from the slope of the indentation curve during unloading. Localized surface elasticity maps can be obtained using dynamic force microscopy in which an oscillating tip is scanned over the sample surface in contact under steady and oscillating load. The lower-frequency operation modes in the kHz range, such as the force modulation mode [29.52, 54] or the pulsed force mode [29.69], are well suited for soft samples such as polymers. However, if the tip–sample contact stiffness becomes significantly higher than the cantilever stiffness, the sensitivity of these techniques strongly decreases. In this case, the sensitivity of the measurement of stiff materials can be improved by using high-frequency operation modes in

the MHz range with a lateral motion, such as acoustic (ultrasonic) force microscopy, referred to as atomic force acoustic microscopy (AFAM) or contact resonance spectroscopy [29.55, 56, 70]. Inclusion of vibration frequencies other than only the first cantilever flexural or torsional resonance frequency, also allows additional information to be obtained.

In the negative lift mode force modulation technique, height data is recorded during primary scanning in the tapping mode, as described earlier. During interleave scanning, the entire cantilever/tip assembly is moved up and down at the force modulation holder's bimorph resonance frequency (about 24 kHz) at some amplitude, here referred to as the force modulation amplitude, and the Z-direction feedback control for the sample $X–Y–Z$ piezo is deactivated, Fig. 29.10a [29.52, 54, 57]. During this scanning, height information from the primary scan is used to maintain a constant lift scan height. This eliminates the influence of height on the measured signals during the interleave scan. Lift scan height is the mean tip-to-sample distance between the tip and sample during the interleave scan. The lift scan height is set such that the tip is in constant contact with the sample, i. e. a constant static load is applied. (A higher lift scan height gives a closer mean tip-to-sample distance.) In addition, the tip motion caused by the bimorph vibration results in a modulating periodic force. The sample surface resists the oscillations of the tip to a greater or lesser extent depending upon the sample's stiffness. The computer records amplitude (which is a function of the elastic stiffness of the material). Contact analyses can be used to obtain a quantitative measure of localized elasticity of soft surfaces [29.54]. Etched single-crystal silicon cantilevers with integrated tips (force modulation etched Si probe or DI FESP) with a radius of 25–50 nm, a stiffness of 1–5 N/m, and a natural frequency of 60–100 kHz, are commonly used for the measurements. Scanning is normally set to a rate of 0.5 Hz along the fast axis.

In the AFAM technique [29.55, 56, 70], the cantilever/tip assembly is moved either in the normal or lateral mode and the contact stiffness is evaluated by comparing the resonance frequency of the cantilever in contact with the sample surface to those of the free vibrations of the cantilever. Several free resonance frequencies are measured. Based on the shift of the measured frequencies, the contact stiffness is determined by solving the characteristic equation for the tip vibrating in contact with the sample surface. The elastic modulus is calculated from contact stiffness using Hertz analysis for a spherical tip indenting a plane. Contact stiffness is equal to 8× contact radius × reduced shear modulus in the shear mode.

In the lateral mode using the AFAM technique, the sample is glued onto cylindrical pieces of aluminum which serve as ultrasonic delay lines coupled to an ultrasonic shear wave transducer, Fig. 29.10b [29.33, 55, 56]. The transducer is driven with frequency sweeps to generate in-plane lateral sample surface vibrations. These couple to the cantilever via the tip–sample contact. To measure torsional vibrations of the cantilever at frequencies up to 3 MHz, the original electronic circuit of the lateral channel of the AFM (using a low-pass filter with limited bandwidth to a few hundred kHz) was replaced by a high-speed scheme which bypasses the low-pass filter. The high-frequency signal was fed to a lock-in amplifier, digitized using a fast analogue-to-digital (A/D) card and fed into a broadband amplifier, followed by an rms-to-dc converter and read by a computer. Etched single-crystal silicon cantilevers (normal stiffness of 3.8–40 N/m) integrated tips are used.

Viscoelastic Mapping

Another form of dynamic force microscopy, phase contrast microscopy, is used to detect the contrast in viscoelastic (viscous energy dissipation) properties of the different materials across the surface [29.53, 57–60, 71, 72]. In these techniques, both deflection amplitude and phase angle contrasts are measured, which are measure of the relative stiffness and viscoelastic properties, respectively. Two phase measurement techniques – tapping mode and torsional resonance (TR) mode – have been developed, which we now describe.

In the tapping-mode (TM) technique, as described earlier, the cantilever/tip assembly is sinusoidally vibrated at its resonant frequency and the sample X–Y–Z piezo is adjusted using feedback control in the z direction to maintain a constant set point, Fig. 29.3 [29.57, 58]. The feedback signal to the Z-direction sample piezo (to keep the set point constant) is a measure of surface roughness. The extender electronics is used to measure the phase-angle lag between the cantilever piezo drive signal and the cantilever response during sample engagement. As illustrated in Fig. 29.3, the phase-angle lag (at least partially) is a function of the viscoelastic properties of the sample material. A range of tapping amplitudes and set points can be used for measurements. Commercially etched single-crystal silicon tip (DI TESP) used for tapping mode, with a radius of 5–10 nm, a stiffness of 20–100 N/m, and a natural frequency of 350 to 400 kHz, is normally used. Scanning is normally set to a rate of 1 Hz along the fast axis.

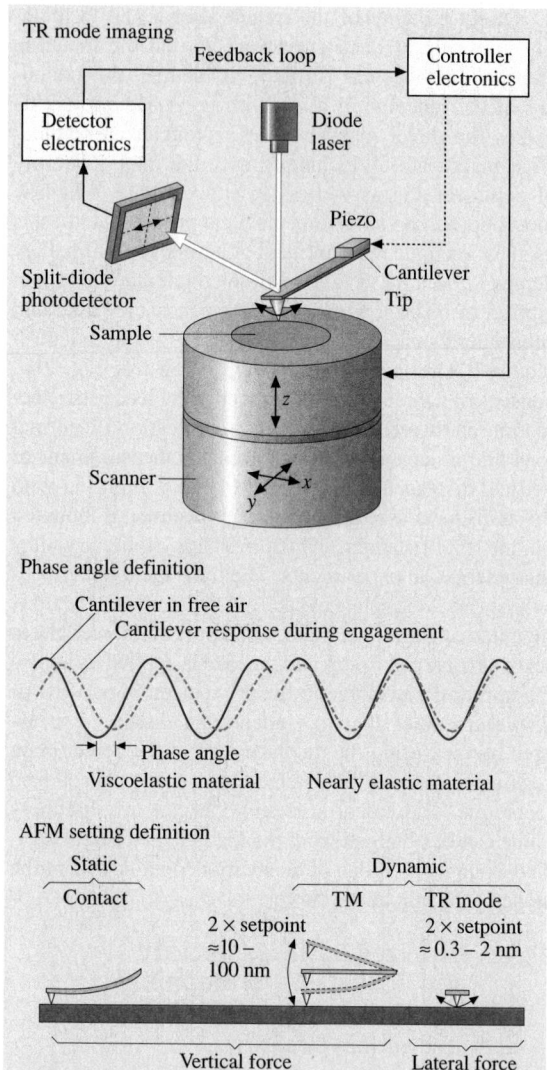

Fig. 29.11 Schematic of torsional resonance mode shown at the top. Two examples of the phase-angle response are shown in the *middle*. One is for materials exhibiting viscoelastic (*left*) and the other nearly elastic properties (*right*). Three AFM settings are compared at the *bottom*: contact, tapping mode (TM), and TR modes. The TR mode is a dynamic approach with a laterally vibrating cantilever tip that can interact with the surface more intensively than other modes. Therefore, more detailed near-surface information is available

In the torsional mode (TR mode), a tip is vibrated in the torsional mode at high frequency at the resonance frequency of the cantilever beam. Etched single-crystal

Nanotribology, Nanomechanics and Materials Characterization | 29.1 Description of AFM/FFM and Various Measurement Techniques   803

Part D | 29.1

silicon cantilever with integrated tip (DI FESP) with a radius of about $5-10$ nm, normal stiffness of $1-5$ N/m, torsional stiffness of about 30 times the normal stiffness and torsional natural frequency of 800 kHz is normally used. A major difference between the TM and the TR modes is the directionality of the applied oscillation: a normal (compressive) amplitude exerted for the TM, and a torsional amplitude for the TR mode. The TR mode is expected to provide good contrast in the tribological and mechanical properties of the near surface region as compared to the TM. Two of the reasons are as follows. (1) In the TM, the interaction is dominated by the vertical properties of the sample, so the tip spends a small fraction of its time in the near-field interaction with the sample. Furthermore, the distance between the tip and the sample changes during the measurements, which changes interaction time and forces, and affects measured data. In the TR mode, the distance remains nearly constant. (2) The lateral stiffness of a cantilever is typically about two orders of magnitude larger than the normal (flexural) stiffness. Therefore, in the TM, if the sample is relatively rigid, and much of the deformation occurs in the cantilever beam, whereas in the TR mode, much of the deformation occurs in the sample. A few comments on the special applications of the TR mode are made next. Since most of the deformation occurs in the sample, the TR mode can be used to measure stiff and hard samples. Furthermore, the properties of thin films can be measured more readily with the TR

mode. For both the TM and TR modes, if the cantilever is driven to vibrate at frequencies above resonance, it would have less motion (high apparent stiffness), leading to higher sample deformation and better contrast. It should be further noted that the TM exerts compressive force, whereas the TR mode exerts torsional force, therefore normal and shear properties are measured in the TM and TR modes, respectively.

In the TR mode, the torsional vibration of the cantilever beam is achieved using a specially designed cantilever holder. It is equipped with a piezo system mounted in a cantilever holder, in which two piezos vibrate out of phase with respect to each other. A tuning process prior to scanning is used to select the torsional vibration frequency. The piezo system excites torsional vibration at the cantilever's resonance frequency. The torsional vibration amplitude of the tip (TR amplitude) is detected by the lateral segments of the split-diode photodetector, Fig. 29.11 [29.59]. The TR mode measures surface roughness and phase angle as follows. During the measurement, the cantilever/tip assembly is first vibrated at its resonance at some amplitude dependent upon the excitation voltage, before the tip engages the sample. Next, the tip engages the sample at some set point. A feedback system coupled to a piezo stage is used to keep a constant TR amplitude during scanning. This is done by controlling the vertical position of the sample using a piezo moving in the Z direction, which changes the degree of tip interaction. The

Table 29.2 Summary of various operating modes of AFM for surface roughness, stiffness, phase angle, and friction

Operating mode	Direction of cantilever vibration	Vibration frequency of cantilever (kHz)	Vibration amplitude (nm)	Feedback control	Data obtained
Contact	n/a			Constant normal load	Surface height, friction
Tapping	Vertical	350–400	10–100	Set point (constant tip amplitude)	Surface height, phase angle (normal viscoelasticity)
Force modulation	Vertical	10–20 (bimorph)	10–100	Constant normal load	Surface height, amplitude (normal stiffness)
Lateral	Lateral (AAFM)	100–3000 (sample)	≈ 5 (sample)	Constant normal load	Shift in contact resonance (normal stiffness, friction)
TR mode I	Torsional	≈ 800	0.3–2	Set point (constant tip amplitude)	Surface height, phase angle (lateral viscoelasticity)
TR mode II	Torsional	≈ 800	0.3–2	Constant normal load	Surface height, amplitude and phase angle (lateral stiffness and lateral viscoelasticity)
TR mode III	Torsional	Higher than 800 in contact	0.3–2	Constant normal load	Shift in contact resonance (friction)

displacement of the sample Z piezo gives a roughness image of the sample. A phase-angle image can be obtained by measuring the phase lag of the cantilever vibration response in the torsional mode during engagement with respect to the cantilever vibration response in free air before engagement. The control feedback of the TR mode is similar to that of tapping, except that the torsional resonance amplitude replaces flexural resonance amplitude [29.59].

Chen and *Bhushan* [29.60] have used a variation to the approach just described (referred to as mode I here). They performed measurements at constant normal cantilever deflection (constant load) (mode II) instead of using constant set point in the *Kasai* et al. [29.59] approach. Their approach overcomes the meniscus adhesion problem present in mode I and reveals true surface properties.

Song and *Bhushan* [29.73] presented a forced torsional vibration model for a tip–cantilever assembly under viscoelastic tip–sample interaction. This model provides a relationship of torsional amplitude and phase shift with lateral contact stiffness and viscosity which can be used to extract in-plane interfacial mechanical properties.

Various operating modes of AFM used for surface roughness, localized surface elasticity, and viscoelastic mapping and friction force measurements (to be discussed later) are summarized in Table 29.2.

29.1.8 Boundary Lubrication Measurements

To study nanoscale boundary lubrication studies, adhesive forces are measured in the force calibration mode, as previously described. The adhesive forces are also calculated from the horizontal intercept of curves of friction versus normal load at zero friction force. For friction measurements, the samples are typically scanned using an Si_3N_4 tip over an area of $2 \times 2 \, \mu m$ at a normal load in the range $5–130 \, nN$. The samples are generally scanned with a scan rate of $0.5 \, Hz$, which results in a scanning speed of $2 \, \mu m/s$. Velocity effects on friction are studied out by changing the scan frequency from 0.1 to $60 \, Hz$, while the scan size is maintained at $2 \times 2 \, \mu m$, which allows the velocity to vary from 0.4 to $240 \, \mu m/s$. To study the durability properties, the friction force and coefficient of friction are monitored during scanning at a normal load of $70 \, nN$ and a scanning speed of $0.8 \, \mu m/s$, for a desired number of cycles [29.39, 40, 42].

29.2 Surface Imaging, Friction and Adhesion

29.2.1 Atomic-Scale Imaging and Friction

Surface height imaging down to atomic resolution of electrically conducing surfaces can be carried out using an STM. An AFM can also be used for surface height imaging and roughness characterization down to the nanoscale. Figure 29.12 shows a sequence of STM images at various scan sizes of solvent-deposited C_{60} film on a 200 nm thick gold coated freshly cleaved mica [29.74]. The film consists of clusters of C_{60} molecules with a diameter of 8 nm. The C_{60} molecules within a cluster appear to pack in a hexagonal array with a spacing of about 1 nm, however, they do not follow any long-range order. The measured cage diameter of the C_{60} molecule is about 0.7 nm, which is very close to the projected diameter of 0.71 nm.

In an AFM measurement during surface imaging, the tip comes into intimate contact with the sample surface and leads to surface deformation with finite tip–sample contact area (typically a few atoms). The finite size of the contact area prevents the imaging of individual point defects, and only the periodicity of the atomic lattice

can be imaged. Figure 29.13a shows the topography image of the freshly cleaved surface of highly oriented pyrolytic graphite (HOPG) [29.30]. The periodicity of the graphite is clearly observed.

To study friction mechanisms on an atomic scale, a freshly cleaved HOPG has been studied by *Mate* et al. [29.22] and *Ruan* and *Bhushan* [29.25]. Figure 29.14a shows the atomic-scale friction force map (raw data) and Fig. 29.13a shows the atomic-scale topography and friction force maps (after 2D spectrum filtering with high-frequency noise truncation) [29.25]. Figure 29.14a also shows a line plot of friction force profiles along some crystallographic direction. The actual shape of the friction profile depends upon the spatial location of axis of tip motion. Note that a portion of the atomic-scale lateral force is conservative. *Mate* et al. [29.22] and *Ruan* and *Bhushan* [29.25] reported that the average friction force increased linearly with normal load and was reversible with load. Friction profiles were similar during sliding of the tip in either direction.

During scanning, the tip moves discontinuously over the sample surface and jumps with discrete steps from

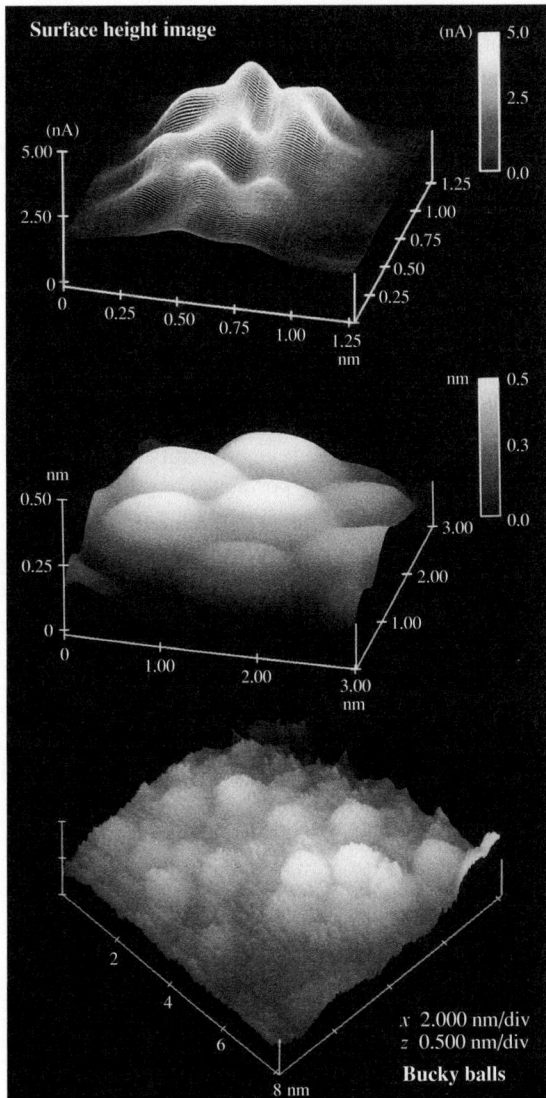

Fig. 29.12 STM images of solvent-deposited C_{60} film on a gold-coated freshly cleaved mica at various scan sizes

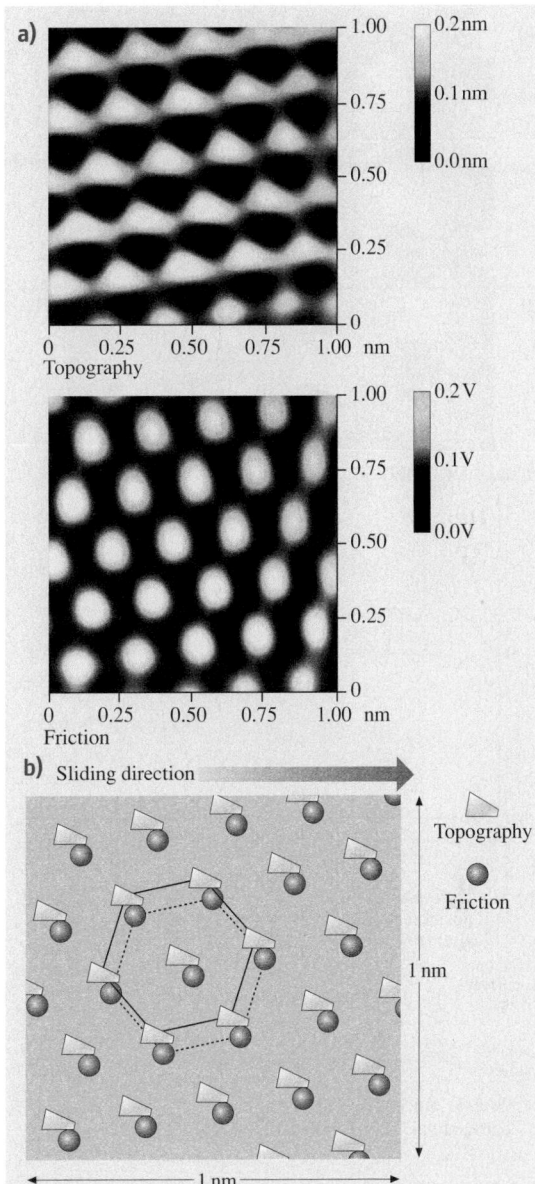

Fig. 29.13 (a) Greyscale plots of surface topography and friction force maps (2D spectrum filtered), measured simultaneously, of a 1 nm × 1 nm area of freshly cleaved HOPG, showing the atomic-scale variation of topography and friction, and (b) schematic of superimposed topography and friction maps from (a); the *symbols* correspond to maxima. Note the spatial shift between the two plots [29.25]

one potential minimum (well) to the next. This leads to a sawtooth-like pattern for the lateral motion (force) with a periodicity of the lattice constant. This motion is called stick–slip movement of the tip [29.5, 10, 22, 25]. The observed friction force includes two components: conservative and periodic, and nonconservative and constant. If the relative motion of the sample and tip were simply that of two rigid collections of atoms, the effective force would be a conservative force oscillating about zero. Slow reversible elastic deformation would

also contribute to the conservative force. The origin of the nonconservative direction-dependent force compo-

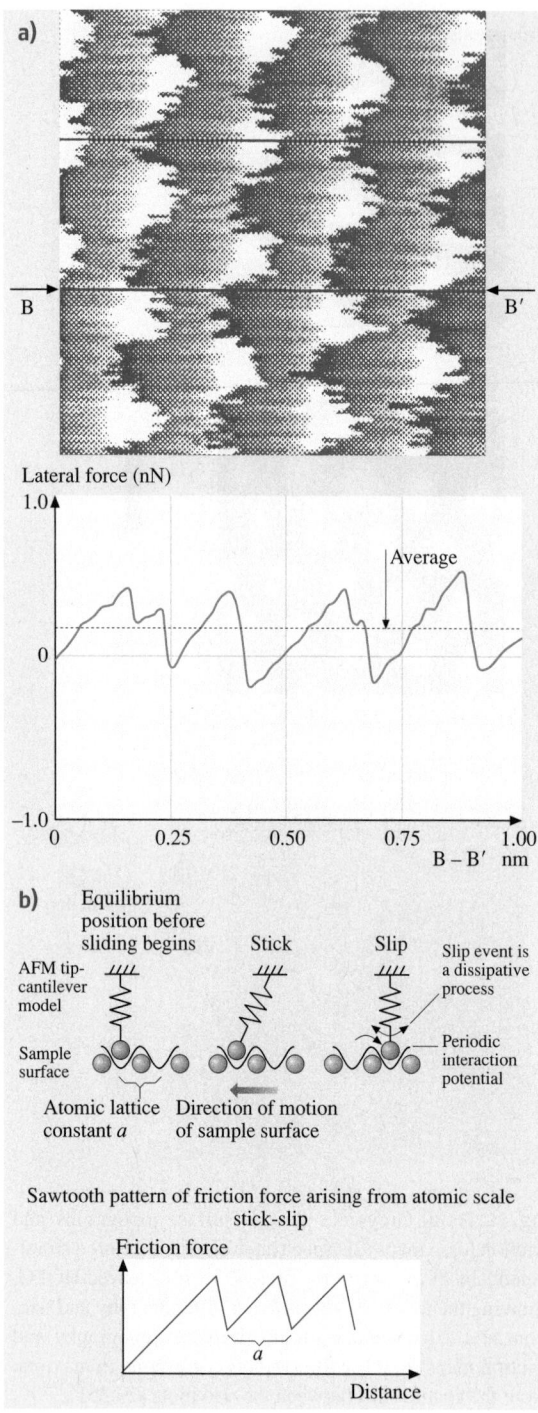

a)

Lateral force (nN)

Average

B – B' nm

b) Equilibrium
    position before
    sliding begins   Stick   Slip
                                                    Slip event is
AFM tip-   a dissipative
cantilever   process
model
                                                    Periodic
Sample   interaction
surface   potential

Atomic lattice   Direction of motion
constant *a*   of sample surface

Sawtooth pattern of friction force arising from atomic scale
stick-slip

Friction force

a

Distance

Fig. 29.14 (a) Greyscale plot of friction force map (raw data) of a $1 \times 1 \, nm^2$ area of freshly cleaved HOPG, showing atomic-scale variation of friction force. High points are shows by lighter color. Also shown is a line plot of friction force profile along the line indicated by *arrows*. The normal load was 25 nN and the cantilever normal stiffness was 0.4 N/m [29.25]. (b) Schematic of a model for a tip atom sliding on an atomically flat periodic surface. The schematic shows the tip jumping from one potential minimum to another, resulting in stick–slip behavior

As just mentioned, stick–slip on the atomic scale is the result of the energy barrier that must be overcome to jump over the atomic corrugations on the sample surface. This corresponds to the energy required for the jump of the tip from a stable equilibrium position on the surface into a neighboring position. The perfect atomic regularity of the surface guarantees the periodicity of the lateral force signal, independent of actual atomic structure of tip apex. Few atoms (based on the magnitude of the friction force, less than 10) on a tip sliding over an array of atoms on the sample are expected to go through the stick–slip. For simplicity, Fig. 29.14b shows a simplified model for one atom on a tip with a one-dimensional spring–mass system. As the sample surface slides against the AFM tip, the tip remains *stuck* initially until it can overcome the energy (potential) barrier, which is illustrated by a sinusoidal interaction potential as experienced by the tip. After some motion, there is enough energy stored in the spring which leads to *slip* into the neighboring stable equilibrium position. During the slip and before attaining stable equilibrium, stored energy is converted into vibrational energy of the surface atoms in the range of 10^{13} Hz (phonon generation) and decays within the range of 10^{-11} s into heat. (A wave of atoms vibrating in concert are termed a phonon.) The stick–slip phenomenon, resulting from irreversible atomic jumps, can be theoretically modeled with classical mechanical models [29.75, 76]. The Tomanek–Zhong–Thomas model [29.76] is the starting point for determining the friction force during atomic-scale stick–slip. The AFM model describes the total potential as the sum of the potential acting on the tip due to interaction with the sample and the elastic energy stored in the cantilever. Thermally activated stick–slip behavior can explain the velocity effects on friction, to be presented later.

Finally, based on Fig. 29.13a, the atomic-scale friction force of HOPG exhibited the same periodicity as that of the corresponding topography, but the peaks in friction and those in topography are displaced rel-

nent would be phonon generation, viscous dissipation, or plastic deformation.

ative to each other (Fig. 29.13b). A Fourier expansion of the interatomic potential was used by *Ruan* and *Bhushan* [29.25] to calculate the conservative interatomic forces between atoms of the FFM tip and those of the graphite surface. Maxima in the interatomic forces in the normal and lateral directions do not occur at the same location, which explains the observed shift between the peaks in the lateral force and those in the corresponding topography.

29.2.2 Microscale Friction

Local variations in the microscale friction of cleaved graphite are observed, Fig. 29.15. These arise from structural changes that occur during the cleaving process [29.26]. The cleaved HOPG surface is largely atomically smooth but exhibits line-shaped regions in which the coefficient of friction is more than an order of magnitude larger. Transmission electron microscopy in-

dicates that the line-shaped regions consist of graphite planes of different orientation, as well as of amorphous carbon. Differences in friction have also been observed for multiphase ceramic materials [29.45]. Figure 29.16 shows the surface roughness and friction force maps of $Al_2O_3 - TiC$ (70-30 wt%). TiC grains have a Knoop hardness of about $2800\,kg/mm^2$ and Al_2O_3 has $2100\,kg/mm^2$, therefore, TiC grains do not polish as much and result in a slightly higher elevation (about 2-3 nm higher than that of Al_2O_3 grains). TiC grains exhibit higher friction force than Al_2O_3 grains. The coefficients of friction of TiC and Al_2O_3 grains are 0.034 and 0.026, respectively, and the coefficient of friction of $Al_2O_3 - TiC$ composite is 0.03. Local variation in

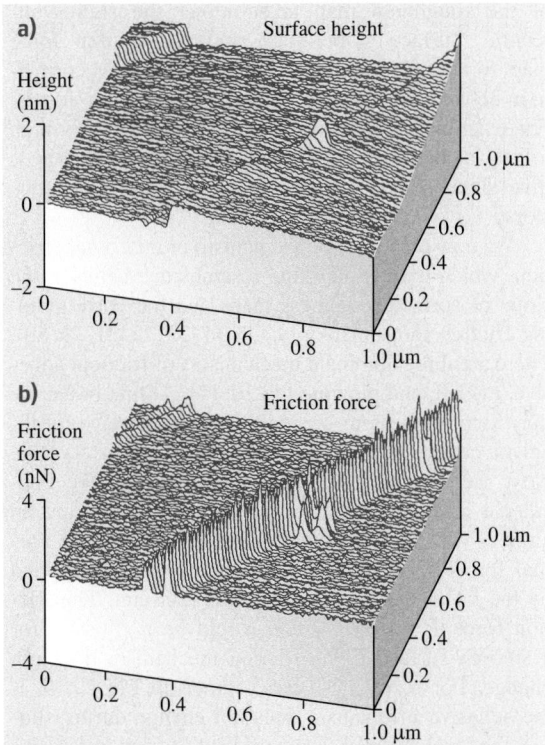

Fig. 29.15 (a) Surface roughness and (b) friction force maps at a normal load of 42 nN of freshly cleaved HOPG surface against an Si_3N_4 FFM tip. Friction in the line-shaped region is over an order of magnitude larger than in the smooth areas [29.25]

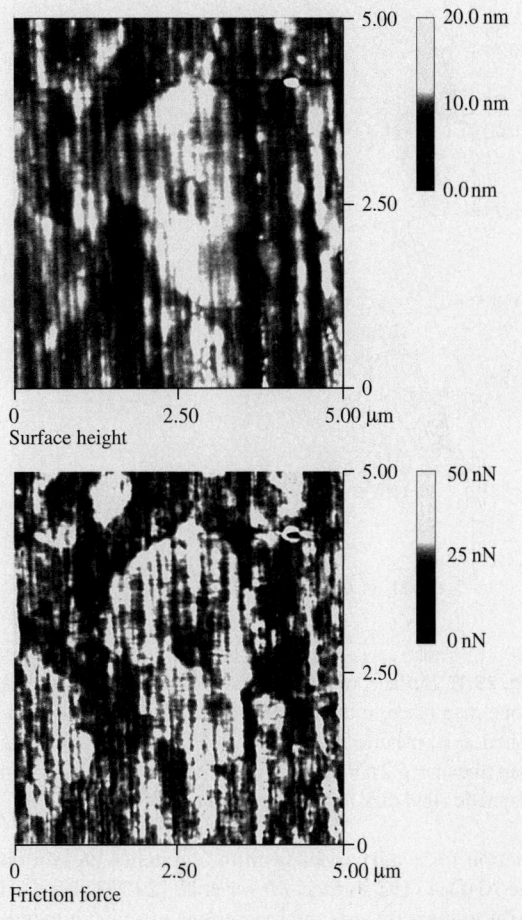

Fig. 29.16 Greyscale surface roughness ($\sigma = 0.80\,nm$) and friction force maps (mean $= 7.0\,nN, \sigma = 0.90\,nN$) for Al_2O_3-TiC (70-30 wt %) at a normal load of 138 nN [29.45]

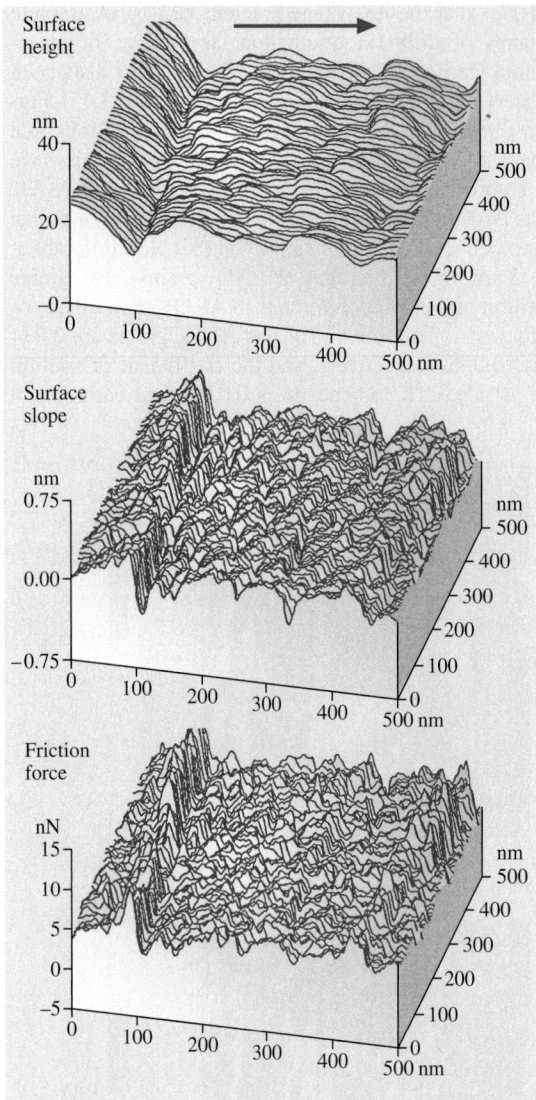

Surface height

Surface slope

Friction force

Fig. 29.17 Surface roughness map ($\sigma = 4.4$ nm), surface slope map taken in the sample sliding direction (the horizontal axis; mean $= 0.023$, $\sigma = 0.197$), and friction force map (mean $= 6.2$ nN, $\sigma = 2.1$ nN) for a lubricated thin-film magnetic rigid disk for a normal load of 160 nN [29.27]

friction force also arises from the scratches present on the $Al_2O_3 - TiC$ surface. *Meyer* et al. [29.77] also used FFM to measure structural variations of organic mono- and multilayer films. All of these measurements suggest that the FFM can be used for structural mapping of the surfaces. FFM measurements can also be used to map chemical variations, as indicated by the use of the FFM

with a modified probe tip to map the spatial arrangement of chemical functional groups in mixed organic monolayer films [29.78]. Here, sample regions that had stronger interactions with the functionalized probe tip exhibited larger friction.

Local variations in the microscale friction of nominally rough, homogeneous-material surfaces can be significant, and are seen to depend on the local surface slope rather than the surface height distribution, Fig. 29.17. This dependence was first reported by *Bhushan* and *Ruan* [29.23], *Bhushan* et al. [29.27], and *Bhushan* [29.65] and later discussed in more detail by *Koinkar* and *Bhushan* [29.79] and *Sundararajan* and *Bhushan* [29.80]. In order to show elegantly any correlation between local values of friction and surface roughness, surface roughness and friction force maps of a gold-coated ruler with somewhat rectangular grids and a silicon grid with square pits were obtained, Fig. 29.18 [29.80]. Figures 29.17 and 29.18 show the surface roughness map, the slopes of the roughness map taken along the sliding direction (surface slope map) and the friction force map for various samples. There is a strong correlation between the surface slopes and friction forces. For example, in Fig. 29.18, the friction force is high locally at the edge of the grids and pits with a positive slope and is low at the edges with a negative slope.

We now examine the mechanism of microscale friction, which may explain the resemblance between the slope of surface roughness maps and the corresponding friction force maps [29.4, 5, 11, 23, 25–27, 79, 80]. There are three dominant mechanisms of friction; adhesive, ratchet, and plowing [29.10, 17]. At first order, we may assume these to be additive. The adhesive mechanism cannot explain the local variation in friction. Next we consider the ratchet mechanism. We consider a small tip sliding over an asperity making an angle θ with the horizontal plane, Fig. 29.19. The normal force W (normal to the general surface) applied by the tip to the sample surface is constant. The friction force F on the sample would be a constant for a smooth surface if the friction mechanism does not change. For a rough surface, shown in Fig. 29.19, if the adhesive mechanism does not change during sliding, the local value of the coefficient of friction remains constant,

$$\mu_0 = S/N \qquad (29.1)$$

where S is the local friction force and N is the local normal force. However, the friction and normal forces are measured with respect to global horizontal and normal axes, respectively. The measured local coefficient of friction μ_1 in the ascending part is

$$\mu_1 = F/W = (\mu_0 + \tan\theta)/(1 - \mu_0\tan\theta)$$
$$\approx \mu_0 + \tan\theta, \text{ for small } \mu_0\tan\theta \qquad (29.2)$$

indicating that, in the ascending part of the asperity, one may simply add the friction force and the asperity slope to one another. Similarly, on the right-hand side (descending part) of the asperity,

$$\mu_2 = (\mu_0 - \tan\theta)/(1 + \mu_0\tan\theta)$$
$$\approx \mu_0 - \tan\theta, \text{ for small } \mu_0\tan\theta \qquad (29.3)$$

For a symmetrical asperity, the average coefficient of friction experienced by the FFM tip traveling across the whole asperity is

$$\mu_{ave} = (\mu_1 + \mu_2)/2$$
$$= \mu_0\left(1 + \tan^2\theta\right)/\left(1 - \mu_0^2\tan^2\theta\right)$$
$$\approx \mu_0\left(1 + \tan^2\theta\right), \text{ for small } \mu_0\tan\theta \qquad (29.4)$$

Finally we consider the plowing component of friction with the tip sliding in either direction, which is [29.10, 17]

$$\mu_p \sim \tan\theta \qquad (29.5)$$

Because in FFM measurements we notice little damage of the sample surface, the contribution by plowing

Fig. 29.18a,b Surface roughness map, surface slope map taken in the sample sliding direction (the horizontal axis), and friction force map for (**a**) a gold-coated ruler (with somewhat rectangular grids with a pitch of 1 μm and a ruling step height of about 70 nm) at a normal load of 25 nN and (**b**) a silicon grid (with 5-μm square pits of depth 180 nm and a pitch of 10 μm) [29.80]

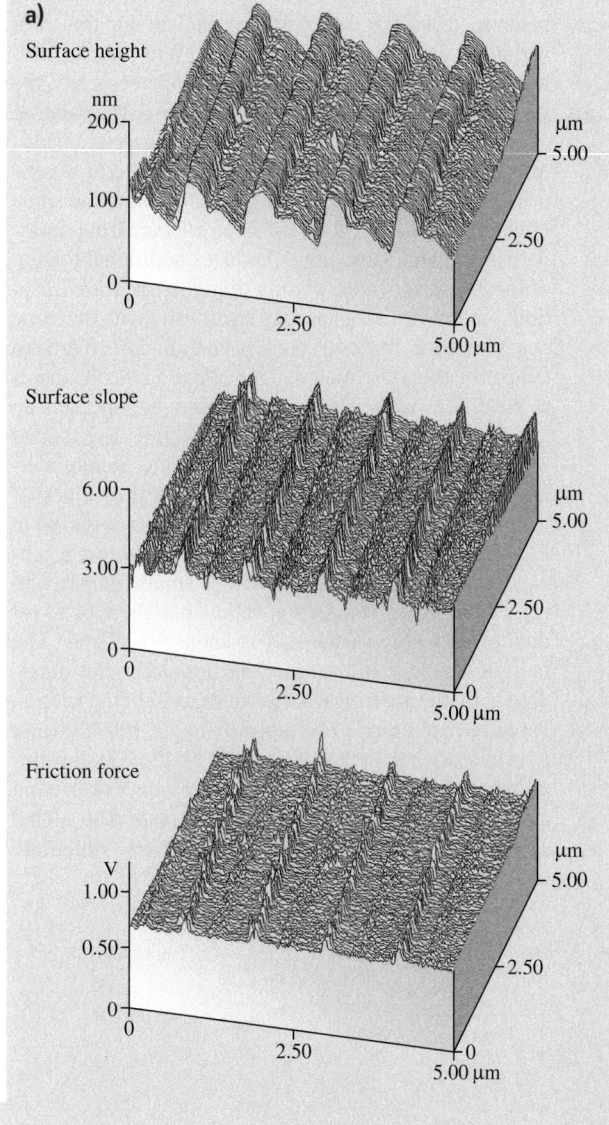

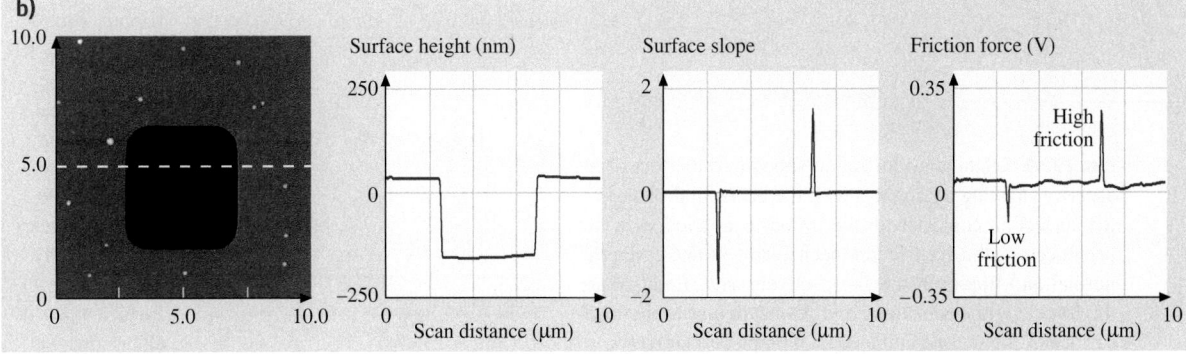

is expected to be small and the ratchet mechanism is believed to be the dominant mechanism for the local variations in the friction force map. With the tip sliding over the leading (ascending) edge of an asperity, the surface slope is positive; it is negative during sliding over the trailing (descending) edge of an asperity. Thus, measured friction is high at the leading edge of asperities and low at the trailing edge. In addition to the slope effect, the collision of the tip when encountering an asperity with a positive slope produces additional torsion of the cantilever beam leading to higher measured friction force. When encountering an asperity with the same negative slope, however, there is no collision effect and hence no effect on torsion. This effect also contributes to the difference in friction forces when the tip scans up and down on the same topography feature. The ratchet mechanism and the collision effects thus semiquantitatively explain the correlation between the slopes of the roughness maps and friction force maps observed in Figs. 29.17 and 29.18. We note that, in the ratchet mechanism, the FFM tip is assumed to be small compared to the size of asperities. This is valid since the typical radius of curvature of the tips is about 10–50 nm. The radii of curvature of the asperities of the samples measured here (the asperities that produce most of the friction variation) are found to be typically about 100–200 nm, which is larger than that of the FFM tip [29.81]. It is important to note that the measured local values of friction and normal forces are measured with respect to global (and not local) horizontal and vertical axes, which are believed to be relevant in applications.

29.2.3 Directionality Effect on Microfriction

During friction measurements, the friction force data from both the forward (trace) and backward (retrace) scans are useful in understanding the origins of the observed friction forces. Magnitudes of material-induced effects are independent of the scanning direction whereas topography-induced effects are different between forward and backward scanning directions. Since the sign of the friction force changes as the scanning direction is reversed (because of the reversal of torque applied to the end of the tip), addition of the friction force data of the forward and backward scan eliminates the material-induced effects while topography-induced effects still remain. Subtraction of the data between forward and backward scans does not eliminate either effect, Fig. 29.20 [29.80].

Owing to the reversal of the sign of the retrace (R) friction force with respect to the trace (T) data; the friction force variations due to topography are in the same

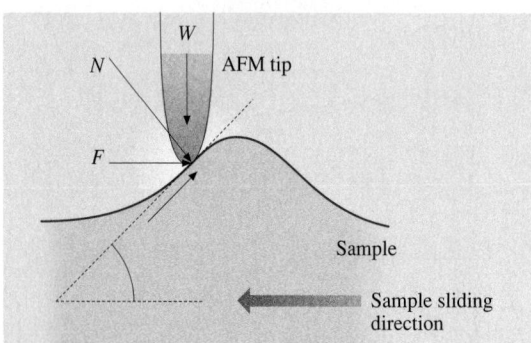

Fig. 29.19 Schematic illustration showing the effect of an asperity (making an angle θ with the horizontal plane) on the surface in contact with the tip on local friction in the presence of adhesive friction mechanism. W and F are the normal and friction forces, respectively, and S and N are the force components along and perpendicular to the local surface of the sample at the contact point, respectively

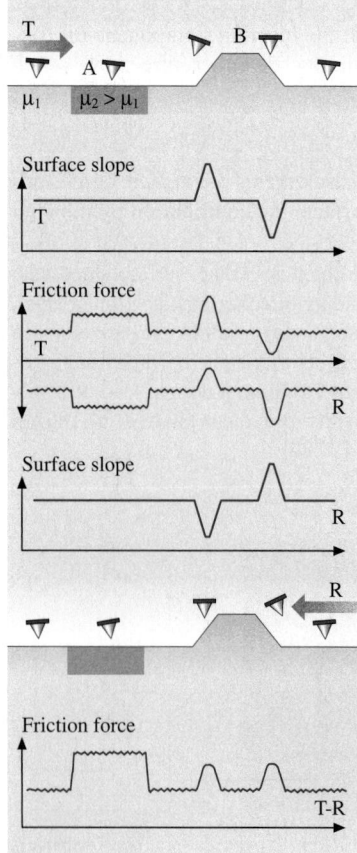

Fig. 29.20 Schematic of friction forces expected when a tip traverses a sample that is composed of different materials and sharp changes in topography. A schematic of surface slope is also shown

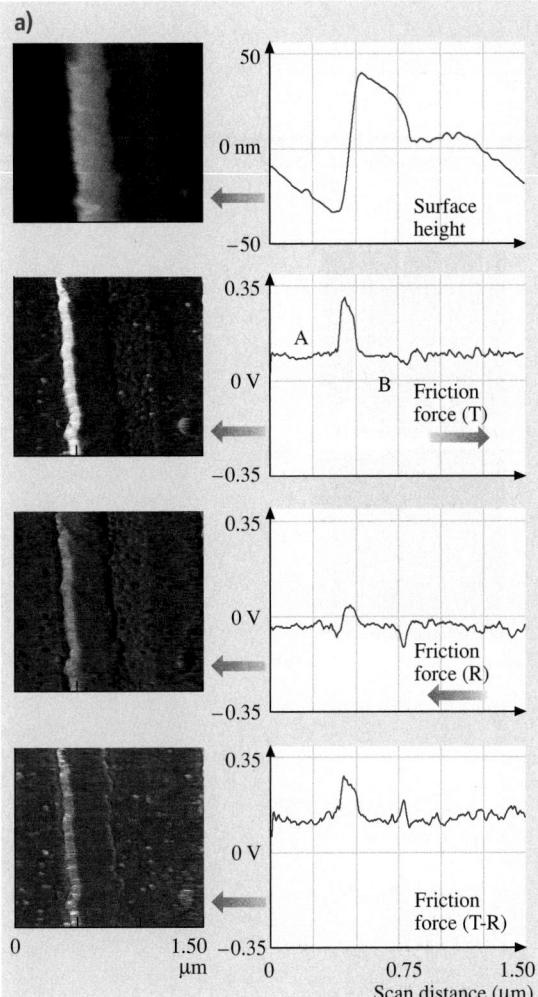

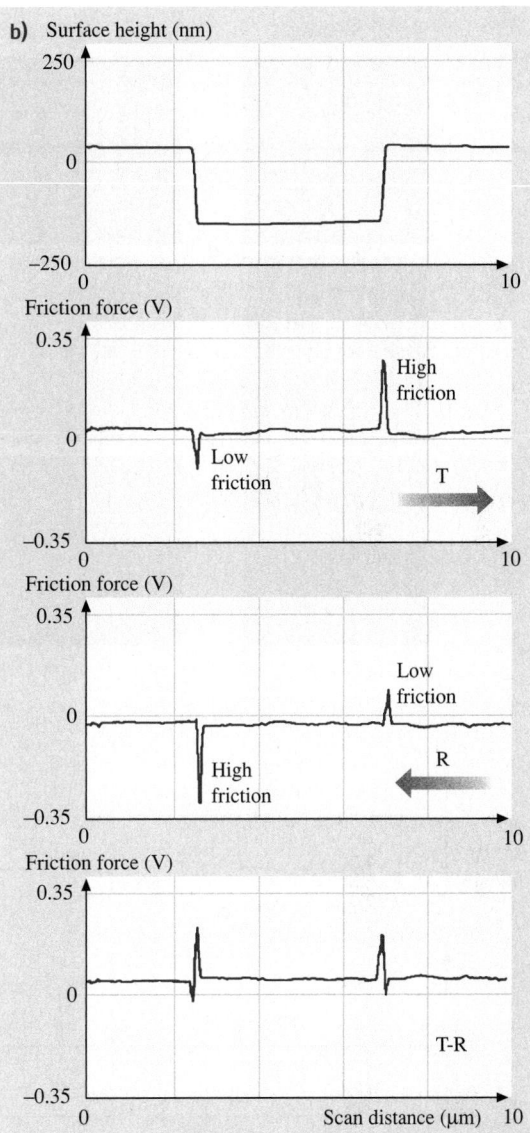

Fig. 29.21 (a) Greyscale images and two-dimensional profiles of surface height and friction forces across a single ruling of the gold-coated ruling, and **(b)** two dimensional profiles of surface height and friction forces across a silicon grid pit. Friction force data in trace and retrace directions, and subtracted force data are presented

direction (peaks in trace correspond to peaks in retrace). However, the magnitudes of the peaks in trace and retrace at a given location are different. An increase in the friction force experienced by the tip when scanning up a sharp change in topography is more than the decrease in the friction force experienced when scanning down the same topography change, partly because of the collision effects discussed earlier. Asperities on engineering

surfaces are asymmetrical, which also affects the magnitude of friction force in the two directions. Asymmetry in tip shape may also have an effect on the directionality effect of friction. We will note later that the magnitude of the surface slopes are virtually identical, therefore, the tip shape asymmetry should not have much effect.

Figure 29.21 shows surface height and friction force data for a gold ruler and a silicon grid in the trace and retrace directions. Subtraction of the two friction data yields a residual peak because of the differences in the magnitudes of the friction forces in the two directions.

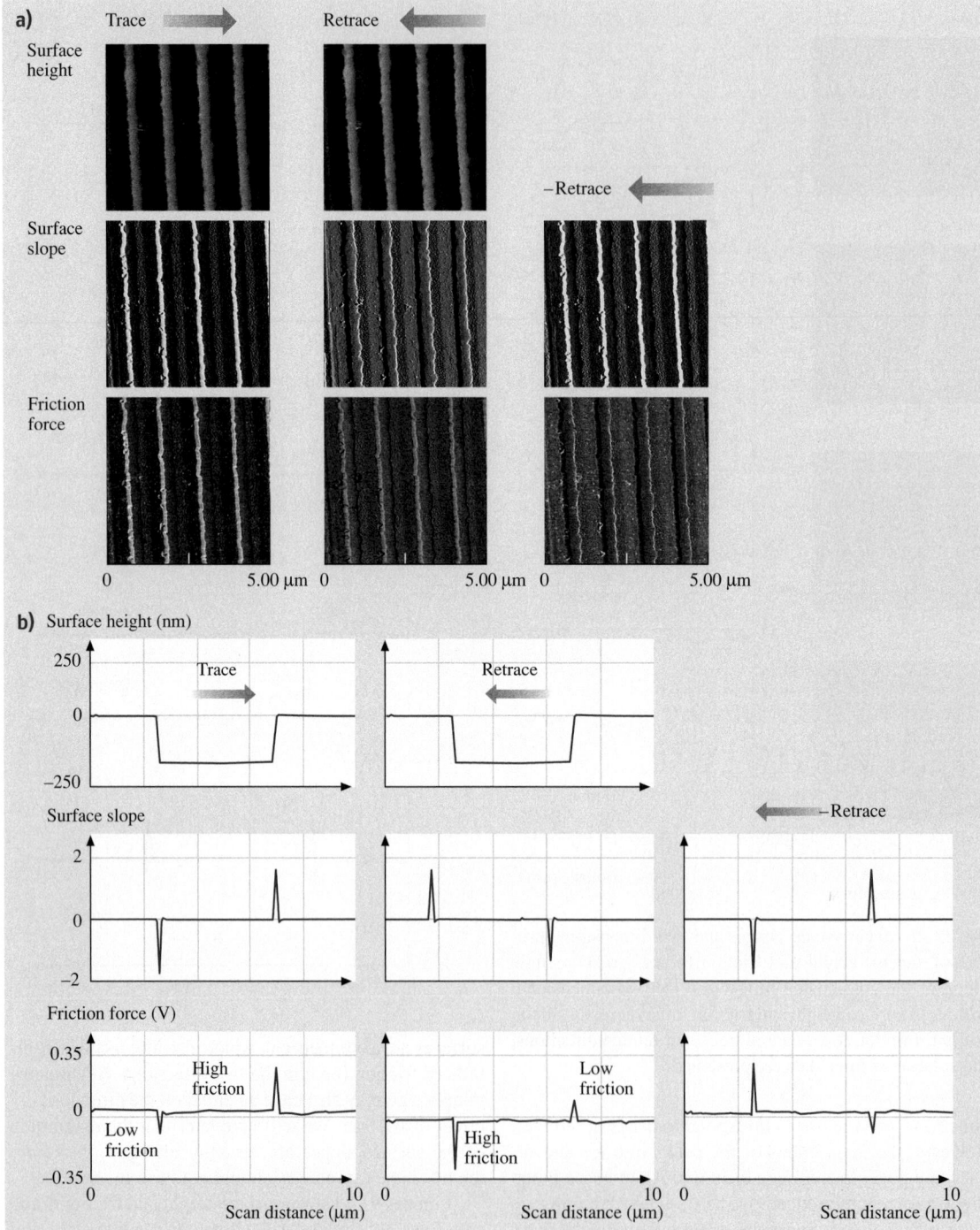

Fig. 29.22 (a) Greyscale images of surface heights, surface slopes and friction forces for scans across a gold-coated ruling, and **(b)** two-dimensional profiles of surface heights, surface slopes and friction forces for scans across the silicon grid pit. *Arrows* indicate the tip sliding direction [29.80]

This effect is observed at all locations where there is significant change in topography.

In order to facilitate comparison of the effect of directionality on friction, it is important to take into account the change of sign of the surface slope and friction force in the trace and retrace directions. Figure 29.22 shows surface height, surface slope, and friction force data for the two samples in the trace and retrace directions. The correlations between surface slope and friction forces are clear. The third column in the figures shows the retrace slope and friction data with an inverted sign (−retrace). Now we can compare the trace data with the −retrace data. It is clear that the friction experienced by the tip is dependent upon the scanning direction because of the surface topography. In addition to the effect of topographical changes discussed earlier, during surface-finishing processes material can be transferred preferentially onto one side of the asperities, which also causes asymmetry and direction dependence. Reduction in local variations and in the directionality of friction properties requires careful optimization of surface-roughness distributions and of surface-finishing processes.

The directionality as a result of the surface asperities effect will also manifest itself in macroscopic friction data, i.e., the coefficient of friction may be different in one sliding direction than that in the other direction. Asymmetrical shape of asperities accentuates this effect. Frictional directionality can also exist in materials with particles having a preferred orientation. The directionality effect in friction on a macroscale is observed in some magnetic tapes. In a macroscale test, a 12.7-mm-wide polymeric magnetic tape was wrapped over an aluminum drum and slid in a reciprocating motion with a normal load of 0.5 N and a sliding speed

of about 60 mm/s [29.3]. The coefficient of friction as a function of sliding distance in either direction is shown in Fig. 29.23. We note that the coefficient of friction on a macroscale for this tape is different in different directions. Directionality in friction is sometimes observed on the macroscale; on the microscale this is the norm [29.4, 15]. On the macroscale, the effect of surface asperities is normally averaged out over a large number of contacting asperities.

29.2.4 Surface-Roughness-Independent Microscale Friction

As just reported, friction contrast in conventional friction measurements is based on interactions that are dependent upon interfacial material properties superimposed with roughness-induced lateral forces, and the cantilever

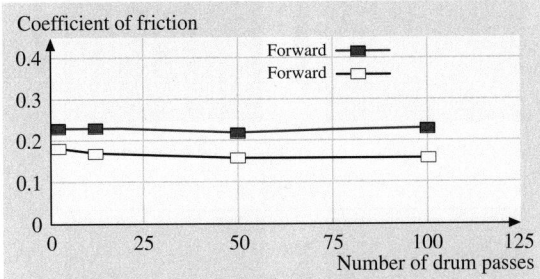

Fig. 29.23 Coefficient of macroscale friction as a function of drum passes for a polymeric magnetic tape sliding over an aluminum drum in a reciprocating mode in both directions. Normal load = 0.5 N over 12.7-mm-wide tape, sliding speed = 60 mm/s [29.65]

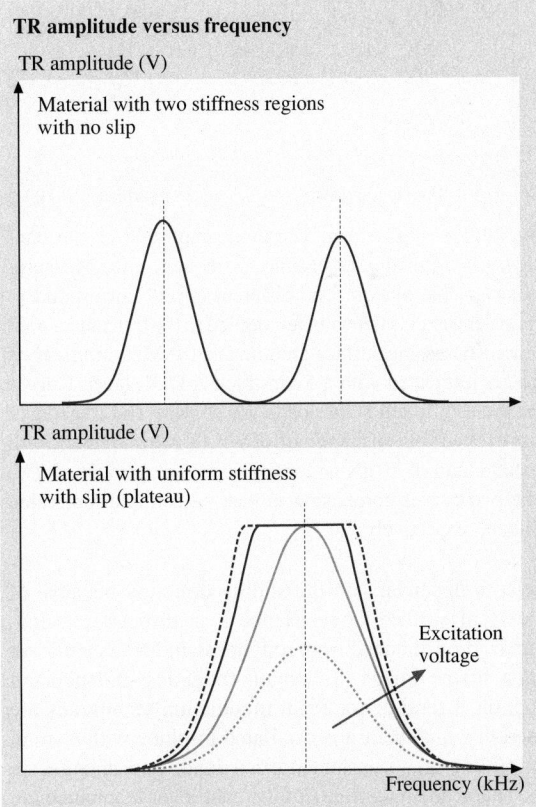

Fig. 29.24 Schematic showing frequency profiles of the TR amplitude for materials with two phases and a single phase. The maximum TR amplitude at the contact resonance frequency of the resonance curve with a flattened top, resulting from slip, can be used for friction force measurement

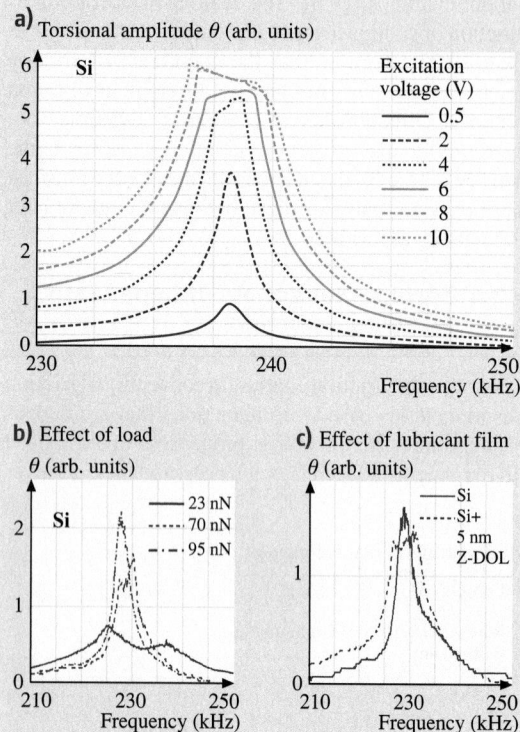

a) Torsional amplitude θ (arb. units)

b) Effect of load
θ (arb. units)

c) Effect of lubricant film
θ (arb. units)

Fig. 29.25a–c Torsional vibration amplitude of the cantilever as a function of excitation frequency. (**a**) Measurement on bare silicon. The different curves correspond to increasing excitation voltages applied to the transducer and, hence, increasing surface amplitudes. (**b**) Measurement on silicon lubricated with a 5-nm-thick Z-DOL layer. Curves for three different static loads are shown. The transducer was excited with an amplitude of 5 V. (**c**) Measurement with a static load of 70 nN and an excitation amplitude of 7 V. The two curves correspond to bare silicon and lubricated silicon, respectively [29.33]

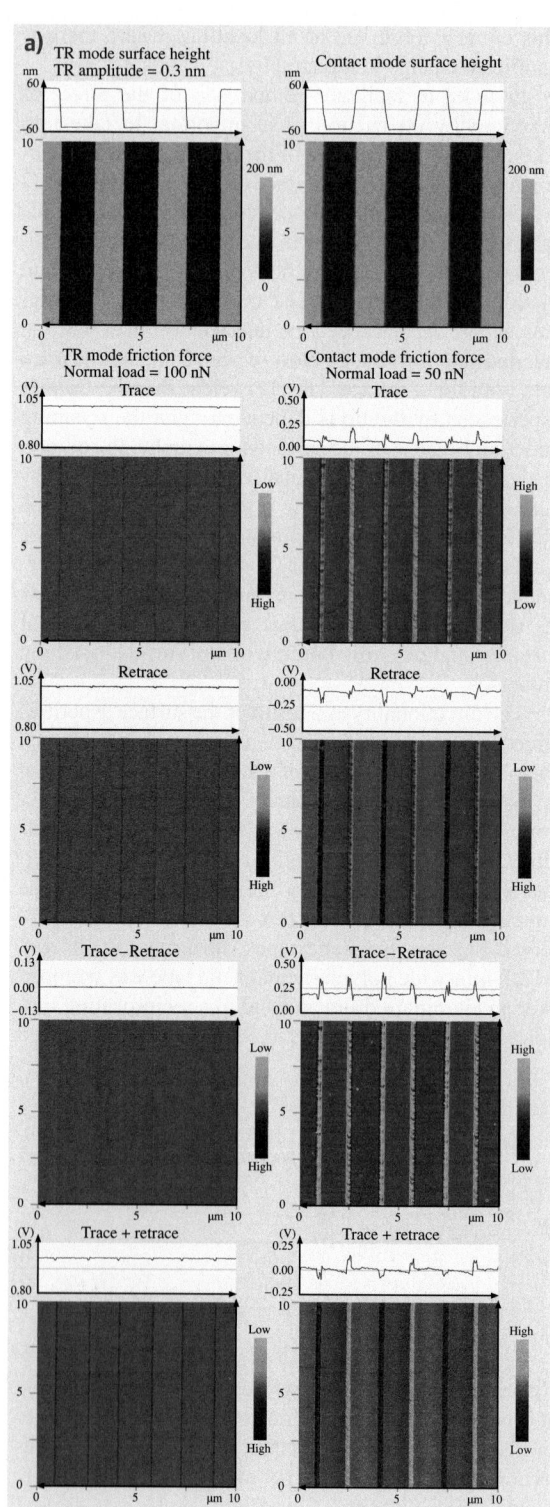

twist is dependent on the sliding direction because of the local surface slope. Hence it is difficult to separate friction-induced from roughness-induced cantilever twist in the image. To obtain roughness-independent friction, lateral or torsional modulation techniques are used in which the tip is oscillated in-plane with a small amplitude at a constant normal load, and changes in the shape and magnitude of the cantilever resonance are used as a measure of the friction force [29.31–36, 82]. These techniques also allow measurements over a very small region (a few nm to a few μm).

Scherer et al. [29.32] and *Reinstaedtler* et al. [29.33, 34] used the lateral mode for friction measurements

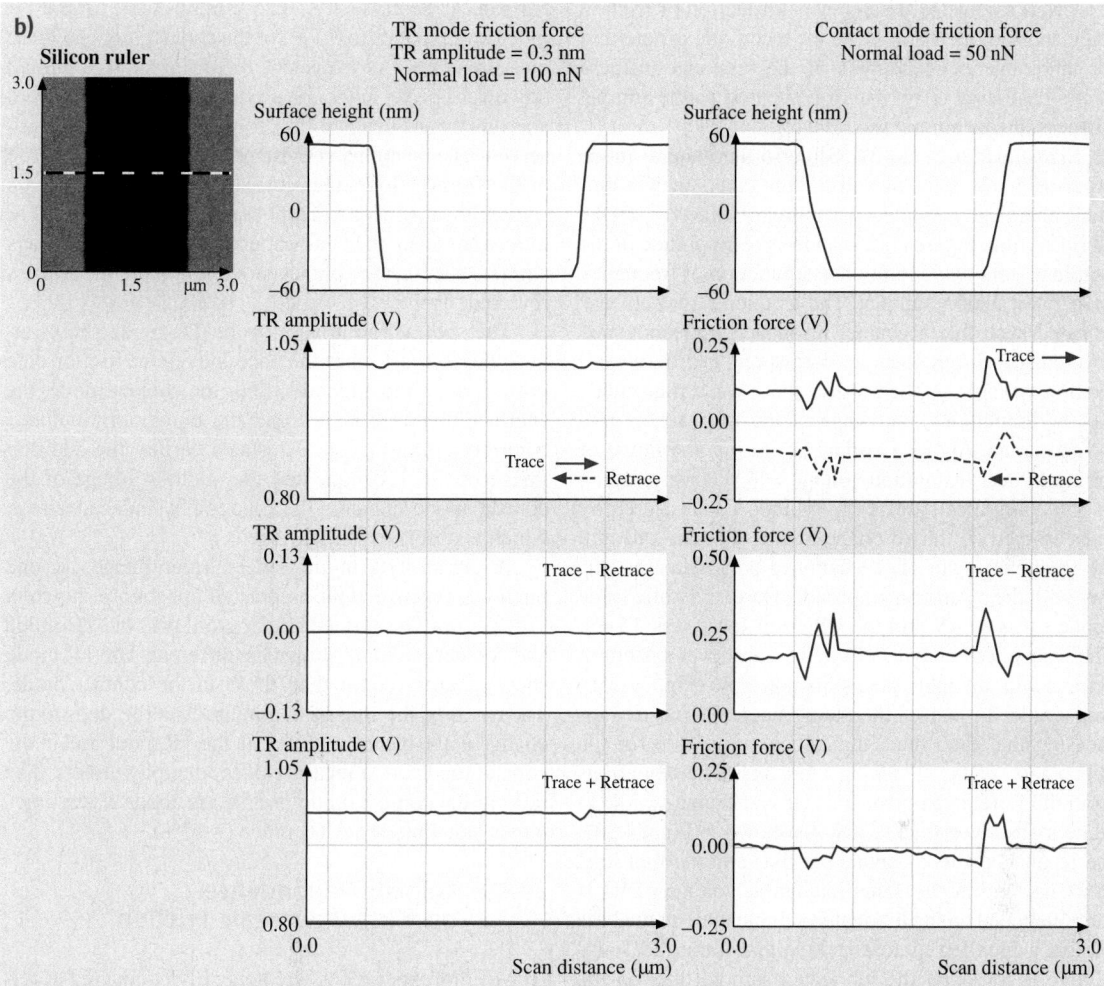

Fig. 29.26 (a) A comparison between the TR-mode friction and contact-mode friction maps together with line scans, on the silicon ruler. The TR-mode surface height and contact-mode surface height images are also shown. ◄
(b) A comparison of the line scans of the TR-mode friction and contact-mode friction on a selected pitch of the silicon ruler [29.36] ▲

(Fig. 29.10b). *Bhushan* and *Kasai* [29.36] used the TR mode for these measurements (Fig. 29.11). Before engagement, the cantilever is driven into torsional motion of the cantilever/tip assembly with a given normal vibration amplitude (the vibration amplitude in free air). After engagement, the vibration amplitude decreases due to the interaction between the tip and the sample, the vibration frequency increases and a phase shift occurs. During scanning, the normal load is kept constant, and the vibration amplitude of the cantilever is measured at the contact frequency.

As mentioned earlier, the shift in the contact resonance frequency in both the lateral and the TR modes is a measure of the contact stiffness, as shown schematically in Fig. 29.24. At excitation voltage above a certain value, as a result of micro-slip at the interface, a flattening of the resonance frequency spectra occurs (Fig. 29.22). At the lowest excitation voltage, the AFM tip sticks to the sample surface and follows the motion like an elastic contact with viscous damping, and the resonance curve is Lorentzian with a well-defined maximum. The excitation voltage should be high enough to initiate micro-slip. The maximum torsional amplitude at a given resonance frequency is a function of friction force and sample stiffness, so the technique is not valid for inhomogeneous samples. If the torsional stiffness of the cantilever is very high compared to the sample stiffness, the technique should work.

Reinstaedtler et al. [29.33] performed lateral-mode experiments on bare Si and Si lubricated with 5-nm-thick chemically bonded perfluoropolyether (Z-DOL) lubricant film. Figure 29.25a shows the amplitude of the cantilever torsional vibration as a function of frequency on a bare silicon sample. The frequency sweep was adjusted such that a contact resonance frequency was covered. The different curves correspond to different excitation voltages applied to the shear-wave transducer. At low amplitudes, the shape of the resonance curve is Lorentzian. Above a critical excitation amplitude of the transducer (excitation voltage $= 4$ V corresponding to ≈ 0.2 nm lateral surface amplitude, as measured by interferometry), the resonance curve flattens out and the frequency span of the flattened part increases further with the excitation amplitude. Here, the static force applied was 47 nN and the adhesion force was 15 nN. The resonance behavior of the tip–cantilever system in contact with the lubricated silicon sample (Fig. 29.25b) was similar to that of the bare silicon sample. By increasing the static load, the critical amplitude for the appearance of the flattening increases. The deviations from the Lorentzian resonance curve became visible at static loads lower than 95 nN. As shown in Fig. 29.25c, the resonance curve obtained at the same normal load of 70 nN and at the same excitation voltage (7 V) is more flattened on the lubricated sample than on the bare silicon, which led us to conclude that the critical amplitude is lower on the lubricated sample than on the bare sample. These experiments clearly demonstrate that torsional vibration of an AFM cantilever at ultrasonic frequencies leads to stick–slip phenomena and sliding friction. Above a critical vibration amplitude, sliding friction sets in.

Bhushan and *Kasai* [29.36] performed friction measurements on a silicon ruler and demonstrated that friction data in the TR mode is essentially independent of surface roughness and sliding direction. Figure 29.26a shows surface height and friction force maps on a silicon ruler obtained using the TR mode and contact-mode techniques. A comparison is made between the TR mode and contact-mode friction force maps. For easy comparison, the line scan profiles near the central area are shown on top of the greyscale maps. The vertical scales of the friction force profiles in the two graphs are selected to cover the same range of friction force so that direct comparison can be made, i. e., 0.25 V in full scale for the TR mode corresponds to 0.5 V for the contact mode in these measurements. As expected, for the trace scan, small downward peaks in the TR mode map and large upward and downward peaks in the contact mode map are observed. The positions of these peaks coincide with those of the surface slope; therefore, the peaks in the friction signals are attributed to a topography-induced effect. For the retrace scan, the peak pattern for the TR mode stays similar, but for the contact mode, the pattern becomes reversed.

The subtraction image for the TR mode shows almost flat contrast, since the trace and retrace friction data profiles are almost identical. For the contact mode, the subtraction image shows that the topography-induced contribution still exists. As stated earlier, the addition image of the TR mode and the addition image of the contact mode enhance the topography-induced effect, which is observed in the figure.

A closer look at the silicon ruler images at one pitch was taken, and the associated images are shown in Fig. 29.26b. The surface height profiles in the TR mode and contact mode are somewhat different. The TR mode shows sharper edges than those in the contact mode. The ratios of the change in amplitude at the steps to the change in the mean amplitude in the TR mode and in the contact mode are a measure of topography effects. The ratio in the contact mode ($\approx 85\%$) is about seven times larger than that of the TR mode ($\approx 12\%$).

29.2.5 Velocity Dependence on Micro/Nanoscale Friction

AFM/FFM experiments are generally conducted at relative velocities as high as about 200 µm/s. To simulate applications, it is of interest to conduct friction experiments at higher velocities. Furthermore, high-velocity experiments (up to 1 m/s) would be useful to study the dependence of friction and wear on velocity. One approach has been to mount samples on a shear-wave transducer (ultrasonic transducer) and then drive it at very high frequencies (in the MHz range) as reported earlier, see Fig. 29.10 [29.31–35, 82, 83]. The coefficient of friction is estimated based on the contact resonance frequency and requires the solution of the characteristic equations for the tip vibrating in contact with the sample surface. The approach is complex and depends on various assumptions.

An alternative approach is to utilize piezo stages with a large amplitude (≈ 100 µm) and a relatively low resonance frequency (a few kHz) and measure the fric-

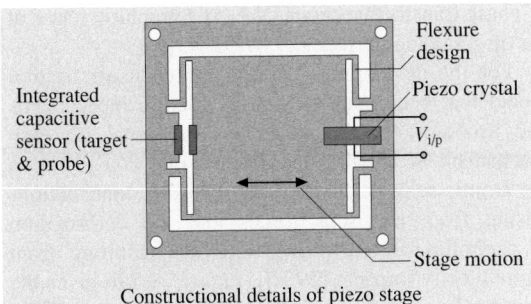

Fig. 29.27 Schematic of cross-sectional view showing construction details of the piezo stage. The integrated capacitive sensors are used as feedback sensors to drive the piezo. The piezo stage is mounted on the standard motorized AFM base and operated using independent amplifier and controller units driven by a frequency generator (not shown in the schematic) [29.37]

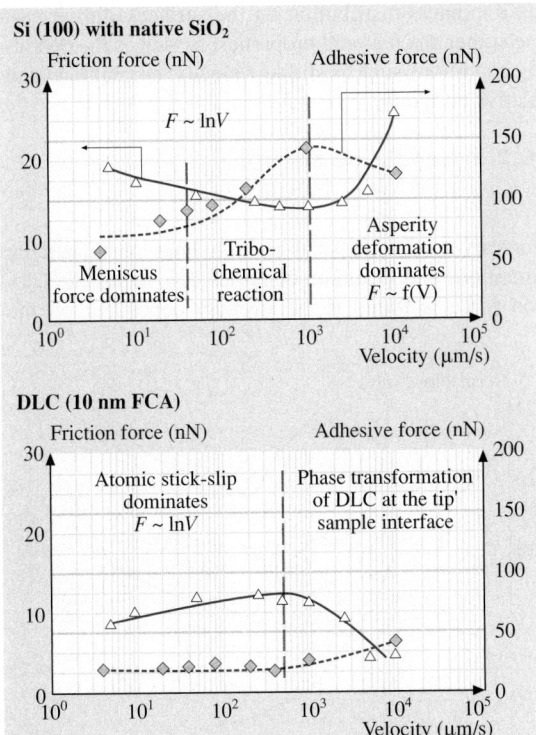

Fig. 29.28 Effect of velocity on friction force at a normal load of 70 nN and adhesive force over a wide range of velocities between 1 μm/s and 10 mm/s measured over a 25-μm scan size on Si(100) and DLC samples. The dominant friction mechanisms acting at different velocities are marked on the figure [29.84]

tion force directly using the FFM signal without any analysis, with the assumptions used in the previous approaches using shear-wave transducers. The commercial AFM set-up modified with this approach yields sliding velocities up to 10 mm/s [29.37]. In the modified set-up, the single-axis piezo stage is oriented such that the scanning axis is perpendicular to the long axis of the AFM cantilever (which corresponds to the 90° scan-angle mode of the commercial AFM; see Fig. 29.27). The displacement is monitored using an integrated capacitive feedback sensor, located diametrically opposite to the piezo crystal, as shown in Fig. 29.27. The capacitive change, corresponding to the stage displacement, gives a measure of the amount of displacement and can be used as feedback to the piezo controller for better guiding and tracking accuracy during scanning. The closed-loop position control of the piezoelectric-driven stages using capacitive feedback sensors provides linearity of motion better than 0.01% with nanometer resolution and a stable drift-free motion [29.37].

Figure 29.28 shows the friction force and adhesive force dependence on sliding velocity for single-crystal silicon, Si(100), and diamond-like carbon (DLC) with 10-nm thickness deposited using the filtered cathode arc (FCA) deposition technique [29.84–87]. The friction force and adhesive force are seen to vary with a change in velocity and both materials exhibit a reversal in friction behavior at certain critical sliding velocities. These reversals correspond to definitive transitions between different dominant friction mechanisms. For Si(100), which is hydrophilic, meniscus forces are dominant. The initial decrease in friction force with velocity corresponds to diminishing meniscus contributions. Beyond a certain critical velocity the residence time of the tip is not sufficient to form meniscus bridges at the sliding interface and the meniscus force contribution to the friction force drops out. At moderate velocities, tribochemical reactions at the tip–sample interface, in which a low-shear-strength $Si(OH)_4$ layer is formed [29.42, 88], also reduce friction. At high sliding velocities, deformation of asperities resulting from the high-velocity impacts becomes dominant and governs the friction behavior (see Fig. 29.29a). For DLC, being partially hydrophobic, meniscus forces are not dominant. The increase in friction force with velocity arises from atomic-scale stick–slip. Based on a thermally-activated Eyring model incorporated by *Bouhacina* et al. [29.89], the potential barrier divided by absolute temperature required for making jumps during stick–slip follows the logarithm of the velocity, and is responsible for the logarithmic dependence on velocity shown by

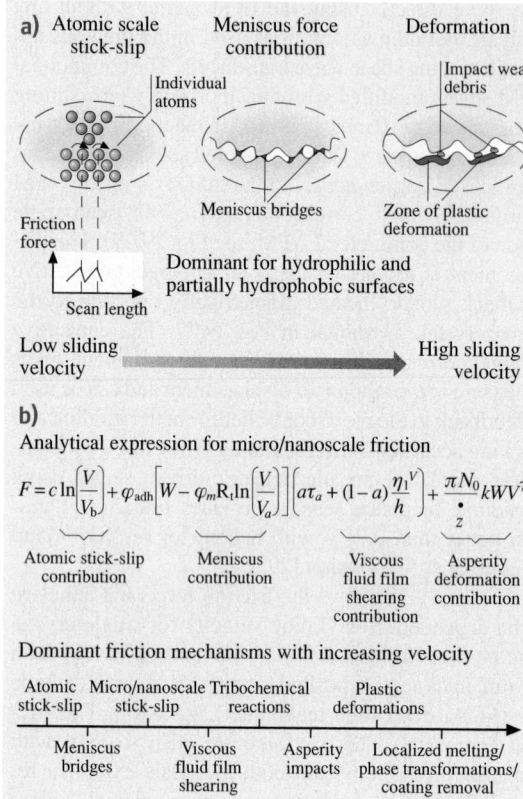

Fig. 29.29 (a) Schematic illustrating various dominant regimes of friction force at different relative sliding velocities from atomic scale stick–slip at low velocities to meniscus force contribution at mid velocities and deformation-related energy dissipation at high velocities, and **(b)** comprehensive analytical expression for velocity dependence of nanoscale friction with the dominant friction mechanisms [29.85]

friction force. Further work by *Gnecco* et al. [29.90] showed that this increase continues up to a certain sliding velocity and then levels off. At high velocities, the large frictional-energy dissipation results in a phase transformation of DLC to a graphite-like phase which is responsible for low friction.

For Si(100), the adhesive force is constant initially but then starts to increase beyond a certain velocity. The increase in the adhesive force for Si(100) is believed to be the result of a tribochemical reaction. This layer gets replenished continuously during sliding and results in a higher adhesive force between the tip and sample surface. For DLC, the adhesive force starts to increase beyond a critical velocity and is believed to be the result

of phase transformation of DLC to a graphitic phase at the tip–sample interface [29.87, 92, 93].

For the development of a comprehensive friction model that encompasses various friction mechanisms and accounts for the velocity dependence of each mechanism, it is necessary to take into consideration the atomic-scale stick–slip, the adhesive contributions arising from meniscus forces and the deformation of asperities at the sliding interface resulting from high-velocity impacts [29.85]. Figure 29.29b gives the analytical expression with different components of friction force. The nonlinear nature of micro/nanoscale friction force and its velocity dependence is apparent from this expression. The different terms representing different friction mechanisms are marked on the analytical expression in Fig. 29.29b and their relative order of precedence with respect to sliding velocity is illustrated schematically. Each of these mechanisms will be dominant depending on the specific sliding interface, the roughness distribution on the surfaces sliding past each other and material properties, as well as the operating conditions such as sliding velocity, normal load and relative humidity.

29.2.6 Nanoscale Friction and Wear Mapping

Contrary to the classical friction laws postulated by Amontons and Coulomb centuries ago, nanoscale friction force is found to be strongly dependent on the

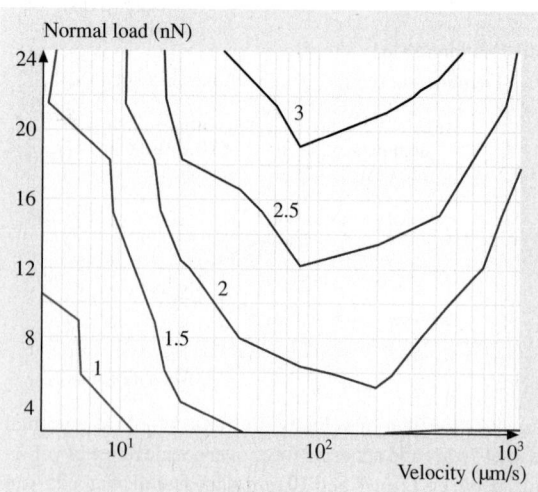

Fig. 29.30 Contour map showing friction force dependence on normal load and sliding velocity for DLC. Contour lines are constant friction force lines [29.91]

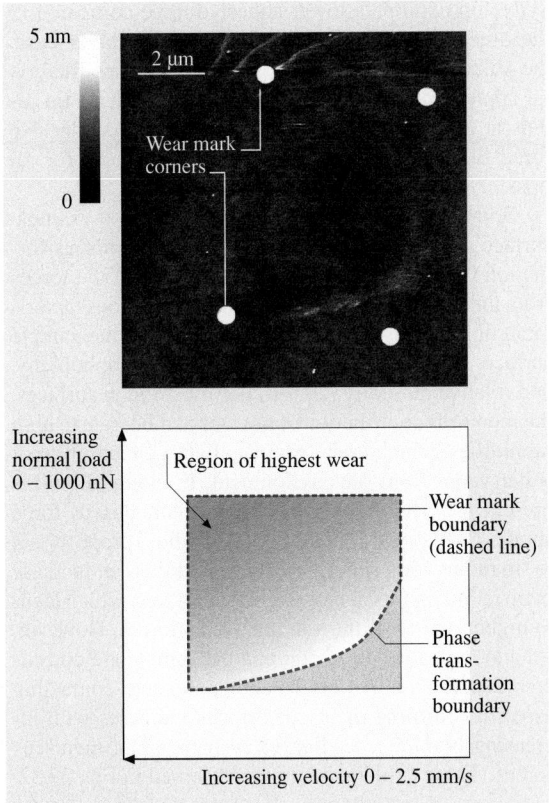

Fig. 29.31 Nanowear map (AFM image and schematic) illustrating the effect of sliding velocity and normal load on the wear of DLC resulting from phase transformation. The *curved area* shows debris lining and is indicative of the minimum frictional energy needed for phase transformation. For clarity, the wear mark corners are indicated by *white dots* in the upper AFM image and the various zones of interest over the entire wear mark are schematically illustrated in the lower image [29.94]

normal load and sliding velocity. Many materials, coatings and lubricants that have wide applications show reversals in friction behavior corresponding to transitions between different friction mechanisms [29.37, 84–86]. Most of the analytical models developed for explaining nanoscale friction behavior have remained limited in their focus and have left investigators shorthanded when trying to explain friction behavior scaling multiple regimes. Nanoscale friction maps provide fundamental insights into friction behavior. They help identify and classify the dominant friction mechanisms, as well as determine the critical operating parameters that influence transitions between different mecha-

nisms [29.85, 86]. Figure 29.30 shows a nanoscale friction map for DLC with friction mapped as a function of the normal load and the sliding velocity [29.91]. The contours represent lines of constant friction force. The friction force is seen to increase with normal load as well as velocity. The increase in friction force with velocity is the result of atomic-scale stick–slip and is a result of thermal activation of the irreversible jumps of the AFM tip that arise from overcoming of the energy barrier between adjacent atomic positions, as described earlier. The concentric contour lines, corresponding to constant friction force predicts a peak point, a point where the friction force reaches a maxima and beyond which any further increase in the normal load or the sliding velocity results in a decrease in the friction force. This characteristic behavior for DLC is the result of a phase transformation of DLC into a graphite-like phase by sp^3-to-sp^2 phase transition, as described earlier. During the AFM experiments, the Si_3N_4 tip gives rise to contact pressures in the range of $1.8–4.4\,GPa$ for DLC for normal loads of $10–150\,nN$ [29.87]. A combination of the high contact pressures that are encountered on the nanoscale and the high frictional-energy dissipation arising from the asperity impacts at the tip–sample interface due to the high sliding velocities accelerates a phase-transition process whereby a low-shear-strength graphite-like layer is formed at the sliding interface.

Similar to friction mapping, one way of exploring the broader wear patterns is to construct wear mechanism maps that summarize data and models for wear, thereby showing mechanisms for any given set of conditions to be identified [29.94–96]. Wear of sliding surfaces can occur by one or more wear mechanisms, including adhesive, abrasive, fatigue, impact, corrosive, and fretting [29.5, 10]. *Tambe* and *Bhushan* [29.87, 94] performed AFM experiments to develop nanoscale wear maps. Figure 29.31 shows a nanowear map generated for a DLC sample by simultaneously varying the normal load and the sliding velocity over the entire scan area. The wear map was generated for a normal load range of $0–1000\,nN$ and sliding velocity range of $0–2.5\,mm/s$. Wear debris, believed to result from phase transformation of DLC by an sp^3-to-sp^2 phase transition, was seen to form only for high values of sliding velocities times normal loads, i.e., only beyond certain threshold friction-energy dissipation [29.87, 94]. Hence the wear region exhibits a transition line indicating that for low velocities and low normal loads there is no phase transformation. For clarity, the wear mark corners are indicated by white dots in the AFM image (top) and the two zones of interest over the entire wear mark are illustrated schematically in Fig. 29.31 (top).

Nanoscale friction and wear mapping are novel techniques for investigating friction force and wear behavior on the nanoscale over a range of operating parameters. By simultaneously varying the sliding velocity and normal load over a large range of values, nanoscale friction and wear behavior can be mapped and the transitions between different wear mechanisms can be investigated. These maps help identify and demarcate critical operating parameters for different wear mechanisms and are very important tools in the process of design and selection of materials/coatings.

29.2.7 Adhesion and Friction in a Wet Environment

Experimental Observations

The tip radius and relative humidity affect adhesion and friction. The relative humidity affects adhesion and friction for dry and lubricated surfaces [29.30, 97, 98]. Figure 29.32 shows the variation of single-point adhesive force measurements as a function of tip radius on a Si(100) sample for several humidities. The adhesive force data are also plotted as a function of relative humidity for several tip radii. The general trend at humidities up to the ambient is that a 50-nm-radius

Si_3N_4 tip exhibits a lower adhesive force compared to the other microtips of larger radii; in the latter case, the values are similar. Thus for the microtips there is no appreciable variation in adhesive force with tip radius at a given humidity up to ambient. The adhesive force increases as relative humidity increases for all tips.

Sources of adhesive force between a tip and a sample surface are van der Waals attraction and meniscus formation [29.5, 10, 98]. Relative magnitudes of the forces from the two sources are dependent upon various factors, including the distance between the tip and the sample surface, their surface roughness, their hydrophobicity, and relative humidity [29.99]. For most rough surfaces, the meniscus contribution dominates at moderate to high humidities, which arise from capillary condensation of water vapor from the environment. If enough liquid is present to form a meniscus bridge, the meniscus force should increase with increasing tip radius (proportional to tip radius for a spherical tip). In addition, an increase in tip radius results in increased contact area, which leads to higher values of the van der Waals forces. However, if nano-asperities on the tip and the sample are considered then the number of contacting and near-contacting asperities forming meniscus bridges increases with increasing humidity, leading to an increase in meniscus forces. These explain the trends observed in Fig. 29.32.

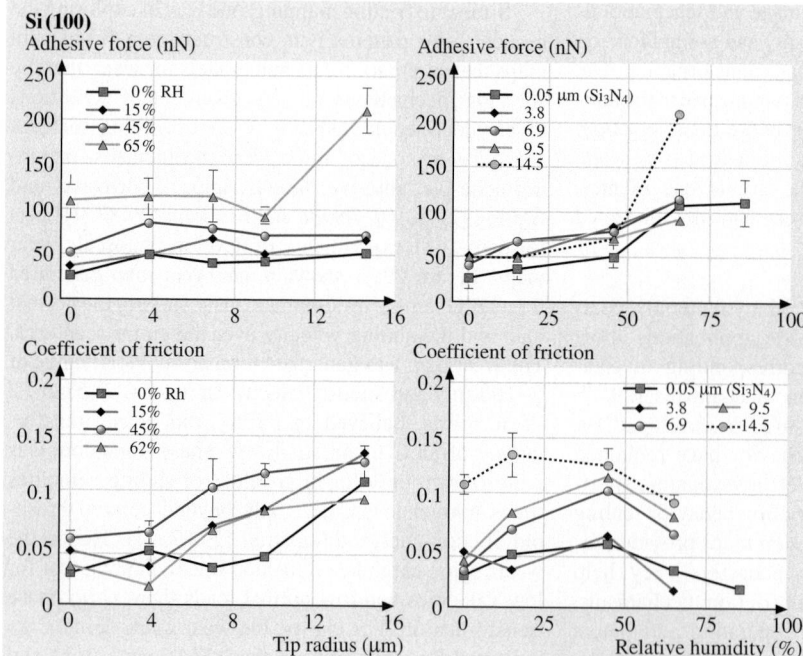

Fig. 29.32 Adhesive force and coefficient of friction as a function of tip radius at several humidities and as a function of relative humidity at several tip radii on Si(100) [29.30]

From the data, the tip radius has little effect on the adhesive forces at low humidities but increases with tip radius at high humidity. Adhesive force also increases with increasing humidity for all tips. This observation suggests that the thickness of the liquid film at low humidities is insufficient to form continuous meniscus bridges that would affect adhesive forces in the case of all tips.

Figure 29.32 also shows the variation in the coefficient of friction as a function of tip radius at given humidity, and as a function of relative humidity for a given tip radius for Si(100). It can be observed that, for 0% RH, the coefficient of friction is about the same for the tip radii except for the largest tip, which shows a higher value. At all other humidities, the trend consistently shows that the coefficient of friction increases with tip radius. An increase in friction with tip radius at low to moderate humidities arises from increased contact area (higher van der Waals forces) and higher values of shear forces required for larger contact area. At high humidities, similarly to the adhesive force data, an increase with tip radius occurs because of both contact-area and meniscus effects. Although AFM/FFM measurements are able to measure the combined effect of the contribution of van der Waals and meniscus forces towards friction force or adhesive force, it is difficult to measure their individual contributions separately. It can be seen that, for all tips, the coefficient of friction increases with humidity to about ambient, beyond which it starts to decrease. The initial increase in the coefficient of friction with humidity arises from the fact that the thickness of the water film increases with increasing humidity, which results in a larger number of nano-asperities forming meniscus bridges and leads to higher friction (larger shear force). The same trend is expected with the microtips beyond 65% RH. This is attributed to the fact that, at higher humidities, the adsorbed water film on the surface acts as a lubricant between the two surfaces. Thus the interface is changed at higher humidities, resulting in lower shear strength and hence a lower friction force and coefficient of friction.

Adhesion and Friction Force Expressions for a Single-Asperity Contact

We now obtain the expressions for the adhesive force and coefficient of friction for a single-asperity contact with a meniscus formed at the interface, Fig. 29.33. For a spherical asperity of radius R in contact with a flat and smooth surface with a composite modulus of elasticity E^* and in the presence of liquid with a concave meniscus, the attractive meniscus force (adhesive force),

designated as F_m or W_{ad} is given by [29.6, 10]:

$$W_{ad} = 2\pi R\gamma \left(\cos\theta_1 + \cos\theta_2\right) \quad (29.6)$$

where γ is the surface tension of the liquid and θ_1 and θ_2 are the contact angles of the liquid with surfaces 1 and 2, respectively. For an elastic contact for both extrinsic (W) and intrinsic (W_{ad}) normal load, the friction force is given as,

$$F_e = \pi\tau \left[\frac{3\left(W + W_{ad}\right)R}{4E^*}\right]^{2/3} \quad (29.7)$$

where W is the external load, and τ is the average shear strength of the contacts. (Surface energy effects are not considered here.) Note that the adhesive force increases linearly with increasing the tip radius, and the friction force increases with increasing tip radius as $R^{2/3}$ and with normal load as $(W + W_{ad})^{2/3}$. The experimental data in support of the $W^{2/3}$ dependence on the friction force can be found in various references (see e.g., *Schwarz* et al. [29.100]). The coefficient of friction μ_e is obtained from (29.7) as

$$\mu_e = \frac{F_e}{(W + W_{ad})} = \pi\tau \left[\frac{3R}{4E^*}\right]^{2/3} \frac{1}{(W + W_{ad})^{1/3}} \quad (29.8)$$

In the plastic contact regime [29.6], the coefficient of friction μ_p is obtained as

$$\mu_p = \frac{F_p}{(W + W_{ad})} = \frac{\tau}{H_s} \quad (29.9)$$

where H_s is the hardness of the softer material. Note that, in the plastic contact regime, the coefficient of friction is independent of the external load, adhesive contributions and surface geometry.

For comparisons, for multiple-asperity contacts in the elastic contact regime the total adhesive force W_{ad}

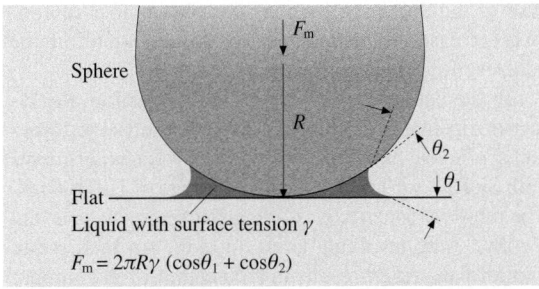

Fig. 29.33 Meniscus formation from a liquid condensate at the interface for a sphere in contact with a plane surface

is the summation of adhesive forces at n individual contacts,

$$W_{ad} = \sum_{i=1}^{n}(W_{ad})_i \quad \text{and}$$

$$\mu_e \approx \frac{3.2\tau}{E^*\left(\sigma_p/R_p\right)^{1/2}+(W_{ad}/W)} \tag{29.10}$$

where σ_p and R_p are the standard deviation of the summit heights and average summit radius, respectively. Note that the coefficient of friction depends upon the surface roughness. In the plastic contact regime, the expression for μ_p in (29.9) does not change.

The source of the adhesive force, in a wet contact in the AFM experiments being performed in an ambient environment, includes mainly attractive meniscus force due to capillary condensation of water vapor from the environment. The meniscus force for a single contact increases with increasing tip radius. A sharp AFM tip in contact with a smooth surface at low loads (on the order of a few nN) for most materials can be simulated as a single-asperity contact. At higher loads, for rough and soft surfaces, multiple contacts would occur. Furthermore, at low loads (nN range) for most materials, the local deformation would be primarily elastic. Assuming that the shear strength of contacts does not change, the adhesive force for smooth and hard surfaces at low normal load (on the order of a few nN) (for a single-asperity contact in the elastic contact regime) would increase with increasing tip radius, and the coefficient of friction would decrease with increasing total normal load as $(W+W_{ad})^{-1/3}$ and would increase with increasing tip radius as $R^{2/3}$. In this case, the Amontons law of friction, which states that the coefficient of friction is independent of normal load and is independent of apparent area of contact, does not hold. For a single-asperity plastic contact and multiple-asperity plastic contacts, neither the normal load nor the tip radius come into play in the calculation of the coefficient of friction. In the case of multiple-asperity contacts, the number of contacts increase with increasing normal load; therefore the adhesive force increases with increasing load.

In the data presented earlier in this section, the effect of tip radius and humidity on the adhesive forces and coefficient of friction is investigated for experiments with Si(100) surface at loads in the range of 10–100 nN. The multiple-asperity elastic-contact regime is relevant for this study involving large tip radii. An increase in humidity generally results in an increase of the number of meniscus bridges, which would increase the adhesive force. As was suggested earlier, this increase in humidity

may also decrease the shear strength of contacts. A combination of an increase in adhesive force and a decrease in shear strength would affect the coefficient of friction. An increase in tip radius would increase the meniscus force (adhesive force). A substantial increase in the tip radius may also increase interatomic forces. These effects influence the coefficient of friction with increasing tip radius.

Roughness Optimization for Superhydrophobic Surfaces

One of the crucial surface properties for surfaces in wet environments is non-wetting or hydrophobicity. It is usually desirable to reduce wetting in fluid flow applications and some conventional applications, such as glass windows and automotive windshields, in order for liquid to flow away along their surfaces. Reduction of wetting is also important in reducing meniscus formation, consequently reducing stiction, friction, and wear. Wetting is characterized by the contact angle, which is the angle between the solid and liquid surfaces. If the liquid wets the surface (referred to as a wetting li-

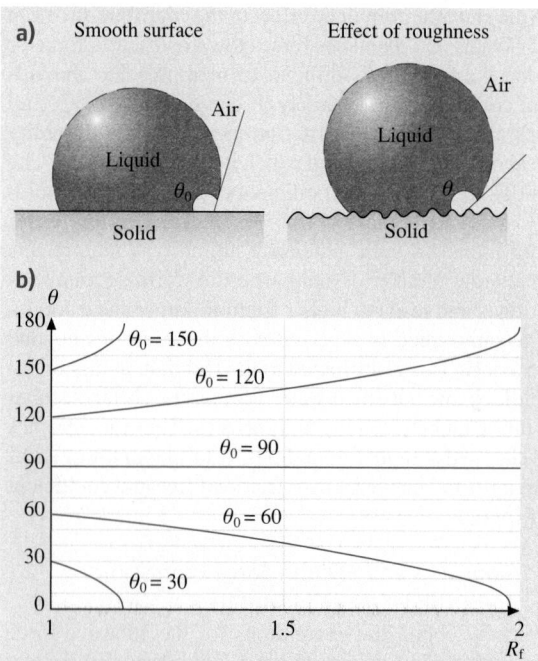

Fig. 29.34 (a) Droplet of liquid in contact with a smooth solid surface (contact angle θ_0) and rough solid surface (contact angle θ), and (b) contact angle for rough surface (θ) as a function of the roughness factor (R_f) for various contact angles for smooth surface (θ_0) [29.101]

quid or hydrophilic surface), the value of the contact angle is $0° \leq \theta \leq 90°$, whereas if the liquid does not wet the surface (referred to as a non-wetting liquid or hydrophobic surface), the value of the contact angle is $90° < \theta \leq 180°$. Superhydrophobic surfaces should also have very low water contact angle hysteresis. A surface is considered superhydrophobic if θ is close to 180°. One of the ways to increase the hydrophobic or hydrophilic properties of the surface is to increase surface roughness. It has been demonstrated experimentally that roughness changes contact angle. Some natural surfaces, including leaves of water-repellent plants such as lotus, are known to be superhydrophobic due to high roughness and the presence of a wax coating. This phenomenon is called in the literature the *lotus effect* [29.102].

If a droplet of liquid is placed on a smooth surface, the liquid and solid surfaces come together under equilibrium at a characteristic angle called the static contact angle θ_0, Fig. 29.34a. The contact angle can be determined from the condition of the total energy of the system being minimized. It can be shown that

$$\cos \theta_0 = \mathrm{d}A_{\mathrm{LA}}/\mathrm{d}A_{\mathrm{SL}} \tag{29.11}$$

where θ_0 is the contact angle for smooth surface, A_{SL} and A_{LA} are the solid–liquid and liquid–air contact areas. Next, let us consider a rough solid surface with a typical size of roughness details smaller than the size of the droplet, Fig. 29.34. For a rough surface, the roughness affects the contact angle due to an increased contact area A_{SL}. For a droplet in contact with a rough surface without air pockets, referred to as a homogeneous interface, based on the minimization of the total surface energy of the system, the contact angle is given as [29.103]

$$\cos \theta = \mathrm{d}A_{\mathrm{LA}}/\mathrm{d}A_{\mathrm{F}}$$
$$= \left(\frac{A_{\mathrm{SL}}}{A_{\mathrm{F}}}\right)(\mathrm{d}A_{\mathrm{LA}}/\mathrm{d}A_{\mathrm{SL}}) = R_{\mathrm{f}} \cos \theta_0 \tag{29.12}$$

where A_{F} is the flat solid–liquid contact area (or a projection of the solid–liquid area A_{SL} onto the horizontal plane). R_{f} is a roughness factor defined as

$$R_{\mathrm{f}} = \frac{A_{\mathrm{SL}}}{A_{\mathrm{F}}} \tag{29.13}$$

Equation (29.13) shows that, if the liquid wets a surface ($\cos \theta_0 > 0$), it will also wet the rough surface with a contact angle $\theta < \theta_0$, and for non-wetting liquids ($\cos \theta_0 < 0$), the contact angle with a rough surface will be greater than that with the flat surface, $\theta > \theta_0$. The dependence of the contact angle on the roughness factor is

presented in Fig. 29.34b for different values of θ_0, based on (29.12). It should be noted that (29.12) is valid only for moderate roughness, when $R_f \cos \theta_0 < 1$ [29.102].

For high roughness, air pockets (composite solid–liquid–air interface) will be formed in the cavities of the surface [29.104]. In the case of partial contact, the contact angle is given by

$$\cos \theta = R_{\mathrm{f}} f_{\mathrm{SL}} \cos \theta_0 - f_{\mathrm{LA}} \tag{29.14}$$

where f_{SL} and f_{LA} are fractional solid–liquid and liquid–air contact areas. The homogeneous and composite interfaces are two metastable states of the system. In reality, some cavities will be filled with liquid, and others with air, and the value of the contact angle is between the values predicted by (29.12) and (29.14). If the distance is large between the asperities or if the slope changes slowly, the liquid–air interface can easily be destabilized due to imperfectness of the profile shape or due to the dynamic effects, such as surface waves (Fig. 29.35). *Nosonovsky* and *Bhushan* [29.101] proposed a stochastic model, which relates the contact angle to the roughness and takes into account the possibility of destabilization of the composite interface due to imperfectness of the shape of the liquid–air interface, caused by effects such as capillary waves.

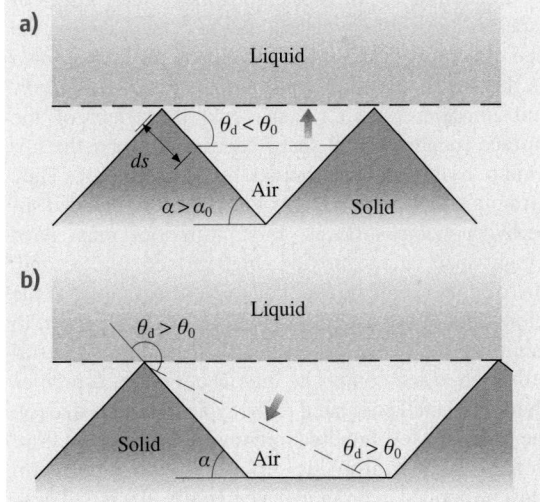

Fig. 29.35 (a) Formation of a composite solid–liquid–air interface for sawtooth and smooth profiles, and **(b)** destabilization of the composite interface for the sawtooth and smooth profiles due to dynamic effects. Dynamic contact angle $\theta_d > \theta_0$ corresponds to advancing liquid–air interface, whereas $\theta_d < \theta_0$ corresponds to the receding interface [29.101]

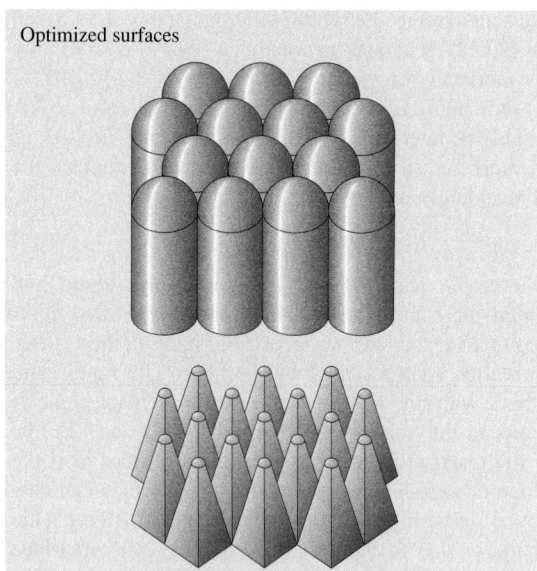

Fig. 29.36 Optimized roughness distribution – hemispherically topped cylindrical asperities (*upper*) and pyramidal asperities (*lower*) with square foundation and rounded tops. Square bases provide higher packing density but introduces undesirable sharp edges [29.102]

In addition to the surface roughness, sharp edges of asperities may affect wetting, because they can pin the line of contact of the solid, liquid, and air (also known as the *triple line*) and resist liquid flow. *Nosonovsky* and *Bhushan* [29.102] considered the effect of the surface roughness and sharp edges and found the optimum roughness distribution for non-wetting. They formulated five requirements for roughness-induced superhydrophobic surfaces. First, asperities must have a high aspect ratio to provide a high surface area. Second, sharp edges should be avoided, to prevent pinning of the triple line. Third, asperities should be tightly packed to minimize the distance between them and avoid destabilization of the composite interface. Fourth, asperities should be small compared to the typical droplet size (on the order of a few hundred microns or larger). And fifth, in the case of hydrophilic surface, a hydrophobic film must be applied in order to have initially $\theta > 90°$. These recommendations can be utilized for producing superhydrophobic surfaces. Remarkably, all these conditions are satisfied by biological water-repellent surfaces, such as some leaves: they have tightly packed hemispherically topped papillae with high (on the order of unity) aspect ratios and a wax coating [29.102]. Figure 29.36 shows two recommended geometries which use either hemi-

spherically topped asperities with a hexagonal packing pattern or pyramidal asperities with a rounded top. These geometries can be used for producing superhydrophobic surfaces.

Burton and *Bhushan* [29.105] have provided indirect evidence of an increase in contact angle and a decrease in adhesive force with the presence of discrete asperities with high aspect ratios based on measurements on smooth hydrophobic films ($\cos \theta_0 < 0$) and those with discrete asperities in a humid environment.

29.2.8 Separation Distance Dependence of Meniscus and van der Waals Forces

When two surfaces are in close proximity, sources of adhesive forces are weak van der Waals attraction and meniscus formation. The relative magnitudes of the forces from the two sources are dependent upon various factors including the interplanar separation, their surface roughness, their hydrophobicity, and relative humidity (liquid volume) [29.99]. The meniscus contribution dominates at moderate to high humidities and van der Waals forces dominate at asperities a few nm apart. In some micro/nanocomponents, it is important to know the relative contribution of the two sources as a function of a given interplanar separation to design an interface for low adhesion. For example, if two ultrasmooth surfaces come in close proximity with interplanar separation on the order of a nm, van der Waals forces may dominate and their magnitude may be reduced by creating bumps on one of the interfaces. This analysis is also of interest in AFM studies to understand the distance dependence of adhesive forces as the tip goes in and out of contact.

Stifter et al. [29.99] modeled the contact of a parabolic-shaped tip and a flat, smooth sample surface. The tip may represent a surface asperity on an interface or an AFM tip in an AFM experiment. They calculated van der Waals and meniscus forces as a function of different parameters, namely, tip geometry, tip–sample starting distance, relative humidity, surface tension, and contact angles. They compared the meniscus forces with van der Waals forces to understand their relative importance in various operating conditions.

The interacting force between the tip and sample in dry conditions is the Lennard–Jones force derived from Lennard–Jones potential. The Lennard–Jones potential is composed of two interactions – the van der Waals attraction and the Pauli repulsion. Van der Waals forces are significant because they are always present.

For a parabolic tip above a half plane with a distance D between the tip and plane, the Lennard–Jones potential is obtained by integrating the atomic potential over the volume of the tip and sample. It is given as [29.99]

$$V(D) = \frac{c}{12}\left(-\frac{A}{D} + \frac{B}{210D^7}\right) \qquad (29.15)$$

where c is the width of the parabolic tip (the diameter in the case of a spherical tip), A and B are two potential parameters where A is Hamakar constant. This equation provides expressions for attractive and repulsive parts. The calculations were made for Lennard–Jones force (total) and van der Waals force (attractive part) for two Hamaker constants 0.04×10^{-19} J (representative of polymers) and 3.0×10^{-19} J (representative of ceramics) and the meniscus force for a water film ($\gamma_\ell = 72.5$ N/m). Figure 29.37 shows various forces as a function of sep-

aration distance. The effect of two relative humidities and three tip radii was also studied which affect meniscus forces. The two dashed curves indicate the spread of possible van der Waals forces for two Hamaker constants. The figure shows that meniscus forces exhibit weaker distance dependence. The meniscus forces can be stronger or weaker than van der Waals forces for distances smaller than about 0.5 nm. For longer distances, the meniscus forces are stronger than the van der Waals forces. van der Waals forces must be considered for a tip–sample distance up to a few nm ($D < 5$ nm). The meniscus forces operate until the meniscus breaks, in the range 5–20 nm [29.99].

29.2.9 Scale Dependence in Friction

Table 29.3 presents adhesive force and coefficient of friction data obtained on the nanoscale and microscale [29.24, 106, 107]. Adhesive force and co-efficient of friction values on the nanoscale are about half to one order of magnitude lower than that on the microscale. Scale dependence is clearly observed in this data. As a further evidence of scale dependence, Table 29.4 shows the coefficient of friction measured for Si(100), HOPG, natural diamond, and DLC on the nanoscale and microscales. It is clearly observed that friction values are scale dependent.

To estimate the scale length, apparent contact radius at test loads are calculated and presented in the table. Mean apparent pressures are also calculated and presented. For nanoscale AFM experiments, it is assumed that an AFM tip coming into contact with a flat surface represents a single asperity and elastic contact, and Hertz analysis was used for the calculations. In the microscale experiments, a ball coming into contact with a flat surface represents multiple-asperity contacts due to the roughness, and the contact pressure of

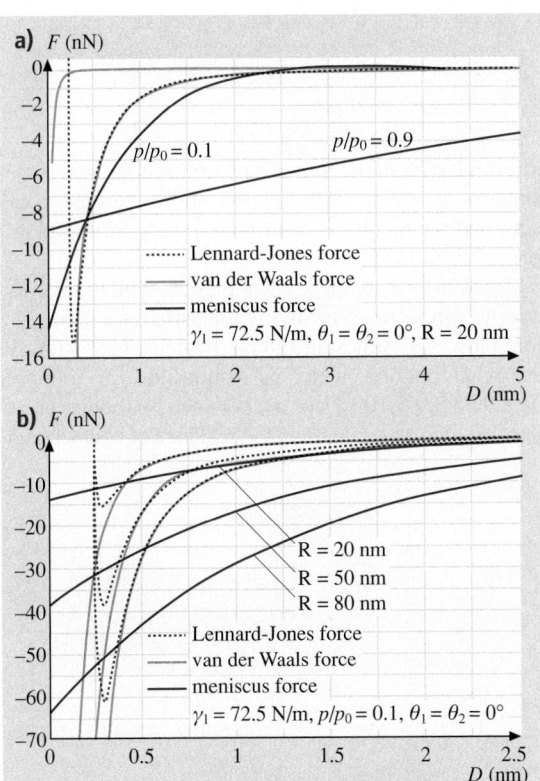

Fig. 29.37a,b Relative contribution of meniscus, van der Waals and Lennard–Jones forces (F) as a function of separation distance (D) and at (**a**) two values of relative humidity (p/p_0) for tip radius of 20 nm and Hamakar constants of 0.04×10^{-19} J and 3.0×10^{-19} J, and (**b**) three tip radii (R) and Hamakar constant of 3.0×10^{-19} J [29.99]

Table 29.3 Micro- and nanoscale values of adhesive force and co-efficient of friction in micro- and nanoscale measurements [29.106]

| Sample | Adhesive force | | Coefficient of friction | |
	Microscale[a] (μN)	Nanoscale[b] (nN)	Micro-scale[a]	Nano-scale[b]
Si(100)	685	52	0.47	0.06
DLC	325	44	0.19	0.03
Z-DOL	315	35	0.23	0.04
HDT	180	14	0.15	0.006

[a] Versus 500-μm-radius Si(100) ball,
[b] versus 50-nm-radius Si_3N_4 tip

Table 29.4 Micro- and nanoscale values of the coefficient of friction, typical physical properties of specimen, and calculated apparent contact radii and apparent contact pressures at loads used in micro- and nanoscale measurements. For calculation purposes it is assumed that contacts on micro- and nanoscale are single-asperity elastic contacts [29.114]

| Sample | Coefficient of friction | | Elastic modulus (GPa) | Poisson's ratio | Hardness (GPa) | Apparent contact radius at test load for | | Mean apparent pressure at test load for | |
	Micro-scale	Nano-scale				Microscale (μm) (upper limit)	Nano-scale (nm)	Microscale (GPa) (lower limit)	Nanoscale (GPa)
Si(100) wafer	0.47[a]	0.06[c]	130[e,f]	0.28[f]	9–10[e,f]	0.8–2.2[a]	1.6–3.4[c]	0.05–0.13[a]	1.3–2.8[c]
Graphite (HOPG)	0.1[b]	0.006[c]	9–15[g] (9)	– (0.25)	0.01[j]	62[b]	3.4–7.4[c]	0.082[b]	0.27–0.58[c]
Natural diamond	0.2[b]	0.05[c]	1140[h]	0.07[h]	80–104[g,h]	21[b]	1.1–2.5[c]	0.74[b]	2.5–5.3[c]
DLC film	0.19[a]	0.03[d]	280[i]	0.25[i]	20–30[i]	0.7–2.0[a]	1.3–2.9[d]	0.06–0.16[a]	1.8–3.8[d]

[a] 500-μm-radius Si(100) ball at 100–2000 μN and 720 μm/s in dry air [29.106]

[b] 3-mm-radius Si$_3$N$_4$ ball (elastic modulus 310 GPa, Poisson's ratio 0.22 [29.108]) at 1 N and 800 μm/s [29.24]

[c] 50-nm-radius Si$_3$N$_4$ tip at load range from 10–100 nN and 0.5 nm/s, in dry air [29.24]

[d] 50-nm-radius Si$_3$N$_4$ tip at load range from 10–100 nN in dry air [29.106]

[e] [29.109] [f] [29.110] [g] [29.108] [h] [29.111] [i] [29.112] [j] [29.113]

the asperity contacts is higher than the apparent pressure. For calculation of a characteristic scale length for multiple-asperity contacts, which is equal to the apparent length of contact, Hertz analysis was also used. This analysis provide an upper limit on the apparent radius and a lower limit on the mean contact pressure.

There are several factors responsible for the differences in the coefficients of friction at the micro- and nanoscale. Among them are the contributions from wear and contaminant particles, transition from elasticity to plasticity, and meniscus effect. The contribution of wear and contaminant particles is more significant at the macro/microscale because of the larger number of trapped particles, referred to as the third-body contribution. It can be argued that, for the nanoscale AFM experiments, the asperity contacts are predominantly elastic (with average real pressure less than the hardness of the softer material) and adhesion is the main contribution to the friction, whereas for the microscale experiments the asperity contact are predominantly plastic and deformation is an important factor. It will be shown later that hardness has a scale effect; it increases with decreasing scale and is responsible for less deformation on a smaller scale. The meniscus effect results in an increase of friction with increasing tip radius (Fig. 29.32). Therefore, the third-body contribution, scale-dependent hardness and other prop-

erties, the transition from elastic contacts in nanoscale contacts to plastic deformation in microscale contacts, and the meniscus contribution, play an important role [29.114–116].

Friction is a complex phenomenon, which involves asperity interactions involving adhesion and deformation (plowing). Adhesion and plastic deformation imply energy dissipation, which is responsible for friction, Fig. 29.38 [29.5,10]. A contact between two bodies takes place on high asperities, and the real area of contact (A_r)

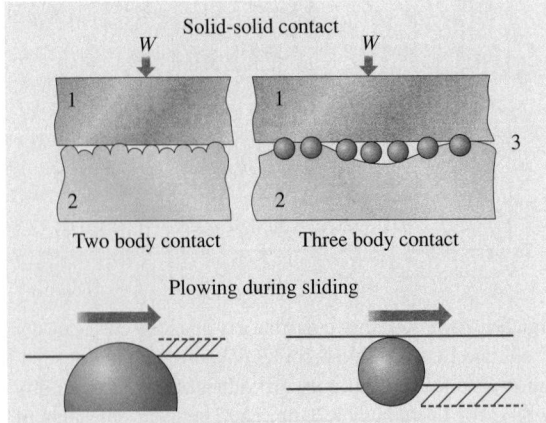

Fig. 29.38 Schematic of two-bodies and three-bodies during dry contact of rough surfaces

is a small fraction of the apparent area of contact. During contact of two asperities, a lateral force may be required for asperities of a given slope to climb against each other. This mechanism is known as the ratchet mechanism, and it also contributes to friction. Wear and contaminant particles present at the interface, referred to as the *third body*, also contribute to the friction, Fig. 29.38. In addition, during contact even at low humidity, a meniscus is formed (Fig. 29.33). Generally any liquid that wets or has a small contact angle on surfaces will condense from vapor into cracks and pores on the surfaces as bulk liquid and in the form of annular-shaped capillary condensate in the contact zone. A quantitative theory of scale effects in friction should consider the effect of scale on physical properties relevant to these various contributions.

According to the adhesion and deformation model of friction, the coefficient of dry friction μ is a sum of an adhesion component μ_a and a deformation (plowing) component μ_d. The later, in the presence of particles, is the sum of the asperity-summit deformation component μ_{ds} and the particles deformation component μ_{dp} so that the total coefficient of friction is [29.115]

$$\mu = \mu_a + \mu_{ds} + \mu_{dp} = \frac{F_a + F_{ds} + F_{dp}}{W}$$
$$= \frac{A_{ra}\tau_a + A_{ds}\tau_{ds} + A_{dp}\tau_{dp}}{W} \qquad (29.16)$$

where W is the normal load, F is the friction force, A_{ra}, A_{ds}, A_{dp} are the real areas of contact during adhesion, two-body deformation and with particles, respectively, and τ is the shear strength. The subscripts a, ds, and dp correspond to adhesion, summit deformation, and particle deformation, respectively.

The adhesional component of friction depends on the real area of contact and adhesion shear strength. The real area of contact is scale dependent due to the scale dependence of surface roughness (for elastic and plastic contact) and due to the scale dependence of hardness (for plastic contact) [29.115]. We limit the analysis here to multiple-asperity contact. For this case, the scale L is defined as the apparent size of the contact between two bodies. (For completeness, for single-asperity contact, the scale is defined as the contact diameter.) It is suggested by *Bhushan* and *Nosonovsky* [29.117] that, for many materials, dislocation-assisted sliding (microslip) is the main mechanism responsible for the shear strength. They considered dislocation-assisted sliding based on the assumption that contributing dislocations are located in a subsurface volume. The thickness of this volume is limited by the distance which dislocations can climb ℓ_s (material parameter) and by the radius of con-

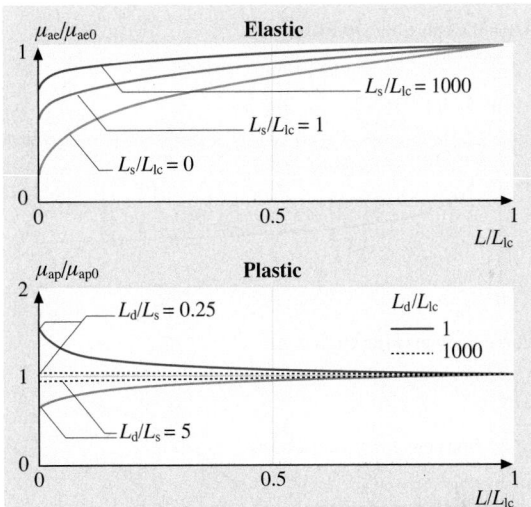

Fig. 29.39 Normalized results for the adhesional component of coefficient of friction, as a function of L/L_{ℓ_c} for multiple-asperity elastic contact. Data are presented for $m = 0.5, n = 0.2$. For multiple-asperity plastic contact, data are presented for two values of L_d/L_{ℓ_c} [29.115]

tact a. They showed that τ_a is scale dependent. Based on this, the adhesional components of the coefficient of friction in the case of elastic contact μ_{ae} and in the case of plastic contact μ_{ap} are given as [29.117]

$$\mu_{ae} = \frac{\mu_{ae0}}{\sqrt{\ell + (\ell_s/a_0)}} \left(\frac{L}{L_{\ell_c}}\right)^{m-n} \sqrt{1 + (L_s/L)^m},$$
$$L < L_{\ell_c} \qquad (29.17)$$

$$\mu_{ap} = \mu_{ap0} \sqrt{\frac{1 + (\ell_d/\overline{a_0})}{1 + (\ell_s/\overline{a_0})}} \sqrt{\frac{1 + (L_s/L)^m}{1 + (L_d/L)^m}},$$
$$L < L_{\ell_c} \qquad (29.18)$$

where μ_{ae0} and μ_{ap0} are values of the coefficient of friction at the macroscale $(L \geq L_{\ell_c})$, m and n are indices that characterize the scale dependence of surface parameters, $\overline{a_0}$ is the macroscale value of the mean contact radius, L_{ℓ_c} is the long-wavelength limit for scale dependence of the contact parameters, ℓ_s and ℓ_d are material-specific characteristic-length parameters, and L_s and L_d are length parameters related to ℓ_s and ℓ_d. The scale dependence of the adhesional component of the coefficient of friction is presented in Fig. 29.39, based on (29.17) and (29.18).

Based on the assumption that multiple asperities of two rough surfaces in contact have a conical shape, the two-body deformation component of friction can

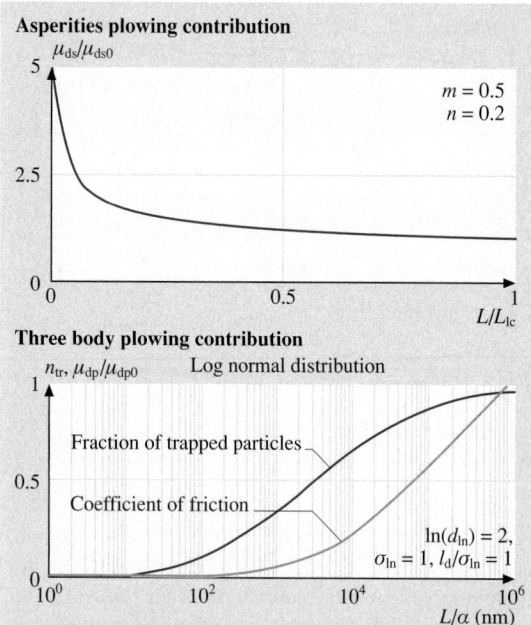

Fig. 29.40 (**a**) Normalized results for the two-body deformation component of the coefficient of friction, and (**b**) the number of trapped particles divided by the total number of particles and three-body deformation component of the coefficient of friction, normalized by the macroscale value for log-normal distribution of debris size, where α is the probability of a particle in the border zone to leave the contact region. Various constants given in the figure correspond to the log-normal distribution [29.115]

be determined as [29.5, 10]

$$\mu_{ds} = \frac{2\tan\theta_r}{\pi} \tag{29.19}$$

where θ_r is the roughness angle (or attack angle) of a conical asperity. Mechanical properties affect the real area of contact and shear strength and these cancel out in (29.16) [29.115]. Based on statistical analysis of a random Gaussian surface [29.115]

$$\mu_{ds} = \frac{2\sigma_0}{\pi\beta_0^*}\left(\frac{L}{L_{\ell_c}}\right)^{n-m} = \mu_{ds0}\left(\frac{L}{L_{\ell_c}}\right)^{n-m}, \quad L < L_{\ell c} \tag{29.20}$$

Fig. 29.42 Coefficient of friction as a function of normal load and for Si(111), SiO$_2$ coating and natural diamond. Inflections in the curves for silicon and SiO$_2$ correspond to the contact stresses equal to the hardness of these materials [29.28] ▶

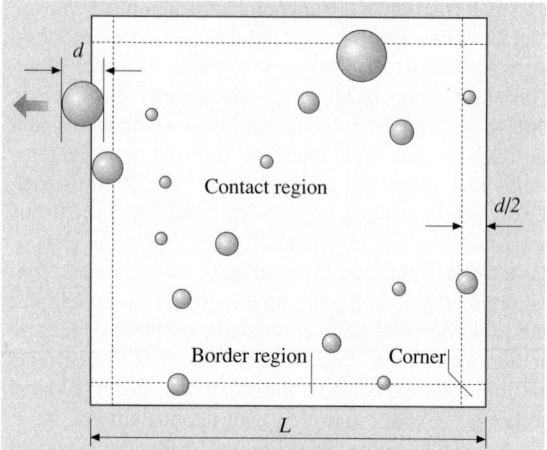

Fig. 29.41 Schematics of debris at the contact zone and at its border region. A particle of diameter d in the border region of $d/2$ is likely to leave the contact zone [29.115]

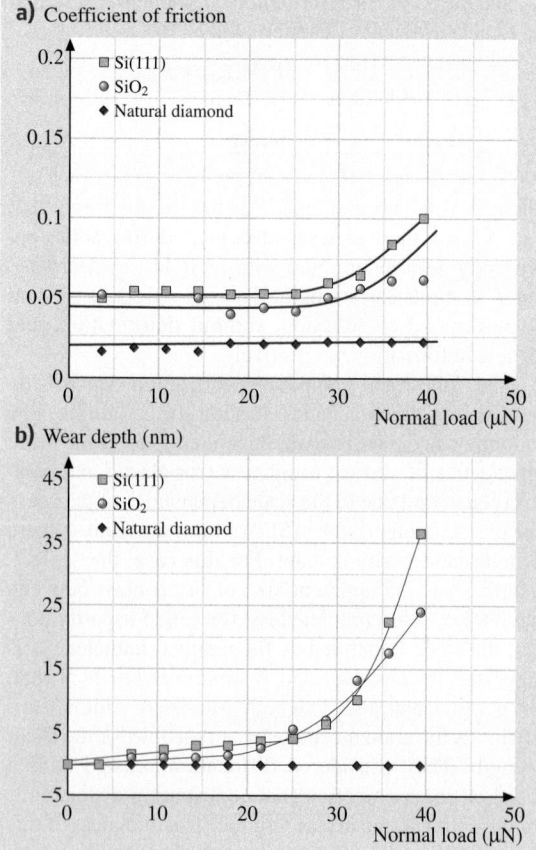

where μ_{ds0} is the value of the coefficient of the summit-deformation component of the coefficient of friction at the macroscale $\left(L \geq L_{\ell_c}\right)$, and σ_0 and β_0^* are the macroscale values of the standard deviation of surface heights and correlation length, respectively, for a Gaussian surface. The scale dependence for the two-body deformation component of the coefficient of friction is presented in Fig. 29.40a for $m = 0.5$, $n = 0.2$, based on (29.20). The coefficient of friction increases with decreasing scale, according to (29.20). This effect is a consequence of increasing average slope or roughness angle.

For three-body deformation, it is assumed that wear and contaminant particles at the borders of the contact region are likely to leave the contact region, while the particles in the center are likely to stay (Fig. 29.41). The plowing three-body deformation is plastic and, assuming that particles are harder than the bodies, the shear strength τ_{dp} is equal to the shear yield strength of the softer body τ_Y, the three-body deformation component of the coefficient of friction is given by [29.116]

$$\mu_{dp} = \mu_{dp0} n_{tr} \frac{\overline{d^2}}{d_0^2} \frac{\sqrt{1 + 2\ell_d/\overline{d}}}{\sqrt{1 + 2\ell_d/\overline{d_0}}} \qquad (29.21)$$

where $\overline{d}$ is the mean particle diameter, $\overline{d_0}$ is the macroscale value of the mean particle diameter, n_{tr} is the number of trapped particles divided by the total number of particles, and μ_{dp0} is the macroscale $(L \to \infty, n_{tr} \to 1)$ value of the third-body deformation component of the coefficient of friction. The scale dependence of μ_{dp} is shown in Fig. 29.40 based on (29.21). Based on the scale-effect predictions presented in Figs. 29.39 and 29.40, trends in the experimental results in Table 29.3 can be explained.

The scale dependence of meniscus effects in friction, wear and interface temperature can be analyzed in a similar way [29.116].

To demonstrate the load dependence of friction at the nano/microscale, the coefficient of friction as a function of normal load is presented in Fig. 29.42. The coefficient of friction was measured by *Bhushan* and *Kulkarni* [29.28, 29] for a Si_3N_4 tip versus Si, SiO_2, and natural diamond using an AFM. They reported that, for low loads, the coefficient of friction is independent of load and increases with increasing load after a certain load. It is noted that the critical value of loads for Si and SiO_2 corresponds to stresses equal to their hardness values, which suggests that the transition to plasticity plays a role in this effect. The friction values at higher loads for Si and SiO_2 approach that of macroscale values.

29.3 Wear, Scratching, Local Deformation, and Fabrication/Machining

29.3.1 Nanoscale Wear

Bhushan and *Ruan* [29.23] conducted nanoscale wear tests on polymeric magnetic tapes using conventional silicon nitride tips, at two different loads of 10 and 100 nN, Fig. 29.43. For a low, normal load of 10 nN, measurements were made twice. There was no discernible difference between consecutive measurements for this load. However, as the load was increased from 10 nN to 100 nN, topographical changes were observed during subsequent scanning at a normal load of 10 nN; material was pushed in the sliding direction of the AFM tip relative to the sample. The material movement is believed to occur as a result of plastic deformation of the tape surface. Thus, deformation and movement of the soft materials on a nanoscale can be observed.

29.3.2 Microscale Scratching

The AFM can be used to investigate how surface materials can be moved or removed on the micro- to nanoscales, for example, in scratching and wear [29.4, 11] (where these things are undesirable), and nanofabrication/nanomachining (where they are desirable). Figure 29.44a shows microscratches made on Si(111) at various loads and a scanning velocity of $2 \, \mu m/s$ after 10 cycles [29.27]. As expected, the scratch depth increases linearly with load. Such microscratching measurements can be used to study failure mechanisms on the microscale and to evaluate the mechanical integrity (scratch resistance) of ultra-thin films at low loads.

To study the effect of scanning velocity, unidirectional scratches, $5 \, \mu m$ in length, were generated at scanning velocities ranging from 1 to $100 \, \mu m/s$ at various normal loads ranging from 40 to $140 \, \mu N$. There is no effect of scanning velocity obtained at a given normal load. For representative scratch profiles at $80 \, \mu N$, see Fig. 29.44b. This may be because of a small effect of frictional heating with the change in scanning velocity used here. Furthermore, for a small change in interface temperature, there is a large underlying volume to dissipate the heat generated during scratching.

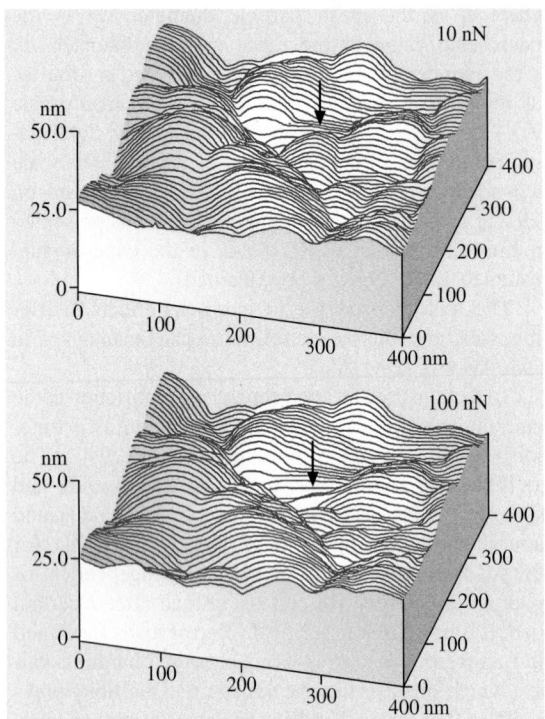

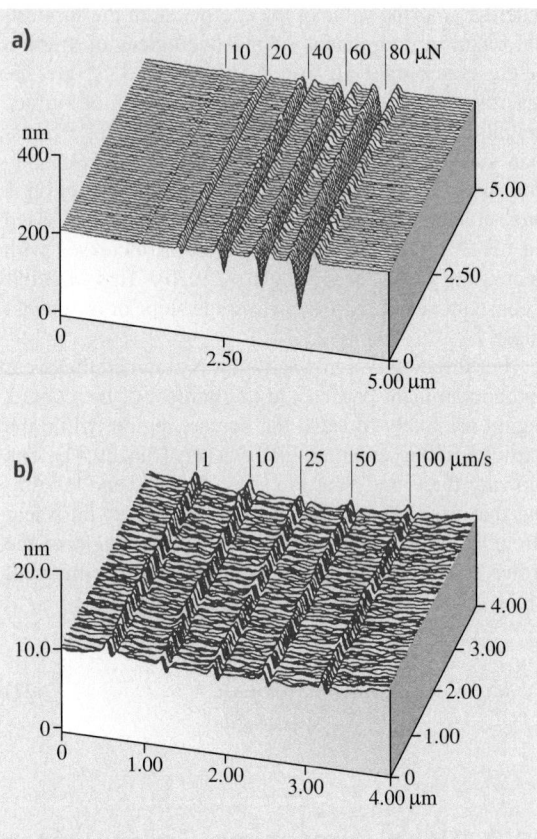

Fig. 29.43 Surface roughness maps of a polymeric magnetic tape at the applied normal load of 10 nN and 100 nN. Location of the change in surface topography as a result of nanowear is indicated by the *arrows* [29.23]

Fig. 29.44a,b Surface plots of (a) Si(111) scratched for ten cycles at various loads and a scanning velocity of 2 μm/s. Note that x and y axes are in μm and the z axis is in nm, and (b) Si(100) scratched in one unidirectional scan cycle at a normal force of 80 μN and different scanning velocities

Scratching can be performed under ramped loading to determine the scratch resistance of materials and coatings. The coefficient of friction is measured during scratching and the load at which the coefficient of friction increases rapidly is known as the *critical load*, which is a measure of scratch resistance. In addition, post-scratch imaging can be performed in situ with the AFM in tapping mode to study failure mechanisms. Figure 29.45 shows data from a scratch test on Si(100) with a scratch length of 25 μm and a scratching velocity of 0.5 μm/s. At the beginning of the scratch, the coefficient of friction is 0.04, which indicates a typical value for silicon. At about 35 μN (indicated by the arrow in the figure), there is a sharp increase in the coefficient of friction, which indicates the critical load. Beyond the critical load, the coefficient of friction continues to increase steadily. In the post-scratch image, we note that at the critical load a clear groove starts to form. This implies that Si(100) was damaged by plowing at the critical load, associated with the plastic flow of the material. At and after the crit-

ical load, small and uniform debris is observed and the amount of debris increases with increasing normal load. *Sundararajan* and *Bhushan* [29.47] have also used this technique to measure the scratch resistance of diamond-like-carbon coatings with thicknesses of 3.5–20 nm.

29.3.3 Microscale Wear

By scanning the sample in two dimensions with the AFM, wear scars are generated on the surface. Figure 29.46 shows the effect of normal load on wear depth. We note that wear depth is very small below 20 μN of normal load [29.118, 119]. A normal load of 20 μN corresponds to contact stresses comparable to the hardness

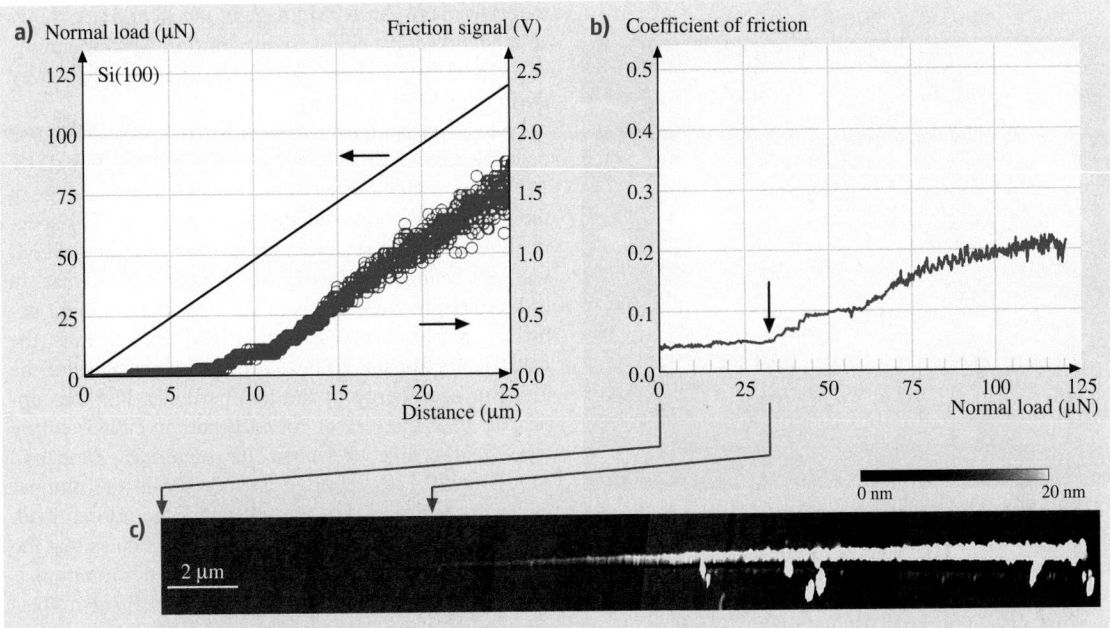

Fig. 29.45 (a) Applied normal load and friction signal measured during the microscratch experiment on Si(100) as a function of scratch distance, **(b)** friction data plotted in the form of coefficient of friction as a function of normal load, and **(c)** AFM surface height image of scratch obtained in tapping mode [29.47]

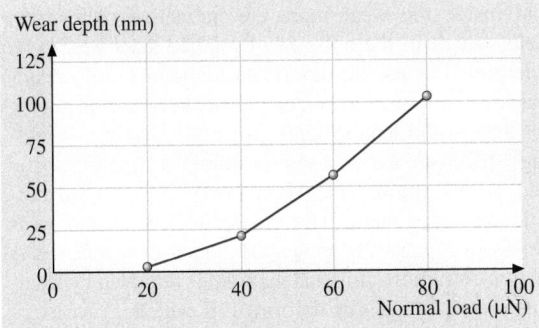

Fig. 29.46 Wear depth as a function of normal load for Si(100) after one cycle [29.119]

of the silicon. Primarily, elastic deformation at loads below 20 μN is responsible for low wear [29.28, 29].

A typical wear mark, of the size 2 μm × 2 μm, generated at a normal load of 40 μN for one scan cycle and imaged using AFM with a scan size of 4 μm × 4 μm at 300 nN load, is shown in Fig. 29.47a [29.118]. The

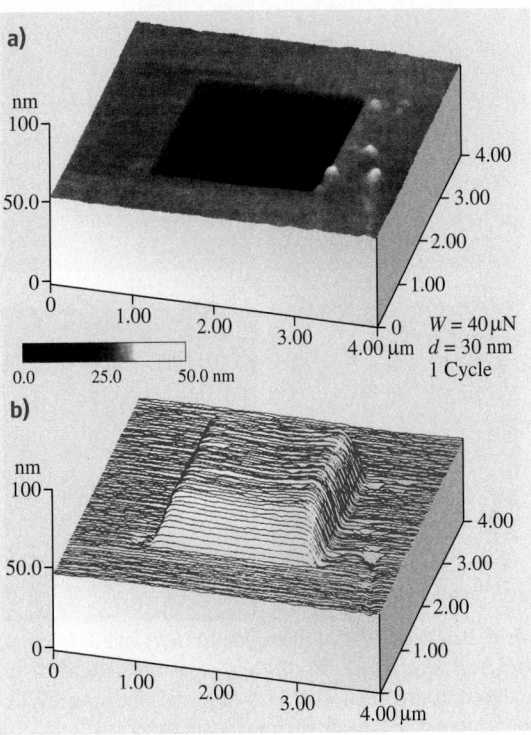

Fig. 29.47 (a) Typical greyscale and **(b)** inverted AFM images of wear mark created using a diamond tip at a normal load of 40 μN and one scan cycle on a Si(100) surface ▶

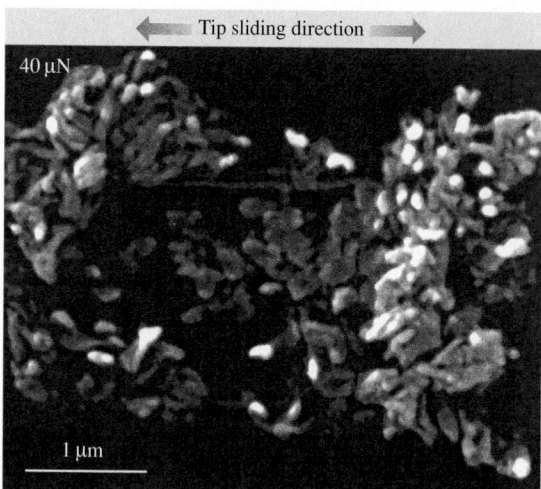

Fig. 29.48 Secondary electron image of wear mark and debris for Si(100) produced at a normal load of 40 μN and one scan cycle

inverted map of wear marks shown in Fig. 29.47b indicates the uniform material removal at the bottom of

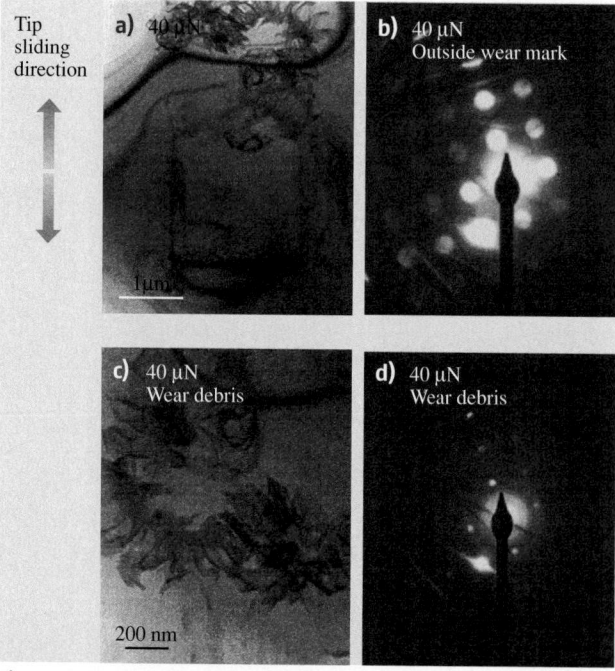

Fig. 29.49a–d Bright-field TEM micrographs (*left*) and diffraction patterns (*right*) of wear mark (**a**), (**b**) and wear debris (**c**), (**d**) in Si(100) produced at a normal load of 40 μN and one scan cycle. Bend contours around and inside wear mark are observed

the wear mark. An AFM image of the wear mark shows small debris at the edges, swiped during AFM scanning. Thus the debris is loose (not sticky) and can be removed during the AFM scanning.

Next we examine the mechanism of material removal on a microscale in AFM wear experiments [29.30, 118, 119]. Figure 29.48 shows a secondary-electron image of the wear mark and associated wear particles. The specimen used for the scanning electron microscope (SEM) was not scanned with the AFM after initial wear, in order to retain wear debris in the wear region. Wear debris is clearly observed. In the SEM micrographs, the wear debris appears to be agglomerated because of the high surface energy of the fine particles. Particles appear to be a mixture of rounded and so-called cutting type (feather-like or ribbon-like material). *Zhao* and *Bhushan* [29.119] reported an increase in the number and size of cutting-type particles with the normal load. The presence of cutting-type particles indicates that the material is removed primarily by plastic deformation.

To understand the material removal mechanisms better, transmission electron microscopy (TEM) has been used. The TEM micrograph of the worn region and associated diffraction pattern are shown in Fig. 29.49a,b. The bend contours are observed to pass through the wear mark in the micrograph. The bend contours around and inside the wear mark are indicative of a strain field, which in the absence of applied stresses can be interpreted as plastic deformation and/or elastic residual stresses. Often, localized plastic deformation during loading would lead to residual stresses during unloading; therefore, bend contours reflect a mix of elastic and plastic strains. The wear debris is observed outside the wear mark. The enlarged view of the wear debris in Fig. 29.49c shows that much of the debris is ribbon-like, indicating that material is removed by a cutting process via plastic deformation, which is consistent with the SEM observations. The diffraction pattern from inside the wear mark is similar to that of virgin silicon, showing no evidence of any phase transformation (amorphization) during wear. A selected area diffraction pattern of the wear debris shows some diffuse rings, which indicates the existence of amorphous material in the wear debris, confirmed as silicon oxide products from chemical analysis. It is known that plastic deformation occurs by generation and propagation of dislocations. No dislocation activity or cracking was observed at 40 μN. However, dislocation arrays could be observed at 80 μN. Figure 29.50 shows the TEM micrographs of the worn region at 80 μN; for better observation of the worn surface, wear debris was moved out

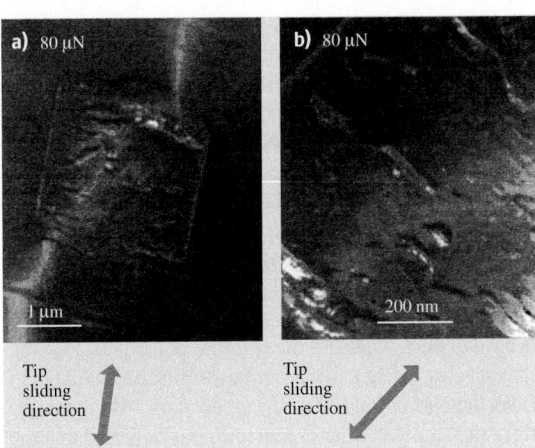

Fig. 29.50 (a) Bright-field and (b) weak-beam TEM micrographs of wear mark in Si(100) produced at a normal load of $80\,\mu N$ and one scan cycle showing bend contours and dislocations [29.119]

of the wear mark by using AFM with a large-area scan at 300 nm after the wear test. The existence of dislocation arrays confirms that material removal occurs by plastic deformation. This corroborates the observations made in scratch tests with a ramped load in the previous section. It is concluded that the material on the microscale at high loads is removed by plastic deformation with a small contribution from elastic fracture [29.119].

To understand wear mechanisms, the evolution of wear can be studied using AFM. Figure 29.51 shows evolution of wear marks of a DLC-coated disk sample. The data illustrate how the microwear profile for a load of $20\,\mu N$ develops as a function of the number of scanning cycles [29.27]. Wear is not uniform, but is initiated at the nanoscratches. Surface defects (with high surface energy) present at nanoscratches act as initiation sites for wear. Coating deposition may also not be uniform on and near nanoscratches, which may lead to coating delamination. Thus, scratch-free surfaces will be relatively resistant to wear.

Wear precursors (precursors to measurable wear) can be studied by making surface potential measurements [29.66–68]. The contact potential difference or simply the surface potential between two surfaces depends on a variety of parameters such as the electronic work function, adsorption, and oxide layers. The surface potential map of an interface gives a measure of changes in the work function which, is sensitive to both physical and chemical conditions of the surfaces including

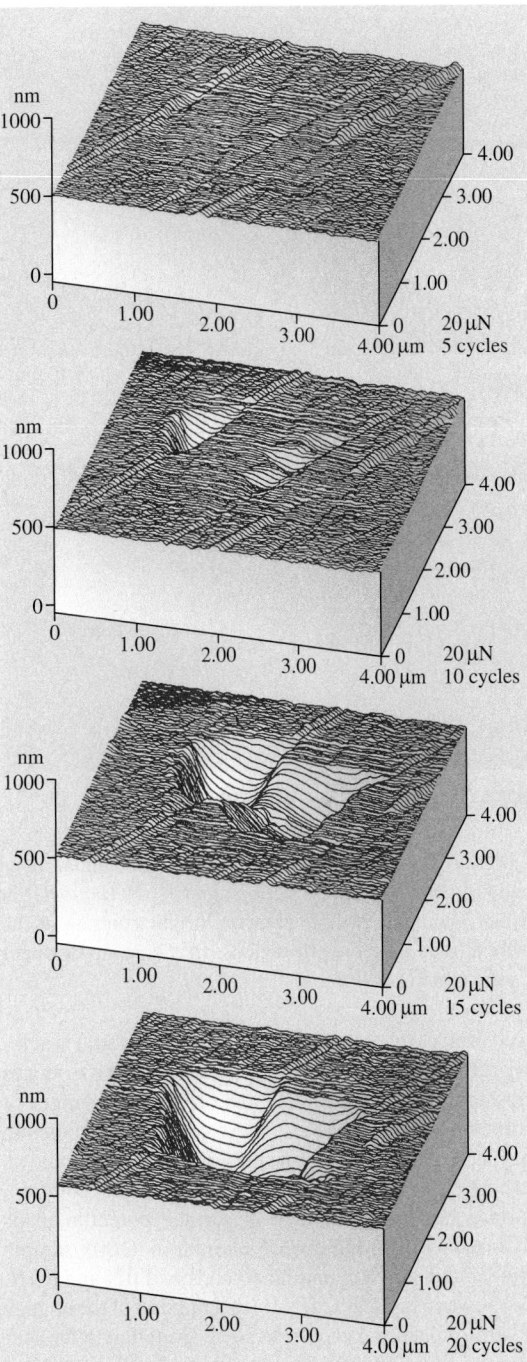

Fig. 29.51 Surface plots of diamond-like carbon-coated thin-film disk showing the worn region; the normal load and number of test cycles are indicated [29.27]

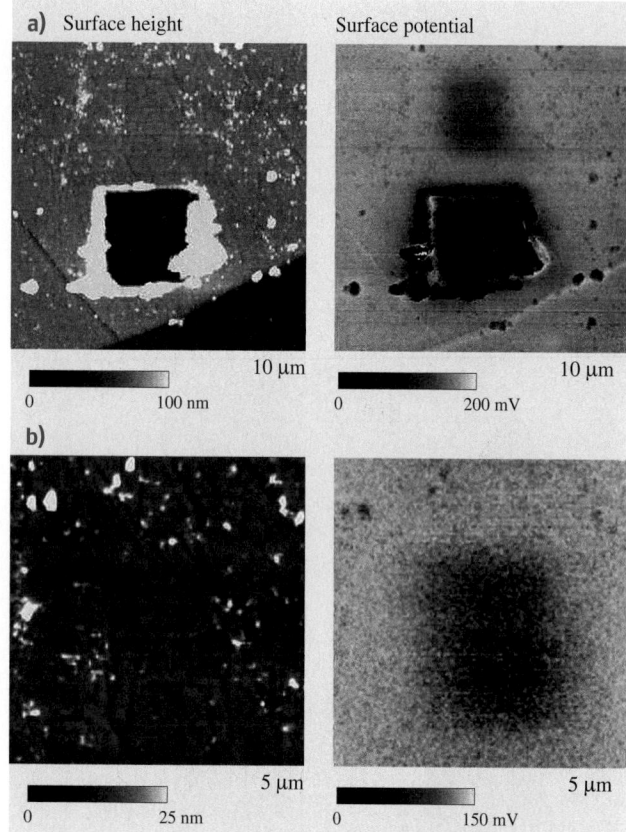

a) Surface height   Surface potential

10 μm   10 μm

0   100 nm   0   200 mV

b)

5 μm   5 μm

0   25 nm   0   150 mV

Fig. 29.52 (a) Surface height and change in surface potential maps of wear regions generated at 1 μN (*top*) and 9 μN (*bottom*) on a single-crystal aluminum sample showing bright contrast in the surface potential map on the worn regions. (b) Close-up of upper (low-load) wear region [29.66]

structural and chemical changes. Before material is actually removed in a wear process, the surface experiences stresses that result in surface and subsurface changes of structure and/or chemistry. These can cause changes in the measured potential of a surface. An AFM tip allows mapping of surface potential with nanoscale resolution. Surface height and change in surface potential maps of a polished single-crystal aluminum (100) sample abraded using a diamond tip at loads of 1 μN and 9 μN, are shown in Fig. 29.52a [Note that the sign of the change in surface potential is reversed here from that in DeVecchio and Bhushan [29.66]]. It is evident that both abraded regions show a large potential contrast (≈ 0.17 V), with respect to the non-abraded area. The black region in the lower right-hand part of the topography scan shows a step that was created during the polishing phase. There

is no potential contrast between the high region and the low region of the sample, indicating that the technique is independent of surface height. Figure 29.52b shows a close-up scan of the upper (low-load) wear region in Fig. 29.52a. Notice that, while there is no detectable change in the surface topography, there is nonetheless a large change in the potential of the surface in the worn region. Indeed, the wear mark of Fig. 29.52b might not be visible at all in the topography map were it not for the noted absence of wear debris generated nearby and then swept off during the low-load scan. Thus, even in the case of zero wear (no measurable deformation of the surface using AFM), there can be a significant change in the surface potential inside the wear mark, which is useful for the study of wear precursors. It is believed that the removal of the thin contaminant layer including the natural oxide layer gives rise to the initial change in surface potential. The structural changes, which precede generation of wear debris and/or measurable wear scars, occur under ultra-low loads in the top few nanometers of the sample, and are primarily responsible for the subsequent changes in surface potential.

29.3.4 In Situ Characterization of Local Deformation

In situ surface characterization of local deformation of materials and thin films is carried out using a tensile stage inside an AFM. Failure mechanisms of polymeric thin films under tensile load were studied by *Bobji* and *Bhushan* [29.61, 62]. The specimens were strained at a rate of 4×10^{-3}% per second and AFM images were captured at different strains up to about 10% to monitor generation and propagation of cracks and deformation bands.

Bobji and *Bhushan* [29.61, 62] studied three magnetic tapes with thickness ranging from 7 to 8.5 μm. One of these was with acicular-shaped metal particle (MP) coating and the other two with metal-evaporated (ME) coating and with and without a thin diamond-like carbon (DLC) overcoat both on a polymeric substrate and all with particulate back-coating [29.15]. They also studied a polyethylene terephthalate (PET) substrate with a thickness of 6 μm. They reported that cracking of the coatings started at about 1% strain for all tapes, well before the substrate starts to yield at about 2% strain. Figure 29.53 shows the topographical images of the MP tape at different strains. At 0.83% strain, a crack can be seen, originating at the marked point. As the tape is further stretched along the direction, as shown in Fig. 29.53, the crack propagates along the shorter boundary of the

Fig. 29.53 Topographical images of the MP magnetic tape at different strains [29.61] ▶

ellipsoidal particle. However, the general direction of the crack propagation remains perpendicular to the direction of the stretching. The length, width, and depth of the cracks increase with strain, and at the same time newer cracks keep on nucleating and propagating with reduced crack spacing. At 3.75% strain, another crack can be seen nucleating. This crack continues to grow parallel to the first one. When the tape is unloaded after stretching up to a strain of about 2%, i. e. within the elastic limit of the substrate, the cracks rejoin perfectly and it is impossible to determine the difference from the unstrained tape.

Figure 29.54 shows topographical images of the three magnetic tapes and the PET substrate after being strained to 3.75%, which is well beyond the elastic limit of the substrate. MP tape develops numerous short cracks perpendicular to the direction of loading. In tapes with metallic coating, the cracks extend throughout the tape width. In ME tape with DLC coating, there is a bulge in the coating around the primary cracks that are initiated when the substrate is still elastic, like crack A in the figure. The white band on the right-hand side of the figure is the bulge of another crack. The secondary cracks, such as B and C, are generated at higher strains and are straighter compared to the primary cracks. In ME tape which has a $Co - O$ film on a PET substrate, with a thickness ratio of 0.03, both with and without DLC coating, no difference is observed in the rate of growth between primary and secondary cracks. The failure is cohesive with no bulging of the coating. This seems to suggest that the DLC coating has residual stresses that relax when the coating cracks, causing delamination. Since the stresses are already relaxed, the secondary crack does not result in delamination. The presence of the residual stress is confirmed by the fact that a free-standing ME tape curls up (in a cylindrical form with its axis perpendicular to the tape length) with a radius of curvature of about 6 mm and the ME tape without the DLC does not curl. The magnetic coating side of the PET substrate is much smoother at shorter scan lengths. However, in 20-μm scans it has a lot of bulging out, which appears as white spots in the figure. These spots change shape even while scanning the samples in tapping mode at very low contact forces.

Fig. 29.54 Comparison of crack morphologies at 3.75% strain in three magnetic tapes and PET substrate. Cracks B and C, nucleated at higher strains, are more linear than crack A [29.62] ▶

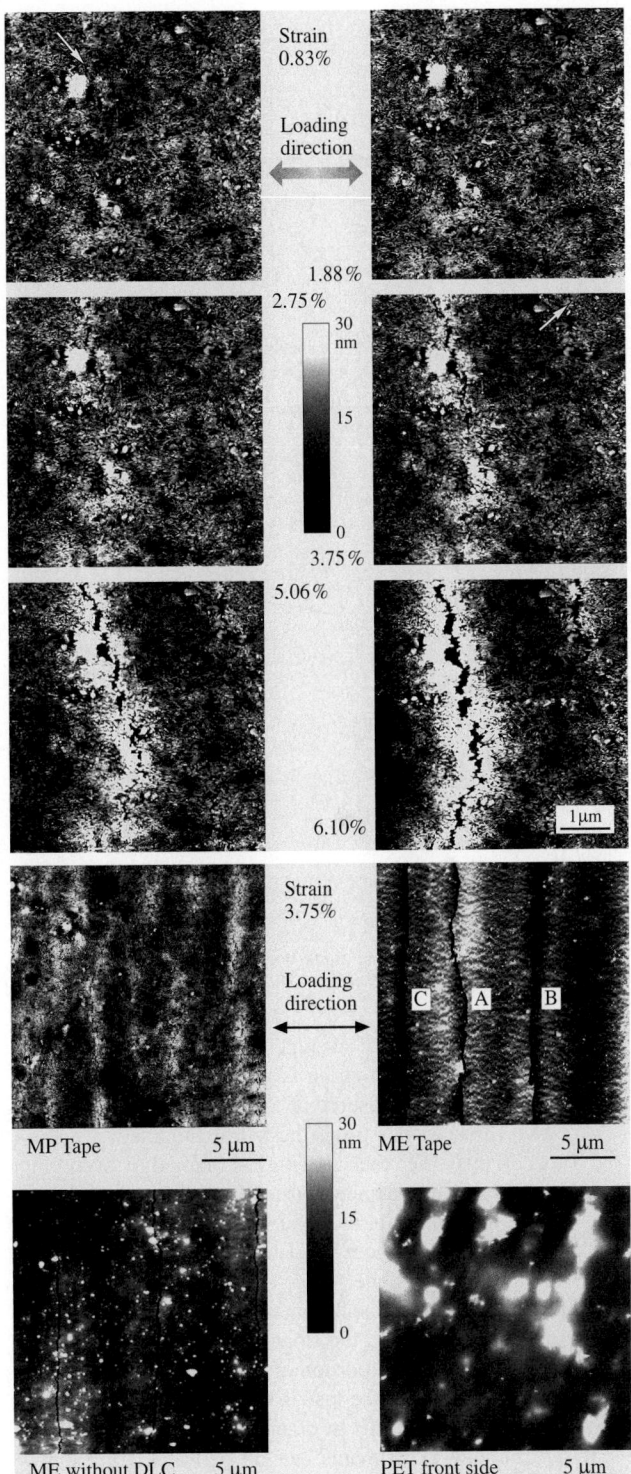

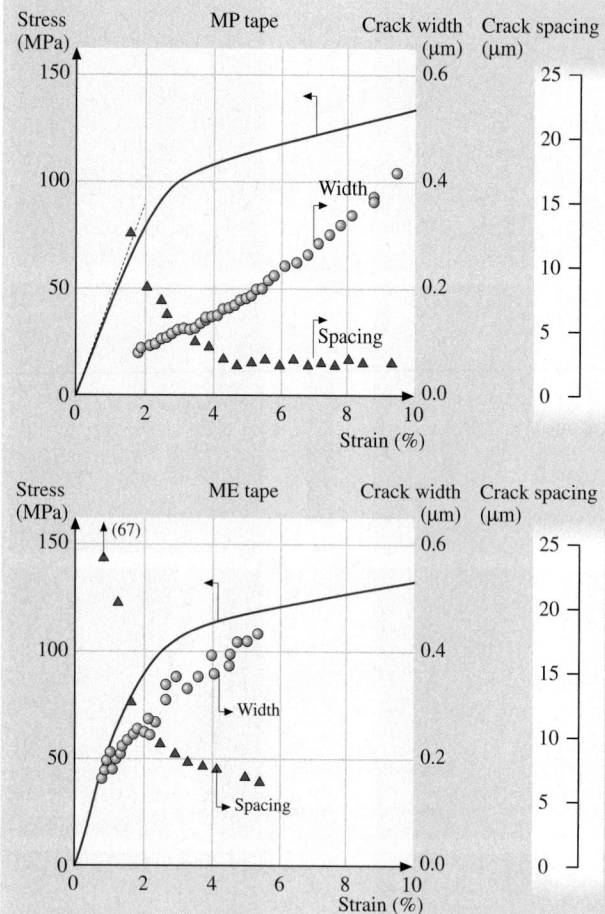

Fig. 29.55 Variation of stress, crack width, and crack spacing with strain in two magnetic tapes [29.61]

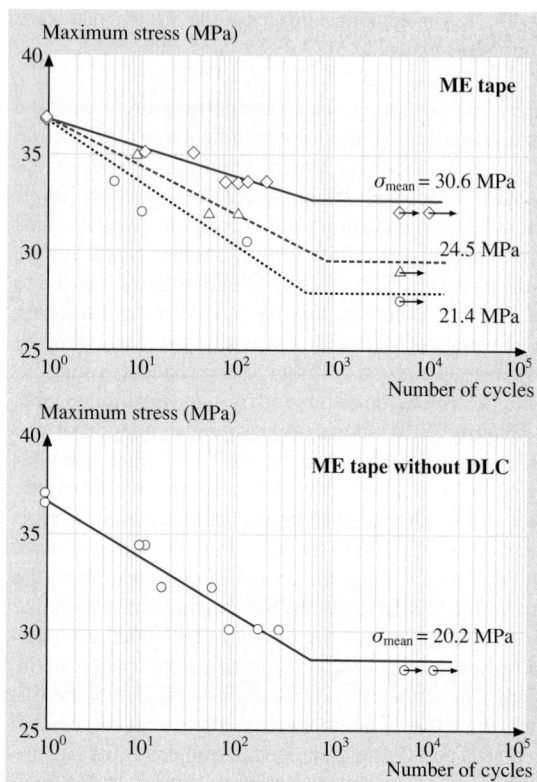

Fig. 29.56 S–N curve for two magnetic tapes with maximum stress plotted on the ordinate and number of cycles to failure on the abscissa. The data points marked with *arrows* indicate tests for which no failure (cracking) was observed in the scan area, even after a large number of cycles (10 000)

The variation of average crack width and average crack spacing with strain is plotted in Fig. 29.55. The crack width is measured at a spot along a given crack over a distance of 1 μm in the 5-μm scan image at different strains. The crack spacing is obtained by averaging the inter-crack distance measured in five separate 50-μm scans at each strain. It can be seen that the cracks nucleate at a strain of about 0.7–1.0%, well within the elastic limit of the substrate. There is a definite change in the slope of the load–displacement curve at the strain where cracks nucleate and the slope after that is closer to the slope of the elastic portion of the substrate. This would mean that most of the load is supported by the substrate once the coating fails by cracking.

Fatigue experiments can be performed by applying a cyclic stress amplitude with a certain mean

stress [29.63]. Fatigue life was determined by the first occurrence of cracks. Experiments were performed at various constant mean stresses and with a range of cyclic stress amplitudes for each mean stress value for various magnetic tapes. The number of cycles to failure were plotted as a function of stress state to obtain a so-called S–N (stress–life) diagram. As the stress is decreased, there is a stress value for which no failure occurs. This stress is termed the endurance limit or simply the fatigue limit. Figure 29.56 shows the S–N curve for an ME tape and an ME tape without DLC. For the ME tape, the endurance limit is seen to go down with decreasing mean stress. This is consistent with the literature and is because for lower mean stress the corresponding stress amplitude is relatively high and this causes failure. The endurance limit is found to be almost the same for all three mean stresses. In the case of ME tape without DLC, the critical number of cycles is also found to be in the same range.

In situ surface characterization of unstretched and stretched films has been used to measure the Poisson's ratio of polymeric thin films by *Bhushan* [29.120]. Uniaxial tension is applied by the tensile stage. Surface height profiles obtained from the AFM images of unstretched and stretched samples are used to monitor the changes in displacements of the polymer films in the longitudinal and lateral directions simultaneously.

29.3.5 Nanofabrication/Nanomachining

An AFM can be used for nanofabrication/nanomachining by extending the microscale scratching operation [29.4, 13, 27, 65]. Figure 29.57 shows two examples of nanofabrication. The patterns were created on a single-crystal silicon (100) wafer by scratching the sample surface with a diamond tip at specified locations and scratching angles. Each line is scribed manually at a normal load of 15 μN and a writing speed of 0.5 μm/s. The separation between lines is about 50 nm and the variation in line width is due to the tip asymmetry. Nanofabrication parameters – normal load, scanning speed, and tip geometry – can be controlled precisely to control depth and length of the devices.

Nanofabrication using mechanical scratching has several advantages over other techniques. Better control over the applied normal load, scan size, and scanning speed can be used for nanofabrication of devices. Using the technique, nanofabrication can be performed on any engineering surface. Use of chemical etching or reactions is not required and this dry nanofabrication process can be used where the use of chemicals and electric fields is prohibited. One disadvantage of this technique is the formation of debris during scratching. At light loads, debris formation is not a problem compared to high-load scratching. However, debris can be removed easily from the scan area at light loads during scanning.

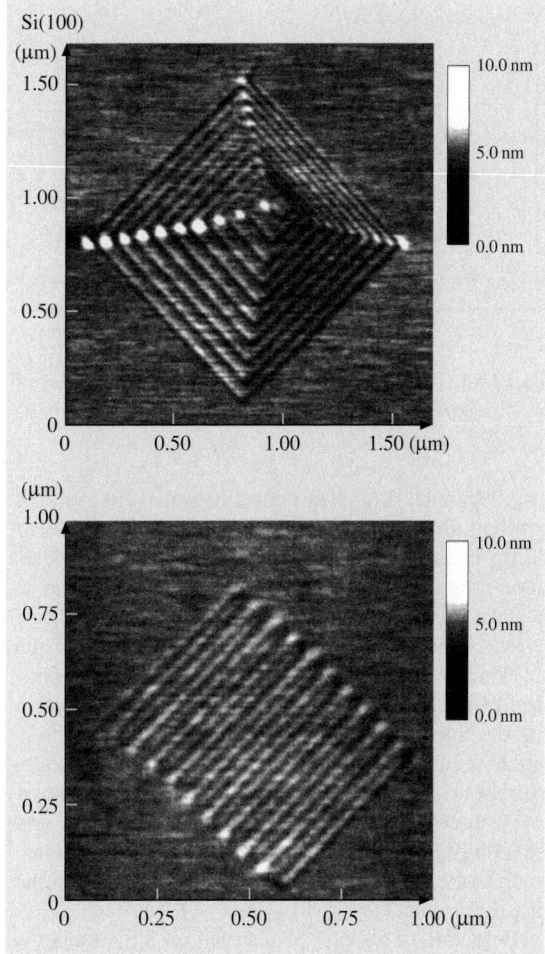

Fig. 29.57 (a) Trim and **(b)** spiral patterns generated by scratching a Si(100) surface using a diamond tip at a normal load of 15 μN and writing speed of 0.5 μm/s

29.4 Indentation

Mechanical properties on relevant scales are needed for the analysis of friction and wear mechanisms. Mechanical properties, such as hardness and Young's modulus of elasticity can be determined on the micro- to picoscales using the AFM [29.23, 27, 44, 50] and a depth-sensing indentation system used in conjunction with an AFM [29.49, 121–123].

29.4.1 Picoindentation

Indentability on the sub-nanometer scale of soft samples can be studied in the force calibration mode (Fig. 29.6) by monitoring the slope of cantilever deflection as a function of sample traveling distance after the tip is engaged and the sample is pushed against the tip. For

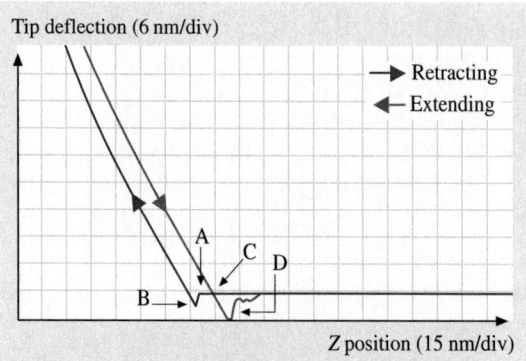

Fig. 29.58 Tip deflection (normal load) as a function of the Z (separation distance) curve for a polymeric magnetic tape [29.23]

a rigid sample, cantilever deflection equals the sample traveling distance, but the former quantity is smaller if the tip indents the sample. In an example for a polymeric magnetic tape shown in Fig. 29.58, the line in the left portion of the figure is curved with a slope of less than 1 shortly after the sample touches the tip, which suggests that the tip has indented the sample [29.23]. Later, the slope is equal to 1, suggesting that the tip no longer indents the sample. This observation indicates that the tape surface is soft locally (polymer-rich) but hard (as a result of magnetic particles) underneath. Since the curves in extending and retracting modes are identical, the indentation is elastic up to the maximum load of about 22 nN used in the measurements.

Detection of transfer of material on a nanoscale is possible with the AFM. Indentation of C_{60}-rich fullerene films with an AFM tip has been shown [29.48] to result in the transfer of fullerene molecules to the AFM tip, as indicated by discontinuities in the cantilever deflection as a function of sample traveling distance in subsequent indentation studies.

29.4.2 Nanoscale Indentation

The indentation hardness of surface films with an indentation depth as small as about 1 nm can be measured using an AFM [29.13, 49, 50]. Figure 29.59 shows the greyscale plots of indentation marks made on Si(111) at normal loads of 60, 65, 70 and 100 μN. Triangular indents can be clearly observed with very shallow depths. At a normal load of 60 μN, indents are observed and the depth of penetration is about 1 nm. As the normal load is increased, the indents become clearer and the

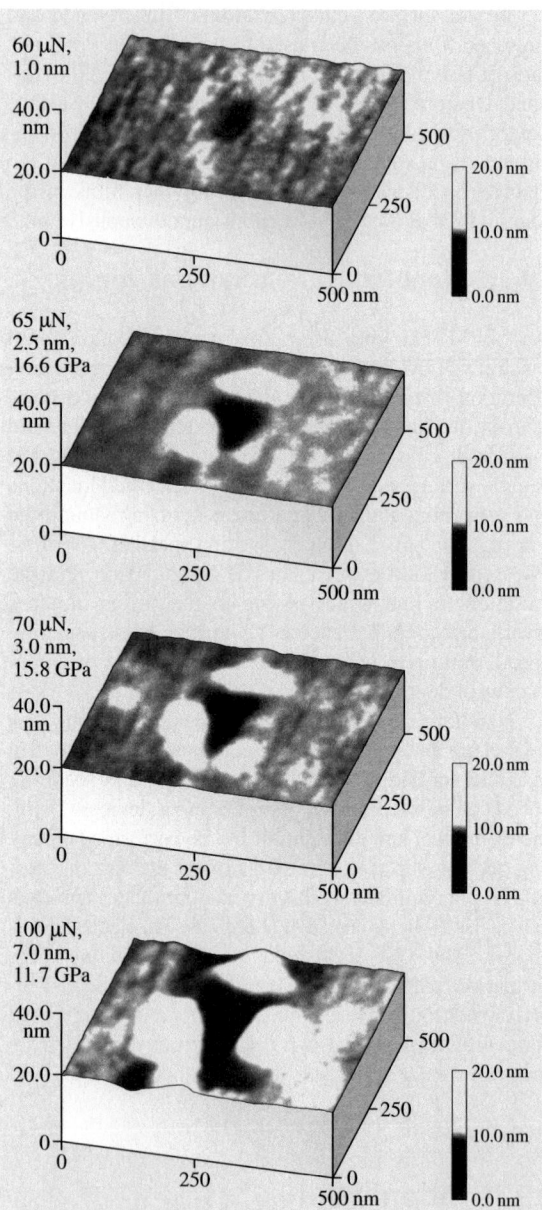

Fig. 29.59 Greyscale plots of indentation marks on the Si(111) sample at various indentation loads. Loads, indentation depths and hardness values are listed in the figure [29.50]

indentation depth increases. For the case of hardness measurements at shallow depths on the same order as variations in surface roughness, it is desirable to subtract the original (unindented) map from the indent map

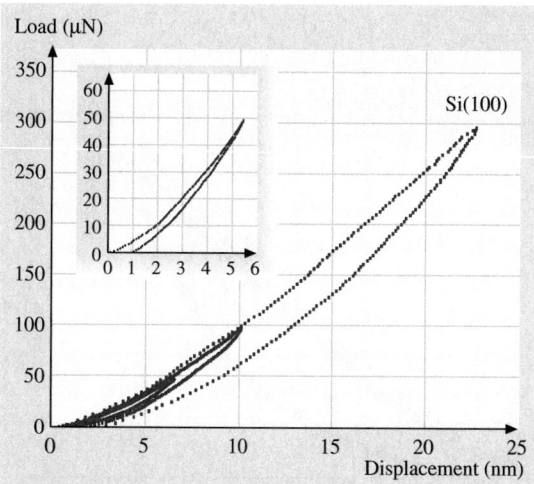

Fig. 29.60 Load–displacement curves at various peak loads for Si(100) [29.49]

for accurate measurement of the indentation size and depth [29.27].

To make accurate measurements of hardness at shallow depths, a depth-sensing nano/picoindentation system (Fig. 29.9) is used [29.49]. Figure 29.60 shows the load–displacement curves at different peak loads for Si(100). Loading/unloading curves often exhibit sharp discontinuities, particularly at high loads. Discontinuities, also referred to as pop-ins, during the initial part of the loading part of the curve mark a sharp transition from pure elastic loading to a plastic deformation of the specimen surface, correspond to an initial yield point. The sharp discontinuities in the unloading part of the curves are believed to be due to the formation of lateral cracks which form at the base of the median crack, which results in the surface of the specimen being thrust upward. Load–displacement data at residual depths as low as about 1 nm can be obtained. The indentation hardness of surface films has been measured for various materials including Si(100) up to a peak load of 500 μN and Al(100) up to a peak load of 2000 μN by *Bhushan* et al. [29.49] and *Kulkarni* and *Bhushan* [29.121–123]. The hardnesses of single-crystal of silicon and single-crystal aluminum on a nanoscale are found to be higher than on a microscale, Fig. 29.61. Microhardness has also been reported to be higher than that on the millimeter scale by several investigators. The data reported to date show that hardness exhibits the scale (size) effect.

During loading, generation and propagation of dislocations is responsible for plastic deformation. A strain gradient plasticity theory has been developed for mi-

cro/nanoscale deformations, and is based on randomly created statistically stored and geometrically necessary dislocations [29.124, 125]. Large strain gradients inherent to small indentations lead to the accumulation of geometrically necessary dislocations, located in a certain subsurface volume, for strain compatibility reasons that cause enhanced hardening. The large strain gradients in small indentations require these dislocations to account for the large slope at the indented surface. These are a function of strain gradient, whereas statistically, stored dislocations are a function of strain. Based on this theory, scale-dependent hardness is given as

$$H = H_0\sqrt{1 + \ell_\mathrm{d} a} \qquad (29.22)$$

where H_0 is the hardness in the absence of strain gradient or macrohardness, ℓ_d is the material-specific characteristic-length parameter, and a is the contact radius. In addition to the role of the strain gradient plasticity theory, an increase in hardness with a decrease in indentation depth can possibly be rationalized on the basis that, as the volume of deformed material decreases, there is a lower probability of encountering material defects.

Bhushan and *Koinkar* [29.44] have used AFM measurements to show that ion implantation of silicon surfaces increases their hardness and thus their wear resistance. Formation of surface alloy films with improved mechanical properties by ion implantation is of growing technological importance as a means of improving the mechanical properties of materials. Hardness of 20-nm-thick DLC films have been measured by *Kulkarni* and *Bhushan* [29.123].

The creep and strain-rate effects (viscoelastic effects) of ceramics can be studied using a depth-sensing

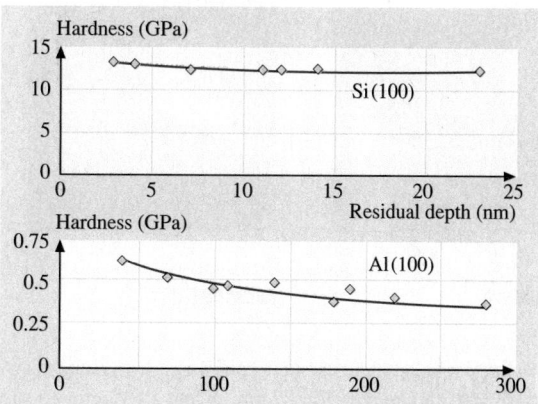

Fig. 29.61 Indentation hardness as a function of residual indentation depth for Si(100) [29.49], and Al(100) [29.121]

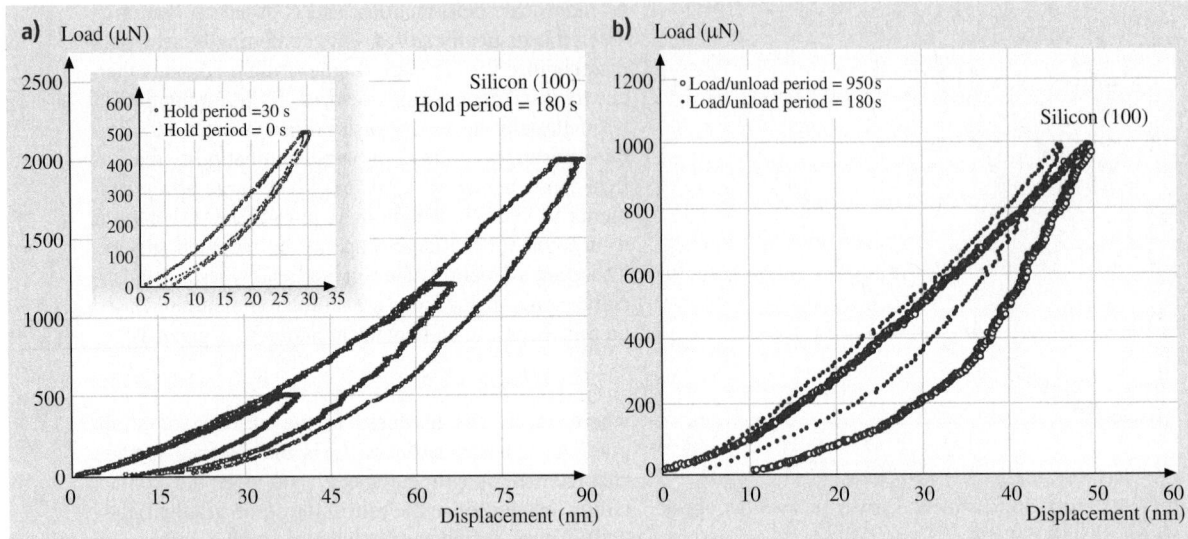

Fig. 29.62 (a) Creep behavior and **(b)** strain-rate sensitivity of Si(100) [29.49]

indentation system. *Bhushan* et al. [29.49] and *Kulkarni and Bhushan* [29.121–123] have reported that ceramics (single-crystal silicon and diamond-like carbon) exhibit significant plasticity and creep on a nanoscale. Figure 29.62a shows the load–displacement curves for single-crystal silicon at various peak loads held for 180 s. To demonstrate the creep effects, the load–displacement curves for a 500 μN peak load held for 0 and 30 s are also shown as an inset. Note that significant creep occurs at room temperature. Nanoindenter experiments conducted by *Li* et al. [29.126] exhibited significant creep only at high temperatures (greater than or equal to 0.25 times the melting point of silicon). The mechanism of dislocation glide plasticity is believed to dominate the indentation creep process on the macroscale. To study the strain-rate sensitivity of silicon, data at two different (constant) rates of loading are presented in Fig. 29.62b. Note that a change in the loading rate by a factor of about five results in a significant change in the load–displacement data. The viscoelastic effects observed here for silicon at ambient temperature could arise from the size effects mentioned earlier. Most likely, creep and strain-rate experiments are being conducted on the hydrated films present on the silicon surface in ambient environment, and these films are expected to be viscoelastic.

29.4.3 Localized Surface Elasticity and Viscoelasticity Mapping

The Young's modulus of elasticity can be calculated from the slope of the indentation curve during unloading. However, these measurements provide a single-point measurement. By using the force modulation technique, it is possible to get localized elasticity maps of soft and compliant materials of near-surface regions with nanoscale lateral resolution. This technique has been successfully used for polymeric magnetic tapes, which consist of magnetic and nonmagnetic ceramic particles in a polymeric matrix. Elasticity maps of a tape can

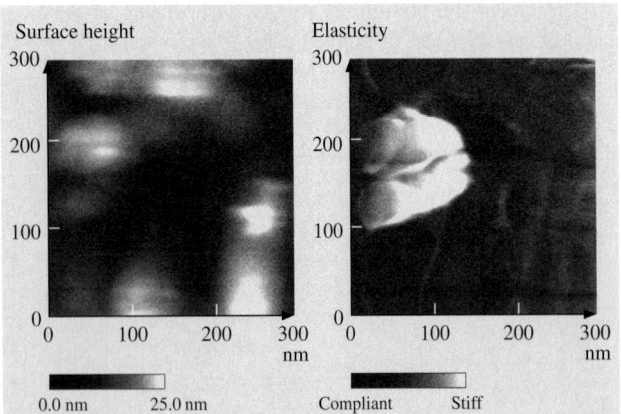

Fig. 29.63 Surface height and elasticity maps on a polymeric magnetic tape ($\sigma = 6.7$ nm and P–V = 32 nm; σ and P–V refer to the standard deviation of surface heights and peak-to-valley distance, respectively). The greyscale on the elasticity map is arbitrary [29.54]

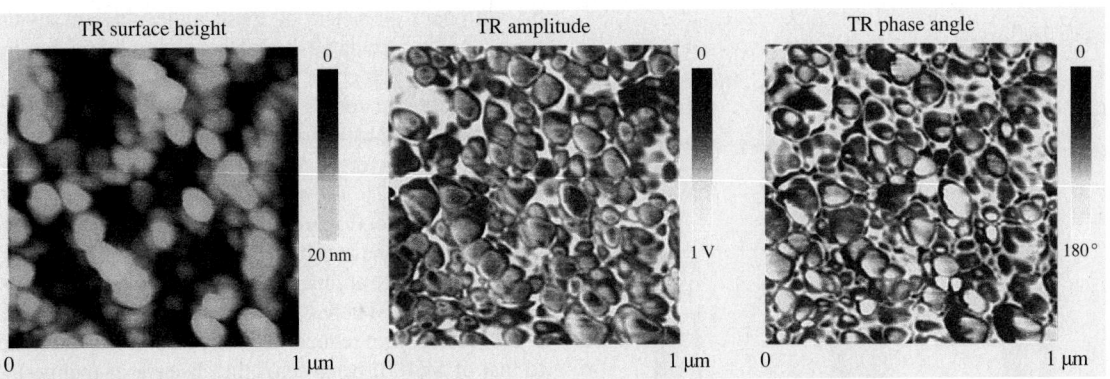

Fig. 29.64 Images of an MP tape obtained with TR mode II (constant deflection). TR mode II amplitude and phase angle images have the largest contrast among tapping, TR mode I and TR mode II techniques [29.60]

be used to identify relative distribution of hard magnetic and nonmagnetic ceramic particles on the tape surface, which has an effect on friction and stiction at the head–tape interface [29.15]. Figure 29.63 shows surface height and elasticity maps on a polymeric magnetic tape [29.54]. The elasticity image reveals sharp variations in surface elasticity due to the composite nature of the film. As can be clearly seen, regions of high elasticity do not always correspond to high or low topography. Based on a Hertzian elastic-contact analysis, the static indentation depth of these samples during the force modulation scan is estimated to be about 1 nm. We conclude that the contrast seen is influenced most strongly by material properties in the top few nanometers, independent of the composite structure beneath the surface layer.

By using phase contrast microscopy, it is possible to get phase contrast maps or the contrast in viscoelastic properties of near-surface regions with nanoscale lateral resolution. This technique has been successfully used for polymeric films and magnetic tapes which consist of ceramic particles in a polymeric matrix [29.57–60].

Figure 29.64 shows typical surface height, TR amplitude and TR phase-angle images for a MP tape

using the TR mode II, described earlier. TR amplitude image provides contrast in lateral stiffness and TR phase-angle image provides contrast in viscoelastic properties. In TR amplitude and phase-angle images, the distribution of magnetic particles can be clearly seen which have better contrast than that in TR surface height image. MP tape samples show granular structure with elliptical shape magnetic particle aggregates (50–100 nm in diameter). Studies by *Scott* and *Bhushan* [29.57], *Bhushan* and *Qi* [29.58], and *Kasai* et al. [29.59] have indicated that the phase shift can be related to the energy dissipation through the viscoelastic deformation process between the tip and the sample. Recent theoretical analysis has established a quantitative correlation between the lateral surface properties (stiffness and viscoelasticity) of materials and amplitude/phase-angle shift in TR measurements [29.73]. The contrast in the TR amplitude and phase-angle images is due to the in-plane (lateral) heterogeneity of the surface. Based on the TR amplitude and phase-angle images, the lateral surface properties (lateral stiffness and viscoelasticity) mapping of materials can be obtained.

29.5 Boundary Lubrication

29.5.1 Perfluoropolyether Lubricants

The classical approach to lubrication uses freely supported multimolecular layers of liquid lubricants [29.5, 10, 15, 127]. The liquid lubricants are sometimes chemically bonded to improve their wear resistance [29.5, 10, 15]. Partially chemically bonded, molecularly thick

perfluoropolyether (PFPE) films are used for lubrication of magnetic storage media because of their thermal stability and extremely low vapor pressure [29.15]. These are considered as potential candidate lubricants for MEMS/NEMS. Molecularly thick PFPEs are well suited because of the following properties: low surface tension and low contact angle, which allow easy spread-

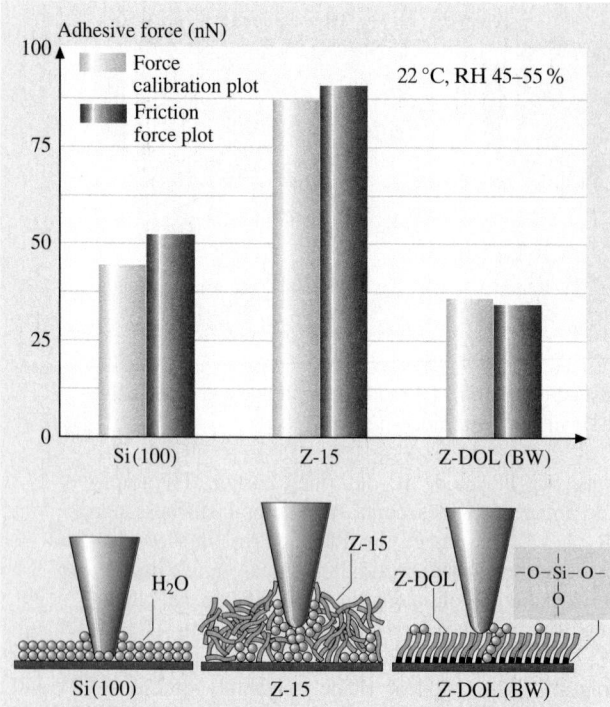

Fig. 29.65 Summary of the adhesive forces of Si(100) and Z-15 and Z-DOL (BW) films measured by force calibration plots and friction force versus normal-load plots in ambient air. The schematic (*bottom*) showing the effect of meniscus, formed between AFM tip and the surface sample, on the adhesive and friction forces [29.42]

ing on surfaces and provide hydrophobic properties; chemical and thermal stability which minimize degradation under use; low vapor pressure, which provides low out-gassing; high adhesion to substrate via organic functional bonds; and good lubricity, which reduces contact surface wear.

For boundary lubrication studies, friction, adhesion and durability experiments have been performed on virgin Si(100) surfaces and silicon surfaces lubricated with various PFPE lubricants [29.39, 40, 42, 128]. Results of two of the PFPE lubricants will be presented here: Z-15 (with $-CF_3$ nonpolar end groups), $CF_3 - O - (CF_2 - CF_2 - O)_m - (CF_2 - O)_n - CF_3$ ($m/n \approx 2/3$) and Z-DOL (with $-OH$ polar end groups), $HO - CH_2 - CF_2 - O - (CF_2 - CF_2 - O)_m - (CF_2 - O)_n - CF_2 - CH_2 - OH$ ($m/n \approx 2/3$). Z-DOL film was thermally bonded at 150 °C for 30 minutes and the unbonded fraction was removed by a solvent (bonded washed or BW) [29.15]. The thicknesses of Z-15 and Z-DOL films were 2.8 nm and 2.3 nm, respectively. Lu-

bricant chain diameters of these molecules are about 0.6 nm and molecularly thick films generally lie flat on surfaces with high coverage.

The adhesive forces of Si(100), Z-15 and Z-DOL (BW) measured by a force calibration plot and plots of friction force versus normal load are summarized in Fig. 29.65 [29.42]. The results measured by these two methods are in good agreement. Figure 29.65 shows that the presence of the mobile Z-15 lubricant film increases the adhesive force compared to that of Si(100) by meniscus formation. Whereas, the presence of solid-phase Z-DOL (BW) film reduces the adhesive force compared to that of Si(100) because of the absence of mobile liquid. The schematic (bottom) in Fig. 29.65 shows the relative size and sources of meniscus. It is well known that the native oxide layer (SiO_2) on top of a Si(100) wafer exhibits hydrophilic properties, and some water molecules can be adsorbed onto this surface. The condensed water will form meniscus as the tip approaches the sample surface. The larger adhesive force in Z-15 is not only caused by the Z-15 meniscus, the nonpolarized Z-15 liquid does not have good wettability and strong bonding with Si(100). Consequently, in the ambient environment, the condensed water molecules from the environment will permeate through the liquid Z-15 lubricant film and compete with the lubricant molecules present on the substrate. The interaction of the liquid lubricant with the substrate is weakened, and a boundary layer of the liquid lubricant forms puddles [29.39, 40]. This dewetting allows water molecules to be adsorbed onto the Si(100) surface as aggregates along with Z-15 molecules. And both of them can form a meniscus while the tip approaches to the surface. Thus the dewetting of liquid Z-15 film results in higher adhesive force and poorer lubrication performance. In addition, as the Z-15 film is quite soft compared to the solid Si(100) surface, and penetration of the tip in the film occurs while pushing the tip down. This leads to the large area of the tip involved to form the meniscus at the tip–liquid (mixture of Z-15 and water) interface. It should also be noted that Z-15 has a higher viscosity than water, therefore Z-15 film provides higher resistance to motion and coefficient of friction. In the case of Z-DOL (BW) film, both of the active groups of Z-DOL molecules are mostly bonded on Si(100) substrate, thus the Z-DOL (BW) film has low free surface energy and cannot be displaced readily by water molecules or readily adsorb water molecules. Thus, the use of Z-DOL (BW) can reduce the adhesive force.

To study the velocity effect on friction and adhesion, the variation of friction force, adhesive force,

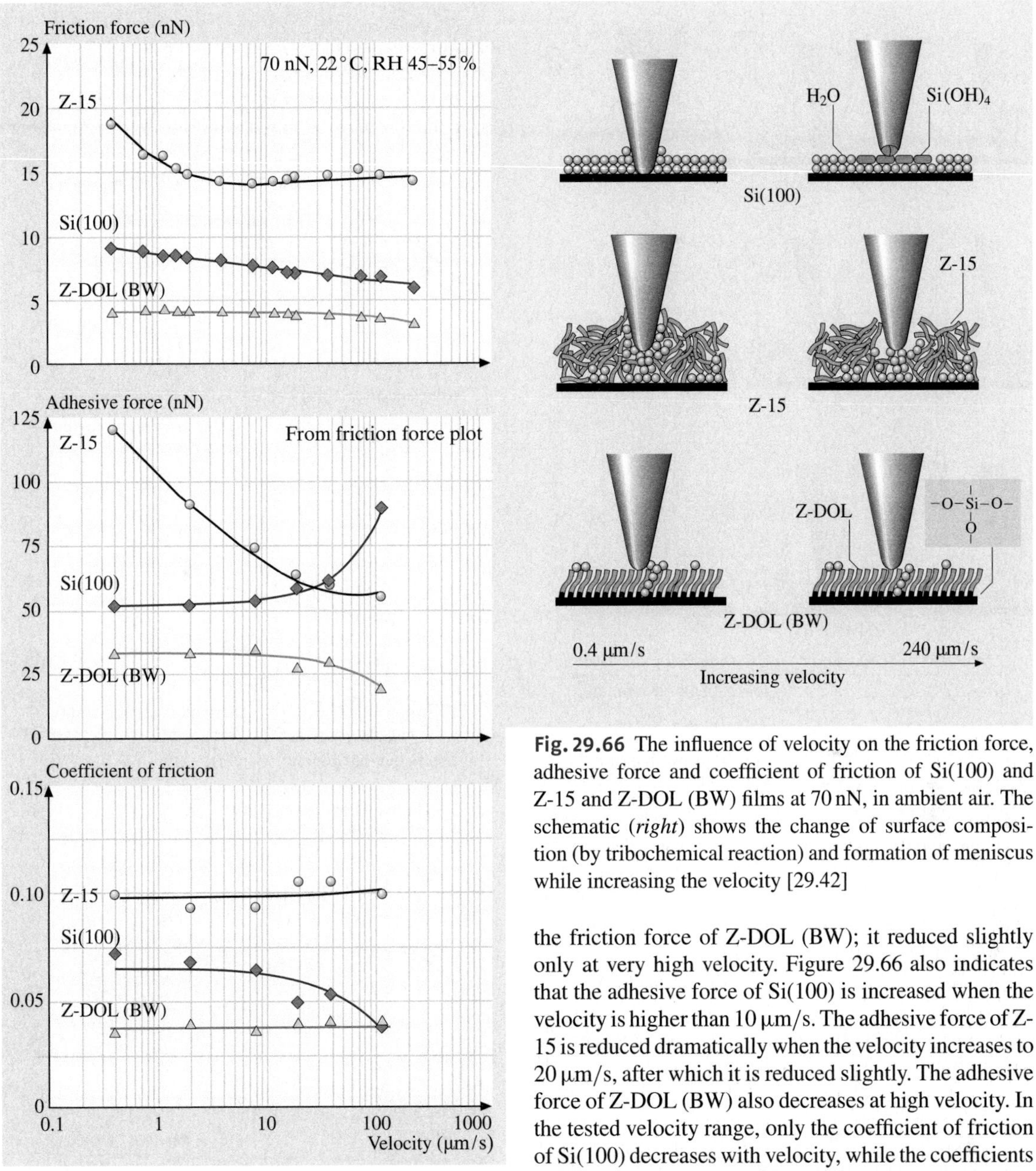

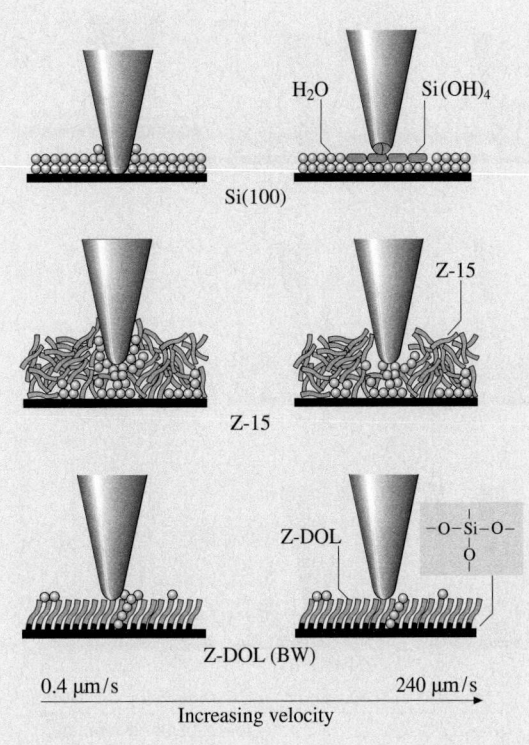

Fig. 29.66 The influence of velocity on the friction force, adhesive force and coefficient of friction of Si(100) and Z-15 and Z-DOL (BW) films at 70 nN, in ambient air. The schematic (*right*) shows the change of surface composition (by tribochemical reaction) and formation of meniscus while increasing the velocity [29.42]

the friction force of Z-DOL (BW); it reduced slightly only at very high velocity. Figure 29.66 also indicates that the adhesive force of Si(100) is increased when the velocity is higher than 10 μm/s. The adhesive force of Z-15 is reduced dramatically when the velocity increases to 20 μm/s, after which it is reduced slightly. The adhesive force of Z-DOL (BW) also decreases at high velocity. In the tested velocity range, only the coefficient of friction of Si(100) decreases with velocity, while the coefficients of friction of Z-15 and Z-DOL (BW) almost remain constant. This implies that the friction mechanisms of Z-15 and Z-DOL (BW) do not change with the variation of velocity.

The mechanisms of the effect of velocity on the adhesion and friction are explained based on the schematics shown in Fig. 29.66 (right) [29.42]. For Si(100), tribochemical reactions play a major role. Although, at

and coefficient of friction of Si(100), Z-15 and Z-DOL(BW) as a function of velocity are summarized in Fig. 29.66 [29.42]. It indicates that, for silicon wafers, the friction force decreases logarithmically with increasing velocity. For Z-15, the friction force decreases with increasing velocity up to 10 μm/s, after which it remains almost constant. The velocity has a very small effect on

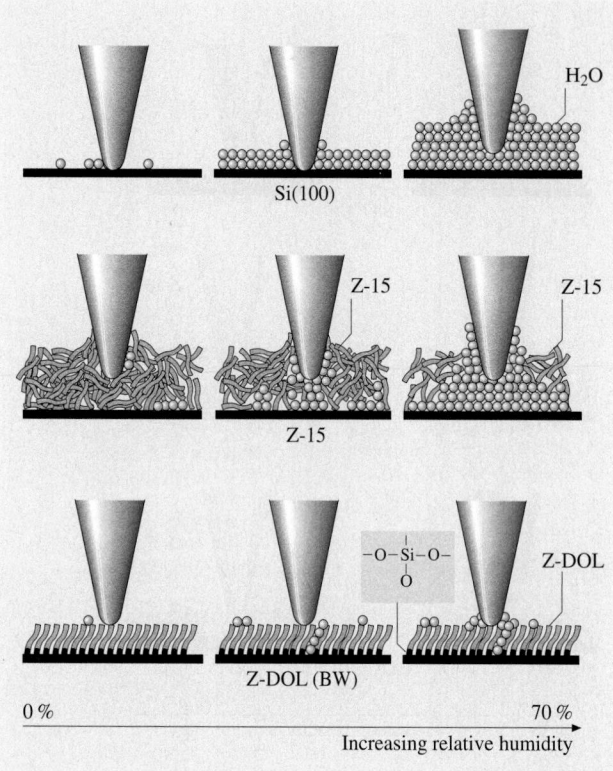

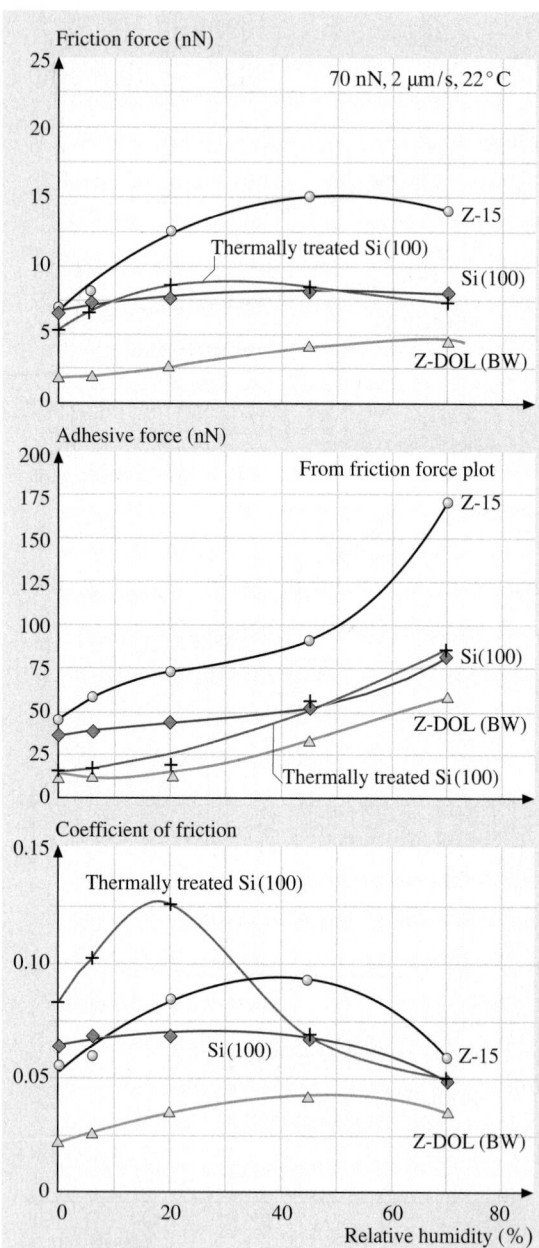

Fig. 29.67 The influence of relative humidity on the friction force, adhesive force, and coefficient of friction of Si(100) and Z-15 and Z-DOL (BW) films at 70 nN, 2 μm/s, and in 22 °C air. The schematic (*left*) shows the change of meniscus while increasing the relative humidity. In this figure, the thermally treated Si(100) represents the Si(100) wafer that was baked at 150 °C for 1 hour in an oven (in order to remove the adsorbed water) just before it was placed in the 0% RH chamber [29.42]

high velocity, the meniscus is broken and does not have enough time to rebuild, the contact stresses and high velocity lead to tribochemical reactions of the Si(100) wafer (which has native oxide (SiO_2)), and Si_3N_4 tip with water molecules, forming $Si(OH)_4$. The $Si(OH)_4$ is removed and continuously replenished during sliding. The $Si(OH)_4$ layer between the tip and Si(100) surface is known to be of low shear strength and causes a decrease in friction force and coefficient of friction [29.10, 17]. The chemical bonds of $Si-OH$ between the tip and the Si(100) surface induce large adhesive force. For Z-15 film, at high velocity the meniscus formed by condensed water and Z-15 molecules is broken and does not have enough time to rebuild; therefore, the adhesive force

and consequently friction force is reduced. The friction mechanisms for the Z-15 film still is shearing the same viscous liquid even at high velocity range, thus the coefficient of friction of Z-15 does not change with velocity. For Z-DOL (BW) film, the surface can adsorb few water molecules in ambient condition, and at high velocity these molecules are displaced, which is respon-

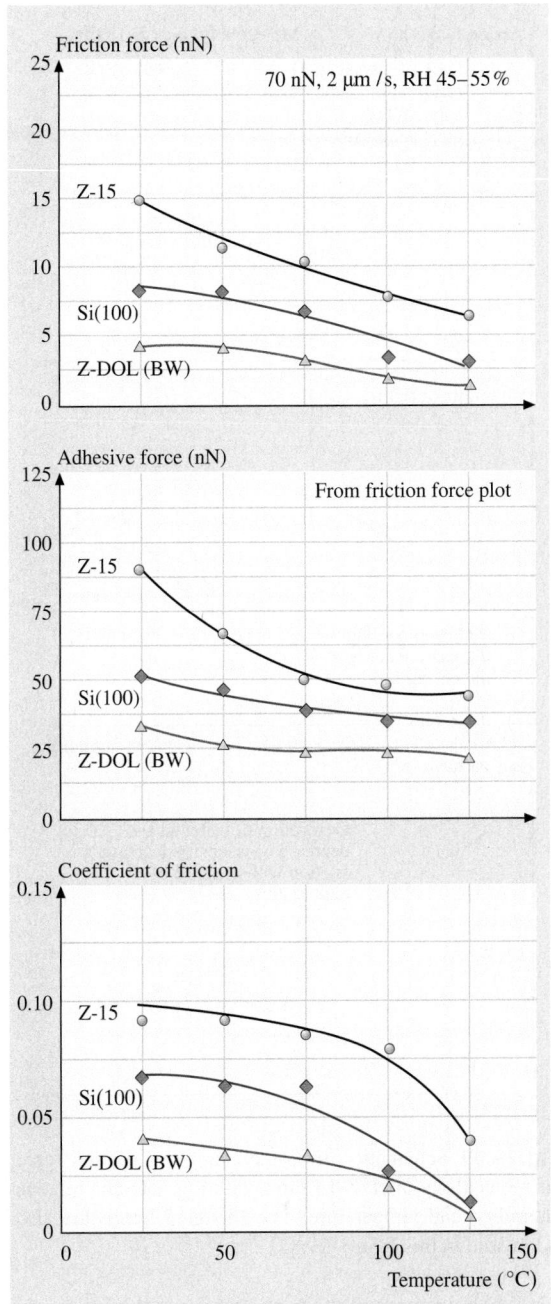

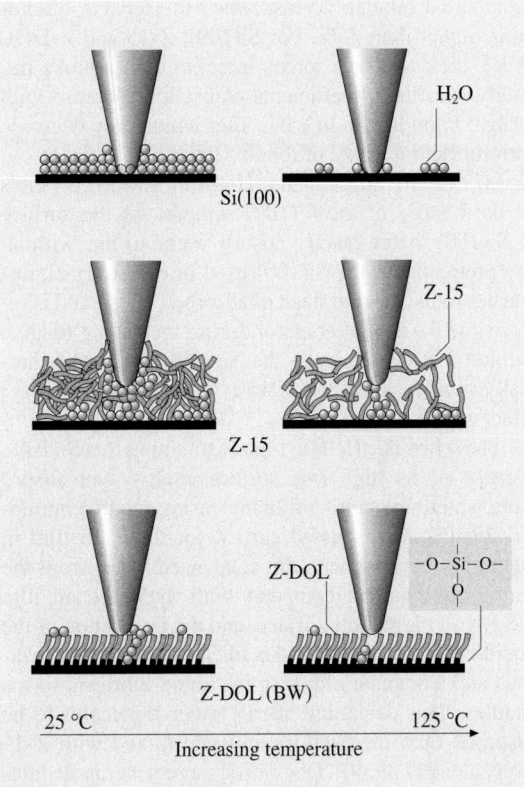

Fig. 29.68 The influence of temperature on the friction force, adhesive force, and coefficient of friction of Si(100) and Z-15 and Z-DOL (BW) films at 70 nN, at 2 μm/s, and in RH 40–50% air. The schematic (*right*) shows that, at high temperature, desorption of water decreases the adhesive forces. And the reduced viscosity of Z-15 leads to the decrease of coefficient of friction. High temperature facilitates orientation of molecules in Z-DOL (BW) film, which results in a lower coefficient of friction [29.42]

in friction force with increasing scanning velocity. This could be another reason for the decrease in friction force for Si(100) and Z-15 film with velocity in this study.

To study the effect of relative humidity on friction and adhesion, the variation of friction force, adhesive force, and coefficient of friction of Si(100), Z-15, and Z-DOL (BW) as a function of relative humidity is shown in Fig. 29.67 [29.42]. It shows that, for Si(100) and Z-15 film, the friction force increases with relative humidity up to 45%, and then shows a slight decrease with further increases in the relative humidity. Z-DOL (BW) has a smaller friction force than Si(100) and Z-15 in the whole testing range and its friction force shows

sible for a slight decrease in friction force and adhesive force. *Koinkar* and *Bhushan* [29.40, 40] have suggested that, in the case of samples with mobile films, such as condensed water and Z-15 films, alignment of liquid molecules (shear thinning) is responsible for the drop

an apparent relative ncrease when the relative humidity is higher than 45%. For Si(100), Z-15 and Z-DOL (BW), their adhesive forces increase with relative humidity, and their coefficients of friction increase with relative humidity up to 45%, after which they decrease with further increases of relative humidity. It is also observed that the effect of humidity on Si(100) depends on the history of the Si(100) sample. As the surface of Si(100) wafer readily adsorb water in air, without any pretreatment the Si(100) used in our study almost reaches to its saturate stage of adsorbed water, and is responsible for a smaller effect during increasing relative humidity. However, once the Si(100) wafer was thermally treated by baking at $150\,^{\circ}$C for 1 hour, a bigger effect was observed.

The schematic (left) in Fig. 29.67 shows that Si(100), because of its high free surface energy, can adsorb more water molecules while increasing relative humidity [29.42]. As discussed earlier, for the Z-15 film in the humid environment, the condensed water from the humid environment competes with the lubricant film present on the sample surface, and the interaction of the liquid lubricant film with the silicon substrate is weakened and a boundary layer of the liquid lubricant forms puddles. This dewetting allows water molecules to be adsorbed onto the Si(100) substrate mixed with Z-15 molecules [29.39, 40]. Obviously, more water molecules can be adsorbed onto the Z-15 surface with increasing relative humidity. The higher number of adsorbed water molecules in the case of Si(100), along with the lubricant molecules in the Z-15 film case, form a larger water meniscus, which leads to an increase of friction force, adhesive force, and coefficient of friction of Si(100) and Z-15 with humidity. However, at the very high humidity of 70%, large quantities of adsorbed water can form a continues water layer that separates the tip and sample surface and acts as a kind of lubricant, which causes a decrease in the friction force and coefficient of friction. For the Z-DOL (BW) film, because of its hydrophobic surface properties, water molecules can be adsorbed at humidity higher than 45%, and this causes an increase in the adhesive force and friction force.

To study the effect of temperature on friction and adhesion, the variation of friction force, adhesive force, and coefficient of friction of Si(100), Z-15 and Z-DOL (BW) as a function of temperature are summarized in Fig. 29.68 [29.42]. It shows that increasing temperature causes a decrease of the friction force, adhesive force and coefficient of friction of Si(100), Z-15 and Z-DOL (BW). The schematic (right) in Fig. 29.68 indicates that, at high temperature, desorption of water

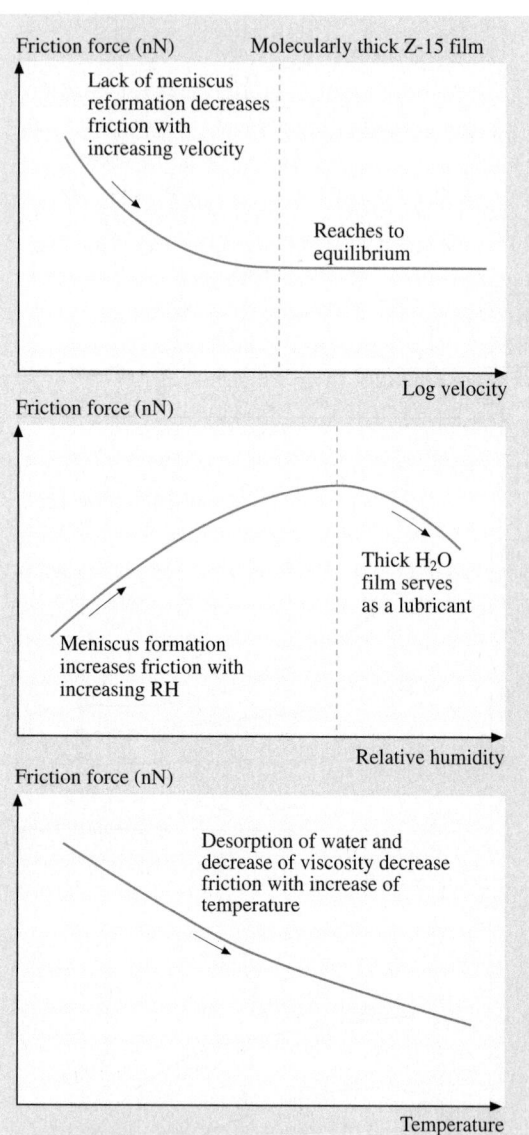

Fig. 29.69 Schematic shows the change of friction force of molecularly thick Z-15 films with log velocity, relative humidity, and temperature. The changing trends are also addressed in this figure [29.42]

leads to a decrease of the friction force, adhesive forces and coefficient of friction for all of the samples. For Z-15 film, the reduction of viscosity at high temperature also makes a contribution to the decrease of friction force and coefficient of friction. In the case of the Z-DOL (BW) film, molecules are more easily oriented at high temperatures, which may be partly responsible for the low friction force and coefficient of friction.

As a brief summary, the influence of velocity, relative humidity, and temperature on the friction force of mobile Z-15 film is presented in Fig. 29.69 [29.42]. The changing trends are also addressed in this figure.

To study the durability of lubricant films at nanoscale, the friction of Si(100), Z-15, and Z-DOL (BW) as a function of the number of scanning cycles are shown in Fig. 29.70 [29.42]. As observed earlier, the friction force and coefficient of friction of Z-15 are higher than that of Si(100), with the lowest values for Z-DOL(BW). During cycling, the friction force and coefficient of friction of Si(100) show a slight decrease during the initial few cycles and then remain constant. This is related to the removal of the top adsorbed layer. In the case of the Z-15 film, the friction force and coefficient of friction show an increase during the initial few cycles and then approach higher, stable values. This is believed to be caused by the attachment of Z-15 molecules to the tip. The molecular interaction between these attached molecules on the tip and molecules on the film surface is responsible for an increase in the friction. However, after several scans, this molecular interaction reaches an equilibrium and the friction force and coefficient of friction then remain constant. In the case of the Z-DOL (BW) film, the friction force and coefficient of friction start low and remain low during the entire test for 100 cycles. This suggests that Z-DOL (BW) molecules do not become attached or displaced as readily as Z-15.

29.5.2 Self-Assembled Monolayers

For lubrication of MEMS/NEMS, another effective approach involves the deposition of organized and dense molecular layers of long-chain molecules. Two common methods to produce monolayers and thin films are the Langmuir–Blodgett (LB) deposition and self-assembled monolayers (SAMs) by chemical grafting of molecules. LB films are physically bonded to the substrate by weak van der Waals attraction, while SAMs are chemically bonded via covalent bonds to the substrate. Because of the choice of chain length and terminal linking group that SAMs offer, they hold great promise for boundary lubrication of MEMS/NEMS. A number of studies have been conducted to study tribological properties of various SAMs [29.38, 41, 43, 129–135].

Bhushan and *Liu* [29.41] studied the effect of film compliance on adhesion and friction. They used hexadecane thiol (HDT), 1,1,biphenyl-4-thiol (BPT), and

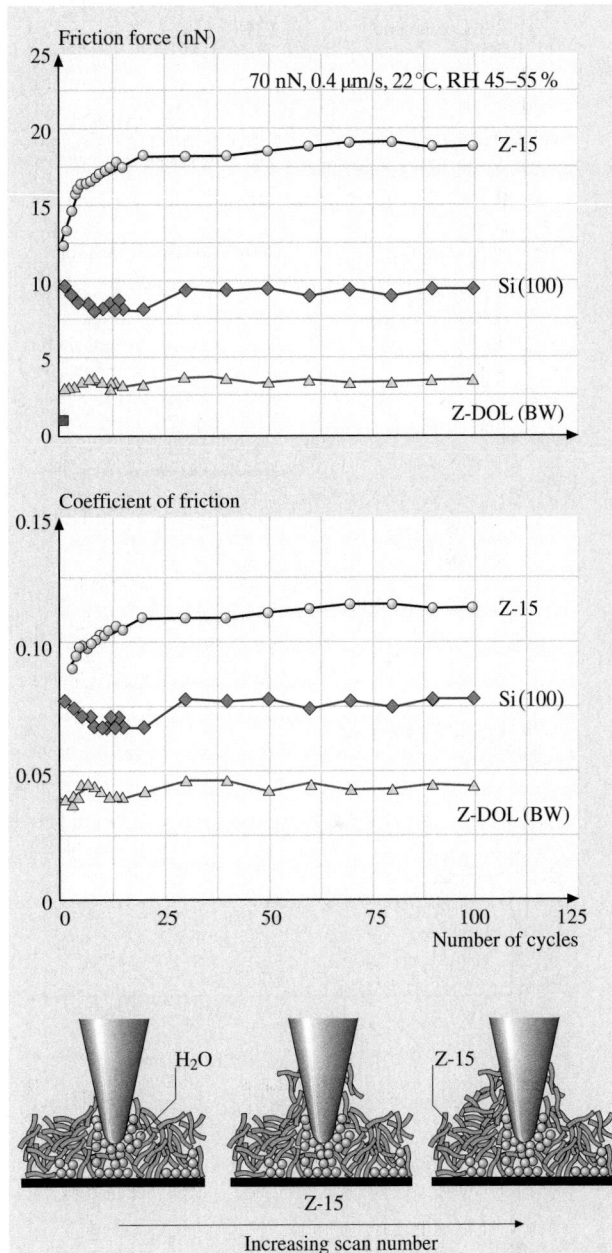

Fig. 29.70 Friction force and coefficient of friction versus number of sliding cycles for Si(100) and Z-15 and Z-DOL (BW) films at 70 nN, 0.8 μm/s, and in ambient air. The schematic (*bottom*) shows that some liquid Z-15 molecules can be attached onto the tip. The molecular interaction between the attached molecules onto the tip with the Z-15 molecules in the film results in an increase of the friction force with multiple scanning [29.42]

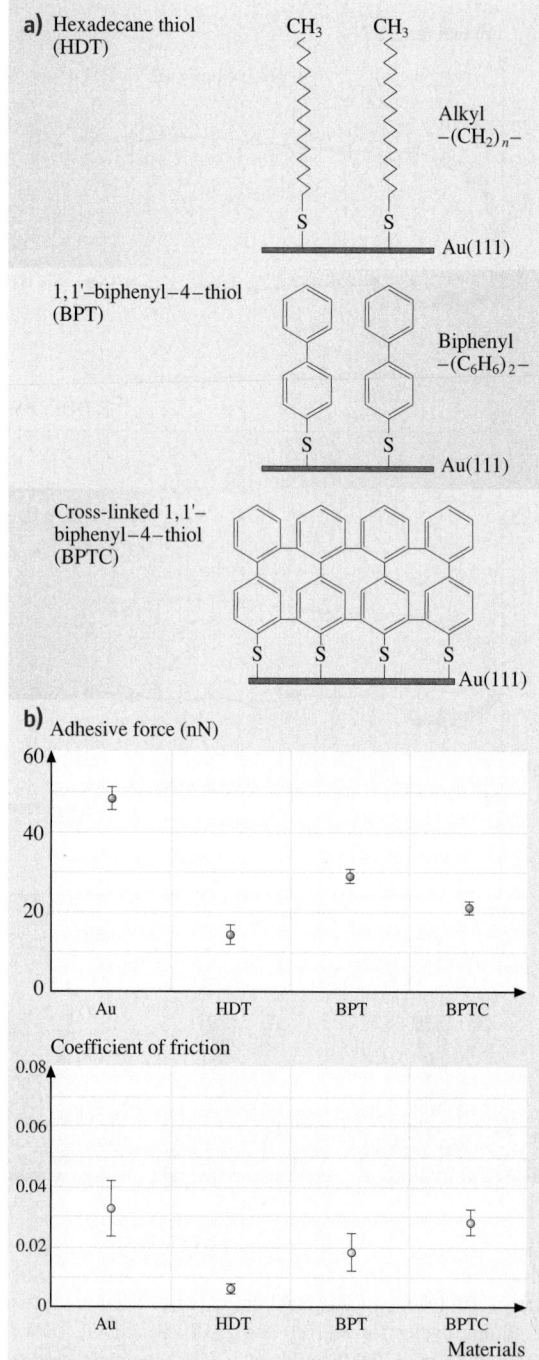

Fig. 29.71 **(a)** Schematics of the structures of hexadecane thiol and biphenyl thiol SAMs on Au(111) substrates, and **(b)** adhesive force and coefficient of friction of Au(111) substrate and various SAMs

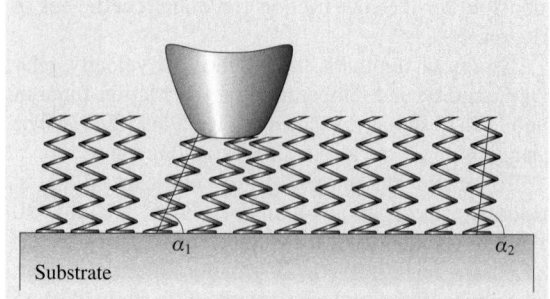

Fig. 29.72 Molecular spring model of SAMs. In this figure, $\alpha_1 < \alpha_2$, which is caused by the further orientation under the normal load applied by an asperity tip [29.41]

crosslinked BPT (BPTC) solvent deposited on Au(111) substrate, Fig. 29.71a. The average values and standard duration of the adhesive force and coefficient of friction are presented in Fig. 29.71b. Based on the data, the adhesive force and coefficient of frictions of SAMs are less than those of corresponding substrates. Among various films, HDT exhibits the lowest values. Based on stiffness measurements of various SAMs, HDT was the most compliant, followed by BPT and BPTC. Based on friction and stiffness measurements, SAMs with high-compliance long carbon chains exhibit low friction; chain compliance is desirable for low friction. Friction mechanism of SAMs is explained by a so-called *molecular spring* model (Fig. 29.72). According to this model, the chemically adsorbed self-assembled molecules on a substrate are just like assembled mo-

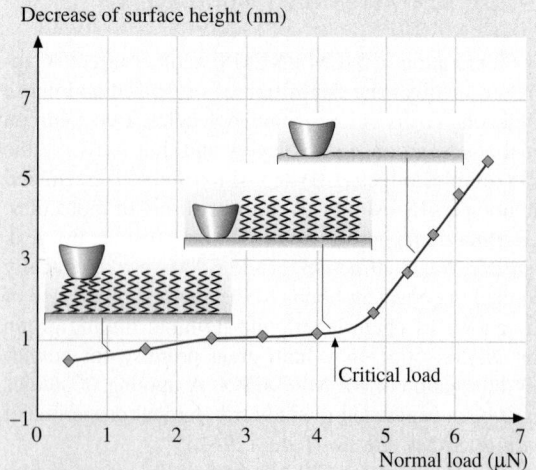

Fig. 29.73 Illustration of the wear mechanism of SAMs with increasing normal load [29.130]

lecular springs anchored to the substrate. An asperity sliding on the surface of SAMs is like a tip sliding on the top of *molecular springs* or a *brush*. The molecular spring assembly has compliant features and can experience orientation and compression under load. The orientation of the molecular springs or brush under normal load reduces the shearing force at the interface, which in turn reduces the friction force. The orientation is determined by the spring constant of a single molecule as well as the interaction between the neighboring molecules, which can be reflected by packing density or packing energy. It should be noted that the orientation can lead to conformational defects along the molecular chains, which lead to energy dissipation.

The SAMs with high-compliance long carbon chains also exhibit the best wear resistance [29.41,130]. In wear experiments, the wear depth as a function of normal load curves show a critical normal load. A representative curve is shown in Fig. 29.73. Below the critical normal load, SAMs undergo orientation; at the critical load SAMs wear away from the substrate due to weak interface bond strengths, while above the critical normal load severe wear takes place on the substrate.

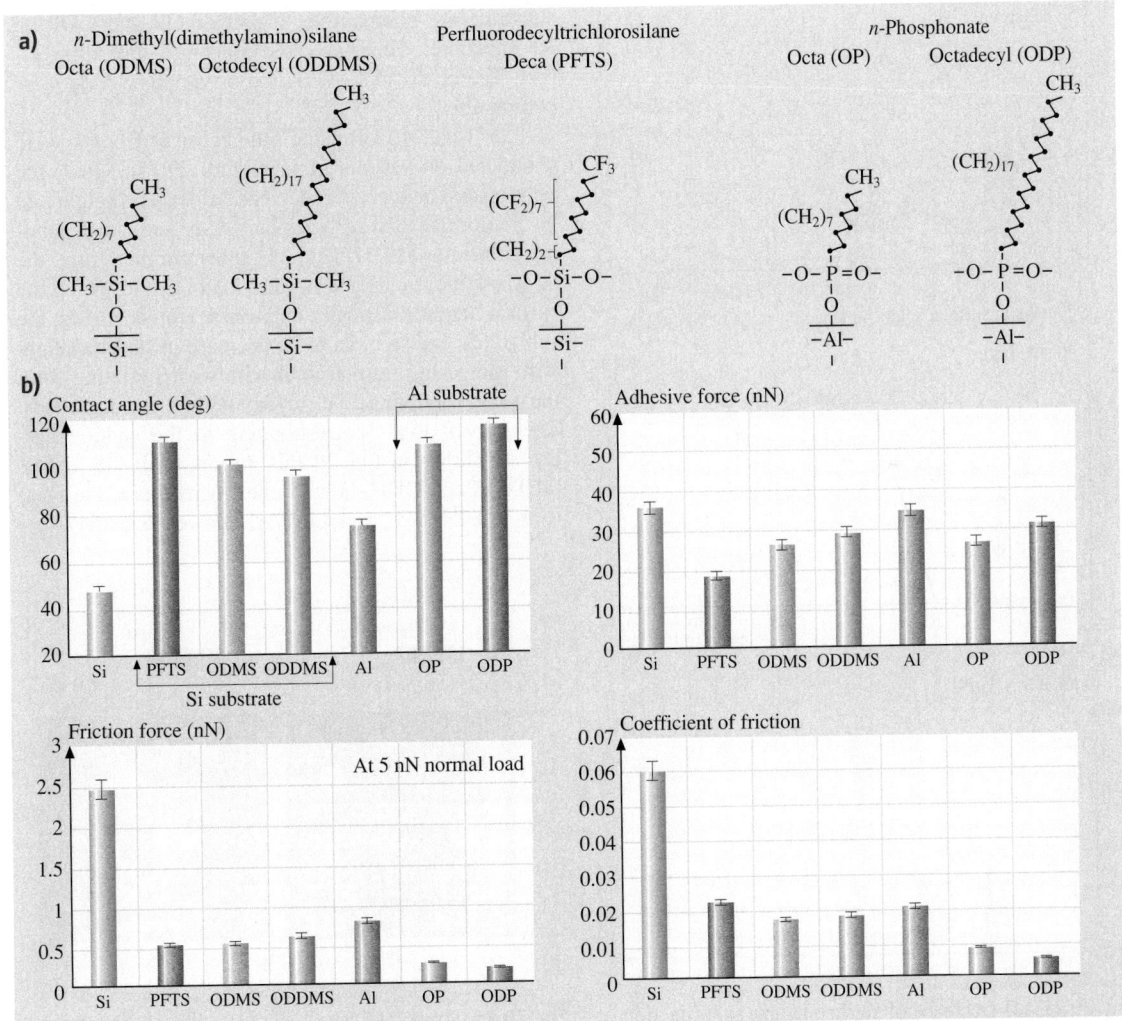

Fig. 29.74 (a) Schematics of structures of perfluoroalkylsilane and alkylsilane SAMs on Si with native oxide substrates, and alkylphosphonate SAMs on Al with native oxide, and (b) contact angle, adhesive force, friction force, and coefficient of friction of Si with native oxide and Al with native oxide substrates and with various SAMs

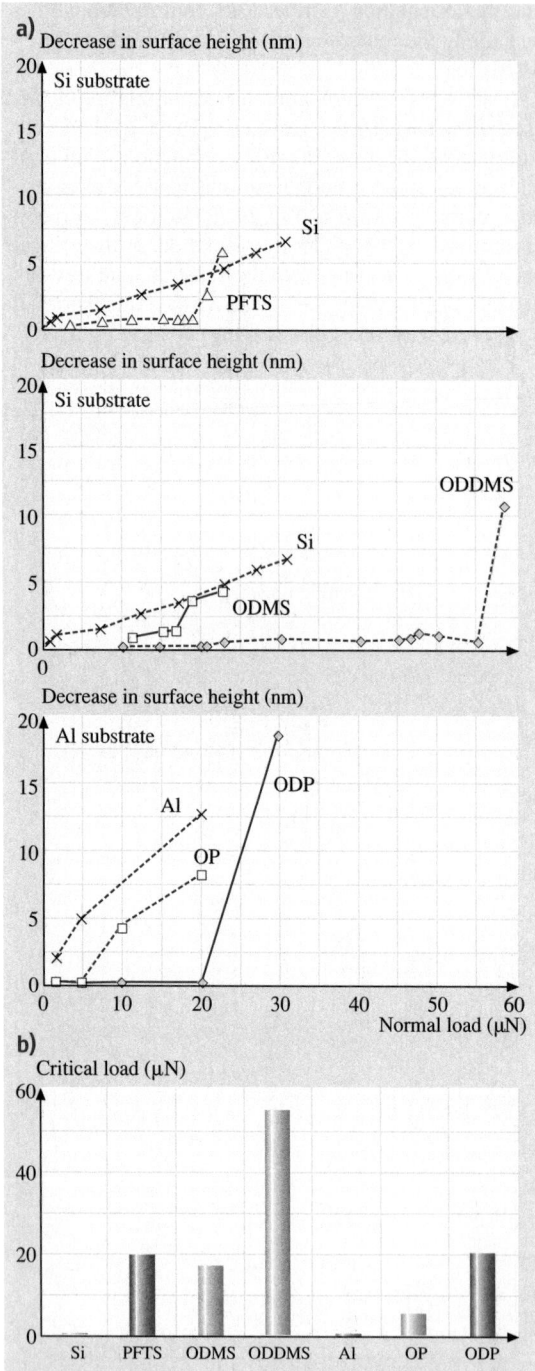

Fig. 29.75 (a) Decrease of surface height as a function of normal load after one scan cycle for various SAMs on Si and Al substrates, and (b) comparison of critical loads for failure during wear tests for various SAMs

Bhushan et al. [29.43], *Kasai* et al. [29.131], and *Tambe* and *Bhushan* [29.134] studied perfluorodecyltricholorosilane (PFTS), *n*-octyldimethyl (dimethylamino) silane (ODMS) ($n = 7$), and *n*-octadecylmethyl (dimethylamino) silane (ODDMS) ($n = 17$) vapor deposited on Si substrate, and octylphosphonate (OP) and octadecylphosphonate (ODP) on Al substrate, Fig. 29.74a. Figure 29.74b presents the contact angle, adhesive force, friction force, and coefficient of friction of two substrates and with various SAMs. Based on the data, PFTS/Si exhibits higher contact angle and lower adhesive force as compared to that of ODMS/Si and ODDMS/Si. Data of ODMS and ODDMS on Si substrate are comparable to that of OP and ODP on Al substrate. Therefore, the substrate had little effect. The coefficient of friction of various SAMs were comparable.

For wear performance studies, experiments were conducted on various films. Figure 29.75a shows the relationship between the decrease of surface height and in the normal load for various SAMs and corresponding substrates [29.37, 131]. As shown in the figure, the SAMs exhibit a critical normal load beyond which the surface height drastically decreases. Unlike SAMs, the substrates show a monotonic decrease in surface height with increasing normal load with wear initiating from the very beginning, i.e., even for low normal loads. The critical loads corresponding to the sudden failure are shown in Fig. 29.75b. Amongst all the SAMs, ODDMS and ODP show the best performance in wear tests. ODDMS/Si and ODP/Al showed a better wear

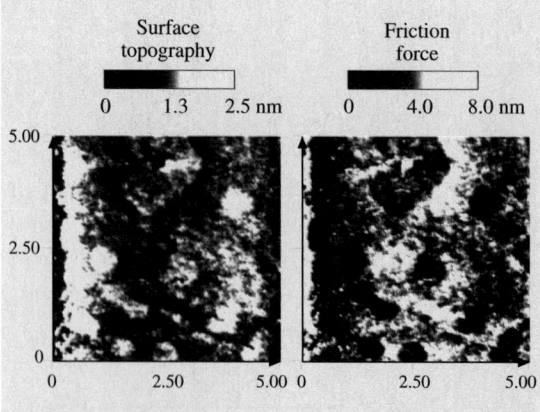

Fig. 29.76 Greyscale plots of the surface topography and friction force obtained simultaneously for unbonded Demnum-type perfluoropolyether lubricant film on silicon [29.39]

Fig. 29.77 Greyscale plots of the adhesive force distribution of a uniformly-coated, 3.5-nm-thick unbonded Z-DOL film on silicon and 3- to 10-nm-thick unbonded Z-DOL film on silicon that was deliberately coated nonuniformly by vibrating the sample during the coating process [29.97]

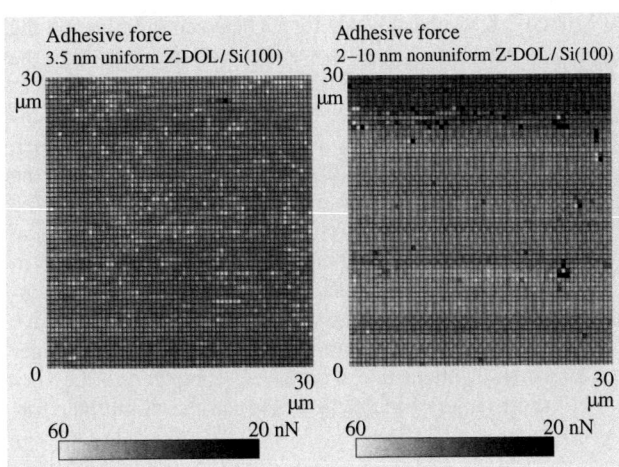

resistance than ODMS/Si and OD/Al due to the chain-length effect. Wear behavior of the SAMs is reported to be mostly determined by the molecule–substrate bond strengths.

29.5.3 Liquid Film Thickness Measurements

Liquid film thickness mapping of ultra-thin films (on the order of couple of 2 nm) can be obtained using friction force microscopy [29.39] and adhesive force mapping [29.97]. Figure 29.76 shows greyscale plots of the surface topography and friction force obtained simultaneously for unbonded Demnum S-100 type PFPE lubricant film on silicon. The friction force plot shows well-distinguished low- and high-friction regions roughly corresponding to high and low regions in the surface topography (thick and thin lubricant regions). A uniformly lubricated sample does not show such a variation in the friction. Friction force imaging can thus be used to measure the lubricant uniformity on the sample surface, which cannot be identified by surface topography alone. Figure 29.77 shows the greyscale plots of the adhesive force distribution for silicon samples coated uniformly and nonuniformly with Z-DOL-type PFPE lubricant. It can be clearly seen that there exists a region that has adhesive force that are distinctly different from the other region for the nonuniformly coated sample. This implies that the liquid film thickness is nonuniform, which gives rise to a difference in the meniscus forces.

Quantitative measurements of liquid film thickness of thin lubricant films (on the order of a few nm) with nanometer lateral resolution can be made with the AFM [29.4, 11, 60, 81]. The liquid film thickness is obtained by measuring the force on the tip as it approaches, contacts and pushes through the liquid film and ultimately contacts the substrate. The distance between the sharp snap-in (owing to the formation of a liquid menis-

Fig. 29.78 Forces between tip and hair surface as a function of tip–sample separation for virgin hair and conditioner-treated hair. A schematic of measurement for the localized conditioner thickness is shown in the inset at the top. The expanded scale view of the force curve at small separation is shown (*bottom*) [29.60]

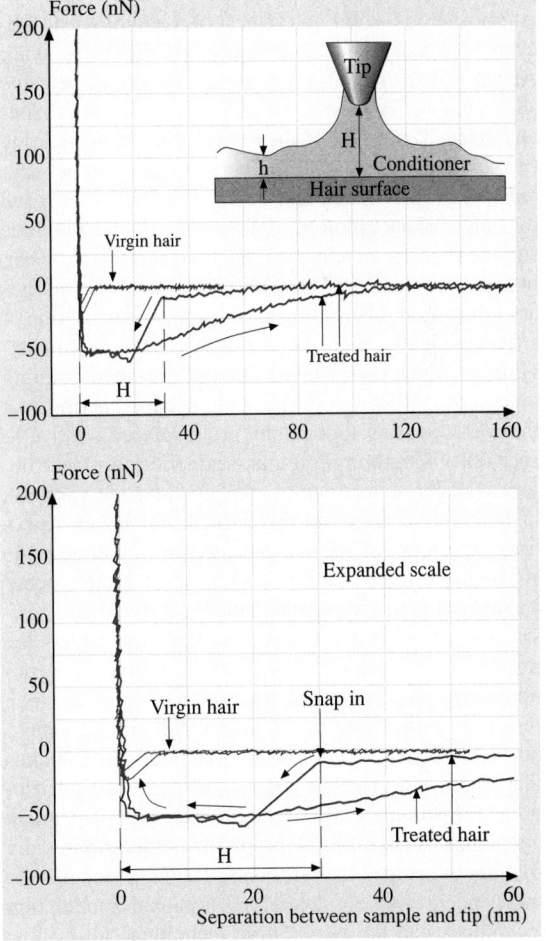

cus and van der Waals forces between the film and the tip) at the liquid surface and the hard repulsion at the substrate surface is a measure of the liquid film thickness. Figure 29.78 shows a plot of the forces between the tip and virgin hair, and hair treated with hair conditioner. The hair sample was first brought into contact with the tip and then pulled away at a velocity of 400 nm/s. The zero tip–sample separation is defined to be the position where the force on the tip is zero and the tip is not in contact with the sample. As the tip approaches the sample, a negative force exists, which indicates an attractive force. The treated hair surface shows much a longer range of interaction with the tip compared to the very short range of interaction between the virgin hair sur-

faces and the tip. Typically, the tip suddenly snaps into contact with the conditioner layer at a finite separation H (about 30 nm), which is proportional to the conditioner thickness h. As the tip contacts the substrate, the tip travels with the sample. When the sample is withdrawn, the forces on the tip slowly decrease to zero once the meniscus of liquid is drawn out from the hair surface. It should be noted that the distance H between the sharp snap-in at the liquid surface and the hard wall contact with the substrate is not the real conditioner thickness h. Due to the interaction of the liquid with the tip at some spacing distance, H tends to be thicker than the actual film thickness, but can still provide an estimate of the actual film thickness and an upper limit on the thickness.

29.6 Closure

At most solid–solid interfaces of technological relevance, contact occurs at many asperities. A sharp AFM/FFM tip sliding on a surface simulates just one such contact. However, asperities come in all shapes and sizes. The effect of the radius of a single asperity (tip) on the friction/adhesion performance can be studied using tips of different radii. AFM/FFM are used to study various tribological phenomena, which include surface roughness, adhesion, friction, scratching, wear, indentation, detection of material transfer, and boundary lubrication. Measurement of atomic-scale friction of a freshly cleaved highly oriented pyrolytic graphite exhibits the same periodicity as that of the corresponding topography. However, the peaks in friction and those in the corresponding topography are displaced relative to each other. Variations in atomic-scale friction and the observed displacement can be explained by the variation in interatomic forces in the normal and lateral directions; the relevant friction mechanism is atomic-scale stick–slip. Local variations in microscale friction occur and are found to correspond to the local slopes, suggesting that a ratchet mechanism and collision effects are responsible for this variation. Directionality in the friction is observed on both micro- and macroscales, which results from the surface roughness and surface preparation. Anisotropy in surface roughness accentuates this effect. The friction contrast in conventional frictional measurements is based on interactions dependent upon interfacial material properties superimposed by roughness-induced lateral forces. To obtain roughness-independent friction, lateral or torsional modulation techniques can be used. These techniques also allow

measurements over a small region. AFM/FFM experiments are generally conducted at relative velocities up to about 200 μm/s. High-velocity experiments can be performed using either by mounting a sample on a shear-wave transducer driven at very high frequencies or by mounting it on a high-velocity piezo stage. By using these techniques, friction and wear experiments can be performed at a range of sliding velocities as well as normal loads, and the data have been used to develop nanoscale friction and wear maps. Relevant friction mechanisms are different for different ranges of sliding velocities and normal loads.

Adhesion and friction in wet environments depends on the tip radius, surface roughness, and relative humidity. Superhydrophobic surfaces can be designed by roughness optimization.

Nanoscale friction is generally found to be smaller than the microscale friction. There are several factors that are responsible for the differences, including wear and contaminant particles, the transition from elasticity to plasticity, scale-dependent roughness and mechanical properties, and meniscus effects. Nanoscale friction values increase with an increase in the normal load above a certain critical load (stress), approaching to the macroscale friction. The critical contact stress corresponds to the hardness of the softer material.

The mechanism of material removal on the microscale is studied. Wear rate for single-crystal silicon is negligible below 20 μN and is much higher and remains approximately constant at higher loads. Elastic deformation at low loads is responsible for negligible wear. Most of the wear debris is loose. SEM and TEM studies of the wear region suggest that the material on the microscale is removed by plastic deformation with a small contribution from elastic fracture; this observation corroborates the scratch data. Evolution of wear has also been studied using AFM. Wear is found to be initiated at nanoscratches. For a sliding interface requiring near-zero friction and wear, contact stresses should be below the hardness of the softer material to minimize plastic deformation and surfaces should be free of nanoscratches. Further, wear precursors can be detected at early stages of wear by using surface potential measurements. It is found that, even in the case of zero wear (no measurable deformation of the surface using AFM), there can be a significant change in the surface potential inside the wear mark, which is useful for study of wear precursors. Detection of material transfer on the nanoscale is possible with AFM.

In situ surface characterization of local deformation of materials and thin coatings can be carried out using a tensile stage inside an AFM. An AFM can also be used for nanofabrication/nanomachining.

Modified AFM can be used to obtain load–displacement curves and for the measurement of nanoindentation hardness and Young's modulus of elasticity, with an indentation depth as low as 1 nm. The hardness of ceramics on the nanoscale is found to be higher than that on the microscale. Ceramics exhibit significant plasticity and creep on a nanoscale. By using the force modulation technique, localized surface elasticity maps of composite materials with penetration depths as low as 1 nm can be obtained. By using phase contrast microscopy in tapping or torsional mode, it is possible to get phase contrast maps or the contrast in viscoelastic properties of near surface regions. Scratching and indentation on the nanoscale are powerful ways to screen for adhesion and resistance to deformation of ultrathin films.

Boundary lubrication studies and measurement of lubricant-film thickness with a lateral resolution on the nanoscale can be conducted using AFM. Chemically bonded lubricant films and self-assembled monolayers are superior in friction and wear resistance. For chemically bonded lubricant films, the adsorption of water, the formation of meniscus and its change during sliding, and surface properties play an important role on the adhesion, friction, and durability of these films. Sliding velocity, relative humidity and temperature affect adhesion and friction. For SAMs, their friction mechanism is explained by a so-called *molecular spring* model. Films with high-compliance long carbon chains exhibit low friction and wear. Also perfluoroalkylsilane SAMs on Si appear to be more hydrophobic with lower adhesion than alkylsilane SAMs on Si.

Investigations of adhesion, friction, wear, scratching and indentation on the nanoscale using the AFM can provide insights into failure mechanisms of materials. Coefficients of friction, wear rates and mechanical properties such as hardness have been found to be different on the nanoscale than on the macroscale; generally, coefficients of friction and wear rates on the micro- and nanoscales are smaller, whereas hardness is greater. Therefore, micro/nanotribological studies may help define the regimes for ultra-low friction and near-zero wear. These studies also provide insight into the atomic origins of adhesion, friction, wear, and lubrication mechanisms.

References

29.1   I. L. Singer, H. M. Pollock: *Fundamentals of Friction: Macroscopic and Microscopic Processes* (Kluwer Academic, Dordrecht 1992) p. 220

29.2   B. N. J. Persson, E. Tosatti: *Physics of Sliding Friction* (Kluwer Academic, Dordrecht 1996) p. E311

29.3   B. Bhushan: *Micro/Nanotribology and its Applications*, Vol. E (Kluwer Academic, Dordrecht 1997) p. 330

29.4   B. Bhushan: *Handbook of Micro/Nanotribology*, Vol. 2nd (CRC, Boca Raton 1999)

29.5   B. Bhushan: *Principles and Applications of Tribology* (Wiley, New York 1999)

29.6   B. Bhushan: Nanoscale tribophysics and tribomechanics, Wear **225–229**, 465–492 (1999)

29.7   B. Bhushan: *Modern Tribology Handbook, Vol. 1: Principles of Tribology* (CRC, Boca Raton 2001)

29.8   B. Bhushan: *Fundamentals of Tribology and Bridging the Gap Between the Macro- and Micro/Nanoscales*, NATO Science Series II, Vol. 10 (Kluwer Academic, Dordrecht 2001)

29.9   B. Bhushan: Nano- to microscale wear and mechanical characterization studies using scanning probe microscopy, Wear **251**, 1105–1123 (2001)

29.10   B. Bhushan: *Introduction to Tribology* (Wiley, New York 2002)

29.11   B. Bhushan: *Nanotribology and Nanomechanics– An Introduction* (Springer, Berlin, Heidelberg 2005)

29.12 B. Bhushan: Nanotribology and nanomechanics, Wear **259**, 1507–1531 (2005)

29.13 B. Bhushan, J. N. Israelachvili, U. Landman: Nanotribology: Friction, wear and lubrication at the atomic scale, Nature **374**, 607–616 (1995)

29.14 H. J. Guntherodt, D. Anselmetti: *Forces in Scanning Probe Methods* (Kluwer Academic, Dordrecht 1995) p. E286

29.15 B. Bhushan: *Tribology and Mechanics of Magnetic Storage Devices*, 2nd edn. (Springer, New York 1996)

29.16 B. Bhushan: *Tribology Issues and Opportunities in MEMS* (Kluwer Academic, Dordrecht 1998)

29.17 B. Bhushan: Wear and mechanical characterisation on micro- to picoscales using AFM, Int. Mater. Rev. **44**, 105–117 (1999)

29.18 B. Bhushan, H. Fuchs, S. Hosaka: *Applied Scanning Probe Methods* (Springer, Berlin, Heidelberg 2004)

29.19 B. Bhushan, H. Fuchs: *Applied Scanning Probe Methods II* (Springer, Berlin, Heidelberg 2006)

29.20 G. Binnig, C. F. Quate, Ch. Gerber: Atomic force microscopy, Phys. Rev. Lett. **56**, 930–933 (1986)

29.21 G. Binnig, Ch. Gerber, E. Stoll, T. R. Albrecht, C. F. Quate: Atomic resolution with atomic force microscope, Europhys. Lett. **3**, 1281–1286 (1987)

29.22 C. M. Mate, G. M. McClelland, R. Erlandsson, S. Chiang: Atomic-scale friction of a tungsten tip on a graphite surface, Phys. Rev. Lett. **59**, 1942–1945 (1987)

29.23 B. Bhushan, J. Ruan: Atomic-scale friction measurements using friction force microscopy: Part II—application to magnetic media, ASME J. Trib. **116**, 389–396 (1994)

29.24 J. Ruan, B. Bhushan: Atomic-scale friction measurements using friction force microscopy: Part I—general principles and new measurement techniques, ASME J. Tribol. **116**, 378–388 (1994)

29.25 J. Ruan, B. Bhushan: Atomic-scale and microscale friction of graphite and diamond using friction force microscopy, J. Appl. Phys. **76**, 5022–5035 (1994)

29.26 J. Ruan, B. Bhushan: Frictional behavior of highly oriented pyrolytic graphite, J. Appl. Phys. **76**, 8117–8120 (1994)

29.27 B. Bhushan, V. N. Koinkar, J. Ruan: Microtribology of magnetic media, Proc. Inst. Mech. Eng., Part J: J. Eng. Tribol. **208**, 17–29 (1994)

29.28 B. Bhushan, A. V. Kulkarni: Effect of normal load on microscale friction measurements, Thin Solid Films **278**, 49–56 (1996)

29.29 B. Bhushan, A. V. Kulkarni: Effect of normal load on microscale friction measurements, Thin Solid Films **293**, 333 (1996)

29.30 B. Bhushan, S. Sundararajan: Micro/nanoscale friction and wear mechanisms of thin films using atomic force and friction force microscopy, Acta Mater. **46**, 3793–3804 (1998)

29.31 V. Scherer, W. Arnold, B. Bhushan: Active Friction Control Using Ultrasonic Vibration. In: *Tribology Issues and Opportunities in MEMS*, ed. by B. Bhushan (Kluwer Academic, Dordrecht 1998) pp. 463–469

29.32 V. Scherer, W. Arnold, B. Bhushan: Lateral force microscopy using acoustic friction force microscopy, Surf. Interface Anal. **27**, 578–587 (1999)

29.33 M. Reinstaedtler, U. Rabe, V. Scherer, U. Hartmann, A. Goldade, B. Bhushan, W. Arnold: On the nanoscale measurement of friction using atomic-force microscope cantilever torsional resonances, Appl. Phys. Lett. **82**, 2604–2606 (2003)

29.34 M. Reinstaedtler, U. Rabe, A. Goldade, B. Bhushan, W. Arnold: Investigating ultra-thin lubricant layers using resonant friction force microscopy, Tribol. Int. **38**, 533–541 (2005)

29.35 M. Reinstaedtler, T. Kasai, U. Rabe, B. Bhushan, W. Arnold: Imaging and measurement of elasticity and friction using the TR mode, J. Phys. D: Appl. Phys. **38**, R269–R282 (2005)

29.36 B. Bhushan, T. Kasai: A surface topography-independent friction measurement technique using torsional resonance mode in an AFM, Nanotechnology **15**, 923–935 (2004)

29.37 N. S. Tambe, B. Bhushan: A new atomic force microscopy based technique for studying nanoscale friction at high sliding velocities, J. Phys. D: Appl. Phys. **38**, 764–773 (2005)

29.38 B. Bhushan, A. V. Kulkarni, V. N. Koinkar, M. Boehm, L. Odoni, C. Martelet, M. Belin: Microtribological characterization of self-assembled and Langmuir–Blodgett monolayers by atomic and friction force microscopy, Langmuir **11**, 3189–3198 (1995)

29.39 V. N. Koinkar, B. Bhushan: Micro/nanoscale studies of boundary layers of liquid lubricants for magnetic disks, J. Appl. Phys. **79**, 8071–8075 (1996)

29.40 V. N. Koinkar, B. Bhushan: Microtribological studies of unlubricated and lubricated surfaces using atomic force/friction force microscopy, J. Vac. Sci. Technol. **14**, 2378–2391 (1996)

29.41 B. Bhushan, H. Liu: Nanotribological properties and mechanisms of alkylthiol and biphenyl thiol self-assembled monolayers studied by AFM, Phys. Rev. B **63**, 245412–1–245412–11 (2001)

29.42 H. Liu, B. Bhushan: Nanotribological characterization of molecularly-thick lubricant films for applications to MEMS/NEMS by AFM, Ultramicroscopy **97**, 321–340 (2003)

29.43 B. Bhushan, T. Kasai, G. Kulik, L. Barbieri, P. Hoffmann: AFM study of perfluorosilane and alkylsilane self-assembled monolayers for anti-stiction in MEMS/NEMS, Ultramicroscopy **105**, 176–188 (2005)

29.44 B. Bhushan, V. N. Koinkar: Tribological studies of silicon for magnetic recording applications, J. Appl. Phys. **75**, 5741–5746 (1994)

29.45   V. N. Koinkar, B. Bhushan: Microtribological studies of $Al_2O_3 - TiC$, polycrystalline and single-crystal $Mn - Zn$ ferrite and SiC head slider materials, Wear **202**, 110–122 (1996)

29.46   V. N. Koinkar, B. Bhushan: Microtribological properties of hard amorphous carbon protective coatings for thin film magnetic disks and heads, Proc. Inst. Mech. Eng. Part J: J. Eng. Tribol. **211**, 365–372 (1997)

29.47   S. Sundararajan, B. Bhushan: Development of a continuous microscratch technique in an atomic force microscope and its application to study scratch resistance of ultra-thin hard amorphous carbon coatings, J. Mater. Res. **16**, 75–84 (2001)

29.48   J. Ruan, B. Bhushan: Nanoindentation studies of fullerene films using atomic force microscopy, J. Mater. Res. **8**, 3019–3022 (1993)

29.49   B. Bhushan, A. V. Kulkarni, W. Bonin, J. T. Wyrobek: Nano/picoindentation measurement using a capacitance transducer system in atomic force microscopy, Philos. Mag. **74**, 1117–1128 (1996)

29.50   B. Bhushan, V. N. Koinkar: Nanoindentation hardness measurements using atomic force microscopy, Appl. Phys. Lett. **64**, 1653–1655 (1994)

29.51   B. Bhushan, X. Li: Nanomechanical characterisation of solid surfaces and thin films (invited), Intern. Mater. Rev. **48**, 125–164 (2003)

29.52   P. Maivald, H. J. Butt, S. A. C. Gould, C. B. Prater, B. Drake, J. A. Gurley, V. B. Elings, P. K. Hansma: Using force modulation to image surface elasticities with the atomic force microscope, Nanotechnology **2**, 103–106 (1991)

29.53   B. Anczykowski, D. Kruger, K. L. Babcock, H. Fuchs: Basic properties of dynamic force microscopy with the scanning force microscope in experiment and simulation, Ultramicroscopy **66**, 251–259 (1996)

29.54   D. DeVecchio, B. Bhushan: Localized surface elasticity measurements using an atomic force microscope, Rev. Sci. Instrum. **68**, 4498–4505 (1997)

29.55   V. Scherer, B. Bhushan, U. Rabe, W. Arnold: Local elasticity and lubrication measurements using atomic force and friction force microscopy at ultrasonic frequencies, IEEE Trans. Magn. **33**, 4077–4079 (1997)

29.56   S. Amelio, A. V. Goldade, U. Rabe, V. Scherer, B. Bhushan, W. Arnold: Measurements of elastic properties of ultra-thin diamond-like carbon coatings using atomic force acoustic microscopy, Thin Solid Films **392**, 75–84 (2001)

29.57   W. W. Scott, B. Bhushan: Use of phase imaging in atomic force microscopy for measurement of viscoelastic contrast in polymer nanocomposites and molecularly-thick lubricant films, Ultramicroscopy **97**, 151–169 (2003)

29.58   B. Bhushan, J. Qi: Phase contrast imaging of nanocomposites and molecularly-thick lubricant films in magnetic media, Nanotechnology **14**, 886–895 (2003)

29.59   T. Kasai, B. Bhushan, L. Huang, C. Su: Topography and phase imaging using the torsional resonance mode, Nanotechnology **15**, 731–742 (2004)

29.60   N. Chen, B. Bhushan: Morphological, nanomechanical and cellular structural characterization of human hair and conditioner distribution using torsional resonance mode in an AFM, J. Micros. **220**, 96–112 (2005)

29.61   M. S. Bobji, B. Bhushan: Atomic force microscopic study of the micro-cracking of magnetic thin films under tension, Scripta Mater. **44**, 37–42 (2001)

29.62   M. S. Bobji, B. Bhushan: In-situ microscopic surface characterization studies of polymeric thin films during tensile deformation using atomic force microscopy, J. Mater. Res. **16**, 844–855 (2001)

29.63   N. Tambe, B. Bhushan: In situ study of nanocracking of multilayered magnetic tapes under monotonic and fatigue loading using an AFM, Ultramicroscopy **100**, 359–373 (2004)

29.64   B. Bhushan, T. Kasai, C. V. Nguyen, M. Meyyappan: Multiwalled carbon nanotube AFM probes for surface characterization of micro/nanostructures, Microsys. Technol. **10**, 633–639 (2004)

29.65   B. Bhushan: Micro/nanotribology and its applications to magnetic storage devices and MEMS, Tribol. Int. **28**, 85–95 (1995)

29.66   D. DeVecchio, B. Bhushan: Use of a nanoscale Kelvin probe for detecting wear precursors, Rev. Sci. Instrum. **69**, 3618–3624 (1998)

29.67   B. Bhushan, A. V. Goldade: Measurements and analysis of surface potential change during wear of single crystal silicon (100) at ultralow loads using Kelvin probe microscopy, Appl. Surf. Sci **157**, 373–381 (2000)

29.68   B. Bhushan, A. V. Goldade: Kelvin probe microscopy measurements of surface potential change under wear at low loads, Wear **244**, 104–117 (2000)

29.69   H. U. Krotil, T. Stifter, H. Waschipky, K. Weishaupt, S. Hild, O. Marti: Pulse force mode: A new method for the investigation of surface properties, Surf. Interface Anal. **27**, 336–340 (1999)

29.70   U. Rabe, K. Janser, W. Arnold: Vibrations of free and surface-coupled atomic force microscope cantilevers: Theory and experiment, Rev. Sci. Instrum. **67**, 3281–3293 (1996)

29.71   J. Tamayo, R. Garcia: Deformation, contact time, and phase contrast in tapping mode scanning force microscopy, Langmuir **12**, 4430–4435 (1996)

29.72   R. Garcia, J. Tamayo, M. Calleja, F. Garcia: Phase contrast in tapping-mode scanning force microscopy, Appl. Phys. A **66**, 309–312 (1998)

29.73   Y. Song, B. Bhushan: Quantitative extraction of in-plane surface properties using torsional resonance mode in atomic force microscopy, J. Appl. Phys. **87**, 83533 (2005)

29.74 B. Bhushan, J. Ruan, B. K. Gupta: A scanning tunnelling microscopy study of Fullerene films, J. Phys. D: Appl. Phys. **26**, 1319–1322 (1993)

29.75 G. A. Tomlinson: A molecular theory of friction, Phil. Mag. Ser. **7**, 905–939 (1929)

29.76 D. Tomanek, W. Zhong, H. Thomas: Calculation of an atomically modulated friction force in atomic force microscopy, Europhys. Lett. **15**, 887–892 (1991)

29.77 E. Meyer, R. Overney, R. Luthi, D. Brodbeck, L. Howald, J. Frommer, H. J. Guntherodt, O. Wolter, M. Fujihira, T. Takano, Y. Gotoh: Friction force microscopy of mixed Langmuir–Blodgett films, Thin Solid Films **220**, 132–137 (1992)

29.78 C. D. Frisbie, L. F. Rozsnyai, A. Noy, M. S. Wrighton, C. M. Lieber: Functional group imaging by chemical force microscopy, Science **265**, 2071–2074 (1994)

29.79 V. N. Koinkar, B. Bhushan: Effect of scan size and surface roughness on microscale friction measurements, J. Appl. Phys. **81**, 2472–2479 (1997)

29.80 S. Sundararajan, B. Bhushan: Topography-induced contributions to friction forces measured using an atomic force/friction force microscope, J. Appl. Phys. **88**, 4825–4831 (2000)

29.81 B. Bhushan, G. S. Blackman: Atomic force microscopy of magnetic rigid disks and sliders and its applications to tribology, ASME J. Tribol. **113**, 452–458 (1991)

29.82 K. Yamanaka, E. Tomita: Lateral force modulation atomic force microscope for selective imaging of friction forces, Jpn. J. Appl. Phys. **34**, 2879–2882 (1995)

29.83 O. Marti, H.-U. Krotil: Dynamic Friction Measurement With the Scanning Force Microscope. In: *Fundamentals of Tribology and Bridging the Gap Between the Macro- and Micro/Nanoscales*, ed. by B. Bhushan (Kluwer, Dordrecht 2001) pp. 121–135

29.84 N. S. Tambe, B. Bhushan: Scale dependence of micro/nano-friction and adhesion of MEMS/NEMS materials, coatings and lubricants, Nanotechnology **15**, 1561–1570 (2004)

29.85 N. S. Tambe, B. Bhushan: Friction model for the velocity dependence of nanoscale friction, Nanotechnology **16**, 2309–2324 (2005)

29.86 N. S. Tambe, B. Bhushan: Durability studies of micro/nanoelectromechanical system materials, coatings, and lubricants at high sliding velocities (up to 10 mm/s) using a modified atomic force microscope, J. Vac. Sci. Technol. A **23**, 830–835 (2005)

29.87 N. S. Tambe, B. Bhushan: Nanoscale friction-induced phase transformation of diamond-like carbon, Scripta Materiala **52**, 751–755 (2005)

29.88 K. Mizuhara, S. M. Hsu: Tribochemical Reaction of Oxygen and Water on Silicon Surfaces. In: *Wear Particles*, ed. by D. Dowson (Elsevier Science, Amsterdam 1992) pp. 323–328

29.89 T. Bouhacina, J. P. Aime, S. Gauthier, D. Michel: Tribological behavior of a polymer grafted on silanized silica probed with a nanotip, Phys. Rev. B. **56**, 7694–7703 (1997)

29.90 E. Gnecco, R. Bennewitz, T. Gyalog, Ch. Loppacher, M. Bammerlin, E. Meyer, H. -J. Guntherodt: Velocity dependence of atomic friction, Phys. Rev. Lett. **84**, 1172–1175 (2000)

29.91 N. S. Tambe, B. Bhushan: Nanoscale friction mapping, Appl. Phys. Lett. **86**, 193102-1–193102-3 (2005)

29.92 A. Grill: Tribology of diamondlike carbon and related materials: An updated review, Surf. Coat. Technol. **94-95**, 507–513 (1997)

29.93 N. S. Tambe, B. Bhushan: Identifying materials with low friction and adhesion for nanotechnology applications, Appl. Phys. Lett **86**, 061906-1–061906-3 (2005)

29.94 N. S. Tambe, B. Bhushan: Nanowear mapping: A novel atomic force microscopy based approach for studying nanoscale wear at high sliding velocities, Tribol. Lett. **20**, 83–90 (2005)

29.95 S. C. Lim, M. F. Ashby: Wear mechanism maps, Acta Metall. **35**, 1–24 (1987)

29.96 S. C. Lim, M. F. Ashby, J. H. Brunton: Wear-rate transitions and their relationship to wear mechanisms, Acta Metall. **35**, 1343–1348 (1987)

29.97 B. Bhushan, C. Dandavate: Thin-film friction and adhesion studies using atomic force microscop, J. Appl. Phys. **87**, 1201–1210 (2000)

29.98 B. Bhushan: Adhesion and stiction: Mechanisms, measurement techniques, and methods for reduction, (invited), J. Vac. Sci. Technol. B **21**, 2262–2296 (2003)

29.99 T. Stifter, O. Marti, B. Bhushan: Theoretical investigation of the distance dependence of capillary and van der Waals forces in scanning probe microscopy, Phys. Rev. B **62**, 13667–13673 (2000)

29.100 U. D. Schwarz, O. Zwoerner, P. Koester, R. Wiesendanger: Friction Force Spectroscopy in the Low-load Regime with Well-defined Tips. In: *Micro/Nanotribology and Its Applications*, ed. by B. Bhushan (Kluwer Academic, Dordrecht 1997) pp. 233–238

29.101 M. Nosonovsky, B. Bhushan: Stochastic model for metastable wetting of roughness-induced superhydrophobic surfaces, Microsyst. Technol. **12**, 231–237 (2005)

29.102 M. Nosonovsky, B. Bhushan: Roughness optimization for biomimetic superhydrophobic surfaces, Microsyst. Technol. **11**, 535–549 (2005)

29.103 R. N. Wenzel: Resistance of solid surfaces to wetting by water, Indus. Eng. Chem. **28**, 988–994 (1936)

29.104 A. Cassie, S. Baxter: Wetting of porous surfaces, Trans. Faraday Soc. **40**, 546–551 (1944)

29.105 Z. Burton, B. Bhushan: Hydrophobicity, adhesion and friction properties with nanopatterned roughness and scale dependence, Nano Lett. **5**, 1607–1613 (2005)

29.106 B. Bhushan, H. Liu, S. M. Hsu: Adhesion and friction studies of silicon and hydrophobic and low friction films and investigation of scale effects, ASME J. Tribol. **126**, 583–590 (2004)

29.107 H. Liu, B. Bhushan: Adhesion and friction studies of microelectromechanical systems/nanoelectromechanical systems materials using a novel microtriboapparatus, J. Vac. Sci. Technol. A **21**, 1528–1538 (2003)

29.108 B. Bhushan, B. K. Gupta: *Handbook of Tribology: Materials, Coatings and Surface Treatments* (McGraw-Hill, New York 1991) reprinted Krieger, Malabar Florida, 1997

29.109 B. Bhushan, S. Venkatesan: Mechanical and tribological properties of silicon for micromechanical applications: A Review, Adv. Info. Storage Sys. **5**, 211–239 (1993)

29.110 Anonymous: *Properties of Silicon*, EMIS Data Reviews Series No. 4. INSPEC, Institution of Electrical Engineers, London. See also Anonymous, MEMS Materials Database, http://www.memsnet.org/material/ (2002)

29.111 J. E. Field: *The properties of natural and synthetic diamond* (Academic, London 1992)

29.112 B. Bhushan: Chemical, mechanical and tribological characterization of ultra-thin and hard amorphous carbon coatings as thin as 3.5 nm: Recent developments, Diamond and Related Materials **8**, 1985–2015 (1999)

29.113 Anonymous: *The Industrial Graphite Engineering Handbook* (National Carbon Company, New York 1959)

29.114 M. Nosonovsky, B. Bhushan: Scale effects in dry friction during multiple-asperity contact, ASME J. Tribol. **127**, 37–46 (2005)

29.115 B. Bhushan, M. Nosonovsky: Comprehensive model for scale effects in friction due to adhesion and two- and three-body deformation (plowing), Acta Mater. **52**, 2461–2474 (2004)

29.116 B. Bhushan, M. Nosonovsky: Scale effects in dry and wet friction, wear, and interface temperature, Nanotechnology **15**, 749–761 (2004)

29.117 B. Bhushan, M. Nosonovsky: Scale effects in friction using strain gradient plasticity and dislocation-assisted sliding (microslip), Acta Mater. **51**, 4331–4345 (2003)

29.118 V. N. Koinkar, B. Bhushan: Scanning and transmission electron microscopies of single-crystal silicon microworn/machined using atomic force microscopy, J. Mater. Res. **12**, 3219–3224 (1997)

29.119 X. Zhao, B. Bhushan: Material removal mechanism of single-crystal silicon on nanoscale and at ultralow loads, Wear **223**, 66–78 (1998)

29.120 B. Bhushan, P. S. Mokashi, T. Ma: A new technique to measure Poisson's ratio of ultrathin polymeric films using atomic force microscopy, Rev. Sci. Instrum. **74**, 1043–1047 (2003)

29.121 A. V. Kulkarni, B. Bhushan: Nanoscale mechanical property measurements using modified atomic force microscopy, Thin Solid Films **290–291**, 206–210 (1996)

29.122 A. V. Kulkarni, B. Bhushan: Nano/picoindentation measurements on single-crystal aluminum using modified atomic force microscopy, Mater. Lett. **29**, 221–227 (1996)

29.123 A. V. Kulkarni, B. Bhushan: Nanoindentation measurement of amorphous carbon coatings, J. Mater. Res. **12**, 2707–2714 (1997)

29.124 N. A. Fleck, G. M. Muller, M. F. Ashby, J. W. Hutchinson: Strain gradient plasticity: Theory and experiment, Acta Metall. Mater. **42**, 475–487 (1994)

29.125 W. D. Nix, H. Gao: Indentation size effects in crystalline materials: A law for strain gradient plasticity, J. Mech. Phys. Solids **46**, 411–425 (1998)

29.126 W. B. Li, J. L. Henshall, R. M. Hooper, K. E. Easterling: The mechanism of indentation creep, Acta Metall. Mater. **39**, 3099–3110 (1991)

29.127 F. P. Bowden, D. Tabor: *The Friction and Lubrication of Solids* (Clarendon, Oxford 1950)

29.128 Z. Tao, B. Bhushan: Bonding, degradation, and environmental effects on novel perfluoropolyether lubricants, Wear **259**, 1352–1361 (2005)

29.129 H. Liu, B. Bhushan, W. Eck, V. Staedtler: Investigation of the adhesion, friction, and wear properties of biphenyl thiol self-assembled monolayers by atomic force microscopy, J. Vac. Sci. Technol. A **19**, 1234–1240 (2001)

29.130 H. Liu, B. Bhushan: Investigation of nanotribological properties of self-assembled monolayers with alkyl and biphenyl spacer chains, Ultramicroscopy **91**, 185–202 (2002)

29.131 T. Kasai, B. Bhushan, G. Kulik, L. Barbieri, P. Hoffman: Nanotribological study of perfluorosilane SAMs for anti-stiction and low wear, J. Vac. Sci. Technol. B **23**, 995–1003 (2005)

29.132 K. K. Lee, B. Bhushan, D. Hansford: Nanotribological characterization of perfluoropolymer thin films for biomedical micro/nanoelectrone chemical systems applications, J. Vac. Sci. Technol. A **23**, 804–810 (2005)

29.133 B. Bhushan, D. Hansford, K. K. Lee: Surface modification of silicon and polymethylsiloxane surfaces with vapor-phase-deposited ultrathin fluorosilane films for biomedical nanodevices, J. Vac. Sci. Technol. A **24**, 1197–1202 (2006)

29.134 N. S. Tambe, B. Bhushan: Nanotribological characterization of self assembled monolayers deposited on silicon and aluminum substrates, Nanotechnology **16**, 1549–1558 (2005)

29.135 Z. Tao, B. Bhushan: Degradation mechanisms and environmental effects on perfluoropolyether, self assembled monolayers, and diamondlike carbon films, Langmuir **21**, 2391–2399 (2005)

30. Surface Forces and Nanorheology of Molecularly Thin Films

In this chapter, we describe the static and dynamic normal forces that occur between surfaces in vacuum or liquids and the different modes of friction that can be observed between: (i) bare surfaces in contact (dry or interfacial friction), (ii) surfaces separated by a thin liquid film (lubricated friction), and (iii) surfaces coated with organic monolayers (boundary friction).

Experimental methods suitable for measuring normal surface forces, adhesion and friction (lateral or shear) forces of different magnitude at the molecular level are described. We explain the molecular origin of van der Waals, electrostatic, solvation and polymer-mediated interactions, and basic models for the contact mechanics of adhesive and nonadhesive elastically deforming bodies. The effects of interaction forces, molecular shape, surface structure and roughness on adhesion and friction are discussed.

Simple models for the contributions of the adhesion force and external load to interfacial friction are illustrated with experimental data on both unlubricated and lubricated systems, as measured with the surface forces apparatus. We discuss rate-dependent adhesion (adhesion hysteresis) and how this is related to friction. Some examples of the transition from wearless friction to friction with wear are shown.

Lubrication in different lubricant thickness regimes is described together with explanations of nanorheological concepts. The occurrence of and transitions between smooth and stick–slip sliding in various types of dry (unlubricated and solid boundary lubricated) and liquid lubricated systems are discussed based on recent experimental results and models for stick–slip involving memory distance and dilatancy.

30.1 Introduction: Types of Surface Forces

In this chapter, we discuss the most important types of surface forces and the relevant equations for the force and friction laws. Several different attractive and repulsive forces operate between surfaces and particles. Some forces occur in vacuum, for example, attractive van der Waals and repulsive hard-core interactions. Other types of forces can arise only when the interacting surfaces are separated by another condensed phase, which is usually a liquid. The most common types of surface forces and their main characteristics are listed in Table 30.1.

In *vacuum*, the two main long-range interactions are the attractive van der Waals and electrostatic (Coulomb) forces. At smaller surface separations (corresponding to molecular contact at surface separations of $D \approx 0.2\,\text{nm}$), additional attractive interactions can be found such as covalent or metallic bonding forces. These attractive forces are stabilized by the hard-core repulsion. Together they determine the surface and interfacial energies of planar surfaces, as well as the strengths of materials and adhesive junctions. Adhesion forces are often strong enough to elastically or plastically deform bodies or particles when they come into contact.

In *vapors* (e.g., atmospheric air containing water and organic molecules), solid surfaces in, or close to, contact will generally have a surface layer of chemisorbed or physisorbed molecules, or a capillary condensed liquid bridge between them. A surface layer usually causes the adhesion to decrease, but in the case of capillary condensation, the additional Laplace pressure or attractive "capillary" force may make the adhesion between the surfaces stronger than in an inert gas or vacuum.

When totally immersed in a *liquid*, the force between particles or surfaces is completely modified from that in vacuum or air (vapor). The van der Waals attraction is generally reduced, but other forces can now arise that can qualitatively change both the range and even the sign of the interaction. The attractive force in such a system can be either stronger or weaker than in the absence of the intervening liquid. For example, the overall attraction can be stronger in the case of two hydrophobic surfaces separated by water, but weaker for two hydrophilic surfaces. Depending on the different forces that may be operating simultaneously in solution, the overall force law is not generally monotonically attractive even at long range; it can be repulsive, or the force can change sign at some finite surface separation. In such cases, the potential-energy minimum, which determines the adhesion force or energy, does not occur at the true molecular contact between the surfaces, but at some small distance further out.

The forces between surfaces in a liquid medium can be particularly complex at *short range*, i.e., at surface separations below a few nanometers or 4–10 molecular diameters. This is partly because, with increasing confinement, a liquid ceases to behave as a structureless continuum with bulk properties; instead, the size and shape of its molecules begin to determine the overall interaction. In addition, the surfaces themselves can no longer be treated as inert and structureless walls (i.e., mathematically flat) and their physical and chemical properties at the atomic scale must now be taken into account. The force laws will then depend on whether the surfaces are amorphous or crystalline (and whether the lattices of crystalline surfaces are matched or not), rough or smooth, rigid or soft (fluid-like), and hydrophobic or hydrophilic.

It is also important to distinguish between *static* (i.e., equilibrium) interactions and *dynamic* (i.e., nonequilibrium) forces such as viscous and friction forces. For example, certain liquid films confined between two contacting surfaces may take a surprisingly long time to equilibrate, as may the surfaces themselves, so that the short-range and adhesion forces appear to be time-dependent, resulting in "aging" effects.

Table 30.1 Types of surface forces in vacuum versus in liquid (colloidal forces)

Type of force	Subclasses or alternative names	Main characteristics
	Attractive forces	
van der Waals	Debye induced dipole force (v & s) London dispersion force (v & s) Casimir force (v & s)	Ubiquitous, occurs both in vacuum and in liquids
Electrostatic	Ionic bond (v) Coulombic force (v & s) Hydrogen bond (v) Charge-exchange interaction (v & s) Acid–base interaction (s) "Harpooning" interaction (v)	Strong, long-range, arises in polar solvents; requires surface charging or charge-separation mechanism
Ion correlation	van der Waals force of polarizable ions (s)	Requires mobile charges on surfaces in a polar solvent
Quantum mechanical	Covalent bond (v) Metallic bond (v) Exchange interaction (v)	Strong, short-range, responsible for contact binding of crystalline surfaces
Solvation	Oscillatory force (s) Depletion force (s)	Mainly entropic in origin, the oscillatory force alternates between attraction and repulsion
Hydrophobic	Attractive hydration force (s)	Strong, apparently long-range; origin not yet understood
Specific binding	"Lock-and-key" or complementary binding (v & s) Receptor–ligand interaction (s) Antibody–antigen interaction (s)	Subtle combination of different non-covalent forces giving rise to highly specific binding; main recognition mechanism of biological systems
	Repulsive forces	
van der Waals	van der Waals disjoining pressure (s)	Arises only between dissimilar bodies interacting in a medium
Electrostatic	Coulombic force (v & s)	Arises only for certain constrained surface charge distributions
Quantum mechanical	Hard-core or steric repulsion (v) Born repulsion (v)	Short-range, stabilizing attractive covalent and ionic binding forces, effectively determine molecular size and shape
Solvation	Oscillatory solvation force (s) Structural force (s) Hydration force (s)	Monotonically repulsive forces, believed to arise when solvent molecules bind strongly to surfaces
Entropic	Osmotic repulsion (s) Double-layer force (s) Thermal fluctuation force (s) Steric polymer repulsion (s) Undulation force (s) Protrusion force (s)	Due to confinement of molecular or ionic species; requires mechanism that keeps trapped species between the surfaces
	Dynamic interactions	
Non-equilibrium	Hydrodynamic forces (s) Viscous forces (s) Friction forces (v & s) Lubrication forces (s)	Energy-dissipating forces occurring during relative motion of surfaces or bodies

Note: (v) applies only to interactions in vacuum, (s) applies only to interactions in solution (or to surfaces separated by a liquid), and (v & s) applies to interactions occurring both in vacuum and in solution.

Part D | 30.1

30.2 Methods Used to Study Surface Forces

30.2.1 Force Laws

The full force law $F(D)$ between two surfaces, i.e., the force F as a function of surface separation D, can be measured in a number of ways [30.2–6]. The simplest is to move the base of a spring by a known amount, ΔD_0. Figure 30.1 illustrates this method when applied to the interaction of two magnets. However, the method is also applicable at the microscopic or molecular level, and it forms the basis of all direct force-measuring apparatuses such as the surface forces apparatus (SFA; [30.3, 7]) and the atomic force microscope (AFM; [30.8–10]). If

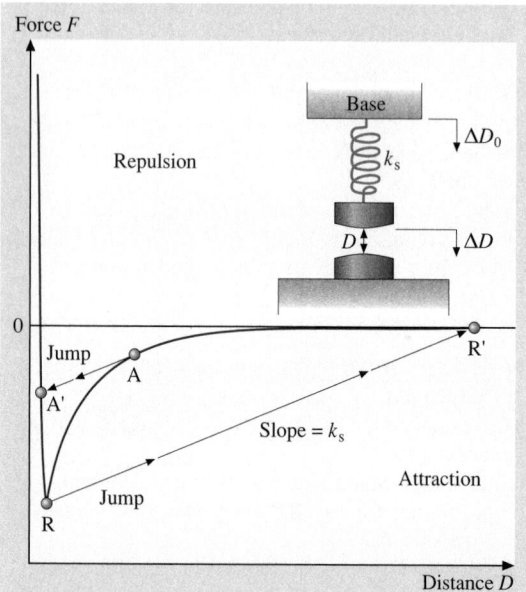

Fig. 30.1 Schematic attractive force law between two macroscopic objects such as two magnets, or between two microscopic objects such as the van der Waals force between a metal tip and a surface. On lowering the base supporting the spring, the latter will expand or contract such that, at any equilibrium separation D, the attractive force balances the elastic spring restoring force. If the gradient of the attractive force dF/dD exceeds the gradient of the spring's restoring force (defined by the spring constant k_s), the upper surface will jump from A into contact at A′ (A for "advancing"). On separating the surfaces by raising the base, the two surfaces will jump apart from R to R′ (R for "receding"). The distance R−R′ multiplied by k_s gives the adhesion force, i.e., the value of F at the point R (after [30.1] with permission)

there is a detectable force between the surfaces, this will cause the force-measuring spring to deflect by ΔD_s, while the surface separation changes by ΔD. These three displacements are related by

$$\Delta D_s = \Delta D_0 - \Delta D . \tag{30.1}$$

The difference in force, ΔF, between the initial and final separations is given by

$$\Delta F = k_s \Delta D_s , \tag{30.2}$$

where k_s is the spring constant. The equations above provide the basis for measurements of the force difference between any two surface separations. For example, if a force-measuring apparatus with a known k_s can measure D (and thus ΔD), ΔD_0, and ΔD_s, the force difference ΔF can be measured between a large initial or reference separation D, where the force is zero ($F = 0$), and another separation $D - \Delta D$. By working one's way in increasing increments of $\Delta D = \Delta D_0 - \Delta D_s$, the full force law $F(D)$ can be constructed over any desired distance regime.

In order to measure an equilibrium force law, it is essential to establish that the two surfaces have stopped moving before the displacements are measured. When displacements are measured while two surfaces are still in relative motion, one also measures a viscous or frictional contribution to the total force. Such dynamic force measurements have enabled the viscosities of liquids near surfaces and in thin films to be accurately determined [30.11–13].

In practice, it is difficult to measure the forces between two perfectly flat surfaces, because of the stringent requirement of perfect alignment for making reliable measurements at distances of a few tenths of a nanometer. It is far easier to measure the forces between curved surfaces, e.g., two spheres, a sphere and a flat surface, or two crossed cylinders. Furthermore, the force $F(D)$ measured between two curved surfaces can be directly related to the energy per unit area $E(D)$ between two flat surfaces at the same separation, D, by the so-called Derjaguin approximation [30.14]:

$$E(D) = \frac{F(D)}{2\pi R} , \tag{30.3}$$

where R is the radius of the sphere (for a sphere and a flat surface) or the radii of the cylinders (for two crossed cylinders, cf. Table 30.2).

30.2.2 Adhesion Forces

The most direct way to measure the adhesion of two solid surfaces (such as two spheres or a sphere on a flat) is to suspend one of them on a spring and measure the adhesion or "pull-off" force needed to separate the two bodies, using the deflection of the spring. If k_s is the stiffness of the force-measuring spring and ΔD is the distance the two surfaces jump apart when they separate, then the adhesion force F_{ad} is given by

$$F_{ad} = F_{max} = k_s \Delta D, \qquad (30.4)$$

where we note that, in liquids, the maximum or minimum in the force may occur at some nonzero surface separation (see Fig. 30.7). From F_{ad} and a known surface geometry, and assuming that the surfaces were everywhere in molecular contact, one may also calculate the surface or interfacial energy γ. For an elastically deformable sphere of radius R on a flat surface, or for two crossed cylinders of radius R, we have [30.3, 15]

$$\gamma = \frac{F_{ad}}{3\pi R}, \qquad (30.5)$$

while for two spheres of radii R_1 and R_2

$$\gamma = \frac{F_{ad}}{3\pi} \left(\frac{1}{R_1} + \frac{1}{R_2} \right), \qquad (30.6)$$

where γ is in units of $J\,m^{-2}$ (see Sect. 30.5.2).

30.2.3 The SFA and AFM

In a typical force-measuring experiment, at least two of the above displacement parameters – ΔD_0, ΔD, and ΔD_s – are directly or indirectly measured, and from these the third displacement and the resulting force law $F(D)$ are deduced using (30.1) and (30.2) together with a measured value of k_s. For example, in SFA experiments, ΔD_0 is changed by expanding or contracting a piezoelectric crystal by a known amount or by moving the base of the spring with sensitive motor-driven mechanical stages. The resulting change in surface separation ΔD is measured optically, and the spring deflection ΔD_s can then be obtained according to (30.1). In AFM experiments, ΔD_0 and ΔD_s are measured using a combination of piezoelectric, optical, capacitance or magnetic techniques, from which the change in surface separation ΔD is deduced. Once a force law is established, the geometry of the two surfaces (e.g., their radii) must also be known before the results can be compared with theory or with other experiments.

The SFA (Fig. 30.2) is used for measurements of adhesion and force laws between two curved molecularly smooth surfaces immersed in liquids or controlled vapors [30.3, 7, 16]. The surface separation is measured by multiple-beam interferometry with an accuracy of ± 0.1 nm. From the shape of the interference fringes one also obtains the radius of the surfaces, R, and any surface deformation that arises during an interaction [30.17–19]. The resolution in the lateral direction is about 1 μm. The surface separation can be independently controlled to within 0.1 nm, and the force sensitivity is about 10^{-8} N. For a typical surface radius of $R \approx 1$ cm, γ values can be measured to an accuracy of about 10^{-3} mJ m^{-2}.

Several different materials have been used to form the surfaces in the SFA, including mica [30.20, 21], silica [30.22], sapphire [30.23], and polymer sheets [30.24]. These materials can also be used as supporting substrates in experiments on the forces between adsorbed or chemically bound polymer layers [30.13, 25–30], surfactant and lipid monolayers, and bilayers [30.31–34], and metal and metal oxide layers [30.35–42]. The range of liquids and vapors that can be used is almost endless, and have thus far included aqueous solutions, organic liquids and solvents, polymer melts, various petroleum oils and lubricant liquids, dyes, and liquid crystals.

Friction attachments for the SFA [30.43–48] allow for the two surfaces to be sheared laterally past each other at varying sliding speeds or oscillating frequencies, while simultaneously measuring both the transverse (frictional or shear) force and the normal force (load) between them. The ranges of friction forces and sliding speeds that can be studied with such methods are currently 10^{-7}–10^{-1} N and 10^{-13}–10^{-2} m s^{-1}, respectively [30.49]. The externally applied load, L, can be varied continuously, and both positive and negative loads can be applied. The distance between the surfaces, D, their true molecular contact area, their elastic (or viscoelastic or elasto-hydrodynamic) deformation, and their lateral motion can all be monitored simultaneously by recording the moving interference-fringe pattern. Equipment for dynamic measurements of normal forces has also been developed. Such measurements give information on the viscosity of the medium and the location of the shear or slip planes relative to the surfaces [30.11–13, 50, 51].

In the atomic force microscope (Fig. 30.3), the force is measured by monitoring the deflection of a soft cantilever supporting a sub-microscopic tip ($R \approx 10$–200 nm) as this interacts with a flat, macroscopic surface [30.8, 52, 53]. The measurements can be

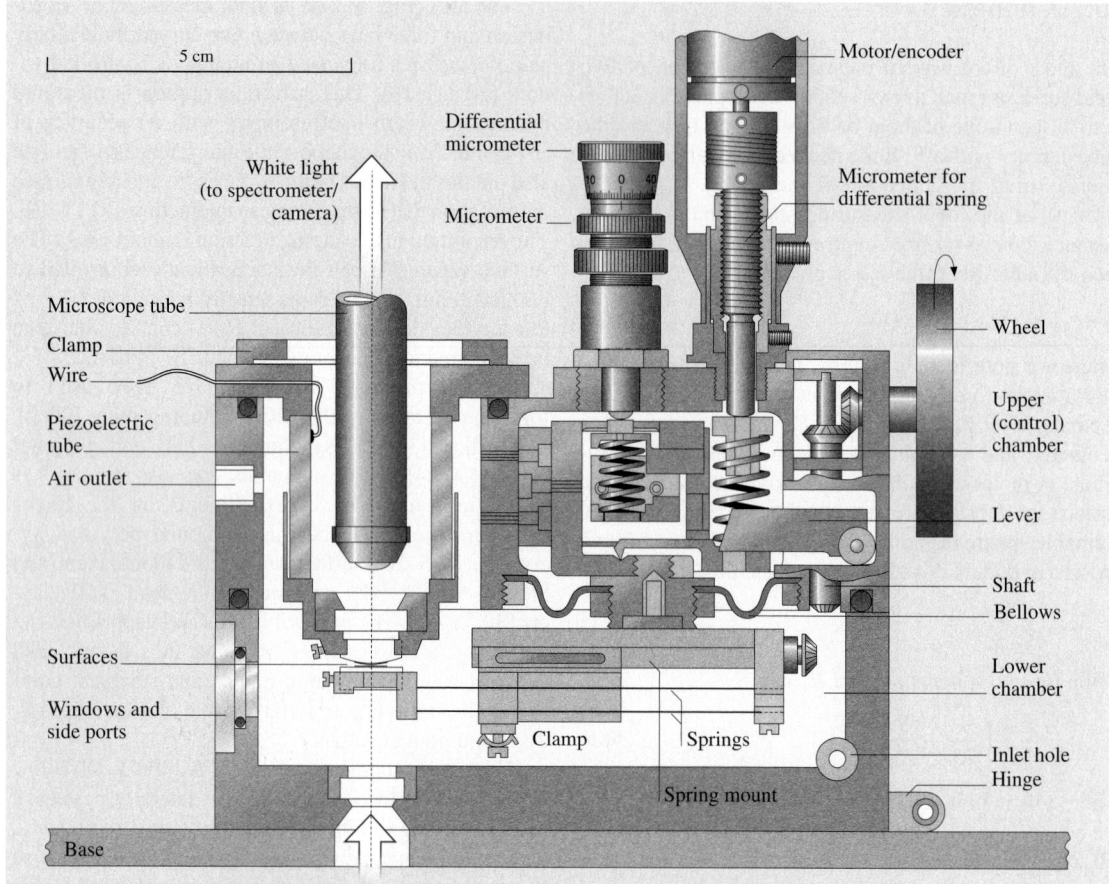

Fig. 30.2 A surface forces apparatus (SFA) where the intermolecular forces between two macroscopic, cylindrical surfaces of local radius R can be directly measured as a function of surface separation over a large distance regime from tenths of a nanometer to micrometers. Local or transient surface deformations can be detected optically. Various attachments for moving one surface laterally with respect to the other have been developed for friction measurements in different regimes of sliding velocity and sliding distance (after [30.16] with permission)

done in a vapor or liquid. The normal (bending) spring stiffness of the cantilever can be as small as $0.01\,\mathrm{N\,m^{-1}}$, allowing measurements of normal forces as small as $1\,\mathrm{pN}\,(10^{-12}\,\mathrm{N})$, which corresponds to the bond strength of single molecules [30.54–57]. Distances can be inferred with an accuracy of about 1 nm, and changes in distance can be measured to about 0.1 nm. Since the contact area can be small when using sharp tips, different interaction regimes can be resolved on samples with a heterogeneous composition on lateral scales of a few nanometers. Height differences and the roughness of the sample can be measured directly from the cantilever deflection or, alternatively, by using a feedback system to raise or lower the sample so that the deflection

(the normal force) is kept constant during a scan over the area of interest. Weak interaction forces and larger (microscopic) interaction areas can be investigated by replacing the tip with a micrometer-sized sphere to form a "colloidal probe" [30.9].

The atomic force microscope can also be used for friction measurements (lateral force microscopy, LFM, or friction force microscopy, FFM) by monitoring the torsion of the cantilever as the sample is scanned in the direction perpendicular to the long axis of the cantilever [30.10, 53, 58, 59]. Typically, the stiffness of the cantilever to lateral bending is much larger than to bending in the normal direction and to torsion, so that these signals are decoupled and height and friction can be de-

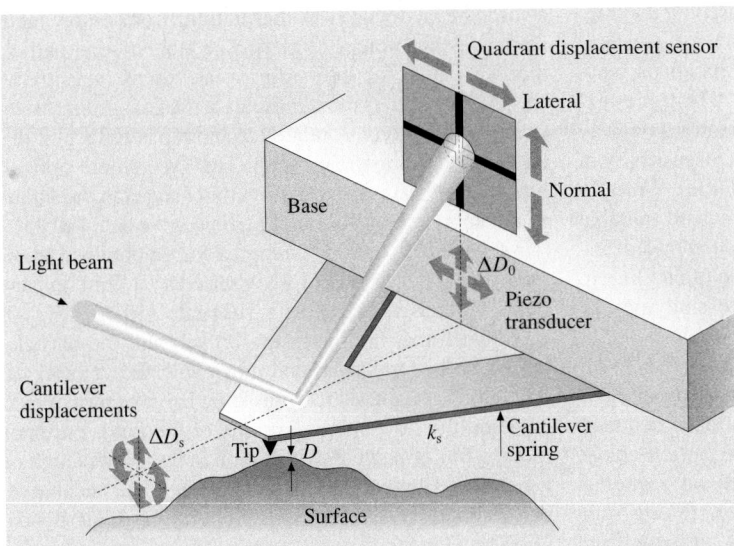

Fig. 30.3 Schematic drawing of an atomic force microscope (AFM) tip supported on a triangular cantilever interacting with an arbitrary solid surface. The normal force and topology are measured by monitoring the calibrated deflection of the cantilever as the tip is moved across the surface by means of a piezoelectric transducer. Various designs have been developed that move either the sample or the cantilever during the scan. Friction forces can be measured from the torsion of the cantilever when the scanning is in the direction perpendicular to its long axis (after [30.60] with permission)

tected simultaneously. The torsional spring constant can be as low as $0.1\,\mathrm{N\,m^{-1}}$, giving a lateral (friction) force sensitivity of $10^{-11}\,\mathrm{N}$.

Rapid technical developments have facilitated the calibrations of the normal [30.61, 62] and lateral spring constants [30.59, 63–65], as well as in situ measurements of the macroscopic tip radius [30.66, 67]. Cantilevers of different shapes with a large range of spring constants, tip radii, and surface treatments (inorganic or organic coatings) are commercially available. The flat surface, and also the particle in the colloidal probe technique, can be any material of interest. However, remaining difficulties with this technique are that the distance between the tip and the substrate, D, and the deformations of the tip and sample, are not directly measurable. Another important difference between the AFM/LFM and SFA techniques is the different size of the contact area, and the related observation that, even when a cantilever with a very low spring constant is used in the AFM, the pressure in the contact zone is typically much higher than in the SFA. Hydrodynamic effects in liquids also affect the measurements of normal forces differently on certain time scales [30.68–71].

30.2.4 Some Other Force-Measuring Techniques

A large number of other techniques are available for the measurements of the normal forces between solid or fluid surfaces (see [30.5, 60]). The techniques discussed in this section are not used for lateral (friction) force measurements, but are commonly used to study normal forces, particularly in biological systems.

Micropipette aspiration is used to measure the forces between cells or vesicles, or between a cell or vesicle and another surface [30.72–74]. The cell or vesicle is held by suction at the tip of a glass micropipette and deforms elastically in response to the net interactions with another surface and to the applied suction. The shape of the deformed surface (cell membrane) is measured and used to deduce the force between the surfaces and the membrane tension [30.73]. The membrane tension, and thus the stiffness of the cell or vesicle, is regulated by applying different hydrostatic pressures. Forces can be measured in the range of $0.1\,\mathrm{pN}$ to $1\,\mathrm{nN}$, and the distance resolution is a few nanometers. The interactions between a colloidal particle and another surface can be studied by attaching the particle to the cell membrane [30.75].

In the osmotic stress technique, pressures are measured between colloidal particles in aqueous solution, membranes or bilayers, or other ordered colloidal structures (viruses, DNA). The separation between the particle surfaces and the magnitude of membrane undulations are measured by X-ray or neutron scattering techniques. This is combined with a measurement of the osmotic pressure of the solution [30.76–79]. The technique has been used to measure repulsive forces, such as Derjaguin–Landau–Verwey–Overbeek (DLVO) interactions, steric forces, and hydration forces [30.80]. The sensitivity in pressure is $0.1\,\mathrm{mN\,m^{-2}}$, and distances can be resolved to $0.1\,\mathrm{nm}$.

The optical tweezers technique is based on the trapping of dielectric particles at the center of a focused laser beam by restoring forces arising from radiation pressure and light-intensity gradients [30.81, 82]. The forces experienced by particles as they are moved toward or away from one another can be measured with a sensitivity in the pN range. Small biological molecules are typically attached to a larger bead of a material with suitable refractive properties. Recent development allows determinations of position with nanometer resolution [30.83], which makes this technique useful for studying the forces during the extension of single molecules.

In total internal reflection microscopy (TIRM), the potential energy between a micrometer-sized colloidal particle and flat surface in aqueous solution is deduced from the average equilibrium height of the particle above the surface, measured from the intensity of scattered light. The average height ($D \approx 10-100$ nm) results from a balance of gravitational force, radiation pressure from a laser beam focused at the particle from below, and intermolecular forces [30.84]. The technique is particularly suitable for measuring weak forces (sensitivity ca. 10^{-14} N), but is more difficult to use for systems with strong interactions. A related technique is reflection interference contrast microscopy (RICM), where optical interference is used to also monitor changes in the shape of the approaching colloidal particle or vesicle [30.85].

An estimate of bond strengths can be obtained from the hydrodynamic shear force exerted by a fluid on particles or cells attached to a substrate [30.86, 87]. At a critical force, the bonds are broken and the particle or cell will be detached and move with the velocity of the fluid. This method requires knowledge of the contact area and the flow-velocity profile of the fluid. Furthermore, a uniform stress distribution in the contact area is generally assumed. At low bond density, this technique can be used to determine the strength of single bonds (1 pN).

30.3 Normal Forces Between Dry (Unlubricated) Surfaces

30.3.1 Van der Waals Forces in Vacuum and Inert Vapors

Forces between macroscopic bodies (such as colloidal particles) across vacuum arise from interactions between the constituent atoms or molecules of each body across the gap separating them. These intermolecular interactions are electromagnetic forces between permanent or induced dipoles (van der Waals forces), and between ions (electrostatic forces). In this section, we describe the van der Waals forces, which occur between all atoms and molecules and between all macroscopic bodies (see [30.3]).

The interaction between two permanent dipoles with a fixed relative orientation can be attractive or repulsive. For the specific case of two freely rotating permanent dipoles in a liquid or vapor (orientational or Keesom interaction), and for a permanent dipole and an induced dipole in an atom or polar or nonpolar molecule (induction or Debye interaction), the interaction is on average always attractive. The third type of van der Waals interaction, the fluctuation or London dispersion interaction, arises from instantaneous polarization of one nonpolar or polar molecule due to fluctuations in the charge distribution of a neighboring nonpolar or polar molecule (Fig. 30.4a). Correlation between these fluctuating induced dipole moments gives an attraction that is present between any two molecules or surfaces across vacuum. At very small separations, the interaction will ultimately be repulsive as the electron clouds of atoms and molecules begin to overlap. The total interaction is thus a combination of a short-range repulsion and a relatively long-range attraction.

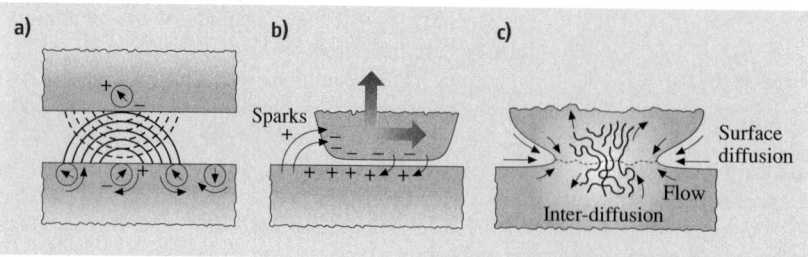

Fig. 30.4a–c Schematic representation of (**a**) van der Waals interaction (dipole–induced dipole interaction), (**b**) charge exchange, which acts to increase adhesion and friction forces, and (**c**) sintering between two surfaces

Except for in highly polar materials such as water, London dispersion interactions give the largest contribution (70–100 %) to the van der Waals attraction. The interaction energy of the van der Waals force between atoms or molecules depends on the separation r as

$$E(D) = \frac{-C_{vdW}}{r^6} , \qquad (30.7)$$

where the constant C_{vdW} depends on the dipole moments and polarizabilities of the molecules. At large separations ($> 10\,\text{nm}$), the London interaction is weakened by a randomizing effect caused by the rapid fluctuations. That is, the induced temporary dipole moment of one molecule may have changed during the time needed for the transmission of the electromagnetic wave (photon) generated by its fluctuating charge density to another molecule and the return of the photon generated by the induced fluctuation in this second molecule. This phenomenon is called retardation and causes the interaction energy to decay as r^{-7} at large separations [30.88].

Dispersion interactions are to a first approximation additive, and their contribution to the interaction energy between two macroscopic bodies (such as colloidal particles) across vacuum can be found by summing the pairwise interactions [30.89]. The interaction is generally described in terms of the Hamaker constant, A_H. Another approach is to treat the interacting bodies and an intervening medium as continuous phases and determine the strength of the interaction from bulk dielectric properties of the materials [30.90, 91]. Unlike the pairwise summation, this method takes into account the screening of the interactions between molecules inside the bodies by the molecules closer to the surfaces and the effects of the intervening medium. For the interaction between material 1 and material 3 across material 2, the non-retarded Hamaker constant given by the Lifshitz theory is approximately [30.3]:

$$A_{H,123} = A_{H,v=0} + A_{H,v>0}$$

$$\approx \tfrac{3}{4}k_B T \left(\frac{\varepsilon_1 - \varepsilon_2}{\varepsilon_1 + \varepsilon_2}\right)\left(\frac{\varepsilon_3 - \varepsilon_2}{\varepsilon_3 + \varepsilon_2}\right)$$

$$+ \frac{3h\nu_e}{8\sqrt{2}} \frac{(n_1^2 - n_2^2)(n_3^2 - n_2^2)}{\sqrt{(n_1^2 + n_2^2)}\sqrt{(n_3^2 + n_2^2)}\left(\sqrt{(n_1^2 + n_2^2)} + \sqrt{(n_3^2 + n_2^2)}\right)} , \qquad (30.8)$$

where the first term ($v = 0$) represents the permanent dipole and dipole–induced dipole interactions and the second ($v > 0$) the London (dispersion) interaction. ε_i and n_i are the static dielectric constants and refractive indexes of the materials, respectively. ν_e is

the frequency of the lowest electron transition (around $3 \times 10^{15}\,\text{s}^{-1}$). Either one of the materials 1, 2, or 3 in (30.8) can be vacuum or air ($\varepsilon = n = 1$). A_H is typically $10^{-20} - 10^{-19}\,\text{J}$ (the higher values are found for metals) for interactions between solids and liquids across vacuum or air.

The interaction energy between two macroscopic bodies is dependent on the geometry and is always attractive between two bodies of the same material [A_H positive, see (30.8)]. The van der Waals interaction energy and force laws ($F = -dE(D)/dD$) for some common geometries are given in Table 30.2. Because of the retardation effect, the equations in Table 30.2 will lead to an overestimation of the dispersion force at large separations. It is, however, apparent that the interaction energy between macroscopic bodies decays more slowly with separation (i. e., has a longer range) than between two molecules.

For inert nonpolar surfaces, e.g., consisting of hydrocarbons or van der Waals solids and liquids, the Lifshitz theory has been found to apply even at molecular contact, where it can be used to predict the surface energies (surface tensions) of such solids and liquids. For example, for hydrocarbon surfaces, $A_H = 5 \times 10^{-20}\,\text{J}$. Inserting this value into the equation for two flat surfaces (Table 30.2) and using a "cut-off" distance of $D_0 \approx 0.165\,\text{nm}$ as an effective separation when the surfaces are in contact [30.3], we obtain for the surface energy γ (which is defined as half the interaction energy)

$$\gamma = \frac{E}{2} = \frac{A_H}{24\pi D_0^2} \approx 24\,\text{mJ}\,\text{m}^{-2} , \qquad (30.9)$$

a value that is typical for hydrocarbon solids and liquids [30.92].

If the adhesion force is measured between two crossed-cylindrical surfaces of $R = 1\,\text{cm}$ (a geometry equivalent to a sphere with $R = 1\,\text{cm}$ interacting with a flat surface, cf. Table 30.2) using an SFA, we expect the adhesion force to be (see Table 30.2) $F = A_H R/(6D_0^2) = 4\pi R\gamma \approx 3.0\,\text{mN}$. Using a spring constant of $k_s = 100\,\text{N}\,\text{m}^{-1}$, such an adhesive force will cause the two surfaces to jump apart by $\Delta D = F/k_s = 30\,\mu\text{m}$, which can be accurately measured. (For elastic bodies that deform in adhesive contact, R changes during the interaction and the measured adhesion force is 25% lower, see Sect. 30.5.2). Surface energies of solids can thus be directly measured with the SFA and, in principle, with the AFM if the contact geometry can be quantified. The measured values are in good agreement with calculated values based on the known surface energies γ of the materials, and

Table 30.2 Van der Waals interaction energy and force between macroscopic bodies of different geometries

Geometry of bodies with surfaces D apart ($D \ll R$)		van der Waals interaction	
		Energy, E	Force, F
Two atoms or small molecules	$r \gtrsim \sigma$	$\dfrac{-C_{vdW}}{r^6}$	$\dfrac{-6C_{vdW}}{r^7}$
Two flat surfaces (per unit area)	$r \gg D$	$\dfrac{-A_H}{12\pi D^2}$	$\dfrac{-A_H}{6\pi D^3}$
Two spheres or macromolecules of radii R_1 and R_2	$R_1, R_2 \gg D$	$\dfrac{-A_H}{6D}\left(\dfrac{R_1 R_2}{R_1 + R_2}\right)$	$\dfrac{-A_H}{6D^2}\left(\dfrac{R_1 R_2}{R_1 + R_2}\right)$
Sphere or macromolecule of radius R near a flat surface	$R \gg D$	$\dfrac{-A_H R}{6D}$	$\dfrac{-A_H R}{6D^2}$
Two parallel cylinders or rods of radii R_1 and R_2 (per unit length)	$R_1, R_2 \gg D$	$\dfrac{-A_H}{12\sqrt{2}\, D^{3/2}}\left(\dfrac{R_1 R_2}{R_1 + R_2}\right)^{1/2}$	$\dfrac{-A_H}{8\sqrt{2}\, D^{5/2}}\left(\dfrac{R_1 R_2}{R_1 + R_2}\right)^{1/2}$
Cylinder of radius R near a flat surface (per unit length)	$R \gg D$	$\dfrac{-A_H \sqrt{R}}{12\sqrt{2}\, D^{3/2}}$	$\dfrac{-A_H \sqrt{R}}{8\sqrt{2}\, D^{5/2}}$
Two cylinders or filaments of radii R_1 and R_2 crossed at 90°	$R_1, R_2 \gg D$	$\dfrac{-A_H \sqrt{R_1 R_2}}{6D}$	$\dfrac{-A_H \sqrt{R_1 R_2}}{6D^2}$

A negative force (A_H positive) implies attraction, a positive force means repulsion (A_H negative)
(after [30.60], with permission)

for nonpolar low-energy solids they are well accounted for by the Lifshitz theory [30.3].

30.3.2 Charge-Exchange Interactions

Electrostatic interactions are present between ions (Coulomb interactions), between ions and permanent dipoles, and between ions and nonpolar molecules in which a charge induces a dipole moment. The interaction energy between ions or between a charge and a fixed permanent dipole can be attractive or repulsive. For an induced dipole or a freely rotating permanent dipole in vacuum or air, the interaction energy with a charge is always attractive.

Spontaneous charge transfer may occur between two dissimilar materials in contact [30.93–97]. The phenomenon, called contact electrification, is especially prominent in contact between a metal and a material with low conductivity (including organic liquids) [30.95,97], but is also observed, for example, between two different polymer layers. It is believed that when two different materials are in static contact, charge transfer might occur due to quantum tunneling of electrons or, in some cases, transfer of surface ions. The charging is generally seen to be stronger with increasing difference in work function (or electron affinity) between the two materials [30.95, 97]. During separation, rolling or sliding of one body over the other, the surfaces experience both charge transition from one surface to the other and charge transfer (conductance) along each surface (Fig. 30.4b). The latter process is typically slower, and, as a result, charges remain on the surfaces as they are separated in vacuum or dry nitrogen gas. The charging gives rise to a strong adhesion with adhesion energies of over $1000 \, \mathrm{mJ \, m^{-2}}$, similar to fracture or cohesion energies of the solid bodies themselves [30.93, 94, 98]. Upon separating the surfaces further apart, a strong, long-range electrostatic attraction is observed. The charging can be decreased through discharges across the gap between the surfaces (which requires a high charging) or through conduction in the solids. The discharge may give rise to triboluminescence [30.99], but can also cause sparks that may ignite combustible materials [30.100]. It has been suggested that charge-exchange interactions are particularly important in rolling friction between dry surfaces (which can simplistically be thought of as an adhesion–separation process), where the distance dependence of forces acting normally to the surfaces plays a larger role than in sliding friction. In the case of sliding friction, charge transfer is also observed between identical materials [30.98, 101]. Mechanisms such as bond formation and breakage (polymer scission), slip at the wall between a flowing liquid and a solid [30.102], or material transfer and the creation of wear particles have been suggested. However, friction electrification or triboelectrification also occurs during wearless sliding, i. e., when the surfaces are not damaged. Other explanations such as the creation or translation of defects on or near the surface have been put forward [30.98]. Attempts have been made to correlate the amount of charging with the normal force or load and with the polarizability of the sliding materials [30.96, 103]. Recent experiments on the sliding friction between metal–insulator surfaces indicate that stick–slip would be accompanied by charge-transfer events [30.104, 105].

Photoinduced charge transfer, or harpooning, involves the transfer of an electron between an atom in a molecular beam or at a solid surface (typically an alkali or transition metal) to an atom or molecule in a gas (typically a halide) to form a negatively charged molecular ion in a highly excited vibrational state. This transfer process can occur at atomic distances of $0.5-0.7 \, \mathrm{nm}$, which is far from molecular contact. The formed molecular ion is attracted to the surface and chemisorbs onto it. Photoinduced charge-transfer processes also occur in the photosynthesis in green plants and in photo-electrochemical cells (solar cells) at the junction between two semiconductors or between a semiconductor and an electrolyte solution [30.106].

30.3.3 Sintering and Cold Welding

When macroscopic particles in a powder or in a suspension come into molecular contact, they can bond together to form a network or solid body with very different density and shear strength compared to the powder (a typical example is porcelain). The rate of bonding is dependent on the surface energy (causing a stress at the edge of the contact) and the atomic mobility (diffusion rate) of the contacting materials. To increase the diffusion rate, objects formed from powders are heated to about one half of the melting temperature of the components in a process called sintering, which can be done in various atmospheres or in a liquid.

In the sintering process, the surface energy of the system is lowered due to the reduction of total surface area (Fig. 30.4c). In metal and ceramic systems, the most important mechanism is solid-state diffusion, initially surface diffusion. As the surface area decreases and the grain boundaries increase at the contacts, grain boundary diffusion and diffusion through the crystal lattice become more important [30.107]. The grain boundaries will eventually migrate, so that larger particles are formed (coarsening). Mass can also be transferred through evaporation and condensation, and through viscous and plastic flow. In liquid-phase sintering, the materials can melt, which increases the mass transport. Amorphous materials like polymers and glasses do not have real grain boundaries and sinter by viscous flow [30.108].

Some of these mechanisms (surface diffusion and evaporation–condensation) reduce the surface area and increase the grain size (coarsening) without densification, in contrast to bulk transport mechanisms such as grain-boundary diffusion and plastic and viscous flow. As the material becomes denser, elongated pores collapse to form smaller, spherical pores with a lower surface energy. Models for sintering typically consider the size and growth rate of the grain boundary (the "neck") formed between two spherical particles. At a high stage of densification, the sintering stress σ at the curved neck between two particles is given by [30.108]

$$\sigma = \frac{2\gamma_{SS}}{G} + \frac{2\gamma_{SV}}{r_p} , \qquad (30.10)$$

where γ_{SS} is the solid–solid grain boundary energy, γ_{SV} is the solid–vapor surface energy, G is the grain size, and r_p is the radius of the pore.

A related phenomenon is cold welding, which is the spontaneous formation of strong junctions between clean (unoxidized) metal surfaces with a mutual solubility when they are brought into contact, with or without an applied pressure. The plastic deformations accompanying the formation and breaking of such contacts on a molecular scale during motion of one surface normally (see Fig. 30.10c,d) or laterally (shearing) with respect to the other have been studied both experimentally [30.109, 110] and theoretically [30.111–116]. The breaking of a cold-welded contact is generally associated with damage or deformation of the surface structure.

30.4 Normal Forces Between Surfaces in Liquids

30.4.1 Van der Waals Forces in Liquids

The dispersion interaction in a medium will be significantly lower than in vacuum, since the attractive interaction between two solute molecules in a medium (solvent) involves displacement and reorientation of the nearest-neighbor solvent molecules. Even though the surrounding medium may change the dipole moment and polarizability from that in vacuum, the interaction between two identical molecules remains attractive in a binary mixture. The extension of the interactions to the case of two macroscopic bodies is the same as described in Sect. 30.3.1. Typically, the Hamaker constants for interactions in a medium are an order of magnitude lower than in vacuum. Between macroscopic surfaces in liquids, van der Waals forces become important at distances below 10–15 nm and may at these distances start to dominate interactions of different origin that have been observed at larger separations.

Figure 30.5 shows the measured van der Waals forces between two crossed-cylindrical mica surfaces in water and various salt solutions. Good agreement is obtained between experiment and theory. At larger surface separations, above about 5 nm, the measured forces fall off more rapidly than D^{-2}. This retardation effect (see Sect. 30.3.1) is also predicted by the Lifshitz theory and is due to the time needed for propagation of the induced dipole moments over large distances.

From Fig. 30.5, we may conclude that, at separations above about 2 nm, or 8 molecular diameters of water, the *continuum* Lifshitz theory is valid. This would mean that water films as thin as 2 nm may be expected to have bulk-like properties, at least as far as their interaction forces

are concerned. Similar results have been obtained with other liquids, where in general continuum properties are manifested, both as regards their interactions and other properties such as viscosity, at a film thickness larger than five or ten molecular diameters. In the absence of a solvent (in vacuum), the agreement of measured van

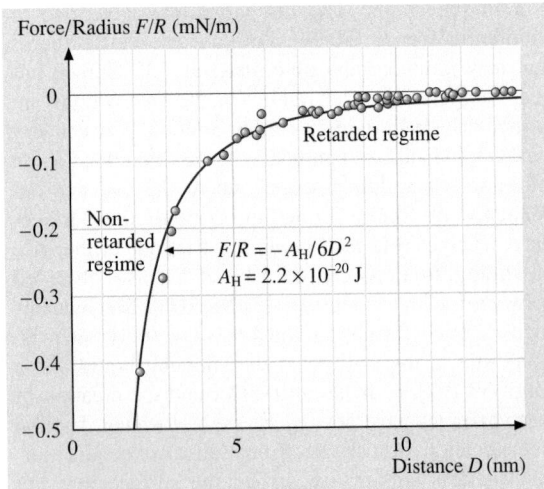

Fig. 30.5 Attractive van der Waals force F between two curved mica surfaces of radius $R \approx 1$ cm measured in water and various aqueous electrolyte solutions. The electrostatic interaction has been subtracted from the total measured force. The measured non-retarded Hamaker constant is $A_H = 2.2 \times 10^{-20}$ J. Retardation effects are apparent at distances larger than 5 nm, as expected theoretically. (After [30.3]. Copyright 1991, with permission from Elsevier Science)

der Waals forces with the continuum Lifshitz theory is generally good at all separations down to molecular contact ($D = D_0$).

Van der Waals interactions in a system of three or more different materials (see (30.8)) can be attractive or repulsive, depending on their dielectric properties. Numerous experimental studies show the attractive van der Waals forces in various systems [30.3], and repulsive van der Waals forces have also been measured directly [30.117, 118]. A practical consequence of the repulsive interaction obtained across a medium with intermediate dielectric properties is that the van der Waals forces will give rise to preferential, nonspecific adsorption of molecules with an intermediate dielectric constant. This is commonly seen as adsorption of vapors or solutes to a solid surface. It is also possible to diminish the attractive interaction between dispersed colloidal particles by adsorption of a thin layer of material with dielectric properties close to those of the surrounding medium (matching of refractive index), or by adsorption of a polymer that gives a steric repulsive force that keeps the particles separated at a distance where the magnitude of the van der Waals attraction is negligible. Thermal motion will then keep the particles dispersed.

30.4.2 Electrostatic and Ion Correlation Forces

Most surfaces in contact with a highly polar liquid (such as water) acquire a surface charge, either by dissociation of ions from the surface into the solution or by preferential adsorption of certain ions from the solution. The surface charge is balanced by a layer of oppositely charged ions (counterions) in the solution at some small distance from the surface (see [30.3]). In dilute solution, this distance is the Debye length, κ^{-1}, which is purely a property of the electrolyte *solution*. The Debye length falls with increasing ionic strength (i. e., with the molar concentration M_i and valency z_i) of the ions in solution:

$$\kappa^{-1} = \left(\frac{\varepsilon \varepsilon_0 k_B T}{e^2 N_A \sum_i z_i^2 M_i} \right)^{1/2} , \qquad (30.11)$$

where e is the electronic charge. For example, for 1:1 electrolytes at 25 °C, $\kappa^{-1} = 0.304\,\text{nm}/\sqrt{M_{1:1}}$, where M_i is given in M (mol dm^{-3}). κ^{-1} is thus about 10 nm in a 1 mM NaCl solution and 0.3 nm in a 1 M solution. In totally pure water at pH 7, where $M_i = 10^{-7}\,\text{M}$, κ^{-1} is 960 nm, or about 1 μm. The Debye length also relates the surface charge density σ of a surface to

the electrostatic surface potential ψ_0 via the Grahame equation, which for 1:1 electrolytes can be expressed as:

$$\sigma = \sqrt{8\varepsilon\varepsilon_0 k_B T} \sinh\left(e\psi_0/2k_B T\right) \times \sqrt{M_{1:1}} . \quad (30.12)$$

Since the Debye length is a measure of the thickness of the diffuse atmosphere of counterions near a charged surface, it also determines the range of the electrostatic "double-layer" interaction between two charged surfaces. The electrostatic double-layer interaction is an entropic effect that arises upon decreasing the thickness of the liquid film containing the dissolved ions. Because of the attractive force between the dissolved ions and opposite charges on the surfaces, the ions stay between the surfaces, but an osmotic repulsion arises as their concentration increases. The long-range electrostatic interaction energy at large separations (weak overlap) between two similarly charged molecules or surfaces is typically repulsive and is roughly an exponentially decaying function of D:

$$E(D) \approx +C_{ES}\,e^{-\kappa D} , \qquad (30.13)$$

where C_{ES} is a constant that depends on the geometry of the interacting surfaces, on their surface charge density, and the solution conditions (Table 30.3). We see that the Debye length is the decay length of the interaction energy between two surfaces (and of the mean potential away from one surface). C_{ES} can be determined by solving the so-called Poisson–Boltzmann equation or by using other theories [30.119–123]. The equations in Table 30.3 are expressed in terms of a constant, Z, defined as

$$Z = 64\pi\varepsilon\varepsilon_0(k_B T/e)^2 \tanh^2\left[z e\psi_0/(4k_B T)\right] , \qquad (30.14)$$

which depends only on the properties of the *surfaces*.

The above approximate expressions are accurate only for surface separations larger than about one Debye length. At smaller separations one must use numerical solutions of the Poisson–Boltzmann equation to obtain the exact interaction potential, for which there are no simple expressions. In the limit of small D, it can be shown that the interaction energy depends on whether the surfaces remain at constant potential ψ_0 (as assumed in the above equations) or at constant charge σ (when the repulsion exceeds that predicted by the above equations), or somewhere between these limits. In the "constant charge limit" the total *number* of counterions in the compressed film does not change as D is decreased, whereas at constant potential, the *concentration* of counterions is constant. The limiting pressure (or force per unit area) at

Table 30.3 Electrical double-layer interaction energy $E(D)$ and force $(F = -\mathrm{d}E/\mathrm{d}D)$ between macroscopic bodies

Geometry of bodies with surfaces D apart $(D \ll R)$			Electric double-layer interaction	
			Energy E	Force F
Two ions or small molecules		$r \geqslant \sigma$	$+z_1 z_2 e^2 \dfrac{\mathrm{e}^{-\kappa(r-\sigma)}}{4\pi\varepsilon\varepsilon_0 r (1+\kappa\sigma)}$	$+z_1 z_2 e^2 \dfrac{(1+\kappa r)}{4\pi\varepsilon\varepsilon_0 r^2 (1+\kappa\sigma)} \mathrm{e}^{-\kappa(r-\sigma)}$
Two flat surfaces (per unit area)		$r \gg D$	$(\kappa/2\pi)\, Z\mathrm{e}^{-\kappa D}$	$\left(\kappa^2/2\pi\right) Z\mathrm{e}^{-\kappa D}$
Two spheres or macromolecules of radii R_1 and R_2		$R_1, R_2 \gg D$	$\left(\dfrac{R_1 R_2}{R_1+R_2}\right) Z\mathrm{e}^{-\kappa D}$	$\kappa \left(\dfrac{R_1 R_2}{R_1+R_2}\right) Z\mathrm{e}^{-\kappa D}$
Sphere or macro-molecule of radius R near a flat surface		$R \gg D$	$RZ\mathrm{e}^{-\kappa D}$	$\kappa RZ\mathrm{e}^{-\kappa D}$
Two parallel cylinders or rods of radii R_1 and R_2 (per unit length)		$R_1, R_2 \gg D$	$\dfrac{\kappa^{1/2}}{\sqrt{2\pi}}\left(\dfrac{R_1 R_2}{R_1+R_2}\right)^{1/2} Z\mathrm{e}^{-\kappa D}$	$\dfrac{\kappa^{3/2}}{\sqrt{2\pi}}\left(\dfrac{R_1 R_2}{R_1+R_2}\right)^{1/2} Z\mathrm{e}^{-\kappa D}$
Cylinder of radius R near a flat surface (per unit length)		$R \gg D$	$\kappa^{1/2}\sqrt{\dfrac{R}{2\pi}}\, Z\mathrm{e}^{-\kappa D}$	$\kappa^{3/2}\sqrt{\dfrac{R}{2\pi}}\, Z\mathrm{e}^{-\kappa D}$
Two cylinders or filaments of radii R_1 and R_2 crossed at 90°		$R_1, R_2 \gg D$	$\sqrt{R_1 R_2}\, Z\mathrm{e}^{-\kappa D}$	$\kappa\sqrt{R_1 R_2}\, Z\mathrm{e}^{-\kappa D}$

The interaction energy and force for bodies of different geometries is based on the Poisson–Boltzmann equation (a continuum, mean-field theory). Equation (30.14) gives the interaction constant Z (in terms of the surface potential ψ_0) for the interaction between similarly charged (ionized) surfaces in aqueous solutions of monovalent electrolyte. It can also be expressed in terms of the surface charge density σ by applying the Grahame equation (30.12) (after [30.60], with permission)

constant charge is the osmotic pressure of the confined ions:

$$F = k_B T \times \text{ion number density}$$
$$= 2\sigma k_B T/(zeD), \quad \text{for } D \ll \kappa^{-1}. \tag{30.15}$$

That is, as $D \to 0$ the double-layer pressure at constant surface charge becomes infinitely repulsive and independent of the salt concentration (at constant potential the force instead becomes a constant at small D). However, at small separations, the van der Waals attraction (which goes as D^{-2} between two spheres or as D^{-3}

between two planar surfaces, see Table 30.2) wins out over the double-layer repulsion, unless some other short-range interaction becomes dominant (see Sect. 30.4.4). This is the theoretical prediction that forms the basis of the so-called Derjaguin–Landau–Verwey–Overbeek (DLVO) theory [30.119, 124], illustrated in Fig. 30.6.

Because of the different distance dependence of the van der Waals and electrostatic interactions, the total force law, as described by the DLVO theory, can show several minima and maxima. Typically, the depth of the outer (secondary) minimum is a few $k_B T$, which is enough to cause reversible flocculation of particles from an aqueous dispersion. If the force barrier between the secondary and primary minimum is lowered, for example, by increasing the electrolyte concentration, particles can be irreversibly coagulated in the primary minimum. In practice, other forces (described in the following sections) often appear at very small separations, so that the full force law between two surfaces or colloidal particles

in solution can be more complex than might be expected from the DLVO theory.

There are situations when the double-layer interaction can be attractive at short range even between surfaces of similar charge, especially in systems with charge regulation due to dissociation of chargeable groups on the surfaces [30.123, 125]; ion condensation [30.126], which may lower the effective surface charge density in systems containing di-and trivalent counterions; or ion correlation, which is an additional van der Waals-like attraction due to mobile and therefore highly polarizable counterions located at the surface [30.127]. The ion correlation (or charge fluctuation) force becomes significant at separations below 4 nm and increases with the surface charge density σ and the valency z of the counterions. Computer simulations have shown that, at high charge density and for monovalent counterions, the ion correlation force can reduce the effective double-layer repulsion by 10–15 %. With divalent counterions, the ion correlation force was found to exceed the double-layer repulsion and the total force then became attractive at a separation below 2 nm even in dilute electrolyte solution [30.128]. Experimentally, such short-range attractive forces have been found between charged bilayers [30.129, 130] and also in other systems [30.131].

30.4.3 Solvation and Structural Forces

When a liquid is confined within a restricted space, for example, a very thin film between two surfaces, it ceases to behave as a structureless continuum. At small surface separations (below about ten molecular diameters), the van der Waals force between two surfaces or even two solute molecules in a liquid (solvent) is no longer a smoothly varying attraction. Instead, there arises an additional "solvation" force that generally oscillates between attraction and repulsion with distance, with a periodicity equal to some mean dimension σ of the liquid molecules [30.132]. Figure 30.7a shows the force law between two smooth mica surfaces across the hydrocarbon liquid tetradecane, whose inert, chain-like molecules have a width of $\sigma \approx 0.4$ nm.

The short-range oscillatory force law is related to the "density distribution function" and "potential of mean force" characteristic of intermolecular interactions in liquids. These forces arise from the confining effects that the two surfaces have on liquid molecules, forcing them to order into quasi-discrete layers. Such layers are energetically or entropically favored and correspond to the minima in the free energy, whereas fractional

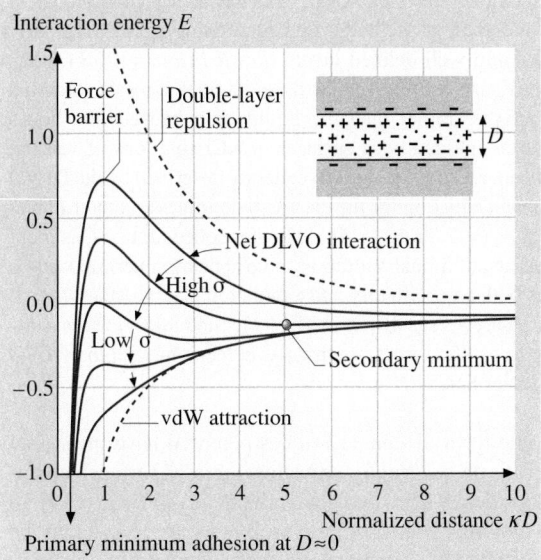

Fig. 30.6 Schematic plots of the DLVO interaction potential energy E between two flat, charged surfaces [or, according to the Derjaguin approximation, (30.3), the force F between two curved surfaces] as a function of the surface separation normalized by the Debye length, κ^{-1}. The van der Waals attraction (inverse power-law dependence on D) together with the repulsive electrostatic "double-layer" force (roughly exponential) at different surface charge σ [or potential, see (30.12)] determine the net interaction potential in aqueous electrolyte solution (after [30.60] with permission)

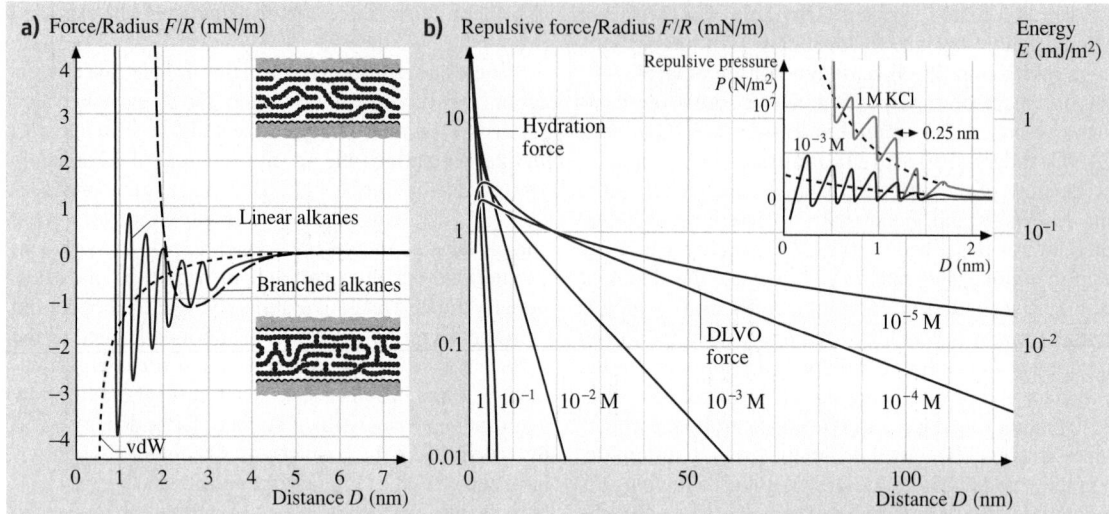

Fig. 30.7 (a) *Solid curve*: Forces measured between two mica surfaces across saturated linear chain alkanes such as *n*-tetradecane and *n*-hexadecane [30.133, 134]. The 0.4-nm periodicity of the oscillations indicates that the molecules are preferentially oriented parallel to the surfaces, as shown schematically in the *upper insert*. The theoretical continuum van der Waals attraction is shown as a *dotted curve*. *Dashed curve*: Smooth, non-oscillatory force law exhibited by irregularly shaped alkanes (such as 2-methyloctadecane) that cannot order into well-defined layers (*lower insert*) [30.134, 135]. Similar non-oscillatory forces are also observed between "rough" surfaces, even when these interact across a saturated linear chain liquid. This is because the irregularly shaped surfaces (rather than the liquid) now prevent the liquid molecules from ordering in the gap. **(b)** Forces measured between charged mica surfaces in KCl solutions of varying concentrations [30.20]. In dilute solutions (10^{-5} and 10^{-4} M), the measured forces are excellently described by the DLVO theory, based on exact solutions to the nonlinear Poisson–Boltzmann equation for the electrostatic forces and the Lifshitz theory for the van der Waals forces (using a Hamaker constant of $A_H = 2.2 \times 10^{-20}$ J). At higher concentrations, as more hydrated K^+ cations adsorb onto the negatively charged surfaces, an additional hydration force appears superimposed on the DLVO interaction at distances below 3–4 nm. This force has both an oscillatory and a monotonic component. *Insert*: Short-range hydration forces between mica surfaces shown as pressure versus distance. The lower and upper curves show surfaces 40% and 95% saturated with K^+ ions. At larger separations, the forces are in good agreement with the DLVO theory. (After [30.3]. Copyright 1991, with permission from Elsevier Science)

layers are disfavored (energy maxima). This effect is quite general and arises in all simple liquids when they are confined between two smooth, rigid surfaces, both flat and curved. Oscillatory forces do not require any attractive liquid–liquid or liquid–wall interaction, only two hard walls confining molecules whose shape is not too irregular and that are free to exchange with molecules in a bulk liquid reservoir. In the absence of any attractive pressure between the molecules, the bulk liquid density could be maintained by an external hydrostatic pressure – in real liquids attractive van der Waals forces play the role of such an external pressure.

Oscillatory forces are now well understood theoretically, at least for simple liquids, and a number of theoretical studies and computer simulations of various confined liquids (including water) that interact via

some form of Lennard–Jones potential have invariably led to an oscillatory solvation force at surface separations below a few molecular diameters [30.136–144]. In a first approximation, the oscillatory force law may be described by an exponentially decaying cosine function of the form

$$E \approx E_0 \cos(2\pi D/\sigma)\mathrm{e}^{-D/\sigma} , \qquad (30.16)$$

where both theory and experiments show that the oscillatory period and the characteristic decay length of the envelope are close to σ.

Once the solvation zones of the two surfaces overlap, the mean liquid density in the gap is no longer the same as in the bulk liquid. Since the van der Waals interaction depends on the optical properties of the liquid, which in turn depends on the density, the van der Waals and the oscillatory solvation forces are not strictly additive. It is more correct to think of the solvation force as *the* van der Waals force at small separations with the molecular properties and density variations of the medium taken into account. It is also important to appreciate that solvation forces do not arise simply because liquid molecules tend to structure into semi-ordered layers at surfaces. They arise because of the disruption or *change* of this ordering during the approach of a second surface. The two effects are related; the greater the tendency toward structuring at an isolated surface the greater the solvation force between two such surfaces, but there is a real distinction between the two phenomena that should be borne in mind.

Oscillatory forces lead to different adhesion values depending on the energy minimum from which two surfaces are being separated. For an interaction energy described by (30.16), "quantized" adhesion energies will be E_0 at $D = 0$ (primary minimum), E_0/e at $D = \sigma$, E_0/e^2 at $D = 2\sigma$, etc. E_0 can be thought of as a depletion force (see Sect. 30.4.5) that is approximately given by the osmotic limit $E_0 \approx -k_B T/\sigma^2$, which can exceed the contribution to the adhesion energy in contact from the van der Waals forces (at $D_0 \approx 0.15–0.20$ nm, as discussed in Sect. 30.3.1, keeping in mind that the Lifshitz theory fails to describe the force law at *intermediate* distances). Such multivalued adhesion forces have been observed in a number of systems, including the interactions of fibers.

Measurements of oscillatory forces between different surfaces across both aqueous and nonaqueous liquids have revealed their richness of properties [30.145–149], for example, their great sensitivity to the shape and rigidity of the solvent molecules, to the presence of other components, and to the structure of the confining surfaces (see Sects. 30.5.3 and 30.9). In particular, the oscillations can be smeared out if the molecules are irregularly shaped (e.g., branched) and therefore unable to pack into ordered layers, or when the interacting surfaces are rough or fluid-like (see Sect. 30.4.6).

It is easy to understand how oscillatory forces arise between two flat, plane parallel surfaces. Between two curved surfaces, e.g., two spheres, one might imagine the molecular ordering and oscillatory forces to be smeared out in the same way that they are smeared out between two randomly rough surfaces (see Sect. 30.5.3); however, this is not the case. Ordering can occur as long as the curvature or roughness is itself regular or uniform, i. e., not random. This is due to the Derjaguin approximation (30.3). If the energy between two flat surfaces is given by a decaying oscillatory function (for example, a co-

sine function as in (30.16)), then the force (and energy) between two curved surfaces will also be an oscillatory function of distance with some phase shift. Likewise, two surfaces with regularly curved regions will also retain their oscillatory force profile, albeit modified, as long as the corrugations are truly regular, i. e., periodic. On the other hand, surface roughness, even on the nanometer scale, can smear out oscillations if the roughness is random and the confined molecules are smaller than the size of the surface asperities [30.150, 151]. If an organic liquid contains small amounts of water, the expected oscillatory force can be replaced by a strongly attractive capillary force (see Sect. 30.5.1).

30.4.4 Hydration and Hydrophobic Forces

The forces occurring in water and electrolyte solutions are more complex than those occurring in nonpolar liquids. According to continuum theories, the attractive van der Waals force is always expected to win over the repulsive electrostatic "double-layer" force at small surface separations (Fig. 30.6). However, certain surfaces (usually oxide or hydroxide surfaces such as clays or silica) swell spontaneously or repel each other in aqueous solution, even at high salt concentrations. Yet in all these systems one would expect the surfaces or particles to remain in strong adhesive contact or coagulate in a primary minimum if the only forces operating were DLVO forces.

There are many other aqueous systems in which the DLVO theory fails and where there is an additional short-range force that is not oscillatory but monotonic. Between hydrophilic surfaces this force is exponentially repulsive and is commonly referred to as the *hydration*, or *structural*, force. The origin and nature of this force has long been controversial, especially in the colloidal and biological literature. Repulsive hydration forces are believed to arise from strongly hydrogen-bonding surface groups, such as hydrated ions or hydroxyl (−OH) groups, which modify the hydrogen-bonding network of liquid water adjacent to them. Because this network is quite extensive in range [30.152], the resulting interaction force is also of relatively long range.

Repulsive hydration forces were first extensively studied between clay surfaces [30.153]. More recently, they have been measured in detail between mica and silica surfaces [30.20–22,154], where they have been found to decay exponentially with decay lengths of about 1 nm. Their effective range is 3–5 nm, which is about twice the range of the oscillatory solvation force in water. Empirically, the hydration repulsion between two hydrophilic surfaces appears to follow the simple equation

$$E = E_0 e^{-D/\lambda_0} , \qquad (30.17)$$

where $\lambda_0 \approx 0.6-1.1$ nm for 1:1 electrolytes and $E_0 = 3-30$ mJ m^{-2} depending on the hydration (hydrophilicity) of the surfaces, higher E_0 values generally being associated with lower λ_0 values.

The interactions between molecularly smooth mica surfaces in dilute electrolyte solutions obey the DLVO theory (Fig. 30.7b). However, at higher salt concentrations, specific to each electrolyte, hydrated cations bind to the negatively charged surfaces and give rise to a repulsive hydration force [30.20, 21]. This is believed to be due to the energy needed to dehydrate the bound cations, which presumably retain some of their water of hydration on binding. This conclusion was arrived at after noting that the strength and range of the hydration forces increase with the known hydration numbers of the electrolyte cations in the order: $Mg^{2+} > Ca^{2+} > Li^+ \sim Na^+ > K^+ > Cs^+$. Similar trends are observed with other negatively charged colloidal surfaces.

While the hydration force between two mica surfaces is overall repulsive below a distance of 4 nm, it is not always monotonic below about 1.5 nm but exhibits oscillations of mean periodicity of 0.25 ± 0.03 nm, roughly equal to the diameter of the water molecule. This is shown in the insert in Fig. 30.7b, where we may note that the first three minima at $D = 0, 0.28$, and 0.56 nm occur at negative energies, a result that rationalizes observations on certain colloidal systems. For example, clay platelets such as montmorillonite often repel each other increasingly strongly as they come closer together, but they are also known to stack into stable aggregates with water interlayers of typical thickness 0.25 and 0.55 nm between them [30.155, 156], suggestive of a turnabout in the force law from a monotonic repulsion to discretized attraction. In chemistry we would refer to such structures as stable hydrates of fixed stoichiometry, whereas in physics we may think of them as experiencing an oscillatory force.

Both surface force and clay swelling experiments have shown that hydration forces can be modified or "regulated" by exchanging ions of different hydration on surfaces, an effect that has important practical applications in controlling the stability of colloidal dispersions. It has long been known that colloidal particles can be precipitated (coagulated or flocculated) by increasing the electrolyte concentration, an effect that was traditionally attributed to the reduced screening of the electrostatic double-layer repulsion between the particles due to the reduced Debye length. However, there are many examples where colloids are stabilized at high salt concentrations, not at low concentrations. This effect is now recognized as being due to the increased hydration repulsion experienced by certain surfaces when they bind highly hydrated ions at higher salt concentrations. Hydration regulation of adhesion and interparticle forces is an important practical method for controlling various processes such as clay swelling [30.155, 156], ceramic processing and rheology [30.157, 158], material fracture [30.157], and colloidal particle and bubble coalescence [30.159].

Water appears to be unique in having a solvation (hydration) force that exhibits both a monotonic and an oscillatory component. Between hydrophilic surfaces the monotonic component is repulsive (Fig. 30.7b), but between hydrophobic surfaces it is attractive and the final adhesion is much greater than expected from the Lifshitz theory.

A hydrophobic surface is one that is inert to water in the sense that it cannot bind to water molecules via ionic or hydrogen bonds. Hydrocarbons and fluorocarbons are hydrophobic, as is air, and the strongly attractive hydrophobic force has many important manifestations and consequences such as the low solubility or miscibility of water and oil molecules, micellization, protein folding, strong adhesion and rapid coagulation of hydrophobic surfaces, non-wetting of water on hydrophobic surfaces, and hydrophobic particle attachment to rising air bubbles (the basic principle of froth flotation).

In recent years, there has been a steady accumulation of experimental data on the force laws between various hydrophobic surfaces in aqueous solution [30.160–178]. These studies have found that the force law between two macroscopic hydrophobic surfaces is of surprisingly long range, decaying exponentially with a characteristic decay length of 1–2 nm in the separation range of 0–10 nm, and then more gradually further out. The hydrophobic force can be far stronger than the van der Waals attraction, especially between hydrocarbon surfaces in water, for which the Hamaker constant is quite small. The magnitude of the hydrophobic attraction has been found to decrease with the decreasing hydrophobicity (increasing hydrophilicity) of lecithin lipid bilayer surfaces [30.31] and silanated surfaces [30.168], whereas examples of the opposite trend have been shown for some Langmuir–Blodgett-deposited monolayers [30.179]. An apparent correlation has been found between high stability of the hydrophobic surface (as measured by its contact angle hysteresis) and the absence of a long-range part of the attractive force [30.180].

For two surfaces in water the purely hydrophobic interaction energy (ignoring DLVO and oscillatory forces) in the range 0–10 nm is given by

$$E = -2\gamma\, e^{-D/\lambda_0} , \qquad (30.18)$$

where typically $\lambda_0 = 1$–2 nm, and $\gamma = 10$–50 mJ m^{-2}. The higher value corresponds to the interfacial energy of a pure hydrocarbon–water interface.

At a separation below 10 nm, the hydrophobic force appears to be insensitive or only weakly sensitive to changes in the type and concentration of electrolyte ions in the solution. The absence of a "screening" effect by ions attests to the non-electrostatic origin of this interaction. In contrast, some experiments have shown that, at separations greater than 10 nm, the attraction does depend on the intervening electrolyte, and that in dilute solutions, or solutions containing divalent ions, it can continue to exceed the van der Waals attraction out to separations of 80 nm [30.165, 181]. Recent research suggests that the interactions at very long range might not be a "hydrophobic" force since they are influenced by the presence of dissolved gas in the solution [30.176, 177], the stability of the hydrophobic surface [30.178, 180], and, on some types of surfaces, bridging submicroscopic bubbles [30.172–174].

The long-range nature of the hydrophobic interaction has a number of important consequences. It accounts for the rapid coagulation of hydrophobic particles in water and may also account for the rapid folding of proteins. It also explains the ease with which water films rupture on hydrophobic surfaces. In this case, the van der Waals force across the water film is repulsive and therefore favors wetting, but this is more than offset by the attractive hydrophobic interaction acting between the two hydrophobic phases across water. Hydrophobic forces are increasingly being implicated in the adhesion and fusion of biological membranes and cells. It is known that both osmotic and electric-field stresses enhance membrane fusion, an effect that may be due to the concomitant increase in the hydrophobic area exposed between two adjacent surfaces.

From the previous discussion we can infer that hydration and hydrophobic forces are not of a simple nature. These interactions are probably the most important, yet the least understood of all the forces in aqueous solutions. The unusual properties of water and the nature of the surfaces (including their homogeneity and stability) appear to be equally important. Some particle surfaces can have their hydration forces regulated, for example, by ion exchange. Others appear to be intrinsically hydrophilic (e.g., silica) and cannot be coagulated by changing the ionic condition, but can be rendered hydrophobic by chemically modifying their surface groups. For example, on heating silica to above 600 °C, two adjacent surface silanol (−OH) groups release a water molecule and form a hydrophobic siloxane (−O−) group, whence the repulsive hydration force changes into an attractive hydrophobic force.

How do these exponentially decaying repulsive or attractive forces arise? Theoretical work and computer simulations [30.138, 140, 182, 183] suggest that the solvation forces in water should be purely oscillatory, whereas other theoretical studies [30.184–191] suggest a monotonically exponential repulsion or attraction, possibly superimposed on an oscillatory force. The latter is consistent with experimental findings, as shown in the inset to Fig. 30.7b, where it appears that the oscillatory force is simply additive with the monotonic hydration and DLVO forces, suggesting that these arise from essentially different mechanisms. It has been suggested that for a sufficiently solvophilic surface, there could be "hydration"-like forces also in nonaqueous systems [30.190].

It is probable that the short-range hydration force between all smooth, rigid, or crystalline surfaces (e.g., mineral surfaces such as mica) has an oscillatory component. This may or may not be superimposed on a monotonic force due to image interactions [30.186], dipole–dipole interactions [30.191], and/or structural or hydrogen-bonding interactions [30.184, 185].

Like the repulsive hydration force, the origin of the hydrophobic force is still unknown. *Luzar* et al. [30.188] carried out a Monte Carlo simulation of the interaction between two hydrophobic surfaces across water at separations below 1.5 nm. They obtained a decaying oscillatory force superimposed on a monotonically attractive curve. In more recent computational and experimental work [30.192–195], it has been suggested that hydrophobic surfaces generate a depleted region of water around them, and that a long-range attractive force due to depletion arises between two such surfaces. Such a difference in density might also cause boundary slip of water at hydrophic surfaces [30.51, 196, 197].

It is questionable whether the hydration or hydrophobic force should be viewed as an ordinary type of solvation or structural force that reflects the packing of water molecules. The energy (or entropy) associated with the hydrogen-bonding network, which extends over a much larger region of space than the molecular correlations, is probably at the root of the long-range interactions of water. The situation in water appears to be governed by much more than the molecular packing effects that dominate the interactions in simpler liquids.

30.4.5 Polymer–Mediated Forces

Polymers or macromolecules are chain-like molecules consisting of many identical segments (monomers or repeating units) held together by covalent bonds. The size of a polymer coil in solution or in the melt is determined by a balance between van der Waals attraction (and hydrogen bonding, if present) between polymer segments, and the entropy of mixing, which causes the polymer coil to expand. In polymer melts above the glass transition temperature, and at certain conditions in solution, the attraction between polymer segments is exactly balanced by the entropy effect. The polymer solution will then behave virtually ideally, and the density distribution of segments in the coil is Gaussian. This is called the theta (θ) condition, and it occurs at the theta or Flory temperature for a particular combination of polymer and solvent or solvent mixture. At lower temperatures (in a poor or bad solvent), the polymer–polymer interactions dominate over the entropic, and the coil will shrink or precipitate. At higher temperatures (good solvent conditions), the polymer coil will be expanded.

High-molecular-weight polymers form large coils, which significantly affect the properties of a solution even when the total mass of polymer is very low. The radius of the polymer coil is proportional to the segment length, a, and the number of segments, n. At theta conditions, the hydrodynamic radius of the polymer coil (the root-mean-square separation of the ends of one polymer chain) is theoretically given by $R_h = a\,n^{1/2}$, and the unperturbed radius of gyration (the average root-mean-square distance of a segment from the center of mass of the molecule) is $R_g = a\,(n/6)^{1/2}$. In a good solvent the perturbed size of the polymer coil, the Flory radius R_F, is proportional to $n^{3/5}$.

Polymers interact with surfaces mainly through van der Waals and electrostatic interactions. The physisorption of polymers containing only one type of segment is reversible and highly dynamic, but the rate of exchange of adsorbed chains with free chains in the solution is low, since the polymer remains bound to the surface as long as one segment along the chain is adsorbed. The adsorption energy per segment is on the order of $k_B T$. In a good solvent, the conformation of a polymer on a surface is very different from the coil conformation in bulk solution. Polymers adsorb in "trains", separated

by "loops" extending into solution and dangling "tails" (the ends of the chain). Compared to adsorption at lower temperatures, good solvent conditions favor more of the polymer chain being in the solvent, where it can attain its optimum conformation. As a result, the extension of the polymer is longer, even though the total amount of adsorbed polymer is lower. In a good solvent, the polymer chains can also be effectively repelled from a surface, if the loss in conformational entropy close to the surface is not compensated for by a gain in enthalpy from adsorption of segments. In this case, there will be a layer of solution (thickness $\approx R_g$) close to the surfaces that is depleted of polymer.

The interaction forces between two surfaces across a polymer solution will depend on whether the polymer adsorbs onto the surfaces or is repelled from them, and also on whether the interaction occurs at "true" or "restricted" thermodynamic equilibrium. At true or full equilibrium, the polymer between the surfaces can equilibrate (exchange) with polymer in the bulk solution at all surface separations. Some theories [30.198, 199] predict that, at full equilibrium, the polymer chains would move from the confined gap into the bulk solution where they could attain entropically more favorable conformations, and that a monotonic attraction at all distances would result from bridging and depletion interactions (which will be discussed below). Other theories suggest that the interaction at small separations would be ultimately repulsive, since some polymer chains would remain in the gap due to their attractive interactions with many sites on the surface (enthalpic) – more sites would be available to the remaining polymer chains if some others desorbed and diffused out from the gap [30.73, 200–202].

At restricted equilibrium, the polymer is kinetically trapped, and the adsorbed amount is thus constant as the surfaces are brought toward each other, but the chains can still rearrange on the surfaces and in the gap. Experimentally, the true equilibrium situation is very difficult to attain, and most experiments are done at restricted equilibrium conditions. Even the equilibration of conformations assumed in theoretical models for restricted equilibrium conditions can be so slow that this condition is difficult to reach experimentally.

In systems of adsorbing polymer, bridging of chains from one surface to the other can give rise to a long-range attraction, since the bridging chains would gain conformational entropy if the surfaces were closer together. In poor solvents, both bridging and intersegment interactions contribute to an attraction [30.26]. However, regardless of solvent and equilibrium conditions, a strong repulsion due to the osmotic interactions is

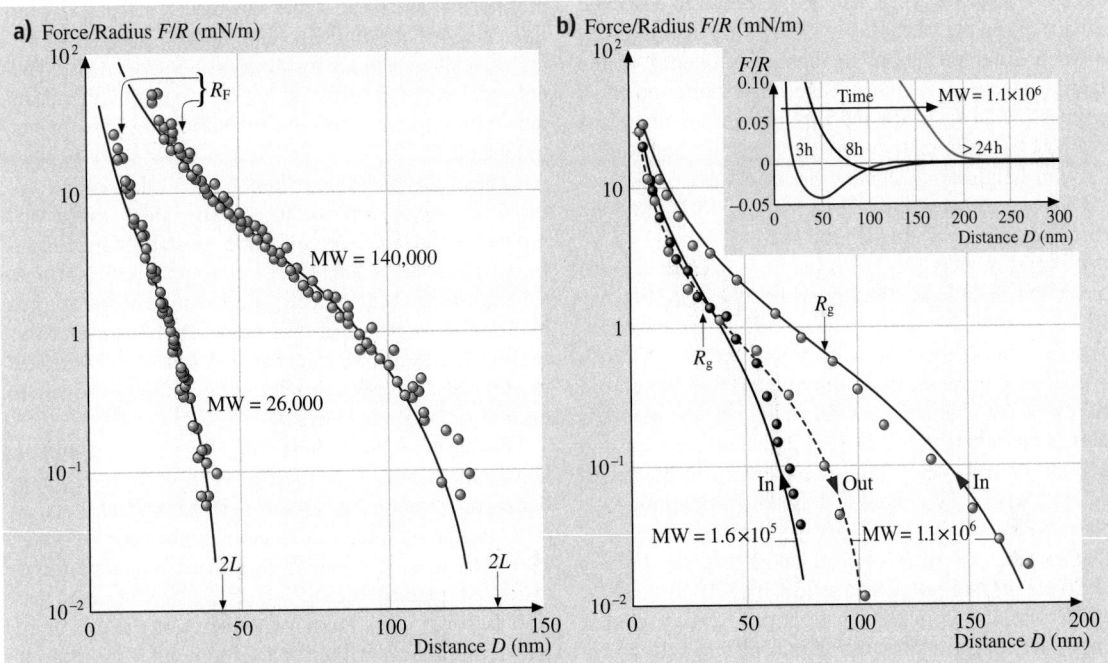

Fig. 30.8a,b Experimentally determined forces in systems of two interacting polymer layers: (**a**) Polystyrene brush layers grafted via an adsorbing chain-end group onto mica surfaces in toluene (a good solvent for polystyrene). *Left curve*: MW = 26 000 g/mol, R_F = 12 nm. *Right curve*: MW = 140 000 g/mol, R_F = 32 nm. Both force curves were reversible on approach and separation. The *solid curves* are theoretical fits using the Alexander–de Gennes theory with the following measured parameters: spacing between attachments sites: s = 8.5 nm, brush thickness: L = 22.5 nm and 65 nm, respectively. (Adapted from [30.203]). (**b**) Polyethylene oxide layers physisorbed onto mica from 150 µg/ml solution in aqueous 0.1 M KNO$_3$ (a good solvent for polyethylene oxide). *Main figure*: Equilibrium forces at full coverage after ∼ 16 h adsorption time. *Left curve*: MW = 160 000 g/mol, R_g = 32 nm. *Right curve:* MW = 1 100 000 g/mol, R_g = 86 nm. Note the hysteresis (irreversibility) on approach and separation for this *physisorbed* layer, in contrast to the absence of hysteresis with *grafted* chains in case (**a**). The *solid curves* are based on a modified form of the Alexander–de Gennes theory. *Inset* in (**b**): evolution of the forces with the time allowed for the higher MW polymer to adsorb from solution. Note the gradual reduction in the attractive bridging component. (Adapted from [30.204–206]. After [30.3]. Copyright 1991, with permission from Elsevier Science)

seen at small surface separations in systems of adsorbing polymers at restricted equilibrium.

In systems containing high concentrations of non-adsorbing polymer, the difference in solute concentration in the bulk and between the surfaces at separations smaller than the approximate polymer coil diameter ($2R_g$, i. e., when the polymer has been squeezed out from the gap between the surfaces) may give rise to an attractive osmotic force (the "depletion attraction") [30.207–212]. In addition, if the polymer coils become initially compressed as the surfaces approach each other, this can give rise to a repulsion ("depletion stabilization") at large separations [30.210]. For a system of two cylindrical surfaces or radius R, the

maximum depletion force, F_{dep}, is expected to occur when the surfaces are in contact and is given by multiplying the depletion (osmotic) pressure, $P_{dep} = \rho k_B T$, by the contact area πr^2, where r is given by the chord theorem: $r^2 = (2R - R_g)R_g \approx 2RR_g$ [30.3]:

$$F_{dep}/R = -2\pi R_g \rho k_B T , \qquad (30.19)$$

where ρ is the number density of the polymer in the bulk solution.

If a part of the polymer (typically an end group) is different from the rest of the chain, this part may preferentially adsorb to the surface. End-adsorbed polymer is attached to the surface at only one point, and the extension of the chain is dependent on the grafting density, i.e., the average distance, s, between adsorbed end groups on the surface (Fig. 30.8). One distinguishes between different regions of increasing overlap of the chains (stretching) called pancake, mushroom, and brush regimes [30.213]. In the mushroom regime, where the coverage is sufficiently low that there is no overlap between neighboring chains, the thickness of the adsorbed layer is proportional to $n^{1/2}$ (i.e., to R_g) at theta conditions and to $n^{3/5}$ in a good solvent.

Several models [30.213–218] have been developed for the extension and interactions between two brushes (strongly stretched grafted chains). They are based on a balance between osmotic pressure within the brush layers (uncompressed and compressed) and the elastic energy of the chains and differ mainly in the assumptions of the segment density profile, which can be a step function or parabolic. At high coverage (in the brush regime), where the chains will avoid overlapping each other, the thickness of the layer is proportional to n.

Experimental work on both monodisperse [30.27, 28, 203, 219] and polydisperse [30.30, 220] systems at different solvent conditions has confirmed the expected range and magnitude of the repulsive interactions resulting from compression of densely packed grafted layers.

30.4.6 Thermal Fluctuation Forces

If a surface is not rigid but very soft or even fluid-like, this can act to smear out any oscillatory solvation force. This is because the thermal fluctuations of such interfaces make them dynamically "rough" at any instant, even though they may be perfectly smooth on a time average. The types of surfaces that fall into this category are fluid-like amphiphilic surfaces of micelles, bilayers, emulsions, soap films, etc., but also solid colloidal particle surfaces that are coated with surfactant monolayers, as occurs in lubricating oils, paints, toners, etc.

These thermal fluctuation forces (also called entropic or steric forces) are usually short range and repulsive and are very effective at stabilizing the attractive van der Waals forces at some small but finite separation. This can reduce the adhesion energy or force by up to three orders of magnitude. It is mainly for this reason that fluid-like micelles and bilayers, biological membranes, emulsion droplets, or gas bubbles adhere to each other only very weakly.

Because of their short range it was, and still is, commonly believed that these forces arise from water ordering or "structuring" effects at surfaces, and that they reflect some unique or characteristic property of water. However, it is now known that these repulsive forces also exist in other liquids [30.221, 222]. Moreover, they appear to become stronger with increasing temperature, which is unlikely if the force originated from molecular ordering effects at surfaces. Recent experiments, theory, and computer simulations [30.223–226] have shown that these repulsive forces have an entropic origin arising from the osmotic repulsion between exposed thermally mobile surface groups once these overlap in a liquid. These phenomena include undulating and peristaltic forces between membranes or bilayers, and, on the molecular scale, protrusion and head-group overlap forces where the interactions are also influenced by hydration forces.

30.5 Adhesion and Capillary Forces

30.5.1 Capillary Forces

When considering the adhesion of two solid surfaces or particles in air or in a liquid, it is easy to overlook or underestimate the important role of capillary forces, i.e., forces arising from the Laplace pressure of curved menisci formed by condensation of a liquid between and around two adhering surfaces (Fig. 30.9).

The adhesion force between a non-deformable spherical particle of radius R and a flat surface in an inert atmosphere (Fig. 30.9a) is

$$F_{ad} = 4\pi R\gamma_{SV} . \tag{30.20}$$

But in an atmosphere containing a condensable vapor, the expression above is replaced by

$$F_{ad} = 4\pi R(\gamma_{LV}\cos\theta + \gamma_{SL}) , \tag{30.21}$$

where the first term is due to the Laplace pressure of the meniscus and the second is due to the direct adhesion of the two contacting solids within the liquid. Note that the above equation does not contain the radius of curvature, r, of the liquid meniscus (see Fig. 30.9b). This is because for smaller r the Laplace pressure γ_{LV}/r increases, but the area over which it acts decreases by the same amount, so the two effects cancel out. Experiments with inert liquids, such as hydrocarbons, condensing between two mica surfaces indicate that (30.21) is valid for values of r as small as $1-2$ nm, corresponding to vapor pressures as low as 40% of saturation [30.148,227,228]. Capillary condensation also occurs in binary liquid systems, e.g., when water dissolved in hydrocarbon liquids condenses around two contacting hydrophilic surfaces or when a vapor cavity forms in water around two hydrophobic surfaces. In the case of water condensing from vapor or from oil, it also appears that the bulk value of γ_{LV} is applicable for meniscus radii as small as 2 nm.

The capillary condensation of liquids, especially water, from vapor can have additional effects on the physical state of the contact zone. For example, if the surfaces contain ions, these will diffuse and build up within the liquid bridge, thereby changing the chemical composition of the contact zone, as well as influencing the adhesion. In the case of surfaces covered with surfactant or polymer molecules (amphiphilic surfaces), the molecules can turn over on exposure to humid air, so that the surface nonpolar groups become replaced by polar groups, which renders the surfaces hydrophilic. When two such surfaces come into contact, water will condense around the contact zone and the adhesion force will also be affected – generally increasing well above the value expected for inert hydrophobic surfaces. It is apparent that adhesion in vapor or a solvent is often largely determined by capillary forces arising from the condensation of liquid that may be present only in very small quantities, e.g., $10-20\%$ of saturation in the vapor, or 20 ppm in the solvent.

30.5.2 Adhesion Mechanics

Two bodies in contact deform as a result of surface forces and/or applied normal forces. For the simplest case of two interacting elastic spheres (a model that is easily extended to an elastic sphere interacting with an undeformable surface, or vice versa) and in the absence of attractive surface forces, the vertical central displacement (compression) was derived by *Hertz* [30.229] (Fig. 30.9c). In this model, the displacement and the contact area are equal to zero when no external force (load) is applied, i.e., at the points of contact and of separation. The contact area A increases with normal force or load as $L^{2/3}$.

In systems where attractive surface forces are present between the surfaces, the deformations are more complicated. Modern theories of the adhesion mechanics of two contacting solid surfaces are based on the Johnson–Kendall–Roberts (JKR) theory [30.15, 230], or on the Derjaguin–Muller–Toporov (DMT) theory [30.231–

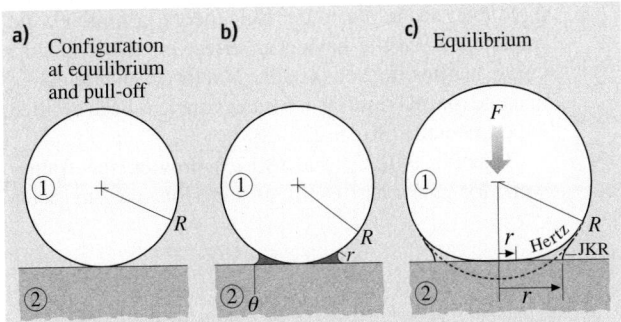

Fig. 30.9a–c Adhesion and capillary forces: (**a**) a non-deforming sphere on a rigid, flat surface in an inert atmosphere and (**b**) in a vapor that can "capillary condense" around the contact zone. At equilibrium, the concave radius, r, of the liquid meniscus is given by the Kelvin equation. For a concave meniscus to form, the contact angle θ has to be less than $90°$. In the case of hydrophobic surfaces surrounded by water, a vapor cavity can form between the surfaces. As long as the surfaces are perfectly smooth, the contribution of the meniscus to the adhesion force is independent of r. (After [30.1] with permission.) (**c**) Elastically deformable sphere on a rigid flat surface in the absence (Hertz) and presence (JKR) of adhesion [(**a**) and (**c**) after [30.3]. Copyright 1991, with permission from Elsevier Science]

233]. The JKR theory is applicable to easily deformable, large bodies with high surface energy, whereas the DMT theory better describes very small and hard bodies with low surface energy [30.234]. The intermediate regime has also been described [30.235].

In the JKR theory, two spheres of radii R_1 and R_2, bulk elastic modulus K, and surface energy γ will flatten due to attractive surface forces when in contact at no external load. The contact area will increase under an external load L or normal force F, such that at mechanical equilibrium the radius of the contact area, r, is given by

$$r^3 = \frac{R}{K}\left[F + 6\pi R\gamma + \sqrt{12\pi R\gamma F + (6\pi R\gamma)^2}\right],$$

(30.22)

where $R = R_1 R_2/(R_1 + R_2)$. In the absence of surface energy, γ, equation (30.22) is reduced to the expression for the radius of the contact area in the Hertz model. Another important result of the JKR theory gives the adhesion force or "pull-off" force:

$$F_{ad} = -3\pi R\gamma_S,$$

(30.23)

where the surface energy, γ_S, is defined through $W = 2\gamma_S$, where W is the reversible work of adhesion.

Note that, according to the JKR theory, a finite elastic modulus, K, while having an effect on the load–area curve, has no effect on the adhesion force, an interesting and unexpected result that has nevertheless been verified experimentally [30.15, 236–238].

Equation (30.22) and (30.23) provide the framework for analyzing results of adhesion measurements

(Fig. 30.10) of contacting solids, known as contact mechanics [30.230, 239], and for studying the effects of surface conditions and time on adhesion energy hysteresis (see Sect. 30.5.4).

The JKR theory has been extended [30.240, 241] to consider rigid or elastic substrates separated by thin compliant layers of very different elastic moduli,

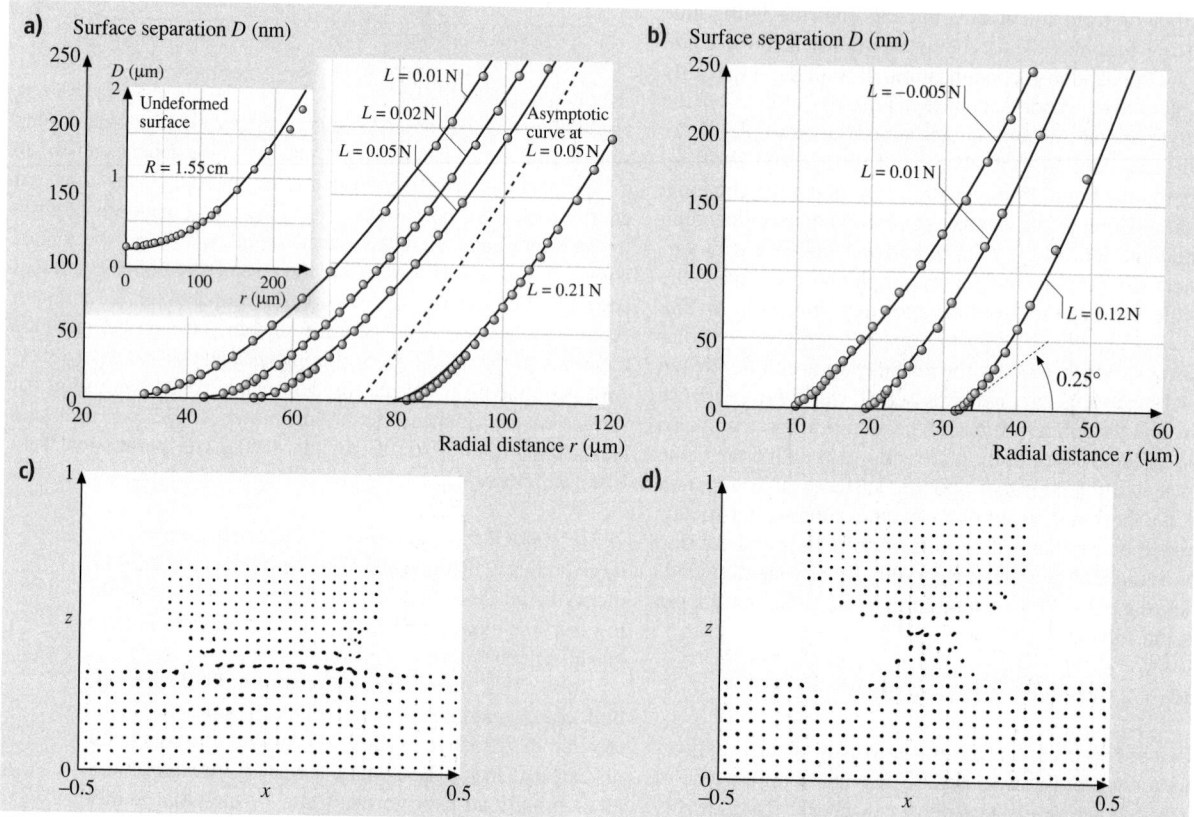

Fig. 30.10a–d Experimental and computer simulation data on contact mechanics for ideal Hertz and JKR contacts. (**a**) Measured profiles of surfaces in nonadhesive contact (*circles*) compared with Hertz profiles (*continuous curves*). The system was mica surfaces in a concentrated KCl solution in which they do not adhere. When not in contact, the surface shape is accurately described by a sphere of radius $R = 1.55$ cm (*inset*). The applied loads were 0.01, 0.02, 0.05, and 0.21 N. The last profile was measured in a different region of the surfaces where the local radius of curvature was 1.45 cm. The Hertz profiles correspond to central displacements of $\delta = 66.5$, 124, 173, and 441 nm. The *dashed line* shows the shape of the undeformed sphere corresponding to the curve at a load of 0.05 N; it fits the experimental points at larger distances (not shown). (**b**) Surface profiles measured with adhesive contact (mica surfaces adhering in dry nitrogen gas) at applied loads of -0.005, 0.01, and 0.12 N. The continuous lines are JKR profiles obtained by adjusting the central displacement in each case to get the best fit to points at larger distances. The values are $\delta = -4.2$, 75.6, and 256 nm. Note that the scales of this figure exaggerate the apparent angle at the junction of the surfaces. This angle, which is insensitive to load, is only about 0.25°. (**c**) and (**d**) Molecular dynamics simulation illustrating the formation of a connective neck between an Ni tip (topmost eight layers) and an Au substrate. The figures show the atomic configuration in a slice through the system at indentation (**c**) and during separation (**d**). Note the crystalline structure of the neck. Distances are given in units of x and z, where $x = 1$ and $z = 1$ correspond to 6.12 nm. [(**a**) and (**b**) after [30.236]. Copyright 1987, with permission from Elsevier Science. (**c**) and (**d**) after [30.112], with kind permission from Kluwer Academic Publishers]

a situation commonly encountered in SFA and AFM experiments. The deformation of the system is then strongly dependant on the ratio of r to the thickness of the confined layer. At small r (low L), the deformation occurs mostly in the thin confined layer, whereas at large r (large L), it occurs mainly in the substrates. Because of the changing distribution of traction across the contact, the adhesion force in a layered system is also modified from that of isotropic systems (30.23) so that it is no longer independent of the elastic moduli.

30.5.3 Effects of Surface Structure, Roughness, and Lattice Mismatch

In a contact between two rough surfaces, the real area of contact varies with the applied load in a different manner than between smooth surfaces [30.243, 244]. For non-adhering surfaces exhibiting an exponential distribution of *elastically* deforming asperities (spherical caps of equal radius), it has been shown that the contact area for rough surfaces increases approximately linearly with the applied normal force (load), L, instead of as $L^{2/3}$ for smooth surfaces [30.243]. It has also been shown that for *plastically* deforming metal microcontacts the real contact area increases with load as $A \propto L$ [30.245, 246]. In systems with attractive surface forces, there is a competition between this attraction and repulsive forces arising from compression of high asperities. As a result, the adhesion in such systems can be very low, especially if the surfaces are not easily deformed [30.247–249]. The opposite is possible for soft (viscoelastic) surfaces where the real (molecular) contact area might be larger than for two perfectly smooth surfaces [30.250]. The size of the real contact area at a given normal force is also an important issue in studies of nanoscale friction, both of single-asperity contacts (cf. Sect. 30.7) and of contacts between rough surfaces (cf. Sect. 30.9.2).

Adhesion forces may also vary depending on the commensurability of the crystallographic lattices of the interacting surfaces. *McGuiggan* and *Israelachvili* [30.251] measured the adhesion between two mica surfaces as a function of the orientation (twist angle) of their surface lattices. The forces were measured in air, water, and an aqueous salt solution where oscillatory structural forces were present. In humid air, the adhesion was found to be relatively independent of the twist angle θ due to the adsorption of a 0.4 nm thick amorphous layer of organics and water at the interface. In contrast, in water, sharp adhesion peaks (energy minima) occurred at $\theta = 0°$, $\pm 60°$, $\pm 120°$ and $180°$, corresponding to the "coincidence" angles of the surface lattices

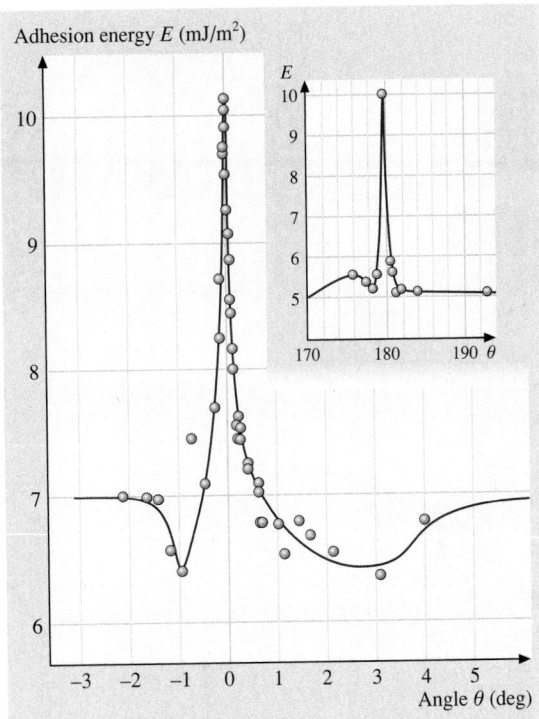

Fig. 30.11 Adhesion energy for two mica surfaces in contact in water (in the primary minimum of an oscillatory force curve) as a function of the mismatch angle θ about $\theta = 0°$ and $180°$ between the mica surface lattices. (After [30.242] with permission)

(Fig. 30.11). As little as $\pm 1°$ away from these peaks, the energy decreased by 50%. In aqueous KCl solution, due to potassium ion adsorption the water between the surfaces becomes ordered, resulting in an oscillatory force profile where the adhesive minima occur at discrete separations of about 0.25 nm, corresponding to integral numbers of water layers. The whole interaction potential was now found to depend on the orientation of the surface lattices, and the effect extended at least four molecular layers.

It has also been appreciated that the structure of the confining surfaces is just as important as the nature of the liquid for determining the solvation forces [30.111, 150, 151, 252–256]. Between two surfaces that are completely flat but "unstructured", the liquid molecules will order into layers, but there will be no lateral ordering within the layers. In other words, there will be positional ordering normal but not parallel to the surfaces. If the surfaces have a crystalline (periodic) lattice, this may induce ordering parallel to the surfaces, as well, and

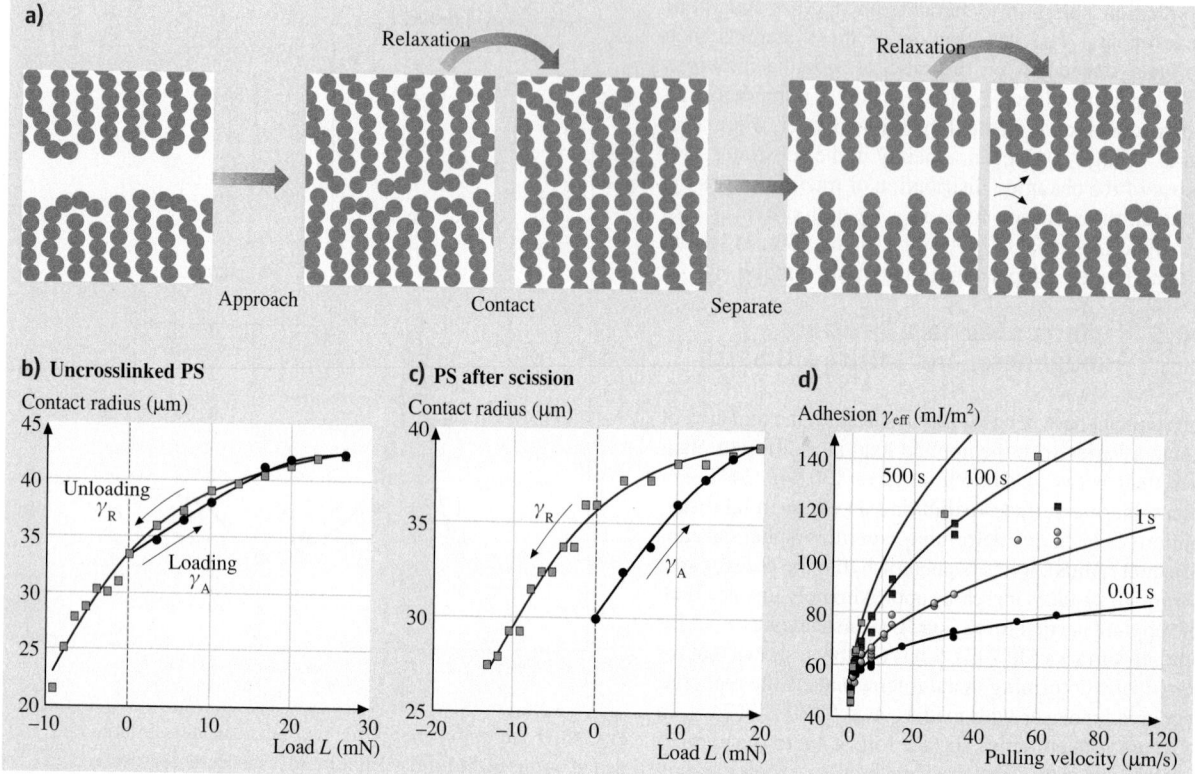

Fig. 30.12 (a) Schematic representation of interpenetrating chains. **(b)** and **(c)**: JKR plots (contact radius r as a function of applied load L) showing small adhesion hysteresis for un-crosslinked polystyrene and larger adhesion hysteresis after chain scission at the surfaces after 18 h irradiation with ultraviolet light in an oxygen atmosphere. The adhesion hysteresis continues to increase with the irradiation time. **(d)** Rate-dependent adhesion of hexadecyl trimethyl ammonium bromide (CTAB) surfactant monolayers. The *solid curves* [30.259] are fits to experimental data on CTAB adhesion after different contact times [30.260] using an approximate analytical solution for a JKR model, including crack tip dissipation. Due to the limited range of validity of the approximation, the fits rely on the part of the experimental data with low effective adhesion energy only. From the fits one can determine the thermodynamic adhesion energy, the characteristic dissipation velocity, and the intrinsic dissipation exponent of the model. [**(a)** after [30.261]. Copyright 1993 American Chemical Society. **(b)** and **(c)** after [30.262]. Copyright 2002 American Association for the Advancement of Science. **(d)** after [30.259]. Copyright 2000 American Chemical Society]

the oscillatory force then also depends on the structure of the surface lattices. Further, if the two lattices have different dimensions ("mismatched" or "incommensurate" lattices), or if the lattices are similar but are not in register relative to each other, the oscillatory force law is further modified [30.251, 257] and the tribological properties of the film are also influenced, as discussed in Sect. 30.9 [30.257, 258].

As shown by the experiments, these effects can alter the magnitude of the adhesive minima found at a given separation within the last one or two nanometers of a thin film by a factor of two. The force barriers (maxima) may also depend on orientation. This could be even more im-

portant than the effects on the minima. A high barrier could prevent two surfaces from coming closer together into a much deeper adhesive well. Thus the maxima can effectively contribute to determining not only the final separation of two surfaces, but also their final adhesion. Such considerations should be particularly important for determining the thickness and strength of intergranular spaces in ceramics, the adhesion forces between colloidal particles in concentrated electrolyte solution, and the forces between two surfaces in a crack containing capillary condensed water.

For surfaces that are *randomly* rough, oscillatory forces become smoothed out and disappear altogether, to be replaced by a purely monotonic solvation force [30.134, 150, 151]. This occurs even if the liquid molecules themselves are perfectly capable of ordering into layers. The situation of *symmetric* liquid molecules confined between *rough* surfaces is therefore not unlike that of *asymmetric* molecules between *smooth* surfaces (see Sect. 30.4.3 and Fig. 30.7a). To summarize, for there to be an oscillatory solvation force, the liquid molecules must be able to be correlated over a reasonably long range. This requires that both the liquid molecules and the surfaces have a high degree of order or symmetry. If either is missing, so will the oscillations. Depending on the size of the molecules to be confined, a roughness of only a few tenths of a nanometer is often sufficient to eliminate any oscillatory component of the force law [30.42, 150].

30.5.4 Nonequilibrium and Rate-Dependent Interactions: Adhesion Hysteresis

Under ideal conditions the adhesion energy is a well-defined thermodynamic quantity. It is normally denoted by E or W (the work of adhesion) or γ (the surface tension, where $W = 2\gamma$) and gives the reversible work done on bringing two surfaces together or the work needed to separate two surfaces from contact. Under ideal, equilibrium conditions these two quantities are the same, but under most realistic conditions they are not; the work needed to separate two surfaces is always greater than that originally gained by bringing them together. An understanding of the molecular mechanisms underlying this phenomenon is essential for understanding many adhesion phenomena, energy dissipation during loading–unloading cycles, contact angle hysteresis, and the molecular mechanisms associated with many frictional processes.

It is wrong to think that hysteresis arises because of some imperfection in the system such as rough or chemically heterogeneous surfaces, or because the supporting material is viscoelastic. Adhesion hysteresis can arise even between perfectly smooth and chemically homogenous surfaces supported by perfectly elastic materials. It can be responsible for such phenomena as rolling friction and elastoplastic adhesive contacts [30.239, 263–266] during loading–unloading and adhesion–decohesion cycles.

Adhesion hysteresis may be thought of as being due to mechanical effects such as instabilities, or chemical effects such as interdiffusion, interdigitation, molecular reorientations and exchange processes occurring at an interface after contact, as illustrated in Fig. 30.12. Such processes induce roughness and chemical heterogeneity even though initially (and after separation and re-equilibration) both surfaces are perfectly smooth and chemically homogeneous. In general, if the energy change, or work done, on separating two surfaces from adhesive contact is not fully recoverable on bringing the two surfaces back into contact again, the adhesion hysteresis may be expressed as

$$W_R \quad > \quad W_A$$
Receding   Advancing

or

$$\Delta W = (W_R - W_A) > 0 , \tag{30.24}$$

where W_R and W_A are the adhesion or surface energies for receding (separating) and advancing (approaching) two solid surfaces, respectively.

Hysteresis effects are also commonly observed in wetting/dewetting phenomena [30.267]. For example, when a liquid spreads and then retracts from a surface the advancing contact angle θ_A is generally larger than the receding angle θ_R. Since the contact angle, θ, is related to the liquid–vapor surface tension, γ_L, and the solid–liquid adhesion energy, W, by the Dupré equation,

$$(1 + \cos\theta)\gamma_L = W , \tag{30.25}$$

we see that *wetting hysteresis* or *contact angle hysteresis* ($\theta_A > \theta_R$) actually implies adhesion hysteresis, $W_R > W_A$, as given by (30.24).

Energy-dissipating processes such as adhesion and contact angle hysteresis arise because of practical constraints of the *finite time* of measurements and the *finite elasticity* of materials. This prevents many loading–unloading or approach–separation cycles from being thermodynamically reversible, even though they would be if carried out infinitely slowly. By thermodynamically irreversible one simply means that one cannot go through the approach–separation cycle via a continuous series of equilibrium states, because some of these are connected via spontaneous – and therefore thermodynamically irreversible – instabilities or transitions where energy is liberated and therefore "lost" via heat or phonon release [30.268]. This is an area of much current interest and activity, especially regarding the fundamental molecular origins of adhesion and friction in polymer and surfactant systems, and the relationships between them [30.239, 259, 260, 262, 264, 269–272].

30.6 Introduction: Different Modes of Friction and the Limits of Continuum Models

Most frictional processes occur with the sliding surfaces becoming damaged in one form or another [30.263]. This may be referred to as "normal" friction. In the case of brittle materials, the damaged surfaces slide past each other while separated by relatively large, micron-sized wear particles. With more ductile surfaces, the damage remains localized to nanometer-sized, plastically de-

formed asperities. Some features of the friction between damaged surfaces will be described in Sect. 30.7.4.

There are also situations in which sliding can occur between two perfectly smooth, undamaged surfaces. This may be referred to as "interfacial" sliding or "boundary" friction and is the focus of the following sections. The term "boundary lubrication" is more com-

Table 30.4 The three main tribological regimes characterizing the changing properties of liquids subjected to increasing confinement between two solid surfaces[a]

Regime	Conditions for getting into this regime	Static/equilibrium properties[b]	Dynamic properties[c]
Bulk	• Thick films (> 10 molecular diameters, $\gg R_g$ for polymers) • Low or zero loads • High shear rates	Bulk (continuum) properties: • Bulk liquid density • No long-range order	Bulk (continuum) properties: • Newtonian viscosity • Fast relaxation times • No glass temperature • No yield point • Elastohydrodynamic lubrication
Intermediate mixed	• Intermediately thick films (4–10 molecular diameters, $\sim R_g$ for polymers) • Low loads or pressure	Modified fluid properties include: • Modified positional and orientational order[a] • Medium- to long-range molecular correlations • Highly entangled states	Modified rheological properties include: • Non-Newtonian flow • Glassy states • Long relaxation times • Mixed lubrication
Boundary	• Molecularly thin films (< 4 molecular diameters) • High loads or pressure • Low shear rates • Smooth surfaces or asperities	Onset of non-fluidlike properties: • Liquid-like to solid-like phase transitions • Appearance of new liquid-crystalline states • Epitaxially induced long-range ordering	Onset of tribological properties: • No flow until yield point or critical shear stress reached • Solid-like film behavior characterized by defect diffusion, dislocation motion, shear melting • Boundary lubrication

Based on work by *Granick* [30.273], *Hu* and *Granick* [30.274], and others [30.38, 261, 275] on the dynamic properties of short chain molecules such as alkanes and polymer melts confined between surfaces

[a] Confinement can lead to an increased or decreased order in a film, depending both on the surface lattice structure and the geometry of the confining cavity

[b] In each regime both the static and dynamic properties change. The static properties include the film density, the density distribution function, the potential of mean force, and various positional and orientational order parameters

[c] Dynamic properties include viscosity, viscoelastic constants, and tribological yield points such as the friction coefficient and critical shear stress

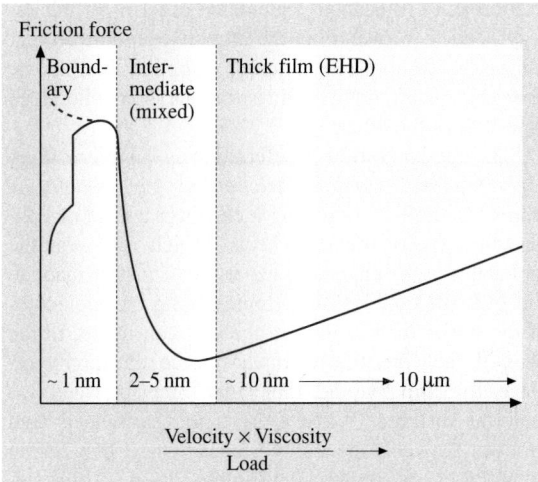

Friction force

Boundary

Intermediate (mixed)

Thick film (EHD)

~ 1 nm   2–5 nm   ~ 10 nm ⟶ 10 μm ⟶

$$\frac{\text{Velocity} \times \text{Viscosity}}{\text{Load}} \longrightarrow$$

Fig. 30.13 Stribeck curve: an empirical curve giving the trend generally observed in the friction forces or friction coefficients as a function of sliding velocity, the bulk viscosity of the lubricating fluid, and the applied load (normal force). The three friction/lubrication regimes are known as the boundary lubrication regime (see Sect. 30.7), the intermediate or mixed lubrication regime (Sect. 30.8.2), and thick film or elasto-hydrodynamic (EHD) lubrication regime (Sect. 30.8.1). The film thicknesses believed to correspond to each of these regimes are also shown. For thick films, the friction force is purely viscous, e.g., Couette flow at low shear rates, but may become complicated at higher shear rates where EHD deformations of surfaces can occur during sliding. (After [30.1] with permission)

monly used to denote the friction of surfaces that contain a thin protective lubricating layer such as a surfactant monolayer, but here we shall use the term more broadly to include any molecularly thin solid, liquid, surfactant, or polymer film.

Experiments have shown that, as a liquid film becomes progressively thinner, its physical properties change, at first quantitatively and then qualitatively [30.44, 47, 273, 274, 276, 277]. The quantitative changes are manifested by an increased viscosity, non-Newtonian flow behavior, and the replacement of normal melting by a glass transition, but the film remains recognizable as a liquid (Fig. 30.13). In tribology, this regime

is commonly known as the "mixed lubrication" regime, where the rheological properties of a film are intermediate between the bulk and boundary properties. One may also refer to it as the "intermediate" regime (Table 30.4).

For even thinner films, the changes in behavior are more dramatic, resulting in a qualitative change in properties. Thus first-order phase transitions can now occur to solid or liquid-crystalline phases [30.46,255,261,275, 278–281], whose properties can no longer be characterized even qualitatively in terms of bulk or continuum liquid properties such as viscosity. These films now exhibit yield points (characteristic of fracture in solids) and their molecular diffusion and relaxation times can be ten orders of magnitude longer than in the bulk liquid or even in films that are just slightly thicker. The three friction regimes are summarized in Table 30.4.

30.7 Relationship Between Adhesion and Friction Between Dry (Unlubricated and Solid Boundary Lubricated) Surfaces

30.7.1 Amontons' Law and Deviations from It Due to Adhesion: The Cobblestone Model

Early theories and mechanisms for the dependence of friction on the applied normal force or load, L, were developed by *da Vinci*, *Amontons*, *Coulomb* and *Euler* [30.282]. For the macroscopic objects investigated, the friction was found to be directly proportional to the load, with no dependence on the contact area. This is described by the so-called Amontons' law:

$$F = \mu L \,, \tag{30.26}$$

where F is the shear or friction force and μ is a constant defined as the coefficient of friction. This friction law

has a broad range of applicability and is still the principal means of quantitatively describing the friction between surfaces. However, particularly in the case of adhering surfaces, Amontons' law does not adequately describe the friction behavior with load, because of the finite friction force measured at zero and even negative applied loads.

When a lateral force, or shear stress, is applied to two surfaces in adhesive contact, the surfaces initially remain "pinned" to each other until some critical shear force is reached. At this point, the surfaces begin to slide past each other either smoothly or in jerks. The frictional force needed to initiate sliding from rest is known as the *static* friction force, denoted by F_s, while the force needed to maintain smooth sliding is referred to as the *kinetic* or *dynamic* friction force, denoted by F_k. In general, $F_s > F_k$. Two sliding surfaces may also move in regular jerks, known as stick–slip sliding, which is discussed in more detail in Sect. 30.8.3. Such friction forces cannot be described by models used for thick films that are viscous (see Sect. 30.8.1) and, therefore, shear as soon as the smallest shear force is applied.

In Sects. 30.7 and 30.8 we will be concerned mainly with single-asperity contacts. Experimentally, it has been found that during both smooth and stick–slip sliding at small film thicknesses the local geometry of the contact zone remains largely unchanged from the static geometry [30.45]. In an adhesive contact, the contact area as a function of load is thus generally well described by the JKR equation, (30.22). The friction force between two molecularly smooth surfaces sliding in *adhesive* contact is not simply proportional to the applied load, L, as might be expected from Amontons' law. There is an additional adhesion contribution that is proportional to the area of contact, A. Thus, in general, the interfacial friction force of dry, unlubricated surfaces sliding smoothly past each other in adhesive contact is given by

$$F = F_k = S_c A + \mu L \,, \qquad (30.27)$$

where S_c is the "critical shear stress" (assumed to be constant), $A = \pi r^2$ is the contact area of radius r given by (30.22), and μ is the coefficient of friction. For low loads we have:

$$F = S_c A = S_c \pi r^2$$
$$= S_c \pi \left[\frac{R}{K} \left(L + 6\pi R \gamma + \sqrt{12\pi R \gamma L + (6\pi R \gamma)^2} \right) \right]^{2/3} ; \qquad (30.28)$$

whereas for high loads (or high μ), or when γ is very low [30.283–287], (30.27) reduces to Amontons' law: $F = \mu L$. Depending on whether the friction force in (30.27) is dominated by the first or second term, one may refer to the friction as *adhesion-controlled* or *load-controlled*, respectively.

The following friction model, first proposed by *Tabor* [30.288] and developed further by *Sutcliffe* et al. [30.289], *McClelland* [30.290], and *Homola* et al. [30.45], has been quite successful at explaining the interfacial and boundary friction of two solid crystalline surfaces sliding past each other in the absence of wear. The surfaces may be unlubricated, or they may be separated by a monolayer or more of some boundary lubricant or liquid molecules. In this model, the values of the critical shear stress S_c, and the coefficient of friction μ, in (30.27) are calculated in terms of the energy needed to overcome the attractive intermolecular forces and compressive externally applied load as one surface is raised and then slid across the molecular-sized asperities of the other.

This model (variously referred to as the *interlocking asperity model*, *Coulomb friction*, or the *cobblestone model*) is similar to pushing a cart over a road of cobblestones where the cartwheels (which represent the molecules of the upper surface or film) must be made to roll over the cobblestones (representing the molecules of the lower surface) before the cart can move. In the case of the cart, the downward force of gravity replaces the attractive intermolecular forces between two material surfaces. When at rest, the cartwheels find grooves between the cobblestones where they sit in potential-energy minima, and so the cart is at some stable mechanical equilibrium. A certain lateral force (the "push") is required to raise the cartwheels against the force of gravity in order to initiate motion. Motion will continue as long as the cart is pushed, and rapidly stops once it is no longer pushed. Energy is dissipated by the liberation of heat (phonons, acoustic waves, etc.) every time a wheel hits the next cobblestone. The cobblestone model is not unlike the *Coulomb* and *interlocking asperity* models of friction [30.282] except that it is being applied at the molecular level and for a situation where the external load is augmented by attractive intermolecular forces.

There are thus two contributions to the force pulling two surfaces together: the externally applied load or pressure, and the (internal) attractive intermolecular forces that determine the adhesion between the two surfaces. Each of these contributions affects the friction force in a different way, which we will discuss in more detail below.

30.7.2 Adhesion Force and Load Contribution to Interfacial Friction

Adhesion Force Contribution
Consider the case of two surfaces sliding past each other, as shown in Fig. 30.14a. When the two surfaces are initially in adhesive contact, the surface molecules will adjust themselves to fit snugly together [30.291], in an analogous manner to the self-positioning of the cartwheels on the cobblestone road. A small tangential force applied to one surface will therefore not result in the sliding of that surface relative to the other. The attractive van der Waals forces between the surfaces must

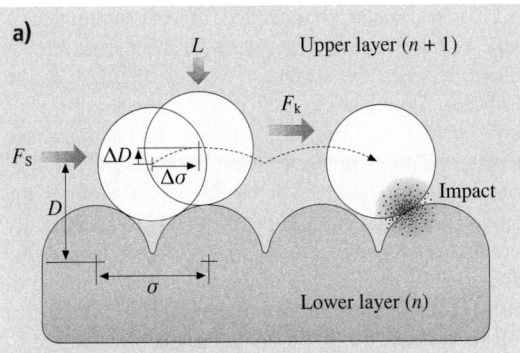

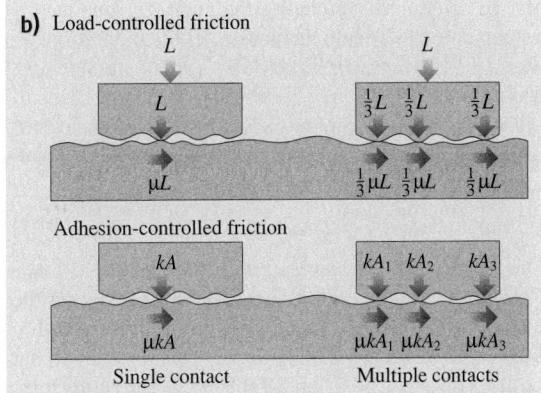

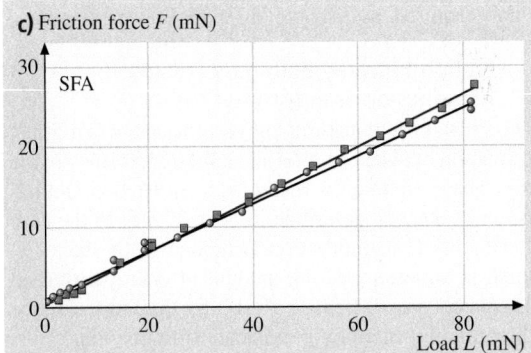

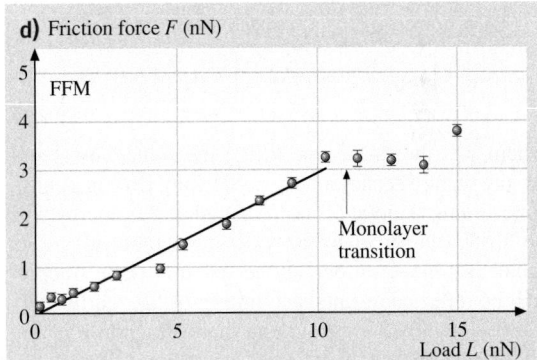

Fig. 30.14 (**a**) Schematic illustration of how one molecularly smooth surface moves over another when a lateral force F is applied (the "cobblestone model"). As the upper surface moves laterally by some fraction of the lattice dimension $\Delta\sigma$, it must also move up by some fraction of an atomic or molecular dimension ΔD before it can slide across the lower surface. On impact, some fraction ε of the kinetic energy is "transmitted" to the lower surface, the rest being "reflected" back to the colliding molecule (upper surface). (After [30.292], with permission) (**b**) Difference in the local distribution of the total applied external load or normal adhesive force between load-controlled non-adhering surfaces and adhesion-controlled surfaces. In the former case, the total friction force F is given either by $F = \mu L$ for one contact point (*left side*) or by $F = \frac{1}{3}\mu L + \frac{1}{3}\mu L + \frac{1}{3}\mu L = \mu L$ for three contact points (*right side*). Thus the load-controlled friction is always proportional to the applied load, independently of the number of contacts and of their geometry. In the case of adhering surfaces, the effective "internal" load is given by kA, where A is the real local contact area, which is proportional to the number of intermolecular bonds being made and broken across each single contact point. The total friction force is now given by $F = \mu kA$ for one contact point (*left side*), and $F = \mu(kA_1 + kA_2 + kA_3) = \mu kA_{tot}$ for three contact points (*right side*). Thus, for adhesion-controlled friction, the friction is proportional to the real contact area, at least when no additional external load is applied to the system. (After [30.287], with permission. Copyright 2004 American Chemical Society) (**c**),(**d**) Friction force between benzyltrichlorosilane monolayers chemically bound to glass or Si, measured in ethanol ($\gamma < 1\,\text{mJ/m}^2$). (**c**) SFA measurements where both glass surfaces were covered with a monolayer. *Circles* and *squares* show two different experiments: one with $R = 2.6\,\text{cm}$, $v = 0.15\,\mu\text{m/s}$, giving $\mu = 0.33 \pm 0.01$; the other with $R = 1.6\,\text{cm}$, $v = 0.5\,\mu\text{m/s}$, giving $\mu = 0.30 \pm 0.01$. (**d**) Friction force microscopy (FFM) measurements of a monolayer-functionalized Si tip ($R = 11\,\text{nm}$) sliding on a monolayer-covered glass surface at $v = 0.15\,\mu\text{m/s}$, giving $\mu = 0.30 \pm 0.01$. Note the different scales in (**c**) and (**d**). (After [30.286], with permission. Copyright 2003 American Chemical Society)

first be overcome by having the surfaces separate by a small amount. To initiate motion, let the separation between the two surfaces increase by a small amount ΔD, while the lateral distance moved is $\Delta\sigma$. These two values will be related via the geometry of the two surface lattices. The energy put into the system by the force F acting over a lateral distance $\Delta\sigma$ is

Input energy: $F \times \Delta\sigma$. $\qquad$ (30.29)

This energy may be equated with the change in interfacial or surface energy associated with separating the surfaces by ΔD, i.e., from the equilibrium separation $D = D_0$ to $D = (D_0 + \Delta D)$. Since $\gamma \propto D^{-2}$ for two flat surfaces (cf. Sect. 30.3.1 and Table 30.2), the energy cost may be approximated by:

Surface energy change × area:

$$2\gamma A\left[1 - D_0^2/(D_0 + \Delta D)^2\right] \approx 4\gamma A(\Delta D/D_0) ,$$
$$(30.30)$$

where γ is the surface energy, A the contact area, and D_0 the surface separation at equilibrium. During steady-state sliding (kinetic friction), not all of this energy will be "lost" or absorbed by the lattice every time the surface molecules move by one lattice spacing: some fraction will be reflected during each impact of the "cartwheel" molecules [30.290]. Assuming that a fraction ε of the above surface energy is "lost" every time the surfaces move across the characteristic length $\Delta\sigma$ (Fig. 30.14a), we obtain after equating (30.29) and (30.30)

$$S_c = \frac{F}{A} = \frac{4\gamma\varepsilon\Delta D}{D_0\Delta\sigma} .$$
$$(30.31)$$

For a typical hydrocarbon or a van der Waals surface, $\gamma \approx 25\,\mathrm{mJ\,m^{-2}}$. Other typical values would be: $\Delta D \approx 0.05\,\mathrm{nm}$, $D_0 \approx 0.2\,\mathrm{nm}$, $\Delta\sigma \approx 0.1\,\mathrm{nm}$, and $\varepsilon \approx 0.1\text{--}0.5$. Using the above parameters, (30.31) predicts $S_c \approx (2.5\text{--}12.5) \times 10^7\,\mathrm{N\,m^{-2}}$ for van der Waals surfaces. This range of values compares very well with typical experimental values of $2 \times 10^7\,\mathrm{N\,m^{-2}}$ for hydrocarbon or mica surfaces sliding in air (see Fig. 30.16) or separated by one molecular layer of cyclohexane [30.45].

The above model suggests that all interfaces, whether dry or lubricated, dilate just before they shear or slip. This is a small but important effect: the dilation provides the crucial extra space needed for the molecules to slide across each other or flow. This dilation is known to occur in macroscopic systems [30.293, 294] and for nanoscopic systems it has been computed by *Thompson* and *Robbins* [30.255] and *Zaloj* et al. [30.295] and measured by *Dhinojwala* et al. [30.296].

This model may be extended, at least semiquantitatively, to lubricated sliding, where a thin liquid film is present between the surfaces. With an increase in the number of liquid layers between the surfaces, D_0 increases while ΔD decreases, hence the friction force decreases. This is precisely what is observed, but with more than one liquid layer between two surfaces the situation becomes too complex to analyze analytically (actually, even with one or no interfacial layers, the calculation of the fraction of energy dissipated per molecular collision ε is not a simple matter). Furthermore, even in systems as simple as linear alkanes, interdigitation and interdiffusion have been found to contribute strongly to the properties of the system [30.143, 297]. Sophisticated modeling based on computer simulations is now required, as discussed in the following section.

Relation Between Boundary Friction and Adhesion Energy Hysteresis

While the above equations suggest that there is a direct correlation between friction and adhesion, this is not the case. The correlation is really between friction and adhesion hysteresis, described in Sect. 30.5.4. In the case of friction, this subtle point is hidden in the factor ε, which is a measure of the amount of energy absorbed (dissipated, transferred, or "lost") by the lower surface when it is impacted by a molecule from the upper surface. If $\varepsilon = 0$, all the energy is reflected, and there will be no kinetic friction force or any adhesion hysteresis, but the absolute magnitude of the adhesion force or energy will remain finite and unchanged. This is illustrated in Figs. 30.17 and 30.19.

The following simple model shows how adhesion hysteresis and friction may be quantitatively related. Let $\Delta\gamma = \gamma_R - \gamma_A$ be the adhesion energy hysteresis per unit area, as measured during a typical loading–unloading cycle (see Figs. 30.17a and 30.19c,d). Now consider the same two surfaces sliding past each other and assume that frictional energy dissipation occurs through the same mechanism as adhesion energy dissipation, and that both occur over the same characteristic molecular length scale σ. Thus, when the two surfaces (of contact area $A = \pi r^2$) move a distance σ, equating the frictional energy ($F \times \sigma$) to the dissipated adhesion energy ($A \times \Delta\gamma$), we obtain

$$\text{Friction force: } F = \frac{A \times \Delta\gamma}{\sigma} = \frac{\pi r^2}{\sigma}(\gamma_R - \gamma_A) ,$$
$$(30.32)$$

or   Critical shear stress: $S_c = F/A = \Delta\gamma/\sigma$,
$$(30.33)$$

which is the desired expression and has been found to give order-of-magnitude agreement between measured friction forces and adhesion energy hysteresis [30.261]. If we equate (30.33) with (30.31), since $4\Delta D/(D_0\Delta\sigma) \approx 1/\sigma$, we obtain the intuitive relation

$$\varepsilon = \frac{\Delta\gamma}{\gamma} .\tag{30.34}$$

External Load Contribution to Interfacial Friction

When there is no interfacial adhesion, S_c is zero. Thus, in the absence of any adhesive forces between two surfaces, the only "attractive" force that needs to be overcome for sliding to occur is the externally applied load or pressure, as shown in Fig. 30.14b.

For a preliminary discussion of this question, it is instructive to compare the magnitudes of the *externally* applied pressure to the *internal* van der Waals pressure between two smooth surfaces. The internal van der Waals pressure between two flat surfaces is given (see Table 30.2) by $P = A_H/6\pi D_0^3 \approx 1$ GPa $(10^4$ atm), using a typical Hamaker constant of $A_H = 10^{-19}$ J, and assuming $D_0 \approx 2$ nm for the equilibrium interatomic spacing. This implies that we should not expect the externally applied load to affect the interfacial friction force F, as defined by (30.27), until the externally applied pressure L/A begins to exceed ~ 100 MPa $(10^3$ atm). This is in agreement with experimental data [30.298] where the effect of load became dominant at pressures in excess of 10^3 atm.

For a more general semiquantitative analysis, again consider the cobblestone model used to derive (30.31), but now include an additional contribution to the surface-energy change of (30.30) due to the work done against the external load or pressure, $L\Delta D = P_{ext}A\Delta D$ (this is equivalent to the work done against gravity in the case of a cart being pushed over cobblestones). Thus:

$$S_c = \frac{F}{A} = \frac{4\gamma\varepsilon\Delta D}{D_0\Delta\sigma} + \frac{P_{ext}\varepsilon\Delta D}{\Delta\sigma} ,\tag{30.35}$$

which gives the more general relation

$$S_c = F/A = C_1 + C_2 P_{ext} ,\tag{30.36}$$

where $P_{ext} = L/A$ and C_1 and C_2 are characteristic of the surfaces and sliding conditions. The constant $C_1 = 4\gamma\varepsilon\Delta D/(D_0\Delta\sigma)$ depends on the mutual adhesion of the two surfaces, while both C_1 and $C_2 = \varepsilon\Delta D/\Delta\sigma$ depend on the topography or atomic bumpiness of the surface groups (Fig. 30.14a). The smoother the surface groups the smaller the ratio $\Delta D/\Delta\sigma$ and hence the lower

the value of C_2. In addition, both C_1 and C_2 depend on ε (the fraction of energy dissipated per collision), which depends on the relative masses of the shearing molecules, the sliding velocity, the temperature, and the characteristic molecular relaxation processes of the surfaces. This is by far the most difficult parameter to compute, and yet it is the most important since it represents the energy-transfer mechanism in any friction process, and since ε can vary between 1 and 0, it determines whether a particular friction force will be large or close to zero. Molecular simulations offer the best way to understand and predict the magnitude of ε, but the complex multibody nature of the problem makes simple conclusions difficult to draw [30.299–302]. Some of the basic physics of the energy transfer and dissipation of the molecular collisions can be drawn from simplified models such as a 1D three-body system [30.268].

Finally, the above equation may also be expressed in terms of the friction force F:

$$F = S_c A = C_1 A + C_2 L .\tag{30.37}$$

Equations similar to (30.36) and (30.37) were previously derived by *Derjaguin* [30.303, 304] and by *Briscoe* and *Evans* [30.305], where the constants C_1 and C_2 were interpreted somewhat differently than in this model.

In the absence of any attractive interfacial force, we have $C_1 \approx 0$, and the second term in (30.36) and (30.37) should dominate (Fig. 30.15). Such situations typically arise when surfaces repel each other across the lubricating liquid film. In such cases, the total frictional force should be low and should increase *linearly* with the external load according to

$$F = C_2 L .\tag{30.38}$$

An example of such lubricated sliding occurs when two mica surfaces slide in water or in salt solution (see Fig. 30.20a), where the short-range "hydration" forces between the surfaces are repulsive. Thus, for sliding in 0.5 M KCl it was found that $C_2 = 0.015$ [30.283]. Another case where repulsive surfaces eliminate the adhesive contribution to friction is for polymer chains attached to surfaces at one end and swollen by a good solvent [30.219]. For this class of systems, $C_2 < 0.001$ for a finite range of polymer layer compressions (normal loads, L). The low friction between the surfaces in this regime is attributed to the entropic repulsion between the opposing brush layers with a minimum of entanglement between the two layers. However, with higher normal loads, the brush layers become compressed and begin to entangle, which results in higher friction (see [30.306]).

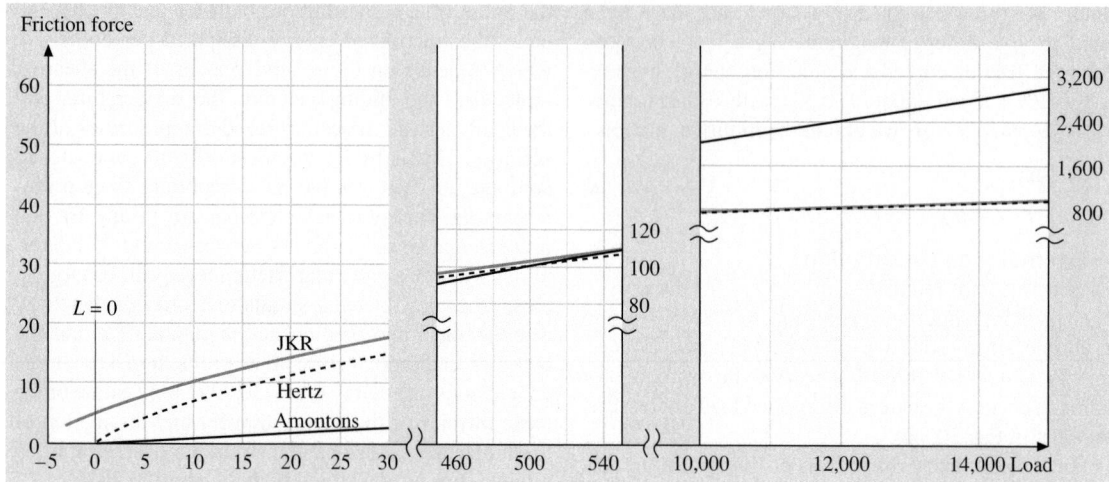

Fig. 30.15 Friction as a function of load for smooth surfaces. At low loads, the friction is dominated by the $C_1 A$ term of (30.37). The adhesion contribution (JKR curve) is most prominent near zero load where the Hertzian and Amontons' contributions to the friction are minimal. As the load increases, the adhesion contribution becomes smaller as the JKR and Hertz curves converge. In this range of loads, the linear $C_2 L$ contribution surpasses the area contribution to the friction. At much higher loads the explicit load dependence of the friction dominates the interactions, and the observed behavior approaches Amontons' law. It is interesting to note that for smooth surfaces the pressure over the contact area does not increase as rapidly as the load. This is because as the load is increased, the surfaces deform to increase the surface area and thus moderate the contact pressure. (After [30.307], with permission of Kluwer Academic Publishers)

It is important to note that (30.38) has exactly the same form as Amontons' Law

$$F = \mu L , \tag{30.39}$$

where μ is the coefficient of friction.

Figure 30.14c,d shows the kinetic friction force measured with both SFA and FFM (friction force microscopy, using AFM) on a system where both surfaces were covered with a chemically bound benzyl-trichlorosilane monolayer [30.286]. When immersed in ethanol, the adhesion in this system is low, and very different contact areas and loads give a linear dependence of F on L with the same friction coefficients, and $F \to 0$ as $L \to 0$. In the FFM measurements (Fig. 30.14d), the plateau in the data at higher loads suggest a transition in the monolayers, similar to previous observations on other monolayer systems. The pressure in the contact region in the SFA is much lower than in the FFM, and no transitions in the friction forces or in the thickness of the confined monolayers were observed in the SFA experiments (and no damage to the monolayers or the underlying substrates was observed during the experiments, indicating that the friction was "wearless"). Despite the difference of more than six orders of magnitude in the contact radii, pressure, loads, and friction

forces, the measured friction coefficients are practically the same.

At the molecular level a thermodynamic analog of the Coulomb or cobblestone models (see Sect. 30.7.1) based on the "contact value theorem" [30.3, 283, 307] can explain why $F \propto L$ also holds at the microscopic or molecular level. In this analysis we consider the surface molecular groups as being momentarily compressed and decompressed as the surfaces move along. Under irreversible conditions, which always occur when a cycle is completed in a finite amount of time, the energy "lost" in the compression–decompression cycle is dissipated as heat. For two non-adhering surfaces, the stabilizing pressure P_i acting locally between any two elemental contact points i of the surfaces may be expressed by the contact value theorem [30.3]:

$$P_i = \rho_i k_B T = k_B T / V_i , \tag{30.40}$$

where $\rho_i = V_i^{-1}$ is the local number density (per unit volume) or activity of the interacting entities, be they molecules, atoms, ions or the electron clouds of atoms. This equation is essentially the osmotic or entropic pressure of a gas of confined molecules. As one surface moves across the other, local regions become compressed and decompressed by a volume ΔV_i. The work done per cycle can be written as $\varepsilon P_i \Delta V_i$, where ε ($\varepsilon \leq 1$) is the fraction of energy per cycle "lost" as heat, as defined earlier. The energy balance shows that, for each compression–decompression cycle, the dissipated energy is related to the friction force by

$$F_i x_i = \varepsilon P_i \Delta V_i , \qquad (30.41)$$

where x_i is the lateral distance moved per cycle, which can be the distance between asperities or the distance between surface lattice sites. The pressure at each contact junction can be expressed in terms of the local normal load L_i and local area of contact A_i as $P_i = L_i / A_i$. The volume change over a cycle can thus be expressed as $\Delta V_i = A_i z_i$, where z_i is the vertical distance of confinement. Inserting these into (30.41), we get

$$F_i = \varepsilon L_i (z_i / x_i) , \qquad (30.42)$$

which is independent of the local contact area A_i. The total friction force is thus

$$F = \sum F_i = \sum \varepsilon L_i (z_i / x_i)$$
$$= \varepsilon \langle z_i / x_i \rangle \sum L_i = \mu L , \qquad (30.43)$$

where it is assumed that on average the local values of L_i and P_i are independent of the local *slope* z_i / x_i. Therefore, the friction coefficient μ is a function only of the average surface topography and the sliding velocity, but is independent of the local (real) or macroscopic (apparent) contact areas.

While this analysis explains non-adhering surfaces, there is still an additional explicit contact area contribution for the case of adhering surfaces, as in (30.37). The distinction between the two cases arises because the initial assumption of the contact value theorem, (30.40), is incomplete for adhering systems. A more appropriate starting equation would reflect the full intermolecular interaction potential, including the attractive interactions, in addition to the purely repulsive contributions of (30.40), much as the van der Waals equation of state modifies the ideal gas law.

30.7.3 Examples of Experimentally Observed Friction of Dry Surfaces

Numerous model systems have been studied with a surface forces apparatus (SFA) modified for friction experiments (see Sect. 30.2.3). The apparatus allows for control of load (normal force) and sliding speed, and simultaneous measurement of surface separation, surface shape, true (molecular) area of contact between smooth

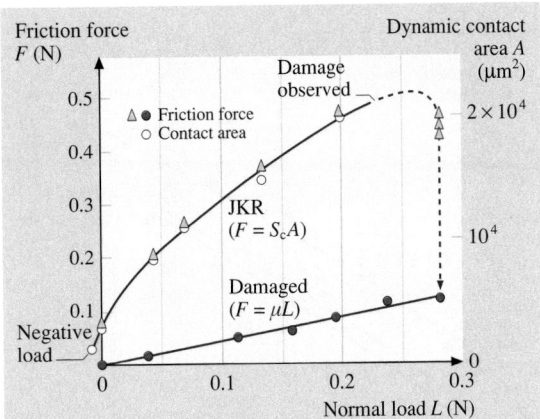

Fig. 30.16 Friction force F and contact area A versus load L for two mica surfaces sliding in adhesive contact in dry air. The contact area is well described by the JKR theory, (30.22), even during sliding, and the friction force is found to be directly proportional to this area, (30.28). The *vertical dashed line* and *arrow* show the transition from interfacial to normal friction with the onset of wear (*lower curve*). The sliding velocity is $0.2\,\mu\text{m s}^{-1}$. (After [30.45], with permission, copyright 1989 American Society of Mechanical Engineers)

surfaces, and friction forces. A variety of both unlubricated and solid- and liquid-lubricated surfaces have been studied both as smooth single-asperity contacts and after they have been roughened by shear-induced damage.

Figure 30.16 shows the contact area, A, and friction force, F, both plotted against the applied load, L, in an experiment in which two molecularly smooth surfaces of mica in adhesive contact were slid past each other in an atmosphere of dry nitrogen gas. This is an example of the low-load adhesion-controlled limit, which is excellently described by (30.28). In a number of different experiments, S_c was measured to be $2.5 \times 10^7\,\text{N m}^{-2}$ and to be independent of the sliding velocity [30.45,308]. Note that there is a friction force even at negative loads, where the surfaces are still sliding in adhesive contact.

Figure 30.17 shows the correlation between adhesion hysteresis and friction for two surfaces consisting of silica films deposited on mica substrates [30.41]. The friction between undamaged hydrophobic silica surfaces showed stick–slip both at dry conditions and at 100% relative humidity. Similar to the mica surfaces in Figs. 30.16, 30.18, and 30.20a, the friction of damaged silica surfaces obeyed Amontons' law with a friction coefficient of 0.25–0.3 both at dry conditions and at 55% relative humidity.

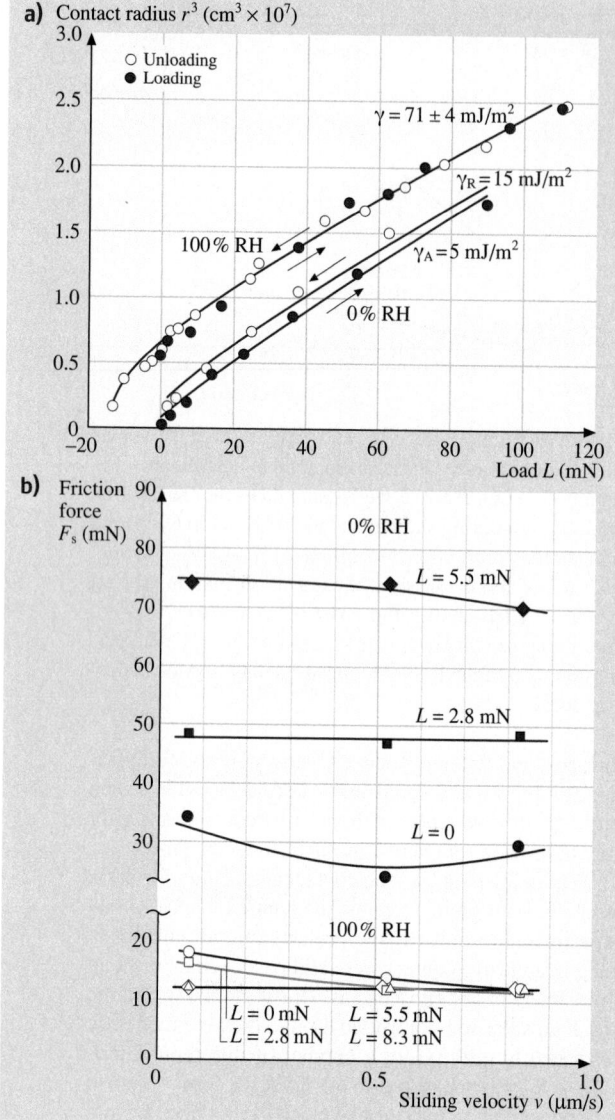

The high friction force of unlubricated sliding can often be reduced by treating the solid surface with a boundary layer of some other solid material that exhibits lower friction, such as a surfactant monolayer, or by ensuring that during sliding a thin liquid film remains between the surfaces (as will be discussed in Sect. 30.8). The effectiveness of a solid boundary lubricant layer on reducing the forces of friction is illustrated in Fig. 30.18. Comparing this with the friction of the unlubricated/untreated surfaces (Fig. 30.16) shows that the critical shear stress has been reduced by a factor of about ten: from 2.5×10^7 to $3.5 \times 10^6 \, \mathrm{N \, m^{-2}}$. At much higher applied loads or pressures, the friction force is proportional to the load, rather than the area of contact [30.298], as expected from (30.27).

Yamada and *Israelachvili* [30.309] studied the adhesion and friction of fluorocarbon surfaces (surfactant-

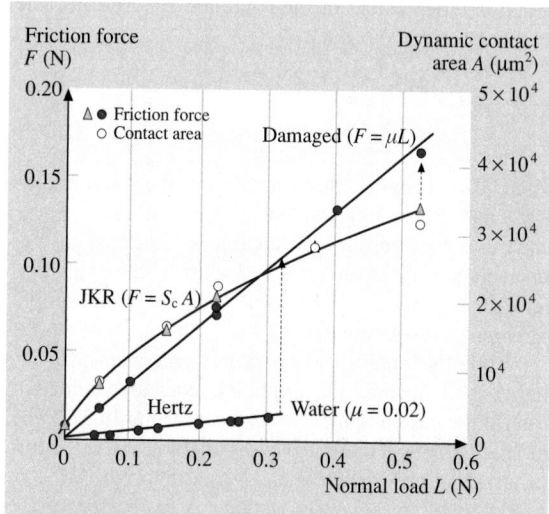

Fig. 30.18 Sliding of mica surfaces, each coated with a 2.5 nm thick monolayer of calcium stearate surfactant, in the absence of damage (obeying JKR-type boundary friction) and in the presence of damage (obeying Amontons-type normal friction). Note that both for this system and for the bare mica in Figs. 30.16 and 30.20a, the friction force obeys Amontons' law with a friction coefficient of $\mu \approx 0.3$ after damage occurs. At much higher applied loads, the undamaged surfaces also follow Amontons-type sliding, but for a different reason: the dependence on adhesion becomes smaller. *Lower line:* interfacial sliding with a monolayer of water between the mica surfaces (load-controlled friction, cf. Fig. 30.20a), shown for comparison. (After [30.308]. Copyright 1990, with permission from Elsevier Science)

Fig. 30.17 (a) Contact radius r versus externally applied load L for loading and unloading of two hydrophilic silica surfaces exposed to dry and humid atmospheres. Note that, while the adhesion is higher in humid air, the *hysteresis* in the adhesion is higher in dry air. **(b)** Effect of velocity on the static friction force F_s for hydrophobic (heat-treated electron-beam-evaporated) silica in dry and humid air. The effects of humidity, load, and sliding velocity on the friction forces, as well as the stick–slip friction of the hydrophobic surfaces, are qualitatively consistent with a "friction" phase diagram representation as in Fig. 30.28. (After [30.41]. Copyright 1994, with permission from Elsevier Science)

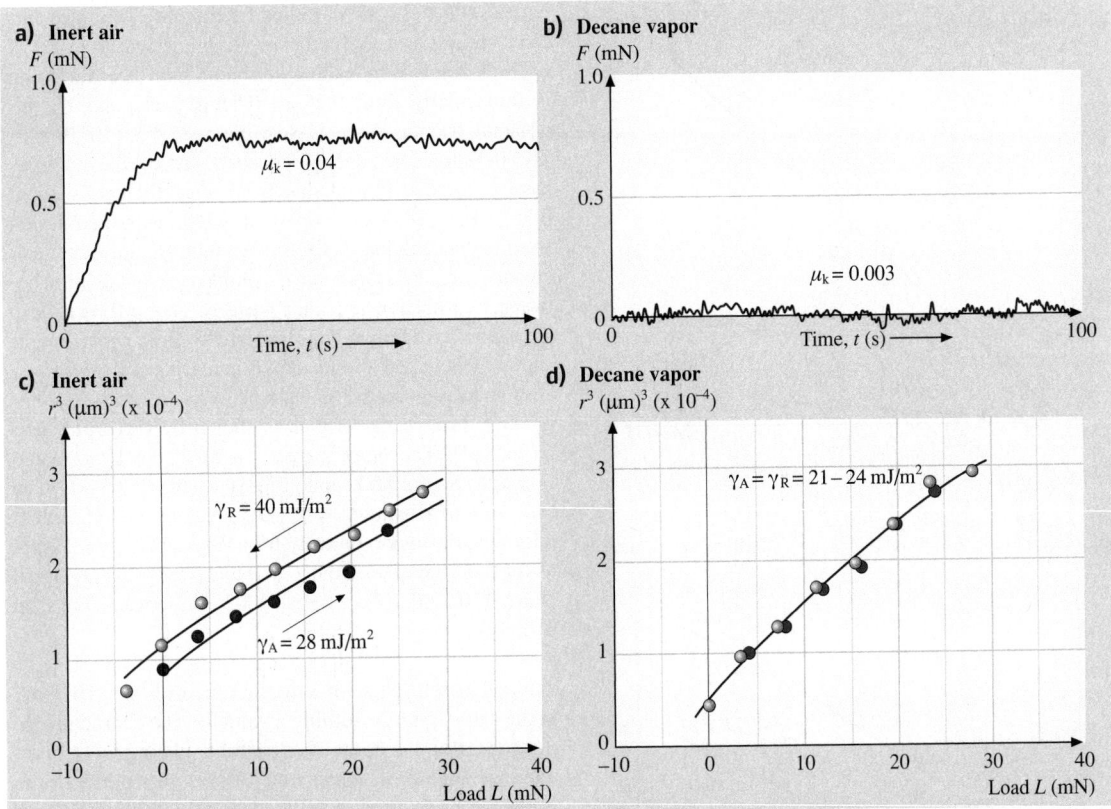

Fig. 30.19a–d *Top*: friction traces for two fluid-like calcium alkylbenzene sulfonate monolayer-coated surfaces at 25 °C showing that the friction force is much higher between dry monolayers (**a**) than between monolayers whose fluidity has been enhanced by hydrocarbon penetration from vapor (**b**). *Bottom*: Contact radius vs. load (r^3 vs. L) data measured for the same two surfaces as above and fitted to the JKR equation (30.22), shown by the *solid curves*. For dry monolayers (**c**) the adhesion energy on unloading ($\gamma_R = 40 \text{ mJ m}^{-2}$) is greater than that on loading ($\gamma_R = 28 \text{ mJ m}^{-2}$), which is indicative of an adhesion energy hysteresis of $\Delta\gamma = \gamma_R - \gamma_A = 12 \text{ mJ m}^{-2}$. For monolayers exposed to saturated decane vapor (**d**) their adhesion hysteresis is zero ($\gamma_A = \gamma_R$), and both the loading and unloading data are well fitted by the thermodynamic value of the surface energy of fluid hydrocarbon chains, $\gamma = 24 \text{ mJ m}^{-2}$. (After [30.261], with permission. Copyright 1993 American Chemical Society)

coated boundary lubricant layers), which were compared to those of hydrocarbon surfaces. They concluded that well-ordered fluorocarbon surfaces have high friction, in spite of their lower adhesion energy (in agreement with previous findings). The low friction coefficient of Teflon (polytetrafluoroethylene, PTFE) must, therefore, be due to some effect other than low adhesion. For example, the softness of PTFE, which allows material to flow at the interface, which thus behaves like a fluid lubricant. On a related issue, *Luengo* et al. [30.310] found that C_{60} surfaces also exhibited low adhesion but high friction. In both cases the high friction appears to arise from the bulky surface groups – fluorocarbon compared to hydro-

carbon groups in the former, large fullerene spheres in the latter. Apparently, the fact that C_{60} molecules rotate *in their lattice* does not make them a good lubricant: the molecules of the opposing surface must still climb over them in order to slide, and this requires energy that is independent of whether the surface molecules are fixed or freely rotating. Larger particles such as $\sim 25 \text{ nm}$ sized nanoparticles (also known as "inorganic fullerenes") do appear to produce low friction by behaving like molecular ball bearings, but the potential of this promising new class of solid lubricant has still to be explored [30.311].

Figure 30.19 illustrates the relationship between adhesion hysteresis and friction for surfactant-coated

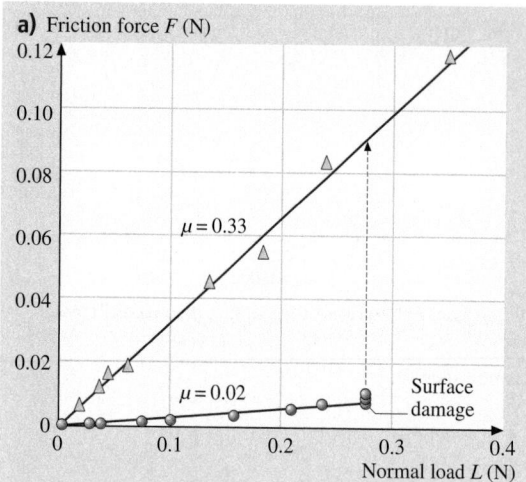

a) Friction force F (N)

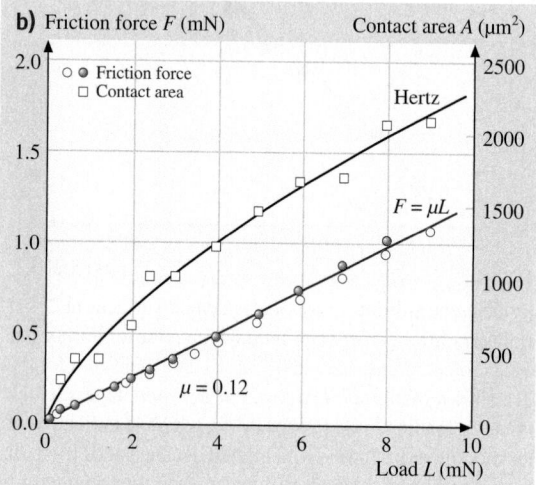

b) Friction force F (mN)   Contact area A (µm²)

Fig. 30.20a,b Load-controlled friction. (a) Two mica surfaces sliding past each other while immersed in a 0.01 M KCl salt solution (nonadhesive conditions). The water film is molecularly thin, 0.25 to 0.5 nm thick, and the interfacial friction force is very low: $S_c \approx 5 \times 10^5 \, \text{N m}^{-2}$, $\mu \approx 0.02$ (before damage occurs). After the surfaces have become damaged, the friction coefficient is about 0.3. (After [30.308]. Copyright 1990, with permission from Elsevier Science) (b) Steady-state friction force and contact area measured on a confined squalane film between two undamaged mica surfaces at $v = 0.6 \, \mu\text{m/s}$ in the smooth sliding regime (no stick–slip). *Open circles* show F obtained on loading (increasing L), *solid circles* show unloading. Both data sets are *straight lines* passing through the origin, as shown by the *brown line* ($\mu = 0.12$). The *black curve* is a fit of the Hertz equation (cf. Sect. 30.5.2 and [30.3]) to the A vs. L data (*open squares*) using $K = 10^{10} \, \text{N/m}^2$, $R = 2$ cm. The thickness D varies monotonically from $D = 2.5$ to $D = 1.7$ nm as the load increases from $L = 0$ to $L = 10$ mN. (Adapted from [30.285]. Copyright 2003 American Physical Society)

$\Delta \gamma = 10 \, \text{mJ m}^{-2}$, with a contact area of $A \approx 10^{-8} \, \text{m}^2$ at the same load. Assuming a value for the characteristic distance σ on the order of one lattice spacing, $\sigma \approx 1$ nm, and inserting these values into (30.32), the friction force is predicted to be $F \approx 100$ mN for the kinetic friction force, which is close to the measured value of 75 mN. Alternatively, we may conclude that the dissipation factor is $\varepsilon = 0.75$, i. e., that almost all the energy is dissipated as heat at each molecular collision.

A liquid lubricant film (Sect. 30.8.3) is usually much more effective at lowering the friction of two surfaces than a solid boundary lubricant layer. However, to use a liquid lubricant successfully, it must "wet" the surfaces, that is, it should have a high affinity for the surfaces, so that not all the liquid molecules become squeezed out when the surfaces come close together, even under a large compressive load. Another important requirement is that the liquid film remains a liquid under tribological conditions, i. e., that it does not epitaxially solidify between the surfaces.

Effective lubrication usually requires that the lubricant be injected between the surfaces, but in some cases the liquid can be made to condense from the vapor. This is illustrated in Fig. 30.20a for two untreated mica surfaces sliding with a thin layer of water between them. A monomolecular film of water (of thickness 0.25 nm per surface) has reduced S_c from its value for dry surfaces (Fig. 30.16) by a factor of more than 30, which

surfaces under different conditions. This effect, however, is much more general and has been shown to hold for other surfaces as well [30.41, 262, 292, 312].

Direct comparisons between absolute adhesion energies and friction forces show little correlation. In some cases, higher adhesion energies for the same system under different conditions correspond to lower friction forces. For example, for hydrophilic silica surfaces (Fig. 30.17) it was found that with increasing relative humidity the adhesion energy *increases*, but the adhesion energy hysteresis measured in a loading–unloading cycle *decreases*, as does the friction force [30.41]. For hydrophobic silica surfaces under dry conditions, the friction at load $L = 5.5$ mN was $F = 75$ mN. For the same sample, the adhesion energy hysteresis was

may be compared with the factor of ten attained with the boundary lubricant layer (of thickness 2.5 nm per surface) in Fig. 30.18. Water appears to have unusual lubricating properties and usually gives wearless friction with no stick–slip [30.313].

The effectiveness of a water film only 0.25 nm thick to lower the friction force by more than an order of magnitude is attributed to the "hydrophilicity" of the mica surface (mica is "wetted" by water) and to the existence of a strongly repulsive short-range hydration force between such surfaces in aqueous solutions, which effectively removes the adhesion-controlled contribution to the friction force [30.283]. It is also interesting that a 0.25 nm thick water film between two mica surfaces is sufficient to bring the coefficient of friction down to 0.01–0.02, a value that corresponds to the unusually low friction of ice. Clearly, a single monolayer of water can be a very good lubricant – much better than most other monomolecular liquid films – for reasons that will be discussed in Sect. 30.9. A linear dependence of F on L has also been observed for mica surfaces separated by certain hydrocarbon liquids [30.275, 285]. Figure 30.20b shows the kinetic friction forces measured at a high velocity across thin films of squalane, a branched hydrocarbon liquid ($C_{30}H_{62}$), which is a model for lubricating oils. Very low adhesive forces are measured between mica surfaces across this liquid [30.285] and the film thickness decreased monotonically with load. The friction force at a given load was found to be velocity-dependent, whereas the contact area was not [30.285].

Dry polymer layers (Fig. 30.21) typically show a high initial static friction ("stiction") as sliding commences from rest in adhesive contact. The development of the friction force after a change in sliding direction, a gradual transition from stick–slip to smooth sliding, is shown in Fig. 30.21. A correlation between adhesion hysteresis and friction similar to that observed for silica surfaces in Fig. 30.17 can be seen for dry polymer layers below their glass-transition temperature. As shown in Fig. 30.12b,c, the adhesion hysteresis for polystyrene surfaces can be increased by irradiation to induce scission of chains, and it has been found that the steady-state friction force (kinetic friction) shows a similar increase with irradiation time [30.262].

Figure 30.22 shows an example of a computer simulation of the sliding of two unlubricated silicon surfaces (modeled as a tip sliding over a planar surface) [30.112]. The sliding proceeds through a series of stick–slip events, and information on the friction force and the local order of the initially crystalline surfaces

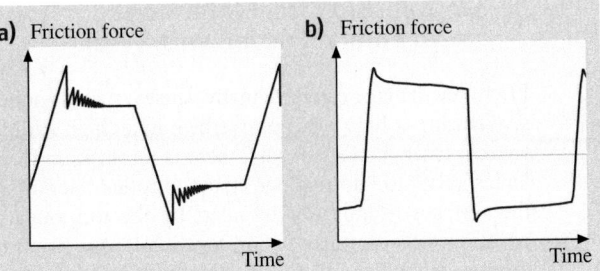

Fig. 30.21a,b Typical friction traces showing how the friction force varies with the sliding time for two symmetric, glassy polymer films under dry conditions. Qualitative features that are common to both polystyrene and polyvinyl benzyl chloride: (**a**) Decaying stick–slip motion is observed until smooth sliding is attained if the motion continues for a sufficiently long distance. (**b**) Smooth sliding observed at sufficiently high speeds. Similar observations have been made by *Berthoud* et al. [30.314] in measurements on polymethyl methacrylate. (After [30.262], with permission. Copyright 2002 American Association for the Advancement of Science)

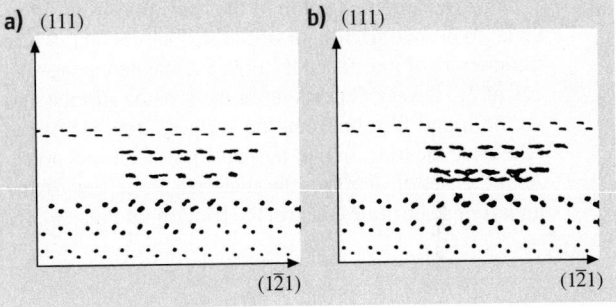

Fig. 30.22a,b Computer simulation of the sliding of two contacting Si surfaces (a tip and a flat surface). Shown are particle trajectories in a constant-force simulation, $F_{z,external} = -2.15 \times 10^{-8}$ N, viewed along the $(10\bar{1})$ direction just before (**a**) and after (**b**) a stick–slip event for a large, initally ordered, dynamic tip. (After [30.112] with permission of Kluwer Academic Publishers)

can be obtained. Similar studies for cold-welding systems [30.112] have demonstrated the occurrence of shear or friction damage within the sliding surface (tip) as the lowest layer of it adheres to the bottom surface. Recent computer simulations have addressed many of the phenomena seen experimentally, including the differences between adhesive and non-adhesive systems, the issue of the dependence of the observed friction on the real contact area (a parameter that is difficult to define or measure at the nanoscale), and the molecular origin of friction responses that follow Amontons' law [30.287, 302, 315–317].

30.7.4 Transition from Interfacial to Normal Friction with Wear

Frictional damage can have many causes, such as adhesive tearing at high loads or overheating at high sliding speeds. Once damage occurs, there is a transition from "interfacial" to "normal" or load-controlled friction as the surfaces become forced apart by the torn-out asperities (wear particles). For low loads, the friction changes from obeying $F = S_c A$ to obeying Amontons' law, $F = \mu L$, as shown in Figs. 30.16 and 30.18, and sliding now proceeds smoothly with the surfaces separated by a 10–100 nm forest of wear debris (in this case, mica flakes). The wear particles keep the surfaces apart over an area that is much greater than their size, so that even one submicroscopic particle or asperity can cause a significant reduction in the area of contact and, therefore, in the friction [30.308]. For this type of frictional sliding, one can no longer talk of the molecular contact area of the two surfaces, although the macroscopic or "apparent" area is still a useful parameter.

One remarkable feature of the transition from interfacial to normal friction of brittle surfaces is that, while the strength of interfacial friction, as reflected in the values of S_c, is very dependent on the type of surface and on the liquid film between the surfaces, this is not the case once the transition to normal friction has occurred. At the onset of damage, the material properties of the underlying substrates control the friction. In Figs. 30.16, 30.18, and 30.20a the friction for the damaged surfaces is that of any damaged mica–mica system, $\mu \approx 0.3$, *independent of the initial surface coatings or liquid films between the surfaces*. A similar friction coefficient was found for damaged silica surfaces [30.41].

In order to modify the frictional behavior of such brittle materials practically, it is important to use coatings that will both alter the interfacial tribological character and remain intact and protect the surfaces from damage during sliding. *Berman* et al. [30.318] found that the friction of a strongly bound octadecyl phosphonic acid monolayer on alumina surfaces was higher than for untreated, undamaged α-alumina surfaces, but the bare surfaces easily became damaged upon sliding, resulting in an ultimately higher friction system with greater wear rates than the more robust monolayer-coated surfaces.

Clearly, the mechanism and factors that determine *normal* friction are quite different from those that govern *interfacial* friction (Sects. 30.7.1–30.7.2). This effect is not general and may only apply to brittle materials. For example, the friction of ductile surfaces is totally different and involves the continuous plastic deformation of contacting surface asperities during sliding, rather than the rolling of two surfaces on hard wear particles [30.263]. Furthermore, in the case of ductile surfaces, water and other surface-active components do have an effect on the friction coefficients under "normal" sliding conditions.

30.8 Liquid Lubricated Surfaces

30.8.1 Viscous Forces and Friction of Thick Films: Continuum Regime

Experimentally, it is usually difficult to unambiguously establish which type of sliding mode is occurring, but an empirical criterion, based on the Stribeck curve (Fig. 30.13), is often used as an indicator. This curve shows how the friction force or the coefficient of friction is expected to vary with sliding speed, depending on which type of friction regime is operating. For thick liquid lubricant films whose behavior can be described by bulk (continuum) properties, the friction forces are essentially the hydrodynamic or viscous drag forces. For example, for two plane parallel surfaces of area A separated by a distance D and moving laterally relative to each other with velocity v, if the intervening liquid is *Newtonian*, i.e., if its viscosity η is independent of the shear rate, the frictional force experienced by the surfaces is given by

$$F = \frac{\eta A v}{D} \,, \tag{30.44}$$

where the shear rate $\dot{\gamma}$ is defined by

$$\dot{\gamma} = \frac{v}{D} \,. \tag{30.45}$$

At higher shear rates, two additional effects often come into play. First, certain properties of liquids may change at high $\dot{\gamma}$ values. In particular, the effective viscosity may become non-Newtonian, one form given by

$$\eta \propto \dot{\gamma}^n \,, \tag{30.46}$$

where $n = 0$ (i. e., $\eta_{\mathrm{eff}} = \mathrm{constant}$) for Newtonian fluids, $n > 0$ for shear-thickening (dilatant) fluids, and $n < 0$

for shear-thinning (pseudoplastic) fluids (the latter become less viscous, i. e., flow more easily, with increasing shear rate). An additional effect on η can arise from the higher local stresses (pressures) experienced by the liquid film as $\dot{\gamma}$ increases. Since the viscosity is generally also sensitive to the pressure (usually increasing with P), this effect also acts to increase η_{eff} and thus the friction force.

A second effect that occurs at high shear rates is surface deformation, arising from the large hydrodynamic forces acting on the sliding surfaces. For example, Fig. 30.23 shows how two surfaces deform elastically when the sliding speed increases to a high value. These deformations alter the hydrodynamic friction forces, and this type of friction is often referred to as *elasto-hydrodynamic lubrication* (EHD or EHL), as mentioned in Table 30.4.

How thin can a liquid film be before its dynamic, e.g., viscous flow, behavior ceases to be described by bulk properties and continuum models? Concerning the static properties, we have already seen in Sect. 30.4.3 that films composed of simple liquids display continuum behavior down to thicknesses of 4–10 molecular diameters. Similar effects have been found to apply to the dynamic properties, such as the viscosity, of simple liquids in thin films. Concerning viscosity measurements, a number of dynamic techniques were recently developed [30.11–13,43,51,319,320] for directly measuring the viscosity as a function of film thickness and shear rate across very thin liquid films between two surfaces. By comparing the results with theoretical predictions of fluid flow in thin films, one can determine the effective positions of the shear planes and the onset of non-Newtonian behavior in very thin films.

The results show that, for simple liquids including linear chain molecules such as alkanes, the viscosity in thin films is the same, within 10%, as the bulk even for films as thin as 10 molecular diameters (or segment widths) [30.11–13,319,320]. This implies that the shear plane is effectively located within one molecular diameter of the solid–liquid interface, and these conclusions were found to remain valid even at the highest shear rates studied (of $\sim 2 \times 10^5 \, \text{s}^{-1}$). With water between two mica or silica surfaces [30.22, 313, 319–321] this has been found to be the case (to within $\pm 10\%$) down to surface separations as small as 2 nm, implying that the shear planes must also be within a few tenths of a nanometer of the solid–liquid interfaces. These results appear to be independent of the existence of electrostatic "double-layer" or "hydration" forces. For the case of the simple liquid toluene confined between surfaces with

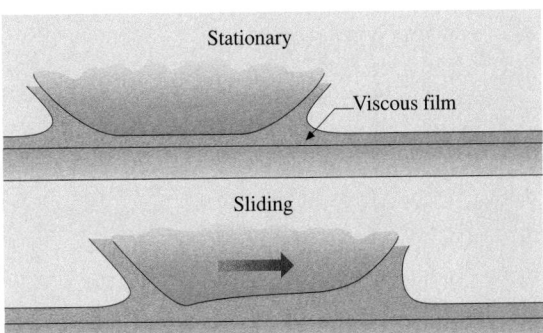

Fig. 30.23 *Top:* Stationary surfaces (one more deformable and one rigid) separated by a thick liquid film. *Bottom:* Elasto-hydrodynamic deformation of the upper surface during sliding. (After [30.1], with permission)

adsorbed layers of C_{60} molecules, this type of viscosity measurement has shown that the traditional no-slip assumption for flow at a solid interface does not always hold [30.322]. The C_{60} layer at the mica–toluene interface results in a "full-slip" boundary, which dramatically lowers the viscous drag or effective viscosity for regular Couette or Poiseuille flow.

With polymeric liquids (polymer melts) such as polydimethylsiloxanes (PDMS) and polybutadienes (PBD), or with polystyrene (PS) adsorbed onto surfaces from solution, the far-field viscosity is again equal to the bulk value, but with the non-slip plane (hydrodynamic layer thickness) being located at $D = 1-2R_g$ away from each surface [30.11, 47], or at $D = L$ or less for polymer brush layers of thickness L per surface [30.13,323]. In contrast, the same technique was used to show that, for non-adsorbing polymers in solution, there is actually a depletion layer of nearly pure solvent that exists at the surfaces that affects the confined solution flow properties [30.321]. These effects are observed from near contact to surface separations in excess of 200 nm.

Further experiments with surfaces closer than a few molecular diameters ($D < 2-4$ nm for simple liquids, or $D < 2-4R_g$ for polymer fluids) indicate that large deviations occur for thinner films, described below. One important conclusion from these studies is, therefore, that the dynamic properties of simple liquids, including water, near an *isolated* surface are similar to those of the bulk liquid *already within the first layer of molecules adjacent to the surface*, only changing when another surface approaches the first. In other words, the viscosity and position of the shear plane near a surface are not simply a property of that surface, but of how far that surface is from another surface. The reason for this

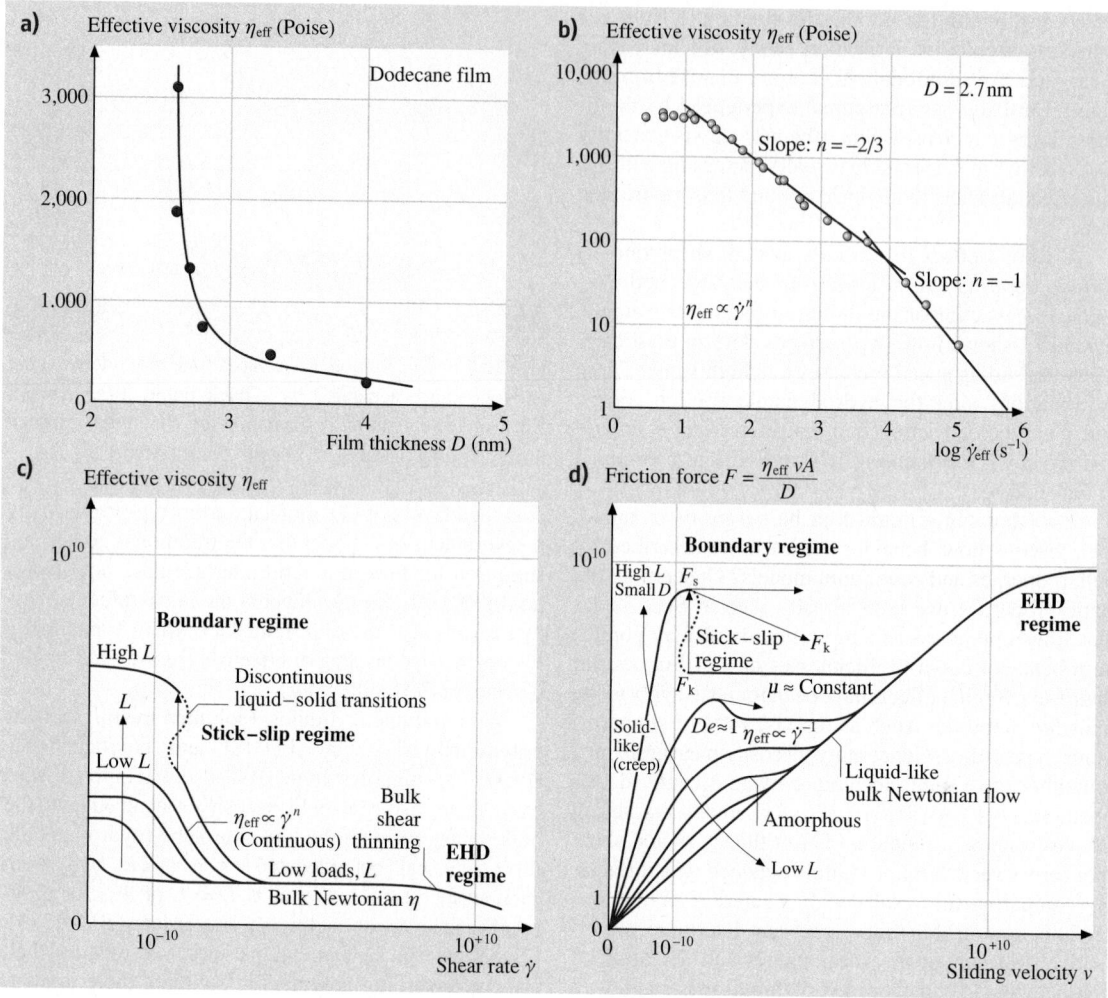

is that, when two surfaces are close together, the constraining effects on the liquid molecules between them are much more severe than when there is only one surface. Another obvious consequence of the above is that one should not make measurements on a single, isolated solid–liquid interface and then draw conclusions about the state of the liquid or its interactions in a thin film *between* two surfaces.

30.8.2 Friction of Intermediate Thickness Films

For liquid films in the thickness range between 4 and 10 molecular diameters, the properties can be significantly different from those of bulk films. Still, the fluids remain recognizable as fluids; in other words, they do not undergo a phase transition into a solid or liquid-crystalline phase. This regime has recently been studied by *Granick* et al. [30.44, 273, 274, 276, 277], who used a friction attachment [30.43, 44] to the SFA where a sinusoidal input signal to a piezoelectric device makes the two surfaces slide back and forth laterally past each other at small amplitudes. This method provides information on the real and imaginary parts (elastic and dissipative components, respectively) of the shear modulus of thin films at different shear rates and film thickness. *Granick* [30.273] and *Hu* et al. [30.277] found that films of simple liquids become non-Newtonian in the 2.5–5 nm regime (about 10 molecular diameters, see Fig. 30.24). Polymer melts become non-Newtonian at much larger film thicknesses, depending on their molecular weight [30.47].

Part D | 30.8

◀ **Fig. 30.24a–d** Typical rheological behavior of liquid films in the mixed lubrication regime. (a) Increase in effective viscosity of dodecane film between two mica surfaces with decreasing film thickness. At distances larger than 4–5 nm, the effective viscosity η_{eff} approaches the bulk value η_{bulk} and does not depend on the shear rate $\dot{\gamma}$. (After [30.273]. Copyright 1991 American Association for the Advancement of Science.) (b) Non-Newtonian variation of η_{eff} with shear rate of a 2.7-nm-thick dodecane film at a net normal pressure of 0.12 MPa and at 28 °C. The effective viscosity decays as a power law, as in (30.46). In this example, $n = 0$ at the lowest $\dot{\gamma}$ and changes to $n = -2/3$ and -1 at higher $\dot{\gamma}$. For films of bulk thickness, dodecane is a low-viscosity Newtonian fluid ($n = 0$). (c) Proposed general friction map of effective viscosity η_{eff} (arbitrary units) as a function of effective shear rate $\dot{\gamma}$ (arbitrary units) on logarithmic scales. Three main classes of behavior emerge: (i) Thick films: elasto-hydrodynamic sliding. At $L = 0$, approximating bulk conditions, η_{eff} is independent of shear rate except when shear thinning might occur at sufficiently large $\dot{\gamma}$. (ii) Boundary layer films, intermediate regime. A Newtonian regime is again observed [$\eta_{eff} = $ constant, $n = 0$ in (30.46)] at low loads and low shear rates, but η_{eff} is much higher than the bulk value. As the shear rate $\dot{\gamma}$ increases beyond $\dot{\gamma}_{min}$, the effective viscosity starts to drop with a power-law dependence on the shear rate [see panel (b)], with n in the range $-1/2$ to -1 most commonly observed. As the shear rate $\dot{\gamma}$ increases still more, beyond $\dot{\gamma}_{max}$, a second Newtonian plateau is encountered. (iii) Boundary layer films, high load. The η_{eff} continues to grow with load and to be Newtonian provided that the shear rate is sufficiently low. Transition to sliding at high velocity is discontinuous ($n < -1$) and usually of the stick–slip variety. (d) Proposed friction map of friction force as a function of sliding velocity in various tribological regimes. With increasing load, Newtonian flow in the elasto-hydrodynamic (EHD) regimes crosses into the boundary regime of lubrication. Note that even EHD lubrication changes, at the highest velocities, to limiting shear stress response. At the highest loads (L) and smallest film thickness (D), the friction force goes through a maximum (the static friction, F_s), followed by a regime where the friction coefficient (μ) is roughly constant with increasing velocity (meaning that the kinetic friction, F_k, is roughly constant). Non-Newtonian shear thinning is observed at somewhat smaller load and larger film thickness; the friction force passes through a maximum at the point where $De = 1$. De, the Deborah number, is the point at which the applied shear rate exceeds the natural relaxation time of the boundary layer film. The velocity axis from 10^{-10} to 10^{10} (arbitrary units) indicates a large span. (Panels (b)–(d) after [30.325]. Copyright 1996, with permission from Elsevier Science)

Klein and *Kumacheva* [30.46, 280, 324] studied the interaction forces and friction of small quasi-spherical liquid molecules such as cyclohexane between molecularly smooth mica surfaces. They concluded that surface epitaxial effects can cause the liquid film to solidify already at six molecular diameters, resulting in a sudden (discontinuous) jump to high friction at low shear rates. Such dynamic first-order transitions, however, may depend on the shear rate.

A generalized friction map (Fig. 30.24c,d) has been proposed by *Luengo* et al. [30.325] that illustrates the changes in η_{eff} from bulk Newtonian behavior ($n = 0$, $\eta_{eff} = \eta_{bulk}$) through the transition regime where n reaches a minimum of -1 with decreasing shear rate to the solid-like creep regime at very low $\dot{\gamma}$ where n returns to 0. A number of results from experimental, theoretical, and computer simulation work have shown values of n from $-1/2$ to -1 for this transition regime for a variety of systems and assumptions [30.273, 274, 299, 326–332].

The intermediate regime appears to extend over a narrow range of film thickness, from about 4 to 10 molecular diameters or polymer radii of gyration. Thinner films begin to adopt boundary or interfacial friction

properties (described below, see also Table 30.5). Note that the intermediate regime is actually a very narrow one when defined in terms of film thickness, for example, varying from about $D = 2$ to 4 nm for hexadecane films [30.273].

A fluid's effective viscosity η_{eff} in the intermediate regime is usually higher than in the bulk, but η_{eff} usually *decreases* with increasing sliding velocity, v (known as *shear thinning*). When two surfaces slide in the intermediate regime, the motion tends to thicken the film (dilatancy). This sends the system into the bulk EHD regime where, as indicated by (30.44), the friction force now *increases* with velocity. This initial decrease, followed by an increase, in the frictional forces of many lubricant systems is the basis for the empirical Stribeck curve of Fig. 30.13. In the transition from bulk to boundary behavior there is first a quantitative change in the material properties (viscosity and elasticity), which can be continuous, to discontinuous qualitative changes that result in yield stresses and non-liquidlike behavior.

The rest of this section is devoted to friction in the boundary regime. Boundary friction may be thought of as applying to the case where a lubricant film is present, but where this film is of molecular dimensions – a few molecular layers or less.

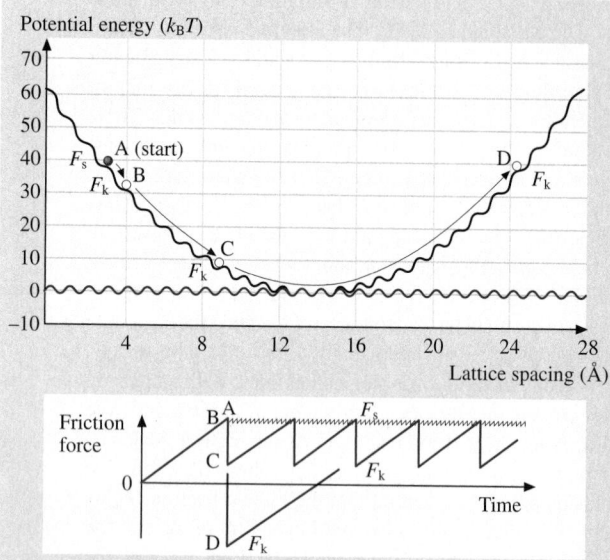

Fig. 30.25 Simple schematic illustration of the most common molecular mechanism leading from smooth to stick–slip sliding in terms of the efficiency of the energy transfer from mechanical to kinetic to phonons. The potential energy of the corrugated surface lattice is shown by the horizontal sine wave. Let the depth of each minimum be ε which is typically $> k_B T$. At equilibrium, a molecule will 'sit' at one of these minima. When the molecule is connected to a horizontal spring, a smooth parabolic curve must be added to the horizontal curve. If this spring is now pushed or pulled laterally at a constant velocity v, the sine curve will move like a wave along the parabola carrying the molecule up with it (towards point A). When the point of inflection at A is reached the molecule will drop and acquire a kinetic energy greater than ε even before it reaches the next lattice site. This energy can be "released" at the next lattice site (i.e., on the first collision), in which case the processes will now be repeated – each time the molecule reaches point A it will fall to point B. This type of motion will give rise to periodic changes in temperature at the interface, as predicted by computer simulations [30.333]. The stick–slip here will have a magnitude of the lattice dimension and, except for AFM measurements that can detect such small atomic-scale jumps [30.59, 334], the measured macro- and microscopic friction forces will be smooth and independent of v. If the energy dissipation (or "transfer") mechanism is less than 100% efficient on each collision, the molecule will move further before it stops. In this case the stick–slip amplitude can be large (point C), and the kinetic friction F_k can even be negative in the case of an overshoot (point D). (After [30.287], with permission. Copyright 2004 American Chemical Society)

30.8.3 Boundary Lubrication of Molecularly Thin Films: Nanorheology

When a liquid is confined between two surfaces or within any narrow space whose dimensions are less than 4 to 10 molecular diameters, both the static (equilibrium) and dynamic properties of the liquid, such as its compressibility and viscosity, can no longer be described even qualitatively in terms of the bulk properties. The molecules confined within such molecularly thin films become ordered into layers ("out-of-plane" ordering), and within each layer they can also have lateral order ("in-plane" ordering). Such films may be thought of as behaving more like a liquid crystal or a solid than a liquid.

As described in Sect. 30.4.3, the measured *normal* forces between two solid surfaces across molecularly thin films exhibit exponentially decaying oscillations, varying between attraction and repulsion with a periodicity equal to some molecular dimension of the solvent molecules. Thus most liquid films can sustain a finite normal stress, and the adhesion force between two surfaces across such films is "quantized", depending on the thickness (or number of liquid layers) between the surfaces. The structuring of molecules in thin films and the oscillatory forces it gives rise to are now reasonably well understood, both experimentally and theoretically, at least for simple liquids.

Work has also recently been done on the dynamic, e.g., viscous or shear, forces associated with molecularly thin films. Both experiments [30.38, 46, 257, 275, 280, 281, 335, 336] and theory [30.254, 255, 326, 337] indicate that, even when two surfaces are in steady-state sliding, they still prefer to remain in one of their stable potential-energy minima, i.e., a sheared film of liquid can retain its basic layered structure. Thus even during motion the film does not become totally liquid-like. Indeed, if there is some "in-plane" ordering within a film, it will exhibit a yield point before it begins to flow. Such films can therefore sustain a finite shear stress, in addition to a finite normal stress. The value of the yield stress depends on the number of layers comprising the film and represents another "quantized" property of molecularly thin films [30.254].

The dynamic properties of a liquid film undergoing shear are very complex. Depending on whether the film is more liquid-like or solid-like, the motion will be smooth or of the stick–slip type illustrated schematically in Fig. 30.25. During sliding, transitions can occur between n layers and $(n-1)$ or $(n+1)$ layers (see

Fig. 30.27). The details of the motion depend critically on the externally applied load, the temperature, the sliding velocity, the twist angle between the two surface lattices, and the sliding direction relative to the lattices.

Smooth and Stick–Slip Sliding

Recent advances in friction-measuring techniques have enabled the interfacial friction of molecularly thin films to be measured with great accuracy. Some of these advances have involved the surface forces apparatus technique [30.38,44–47,274,275,280,281,285,286,296, 297,308,313,335,336,338] while others have involved the atomic force microscope [30.10, 58, 59, 284, 290, 339, 340]. In addition, computer simulations [30.111, 151,254,255,287,295,299–302,315–317,333,337,341] have become sufficiently sophisticated to enable fairly complex tribological systems to be studied. All these advances are necessary if one is to probe such subtle effects as smooth or stick–slip friction, transient and memory effects, and ultralow friction mechanisms at the molecular level.

The theoretical models presented in this section will be concerned with a situation commonly observed experimentally: stick–slip occurs between a static state with high friction and a low-friction kinetic state, and a transition from this sliding regime to smooth sliding can be induced by an increase in velocity. Experimental data on various systems showing this behavior are shown in Figs. 30.27, 30.30b, and 30.31a. Recent studies on adhesive systems have revealed the possibility of other dynamic responses such as inverted stick–slip between two kinetic states of higher and lower friction and with a transition from smooth sliding to stick–slip with increasing velocity, as shown in Fig. 30.30c [30.342]. Similar friction responses have also been seen in computer simulations [30.343].

With the added insights provided by computer simulations, a number of distinct molecular processes have been identified during smooth and stick–slip sliding in model systems for the more familiar static-to-kinetic stick–slip and transition from stick–slip to smooth sliding. These are shown schematically in Fig. 30.26 for the case of spherical liquid molecules between two solid crystalline surfaces. The following regimes may be identified:

Surfaces at rest (Fig. 30.26a): even with no externally applied load, solvent–surface epitaxial interactions can cause the liquid molecules in the film to attain a solid-like state. Thus at rest the surfaces are stuck to each other through the film.

Sticking regime (frozen, solid-like film) (Fig. 30.26b): a progressively increasing lateral shear stress is applied. The solid-like film responds elastically with a small lateral displacement and a small increase or dilatancy in film thickness (less than a lattice spacing or molecular dimension, σ). In this regime the film retains its frozen, solid-like state: all the strains are elastic and reversible, and the surfaces remain effectively stuck to each other. However, slow creep may occur over long time periods.

Slipping and sliding regimes (molten, liquid-like film) (Fig. 30.26c,c′,c″): when the applied shear stress or force has reached a certain critical value, the *static* friction force, F_s, the film suddenly melts (known as "shear melting") or rearranges to allow for wall slip or slip within the film to occur at which point the two surfaces begin to slip rapidly past each other. If the applied stress is kept at a high value, the upper surface will continue to slide indefinitely.

Refreezing regime (resolidification of film) (Fig. 30.26d): In many practical cases, the rapid slip of the upper surface relieves some of the applied force, which

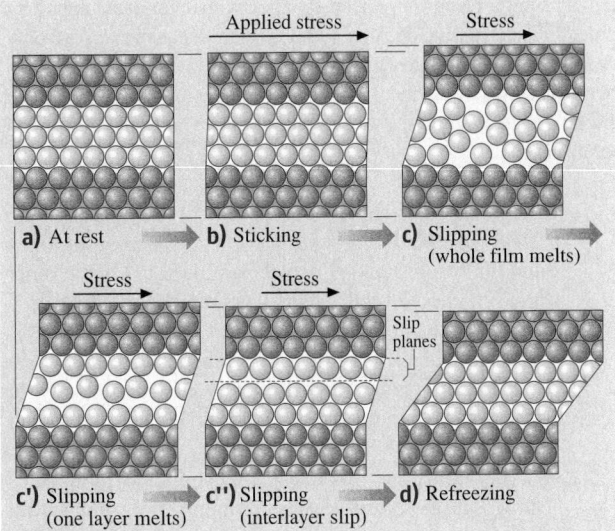

Fig. 30.26a–d Idealized schematic illustration of molecular rearrangements occurring in a molecularly thin film of spherical or simple chain molecules between two solid surfaces during shear. Depending on the system, a number of different molecular configurations within the film are possible during slipping and sliding, shown here as stages (**c**): total disorder as the whole film melts; (**c′**): partial disorder; and (**c″**): order persists even during sliding with slip occurring at a single slip plane either within the film or at the walls. A dilation is predicted in the direction normal to the surfaces. (After [30.278], with permission)

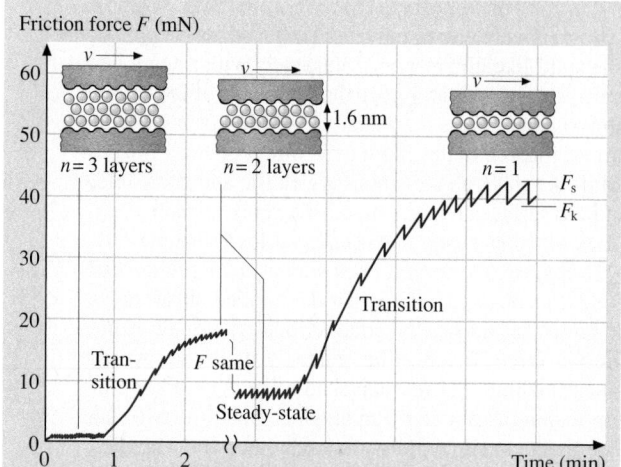

Fig. 30.27 Measured change in friction during interlayer transitions of the silicone liquid octamethylcyclotetrasiloxane (OMCTS), an inert liquid whose quasi-spherical molecules have a diameter of 0.8 nm. In this system, the shear stress $S_c = F/A$ was found to be constant as long as the number of layers, n, remained constant. Qualitatively similar results have been obtained with other quasi-spherical molecules such as cyclohexane [30.335]. The shear stresses are only weakly dependent on the sliding velocity v. However, for sliding velocities above some critical value, v_c, the stick–slip disappears and sliding proceeds smoothly at the kinetic value. (After [30.275], with permission)

eventually falls below another critical value, the *kinetic* friction force F_k, at which point the film resolidifies and the whole stick–slip cycle is repeated. On the other hand, if the slip rate is smaller than the rate at which the external stress is applied, the surfaces will continue to slide smoothly in the kinetic state and there will be no more stick–slip. The critical velocity at which stick–slip disappears is discussed in more detail in Sect. 30.8.3.

Experiments with linear chain (alkane) molecules show that the film thickness remains quantized during sliding, so that the structure of such films is probably more like that of a nematic liquid crystal where the liquid molecules have become shear-aligned in some direction, enabling shear motion to occur while retaining some order within the film [30.344]. Experiments on the friction of two molecularly smooth mica surfaces separated by three molecular layers of the liquid octamethylcyclotetrasiloxane (OMCTS, see Fig. 30.27) show how the friction increases to higher values in a quantized way when the number of layers falls from $n = 3$ to $n = 2$ and then to $n = 1$.

Computer simulations for simple spherical molecules [30.255] further indicate that during slip the film thickness is roughly 15% higher than at rest (i.e., the film density falls), and that the order parameter within the film drops from 0.85 to about 0.25. Such dilatancy has been investigated both experimentally [30.296] and in further computer simulations [30.295]. The changes in thickness and in the order parameter are consistent with a disorganized state for the whole film during the slip [30.337], as illustrated schematically in Fig. 30.26c. At this stage, we can only speculate on other possible configurations of molecules in the slipping and sliding regimes. This probably depends on the shapes of the molecules (e.g., whether they are spherical or linear or branched), on the atomic structure of the surfaces, on the sliding velocity, etc. [30.345]. Figure 30.26c,c′,c″ shows three possible sliding modes wherein the shearing film either totally melts, or where the molecules retain their layered structure and where slip occurs between two or more layers. Other sliding modes, for example, involving the movement of dislocations or disclinations are also possible, and it is unlikely that one single mechanism applies in all cases.

Both friction and adhesion hysteresis vary nonlinearly with temperature, often peaking at some particular temperature, T_0. The temperature dependence of these forces can, therefore, be represented on a friction phase diagram such as the one shown in Fig. 30.28. Experi-

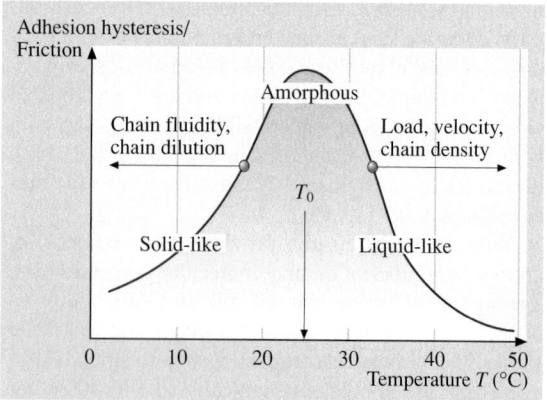

Fig. 30.28 Schematic friction phase diagram representing the trends observed in the boundary friction of a variety of different surfactant monolayers. The characteristic bell-shaped curve also correlates with the adhesion energy hysteresis of the monolayers. The *arrows* indicate the direction in which the whole curve is dragged when the load, velocity, etc., is increased. (After [30.292], with permission)

ments have shown that T_0, and the whole bell-shaped curve, are shifted along the temperature axis (as well as in the vertical direction) in a systematic way when the load, sliding velocity, etc., are varied. These shifts also appear to be highly correlated with one another, for example, an increase in temperature produces effects that are similar to *decreasing* the sliding speed or load.

Such effects are also commonly observed in other energy-dissipating phenomena such as polymer viscoelasticity [30.346], and it is likely that a similar physical mechanism is at the heart of all such phenomena. A possible molecular process underlying the energy dissipation of chain molecules during boundary-layer sliding is illustrated in Fig. 30.29, which shows the three main dynamic phase states of boundary monolayers.

In contrast to the characteristic relaxation time associated with fluid lubricants, it has been established that for unlubricated (dry, solid, rough) surfaces, there is a characteristic memory distance that must be exceeded before the system loses all memory of its initial state (original surface topography). The underlying mechanism for a characteristic distance was first used to successfully explain rock mechanics and earthquake faults [30.347] and, more recently, the tribological behavior of unlubricated surfaces of ceramics, paper and elastomeric polymers [30.314, 348]. Recent experiments [30.285, 344, 345, 349] suggest that fluid lubricants composed of complex branched-chained or polymer molecules may also have characteristic distances (in addition to characteristic relaxation times) associated with their tribological behavior – the characteristic distance being the total sliding distance that must be exceeded before the system reaches its steady-state tribological conditions (see Sect. 30.8.3).

Abrupt versus Continuous Transitions Between Smooth and Stick–Slip Sliding

An understanding of stick–slip is of great practical importance in tribology [30.350], since these spikes are the major cause of damage and wear of moving parts. Stick–slip motion is a very common phenomenon and is also the cause of sound generation (the sound of a violin string, a squeaking door, or the chatter of machinery), sensory perception (taste texture and feel), earthquakes, granular flow, nonuniform fluid flow such as the spurting flow of polymeric liquids, etc. In the previous section, the stick–slip motion arising from freezing–melting transitions in thin interfacial films was described. There are other mechanisms that can give rise to stick–slip friction, which will now be considered.

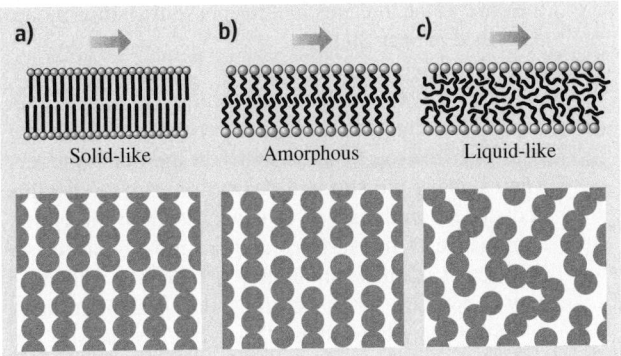

Fig. 30.29a–c Different dynamic phase states of boundary monolayers during adhesive contact and/or frictional sliding. Solid-like (**a**) and liquid-like monolayers (**c**) exhibit low adhesion hysteresis and friction. Increasing the temperature generally shifts a system from the left to the right. Changing the load, sliding velocity, or other experimental conditions can also change the dynamic phase state of surface layers, as shown in Fig. 30.28. (After [30.292], with permission)

However, before proceeding with this, it is important to clarify exactly what one is measuring during a friction experiment.

Most tribological systems and experiments can be described in terms of an equivalent mechanical circuit with certain characteristics. The friction force F_0, which is generated at the surfaces, is generally measured as F at some other place in the set-up. The mechanical coupling between the two may be described in terms of a simple elastic stiffness or compliance, K, or as more complex nonelastic coefficients, depending on the system. The distinction between F and F_0 is important because, in almost all practical cases, the applied, measured, or detected force, F, is *not* the same as the "real" or "intrinsic" friction force, F_0, generated at the surfaces. F and F_0 are coupled in a way that depends on the mechanical construction of the system, for example, the axle of a car wheel that connects it to the engine. This coupling can be modeled as an elastic spring of stiffness K and mass m. This is the simplest type of mechanical coupling and is also the same as in SFA- and AFM-type experiments. More complicated real systems can be reduced to a system of springs and dashpots, as described by *Peachey* et al. [30.351] and *Luengo* et al. [30.47].

We now consider four different models of stick–slip friction, where the mechanical couplings are assumed to be of the simple elastic spring type. The first three mechanisms may be considered traditional or classical mechanisms or models [30.350], the fourth is essentially the same as the freezing–melting phase-transition model described in Sect. 30.8.3.

Rough Surfaces or Surface Topology Model. Rapid slips can occur whenever an asperity on one surface goes over the top of an asperity on the other surface. The extent of the slip will depend on asperity heights and slopes, on the speed of sliding, and on the elastic compliance of the surfaces and the moving stage. As in all cases of stick–slip motion, the driving velocity v may be constant, but the resulting motion at the surfaces v_0 will display large slips. This type of stick–slip has been described by *Rabinowicz* [30.350]. It will not be of much concern here since it is essentially a noise-type fluctuation, resulting from surface imperfections rather than from the intrinsic interaction between two surfaces. Actually, at the atomic level, the regular atomic-scale corrugations of surfaces can lead to periodic stick–slip motion of the type shown here. This is what is sometimes measured by AFM tips [30.10, 58, 59, 290, 339, 340].

Distance-Dependent or Creep Model. Another theory of stick–slip, observed in solid-on-solid sliding, is one that involves a characteristic *distance*, but also a characteristic time, τ_s, this being the characteristic time required for two asperities to increase their adhesion strength after coming into contact. Originally proposed by *Rabinowicz* [30.350, 352], this model suggests that two rough macroscopic surfaces adhere through their microscopic asperities of a characteristic length. During shearing, each surface must first creep this distance (the size of the contacting junctions) after which the surfaces continue to slide, but with a lower (kinetic) friction force than the original (static) value. The reason for the decrease in the friction force is that even though, on average, new asperity junctions should form as rapidly as the old ones break, the time-dependent adhesion and friction of the new ones will be lower than the old ones.

The friction force, therefore, remains high during the creep stage of the slip. However, once the surfaces have moved the characteristic distance, the friction rapidly drops to the kinetic value. For any system where the kinetic friction is less than the static force (or one that has a negative slope over some part of its curve of F_0 versus v_0) will exhibit regular stick–slip sliding motion for certain values of K, m, and driving velocity, v.

This type of friction has been observed in a variety of dry (unlubricated) systems such as paper-on-paper [30.353, 354] and steel-on-steel [30.352, 355, 356]. This model is also used extensively in geologic systems

to analyze rock-on-rock sliding [30.357, 358]. While originally described for adhering macroscopic asperity junctions, the distance-dependent model may also apply to molecularly smooth surfaces. For example, for polymer lubricant films, the characteristic length would now be the chain–chain entanglement length, which could be much larger in a confined geometry than in the bulk.

Velocity-Dependent Friction Model. In contrast to the two friction models mentioned above, which apply mainly to unlubricated, solid-on-solid contacts, the stick–slip of surfaces with thin liquid films between them is better described by other mechanisms. The velocity-dependent friction model is the most studied mechanism of stick–slip and, until recently, was considered to be the only cause of intrinsic stick–slip. If a friction force decreases with increasing sliding velocity, as occurs with boundary films exhibiting shear thinning, the force (F_s) needed to initiate motion will be higher than the force (F_k) needed to maintain motion.

A decreasing intrinsic friction force F_0 with sliding velocity v_0 results in the sliding surface or stage moving in a periodic fashion, where during each cycle rapid acceleration is followed by rapid deceleration. As long as the drive continues to move at a fixed velocity v, the surfaces will continue to move in a periodic fashion punctuated by abrupt stops and starts whose frequency and amplitude depend not only on the function $F_0(v_0)$, but also on the stiffness K and mass m of the moving stage, and on the starting conditions at $t = 0$.

More precisely, the motion of the sliding surface or stage can be determined by solving the following differential equation:

$$m\ddot{x} = (F_0 - F) = F_0 - (x_0 - x)K$$
$$\text{or}\quad m\ddot{x} + (x_0 - x)K - F_0 = 0\,, \tag{30.47}$$

where $F_0 = F_0(x_0, v_0, t)$ is the intrinsic or "real" friction force at the shearing surfaces, F is the force on the spring (the externally applied or measured force), and $(F_0 - F)$ is the force on the stage. To solve (30.47) fully, one must also know the initial (starting) conditions at $t = 0$, and the driving or steady-state conditions at finite t. Commonly, the driving condition is: $x = 0$ for $t < 0$ and $x = vt$ for $t > 0$, where $v = $ constant. In other systems, the appropriate driving condition may be $F = $ constant.

Various, mainly phenomenological, forms for $F_0 = F_0(x_0, v_0, t)$ have been proposed to explain various kinds of stick–slip phenomena. These models generally assume a particular functional form for the friction as a function of velocity only, $F_0 = F_0(v_0)$, and they

Fig. 30.30 (a) "Phase transitions" model of stick–slip where a thin liquid film alternately freezes and melts as it shears, shown here for 22 spherical molecules confined between two solid crystalline surfaces. In contrast to the velocity-dependent friction model, the intrinsic friction force is assumed to change abruptly (at the transitions), rather than smooth or continuously. The resulting stick–slip is also different, for example, the peaks are sharper and the stick–slip disappears above some critical velocity v_c. Note that, while the slip displacement is here shown to be only two lattice spacings, in most practical situations it is much larger, and that freezing and melting transitions at surfaces or in thin films may not be the same as freezing or melting transitions between the bulk solid and liquid phases. **(b)** Exact reproduction of a chart-recorder trace of the friction force for hexadecane between two untreated mica surfaces at increasing sliding velocity v, plotted as a function of time. In general, with increasing sliding speed, the stick–slip spikes increase in frequency and decrease in magnitude. As the critical sliding velocity v_c is approached, the spikes become erratic, eventually disappearing altogether at $v = v_c$. At higher sliding velocities the sliding continues smoothly in the kinetic state. Such friction traces are fairly typical for simple liquid lubricants and dry boundary lubricant systems (see Fig. 30.31a) and may be referred to as the "conventional" type of static-kinetic friction (in contrast to panel (**c**)). Experimental conditions: contact area $A = 4 \times 10^{-9}\,\text{m}^2$, load $L = 10\,\text{mN}$, film thickness $D = 0.4$–$0.8\,\text{nm}$, $v = 0.08$–$0.4\,\mu\text{m s}^{-1}$, $v_c \approx 0.3\,\mu\text{m s}^{-1}$, atmosphere: dry N_2 gas, $T = 18\,°\text{C}$. [(**a**) and (**b**) after [30.362] with permission. Copyright 1993 American Chemical Society.] (**c**) Transition from smooth sliding to "inverted" stick–slip and to a second smooth-sliding regime with increasing driving velocity during shear of two adsorbed surfactant monolayers in aqueous solution at a load of $L = 4.5\,\text{mN}$ and $T = 20\,°\text{C}$. The smooth sliding (*open circles*) to inverted stick–slip (*squares*) transition occurs at $v_c \sim 0.3\,\mu\text{m/s}$. Prior to the transition, the kinetic stress levels off at after a logarithmic dependence on velocity. The quasi-smooth regime persists up to the transition at v_c. At high driving velocities (*filled circles*), a new transition to a smooth sliding regime is observed between 14 and 17 $\mu\text{m/s}$. (After [30.342] with permission) (**d**) Friction response of a thin squalane (a branched hydrocarbon) film at different loads and a constant sliding velocity $v = 0.08\,\mu\text{m s}^{-1}$, slightly above the critical velocity for this system at low loads. Initially, with increasing load, the stick–slip amplitude and the mean friction force decrease with sliding time or sliding distance. However, at high loads or pressures, the mean friction force increases with time, and the stick–slip takes on a more symmetrical,

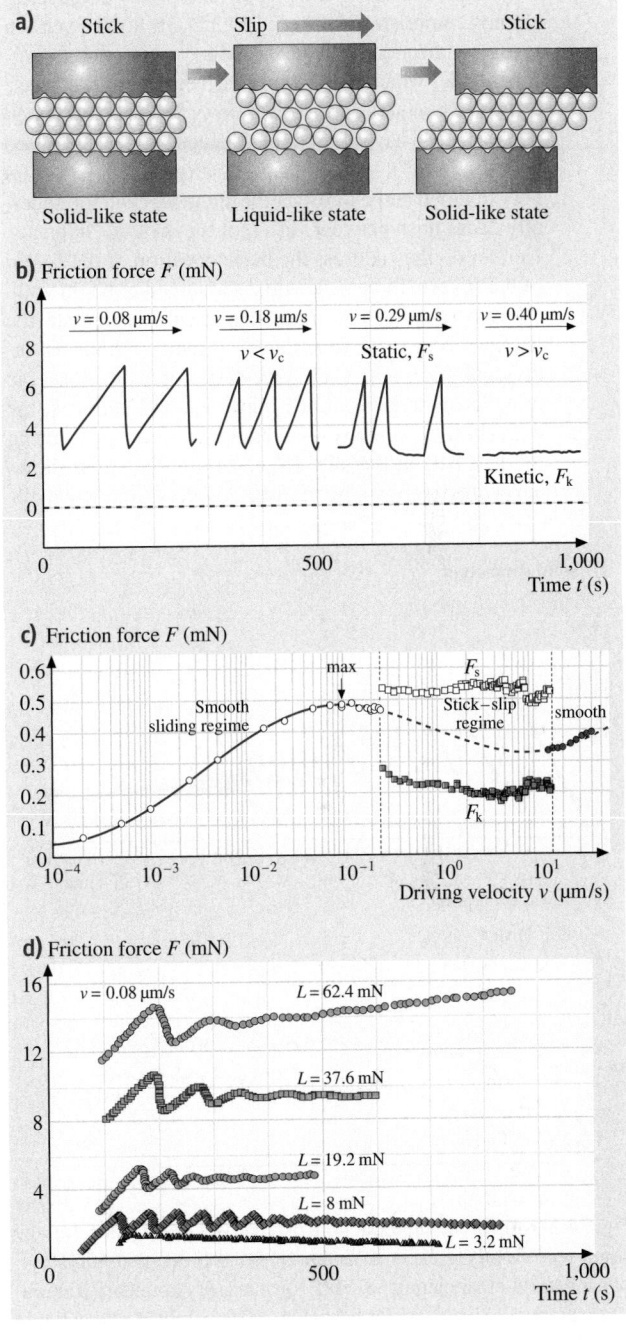

sinusoidal shape. At all loads investigated, the stick–slip component gradually decayed as the friction proceeded towards smooth sliding. (After [30.285] with permission. Copyright 2003 American Physical Society)

may also contain a number of mechanically coupled elements comprising the stage [30.359, 360]. One version is a two-state model characterized by two friction forces, F_s and F_k, which is a simplified version of the phase transitions model described below. More complicated versions can have a rich $F–v$ spectrum, as proposed by *Persson* [30.361]. Unless the experimental data is very detailed and extensive, these models cannot generally distinguish between different types of mechanisms. Neither do they address the basic question of the *origin* of the friction force, since this is assumed to begin with.

Experimental data has been used to calculate the friction force as a function of velocity *within* an individual stick–slip cycle [30.363]. For a macroscopic granular material confined between solid surfaces, the data shows a velocity-weakening friction force during the first half of the slip. However, the data also shows a hysteresis loop in the friction–velocity plot, with a dif-

ferent behavior in the deceleration half of the slip phase. Similar results were observed for a 1–2 nm liquid lubricant film between mica surfaces [30.345]. These results indicate that a purely velocity-dependent friction law is insufficient to describe such systems, and an additional element such as the *state* of the confined material must be considered.

Phase Transitions Model. In recent molecular dynamics computer simulations it has been found that thin interfacial films undergo first-order phase transitions between solid-like and liquid-like states during sliding [30.255, 364], as illustrated in Fig. 30.30. It has been suggested that this is responsible for the observed stick–slip behavior of simple isotropic liquids between two solid crystalline surfaces. With this interpretation, stick–slip is seen to arise because of the abrupt change in the flow properties of a film at a transition [30.278, 279, 326], rather than the gradual or continuous change, as occurs in the previous example. Other computer simulations indicate that it is the stick–slip that induces a disorder ("shear melting") in the film, not the other way around [30.365].

An interpretation of the well-known phenomenon of decreasing coefficient of friction with increasing sliding velocity has been proposed by *Thompson* and *Robbins* [30.255] based on their computer simulation. This postulates that it is not the friction that changes with sliding speed v, but rather the time various parts of the system spend in the sticking and sliding modes. In other words, at any instant during sliding, the friction at any local region is always F_s or F_k, corresponding to the "static" or "kinetic" values. The measured frictional force, however, is the sum of all these discrete values averaged over the whole contact area. Since as v increases each local region spends more time in the sliding regime (F_k) and less in the sticking regime (F_s), the overall friction coefficient falls. One may note that this interpretation reverses the traditional way that stick–slip has been explained. Rather than invoking a decreasing friction with velocity to explain stick–slip, it is now the more fundamental stick–slip phenomenon that is producing the apparent decrease in the friction force with increasing sliding velocity. This approach has been studied analytically by *Carlson* and *Batista* [30.366], with a comprehensive rate- and state-dependent friction force law. This model includes an analytic description of the freezing–melting transitions of a film, resulting in a friction force that is a function of sliding velocity in a natural way. This model predicts a full range of stick–slip behavior observed experimentally.

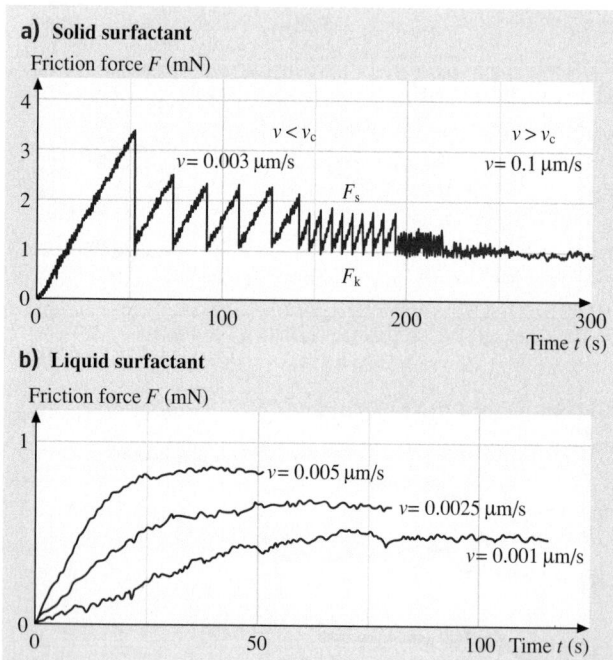

Fig. 30.31 (a) Exact reproduction of chart-recorder trace for the friction of closely packed surfactant monolayers (L-α-dimirystoyl-phosphatidyl-ethanolamine, DMPE) on mica (dry boundary friction) showing qualitatively similar behavior to that obtained with a liquid hexadecane film (Fig. 30.30b). In this case, $L = 0$, $v_c \approx 0.1\,\mu m\,s^{-1}$, atmosphere: dry N_2 gas, $T = 25\,°C$. **(b)** Sliding typical of liquid-like monolayers, here shown for calcium alkylbenzene sulfonate in dry N_2 gas at $T = 25\,°C$ and $L = 0$. (After [30.261], with permission. Copyright 1993 American Chemical Society)

An example of the rate- and state-dependent model is observed when shearing thin films of OMCTS between mica surfaces [30.367, 368]. In this case, the static friction between the surfaces is dependent on the time that the surfaces are at rest with respect to each other, while the intrinsic kinetic friction $F_{k,0}$ is relatively constant over the range of velocities. At slow driving velocities, the system responds with stick–slip sliding with the surfaces reaching maximum static friction before each slip event, and the amplitude of the stick–slip, $F_s - F_k$, is relatively constant. As the driving velocity increases, the static friction decreases as the time at relative rest becomes shorter with respect to the characteristic time of the lubricant film. As the static friction decreases with increasing drive velocity, it eventually equals the intrinsic kinetic friction $F_{k,0}$, which defines the critical velocity v_c, above which the surfaces slide smoothly without the jerky stick–slip motion.

The above classifications of stick–slip are not exclusive, and molecular mechanisms of real systems may exhibit aspects of different models simultaneously. They do, however, provide a convenient classification of existing models and indicate which experimental parameters should be varied to test the different models.

Critical Velocity for Stick–Slip. For any given set of conditions, stick–slip disappears above some critical sliding velocity v_c, above which the motion continues smoothly in the liquid-like or kinetic state [30.261, 285, 342, 345, 362]. The critical velocity is well described by two simple equations. Both are based on the phase transition model, and both include some parameter associated with the inertia of the measuring instrument. The first equation is based on both experiments and simple theoretical modeling [30.362]:

$$v_c \approx \frac{(F_s - F_k)}{5K\tau_0} , \qquad (30.48)$$

where τ_0 is the *characteristic nucleation time* or freezing time of the film. For example, inserting the following typically measured values for a ~ 1 nm thick hexadecane film between mica: $(F_s - F_k) \approx 5$ mN, spring constant $K \approx 500\,\mathrm{N\,m^{-1}}$, and nucleation time [30.362] $\tau_0 \approx 5$ s, we obtain $v_c \approx 0.4\,\mathrm{\mu m\,s^{-1}}$, which is close to typically measured values (Fig. 30.30b).

The second equation is based on computer simulations [30.364]:

$$v_c \approx 0.1\sqrt{\frac{F_s\sigma}{m}} , \qquad (30.49)$$

where σ is a molecular dimension and m is the mass of the stage. Again, inserting typical experimental values into this equation, viz., $m \approx 20$ g, $\sigma \approx 0.5$ nm, and $(F_s - F_k) \approx 5$ mN as before, we obtain $v_c \approx 0.3\,\mathrm{\mu m\,s^{-1}}$, which is also close to measured values.

Stick–slip also disappears above some critical temperature T_c, which is not the same as the melting temperature of the bulk fluid [30.285]. Certain correlations have been found between v_c and T_c and between various other tribological parameters that appear to be consistent with the principle of time–temperature superposition (see Sect. 30.8.3), similar to that occurring in viscoelastic polymer fluids [30.346, 369, 370].

Recent work on the coupling between the mechanical resonances of the sliding system and molecular-scale relaxations [30.295, 338, 341, 371] has resulted in a better understanding of a phenomenon previously noted in various engineering applications: the vibration of one of the sliding surfaces perpendicularly to the sliding direction can lead to a significant reduction of the friction. At certain oscillation amplitudes and a frequency higher than the molecular-scale relaxation frequency, stick–slip friction can be eliminated and replaced by an ultralow kinetic-friction state.

30.9 Effects of Nanoscale Texture on Friction

The above scenario is already quite complicated, and yet this is the situation for the simplest type of experimental system. The factors that appear to determine the critical velocity v_c depend on the type of liquid between the surfaces, as well as on the surface lattice structure.

30.9.1 Role of the Shape of Confined Molecules

Small spherical molecules such as cyclohexane and OMCTS have been found to have very high v_c, which indicates that these molecules can rearrange relatively quickly in thin films. Chain molecules and especially branched-chain molecules have been found to have much lower v_c, which is to be expected, and such liquids tend to slide smoothly or with erratic stick–slip [30.345], rather than in a stick–slip fashion (see Table 30.5). With highly asymmetric molecules, such as multiply branched isoparaffins and polymer melts, no regular spikes or stick–slip behavior occurs at any speed, since these molecules can never order themselves sufficiently to solidify. Examples of such liquids are some perfluoropolyethers and polydimethylsiloxanes (PDMS).

Recent computer simulations [30.144, 151, 287, 315, 372] of the structure, interaction forces, and tribological behavior of chain molecules between two shearing surfaces indicate that both linear *and* singly or doubly branched-chain molecules order between two flat surfaces by aligning into discrete layers parallel to the surfaces. However, in the case of the weakly branched molecules, the expected oscillatory forces do not appear because of a complex cancelation of entropic and enthalpic contributions to the interaction free energy, which results in a monotonically smooth interaction, exhibiting a weak energy minimum rather than the os-

cillatory force profile that is characteristic of linear molecules. During sliding, however, these molecules can be induced to further align, which can result in a transition from smooth to stick–slip sliding.

Table 30.5 shows the trends observed with some organic and polymeric liquid between smooth mica surfaces. Also listed are the bulk viscosities of the liquids. From the data of Table 30.5 it appears that there is a direct correlation between the shapes of molecules and their coefficient of friction or effectiveness as lubricants (at least at low shear rates). Small spherical or chain molecules have high friction with stick–slip, because they can pack into ordered solid-like layers. In contrast, longer chained and irregularly shaped molecules remain in an entangled, disordered, fluid-like state even in very thin films, and these give low friction and smoother sliding. It is probably for this reason that irregularly shaped branched chain molecules are usually better lubricants. It is interesting to note that the friction coefficient generally decreases as the bulk viscosity of the liquids *increases*. This unexpected trend occurs because the factors that are conducive to low friction are generally conducive to high viscosity. Thus molecules with side groups such as branched alkanes and polymer melts

Table 30.5 Effect of molecular shape and short-range forces on tribological properties[a]

Liquid	Short-range force	Type of friction	Friction coefficient	Bulk liquid viscosity (cP)
Organic (water-free)				
Cyclohexane	Oscillatory	Quantized stick–slip	$\gg 1$	0.6
OMCTS[b]	Oscillatory	Quantized stick–slip	$\gg 1$	2.3
Octane	Oscillatory	Quantized stick–slip	1.5	0.5
Tetradecane	Oscillatory $\leftrightarrow$ smooth	stick–slip $\leftrightarrow$ smooth	1.0	2.3
Octadecane (branched)	Oscillatory $\leftrightarrow$ smooth	stick–slip $\leftrightarrow$ smooth	0.3	5.5
PDMS[b] ($M = 3700\,\mathrm{g\,mol}^{-1}$, melt)	Oscillatory $\leftrightarrow$ smooth	Smooth	0.4	50
PBD[b] ($M = 3500\,\mathrm{g\,mol}^{-1}$, branched)	Smooth	Smooth	0.03	800
Water				
Water (KCl solution)	Smooth	Smooth	0.01–0.03	1.0

[a] For molecularly thin liquid films between two shearing mica surfaces at 20 °C

[b] OMCTS: Octamethylcyclotetrasiloxane, PDMS: Polydimethylsiloxane, PBD: Polybutadiene

usually have higher bulk viscosities than their linear homologues for obvious reasons. However, in thin films the linear molecules have higher shear stresses, because of their ability to become ordered. The only exception to the above correlations is water, which has been found to exhibit both low viscosity *and* low friction (see Fig. 30.20a, and Sect. 30.7.3). In addition, the presence of water can drastically lower the friction and eliminate the stick–slip of hydrocarbon liquids when the sliding surfaces are hydrophilic.

If an "effective" viscosity, η_{eff}, were to be calculated for the liquids of Table 30.5, the values would be many orders of magnitude higher than those of the bulk liquids. This can be demonstrated by the following simple calculation based on the usual equation for Couette flow (see (30.44)):

$$\eta_{eff} = F_k D/Av , \qquad (30.50)$$

where F_k is the kinetic friction force, D is the film thickness, A the contact area, and v the sliding velocity. Using typical values for experiments with hexadecane [30.362]: $F_k = 5\,\mathrm{mN}$, $D = 1\,\mathrm{nm}$, $A = 3 \times 10^{-9}\,\mathrm{m}^2$, and $v = 1\,\mu\mathrm{m\,s}^{-1}$, one gets $\eta_{eff} \approx 2000\,\mathrm{Ns\,m}^{-2}$, or 20 000 Poise, which is $\sim 10^6$ times higher than the bulk viscosity, η_{bulk}, of the liquid. It is instructive to consider that this very high effective viscosity nevertheless still produces a low friction force or friction coefficient μ of about 0.25. It is interesting to speculate that, if a 1 nm film were to exhibit bulk viscous behavior, the friction coefficient under the same sliding conditions would be as low as 0.000001. While such a low value has never been reported for any tribological system, one may consider it a theoretical lower limit that could conceivably be attained under certain experimental conditions.

30.9.2 Effects of Surface Structure

Various studies [30.44, 273, 274, 276, 284–286] have shown that confinement and load generally increase the effective viscosity and/or relaxation times of molecules, suggestive of an increased glassiness or solid-like behavior (Figs. 30.32 and 30.33). This is in marked contrast to studies of liquids in small confining capillaries where the opposite effects have been observed [30.373, 374]. The reason for this is probably because the two modes of confinement are different. In the former case (confinement of molecules between two structured solid surfaces), there is generally little opposition to any lateral or vertical displacement of the two surface lattices relative to each other. This means that the two lattices can shift in the x–y–z planes (Fig. 30.32a)

to accommodate the trapped molecules in the most crystallographically commensurate or epitaxial way, which would favor an ordered, solid-like state. In contrast, the walls of capillaries are rigid and cannot easily move or adjust to accommodate the confined molecules (Fig. 30.32b), which will therefore be forced into a more disordered, liquid-like state (unless the capillary wall geometry and lattice are *exactly* commensurate with the liquid molecules, as occurs in certain zeolites [30.374]).

Experiments have demonstrated the effects of surface lattice mismatch on the friction between surfaces [30.257, 258, 375]. Similar to the effects of lattice mismatch on adhesion (Fig. 30.11), the static friction of a confined liquid film is maximum when the lattices of the confining surfaces are aligned. For OMCTS confined between mica surfaces [30.258] the static friction was found to vary by more than a factor of 4, while for bare mica surfaces the variation was by a factor of 3.5 [30.375]. In contrast to the sharp variations

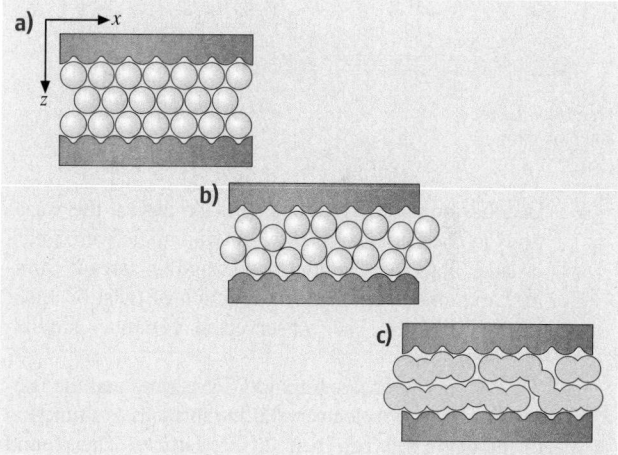

Fig. 30.32a–c Schematic view of interfacial film composed of spherical molecules under a compressive pressure between two solid crystalline surfaces. (**a**) If the two surface lattices are free to move in the x–y–z directions, so as to attain the lowest energy state, they could equilibrate at values of x, y, and z, which induce the trapped molecules to become "epitaxially" ordered into a "solid-like" film. (**b**) Similar view of trapped molecules between two solid surfaces that are not free to adjust their positions, for example, as occurs in capillary pores or in brittle cracks. (**c**) Similar to (**a**), but with chain molecules replacing the spherical molecules in the gap. These may not be able to order as easily as do spherical molecules even if x, y, and z can adjust, resulting in a situation that is more akin to (**b**). (After [30.362] with permission. Copyright 1993 American Chemical Society)

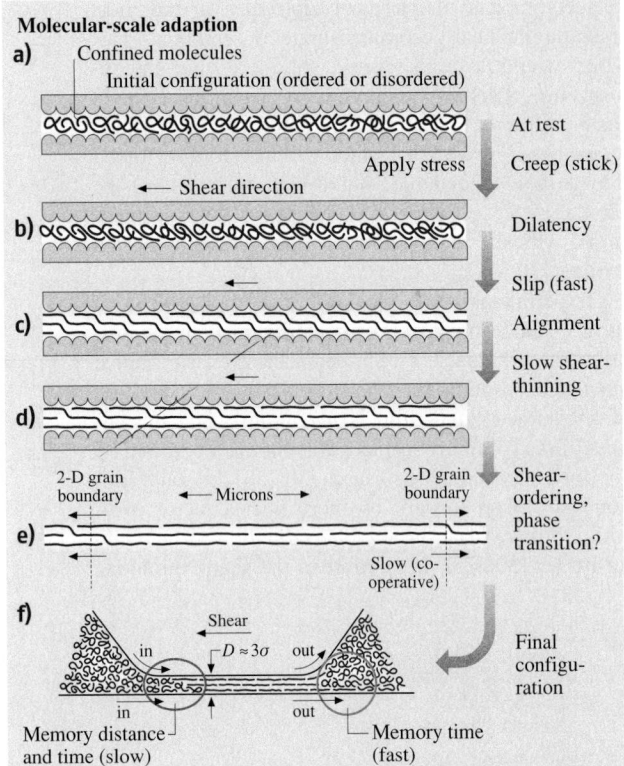

Molecular-scale adaption

a) Confined molecules
Initial configuration (ordered or disordered)

At rest

Shear direction

Apply stress   Creep (stick)

b)

Dilatency

Slip (fast)

c)

Alignment

Slow shear-thinning

d)

2-D grain boundary   ←— Microns —→   2-D grain boundary

e)

Shear-ordering, phase transition?

Slow (co-operative)

f)

Shear

in $D \approx 3\sigma$ out

in   out

Memory distance and time (slow)   Memory time (fast)

Final configuration

Fig. 30.33a–f Schematic representation of the film under shear. (**a**) The lubricant molecules are just confined, but not oriented in any particular direction. Because of the need to shear, the film dilates (**b**). The molecules disentangle (**c**) and get oriented in a certain direction related to the shear direction (**d**). (**e**) Slowly evolving domains grow inside the contact region. These macroscopic domains are responsible for the long relaxation times. (**f**) At the steady-state, a continuous gradient of confinement time and molecular order is established in the contact region, which is different for molecules adsorbed on the upper and lower surfaces. Molecules entering into the contact are not oriented or ordered. The required sliding distance to modify their state defines a characteristic distance. Molecules leaving the contact region need some (short) characteristic time to regain their bulk, unconfined configuration. (After [30.344], with permission. Copyright 2000 American Chemical Society)

in adhesion energy over small twist angles, the variations in friction as a function of twist angle were much broader both in magnitude and angular spread. Similar variations in friction as a function of twist or misfit angles have also been observed in computer simulations [30.376].

Robbins and coworkers [30.315] computed the friction forces of two clean crystalline surfaces as a function of the angle between their surface lattices. They found that, for all non-zero angles (finite "twist" angles), the friction forces fell to zero due to incommensurability effects. They further found that sub-monolayer amounts of organic or other impurities trapped between two incommensurate surfaces can generate a finite friction force. They therefore concluded that any finite friction force measured between incommensurate surfaces is probably due to such "third-body" effects.

The reason why surface texture (lattice structure, roughness, granularity, topography, etc.) has a larger ef-

fect on the lateral (shear or friction) forces between two surfaces than on their normal (adhesion) forces is because friction is proportional to the adhesion hysteresis (Sect. 30.7.2), which can be low even when the adhesion force is high. It is also important to recognize that a system might be defined by more than one length scale. Some systems have well-defined dimensions or size (e.g., a perfect lattice, monodisperse nanoparticles), while others have different lateral and vertical dimensions and macroscopic curvature [30.249]. Furthermore, the morphology or texture of many systems, such as asperities that are randomly distributed over a surface, affects adhesion and tribological properties [30.244, 249, 287, 317, 377–379].

With rough surfaces, i.e., those that have *random* protrusions rather than being periodically structured, we expect a smearing out of the correlated intermolecular interactions that are involved in film freezing and melting (and in phase transitions in general). This should effectively eliminate the highly regular stick–slip and may also affect the location of the slipping planes [30.151, 287, 314, 348]. The stick–slip friction of "real" surfaces, which are generally rough, may, therefore, be quite different from those of perfectly smooth surfaces composed of the same material. We should note, however, that even between rough surfaces, most of the contacts occur between the tips of microscopic asperities, which may be smooth over their microscopic contact area [30.380].

References

30.1   J. N. Israelachvili: Surface Forces and Microrheology of Molecularly Thin Liquid Films. In: *Handbook of Micro/Nanotribology*, ed. by B. Bhushan (CRC, Boca Raton 1995) pp. 267–319

30.2   K. B. Lodge: Techniques for the measurement of forces between solids, Adv. Colloid Interface Sci. **19**, 27–73 (1983)

30.3   J. N. Israelachvili: *Intermolecular and Surface Forces*, 2nd edn. (Academic, London 1991)

30.4   P. F. Luckham, B. A. de L. Costello: Recent developments in the measurement of interparticle forces, Adv. Colloid Interface Sci. **44**, 183–240 (1993)

30.5   P. M. Claesson, T. Ederth, V. Bergeron, M. W. Rutland: Techniques for measuring surface forces, Adv. Colloid Interface Sci. **67**, 119–183 (1996)

30.6   V. S. J. Craig: An historical review of surface force measurement techniques, Colloids Surf. A **129–130**, 75–94 (1997)

30.7   J. N. Israelachvili, G. E. Adams: Measurements of forces between two mica surfaces in aqueous electrolyte solutions in the range 0–100 nm, J. Chem. Soc. Faraday Trans. I **74**, 975–1001 (1978)

30.8   G. Binnig, C. F. Quate, C. Gerber: Atomic force microscope, Phys. Rev. Lett. **56**, 930–933 (1986)

30.9   W. A. Ducker, T. J. Senden, R. M. Pashley: Direct measurement of colloidal forces using an atomic force microscope, Nature **353**, 239–241 (1991)

30.10   E. Meyer, R. M. Overney, K. Dransfeld, T. Gyalog: *Nanoscience: Friction and Rheology on the Nanometer Scale* (World Scientific, Singapore 1998)

30.11   J. N. Israelachvili: Measurements of the viscosity of thin fluid films between two surfaces with and without adsorbed polymers, Colloid Polym. Sci. **264**, 1060–1065 (1986)

30.12   J. P. Montfort, G. Hadziioannou: Equilibrium and dynamic behavior of thin films of a perfluorinated polyether, J. Chem. Phys. **88**, 7187–7196 (1988)

30.13   A. Dhinojwala, S. Granick: Surface forces in the tapping mode: Solvent permeability and hydrodynamic thickness of adsorbed polymer brushes, Macromolecules **30**, 1079–1085 (1997)

30.14   B. V. Derjaguin: Untersuchungen über die Reibung und Adhäsion, IV. Theorie des Anhaftens kleiner Teilchen, Kolloid Z. **69**, 155–164 (1934)

30.15   K. L. Johnson, K. Kendall, A. D. Roberts: Surface energy and the contact of elastic solids, Proc. R. Soc. London A **324**, 301–313 (1971)

30.16   J. N. Israelachvili, P. M. McGuiggan: Adhesion and short-range forces between surfaces. Part 1: New apparatus for surface force measurements, J. Mater. Res. **5**, 2223–2231 (1990)

30.17   J. N. Israelachvili: Thin film studies using multiple-beam interferometry, J. Colloid Interface Sci. **44**, 259–272 (1973)

30.18   Y. L. Chen, T. Kuhl, J. Israelachvili: Mechanism of cavitation damage in thin liquid films: Collapse damage vs. inception damage, Wear **153**, 31–51 (1992)

30.19   M. Heuberger, G. Luengo, J. Israelachvili: Topographic information from multiple beam interferometry in the surface forces apparatus, Langmuir **13**, 3839–3848 (1997)

30.20   R. M. Pashley: DLVO and hydration forces between mica surfaces in Li$^+$, Na$^+$, K$^+$, and Cs$^+$ electrolyte solutions: A correlation of double-layer and hydration forces with surface cation exchange properties, J. Colloid Interface Sci. **83**, 531–546 (1981)

30.21   R. M. Pashley: Hydration forces between mica surfaces in electrolyte solution, Adv. Colloid Interface Sci. **16**, 57–62 (1982)

30.22   R. G. Horn, D. T. Smith, W. Haller: Surface forces and viscosity of water measured between silica sheets, Chem. Phys. Lett. **162**, 404–408 (1989)

30.23   R. G. Horn, D. R. Clarke, M. T. Clarkson: Direct measurements of surface forces between sapphire crystals in aqueous solutions, J. Mater. Res. **3**, 413–416 (1988)

30.24   W. W. Merrill, A. V. Pocius, B. V. Thakker, M. Tirrell: Direct measurement of molecular level adhesion forces between biaxially oriented solid polymer films, Langmuir **7**, 1975–1980 (1991)

30.25   J. Klein: Forces between mica surfaces bearing adsorbed macromolecules in liquid media, J. Chem. Soc. Faraday Trans. I **79**, 99–118 (1983)

30.26   S. S. Patel, M. Tirrell: Measurement of forces between surfaces in polymer fluids, Annu. Rev. Phys. Chem. **40**, 597–635 (1989)

30.27   H. Watanabe, M. Tirrell: Measurements of forces in symmetric and asymmetric interactions between diblock copolymer layers adsorbed on mica, Macromolecules **26**, 6455–6466 (1993)

30.28   T. L. Kuhl, D. E. Leckband, D. Lasic, J. N. Israelachvili: Modulation and modeling of interaction forces between lipid bilayers exposing terminally grafted polymer chains. In: *Stealth Liposomes*, ed. by D. Lasic, F. Martin (CRC, Boca Raton 1995) pp. 73–91

30.29   J. Klein: Shear, friction, and lubrication forces between polymer-bearing surfaces, Annu. Rev. Mater. Sci. **26**, 581–612 (1996)

30.30   M. Ruths, D. Johannsmann, J. Rühe, W. Knoll: Repulsive forces and relaxation on compression of entangled, polydisperse polystyrene brushes, Macromolecules **33**, 3860–3870 (2000)

30.31   C. A. Helm, J. N. Israelachvili, P. M. McGuiggan: Molecular mechanisms and forces involved in the adhesion and fusion of amphiphilic bilayers, Science **246**, 919–922 (1989)

Part D | 30

30.32 Y. L. Chen, C. A. Helm, J. N. Israelachvili: Molecular mechanisms associated with adhesion and contact angle hysteresis of monolayer surfaces, J. Phys. Chem. **95**, 10736–10747 (1991)

30.33 D. E. Leckband, J. N. Israelachvili, F.-J. Schmitt, W. Knoll: Long-range attraction and molecular rearrangements in receptor–ligand interactions, Science **255**, 1419–1421 (1992)

30.34 J. Peanasky, H. M. Schneider, S. Granick, C. R. Kessel: Self-assembled monolayers on mica for experiments utilizing the surface forces apparatus, Langmuir **11**, 953–962 (1995)

30.35 C. J. Coakley, D. Tabor: Direct measurement of van der Waals forces between solids in air, J. Phys. D **11**, L77–L82 (1978)

30.36 J. L. Parker, H. K. Christenson: Measurements of the forces between a metal surface and mica across liquids, J. Chem. Phys. **88**, 8013–8014 (1988)

30.37 C. P. Smith, M. Maeda, L. Atanasoska, H. S. White, D. J. McClure: Ultrathin platinum films on mica and the measurement of forces at the platinum/water interface, J. Phys. Chem. **92**, 199–205 (1988)

30.38 S. J. Hirz, A. M. Homola, G. Hadziioannou, C. W. Frank: Effect of substrate on shearing properties of ultrathin polymer films, Langmuir **8**, 328–333 (1992)

30.39 J. M. Levins, T. K. Vanderlick: Reduction of the roughness of silver films by the controlled application of surface forces, J. Phys. Chem. **96**, 10405–10411 (1992)

30.40 S. Steinberg, W. Ducker, G. Vigil, C. Hyukjin, C. Frank, M. Z. Tseng, D. R. Clarke, J. N. Israelachvili: Van der Waals epitaxial growth of α-alumina nanocrystals on mica, Science **260**, 656–659 (1993)

30.41 G. Vigil, Z. Xu, S. Steinberg, J. Israelachvili: Interactions of silica surfaces, J. Colloid Interface Sci. **165**, 367–385 (1994)

30.42 M. Ruths, M. Heuberger, V. Scheumann, J. Hu, W. Knoll: Confinement-induced film thickness transitions in liquid crystals between two alkanethiol monolayers on gold, Langmuir **17**, 6213–6219 (2001)

30.43 J. Van Alsten, S. Granick: Molecular tribometry of ultrathin liquid films, Phys. Rev. Lett. **61**, 2570–2573 (1988)

30.44 J. Van Alsten, S. Granick: Shear rheology in a confined geometry: Polysiloxane melts, Macromolecules **23**, 4856–4862 (1990)

30.45 A. M. Homola, J. N. Israelachvili, M. L. Gee, P. M. McGuiggan: Measurements of and relation between the adhesion and friction of two surfaces separated by molecularly thin liquid films, J. Tribol. **111**, 675–682 (1989)

30.46 J. Klein, E. Kumacheva: Simple liquids confined to molecularly thin layers. I. Confinement-induced liquid-to-solid phase transitions, J. Chem. Phys. **108**, 6996–7009 (1998)

30.47 G. Luengo, F.-J. Schmitt, R. Hill, J. Israelachvili: Thin film rheology and tribology of confined polymer melts: Contrasts with bulk properties, Macromolecules **30**, 2482–2494 (1997)

30.48 L. Qian, G. Luengo, D. Douillet, M. Charlot, X. Dollat, E. Perez: New two-dimensional friction force apparatus design for measuring shear forces at the nanometer scale, Rev. Sci. Instrum. **72**, 4171–4177 (2001)

30.49 E. Kumacheva: Interfacial friction measurements in surface force apparatus, Prog. Surf. Sci. **58**, 75–120 (1998)

30.50 F. Restagno, J. Crassous, E. Charlaix, C. Cottin-Bizonne, M. Monchanin: A new surface forces apparatus for nanorheology, Rev. Sci. Instrum. **73**, 2292–2297 (2002)

30.51 C. Cottin-Bizonne, B. Cross, A. Steinberger, E. Charlaix: Boundary slip on smooth hydrophobic surfaces: Intrinsic effects and possible artifacts, Phys. Rev. Lett. **94**, 056102/1–4 (2005)

30.52 A. L. Weisenhorn, P. K. Hansma, T. R. Albrecht, C. F. Quate: Forces in atomic force microscopy in air and water, Appl. Phys. Lett. **54**, 2651–2653 (1989)

30.53 G. Meyer, N. M. Amer: Simultaneous measurement of lateral and normal forces with an optical-beam-deflection atomic force microscope, Appl. Phys. Lett. **57**, 2089–2091 (1990)

30.54 E. L. Florin, V. T. Moy, H. E. Gaub: Adhesion forces between individual ligand–receptor pairs, Science **264**, 415–417 (1994)

30.55 G. U. Lee, D. A. Kidwell, R. J. Colton: Sensing discrete streptavidin–biotin interactions with atomic force microscopy, Langmuir **10**, 354–357 (1994)

30.56 H. Skulason, C. D. Frisbie: Detection of discrete interactions upon rupture of Au microcontacts to self-assembled monolayers terminated with $-S(CO)CH_3$ or $-SH$, J. Am. Chem. Soc. **122**, 9750–9760 (2000)

30.57 M. Carrion-Vazquez, A. F. Oberhauser, S. B. Fowler, P. E. Marszalek, S. E. Broedel, J. Clarke, J. M. Fernandez: Mechanical and chemical unfolding of a single protein: A comparison, Proc. Nat. Acad. Sci. USA **96**, 3694–3699 (1999)

30.58 C. M. Mate, G. M. McClelland, R. Erlandsson, S. Chiang: Atomic-scale friction of a tungsten tip on a graphite surface, Phys. Rev. Lett. **59**, 1942–1945 (1987)

30.59 R. W. Carpick, M. Salmeron: Scratching the surface: Fundamental investigations of tribology with atomic force microscopy, Chem. Rev. **97**, 1163–1194 (1997)

30.60 D. Leckband, J. Israelachvili: Intermolecular forces in biology, Quart. Rev. Biophys. **34**, 105–267 (2001)

30.61 J. P. Cleveland, S. Manne, D. Bocek, P. K. Hansma: A nondestructive method for determining the spring constant of cantilevers for scanning force microscopy, Rev. Sci. Instrum. **64**, 403–405 (1993)

Part D | 30

30.62 J. E. Sader, J. W. M. Chon, P. Mulvaney: Calibration of rectangular atomic force microscope cantilevers, Rev. Sci. Instrum. **70**, 3967–3969 (1999)

30.63 Y. Liu, T. Wu, D. F. Evans: Lateral force microscopy study on the shear properties of self-assembled monolayers of dialkylammonium surfactant on mica, Langmuir **10**, 2241–2245 (1994)

30.64 A. Feiler, P. Attard, I. Larson: Calibration of the torsional spring constant and the lateral photodiode response of frictional force microscopes, Rev. Sci. Instrum. **71**, 2746–2750 (2000)

30.65 C. P. Green, H. Lioe, J. P. Cleveland, R. Proksch, P. Mulvaney, J. E. Sader: Normal and torsional spring constants of atomic force microscope cantilevers, Rev. Sci. Instrum. **75**, 1988–1996 (2004)

30.66 C. Neto, V. S. J. Craig: Colloid probe characterization: Radius and roughness determination, Langmuir **17**, 2097–2099 (2001)

30.67 G. M. Sacha, A. Verdaguer, J. Martinez, J. J. Saenz, D. F. Ogletree, M. Salmeron: Effective tip radius in electrostatic force microscopy, Appl. Phys. Lett. **86**, 123101/1–3 (2005)

30.68 R. G. Horn, J. N. Israelachvili: Molecular organization and viscosity of a thin film of molten polymer between two surfaces as probed by force measurements, Macromolecules **21**, 2836–2841 (1988)

30.69 R. G. Horn, S. J. Hirz, G. Hadziioannou, C. W. Frank, J. M. Catala: A reevaluation of forces measured across thin polymer films: Nonequilibrium and pinning effects, J. Chem. Phys. **90**, 6767–6774 (1989)

30.70 O. I. Vinogradova, H.-J. Butt, G. E. Yakubov, F. Feuillebois: Dynamic effects on force measurements. I. Viscous drag on the atomic force microscope cantilever, Rev. Sci. Instrum. **72**, 2330–2339 (2001)

30.71 V. S. J. Craig, C. Neto: In situ calibration of colloid probe cantilevers in force microscopy: Hydrodynamic drag on a sphere approaching a wall, Langmuir **17**, 6018–6022 (2001)

30.72 E. Evans, D. Needham: Physical properties of surfactant bilayer membranes: Thermal transitions, elasticity, rigidity, cohesion, and colloidal interactions, J. Phys. Chem. **91**, 4219–4228 (1987)

30.73 E. Evans, D. Needham: Attraction between lipid bilayer membranes in concentrated solutions of nonadsorbing polymers: Comparison of mean-field theory with measurements of adhesion energy, Macromolecules **21**, 1822–1831 (1988)

30.74 S. E. Chesla, P. Selvaraj, C. Zhu: Measuring two-dimensional receptor-ligand binding kinetics by micropipette, Biophys. J. **75**, 1553–1572 (1998)

30.75 E. Evans, K. Ritchie, R. Merkel: Sensitive force technique to probe molecular adhesion and structural linkages at biological interfaces, Biophys. J. **68**, 2580–2587 (1995)

30.76 D. M. LeNeveu, R. P. Rand, V. A. Parsegian: Measurements of forces between lecithin bilayers, Nature **259**, 601–603 (1976)

30.77 A. Homola, A. A. Robertson: A compression method for measuring forces between colloidal particles, J. Colloid Interface Sci. **54**, 286–297 (1976)

30.78 V. A. Parsegian, N. Fuller, R. P. Rand: Measured work of deformation and repulsion of lecithin bilayers, Proc. Nat. Acad. Sci. USA **76**, 2750–2754 (1979)

30.79 R. P. Rand, V. A. Parsegian: Hydration forces between phospholipid bilayers, Biochim. Biophys. Acta **988**, 351–376 (1989)

30.80 S. Leikin, V. A. Parsegian, D. C. Rau, R. P. Rand: Hydration forces, Annu. Rev. Phys. Chem. **44**, 369–395 (1993)

30.81 S. Chu, J. E. Bjorkholm, A. Ashkin, A. Cable: Experimental observation of optically trapped atoms, Phys. Rev. Lett. **57**, 314–317 (1986)

30.82 A. Ashkin: Optical trapping and manipulation of neutral particles using lasers, Proc. Nat. Acad. Sci. USA **94**, 4853–4860 (1997)

30.83 K. Visscher, S. P. Gross, S. M. Block: Construction of multiple-beam optical traps with nanometer-resolution positioning sensing, IEEE J. Sel. Top. Quantum Electron. **2**, 1066–1076 (1996)

30.84 D. C. Prieve, N. A. Frej: Total internal reflection microscopy: A quantitative tool for the measurement of colloidal forces, Langmuir **6**, 396–403 (1990)

30.85 J. Rädler, E. Sackmann: On the measurement of weak repulsive and frictional colloidal forces by reflection interference contrast microscopy, Langmuir **8**, 848–853 (1992)

30.86 P. Bongrand, C. Capo, J.-L. Mege, A.-M. Benoliel: Use of hydrodynamic flows to study cell adhesion. In: *Physical Basis of Cell–Cell Adhesion*, ed. by P. Bongrand (CRC, Boca Raton 1988) pp. 125–156

30.87 G. Kaplanski, C. Farnarier, O. Tissot, A. Pierres, A.-M. Benoliel, A.-C. Alessi, S. Kaplanski, P. Bongrand: Granulocyte–endothelium initial adhesion: Analysis of transient binding events mediated by E-selectin in a laminar shear flow, Biophys. J. **64**, 1922–1933 (1993)

30.88 H. B. G. Casimir, D. Polder: The influence of retardation on the London–van der Waals forces, Phys. Rev. **73**, 360–372 (1948)

30.89 H. C. Hamaker: The London–van der Waals attraction between spherical particles, Physica **4**, 1058–1072 (1937)

30.90 E. M. Lifshitz: The theory of molecular attraction forces between solid bodies, Sov. Phys. JETP (English translation) **2**, 73–83 (1956)

30.91 I. E. Dzyaloshinskii, E. M. Lifshitz, L. P. Pitaevskii: The general theory of van der Waals forces, Adv. Phys. **10**, 165–209 (1961)

30.92 H. W. Fox, W. A. Zisman: The spreading of liquids on low-energy surfaces. III. Hydrocarbon surfaces, J. Colloid Sci. **7**, 428–442 (1952)

Part D | 30

30.93   B. V. Derjaguin, V. P. Smilga: Electrostatic compo-
        nent of the rolling friction force moment, Wear **7**,
        270–281 (1964)

30.94   R. G. Horn, D. T. Smith: Contact electrification and
        adhesion between dissimilar materials, Science
        256, 362–364 (1992)

30.95   W. R. Harper: *Contact and frictional electrification*
        (Laplacian, Morgan Hill 1998)

30.96   J. A. Wiles, B. A. Grzybowski, A. Winkleman,
        G. M. Whitesides: A tool for studying contact
        electrification in systems comprising metals and
        insulating polymers, Anal. Chem. **75**, 4859–4867
        (2003)

30.97   J. Lowell, A. C. Rose-Innes: Contact electrification,
        Adv. Phys. **29**, 947–1023 (1980)

30.98   C. Guerret-Piecourt, S. Bec, F. Segualt, D. Juve,
        D. Treheux, A. Tonck: Adhesion forces due to nano-
        triboelectrification between similar materials, Eur.
        Phys. J.: Appl. Phys. **28**, 65–72 (2004)

30.99   T. Miura, N. Hirokawa, K. Enokido, I. Arakawa: Spa-
        tially resolved spectroscopy of gas discharge during
        sliding friction between diamond and quartz in N_2
        gas, Appl. Surf. Sci. **235**, 114–118 (2004)

30.100 T. E. Fischer: Tribochemistry, Annu. Rev. Mater. Sci.
        18, 303–323 (1988)

30.101 J. Lowell, W. S. Truscott: Triboelectrification of
        identical insulators. I. An experimental investiga-
        tion, J. Phys. D: Appl. Phys. **19**, 1273–80 (1986)

30.102 L. Perez-Trejo, J. Perez-Gonzalez, L. de Vargas,
        E. Moreno: Triboelectrification of molten linear
        low-density polyethylene under continuous extru-
        sion, Wear **257**, 329–337 (2004)

30.103 K. Ohara, T. Tonouchi, S. Uchiyama: Frictional
        electrification of thin films deposited by the
        Langmuir–Blodgett method, J. Phys. D: Appl. Phys.
        23, 1092–1096 (1990)

30.104 R. Budakian, S. J. Putterman: Correlation between
        charge transfer and stick–slip friction at a metal–
        insulator interface, Phys. Rev. Lett. **85**, 1000–1003
        (2000)

30.105 J. V. Wasem, P. Upadhyaya, S. C. Langford,
        J. T. Dickinson: Transient current generation dur-
        ing wear of high-density polyethylene by a
        stainless steel stylus, J. Appl. Phys. **93**, 719–730
        (2003)

30.106 M. Grätzel: Photoelectrochemical cells, Nature **414**,
        338–344 (2001)

30.107 X. Xu, Y. Liu, R. M. German: Reconciliation of sin-
        tering theory with sintering practice, Adv. Powder
        Metallurgy Particulate Mater. **5**, 67–78 (2000)

30.108 R. M. German: *Sintering Theory and Practice* (Wiley,
        New York 1996)

30.109 R. Budakian, S. J. Putterman: Time scales for cold
        welding and the origins of stick–slip friction, Phys.
        Rev. B **65**, 235429/1–5 (2002)

30.110 D. H. Buckley: Influence of various physical prop-
        erties of metals on their friction and wear behavior
        in vacuum, Metals Eng. Quart. **7**, 44–53 (1967)

30.111 U. Landman, W. D. Luedtke, N. A. Burnham,
        R. J. Colton: Atomistic mechanisms and dynamics of
        adhesion, nanoindentation, and fracture, Science
        248, 454–461 (1990)

30.112 U. Landman, W. D. Luedtke, E. M. Ringer: Molecular
        dynamics simulations of adhesive contact forma-
        tion and friction, NATO Sci. Ser. E **220**, 463–510
        (1992)

30.113 W. D. Luedtke, U. Landman: Solid and liquid junc-
        tions, Comput. Mater. Sci. **1**, 1–24 (1992)

30.114 U. Landman, W. D. Luedtke: Interfacial junctions
        and cavitation, MRS Bull. **18**, 36–44 (1993)

30.115 B. Bhushan, J. N. Israelachvili, U. Landman: Nano-
        tribology: Friction, wear and lubrication at the
        atomic   scale,   Nature   **374**,   607–
        616 (1995)

30.116 M. R. Sørensen, K. W. Jacobsen, P. Stoltze: Simula-
        tions of atomic-scale sliding friction, Phys. Rev. B
        53, 2101–2113 (1996)

30.117 A. Meurk, P. F. Luckham, L. Bergström: Direct meas-
        urement of repulsive and attractive van der Waals
        forces between inorganic materials, Langmuir **13**,
        3896–3899 (1997)

30.118 S.-w. Lee, W. M. Sigmund: AFM study of repulsive
        van der Waals forces between Teflon AF thin film
        and silica or alumina, Colloids Surf. A **204**, 43–50
        (2002)

30.119 E. J. W. Verwey, J. T. G. Overbeek: *Theory of the
        Stability of Lyophobic Colloids*, 1st edn. (Elsevier,
        Amsterdam 1948)

30.120 D. Y. C. Chan, R. M. Pashley, L. R. White: A simple
        algorithm for the calculation of the electrostatic
        repulsion between identical charged surfaces in
        electrolyte, J. Colloid Interface Sci. **77**, 283–285
        (1980)

30.121 J. E. Sader, S. L. Carnie, D. Y. C. Chan: Accurate an-
        alytic formulas for the double-layer interaction
        between spheres, J. Colloid Interface Sci. **171**, 46–54
        (1995)

30.122 D. Harries: Solving the Poisson–Boltzmann equa-
        tion for two parallel cylinders, Langmuir **14**,
        3149–3152 (1998)

30.123 P. Attard: Recent advances in the electric double
        layer in colloid science, Curr. Opin. Colloid Interface
        Sci. **6**, 366–371 (2001)

30.124 B. Derjaguin, L. Landau: Theory of the stability of
        strongly charged lyophobic sols and of the adhe-
        sion of strongly charged particles in solutions of
        electrolytes, Acta Physicochim. URSS **14**, 633–662
        (1941)

30.125 D. Chan, T. W. Healy, L. R. White: Electrical dou-
        ble layer interactions under regulation by surface
        ionization equilibriums – dissimilar amphoteric
        surfaces, J. Chem. Soc. Faraday Trans. 1 **72**, 2844–
        2865 (1976)

30.126 G. S. Manning: Limiting laws and counterion con-
        densation in polyelectrolyte solutions. I. Colligative
        properties, J. Chem. Phys. **51**, 924–933 (1969)

30.127 L. Guldbrand, V. Jönsson, H. Wennerström, P. Linse: Electrical double-layer forces: A Monte Carlo study, J. Chem. Phys. **80**, 2221–2228 (1984)

30.128 H. Wennerström, B. Jönsson, P. Linse: The cell model for polyelectrolyte systems. Exact statistical mechanical relations, Monte Carlo simulations, and the Poisson–Boltzmann approximation, J. Chem. Phys. **76**, 4665–4670 (1982)

30.129 J. Marra: Effects of counterion specificity on the interactions between quaternary ammonium surfactants in monolayers and bilayers, J. Phys. Chem. **90**, 2145–2150 (1986)

30.130 J. Marra: Direct measurement of the interaction between phosphatidylglycerol bilayers in aqueous-electrolyte solutions, Biophys. J. **50**, 815–825 (1986)

30.131 B. Jönsson, H. Wennerström: Ion–ion correlations in liquid dispersions, J. Adh. **80**, 339–364 (2004)

30.132 R. G. Horn, J. N. Israelachvili: Direct measurement of structural forces between two surfaces in a nonpolar liquid, J. Chem. Phys. **75**, 1400–1411 (1981)

30.133 H. K. Christenson, D. W. R. Gruen, R. G. Horn, J. N. Israelachvili: Structuring in liquid alkanes between solid surfaces: Force measurements and mean-field theory, J. Chem. Phys. **87**, 1834–1841 (1987)

30.134 M. L. Gee, J. N. Israelachvili: Interactions of surfactant monolayers across hydrocarbon liquids, J. Chem. Soc. Faraday Trans. **86**, 4049–4058 (1990)

30.135 J. N. Israelachvili, S. J. Kott, M. L. Gee, T. A. Witten: Forces between mica surfaces across hydrocarbon liquids: Effects of branching and polydispersity, Macromolecules **22**, 4247–4253 (1989)

30.136 I. K. Snook, W. van Megen: Solvation forces in simple dense fluids I, J. Chem. Phys. **72**, 2907–2913 (1980)

30.137 W. J. van Megen, I. K. Snook: Solvation forces in simple dense fluids II. Effect of chemical potential, J. Chem. Phys. **74**, 1409–1411 (1981)

30.138 R. Kjellander, S. Marcelja: Perturbation of hydrogen bonding in water near polar surfaces, Chem. Phys. Lett. **120**, 393–396 (1985)

30.139 P. Tarazona, L. Vicente: A model for the density oscillations in liquids between solid walls, Mol. Phys. **56**, 557–572 (1985)

30.140 D. Henderson, M. Lozada-Cassou: A simple theory for the force between spheres immersed in a fluid, J. Colloid Interface Sci. **114**, 180–183 (1986)

30.141 J. E. Curry, J. H. Cushman: Structure in confined fluids: Phase separation of binary simple liquid mixtures, Tribol. Lett. **4**, 129–136 (1998)

30.142 M. Schoen, T. Gruhn, D. J. Diestler: Solvation forces in thin films confined between macroscopically curved substrates, J. Chem. Phys. **109**, 301–311 (1998)

30.143 F. Porcheron, B. Rousseau, M. Schoen, A. H. Fuchs: Structure and solvation forces in confined alkane films, Phys. Chem. Chem. Phys. **3**, 1155–1159 (2001)

30.144 J. Gao, W. D. Luedtke, U. Landman: Layering transitions and dynamics of confined liquid films, Phys. Rev. Lett. **79**, 705–708 (1997)

30.145 H. K. Christenson: Forces between solid surfaces in a binary mixture of non-polar liquids, Chem. Phys. Lett. **118**, 455–458 (1985)

30.146 H. K. Christenson, R. G. Horn: Solvation forces measured in non-aqueous liquids, Chem. Scr. **25**, 37–41 (1985)

30.147 J. Israelachvili: Solvation forces and liquid structure, as probed by direct force measurements, Acc. Chem. Res. **20**, 415–421 (1987)

30.148 H. K. Christenson: Non-DLVO forces between surfaces – solvation, hydration and capillary effects, J. Disp. Sci. Technol. **9**, 171–206 (1988)

30.149 V. Franz, H.-J. Butt: Confined liquids: Solvation forces in liquid alcohols between solid surfaces, J. Phys. Chem. B **106**, 1703–1708 (2002)

30.150 L. J. D. Frink, F. van Swol: Solvation forces between rough surfaces, J. Chem. Phys. **108**, 5588–5598 (1998)

30.151 J. Gao, W. D. Luedtke, U. Landman: Structures, solvation forces and shear of molecular films in a rough nano-confinement, Tribol. Lett. **9**, 3–13 (2000)

30.152 H. E. Stanley, J. Teixeira: Interpretation of the unusual behavior of H_2O and D_2O at low temperatures: Tests of a percolation model, J. Chem. Phys. **73**, 3404–3422 (1980)

30.153 H. van Olphen: *An Introduction to Clay Colloid Chemistry*, 2nd edn. (Wiley, New York 1977) Chap. 10

30.154 N. Alcantar, J. Israelachvili, J. Boles: Forces and ionic transport between mica surfaces: Implications for pressure solution, Geochimica et Cosmochimica Acta **67**, 1289–1304 (2003)

30.155 U. Del Pennino, E. Mazzega, S. Valeri, A. Alietti, M. F. Brigatti, L. Poppi: Interlayer water and swelling properties of monoionic montmorillonites, J. Colloid Interface Sci. **84**, 301–309 (1981)

30.156 B. E. Viani, P. F. Low, C. B. Roth: Direct measurement of the relation between interlayer force and interlayer distance in the swelling of montmorillonite, J. Colloid Interface Sci. **96**, 229–244 (1983)

30.157 R. G. Horn: Surface forces and their action in ceramic materials, J. Am. Ceram. Soc. **73**, 1117–1135 (1990)

30.158 B. V. Velamakanni, J. C. Chang, F. F. Lange, D. S. Pearson: New method for efficient colloidal particle packing via modulation of repulsive lubricating hydration forces, Langmuir **6**, 1323–1325 (1990)

30.159 R. R. Lessard, S. A. Zieminski: Bubble coalescence and gas transfer in aqueous electrolytic solutions, Ind. Eng. Chem. Fundam. **10**, 260–269 (1971)

30.160 J. Israelachvili, R. Pashley: The hydrophobic interaction is long range, decaying exponentially with distance, Nature **300**, 341–342 (1982)

30.161 R. M. Pashley, P. M. McGuiggan, B. W. Ninham, D. F. Evans: Attractive forces between uncharged hydrophobic surfaces: Direct measurements in aqueous solutions, Science **229**, 1088–1089 (1985)

30.162 P. M. Claesson, C. E. Blom, P. C. Herder, B. W. Ninham: Interactions between water-stable hydrophobic Langmuir–Blodgett monolayers on mica, J. Colloid Interface Sci. **114**, 234–242 (1986)

30.163 Ya. I. Rabinovich, B. V. Derjaguin: Interaction of hydrophobized filaments in aqueous electrolyte solutions, Colloids Surf. **30**, 243–251 (1988)

30.164 J. L. Parker, D. L. Cho, P. M. Claesson: Plasma modification of mica: Forces between fluorocarbon surfaces in water and a nonpolar liquid, J. Phys. Chem. **93**, 6121–6125 (1989)

30.165 H. K. Christenson, J. Fang, B. W. Ninham, J. L. Parker: Effect of divalent electrolyte on the hydrophobic attraction, J. Phys. Chem. **94**, 8004–8006 (1990)

30.166 K. Kurihara, S. Kato, T. Kunitake: Very strong long range attractive forces between stable hydrophobic monolayers of a polymerized ammonium surfactant, Chem. Lett. **19**, 1555–1558 (1990)

30.167 Y. H. Tsao, D. F. Evans, H. Wennerstrom: Long-range attractive force between hydrophobic surfaces observed by atomic force microscopy, Science **262**, 547–550 (1993)

30.168 Ya. I. Rabinovich, R.-H. Yoon: Use of atomic force microscope for the measurements of hydrophobic forces between silanated silica plate and glass sphere, Langmuir **10**, 1903–1909 (1994)

30.169 V. S. J. Craig, B. W. Ninham, R. M. Pashley: Study of the long-range hydrophobic attraction in concentrated salt solutions and its implications for electrostatic models, Langmuir **14**, 3326–3332 (1998)

30.170 P. Kékicheff, O. Spalla: Long-range electrostatic attraction between similar, charge-neutral walls, Phys. Rev. Lett. **75**, 1851–1854 (1995)

30.171 H. K. Christenson, P. M. Claesson: Direct measurements of the force between hydrophobic surfaces in water, Adv. Colloid Interface Sci. **91**, 391–436 (2001)

30.172 P. Attard: Nanobubbles and the hydrophobic attraction, Adv. Colloid Interface Sci. **104**, 75–91 (2003)

30.173 J. L. Parker, P. M. Claesson, P. Attard: Bubbles, cavities, and the long-ranged attraction between hydrophobic surfaces, J. Phys. Chem. **98**, 8468–8480 (1994)

30.174 T. Ederth, B. Liedberg: Influence of wetting properties on the long-range "hydrophobic" interaction between self-assembled alkylthiolate monolayers, Langmuir **16**, 2177–2184 (2000)

30.175 S. Ohnishi, V. V. Yaminsky, H. K. Christenson: Measurements of the force between fluorocarbon monolayer surfaces in air and water, Langmuir **16**, 8360–8367 (2000)

30.176 Q. Lin, E. E. Meyer, M. Tadmor, J. N. Israelachvili, T. L. Kuhl: Measurement of the long- and short-range hydrophobic attraction between surfactant-coated surfaces, Langmuir **21**, 251–255 (2005)

30.177 E. E. Meyer, Q. Lin, J. N. Israelachvili: Effects of dissolved gas on the hydrophobic attraction between surfactant-coated surfaces, Langmuir **21**, 256–259 (2005)

30.178 E. E. Meyer, Q. Lin, T. Hassenkam, E. Oroudjev, J. Israelachvili: Origin of the long-range attraction between surfactant-coated surfaces, Proc. Nat. Acad. Sci. USA **102**, 6839–6842 (2005)

30.179 M. Hato: Attractive forces between surfaces of controlled "hydrophobicity" across water: A possible range of "hydrophobic interactions" between macroscopic hydrophobic surfaces across water, J. Phys. Chem. **100**, 18530–18538 (1996)

30.180 H. K. Christenson, V. V. Yaminsky: Is long-range hydrophobic attraction related to the mobility of hydrophobic surface groups?, Colloids Surf. A **129–130**, 67–74 (1997)

30.181 H. K. Christenson, P. M. Claesson, J. Berg, P. C. Herder: Forces between fluorocarbon surfactant monolayers: Salt effects on the hydrophobic interaction, J. Phys. Chem. **93**, 1472–1478 (1989)

30.182 N. I. Christou, J. S. Whitehouse, D. Nicholson, N. G. Parsonage: A Monte Carlo study of fluid water in contact with structureless walls, Faraday Symp. Chem. Soc. **16**, 139–149 (1981)

30.183 B. Jönsson: Monte Carlo simulations of liquid water between two rigid walls, Chem. Phys. Lett. **82**, 520–525 (1981)

30.184 S. Marcelja, D. J. Mitchell, B. W. Ninham, M. J. Sculley: Role of solvent structure in solution theory, J. Chem. Soc. Faraday Trans. II **73**, 630–648 (1977)

30.185 D. W. R. Gruen, S. Marcelja: Spatially varying polarization in water: A model for the electric double layer and the hydration force, J. Chem. Soc. Faraday Trans. 2 **79**, 225–242 (1983)

30.186 B. Jönsson, H. Wennerström: Image-charge forces in phospholipid bilayer systems, J. Chem. Soc. Faraday Trans. 2 **79**, 19–35 (1983)

30.187 D. Schiby, E. Ruckenstein: The role of the polarization layers in hydration forces, Chem. Phys. Lett. **95**, 435–438 (1983)

30.188 A. Luzar, D. Bratko, L. Blum: Monte Carlo simulation of hydrophobic interaction, J. Chem. Phys. **86**, 2955–2959 (1987)

30.189 P. Attard, M. T. Batchelor: A mechanism for the hydration force demonstrated in a model system, Chem. Phys. Lett. **149**, 206–211 (1988)

30.190 J. Forsman, C. E. Woodward, B. Jönsson: Repulsive hydration forces and attractive hydrophobic forces in a unified picture, J. Colloid Interface Sci. **195**, 264–266 (1997)

30.191 E. Ruckenstein, M. Manciu: The coupling between the hydration and double layer interactions, Langmuir **18**, 7584–7593 (2002)

30.192 K. Leung, A. Luzar: Dynamics of capillary evaporation. II. Free energy barriers, J. Chem. Phys. **113**, 5845–5852 (2000)

30.193 D. Bratko, R. A. Curtis, H. W. Blanch, J. M. Prausnitz: Interaction between hydrophobic surfaces with metastable intervening liquid, J. Chem. Phys. **115**, 3873–3877 (2001)

30.194 J. R. Grigera, S. G. Kalko, J. Fischbarg: Wall–water interface. A molecular dynamics study, Langmuir **12**, 154–158 (1996)

30.195 M. Mao, J. Zhang, R.-H. Yoon, W. A. Ducker: Is there a thin film of air at the interface between water and smooth hydrophobic solids?, Langmuir **20**, 1843–1849 (2004). Erratum: Langmuir **20** 4310 (2004)

30.196 Y. Zhu, S. Granick: Rate-dependent slip of Newtonian liquid at smooth surfaces, Phys. Rev. Lett. **87**, 096105/1–4 (2001)

30.197 D. Andrienko, B. Dunweg, O. I. Vinogradova: Boundary slip as a result of a prewetting transition, J. Chem. Phys. **119**, 13106–13112 (2003)

30.198 P. G. de Gennes: Polymers at an interface. 2. Interaction between two plates carrying adsorbed polymer layers, Macromolecules **15**, 492–500 (1982)

30.199 J. M. H. M. Scheutjens, G. J. Fleer: Interaction between two adsorbed polymer layers, Macromolecules **18**, 1882–1900 (1985)

30.200 E. A. Evans: Force between surfaces that confine a polymer solution: Derivation from self-consistent field theories, Macromolecules **22**, 2277–2286 (1989)

30.201 H. J. Ploehn: Compression of polymer interphases, Macromolecules **27**, 1627–1636 (1994)

30.202 J. Ennis, B. Jönsson: Interactions between surfaces in the presence of ideal adsorbing block copolymers, J. Phys. Chem. B **103**, 2248–2255 (1999)

30.203 H. J. Taunton, C. Toprakcioglu, L. J. Fetters, J. Klein: Interactions between surfaces bearing end-adsorbed chains in a good solvent, Macromolecules **23**, 571–580 (1990)

30.204 J. Klein, P. Luckham: Forces between two adsorbed poly(ethylene oxide) layers immersed in a good aqueous solvent, Nature **300**, 429–431 (1982)

30.205 J. Klein, P. F. Luckham: Long-range attractive forces between two mica surfaces in an aqueous polymer solution, Nature **308**, 836–837 (1984)

30.206 P. F. Luckham, J. Klein: Forces between mica surfaces bearing adsorbed homopolymers in good solvents, J. Chem. Soc. Faraday Trans. **86**, 1363–1368 (1990)

30.207 S. Asakura, F. Oosawa: Interaction between particles suspended in solutions of macromolecules, J. Polym. Sci. **33**, 183–192 (1958)

30.208 J. F. Joanny, L. Leibler, P. G. de Gennes: Effects of polymer solutions on colloid stability, J. Polym. Sci. Polym. Phys. **17**, 1073–1084 (1979)

30.209 B. Vincent, P. F. Luckham, F. A. Waite: The effect of free polymer on the stability of sterically stabilized dispersions, J. Colloid Interface Sci. **73**, 508–521 (1980)

30.210 R. I. Feigin, D. H. Napper: Stabilization of colloids by free polymer, J. Colloid Interface Sci. **74**, 567–571 (1980)

30.211 P. G. de Gennes: Polymer solutions near an interface. 1. Adsorption and depletion layers, Macromolecules **14**, 1637–1644 (1981)

30.212 G. J. Fleer, R. Tuinier: Concentration and solvency effects on polymer depletion and the resulting pair interaction of colloidal particles in a solution of non-adsorbing polymer, Polymer Prepr. **46**, 366 (2005)

30.213 P. G. de Gennes: Polymers at an interface; a simplified view, Adv. Colloid Interface Sci. **27**, 189–209 (1987)

30.214 S. Alexander: Adsorption of chain molecules with a polar head. A scaling description, J. Phys. (France) **38**, 983–987 (1977)

30.215 P. G. de Gennes: Conformations of polymers attached to an interface, Macromolecules **13**, 1069–1075 (1980)

30.216 S. T. Milner, T. A. Witten, M. E. Cates: Theory of the grafted polymer brush, Macromolecules **21**, 2610–2619 (1988)

30.217 S. T. Milner, T. A. Witten, M. E. Cates: Effects of polydispersity in the end-grafted polymer brush, Macromolecules **22**, 853–861 (1989)

30.218 E. B. Zhulina, O. V. Borisov, V. A. Priamitsyn: Theory of steric stabilization of colloid dispersions by grafted polymers, J. Colloid Interface Sci. **137**, 495–511 (1990)

30.219 J. Klein, E. Kumacheva, D. Mahalu, D. Perahia, L. J. Fetters: Reduction of frictional forces between solid surfaces bearing polymer brushes, Nature **370**, 634–636 (1994)

30.220 D. Goodman, J. N. Kizhakkedathu, D. E. Brooks: Evaluation of an atomic force microscopy pull-off method for measuring molecular weight and polydispersity of polymer brushes: Effect of grafting density, Langmuir **20**, 6238–6245 (2004)

30.221 T. J. McIntosh, A. D. Magid, S. A. Simon: Range of the solvation pressure between lipid membranes: Dependence on the packing density of solvent molecules, Biochemistry **28**, 7904–7912 (1989)

30.222 P. K. T. Persson, B. A. Bergenståhl: Repulsive forces in lecithin glycol lamellar phases, Biophys. J. **47**, 743–746 (1985)

30.223 J. N. Israelachvili, H. Wennerström: Hydration or steric forces between amphiphilic surfaces?, Langmuir **6**, 873–876 (1990)

30.224 J. N. Israelachvili, H. Wennerström: Entropic forces between amphiphilic surfaces in liquids, J. Phys. Chem. **96**, 520–531 (1992)

30.225 M. K. Granfeldt, S. J. Miklavic: A simulation study of flexible zwitterionic monolayers. Interlayer interaction and headgroup conformation, J. Phys. Chem. **95**, 6351–6360 (1991)

30.226 G. Pabst, J. Katsaras, V. A. Raghunathan: Enhancement of steric repulsion with temperature in oriented lipid multilayers, Phys. Rev. Lett. **88**, 128101/1–4 (2002)

30.227 L. R. Fisher, J. N. Israelachvili: Direct measurements of the effect of meniscus forces on adhesion: A study of the applicability of macroscopic thermodynamics to microscopic liquid interfaces, Colloids Surf. **3**, 303–319 (1981)

30.228 H. K. Christenson: Adhesion between surfaces in unsaturated vapors – a reexamination of the influence of meniscus curvature and surface forces, J. Colloid Interface Sci. **121**, 170–178 (1988)

30.229 H. Hertz: Über die Berührung fester elastischer Körper, J. Reine Angew. Math. **92**, 156–171 (1881)

30.230 H. M. Pollock, D. Maugis, M. Barquins: The force of adhesion between solid surfaces in contact, Appl. Phys. Lett. **33**, 798–799 (1978)

30.231 B. V. Derjaguin, V. M. Muller, Yu. P. Toporov: Effect of contact deformations on the adhesion of particles, J. Colloid Interface Sci. **53**, 314–326 (1975)

30.232 V. M. Muller, V. S. Yushchenko, B. V. Derjaguin: On the influence of molecular forces on the deformation of an elastic sphere and its sticking to a rigid plane, J. Colloid Interface Sci. **77**, 91–101 (1980)

30.233 V. M. Muller, B. V. Derjaguin, Y. P. Toporov: On 2 methods of calculation of the force of sticking of an elastic sphere to a rigid plane, Colloids Surf. **7**, 251–259 (1983)

30.234 D. Tabor: Surface forces and surface interactions, J. Colloid Interface Sci. **58**, 2–13 (1977)

30.235 D. Maugis: Adhesion of spheres: The JKR–DMT transition using a Dugdale model, J. Colloid Interface Sci. **150**, 243–269 (1992)

30.236 R. G. Horn, J. N. Israelachvili, F. Pribac: Measurement of the deformation and adhesion of solids in contact, J. Colloid Interface Sci. **115**, 480–492 (1987)

30.237 V. Mangipudi, M. Tirrell, A. V. Pocius: Direct measurement of molecular level adhesion between poly(ethylene terephthalate) and polyethylene films: Determination of surface and interfacial energies, J. Adh. Sci. Technol. **8**, 1251–1270 (1994)

30.238 H. K. Christenson: Surface deformations in direct force measurements, Langmuir **12**, 1404–1405 (1996)

30.239 M. Barquins, D. Maugis: Fracture mechanics and the adherence of viscoelastic bodies, J. Phys. D: Appl. Phys. **11**, 1989–2023 (1978)

30.240 I. Sridhar, K. L. Johnson, N. A. Fleck: Adhesion mechanics of the surface force apparatus, J. Phys. D: Appl. Phys. **30**, 1710–1719 (1997)

30.241 I. Sridhar, Z. W. Zheng, K. L. Johnson: A detailed analysis of adhesion mechanics between a compliant elastic coating and a spherical probe, J. Phys. D: Appl. Phys. **37**, 2886–2895 (2004)

30.242 P. M. McGuiggan, J. Israelachvili: Measurements of the effects of angular lattice mismatch on the adhesion energy between two mica surfaces in water, Mat. Res. Soc. Symp. Proc. **138**, 349–360 (1989)

30.243 J. A. Greenwood, J. B. P. Williamson: Contact of nominally flat surfaces, Proc. R. Soc. London A **295**, 300–319 (1966)

30.244 B. N. J. Persson: Elastoplastic contact between randomly rough surfaces, Phys. Rev. Lett. **87**, 116101/1–4 (2001)

30.245 F. P. Bowden, D. Tabor: *An Introduction to Tribology* (Anchor/Doubleday, Garden City 1973)

30.246 D. Maugis, H. M. Pollock: Surface forces, deformation and adherence at metal microcontacts, Acta Metallurgica **32**, 1323–1334 (1984)

30.247 K. N. G. Fuller, D. Tabor: The effect of surface roughness on the adhesion of elastic solids, Proc. R. Soc. London A **345**, 327–342 (1975)

30.248 D. Maugis: On the contact and adhesion of rough surfaces, J. Adh. Sci. Technol. **10**, 161–175 (1996)

30.249 B. N. J. Persson, E. Tosatti: The effect of surface roughness on the adhesion of elastic solids, J. Chem. Phys. **115**, 5597–5610 (2001)

30.250 H.-C. Kim, T. P. Russell: Contact of elastic solids with rough surfaces, J. Polym. Sci. Polym. Phys. **39**, 1848–1854 (2001)

30.251 P. M. McGuiggan, J. N. Israelachvili: Adhesion and short-range forces between surfaces. Part II: Effects of surface lattice mismatch, J. Mater. Res. **5**, 2232–2243 (1990)

30.252 C. L. Rhykerd, Jr., M. Schoen, D. J. Diestler, J. H. Cushman: Epitaxy in simple classical fluids in micropores and near-solid surfaces, Nature **330**, 461–463 (1987)

30.253 M. Schoen, D. J. Diestler, J. H. Cushman: Fluids in micropores. I. Structure of a simple classical fluid in a slit-pore, J. Chem. Phys. **87**, 5464–5476 (1987)

30.254 M. Schoen, C. L. Rhykerd, Jr., D. J. Diestler, J. H. Cushman: Shear forces in molecularly thin films, Science **245**, 1223–1225 (1989)

30.255 P. A. Thompson, M. O. Robbins: Origin of stick–slip motion in boundary lubrication, Science **250**, 792–794 (1990)

30.256 K. K. Han, J. H. Cushman, D. J. Diestler: Grand canonical Monte Carlo simulations of a Stockmayer fluid in a slit micropore, Mol. Phys. **79**, 537–545 (1993)

30.257 M. Ruths, S. Granick: Influence of alignment of crystalline confining surfaces on static forces and shear in a liquid crystal, 4′-n-pentyl-4-cyanobiphenyl (5CB), Langmuir **16**, 8368–8376 (2000)

30.258 A. D. Berman: Dynamics of molecules at surfaces. Ph.D. Thesis (Univ. of California, Santa Barbara 1996)

30.259 E. Barthel, S. Roux: Velocity dependent adherence: An analytical approach for the JKR and DMT models, Langmuir **16**, 8134–8138 (2000)

30.260 M. Ruths, S. Granick: Rate-dependent adhesion between polymer and surfactant monolayers on elastic substrates, Langmuir **14**, 1804–1814 (1998)

30.261 H. Yoshizawa, Y. L. Chen, J. Israelachvili: Fundamental mechanisms of interfacial friction. 1: Relation between adhesion and friction, J. Phys. Chem. **97**, 4128–4140 (1993)

30.262 N. Maeda, N. Chen, M. Tirrell, J. N. Israelachvili: Adhesion and friction mechanisms of polymer-on-polymer surfaces, Science **297**, 379–382 (2002)

30.263 F. P. Bowden, D. Tabor: *The Friction and Lubrication of Solids* (Clarendon, London 1971)

30.264 J. A. Greenwood, K. L. Johnson: The mechanics of adhesion of viscoelastic solids, Phil. Mag. A **43**, 697–711 (1981)

30.265 D. Maugis: Subcritical crack growth, surface energy, fracture toughness, stick–slip and embrittlement, J. Mater. Sci. **20**, 3041–3073 (1985)

30.266 F. Michel, M. E. R. Shanahan: Kinetics of the JKR experiment, C. R. Acad. Sci. II (Paris) **310**, 17–20 (1990)

30.267 C. A. Miller, P. Neogi: *Interfacial Phenomena: Equilibrium and Dynamic Effects* (Dekker, New York 1985)

30.268 J. Israelachvili, A. Berman: Irreversibility, energy dissipation, and time effects in intermolecular and surface interactions, Israel J. Chem. **35**, 85–91 (1995)

30.269 A. N. Gent, A. J. Kinloch: Adhesion of viscoelastic materials to rigid substrates. III. Energy criterion for failure, J. Polym. Sci. A-2 **9**, 659–668 (1971)

30.270 A. N. Gent: Adhesion and strength of viscoelastic solids. Is there a relationship between adhesion and bulk properties?, Langmuir **12**, 4492–4496 (1996)

30.271 H. R. Brown: The adhesion between polymers, Annu. Rev. Mater. Sci. **21**, 463–489 (1991)

30.272 M. Deruelle, M. Tirrell, Y. Marciano, H. Hervet, L. Léger: Adhesion energy between polymer networks and solid surfaces modified by polymer attachment, Faraday Discuss. **98**, 55–65 (1995)

30.273 S. Granick: Motions and relaxations of confined liquids, Science **253**, 1374–1379 (1991)

30.274 H. W. Hu, S. Granick: Viscoelastic dynamics of confined polymer melts, Science **258**, 1339–1342 (1992)

30.275 M. L. Gee, P. M. McGuiggan, J. N. Israelachvili, A. M. Homola: Liquid to solidlike transitions of molecularly thin films under shear, J. Chem. Phys. **93**, 1895–1906 (1990)

30.276 J. Van Alsten, S. Granick: The origin of static friction in ultrathin liquid films, Langmuir **6**, 876–880 (1990)

30.277 H.-W. Hu, G. A. Carson, S. Granick: Relaxation time of confined liquids under shear, Phys. Rev. Lett. **66**, 2758–2761 (1991)

30.278 J. Israelachvili, M. Gee, P. McGuiggan, P. Thompson, M. Robbins: Melting–freezing transitions in molecularly thin liquid films during shear, Fall Meeting of the Mater. Res. Soc., Boston, MA. 1990, ed. by J. M. Drake, J. Klafter, R. Kopelman (MRS Publications, Pittsburgh, PA1990) 3–6

30.279 J. Israelachvili, P. McGuiggan, M. Gee, A. Homola, M. Robbins, P. Thompson: Liquid dynamics in molecularly thin films, J. Phys.: Condens. Matter **2**, SA89–SA98 (1990)

30.280 J. Klein, E. Kumacheva: Confinement-induced phase transitions in simple liquids, Science **269**, 816–819 (1995)

30.281 A. L. Demirel, S. Granick: Glasslike transition of a confined simple fluid, Phys. Rev. Lett. **77**, 2261–2264 (1996)

30.282 D. Dowson: *History of Tribology*, 2nd edn. (Professional Engineering Publishing, London 1998)

30.283 A. Berman, C. Drummond, J. Israelachvili: Amontons' law at the molecular level, Tribol. Lett. **4**, 95–101 (1998)

30.284 M. Ruths: Boundary friction of aromatic self-assembled monolayers: Comparison of systems with one or both sliding surfaces covered with a thiol monolayer, Langmuir **19**, 6788–6795 (2003)

30.285 D. Gourdon, J. N. Israelachvili: Transitions between smooth and complex stick–slip sliding of surfaces, Phys. Rev. E **68**, 021602/1–10 (2003)

30.286 M. Ruths, N. A. Alcantar, J. N. Israelachvili: Boundary friction of aromatic silane self-assembled monolayers measured with the surface forces apparatus and friction force microscopy, J. Phys. Chem. B **107**, 11149–11157 (2003)

30.287 J. Gao, W. D. Luedtke, D. Gourdon, M. Ruths, J. N. Israelachvili, U. Landman: Frictional forces and Amontons' law: From the molecular to the macroscopic scale, J. Phys. Chem. B **108**, 3410–3425 (2004)

30.288 D. Tabor: The role of surface and intermolecular forces in thin film lubrication, Tribol. Ser. **7**, 651–682 (1982)

30.289 M. J. Sutcliffe, S. R. Taylor, A. Cameron: Molecular asperity theory of boundary friction, Wear **51**, 181–192 (1978)

30.290 G. M. McClelland: Friction at weakly interacting interfaces. In: *Adhesion and Friction*, ed. by M. Grunze, H. J. Kreuzer (Springer, Berlin, Heidelberg 1989) pp.1–16

30.291 D. H. Buckley: The metal-to-metal interface and its effect on adhesion and friction, J. Colloid Interface Sci. **58**, 36–53 (1977)

30.292 J. N. Israelachvili, Y.-L. Chen, H. Yoshizawa: Relationship between adhesion and friction forces, J. Adh. Sci. Technol. **8**, 1231–1249 (1994)

30.293 J.-C. Géminard, W. Losert, J. P. Gollub: Frictional mechanics of wet granular material, Phys. Rev. E **59**, 5881–5890 (1999)

30.294 R. G. Cain, N. W. Page, S. Biggs: Microscopic and macroscopic aspects of stick–slip motion in granular shear, Phys. Rev. E **64**, 016413/1–8 (2001)

30.295 V. Zaloj, M. Urbakh, J. Klafter: Modifying friction by manipulating normal response to lateral motion, Phys. Rev. Lett. **82**, 4823–4826 (1999)

30.296 A. Dhinojwala, S. C. Bae, S. Granick: Shear-induced dilation of confined liquid films, Tribol. Lett. **9**, 55–62 (2000)

30.297 L. M. Qian, G. Luengo, E. Perez: Thermally activated lubrication with alkanes: The effect of chain length, Europhys. Lett. **61**, 268–274 (2003)

30.298 B. J. Briscoe, D. C. B. Evans, D. Tabor: The influence of contact pressure and saponification on the sliding behavior of steric acid monolayers, J. Colloid Interface Sci. **61**, 9–13 (1977)

30.299 M. Urbakh, L. Daikhin, J. Klafter: Dynamics of confined liquids under shear, Phys. Rev. E **51**, 2137–2141 (1995)

30.300 M. G. Rozman, M. Urbakh, J. Klafter: Origin of stick-slip motion in a driven two-wave potential, Phys. Rev. E **54**, 6485–6494 (1996)

30.301 M. G. Rozman, M. Urbakh, J. Klafter: Stick–slip dynamics as a probe of frictional forces, Europhys. Lett. **39**, 183–188 (1997)

30.302 M. Urbakh, J. Klafter, D. Gourdon, J. Israelachvili: The nonlinear nature of friction, Nature **430**, 525–528 (2004)

30.303 B. V. Derjaguin: Molekulartheorie der äußeren Reibung, Z. Physik **88**, 661–675 (1934)

30.304 B. V. Derjaguin: Mechanical properties of the boundary lubrication layer, Wear **128**, 19–27 (1988)

30.305 B. J. Briscoe, D. C. B. Evans: The shear properties of Langmuir–Blodgett layers, Proc. R. Soc. London A **380**, 389–407 (1982)

30.306 P. F. Luckham, S. Manimaaran: Investigating adsorbed polymer layer behaviour using dynamic surface force apparatuses – a review, Adv. Colloid Interface Sci. **73**, 1–46 (1997)

30.307 A. Berman, J. Israelachvili: Control and minimization of friction via surface modification, NATO ASI Ser. E: Appl. Sci. **330**, 317–329 (1997)

30.308 A. M. Homola, J. N. Israelachvili, P. M. McGuiggan, M. L. Gee: Fundamental experimental studies in tribology: The transition from "interfacial" friction of undamaged molecularly smooth surfaces to "normal" friction with wear, Wear **136**, 65–83 (1990)

30.309 S. Yamada, J. Israelachvili: Friction and adhesion hysteresis of fluorocarbon surfactant monolayer-coated surfaces measured with the surface forces apparatus, J. Phys. Chem. B **102**, 234–244 (1998)

30.310 G. Luengo, S. E. Campbell, V. I. Srdanov, F. Wudl, J. N. Israelachvili: Direct measurement of the adhesion and friction of smooth C_{60} surfaces, Chem. Mater. **9**, 1166–1171 (1997)

30.311 L. Rapoport, Y. Bilik, Y. Feldman, M. Homyonfer, S. R. Cohen, R. Tenne: Hollow nanoparticles of WS$_2$ as potential solid-state lubricants, Nature **387**, 791–793 (1997)

30.312 J. Israelachvili, Y.-L. Chen, H. Yoshizawa: Relationship between adhesion and friction forces. In: *Fundamentals of Adhesion and Interfaces*, ed. by D. S. Rimai, L. P. DeMejo, K. L. Mittal (VSP, Utrecht, The Netherlands 1995) pp. 261–279

30.313 U. Raviv, S. Perkin, P. Laurat, J. Klein: Fluidity of water confined down to subnanometer films, Langmuir **20**, 5322–5332 (2004)

30.314 P. Berthoud, T. Baumberger, C. G'Sell, J. M. Hiver: Physical analysis of the state- and rate-dependent friction law: Static friction, Phys. Rev. B **59**, 14313–14327 (1999)

30.315 G. He, M. Müser, M. Robbins: Adsorbed layers and the origin of static friction, Science **284**, 1650–1652 (1999)

30.316 M. H. Müser, L. Wenning, M. O. Robbins: Simple microscopic theory of Amontons's laws for static friction, Phys. Rev. Lett. **86**, 1295–1298 (2001)

30.317 B. Luan, M. O. Robbins: The breakdown of continuum models for mechanical contacts, Nature **435**, 929–932 (2005)

30.318 A. Berman, S. Steinberg, S. Campbell, A. Ulman, J. Israelachvili: Controlled microtribology of a metal oxide surface, Tribol. Lett. **4**, 43–48 (1998)

30.319 D. Y. C. Chan, R. G. Horn: The drainage of thin liquid films between solid surfaces, J. Chem. Phys. **83**, 5311–5324 (1985)

30.320 J. N. Israelachvili, S. J. Kott: Shear properties and structure of simple liquids in molecularly thin films: The transition from bulk (continuum) to molecular behavior with decreasing film thickness, J. Colloid Interface Sci. **129**, 461–467 (1989)

30.321 T. L. Kuhl, A. D. Berman, S. W. Hui, J. N. Israelachvili: Part 1: Direct measurement of depletion attraction and thin film viscosity between lipid bilayers in aqueous polyethylene glycol solutions, Macromolecules **31**, 8250–8257 (1998)

30.322 S. E. Campbell, G. Luengo, V. I. Srdanov, F. Wudl, J. N. Israelachvili: Very low viscosity at the solid–liquid interface induced by adsorbed C_{60} monolayers, Nature **382**, 520–522 (1996)

30.323 J. Klein, Y. Kamiyama, H. Yoshizawa, J. N. Israelachvili, G. H. Fredrickson, P. Pincus, L. J. Fetters: Lubrication forces between surfaces bearing polymer brushes, Macromolecules **26**, 5552–5560 (1993)

30.324 E. Kumacheva, J. Klein: Simple liquids confined to molecularly thin layers. II. Shear and frictional behavior of solidified films, J. Chem. Phys. **108**, 7010–7022 (1998)

30.325 G. Luengo, J. Israelachvili, A. Dhinojwala, S. Granick: Generalized effects in confined fluids: New friction map for boundary lubrication, Wear **200**, 328–335 (1996) Erratum. Wear **205** (1997) 246

30.326 P. A. Thompson, G. S. Grest, M. O. Robbins: Phase transitions and universal dynamics in confined films, Phys. Rev. Lett. **68**, 3448–3451 (1992)

30.327 Y. Rabin, I. Hersht: Thin liquid layers in shear: Non-Newtonian effects, Physica A **200**, 708–712 (1993)

30.328 P. A. Thompson, M. O. Robbins, G. S. Grest: Structure and shear response in nanometer-thick films, Israel J. Chem. **35**, 93–106 (1995)

30.329 M. Urbakh, L. Daikhin, J. Klafter: Sheared liquids in the nanoscale range, J. Chem. Phys. **103**, 10707–10713 (1995)

30.330 A. Subbotin, A. Semenov, E. Manias, G. Hadziioannou, G. ten Brinke: Rheology of confined polymer melts under shear flow: Strong adsorption limit, Macromolecules **28**, 1511–1515 (1995)

30.331 A. Subbotin, A. Semenov, E. Manias, G. Hadziioannou, G. ten Brinke: Rheology of confined polymer melts under shear flow: Weak adsorption limit, Macromolecules **28**, 3901–3903 (1995)

30.332 J. Huh, A. Balazs: Behavior of confined telechelic chains under shear, J. Chem. Phys. **113**, 2025–2031 (2000)

30.333 H. Xie, K. Song, D. J. Mann, W. L. Hase: Temperature gradients and frictional energy dissipation in the sliding of hydroxylated α-alumina surfaces, Phys. Chem. Chem. Phys. **4**, 5377–5385 (2002)

30.334 U. Landman, W. D. Luedtke, A. Nitzan: Dynamics of tip-substrate interactions in atomic force microscopy, Surf. Sci. **210**, L177–L184 (1989)

30.335 J. N. Israelachvili, P. M. McGuiggan, A. M. Homola: Dynamic properties of molecularly thin liquid films, Science **240**, 189–191 (1988)

30.336 A. M. Homola, H. V. Nguyen, G. Hadziioannou: Influence of monomer architecture on the shear properties of molecularly thin polymer melts, J. Chem. Phys. **94**, 2346–2351 (1991)

30.337 M. Schoen, S. Hess, D. J. Diestler: Rheological properties of confined thin films, Phys. Rev. E **52**, 2587–2602 (1995)

30.338 M. Heuberger, C. Drummond, J. Israelachvili: Coupling of normal and transverse motions during frictional sliding, J. Phys. Chem. B **102**, 5038–5041 (1998)

30.339 G. M. McClelland, S. R. Cohen: *Chemistry and Physics of Solid Surfaces VIII* (Springer, Berlin, Heidelberg 1990) pp. 419–445

30.340 E. Gnecco, R. Bennewitz, T. Gyalog, E. Meyer: Friction experiments on the nanometre scale, J. Phys. Cond. Matter **13**, R619–R642 (2001)

30.341 J. Gao, W. D. Luedtke, U. Landman: Friction control in thin-film lubrication, J. Phys. Chem. B **102**, 5033–5037 (1998)

30.342 C. Drummond, J. Elezgaray, P. Richetti: Behavior of adhesive boundary lubricated surfaces under shear: A new dynamic transition, Europhys. Lett. **58**, 503–509 (2002)

30.343 A. E. Filippov, J. Klafter, M. Urbakh: Inverted stick–slip friction: What is the mechanism?, J. Chem. Phys. **116**, 6871–6874 (2002)

30.344 C. Drummond, J. Israelachvili: Dynamic behavior of confined branched hydrocarbon lubricant fluids under shear, Macromolecules **33**, 4910–4920 (2000)

30.345 C. Drummond, J. Israelachvili: Dynamic phase transitions in confined lubricant fluids under shear, Phys. Rev. E. **63**, 041506/1–11 (2001)

30.346 J. D. Ferry: *Viscoelastic Properties of Polymers*, 3rd edn. (Wiley, New York 1980)

30.347 A. Ruina: Slip instability and state variable friction laws, J. Geophys. Res. **88**, 10359–10370 (1983)

30.348 T. Baumberger, P. Berthoud, C. Caroli: Physical analysis of the state- and rate-dependent friction law. II. Dynamic friction, Phys. Rev. B **60**, 3928–3939 (1999)

30.349 J. Israelachvili, S. Giasson, T. Kuhl, C. Drummond, A. Berman, G. Luengo, J.-M. Pan, M. Heuberger, W. Ducker, N. Alcantar: Some fundamental differences in the adhesion and friction of rough versus smooth surfaces, Tribol. Ser. **38**, 3–12 (2000)

30.350 E. Rabinowicz: *Friction and Wear of Materials*, 2nd edn. (Wiley, New York 1995) Chap. 4

30.351 J. Peachey, J. Van Alsten, S. Granick: Design of an apparatus to measure the shear response of ultrathin liquid films, Rev. Sci. Instrum. **62**, 463–473 (1991)

30.352 E. Rabinowicz: The intrinsic variables affecting the stick–slip process, Proc. Phys. Soc. **71**, 668–675 (1958)

30.353 T. Baumberger, F. Heslot, B. Perrin: Crossover from creep to inertial motion in friction dynamics, Nature **367**, 544–546 (1994)

30.354 F. Heslot, T. Baumberger, B. Perrin, B. Caroli, C. Caroli: Creep, stick–slip, and dry-friction dynamics: Experiments and a heuristic model, Phys. Rev. E **49**, 4973–4988 (1994)

30.355 J. Sampson, F. Morgan, D. Reed, M. Muskat: Friction behavior during the slip portion of the stick–slip process, J. Appl. Phys. **14**, 689–700 (1943)

30.356 F. Heymann, E. Rabinowicz, B. Rightmire: Friction apparatus for very low-speed sliding studies, Rev. Sci. Instrum. **26**, 56–58 (1954)

30.357 J. H. Dieterich: Time-dependent friction and the mechanics of stick–slip, Pure Appl. Geophys. **116**, 790–806 (1978)

30.358 J. H. Dieterich: Modeling of rock friction. 1. Experimental results and constitutive equations, J. Geophys. Res. **84**, 2162–2168 (1979)

30.359 G. A. Tomlinson: A molecular theory of friction, Phil. Mag. **7**, 905–939 (1929)

30.360 J. M. Carlson, J. S. Langer: Mechanical model of an earthquake fault, Phys. Rev. A **40**, 6470–6484 (1989)

30.361 B. N. J. Persson: Theory of friction: The role of elasticity in boundary lubrication, Phys. Rev. B **50**, 4771–4786 (1994)

30.362 H. Yoshizawa, J. Israelachvili: Fundamental mechanisms of interfacial friction. 2: Stick–slip friction of spherical and chain molecules, J. Phys. Chem. **97**, 11300–11313 (1993)

30.363 S. Nasuno, A. Kudrolli, J. P. Gollub: Friction in granular layers: Hysteresis and precursors, Phys. Rev. Lett. **79**, 949–952 (1997)

30.364 M. O. Robbins, P. A. Thompson: Critical velocity of stick–slip motion, Science **253**, 916 (1991)

30.365 P. Bordarier, M. Schoen, A. Fuchs: Stick–slip phase transitions in confined solidlike films from an equilibrium perspective, Phys. Rev. E **57**, 1621–1635 (1998)

30.366 J. M. Carlson, A. A. Batista: Constitutive relation for the friction between lubricated surfaces, Phys. Rev. E **53**, 4153–4165 (1996)

30.367 A. D. Berman, W. A. Ducker, J. N. Israelachvili: Origin and characterization of different stick–slip friction mechanisms, Langmuir **12**, 4559–4563 (1996)

30.368 A. D. Berman, W. A. Ducker, J. N. Israelachvili: Experimental and theoretical investigations of stick–slip friction mechanisms, NATO ASI Ser. E: Appl. Sci. **311**, 51–67 (1996)

30.369 K G. McLaren, D. Tabor: Viscoelastic properties and the friction of solids. Friction of polymers and influence of speed and temperature, Nature **197**, 856–858 (1963)

30.370 K. A. Grosch: Viscoelastic properties and friction of solids. Relation between friction and viscoelastic properties of rubber, Nature **197**, 858–859 (1963)

30.371 L. Bureau, T. Baumberger, C. Caroli: Shear response of a frictional interface to a normal load modulation, Phys. Rev. E **62**, 6810–6820 (2000)

30.372 J. P. Gao, W. D. Luedtke, U. Landman: Structure and solvation forces in confined films: Linear and branched alkanes, J. Chem. Phys. **106**, 4309–4318 (1997)

30.373 J. Warnock, D. D. Awschalom, M. W. Shafer: Orientational behavior of molecular liquids in restricted geometries, Phys. Rev. B **34**, 475–478 (1986)

30.374 D. D. Awschalom, J. Warnock: Supercooled liquids and solids in porous glass, Phys. Rev. B **35**, 6779–6785 (1987)

30.375 M. Hirano, K. Shinjo, R. Kaneko, Y. Murata: Anisotropy of frictional forces in muscovite mica, Phys. Rev. Lett. **67**, 2642–2645 (1991)

30.376 T. Gyalog, H. Thomas: Friction between atomically flat surfaces, Europhys. Lett. **37**, 195–200 (1997)

30.377 B. N. J. Persson, F. Bucher, B. Chiaia: Elastic contact between randomly rough surfaces: comparison of theory with numerical results, Phys. Rev. B **65**, 184106/1–7 (2002)

30.378 S. Hyun, L. Pei, J.-F. Molinari, M. O. Robbins: Finite-element analysis of contact between elastic self-affine surfaces, Phys. Rev. E **70**, 026117/1–12 (2004)

30.379 J. Israelachvili, N. Maeda, K. J. Rosenberg, M. Akbulut: Effects of sub-ångstrom (pico-scale) structure of surfaces on adhesion, friction and bulk mechanical properties, J. Mater. Res. **20**, 1952–1972 (2005)

30.380 T. R. Thomas: *Rough Surfaces*, 2nd edn. (Imperial College Press, London 1999)

31. Interfacial Forces and Spectroscopic Study of Confined Fluids

In this chapter we discuss three specific issues which are relevant for liquids in intimate contact with solid surfaces. (1) Studies of the hydrodynamic flow of simple and complex fluids within ultra-narrow channels show the effects of flow rate, surface roughness and fluid–surface interaction on the determination of the boundary condition. We draw attention to the importance of the microscopic particulars to the discovery of what boundary condition is appropriate for solving continuum equations and the potential to capitalize on *slip at the wall* for purposes of materials engineering. (2) We address the long-standing question of the structure of aqueous films near a hydrophobic surface. When water was confined between adjoining hydrophobic and hydrophilic surfaces (a Janus interface), giant fluctuations in shear responses were observed, which implies some kind of flickering, fluctuating complex at the water–hydrophobic interface. (3) Finally we discuss recent experiments that augment friction studies by measurement of diffusion, using fluorescence correlation spectroscopy (FCS). Here spatially resolved measurements showed that translation diffusion slows exponentially with increasing mechanical pressure from the edges of a Hertzian contact toward the center, accompanied by increasingly heterogeneous dynamical responses. This dynamical probe of how liquids order in molecularly thin films fails to support the hypothesis that shear produces a melting transition.

Confinement of fluids, ranging from water, hydrocarbon oil, polymer melts and solutions, to DNA, protein and other bio-macromolecules, can strongly affect the structures and dynamics of the fluid molecules, particularly when the thickness of fluid films becomes comparable to the length scale of the molecular dimension. Phase transitions, such as solidification or glass transition, can be induced, and confinement can not only result in structural changes in terms of molecular orientation and packing, but changes of dynamical processes, such as molecular translational as well as rotational diffusion. These questions about confined fluid structure and dynamics involve deep scientific puzzles. On the applied side, they also pertain directly to the understanding of many fundamental issues in nature and technology, including various applications in adhesion, colloidal stabilization, lubrication, micro/nanofluidics, and micro/nano-electromechanical systems (MEMS/NEMS) devices, among others.

The techniques that have been used to study confined fluids are mainly surface forces apparatus (SFA) [31.1–4] and recently atomic force microscopy (AFM) and its derivatives [31.5,6]. SFA, which was originally designed to directly measure the van der Waals and Derjaguin–Landau–Verwey–Overbeek (DLVO) forces in simple liquids and in colloidal systems, was later modified by several groups to study the frictional forces of molecularly thin films confined between rigid surfaces [31.2–4]. AFM is an alternative method to study surface force, friction, adhesion and lubrication of thin film. With these two techniques many new phenomena about confined fluids, such as layering structure, solidification, stick–slip motion, etc. have been experimentally observed and interpreted.

The previous Chapt. 30 reviewed the advances in the studies of surface forces, adhesion and shear interaction of confined liquids. In this chapter, we review recent studies on some specific interfacial forces of thin liquid films by using modified SFA devices and the results from an integrated experimental platform, which enables direct fluorescence-based observation of the dynamics of individual molecules under confinement.

This chapter is organized as follows. This introductory section is followed by a discussion of the validity of the no-slip boundary for fluid molecules moving past a solid surface, determined by measuring the hydrodynamic force in flow geometry. A specific interfacial force – the long-range hydrophobic interaction – and its related hydrophobic effect on the behavior of water molecules near a hydrophobic interface are discussed in Sect. 31.2. In Sects. 31.3–31.5, recent developments in combining ultrafast spectroscopy with the SFA to investigate the dynamic behavior of individual confined molecules, instead of their ensemble-averaged behavior, are reviewed.

31.1 Hydrodynamic Force of Fluids Flowing in Micro- to Nanofluidics: A Question About No-Slip Boundary Condition

Viscous flow is familiar and useful, yet the underlying physics is surprisingly subtle and complex. A tenet of textbook continuum fluid dynamics is the *no-slip* boundary condition, which means that fluid molecules immediately adjacent to the surface of a solid move with exactly the same velocity as that solid. It stands at the bedrock of our understanding of the flow of simple low-viscosity fluids and comprises a springboard for much sophisticated calculation. While it is true that at some level of detail this continuum description must fail and demand description at the molecular level, it is tremendously successful as the basis for continuum-based calculations. This section is adapted from discussions in some previous primary accounts [31.7–9].

As expressed in a prominent fluid dynamics textbook [31.10]:

"In other words there is no relative motion between the fluid and the solid. This fact may seem surprising but it is undoubtedly true. No matter how smooth the solid surface or how small the viscosity of the fluid, the particles immediately adjacent to the surface do not move relative to it. It is perhaps not without interest that Newton's term for viscosity was 'defectus lubricitatis' – 'lack of slipperiness'. Even for a fluid that does not 'wet' the surface this rule is not violated."

Feynman noted in his lectures that the no-slip boundary condition explains why large particles are easy to remove by blowing past a surface, but small particles are not.

However, for many years it was observed that the *no-slip* boundary condition is not intuitively obviously and has been controversial for centuries. A pantheon of great scientists – among them, Bernoulli, Coulomb, Navier, Poisson, Stokes, Taylor, Debye, de Gennes – has worried about it. The compelling rational arguments – for and against – were divided by the pragmatic observation that predictions agree with experiments. The possibility of slip was discussed until recently, in mainstream literature, only in the context of the flow of polymer melts [31.11,12], though over the years persistent doubts were expressed [31.13]. *No-slip* contrasts with *slip* characteristic of highly viscous polymers [31.11, 12], monolayers of gas condensed on vibrated solids [31.14], superfluid helium [31.15], moving contact lines of liquid droplets on solids [31.16], and kinetic friction of liquid films less than 5–10 molecular dimensions thick [31.17, 18]. However, experimental capability and technical needs have changed – especially so with the emerging interest in microfluidics and microelectromechanical system (MEMS)-based devices.

The situation has changed but the enormous and enduring success of the no-slip assumption for modeling must be emphasized. It works beautifully provided that certain assumptions are met: a single-component fluid, a wetted surface, and low levels of shear stress. Then careful experiments imply that the fluid comes to rest within 1−2 molecular diameters of the surface [31.19–21]. But the necessary assumptions are more restrictive, and their applicability is more susceptible to intentional control than is widely appreciated. Generally, one may argue from the fact that fluid molecules are ("obviously") stuck to walls by intermolecular forces. The traditional explanation is that, since most surfaces are rough, the viscous dissipation as fluid flows past surface irregularities brings it to rest, regardless of how weakly or strongly molecules are attracted to the surface [31.7, 8, 22–24]. This has been challenged by accumulating evidence that, if molecularly smooth surfaces are wet only partially, hydrodynamic models work better when one uses instead *partial slip* boundary conditions [31.13, 21, 25–30]. Then the main issue would be whether the fluid molecules attract the surface or the fluid more strongly [31.9, 13, 27–34].

31.1.1 How to Quantify the Amount of Slip

To know what boundary condition is appropriate in solving continuum equations requires inquiry into microscopic particulars. Attention is drawn to unresolved topics of investigation and to the potential to purposely capitalize on *slip at the wall* for purposes of materials engineering.

The formal idea of a *slip length* is common. *Slip* signifies that, in the continuum model of flow, the fluid velocity at the surface is finite (the slip velocity, v_s) and increases linearly with distance normal to it. The slip velocity is assumed to be proportional to the shear stress at the surface, σ_s in the relation:

$$\eta v_s = b\sigma_s \, , \tag{31.1}$$

where η is the viscosity and b, the slip length, is the notional distance inside the surface at which the velocity equals zero, if the velocity profile is extrapolated inside the surface until it reaches zero. In much of the literature the slip length has been assumed to be a constant that characterizes the material response of a given fluid–surface pair, but evidence of additional dependence on velocity is discussed below.

It is unreasonable to expect this continuum description to yield microscopic information. One example of this was already given – the appearance of no-slip conditions when the microscopic reason is that flow irregularities pin the fluid to the wall [31.7, 8, 22–24]. Another example is apparent slip when a low-viscosity component in the fluid facilitates flow because it seg-

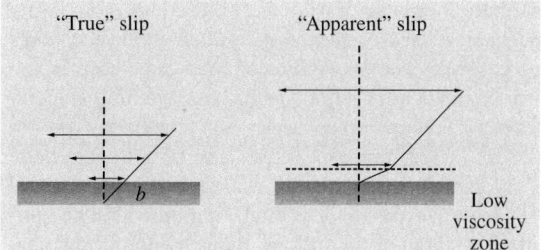

Fig. 31.1 Schematic illustration of the distinction between true slip (*left*) and apparent slip (*right*) in oscillatory flow, although for both cases the velocity of the moving fluid extrapolates to zero at a notional distance inside the wall and is finite where it crosses the wall. For true slip, this is literally so. It may also happen that a low-viscosity component in the fluid facilitates flow because it segregates near the surface. The velocity gradient is then larger nearest the surface because the viscosity is smaller. When specific real systems are investigated, structural and chemical materials analysis at the microscopic scale are needed to distinguish between these possibilities. After [31.7] with permission

regates near the surface [31.35, 36]. When conventional continuum mechanics contends with situations that are more complex than the model allows, one should resist the temptation to interpret literally the parameters in which the continuum mechanics model is couched. These examples emphasize instances where the notions of *stick* and *slip* are numerical conveniences not to be interpreted literally in terms of molecular mechanism. The appearance of slip owing to surface segregation of a low-viscosity component is illustrated in Fig. 31.1.

31.1.2 The Mechanisms that Control Slip in Low–Viscosity Fluids

Partial slip of so-called Newtonian fluids such as alkanes and water is predicted by an increasing number of computer simulations [31.25–27, 37–39] and, in the laboratory when forces are measured, systematic deviations from the predictions based on *stick* are found [31.7–9, 28–34, 40, 41]. Some sense of urgency comes from potential practical applications. Typical magnitudes of the slip length reported in the literature are submicron, so small that the practical consequences of slip would be minimal for flow in channels whose size is macroscopic. But if the channel size is very small, there are potential ramifications for microfluidics.

The simulations must be believed because they are buttressed by direct measurements. In the past, all laboratory reports of slip were based on comparing mechanical measurements of force to fluid mechanics models and hence were indirect inferences. Recently, optical methods were introduced to measure fluid velocity directly. For example, *Léger* and coworkers photo-bleached tracer fluorescent dyes and from the time rate of fluorescence recovery, measured in attenuated total reflection in order to focus on the region within an optical wavelength of the surface, the velocity of flow near the surface was inferred [31.28]. They reported slip of hexadecane near an oleophobic surface provided that it was smooth, but not when it was rough. *Tretheway* and *Meinhart* used laser particle image velocimetry of tracer latex particles to infer the velocity of water flow in microchannels [31.34]. They reported slip when the surface was hydrophobic but stick when it was hydrophilic. *Callahan* and coworkers demonstrated the feasibility of using nuclear magnetic resonance (NMR) velocity imaging, though this method has been used to date only for multicomponent fluids [31.42].

An important hint about the mechanism comes from the repeated observation that the amount of slip depends on the flow rate, in measurements using not only the surface forces apparatus (SFA) [31.8, 9, 31, 32] but also atomic force microscopy (AFM) [31.29, 41]. The main idea of all of these experiments is that two solids of mean radius of curvature R, at spacing D, experience a hydrodynamic force, F_H as they approach one another (or retreat from one another) in a liquid medium, thereby squeezing fluid out of (or into) the intervening gap. This force F_H is proportional to the rate at which the spacing changes, dD/dt (where t denotes time), is proportional to the viscosity, η (assumed to be constant), and is inversely proportional to D. The no-slip boundary condition combined with the Navier–Stokes equations gives to first order the following expression, known as the Reynolds equation:

$$F_H = f \frac{6\pi R^2 \eta}{D} \frac{dD}{dt} . \tag{31.2}$$

Higher-order solutions of the Navier–Stokes equations confirm that the lowest-order term is enough. The deviation of the dimensionless number f^* from unity quantifies the deviation from the classical no-slip prediction. The classical prediction is analogous when the surface spacing is vibrated. In that case a sinusoidal oscillatory drive generates an oscillatory hydrodynamic force whose peak we denote as $F_{H,peak}$. The peak velocity of vibration is $v_{peak} = d \cdot \omega$, where d is the vibration amplitude and ω the angular frequency of vibration. Studies show that, when the frequency and amplitude of oscillatory flow are varied, results depend on their product, v_{peak} and the deviations from (31.2) depend on v_{peak}/D. This ratio, the flow rate, is the ratio suggested by the form of (31.2).

Without necessarily assigning physical meaning to this quantity, it can be used as an alternative expression of the same data. Mathematical manipulation [31.43] shows that f^* and the slip length, b are related as:

$$f^* = 2 \frac{D}{6b} \left[\left(1 + \frac{D}{6b} \right) \ln \left(1 + \frac{6b}{D} \right) - 1 \right] . \tag{31.3}$$

31.1.3 Experimental

The slippery question was studied experimentally by testing the limits of ideas about interfacial force and surface roughness. Figure 31.2 sketches the experimental strategy and shows AFM (atomic force microscopy) images of some of the surfaces studied. Three strategies were used to vary surface roughness systematically. The first was based on using collapsed polymers. Narrow-distribution samples of diblock copolymers of polystyrene ($M_w = 55\,400$) and polyvinylpyridine ($M_w = 28\,200$), PS/PVP, were allowed to adsorb for a limited time from dilute (5×10^3 μg/ml) toluene solution onto freshly cleaved muscovite mica. They appeared to aggregate during the drying process. The remaining bare regions of mica were then coated with an organic monolayer of condensed octadecyltriethoxysilane (OTE) [31.44]. The contact angles against water and tetradecane were stable in time, which shows that the

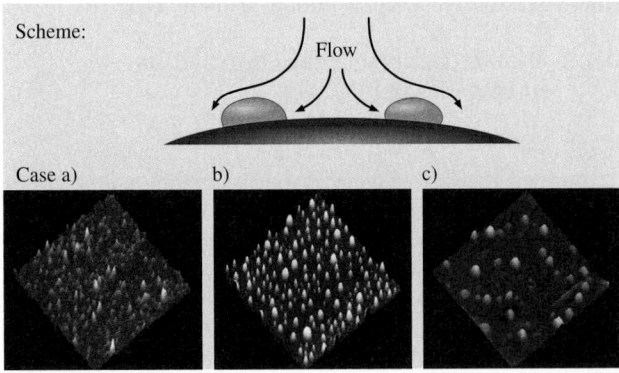

Fig. 31.2a–c The scheme of flow over a rough surface is shown schematically in the top portion. In the bottom panel, AFM images are shown of the following case: (**a**) self-assembled OTS layers; (**b**) PS/PVP-OTE layers (surface coverage $\approx 80\%$); (**c**) PS/PVP-OTE layers (surface coverage $\approx 20\%$). Each AFM image concerns an area $3\,\mu m \times 3\,\mu m$. After [31.8] with permission

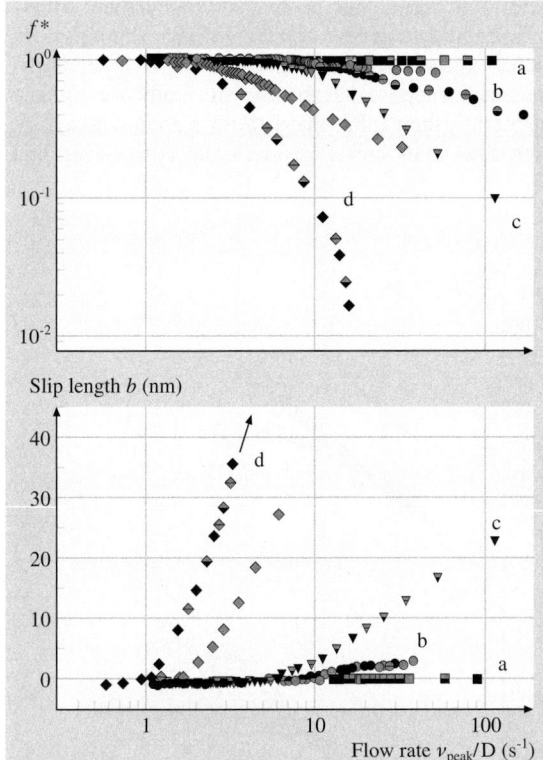

Fig. 31.3 f^* as a function of logarithmic flow rate v_{peak}/D (*top panel*) where f^* is defined in (31.2), and the equivalent slip length (*bottom panel*), for deionized water (*filled symbols*) and tetradecane (*open symbols*) between surfaces of different levels of rms surface roughness, specifically: 6 nm (case a; *squares*), 3.5 nm (case b; *circles*), 2 nm (case c; *down triangles*), 1.2 nm (case f; *hexagons*), 0.6 nm (case e, *upper triangles*) and molecularly-smooth (case d; *diamonds*). The data, taken at different amplitudes in the range of 0.3–1.5 nm and frequencies in the range 6.3–250 rad/s, are mostly not distinguished in order to avoid clutter. To illustrate the similarly successful collapse at these rough surfaces, data taken at the two frequencies 6.3 rad/s (*cross filled symbols*) and 31 rad/s (*semi-filled symbols*) for water are included explicitly. After [31.8] with permission

liquids did not penetrate them. AFM images in Fig. 31.2 quantify the roughness. To produce larger roughness values, mica was coated with condensed octadecyltrichlorosilane (OTS) after first saturating the cyclohexane deposition solution with H_2O to encourage partial polymerization in solution before surface attachment. In the third method, self-assembled monolayers of alkane thiols were formed on silver whose roughness was con-

trolled by a direct-current (DC) bias applied during sputter deposition. The silver was coated with octadecanethiol deposited from ethanol solution and placed in opposition with a molecularly smooth OTE surface.

According to (31.2), the *stick* prediction ($f^* = 1$) corresponds to a horizontal line and one observes in Fig. 31.3 that deviations from this prediction decreased systematically as the surface roughness increased. In addition, deviations from the predictions of the no-slip boundary condition are alternatively often represented as the *slip length* discussed above, the fictitious distance inside the solid at which the no-slip flow boundary condition would hold; the equivalent representation of this data in terms of the slip length is included in Fig. 31.3. While it was known previously that a very large amount of roughness is sufficient to generate *no-slip* [31.22–24], an experimental study in which roughness was varied systematically [31.7,8] suc-

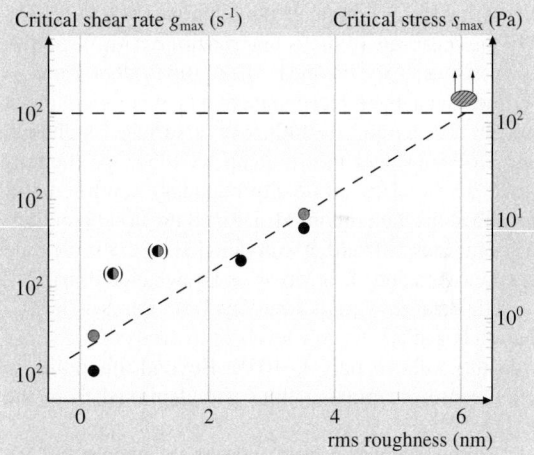

Fig. 31.4 Illustration that deviations from the traditional no-slip boundary condition depend systematically on surface roughness. Here the critical shear rate for onset of slip (*left ordinate*) and critical shear stress (*right ordinate*) are plotted semilogarithmically against rms surface roughness for flow of deionized water (*solid symbols*) and tetradecane (*open symbols and semi-filled symbols as identified in* [31.32]. The data in parentheses indicate the asymmetric situation – one surface was rough and the opposed surface was atomically smooth, as discussed in [31.32]. Maximum shear rate and shear stress on the crossed cylinders were calculated using known relations based on continuum hydrodynamics from [31.13, 42]. Specifically, $\gamma_{max} = A\sqrt{\dfrac{R}{D}\dfrac{v_{peak}}{D}}$ and $\sigma_{peak} = 1.378\eta R^{1/2}\dfrac{v_{peak}}{D^{3/2}}$. After [31.7] with permission

ceeded in quantifying how much actual roughness was needed in an actual system. The critical level of 6 nm considerably exceeded the size of the fluid molecules. As methods are known to achieve greater smoothness in MEMS devices, and potentially in microfluidic devices, this offers the promise of practical applications.

Observation of rate-dependent slip suggests that fluid is pinned up to some critical shear stress, beyond which it slips. However, some laboratories report slip *regardless* of flow rate [31.27, 29, 32–39]. Perhaps, the essential difference is that the magnitude of shear stress is larger in the latter experiments [31.45]. But in cases where slip is rate dependent, this affords a potential strategy by which to effect purposeful mixing in a microfluidic device. The idea would be simply to make some patches on the surface hydrophobic and other patches hydrophilic, so that when flow was slow enough it would be smooth, but when it was fast enough, mixing would result from jerkiness at the hydrophobic patches.

While it is true that slip at smooth surfaces is predicted based on computer simulations [31.25–27, 37–39], the shear rate of molecular dynamics (MD) simulations so much exceeds shear rate in those laboratory experiments that the direct connection to experiment is not evident. To quantify the influence of surface roughness, Fig. 31.4 considers the limit up to which predictions based on the classical no-slip boundary condition still described the data in Fig. 31.3. Since the no-slip boundary condition still held, it was valid [31.7–9] to calculate the shear rate and shear stress by known equations.

The data show that deviations from the no-slip prediction began at very low levels of hydrodynamic stress – on the order of only 1–10 Pa. Beyond this point, in some sense the moving fluid was depinned from the surface.

Slip need not necessarily be predicated on having surfaces coated with self-assembled monolayers to render them partially wetted, though this was the case in most of the studies cited so far. The no-slip boundary condition switches to partial slip when the fluid contains a small amount of adsorbing surfactant [31.30, 32].

31.1.4 Slip Can Be Modulated by Dissolved Gas

When experimental observations deviate from expectations based on the *stick* boundary condition, there are at least two alternative scenarios with microscopic interpretations. One might argue that the fluid viscosity depends on distance from the wall, but for Newtonian

fluids this would not be realistic. Why then do experimental data appear to undergo shear thinning with increasing values of the parameter v_{peak}/D, if it is unreasonable to suppose that the viscosity really diminished? Inspection shows that the data for smooth surfaces at high flow rates are consistent with a two-layer-fluid

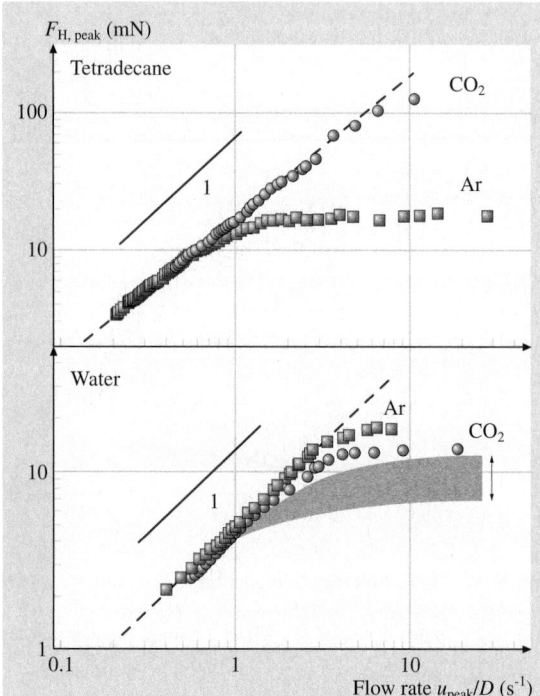

Fig. 31.5 Illustration that the onset of slip depends on dissolved gas, when simple Newtonian fluids flow past atomically smooth surfaces, either wetted or partially wetted. On log–log scales, the hydrodynamic force $F_{H,peak}$ is plotted against reduced flow rate, v_{peak}/D, such that a straight line of slope unity would indicate the *no-slip* condition assumed by (31.1). The vibration frequency is 9 Hz. *Top panel*: tetradecane flowing between the asymmetric case of a wetted mica surface on one side, a partially wetted surface of methyl-terminated self-assembled monolayer on the other side, prepared as described elsewhere [31.33]. *Filled symbols*, tetradecane saturated with carbon dioxide; *open symbols*, tetradecane saturated with argon. *Bottom panel*: deionized water flowing between mica surfaces that are wetted by this fluid. The hatched region of the graph shows the range of irreproducible results obtained when the gas dissolved in the water was not controlled. *Filled symbols*, water saturated with carbon dioxide; *open symbols*, water saturated with argon. After [31.7] with permission

model in which a layer $< 1\,nm$ thick, but with viscosity $10–20$ times *less* than the bulk fluid, adjoins each solid surface [31.9]. A possible mechanism to explain its genesis was proposed by *de Gennes* [31.46], who conjectured that shear may induce nucleation of vapor bubbles; once the nucleation barrier is exceeded the bubbles grow to cover the surface, and flow of liquid is over this thin gas film rather the solid surface itself. Indeed, it is likely that incomplete air removal from the solid surfaces can profoundly influence findings in these situations where surface roughness is so low. This has been identified by recent research as a likely source of the misnamed *long-range hydrophobic attraction* [31.47, 48]; gases also appear to influence the sedimentation rate of small particles in liquid [31.49].

Accordingly, similar experiments were performed, in which the surface forces apparatus was used to measure hydrodynamic forces of Newtonian fluids that had been purged with various gases. Dissolved gas strongly influences hydrodynamic forces, in spite of the fact that gas solubility is low.

Figure 31.5 (top panel) illustrates experiments in which a simple nonpolar fluid (tetradecane) was placed between a wetted mica surface on one side, and a partially wetted methyl-terminated surface on the other, using methods described in detail elsewhere [31.8, 9]. The surface–surface spacing of $10–100\,nm$ substantially exceeded the size of the fluid molecules. The spacings were vibrated with small amplitude at these spacings where the fluid responded as a continuum, and the magnitude of hydrodynamic force was measured as a function of the ratio v_{peak}/D suggested by (31.2). The experiments showed that, whereas textbook behavior [31.10] was nearly followed when the tetradecane had been saturated with carbon dioxide gas, massive deviations from this prediction were found when the tetradecane was saturated with argon. This makes it seem likely that argon segregated to the solid walls, creating a low-viscosity boundary layer, in this way greasing the flow of fluid past that surface. Presumably, the amount of segregation is a material property of the fluid, the chemical makeup of the surface, and the chemical identity of the dissolved gas. In this example, the fact that argon possesses low solubility in tetradecane may have made it more prone to segregate to the surfaces.

Indeed, when a solid wall is hydrophobic and immersed in water, the ideas of *Chandler* and coworkers [31.50] suggest that thermodynamics may assist the formation of a vapor phase near the wall. Recent force measurements support this idea [31.51, 52].

31.1.5 Slip Past Wetted Surfaces

The influence of dissolved gas (just discussed) casts doubt on a traditional assumption of work in this field, which is that slip arises because fluid–fluid intermolecular interactions are stronger than those between fluid and surface, i.e. that the surface must be wetted only partially. Yet for several years, there have been prominent counterexamples [31.28, 41]. Recent experiments show that dissolved gases can mediate apparent slip even for solid surfaces that are fully wetted by the flowing fluid.

Figure 31.5 (bottom panel) summarizes experiments in which deionized water was placed between wetted surfaces of mica and the surface–surface spacing of $10–100\,nm$ was vibrated with small amplitude in the manner described previously [31.7–9, 31–34]. Hydrodynamic force is plotted against the ratio, v_{peak}/D. It is obvious that the prediction based on (31.2), a straight line on the log–log plot with a slope of unity, was violated systematically when the hydrodynamic force reached a critical level. An intriguing point is that initial findings were found to be irreproducible (they varied within the range marked by the hatched lines in the graph) but became reproducible when the water was first deliberately saturated with gas. One observes that water saturated with argon appeared to *slip* at a slightly higher level of shear stress than water saturated with carbon dioxide, and that in both cases the limiting hydrodynamic force was larger than when the nature of the dissolved gas was not controlled.

This rich and complex sensitivity to the detailed materials chemistry of the system disappears, unfortunately, when surfaces are so rough that the *stick* boundary condition is produced trivially by the influence of surface roughness [31.7, 8, 22–24]. Therefore for scientific and practical reasons alike, these issues of flow past nearly smooth surfaces comprise fertile ground for future work.

31.1.6 The Purposeful Generation of Slip

Inspired by these ideas to design new engineering structures, one might strive to "grease" the flow of liquids past solid surfaces by altering the boundary condition. One strategy is to make the surfaces ultra-smooth [31.7–9]. Another (also mentioned already) is to add processing aids that segregate to the surface [31.30, 32, 36]. A third way is to purposefully use multicomponent fluids to generate concentration gradients and differential wetting to generate slip, as can occur even if there is no velocity gradient in the fluid [31.38]. These methods could potentially be used in nanomotors or nanopumps.

Alternatively, one may seek to maximize contact with air, which is exceedingly solvophobic. Readers will have noticed that water ubiquitously beads up on the leaves of plants. Some plants can display a contact angle that approaches 180°, even though water at a smooth surface of the same chemical makeup displays a much lower contact angle. A beautiful recent series of experiments from the Kao Corporation in Japan provided insight into why [31.53] – the surfaces are rough on many length scales [31.54, 55] and trap air beneath them. Readers will have noticed that, if one tilts a leaf, a drop of water on it rolls smoothly, because it rides mainly on a cushion of air, whose effect will be further discussed in the next section. It is the principle of an ingenious method introduced recently to lower the viscous drag when fluids [31.56] are caused to flow through pipes whose diameter is macroscopic. Of course, given a long enough period of time it is likely that the trapped gas would dissolve into the flowing fluid, but perhaps this effect could be enhanced by placing air nozzles along the walls of the tube and replenishing the trapped gas with a stream of inlet air.

A final method by which flow of a Newtonian fluid past surfaces may be facilitated is to *ciliate* the surfaces by coating with chain molecules – polymers, proteins, or sugars. Recent experiments using a surface forces apparatus (SFA) suggest a similar (but less dramatic) rate-dependent slip in this case also [31.31]. This is possibly related to fluid flow in biological organs whose surfaces are also extensively ciliated, such as blood vessels and the kidney [31.57].

31.1.7 Outlook

The textbook presentation of engineering fluid mechanics is often of a subject thoroughly understood, but recent experiments and simulations using smooth surfaces show behavior that is richer and more complex than had been supposed. The correct boundary condition appears to depend on physical chemical properties of the solid–liquid interface that are susceptible to rational control.

31.2 Hydrophobic Interaction and Water at a Hydrophobicity Interface

The role of water in physical situations from biology to geology is almost universally thought to be important but the details are disputed [31.1, 48, 50–52, 58–72]. For example, as concerns proteins, the side-chains of roughly half the amino acids are polar while the other half are hydrophobic; the non-mixing of the two is a major mechanism steering the folding of proteins and other self-assembly processes. As a second example, it is an everyday occurrence to observe beading-up of raindrops, on raincoats or leaves of plants. Moreover, it is observed theoretically and experimentally that, when the gap between two hydrophobic surfaces becomes critically small, water is ejected spontaneously [31.50, 51, 70–72] whereas water films confined between symmetric hydrophilic surfaces are stable [31.1]. Despite its importance, water exhibits many anomalous behaviors when compared to other fluids. Particularly, it presents some even more puzzling behaviors near hydrophobic surfaces. This section is adapted from discussions in several primary accounts published previously [31.51, 52].

In its liquid form, water consists of an ever-changing three-dimensional network of hydrogen bonds. Hydrophobic surfaces cannot form hydrogen bonds, and the hydrogen-bonding network must be disrupted. So what happens when water is compelled to be close to a hydrophobic surface? Energetically, it is expected that the system forms as many hydrogen bonds as possible, resulting in a preferential ordering of the water. Entropically, it is expected that the system orients randomly and thus samples the maximum number of states. Which of these two competing interactions dominates? What effect does the competition have on the dynamic and equilibrium properties of the system? The answers to these questions are still hotly debated. To help resolve this debate, static and dynamic interaction of water confined to a hydrophobic surface is studied by SFA.

31.2.1 Experimental

The atomically smooth clay surfaces used in this study, muscovite mica (hydrophilic) and muscovite mica blanketed with a methyl-terminated organic self-assembled monolayer (hydrophobic), allowed the surface separation to be measured, by multiple beam interferometry, with a resolution of $\pm 2-5$ Å. Surfaces were brought to the spacings described below using a surface forces apparatus modified for dynamic oscillatory shear [31.44, 73]. A droplet of water was placed between the two surfaces oriented in a crossed cylinder geometry. Piezo-electric bimorphs were used to produce and detect controlled shear motions. The deionized water was previously passed through a purification system, Barnstead Nanopure II (control experiments with water containing dissolved salt were similar). In experiments using degassed water, the water was either first boiled, then cooled in a sealed container, or vacuumed for 5–10 h in an oven at room temperature. The temperature of measurements was 25 °C.

In order to determine firmly that findings did not depend on details of surface preparation, three methods were used to render one surface hydrophobic. In order of increasing complexity, these were: (a) atomically smooth mica coated with a self-assembled monolayer of condensed octadecyltriethoxysilane (OTE), using methods described previously [31.44]; (b) mica coated using Langmuir–Blodgett methods with a monolayer of condensed octadecyltriethoxysilane; and (c) a thin film of silver sputtered onto atomically smooth mica, then coated with a self-assembled thiol monolayer. In method (a), the monolayer quality was improved by distilling the OTE before self-assembly. In method (b), OTE was spread onto aqueous HCl (pH= 2.5), 0.5 h was allowed for hydrolysis, the film was slowly compressed to the surface pressure $\pi = 20$ mN/m (3–4 h), and the close-packed film was transferred onto mica by the Langmuir–Blodgett technique at a creep-up speed of 2 mm/min. Finally the transferred films were vacuum baked at 120 °C for 2 h. In method (c), 650 Å of silver were sputtered at 120 V (1 Å/s) onto mica that was held at room temperature, and then octadecanethiol was deposited from 0.5 mM ethanol solution. In this case, AFM (atomic force microscopy; Nanoscope II) showed the rms (root mean square) roughness to be 0.5 nm. All three methods led to the same conclusions, summarized below.

31.2.2 Hydrophobic Interaction

A puzzling aspect of the hydrophobic attraction is that its intensity and range appear to be qualitatively different as concerns extended surfaces of large area, and small molecules of modest size [31.50, 60, 67, 74]. One difference is fundamental: the hydrogen-bond network of water is believed for theoretical reasons to be less disrupted near a single alkane molecule than near an extended surface [31.50, 60, 67, 74]. A second difference is phenomenological: direct measurement shows attractive

forces between extended surfaces starting at separations too large to be reasonably explained by disruption of the hydrogen-bond network. This conclusion comes from 20 years of research using the surface forces apparatus (SFA) and, more recently, atomic force microscopy (AFM). The onset of attraction, ≈ 10 nm in the first experiments [31.69, 75–78], soon increased by nearly an order of magnitude [31.79–81] and has been reported, in the most recent work, to begin at separations as large as 500 nm [31.71]. This has engendered much speculation because it is unreasonably large compared to the size of the water molecule (≈ 0.25 nm). The range of interaction decreases if the system (water and the hydrophobic surfaces) are carefully degassed [31.48, 82–86]. Water in usual laboratory experiments is not degassed, however, so it is relevant to understand the origin of long-range

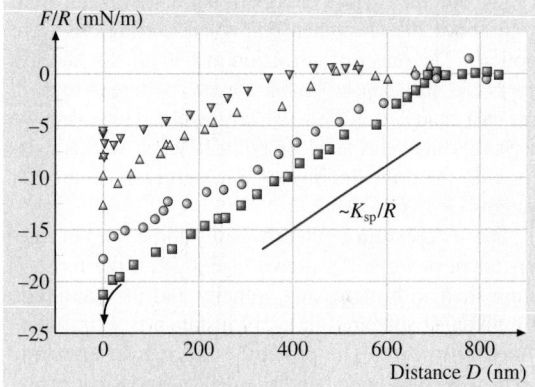

Fig. 31.6 Force–distance profiles of deionized water between hydrophobic surfaces (OTE monolayers on mica). Force F, normalized by the mean radius of curvature ($R \approx 2$ cm) of the crossed cylinders, is plotted against surface separation. Forces were measured during approach from static deflection of the force-measuring spring, while simultaneously applying small-amplitude harmonic oscillations in the normal direction with peak velocity $v_{\text{peak}} = d \times 2\pi f$ where d denotes displacement amplitude and f denotes frequency. This velocity was zero (*solid squares*), 7.6 nm/s ($d = 1.6$ nm, $f = 0.76$ Hz; *circles*), 26 nm/s ($d = 3.2$ nm, $f = 1.3$ Hz; *up triangles*), 52 nm/s ($d = 3.2$ nm, $f = 2.6$ Hz; *down triangles*). The pull-off adhesion forces ("jump out"), measured at rest and with oscillation, are indicated by the respective *semi-filled symbols*. The approach data follow the straight line with slope K_{sp}/R (*drawn separately as a guide to the eye*), indicating that they represented a spring instability ("jump in") such that the gradient of the attractive force exceeded the spring constant (K_{sp}), 930 N/m. After [31.51] with permission

attraction in that environment. A recent review summarizes the experimental and theoretical situation [31.81].

In the course of experiments intended to probe the predicted slip of water over hydrophobic surfaces [31.9, 45] (see the previous section), weakening of the long-range hydrophobic force to the point of vanishing was observed when the solid surfaces experienced low-level vibrations around a mean static separation.

The attraction recorded during the approach of OTE surfaces with a droplet of deionized water in between is plotted in Fig. 31.6 as a function of surface–surface separation (D). $D = 0$ here refers to a monolayer–monolayer contact in air. In water, the surfaces jumped into adhesive contact at 0 ± 2 Å. This *jump in* was very slow to develop, however. The pull-off force to separate the surfaces from contact at rest (113 mN/m in Fig. 31.6) implies, from the Johnson–Kendall–Roberts (JKR) theory [31.76], the surface energy of about $12 \, \text{mJ/m}^2$ (and up to about 30% less than this when oscillations were applied). The onset of attraction at 650 nm for the hydrophobic surfaces at rest is somewhat larger than in any past study of which we are aware. However, we emphasize that the level of pull-off force was consistent with the prior findings of other groups using other systems [31.1, 60–72].

These observations clearly imply some kind of rate-dependent process. As shown in Fig. 31.7, the force F diminished with increasing velocity and its magnitude at a given D appeared in every instance to extrapolate smoothly to zero. The possible role of hydrodynamic forces was considered but discarded as a possible explanation. Similar results (not shown) were also obtained when the surfaces were vibrated parallel to one another rather than in the normal direction. Some precedence is found in a recent AFM study that reported weakened hydrophobic adhesion force with increasing approach rate [31.85].

These observations remove some of the discrepancy between the range of hydrophobic forces between extended surfaces of macroscopic size [31.1, 60–72] and the range that is expected theoretically [31.50,60,67,74]. A tentative explanation is based on the frequent suggestion that the long-range hydrophobic attraction between extended surfaces stems from the action of microscopic or submicron-sized bubbles that arise either from spontaneous cavitation or the presence of adventitious air droplets that form bridges between the opposed surfaces [31.48,81,83–86]. The experiments reported here show that this effect required time to develop. Hydrophobic attraction at long range was softened to the point of vanishing when the solid surfaces were not stationary.

31.2.3 Hydrophobicity at a Janus Interface

Due to the strong propensity to repel water completely out of two hydrophobic surfaces, it is then interesting to consider the antisymmetric situation, a hydrophilic surface on one side (A) and a hydrophobic surface (B) on the other. The surface A prevents water from being expelled, to successfully retain a stabilized aqueous thin film at intimately close contact. The surface B introduces the hydrophobic effect. This Janus situation is shown in the cartoon of Fig. 31.8.

Similar to the force between two OTE surfaces, long-range attraction was also observed at the Janus interface, as shown in the inset of Fig. 31.8. The opposed surfaces ultimately sprang into contact from $D \approx 5$ nm and upon pulling the surfaces apart, an attractive minimum was observed at $D = 5.4$ nm. The surfaces could be squeezed to a lower thickness, $D \approx 2.0$ nm. Knowing that the linear dimension of a water molecule is ≈ 0.25 nm [31.1], the thickness of the resulting aqueous films amounted to the order of 5–20 water molecules, although (see below) it is not clear that molecules were distributed evenly across this space. Below we discuss shear results.

In the shear measurements, the sinusoidal shear deformations were gentle – the significance of the resulting linear response being that the act of measurement did not perturb their equilibrium structure (linear responses were verified from the absence of harmonics in the time dependence of shear motion). Using techniques that are

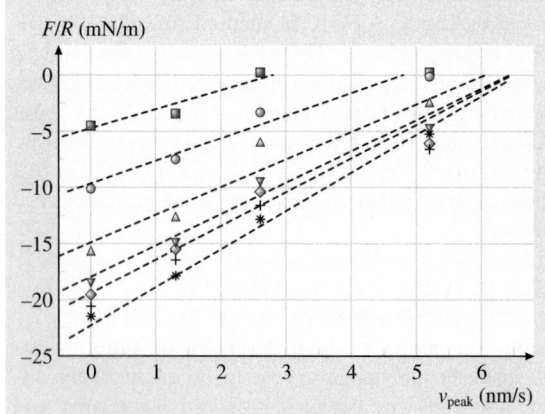

Fig. 31.7 The attractive force (F/R) at seven different surface separations (D) is plotted against peak velocity. The film thickness was $D = 720$ nm (*squares*), 540 nm (*circles*), 228 nm (*up triangles*), 116 nm (*down triangles*), 63 nm (*diamonds*), 17 nm (*crosses*), 5 nm (*stars*). After [31.51] with permission

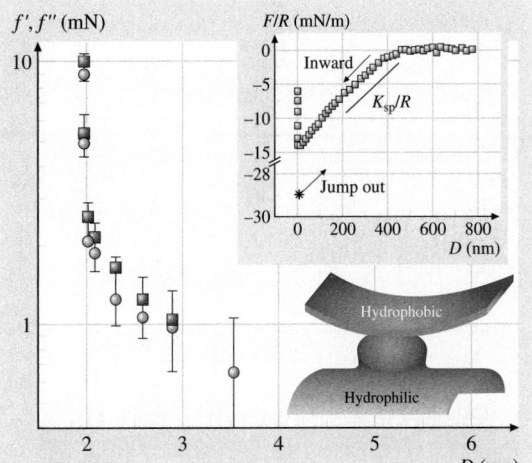

Fig. 31.8 Deionized water confined between a hydrophilic surface on one side and a hydrophobic surface on the other (*cartoon*). The cartoon is not to scale because the gap thickness is nanometers at closest approach and the droplet size (≈ 2 mm on a side) vastly exceeds the contact zone ($\approx 10\,\mu$m on a side). The main figure shows the time-averaged viscous (*circles*) and elastic (*squares*) shear forces measured at 1.3 Hz and 0.3-nm deflection, plotted semilogarithmically against surface separation for deionized water confined between OTE deposited onto mica using the Langmuir–Blodgett (LB) technique (shear impulses were applied to this hydrophobic surface). The *inset* shows the static force–distance relations. Force, normalized by the mean radius of curvature ($R \approx 2$ cm) of the crossed cylinders, is plotted against the thickness of the water film ($D = 0$ refers to contact in air). The pull-off adhesion at $D \approx 5.4$ nm is indicated by a *star*. The *straight line* with slope K_{sp}/R indicates the onset of a spring instability where the gradient of attractive force exceeds the spring constant (K_{sp}), 930 N/m. Following this *jump* into contact, films of stable thickness resulted, whose thickness could be varied in the range $D = 1$–4 nm with application of compressive force. After [31.52] with permission

well known in rheology, from the phase lag and amplitude during oscillatory excitation the responses to shear excitation were decomposed into one in-phase component (the elastic force, f') and one out-of-phase component (the viscous force, f'') [31.87]. Figure 31.8 (main portion) illustrates responses at a single frequency and variable thickness. The shear forces stiffened by more than one order of magnitude as the films were squeezed. It is noteworthy that, when molecularly thin aqueous films are confined between clay surfaces that are

symmetrically hydrophilic, deviations from the response of bulk water appear only at smaller separations [31.88]; evidently the physical origin is different here. Moreover, at each separation the elastic and viscous forces were nearly identical. The equality of elastic and viscous forces proved to be general, not an accident of the shear frequency chosen. Again this contrasts with recent studies of molecularly thin water films between surfaces that are symmetrically hydrophilic [31.88]. The magnitudes of the shear moduli in Fig. 31.8 are "soft" – something like those of agar or jelly. They were considerably softer than for molecularly thin aqueous films between symmetrically hydrophilic surfaces. This again emphasizes the different physical origin of shear forces in the present Janus situation.

Figure 31.9 illustrates the unusual result that the shear forces scaled in magnitude with the *same* power law, the square root of excitation frequency. This behavior, which is intermediate between solid and liquid, is often associated in other systems with dynamical heterogeneity [31.89, 90]. By known arguments it indicates a broad distribution of relaxation times rather than any single dominant one [31.87]. The slope of $1/2$ is required mathematically by the Kramers–Kronig relations if $G'(\omega) = G''(\omega)$ [31.87]; its observation lends credibility to the measurements. Figure 31.9 (main panel) illustrates this scaling for an experiment in which data were averaged over a long time. The inset shows that the same was observed using other methods to prepare a hydrophobic surface. In all of these instances, $\omega^{1/2}$ scaling was observed regardless of the method of rendering the surface hydrophobic. But to observe clean $\omega^{1/2}$ scaling required extensive time averaging – see later.

In repeated measurements at the same frequency, we observed giant fluctuations (± 30–40%) around a definite mean, as illustrated in Fig. 31.10, although water confined between symmetric hydrophilic–hydrophilic surfaces (bottom panel) did not display this. It is extraordinary that fluctuations did not average out over the large contact area ($\approx 10\,\mu$m on a side) that far exceeded any molecular size. The structural implication is that the confined water film comprised some kind of fluctuating complex – seeking momentarily to dewet the hydrophobic side by a thermal fluctuation in one direction, but unable to because of the nearby hydrophilic side; seeking next to dewet the hydrophobic side by a thwarted fluctuation in another direction; and so on. Apparently, nearby hydrophobic and hydrophilic surfaces may produce a quintessential instance of competing terms in the free energy, to satisfy which

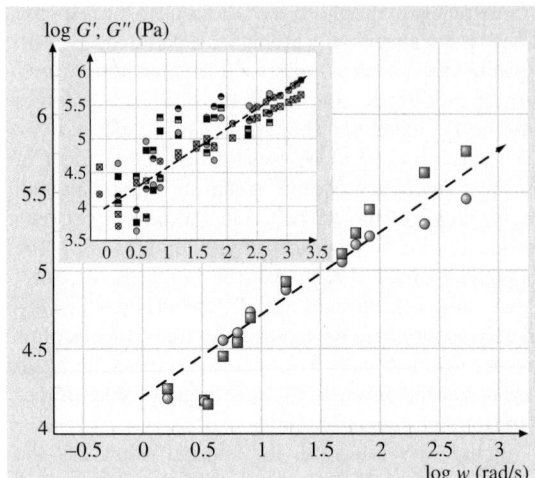

Fig. 31.9a–c The frequency dependence of the momentum transfer between the moving surface (hydrophobic) and the aqueous film with an adjoining hydrophilic surface is plotted on log–log scales. Time-averaged quantities are plotted. On the *right-hand ordinate* scale are the viscous, g'' (*circles*), and elastic, g' (*squares*) spring constants. The equivalent loss moduli, G''_{eff} (*circles*), and elastic moduli, G'_{eff} (*squares*), are on the *left-hand ordinate*. All measurements were made just after the jump into contact shown in Fig. 31.6, i. e. at nearly the same compressive pressure ≈ 3 MPa. The main panel, representing LB-deposited OTE, shows $\omega^{1/2}$ scaling after long-time averaging, 0.5–1 h per datum. The *inset* shows comparisons with using other methods to produce a hydrophobic surface. In those experiments the thickness was generally $D = 1.5$–1.6 nm but occasionally as large as 2.5 nm when dealing with octadecanethiol monolayers. Symbols in the *inset* show data averaged over only 5–10 min with hydrophobic surfaces prepared with (**a**) a self-assembled monolayer of OTE (*half-filled symbols*), (**b**) Langmuir–Blodgett deposition of OTE (*crossed symbols*), or (**c**) deposition of octadecanethiol on Ag (*open symbols*). As in the main figure, *circles* denote viscous forces, *squares* denote elastic forces. Scatter in this data reflects shorter averaging times than in the main part of this figure (Fig. 31.8). After [31.52] with permission

there may be many metastable states that are equally bad (or almost equally bad) compromises [31.91, 92]. This suggests the physical picture of flickering capillary-type waves, sketched hypothetically in Fig. 31.11. These proposed long-wavelength capillary fluctuations would differ profoundly from those at the free liquid–gas interface because they would be constricted by the nearby solid surface.

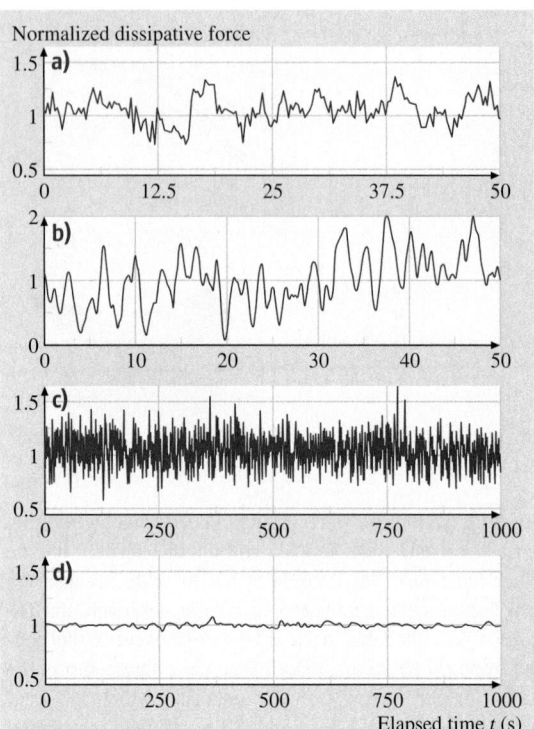

Fig. 31.10a–d The prominence of fluctuations for water confined in a Janus interface is illustrated. In panels (**a**)–(**c**), the viscous forces, normalized to the mean (at 1.3 Hz), are plotted against time elapsed. In panel (**a**), the surfaces were first wetted with ethanol to remove adsorbed gas, then flushed with degassed, deionized water. In panel (**b**), the ethanol rinse was omitted. Panel (**c**) shows a long-time trace for data taken under the same conditions as for panel (**b**). Panel (**d**) illustrates that water confined between symmetric hydrophilic–hydrophilic surfaces (panel (**d**)) did not display noisy responses. Confined ethanol films likewise failed to display noisy responses (not shown). After [31.52] with permission

The power spectrum is the decomposition of the traces into their Fourier components whose squared amplitudes are plotted, on log–log scales, against Fourier frequency in Fig. 31.11. In the power spectrum one observes, provided the Fourier frequency is sufficiently low, a high level of "white" frequency-independent noise. But the amplitude began to decrease beyond a threshold Fourier frequency ($f \approx 0.001$ Hz), about 10^3 times less than the drive frequency. Other experiments (not shown) showed the same threshold Fourier frequency when the drive frequency was raised from 1.3

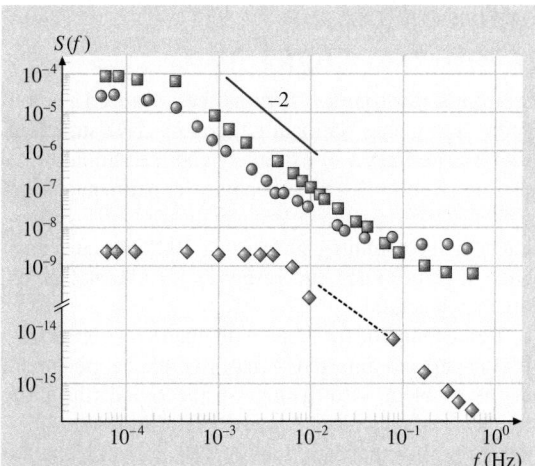

Fig. 31.11 The power spectrum for deionized water and the hydrophobized surface comprised of OTE deposited by the LB technique (*squares*); degassed deionized water and the hydrophobized surface comprised of an octadecanethiol monolayer (*circles*); and water containing 25 mM NaCl confined between symmetric hydrophilic–hydrophilic surfaces (*diamonds*). To calculate the power spectra the time elapsed was at least 10^5 s (the complete time series is not shown). The power-law exponent is close to -2. After [31.52] with permission

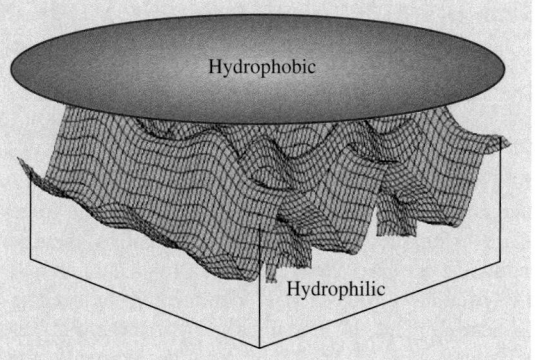

Fig. 31.12 Schematic illustration of the capillary waves of water meeting the hydrophobic surface with a flickering vapor phase in between

to 80 Hz, and therefore it appears to be a characteristic feature of the system. It defines a characteristic time for rearrangement of some kind of structure, $\approx 10^3/2\pi$s; we tentatively identify this with the lifetime of bubbles or vapor (see below). The subsequent decay of the power spectrum as roughly $1/f^2$ suggests that these fluctuations reflect discrete entities, as smooth variations would decay more rapidly. Noise again appeared to become "white" but with an amplitude 10^4 times smaller at a Fourier frequency of $f > 0.1$ Hz but the physical origin of this is not evident at this time.

The possible role of dissolved gas is clear in the context of our proposed physical explanation. Indeed, submicron-sized bubbles resulting from dissolved air have been proposed to explain the anomalously long range of the hydrophobic attraction observed between extended surfaces [31.48, 51, 70–72]. To test this idea, we performed control experiments using degassed water. The power spectrum, shown in Fig. 31.12, was nearly the same. Although we cannot exclude that a small amount

of residual dissolved gas was responsible, this method of degassing is reported to remove long-range hydrophobic attraction [31.71], whereas the comparison in Fig. 31.12 shows the consequence in the present situation to be minor. We conclude that the effects reported in this chapter do not appear to stem from the presence of dissolved gas.

From recent theoretical analysis of the hydrophobic interaction the expectation of dewetting emerges – it is predicted that an ultrathin gas gap, with a thickness on the order of 1 nm, forms spontaneously when an extended hydrophobic surface is immersed in water [31.50,61,67]. The resulting capillary-wave spectrum does not appear to have yet been considered theoretically, but for the related case of the free water–vapor interface, measurements confirm that capillary waves with a broad spectrum of wavelengths up to micrometers in size contribute to its width [31.93]. On physical grounds, the thin gas gap suggested by our measurements should also be expected to possess soft modes with fluctuations whose wavelength ranges from small to large. From this perspective, we then expect that the experimental geometry of a Janus-type water film, selected for experimental convenience, was incidental to the main physical effect.

This has evident connections to understanding the long-standing question of the structure of aqueous films near a hydrophobic surface and may have a bearing on understanding the structure of water films near the patchy hydrophilic–hydrophobic surfaces that are so ubiquitous in proteins.

31.3 Ultrafast Spectroscopic Study of Confined Fluids: Combining Ultra-Fast Spectroscopy with Force Apparatus

The surface forces apparatus (SFA) modified to measure interfacial rheology has been used widely in last few years to study the viscoelasticity of molecularly thin fluid films [31.1–4, 94–98]. A recent application of this technique is described in Sect. 31.2. Though the force-based techniques are powerful and sensitive, they are indirect. The observation of structure-related transitions, e.g., oscillatory forces [31.1], confinement-induced solidification [31.4, 94, 95], and stick–slip motion in SFA experiments [31.17, 96, 97] have not been directly verified experimentally using an independent technique.

The power of scattering, microscopy and spectroscopic techniques in the studies of confined fluids has been of speculative interest for a long time. However, there are only a few reported successes, primarily because of technical difficulties of combining these forms of techniques with SFA. Neutron and X-ray diffraction methods are very powerful for direct determination of structures at the nanoscale. Recently developed X-ray surface forces apparatus permits simultaneous X-ray diffraction and direct normal and lateral forces measurements of complex fluids under shear and confinement [31.98]. *Safinya* et al. have investigated the structure of thin liquid crystal films under confinement using this apparatus [31.99]. The deep penetrating power of neutrons and the ability to substitute hydrogen with deuterium in many liquid systems can be exploited to measure the molecular density and orientation of confined fluids by using neutron diffraction [31.100]. The structure of end-grafted polymer brush layers have been investigated in this manner. Successful utilization of this method requires one to devise an apparatus that can keep single-crystal substrates of quartz or sapphire with areas up to tens of square centimeters parallel at controlled and well-defined separations [31.101]. So far, both neutron and X-ray confinement cells are limited to confining gaps ranging from several hundred angstroms to millimeters and are not capable of studying ultrathin ($\approx$ nm) liquid films. This intermediate length scale is more suited to study complex fluids, e.g., long polymer chains, colloidal particles and biological cells, where self-organized structures of larger scale come into play. For simple fluids, the difficulty arises because, as the film thickness decreases, the total number of scatterers decreases and the signal-to-noise ratio presents severe limitations. It is difficult to distinguish with sufficient confidence the structure of a molecularly thin fluid film from that of the thicker solids that envelop it. Synchrotron X-ray sources, such as the advanced photon source at Argonne National Laboratory have sufficient flux for experiments of extremely confined liquid films possible. Recent X-ray reflectivity experiments have confirmed the expected layering in the direction perpendicular to the confining surfaces[31.102], but questions about in-plane order and responses to external fields remain conjectural.

The interaction of light with matter – for example Raman and infrared – has impressive potential, but the problem is to distinguish the signal (the fluid monolayer) from all the noise (the sliding surfaces and the solids beneath them). The microrheometer developed by *Dhinojwala* et al. can readily be combined with spectroscopy (Fourier-transform infrared spectroscopy and dielectric spectroscopy) or scattering (X-ray and neutron) techniques [31.103]. It uses two parallel optically flat windows plates whose separation can be controlled from a few tens of nm to tens of μm, but is more suited to thicker (0.1–10 μm) films. It has been used for in situ study of shear-induced molecular orientation of nematic liquid crystals by using Fourier-transform infrared time-resolved spectroscopy (FTIR-TRS) synchronized with the shear motion [31.104]. By replacing one of the plates with a prism, recently it has been shown that this rheometer can be combined with the surface-sensitive technique of infrared-visible sum-frequency generation (SFG) in the total internal reflection (TIR) geometry [31.105]. This combination can be used to probe the orientation, alignment and relaxation modes of organic molecules at the buried interface in a condition of flow or shear. Some years back it was shown that SFG can be combined with the surface forces apparatus to study nanometer-thick films of self-assembled monolayer confined between atomically smooth mica surfaces [31.106], but implementation of this approach to other experimental situations, such as confined fluids still presents significant challenges.

Another problematic issue arises in interpreting experiments that measure mechanical properties, such as the SFA or AFM. The measurement generates a single number, the force, but although the surface separations are molecular, the areas of interaction are macroscopic. So this force is the result of the average response of a large collection of molecules. Recent advances in optical spectroscopy and microscopy have made it possible not only to detect and image single molecules, but to conduct spectroscopic measurements and monitor dynamic processes as well at the level of single (or a handful of) molecules [31.107, 108]. These studies illustrate their utility to dissect the distributions around the average for processes such as diffusion or chemical reactions. In many of these experiments, a fluorescent molecule is doped into the sample, which acts as a probe of its local environment [31.109]. Monitoring motions of the probe over time and in the presence of external fields can offer insights into changes in this local environment within which the dye molecule is embedded. However, to integrate force measurements using SFA with concurrent measurements using fluorescence spectroscopy required significant modification of the usual methods [31.110]. In the following we discuss the challenges of combining SFA with single-molecule-sensitive spectroscopy techniques. This section is adapted from discussions in several primary accounts published previously [31.109, 110].

31.3.1 Challenges

One of the major difficulties is to detect and collect fluorescence efficiently and to separate it from background noise. Background originates from many sources: Raman and Rayleigh scattering, fluorescence owing to impurities in the solvent, and from the substrates, which includes the lens, glue and mica (the glue attaches a cleaved mica sheet onto the supporting cylindrical lens in SFA experiments). Typical background counts can far exceed those from a dilute concentration of fluorophore molecules doped inside the sample of interest.

Another type of challenge comes from the limited photochemical lifetime of a fluorophore. A common fluorophore photobleaches irreversibly after emitting a finite number of photons (10^5–10^6). This problem becomes severe in ultrathin films, where the dynamics can become slower and a dye molecule resides for a long time within the laser focus.

A third difficulty is the necessity to perform spectroscopy at the same time as multiple-beam interferometry (MBI) to determine the film thickness. Traditionally a silver coating of thickness ≈ 63 nm is used at the back side of mica for the purpose of MBI, but the high reflectivity of silver from the infrared to UV regime excludes this possibility here.

The final challenge is to incorporate the SFA and the needed optics. As the signal must be as large as possible, the maximum possible amount of fluorescence from the fluorophore of interest should be detected for a successful experiment. A high numerical aperture (N.A.) objective is desirable but such objectives have a very small working distance (≈ 1–2 mm). This requires significant modification of the traditional SFA in order to make it possible to focus the laser beam on the sample.

We recently succeeded in implementing the technique of fluorescence correlation spectroscopy (FCS), which can measure the translational diffusion with surface forces measurement and friction studies within the SFA [31.111]. The scientific objective of building this integrated platform was to answer questions such as: how is the rate of molecular probe diffusion, within a confined fluid, related to the stress relaxation time? Is there significant collective molecular motion or dynamical heterogeneity? What happens to the molecules during the stick–slip motion? The principle of the FCS technique and the experimental set-up of the combined platform are described below.

31.3.2 Principles of FCS Measurement

Fluorescence correlation spectroscopy (FCS) is an experimental method to extract information on dynamical processes from the fluctuation of fluorescence intensity [31.112]. The technique has enjoyed widespread application recently in the field of chemical biology because of its ability to access to a multitude of parameters with biological relevance [31.113, 114]. The fluctuation of fluorescence, when dye molecules are dilute, can in principle result from diffusion, aggregation, or chemical reaction. Compared to other techniques for studying diffusion problems, such as quasi-elastic light scattering (QELS), fluorescence recovery after photobleaching (FRAP), and laser-induced transient grating spectroscopy, FCS presents the unique capability of measuring extremely dilute systems with high spatial resolution (down to the optical diffraction limit). On the average there can be as few as 1–5 dye molecules within the ≈ 1 fl volume element of the focused laser beam. However, these dye molecules move in and out due to Brownian motion, causing intensity fluctuations which can be observed as low-frequency noise on the mean fluorescence signal (Fig. 31.13). By inspecting the autocorrelation function of this fluctuation,

$$G(\tau) \equiv \langle \delta I(t) \delta I(t+\tau) \rangle / \langle \delta I(t) \rangle^2 , \qquad (31.4)$$

(here I denotes fluorescence intensity and t is the time variable), and by choosing a suitable model to analyze it, the rate of dynamic process is obtained [31.112].

If the primary reason for fluctuation is translational diffusion, and assuming that the fluorescence characteristics of the diffusing molecules do not change while traversing the laser volume, one can use Fick's second law to calculate the translational diffusion coefficient (D) from the autocorrelation function by using the relation [31.115],

$$G(\tau) = G(0)/\left(1 + 8D\tau/\omega_0^2\right). \tag{31.5}$$

This result follows from the convolution of the concentration correlation with the spatial profile of the laser focus, which has been assumed to be a two-dimensional Gaussian of width ω_0. The magnitude of the autocorrelation function at time zero, $G(0)$, is related to the average number of fluorophores (N) in the observation volume by the relation [31.116]

$$G(0) = 1/(2\sqrt{2}N). \tag{31.6}$$

Molecular mobilities can be measured over a wide range of characteristic time constants from $\approx 10^{-3}$ to 10^3 ms by using this technique.

Fluctuation analysis is best performed if the system under observation is restricted to very small ensembles and if the background is efficiently suppressed. These can be accomplished by a combination of very low sample concentrations ($\approx$ nanomolar) with extremely small measurements volumes ($\approx$femtoliter). The excitation of the fluorophores can be performed either with two photons using a pulsed laser or with one photon using continuous-wave lasers [31.115]. In one-photon FCS, spatial resolution is obtained with a confocal set-up, in which a small pinhole inserted into the image plane can reject the out-of-focus fluorescence. For two-photon excitation on the other hand, simultaneous (within $\approx 10^{-15}$ s) absorption of two lower-energy photons of approximately twice the wavelength is required for a transition to the excited state. Mode-locked lasers providing short pulses ($\approx 10^{-13}$ s) with a high repetition rate (10^8 Hz) can provide sufficient photon flux densities for two-photon processes. As the excitation probability is proportional to the mean square of the intensity, it results in inherent depth discrimination. Additionally, two-photon excitation improves the signal-to-background ratio considerably. As the most prominent scattering came from the incident light, which is well separated in wavelength from the induced fluorescence, this makes it easy to separate the fluorescence emission from the excitation light and the scattered light. However, the photobleaching rates with two-photon excitation are significantly enhanced with respect to

one-photon excitation at comparable photon-emission yields [31.117].

31.3.3 Experimental Set-up

A schematic diagram of the method of combing fluorescence correlation spectroscopy with the surface forces apparatus is shown in Fig. 31.12. The FCS portion of the set-up consists of three major parts: light source, microscope and data acquisition [31.110]. A femtosecond Ti:sapphire laser, which typically generates laser pulses with full width at half maximum (FWHM) of 100 fs at a repetition rate of 80 MHz can be used for the two-photon excitation of the fluorophores. An inverted microscope serves as the operational platform for the whole experiment. The excitation light is focused onto the sample with an objective lens of high N.A. and the emitted light is collected through the same objective and is detected by a photomultiplier tube (PMT) or avalanche photodiode (APD). The photon counting output is recorded by an integrated FCS data-acquisition board and data analysis can be performed with commercial or home-written software. By introducing the laser beam through the objective lens, a small excitation volume (≈ 1 fl) is generated within the sample. The lateral dimension of the excitation spot is about $\approx 0.5\,\mu$m, which can be determined by a calibration experiment using widely used dyes, such as fluorescein, whose diffusion coefficient in water is known to be $\approx 300\,\mu\text{m}^2/\text{s}$. The excitation power at the sample needs to be less than 1 mW to avoid photobleaching and heating effects of the sample.

The modified surface forces apparatus sat directly on the microscope stage. The traditional crossed-cylindrical geometry produced a circular contact of parallel plates when the crossed cylinders were squeezed together such that they flattened at the apex. Using an inchworm motor, separation of the surfaces can be controlled from nanometers to millimeters. To determine the separation between the surfaces, the traditional silver sheets for interferometric measurements of surface spacing in the SFA need to be replaced by multilayer dielectric coatings [31.118]. These multilayers can be produced by successive evaporation of layers (typically 13 or 15) of TiO_x and Al_2O_3 by electron-beam evaporation. The optical thickness of each layer was approximately $\lambda/4$ ($\lambda \approx 650$ nm), as determined by the optical monitor within the coating chamber. The thickness of each coating determines the windows of reflectivity and translucency. This approach can produce high reflectivity in the region $600-700$ nm, as well as translucent

Fig. 31.13a–d Schematic illustration of the utility of fluorescence correlation spectroscopy in a confined geometry: (**a**) A fluorescent molecule is doped within an ultrathin film of fluids (e.g., simple alkanes, polymers, colloidal particles) confined between two solid surfaces. Photon emission counts can fluctuate with time (**c**) resulting from the diffusion of fluorophores through the laser focus (**b**). From the autocorrelation function of this fluctuation $G(\tau)$, the rate of dynamic process can be obtained (**d**). Calculated $G(\tau)$ for pure Brownian diffusion (*dashed curve*) and flow superimposed with diffusion (*solid curve*) are shown

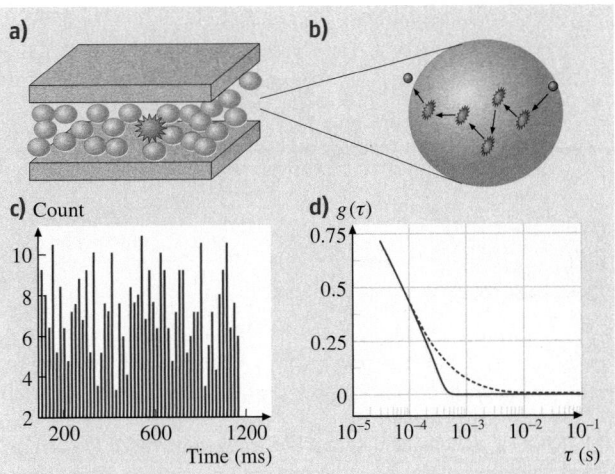

windows in the region $\approx 800\,\text{nm}$ (to allow fluorescence excitation) and $400\text{–}550\,\text{nm}$ (to detect fluorescence). The reflectivity as a function of wavelength is shown in Fig. 31.13 for the bare mica surface and for surfaces with different numbers of multilayers.

The same set-up (Fig. 31.14) with some modification can be used to probe molecular rotational diffusion. In the ground state fluorophores are all randomly oriented. When excited by polarized light, only those fluorophores that have their dipole moments oriented along the electric vector of the incident light are preferentially excited. So the excited-state population is not randomly oriented, instead there is a larger number of excited molecules having their transition moments oriented along the polarization direction of the incident light. This anisotropy of orientation can be determined by measuring the intensity of light polarized parallel to the incident light and perpendicular to the incident light. This preferential

anisotropy at time zero decays due to the rotational diffusion of the dye molecule, following an exponential law with a *characteristic rotation time* (τ). As the typical rotational time ranges from picoseconds to nanoseconds, the dynamics on much smaller length scales, that of only one or two nm, can be investigated by this method. Translational diffusion experiments by FCS, on the other hand, (as discussed in the following sections) involves a large distance – the probe molecules travel hundreds of nm into and out of the laser spot – to produce a signal. Therefore, it is more sensitive to the *global* environment of the molecules.

31.4 Contrasting Friction with Diffusion in Molecularly Thin Films

In FCS experiments, the magnitude of the fluctuation autocorrelation function scales inversely with the number of fluorescent molecules in the observation volume (31.6). Large fluorophore concentrations more than micromolar are not efficient in FCS, because $G(0)$ eventually becomes too small for fluctuation analysis. Typical dye concentration for confined fluid experiments is kept at $50\,\text{nM}$. A key point for these experiments is to find systems in which adsorption of the fluorophore would not complicate the situation. In other words, the fluids themselves, not the fluorophores, should be attracted preferentially to the confining solid surfaces [31.111]. This point can be verified by scanning the laser focus vertically from within the mica, through the surface, into the bulk fluids, and observing that there is no jump in fluorescence counts as the surface is crossed. Finally, one needs to be sensitive to the concern

that, when using fluorophores to probe local micro-environments, micro-environments might be perturbed by their presence. Therefore, it is essential to perform normal and shear force experiments with and without the presence of dye molecules to verify that these are not affected. This section is adapted from discussions in several primary accounts published previously [31.111].

In SFA experiments, a drop of the fluid for study is placed between the two mica sheets, oriented as crossed cylinders so that in projection the geometry os a sphere against a flat (Fig. 31.16). In the study of surface forces, it is well known that, as surfaces separated by fluid are brought together, fluid drains smoothly until a thickness of $5\text{–}10$ molecular dimensions, at which point the fluid supports stress owing to packing of molecules at the surface [31.1]. When rounded surfaces of this kind are pressed together, separated by

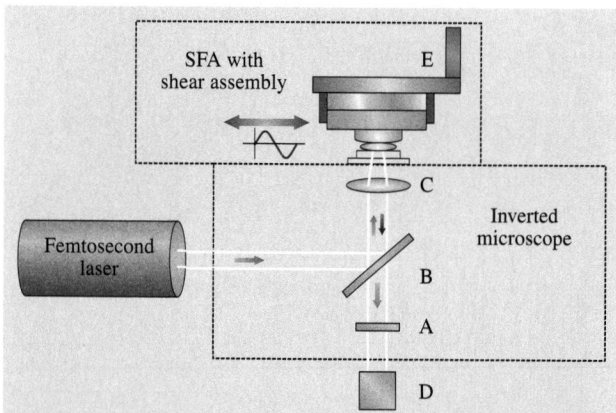

Fig. 31.14a–e Schematic diagram of the assembly used to perform fluorescence correlation spectroscopy within a surface forces apparatus equipped for shear experiments. A miniaturized surface forces apparatus sits on a microscope stage. A femtosecond pulsed laser excites fluorescent dye molecules within a molecularly thin liquid film contained between two opposed surfaces of muscovite mica. (a) colored filter to remove the residual excitation light ($\lambda = 800$ nm) from the fluorescence light (400–550 nm) which is collected by the single-photon counting module (d). (b) Dichroic mirror and (c) the objective lens used to focus and collect the light. An inchworm motor (e) controls the separation of the surfaces from nanometers up to several micrometers

fluid, the curved surfaces flatten at the apex to form parallel plates. The resulting inhomogeneous pressure distribution over the contact region is well known in the field of tribology. It is approximately *Hertzian*; zero at the edges of the contact zone and, at the center, 3/2 the mean value [31.119]. The Hertzian model is generally a good approximation in the absence of strongly attractive forces.

Figure 31.15 shows results for two different fluid systems: (a) propane diol containing ≈ 50 nM rhodamine 123, and (b) octamethylcyclotetrasiloxane (OMCTS), containing ≈ 50 nM coumarin. Propane diol is a low-viscosity fluid (≈ 0.4 Poise) with a glass-transition temperature far below room temperature ($T_g = -105\,°C$). The OMCTS molecule is ring-shaped; it is the cyclic tetramer of dimethylsiloxane. It has a viscosity much like water (≈ 0.002 Poise) and possesses the intriguing feature that it crystallizes at 1 atm near room temperature ($T_m = 17\,°C$), thus enhancing the possibility that a confinement-induced elevation of the melting temperature might be detected. There is a long tradition of considering it to constitute a model system when studying friction and surface-induced structure

of nonpolar confined fluids because numerous computer simulations designed to model lubrication have concerned particles of spherical shape [31.2, 95, 120]. As the typical size of the contact area is $\approx 50\,\mu$m and the size of the laser spot is $\approx 0.5\,\mu$m, it is possible to scan the laser focus laterally, as sketched in Fig. 31.14. From time series of fluctuations of the fluorescence intensity, the intensity–intensity autocorrelation function can be calculated and is plotted against logarithmic time lag. From Fig. 31.15 it is obvious that the characteristic diffusion time increased with increasing proximity to the center of the contact. Their physical meaning is to describe the time to diffuse through the spot of calibrated diameter, $\approx 0.5\,\mu$m, at which the interrogatory laser beam was focused. The time scale of these processes, which can be estimated as the time at which the autocorrelation function decayed to one half of its initial value, slowed for example from 2 ms (rhodamine 123 in bulk propane diol) to 800 ms (the slowest curve shown in Fig. 31.15a).

For quantitative analysis, one can fit these curves to the standard model for two-dimensional Fickian diffusion and assume a single diffusion process; the imperfect fits towards the center of the contact zone are

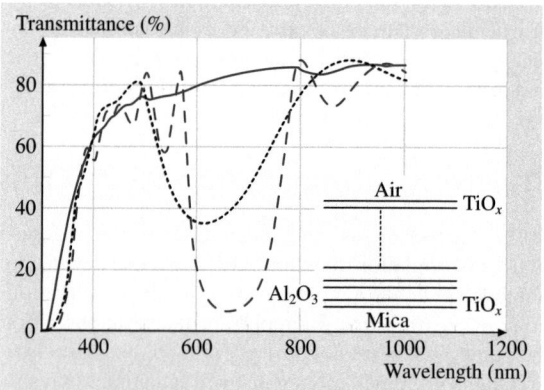

Fig. 31.15 Optical transmittance of mica coated with a multilayer dielectric, showing the feasibility of performing optical spectroscopies in the region 400–600 nm while also measuring surface–surface separation by multiple beam interferometry in the region 650–750 nm. Bare mica (*solid line*), mica coated with 7 layers of coating (*dotted line*) and with 13 layers of coating (*dash-dot line*). The window of optical transmission is adjustable through variations of multilayer dielectric. (*Inset*) The schematic diagram of the dielectric coating, composed of alternate layers of TiO_x and Al_2O_3

discussed below. Figure 31.18 shows D/D_{bulk}, the relative diffusion coefficient, plotted against position within the contact on semi-logarithmic scales; the logarithmic scale is needed because the dependence was so strong. Equivalently, assuming a Hertzian contact-pressure distribution, D/D_{bulk} is plotted against relative pressure within the contact in Fig. 31.18 (inset). A separate control experiment in which the bulk pressure was varied showed the diffusion in propane diol slowed by a much smaller amount, a factor of only 1.5, when the pressure was raised to 7 MPa. Evidently the findings described in this report are significantly larger than those produced by compressibility of the bulk fluid under isotropic pressure, and a different sort of explanation is needed.

Postulating that D_{eff} was proportional to a Boltzmann factor in energy ($\Delta E/kT$, where E is the energy, k the Boltzmann constant, and T the absolute temperature, which was constant), ΔE can be regarded as the net differential normal pressure, Δp, times an activation volume, ΔV_{act}. Figure 31.18 (inset) shows that the data are consistent with the implied exponential decrease of D_{eff} with p, in spite of the fact that p is mechanical pressure squeezing the confined fluid, not the usual isotropic pressure. From the slope in Fig. 31.18, one deduces that $\Delta V_{\text{act}} = 15–20\,\text{nm}^3$. It is intriguingly close to the activation volume obtained some years ago from independent friction measurements [31.121]. In the bulk, by contrast, the activation volume for diffusion is only $\approx 0.2\,\text{nm}^3$, the size of a fluid molecule. This analysis highlights one of the key conclusions that diffusion appeared to involve cooperative rearrangements of many molecules. But this concept assumes a single reaction coordinate and a fully equilibrated homogeneous system. Therefore, it may not seem physically meaningful to identify the deduced activation volume with the lateral size of cooperatively rearranging regions within the confined films.

The inflections in the intensity–intensity autocorrelation functions of Fig. 31.17 are quantitatively reproducible on different days with fresh samples. They refer to the same data-acquisition time, ≈ 45 min. Near the edges of the contact this time is enough to produce an autocorrelation curve that conformed well to expectations for a diffusion process with a single diffusion rate. But in the same system, at the same film thickness, at the same temperature, and in the same experiment, the autocorrelation curves deviated from this expectation more and more, the higher the local mechanical pressure. The results suggest that the system became increasingly heterogeneous with increasing pressure, so much so that the spatial resolution of this experiment sampled zones

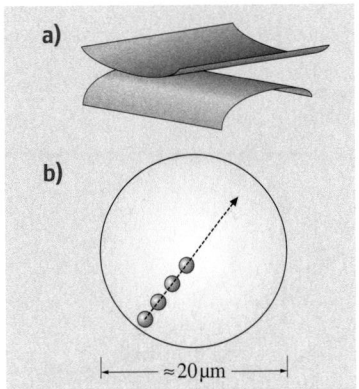

Fig. 31.16a,b Experimental scheme. (**a**) Crossed-cylinder configuration. Droplets of fluids were placed between crossed cylinders of mica as a drop and force is applied in the normal direction, causing the formation of a circular area of flattened contact. (**b**) Fluorescence correlation spectroscopy (FCS) is performed after compressing the films with a mean normal pressure of 2–3 MPa. The sequence of smaller circles illustrate that the focus of the laser beam (diameter $\approx 0.5\,\mu\text{m}$) was scanned across the contact (radius $\approx 10\,\mu\text{m}$), enabling spatially resolved measurements. After [31.111] with permission

of slightly different dynamical response. Molecules diffused at one rate through a certain micro-environment, then entered another. Physically, this signifies some kind of long-lived but quantitatively reproducible heterogeneity, which becomes increasingly intense when moving from the edges of the contact towards the center. To make this more precise, it is worth remembering that molecules in confined fluid films are known to organize into layers parallel to the confining surfaces, in which local density differs. In this scenario, the multiple processes suggested in the autocorrelation curves signify that translational diffusion rates differed in these layers. For example, for OMCTS the slowest curve in Fig. 31.17b can be decomposed into sub-processes with $D_{\text{eff}} = 0.1, 0.7, 2.9,$ and $40\,\mu\text{m}^2/\text{s}$, the average being $5.5\,\mu\text{m}^2/\text{s}$ because the slowest two processes had the smallest amplitude and contributed least. This interpretation is consistent with the observation that the heterogeneity was more pronounced for OMCTS than for 1,2-propane diol, which is not expected to organize so definitively into distinct layers. From another perspective, these data are qualitatively reminiscent of the 'cage' slowing down observed in autocorrelation curves from dynamic light scattering (DLS) studies of colloidal glasses – some kind of incipient but incomplete solidifi-

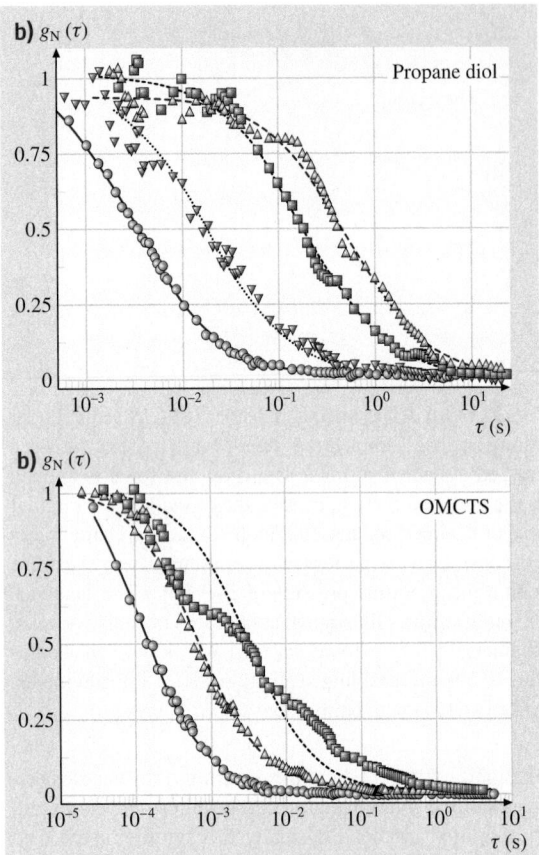

Fig. 31.17 Fluorescence autocorrelation functions, normalized to unity, are plotted against logarithmic time for rhodamine 123 in 1,2-propane diol (a) and coumarin 153 in OMCTS (b). The focus was at distance a from the center of the contact and the ratio $x \equiv a/r$ was considered, where r is the contact radius. In panel c, $x \approx 0.95$ (*circles*), 0.82 (*down triangles*), 0.7 (*squares*), and 0.6 (*up triangles*). In panel , $x \approx 0.95$ (*circles*), 0.8 (*up triangles*), and 0.7 (*squares*). Lines through the autocorrelation curves are the least-squares fit to a single Fickian diffusion process. After [31.111] with permission

ios, but one conclusion is firm: the scale of heterogeneity must have been impressively large, when one considers that these heterogeneities did not average out in spite of the long averaging time and the fact that the laser beam spot ($\approx 500 - \text{nm}$ diameter) exceeded the size of the diffusing molecules by so much.

Concerning the question of how the diffusion of individual molecules is related to friction, these results suggest that the slit-averaged retardation of diffusion is much less than the confinement-induced enhancement of effective viscosity, which is known to be at least 6–7 orders of magnitude [31.94,95]. However, the generality of the observation has been cast into doubt recently by mechanical experiments, which showed that the solidity of molecularly thin films using mica as substrates depends on a particular method of surface preparation [31.124]. Mica recleaved immediately before an experiment to minimize potential exposure to airborne contaminants, is reported to produce very low friction [31.125]. These results are consistent with the spectroscopy experiments, with molecular dynamics simulation and dielectric measurements, none of which had observed divergence of relaxation times, or confinement induced solidification.

cation [31.122]. From the available data it is difficult so far to assess the relative importance of these two scenar-

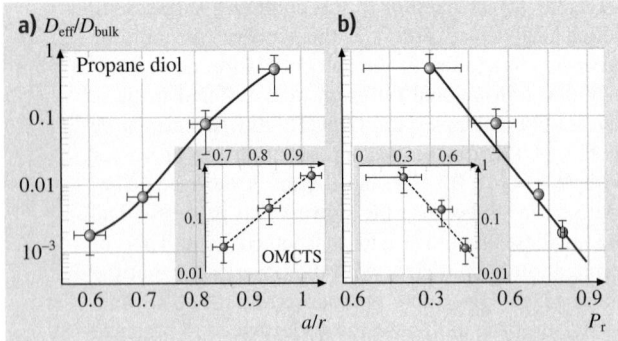

Fig. 31.18 Quantification of the effective diffusion coefficients (D_{eff}) inferred from the raw data in the previous figure. (**a**) The ratio, $D_{\text{eff}}/D_{\text{bulk}}$, is plotted against focus position, a/r. (**b**) The ratio, $D_{\text{eff}}/D_{\text{bulk}}$, is plotted against relative pressure squeezing the surfaces together, $P_{\text{r}} \equiv P/P_{\text{max}}(P_{\text{max}} \approx 4\,\text{MPa})$ assuming a Hertzian pressure distribution. *Open circles* (main figures) denote rhodamine 123 in 1,2-propane diol ($D_{\text{bulk}} \approx 8\,\mu\text{m}^2/\text{s}$). *Filled circles* (insets) denote coumarin 153 in OMCTS ($D_{\text{bulk}} \approx 190\,\mu\text{m}^2/\text{s}$). After [31.123] with permission

31.5 Diffusion of Confined Molecules During Shear

In most previous friction experiments using the traditional mica cleaving procedure, static friction (a mechanically solid state) was found to give way, under sliding, to kinetic friction (a mechanically molten state) [31.126]. The transition from mechanical *solidity* to *fluidity* has been interpreted to suggest a shear-induced phase transition. The premise was that, if the hypothesis of shear-melting held, shear should induce considerable quickening of the intensity–intensity autocorrelation function time scale of the dye molecule. Shear forces in SFA experiments are generated by sliding the top cylindrical lens, which is suspended from the upper portion of the apparatus by two piezoelectric bimorphs, mounted symmetrically to the two ends of the mount (Fig. 31.14). Shear forces were generated by one of the piezoelectric bimorphs (the "sender"), and the response of the device induced a voltage across the right-hand bimorph (the

"receiver"). A typical frequency range is 0.1–256 Hz, with shear displacement amplitude of 0.1–10 μm. These forces were usually sinusoidal in time but sometimes were a triangular ramp of constant slope, for better comparison with nanotribology measurements and simulations that employed a constant sliding velocity. The triangular ramp produced nearly constant velocity except for a slight acceleration to prevent stick–slip. This section is adapted from discussions in several primary accounts published previously [31.123].

Figure 31.19 shows representative autocorrelation curves of the fluorescence fluctuations. In Fig. 31.20, the ratio of D_{eff} during sliding to D_{eff} at rest is plotted against the shear rate; the main figure shows the entire range of shear rate and the inset magnifies the regime of small shear rate. One sees that shear speeded up this measure of the time scale of the autocorrelation function by less than a factor of 5 in every case [31.123]. The significance is that this affords a direct (negative) test of the common hypothesis that the transition from rest to sliding reflects shear melting that is analogous to the melting transition of a solid. The reason that the autocorrelation functions were unaffected by motion is believed to be that the fluid undergoes partial slip when the moving surface is smooth; the same conclusion follows from friction measurements in similar systems.

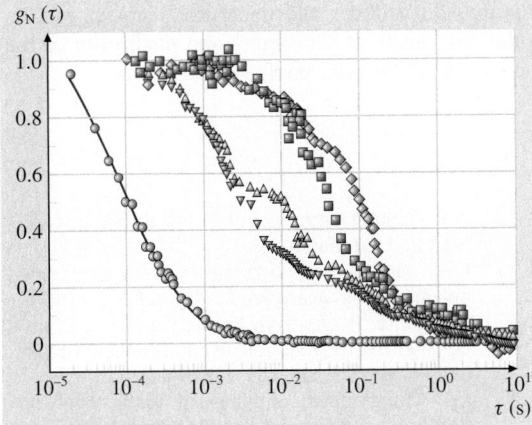

Fig. 31.19 Illustrative fluorescence intensity–intensity autocorrelation functions, $g_N(\tau)$, normalized to unity at short time and plotted against logarithmic time lag τ. The thickness of the liquid film was 3.0 ± 0.4 nm relative to air, corresponding to ≈ 3–4 molecular layers. The curves compare diffusion at rest in the unconfined bulk (*circles*; $D = 180 \, \mu m^2/s$) and at two radial positions ("a") within the contact zone of radius $r \approx 25 \, \mu m$; $a/r = 0.7$ (*triangles*) and $a/r = 0.5$ (*squares*). These data are taken at rest (*open symbols*) and while sliding (*filled symbols*) at shear rates of 10^4 and 10^2 s^{-1}, respectively. Sliding was performed at 1–256 Hz such that it was unidirectional for half the period then reversed direction, and so on repetitively; for the data shown here, it was 256 and 32 Hz, respectively. Lines through the data for the bulk are fits to a single diffusion process. After [31.111] with permission

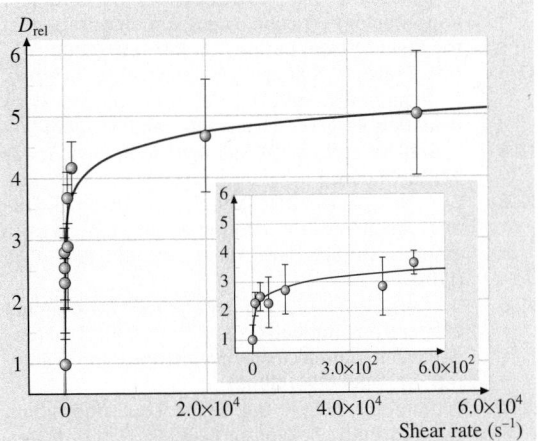

Fig. 31.20 The ratio (D_{rel}) of the effective diffusion coefficient during sliding (defined in text) to that at rest, plotted against peak shear rate. The data represent the average of more than 30 experiments, with error bars indicated. The *inset* shows the low-shear part of this data. After [31.123] with permission

Taken together, these ultrafast spectroscopy experiments show that these complex molecularly thin systems retain a high degree of fluidity at the molecular level. While it is true that the structure factor of in-plane correlation is predicted from computer simulations to be enhanced relative to the bulk and that the activation volume for diffusion exceeds that in the bulk liquid by about three orders of magnitude, indicating a higher degree of order than in the bulk fluid, shear does not appear to substantially modify the degree of fluidity inferred by direct measurements of the diffusion of individual molecules. An agenda for future investigation will be to understand better the relationship between the mechanical friction response, which is an ensemble average, and measurements such as those presented here, which refer to the motion of individual molecules.

31.6 Summary

We have reviewed some recent advances in the experimental study of confined fluids. Some important questions involving thin liquid films with broad applications from biology to tribology have been addressed with complementary approaches. Surface and interfacial forces of liquids in intimate contact with surfaces can be directly measured by SFA, AFM and other force-based techniques. Meanwhile, scattering, microscopic and spectroscopic techniques combined with the surface forces apparatus have been developed to provide unique experimental platforms in order to understand the structure, the phase behavior, and the dynamical responses of fluids confined between surfaces whose spacing becomes comparable to the correlation length of short-range packing, the size of supramolecular structures, and/or the size of the molecules. Taking all the efforts together, the study of fluid molecules in close proximity to surfaces and under confinement gets us to the nitty-gritty of the beauty and distinction of the nano-world and nanotechnology.

References

31.1   J. N. Israelachvili: *Intermolecular and Surface Forces*, 2nd edn. (Academic, New York 1991)

31.2   B. Bhushan, J. N. Israelachvili, U. Landman: Nanotribology—Friction, wear and lubrication at the atomic-scale, Nature **374**, 607–616 (1995)

31.3   J. M. Drake, J. Klafter, P. E. Levitz, R. M. Overney, M. Urbakh: *Dynamics in Small Confining Systems V* (Materials Research Society, Warrendale 2000)

31.4   S. Granick: Soft matter in a tight spot, Phys. Today **52**, 26–31 (1999)

31.5   C. M. Mate, G. M. McClelland, R. Erlandsson, S. Chiang: Atomic-scale friction of a tungsten tip on a graphite surface, Phys. Rev. Lett. **59**, 1942–1945 (1987)

31.6   G. Meyer, N. M. Amer: Simultaneous measurement of lateral and normal forces with an optical-beam-deflection atomic force microscope, Appl. Phys. Lett. **57**, 2089–2091 (1990)

31.7   S. Granick, Y. Zhu, H. Lee: Slippery questions about complex fluids flowing past solids, Nature Mater. **2**, 221–227 (2003)

31.8   Y. Zhu, S. Granick: Limits of the hydrodynamic no-slip boundary condition, Phys. Rev. Lett. **88**, 106102–(1–4) (2002)

31.9   Y. Zhu, S. Granick: Rate-dependent slip of Newtonian liquid at smooth surfaces, Phys. Rev. Lett. **87**, 096105–(1–4) (2001)

31.10 B. S. Massey: *Mechanics of Fluids*, 6th edn. (Chapman Hall, London 1989)

31.11 P.-G. de Gennes: Viscometric flows of tangled polymers, C. R. Acad. Sci. B. Phys. **288**, 219 (1979)

31.12 L. Léger, E. Raphael, H. Hervet: Surface-anchored polymer chains: Their role in adhesion and friction, Adv. Polym. Sci. **138**, 185–225 (1999)

31.13 O. I. Vinogradova: Slippage of water over hydrophobic surfaces, Int. J. Miner. Process **56**, 31–60 (1999)

31.14 C. Mak, J. Krim: Quartz–crystal microbalance studies of the velocity dependence of interfacial friction, Phys. Rev. B **58**, 5157–5179 (1998)

31.15 S. M. Tholen, J. M. Parpia: Slip and the effect of He-4 at the He-3–silicon interface, Phys. Rev. Lett. **67**, 334–337 (1991)

31.16 C. Huh, L. E. Scriven: Hydrodynamic model of steady movement of a solid/liquid/fluid contact line, J. Colloid Interface Sci. **35**, 85–101 (1971)

31.17 G. Reiter, A. L. Demirel, S. Granick: From static to kinetic friction in confined liquid-films, Science **263**, 1741–1744 (1994)

31.18 G. Reiter, A. L. Demirel, J. S. Peanasky, L. Cai, S. Granick: Stick to slip transition and adhesion of lubricated surfaces in moving contact, J. Chem. Phys. **101**, 2606–2615 (1994)

31.19 N. V. Churaev, V. D. Sobolev, A. N. Somov: Slippage of liquids over lyophobic solid surfaces, J. Colloid Interface Sci. **97**, 574–581 (1984)

31.20 D. Y. C. Chan, R. G. Horn: The drainage of thin liquid films between solid surfaces, J. Chem. Phys. **83**, 5311–5324 (1985)

31.21 J. N. Israelachvili: Measurement of the viscosity of liquids in very thin films, J. Colloid Interface Sci. **110**, 263–271 (1986)

31.22 J. F. Nye: A calculation on the sliding of ice over a wavy surface using a Newtonian viscous approximation, Proc. Roy. Soc. A **311**, 445–467 (1969)

31.23 S. Richardson: On the no-slip boundary condition, J. Fluid Mech. **59**, 707–719 (1973)

31.24 K. M. Jansons: Determination of the macroscopic (partial) slip boundary condition for a viscous flow over a randomly rough surface with a perfect slip microscopic boundary condition, Phys. Fluids **31**, 15–17 (1988)

31.25 P. A. Thompson, M. O. Robbins: Shear flow near solids: epitaxial order and flow boundary condition, Phys. Rev. A **41**, 6830–6839 (1990)

31.26 P. A. Thompson, S. Troian: A general boundary condition for liquid flow at solid surfaces, Nature **389**, 360–362 (1997)

31.27 J.-L. Barrat, L. Bocquet: Large slip effect at a non-wetting fluid–solid interface, Phys. Rev. Lett. **82**, 4671–4674 (1999)

31.28 R. Pit, H. Hervet, L. Léger: Direct experimental evidence of slip in hexadecane–solid interfaces, Phys. Rev. Lett. **85**, 980–983 (2000)

31.29 V. S. J. Craig, C. Neto, D. R. M. Williams: Shear-dependent boundary slip in aqueous Newtonian liquid, Phys. Rev. Lett. **87**, 54504–(1–4) (2001)

31.30 O. A. Kiseleva, V. D. Sobolev, N. V. Churaev: Slippage of the aqueous solutions of cetyltriimethylammonium bromide during flow in thin quartz capillaries, Colloid J. **61**, 263–264 (1999)

31.31 Y. Zhu, S. Granick: Apparent slip of Newtonian fluids past adsorbed polymer layers, Macromolecules **36**, 4658–4663 (2002)

31.32 Y. Zhu, S. Granick: The no slip boundary condition switches to partial slip when the fluid contains surfactant, Langmuir **18**, 10058–10063 (2002)

31.33 J. Baudry, E. Charlaix, A. Tonck, D. Mazuyer: Experimental evidence of a large slip effect at a nonwetting fluid–solid interface, Langmuir **17**, 5232–5236 (2002)

31.34 D. C. Tretheway, C. D. Meinhart: Apparent fluid slip at hydrophobic microchannel walls, Phys. Fluids **14**, L9–L12 (2002)

31.35 H. A. Barnes: A review of the slip (wall depletion) of polymer solutions, emulsions and particle suspensions in viscometers: Its cause, character, and cure, J. Non-Newtonian Fluid Mech. **56**, 221–251 (1995)

31.36 E. C. Achilleos, G. Georgiou, S. G. Hatzikiriakos: Role of processing aids in the extrusion of molten polymers, J. Vinyl Additive Technol. **8**, 7–24 (2002)

31.37 H. Brenner, V. Ganesan: Molecular wall effects: Are conditions at a boundary 'boundary conditions'?, Phys. Rev. E. **61**, 6879–6897 (2000)

31.38 J. Gao, W. D. Luedtke, U. Landman: Structures, solvation forces and shear of molecular films in a rough nano-confinement, Tribology Lett. **9**, 3–134 (2000)

31.39 C. Denniston, M. O. Robbins: Molecular and continuum boundary conditions for a miscible binary fluid, Phys. Rev. Lett. **87**, 178302(1–4) (2001)

31.40 S. E. Campbell, G. Luengo, V. I. Srdanov, F. Wudl, J. N. Israelachvili: Very low viscosity at the solid–liquid interface induced by adsorbed C-60 monolayers, Nature **382**, 520–522 (1996)

31.41 E. Bonaccurso, M. Kappl, H.-J. Butt: Hydrodynamic force measurements: Boundary slip of water on hydrophilic surfaces and electrokinetic effects, Phys. Rev. Lett. **88**, 076103(1–4) (2002)

31.42 M. M. Britton, P. T. Callaghan: Two-phase shear band structures at uniform stress, Phys. Rev. Lett. **78**, 4930–4933 (1997)

31.43 O. I. Vinogradova: Drainage of a thin liquid-film confined between hydrophobic surfaces, Langmuir **11**, 2213–2220 (1995)

31.44 J. S. Peanasky, H. M. Schneider, S. Granick, C. R. Kessel: Self-assembled monolayers on mica for experiments utilizing the surface forces apparatus, Langmuir **11**, 953–962 (1995)

31.45 H. A. Spikes: The half-wetted bearing. Part 2: Potential application to low load contacts, Proc. Inst. Mech. Eng. Part J **217**, 15–26 (2003)

31.46 P.-G. de Gennes: On fluid/wall slippage, Langmuir **18**, 3413–3414 (2002)

31.47 J. W. G. Tyrrell, P. Attard: Atomic force microscope images of nanobubbles on a hydrophobic surface and corresponding force–separation data, Langmuir **18**, 160–167 (2002)

31.48 N. Ishida, T. Inoue, N. Miyahara, K. Higashitani: Nano bubbles on a hydrophobic surface in water observed by tapping-mode atomic force microscopy, Langmuir **16**, 6377–6380 (2000)

31.49 U. C. Boehnke, T. Remmler, H. Motschmann, S. Wurlitzer, J. Hauwede, T. M. Fischer: Partial air wetting on solvophobic surfaces in polar liquids, J. Coll. Int. Sci. **211**, 243–251 (1999)

31.50 K. Lum, D. Chandler, J. D. Weeks: Hydrophobicity at small and large length scales, J. Phys. Chem. B **103**, 4570–4577 (1999)

31.51 X. Zhang, Y. Zhu, S. Granick: Softened hydrophobic attraction between macroscopic surfaces in relative motion, J. Am. Chem. Soc. **123**, 6736–6737 (2001)

31.52 X. Zhang, Y. Zhu, S. Granick: Hydrophobicity at a Janus Interface, Science **295**, 663–666 (2002)

31.53 T. Onda, S. Shibuichi, N. Satoh, K. Tsuji: Super-water-repellent fractal surfaces, Langmuir **12**, 2125–2127 (1996)

31.54 J. Bico, C. Marzolin, D. Quéré: Pearl drops, Europhys. Lett. **47**, 220–226 (1999)

Part D | 31

31.55   S. Herminghaus: Roughness-induced non-wetting, Europhys. Lett. **52**, 165–170 (2000)

31.56   K. Watanabe, Y. Udagawa, H. Udagawa: Drag reduction of Newtonian fluid in a circular pipe with a highly water-repellent wall, J. Fluid Mech. **381**, 225–238 (1999)

31.57   D. W. Bechert, M. Bruse, W. Hage, R. Meyer: Fluid mechanics of biological surfaces and their technological application, Naturwiss. **87**, 157–171 (2000)

31.58   W. Kauzmann: Some forces in the interpretation of protein denaturation, Adv. Prot. Chem. **14**, 1 (1959)

31.59   C. Tanford: *The Hydrophobic Effect—Formation of Micelles and Biological Membranes* (Wiley-Interscience, New York 1973)

31.60   F. H. Stillinger: Structure in aqueous solutions of nonpolar solutes from the standpont of scaled-particle theory, J. Solution Chem. **2**, 141 (1973)

31.61   E. Ruckinstein, P. Rajora: On the no-slip boundary-condition of hydrodynamics, J. Colloid Interface Sci. **96**, 488–491 (1983)

31.62   L. R. Pratt, D. Chandler: Theory of hydrophobic effect, J. Chem. Phys. **67**, 3683–3704 (1977)

31.63   A. Ben-Naim: *Hydrophobic Interaction* (Kluwer, New York 1980)

31.64   A. Wallqvist, B. J. Berne: Computer-simulation of hydrophobic hydration forces stacked plates at short-range, J. Phys. Chem. **99**, 2893–2899 (1995)

31.65   G. Hummer, S. Garde, A. E. Garcia, A. Pohorille, L. R. Pratt: An information theory model of hydrophobic interactions, Proc. Nat. Acad. Sci. USA **93**, 8951–8955 (1996)

31.66   Y. K. Cheng, P. J. Rossky: The effect of vicinal polar and charged groups on hydrophobic hydration, Biopolymers **50**, 742–750 (1999)

31.67   D. M. Huang, D. Chandler: Temperature and length scale dependence of hydrophobic effects and their possible implications for protein folding, Proc. Nat. Acad. Sci. USA **97**, 8324–8327 (2000)

31.68   G. Hummer, S. Garde, A. E. Garcia, L. R. Pratt: New perspectives on hydrophobic effects, Chem. Phys. **258**, 349–370 (2000)

31.69   D. Bratko, R. A. Curtis, H. W. Blanch, J. M. Prausnitz: Interaction between hydrophobic surfaces with metastable intervening liquid, J. Chem. Phys. **115**, 3873–3877 (2001)

31.70   Y.-H. Tsao, D. F. Evans, H. Wennerstöm: Long-range attractive force between hydrophobic surfaces observed by atomic force microscopy, Science **262**, 547–550 (1993) and references therein

31.71   R. F. Considine, C. J. Drummond: Long-range force of attraction between solvophobic surfaces in water and organic liquids containing dissolved air, Langmuir **16**, 631–635 (2000)

31.72   J. W. G. Tyrrell, P. Attard: Images of nanobubbles on hydrophobic surfaces and their interactions, Phys. Rev. Lett. **87**, 176104 (2001)

31.73   J. Peachey, J. Van Alsten, S. Granick: Design of an apparatus to measure the shear response of ul-

31.74   C. Y. Lee, J. A. McCammon, P. J. Rossky: The structure of liquid water at an extended hydrophobic surface, J. Chem. Phys. **80**, 4448–4455 (1984)

31.75   J. N. Israelachvili, R. M. Pashley: The hydrophobic interaction is long-range, decaying exponentially with distance, Nature **300**, 341–342 (1982)

31.76   J. N. Israelachvili, R. M. Pashley: Measure of the hydrophobic interaction between 2 hydrophobic surfaces in aqueous-electrolyte solutions, J. Colloid Interface Sci. **98**, 500–514 (1984)

31.77   R. M. Pashley, P. M. McGuiggan, B. W. Ninham, D. F. Evans: Attractive forces between uncharged hydrophobic surfaces-direct measurement in aqueous-solution, Science **229**, 1088–1089 (1985)

31.78   P. M. Claesson, C. E. Blom, P. C. Herder, B. W. Ninham: Interactions between water-stable hydrophobic Langmuir–Blodgett monolayers on mica, J. Colloid Interface Sci. **114**, 234–242 (1986)

31.79   P. M. Claesson, H. K. Christenson: Very long-range attractive forces between uncharged hydrocarbon and fluorocarbon surfaces in water, J. Phys. Chem. **92**, 1650–1655 (1988)

31.80   H. K. Christenson, P. M. Claesson, J. Berg, P. C. Herder: Forces between fluorocarbon surfactant monolayers—salt effects on the hydrophobic interact, J. Phys. Chem. **93**, 1472–1478 (1989)

31.81   O. Spalla: Long-range attraction between surfaces: Existence and amplitude?, Curr. Opin. Colloid Interface Sci. **5**, 5–12 (2000) and references therein

31.82   J. Wood, R. Sharma: How long is the long-range hydrophobic attraction?, Langmuir **11**, 4797–4802 (1995)

31.83   J. L. Parker, P. M. Claesson, P. Attard: Bubbles, cavities, and the long-range attraction between hydrophobic surfaces, J. Phys. Chem. **98**, 8468–8480 (1994)

31.84   A. Carambassis, L. C. Jonker, P. Attard, M. W. Rutland: Forces measured between hydrophobic surfaces due to a submicroscopic bridging bubble, Phys. Rev. Lett. **80**, 5357–5360 (1998)

31.85   V. S. J. Craig, B. W. Ninham, R. M. Pashley: Direct measurement of hydrophobic forces: A study of dissolved gas, approach rate, and neutron irradiation, Langmuir **15**, 1562–1569 (1999)

31.86   R. F. Considine, R. A. Hayes, R. G. Horn: Forces measured between latex spheres in aqueous electrolyte: Non-DLVO behavior and sensitivity to dissolved gas, Langmuir **15**, 1657–1659 (1999)

31.87   J. D. Ferry: *Viscoelastic Properties of Polymers*, 3rd edn. (Wiley, New York 1982)

31.88   Y. Zhu, S. Granick: Viscosity of interfacial water, Phys. Rev. Lett. **87**, 096104 (2001)

31.89   H. H. Winter, F. Chambon: Analysis of linear viscoelasticity of a cross-linking polymer at the gel point, J. Rheol. **30**, 367–382 (1986)

trathin liquid-films, Rev. Sci. Instrum. **62**, 463–473 (1991)

31.90 R. Yamamoto, A. Onuki: Dynamics of highly supercooled liquids: Heterogeneity, rheology, and
diffusion, Phys. Rev. E **58**, 3515–3529 (1998) and
references therein

31.91 A. O. Parry, R. Evans: Influence of wetting on
phase-equilibra–a novel mechanism for critical-
point shifts in films, Phys. Rev. Lett. **64**, 439–442
(1990)

31.92 K. Binder, D. P. Landau, A. M. Ferrenberg: Thin ising
films with completing walls–a Monte Carlo study,
Phys. Rev. E **51**, 2823–2838 (1995)

31.93 D. K. Schwartz, M. L. Schlossman, E. H. Kawamoto,
G. J. Kellog, P. S. Perhan, B. M. Ocko: Thermal diffuse X-ray-scattering studies of the water–vapor
interface, Phys. Rev. A **41**, 5687–5690 (2000)

31.94 S. Granick: Motions and relaxations of confined
liquids, Science **253**, 1374–1379 (1991)

31.95 J. Klein, E. Kumacheva: Simple liquids confined to
molecularly thin layers. I. Confinement-induced
liquid-to-solid phase transitions, J. Chem. Phys.
108, 6996–7009 (1998)

31.96 E. Kumacheva, J. Klein: Simple liquids confined
to molecularly thin layers. II. Shear and frictional
behavior of solidified films, J. Chem. Phys. **108**,
7010–7022 (1998)

31.97 C. Drummond, J. Israelachvili: Dynamic phase transitions in confined lubricant fluids under shear,
Phys. Rev. E **63**, 041506–1–11 (2001)

31.98 Y. Golan, M. Seitz, C. Luo, A. Martin-Herranz,
M. Yasa, Y. L. Li, C. R. Safinya, J. Israelachvili: The
X-ray surface forces apparatus for simultaneous X-
ray diffraction and direct normal and lateral force
measurements, Rev. Sci. Instrum. **73**, 2486–248
(2002)

31.99 Y. Golan, A. Martin-Herranz, Y. Li, C. R. Safinya,
J. Israelachvili: Direct observation of shear-
induced orientational phase coexistence in a
lyotropic system using a modified X-ray surface
forces apparatus, Phys. Rev. Lett. **86**, 1263–1266
(2001)

31.100 S. M. Baker, G. Smith, R. Pynn, P. Butler, J. Hayter,
W. Hamilton, L. Magid: Shear cell for the study of
liquid–solid interfaces by neutron scattering, Rev.
Sci. Instrum. **65**, 412–416 (1994)

31.101 T. L. Kuhl, G. S. Smith, J. N. Israelachvili, J. Ma-
jewski, W. Hamilton: Neutron confinement cell for
investigating complex fluids, Rev. Sci. Instrum. **72**,
1715–1720 (2001)

31.102 O. H. Seeck, H. Kim, D. R. Lee, D. Shu, I. D. Kaendler,
J. K. Basu, S. K. Sinha: Observation of thickness
quantization in liquid films confined to molecular
dimension, Europhys. Lett. **60**, 376–382 (2002)

31.103 A. Dhinojwala, S. Granick: Micron-gap rheo-optics
with parallel plates, J. Chem. Phys. **107**, 8664–8667
(1998)

31.104 I. Soga, A. Dhinojwala, S. Granick: Optorheological
studies of sheared confined fluids with mesoscopic
thickness, Langmuir **4**, 1156–1161 (1998)

31.105 S. Mamedov, A. D. Schwab, A. Dhinojwala: A device for surface study of confined micron thin films
in a total internal reflection geometry, Rev. Sci.
Instrum. **73**, 2321–2324 (2002)

31.106 P. Frantz, F. Wolf, X. D. Xiao, Y. Chen, S. Bosch,
M. Salmeron: Design of surface forces apparatus
for tribolgy studies combined with nonlinear optical spectroscopy, Rev. Sci. Instrum. **68**, 2499–2504
(1997)

31.107 X. S. Xie, J. K. Trautman: Optical studies of single
molecules at room temperature, Annu. Rev. Phys.
Chem. **49**, 441–480 (1998)

31.108 W. E. Moerner, M. Orritt: Illuminating single
molecules, Science **283**, 670–1676 (1999)

31.109 L. A. Deschenes, D. A. Vanden Bout: Single molecule
studies of heterogeneous dynamics in polymer
melts near the glass transition, Science **292**, 255–
258 (2001)

31.110 A. Mukhopadhyay, S. Granick: An integrated
platform for surface force measurements and
fluorescence correlation spectroscopy, Rev. Sci. Instrum. **74**, 3067–3072 (2003)

31.111 A. Mukhopadhyay, J. Zhao, S. C. Bae, S. Granick:
Contrasting friction and diffusion in molecularly-
thin films, Phys. Rev. Lett. **89**, 136103 (2002)

31.112 K. M. Berland, P. T. C. So, E. Gratton: 2-Photon fluorescence correlation spectroscopy–method and
application to the intracellular environment, Biophys. J. **68**, 694–701 (1995)

31.113 U. Kettling, A. Koltermann, P. Schwille, M. Eigen:
Real-time enzyme kinetics monitored by dual-
color fluorescence cross-correlation spectroscopy,
Proc. Natl. Acad. Sci. USA **95**, 1416–1420 (1998)

31.114 A. M. Lieto, R. C. Cush, N. L. Thompson: Ligand-
receptor kinetics measured by total internal
reflection with fluorescence correlation spectroscopy, Biophys. J. **85**, 3294–3302 (2003)

31.115 P. Schwille, U. Haupts, S. Maiti, W. W. Webb:
Molecular dynamics in living cells observed by fluorescence correlation spectroscopy with one- and
two-photon excitation, Biophys. J **77**, 2251–2265
(1999)

31.116 W. W. Webb: Fluorescence correlation spectroscopy: inception, biophysical experimenta-
tions, and prospectus, Appl. Opt. **40**, 3969–3983
(2001)

31.117 P. S. Dittrich, P. Schwill: Photobleaching and stabilization of fluorophores used for single-molecule
analysis with one- and two-photon excitation,
Appl. Phys. B **73**, 829–837 (2001)

31.118 M. Born, E. Wolf: *Principles of Optics* (Cambridge
Univ. Press, Cambridge 1999) p. 7

31.119 I. Sridhar, K. L. Johnson, N. A. Fleck: Adhesion mechanics of the surface force apparatus, J. Appl.
Phys. D **30**, 1710–1719 (1997)

31.120 Y. Zhu, S. Granick: Reassessment of solidification
in fluids confined between mica sheets, Langmuir
19, 8148–8151 (2003)

31.121 H.-W. Hu, G. Carson, S. Granick: Relaxation-time of confined liquids under shear, Phys. Rev. Lett. **66**, 2758–2761 (1991)

31.122 K. N. Pham, A. M. Puertas, J. Bergenholtz, S. U. Egelhaaf, A. Moussaid, P. N. Pusey, B. Schofield, M. E. Cates, M. Fuchs, W. C. K. Poon: Multiple glassy states in a simple model system, Science **296**, 104–106 (2002)

31.123 A. Mukhopadhyay, S. C. Bae, J. Zhao, S. Granick: How confined lubricants diffuse during shear, Phys. Rev. Lett. **93**, 236105 (2004)

31.124 Z. Q. Lin, S. Granick: Platinum nanoparticles at mica surfaces, Langmuir **19**, 7061–7070 (2003)

31.125 Y. Zhu, S. Granick: Superlubricity: A paradox about confined fluids resolved, Phys. Rev. Lett. **93**, 096101 (2004)

31.126 M. Urbakh, J. Klafteer, D. Gourdon, J. Israelachvili: The nonlinear nature of friction, Nature **430**, 525–528 (2004)

32. Scanning Probe Studies of Nanoscale Adhesion Between Solids in the Presence of Liquids and Monolayer Films

Adhesion between solids is a ubiquitous phenomenon whose importance is magnified at the micrometer and nanometer scales, where the surface-to-volume ratio diverges as we approach the size of a single atom.

Numerous techniques for measuring adhesion at the atomic scale have been developed, but significant limitations exist. Instrumental improvements and reliable quantification are still needed. Recent studies have highlighted the unique and important effect of liquid capillaries, particularly water, at the nanometer scale. The results demonstrate that macroscopic considerations of classic meniscus theory must be modified to take into account new scaling and geometric relationships unique to the nanometer scale. More generally, a molecular scale description of wetting and capillary condensation as it applies to nanoscale interfaces is clearly desirable, but remains an important challenge.

The measurement of adhesion between self-assembled monolayers has proven to be a reliable way to probe the influence of surface chemistry and local environment on adhesion. To date, however, few of these systems have been investigated in detail quantitatively. The molecular origins of adhesion down to the single-bond level still need to be fully investigated. The most recent studies illustrate that, while new information about adhesion in these systems has been revealed, further enhancements of current techniques as well as the development of new methodologies coupled with accurate theoretical modeling are required to adequately tackle these complex measurements.

32.1 The Importance of Adhesion at the Nanoscale

The mutual attraction and bonding of two surfaces, which can occur with or without an intervening medium, are common phenomena with far-reaching manifestations and applications in society. The adherence between a raindrop and a window pane, the climbing of a gecko up a vine, the sticking of multiple adhesive note pads to a professor's wall, the force required to separate hook-and-eye (Velcro) strips, the building of a sand castle, and the book page turned by a wetted finger are all scenarios where adhesion is important. Within the complexity of these examples and others lies a central theme: that the mechanical forces between a pair of materials can be fundamentally affected by not just the macroscopic or microscopic structures of the surfaces, but also by the interatomic and intermolecular forces that exist between them.

Adhesion and intermolecular forces have been studied for many years dating back to ancient times [32.1], and active research continues today for topics as broad as insect and reptile locomotion [32.2], interactions between cells in the body [32.3], and the design of self-healing composites [32.4], to name but a few examples. While adhesion is clearly of interest for a wide range of macroscopic applications, the importance of adhesion becomes dominant at micrometer and nanometer scales. This is primarily due to the dramatic increase in the surface-to-volume ratio of materials at these scales, an effect which renders friction and interfacial wear at such length scales critical phenomena too [32.5]. For example, the dominating effect of adhesion at this scale has affected the development of microelectromechanical systems (MEMS), where interfacial forces can prevent devices from functioning properly since the small flexible parts often emerge from the fabrication line stuck together. Studies of adhesion in MEMS are ongoing [32.6, 7], and MEMS devices that are commercially deployed rely upon sophisticated surface treatments to reduce and control adhesion [32.8–10].

Detailed control of adhesion at the molecular level will be essential for the design of even smaller nanoelectromechanical systems (NEMS). Much has been written about the possibilities for nanoscale machines, sensors, actuators, and so on. It is crucial to understand that a molecule is *all* surface, and therefore molecular and nanoscale devices cannot be properly designed, implemented, or characterized unless an understanding of atomic-scale adhesion is thoroughly presented, partic-

ularly if these devices are to involve any moving parts that will come into and out of contact. Studying adhesion at the nanoscale is important for other reasons. The protruding asperities in most MEMS materials are often nanoscale in dimension, and therefore a complete understanding of adhesion in MEMS requires investigation of the adhesive properties of the individual nanoscale asperities. In addition, the experimental study of adhesion at the nanoscale is required for the development of detailed atomic-scale models of adhesion. Such progress requires a close collaboration between experiment and theory, which is essential for the eventual success of such an endeavor.

There has been significant progress in the experimental study of adhesion at the nanometer scale using scanning probe methods, but numerous challenges exist. A discussion of solid–solid adhesion without an intervening medium is provided elsewhere in this book. This chapter focuses on how adhesion is affected by the ubiquitous presence of water, and how it can be controlled through the application of self-assembled monolayer (SAM) coatings, again in the presence of a liquid medium. In addition, we discuss specific instrumental challenges that are inherent to adhesion measurements. We do not delve into the realm of atomic-scale modeling of adhesion, nor do we discuss the role of more complex coatings such as polymer brushes and blends. Rather, our focus will be on critically evaluating the relevant experimental techniques, and critically reviewing recent results from studies of water and SAM films, which are perhaps the two most commonly encountered media in nanoscale adhesion applications.

32.2 Techniques for Measuring Adhesion

The experimental study of adhesion at the nanoscale experienced two renaissances in the previous century. The first occurred with the development of the surface forces apparatus (SFA) [32.11, 12], and the second with the later development of the AFM [32.13] and other related scanning force techniques.

SFA experiments have contributed profoundly to our understanding of adhesion. The SFA consists of a pair of atomically smooth surfaces, usually mica sheets, mounted on crossed cylinders that can be pressed together to form a controlled circular contact. The applied load, normal displacement, surface separation, contact area, and shear force (if applied) can all be controlled

and/or measured [32.11, 14–16]. The SFA can be operated in air, a controlled environment, or under liquid conditions. The surfaces are often treated to attach molecules whose behavior under confinement can be studied. Alternately, the behavior of a confined fluid layer can be observed. The surface separation can be measured and controlled in the Ångstrom (Å) regime. The lateral resolution is limited to the range of several tens of micrometers. However, the true contact area between the interfaces can be directly measured, which is a key advantage since it allows an adhesive force to be converted to a force (or energy) per unit area (or per molecule), thus separating geometrical contributions to adhesion from chemical contributions. In this chapter, we will refer to relevant results in the context of the scanning probe experiments that we discuss below. As a very brief summary, some of the most important results pertaining to adhesion include the observation of capillary effects on adhesion [32.17, 18], the presence of hysteresis in adhesion due to pressure-induced restructuring of the interface [32.19], and numerous studies of how interfacial chemistry (hydrophilicity, surface charge, specific chemical groups, and so forth) affect adhesive forces [32.19–21].

AFM instrumentation is described elsewhere in this book. Significant ways in which the AFM differs from the SFA are: (1) the contact radius is nanometers, not microns, due to the fact that the tip is usually $< 100\,\text{nm}$ in radius, and the contact area at low applied loads will be a fraction of this radius [32.22]; (2) the force resolution in standard commercial AFMs is typically $10^{-10}\,\text{N}$, and high-resolution systems have been developed that measure forces orders of magnitudes smaller, as opposed to $\approx 10^{-6}\,\text{N}$ with the SFA; (3) the contact area is not directly observable, which is a key disadvantage, although it may be determined or inferred by the measurement of related quantities such as contact conductance [32.23, 24] or contact stiffness [32.25, 26]; (4) the actual separation between the tip and sample are not directly observable, which is a key disadvantage; (5) the relative separation between the sample and the *cantilever* (not the tip) is controlled in the 0.1 nm range or better; (6) the measurement bandwidth is typically in the kHz regime, but can extend into the MHz regime depending on the data acquisition technique; (7) the operating environment includes ultrahigh vacuum (UHV), and cryogenic to elevated temperatures; (8) there is virtually no limit to the range of sample materials that can be probed by the AFM, provided the sample is not overly rough; and (9) half of the interface (the tip) is essentially unknown or uncontrolled without devoting particular effort, whereas with the SFA both surfaces are well-defined. This is another key challenge for AFM that has yet to be consistently addressed.

The general set-up of the AFM is as follows. A small sharp tip (with a radius of typically $10-100\,\text{nm}$) is attached to the end of a compliant cantilever (Fig. 32.1). The tip is brought in close proximity to a sample surface. Forces acting between the tip and the sample's result in deflections of the cantilever (Fig. 32.2a). The cantilever bends vertically (toward or away from the sample) in response to attractive or repulsive forces acting on the tip. The deflection of the cantilever from its equilibrium position is proportional to the normal load applied to the tip

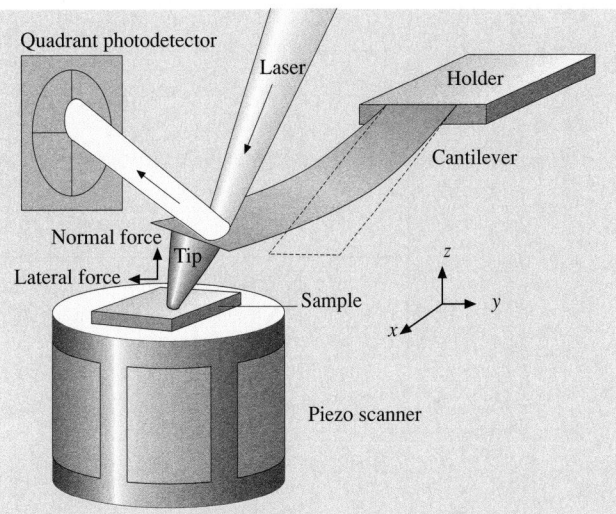

Fig. 32.1 Diagram of the AFM set-up for the optical beam deflection method. The tip is in contact with a sample surface. A laser beam is focused on the back of the cantilever and reflects into a four-quadrant photodetector. Normal forces deflect the cantilever up or down, while lateral forces twist the cantilever left and right. These deflections are simultaneously and independently measured by monitoring the deflection of the reflected laser beam

by the cantilever. Lateral forces result in a twisting of the cantilever about its long axis (torsion). These measurements can be performed in a variety of environments: ambient air, controlled atmosphere, liquids [32.27], or ultrahigh vacuum (UHV) [32.28–30].

While adhesion is often quantified in terms of a measured force, there is actually a force–displacement characteristic associated with an adhesive interface. The *work of adhesion*, as determined by fracture mechanics experiments for example, is the mechanical work per unit area needed to completely separate an interface from its initial equilibrium separation. That is, the work of adhesion is the *integral* of the interfacial force-displacement curve, normalized per unit interfacial area in contact at equilibrium. We must be careful to distinguish between the AFM tip-sample force–displacement curve and the interfacial force-displacement curve, because the former pertains to a tip on a flat surface, while the latter pertains to two flat surfaces. As we discuss in Sect. 32.5.2, if the tip radius is known it is possible to obtain the work of adhesion from an AFM pull-off force measurement.

The simplest way (perhaps deceptively so) to measure the effect of adhesion with the AFM is through so-called force–displacement plots (also referred to

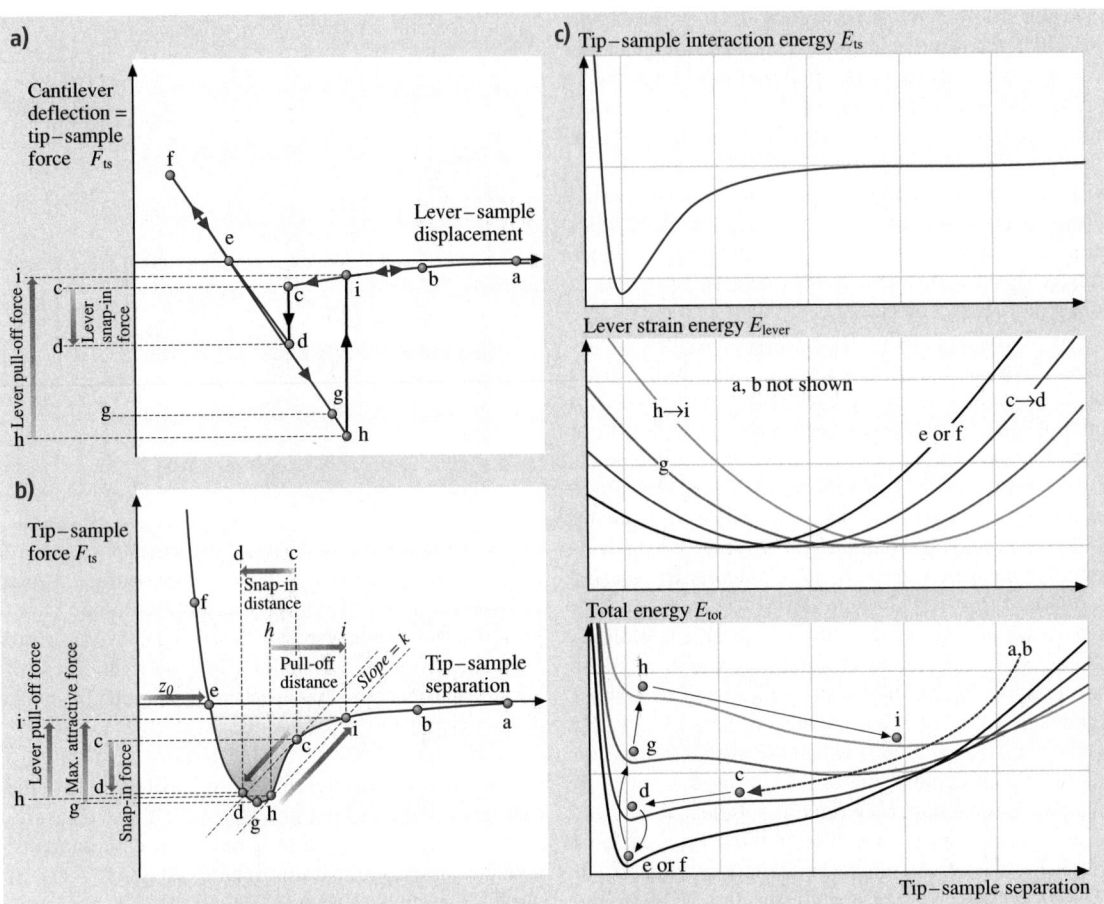

Fig. 32.2 (**a**) A "force–displacement" curve displays the vertical cantilever deflection versus the lever–sample displacement. This displacement is measured between the sample and the rigidly-held back end of the cantilever (as opposed to the front end with the tip which will bend in response to interaction forces) (**b**) The true tip–sample interaction as a function of *tip–sample* separation (**c**) Tip–sample interaction energy, lever strain energy, and total energy as a function of tip–sample separation. Note that these schematics correspond to the case of weak adhesion, when no frakture-like processes occur at the interface upon retraction. The case of strong adhesion with interfacial rupture is discussed further in the text

in the literature as force curves and force–distance curves or plots). A force-displacement plot, quite distinct from the tip-sample force-displacement curve, displays the cantilever's normal deflection versus the lever-sample displacement. The cantilever deflection can be calibrated to give the cantilever force F_{lever} (discussed further below). If we consider only the vertical forces acting on the tip to be the tip–sample interaction force F_{ts} and the cantilever force, then by force equilibrium, $F_{lever} = -F_{ts}$; in other words, these two forces are equal in magnitude. Thus, very simply, the calibrated signal from the photodiode does

indeed represent the force the sample is exerting on the tip.

The lever–sample displacement is measured between the sample and the rigidly-held back end of the cantilever (as opposed to the front end with the tip, which will bend in response to interaction forces). This displacement is altered by varying the vertical position of the piezo tube which, depending on the type of AFM, displaces either the tip or the sample in the direction normal to the sample. Referring to Fig. 32.2a, the stages of acquisition of a force–displacement plot are as follows: (*a*) The lever and sample are initially far apart and no forces act. (*b*) The lever is brought closer to the sample, and the tip senses attractive forces which cause the tip end of the lever to bend downward, thus signifying a negative (attractive) force. These forces may be of electrostatic, van der Waals, or of other origin. (*c*) At this point, the attractive force gradient (slope) of the tip–sample interaction force exceeds the normal spring constant k of the lever, and this causes an instability whereby the tip snaps into contact (*d*) with the sample. The lever–sample displacement can continue to decrease, and it eventually crosses the force axis (*e*), which corresponds to zero externally applied load. As this tip is in repulsive contact with the sample, the front end of the lever is pushed further upward, and the force corresponds to the externally applied load. (*f*) The lever-sample displacement direction is reversed at a point chosen by the user. As the lever–sample distance is reduced, the force becomes negative. Adhesion between the tip and sample maintains the contact although there is now a net negative (tensile) load. Eventually the tip passes through the point of maximum adhesion (*g*). Now the attractive force between the tip and sample starts to decrease, and so in principle the cantilever deflection will decrease. However, close to here, an unstable point is reached (*h*) where the adhesive bond ruptures. For a weak adhesive interaction, such as that due purely to van der Waals interactions in a liquid, this will occur when the cantilever stiffness exceeds the force gradient of the tip-sample interaction, and the tip snaps out of contact with the sample, as shown in (*h*)-(*i*). However, for stronger solid-solid adhesive contacts, the situation is different than that shown. The instability can be thought of as a fracture process. The rupture occurs due to high tensile stresses at the edge of the contact zone that exceed the bond strength. Therefore, pull-off will occur at (*g*). The difference in tip-sample separation between points (*g*) and (*h*) as represented in this plot is purely schematic, and in practice, fluctuations due to vibrations will cause the pull-off instability to occur before any reduction in force is observed. In any event, the resulting change in force relaxation is usually called the *pull-off force*. Note that the forces and distances are not drawn to scale; in particular, the attractive part of the interaction is exaggerated beyond that which often occurs for inert, neutral surfaces.

The cantilever force–displacement plot can be measured at a single location on the sample, or a series of measurements can be carried out over an area of interest. These so-called adhesion mapping techniques allow for spatially resolved adhesion measurements to be correlated with other sample properties such as fric-

tion, chemical termination, and other types of material heterogeneity.

To properly derive information about the tip-sample adhesion force from the AFM, it is critical to understand the nature of the mechanical instability in both snap-in and pull-off processes. The fundamental point to comprehend is that *the pull-off force is <u>not</u> the adhesive bond strength*. In other words, *it is not a direct measure of the actual adhesive forces that were acting between the tip and sample in the absence of applied load*. This important point is often overlooked or misunderstood, and so we discuss it in some detail here.

A cartoon of the interaction force F_{ts} between the tip and sample is sketched (not to any particular scale) as a function of the true tip–sample separation in Fig. 32.2b. Superimposed on this is the force–distance relation of the cantilever: a straight line with slope k (N/m), the cantilever stiffness (diagonal dashed lines). Points of instability are shown as gray dots, labels and arrows, as opposed to stable points shown as black dots, labels, and arrows. Snap-in occurs when the attractive force gradient dF_{ts}/dz just exceeds k; in other words, when the dashed line is tangent to the tip–sample force curve, as shown for point (*c*). This instability is a direct consequence of Newton's Second Law and is explained further below. Similarly, the pull-off occurs when during retraction k finally just exceeds dF_{ts}/dz, as shown for point (*h*), where once again, the diagonal line is just tangent to the tip–sample force curve. By definition, since k is finite, this point *cannot* correspond to the point of maximum attractive force (or adhesive force), (*g*). Thus, the force at which pull-off occurs is *not* precisely equal to the true adhesion force. How significant this deviation will be is discussed below.

This can also be illustrated from an energetic perspective. In Fig. 32.2c, we show separately the tip–sample interaction energy, E_{ts}, as a function of tip–sample separation (top), followed by the quadratic elastic strain energy, E_{lever}, of the cantilever (middle) and the sum, E_{total}, of the two (bottom). As the lever–sample displacement is varied, the elastic energy curve is shifted to the appropriate position which, for this example, is represented on the tip–sample separation axis. Far away from the sample, the system resides in a deep minimum (points (*a*) and (*b*), not shown). As the tip–sample separation is reduced, this minimum becomes increasingly shallow (evolving along the dashed arrow). At the snap-in point (*c*) the minimum is eliminated since the attractive energy of the tip–sample interaction has overwhelmed the energy minimum of the cantilever. Mathematically, this is described as an inflection point, where:

$$\frac{d^2 E_{tot}}{dz^2} = 0 .$$ (32.1a)

The total energy is given by

$$E_{tot} = E_{ts} + E_{lever}$$
$$= E_{ts} + \frac{1}{2}k(z - z_0)^2$$ (32.1b)

Therefore, snap-in occurs when

$$\frac{d^2 E_{ts}}{dz^2} = -\frac{dF_{ts}}{dz} = -k .$$ (32.1c)

This explains why the instability is described by the line of slope k being tangent to the tip–sample force curve as shown in Fig. 32.2b. The system now dynamically finds the new minimum, by following the gray arrow to point d where a mechanical contact between the tip and sample is now formed. There will, of course, be dynamics associated with this transition which can result in a damped oscillation as the system settles into its new stable equilibrium.

Note that once the contact is formed, the tip and sample elastically deform and so, strictly speaking, the tip–sample potential will be distinct from the potential shown before contact occurs. For simplicity we have left out this change, and instead have drawn the potential as a single-valued function. However, such changes in the energy landscape should be considered in a complete description of the problem [32.31].

Nevertheless, it will also hold true that upon retracting the tip from the surface, the stable minimum seen at (e), (f), and then (g) becomes more shallow and eventually disappears as it is overwhelmed by the strain energy of the cantilever. Again, an inflection point is created when:

$$\frac{d^2 E_{ts}}{dz^2} = -k ,$$ (32.2)

in other words, when again $\frac{dF_{ts}}{dz} = k .$ (32.3)

This second instability occurs at point (h), the pull-off point. The system then follows the gray arrow to point (i), where the tip is now out of contact with the sample (but experiencing a small amount of attractive force due to whatever long-range attractive forces exist). If the spring is sufficiently compliant (low k), or the potential sufficiently curved (large $\frac{dF_{ts}}{dz}$ shortly past the minimum at (g)), and if the long-range attractive force at (i) is small, then the pull-off force *does* nearly correspond to the *maximum attractive force*, as indicated in

Fig. 32.2b. In other words, point (h) would occur very close to point (g), and point (i) is close to a force of zero. The value of the force at (i) can be determined by retracting the cantilever sufficiently far from the surface, which is often practical except for strong, long-range forces that may occur, for example, when charge is present on the surface or the tip. Of more fundamental concern is that if the cantilever is somewhat stiff, or if the potential is rather compliant (which may be the case for organic, polymeric, biological or liquid systems), then the pull-off force may differ substantially from the maximum attractive force. For such cases, the distinction between the pull-off force and the maximum attractive force is important, and the limitations imposed by the AFM's intrinsic instabilities become apparent. Of course, if a cantilever with a stiffness that exceeds the attractive force gradient at all points is used, one will avoid the instabilities. However, since it is the deflection of the cantilever that is used to sense the force, one would have to trade off force sensitivity with stability, which is often an unwanted compromise.

In addition, the discussion above is entirely predicated upon the notion that there is a unique (i. e. non-hysteretic) force-displacement interaction between the tip and sample. In fact, as mentioned above, local rupture of adhesive bonds between the tip and the sample during retraction can occur because of the high tensile stresses. Therefore, for the same tip position, more than one metastable configuration of the atoms is possible. One case could be where the atoms have separated, and only weak van der Waals interactions occur. Another case would be where the atoms remain chemically bonded and strained in tension. The rupture of the tip-sample contact will occur instead at or very close to point (g) in Figs. 32.2a and 32.2b, regardless of the spring constant of the cantilever.

When a compliant spring or holder is used to manipulate a probe, as is the case for AFM, the technique is generally referred to as being "load-controlled", since the load can be prescribed, but the actual displacement of the probe with respect to the sample cannot (as illustrated by the jump in displacement that occurs during snap-in or pull-off).

In contrast, "displacement-controlled" techniques avoid this instability by effectively eliminating the compliance of the spring, thereby directly probing the tip-sample interaction. This has been carried out for decades in the mechanical testing community, and displacement-controlled scanning probes have been developed over the past ten years. This is accomplished by displacing the tip by direct application of a force to the tip itself. Pethica and coworkers [32.32, 33] use a magnetic coating on the tip and external coils to apply forces to the tip. They refer to the instrument as a "force-controlled microscope". *Houston* and coworkers [32.34, 35] control the force electrostatically and refer to the instrument as an "interfacial force microscope" (IFM). Lieber and coworkers use a variation on Pethica's method, where a magnetic coil is used to apply a force to the cantilever [32.36], and adapted it to work in solution.

An example of an adhesion measurement with the IFM is shown in [32.37], which shows that the instrument is able to measure the entire interaction force curve without instabilities. Thus, this provides a *direct* measurement of the minimum interaction force (as well as forces at all other tip–sample separations) and is a more reliable measure of adhesion. The disadvantage with these techniques is primarily one of inconvenience: the probes require extra manufacturing steps and control electronics. However, given the importance of adhesion in nanoscale science and technology, the extra information gained makes these techniques clearly worth the effort.

Pull-off instabilities can occur even in a displacement-controlled experiment. In this case, the instability is an "intrinsic" instability whereby the adhesive force gradient competes with the stiffness of the *contact itself* [32.38, 39]. Adhesive materials with low stiffness, such as polymers, may show this behavior.

Another way to avoid these instabilities while maintaining, or in fact, enhancing the force sensitivity, is to use dynamic AFM techniques [32.40]. In this case, the inertia of the cantilever, driven at or near its resonance, prevents the instabilities from occurring. The resonance frequency shift of the cantilever is sensitively measured, and this can be related to the integral of the force the tip experiences during its oscillation cycle. Thus, the force–displacement curve is mapped out by interpolating the data [32.41]. Uncertainty can be introduced by the interpolation scheme, since significant attraction is only experienced at the very bottom of the tip's oscillation cycle. Using small oscillations with high sensitivity force detection avoids this difficulty [32.42].

Finally, the pull-off force may show a time dependence that arises from intrinsic viscoelasticity of the tip or sample materials [32.43], or kinetic effects due to adsorption or reaction of materials at the interface [32.44]. These regimes remain relatively unexplored but certainly, to compare adhesion measurements between labs, the velocity of approach and retraction, as well as the time in contact, ought to be reported for any published experimental results.

A quantitative and reliable examination of adhesion therefore requires careful consideration of the mechanics of the contact and the cantilever. For an AFM experiment where instabilities occur, one can conclude that the pull-off force is a good measure of adhesion only if (1) the materials are fully elastic with little or no viscoelastic character, (2) the interface is chemically stable, (3) the cantilever stiffness is sufficiently low compared with the adhesive force gradient, and (4) the contact stiffness is sufficiently high compared with the adhesive force gradient. Otherwise, a more thorough investigation is required.

32.3 Calibration of Forces, Displacements, and Tips

Whether the forces are to be measured with load-controlled or displacement-controlled techniques, measurements cannot be compared between laboratories unless the forces are properly calibrated. Unfortunately this can be a rather involved task, and adoption of standards has yet to become widespread and robust. Here we provide a summary of the pertinent issues with references to other works for further reading; these issues are also discussed elsewhere in this book.

32.3.1 Force Calibration

Commercially available AFM cantilevers often come in two common forms: V-shaped and rectangular. Silicon and silicon nitride are the typical materials, but other choices are becoming popular. Reflective coatings are often applied to enhance the reflectivity to collect more laser light in the photodiode. The normal force constant of a monolithic rectangular cantilever beam requires knowledge of all lever dimensions and its modulus [32.45]. For a rectangular beam of length L, width w, thickness t and Young's modulus E, the normal bending stiffness k is given by:

$$k = \frac{Ewt^3}{4L^3} . \tag{32.4}$$

The cubic dependence on thickness is particularly problematic since in microfabrication processes t is usually determined by etching processes that are not precisely controlled, and the thickness is difficult and cumbersome to measure experimentally. Variations in E can also occur depending on the type of cantilever, particularly for silicon nitride cantilevers, although the dependence on E is clearly not as critical. If a bulk value for E is used, the thickness can be determined by measuring the resonance frequency of the cantilever [32.46]. However, this method only works for uncoated monolithic cantilevers, since the metal coatings applied to enhance reflectivity or for other purposes will also alter the spring constant substantially [32.47]. In addition, formulae for V-shaped levers are substantially more complicated. Uncoated, single-crystal silicon cantilevers are perhaps the only ones for which the force constant can be reliably determined by using (32.4) and the resonance frequency [32.48]. Otherwise, an experimental, preferably in situ, calibration method is strongly encouraged.

Experimental methods used to calibrate the force constant of AFM cantilevers have been extensively discussed [32.46,49–59]. It has been repeatedly shown that the manufacturer's quoted spring constants can be in error by large factors and should not be used in any quantitative research effort. We will not delve into the details of these calibration methods, but we do make note of one particularly recent method proposed by *Sader* et al. [32.60] which appears to be reliable and simple to perform for rectangular levers. It relies on measuring the resonance frequency and the quality factor of the cantilever in air. Use of the hydrodynamic function relates the damping effect of air to the quality factor and resonance frequency, and the dependences on E and t are eliminated from the resulting formula for the force constant. The measurement and calibration method can be carried out in a matter of minutes, particularly if the AFM software can find the resonance and then calculate the quality factor and resonance frequency. Care should be taken to account for background signal and filtering by the detection instrumentation at high frequencies.

32.3.2 Probe Tip Characterization

A problem of quite a different nature is that the geometry of the contact formed between the AFM tip and sample surface is not defined if the tip shape and composition is not known. This issue is of crucial importance since one is trying to understand the properties of an *interface*, and the tip is half of that interface.

The adhesion force between the tip and sample is a meaningless quantity in the absence of any knowledge of the tip shape [32.22]. The only use for such measure-

ments is in cases where the *same* tip is used to compare different samples or different conditions, and verification by cyclic repetition of the experiments is carried out to ensure that the tip itself did not change during the experiment.

There are several in situ methods that are used to characterize the tip shape. A topographic AFM image is actually a convolution of the tip and the sample geometry [32.61]. Separation of the tip and sample contributions by contact imaging of known, or at least sharp, sample features allows the tip shape to be determined to a significant extent [32.62–73]. Ex situ tip imaging by transmission electron microscopy has also been performed and can produce images with nanometer-scale resolution [32.72,74]. Some of these measurements have revealed that a large fraction of microfabricated cantilevers possess double tips and other unsuitable tip structures [32.63, 67, 73]. This convincingly demonstrates that tip characterization is absolutely necessary for useful nanotribological measurements with AFM. Thin film coatings applied to the microfabricated levers can provide robust, smooth and even conductive coatings [32.74–76]. Further work in this direction would be useful, so as to provide a wider array of dependable tip structures and materials.

In addition to the shape of the tip, the chemical composition of the tip is equally important, but is also challenging to determine or control. *Xiao* and coworkers have shown that the AFM tip is readily chemically modified when scanned in contact with various materials [32.77], even tips that have been coated with self-assembled monolayers in order to control their chemistry. They recommend "running in" the tip with a standard sample to give reproducible results. The stresses that take place in a nanocontact can be very large [32.5], and so modification of both the chemistry and structure of the tip is important to consider.

One class of experiments where the tip shape and chemistry is not as critical is in cases where a molecule or nanostructure is tethered to the end of the tip and specific interactions are probed [32.78–82]. Another alternative method is to use colloid probes [32.83, 84], whereby colloidal particles are attached to the cantilever on top of (or in place of) the tip. This method requires a unique calibration procedure [32.85], but provides particles whose structure and chemistry can be measured and perhaps controlled prior to attachment to the lever. The nanoscale roughness of these probes does need to be considered carefully though, as it will affect the interfacial properties.

32.3.3 Displacement Calibration

Proper signal and spatial calibration also requires knowledge of the sensitivity of the piezoelectric scanning elements. This can involve complications due to instrumental drift [32.87] and inherent piezoelectric effects, namely nonlinearity, hysteresis, creep, and variations of sensitivity with applied voltage [32.88–92]. Caution must be exercised when determining and relying upon these parameters. Techniques such as laser interferometry [32.93], scanning sloped samples [32.53, 89], scanning known surface step heights [32.94] or the use of precalibrated piezoelectrics [32.95] can facilitate piezo calibration.

32.3.4 Cantilever Tilt

One final issue of concern is that the cantilever itself is tilted by typically 10–20 °C in most AFMs (Fig. 32.3). This is to ensure that the tip makes contact with the sample before any other component, such as the nearby sides of the cantilever chip, but it introduces coupling between the mechanical forces of interest. Consider an experiment where the load is varied by moving the fixed end of the lever relative to the sample along the z-axis. Because of the tilt, the tip end of the lever displaces in the x-direction in Fig. 32.3b. Thus, with increasing load (with decreasing separation between the fixed end of the lever and the sample surface), the tip end of the lever moves in the $+x$-direction. Similarly, the tip retraces this path when the lever retracts from the surface. This issue was highlighted some time ago by *Overney* et al. who discussed the effect of in-plane displacement on elastic compliance measurements and accounted for it in their experiments [32.96]. In addition, *Marcus* et al. and *D'Amato*

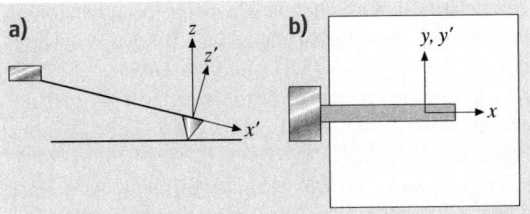

Fig. 32.3 (a) *Side view* and (b) *top view* of the lever–sample system in an atomic force microscope. The x-axis corresponds to the projection of the cantilever onto the sample surface. Tip displacement (or motion of the tip end of the lever) versus load occurs along this axis. Load is varied by moving the fixed end of the lever relative to the sample along the z-axis. After [32.86]

et al. addressed consequences of the tilt angle in AFM in relation to phase contrast imaging in intermittent-contact AFM [32.97, 98], and *Cannara* et al. [32.86] discussed a methodology for correcting the effect when acquiring friction measurements. Both *Overney* et al. and *Cannara* et al. propose moving the fixed end of the lever relative to the sample along the z'-axis shown in Fig. 32.3b, instead of the z-axis. This is also useful for adhesion measurements, as it will restrict the range of motion of the tip across the sample as the load is varied.

Heim et al. [32.99] pointed out that pull-off forces or any other measured normal force needs to be corrected for the effect of tilt. They propose a specific correction factor and show that the correction can be as large as 20-30% for tilt angles of 20°. *Hutter* [32.100] has proposed a modification to their equation. Note that issues of tilt are absent for the IFM described above, since the tip and force sensor are oriented directly normal to the sample surface.

32.4 The Effect of Liquid Capillaries on Adhesion

32.4.1 Theoretical Background and Approximations

The adsorption of water and other liquids onto surfaces, and their subsequent behavior at interfaces, continues to be a vibrant area of research. The importance of liquid–solid interface behavior is massive, encompassing topics as broad as paints, textiles, lubricants, geology and environmental chemistry, and covering all corners of biology.

The ability to measure forces at the nanoscale using scanning probe microscopy has generated much interest in these fields. For any force measurement carried out in ambient laboratory conditions, the possibility of a capillary neck forming between the tip and sample must be considered. The study of such necks may in turn provide insight into the behavior of the liquid, which is discussed in other chapters in this book. In order to provide background for these emerging areas, we consider here the fundamental mechanical and chemical aspects of adhesion in the presence of a liquid meniscus. By way of introduction, *Israelachvili* [32.1] and *de Gennes* [32.101] provide rigorous coverage of the terminology, physics, and chemistry of liquid films and their wetting properties.

Water readily adsorbs at many surfaces. At a crack or sharp corner, it can condense to form a meniscus if its contact angle is sufficiently small. The small gap between an AFM tip and a surface is therefore an ideal occasion for such condensation. Early on it was realized that liquid condensation plays a significant role in tip–sample interactions [32.103, 104].

The AFM literature has so far idealized the tip as a sphere of radius R and applied the classic theory of capillary condensation between a sphere and a plane, which as based on the thermodynamics of capillary formation. Within this theory, geometrical assumptions are often made. In particular, when the radius of the sphere is large with respect to the size of the capillary, one possible approximation is known as the "circle approximation", in which the meniscus radii of curvature are taken to be constant. The geometry of the capillary meniscus using the circle approximation is shown in Fig. 32.4, which is adapted from [32.102]. Hydrophobic surfaces are sketched on the left half ($r_1 > 0$), and hydrophilic surfaces sketched on the right half ($r_1 < 0$). The water contact angles with the sample and tip respectively are θ_1 and θ_2. D represents the separation of the tip and sample. The angle ϕ is referred to as the "filling angle". The pressure difference, or Laplace pressure, across a curved interface is given by the Young–Laplace

equation [32.1, 102]:

$$\Delta p = \gamma \left(\frac{1}{r_1} + \frac{1}{r_2} \right) \tag{32.5}$$

where r_1 and r_2 are defined in Fig. 32.4 and γ is the surface tension of the liquid, which in this case is water. The resulting force is attractive if $\Delta p < 0$. Note that $r_1 > 0$ and $r_2 < 0$. If the capillary formation is isothermal, then one can derive the Kelvin equation:

$$R_M T \ln \left(\frac{p}{p_0} \right) = \gamma V \left(\frac{1}{r_1} + \frac{1}{r_2} \right), \tag{32.6}$$

where p_0 is the saturation pressure of the liquid, V is the molar volume of the liquid, T is the temperature, and R_M the molar gas constant. The ratio p/p_0 simply corresponds to the relative vapor pressure of the liquid, which in the case of water is just the relative humidity (RH). This is often rewritten in terms of the Kelvin radius r_K:

$$r_K^{-1} = \left(\frac{1}{r_1} + \frac{1}{r_2} \right) = \frac{R_M T}{\gamma V} \ln \left(\frac{p}{p_0} \right). \tag{32.7}$$

The Kelvin radius varies logarithmically with the partial pressure of the liquid and is by definition less than zero (negative values correspond to convex curvature). For water at $20\,°C$, $\gamma V / R_M T = 0.54$ nm. A graph of r_K versus p/p_0 is shown in Fig. 32.5. Of particular note is the fact that starting from $p/p_0 = 0.75$ and lower, we find that $|r_K| < 2$ nm. Although widely used, Eqs. (32.5–32.7) are approximate because r_1 and r_2 are not normal everywhere to the meniscus surface and hence the assumed meniscus shape is not isobaric. This point will be elaborated in Sect. 32.4.3.

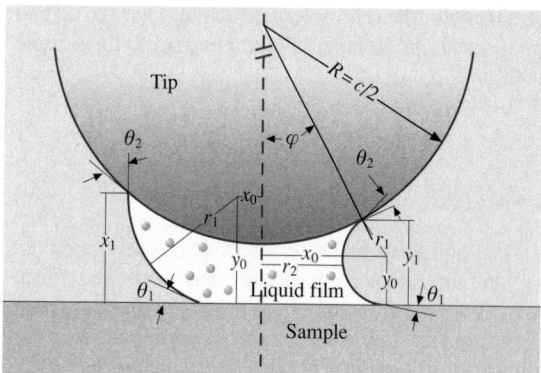

Fig. 32.4 The AFM tip, considered as a sphere at a distance D from the sample. The liquid film in-between may form a concave (*right*) or convex (*left*) meniscus in the plane shown. Figure is based on [32.102] but with significant changes

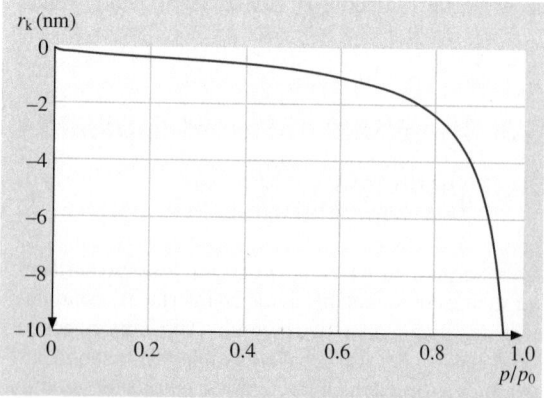

Fig. 32.5 The Kelvin radius of water at $20\,°C$ plotted as a function of the relative humidity

A simple equation for the maximum attractive capillary force F_c between the tip and sample that is commonly used is

$$F_c = -4\pi\gamma R \cos\theta. \tag{32.8}$$

This equation is calculated by considering the Laplace pressure only. The adhesion force is predicted to be independent of RH. Notice that an immediate problem arises, which is that it predicts a finite force even at 0% RH, where there can be no capillary formation. Derivations of (32.8) are presented in several other publications [32.1, 102, 105, 106]. Significantly, this equation and its use with AFM experiments contains several assumptions, *many of which may not be valid for an AFM experiment*:

1. $|r_1| \ll R$, which is equivalent to saying ϕ is small;
2. $|r_1| \ll |r_2|$;
3. $\theta_1 = \theta_2$;
4. $D \ll |r_1|$, $|r_2|$;
5. The tip is shaped like a perfect sphere.;
6. The effect of solid–solid adhesion is negligible with respect to the meniscus force;
7. The meniscus cross-sections are perfect circular arcs;
8. The force from the Laplace pressure dominates the force due to the resolved surface tension of the meniscus;
9. The surface tension γ is independent of the meniscus size;
10. The meniscus volume remains constant as the tip is retracted;
11. The maximum force of attraction is equal to the pull-off force;
12. The tip and sample are perfectly rigid.

Assumption (1) may not be true, since the tip radius may indeed be small and comparable to the meniscus curvature radii. Assumption (2) may not be true since the small tip geometry may cause the two meniscus curvature radii to be similar. Assumption (3) is by no means true if the tip and sample are made of different materials. Assumption (4) may not be true, since both the separation of the tip and sample as well as the radii r_1 and r_2 may be in the nanometer range. Assumption (5) may be slightly or grossly in error, and is a particularly dubious assumption in the absence of tip characterization. Also, for large menisci, the capillary will grow beyond the end of the tip and start climbing up its shank which may be pyramidal or conical in shape. Assumption (6) may also be inaccurate if van der Waals or other adhesive forces are significant, and this is discussed further in

the next section. Assumption (7) is not correct [32.106], and the regimes where it is a reasonable approximation require close scrutiny. Assumption (8) will be inaccurate at high relative vapor pressures [32.1, 105]. The nature of assumptions (1)-(8), and corrections to the theory to account for their violation, are presented in the work of *Orr* and *Scriven* [32.106]. This theory remains within the bounds of the classical picture of capillary formation. Results from numerical calculations in which this theory is applied are given later in this section. It will be seen that for (32.8) the geometrical assumptions alone do not severely restrict it.

Assumption (9) concerns an important scientific question that has not been fully resolved and represents the possible violation of the classical framework by molecular effects at the nanoscale. SFA measurements by *Israelachvili* have indicated that for cyclohexane and other inert organic liquids, γ remains nearly equal to its macroscopic value even for Kelvin radii that, remarkably, correspond to one or two molecular diameters. However, for water, the adhesion force comes to within 10% of the bulk prediction only for $p/p_0 > 0.9$, which corresponds to Kelvin radii greater than ≈ 5 nm in magnitude [32.17]. Later, *Christenson* [32.107] improved the symmetry of the SFA leaf spring and found that (32.8) holds for cyclohexane, n-hexane and water for $p/p_0 > 0.7$. Nonetheless, these deviations from macroscopic thermodynamic predictions alone calls the use of (32.8) into serious question for AFM measurements. Rather, the exploration of this deviation at the molecular scale presents a unique opportunity for scanning probe measurements.

Assumptions (10) and (11) are not assumptions of (32.8) itself, but rather assumptions that are often used when applying (32.8) to AFM measurements. Equation (32.8) simply gives the maximum force of attraction between the tip and sample. As discussed above, an AFM does *not* measure this quantity. Rather, it measures the force at which an instability occurs. If a capillary has formed between the tip and sample, then the force as a function of distance can be calculated. Calculating this force requires making one of two assumptions: either the volume of the capillary is conserved (due to the rate of displacement being large with respect to the adsorption or desorption kinetics of the liquid) or the Kelvin radius is conserved (the rate of displacement is slow with respect to the adsorption or desorption kinetics of the liquid, and so the capillary remains in equilibrium). The constant volume assumption (10) has been used in every paper we have reviewed. *Israelachvili*, however, pointed out the difference between these two approaches in his book [32.1], and left the solution of the problem as an exercise to the reader. The force is the same at $D = 0$, but the reduction in force with displacement is more rapid and linear for the constant Kelvin radius case. With the assumptions listed above for the constant volume case

$$F(D) = 4\pi R\gamma_l \cos\theta \left(1 - \frac{D}{\sqrt{4r_K^2 \cos^2\theta + D^2}}\right),$$

(32.9)

while for the constant Kelvin radius case,

$$F(D) = 4\pi R\gamma_l \cos\theta \left(1 - \frac{D}{2\,|r_K|\cos\theta}\right). \quad (32.10)$$

As with the problem of scale-dependent surface tension mentioned above, the kinetics of capillary formation and dissolution is a relatively unexplored problem and is therefore worthy of further investigation. A recent study of the humidity dependence of friction as a function of sliding speed is an example where this issue is raised [32.108].

Once an assumption about how the meniscus changes with displacement has been made, one still needs to consider the nature of the instability in order to relate the AFM pull-off measurement to the capillary's properties. As stated above and shown in Fig. 32.2, a low lever stiffness k or a strongly varying adhesive force will lead to a pull-off force that is nearly equal to the adhesive force, and so assumption (11) would be valid. However, if k is sufficiently large, or the capillary stiffness sufficiently weak, this assumption will fail. As we shall see below, experimental efforts to investigate this point are yet to be carried out.

Finally, assumption (12), if violated, requires a substantially more complex analysis to deal with it. The question has been addressed independently by *Maugis* [32.109] and *Fogden* and *White* [32.110]. Both papers provide a nondimensional parameter that allows one to determine the severity of the effect. In the limit of small tips, stiff materials, large (in magnitude) Kelvin radii, and low surface tensions, the effect of elastic deformation is negligible. However, for relatively compliant materials, large tips, and small Kelvin radii, the meniscus can appreciably deform the contact in the immediate vicinity of the meniscus. This can substantially alter the mechanics of adhesion as well as significantly affecting the stresses. The dependence on the Kelvin radius is particularly critical. This effect may be of particular concern with soft materials like polymers or biological specimens. According to

Maugis, the problem becomes analogous to the adhesive contact problem for solids, discussed by *Johnson* et al. [32.111], and further studied in many papers since [32.31, 112, 113].

32.4.2 Experimental Studies of Capillary Formation with Scanning Probes

There have been several experimental and theoretical investigations of how pull-off forces are affected by liquid capillaries. These studies have mostly focused on the effect of relative humidity. We do not present an exhaustive review here, but rather we summarize a few key results that highlight the important trends observed and the outstanding questions that remain.

Early on, it was realized that water capillary formation occurred readily if at least one of the two surfaces in contact was hydrophilic. Higher adhesion will lead to higher contact forces and therefore larger elastic contact areas, and this can degrade the lateral spatial resolution of the AFM, as was observed by *Thundat* and coworkers [32.114]. Furthermore, several observations have confirmed the expected result that capillary formation was readily prevented for hydrophobic surfaces. For example, *Binggeli* and *Mate* [32.104, 115] showed that tungsten tips in contact with clean silicon wafers with a hydrophilic native oxide exhibited strong adhesion and long pull-off lengths, whereas surfaces treated with a hydrophobic perfluoropolyether (such as Z-DOL) showed pull-off forces reduced by a factor of 2–3. These results were confirmed by *Bhushan* and *Sundararajan* [32.116],

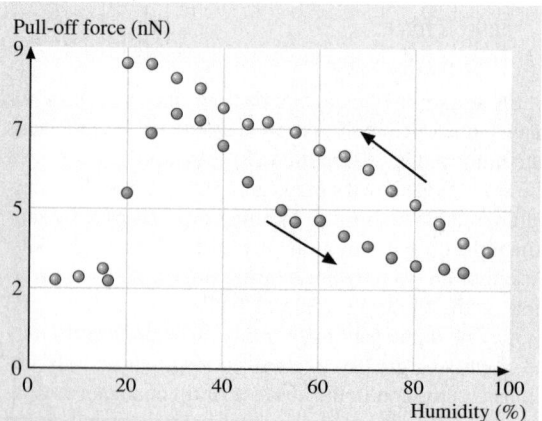

Fig. 32.6 Pull-off force between a silicon nitride tip and the muscovite mica surface as a function of RH. Two sets of data are shown; increasing (*open circles*) and decreasing (*closed circles*) humidity. After [32.44]

who investigated the pull-off force between silicon nitride tips and Si(100) with and without Z-DOL coatings for a range of relative humidities. As another example, MoO_3 films, likely hydrophilic, showed a sixfold increase in pull-off force measured with silicon tips from 0 to 50% RH [32.117].

While these examples illustrate some basic trends, the need for more finely resolved measurements as a function of humidity has been addressed only quite recently. For example, *Xu* et al. measured adhesion between a silicon nitride tip and (hydrophilic) muscovite mica in ≈ 5% RH increments. The result is shown in Fig. 32.6. There are three identifiable regions: constant adhesion at low RH (<20%), increasing adhesion, and then decreasing adhesion. Some hysteresis is seen between experiments conducted with increasing and decreasing RH, but the overall trend is preserved. *Xu* et al. correlated their measurements with detailed studies of the growth of molecular water films on the mica surface which they could image directly using scanning polarization force microscopy (Fig. 32.7). They proposed that below 20%, capillary condensation does not occur, and indeed, they see no evidence of a water film at these low humidities. Above 20% RH, a strongly bound molecular water layer is formed on the bare mica surface (Fig. 32.7). In the presence of the tip, a capillary meniscus can condense. Above 40% RH, the pull-off force decreases. Recognizing one of the limitations of (32.8), they attribute this to the violation of assumption (5) listed above. They argue that for a pyramidal AFM tip, r_1 and r_2 become comparable in magnitude (and remain opposite in sign) once the capillary reaches the shank of the tip, leading to a near cancellation of the Laplace

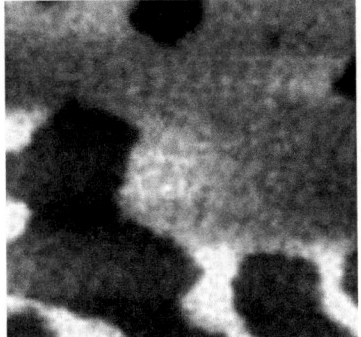

Fig. 32.7 Scanning polarization force microscopy image of water structures on the muscovite mica surface. A degree of polygonal shape to the boundaries can be seen. The signal represents two distinct phases of water that are present as a molecular film below 45% RH After [32.44]

pressure given in (32.5). This argument is certainly plausible, although a rigorous proof is not provided, and the other limitations of (32.8) are not discussed in relation to this issue. Nevertheless, the correlation between the onset of adhesion increase and the formation of the molecular water film as seen directly in their dramatic images is an extremely convincing case where the classical assumptions of (32.8) must be modified to account for the molecular structure of the water film.

Further considerations of the limitations of (32.8) were measured, discussed, and modeled in a detailed paper by *Xiao* and *Qian* [32.105]. Adhesion measurements, collected in large numbers for good statistics, were carried out with the same silicon nitride tip on two different surfaces: hydrophilic SiO_2, and a hydrophobic layer of *n*-octadecyltrimethoxysilane (OTE) on SiO_2. Contact angle measurements to confirm the assertion of hydrophilicity were not presented; however, the OTE surface was confirmed to be hydrophobic with a water contact angle of 108°. The results are shown in Fig. 32.8. The hydrophobic surface shows no dependence on RH, whereas the hydrophilic bare SiO_2 surface shows three regimes similar to the result of *Xu* et al.: constant pull-off force (≳30% RH), increasing pull-off force (30%–70% RH) and then decreasing pull-off force. *Xiao* and *Qian* discuss the limitations of (32.8) in substantial detail. In particular, they consider the violation of assumptions (1) through (8) listed above. This includes a treatment of van der Waals adhesion, with the effect of electrostatic screening of this force by the water itself taken into consideration. Equations for this more general case are presented. With all these aspects taken into account, they are only able to fit the model to their data qualitatively (Fig. 32.9). At low RH (< 10%), the van der Waals force dominates and the adhesion is initially constant. At intermediate values, the Laplace pressure contribution increases and then begins to saturate. The contribution from the resolved surface tension becomes significant at high RH (above ≈ 80%), and the contribution from the Laplace pressure begins to drop strongly around this same point. It should be noted that their capillary model still assumes constant values of r_1 and r_2 for the entire meniscus, and while it may be an improvement on (32.8), it is not precise. Qualitatively, this reproduces their results (and the aforementioned ones) by producing regimes of constant, increasing, then decreasing adhesion. However, the humidities at which the transitions occur, and the relative changes in adhesion, do not match the data. Somewhat better agreement at high RH (> 70%) is found by considering alternate (blunt) tip

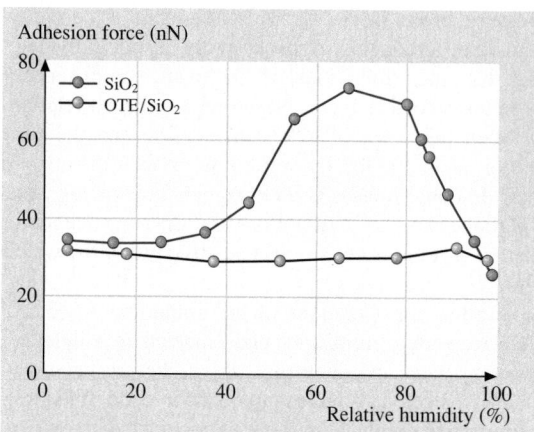

Fig. 32.8 Pull-off force as a function of RH for adhesion between a silicon nitride tip and SiO$_2$ (*filled circles*) and OTE/SiO$_2$ (*open circles*) samples. After [32.105]

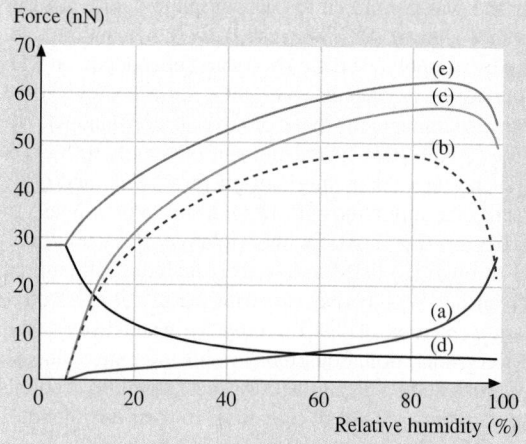

Fig. 32.9a–e Contributions to the adhesion force as a function of humidity: (**a**) resolved surface tension force; (**b**) Laplace pressure force; (**c**) total capillary force (Laplace + surface tension); (**d**) van der Waals force; (**e**) total adhesion force. After [32.105]

shapes. Nevertheless, the most significant discrepancy occurs for low RH, where the extent of the constant adhesion force is underestimated by the classical theory. The authors attribute this to a failure of the classical continuum theory to properly describe the properties of a molecular-scale meniscus, as earlier concluded using the SFA [32.107]. Another interesting point of this study is that at low RH, the adhesion is very similar for both samples, a fact which the authors attribute to the dominance of the van der Waals force for both samples (which

is largely determined by the substrate and not affected significantly by the OTE film).

Slightly more recently, *He* et al. have studied capillary forces for a variety of tip–sample pairs [32.118]. Hydrophilic tips (silicon and silicon nitride with no surface treatment) and hydrophobic tips (coated with *n*−octadecyltrichlorosilane) were used. The hydrophobic character of the tip was asserted based on a water contact angle measurement of 105.5°, presumably taken on a different region of the cantilever chip. Solvent-cleaned silicon samples and calcium fluoride films were used as hydrophilic substrates. As with the measurements of *Xiao* and *Qian*, contact angle measurements for the hydrophilic samples and tips were not presented. Results for hydrophobic and hydrophilic tips are shown in Fig. 32.10.

These results, which were carried out independently and without knowledge of *Xiao* and *Qian's* work, show impressive agreement. This is particularly interesting given that here the *tip* was varied from hydrophilic to hydrophobic (while the sample stayed hydrophilic), whereas in *Xiao* and *Qian's* work, the tip presumably remained hydrophilic and the sample was varied from hydrophilic to hydrophobic. For the hydrophobic tip, the pull-off force remains constant, indicating once again that capillary formation was suppressed. However, for the hydrophilic tip, three regimes of adhesion are found Fig. 32.11. Similar results are found for a hydrophilic glass microsphere used as a tip. The authors refer to these three regimes as the van der Waals regime, the mixed van der Waals–capillary regime, and the capillary regime. In agreement with *Xiao* and *Qian's* assessment, the authors propose that at low RH ($\lesssim 35\%$ in this case), the formation of the water meniscus is suppressed and the adhesion is dominated by solid–solid (presumably van der Waals) interactions. They propose, based on the work of *de Gennes* on the theory of spreading [32.101], that a minimum precursor film thickness is required to form the meniscus. The authors also present a calculation of the adhesion force when assumptions (1) and (2) are relaxed.

32.4.3 Theoretical Issues Revisited

As detailed in Sect. 32.4.1, there are many approximations leading to (32.8). However, the geometrical assumptions (1)-(4), (7) and (8) are addressed by the numerical treatment of *Orr* et al. [32.106]. In this approach, the circle approximation is removed. Boundary conditions determined by the geometry are applied to the axisymmetric Young–Laplace differential equation,

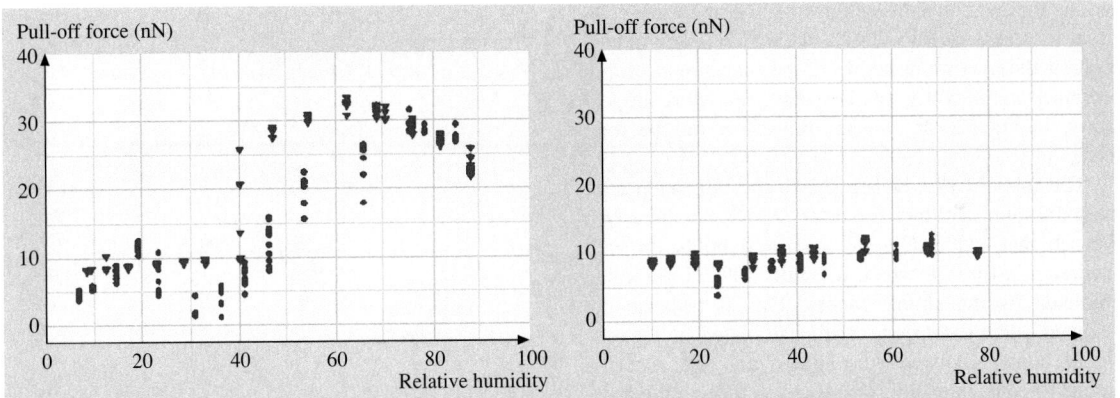

Fig. 32.10 (**a**) Pull-off force versus RH measured between a hydrophilic tip and a flat silicon sample. *Circles*: measured when increasing RH. *Triangles*: measured when decreasing RH. (**b**) Pull-off force versus RH measured between a sharp SFM tip coated with OTS and a flat silicon sample. The pull-off force is independent of humidity. After [32.118]

which is then solved exactly in terms of incomplete elliptic integrals. The total force is calculated from the Laplace pressure and the resolved surface tension force contributions, and is independent of the z-plane at which it is calculated. Let us qualitatively consider Fig. 32.12a, which is a revised form of Fig. 32.3 showing the "pendular bridge" geometry. Now we have replaced the commonly used radii r_1 and r_2 with the *principal radii of curvature*, r_a (the azimuthal radius) and r_m (the meridional radius). The principal radii are normal everywhere to the meniscus surface and are contained in two orthogonal planes. Each of these orthogonal planes contain the surface normal. Although any two orthogonal planes can be used to find the surface curvature, the principal radii are in the planes oriented such that they contain the minimum and maximum surface radii. Note

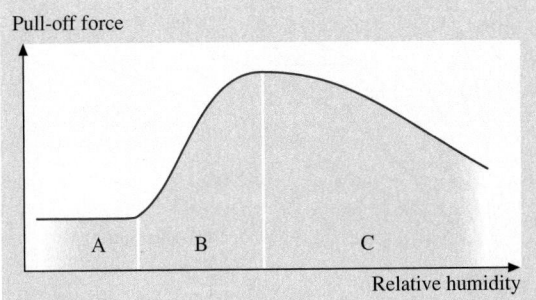

Fig. 32.11 Figure illustrating the distinct regimes of the pull-off force as a function of RH. Regimes I, II, and III are referred to as the van der Waals regime, mixed van der Waals–capillary regime, and capillary regime respectively. After [32.118]

that these quantities *are not constant – they have distinct local values at different locations on the meniscus surface*. The particular radii drawn refer to the point A on the meniscus surface. The radii drawn in Fig. 32.12a refer to, where the direction indicates the normal to the meniscus surface at A. Being outside the pendular ring, the meriodonal radius r_m is negative, and its magnitude is equal to the radius of curvature of the arc formed by the intersection of the plane of the page and the meniscus surface. Being inside the pendular ring, the azimuthal radius r_a is positive, and its magnitude is equal to the radius of curvature of the corresponding arc normal to the plane of the page.

Using r_a and r_m, the mensiscus is now isobaric. Equation (32.7) is no longer approximate, that is:

$$r_K^{-1} = \left(\frac{1}{r_m} + \frac{1}{r_a} \right) = \frac{RT}{\gamma V} \ln \left(\frac{p}{p_0} \right) . \qquad (32.11)$$

Let us take R to be small, say 20 nm, of the order of an AFM tip. At very low partial pressure p/p_0, the geometrical approximations leading to (32.8) hold, and (32.9) for the constant volume case and (32.10) for the constant Kelvin radius case still hold. However, for $D = 0$, as p/p_0 begins to increase, we see from Fig. 32.5 that $|r_K|$ begins to approach R. Recall that the azimuthal radius r_a is positive while the meridional radius r_m is negative. As p/p_0 increases, r_a begins to approach $|r_m|$ in magnitude because of the high tip curvature. Hence, $|r_m|$ must decrease to keep r_K (a negative quantity) constant. This significantly changes the meniscus profiles.

In Fig. 32.12b, we show our own calculated results following the numerical approach with $R = 20$ nm, and assuming constant volume of the meniscus. As p/p_0 increases in Fig. 32.12b, with $D = 0$, we indeed see that the maximum height in the meridional profile becomes significantly smaller than r_K. More importantly, in Fig. 32.12c, we see the effect on the total force.

The inset in Fig. 32.12c shows the dependence of the maximum capillary force (at $D = 0$) versus p/p_0. We see that {eq08-d32-eq8 at $D = 0$ holds up well even at $p/p0 = 0.9$, where it is still 80% of the value predicted by the simple theory. This is because the resolved surface tension, when calculated at the circle of contact between the sphere and the asperity, begins to contribute significantly even though the area-dependent contribution from the Laplace pressure is reduced.

These exact numerical results substantiate the conclusions drawn earlier [32.44], that the strong decrease in pull-off force with increasing p/p_0 is not due to a decrease in the capillary adhesion. However, besides the tip shape argument already mentioned [32.44, 105], another explanation for the reduction in force at high RH has been proposed [32.104, 118]. From thermodynamics, the component of the attractive force acting on the

Fig. 32.12 (a) The general pendular ring geometry for a meniscus between a sphere of radius R and a flat substrate. The contact angles are θ_1 and θ_2 for the sphere and plane respectively. z represents the separation between the sphere and the flat surface. The local meridional radius of curvature, r_m (<0), and local azimuthal radius of curvature, r_a (>0), are drawn for the point A. **(b)** Geometric shape of a water meniscus for a range of RH values, from the numerical solution to the Young-Laplace equation for a 20 nm radius sphere at zero separation (contact) for perfectly hydrophilic surfaces. Clearly, the size of the meniscus starts to approach the size of the sphere itself at high RH. The lack of applicability of the circle approximation is also evident. **(c)** The meniscus force between the tip and the surface as a function of separation for a range of RH values from 0.3 to 0.95. The unstable part of the force–separation curve is indicated with a *dotted line*. *Dashed lines* represent the force–displacement relation for a cantilever with force constants of $k =0.5$, 2.5, and 50 N/m. Instabilities will occur when the gradient of the meniscus force exceeds k. A force constant of 50 N/m will remain stable for all RH, and one of 0.5 N/m will pull off immediately upon applying any retraction. At 2.5 N/m, metastable points may be reached depending on the RH. The *solid line* at the bottom represents the prediction of (32.8). The *inset* shows the maximum attractive force (at zero separation) as a function of RH

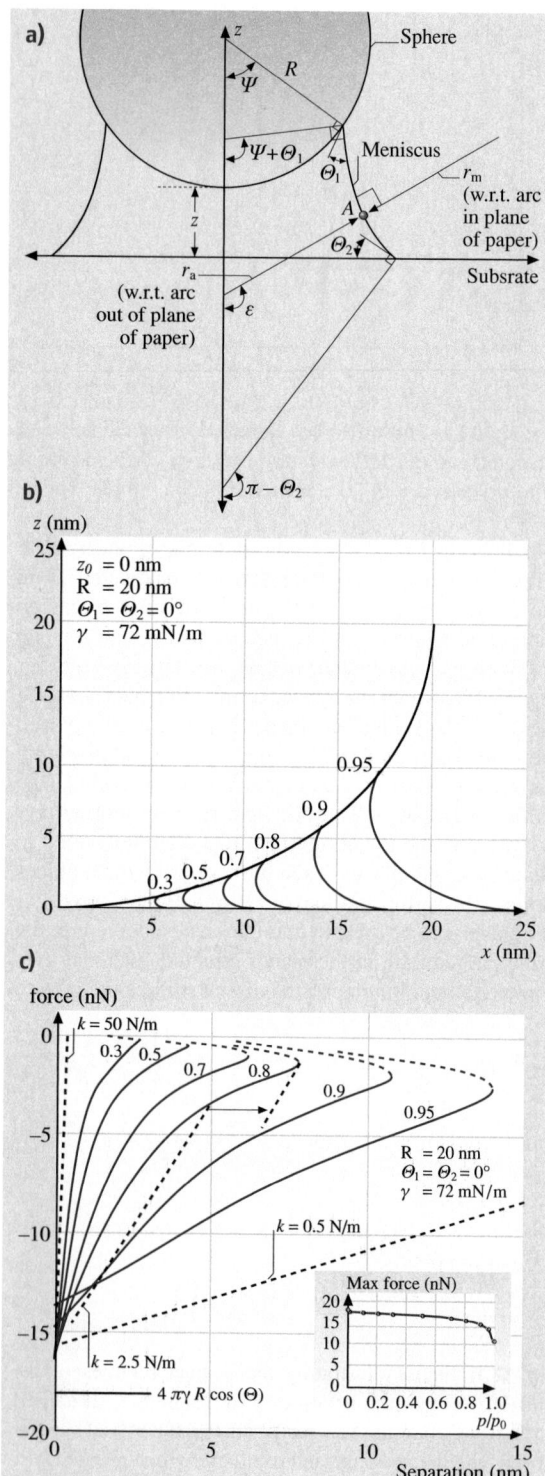

tip from the liquid in the gaps is given by

$$F_{\text{chem}} = -\frac{\partial G}{\partial z} = -\frac{A}{v}kT \ln\left(\frac{p}{p_0}\right) , \qquad (32.12)$$

where G is the Gibbs free energy, A is the area of the liquid film in the gap, and v is the molar volume. This term corresponds to a positive (repulsive) force opposing the negative (attractive) capillary force. For a small radius R, the relative importance of A increases, and hence this term can become important.

In addition, the force–displacement curves in Fig. 32.12c exhibit a "nose". Points below this nose (indicated by the solid lines) are stable capillary shapes, while points above it (indicated by the dotted lines) are unstable, and will not be accessed by an experiment. The dashed lines in Fig. 32.12c show the cantilever load lines for three different stiffnesses and are drawn for the case $p/p_0 = 0.8$. If the AFM cantilever is stiff, the full force–displacement curve, up to the tip of the nose, will be sampled. For example, *Binggeli* and *Mate* [32.104] used a spring stiffness of $\approx 50\,\text{N/m}$, and their results exhibit the full capillary force–displacement curves (they estimate $R = 100\,\text{nm}$). On the other hand, *Xiao* and *Qian* [32.105] used a nominal spring stiffness of $\approx 0.5\,\text{N/m}$. That stiffness would only measure the maximum force in this situation. The stiffnesses used by *He* et al. [32.118] were nominally $0.5\,\text{N/m}$ and below. Indeed, they saw no dependence of the pull-off force on the spring constant. However, if the spring constant were $2.5\,\text{N/m}$ with $p/p_0 = 0.8$, we would expect a jump from the maximum capillary force to another stable point. Beyond this point, the experiment would sample the remainder of the curve.

This pendular ring treatment has not considered that a thin film of liquid is present on the surface according to the BET adsorption isotherm [32.119], nor has it considered the associated effect of disjoining pressure. Disjoining pressure can be thought of as the force per unit area experienced by surface molecules on a solid surface relative to molecules on the bulk liquid [32.120]. *Gao* [32.121] has shown that the effect of disjoining pressure is small when $r_a \ll R$ and $D \ll D_{\text{max}}$ (where D_{max} is the distance corresponding to rupture). Clearly these conditions are not always met by the pendular ring, and therefore experimental work to quantify the effect of disjoining pressure on the true capillary force-displacement curves of AFM-sized tips would be welcome.

There is a large body of literature exploring the numerical solution to the axisymmetric Young–Laplace differential equation. For example, *Lian* et al. [32.122] examined the stability of the curves in detail. *Willett* et al. [32.123] provide an excellent summary of this literature and of the macroscale experiments that have been conducted. They also perform experiments that match the theoretical curves using millimeter-sized spheres, and further give analytical expressions derived from curve fitting procedures to approximate the numerical results.

32.4.4 Future Directions

The results and modeling so far indicate the possibility of two trends: hydrophobic surfaces will exhibit little dependence of adhesion on RH, whereas if one surface is hydrophilic, three regimes of behavior occur: constant solid–solid adhesion at low RH, increasing adhesion at intermediate RH, and decreasing adhesion at the highest RH. Some insight into the physical mechanisms behind these regimes has been presented, but it would be desirable to pursue further work in this area.

Perhaps most critically needed is an atomic-scale picture of the menisci and water films present under low partial pressures. This could address the question of why adhesion is initially independent of RH, and when the meniscus itself would start to form. Further theoretical developments that address the assumptions laid out above would also help to clarify the picture. It is also important to extend these studies beyond simply the case of water, as the properties of other liquids are also of great interest, and could be compared to previous experiments with the SFA and other tools.

Studies that clarify the kinetics of meniscus formation are also needed. There should be a noticeable transition in behavior once the rate of displacement becomes comparable to the appropriate kinetic rates, and this could provide valuable information about these kinetic processes at the nanometer scale.

Finally, as mentioned before, there continues to be a gap in reproducibility and comparability between laboratories that will only be bridged when standard techniques for tip characterization and force and displacement calibration are addressed. Efforts that take these considerations into account are worthy of further support.

Part D | 32.4

32.5 Self-Assembled Monolayers

32.5.1 Adhesion at SAM Interfaces

When thinking about adhesion and the related phenomenon of friction, it is important to realize that the interfaces of real surfaces in contact are rarely atomically smooth. Surfaces that appear smooth on the macroscopic scale are, upon closer inspection, found to consist of nanometer-scaled asperities (typically on the order of 10 nm) whose intentional or accidental interactions ultimately control adhesion, friction and wear at contacts [32.124–128]. The sizes of these asperities become particularly important when one considers that the true contact area between interfaces for the distribution of load is localized through these asperity–asperity interactions where extremely high pressures can be produced, resulting in sharply increasing local stress fields that can cause materials to yield and shear as they encounter each other during sliding and intermittent contact. In addition to load distribution at nanoscale asperity–asperity contacts, their size will influence surface wetting and adhesion due to capillary forces localized at the contacts [32.126–128]. The structures of applied lubricant films at such asperities will be highly dependent upon asperity curvature, and defects in lubricant film structure may form more readily here than on atomically flat surfaces.

The minimization of adhesion at such asperity–asperity contacts is a critical issue in MEMS devices [32.125–128]. In fact, the intentional introduction of surface roughness (on the order of 10 nm RMS) can be employed to lead to reduced stiction during post-processing feature release. These same asperities, however, must later resist wear during controlled or accidental contact during device operation. Thus, the specific details of adhesion and energy dissipation at such asperity–asperity contacts are required for the rational design of such systems.

To function as a protective lubricant layer in such systems, self-assembled monolayers (SAMs) of alkylsilane and fluorosilane compounds with chain lengths ranging from C_{10} to C_{18} have been shown to be useful in the reduction of friction and adhesion in MEMS [32.125–128]. Such direct applications of SAMs as lubricant films, combined with the ease of sample preparation and the ability to generate model surfaces with well defined film chemistry and structures have made nanotribological and adhesion studies of SAMs a rich area of research [32.5]. Many of these nanotribological studies have used AFM to examine either SAMs of alkylsilane films on atomically smooth Si wafers, glass or mica surfaces. Alternatively, many researchers have examined alkanethiol films on atomically smooth Au(111) surfaces. Using this approach many molecular level details, such as the influence of film chemistry and molecular organization on friction, adhesion and wear of SAMs can be obtained. Developing a clear understanding of the details of adhesion at SAM-modified surfaces allows for the complex link between surface chemistry and adhesion and friction to be understood at the molecular level. In this section we overview AFM studies of adhesion on SAM-modified surfaces.

32.5.2 Chemical Force Microscopy: General Methodology

In order to probe adhesion between chemical modified surfaces using AFM, the probe tips and sample surfaces are typically modified via self-assembly of monolayers using organosilanes on surfaces such as mica, glass or oxidized Si or Si_3N_4 (the latter two being the typical materials of which AFM tips are made) or formed from thiols on Au-coated AFM tips and surfaces (Fig. 32.12) [32.129, 130]. While this has been shown to be a facile method for the modification of AFM tips for chemical force measurements, it should be noted that the details of the packing densities of the monolayers formed on the AFM tips are in general not known. This lack of detail regarding molecular overlayers on AFM probe tips can be a problem which requires careful consideration when using such chemically modified tips for the quantitative determination of adhesion forces and molecular interactions, as the number density of species in the tip–sample contact is related to the measured adhesion.

Details of the environment in which the adhesion measurements are carried out are another important consideration. Under ambient environmental lab conditions, surfaces are contaminated with organic compounds from the air as well as a layer of condensed water vapor, which varies with humidity. The condensed water layer can form a contact meniscus between the tip and sample introducing a capillary force into the measured adhesion [32.131]. The presence of this capillary force can overwhelm the details of the adhesion from the SAM-terminated surfaces to be probed. To avoid this, many studies are performed in liquid environments or ultrahigh vacuum to eliminate capillary forces. In liquid, the nature of the solvent will of course impact on the measured adhesion for a given pair of interacting surfaces, as solvent exclusion plays an important role. Also, in the case of water, the pH and ionic strength of the water environment can also influence the measured adhesion in the presence of any surface-bound charges.

When quantifying adhesion energies from AFM measurements, the contact mechanics model developed by *Johnson*, *Kendall* and *Roberts* (JKR) [32.22, 111] is often employed in the analysis of the adhesion data acquired by force–distance spectroscopy, whereby the number of interacting species (and consequently the average 'unit' interaction force or energy) can be derived from the estimated contact area and the average molecular packing density. Using the JKR model, the force of adhesion (AFM pull-off force) is related to the work of adhesion, W_{adh}, and the reduced radius, R, of the tip–surface contact:

$$F_{adh} = -\frac{3}{2}\pi R W_{adh} \,. \qquad (32.13)$$

The work of adhesion is a combination of the tip–surface (γ_{ts}), tip–solvent (γ_{tl}) and surface–solvent (γ_{sl}) interfacial energies ($W_{adh} = \gamma_{sl} + \gamma_{tl} - \gamma_{ts}$), and for tip–surface combinations that have the same chemical composition, the surface energy may be estimated directly from the adhesion measurement as W_{adh}. The effective contact radius at separation, r_s, from the JKR model is given as:

$$r_S = \left(\frac{3\pi W_{adh} R^2}{2K}\right)^{\frac{1}{3}} , \qquad (32.14)$$

where K is the reduced elastic modulus of the tip and surface. Using the contact area at separation and the assumed packing density of the molecules at the surfaces in contact, an estimate of the adhesion force or interaction energy can be made on a per molecule basis.

The accuracy of the interfacial energies and per molecule values obtained using this approach, however, must be considered carefully due to the accumulation of error carried though by the imprecise knowledge of the contact, including the tip radius, molecular packing densities of the modified surfaces, as well as the associated elastic properties of the contact at the monolayer level. As the details of the elastic properties of self-assembled monolayers are generally not known, the elastic properties of the contacts are typically assumed to be dominated by those of the underlying substrate, and the bulk values of the surface and/or tip materials (Au, Si, mica, Si_3N_4, SiO_2) are often employed in these calculations. Moreover, as mentioned above, if the molecular packing densities of the monolayers being evaluated are not known (as is the case with a typical AFM tip), then estimations must be used. For contact areas at pull-off approaching $1\,nm^2$, if the error in packing density is as much as one molecule per nm^2, for a typical alkanethiol this can lead to an error as high as 25% or more in the reported per molecule adhesion force, and such details should be taken into consideration when describing quantitative measurements.

An alternative approach to the measurement of adhesive interactions based on Poisson statistics has been promoted by *Beebe* and coworkers for the statistical evaluation of single bond forces without a priori knowledge of tip–surface contact details involved [32.132–134]. A main limitation of this approach, however, is that a completely homogeneous chemical system is assumed, so that there is only one type of discrete interaction present that gives rise to the observed adhesion. Unfortunately, for many solution phase systems a number of different interactions are typically operating, including energetic exchange with and reorganization of solvent molecules, depending on the solution conditions, issues that have never been thoroughly addressed in any molecular level measurement of adhesion.

32.5.3 Adhesion at SAM-Modified Surfaces in Liquids

A number of researchers have used AFM to probe interfacial adhesion for a variety of different chemical systems. Most notably, *Lieber* and coworkers promoted the use of chemically modified substrates and AFM tips to study selective molecular interactions [32.129, 135–139]. A number of other researchers have adopted a similar methodology leading to the measurement of adhesion forces and interfacial energies for the interactions of a variety of molecular functional groups in various environments, although to date few specific types of interactions have been thoroughly investigated (Table 32.1) [32.36, 129, 132, 135, 136, 140–149].

Measurements in air suffer from issues due to water vapor, as described above, so the adhesion of many SAM-terminated surfaces have been evaluated under liquid and it is these systems that we shall focus on here. The value of the adhesion is of course modified depending on the solvent. For example, the interaction of methyl-terminated interfaces is much stronger in water than in nonpolar solvents, in agreement with the general concept that upon separation the generation of hydrophobic interfaces in a polar solvent is highly energetically unfavorable. The impact of solvent on adhesion can be addressed basically as a variation in the Hamaker constant.

When ionizable end-groups are studied in water, the details of the adhesion measurements and results become more complicated, as the chemical natures of the surfaces are now dependent on the pH and ionic strength of the solution. In these circumstances, multiple interactions including ionic, van der Waals and double-layer forces come into play simultaneously. Under these conditions, the general form for the JKR adhesion force may be modified to include these additional forces as follows:

$$F_{adh} = -\frac{3}{2}\pi R W_{adh} - \frac{AR}{6D} + 6\pi R W_{dl} , \qquad (32.15)$$

where the second term now includes the attractive van der Waals component and the third term is the repulsive double-layer.

Table 32.1 List of various interactions evaluated by atomic force microscopy adhesion measurements

Chemical Contacts	References
$-CH_3/-CH_3$	[32.133, 135–137, 143]
	[32.146–148, 150]
$-CH_3/-COOH$	[32.135, 136, 150]
$-CH_3/-CONH_2$	[32.150]
$-CH_3/-NH_2$	[32.150]
$-CH_3/-OH$	[32.150]
$-CF_3/-CF_3$	[32.147]
$-CF_3/-CH_3$	[32.147]
$-OH/-OH$	[32.36, 133, 134, 137]
	[32.141, 148, 150]
$-OH/NH_2$	[32.150]
$-OH/-COOH$	[32.137, 150]
$-OH/-CONH_2$	[32.150]
$-COOH/-COOH$	[32.36, 132, 136, 137]
	[32.143, 148, 150]
$-CONH_2/-CONH_2$	[32.150]
$-CONH_2/-COOH$	[32.150]
$-NH_2/-NH_2$	[32.137, 150, 151]
$-CH_3/-NH_2$	[32.150, 151]
$-NH_2/-COOH$	[32.150]
$-NH_2/-CONH_2$	[32.150]
$-SO_3H/-SO_3H$	[32.151]
$-CH_3/-SO_3H$	[32.151]
$-NH_2/-SO_3H$	[32.151]
$-PO_3H_2/-PO_3H_2$	[32.152, 153]
$-SH/-SH$	[32.134]
$Au-Au$	[32.154]

Several groups have utilized the ability of AFM to function as a local probe of ionicity and to carry out local force titration measurements on several functional groups including $-COOH$ [32.36, 132, 136, 137, 143, 148, 150], $-NH_2$ [32.137, 150, 151], $-SO_3H$ [32.151] and $-PO_3H_2$ [32.152, 153] Table 32.1. The force–distance curves themselves can give a general sense of the local chemical state based on whether the approach curve is attractive or repulsive. In force titrations, the adhesion is measured as a function of pH, and the change in force is dependent upon the equilibrium mixture of charged species within the tip–sample contact at the time of measurement. Peaks in the adhesion force versus pH permit the determination of the local pK values for the ionized species, which are shown to be dependent on ionic strength. Shown in Fig. 32.13 is a force titration of a diprotic acid (11-thioundecyl-1-phosphonic acid). Shifts in the pK values from those measured in free solution have been ascribed to a variety of factors including build-up of excess surface charge and solvation effects of the surface-bound ionic species. However, the localization of charge and change migration are effects that are yet to be fully explored in these systems.

32.5.4 Impact of Intra- and Interchain Interactions on Adhesion

Several studies of the effect of chain length on friction and adhesion have found that in general, adhesion decreases with increasing chain length due to increased stability from lateral chain–chain van der Waals interactions within the film, which increases the overall film stiffness. The increased stiffness of the films acts to reduce the effective contact area that develops under compression, consequently reducing adhesion and friction [32.142, 144, 155]. However, in alkyl chain-based monolayers, the end-group orientation is also dependent on chain length. This results in an "odd–even" effect on the measured adhesion/friction of the monolayer. With $-CH_3$ terminated films, the methyl group orientation differs for odd- and even-length molecules. This impacts the orientation of the methyl group net dipole and hence the local surface free energy [32.156]. In circumstances where interchain hydrogen bonding can occur, additional film stability can be introduced, also yielding an "odd–even" effect, as observed by *Houston* and coworkers [32.140, 143].

AFM studies of alkanethiol films on Au surfaces have shown that, depending upon the tip size, under varying loads, the tip can readily penetrate the SAM film, displacing the thiol layer from the tip–sample con-

Fig. 32.13 The modification of AFM tips for chemical force microscopy is frequently carried out via chemical functionalization by alkylsilane monolayers on the oxidized surfaces of Si- or Si_3N_4-based tips or with alkylthiol monolayers assembled on Au-coated ($\approx 50\,nm$) AFM tips. The Au tips often also have a Cr binding layer ($\approx 5\,nm$) placed on before Au coating

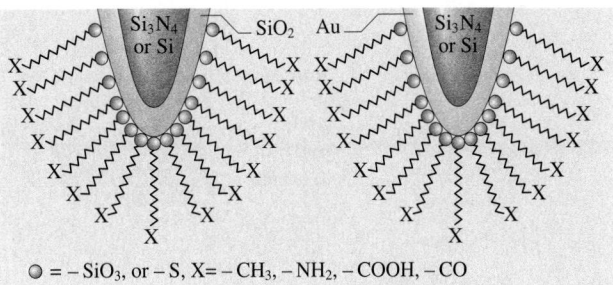

$\circ = -SiO_3$, or $-S$, $X = -CH_3, -NH_2, -COOH, -CO$

tact [32.157–159]. Upon reduction of the force, the tip again moves out of the SAM film and the surface structure is returned to its original condition. Of key importance in such studies is the mechanism by which the film is displaced. Recently, this has been modeled using molecular dynamics simulations by *Harrison* [32.160]. These studies have clearly demonstrated that in the initial stages of film compression and penetration by an asperity, *gauche* defects within the typically all-*trans* configurations of molecules in the SAM layers appear and propagate. The introduction of such defects is the catalyst for the weakening of the chain–chain lateral interactions that help stabilize and maintain film integrity. As this is lost, the asperity can rapidly penetrate the film and alkanethiol displacement can occur either via chain collapse or bond scission from the surface. *Salmeron* and coworkers demonstrated that the prevalence of molecular displacement versus film compression depends heavily on the AFM tip size, with sharp tips readily penetrating and displacing surface-bound thiols, while large tips spread the load over more molecules and induce compression over displacement (Fig. 32.14).

The inherent stability of the film structure has also been confirmed by sum-frequency generation spectroscopic studies which have indicated that without the presence of lateral chain–chain interactions, *gauche* defects appear within the chain structure that reduce overall order and lateral interactions within films [32.73, 161]. The ability to form these defects through poor film order will consequently increase adhesion, friction and wear of the film. This same spectroscopic study further demonstrated that the appearance of *gauche* defects can be induced by controlling the local environment (with the presence of water), causing chains to collapse back upon themselves.

As the molecular structure in SAMs moves away from ideally organized layers, chain–chain interactions

Fig. 32.14a,b Adhesion versus pH for 11-thioundecyl-1-phosphonic acid in buffer illustrating the impact of pH and ionic strength, (**a**) 10^{-4}M and (**b**) 10^{-1} M, on the measured adhesion. The peaks in adhesion provide local measures of the surface pK. After [32.153]

between the contacting surfaces can also result in entanglements. A recent study of the adhesive interactions of Au(111) surfaces modified with dialkylsulfides $[CH_3(CH_2)_n - S - (CH_2)_9CH_3$; $n = 9, 11, 13, 15, 17]$ with varying chain arm lengths probed the combined effect of chain length, solvent and intersurface chain entanglement on friction and adhesion using simultaneously modified surfaces and AFM tips [32.162]. This study found that chain–chain interpenetration produced the reverse dependence on chain length for the meas-

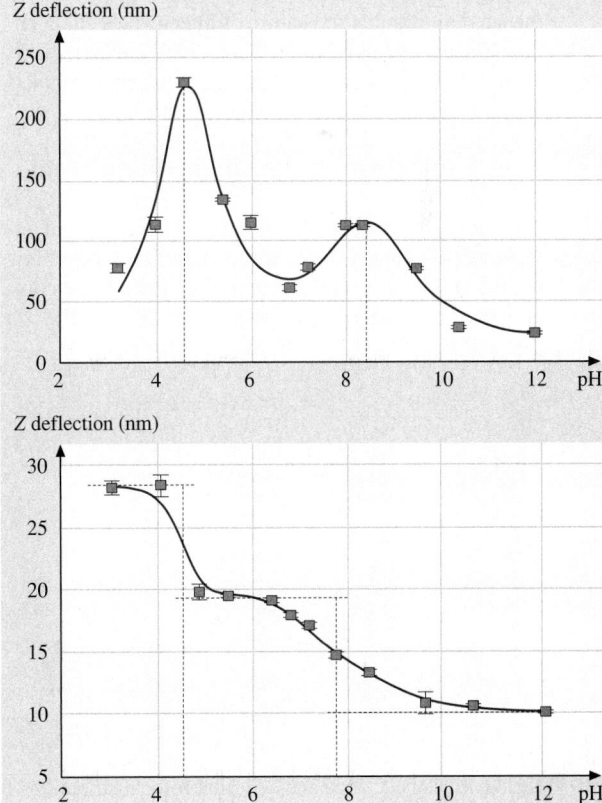

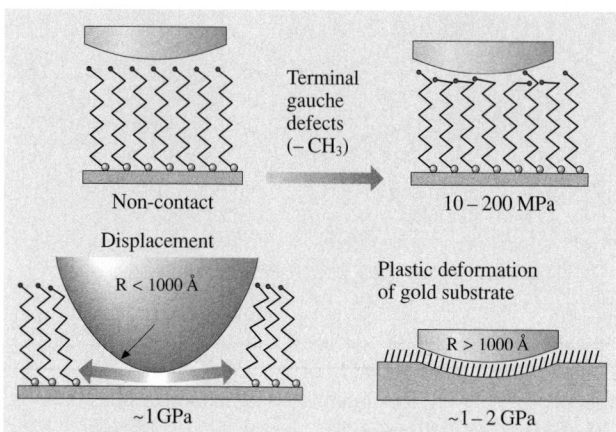

Fig. 32.15 Schematic illustrating the compression of a model lubricant layer under an AMF tip. At low pressures, *gauche* defects can form at the tip–sample junction but the molecules remain in place. Depending on the tip size and load, displacement (for sharp tips) or trapping of monolayer molecules (for blunt tips) can occur. After [32.5]

ured adhesion mentioned above. This work points to the need to examine the nature of intersurface chain entanglements in nanoscale systems. Such entanglements should be more prevalent in asperity–asperity contacts where ideal film structure will not be feasible.

32.5.5 Adhesion at the Single-Bond Level

The ability to resolve the discrete components of interactions is highly desirable. There have been few reports of the direct observation of discrete force components observed with the separation of an AFM tip

from a surface. *Beebe* and coworkers have utilized a statistical method (as described above) for the direct determination of single-bond forces for a variety of interactions [32.132–134], including biological systems such as biotin–avidin [32.163, 164]. The first report of quantized force measurements was described by *Hoh* et al., which lead to the estimation of single hydrogen bonding forces from studies of glass surfaces in water as being on the order of 10 pN [32.165]. The use of AFM for the study of the energetics of true single chemical bond cleavage has also received little attention. One previous report described discrete covalent bond scission using AFM. In that case it was proposed that the jumps in the observed pull-off curves were due to sequential scission of chemical bonds contained in a large multifunctional polymeric species as it progressively detached from the substrate [32.166]. However, the identification of the relevant chemical bonds involved at each stage was largely based on the known (gas phase) bond strengths of the potentially active functional groups, such that solvation effects on the bond energies were ignored – a simplification that profoundly affects the estimated energetics, as energetic exchange with the solvent must be included. More recently, the measurement of discrete bond scission was reported by *Frisbie* and coworkers for Au-thiol complexes. Here the details of Au atom abstraction were reported with a quantized value of 100 pN (estimated at $\approx$ 10 kJ/mol, based on an assumed bond rupture length of 1 Å) [32.149]. These studies have demonstrated the feasibility of probing local single-bond energetics, and have suggested some general requirements for the measurement of adhesion quantization in SAM layers, including the need for a significant negative tip–surface interfacial energy coupled with minimal

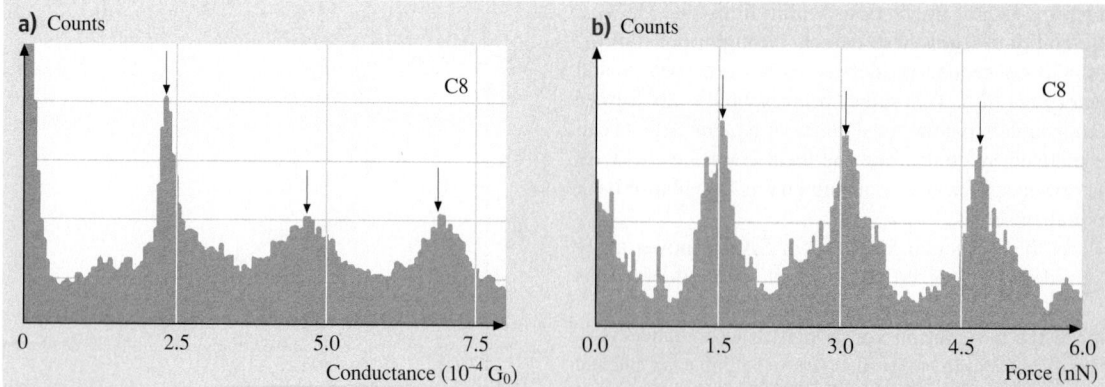

Fig. 32.16 Conductance and adhesion force distributions for 1,8′-octanedithiol, illustrating the quantization of the both the conductance and associated forces within the junction. After [32.154]

solvent surface tension [32.149, 167]. More recently, discrete bond forces and the associated quantized changes in through-molecule conductance have been reported by *Tao* and coworkers (Fig. 32.16) [32.154]. These measurements have shown how the electrical and mechanical properties of bonds are linked within molecules. Here quantized values for Au–Au bond scission were reported to be on the order of $\approx 1.5\,\text{nN}$ for $1,8'$-octanedithiol. These measurements offer the ability to quantify not only molecular forces and bonds at the single molecule level, but also to determine how charge transport within molecules is impacted by the mechanical deformation of the molecules within the junction, a key element in developing measurements of single-molecule conduction critical to molecular-based electronic devices.

32.5.6 Future Directions

Extending AFM adhesion measurements to reactive systems where chemical bonds can form between the tip and surface affords an expansion of chemical reaction dynamics to solution-based chemistries, whereby the energetic details of single reaction events previously only accessible for gas phase scattering experiments may be obtained. Studies of such complex heterogeneous systems will open the door to evaluations of the energetic pathways of solution-based chemistries for any system where the appropriate functionalization of surfaces can be exploited. Adhesion has already been demonstrated as a reasonable local probe of surface reaction kinetics whereby the local changes in the chemical forces may be followed as a function of time during surface chemical reactions [32.168, 169]. In addition to advances in measurements of reactive systems by AFM, complete insight into the operative molecular mechanisms can only be gained when combined with a detailed theory that takes into account not only the specific types of interactions present between the surfaces, but also the requisite energetic exchange with local solvent molecules. Advances in computer technologies and in computational theory have made this realistic [32.145].

To advance the field of adhesion measurements at the molecular level, energetic barriers for specific interactions also need to be evaluated, with attention given to the nature of the molecular interactions being probed, importantly including the details of energetic exchange with the solvent surrounding the interacting molecules. Studies by AFM of molecular interactions within sharply confined geometries ($\approx 1\,\text{nm}^2$ contact area) provide an opportunity to evaluate such contributions with molecular detail. Here again, *Lieber* and coworkers have been advancing the approach of chemically functionalizing carbon nanotubes to reduce both the type and number of specific interacting species [32.138]. This approach may hold some promise for probing well-defined specific chemical interactions and/or reactions as long as the nanotubes can be sufficiently stabilized against the buckling that is predicted by recent theoretical studies [32.160].

In addition to the modification of probe geometries to improve the localization of interactions for adhesion measurements, one of the principal difficulties with performing AFM measurements of adhesion at SAM surfaces is the unavoidable snap to contact. This makes details about the long-range interaction potential almost completely inaccessible. To address this issue, *Lieber* and coworkers have developed a modified AFM system in which the cantilever is magnetized, providing an additional feedback mechanism to help avoid snap-in [32.36]. When used, this approach provides a smooth approach and retract curve. Similarly, the capacitive coupling feedback mechanism of the interfacial force microscope (IFM) also affords measurements of this transition from out of contact without snap-in [32.140].

32.6 Concluding Remarks

Scanning probes are powerful tools for determining the fundamental molecular basis of adhesion. Continuum models of adhesion and capillary condensations are useful, but careful attention must be paid to their limits and assumptions. Further progress in these areas requires detailed analysis of the structure and chemistry of both the tip and sample, as well as their environment (solvent, humidity, and so on).

The ability to determine the effects of molecular-scale water menisci or single-bond energetics are truly spectacular accomplishments that continue to inspire researchers worldwide to pursue these measurements. With attention paid to previous work and a consideration of the issues raised in this chapter, many more discoveries are sure to be made.

References

32.1   J. N. Israelachvili: *Intermolecular and Surface Forces* (Academic Press, London 1992)

32.2   K. Autumn, Y. A. Liang, S. T. Hsieh, W. Zesch, C. Wai Pang, T. W. Kenny, R. Fearing, R. J. Full: Adhesive force of a single gecko foot-hair, Nature **405**, 681–685 (2000)

32.3   A. L. Baldwin, G. Thurston: Mechanics of endothelial cell architecture and vascular permeability, Crit. Rev. Biomed. Eng. **29**, 247–278 (2001)

32.4   S. R. White, N. R. Sottos, P. H. Guebelle, J. S. Moore, M. R. Kessler, S. R. Sriram, E. N. Brown, S. Viswanathan: Autonomic healing of polymer composites, Nature **409**, 794–797 (2001)

32.5   R. W. Carpick, M. Salmeron: Scratching the surface: Fundamental investigations of tribology with atomic force microscopy, Chem. Rev. **97**, 1163–1194 (1997)

32.6   M. P. De boer, T. A. Michalske: Accurate method for determining adhesion of cantilever beams, J. Appl. Phys. **86**, 817–827 (1999)

32.7   M. P. De boer, J. A. Knapp, T. A. Michalske, U. Srinivasan, R. Maboudian: Adhesion hysteresis of silane-coated microcantilevers, Acta Mater. **48**, 4531–4541 (2000)

32.8   R. Maboudian, R. T. Howe: Critical review: Adhesion in surface micromechanical structures, J. Vacuum Sci. Technol. **15**, 1–20 (1997)

32.9   R. Maboudian, W. R. Ashurst, C. Carraro: Tribological challenges in micromechanical systems, Tribol. Lett. **12**, 95–100 (2002)

32.10   R. Maboudian: Adhesion and friction issues associated with reliable operation of MEMS, MRS Bull. **23**, 47–51 (1998)

32.11   J. N. Israelachvili: Thin film studies using multiple-beam interferometry, J. Colloid Interf. Sci. **44**, 259–272 (1973)

32.12   J. N. Israelachvili, D. Tabor: The measurement of van der Waals dispersion forces in the range of 1.5 to 130 nm, Proc. R. Soc. Lond. A **331**, 19–38 (1972)

32.13   G. Binnig, C. F. Quate, C. Gerber: Atomic force microscope, Phys. Rev. Lett. **56**, 930–933 (1986)

32.14   J. N. Israelachvili, P. M. Mcguiggan, A. M. Homola: Dynamic properties of molecularly thin liquid films, Science **240**, 189–191 (1988)

32.15   J. Peachey, J. Van Alsten, S. Granick: Design of an apparatus to measure the shear response of ultrathin liquid, Rev. Sci. Instrum. **62**, 463–473 (1991)

32.16   P. Frantz, N. Agraït, M. Salmeron: Use of capacitance to measure surface forces. 1. Measuring distance of separation with enhanced spatial and time resolution, Langmuir **12**, 3289–3294 (1996)

32.17   L. R. Fisher, J. N. Israelachvili: Direct measurement of the effect of meniscus forces on adhesion: a study of the applicability of macroscopic thermodynamics to microscopic liquid interfaces, Colloids and Surfaces **3**, 303–319 (1981)

32.18   L. R. Fisher, J. N. Israelachvili: Experimental studies on the applicability of the Kelvin equation to highly curved concave menisci, J. Colloid Interf. Sci. **80**, 528–541 (1981)

32.19   J. N. Israelachvili: Adhesion, friction and lubrication of molecularly smooth surfaces. In: *Fundamentals of Friction*, ed. by I. L. Singer, H. M. Pollock (Kluwer, Dordrecht 1992) pp. 351–385

32.20   G. Reiter, A. L. Demirel, J. Peanasky, L. L. Cai, S. Granick: Stick to slip transition and adhesion of lubricated surfaces in moving contact, J. Chem. Phys. **101**, 2606–2615 (1994)

32.21   S. Granick: Molecular tribology of fluids. In: *Fundamentals of Friction*, ed. by I. L. Singer, H. M. Pollock (Kluwer, Dordrecht 1992) p. 387

32.22   K. L. Johnson: *Contact Mechanics* (University Press, Cambridge 1987)

32.23   M. A. Lantz, S. J. O'shea, M. E. Welland: Simultaneous force and conduction measurements in atomic force microscopy, Phys. Rev. B **56**, 15345–15352 (1997)

32.24   M. Enachescu, R. J. A. Van Den Oetelaar, R. W. Carpick, D. F. Ogletree, C. F. J. Flipse, M. Salmeron: An AFM study of an ideally hard contact: The diamond(111)/tungsten-carbide interface, Phys. Rev. Lett. **81**, 1877–1880 (1998)

32.25   M. A. Lantz, S. J. O'shea, A. C. F. Hoole, M. E. Welland: Lateral stiffness of the tip and tip–sample contact in frictional force microscopy, Appl. Phys. Lett. **70**, 970–972 (1997)

32.26   R. W. Carpick, D. F. Ogletree, M. Salmeron: Lateral stiffness: A new nanomechanical measurement with friction force microscopy, Appl. Phys. Lett. **70**, 1548–1550 (1997)

32.27   O. Marti, B. Drake, P. K. Hansma: Atomic force microscopy of liquid-covered surfaces: atomic resolution images, Appl. Phys. Lett. **51**, 484–486 (1987)

32.28   G. J. Germann, S. R. Cohen, G. Neubauer, G. M. Mcclelland, H. Seki, D. Coulman: Atomic scale friction of a diamond tip on diamond (100) and (111) surfaces, J. Appl. Phys. **73**, 163–167 (1993)

32.29   L. Howald, E. Meyer, R. Lüthi, H. Haefke, R. Overney, H. Rudin, H.-J. Güntherodt: Multifunctional probe microscope for facile operation in ultrahigh vacuum, Appl. Phys. Lett. **63**, 117–119 (1993)

32.30   M. Kageshima, H. Yamada, K. Nakayama, H. Sakama, A. Kawau, T. Fujii, M. Suzuki: Development of an ultrahigh vacuum atomic force microscope for investigations of semiconductor surfaces, J. Vacuum Sci. Technol. B **11**, 1987–1991 (1993)

32.31   J. A. Greenwood: Adhesion of elastic spheres, Proc. R. Soc. Lond. A **453**, 1277–1297 (1997)

32.32   S. P. Jarvis, A. Oral, T. P. Weihs, J. B. Pethica: A novel force microscope and point contact probe, Rev. Sci. Instrum. **64**, 3515–3520 (1993)

32.33   S. P. Jarvis, H. Yamada, S.-I. Yamamoto, H. Tokumoto: A new force controlled atomic force microscope for use in ultrahigh vacuum, Rev. Sci. Instrum. **67**, 2281–2285 (1996)

32.34   S. A. Joyce, J. E. Houston: A new force sensor incorporating force-feedback control for interfacial force microscopy, Rev. Sci. Instrum. **62**, 710–715 (1991)

32.35   S. A. Joyce, J. E. Houston, T. A. Michalske: Differentiation of topographical and chemical structures using an interfacial force microscope, Appl. Phys. Lett. **60**, 1175 (1992)

32.36   P. D. Ashby, L. W. Chen, C. M. Lieber: Probing intermolecular forces and potentials with magnetic feedback chemical force microscopy, J. Am. Chem. Soc. **122**, 9467–9472 (2000)

32.37   H. I. Kim, V. Boiadjiev, J. E. Houston, X. Y. Zhu, J. D. Kiely: Tribological properties of self-assembled monolayers on Au, SiO$_x$, Si surfaces, Tribol. Lett. **10**, 97–101 (2001)

32.38   P. A. Taylor, J. S. Nelson, B. W. Dodson: Adhesion between atomically flat metallic surfaces, Phys. Rev. B **44**, 5834–5841 (1991)

32.39   J. S. Nelson, B. W. Dodson, P. A. Taylor: Adhesive avalanche in covalently bonded materials, Phys. Rev. B **45**, 4439–4444 (1992)

32.40   R. Garcia, R. Perez: Dynamic atomic force microscopy methods, Surf. Sci. Rep. **47**, 197–301 (2002)

32.41   F. J. Giessibl: Forces and frequency shifts in atomic-resolution dynamic-force microscopy, Phys. Rev. B **56**, 16010–15 (1997)

32.42   P. M. Hoffmann, A. Oral, R. A. Grimble, H. O. Ozer, S. Jeffery, J. B. Pethica: Direct measurement of interatomic force gradients using an ultra-low-amplitude atomic force microscope, Proc. R. Soc. Lond. A **457**, 1161–1174 (2001)

32.43   K. R. Shull: Contact mechanics and the adhesion of soft solids, Mater. Sci. Eng. R **R36**, 1–45 (2002)

32.44   L. Xu, A. Lio, J. Hu, D. F. Ogletree, M. Salmeron: Wetting and capillary phenomena of water on mica, J. Phys. Chem. B **102**, 540–548 (1998)

32.45   S. P. Timoshenko, J. N. Goodier: *Theory of Elasticity* (McGraw-Hill, New York 1987)

32.46   J. P. Cleveland, S. Manne, D. Bocek, P. K. Hansma: A nondestructive method for determining the spring constant of cantilevers for scanning force microscopy, Rev. Sci. Instrum. **64**, 403–405 (1993)

32.47   J. E. Sader, I. Larson, P. Mulvaney, L. R. White: Method for the calibration of atomic force microscope cantilevers, Rev. Sci. Instrum. **66**, 3789–3798 (1995)

32.48   M. Tortonese, M. Kirk: Characterization of application specific probes for SPMs, Proc. SPIE **3009**, 53–60 (1997)

32.49   T. R. Albrecht, S. Akamine, T. E. Carver, C. F. Quate: Microfabrication of cantilever styli for the atomic force microscope, J. Vacuum Sci. Technol. A **8**, 3386–96 (1990)

32.50   H.-J. Butt, P. Siedle, K. Seifert, K. Fendler, T. Seeger, E. Bamberg, A. L. Weisenhorn, K. Goldie, A. Engel: Scan speed limit in atomic force microscopy, J. Microsc. **169**, 75–84 (1993)

32.51   J. M. Neumeister, W. A. Ducker: Lateral, normal, and longitudinal spring constants of atomic force microscopy cantilevers, Rev. Sci. Instrum. **65**, 2527–2531 (1994)

32.52   J. E. Sader: Parallel beam approximation for V-shaped atomic force microscope cantilevers, Rev. Sci. Instrum. **66**, 4583–4587 (1995)

32.53   D. F. Ogletree, R. W. Carpick, M. Salmeron: Calibration of frictional forces in atomic force microscopy, Rev. Sci. Instrum. **67**, 3298–3306 (1996)

32.54   R. Lüthi, E. Meyer, H. Haefke, L. Howald, W. Gutmannsbauer, M. Guggisberg, M. Bammerlin, H.-J. Güntherodt: Nanotribology: An UHV-SFM study on thin films of C$_{60}$ and AgBr, Surf. Sci. **338**, 247–260 (1995)

32.55   U. D. Schwarz, P. Koster, R. Wiesendanger: Quantitative analysis of lateral force microscopy experiments, Rev. Sci. Instrum. **67**, 2560–2567 (1996)

32.56   T. J. Senden, W. A. Ducker: Experimental determination of spring constants in atomic force microscopy, Langmuir **10**, 1003–1004 (1994)

32.57   A. Torii, M. Sasaki, K. Hane, S. Okuma: A method for determining the spring constant of cantilevers for atomic force microscopy, Meas. Sci. Technol. **7**, 179–184 (1996)

32.58   J. A. Ruan, B. Bhushan: Atomic-scale friction measurements using friction force microscopy: Part I – general principles and new measurement techniques, Trans. ASME J. Tribol. **116**, 378–388 (1994)

32.59   Y. Q. Li, N. J. Tao, J. Pan, A. A. Garcia, S. M. Lindsay: Direct measurement of interaction forces between colloidal particles using the scanning force microscope, Langmuir **9**, 637–641 (1993)

32.60   J. E. Sader, J. W. M. Chon, P. Mulvaney: Calibration of rectangular atomic force microscope cantilevers, Rev. Sci. Instrum. **70**, 3967–3969 (1999)

32.61   J. S. Villarrubia: Morphological estimation of tip geometry for scanned probe microscopy, Surf. Sci. **321**, 287–300 (1994)

32.62   J. S. Villarrubia: Algorithms for scanned probe microscope image simulation, surface reconstruction, and tip estimation, J. Res. Natl. Inst. Stand. Technol. (USA) **102**, 425–454 (1997)

32.63   L. S. Dongmo, J. S. Villarrubia, S. N. Jones, T. B. Renegar, M. T. Postek, J. F. Song: Experimental test of blind tip reconstruction for scanning probe microscopy, Ultramicroscopy **85**, 141–153 (2000)

Part D | 32

32.64   R. W. Carpick, N. Agraït, D. F. Ogletree, M. Salmeron: Measurement of interfacial shear (friction) with an ultrahigh vacuum atomic force microscope, J. Vacuum Sci. Technol. B **14**, 1289–1295 (1996)

32.65   K. F. Jarausch, T. J. Stark, P. E. Russell: Silicon structures for in situ characterization of atomic force microscope probe geometry, J. Vacuum Sci. Technol. B **14**, 3425–3430 (1996)

32.66   F. Atamny, A. Baiker: Direct imaging of the tip shape by AFM, Surf. Sci. **323**, L314–L318 (1995)

32.67   S. S. Sheiko, M. Moller, E. M. C. M. Reuvekamp, H. W. Zandbergen: Evaluation of the probing profile of scanning force microscopy tips, Ultramicroscopy **53**, 371–380 (1994)

32.68   C. Odin, J. P. Aimé, Z. El Kaakour, T. Bouhacina: Tip's finite size effects on atomic force microscopy in the contact mode: Simple geometrical considerations for rapid estimation of apex radius, tip angle based on the study of polystyrene latex balls, Surf. Sci. **317**, 321–340 (1994)

32.69   K. L. Westra, D. J. Thomson: Atomic force microscope tip radius needed for accurate imaging of thin film surfaces, J. Vacuum Sci. Technol. B **12**, 3176–3181 (1994)

32.70   P. Markiewicz, M. C. Goh: Atomic force microscope tip deconvolution using calibration arrays, Rev. Sci. Instrum. **66**, 3186–3190 (1995)

32.71   R. Dixson, J. Schneir, T. Mcwaid, N. Sullivan, V. W. Tsai, S. H. Zaidi, S. R. J. Brueck: Toward accurate linewidth metrology using atomic force microscopy and tip characterization, Proc. SPIE **2725**, 589–607 (1996)

32.72   P. Siedle, H.-J. Butt, E. Bamberg, D. N. Wang, W. Kuhlbrandt, J. Zach, M. Haider: Determining the form of atomic force microscope tips. In: *X-Ray Optics and Microanalysis 1992*, Proceedings of the Thirteenth International Congress, Manchester, UK, ed. by P. B. Kenway, P. J. Duke, G. W. Lorimer, T. Mulvey, J. W. Drummond, G. Love, A. G. Michette, M. Stedman (IOP, Bristol 1992)

32.73   S. Xu, M. F. Arnsdorf: Calibration of the scanning (atomic) force microscope with gold particles, J. Microsc. **3**, 199–210 (1994)

32.74   U. D. Schwarz, O. Zwörner, P. Köster, R. Wiesendanger: Friction force spectroscopy in the low-load regime with well-defined tips. In: *Micro/Nanotribology and Its Applications*, ed. by B. Bhushan (Kluwer, Dordrecht 1997)

32.75   S. J. O'Shea, R. N. Atta, M. E. Welland: Characterization of tips for conducting atomic force microscopy, Rev. Sci. Instrum. **66**, 2508–2512 (1995)

32.76   P. Niedermann, W. Hanni, N. Blanc, R. Christoph, J. Burger: Chemical vapor deposition diamond for tips in nanoprobe experiments, J. Vacuum Sci. Technol. A **14**, 1233–1236 (1995)

32.77   L. M. Qian, X. D. Xiao, S. Z. Wen: Tip in situ chemical modification and its effects on tribological measurements, Langmuir **16**, 662–670 (2000)

32.78   E. L. Florin, V. T. Moy, H. E. Gaub: Adhesion forces between individual ligand–receptor pairs, Science **264**, 415–417 (1994)

32.79   V. T. Moy, E. L. Florin, H. E. Gaub: Intermolecular forces and energies between ligands and receptors, Science **266**, 257–259 (1994)

32.80   G. U. Lee, L. A. Chrisey, R. J. Colton: Direct measurement of the forces between complementary strands of DNA, Science **266**, 771–773 (1994)

32.81   S. S. Wong, E. Joselevich, A. T. Woolley, C. Chin Li, C. M. Lieber: Covalently functionalized nanotubes as nanometre-sized probes in chemistry and biology, Nature **394**, 52–55 (1998)

32.82   O. H. Willemsen, M. M. E. Snel, K. O. Van Der Werf, B. G. De Grooth, J. Greve, P. Hinterdorfer, H. J. Gruber, H. Schindler, Y. Van Kooyk, C. G. Figdor: Simultaneous height and adhesion imaging of antibody-antigen interactions by atomic force microscopy, Biophys. J. **75**, 2220–2228 (1998)

32.83   W. A. Ducker, T. J. Senden, R. M. Pashley: Direct measurement of colloidal forces using an atomic force microscope, Nature **353**, 239–241 (1991)

32.84   H. J. Butt: Measuring electrostatic, van der Waals, and hydration forces in electrolyte solutions with an atomic force microscope, Biophys. J. **60**, 1438–1444 (1991)

32.85   V. S. J. Craig, C. Neto: In situ calibration of colloid probe cantilevers in force microscopy: hydrodynamic drag on a sphere approaching a wall, Langmuir **17**, 6018–6022 (2001)

32.86   R. J. Cannara, M. J. Brukman, R. W. Carpick: Cantilever tilt compensation for variable-load atomic force microscopy, Rev. Sci. Instrum. **76**, 53706 (2005)

32.87   R. Staub, D. Alliata, C. Nicolini: Drift elimination in the calibration of scanning probe microscopes, Rev. Sci. Instrum. **66**, 2513–2516 (1995)

32.88   S. M. Hues, C. F. Draper, K. P. Lee, R. J. Colton: Effect of PZT and PMN actuator hysteresis and creep on nanoindentation measurements using force microscopy, Rev. Sci. Instrum. **65**, 1561–1565 (1994)

32.89   J. Fu: In situ testing and calibrating of z-piezo of an atomic force microscope, Rev. Sci. Instrum. **66**, 3785–3788 (1995)

32.90   J. Garnaes, L. Nielsen, K. Dirscherl, J. F. Jorgensen, J. B. Rasmussen, P. E. Lindelof, C. B. Sorensen: Two-dimensional nanometer-scale calibration based on one-dimensional gratings, Appl. Phys. A **66**, 831–5 (1998)

32.91   J. F. Jorgensen, K. Carneiro, L. L. Madsen, K. Conradsen: Hysteresis correction of scanning tunneling microscope images, J. Vacuum Sci. Technol. B **1**, 1702–1704 (1994)

32.92   J. F. Jorgensen, L. L. Madsen, J. Garnaes, K. Carneiro, K. Schaumburg: Calibration, drift elimination, and molecular structure analysis, J. Vacuum Sci. Technol. B **12**, 1698–1701 (1994)

32.93   M. Jaschke, H. J. Butt: Height calibration of optical lever atomic force microscopes by simple laser interferometry, Rev. Sci. Instrum. **66**, 1258–1259 (1995)

32.94   L. A. Nagahara, K. Hashimoto, A. Fujishima, D. Snowden-Ifft, P. B. Price: Mica etch pits as a height calibration source for atomic force microscopy, J. Vacuum Sci. Technol. B **12**, 1694–7 (1993)

32.95   H. M. Brodowsky, U.-C. Boehnke, F. Kremer: Wide range standard for scanning probe microscopy height calibration, Rev. Sci. Instrum. **67**, 4198–4200 (1996)

32.96   R. M. Overney, H. Takano, M. Fujihira: Elastic compliances measured by atomic force microscopy, Europhys. Lett. **26**, 443–447 (1994)

32.97   M. S. Marcus, R. W. Carpick, D. Y. Sasaki, M. A. Eriksson: Material anisotropy revealed by phase contrast in intermittent contact atomic force microscopy, Phys. Rev. Lett. **88**, 226103 (2002)

32.98   M. J. D'amato, M. S. Marcus, M. A. Eriksson, R. W. Carpick: Phase imaging and the lever-sample tilt angle in dynamic atomic force microscopy, Appl. Phys. Lett. **85**, 4378–4380 (2004)

32.99   L.-O. Heim, M. Kappl, H.-J. Butt: Tilt of atomic force microscope cantilevers: Effect on spring constant and adhesion measurements, Langmuir **20**, 2760–2764 (2004)

32.100   J. L. Hutter: Comment on tilt of atomic force microscope cantilevers: Effect on spring constant and adhesion measurements, Langmuir **21**, 2630–2632 (2005)

32.101   P. G. De Gennes: Wetting: statistics and dynamics, Rev. Mod. Phys. **57**, 827–863 (1985)

32.102   T. Stifter, O. Marti, B. Bhushan: Theoretical investigation of the distance dependence of capillary and van der Waals forces in scanning force microscopy, Phys. Rev. B **62**, 13667–13673 (2000)

32.103   Y. Sugawara, M. Ohta, T. Konishi, S. Morita, M. Suzuki, Y. Enomoto: Effects of humidity and tip radius on the adhesive force measured with atomic force microscopy, Wear **168**, 13–16 (1993)

32.104   M. Binggeli, C. M. Mate: Influence of capillary condensation of water on nanotribology studied by force microscopy, Appl. Phys. Lett. **65**, 415–417 (1994)

32.105   X. Xiao, Q. Linmao: Investigation of humidity-dependent capillary force, Langmuir **16**, 8153–8158 (2000)

32.106   F. M. Orr, L. E. Scriven, A. P. Rivas: Pendular rings between solids: meniscus properties and capillary force, J. Fluid Mech. **67**, 723–742 (1975)

32.107   H. K. Christenson: Adhesion between surfaces in undersaturated vapors – a re-examination of the influence of meniscus curvature and surface forces, J. Colloid Interf. Sci. **121**, 170–178 (1988)

32.108   E. Riedo, F. Levy, H. Brune: Kinetics of capillary condensation in nanoscopic sliding friction, Phys. Rev. Lett. **88**, 185505/1–4 (2002)

32.109   D. Maugis, B. Gauthiermanuel: JKR–DMT transition in the presence of a liquid meniscus, J. Adhes. Sci. Technol. **8**, 1311–1322 (1994)

32.110   A. Fogden, L. R. White: Contact elasticity in the presence of capillary condensation. 1. The nonadhesive Hertz problem, J. Colloid Interf. Sci. **138**, 414–430 (1990)

32.111   K. L. Johnson, K. Kendall, A. D. Roberts: Surface energy and the contact of elastic solids, Proc. R. Soc. Lond. A **324**, 301–313 (1971)

32.112   K. L. Johnson: Adhesion and friction between a smooth elastic asperity and a plane surface, Proc. R. Soc. Lond. A **453**, 163–179 (1997)

32.113   K. Johnson, J. Greenwood: An adhesion map for the contact of elastic spheres, J. Colloid Interf. Sci. **192**, 326–333 (1997)

32.114   T. Thundat, X. Y. Zheng, G. Y. Chen, R. J. Warmack: Role of relative humidity in atomic force microscopy imaging, Surf. Sci. **294**, L939–L943 (1993)

32.115   M. Binggeli, R. Christoph, H.-E. Hintermann: Observation of controlled, electrochemically induced friction force modulations in the nano-Newton range, Tribol. Lett. **1**, 13–21 (1995)

32.116   B. Bhushan, S. Sundararajan: Micro/nanoscale friction and wear mechanisms of thin films using atomic force and friction force microscopy, Acta Mater. **46**, 3793–3804 (1998)

32.117   W. Gulbinski, D. Pailharey, T. Suszko, Y. Mathey: Study of the influence of adsorbed water on AFM friction measurements on molybdenum trioxide thin films, Surf. Sci. **475**, 149–158 (2001)

32.118   M. He, A. S. Blum, D. E. Aston, C. Buenviaje, R. M. Overney, R. Luginbuhl: Critical phenomena of water bridges in nanoasperity contacts, J. Chem. Phys. **114**, 1355–1360 (2001)

32.119   A. W. Adamson: *Physical Chemistry of Surfaces*, 5th edn. (Wiley, 1990) Chap. 16, pp. 591–615

32.120   C. M. Mate: Application of disjoining and capillary pressure to liquid lubricant films in magnetic recording, J. Appl. Phys. **72**(2), 3084–3090 (1992)

32.121   C. Gao: Theory of menisci and its applications, Appl. Phys. Lett. **71**(13), 1801–1803 (1997)

32.122   G. Lian, C. Thornton, M. J. Adams: A theoretical study of the liquid bridge forces between two rigid spherical bodies, J. Colloid Interf. Sci. **161**, 138–147 (1993)

32.123   C. D. Willett, M. J. Adams, S. A. Johnson, J. P. K. Seville: Capillary bridges between two spherical bodies, Langmuir **16**, 9396–9405 (2000)

32.124   J. A. Greenwood, J. B. P. Williamson: Contact of nominally flat surfaces, Proc. R. Soc. Lond. A **295**, 300–319 (1966)

32.125   R. Maboudian: Surface processes in MEMS technology, Surf. Sci. Rep. **30**, 207–270 (1998)

32.126   R. Maboudian, W. R. Ashurst, C. Carraro: Self-assembled monolayers as anti-stiction coatings for MEMS: Characteristics and recent developments, Sensors Actuat. A **82**, 219–223 (2000)

Part D | 32

32.127 R. Maboudian, W. R. Ashurst, C. Carraro: Tribological challenges in micromechanical systems, Tribol. Lett. **12**, 95–100 (2002)

32.128 K. Komvopoulous: Surface engineering and microtribology for microelectromechanical systems, Wear **200**, 305–327 (1996)

32.129 A. Noy, D. V. Vezenov, C. M. Lieber: Chemical force microscopy, Annu. Rev. Mater. Sci **27**, 381–421 (1997)

32.130 H. Takano, J. R. Kenseth, S.-S. Wong, J. C. O'brien, M. D. Porter: Chemical and biochemical analysis using scanning force microscopy, Chem. Rev. **99**, 2845–2890 (1999)

32.131 D. L. Sedin, K. L. Rowlen: Adhesion forces measured by atomic force microscopy in humid air, Anal. Chem. **72**, 2183–2189 (2000)

32.132 T. Han, J. M. Williams, T. P. Beebe: Chemical bonds studied with functionalized atomic force microscopy tips, Anal. Chim. Acta **307**, 365–376 (1995)

32.133 L. A. Wenzler, G. L. Moyes, L. G. Olson, J. M. Harris, T. P. Beebe: Single-molecule bond rupture force analysis of interactions between AFM tips and substrates modified with organosilanes, Anal. Chem. **69**, 2855–2861 (1997)

32.134 L. A. Wenzler, G. L. Moyes, G. N. Raikar, R. L. Hansen, J. M. Harris, T. P. B.. Jr: Measurements of single-molecule bond rupture forces between self-assembled monolayers of organosilanes with the atomic force microscope, Langmuir **13**, 3761–3768 (1997)

32.135 C. D. Frisbie, L. F. Rozsnyai, A. Noy, M. S. Wrighton, C. M. Lieber: Functional group imaging by chemical force microscopy, Science **265**, 2071–2074 (1994)

32.136 A. Noy, C. D. Frisbie, L. F. Rozsnyai, M. S. Wrighton, C. M. Lieber: Chemical force microscopy: Exploiting chemically-modified tips to quantify adhesion, friction and functional group distributions in molecular assemblies, J. Am. Chem. Soc. **117**, 7943–7951 (1995)

32.137 D. V. Vezenov, A. Noy, L. F. Rozsnyai, C. M. Lieber: Force titrations, ionization state sensitive imaging of functional group distributions in molecular assemblies, J. Am. Chem. Soc. **119**, 2006–2015 (1997)

32.138 S. S. Wong, A. T. Woolley, E. Joselevich, C. L. Cheung, C. M. Lieber: Covalently-functionalized single-walled carbon nanotube tips for chemical force microscopy, J. Am. Chem. Soc. **120**, 8557–8558 (1998)

32.139 D. V. Vezenov, A. V. Zhuk, G. M. Whitesides, C. M. Likeber: Chemical force spectroscopy in heterogeneous systems: Intermolecular interactions involving epoxy polymer, mixed monolayers and polar solvents, J. Am. Chem. Soc. **124**, 10578–10588 (2002)

32.140 J. E. Houston, H. I. Kim: Adhesion, friction, and mechanical properties of functionalized alkanethiol self-assembled monolayers, Acc. Chem. Res. **35**, 547–553 (2002)

32.141 T. Ito, M. Namba, P. Buhlmann, Y. Umezawa: Modification of silicon nitride tips with trichlorosilane self-assembled monolayers (SAMs) for chemical force microscopy, Langmuir **13**, 4323–4332 (1997)

32.142 H. I. Kim, M. Graupe, O. Oloba, T. Koini, S. Imaduddin, T. R. Lee, S. S. Perry: Molecularly specific studies of the frictional properties of monolayer films: A systematic comparison of CF_3-, (CH3) (32.2) CH-, CH_3-terminated films, Langmuir **15**, 3179–3185 (1999)

32.143 H. I. Kim, J. E. Houston: Separating mechanical and chemical contributions to molecular-level friction, J. Am. Chem. Soc. **122**, 12045–12046 (2000)

32.144 S. Lee, Y. S. Shon, R. Colorado, R. L. Guenard, T. R. Lee, S. S. Perry: The influence of packing densities and surface order on the frictional properties of alkanethiol self-assembled monolayers (SAMs) on gold: A comparison of SAMs derived from normal and spiroalkanedithiols, Langmuir **16**, 2220–2224 (2000)

32.145 Y. Leng, S. Jiang: Dynamic simulations of adhesion and friction in chemical force microscopy, J. Am. Chem. Soc. **124**, 11764–11770 (2002)

32.146 T. Nakagawa, K. Ogawa, T. Kurumizawa: Discriminating molecular length of chemically adsorbed molecules using an atomic force microscope having a tip covered with sensor molecules (an atomic force microscope having chemical sensing function), Jpn. J. Appl. Phys. **32**, 294–296 (1993)

32.147 T. Nakagawa, K. Ogawa, T. Kurumizawa: Atomic force microscope for chemical sensing, J. Vacuum Sci. Tech. B **12**, 2215–2218 (1994)

32.148 S. K. Sinniah, A. B. Steel, C. J. Miller, J. E. Reutt-Robey: Solvent exclusion and chemical contrast in scanning force microscopy, J. Am. Chem. Soc. **118**, 8925–8931 (1996)

32.149 H. Skulason, C. D. Frisbie: Detection of discrete interactiuons upon rupture of Au microcontacts to self-assembled monolayers terminated with −S(CO)CH$_3$ or −SH, J. Am. Chem. Soc. **122**, 9750–9760 (2000)

32.150 E. W. V. D. Vegte, G. Hadziioannou: Scanning force microscopy with chemical specificity: An extensive study of chemically specific tip–surface interactions and the chemical imaging of surface functional groups, Langmuir **13**, 4357–4368 (1997)

32.151 V. Tsukruk, V. N. Bliznyuk: Adhesive and friction forces between chemically modified silicon and silicon nitride surfaces, Langmuir **14**, 446–455 (1998)

32.152 E. W. V. D. Vegte, G. Hadziioannou: Acid-base properties and the chemical imaging of surface-bound functional groups with scanning force microscopy, J. Phys. Chem. B **101**, 9563–9569 (1997)

32.153 J. Zhang, J. Kirkham, C. Robinson, M. L. Wallwork, D. A. Smith, A. Marsh, M. Wong: Determination of the ionization state of 11-thioundecyl-1-phosphonic acid in self-assembled monolayers by chemical force microscopy, Anal. Chem. **72**, 1973–1978 (2000)

32.154 B. Xu, X. Xiao, N. J. Tao: Measurements of single-molecule electromechanical properties, J. Am. Chem. Soc. **125**, 16164–16165 (2003)

32.155 A. Lio, C. Morant, D. F. Ogletree, M. Salmeron: Atomic force microscopy study of the pressure-dependent structural and frictional properties of n-alkanethiols on gold, J. Phys. Chem. B **101**, 4767–4773 (1997)

32.156 S.-S. Wong, H. Takano, M. D. Porter: Mapping orientation differences of terminal functional groups by friction force microscopy, Anal. Chem. **70**, 5209–5212 (1998)

32.157 A. Lio, D. H. Charych, M. Salmeron: Comparative atomic force microscopy study of the chain length dependence of frictional properties of alkanethiols on gold and alkysilanes on mica, J. Phys. Chem. B **101**, 3800–3805 (1997)

32.158 X. Xiao, J. Hu, D. H. Charych, M. Salmeron: Chain length dependence of the frictional properties of alkylsilane molecules self-assembled on mica studied by atomic force microscopy, Langmuir **12**, 235–237 (1996)

32.159 G.-Y. Liu, M. Salmeron: Reversible displacement of chemisorbed n-alkane thiol molecules on Au(111) surface: An atomic force microscopy study, Langmuir **10**, 367–370 (1994)

32.160 A. B. Tutein, S. J. Stuart, J. A. Harrison: Indentation analysis of linear-chain hydrocarbon monolayers anchored to diamond, J. Phys. Chem. B **103**, 11357–11365 (1999)

32.161 R. L. Pizzolatto, Y. J. Yang, L. K. Wolf, M. C. Messmer: Conformational aspects of modal chromatographic surfaces studied by sum-frequency generation, Anal. Chem. Acta **397**, 81–92 (1999)

32.162 E. W. V. D. Vegte, A. Subbotin, G. Hadziioannou: Nanotribological properties of unsymmetrical n-dialkyl sulfide monolayers on gold: Effect of chain length on adhesion, friction and imaging, Langmuir **16**, 3249–3256 (2000)

32.163 Y.-S. Lo, J. Simons, T. P. B. Jr: Temperature dependence of the biotin–avidin bond rupture force studied by atomic force microscopy, J. Phys. Chem. B **106**, 9847–9857 (2002)

32.164 Y.-S. Lo, N. D. Huefner, W. S. Chan, F. Stevebs, J. M. Harris, J. T. P. Beebe: Specific interactions between biotin and avidin studies by atomic force microscopy using the Poisson statistical analysis method, Langmuir **15**, 1373–1382 (1999)

32.165 J. H. Hoh, J. P. Cleavland, C. B. Prater, J.-P. Revel, P. K. Hansma: Quantitized adhesion detected with the atomic force microscope, J. Am. Chem. Soc. **114**, 4917–4918 (1992)

32.166 M. Grandbois, M. Beyer, M. Rief, H. Clausen-Schaumann, H. E. Gaub: How strong is a covalent bond?, Science **283**, 1727–1730 (1999)

32.167 H. Skulason, C. D. Frisbie: Contact mechanics modeling of pull-off measurements: Effect of solvent, probe radius, and chemical binding probability on the detection of single-bond rupture forces by atomic force microscopy, Anal. Chem. **74**, 3096–3104 (2002)

32.168 H. Schonherr, V. Chechik, C. J. M. Stirling, G. J. Vancso: Monitoring surface reactions at an AFM tip: An approach to following reaction kinetics in self-assembled monolayers on the nanometer scale, J. Am. Chem. Soc. **122**, 3679–3687 (2000)

32.169 M. P. L. Werts, E. W. V. D. Vegte, G. Hadziioannou: Surface chemical reactions probed with scanning force microscopy, Langmuir **13**, 4939–4942 (1997)

33. Friction and Wear on the Atomic Scale

Friction has long been the subject of research: the empirical da Vinci–Amontons friction laws have been common knowledge for centuries. Macroscopic experiments performed by the school of *Bowden* and *Tabor* revealed that macroscopic friction can be related to the collective action of small asperities. Over the last 15 years, experiments performed with the atomic force microscope have provided new insights into the physics of single asperities sliding over surfaces. This development, together with the results from complementary experiments using surface force apparatus and the quartz microbalance, have led to the new field of *nanotribology*. At the same time, increasing computing power has permitted the simulation of processes that occur during sliding contact involving several hundreds of atoms. It has become clear that atomic processes cannot be neglected when interpreting nanotribology experiments. Even on well-defined surfaces, experiments have revealed that atomic structure is directly linked to friction force. This chapter will describe friction force microscopy experiments that reveal, more or less directly, atomic processes during sliding contact.

We will begin by introducing friction force microscopy, including the calibration of cantilever force sensors and special aspects of the ultrahigh vacuum environment. The empirical Tomlinson model often used to describe atomic stick–slip results is therefore presented in detail. We review experimental results regarding atomic friction, including thermal activation, velocity dependence and temperature dependence. The geometry of the contact is crucial to the interpretation of experimental results, such as the calculation of the lateral contact stiffness, as we shall see. The onset of wear on the atomic scale has recently been sudied experimentally and it is described here. In order to compare results, we present molecular dynamics simulations that are directly related to atomic friction experiments. The chapter ends with a discussion of dissipation measurements performed in noncontact force microscopy, which

may become an important complementary tool for the study of mechanical dissipation in nanoscopic devices.

33.1 Friction Force Microscopy in Ultrahigh Vacuum

The *friction force microscope* (FFM, also called the lateral force microscope, LFM) exploits the interaction of a sharp tip sliding on a surface in order to quantify dissipative processes down to the atomic scale (Fig. 33.1).

33.1.1 Friction Force Microscopy

The relative motion of tip and surface is realized by a *scanner* created from piezoelectric elements, which moves the surface perpendicularly to the tip with a certain periodicity. The scanner can be also extended or retracted in order to vary the normal force (F_N) that is applied to the surface. This force is responsible for the deflection of the *cantilever* that supports the tip. If the normal force F_N increases while scanning due to the local slope of the surface, the scanner is retracted by a feedback loop. On the other hand, if F_N decreases, the surface is brought closer to the tip by extending the scanner. In this way, the surface topography can be determined line-by-line from the vertical displacement of the scanner. Accurate control of such vertical movement is made possible by a light beam reflected from the rear of the lever into a photodetector. When the cantilever bends, the light spot on the detector moves up or down and causes the photocurrent to vary, when in turn triggers a corresponding change in the normal force F_N applied.

The relative sliding of tip and surface is usually also accompanied by *friction*. A lateral force (F_L), which acts in the opposite direction to the scan velocity, v, hinders the motion of the tip. This force causes torsion in the cantilever, which can be observed along with the topography if the photodetector can detect not only the normal deflection but also the lateral movement of the lever while scanning. In practice this is achieved using a four-quadrant photodetectors, as shown in Fig. 33.1. We should note that friction forces also cause lateral bending of the cantilever, but this effect is negligible if the thickness of the lever is much less than the width.

The FFM was first used by *Mate* et al. in 1987 to study the friction associated with atomic features [33.1] (just one year after *Binnig* et al. introduced the atomic force microscope [33.2]). In their experiment, Mate used a tungsten wire and a slightly different technique to that described above to detect lateral forces (nonfiber interferometry). Optical beam deflection was introduced later by *Marti* et al. and *Meyer* et al. [33.3, 4]. Other methods of measuring the forces between tip and surface include capacitance detection [33.5], dual fiber interfer-

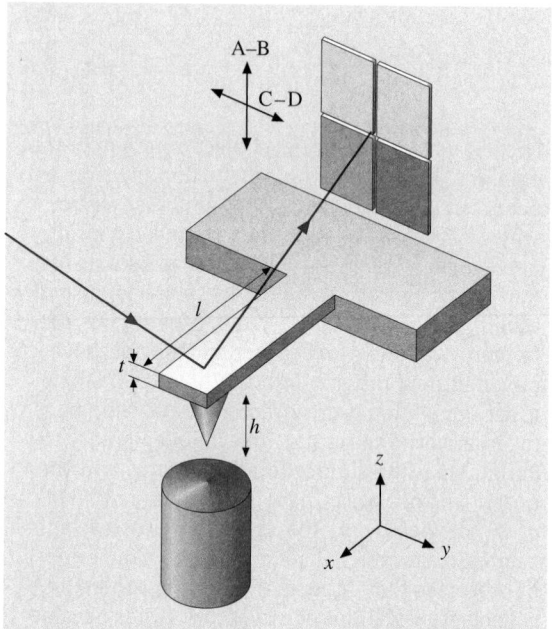

Fig. 33.1 Schematic diagram of a beam-deflection friction force microscope

ometry [33.6] and piezolevers [33.7]. In the first method, two plates close to the cantilever reveal the capacitance while scanning. The second technique uses two optical fibers to detect the cantilever deflection along two orthogonal directions aligned 45° with respect to the surface normal. Finally, in the third method, cantilevers with two Wheatstone bridges at their bases reveal normal and lateral forces, which are respectively proportional to the sum and the difference of both bridge signals.

33.1.2 Force Calibration

Force calibration is relatively simple if rectangular cantilevers are used. Due to possible discrepancies with the geometric values provided by manufacturers, one should use optical and electron microscopes to determine the width, thickness and length of the cantilever (w, t, l), the tip height h and the position of the tip with respect to the cantilever. The thickness of the cantilever can also be determined from the resonance frequency of the lever, f_0, using the relation [33.8]:

$$ t = \frac{2\sqrt{12}\pi}{1.875^2}\sqrt{\frac{\rho}{E}}\, f_0 l^2 \,, \tag{33.1} $$

Here ρ is the density of the cantilever and E is its Young's modulus. The normal spring constant (c_N) and the lateral spring constant (c_L) of the lever are given by

$$c_N = \frac{Ewt^3}{4l^3}, \quad c_L = \frac{Gwt^3}{3h^2l}, \quad (33.2)$$

where G is the shear modulus. Figure 33.2 shows some SEM images of rectangular silicon cantilevers used for FFM. In the case of silicon, $\rho = 2.33 \times 10^3 \, \text{kg/m}^3$, $E = 1.69 \times 10^{11} \, \text{N/m}^2$ and $G = 0.5 \times 10^{11} \, \text{N/m}^2$. Thus, for the cantilever shown in Fig. 33.2, $c_N = 1.9 \, \text{N/m}$ and $c_L = 675 \, \text{N/m}$.

The next force calibration step consists of measuring the sensitivity of the photodetector, S_z (nm/V). For beam-deflection FFMs, the sensitivity (S_z) can be determined by force vs. distance curves measured on hard surfaces (such as Al_2O_3), where elastic deformations are negligible and the vertical movement of the scanner equals the deflection of the cantilever. A typical relation between the difference between the vertical signals on the four-quadrant detector (V_N) and the distance from the surface (z) is sketched in Fig. 33.3. When the tip is approached, no signal is revealed until the tip jumps into contact at $z = z_1$. Further extension or retraction of the scanner results in elastic behavior until the tip jumps out of contact again at a distance $z_2 > z_1$. The slope of the elastic part of the curve gives the required sensitivity, S_z.

The normal and lateral forces are related to the voltage V_N, and the difference between the horizontal signals, V_L, as follows:

$$F_N = c_N S_z V_N, \quad F_L = \frac{3}{2} c_L \frac{h}{l} S_z V_L, \quad (33.3)$$

It is assumed here that the light beam is positioned above the probing tip.

The normal spring constant c_N can also be calibrated using other methods. *Cleveland* et al. [33.10] attached

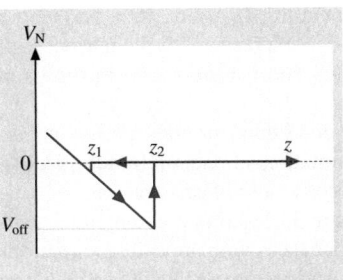

Fig. 33.3 Sketch of a typical force vs. distance curve

tungsten spheres to the tip, which changes the resonance frequency f_0 according to the formula

$$f_0 = \frac{1}{2\pi} \sqrt{\frac{c_N}{M + m^*}}. \quad (33.4)$$

M is the mass of the added object, and m^* is an effective mass of the cantilever, which depends on its geometry [33.10]. The spring constant can be extrapolated from the frequency shifts corresponding to the different masses attached.

As an alternative, *Hutter* et al. observed that the spring constant c_N can be related to the area of the power spectrum of the thermal fluctuations of the cantilever, P [33.11]. The correct relation is $c_N = 4k_B T/3P$, where $k_B = 1.38 \times 10^{-3} \, \text{J/K}$ is Boltzmann's constant and T is the temperature [33.12].

Cantilevers with different shapes require finite element analysis, although analytical formulas can be derived in a few cases. For V-shaped cantilevers, *Neumeister* et al. derived the following approximation for the lateral spring constant c_L [33.13]:

$$c_L = \frac{Et^3}{3(1+\nu)h^2}$$
$$\times \left(\frac{1}{\tan\alpha} \ln \frac{w}{d \sin\alpha} + \frac{L\cos\alpha}{w} - \frac{3\sin 2\alpha}{8} \right)^{-1}, \quad (33.5)$$

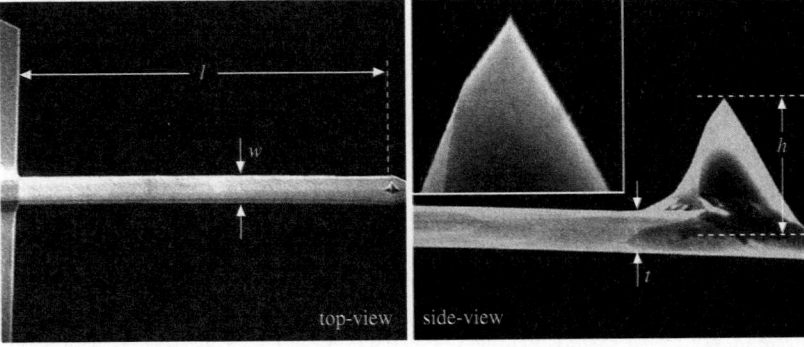

Fig. 33.2 SEM images of a rectangular cantilever. The relevant dimensions are $l = 445 \, \mu m$, $w = 43 \, \mu m$, $t = 4.5 \, \mu m$, $h = 14.75 \, \mu m$. Note that h is given by the sum of the tip height and half of the cantilever thickness. (After [33.9])

Part D | 33.1

The geometrical quantities L, w, α, d, t and h are defined in Fig. 33.4. The expression for the normal constant is more complex and can be found in the cited reference.

Surfaces with well-defined profiles permit an alternative in situ calibration of lateral forces [33.14]. We present a slightly modified version of the method [33.15]. Figure 33.5 shows a commercial grating formed by alternate faces with opposite inclinations with respect to the scan direction. When the tip slides on the inclined planes, the normal force (F_N) and the lateral force (F_L) with respect to the surface are different from the two components $F_\perp$ and $F_\parallel$, which are separated by the photodiode (see Fig. 33.6a).

If the linear relation $F_L = \mu F_N$ holds (see Sect. 33.5), the component $F_\parallel$ can be expressed in terms of $F_\perp$:

$$F_\parallel = \frac{\mu + \tan\theta}{1 - \mu\tan\theta} F_\perp . \qquad (33.6)$$

Fig. 33.5 Silicon grating formed by alternated faces angled at $\pm 55°$ from the surface (courtesy Silicon-MDT Ltd., Moscow, Russia)

10 μm

The component $F_\perp$ is kept constant by the feedback loop. The sum of and the difference between the $F_\parallel$ values for the two planes (1) and (2) are given by

$$F_+ \equiv F_\parallel^{(1)} + F_\parallel^{(2)} = \frac{2\mu\left(1 + \tan^2\theta\right)}{1 - \mu^2\tan^2\theta} F_\perp$$

$$F_- \equiv F_\parallel^{(1)} - F_\parallel^{(2)} = \frac{2\left(1 + \mu^2\right)\tan\theta}{1 - \mu^2\tan^2\theta} F_\perp . \qquad (33.7)$$

The values of F_+ and F_- (in volts) can be measured by scanning the profile back and forth (Fig. 33.6b). If F_+ and F_- are recorded with different values of $F_\perp$, one can determine the conversion ratio between volts and nanonewtons as well as the coefficient of friction, μ.

An accurate error analysis of lateral force calibration was provided by *Schwarz* et al., who revealed the importance of the cantilever oscillations induced by the feedback loop and the so-called "pull-off force" (Sect. 33.5) in friction measurements, aside from the geometrical positioning of the cantilevers and laser beams [33.16].

An adequate estimation of the radius of curvature of the tip, R, is also important for some applications (Sect. 33.5.2). This quantity can be evaluated with a scanning electron microscope. This allows well-

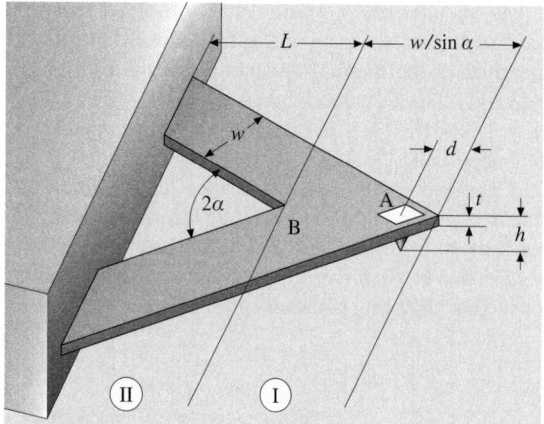

Fig. 33.4 Geometry of a V-shaped cantilever. (After [33.13])

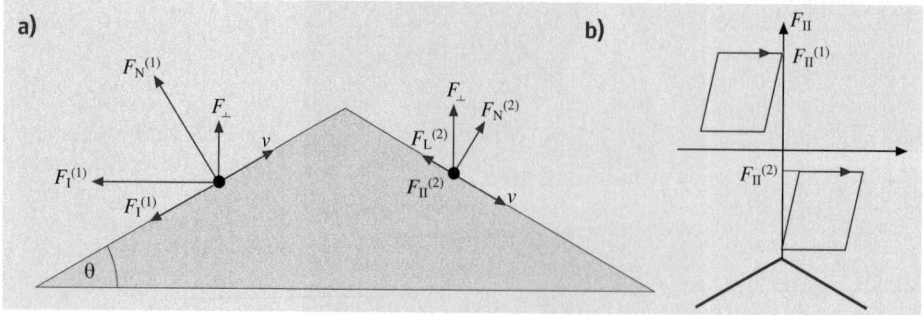

Fig. 33.6 (a) Forces acting on a FFM tip sliding on the grating shown in Fig. 33.5; (b) Friction loops acquired on the two faces

defined structures such as step sites [33.17, 18] or whiskers [33.19] to be imaged. Images of these high aspect ratio structures are convolutions with the tip structure. A deconvolution algorithm that allows for the extraction of the probe tip's radius of curvature was suggested by *Villarrubia* [33.20].

33.1.3 The Ultrahigh Vacuum Environment

Atomic friction studies require well-defined surfaces and – whenever possible – tips. For the surfaces, established methods of surface science performed in ultra-high vacuum (UHV) can be employed. Ionic crystals such as NaCl have become standard materials for friction force microscopy on the atomic scale. Atomically clean and flat surfaces can be prepared by cleavage in UHV. The crystal has to be heated to about 150 °C for 1 h in order to remove charge after the cleavage process. Metal surfaces can be cleaned and flattened by cycles of sputtering with argon ions and annealing. Even surfaces prepared in air or liquids, such as self-assembled molecular monolayers, can be transferred into the vacuum and studied after careful heating procedures that remove water layers.

Tip preparation in UHV is more difficult. Most force sensors for friction studies have silicon nitride or pure silicon tips. Tips can be cleaned and oxide layers removed by sputtering with argon ions. However, the sharpness of the tip is normally reduced by sputtering. As an alternative, tips can be etched in fluoric acid directly before transfer to the UHV. The significance of tip preparation is limited by the fact that the chemical and geometrical structure of the tip can undergo significant changes when sliding over the surface.

Using the friction force microscope in UHV conditions requires some additional effort. First of all, only materials with low vapor pressures can be used, which excludes most plastics and lubricants. Beam-deflection force microscopes employ either a light source in the vacuum chamber or an optical fiber guiding the light into the chamber. The positioning of the light beam on the cantilever and the reflected beam on the position-sensitive detector is achieved by motorized mirrors [33.21] or by moving the light source or detector [33.22]. Furthermore, a motorized sample approach must be realized.

The quality of the force sensor's electrical signal can seriously deteriorate when it is transferred out of the vacuum chamber. Low noise and high bandwidth can be preserved using a preamplifier in the vacuum. Again, the choice of materials for printing and devices is limited by the need for low vapor pressure. Stronger heating

of the electrical circuitry in vacuum, therefore, may be needed.

33.1.4 A Typical Microscope Operated in UHV

A typical AFM used in UHV is shown in Fig. 33.7. The housing (1) contains the light source and a set of lenses that focus the light onto the cantilever. Alternatively, the light can be guided via an optical fiber into the vacuum. By using light emitting diodes with low coherency it is possible to avoid interference effects often found in instruments that use a laser as the light source. A plane mirror fixed on the spherical rotor of a first stepping motor (2) can be rotated around vertical and horizontal axes in order to guide the light beam onto the rear of the cantilever, which is mounted on a removable carrier plate (3). The light is reflected off the cantilever toward a second motorized mirror (4) that guides the beam to the center of the quadrant photodiode (5), where the light is then converted into four photocurrents. Four preamplifiers in close vicinity to the photodiode allow low-noise measurements with a bandwidth of 3 MHz.

The two motors with spherical rotors, used to realign the light path after the cantilever has been exchanged,

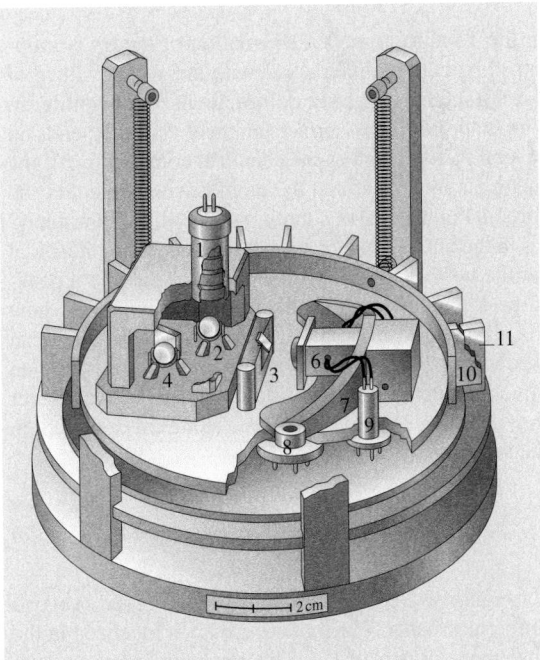

Fig. 33.7 Schematic view of the UHV-AVM realized at the University of Basel. (After [33.21])

work as *inertial stepping motors*: the sphere rests on three piezoelectric legs that can be moved in small amounts tangentially to the sphere. Each step of the motor consists of the slow forward motion of two legs followed by an abrupt jump backwards. During the slow forward motion, the sphere follows the legs due to friction, whereas it cannot follow the sudden jump due to its inertia. A series of these tiny steps rotates the sphere macroscopically.

The sample, which is also placed on an exchangeable carrier plate, is mounted at the end of a tube scanner (6), which can move the sample in three dimensions over several micrometers. The whole scanning head (7) is the slider of a third inertial stepping motor

for coarse positioning of the sample. It rests with its flat and polished bottom on three supports. Two of them are symmetrically placed piezoelectric legs (8), whereas the third central support is passive. The slider (7) can be moved in two dimensions and rotated about a vertical axis by several millimeters (rotation is achieved by antiparallel operation of the two legs). The slider is held down by two magnets, close to the active supports, and its travel is limited by two fixed posts (9) that also serve as cable attachments. The whole platform is suspended by four springs. A ring of radial copper lamellae (10), floating between a ring of permanent magnets (11) on the base flange, acts to efficiently damp eddy currents.

33.2 The Tomlinson Model

In Sect. 33.3, we show that the FFM can reveal friction forces down to the atomic scale, which are characterized by a typical sawtooth pattern. This phenomenon can be seen as a consequence of a *stick-slip* mechanism, first discussed by *Tomlinson* in 1929 [33.23].

33.2.1 One-Dimensional Tomlinson Model

In the Tomlinson model, the motion of the tip is influenced by both the interaction with the atomic lattice of the surface and the elastic deformations of the cantilever. The shape of the tip–surface potential, $V(r)$, depends on several factors, such as the chemical composition of the materials in contact and the atomic arrangement at the tip end. For the sake of simplicity, we will start the analysis in the one-dimensional case considering a sinusoidal profile with an atomic lattice periodicity a and a peak-to-peak amplitude E_0. In Sect. 33.5, we will show how the elasticity of the cantilever and the contact area can be described in a unique framework by introducing an effective lateral spring constant, k_{eff}. If the cantilever moves with a constant velocity v along the x-direction, the total energy of the system is

$$E_{tot}(x, t) = -\frac{E_0}{2} \cos \frac{2\pi x}{a} + \frac{1}{2} k_{eff}(vt - x)^2 . \quad (33.8)$$

Figure 33.8 shows the energy profile $E_{tot}(x, t)$ at two different instants. When $t = 0$, the tip is localized in the absolute minimum of E_{tot}. This minimum increases with time due to the cantilever motion, until the tip position becomes unstable when $t = t^*$.

At a given time t, the position of the tip can be determined by equating the first derivative of $E_{tot}(x, t)$ with respect to x to zero:

$$\frac{\partial E_{tot}}{\partial x} = \frac{\pi E_0}{a} \sin \frac{2\pi x}{a} - k_{eff}(vt - x) = 0 . \quad (33.9)$$

The critical position x^* corresponding to $t = t^*$ is determined by equating the second derivative $\partial^2 E_{tot}(x, t)/\partial x^2$ to zero, which gives

$$x^* = \frac{a}{4} \arccos \left(-\frac{1}{\gamma} \right), \quad \gamma = \frac{2\pi^2 E_0}{k_{eff} a^2} . \quad (33.10)$$

The coefficient γ compares the strength of the interaction between the tip and the surface with the stiffness of the system. When $t = t^*$ the tip suddenly *jumps* into the next minimum of the potential profile. The lateral force $F^* = k_{eff}(vt - x^*)$ that induces the jump can be

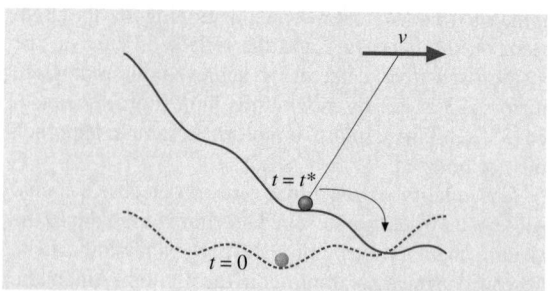

Fig. 33.8 Energy profile experienced by the FFM tip (*black circle*) at $t = 0$ (*dotted line*) and $t = t^*$ (*continuous line*)

evaluated from (33.9) and (33.10):

$$F^* = \frac{k_{\text{eff}} a}{2\pi} \sqrt{\gamma^2 - 1} \,. \qquad (33.11)$$

Thus the stick-slip is observed only if $\gamma > 1$: when the system is not too stiff or when the tip–surface interaction is strong enough. Figure 33.9 shows the lateral force F_L as a function of the cantilever position, X. When the cantilever is moved to the right, the lower part of the curve in Fig. 33.9 is obtained. If, at a certain point, the cantilever's direction of motion is suddenly inverted, the force has the profile shown in the upper part of the curve. The area of the *friction loop* obtained by scanning back and forth gives the total energy dissipated.

On the other hand, when $\gamma < 1$, the stick-slip is suppressed. The tip slides in a continuous way on the surface and the lateral force oscillates between negative and positive values. Instabilities vanish in this regime, which leads to the disappearance of lateral force hysteresis and correspondingly negligible dissipation losses.

33.2.2 Two-Dimensional Tomlinson Model

In two dimensions, the energy of our system is given by

$$E_{\text{tot}}(\boldsymbol{r}, t) = U(\boldsymbol{r}) + \frac{k_{\text{eff}}}{2}(\boldsymbol{v}t - \boldsymbol{r})^2 \,, \qquad (33.12)$$

where $\boldsymbol{r} \equiv (x, y)$ and $\boldsymbol{v}$ is arbitrarily oriented on the surface (note that $\boldsymbol{v} \neq \mathrm{d}\boldsymbol{r}/\mathrm{d}t$!). Figure 33.10 shows the total energy corresponding to a periodic potential of the form

$$U(x, y, t) = -\frac{E_0}{2}\left(\cos\frac{2\pi x}{a} + \cos\frac{2\pi y}{a}\right)$$
$$+ E_1 \cos\frac{2\pi x}{a}\cos\frac{2\pi y}{a} \,. \qquad (33.13)$$

The equilibrium condition becomes

$$\nabla E_{\text{tot}}(\boldsymbol{r}, t) = \nabla U(\boldsymbol{r}) + k_{\text{eff}}(\boldsymbol{r} - \boldsymbol{v}t) = 0 \,. \qquad (33.14)$$

The stability of the equilibrium can be described by introducing the Hessian matrix

$$H = \begin{pmatrix} \dfrac{\partial^2 U}{\partial x^2} + k_{\text{eff}} & \dfrac{\partial^2 U}{\partial x \partial y} \\ \dfrac{\partial^2 U}{\partial y \partial x} & \dfrac{\partial^2 U}{\partial y^2} + k_{\text{eff}} \end{pmatrix} \,. \qquad (33.15)$$

When both eigenvalues $\lambda_{1,2}$ of the Hessian are positive, the position of the tip is stable. Figure 33.11 shows these regions for a potential of the form (33.13). The tip follows the cantilever adiabatically as long as it remains in the $(++)$-region. When the tip is dragged to the border of the region, it suddenly jumps into the next $(++)$-region. A comparison between a theoretical friction map deduced from the 2-D Tomlinson model and an experimental map acquired by UHV-FFM is given in the next section.

33.2.3 Friction Between Atomically Flat Surfaces

So far we have implicitly assumed that the tip is terminated by only one atom. It is also instructive to consider the case of a periodic surface sliding on another periodic surface. In the Frenkel–Kontorova–Tomlinson (FKT) model, the atoms of one surface are harmonically coupled with their nearest neighbors. We will restrict ourselves to the case of quadratic symmetries, with lattice constants a_1 and a_2 for the upper and lower surfaces, respectively (Fig. 33.12). In this context, the role of *commensurability* is essential. It is well known that any real number z can be represented as a continued

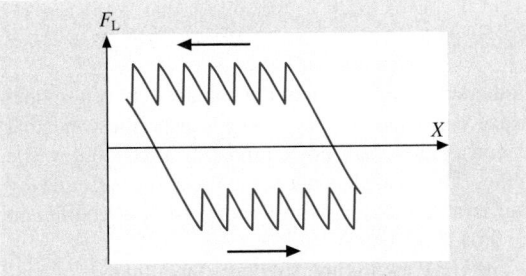

Fig. 33.9 Friction loop obtained by scanning back and forth in the 1-D Tomlinson model. The effective spring constant k_{eff} is the slope of the sticking part of the loop (if $\gamma \gg 1$)

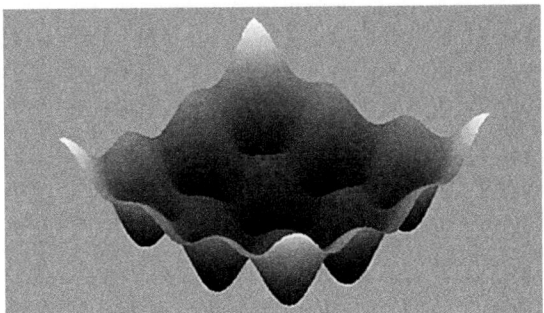

Fig. 33.10 Energy landscape experienced by the FFM tip in 2-D

Part D | 33.2

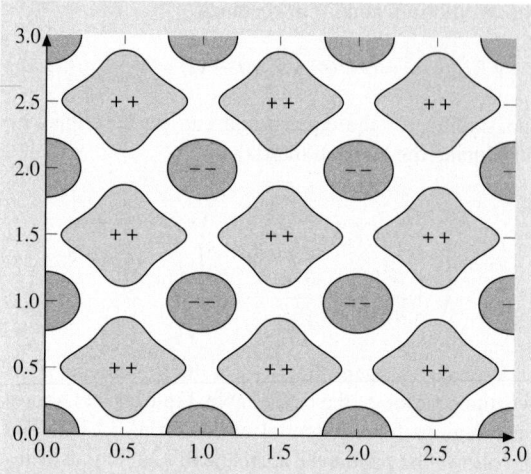

Fig. 33.11 Regions on the tip plane labeled according to the signs of the eigenvalues of the Hessian matrix. (After [33.24])

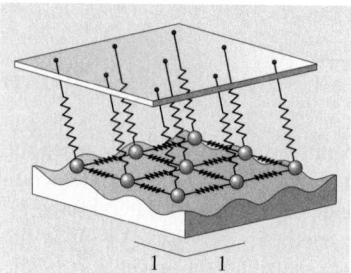

Fig. 33.12 The FKT model in 2-D. (After [33.25])

fraction:

$$z = N_0 + \cfrac{1}{N_1 + \cfrac{1}{N_2 + \dots}} . \qquad (33.16)$$

The sequence that converges most slowly is obtained when all $N_i = 1$, which corresponds to the *golden mean* $\bar{z} = (\sqrt{5} - 1)/2$. In 1-D, *Weiss* and *Elmer* predicted that friction should decrease with decreasing commensurability, the minimum friction being reached when $a_1/a_2 = \bar{z}$ [33.26].

In 2-D, *Gyalog* and *Thomas* studied the case $a_1 = a_2$, with a misalignment between the two lattices given by an angle θ [33.25]. When the sliding direction changes, friction also varies from a minimum value (corresponding to

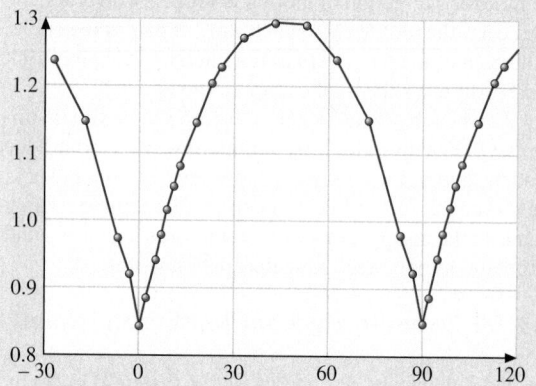

Fig. 33.13 Friction as a function of the sliding angle φ in the 2-D FKT model. (After [33.25])

the sliding angle $\varphi = \theta/2$) to a maximum value (which is reached when $\varphi = \theta/2 + \pi/4$; see Fig. 33.13). The misfit angle θ is related to the commensurability. Since the misfit angles that give rise to commensurate structure form a dense subset, the dependence of friction on θ should be discontinuous. The numerical simulations performed by Gyalog are in agreement with this conclusion.

33.3 Friction Experiments on the Atomic Scale

Figure 33.14 shows the first atomic-scale friction map, as observed by Mate. The periodicity of the lateral force is the same as that of the atomic lattice of graphite. The series of friction loops in Fig. 33.15 reveals the stick-slip effect discussed in the previous section. The applied loads are in the range of tens of µN. According to the continuum models discussed in Sect. 33.5, these values correspond to contact diameters of 100 nm. A possible explanation for the atomic features observed at such high loads is that graphite flakes may have detached from the surface and adhered to the tip [33.27].

Another explanation is that the contact between tip and surface consisted of few nm-scale asperities and that the corrugation was not entirely averaged out while sliding. The load dependence of friction as found by Mate is rather linear, with a small friction coefficient $\mu = 0.01$ (Fig. 33.16).

The UHV environment reduces the influence of contaminants on the surface and leads to more precise and reproducible results. *Meyer* et al. [33.28] obtained a series of interesting results on ionic crystals using the UHV-FFM apparatus described in Sect. 33.1.4. By

Fig. 33.14 First atomic friction map acquired on graphite with a normal force $F_N = 56\,\mu N$. Frame size: 2 nm. (After [33.1])

applying subnanonewton loads to a NaCl surface, *Socoliuc* et al. observed the transition from stick-slip to continuous sliding discussed in Sect. 33.2.1 [33.29]. In Fig. 33.17, a friction map recorded on KBr(100) is compared with a theoretical map obtained with the 2-D Tomlinson model. The periodicity $a = 0.47$ nm corresponds to the spacing between equally charged ions. No individual defects were observed. One possible reason

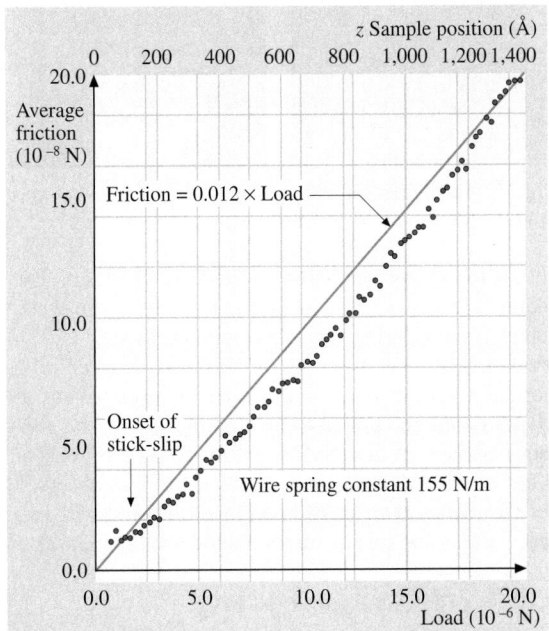

Fig. 33.16 Load dependence of friction on graphite. (After [33.1])

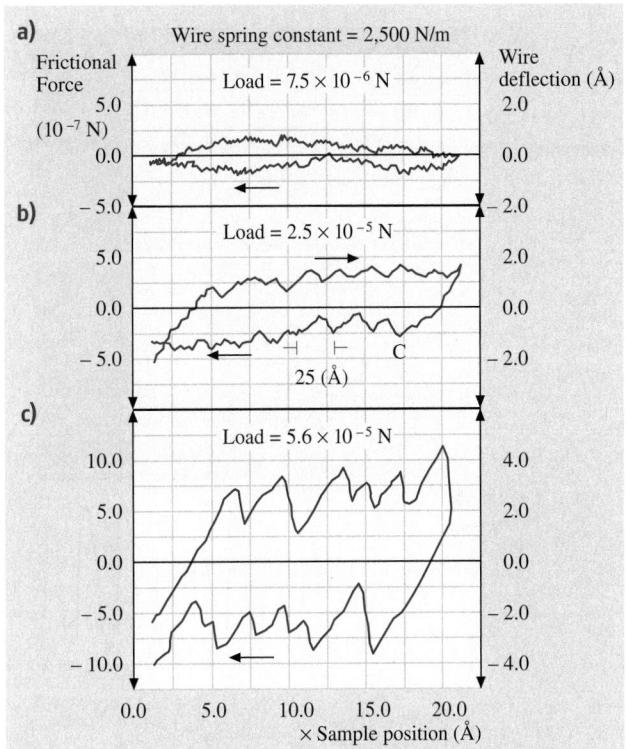

Fig. 33.15a–c Friction loops on graphite acquired with (**a**) $F_N = 7.5\,\mu N$, (**b**) 24 μN and (**c**) 75 μN. (After [33.1])

is that the contact realized by the FFM tip is always formed by many atoms, which superimpose and average their effects. Molecular dynamics (MD) calculations (Sect. 33.7) show that even single-atom contact may cause rather large stresses in the sample, which lead to the motion of defects far away from the contact area. In a rather picturesque analogy, we can say that "defects behave like dolphins that swim away in front of an ocean cruiser" [33.28].

Lüthi et al. [33.30] even detected atomic-scale friction on a reconstructed Si(111)7×7 surface. However, uncoated Si tips and tips coated with Pt, Au, Ag, Cr and Pt/C damaged the sample irreversibly, and the observation of atomic features was achieved only after coating the tips with polytetrafluoroethylene (PTFE), which has lubricant properties and does not react with the dangling bonds of Si(111)7×7 (Fig. 33.18).

Recently friction was even resolved on the atomic scale on metallic surfaces in UHV [33.31]. In Fig. 33.19a a reproducible stick-slip process on Cu(111) is shown. Sliding on the (100) surface of copper produced irregular patterns, although atomic features were recognized even

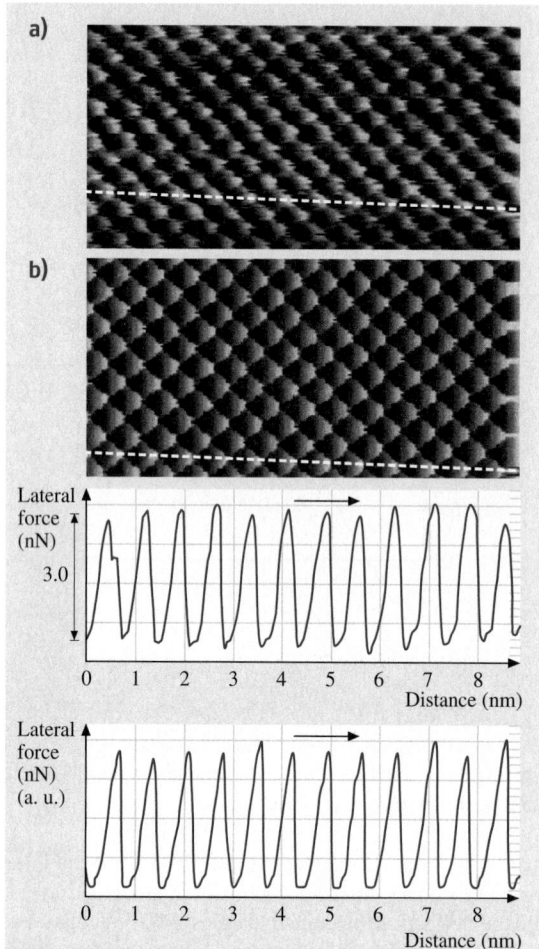

Fig. 33.17 (a) Measured and (b) theoretical friction map on KBr(100). (After [33.32])

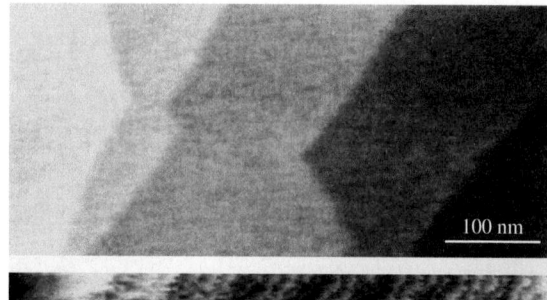

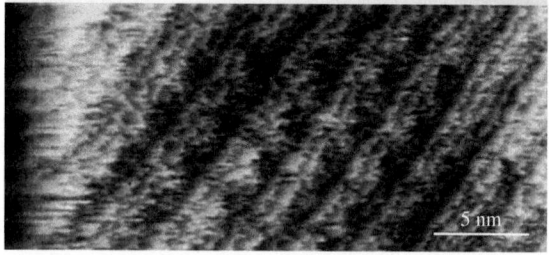

Fig. 33.18 (a) Topography and (b) friction image of Si(111)7×7 measured with a PTFE-coated Si tip. (After [33.30])

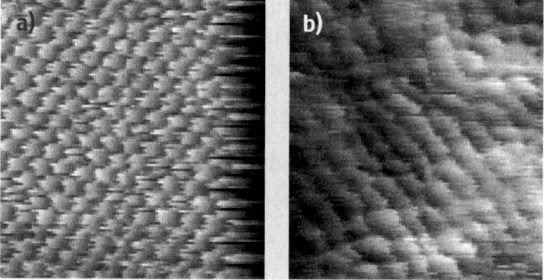

Fig. 33.19a,b Friction images of (a) Cu(111) and (b) Cu(100). Frame size: 3 nm. (After [33.35])

in this case (Fig. 33.19b). Molecular dynamics suggests that wear should occur more easily on the Cu(100) surface than on the close-packed Cu(111) (Sect. 33.7). This conclusion was achieved by adopting copper tips in computer simulations. The assumption that the FFM tip used in the experiments was covered by copper is supported by current measurements performed at the same time.

Atomic stick-slip on diamond was observed by *Germann* et al. with an apposite diamond tip prepared by chemical vapor deposition [33.33] and, a few years later, by *van der Oetelaar* et al. [33.34] with standard silicon tips. The values of friction vary dramatically depending on the presence or absence of hydrogen on the surface.

Fujisawa et al. [33.36] measured friction on mica and on MoS$_2$ with a 2-D FFM apparatus that could also

reveal forces perpendicular to the scan direction. The features in Fig. 33.20 correspond to a zigzag tip walk, which is predicted by the 2-D Tomlinson model [33.37]. Two-dimensional stick-slip on NaF was detected with normal forces below 14 nN, whereas loads of up to 10 μN could be applied to layered materials. The contact between tip and NaF was thus formed by one or a few atoms. A zigzag walk on mica was also observed by *Kawakatsu* et al. using an original 2-D FFM with two laser beams and two quadrant photodetectors [33.38].

33.3.1 Anisotropy of Friction

The importance of the misfit angle in the reciprocal sliding of two flat surfaces was first observed experimentally

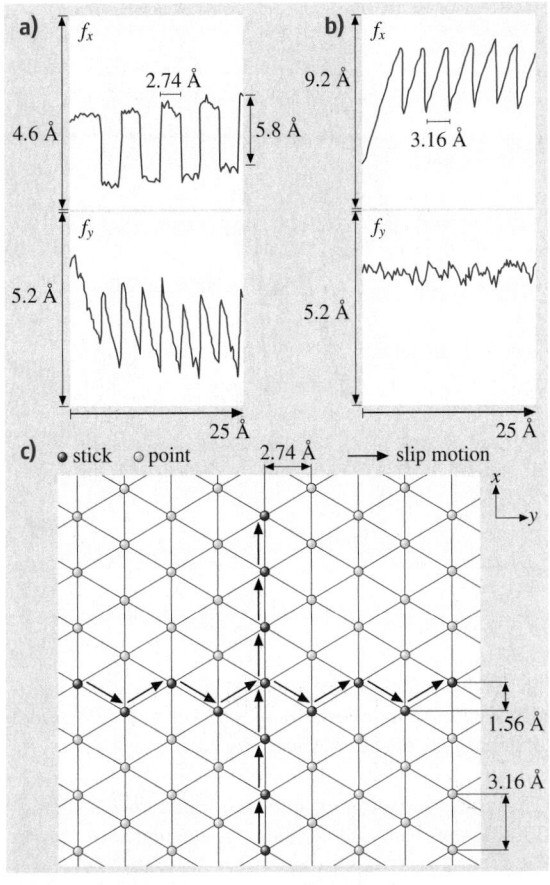

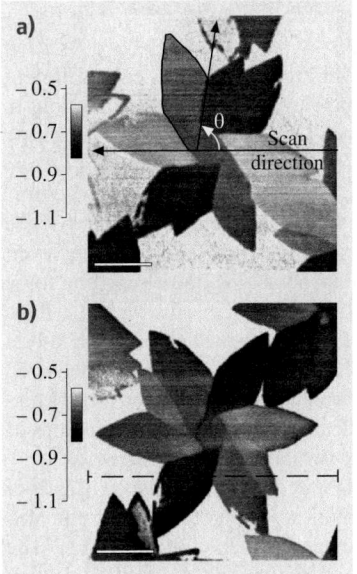

Fig. 33.20 (a) Friction force on MoS_2 acquired by scanning along the cantilever and (b) across the cantilever. (c) Motion of the tip on the sample. (After [33.36])

Fig. 33.21 Friction images of a thiolipid monolayer on a mica surface. (After [33.41])

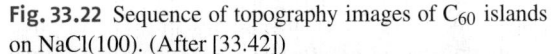

Fig. 33.22 Sequence of topography images of C_{60} islands on NaCl(100). (After [33.42])

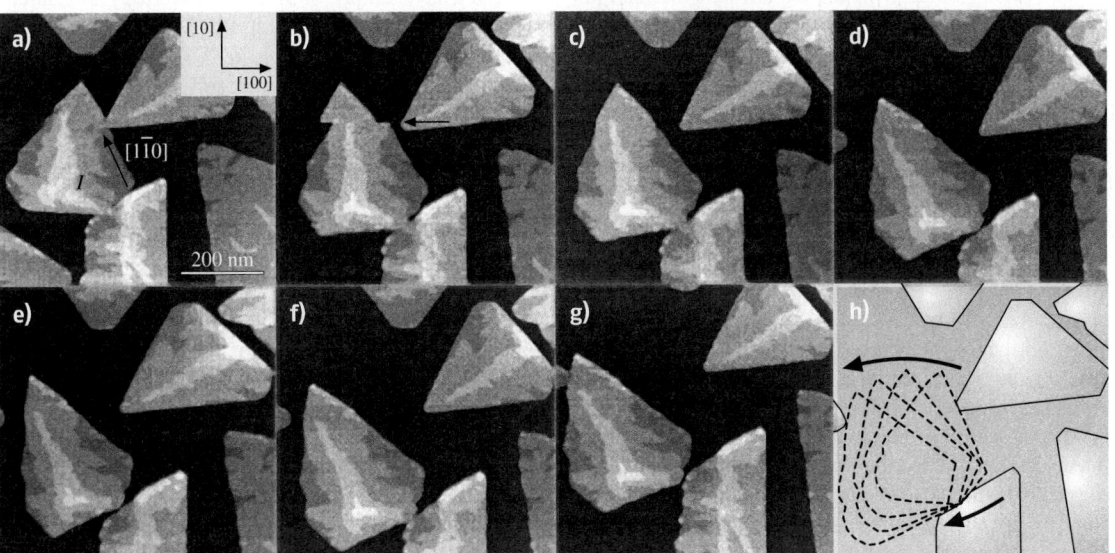

by *Hirano* et al. in the contact of two mica sheets [33.39]. The friction increased when the two surfaces formed commensurate structures, in agreement with the discussion in Sect. 33.2.3. In more recent measurements with a monocrystalline tungsten tip on Si(001), *Hirano* et al. observed *superlubricity* in the incommensurate case [33.40].

Overney et al. [33.44] studied the effects of friction anisotropy on a bilayer lipid film. In this case, different molecular alignments resulted in significant variations in the friction. Other measurements of friction anisotropy on single crystals of stearic acid were reported by *Takano* and *Fujihira* [33.45]. An impressive confirmation of this effect recently came from a dedicated force microscope developed by Frenken and coworkers, the Tribolever, which allows quantitative tracking of the scanning force in three dimensions [33.46]. With this instrument, a flake from a graphite surface was picked up and the lateral forces between the flake and the surface were measured at different angles of rotation. Stick-slip and energy dissipation were only clearly revealed at rotation angles of $0°$ and $60°$, when the two lattices are in registry.

Liley et al. [33.41] observed flower-shaped islands of a lipid monolayer on mica, which consisted of domains with different molecular orientations (Fig. 33.21). The angular dependence of friction reflects the tilt direction of the alkyl chains of the monolayer, as revealed by other techniques.

Lüthi et al. [33.42] used the FFM tip to move C_{60} islands, which slide on sodium chloride in UHV without disruption (Fig. 33.22). In this experiment the friction was found to be independent of the sliding direction. This was not the case in other experiments performed by *Sheehan* and *Lieber*, who observed that the misfit angle is relevant when MoO_3 islands are dragged on the MoS_2 surface [33.47]. In these experiments, sliding was possible only along low index directions.

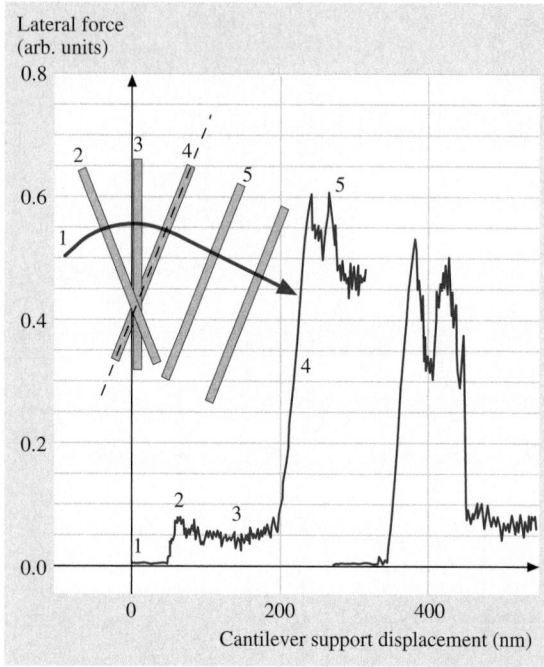

Fig. 33.23 Friction force experienced as a carbon nanotube is rotated into (*left trace*) and out of (*right trace*) commensurate contact. (After [33.43])

The weak orientation dependence found by *Lüthi* et al. [33.42] is probably due to the large mismatch of C_{60} on NaCl.

A recent example of friction anisotropy is related to carbon nanotubes. *Falvo* et al. [33.43] manipulated nanotubes on graphite using a FFM tip (Fig. 33.23). A dramatic increase in the lateral force was found in directions corresponding to commensurate contact. At the same time, the nanotube motion changed from sliding/rotating to stick-roll.

33.4 Thermal Effects on Atomic Friction

Although the Tomlinson model gives a good interpretation of the basic mechanism of the atomic stick-slip discussed in Sect. 33.2, it cannot explain some minor features observed in the atomic friction. For example, Fig. 33.24 shows a friction loop acquired on NaCl(100). The peaks in the sawtooth profile have different heights, which is in contrast to the result in Fig. 33.9. Another effect is observed if the scan velocity v is varied: the mean friction force increases with the logarithm of v

(Fig. 33.25). This effect cannot be interpreted within the mechanical approach in Sect. 33.2 without further assumptions.

33.4.1 The Tomlinson Model at Finite Temperature

Let us focus again on the energy profile discussed in Sect. 33.2.1. For the sake of simplicity, we will as-

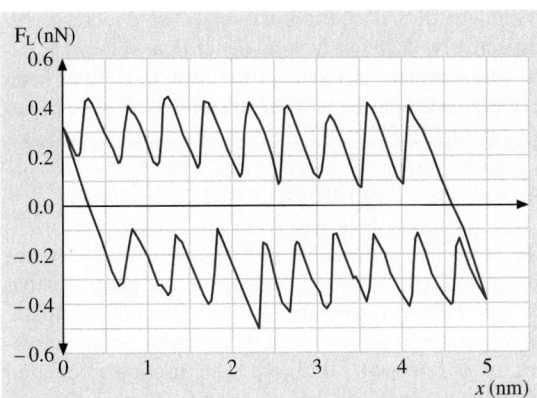

Fig. 33.24 Friction loop on NaCl(100). (After [33.48])

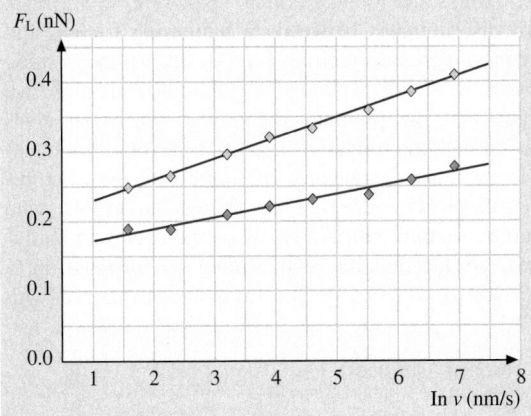

Fig. 33.25 Mean friction force vs. scanning velocity on NaCl(100) at $F_N = 0.44$nN (+) and $F_N = 0.65$nN (×). (After [33.48])

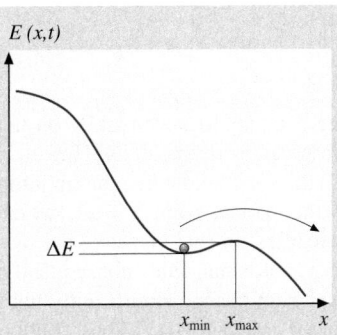

Fig. 33.26 Energy barrier that hinders the tip jump in the Tomlinson model

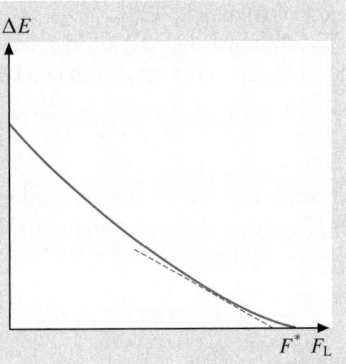

Fig. 33.27 Energy barrier ΔE as a function of the lateral force F_L. The *dashed line* close to the critical value corresponds to the linear approximation (33.17)

sume that $\gamma \gg 1$. At a given time $t < t^*$, the tip jump is prevented by the energy barrier $\Delta E = E(x_{max}, t) - E(x_{min}, t)$, where x_{max} corresponds to the first maximum observed in the energy profile and x_{min} is the actual position of the tip (Fig. 33.26). The quantity ΔE decreases with time or, equivalently, with the frictional force F_L until it vanishes when $F_L = F^*$ (Fig. 33.27). Close to the critical point, the energy barrier can be written approximately as

$$\Delta E = \lambda(\tilde{F} - F_L), \tag{33.17}$$

where $\tilde{F}$ is close to the critical value $F^* = \pi E_0/a$.

At finite temperature T, the lateral force required to induce a jump is lower than F^*. To estimate the most probable value of F_L at this point, we first consider the probability p that the tip does *not* jump. The proba-

bility p changes with time t according to the master equation

$$\frac{dp(t)}{dt} = -f_0 \exp\left(-\frac{\Delta E(t)}{k_B T}\right) p(t), \tag{33.18}$$

where f_0 is a characteristic frequency of the system. The physical meaning of this frequency is discussed in Sect. 33.4.2. We should note that the probability of a reverse jump is neglected, since in this case the energy barrier that must be overcome is much higher than ΔE. If time is replaced by the corresponding lateral force, the master equation becomes

$$\frac{dp(F_L)}{dF_L} = -f_0 \exp\left(-\frac{\Delta E(F_L)}{k_B T}\right)\left(\frac{dF_L}{dt}\right)^{-1} p(F_L). \tag{33.19}$$

At this point, we substitute

$$\frac{dF_L}{dt} = \frac{dF_L}{dX}\frac{dX}{dt} = k_{eff}v \tag{33.20}$$

and use the approximation (33.17). The maximum probability transition condition $d^2 p(F)/dF^2 = 0$ then yields

$$F_L(v) = F^* - \frac{k_B T}{\lambda} \ln \frac{v_c}{v} \tag{33.21}$$

with

$$v_c = \frac{f_0 k_B T}{k_{eff} \lambda} .$$ (33.22)

Thus, the lateral force depends logarithmically on the sliding velocity, as observed experimentally. However, approximation (33.17) does not hold when the tip jump occurs very close to the critical point $x = x^*$, which is the case at high velocities. In this instance, the factor $(dF_L \, dt)^{-1}$ in (33.19) is small and, consequently, the probability $p(t)$ does not change significantly until it suddenly approaches 1 when $t \to t^*$. Thus friction is constant at high velocities, in agreement with the classical Coulomb's law of friction [33.28].

Sang et al. [33.50] observed that the energy barrier close to the critical point is better approximated by a relation like

$$\Delta E = \mu (F^* - F_L)^{3/2} .$$ (33.23)

The same analysis performed using approximation (33.23) instead of (33.17) leads to the expression [33.51]

$$\frac{\mu (F^* - F_L)^{3/2}}{k_B T} = \ln \frac{v_c}{v} - \ln \sqrt{1 - \frac{F^*}{F_L}} ,$$ (33.24)

where the critical velocity v_c is now

$$v_c = \frac{\pi \sqrt{2}}{2} \frac{f_0 k_B T}{k_{eff} a} .$$ (33.25)

The velocity v_c discriminates between two different regimes. If $v \ll v_c$, the second logarithm in (33.24) can be neglected, which leads to the logarithmic dependence

$$F_L(v) = F^* - \left(\frac{k_B T}{\mu}\right)^{2/3} \left(\ln \frac{v_c}{v}\right)^{2/3} .$$ (33.26)

In the opposite case, $v \gg v_c$, the term on the left in (33.23) is negligible and

$$F_L(v) = F^* \left(1 - \frac{v_c}{v}\right)^2 .$$ (33.27)

In such a case, the lateral force F_L tends to F^*, as expected.

In a recent work, *Reimann* et al. distinguished between the dissipation that occurs in the tip apex and that in the substrate volume in contact with the tip [33.52]. After the initial logarithmic increase, the velocity dependence of friction changes in different ways, depending on the relative contribution of the tip apex to the total dissipation. A 'friction plateau' is only predicted when $\theta \approx 0.5$ over a limited velocity range. At lower

or higher values of θ, friction is expected to increase, or, respectively, decrease beyond the critical velocity v_c.

The thermally activated Tomlinson model has been recently extended to two dimensions by *Fasolino* and coworkers [33.53].

33.4.2 Velocity Dependence of Friction

The velocity dependence of friction was only recently studied by FFM. *Zwörner* et al. observed that friction between silicon tips and diamond, graphite or amorphous carbon is constant with scan velocities of a few μm/s [33.54]. The friction decreased when v was reduced below 1 μm/s. In their experiment on lipid films on mica (Sect. 33.3.1), *Gourdon* et al. [33.49] explored a range of velocities from 0.01 to 50 μm/s and found a critical velocity $v_c = 3.5$ μm/s that discriminates between an increasing friction and a constant friction regime (Fig. 33.28). Although these results were not explained by thermal activation, we argue that the previous theoretical discussion gives the correct interpretative key. A clear observation of a logarithmic dependence of friction on the micrometer scale was reported by *Bouhacina* et al., who studied friction on triethoxysilane molecules and polymers grafted on silica with sliding velocities of up to $v = 300$ μm/s [33.55]. The result was explained with

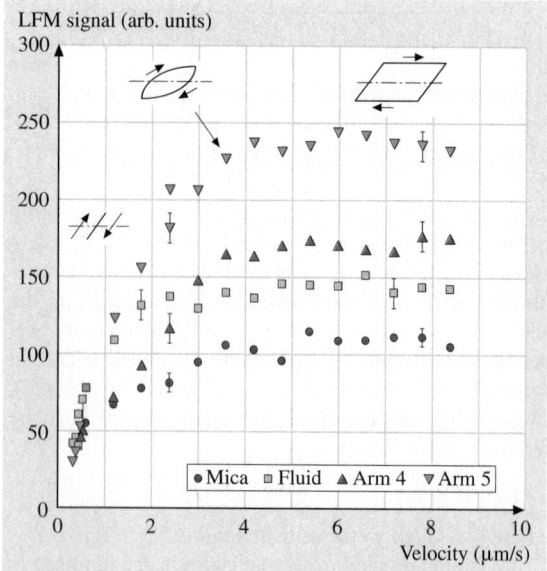

Fig. 33.28 Velocity dependence of friction on mica and on lipid films with different orientations (arms 4 and 5) and in a fluid phase. (After [33.49])

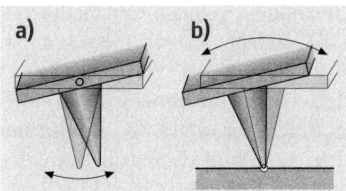

Fig. 33.29a,b Torsional modes of cantilever oscillation (**a**) when the tip is free and (**b**) when the tip is in contact with a surface. (After [33.58])

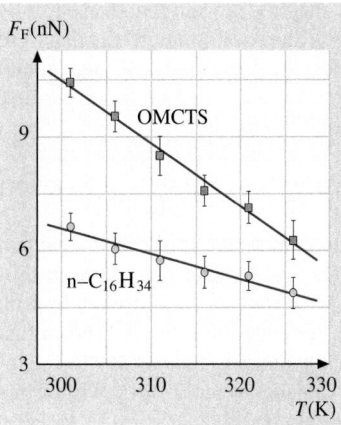

Fig. 33.31 Temperature dependence of friction on *n*-hexadecane and octamethylcyclotetrasiloxane. (After [33.60])

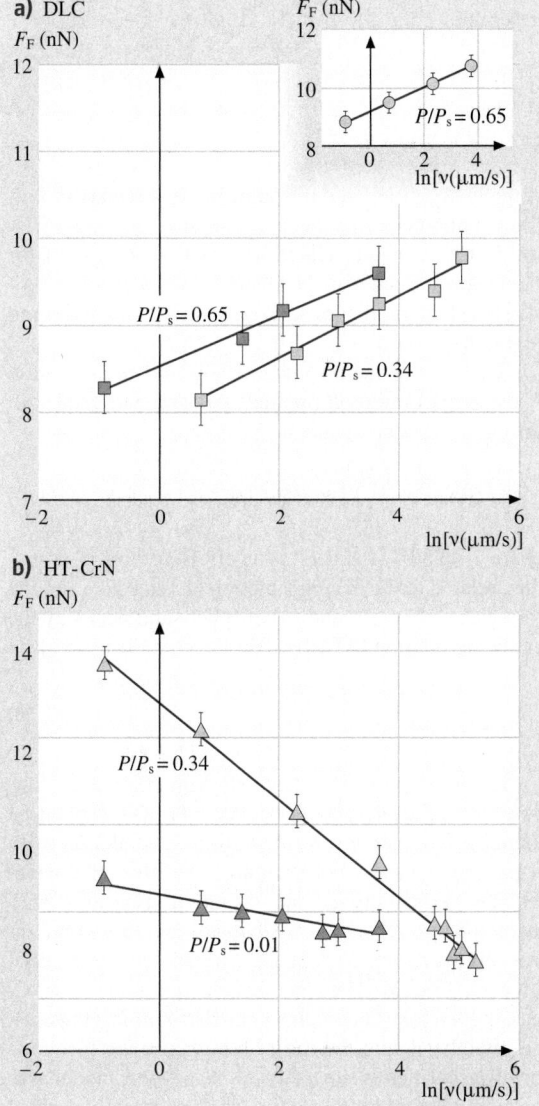

Fig. 33.30a,b Friction vs. sliding velocity (**a**) on hydrophobic surfaces and (**b**) on hydrophilic surfaces. (After [33.59])

a thermally activated Eyring model, which does not differ significantly from the model discussed in the previous subsection [33.56, 57].

The first measurements on the atomic scale were performed by *Bennewitz* et al. on copper and sodium chloride [33.31, 48]; in both cases a logarithmic dependence of friction was revealed up to $v < 1 \,\mu\text{m/s}$ (Fig. 33.25), in agreement with (33.21). Higher values of velocities were not explored, due to the limited range of the scan frequencies possible with FFM on the atomic scale. The same limitation does not allow a clear distinction between (33.21) and (33.26) when interpreting the experimental results.

At this point we would like to discuss the physical meaning of the characteristic frequency f_0. With a lattice constant a of a few angstroms and an effective spring constant k_{eff} of about 1 N/m, which are typical of FFM experiments, (33.25) gives a value of a few hundred kHz for f_0. This is the characteristic range in which the torsional eigenfrequencies of the cantilevers are located in both contact and noncontact modes (Fig. 33.29). Future work may clarify whether or not f_0 must be identified with these frequencies.

To conclude this section, we should emphasize that the increase in friction with increasing velocity is ultimately related to the materials and the environment in which the measurements are realized. In a humid environment, *Riedo* et al. observed that the surface wettability plays an important role [33.59]. Friction *decreases* with increasing velocity on hydrophilic surfaces, and the rate of this decrease depends drastically on humidity. A logarithmic increase is again found on partially hydrophobic surfaces (Fig. 33.30). These results were interpreted considering the thermally activated nucleation of water bridges between tip and sample asperities, as discussed in the cited reference.

33.4.3 Temperature Dependence of Friction

Thus far we have used thermal activation to explain the velocity dependence of friction. The same mechanism also predicts that friction should change with temperature. The master equation ((33.18)) shows that the probability of a tip jump is reduced at low temperatures T until it vanishes when $T = 0$. Within this limit case, thermal activation is excluded, and the lateral force F_L is equal to F^*, independent of the scanning velocity v.

To our knowledge, stick-slip processes at low temperatures have not been reported. A significant increase in the mean friction with decreasing temperature was recently measured by *He* et al. [33.60] by FFM (Fig. 33.31). Neglecting logarithmic contributions, (33.21) and (33.26) predict that $(F^* - F_L) \approx T$ and $(F^* - F_L) \approx T^{2/3}$ respectively for the temperature dependence of friction. Although He et al. applied a linear fit to their data, the 30 K range that they considered is again not large enough to prove that a $T^{2/3}$ fit would be preferable.

33.5 Geometry Effects in Nanocontacts

Friction is ultimately related to the real shape of the contact between the sliding surfaces. On the macroscopic scale, the contact between two bodies is studied within the context of continuum mechanics, which is based on the elasticity theory developed by Hertz in the nineteenth century. Various FFM experiments have shown that continuum mechanics is still valid down to contact areas just a few nanometers in size. Only when contact is reduced to few atoms does the continuum frame become unsuitable, and other approaches like molecular dynamics become necessary. This section will deal with continuum mechanics theory; molecular dynamics will be discussed in Sect. 33.7.

33.5.1 Continuum Mechanics of Single Asperities

The lateral force F_L between two surfaces in reciprocal motion depends on the size of the real area of contact, A, which can be a few orders of magnitude smaller than the apparent area of contact. The simplest assumption is that friction is proportional to A; the proportionality factor is called the *shear strength* σ [33.61]:

$$F_L = \sigma A . \tag{33.28}$$

For plastic deformation, the asperities are compressed until the pressure (p) equals a certain yield value, p^*. The resulting contact area is thus $A = F_N/p^*$, and the well-known Amontons' law is obtained: $F_L = \mu F_N$, where $\mu = \sigma/p^*$ is the *coefficient of friction*. The same idea can be extended to contacts formed by many asperities, and it leads again to Amontons' law. The simplicity of this analysis explains why most friction processes were related to plastic deformation for a long time. Such a mechanism, however, should provoke quick disruption of surfaces, which is not observed in practice.

Elastic deformation can be easily studied in the case of a sphere of radius R pressed against a flat surface. In this case, the contact area is

$$A(F_N) = \pi \left(\frac{R}{K} \right)^{2/3} F_N^{2/3} , \tag{33.29}$$

where $K = 3E^*/4$ and E^* is an effective Young's modulus, related to the Young's moduli (E_1 and E_2) and the Poisson numbers (ν_1 and ν_2) of sphere and plane, by the following relation [33.62]:

$$\frac{1}{E^*} = \frac{1 - \nu_1^2}{E_1} + \frac{1 - \nu_2^2}{E_2} . \tag{33.30}$$

The result $A \propto F_N^{2/3}$ contrasts with Amontons' law. However, a linear relation between F_L and F_N can be obtained for contacts formed from several asperities in particular cases. For example, the area of contact between a flat surface and a set of asperities with an exponential height distribution and the same radius of curvature R depends linearly on the normal force F_N [33.63]. The same conclusion holds approximately even for a Gaussian height distribution. However, the hypothesis that the radii of curvature are the same for all asperities is not realistic. A general model was recently proposed by *Persson*, who analytically derived the proportionality between contact area and load for a large variety of elastoplastic contacts formed by surfaces with arbitrary roughnesses [33.64]. However, this discussion is not straightforward and goes beyond the purposes of this section.

Further effects are observed if adhesive forces between the asperities are taken into account. If the range of action of these forces is smaller than the elastic deformation, (33.29) is extended to the Johnson–Kendall–Roberts (JKR) relation

$$A(F_N) = \pi \left(\frac{R}{K}\right)^{2/3}$$ (33.31)

$$\times \left(F_N + 3\pi\gamma R + \sqrt{6\pi\gamma R F_N + (3\pi\gamma R)^2}\right)^{2/3},$$

where γ is the surface tension of the sphere and plane [33.66]. The real contact area at zero load is finite and the sphere can be detached only by pulling it away with a certain force. This is also true in the opposite case, in which the range of action of adhesive forces is larger than the elastic deformation. In this case, the relation between contact area and load takes the simple form

$$A(F_N) = \pi \left(\frac{R}{K}\right)^{2/3} (F_N - F_{off})^{2/3},$$ (33.32)

where F_{off} is the negative load required to break the contact. The Hertz-plus-offset relation (33.32) can be derived from the Derjaguin–Muller–Toporov (DMT) model [33.67]. To discriminate between the JKR or DMT models, *Tabor* introduced a nondimensional parameter

$$\Phi = \left(\frac{9R\gamma^2}{4K^2 z_0^3}\right)^{1/3},$$ (33.33)

where z_0 is the equilibrium distance during contact. The JKR model can be applied if $\Phi > 5$; the DMT model holds when $\Phi < 0.1$ [33.68]. For intermediate values of Φ, the Maugis–Dugdale model [33.69] could reasonably explain experimental results (Sect. 33.5.3).

33.5.2 Dependence of Friction on Load

The FFM tip represents a single asperity sliding on a surface. The previous discussion suggests a nonlinear dependence of friction on the applied load, provided that continuum mechanics is applicable. Schwarz et al. ob-

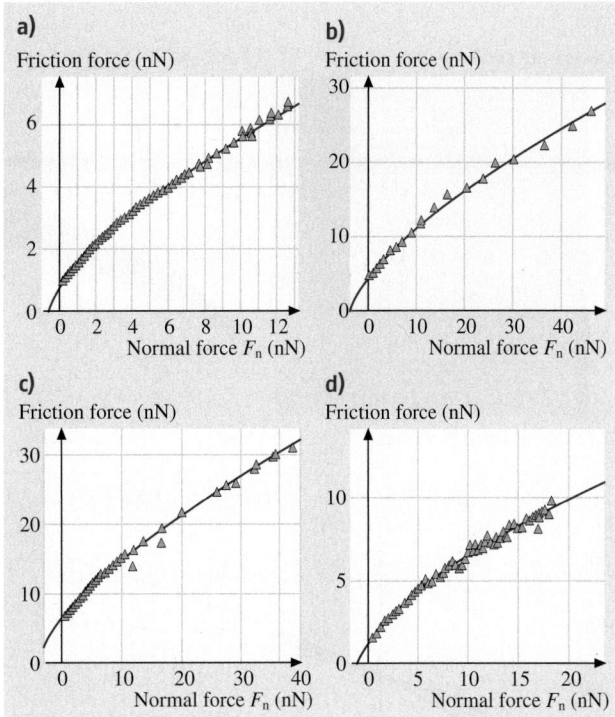

Fig. 33.32a–d Friction vs. load curve on amorphous carbon in argon atmosphere. Curves (a)–(d) refer to tips with different radii of curvature. (After [33.65])

served the Hertz-plus-offset relation (33.32) on graphite, diamond, amorphous carbon and C_{60} in an argon atmosphere (Fig. 33.32). In their measurements, they used well-defined spherical tips with radii of curvature of tens of nanometers, obtained by contaminating silicon tips with amorphous carbon in a transmission electron microscope. In order to compare the tribological behavior of different materials, *Schwarz* et al. suggested the introduction of an effective coefficient of friction, $\tilde{C}$, which is independent of the tip curvature [33.65].

Meyer et al., *Carpick* et al., and *Polaczyc* et al. performed friction measurements in UHV in agreement with JKR theory [33.17,70,71]. Different materials were considered (ionic crystals, mica and metals) in these experiments. In order to correlate the lateral and normal forces with improved statistics, Meyer et al. applied an original 2-D histogram technique (Fig. 33.33). Carpick et al. extended the JKR relation (33.32) to include nonspherical tips. In the case of an axisymmetric tip profile $z \propto r^{2n}$ ($n > 1$), it can be proven analytically that the increase in the friction becomes less pronounced with increasing n (Fig. 33.34).

a)
Decreasing normal load

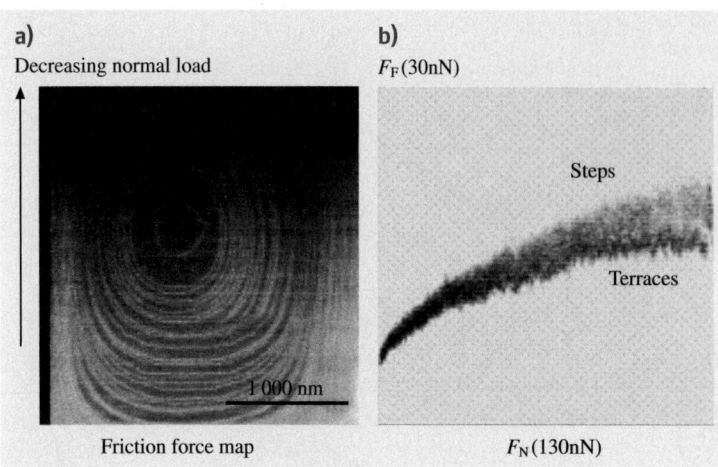

Friction force map

b)
F_F (30nN)

Steps

Terraces

F_N (130nN)

Fig. 33.33 (a) Friction force map on NaCl(100). The load is decreased from 140 to 0nN (jump-off point) during imaging. **(b)** 2-D-histogram of **(a)**. (After [33.17])

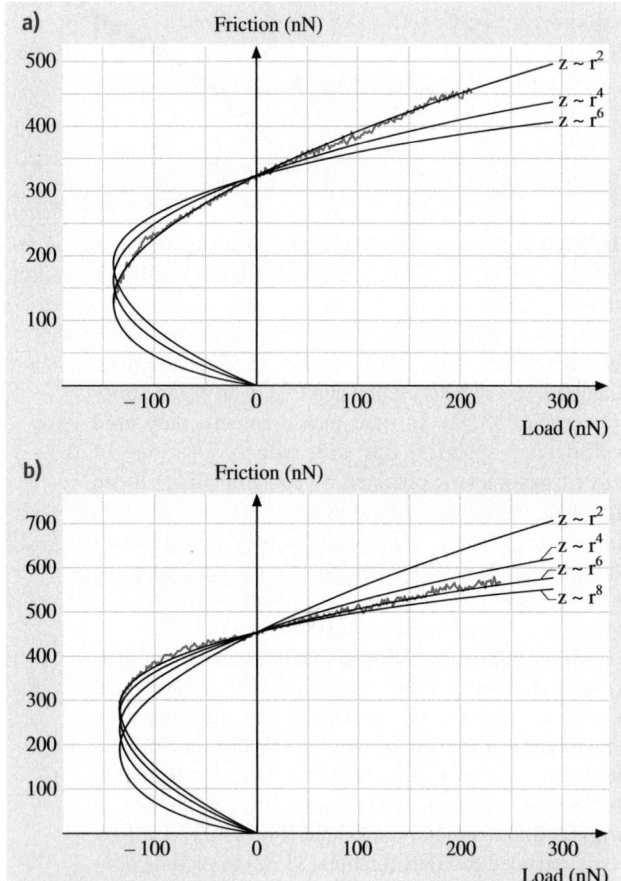

Fig. 33.34a,b Friction vs. load curves **(a)** for a spherical tip and **(b)** for a blunted tip. The *solid curves* are determined using the JKR theory. (After [33.70])

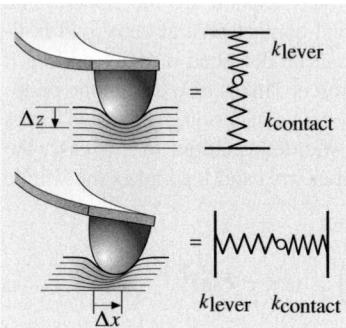

Fig. 33.35 Sketch of normal and lateral stiffness of the contact between tip and surface. (After [33.72])

33.5.3 Estimation of the Contact Area

In contrast to other tribological instruments, such as the surface force apparatus [33.73], the contact area cannot be measured directly by FFM. Indirect methods are provided by contact stiffness measurements. The contact between the FFM tip and the sample can be modeled by a series of two springs (Fig. 33.35). The effective constant k_{eff}^z of the series is given by

$$\frac{1}{k_{eff}^z} = \frac{1}{k_{contact}^z} + \frac{1}{c_N} , \qquad (33.34)$$

where c_N is the normal spring constant of the cantilever and $k_{contact}^z$ is the normal stiffness of the contact. This quantity is related to the radius of the contact area (a) by the simple relation

$$k_{contact}^z = 2aE^* , \qquad (33.35)$$

where E^* is the effective Young's modulus introduced previously [33.74]. Typical values of $k_{contact}^z$ are an order of magnitude larger than c_N, however, and practical application of (33.34) is not possible.

Carpick et al. independently suggested an alternative method [33.72, 75]. According to various models, the *lateral* contact stiffness of the contact between a sphere and a flat surface is [33.76]

$$k_{\text{contact}}^x = 8aG^* , \qquad (33.36)$$

where the effective shear stress G^* is defined by

$$\frac{1}{G^*} = \frac{2 - v_1^2}{G_1} + \frac{2 - v_2^2}{G_2} \qquad (33.37)$$

G_1, G_2 are the shear moduli of the sphere and the plane, respectively. The contact between the FFM tip and the sample can again be modeled by a series of springs (Fig. 33.35). The effective constant k_{eff}^x of the series is given by

$$\frac{1}{k_{\text{eff}}^x} = \frac{1}{k_{\text{contact}}^x} + \frac{1}{k_{\text{tip}}^x} + \frac{1}{c_L} , \qquad (33.38)$$

where c_L is the lateral spring constant of the cantilever and k_{contact}^x is the lateral stiffness of the contact. As

suggested by Lantz, (33.38) also includes the lateral stiffness of the tip, k_{tip}^x, which can be comparable to the lateral spring constant. The effective spring constant k_{eff}^x is simply given by the slope dF_L/dx of the friction loop (Sect. 33.2.1). Once k_{contact}^x is determined, the contact radius a is easily estimated by (33.36).

The lateral stiffness method was applied to contacts between silicon nitride and muscovite mica in air and between NbSe2 and graphite in UHV. The dependences of both the spring constant k_{eff}^x and the lateral force F_L on the load F_N were explained within the same models (JKR and Maugis–Dugdale, respectively), which confirms that friction is proportional to the contact area for the range of loads applied (up to $F_N = 40$ nN in both experiments).

Enachescu et al. estimated the contact area by measuring the contact conductance on diamond as a function of the applied load [33.77, 78]. Their experimental data were fitted with the DMT model, which was also used to explain the dependence of friction on load. Since the contact conductance is proportional to the contact area, the validity of the hypothesis (33.28) was confirmed again.

33.6 Wear on the Atomic Scale

If the normal force F_N applied by the FFM exceeds a critical value, which depends on the tip shape and on the material under investigation, the surface topography is permanently modified. In some cases wear is exploited to create patterns with well-defined shapes. Here we will focus on the mechanisms that act at the nanometer scale, where recent experiments have demonstrated the unique ability of the FFM to both scratch and image surfaces down to the atomic scale.

33.6.1 Abrasive Wear on the Atomic Scale

Lüthi et al. observed the appearance of wear at very low loads, i.e. $F_N = 3$ nN, for ionic crystals [33.32]. Atomically resolved images of the damage produced by scratching the FFM tip area on potassium bromide were obtained very recently by *Gnecco* et al. [33.79]. In Fig. 33.36, a small mound that has piled up at the end of a groove on KBr(100) is shown at different

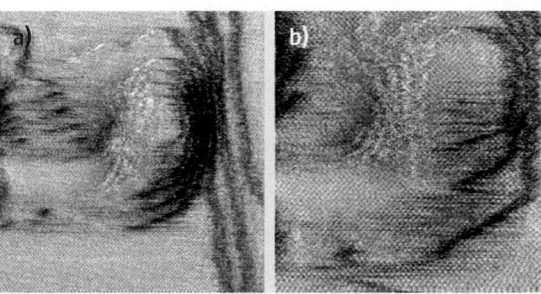

Fig. 33.36a,b Lateral force images acquired at the end of a groove scratched 256 times with a normal force $F_N = 21$ nN. Frame sizes: (a) 39 nm, (b) 25 nm

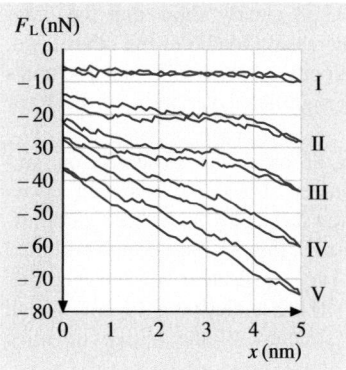

Fig. 33.37 Friction loops acquired while scratching the KBr surface on 5 nm long lines with different loads $F_N = 5.7$ to 22.8 nN. (After [33.79])

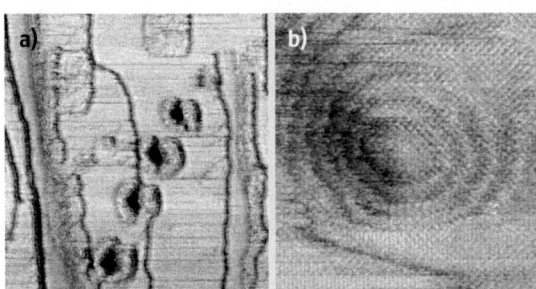

Fig. 33.38 (a) Lateral force images of the pits produced with $F_N = 5.7$ to 22.8 nN. Frame size: 150 nm; **(b)** Detailed image of the fourth pit from the top with pseudo-atomic resolution. Frame size: 20 nm

magnifications. The groove was created a few minutes before imaging by repeatedly scanning with the normal force $F_N = 21$ nN. The image shows a lateral force map acquired with a load of about 1 nN; no atomic features were observed in the corresponding topographic signal. Figure 33.36a,b shows that the debris extracted from the groove recrystallized with the same atomic arrangement of the undamaged surface, which suggests that the wear process occurred in a similar way to epitaxial growth, assisted by the microscope tip.

Although it is not that easy to understand how wear is initiated and how the tip transports the debris, important indications are given by the profile of the lateral force F_L recorded while scratching. Figure 33.37 shows some friction loops acquired when the tip was scanned laterally on areas of size 5×5 nm^2. The mean lateral force multiplied by the scanned length gives the total energy dissipated in the process. The tip movement produces the pits shown in Fig. 33.38a. Thanks to the pseudo-atomic resolution obtained by FFM (Fig. 33.38b), the number of removed atoms can be determined from lateral force images, which allow us to estimate that 70% of the dissipated energy went into wearless friction [33.79]. Figures 33.37 and 33.38 clearly show that the damage increases with increasing load. On the other hand, changing the scan velocity v between 25 and 100 nm/s did not produce any significant variation in the wear process.

A different kind of wear was observed on layered materials. *Kopta* et al. [33.80] removed layers from a muscovite mica surface by scratching with normal force $F_N = 230$ nN (Fig. 33.39a). Fourier-filtered images acquired on very small areas revealed the different periodicities of the underlying layers, which reflect the complex structure of the muscovite mica (Fig. 33.39b,c).

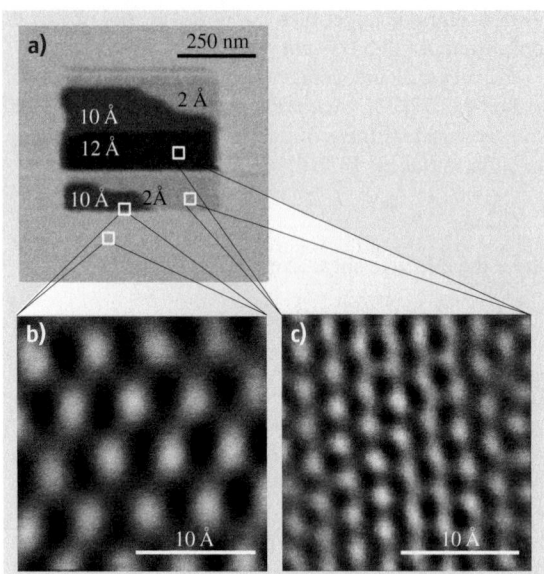

Fig. 33.39 (a) Topography image of an area scratched on muscovite mica with $F_N = 230$ nN; **(b),(c)** Fourier-filtered images of different regions. (After [33.80])

33.6.2 Contribution of Wear to Friction

The mean lateral force detected while scratching a KBr(100) surface with a fixed load $F_N = 11$ nN is shown in Fig. 33.40. A rather continuous increase in "friction" with the number of scratches N is observed, which can be approximated with the following exponential law:

$$F_L = F_0 \, e^{-N/N_0} + F_\infty \left(1 - e^{-N/N_0}\right) . \qquad (33.39)$$

Equation (33.39) is easily interpreted by assuming that friction is proportional to contact area $A(N)$, and

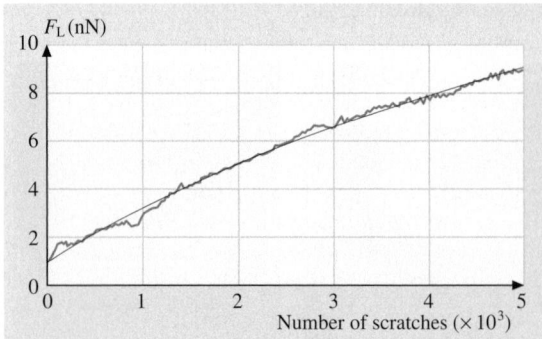

Fig. 33.40 Mean value of the lateral force during repeated scratching with $F_N = 11$ nN on a 500 nm line. (After [33.79])

Fig. 33.41 Friction vs. load curve during the creation of a hole in the muscovite mica. (After [33.80])

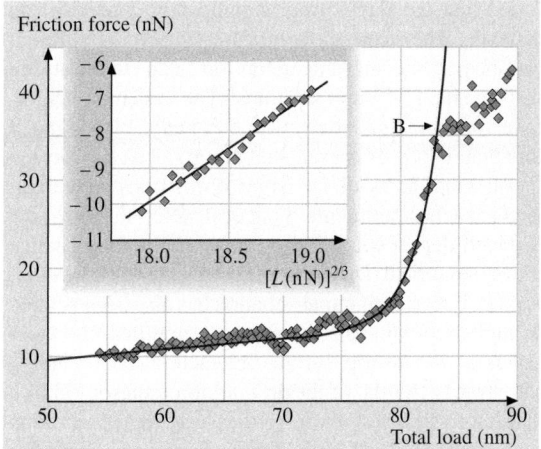

that time evolution of $A(N)$ can be described by

$$\frac{dA}{dN} = \frac{A_\infty - A(N)}{N_0} , \qquad (33.40)$$

Here A_∞ is the limit area in which the applied load can be balanced without scratching.

To interpret their experiment on mica, Kopta et al. assumed that wear is initiated by atomic defects. When the defects accumulate beyond a critical concentration, they grow to form the scars shown in Fig. 33.39. Such a process was once again related to thermal activation. The number of defects created in the contact area $A(F_N)$ is

$$N_{\text{def}}(F_N) = t_{\text{res}} n_0 A(F_N) f_0 \exp\left(-\frac{\Delta E}{k_B T}\right) , \qquad (33.41)$$

where t_{res} is the residence time of the tip, n_0 is the surface density of atoms, and f_0 is the frequency of attempts to overcome the energy barrier ΔE to break a Si$-$O bond, which depends on the applied load. When the defect density reaches a critical value, a hole is nucleated. The friction force during the creation of a hole was also estimated via thermal activation by Kopta et al., who

derived the formula

$$F_L = c(F_N - F_{\text{off}})^{2/3} + \gamma F_N^{2/3} \exp\left(B_0 F_N^{2/3}\right) . \qquad (33.42)$$

The first term on the right gives the wearless dependence of friction in the Hertz-plus-offset model (Sect. 33.5.1); the second term is the contribution of the defect production. The agreement between (33.42) and experiment can be observed in Fig. 33.41.

33.7 Molecular Dynamics Simulations of Atomic Friction and Wear

Section 33.5 mentioned that small sliding contacts can be modeled by continuum mechanics. This modeling has several limitations. The first and most obvious is that continuum mechanics cannot account for atomic-scale processes like atomic stick-slip. While this limit can be overcome by semiclassical descriptions like the Tomlinson model, one definite limit is the determination of contact stiffness for contacts with a radius of a few nanometers. Interpreting experimental results with the methods introduced in Sect. 33.5.3 regularly yields contact radii of atomic or even smaller size, in clear contradiction to the minimal contact size given by adhesion forces. Macroscopic quantities such as shear modulus or pressure fail to describe the mechanical behavior of these contacts. Microscopic modeling that includes the atomic structure of the contact is therefore required. This is usually achieved through a *molecular dynamics* (MD) simulation of the contact. In such simulations, the sliding contact is set up by boundaries of fixed atoms in relative motion and the atoms of the contact, which are

allowed to relax their positions according to interactions between each pair of atoms. Methods of computer simulation used in tribology are discussed elsewhere in this book. In this section we will discuss simulations that can be directly compared to experimental results showing atomic friction processes. The major outcome of the simulations beyond the inclusion of the atomic structure is the importance of including displacement of atoms in order to correctly predict forces. Then we present simulation studies that include wear of the tip or the surface.

33.7.1 Molecular Dynamics Simulations of Friction Processes

The first experiments that exhibited the features of atomic friction were performed on layered materials, often graphite. A theoretical study of forces between an atomically sharp diamond tip and the graphite surface has been reported by *Tang* et al. [33.81]. The authors found that the forces were significantly dependent on distance. The strongest contrasts appeared at different distances for normal and lateral forces due to the strong displacement of surface atoms. The order of magnitude found in this study was one nanonewton, much less than in most experimental reports, which indicated that contact areas of far larger dimensions were realized in such experiments. Tang et al. determined that the distance dependence of the forces could even change the symmetrical appearance of the lateral forces observed. The experimental situation has also been studied in numerical simulations using a simplified one-atom potential for the tip–surface interaction but including the spring potential of the probing force sensor [33.37]. The motivation for these studies was the observation of a hexagonal pattern in the friction force, while the surface atoms of graphite are ordered in a honeycomb structure. The simulations revealed how the jump path of the tip under lateral force is dependent on the force constant of the probing force sensor.

Surfaces of ionic crystals have become model systems for studies in atomic friction. Atomic stick-slip behavior has been observed by several groups with a lateral force modulation of the order of 1 nN. Pioneering work in atomistic simulation of sliding contacts has been done by *Landman* et al. The first ionic system studied was a CaF$_2$ tip sliding over a CaF$_2$(111) surface [33.82]. In MD calculations with controlled temperature, the tip was first moved toward the surface up to the point at which an attractive normal force of -3 nN acted on the tip. Then the tip was moved laterally, and the lateral force determined. An oscillation with a periodicity corresponding to the atomic periodicity of the surface and with an amplitude decreasing from 8 nN was found. Inspection of the atomic positions revealed a wear process from shear cleavage of the tip. This transfer of atoms between tip and surface plays a crucial role in atomic friction studies, as was shown by *Shluger* et al. [33.83]. These authors simulated a MgO tip scanning laterally over a LiF(100) surface. Initially an irregular oscillation of the system's energy is found together with transfer of atoms between surface and tip. After a while, the tip apex structure is changed by adsorption of Li and F ions in such a way that nondestructive sliding with perfectly regular energy oscillations correlating with the periodicity of the surface was observed. The authors called this effect self-lubrication and speculate that, in general, dynamic self-organization of the surface material on the tip might promote the observation of periodic forces. In a less costly molecular mechanics study, in which the forces were calculated for each fixed tip–sample configuration, *Tang* et al. produced lateral and normal force maps for a diamond tip over a NaCl(100) surface, including such defects as vacancies and a step [33.84]. As with the studies mentioned before, they found that significant atomic force contrast can be expected for tip–sample distances of less than 0.35 nm, while distances below 0.15 nm result in destructive forces. For the idealized conditions of scanning at constant height in this regime, the authors predict that even atomic-sized defects could be imaged. Experimentally, these conditions cannot be stabilized in the static modes used so far in lateral force measurements. However, dynamic modes of force microscopy have given atomic resolution of defects within the distance regime of 0.2 and 0.4 nm [33.85]. Recent experimental progress in atomic friction studies of surfaces of ionic crystals include the velocity dependence of lateral forces and atomic-scale wear processes. Such phenomena are not yet accessible by MD studies: the experimental scanning timescale is too far from the atomic relaxation timescales that govern MD simulations. Furthermore, the number of freely transferable atoms that can be included in a simulation is simply limited by meaningful calculation time.

Landman et al. also simulated a system of high reactivity, namely a silicon tip sliding over a silicon surface [33.86]. A clear stick-slip variation in the lateral force was observed for this situation. Strong atom displacements created an interstitial atom under the influence of the tip, which was annealed as the tip moved on. Permanent damage was predicted, however, when the tip enters the repulsive force regime. Although the simulated Si(111) surface is not experimentally accessible, it should be mentioned that the tip had to be passivated by a Teflon layer on the Si(111)7×7 reconstructed surface before nondestructive contact mode measurements became possible (Sect. 33.3). It is worth noting that the simulations for the Cu(111) surface revealed a linear relation between contact area and mean lateral force, similar to classical macroscopic laws.

Wear processes are predicted by several MD studies of metallic sliding over metallic surfaces, which will be discussed in the following section. For a (111)-terminated copper tip sliding over a Cu(111) surface, however, Sørensen et al. found that nondestructive sliding is possible while the lateral force exhibits the sawtooth-like shape characteristic of atomic stick-slip (Fig. 33.42). In contrast, a Cu(100) surface would be disordered by a sliding contact (Fig. 33.43). This difference between the (100) and (111) plane (as well as the absolute lateral forces) has been confirmed experimentally (Sect. 33.3).

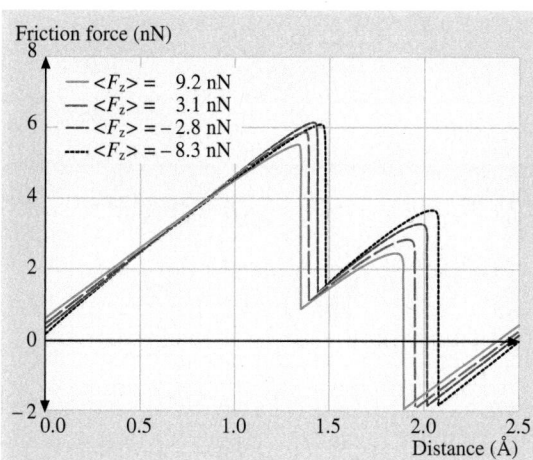

Fig. 33.42 Lateral force acting on a Cu(111) tip in matching contact with a Cu(111) substrate as a function of the sliding distance at different loads. (After [33.87])

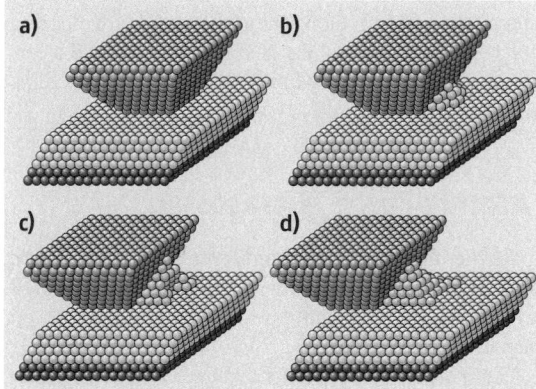

Fig. 33.43a–d Snapshot of a Cu(100) tip on a Cu(100) substrate during sliding. (**a**) Starting configuration; (**b**)–(**d**) snapshots after two, four, and six slips. (After [33.87])

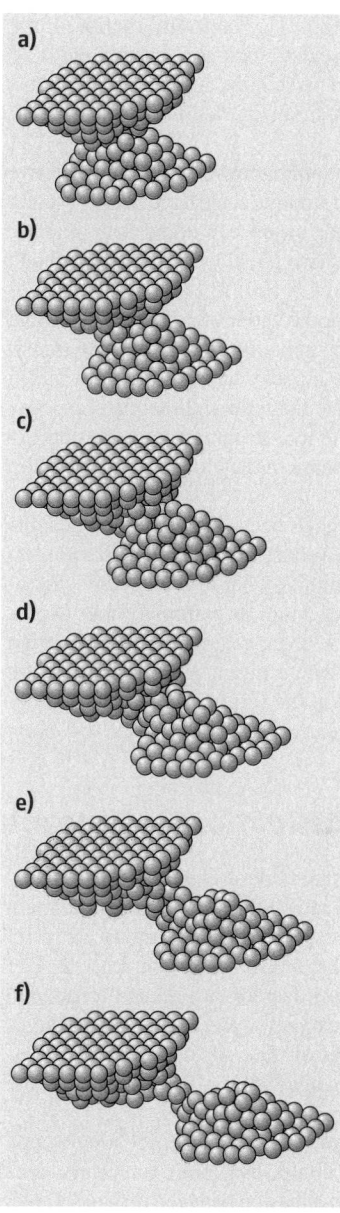

Fig. 33.44a–f
Snapshot of a Cu(100) neck during shearing starting from configuration (**a**). The upper substrate was displaced 4.2 Å between subsequent pictures. (After [33.87])

33.7.2 Molecular Dynamics Simulations of Abrasive Wear

The long timescales characteristic of wear processes and the large amount of material involved make any attempt to simulate these mechanisms on a computer a tremendous challenge. Despite this, MD can provide useful insights on the mechanisms of removal and deposition of single atoms by the FFM tip, which is not the kind of information directly observable experimentally. Complex processes like abrasive wear and nanolithography

can be investigated only within approximate classical mechanics.

The observation made by Livshits and Shluger, that the FFM tip undergoes a process of self-lubrication when scanning ionic surfaces (Sect. 33.7.1), proves that friction and wear are strictly related phenomena. In their MD simulations on copper, Sørensen et al. considered not only ordered (111)- and (100)-terminated tips, but also amorphous structures obtained by "heating" the tip

Part D | 33.7

to high temperatures [33.87]. The lateral motion of the neck thus formed revealed stick-slip behavior due to combined sliding and stretching, as well as ruptures, accompanied by deposition of debris on the surface (Fig. 33.44).

To our knowledge, only a few examples of abrasive wear simulations on the atomic scale have been reported. *Buldum* and *Ciraci*, for instance, studied nanoindentation and sliding of a sharp Ni(111) tip on Cu(110) and a blunt Ni(001) tip on Cu(100) [33.89]. In the case of the sharp tip, quasiperiodic variations of the lateral force were observed, due to stick-slip involving phase transition. One layer of the asperity was deformed to match the substrate during the first slip and then two asperity layers merged into one in a structural transition during the second slip. In the case of the blunt tip, the stick-slip was less regular.

Different results have been reported in which the tip is harder than the underlying sample. *Komanduri* et al. considered an infinitely hard Ni indenter scratching single crystal aluminium at extremely low depths (Fig. 33.45) [33.88]. A linear relation between friction and load was found, with a high coefficient of friction $\mu = 0.6$, independent of the scratch depth. Nanolithography simulations were recently performed by *Fang*

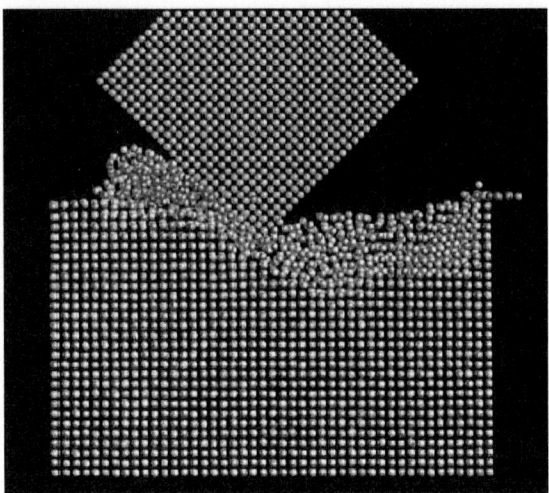

Fig. 33.45 MD simulation of a scratch realized with an infinitely hard tool. (After [33.88])

et al. [33.90], who investigated the role of the displacement of the FFM tip along the direction of slow motion between a scan line and the next one. They found a certain correlation with FFM experiments on silicon films coated with aluminium.

33.8 Energy Dissipation in Noncontact Atomic Force Microscopy

Historically, the measurement of energy dissipation induced by tip–sample interaction has been the domain of friction force microscopy, where the sharp AFM tip slides over a sample that it is in gentle contact with. The origins of dissipation in friction are related to phonon excitation, electronic excitation and irreversible changes of the surface. In a typical stick-slip experiment, the energy dissipated in a single atomic slip event is of the order of 1 eV.

However, the lateral resolution of force microscopy in the contact mode is limited by a minimum contact area of several atoms due to adhesion between tip and sample.

This problem has been overcome in noncontact dynamic force microscopy. In the dynamic mode, the tip oscillates with a constant amplitude A of typically 1–20 nm at the eigenfrequency f of the cantilever, which shifts by Δf due to interaction forces between tip and sample. This technique is described in detail in Part B of this book.

Dissipation also occurs in the noncontact mode of force microscopy, where the atomic structure of tip and sample are reliably preserved. In this dynamic

mode, the damping of the cantilever oscillation can be deduced from the excitation amplitude A_{exc} required to maintain the constant tip oscillation amplitude on resonance.

Compared to friction force microscopy, the interpretation of noncontact AFM (nc-AFM) experiments is complicated due to the perpendicular oscillation of the tip, typically with an amplitude that is large compared to the minimum tip–sample separation. Another problem is to relate the measured damping of the cantilever to the different origins of dissipation.

In all dynamic force microscopy measurements, a power dissipation P_0 caused by internal friction of the freely oscillating cantilever is observed, which is proportional to the eigenfrequency ω_0 and to the square of the amplitude A and is inversely proportional to the known Q value of the cantilever. When the tip–sample distance is reduced, the tip interacts with the sample and therefore additional damping of the oscillation is encountered. This extra dissipation P_{ts} caused by the tip–sample interaction can be calculated from the excitation signal A_{exc} [33.91].

The observed energy losses per oscillation cycle (100 meV) [33.92] are roughly similar to the 1 eV energy loss in the contact slip process. When estimating the contact area in the contact mode for a few atoms, the energy dissipation per atom that can be associated with a bond being broken and reformed is also around 100 meV.

The idea of relating the additional damping of the tip oscillation to dissipative tip–sample interactions has recently attracted much attention [33.93]. The origins of this additional dissipation are manifold: one may distinguish between apparent energy dissipation (for example from inharmonic cantilever motion, artefacts from the phase controller, or slow fluctuations round the steady state solution [33.93, 94]), velocity-dependent dissipation (for example electric and magneticfield-mediated Joule dissipation [33.95, 96]) and hysteresis-related dissipation (due to atomic instabilities [33.97, 98] or hysteresis due to adhesion [33.99]).

Forces and dissipation can be measured by recording Δf and A_{exc} simultaneously during a typical AFM experiment. Many experiments show true atomic contrast in topography (controlled by Δf) and in the dissipation signal A_{exc} [33.100]; however, the origin of the atomic energy dissipation process is still not completely resolved.

To prove that the observed atomic-scale variation in the damping is indeed due to atomic-scale energy dissipation and not an artefact of the distance feedback, *Loppacher* et al. [33.92] carried out a nc-AFM experiment on Si(111)-7×7 at constant height (with distance feedback stopped). Frequency-shift and dissipation exhibit atomic-scale contrast, demonstrating true atomic-scale variations in force and dissipation.

Strong atomic-scale dissipation contrast at step edges has been demonstrated in a few experiments (NaCl on Cu [33.85] or measurements on KBr [33.101]). In Fig. 33.46, ultrathin NaCl islands grown on Cu(111) are shown. As shown in Fig. 33.46a, the island edges have a higher contrast than the NaCl terrace and they show atomically resolved corrugation. The strongly enhanced contrast of the step edges and kink sites could be attributed to a slower decay of the electric field and to easier relaxation of the positions of the ions at these locations. The dissipation image shown in Fig. 33.46b was recorded at the same time. To establish a direct spatial correspondence between the excitation and the topography signal, the match between topography and A_{exc} has been studied on many images. Sometimes the topography and A_{exc} are in phase, sometimes they are shifted a little bit, and sometimes A_{exc} is at a minimum when topography is at a maximum. The local contrast formation thus depends strongly on the atomic tip structure. In fact, the strong dependence of the dissipation contrast on the atomic state of the tip apex is impressively confirmed by the tip change observed in the experimental images shown in Fig. 33.46b. The dissipation contrast is seriously enhanced, while the topography contrast remains almost unchanged. The dissipation clearly depends strongly on the state of the tip and exhibits more short-range character than the frequency shift.

More directly related to friction measurements, where the tip is sliding in contact with the sample, are nc-AFM experiments, where the tip is oscillating parallel to the surface. *Stowe* et al. [33.102] oriented cantilever beams with in-plane tips perpendicular to the surface, so that the tip motion was approximately parallel to the surface. The noncontact damping of the lever was used to measure localized electrical Joule dissipation. They were able to image the dopant density for n- and p-type silicon samples with 150 nm spatial resolution. A dependence of U_{ts}^2 on the tip–sample voltage was found for the dissipation, as proposed by *Denk* and *Pohl* [33.95] for electric field Joule dissipation.

Stipe et al. [33.103] measured the noncontact friction between a Au(111) surface and a gold-coated tip with the same set-up. They observed the same U_{ts}^2 dependence of the bias voltage and a distance dependence that follows the power law $1/d^n$, where n is between 1.3 and 3 [33.103, 104]. A substantial electric-field is present even when the external bias voltage is zero. The presence of inhomogeneous tip–sample electric fields is

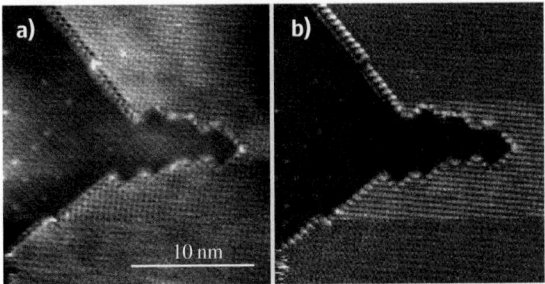

Fig. 33.46 (a) Topography and **(b)** A_{exc} images of a NaCl island on Cu(111). The tip changes after 1/4 of the scan, thereby changing the contrast in the topography and enhancing the contrast in A_{exc}. After 2/3 of the scan, the contrast from the *lower part* of the image is reproduced, indicating that the tip change was reversible. (After [33.85])

difficult to avoid, even under the best experimental conditions. Although this dissipation is electrical in origin, the detailed mechanism is not totally clear. The most straightforward mechanism is to assume that inhomogeneous fields emanating from the tip and the sample induce surface charges in the nearby metallic sample. When the tip moves, currents are induced, causing ohmic dissipation [33.95, 102]. But in metals with good electrical conductivity, ohmic dissipation is insufficient to account for the observed effect [33.105]. Thus the tip–sample electric field must have an additional effect, such as driving the motions of adsorbates and surface defects.

When exciting the torsional oscillation of commercial, rectangular AFM cantilevers, the tip is oscillating approximately parallel to the surface. In this mode, it was possible to measure lateral forces acting on the tip at step edges and near impurities quantitatively [33.58]. Enhanced energy dissipation was also observed at the impurities. When the tip is moved further toward the sample, contact formation transforms the nearly free torsional oscillation of the cantilever into a different mode, with the tip–sample contact acting as a hinge. When this contact is formed, a rapid increase in the power required to maintain a constant tip oscillation amplitude and a positive frequency shift are found. The onsets of the simultaneously recorded damping and

positive frequency shift are sharp and essentially coincide. It is assumed that these changes indicate the formation of a tip–sample contact. Two recent studies [33.106, 107] report on the use of the torsional eigenmode to measure the elastic properties of the tip–sample contact, where the tip is in contact with the sample and the shear stiffness depends on the normal load.

Kawagishi et al. [33.108] scanned with lateral amplitudes of the order of 10 pm to 3 nm; their imaging technique showed up contrast between graphite terraces, silicon and silicon dioxide, graphite and mica. Torsional self-excitation showed nanometric features of self-assembled monolayer islands due to different lateral dissipations.

Giessibl et al. [33.109] recently established true atomic resolution of both conservative and dissipative forces by lateral force microscopy. The interaction between a single tip atom oscillated parallel to an Si(111)-7×7 surface was measured. A dissipation energy of up to 4 eV per oscillation cycle was found, which is explained by the plucking action of one atom onto the other, as described by *Tomlinson* in 1929 [33.23].

A detailed review of dissipation phenomena in noncontact force microscopy has been given by *Hug* [33.110].

33.9 Conclusion

Over the last 15 years, two instrumental developments have stimulated scientific activities in the field of nanotribology. On the one hand, the invention and development of friction force microscopy has allowed us to quantitatively study single-asperity friction. As we have discussed in this chapter, atomic processes are observed using forces of around 1 nN (forces related to single chemical bonds). On the other hand, the enormous increase in achievable computing power has provided the basis for molecular dynamics simulations of systems containing several hundreds of atoms. These methods allow the development of the atomic structure in a sliding contact to be analyzed and the forces to be predicted.

The most prominent observation of atomic friction is stick-slip behavior with the periodicity of the surface atomic lattice. Semiclassical models can explain experimental findings, including the velocity dependence, which is a consequence of the thermal activation of slip events. Classical continuum mechanics can also describe the load dependence of friction in contacts with

an extension of several tens of nanometers. However, when we try to apply continuum mechanics to contacts formed at just ten atoms, obviously wrong numbers result (for the contact radius for instance). Only comparison with atomistic simulations can provide a full, meaningful picture of the physical parameters of such sliding contacts. These simulations predict a close connection between wear and friction, in particular the transfer of atoms between surface and tip, which in some cases can even lower the friction in a process of self-lubrication.

First experiments have succeeded in studying the onset of wear with atomic resolution. Research into microscopic wear processes will certainly grow in importance as nanostructures are produced and their mechanical properties exploited. Simulations of such processes involving the transfer of thousands of atoms will become feasible with further increases in computing power. Another aspect of nanotribology is the expansion of atomic friction experiments toward surfaces with well-defined roughnesses. In general, the problem of bridging the gap between single-asperity experiments on well-defined surfaces and macroscopic friction should be approached, both experimentally and via modeling.

References

33.1 C. M. Mate, G. M. McClelland, R. Erlandsson, S. Chiang: Atomic-scale friction of a tungsten tip on a graphite surface, Phys. Rev. Lett. **59**, 1942–1945 (1987)

33.2 G. Binnig, C. F. Quate, Ch. Gerber: Atomic force microscope, Phys. Rev. Lett. **56**, 930–933 (1986)

33.3 O. Marti, J. Colchero, J. Mlynek: Combined scanning force and friction microscopy of mica, Nanotechnology **1**, 141–144 (1990)

33.4 G. Meyer, N. Amer: Simultaneous measurement of lateral and normal forces with an optical-beam-deflection atomic force microscope, Appl. Phys. Lett. **57**, 2089–2091 (1990)

33.5 G. Neubauer, S. R. Cohen, G. M. McClelland, D. E. Horn, C. M. Mate: Force microscopy with a bidirectional capacitance sensor, Rev. Sci. Instrum. **61**, 2296–2308 (1990)

33.6 G. M. McClelland, J. N. Glosli: Friction at the atomic scale. In: *NATO ASI Series E*, Vol. 220, ed. by L. Singer, H. M. Pollock (Kluwer, Dordrecht 1992) pp. 405–425

33.7 R. Linnemann, T. Gotszalk, I. W. Rangelow, P. Dumania, E. Oesterschulze: Atomic force microscopy and lateral force microscopy using piezoresistive cantilevers, J. Vacuum Sci. Technol. B **14**, 856–860 (1996)

33.8 M. Nonnenmacher, J. Greschner, O. Wolter, R. Kassing: Scanning force microscopy with micromachined silicon sensors, J. Vacuum Sci. Technol. B **9**, 1358–1362 (1991)

33.9 R. Lüthi: Untersuchungen zur Nanotribologie und zur Auflösungsgrenze im Ultrahochvakuum mittels Rasterkraftmikroskopie. Ph.D. Thesis (Univ. of Basel, Basel 1996)

33.10 J. Cleveland, S. Manne, D. Bocek, P. K. Hansma: A nondestructive method for determining the spring constant of cantilevers for scanning force microscopy, Rev. Sci. Instrum. **64**, 403–405 (1993)

33.11 J. L. Hutter, J. Bechhoefer: Calibration of atomic-force microscope tips, Rev. Sci. Instrum. **64**, 1868–1873 (1993)

33.12 H. J. Butt, M. Jaschke: Calculation of thermal noise in atomic-force microscopy, Nanotechnology **6**, 1–7 (1995)

33.13 J. M. Neumeister, W. A. Ducker: Lateral, normal, and longitudinal spring constants of atomic-force microscopy cantilevers, Rev. Sci. Instrum. **65**, 2527–2531 (1994)

33.14 D. F. Ogletree, R. W. Carpick, M. Salmeron: Calibration of frictional forces in atomic force microscopy, Rev. Sci. Instrum. **67**, 3298–3306 (1996)

33.15 E. Gnecco: AFM study of friction phenomena on the nanometer scale. Ph.D. Thesis (Univ. of Genova, Genova 2001)

33.16 U. D. Schwarz, P. Köster, R. Wiesendanger: Quantitative analysis of lateral force microscopy experiments, Rev. Sci. Instrum. **67**, 2560–2567 (1996)

33.17 E. Meyer, R. Lüthi, L. Howald, M. Bammerlin, M. Guggisberg, H.-J. Güntherodt: Site-specific friction force spectroscopy, J. Vacuum Sci. Technol. B **14**, 1285–1288 (1996)

33.18 S. S. Sheiko, M. Möller, E. M. C. M. Reuvekamp, H. W. Zandberger: Calibration and evaluation of scanning-force microscopy probes, Phys. Rev. B **48**, 5675 (1993)

33.19 F. Atamny, A. Baiker: Direct imaging of the tip shape by AFM, Surf. Sci. **323**, L314 (1995)

33.20 J. S. Villarrubia: Algorithms for scanned probe microscope image simulation, surface reconstruction, and tip estimation, J. Res. Natl. Inst. Stand. Technol. **102**, 425–454 (1997)

33.21 L. Howald, E. Meyer, R. Lüthi, H. Haefke, R. Overney, H. Rudin, H.-J. Güntherodt: Multifunctional probe microscope for facile operation in ultrahigh vacuum, Appl. Phys. Lett. **63**, 117–119 (1993)

33.22 Q. Dai, R. Vollmer, R. W. Carpick, D. F. Ogletree, M. Salmeron: A variable temperature ultrahigh vacuum atomic force microscope, Rev. Sci. Instrum. **66**, 5266–5271 (1995)

33.23 G. A. Tomlinson: A molecular theory of friction, Philos. Mag. Ser. **7**, 905 (1929)

33.24 T. Gyalog, M. Bammerlin, R. Lüthi, E. Meyer, H. Thomas: Mechanism of atomic friction, Europhys. Lett. **31**, 269–274 (1995)

33.25 T. Gyalog, H. Thomas: Friction between atomically flat surfaces, Europhys. Lett. **37**, 195–200 (1997)

33.26 M. Weiss, F. J. Elmer: Dry friction in the Frenkel–Kontorova–Tomlinson model: Static properties, Phys. Rev. B **53**, 7539–7549 (1996)

33.27 J. B. Pethica: Comment on "Interatomic forces in scanning tunneling microscopy: Giant corrugations of the graphite surface", Phys. Rev. Lett. **57**, 3235 (1986)

33.28 E. Meyer, R. M. Overney, K. Dransfeld, T. Gyalog: *Nanoscience, Friction and Rheology on the Nanometer Scale* (World Scientific, Singapore 1998)

33.29   A. Socoliuc, R. Bennewitz, E. Gnecco, E. Meyer: Transition from stick-slip to continuous sliding in atomic friction: Entering a new regime of ultralow friction, Phys. Rev. Lett. **92**, 134301 (2004)

33.30   L. Howald, R. Lüthi, E. Meyer, H.-J. Güntherodt: Atomic-force microscopy on the Si(111)7×7 surface, Phys. Rev. B **51**, 5484–5487 (1995)

33.31   R. Bennewitz, T. Gyalog, M. Guggisberg, M. Bammerlin, E. Meyer, H.-J. Güntherodt: Atomic-scale stick-slip processes on Cu(111), Phys. Rev. B **60**, R11301–R11304 (1999)

33.32   R. Lüthi, E. Meyer, M. Bammerlin, L. Howald, H. Haefke, T. Lehmann, C. Loppacher, H.-J. Güntherodt, T. Gyalog, H. Thomas: Friction on the atomic scale: An ultrahigh vacuum atomic force microscopy study on ionic crystals, J. Vacuum Sci. Technol. B **14**, 1280–1284 (1996)

33.33   G. J. Germann, S. R. Cohen, G. Neubauer, G. M. McClelland, H. Seki: Atomic-scale friction of a diamond tip on diamond (100) and (111) surfaces, J. Appl. Phys. **73**, 163–167 (1993)

33.34   R. J. A. van den Oetelaar, C. F. J. Flipse: Atomic-scale friction on diamond(111) studied by ultrahigh vacuum atomic force microscopy, Surf. Sci. **384**, L828–L835 (1997)

33.35   R. Bennewitz, E. Gnecco, T. Gyalog, E. Meyer: Atomic friction studies on well-defined surfaces, Tribol. Lett. **10**, 51–56 (2001)

33.36   S. Fujisawa, E. Kishi, Y. Sugawara, S. Morita: Atomic-scale friction observed with a two-dimensional frictional-force microscope, Phys. Rev. B **51**, 7849–7857 (1995)

33.37   N. Sasaki, M. Kobayashi, M. Tsukada: Atomic-scale friction image of graphite in atomic-force microscopy, Phys. Rev. B **54**, 2138–2149 (1996)

33.38   H. Kawakatsu, T. Saito: Scanning force microscopy with two optical levers for detection of deformations of the cantilever, J. Vacuum Sci. Technol. B **14**, 872–876 (1996)

33.39   M. Hirano, K. Shinjo, R. Kaneko, Y. Murata: Anisotropy of frictional forces in muscovite mica, Phys. Rev. Lett. **67**, 2642–2645 (1991)

33.40   M. Hirano, K. Shinjo, R. Kaneko, Y. Murata: Observation of superlubricity by scanning tunneling microscopy, Phys. Rev. Lett. **78**, 1448–1451 (1997)

33.41   M. Liley, D. Gourdon, D. Stamou, U. Meseth, T. M. Fischer, C. Lautz, H. Stahlberg, H. Vogel, N. A. Burnham, C. Duschl: Friction anisotropy and asymmetry of a compliant monolayer induced by a small molecular tilt, Science **280**, 273–275 (1998)

33.42   R. Lüthi, E. Meyer, H. Haefke, L. Howald, W. Gutmannsbauer, H.-J. Güntherodt: Sled-type motion on the nanometer scale: Determination of dissipation and cohesive energies of C_{60}, Science **266**, 1979–1981 (1994)

33.43   M. R. Falvo, J. Steele, R. M. Taylor, R. Superfine: Evidence of commensurate contact and rolling motion: AFM manipulation studies of carbon nanotubes on HOPG, Tribol. Lett. **9**, 73–76 (2000)

33.44   R. M. Overney, H. Takano, M. Fujihira, W. Paulus, H. Ringsdorf: Anisotropy in friction and molecular stick-slip motion, Phys. Rev. Lett. **72**, 3546–3549 (1994)

33.45   H. Takano, M. Fujihira: Study of molecular scale friction on stearic acid crystals by friction force microscopy, J. Vacuum Sci. Technol. B **14**, 1272–1275 (1996)

33.46   M. Dienwiebel, G. Verhoeven, N. Pradeep, J. Frenken, J. Heimberg, H. Zandbergen: Superlubricity of graphite, Phys. Rev. Lett. **92**, 126101 (2004)

33.47   P. E. Sheehan, C. M. Lieber: Nanotribology and nanofabrication of MoO_3 structures by atomic force microscopy, Science **272**, 1158–1161 (1996)

33.48   E. Gnecco, R. Bennewitz, T. Gyalog, Ch. Loppacher, M. Bammerlin, E. Meyer, H.-J. Güntherodt: Velocity dependence of atomic friction, Phys. Rev. Lett. **84**, 1172–1175 (2000)

33.49   D. Gourdon, N. A. Burnham, A. Kulik, E. Dupas, F. Oulevey, G. Gremaud, D. Stamou, M. Liley, Z. Dienes, H. Vogel, C. Duschl: The dependence of friction anisotropies on the molecular organization of LB films as observed by AFM, Tribol. Lett. **3**, 317–324 (1997)

33.50   Y. Sang, M. Dubé, M. Grant: Thermal effects on atomic friction, Phys. Rev. Lett. **87**, 174301 (2001)

33.51   E. Riedo, E. Gnecco, R. Bennewitz, E. Meyer, H. Brune: Interaction potential and hopping dynamics governing sliding friction, Phys. Rev. Lett. **91**, 084502 (2003)

33.52   P. Reimann, M. Evstigneev: Nonmonotonic velocity dependence of atomic friction, Phys. Rev. Lett. **93**, 230802 (2004)

33.53   C. Fusco, A. Fasolino: Velocity dependence of atomic-scale friction: A comparative study of the one- and two-dimensional Tomlinson model, Phys. Rev. B **71**, 45413 (2005)

33.54   O. Zwörner, H. Hölscher, U. D. Schwarz, R. Wiesendanger: The velocity dependence of frictional forces in point-contact friction, Appl. Phys. A **66**, 263–267 (1998)

33.55   T. Bouhacina, J. P. Aimé, S. Gauthier, D. Michel, V. Heroguez: Tribological behavior of a polymer grafted on silanized silica probed with a nanotip, Phys. Rev. B **56**, 7694–7703 (1997)

33.56   H. J. Eyring: The activated complex in chemical reactions, J. Chem. Phys. **3**, 107 (1937)

33.57   J. N. Glosli, G. M. McClelland: Molecular dynamics study of sliding friction of ordered organic monolayers, Phys. Rev. Lett. **70**, 1960–1963 (1993)

33.58   O. Pfeiffer, R. Bennewitz, A. Baratoff, E. Meyer, P. Grütter: Lateral-force measurements in dynamic force microscopy, Phys. Rev. B **65**, 161403 (2002)

33.59 E. Riedo, F. Lévy, H. Brune: Kinetics of capillary condensation in nanoscopic sliding friction, Phys. Rev. Lett. **88**, 185505 (2002)

33.60 M. He, A. S. Blum, G. Overney, R. M. Overney: Effect of interfacial liquid structuring on the coherence length in nanolubrucation, Phys. Rev. Lett. **88**, 154302 (2002)

33.61 F. P. Bowden, F. P. Tabor: *The Friction and Lubrication of Solids* (Oxford Univ. Press, Oxford 1950)

33.62 L. D. Landau, E. M. Lifshitz: *Introduction to Theoretical Physics* (Nauka, Moscow 1998) Vol. 7

33.63 J. A. Greenwood, J. B. P. Williamson: Contact of nominally flat surfaces, Proc. R. Soc. Lond. A **295**, 300 (1966)

33.64 B. N. J. Persson: Elastoplastic contact between randomly rough surfaces, Phys. Rev. Lett. **87**, 116101 (2001)

33.65 U. D. Schwarz, O. Zwörner, P. Köster, R. Wiesendanger: Quantitative analysis of the frictional properties of solid materials at low loads, Phys. Rev. B **56**, 6987–6996 (1997)

33.66 K. L. Johnson, K. Kendall, A. D. Roberts: Surface energy and contact of elastic solids, Proc. R. Soc. Lond. A **324**, 301 (1971)

33.67 B. V. Derjaguin, V. M. Muller, Y. P. Toporov: Effect of contact deformations on adhesion of particles, J. Colloid Interf. Sci. **53**, 314–326 (1975)

33.68 D. Tabor: Surface forces and surface interactions, J. Colloid Interf. Sci. **58**, 2–13 (1977)

33.69 D. Maugis: Adhesion of spheres: the JKR–DMT transition using a Dugdale model, J. Colloid Interf. Sci. **150**, 243–269 (1992)

33.70 R. W. Carpick, N. Agraït, D. F. Ogletree, M. Salmeron: Measurement of interfacial shear (friction) with an ultrahigh vacuum atomic force microscope, J. Vacuum Sci. Technol. B **14**, 1289–1295 (1996)

33.71 C. Polaczyk, T. Schneider, J. Schöfer, E. Santner: Microtribological behavior of Au(001) studied by AFM/FFM, Surf. Sci. **402**, 454–458 (1998)

33.72 R. W. Carpick, D. F. Ogletree, M. Salmeron: Lateral stiffness: A new nanomechanical measurement for the determination of shear strengths with friction force microscopy, Appl. Phys. Lett. **70**, 1548–1550 (1997)

33.73 J. N. Israelachvili, D. Tabor: Measurement of van der Waals dispersion forces in range 1.5 to 130 nm, Proc. R. Soc. Lond. A **331**, 19 (1972)

33.74 S. P. Jarvis, A. Oral, T. P. Weihs, J. B. Pethica: A novel force microscope and point-contact probe, Rev. Sci. Instrum. **64**, 3515–3520 (1993)

33.75 M. A. Lantz, S. J. O'Shea, M. E. Welland, K. L. Johnson: Atomic-force-microscope study of contact area and friction on NbSe$_2$, Phys. Rev. B **55**, 10776–10785 (1997)

33.76 K. L. Johnson: *Contact Mechanics* (Cambridge Univ. Press, Cambridge 1985)

33.77 M. Enachescu, R. J. A. van den Oetelaar, R. W. Carpick, D. F. Ogletree, C. F. J. Flipse, M. Salmeron: Atomic force microscopy study of an ideally hard contact: the diamond(111)/tungsten carbide interface, Phys. Rev. Lett. **81**, 1877–1880 (1998)

33.78 M. Enachescu, R. J. A. van den Oetelaar, R. W. Carpick, D. F. Ogletree, C. F. J. Flipse, M. Salmeron: Observation of proportionality between friction and contact area at the nanometer scale, Tribol. Lett. **7**, 73–78 (1999)

33.79 E. Gnecco, R. Bennewitz, E. Meyer: Abrasive wear on the atomic scale, Phys. Rev. Lett. **88**, 215501 (2002)

33.80 S. Kopta, M. Salmeron: The atomic scale origin of wear on mica and its contribution to friction, J. Chem. Phys. **113**, 8249–8252 (2000)

33.81 H. Tang, C. Joachim, J. Devillers: Interpretation of AFM images – the graphite surface with a diamond tip, Surf. Sci. **291**, 439–450 (1993)

33.82 U. Landman, W. D. Luedtke, E. M. Ringer: Atomistic mechanisms of adhesive contact formation and interfacial processes, Wear **153**, 3–30 (1992)

33.83 A. I. Livshits, A. L. Shluger: Self-lubrication in scanning force microscope image formation on ionic surfaces, Phys. Rev. B **56**, 12482–12489 (1997)

33.84 H. Tang, X. Bouju, C. Joachim, C. Girard, J. Devillers: Theoretical study of the atomic-force microscopy imaging process on the NaCl(100) surface, J. Chem. Phys. **108**, 359–367 (1998)

33.85 R. Bennewitz, A. S. Foster, L. N. Kantorovich, M. Bammerlin, Ch. Loppacher, S. Schär, M. Guggisberg, E. Meyer, A. L. Shluger: Atomically resolved edges and kinks of NaCl islands on Cu(111): Experiment and theory, Phys. Rev. B **62**, 2074–2084 (2000)

33.86 U. Landman, W. D. Luetke, M. W. Ribarsky: Structural and dynamical consequences of interactions in interfacial systems, J. Vacuum Sci. Technol. A **7**, 2829–2839 (1989)

33.87 M. R. Sørensen, K. W. Jacobsen, P. Stoltze: Simulations of atomic-scale sliding friction, Phys. Rev. B **53**, 2101–2113 (1996)

33.88 R. Komanduri, N. Chandrasekaran, L. M. Raff: Molecular dynamics simulation of atomic-scale friction, Phys. Rev. B **61**, 14007–14019 (2000)

33.89 A. Buldum, C. Ciraci: Contact, nanoindentation and sliding friction, Phys. Rev. B **57**, 2468–2476 (1998)

33.90 T. H. Fang, C. I. Weng, J. G. Chang: Molecular dynamics simulation of a nanolithography process using atomic force microscopy, Surf. Sci. **501**, 138–147 (2002)

33.91 B. Gotsmann, C. Seidel, B. Anczykowski, H. Fuchs: Conservative and dissipative tip–sample interaction forces probed with dynamic AFM, Phys. Rev. B **60**, 11051–11061 (1999)

33.92 C. Loppacher, R. Bennewitz, O. Pfeiffer, M. Guggisberg, M. Bammerlin, S. Schär, V. Barwich,

Part D | 33

A. Baratoff, E. Meyer: Experimental aspects of dissipation force microscopy, Phys. Rev. B **62**, 13674–13679 (2000)

33.93   M. Gauthier, M. Tsukada: Theory of noncontact dissipation force microscopy, Phys. Rev. B **60**, 11716–11722 (1999)

33.94   J. P. Aimé, R. Boisgard, L. Nony, G. Couturier: Nonlinear dynamic behavior of an oscillating tip-microlever system and contrast at the atomic scale, Phys. Rev. Lett. **82**, 3388–3391 (1999)

33.95   W. Denk, D. W. Pohl: Local electrical dissipation imaged by scanning force microscopy, Appl. Phys. Lett. **59**, 2171–2173 (1991)

33.96   S. Hirsekorn, U. Rabe, A. Boub, W. Arnold: On the contrast in eddy current microscopy using atomic force microscopes, Surf. Interf. Anal. **27**, 474–481 (1999)

33.97   U. Dürig: Atomic-Scale Metal Adhesion. In: *Forces in Scanning Probe Methods*, NATO ASI, Ser. E, Vol. 286, ed. by H. J. Güntherodt, D. Anselmetti, E. Meyer (Kluwer, Dordrecht 1995) pp. 191–234

33.98   N. Sasaki, M. Tsukada: Effect of microscopic non-conservative process on noncontact atomic force microscopy, Jpn. J. Appl. Phys. **39**, L1334–L1337 (2000)

33.99   B. Gotsmann, H. Fuchs: The measurement of hysteretic forces by dynamic AFM, Appl. Phys. A **72**, 55–58 (2001)

33.100   M. Guggisberg, M. Bammerlin, A. Baratoff, R. Lüthi, C. Loppacher, F. M. Battiston, J. Lü, R. Bennewitz, E. Meyer, H. J. Güntherodt: Dynamic force microscopy across steps on the Si(111)-(7 × 7) surface, Surf. Sci. **461**, 255–265 (2000)

33.101   R. Bennewitz, S. Schär, V. Barwich, O. Pfeiffer, E. Meyer, F. Krok, B. Such, J. Kolodzej, M. Szymon-ski: Atomic-resolution images of radiation damage in KBr, Surf. Sci. **474**, 197–202 (2001)

33.102   T. D. Stowe, T. W. Kenny, J. Thomson, D. Rugar: Silicon dopant imaging by dissipation force microscopy, Appl. Phys. Lett. **75**, 2785–2787 (1999)

33.103   B. C. Stipe, H. J. Mamin, T. D. Stowe, T. W. Kenny, D. Rugar: Noncontact friction and force fluctuations between closely spaced bodies, Phys. Rev. Lett. **87**, 96801 (2001)

33.104   B. Gotsmann, H. Fuchs: Dynamic force spectroscopy of conservative and dissipative forces in an Al-Au(111) tip–sample system, Phys. Rev. Lett. **86**, 2597–2600 (2001)

33.105   B. N. J. Persson, A. I. Volokitin: Comment on "Brownian motion of microscopic solids under the action of fluctuating electromagnetic fields", Phys. Rev. Lett. **84**, 3504 (2000)

33.106   K. Yamanaka, A. Noguchi, T. Tsuji, T. Koike, T. Goto: Quantitative material characterization by ultrasonic AFM, Surf. Interf. Anal. **27**, 600–606 (1999)

33.107   T. Drobek, R. W. Stark, W. M. Heckl: Determination of shear stiffness based on thermal noise analysis in atomic force microscopy: Passive overtone microscopy, Phys. Rev. B **64**, 045401 (2001)

33.108   T. Kawagishi, A. Kato, Y. Hoshi, H. Kawakatsu: Mapping of lateral vibration of the tip in atomic force microscopy at the torsional resonance of the cantilever, Ultramicroscopy **91**, 37–48 (2002)

33.109   F. J. Giessibl, M. Herz, J. Mannhart: Friction traced to the single atom, Proc. Natl. Acad. Sci. USA **99**, 12006–12010 (2002)

33.110   H.-J. Hug, A. Baratoff: Measurement of dissipation induced by tip–sample interactions. In: *Noncontact Atomic Force Microscopy*, ed. by S. Morita, R. Wiesendanger, E. Meyer (Springer, Berlin, Heidelberg 2002) p. 395

34. Velocity Dependence of Nanoscale Friction, Adhesion and Wear

The advent of micro/nanostructures and the subsequent miniaturization of moving components for various nanotechnology applications, such as micro/nanoelectromechanical systems (MEMS/NEMS), have ascribed paramount importance to tribology and mechanics on the nanoscale. Most of these micro/nanodevices and components operate at very high sliding velocities (of the order of tens of mm/s to few m/s). Atomic force microscopy (AFM) studies into potential materials, coatings and lubricants for these devices have been rendered inadequate due to the inherent limitations on the highest sliding velocities achievable with commercial AFMs ($< 250\,\mu$m/s). The development of a new AFM-based technique has allowed nanotribological investigations to be performed over a wide range of velocities (up to 10 mm/s). Research conducted on various materials, coatings and lubricants reveals a strong velocity dependence of friction, adhesion and wear on the nanoscale. Based on the experimental evidence, theoretical formulations have been realized for nanoscale friction behavior in order to design a comprehensive analytical model that explains the velocity dependence. The model takes into consideration the contributions of adhesion at the tip–sample interface, high impact velocity related deformations at the contacting asperities, and atomic-scale stick-slip. Dominant friction mechanisms are identified and critical operating parameters for their transitions are defined. Wear studies are conducted at high sliding velocities to elucidate the primary failure mechanisms. A novel AFM-based nanowear mapping technique to map wear on the nanoscale is developed, and the interdependence of normal load and sliding velocity on sample surface wear is studied. This technique helps identify and classify wear mechanisms and determine the critical parameters responsible for their transitions. The interdependence of mechanical and tribological properties for various

materials is explored and tribologically ideal materials with low adhesion and friction, suitable for nanotechnology applications, are identified.

34.1 Bridging Science and Engineering for Nanotribological Investigations

Nanotechnology, defined literally as any technology performed on the nanoscale that has applications in the real world [34.5], has spurred the development of innovative micro/nanosystems with the discovery of novel materials, processes and phenomena on the micro/nanoscale, and has led to the rapid advancement of micro/nanoelectromechanical systems (MEMS/NEMS). Many applications of MEMS/NEMS already available on the commercial market – such as the 'lab-on-a-chip' sensors used for drug delivery, accelerometers used for automobile airbag deployment and digital micromirror devices (DMDs) used in high-definition TVs and video projectors in homes and theaters – have been described in greater detail in previous chapters in this handbook. These still form only the tip of the iceberg when considering the plethora of potential applications of nanotechnology. In fact, MEMS/NEMS are now believed to be the next logical step in the "silicon revolution". Visionaries and leading scientists and researchers, presenting at the National Nanotechnology Initiatives (NNI) workshop on Nanotechnology in Space Exploration held in Palo Alto, CA (USA) in Aug 2004, have slated the emerging field of

nanotechnology to be the next disruptive technology that will have major impact in the next one to three decades.

Despite the increasing popularity and technological advances of MEMS/NEMS applications, severe tribological (friction, adhesion and wear) problems tend to undermine their performance and reliability. Several studies have shown that tribology and mechanics are the limiting factors to the imminent broad-based impact of nanotechnology on our everyday lives [34.6–10]. Miniaturization and the subsequent development of MEMS/NEMS require better tribological performance of the system components and a fundamental understanding of basic phenomena underlying friction, wear and lubrication on the micro/nanoscale [34.11–15]. The components used in micro/nanostructures are very light (on the order of a few micrograms) and operate under very light loads (on the order of few micrograms to a few milligrams). As we move from macro- to nanoscale the surface-to-volume ratio increases considerably and becomes a cause of serious concern from a tribological point of view. On the microscale, surface forces (such as friction, adhesion, meniscus forces, viscous drag and surface tension) that are proportional to area significantly increase and limit the life and reliability of MEMS/NEMS. As a result, friction, adhesion and wear of lightly loaded micro/nanostructures are highly dependent on surface interactions.

Further, many MEMS/NEMS devices are designed to operate at very high sliding velocities (Fig. 34.1). These devices often have components that operate at rotational speeds ranging well above 100 000 rpm, as in the case of micromotors [34.10], to well above 1 million rpm, as in some applications such as microgas turbines [34.16]. Given the sizes of these devices (the micromotor has a rotor diameter of 120 μm), the relative sliding velocities that they operate at can reach several m/s. Continuous sliding at such high relative velocities eventually leads to high wear and subsequent device failure. Wear is a known issue in the failure of micro/nanoscale devices and components [34.17, 18] and has been shown to result from high frictional energy dissipation due to the impacts of contacting asperities at high sliding velocities [34.19–28]. Wear also arises from sustained operation over a period of time as a result of fatigue. Commercial applications such as the DMD are designed for a lifetime operation of over 100 000 h without loss of image quality. Such stringent operating conditions mean that the oscillating micromirrors in

Electrostatic micromotor [83]

Microturbine bladed rotor and nozzle guide vanes on the stator [84]

Six gear chain (www.sandia.gov)

Ni-Fe Wolfrom-type gear system by LIGA [88]

Fig. 34.1a–d Examples of applications of nanotechnology that operate at high sliding velocities: (**a**) electrostatic micromotor [34.1], (**b**) microturbine [34.2], (**c**) microgear drive (from [34.3]) and (**d**) planetary gear system [34.4]

Table 34.1 Brief description of sample preparation

Material	Details
Si (100)	Wafer World Inc. (West Palm Beach, Florida, USA)
DLC [a]	Filtered cathodic arc deposition on Si (100) substrate, 10 nm thick film
Al [b]	250 nm thick Al alloy film on Si substrate (film consists of an alloy with >90% Al and traces of Ti and Si)
PMMA [c]	10% by weight of PMMA dissolved in an organic solvent (anisole) is spin coated on a glass substrate, 1–2 μm thick film
PDMS [c]	Mixture of a translucent base in a curing agent (10:1 ratio) is vacuum dried and cured for 48 h at room temperature
HDT [a]	24 h immersion of Au (111)/Si (100) substrates in 1 mM HDT solution in ethanol, 1.89 nm thick film
Z-15 [d]	2–3 nm thick film deposited on Si (100) substrate by dip coating technique
Z-DOL [d]	2–3 nm thick film deposited on Si (100) substrate by dip coating technique

[a][34.30], [b][34.31], [c][34.32], [d][34.33]

the DMD can be subject to well over 2.5 trillion cycles during their lifetime. Wear during contacts between spring tips and the landing sites, hinge fatigue, shock and vibration-related failures are some of the important issues known to affect the reliable operation of the DMD [34.29]. Wear in many such devices can occur by one or more wear mechanisms, including adhesive, abrasive, fatigue, impact, corrosive and fretting mechanisms, and can seriously undermine the reliability of these devices (Fig. 34.2 shows wear of a microgear set component).

Controlling friction and wear is important in all machine components requiring relative motion. Two basic laws of dry (or conventional) friction are generally obeyed over a wide range of applications. These laws are often referred to as Amontons' laws, after the French physicist *Amontons* who rediscovered them in 1699 [34.34] after Leonardo Da Vinci first described them some 200 years earlier. Amontons' first law states that the friction force is proportional to the normal load. The second law states that friction force (or the coefficient of friction) is independent of the apparent area of contact between the contacting bodies. A third law is sometimes added to these two laws, which is often attributed to *Coulomb* [34.35]. It states that the kinetic friction force (or coefficient of friction) is independent of the sliding velocity once motion starts. Coulomb also made a clear distinction between static and kinetic friction. Over the years, these laws have been found to fall down in many cases [34.13, 36–39]. Various studies on the micro/nanoscale have indicated a strong normal load and sliding velocity dependence of friction force [34.19–28, 30, 40–46].

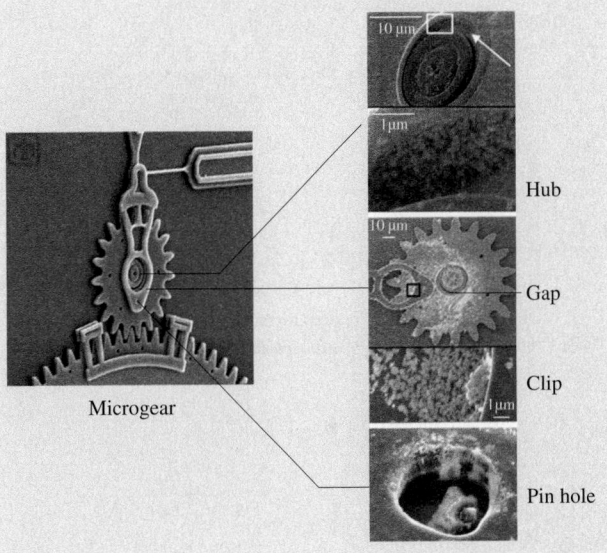

Fig. 34.2 SEM images showing a microgear set and wear of different components such as a hub, a clip and a pin-hole from high sliding velocity [34.3]

As the dimensions of the components and loads used continue to decrease, tribological and mechanical properties on the micro- to nanoscale become very important. It is critical to evaluate MEMS/NEMS component materials and coatings for failures resulting from high static friction and adhesion (stiction). A fundamental understanding of nanoscale friction is of paramount importance, not only for the design of reliable industrial and scientific applications but also to provide a bridge between science and engineering on the micro/nanoscale.

Part D | 34.1

34.2 Instrumentation

34.2.1 Extending the Capabilities of AFM for High Sliding Velocity Studies

As extensive as the research efforts have been to characterize and understand the velocity dependence of friction, inherent instrument limitations on the highest sliding velocities achievable with commercial AFMs ($< 250\,\mu m/s$) have stymied research pursuits geared towards obtaining a fundamental understanding of failures resulting from high relative sliding velocities found in

Modified AFM setup

Construction details of piezo stage

many real world applications (tens of mm/s to few m/s). With the lack of sufficient experimental evidence, theoretical formulations have been limited to low sliding velocity regimes. Studies into the velocity dependence of friction and adhesion for sliding velocity ranges of engineering importance are crucial to the future design and development of micro/nano structures and devices for various nanotechnology applications.

To achieve higher sliding velocities between the AFM cantilever tip and the sample surface, the primary approach has been to incorporate a customized stage that is capable of providing high sliding velocity. To date, various techniques have been developed that modify the basic AFM set-up in order to enhance its measurement capabilities and to broaden the scope of tribological studies into higher sliding velocity regimes. One such approach has been to mount samples on a shear wave transducer and then drive it at very high frequencies (in the MHz range) [34.47–50]. The main challenges associated with modifying the AFM set-up to achieve higher velocities are maintaining a constant velocity profile during the entire scan duration while also scanning larger surface areas to get a better understanding of nanotribological properties. In techniques employed by researchers using a shear wave transducer, the modulation amplitude was very small, in the nanometer range. Moreover, friction force measurements made by these researchers did not directly provide a fundamental understanding of nanoscale friction since they used a combination of oscillators, one to achieve high velocity and another for scanning. This does not provide a good estimate of the nanoscale friction resulting from actual relative sliding between two components since the oscillation of the tip (or the sample) while scanning changes the adhesive force at the tip–sample interface and influences stick-slip behavior.

An alternative approach to accomplishing such a modification is to use either a motorized stage or

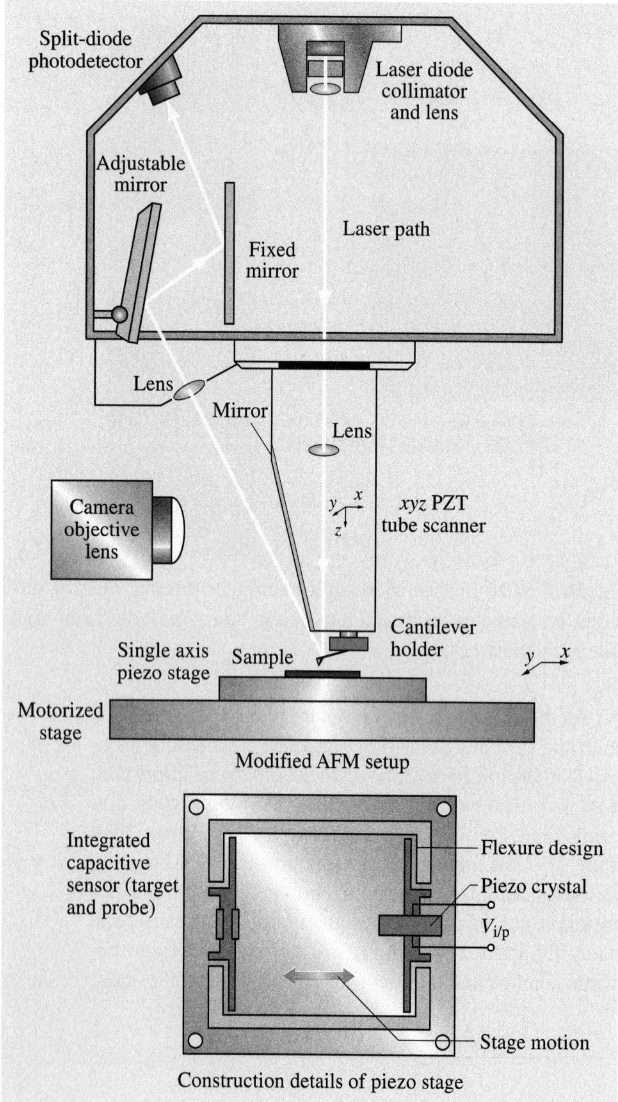

Fig. 34.3 Schematic showing modifications made to a commercial AFM set-up using the single-axis piezo stage, and a cross-sectional view showing constructional details of the piezo stage. The integrated capacitive sensors are used as feedback sensors to drive the piezo. The piezo stage is mounted on the standard motorized AFM base and operated using independent amplifier and controller units driven by a frequency generator (not shown in the schematic) [34.20]

a piezo stage with large amplitude ($\approx 100\,\mu m$) and relatively low resonant frequency (a few kHz). *Tambe* and *Bhushan* [34.20] modified a commercial AFM set-up (D3100, Nanoscope IIIa controller, Digital Instruments, Santa Barbara, CA, USA) by incorporating a custom calibrated piezo stage (P621.1CL, HERA Nanopositioner with capacitive feedback, Polytec PI, Karlsruhe, Germany). This piezo stage has an unloaded resonance frequency of 800 Hz and a maximum scan length of $100\,\mu m$. In the modified set-up (Fig. 34.3), the single-axis piezo stage is oriented such that the scanning axis is perpendicular to the long axis of the AFM cantilever (this corresponds to the 90° scan angle mode of the commercial AFM). Scanning is achieved by providing a triangular voltage pulse to the piezo amplifier. The displacement is monitored using an integrated capacitive feedback sensor, located diametrically opposite to the piezo crystal as shown in Fig. 34.3. The capacitive sensor has a stationary target element mounted on the stage block and a moving probe element mounted on the flexure. The capacitance change, corresponding to the stage displacement, gives an indication of the amount of displacement. The stage can be operated in both open loop and closed loop operational modes. In the closed loop mode, the capacitive sensor signals are used as feedback by the piezo controller to provide better guiding and tracking accuracy. The closed loop position control of piezoelectric driven stages using capacitive feedback sensors provides a linearity of motion of better than 0.01% with nanometer resolution and stable drift-free motion [34.51, 52]. The piezo stage drive units, based on P-885 PICMA® low-voltage multilayer piezo ceramics, have an extremely stiff friction-free flexure system with excellent guiding accuracy (typically less than $5\,\mu rad$ pitch/yaw over the entire travel range). The closed loop position control of piezoelectric driven stages using capacitive feedback sensors provides linearity of motion better than 0.01% with nanometer resolution and a stable drift free and hysteresis free motion.

The output of the capacitive sensor was calibrated to get the actual scan size in μm. A capacitive sensor output voltage of 1 V corresponds to a scan size of $10\,\mu m$. This calibration data was obtained from the stage manufacturer, Physik Instrumente (PI), Karlsruhe, Germany. To obtain the velocity, the stage displacement was calibrated for different scanning frequencies. The velocity then is the product of displacement and the scanning frequency. The instantaneous velocity of the stage can be obtained by differentiating the displacement signals.

34.2.2 Friction Force and Adhesive Force Measurements

During the experiments, the AFM cantilever was held stationary by maintaining a scan size of zero. The AFM controller feedback functioned in the conventional manner and maintained a constant normal load by adjusting the vertical deflection of the cantilever. As the mounted sample is scanned below the AFM tip, the torsional deflections of the tip are monitored by the photodiode detector. The raw signals from the optical detection system are directly routed to a high-speed data acquisition A/D board (NI PCI-6040E (PCI-MIO-16E-4), 12-bit, 16 Analog Input Multifunction DAQ, National Instruments, Austin, TX, USA). The photodiode signals are a direct indication of the friction force encountered by the tip during sliding motion. The data acquisition system is synchronized to collect friction signals, with the input voltage pulse acting as the trigger. Raw friction data is processed using high sampling rates of up to 25 kilosamples/s/channel.

Measurements for adhesive force can be done in two different ways. In one method, the friction force is measured for various normal loads and adhesive force is obtained from the negative intercept of the friction force vs. normal load curve on the normal load axis. The other method of calculating adhesive force is to use the 'force calibration plot' technique [34.12]. Both methods have been shown to yield similar adhesive force values [34.30, 52].

The velocity dependence experiments were performed using square-pyramidal Si_3N_4 tips with a nominal radius of 30–50 nm and a nominal spring stiffness of 0.58 N/m. The friction force measurements were conducted over a range of velocities, between $1\,\mu m/s$ and 10 mm/s, for scan sizes of $25\,\mu m$ and at a constant normal load of 70 nN. The friction measurements at each velocity were carried out at different locations on the samples to prevent any bias in the measurements. All measurements were conducted in a controlled environment of $20 \pm 2\,°C$ and $50 \pm 5\%$ RH.

34.2.3 Wear

Effect of Continuous Sliding on Friction
Wear initiation was studied by continuously scanning on the same sample location and monitoring the change in friction force with sliding distance. Experiments were performed using square-pyramidal Si_3N_4 tips and triangular pyramidal tips of single-crystal natural diamond. The Si_3N_4 tips had a nominal radius of 30–50 nm and

were mounted on gold-coated triangular Si_3N_4 cantilevers with a nominal spring stiffness of 0.58 N/m. Experiments with Si_3N_4 tips were performed at normal loads of 70 nN and 150 nN and for sliding velocities of between 1 μm/s and 10 mm/s. The diamond tip, ground to the shape of a three-sided pyramid with an apex angle of 80° and a tip radius of about 50 nm, was bonded using an epoxy to a gold-coated stainless steel cantilever beam (20 μm thick, 0.2 mm wide and approximately 2 mm long). The stiffness of the beam was 10 N/m and was calculated using the analysis for an end-loaded cantilever beam with a rectangular cross-section, as described in [34.12].

Nanowear Mapping Technique

For the nanowear mapping experiments, the sample was oscillated using the piezo stage and the AFM tip was simultaneously dragged perpendicular to the direction of motion of the sample (schematic in Fig. 34.4). The sample oscillation frequency (the scan speed), the AFM tip velocity and the normal load were controlled to achieve appropriate relative sliding velocities, normal loads and a specific number of sliding cycles. The relative sliding velocity was varied by changing the scan speed while the number of sliding cycles was varied by varying the rate of movement of the AFM tip (the velocity of the AFM tip while sliding perpendicular to the direction of motion of the sample). This varying rate of movement results in the AFM tip residing for different time intervals on the sample surface and so the number of sliding cycles obtained at each location on the sample surface is different.

Nanoscale wear maps were generated on a scan size of 5 μm using Si tips (coated with Ti/Pt) with a nominal cantilever stiffness of 3.5 N/m and nominal tip radii

of 40 nm. After the wear tests, the sample surface was imaged using the same AFM tip but at relatively low normal loads. Wear maps were first obtained by varying only one parameter out of the following three at a time: normal load, sliding velocity and number of sliding cycles. To study the effect of increasing normal load and increasing number of sliding cycles at constant sliding velocity, the AFM tip was programmed to make controlled movements at desired normal loads and velocities. This controlled motion of the AFM tip was achieved using custom software code written in NanoScript[TM] [34.53]. Next, wear maps were generated to study the effect of normal load and sliding velocity simultaneously. To vary the sliding velocity across the scan area, the input voltage pulse for the piezo stage was slightly modified. In normal operation a triangular voltage pulse is provided to the piezo stage to achieve scanning operation and to obtain a constant sliding velocity. For the wear mapping experiments, a parabolic voltage pulse was used as input to drive the piezo. This resulted in a steady increase in sliding velocity across the scan area. The synchronized and controlled movement of the AFM tip and the sample is a novel approach to obtaining nanoscale wear maps and one that helps generate a visual representation of the sample surface wear as a function of sliding velocity, applied normal load and number of sliding cycles [34.27]. Using this approach, a wear map can be generated in a single experiment, compared to the rather cumbersome approach where researchers conduct multiple experiments at different normal loads, sliding velocities and for different numbers of sliding cycles and then generate a contour map for the sample wear based on individual data points obtained from each experiment.

34.2.4 Materials, Coatings and Lubricants for Nanotechnology Applications

Traditionally, most micro/nanoscale devices have been built using silicon as the primary structural material, not only owing to the well-developed state of silicon micromachining technologies but also due to its good mechanical properties [34.10, 54, 55]. Despite its dominance though, many other structural materials are currently in use. Aluminium is used for micromirrors in the digital micromirror device (DMD) [34.10,56,57] and gold is used mainly as a reflective material for optical MEMS/NEMS, microswitches and microrelays [34.10]. SiC, due to its excellent mechanical properties, thermal dissipative characteristics, chemical inertness and optical transparency, is particularly suited for applications

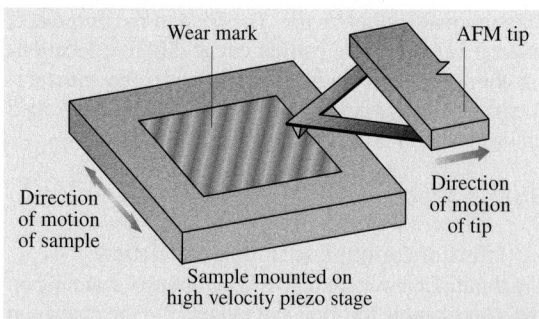

Fig. 34.4 Novel technique for wear mapping, achieved through controlled movement of the AFM tip while the sample is sliding perpendicular to the direction of the AFM tip movement [34.24]

involving harsh environments [34.58, 59]. Diamond and hard amorphous carbon coatings, commonly called diamondlike carbon (DLC), exhibit low friction and wear and are potential materials for various industrial applications [34.60, 61].

These materials pose a serious challenge when making devices with moving parts due to their inherent high stiffness. Also, it is not clear whether they should be the first choice materials for applications in biology and chemistry that require fluidic control [34.62]. Intensive research, focused on exploring fluid applications, has led to the introduction of alternative polymeric materials and their related fabrication technologies [34.63, 64]. The wide range of available polymer materials allows manufacturers to choose materials with properties suitable for their specific application [34.10, 62, 65].

Not all metal/polymer materials or coatings are suitable for all nanotechnology applications, and material selection is strongly governed by the operating and environmental parameters that the micro/nanostructures and devices are designed to work in. The efficiency,

power output and steady state operation of micro/nanostructures and devices can be critically influenced by adhesion, friction and wear [34.7–10, 29]. This has necessitated the application of low-friction and low-adhesion ultrathin lubricant films for the protection of contacting surfaces in micro/nanostructures. One of the lubricant systems used for this purpose consists of self-assembled monolayers (SAMs), monomolecular thick films deposited on solid substrates. SAMs consist of three building groups: a head group that reacts with a substrate, a tail group that interacts with the outer surface of the film, and a spacer (backbone) chain group that connects the head and tail groups [34.66, 67]. The appropriate choice of the three groups will contribute to the optimal design of the SAMs. DMDs used commercially are coated with a self-assembled monolayer of perfluorinated n-alkanoic acid ($C_nF_{2n-1}O_2H$) in a vapor phase deposition process [34.68–74].

Table 34.1 lists the samples studied along with brief descriptions and preparation details.

34.3 Velocity Dependence of Nanoscale Friction and Adhesion

Various researchers have investigated the dominant friction mechanisms on the micro/nanoscale and presented analytical models to corroborate their studies [34.9, 44, 46, 75–77]. Most of these models, however, remain limited in their focus and researchers are left short-handed when trying to explain friction behavior scaling multiple regimes. The new AFM-based technique developed by *Tambe* and *Bhushan* [34.20] promises high sliding velocities while enabling friction measurements over large scans. It is ideal for investigating friction mechanisms at different sliding velocities (up to 10 mm/s) and for developing a comprehensive analytical model to explain the velocity dependence of nanoscale friction for both dry (unlubricated) and lubricated interfaces with both hydrophilic and hydrophobic properties, as well as for studying the transitions between different friction mechanisms.

Figures 34.5a and b give the velocity dependence of friction force for the samples studied over a wide range of velocities, from a few μm/s to 10 mm/s. For Si(100) with a native oxide layer and Z-15, the friction force was found to initially decrease with velocity, reach minima, and then start to increase with velocity. The initial decrease is logarithmic and has been known to result from diminishing meniscus force contributions to the overall friction force with velocity [34.9, 19, 21, 30, 40]. For

the hydrophilic Si(100) sample, the meniscus force contribution arises from condensed water molecules, while for Z-15, the meniscus contribution arises from the Z-15 molecules as well as the condensed water molecules. Beyond a certain critical velocity, the residence time of the tip at the sample surface is not sufficient to form meniscus bridges and the meniscus force contribution to the friction force drops out. *Koinkar* and *Bhushan* [34.40] have suggested that in the case of samples with mobile films, such as condensed water and Z-15 films, alignment of liquid molecules (shear thinning) is responsible for the drop of friction force with an increase in scanning velocity. This could be another reason for the decrease in friction force for Si(100) and Z-15 films.

Unlike the results obtained for Si(100), experiments run on the DLC sample show an initial logarithmic increase in friction force followed by a near-logarithmic decrease with velocity. The DLC film is partially hydrophobic and can absorb only a few water molecules in ambient conditions [34.19, 33]. Surprisingly and interestingly, beyond a critical velocity, the friction force is found to drop nearly logarithmically with any further increase in velocity. The decrease in friction force is the result of phase transformation of the sample surface due to higher energy dissipation from high velocity impacts [34.24]. These findings are discussed in more

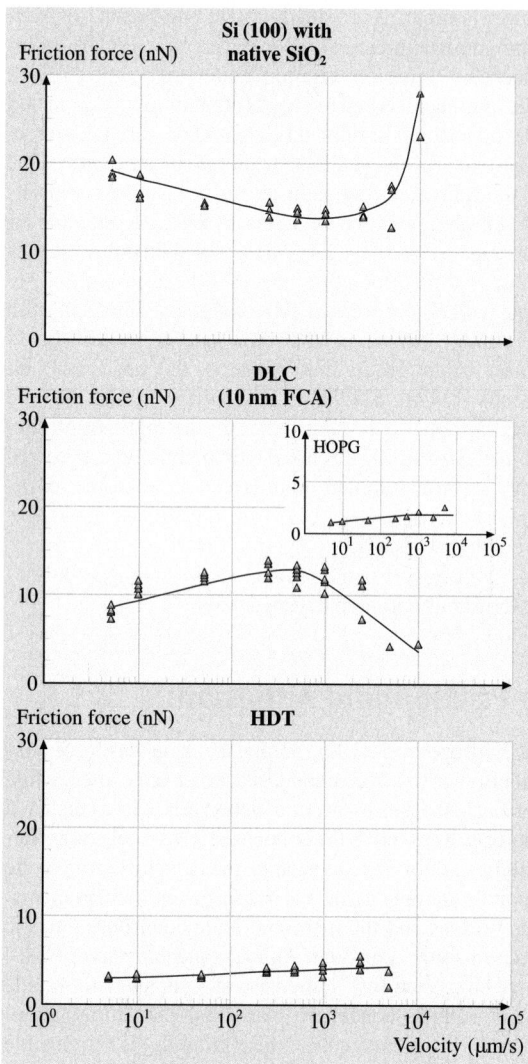

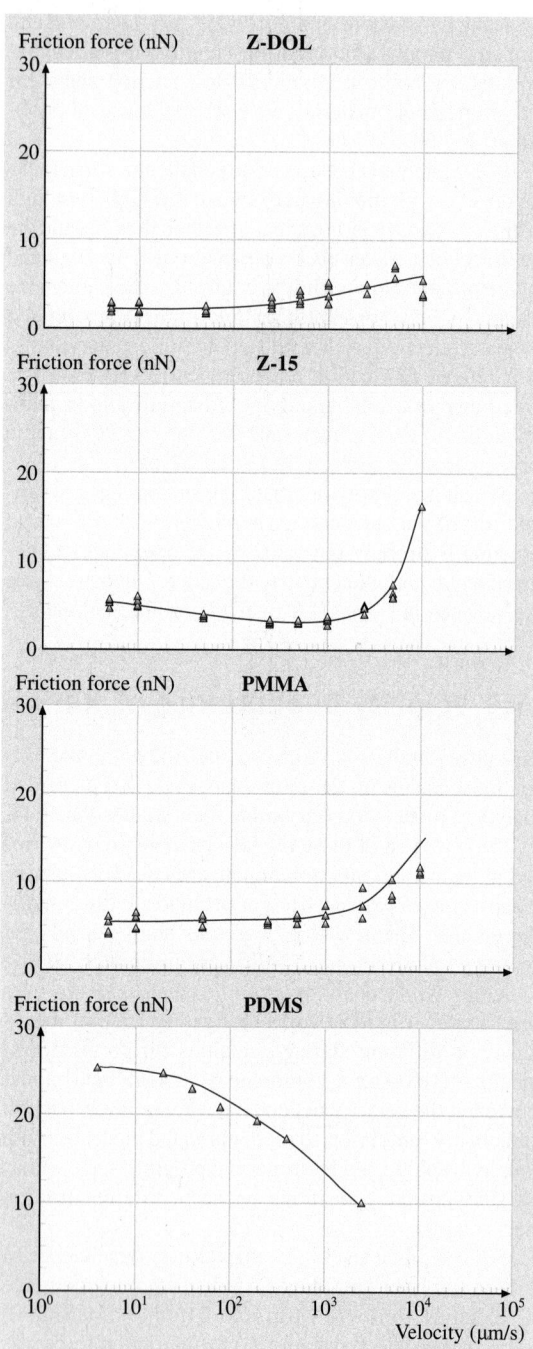

Fig. 34.5a,b Velocity dependence of friction force measured using a Si_3N_4 tip of 30–50 nm tip radius at a normal load of 70 nN and for velocities ranging between 1 μm/s and 10 mm/s at $20 \pm 2\,°C$ and $50 \pm 5\%$ RH for (**a**) Si(100) with native oxide, DLC, and HDT, and (**b**) Z-DOL, Z-15, PMMA and PDMS

detail in a later section. In the case of HDT, the friction force is seen to increase with velocity over the entire range of velocities studied. A similar increase has been reported before at lower velocities by *Liu* and *Bhushan* [34.30]. However, they observed different behavior at higher normal loads and instead found that the friction force initially increases linearly with

velocity and then reaches a plateau after a certain critical velocity. This trend is believed to be followed even for lower normal loads, but the transition is not as distinct. Moreover, the increase is believed to have

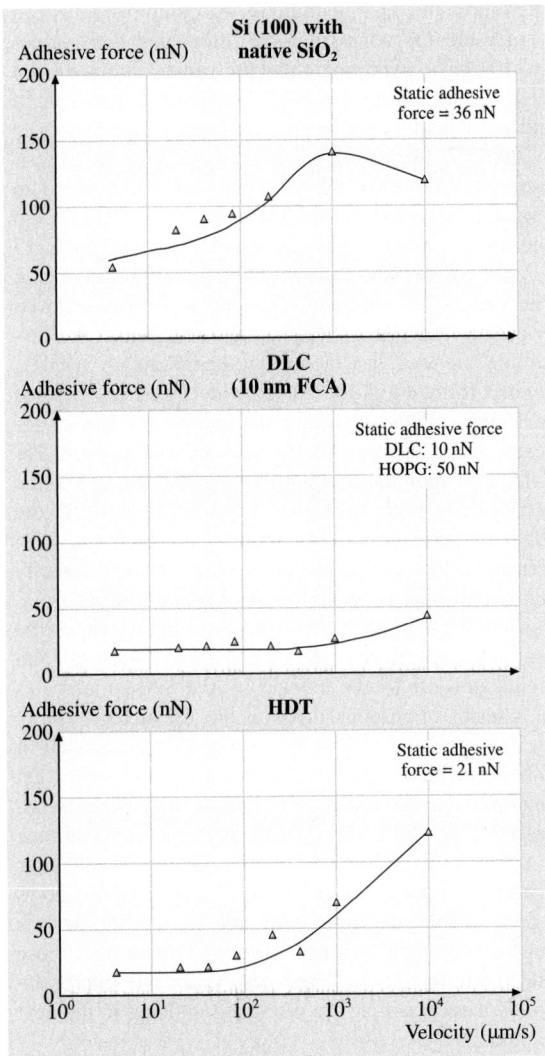

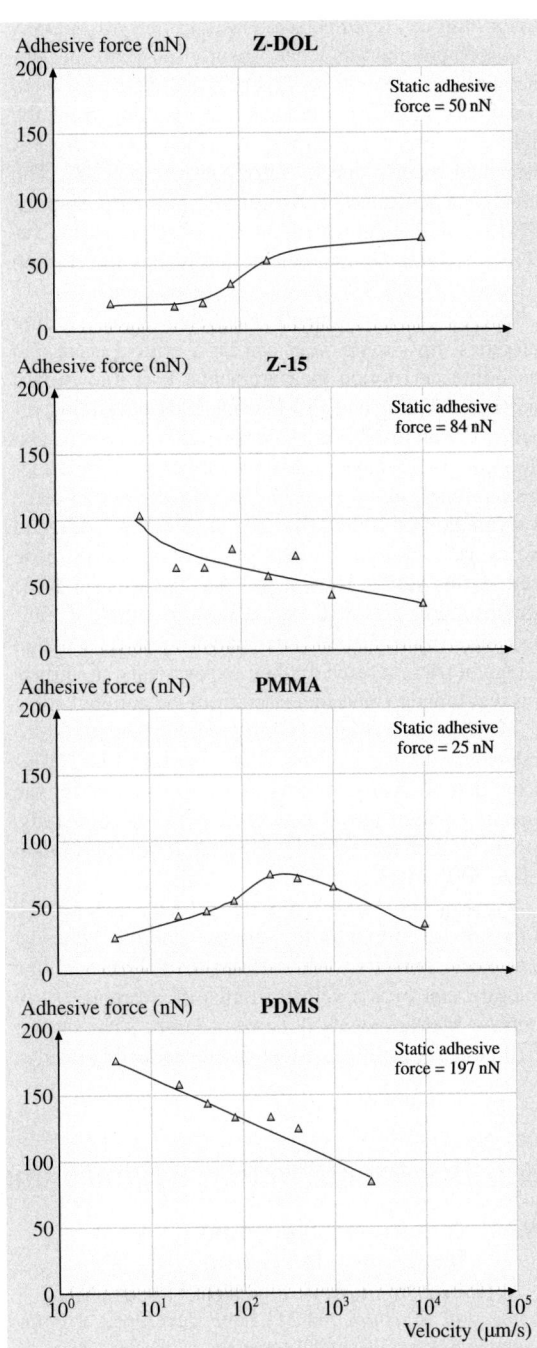

Fig. 34.6a,b Velocity dependence of adhesive force measured using a Si_3N_4 tip of $30–50$ nm tip radius at a normal load of 70 nN and for velocities ranging between $1 \mu m/s$ and 10 mm/s at $20 \pm 2\,°C$ and $50 \pm 5\%$ RH for (**a**) Si(100) with native oxide, DLC, and HDT, and (**b**) Z-DOL, Z-15, PMMA and PDMS

a linear trend, not a logarithmic trend, for all normal loads.

Z-DOL and PMMA have partially hydrophobic surfaces while PDMS is hydrophobic [34.19, 78] and meniscus force is not dominant. While Z-DOL and PMMA exhibit very little velocity dependence at low velocities, beyond a critical velocity they do show

a nonlinear increase in friction force with velocity owing to higher energy dissipation from asperity impacts. The fully bonded Z-DOL behaves more like a solid-like film and its velocity-dependent friction behavior has been reported previously [34.30, 40]. PDMS is a soft polymer and can absorb most of the impacts at high velocities. The primary contribution to friction for PDMS is the stick-slip resulting from the high adhesion at the sliding interface and energy is dissipated mainly due to deformation hysteresis. The velocity dependence of viscoelastic polymers such as PDMS is well documented [34.38]. With increasing velocity, the stick-slip component diminishes and friction reduces.

For friction experiments conducted at high sliding velocities, tip/sample wear can be a critical issue and can influence friction measurements. It is known that continuous sliding and high normal loads and sliding velocities can all result in considerable tip wear. *Tambe* and *Bhushan* [34.22] have studied the influence of these factors on sample and tip wear by measuring friction force as a function of sliding distance at different loads and sliding velocities for various samples including those used in this study. They found that the friction force remains fairly constant over a large number of sliding cycles (up to 50 000) after which there is a sudden change in friction force. In their experiments, the tip radius was found to have increased from the nominal value of 30–50 nm to as high as 60–70 nm. Further studies to understand the wear of the sample surface [34.27] indicate that no visible wear is expected to occur for the operating parameters that were selected in this study, in which all of the experiments were terminated much before 500 sliding cycles.

Figures 34.6a and b give the velocity dependence of the adhesive force for the samples studied. Adhesive force measurements were made on a smaller scan size of 2 μm, and over a velocity range of a few μm/s to 1 mm/s. Measurements done on a larger scan size of 25 μm did not yield consistent and reproducible results.

Therefore, the FCP technique was employed instead to measure the adhesive force after each high velocity (10 mm/s) experiment and the values obtained from these tests are plotted in Fig. 34.6a and b. Figures 34.6a and b also give the static adhesive force values measured by the FCP technique before the high velocity experiments. For Si(100), the adhesive force was found to remain fairly constant initially, then increase rapidly and later drop again at very high velocity, whereas for Z-15 it rapidly decreases and then remains almost constant. The increase in adhesive force for Si(100) is believed to be the result of a tribochemical reaction at the tip–sample interface, in which a low shear strength $Si(OH)_4$ layer is formed [34.30, 79]. This layer gets replenished continuously during sliding and results in a higher adhesive force between the tip and sample surface. For Z-15, with no meniscus contributions, the adhesive force decreases at high velocities. For DLC and HDT, the adhesive force is constant initially but then starts increasing. The increase is rapid beyond a critical velocity and is believed to be the result of a phase transformation via formation of a low shear strength layer at the tip–sample interface in the case of DLC [34.24, 60] and higher pull-off forces exerted by the SAM molecules as a result of viscous drag on the tip at high velocity in the case of HDT [34.80]. For Z-DOL as well, it is believed that at high velocities some molecules get displaced and become mobile thereby increasing the adhesive force on the tip. PMMA shows a classical peak in adhesive force behavior with increasing velocity, as is known to occur for polymers [34.38]. PDMS shows a drop in adhesive force with velocity and the peak is believed to occur at much lower velocities than those selected in this study. These results and the interpretations for each sample are discussed in detail in the next section.

34.4 Dominant Friction Regimes and Mechanisms

34.4.1 Comprehensive Model for Nanoscale Friction

Tambe and *Bhushan* [34.21] have developed a comprehensive friction model starting from the classical theory of friction [34.13, 36, 37, 39] by identifying the primary sources of nanoscale friction. The nanoscale friction force between two contacting surfaces is a result of three components: interfacial adhesion between contacting asperities, energy required for deforma-

tion of contacting asperities during relative motion, and stick-slip. The stick-slip effect can arise on both atomic and micro/nanoscales and is known to be velocity-dependent [34.12, 42, 44, 76, 81–83]. Assuming negligible interaction between the adhesion and deformation processes and the stick-slip during sliding, we can add them [34.13]. The total friction force, F, is therefore,

$$F = F_{\text{stick–slip}} + F_{\text{adh}} + F_{\text{def}} , \tag{34.1}$$

where

$$F_{\text{stick–slip}} = \text{friction force due to atomic scale}$$
$$\text{stick-slip between contacting}$$
$$\text{surfaces,}$$

$$F_{\text{adh}} = \text{friction force due to adhesive}$$
$$\text{interaction at the contacting surfaces ,}$$

$$F_{\text{def}} = \text{friction force due to deformation at the}$$
$$\text{interface of the contacting surfaces.}$$

We now discuss each of these three contributing factors and their velocity dependences in detail.

Atomic Scale Stick–Slip Contribution to Friction Force

The term stick-slip, first coined by *Bowden* and *Leben* [34.85], corresponds to the build-up of the friction force to a certain value, the static friction force, followed by slip at the interface once this force is overcome. The analysis of stick-slip is of particular importance in tribology because of its potential to cause damage and wear of moving parts, and various models have been proposed to explain this behavior. *Rabinowicz* [34.86] uses a model for rough surfaces which produces irregular stick-slip. *Bhushan* [34.13] uses a spring-dashpot mechanism to model the displacement of a block as a function of time during stick-slip behavior. *Israelachvili* et al. [34.83] use the phase transition model to explain how thin liquid films alternately freeze and melt as they are sheared.

Atomic scale stick-slip on highly oriented pyrolytic graphite (HOPG) has been reported by *Mate* et al. [34.81] and *Ruan* and *Bhushan* [34.84]. Figure 34.7a shows AFM images for surface roughness and friction force measured on a HOPG sample and a 2-D cross-sectional profile of the friction force measured in the forward scanning direction [34.84]. Stick-slip on the atomic scale is the result of the energy barrier that needs to be overcome to jump over atomic corrugations on the sample surface. It corresponds to the energy required for the jump of the tip from a stable equilibrium position on the surface into a neighboring position. During actual sliding between the tip and the sample surface, a few

hundred atoms on the tip are expected to go through the stick-slip process. Figure 34.7b shows a simplified version of the AFM where the tip is modeled as a one-dimensional spring mass system. As the sample surface slides against the AFM tip, the tip remains "stuck" initially until it can overcome the energy barrier, illustrated by a sinusoidal interaction potential as experienced by the tip. After some initial motion there is enough energy stored in the spring that leads to "slip" into the neighboring stable equilibrium position. After the slip and before attaining stable equilibrium, stored energy is converted into vibrational energy of surface atoms and dissipated as heat through phonon generation. The stick-slip phenomenon resulting from the irreversible atomic jumps between neighboring equilibrium positions on the sam-

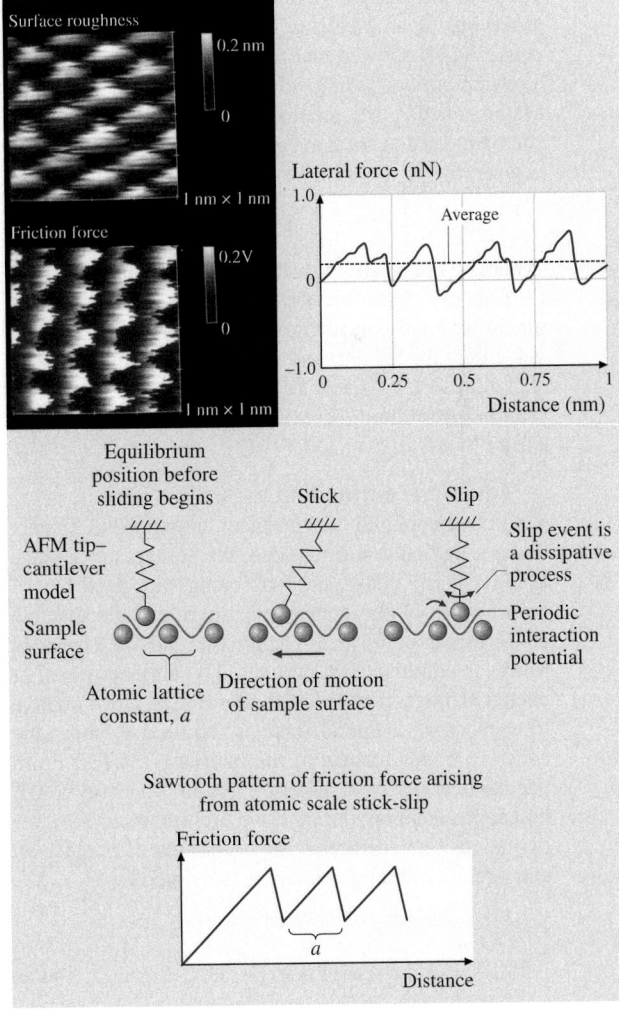

Fig. 34.7 (a) AFM images showing topography and friction data measured on a freshly cleaved sample of HOPG and 2-D cross-sectional profile of friction force at a particular scan location revealing atomic-scale stick-slip [34.84], and **(b)** simplified model of an AFM illustrating the atomic-scale stick-slip process and the corresponding sawtooth pattern obtained during friction force measurements

ple surface can be theoretically modeled with classical mechanical models [34.87, 88]. The Tomanek–Zhong–Thomas AFM model [34.88] is the starting point for determining friction force during atomic scale stick-slip. The AFM model describes the total potential as the sum of the potential acting on the tip due to the interaction with the sample and the elastic energy stored in the cantilever.

A stress-modified thermally activated Eyring model has been incorporated by *Bouhacina* et al. [34.42] to describe friction on the atomic scale. The potential barrier required to make jumps during stick-slip follows the logarithm of the sliding velocity and is responsible for the logarithmic dependence on velocity shown by the friction force. *Gnecco* et al. [34.76] have derived an expression for the lateral force (friction force) based upon the thermally activated jumps of the atoms in the proximity of an effective interaction potential. *Riedo* et al. [34.44] showed that this increase continues up to a certain critical sliding velocity v_b and then levels off. Mathematically, the relation between the atomic scale stick-slip-related friction force and the sliding velocity is given by

$$F_{\text{stick–slip}} = c_1 \ln \left(\frac{v}{v_b} \right) + c_2 , \qquad (34.2)$$

where c_1 and c_2 are characteristic constants.

For dry (unlubricated) samples (Fig. 34.5), the meniscus force contribution dominates (at low sliding velocities) for Si(100) and so we do not expect atomic stick-slip to play a role for Si(100). For DLC, we do see the logarithmic increase, as has also been reported earlier by *Riedo* et al. [34.77].

Adhesion Contribution to Friction Force

When two nominally flat surfaces are placed in contact under a particular normal load, the contact takes place at the asperity peaks, the load being supported by the deformation of the contacting asperities, and discrete contact spots are formed. The proximity of the asperities results in adhesive contacts caused by either physical or chemical interaction. When these surfaces move relative to each other, a lateral force is required to shear the adhesive bonds formed at the interface [34.13]. From the classical theory of adhesion [34.36], to a very rough first approximation, the adhesive friction force, F_{adh}, for two contacting surfaces with a real area of contact A_r and a relative sliding velocity v, is defined by

$$F_{\text{adh}} = A_r \left[\alpha \tau_a + (1 - \alpha) \tau_l \right]$$
$$= A_r \left(\alpha \tau_a + (1 - \alpha) \frac{\eta_l v}{h} \right) , \qquad (34.3)$$

where, $\alpha =$ fraction of dry contact ($0 \le \alpha \le 1$), $\tau_a =$ average shear strength of dry contact, $\tau_l =$ average shear strength of liquid (water or lubricant) film, $\eta_l =$ viscosity of liquid (water or lubricant) film, $h =$ thickness of liquid (water or lubricant) film.

The contacts at the asperities can be either elastic or plastic, depending primarily on the surface roughness and mechanical properties of the mating surfaces (and on velocity, as shown later). The real area of contact is therefore given by following equations [34.13]:

$$\text{For elastic contacts, } A_r \sim \frac{3.2 F_N R_p^{1/2}}{E^* \sigma_p^{1/2}} ,$$

$$\text{and for plastic contacts, } A_r \sim \frac{F_N}{H} , \qquad (34.4)$$

Here $F_N =$ total normal load (normal load of tip + meniscus forces, if any), $E^* =$ composite Young's modulus for the two surfaces in contact, $R_p =$ radius (average value) of asperity peaks, $\sigma_p =$ standard deviation of asperity peak heights, $H =$ hardness of softer sample.

With the presence of a thin liquid film such as a lubricant or an adsorbed water layer at the contact interface, menisci form around the contacting and near-contacting asperities due to surface energy effects. The attractive meniscus force arises from the negative Laplace pressure inside the curved menisci and is given by the product of this pressure difference and the immersed surface area of the asperity. This intrinsic attractive force may result in high friction and wear. For nanoscale contacts, the meniscus force contribution F_m becomes comparable to the normal load W [34.9] and so the total normal load can be written as

$$F_N = W + F_m . \qquad (34.5)$$

The total meniscus force at the sliding interface is obtained by summing up the meniscus forces from all individual contacting and noncontacting asperities where menisci are formed [34.9] and is given by the expression

$$F_m = 2\pi R_t \gamma \left[\cos(\theta_1) + \cos(\theta_2) \right] N(t) , \qquad (34.6)$$

where $R_t =$ radius of the contacting tip, $\gamma =$ surface tension of the liquid film, θ_1 and $\theta_2 =$ contact angles for the sample and the tip (solid–liquid interface), $N(t) =$ number of contacting and near-contacting asperities where menisci build up and is a function of the rest time (or residence time) of the tip at the tip–sample interface.

To derive the velocity dependence of the friction force arising from meniscus force contributions, we first determine the number of meniscus bridges that are formed at the interface as a function of time. There are two approaches to determining $N(t)$. In one approach, a thin water film is assumed to already exist at the interface [34.75], while in the other a thermal activation process is considered where water molecules condense preferentially at interstitial locations [34.89].

In the first approach the model considers liquid flow towards the contact zone until equilibrium is attained. The driving forces for such a flow are the Laplace or capillary pressure, which is the pressure gradient due to the curved liquid–air interface, and the disjoining pressure, which is the force per unit area that the molecules on the surface of a liquid film experience relative to that experienced by the molecules on the surface of the bulk liquid. This model is particularly developed for a liquid film and it takes into consideration only those asperities which are fully immersed while estimating the total meniscus force. Meniscus bridge formation is, however, not restricted to interfaces that are fully immersed in a liquid film. Meniscus bridges can form through condensation of liquid at preferential interstitial locations. An alternate approach to modeling meniscus bridge formation is the one based on the thermal activation process. The meniscus bridge formation is assumed to be the result of preferential condensation of liquid at the interface by overcoming an energy barrier that is required for the coating films to fully grow and coalesce and form menisci at the interface. The expression for the total number of asperities where meniscus bridges are formed is derived starting from the Kelvin relation and by determining the total energy required for nucleating meniscus bridges by condensation of a certain volume of water [34.90] at the tip–sample interface.

For the condensation of a meniscus bridge by thermal activation, the spatial extent of the meniscus bridge is controlled by the radius of curvature, r_{eq}, of the liquid interface, which, at the liquid–vapor equilibrium, is fixed by the Kelvin relation [34.90]

$$\frac{1}{r_{eq}} = \frac{\rho k_B T}{\gamma} \ln (p_s/p) , \qquad (34.7)$$

where ρ = density of the liquid, k_B = Boltzmann constant, T = absolute ambient temperature, p_s/p = ratio of ambient pressure to the saturation pressure at that corresponding ambient temperature (equal to relative humidity).

Under ambient conditions, this expression yields a value of r_{eq} of nanometer dimensions, thereby implying that liquid bridges are able to form only in nanometer-scale interstitial locations. For two surfaces in close contact with an asperity spacing that is less than

a critical distance of the order of the Kelvin radius r_{eq}, capillary condensation should occur [34.90]. However, an energy barrier has to be overcome, as the coating films have to grow and coalesce in order to fill the gap between the surfaces. Along this path, the free energy of the system increases to a maximum as the films are about to merge. The energy barrier is therefore the free energy required to condense a certain water volume from the corresponding undersaturated vapor phase [34.89] and

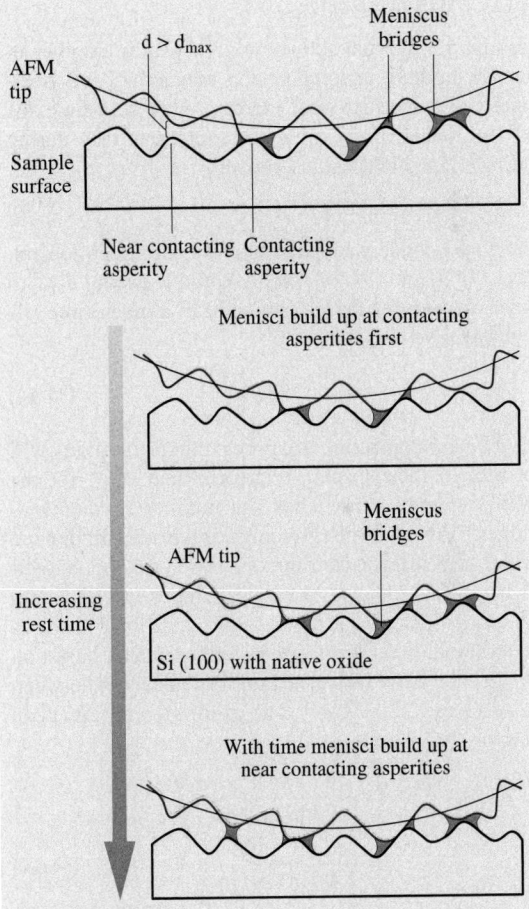

Fig. 34.8 (a) Schematic showing meniscus bridge formation at contacting and near-contacting asperities. Menisci bridges cannot be formed at near-contacting asperities if the spacing d is greater than the maximum spacing d_{max}, corresponding to a particular given residence time of the tip at the tip–sample interface [34.21], and (b) schematic illustrating the effect of increasing residence time on the build-up of meniscus bridges between the contacting and near-contacting asperities at the tip–sample interface [34.20]

is given by the expression

$$\Delta E(d) = \frac{1}{r_{eq}} \gamma V_1 = \rho k_B T \ln (p_s/p) V_1 , \qquad (34.8)$$

where $V_1 =$ liquid volume needed to nucleate the liquid bridge.

For a meniscus bridge with cross-sectional area A and a spacing d between the surfaces at the nucleating site, the expression for the free threshold energy becomes

$$\Delta E(d) = \rho k_B T \ln (p_s/p) \, dA . \qquad (34.9)$$

Assuming a thermal activation process, the number of menisci bridges forming at the contacting and near-contacting asperities will increase with rest time (or residence time of the tip at the sample surface during sliding) (Fig. 34.8a) and is given by

$$t(d) = t_a \exp[\Delta E(d)/k_B T] , \qquad (34.10)$$

where $t_a =$ condensation time for one liquid monolayer. From (34.9) and (34.10), the maximum spacing d_{max} to which a meniscus bridge can build in a given time t is therefore given by

$$d_{max}(t) = \frac{1}{A\rho} \frac{1}{\ln (p_s/p)} \ln \left(\frac{t}{t_a}\right) . \qquad (34.11)$$

For the near-contacting asperities, menisci bridges will not form if their spacing is greater than d_{max}, for the given residence time. Thus the number of meniscus bridges forming at the tip–sample interface will depend on the peak height distribution. Based on the statistical height distribution, the total number of contacting and near-contacting asperities that contribute towards the total meniscus force can be approximated based on the typical width of the distribution of distances between the surfaces λ [34.77, 89]. The expression for $N(t)$ can therefore be given by

$$N(t) = \frac{d_{max}}{\lambda} = \frac{1}{\lambda A\rho} \frac{1}{\ln (p_s/p)} \ln \left(\frac{t}{t_a}\right)$$
$$= \frac{1}{\lambda A\rho} \frac{1}{\ln (p_s/p)} \ln \left(\frac{v_a}{v}\right) , \qquad (34.12)$$

where $v_a =$ critical velocity corresponding to the condensation time of one liquid monolayer. Having determined the total number of meniscus bridges formed for a specific sliding velocity we can now determine the total meniscus force from (34.12) and (34.6), and we get

$$F_m = \frac{2\pi R_t \gamma \, [\cos(\theta_1) + \cos(\theta_2)]}{\lambda A\rho \ln (p_s/p)} \ln \left(\frac{v_a}{v}\right)$$
$$= -\phi_m R_t \ln \left(\frac{v}{v_a}\right) ,$$

where

$$\phi_m = \frac{2\pi\gamma \, [\cos(\theta_1) + \cos(\theta_2)]}{\lambda A\rho \ln (p_s/p)} . \qquad (34.13)$$

Then from (34.3, 5, 6) and (34.13) we have

$$F_{adh} = \phi_{adh} \left(W - \phi_m R_t \ln \left(\frac{v}{v_a}\right)\right)$$
$$\times \left(\alpha\tau_a + (1-\alpha)\frac{\eta_l v}{h}\right) \qquad (34.14)$$

where $\phi_{adh} = \frac{3.2 R_p^{1/2}}{E^* \sigma_p^{1/2}}$ for elastic contacts and $\phi_{adh} = \frac{1}{H}$ for plastic contacts.

As velocity increases, the time available for stable menisci formation reduces and so menisci do not form at all of the contacting or near-contacting asperities at the tip–sample interface. Figure 34.8b illustrates the building-up of meniscus bridges relative to the residence time. The maximum height that a meniscus bridge can build up to decreases logarithmically with decreasing residence time or with increasing sliding velocity, as is evident from (34.11). Moreover, the contribution of viscous shearing of the water film at contacting and near-contacting asperities, where menisci are formed, to the adhesive friction force is very small and can be neglected in most instances [34.13]. For such a case, (34.14) reduces to

$$F_{adh} \approx \mu_{adh} \left(W - \phi_m R_t \ln \left(\frac{v}{v_a}\right)\right) ,$$

where

$$\mu_{adh} = \phi_{adh} \, (\alpha\tau_a) . \qquad (34.15)$$

As v approaches v_a, the meniscus contribution decreases and friction force drops until a critical velocity $v = v_a$, when the sliding velocity is too high for stable menisci to form anywhere at the interface and this is when the friction force levels off. At this point the contribution of adhesive friction force to the overall friction force diminishes and friction is now dominated only by deformation-related friction force and stick-slip.

In our experiments, meniscus force contribution needs to be considered for Si(100) and Z-15 samples. In the case of Si(100), the meniscus force contribution is solely the result of condensed water molecules and (34.15) holds. For Z-15, however, meniscus force contribution arises from condensed water molecules as well as the liquid Z-15 film itself. For such a case, a combination of both the kinetic meniscus model suggested by *Chilamakuri* and *Bhushan* [34.75] and the thermal activation model suggested by *Bouquet* et al. [34.89] needs to be taken into consideration while determining the velocity dependence of friction force. Both of these models, however, predict a logarithmic decrease in friction force with velocity. From our results, we find that the friction force decreases logarithmically with velocity and follows (34.15) pretty closely for both Si(100) and Z-15 samples.

Contribution of Asperity Deformation to Friction Force

Now we address the second contribution to the nanoscale friction, which results from energy dissipation due to deformation of the contacting asperities (Fig. 34.9). During any relative motion, adhesion and asperity interactions are always present. Their contribution, though, may or may not be significant, and this depends on surface roughness, relative hardness of the two surfaces in contact, the normal load and sliding velocity. Low sliding velocity means less energy is available for deformation. As the relative sliding velocity increases, the impacts between surface asperities result in higher energy dissipation. The deformation-related friction force can be written as

$$F_{def}$$
$$= \frac{\text{work done (frictional energy dissipated)}}{\text{sliding distance over which energy is expended}}$$
$$= (\text{force to deform one asperity})$$
$$\quad \times (\text{number of asperities deformed}) \qquad (34.16)$$

We derive a relation between sliding velocity and the deformation-related friction force, starting from the Archard equation for wear of asperities of contacting surfaces. The amount of area worn away depends on the roughness of the sample surfaces, the sliding velocity (energy available from the relative sliding motion) and the relative hardness of the contacting surfaces. For a surface with a random peak height distribution $\varphi(z)$ and the number of peaks in the nominal contact area N_0, the total force corresponding to the energy dissipated during deformation of asperities can be written as [34.13]

$$F_{def} = \int_0^{z_0} \left(\pi a^2 H \right) N_0 \varphi(z) \, dz$$

for plastic contacts

and,

$$F_{def} = \int_0^{z_0} \left[\pi a^2 E^* (\sigma_p / R_p)^{1/2} \right] N_0 \varphi(z) \, dz$$

for elastic contacts , $\qquad (34.17)$

where $z_0 =$ maximum nominal depth over which wear can occur for a given sliding velocity.

The deformation-related contribution to friction can occur for elastic as well as plastic contacts. In elastic contacts, strong adhesion of some contacts can lead to the generation of wear particles [34.13]. This is particularly significant for elastic contacts that occur in interfaces where one of the materials has a low modulus of elasticity or for very smooth surfaces (such as in magnetic recording interfaces and nanotechnology applications [34.10, 91]). Repeated elastic contacts can also fail by surface/subsurface fatigue [34.13].

Now consider a single asperity interaction at the tip–sample interface, as illustrated in Fig. 34.9. Assuming the profile of the asperity on the sample surface to be

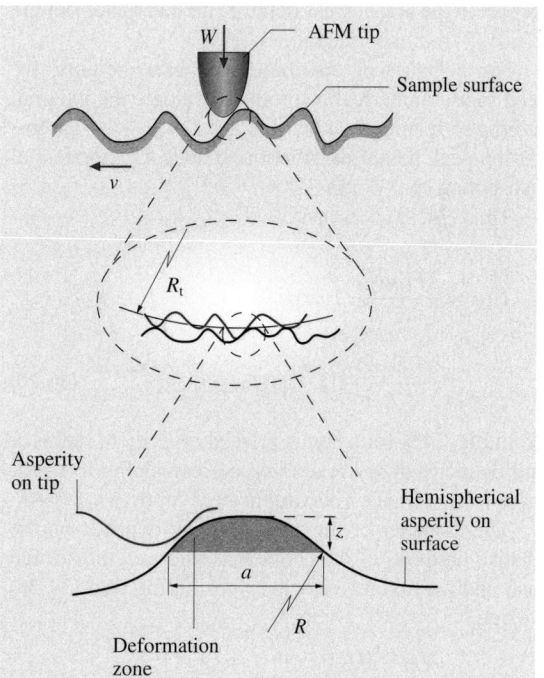

Fig. 34.9 Schematic of multiasperity interaction at the AFM tip–sample interface resulting from asperity impacts. The normal load is W, the tip radius is R_t, and for a hemispherical asperity of radius R, the impact results in a deformation zone of diameter a and height z [34.21]

hemispherical, we then have, for the two surfaces moving with relative sliding velocity v, a wear depth of z and a worn area of diameter a. The rate of wear depth is given by [34.13]

$$\dot{z} = \frac{kF_N V}{\pi a^2 H} \quad \text{for plastic contacts and,}$$

$$\dot{z} = \frac{kF_N V}{\pi a^2 E^* (\sigma_p/R_p)^{1/2}} \quad \text{for elastic contacts,} \quad (34.18)$$

where k = nondimensional wear coefficient that depends on the materials in contact. Substituting (34.18) into (34.17) we obtain

$$F_{def} = \int_0^{z_0} \frac{kF_N v}{\dot{z}} N_0 \varphi(z) \, dz$$

$$= kF_N v N_0 \int_0^{z_0} \frac{1}{\dot{z}} \varphi(z) \, dz \,. \quad (34.19)$$

Assuming that the rate of wear depth, $\dot{z}$, is constant with respect to the actual wear depth z, we can factor out the term $1/\dot{z}$ from the integral.

For a Gaussian distribution of peak heights, we need to use numerical methods to evaluate the integral. A simpler relation can be obtained for F_{def} by assuming the peak height distribution to have an exponential distribution $\varphi(z) = \exp(-z)$.

Thus (34.19) becomes

$$F_{def} = \frac{kF_N v N_0}{\dot{z}} \int_0^{z_0} \exp(-z) \, dz$$

$$= \frac{\pi k F_N v N_0}{\dot{z}} [1 - \exp(-z_0)] \,. \quad (34.20)$$

From Fig. 34.9, for a hemispherical asperity of radius R and diameter of worn area a_0, corresponding to maximum wear, we have a maximum wear depth $z_0 = a_0^2/8R$.

The diameter of the worn asperity depends on the relative hardness of the contacting surfaces, the normal load and the sliding velocity. Substituting into (34.20) we get,

$$F_{def} = \frac{\pi k F_N v N_0}{\dot{z}} \left[1 - \exp\left(-a_0^2/8R\right)\right] \,. \quad (34.21)$$

Now, the energy required for deformation comes from the kinetic energy of the tip impacting on the sample surface. It is the product of the force required to deform one asperity and the total sliding distance, and can be written as

$$\varepsilon \frac{mv^2}{2} = \left(H\pi a_0^2\right) a_0$$

for plastic contacts

and

$$\varepsilon \frac{mv^2}{2} = \left[E^* (\sigma_p/R_p)^{1/2} \pi a_0^2\right] a_0$$

for elastic contacts , $\quad (34.22)$

where m = equivalent mass of the tip and ε = fraction of kinetic energy due to relative sliding motion that is expended during deformation.

Thus, for given contacting surfaces and a given normal load, wear will initiate only after sufficient energy is available for deformation; in other words once the sliding velocity is high enough. Below a certain critical velocity there will be no frictional losses due to deformation.

Once wear starts, from (34.22) we have $a_0 \sim v^{2/3}$. Substituting into (34.21), we can write the deformation-related friction force contribution as,

$$F_{def} = \frac{\pi N_0}{\dot{z}} k F_N v \left[1 - \exp\left(-\phi_{def} v^{4/3}\right)\right] \,, \quad (34.23)$$

where

$$\phi_{def} = \frac{1}{8R} \left(\frac{\varepsilon m}{2\pi H}\right)^{2/3}$$

for plastic contacts
and

$$\phi_{def} = \frac{1}{8R} \left(\frac{\varepsilon m}{2\pi E^* (\sigma_p/R_p)^{1/2}}\right)^{2/3}$$

for elastic contacts . $\quad (34.24)$

If the normal load is small – in other words the normal load itself does not cause plastic deformation – then the friction force due to deformation will become dominant only at higher sliding velocities, when the energy from impacts is higher. In such a case, there can be a clear distinction between the adhesive friction force regime and the deformation-related friction force regime. We can then replace the F_N term in (34.23) by W and further simplify it by using Taylor series expansion of the exponential term and by neglecting higher order terms. Thus, the (34.21) can be written as

$$F_{def} = \mu_{def} W$$

where

$$\mu_{\text{def}} = \frac{\pi N_0}{\dot{z}} \phi_{\text{def}} k v^{7/3} \qquad (34.25)$$

The deformation-related friction force increases monotonically with sliding velocity. This increase is similar to our experimental data for a Si(100) sample (Fig. 34.5a). It should be noted, though, that the increase in the deformation-related friction force will not continue forever as the sliding velocity is increased. At very high sliding velocities there will be larger energy dissipation at the contacting asperities, creating very high flash temperatures. Localized melting will set in once the temperatures increase beyond a certain value. The sliding mechanism at this point will become that of viscous shearing of the melted contact zone.

Total Contribution to Friction Force

Based on the analytical theory developed so far, taking into consideration the various factors contributing to nanoscale friction, we can now integrate all the results together to yield a single mathematical expression relaying the velocity dependence of friction force from atomic scales to micro/nanoscales for dry (unlubricated) and lubricated interfaces as well as for interfaces that are hydrophobic and hydrophilic. The total friction force is therefore

$$
\begin{aligned}
F &= F_{\text{stick-slip}} + F_{\text{adh}} + F_{\text{def}} \\
&= c_1 \ln\left(\frac{v}{v_{\text{b}}}\right) + c_2 + \phi_{\text{adh}}\left[W - \phi_{\text{m}} R_{\text{t}} \ln\left(\frac{v}{v_{\text{a}}}\right)\right] \\
&\quad \times \left[\alpha \tau_{\text{a}} + (1-\alpha)\frac{\eta_{\text{l}} v}{h}\right] \\
&\quad + \frac{\pi N_0}{\dot{z}} k F_{\text{N}} v \left[1 - \exp\left(-\phi_{\text{def}} v^{4/3}\right)\right] . \qquad (34.26)
\end{aligned}
$$

$$
F = c \ln\left(\frac{v}{v_{\text{b}}}\right) + \phi_{\text{adh}}\left(W - \phi_{\text{m}} R_{\text{t}} \ln\left(\frac{v}{v_{\text{a}}}\right)\right)\left(\alpha \tau_{\text{a}} + (1-\alpha)\frac{\eta_{\text{l}} v}{h}\right) + \frac{\pi N_0}{\dot{z}} kWv^{7/3}
$$

Atomic stick-slip contribution	Meniscus contribution	Viscous fluid film shearing contribution	Asperity deformation contribution

Dominant friction mechanisms with increasing velocity

Atomic stick-slip	Micro/nanoscale stick-slip	Tribochemical reactions	Plastic deformations
Meniscus bridges	Viscous fluid film shearing	Asperity impacts	Localized melting/phase trasnformations/coating removal

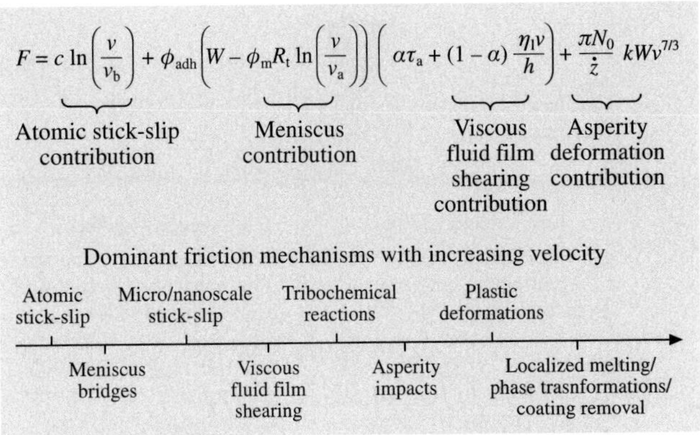

Fig. 34.11 Comprehensive analytical expression for velocity dependence of nanoscale friction with the dominant friction mechanisms [34.21]

In Fig. 34.10, the dominant mechanisms of friction force at different relative sliding velocities are shown schematically for a dry hydrophobic sample surface. Figure 34.11 gives the comprehensive analytical expression for nanoscale friction and its velocity dependence. Various terms indicating the dominant friction mechanisms are shown and their relative order of precedence with respect to sliding velocity is illustrated. It should be noted that this precedence order is not necessarily valid for all cases and depends upon the nature of the sliding interface.

34.4.2 Molecular Spring Model for Compliant SAMs

The velocity dependence of friction force for compliant SAMs cannot be explained using the analytical model given by (34.26) above. For the HDT samples used in

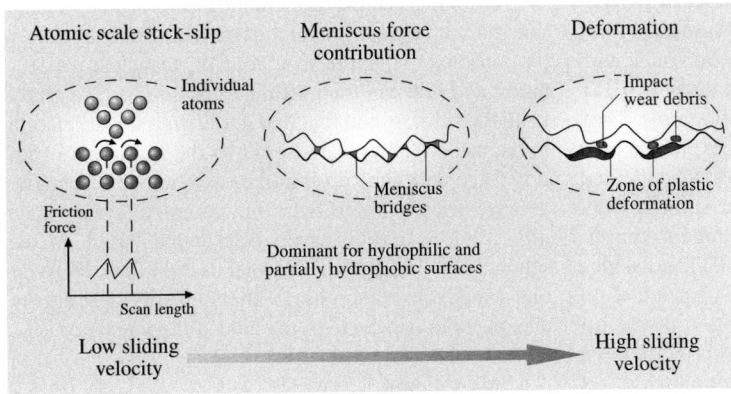

Fig. 34.10 Schematic illustrating various dominant regimes of friction force at different relative sliding velocities from atomic-scale stick-slip at low velocities to deformation-related energy dissipation at high velocities [34.21]

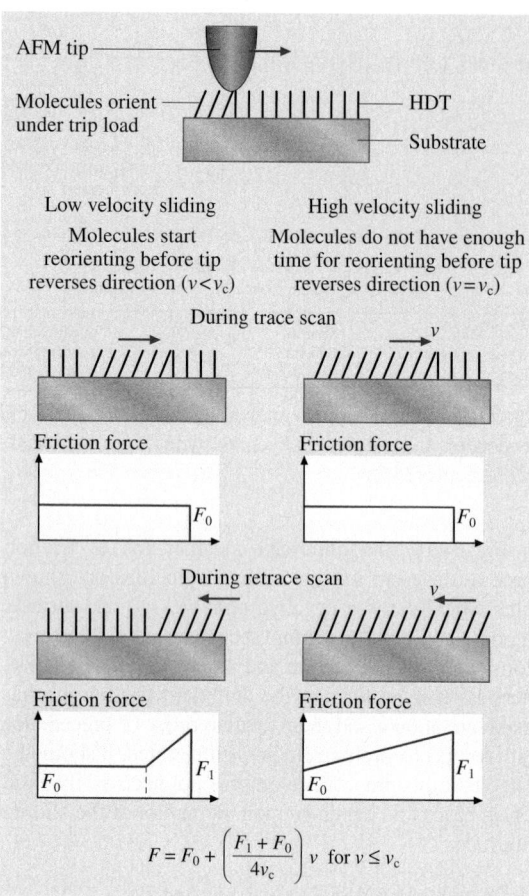

Fig. 34.12 Molecular spring model for compliant SAMs showing friction force variation with velocity during trace and retrace scans, corresponding to molecular reorientation after relaxation of tip normal load (initial molecular tilt is not shown in order to avoid complexity) [34.21]

this study, friction is governed by the viscoelastic behavior of the SAM molecules. *Liu* and *Bhushan* [34.45] have shown that there is an increase in friction force with velocity for compliant SAM molecules such as HDT and MHA (16-mercaptohexadecanoic acid thiol). They proposed a molecular spring model that explained how SAMs reorient under the tip normal load. *Tambe* and *Bhushan* [34.21] extended the molecular spring model to explain the velocity dependence of friction for compliant SAMs. As shown in Fig. 34.12, the HDT molecules orient by tilting through a certain angle, dependant on the normal load, in the direction of motion of the tip (the initial orientation of the molecules is not shown in the schematic in Fig. 34.12 to avoid complexity). As

the tip proceeds along the scan direction, the molecules are 'relieved' from the tip load and reorient to their initial position. Monte Carlo simulations of the mechanical relaxation of a $CH_3(CH_2)_{15}SH$ SAM performed by *Siepmann* and *McDonald* [34.92] indicated that SAMs respond almost elastically to microindentation by an AFM tip under a critical normal load. They suggested that the monolayer could be compressed, leading to a major change in the mean molecular tilt (orientation), but that the original structure is recovered as the normal load is removed. *Garcia-Parajo* et al. [34.93] observed the compression and relaxation of octadecyltrichlorosilane (OTS) film in their tests while *Liu* and *Bhushan* [34.45] confirmed similar behavior on various SAMs including the HDT that we used in this study. *Gourdon* et al. [34.94] have reported a strong dependence of the molecular organization of Langmuir–Blodgett films physisorbed on mica substrates on the friction force measured during forward and reverse scans.

In the experiments, run over a range of velocities between 1 μm/s and 10 mm/s over a scan size of 25 μm, the maximum time available for reorientation of the most distant molecule on the scan path before the tip returns to that same location while scanning will correspondingly vary between the two extremes of 50 s and 5 ms. There will be a critical time taken by the SAM molecules to reorient and this time will be a function of the packing density, molecular complexity, and the angle through which the molecules have to spring back [34.95]. When more time is available for reorientation before the tip reaches the same location on the sample surface again, most of the molecules in the wake of the tip scan direction will spring back. However, as the velocity increases, fewer molecules get sufficient time to reorient and these molecules now oppose the tip motion. Thus the friction force is higher when the tip is returning at higher velocities.

Assuming that the reorientation of molecules occurs linearly with time, the average friction force is inversely proportional to the time available for molecular reorientation and one can then estimate friction force directly from the velocity and find that it follows a linear relation to the velocity. Figure 34.12 shows the variation of friction force with time during trace and retrace scans. The average friction force over one complete cycle (trace plus retrace) can be easily determined based on the schematic in Fig. 34.12. The total friction force (F) over one scan cycle is the ratio of the total frictional energy dissipated in one cycle to the total sliding distance (s):

$$F = \frac{1}{2}\left(F_{\text{trace}} + F_{\text{retrace}}\right) \tag{34.27}$$

Here F_{trace} and F_{retrace} are the total friction forces during the trace and retrace scans. For a critical sliding distance s_c, over which the molecules do not get sufficient time for reorientation, we can determine the total friction force as

$$F = F_0 + \frac{s}{4s_c}(F_1 - F_0), \text{ for } s \leq s_c, \quad (34.28)$$

where F_0 and F_1 are the maximum and minimum values of the friction force, as shown in the schematic in Fig. 34.12.

The friction force can be expressed as a function of the ratio of the actual sliding velocity v to the critical sliding velocity v_c, corresponding to the maximum time required by the farthest molecule on the scan path to return to its initial orientation, and we have

$$F = F_0 + \left(\frac{F_1 - F_0}{4v_c}\right)v \text{ for } v \leq v_c,$$
$$F = \left(\frac{F_1 + 3F_0}{4}\right) \text{ for } v > v_c. \quad (34.29)$$

Once the critical velocity is reached, no molecules over the entire tip sliding distance have sufficient time to reorient and the friction force will then stabilize and reach a plateau. We therefore see a linear increase in friction force with velocity initially, and for higher velocities the friction force levels off. This is similar to the data presented by *Liu* and *Bhushan* [34.45]. Their data at high normal loads clearly shows a linear increase in friction force followed by a distinct plateau beyond a certain critical velocity.

It should be noted that it is assumed when deriving these equations that the molecules have no initial tilt angle. Any initial tilt angle will change the friction forces measured while sliding in the direction of the molecular tilt and while sliding in a direction opposite to that. The smaller the tilt angle, larger will be the difference in the friction force measured over one scan cycle. The molecular reorientation will however always follow the same rule at different velocities and thus the molecular spring model used to derive the friction force above in (34.29) remains unaffected.

34.4.3 Adhesion and Deformation Hysteresis Model for Polymers

Polymers form a special class of materials and, owing to their viscoelastic nature, friction force is largely affected by viscoelastic recovery during sliding, deformation and adhesion hysteresis [34.13, 38]. *Tabor* [34.96] extended the concept that the coefficient of friction includes

both adhesion and deformation terms to polymers and demonstrated that the latter becomes significant for a high-hysteresis polymer. At any given operating temperature, the increase in sliding velocity produces an adhesion peak at creep velocities and a hysteresis peak at very high velocities [34.38]. The adhesion component of the friction force F_{adh} for any polymer may be attributed to the molecular bonding of exposed surface atoms in both surfaces at the sliding interface, according to a stretch, break and relaxation cycle of events. The deformation term F_{def} is due to a delayed recovery of the polymer after indentation by a particular asperity and gives rise to what is generally called the hysteresis component of friction F_{hyst}. Modern theories of adhesion in a given velocity and temperature range have described adhesion as a thermally activated molecular stick-slip process [34.38]. Assuming negligible interaction between the adhesion and deformation processes and the stick-slip during sliding, we can add them (Fig. 34.13) [34.13, 38]. The total friction force F is therefore,

$$F = F_{\text{adh}} + F_{\text{def}} = F_{\text{adh}} + F_{\text{hyst}}. \quad (34.30)$$

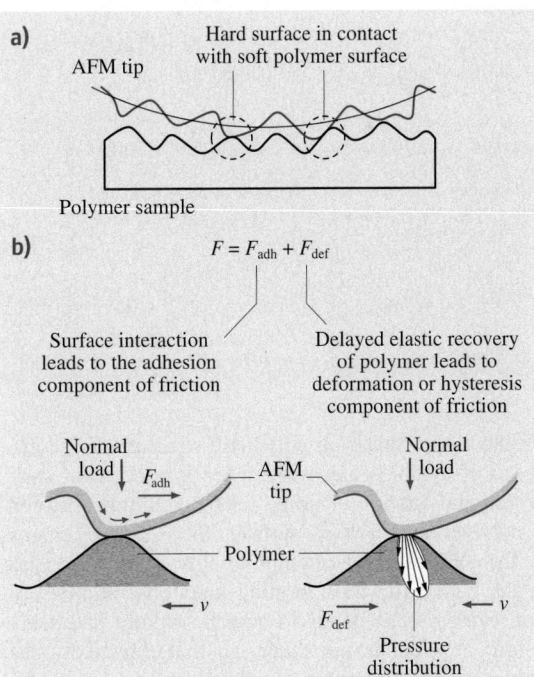

Fig. 34.13 (a) Schematic illustrating multiasperity interaction at the tip–sample interface for a soft polymer sample. (b) The principal components of friction, adhesion and deformation (hysteresis) and their causes

Table 34.2 Analytical expressions for friction force at different sliding velocities and the corresponding dominant friction regimes for various samples

Sample	Analytical expressions	
	Low velocity regime	**High velocity regime**
Si(100) (unlubricated, hydrophilic)	$F = \mu_{adh}\left[W - \phi_m R \ln\left(\frac{v}{v_a}\right)\right]$ or $F = F_0 - F_1 \ln(v)$ (meniscus force contributions)	$F = \frac{\pi N_0}{\dot{z}} k W v^{7/3}$ or $F = F_2 v^{7/3}$ (asperity impact deformations)
DLC (hydrophobic)	$F = \mu_{adh} W + c \ln\left(\frac{v}{v_b}\right)$ or $F = F_0 + F_1 \ln(v)$ (atomic stick–slip)	$F = \mu_{adh} W + \frac{\pi N_0}{\dot{z}} k F_N V \left[1 - \exp\left(-\phi_{def} v^{4/3}\right)\right]$ Phase transformation
HDT (compliant SAM)	$F = F_0 + \left(\frac{F_1 - F_0}{4 v_c}\right) v$ for $v \le v_c$ or $F = F_0' + F_1' v$ (molecular reorientation)	$F = \left(\frac{F_1 + 3 F_0}{4}\right)$ for $v > v_c$ or $F = $ constant (molecules do not have sufficient time to reorient at high velocities)
Z-15 (liquid, PFPE lubricant)	$F = \mu_{adh}\left[W - \phi_m R \ln\left(\frac{v}{v_a}\right)\right]$ or $F = F_0 - F_1 \ln(v)$ (meniscus contributions from condensed water molecules and mobile Z-15 molecules)	$F = \phi_{adh} W \frac{\eta_1 v}{h}$ or $F = F_2 v$ (viscous shearing of mobile lubricant film)
Z-DOL (solid-like PFPE lubricant)	$F = \mu_{adh} W + c \ln\left(\frac{v}{v_b}\right)$ or $F = F_0 + F_1 \ln(v)$ (stick–slip)	$F = \frac{\pi N_0}{\dot{z}} k \phi_{def} W v^{7/3}$ or $F = F_2 v^{7/3}$ (tip impacts solid-like phase dissipating frictional energy)
PMMA (polymer)	$F = \mu_{adh} W + c \ln\left(\frac{v}{v_b}\right)$ or $F = F_0 + F_1 \ln(V)$ (stick–slip)	$F = \frac{\pi N_0}{\dot{z}} k \phi_{def} W v^{7/3}$ or $F = F_2 v^{7/3}$ (tip impacts solid-like phase dissipating frictional energy)
PDMS (polymer)	$F = A_r \tau_a = A_r\left(\tau_0 + \alpha' p_r\right) = \left(\frac{\tau_0}{p_r} + \alpha'\right) W = \mu W$ $\mu = \left(\frac{\tau_0}{p_r} + \alpha'\right) = K\left[\left(\frac{E'}{p^r}\right) + K'\left(\frac{p}{E'}\right)^n\right]\tan\delta$	

Unlike a hard material, polymeric structures are composed of flexible chains which are in a constant state of thermal motion. During relative sliding between a polymer and a hard surface, the separate chains in the surface layer attempt to link with molecules in the hard base, thus forming junctions. Sliding action causes these bonds to stretch, rupture and relax before new bonds are made, so that effectively the polymer molecules jump a molecular distance to their new equilibrium position. Thus, a dissipative stick-slip molecular level process is fundamentally responsible for adhesion, and several theories exist to explain this phenomenon [34.38].

On a macroscopic level, both adhesion and hysteresis can be attributed to the viscoelastic properties of the polymer. If we define the complex modulus E^0 as the ratio of stress to strain for a viscoelastic body, it can be shown from viscoelastic theory that

$$E^0 = E' + i E'', \tag{34.31}$$

where E' is the storage modulus or stress:strain ratio for the component of strain in phase with the applied stress, and E'' is the loss modulus or stress:strain ratio for the component of strain 90° out of phase with the applied stress. The ratio of the energy dissipated (whether in the stretch and rupture cycle of individual bonds in the molecular adhesion model, or in the deformation and recovery cycle associated with hysteresis) to energy stored per cycle is defined as the tangent modulus and is given by

$$\tan \delta = \frac{E''}{E'} \, . \tag{34.32}$$

The adhesional and hysteresis components of friction are given by [34.38]

$$F_{adh} = K_2 \frac{E'}{p^r} W \tan \delta \, , \text{ and}$$

$$F_{hyst} = K_3 \left(\frac{p}{E'} \right)^n W \tan \delta \, , \text{ where } n \geq 1 \, , \tag{34.33}$$

where p is the nominal pressure, r is an exponent with a value in the neighborhood of 0.2, and K_2 and K_3 are constants that depend upon the particular sliding combination. Combining the above two equations ((34.32) and (34.33)), we obtain

$$F = F_{adh} + F_{hyst} = K_2 \left[\frac{E'}{p^r} + K_4 \left(\frac{p}{E'} \right)^n \right] W \tan \delta \, . \tag{34.34}$$

Though the velocity dependence of friction force is not directly evident from (34.34), the terms E' and $\tan \delta$ have to be measured at the frequency of deformation and are thus velocity-dependent [34.13].

34.4.4 Dominant Friction Force Regimes for Various Samples

The analytical and molecular spring models developed to explain nanoscale friction behavior can be used to explain experimental results for various samples. The experimental data suggests two distinct regimes for all of the samples studied and reversals in friction force are seen to occur close to the $100-500 \, \mu m/s$ velocity range. However, each has a different friction mechanism acting at different relative sliding velocities. Table 34.2 gives the analytical expressions relating the friction force to the sliding velocity for all of the samples.

Dry (Unlubricated) Hydrophilic Solids
For hydrophilic interfaces, meniscus forces dominate at low velocities and hence the contributions of atomic stick-slip can be neglected. Moreover, the contribution of viscous shearing of the water film (at contacting and near-contacting asperities where menisci are formed) to the adhesive friction force is very small and can be neglected in most instances [34.13]. For such a case,

(34.14) reduces to

$$F_{adh} = \mu_{adh} \left[W - \phi_m R_t \ln \left(\frac{v}{v_a} \right) \right] ,$$

$$\text{where, } \mu_{adh} = \phi_{adh} \tau_a \, . \tag{34.35}$$

As v approaches v_a, the meniscus contribution decreases and friction force drops until a critical velocity $v = v_a$ where the sliding velocity is too high for stable menisci to form anywhere at the interface, and this is when the friction force levels off. At this point the adhesive friction force contribution to the overall friction force diminishes and friction is now dominated only by deformation-related friction force and stick-slip.

If the normal load is small (in other words the normal load itself does not cause plastic deformation), then the friction force due to deformation will become dominant only at higher sliding velocities, when the energy from impacts is higher. In such a case there can be a clear distinction between the adhesive friction force regime and the deformation-related friction force regime. Thus the expression for nanoscale friction for a dry (unlubricated) solid material is

$$F = \mu_{adh} \left[W - \phi_m R_t \ln \left(\frac{v}{v_a} \right) \right] + \frac{\pi N_0}{\overset{\bullet}{z}} k F_N v^{7/3} \, . \tag{34.36}$$

This expression is valid for Si(100), which is an unlubricated hydrophilic solid. For Si(100), meniscus forces dominate only at low velocities, when there is sufficient time available for stable menisci formation at the tip–sample interface. At higher velocities, Si(100) shows a nonlinear increase in friction with velocity and this is attributable to the energy dissipation by asperity deformations at the tip–sample interface. For moderately high velocities, a tribochemical reaction has been reported to occur, wherein a low shear strength layer of Si(OH)$_4$ is formed at the tip–sample interface [34.30, 79]. This Si(OH)$_4$ layer is continuously removed and replenished, causing a drop in friction force with velocity, similar to the one resulting from the diminishing meniscus force contributions. However, formation of Si(OH)$_4$ also corresponds to an increase in the adhesive force and can thus be easily distinguished. At higher velocities, deformation and wear of the interface resulting from impacting asperities causes the friction force to increase again. The dominant regimes of the velocity-dependent friction force acting at the tip sample interface for Si(100) are schematically illustrated in Fig. 34.14a.

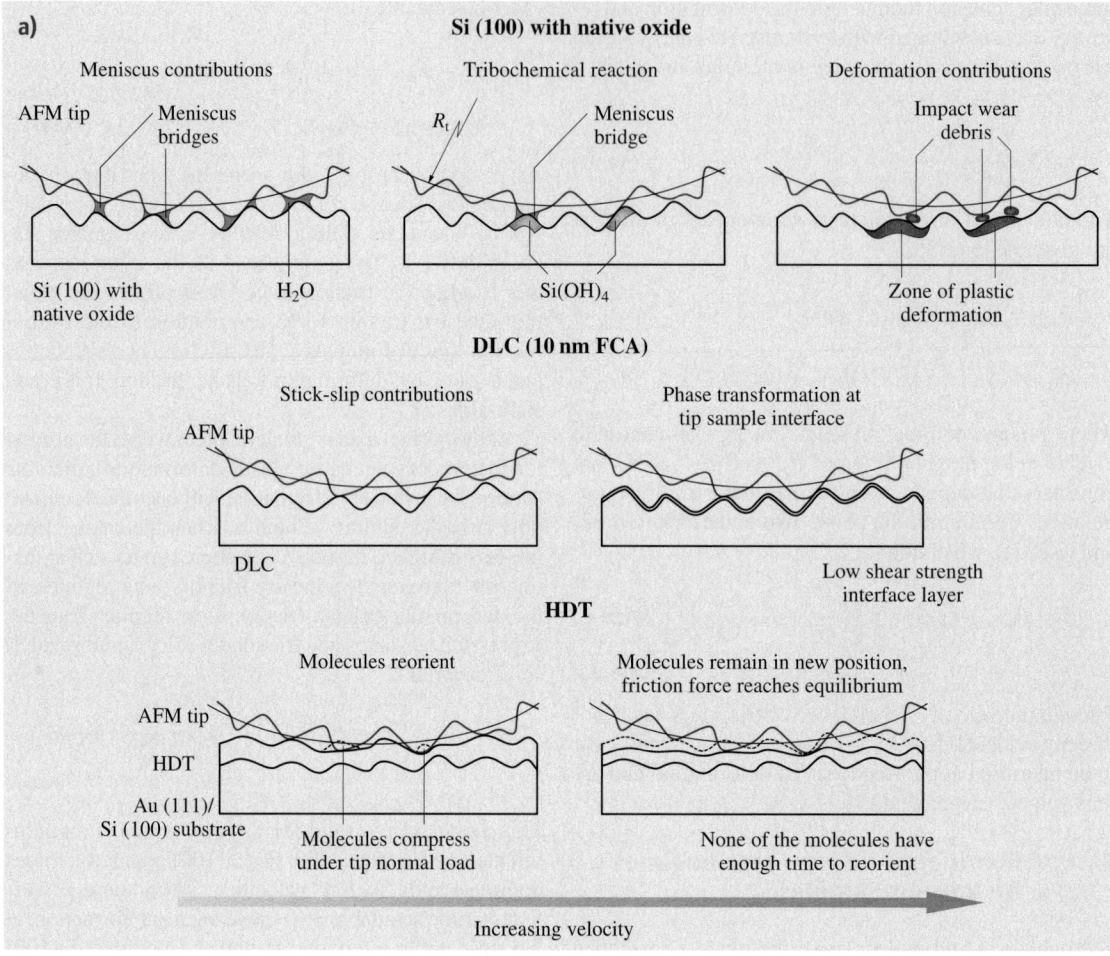

Wet (Water Film or Mobile Lubricant) Solids

For wet surfaces (those having either a water film or mobile lubricant film), the deformation component of friction force is not dominant and can be neglected. In this case, meniscus force dominates at low velocities and viscous fluid film shearing is the dominant mechanism at high velocities. The expression for nanoscale friction thus becomes

$$F = \mu_{adh} \left[W - \phi_m R_t \ln \left(\frac{v}{v_a} \right) \right] + \mu'_{adh} W \frac{\eta_l v}{h} , \tag{34.37}$$

where $\mu_{adh} = \phi_{adh} \tau_a$ and $\mu'_{adh} = \phi_{adh}$.

This expression is valid for Z-15, which is a mobile lubricant. The friction force for Z-15 is seen to once again follow a similar trend to that seen with Si(100) at low velocities. There is a logarithmic decrease in friction

force with velocity initially and that is attributable to diminishing meniscus force contributions to the friction force. For Z-15, menisci are formed by condensed water and the Z-15 molecules. As velocity increases there isn't sufficient time available for formation of stable menisci and consequently the friction drops. At high velocities, no menisci are formed but as the tip slides over the interface there is a pure viscous film shearing effect that dictates the friction force behavior. Thus we see a linear increase in friction at high velocities. The two dominant mechanisms are schematically illustrated in Fig. 34.14b.

Dry (Unlubricated) Hydrophobic Solids

For hydrophobic surfaces, meniscus forces are not the dominant friction mechanism and atomic stick-slip is more significant at low velocities. At high velocities, deformation is the dominant mechanism for solids and

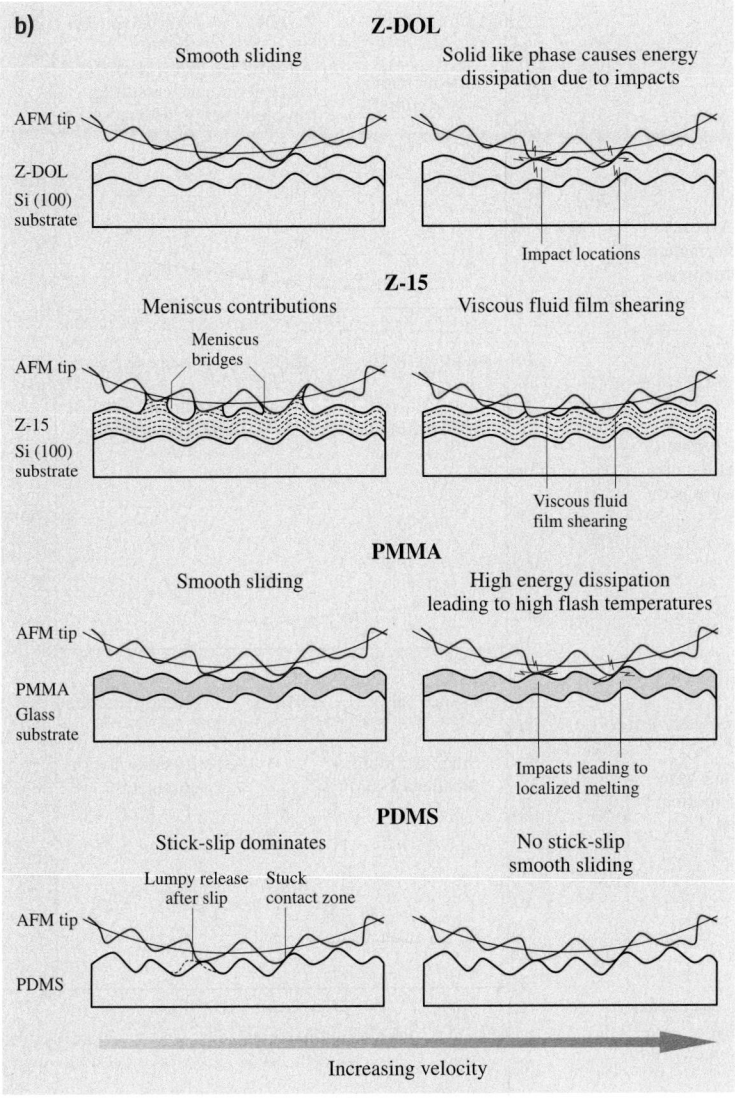

b)

Fig. 34.14a,b Schematic illustration of tip–sample interaction at different sliding velocities depicting the various friction mechanisms and the corresponding effect on the sliding interface for **(a)** ◄ Si(100) with native oxide, DLC, and HDT, and **(b)** Z-DOL, Z-15, PMMA and PDMS

the expression for nanoscale friction can be reduced to the following form

$$F = c_1 \ln \left(\frac{v}{v_b} \right) + \mu_{adh} W + \frac{\pi N_0}{\dot{z}} k F_N v^{7/3} \; . \quad (34.38)$$

This expression is valid for DLC, which is an unlubricated hydrophobic solid. Unlike the results obtained for Si(100), experiments run on the DLC sample show an initial logarithmic increase in friction force followed by a near-logarithmic decrease with velocity. The DLC film is partially hydrophobic and can absorb only a few water molecules in ambient conditions [34.33].

Previous researchers [34.77] have shown that friction force increases logarithmically with sliding velocity for DLC. In fact we find that their predicted value for the critical velocity of 400 μm/s is very close to the velocity at which we see a reversal in friction force and a corresponding increase in the adhesive force in our experimental results. Table 34.2 gives the analytical relation for the velocity dependence of friction force at low velocities. The difference between the expression for Si(100) and DLC is that while the slope is negative for Si(100), the slope is positive for DLC (F_1 in Table 34.2).

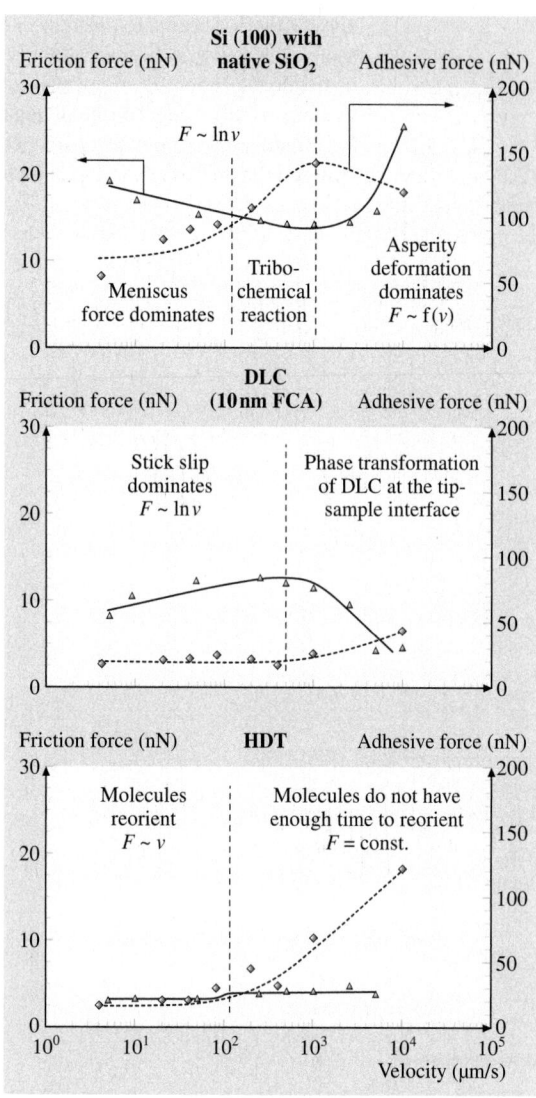

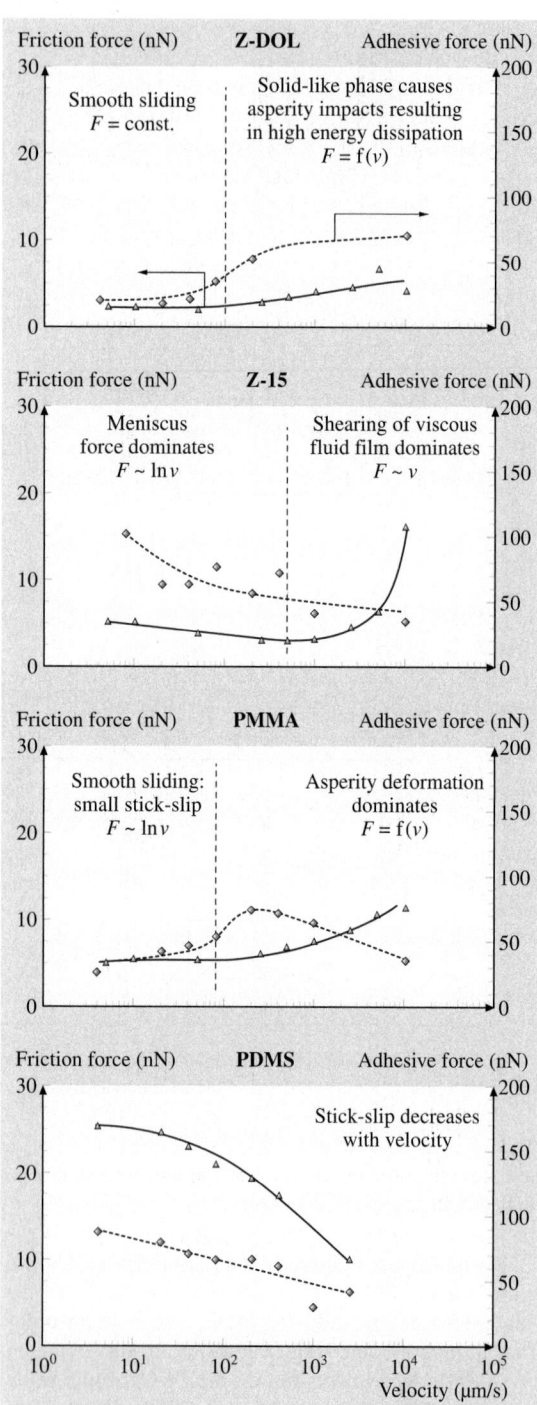

Fig. 34.15a,b Velocity dependence of friction and adhesion with dominant friction mechanisms at different sliding velocities for (**a**) Si(100) with native oxide, DLC, and HDT, and (**b**) Z-DOL, Z-15, PMMA and PDMS

Beyond the critical velocity, the friction force drops nearly logarithmically with any further increase in velocity. The decrease in the friction force with velocity for DLC is the result of the formation of an interfacial layer of low shear strength. Friction-induced phase transformation of DLC to graphite by a sp³ to sp² phase transition has been shown to occur by *Voevodin* et al. [34.97] and *Tambe* and *Bhushan* [34.24].

For DLC, the tribological behavior appears to be controlled by a low shear strength layer formed during sliding of the surfaces in contact in most cases. Experiments with DLC have shown a decrease in the coefficient of friction with increasing load and speed, and this decrease has been attributed to an increase in the thickness of the low shear strength layer [34.24, 60]. At high velocities, the high contact pressures combined with the high frictional energy dissipation due to asperity impacts at the contacting asperities on the tip–sample interface accelerate the phase transformation process.

Experiments run to study the velocity dependence of friction force for highly oriented pyrolitic graphite (HOPG) (Fig. 34.5a, inset) reveal that, for HOPG, friction force initially increases with velocity as a result of atomic stick-slip, but then levels off beyond a certain critical velocity. The friction force values measured for DLC and HOPG at the highest velocity of 10 mm/s are almost the same. Moreover, in our experiments for DLC we also find that the decrease in friction force is followed by an increase in adhesive force. The adhesive force increased from about 20 nN at 10 μm/s to around 45 nN at 10 mm/s. Adhesive force measurements done on HOPG using the FCP technique showed that HOPG has an adhesive force of 50–60 nN. We therefore believe that the decrease in friction force beyond a certain critical velocity in our experiments is the result of a phase transformation of DLC that yields a low shear strength layer. Figure 34.14a schematically shows such a layer at the tip–sample interface.

Compliant SAMs: Molecular Spring Model

Based on the molecular spring model developed for compliant SAMs, the friction data for HDT can be demarcated into two distinct friction force regimes, one where the friction force increases linearly with the velocity and the other where friction force is independent of the sliding velocity. Previously published data [34.30] showed a logarithmic increase in friction force with velocity at low loads, but at higher normal loads a linear increase followed by a constant friction force regime was distinctly seen. The adhesive force data presented here clearly shows a change in friction behavior beyond a certain critical velocity. We therefore believe that the friction force behavior can be explained by the molecular spring model even though the transition between the friction force regimes is not distinct at low normal loads. Previous studies by *Clear* and *Nealey* [34.80] suggest that a viscous drag acts on the tip and plays a significant role in determining the overall friction behavior at high velocities for SAMs. The rapid increase in adhesive force beyond a critical velocity observed in our experiments must be the result of higher pull-off forces due to such a viscous drag.

Based on the experimental data in Figs. 34.5 and 34.6, the analytical models developed in the previous section and with the help of the illustrations in Figs. 34.14a and b showing the various dominant regimes for the samples studied, the velocity dependence of friction force for all of the samples can now be summarized as shown in Figs. 34.15a and b.

34.5 Nanoscale Friction Mapping

Many materials, coatings and lubricants are found to show a strong dependence of friction force and adhesive force on the sliding velocity. The comprehensive analytical model developed in order to understand nanoscale friction behavior reveals the nonlinear nature of nanoscale friction. A fundamental understanding of the normal load and velocity dependence of nanoscale friction is of paramount importance, not only for the design of reliable industrial and scientific applications but also to provide a bridge between science and engineering on the micro/nanoscale.

Tambe and *Bhushan* [34.28] have studied nanoscale friction as a function of normal load and sliding velocity and used a novel mapping technique that helps identify, demarcate and classify different friction mechanisms. Figure 34.16 shows the nanoscale friction maps obtained in order to study the dependence of friction force on normal load and sliding velocity. The contour maps were generated from friction force data collected at different normal loads and sliding velocities. The contours represent constant friction force lines and are marked by the value of the friction force in nN. The contours for each material are characteristic of the friction behavior exhibited by that particular material even though there are some features that can be classified as universal irrespective of the material and these are discussed next. Based on the specific arrangement of contour lines, the dominant friction mechanisms can be identified. At the bottom half of Fig. 34.16, the most commonly observed features in the contours (or the characteristic contours) as found from our study are summarized and the significance of each is stated.

Fig. 34.16 Contour maps showing dependence of friction force on normal load and sliding velocity and summary of characteristic contour patterns and their significance [34.28]

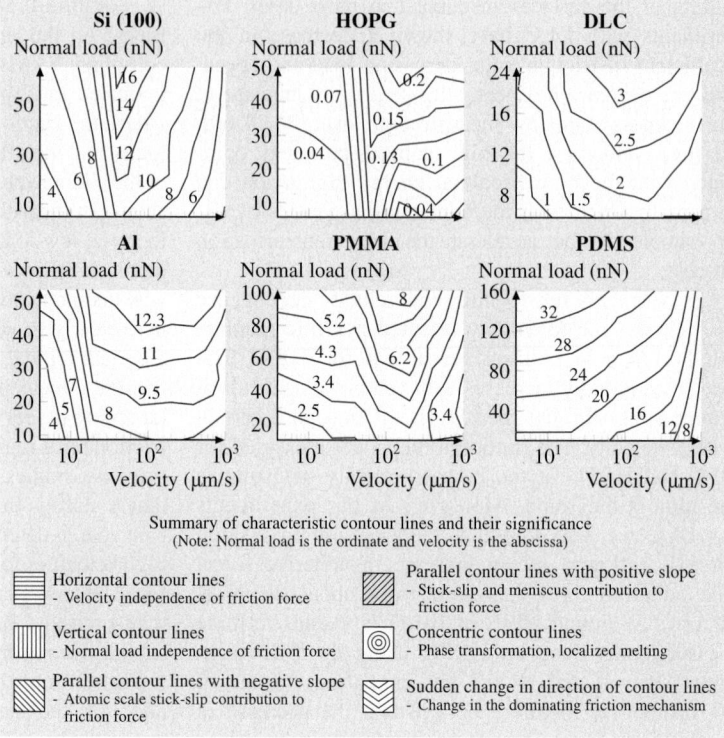

Horizontal contour lines indicate the velocity-independent nature of friction force. This behavior is found at high velocities for HOPG, at moderately high velocities for Al and at low velocities for PMMA. Studies on the dependence of friction force on velocity at the nanoscale [34.44] indicate that friction force becomes constant relative to sliding velocity when the atomic-scale stick-slip that occurs at low sliding velocities loses its dominance. A constant friction force with respect to sliding velocity would appear as a horizontal contour on a friction map, and is the reason, for example, for the horizontal contours seen for HOPG at high velocities. Vertical contour lines indicate a normal load independence of friction force. This behavior is not seen in any of the samples studied, although the steep contour lines for Si(100), HOPG and Al indicate that there is a very small normal load dependence on friction at low velocities. For all practical instances it would be impossible to find a material that shows normal load independence of friction force.

Some other characteristic contours are those with slanting lines with either a positive or negative slope. These friction contours arise from the microscale stick-slip-related contributions or from the formation of

meniscus bridges by preferential condensation of liquid films at the sliding interface, particularly for hydrophilic interfaces. Researchers have shown that friction force increases with velocity due to atomic-scale stick-slip [34.44, 46]. On a friction map, this increase would be seen as slanted contours with a positive slope; an increase in friction force as one moves from left to right on the map. In this study, HOPG, DLC and Al showed this behavior. Stick-slip can also originate as a result of other mechanisms and results in a decrease in friction force with an increase in sliding velocity [34.83]. This behavior would result in slanted contours with a negative slope on the friction map, such as seen for PMMA at high sliding velocities and for PDMS. The friction force arising from meniscus contributions, found for hydrophilic surfaces such as Si (100), results in a drop in friction force with increasing sliding velocity. A minimum threshold equilibrium time is necessary for the formation of stable meniscus bridges at contacting and near-contacting asperities for a sliding interface [34.21, 89]. With increasing velocity fewer meniscus bridges build up at the interface and thus the overall contribution to friction force drops with increasing velocity. This behavior would manifest itself in the form of slanted contour lines with negative slopes on the friction map.

Contour maps can also consist of concentric contour lines. DLC appears to show this kind of behavior. The concentric contour map implies a peak friction force corresponding to a critical normal load and sliding velocity. Beyond that point any further increase in either the normal load or the sliding velocity would result in a decrease in the friction force. This kind of behavior would typically imply a phase transformation by the formation of a low friction phase at the interface or localized melting at the contact zone. Phase transformation has been known to occur for DLC, resulting in a low-friction graphite-like layer via a sp^3 to sp^2 phase transition [34.24, 60]. Localized melting would arise from very high frictional energy dissipation and is particularly expected for polymer materials. In this study, PMMA appeared to show concentric lines at moderately high velocities, but for the given range of normal loads and sliding velocities, the experimental evidence is not sufficient to support this hypothesis.

Another characteristic contour is the one where contour lines change direction suddenly. This implies a sudden change in the dominant friction mechanism. Si(100) and PMMA showed such behavior. In the case of silicon, researchers have reported formation of a $Si(OH)_4$ layer at the sliding interface at high sliding velocities [34.79]. It is believed, based on the contour maps, that this effect is initiated at a particular sliding velocity and that this is the reason for the sudden change in friction force. For Si(100), the meniscus contribution to the friction force is also a dominant mechanism; however, both of these mechanisms are known to coexist in tandem and result in a decrease in friction force with velocity [34.19, 21, 30].

The nanoscale friction maps provide fundamental insights into the dependence of friction force on normal load and sliding velocity. They help to identify and classify the dominant friction mechanisms as well as to determine the critical operating parameters that influence transitions between different mechanisms. Other than normal load and sliding velocity, the number of sliding cycles or the sliding duration are also vital to understanding the evolution of friction and in particular wear. Next we discuss investigations into the wear initiation process and the wear mechanisms involved. Later on, the utility of nanoscale friction mapping, in conjunction with two other novel techniques, nanoscale wear mapping and material mapping, are discussed.

34.6 Wear Studies at High Sliding Velocities

The effects of operating parameters on friction and adhesion were presented in the previous section. Nanoscale friction force was seen to strongly depend on the sliding velocity for various materials, coatings and lubricants. At high sliding velocities, friction behavior is found to be governed by asperity impacts and the resulting deformations of the contacting asperities due to high frictional energy dissipation. Next we discuss the effect of continuous sliding on friction force in order to understand the wear initiation process and then we investigate the wear mechanisms involved for the specific tip–sample interface.

34.6.1 Effect of Sliding Velocity on Friction and Wear Initiation

At high velocities, higher energy dissipation due to asperity impacts is expected to have an influence on wear initiation. *Tambe* and *Bhushan* [34.22] have investigated the effect of high velocity on the wear initiation for Si(100) and DLC by monitoring friction force as a function of sliding distance at two different velocities (Fig. 34.17). For Si(100), experiments were performed at $10 \mu m/s$ and $10 mm/s$ (friction was high for both of these velocities Fig. 34.5a). The friction force increased drastically during high velocity tests ($10 mm/s$) whereas it slightly decreased and leveled off to a value lower than that at the start during the low velocity tests ($10 \mu m/s$). Similar behavior was observed at normal loads of 70 and 150 nN; however, as expected, the increase in friction force was higher at 150 nN. The drastic increase is believed to be the result of a combination of factors: deformation resulting from impacts of the contacting asperities and the corresponding three-body wear due to debris particles generated at the sliding interface, and a tribochemical reaction which is accelerated by high velocities [34.79]. Previous researchers [34.98] have proposed a mechanism of failure of silicon-based microfilms that arises due to high stress cycles and subsequent wear through a process of sequential mechanically induced oxidation and environmentally assisted cracking of the surface layers.

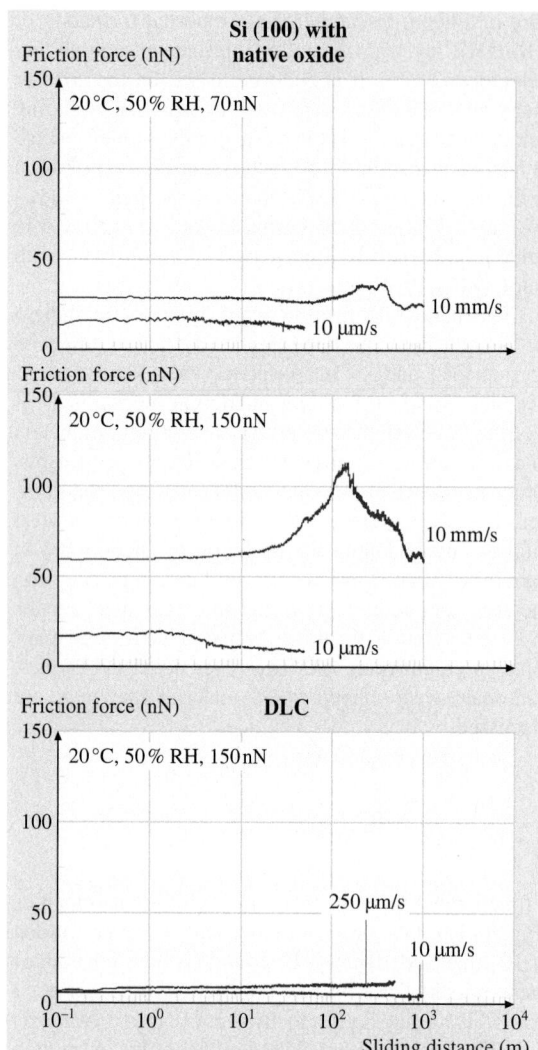

Fig. 34.17 Effect of sliding velocity on durability of Si(100) and DLC measured at 70 nN and 150 nN normal load using Si$_3$N$_4$ tips at 20 °C and 50% RH [34.22]

For DLC the experiments were conducted at 250 μm/s and 10 mm/s. The friction force remained fairly constant with sliding distance for the test at 250 μm/s; however, it decreased with sliding distance for the test at 10 mm/s. These results are consistent with those reported before for macroscale pin-on-disk experiments [34.60]. The ultralow wear rates and the reduction in the friction force in long duration tests have been attributed to the formation of a low shear strength layer at the sliding interface. Repeated friction is believed to

accelerate the sp^3 to sp^2 phase transition of some DLC into a graphite-like phase [34.24, 60, 97].

The effect of continuous sliding on the tip wear was studied by estimating the tip radii before and after the tests using tip characterization software [34.99]. In all cases the tip radius was found to have increased from the initial nominal value of 30–50 nm to around 60–75 nm. However, no specific trend was found that could suggest more tip wear for high velocity than for high normal load.

34.6.2 Effect of Continuous Sliding on Friction at High Sliding Velocities

The strong dependence of sliding velocity on the wear initiation process motivates a comparative study for different samples. Figures 34.18 and 34.19 show the results from tests run using Si$_3$N$_4$ and diamond tips at sliding velocities of 10 mm/s and normal loads of 150 nN and 1 μN respectively. For wear studies on Si(100) and DLC, it has been shown that higher normal loads (>10 μN) can lead to high tip wear [34.100], while high normal loads (>5 μN) can wear off HDT films from substrate [34.45]. To determine the effect of sliding velocity, therefore, lower normal loads were used.

Si(100) had the lowest durability; in other words, wear initiated the earliest for Si(100) amongst all samples. The friction force increased drastically after only a short sliding distance due to high deformation and wear of the contacting asperities from impacts. For Z-15, the friction force was constant initially but then increased drastically and reached a value close to that for Si(100). A similar behavior has been reported previously [34.40] and the reason for this is believed to be the displacement of the mobile Z-15 molecules from under the scanning tip at the sliding interface. The depletion of lubricant molecules at the sliding interface exposes the Si(100) substrate and results in higher friction.

For DLC, phase transformation is believed to occur due to high contact pressures at the sliding interface, as noted in the previous section. However, the results show that while the friction force decreased with sliding distance at low normal load with a Si$_3$N$_4$ tip, it increased at a higher normal load with diamond tip. Diamond is much harder than Si$_3$N$_4$ and the contact pressures are much higher for the diamond tip (Table 34.3). The increase in friction force thus suggests that the graphite-like debris particles get ejected from the interface during sliding instead of forming a thin film at the interface [34.22].

For HDT, the friction force remained fairly constant over a large sliding distance. A slight drop in friction

Fig. 34.18 Friction force as a function of sliding distance measured using Si$_3$N$_4$ tips at a normal load of 150 nN and a sliding velocity of 10 mm/s at 20 °C and 50% RH [34.22]

was seen towards the end of the test at low normal load and this was probably due to orientation and relocation of the SAM molecules due to continuous scanning. *Liu and Bhushan* [34.101] have reported that friction force can significantly reduce after several scans for compliant SAMs. At high normal load, however, a slight increase was seen in the friction force. In the experiments, the combination of high velocity and high normal load is believed to result in the detachment and removal of some SAM molecules from the interface. This eventually exposes the underlying Au (111) substrate, which has a higher coefficient of friction [34.102] and is the reason for the increase in friction.

For Z-DOL, the friction force dropped after large sliding distances for both normal loads, and similar to HDT this drop was most probably the result of some of the bonded Z-DOL molecules breaking off from the substrate and acting as a mobile lubricant fraction. The results therefore indicate that higher sliding velocities can have a similar effect on the wear initiation process to higher normal loads.

At low normal load, PMMA showed a constant friction force up to a certain sliding distance followed by a drastic increase. PMMA is a compliant polymer (Table 34.4) with a low melting point in the range of 85–100 °C [34.65]. The continuous sliding and the high energy dissipation from asperity impacts will subsequently lead to localized melting at the contact zone. That in turn will result in an increase in the friction force. At higher normal loads, friction is higher for PMMA due to larger energy dissipation. PDMS, which is more compliant than PMMA (Table 34.4) can, on the other hand, absorb impacts even at high velocities. Consequently friction force remains fairly constant for PDMS over the entire test duration. Some localized melting may still take place due to the continuous sliding and result in an increase in friction force towards the end of the test.

Adhesive force was measured before and after the wear initiation studies with Si$_3$N$_4$ tips to get a measure of the change in surface properties of the sample and/or tip. The adhesive force was seen to increase for all of the samples studied (Fig. 34.20). For Si(100) the increase is the result of formation of wear debris

Fig. 34.19 Friction force as a function of sliding distance measured using diamond tips at a normal load of 1 µN and a sliding velocity of 10 mm/s at 20 °C and 50% RH [34.22]

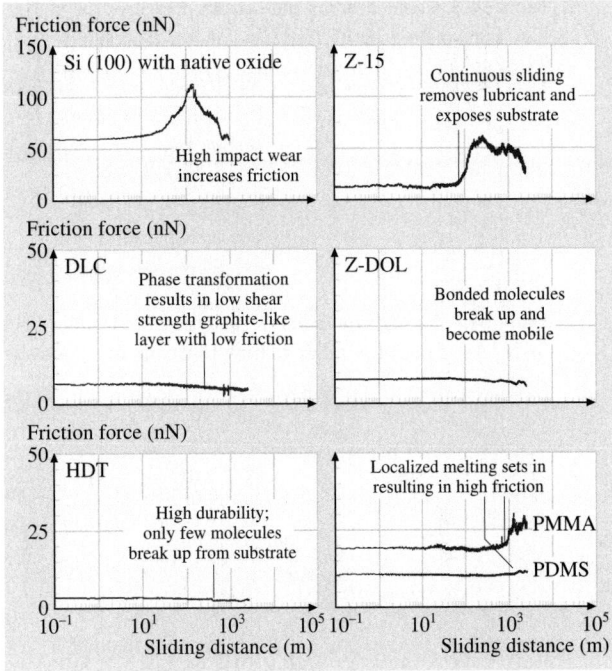

particles [34.22, 27] and a tribochemical reaction that occurs at the interface [34.79] along with tip wear, as noted in the previous section. DLC, on the other hand, is partially hydrophobic and tip wear does not lead to

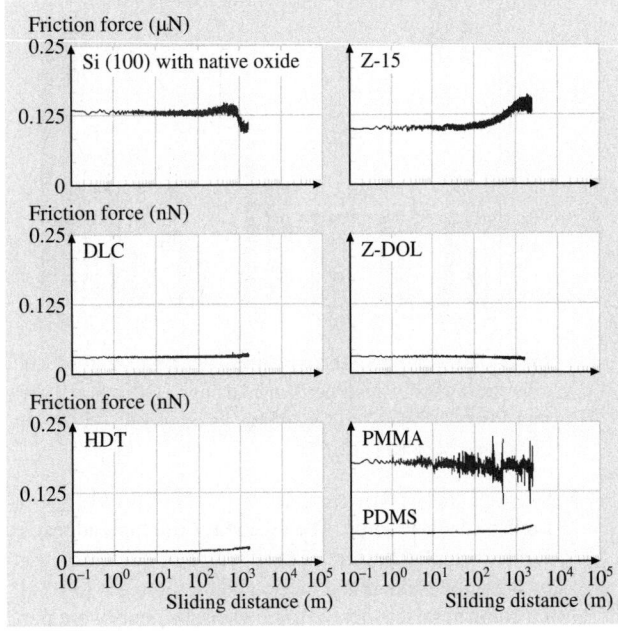

Table 34.3 Contact radius and contact pressures for Si(100) and DLC samples used in the experiments. (Calculations are based on single-asperity Hertzian contact analysis)

	Tip used	Si₃N₄	Diamond
	Nominal tip radius (nm)	30–50	50
	Normal load range	10–150 nN	0.5–10 μN
Si (100)	Contact radius (nm)	1.6–3.9	5.3–14.4
	Contact pressure (GPa)	1.24–3.14	5.67–15.4
DLC	Contact radius (nm)	1.3–3.3	4.3–11.7
	Contact pressure (GPa)	1.8–4.4	8.6–23.3

Table 34.4 Young's modulus (E) and measured values of coefficient of friction. All measurements were performed using Si₃N₄ AFM tips of nominal radius 30–50 nm in a controlled environment of $20 \pm 2\,°C$ and 50% RH

Material	E (GPa)	Ref.	Coefficient of friction	RMS roughness (nm)	Ref.
Diamond	1140	[34.103]	0.05	2.3	[34.41]
SiC	395	[34.59]	0.02	0.89	[34.59]
DLC	280	[34.61]	0.03	0.14	
Si (111)	188	[34.104, 105]	0.04	0.14	[34.41]
Polysilicon	167	[34.59]	0.04	0.86	[34.59]
Si (100)	130	[34.104, 105]	0.05	0.14	
Al alloy film	91	[34.31]	0.06	1.6	
Au (111)	77	[34.106]	0.035	0.37	[34.102]
SiO₂	73	[34.107]	0.05	0.14	[34.41]
HOPG	9–15	[34.103]	0.008	0.09	
PMMA	7.7	[34.108]	0.07	0.98	
PDMS	0.36–0.87	[34.78]	0.1		

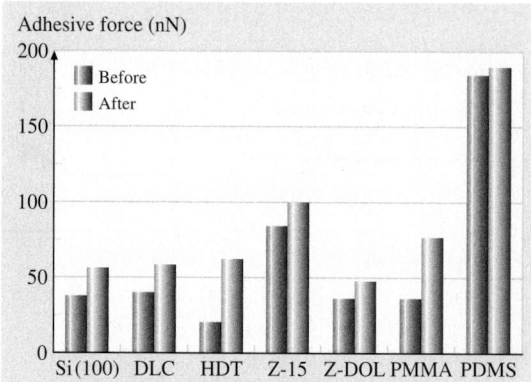

Fig. 34.20 Adhesive force values measured before and after the durability tests performed using Si₃N₄ tips at a normal load of 150 nN and a sliding velocity of 10 mm/s at 20 °C and 50% RH [34.22]

formed on highly oriented pyrolytic graphite (HOPG) in an ambient environment of 20 °C and 50% RH, and a value of 50–60 nN was obtained [34.21, 24]. As the adhesive force for DLC after the test was in this range, the increase is believed to be due to phase transformation. The increases in adhesive force for HDT, Z-15 and Z-DOL are believed to be the result of some of the unbonded or displaced molecules becoming attached to the tip [34.22]. For PMMA and PDMS, continuous sliding leads to localized melting at the contacting asperities. This results in softening of the contact zone and is the reason for the increase in the adhesive force. The increase, however, is only marginal in the case of PDMS.

Figure 34.21 summarizes the wear initiation mechanisms for various materials, coatings and lubricants at high sliding velocities. The transformations of the contact zone at the tip–sample interface are illustrated schematically.

higher adhesive force. The reason for the high adhesive force, as stated before, is believed to be the formation of a graphite-like phase at the sliding interface [34.24]. To confirm this, adhesive force measurements were per-

34.6.3 Mapping Wear on the Nanoscale

An AFM-based 'continuous microscratch' technique has been developed by *Sundararajan* and *Bhushan* for

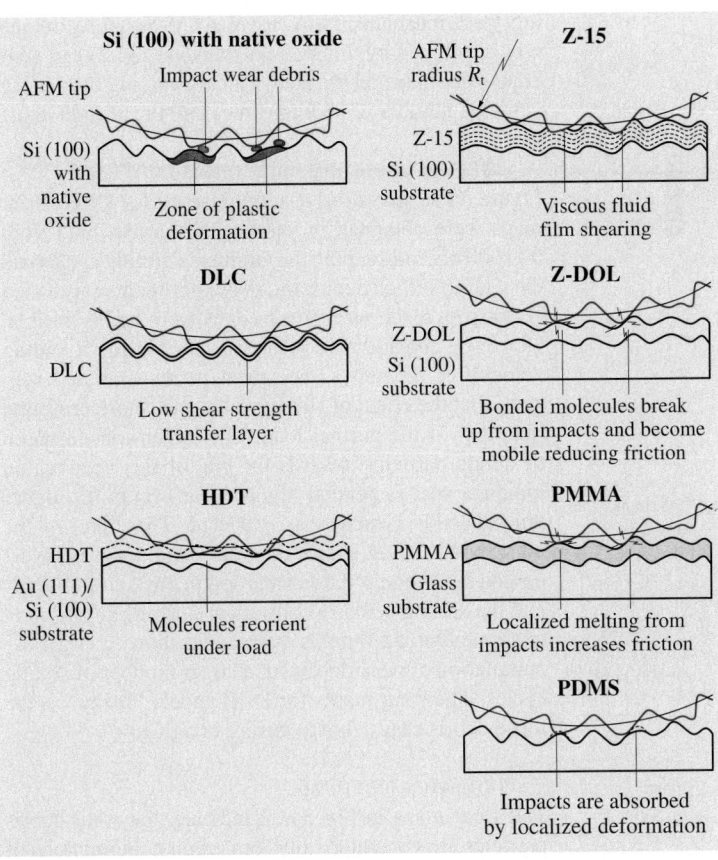

Fig. 34.21 Schematic illustrating the wear mechanisms responsible for different materials, coatings and lubricants [34.22]

wear studies [34.100]. That technique gave a direct dependence of the scratch depth on the applied normal load, thereby giving the critical normal load for failure for materials and coatings. This technique has been used by *Liu* and *Bhushan* [34.45] to study the wear behavior of various SAM lubricants as well as substrates. The microscratch technique, however, doesn't provide the dependence of normal load, sliding veloc-

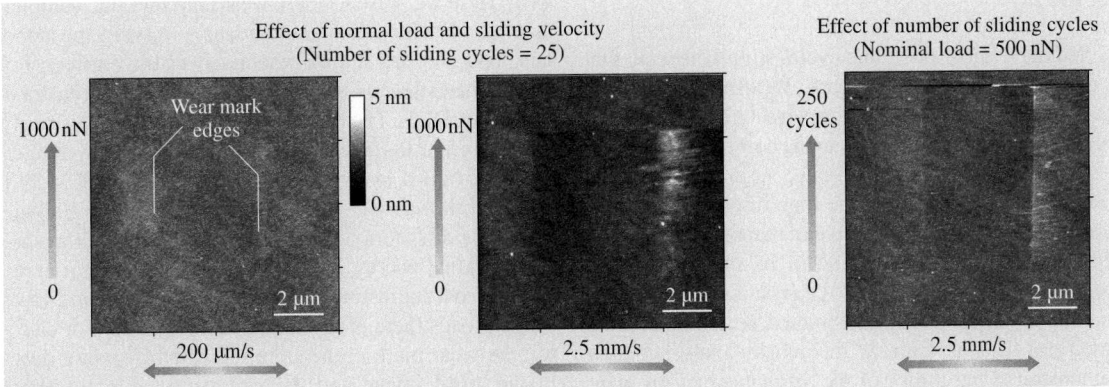

Fig. 34.22 Wear maps showing effect of normal load and sliding velocity, as well as effect of number of sliding cycles for DLC [34.24]

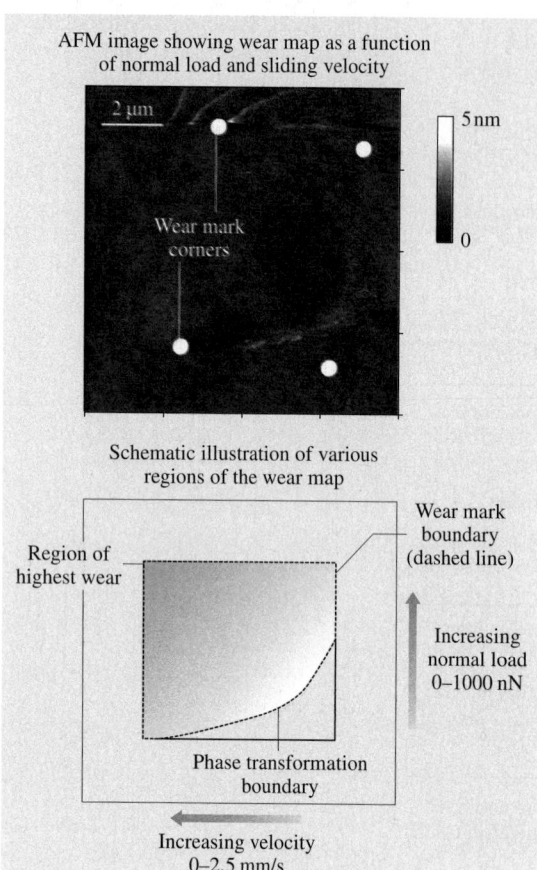

AFM image showing wear map as a function
of normal load and sliding velocity

2 µm

5 nm

Wear mark
corners

0

Schematic illustration of various
regions of the wear map

Region of
highest wear

Wear mark
boundary
(dashed line)

Increasing
normal load
0–1000 nN

Phase transformation
boundary

Increasing velocity
0–2,5 mm/s

Fig. 34.23 Nanowear map (AFM image and schematic) illustrating the influence of sliding velocity and normal load on the wear of DLC resulting from phase transformation. *Curved area* shows debris lining and is indicative of the minimum frictional energy needed for phase transformation [34.27]

ity and the number of sliding cycles simultaneously that is needed to generate wear maps. Wear maps give a direct representation of various operating conditions and can help identify and demarcate the corresponding wear mechanisms. On the macroscale, various researchers have used an approach where they run a multitude of experiments and then plot wear maps based on the failure data and rate of removal of material during wear [34.13,36,37,39,109,110]. *Lim* and *Ashby* [34.109] and *Lim* et al. [34.110] constructed wear maps using empirical data as well as theoretical analysis. They demonstrated the utility of the wear mechanism mapping method as a way of classifying and ordering wear data and of showing the relationships between compet-

ing wear mechanisms. A novel AFM-based technique was developed by *Tambe* and *Bhushan* [34.27] to generate wear maps on the nanoscale by varying the sliding velocity, number of sliding cycles and the normal load.

Effect of Operating Parameters on Wear

Figure 34.22 gives the results obtained for DLC. Wear maps were obtained by varying the normal load from 0–1000 nN and keeping the number of sliding cycles and the sliding velocity constant. For experiments conducted at 200 µm/s, the wear mark edges were barely visible. However, considerable wear was visible for a sliding velocity of 2.5 mm/s. The wear mark generated suggests that the effect of sliding velocity is more profound than that of the normal load. The larger concentration of debris particles towards the end of the wear region indicates that in general higher wear occurs for higher normal loads, which is as expected. The effect of the number of sliding cycles on wear behavior was investigated by keeping the normal load constant at 500 nN and the sliding velocity constant at 2.5 mm/s. The wear mark generated from this experiment shows a larger accumulation of wear debris for a larger number of sliding cycles. The wear marks for DLC appear "fuzzy" as the loose debris easily moves during imaging.

Nanowear Mapping

The wear maps in Fig. 34.22 indicate that wear debris particles are generated only for certain combinations of sliding velocities, normal loads and number of sliding cycles. In these wear maps only one operating parameter was varied at a time. To obtain true wear maps that can allow investigation of different wear mechanisms simultaneously, it is necessary to vary both normal load and sliding velocity and investigate the resulting wear. Here we present a nanowear map for DLC sample that was obtained by simultaneously varying the normal load and the sliding velocity over the entire scan area to investigate wear resulting from phase transformation of DLC. The wear map generated for a normal load range of 0–1000 nN and a sliding velocity range of 0–2.5 mm/s in this fashion is shown in Fig. 34.23. Wear debris was seen to form only for particular sliding velocities and normal loads (beyond a certain threshold frictional energy dissipation). Hence the wear area was curved, indicating that for low velocities and low normal loads there is no phase transformation. For clarity, the wear mark corners are indicated by white dots in the AFM image and the various zones of interest over the entire wear mark are schematically illustrated in Fig. 34.23.

34.7 Identifying Materials with Low Friction and Adhesion for Nanotechnological Applications

Nanotribological investigations have been discussed for various materials, coatings and lubricants that are considered potential solutions for nanotechnology applications at high sliding velocities. They have encompassed velocity regimes of scientific as well as engineering interest, thereby enabling development of a broader and more fundamental understanding of nanoscale friction behavior. Other than tribology, an understanding of mechanics on the nanoscale has been shown to be crucial to the design of reliable and failure-proof micro/nanodevices. The interdependence between friction and material properties on the nanoscale is of significant interest when selecting materials that would be ideal from tribology point of view: materials with low friction and adhesion. Scientific studies indicate that mechanical properties can strongly affect the tribological performance [34.13]. Efforts to explicitly characterize the nanoscale friction and adhesion of various materials on the basis of their mechanical properties remain limited though. Recently, *Tambe* and *Bhushan* [34.25] have established a vital link between the Young's moduli of materials and their coefficients of friction and adhesive forces over a range of sliding velocities.

Table 34.4 lists the materials used in this study along with their Young's moduli and the corresponding coefficients of friction. Also listed are the materials studied by previous researchers under identical experimental and environmental conditions. For materials where a range

of values have been reported for the coefficient of friction, the average values are listed. Hence forward, in this section, materials with lower Young's modulus are referred to as low-E materials and those with higher Young's modulus are referred to as high-E materials.

A clear trend is observed for the dependence of the coefficient of friction on the Young's modulus (Fig. 34.24). (It should be noted that the sliding interface will only undergo elastic deformations under the very low normal loads used in the experiments.) Low-E materials show higher coefficients of friction than high-E materials. This result can be intuitively inferred from the classical theories of friction [34.13,37,111]. An approximate relation can be developed between the coefficient of friction and the Young's modulus by assuming a single-asperity elastic contact and applying Hertzian contact analysis. For most sliding interfaces though, the contact is often a multi-asperity contact and no closed-form analytical solutions exist. Numerical methods must be employed to solve problems dealing with multi-asperity contact [34.13]. Moreover, nanoscale friction and adhesion is largely dependent on sample surface roughness and the shear strength of the sliding interface [34.13,111]. Table 34.4 indicates that the roughness values of the samples are not the same, although they are comparable. In light of this limitation on analytical formulations and the inherent complexity involved in relating the nanoscale friction and adhesion to the material properties, the near-logarithmic dependence of the coefficient of friction on the Young's modulus shown by a wide variety of materials in the experimental data is extremely interesting. Only highly-oriented pyrolytic graphite (HOPG) stands out as an anomaly in this trend, and this is because of its extremely low shear strength.

Figure 34.25 shows the velocity dependence of adhesion and the coefficient of friction over a range of velocities for different materials. The contour maps (Fig. 34.25) give the Young's modulus dependences of these two quantities for the same materials. Some very interesting trends are revealed by the contour maps. Adhesion is high for low-E materials and at low sliding velocities and it gradually decreases with an increase in velocity. The dependence of friction and adhesion on velocity for viscoelastic polymers (such as PDMS used in this study) is well known and it is defined by a definite peak occurring at a specific sliding velocity [34.13,38,111]. Similar behavior was found even for materials with higher Young's moduli where the adhe-

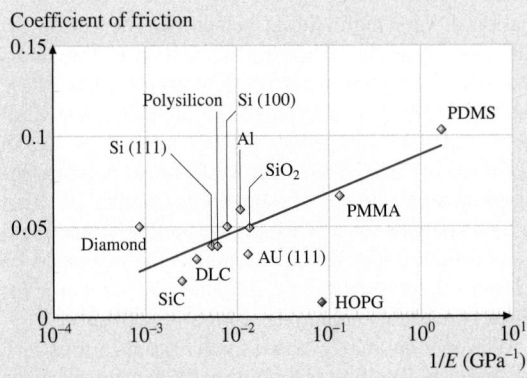

Fig. 34.24 Dependence of coefficient of friction on the Young's modulus for various materials. Measurements were made using Si_3N_4 AFM tips of nominal radius 30–50 nm in a controlled environment of $20 \pm 2\,°C$ and 50% RH

Fig. 34.25 Dependence of adhesive force and coefficient of friction on velocity for various materials, and contour plots giving these dependences as a function of Young's modulus [34.25]

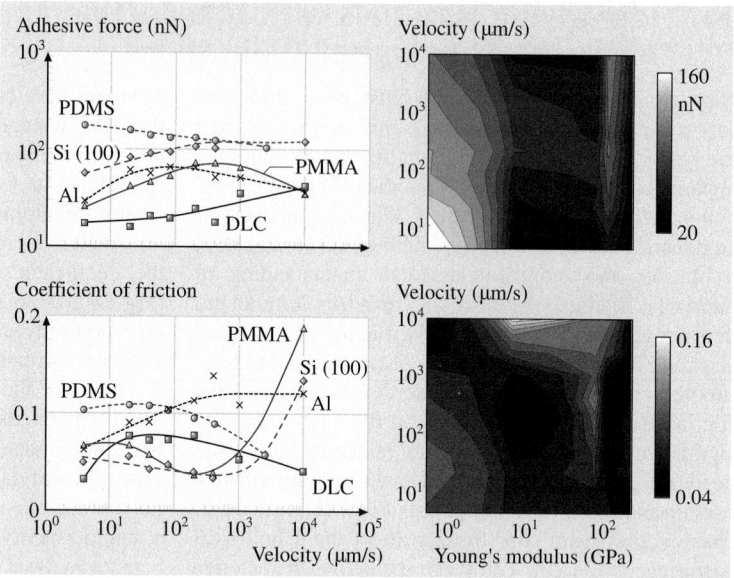

sive force increased with velocity and reached a peak. Moreover, the peak was attained at higher velocities for materials with higher E values. The contour map for the coefficient of friction also reveals peculiar trends.

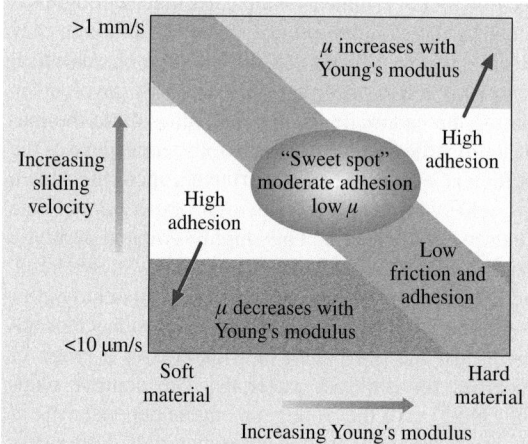

Fig. 34.26 Map for identifying materials with low friction and adhesion. At low velocities ($< 10\,\mu m/s$), the coefficient of friction (μ) decreases logarithmically with Young's modulus. At high velocities ($> 1\,m/s$), μ increases with increasing Young's modulus. The 'sweet spot' corresponds to materials with low μ and moderate adhesion. The ideal zone for material selection is, however, the one where both μ and adhesion are either low or moderate (the *shaded diagonal portion*) [34.25]

At low velocities, the coefficient of friction decreases with increasing Young's modulus and this decrease is found to be nearly logarithmic in nature (seen from Fig. 34.24). However, at high velocities, this trend is reversed and the coefficient of friction is found to increase with increasing Young's modulus. At high velocities, friction is primarily governed by impact deformations and plowing effects [34.13, 21]. Thus, while low-E materials are able to absorb most of the impacts during sliding, for the high-E materials, the impacts during sliding result in high friction. DLC is the only high-E material which shows a decrease in the coefficient of friction at very high sliding velocities. The reason for this anomaly is the phase transformation of the amorphous DLC to a low shear strength graphite-like phase [34.24, 60, 97]. The sharp Si_3N_4 tip used in AFM studies (30–50 nm radius) gives rise to contact pressures in the range of 1.8–3.8 GPa for DLC films, corresponding to normal loads of 10–100 nN (assuming Hertzian contact analysis for elastic single-asperity contact) and *Voevodin* et al. [34.97] have reported the formation of debris with polycrystalline graphite-like structure for contact pressures in the range of 0.8–1.1 GPa.

The contour map reveals a small central zone of very low friction. The corresponding adhesive force values for this zone are also moderate. This can be considered to be a 'sweet spot' and corresponds to an ideal material that a tribologist would like to choose. HOPG falls into this zone. Various other zones of interest are summarized in Fig. 34.26. They indicate the dependence of

Young's modulus on the operating parameters and thus can be used as a guide for selecting materials for various nanotechnological applications. We find that, from the tribology point of view, low-E materials show promise for high sliding velocity applications, while high-E materials are more suitable for lower sliding velocities.

34.8 Closure

A fundamental understanding of tribology and mechanics on the micro/nanoscale is central to the mitigation of various issues hindering the broad-based impact of nanotechnological applications on everyday lives. Further, it is imperative that these investigations be conducted at sliding velocities that are of engineering interest, those that are experienced by most real world micro/nanodevices. Nanotribological investigations conducted for various materials, coatings and lubricants that are considered potential solutions for nanotechnological applications have encompassed velocity regimes of scientific as well as engineering interest, thereby enabling the development of a broader and more fundamental understanding of nanoscale friction behavior.

A strong dependence of friction on velocity was found for all materials, coatings and lubricants. The comprehensive analytical friction model explains friction behavior scaling multiple regimes and also helps to identify the critical parameters corresponding to transitions between various friction regimes. At high velocities, deformation of contacting asperities from the high-velocity impacts was found to result in high friction for most of the samples. Si(100) showed the poorest performance and this is believed to be the result of a combination of deformation resulting from impacts of the contacting asperities and the corresponding three-body wear due to debris particles generated at the sliding interface and a tribochemical reaction which is accelerated by high velocities. For DLC, the high-energy dissipation results in phase transformation to a low shear strength graphite-like layer at the sliding interface which has low friction, so DLC shows better performance. At higher normal loads, however, the graphite-like debris can get ejected from the interface and increase friction. For the lubricants HDT, Z-15 and Z-DOL, performance was governed by the displacement of the mobile molecules or the removal of the bonded molecules from the sliding interface. For the polymers PMMA and PDMS, performance was governed by localized melting of the contact zone due to high energy dissipation.

The novel AFM-based approach based on simultaneous controlled movements of the sample and AFM tip provides a fast and efficient means of mapping the sample surface wear as a function of the sliding velocity, the applied normal load and the number of sliding cycles. For DLC wear, debris was believed to be generated as a result of the phase transformation of DLC to a graphite-like phase. The phase transformation process required a minimum threshold frictional energy dissipation and was found to occur only for certain combinations of sliding velocities, normal loads and number of sliding cycles. A nanowear map generated on DLC by simultaneously varying the sliding velocity and the normal load revealed this transition very well. The ability to achieve high sliding velocities that are close to those of engineering interest in conjunction with the novel approach developed for studying nanoscale wear can be adapted very easily to study any materials, coatings and lubricants. By selecting operating parameters appropriately, the nanoscale wear can be mapped in a fast and efficient manner and the competing wear mechanisms and their transitions can be studied.

Material maps created by plotting the coefficient of friction and adhesion as a function of the Young's modulus reveal various interesting facets of the behavior of nanoscale friction. For example, the coefficient of friction decreased with increasing velocity for materials with low Young's modulus, but the reverse was true for materials with high Young's modulus. The map shows that if the sliding velocity is likely to be high, a compliant material would perform better than a stiffer material.

Friction, adhesion and wear are important aspects of a nanotechnology with moving components. Finding a slippery surface is seen to be a subtle business. The performance and robustness of a MEMS/NEMS device might hinge on its frictional properties. The dependence of friction, adhesion and wear on velocity at the nanoscale implies that it is best to adapt materials to a particular nanotechnological application.

References

34.1   Y. C. Tai, L. S. Fan, R. S. Muller: IC–processed micro-motors: Design, technology and testing, Proc. IEEE MEMS **20**, 1–6 (1989)

34.2   S. M. Spearing, K. S. Chen: Micro gas turbine engine materials and structures, Ceram. Eng. Sci. Proc. **18**, 11–18 (1997)

34.3   Sandia Corp.: Sandia National Laboratories website, www.sandia.gov (2005)

34.4   H. Lehr, S. Abel, J. Doppler, W. Ehrfeld, B. Hagemann, K. P. Kamper, F. Michel, Ch. Schulz, Ch. Thurigeen: Microactuators as driving units for microrobotic systems. In: *Proc. Microrobotics: Components and Applications*, Vol. 2906, ed. by A. Sulzmann (SPIE, Lausanne 1996) pp. 202–210

34.5   R. P. Feynman: There's plenty of room at the bottom, Eng. Sci. **23**, 22–36 (1960)

34.6   K. Komvopoulos: Adhesion and friction forces in microelectromechanical systems: mechanisms, measurement, surface modification techniques and adhesion theory, J. Adhes. Sci. Technol. **17**, 477–517 (2003)

34.7   R. Maboudian, R. T. Howe: Critical review: Adhesion in surface micromachined structures, J. Vacuum Sci. Technol. B **15**, 1–20 (1997)

34.8   B. Bhushan: *Tribology Issues and Opportunities in MEMS* (Kluwer, Dordrecht 1998)

34.9   B. Bhushan: Adhesion and stiction: Mechanisms, measurement techniques, J. Vacuum Sci. Technol. B **21**, 2262–2296 (2003)

34.10   B. Bhushan (eds.): *Springer Handbook of Nanotechnology* (Springer, Berlin, Heidelberg 2004)

34.11   B. Bhushan: *Micro/Nanotribology and its Applications*, NATO ASI Series E: Applied Sciences, Vol. 330 (Kluwer, Dordrecht 1997)

34.12   B. Bhushan: *Handbook of Micro/Nanotribology*, 2nd edn. (CRC, Boca Raton 1999)

34.13   B. Bhushan: *Principles and Applications of Tribology* (Wiley, New York 1999)

34.14   B. Bhushan (ed): *Modern Tribology Handbook, Vol 2: Materials, Coatings and Industrial Applications* (CRC, Boca Raton 2001)

34.15   B. Bhushan: *Fundamentals of Tribology and Bridging the Gap between the Macro- and Micro/Nanoscales*, NATO Science Series II, Vol. 10 (Kluwer, Dordrecht 2001)

34.16   L. G. Frechette, S. A. Jacobson, K. S. Breuer, F. F. Ehrich, R. Ghodssi, R. Khanna, C. W. Wong, X. Zhang, M. A. Schmidt, A. Epstein: High-speed microfabricated silicon tribomachinery and fluid film bearings, J. Microelectromech. Sys. **14**, 141–152 (2005)

34.17   C. Muhlstein, S. Brown: Reliability and fatigue testing of MEMS. In: *Tribology Issues and Opportunities in MEMS*, ed. by B. Bhushan (Kluwer, Dordrecht 1997) pp. 519–528

34.18   D. M. Tanner, N. F. Smith, L. W. Irwin: *MEMS Reliability: Infrastructure, Test Structures, Experiments and Failure Modes* (Sandia National Laboratories, Alberquerque 2000) pp. 2000–2091

34.19   N. S. Tambe, B. Bhushan: Scale dependence of micro/nanofriction and -adhesion of MEMS/NEMS materials, coatings and lubricants, Nanotechnology **15**, 1561–1570 (2004)

34.20   N. S. Tambe, B. Bhushan: A new atomic force microscopy based technique for studying nanoscale friction at high sliding velocities, J. Phys. D **38**, 764–773 (2005)

34.21   N. S. Tambe, B. Bhushan: Friction model for velocity dependence of friction force, Nanotechnology **16**, 2309–2324 (2005)

34.22   N. S. Tambe, B. Bhushan: Durability studies of micro/nanoelectromechanical systems materials, coatings and lubricants at high sliding velocities (up to 10 mm/s) using a modified atomic force microscope, J. Vacuum Sci. Technol. A **23**, 830–835 (2005)

34.23   N. S. Tambe, B. Bhushan: Micro/nanotribological characterization of PDMS and PMMA used for BioMEMS/NEMS applications, Ultramicroscopy **105**, 238–247 (2005)

34.24   N. S. Tambe, B. Bhushan: Nanoscale friction induced phase transformation of diamondlike carbon, Scripta Mater. **52**, 751–755 (2005)

34.25   N. S. Tambe, B. Bhushan: Identifying materials with low friction and adhesion for nanotechnology applications, Appl. Phys. Lett. **86**, 061906:1–3 (2005)

34.26   N. S. Tambe, B. Bhushan: Nanotribological characterization of self-assembled monolayers deposited on silicon and aluminum substrates, Nanotechnology **16**, 1549–1558 (2005)

34.27   N. S. Tambe, B. Bhushan: Nanowear mapping: A novel atomic force microscopy based approach for mapping nanoscale wear at high sliding velocities, Tribol. Lett. **20**, 83–90 (2005)

34.28   N. S. Tambe, B. Bhushan: Nanoscale friction mapping, Appl. Phys. Lett. **86**, 193102 (2005)

34.29   H. Liu, B. Bhushan: Investigation of nanotribological and nanomechanical properties of DMD by an AFM, J. Vacuum Sci. Technol. A **22**, 1388–1396 (2004)

34.30   H. Liu, B. Bhushan: Nanotribological characterization of molecularly thick lubricant films for applications to MEMS/NEMS by AFM, Ultramicroscopy **97**, 321–340 (2003)

34.31   G. Wei, B. Bhushan, S. J. Jacobs: Nanomechanical characterization of multilayered thin film structures for digital micromirror devices, Ultramicroscopy **100**, 375–389 (2004)

34.32   B. Bhushan, Z. Burton: Adhesion and friction properties of polymers in microfluidic devices, Nanotechnology **16**, 467–478 (2005)

34.33   H. Liu, B. Bhushan: Adhesion and friction studies of microelectromechanical systems/nanoelectromechanical systems materials using a novel microtriboapparatus, J. Vacuum Sci. Technol. A **24**, 1528–1538 (2003)

34.34   G. Amontons: De la resistance cause dans les Machines, Mem. Acad. Roy. A, 257–282 (1699)

34.35   C. A. Coulomb: Theorie des Machines Simples, en ayant regard au Frottement de leurs Parties, et a la Roideur des Cordages, Mem. Math. Phys. **X**, 161–342 (1785) X, Paris

34.36   F. P. Bowden, D. Tabor: *The Friction and Lubrication of Solids, Part I* (Clarendon, Oxford 1950)

34.37   F. P. Bowden, D. Tabor: *The Friction and Lubrication of Solids, Part II* (Clarendon, Oxford 1964)

34.38   D. F. Moore: *The Friction and Lubrication of Elastomers*, 1st edn. (Pergamon, New York 1972)

34.39   I. L. Singer, H. M. Pollock: *Fundamentals of Friction: Macroscopic and Microscopic Processes*, NATO Series E: Applied Sciences (Kluwer, Boston 1992) p. 220

34.40   V. Koinkar, B. Bhushan: Microtribological studies of unlubricated and lubricated surfaces using atomic force/friction force microscopy, J. Vac. Sci. Technol. A **14**, 2378–2391 (1996)

34.41   B. Bhushan, A. V. Kulkarni: Effect of normal load on microscale friction measurements, Thin Solid Films **278**, 49–56 (1996)

34.42   T. Bouhacina, J. P. Aime, S. Gauthier, D. Michel: Tribological behavior of a polymer grafter on silanized silica probed with a nanotip, Phys. Rev. B **56**, 7694–7703 (1997)

34.43   T. Baumberger, P. Berthoud, C. Caroli: Physical analysis of the state- and rate-dependent friction law. II. Dynamic friction, Phys. Rev. B **60**, 3928–3939 (1999)

34.44   E. Riedo, E. Gnecco, R. Bennewitz, E. Meyer, H. Brune: Interaction potential and hopping dynamics governing sliding friction, Phys. Rev. Lett. **91**, 084502-1 (2003)

34.45   H. Liu, B. Bhushan: Investigation of nanotribological properties of self-assembled monolayers with alkyl and biphenyl spacer chains (invited), Ultramicroscopy **91**, 185–202 (2002)

34.46   E. Gnecco, R. Bennewitz, O. Pfeiffer, A. Socoliuc, E. Meyer: Friction and Wear on the Atomic Scale. In: *Springer Handbook of Nanotechnology*, ed. by B. Bhushan (Springer, Berlin, Heidelberg 2004) pp. 631–660

34.47   K. Yamanaka, E. Tomita: Lateral force modulation atomic force microscope for selective imaging of friction forces, Jpn. J. Appl. Phys. **34**, 2879–2882 (1995)

34.48   V. Scherer, B. Bhushan, W. Arnold: Lateral force microscopy using acoustic friction force microscopy, Surf. Interf. Anal. **27**, 578–586 (1999)

34.49   O. Marti, H.-U. Krotil: Dynamic friction measurement with the scanning force microscope. In: *Fundamentals of Tribology and Bridging the Gap between the Macro- and Micro/Nanoscales*, ed. by B. Bhushan (Kluwer, Dordrecht 2001) pp. 121–135

34.50   M. Reinstadler, U. Rabe, V. Scherer, U. Hartmann, A. Goldade, B. Bhushan, W. Arnold: On the nanoscale measurement of friction using atomic-force microscope cantilever torsional resonances, Appl. Phys. Lett. **82**, 2604–2606 (2003)

34.51   Anonymous: User manuals (PZ 106E, PZ-62E) and www.polytecpi.com, (2003)

34.52   N. S. Tambe: Nanotribological investigations of materials, coatings and lubricants for nanotechnology applications at high sliding velocities, Ph.D. dissertation, The Ohio State University; available from http://www.ohiolink.edu/etd/send-pdf.cgi?osu1109949835, (2005)

34.53   Digital Instruments, Inc.: Appendix D: Lithography. In: *Nanoscope® command reference manual, Version 4.42* (Digital Instruments, Inc., Santa Barbara 1999)

34.54   S. D. Senturia: *Microsystem Design* (Kluwer, Boston 2001)

34.55   M. Madou: *Fundamentals of Microfabrication: The Science of Miniaturization*, 2nd edn. (CRC, Boca Raton 2002)

34.56   L. J. Hornbeck, W. E. Nelson: Bistable deformable mirror device. In: *Spatial Light Modulators and Applications*, OSA Technical Digest Series, Vol. 8 (Optical Society of America, Washington DC, Washington DC 1988) p. 107

34.57   L. J. Hornbeck: The DMD™ projection display chip: A MEMS-based technology, MRS Bull. **26**, 325–328 (2001)

34.58   M. Mehregany, C. A. Zorman: SiC MEMS: Opportunities and challenges for applications in harsh environments, Thin Solid Films **355-356**, 518–524 (1999)

34.59   S. Sundararajan, B. Bhushan: Micro/nanotribological studies of polysilicon and SiC films for MEMS applications, Wear **217**, 251–261 (1998)

34.60   A. Grill: Tribology of diamondlike carbon and related materials: an updated review, Surf. Coat. Technol. **94-95**, 507–513 (1997)

34.61   B. Bhushan: Chemical, mechanical and tribological characterization of ultrathin and hard amorphous carbon coatings as thin as 3.5 nm: recent developments, Diamond Relat. Mater. **8**, 1985–2015 (1999)

34.62   S. R. Quake, A. Scherer: From micro- to nanofabrication with soft materials, Science **290**, 1536–1540 (2000)

34.63   W. C. Tang, A. P. Lee: Defense applications of MEMS, MRS Bull. **26**, 318–319 (2001)

Part D | 34

34.64   J. C. McDonald, G. M. Whitesides: Poly(dimethyl-siloxane) as a material for fabricating microfluidic devices, Acc. Chem. Res. **35**, 491–499 (2000)

34.65   H. Becker, L. E. Locascio: Polymer microfluidic devices, Talanta **56**, 267–287 (2002)

34.66   A. Ulman: *An Introduction to Ultrathin Organic Films: From Langmuir-Blodgett to Self-Assembly* (Academic, San Diego 1991)

34.67   A. Ulman: Formation and structure of self-assembled monolayers, Chem. Rev. **96**, 1533–1554 (1996)

34.68   L. J. Hornbeck: Low surface energy passivation layer for micromachined devices, US Patent No. 5,602,671, February 11, 1997

34.69   R. M. Wallace, S. A. Henck, D. Webb: A PFPE coating for micro-mechanical devices, US Patent No. 5,512,374, April 30, 1996

34.70   S. A. Henck: Lubrication of digital micromirror devices®, Tribol. Lett. **3**, 239–247 (1997)

34.71   S. H. Lee, M. J. Kwon, J. G. Park, Y. K. Kim, H. J. Shin: Preparation characterization of perfluoro-organic thin films on aluminium, Surf. Coat. Technol. **112**, 48–51 (1999)

34.72   K. K. Lee, N. G. Cha, J. S. Kim, J. G. Park, H. J. Shin: Chemical, optical and tribological characterization of perfluoropolymer films as an anti-stiction layer in micromirror arrays, Thin Solid Films **377-378**, 727–732 (2000)

34.73   C. S. Gudeman: Vapor phase low molecular weight lubricants, US Patent No. 6,251,842, B1, June 26, 2001

34.74   R. A. Robbins, S. J. Jacobs: Lubricant delivery for micro-mechanical devices, US Patent No. 6,300,294 B1, October 9, 2001

34.75   S. Chilamakuri, B. Bhushan: A comprehensive kinetic meniscus model for prediction of long-term static friction, J. Appl. Phys. **86**, 4649–4656 (1999)

34.76   E. Gnecco, R. Bennewitz, T. Gyalog, Ch. Loppacher, M. Bammerlin, E. Meyer, H.-J. Guntherodt: Velocity dependence of atomic friction, Phys. Rev. Lett. **84**, 1172–1175 (2000)

34.77   E. Riedo, F. Levy, H. Brune: Kinetics of capillary condensation in nanoscopic sliding friction, Phys. Rev. Lett. **88**, 185505–1 (2002)

34.78   C. Livermore, J. Voldman: Material properties database, http://www.mit.edu/~6.777/matprops/matprops.htm, (2004)

34.79   K. Mizuhara, S. M. Hsu: *Wear Particles*, ed. by D. Dowson eds., C. M. Taylor, T. H. C. Childs, M. Godet, G. Dalmaz (Elsevier, Amsterdam 1992) pp. 323–328

34.80   S. C. Clear, P. F. Nealey: Lateral force microscopy study of the frictional behavior of self-assembled monolayers of octadecyltrichlorosilane on silicon/silicon dioxide immersed in n-alcohols, Langmuir **17**, 720–732 (2001)

34.81   C. M. Mate, G. M. McClelland, R. Erlandsson, S. Chiang: Atomic-scale friction of a tungsten tip on a graphite surface, Phys. Rev. Lett. **59**, 1942–1945 (1987)

34.82   R. Bennewitz, E. Meyer, M. Bammerlin, T. Gyalog, E. Gnecco: Atomic-scale stick slip. In: *Fundamentals of Tribology and Bridging the Gap between the Macro- and Micro/Nanoscales*, ed. by B. Bhushan (Kluwer, Dordrecht 2001) pp. 53–66

34.83   J. N. Israelachvili, Y.-L. Chen, H. Yoshizawa: Relationship between adhesion and friction forces, J. Adhes. Sci. Technol. **8**, 1231–1249 (1994)

34.84   J. Ruan, B. Bhushan: Atomic scale friction measurements using friction force microscopy: Part I. General principles and new measurement technique, ASME J. Tribol. **116**, 378–388 (1994)

34.85   F. P. Bowden, L. Leben: The nature of sliding and analysis of friction, Proc. R. Soc. Lond. A **169**, 371–379 (1939)

34.86   E. Rabinowicz: *Friction and Wear of Materials*, 2nd edn. (Wiley, New York 1995)

34.87   G. A. Tomlinson: A molecular theory of friction, Philos. Mag. Ser. **7**, 905–939 (1929)

34.88   D. Tománek, W. Zhong, H. Thomas: Calculation of an atomically modulated friction force in atomic-force microscopy, Europhys. Lett. **15**, 887–892 (1991)

34.89   L. Bouquet, E. Charlaix, S. Ciliberto, J. Crassous: Moisture-induced ageing in granular media and the kinetics of capillary condensation, Nature **396**, 735–737 (1998)

34.90   R. Evans: *Liquids at Interfaces*, ed. by J. Charvolin, J. F. Joanny, J. Zinn-Justin (Elsevier, New York 1989) pp. 3–98

34.91   B. Bhushan: *Tribology and Mechanics of Magnetic Storage Devices*, 2nd edn. (Springer, Berlin, Heidelberg 1996)

34.92   J. I. Siepmann, I. R. McDonald: Monte Carlo simulation of the mechanical relaxation of a self-assembled monolayer, Phys. Rev. Lett. **70**, 453–456 (1993)

34.93   M. Garcia-Parajo, C. Longo, J. Servat, P. Gorostiza, F. Sanz: Nanotribological properties of octadecyltrichlorosilane self-assembled ultrathin films studied by atomic force microscopy: contact and tapping modes, Langmuir **13**, 2333–2339 (1997)

34.94   D. Gourdon, N. A. Burnham, A. Kulik, E. Dupas, F. Oulevey, G. Gremaud: The dependence of friction anisotropies on the molecular organization of LB films as observed by AFM, Tribol. Lett. **3**, 317–324 (1997)

34.95   S. A. Joyce, R. C. Thomas, J. E. Houston, T. A. Michalske, R. M. Crooks: Mechanical relaxation of organic monolayer films measured by force microscopy, Phys. Rev. Lett. **68**, 2790–2793 (1992)

34.96   D. Tabor: Proc. First Int. Skid Prevention Conf., Part 1, (1958)

34.97   A. A. Voevodin, A. W. Phelps, J. S. Zabinski, M. S. Donley: Friction-induced phase transformation of pulsed laser deposited diamond-like carbon, Diamond Relat. Mat. **5**, 1264–1269 (1996)

34.98   C. L. Muhlstein, S. B. Brown, R. O. Ritchie: High-cycle fatigue of single crystal silicon thin films, J. Microelectromech. Syst. **10**, 593–600 (2001)

34.99   Image Metrology A/S: Scanning Probe Image Processor, Version 3.2.2.0. (Image Metrology A/S, Lyngby 2004)

34.100 S. Sundararajan, B. Bhushan: Development of a continuous microscratch technique in an atomic force microscope and its applications to study scratch resistance of ultra-thin hard amorphous carbon coatings, J. Mater. Res. **16**, 437–445 (2001)

34.101 H. Liu, B. Bhushan: Orientation and relocation of biphenyl thiol self-assembled monolayers under sliding, Ultramicroscopy **91**, 177–183 (2002)

34.102 B. Bhushan, H. Liu: Nanotribological properties and mechanisms of alkylthiol and biphenyl thiol self-assembled monolayers studied by AFM, Phys. Rev. B **63**, 245412-1 (2001)

34.103 J. E. Field: *The Properties of Natural and Synthetic Diamond* (Academic, London 1992)

34.104 Institution of Electrical Engineers: Properties of Silicon, EMIS Data Reviews Series No. 4, (INSPEC, London 1988)

34.105 Anonymous: MEMS Materials Database, http://www.memsnet.org/material/ (2002)

34.106 Anonymous: MatWeb (Material Property Data), http://www.matweb.com (2004)

34.107 B. Bhushan, B. K. Gupta: *Handbook of Tribology: Materials, Coatings and Surface Treatments* (McGraw-Hill, New York 1997)

34.108 G. Wei: The Young's modulus and hardness were obtained on NanoIndentor II (MTS Systems Corporation) using continuous stiffness measurement technique. The indentation depth was 200 nm with oscillation frequency of 45 Hz, oscillation amplitude of 1 nm (2004)

34.109 S. C. Lim, M. F. Ashby: Wear-mechanism maps, Acta Metall. **35**, 1–24 (1987)

34.110 S. C. Lim, M. F. Ashby, J. H. Brunton: Wear-rate transitions and their relationship to wear mechanisms, Acta Metall. **35**, 1343–1348 (1987)

34.111 B. N. J. Persson: *Sliding friction: Physical principles and applications*, 2nd edn. (Springer, Berlin, Heidelberg 2000)

Part D | 34

35. Computer Simulations of Nanometer-Scale Indentation and Friction

Engines and other machines with moving parts are often limited in their design and operational lifetime by friction and wear. This limitation has motivated the study of fundamental tribological processes with the ultimate aim of controlling and minimizing their impact. The recent development of miniature apparatus, such as microelectromechanical systems (MEMS) and nanometer-scale devices, has increased interest in atomic-scale friction, which has been found to, in some cases, be due to mechanisms that are distinct from the mechanisms that dominate in macroscale friction.

Presented in this chapter is a review of computational studies of tribological processes at the atomic and nanometer scale. In particular, a review of the findings of computational studies of nanometer-scale indentation, friction and lubrication is presented, along with a review of the salient computational methods that are used in these studies, and the conditions under which they are best applied.

Engines and other machines with moving parts are often limited in their design and operational lifetime by friction and wear. This limitation has motivated the study of tribological processes with the aim of controlling and minimizing the impact of these processes. There are numerous historical examples that illustrate the importance of friction to the development of civilizations, including the ancient Egyptians who invented technologies to move the stones used to build the pyramids [35.1]; Coulomb, who was motivated to study friction by the need to move ships easily and without wear from land to the water [35.1]; and *Johnson* et al. [35.2], who developed an improved understanding of contact mechanics and surface energies through the study of automobile windshield wipers. At present, substantial research and development is aimed at microscale and nanoscale machines with moving parts that at times challenge our fundamental understanding of friction and wear. This has motivated the study of atomic-scale friction and has, consequently, led to new discoveries such as self-lubricating surfaces and wear-resistant materials. While there are similarities between friction at the macroscale and the atomic scale, in many instances the mechanisms that lead to friction at these two scales are quite different. Thus, as devices such as magnetic storage disks and microelectromechanical systems (MEMS) [35.3] continue to shrink in size, it is expected that new phenomena associated with atomic-scale friction, adhesion and wear will dominate the functioning of these devices.

The last two decades have seen considerable scientific effort expended on the study of atomic-scale friction [35.4–17]. This effort has been facilitated by the development of new advanced experimental tools to measure friction over nanometer-scale distances at low loads, rapid improvements in computer power, and the maturation of computational methodologies for the modeling of materials at the atomic scale. For example, friction-force and atomic-force microscopes (FFM and AFM) allow the frictional properties of solids to be characterized with atomic-scale resolution under single-asperity indentation and sliding conditions [35.18–21]. In addition, the surface force apparatus (SFA) provides data about the tribological and lubrication responses of many liquid and solid systems with atomic resolution [35.22], and the quartz crystal microbalance (QCM) provides information about the atomic-scale origins of friction [35.4, 23]. These and related experimental methods allow researchers to study sliding surfaces at the atomic scale and relate the observed phenomena to macroscopically observed friction, lubrication and wear.

Analytic models and computational simulations have played an important role in characterizing and understanding friction. They can, for example, assist in the interpretation of experimental data or provide predictions that subsequent experiments can confirm or refute. Analytic models have long been used to study friction, including early studies by *Tomlinson* [35.24] and *Frenkel* and *Kontorova* [35.25] and more recent studies by *McClelland* et al. [35.26], *Sokoloff* [35.13, 27–33], *Persson* [35.34–37] and others [35.38–44]. Most of these idealized models divide the complex motions that create friction into more fundamental components defined by quantities such as spring constants, the curvature and magnitude of potential wells, and bulk phonon frequencies. While these simplifications provide these approaches with some predictive capabilities, many assumptions must be made in order to be able to apply these models to study friction, which may lead to incorrect or incomplete results.

In atomic-scale molecular dynamics (MD) simulations, atom trajectories are calculated by numerically integrating coupled classical equations of motion. Interatomic forces that enter these equations are typically calculated either from total energy methods that include electronic degrees of freedom, or from simplified mathematical expressions that give the potential energy as a function of interatomic displacements. MD simulations can be considered numerical experiments that provide a link between analytic models and experiments. The main strength of MD simulations is that they can reveal unanticipated phenomena or unexpected mechanisms for well-known observations. Weaknesses include a lack of quantum effects in classical atomistic dynamics, and perhaps more importantly, the fact that meaningless results can be obtained if the simulation conditions are chosen incorrectly. The next section contains a review of MD simulations, including the approximations that are inherent in their application to the study of friction, and the conditions under which they should and should not be applied.

35.1 Computational Details

Molecular dynamics simulations are straightforward to describe: given a set of initial conditions and a way of mathematically modeling interatomic forces, Newton's (or equivalent) classical equation of motion is numerically integrated [35.45]

$$F = ma \,, \tag{35.1a}$$

$$-\nabla E = m(\partial^2 r / \partial t^2) \,, \tag{35.1b}$$

where F is the force on each atom, m is the atomic mass, a is the atomic acceleration, E is the potential energy felt by each atom, r is the atomic position, and t is time. The forces acting on any given atom are calculated, and then the atoms move a short increment ∂t (called a time step) forward in time in response to these applied forces. This is accompanied by a change in atomic positions, velocities and accelerations. The process is then repeated for some specified number of time steps.

The output of these simulations includes new atomic positions, velocities, and forces that allow additional quantities such as temperature and pressure to be determined. As the size of the system increases, it is useful to render the atomic positions in animated movies that reveal the responses of the system in a qualitative manner. Quantitative data can be obtained by analyzing the numerical output directly.

The following sections review the way in which energies and forces are calculated in MD simulations and the important approximations that are used to realistically model the friction that occurs in experiments with smaller systems of only a few tens of thousands of atoms in simulations. The reader is referred to additional sources [35.46–52] for a more comprehensive overview of MD simulations (including computer algorithms) and the potentials that are used to calculate energies and forces in MD simulations.

35.1.1 Energies and Forces

There are several different approaches by which interatomic energies and forces are determined in MD simulations. The most theoretically rigorous methods are those that are classified as *ab initio* or first principles. These techniques, which include density functional theory [35.53, 54] and quantum chemical *ab initio* [35.55] methods, are derived from quantum mechanical principles and are generally both the most accurate and the most computationally intensive. They are therefore limited to a small number of atoms (< 500), which has limited their use in the study of friction. Alternatively, empirical methods are functions containing parameters that are determined by fitting to experimental data or the results of *ab initio* calculations [35.50]. These techniques can usually be relied on to correctly describe qualitative trends and are often the only choice available for modeling systems containing tens of thousands, millions, or billions of atoms. Empirical methods have therefore been widely used in studies of friction. Semi-empirical methods, including tight-binding methods, include some elements of both empirical methods and *ab initio* methods. For instance, they require quantum mechanical information in the form of, for example, on-site and hopping matrix elements, and include fits to experimental data [35.56].

Empirical methods simplify the modeling of materials by treating the atoms as spheres that interact with each other via repulsive and attractive terms that can be either pairwise additive or many-body in nature. In this approach, electrons are not treated explicitly, although it is understood that the interatomic interactions are ultimately dependent on them. As discussed in this section, some empirical methods explicitly include charge through classical electrostatic interactions, although most methods assume charge-neutral systems. The repulsive and attractive functional forms generally depend on interatomic distances and/or angles and contain adjustable parameters that are fit to *ab initio* results and/or experimental data.

The main strength of empirical potentials is their computational speed. Recent simulations with these approaches have modeled billions of atoms [35.57], something that is not possible with *ab initio* or semi-empirical approaches at this time. The main weakness of empirical potentials is their lack of quantitative accuracy, especially if they are poorly formulated or applied to systems that are too far removed from the fitting database used in their construction. Furthermore, because of the differences in the nature of chemical bonding in various materials, such as covalent bonding in carbon versus metallic bonding in gold, empirical methods have been historically derived for particular classes of materials. They are therefore generally nontransferable, although some methods have been shown to be theoretically equivalent [35.51, 58], and in recent years there has been progress towards the development of empirical methods that can model heterogeneous material systems [35.59–64].

Several of the most important and common general classes of empirical methods used for calculating interatomic energies and forces in materials, the so-called potentials, are reviewed here. The first to be considered are the potentials that are used to model covalently bound materials, including the bond-order potential and the Stillinger-Weber potential.

The bond-order potential was first formulated by *Abell* [35.65] and subsequently developed and parameterized by *Tersoff* for silicon and germanium [35.66,67], *Brenner* and coworkers for hydrocarbons [35.52, 68, 69], *Dyson* and *Smith* for carbon–silicon–hydrogen systems [35.70], *Sinnott* and coworkers for carbon–oxygen–hydrogen systems [35.71], and *Graves* and coworkers [35.72] and *Sinnott* and coworkers [35.73] for fluorocarbons.

The bond-order potential has the general functional form

$$E = \sum_i \sum_{j(>i)} [V_R(r_{ij}) - b_{ij} V_A(r_{ij})] \qquad (35.2)$$

where $V_R(r)$ and $V_A(r)$ are pair-additive interactions that model the interatomic repulsion and electron–nuclear attraction, respectively. The quantity r_{ij} is the distance between pairs of nearest-neighbor atoms i and j, and b_{ij} is a bond-order term that takes into account the many-body interactions between atoms i and j, including those due to nearest neighbors and angle effects. The potential is short-ranged and only considers nearest neighbor bonds. To model long-range nonbonded interactions, the bond-order potential is combined with pair-wise potentials either directly through splines [35.74] or indirectly with more sophisticated functions [35.75].

The *Stillinger-Weber* potential [35.76] potential was formulated to model silicon, with a particular emphasis on the liquid phases of silicon. It includes many-body interactions in the form of a sum of two- and three-body interactions

$$E = \sum_{ij} V_{ij}^2(r_{ij}) + \sum_{jik} V_{jik}^3(r_{ij}, r_{ik}) , \qquad (35.3)$$

where V^2 is a pair-additive interaction and V^3 is a three-body term. The three-body term includes an angular interaction that minimizes the potential energy for tetrahedral angles. This term favors the formation of open structures, such as the diamond cubic crystal structure.

The second potential is the embedded atom method (EAM) approach [35.77, 78] and related methods [35.79], which were initially developed for modeling metals and alloys. The functional form in the EAM is

$$E = \sum_i F(\rho_i) + \sum_{i>j} \Phi(r_{ij}) \, , \qquad (35.4)$$

where F is called the embedding energy. This term models the energy due to embedding an atom into a uniform electron gas with a uniform compensating positive background (jellium) of density ρ_i that is equal to the actual electron density of the system. The term $\Phi(r_{ij})$ is a pairwise functional form that corrects for the jellium approximation. Several parameterizations of the EAM exist (see, for example, [35.77, 78, 80–82]) and it has recently been extended to model nonmetallic systems. For example, the modified EAM (MEAM) approach [35.64, 83, 84] was developed so that EAM could be applied to metal oxides [35.60] and covalently bound materials [35.84].

The third method is the general class of Coulomb or multipole interaction potentials used to model charged ionic materials or molecules [35.85]. In this formalism, an energy term is given as

$$E = \sum_i \sum_{j(>i)} [(q(r_i)q(r_j)/r_{ij})] \, , \qquad (35.5)$$

where $q(r_i)$ is the charge on atom i and r_{ij} is the distance between atoms i and j. More complex formalisms that take into account, say, the Madelung constant in the case of ionic crystals, are used in practice. In general, the charges are held fixed, but methods that allow charge to vary in a realistic manner have been developed [35.61, 86].

Lastly, long-range van der Waals or related forces are typically modeled with pairwise additive potentials. A widely used approximation is the Lennard-Jones (LJ) potential [35.87], which has the following functional form:

$$E = 4\varepsilon \sum_i \sum_{j(>i)} \left[(\sigma/r_{ij})^{12} - (\sigma/r_{ij})^6 \right] \, . \qquad (35.6)$$

In this approach ε and σ are parameters and r_{ij} is the distance between atoms i and j.

All of these potentials are widely used in MD simulations of materials, including studies of friction, lubrication, and wear.

35.1.2 Important Approximations

Several approximations are typically used in MD simulations of friction. The first is the use of periodic boundary conditions (PBCs) and the minimum image convention for interatomic interactions [35.48]. In both cases the simulation supercell is surrounded by replicas of itself so that atoms (or phonons, etc.) that exit one side of the supercell remerge into the simulation through the opposite side of the supercell. In the minimum image convention an atom interacts either with another atom in the supercell or its equivalent atom in a surrounding cell depending on which distance to the atom is shortest. This process is illustrated in Fig. 35.1. In this convention supercells must be large enough that atoms do not interact with themselves over the periodic boundaries. In computational studies of friction and wear, PBCs are usually applied in the two dimensions within the plane(s) of the sliding surface(s). The strength of this approach is that it allows a finite number of atoms to model an infinite system. However, the influence of boundaries on system dynamics is not completely eliminated; for example, phonon scattering due to the periodic boundaries can influence heat transport and therefore frictional properties of sliding interfaces.

Another important tool that is often used in MD simulations of friction is thermostats to regulate system temperature. In macroscopic systems, heat that is generated from friction is dissipated rapidly from the surface to the bulk phonon modes. Because atomistic computer simulations are limited systems that are many orders of magnitude smaller than systems that are generally studied experimentally, thermostats are needed to prevent the

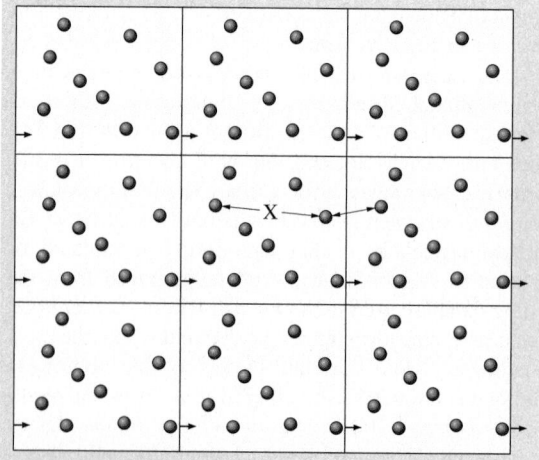

Fig. 35.1 Illustration of periodic boundary conditions consisting of a central simulation cell surrounded by replica systems. The *solid arrows* indicate an atom leaving the central box and re-entering on the opposite side. The *dotted arrows* illustrate the minimum image convention

system temperature from rising in a nonphysical manner. Typically in simulations of indentation or friction, the thermostat is applied to a region of the simulation cell that is well removed from the interface where friction and indentation is taking place. In this way, local heating of the interface that occurs as work is done on the system, but excess heat is efficiently dissipated from the system as a whole. In this manner the adjustment of atomic temperatures occurs away from the processes of interest, and simplified approximations for the friction term can be used without unduly influencing the dynamics produced by the interatomic forces.

There are several different formalisms for atomistic thermostats. The simplest of these controls the temperature by intermittently rescaling the atomic velocities to values corresponding to the desired temperature [35.88] such that

$$\left(\frac{v_{\text{new}}}{v_{\text{old}}}\right)^2 = \frac{T}{T_{\text{ins}}}, \tag{35.7}$$

where v_{new} is the rescaled velocity, and v_{old} is the velocity before the rescaling. This approach, which is called the velocity rescaling method, is both simple to implement and effective at maintaining a given temperature over the course of an MD simulation. It was consequently widely used in early MD simulations. The velocity rescaling approach does have some significant disadvantages, however. First, there is little theoretical basis for the adjustment of atomic velocities, and the system dynamics are not time-reversible, which is inconsistent with classical mechanics. Second, the rate and mode of heat dissipation are disconnected from system properties, which may affect system dynamics. Lastly, for typical MD simulation system sizes, the averaged quantities that are obtained, such as pressure for instance, do not correspond to values in any thermodynamic ensemble.

For these reasons, more sophisticated methods for maintaining system temperatures in MD simulations have been developed. The Langevin dynamics approach [35.48], which was originally developed from the theory of Brownian motion, falls into this category. In this approach, terms are added to the interatomic forces that correspond to a random force and a frictional term [35.46, 89, 90]. Therefore, Newton's equation of motion for atoms subjected to Langevin thermostats is given by the following equation rather than (35.1a, 1b):

$$m\mathbf{a} = \mathbf{F} - m\xi\mathbf{v} + R(t), \tag{35.8}$$

where $\mathbf{F}$ are the forces due to the interatomic potential, ξ is a friction coefficient, m and $\mathbf{v}$ are the particle's mass and velocity, respectively, and $R(t)$ is a random force that acts as "white noise". The friction term can be formulated in terms of a memory kernal, typically for harmonic solids [35.91–93], or a friction coefficient can be approximated using the Debye frequency. The random force can be given by a Gaussian distribution where the width, which is chosen to satisfy the fluctuation-dissipation theorem, is determined from the equation

$$\langle R(0)R(t)\rangle = 2mk_{\text{B}}T\xi\delta(t). \tag{35.9}$$

Here, the function R is the random force in (35.8), m is the particle mass, T is the desired temperature, k_{B} is Boltzmann's constant, t is time, and ξ is the friction coefficient. It should be noted that the random forces are uncoupled from those at previous steps, which is denoted by the delta function. Additionally, the width of the Gaussian distribution from which the random force is obtained varies with temperature. Thus, the Langevin approach does not require any feedback from the current temperature of the system as the random forces are determined solely from (35.9).

In the early 1980s, *Nosé* developed a new thermostat that corresponds directly to a canonical ensemble (system with constant temperature, volume and number of atoms) [35.94, 95], which is a significant advance from the methods described so far. In this approach, Nosé introduces a degree of freedom s that corresponds to the heat bath and acts as a time scaling factor, and adds a parameter Q that may be regarded as the heat bath "mass". A simplified form of Nosé's method was subsequently implemented by *Hoover* [35.46] that eliminated the time scaling factor whilst introducing a thermodynamic friction coefficient ζ. Hoover's formulation of Nosé's method is therefore easy to use and is commonly referred to as the Nosé–Hoover thermostat.

When this thermostat is applied to a system containing N atoms, the equations of motion are written as (dots denote time derivatives):

$$\dot{\mathbf{r}}_i = \frac{\mathbf{p}_i}{m_i},$$
$$\dot{\mathbf{p}}_i = \mathbf{F}_i - \zeta\mathbf{p}_i,$$
$$\dot{\zeta} = \frac{1}{Q}\left(\sum_{i=1}^{N}\frac{p_i^2}{m_i} - N_f k_{\text{B}}T\right), \tag{35.10}$$

where $\mathbf{r}_i$ is the position of atom i, $\mathbf{p}_i$ is the momentum and $\mathbf{F}_i$ is the force applied to each atom. The last equation in (35.10) contains the temperature control mechanism in the Nosé–Hoover thermostat. In particular, the term between the parentheses on the right-hand side of this equation is the difference between the system's instantaneous kinetic energy and the kinetic energy at the desired temperature. If the instantaneous value is higher than the desired one, the friction force will increase to lower it and vice versa.

It should be pointed out that the choice of the heat bath "mass" Q is arbitrary but crucial to the successful performance of the thermostat. For example, a small value of Q leads to rapid temperature fluctuation while large Q values result in inefficient sampling of phase space. Nosé recommended that Q should be proportional to $N_f k_B T$ and should allow the added degree of freedom s to oscillate around its averaged value at a frequency of the same order as the characteristic frequency of the physical system [35.94, 95]. If ergodic dynamic behavior is assumed, the Nosé–Hoover thermostat will maintain a well-defined canonical distribution in both momentum and coordinate space. However, for small systems where the dynamic is not ergodic, the Nosé–Hoover thermostat fails to generate a canonical distribution. Therefore, more sophisticated algorithms based on the Nosé–Hoover thermostat have been proposed to fix its ergodicity problem; for example, the "Nosé–Hoover chain" method of *Martyna* et al. [35.96]. However, these complex thermostats are not as easy to apply as the Nosé–Hoover thermostat due to the difficult evaluation of the coupling parameters for each different case and the significant computational cost [35.97]. From a practical point of view, if the molecular system is large enough that the movements of the atoms are sufficiently chaotic, ergodicity is guaranteed and the performance of the Nosé–Hoover thermostat is satisfactory [35.25].

In an alternative approach, *Schall* et al. recently introduced a hybrid continuum-atomistic thermostat [35.98]. In this method, an MD system is divided into grid regions, and the average kinetic energy in the atomistic simulation is used to define a temperature for each region. A continuum heat transfer equation is then solved stepwise on the grid using a finite difference approximation, and the velocities of the atoms in each grid region are scaled to match the solution of the continuum equation. To help account for a time lag in the transfer of kinetic to potential energy, Hoover constraining forces are added to those from the interatomic potential. This process is continued, leading to an ad hoc feedback between the continuum and atomistic simulations. The main advantage of this approach is that the experimental thermal diffusivity can be used in the continuum expression, leading to heat transfer behavior that matches

experimental data. For example, in metals the majority of the thermal properties at room temperature arise from electronic degrees of freedom that are neglected with strictly classical potentials. This thermostat is relatively straightforward to implement, and requires only the interatomic potential and the bulk thermal diffusivity as input. It is also appropriate for nonequilibrium heat transfer, such as occurs as heat is dissipated from sliding surfaces moving at high relative velocities.

Cushman et al. [35.99, 100] developed a unique alternative to the grand canonical ensemble by performing a series of grand canonical Monte Carlo simulations [35.48, 101] at various points along a hypothetical sliding trajectory. The results from these simulations are then used to calculate the correct particle numbers at a fixed chemical potential, which are then used as inputs to nonsliding, constant-*NVE* MD simulations at each of the chosen trajectory points. The sliding speed can be assumed to be infinitely slow because the system is fully equilibrated at each step along the sliding trajectory. This approach offers a useful alternative to continuous MD simulations that are restricted to sliding speeds that are orders of magnitude larger than most experimental studies (about 1 m/s or greater).

To summarize, this section provides a brief review and description of components that are used in atomistic, molecular dynamics simulation of many of the processes related to friction, such as indentation, sliding, and wear. The components discussed here include the potential energy expression used to calculate energies and forces in the simulations, periodic boundary conditions and thermostats. Each of these components has their own strengths and weaknesses that should be well-understood both prior to their use and in the interpretation of results. For example, general principles related to liquid lubrication in confined areas may be most easily understood and generalized from simulations that use pair potentials and may not require a thermostat. On the other hand, if one wants to study the wear or indentation of a surface of a particular metal, then EAM or other semiempirical potentials, together with a thermostat, would be expected to yield more reliable results. If one requires information on electronic effects, *ab initio* or semi-empirical approaches that include the evaluation of electronic degrees of freedom must be used. Thus, the best combination of components for a particular study depends on the chemical nature of the system of interest, the processes being simulated, the type of information desired, and the available computational resources.

35.2 Indentation

It is critical to understand the nanometer-scale properties of materials that are being considered for use as new coatings with specific friction and wear behavior. Experimental determination of these properties is most frequently done with the AFM, which provides a variety of data related to the interaction of the microscope tips with the sample surface [35.102–104]. In AFM experiments, the tip has a radius of about 1–100 nm and is pressed against the surface under ambient conditions (in air), ultrahigh vacuum (UHV) conditions, or in a liquid. The microscope tip can either move in the direction normal to the surface, which is the case in nanoindentation studies, or raster across the surface, which is the case in surface imaging or friction studies. Sliding rates of 1 nm/s–1 μ/s are typically used, which are many orders of magnitude slower than the rates used in MD simulations of sliding or indentation of around 1–100 m/s. As discussed in the previous section, the higher rates used in computational simulations are a consequence of modeling full atomic motion, which occurs on a femto- to picosecond timescale, and the stepwise solution of the classical equations of motion, which makes the large number of simulation steps needed to reach experimental timescales computationally impossible with current processor speeds.

As the tip moves either normal to or across the surface, the forces acting upon it as a result of its interactions with the surface are measured. When the tip is moved in the surface normal direction, it can penetrate the surface on the nanometer scale and provide information on the nanometer-scale mechanical properties of the surface [35.105, 106]. The indentation process also causes the force on the tip to increase, and the rate of increase is related to both the depth of indentation and the properties of the surface. The region of the force curve that reflects this high force is known as the repulsive wall region [35.102], or, when considered without any lateral motion of the tip, an indentation curve. When the tip is retracted after indentation, enhanced adhesion between the tip and surface relative to the initial contact can result. This phenomenon is indicated by hysteresis in the force curve.

Tip–surface adhesion can result from the formation of chemical bonds between the tip and the sample, or from the formation of liquid capillaries between the microscope tip and the surface caused by the interaction of the tip with a layer of liquid contamination on the surface. The latter case is especially prevalent in AFM studies conducted in ambient environments. In the case of clean metallic systems, the sample can wet the tip or the tip can wet the sample in the form of a connective "neck" of metal atoms between the surface and the tip that can lead to adhesion. In the case of polymeric or molecular systems, entanglement of molecules that are anchored on the tip with molecules anchored on the sample can be responsible for force curve hysteresis.

In the case of horizontal rastering of AFM tips across surfaces, the force curve data provide a map of the surface that is indicative of the surface topography [35.107]. If the deflection of the tip in the lateral direction is recorded while the tip is being rastered, a friction map of the surface [35.20] is produced.

The rest of this section discusses some of the important insights and findings that have been obtained from MD simulations of nanoindentation. These studies have not only provided insight into the physical phenomena responsible for the qualitative shapes of AFM force curves, they have also revealed a wealth of atomic-scale phenomena that occur during nanoindentation that was not previously known.

35.2.1 Surfaces

The nature of adhesive interactions between clean, deformable metal tips indenting metal surfaces have been identified and clarified over the course of the last decade through the use of MD simulations [35.104, 108–112]. In particular, the high surface energies associated with clean metal surfaces can lead to strongly attractive interactions between surfaces in contact. The strength of this attraction can be so large that when the tip gets close enough to the surface to interact with it, surface atoms "jump" upwards to wet the tip in a phenomena known as jump-to-contact (JC). This wetting mechanism was first discovered in MD simulations [35.111] and has been confirmed experimentally [35.105, 113–115] using the AFM, as shown in Fig. 35.2.

The MD simulations of *Landman* et al. [35.111, 116–118] using EAM potentials revealed that the JC phenomenon in metallic systems is driven by the need of the atoms at the tip-surface interface to optimize their interaction energies while maintaining their individual material cohesive binding. When the tip advances past the JC point it indents the surface, which causes the force to increase. This behavior is indicated in Fig. 35.3, points D to M. This region of the computer-generated force curve has a maximum not present in the force curve

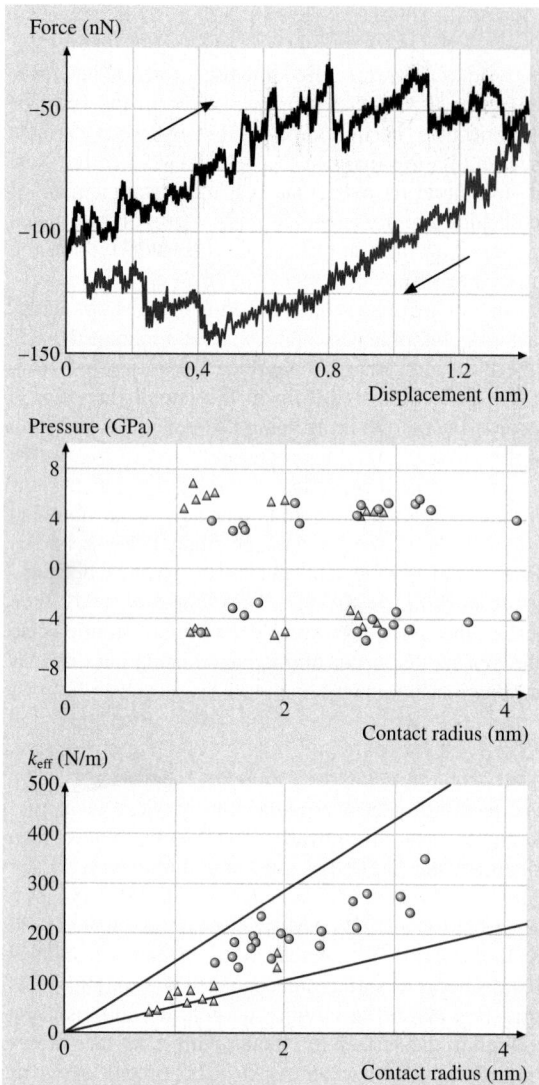

Fig. 35.2 *Top*: The experimental values for the force between a tip and a surface that have a connective neck between them. The neck contracts and extends without breaking on the scales shown. *Bottom*: The effective spring constant k_{eff} determined experimentally for the connective necks and corresponding maximum pressures, versus contact radius of the tip. The *triangles* indicate measurements taken at room temperature; the *circles* are the measurements taken at liquid He temperatures. After [35.115], with permission of the ACS (1996)

surface that causes "pile-up" of the surface atoms around the edges of the indenter. Hysteresis on the withdrawal of the tip, shown in Fig. 35.3, points M to X, is present due to adhesion between the tip and the substrate. In particular, as the tip retracts from the sample, a connective "neck" or nanowire of atoms forms between the tip and the substrate that is primarily composed of metal atoms from the surface with some atoms from the metal indenter that have diffused into the structure. A snapshot from the MD simulations that illustrates this behavior is shown in Fig. 35.4.

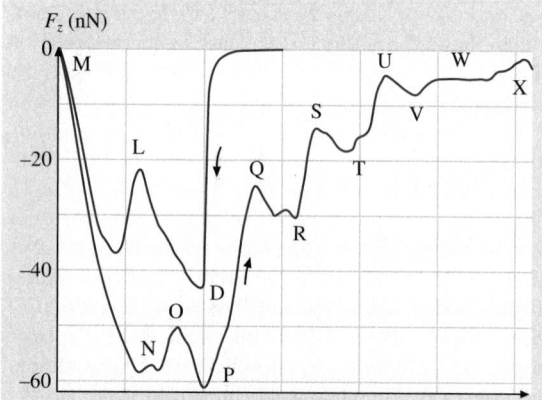

Fig. 35.3 Computationally derived force F_z versus tip-to-sample distance d_{hs} curves for approach, contact, indentation, then separation using the same tip–sample system shown in Fig. 35.4. These data were calculated from an MD simulation. After [35.111] with the permission of the AAAS (1990)

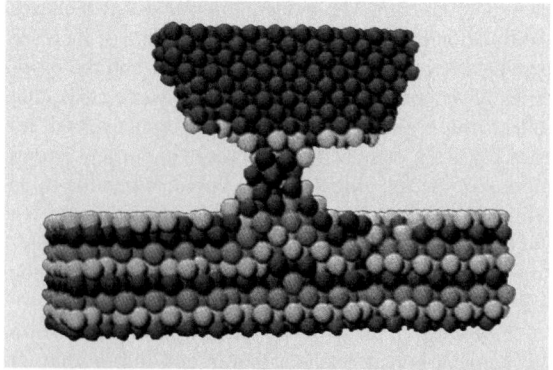

Fig. 35.4 Illustration of atoms in the MD simulation of a Ni tip being pulled back from an Au substrate. This causes the formation of a connective neck of atoms between the tip and the surface. After [35.111] with the permission of the AAAS (1990)

generated from experimental data (Fig. 35.3, point L). This is due to tip-induced flow of the metal atoms in the

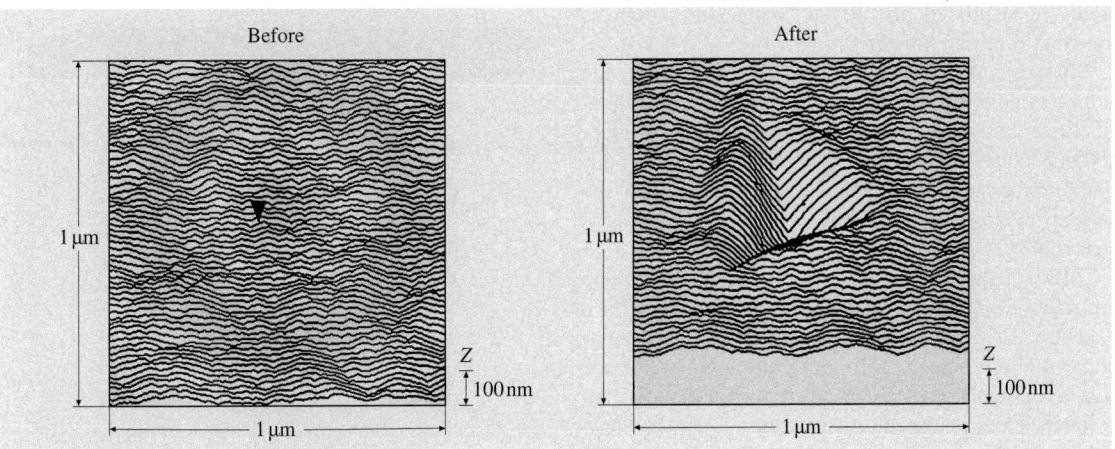

Fig. 35.5 Images of a gold surface before and after being indented with a pyramidal shaped diamond tip in air. The indentation created a surface crater. Note the pile-up around the crater edges. After [35.120] with the permission of Elsevier (1993)

As the tip is withdrawn farther, the magnitude of the force increases (becomes more negative) until, at a critical force, the atoms in adjacent layers of the connective nanowire rearrange so that an additional row of atoms is created. This process causes elongation of the connective nanowire and is responsible for the fine structure (apparent as a series of maxima) present in the retraction portion of the force curve. These elongation and rearrangement steps are repeated until the connection between the tip and the surface is broken. Similar elongation events have been observed experimentally. For example, scanning tunneling microscopy (STM) experiments demonstrate that the metal nanowires between metal tips and surfaces can elongate approximately 2500 Å without breaking [35.119].

The JC process has been shown to affect the temperature at the tip–surface interface. For instance, the constant-energy MD simulations of *Tomagnini* et al. [35.121] predicted that the energy released due to the wetting of the tip by surface atoms increases the temperature of the tip by about 15 K at room temperature and is accompanied by significant structural rearrangement. At temperatures high enough to cause the first few metal surface layers to be liquid, the distance at which the JC occurs increases, as does the contact area between the tip and the surface and the amount of nanowire elongation prior to breakage.

Simulations by *Komvopoulos* and *Yan* [35.122] using LJ potentials showed how metallic surfaces respond to single and repeated indentation by metallic, or covalently bound, rigid tips. The simulations predicted

that a single indentation event produces hysteresis in the force curve as a result of surface plastic deformation and heating. The repulsive force decreases abruptly during surface penetration by the tip and surface plastic deformation. Repeated indentation results in the continuous decrease of the elastic stiffness, surface heating, and mean contact pressures at maximum penetration depths to produce behaviors that are similar to cyclic work hardening and softening by annealing observed in metals at the macroscale.

When the tip is much stiffer than the surface, pile-up of surface atoms around the tip occurs to relieve the stresses induced by nanoindentation. In contrast, when the surface is much stiffer than the tip, the tip can be damaged or destroyed. Simulations by *Belak* et al. [35.123] using perfectly rigid tips showed the mechanism by which the surface yields plastically after its elastic threshold is exceeded. The simulations showed how nanoindentation causes surface atoms to move on to the surface but under the tip and thus cause atomic pile-up. In this study, variations in the indentation rate reveal that point defects created as a result of nanoindentation relax by moving through the surface if the rate of indentation is slow enough. If the indentation rate is too high, there is no time for the point defects to relax and move away from the indentation area and so strain builds up more rapidly. The rigid indenters considered in these MD simulations are analogous to experiments that use surface passivation to prevent JC between the tip and the surface [35.120, 124], the results of which agree with the predicted results of pile-up and crater formation, as shown in Fig. 35.5 [35.120].

In short, MD simulations are able to explain the atomic-scale mechanisms behind measured experimental force curves produced when metal tips indent homogeneous metal surfaces to nanometer-scale depths. This preliminary work has spawned much of the current interest in using the JC to produce metal nanowires [35.126–128].

MD simulations have also been used to examine the relationship between nanoindentation and surface structure. This is most apparent in a series of computational studies that consider the indentation of a surface with a "virtual" hard-sphere indenter in a manner that is independent of the rate of indentation, as shown in Fig. 35.6. The virtual indenter is modeled through the application of a repulsive force to the surface rather than through the presence of an actual atomic tip. *Kelchner* et al. [35.129], rather than use MD, pushed the indenter against the surface a short distance and then allowed the system to relax using standard energy minimization methods in combination with EAM potentials. The system is fully relaxed when the energy of the surface system is minimized. After relaxation, the tip is pushed further into the surface and the process is repeated. As the tip generates more stresses in the surface, dislocations are generated and plastic deformation occurs. If the tip is pulled back after indenting less than a specific critical value, the atoms that were plastically deformed are healed during the retraction and the surface recovers its original structure. In contrast, if the tip is indented past the critical depth, additional dislocations are created that interfere with the surface healing process on tip withdrawal. In this case, a surface crater is left on the surface following nanoindentation.

Fig. 35.7 Snapshot of two partial dislocations separated by a stacking fault. The *dark spheres in the center* of the structure indicate atoms in perfect crystal positions after both partial dislocations have passed. After [35.125] with the permission of Elsevier (1993)

Fig. 35.8 Snapshot of the high-energy atoms only after a load drop caused by dislocation generation during the nanoindentation of gold near a grain boundary. After [35.125] with the permission of Elsevier (1993)

A similar study by *Lilleodden* et al. [35.125] considered the generation of dislocations in perfect crystals and near grain boundaries in gold. Analysis of the relationship between the load and the tip displacement in the perfect crystal shows discrete load drops followed by elastic behavior. These load drops are shown to correspond to the homogeneous nucleation of dislocations, as illustrated in Fig. 35.7, which is a snapshot taken just

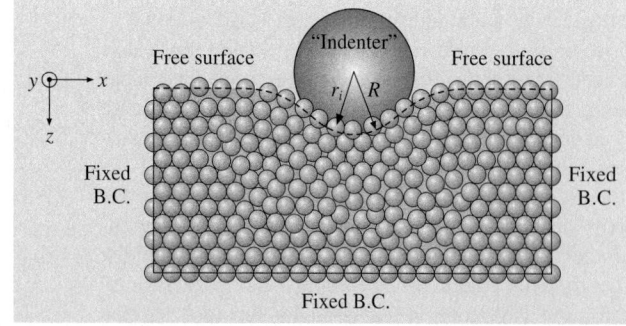

Fig. 35.6 A schematic of a spherical, virtual tip indenting a metal surface. After [35.125] with the permission of Elsevier (1993)

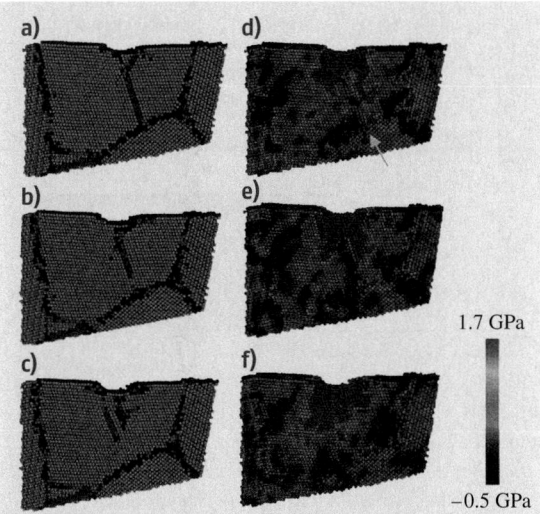

Fig. 35.9a–f Snapshots showing the atomic stress distribution and atomic structures in a gold surface. Figures (**a**)–(**c**) show the atomic structure at indentation depths of 7.9, 8.6, and 9.6 Å, respectively, with a virtual spherical indenter. A dislocation is represented by the two parallel {111} planes (*two dark lines*) that show the stacking fault left behind after the leading partial dislocation has passed. Figures (**d**)–(**f**) show the atomic stress distribution of the same system at the same indentation depths. Here the *dark color* indicates compressive hydrostatic pressures of 1.7 GPa and higher while the *gray* color indicates tensile pressures of − 0.5 GPa and lower. The *arrow* in (**d**)shows the region of the system where a dislocation interacts with a grain boundary. After [35.130] with the permission of Elsevier (2004)

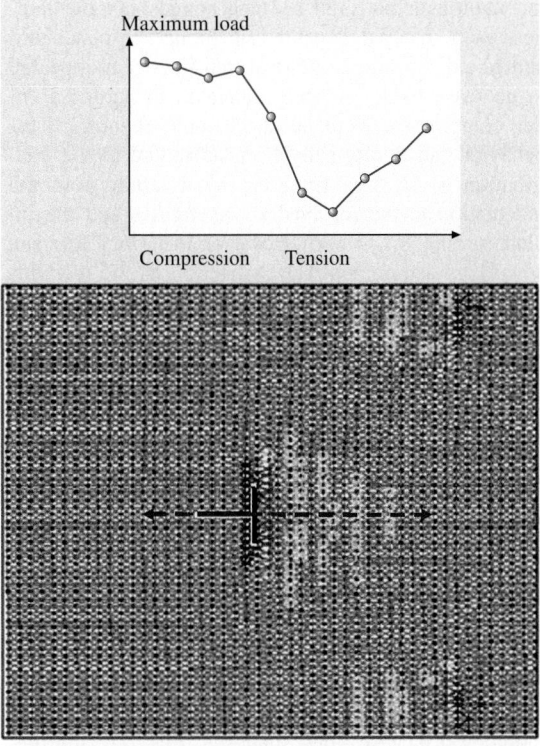

Fig. 35.10 Data and system illustration from a simulation of a gold surface containing a dislocation. *Top*: Maximum load for simulated shallow indentation at several points along the *dotted line* in the *bottom* illustration. *Bottom*: Top view of the simulated surface. The dislocation is denoted by the *solid black lines*

after the load drop. When nanoindentation occurs close to a grain boundary, similar relationships between the load and tip displacement are predicted to occur as were seen for the perfect crystal. However, the dislocations responsible for the load drop are preferentially emitted from the grain boundaries, as illustrated in Fig. 35.8.

Simulations can also show how atomic structure and stresses are affected by nanoindentation. For instance, MD simulations with a virtual indenter by *Hasnaoui* et al. [35.130] using semi-empirical tight-binding methods showed the interaction between the grain boundaries under the indenter and the dislocations generated by the indentation, as illustrated in Fig. 35.9. This study shows that if the size of the indenter is smaller than the grain size, the grain boundaries can emit, absorb, and reflect the dislocations in a manner that depends on atomic structure and the distribution of stresses.

Zimmerman et al. considered the indentation of a single-crystal gold substrate both near and far from a surface step [35.131]. The results of these simulations, which used EAM potentials, showed that the onset of plastic deformation depends to a significant degree on the distance of indentation from the step, and whether the indentation is on the plane above or below the step. In a related set of simulations, *Shenderova* et al. [35.132] examined whether ultrashallow elastic nanoindentation can nondestructively probe surface stress distributions associated with surface structures such as a trench and a dislocation intersecting a surface. The simulations carried out the nanoindentation to a constant depth. They predicted maximum loads that reflect the in-plane stresses at the point of contact between the indenter and the substrate, as illustrated in Fig. 35.10.

Since the 1930s, studies have been performed using hardness measurement techniques [35.133–136] and indentation methods [35.137] that suggest that the hardness of a material depends on applied in-plane uni- and bi-axial strain. In general, tensile strain appeared to decrease hardness while increases in hardness under compressive in-plane strain were reported. This behavior had traditionally been attributed to the contribution of stresses from the local strain from the indentation to the resolved shear stresses and the in-plane strain [35.134, 136]. However, in 1996, *Pharr* and coworkers determined that changes in elastic modulus determined from unloading curves of strained substrates using contact areas estimated via an elastic model are too large to have physical significance, a result that brought into question the interpretation of prior hardness data [35.137, 138]. They hypothesized that the apparent change in modulus (and hardness) with in-plane strain is mainly due to changes in contact area that are not typically taken into consideration in elastic half-space models. This hypothesis was based on experimental nanoindentation studies of a strained polycrystalline aluminium alloy and finite element calculations on an isotropic solid [35.137, 138]. They further suggested that in-plane compression increases pile-up around the indenter that, when not taken into account in the analysis of unloading curves, implies a nonphysical increase in modulus. Likewise, they suggested that in-plane tensile strain reduces the amount of material that is piled up around an indenter, which leads to a corresponding reduced (nonphysical) modulus when interpreting unloading curves using elastic models.

To explore in more detail the issue of pile-up and its influence on the interpretation of loading curves, *Schall* and *Brenner* used MD simulations and EAM potentials to model the plastic nanoindentation of a single-crystal gold surface under an applied in-plane strain [35.139]. These simulations predicted that the mean pressure, calculated from true contact areas that take into account plastic pile-up around the indenter, varies only slightly with applied pre-stress. They also predicted that the higher values occur in compression rather than in tension, and that the modulus calculated from the true contact area is essentially independent of the pre-stress level in the substrate. In contrast, if the contact area is estimated from approximate elastic formulae, the contact area is underestimated, which leads to a strong, incorrect dependence of apparent modulus on the pre-stress level. The simulations also showed larger pile-up in compression than in tension, in agreement with the Pharr model, and both regimes produced contact areas larger than those typically assumed in elastic analyses. These findings are illustrated in Fig. 35.11.

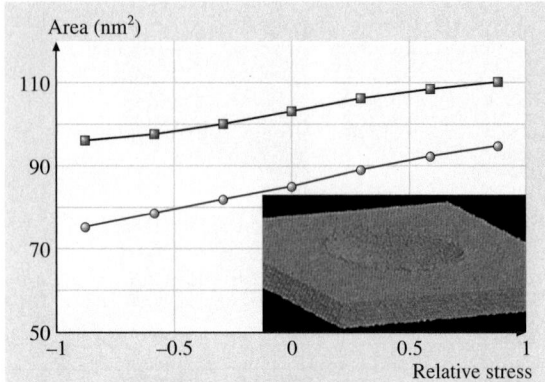

Fig. 35.11 Contact area projected in the plane at a maximum load for simulated indention of a gold surface as a function of in-plane biaxial stress. The stress is normalized to the theoretical yield stress. The *top curve* is from an atomistic simulation; the *bottom curve* is from an elastic model. *Inset*: Illustration of the region near the indention from the simulation. The tip is not shown for clarity. Initial formation of pile-up around the edge of the indentation is apparent

Nanometer-scale indentation of ceramic systems has also been investigated with MD simulations. Ceramics are stiffer and more brittle than metals at the macroscale and examining the nanoindentation of ceramic surfaces provides information about the nanometer-scale properties. They also reveal the manner by which defects form in covalent and ionic materials. For example, *Landman* et al. [35.110, 140] considered the interaction of a CaF_2 tip with a CaF_2 substrate in MD simulations using empirical potentials. As the tip approaches the surface, the attractive force between them steadily increases. This attractive force increases dramatically at the critical distance of 2.3 Å as the interlayer spacing of the tip increases (the tip is elongated) in a process that is similar to the JC phenomenon observed in metals. An important difference, however, is the amount of elongation, which is 0.35 Å in the case of the ionic ceramics and several angstroms in the case of metals. As the distance between the tip and the surface decreases further, the attractive nature of their interaction increases until a maximum value is reached. Indentation beyond this point results in a repulsive tip–substrate interaction, compression of the tip, and ionic bonding between the tip and substrate. These bonds are responsible for the hysteresis predicted to occur in the force curve on retraction, which ultimately leads to plastic deformation of the tip followed by fracture.

The responses of covalently bound ceramics such as diamond and silicon to nanoindentation have been heavily studied with MD simulations. One of the first of these computational studies was carried out by *Kallman* et al. who used the Stillinger-Weber potential to examine the indentation of amorphous and crystalline silicon [35.141]. The motivation for this study came from experimental data that indicated a large change in electrical resistivity during indentation of silicon, which led to the suggestion of a load-induced phase transition below the indenter. *Clarke* et al., for example, reported forming an Ohmic contact under load, and using transmission electron microscopy they observed an amorphous phase at the point of contact after indentation [35.142]. Using micro-Raman microscopy, *Kailer* et al. identified a metallic β-Sn phase in silicon near the interface of a diamond indenter during hardness loading [35.143]. Furthermore, upon rapid unloading they detected amorphous silicon as in the *Clarke* et al. [35.142] experiments, while slow unloading resulted in a mixture of high-pressure polymorphs near the indent point. At the highest indentation rate and the lowest temperature, the simulations by *Kallman* et al. [35.141] showed that amorphous and crystalline silicon have similar yield strengths of 138 and 179 kbar, respectively. In contrast, at temperatures near the melting temperature and at the slowest indentation rate, both amorphous and crystalline silicon are predicted to have lower yield strengths of 30 kbar. The simulations thus show how the predicted yield strength of silicon at the nanometer scale depends on structure, rate of deformation, and surface temperature.

Interestingly, *Kallman* et al. [35.141] found that amorphous silicon does not crystallize upon indentation, but indentation of crystalline silicon at temperatures near the melting point transforms the surface structure near the indenter to the amorphous phase. The simulations do not predict transformation to the β-Sn structure under any of the conditions considered. These results agree with the outcomes of scratching experiments [35.144] that showed that amorphous silicon emerges from room-temperature scratching of crystalline silicon.

Kaxiras and coworkers revisited the silicon nanoindentation issue using a quasi-continuum model that couples interatomic forces from the Stillinger-Weber potential to a finite element grid [35.145]. They report good agreement between simulated loading curves and experiment provided that the curves are scaled by the indenter size. Rather than the β-Sn structure, however, atomic displacements suggest formation of a metallic structure with fivefold coordination below the indenter upon loading, and a residual simple cubic phase near the indentation site after the load is released rather than the mix of high-pressure phases characterized experimentally. *Smith* et al. attribute this discrepancy to shortcomings of the Stillinger-Weber potential in adequately describing the high-pressure phases of silicon. They also used a simple model for changes in electrical resistivity with loading involving contributions from both a Schottky barrier and spreading resistance. Simulated resistance-versus-loading curves agree well with experiment despite possible discrepancies between the high-pressure phases under the indenter, suggesting that

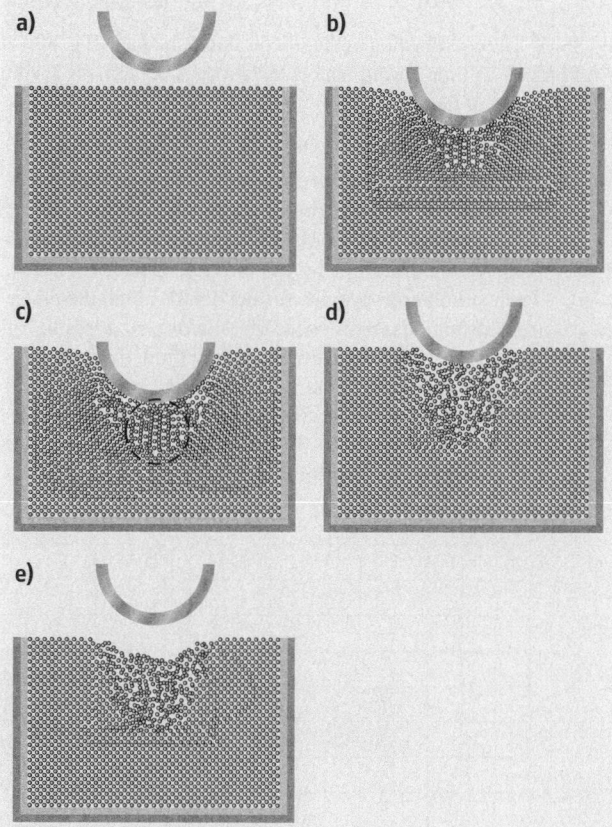

Fig. 35.12a–e Snapshots of a silicon sample during indentation. The *smaller dots* are diamond atoms. (**a**) Crystalline silicon prior to indentation. (**b**) Atoms beneath the indenter are displaced as a result of indentation. (**c**) The system at maximum indentation. Some of the atoms are in a crystalline arrangement (*circled region*) that is different from the diamond structure. (**d**) The surface structure is largely amorphous as the tip is withdrawn. (**e**) The surface after indentation. Note the amorphous region at the site of the indentation process. After [35.130] with the permission of the IOP (2000)

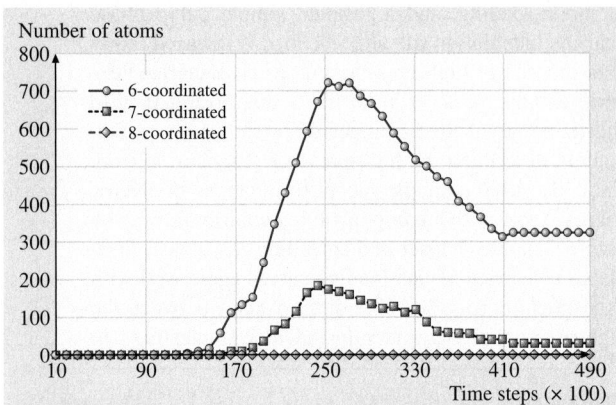

Fig. 35.13 The coordination of the silicon atoms shown in Fig. 35.12 as a function of time during nanoindentation. After [35.146] with the permission of the IOP (2000)

the salient features of the experiment are not dependent on the details of the high-pressure phases produced.

Additional MD simulations of the indentation of silicon were carried out by *Cheong* and *Zhang* [35.146]. Their simulations provide further details about the phase transformations that occur in silicon as a result of nanoindentation. In particular, they find that the diamond cubic silicon is transformed into a body-centered tetragonal structure (β-Si) upon loading of the indenter, as illustrated in Fig. 35.12. Figure 35.13 shows that the coordination numbers of silicon atoms also coin-

cide with that of the theoretical β-Si structure. The body-centered tetragonal structure is transformed into amorphous silicon during the unloading stage. A second indentation simulation again predicted that that this is a reversible process. Atomistic simulations by *Sanz-Navarro* et al. [35.147] shows the relation between the indentation of silicon and the hydrostatic pressure on surface cells due to the nanoindentation, as illustrated in Fig. 35.14. These simulations further predict that the transformation of diamond silicon into the β-Si structure can occur if the hydrostatic pressure is somewhat over 12 GPa.

Multimillion atom simulations of the indentation of silicon nitride were recently carried out by *Walsh* et al. [35.148]. The elastic modulus and hardness of the surface was calculated using load–displacement relationships. Snapshots from the simulations, illustrated in Fig. 35.15, show that pile-up occurs on the surface along the edges of the tip. Plastic deformation of the surface is predicted to extend a significant distance beyond the actual contact area of the indenter, as illustrated in Fig. 35.15.

The indentation of bare and hydrogen-terminated diamond (111) surfaces beyond the elastic limit was investigated by *Harrison* et al. [35.149] using a hydrogen-terminated sp^3-bonded tip in MD simulations that utilized bond-order potentials. The simulations identified the depth and applied force at which the diamond (111) substrate incurred plastic deformation due to indentation. At low indentation forces, the tip–

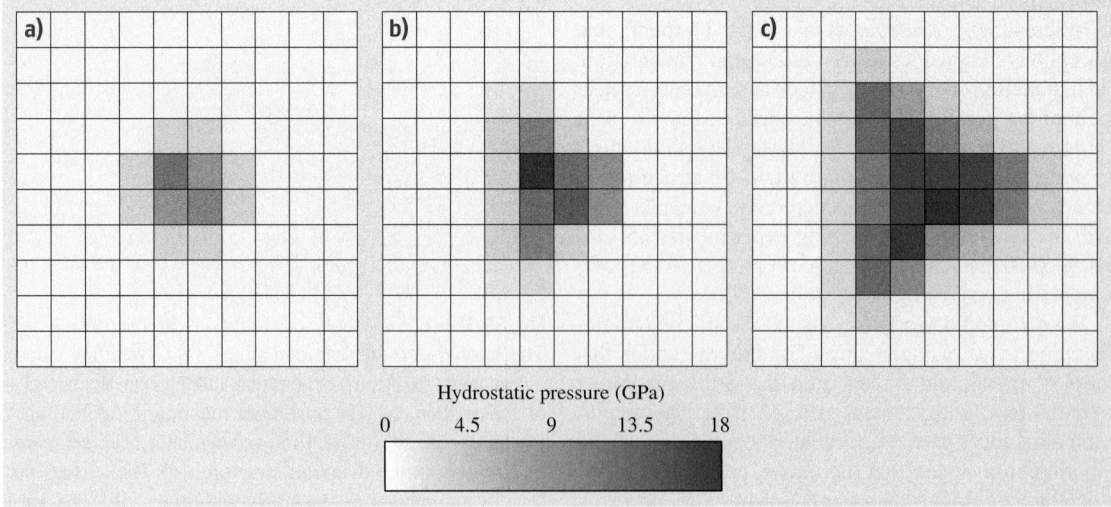

Fig. 35.14a–c Calculated hydrostatic pressure of surface cells at indentation depths of (**a**) 8.9 Å, (**b**) 15.7 Å, and (**c**) 25.3 Å. After [35.147] with the permission of the IOP (2000)

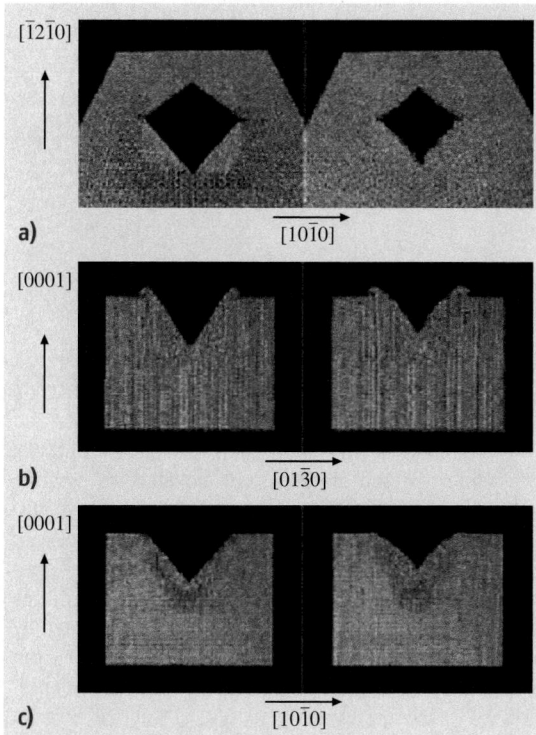

Fig. 35.15a–c Snapshots of the silicon nitride (**a**) surface, (**b**) slide parallel to the edges of the indenter, and (**c**) slide across the indenter diagonal. The *left-hand side* shows the surface when it is fully loaded, while the *right-hand side* shows the surface after the tip has been withdrawn. After [35.148] with the permission of the AIOP (2003)

surface interaction is purely elastic, as illustrated in Fig. 35.16. This finding agrees with the findings of *Cho* et al. [35.150], who examined the atomic-scale mechanical hysteresis experienced by an AFM tip indenting Si(100) with density functional theory. The calculations predicted that at low rates it is possible to cycle repeatedly between two buckled configurations of the surface without adhesion.

Fig. 35.17 Illustration of atoms in the MD simulation of the indentation of a hydrogen-terminated diamond (111) substrate with a hydrogen-terminated, sp^3-hybridized tip at selected time intervals. The figure illustrates the tip–substrate system as the tip was being withdrawn from the sample. *Large* and *small* spheres represent carbon and hydrogen atoms, respectively. After [35.149] with the permission of Elsevier (1992)

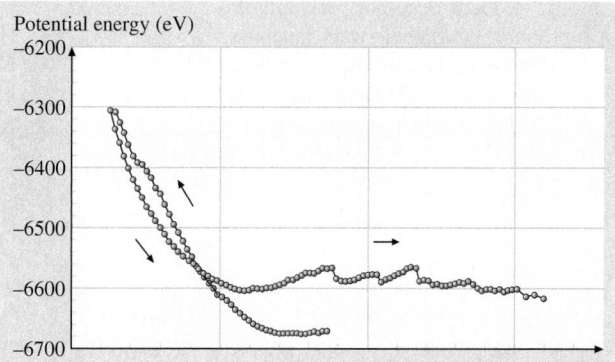

Fig. 35.16 Potential energy as a function of rigid-layer separation generated from an MD simulation of an elastic (nonadhesive) indentation of a hydrogen-terminated diamond (111) surface using a hydrogen-terminated, sp^3-hybridized tip. After [35.149] with the permission of Elsevier (1992)

When the nanoindentation process of diamond (111) is plastic, connective strings of atoms are formed between the tip and the surface, as illustrated in Fig. 35.17. These strings break as the distance between the tip and crystal increases and each break is accompanied

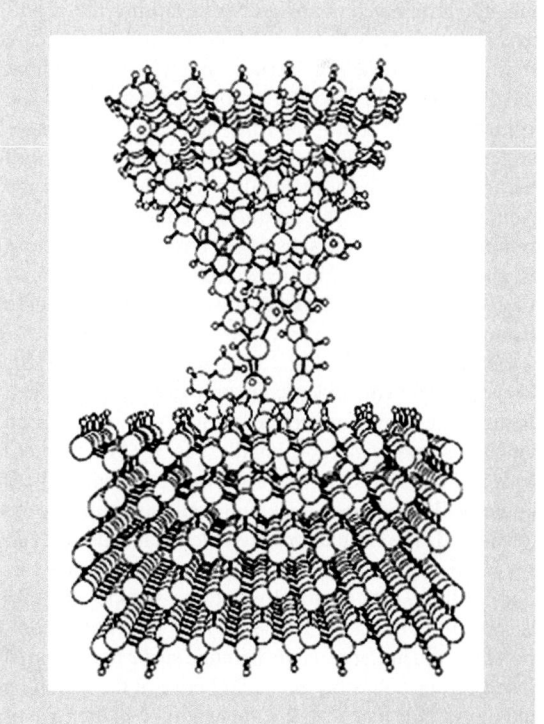

Fig. 35.18 Snapshots of the indentation of a single-wall nanotube (*left-hand image*) and a bundle of nanotubes (*right-hand image*) on hydrogen-terminated diamond (111) ▶

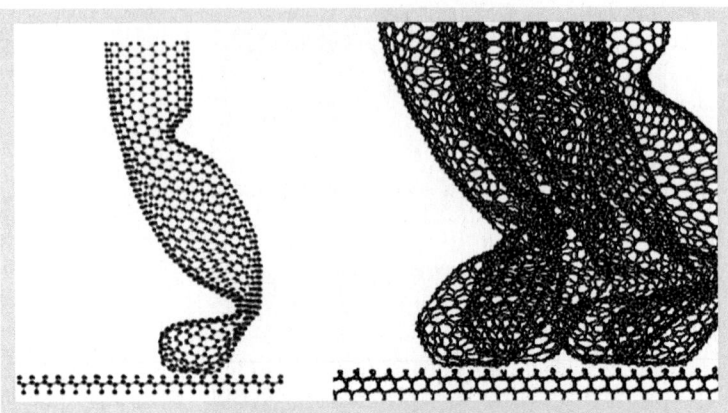

by a sudden drop in the potential energy at large positive values of tip–substrate separation. The simulations further predict that the tip end twists to minimize interatomic repulsive interactions between the hydrogen atoms on the surface and the tip. This behavior is predicted to lead to new covalent bond formation between the tip and the carbon atoms below the first layer of the surface and connective strings of atoms between the tip and the surface when the tip is retracted. Not surprisingly, when the surface is bare and not terminated with hydrogen atoms, the repulsive interactions between the tip and the surface are minimized and the tip indents the substrate without twisting [35.149]. Because carbon–carbon bonds are formed between the tip and the first layer of the substrate, the indentation is ordered (the surface is not disrupted as much by interacting with the tip) and the eventual fracture of the tip during retraction results in minimal damage to the substrate. The concerted fracture of all bonds in the tip gives rise to a single maximum in the potential versus distance curve at large distances.

Harrison et al. [35.152, 153] and *Garg* et al. [35.151, 154] considered the indentation of hydrogen-terminated diamond and graphene surfaces with AFM tips of carbon nanotubes and nanotube bundles using MD simulations and bond-order potentials. Tips consisting of both single-wall nanotubes and multiwall nanotubes were considered. The simulations predicted that nanotubes do not plastically deform during tip crashes on these surfaces. Rather, they elastically deform, buckle, and slip as shown in Fig. 35.18. However, as is the case for diamond tips indenting reactive diamond surfaces discussed above, strong adhesion can occur between the nanotube and the surface that destroys the nanotube in the case of highly reactive surfaces, as illustrated in Fig. 35.19.

To summarize, MD simulations reveal the properties of ceramic tips and surfaces with covalent or ceramic bonding that are most important for nanometer-scale indentation. They predict that brittle fracture of the tip can occur that is sometimes accompanied by strong adhesion with the surface. They also reveal the conditions under which neither the tip nor the surface is affected by the nanoindentation process. The insight gained from these simulations helps in the interpretation of exper-

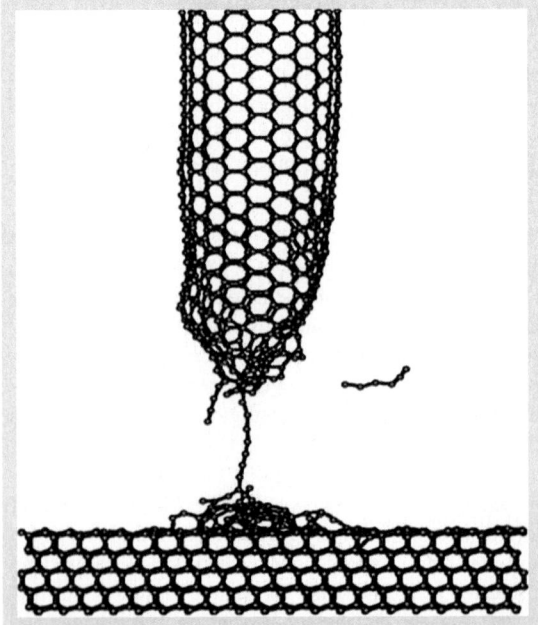

Fig. 35.19 Snapshot of a single-wall carbon nanotube as it is withdrawn following indentation on a bare diamond (111) surface. After [35.151] with the permission of APS (1999)

imental data, and it also reveals the nanometer-scale mechanisms by which, for example, tip buckling and permanent modification of the surface occur.

35.2.2 Thin Films

In many instances, surfaces are covered with thin films that can range in thickness from a few atomic layers to several μm. These films are more likely to have properties that differ from the properties of bulk materials of similar composition, and the likelihood of this increases as the film thickness decreases. Nanoindentation is one of the best approaches to determining the properties of these films. Consequently, numerous computational simulations of this process have been carried out.

For example, MD simulations have been used to study the indentation of metal surfaces covered with liquid *n*-hexadecane films, as illustrated in Fig. 35.20. As the metal tip touches the film, some of the molecules from the surface transfer to the tip and this causes the film to "swell". As the tip continues to push against the surface, the hydrocarbon film wets the side of the tip. The simulations show how the hydrocarbon film passivates the surface and prevents the strong attractive interactions discussed above for clean metal surfaces and tips from occurring.

In a series of MD simulations, *Tupper* and *Brenner* modeled the compression of a thiol self-assembled monolayer (SAM) on a rigid gold surface using both a smooth compressing surface [35.155] and a compressing surface with an asperity [35.156]. These simulations showed that compression with the smooth surface produced a compression-induced structural change that led to a change in slope of the simulated force versus compression curve. This transition is reversible and involves a change in the ordered arrangement of the sulfur head groups on the gold surface. A similar change in slope seemed to be present in the experimental indentation curves of *Houston* and coworkers [35.157], but was not discussed by the authors. The simulations with the asperity showed that the asperity is able to penetrate the tail groups of the SAM, as illustrated in Fig. 35.21, before an appreciable load is apparent on the compressing surface. This result indicates that it is possible to image the head groups of a thiol self-assembled monolayer that are adsorbed onto the surface of a gold substrate using STM, and consequently ordered images of these systems may not be indicative of the arrangement of the tail groups.

Zhang et al. [35.158] used a hybrid MD simulation approach, where a dynamic element model for the AFM

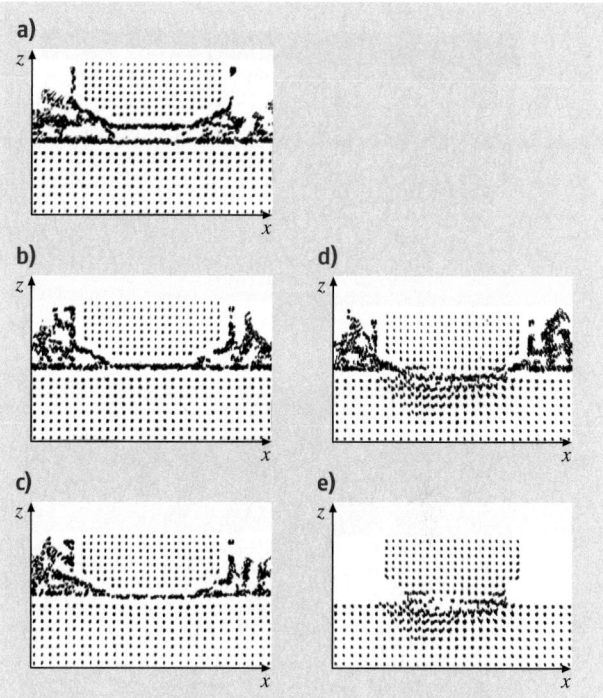

Fig. 35.20a–e Cutaways of the side view from molecular dynamics simulations of a Ni tip indenting a Au (001) surface covered with a hexadecane film. In (e) only the metal atoms are shown. Note how the hexadecane is forced out from between the metal surfaces. After [35.110] with the permission of Elsevier (1995)

cantilever was merged with a MD relaxation approach for the rest of the system, to study the frictional properties of alkanethiol SAMs on gold. They investigated the effect of several variables like chain length, terminal group, scan direction, and scan velocity. Their results show that friction forces decrease as the chain length of the SAMs increase. In the case of shorter chains such as C_7CH_3, the SAMs near the tip can be deformed by indentation, as illustrated in Fig. 35.22. This behavior is predicted to be the cause of higher friction that occurs for the short-length chains.

Harrison and coworkers have used classical MD simulations [35.153, 159] to examine the indentation of monolayers composed of linear hydrocarbon chains that are chemically bound (or anchored) to a diamond substrate. Both flexible and rigid single-wall, capped nanotubes were used as tips. The simulations showed that indentation causes the ordering of the monolayer to be disrupted regardless of the type of tip used. Indentation results in the formation of gauche defects within the monolayer and, for deep indents, results in the pinning of

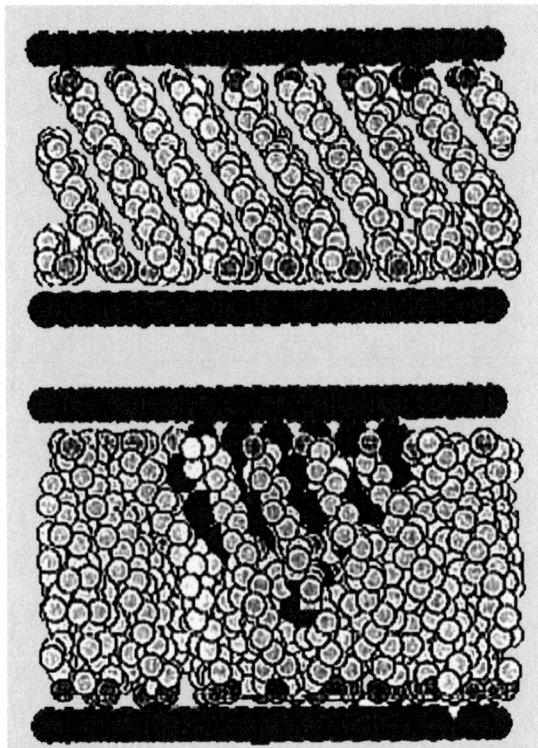

Fig. 35.21 Snapshots illustrating the compression of a self-assembled thiol film on gold for a smooth surface (*top*) and a surface containing an asperity (*bottom*). The asperity can penetrate and disorder the film tail groups before appreciable load occurs

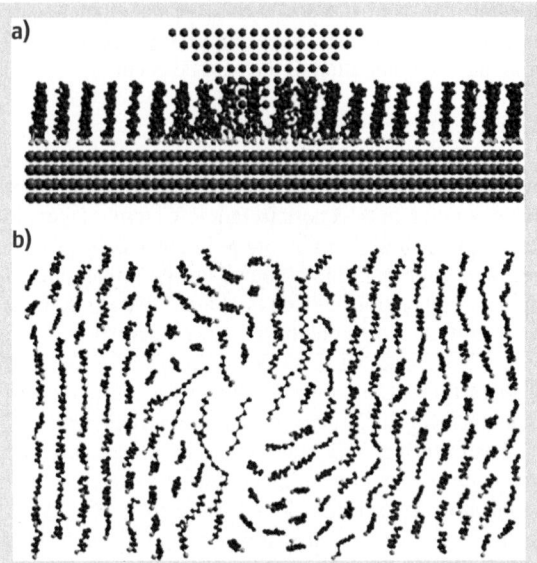

Fig. 35.22 (**a**) Side and (**b**) top views of the final configuration of a C_7CH_3 self-assembled monolayer on Au (111) under a high normal load of 1.2 nN at 300 K. The tip is not shown in *(b)* for clarity. After [35.158] with the permission of the ACS (2003)

selected hydrocarbon chains beneath the tube. Flexible nanotubes tilt slightly as they begin to indent the softer monolayers. This small distortion is due to the fact that nanotubes are stiff along their axial direction and more flexible in the transverse direction. In contrast, when the nanotubes encounter the hard diamond substrate, after "pushing" through the monolayer, they buckle. This process is illustrated in Fig. 35.23 and the force curves are shown in Fig. 35.24. The buckling of the nanotube was previously observed when single-wall, capped nanotubes were brought into contact with hydrogen-terminated diamond (111) surfaces [35.151, 152]. In the absence of the monolayer, the nanotube tips encounter the hard substrate in an almost vertical position. This interaction with the diamond substrate causes the cap of the nanotubes to be "pushed" inside the nanotube (they invert). Increasing the load on the nanotubes causes the walls of the tube to buckle. Both the cap inversion and

the buckling are reversible processes. That is, when the load on the tube is removed, it recovers its original shape.

Deep indents of the hydrocarbon monolayers using rigid nanotubes result in rupture of chemical bonds. The simulations also show that the number of gauche defects generated by the indentation is a linear function of penetration depth and equal for C_{13} and C_{22} monolayers. Thus, it is the tip that governs the number of gauche defects generated.

Leng and *Jiang* [35.160] investigated the effect of using tips coated with SAMs containing hydrophobic methyl (CH_3) or hydrophilic hydroxyl (OH) terminal groups to nanoindent gold surfaces that also are covered with SAMs with the identical terminal groups as the tip. Figure 35.25 contains snapshots for the indentation process predicted to occur for terminal OH/OH interactions during compression and the pull-off. The adhesion force of OH/OH pairs is calculated to be about four times larger than that of CH_3/CH_3 pairs, as shown in Fig. 35.27. This is due to the formation of hydrogen bonding between OH/OH pairs. This interaction is also expected to increase the frictional force between monolayers with OH terminations.

Related MD simulations by *Mate* predict that the end groups on polymer lubricants have a significant influ-

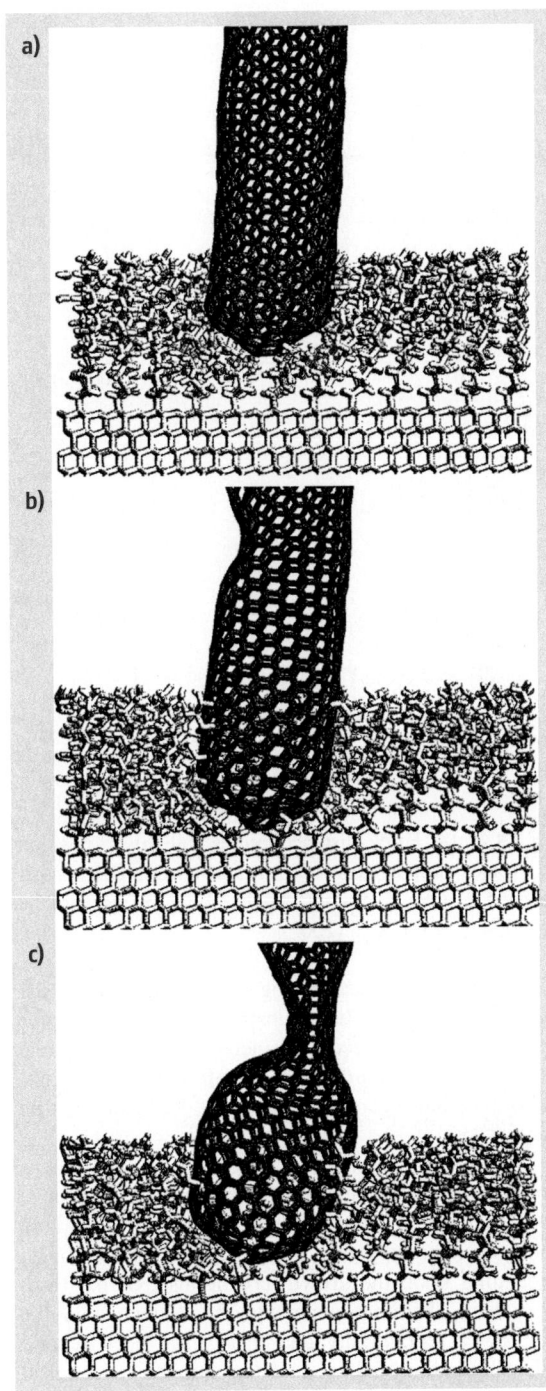

a)

b)

c)

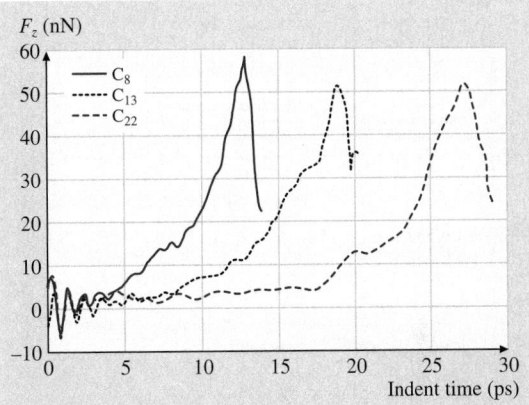

Fig. 35.23a–c Snapshots from the simulation of the interaction of a flexible single-walled carbon nanotube with a monolayer of C_{13} chains on diamond. The loads are (a) 19.8 nN, (b) 41.2 nN, and (c) 36.0 nN. After [35.159] with the permission of the ACS (2003)

Fig. 35.24 The load on the upper two layers of the flexible carbon nanotube indenter shown in Fig. 35.23 as a function of indentation time for the nanoindentation of the indicated hydrocarbon monolayes on diamond. After [35.159] with the permission of the ACS (1999)

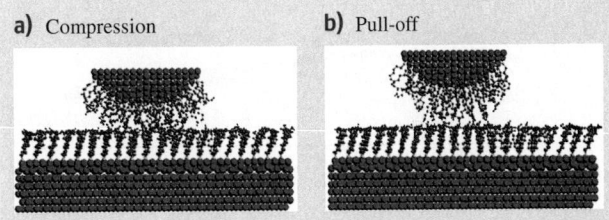

a) Compression   b) Pull-off

Fig. 35.25a,b Snapshots from the OH/OH pair interaction during (a) compression and (b) pull-off. After [35.160] with permission of the ACS (2002)

less reactive than regular alcohol end groups. When fluorinated films are indented, the normal force becomes more attractive as the distance between the tip and film decreases until the hard wall limit is reached and the interactions become repulsive. In contrast, when AFM tips indent hydrogenated films, the forces become increasingly repulsive as the distance between them decreases, as shown in Fig. 35.26 and Fig. 35.28. This predicted behavior is due to the compression of the end group beneath the tip. For the lubricant molecules to be squeezed out from between the tip and the surface, the hydrogen bonding between the two must first be broken and this

ence on the lubrication properties of polymers [35.161]. For instance, fluorinated end groups are predicted to be

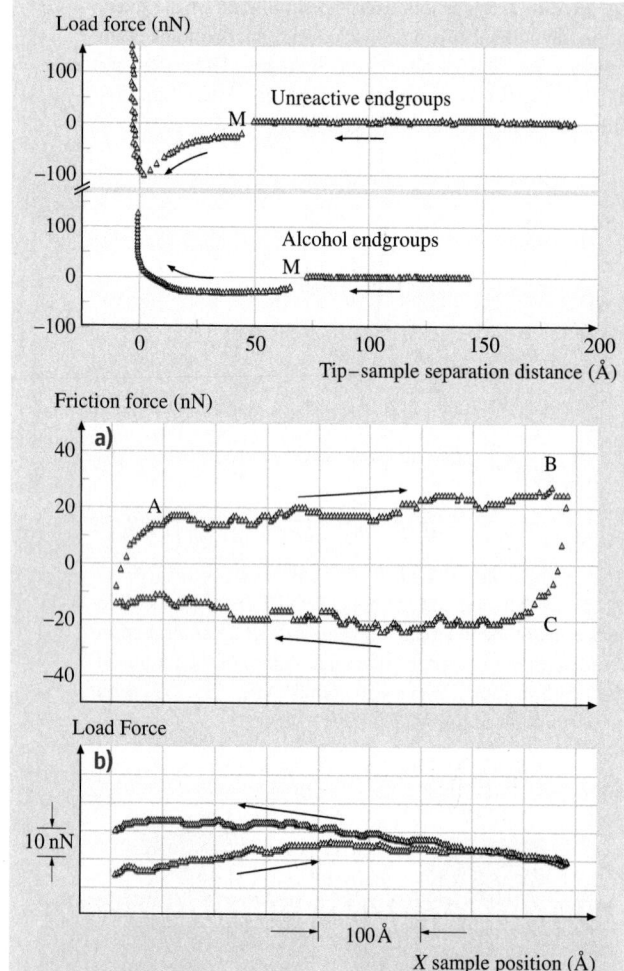

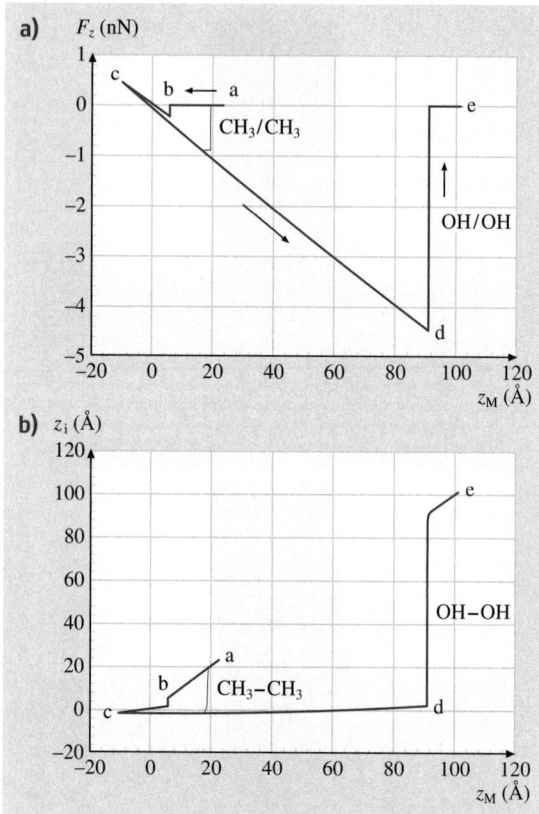

Fig. 35.27 (a) Force–distance curves and (b) tip position (z_i) versus support position (z_M) for the OH/OH contact pair and the CH_3/CH_3 contact pair. After [35.160] with the permission of the ACS (2002)

Fig. 35.26 *Top*: The force versus distance curve (indentation part only) for unbonded perfluoropolyether on Si (100). The unreactive end groups were from a 10 Å-thick film; the reactive alcohol end groups were from a 30 Å-thick film. The negative forces represent attractive interactions between the tip and the surface. *Bottom*: Measured plots of friction and load forces of the tip as it slides over the sample with the alcohol end groups. After [35.161] with the permission of the APS (1992)

increases the force needed to indent the system. As a result, a major effect of the presence of alcohol end groups is to dramatically increase the load that a liquid lubricant can support before failure (solid–solid contact) occurs.

When atomically sharp tips are used to indent solid-state thin films where there is a large mismatch in the mechanical properties of the film and the substrate, it is difficult to determine the true contact area between

the tip and the surface during nanoindentation. In the case of soft films on hard substrates, pile-up can occur around the tip that effectively increases the contact area. In contrast, with hard films on soft substrates, "sink-in" is experienced around the tip that decreases the true contact area.

A class of coatings that has received much attention is diamond-like amorphous carbon (DLC) coatings. DLC coatings are almost as hard as crystalline diamond and may have very low friction coefficients (< 0.01) depending upon the growth conditions [35.162–165]. They have therefore generated much interest in the tribological community and there have been several MD simulation studies to determine the mechanical and atomic-scale frictional properties of DLC coatings. MD simulations with bond-order potentials by *Sinnott* et al. [35.166] examined the differences in indentation behavior of a hydrogen-terminated diamond

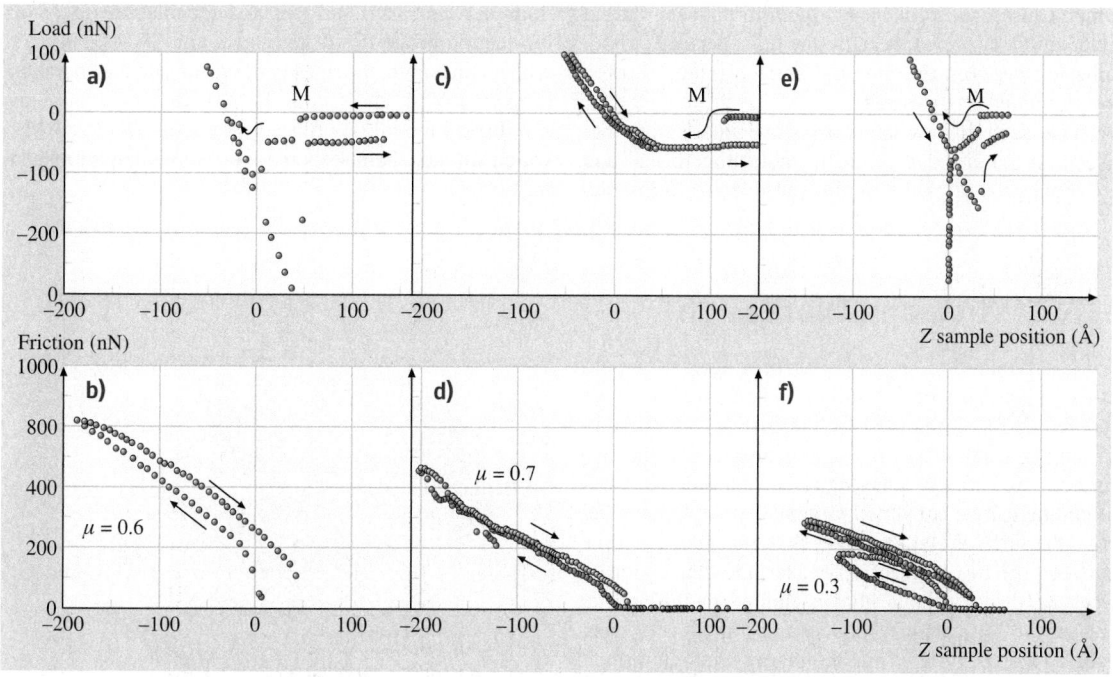

Fig. 35.28a–f Measured values for friction and load as an atomic-force microscope tip is scanned across a 30 Å-thick sample of perfluoropolyether on Si(100). (**a**) and (**b**) The unbonded polymer with unreactive end groups. (**c**) and (**d**) The unbonded polymer with alcohol end groups. (**e**) and (**f**) A bonded polymer. After [35.161] with the permission of the APS (1992)

tip on hydrogen-terminated single-crystal diamond surfaces and diamond surfaces covered with DLC. In the former case, the tip goes through shear and twist deformations at low loads that change to plastic deformation and adhesion with the surface at high loads. When the surface is covered with the DLC film, the tip easily penetrates the film, as illustrated in Fig. 35.29, which "heals" easily when the tip is retracted so that no crater or other evidence of the indentation is left behind.

MD simulations by *Glosli* et al. [35.167] of the indentation of DLC films that are about 20 nm-thick give similar results. In this case a larger, rigid diamond tip was used in the indentations and was also slid across the surface. During sliding, the tip plows the surface,

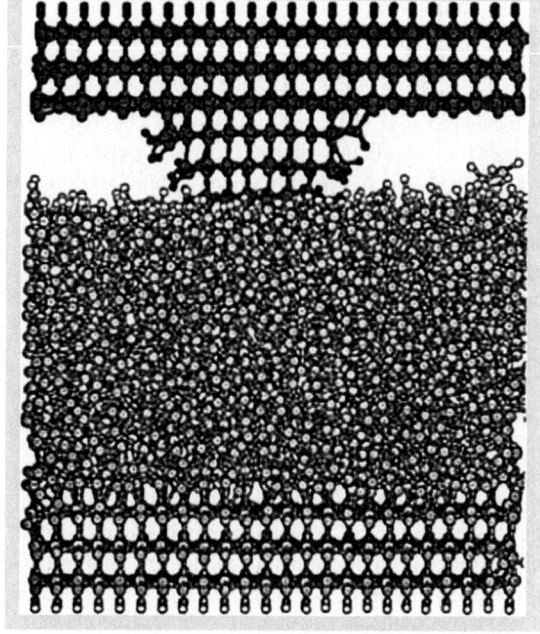

Fig. 35.29 Snapshot from a molecular dynamics simulation where a pyramidal diamond tip indented an amorphous carbon thin film that is 20 layers thick. The simulation took place at room temperature and the carbon atoms in the film were 21% sp^3-hybridized and 58% sp^2-hybridized (the remaining atoms were on the surface and were not counted). After [35.166] with the permission of the AIP (1997)

which causes some changes to the film not seen during indentation. However, because the tip is perfectly rigid, adhesion between the film and surface is not allowed which influences the results.

This section shows that repulsive interactions between surfaces covered with molecular films and proximal probe tips are minimized relative to interactions between bare surfaces and indentation tips. The lubrication properties of polymers and SAMs can vary with chain length, the rigidity of the tip, and the chemical properties of the end groups. In some cases, indentations can disrupt the initial ordering of polymers and SAMs, which affects their responses to nanoindentation and friction.

35.3 Friction and Lubrication

Work is required to slide two surfaces against one another. When the work of sliding is converted to a less ordered form, as required by the first law of thermodynamics, friction will occur. For instance, if the two surfaces are strongly adhering to one another, the work of sliding can be converted to damage that extends beyond the surfaces and into the bulk. If the adhesive force between the two surfaces is weaker, the conversion of work results in damage that is limited to the area at or near the surface and produces transfer films or wear debris [35.169, 170]. While the thermodynamic principles of the conversion of work to heat are well known, the mechanisms by which this takes place at sliding surfaces are much less well established despite their obvious importance for a wide variety of technological applications.

Atomic-scale simulations of friction are therefore important tools for achieving this understanding. They have consequently been applied to numerous materials in a wide variety of structures and configurations, including atomically flat and atomically rough diamond surfaces [35.171–173], rigid substrates covered with monolayers of alkane chains [35.174], perfluorocarboxylic acid and hydrocarboxylic Langmuir–Blodgett (LB) monolayers [35.175], between contacting copper surfaces [35.168, 176], between a silicon tip and a silicon substrate [35.116, 140], and between contacting diamond surfaces that have organic molecules absorbed on them [35.177]. These and several other studies are discussed below.

35.3.1 Bare Surfaces

Sliding friction that takes place between two surfaces in the absence of lubricant is termed "dry" friction even if the process occurs in an ambient environment. Simple models have been developed to model dry sliding friction that, for example, consider the motion of a single atom over a monoatomic chain [35.178]. Results from

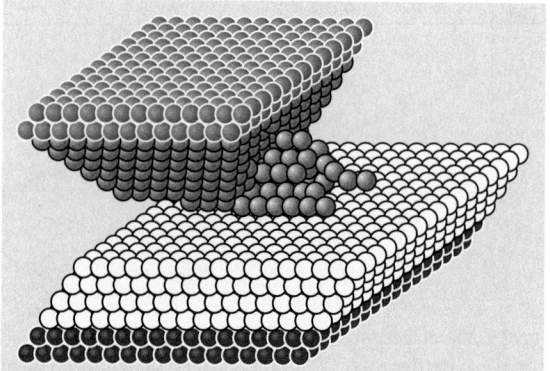

Fig. 35.30 Snapshot from a molecular dynamics simulation of a copper tip sliding across a Cu(100) surface. A connective neck between the two is sheared during the sliding, leading to wear of the tip. The simulation was performed at a temperature of 0 K. After [35.168] with the permission of the APS (1996)

these models reveal how elastic deformation of the substrate from the sliding atom affects energy dissipation and how the average frictional force varies with changes in the force constant of the substrate in the direction normal to the scan direction. Much of the correct behavior involved in dry sliding friction is captured by these types of simple models. However, more detailed models and simulations, such as MD simulations, are required to provide information about more complex phenomena.

MD simulations have been used to study the sliding of metal tips across clean metal surfaces by numerous groups [35.168, 179–183]. An illustrative case is shown in Fig. 35.30 for a copper tip sliding across a copper surface [35.168]. Adhesion and wear occur when the attractive force between the atoms on the tip and the atoms at the surface becomes greater than the attractive forces within the tip itself. Atomic-scale stick and slip can occur through nucleation and subsequent motion of

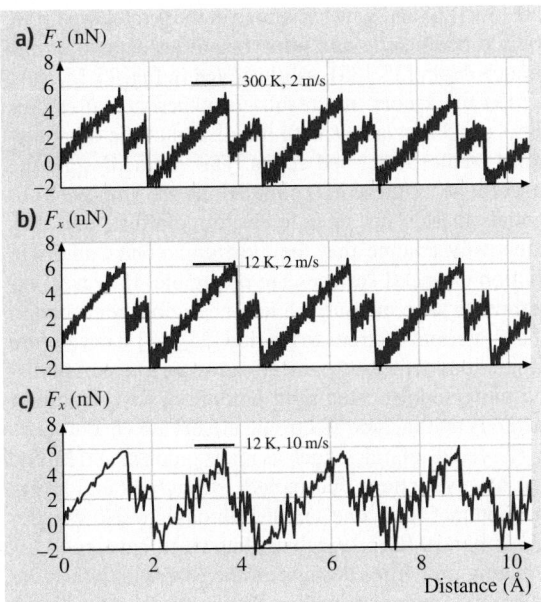

Fig. 35.31 Plots of the lateral force versus distance from a simulation similar to that shown in Fig. 35.30. The plots illustrate the dependence of the force on temperature and sliding velocity. After [35.168] with the permission of the APS (1996)

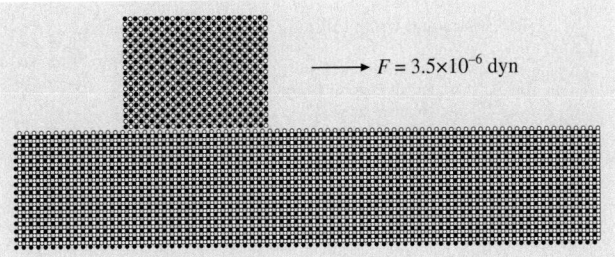

Fig. 35.32 Starting configuration of sliding NiAl tip on a NiAl surface. After [35.184] with the permission of the AIP (2001)

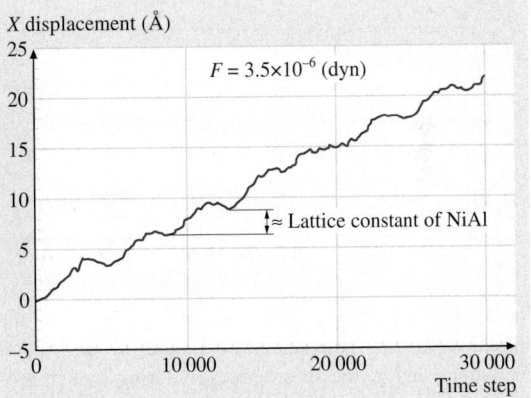

Fig. 35.33 A structured curve of frictional dynamics of an atom in the *upper right corner* that is indicative of stick-slip. After [35.184] with the permission of the AIP (2001)

dislocations, and wear can occur if part of the tip gets left behind on the surface (Fig. 35.30). The simulations can further provide data on how the characteristic 'stick-slip' friction motion can depend on the area of contact, the rate of sliding, and the sliding direction (Fig. 35.31).

An additional study of stick-slip in the sliding of much larger, square-shaped metal tips across metal surfaces was carried out by *Li* et al. [35.184] using EAM potentials. The initial structure of a NiAl tip and surface system is shown in Fig. 35.32. This study predicted that collective elastic deformation of the surface layers in response to sliding is the main cause of the stick-slip behavior shown in Fig. 35.33. The simulations also predicted that stick-slip produces phonons that propagate through the surface slab.

Large-scale simulations using pairwise Morse potentials that are similar in form to (35.6) were used to study the wear of metal surfaces caused by metal tips that plow the surface, as illustrated in Fig. 35.34. They provide insight into the wear track dependence of the sliding

Fig. 35.34 Snapshots of the scratching of an aluminium surface with a rigid tip at a depth of 0.8 nm. After [35.185] with the permission of the APS (2000)

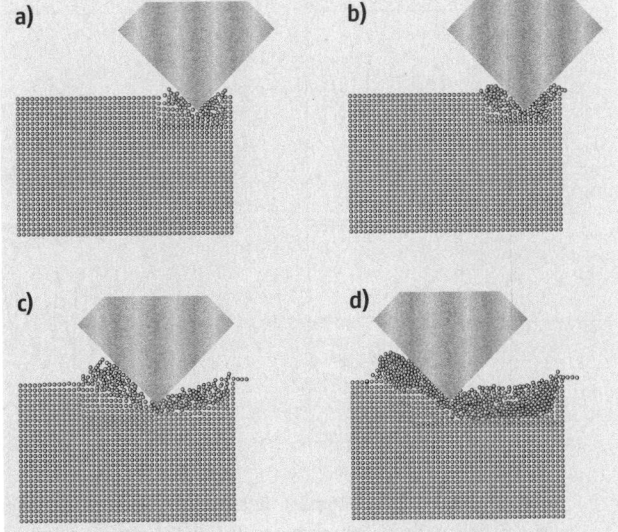

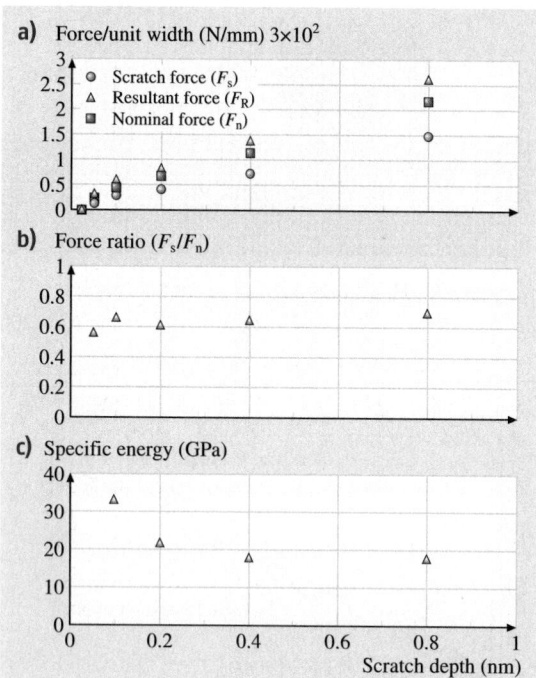

Fig. 35.35a–c Variation in (**a**) the scratching force, the normal force, and the resultant force, and (**b**) the friction coefficient, and (**c**) the specific energy during scratch processes similar to those shown in Fig. 35.34 at scratch depths ranging from 0.8 nm to almost 0 nm. After [35.185] with the permission of the APS (2000)

rate [35.186] and how variations in the scratching force, friction coefficient, and other quantities depend on the scratch depth [35.185], as illustrated in Fig. 35.35.

On the whole, the results of experimental studies show good agreement with the results of the computational studies described above. This is true despite the fact that all of these MD simulations use empirical potentials that do not include electronic effects and thus effectively assume that the electronic contributions to friction on metal surfaces are negligible. However, experiments have measured a non-negligible contribution of conduction electrons to friction [35.188]. Thus, future simulations of metal-tip–metal-substrate interactions using more sophisticated tight-binding or first principles methods that include electronic effects are encouraged.

Layered ceramics, such as mica, graphite and MoS_2, that have structures that include strongly bound layers that interact with one another through weak van der Waals bonds, have long been known to have good lubricating properties because of the ease with which the layers slide over one another. They have, therefore, been the focus of some of the earliest experimental studies of nanometer-scale friction [35.19, 189]. The results of these early studies lead researchers to hypothesize that at high loads measured friction forces were related to "incipient sliding" [35.190, 191] caused by a small flake from the surface becoming attached to the end of the tip. If true, this would mean that all measured interactions were between the surface and the flake, which has a larger contact area than the clean tip. However, sub-

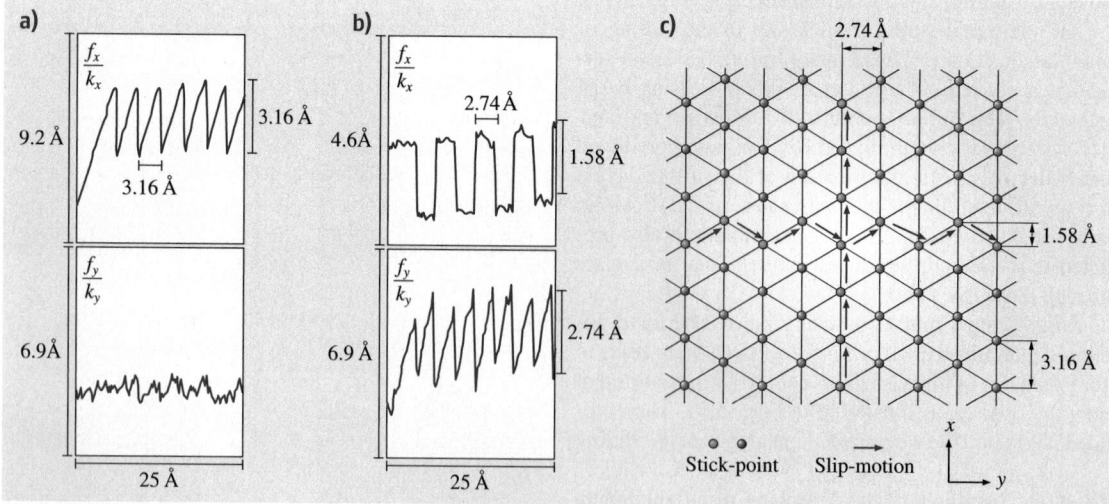

Fig. 35.36a–c Displacement data from a scan across a MoS_2(001) surface. The data in (**a**) and (**b**) are form scans along the x-and y-directions, respectively, on the surface shown in (**c**). After [35.187] with the permission of the APS (1996)

sequent simulations of constant force AFM images of graphite by *Tang* et al. [35.192] showed that there is no need for the assumption of a graphite flake under the tip to reproduce the experimental images of a graphite surface.

Surprisingly strong localized fluctuations in atomic-scale friction are displayed by layered ceramics [35.187, 194–196]. For instance, square-well signals with sub-angstrom lateral width are obtained in FFM scans on $MoS_2(001)$ in the direction across the scan direction,

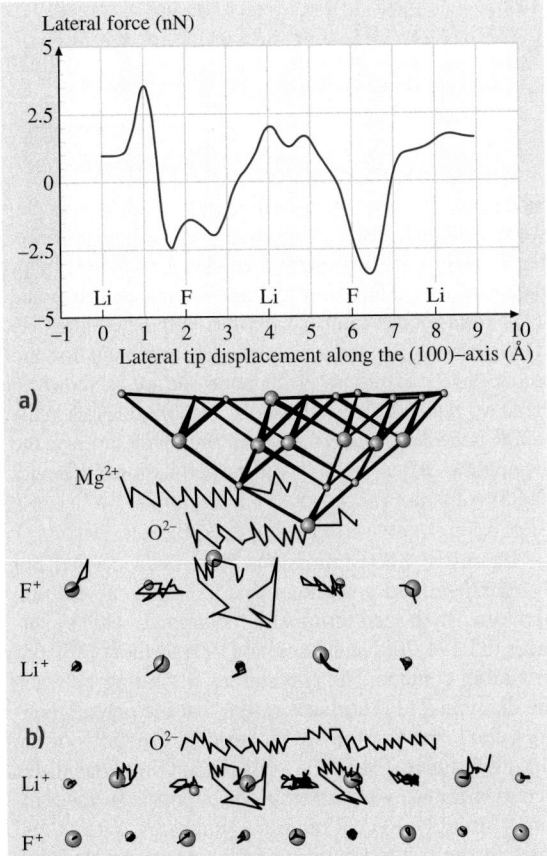

Fig. 35.37a,b *Top*: The lateral force calculated for a MgO tip scanning in the ⟨001⟩-direction on LiF(001). *Bottom*: A view of the side of the surface plane along the scan direction. The surface Li^+ and F^- atoms are seen to relax to relieve the frictional energy and this relaxation motion is indicated in the figure by the *category lines*. (**a**) How a F^- ion on the surface can be moved into an interstitial site by the tip and then it returns to its original position. (**b**) How the relaxation of the surface atoms is reversible. After [35.193] with the permission of Elsevier (1995)

while sawtooth signals are detected along the scan direction, as shown in Fig. 35.36. This finding can be explained by a stick-slip model by *Mate* [35.19] and *Erlandsson* [35.189] that assumes that the tip does a zigzag walk along the scan. Measured variations in the frictional force with the periodicity of cleavage planes [35.189] are consistent with the results of this simple model. However, additional experiments indicate a more complex tip–surface interaction, such as changes in the intrinsic lateral force between the substrate and the AFM tip [35.197] or sliding-induced chemistry between the tip and the surface [35.198, 199].

Crystalline ceramics differ from layered ceramics in that they are held together by relatively strong covalent or ionic bonds. In the case of ionic systems, *Shluger* et al. [35.193] used a mixture of atomistic and macroscopic modeling methods to study the interaction of a MgO tip and a LiF surface. In particular, the tip–surface interaction was treated atomistically and the cantilever deflection was treated with a macroscopic approach. The results, shown in Fig. 35.37, show that if the tip is charged and in hard contact with the surface, tip and surface distortions are possible that can lead to motion of the surface ions within the surface plane and the transfer of some of the ions onto the tip.

In the case of covalently bound ceramics, there is extensive literature related to friction of diamond [35.200, 201] because, while it is the hardest material known, it also exhibits relatively low friction. The "ratchet mechanism" has been proposed for energy dissipation during friction on the macroscale in diamond, where energy is released by the transfer of normal force from one surface asperity to another. The elastic mechanism is another mechanism that has been proposed, where the released energy comes from elastic strain in an asperity. Atomic-scale friction has been measured experimentally [35.20] for diamond tips with near atomic-scale radii sliding over hydrogen-terminated diamond surfaces. These experiments are sensitive enough to detect the 2×1 reconstruction on the diamond (100) surface. Furthermore, the average friction coefficient determined with an AFM on H-terminated diamond (111) surfaces is about two orders of magnitude smaller than the value measured on bare, 2×1 diamond (111) surfaces, indicating greater adhesion in the latter case [35.202]. More recently, the friction between a tungsten carbide tip and hydrogen-terminated diamond (111) was examined with AFM in UHV by *Enachescu* et al. [35.203]. The friction between these two hard surfaces was shown to obey Derjaguin–Muller–Toporov or DMT [35.204] contact

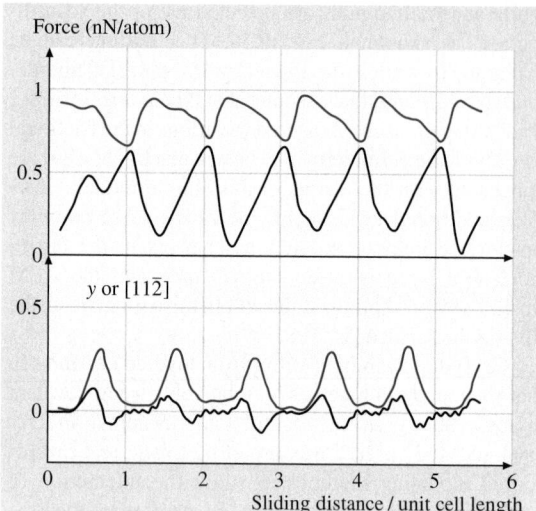

Fig. 35.38 Calculated frictional force (*lower lines*) and normal force (*upper lines*) felt by a hydrogen-terminated (111) surface as it slides against another hydrogen-terminated diamond (111) surface in a MD simulation. The sliding direction is given in the legend. The sliding speed is 1 Å/ps and the simulation temp is 300 K. The two plots show how the simulated stick-slip motion changes as a function of the applied load. The load is high and low in the *upper* and *lower* panels, respectively. After [35.205] with the permission of the ACS (1995)

mechanics and the shear strength of the interface was determined to be 246 MPa.

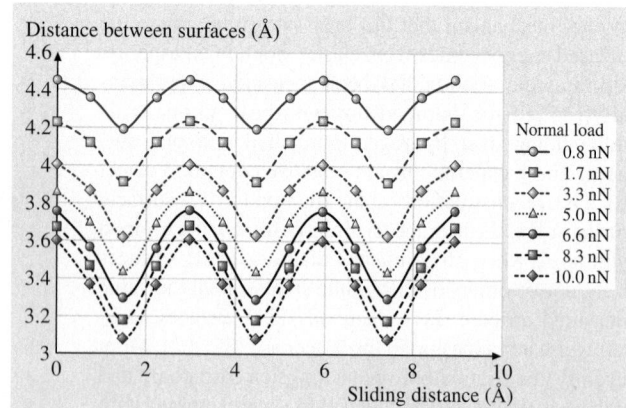

Fig. 35.39 Distance between hydrogen-terminated (111) crystals as a function of sliding distance. After [35.206] with the permission of the ACS (2001)

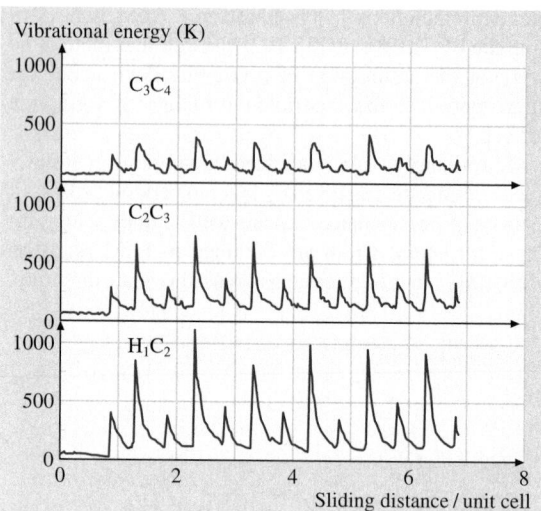

Fig. 35.40 Average vibrational energy of oscillators between diamond layers as a function of sliding distance. These energies are derived from a molecular dynamics simulation of the sliding of a hydrogen-terminated diamond (111) surface over another hydrogen-terminated diamond (111) surface. The vibrational energy between the first and second layers of the lower diamond surface is shown in the *lower panel*, between the second and third layers in the *middle panel*, and between the third and fourth layers in the *upper panel*. After [35.207] with the permission of Elsevier (1995)

Extensive MD simulations have been carried out by *Harrison* and coworkers that examine the friction between hydrogen-terminated diamond (111) surfaces [35.171,208] and diamond (100) surfaces [35.205] in sliding contact. The simulations of sliding between the diamond (111) surfaces reveal that the potential energy, load, and friction are all periodic functions of the sliding distance (Fig. 35.38). Maxima in these quantities occur when the hydrogen atoms on opposing surfaces interact strongly. Recent *ab initio* studies by *Neitola* and *Park* of the friction between hydrogen-terminated diamond (111) surfaces also show that the potential energy is periodic with sliding distance (Fig. 35.39) [35.206]. Because the results of the *ab initio* studies and the MD simulations are in good agreement, Neitola and Park conclude that the potential model used in the MD studies is accurate.

As mentioned previously, the maxima in the load and the friction values during sliding are caused by the interactions of hydrogen atoms on opposing surfaces. When sliding in the [11$\bar{2}$] direction, the H atoms "revolve" around one another, thus decreasing the repulsive interaction between the sliding surfaces because the hydrogen atoms are not forced to pass directly over one another [35.171]. Increasing the load causes increased stress at the interface. The opposing hydrogen atoms become "stuck". Once the stress at the interface becomes large enough to overcome the hydrogen–hydrogen interaction between opposing surfaces, the hydrogen atoms "slip" past one another with the same "revolving" motion observed at low loads. This phenomenon is known as atomic-scale stick-slip and has the periodicity of the diamond lattice. It should be noted that due to the alignment of the opposing surfaces, the hydrogen atoms are directly in line with each other when sliding in the [11$\bar{2}$] direction. However, the hydrogen atoms are not "aligned" with each other when sliding in the [11$\bar{0}$], so the friction in this direction is lower than in the [11$\bar{2}$] direction. It should be noted, however, that experimentally all initial alignments are likely to be probed.

Harrison and coworkers have further shown that the peaks in the frictional force are correlated with peaks in the temperature of the atoms at the interface when two hydrogen-terminated diamond (111) surfaces are in sliding contact [35.208]. Figure 35.40 shows the vibrational energy (or temperature) between diamond layers as a function of sliding distance. These data clearly show that layers close to the sliding interface can be vibrationally excited during sliding. When the hydrogen atoms are "stuck" or interacting with each other strongly, the stress and friction force at the interface build up. When the hydrogen atoms "slip" past one another, the stress at the interface is relieved and the energy is transferred to the diamond in the form of vibration or heat. Thus, the peaks in the temperature occur slightly after the peaks in the frictional force.

It should be noted that atomic-scale stick slip is observed in other systems. *Harrison* and coworkers used MD simulations to demonstrate that two hydrogen-terminated diamond (100) (2 × 1) surfaces in sliding contact also exhibit stick-slip [35.205]. In addition, it was shown that the shape of the friction versus sliding distance curves is influenced slightly by the speed of the sliding, with features in the curves becoming more pronounced at slower speeds. Stick-slip behavior was also observed in AFM studies of diamond (100) (2 × 1) surfaces [35.202]. However, in this case the stick-slip was over a much longer length scale and may be due to the fact that the surfaces were not hydrogen-terminated.

Mulliah et al. [35.210] used MD simulations with bond-order potentials [35.211] to model interactions between indenter atoms, EAM potentials [35.212] to

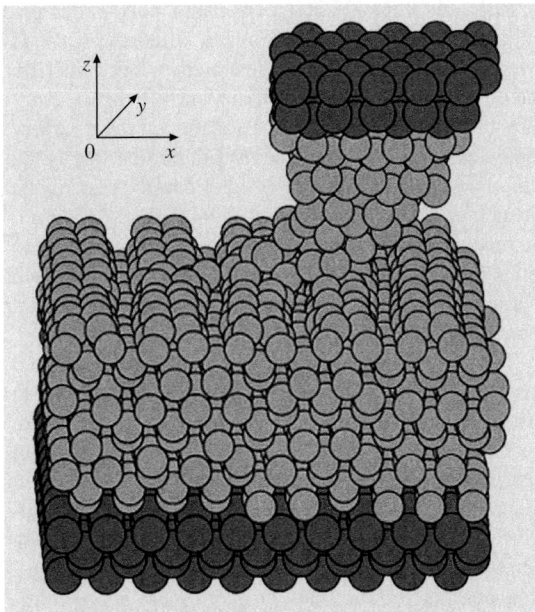

Fig. 35.41 Snapshots of a Si(111) tip interacting with a Si(001) 2 × 1 surface. The tip is rastering along the surface in the x-direction and starts off at a distance of 9 Å from the surface. After [35.209] with the permission of the CCLRC (2002)

model interactions between substrate atoms, and the Ziegler–Biersack–Littmack potential [35.213] to model interactions between indenter and substrate atoms to study the atomic-scale stick-slip phenomenon of a pyramidal diamond tip interacting with a silver surface at several sliding rates and vertical support displacements. These simulations showed that dislocations are related to the stick events emitting a dislocation in the substrate near the tip. The scratch in the substrate is discrete due to the tip jumping over the surface in the case of small vertical displacements. In contrast, large displacements of 15 Å or more result in a continuous scratch. These simulations also showed how the dynamic friction coefficient and the static friction coefficient increase with increasing tip depth. The tip moves continuously through a stick and slip motion at large depths, whereas it comes to a halt in the case of shallow indents. Although the sliding rate can change the exact points of stick and slip, the range of sliding rates over the range of values considered in this study (1.0 to 5.0 m/s) has no influence on the damage to the substrate, the atomistic stick-slip mechanisms, or the calculated friction coefficients.

The effect of the way in which the tip is rastered across the surface in MD simulations was considered by *Cai* and *Wang* [35.209, 214] using bond-order potentials. In particular, they dragged silicon tips across several silicon surfaces, as illustrated in Fig. 35.41, in two different ways. In the first, they moved the tip every MD step while in the other they advanced the tip every 1000 steps. In both cases, the overall sliding rate is the same and equals 1.67 m/s. In both cases, wear of the tip such as is illustrated in Fig. 35.41 occurs. However, the mechanisms by which the wear occurs are found to depend on the approach used, and the latter approach is found to be in better agreement with experimental data.

In many studies, diamond tips or diamond-decorated tips are used in friction measurements. Diamond is an attractive material for an FFM tip because of its high mechanical strength and the belief that such tips are wear-resistant. However, diamond tips that were used to scratch diamond and silicon surfaces and then imaged showed significant wear that increased with the increasing hardness of the tested material [35.215, 216]. This wear altered the shape of the tip and hence influenced the contact area that is used to determine friction coefficients.

In summary, MD simulations provide insight into dry sliding friction and the sliding of metal tips across clean metal, crystalline ceramics, and layered ceramics surfaces. Stick-slip friction or wear can occur depending on the sliding conditions. The good lubricating properties of layered ceramics are observed in the simulations along with localized fluctuations in atomic-scale friction. Crystalline ceramics, such as diamond, exhibit relatively low friction and the simulations show how stick-slip atomic-scale motion changes with the conditions of sliding and the way in which the simulation is performed.

35.3.2 Decorated Surfaces

While dry sliding friction in vacuum assumes that ambient gas particles have no direct effect on the results, MD simulations show that free particles between two surfaces in sliding contact influence friction to a surprisingly large degree. These so-called third-body molecules have been studied extensively by *Perry* et al. [35.177, 217, 218] using MD simulations with bond-order and LJ potentials. These simulations focus on the effect of trapped small hydrocarbon molecules on the atomic-scale friction of two (111) crystal faces of diamond with hydrogen termination. These molecules might represent hydrocarbon contamination trapped between contacting surfaces prior to a sliding experiment in dry friction, or hydrocarbon debris formed during sliding.

In particular, the effects on friction of methane (CH_4), ethane (C_2H_6), and isobutane (CH_3)$_3CH$ trapped between diamond (111) surfaces in sliding contact were examined in separate studies (Fig. 35.42). The frictional force for all these systems generally increases as the load increases, as illustrated in Fig. 35.43. The simulations predict that the third-body molecules markedly reduce the average frictional force compared to the results for pristine hydrogen-terminated surfaces. This is particularly true at high loads, where the third-body molecules act as a boundary layer between the two diamond surfaces. That is, the third-body molecules reduce the interaction of hydrogen atoms on opposing surfaces [35.218]. This is demonstrated by examining the vibrational energy excited in the diamond lattice during the sliding (Fig. 35.44). Significant vibrational excitation of the diamond outer layer (C–H) occurs in the absence of the methane molecules. Thus, the friction is approximately 3.5 times larger when methane is not present. The application of load to the diamond surfaces causes the normal mode vibrations of the trapped

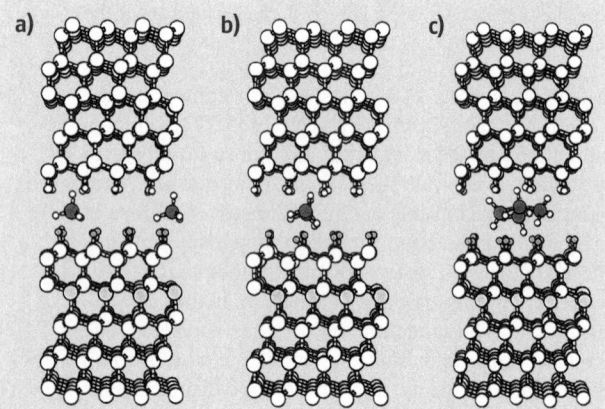

Fig. 35.42a–c Initial configuration at low load for the diamond plus third-body molecule systems. These systems are composed of two diamond surfaces, viewed along the [$\bar{1}10$] direction, and two methane molecules in (**a**), one ethane molecule in (**b**), and one isobutane molecule in (**c**). Large *white* and *dark gray spheres* represent carbon atoms of the diamond surfaces and the third-body molecules, respectively. *Small gray spheres* represent hydrogen atoms of the lower diamond surface. Hydrogen atoms of the upper diamond surface and the third-body molecules are both represented by *small white spheres*. Sliding is achieved by moving the rigid layers of the upper surface from *left to right* in the figure. After [35.177] with the permission of the ACS (1997)

methane molecules to change. Power spectra calculated from MD simulations [35.217,218] show that even under low loads, the peaks in the power spectra are significantly broadened. Peaks in the low-energy region of the spectrum almost disappear with the additional application of load.

The size of the methane molecules allows them to be "pushed" in-between hydrogen atoms on the diamond surfaces while sliding [35.218]. However, steric considerations cause the larger ethane and isobutane molecules to change orientation during sliding. Conformations that lead to increased interactions with the diamond surfaces increase the average frictional force. Thus, despite the fact that the two diamond surfaces are

farther apart when ethane and isobutane are present compared to when methane is present, the friction is larger because these molecules do not "fit" nicely into potential energy valleys between hydrogen atoms when sliding.

When similar hydrocarbon molecules (methyl, ethyl, and n-propyl groups) are chemisorbed to one of the sliding diamond surfaces, instead of trapped between the surfaces, different behavior is observed by *Harrison* et al. [35.172,173,208,219]. Simulations show that methyl-termination does not decrease friction significantly but results in frictional forces that are nearly the same as they are for hydrogen-terminated diamond surfaces [35.207]. While the methane third-body molecules decrease the frictional force to a greater extent than the chemisorbed methyl groups, friction as a function of load is comparable for the ethyl-terminated and ethane systems, with the former giving slightly higher frictional forces. Attaching the hydrocarbon groups to the diamond surfaces causes them to have less freedom to move between hydrogen atoms on opposing diamond surfaces during sliding. This generally increases their repulsive interaction with the diamond counterface.

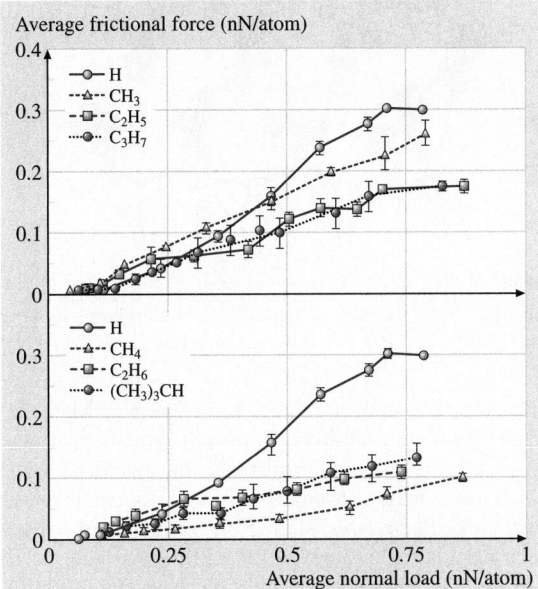

Fig. 35.43 Average frictional force per rigid-layer atom as a function of average normal load per rigid-layer atom for sliding the upper diamond surface in the [11$\bar{2}$] crystallographic direction. Data for the methane (CH$_4$) system (*open triangles*), the ethane (C$_2$H$_6$) system (*open squares*), the isobutane (CH$_3$)$_3$CH system (*filled circles*), and diamond surfaces in the absence of third-body molecules (*open circles*) are shown in the lower panel. Data for the methyl-terminated −CH$_3$ system (*open triangles*), the ethyl-terminated (−C$_2$H$_5$) system (*open squares*), the n-propyl-terminated (−C$_3$H$_7$) system (*filled circles*), and diamond surfaces in the absence of third-body molecules (*open circles*) are shown in the *upper panel. Lines* have been drawn to aid the eye. After [35.177] with the permission of the ACS (1997)

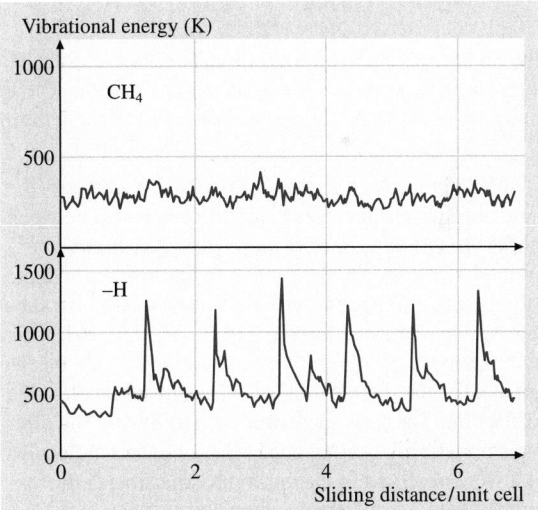

Fig. 35.44 Average vibrational energy between the (C–H) bonds of the upper diamond (111) surface versus sliding distance for hydrogen-terminated diamond (111) surfaces, with (CH$_4$) and without methane (H), trapped between them. The average normal load is approximately the same in both simulations and is in the range 0.8–0.85 nN/atom. The average frictional force on the upper surface is about 3.5 times smaller in the presence of the methane third-body molecules. After [35.218] with the permission of Elsevier (1996)

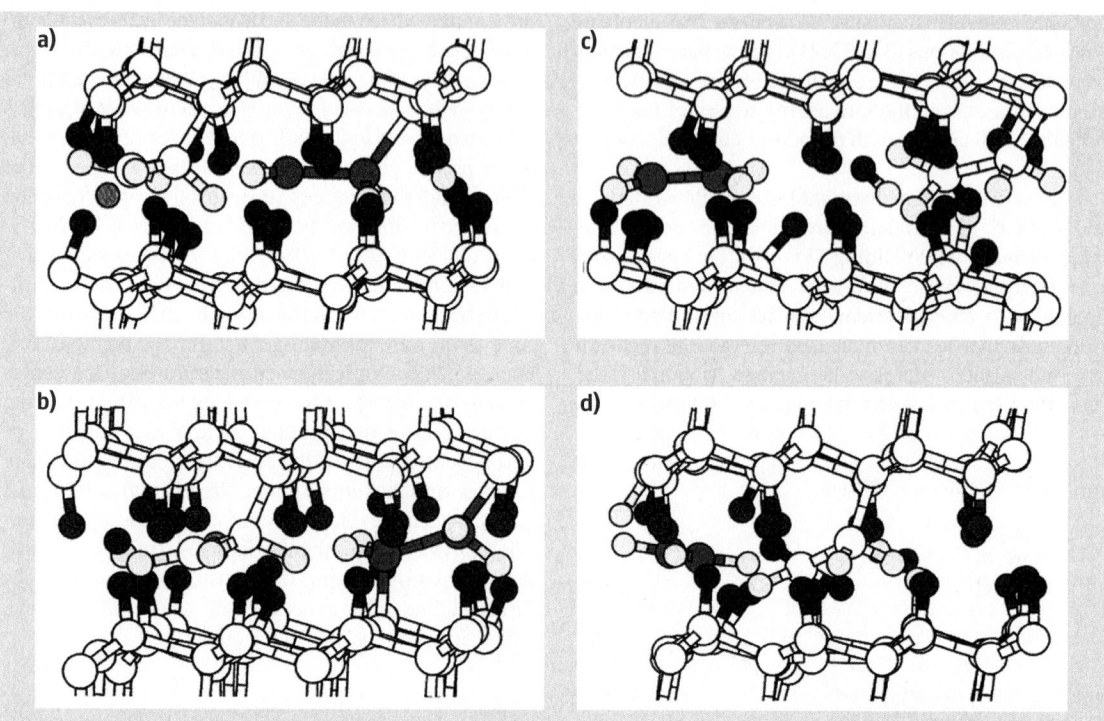

Fig. 35.45 Snapshots from a molecular dynamics simulation of the sliding of two hydrogen-terminated diamond (111) surfaces against one another in the [11$\overline{2}$] direction. The upper surface has two ethyl fragments chemisorbed to it. The simulation shows how sliding can induce chemistry at the interface. After [35.220] with the permission of the ACS (1994)

MD simulations can also provide insight into the rich, nonequilibrium tribochemistry that occurs between surfaces in sliding contact. *Harrison* and *Brenner* examined the tribochemistry that occurs when ethane molecules are trapped between diamond surfaces in sliding contact, as illustrated in Fig. 35.45 [35.220]. This simulation was the first to show the atomic-scale mechanisms for the degradation of lubricant molecules due to friction. The type of debris formed during the sliding simulation is similar to the types of debris molecules that were observed in macroscopic experiments that examined the friction between diamond surfaces [35.221].

In the case of sliding metal surfaces, impurity molecules or atoms (both adsorbed and absorbed) on thin metal films can be expected to affect the film's properties. For example, calculations have shown that resistivity changes in the metal are strongly dependent on the nature of the adsorption bond [35.222]. When this result is used to interpret atomic-scale friction results obtained with the QCM, the sliding of adsorbate structures on metal surfaces are shown to be a combination of electron excitation and lattice vibrations. Additionally, other

interesting quantum effects can come into play when the adsorbate is very different chemically from the surface on which it is sliding. For instance, the electronic frictional forces acting on small, inert atoms and molecules, such as C_2H_6 and Xe, sliding on metal surfaces have been calculated by *Persson* [35.223], where the metal surface was approximated by a electron gas (jellium) model. The calculations showed that the Pauli repulsive and attractive van der Waals forces are of similar magnitudes. In addition, the calculated electronic friction contributions agree well with the values derived from surface resistivity by *Grabhorn* et al. [35.224] and QCM measurements. These studies showed that parallel friction is mainly due to electronic effects while perpendicular friction is phononic in nature in this system.

In summary, MD simulations show that the average frictional force decreases significantly in systems with third-body molecules, especially at high loads. Simulations also provide information about the details of tribochemical interactions that can occur between lubricants and sliding surfaces. Additionally, the effect of the presence of small molecules on thin metal films can influence film properties, such as resistivity.

35.3.3 Thin Films

As discussed at the beginning of this section, the conversion of the work of sliding into some other less ordered form is responsible for friction at sliding solid interfaces. In the case of adhering systems, the work of sliding may be converted into damage within the bulk (plastic deformation), while in the case of weakly adhesive forces, friction can occur through the conversion of work to heat at the interface that causes no permanent damage to the surface (wearless friction). The latter case, when it is achieved through the presence of lubricating thin films, is the topic of this section.

There are several types of lubricating thin films, the simplest of which consist of small molecules that are analogous to wear debris that can "roll" between the sliding surfaces or that represent very short-chain bonded lubricants. These thin films were discussed in the previous subsection. The rest of this subsection will, therefore, focus on the effects of liquids, larger nanoparticles, self-assembled monolayers and solid thin films on lubrication and friction.

Liquids
Liquids are common lubricants and so they have been studied in great depth at the macroscale. At the nanoscale, the tribological response of spherical liquid molecules has been well-characterized experimentally using the SFA and computationally with MD simulations by *Berman* et al. [35.225]. The SFA experiments considered one to three liquid layers and the stick-slip motion at the interface is found to increase in a quantized fashion as the number of lubricant layers decrease. When no external forces are applied to the system, the sliding stops and the solid–lubricant interactions are strong enough to force the liquid molecules to form a close-packed structure that is ordered. The transformation of the liquid into this solid-like structure causes the two surfaces to effectively bond to each other through the lubricant. When the surfaces start to slide again, lateral shear forces are introduced that steadily increase, which causes the molecules in the liquid to undergo small lateral displacements that change the film thickness. If these shear forces become greater than a critical value, the film disorders in a manner that is analogous to melting. This allows the surfaces to slide easily past each other in a manner that is still quantized. This sequence of events is nicely illustrated

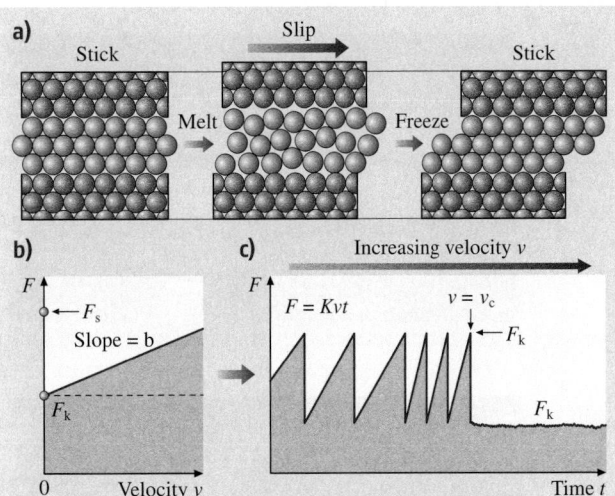

Fig. 35.46 The stick/slip transition that occurs for thin films of liquid between two sliding solid surfaces. F is the intrinsic friction and F_s is the friction where the liquid is in the rigid state; F_k is the friction where the liquid is in the liquid-like state. After [35.225] with the permission of the ACS (1996)

in Fig. 35.46 and can be reproduced multiple times for the same system.

Persson et al. [35.226] used MD simulations with pairwise potentials similar to those in (35.6) to examine the mechanism by which this sharp transition occurs. They find that in the case of sliding on insulating crystal surfaces, the solid-state lubricant may be in a "superlubric" state where the friction is negligible. It is clear from the simulations, however, that any surface defects, even in low concentrations, will disrupt this state and transform the lubricant back into a fluid. In addition, when sliding occurs on metallic surfaces above cryogenic temperatures, the electronic contributions to friction are no longer zero and no superlubric state is possible.

High applied pressures can force the fluid molecules out from between the two confining surfaces [35.227]. The fact that liquid molecules close to a stiff surface are strongly layered in the direction perpendicular to the surface explains the experimental observation of a $(n \rightarrow n-1)$ layer transition, where n is number of monolayers, that is observed as the normal load increases [35.228]. Nucleation theory is used to calculate the critical pressure and determine the spreading dynamics of the $(n-1)$ island.

The reactivity of the liquid molecules are also critically important to boundary layer friction. MD studies by *Persson* et al. [35.229] show that inert molecules in-

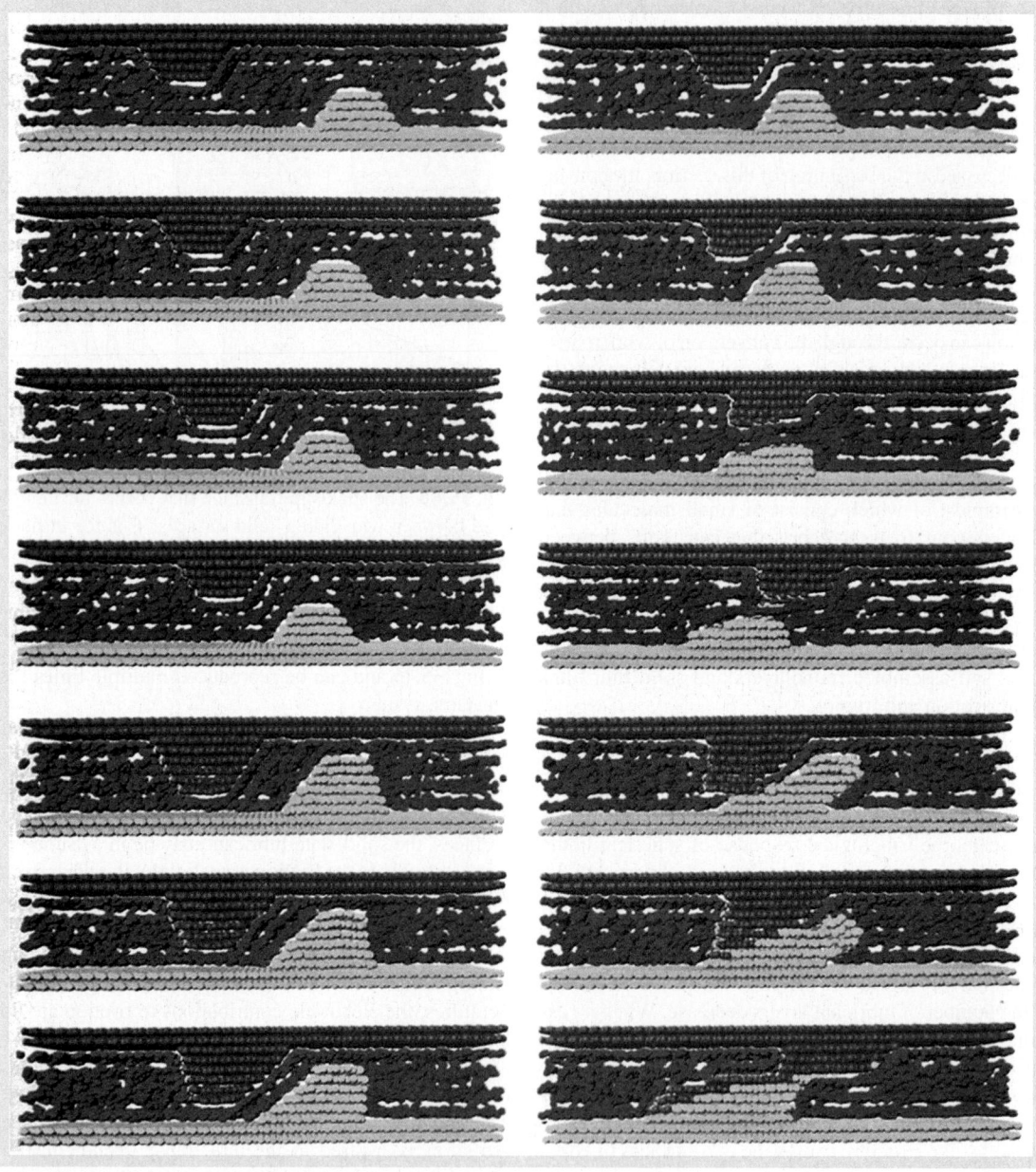

Fig. 35.47 Stills from a molecular dynamics simulation where Au (111) surfaces with surface roughness slide over one another while separated by hexadecane molecules. The scanning velocity is 10 m/s. Layering of the lubricant and asperity deformation occurs as the sliding continues. The *top three rows* show the results when the asperity heights are separated by 4.6 Å. The *bottom three rows* show the results when the asperity heights are separated by -6.7 Å. After [35.230] with the permission of the ACS (1996)

teract weakly with sliding surfaces. Consequently, as the rate of sliding increases, the molecular conversion from the solid state to the liquid state occurs in an abrupt manner. However, when the molecules interact strongly with

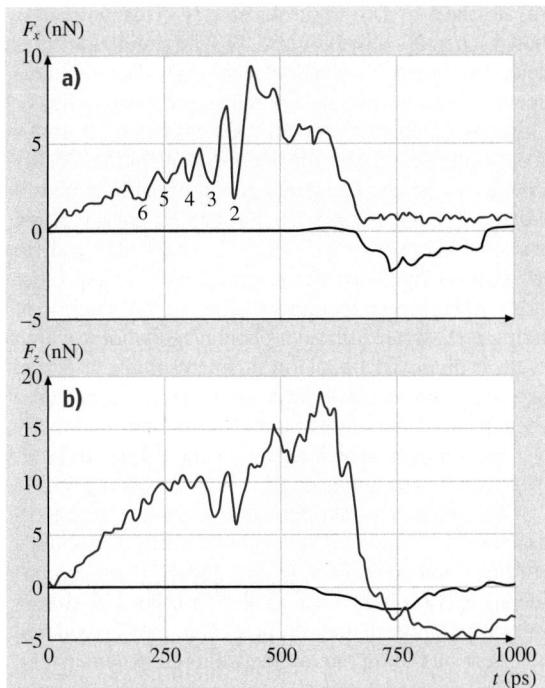

Fig. 35.48 The lateral force (F_x) and normal force (F_z) from the molecular dynamics simulations shown in Fig. 35.35 as a function of time. The forces between the two metal surfaces are shown by the *dashed line*. The force oscillations correspond to the structural changes of the lubricant in Fig. 35.35. After [35.230] with the permission of the ACS (1996)

the surfaces, they undergo a more gradual transition from the solid to the liquid state. *Persson* et al. [35.34] also considered systems where the molecules are attached to one of the surfaces, which causes the transitions to be abrupt. This is especially true if there are large separations between the chains.

While the studies discussed so far have focused on spherical liquids, most widely used liquid lubricants consist of long-chain hydrocarbons. Nonspherical liquid molecules have more difficulty aligning and solidifying. This is borne out in MD simulations by *Thompson* et al. [35.231] that show that spherical molecules have higher critical velocities than branched molecules. In particular, the simulations show that when the molecules are branched, the amount of time various parts of the system spend in the sticking and sliding modes changes with sliding rate. The critical velocity can also depend on the number of liquid layers in the film, the structure and relative orientation of the two slid-

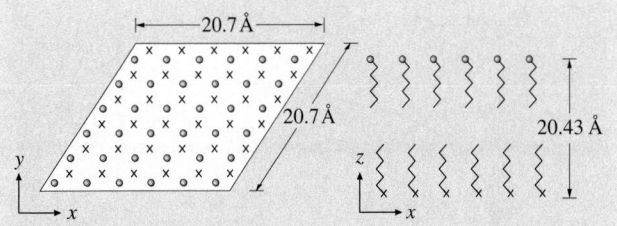

Fig. 35.49 Top and side views of the alkane chains attached to surfaces that are sliding against each other. After [35.174] with the permission of the APS (1993)

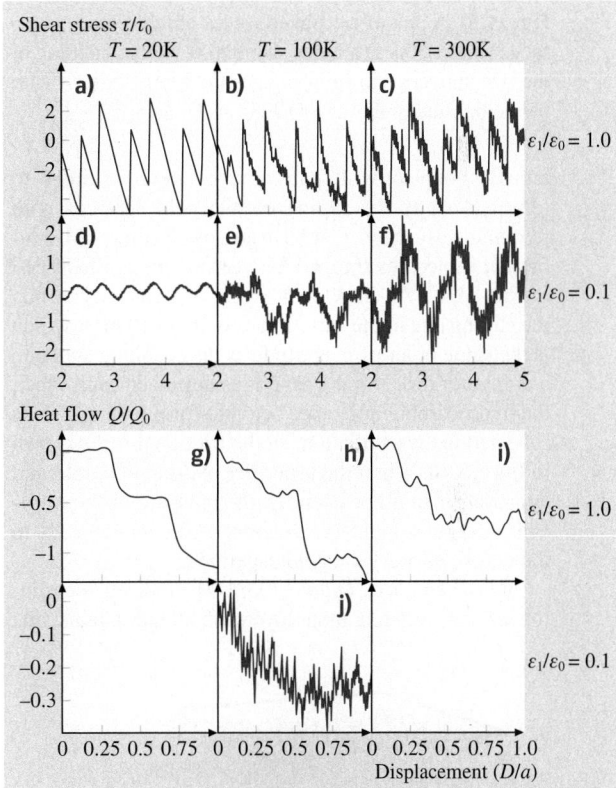

Fig. 35.50a–j Data from molecular dynamics simulations of the sliding of the surfaces shown in Fig. 35.32. (**a**) – (**f**)The shear stress and (**b**) – (**j**) the heat flow as a function of sliding for normal and reduced interfacial strengths. The *plots* show how the calculated values change with system temperature. After [35.174] with the permission of the APS (1993)

ing surfaces, the applied load and the stiffness of the surfaces.

Additional studies by *Landman* et al. [35.230] used MD simulations with bond-order and EAM po-

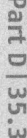

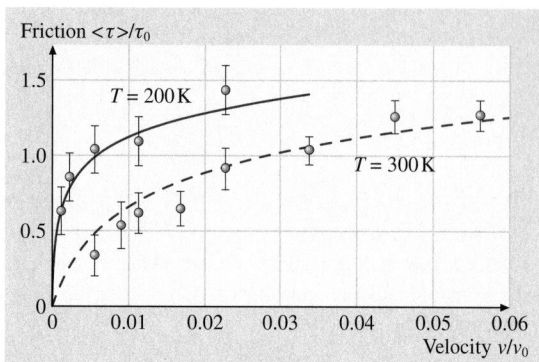

Fig. 35.51 A plot of calculated values of the average interfacial shear stress as a function of the velocity of sliding of the two surfaces shown in Fig. 35.33. After [35.174] with the permission of the APS (1993)

tentials coupled with pair-wise potentials similar to (35.6) to study the sliding of two gold surfaces with pyramidal asperities that have straight chain $C_{16}H_{34}$ lubricant molecules trapped between them, as illustrated in Fig. 35.47. An important aspect of this study is that the sliding rate in the simulations is about $10\,\text{m/s}$, which is the same order of magnitude as the scanning speed in a computer disk. As the asperities approach each other, the hydrocarbon molecules begin to form layers. This is reflected in the oscillations in the frictional force shown in Fig. 35.48. When the asperities overlap in height and approach each other laterally, the pressure of the lubricant molecules increases to about $4\,\text{GPa}$ which leads to the deformation of the gold asperities.

Glosli and *McClelland* [35.174] modeled the sliding of two ordered monolayers of alkane chains that are attached to two rigid substrates. This system is shown schematically in Fig. 35.49. The simulations predicted that energy dissipation occurs by a discontinuous plucking mechanism (sudden release of shear strain) or a viscous mechanism (continuous collisions of atoms of opposite films). The specific mechanism that occurs depends on the interfacial interaction strength. In particular, the "pluck" occurs when mechanical energy stored as strain is converted into thermal energy that leads to low friction forces at low temperatures. On the other hand, at higher temperatures some of the energy of sliding is dissipated through phonon excitations, which results in higher frictional forces. Interestingly, this trend reverses again at the highest temperatures considered when the molecules move so much that they slide easily over the surfaces, which decreases the frictional force. These results are summarized in Figs. 35.50 and 35.51.

Other studies of sliding surfaces with attached organic chains include MD simulations with LJ potentials by *Müser* and coworkers, [35.41, 232, 233] which considered friction between polymer "brushes" in sliding contact with one another. In particular, they considered the effect of sliding rate on the tilting of polymers and the effect of steady-state sliding versus non-steady-state ("transient") sliding. The simulations find that shear forces are lower for chains that tilt in a direction that is parallel to the shear direction. This tilting effect is significant for grafted polymers, as illustrated in Fig. 35.52, and less significant for absorbed polymers. This is due to the decrease in the differential frictional coefficient for the grafted polymers as well as the increase in the friction coefficient for absorbed polymers under shear. The tilting is also affected by the rate of sliding and is much larger at high sliding rates than small rates, as in-

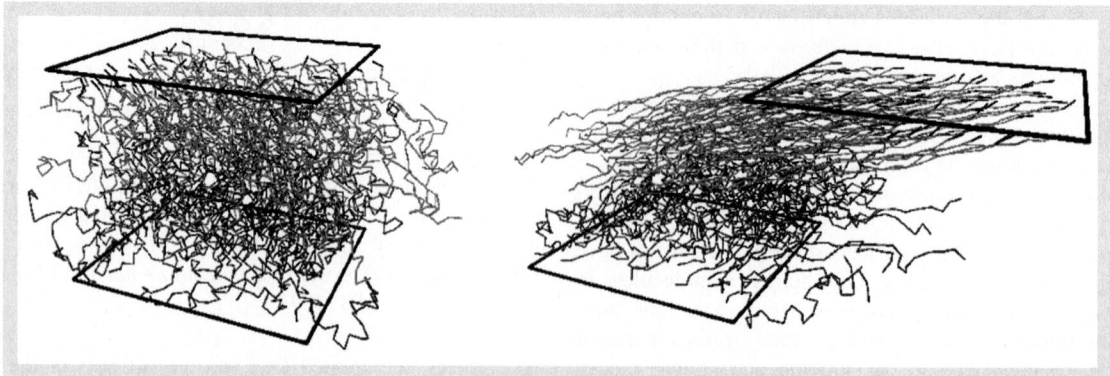

Fig. 35.52 Snapshots of sliding walls with attached polymers in a solvent. *Right-hand figure* illustrates the sliding process at low sliding rates while the *left-hand figure* illustrates the sliding process at high sliding rates. After [35.41] with the permission of the CCLR (2002)

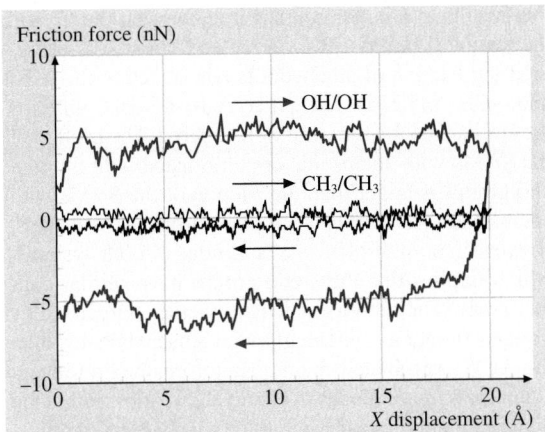

Fig. 35.53 Typical friction loops for the systems shown in Fig. 35.25 for CH_3/CH_3 and OH/OH pairs under a contact load of 0.2 nN. After [35.160] with the permission of the ACS (2002)

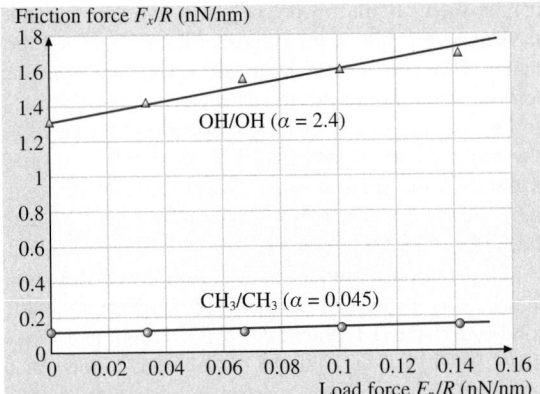

Fig. 35.54 Friction force versus contact load from the systems shown in Fig. 35.25 for CH_3/CH_3 and OH/OH. After [35.160] with the permission of the ACS (2002)

dicated in Fig. 35.52. The simulations further show that the inclination angle of the chains decreases much more slowly than the shear stress, and the shear stress maximum is more pronounced if there is hysteresis in the chain orientations.

Typical friction loops for tips that are functionalized and sliding against surfaces that are functionalized in the same manner as illustrated in Fig. 35.25 are shown in Fig. 35.53. The friction force between the OH/OH pairs is significantly larger than the friction force between the CH_3/CH_3 pairs. This is due to the formation and breaking of hydrogen bonds during the shearing for the OH/OH pairs. The mean forces vs. load forces for the

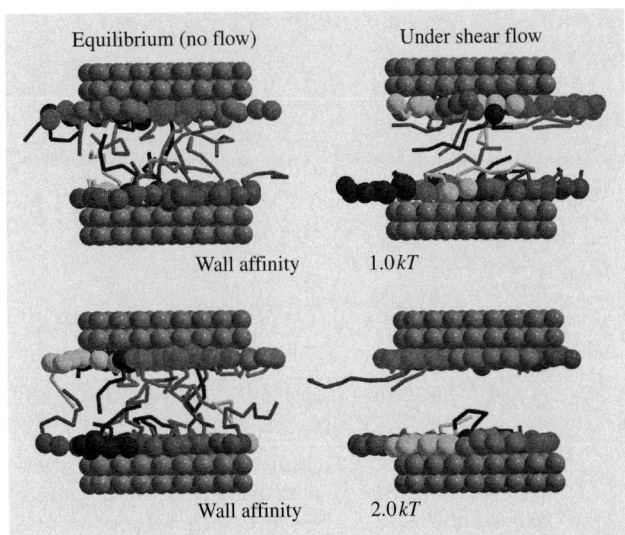

Fig. 35.55 Changes in the conformation of adsorbed hydrocarbon chains on weakly (*top*) and strongly (*bottom*) physisorbing surfaces at equilibrium and under shear. After [35.234] with the permission of the ACS (1996)

OH/OH and CH_3/CH_3 pairs given in Fig. 35.54 are reduced by the tip radius.

MD simulations by *Manias* et al. considered the shearing of entangled oligomer chains that are attached to sliding surfaces, as illustrated in Fig. 35.55 [35.234]. They find that slip takes place within the film and that this occurs through changes in the chain conformations. Increased viscosity is predicted at the film–surface interface compared to the middle of the film, which results in a range of viscosities across the film as one moves away from the points of sliding contact.

To summarize this section, experiments and MD simulations show similar stick/slip transitions that occur for thin films of liquid between two sliding solid surfaces. Frictional properties are found to depend to a significant degree on molecular shape, whether the molecules are grafted on the surfaces or merely absorbed on them, and the degree of tilting in the case of molecular chains. In the case of long-chain molecules, temperature is found to affect the frictional force because the mechanical energy stored in long-chains can be converted into thermal energy by friction.

Self-Assembled Structures

There have been numerous experimental studies of friction on SAMs on solid surfaces with AFM and FFM. The experimental results reveal relationships among

Part D | 35.3

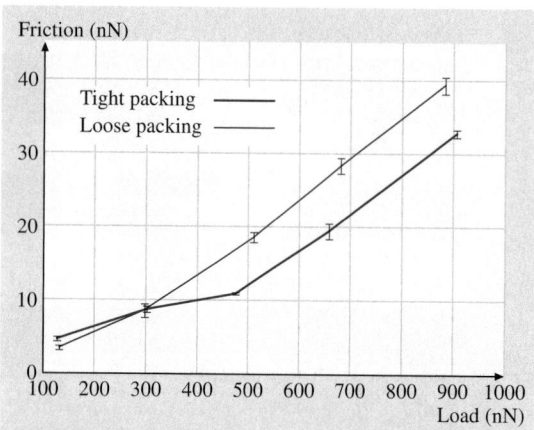

Fig. 35.56 Friction as a function of load when a hydrogen-terminated counterface is in sliding contact with C_{18} alkane monolayers. After [35.239] with the permission of the ACS (2001)

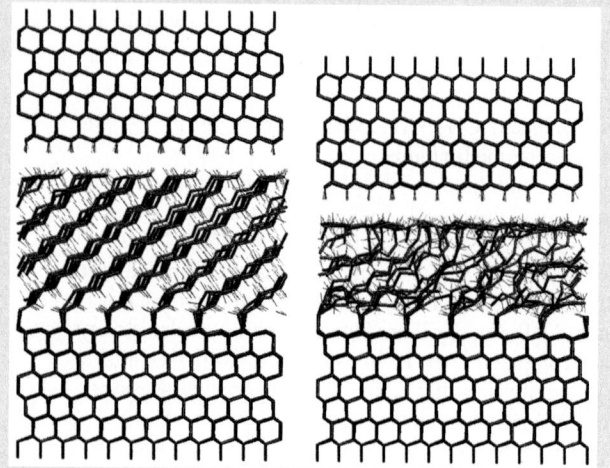

Fig. 35.57 Snapshots of tightly packed C_{18} alkane monolayers on the *left*, and loosely packed monolayers on the *right* under a load of about 500 nN. The chains in both systems are arranged in a (2×2) arrangement on diamond (111). The loosely packed system has 30% of the chains randomly removed. The sliding direction is from *left to right*. After [35.239] with the permission of the ACS (2001)

elastic compliance, topography and friction on thin LB films [35.235]. For example, they have detected differences in the adhesive interactions between the microscope tips and CH_3 and CF_3 end groups [35.106]. Fluorocarbon domains generally exhibit higher friction than the hydrocarbon films, which the authors attribute to the lower elasticity modulus of the fluorocarbon films

that results in a larger contact area between the tip and the sample [35.235–237]. *Perry* and coworkers examined the friction of alkanethiols terminated with $-CF_3$ and $-CH_3$ [35.238]. The lattice constants for both films are similar and the films are well-ordered. The friction of the SAMs with chains that are terminated with fluorine end groups is larger than the friction of the SAMs with chains that are terminated with hydrogen end groups. However, the pull-off force is similar in both systems, which implies that these end groups have similar contact areas. The authors speculate that the larger $-CF_3$ groups interact more strongly with adjacent chains than the $-CH_3$-terminated chains. Therefore, the fluorinated chains have more modes of energy dissipation within the plane of the monolayer and, thus, have larger friction.

Molecular disorder of the alkyl chains at the surface can also affect the frictional properties of self-assembled films if the layers are not packed too closely together [35.240]. Indentation can induce disorder in the chains that then compress as the tip continues to press against them. If the tip presses hard enough, the film hardens as a result of the repulsive forces between the chains. However, if the chains are tilted, they bend or deform when the tip pushes on them in a mostly elastic fashion that produces long lubrication lifetimes. At low contact loads of about 10^{-8} N, wear usually occurs at defect sites, such as steps. Wear can also occur if there are strong adhesive forces between the film and the surface [35.241].

The friction of model SAMs composed of alkane chains was examined using MD simulations with bond-order and LJ potentials by *Mikulski* and *Harrison* [35.239, 242]. These simulations show that periodicities observed in a number of system quantities are the result of the synchronized motion of the chains when they are in sliding contact with the diamond counterface. The tight packing of the monolayer and commensurability of the counterface are both needed to achieve synchronized motion when sliding in the direction of chain tilt. The tightly packed monolayer is composed of alkane chains attached to diamond (111) in the (2×2) arrangement and the loosely packed system has approximately 30% fewer chains. The average friction at low loads is similar in both the tightly and loosely packed systems at low loads. Increasing the load, however, causes the tightly packed monolayer to have significantly lower friction than the loosely packed monolayer (Fig. 35.56). While the movement of chains is somewhat restricted in both systems, the tightly packed monolayer under high loads is more constrained with respect to the movement of individual chains than the loosely packed monolayer, as illustrated in Fig. 35.57. Therefore, sliding initiates larger bond-length fluctuations in the loosely packed system, which ultimately lead to more energy dissipation via vibration and, thus, higher friction. Thus, the efficient packing of the chains is responsible for the lower friction observed for tightly packed monolayers under high loads.

Several AFM experiments have examined the friction of SAMs composed of chains of mixed lengths [35.243]. For example, the friction of SAMs composed of spiroalkanedithiols was examined by *Perry* and coworkers [35.244]. The effects of crystalline order at the sliding interface were examined by systematically shortening some of the chains. The resulting increase in disorder at the sliding interface causes an increase in friction.

The link between friction and disorder in monolayers composed of *n*-alkane chains was recently examined using MD simulations by *Harrison* and coworkers [35.245]. The tribological behavior of monolayers of 14 carbon atom-containing alkane chains, or pure monolayers, was compared to monolayers that randomly combine equal amounts of 12 and 16 carbon-atom chains, or mixed monolayers. Pure monolayers consistently show lower friction than mixed monolayers when sliding under repulsive (positive) loads in the direction of chain tilt. These MD simulations reproduce trends observed in AFM experiments of mixed-length alkanethiols [35.243] and spiroalkanedithiols on Au [35.246].

Because the force on individual atoms is known as a function of time in MD simulations, it is possible to calculate the contact forces between individual monolayer chain groups and the tip, where contact force is defined as the force between the tip and a $-CH_3$ or a $-CH_2$-group in the alkane chains. The distribution of contact forces between individual monolayer chain groups and the tip are shown in Fig. 35.58. It is clear from these contact force data that the magnitude, or scale of the forces, is similar in both the pure and the mixed monolayers. In addition, it is also apparent that the pure and mixed monolayers resist tip motion in the same way. That is, the shape of the histograms in the positive force intervals is similar. In contrast, the contact forces "pushing" the tip along differ in the two monolayers. The pure monolayers exhibit a high level of symmetry between resisting and pushing forces. Because the net friction is the sum of the resisting and pushing forces, the symmetry in these distributions of the pure monolayers results in a lower net friction than the mixed monolayers. Thus, the ordered, densely packed nature of the pure mono-

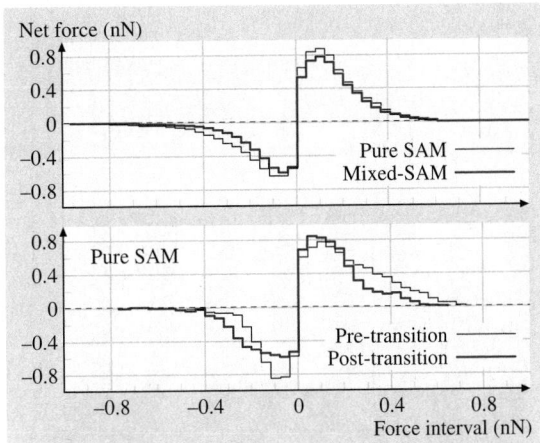

Fig. 35.58 The distribution of contact forces along the sliding direction (friction force). In the *upper panel*, the forces for the mixed and pure system sliding in the direction of chain tilt are shown. The forces for the pure system sliding in the transverse direction to the chain tilt are shown in the *lower panel*. Positive force intervals correspond to chain groups that resist tip motion while negative intervals correspond to chain groups that "push" the tip in the sliding direction. Forces from four runs with independent starting configurations are binned for all sets of data

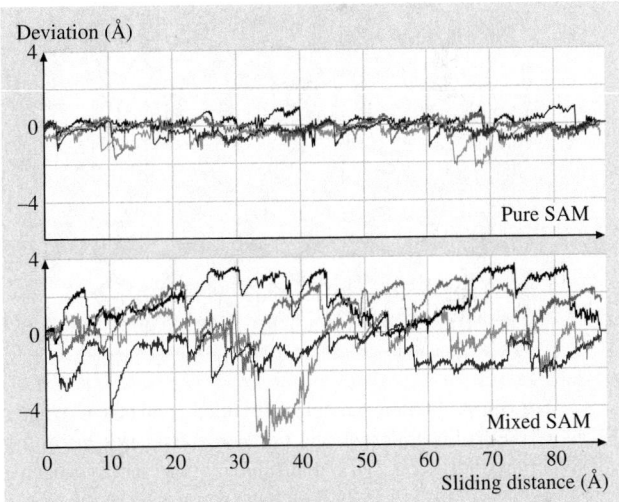

Fig. 35.59 Trajectories of individual chain groups that generate the largest contact forces when sliding in the direction of chain tilt for both the pure and mixed monolayer systems. The deviation is defined as the change in position along the sliding direction relative to the chain group's starting position. (The positions are averaged over 2000 simulation steps)

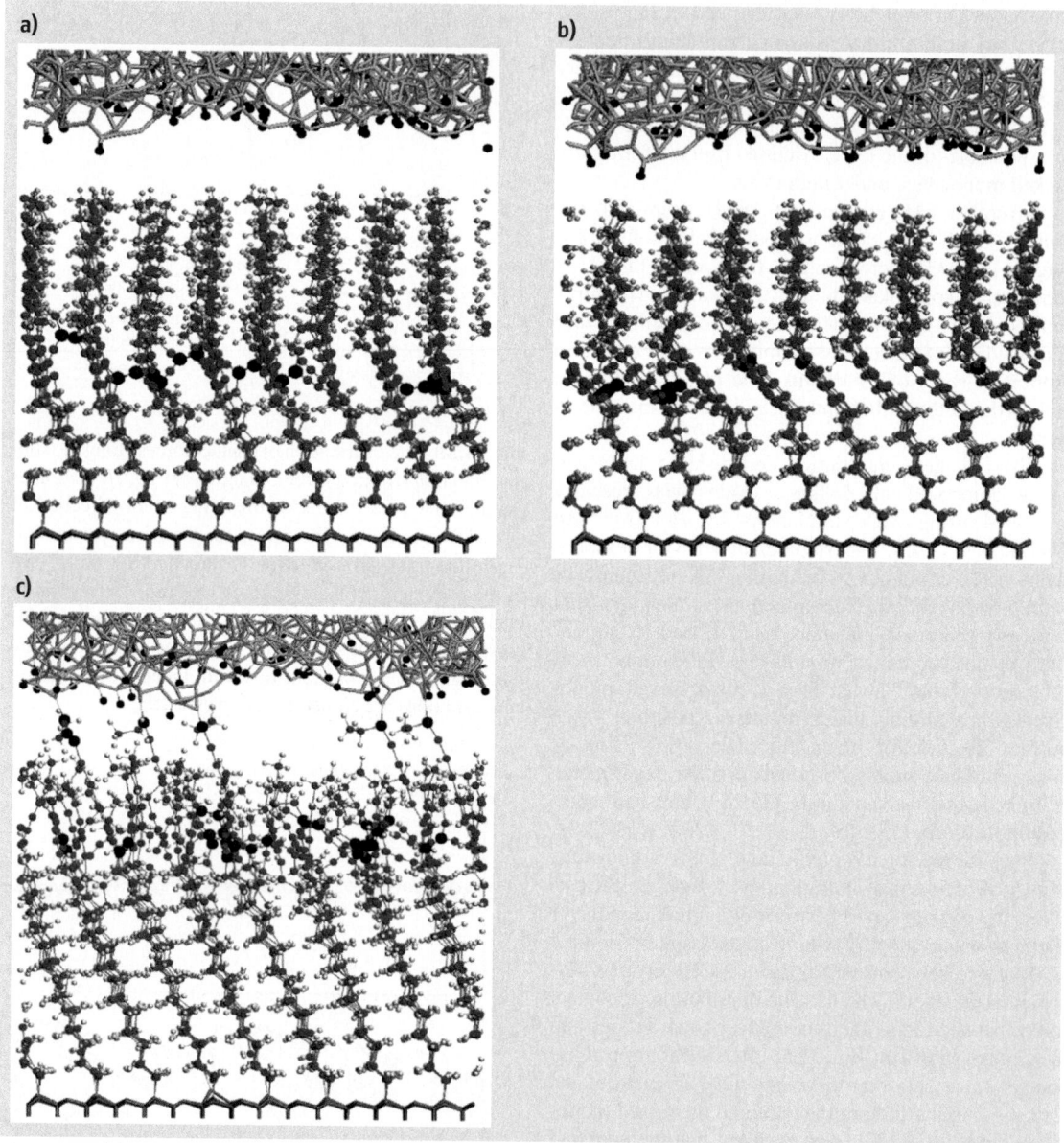

Fig. 35.60 (a) Perpendicular-, (b) tilted-, and (c) end-chain monolayer systems after compression to 200 nN and pull-back of the hydrogen-terminated tip. *Large, dark spheres* in the hydrocarbon monolayers represent cross-linked atoms with sp² hybridization. *Dark, small spheres* represent hydrogen atoms that are initially on the hydrogen-terminated amorphous carbon tip. After [35.247] with the permission of the ACS (2004)

layers allows the energy stored when the monolayer is resisting tip motion (positive forces) to be regained efficiently when the monolayer "pushes" on the tip (negative forces). The distribution of negative contact forces in the mixed monolayers is different from the distribution of the positive forces. For this reason, mechanical energy is not efficiently channeled back into the mixed monolayer as the tip passes over the chains and, as a result, the friction is higher. The range of motion of the chains is monitored by computing the deviation in a chain group's position compared to its starting position, as illustrated in Fig. 35.59. It is clear from analyzing these data that the increased range of motion is linked to large contact forces. The increased range of motion of the protruding tails in the mixed system prevents the efficient recovery of energy during sliding (negative contact force distribution) and allows for the dissipation of energy.

The pure monolayers exhibit marked friction anisotropy. The contact force distribution changes dramatically as a result of the change in sliding direction, resulting in an increase in friction (Fig. 35.58). Sliding in the direction perpendicular to chain tilt can cause both types of monolayers to transition to a state where the chains are primarily tilted along the sliding direction. This transition is accompanied by a large change in the distribution of contact forces and a reduction in friction.

Recently, the response of monolayers composed of alkyne chains, which contain diacetylene moieties, to compression and shear [35.247] was examined using MD simulations. These are the only simulations to date that show that compression and shear can result in crosslinking, or polymerization, between chains. The vertical positioning of the diacetylene moieties within the alkyne chains (spacer length) and the sliding direction both have an influence on the pattern of cross-linking and friction. Compression and shear cause irregular polymerization patterns to be formed among the carbon backbones, as illustrated in Fig. 35.60. When diacetylene moieties are located at the ends of the chains closest to the tip, chemical reactions between the chains of the monolayer and the amorphous carbon tip occur causing the friction to increase 100 times, as indicated in Fig. 35.61. The friction between the amorphous carbon tip and all of the diacetylene-containing chains is larger than the friction between a hydrogen-terminated diamond counterface and tightly packed monolayers composed of n-alkane chains [35.239]. This is attributed to the disorder at the interface caused by the irregular counterface.

Zhang et al. [35.248] used MD simulations to study the effect of confined water between alkyl monolayers terminated with $-CH_3$ (hydrophobic) and $-OH$ (hydrophilic) groups on Si(111), as illustrated in Fig. 35.62. For the hydrophobic molecules, the friction coefficient is almost constant independent of the number of water molecules. For the hydrophilic molecules, the friction coefficient decreases rapidly with an increase in the number of water molecules, as shown in Fig. 35.63. These results are in good agreement with surface force microscopy (SFM) experimental results. *Zhang* et al. [35.249] also studied the friction of alkanethiol SAMs on gold using hybrid molecular simulations at the

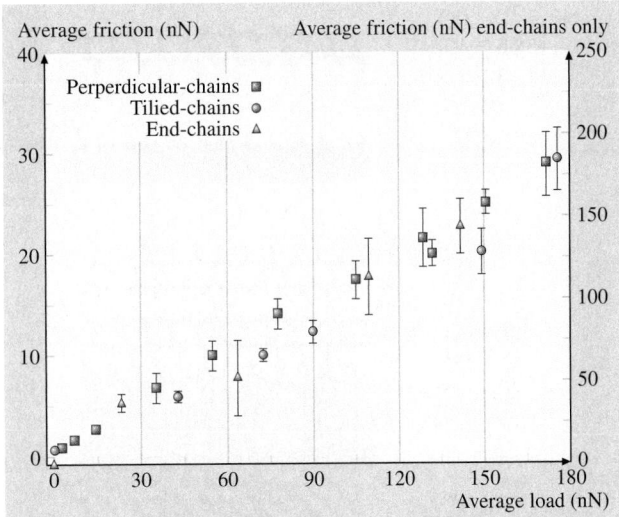

Fig. 35.61 Average friction on the tip as a function of load for the monolayer systems shown in Fig. 35.60. The *scale* for the average friction in the end-chain system is shown on the *right-hand side* of the figure. After [35.247] with the permission of the ACS (2004)

same time scales as are used in AFM and FFM experiments. Various quantities were varied in the simulations, including chain length, terminal group, scan direction and scan velocity. The simulations showed that the frictional force decreases as the chain length increases and is smallest when scanned along the tilt direction. They also predicted a maximum friction coefficient for hydrophobic $-CH_3$-terminated SAMs and low friction coefficients for hydrophilic $-OH$-terminated SAMs as the scan velocity increases. The simulations further predicted a saturated constant value at high scan rates for both surfaces. These results are summarized in Figs. 35.64 and 35.65.

The work of *Chandross* et al. [35.250, 251] illustrates the effects of chain length on friction and stick-slip behavior between two ordered SAMs consisting of alkylsilane chains over a range of shear rates at various separation distances or pressure, as illustrated in Fig. 35.66. The adhesion forces between the two SAMs at the same separation distance decrease as the chain length increases from 6 to 18 carbon atoms. However, the friction forces are independent of the chain length and the shear velocity. The system size is shown to have an effect on the sharpness of the slip transitions but not on the dynamical events, as shown in Fig. 35.67.

In short, atomic-scale simulations show the relationship between elastic properties, degree of molecular

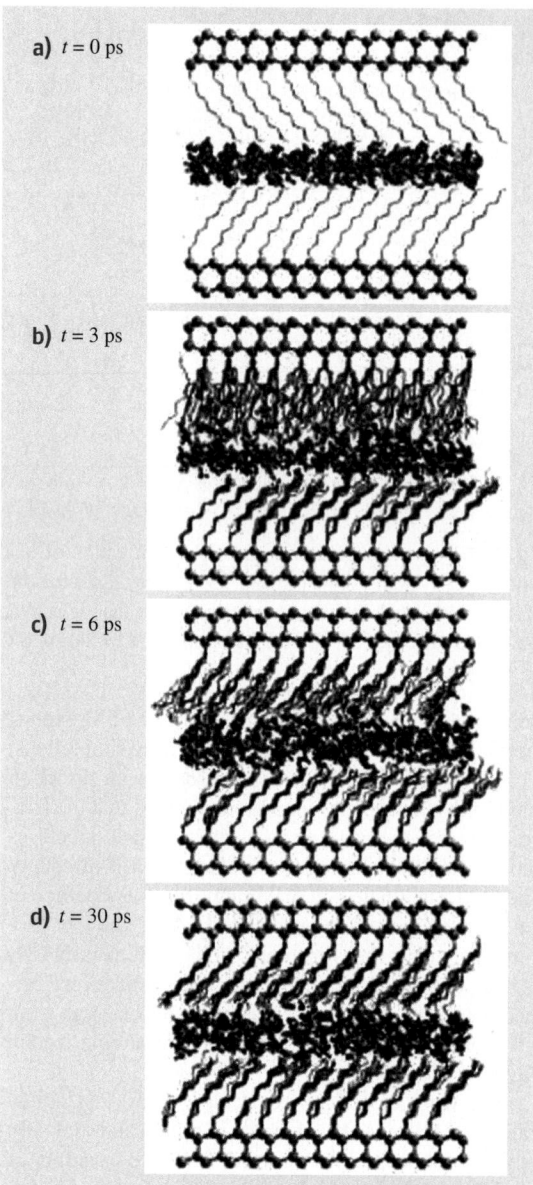

Fig. 35.62 Snapshots of hydrophilic monolayers and confined water molecules from MD simulations at 300 K. The tilt direction of monolayers on the top plate changed after $t = 10.0$ ps. After [35.248] with the permission of the AIP (2005)

Nanoparticles

Nanoparticles are being considered for a wide variety of applications, including as fillers for nanocomposite materials, novel catalysts or catalytic supports, and components for nanometer-scale electronic devices [35.252]. They have also generated considerable interest as possible new lubricant materials that have the potential to function as "nanoballbearings" with exceptionally low friction coefficients. The nanoparticles

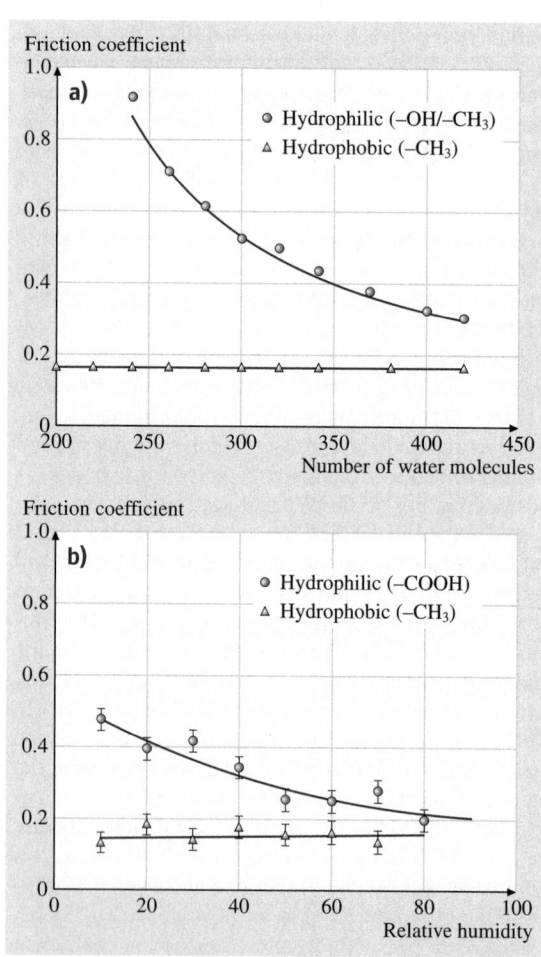

Fig. 35.63 (a) Friction coefficients for hydrophobic ($-CH_3$) and hydrophilic (50% mixed $-CH_3/-OH$) monolayers as a function of water molecules from MD simulations at 300 K ($H = 6.0$ Å), and **(b)** scanning force microscopy measurements of frictional forces of difference surfaces under various relative humidities. After [35.248] with the permission of the AIP (2005)

disorder and friction of self-assembled thin films that illuminates the origin of the properties that are measured experimentally.

Fig. 35.64 Schematic illustration of the chain tilt and scan directions on alkanethiol SAMs/Au(111) in hybrid molecular simulations; θ is the angle between the tip moving direction and the chain tilt direction. The *larger spheres* represent substrate Au atoms, smaller spheres sulfur atoms in molecular chains, and *zigzag lines* molecular chains. After [35.249] with the permission of the ACS (2003)

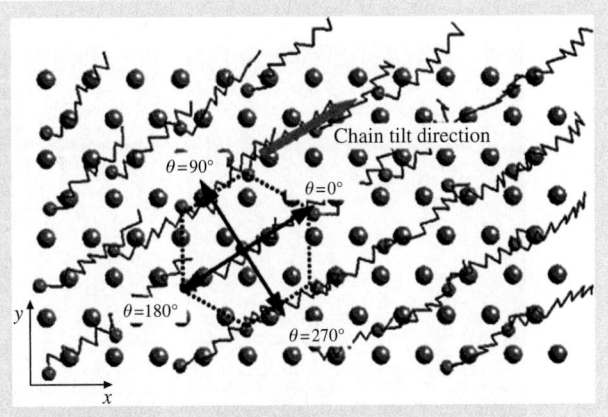

of most interest for tribological applications include C_{60} [35.253–265], carbon nanotubes [35.266–273], and MoO_3 nanoparticles [35.274, 275], among others [35.276].

The experiments report wide variations in frictional coefficients (for instance, values of 0.06 to 0.9 have been measured for C_{60}) that may be caused by differences in the experimental methods used, the thickness of the nanoparticle layer or island, the atmosphere (argon versus air, levels of humidity) used, and the transfer of nanoparticles to the FFM tips. As a result, there is much that remains to be clarified about the tribological behavior of nanoparticle films.

In the case of C_{60}, the mechanistic response to applied shear forces has not been definitively determined. For example, some experimental studies show evidence of C_{60} molecules rolling against the substrate, each other, or the sliding surfaces [35.253, 258, 260, 263, 265] while others hypothesize that the low friction of C_{60} films is due in part to blunting of the tip by transfer of fullerene molecules to the tip apex. Fullerene films are found experimentally to have dissipation energies and shear strengths that are a full order of magnitude lower than the values that are typical for boundary lubricants [35.277]. Experimental testing of the frictional properties of fullerenes reveal low mechanical stability accompanied by progressive wear and transfer of fullerene materials when they are only physisorbed on a solid surface [35.278]. Furthermore, measurements with a FFM show that under certain conditions, adsorbed fullerene films deteriorate at pressures as low as about 0.1 GPa [35.279]. The challenge is therefore to obtain mechanically stable, ordered molecular films of fullerenes firmly attached to a solid substrate.

Fig. 35.65a,b Frictional force as a function of scan direction from hybrid simulations for $C_{11}CH_3$ SAMs on Au(111) at (**a**) 300 K and (**b**) 1.0 K. Frictional force is the smallest when scanned along the tilt direction, the largest when scanned against the tilt direction, and between when scanned perpendicular to the tilt direction at both temperatures. After [35.249] with the permission of the ACS (2003)

There have been several MD simulation studies to investigate the tribological properties of fullerenes. A representative study by *Legoas* et al. [35.280] investigated the experimentally observed low-friction system

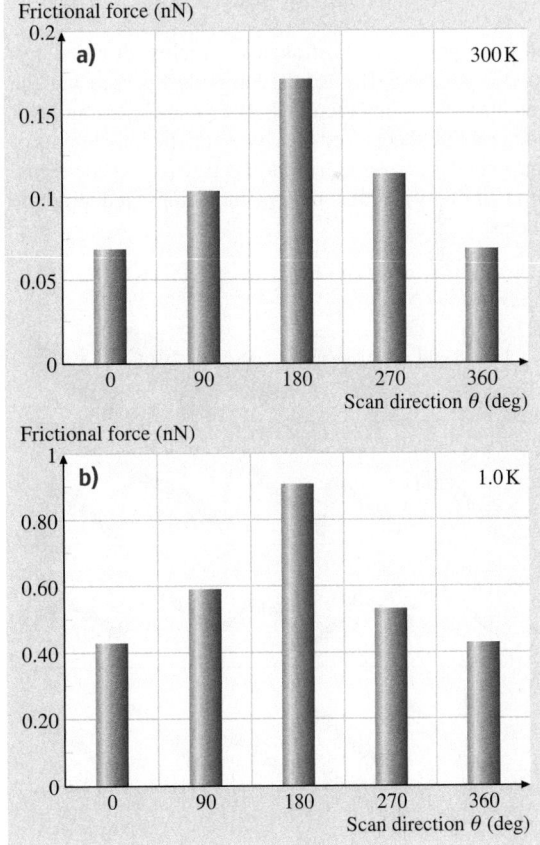

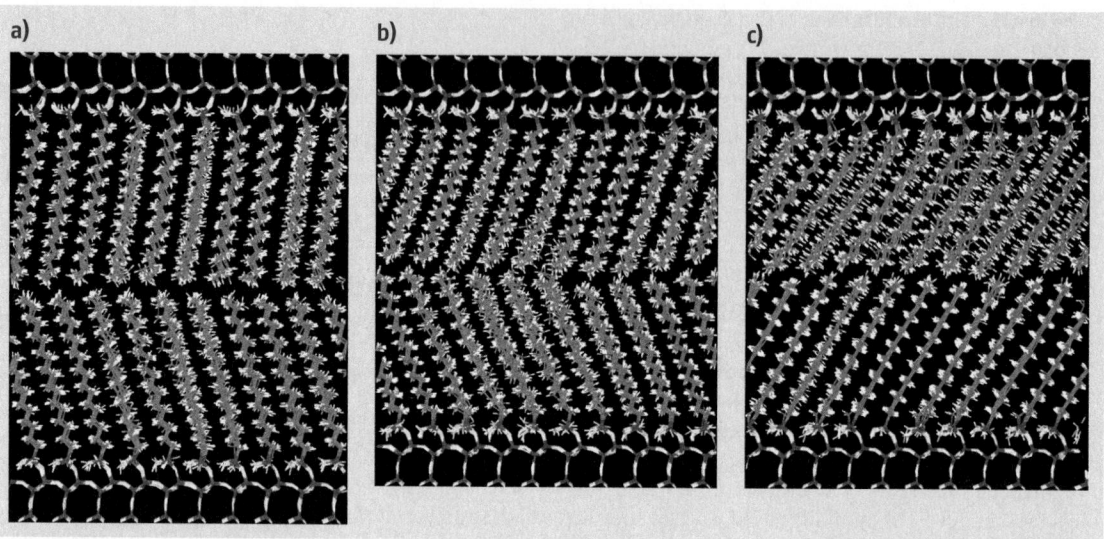

Fig. 35.66a–c Wireframe images of $n = 18$ SAMs at fixed separations of (**a**) $d = -5.2$ Å (low pressure, under compression only) (**b**) $d = -10.2$ Å (high pressure, under compression only) and (**c**) $d = -10.2$ Å (high pressure, under shear). After [35.251] with the permission of the ACS (2005)

of C_{60} molecules positioned on highly oriented pyrolytic graphite. The results show that decreasing the van der Waals interaction between a C_{60} monolayer and graphite sheets, and the characteristic movements of graphite flakes over C_{60} monolayers, explains the measured ultralow friction of C_{60} molecules and graphite sheets.

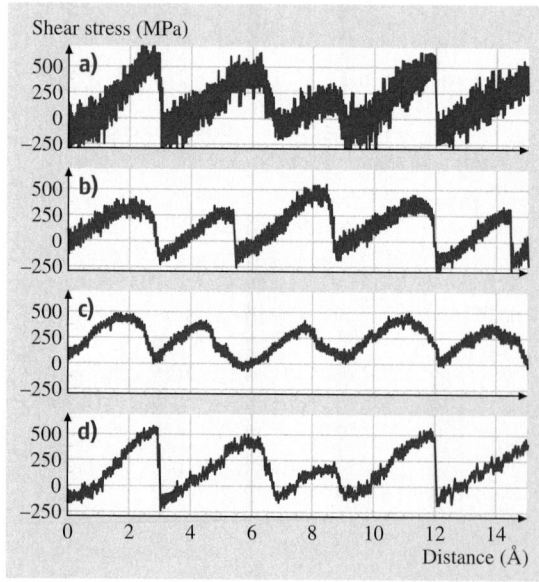

Fig. 35.67a–d Shear stress σ_s as a function of system size for $n = 6$ SAMs corresponding to a pressure of 200 MPa at $v = 1.0$ m/s: (**a**) 100 chains per surface, (**b**) 400 chains per surface, (**c**) 1600 chains per surface, and (**d**) 16 point box average of system with 100 chains per surface. After [35.251] with the permission of the ACS (2002)

Several MD simulation studies have also been carried out on the tribological properties of carbon nanotubes. For example, simulations by *Buldum* et al. and *Schall* et al. [35.266, 269] indicate that single-wall carbon nanotubes roll when their honeycomb lattice is "in registry" with the honeycomb lattice of the graphite. If this registry is not present, the carbon nanotubes respond to applied forces from an AFM by sliding. This behavior is nicely summarized in Fig. 35.68. These MD simulation findings were simultaneously confirmed in experimental studies by *Falvo* et al. [35.268]. Experimental studies of multiwall carbon nanotubes on graphite [35.272] show similar evidence of nanotube rolling when the outer tube is pushed.

The tribological properties of nanotube bundles are important, as it is well-known that carbon nanotubes agglomerate together very readily to form bundles and are often grown in bundle form [35.252]. An experimental study by *Miura* et al. [35.273] of carbon nanotube bundles being pushed around on a KCl surface with an AFM tip indicates that bundles of single-wall carbon

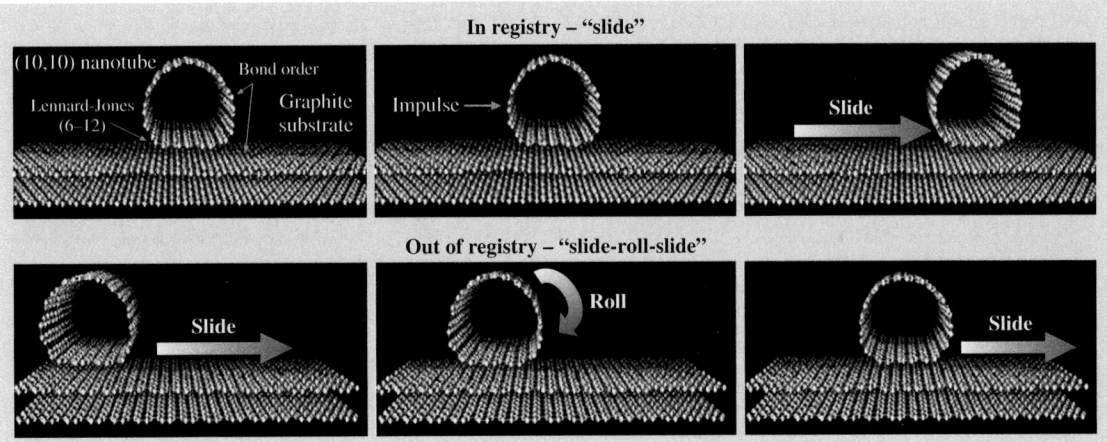

Fig. 35.68 Dynamics of a nanotube on a graphite surface. When the nanotube and graphite plane are out of registry, the nanotube slides as it slows down from an initial impulse (*upper right panel*). When the nanotube is oriented such that it is in registry with the graphite, it slows by a combination of rolling and sliding

nanotubes can be induced to roll in a manner that is similar to the rolling observed for multiwall nanotubes.

MD simulations by *Ni* et al. [35.270, 271] considered the responses of horizontally and vertically aligned single-wall carbon nanotubes between two hydrogen-terminated diamond surfaces, where the top surface is slid relative to the bottom surface. The movement of the carbon nanotubes in response to the shear forces was predicted to be simple sliding for both orientations. Interestingly, the simulations do not predict rolling of the horizontally arranged carbon nanotubes even when they are aligned with each other in two-layer and three-layer structures. Instead, at low compressive forces, illustrated in Fig. 35.69, the nanotube bundles slide as a single unit, and at high compressive forces, also illustrated in Fig. 35.69, the deformed carbon nanotubes closest to the topmost moving diamond surface start to slide in a motion reminiscent of the movement of a tank or bulldozer wheel belt. However, when these moving carbon nanotube atoms would have turned the first corner at the top of the ellipse, they encounter the neighboring nanotube and cannot slide past it. This causes them to deform even further, form cross-links with one another, and, in some cases, move in the reverse direction to the sliding motion of the diamond surface. This causes the large oscillations in the normal and lateral forces plotted in Fig. 35.69.

The responses of the horizontally arranged carbon nanotubes are substantially different from the responses of the vertically arranged nanotubes at high compression, as can be seen by comparing Figs. 35.69 and 35.70.

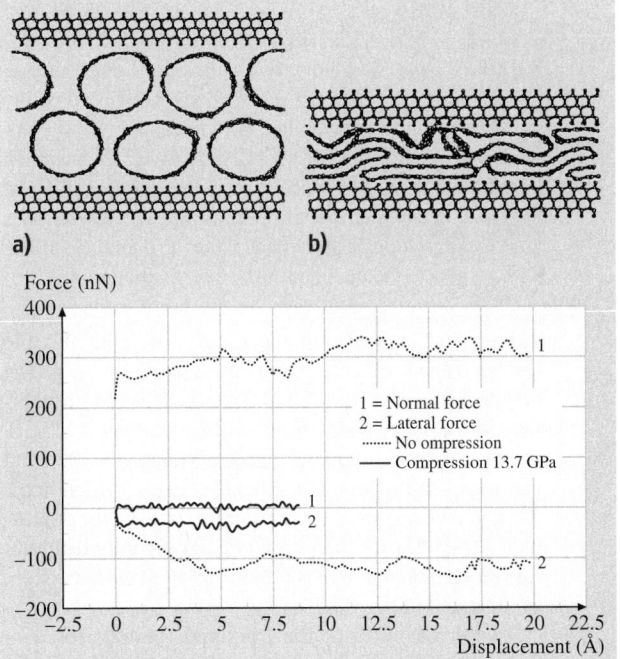

Fig. 35.69a,b *Left*: Snapshots from simulations that examine the sliding of the topmost diamond surface on horizontally arranged nanotubes at different compressions; (**a**) is at a pressure of $\approx 0\,\mathrm{GPa}$; (**b**) is with a pressure of 13.7 GPa. *Right*: Plots of the normal and lateral components of force during sliding of the top diamond's surface on horizontally arranged nanotubes as a function of the displacement of the top diamond surface with respect to the diamond surface on the bottom

Part D | 35.3

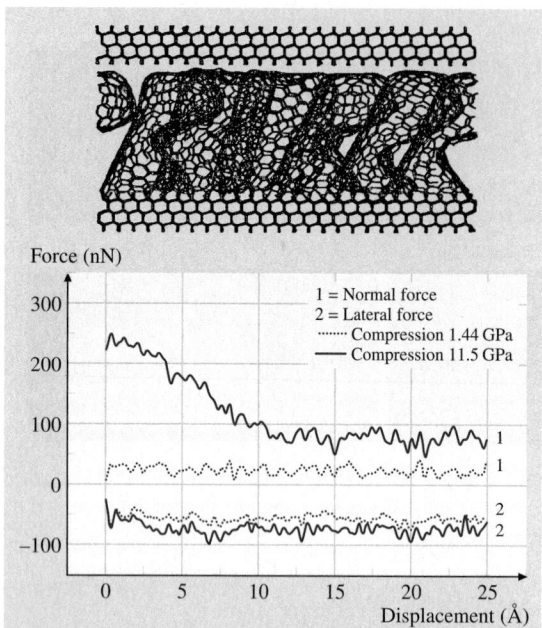

Force (nN)

1 = Normal force
2 = Lateral force
········ Compression 1.44 GPa
——— Compression 11.5 GPa

Fig. 35.70 *Top*: Snapshots from simulations that examine the sliding of the topmost diamond surface on vertically arranged nanotubes with one set of capped ends compressed at a pressure of 11.5 GPa. *Bottom*: Plots of the normal and lateral components of force during sliding of the top diamond's surface on vertically arranged nanotubes as a function of the displacement of the top diamond surface with respect to the diamond surface on the bottom

The vertical, capped carbon nanotubes are quite flexible and bend and buckle in response to applied forces. As the buckle is forming, the normal force decreases then stabilizes in the buckled structure, as illustrated in Fig. 35.70. As the topmost diamond surface slides, the buckled nanotubes swing around the buckle "neck" which helps dissipate the applied stresses. For this reason, the magnitudes of the lateral forces are not significantly different for the vertical nanotubes at low and high compression, as indicated in Fig. 35.70.

When the ratio of the frictional (lateral) force to the normal force is taken to calculate friction coefficients for these systems, high, nonintuitive values were obtained. As outlined by *Ni* et al. [35.270], this is because the actual contact area of the nanotubes is not proportional to the sliding force. In the case of the horizontal nanotube bundles, the tubes are able to deform and significantly change their contact area with the sliding surface with minimal change in the normal force, as shown in Fig. 35.69. In the case of the vertical nano-

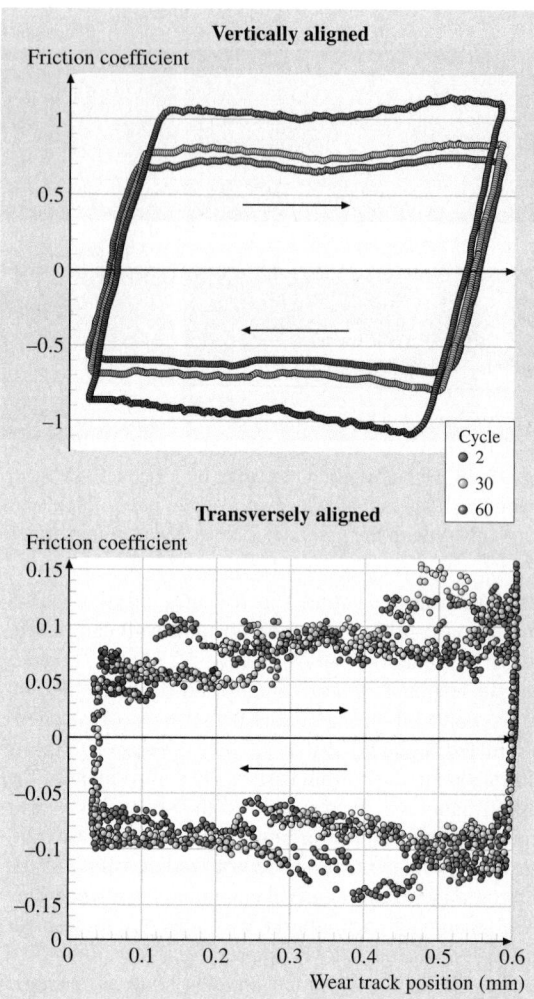

Fig. 35.71 Coefficient of friction data versus track position collected for one full cycle of reciprocating sliding for nanotubes that are vertically and transversely aligned. After [35.281]

tubes, the contact area remains approximately the same regardless of the initial loading force because of the flexibility of the nanotubes. This causes the lateral forces to change only slightly with significant changes in the normal force, as shown in Fig. 35.70. This analysis indicates that care must be taken in calculating friction coefficients for nanotube systems. Recent experiments by *Dickrell* et al. [35.281] show good agreement with these predictions, as shown in Fig. 35.71.

To summarize, this section shows that nanoparticles show some promise as lubricating materials due to their exceptionally low friction coefficients in ex-

Fig. 35.72a–d A series of chemical reactions induced by sliding of the counterface over the thin film under an average load of 300 nN. (**a**) The sliding causes the rupture of a carbon–hydrogen bond in the counterface. (**b**) The hydrogen atom is incorporated into the film and forms a bond to a carbon atom in the film. (**c**) A bond is formed between the unsaturated carbon atoms in the film and the carbon that suffered the bond rupture in the counterface, and continued sliding causes this carbon to be transferred into the film. (**d**) The transferred carbon forms a bond with another carbon in the counterface. The counterface has slid 0.0 (a), 15.9 (b), 26.1 (c), and 30.5 Å (d). After [35.282] with the permission of the ACS (2002)

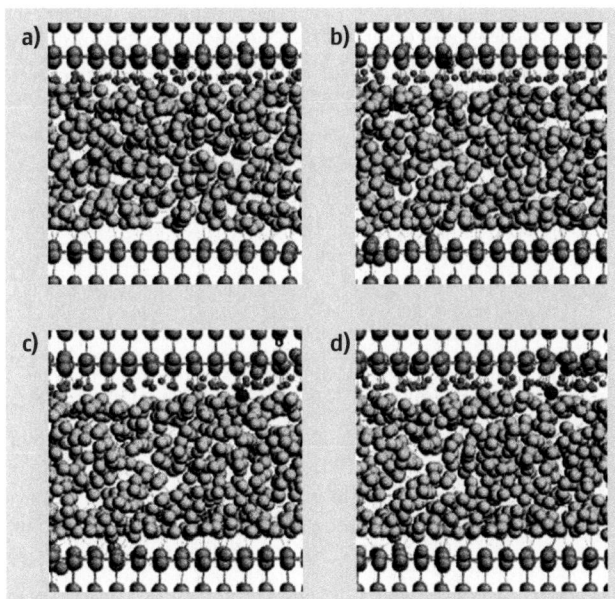

periments and simulations. Some nanopaticles show lattice-directed sliding on substrates due to their unique atomic structures. However, there is much that remains to be done before the nanometer-scale friction of these materials is well understood.

Solid State

Surfaces are able to slide over each other at high loads with a minimum of resistance from friction in the presence of liquid lubricants. Some solid thin films can also fulfill these functions and, when they do, are termed solid lubricants. Solid lubricants are generally defined as having friction coefficients of 0.3 or less and low wear.

Bowden and *Tabor* showed how thin solid films can reduce friction as follows [35.283]. The total friction force F_f is given as

$$F_f = A F_S + F_p \qquad (35.11)$$

where F_p is the plowing term, A is the area of contact and F_S is the shear strength of the interface. If the surfaces are soft, F_S will be reduced while the other parameters will increase. However, if the surfaces under the solid film are very stiff, A and F_p will decrease thereby decreasing friction. The properties specific to the film will also have an effect on friction. For instance, if the films are less than 1 μm thick, the surface asperities will be able to break through the film to eventually cause wear between the surfaces under normal circumstances. On the other hand, if the lubricant film is too thick, there will be increased plowing and wear that causes the frictional forces to increase. It is important that the lubricant not delaminate in response to the frictional forces, so strong bonds between the lubricant and the surface are required for a solid state lubricant to be effective.

The most common materials used as solid lubricants have layered structures like graphite or MoS_2, that, as

discussed above, experience low friction. It is not necessary for the lubricant film to have a layered structure to give low friction. For example, diamond-like carbon has some of the lowest coefficients of friction measured and yet does not have a layered structure. Similarly, not all layered structures are lubricants. For instance, mica gives a relatively high coefficient of friction (> 1).

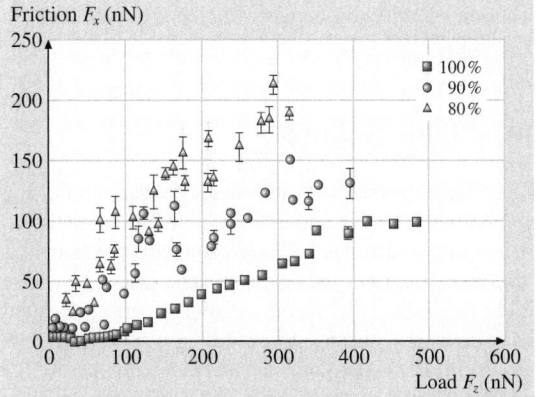

Fig. 35.73 Friction curves for the thin film system with a counterface that is 100% hydrogen-terminated (*open squares*), 90% hydrogen-terminated (*filled squares*), and 80% hydrogen-terminated (*open circles*). After [35.282] with the permission of the ACS (2002)

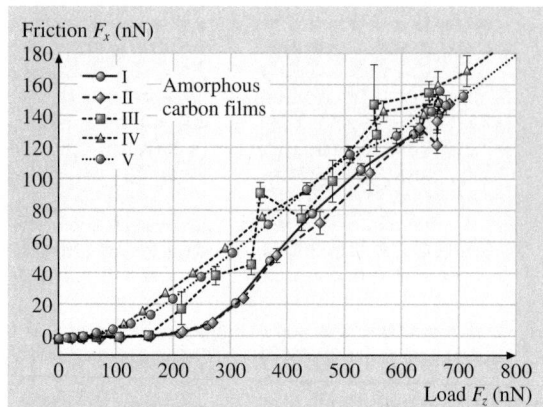

Fig. 35.74 Average friction versus load for five amorphous carbon films. Films I–III are hydrogen-free and contain various ratios of sp^2-to-sp^3 carbon. Films IV and V are both over 90% sp^3 carbon and have surface hydrogenation

The atomic-scale tribological behavior that occurs when a hydrogen-terminated diamond (111) counterface is in sliding contact with amorphous, hydrogen-free, DLC films was examined using MD simulations by *Gao* et al. [35.282]. Two films, with approximately the same ratio of sp^3–sp^2 carbon but different thicknesses, were examined. Similar average friction was obtained from both films in the load range examined. A series of tribochemical reactions occur above a critical load that result in a significant restructuring of the film, which is analogous to the "run-in" observed in macroscopic friction experiments, and reduces the friction. The contribution of adhesion between the counterface and the sample to friction is examined by varying the saturation of the counterface. The friction increases when the degree of saturation of the diamond counterface is reduced by randomly removing hydrogen atoms. Lastly, two potential energy functions that differ only in their long-range forces are used to examine the contribution of long-range interactions to friction in the same system (as illustrated in Figs. 35.72 and 35.73).

MD simulations were also recently used by *Gao* et al. [35.284] to examine the effects of the sp^2–sp^3 carbon ratio and surface hydrogen on the mechanical and tribological properties of amorphous carbon films. This work showed that, in addition to the sp^2–sp^3 ratio of carbon, the three-dimensional structures of the films are important when determining the mechanical properties of the films. For example, it is possible to have high sp^2-carbon content, which is normally associated with softer films, and large elastic constants. This occurs when sp^2-ringlike structures are oriented perpendicular to the compression direction. The layered nature of the amorphous films examined leads to novel mechanical behavior that influences the shape of the friction versus load data, as illustrated in Fig. 35.74. When load is applied to the films, the film layer closest to the interface is compressed. This results in the very low friction of films I and II up to approximately 300 nN and the response of films IV and V up to 100 nN. Once the outer film layers have been compressed, additional application of load causes an almost linear increase in friction for films I and II as well as IV and V. Film III has an erratic friction versus load response due to the early onset of tribochemical reactions between the tip and the film.

35.4 Conclusions

This chapter provides a wide-ranging discussion of the background of MD and related simulation methods, their role in the study of nanometer-scale indentation and friction, and their contributions to these fields. Specific, illustrative examples are presented that show how these approaches are providing new and exciting insights into mechanisms responsible for nanoindentation, atomic-scale friction, wear, and related atomic-scale and molecular scale processes. The examples also illustrate how the results from MD and related simulations are complementary to experimental studies, serve to guide experimental work, and assist in the interpretation of experimental data. The ability of these simulations and experimental techniques such as the surface force apparatus and proximal probe microscopes to study nanometer-scale indentation and friction at approximately the same scale is revolutionizing our understanding of the origin of friction at its most fundamental atomic level.

References

35.1 D. Dowson: *History of Tribology* (Longman, London 1979)

35.2 K. L. Johnson, K. Kendell, A. D. Roberts: Surface energy and the contact of elastic solids, Proc. R. Soc. Lond. A **324**, 301–313 (1971)

35.3 M. Gad-el-Hak (Ed.): *The MEMS Handbook*, Mech. Eng. Handbook Ser. (CRC, Boca Raton 2002)

35.4 J. Krim: Friction at the atomic scale, Sci. Am. **275**, 74–80 (1996)

35.5 J. Krim: Atomic-scale origins of friction, Langmuir **12**, 4564–4566 (1996)

35.6 J. Krim: Progress in nanotribology: experimental probes of atomic-scale friction, Comment Cond. Mat. Phys. **17**, 263–280 (1995)

35.7 A. P. Sutton: Deformation mechanisms, electronic conductance and friction of metallic nanocontacts, Curr. Opin. Sol. St. Mater. Sci. **1**, 827–833 (1996)

35.8 C. M. Mate: Force microscopy studies of the molecular origins of friction and lubrication, IBM J. Res. Dev. **39**, 617–627 (1995)

35.9 A. M. Stoneham, M. M. D. Ramos, A. P. Sutton: How do they stick together – the statics and dynamics of interfaces, Philos. Mag. A **67**, 797–811 (1993)

35.10 I. L. Singer: Friction and energy dissipation at the atomic scale: A review, J. Vacuum Sci. Technol. A **12**, 2605–2616 (1994)

35.11 B. Bhushan, J. N. Israelachvili, U. Landman: Nanotribology – friction, wear and lubrication at the atomic scale, Nature **374**, 607–616 (1995)

35.12 J. A. Harrison, D. W. Brenner: *Handbook of Micro/Nanotechnology*, ed. by B. Bhushan (CRC, Boca Raton 1995)

35.13 J. B. Sokoloff: Theory of atomic level sliding friction between ideal crystal interfaces, J. Appl. Phys. **72**, 1262–1270 (1992)

35.14 W. Zhong, G. Overney, D. Tomanek: Theory of atomic force microscopy on elastic surfaces. In: *The Structure of Surfaces III: Proc. 3rd Int. Conf. on the Structure of Surfaces*, Vol. 24, ed. by S. Y. Tong, M. A. V. Hove, X. Xide, K. Takayanagi (Springer, Berlin, Heidelberg 1991) p. 243

35.15 J. N. Israelachvili: Adhesion, friction and lubrication of molecularly smooth surfaces. In: *Fundamentals of Friction: Macroscopic and Microscopic processes*, ed. by I. L. Singer, H. M. Pollock (Kluwer, Dordrecht 1992) pp. 351–385

35.16 S. B. Sinnott: Theory of atomic-scale friction. In: *Handbook of Nanostructured Materials and Nanotechnology*, Vol. 2, ed. by H. Nalwa (Academic, San Diego 2000) pp. 571–618

35.17 S.-J. Heo, S. B. Sinnott, D. W. Brenner, J. A. Harrison: Computational modeling of nanometer-scale tribology. In: *Nanotribology and Nanomechanics*, ed. by B. Bhushan (Springer, Berlin, Heidelberg 2005)

35.18 G. Binnig, C. F. Quate, C. Gerber: Atomic force microscope, Phys. Rev. Lett. **56**, 930–933 (1986)

35.19 C. M. Mate, G. M. Mcclelland, R. Erlandsson, S. Chiang: Atomic-scale friction of a tungsten tip on a graphite surface, Phys. Rev. Lett. **59**, 1942–1945 (1987)

35.20 G. J. Germann, S. R. Cohen, G. Neubauer, G. M. Mcclelland, H. Seki, D. Coulman: Atomic-scale friction of a diamond tip on diamond (100) surface and (111) surface, J. Appl. Phys. **73**, 163–167 (1993)

35.21 R. W. Carpick, M. Salmeron: Scratching the surface: Fundamental investigations of tribology with atomic force microscopy, Chem. Rev. **97**, 1163–1194 (1997)

35.22 J. N. Israelachvili: *Intermolecular and surface forces: With applications to colloidal and biological systems* (Academic, London 1992)

35.23 J. Krim, D. H. Solina, R. Chiarello: Nanotribology of a Kr monolayer – a quartz crystal microbalance study of atomic-scale friction, Phys. Rev. Lett. **66**, 181–184 (1991)

35.24 G. A. Tomlinson: A molecular theory of friction, Philos. Mag. Ser. 7 **7**, 905–939 (1929)

35.25 F. C. Frenkel, T. Kontorova: On the theory of plastic deformation and twinning, Zh. Eksp. Teor. Fiz. **8**, 1340 (1938)

35.26 G. M. McClelland, J. N. Glosli: Friction at the atomic scale. In: *Fundamentals of friction: Macroscopic and microscopic processes*, ed. by I. L. Singer, H. M. Pollock (Kluwer, Dordrecht 1992) pp. 405–422

35.27 J. B. Sokoloff: Theory of dynamical friction between idealized sliding surfaces, Surf. Sci. **144**, 267–272 (1984)

35.28 J. B. Sokoloff: Theory of energy dissipation in sliding crystal surfaces, Phys. Rev. B **42**, 760–765 (1990)

35.29 J. B. Sokoloff: Possible nearly frictionless sliding for mesoscopic solids, Phys. Rev. Lett. **71**, 3450–3453 (1993)

35.30 J. B. Sokoloff: Microscopic mechanisms for kinetic friction: Nearly frictionless sliding for small solids, Phys. Rev. B **52**, 7205–7214 (1995)

35.31 J. B. Sokoloff: Theory of electron and phonon contributions to sliding friction. In: *Physics of Sliding Friction*, ed. by B. N. J. Persson, E. Tosatti (Kluwer, Dordrecht 1996) pp. 217–229

35.32 J. B. Sokoloff: Static friction between elastic solids due to random asperities, Phys. Rev. Lett. **86**, 3312–3315 (2001)

35.33 J. B. Sokoloff: Possible microscopic explanation of the virtually universal occurrence of static friction, Phys. Rev. B **65**, 115415 (2002)

35.34 B. N. J. Persson, D. Schumacher, A. Otto: Surface resistivity and vibrational damping in adsorbed layers, Chem. Phys. Lett. **178**, 204–212 (1991)

Part D | 35

35.35   A. I. Volokitin, B. N. J. Persson: Resonant photon tunneling enhancement of the van der Waals friction, Phys. Rev. Lett. **91**, 106101 (2003)

35.36   A. I. Volokitin, B. N. J. Persson: Noncontact friction between nanostructures, Phys. Rev. B **68**, 155420 (2003)

35.37   A. I. Volokitin, B. N. J. Persson: Adsorbate-induced enhancement of electrostatic noncontact friction, Phys. Rev. Lett. **94**, 086104 (2005)

35.38   J. S. Helman, W. Baltensperger, J. A. Holyst: Simple model for dry friction, Phys. Rev. B **49**, 3831–3838 (1994)

35.39   T. Kawaguchi, H. Matsukawa: Dynamical frictional phenomena in an incommensurate two-chain model, Phys. Rev. B **56**, 13932–13942 (1997)

35.40   M. H. Müser: Nature of mechanical instabilities and their effect on kinetic friction, Phys. Rev. Lett. **89**, 224301 (2002)

35.41   M. H. Müser: Towards an atomistic understanding of solid friction by computer simulations, Comput. Phys. Commun. **146**, 54–62 (2002)

35.42   P. Reimann, M. Evstigneev: Nonmonotonic velocity dependence of atomic friction, Phys. Rev. Lett. **93**, 230802 (2004)

35.43   C. Ritter, M. Heyde, B. Stegemann, K. Rademann, U. D. Schwarz: Contact area dependence of frictional forces: Moving adsorbed antimony nanoparticles, Phys. Rev. B **71**, 085405 (2005)

35.44   C. Fusco, A. Fasolino: Velocity dependence of atomic-scale friction: A comparative study of the one- and two-dimensional Tomlinson model, Phys. Rev. B **71**, 045413 (2005)

35.45   C. W. Gear: *Numerical Initial Value Problems in Ordinary Differential Equations* (Prentice-Hall, Englewood Cliffs 1971)

35.46   W. G. Hoover: *Molecular Dynamics* (Springer, Berlin, Heidelberg 1986)

35.47   D. W. Heermann: *Computer Simulation Methods in Theoretical Physics* (Springer, Berlin, Heidelberg 1986)

35.48   M. P. Allen, D. J. Tildesley: *Computer Simulation of Liquids* (Clarendon, Oxford 1987)

35.49   J. M. Haile: *Molecular Dynamics Simulation: Elementary Methods* (Wiley, New York 1992)

35.50   M. Finnis: *Interatomic Forces in Condensed Matter* (Oxford University Press, Oxford 2003)

35.51   D. W. Brenner: Relationship between the embedded-atom method and Tersoff potentials, Phys. Rev. Lett. **63**, 1022–1022 (1989)

35.52   D. W. Brenner: The art and science of an analytic potential, Phys. Stat. Sol. B **217**, 23–40 (2000)

35.53   R. G. Parr, W. Yang: *Density Functional Theory of Atoms and Molecules* (Oxford Univ. Press, New York 1989)

35.54   R. Car, M. Parrinello: Unified approach for molecular dynamics and density functional theory, Phys. Rev. Lett. **55**, 2471–2474 (1985)

35.55   C. Cramer: *Essentials of Computational Chemistry, Theories and Models* (Wiley, Chichester 2004)

35.56   A. P. Sutton: *Electronic Structure of Materials* (Clarendon, Oxford 1993)

35.57   K. Kadau, T. C. Germann, P. S. Lomdahl: Large-scale molecular dynamics simulation of 19 billion particles, Int. J. Mod. Phys. C **15**, 193–201 (2004)

35.58   B. J. Thijsse: Relationship between the modified embedded-atom method and Stillinger–Weber potentials in calculating the structure of silicon, Phys. Rev. B **65**, 195207 (2002)

35.59   M. I. Baskes, J. S. Nelson, A. F. Wright: Semiempirical modified embedded atom potentials for silicon and germanium, Phys. Rev. B **40**, 6085–6100 (1989)

35.60   T. Ohira, Y. Inoue, K. Murata, J. Murayama: Magnetite scale cluster adhesion on metal oxide surfaces: Atomistic simulation study, Appl. Surf. Sci. **171**, 175–188 (2001)

35.61   F. H. Streitz, J. W. Mintmire: Electrostatic potentials for metal oxide surfaces and interfaces, Phys. Rev. B **50**, 11996–12003 (1994)

35.62   A. Yasukawa: Using an extended Tersoff interatomic potential to analyze the static fatigue strength of SiO_2 under atmospheric influence, JSME Int. J. A **39**, 313–320 (1996)

35.63   T. Iwasaki, H. Miura: Molecular dynamics analysis of adhesion strength of interfaces between thin films, J. Mater. Res. **16**, 1789–1794 (2001)

35.64   B.-J. Lee, M. I. Baskes: Second nearest-neighbor modified embedded-atom method potential, Phys. Rev. B **62**, 8564–8567 (2000)

35.65   G. C. Abell: Empirical chemical pseudopotential theory of molecular and metallic bonding, Phys. Rev. B **31**, 6184–6196 (1985)

35.66   J. Tersoff: New empirical approach for the structure and energy of covalent systems, Phys. Rev. B **37**, 6991–7000 (1988)

35.67   J. Tersoff: Modeling solid-state chemistry: Interatomic potentials for multicomponent systems, Phys. Rev. B **39**, 5566–5569 (1989)

35.68   D. W. Brenner: Empirical potential for hydrocarbons for use in simulating the chemical vapor deposition of diamond films, Phys. Rev. B **42**, 9458–9471 (1990)

35.69   D. W. Brenner, O. A. Shenderova, J. A. Harrison, S. J. Stuart, B. Ni, S. B. Sinnott: Second generation reactive empirical bond order (REBO) potential energy expression for hydrocarbons, J. Phys. C **14**, 783–802 (2002)

35.70   A. J. Dyson, P. V. Smith: Extension of the Brenner empirical interactomic potential to C–Si–H, Surf. Sci. **355**, 140–150 (1996)

35.71   B. Ni, K.-H. Lee, S. B. Sinnott: Development of a reactive empirical bond order potential for hydrocarbon-oxygen interactions, J. Phys. C **16**, 7261–7275 (2004)

35.72 J. Tanaka, C.F. Abrams, D.B. Graves: New C-F interatomic potential for molecular dynamics simulation of fluorocarbon film formation, Nucl. Instrum. Meth. B **18**, 938–945 (2000)

35.73 I. Jang, S.B. Sinnott: Molecular dynamics simulations of the chemical modification of polystyrene through $C_xF_y^+$ beam deposition, J. Phys. Chem. B **108**, 9656–9664 (2004)

35.74 S.B. Sinnott, O.A. Shenderova, C.T. White, D.W. Brenner: Mechanical properties of nanotubule fibers and composites determined from theoretical calculations and simulations, Carbon **36**, 1–9 (1998)

35.75 S.J. Stuart, A.B. Tutein, J.A. Harrison: A reactive potential for hydrocarbons with intermolecular interactions, J. Chem. Phys. **112**, 6472–6486 (2000)

35.76 F.H. Stillinger, T.A. Weber: Computer simulation of local order in condensed phases of silicon, Phys. Rev. B **31**, 5262–5271 (1985)

35.77 S.M. Foiles: Application of the embedded-atom method to liquid transition metals, Phys. Rev. B **32**, 3409–3415 (1985)

35.78 M.S. Daw, M.I. Baskes: Semiempirical, quantum mechanical calculation of hydrogen embrittlement in metals, Phys. Rev. Lett. **50**, 1285–1288 (1983)

35.79 T.J. Raeker, A.E. Depristo: Theory of chemical bonding based on the atom-homogeneous electron gas system, Int. Rev. Phys. Chem. **10**, 1–54 (1991)

35.80 R.W. Smith, G.S. Was: Application of molecular dynamics to the study of hydrogen embrittlement in Ni-Cr-Fe alloys, Phys. Rev. B **40**, 10322–10336 (1989)

35.81 R. Pasianot, D. Farkas, E.J. Savino: Empirical many-body interatomic potential for bcc transition metals, Phys. Rev. B **43**, 6952–6961 (1991)

35.82 R. Pasianot, E.J. Savino: Embedded-atom method interatomic potentials for hcp metals, Phys. Rev. B **45**, 12704–12710 (1992)

35.83 M.I. Baskes, J.S. Nelson, A.F. Wright: Semiempirical modified embedded-atom potentials for silicon and germanium, Phys. Rev. B **40**, 6085–6100 (1989)

35.84 M.I. Baskes: Modified embedded-atom potentials for cubic materials and impurities, Phys. Rev. B **46**, 2727–2742 (1992)

35.85 K. Ohno, K. Esfarjani, Y. Kawazoe: *Computational Materials Science from Ab Initio to Monte Carlo Methods* (Springer, Berlin, Heidelberg 1999)

35.86 A.K. Rappe, W.A. Goddard III: Charge equilibration for molecular dynamics simulations, J. Phys. Chem. **95**, 3358–3363 (1991)

35.87 D. Frenkel, B. Smit: *Understanding Molecular Simulation: From Algorithms to Applications* (Academic, San Diego 1996)

35.88 L.V. Woodcock: Isothermal molecular dynamics calculations for liquid salts, Chem. Phys. Lett. **10**, 257–261 (1971)

35.89 T. Schneider, E. Stoll: Molecular dynamics study of a three-dimensional one-component model for distortive phase transitions, Phys. Rev. B **17**, 1302–1322 (1978)

35.90 K. Kremer, G.S. Grest: Dynamics of entangled linear polymer melts – a molecular dynamics simulation, J. Chem. Phys. **92**, 5057–5086 (1990)

35.91 S.A. Adelman, J.D. Doll: Generalized Langevin equation approach for atom-solid-surface scattering – general formulation for classical scattering off harmonic solids, J. Chem. Phys. **64**, 2375–2388 (1976)

35.92 S.A. Adelman: Generalized Langevin equations and many-body problems in chemical dynamics, Adv. Chem. Phys. **44**, 143–253 (1980)

35.93 J.C. Tully: Dynamics of gas-surface interactions – 3D generalized Langevin model applied to fcc and bcc surfaces, J. Chem. Phys. **73**, 1975–1985 (1980)

35.94 S. Nosé: A unified formulation of the constant temperature molecular dynamics methods, J. Chem. Phys. **81**, 511–519 (1984)

35.95 S. Nosé: A molecular dynamics method for simulations in the canonical ensemble, Mol. Phys. **52**, 255–268 (1984)

35.96 G.J. Martyna, M.L. Klein, M. Tuckerman: Nose-Hoover chains – the canonical ensemble via continuous dynamics, J. Chem. Phys. **97**, 2635–2643 (1992)

35.97 M. D'Alessandro, M. D'Abramo, G. Brancato, A. Di Nola, A. Amadei: Statistical mechanics and thermodynamics of simulated ionic solutions, J. Phys. Chem. B **106**, 11843–11848 (2002)

35.98 J.D. Schall, C.W. Padgett, D.W. Brenner: Ad hoc continuum-atomistic thermostat for modeling heat flow in molecular dynamics simulations, Mol. Simul. **31**, 283–288 (2005)

35.99 M. Schoen, C.L. Rhykerd, D.J. Diestler, J.H. Cushman: Shear forces in molecularly thin films, Science **245**, 1223–1225 (1989)

35.100 J.E. Curry, F.S. Zhang, J.H. Cushman, M. Schoen, D.J. Diestler: Transient coexisting nanophases in ultrathin films confined between corrugated walls, J. Chem. Phys. **101**, 10824–10832 (1994)

35.101 D.J. Adams: Grand canonical ensemble Monte Carlo for a Lennard-Jones fluid, Mol. Phys. **29**, 307–311 (1975)

35.102 N.A.a.C., R.J. Burnham: Force microscopy. In: *Scanning Tunneling Microscopy and Spectroscopy: Theory, Techniques, and Applications*, ed. by D.A. Bonnell (VCH, New York 1993) pp.191–249

35.103 E. Meyer: *Nanoscience: Friction and Rheology on the Nanometer Scale* (World Scientific, Hackensack 1998)

35.104 G.E. Totten, H. Liang: *Mechanical Tribology: Materials Characterization and Applications* (Marcel Dekker, New York 2004)

35.105 N.A. Burnham, R.J. Colton: Measuring the nanomechanical properties and surface forces of materials using an atomic force microscope, J. Vacuum Sci. Technol. A **7**, 2906–2913 (1989)

35.106   N. A. Burnham, D. D. Dominguez, R. L. Mowery, R. J. Colton: Probing the surface forces of monolayer films with an atomic force microscope, Phys. Rev. Lett. **64**, 1931–1934 (1990)

35.107   E. Meyer, R. Overney, D. Brodbeck, L. Howald, R. Luthi, J. Frommer, H. J. Guntherodt: Friction and wear of Langmuir–Blodgett films observed by friction force microscopy, Phys. Rev. Lett. **69**, 1777–1780 (1992)

35.108   A. P. Sutton, J. B. Pethica, H. Rafii-Tabar, J. A. Nieminen: Mechanical properties of metals at the nanometer scale. In: *Electron Theory in Alloy Design*, ed. by D. G. Pettifor, A. H. Cottrell (Institute of Materials, London 1992) pp. 191–233

35.109   H. Raffi-Tabar, A. P. Sutton: Long-range Finnis–Sinclair potentials for fcc metallic alloys, Philos. Mag. Lett. **63**, 217–224 (1991)

35.110   U. Landman, W. D. Luedtke, E. M. Ringer: Atomistic mechanisms of adhesive contact formation and interfacial processes, Wear **153**, 3–30 (1992)

35.111   U. Landman, W. D. Luedtke, N. A. Burnham, R. J. Colton: Atomistic mechanisms and dynamics of adhesion, nanoindentation and fracture, Science **248**, 454–461 (1990)

35.112   O. Tomagnini, F. Ercolessi, E. Tosatti: Microscopic interaction between a gold tip and a Pb(110) surface, Surf. Sci. **287/288**, 1041–1045 (1991)

35.113   N. Ohmae: Field ion microscopy of microdeformation induced by metallic contacts, Philos. Mag. A **74**, 1319–1327 (1996)

35.114   N. A. Burnham, R. J. Colton, H. M. Pollock: Interpretation of force curves in force microscopy, Nanotechnology **4**, 64–80 (1993)

35.115   N. Agrait, G. Rubio, S. Vieira: Plastic deformation in nanometer-scale contacts, Langmuir **12**, 4505–4509 (1996)

35.116   U. Landman, W. D. Luedtke, A. Nitzan: Dynamics of tip-substrate interactions in atomic force microscopy, Surf. Sci. **210**, L177–L182 (1989)

35.117   U. Landman, W. D. Luedtke: Nanomechanics and dynamics of tip substrate interactions, J. Vacuum Sci. Technol. B **9**, 414–423 (1991)

35.118   U. Landman, W. D. Luedtke, J. Ouyang, T. K. Xia: Nanotribology and the stability of nanostructures, Jpn. J. Appl. Phys. Pt. 1 **32**, 1444–1462 (1993)

35.119   J. W. M. Frenken, H. M. Vanpinxteren, L. Kuipers: New views on surface melting obtained with STM and ion scattering, Surf. Sci. **283**, 283–289 (1993)

35.120   T. Yokohata, K. Kato: Mechanism of nanoscale indentation, Wear **168**, 109–114 (1993)

35.121   O. Tomagnini, F. Ercolessi, E. Tosatti: Microscopic interaction between a gold tip and a Pb(110) surface, Surf. Sci. **287**, 1041–1045 (1993)

35.122   K. Komvopoulos, W. Yan: Molecular dynamics simulation of single and repeated indentation, J. Appl. Phys. **82**, 4823–4830 (1997)

35.123   J. Belak, I. F. Stowers: *A Molecular Dynamics Model of the Orthogonal Cutting Process* (Proc. Am. Soc. Precision Eng. Annu. Conf., 1990) pp. 76–79

35.124   M. Fournel, E. Lacaze, M. Schott: Tip-surface interactions in STM experiments on Au(111): Atomic-scale metal friction, Europhys. Lett. **34**, 489–494 (1996)

35.125   E. T. Lilleodden, J. A. Zimmerman, S. M. Foiles, W. D. Nix: Atomistic simulations of elastic deformation and dislocation nucleation during nanoindentation, J. Mech. Phys. Sol. **51**, 901–920 (2003)

35.126   J. L. Costakramer, N. Garcia, P. Garciamochales, P. A. Serena: Nanowire formation in macroscopic metallic contacts – quantum-mechanical conductance tapping a table top, Surf. Sci. **342**, L1144–L1149 (1995)

35.127   A. I. Yanson, J. M. van Ruitenbeek, I. K. Yanson: Shell effects in alkali metal nanowires, Low Temp. Phys. **27**, 807–820 (2001)

35.128   A. I. Yanson, I. K. Yanson, J. M. van Ruitenbeek: Crossover from electronic to atomic shell structure in alkali metal nanowires, Phys. Rev. Lett. **8721**, 216805 (2001)

35.129   C. L. Kelchner, S. J. Plimpton, J. C. Hamilton: Dislocation nucleation and defect structure during surface indentation, Phys. Rev. B **58**, 11085–11088 (1998)

35.130   A. Hasnaoui, P. M. Derlet, H. V. Swygenhoven: Interaction between dislocations, grain boundaries under an indenter – a molecular dynamics simulation, Acta Mater. **52**, 2251–2258 (2004)

35.131   O. R. de la Fuente, J. A. Zimmerman, M. A. Gonzalez, J. de la Figuera, J. C. Hamilton, W. W. Pai, J. M. Rojo: Dislocation emission around nanoindentations on a (001) fcc metal surface studied by scanning tunneling microscopy and atomistic simulations, Phys. Rev. Lett. **88**, 036101 (2002)

35.132   O. A. Shenderova, J. P. Mewkill, D. W. Brenner: Nanoindentation as a probe of nanoscale residual stresses, Mol. Simul. **25**, 81–92 (2000)

35.133   S. Kokubo: On the change in hardness of a plate caused by bending, Sci. Rep. Tohoku Imperial University **21**, 256–267 (1932)

35.134   G. Sines, R. Calson: Hardness measurements for determination of residual stresses, ASTM Bull. **180**, 35–37 (1952)

35.135   G. U. Oppel: Biaxial elasto-plastic analysis of load and residual stresses, Exp. Mech. **21**, 135–140 (1964)

35.136   T. R. Simes, S. G. Mellor, D. A. Hills: A note on the influence of residual stress on measured hardness, J. Strain Anal. Eng. Des. **19**, 135–137 (1984)

35.137   T. Y. Tsui, G. M. Pharr, W. C. Oliver, C. S. Bhatia, C. T. White, S. Anders, A. Anders, I. G. Brown: Nanoindentation and nanoscratching of hard carbon coatings for magnetic disks, Mat. Res. Soc. Symp. Proc. **383**, 447–452 (1995)

35.138   A. Bolshakov, W. C. Oliver, G. M. Pharr: Influences of stress on the measurement of mechanical properties using nanoindentation. 2. Finite element simulations, J. Mater. Res. **11**, 760–768 (1996)

35.139 J. D. Schall, D. W. Brenner: Atomistic simulation of the influence of pre-existing stress on the interpretation of nanoindentation data, J. Mater. Res. **19**, 3172–3180 (2004)

35.140 U. Landman, W. D. Luedtke, M. W. Ribarsky: Structural and dynamical consequences of interactions in interfacial systems, J. Vacuum Sci. Technol. A **7**, 2829–2839 (1989)

35.141 J. S. Kallman, W. G. Hoover, C. G. Hoover, A. J. Degroot, S. M. Lee, F. Wooten: Molecular-dynamics of silicon indentation, Phys. Rev. B **47**, 7705–7709 (1993)

35.142 D. R. Clarke, M. C. Kroll, P. D. Kirchner, R. F. Cook, B. J. Hockey: Amorphization and conductivity of silicon and germanium induced by indentation, Phys. Rev. Lett. **60**, 2156–2159 (1988)

35.143 A. Kailer, K. G. Nickel, Y. G. Gogotsi: Raman microspectroscopy of nanocrystalline and amorphous phases in hardness indentations, J. Raman Spec. **30**, 939–961 (1999)

35.144 K. Minowa, K. Sumino: Stress-induced amorphization of a silicon crystal by mechanical scratching, Phys. Rev. Lett. **69**, 320–322 (1992)

35.145 G. S. Smith, E. B. Tadmor, E. Kaxiras: Multiscale simulation of loading and electrical resistance in silicon nanoindentation, Phys. Rev. Lett. **84**, 1260–1263 (2000)

35.146 W. C. D. Cheong, L. C. Zhang: Molecular dynamics simulation of phase transformations in silicon monocrystals due to nano-indentation, Nanotechnology **11**, 173–180 (2000)

35.147 C. F. Sanz-Navarro, S. D. Kenny, R. Smith: Atomistic simulations of structural transformations, Nanotechnology **15**, 692–697 (2004)

35.148 P. Walsh, A. Omeltchenko, R. K. Kalia, A. Nakano, P. Vashishta, S. Saini: Nanoindentation of silicon nitride: A multimillion-atom molecular dynamics study, Appl. Phys. Lett. **82**, 118–120 (2003)

35.149 J. A. Harrison, C. T. White, R. J. Colton, D. W. Brenner: Nanoscale investigation of indentation, adhesion and fracture of diamond (111) surfaces, Surf. Sci. **271**, 57–67 (1992)

35.150 K. Cho, J. D. Joannopoulos: Mechanical hysteresis on an atomic-scale, Surf. Sci. **328**, 320–324 (1995)

35.151 A. Garg, J. Han, S. B. Sinnott: Interactions of carbon-nanotubule proximal probe tips with diamond and graphene, Phys. Rev. Lett. **81**, 2260–2263 (1998)

35.152 J. A. Harrison, S. J. Stuart, D. H. Robertson, C. T. White: Properties of capped nanotubes when used as SPM tips, J. Phys. Chem. B **101**, 9682–9685 (1997)

35.153 J. A. Harrison, S. J. Stuart, A. B. Tutein: A new, reactive potential energy function to study indentation and friction of C_{13} n-alkane monolayers. In: *Interfacial Properties on the Submicron Scale*, ed. by

J. E. Frommer, R. Overney (ACS, Washington 2001) pp. 216–229

35.154 A. Garg, S. B. Sinnott: Molecular dynamics of carbon nanotube proximal probe tip-surface contacts, Phys. Rev. B **60**, 13786–13791 (1999)

35.155 K. J. Tupper, D. W. Brenner: Compression-induced structural transition in a self-assembled monolayer, Langmuir **10**, 2335–2338 (1994)

35.156 K. J. Tupper, R. J. Colton, D. W. Brenner: Simulations of self-assembled monolayers under compression – effect of surface asperities, Langmuir **10**, 2041–2043 (1994)

35.157 S. A. Joyce, R. C. Thomas, J. E. Houston, T. A. Michalske, R. M. Crooks: Mechanical relaxation of organic monolayer films measured by force microscopy, Phys. Rev. Lett. **68**, 2790–2793 (1992)

35.158 L. Zhang, Y. Leng, S. Jiang: Tip-based hybrid simulation study of frictional properties of self-assembled monolayers: Effects of chain length, terminal group, and scan direction, scan velocity, Langmuir **19**, 9742–9747 (2003)

35.159 A. B. Tutein, S. J. Stuart, J. A. Harrison: Indentation analysis of linear-chain hydrocarbon monolayers anchored to diamond, J. Phys. Chem. B **103**, 11357–11365 (1999)

35.160 Y. Leng, S. Jiang: Dynamic simulations of adhesion and friction in chemical force microscopy, J. Am. Chem. Soc. **124**, 11764–11770 (2002)

35.161 C. M. Mate: Atomic force microscope study of polymer lubricants on silicon surfaces, Phys. Rev. Lett. **68**, 3323–3326 (1992)

35.162 K. Enke, H. Dimigen, H. Hubsch: Frictional properties of diamond-like carbon layers, Appl. Phys. Lett. **36**, 291–292 (1980)

35.163 K. Enke: Some new results on the fabrication of and the mechanical, electrical, optical properties of I-carbon layers, Thin Solid Films **80**, 227–234 (1981)

35.164 S. Miyake, S. Takahashi, I. Watanabe, H. Yoshihara: Friction and wear behavior of hard carbon films, ASLE Trans. **30**, 121–127 (1987)

35.165 A. Erdemir, C. Donnet: Tribology of diamond, diamond-like carbon, and related films. In: *Modern Tribology Handbook*, Vol. II, ed. by B. Bhushan (CRC, Boca Raton 2000) pp. 871–908

35.166 S. B. Sinnott, R. J. Colton, C. T. White, O. A. Shenderova, D. W. Brenner, J. A. Harrison: Atomistic simulations of the nanometer-scale indentation of amorphous carbon thin films, J. Vacuum Sci. Technol. A **15**, 936–940 (1997)

35.167 J. N. Glosli, M. R. Philpott, G. M. McClelland: Molecular dynamics simulation of mechanical deformation of ultra-thin amorphous carbon films, Mater. Res. Soc. Symp. Proc. **383**, 431–435 (1995)

35.168 M. R. Sorensen, K. W. Jacobsen, P. Stoltze: Simulations of atomic-scale sliding friction, Phys. Rev. B **53**, 2101–2113 (1996)

35.169 I. L. Singer: A thermochemical model for analyzing low wear-rate materials, Surf. Coat. Technol. **49**, 474–481 (1991)

35.170 I. L. Singer, S. Fayeulle, P. D. Ehni: Friction and wear behavior of tin in air – the chemistry of transfer films and debris formation, Wear **149**, 375–394 (1991)

35.171 J. A. Harrison, C. T. White, R. J. Colton, D. W. Brenner: Molecular dynamics simulations of atomic-scale friction of diamond surfaces, Phys. Rev. B **46**, 9700–9708 (1992)

35.172 J. A. Harrison, R. J. Colton, C. T. White, D. W. Brenner: Effect of atomic-scale surface roughness on friction – a molecular dynamics study of diamond surfaces, Wear **168**, 127–133 (1993)

35.173 J. A. Harrison, C. T. White, R. J. Colton, D. W. Brenner: Atomistic simulations of friction at sliding diamond interfaces, MRS Bull. **18**, 50–53 (1993)

35.174 J. N. Glosli, G. M. Mcclelland: Molecular dynamics study of sliding friction of ordered organic monolayers, Phys. Rev. Lett. **70**, 1960–1963 (1993)

35.175 A. Koike, M. Yoneya: Molecular dynamics simulations of sliding friction of Langmuir-Blodgett monolayers, J. Chem. Phys. **105**, 6060–6067 (1996)

35.176 J. E. Hammerberg, B. L. Holian, S. J. Zhou: Studies of sliding friction in compressed copper, Conference of the American Physical Society Topical Group on Shock Compression of Condensed Matter, Seattle, WA 1995, ed. by S. C. Schmidt, W. C. Tao (AIP, New York 1995) 370

35.177 M. D. Perry, J. A. Harrison: Friction between diamond surfaces in the presence of small third-body molecules, J. Phys. Chem. B **101**, 1364–1373 (1997)

35.178 A. Buldum, S. Ciraci: Atomic-scale study of dry sliding friction, Phys. Rev. B **55**, 2606–2611 (1997)

35.179 A. P. Sutton, J. B. Pithica: Inelastic flow processes in nanometre volumes of solids, J. Phys. Cond. Matter **2**, 5317–5326 (1990)

35.180 S. Akamine, R. C. Barrett, C. F. Quate: Improved atomic force microscope images using microcantilevers with sharp tips, Appl. Phys. Lett. **57**, 316–318 (1990)

35.181 J. A. Nieminen, A. P. Sutton, J. B. Pethica: Static junction growth during frictional sliding of metals, Acta Metall. Mater. **40**, 2503–2509 (1992)

35.182 J. A. Niemienen, A. P. Sutton, J. B. Pethica, K. Kaski: Mechanism of lubrication by a thin solid film on a metal surface, Model. Simul. Mater. Sci. Eng **1**, 83–90 (1992)

35.183 V. V. Pokropivny, V. V. Skorokhod, A. V. Pokropivny: Atomistic mechanism of adhesive wear during friction of atomic sharp tungsten asperity over (114) bcc-iron surface, Mater. Lett. **31**, 49–54 (1997)

35.184 B. Li, P. C. Clapp, J. A. Rifkin, X. M. Zhang: Molecular dynamics simulation of stick-slip, J. Appl. Phys. **90**, 3090–3094 (2001)

35.185 R. Komanduri, N. Chandrasekaran: Molecular dynamics simulation of atomic-scale friction, Phys. Rev. B **61**, 14007–14019 (2000)

35.186 T.-H. Fang, C.-I. Weng, J.-G. Chang: Molecular dynamics simulation of nanolithography process using atomic force microscopy, Surf. Sci. **501**, 138–147 (2002)

35.187 S. Morita, S. Fujisawa, Y. Sugawara: Spatially quantized friction with a lattice periodicity, Surf. Sci. Rep. **23**, 1–41 (1996)

35.188 A. Dayo, W. Alnasrallah, J. Krim: Superconductivity-dependent sliding friction, Phys. Rev. Lett. **80**, 1690–1693 (1998)

35.189 R. Erlandsson, G. Hadziioannou, C. M. Mate, G. M. Mcclelland, S. Chiang: Atomic scale friction between the muscovite mica cleavage plane and a tungsten tip, J. Chem. Phys. **89**, 5190–5193 (1988)

35.190 K. L. Johnson: *Contact Mechanics* (Cambridge Univ. Press, Cambridge 1985)

35.191 J. B. Pethica: Interatomic forces in scanning tunneling microscopy – giant corrugations of the graphite surface – comment, Phys. Rev. Lett. **57**, 3235–3235 (1986)

35.192 H. Tang, C. Joachim, J. Devillers: Interpretation of AFM images – the graphite surface with a diamond tip, Surf. Sci. **291**, 439–450 (1993)

35.193 A. L. Shluger, R. T. Williams, A. L. Rohl: Lateral and friction forces originating during force microscope scanning of ionic surfaces, Surf. Sci. **343**, 273–287 (1995)

35.194 S. Fujisawa, Y. Sugawara, S. Morita: Localized fluctuation of a two-dimensional atomic-scale friction, Jpn. J. Appl. Phys. Pt. 1 **35**, 5909–5913 (1996)

35.195 S. Fujisawa, Y. Sugawara, S. Ito, S. Mishima, T. Okada, S. Morita: The two-dimensional stick-slip phenomenon with atomic resolution, Nanotechnology **4**, 138–142 (1993)

35.196 S. Fujisawa, Y. Sugawara, S. Morita, S. Ito, S. Mishima, T. Okada: Study on the stick-slip phenomenon on a cleaved surface of the muscovite mica using an atomic-force lateral force microscope, J. Vacuum Sci. Technol. B **12**, 1635–1637 (1994)

35.197 J. A. Ruan, B. Bhushan: Atomic-scale and microscale friction studies of graphite and diamond using friction force microscopy, J. Appl. Phys. **76**, 5022–5035 (1994)

35.198 R. W. Carpick, N. Agrait, D. F. Ogletree, M. Salmeron: Variation of the interfacial shear strength and adhesion of a nanometer-sized contact, Langmuir **12**, 3334–3340 (1996)

35.199 R. W. Carpick, N. Agrait, D. F. Ogletree, M. Salmeron: Measurement of interfacial shear (friction) with an ultrahigh vacuum atomic force microscope, J. Vacuum Sci. Technol. B **14**, 1289,2772 (1996)

35.200 B. Samuels, J. Wilks: The friction of diamond sliding on diamond, J. Mater. Sci. **23**, 2846–2864 (1988)

35.201 T. Cagin, J.W. Che, M.N. Gardos, A. Fijany, W.A. Goddard: Simulation and experiments on friction, wear of diamond: A material for MEMS and NEMS application, Nanotechnology **10**, 278–284 (1999)

35.202 R.J.A. van den Oetelaar, C.F.J. Flipse: Atomic-scale friction on diamond(111) studied by ultra-high vacuum atomic force microscopy, Surf. Sci. **384**, L828–L835 (1997)

35.203 M. Enachescu, R.J.A. van den Oetelaar, R.W. Carpick, D.F. Ogletree, C.F.J. Flipse, M. Salmeron: Atomic force microscopy study of an ideally hard contact: The diamond(111) tungsten carbide interface, Phys. Rev. Lett. **81**, 1877–1880 (1998)

35.204 B.V. Derjaguin, V.M. Muller, Y. Toporov: Effect of contact deformations on adhesion of particles, J. Colloid Interf. Sci. **53**, 314–326 (1975)

35.205 M.D. Perry, J.A. Harrison: Universal aspects of the atomic-scale friction of diamond surfaces, J. Phys. Chem. B **99**, 9960–9965 (1995)

35.206 R. Neitola, T.A. Pakannen: Ab initio studies on the atomic-scale origin of friction between diamond (111) surfaces, J. Phys. Chem. B **105**, 1338–1343 (2001)

35.207 J.A. Harrison, R.J. Colton, C.T. White, D.W. Brenner: Atomistic simulation of the nanoindentation of diamond and graphite surfaces, Mater. Res. Soc. Sym. Proc. **239**, 573–578 (1992)

35.208 J.A. Harrison, C.T. White, R.J. Colton, D.W. Brenner: Investigation of the atomic-scale friction and energy dissipation in diamond using molecular dynamics, Thin Solid Films **260**, 205–211 (1995)

35.209 J. Cai, J.-S. Wang: Friction between Si tip and (001)–2×1 surface: A molecular dynamics simulation, Comput. Phys. Commun. **147**, 145–148 (2002)

35.210 D. Mulliah, S.D. Kenny, R. Smith: Modeling of stick-slip phenomena using molecular dynamics, Phys. Rev. B **69**, 205407 (2004)

35.211 D.W. Brenner: Empirical potential for hydrocarbons for use in simulating the chemical vapor deposition of diamond films, Phys. Rev. B **42**, 9458–9471 (1990)

35.212 G.J. Ackland, G. Tichy, V. Vitek, M.W. Finnis: Simple n-body potentials for the noble metals and nickel, Philos. Mag. A **56**, 735–756 (1987)

35.213 J.P. Biersack, J. Ziegler, U. Littmack: *The Stopping and Range of Ions in Solids* (Pergamon, Oxford 1985)

35.214 J. Cai, J.S. Wang: Friction between a Ge tip and the (001)–2×1 surface: A molecular dynamics simulation, Phys. Rev. B **64**, 113313 (2001)

35.215 A.G. Khurshudov, K. Kato, H. Koide: Nano-wear of the diamond AFM probing tip under scratching of silicon, studied by AFM, Tribol. Lett. **2**, 345–354 (1996)

35.216 A. Khurshudov, K. Kato: Volume increase phenomena in reciprocal scratching of polycarbonate studied by atomic-force microscopy, J. Vacuum Sci. Technol. B **13**, 1938–1944 (1995)

35.217 M.D. Perry, J.A. Harrison: Molecular dynamics studies of the frictional properties of hydrocarbon materials, Langmuir **12**, 4552–4556 (1996)

35.218 M.D. Perry, J.A. Harrison: Molecular dynamics investigations of the effects of debris molecules on the friction and wear of diamond, Thin Solid Films **291**, 211–215 (1996)

35.219 J.A. Harrison, C.T. White, R.J. Colton, D.W. Brenner: Effects of chemically-bound, flexible hydrocarbon species on the frictional properties of diamond surfaces, J. Phys. Chem. **97**, 6573–6576 (1993)

35.220 J.A. Harrison, D.W. Brenner: Simulated tribochemistry – an atomic-scale view of the wear of diamond, J. Am. Chem. Soc. **116**, 10399–10402 (1994)

35.221 Z. Feng, J.E. Field: Friction of diamond on diamond and chemical vapor deposition diamond coatings, Surf. Coat. Technol. **47**, 631–645 (1991)

35.222 B.N.J. Persson: Applications of surface resistivity to atomic scale friction, to the migration of hot adatoms, and to electrochemistry, J. Chem. Phys. **98**, 1659–1672 (1993)

35.223 B.N.J. Persson, A.I. Volokitin: Electronic friction of physisorbed molecules, J. Chem. Phys. **103**, 8679–8683 (1995)

35.224 H. Grabhorn, A. Otto, D. Schumacher, B.N.J. Persson: Variation of the dc-resistance of smooth and atomically rough silver films during exposure to C_2H_6 and C_2H_4, Surf. Sci. **264**, 327–340 (1992)

35.225 A.D. Berman, W.A. Ducker, J.N. Israelachvili: Origin and characterization of different stick-slip friction mechanisms, Langmuir **12**, 4559–4563 (1996)

35.226 B.N.J. Persson: Theory of friction – dynamical phase transitions in adsorbed layers, J. Chem. Phys. **103**, 3849–3860 (1995)

35.227 B.N.J. Persson, E. Tosatti: Layering transition in confined molecular thin films – nucleation and growth, Phys. Rev. B **50**, 5590–5599 (1994)

35.228 H. Yoshizawa, J. Israelachvili: Fundamental mechanisms of interfacial friction. 2. Stick-slip friction of spherical and chain molecules, J. Phys. Chem. **97**, 11300–11313 (1993)

35.229 B.N.J. Persson: Theory of friction: Friction dynamics for boundary lubricated surfaces, Phys. Rev. B **55**, 8004–8012 (1997)

35.230 U. Landman, W.D. Luedtke, J.P. Gao: Atomic-scale issues in tribology: Interfacial junctions and nano-elastohydrodynamics, Langmuir **12**, 4514–4528 (1996)

35.231 P.A. Thompson, M.O. Robbins: Origin of stick-slip motion in boundary lubrication, Science **250**, 792–794 (1990)

35.232 T. Kreer, M.H. Müser, K. Binder, J. Klein: Frictional drag mechanisms between polymer-bearing surfaces, Langmuir **17**, 7804–7813 (2001)

35.233 T. Kreer, K. Binder, M.H. Müser: Friction between polymer brushes in good solvent conditions: Steady-state sliding versus transient behavior, Langmuir **19**, 7551–7559 (2003)

35.234 E. Manias, G. Hadziioannou, G. ten Brinke: Inhomogeneities in sheared ultrathin lubricating films, Langmuir **12**, 4587–4593 (1996)

35.235 R.M. Overney, T. Bonner, E. Meyer, M. Reutschi, R. Luthi, L. Howald, J. Frommer, H.J. Guntherodt, M. Fujihara, H. Takano: Elasticity, wear, and friction properties of thin organic films observed with atomic-force microscopy, J. Vacuum Sci. Technol. B **12**, 1973–1976 (1994)

35.236 R.M. Overney, E. Meyer, J. Frommer, D. Brodbeck, R. Luthi, L. Howald, H.J. Guntherodt, M. Fujihira, H. Takano, Y. Gotoh: Friction measurements on phase-separated thin-films with a modified atomic force microscope, Nature **359**, 133–135 (1992)

35.237 R.M. Overney, E. Meyer, J. Frommer, H.J. Guntherodt, M. Fujihira, H. Takano, Y. Gotoh: Force microscopy study of friction and elastic compliance of phase-separated organic thin-films, Langmuir **10**, 1281–1286 (1994)

35.238 H.I. Kim, T. Koini, T.R. Lee, S.S. Perry: Systematic studies of the frictional properties of fluorinated monolayers with atomic force microscopy: Comparison of CF_3- and CH_3-terminated films, Langmuir **13**, 7192–7196 (1997)

35.239 P.T. Mikulski, J.A. Harrison: Packing density effects on the friction of n-alkane monolayers, J. Am. Chem. Soc. **123**, 6873–6881 (2001)

35.240 M. GarciaParajo, C. Longo, J. Servat, P. Gorostiza, F. Sanz: Nanotribological properties of octadecyltrichlorosilane self-assembled ultrathin films studied by atomic force microscopy: Contact and tapping modes, Langmuir **13**, 2333–2339 (1997)

35.241 R.M. Overney, H. Takano, M. Fujihira, E. Meyer, H.J. Guntherodt: Wear, friction and sliding speed correlations on Langmuir-Blodgett films observed by atomic force microscopy, Thin Solid Films **240**, 105–109 (1994)

35.242 P.T. Mikulski, J.A. Harrison: Periodicities in the properties associated with the friction of model self-assembled monolayers, Tribol. Lett. **10**, 29–35 (2001)

35.243 E. Barrena, C. Ocal, M. Salmeron: A comparative AFM study of the structural and frictional properties of mixed and single component films of alkanethiols on Au(111), Surf. Sci. **482**, 1216–1221 (2001)

35.244 Y.-S. Shon, S. Lee, R. Colorado, S.S. Perry, T.R. Lee: Spiroalkanedithiol-based SAMS reveal unique insight into the wettabilities and frictional properties of organic thin films, J. Am. Chem. Soc. **122**, 7556–7563 (2000)

35.245 P.T. Mikulski, G. Gao, G.M. Chateauneuf, J.A. Harrison: Contact forces at the sliding interface: Mixed versus pure model alkane monolayers, J. Chem. Phys. **122**, 024701 (2005)

35.246 S. Lee, Y.S. Shon, R. Colorado, R.L. Guenard, T.R. Lee, S.S. Perry: The influence of packing densities, surface order on the frictional properties of alkanethiol self-assembled monolayers (SAMs) on gold: A comparison of SAMs derived from normal and spiroalkanedithiols, Langmuir **16**, 2220–2224 (2000)

35.247 G.M. Chateauneuf, P.T. Mikulski, G.T. Gao, J.A. Harrison: Compression- and shear-induced polymerization in model diacetylene-containing monolayers, J. Phys. Chem. B **108**, 16626–16635 (2004)

35.248 L. Zhang, S. Jiang: Molecular simulation study of nanoscale friction for alkyl monolayers on Si(111), J. Chem. Phys. **117**, 1804–1811 (2002)

35.249 L.Z. Zhang, Y.S. Leng, S.Y. Jiang: Tip-based hybrid simulation study of frictional properties of self-assembled monolayers: Effects of chain length, terminal group, scan direction, and scan velocity, Langmuir **19**, 9742–9747 (2003)

35.250 M. Chandross, E.B.W. III, M.J. Stevens, G.S. Grest: Systematic study of the effect of disorder on nanotribology of self-assembled monolayers, Phys. Rev. Lett. **93**, 166103 (2004)

35.251 M. Chandross, G.S. Grest, M.J. Stevens: Friction between alkylsilane monolayers: Molecular simulation of ordered monolayers, Langmuir **18**, 8392–8399 (2002)

35.252 S.B. Sinnott, R. Andrews: Carbon nanotubes: Synthesis, properties and applications, Crit. Rev. Sol. St. Mater. Sci. **26**, 145–249 (2001)

35.253 B. Bhushan, B.K. Gupta, G.W. Van Cleef, C. Capp, J.V. Coe: Sublimed C_{60} films for tribology, Appl. Phys. Lett. **62**, 3253–3255 (1993)

35.254 T. Thundat, R.J. Warmack, D. Ding, R.N. Compton: Atomic force microscope investigation of C_{60} adsorbed on silicon and mica, Appl. Phys. Lett. **63**, 891–893 (1993)

35.255 C.M. Mate: Nanotribology studies of carbon surfaces by force microscopy, Wear **168**, 17–20 (1993)

35.256 R. Lüthi, E. Meyer, H. Haefke: Sled-type motion on the nanometer scale: Determination of dissipation and cohesive energies of C_{60}, Science **266**, 1979–1981 (1993)

35.257 R. Lüthi, H. Haefke, E. Meyer, L. Howald, H.-P. Lang, G. Gerth, H.J. Güntherodt: Frictional and atomic-scale study of C_{60} thin films by scanning force microscopy, Z. Phys. B **95**, 1–3 (1994)

35.258 Q.-J. Xue, X.-S. Zhang, F.-Y. Yan: Study of the structural transformations of C_{60}/C_{70} crystals during friction, Chin. Sci. Bull. **39**, 819–822 (1994)

35.259 W. Allers, U.D. Schwarz, G. Gensterblum, R. Wiesendanger: Low-load friction behavior of epitaxial C_{60} monolayers, Z. Phys. B **99**, 1–2 (1995)

35.260 U.D. Schwarz, W. Allers, G. Gensterblum, R. Wiesendanger: Low-load friction behavior of epitaxial C_{60} monolayers under Hertzian contact, Phys. Rev. B **52**, 14976–14984 (1995)

35.261 J. Ruan, B. Bhushan: Nanoindentation studies of sublimed fullerene films using atomic force microscopy, J. Mater. Res. **8**, 3019–3022 (1996)

35.262 U. D. Schwarz, O. Zworner, P. Koster, R. Wiesendanger: Quantitative analysis of the frictional properties of solid materials at low loads. I. Carbon compounds, Phys. Rev. B **56**, 6987–6996 (1997)

35.263 S. Okita, M. Ishikawa, K. Miura: Nanotribological behavior of C_{60} films at an extremely low load, Surf. Sci. **442**, L959–L963 (1999)

35.264 S. Okita, K. Miura: Molecular arrangement in C_{60} and C_{70} films on graphite and their nanotribological behavior, Nano Lett. **1**, 101–103 (2001)

35.265 K. Miura, S. Kamiya, N. Sasaki: C_{60} molecular bearings, Phys. Rev. Lett. **90**, 055509 (2003)

35.266 A. Buldum, J. P. Lu: Atomic scale sliding and rolling of carbon nanotubes, Phys. Rev. Lett. **83**, 5050–5053 (1999)

35.267 M. R. Falvo, R. M. Taylor, A. Helser, V. Chi, F. P. Brooks, S. Washburn, R. Superfine: Nanometer-scale rolling and sliding of carbon nanotubes, Nature **397**, 236–238 (1999)

35.268 M. R. Falvo, J. Steele, R.M.T. II, R. Superfine: Gear-like rolling motion mediated by commensurate contact: Carbon nanotubes on HOPG, Phys. Rev. B **62**, R10664–R10667 (2000)

35.269 J. D. Schall, D. W. Brenner: Molecular dynamics simulations of carbon nanotube rolling and sliding on graphite, Mol. Simul. **25**, 73–80 (2000)

35.270 B. Ni, S. B. Sinnott: Tribological properties of carbon nanotube bundles, Surf. Sci. **487**, 87–96 (2001)

35.271 B. Ni, S. B. Sinnott: Mechanical and tribological properties of carbon nanotubes investigated with atomistic simulations, *Nanotubes and related materials*. In: *Nanotubes and Related Materials*, MRS Proceedings, Vol. 633 (Materials Research Society, Pittsburgh, PA 2001) pp. A17.13.11–A17.13.15

35.272 K. Miura, T. Takagi, S. Kamiya, T. Sahashi, M. Yamauchi: Natural rolling of zigzag multiwalled carbon nanotubes on graphite, Nano Lett. **1**, 161–163 (2001)

35.273 K. Miura, M. Ishikawa, R. Kitanishi, M. Yoshimura, K. Ueda, Y. Tatsumi, N. Minami: Bundle structure and sliding of single-walled carbon nanotubes observed by friction-force microscopy, Appl. Phys. Lett. **78**, 832–834 (2001)

35.274 P. E. Sheehan, C. M. Lieber: Nanotribology and nanofabrication of MoO_3 structures by atomic force microscopy, Science **272**, 1158–1161 (1996)

35.275 J. Wang, K. C. Rose, C. M. Lieber: Load-independent friction: MoO_3 nanocrystal lubricants, J. Phys. Chem. B **103**, 8405–8408 (1999)

35.276 Q. Ouyang, K. Okada: Nanoballbearing effect of ultra-fine particles of cluster diamond, Appl. Surf. Sci. **78**, 309–313 (1994)

35.277 R. Luthi, E. Meyer, H. Haefke, L. Howald, W. Gutmannsbauer, H. J. Guntherodt: Sled-type motion on the nanometer-scale – determination of dissipation and cohesive energies of C_{60}, Science **266**, 1979–1981 (1994)

35.278 B. Bhushan, B. K. Gupta, G. W. Vancleef, C. Capp, J. V. Coe: Fullerene (C_{60}) films for solid lubrication, Tribol. Trans. **36**, 573–580 (1993)

35.279 U. D. Schwarz, W. Allers, G. Gensterblum, R. Wiesendanger: Low-load friction behavior of epitaxial C_{60} monolayers under Hertzian contact, Phys. Rev. B **52**, 14976–14984 (1995)

35.280 S. B. Legoas, R. Giro, D. S. Galvao: Molecular dynamics simulations of C_{60} nanobearings, Chem. Phys. Lett. **386**, 425–429 (2004)

35.281 P. L. Dickrell, S. B. Sinnott, D. W. Hahn, N. R. Raravikar, L. S. Schadler, P. M. Ajayan, W. G. Sawyer: Frictional anisotropy of oriented carbon nanotube surfaces, Tribol. Lett. **18**, 59–62 (2005)

35.282 G. T. Gao, P. T. Mikulski, J. A. Harrison: Molecular-scale tribology of amorphous carbon coatings: Effects of film thickness, adhesion, and long-range interactions, J. Am. Chem. Soc. **124**, 7202–7209 (2002)

35.283 F. P. Bowden, D. Tabor: *The Friction and Lubrication of Solids, Part 2* (Clarenden, Oxford 1964)

35.284 G. T. Gao, P. T. Mikulski, G. M. Chateauneuf, J. A. Harrison: The effects of film structure and surface hydrogen on the properties of amorphous carbon films, J. Phys. Chem. B **107**, 11082–11090 (2003)

36. Nanoscale Mechanical Properties – Measuring Techniques and Applications

This chapter describes local measurements and applications of nanoscale mechanical properties. It includes detailed state of the art presentation and in-depth analysis of experimental techniques, results and interpretations.

After a short introduction, the second part of the chapter explores the application of local mechanical spectroscopy using coupled atomic force microscopy and ultrasound. This technique allows us to rapidly map elasticity and anelastic properties. Semiquantitative measurements can be taken as a function of temperature at specific point in the sample. The results obtained on the nanoscale are similar to those from bulk measurements, with interpretable differences. The local elasticity and damping were measured during phase transitions of polymer samples and shape-memory alloys.

The third part describes the "nano-Swiss cheese" method of measuring the elastic properties of nanosized tubular objects such as carbon nanotubes and microtubules. It is probably the only experiment in which the properties of single-wall nanotube ropes have been measured as a function of the rope diameter. We extended this idea to biological objects, microtubules, and successfully solved major experimental difficulties. We not only measured the temperature dependency of the microtubule modulus under pseudo-physiological conditions, but we also estimated the shear modulus using the same microtubule with several lengths of suspended segments.

The fourth section describes the scanning nanoindentation technique as applied to human bone tissue. This instrument allows topography scans and indentation tests to be performed using an identical tip. The technique allows the indenter tip to be positioned on structures of interest with great precision during suface scans. For very inhomogeneous samples, such as bone tissue, this tool provides a way to detect local variations in mechanical properties. The indentation test allows

us to study quantitative parameters like elastic modulus and hardness at the submicron level. Using this, local mechanical properties of compact and trabecular bone lamellae were tested under both dry and pseudo-physiological conditions.

The chapter ends with a discussion of future prospects for the field and some conclusions.

Part D | 36

Measuring mechanical properties at the nanoscale is the key to understanding mesoscopic materials and inhomogeneous materials.

The most prominent technique used to perform such measurements uses an atomic force microscope in static mode to obtain force–distance (f-d) curves. Ideally, the force applied to the tip and the tip–sample distance are measured, which allows the local (nanoscale) reduced Young's modulus to be determined. In practice, however, these values are not measured directly. Instead, the applied force is offset by adhesive and capillary forces. Furthermore, the tip–sample distance must be obtained by subtracting the sample (Z-piezo) displacement and the deflection of the cantilever, neither of which are calibrated on most commercial instruments. Often only the voltage applied to the nonlinear and hysteretic Z-piezo and the uncalibrated cantilever deflection is available. These data must therefore be tediously calibrated and converted to real displacements and forces. The calibration procedure given by *Radmacher* [36.1] works well, but only when applied to very compliant materials such as biological or polymer samples. An additional difficulty is that since one end of the cantilever is fixed, neither the orientation of the tip nor its position on the sample surface remains constant during f-d curve acquisition. Therefore, movement of the tip on the sample surface occurs not only vertically but also along the cantilever axis. As a result, applied contact mechanics may not be valid. Finally, uncertainty about the exact tip shape and radius complicates the procedure even further. The instrumental challenges involved are so great that applying static AFM in order to obtain absolute values of the elastic modulus using commercial instruments is difficult at best, especially on stiff surfaces, and this provided the motivation to develop other techniques.

Although absolute values of mechanical properties are difficult to measure, relative measurements or maps (images) of these properties are still feasible and of great interest. Acoustic vibrations of the AFM sample surface can be used to access local elastic and anelastic properties. The first part of this chapter describes these techniques and applications.

Studies of mesoscopic specimens are even more challenging – it is difficult and time-consuming to locate a sample, especially one that is positioned properly. In the second part of the chapter, a technique based on the use of 'nano-Swiss cheese' is introduced as a way to investigate the axial and shear moduli of several species of carbon nanotubes. This method has been extended for use with biological microtubules, where experiments are performed in liquid and the properties are studied as a function of the temperature.

Reliable mechanical measurements can be obtained by combining the best features of both the AFM and the nanoindenter. Classical nanoindenters use optical tip positioning, which is inadequate on the nanoscale. The scanning nanoindenter uses the same diamond tip for imaging (with micronewton forces) and measuring. Its application to human bones in liquid at 37 °C is described in the third part of this chapter. We then discuss how the same technique can be used to study the nanomechanical properties of human hair.

One should underline the importance of AFM linearity and stability in such experiments – good design can minimize the frustration of and the experimental time needed by courageous PhD students, and lot of improvements are still necessary in this area.

Finally, a short list of hopes and prospects for this field conclude this chapter.

36.1 Local Mechanical Spectroscopy via Dynamic Contact AFM

Mechanical properties of solids (elasticity, anelasticity, plasticity) are generally measured on macroscopic samples. However, many phenomena in materials science require measurements of the mechanical properties at the surface of a material, or at the interfaces between thin layers deposited on a surface. High spatial resolution is also important; for example for multiphased materials or composite materials, phase transitions, lattice softening in shape-memory alloys, precipitation in light alloys and glass transitions of amorphous materials.

For multiphased materials, such as nanomaterials, composites, alloys or polymer blends, the location of the

dissipative mechanism in one phase must be determined either through modeling or by separately studying each phase, without changing its behavior if possible. The latter is only possible in a limited number of cases due to the interactions between the different phases within a material. To give an example, the global behavior of a composite is mostly driven by the stress transfer properties between the reinforcement and the matrix, which are controlled by the local mechanical properties in the interface region, and in particular the dynamics of the structural defects in this area. It is obviously impossible to prepare a sample only composed of interface regions. Therefore, a method that may be used to study the dynamics of structural defects locally would thus help us to make important strides in the understanding and enhancement of such materials.

Different techniques based on *scanning probe microscopy* (SPM) have been developed in order to probe the elastic and anelastic properties of surfaces, interfaces or phases of inhomogeneous materials at micrometer and the nanometer scales. *Scanning acoustic microscopy* (SAM), which was first developed in the mid-1970s, allows one to study the properties of materials at the micrometer scale.

Amongst the various techniques used to study the local mechanical properties of materials, several groups have recently used methods based on *scanning force microscopy* (SFM) [36.2]. In most of these, the focus has been on studying "elasticity", using the so-called "force modulation mode" (FMM) at low frequencies [36.3]. The force modulation mode generally uses a large amplitude (greater than 10 nm), low frequency (a few kHz) vibration of the sample beneath the SFM tip. The component of the tip motion at the excitation frequency and the mean position of the tip are recorded simultaneously, giving several images of the sample surface. In particular, the in-phase and out-of-phase components of force modulation mode at room temperature have been interpreted in terms of stiffness ("elasticity") and damping ("viscoelasticity") [36.4, 5]. However, it has been recently shown by *Mazeran* et al. [36.6] that the contrast observed with the force modulation mode is dominated by friction properties, giving only minor information on the elasticity. Consequently, some care has to be taken over interpretations of these low-frequency studies. One way to suppress the influence of friction on the contrast is to use smaller amplitudes (a few Å) at higher frequencies [36.7]. *Scanning local-acceleration microscopy* (SLAM) implements this idea [36.8]: SLAM is a modification of contact mode scanning force microscopy. In SLAM, the sample is vibrated at a frequency just above the resonance of the tip–sample system. In this case, the inertia of the tip prevents it from completely following the imposed high-frequency displacement, which induces non-negligible forces and gives rise to elastic deformation of the sample. The contact stiffness is obtained by measuring the residual displacement of the tip. Mapping the contact stiffness at different temperatures with SLAM [36.9] allows local mechanical spectroscopy to be performed. Some other techniques also use high frequencies, but with different approaches to imaging elasticity at room temperature [36.7, 10, 11]. Each of these high-frequency techniques are capable

of mapping properties such as stiffness or adhesion at constant temperature with very high lateral resolution.

36.1.1 Variable Temperature SLAM (T–SLAM)

The technique described here combines the lateral resolution of scanning force microscopy with the physical information available from temperature ramps. It called *variable temperature* SLAM (T–SLAM) [36.12, 13]. Ramping the temperature of the sample during a SLAM measurement and acquiring both the amplitude and the phase of the tip's motion allows one to obtain local mechanical spectroscopic data. A simple model enables the measurement to be interpreted in terms of material properties.

T-SLAM is basically SLAM [36.8, 14], performed with a variable temperature sample holder [36.9, 15]. An ultrasonic transducer is placed beneath the sample in a commercial scanning force microscope (Fig. 36.1) and

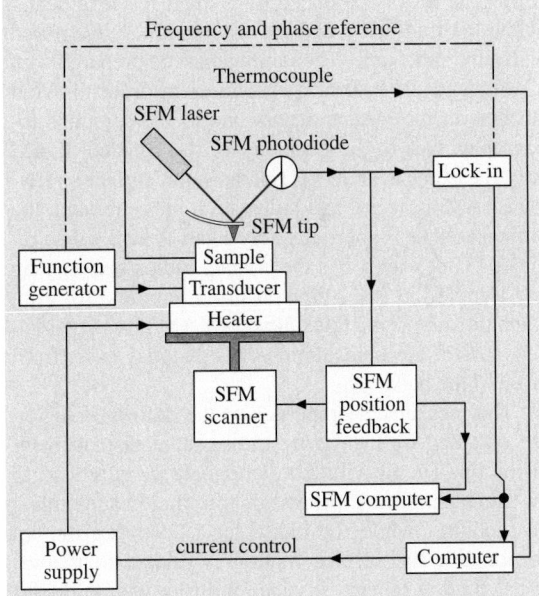

Fig. 36.1 Schematic diagram of T-SLAM. Ultrasound is generated by a transducer (connected to a function generator) that is placed beneath the sample. The motion of the scanning force microscope tip is detected optically. The detected signal is fed to a lock-in amplifier, which extracts the amplitude and phase relative to the motion of the transducer. The temperature is controlled using a small resistive heater and is measured with a thermocouple. The rest of the microscope head (to which the tip, the laser and the photodiode are attached) is not represented in this diagram

excited by using a function generator. The ultrasound is transmitted through the sample, forcing periodic local deformation of the sample's surface underneath the tip. The motion of the scanning force microscopy tip is detected optically by laser beam deflection from the backside of the cantilever. The detection signal is then fed to a lock-in amplifier, which extracts the tip's amplitude (related to the elasticity) and the phase (related to the viscoelasticity) relative to the transducer's motion. The typical frequency of the transducer is 825 kHz. This system can be operated in two ways, where either the amplitude and phase of the signal are mapped as a function of position at fixed temperature, or the amplitude and phase are recorded as a function of temperature at a "fixed" location. The first method is known as "SLAM imaging", where the output signal of the lock-in is fed into an auxiliary data acquisition channel of the microscope. An extra computer is used to store local mechanical spectroscopic data as a function of the temperature at a fixed location and to control temperature. The heat is produced by a small resistive heater below the transducer and the temperature is measured with a thermocouple. The sample must be prepared with a sufficiently low surface roughness in order to avoid artifacts in the measurements due to the sample's topography. Due to the geometry of the contact, SLAM only measures a small volume near the surface. Viewing SLAM as a very fast indentation measurement, the probed volume is approximately a half-sphere with a radius of $10a$, where a is the contact radius between tip and sample [36.16]. a is typically a few nm. So, even when the sample is a thick film, T-SLAM gives access to near-surface mechanical properties that may differ from those of the bulk.

The mechanical properties of the deformed region are obtained by measuring the residual displacement of the tip. The tip vibration amplitude d_1 is related to the contact stiffness, proportional to the dynamic elastic modulus, while the phase lag φ between the tip motion and the surface motion is related to the internal friction (energy dissipation inside the deformed volume). Mapping d_1 and φ at different temperatures using T-SLAM allows the homogeneity of mechanical relaxations or phase transitions to be studied. By recording d_1 and φ as a function of temperature at a fixed location, local mechanical spectroscopy can be performed.

When used just above the first resonance of the tip–sample system, the SLAM system can be described and analyzed using a point mass rheological model [36.9, 12], which allows one to obtain an equation relating the

damping (loss factor η') to the measured parameters:

$$\eta' = \left(\frac{1}{2} \frac{k_e}{k_c - m\omega^2} \right) \frac{\sin \varphi}{d_1/z_1} , \qquad (36.1)$$

Here ω is the measurement frequency, k_c and k_e are parameters related to the elastic modulus of the AFM cantilever and the elastic modulus of the sample, respectively, z_1 is the transducer vibration amplitude, and $m = k_c/\omega_c^2$ is the equivalent point mass of the tip, where ω_c is the free resonant frequency of the cantilever.

The principal limitation of (36.1) comes from the model's assumption that the cantilever is a point mass restricted to vertical displacements. Due to possible lateral displacement of the tip [36.6] or the existence of other vibrational modes of the cantilever, which are not described by the point mass model, (36.1) is only reliable for measuring the damping just above the first resonance of the tip–sample system. In order to quantitatively measure this quantity over a larger frequency spectrum, it is necessary to develop a more realistic model of the cantilever interacting with the sample surface, for example the model presented by *Dupas* et al. [36.17], in which the cantilever is modeled as a beam.

36.1.2 Example 1: Local Mechanical Spectroscopy of Polymers

Figure 36.2 shows local mechanical spectra [36.13, 18] obtained for bulk technical PVC (including plasticizer and pigments, and taken "off the shelf") as a function of temperature (a), and differential scanning calorimetry (DSC) measurements for the same sample (b). The mechanical measurement (a) displays the amplitude of vibration of the SLAM tip (thin line) and its phase lag (thick line) as a function of temperature. Four temperature domains can be identified. At lower temperatures (1), the first domain shows a phase lag peak associated with a decrease in amplitude. The second temperature (2) domain corresponds to a zone where the vibration amplitude increases without any variation in the phase lag. The third temperature domain (3) is characterized by a large phase lag peak and a significant decrease in vibration amplitude. The last temperature domain (4) shows a slow increase in the phase lag without any variation in the vibration amplitude.

The calorimetry curves (b) display the heat flow as a function of temperature for the same material, but on a larger scale. For clarity, the same temperature domains have been reported on the graph. The dashed line displays the first heating, where both reversible and ir-

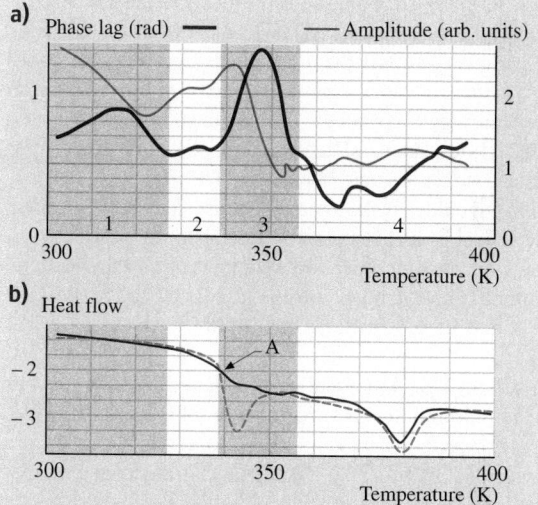

Fig. 36.2 (a) Local mechanical spectra as a function of temperature for a technical PVC. The vibration amplitude is displayed as a *thin line* and the phase lag as a *thick line*. Four temperature domains can be distinguished. From *left* to *right*, a small decrease in vibration amplitude is associated with the first phase lag peak (1); vibration amplitude increases in the next domain (2); then a large decrease in vibration amplitude is associated with a large phase lag peak (3); finally the phase lag slowly increases (4). **(b)** DSC measurements of the same sample. For clarity, the temperature domains observed in Fig. 36.5a have been reported. The graph displays the first (*dashed line*) and the second (*solid line*) heatings. The glass transition occurs around 340 K (labeled *A*), slightly below the temperature range of domain 3. An irreversible endothermic relaxation takes place within the same temperature range (only visible on the first run). A reversible endothermic event occurs around 380 K

reversible events are present; the solid line shows the second heating, where only reversible events are present. An endothermic event (labeled A) can be observed on the solid line near the border between temperature domains 2 and 3. An irreversible endothermic relaxation is superimposed on (A) on the first heating. A reversible endothermic peak is observed around 380 K, in domain 4. The endothermic event (A) has the characteristics of a glass transition: the specific heat changes from one value to another without a peak [36.19]. The glass transition temperature of this PVC is therefore approximately 340 K. The irreversible relaxation that occurs in the same temperature range is associated with physical aging or structural relaxation. The reversible peak at

higher temperature is associated with the melting of the small crystalline volume fraction.

These results from local mechanical spectroscopy are in good qualitative agreement with macroscopic global measurements [36.13,18]. They yield the same information as the macroscopic global measurements, but from a much smaller volume, and therefore allow the different mechanisms involved to be elucidated. Both measurements (local and global) show a phase peak for the primary and secondary relaxations, connected with a large decrease in the stiffness for the primary relaxation and a much smaller one for the secondary relaxation. Plasticity induces an increase in the phase lag in both cases. The temperatures observed for the relaxation related to the glass transition compare well with the calorimetric data. Based on all of this evidence, there is no doubt that the amplitude and phase lag of T-SLAM are functions of the stiffness and damping. In this regard, T-SLAM provides an extension of the global method, allowing dissipative phenomena to be located and spatial homogeneities of phase transitions or of relaxations to be studied. This will allow new insight into the field of inhomogeneous or composite materials. In addition, T-SLAM only measures a small volume near the surface. Therefore, even if the sample is a thick film, T-SLAM gives access to the near-surface mechanical properties, which may differ from bulk properties. This will bring interesting new perspectives to the field of surface dynamics, which is still the source of much debate.

36.1.3 Example 2: Local Mechanical Spectroscopy of NiTi

Near-stoichiometric NiTi alloys exhibit a martensitic phase transformation between a low-temperature monoclinic phase, called martensite, and a high temperature cubic phase with B2 structure, called austenite (1). This transformation is responsible for the shape-memory and pseudo-elastic effects in deformed NiTi alloys. Upon transformation to the martensitic phase, an intermediate rhombohedral (R) phase can form [36.20, 21]. Although *Bataillard* [36.21] demonstrated that the R phase tends to be finely dispersed inside the material, there is still controversy about the spatial scale at which the decomposition of the transformation occurs. Another puzzling question concerns the transformation itself. Optical microscopy observations suggest that the transformation occurs very suddenly inside an austenite grain. This has led to the concept of "military transformation". The width of the globally measured

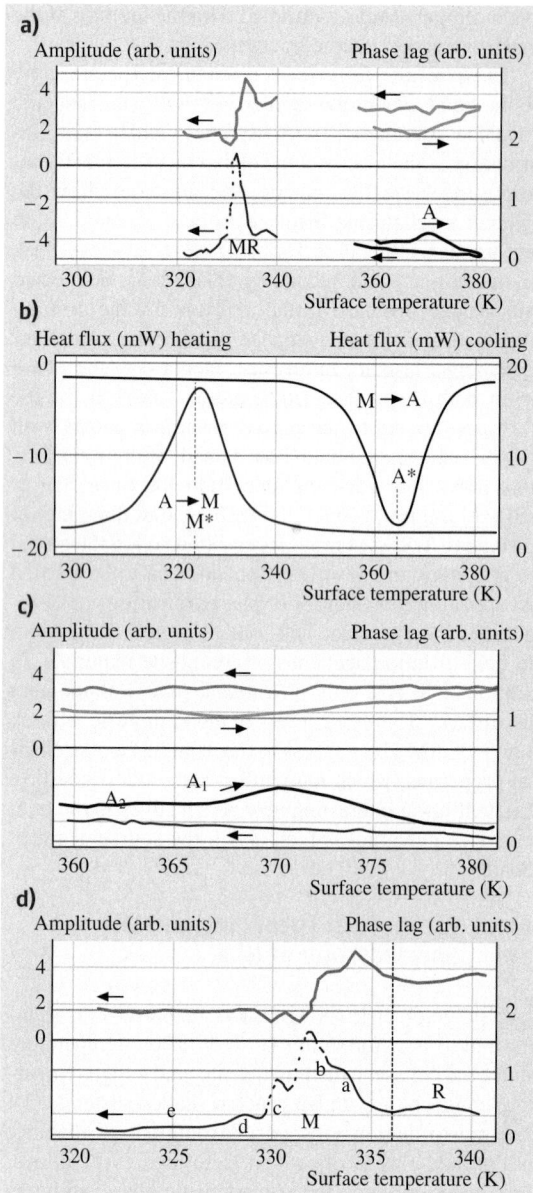

a)

Amplitude (arb. units)   Phase lag (arb. units)

b)

Heat flux (mW) heating   Heat flux (mW) cooling

c)

Amplitude (arb. units)   Phase lag (arb. units)

d)

Amplitude (arb. units)   Phase lag (arb. units)

Fig. 36.3 (a)–(b) Comparison of local mechanical spectro-scopic measurements of NiTi with those from a calorimetric experiment. **(a)** Both the reverse and direct transformations are associated with a phase lag peak and a variation in modulus. **(b)** Calorimetric measurement of the same sample. **(c)–(d)** Zoom of temperature ranges of the reverse **(c)** and direct **(d)** transformations. The peak (A1) may exhibit a shoulder on the low temperature side (A2). The direct transformation is characterized by a recovery of the original vibration amplitude and a complex phase lag spectrum, formed from two main maxima, denoted R and M. Peaks a, b, c, d, e are substructures of peak M

son that T-SLAM local mechanical spectroscopy has been applied to the martensitic phase transition of NiTi [36.12, 22]. The results from these studies are presented in Fig. 36.3. The global transformation behavior of NiTi has been defined with calorimetry spectra (Fig. 36.3b, temperature scanning rate 10 K/min.).

The vibration amplitude ("elasticity") and the phase lag ("internal friction") are shown in the top and bottom curves, respectively. The presented data is incomplete due to experimental limitations. A phase lag peak can be observed upon heating at approximately 370 K, associated with a change in the vibration amplitude. This event can be correlated with the phase transformation from martensite to austenite observed using calorimetry (Fig. 36.3b). Upon cooling, no event is observed around 370 K; the vibration amplitude is stable and the phase lag curve does not exhibit any peak. However, an intense phase lag peak can be observed upon cooling near 330 K, associated with a restoration of the vibration amplitude characteristic of the martensite. This event correlates with the phase transformation from austenite to martensite observed via calorimetry. Changes in the vibration amplitude spectrum are correlated with events in the phase lag spectrum. Two main features should be noted. First, the transformation peaks observed with the SLAM method are narrower than those observed in the (global) calorimetry measurements. Secondly, the measured temperatures for the local peak are located on the high-temperature sides of the peaks measured by calorimetry. The two transformation peaks are displayed in detail in Figs. 36.3c and 36.3d. The reverse transformation (Fig. 36.3c) is characterized by a change in the vibration amplitude level and a phase lag peak. This peak (A1) may have a shoulder on the low-temperature side (A2). A small decrease in the vibration amplitude may precede this increase. However, the magnitude of this softening and the very low intensity of peak A2 are too small to exclude experimental artifacts. The direct

transformation would then be the sum of various narrow contributions deriving from different places inside the sample. This image is, however, not universally accepted.

Measuring inside a single grain of a polycrystal would be a way to address these questions. Both the spatial scale of the R phase distribution and the "military" character of the transformation will have an effect on the result of such a measurement. This is the rea-

transformation (Fig. 36.3d) displays a recovery of the original vibration amplitude and a complicated phase lag maximum.

Comparing the transition temperatures measured by calorimetry with the results from local mechanical spectroscopy leads to the conclusion that the peaks observed in the phase lag and the change in the vibration amplitude originate from the martensitic phase transition. At the scale of observation, the martensite transforms into austenite around 370 K (A1 peak) and the austenite transforms into martensite close to 332 K (M peak), with the formation of the rhombohedral R phase at 337 K (R peak). The peaks A2 and e could be linked to the presence of two different types of martensite, as already observed elsewhere [36.20, 21]. It is striking to note that all of the events observed in macroscopic experiments seem to be reproducible at the scale of observation, namely less than 10^{-3} μm^3. In particular, this would confirm that the R phase precipitates are very finely distributed in the austenite matrix, as observed by *Bataillard* [36.21]. Otherwise, the R peak could not be observed reproducibly using such a technique. Local measurements differ from global ones in at least two aspects: first, the width of the transformation temper-

ature range is smaller; secondly, the M peak exhibits substructure.

Global internal friction measurements on these materials have produced spectra with peaks that have a similar breadth to the calorimetry peaks [36.23], whereas the peaks measured locally are narrower. This is easy to understand if the martensitic transformation occurs inhomogeneously inside the sample. Grains tend to transform at different temperatures depending on their stress state. As local measurements have sufficient spatial resolution to probe a single grain inside the material, it is logical that the transformation occurs over a narrower temperature range than in a global experiment probing a large number of grains.

The substructure in the M peak could have three possible origins: (i) the technique is sensitive to the mechanical relaxation inside newly formed plates, such as stress relaxation by twin motion; (ii) the probed volume contains several martensite plates that grow one-by-one, each with its own, distinct "elementary" peak (a, b, ...); (iii) the region of the sample surface analyzed changes due to thermal drifts, so the transformations of several growing martensite plates are probed, leading to multiple "elementary" peaks.

36.2 Static Methods – Mesoscopic Samples, Shear and Young's Modulus

36.2.1 Carbon Nanotubes – Introduction to Basic Morphologies and Production Methods

Carbon nanotubes (CNTs) are the newest forms of carbon, found in 1991 by *Iijima* [36.24]. Owing to their remarkable properties, this discovery has opened up whole new fields of study in physics, chemistry and material science. They possess a unique combination of small size (diameters ranging from ≈ 1 to 50 nm with lengths of up to ≈ 1 cm), low density (similar to that of graphite), high stiffness, high strength and a broad range of electronic properties from metallic to semiconducting. Their range of potential applications is immense, including reinforcing elements in high-strength composites, electron sources in field emission displays and small X-ray sources, ultrasharp and resistant AFM tips with high aspect ratios, gas sensors, and as components in anticipated nanoscale electronics. In addition, they provide a very popular system for studying fundamental physical phenomena on the mesoscopic scale. With advances in manufacturing and processing, it is likely they

will come to play an integral part in many devices that we use in our everyday life.

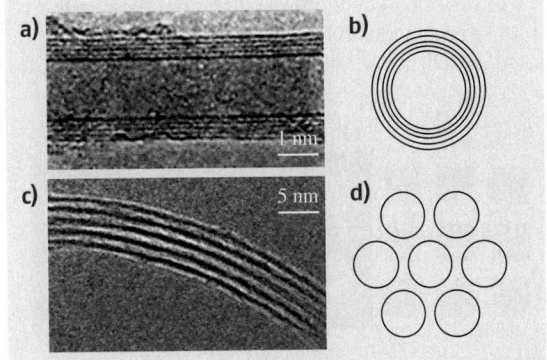

Fig. 36.4a–d TEM images and schematic drawings of cross-sections of different morphologies of carbon nanotubes. (**a**) and (**b**) Multiwalled carbon nanotubes, consisting of concentric, nested tubes. (**c**) and (**d**) Single-walled carbon nanotubes bundled up into ropes and held together by van der Waals interactions

From a structural point of view, carbon nanotubes can be thought of as rolled-up sheets of graphite (*graphene*). They can be divided into two distinct groups. Multiwalled carbon nanotubes (MWNT) exhibit a Russian doll-like structure of nested, concentric tubes, see Fig. 36.4a and b. They were also the first CNTs to be discovered experimentally. The interlayer spacing can range from 0.342 to 0.375 nm, depending on the diameters and the number of shells comprising the tube [36.25]. For comparison, the interlayer spacing in graphite is 0.335 nm, suggesting relatively weak interaction between individual shells. This fact has been corroborated by studies of mechanical and electronic properties of CNTs.

The second type of carbon nanotubes takes the form of a single rolled-up graphitic sheet – a single-walled CNT (SWNT). When produced, their diameter distribution is relatively narrow, so they often bundle-up in the form of crystalline "ropes" [36.27], Fig. 36.4c and d, where single tubes are held together by van der Waals interactions.

There are several distinct classes of production methods. The earliest known is based on the cooling of carbon plasma. When voltage is applied between two graphitic electrodes in an inert atmosphere, they gradually evaporate and form plasma. Upon cooling, soot containing multiwalled nanotubes is formed [36.24]. If the anode is filled with catalysts, for example cobalt or iron particles, SWNT ropes form. Another way of producing CNTs from carbon plasma is by laser ablation of a graphitic target [36.27]. It is generally considered that these methods produce CNTs of higher quality, albeit in very small quantities and without the possibility of scale-up to industrial production. Other methods are based on chemical vapor deposition (CVD), the cat-

alytic decomposition of various hydrocarbons such as methane or acetylene mixed with nitrogen or hydrogen in the presence of catalysts [36.28]. This method offers the ability to control the growth of nanotubes by patterning the catalyst [36.29], and is therefore more suited to producing nanoscale structures with integrated CNTs. This method is also capable of producing CNTs in industrial quantities. Its main disadvantage is the higher concentration of defects that diminish their mechanical properties.

36.2.2 Measuring the Mechanical Properties of Carbon Nanotubes by SPM

Mechanical measurements of CNTs performed using AFM have confirmed theoretical expectations [36.30] of their superior mechanical properties. These experiments involve measuring the deformation caused by applying controlled force, resulting in the bending of immobilized carbon nanotubes in either the lateral [36.26] or the normal direction [36.31], and also the tensile stretching of CNTs fixed at both ends to AFM tips, observed using a scanning electron microscope [36.32].

It is not obvious that continuum mechanics and its associated concepts like the Young's and shear moduli and tensile strength should work on the nanoscale. In order to apply them, one should also define the "thickness" of the nanotube walls, in the frame of the continuous beam approximation. Most scientists working in this field use the value 0.34 nm, close to the interlayer sepa-

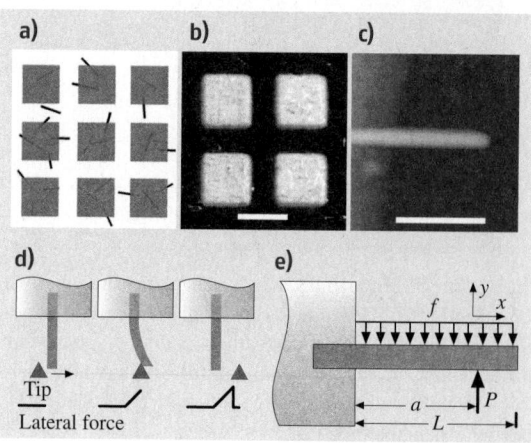

Fig. 36.5a–e Schematic of the experiment performed by *Wong* et al. [36.26]. (**a**) Nanotubes were dispersed on a substrate and pinned down by depositing SiO pads. (**b**) Optical micrograph showing the pads (*light*) and MoS$_2$ substrate (*dark*). The scale bar is 8 μm long. (**c**) AFM image of a SiC nanorod protruding from the pad. The scale bar is 500 nm long. The same experimental set-up was used when elastically deforming MWNTs. (**d**) The tip, shown as a *triangle*, moves in the direction of the arrow. The lateral force is indicated by the trace at the *bottom*. Before the tip contacts the beam, the lateral force remains constant and equal to the friction force. During bending, a linear increase in the lateral force with deflection is measured. After the tip has passed over the beam, the lateral force drops to its initial value and the beam snaps back to its equilibrium position. (**e**) Schematic of a pinned beam with a free end. The beam of length L is subjected to a point load P at $x = a$ and a friction force f (reproduced with permission from [36.26], © American Association for the Advancement of Science)

ration in graphite, as the thickness of a nanotube. When comparing different results, one must however bear in mind that in order to convert relatively precise force–displacement measurements into macroscopic quantities like the Young's and shear moduli, one has to introduce various geometrical factors, namely diameter and length. Even a small imprecision in their determination is very unforgiving, because the diameter d enters into equations for beam deformation as d^4 and length l as l^3, leading to large uncertainties in the final result.

The first quantitative measurement of the Young's modulus of carbon nanotubes was reported by *Wong* et al. [36.26]. They deposited MWNTs and SiC nanorods (similar in dimensions to MWNTs) on flat surfaces and bent them laterally. The MWNTs were first randomly dispersed on a flat surface of MoS_2 single crystals that were used because of their low friction coefficient and exceedingly flat surfaces. Friction between the tubes and substrate was further reduced by performing the measurements in water. The tubes were pinned on one side to this substrate by depositing an array of square pads through a shadow mask, Fig. 36.5a–c. AFM was used to find and characterize the dimensions of protruding tubes. The beam was deformed laterally by the AFM tip, until a certain deformation was reached where the tip would pass over the tube, allowing the tube to snap back to its relaxed position. Force–distance curves were acquired at different positions along the chosen beam, see Fig. 36.5d and e. The maximum deflection of the nanobeam was controlled to a certain degree by the applied normal load, and in this way tube breaking could be avoided or achieved in a controlled manner. The applied lateral load P is given in terms of lateral displacement y at the position x along the beam by the equation:

$$P(x, y) = 3EI\frac{y}{x^3} + \frac{f}{8}\left(x - 4L - 6\frac{L^2}{x}\right), \quad (36.2)$$

where E is the Young's modulus of the beam, I is the second moment of the cross-section, equal to $\pi d^4/64$ for a solid cylinder of diameter d, and f is the unknown friction force, presumably small due to the experimental design. The lateral force (Fig. 36.6a) was only known up to a factor of proportionality, because the AFM lever's lateral force constant wasn't calibrated for these measurements. This uncertainty and the effect of friction were eliminated by calculating the lateral force constant of the nanobeam:

$$\frac{dP}{dy} \equiv k = \frac{3\pi d^4}{64x^3}E \quad (36.3)$$

This is shown in Fig. 36.6b.

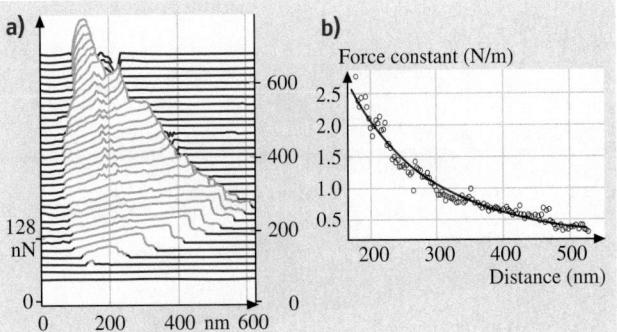

Fig. 36.6 (a) A series of lateral force–distance curves acquired by *Wong* et al. [36.26] for different positions along a MWNT. (b) The lateral spring constant as a function of position on the beam. The curve is a fit to equation (reproduced with permission from [36.26], © American Association for the Advancement of Science)

The mean value of the Young's modulus for the MWNTs was $E = 1.3\pm0.6$ TPa, similar to that of diamond ($E = 1.2$ TPa). Discontinuities in the bending curves were also observed for larger deformations, which were attributed to elastic buckling of the nanotubes [36.33].

In another series of experiments, *Salvetat* et al. measured the Young's modulus of isolated SWNTs and SWNT ropes [36.31], MWNTs produced using different methods [36.34], and the shear modulus of SWNT ropes [36.35]. The experimental set-up that enabled them to perform measurements on such a wide range of CNT morphologies involved measuring the vertical deflections of nanotubes bridging holes in a porous membrane.

In their measurement method, the CNTs were suspended in ethanol and deposited on the surface of a well-polished alumina (Al_2O_3) ultrafiltration membrane. Tubes adhered to the surface due to the van der Waals interaction, occasionally spanning holes, Fig. 36.7a. After a suitable nanotube had been found, a series of AFM images was taken under different loads. Each image corresponded to the surface (and the tube) under a given normal load. Linescans across the tube revealed the vertical deformation, Fig. 36.7b. For the range of normal loads applied, the deflection of a thin, long nanotube at the midpoint δ as a function of the normal force F can be fitted using the clamped beam formula [36.36]:

$$\delta = \frac{FL^3}{192EI}, \quad (36.4)$$

where L is the suspended length.

a)

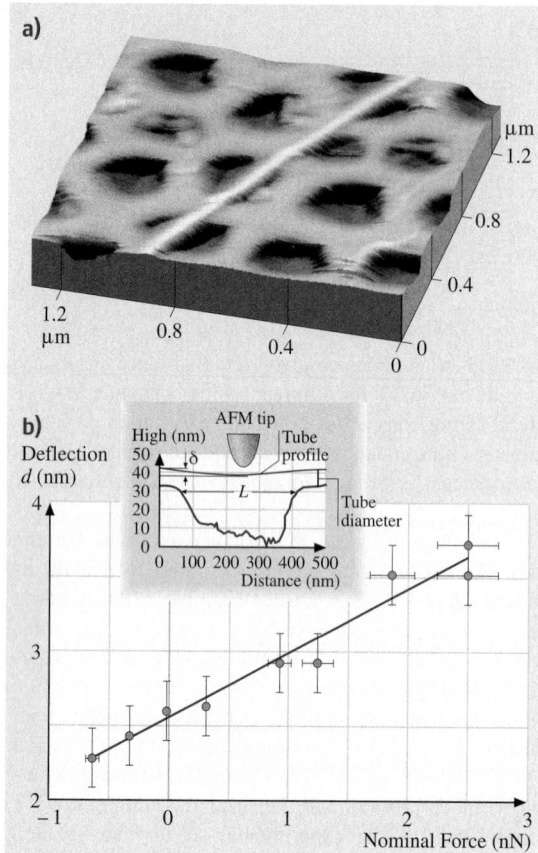

b)

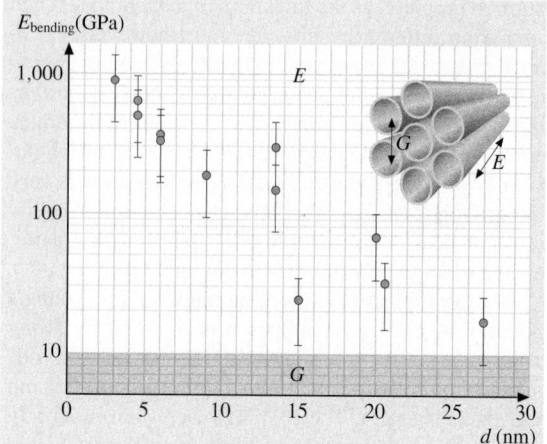

$E_{bending}(GPa)$

Fig. 36.8 Shear moduli of 12 SWNT ropes with different diameters. The measured $E_{bending}$ of thin ropes corresponds to E, while for thick ropes one obtains the value of the shear modulus G (reproduced with permission from [36.35], © American Physical Society 1999)

Fig. 36.7 (a) An AFM image of a CNT lying on a porous alumina filter [36.35]. **(b)** Dependence of the vertical deflection on the applied nominal force. The *inset* shows a comparison between linescans taken on the tube and over a hole [36.34] (reproduced with permission **(a)** from [36.35], © American Physical Society 1999, **(b)** from [36.34], © Wiley 1999)

The fitted line does not pass through the origin because the force acting on the nanotube is not equal to the nominal force alone; it contains a constant term arising from the attractive force between the AFM tip and the tube. The tube's deflection should also contain a term corresponding to the interaction between the tube and the substrate, but this is generally regarded as negligible.

This variable load imaging technique is advantageous when obtaining quantitative information, as one is then assured that the AFM tip is in the desired location when deforming the tube. Equation (36.4) is only valid if the tubes adhere well to the substrate, confirmed by the fact that the images reveal no dis-

placement of the parts of the tube in contact with the membrane.

Using this technique, a Young's modulus of 1 TPa was found for SWNTs. Values for MWNTs showed a strong dependence on the amount of disorder in the graphitic layers – an average value of $E = 870\,GPa$ was found for the arc discharge-grown tubes, while catalytically grown MWNTs, known to include a high concentration of defects, can have a Young's modulus as low as 12 GPa [36.34, 37].

An additional term must be added to the bending formula for the deflection of SWNT ropes, due to the influence of shearing between the tubes comprising the rope. The CNTs are held together via weak van der Waals interactions. Therefore, the ropes behave as an assembly of individual tubes rather than as a thick beam. The deflection can be modeled as a sum of deflections due to bending and shearing [36.36]:

$$\delta = \delta_{bending} + \delta_{shearing} = \frac{FL^3}{192EI} + f_s \frac{FL}{GA}$$
$$= \frac{FL^3}{192E_{bending}I} \,, \tag{36.5}$$

where f_s is a shape factor (equal to 10/9 for a cylinder), G the shear modulus and A is the beam's cross-sectional area. $E_{bending}$ is the effective bending modulus, equal to Young's modulus when the influence of shearing can be neglected (for thin, long ropes).

The Young's and the intertube shear moduli can thus be extrapolated by measuring the $E_{bending}$ of an ensem-

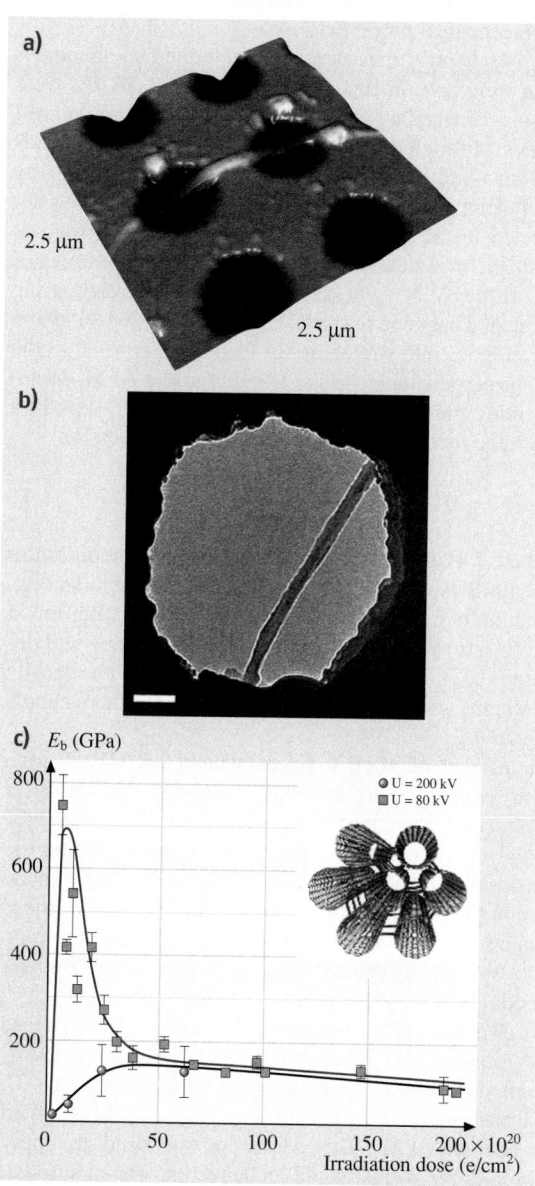

Fig. 36.10 (a) Three-dimensional rendering based on an AFM image of a SWNT bundle spanning two holes on silicon nitride membrane with a microfabricated hole array. Measurements were performed on the hole on the right. **(b)** TEM image of the same nanotube bundle. Scale bar 100 nm. **(c)** Behavior of the bending modulus of the SWNT bundle as a function of received e-beam dose. (After [36.42])

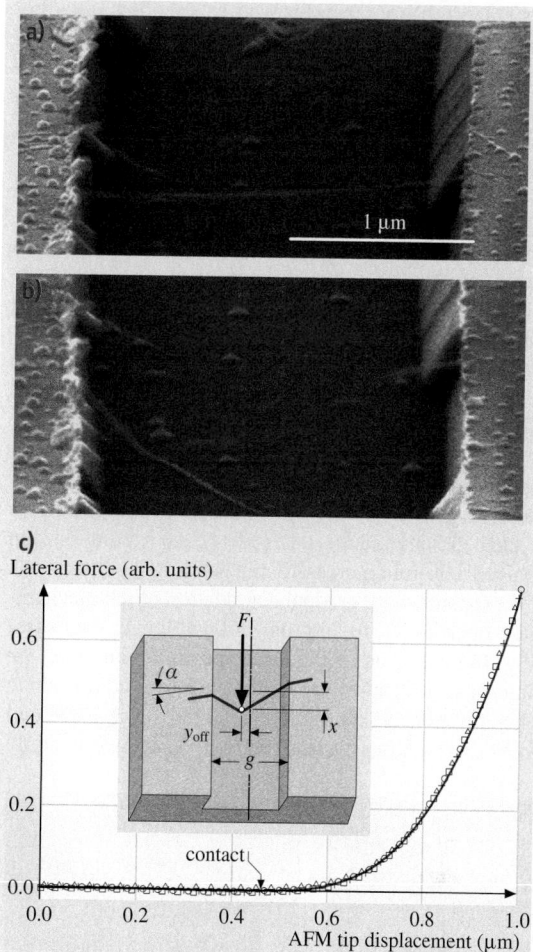

Fig. 36.9 (a) SEM image of a SWNT rope suspended over a trench in silicon before and **(b)** after being deformed past its elastic limit. **(c)** Lateral force on a single-wall nanotube rope as a function of AFM tip displacement (reproduced with permission from [36.41], © American Institute of Physics)

ble of ropes with different diameter-to-length ratios – for thin ropes one obtains the value of the Young's modulus, while for thick ones, $E_{bending}$ approaches the value of G, on the order of 10 GPa (see Fig. 36.8). This phenomenon of easy intertube sliding is also encountered in macroscopic objects, such as ribbons [36.38], cables [36.39] and centimeter-long strands [36.40], composed entirely of carbon nanotubes. These structures have mechanical properties that are much poorer than those of carbon nanotubes due to the weak connections between the individual nanotubes in them.

This problem can be solved by creating links between the nanotubes. In the ideal case, these links should be covalent carbon–carbon bonds, because of their high strength. *Kis* et al. demonstrated that stable links can form between nanotubes in bundles during electron beam irradiation [36.42]. SWNT bundles were deposited on silicon nitride membranes compatible with transmission electron microscopy imaging and containing microfabricated hole arrays. The use of silicon nitride membranes allowed mechanical AFM measurements, TEM observations and irradiation to be combined for the first time on the same, isolated SWNT bundle (Fig. 36.10a and b). Mechanical measurements were performed using an AFM by deflecting nanotube bundles deposited over holes in the membrane. The same nanotube bundle was placed in a transmission electron microscope (TEM), imaged with high resolution and irradiated with a focused electron beam. Measurements of the bending modulus taken after each irradiation step showed a 30-fold increase in the bending modulus upon low exposure, with gradual degradation of the mechanical properties due to amorphization at longer exposures (Fig. 36.10c). This method of depositing nanotubes on patterned, electron-transparent membranes could enable a host of related experiments in which TEM, SEM and AFM observations on nanotubes and nanowires could

be combined. Fabricating additional electrical contacts on these electron-transparent membranes would allow in situ measurements of electrical transport and electromechanical properties.

Mechanical measurements on individual nanotubes that were constituents of nanomechanical devices were also performed by several groups. *Walters* et al. pinned ropes of single-walled nanotubes beneath metal pads on an oxidized silicon surface and released them by wet etching, see Fig. 36.9a [36.41]. The SWNT was deflected in the lateral direction using an AFM tip, see Fig. 36.9b. As the suspended length is on the order of 1 μm, the SWNT rope can be modeled as an elastic string stretched between the pads. Upon deformation, all of the strain goes into stretching. In the simple case of a tube lying perpendicular to the trench and the AFM tip deforming the tube in the middle, the force F exerted on the tube by the AFM tip is given by the expression:

$$F = 2T \sin\theta = 2T \frac{2x}{L} \approx \frac{8kx^3}{L_0} \, ,$$

where T is the tension in the string, L_0 is its equilibrium length, k is the spring constant, and x is the lateral deflection in the middle. Using this set-up, they deformed SWNT ropes to a maximal strain of 5.8±0.9% and determined a lower bound of 45±7 GPa for the tensile strength, assuming a value of 1.2 TPa for the Young's modulus.

Kim et al. used a set-up where the SWNT rope was embedded in metallic electrodes deposited on a silicon substrate coated with a sacrificial layer of poly(methylmethacrylate) (PMMA), thus avoiding the exposure of SWNTs to wet etchants. In their experiment, the tube can also be modeled as an elastic string. Using an AFM tip, they vertically deformed the rope and obtained an estimate of $E = 0.4$ TPa for the Young's modulus of a SWNT rope.

Williams et al. measured the mechanical response of torsional or "paddle" oscillators containing multi-walled carbon nanotubes as spring elements [36.43]. Measurements were performed using an AFM mounted inside an SEM. The AFM tip was used to apply force to the paddle while its deflection was monitored, Fig. 36.11a–c, resulting in force–distance curves like the one in Fig. 36.11d. Values of single-shell shear moduli were in the 210–830 GPa range. Repeated mechanical stressing of the devices also showed a significant increase in stiffness followed by saturation after several hundred deflections.

Finally, the first direct measurements of the elastic properties of CNTs that did not rely on the

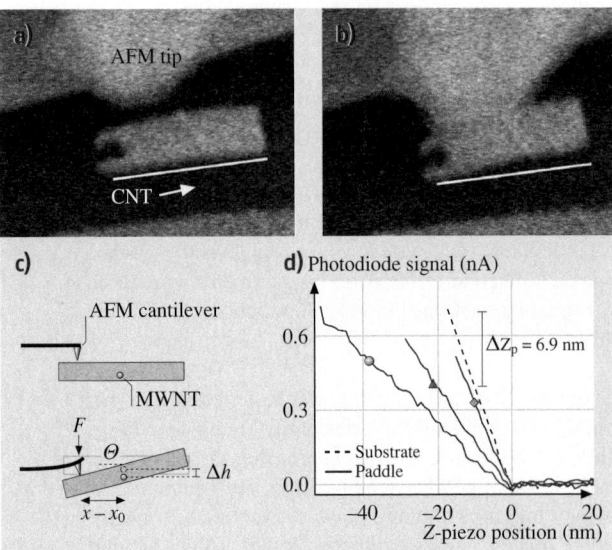

Fig. 36.11 (a) SEM image of an AFM tip close to the paddle of the torsional oscillator. **(b)** The device during measurement. **(c)** A schematic of the measurement procedure. **(d)** Force–distance curves on the substrate and at different positions along the paddle. (After [36.43])

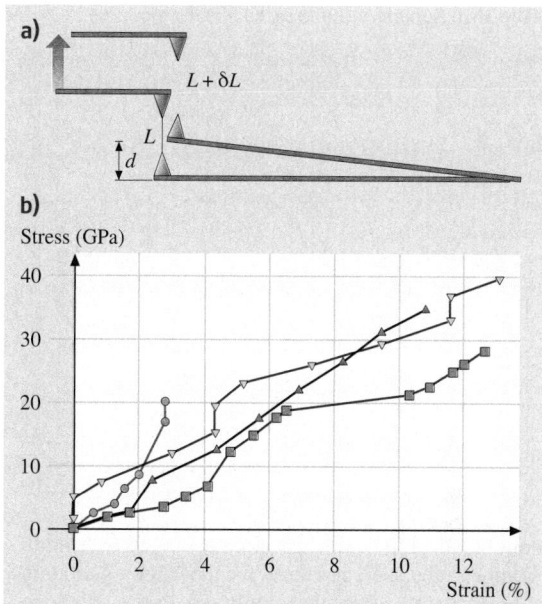

Fig. 36.12 (**a**) SEM image of a MWNT mounted between two opposing AFM tips. (**b**) A close-up of the tip region (reproduced with permission from [36.44] © American Association for the Advancement of Science 1999)

beam or stretching string set-up involved deforming MWNTs [36.44] and SWNT ropes [36.32] under axial strain. This was achieved by identifying and attaching opposite ends of MWNTs or SWNT ropes to two AFM tips, all inside a SEM. One tip was integrated with a rigid cantilever with a spring constant of more than 20 N/m and the other was compliant (0.1 N/m), Fig. 36.12. The rigid lever was then driven using a linear piezo motor. At the other end, the compliant lever bent under the applied tensile load. The deflection of the compliant cantilever – corresponding to the force applied to the nanotube – and the strain on the nanotube were measured simultaneously. The force F is calculated as $F = kd$, where k is the spring constant of the flexible AFM lever and d its displacement in the vertical direction. The strain of the nanotube is $\delta L/L$, Fig. 36.13. From the stress–strain curves obtained in this fashion, Fig. 36.13b, Young's moduli ranging from 270–950 GPa were found. Examinations of the same broken tubes inside a TEM revealed that nanotubes break with the "sword in sheath" mechanism, where only the outer layer appears to carry the load. After it breaks, the inner shells are pulled out. An average bending strength of 14 GPa and axial strengths of up to 63 GPa were found.

Firm attachment of the nanotubes to AFM tips was ensured by the deposition of carbonaceous material achieved by concentrating the electron beam in the contact area [36.45].

These measurements of the elastic properties of carbon nanotubes can be summarized in Table 36.1.

Before comparing them, it should be noted that the absolute values of these mechanical constants have relatively large uncertainties, because of the huge influence of the precision of determining tube diameters and lengths on the final result. Also, the cited values

Fig. 36.13 (**a**) The principle of the experiment performed by *Yu* et al. [36.32, 44]. As the rigid cantilever is driven upward, the lower, compliant cantilever bends by an amount d and the nanotube is stretched by δL. As a result, the nanotube is strained by $\delta L/L$ under the action of force $F = kd$, where k is the elastic constant of the lower AFM lever. (**b**) Stress versus strain curves for different individual MWNTs (reproduced with permission from [36.44] © American Association for the Advancement of Science 1999)

represent averages obtained for several tubes, with the exceptions of the lowest value given for catalytically grown MWNTs by *Salvetat* et al. [36.31], a single value for an individual SWNT rope from *Kim* et al. [36.46], and the lowest and highest values from *Yu* et al. [36.32] and *Williams* [36.43]. One also has to bear in mind that concepts like Young's and shear moduli and tensile strength were introduced to describe macroscopic and continuous structures. Therefore, applying them to describe mesoscopic objects like nanotubes has its limitations.

The methods presented above are all "single-molecule" methods, in the sense that they measure the properties of individual objects. Results from a single experiment therefore represent the properties of a particular object and will differ from case to case (due to, for example, defects coming from production and purification, or more prosaic reasons like experimental errors). In order to perform comparisons it is therefore

Table 36.1 Summary of the mechanical properties of carbon nanotubes, measured using SPM methods

Young's modulus E (GPa)	Shear modulus G (GPa)	Tensile strength σ (GPa)	Nanotube type	Deformation method	Reference
1300±600	–	–	MWNT arc grown	Lateral bending	[36.26]
1000±600	–	–	SWNT	Normal bending	[36.31]
1000±600	≈ 1 GPa (intertube)	–	SWNT ropes	Normal bending	[36.35]
870±400	–	–	MWNT arc-grown	Normal bending	[36.31]
12±6	–	–	MWNT catalytic	Normal bending	[36.31]
–	≈ 1 GPa (intertube)	–	SWNT	Normal bending	[36.42]
–	–	45±7	SWNT rope	Lateral bending	[36.41]
400	–	–	SWNT rope	Normal bending	[36.46]
–	210–830	–	MWNT arc-grown	Torsional	[36.43]
1020	–	30	SWNT ropes	Tensile loading	[36.32]
270–950	–	11–63	MWNT arc-grown	Tensile loading	[36.44]

more practical to deal with values averaged over multiple tubes. The average values for the Young's modulus of high-quality tubes are, within the limits of experimental error, all around 1 TPa, close to that of diamond (1.2 TPa), while the tensile strength can be 30 times higher than that of steel (1.9 GPa). Catalytically grown MWNTs have Young's moduli that can be as low as 12±6 GPa, which is definitely disappointing and shows that the production method plays an important role in the quality of carbon nanotubes from the point of view of their mechanical properties.

Future improvements in large-scale production and processing are therefore necessary before they are applied as the ultimate reinforcement fibers. Even so, progress in measuring the mechanical properties of CNTs has and will continue to be closely related to, and often motivate progress in, nanoscale manipulation in general.

36.2.3 Microtubules and Their Elastic Properties

Microtubules are a vital biological nanostructure, similar in dimension and shape to carbon nanotubes. In fact, the initial name given to carbon nanotubes by their discoverer (S. Iijima) was "microtubules of graphitic carbon". From a structural point of view, they are much more complicated. They self-assemble in buffers maintained at a physiological temperature of 37 °C from protein subunits, α- and β-tubulin, each of which has a molecular weight of 40 kDa. These protein subunits are bound laterally into protofilaments. These protofilaments are in turn arranged into the shape of a hollow cylinder with an

external diameter of 25 nm and an internal diameter of 15 nm. Microtubules are remarkable: inside living cells, their length constantly fluctuates – they can even completely disassemble, and consume energy in the form of GTP.

Together with actin and intermediate filaments, they constitute the cellular cytoskeleton. In addition, they perform various unique vital functions: they act as tracks along which molecular motors move, they help pull apart chromosomes during cell division, and form bundles that propel sperms and some bacteria. All of these roles are determined by their structure and mechanical properties. However, even after more than a decade of research, there are still large discrepancies between values of Young's modulus reported for them in the literature. Several methods have been applied, like bending or buckling the microtubules using optical tweezers [36.47], hydrodynamic flow [36.48], thermally induced vibrations or shape fluctuations [36.49], buckling in vesicles [36.50], or squashing with an AFM tip [36.51, 52] – yielding results ranging from 1 [36.52] to several GPa [36.47].

Since microtubules are geometrically similar to nanotubes, *Kis* et al. [36.53] used the suspended tube configuration. Microtubules were deposited on porous membranes and AFM images were acquired under different nominal normal loads. All of the measurements were performed in liquid and at controlled temperatures in order to prevent the degradation of proteins, so the substrate had to be functionalized in order to ensure good adhesion between the tubes and the support, silicon with a layer of PMMA. Slits were prepared in PMMA using electron beam lithography, providing the ability to measure the elastic responses of individual microtubules lying over four different-sized holes 80–200 nm across and 400 nm deep, see Fig. 36.14.

Clearly, the results are dependent on the hole diameter, which could only result from a shear component within the microtubules. Simplifying (36.5), the deformation of microtubules can be modeled as:

$$\frac{1}{E_{\mathrm{bending}}} = \frac{1}{E} + \frac{10}{3} \frac{D_{\mathrm{ext}}^2 + D_{\mathrm{int}}^2}{L^2} \frac{1}{G} \,,$$

where and D_{int} represent the external and internal MT diameters. The shear modulus can be calculated from the slope of the fit to the plot of $1/E_{\mathrm{bending}}$ versus the $(D_{\mathrm{ext}}^2 + D_{\mathrm{int}}^2)/L^2$, and the inverse Young's modulus will be equal to the intercept on the y-axis. In this way, a shear modulus of $G = 1.4 \pm 0.3$ GPa and a lower limit of $E = 150$ MPa were obtained simultaneously from measurements on an individual microtubule.

This anisotropy comes from the fact that microtubules are constructed from monomers that are strongly bound in the longitudinal direction, along single protofilaments. The link between neighboring protofilaments is much weaker. Therefore, microtubules must be considered to be anisotropic beams, with an E_{bending} that depends on the scale on which they are deformed [36.53]. This is, in all respects, analogous to the situation with SWNT ropes that are built of stiff individual SWNTs, held together in bundles by weak van der Waals interactions. This conclusion may explain the large discrepancies between the values of Young's moduli reported in the literature over the past ten years by inadequate modeling.

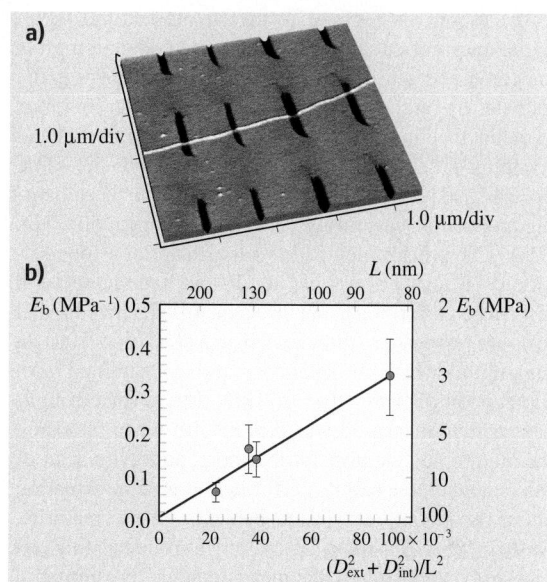

Fig. 36.14 (a) A pseudo 3-D rendering of a single microtubule suspended over four different-sized holes in PMMA. **(b)** The variation of the E_{bending} with length was used to determine the shear and Young's moduli for the microtubule displayed in **(a)** (reproduced with permission from [36.53] © American Physical Society 2002)

As for carbon nanotubes, measuring the mechanical properties of microtubules can provide valuable insights into their structure and provide a deeper understanding of the functioning of these remarkable structures.

36.3 Scanning Nanoindentation as a Tool to Determine Nanomechanical Properties of Biological Tissue Under Dry and Wet Conditions

In the following section, nanoindentation is discussed as a tool for determining nanomechanical properties. This method was developed in the early 1980s, and it evolved from traditional Vicker hardness-testing devices. The latter is based on the concept that a pyramidal tip is pushed into a material with a known force. After the test, the size of the remaining imprint is measured under an optical microscope. The ratio of the force employed to the imprint area after the load is removed is defined as the hardness, which represents the mean pressure that the material can resist. Unfortunately, this mechanical parameter is a complex combination of elastic and postyield properties and can't be easily interpreted at the level of continuum mechanics because the nature of the

postyield deformation mechanisms depend on the tested material. This point raised important concerns about determining elastic constants such as the Youngs modulus from an indentation test. The integration of a displacement transducer with high sensitivity was necessary to provide continuous acquisition of the applied load and the resulting indentation depth. Nanoindentation represents the state of the art of this development allowing indentation tests on the nanometer scale.

36.3.1 Scanning Nanoindentation

Using this continuous force-displacement data, it is possible to quantify an elastic modulus for material volumes in the submicron regime. To investigate features down to the micrometer scale, classical nanoindentation tools are combined with an optical microscope to position the indenter tip on the region of interest. However, in order to defeat the limits of optical resolution, indentation and scanning probe microscopy were combined; the same tip was used to make nanoscale control of the indenting tip position easier (Hysitron, Inc., Minneapolis, MN USA). This instrument allows the material's topography to be scanned in AFM mode and nanoindentation tests to be performed using the same tip. The mechanical tests are restricted to the scanned area ($100\,\mu m \times 100\,\mu m$ maximum) where the indenter can be positioned with a precision of better than 100 nm. The surface roughness can be measured which is helpful when choosing the area to be indented. Furthermore, an in situ scan of the indented area can be performed to give information about the pile-up or sink-in behavior of the material. *Kulkarni* and *Bhushan* [36.54, 55] introduced this device and demonstrated that measurements on aluminum and silicon were similar to the results from nonscanning indentation systems.

36.3.2 Application of Scanning Nanoindentation

The development of the scanning nanoindentation (SN) technique has led to a variety of studies, many of them characterizing thin layers. Applications range from the mechanical characterization of corrosion-free film grown on single-crystalline iron [36.56] to indentation and microscratch tests on Fe-N/Ti-N multilayers [36.57]. One author included nanoindentation when discussing two different techniques for electrochemically depositing thin Ni-P layers on pure Ni [36.58]. *Rar* et al. [36.59] reported studies on the growth of thin gold layers on native oxides of silicon while *Wei* et al. [36.60] tested the nanomechanical properties of thin CrN_x films deposited on silicon.

Ott et al. [36.61] conducted investigations focusing on wear-resistant Ti-B-N coatings.

Schiffmann [36.62] determined the friction coefficient and the plastic properties of diamondlike carbon coatings deposited on glass using indents.

Zhao [36.63] tested a set of single crystalline and polycrystalline metallic samples and discussed indentation size effects. They concluded that indentation size depended on characteristic material properties, like the Burgers vector, the shear modulus and an intrinsic material length scale.

Wang [36.64] discussed the pile-up behavior of the material around the indented area. This study was performed on copper single crystals and includes finite element simulations.

Studies of heterogeneous materials clearly demonstrate the advantages of the surface scanning available. *Malkow* and *Bull* [36.65] investigated the elastic/plastic behavior of carbon nitride films deposited on silicon. They determined the load-on and load-off hardness, the latter accessed by scanning the indent impression. *Shima* et al. [36.66] studied silicon oxynitride films on pure silicon and found the hardness as a function of the deposition temperature. *Göken* et al. [36.67] used the high positioning precision to characterize individual lamellae of a TiAl alloy that consists of a two-phase structure, and to study the mechanical properties of nanosized precipitates of nickel-based super-alloys [36.68]. Performing in situ electrochemical treatment of iron single crystals, *Seo* et al. [36.69] used SN to study the variation in the remaining imprint shape with time.

In the field of biomechanics, work was done on bone tissue in wild-type and gene-mutated zebra fish. Characterization of the residual indentations in AFM mode supported the statement that gene mutation can change bone brittleness [36.70]. Other investigations focused on the demineralizing effects of soft drinks on tooth enamel by studying changes in elastic properties and surface topography [36.71]. *Habelitz* et al. [36.72] and *Marshall* et al. [36.73] characterized the junction between human tooth enamel and the mechanically different dentin. *Chang* [36.74] determined the nanomechanical properties of pig alveolar bone surrounding a titan implant. After one month healing they tested the nanomechanics of the bone as a function of distance from the implant–bone interface.

Hengsberger et al. [36.75] took advantage of the nanometer positioning capability of the SN to investigate the elastic and plastic properties of individual human bone lamellae. Related work is presented in greater detail below to demonstrate the use and benefits of the SN technique in these examples.

36.3.3 Example 1: Study of the Mechanical Properties of Bone Lamellae using SN

Before we discuss the nanomechanical properties of bone lamellae it is useful to review the structure of them on the macroscopic level. Figure 36.15 presents the hierarchy of human bone tissue for the example of the femoral neck. The outer shell consists of cortical bone while the porous trabecular bone structure gives inner support. At the next lower level, both bone types show Bone Structural Units (BSU [36.76]) that consist of some tens of lamellae. In compact bone, the lamellae are arranged concentrically, while in trabecular bone these lamellae are parallel. The nomenclature "BSU" was motivated by its underlying cellular process. The formation of such an individual structural unit occurs within a spatially and timely coordinated cellular process. The optical contrast shows two alternating types of lamellae, thin lamellae that appear dark and thick lamellae that appear bright. These bone lamellae show widths ranging from 1 to 3 μm for thin and 2 to 4 μm for thick lamellae [36.75].

At the next level, the bone matrix consists of a complex collagen and mineral nanostructure. It is still not entirely understood what type of variations in the collagen/mineral meshwork are responsible for the lamellar structure.

The mechanical properties at these different levels of bone hierarchy are increasingly well understood. However, little is known about how the macroscopic mechanical properties of the whole bone relate to its nanomechanical properties. The possibility that factors such as fracture risk might be better understood with an analysis of small bone volumes motivates the application of nanoindentation in this field.

A set of nanoindentation studies is available (mainly using the SN device combined with an optical microscope) that presents the intrinsic mechanical properties of bone tissue. Among recent work, differences were reported between donors, anatomical sites, BSU and thin and thick lamellae [36.75, 77–80]. Due to the technical constraints of this sensitive nanomechanical device, the majority of these studies present dehydrated or dried tissue properties measured at ambient temperature [36.81–84]. However, removal of the water content may lead to anisotropic shrinking of the matrix that creates microcracks and alters the mechanical properties of the bone constituents. For an accurate characterization of in vivo properties, the nanomechanical tests should therefore be done under physiological conditions.

There have been a few attempts to characterize bone properties under physiological conditions [36.77,79,85]. This was achieved by studying fresh bone samples that are kept moist with a thin layer of liquid (less than a hundred microns) on the surface, or with subsurface water irrigation. However, local evaporation of the thin liquid layer may have led to indents at areas that were partially dried out during the test. One possible solution to such

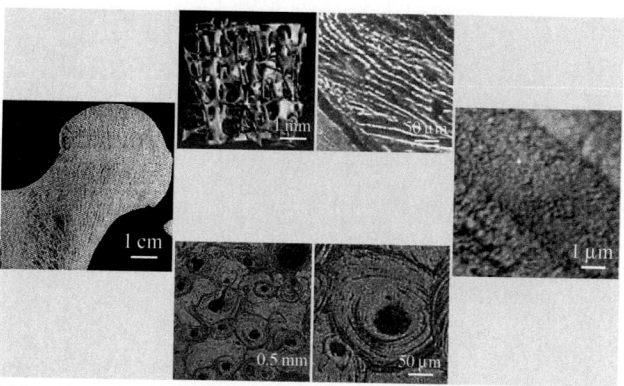

Fig. 36.15 Hierarchy of the human femoral neck. (From *left* to *right*.) The *far left figure* shows a cut through a frontal plane of the femoral neck. The outer shell consists of compact bone while the inside is made of the spongier trabecular bone. The *left pair* of images show the trabecular structure (*top*) and a transverse cut of the compact shell that shows vascular canals (*bottom*).
The *right pair* of images shows packets of trabecular bone lamellae (*top*) and a single osteon (*bottom*), a vascular canal surrounded by concentric lamellae. Packets of trabecular bone lamellae and osteons represent a structural unit (BSU). Note the alternate bright (*thick*) and dark (*thin*) lamellae.
The figure *far right* shows three bone lamellae, structures that are similar for trabecular and compact bone. The bone matrix within the lamellae mainly consists of collagen fibers and hydroxyapatite crystals

local drying might be to conduct measurements where the indentation tip and tissue sample are both fully immersed in a liquid cell and simultaneously heated up to body temperature.

However, practical limitations on temperature stability and inaccuracy in contact force detection due to liquid on the surface make such measurements extraordinarily difficult. Therefore, for statistically powerful studies, drying the samples seems unavoidable.

The objective of this study was therefore to use SN to determine the effect of drying on the stiffness of single bone lamellae. The goal was to determine a conversion factor that allows dry tissue properties to be recalculated into their in vivo properties.

For this purpose, we measured an identical set of lamellae selected from human trabecular and compact bone, first under physiological and then under dry conditions.

Experimental Set-Up and Technical Features

Figure 36.16 demonstrates an optical picture and a sketch of the combined AFM and nanoindenter de-

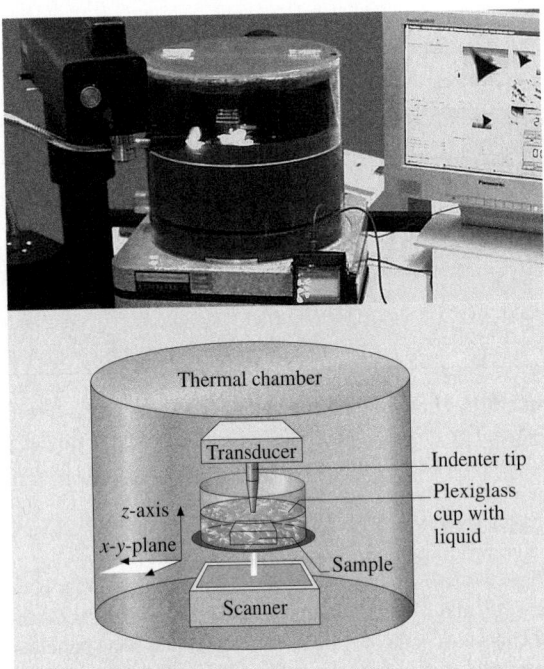

Fig. 36.16 Optical picture (*top*) and sketch (*bottom*) of the scanning nanoindentation device for measurements performed under physiological conditions. The sample was installed in a plexiglass cup so that liquid could be added. The entire instrument was heated in a custom-made thermal chamber

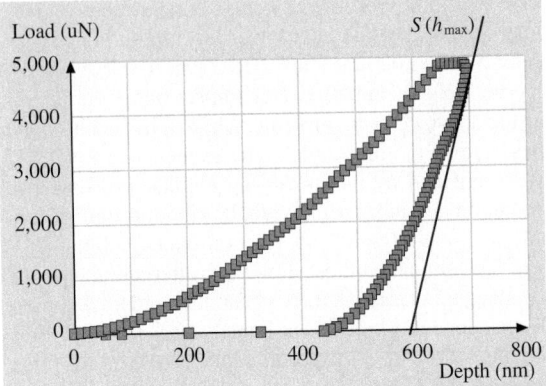

Fig. 36.17 Force–displacement curve obtained during a nanoindentation test consisting of three parts (see text). Hardness was determined at maximum load. The elastic indentation modulus was calculated using the slope at the point of initial unloading $S(h_{max})$

vice (Hysitron Inc., Minneapolis, MN). The transducer consists of a three-plate capacitor on whose central plate a Berkovich (three-sided pyramid) diamond tip is mounted. The transducer provides a contact force feedback between the tip and the sample surface. The sample is mounted on a piezoelectric scanner that allows movement in the x, y and z-directions. During x, y-surface scanning, the piezoelectric scanner keeps this feedback signal constant by correcting the z-height. The movement of the piezoelectric scanner therefore describes the negative surface of the constant contact force (this is commonly called an AFM scan). For the liquid cell tests, the sample was glued into a Plexiglas cup for the addition of a liquid layer several millimeters high. A commercially available liquid cell tip (Hysitron Inc.) was used, which contains an additional shaft (approximately 300 µm in diameter and 5 mm in length) used to protect the transducer from the fluid. A small layer of latex was placed between the sample holder and the magnetic stick of the piezoelectric scanner to protect the latter from liquid. The additional shaft of the inden-

ter tip and the latex layer represent elastic components that increase the machine compliance ($C_m = 7$ nm/mN instead of 3 nm/mN). This variable corrects for the deformation of all machine components while indentation data are recorded and can be determined from a tip shape calibration curve (Hysitron Inc.). The nanoindentation device was installed in a custom-made thermal chamber to allow the sample to be heated to 37 °C. Note that the increased temperature and humidity also changes the value of the electrostatic force constant (EFC). The latter corrects the force due to the springs that support the central plate on which the indenter tip is mounted. The EFC can be calibrated by performing out-of-contact indents and varying this value until a zero line in the force–displacement curve is measured.

Using an image acquired in AFM mode, the indentation area can be selected with a high spatial resolution (< 0.1 µm). Figure 36.17 shows force–displacement data for a typical nanoindentation curve. In the first step, the tip is pressed into the material, resulting in indistinguishable elastic and plastic deformation. Then the tip is held at maximum force, resulting in creep of the material under the tip. In a third step, unloading is performed, which leads to elastic recovery of the material. Typically each indent requires between 15 s and several minutes, representing a compromise between a desired quasistatic strain rate and the thermal drift of the instrument (possibly below 0.1 nm/s). This device is load-controlled and linearly increasing and decreasing loading protocols were therefore applied. The loading rate corresponds to

$$\frac{dP}{dt} = \frac{P_{max}}{T} \qquad (36.6)$$

where P_{max} is the maximum load and T is the (un)loading time.

The elastic constants of the sample were determined using the unloading part (step 3) of the nanoindentation curve. Based on analytical work by *Sneddon* [36.86], *Oliver* and *Pharr* [36.87,88] derived the following equation for an indenter of revolution that is pressed into an isotropic elastic material:

$$S(h_{max}) = \frac{dP}{dh}(h_{max}) = \frac{2}{\sqrt{\pi}} E_r \sqrt{A_c(h_{max})} \,.$$

$$(36.7)$$

Pharr et al. [36.88] showed that this equation is a good approximation for a Berkovich indenter tip. P represents the applied load, $S(h_{max})$ is the derivative of the unloading curve at the point of initial unloading h_{max}. This variable is determined by fitting typically 40% to 95% of the unloading curve to avoid the influence of viscoelastic effects at initial unloading when a singularity in the strain rate occurs. $A_c(h)$ is the contact area over which the material and the indenter are in instantaneous contact. The latter function must be calibrated by performing indents with increasing depth in a standard material, typically fused silica. The latter has a known reduced modulus of $E_r = 69.9$ GPa that allows the contact area $A_c(h)$ to be calibrated using the measured contact stiffness $S(h)$ and (36.7). The reduced modulus E_r combines the deformation of the material and the diamond tip as follows [36.89]:

$$\frac{1}{E_r} = \frac{1 - \nu_{specimen}^2}{E_{specimen}} + \frac{1 - \nu_{tip}^2}{E_{tip}} \,. \qquad (36.8)$$

We use the indentation modulus

$$E_{ind} = \left(\frac{1}{E_r} - \frac{1 - \nu_{tip}^2}{E_{tip}}\right)^{-1} \qquad (36.9)$$

that can be calculated with the reduced modulus and the elastic properties of the diamond indenter tip $\nu_{tip} = 0.07$ and $E_{tip} = 1140$ GPa. This variable is combined with

$$E_{ind} = \frac{E_{specimen}}{1 - \nu_{specimen}^2}, \qquad (36.10)$$

the local Young's modulus $E_{specimen}$ and the Poisson ratio $\nu_{specimen}$ of the specimen.

It is important to note that this theory assumes an isotropic material. For an anisotropic material the indentation modulus represents an average of the elastic properties in all directions. The latter strongly depends on the geometry of the indenter. For indents with a blunt indenter (such as the Berkovich, with an opening angle of 143°), non-negligible deformations will occur in the plane perpendicular to the loading direction. In this context it is useful to present an approximation of the volume that is tested by the indentation test.

The stress field generated by the indentation process is heterogeneous and leads to plastic deformation and damage in the vicinity of the tip. Using Hertz's theory [36.89], the spatial dependence of the stress components during indentation can be estimated by considering the elastic contact of a spherical indenter with a semi-infinite half space. In the direction of loading (z-axis), the stress component below the indenter decreases according to

$$\frac{\sigma_{zz}}{p_0} = -\left(1 + \frac{z^2}{a^2}\right)^{-1}, \qquad (36.11)$$

where p_0 indicates the maximum pressure below the indenter and a the contact radius.

In a horizontal plane at the surface ($z = 0$), the radial and circumferential components of the stress field next to the contact area obey

$$\frac{\sigma_{rr}}{p_0} = -\frac{\sigma_{\theta\theta}}{p_0} = \frac{(1 - 2\nu)\, a^2}{3r^2} \qquad (36.12)$$

where θ and r are the cylindrical coordinates of the periphery of the indenter and ν is the Poisson ratio. For $\nu = 0.3$, the stress field components reach their 10%-boundary, defined by $\sigma_{zz} = -0.1 p_0$ and $\sigma_{rr} = -\sigma_{\theta\theta} = 0.1 p_0$ at a depth of $z \cong 3a$ or in the lateral direction at $r \cong \frac{2}{\sqrt{3}} a$. For a Berkovich tip, the ratio between maximum indentation depth and contact radius is approximately $a \cong 3h_{max}$. The mechanical properties measured by nanoindentation therefore correspond to a semiellipsoidal volume extending to about nine times the indentation depth ($z \cong 3a \cong 9h_{max}$) employed in the vertical direction (z) and about seven times this same depth ($2r \cong \frac{4}{\sqrt{3}} a \cong 7h_{max}$) in the radial direction (r).

The load-on hardness is determined with

$$H = \frac{P(h_{max})}{A_c(h_{max})} \qquad (36.13)$$

as the ratio of the maximum load to the ("load-on") contact area at maximum load. This is different from the Vickers hardness, where the contact area is characterized by the remaining imprint after load removal. Differences between nanoindentation and Vickers therefore occur for materials with non-negligible elastic recovery. After unloading, such materials may expose an imprint that is smaller than the area of contact at maximum load. Since the SN tool can be used to image the remaining imprint after unloading, hardness values obtained using both definitions may thus be compared.

For fused silica, the literature provides a hardness range of between 8.3 and 9.5 GPa, while an intercomparison of SN users showed an average of 8.96 GPa (Surface, Hückelhoven). Conveniently, the hardness value also provides a way to check the area function of the tip, which is based on the reduced modulus of the calibration material.

Tests of Bone Tissue
Under Pseudo-Physiological Conditions

Samples containing trabecular and compact bone lamellae were dissected from the medial part of the femoral neck of an 86 year old female. After embedding in PMMA, the samples were polished with successive grades of carbide paper and finished with 0.05 μm alumina solution. Polishing represents an important preparation step since the here-employed theory assumes an infinitely flat surface. The mean surface roughness of the indented area should therefore be much less than the indentation depth employed. Unfortunately, no objective criteria that determines the maximum allowable surface roughness as a function of the indentation depth has been formulated so far.

Thick and thin lamellae of trabecular and compact bone were first characterized under physiological conditions: fully wet and at 37 °C. Then the specimens were dried for 24 h at 50 °C and identical tests were carried out again but under dry conditions.

In both cases, 16 indentation tests were performed to maximum depths of 100 nm and 500 nm. Each test consisted of 10 s loading, 10 s holding and 10 s unloading. The maximum allowable thermal drift was set 0.1 nm/s. To avoid proximity effects of neighboring indents, adjacent testing areas in identical lamellae were chosen after changing the testing conditions.

The tests performed under fully wet conditions at body temperature were found to be very sensitive to thermal stability. The nanoindentation device was heated for several days to reach the stable thermal conditions of the instrument's components before the sample could be installed. The electrostatic force constant (EFC) was checked daily before beginning data acquisition. The approach of the indenter tip in the liquid environment was performed using a contact force of 7 μN. However, this value is offset by approximately 1 μN per mm of water penetration due to the Archimedes force acting on

Fig. 36.18 Surface topography of a trabecular bone structural unit which shows the lamellar structure. Thick, (*bright*) lamellae correspond to the tops in the topography. Note that the two holes are not indents; they are ellipsoidal lacunae that embed bone cells

the fluid cell tip. Further effects such as water surface tension may also exert a repulsive force on the tip.

AFM scans (as in Fig. 36.18, which shows the topography of trabecular bone lamellae under dry conditions) allowed the two lamella types to be identified. Thick lamellae correspond to the tops (bright contrast) and thin lamellae to the valleys (dark contrast), where the surface relief results from preferential polishing.

Indentation Modulus of Bone Tissue

Under wet conditions, trabecular bone showed a mean indentation modulus of 12.5±4.0 GPa when the data for both lamella types and indentation depths were combined. Under dry conditions, the results showed a mean of 19.6±2.6 GPa, an increase of 57%. The mean stiffness of lamellae of compact bone increased by 76.5% from 14.9±4.5 GPa to 26.4±3.8 GPa.

These conversion factors between dried and fully wet tissue properties should be compared with other studies. *Rho* [36.85] reported an increase in the indentation modulus of bovine compact bone of 15.8%. Our study gives a change of +76.5% for human compact bone. This high discrepancy may be attributed to different preparation and testing protocols. Rho tested the bone samples at ambient temperature while they were kept moist by a thin film of deionized water. In our study the tissue properties were determined under fully wet conditions and at body temperature. Furthermore, Rho dried for 14 d at ambient temperature, while in our study the drying process occurred for 24 h at 50°. These points may explain the different relative changes in mechanical properties we detected compared to Rho.

Table 36.2 Absolute and relative changes in indentation moduli with respect to initial values measured under wet conditions. Note the greater increase for the thin lamellae after drying. Indents to a depth of 100 nm represent properties of single lamellae while 500 nm indents are also influenced by neighboring lemellae

Indentation modulus	Lamella	Indentation depth (nm)	Wet (GPa)	Dry (GPa)	Relative change (%)
Trabecular bone	Thin	100	11.6±4.1	20.5±4.3	+76
		500	9.8±1.9	18.1±3.2	+84
	Thick	100	12.4±5	19.1±1.11	+54
		500	15.6±1.3	19.0±1.1	+22
Compact bone	Thin	100	10.4±0.2	23.4±3.6	+124
		500	13.9±2.0	27.1±3.1	+95
	Thick	100	19.9±4.8	27.9±4.6	+40
		500	15.5±2.6	23.0±0.8	+49

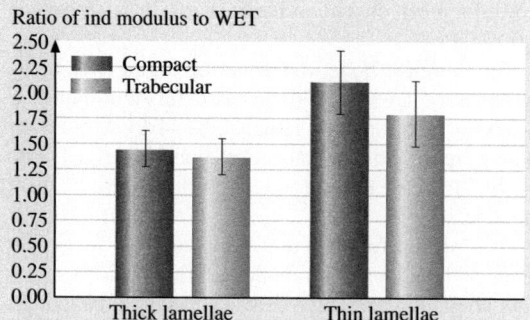

Fig. 36.19 Relative increase in indentation modulus for compact and trabecular bone lamellae after drying. The results are normalized with respect to their initial values under wet conditions. Note that the thin lamellae are more affected by drying for both types of bone

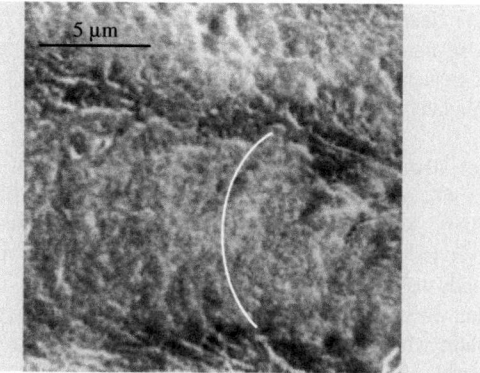

Fig. 36.20 SEM scan of bone lamellae adjacent to the spot characterized by nanoindentation. This SEM scan lends support to the bone lamellation model that is based on a smooth change in the orientation of the collagen fibers [36.90]

Figure 36.19 presents the indentation moduli (combining both depths) normalized with respect to their initial wet values under physiological conditions. It is interesting that the increase in this elastic parameter after drying was significantly ($p < 0.00001$) higher for thin than for thick lamellae. The relative change in stiffness due to drying was +44% for thick lamellae and +109% for thin lamellae of compact bone. For trabecular bone, the corresponding values were +37% and +78% for thick and thin lamellae, respectively.

Table 36.2 shows the results for both indentation depths. The differences are likely related to the volume sampled during the indentation, a semiellipsoid following the approximations of (36.11) and (36.12) with 0.7 μm diameter and 0.9 μm height for 100 nm indents (and 3.5 μm × 4.5 μm for 500 nm indents). Given typical lamellae dimensions of 1–4 μm, the shallow indentation measurements therefore represent properties of single lamellae whereas the deeper indents include

neighboring lamellae. It is also worth noting that thin lamellae showed a greater influence from drying when only the shallow indents were considered.

This important result should be discussed in relation to published models that address the phenomenon of bone lamellation. *Marotti* [36.91] proposed that bone lamellae are due to alternating changes of the collagen fiber density. According to his SEM results, the density of collagen is higher in thin lamellae than in thick lamellae. Collagen fibers are long chains of proteins and contain adsorption sites for polar water molecules. The higher density of the collagen fibers in the thinner lamellae results in a higher water binding capacity, which may explain the higher relative influence of drying.

Giraud-Guille [36.90], on the other hand, proposed a nested arc structure for bone lamellae, with smooth orientation changes between adjacent colla-

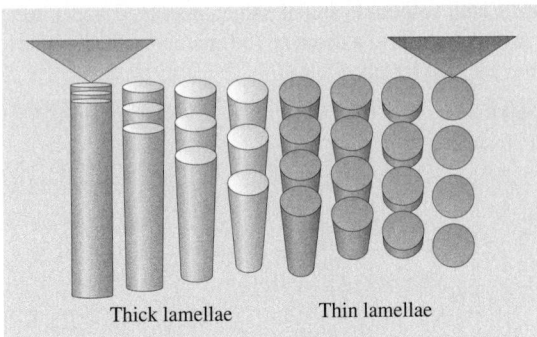

Fig. 36.21 Bone lamellation model proposed by this study. This model represents a combination of a smooth orientation change and density variations of the collagen fibers. Also note the spots where the indentations were made. A greater change in mechanical properties was detected for thin lamellae after drying

gen fibers. According to this model, indents on thick lamellae load in the longitudinal direction of the fibers, whereas the load is perpendicular to the long axis for thin lamellae. Removal of the liquid phase leads to packaging of the collagen fibers, an effect that is intuitively anisotropic and one that may explain why the effect of drying was different for both lamella types. Additionally, drying leads to microfracture initiation within thick lamellae [36.75], possibly explaining the diminished effect of drying on these structures.

Figure 36.20 presents a SEM image from a bone sample used in this study. This scan confirms the nested arc structure Giraud-Guille observed with TEM.

Our results are therefore in agreement with both the Marotti and the Giraud-Guille models. This implies a model that combines smooth orientation changes in the collagen fibers with changes in density. This architecture is sketched in Fig. 36.21. Also note the spots where the indentations were made.

Hardness of Bone Tissue

Compact bone revealed a mean hardness of 0.46 GPa under wet conditions, which increased by 74% to 0.80 GPa after drying. Upon drying, trabecular bone lamellae showed a mean change in hardness of 76% (from 0.41 to 0.72 GPa).

For compact bone, the effect of drying was again significantly ($p = 0.0006$) greater for thin lamellae (+108%) than for thick lamellae (+44%). Similar results were found for trabecular bone lamellae, with $a + 99\%$ increase for thin and $a + 56\%$ increase for thick lamellae.

The results for hardness lead therefore to similar trends as for indentation modulus. This leads us to another point, the correlation between indentation modulus and hardness. These mechanical parameters show an empirically demonstrated relationship [36.92] that may change upon drying. Therefore, we related the data for the indentation modulus from *Hengsberger* et al. [36.75] to the corresponding unpublished hardness results. The

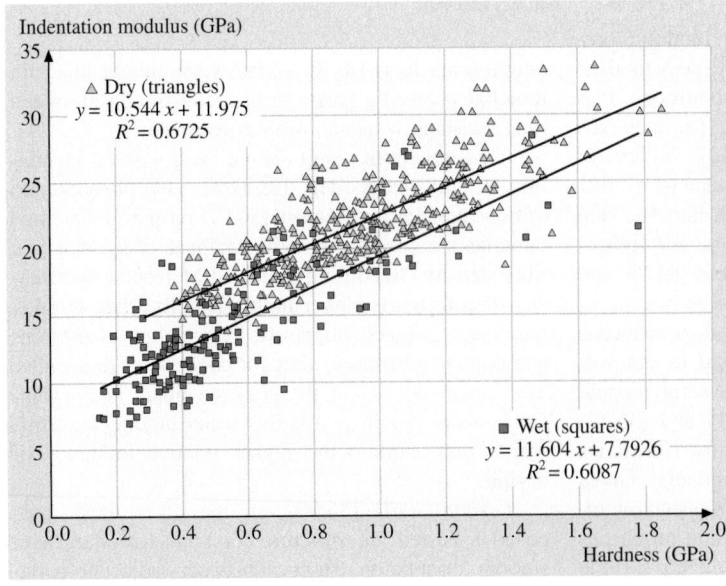

Fig. 36.22 Indentation modulus and hardness under dry and wet conditions using raw data from *Hengsberger* et al. [36.75]. Drying does not change the correlation coefficient and it changes the slope of the regression curve slightly but significantly

correlation between hardness and indentation modulus is similar under wet ($R^2 = 0.61$) and dry ($R^2 = 0.67$) conditions (Fig. 36.22). However, the slopes of the linear regressions show close but significantly different values ($p < 0.0001$) for the wet and dry samples. Hardness and indentation modulus therefore show a similar but significantly different relative shift upon drying.

36.3.4 Example 2: Nanoindentation of Human Dry and Wet Hair: Indentation Depth Issues

Hair is another example of a composite biological structure for which the macroscopic mechanical response represents some weighted average of the mechanical properties of its constituents. In the following section we address the hierarchical structure of human hair.

The structure of hair can be divided into three layers: the medulla, the cortex and the cuticle cells. Figure 36.23a shows these three layers on human hair (of Afro-American origin) sliced transversely; the hair has an elliptical shape with a minor diameter of $55\pm2\,\mu m$ and a major diameter of $98\pm2\,\mu m$ [36.93].

The *medulla* is a central channel of hair that is aligned with the longitudinal axis of the hair and composed of honeycomb-like cells with empty spaces. The medulla is not necessarily a continuous chan-

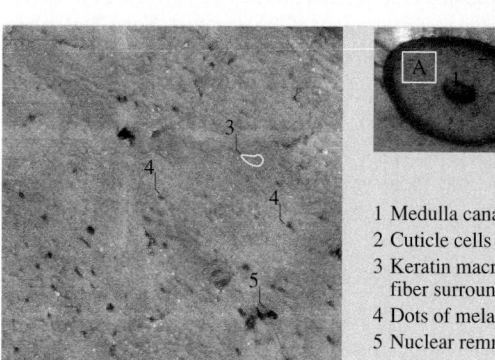

1	Medulla canal
2	Cuticle cells
3	Keratin macro-fiber surrounded
4	Dots of melanin
5	Nuclear remnant

Fig. 36.23 (a) Transverse slice of hair showing an elliptical shape with a large diameter of approximately 98 μm and an ellipticity of 57%. A layer of cuticle cells surrounds the internal structure of hair. The *square* indicates the approximate size of the AFM scan on the large figure. **(b)** AFM scan of an area approximately 20 μm×20 μm in size. The internal structure of hair consists of keratin macrofibrils. Melanin granules appear as dots on this scan. The *larger black areas* are nuclear remnants, residuals of cortical cell nuclei

nel from the root to the distal end, and is sometimes absent.

The *cortex* is the core of the hair (ca. 80%). During the "keratinization" process, the cells at the root of hair start to elongate and produce highly oriented keratin structures that form roughly cylindrical fibers. The cells die as a result of the keratinization, and become units called cortical cells. The cortical cells ($\approx100\,\mu m$ in length and $\approx1-6\,\mu m$ in diameter) are aligned with the long axis of the hair fiber and filled with partially crystallized keratin. Keratin is a highly organized helical polypeptide. Serine, glutamic acid, proline, threonine, leucine, n-acetyl serine and cystine are the main aminoacids of keratin. The chains are stabilized by many inter- and intramolecular interactions that are of electrostatic, van der Waals, hydrogen and covalent (disulfide bonds) nature. Keratin forms basic filament structures ($\approx7.5\,nm$ in diameter) known as microfibrils or the intermediate filaments (IF). Microfibrils are bundled up together, embedded in an amorphous high sulfur keratin matrix and bundled up forming the macrofibrils ($\approx100-400\,nm$ in diameter) filling up the cortical cell. The Melanin granules are inside the cortex.

Melanin, which is responsible for hair color, forms spherical granules ca. 200–800 nm in size and is distributed randomly in the cortex.

The *cuticle* forms the outer surface of the hair, protecting the cortex with a tight structure of overlapping cells.

All of these substructures, strongly anisotropic, are responsible of hair's composite biological structure that enable it to exhibit particular macroscopic properties like high rupture stress and strain, high elastic modulus, and fast and total recovery at high strains.

Figure 36.23b shows a typical topographic image of a hair cut obtained by conventional AFM. The brighter areas are keratin macrofibrils with a diameter of approximately 0.4 μm. The black dots are melanin granules. The larger black areas are remnants of nuclei that are what remain of the cells that died during the keratinisation process.

The macroscopic elastic modulus of dry human hair was characterized experimentally. The traction results, presented by *Franbourg* et al. [36.93], show a macroscopic elastic modulus of approximately 3.3 GPa under dry conditions. The change in macroscopic elastic modulus after wetting was analyzed by *Gamez-Garcia* et al. [36.94]. Their quasi-static tensile tests showed that the Young's modulus (defined as the ratio of stress to strain at 1% relative deformation) dropped to 30% of the initial value after immersion in deionized water. This large decrease in macroscopic elastic modulus, corresponding to a factor of 3.3 change after wetting, motivated the analysis of the nanomechanical properties of hair under wet and dry conditions.

In order to quantify the local mechanical changes due to wetting, virgin human hair of African descent was measured by nanoindentation.

Hair samples were embedded in LR White hard-grade acrylic resin. An internal study by L'Oréal Recherche showed that LR White resin does not penetrate the cuticle cell structure and therefore does not alter the mechanical properties of hair. The embedded sample was cut perpendicular to the long axis of the hair, obtaining transverse slices (see Fig. 36.24). The sample was then polished metallographically using successive grades of carbide paper and finished with 0.05 μm alumina solution. The sample was rinsed in an ultrasound bath between successive polishing grades. After polishing, the sample was dried at room temperature for a day.

Nanoindentation was then performed on the hair samples, at first under dry conditions and then under fully wet conditions, the sample being immersed in pure

Fig. 36.24 Topographic scan of the hair structure employing the scanning nanoindentation tool (5 μm × 5 μm) under dry conditions. The resolution of this scan allows the indenter tip to be positioned on structures like keratin macrofibrils (*bright areas*), melanin (*dots*) and the lipid interface between the keratin macrofibrils. Topography scans performed under wet conditions did not allow the lipid interfaces to be resolved

water and at ambient temperature (Fig. 36.16). The same sample holder used to study bone tissue was employed. Unlike bone tissue, the structure of dead hair does not degrade in vitro and so a pure water solution can be used without time constraints.

The time taken for the hair tissue to absorb the water does not represent a critical factor. *Franbourg et al.* [36.93] have shown that water saturation is achieved after 200 s.

Indents were performed under dry and wet conditions on keratin macrofibrils, on melanin and on the medulla. Additional indents were performed on the lipid interface between adjacent macrofibrils under dry conditions. Since the keratin macrofibrils, the melanin granules and the lipid interfaces are typically hundreds of nanometers in size, the indentation depth represents a critical parameter.

Employing the estimate based on (36.11) and (36.12) for a Berkovich indenter, we can propose a contact depth of

$$h_c \approx \frac{d_{structure}}{7} \,, \tag{36.14}$$

where $d_{structure}$ represents the diameter of the structure to be characterized. Here we approximated the maximum depth h_{max} by the contact depth h_c. If the nanoindenter tool only allows for load-controlled indents, the maximum force must be estimated. To do this, some preliminary first indents need to be performed to determine the mean hardness of the structure. For a Berkovich indenter, this parameter ideally corresponds to

$$H = \frac{P_{max}}{24.5 h_c^2} \,, \tag{36.15}$$

where the ideal area function $A_c(h) = 24.5 h_c^2$ of the Berkovich geometry was used. A little manipulation gives

$$P_{max} = 24.5 H h_c^2 \approx 24.5 H \left(\frac{d_{structure}}{7} \right)^2$$
$$= 0.5 H d_{structure}^2 \,. \tag{36.16}$$

This equation represents a coarse estimate of the maximum allowable load. Since the hardness of biological tissue can vary between regions, it is possible that indents for a given maximum load achieve higher contact depths than allowed by (36.14). Furthermore, the hardness can change dramatically upon moving from dry to wet conditions. Additionally, the indenter shape often varies from the ideal area function for low-depth indents (<50 nm). For low-depth indents the calibrated

Table 36.3 Elastic moduli of different hair substructures under dry and wet conditions

Structure	Elastic modulus when dry	Elastic modulus when wet	Relative change (factor of)
Keratin macrofibrils	5.93 ±1.06 GPa	1.18 ±0.39 GPa	5
Melanin granules	7.02 ±2.84 GPa	1.37 ±1.18 GPa	5.1
Medulla channel	3.90 ±1.40 GPa	1.88 ±1.18 GPa	2.1
Lipid interface	5.26 ±0.39 GPa		

area function should be used in (36.16) instead of the ideal area function.

Note that indents to shallow indentation depths (<50 nm) do not necessarily lead to permanent imprints that can be resolved afterwards via topography scans performed with the identical tip.

For this study, a total of 122 indents were performed, 49 on keratin, 37 on melanin, 22 on the lipid interface under dry conditions and 14 on the medulla. The tests were employed a 10 s loading, 10 s holding and 10 s unloading cycle. The maximum load was adapted to achieve indents to a mean contact depth of 35 nm. When choosing the maximum load, the individual size of the indented structure was taken into account.

Hardness of Hair

The hardness of keratin decreased by factor of 2.5 after wetting (from 0.43 ± 0.18 to 0.17 ± 0.08 GPa). A relative decrease in hardness of a factor of 2.85 (from 0.54 ± 0.32 to 0.19 ± 0.06 GPa) was observed for melanin. The soft cells of the medulla channel showed the smallest decrease in hardness after wetting (from 0.35 ± 0.17 to 0.26 ± 0.12 GPa).

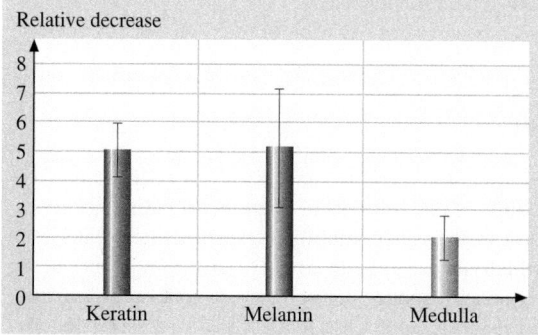

Fig. 36.25 Relative decrease (mean±standard deviation) in the elastic moduli of hair substructures after wetting. Note that the elastic modulus decreases by about a factor of five for the keratin macrofibrils (49 indents) and the melanin granules (37 indents). The medulla (14 indents) only showed a factor of 2.1 change in modulus

The lipid interface of dry hair showed hardness values of approximately 0.38 GPa±0.05 GPa.

Elastic Modulus of Hair

The indentation modulus of the whole hair structure was determined and an elastic modulus E_{specimen} was calculated using (36.10) assuming a Poisson ratio of $\nu = 0.4$ (L'Oréal, internal communication). The elastic modulus (instead of the indentation modulus, which involves the Poisson ratio) was used in this study to allow for a direct comparison with macroscopic tensile tests.

Table 36.3 and Fig. 36.25 show the changes in elastic properties upon moving from dry to wet conditions. Note that the small lipid interfaces separating the keratin macrofibrils could only be characterized under dry conditions, since performing the topography scan under wet conditions did not allow this structure to be identified.

The elastic properties of keratin macrofibrils obtained by this study can be compared with the Young's modulus of native wool (cortex) reported by *Parbhu* et al. [36.95]. Using AFM-based indentation tests, they obtained an elastic modulus of 4.00±1.7 GPa. The hardness of dry keratin obtained in our study compares well with that of wool keratin. Their indentation tests showed a value of 0.4±0.2 GPa.

Our data show that the mechanical properties of hair substructure change dramatically after wetting. The decrease in the elastic modulus of the keratin macrofibrils, the main constituent of hair, is greater than the macroscopic change after wetting (a factor of 3.3).

The melanin granules, which represent the hardest and stiffest hair component under dry conditions, show a factor of 5.1 decrease in mechanical properties after wetting. These granules (mean size: 200 to 800 nm) are embedded in the hair matrix, which is mainly keratin. It is possible that the decrease in the mechanical properties of melanin is less dramatic than observed in our indentations. We cannot exclude the possibility that the melanin granules are pushed into the soft hair matrix as they are indented under wet conditions. This complex point needs to be addressed with further studies.

The medulla exhibits a lower elastic modulus under dry conditions in our study. Furthermore, this structure is less affected by wetting than the other substructures. These observations may be attributed to its empty honeycomb structure, which is very soft under dry conditions. After wetting, the presence of additional water in the empty spaces may compensate for the decrease in the elastic modulus of the protein structure that constitutes the medulla.

Precise interpretation of these results requires an adequate mechanical model that involves the protein phases, their crosslinking, as well as the three main structures: the medulla, the cortex and the cuticle cells. In particular, the role of the cuticle cells in the macroscopic mechanical response should be clarified. A simplified mechanical model for hair was proposed by [36.96, 97]. However, this model does not take into account the complex structure of hair, which involves at least six hierarchical levels [36.98].

Swelling of the tissue presents a further issue in tests performed under wet conditions [36.93, 94]. Unembedded hair swells macroscopically after wetting by between 6% and 10% radially and by around 1% longitudinally. Since the hair tissue must be embedded for nanomechanical studies, this effect can result in additional stresses in the hair matrix.

Another point relates to the viscoelastic properties of hair. The loading sequence employed for the nanomechanical tests represents a compromise between the desired quasi-static strain rate and the thermal drift of the indenter tip. According to macroscopic relaxation tests [36.99], viscoelastic properties of dry hair depend strongly on the applied strain. For a relative strain of 1% and a 1 s holding period, a macroscopic relaxation modulus of approximately 4.8 GPa was found, which dropped to 4.2 GPa after 10 s and finally to 4 GPa after holding

for 500 s. The viscoelastic properties of hair should be taken into account before comparing nanoindentation data with quasistatic tensile tests.

36.3.5 Conclusion

Modern scanning nanoindentation has clearly solved two problems: nanoscale indentations can be performed with high force and displacement resolution, and high (nanometer) lateral control of the indentation position allows small structures of heterogeneous materials to be characterized. Interfaces within composites, variations in local density or composition in chemical products, and biological structures can all therefore be investigated using the SN tool.

Recent theoretical progress [36.100] means that the nanoindentation technique can also be used to characterize anisotropic materials by indenting in different directions. Other parameters can also be determined. For instance, creep behavior is accessible when the indentation load is held constant at a maximum load. The hysteresis of the force–displacement curve represents the energy that is dissipated during indentation, and so it provides information about the post-elastic behavior of a sample. Dynamical variables such as viscoelastic properties are also accessible if sinusoidal oscillations are performed at different frequencies during loading.

Future interest from the nanotechnology community may direct work in this area towards tests on lower levels of structural organization, like the molecular level. This would require significant improvements in transducer sensitivity and indenter tip machining. New ways to reduce the thermal drift are also required for next-generation devices.

36.4 General Summary and Perspectives

We have demonstrated that AFM and scanning nanoindentation are ideal tools for investigating variations in local properties of bulk materials, like bone and hair, and performing physical measurements on individual nanoscale objects, such as carbon nanotubes and protein polymers. These techniques allow us to access previously inaccessible properties of living matter like the Young's and shear moduli. It should be emphasized that shear is omnipresent because biostructures are "composite" materials with strongly anisotropic interactions between their constituents. We have also explored the mechanical properties of proteins as a function of tem-

perature. Interactions can vary remarkably, even over a range of just few tens of degrees, providing deeper insight in the functioning of these structures. One issue (which seems difficult to solve at the moment) is the frequency-dependent responses of nano- and biostructures. In the latter case it is limited to a few kilohertz, mainly because of the surrounding liquid, but one still requires several decades in frequency for a meaningful analysis.

Improved scanning probe tips are needed to provide better resolution, longer lifetimes, and easy functionalization (for sensing different chemical environments).

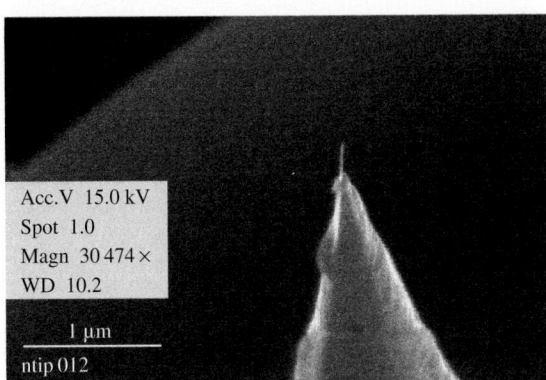

Fig. 36.26 SEM picture of an AFM tip with a multiwalled carbon nanotube attached

This may be achieved through the use of carbon nanotube tips (Fig. 36.26), which have a Young's modulus of 1 TPa, a very well defined and sharp tip structure, and pentagon "defects" at the apex that provide sites for functionalization. It is a general feeling within the scanning probe community that carbon nanotubes will open up new avenues in the study of living matter.

Finally, the development of the photonic force microscope in Heidelberg has allowed biostructures to be imaged with unprecedented three-dimensional resolution; features that are otherwise inaccessible, even to AFM tips, can be seen. In the future, this instrument will certainly provide radically new insights into biological functioning.

References

36.1   M. Radmacher: Measuring the elastic properties of biological samples with the AFM, IEEE Eng. Med. Biol. Mag. **16**(2), 47–57 (1997)

36.2   G. Binnig, C. F. Quate, C. Gerber: Atomic force microscope, Phys. Rev. Lett. **56**, 930–933 (1986)

36.3   P. Maivald, H. J. Butt, S. A. C. Gould, C. B. Prater, B. Drake, J. A. Gurley, V. B. Elings, P. K. Hansma: Using force modulation to image surface elasticities with the atomic force microscope, Nanotechnology **2**, 103–106 (1991)

36.4   T. Kajiyama, K. Takata, I. Ohki, S.-R. Ge, J.-S. Yoon, A. Takahara: Imaging of dynamic viscoelastic properties of a phase-separated polymer by forced oscillation atomic force microscopy, Macromolecules **27**, 7932–7934 (1994)

36.5   B. Nysten, R. Legras, J.-L. Costa: Atomic force microscopy imaging of viscoelastic properties in toughened polypropylene resins, J. Appl. Phys. **78**, 5953–5958 (1995)

36.6   P.-E. Mazeran, J.-L. Loubet: Normal and lateral modulation with a scanning force microscope, an analysis: Implication in quantitative elastic and friction imaging, Tribol. Lett. **3**, 125–129 (1997)

36.7   B. Cretin, F. Sthal: Scanning microdeformation microscopy, Appl. Phys. Lett. **62**, 829–831 (1993)

36.8   N. A. Burnham, A. J. Kulik, G. Gremaud, P.-J. Gallo, F. Oulevey: Scanning local-acceleration microscopy, J. Vacuum Sci. Technol. B **14**, 794–799 (1996)

36.9   F. Oulevey, N. A. Burnham, A. J. Kulik, G. Gremaud, W. Benoit: Mechanical properties studied at the nanoscale using scanning local-acceleration microscopy (SLAM), J. Phys. IV, C8-731–734 (1996)

36.10   U. Rabe, W. Arnold: Acoustic microscopy by atomic force microscopy, Appl. Phys. Lett. **64**, 1493–1495 (1994)

36.11   O. Kolosov, K. Yamanaka: Nonlinear detection of ultrasonic vibrations in an atomic force microscope, Jpn. J. Appl. Phys. **32**, 22–25 (1993)

36.12   F. Oulevey: Cartographie et spectrométrie des propriétés mécaniques à l'échelle nanométrique par microscopie acoustique en champ proche. Ph.D. Thesis (EPFL, Lausanne 1999)

36.13   F. Oulevey, G. Gremaud, A. Semoroz, A. J. Kulik, N. A. Burnham, E. Dupas, D. Gourdon: Local mechanical spectroscopy with nanometer-scale lateral resolution, Rev. Sci. Instrum. **69**, 2085–2094 (1998)

36.14   N. A. Burnham, G. Gremaud, A. J. Kulik, P.-J. Gallo, F. Oulevey: Materials' properties measurements: Choosing the optimal scanning probe microscope configuration, J. Vacuum Sci. Technol. B **14**, 1308–1312 (1996)

36.15   F. Oulevey, G. Gremaud, A. J. Kulik, B. Guisolan: Simple low-drift heating stage for scanning probe microscopes, Rev. Sci. Instrum. **70**, 1889–1890 (1999)

36.16   H. M. Pollock: Nanoindentation. In: *Friction Lubrication and Wear Technology*, AMS Handbook, Vol. 18 (AMS, Ohio 1992) p. 419

36.17   E. Dupas, G. Gremaud, A. Kulik, J.-L. Loubet: High-frequency mechanical spectroscopy with an atomic force microscope, Rev. Sci. Instrum. **72**(10), 3891–3897 (2001)

36.18   F. Oulevey, N. A. Burnham, G. Gremaud, A. J. Kulik, H. M. Pollock, A. Hammiche, M. Reading, M. Song, D. J. Hourston: Dynamic mechanical analysis at the submicron scale, Polymer **41**, 3087–3092 (2000)

36.19   W. Wm. Wendlandt: *Thermal Analysis in Chemical Analysis*, Vol. 19, 3rd edn. (Wiley, New York 1996) p. 360

36.20   D. Mari, D. C. Dunand: NiTi and NiTi–TiC composites: Part I. Transformation and thermal cycling behavior, Metall. Mater. Trans. A **26A**, 2833–2847 (1995)

36.21   L. Bataillard: Transformation martensitique multiple dans un alliage à mémoire de forme Ni–Ti. Ph.D. Thesis (EPF, Lausanne 1996)

36.22   F. Oulevey, G. Gremaud, D. Mari, A. J. Kulik, N. A. Burnham, W. Benoit: Martensitic transformation of NiTi studied at the nanometer scale by local mechanical spectroscopy, Scripta Mater. **42**, 31–36 (2000)

36.23   D. Mari, L. Bataillard, D. C. Dunand, R. Gotthardt: Martensitic transformation of NiTi and NiTi–TiC composites, J. Phys. (France) **IV**, C8–659–664 (1995)

36.24   S. Iijima: Helical microtubules of graphitic carbon, Nature **354**, 56–58 (1991)

36.25   C.-H. Kiang, M. Endo, P. M. Ajayan, G. Dresselhaus, M. S. Dresselhaus: Size effects in carbon nanotubes, Phys. Rev. Lett. **81**, 1869–1872 (1998)

36.26   E. W. Wong, P. E. Sheehan, C. M. Lieber: Nanobeam mechanics: Elasticity, strength and toughness of nanorods and nanotubes, Science **277**, 1971–1975 (1997)

36.27   A. Thess, R. Lee, P. Nikolaev, H. Dai, P. Petit, J. Robert, C. Xu, Y. H. Lee, S. G. Kim, A. G. Rinzler, D. T. Colbert, G. U. Scuseria, D. Tománek, J. E. Fischer, R. E. Smalley: Crystalline ropes of metallic carbon nanotubes, Science **273**, 483–487 (1996)

36.28   W. Z. Li, S. S. Xie, L. X. Qian, B. H. Chang, B. S. Zou, W. Y. Zhou, R. A. Zhao, G. Wang: Large-scale synthesis of aligned carbon nanotubes, Science **274**, 1701–1703 (1996)

36.29   J. Kong, H. T. Soh, A. M. Cassell, C. F. Quate, H. J. Dai: Synthesis of individual single-walled carbon nanotubes on patterned silicon wafers, Nature **395**, 878–881 (1998)

36.30   J. P. Lu: Elastic properties of carbon nanotubes and nanoropes, Phys. Rev. Lett. **79**, 1297–1300 (1997)

36.31   J.-P. Salvetat, J. M. Bonard, N. H. Thomson, A. J. Kulik, L. Forró, W. Benoit, L. Zuppiroli: Mechanical properties of carbon nanotubes, Appl. Phys. A **69**, 255–260 (1999)

36.32   M.-F. Yu, B. S. Files, S. Arepalli, R. S. Ruoff: Tensile loading of ropes of single wall carbon nanotubes and their mechanical properties, Phys. Rev. Lett. **84**, 5552–5555 (2000)

36.33   S. Iijima, C. Brabec, A. Maiti, J. Bernholc: Structural flexibility of carbon nanotubes, J. Chem. Phys. **104**, 2089–2092 (1996)

36.34   J.-P. Salvetat, A. J. Kulik, J.-M. Bonard, G. A. D. Briggs, T. Stöckli, K. Méténier, S. Bonnamy, F. Béguin, N. A. Burnham, L. Forró: Elastic modulus of ordered and disordered multiwalled carbon nanotubes, Adv. Mater. **11**, 161–165 (1999)

36.35   J.-P. Salvetat, G. A. D. Briggs, J.-M. Bonard, R. R. Bacsa, A. J. Kulik, T. Stöckli, N. Burnham, L. Forró: Elastic and shear moduli of single-walled

36.36   J. M. Gere, S. P. Timoshenko: *Mechanics of Materials* (PWS-Kent, Boston 1984)

36.37   B. Lukic, J. W. Seo, E. Couteau, K. Lee, S. Gradecak, R. Berkecz, K. Hernadi, S. Delpeux, T. Cacciaguerra, F. Béguin, A. Fonseca, J. B. Nagy, G. Csányi, A. Kis, A. J. Kulik, L. Forró: Elastic modulus of multi-walled carbon nanotubes produced by catalytic chemical vapour deposition, Appl. Phys. A **80**, 695–700 (2005)

36.38   B. Vigolo, A. Penicaud, C. Coulon, C. Sauder, R. Pailler, C. Journet, P. Bernier, P. Poulin: Macroscopic fibers and ribbons of oriented carbon nanotubes, Science **290**, 1331–1334 (2000)

36.39   A. B. Dalton, S. Collins, E. Muñoz, J. M. Razal, V. H. Ebron, J. P. Ferraris, J. N. Coleman, B. G. Kim, R. H. Baughman: Super-tough carbon nanotube fibres, Nature **423**, 703 (2003)

36.40   H. W. Zhu, C. L. Xu, D. H. Wu, B. Q. Wei, R. Vajtai, P. M. Ajayan: Direct synthesis of long single-walled carbon nanotube strands, Science **296**, 884–886 (2002)

36.41   D. A. Walters, L. M. Ericson, M. J. Casavant, J. Liu, D. T. Colbert, K. A. Smith, R. E. Smalley: Elastic strain of freely suspended single-wall carbon nanotube ropes, Appl. Phys. Lett. **74**, 3803–3805 (1999)

36.42   A. Kis, G. Csanyi, J.-P. Salvetat, T.-N. Lee, E. Couteau, A. J. Kulik, W. Benoit, J. Brugger, L. Forró: Reinforcement of single-walled nanotube bundles by intertube bridging, Nature Mater. **3**, 153–157 (2004)

36.43   P. A. Williams, S. J. Papadakis, A. M. Patel, M. R. Falvo, S. Washburn, R. Superfine: Torsional response and stiffening of individual multiwalled carbon nanotubes, Phys. Rev. Lett. **89**, 255502 (2002)

36.44   M.-F. Yu, O. Lourie, M. J. Dyer, K. Moloni, T. F. Kelly, R. S. Ruoff: Strength and breaking mechanism of multiwalled carbon nanotubes under tensile load, Science **287**, 637–640 (2000)

36.45   T. Fujii, M. Suzuki, M. Miyashita, M. Yamaguchi, T. Onuki, H. Nakamura, T. Matsubara, H. Yamada, K. Nakayama: Micropattern measurement with an atomic force microscope, J. Vacuum Sci. Technol. B **9**, 666–669 (1991)

36.46   G.-T. Kim, G. Gu, U. Waizmann, S. Roth: Simple method to prepare individual suspended nanofibers, Appl. Phys. Lett. **80**, 1815–1817 (2002)

36.47   M. Kurachi, M. Hoshi, H. Tashiro: Buckling of a single microtubule by optical trapping forces: Direct measurement of microtubule rigidity, Cell. Motil. Cyt. **30**, 221–228 (1995)

36.48   R. B. Dye, S. P. Fink, R. C. Williams: Taxol-induced flexibility of microtubules and its reversal by Map-2 and Tau, J. Biol. Chem. **268**, 6847–6850 (1993)

carbon nanotube ropes, Phys. Rev. Lett. **82**, 944–947 (1999)

36.49 F. Gittes, B. Mickey, J. Nettleton, J. Howard: Flexual rigidity of microtubules and actin filaments measured from thermal fluctuations in shape, J. Cell Biol. **120**, 923–934 (1993)

36.50 M. Elbaum, D. K. Fygenson, A. Libchaber: Buckling microtubules in vesicles, Phys. Rev. Lett. **76**, 4078–4081 (1996)

36.51 P. J. D. Pablo, I. A. T. Schaap, F. C. Mackintosh, C. F. Schmidt: Deformation and collapse of microtubules on the nanometer scale, Phys. Rev. Let. **91**, 098101 (2003)

36.52 A. Vinckier, C. Dumortier, Y. Engelborghs, L. Hellemans: Dynamical and mechanical study of immobilized microtubules with atomic force microscopy, J. Vacuum Sci. Technol. B **14**, 1427–1431 (1996)

36.53 A. S. K. Kis, B. Babic, A. J. Kulik, W. Benoît, G. A. D. Briggs, C. Schönenberger, S. Catsicas, L. Forró: Nanomechanics of microtubules, Phys. Rev. Lett. **89**, 248101 (2002)

36.54 A. V. Kulkarni, B. Bhushan: Nanoscale mechanical property measurements using modified atomic force microscopy, Thin Solid Films **290-291**, 206–210 (1996)

36.55 A. V. Kulkarni, B. Bhushan: Nano/picoindentation measurements on single-crystal aluminum using modified atomic force microscopy, Mater. Lett. **29**, 221–227 (1996)

36.56 M. Chiba, M. Seo: Effects of dichromate treatment on mechanical properties of passivated single crystal iron (100) and (110) surfaces, Corros. Sci. **44**, 2379–2391 (2002)

36.57 X. C. Lu, B. Shi, L. K. Y. Li, J. Luo, X. Chang, Z. Tian, J. I. Mou: Nanoindentation and microtribological behavior of Fe-N/Ti-N multilayers with different thickness of Ti-N layers, Wear **251**, 1144–1149 (2001)

36.58 P. Peeters, Gvd. Horn, T. Daenen, A. Kurowski, G. Staikov: Properties of electroless and electroplated Ni-P and its application in microgalvanics, Electrochim. Acta **47**, 161–169 (2001)

36.59 A. Rar, J. N. Zhou, W. J. Liu, J. A. Barnard, A. Bennett, S. C. Street: Dendrimer-mediated growth of very flat ultrathin Au films, Appl. Surf. Sci. **175-176**, 134–139 (2001)

36.60 G. Wei, A. Rar, J. A. Barnard: Composition, structure, and nanomechanical properties of DC-sputtered CrN_x ($0 \leq x \leq 1$) thin films, Thin Solid Films **460-464**, 398–399 (2001)

36.61 R. D. Ott, C. Ruby, F. Huang, M. L. Weaver, J. A. Barnard: Nanotribology and surface chemistry of reactively sputtered Ti-B-N hard coatings, Thin Solid Films **377-378**, 602–606 (2000)

36.62 A. Hieke: Analysis of microwear experiments on thin DLC coatings: Friction, wear and plastic deformation, Wear **254/5-6**, 565–572 (2003)

36.63 M. Zhao, W. S. Slaughter, M. Li, S. X. Mao: Material-length-scale-controlled nanoindentation size ef-

fects due to strain-gradient plasticity, Acta Mater. **51/15**, 4461–4469 (2003)

36.64 D. Raabe, C. Klüber, F. Roters: Orientation dependence of nanoindentation pile-up patterns and of nanoindentation microtextures in copper single crystals, Acta Mater. **52/8**, 2229–2238 (2004)

36.65 T. Malkow, S. J. Bull: Hardness measurements on thin IBAD CN_x films – a comparative study, Surf. Coat. Technol. **137**, 197–204 (2001)

36.66 Y. Shima, H. Hasuyama, T. Kondoh, Y. Imaoka, T. Watari, K. Baba, R. Hatada: Mechanical properties of silicon oxynitride thin films prepared by low energy ion beam assisted deposition, Nucl. Instrum. Meth. B **148**, 599–603 (1999)

36.67 M. Göken, M. Kempf, W. D. Nix: Hardness and modulus of the lamellar microstructure in PST-TiAl studied by nanoindentations and AFM, Acta Mater. **49**, 903–911 (2001)

36.68 M. Göken, M. Kempf: Microstructural properties of superalloys investigated by nanoindentations in an atomic force microscope, Acta Mater. **47**, 1043–1052 (1999)

36.69 M. Seo, M. Chiba, K. Suzuki: Nano-mechano-electrochemistry of the iron (100) surface in solution, J. Electroanal. Chem. **473**, 49–53 (1999)

36.70 Y. Zhang, F. Z. Cui, X. M. Wang, Q. L. Feng, X. D. Zhu: Mechanical properties of skeletal bone in gene-mutated $stöpsel^{dtl28d}$ and wild-type zebrafish (*Danio rerio*) measured by atomic force microscopy-based nanoindentation, Bone **30**, 541–546 (2002)

36.71 M. Finke, J. A. Hughes, D. M. Parker, K. D. Jandt: Mechanical properties of in situ demineralised human enamel measured by AFM nanoindentation, Surf. Sci. **491**, 456–467 (2001)

36.72 S. Habelitz, S. J. Marshall, G. W. Marshall, J. R. Balooch, M. Balooch: The functional width of the dentino–enamel junction determined by AFM-based nanoscratching, J. Struct. Biol. **135**, 294–301 (2001)

36.73 G. W. Marshall Jr., M. Balooch, R. R. Gallagher, S. A. Gansky, S. J. Marshall: Mechanical properties of the dentinoenamel junction: AFM studies of nanohardness, elastic modulus, and fracture, J. Biomed. Mater. Res. **54**, 87–95 (2001)

36.74 M. C. Chang, C. C. Ko, C. C. Liu, W. H. Douglas, R. DeLong, W. J. Seong, J. Hodges, K.-N. An: Elasticity of alveolar bone near dental implant–bone interfaces after one month's healing, J. Biomech. **36/8**, 1209–1214 (2003)

36.75 S. Hengsberger, A. Kulik: Nanoindentation discriminates the elastic properties of individual human bone lamellae under dry and physiological conditions, Bone **30**, 178–184 (2002)

36.76 E. F. Eriksen, D. W. Axelrod, F. M. Melsen: *Bone Histomorphometry*, 1st edn. (Raven, New York 1994)

36.77   C. E. Hoffler, K. E. Moore, K. Kozloff, P. K. Zysset, M. B. Brown, S. A. Goldstein: Heterogeneity of bone lamellar-level elastic moduli, Bone **26**, 603–609 (2000)

36.78   C. E. Hoffler, K. E. Moore, K. Kozloff, P. K. Zysset, S. A. Goldstein: Age, gender, and bone lamellae elastic moduli, J. Orth. Res. **18**, 432–437 (2000)

36.79   P. K. Zysset, X. E. Guo, C. E. Hoffler, K. E. Moore, S. A. Goldstein: Elastic modulus and hardness of cortical and trabecular bone lamellae measured by nanoindentation in the human femur, J. Biomech. **32**, 1005–1012 (1999)

36.80   J. Y. Rho, P. Zioupos, J. D. Currey, G. M. Pharr: Variations in the individual thick lamellar properties within osteons by nanoindentation, Bone **25**, 295–300 (1999)

36.81   J. Y. Rho, M. E. Roy, T. Y. Tsui, G. M. Pharr: Elastic properties of microstructural components of human bone tissue as measured by nanoindentation, J. Biomed. Mater. Res. **45**, 48–54 (1999)

36.82   S. Hengsberger, A. Kulik, P. Zysset: A combined atomic force microscopy and nanoindentation technique to investigate the elastic properties of bone structural units, Eur. Cells Mater. **1**, 12–16 (2001)

36.83   C. H. Turner, J. Y. Rho, Y. Takano, T. Y. Tsui, G. M. Pharr: The elastic properties of trabecular and cortical bone tissues are similar: Results from two microscopic measurement techniques, J. Biomech. **32**, 437–441 (1999)

36.84   M. E. Roy, J. Y. Rho, T. Y. Tsui, N. S. Evans, G. M. Pharr: Mechanical and morphological variation of the human lumbar vertebral cortical and trabecular bone, J. Biomed. Mater. Res. **44**, 191–199 (1999)

36.85   J. Y. Rho, G. M. Pharr: Effects of drying on the mechanical properties of bovine femur measured by nanoindentation, J. Mater. Sci. Mater. M. **10**, 485–488 (1999)

36.86   I. N. Sneddon: The relation between load and penetration in the axisymmetric Boussinesq problem for a punch of arbitrary profile, Int. J. Eng. Sci. **3**, 47–57 (1965)

36.87   W. C. Oliver, G. M. Pharr: An improved technique for determining hardness and elastic modulus using load and displacement sensing indentation experiments, Mater. Res. Soc. **7/6**, 1564–1583 (1992)

36.88   G. M. Pharr, W. C. Oliver, F. R. Brotzen: On the generality of the relationship among contact stiffness, contact area, and elastic modulus during indentation, J. Mater. Res. **7/3**, 613–617 (1992)

36.89   K. L. Johnson: *Contact Mechanics*, 1st edn. (Cambridge Univ. Press, Cambridge 1985) pp. 84–106

36.90   M. M. Giraud-Guille: Plywood structures in nature, Curr. Opin. Sol. St. Mater. Sci. **3**, 221–227 (1998)

36.91   G. Marotti: A new theory of bone lamellation, Calcif. Tissue Int. **53**, 47–56 (1993)

36.92   G. P. Evans, J. C. Behiri, J. D. Currey, W. Bonfield: Microhardness and Young's modulus in cortical bone exhibiting a wide range of mineral volume fractions, and in a bone analogue, J. Mater. Sci. Mater. M. **1**, 38–43 (1990)

36.93   A. Franbourg, P. Hallegot, F. Baltenneck, C. Toutain, F. Leroy: Current research on ethnic hair, J. Am. Acad. Dermatol. **48**, 115–119 (2003)

36.94   M. Gamez-Garcia: Effects of some oils, emulsions, and other aqueous systems on the mechanical properties of hair at small deformations, J. Soc. Cosmet. Chem. **44**, 69–87 (1993)

36.95   A. N. Parbhu, W. G. Bryson, R. Lal: Disulfide bonds in the outer layer of keratin fibers confer higher mechanical rigidity: correlative nano-indentation and elasticity measurement with an AFM, Biochemistry **38**, 11755–11761 (1999)

36.96   M. Feughelman, B. K. Willis: Mechanical extension of human hair and the movement of the cuticle, J. Cosmet. Sci. **52**, 185–193 (2001)

36.97   M. Feughelman: *Mechanical Properties and Structure of α-Keratin Fibers* (UNSW Press, Sydney 1997)

36.98   J. W. S. Hearle: A critical review of the structural mechanics of wool and hair fibres, Int. J. Bio. Macromol. **27**, 123–138 (2000)

36.99   H. A. Barnes, G. P. Robert: The non-linear viscoelastic behaviour of human hair at moderate extensions, Int. J. Cosmet. Sci. **22**, 259–264 (2000)

36.100 J. G. Swadener, G. M. Pharr: Indentation of elastically anisotropic half-spaces by cones and parabolae of revolution, Philos. Mag. A **81**, 447–466 (2001)

37. Nanomechanical Properties of Solid Surfaces and Thin Films

Instrumentation for the testing of mechanical properties on the submicron scale has developed enormously in recent years. This has enabled the mechanical behavior of surfaces, thin films, and coatings to be studied with unprecedented accuracy. In this chapter, the various techniques available for studying nanomechanical properties are reviewed with particular emphasis on nanoindentation. The standard methods for analyzing the raw data obtained using these techniques are described, along with the main sources of error. These include residual stresses, environmental effects, elastic anisotropy, and substrate effects. The methods that have been developed for extracting thin-film mechanical properties from the often convoluted mix of film and substrate properties measured by nanoindentation are discussed. Interpreting the data is frequently difficult, as residual stresses can modify the contact geometry and, hence, invalidate the standard analysis routines. Work hardening in the deformed region can also result in variations in mechanical behavior with indentation depth. A further unavoidable complication stems from the ratio of film to substrate mechanical properties and the depth of indentation in comparison to film thickness. Even very shallow indentations may be influenced by substrate properties if the film is hard and very elastic but the substrate is compliant. Under these circumstances nonstandard methods of analysis must be used. For multilayered systems many different mechanisms affect the nanomechanical behavior, including Orowan strengthening, Hall–Petch behavior, image force effects, coherency and thermal stresses, and composition modulation.

The application of nanoindentation to the study of phase transformations in semiconductors, fracture in brittle materials, and mechanical properties in biological materials are described. Recent developments such as the testing of viscoelasticity using nanoindentation methods are likely to be particularly important in future studies of poly-

mers and biological materials. The importance of using a range of complementary methods such as electron microscopy, in situ AFM imaging, acoustic monitoring, and electrical contact measurements is emphasized. These are especially important on the nanoscale because so many different physical and chemical processes can affect the measured mechanical properties.

When two bodies come into contact their surfaces experience the first and usually largest mechanical loads. Hence, characterizing and understanding the mechanical properties of surfaces is of paramount importance in a wide range of engineering applications. Obvious examples of where surface mechanical properties are important are in wear-resistant coatings on reciprocating surfaces and hard coatings for machine tool bits. This chapter details the current methods for measuring the mechanical properties of surfaces and highlights some of the key experimental results that have been obtained.

The experimental technique that is highlighted in this chapter is nanoindentation. This is for the simple reason

that it is now recognized as the preferred method for testing thin film and surface mechanical properties. Despite this recognition, there are still many pitfalls for the unwary researcher when performing nanoindentation tests. The commercial instruments that are currently available all have attractive, user-friendly software, which makes the performance and analysis of nanoindentation tests easy. Hidden within the software, however, are a myriad of assumptions regarding the tests that are being performed and the material that is being examined. Unless the researcher is aware of these, there is a real danger that the results obtained will say more about the analysis routines than they do about the material being tested.

37.1 Instrumentation

The instruments used to examine nanomechanical properties of surfaces and thin films can be split into those based on point probes and those complimentary methods that can be used separately or in conjunction with point probes. The complimentary methods include a wide variety of techniques ranging from optical tests such as micro-Raman spectroscopy to high-energy diffraction studies using X-rays, neutrons, or electrons to mechanical tests such as bulge or blister testing.

Point-probe methods have developed from two historically different methodologies, namely, scanning probe microscopy [37.1] and microindentation [37.2]. The two converge at a length scale between 10–1000 nm. Point-probe mechanical tests in this range are often referred to as nanoindentation.

37.1.1 AFM and Scanning Probe Microscopy

Atomic force microscopy (AFM) and other scanning probe microscopies are covered in detail elsewhere in this volume, but it is worth briefly highlighting the main features in order to demonstrate the similarities to nanoindentation. There are now a myriad of different variants on the basic scanning probe microscope. All use piezoelectric stacks to move either a probe tip or the sample with subnanometer precision in the lateral and vertical planes. The probe itself can be as simple as a tungsten wire electrochemically polished to give a single atom at the tip, or as complex as an AFM tip that is bio-active with, for instance, antigens attached. A range of scanning probes have been developed with the intention of measuring specific physical properties such as magnetism and heat capacity.

To measure mechanical properties with an AFM, the standard configuration is a hard probe tip (such as silicon nitride or diamond) mounted on a cantilever (see Fig. 37.1). The elastic deflection of the cantilever is monitored either directly or via a feedback mechanism to measure the forces acting on the probe. In general,

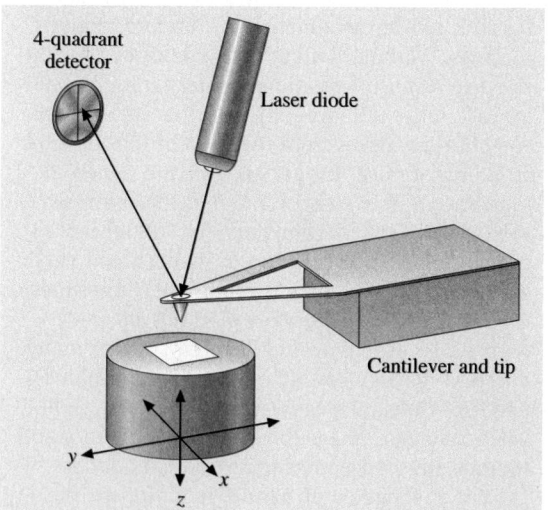

Fig. 37.1 Diagram of a commercial AFM. The AFM tip is mounted on a compliant cantilever, and a laser light is reflected off the back of the cantilever onto a position-sensitive detector (4-quadrant detector). Any movements of the cantilever beam cause a deflection of the laser light that the detector senses. The sample is moved using piezoelectric stack, and forces are calculated from the cantilever's stiffness and the measured deflection

the forces experienced by the probe tip split into attractive or repulsive forces. As the tip approaches the surface, it experiences intermolecular forces that are attractive, although they can be repulsive under certain circumstances [37.3]. Once in contact with the surface the tip usually experiences a combination of attractive intermolecular forces and repulsive elastic forces. Two schools of thought exist regarding the attractive forces when the tip is in contact with the surface. The first is often referred to as the DMT or Bradley model. It holds that attractive forces only act outside the region of contact [37.4–7]. The second theory, usually called the JKR model, assumes that all the forces experienced by the tip, whether attractive or repulsive, act in the region of contact [37.8]. Most real nanoscale contacts lie somewhere between these two theoretical extremes.

37.1.2 Nanoindentation

The fundamental difference between AFM and nanoindentation is that during a nanoindentation experiment an external load is applied to the indenter tip. This load enables the tip to be pushed into the sample, creating a nanoscale impression on the surface, otherwise referred to as a nanoindentation or nanoindent.

Conventional indentation or microindentation tests involve pushing a hard tip of known geometry into the sample surface using a fixed peak load. The area of indentation that is created is then measured, and the mechanical properties of the sample, in particular its hardness, is calculated from the peak load and the indentation area. Various types of indentation testing are used in measuring hardness, including Rockwell, Vickers, and Knoop tests. The geometries and definitions of hardness used in these tests are shown by Fig. 37.2.

When indentations are performed on the nanoscale there is a basic problem in measuring the size of the indents. Standard optical techniques cannot easily be used to image anything smaller than a micron, while electron microscopy is simply impractical due to the time involved in finding and imaging small indents. To overcome these difficulties, nanoindentation methods have been developed that continuously record the load, displacement, time, and contact stiffness throughout the indentation process. This type of continuously recording indentation testing was originally developed in the former Soviet Union [37.9–13] as an extension of microindentation tests. It was applied to nanoscale indentation testing in the early

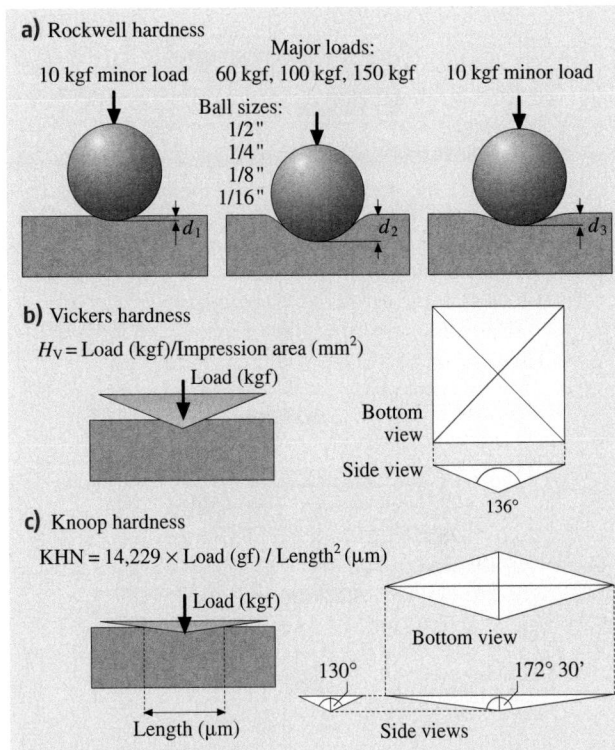

Fig. 37.2 (a) The standard Rockwell hardness test involves pushing a ball into the sample with a minor load, recording the depth, d_1, then applying a major load and recording the depth, d_2, then returning to the lower load and recording the depth, d_3. Using the depths, the hardness is calculated. (b) Vickers hardness testing uses a four-sided pyramid pushed into the sample with a known load. The area of the resulting indentation is measured optically and the hardness calculated as the load is divided by area. (c) Knoop indentation uses the same definition of hardness as the Vickers test, load divided by area, but the indenter geometry has one long diagonal and one short diagonal

1980s [37.14, 15], hence, giving rise to the field of nanoindentation testing.

In general, nanoindentation instruments include a loading system that may be electrostatic, electromagnetic, or mechanical, along with a displacement measuring system that may be capacitive or optical. Schematics of several commercial nanoindentation instruments are shown in Fig. 37.3a–c.

Among the many advantages of nanoindentation over conventional microindentation testing is the ability to measure the elastic, as well as the plastic properties of the test sample. The elastic modulus is obtained from the contact stiffness (S) using the following equation that

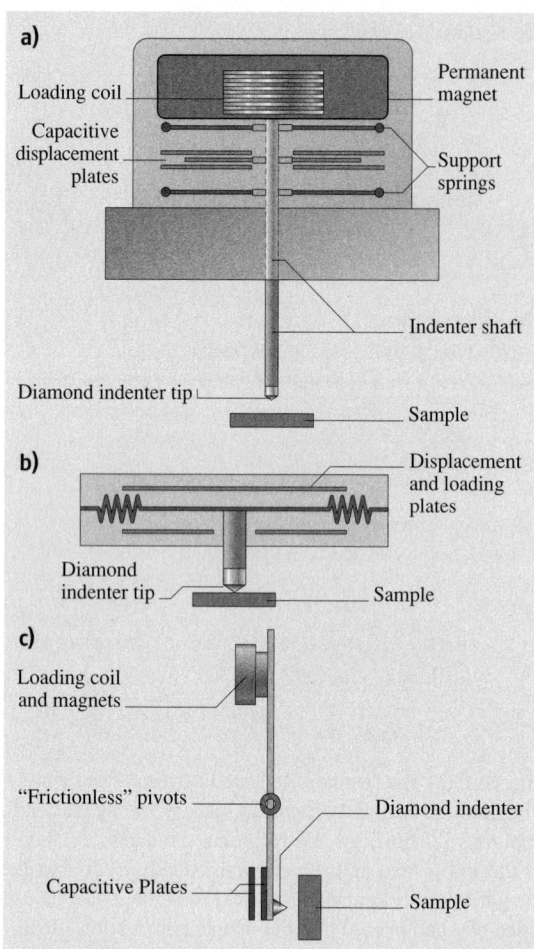

Fig. 37.3a–c Schematics of three commercial nanoindentation devices made by (**a**) MTS Nanoinstruments, Oak Ridge, Tennessee, (**b**) Hysitron Inc., Minneapolis, Minnesota, (**c**) Micro Materials Limited, Wrexham, UK. Instruments (**a**) and (**c**) use electromagnetic loading, while (**b**) uses electrostatic loads

appears to be valid for all elastic contacts [37.16, 17]:

$$S = \frac{2}{\sqrt{\pi}} E_r \sqrt{A} .$$ (37.1)

A is the contact area and E_r is the reduced modulus of the tip and sample as given by:

$$\frac{1}{E_r} = \frac{\left(1 - v_t^2\right)}{E_t} + \frac{\left(1 - v_s^2\right)}{E_s} ,$$ (37.2)

where E_t, v_t and E_s, v_s are the elastic modulus and Poissons ratio of the tip and sample, respectively.

37.1.3 Adaptations of Nanoindentation

Several adaptations to the basic nanoindentation setup have been used to obtain additional information about the processes that occur during nanoindentation testing, for example, in situ measurements of acoustic emissions and contact resistance. Environmental control has also been used to examine the effects of temperature and surface chemistry on the mechanical behavior of nanocontacts. In general, it is fair to say that the more information that can be obtained and the greater the control over the experimental parameters the easier it will be to understand the nanoindentation results. Load-displacement curves provide a lot of information, but they are only part of the story.

During nanoindentation testing discontinuities are frequently seen in the load–displacement curve. These are often called "pop-ins" or "pop-outs", depending on their direction. These sudden changes in the indenter displacement, at a constant load (see Fig. 37.4), can be caused by a wide range of events, including fracture, delamination, dislocation multiplication, or nucleation and phase transformations. To help distinguish between the various sources of discontinuities, acoustic transducers have been placed either in contact with the sample or immediately behind the indenter tip. For example, the results of nanoindentation tests that monitor acoustic emissions have shown that the phase transformations seen in silicon during nanoindentation are not the sudden events that they would appear to be from the load–displacement curve. There is no acoustic emission associated with the pop-out seen in the unloading curve of silicon [37.18]. An acoustic emission would be expected if there were a very rapid phase transformation causing a sudden change in volume. Fracture and delamination of films, however, give very strong acoustic signals [37.19],

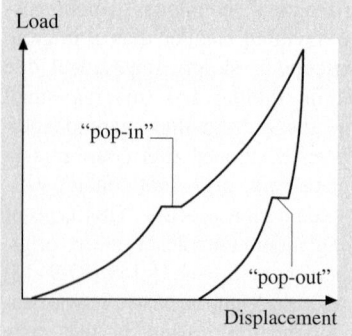

Fig. 37.4 Sketch of a load/displacement curve showing a pop-in and a pop-out

but the exact form of the signal appears to be more closely related to the sample geometry than to the event [37.20].

Additional information about the nature of the deformed region under the nanoindentation can be obtained by performing in situ measurements of contact resistance. The basic setup for this type of testing is shown in Fig. 37.5. An electrically conductive tip is needed to study contact resistance. Consequently, a conventional diamond tip is of limited use. Elastic, hard, and metallically conductive materials such as vanadium carbide can be used as substitutes for diamond [37.21, 22], or a thin conductive film (e.g., Ag) can be deposited on the diamond's surface (such a film is easily transferred to the indented surface so great care must be taken if multiple indents are performed). Measurements of contact resistance have been most useful for examining phase transformations in semiconductors [37.21, 22] and the dielectric breakdown of oxide films under mechanical loading [37.23].

One factor that is all too frequently neglected during nanoindentation testing is the effect of the experimental environment. Two obvious ways in which the environment can affect the results of nanoindentation tests are increases in temperature, which give elevated creep rates, and condensation of water vapor, which modifies the tip-sample interactions. Both of these environmental effects have been shown to significantly affect the measured mechanical properties and the modes of deformation that occur during nanoindentation [37.24–27]. Other environmental effects, for instance, those due to photoplasticity or hydrogen ion absorption, are also possible, but they are generally less troublesome than temperature fluctuations and variations in atmospheric humidity.

37.1.4 Complimentary Techniques

Nanoindentation testing is probably the most important technique for characterizing the mechanical properties of thin films and surfaces, but there are many alternative or additional techniques that can be used. One of the most important alternative methods for measuring the mechanical properties of thin films uses bulge or blister testing [37.28]. Bulge tests are performed on thin films mounted on supporting substrates. A small area of the substrate is removed to give a window of unsupported film. A pressure is then applied to one side of the window causing it to bulge. By measuring the height of the bulge, the stress-strain curve and the residual stress are obtained. The basic configuration for bulge testing is shown in Fig. 37.6.

37.1.5 Bulge Tests

The original bulge tests used circular windows because they are easier to analyze mathematically, but now square and rectangular windows have become common [37.29]. These geometries tend to be easier to fabricate. Unfortunately, there are several sources of errors in bulge testing that can potentially lead to large errors in the measured mechanical properties. These errors at one time led to the belief that multilayer films can show a "super modulus" effect, where the elastic modulus of the multilayer is several times that of its constituent layers [37.30]. It is now accepted that any enhancement

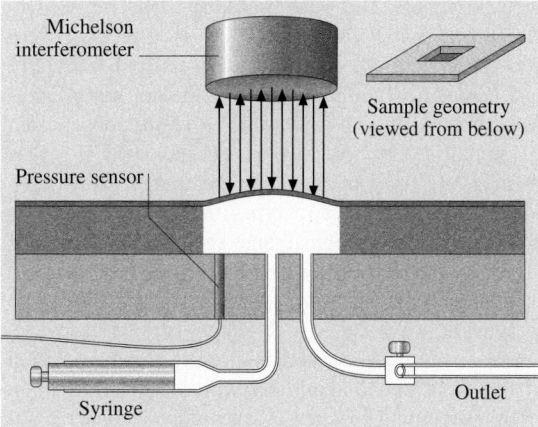

Fig. 37.6 Schematic of the basic setup for bulge testing. The sample is prepared so that it is a thin membrane, and then a pressure is applied to the back of the membrane to make it bulge upwards. The height of the bulge is measured using an interferometer

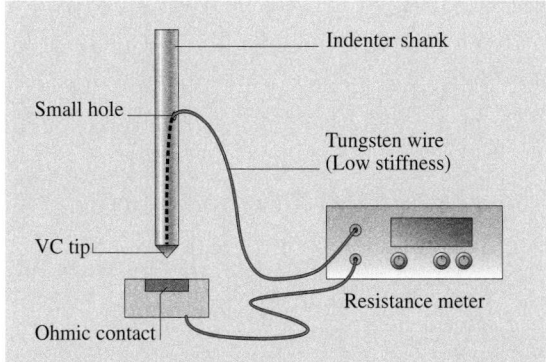

Fig. 37.5 Schematic of the basic setup for making contact resistance measurements during nanoindentation testing

to the elastic modulus in multilayer films is small, on the order of 15% [37.31]. The main sources of error stem from compressive stresses in the film (tensile stress is not a problem), small variations in the dimensions of the window, and uncertainty in the exact height of the bulge. Despite these difficulties, one advantage of bulge testing over nanoindentation testing is that the stress state is biaxial, so that only properties in the plane of the film are measured. In contrast, nanoindentation testing measures a combination of in-plane and out-of-plane properties.

37.1.6 Acoustic Methods

Acoustic and ultrasonic techniques have been used for many years to study the elastic properties of materials. Essentially, these techniques take advantage of the fact that the velocity of sound in a material is dependent on the inter-atomic or inter-molecular forces in the material. These, of course, are directly related to the material's elastic constants. In fact, any nonlinearity of inter-atomic forces enables slight variations in acoustic signals to be used as a measure of residual stress.

An acoustic method ideally suited to studying surfaces is scanning acoustic microscopy (SAM) [37.32]. There are also several other techniques that have been used to study surface films and multilayers, but we will first consider SAM in detail. In a SAM, a lens made of sapphire is used to bring acoustic waves to focus via a coupling fluid on the surface. A small piezo-electric transducer at the top of the lens generates the acoustic signal. The same transducer can be used to detect the signal when the SAM is used in reflection mode. The use of a transducer as both generator and detector, a common imaging mode, necessitates the use of a pulsed rather than a continuous acoustic signal. Continuous waves can be used if phase changes are used to build up the image. The transducer lens generates two types of acoustic waves in the material: longitudinal and shear. The ability of a solid to sustain both types of wave (liquids can only sustain longitudinal waves) gives rise to a third type of acoustic wave called a Rayleigh, or surface, wave. These waves are generated as a result of superposition of the shear and longitudinal waves with a common phase velocity. The stresses and displacements associated with a Rayleigh wave are only of significance to a depth of ≈ 0.6 Rayleigh wavelengths below the solid surface. Hence, using SAM to examine Rayleigh waves in a material is a true surface characterization technique.

Using a SAM in reflection mode gives an image where the contrast is directly related to the Rayleigh wave velocity, which is in turn a function of the material's elastic constants. The resolution of the image depends on the frequency of the transducer used, i. e., for a 2 GHz signal a resolution better than 1 μm is achievable. The contrast in the image results from the interference of two different waves in the coupling fluid. Rayleigh waves that are excited in the surface "leak" into the coupling fluid and interfere with the acoustic signal that is directly reflected back from the surface. It is usually assumed that the properties of the coupling fluid are well characterized. The interference of the two waves gives a characteristic $V(z)$ curve, as illustrated by Fig. 37.7, where z is the separation between the lens and the surface. Analyzing the periodicity of the $V(z)$ curves provides information on the Rayleigh wave velocity. As with other acoustic waves, the Rayleigh velocity is related to the elastic constants of the material. When using the SAM for a material's characterization, the lens is usually held in a fixed position on the surface. By using a lens designed specifically to give a line-focus beam, rather than the standard spherical lens, it is possible to use SAM to look at anisotropy in the wave velocity [37.33] and hence in elastic properties by producing waves with a specific direction.

One advantage of using SAM in conjunction with nanoindentation to characterize a surface is that the measurements obtained with the two methods have a slightly different dependence on the test material's elastic properties, E_s and ν_s (the elastic modulus and Poisson's ratio). As a result, it is possible to use SAM and nanoindentation combined to find both E_s and ν_s,

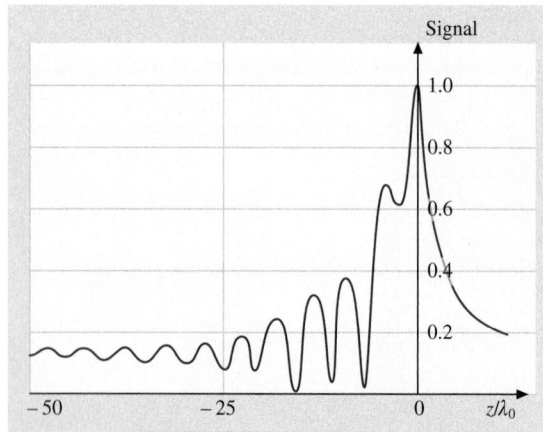

Fig. 37.7 A typical $V(z)$ curve obtained with a SAM when testing fused silica

as illustrated by Fig. 37.8 [37.34]. This is not possible when using only one of the techniques alone.

In addition to measuring surface properties, SAM has been used to study thin films on a surface. However, the Rayleigh wave velocity can be dependent on a complex mix of the film and substrate properties. Other acoustic methods have been utilized to study freestanding films. A freestanding film can be regarded as a plate, and, therefore, it is possible to excite Lamb waves in the film. Using a pulsed laser to generate the waves and a heterodyne interferometer to detect the arrival of the Lamb wave, it is possible to measure the flexural modulus of the film [37.35]. This has been successfully demonstrated for multilayer films with a total thickness $< 10\,\mu m$. In the plate configuration, due to the nonlinearity of elastic properties, it is also possible to measure stress. This has been demonstrated for horizontally polarized shear waves in plates [37.36], but thin plates require very high frequency transducers or laser sources.

37.1.7 Imaging Methods

When measuring the mechanical properties of a surface or thin film using nanoindentation, it is not always easy to visualize what is happening. In many instances there is a risk that the mechanical data can be completely misinterpreted if the geometry of the test is not as expected. To expedite the correct interpretation of the mechanical data, it is generally worthwhile to use optical, electron, or atomic force microscopy to image the nanoindentations. Obviously, optical techniques are only of use for larger indentations, but they will often reveal the presence of median or lateral cracks [37.37]. Electron microscopy and AFM, however, can be used to examine even the smallest nanoindentations. The principle problem with these microscopy techniques is the difficulty in finding the nanoindentations. It is usually necessary to make large, "marker" indentations in the vicinity of the nanoindentations to be examined in order to find them [37.38].

It is possible to see features such as extrusions with a scanning electron microscope (SEM) [37.39], as well as pile-up and sink-in around the nanoindents, though AFM is generally better for this. Transmission elec-

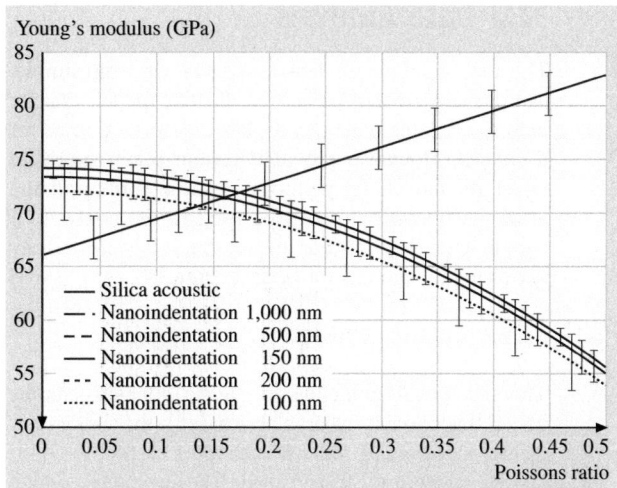

Fig. 37.8 Because SAM and nanoindentation have different dependencies on Young's modulus, E, and Poissons ratio, ν, it is possible to use the two techniques in combination to find E and ν [37.34]. On the graph, the intersection of the curves gives E and ν

tron microscopy (TEM) is useful for examining what has happened subsurface, for instance, the indentation induced dislocations in a metal [37.40] or the phases present under a nanoindent in silicon [37.21]. However, with TEM there is the added difficulty of sample preparation and the associated risk of observing artifacts. Recently, there has been considerable interest in the use of focused ion beams to cut cross sections through nanoindents [37.41]. When used in conjunction with SEM or TEM this provides an excellent means to see what has happened in the subsurface region.

One other technique that has proved to be useful in studying nanoindents is micro-Raman spectroscopy. This involves using a microscope to focus a laser on the sample surface. The same microscope is also used to collect the scattered laser light, which is then fed into a spectroscope. The Raman peaks in the spectrum provide information on the bonding present in a material, while small shifts in the wave number of the peaks can be used as a measure of strain. Micro-Raman has proven to be particularly useful for examining the phases present around nanoindentations in silicon [37.42].

37.2 Data Analysis

The analysis of nanoindentation data is far from simple. This is mostly due to the lack of effective models that are able to combine elastic and plastic deformation under a contact. However, provided certain precautions are taken, the models for perfectly elastic deformation and ideal plastic materials can be used in the analysis of nanoindentation data. For this reason, it is worth briefly reviewing the models for perfect contacts.

37.2.1 Elastic Contacts

The theoretical modeling of elastic contacts can be traced back many years, at least to the late nineteenth century and the work of *Hertz* (1882) [37.43] and *Boussinesq* (1885) [37.44]. These models, which are still widely used today, consider two axisymmetric curved surfaces in contact over an elliptical region (see Fig. 37.9). The contact region is taken to be small in comparison to the radius of curvature of the contacting surfaces, which are treated as elastic half-spaces. For an elastic sphere, radius R, in contact with a flat, elastic half-space, the contact region will be circular and the Hertz model gives the following relationships:

$$a = \sqrt[3]{\frac{3PR}{4E_r}} , \tag{37.3}$$

$$\delta = \sqrt[3]{\frac{9P^2}{16RE_r^2}} , \tag{37.4}$$

$$P_0 = \sqrt[3]{\frac{6PE_r^2}{\pi^3 R^2}} , \tag{37.5}$$

where a is the radius of the contact region, E_r is given by (37.2), δ is the displacement of the sphere into the surface, P is the applied load and P_0 is the maximum pressure under the contact (in this case at the center of the contact).

The work of *Hertz* and *Boussinesq* was extended by *Love* [37.45, 46] and later by *Sneddon* [37.47], who simplified the analysis using Hankel transforms. *Love* showed how *Boussinesq*'s model could be used for a flat-ended cylinder and a conical indenter, while *Sneddon* produced a generalized relationship for any rigid axisymmetric punch pushed into an elastic half-space. *Sneddon* applied his new analysis to punches of various shapes and derived the following relationships between the applied load, P, and displacement, δ, into the elastic half-space for, respectively, a flat-ended cylinder, a cone

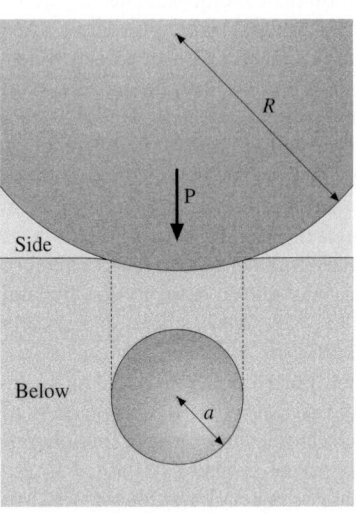

Fig. 37.9 Hertzian contact of a sphere, radius R, on a semi-infinite, flat surface. The contact in this case is a circular region of radius a

of semi-vertical angle ϕ, and a parabola of revolution where $a^2 = 2k\delta$:

$$P = \frac{4\mu a \delta}{1 - \nu} , \tag{37.6}$$

$$P = \frac{4\mu \cot \phi}{\pi (1 - \nu)} \delta^2 , \tag{37.7}$$

$$P = \frac{8\mu}{3(1 - \nu)} \left(2k\delta^3\right)^{1/2} , \tag{37.8}$$

where μ and ν are the shear modulus and Poisson's ratio of the elastic half-space, respectively.

The key point to note about (37.6), (37.7), and (37.8) is that they all have the same basic form, namely:

$$P = \alpha \delta^m , \tag{37.9}$$

where α and m are constants for each geometry.

Equation (37.9) and the relationships developed by Hertz and his successors, (37.3–37.8), form the foundation for much of the current nanoindentation data analysis routines.

37.2.2 Indentation of Ideal Plastic Materials

Plastic deformation during indentation testing is not easy to model. However, the indentation response of ideal plastic metals was considered by *Tabor* in his classic text, "The Hardness of Metals" [37.48]. An ideal plastic material (or more accurately an ideal elastic-plastic material) has a linear stress-strain curve until it reaches its elastic limit and then yields plastically at a yield stress, Y_0, that remains constant even after deformation has commenced. In a 2-D problem, the yielding occurs because the Huber-Mises [37.49] criterion has been reached. In other words, the maximum shear stress acting on the material is around $1.15Y_0/2$.

First, we consider a 2-D flat punch pushed into an ideal plastic material. By using the method of slip lines it is found that the mean pressure, P_m, across the end of the punch is related to the yield stress by:

$$P_m = 3Y_0 . \tag{37.10}$$

If the *Tresca* criterion [37.50] is used, then P_m is closer to $2.6Y_0$. In general, for both 2-D and three-dimensional punches pushed into ideal plastic materials, full plasticity across the entire contact region can be expected when $P_m = 2.6$ to $3.0Y_0$. However, significant deviations from this range can be seen if, for instance, the material undergoes work-hardening during indentation, or the material is a ceramic, or there is friction between the indenter and the surface.

The apparently straightforward relationship between P_m and Y_0 makes the mean pressure a very useful quantity to measure. In fact, P_m is very similar to the Vickers hardness, H_V, of a material:

$$H_V = 0.927P_m . \tag{37.11}$$

During nanoindentation testing it is the convention to take the mean pressure as the nanohardness. Thus, the "nanohardness", H, is defined as the peak load, P, applied during a nanoindentation divided by the projected area, A, of the nanoindentation in the plane of the surface, hence:

$$H = \frac{P}{A} . \tag{37.12}$$

37.2.3 Adhesive Contacts

During microindentation testing and even most nanoindentation testing the effects of intermolecular and surface forces can be neglected. Very small nanoindentations, however, can be influenced by the effects of intermolecular forces between the sample and the tip. These adhesive effects are most readily seen when testing soft polymers, but there is some evidence that forces between the tip and sample may be important in even relatively strong materials [37.51, 52].

Contact adhesion is usually described by either the JKR or DMT model, as discussed earlier in this chap-

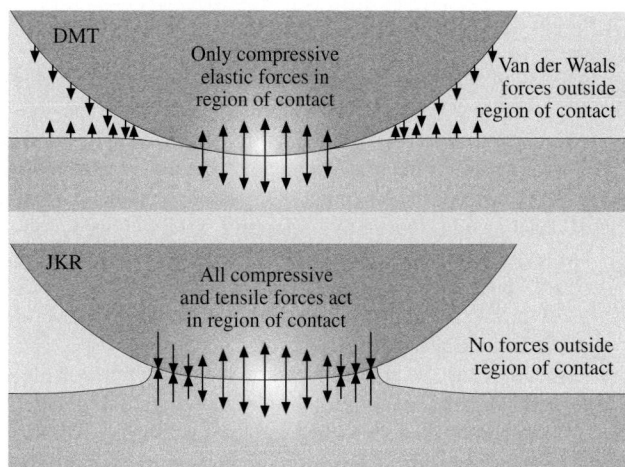

Fig. 37.10 The contact geometry for the DMT and JKR models for adhesive contact. Both models are based on the Hertzian model. In the DMT, model van der Waals forces outside the region of contact introduce an additional load in the Hertz model. But for the JKR model, it is assumed that tensile, as well as compressive stresses can be sustained within the region of contact

ter. Both the models consider totally elastic spherical contacts under the influence of attractive surface forces. The JKR model considers the surface forces in terms of the associated surface energy, whereas the DMT model considers the effects of adding van der Waals forces to the Hertzian contact model. The differences between the two models are illustrated by Fig. 37.10.

For nanoindentation tests conducted in air the condensation of water vapor at the tip-sample interface usually determines the size of the adhesive force acting during unloading. The effects of water vapor on a single nanoasperity contact have been studied using force-controlled AFM techniques [37.53] and, more recently, nanoindentation methods [37.26]. Unsurprisingly, it has also been found that water vapor can affect the deformation of surfaces during nanoindentation testing [37.27].

In addition to water vapor, other surface adsorbates can cause dramatic changes in the nanoscale mechanical behavior. For instance, oxygen on a clean metal surface can cause an increase in the apparent strength of the metal [37.54]. These effects are likely to be related to, firstly, changes in the surface and intermolecular forces acting between the tip and the sample and, secondly, changes in the mechanical stability of surface nanoasperities and ledges. Adsorbates can help stabilize atomic-scale variations in surface morphology, thereby making defect generation at the surface more difficult.

37.2.4 Indenter Geometry

All of the indenter geometries considered up to this point have been axisymmetric, largely because they are easier to deal with theoretically. Unfortunately, fabricating axisymmetric nanoindentation tips is extremely difficult, because shaping a hard tip on the scale of a few nanometers is virtually impossible. Despite these problems, there has been considerable effort put into the use of spherical nanoindentation tips [37.55]. This clearly demonstrates that the spherical geometry can be useful at larger indentation depths.

Because of the problems associated with creating axisymmetric nanoindentation tips, pyramidal indenter geometries have now become standard during nanoindentation testing. The most common geometries are the three-sided Berkovich pyramid and cube-corner (see Fig. 37.11). The Berkovich pyramid is based on the four-sided Vickers pyramid, the opposite sides of which make an 136° angle. For both the Vickers and Berkovich pyramids the cross-sectional area of the pyramid's base, A, is related to the pyramid's height, D, by:

$$A = 24.5D^2 . \tag{37.13}$$

The cube-corner geometry is now widely used for making very small nanoindentations, because it is much sharper than the Berkovich pyramid. This makes it easier to initiate plastic deformation at very light loads, but great care should be taken when using the cube-corner geometry. Sharp cube-corners can wear down quickly and become blunt, hence the cross-sectional area as a function of depth can change over the course of several indentations. There is also a potential problem with the standard analysis routines [37.56], which were developed for much blunter geometries and are based on the elastic contact models outlined earlier. The elastic contact models all assume the displacement into the surface is small compared to the tip radius. For the cube-corner geometry this is probably only the case for nanoindentations that are no more than a few nanometers deep.

37.2.5 Analyzing Load/Displacement Curves

The load/displacement curves obtained during nanoindentation testing are deceptively simple. Most newcomers to the area will see the curves as being somewhat akin to the stress/strain curves obtained during tensile testing. There is also a real temptation just to use the values of hardness, H, and elastic modulus, E, obtained from standard analysis software packages as the "true" values. This may be the case in many instances, but for very shallow nanoindents and tests on thin films the geometry of the contact can differ significantly from the geometry assumed in the analysis routines. Consequently, experimentalists should think very carefully about the test itself before concluding that the values of H and E are correct.

The basic shape of a load/displacement curve can reveal a great deal about the type of material being tested. Figure 37.12 shows some examples of ideal curves for materials with different elastic moduli and yield stresses. Discontinuities in the load/displacement curve can also provide information on such processes as fracture, dislocation nucleation, and phase transformations. Initially, though, we will consider ideal situations such as those illustrated by Fig. 37.12.

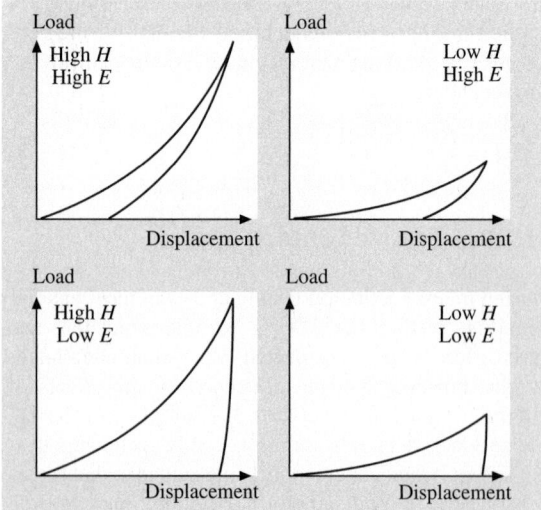

Fig. 37.12 Examples of load/displacement curves for idealized materials with a range of hardness and elastic properties

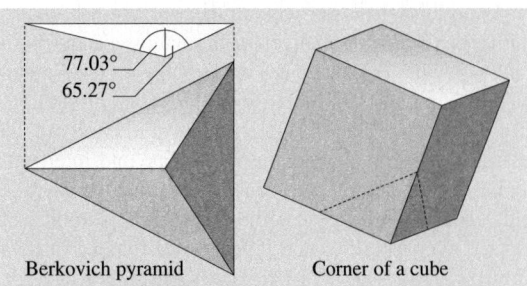

Fig. 37.11 The ideal geometry for the three-sided Berkovich pyramid and cube corner tips

The loading section of the load/displacement curve approximates a parabola [37.57] whose width depends on a combination of the material's elastic and plastic properties. The unloading curve, however, has been shown to follow a more general relationship [37.56] of the form:

$$P = \alpha \, (\delta - \delta_i)^m \; , \tag{37.14}$$

where δ is the total displacement and δ_i is the intercept of the unloading curve with the displacement axis shown in Fig. 37.13.

Equation (37.14) is essentially the same as (37.9) but with the origin displaced. Since (37.9) is obtained by considering purely elastic deformation, it follows that the unloading curve is exhibiting purely elastic behavior. Since the shape of the unloading curve is determined by the elastic recovery of the indented region, it is not entirely surprising that its shape resembles that found for purely elastic deformation. What is fortuitous is that the

elastic analysis used for an elastic half-space seems to be valid for a surface where there is a plastically formed indentation crater present under the contact. However, the validity of this analysis may only hold when the crater is relatively shallow and the geometry of the surface does not differ significantly from that of a flat, elastic half-space. For nanoindentations with a Berkovich pyramid, this is generally the case.

Before *Oliver* and *Pharr* [37.56] proposed their now standard method for analyzing nanoindentation data, the analysis had been based on the observation that the initial part of the unloading curve is almost linear. A linear unloading curve, equivalent to $m = 1$ in (37.14), is expected when a flat punch is used on an elastic half-space. The flat punch approximation for the unloading curve was used in [37.58–60] to analyze nanoindentation data. When Oliver and Pharr looked at a range of materials they found m was typically larger than 1, and that $m = 1.5$, or a paraboloid, was a better approximation than a flat punch. Oliver and Pharr used (37.1) and (37.12) to obtain the values for a material's elastic modulus and hardness. Equation (37.1) relates the contact stiffness during the initial part of the unloading curve (see Fig. 37.13) to the reduced elastic modulus and the contact area at the peak load. Equation (37.12) gives the hardness as the peak load divided by the contact area. It is immediately obvious that the key to measuring the mechanical properties of a material is knowing the contact area at the peak load. This is the single most important factor in analyzing nanoindentation data. Most mistakes in the analysis come from incorrect assumptions about the contact area.

To find the contact area, a function relating the contact area, A_c, to the contact depth, δ_c, is needed. For a perfect Berkovich pyramid this would be the same as (37.13). But since making a perfect nanoindenter tip is impossible, an expanded equation is used:

$$A_c \, (\delta_c) = 24.5\delta_c^2 + \sum_{j=1}^{7} C_j \sqrt[2j]{\delta_c} \; , \tag{37.15}$$

where C_j are calibration constants of the tip.

There is a crucial step in the analysis before A_c can be calculated, namely, finding δ_c. The contact depth is not the same as the indentation depth, because the surface around the indentation will be elastically deflected during loading, as illustrated by Fig. 37.14. Sneddon's analysis [37.47] provides a way to calculate the deflection of the surface at the edge of an axisymmetric contact. Subtracting the deflection from the total indentation depth at peak load gives the contact depth. For a paraboloid, as used by *Oliver* and *Pharr* [37.56] in

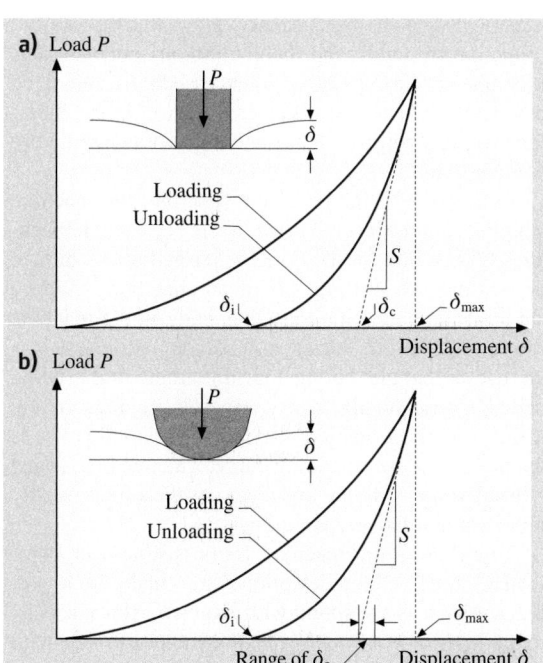

Fig. 37.13a,b Analysis of the load/displacement curve gives the contact stiffness, S, and the contact depth, δ_c. These can then be used to find the hardness, H, and elastic, or Young's modulus, E. (**a**) The first method of analysis [37.58–60] assumed the unloading curve could be approximated by a flat punch on an elastic half-space. (**b**) A more refined analysis [37.56] uses a paraboloid on an elastic half-space

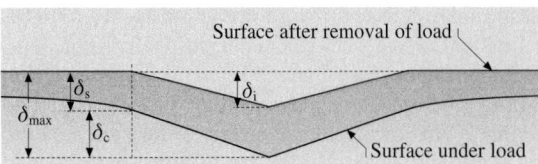

Fig. 37.14 Profile of surface under load and unloaded showing how δ_c compares to δ_i and δ_{max}

their analysis, the elastic deflection at the edge of the contact is given by:

$$\delta_s = \varepsilon \frac{P}{S} = 0.75 \frac{P}{S} \,, \tag{37.16}$$

where S is the contact stiffness and P the peak load. The constant ε is 0.75 for a paraboloid, but ranges between 0.72 (conic indenter) and 1 (flat punch). Figure 37.15 shows how the contact depth depends on the value of ε. The contact depth at the peak load is, therefore:

$$\delta_c = \delta - \delta_s \,. \tag{37.17}$$

Using the load/displacement data from the unloading curve and (37.1), (37.2), (37.12), (37.14–37.17), the hardness and reduced elastic modulus for the test sample can be calculated. To find the elastic modulus of the sample, E_s, it is also necessary to know Poisson's ratio, ν_s, for the sample, as well as the elastic modulus, E_t, and Poisson's ratio, ν_t, of the indenter tip. For diamond these are 1141 GPa and 0.07, respectively.

There also remains the issue of calibrating the tip shape, or finding the values for C_j in (37.15). Knowing the exact expansion of $A_c(\delta_c)$ is vital if the values for E_s and H are to be accurate. Several methods for calibrating the tip shape have been used, including imaging the tip with an electron microscope, measuring the size of nanoindentations using SEM or TEM of negative replicas, and using scanning probes to examine either the tip itself or the nanoindentations made with the tip. There are strengths and weaknesses to each of these methods. In general, however, the accuracy and

usefulness of the methods depends largely on how patient and rigorous the experimentalist is in performing the calibration.

Because of the experimental difficulties and time involved in calibrating the tip shape by these methods, *Oliver* and *Pharr* [37.56] developed a method for calibration based on standard specimens. With a standard specimen that is mechanically isotropic and has a known E and H that does not vary with indentation depth, it should be possible to perform nanoindentations to a range of depths, and then use the analysis routines in reverse to deduce the tip area function, $A_c(\delta_c)$. In other words, if you perform a nanoindentation test, you can find the contact stiffness, S, at the peak load, P, and the contact depth, δ_c, from the unloading curve. Then if you know E a priori, (37.1) can be used to calculate the contact area, A, and, hence, you have a value for A_c at a depth δ_c. Repeating this procedure for a range of depths will give a numerical version of the function $A_c(\delta_c)$. Then, it is simply a case of fitting (37.15) to the numerical data. If the hardness, H, is known and not a function of depth, and the calibration specimen was fully plastic during testing, then essentially the same approach could be used but based on (37.12). Situations where a constant H is used to calibrate the tip are extremely rare.

In addition to the tip shape function, the machine compliance must be calibrated. Basic Newtonian mechanics tells us that for the tip to be pushed into a surface the tip must be pushing off of another body. During nanoindentation testing the other body is the machine frame. As a result, during a nanoindentation test it is not just the sample, but the machine frame that is being loaded. Consequently, a very small elastic deformation of the machine frame contributes to the total stiffness obtained from the unloading curve. The machine frame is usually very stiff, $> 10^6$ N/m, so the effect is only important at relatively large loads.

To calibrate the machine frame stiffness or compliance, large nanoindentations are made in a soft material such as aluminum with a known, isotropic elastic modulus. For very deep nanoindentations made with a Berkovich pyramid, the contact area, $A_c(\delta_c)$, can be reasonably approximated to $24.5\delta_c^2$, thus (37.1) can be used to find the expected contact stiffness for the material. Any difference between the expected value of S and the value measured from the unloading curve will be due to the compliance of the machine frame. Performing a number of deep nanoindentations enables an accurate value for the machine frame compliance to be obtained.

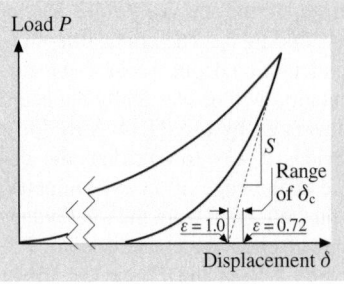

Fig. 37.15 Load-displacement curve showing how δ_c varies with ε

Currently, because of its ready availability and predictable mechanical properties, the most popular calibration material is fused silica ($E = 72\,\text{GPa}$, $\nu = 0.17$), though aluminum is still used occasionally.

37.2.6 Modifications to the Analysis

Since the development of the analysis routines in the early 1990s, it has become apparent that the standard analysis of nanoindentation data is not applicable in all situations, usually because errors occur in the calculated contact depth or contact area. *Pharr* et al. [37.61–64] have used finite element modeling (FEM) to help understand and overcome the limitations of the standard analysis. Two important sources of errors have been identified in this way. The first is residual stress at the sample surface. The second is the change in the shape of nanoindents after elastic recovery.

The effect of residual stresses at a surface on the indentation properties has been the subject of debate for many years [37.65–67]. The perceived effect was that compressive stresses increased hardness, while tensile stresses decreased hardness. Using FEM it is possible to model a pointed nanoindenter being pushed into a model material that is in residual tension or compression. An FEM model of nanoindentation into aluminum alloy 8009 [37.61] has confirmed earlier experimental observations [37.68] indicating that the contact area calculated from the unloading curve is incorrect if there are residual stresses. In the FEM model of an aluminum alloy the mechanical behavior of the material is modeled using a stress-strain curve, which resembles that of an elastic-perfectly-plastic metal with a flow stress of 425.6 MPa. Yielding starts at 353.1 MPa and includes a small amount of work hardening. The FEM model was used to find the contact area directly and using the simulated unloading curve in conjunction with Oliver and Pharr's method. The results as a function of residual stress are illustrated in Fig. 37.16. Note that the differences between the two measured contact areas lead to miscalculations of E and H.

Errors in the calculated contact area stem from incorrect assumptions about the pile-up and sink-in at the edge of the contact, as illustrated by Fig. 37.17. The Oliver and Pharr analysis assumes the geometry of the sample surface is the same as that given by *Sneddon* [37.47] in his analytical model for the indentation of elastic surfaces. Clearly, for materials where there is significant plastic deformation, it is possible that there will be large deviations from the surface geometry found using Sneddon's elastic model. In reality, the error in the contact

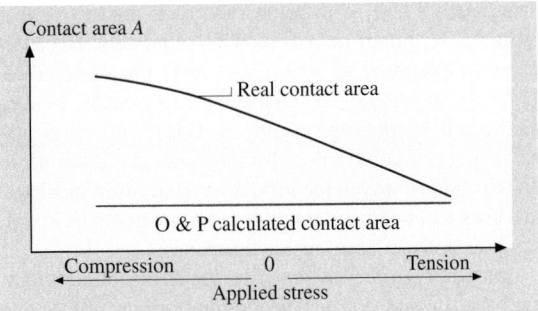

Fig. 37.16 When a surface is in a state of stress there is a significant difference between the contact area calculated using the Oliver and Pharr method and the actual contact area [37.61]. For an aluminum alloy this can lead to significant errors in the calculated hardness and elastic modulus

area depends on how much the geometry of the test sample surface differs from that of the calibration material (typically fused silica). It is possible that a test sample, even without a residual stress, will have a different surface geometry and, hence, contact area at a given depth, when compared to the calibration material. This is often seen for thin films on a substrate (e.g., *Tsui* et al. [37.69, 70]). Residual stresses increase the likelihood that the contact area calculated using Oliver and Pharr's method will be incorrect.

The issue of sink-in and pile-up is always a factor in nanoindentation testing. However, there is still no effective way to deal with these phenomena other than reverting to imaging of the indentations to identify the true contact area. Even this is difficult, as the edge of an indentation is not easy to identify using AFM or electron microscopy. One approach that has

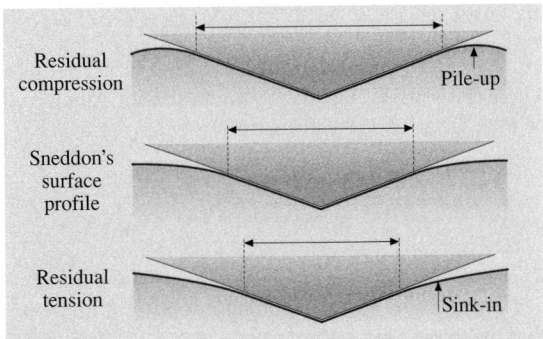

Fig. 37.17 Pile-up and sink-in are affected by residual stresses, and, hence, errors are introduced into standard Oliver and Pharr analysis

been used [37.71] with some success is measuring the ratio E_r^2/H, rather than E_r and H separately. Because E_r is proportional to $1/\sqrt{A}$ and H is proportional to $1/A$, E_r^2/H should be independent of A and, hence, unaffected by pile-up or sink-in. While this does not provide quantitative values for mechanical properties, it does provide a way to identify any variations in mechanical properties with indentation depth or between similar samples with different residual stresses.

Another source of error in the Oliver and Pharr analysis is due to incorrect assumptions about the nanoindentation geometry after unloading [37.63]. Once again, this is due to differences between the test sample and the calibration material. The exact shape of an unloaded nanoindentation on a material exhibiting elastic recovery is not simply an impression of the tip shape; rather, there is some elastic recovery of the nanoindentation sides giving them a slightly convex shape (see Fig. 37.18). The shape actually depends on Poisson's ratio, so the standard Oliver and Pharr analysis will only be valid for a material where $\nu = 0.17$, the value for fused silica, assuming it is used for the calibration.

To deal with the variations in the recovered nanoindentation shape, it has been suggested [37.63] that a modified nanoindenter geometry with a slightly concave side be used in the analysis (see Fig. 37.18). This requires a modification to (37.1):

$$S = \gamma 2 E_r \sqrt{\frac{A}{\pi}} , \qquad (37.18)$$

where γ is a correction term dependent on the tip geometry. For a Berkovich pyramid the best value is:

$$\gamma = \frac{\frac{\pi}{4} + 0.15483073 \cot \Phi \left(\frac{(1-2\nu_s)}{4(1-\nu_s)}\right)}{\left[\frac{\pi}{2} - 0.83119312 \cot \Phi \left(\frac{(1-2\nu_s)}{4(1-\nu_s)}\right)\right]^2} , \qquad (37.19)$$

where $\Phi = 70.32°$. For a cube corner the correction can be even larger and γ is given by:

$$\gamma = 1 + \left(\frac{(1-2\nu_s)}{4(1-\nu_s)\tan \Phi}\right) , \qquad (37.20)$$

where $\Phi = 42.28°$. Figure 37.19 shows how the modified contact area varies with depth for a real diamond Berkovich pyramid.

The validity of the γ-modified geometry is questionable from the perspective of contact mechanics since it relies on assuming an incorrect geometry for the nanoin-

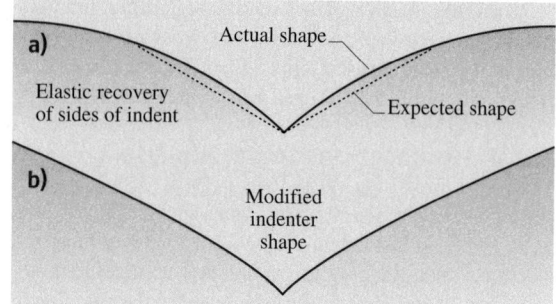

Fig. 37.18 (a) *Hay* et al. [37.63] found from experiments and FEM simulations that the actual shape of an indentation after unloading is not as expected. **(b)** They introduced a γ term to correct for this effect. This assumes the indenter has slightly concave sides

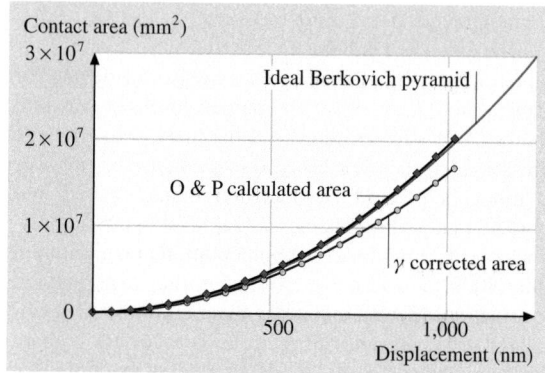

Fig. 37.19 For a real Berkovich tip the γ corrected area [37.63] is less at a given depth than the area calculated using the Oliver and Pharr method

denter tip to correct for an error in the geometry of the nanoindentation impression. The values for E and H obtained using the γ-modification are, however, good and can be significantly different to the values obtained with the standard Oliver and Pharr analysis.

37.2.7 Alternative Methods of Analysis

All of the preceding discussion on the analysis of nanoindentation curves has focused on the unloading curve, virtually ignoring the loading curve data. This is for the simple reason that the unloading curve can in many cases be regarded as purely elastic, whereas the shape of the loading curve is determined by a complex mix of elastic and plastic properties.

It is clear that there is substantially more data in the loading curve if it can be extracted. *Page* et al. [37.57,

72] have explored the possibility of curve fitting to the loading data using a combination of elastic and plastic properties. By a combination of analysis and empirical fitting to experimental data, it was suggested that the loading curve is of the following form:

$$P = E \left(\psi \sqrt{\frac{H}{E}} + \phi \sqrt{\frac{E}{H}} \right)^{-2} \delta^2 , \qquad (37.21)$$

where ψ and ϕ are determined experimentally to be 0.930 and 0.194, respectively. For homogenous samples this equation gives a linear relationship between P and δ^2. Coatings, thin film systems, and samples that strain-harden can give significant deviations from linearity. Analysis of the loading curve has yet to gain popularity as a standard method for examining nanoindentation data, but it should certainly be regarded as a prime area for further investigation.

Another alternative method of analysis is based on the work involved in making an indentation. In essence, the nanoindentation curve is a plot of force against distance indicating integration under the loading curve will give the total work of indentation, or the sum of the elastic strain energy and the plastic work of indentation. Integrating under the unloading curve should give only the elastic strain energy. Thus, the work involved in both elastic and plastic deformation during nanoindentation can be found. *Cheng* and *Cheng* [37.73] combined measurements of the work of indentation with a dimensional analysis that deals with the effects of scaling in a material that work-hardens to estimate H/E_r. They subsequently evaluated H and E using the Oliver and Pharr approach to find the contact area.

37.2.8 Measuring Contact Stiffness

As discussed earlier, it is possible to add a small AC load on top of the DC load used during nanoindentation testing, providing a way to measure the contact stiffness throughout the entire loading and unloading cycle [37.74, 75]. The AC load is typically at a frequency of $\approx 60\,\mathrm{Hz}$ and creates a dynamic system, with the sample acting as a spring with stiffness S (the contact stiffness), and the nanoindentation system acting as a series of springs and dampers. Figure 37.20 illustrates how the small AC load is added to the DC load. Figure 37.21 shows how the resulting dynamic system can be modeled. An analysis of the dynamic system gives the following relationships for S based on the amplitude of the AC displacement oscillation and the phase difference

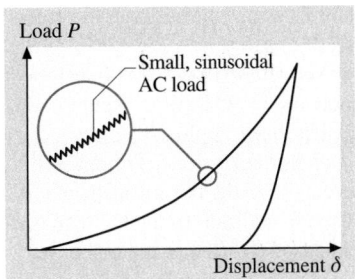

Fig. 37.20 A small AC load can be added to the DC load. This enables the contact stiffness, S, to be calculated throughout the indentation cycle

between the AC load and displacement signals:

$$\left| \frac{P_{\mathrm{os}}}{\delta(\omega)} \right| = \sqrt{\left[\left(S^{-1} + C_{\mathrm{f}} \right)^{-1} + K_{\mathrm{s}} - m\omega^2 \right]^2 + \omega^2 D^2} , \qquad (37.22)$$

$$\tan(\chi) = \frac{\omega D}{\left(S^{-1} + C_{\mathrm{f}} \right)^{-1} + K_{\mathrm{s}} - m\omega^2} , \qquad (37.23)$$

where C_{f} is the load frame compliance (the reciprocal of the load frame stiffness), K_{s} is the stiffness of the support springs (typically in the region of $50\text{--}100\,\mathrm{N/m}$), D is the damping coefficient, P_{os} is the magnitude of the load oscillation, $\delta(\omega)$ is the magnitude of the displacement oscillation, ω is the oscillation frequency, m is the mass of the indenter, and χ is the phase angle between the force and the displacement.

In order to find S using either (37.22) or (37.23), it is necessary to calibrate the dynamic response of the system when the tip is not in contact with a sample ($S^{-1} = 0$). This calibration combined with the standard DC calibrations will provide the values for all of the constants in the two equations. All that needs to be measured in order to obtain S is either $\delta(\omega)$ or χ, both of which are measured by the lock-in amplifier used to generate the AC signal. Since the S obtained is the same as the S in (37.1), it follows that the Oliver and Pharr analysis

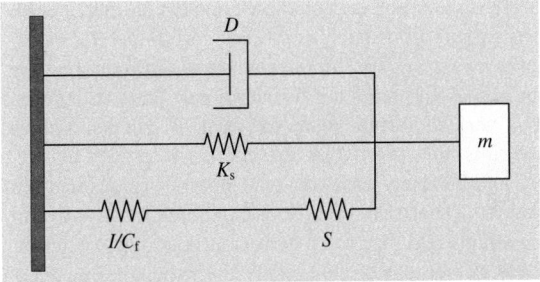

Fig. 37.21 The dynamic model used in the analysis of the AC response of a nanoindentation device

can be applied to obtain E_r and H throughout the entire nanoindentation cycle.

The dynamic analysis detailed here was developed for the MTS Nanoindenter™ (Oakridge, Tennessee), but a similar analysis has been applied to other commercial instruments such as the Hysitron Triboscope™ (Minneapolis, Minnesota) [37.76]. For all instruments, an AC oscillation is used in addition to the DC voltage, and a dynamic model is used to analyze the response.

37.2.9 Measuring Viscoelasticity

Using an AC oscillation in addition to the DC load introduces the possibility of measuring viscoelastic properties during nanoindentation testing. This has recently been the subject of considerable interest with researchers looking at the loss modulus, storage modulus, and loss tangent of various polymeric materials [37.25, 77]. Recording the displacement response to the AC force oscillation enables the complex modulus (including the loss and storage modulus) to be found. If the modulus is complex, it is clear from (37.1) that the stiffness also becomes complex. In fact, the stiffness will have two components: S', the component in phase with the AC force and S'', the component out of phase with the AC force.

The dynamic model illustrated in Fig. 37.21 is no longer appropriate for this situation, as the contact on

the test sample also includes a damping term, shown in Fig. 37.22. Equations (37.22) and (37.23) must also

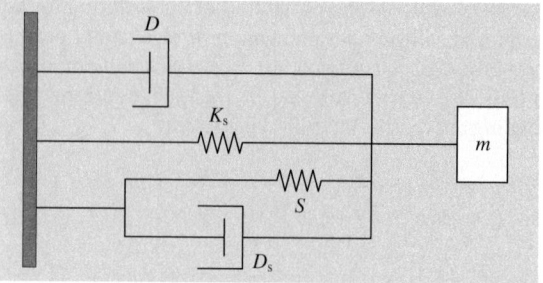

Fig. 37.22 The simplified dynamic model used when the sample is viscoelastic. It is assumed that the load frame compliance is negligible

be revised. Neglecting the load frame compliance, C_f, which in most real situations is negligible, (37.22) and (37.23) when the sample damping, D_s, is included become:

$$\left| \frac{P_{os}}{\delta(\omega)} \right| = \sqrt{\{S + K_s - m\omega^2\}^2 + \omega^2 (D + D_s)^2},$$

(37.24)

$$\tan(\chi) = \frac{\omega(D + D_s)}{S + K_s - m\omega^2}.$$

(37.25)

In order to find the loss modulus and storage modulus, (37.1) is used to relate S' (storage component) and S'' (loss component) to the complex modulus.

This method for measuring viscoelastic properties using nanoindentation has now been proven in principal, but has still only been applied to a very small range of polymers and remains an area of future growth.

37.3 Modes of Deformation

As described earlier, the analysis of nanoindentation data is based firmly on the results of elastic continuum mechanics. In reality, this idealized, purely elastic situation rarely occurs. For very shallow contacts on metals with thin surface films such as oxides, carbon layers, or organic layers [37.78, 79], the contact can initially be very similar to that modeled by Hertz and, later, Sneddon. It is very important to realize that this in itself does not constitute proof that the contact is purely elastic, because in many cases a small number of defects are present. These may be preexisting defects that move in the strain-field generated beneath the contact. Alternatively, defects can be generated either when the contact is first made or during the initial loading [37.52, 80]. When defects such as short lengths of dislocation are present

the curves may still appear to be elastic even though inelastic processes like dislocation glide and cross-slip are taking place.

37.3.1 Defect Nucleation

Nucleation of defects during nanoindentation testing has been the subject of many experimental [37.81, 82] and theoretical studies [37.83, 84]. This is probably because nanoindentation is seen as a way to deform a small, defect-free volume of material to its elastic limit and beyond in a highly controlled geometry. There are, unfortunately, problems in comparing experimental results with theoretical predictions, largely because the kinetic processes involved in defect nucleation are difficult to model. Simulations conducted at 0 K do not permit kinetic processes, and molecular dynamics simulations are too fast (nanoseconds or picoseconds). Real nanoindentation experiments take place at ≈ 293 K and last for seconds or even minutes.

Kinetic effects appear in many forms, for instance, during the initial contact between the indenter tip and the surface when defects can be generated by the combined action of the impact velocity and surface forces [37.51]. A second example of a kinetic effect occurs during hold cycles at large loads when what appears to be an elastic contact can suddenly exhibit a large discontinuity in the displacement data [37.80]. Figure 37.23 shows how these kinetic effects can affect the nanoindentation data and the apparent yield point load.

During the initial formation of a contact, the deformation of surface asperities [37.51] and ledges [37.85] can create either point defects or short lengths of dislocation line. During the subsequent loading, the defects can help in the nucleation and multiplication of dislocations. The large strains present in the region surrounding the contact, coupled with the existence of defects generated on contact, can result in the extremely rapid multiplication of dislocations and, hence, pronounced discontinuities in the load-displacement curve. It is important to realize that the discontinuities are due to the rapid multiplication of dislocations, which may or may not occur at the same time that the first dislocation is nucleated. Dislocations may have been present for some time with the discontinuity only occurring when the existing defects are configured appropriately, as a Frank-Read source, for instance. Even under large strains, the time taken for a dislocation source to form from preexisting defects may be long. It is, therefore, not surprising that large discontinuities can be seen during hold cycles or unloading.

The generation of defects at the surface and the initiation of yielding is a complex process that is extremely dependent on surface asperities and surface forces. These, in turn, are closely related to the surface chemistry. It is not only the magnitude of surface forces, but also their range in comparison to the height of surface asperities that determines whether defects are generated on contact. Small changes in the surface chemistry or the velocity of the indenter tip when it first contacts the surface, can cause a transition from a situation in which defects are generated on contact to one where the contact is purely elastic [37.52].

When the generation of defects during the initial contact is avoided and the deformed region under the

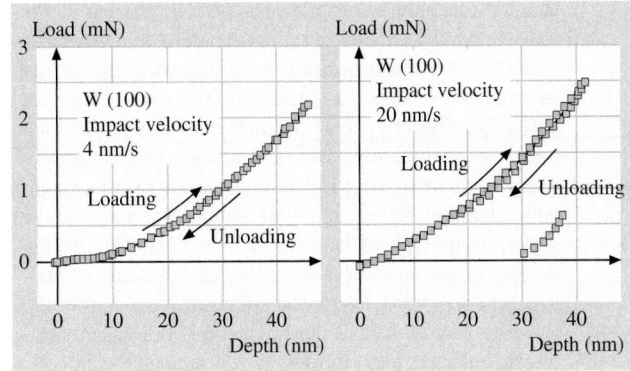

Fig. 37.23a,b Load-displacement curves for W(100) showing how changes in the impact velocity can cause a transition from perfectly elastic behavior to yielding during unloading

contact is truly defect free, then the yielding of the sample should occur at the yield stress of a perfect crystal lattice. The load at which plastic deformation commences under these circumstances becomes very reproducible [37.86]. Unfortunately, nanoindenter tips on the near-atomic scale are not perfectly smooth or axisymmetric. As a result, accurately measuring the yield stress is very difficult. In fact, a slight rotation in the plane of the surface of either the sample or the tip can give a substantial change in the observed yield point load. Coating the surface in a cushioning self-assembled monolayer [37.87] can alleviate some of these variations, but it also introduces a large uncertainty in the contact area. Surface oxide layers, which may be several nanometers thick, have also been found to enhance the elastic behavior seen for very shallow nanoindentations on metallic surfaces [37.78]. Removal of the oxide has been shown to alleviate the initial elastic response.

While nanoindentation testing is ideal for examining the mechanical properties of defect-free volumes and looking at the generation of defects in perfect crystal lattices, it should be clear from the preceding discussion that great care must be taken in examining how the surface properties and the loading rate affect the results, particularly when comparisons are being made to theoretical models for defect generation.

37.3.2 Variations with Depth

Ideal elastic-plastic behavior, as described by *Tabor* [37.48], can be seen during indentation testing, provided the sample has been work-hardened so that the flow stress is a constant. However, it is often the case that the mechanical properties appear to change as the load (or depth) is increased. This apparent change can be a result of several processes, including work-hardening during the test. This is a particularly important effect for soft metals like copper. These metals usually have a high hardness at shallow depths, but it decreases asymptotically with increasing indentation depth to a hardness value that may be less than half that observed at shallow depths. This type of behavior is due to the increasing density of geometrically necessary dislocations at shallow depths [37.88]. Hence the effects of work-hardening are most pronounced at shallow depths. For hard materials the effect is less obvious.

Work-hardening is one of the factors that contribute to the so-called indentation size effect (ISE), whereby at shallow indentation depths the material appears to be harder. The ISE has been widely observed during microindentation testing, with at least part of the effect appearing to result from the increased difficulty in optically measuring the area of an indentation when it is small. During nanoindentation testing the ISE can also be observed, but it is often due to the tip area function, $A_c(\delta_c)$, being incorrectly calibrated. However, there are physical reasons other than work-hardening for expecting an increase in mechanical strength in small volumes. As described in the previous section, small volumes of crystalline materials can have either no defects or only a small number of defects present, making plastic yielding more difficult. Also, because of dislocation line tension, the shear stress required to make a dislocation bow out increases as the radius of the bow decreases. Thus, the shear stress needed to make a dislocation bow out in a small volume is greater than it is in a large volume. These physical reasons for small volumes appearing stronger than large volumes are particularly important in thin film systems, as will be discussed later. Note, however, that these physical reasons for increased hardness do not apply for an amorphous material such as fused silica, which partially explains its value as a calibration material.

37.3.3 Anisotropic Materials

The analysis methods detailed earlier were concerned primarily with the interpretation of data from nanoindentations in isotropic materials where the elastic modulus is assumed to be either independent of direction or a polycrystalline average of a material's elastic constants. Many crystalline materials exhibit considerable anisotropy in their elastic constants, hence, these analysis techniques may not always be appropriate. The

theoretical problem of a rigid indenter pressed into an elastic, anisotropic half-space has been considered by *Vlassak* and *Nix* [37.89]. Their aim was to identify the feasibility of interpreting data from a depth-sensing indentation apparatus for samples with elastic constants that are anisotropic. Nanoindentation experiments [37.90] have shown the validity of the elastic analysis for crystalline zinc, copper, and beta-brass. The observed indentation modulus for zinc, as predicted, varied by as much as a factor of 2 between different orientations. The variations in the observed hardness values for the same materials were smaller, with a maximum variation with orientation of 20% detected in zinc. While these variations are clearly detectable with nanoindentation techniques, the variations are small in comparison to the actual anisotropy of the test material's elastic properties. This is because the indentation modulus is a weighted average of the stiffness in all directions.

At this time the effects of anisotropy on the hardness measured using nanoindentation have not been fully explored. For materials with many active slip planes it is likely that the small anisotropy observed by *Vlassak* and *Nix* is correct once plastic flow has been initiated. It is possible, however, that for defect-free crystalline specimens with a limited number of active slip planes that very shallow nanoindentations may show a much larger anisotropy in the observed hardness and initial yield point load.

37.3.4 Fracture and Delamination

Indentation testing has been widely used to study fracture in brittle materials [37.91], but the lower loads and smaller deformation regions of nanoindentation tests make it harder to initiate cracks and, hence, less useful as a way to evaluate fracture toughness. To overcome these problems the cube corner geometry, which generates larger shear stresses than the Berkovich pyramid, has been used with nanoindentation testing to study fracture [37.92]. These studies have had mixed success, because the cube corner geometry blunts very quickly when used on hard materials. In many cases, brittle materials are very hard.

Depth sensing indentation is better suited to studying delamination of thin films. Recent work extends the research conducted by *Marshall* et al. [37.93, 94], who examined the deformation of residually stressed films by indentation. A schematic of their analysis is given by Fig. 37.24. Their indentations were several microns deep, but the basic analysis is valid for nanoindentations. The analysis has been extended to multilayers [37.95],

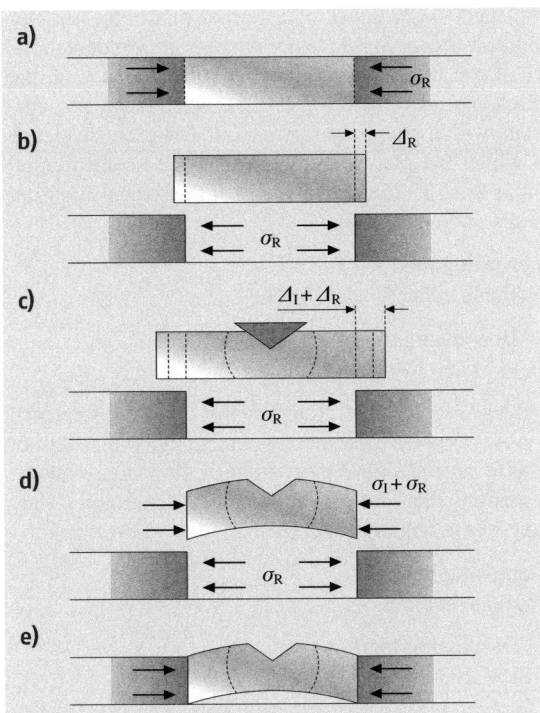

Fig. 37.24a–e To model delamination *Kriese* et al. [37.95] adapted the model developed by *Marshall* and *Evans* [37.93]. The model considers a segment of removed stressed film that is allowed to expand and then indented, thereby expanding it further. Replacing the segment in its original position requires an additional stress, and the segment bulges upwards

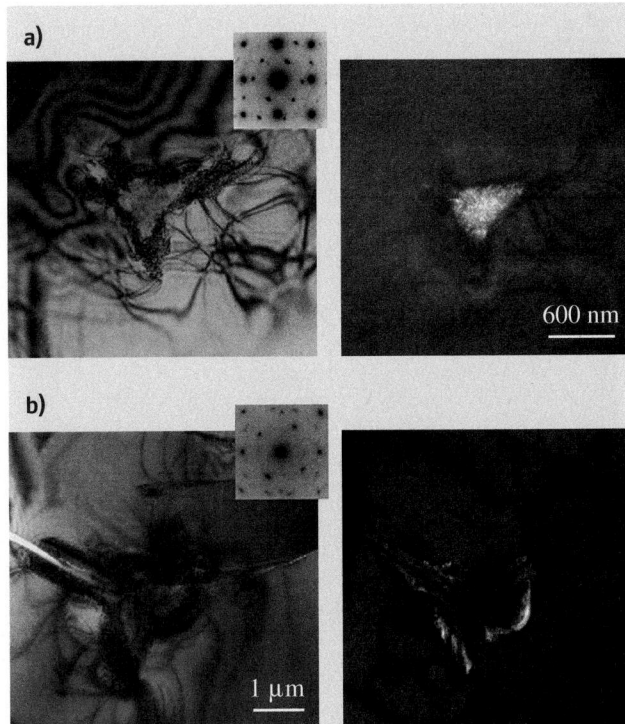

Fig. 37.25a,b Bright-field and dark-field TEM of (**a**) small and (**b**) large nanoindentations in Si. In small nanoindents the metastable phase BC-8 is seen in the center, but for large nanoindents BC-8 is confined to the edge of the indent, while the center is amorphous

which is important since it enables a quantitative assessment of adhesion energy when an additional stressed film has been deposited on top of the film and substrate of interest. The additional film limits the plastic deformation of the film of interest and also applies extra stress that aids in the delamination. After indentation, the area of the delaminated film is measured optically or with an AFM to assess the extent of the delamination. This measurement, coupled with the load-displacement data, enables quantitative assessment of the adhesion energy to be made for metals [37.96] and polymers [37.97].

37.3.5 Phase Transformations

The pressure applied to the surface of a material during indentation testing can be very high. Equation (37.10) indicates that the pressure during plastic yielding is about three times the yield stress. For many materials, high

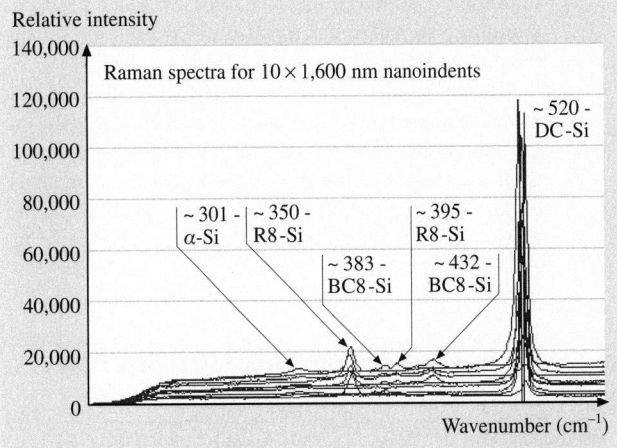

Fig. 37.26 Micro-Raman generally shows the BC-8 and R-8 Si phases that are at the edge of the nanoindents, but the amorphous phase in the center is not easy to detect, as it is often subsurface and the Raman peak is broad

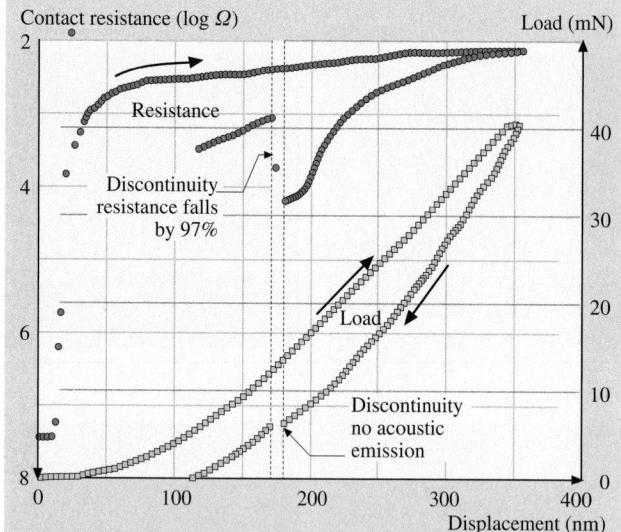

Fig. 37.27 Nanoindentation curves for deep indents on Si show a discontinuity during unloading and simultaneously a large drop in contact resistance

hydrostatic pressures can cause phase transformations, and provided the transformation pressure is less than the pressure required to cause plastic yielding, it is possible during indentation testing to induce a phase transformation. This was first reported for silicon [37.98], but it has also been speculated [37.99] that many other materials may show the same effects. Most studies still focus on silicon because of its enormous technological importance, although there is some evidence that germanium also undergoes a phase transformation during the nanoindentation testing [37.100].

Recent results [37.21, 22, 41, 101, 102] indicate that there are actually multiple phase transformations during the nanoindentation of silicon. TEM of nanoindentations in diamond cubic silicon have shown the presence of amorphous-Si and the body-centered cubic BC-8 phase (see Fig. 37.25). Micro-Raman spectroscopy has

indicated the presence of a further phase, the rhomb-hedral R-8 (see Fig. 37.26). For many nanoindentations on silicon there is a characteristic discontinuity in the unloading curve (see Fig. 37.27), which seems to correlate with a phase transformation. The exact sequence in which the phases form is still highly controversial with some [37.42], suggesting that the sequence during loading and unloading is:

Increasing load →
Diamond cubic Si → β-Sn Si

← **Decreasing load**
BC-8 Si and R-8 Si ← β-Sn Si

Other groups [37.21, 22] suggest that the above sequence is only valid for shallow nanoindentations that do not exhibit an unloading discontinuity. For large nanoindentations that show an unloading discontinuity, they suggest the sequence will be:

Increasing load →
Diamond cubic Si → β-Sn Si

← **Decreasing load**
α-Si ← BC-8 Si and R-8 Si ← β-Sn Si

The disagreement is over the origin of the unloading discontinuity. *Mann* et al. [37.21] suggest it is due to the formation of amorphous silicon, while *Gogotsi* et al. [37.42] believe it is the β-Sn Si to BC-8 or R-8 transformation. Mann et al. argue that the high contact resistance before the discontinuity and the low contact resistance afterwards rule out the discontinuity being the metallic β-Sn Si transforming to the more resistive BC-8 or R-8. The counterargument is that amorphous Si is only seen with micro-Raman spectroscopy when the unloading is very rapid or there is a large nanoindentation with no unloading discontinuity. The importance of unloading rate and cracking in determining the phases present are further complications. The controversy will remain until in situ characterization of the phases present is undertaken.

37.4 Thin Films and Multilayers

In almost all real applications, surfaces are coated with thin films. These may be intentionally added such as hard carbide coatings on a tool bit, or they may simply be native films such as an oxide layer. It is also likely that there will be adsorbed films of water and organic contaminants that can range from a single molecule in thickness up to several nanometers. All of these films, whether native or intentionally placed on the surface, will affect the surface's mechanical behavior on the nanoscale. Adsorbates can have a significant impact on the surface forces [37.3] and, hence, the geometry and stability of asperity contacts. Oxide films can have dramatically different mechanical properties to the bulk and will also modify the surface forces. Some of the effects of native films have been detailed in the earlier sections on dislocation nucleation and adhesive contacts.

The importance of thin films in enhancing the mechanical behavior of surfaces is illustrated by the abundance of publications on thin film mechanical properties (see for instance *Nix* [37.88] or *Cammarata* [37.31] or *Was* and *Foecke* [37.103]). In the following sections, the mechanical properties of films intentionally deposited on the surface will be discussed.

37.4.1 Thin Films

Measuring the mechanical properties of a single thin surface film has always been difficult. Any measurement performed on the whole sample will inevitably be dominated by the bulk substrate. Nanoindentation, since it looks at the mechanical properties of a very small region close to the surface, offers a possible solution to the problem of measuring thin film mechanical properties. However, there are certain inherent problems in using nanoindentation testing to examine the properties of thin films. The problems stem in part from the presence of an interface between the film and substrate. The quality of the interface can be affected by many variables, resulting in a range of effects on the apparent elastic and plastic properties of the film. In particular, when the deformation region around the indent approaches the interface, the indentation curve may exhibit features due to the thin film, the bulk, the interface, or a combination of all three. As a direct consequence of these complications, models for thin-film behavior must attempt to take into account not only the properties of the film and substrate, but also the interface between them.

If, initially, the effect of the interface is neglected, it is possible to divide thin-coated systems into a number of categories that depend on the values of E (elastic modulus) and Y (the yield stress) of the film and substrate. These categories are typically [37.104, 105]:

1. coatings with high E and high Y, substrates with high E and high Y;
2. coatings with high E and high Y, substrates with high or low E and low Y;
3. coatings with high or low E and low Y, substrates with high E and high Y;
4. coatings with high or low E and low Y, substrates with high or low E and low Y.

The reasons for splitting thin film systems into these different categories have been amply demonstrated experimentally by *Whitehead* and *Page* [37.104, 105] and theoretically by *Fabes* et al. [37.106]. Essentially, hard, elastic materials (high E and Y) will possess smaller plastic zones than soft, inelastic (low E and Y) materials. Thus, when different combinations of materials are used as film and substrate, the overall plastic zone will differ significantly. In some cases, the plasticity is confined to the film, and in other cases, it is in both the film and substrate, as shown by Fig. 37.28. If the standard nanoindentation analysis routines are to be used, it is essential that the plastic zone and the elastic strain field are both confined to the film and do not reach the substrate. Clearly, this is difficult to achieve unless extremely shallow nanoindentations are used. There is an often quoted 10% rule, that says nanoindents in a film must have a depth of less than 10% of the film's thickness if only the film properties are to be measured. This has no real validity [37.107]. There are film/substrate combinations for which 10% is very conservative, while for other combinations even 5% may be too deep. The effect of the substrate for different combinations of film and substrate properties has been studied using FEM [37.108], which has shown that the maximum nanoindentation depth to measure film only properties decreases in moving from soft on hard to hard on soft combinations. For a very soft film on a hard substrate, nanoindentations of 50% of the film thickness are alright, but this drops to < 10% for a hard film on a soft substrate. For a very strong film on

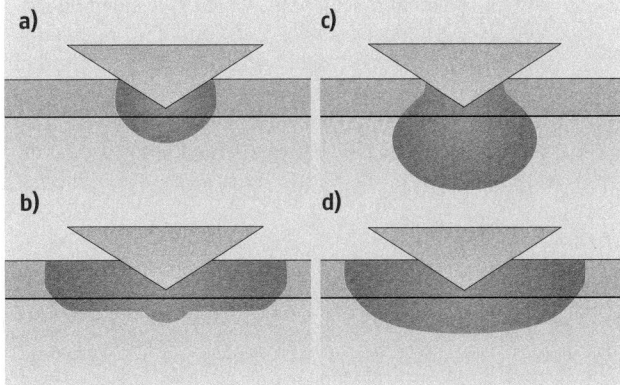

Fig. 37.28a–d Variations in the plastic zone for indents on films and substrates of different properties. (**a**) Film and substrate have high E and Y, (**b**) film has a high E and Y, substrate has a high or low E and low Y, (**c**) film has a high or low E and a low Y, and substrate has a high E and Y, (**d**) film has a high or low E and a low Y, and substrate has a high or low E and a low Y

a soft substrate, the surface film behaves like an elastic membrane or a bending plate.

Theoretical analysis of thin-film mechanical behavior is difficult. One theoretical approach that has been adopted uses the volumes of plastically deformed material in the film and substrate to predict the overall hardness of the system. However, it should be noted that this method is only really appropriate for soft coatings and indentation depths below the thickness of the coating (see cases c and d of Fig. 37.28), otherwise the behavior will be closer to that detailed later and shown by Fig. 37.29.

The technique of combining the mechanical properties of the film and substrate to evaluate the overall hardness of the system is generally referred to as the rule of mixtures. It stems from work by *Burnett* et al. [37.109–111] and *Sargent* [37.112], who derived a weighted average to relate the "composite" hardness (H) to the volumes of plastically deformed material in the film (V_f) and substrate (V_s) and their respective values of hardness, H_f and H_s. Thus,

$$H = \frac{H_f V_f + H_s V_s}{V_{total}} , \qquad (37.26)$$

where V_{total} is $V_f + V_s$.

Equation (37.26) was further developed by *Burnett* and *Page* [37.109] to take into account the indentation size effect. They replaced H_s with $K\delta_c^{n-2}$, where K and n are experimentally determined constants dependent on the indenter and sample, and δ_c is the contact depth. This expression is derived directly from Meyer's law for spherical indentations, which gives the relationship $P = Kd^n$ between load, P, and the indentation dimension, d. Burnett and Page also employed a further refinement to enable the theory to fit experimental results from a specific sample, ion-implanted silicon. This particular modification essentially took into account the different sizes of the plastic zones in the two materials

by multiplying H_s by a dimensionless factor (V_s/V_{total}). While this seems to be a sensible approach, it is mostly empirical, and the physical justification for using this particular factor is not entirely clear. Later, *Burnett* and *Rickerby* [37.110, 111] took this idea further and tried to generalize the equations to take into account all of the possible scenarios. Thus, the following equations were suggested:

$$H = \frac{H_f \left(\Omega^3\right) V_f + H_s V_s}{V_{total}} , \qquad (37.27)$$

$$H = \frac{H_f V_f + H_s \left(\Omega^3\right) V_s}{V_{total}} . \qquad (37.28)$$

The first of these, (37.27), deals with the case of a soft film on a hard substrate, and the second, (37.28), with a hard film on a soft substrate. The Ω term expresses the variation of the total plastic zone from the ideal hemispherical shape. This was taken still further by *Bull* and *Rickerby* [37.113], who derived an approximation for Ω based on the film and substrate zone radii being related to their respective hardness and elastic modulii [37.114, 115]. Hence:

$$\Omega = (E_f H_s / E_s H_f)^l , \qquad (37.29)$$

where l is determined empirically. E_f and H_f and E_s and H_s are the elastic modulus and hardness of the film and substrate, respectively.

Experimental data [37.116] indicate that the effect of the substrate on the elastic modulus of the film can be quite different than the effect on hardness, due to the zones of the elastic and plastic strain fields being different sizes.

Chechechin et al. [37.117] have recently studied the behavior of Al_2O_3 films of various thicknesses on different substrates. Their results indicate that many of the models correctly predict the transition between the properties of the film and those of the substrate, but do not always fit the observed hardness against depth curves. This group have also studied the pop-in behavior of Al_2O_3 films [37.118] and have attempted to model the range of loads and depths at which they occur via a Weibull-type distribution, as utilized in fracture analysis.

A point raised by Burnett and Rickerby should be emphasized. They state that there are two very distinct modes of deformation during nanoindentation testing. The first, referred to as *Tabor*'s [37.48] model for low Y/E materials, involves the buildup of material at the side of the indenter through movement of material at slip lines. The second, for materials with large Y/E does

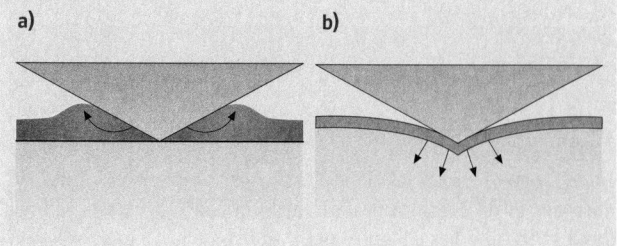

Fig. 37.29a,b Two different modes of deformation during nanoindentation of films. In (**a**) materials move upwards and outwards, while in (**b**) the film acts like a membrane and the substrate deforms

not result in surface pile-up. The displaced material is then accommodated by radial displacements [37.115]. The point is that a thin, strong, and well-bonded surface film can cause a substrate that would normally deform by Tabor's method to behave more like a material with high Y/E (see Fig. 37.29). It should be noted that this only applies as long as the film does not fail.

In recent theoretical and experimental studies the importance of material pile-up and sink-in has been investigated extensively. As discussed in an earlier section, pile-up can be increased by residual compressive stresses, but even in the absence of residual stresses pile-up can introduce a significant error in the calculated contact area. This is most pronounced in materials that do not work-harden [37.62]. For these materials using the Oliver and Pharr method fails to account for the pile-up and results in a large error in the values for E and H. For thin films *Tsui* et al. have used a focused ion beam to section through Knoop indentations in both soft films on hard substrates [37.69] and hard films on soft substrates [37.70]. The soft films, as expected, exhibit pile-up, while the hard film acts more like a membrane and the indentation exhibits sink-in with most of the plasticity in the substrate. Thus, there are three clearly identifiable factors affecting the pile-up and sink-in around nanoindents during testing of thin films:

1. Residual stresses
2. Degree of work-hardening
3. Ratio of film and substrate mechanical properties.

The bonding or adhesion between the film and substrate could also be added to this list. And it should not be forgotten that the depth of the nanoindentation relative to the film thickness also affects pile-up. For a very deep nanoindentation into a thin, soft film on a hard substrate pile-up is reduced, due to the combined constraints on the film of the tip and substrate [37.119]. Due to all of these complications, using nanoindentation to study thin film mechanical properties is fraught with danger. Many unprepared researchers have misguidedly taken the values of E and H obtained during nanoindentation testing to be absolute values only to find out later that the values contain significant errors.

Many of the problems associated with nanoindentation testing are related to incorrectly calculating the contact area, A. The *Joslin* and *Oliver* method [37.71] is one way that A can be removed from the calculations. This approach has been used with some success to look at strained epitaxial II/VI semiconductor films [37.120], but there is evidence that the lattice mismatch in these

films can cause dramatic changes in the mechanical properties of the films [37.121]. This may be due to image forces and the film/substrate interface acting as a barrier to dislocation motion. Recently, it has been shown that using films and substrates with known matching elastic moduli, it is possible to use the assumption of constant elastic modulus with depth to evaluate H [37.122]. In effect, this is using (37.1) to evaluate A from the contact stiffness data, and then substituting the value for A into (37.12). The value of E is measured independently, for instance, using acoustic techniques.

37.4.2 Multilayers

Multilayered materials with individual layers that are a micron or less in thickness, sometimes referred to as superlattices, can exhibit substantial enhancements in hardness or strength. This should be distinguished from the super modulus effect discussed earlier, which has been shown to be largely an artifact. The enhancements in hardness can be as much as 100% when compared to the value expected from the rule of mixtures, which is essentially a weighted average of the hardness for the constituents of the two layers [37.123]. Table 37.1 shows how the properties of isostructural multilayers can show a substantial increase in hardness over that for fully interdiffused layers. The table also shows how there can be a substantial enhancement in hardness for non-isostructural multilayers compared to the values for the same materials when they are homogeneous.

There are many factors that contribute to enhanced hardness in multilayers. These can be summarized as [37.103]:

1. Hall-Petch behavior
2. Orowan strengthening
3. Image effects
4. Coherency and thermal stresses
5. Composition modulation

Hall-Petch behavior is related to dislocations piling-up at grain boundaries. (Note that pile-up is used to describe two distinct effects: One is material building up at the side of an indentation, the other is an accumulation of dislocations on a slip-plane.) The dislocation pile-up at grain boundaries impedes the motion of dislocations. For materials with a fine grain structure there are many grain boundaries, and, hence, dislocations find it hard to move. In polycrystalline multilayers, it is often the case that the size of the grains within a layer scales with the layer thickness so that reducing the layer

Table 37.1 Results for some experimental studies of multilayer hardness

Study	Multilayer	Maximum hardness and multilayer repeat length	Reference hardness value	Range of hardness values for multilayers
Isostructural Knoop hardness [37.124]	Cu/Ni	524 at 11.6 nm	284 (interdiffused)	295–524
Non-isostructural Nanoindentation [37.125]	Mo/NbN	33 GPa at 2 nm	NbN – 17 GPa Mo – 2.7 GPa Wo – 7 GPa	12–33 GPa
	W/NbN	29 GPa at 3 nm	(individual layer materials)	23–29 GPa

thickness reduces the grain size. Thus, the Hall-Petch relationship (below) should be applicable to polycrystalline multilayer films with the grain size, d_g, replaced by the layer thickness.

$$Y = Y_0 + k_{HP} \, d_g^{-0.5} \, , \tag{37.30}$$

where Y is the enhanced yield stress, Y_0 is the yield stress for a single crystal, and k_{HP} is a constant.

There is an ongoing argument about whether Hall-Petch behavior really takes place in nanostructured multilayers. The basic model assumes many dislocations are present in the pile-up, but such large dislocation pile-ups are not seen in small grains [37.126] and are unlikely to be present in multilayers. As a direct consequence, studies have found a range of values, between 0 and −1, for the exponent in (37.30), rather than the −0.5 predicted for Hall-Petch behavior.

Orowan strengthening is due to dislocations in layered materials being effectively pinned at the interfaces. As a result, the dislocations are forced to bow out along the layers. In narrow films, dislocations are pinned at both the top and bottom interfaces of a layer and bow out parallel to the plane of the interface [37.127, 128]. Forcing a dislocation to bow out in a layered material requires an increase in the applied shear stress beyond that required to bow out a dislocation in a homogeneous sample. This additional shear stress would be expected to increase as the film thickness is reduced.

Image effects were suggested by *Koehler* [37.129] as a possible source of enhanced yield stress in multilayered materials. If two metals, A and B, are used to make a laminate and one of them, A, has a high dislocation line energy, but the other, B, has a low dislocation line energy, then there will be an increased resistance to dislocation motion due to image forces.

However, if the individual layers are thick enough that there may be a dislocation source present within the layer, then dislocations could pile-up at the interface. This will create a local stress concentration point and the enhancement to the strength will be very limited. If the layers are thin enough that there will be no dislocation source present, the enhanced mechanical strength may be substantial. In Koehler's model only nearest neighbor layers were taken to contribute to the image forces. However, this was extended to include more layers [37.130] without substantial changes in the results. The consequence on image effects of reducing the thickness of the individual layers in a multilayer is that it prevents dislocation sources from being active within the layer.

For many multilayer systems there is an increase in strength as the bilayer repeat length is reduced, but there is often a critical repeat length (e.g., 3 nm for the W/NbN multilayer of Table 37.1) below which the strength falls. One explanation for the fall in strength involves the effects of coherency and thermal stresses on dislocation energy. Unlike image effects where the energy of dislocations are a maximum or minimum in the center of layers, the energy maxima and minima are at the interfaces for coherency stresses. Combining the effects of varying moduli and coherency stresses shows that the dependence of strength on layer thickness has a peak near the repeat period where coherency strains begin decreasing [37.131].

Another source of deviations in behavior at very small repeat periods is the imperfect nature of interfaces. With the exception of atomically perfect epitaxial films, interfaces are generally not atomically flat and there is some interdiffusion. For the Cu/Ni film of Table 37.1, the effects of interdiffusion on hardness were examined [37.124] by annealing the multilayers. The results were in agreement with a model by *Krzanowski* [37.132] that predicted the variations in hardness would be proportional to the amplitude of the composition modulation.

It is interesting to note that the explanations for enhanced mechanical properties in multilayered materials are all based on dislocation mechanisms. So it would seem natural to assume that multilayered materials that do not contain dislocations will show no enhanced hardness over their rule of mixtures values. This has been verified by studies on amorphous metal multilayers [37.133], which shows that the hardness of the multilayers, firstly, lies between that of the two individual materials and, secondly, has almost no variation with repeat period.

37.5 Developing Areas

Over the past 20 to 30 years, the driving force for studying nanomechanical behavior of surfaces and thin films has been largely, though not exclusively, the microelectronics industry. The importance of electronics to the modern world is only likely to grow in the foreseeable future, but other technological areas may overtake microelectronics as the driving force for research, including the broad fields of biomaterials and nanotechnology. In several places in this chapter, a number of developing areas have been mentioned. These include nanoscale measurements of viscoelasticity and the study of environmental effects (temperature and surface chemistry) on nanomechanical properties. Both of these topics will be vital in the study of biological systems and, as a result, will be increasingly important from a research point of view.

We still have a relatively rudimentary understanding of the nanomechanics of complex biological systems such as bone cells (osteoblasts and osteoclasts) and skin cells (fibroblasts), or even, for that matter, simpler biological structures such as dental enamel. For example, Fig. 37.30 shows how the mechanical properties of dental enamel can vary within a single tooth [37.134]. But this is still a relatively large-scale measure of mechanical behavior. The prismatic structure of enamel means that there are variations in mechanical properties on a range of scales from millimeters down to nanometers.

In terms of data analysis, there remains much to be done. If an analysis method can be developed

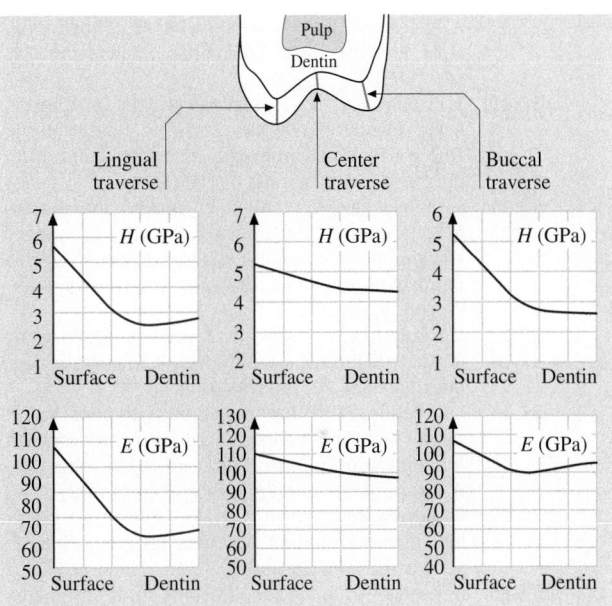

Fig. 37.30 Variations in E and H across human dental enamel. The sample is an upper 2nd molar cut in cross section from the lingual to the buccal side. Nanoindentations are performed across the surface to examine how the mechanical properties vary

that deals with the problems of pile-up and sink-in, the utility of nanoindentation testing will be greatly enhanced.

References

37.1   G. Binnig, C. F. Quate, C. Gerber: Atomic force microscope, Phys. Rev. Lett. **56**, 930–933 (1986)

37.2   *Microindentation Techniques in Materials Science and Engineering*, ed. by P. J. Blau, B. R. Lawn (ASTM, Pennsylvannia 1986)

37.3   J. N. Israelachvili: *Intermolecular and Surface Forces* (Academic, London 1992)

37.4   R. S. Bradley: The cohesive force between solid surfaces and the surface energy of solids, Philos. Mag. **13**, 853–862 (1932)

37.5   B. V. Derjaguin, V. M. Muller, Yu. P. Toporov: Effect of contact deformations on the adhesion of particles, J. Coll. Interface Sci. **53**, 314–326 (1975)

37.6   V. M. Muller, V. S. Yuschenko, B. V. Derjaguin: On the influence of molecular forces on the deformation of an elastic sphere and its sticking to a rigid plane, J. Coll. Interface Sci. **77**, 91–101 (1980)

37.7   V. M. Muller, B. V. Derjaguin, Yu. P. Toporov: On two methods of calculation of the force of sticking of an elastic sphere to a rigid plane, Coll. Surf. **7**, 251–259 (1983)

37.8   K. L. Johnson, K. Kendal, A. D. Roberts: Surface energy and the contact of elastic solids, Proc. R. Soc. A **324**, 301–320 (1971)

37.9   A. P. Ternovskii, V. P. Alekhin, M. Kh. Shorshorov, M. M. Khrushchov, V. N. Skvortsov: Zavod Lab. **39**, 1242 (1973)

37.10   S. I. Bulychev, V. P. Alekhin, M. Kh. Shorshorov, A. P. Ternovskii, G. D. Shnyrev: Determining Young's modulus from the indentor penetration diagram, Zavod Lab. **41**, 1137 (1975)

37.11   S. I. Bulychev, V. P. Alekhin, M. Kh. Shorshorov, A. P. Ternovskii: Mechanical properties of materials studied from kinetic diagrams of load versus depth of impression during microimpression, Prob. Prochn. **9**, 79 (1976)

37.12   S. I. Bulychev, V. P. Alekhin: Zavod Lab. **53**, 76 (1987)

37.13   M. Kh. Shorshorov, S. I. Bulychev, V. P. Alekhin: Sov. Phys. Doklady **26**, 769 (1982)

37.14   J. B. Pethica: Microhardness tests with penetration depths less than ion implanted layer thickness. In: *Ion Implantation into Metals*, ed. by V. Ashworth, W. Grant, R. Procter (Pergamon, Oxford 1982) p. 147

37.15   D. Newey, M. A. Wilkens, H. M. Pollock: An ultra-low-load penetration hardness tester, J. Phys. E: Sci. Instrum. **15**, 119 (1982)

37.16   D. Kendall, D. Tabor: An ultrasonic study of the area of contact between stationary and sliding surfaces, Proc. R. Soc. A **323**, 321–340 (1971)

37.17   G. M. Pharr, W. C. Oliver, F. R. Brotzen: On the generality of the relationship among contact stiffness, contact area and elastic-modulus during indentation, J. Mater. Res. **7**, 613 (1992)

37.18   T. P. Weihs, C. W. Lawrence, B. Derby, C. B. Scruby, J. B. Pethica: Acoustic emissions during indentation tests, MRS Symp. Proc. **239**, 361–366 (1992)

37.19   D. F. Bahr, J. W. Hoehn, N. R. Moody, W. W. Gerberich: Adhesion and acoustic emission analysis of failures in nitride films with 14 metal interlayer, Acta Mater. **45**, 5163 (1997)

37.20   D. F. Bahr, W. W. Gerberich: Relationships between acoustic emission signals and physical phemomena during indentation, J. Mat. Res. **13**, 1065 (1998)

37.21   A. B. Mann, D. van Heerden, J. B. Pethica, T. P. Weihs: Size-dependent phase transformations during point-loading of silicon, J. Mater. Res. **15**, 1754 (2000)

37.22   A. B. Mann, D. van Heerden, J. B. Pethica, P. Bowes, T. P. Weihs: Contact resistance and phase transformations during nanoindentation of silicon, Philos. Mag. A **82**, 1921 (2002)

37.23   S. Jeffery, C. J. Sofield, J. B. Pethica: The influence of mechanical stress on the dielectric breakdown field strength of SiO_2 films, Appl. Phys. Lett. **73**, 172 (1998)

37.24   B. N. Lucas, W. C. Oliver: Indentation power-law creep of high-purity indium, Metall. Trans. A **30**, 601 (1999)

37.25   S. A. Syed Asif: Time dependent micro deformation of materials. Ph.D. Thesis (Oxford Univ., Oxford 1997)

37.26   S. A. Syed Asif, R. J. Colton, K. J. Wahl: Nanoscale Surface Mechanical Property Measurements: Force Modulation Techniques Applied to Nanoindentation. In: *Interfacial Properties on the Submicron Scale*, ed. by J. Frommer, R. Overney (ACS Books, Whashington 2000)

37.27   A. B. Mann, J. B. Pethica: Nanoindentation studies in a liquid environment, Langmuir **12**, 4583 (1996)

37.28   J. W. Beams: Mechanical properties of thin films of gold and silver. In: *Structure and Properties of Thin Films*, ed. by C. A. Neugebauer, J. B. Newkirk, D. A. Vermilyea (Wiley, New York 1959) pp. 183–192

37.29   J. J. Vlassak, W. D. Nix: A new bulge test technique for the determination of Youngs modulus and Poissons ratio of thin-films, J. Mater. Res. **7**, 3242 (1992)

37.30   W. M. C. Yang, T. Tsakalakos, J. E. Hilliard: Enhanced elastic modulus in composition modulated gold-nickel and copper-palladium foils, J. Appl. Phys. **48**, 876 (1977)

37.31   R. C. Cammarata: Mechanical properties of nanocomposite thin-films, Thin Solid Films **240**, 82 (1994)

37.32   G. A. D. Briggs: *Acoustic Microscopy* (Clarendon, Oxford 1992)

37.33   J. Kushibiki, N. Chubachi: Material characterization by line-focus-beam acoustic microscope, IEEE Trans. Sonics Ultrasonics **32**, 189–212 (1985)

37.34   M. J. Bamber, K. E. Cooke, A. B. Mann, B. Derby: Accurate determination of Young's modulus and Poisson's ratio of thin films by a combination of acoustic microscopy and nanoindentation, Thin Solid Films **398**, 299–305 (2001)

37.35   S. E. Bobbin, R. C. Cammarata, J. W. Wagner: Determination of the flexural modulus of thin-films from measurement of the 1st arrival of the symmetrical Lamb wave, Appl. Phys. Lett. **59**, 1544–1546 (1991)

37.36   R. B. King, C. M. Fortunko: Determination of inplane residual-stress state in plates using horizontally polarized shear waves, J. Appl. Phys. **54**, 3027–3035 (1983)

37.37   R. F. Cook, G. M. Pharr: Direct observation and analysis of indentation cracking in glasses and ceramics, J. Am. Ceram. Soc. **73**, 787–817 (1990)

37.38 T. F. Page, W. C. Oliver, C. J. McHargue: The deformation-behavior of ceramic crystals subjected to very low load (nano)indentations, J. Mater. Res. **7**, 450–473 (1992)

37.39 G. M. Pharr, W. C. Oliver, D. S. Harding: New evidence for a pressure-induced phase-transformation during the indentation of silicon, J. Mater. Res. **6**, 1129–1130 (1991)

37.40 C. F. Robertson, M. C. Fivel: The study of submicron indent-induced plastic deformation, J. Mater. Res. **14**, 2251–2258 (1999)

37.41 J. E. Bradby, J. S. Williams, J. Wong-Leung, M. V. Swain, P. Munroe: Transmission electron microscopy observation of deformation microstructure under spherical indentation in silicon, Appl. Phys. Lett. **77**, 3749–3751 (2000)

37.42 Y. G. Gogotsi, V. Domnich, S. N. Dub, A. Kailer, K. G. Nickel: Cyclic nanoindentation and Raman microspectroscopy study of phase transformations in semiconductors, J. Mater. Res. **15**, 871–879 (2000)

37.43 H. Hertz: Über die Berührung fester elastischer Körper, J. reine angew. Math. **92**, 156–171 (1882)

37.44 J. Boussinesq: *Application des potentiels à l'étude de l'équilibre et du mouvement des solides élastiques* (Blanchard, Paris 1885) Reprint (1996)

37.45 A. E. H. Love: The stress produced in a semi-infinite solid by pressure on part of the boundary, Philos. Trans. R. Soc. **228**, 377–420 (1929)

37.46 A. E. H. Love: Boussinesq's problem for a rigid cone, Quarter. J. Math. **10**, 161 (1939)

37.47 I. N. Sneddon: The relationship between load and penetration in the axisymmetric Boussinesq problem for a punch of arbitrary profile, Int. J. Eng. Sci. **3**, 47–57 (1965)

37.48 D. Tabor: *Hardness of Metals* (Oxford Univ. Press, Oxford 1951)

37.49 R. von Mises: Mechanik der festen Körper in plastisch deformablen Zustand, Goettinger Nachr. Math.-Phys. **K1**, 582–592 (1913)

37.50 H. Tresca: Sur l'ecoulement des corps solids soumis s fortes pression, Compt. Rend. **59**, 754 (1864)

37.51 A. B. Mann, J. B. Pethica: The role of atomic-size asperities in the mechanical deformation of nanocontacts, Appl. Phys. Lett. **69**, 907–909 (1996)

37.52 A. B. Mann, J. B. Pethica: The effect of tip momentum on the contact stiffness and yielding during nanoindentation testing, Philos. Mag. A **79**, 577–592 (1999)

37.53 S. P. Jarvis: Atomic force microscopy and tip-surface interactions. Ph.D. Thesis (Oxford Univ., Oxford 1993)

37.54 J. B. Pethica, D. Tabor: Contact of characterised metal surfaces at very low loads: Deformation and adhesion, Surf. Sci. **89**, 182 (1979)

37.55 J. S. Field, M. V. Swain: Determining the mechanical-properties of small volumes of materials from submicrometer spherical indentations, J. Mater. Res. **10**, 101–112 (1995)

37.56 W. C. Oliver, G. M. Pharr: An improved technique for determining hardness and elastic-modulus using load and displacement sensing indentation experiments, J. Mater. Res. **7**, 1564–1583 (1992)

37.57 S. V. Hainsworth, H. W. Chandler, T. F. Page: Analysis of nanoindentation load-displacement loading curves, J. Mater. Res. **11**, 1987–1995 (1996)

37.58 J. L. Loubet, J. M. Georges, O. Marchesini, G. Meille: Vickers indentation curves of magnesium oxide (MgO), Mech. Eng. **105**, 91–92 (1983)

37.59 J. L. Loubet, J. M. Georges, O. Marchesini, G. Meille: Vickers indentation curves of magnesium oxide (MgO), J. Tribol. Trans. ASME **106**, 43–48 (1984)

37.60 M. F. Doerner, W. D. Nix: A method for interpreting the data from depth sensing indentation experiments, J. Mater. Res. **1**, 601–609 (1986)

37.61 A. Bolshakov, W. C. Oliver, G. M. Pharr: Influences of stress on the measurement of mechanical properties using nanoindentation. 2. Finite element simulations, J. Mater. Res. **11**, 760–768 (1996)

37.62 A. Bolshakov, G. M. Pharr: Influences of pileup on the measurement of mechanical properties by load and depth sensing instruments, J. Mater. Res. **13**, 1049–1058 (1998)

37.63 J. C. Hay, A. Bolshakov, G. M. Pharr: A critical examination of the fundamental relations used in the analysis of nanoindentation data, J. Mater. Res. **14**, 2296–2305 (1999)

37.64 G. M. Pharr, T. Y. Tsui, A. Bolshakov, W. C. Oliver: Effects of residual-stress on the measurement of hardness and elastic-modulus using nanoindentation, MRS Symp. Proc. **338**, 127–134 (1994)

37.65 T. R. Simes, S. G. Mellor, D. A. Hills: A note on the influence of residual-stress on measured hardness, J. Strain Anal. Eng. Des. **19**, 135–137 (1984)

37.66 W. R. Lafontaine, B. Yost, C. Y. Li: Effect of residual-stress and adhesion on the hardness of copper-films deposited on silicon, J. Mater. Res. **5**, 776–783 (1990)

37.67 W. R. Lafontaine, C. A. Paszkiet, M. A. Korhonen, C. Y. Li: Residual stress measurements of thin aluminum metallizations by continuous indentation and X-ray stress measurement techniques, J. Mater. Res. **6**, 2084–2090 (1991)

37.68 T. Y. Tsui, W. C. Oliver, G. M. Pharr: Influences of stress on the measurement of mechanical properties using nanoindentation. 1. Experimental studies in an aluminum alloy, J. Mater. Res. **11**, 752–759 (1996)

37.69 T. Y. Tsui, J. Vlassak, W. D. Nix: Indentation plastic displacement field: Part I. The case of soft films on hard substrates, J. Mater. Res. **14**, 2196–2203 (1999)

37.70 T. Y. Tsui, J. Vlassak, W. D. Nix: Indentation plastic displacement field: Part II. The case of hard films on soft substrates, J. Mater. Res. **14**, 2204–2209 (1999)

37.71 D. L. Joslin, W. C. Oliver: A new method for analyzing data from continuous depth-sensing microindentation tests, J. Mater. Res. **5**, 123–126 (1990)

37.72   M. R. McGurk, T. F. Page: Using the P-delta(2) analysis to deconvolute the nanoindentation response of hard-coated systems, J. Mater. Res. **14**, 2283–2295 (1999)

37.73   Y. T. Cheng, C. M. Cheng: Relationships between hardness, elastic modulus, and the work of indentation, Appl. Phys. Lett. **73**, 614–616 (1998)

37.74   J. B. Pethica, W. C. Oliver: Mechanical properties of nanometer volumes of material: Use of the elastic response of small area indentations, MRS Symp. Proc. **130**, 13–23 (1989)

37.75   W. C. Oliver, J. B. Pethica: Method for continuous determination of the elastic stiffness of contact between two bodies, United States Patent Number 4,848,141, (1989)

37.76   S. A. S. Asif, K. J. Wahl, R. J. Colton: Nanoindentation and contact stiffness measurement using force modulation with a capacitive load-displacement transducer, Rev. Sci. Instrum. **70**, 2408–2413 (1999)

37.77   J. L. Loubet, W. C. Oliver, B. N. Lucas: Measurement of the loss tangent of low-density polyethylene with a nanoindentation technique, J. Mater. Res. **15**, 1195–1198 (2000)

37.78   W. W. Gerberich, J. C. Nelson, E. T. Lilleodden, P. Anderson, J. T. Wyrobek: Indentation induced dislocation nucleation: The initial yield point, Acta Mater. **44**, 3585–3598 (1996)

37.79   J. D. Kiely, J. E. Houston: Nanomechanical properties of Au(111), (001), and (110) surfaces, Phys. Rev. B **57**, 12588–12594 (1998)

37.80   D. F. Bahr, D. E. Wilson, D. A. Crowson: Energy considerations regarding yield points during indentation, J. Mater. Res. **14**, 2269–2275 (1999)

37.81   D. E. Kramer, K. B. Yoder, W. W. Gerberich: Surface constrained plasticity: Oxide rupture and the yield point process, Philos. Mag. A **81**, 2033–2058 (2001)

37.82   S. G. Corcoran, R. J. Colton, E. T. Lilleodden, W. W. Gerberich: Anomalous plastic deformation at surfaces: Nanoindentation of gold single crystals, Phys. Rev. B **55**, 16057–16060 (1997)

37.83   E. B. Tadmor, R. Miller, R. Phillips, M. Ortiz: Nanoindentation and incipient plasticity, J. Mater. Res. **14**, 2233–2250 (1999)

37.84   J. A. Zimmerman, C. L. Kelchner, P. A. Klein, J. C. Hamilton, S. M. Foiles: Surface step effects on nanoindentation, Phys. Rev. Lett. **87**, article 165507 (1–4) (2001)

37.85   J. D. Kiely, R. Q. Hwang, J. E. Houston: Effect of surface steps on the plastic threshold in nanoindentation, Phys. Rev. Lett. **81**, 4424–4427 (1998)

37.86   A. B. Mann, P. C. Searson, J. B. Pethica, T. P. Weihs: The relationship between near-surface mechanical properties, loading rate and surface chemistry, Mater. Res. Soc. Symp. Proc. **505**, 307–318 (1998)

37.87   R. C. Thomas, J. E. Houston, T. A. Michalske, R. M. Crooks: The mechanical response of gold substrates passivated by self-assembling monolayer films, Science **259**, 1883–1885 (1993)

37.88   W. D. Nix: Elastic and plastic properties of thin films on substrates: Nanoindentation techniques, Mater. Sci. Eng. A **234**, 37–44 (1997)

37.89   J. J. Vlassak, W. D. Nix: Indentation modulus of elastically anisotropic half-spaces, Philos. Mag. A **67**, 1045–1056 (1993)

37.90   J. J. Vlassak, W. D. Nix: Measuring the elastic properties of anisotropic materials by means of indentation experiments, J. Mech. Phys. Solids **42**, 1223–1245 (1994)

37.91   B. R. Lawn: *Fracture of Brittle Solids* (Cambridge Univ. Press, Cambridge 1993)

37.92   G. M. Pharr: Measurement of mechanical properties by ultra-low load indentation, Mater. Sci. Eng. A **253**, 151–159 (1998)

37.93   D. B. Marshall, A. G. Evans: Measurement of adherence of residually stressed thin-films by indentation. 1. Mechanics of interface delamination, J. Appl. Phys. **56**, 2632–2638 (1984)

37.94   C. Rossington, A. G. Evans, D. B. Marshall, B. T. Khuriyakub: Measurement of adherence of residually stressed thin-films by indentation. 2. Experiments with ZnO/Si, J. Appl. Phys. **56**, 2639–2644 (1984)

37.95   M. D. Kriese, W. W. Gerberich, N. R. Moody: Quantitative adhesion measures of multilayer films: Part I. Indentation mechanics, J. Mater. Res. **14**, 3007–3018 (1999)

37.96   M. D. Kriese, W. W. Gerberich, N. R. Moody: Quantitative adhesion measures of multilayer films: Part II. Indentation of W/Cu, W/W, Cr/W, J. Mater. Res. **14**, 3019–3026 (1999)

37.97   M. Li, C. B. Carter, M. A. Hillmyer, W. W. Gerberich: Adhesion of polymer-inorganic interfaces by nanoindentation, J. Mater. Res. **16**, 3378–3388 (2001)

37.98   D. R. Clarke, M. C. Kroll, P. D. Kirchner, R. F. Cook, B. J. Hockey: Amorphization and conductivity of silicon and germanium induced by indentation, Phys. Rev. Lett. **60**, 2156–2159 (1988)

37.99   J. J. Gilman: Insulator-metal transitions at microindentation, J. Mater. Res. **7**, 535–538 (1992)

37.100   G. M. Pharr, W. C. Oliver, R. F. Cook, P. D. Kirchner, M. C. Kroll, T. R. Dinger, D. R. Clarke: Electrical-resistance of metallic contacts on silicon and germanium during indentation, J. Mater. Res. **7**, 961–972 (1992)

37.101   A. Kailer, Y. G. Gogotsi, K. G. Nickel: Phase transformations of silicon caused by contact loading, J. Appl. Phys. **81**, 3057–3063 (1997)

37.102   J. E. Bradby, J. S. Williams, J. Wong-Leung, M. V. Swain, P. Munroe: Mechanical deformation in silicon by micro-indentation, J. Mater. Res. **16**, 1500–1507 (2000)

37.103 G. S. Was, T. Foecke: Deformation and fracture in microlaminates, Thin Solid Films **286**, 1–31 (1996)

37.104 A. J. Whitehead, T. F. Page: Nanoindentation studies of thin-film coated systems, Thin Solid Films **220**, 277–283 (1992)

37.105 A. J. Whitehead, T. F. Page: Nanoindentation studies of thin-coated systems, NATO ASI Ser. E **233**, 481–488 (1993)

37.106 B. D. Fabes, W. C. Oliver, R. A. McKee, F. J. Walker: The determination of film hardness from the composite response of film and substrate to nanometer scale indentations, J. Mater. Res. **7**, 3056–3064 (1992)

37.107 T. F. Page, S. V. Hainsworth: Using nanoindentation techniques for the characterization of coated systems – a critique, Surface Coat. Technol. **61**, 201–208 (1993)

37.108 X. Chen, J. J. Vlassak: Numerical study on the measurement of thin film mechanical properties by means of nanoindentation, J. Mater. Res. **16**, 2974–2982 (2001)

37.109 P. J. Burnett, T. F. Page: Surface softening in silicon by ion-implantation, J. Mater. Sci. **19**, 845–860 (1984)

37.110 P. J. Burnett, D. S. Rickerby: The mechanical-properties of wear resistant coatings. 1. Modeling of hardness behavior, Thin Solid Films **148**, 41–50 (1987)

37.111 P. J. Burnett, D. S. Rickerby: The mechanical-properties of wear resistant coatings. 2. Experimental studies and interpretation of hardness, Thin Solid Films **148**, 51–65 (1987)

37.112 P. M. Sargent: A better way to present results from a least-squares fit to experimental-data – an example from microhardness testing, J. Test. Eval. **14**, 122–127 (1986)

37.113 S. J. Bull, D. S. Rickerby: Evaluation of coatings, Brit. Ceram. Trans. J. **88**, 177–183 (1989)

37.114 B. R. Lawn, A. G. Evans, D. B. Marshall: Elastic/plastic indentation damage in ceramics: The median/radial crack system, J. Am. Ceram. Soc. **63**, 574–581 (1980)

37.115 R. Hill: *The Mathematical Theory of Plasticity* (Clarendon, Oxford 1950)

37.116 W. C. Oliver, C. J. McHargue, S. J. Zinkle: Thin-film characterization using a mechanical-properties microprobe, Thin Solid Films **153**, 185–196 (1987)

37.117 N. G. Chechechin, J. Bottiger, J. P. Krog: Nanoindentation of amorphous aluminum oxide films. 1. Influence of the substrate on the plastic properties, Thin Solid Films **261**, 219–227 (1995)

37.118 N. G. Chechechin, J. Bottiger, J. P. Krog: Nanoindentation of amorphous aluminum oxide films. 2. Critical parameters for the breakthrough and a membrane effect in thin hard films on soft substrates, Thin Solid Films **261**, 228–235 (1995)

37.119 D. E. Kramer, A. A. Volinsky, N. R. Moody, W. W. Gerberich: Substrate effects on indentation plastic zone development in thin soft films, J. Mater. Res. **16**, 3150–3157 (2001)

37.120 A. B. Mann: Nanomechanical measurements: Surface and environmental effects. Ph.D. Thesis (Oxford Univ., Oxford 1995)

37.121 A. B. Mann, J. B. Pethica, W. D. Nix, S. Tomiya: Nanoindentation of epitaxial films: A study of pop-in events, Mater. Res. Soc. Symp. Proc. **356**, 271–276 (1995)

37.122 R. Saha, W. D. Nix: Effects of the substrate on the determination of thin film mechanical properties by nanoindentation, Acta Mater. **50**, 23–38 (2002)

37.123 S. A. Barnett: Deposition and mechanical properties of superlattice thin films. In: *Physics of Thin Films*, ed. by M. H. Francombe, J. L. Vossen (Academic, New York 1993)

37.124 R. R. Oberle, R. C. Cammarata: Dependence of hardness on modulation amplitude in electrodeposited Cu-Ni compositionally modulated thin-films, Scripta Metall. **32**, 583–588 (1995)

37.125 A. Madan, Y. Y. Wang, S. A. Barnett, C. Engstrom, H. Ljungcrantz, L. Hultman, M. Grimsditch: Enhanced mechanical hardness in epitaxial non-isostructural Mo/NbN and W/NbN superlattices, J. Appl. Phys. **84**, 776–785 (1998)

37.126 R. Venkatraman, J. C. Bravman: Separation of film thickness and grain-boundary strengthening effects in Al thin-films on Si, J. Mater. Res. **7**, 2040–2048 (1992)

37.127 J. D. Embury, J. P. Hirth: On dislocation storage and the mechanical response of fine-scale microstructures, Acta Mater. **42**, 2051–2056 (1994)

37.128 D. J. Srolovitz, S. M. Yalisove, J. C. Bilello: Design of multiscalar metallic multilayer composites for high-strength, high toughness, and low CTE mismatch, Metall. Trans. A **26**, 1805–1813 (1995)

37.129 J. S. Koehler: Attempt to design a strong solid, Phys. Rev. B **2**, 547–551 (1970)

37.130 S. V. Kamat, J. P. Hirth, B. Carnahan: Image forces on screw dislocations in multilayer structures, Scripta Metall. **21**, 1587–1592 (1987)

37.131 M. Shinn, L. Hultman, S. A. Barnett: Growth, structure, and microhardness of epitaxial TiN/NbN superlattices, J. Mater. Res. **7**, 901–911 (1992)

37.132 J. E. Krzanowski: The effect of composition profile on the strength of metallic multilayer structures, Scripta Metall. **25**, 1465–1470 (1991)

37.133 J. B. Vella, R. C. Cammarata, T. P. Weihs, C. L. Chien, A. B. Mann, H. Kung: Nanoindentation study of amorphous metal multilayered thin films, MRS Symp. Proc. **594**, 25–29 (2000)

37.134 J. L. Cuy, A. B. Mann, K. J. Livi, M. F. Teaford, T. P. Weihs: Nanoindentation mapping of the mechanical properties of human molar tooth enamel, Arch. Oral Biol. **47**, 281–291 (2002)

38. Scale Effect in Mechanical Properties and Tribology

A model, which explains scale effects in mechanical properties and tribology is presented. Mechanical properties are scale dependent based on the strain gradient plasticity and the effect of dislocation-assisted sliding. Both single asperity and multiple asperity contacts are considered. The relevant scaling length is the nominal contact length – contact diameter for a single-asperity contact, and scan length for multiple-asperity contacts. For multiple asperity contacts, based on an empirical power-rule for scale dependence of roughness, contact parameters are calculated. The effect of load on the contact parameters and the coefficient of friction is also considered. During sliding, adhesion and two- and three-body deformation, as well as ratchet mechanism, contribute to the dry friction force. These components of the friction force depend on the relevant real areas of contact (dependent on roughness and mechanical properties), average asperity slope, number of trapped particles, and shear strength during sliding. Scale dependence of the components of the coefficient of friction is studied. A scale dependent transition index, which is responsible for transition from predominantly elastic adhesion to plastic deformation has been proposed. Scale dependence of the wet friction, wear, and interface temperature has been also analyzed. The proposed model is used to explain the trends in the experimental data for various materials at nanoscale and microscale, which indicate that nanoscale values of coefficient of friction are lower than the microscale values due to an increase of the three-body deformation and transition from elastic adhesive contact to plastic deformation.

38.1 Nomenclature

$a, \bar{a}, \bar{a}_0, a_{max}, \bar{a}_{max}$: Contact radius, mean contact radius, macroscale value of mean contact radius, maximum contact radius, mean value of maximum contact radius.

$A_a, A_r, A_{ra}, A_{re}, A_{re0}, A_{rp}, A_{rp0}, A_{ds}, A_{dp}$: Apparent area of contact, real area of contact, real area of contact during adhesion, real area of elastic contact, macroscale value of real area of elastic contact, real area of plastic contact, macroscale value of real area of plastic contact, real area of contact during asperity summit deformation, area of contact with particles

b: Burgers vector

c: Constant, specified by crystal structure

C_0: Constant required for normalization of $p(d)$

d, d_e, d_n, d_{ln}, $\bar{d}$, $\bar{d}_0$: Particle diameter, minimum for exponential distribution, mean for normal distribution, exponential of mean of $\ln(d)$ for log-normal distribution, mean trapped particles diameter, macroscale value of mean trapped particles diameter

D: Interface zone thickness

E_1, E_2, E^*: Elastic moduli of contacting bodies, effective elastic modulus

F, F_a, F_d, F_{ae}, F_{ap}, F_a, F_{ds}, F_{dp}, F_m, F_{m0}: Friction force, friction force due to adhesion, friction force due to deformation, friction force during elastic adhesional contact, plastic adhesional contact, summit deformation, particles deformation respectively, meniscus force for wet contact, macroscale value of meniscus force

G: Elastic shear modulus

h: Indentation depth

h_f: Liquid film thickness

H, H_0: Hardness, hardness in absence of strain gradient

k, k_0: Wear coefficient, macroscale value of wear coefficient

l_s, l_d: Material-specific characteristic length parameters

L, L_{lwl}, L_{lc}, L_s, L_d: Length of the nominal contact zone, long wavelength limit for roughness parameters, long wavelength limit for contact parameters, length parameters related to l_s and l_d

L_p: Peclet number

m, n: Indices of exponents for scale-dependence of σ and β^*

n_{tr}: Number of trapped particles divided by the total number of particles

p_a, p_{ac}: Apparent pressure, critical apparent pressure

$p(d)$, $p_{tr}(d)$: Probability density function for particle size distribution, probability density function for trapped particle size distribution

$P(d)$: Cumulative probability distribution for particle size

R, R_p, $\overline{R_p}$, $\overline{R_{p0}}$: Effective radius of summit tips, radius of summit tip, mean radius of summit tips, macroscale value of the mean radius of summit tips

$R(\tau)$: Autocorrelation function

s: Spacing between slip steps on the indentation surface

s_d: Separation distance between reference planes of two surfaces in contact

N, N_0: Total number of contacts, macroscale value of total number of contacts

T, T_0: Maximum flash temperature rise, macroscale value of temperature rise

x: Sliding distance

v: Volume of worn material

V: Sliding velocity

W: Normal load

z, z_{min}, z_{max}: Random variable, minimum and maximum value of z

α: Probability for a particle in the border zone to leave the contact region

β^*, β_0^*: Correlation length, macroscale value of correlation length

γ: Surface tension

Γ: Gamma function

ε: Strain

η: Density of particles per apparent area of contact

η_{int}, η_{cr}: Density of dislocation lines per interface area, critical density of dislocation lines per interface area

κ: Curvature

κ_t: Thermal diffusivity

θ: Contact angle between the liquid and surface

θ_i: Indentation angle

θ_r: Roughness angle

μ, μ_a, μ_{ae}, μ_{ae0}, μ_{ap}, μ_{ap0}, μ_d, μ_{ds}, μ_{ds0}, μ_{dp}, μ_{dp0}, μ_r, μ_{r0}, μ_{re}, μ_{re0}, μ_{rp}, μ_{rp0}, μ_{wet}: Coefficient of friction, coefficient of adhesional friction, coefficient of adhesional elastic friction, macroscale value of coefficient of adhesional elastic friction, coefficient of adhesional plastic friction, macroscale value of coefficient of adhesional plastic friction, coefficient of deformation friction, coefficient of summits deformation friction, macroscale value of coefficient of summits deformation friction, coefficient of particles deformation friction, macroscale value of coefficient of particles deformation friction, ratchet component of the coefficient of friction, macroscale value of ratchet component of the coefficient of friction, ratchet component of the coefficient of elastic friction, macroscale value of ratchet component of the coefficient of elastic friction, ratchet component of the coefficient of plastic friction, macroscale value of ratchet component of the coefficient of plastic friction, and coefficient of wet friction

ν_1, ν_2: Poisson's ratios of contacting bodies

ρc_p: Volumetric specific heat

σ, σ_0, σ_e, σ_n, σ_{ln}: Standard deviation of rough surface profile height, macroscale value of standard deviation of rough surface profile height, standard deviation for the exponential distribution, standard deviation for the normal distributions, standard deviation for $\ln(d)$ of the log normal distribution

ρ, ρ_G, ρ_S: Total density of dislocation lines per volume, density of GND per volume, density of SSD per volume

ϕ, ϕ_0: Transition index, macroscale value of transition index

τ, τ_0: Spatial parameter, value at which the autocorrelation function decays

τ_a, τ_{a0}, τ_Y, τ_{Y0}, τ_{ds}, τ_{ds0}, τ_{dp}, τ_{dp0}, τ_p: Adhesional shear strength during sliding, macroscale value of adhesional shear strength, shear yield strength, shear yield strength in absence of strain gradient, shear strength during summits deformation, macroscale value of shear strength during summits deformation, shear strength during particles deformation, macroscale value of shear strength during particles deformation, Peierls stress

38.2 Introduction

Microscale and nanoscale measurements of tribological properties, which became possible due to the development of the Surface Force Apparatus (SFA), Atomic Force Microscope (AFM), and Friction Force Microscope (FFM) demonstrate scale dependence of adhesion, friction, and wear as well as mechanical properties including hardness [38.1–4]. Advances of micro/nanoelectromechanical systems (MEMS/NEMS) technology in the past decade make understanding of scale effects in adhesion, friction, and wear especially important, since surface to volume ratio grows with miniaturization and surface phenomena dominate. Dimensions of MEMS/NEMS devices range from about 1 mm to few nm.

Experimental studies of scale dependence of tribological phenomena have been conducted recently. AFM experiments provide data on nanoscale [38.5–10] whereas microtriboapparatus [38.11, 12] and SFA [38.13] provide data on microscale. Experimental data indicate that wear mechanisms and wear rates are different at macro- and micro/nanoscales [38.14,15]. During sliding, the effect of operating conditions such as load and velocity on friction and wear are frequently manifestations of the effect of temperature rise on the variable under study. The overall interface temperature rise is a cumulative result of numerous flash temperature rises at individual asperity contacts. The temperature rise at each contact is expected to be scale dependent, since it depends on contact size, which is scale dependent.

Friction is a complex phenomenon, which involves asperity interactions involving adhesion and deformation (plowing), Fig. 38.1. Adhesion and plastic deformation imply energy dissipation, which is responsible for friction. A contact between two bodies takes place on high asperities, and the real area of contact (A_r) is a small fraction of the apparent area of contact [38.16]. During contact of two asperities, a lateral force may be required for asperities of a given slope to climb against each other. This mechanism is known as ratchet mechanism, and it also contributes to the friction. Wear and contaminant particles present at the interface, referred as the "third body", also contribute to friction, Fig. 38.2a. In addition, during contact, even at low humidity, a meniscus is formed. Generally, any liquid that wets or has a small contact angle on surfaces will condense from vapor into cracks and pores on surfaces as bulk liquid and in the form of annular-shaped capil-

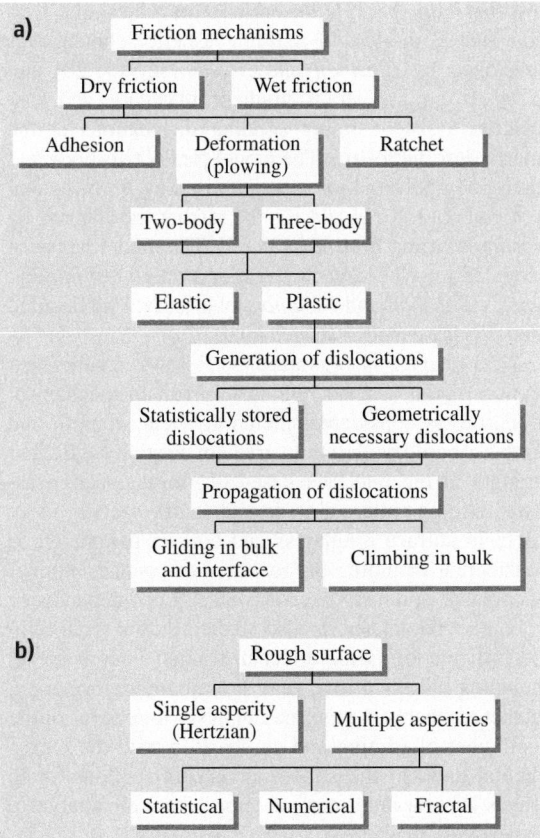

Fig. 38.1 (a) A block diagram showing friction mechanisms and generation and propagation of dislocations during sliding, **(b)** a block diagram of rough contact models

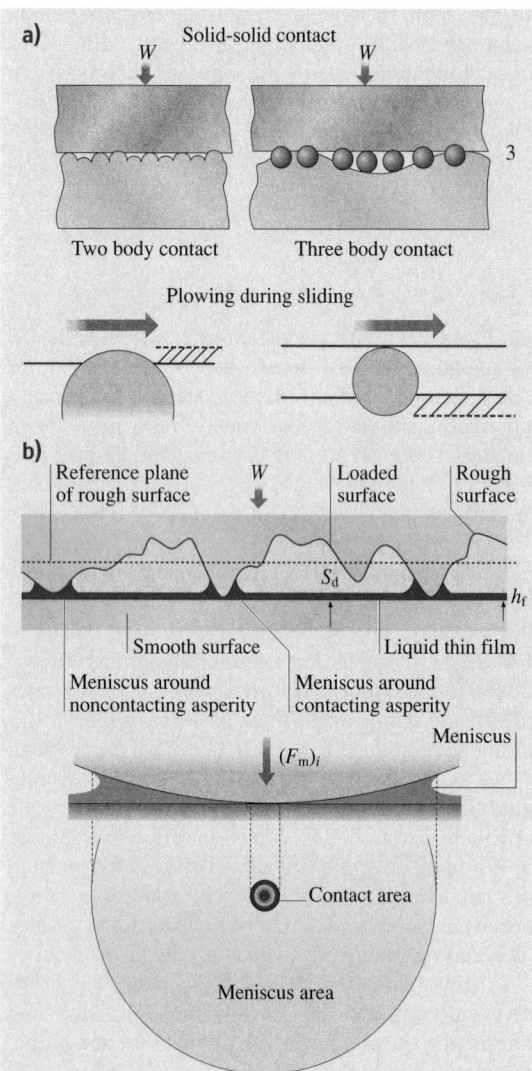

Fig. 38.2a,b Schematics of (**a**) two-bodies and three-bodies during dry contact of rough surfaces, (**b**) formation of menisci during wet contact

A quantitative theory of scale effects in friction should consider scale effect on physical properties relevant to these contributions. However, conventional theories of contact and friction lack characteristic length parameters, which would be responsible for scale effects. The linear elasticity and conventional plasticity theories are scale-invariant and do not include any material length scales. A strain gradient plasticity theory has been developed, for microscale deformations, by *Fleck* et al. [38.17], *Nix* and *Gao* [38.18] and *Hutchinson* [38.19]. Their theory predicts a dependence of mechanical properties on the strain gradient, which is scale dependent: the smaller is the size of the deformed region, the greater is the gradient of plastic strain, and, the greater is the yield strength and hardness.

A comprehensive model of scale effect in friction including adhesion, two- and three-body deformations and the ratchet mechanism, has recently been proposed by *Bhushan* and *Nosonovsky* [38.20–22] and *Nosonovsky* and *Bhushan* [38.23]. The model for adhesional friction during single and multiple asperity contact was developed by *Bhushan* and *Nosonovsky* [38.20] and is based on the strain gradient plasticity and dislocation assisted sliding (gliding dislocations at the interface or microslip). The model for the two-body and three-body deformation was proposed by *Bhushan* and *Nosonovsky* [38.21] and for the ratchet mechanism by *Nosonovsky* and *Bhushan* [38.23]. The model has been extended for wet contacts, wear and interface temperature by *Bhushan* and *Nosonovsky* [38.22]. The detailed model is presented in this chapter.

The chapter is organized as follows. In the next section of this chapter, the scale effect in mechanical properties is considered, including yield strength and hardness based on the strain gradient plasticity and shear strength at the interface based on the dislocation assisted sliding (microslip). In the fourth section, scale effect in surface roughness and contact parameters is considered, including the real area of contact, number of contacts, and mean size of contact. Load dependence of contact parameters is also studied in this section. In the fifth section, scale effect in friction is considered, including adhesion, two- and three-body deformation, ratchet mechanism, meniscus analysis, total value of the coefficient of friction and comparison with the experimental data. In the sixth and seventh sections, scale effects in wear and interface temperature are analyzed, respectively.

lary condensate in the contact zone. Figure 38.2b shows a random rough surface in contact with a smooth surface with a continuous liquid film on the smooth surface. The presence of the liquid film of the condensate or preexisting film of the liquid can significantly increase the adhesion between the solid bodies [38.16]. The effect of meniscus is scale-dependent.

38.3 Scale Effect in Mechanical Properties

In this section, scale dependence of hardness and shear strength at the interface is considered. A strain gradient plasticity theory has been developed, for microscale deformations, by *Fleck* et al. [38.17], *Nix* and *Gao* [38.18], *Hutchinson* [38.19], and others, which is based on statistically stored and geometrically necessary dislocations (to be described later). Their theory predicts a dependence of mechanical properties on the strain gradient, which is scale dependent: the smaller is the size of the deformed region, the greater is the gradient of plastic strain, and, the greater is the yield strength and hardness. *Gao* et al. [38.24] and *Huang* et al. [38.25] proposed a mechanism-based strain gradient (MSG) plasticity theory, which is based on a multiscale framework, linking the microscale (10–100 nm) notion of statistically stored and geometrically necessary dislocations to the mesoscale (1–10 μm) notion of plastic strain and strain gradient. *Bazant* [38.26] analyzed scale effect based on the MSG plasticity theory in the limit of small scale, and found that corresponding nominal stresses in geometrically similar structures of different sizes depend on the size according to a power exponent law.

It was recently suggested also, that relative motion of two contacting bodies during sliding takes place due to dislocation-assisted sliding (microslip), which results in scale-dependent shear strength at the interface [38.20]. Scale effects in mechanical properties (yield strength, hardness, and shear strength at the interface) based on the strain gradient plasticity and dislocation-assisted sliding models are considered in this section.

38.3.1 Yield Strength and Hardness

Plastic deformation occurs during asperity contacts because a small real area of contact results in high contact stresses, which are often beyond the limits of the elasticity. As stated earlier, during loading, generation and propagation of dislocations is responsible for plastic deformation. Because dislocation motion is irreversible, plastic deformation provides a mechanism for energy dissipation during friction. The strain gradient plasticity theories [38.17–19] consider two types of dislocations: randomly created Statistically Stored Dislocations (SSD) and Geometrically Necessary Dislocations (GND). The GND are required for strain compatibility reasons. Randomly created SSD during shear and GND during bending are presented in Fig. 38.3a. The density of the GND (total length of dislocation lines per volume) during bending is proportional

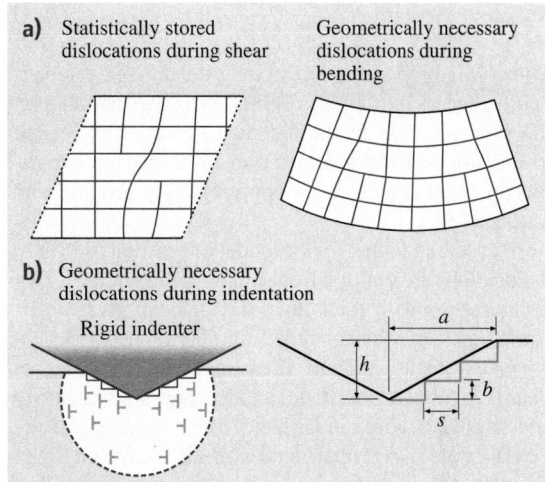

Fig. 38.3 (a) Illustration of statistically stored dislocations during shear and geometrically necessary dislocations during bending, **(b)** geometrically necessary dislocations during indentation

to the curvature κ and to the strain gradient

$$\rho_G = \frac{\kappa}{b} = \frac{1}{b}\frac{\partial \varepsilon}{\partial z} \propto \nabla \varepsilon , \qquad (38.1)$$

where ε is strain, b is the Burgers vector, and $\nabla \varepsilon$ is the strain gradient.

The GND during indentation, Fig. 38.3b, are located in a certain sub-surface volume. The large strain gradients in small indentations require GND to account for the large slope at the indented surface. SSD, not shown here, also would be created and would contribute to deformation resistance, and are function of strain rather than strain gradient. According to *Nix* and *Gao* [38.18], we assume that indentation is accommodated by circular loops of GND with Burgers vector normal to the plane of the surface. If we think of the individual dislocation loops being spaced equally along the surface of the indentation, then the surface slope

$$\tan \theta_i = \frac{h}{a} = \frac{b}{s} , \qquad (38.2)$$

where θ_i is the angle between the surface of the conical indenter and the plane of the surface, a is the contact radius, h is the indentation depth, b is the Burgers vector, and s is the spacing between individual slip steps on the indentation surface (Fig. 38.3b). They reported that for geometrical (strain compatibility) considerations, the

density of the GND is

$$\rho_G = \frac{3}{2bh}\tan^2\theta_i = \frac{3}{2b}\left(\frac{\tan\theta_i}{a}\right) = \frac{3}{2b}\nabla\varepsilon. \quad (38.3)$$

Thus ρ_G is proportional to strain gradient (scale dependent) whereas the density of SSD, ρ_S is dependent upon the average strain in the indentation, which is related to the slope of the indenter ($\tan\theta_i$). Based on experimental observations, ρ_S is approximately proportional to strain [38.17].

According to the Taylor model of plasticity [38.30], dislocations are emitted from Frank–Read sources. Due to interaction with each other, the dislocations may become stuck in what is called the Taylor network, but when externally applied stress reaches the order of Peierls stress for the dislocations, they start to move and the plastic yield is initiated. The magnitude of the Peierls stress τ_p is proportional to the dislocation's Burgers vector b divided by a distance between dislocation lines s [38.30, 31]

$$\tau_p = Gb/(2\pi s), \quad (38.4)$$

where G is the elastic shear modulus. An approximate relation of the shear yield strength τ_Y to the dislocations density at a moment when yield is initiated is given by [38.30]

$$\tau_{Y0} = cGb/s = cGb\sqrt{\rho}, \quad (38.5)$$

where c is a constant on the order of unity, specified by the crystal structure and ρ is the total length of dislocation lines per volume, which is a complicated function of strain ε and strain gradient ($\nabla\varepsilon$)

$$\rho = \rho_S(\varepsilon) + \rho_G(\nabla\varepsilon). \quad (38.6)$$

The shear yield strength τ_Y can be written now as a function of SSD and GND densities [38.30]

$$\tau_Y = cGb\sqrt{\rho_S + \rho_G} = \tau_{Y0}\sqrt{1 + (\rho_G/\rho_S)}, \quad (38.7)$$

where

$$\tau_{Y0} = cGb\sqrt{\rho_S} \quad (38.8)$$

is the shear yield strength value in the limit of small ρ_G/ρ_S ratio (large scale) that would arise from the SSD, in the absence of GND. Note that the ratio of the two densities is defined by the problem geometry and is scale dependent. Based on the relationships for ρ_G (38.3) and ρ_S, the ratio ρ_G/ρ_S is inversely proportional to a and (38.7) reduces to

$$\tau_Y = \tau_{Y0}\sqrt{1 + (l_d/a)}, \quad (38.9)$$

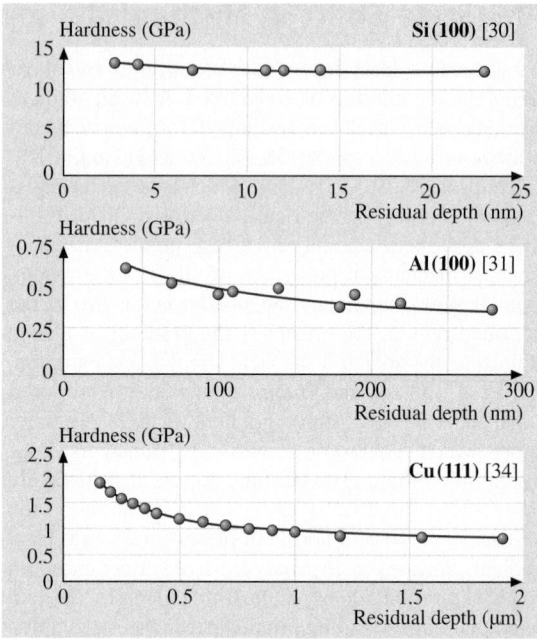

Fig. 38.4 Indentation hardness as a function of residual indentation depth for Si(100) [38.27], Al(100) [38.28], Cu(111) [38.29]

where l_d is a plastic deformation length that characterizes depth dependence on shear yield strength. According to *Hutchinson* [38.19], this length is physically related to an average distance a dislocation travels, which was experimentally determined to be between $0.2\,\mu m$ and $5\,\mu m$ for copper and nickel. Note that l_d is a function of the material and the asperity geometry and is dependent on SSD.

Using von Mises yield criterion, hardness $H = 3\sqrt{3}\tau_Y$. From (38.9) the hardness is also scale-dependent [38.18]

$$H = H_0\sqrt{1 + (l_d/a)}, \quad (38.10)$$

where H_0 is hardness in absence of strain gradient. Equation (38.9) provides dependence of the resistance force to deformation upon the scale in a general case of plastic deformation [38.20].

Scale dependence of yield strength and hardness has been well established experimentally. *Bhushan* and *Koinkar* [38.32] and *Bhushan* et al. [38.27] measured hardness of single-crystal silicon (100) up to a peak load of $500\,\mu N$. *Kulkarni* and *Bhushan* [38.28] measured hardness of single crystal aluminium (100) up to $2000\,\mu N$ and *Nix* and *Gao* [38.18] presented data for single crystal copper; using a three-sided pyramidal

(Berkovich) diamond tip. The hardness on nanoscale is found to be higher than on microscale (Fig. 38.4). Similar results have been reported in other tests, including indentation tests for other materials [38.29, 33, 34], torsion and tension experiments on copper wires [38.17, 19], and bending experiments on silicon and silica beams [38.35].

38.3.2 Shear Strength at the Interface

Mechanism of slip involves motion of large number of dislocations, which is responsible for plastic deformation during sliding. Dislocations are generated and

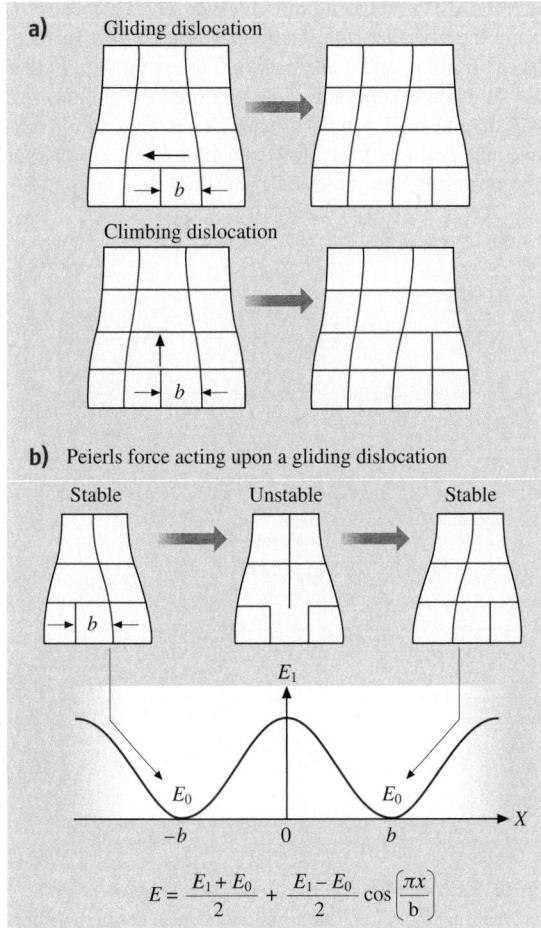

Fig. 38.5 (a) Schematics of gliding and climbing dislocations motion by a unit step of Burgers vector *b*. **(b)** Origin of the periodic force acting upon a gliding dislocation (Peierls force). Gliding dislocation passes locations of high and low potential energy

stored in the body and propagate under load. There are two modes of possible line (or edge) dislocation motion: gliding, when dislocation moves in the direction of its Burgers vector *b* by a unit step of its magnitude, and climbing, when dislocation moves in a direction, perpendicular to its Burgers vector (Fig. 38.5a). Motion of dislocations can take place in the bulk of the body or at the interface. Due to periodicity of the lattice, a gliding dislocation experiences a periodic force, known as the Peierls force [38.31]. The Peierls force is responsible for keeping the dislocation at a central position between symmetric lattice lines and it opposes dislocation's gliding (Fig. 38.5b). Therefore, an external force should be applied to overcome Peierls force resistance against dislocation's motion. *Weertman* [38.36] showed that a dislocation or a group of dislocations can glide uniformly along an interface between two bodies of different elastic properties. In continuum elasticity formulation, this motion is equivalent to a propagating interface slip pulse, however the physical nature of this deformation is plastic, because dislocation motion is irreversible. The local plastic deformation can occur at the interface due to concentration of dislocations even in the predominantly elastic contacts. Gliding of a dislocation along the interface results in a relative displacement of the bodies for a distance equal to the Burgers vector of the dislocation, whereas a propagating set of dislocations effectively results in dislocation-assisted sliding, or microslip (Fig. 38.6).

Several types of microslip are known in the tribology literature [38.16], the dislocation-assisted sliding is one type of microslip, which propagates along the interface. Conventional mechanism of sliding is considered to be concurrent slip with simultaneous breaking of all adhesive bonds. Based on *Johnson* [38.37] and *Bhushan* and *Nosonovsky* [38.20], for contact sizes on the or-

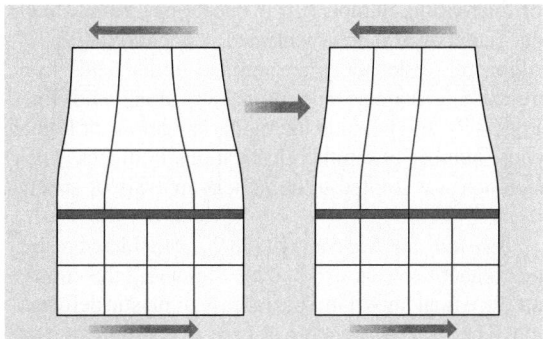

Fig. 38.6 Schematic showing microslip due to gliding dislocations at the interface

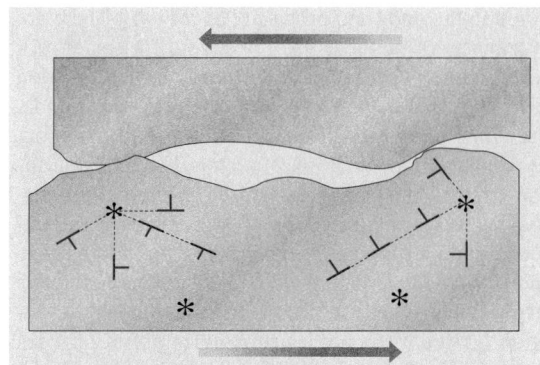

Fig. 38.7 Generation of dislocations from sources (∗) during plowing due to plastic deformation

der of few nm to few μm, dislocation-assisted sliding is more energetically profitable than a concurrent slip. Their argument is based on the fact that experimental measurements with the SFA demonstrated that, for mica, frictional stress is of the same order as Peierls stress, which is required for gliding of dislocations.

Polonsky and *Keer* [38.38] considered the preexisting dislocation sources and carried out a numerical microcontact simulation based on contact plastic deformation representation in terms of discrete dislocations. They found that when the asperity size decreases and becomes comparable with the characteristic length of materials microstructure (distance between dislocation sources), resistance to plastic deformation increases, which supports conclusions drawn from strain gradient plasticity. *Deshpande* et al. [38.39] conducted discrete plasticity modeling of cracks in single crystals and considered dislocation nucleation from Frank–Read sources distributed randomly in the material. Pre-existing sources of dislocations, considered by all of these authors, are believed to be a more realistic reason for increasing number of dislocations during loading, rather than newly nucleated dislocations [38.30]. In general, dislocations are emitted under loads from pre-existing sources and propagate along slip lines (Fig. 38.7). As shown in the figure, in regions of higher loads, number of emitted dislocations is higher. Their approach was limited to numerical analysis of special cases.

Bhushan and *Nosonovsky* [38.20] considered a sliding contact between two bodies. Slip along the contact interface is an important special case of plastic deformation. The local dislocation-assisted microslip can exist even if the contact is predominantly elastic due to concentration of dislocations at the interface. Due to these

dislocations, the stress at which yield occurs at the interface is lower than shear yield strength in the bulk. This means that average shear strength at the interface is lower than in the bulk.

An assumption that all dislocations produced by externally applied forces are distributed randomly throughout the volume would result in vanishing small probability for a dislocation to be exactly at the interface. However, many traveling (gliding and climbing) dislocations will be stuck at the interface as soon as they reach it. As a result of this, a certain number of dislocations will be located at the interface. In order to account for a finite dislocation density at the interface, *Bhushan* and *Nosonovsky* [38.20] assumed, that the interface zone has a finite thickness D. Dislocations within the interface zone may reach the contact surface due to climbing and contribute into the microslip. In the case of a small contact radius a, compared to interface zone thickness D, which is scale dependent, and is approximately equal to a. However, in the case of a large contact radius, the interface zone thickness is approx-

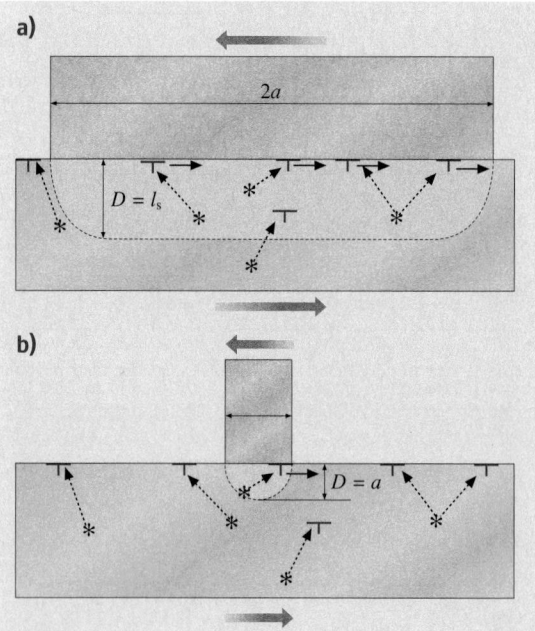

Fig. 38.8a,b Gliding dislocations at the interface generated from sources (∗). Only dislocations generated within the interface zone can reach the interface. (**a**) For a large contact radius a, thickness of this zone D is approximately equal to an average distance dislocations climb l_s. (**b**) For small contact radius a, the thickness of the interface zone is approximately equal to a

imately equal to the average distance dislocations can climb l_s. An illustration of this is provided in Fig. 38.8. The depth of the subsurface volume, from which dislocations have a high chance to reach the interface is limited by l_s and by a, respectively, for the two cases considered here. Based on these geometrical considerations, an approximate relation can be written as

$$D = \frac{al_s}{l_s + a} . \tag{38.11}$$

The interface density of dislocations (total length of dislocation lines per interface area) is related to the volume density as

$$\eta_{int} = \rho D = \rho \left(\frac{al_s}{l_s + a} \right) . \tag{38.12}$$

During sliding, dislocations must be generated at the interface with a certain critical density $\eta_{int} = \eta_{cr}$. The corresponding shear strength during sliding can be written following (38.9) as

$$\tau_a = \tau_{a0} \sqrt{1 + (l_s/a)} , \tag{38.13}$$

where

$$\tau_{a0} = cGb \sqrt{\frac{\eta_{cr}}{l_s}} \tag{38.14}$$

is the shear strength during sliding in the limit of $a \gg l_s$.

Equation (38.13) gives scale-dependence of the shear strength at the interface and is based on the following assumptions. First, it is assumed that only dislocations in the interface zone of thickness D, given by (38.11), contribute into sliding. Second, it is assumed, that a critical density of dislocations at the interface η_{cr} is required for sliding. Third, the shear strength is equal to the Peierls stress, which is related to the volume density of the dislocations $\rho = \eta/D$ according to (38.4), with the typical distance between dislocations $s = 1/\sqrt{\rho}$. It is noted, that proposed scaling rule for the dislocation assisted sliding mechanism (38.13) has a similar form to that for the yield strength (38.9), since both results are consequences of scale dependent generation and propagation of dislocations under load [38.20].

38.4 Scale Effect in Surface Roughness and Contact Parameters

During multiple-asperity contact, scale dependence of surface roughness is a factor which contributes to scale dependence of the real area of contact. Roughness parameters are known to be scale dependent [38.16], which results, during the contact of two bodies, in scale dependence of the real area of contact, number of contacts and mean contact size. The contact parameters also depend on the normal load, and the load dependence is similar to the scale dependence [38.23]. Both effects are analyzed in this section.

38.4.1 Scale Dependence of Roughness and Contact Parameters

A random rough surface with Gaussian height distribution is characterized by the standard deviation of surface height σ and the correlation length β^* [38.16]. The correlation length is a measure of how quickly a random event decays and it is equal to the length, over which the autocorrelation function drops to a small fraction of the value at the origin. The correlation length can be considered as a distance, at which two points on a surface have just reached the condition where they can be regarded as being statistically independent. Thus, σ is a measure of height distribution and β^* is a measure of spatial distribution.

A surface is composed of a large number of length scales of roughness that are superimposed on each other. According to AFM measurements on glass-ceramic disk surface, both σ and β^* initially increase with the scan size and then approach a constant value, at certain scan size (Fig. 38.9). This result suggests that disk roughness has a long wavelength limit, L_{lwl}, which is equal to the scan size at which the roughness values approach a constant value [38.16]. It can be assumed that σ and β^* depend on the scan size according to an empirical power rule

$$\sigma = \sigma_0 \left(\frac{L}{L_{lwl}} \right)^n , \qquad L < L_{lwl} ,$$

$$\beta^* = \beta_0^* \left(\frac{L}{L_{lwl}} \right)^m , \qquad L < L_{lwl} , \tag{38.15}$$

where n and m are indices of corresponding exponents and σ_0 and β_0^* are macroscale values [38.20]. Based on the data, presented in Fig. 38.9, it is noted that for glass-ceramic disk, long-wavelength limit for σ and β^* is about $17 \, \mu m$ and $23 \, \mu m$, respectively. The difference is expected to be due to measurement errors. An average value $L_{lwl} = 20 \, \mu m$ is taken here for calculations. The values of the indices are found as $m = 0.5$, $n = 0.2$, and the macroscale values are $\sigma_0 = 5.3 \, nm$, $\beta_0^* = 0.37 \, \mu m$ [38.23].

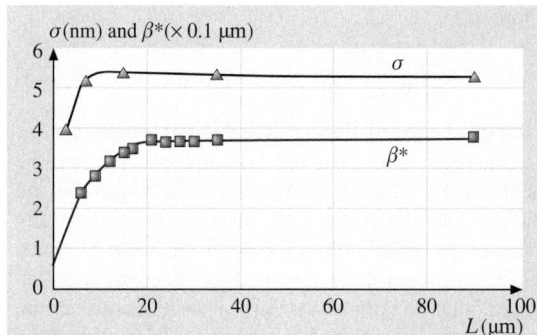

Fig. 38.9 Roughness parameters as a function of scan size for a glass-ceramic disk measured using AFM [38.16]

For two random surfaces in contact, the length of the nominal contact size L defines the characteristic length scale of the problem. The contact problem can be simplified by considering a rough surface with composite roughness parameters in contact with a flat surface. The composite roughness parameters σ and β^* can be obtained based on individual values for the two surfaces [38.16]. For Gaussian surfaces, the contact parameters of interest, to be discussed later, are the real area of contact A_r, number of contacts N, and mean contact radius $\bar{a}$. The long wavelength limit for scale dependence of the contact parameters L_{lc}, which is not necessarily equal to that of the roughness, L_{lwl}, will be used for normalization of length parameters. The scale dependence of the contact parameters exists if $L < L_{lc}$ [38.23].

The mean of surface height distribution corresponds to so-called reference plane of the surface. Separation s_d is a distance between reference planes of two surfaces in contact, normalized by σ. For a given s_d and statistical distribution of surface heights, the total real area of contact (A_r), number of contacts (N), and elastic normal load W_e can be found, using statistical analysis of contacts. The real area of contact, number of contacts and elastic normal load are related to the separation distance s_d [38.40]

$$A_r \propto F_A(s_d) \,,$$

$$N \propto \frac{1}{(\beta^*)^2} F_N(s_d) \,,$$

$$W_e \propto \frac{E^* \sigma}{\beta^*} F_W(s_d) \,, \tag{38.16}$$

where $F_A(s_d)$, $F_N(s_d)$, and $F_W(s_d)$, are integral functions defined by *Onions* and *Archard* [38.40]. It should be noted, that A_r and N as functions of s_d are prescribed by the contact geometry (σ, β^*) and do not depend on

whether the contact is elastic or plastic. Based on *Onions* and *Archard* data, it is observed that the ratio F_W/F_A is almost constant for moderate $s_d < 1.4$ and increases slightly for $s_d > 1.4$. The ratio F_A/F_N decreases rapidly with s_d and becomes almost constant for $s_d > 2.0$. For moderate loads, the contact is expected to occur on the upper parts of the asperities ($s_d > 2.0$), and a linear proportionality of $F_A(s_d)$, $F_N(s_d)$, and $F_W(s_d)$ can be assumed [38.20].

Based on (38.16) and the observation that F_W/F_A is almost constant, for moderate loads, A_{re} (the real area of elastic contact), N, and $\bar{a}$ are related to the roughness, based on the parameter L_{lc}, as

$$A_{re} \propto \frac{\beta^*}{\sigma E^*} W = A_{re0} \left(\frac{L}{L_{lc}} \right)^{m-n} \,, \quad L < L_{lc} \,, \tag{38.17}$$

$$N \propto \frac{W}{\beta^* \sigma E^*} = N_0 \left(\frac{L}{L_{lc}} \right)^{-m-n} \,, \quad L < L_{lc} \,, \tag{38.18}$$

$$\bar{a} \propto \beta^* = \sqrt{\frac{A_r}{N}} = \bar{a}_0 \left(\frac{L}{L_{lc}} \right)^m \,, \quad L < L_{lc} \,. \tag{38.19}$$

The mean radius of summit tips $\overline{R_p}$ is given, according to *Whitehouse* and *Archard* [38.41]

$$\overline{R_p} \propto \frac{(\beta^*)^2}{\sigma} = \overline{R_{p0}} \left(\frac{L}{L_{lwl}} \right)^{2m-n} \,, \quad L < L_{lwl} \,, \tag{38.20}$$

where $\bar{a}_0$, N_0 and $\overline{R_{p0}}$ are macroscale values, E^* is the effective elastic modulus of contacting bodies [38.22], which is related to the elastic moduli E_1, E_2 and Poisson's ratios ν_1, ν_2 of the two bodies as $1/E^* = (1 - \nu_1^2)/E_1 + (1 - \nu_2^2)/E_2$ and which is known to be scale independent, and variables with the subscript "0" are corresponding macroscale values (for $L \geq L_{lc}$).

Dependence of the real area of plastic contact A_{rp} on the load is given by

$$A_{rp} = \frac{W}{H} \,, \tag{38.21}$$

where H is hardness. According to the strain gradient plasticity model [38.17,18], the yield strength τ_Y is given by (38.9) and hardness H is given by (38.10). In the case of plastic contact, the mean contact radius can be determined from (38.19), which is based on the contact geometry and independent of load [38.20]. Assuming the contact radius as its mean value from (38.19) based

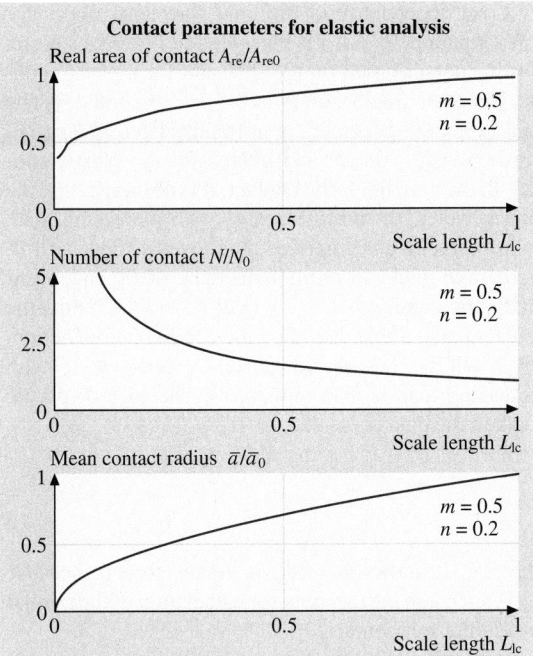

Fig. 38.10a–c Scale length dependence of normalized contact parameters ($m = 0.5$, $n = 0.2$) (**a**) real area of contact, (**b**) number of contacts, and (**c**) mean contact radius

on elastic analysis, and combining (38.10), (38.19) and (38.21), the real area of plastic contact is given as

$$A_{\mathrm{rp}} = \frac{W}{H_0\sqrt{1+(l_{\mathrm{d}}/\bar{a})}}$$
$$= \frac{W}{H_0\sqrt{1+(L_{\mathrm{d}}/L)^m}}, \quad L < L_{\mathrm{lc}}, \quad (38.22)$$

where L_{d} is a characteristic length parameter related to l_{d}, $\bar{a}$, and L_{lc} [38.20]

$$L_{\mathrm{d}} = L_{\mathrm{lc}}\left(\frac{l_{\mathrm{d}}}{\bar{a}_0}\right)^{1/m}. \quad (38.23)$$

The scale dependence of A_{re}, N, and $\bar{a}$ is presented in Fig. 38.10.

38.4.2 Dependence of Contact Parameters on Load

The effect of short and long wavelength details of rough surfaces on contact parameters also depends on the normal load. For low loads, the ratio of real to apparent areas of contact $A_{\mathrm{r}}/A_{\mathrm{a}}$, is small, contact spots are small, and long wavelength details are irrelevant.

For higher $A_{\mathrm{r}}/A_{\mathrm{a}}$, long wavelength details become important, whereas small wavelength details of the surface geometry become irrelevant. The effect of increased load is similar to the effect of increased scale length [38.23].

In the preceding subsections, it was assumed that the roughness parameters are scale-dependent for $L < L_{\mathrm{lwl}}$, whereas the contact parameters are scale-dependent for $L < L_{\mathrm{lc}}$. The upper limit of scale dependence for the contact parameters, L_{lc}, depends on the normal load, and it is reasonable to assume that L_{lc} is a function of $A_{\mathrm{r}}/A_{\mathrm{a}}$, and the contact parameters are scale-dependent when $A_{\mathrm{r}}/A_{\mathrm{a}}$ is below a certain critical value. It is convenient to consider the apparent pressure p_{a}, which is equal to the normal load divided by the apparent area of contact [38.23].

For elastic contact, based on (38.15) and (38.17), this condition can be written as

$$\frac{A_{\mathrm{re}}}{A_{\mathrm{a}}} \propto \frac{\beta^* p_{\mathrm{a}}}{\sigma} = p_{\mathrm{a}}\frac{\beta_0^*}{\sigma_0}\left(\frac{L}{L_{\mathrm{lwl}}}\right)^{m-n} < p_{\mathrm{ac}}, \quad (38.24)$$

where p_{ac} is a critical apparent pressure, below which the scale dependence occurs [38.23]. From (38.24) one can find

$$L < L_{\mathrm{lwl}}\left(\frac{\beta_0^*}{\sigma_0}\frac{p_{\mathrm{a}}}{p_{\mathrm{ac}}}\right)^{1/(n-m)}. \quad (38.25)$$

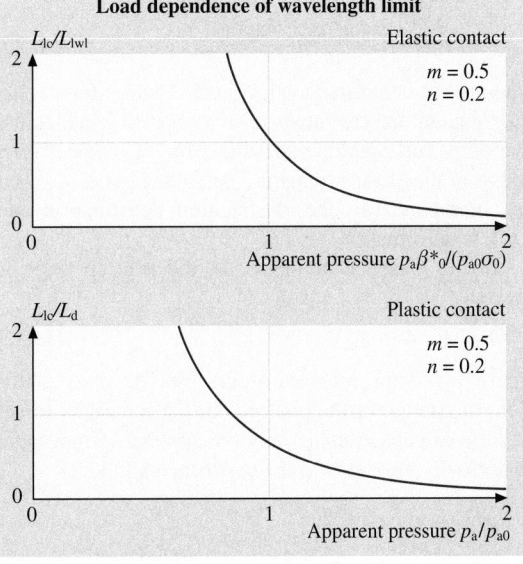

Fig. 38.11 Dependence of the normalized long wavelength limit for contact parameters on load (normalized apparent pressure) for elastic and plastic contacts ($m = 0.5$, $n = 0.2$)

The right-hand expression in (38.24) is defined as L_{lc}

$$L_{lc} = L_{lwl} \left(\frac{\beta_0^*}{\sigma_0} \frac{p_a}{p_{ac}} \right)^{1/(n-m)} . \qquad (38.26)$$

For plastic contact, based on (38.22)

$$\frac{A_{rp}}{A_a} \propto \frac{p_a}{\sqrt{1 + (L_d/L)^m}} < p_{ac} . \qquad (38.27)$$

In a similar manner to the elastic case, (38.27) yields [38.23]

$$L_{lc} = L_d \left[\left(\frac{p_a}{p_{ac}} \right)^2 - 1 \right]^{-1/m} . \qquad (38.28)$$

Load dependence of the long wavelength limit for contact parameters, L_{lc} is presented in Fig. 38.11 for an elastic contact based on (38.28), and for a plastic contact based on (38.28), for $m = 0.5$, $n = 0.2$ [38.23]. The load (apparent pressure) is normalized by $\beta_0^*/(p_{ac}\sigma_0)$ for the elastic contact and by p_{ac} for the plastic contact. In the case of elastic contact, it is observed, that the long wavelength limit decreases with increasing load. For a problem, characterized by a given scale length L, increase of load will result in decrease of L_{lc} and, eventually, the condition $L < L_{lc}$ will be violated; thus the contact parameters, including the coefficient of friction, will reach the macroscale values. Decrease of L_{lc} with increasing load is also observed in the case of plastic contact, the data presented for $p_a/p_{ac} > 1$.

38.5 Scale Effect in Friction

According to the adhesion and deformation model of friction [38.16], the coefficient of dry friction μ can be presented as a sum of adhesion component μ_a and deformation (plowing) component μ_d. The later, in the presence of particles, is a sum of asperity summits deformation component μ_{ds} and particles deformation component μ_{dp}, so that the total coefficient of friction is [38.21]

$$\mu = \mu_a + \mu_{ds} + \mu_{dp} = \frac{F_a + F_{ds} + F_{dp}}{W}$$
$$= \frac{A_{ra}\tau_a + A_{ds}\tau_{ds} + A_{dp}\tau_{dp}}{W} , \qquad (38.29)$$

where W is the normal load, F is the friction force, A_{ra}, A_{ds}, A_{dp} are the real areas of contact during adhesion, two body deformation and with particles, respectively, and τ is the shear strength. The subscripts a, ds, and dp correspond to adhesion, summit deformation and particle deformation.

In the presence of meniscus, the friction force is given by

$$F = \mu (W + F_m) , \qquad (38.30)$$

where F_m is the meniscus force [38.16]. The coefficient of friction in the presence of the meniscus force, μ_{wet}, is calculated using only the applied normal load, as normally measured in the experiments [38.22]

$$\mu_{wet} = \mu \left(1 + \frac{F_m}{W} \right)$$
$$= \frac{A_{ra}\tau_a + A_{ds}\tau_{ds} + A_{dp}\tau_{dp}}{W} \left(1 + \frac{F_m}{W} \right) . \qquad (38.31)$$

The (38.31) shows that μ_{wet} is greater than μ, because F_m is not taken into account for calculation of the normal load in the wet contact.

It was shown by *Greenwood* and *Williamson* [38.42] and by subsequent modifications of their model, that for contacting surfaces with common statistical distributions of asperity heights, the real area of contact is almost linearly proportional to the normal load. This linear dependence, along with (38.29), result in linear dependence of the friction force on the normal load, or coefficient of friction being independent of the normal load. For a review of the numerical analysis of rough surface contacts, see *Bhushan* [38.43,44] and *Bhushan* and *Peng* [38.45]. The statistical and numerical theories of contact involve roughness parameters – e.g. the standard deviation of asperity heights and the correlation length [38.16]. The roughness parameters are scale dependent. In contrast to this, the theory of self-similar (fractal) surfaces solid contact developed by *Majumdar* and *Bhushan* [38.46] does not include length parameters and are scale-invariant in principle. The shear strength of the contacts in (38.29) is also scale dependent. In addition to the adhesional contribution to friction, elastic and plastic deformation on nano- to macroscale contributes to friction [38.16]. The deformations are also scale dependent.

38.5.1 Adhesional Friction

The adhesional component of friction depends on the real area of contact and adhesion shear strength. Here we derive expressions for scale dependence of adhesional friction during single-asperity and multiple-asperity contacts.

Single–Asperity Contact

The scale length during single-asperity contact is the nominal contact length, which is equal to the contact

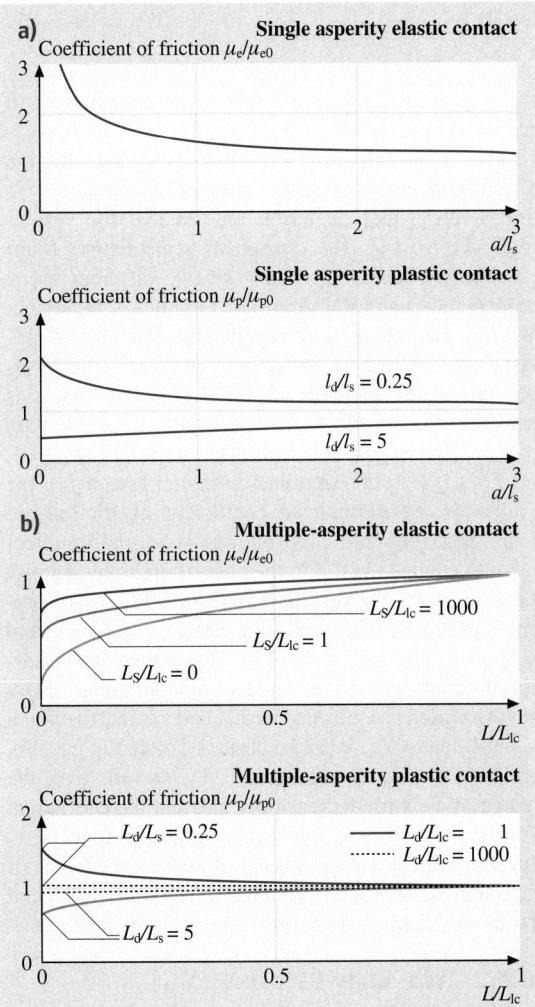

Fig. 38.12 Normalized results for the adhesional component of the coefficient of friction, as a function of scale (a/l_s for single asperity contact and L/L_{lc} for multi-asperity contact). In the case of single asperity plastic contact, data are presented for two values of l_d/l_s. In the case of multi-asperity contact, data are presented for $m = 0.5$, $n = 0.2$. For multi-asperity elastic contact, data are presented for three values of L_S/L_{lc}. For multi-asperity plastic contact, data are presented for two values of L_d/L_s

diameter $2a$. In the case of predominantly elastic contacts, the real area of contact A_{re} depends on the load according to the Hertz analysis [38.47]

$$A_{re} = \pi a^2 , \tag{38.32}$$

and

$$a = \left(\frac{3WR}{4E^*} \right)^{1/3} , \tag{38.33}$$

where R is effective radius of curvature of summit tips, and E^* is the effective elastic modulus of the two bodies. In the case of predominantly plastic contact, the real area of contact A_{rp} is given by (38.21), whereas the hardness is given by (38.10).

Combining (38.10), (38.13), (38.29), and (38.32), the adhesional component of the coefficient of friction can be determined for the predominantly elastic contact as

$$\mu_{ae} = \mu_{ae0}\sqrt{1 + (l_s/a)} \tag{38.34}$$

and for the predominantly plastic contact as

$$\mu_{ap} = \mu_{ap0}\sqrt{\frac{1 + (l_s/a)}{1 + (l_d/a)}} , \tag{38.35}$$

where μ_{ae0} and μ_{ap0} are corresponding values at the macroscale [38.20].

The scale dependence of adhesional friction in single-asperity contact is presented in Fig. 38.12a. In the case of single asperity elastic contact, the coefficient of friction increases with decreasing scale (contact diameter), because of an increase in the adhesion strength, according to (38.34). In the case of single asperity plastic contact, the coefficient of friction can increase or decrease with decreasing scale, because of an increased hardness or increase in adhesional strength. The competition of these two factors is governed by l_d/l_s, according to (38.35). There is no direct way to measure l_d and l_s. We will see later, from experimental data, that the coefficient of friction tends to decrease with decreasing scale, therefore, it must be assumed that $l_d/l_s > 1$ for the data reported in the paper [38.20].

Multiple–Asperity Contact

The adhesional component of friction depends on the real area of contact and adhesion shear strength. Scale dependence of the real area of contact was considered in the preceding section. Here we derive expressions for scale-dependence of the shear strength at the interface during adhesional friction. It is suggested by

Table 38.1 Scaling factors for the coefficient of adhesional friction [38.20]

Single asperity elastic contact	Single asperity plastic contact	Multiple-asperity elastic contact	Multiple-asperity plastic contact
$\mu_e =$ $\mu_{e0}\sqrt{1+(l_s/a)}$	$\mu_e =$ $\mu_{e0}\sqrt{\dfrac{1+(l_s/a)}{1+(l_d/a)}}$	$\mu_e =$ $\mu_{e0}C_E L^{m-n}$ $\times\sqrt{1+(L_s/L)^m}$	$\mu_p =$ $\mu_{p0}C_P\sqrt{\dfrac{1+(L_s/L)^m}{1+(L_d/L)^m}}$

Bhushan and *Nosonovsky* [38.20] that, for many materials, dislocation-assisted sliding (microslip) is the main mechanism, which is responsible for the shear strength. They considered dislocation assisted sliding based on the assumption, that contributing dislocations are located in a subsurface volume. The thickness of this volume is limited by the distance which dislocations can climb l_s (material parameter) and by the radius of contact a. They showed that τ_a is scale dependent according to (38.13). Assuming the contact radii equal to the mean value given by (38.19)

$$\tau_a = \tau_{a0}\sqrt{1+(L_s/L)^m}\,, \quad L < L_{lc}\,, \tag{38.36}$$

where

$$L_s = L_{lc}\left(\frac{l_s}{\bar{a}_0}\right)^{1/m}\,. \tag{38.37}$$

In the case of absence of the microslip (e.g., for an amorphous material), it should be assumed in (38.34–38.36), $L_s = l_s = 0$.

Based on (38.9, 38.17, 38.24, 38.29, 38.36, 38.37), the adhesional component of the coefficient of friction in the case of elastic contact μ_{ae} and in the case of plastic contact μ_{ap}, is given as [38.20]

$$\mu_{ae} = \frac{\tau_a A_{re}}{W}$$

$$= \frac{\tau_{a0} A_{re0}}{W}\left(\frac{L}{L_{lc}}\right)^{m-n}\sqrt{1+(L_s/L)^m}$$

$$= \frac{\mu_{ae0}}{\sqrt{1+(l_s/\bar{a}_0)}}\left(\frac{L}{L_{lc}}\right)^{m-n}$$

$$\times\sqrt{1+(L_s/L)^m}\,, \quad L < L_{lc}\,; \tag{38.38}$$

$$\mu_{ap} = \frac{\tau_{a0}}{H_0}\sqrt{\frac{1+(L_s/L)^m}{1+(L_d/L)^m}}$$

$$= \mu_{ap0}\sqrt{\frac{1+(l_d/\bar{a}_0)}{1+(l_s/\bar{a}_0)}}\sqrt{\frac{1+(L_s/L)^m}{1+(L_d/L)^m}}\,, \quad L < L_{lc}\,, \tag{38.39}$$

where μ_{ae0} and μ_{ap0} are values of the coefficient of friction at macroscale ($L \geq L_{lc}$).

The scale dependence of adhesional friction in multiple-asperity elastic contact is presented in Fig. 38.12b, which is based on (38.38), for various values of L_s/L_{lc}. The change of scale length L affects the coefficient of friction in two different ways: through the change of A_{re} (38.17) and τ_a (38.36) below L_{lc}. Further, τ_a is controlled by the ratio L_s/L. Based on (38.36), for small ratio of L_s/L_{lc}, scale effects on τ_a is insignificant for $L/L_{lc} > 0$. As it is seen from Fig. 38.12b by comparison of the curve with $L_s/L_{lc} = 0$ (insignificant scale effect on τ_a), $L_s/L_{lc} = 1$, and $L_s/L_{lc} = 1000$ (significant scale effect on τ_a), the results for the normalized coefficient of friction are close, thus, the main contribution to the scaling effect is due to change of A_{re}. In the case of multiple-asperity plastic contact, the results, based on (38.39), are presented in Fig. 38.12b for $L_d/L_s = 0.25$, $L_d/L_s = 5$ and $L_d/L_{lc} = 1$ and $L_d/L_{lc} = 1000$. The change of scale affects the coefficient of friction through the change of A_{rp} (38.34), which is controlled by L_d, and τ_a (38.36), which is controlled by L_s. It can be observed from Fig. 38.12b, that for $L_d > L_s$, the change of A_{rp} prevails over the change of τ_a, with decreasing scale, and the coefficient of friction decreases. For $L_d < L_s$, the change of τ_a prevails, with decreasing scale, and the coefficient of friction increases [38.20]. Expressions for the coefficient of adhesional friction are presented in Table 38.1.

38.5.2 Two-Body Deformation

Based on the assumption that multiple asperities of two rough surfaces in contact have conical shape, the two-body deformation component of friction can be determined as

$$\mu_{ds} = \frac{2\tan\theta_r}{\pi}\,, \tag{38.40}$$

where θ_r is the roughness angle (or attack angle) of a conical asperity [38.16, 48]. Mechanical properties af-

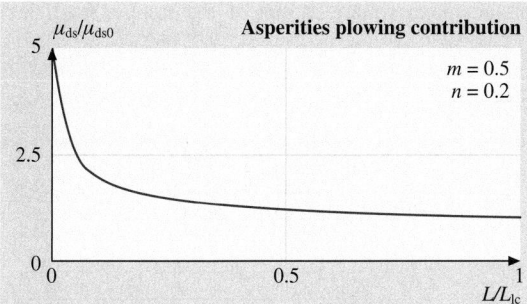

Fig. 38.13 Normalized results for the two-body deformation component of the coefficient of friction

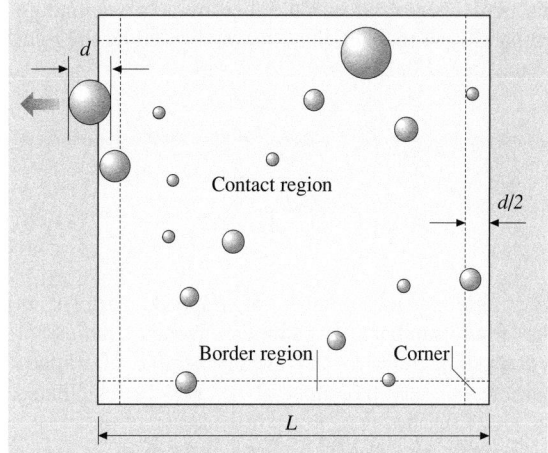

Fig. 38.14 Schematics of debris at the contact zone and at its border region. A particle of diameter d in the border region of $d/2$ is likely to leave the contact zone

fect the real area of contact and shear strength and these cancel out in (38.29).

The roughness angle is scale-dependent and is related to the roughness parameters [38.41]. Based on statistical analysis of a random Gaussian surface,

$$\tan\theta_{\mathrm{r}} \propto \frac{\sigma}{\beta^*} .\qquad(38.41)$$

From (38.40) it can be interpreted that stretching the rough surface in the vertical direction (increasing vertical scale parameter σ) increases $\tan\theta_{\mathrm{r}}$, and stretching in the horizontal direction (increasing vertical scale parameter β^*) decreases $\tan\theta_{\mathrm{r}}$.

Using (38.40) and (38.41), the scale dependence of the two-body deformation component of the coefficient of friction is given as [38.21]

$$\begin{aligned}\mu_{\mathrm{ds}} &= \frac{2\sigma_0}{\pi\beta^*}\left(\frac{L}{L_{\mathrm{lc}}}\right)^{n-m}\\&=\mu_{\mathrm{ds}0}\left(\frac{L}{L_{\mathrm{lc}}}\right)^{n-m} ,\quad L < L_{\mathrm{lc}} ,\end{aligned}\qquad(38.42)$$

where $\mu_{\mathrm{ds}0}$ is the value of the coefficient of summits deformation component of the coefficient of friction at macroscale ($L \geq L_{\mathrm{lc}}$).

The scale dependence for the two-body deformation component of the coefficient of friction is presented in Fig. 38.13 for $m = 0.5$, $n = 0.2$. The coefficient of friction increases with decreasing scale, according to (38.42). This effect is a consequence of increasing average slope or roughness angle.

38.5.3 Three-Body Deformation Friction

In this sections of the paper, size distribution of particles will be idealized according to the exponential, normal, and log normal density functions, since these distributions are the most common in nature and industrial applications (see Sect. 38.A). The probability for a particle of a given size to be trapped at the interface depends on the size of the contact region. Particles at the edge of the region of contact are likely to leave the contact area, whereas those in the middle are likely to be trapped. The ratio of the edge region area to the total apparent area of contact increases with decreasing scale size. Therefore, the probability for a particle to be trapped decreases, as well as the three-body component of the coefficient of friction [38.21].

Let us consider a square region of contact of two rough surfaces with a length L (relevant scale length), with the density of debris of η particles per unit area (Fig. 38.14). We assume that the particles have the spherical form and that $p(d)$ is the probability density function of particles size. It is also assumed that, for a given diameter, particles at the border region of the contact zone of the width $d/2$ are likely to leave the contact zone, with a certain probability α, whereas particles at the center of the contact region are likely to be trapped. It should be noted, that particles in the corners of the contact region can leave in two different directions, therefore, for them the probability to leave is 2α. The total nominal contact area is equal to L^2, the area of the border region, without the corners, is equal to $4(L-d)d/2$, and the area of the corners is equal to d^2.

The probability density of size distribution for the trapped particles $p_{\mathrm{tr}}(d)$ can be calculated by multiplying $p(d)$ by one minus the probability of a particle with diameter d to leave; the later is equal to the ratio of the border region area, multiplied by a corresponding probability of the particle to leave, divided by the total contact area [38.21]

$$p_{tr}(d) = p(d)\left(1 - \frac{2\alpha(L-d)d + 2\alpha d^2}{L^2}\right)$$
$$= p(d)\left(1 - \frac{2\alpha d}{L}\right), \quad d < \frac{L}{2\alpha}. \quad (38.43)$$

The ratio of the number of trapped particles to the total number of particles, average radius of a trapped particle $\overline{d}$, and average square of trapped particles $\overline{d^2}$, as functions of L, can be calculated as

$$n_{tr} = \frac{\int_0^{L/2} p_{tr}(d)\,dd}{\int_0^\infty p(d)\,dd} = \frac{\int_0^{L/2} p(d)\left(1 - \frac{2\alpha d}{L}\right)dd}{\int_0^\infty p(d)\,dd},$$

$$\overline{d} = \frac{\int_0^{L/2} d\,p_{tr}(d)\,dd}{\int_0^{L/2} p_{tr}(d)\,dd},$$

$$\overline{d^2} = \frac{\int_0^{L/2} d^2\,p_{tr}(d)\,dd}{\int_0^{L/2} p_{tr}(d)\,dd}. \quad (38.44)$$

Let us assume an exponential distribution of particles' size (38.A7) with $d_e = 0$. Substituting (38.A7) into (38.44) and integrating yields for the ratio of trapped particles [38.21]

$$n_{tr} = \frac{\int_0^{L/(\alpha 2)} \frac{1}{\sigma_e} \exp\left(-\frac{d}{\sigma_e}\right)\left(1 - \frac{2\alpha d}{L}\right)dd}{\int_0^\infty \frac{1}{\sigma_e} \exp\left(-\frac{d}{\sigma_e}\right)dd}$$

$$= \exp\left(-\frac{d}{\sigma_e}\right)\frac{\sigma_e - L/(2\alpha) + d}{L/(2\alpha)}\Bigg|_0^{L/(2\alpha)}$$

$$= \frac{2\alpha\sigma_e}{L}\left[\exp\left(-\frac{L}{2\alpha\sigma_e}\right) - 1\right] + 1 \quad (38.45)$$

whereas the mean diameter of the trapped particles is

$$\overline{d} = \frac{\int_0^{L/(2\alpha)} d\exp\left(-\frac{d}{\sigma_e}\right)\left(1 - \frac{2\alpha d}{L}\right)dd}{\int_0^{L/(2\alpha)} \exp\left(-\frac{d}{\sigma_e}\right)\left(1 - \frac{2\alpha d}{L}\right)dd}$$

$$= \sigma_e \frac{\exp\left(-\frac{L}{2\alpha\sigma_e}\right)\left(1 + \frac{4\alpha\sigma_e}{L}\right) + 1 - \frac{4\alpha\sigma_e}{L}}{\frac{2\alpha\sigma_e}{L}\left[\exp\left(-\frac{L}{2\alpha\sigma_e}\right) - 1\right] + 1} \quad (38.46)$$

and the mean square radius of the trapped particles is

$$\overline{d^2}$$
$$= \frac{\int_0^{L/(2\alpha)} d^2\exp\left(-\frac{d}{\sigma_e}\right)\left(1 - \frac{2\alpha d}{L}\right)dd}{\int_0^{L/(2\alpha)} \exp\left(-\frac{d}{\sigma_e}\right)\left(1 - \frac{2\alpha d}{L}\right)dd}$$

$$= \sigma_e^2 \frac{\exp\left(-\frac{L}{2\alpha\sigma_e}\right)\left(\frac{L}{2\alpha\sigma_e} + 4 + \frac{12\alpha\sigma_e}{L}\right) + 2 - \frac{12\alpha\sigma_e}{L}}{\frac{2\alpha\sigma_e}{L}\left(\exp\left(-\frac{L}{2\alpha\sigma_e}\right) - 1\right) + 1}. \quad (38.47)$$

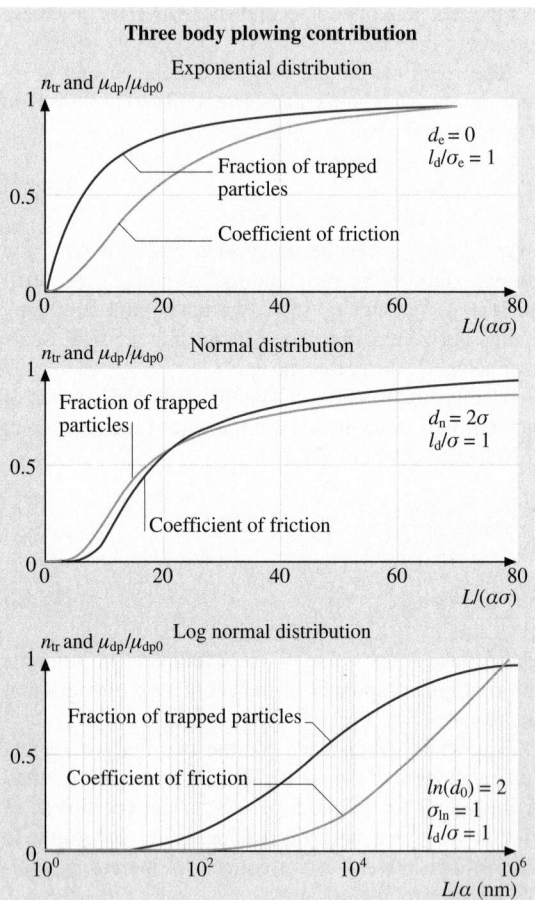

Fig. 38.15a–c The number of trapped particles divided by the total number of particles and three-body deformation component of the coefficient of friction, normalized by the macroscale value, for three different distributions of debris size: (**a**) exponential, (**b**) normal, and (**c**) log-normal distributions

For the normal and log normal distributions, similar calculations can be conducted numerically.

The area, supported by particles can be found as the number of trapped particles $\eta L^2 n_{tr}$ multiplied by average particle contact area

$$A_{dp} = \eta L^2 n_{tr} \frac{\pi \overline{d^2}}{4} , \tag{38.48}$$

where $\overline{d^2}$ is mean square of particle diameter, η is particle density per apparent area of contact (L^2) and n_{tr} is a number of trapped particles divided by the total number of particles [38.21].

The plowing deformation is plastic and, assuming that particles are harder than the bodies, the shear strength τ_{dp} is equal to the shear yield strength of the softer body τ_Y which is given by the (38.9) with $a = \overline{d}/2$. Combining (38.29) with (38.9) and (38.48)

$$
\begin{aligned}
\mu_{dp} &= \frac{A_{dp}\tau_{dp}}{W} = \eta \frac{L^2}{W} \frac{\pi \overline{d^2}}{4} n_{tr}\tau_{Y0}\sqrt{1+2l_d/\overline{d}} \\
&= \mu_{dp0} n_{tr} \frac{\overline{d^2}}{\overline{d_0^2}} \frac{\sqrt{1+2l_d/\overline{d}}}{\sqrt{1+2l_d/\overline{d_0}}} ,
\end{aligned}
\tag{38.49}
$$

where $\overline{d}$ is mean particle diameter, $\overline{d_0}$ is the macroscale value of mean particle diameter, and μ_{dp0} is macroscale ($L \to \infty$, $n_{tr} \to 1$) value of the third-body deformation component of the coefficient of friction given as

$$\mu_{dp0} = \eta \frac{L^2}{W} \frac{\pi \overline{d_0^2}}{4} \tau_{Y0}\sqrt{1+2l_d/\overline{d_0}} . \tag{38.50}$$

Scale dependence of the three-body deformation component of the coefficient of friction is presented in Fig. 38.15, based on (38.49). The number of trapped particles divided by the total number of particles, as well as the three-body deformation component of the coefficient of friction, are presented as a function of scale size divided by α for the exponential, normal, and log normal distributions. The dependence of μ_d/μ_{d0} is shown as a function of $L/(\alpha\sigma_e)$ for the exponential distribution and normal distribution, for $d_n = d_e = 2\sigma_e$ and $l_d/\sigma_e = 1$, whereas for the log normal distribution the results are presented as a function of L/α, for $(\ln d_{ln}) = 2$, $\sigma_{ln} = 1$, and $l_d/\sigma_{ln} = 1$. This component of the coefficient of friction decreases for all of the three distributions. The results are shown for $l_d/\sigma_{ln} = 1$, however, variation of l_d/σ_{ln} in the range between 0.1 and 10 does not change significantly the shape of the curve. The decrease of the three-body deformation friction force with decreasing scale results with this component being small at the nanoscale.

38.5.4 Ratchet Mechanism

Surface roughness can have an appreciable influence on friction during adhesion. If one of the contacting surfaces has asperities of much smaller lateral size, such that a small tip slides over an asperity, having the average angle θ_r (so called ratchet mechanism), the corresponding component of the coefficient of friction is given by

$$\mu_r = \mu_a \tan^2 \theta_r , \tag{38.51}$$

where μ_r is the ratchet mechanism component of friction [38.16]. Combining (38.15, 38.41, 38.38, 38.39) yields for the scale dependence of the ratchet component of the coefficient of friction in the case of elastic, μ_{re}, and plastic contact, μ_{rp}

$$
\begin{aligned}
\mu_{re} &= \mu_{ae}\left[\frac{2\sigma_0}{\pi\beta_0^*}\left(\frac{L}{L_{lc}}\right)^{n-m}\right]^2 \\
&= \frac{\mu_{re0}}{\sqrt{1+(l_s/\overline{a_0})}}\left(\frac{L}{L_{lc}}\right)^{n-m} \\
&\quad \times \sqrt{1+(L_s/L)^m} , \quad L < L_{lc} ,
\end{aligned}
\tag{38.52}
$$

$$
\begin{aligned}
\mu_{rp} &= \mu_{ap}\left[\frac{2\sigma_0}{\pi\beta_0^*}\left(\frac{L}{L_{lc}}\right)^{n-m}\right]^2 \\
&= \mu_{rp0}\left(\frac{L}{L_{lc}}\right)^{2(n-m)}\sqrt{\frac{1+(l_d/\overline{a_0})}{1+(l_s/\overline{a_0})}} \\
&\quad \times \sqrt{\frac{1+(L_s/L)^m}{1+(L_d/L)^m}} , \quad L < L_{lc} ,
\end{aligned}
\tag{38.53}
$$

where μ_{re0} and μ_{rp0} are the macroscale values of the ratchet component of the coefficient of friction for elastic and plastic contact correspondingly [38.23].

Scale dependence of the ratchet component of the coefficient of friction, normalized by the macroscale value, is presented in Fig. 38.16, for scale independent adhesional shear strength, $\tau_a = \text{const}$, ($L_s = 0$) and for scale dependent τ_a ($L_s = 10L_d$), based on (38.51) and (38.53). The ratchet component during adhesional elastic friction, μ_{re}, is presented in Fig. 38.16a. It is observed, that, with decreasing scale, μ_{re} increases. The ratchet component during adhesional plastic friction, μ_{rp}, is presented in Fig. 38.16b. It is observed, that, for $L_s = 0$, with decreasing scale, μ_{rp} increases [38.23].

38.5.5 Meniscus Analysis

During contact, if a liquid is introduced at the point of asperity contact, the surface tension results in a pressure difference across a meniscus surface, referred to

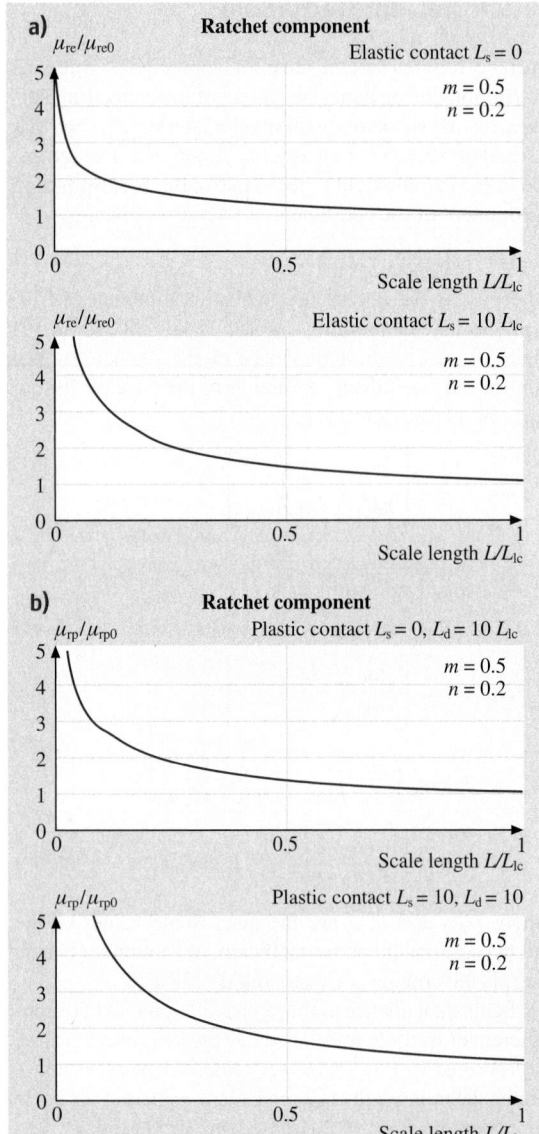

Fig. 38.16a,b Normalized results for the ratchet component of the coefficient of friction, as a function of scale, for scale independent ($L_s = 0$) and scale dependent ($L_s = 10L_{lc}$) shear strength ($m = 0.5, n = 0.2$). (**a**) contact, (**b**) Plastic contact ($L_d = 10L_{lc}$)

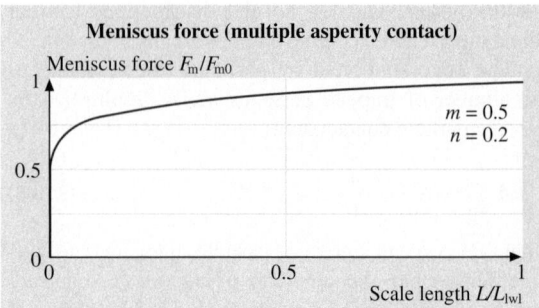

Fig. 38.17 Meniscus force for $m = 0.5, n = 0.2$

a surface with continuous liquid film [38.16]

$$F_m \propto R_p . \qquad (38.54)$$

The case of multiple-asperity contact is shown in Fig. 38.1b. Note, that both contacting and near-contacting asperities wetted by the liquid film contribute to the total meniscus force. A statistical approach can be used to model the contact. In general, given the interplanar separation s_d, the mean peak radius $\overline{R_p}$, the thickness of liquid film h_f, the surface tension γ, liquid contact angle between the liquid and surface θ, and the total number of summits in the nominal contact area N,

$$F_m = 2\pi \overline{R_p}\gamma(1 + \cos\theta)N . \qquad (38.55)$$

In (38.54), γ and θ are material properties, which are not expected to depend on scale, whereas $\overline{R_p}$ and N depend on surface topography, and are scale-dependent, according to (38.18) and (38.20).

$$F_m \propto \overline{R_p}N = F_{m0}\left(\frac{L}{L_{lwl}}\right)^{m-2n} , \quad L < L_{lwl} , \qquad (38.56)$$

where F_{m0} is the macroscale value of the meniscus force ($L \geq L_{lwl}$).

Scale dependence of the meniscus force is presented in Fig. 38.17, based on (38.56) for $m = 0.5, n = 0.2$. It may be observed that, depending on the value of D, the meniscus force may increase or decrease with decreasing scale size.

38.5.6 Total Value of Coefficient of Friction and Transition from Elastic to Plastic Regime

During transition from elastic to plastic regime, contribution of each of the three components of the coefficient

as capillary pressure or Laplace pressure. The attractive force for a sphere in contact with a plane surface is proportional to the sphere radius R_p, for a sphere close to a surface with separation s or for a sphere close to

of friction in (38.29) changes. In the elastic regime, the dominant contribution is expected to be adhesion involving elastic deformation, and in the plastic regime the dominant contribution is expected to be deformation. Therefore, in order to study transition from elastic to plastic regime, the ratios of deformation to adhesion component should be considered. The expression for the total value of the coefficient of friction, which includes meniscus force contribution, based on (38.29) and (38.31) can be rewritten as [38.21]

$$\mu_{\text{wet}} = \mu_a \left(1 + \frac{\mu_{ds}}{\mu_a} + \frac{\mu_{dp}}{\mu_a}\right)\left(1 + \frac{F_m}{W}\right) . \quad (38.57)$$

The ratchet mechanism component is ignored here since it is present only in special cases. Results in the preceding subsection provide us with data about the adhesion and two-body and three-body deformation components of the coefficient of friction, normalized by their values at the macroscale. However, that analysis does not provide any information about their relation to each other or about transition from the elastic to plastic regime. In order to analyze the transition from pure adhesion involving elastic deformation to plastic deformation, a transition index ϕ can be considered [38.21]. The transition index is equal to the ratio of average pressure in the elastic regime (normal load per real area of elastic contact) to hardness or simply the ratio of the real area of plastic contact divided by the real area of elastic contact

$$\phi = \frac{W}{A_{re}H} = \frac{A_{rp}}{A_{re}} . \quad (38.58)$$

Using (38.17) and (38.22), the scale-dependence of ϕ is

$$\phi = \frac{W}{A_{re0}(L/L_{lc})^{m-n}H_0\sqrt{1+(L_s/L)^m}}$$
$$= \phi_0 \frac{\sqrt{1+(l_s/a)}(L/L_{lc})^{n-m}}{\sqrt{1+(L_s/L)^m}} , \quad L < L_{lc} , \quad (38.59)$$

where ϕ_0 is the macroscale value of the transition index [38.21].

With a low value of ϕ close to zero, the contacts are mostly elastic and only adhesion contributes to the coefficient of friction involving elastic deformation. Whereas with increasing ϕ approaching unity, the contacts become predominantly plastic and deformation becomes a dominant contributor. It can be argued that A_{ds}/A_{re} and A_{dp}/A_{re} will also be a direct function of ϕ, and in the paper these will be assumed to have linear relationship.

Next, the ratio of adhesion and deformation components of the coefficient of friction in terms of ϕ is obtained. In this relationship, τ_{ds} and τ_{dp} are equal to

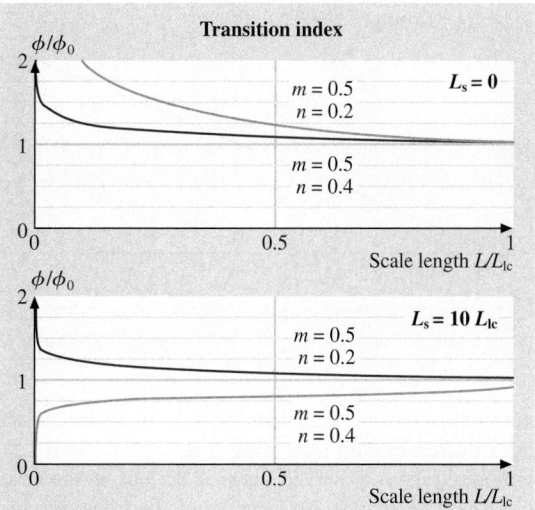

Fig. 38.18 The transition index as a function of scale. Presented for $m = 0.5$, $n = 0.2$ and $m = 0.5$, $n = 0.4$

the shear yield strength, which is proportional to hardness and can be obtained from (38.9), using (38.19) and (38.36)

$$\frac{\mu_{ds}}{\mu_{ae}} = \frac{A_{ds}\tau_{ds}}{A_{re}\tau_a}$$
$$\propto \phi \frac{\tau_{ds}}{\tau_a} = \phi \frac{\tau_{ds0}\sqrt{1+(L_d/L)^m}}{\tau_{a0}\sqrt{1+(L_s/L)^m}} , \quad L < L_{lc} , \quad (38.60)$$

$$\frac{\mu_{dp}}{\mu_{ae}} = \frac{A_{dp}\tau_{dp}}{A_{re}\tau_a}$$
$$\propto \phi \frac{\tau_{dp}}{\tau_a} = \phi \frac{\tau_{Y0}\sqrt{1+2l_d/\bar{d}}}{\tau_{a0}\sqrt{1+(L_s/L)^m}} , \quad L < L_{lc} . \quad (38.61)$$

The sum of adhesion and deformation components [38.21]

$$\mu_{\text{wet}}$$
$$= \mu_{ae}\left[1 + \phi\left(\frac{\tau_{ds0}\sqrt{1+(L_d/L)^m}}{\tau_{a0}\sqrt{1+(L_s/L)^m}} + \frac{\tau_{Y0}\sqrt{1+2l_d/\bar{d}}}{\tau_{a0}\sqrt{1+(L_s/L)^m}}\right)\right]$$
$$\times \left[1 + \frac{F_{m0}}{W}\left(\frac{L}{L_{lwl}}\right)^{m-2n}\right] , \quad L < L_{lc} . \quad (38.62)$$

Note that ϕ itself is a complicated function of L, according to (38.59).

Scale dependence of the transition index, normalized by the macroscale value, is presented in Fig. 38.18, based

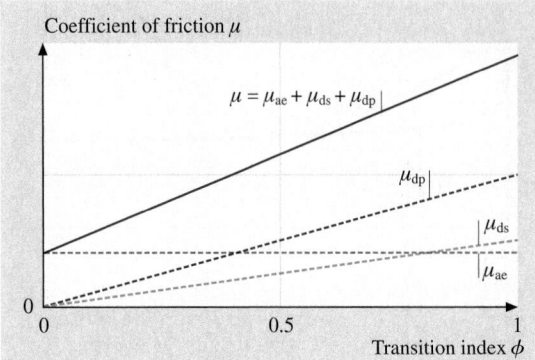

Fig. 38.19 The coefficient of friction (dry contact) as a function of the transition index for given scale length L. With increasing ϕ and onset of plastic deformation, both μ_{ds} and μ_{dp} grow, as a result of this, the total coefficient of friction μ grows as well

rolls with a single contact zone (before onset of wear), Fig. 38.21b. Contacts relevant in these experiments can be considered as single-asperity, predominantly elastic in all of these cases. For a single-asperity elastic contact

on (38.59). It is observed that, for $L_s = 0$, the transition index decreases with increasing scale. For $L_s = 10L_{lc}$, the same trend is observed for $m > 2n$, but, in the case $m < 2n$, ϕ decreases. An increase of the transition index means that the ratio of plastic to elastic real areas of contact increases. With decreasing scale, the mean radius of contact decreases, causing hardness enhancement and decrease of the plastic area of contact. Based on this, the model may predict an increase or decrease of the transition index, depending on whether elastic or plastic area decreases faster.

The dependence of the coefficient of friction on ϕ is illustrated in Fig. 38.19, based on (38.62). It is assumed in the figure that the slope for the dependence of μ_{dp} on ϕ is greater than the slope for the dependence of μ_{ds} on ϕ. For ϕ close to zero, the contact is predominantly elastic, whereas for ϕ approaching unity the contact is predominantly plastic.

38.5.7 Comparison with the Experimental Data

Experimental data on friction at micro- and nanoscale are presented in this subsection and compared with the model. First, a single-asperity predominantly elastic contact is considered [38.20], then transition to plastic deformation involving multiple asperity contacts is analyzed [38.23].

Single–Asperity Predominantly–Elastic Contact
Nanoscale dependence of friction force upon the normal load was studied for Pt-coated AFM tip versus mica in ultra-high vacuum (UHV) by *Carpick* et al. [38.7], for Si tip versus diamond and amorphous carbon by *Schwarz* et al. [38.8] and for Si_3N_4 tip on Si, SiO_2, and diamond by *Bhushan* and *Kulkarni* [38.6] (Fig. 38.20a). *Homola* et al. [38.13] conducted SFA experiments with mica

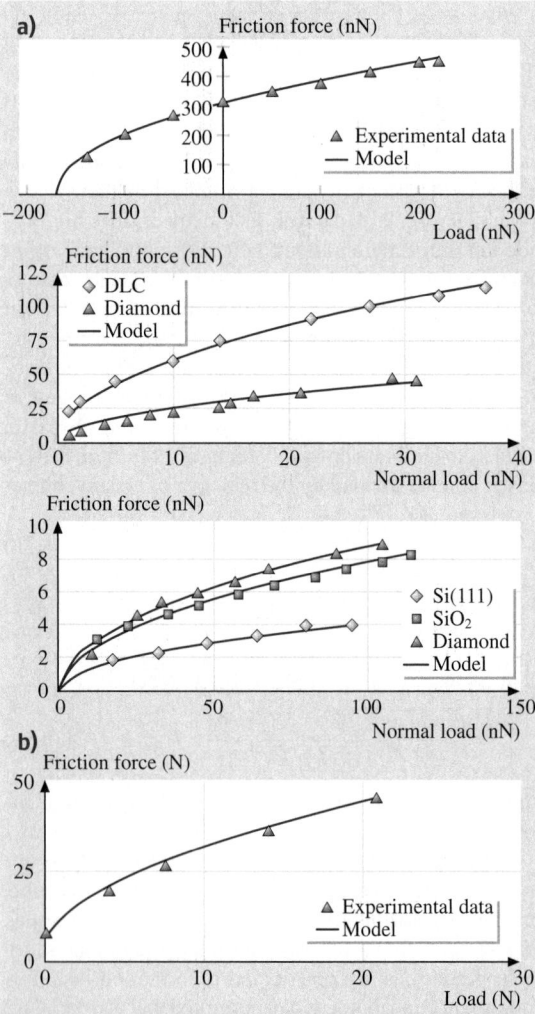

Fig. 38.20a,b Summary of (**a**) AFM data (*upper*: Pt-coated tip on mica in UHV [38.7], *middle*: Si tip on DLC and diamond in UHV [38.8], *lower*: Si_3N_4 tip on various materials [38.6]) and (**b**) SFA data (on mica vs. mica in dry air [38.13]) for friction force as a function of normal load

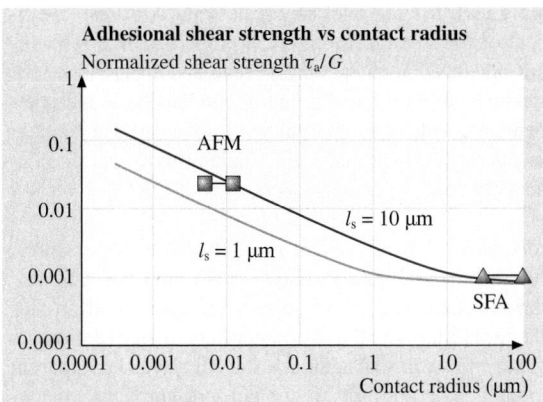

Fig. 38.21 Shear stress as a function of contact radius. Microscale and nanoscale data compared with the model for $l_s = 1\,\mu m$ and $l_s = 10\,\mu m$

of radius a, expression for μ is given by (38.17). For the limit of a small contact radius $a \ll l_s$, the (38.13) combined with the Hertzian dependence of the contact area upon the normal load (38.33) yields

$$F_e \approx \pi a^2 \tau_0 \sqrt{l_s/a} \propto a^{3/2} \propto W^{1/2}\ . \qquad (38.63)$$

If an adhesive pull-off force W_0 is large, (38.63) can be modified as

$$F_e = C_0\sqrt{W + W_0}\ , \qquad (38.64)$$

where C_0 is a constant. Friction force increases with square root of the normal load, opposed to the two third exponent in scale independent analysis.

The results in Fig. 38.20 demonstrate a reasonable agreement of the experimental data with the model. The platinum-coated tip versus mica [38.7] has a relatively high pull-off force and the data fit with $C_0 = 23.7\,(nN)^{1/2}$ and $W_0 = 170\,nN$. For the silicon tip vs. amorphous carbon and natural diamond, the fit is given by $C_0 = 8.0, 19.3\,(nN)^{1/2}$ and small W_0. For the virgin Si(111), SiO$_2$, and natural diamond sliding versus Si$_3$N$_4$ tip [38.8], the fit is given by $C_0 = 0.40, 0.76, 0.86\,(nN)^{1/2}$ for Si(111), SiO$_2$, and diamond, respectively and small W_0. For two mica rolls [38.13], the fit is given by $C_0 = 10\,N^{1/2}$ and $W_0 = 0.5\,N$ [38.20].

AFM experiments provide data on nanoscale, whereas SFA experiments provide data on microscale. Next we study scale dependence on the shear strength based on these data. In the AFM measurements by *Carpick* et al. [38.7], the average shear strength during sliding for Pt–mica interface was reported as 0.86 GPa, whereas the pull-off contact radius was reported as 7 nm.

In the SFA measurements by *Homola* et al. [38.13], the average shear strength during sliding for mica–mica interface was reported as 25 MPa, whereas the contact area during high loads was on the order of $10^{-8}\,m^2$, which corresponds to a contact radius on the order 100 μm. To normalize shear strength, we need shear modulus. The shear modulus for mica is $G_{mica} = 25.7\,GPa$ [38.49] and for Pt is $G_{Pt} = 63.7\,GPa$ [38.50]. For mica–Pt interface, the effective shear modulus is

$$G = 2G_{mica}G_{Pt}/(G_{mica} + G_{Pt}) = 36.6\,GPa\ . \quad (38.65)$$

This yields the value of the shear stress normalized by the shear modulus $\tau_a/G = 2.35 \times 10^{-2}$ for *Carpick* et al. [38.7] AFM data and 9.73×10^{-4} for the SFA data. These values are presented in the Fig. 38.21 together with the values predicted by the model for assumed values of $l_s = 1\,\mu m$ and $10\,\mu m$. It can be seen that the model (38.13) provides an explanation of adhesional shear strength increase with a scale decrease [38.20].

Transition to Predominantly Plastic Deformation Involving Multiple Asperity Contacts

Next, we analyze the effect of transition from predominantly elastic adhesion to predominantly plastic deformation involving multiple asperity contacts [38.23]. The data on nano- and microscale friction for various materials, are presented in Table 38.2, based on *Ruan and Bhushan* [38.5], *Liu and Bhushan* [38.11], and *Bhushan* et al. [38.12], for Si(100), graphite (HOPG), natural diamond, and diamond-like carbon (DLC). There are several factors responsible for the differences in the coefficients of friction at micro- and nanoscale. Among them are the contributions from ratchet mechanism, meniscus effect, wear and contamination particles, and transition from elasticity to plasticity. The ratchet mechanism and meniscus effect result in an increase of friction with decreasing scale and cannot explain the decrease of friction found in the experiments. The contribution of wear and contamination particles is more significant at macro/microscale because of larger number of trapped particles (Fig. 38.15). It can be argued, that for the nanoscale AFM experiments the contacts are predominantly elastic and adhesion is the main contribution to the friction, whereas for the microscale experiments the contacts are predominantly plastic and deformation is an important factor. Therefore, transition from elastic contacts in nanoscale contacts to plastic deformation in microscale contacts is an important effect [38.23].

According to (38.29), the friction force depends on the shear strength and a relevant real area of contact.

Part D | 38.5

Specimen	Coefficient of friction		Elastic modulus (GPa)	Poisson's ratio	Hardness (GPa)	Apparent contact radius at test load for		Mean apparent pressure at test load for	
	Microscale	Nanoscale				Microscale (μm)	Nanoscale (nm)	Microscale (GPa)	Nanoscale (GPa)
Si(100) wafer	0.47[a]	0.06[c]	130[e,f]	0.28[f]	9–10[e,f]	0.8–2.2	1.6–3.4	0.05–0.13[a]	1.3–2.8[c]
Graphite (HOPG)	0.1[b]	0.006[c]	9–15[g] (9)	– (0.25)	0.01[j]	62	3.4–7.4	0.082[b]	0.27–0.58[c]
Natural diamond	0.2[b]	0.05[c]	1140[h]	0.07[h]	80–104[g,h]	21	1.1–2.5	0.74[b]	2.5–5.3[c]
DLC film	0.19[b]	0.03[d]	280[i]	0.25[i]	20–30[i]	0.7–2.0	1.3–2.9	0.06–0.16[a]	1.8–3.8[d]

[a] 500 μm radius Si(100) ball at 100–2000 μN and 720 μm/s in dry air [38.12]
[b] 3 mm radius Si$_3$N$_4$ ball (Elastic modulus 310 GPa, Poisson's ratio 0.22 [38.50]), at 1 N and 800 μm/s [38.5]
[c] 50 nm radius Si$_3$N$_4$ tip at load range from 10–100 nN and 0.5 nm/s, in dry air [38.5]
[d] 50 nm radius Si$_3$N$_4$ tip at load range from 10–100 nN in dry air [38.12]
[e] [38.54], [f] [38.52], [g] [38.50], [h] [38.53], [i] [38.55], [j] [38.51]

Table 38.2 Micro- and nanoscale values of the coefficient of friction, typical physical properties of specimens, and calculated apparent contact radii and apparent contact pressures at loads used in micro- and nanoscale measurements. For calculation purposes it is assumed that contacts on micro- and nanoscale are single-asperity elastic contacts [38.23]

For calculation of contact radii and contact pressures, the elastic modulus, Poisson's ratio, and hardness for various samples, are required and presented in Table 38.2 [38.50–55]. In the nanoscale AFM experiments a sharp tip was slid against a flat sample. The apparent contact size and mean contact pressures are calculated based on the assumption, that the contacts are single asperity, elastic contacts (contact pressures are small compared to hardness). Based on the Hertz equation [38.47], for spherical asperity of radius R in contact with a flat surface, with an effective elastic modulus E^*, under normal load W, the contact radius a and mean apparent contact pressure p_a are given by

$$a = \left(\frac{3WR}{4E^*} \right)^{1/3} , \tag{38.66}$$

$$p_a = \frac{W}{\pi a^2} . \tag{38.67}$$

The surface energy effect [38.16] was neglected in (38.66) and (38.67), because the experimental value of the normal adhesion force was small, compared to W [38.5]. The calculated values of a and p_a for the relevant normal load are presented in Table 38.2 [38.23].

In the microscale experiments, a ball was slid against a nominally flat surface. The contact in this case is multiple-asperity contact due to the roughness, and the contact pressure of the asperity contacts is higher than the apparent pressure. For calculation of a characteristic scale length for the multiple asperity contacts, which is equal to the apparent length of contact, (38.66) was also used. Apparent radius and mean apparent contact pressure for microscale contacts at relevant load ranges are also presented in Table 38.2 [38.23].

A quantitative estimate of the effect of the shear strength and the real area of contact on friction is presented in Table 38.3. The friction force at mean load (average of maximum and minimum loads) is shown, based on the experimental data presented in Table 38.2. For microscale data, the real area of contact was estimated based on the assumption that the contacts are plastic and based on (38.33) for mean loads given in Table 38.2. For nanoscale data, the apparent area of contact was on the order of several square nanome-

Table 38.3 Mean friction force, the real area of contact and lower limit of shear strength [38.23]

Specimen	Friction force at mean load [a]		Upper limit of real area of contact at mean load		Lower limit of mean shear strength (GPa)	
	Microscale (mN)	Nanoscale (nN)	Microscale [b] (μm^2)	Nanoscale [c] (nm^2)	Microscale [d]	Nanoscale [d]
Si(100) wafer	0.49	3.3	0.11	19	4.5	0.17
Graphite (HOPG)	100	0.33	10^5	92	0.001	0.004
Natural diamond	200	2.7	10.9	10	18.4	0.27
DLC film	0.2	1.7	0.042	14	4.8	0.12

[a] Based on the data from Table 38.2. Mean load at microscale is 1050 μN for Si(100) and DLC film and 1 N for HOPG and natural diamond, and 55 nN for all samples at nanoscale

[b] For plastic contact, based on hardness values from Table 38.2. Scale-dependent hardness value will be higher at relevant scale, presented values of the real area of contact are an upper estimate

[c] Upper limit for the real area is given by the apparent area of contact calculated based on the radius of contact data from Table 38.2

[d] Lower limit for the mean shear strength is obtained by dividing the friction force by the upper limit of the real area of contact

ters, and it was assumed that the real area of contact is comparable with the mean apparent area of contact, which can be calculated for the mean apparent contact radius, given in Table 38.2. The estimate provides with the upper limit of the real area of contact. The lower limit of the shear strength is calculated as friction force, divided by the upper limit of the real area of contact, and presented in Table 38.3 [38.23]. Based on the data in Table 38.3, for Si(100), natural diamond and DLC film, the microscale value of shear strength is about two orders of magnitude higher, than the nanoscale value, which indicates, that transition from adhesion to deformation mechanism of friction and the third-body effect are responsible for an increase of friction at mi-

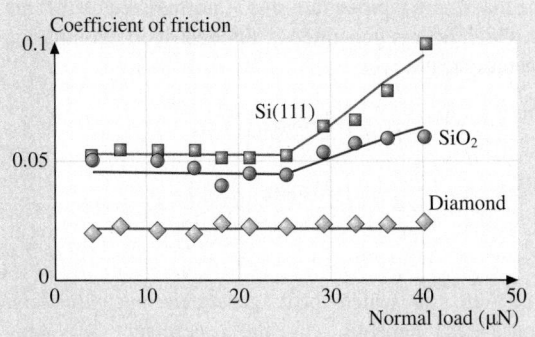

Fig. 38.22 Coefficient of friction as a function of normal load [38.6]

croscale. For graphite, this effect is less pronounced due to molecularly smooth structure of the graphite surface [38.23].

Based on data available in the literature [38.6], load dependence of friction at nano/microscale as a function of normal load is presented in Fig. 38.22. Coefficient of friction was measured for Si_3N_4 tip versus Si, SiO_2, and natural diamond using an AFM. They reported that for low loads the coefficient of friction is independent of load and increases with increasing load after a certain load. It is noted that the critical value of loads for Si and SiO_2 corresponds to stresses equal to their hardness values, which suggests that transition to plasticity plays a role in this effect. The friction values at higher loads for Si and SiO_2 approach to that of macroscale values. This result is consistent with predictions of the model for plastic contact (Fig. 38.11), which states that, with increasing normal load, the long wavelength limit for the contact parameters decreases. This decrease results in violation of the condition $L < L_{lc}$, and the contact parameters and the coefficient of friction reach the macroscale values, as was discussed earlier. It must be noted, that the values of $m = 0.5$ and $n = 0.2$ are taken based on available data for the glass-ceramic disk (Fig. 38.9), these parameters depend on material and on surface preparation and may be different for Si, SiO_2, and natural diamond, however, no experimental data on scale dependence of roughness parameters for the materials of interest are available.

38.6 Scale Effect in Wear

The amount of wear during adhesive or abrasive wear involving plastic deformation is proportional to the load and sliding distance x, divided by hardness [38.16]

$$v = k_0 \frac{Wx}{H} , \tag{38.68}$$

where v is volume of worn material and k_0 is a nondimensional wear coefficient. Using (38.10) and (38.19), the relationships can be obtained for scale dependence of the coefficient of wear in the case of the fractal surface and power-law dependence of roughness parameters

$$v = k \frac{Wx}{H_0} \tag{38.69}$$

and

$$k = \frac{k_0}{\sqrt{1 + (l_\mathrm{d}/a)}} = \frac{k_0}{\sqrt{1 + (L_\mathrm{d}/L)^m}} , \quad L < L_{\mathrm{lwl}} , \tag{38.70}$$

where k is scale-dependent wear coefficient, and k_0 corresponds to the macroscale limit of the value of k [38.22].

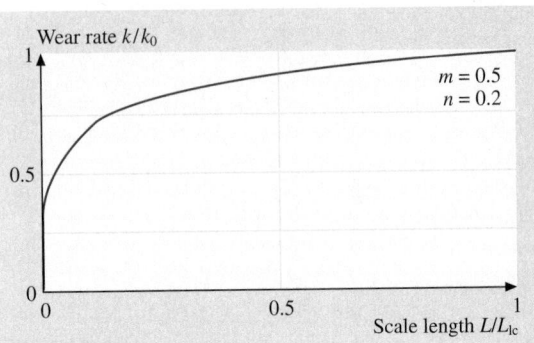

Fig. 38.23 The wear coefficient as a function of scale, presented for $m = 0.5$, $n = 0.2$

Scale dependence of the wear coefficient is presented in Fig. 38.23 for $m = 0.5$ and $n = 0.2$, based on (38.70). It is observed, that the wear coefficient decreases with decreasing scale; this is due to the fact that the hardness increases with decreasing mean contact size.

38.7 Scale Effect in Interface Temperature

Frictional sliding is a dissipative process, and frictional energy is dissipated as heat over asperity contacts. Therefore, a high amount of heat per unit area is generated during sliding. A contact is formed and destroyed as one asperity passes the other at a given velocity. When an asperity comes into contact with another asperity, the real area of contact starts to grow, when the asperities are directly above each other, the area is at maximum, as they move away from each other, the area starts to get smaller. There are number of contacts at a given time during sliding. For each individual asperity contact, a flash temperature rise can be calculated. High temperature rise affects mechanical and physical properties of contacting bodies.

For thermal analysis, a dimensionless Peclet number is used

$$L_\mathrm{p} = \frac{6Va_{\max}}{16\kappa_\mathrm{t}} , \tag{38.71}$$

where V is sliding velocity, $a_{\max}$ is maximum radius of contact for a given contact spot, and κ_t is thermal diffusivity. This parameter indicates whether the sliding is high-speed or low-speed. If $L_\mathrm{p} > 10$, the contact

falls into the category of high speed; if $L_\mathrm{p} < 0.5$, it falls into the category of low speed; if $0.5 \leq L_\mathrm{p} \leq 10$, a transition regime should be considered [38.16]. For high L_p, there is not enough time for the heat to flow to the sides during the lifetime of the contact and the heat flows only in the direction, perpendicular to the sliding surface. Based on the numerical calculations for flash temperature rise of as asperity contact for adhesional contact [38.16], the following relation holds for the maximum temperature rise T, normalized by the rate at which heat is generated q, divided by the volumetric specific heat ρc_p

$$\begin{aligned} \frac{T \rho c_\mathrm{p} V}{q} &= 0.95 \left(\frac{2V a_{\max}}{\kappa_\mathrm{t}} \right)^{1/2} , \quad L_\mathrm{p} > 10 \\ &= 0.33 \left(\frac{2V a_{\max}}{\kappa_\mathrm{t}} \right) , \quad L_\mathrm{p} < 0.5 . \end{aligned} \tag{38.72}$$

The rate at which heat generated per time per unit area depends on the coefficient of friction μ, sliding velocity V, apparent normal pressure p_a, and ratio of the apparent to real areas of

contact (A_a/A_r)

$$q = \mu p_a V \frac{A_a}{A_r} . \qquad (38.73)$$

Based on (38.72) and (38.73),

$$\frac{T\rho c_p}{p_a} = 0.95 \frac{A_r}{A_a} \mu \left(\frac{2Va_{max}}{\kappa_t}\right)^{1/2} , \quad L_p > 10$$

$$= 0.33 \frac{A_r}{A_a} \mu \left(\frac{2Va_{max}}{\kappa_t}\right) , \quad L_p < 0.5 .$$

$$(38.74)$$

For a multiple asperity contact, mean temperature in terms of average of maximum contact size can be written as

$$\frac{\overline{T}\rho c_p}{p_a} = 0.95 \frac{A_r}{A_a} \mu \left(\frac{2V\bar{a}_{max}}{\kappa_t}\right)^{1/2} , \quad L_p > 10$$

$$= 0.33 \frac{A_r}{A_a} \mu \left(\frac{2V\bar{a}_{max}}{\kappa_t}\right) , \quad L_p < 0.5 .$$

$$(38.75)$$

In (38.75) $\bar{a}_{max}$, μ and A_a/A_r are scale dependent parameters. During adhesional contact, the maximum radius $\bar{a}_{max}$ is proportional to the contact radius $\bar{a}$, and the scale dependence for $\bar{a}_{max}$ is given by (38.19), for μ by (38.38–38.39), and for A_{re} and A_{rp} by (38.17) and (38.21). The scale dependence of q, involving μ and A_r, and $\bar{a}_{max}$ in (38.72) can be considered separately and then combined. For the sake of simplicity, we only consider the scale dependence of $\bar{a}_{max}$. For the empirical rule dependence of surface roughness parameters and the fractal model, in the case of high and low velocity, (38.75)

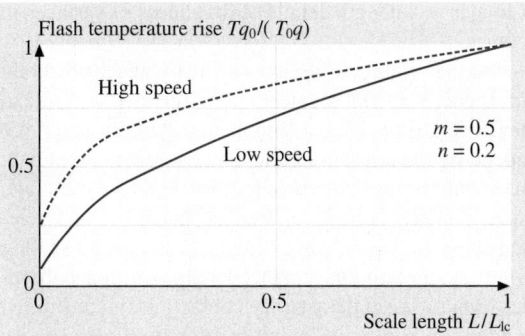

Fig. 38.24 Ratio of the flash temperature rise to the amount of heat generated per unit time per unit area, for a given sliding velocity, as a function of scale. Presented for $m = 0.5$, $n = 0.2$

yields [38.22]

$$\frac{\overline{T}\rho c_p V}{q} = 0.95 \left(\frac{2VC_A L^m}{\kappa}\right)^{1/2} ,$$

$$L < L_{lwl} , L_p > 10$$

$$= 0.33 \left(\frac{2VC_A L^m}{\kappa}\right) ,$$

$$L < L_{lwl} , L_p < 0.5 . \qquad (38.76)$$

Scale dependence for the ratio of the flash temperature rise to the amount of heat generated per unit time per unit area, for a given sliding velocity, as a function of scale, is presented in Fig. 38.24, based on (38.76), for the high-speed and low-speed cases. For the empirical rule dependence of roughness parameters, the results are shown for $m = 0.5$, $n = 0.2$.

38.8 Closure

A model, which explains scale effects in mechanical properties (yield strength, hardness, and shear strength at the interface) and tribology (surface roughness, contact parameters, friction, wear, and interface temperature), has been presented in this chapter.

Both mechanical properties and roughness parameters are scale-dependent. According to the strain gradient plasticity, the scale dependence of the so-called geometrically necessary dislocations causes enhanced yield strength and hardness with decreasing scale. The shear strength at the interface is scale dependent due to the effect of dislocation-assisted sliding. An empirical

rule for scale dependence of the roughness parameters has been proposed, namely, it was assumed, that the standard deviation of surface height and autocorrelation length depend on scale according to a power law when scale is less than the long wavelength limit value.

Both single asperity and multiple asperity contacts were considered. For multiple asperity contacts, based on the empirical power-rule for scale dependence of roughness, contact parameters were calculated. The effect of load on the contact parameters was also studied. The effect of increasing load is similar to that of increasing scale because it results in increased relevance of longer wavelength details of roughness of surfaces in contact.

During sliding, adhesion and two- and three-body deformation, as well as ratchet mechanism, contribute to the friction force. These components of the friction force depend on the relevant real areas of contact (dependent on roughness, mechanical properties, and load), average asperity slope, number of trapped particles, and relevant shear strength during sliding. The relevant scaling length is the nominal contact length – contact diameter ($2a$) for a single-asperity contact, only considered in adhesion, and scan length (L) for multiple-asperity contacts, considered in adhesion and deformation.

For the adhesional component of the coefficient of friction, the shear yield strength and hardness increase with decreasing scale. In the case of elastic contact, the real area of contact is scale independent for single-asperity contact, and may increase or decrease depending on roughness parameters, for multiple-asperity contact. In the case of plastic contact, enhanced hardness results in a decrease in the real area of contact. The adhesional shear strength at the interface may remain constant or increase with decreasing scale, due to dislocation-assisted sliding (or microslip). The model predicts that the adhesional component of the coefficient of friction may increase or decrease with scale, depending on the material parameters and roughness. The coefficient of friction during two-body deformation and the ratchet component depend on the average slope of the rough surface. The average slope increases with scale due to scale dependence of the roughness parameters. As a result, the two-body deformation component of the coefficient of friction increases with decreasing scale. The three-body component of the coefficient of friction depends on the concentrations of particles, trapped at the interface, which decreases with decreasing scale.

The transition index, which is responsible for transition from predominantly elastic adhesional friction to plastic deformation was proposed and was found to change with scale, due to scale dependence of roughness parameters. For the transition index close to zero, the contact is predominantly elastic and the dominant contribution to friction is adhesion involving elastic deformation. The increase of the transition index leads to an increase in plastic deformation with increasing contribution of the deformation component of friction, which results in larger value of the total coefficient of friction.

In presence of the meniscus force, the measured value of the coefficient of friction is greater than the value of the coefficient of dry friction. The difference is especially important for small loads, when the normal load is comparable with the meniscus force. The meniscus force depends on peak radii and may either increase or decrease with scale, depending on the surface parameters.

The wear coefficient and the ratio of the maximum flash temperature rise to the amount of heat generated per unit time per unit area, for a given sliding velocity, as a function of scale, decrease with decreasing scale due to decrease in the mean contact size.

The proposed model is used to explain the trends in the experimental data for various materials at nanoscale and microscale, which indicate that nanoscale values of coefficient of friction are lower than the microscale values (Tables 38.2 and 38.3). The two factors responsible for this trend are the increase of the three-body deformation and transition from elastic adhesive contact to plastic deformation. Experimental data show that the coefficient of friction increases with increasing load after a certain load and reaches the macroscale value. This is due to the onset of plastic deformation with increasing load and the effect of load on contact parameters, which affect the coefficient of friction.

38.A Statistics of Particle Size Distribution

38.A.1 Statistical Models of Particle Size Distribution

Particle size analysis is an important field for different areas of engineering, environmental, and biomedical studies. In general, size distribution of particles depends on how the particles were formed and sorted. Several statistical distributions, which govern distribution of random variables including particle size, have been suggested (Fig. 38.25), [38.56–60]. Statistical distributions commonly used are either the probability density (or frequency) function (PDF), $p(z)$, or cumulative distribution function (CDF), $P(h)$. $P(h)$ associated with random variable $z(x)$, which can take any value between $-\infty$ and $+\infty$ or z_{min} and z_{max}, is defined as the probability of the event $z(x) \leq z'$ and is written as [38.61]

$$P(z) = Prob(z \leq z') \tag{38.A1}$$

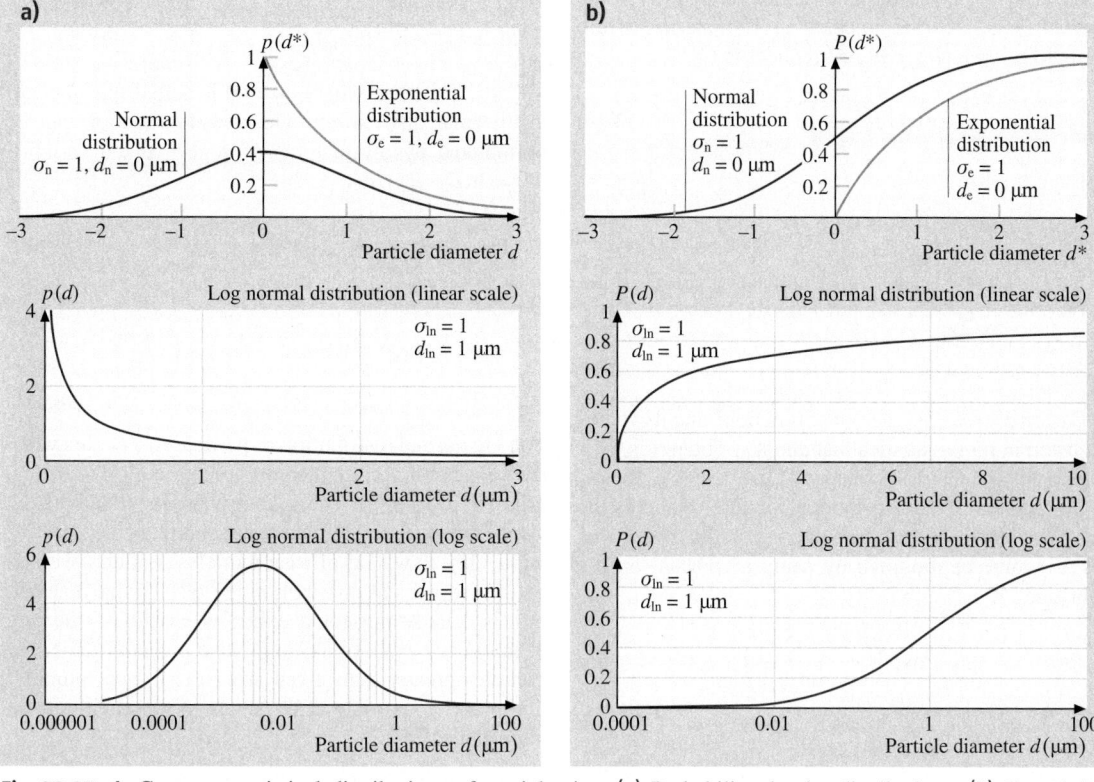

Fig. 38.25a,b Common statistical distributions of particle size. (**a**) Probability density distributions. (**b**) Cumulative distributions

with $P(-\infty) = 0$ and $P(\infty) = 1$.

The pdf is the slope of the CDF given by its derivative

$$P(z) = \frac{\mathrm{d}P(z)}{\mathrm{d}z} \qquad (38.A2)$$

or

$$P(z \leq z') = P(z') = \int_{-\infty}^{z'} p(z)\,\mathrm{d}z \ . \qquad (38.A3)$$

Furthermore, the total area under the probability density function must be unity; that is, it is certain that the value of z at any x must fall somewhere between plus and minus infinity or $z_{\min}$ and $z_{\max}$. The definition of $p(z)$ is phrased as that the random variable $z(x)$ is distributed as $p(z)$.

The probability density (or frequency) function, $p(d)$, in the exponential form is the simplest distribution mathematically

$$p(d) = \frac{1}{\sigma_e} \exp\left(-\frac{d - d_e}{\sigma_e}\right) \ , \qquad d \geq d_0 \ , \qquad (38.A4)$$

where d is particle diameter, σ_e is standard deviation, and d_e is minimum value (for this distribution). For convenience, the density function can be normalized by σ_e in terms of a normalized variable, d^*, equal to $(d - d_e)/\sigma_e$

$$p(d^*) = \exp\left(-d^*\right) \ , \qquad d^* \geq 0 \qquad (38.A5)$$

which has zero minimum and unity standard deviation. The cumulative distribution function $P(d')$ is given as

$$P(d') = P(d^* \leq d') = 1 - \exp\left(-d'\right) \ . \qquad (38.A6)$$

The Gaussian or normal distribution is used to represent data for a wide collection of random physical phenomena in practice such as surface roughness. The probability density and cumulative distribution functions are given as

$$p(d) = \frac{1}{\sqrt{2\pi}\sigma_n} \exp\left(-\frac{(d - d_n)^2}{2\sigma_n^2}\right) \ ,$$
$$-\infty < d < \infty, -\infty < d_n < \infty, \sigma_e > 0 \ ,$$
$$\qquad (38.A7)$$

where d_n is the mean value. The integral of $p(d)$ in the interval $-\infty < d < \infty$ is equal to 1. In terms of the normalized variables, (38.A6) reduces to

$$p(d^*) = \frac{1}{\sqrt{2\pi}} \exp\left(-\frac{d^{*2}}{2}\right) \qquad (38.A8)$$

and

$$P(d') = P(d^* \leq d')$$
$$= \frac{1}{\sqrt{2\pi}} \int_{-\infty}^{d'} \exp\left[-(d^*)^2/2\right] dd^* = \text{erf}(d'),$$
$$(38.A9)$$

where $\text{erf}(d')$ is called the "error function" and its values are listed in most statistical handbooks. The pdf is bell-shaped and the CDF is S-shaped.

For particle size distribution, of interest here, the diameter cannot be less than zero. For this condition, (38.A7) must be modified by using a constant on the right side

$$p(d) = \frac{C_0}{\sqrt{2\pi}\sigma_e} \exp\left(-\frac{(d-d_n)^2}{2\sigma_e^2}\right), \quad 0 \leq d < \infty,$$
$$(38.A10)$$

where

$$C_0 = \left[\frac{1}{\sqrt{2\pi}} \int_{-d_0/\sigma}^{\infty} \exp\left(-\frac{t^2}{2}\right) dt\right]^{-1}.$$

The constant is calculated by integrating $p(d)$ in the interval $0 \leq d \leq \infty$ and equating to one

$$\int_0^\infty p(d)\, dd = 1. \qquad (38.A11)$$

The log normal distribution is commonly used to describe particle size distribution. A variable d is log normally distributed if $\ln d$ is normally distributed. Log normal probability density function for variable d, for which $\ln(d)$ has a Gaussian distribution with a mean $\ln(d)_{ln}$ and standard deviation σ_{ln}, is given as

$$p(d) = \frac{1}{\sqrt{2\pi}\sigma_{ln}} \left(\frac{1}{d}\right) \exp\left(-\frac{[\ln(d/d_{ln})]^2}{2\sigma_{ln}^2}\right),$$
$$0 < d < \infty. \qquad (38.A12)$$

The mean of the log normal distribution is $\exp(\ln d_{ln} + \sigma_{ln}^2/2)$, the standard deviation is

$\exp(2 \ln d_{ln} + \sigma_{ln}^2)[\exp(\sigma_{ln}^2) - 1]$, the skewness is $[\exp(\sigma_{ln})^2 + 2][\exp(\sigma_{ln}^2) - 1]^{1/2}$, and kurtosis is $\exp[4(\sigma_{ln})^2] + 2\exp[3(\sigma_{ln})^2] + 3\exp[2(\sigma_{ln})^2] - 3$ [38.58]. The case where $d_{ln} = 0$ is called the standard log normal distribution. The density function can be normalized by σ_{ln} in terns of a normalized variable, $d^* = (\ln d - d_{ln})/\sigma_{ln}$

$$p(d^*) = \frac{1}{\sqrt{2\pi}} \left(\frac{1}{d^*}\right) \exp\left(-\frac{(d^*)^2}{2}\right) \qquad (38.A13)$$

and

$$P(d') = P(d^* \leq d') = \frac{1}{2}\left[1 + \text{erf}\left(\frac{d'}{\sqrt{2}}\right)\right].$$
$$(38.A14)$$

The log normal distribution of particle size occurs when the dispersion is attained by comminution (milling, grinding, crushing). The size distribution of pulverized silica, granite, calcite, limestone, quartz, soda, ash, alumina, clay, as well as of wear particles is often governed by the log-normal distribution [38.62]. A size distribution is usually presented either as probability density or frequency $p(d)$, or as cumulative percent (percent of particles greater than given size) $P(d)$, or as cumulative mass vs. particle size. All these presentations are interrelated [38.62].

38.A.2 Typical Particle Size Distribution Data

Typical experimental data for size distributions of atmospheric (dust), sand, and abrasive diamond particles are presented in Fig. 38.26a. It can be seen, that the atmospheric particles [38.63] follow the normal distribution function. The dune sand is low in heavy mineral content, so the curve is concaved downward. Micaceous dune sand is sorted by gravity slide on a sharp mountain slope and appears to follow log normal distribution, as many distributions of sediments, which are sorted by gravity. Whereas beach sand distribution curve is concaved upward, due to richness in smaller size component [38.62]. The abrasive diamond particles follow the log normal distribution [38.64].

Size distribution of wear particles has been studied actively since 1970s, when the ferrography was introduced [38.65, 66]. The data for wear particles is presented in Fig. 38.26b. *Xuan* et al. [38.67] studied the size distribution of submicrometer particles during sliding of steel-steel using a Falex 3, pin-on-disk machine, using a laser particle counter for various sliding distances. They found a distribution, which is close to

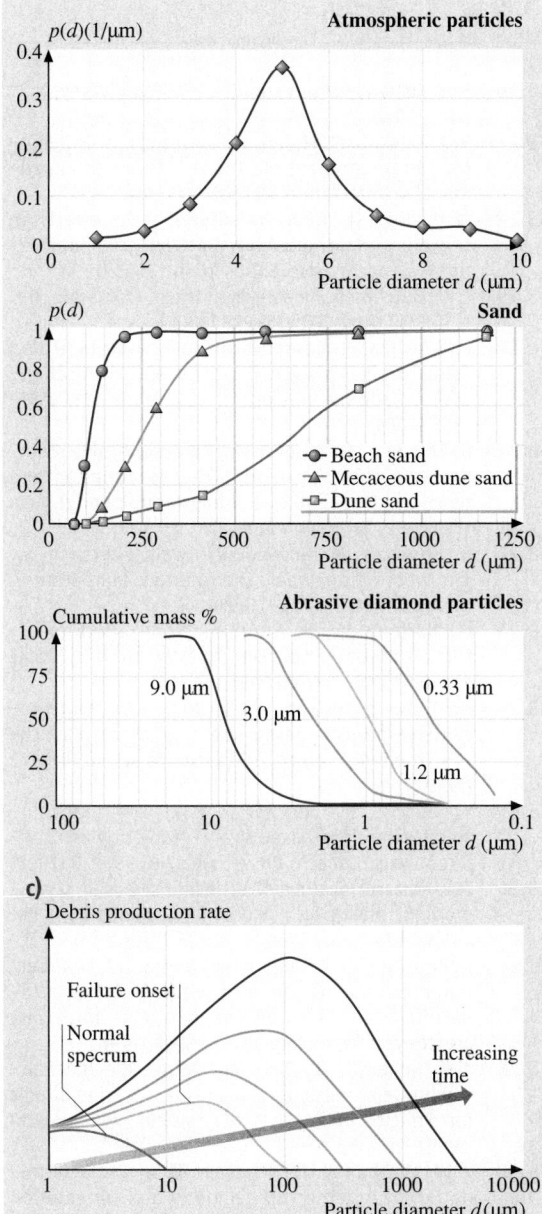

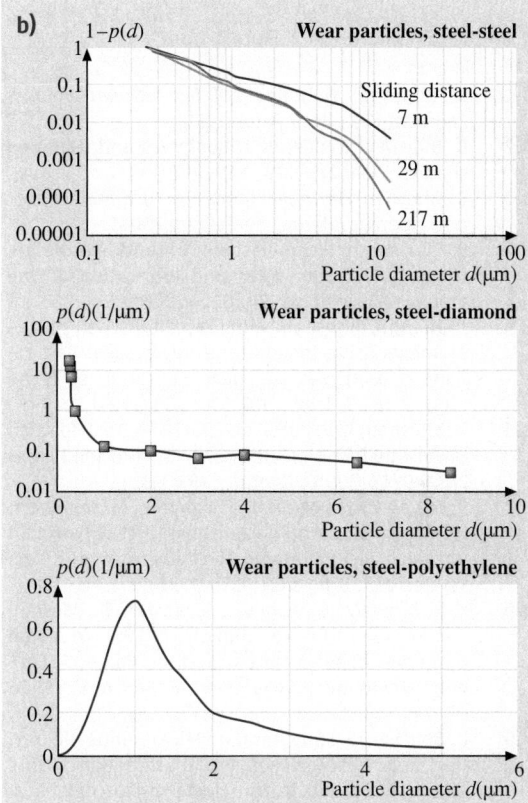

Fig. 38.26 (a) Experimental data for atmospheric [38.63], sand [38.62], and abrasive diamond [38.64] particle size distribution. (b) Experimental data for wear particle size distribution (steel–steel [38.67], steel–diamond [38.68], steel–polyethylene [38.69]). (c) Change with time of wear debris production rate during lubricated sliding as a function of particle size [38.71]

the log normal function. *Mizumoto* and *Kato* [38.68] studied size distribution of particles generated during pin-on-disk test, for diamond, sapphire, silicon carbide, and tungsten carbide pins vs. steel disk, using a laser particle counter. They found that the probability density function is exponential for particles greater than 1 μm diameter, however for smaller particles a linear law was assumed. *Shanbhag* et al. [38.69] studied wear particles for ultrahigh molecular weight polyethylene (UHMWPE) versus titanium in biomedical applications (total knee replacement) using a scanning electron microscope. They found that the distribution is close to that of the normal distribution. Numerous data for wear particles are presented by *Anderson* [38.70]. *Hunt* [38.71] discusses various techniques of debris measurement and analysis in lubricants. A typical change in wear debris generation rate, which occurs with time, is presented in Fig. 38.26c. Change in the size distribution of wear particles in lubricant indicates an onset of mechanical failure.

References

38.1   B. Bhushan: *Handbook of Micro/Nanotribology*, 2nd edn. (CRC, Boca Raton 1999)

38.2   B. Bhushan: Nanoscale tribophysics and tribomechanics, Wear **225–229**, 465–492 (1999)

38.3   B. Bhushan: *Springer Handbook of Nanotechnology* (Springer, Berlin, Heidelberg 2004)

38.4   B. Bhushan, J. N. Israelachvili, U. Landman: Nanotribology: Friction, wear and lubrication at the atomic scale, Nature **374**, 607–616 (1995)

38.5   J. Ruan, B. Bhushan: Atomic-scale friction measurements using friction force microscopy: Part I – General principles and new measurement technique, ASME J. Tribol. **116**, 378–388 (1994)

38.6   B. Bhushan, A. V. Kulkarni: Effect of normal load on microscale friction measurements, Thin Solid Films **278**, 49–56 (1996); Erratum: **293** 333

38.7   R. W. Carpick, N. Agrait, D. F. Ogletree, M. Salmeron: Measurement of interfacial shear (friction) with an ultrahigh vacuum atomic force microscope, J. Vac. Sci. Technol. B **14**, 1289–1295 (1996)

38.8   U. D. Schwarz, O. Zwörner, P. Köster, R. Wiesendanger: Quantitative analysis of the frictional properties of solid materials at low loads. 1. Carbon compounds, Phys. Rev. B **56**, 6987–6996 (1997)

38.9   B. Bhushan, S. Sundararajan: Micro/nanoscale friction and wear mechanisms of thin films using atomic force and friction force microscopy, Acta Mater. **46**, 3793–3804 (1998)

38.10   B. Bhushan, C. Dandavate: Thin-film friction and adhesion studies using atomic force microscopy, J. Appl. Phys. **87**, 1201–1210 (2000)

38.11   H. Liu, B. Bhushan: Adhesion and friction studies of microelectromechanical systems/nanoelectromechanical systems materials using a novel microtriboapparatus, J. Vac. Sci. Technol. A **21**, 1538–1538 (2003)

38.12   B. Bhushan, H. Liu, S. M. Hsu: Adhesion and friction studies of silicon and hydrophobic and low friction films and investigation of scale effects, ASME J. Tribol. **126**, 583–590 (2004)

38.13   A. W. Homola, J. N. Israelachvili, P. M. McGuiggan, M. L. Gee: Fundamental experimental studies in tribology: The transition from interfacial friction of undamaged molecularly smooth surfaces to normal friction with wear, Wear **136**, 65–83 (1990)

38.14   V. N. Koinkar, B. Bhushan: Scanning and transmission electron microscopies of single-crystal silicon microworn/machined using atomic force microscopy, J. Mater. Res. **12**, 3219–3224 (1997)

38.15   X. Zhao, B. Bhushan: Material removal mechanisms of single-crystal silicon on nanoscale and at ultralow loads, Wear **223**, 66–78 (1998)

38.16   B. Bhushan: *Introduction to Tribology* (Wiley, New York 2002)

38.17   N. A. Fleck, G. M. Muller, M. F. Ashby, J. W. Hutchinson: Strain gradient plasticity: Theory and experiment, Acta Metall. Mater. **42**, 475–487 (1994)

38.18   W. D. Nix, H. Gao: Indentation size effects in crystalline materials: A law for strain gradient plasticity, J. Mech. Phys. Solids **46**, 411–425 (1998)

38.19   J. W. Hutchinson: Plasticity at the micron scale, Int. J. Solids Struct. **37**, 225–238 (2000)

38.20   B. Bhushan, M. Nosonovsky: Scale effects in friction using strain gradient plasticity and dislocation-assisted sliding (microslip), Acta Mater. **51**, 4331–4345 (2003)

38.21   B. Bhushan, M. Nosonovsky: Comprehensive model for scale effects in friction due to adhesion and two- and three-body deformation (plowing), Acta Mater. **52**, 2461–2474 (2004)

38.22   B. Bhushan, M. Nosonovsky: Scale effects in dry and wet friction, wear, and interface temperature, Nanotechnol. **15**, 749–761 (2004)

38.23   M. Nosonovsky, B. Bhushan: Scale effect in dry friction during multiple asperity contact, ASME J. Tribol. **127**, 37–46 (2005)

38.24   H. Gao, Y. Huang, W. D. Nix, J. W. Hutchinson: Mechanism-based strain-gradient plasticity – I. theory, J. Mech. Phys. Solids **47**, 1239–1263 (1999)

38.25   Y. Huang, H. Gao, W. D. Nix, J. W. Hutchinson: Mechanism-based strain-gradient plasticity – II. analysis, J. Mech. Phys. Solids **48**, 99–128 (2000)

38.26   Z. P. Bazant: Scaling of dislocation-based strain-gradient plasticity, J. Mech. Phys. Solids **50**, 435–448 (2002)

38.27   B. Bhushan, A. V. Kulkarni, W. Bonin, J. T. Wyrobek: Nano/picoindentation measurement using a capacitive transducer system in atomic force microscopy, Philos. Mag. **74**, 1117–1128 (1996)

38.28   A. V. Kulkarni, B. Bhushan: Nanoscale mechanical property measurements using modified atomic force microscopy, Thin Solid Films **290–291**, 206–210 (1996)

38.29   K. W. McElhaney, J. J. Vlassak, W. D. Nix: Determination of indenter tip geometry and indentation contact area of depth-sensing indentation experiments, J. Mater. Res. **13**, 1300–1306 (1998)

38.30   J. Friedel: *Dislocations* (Pergamon, New York 1964)

38.31   J. Weertman, J. R. Weertman: *Elementary Dislocations Theory* (MacMillan, New York 1966)

38.32   B. Bhushan, A. V. Koinkar: Nanoindentation hardness measurements using atomic force microscopy, Appl. Phys. Lett. **64**, 1653–1655 (1994)

38.33   N. Gane, J. M. Cox: The micro-hardness of metals at very low loads, Philos. Mag. **22**, 881–891 (1970)

38.34 M. A. Stelmashenko, M. G. Walls, L. M. Brown, Y. V. Miman: Microindentation on W and Mo oriented single crystal an SEM study, Acta Met. Mater. **41**, 2855–2865 (1993)

38.35 S. Sundararajan, B. Bhushan: Development of AFM-based techniques to measure mechanical properties of nanoscale structures, Sensors Actuator A **101**, 338–351 (2002)

38.36 J. J. Weertman: Dislocations moving uniformly on the interface between isotropic media of different elastic properties, J. Mech. Phys. Solids **11**, 197–204 (1963)

38.37 K. L. Johnson: Adhesion and friction between a smooth elastic spherical asperity and a plane surface, Proc. R. Soc. London A **453**, 163–179 (1997)

38.38 I. A. Polonsky, L. M. Keer: Scale effects of elastic-plastic behavior of microscopic asperity contact, ASME J. Tribol. **118**, 335–340 (1996)

38.39 V. S. Deshpande, A. Needleman, E. Van der Giessen: Discrete dislocation plasticity modeling of short cracks in single crystals, Acta Mater. **51**, 1–15 (2003)

38.40 R. A. Onions, J. F. Archard: The contact of surfaces having a random structure, J. Phys. D **6**, 289–304 (1973)

38.41 D. J. Whitehouse, J. F. Archard: The properties of random surfaces of significance in their contact, Proc. R. Soc. London A **316**, 97–121 (1970)

38.42 J. A. Greenwood, J. B. P. Williamson: Contact of nominally flat surfaces, Proc. R. Soc. London A **295**, 300–319 (1966)

38.43 B. Bhushan: Contact mechanics of rough surfaces in tribology: Single asperity contact, Appl. Mech. Rev. **49**, 275–298 (1996)

38.44 B. Bhushan: Contact mechanics of rough surfaces in tribology: Multiple asperities contact, Tribol. Lett. **4**, 1–35 (1998)

38.45 B. Bhushan, W. Peng: Contact modeling of multilayered rough surfaces, Appl. Mech. Rev. **55**, 435–480 (2002)

38.46 A. Majumdar, B. Bhushan: Fractal model of elastic-plastic contact between rough surfaces, ASME J. Tribol. **113**, 1–11 (1991)

38.47 K. L. Johnson: *Contact Mechanics* (Clarendon, Oxford 1985)

38.48 E. Rabinowicz: *Friction and Wear of Materials*, 2nd edn. (Wiley, New York 1995)

38.49 H. R. Clauser (Ed.): *The Encyclopedia of Engineering Materials and Processes* (Reinhold, London 1963)

38.50 B. Bhushan, B. K. Gupta: *Handbook of Tribology: Materials, Coatings, and Surface Treatments* (McGraw-Hill, New York 1991; Krieger, Malabar, New York 1997)

38.51 National Carbon Comp.: *The Industrial Graphite Engineering Handbook* (National Carbon Company, New York 1959)

38.52 INSPEC: *Properties of Silicon*, EMIS Data Rev. Ser. No. 4 (INSPEC, Institution of Electrical Engineers, London 2002) See also, MEMS Materials Database, http://www.memsnet.org/material/

38.53 J. E.. Field (Ed.): *The Properties of Natural and Synthetic Diamond* (Academic, London 1992)

38.54 B.. Bhushan, S. Venkatesan: Mechanical and tribological properties of silicon for micromechanical applications: A review, Adv. Info. Storage Syst. **5**, 211–239 (1993)

38.55 B. Bhushan: Chemical, mechanical and tribological characterization of ultra-thin and hard amorphous carbon coatings as thin as 3.5 nm: Recent developments, Diam. Relat. Mater. **8**, 1985–2015 (1999)

38.56 C. Bernhardt: *Particle Size Analysis* (Chapman Hall, London 1994)

38.57 J. L. Devoro: *Probability and Statistics for Engineering and the Sciences* (Duxbury, New York 1995)

38.58 B. S. Everitt: *The Cambridge Dictionary of Statistics* (Cambridge Univ. Press, Cambridge 1998)

38.59 D. Zwillinger, S. Kokoska: *CRC Standard Probability and Statistics Tables and Formulas* (CRC, Boca Raton 2000)

38.60 S. Wolfram: *The Mathematica Book*, 5th edn. (Wolfram Media, Champaign 2003)

38.61 J. S. Bendet, A. G. Piersol: *Engineering Applications of Correlation and Spectral Analysis*, 2nd edn. (Wiley, New York 1986)

38.62 G. Herdan: *Small Particle Statistics* (Butterworth, London 1960)

38.63 R. D. Cadle: *Particle Size – Theory and Industrial Applications* (Reinhold, New York 1965)

38.64 Y. Xie, B. Bhushan: Effect of particle size, polishing pad and contact pressure in free abrasive polishing, Wear **200**, 281–295 (1996)

38.65 W. W. Seifert, V. C. Westcott: A method for the study of wear particles in lubricating oil, Wear **21**, 27–42 (1972)

38.66 D. Scott, V. C. Westcott: Predictive maintenance by ferrography, Wear **44**, 173–182 (1977)

38.67 J. L. Xuan, H. S. Cheng, R. J. Miller: Generation of submicrometer particles in dry sliding, ASME J. Tribol. **112**, 664–691 (1990)

38.68 M. Mizumoto, K. Kato: Size distribution and number of wear particles generated by the abrasive sliding of a model asperity in the SEM-tribosystem. In: *Wear Particles: From the Cradle to the Grave*, ed. by D. Dowson et al. (Elsevier, Amsterdam 1992) pp. 523–530

38.69 A. S. Shanbhag, H. O. Bailey, D. S. Hwang, C. W. Cha, N. G. Eror, H. E. Rubash: Quantitative analysis of ultrahigh molecular weight polyethylene (UHMWPE) wear debris associated with total knee replacements, J. Biomed. Mater. Res. **53**, 100–110 (2000)

38.70 D. P. Anderson: *Wear Particle Atlas*, 2nd edn. (Spectro Inc. Industrial Tribology Systems, Littleton 1991)

38.71 T. M. Hunt: *Handbook of Wear Debris Analysis and Particle Detection in Liquids* (Elsevier Applied Science, London 1993)

39. Mechanics of Biological Nanotechnology

One of the most compelling areas to be touched by nanotechnology is biological science. Indeed, we will argue that there is a fascinating interplay between these two subjects, with biology as a key beneficiary of advances in nanotechnology as a result of a new generation of single-molecule experiments that complement traditional assays involving statistical assemblages of molecules. This interplay runs in both directions, with nanotechnology continually receiving inspiration from biology itself. The goal of this chapter is to highlight some representative examples of the exchange between biology and nanotechnology and to illustrate the role of nanomechanics in this field and how mechanical models have arisen in response to the emergence of this new field. Primary attention will be given to the particular example of the processes that attend the life cycle of bacterial viruses. Viruses feature many of the key lessons of biological nanotechnology, including self assembly, as evidenced in the spontaneous formation of the protein shell (capsid) within which the viral genome is packaged, and a motor-mediated biological process, namely the packaging of DNA in this capsid by a molecular motor that pushes the DNA into the capsid. We argue that these processes in viruses are a compelling real-world example of nature's nanotechnology, and they reveal the nanomechanical challenges that will continue to be confronted at the nanotechnology–biology interface.

Nanotechnology is the seat of a broad variety of interdisciplinary activity, with applications as diverse as optoelectronics, microfluidics, and medicine. However, one of the richest interdisciplinary areas, which is only now beginning to be harvested, is the interface between the biological sciences and nanotechnology. The aim of the present chapter is threefold. First, we wish to illustrate the synergy that exists between biological nanotechnology (such as molecular motors and transmembrane pumps) and biologically inspired nanotechnology (including synthetic proteins, artificial viruses aimed at delivering drugs and biofunctionalized cantilevers). Secondly, through a series of order-of-magnitude estimates and an examination of the units used to describe spatial dimensions, temporal processes, forces at the nanoscale, and the energy budget in nanoscale systems, we aim to build an intuitive sense of the workings of nanotechnology in the biological setting. Finally, through several specific case studies, we hope to illustrate some of the challenges faced when modeling the mechanical processes of biological nanotechnology, and show how such challenges have been met thus far as well as how they might be met in the future.

39.1 Science at the Biology–Nanotechnology Interface

39.1.1 Biological Nanotechnology

Though the innovations leading to the adoption of the expression "nanotechnology" are indeed impressive, the perspective of this chapter is that one need only look inward – at the way that our muscles move, at how we digest and synthesize molecules of dazzling complexity, at the way in which we think the thoughts that permit us to fill the shelves of libraries with scientific journals – to realize that the greatest nanotechnology of all is that which is revealed in the living world. In other words, one of the central thrusts is the idea that the microscopic workings of life offer an inspiring vision of nanotechnology. Clearly, the diversity of the examples of "nanotechnology" seen in the living world can (and do) fill the pages of learned texts. A guided tour of the machinery of the cell can be found in *Alberts* et al. [39.1], while they have also envisioned the cell as an assemblage of "protein machines" [39.2]. The aim of the present section is to provide several cursory examples of the nanotechnological marvels that power the living realm.

39.1.2 Self–Assembly as Biological Nanotechnology

One of the most intriguing nanotechnological tricks of the living world is the central role played by self-assembly in such systems. Whether we consider the spontaneous assembly of viruses, either in test tubes or in the interior of an infected cell, or the fusion of one membrane-bound region to another through vesicle fusion, spontaneous formation of functional "materials" is a key part of the biological repertoire. To be more concrete, we note that self-assembly in biological systems takes place in a number of different guises. First, the self-assembly of linear assemblies is a key part of the cytoskeletal assembly process, with G-actin associating to form actin filaments and, similarly, tubulin monomers joining to form microtubules. This is also the process that is used by certain bacteria such as *Listeria* for locomotion. Both of these examples are described more fully in [39.3]. This same type of process is taken to the next level of sophistication in simple viruses such as tobacco mosaic virus that involve not only the assembly of protein monomers, but also the genetic message in the form of long RNA molecules. A second broad class of self-assembly processes is associated with the formation of containers such as liposomes or viral capsids. In the case of liposomes, lipid molecules such as phosphatidylcholine spontaneously organize in a way that sequesters the hydrophobic tails from the surrounding water. Similarly, protein subunits spontaneously assemble to form viral capsids [39.4]. These structures play a variety of roles in the biological realm, from serving as containers for different macromolecules to providing for concentration gradients of small ions that are maintained and utilized by molecular motors such as ATPsynthase [39.1].

Beyond the simplistic description of self-assembly advanced above, a second key feature of biological self-assembly must also be considered. In particular, a variety of self-assembly processes in the biological realm are templated (or coded). As is well-known, proteins are a hugely versatile class of molecules all based upon the same fundamental building blocks. Interestingly, the enormous diversity of protein action is founded upon 20 distinct amino acid building blocks, and the template for assembling a given protein is carried in the form of messenger RNA, which is then read and used as the basis for protein synthesis by the ribosome. One of the most exciting developments in modern nanotechnology is the attempt to exploit templated self-assembly processes for the purposes of creating new materials. An example in the context of protein-based materials can be found in the article of *van Hest* and *Tirrell* [39.5], where the machinery of the cell has been tricked into incorporating artificial amino acids into synthetic proteins.

39.1.3 Molecular Motors as Biological Nanotechnology

One of the hallmarks of life is change and motion. At the cellular level, such motion is effected through a dizzying variety of mechanisms, most of which when viewed at the molecular level are seen to be the result of the action of molecular motors [39.3]. For example, muscle contraction reflects the coherent action of huge numbers of myosin motor molecules as they march along actin filaments. Similarly, the motion of certain bacteria can be traced to the rotation of a rotary motor embedded in the cell wall, which is attached to filaments known as flagella [39.6]. In a similar vein, just as our modern society is replete with examples of systems aimed at allowing for communication and transport between widely separated geographic locations, so too has the living world had to answer these same challenges. As will be shown in the section of this chapter aimed at making estimates of various scales and processes in nanotechnology, one of the key mechanisms for communication and transport is diffusion. That is, a chemical concentration at one place can make its presence known at a distance x with a characteristic time, $t_{\text{diffusive}} \approx x^2/2D$, where D is the diffusion constant. However, as will be shown later in the chapter, there are many cellular processes that cannot wait as long as $t_{\text{diffusion}}$. As a result, a host of molecular motors and an associated transport system (the elements of the cytoskeleton) permit active transport. For example, kinesin is a motor molecule that transports material along long, relatively rigid, polymeric assemblies known as microtubules [39.1].

Even more incredible are rotary machines like those associated with ATPsynthase and bacterial flagella. ATPsynthase is a membrane-embedded machine that rotates and, in so doing, synthesizes new ATP molecules (adenosine triphosphate, energy currency of the cell) [39.1, 7–9]. Similarly, the bacterial flagellar motor rotates with the result that the attached bacterial flagellum rotates and thereby induces motion of the cell [39.3, 6]. The exquisite details of the construction of these devices are themselves breathtaking. The rotary motors in bacteria are constructed from several components that are much like a rotor and stator and perform periodic motions by deriving energy from the flow of protons across a membrane [39.6]. These rotary motors are very powerful, as evidenced by F_1-ATPsynthase, which can generate a torque large enough to rotate a molecule of actin 100 times its own length [39.10]. In addition, bacterial motors have another layer of sophistication in that stimuli from the external environment, which are sensed by the bacteria through the pores on its membrane, can change the direction of rotation of the motor in a process known as chemotaxis [39.3]. Feedback and signal transmission are implemented in engineering devices by means of complex circuitry, whereas in the living world these functions are often accomplished by means of chemistry and conformational changes in large molecules.

All of these molecular machines involve a rich interplay between chemistry, thermodynamics, and mechanics. From a structural perspective, most molecular motors are proteins with different subunits performing different functions [39.11]. Often there is some region within the motor where chemical energy is derived from the hydrolysis of adenosine triphosphate (ATP). This chemical energy is then converted to mechanical energy through a conformational change in the protein. These examples serve to call the reader's attention to the importance and variety of motor molecules found throughout the living world and which almost any sensible definition of nanotechnology would have to include as particularly sophisticated examples.

39.1.4 Molecular Channels and Pumps as Biological Nanotechnology

From a structural perspective, one of the most intriguing features of cellular systems is their division into a number of separate membrane-bound compartments. We have already touched upon this compartmentalization in the context of self-assembly, and note that the role of membranes and the proteins bound within them is described clearly in *Alberts* et al. [39.1]. The presence of such compartments, often marked by large concentration gradients with respect to the surrounding medium, hints at another nanotechnological wonder of the living world, namely the presence of a wide range of transmembrane channels that mediate the exchange of material between these different compartments. Certain passive versions of these channels are gated by various mechanisms such as the arrival of signaling molecules or tension in the membrane within which they are found [39.12–14]. Once the channel is in the open state, ions pass through passively, by diffusion. Active versions of ion channels that transfer ions such as Na^+ and Ca^{2+} are similarly critical to the functioning of a cell, for both unicellular and multicellular organisms. For example, in the case of Na^+ ions, typical concentrations within the cell can be as much as a factor of 10–20 less than those in the extracellular milieu. Such concentration gradients imply the need for sophisticated active "devices" which can do work against such gradients. One of the most remarkable machines of this type is the Na^+–K^+ pump [39.1]. This machine is powered by the hydrolysis of ATP (the consumption of ATP fuel), and it can pump ions up a potential gradient. In particular, this pump hydrolyzes ATP and pumps Na^+ ions *out* of the cell against a very steep concentration gradient while pumping K^+ ions *in* again against a steep gradient.

Part D | 39.1

The point of this brief discussion has been to illustrate the first of several perspectives that we will bring to bear on the question of the biology–nanotechnology interface. Thus far, we have noted that nature is replete with examples of macromolecules and macromolecular assemblies that perform nanotechnological tasks and in this capacity serve as examples of biological nanotechnology. Next, we wish to examine the ways in which biological phenomena can inspire nanotechnology itself.

39.1.5 Biologically Inspired Nanotechnology

As noted in the beginning of this chapter, biological systems are nanotechnologically relevant for several reasons. First, as shown above, the living world is full of examples of nanotechnological devices. However, a second key point is that nanotechnology has been driven and inspired by the example of biological systems and the need (for example, in medicine) to influence biological systems on the scale of a single cell. In addition, preliminary steps have been taken to harness the nanotechnology of biological systems and use it to perform useful functions.

One compelling example of a proof-of-principle, biologically inspired device emerged from work aimed at exploring the function of ATP synthase, the rotary device already described above. Fluorescently labeled filamentous proteins from the cytoskeleton, known as actin, were attached to the putative rotary component of the F_1 subunit. Such filaments are observable using light microscopy. It was then observed that the long actin filament rotated like a propeller when the ATP synthase performed ATP hydrolysis [39.15].

A second example, also involving ATP synthase, is suggested by a set of beautiful experiments performed by *Racker* and others [39.16, 17]. The idea is illustrated schematically in Fig. 39.1. Two different protein machines are "reconstituted" in an artificial membrane-bound region known as a liposome. One of these devices is known as bacteriorhodopsin and has the capacity to pump hydrogen ions when it is exposed to light. Since both the bacteriorhodopsin and ATPsynthase are embedded in the same membrane, the ATPsynthase can then exploit the light-induced proton gradient to perform ATP synthesis. Again, these experiments were undertaken not for their role as possible devices, but rather to probe the nature of various molecular machines. Nevertheless, we view them as a provocative demonstration of both the manipulation and use of such machines in artificial environments. As such, they provide an inspiring vision of the possibilities for biologically inspired nanotechnology.

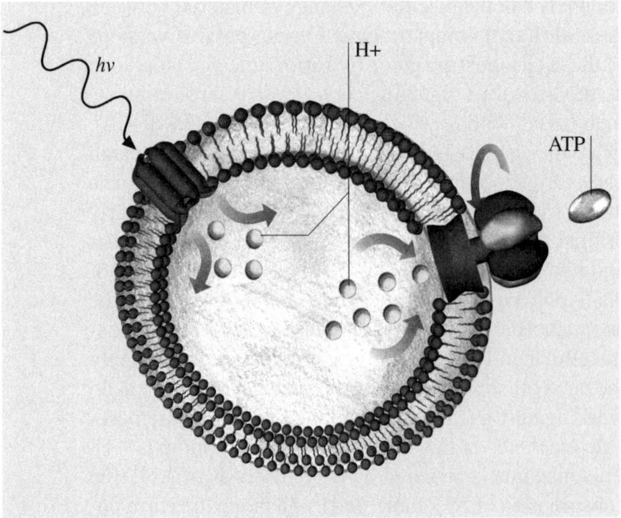

Fig. 39.1 Schematic of the experiment in which bacteriorhodopsin (*top of figure*) and ATP synthase (*right of figure*) are artificially reconstituted in a liposome and act in unison to produce ATP molecules

Fig. 39.2 Schematic of the use of biofunctionalized cantilevers as a tool for detecting molecules of biological interest (figure courtesy of Arun Majumdar, Berkeley)

Another fascinating example of biologically inspired nanotechnology is that of biofunctionalized cantilevers (see Fig. 39.2). A typical example is provided by the experiments of *Wu* et al. [39.18], who have demonstrated the use of a biofunctionalized cantilever as a scheme for detecting small concentrations of biologically interesting molecules such as prostrate-specific antigen and single-stranded DNA. One surface of the cantilever is coated with an antibody, and then it is placed in environments containing different concentrations of the antigen. The key ideas from a mechanical perspective are that: (1) the difference in surface energy between the top and bottom surfaces of the cantilever induces spontaneous bending, and (2) the binding of molecules of interest to target molecules initially present on the surface leads to surface energy differences and bending that can be detected by optical means. In this way it is possible to measure the concentration of the molecules of interest. The specificity and sensitivity of the method makes it viable for use in the laboratory, as well as for commercial purposes. We return to this example in our discussion of the modeling challenges posed by problems at the interface between biology and nanotechnology.

39.1.6 Nanotechnology and Single-Molecule Assays in Biology

One of the key refrains of the nanotechnological era is Feynman's quip that "there is plenty of room at the bottom" [39.20]. The benefits of miniaturization are evident at every turn in applications ranging from our cars to our computers. Associated with the development of the inspiring new techniques made possible by nanotechnology has been the emergence of a host of scientific opportunities. One of the arenas to benefit from these new techniques is biology. As scientists and engineers have taken the plunge to the nanotechnological "bottom" foreseen by Feynman, opportunities have constantly arisen to manipulate biological systems in ways that were previously unimagined, culminating in a new era of single-molecule biology. Single-molecule experiments have, in fact, presented us with a view of Feynman's "room at the bottom" as being filled with very complicated machines whose functioning makes life possible.

Single-molecule assays complement statistical/collective studies involving a large number of molecular actors by revealing the prominent role of fluctuations at the subcellular scale. For example, a photospectrometer measures the optical response of a huge collection of molecules, whereas optical tweezers pull on a single molecule of DNA and enable us to follow the changes in conformation or the breakage of bonds. In fact, experimental methods are so advanced that it is now possible to manipulate a single molecule even as we watch it on a screen as it jiggles around in different conformations. In what follows, we give several examples of how nanotechnology has reached out to help create single-molecule biology and in the process has led to the advent of new quantitative opportunities for investigating biological systems.

Atomic Force Microscopy

One of the tools that has revolutionized nanotechnology, in general, and single-molecule biology, in particular, is the atomic force microscope. The AFM has helped create the field of single-molecule force spectroscopy [39.21]. We note that mechanics has a long tradition of using force–extension data (much like the electrical engineer uses current–voltage data) to probe the inner workings of various materials. It is now possible to apply forces of known magnitude to a macromolecule and then to study how it deforms under the force. This furnishes structural information and provides insights into the energy landscape the molecule needs to navigate as it undergoes force-induced conformational changes. The energy of deformation associated with such molecules is primarily determined by weak forces such as hydrogen bonds, van der Waals contacts and hydration effects. On a more philosophical note, these experiments

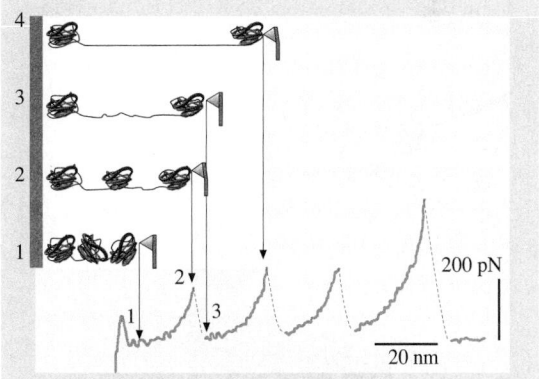

Fig. 39.3 Schematic illustrating the way that the AFM has been harnessed as a nanotechnological analog of the Instron machine for the measurement of the force–extension properties of single molecules. Four snapshots in the life history of a globular protein subject to loading are shown, as well as the measured force–extension curve. Each sawtooth corresponds to the unfolding of a single protein domain. (Figure adapted from [39.19])

force us to think in terms of forces and not energy, complementing the traditional views held in molecular biology, and they can lead to many new insights into the relation between structure and function in proteins, polynucleotides, and other macromolecular entities.

The AFM has been used in a wide variety of single-molecule experiments on many of the key classes of molecules found in the living world, including nucleic acids, proteins, and carbohydrates. One fascinating example is the use of atomic force microscopy to examine the mechanical properties of the muscle protein titin [39.22]. The experiment is illustrated schematically in Fig. 39.3, which shows that there is a series of force–extension signatures (increasing force and extension followed by a precipitous load drop) that correspond to the unbinding of the individual domains that make up this protein.

Optical Tweezers

Another instrument that has been used with great success in the realm of biophysics for the purposes of performing nanomechanical measurements on macromolecules and their assemblies is the optical tweezer [39.23, 24]. While the AFM is relatively stiff and applies large forces (on the order of 100–1000 pN), optical tweezers are compliant and can measure smaller forces (on the order of

0.1–10 pN). Some of the most interesting experiments performed with optical tweezers concern the functioning of molecular motors, which by themselves are marvels of nanotechnology. For example, *Svoboda* et al. [39.25] attached kinesin to an optically trapped bead and observed its movement along a microtubule. An example of the type of data to emerge from such experiments is shown in Fig. 39.4. One of the conclusions to emerge from such experiments is that kinesin can exert forces on the order of 5–7 pN before it stalls. Such experiments also permit an examination of the effect of changing the concentration of ATP on the functioning of kinesin [39.26]. Similar experiments have been performed on RNA polymerase as it advances along DNA to deduce not only the stall force but also its velocity as a function of the constraining force [39.27]. Such measurements provide the mechanochemical basis of biological function and go a long way to revealing connections between chemical kinetics and mechanical processes at the molecular level.

As noted at the start of the chapter, one of the most fascinating examples of the biology–nanotechnology interface is that of bacterial viruses, known as bacteriophages. The life cycle of a large class of bacteriophages is characterized by self-assembly processes that lead to the formation of the protein shell of the virus followed by active packaging of the viral DNA within this shell by a motor. The structure of this so-called "viral portal motor" has recently been solved using X-ray crystallography [39.28]. In a recent experiment, *Smith* et al. used optical tweezers to study the characteristics of the DNA packaging process of the ϕ-29 bacteriophage [39.29]. One of the conclusions of this experiment is that the motor has to act against an increasing resistive force as more and more of the DNA is packed inside the capsid. From a quantitative perspective, this experiment yields the force and the rate of packing as a function of the fraction of the genome packed. It is also important to note that bacteriophages are not only an obscure subject of quiet enquiry, but are also the basis of a huge range of cloning products (see, for example, the lambda ZAP vectors of Stratagene, La Jolla, CA, USA) used in experiments with recombinant DNA, and, more generally, viruses are being explored as the basis of gene therapy.

Our discussion thus far has been aimed at providing a rough overview of the vast landscape that sits at the interface between biology and nanotechnology. It is hoped that the few representative case studies set forth above suffice to illustrate our basic thesis, namely that biological nanotechnology represents nanotechnology at its best.

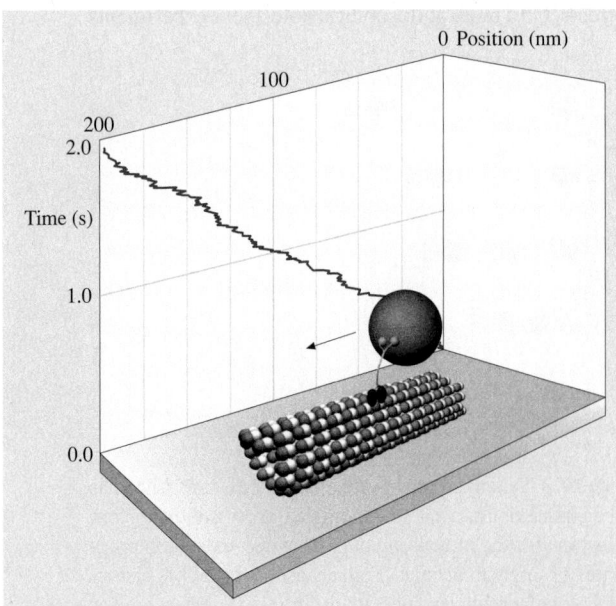

Fig. 39.4 Schematic illustrating the use of optical tweezers to measure the speed of the molecular motor kinesin in its journey along a microtubule

39.1.7 The Challenge of Modeling the Bio–Nano Interface

As highlighted in the previous two subsections, there have been huge advances at the interface between nanotechnology and biology. We have argued that there are two distinct representations of the interface between biology and nanotechnology, and each has its own associated set of modeling challenges. The argument of the present discussion is that another key part of the infrastructure that must attend these developments is that associated with the modeling of these systems. One of the intriguing ways in which modeling at the biology–nanotechnology interface is assuming greater importance is that, with increasing regularity, experimental data on biological systems is of a quantitative character. As a result, the models that are put forth to greet these experiments must similarly be of a quantitative character [39.30].

As an example of the type of modeling challenges that must be faced in contemplating the types of problems described above, we return to the example of biofunctionalized cantilevers as a problem in nanomechanics [39.18, 31]. The basic physics behind the use of biofunctionalized cantilevers as sensors is a competition between elastic bending energy and the surface free energy difference between the upper and lower faces of the cantilever. The face with the lower free energy per unit area tends to increase its area by bending the cantilever. The amount of bending, on the other hand, is limited by the elastic energy cost. The utility of this device derives from the fact that the difference in the surface free energy is affected by specific binding of target molecules to probe molecules that are initially deposited on one side of the cantilever.

To provide a quantitative model of the biofunctionalized cantilever, we construct an energy functional that takes into account the elastic energy of beam bending and the surface energy. Both contributions to the total energy can be written as functionals of $u(x)$, the deflection of the cantilever. Note that in the case in which the two surfaces have the same free energy per unit area, $\gamma_{\text{up}} = \gamma_{\text{down}}$, the equilibrium configuration corresponds to $u(x) = 0$. The case of interest here is that in which the two surfaces have different energies.

We recall that the energy associated with beam bending is of the form [39.32]

$$E_{\text{bend}}[u(x)] = \frac{EI}{2} \int_0^L \left[u''(x)\right]^2 \mathrm{d}x \,, \tag{39.1}$$

where E is Young's modulus, L is the length of the beam, and $I = wt^3/12$ is the areal moment of inertia; t and w are the beam thickness and width. The main approximation we have made in writing (39.1) is that the cross-section of the beam remains unchanged by the bending process. The surface contribution to the total energy is associated with the changes in the areas of the upper and lower surfaces by virtue of the beam deforming. In particular, we have

$$E_{\text{surf}}[u(x)] = \gamma_{\text{up}} w \left[L - \frac{t}{2} \int_0^L u''(x)\,\mathrm{d}x \right]$$
$$+ \gamma_{\text{down}} w \left[L + \frac{t}{2} \int_0^L u''(x)\,\mathrm{d}x \right] \,, \tag{39.2}$$

where the terms in parentheses are the arc lengths along the top and bottom surface of the beam, respectively. The physical content of this functional is the idea that if $u'' > 0$ (concave upward) then the upper surface area will shrink while the lower surface area will increase. Neglecting uninteresting constant terms, the total energy functional is

$$E_{\text{tot}}[u(x)] = \frac{EI}{2} \int_0^L \left[u''(x)\right]^2 \mathrm{d}x$$
$$+ \frac{tw}{2}(\gamma_{\text{down}} - \gamma_{\text{up}}) \int_0^L u''(x)\,\mathrm{d}x \,. \tag{39.3}$$

Our goal is to find the displacement profile $u(x)$ that yields the minimum value for the total energy. The general mathematical framework for effecting this minimization is the calculus of variations. For the functional in (39.3), a more direct route to the result can be obtained by completing the square. Namely, we note that the E_{tot} involves a term quadratic in $u''(x)$ and a second term that is linear in the same function. As a result, the total energy can be rewritten in the form

$$E_{\text{tot}}[u(x)] = \frac{EI}{2} \int_0^L \mathrm{d}x \left[u''(x) + \frac{tw\Delta\gamma}{2EI} \right]^2 \,, \tag{39.4}$$

where we have introduced $\Delta\gamma = \gamma_{\text{down}} - \gamma_{\text{up}}$, and we neglect an uninteresting constant term. Clearly the $u(x)$ that minimizes the energy is one for which the integrand vanishes everywhere, or

$$u''(x) + \frac{tw\Delta\gamma}{2EI} = 0 \,. \tag{39.5}$$

At this point, we are left with a standard differential equation whose solution, given the boundary conditions $u(0) = 0$ and $u'(0) = 0$, is

$$u(x) = -\frac{tw}{4EI}\Delta\gamma x^2 \ . \tag{39.6}$$

The physical meaning of this solution is hinted at by observing that if $\gamma_{\text{down}} > \gamma_{\text{up}}$ then the beam curves downward with the result that the area of the lower surface is reduced.

Measurements of *Wu* et al. [39.31] indicate an upward cantilever deflection when single-stranded DNA (ssDNA), up to 20 nucleotides in length, hybridizes with complementary strands of ssDNA, which were deposited initially so as to functionalize the beam. This effect might be attributed to the different elastic properties of ssDNA and its double-stranded counterpart. Namely, under physiological conditions, ssDNA has a persistence length equal to two nucleotides, while dsDNA is much stiffer (due to hydrogen bonding and base stacking interactions between the two strands) and has a persistence length of 150 nucleotides. Therefore, deposition of the flexible ssDNA molecules initially leads to bending of the cantilever downwardsdue to entropic repulsion between the ssDNA chains. After hybridization, rigid dsDNA strands are formed, there is no longer any entropy to be gained by increasing the area of the top surface, and the beam bends back upwards. Remarkably, *Wu* et al. demonstrate that their biofunctionalized cantilever is sensitive to ssDNA that differ in length by a single nucleotide!

To gain further insight into the physics of the biofunctionalized cantilever, it is instructive to examine some quantitative aspects of the experiment using (39.6). Namely, *Wu* et al. find a deflection of $u(L) = 12$ nm when a 20-nucleotide strand of ssDNA hybridizes with a 20-nucleotide-long complementary target. Using the quoted numbers for the cantilever, $E = 180$ GPa, $L = 200\,\mu$m, $t = 0.5\,\mu$m, $w = 20\,\mu$m, leads to $\Delta\gamma = 4.5$ fJ/μm^2. From this result we can estimate the change in surface free energy due to beam bending,

$$\Delta E_{\text{surf}} \approx wL\Delta\gamma \ . \tag{39.7}$$

We find $\Delta E_{\text{surf}} = -18$ pJ, which, given the quoted areal chain density of 6×10^{12} cm^{-2}, leads to a decrease in free energy of 75 pN nm, or 18 $k_{\text{B}}T$, per chain. This is comparable to the entropy of a 20 nt ssDNA, which is on the order of 10 k_{B}, lending support to the idea that entropic repulsion between ssDNA strands is implicated in cantilever bending.

While this example gives a feel for the way in which quantitative models have been put forth to respond to biologically inspired nanotechnologies, the remainder of the chapter will emphasize attempts to construct nanomechanical models of biological nanotechnology itself. We turn first to a discussion of the various scales that arise when considering the biology–nanotechnology interface, and then conclude with several modeling examples.

39.2 Scales at the Bio–Nano Interface

Every scientific discipline has a preferred set of units that lends itself to building intuition about the system at hand. For example, an astronomer thinks of distances between stars in light years, not kilometers. Though most of us have an intuitive sense of the meaning of a kilometer, by the time we add more than six zeros, all intuition is lost. At terrestrial scales, we talk of distances between cities in terms of the flying time or the driving time between them and usually not in terms of hundreds of miles. Similar choices must be faced in the biological setting. For example, a biologist might characterize the complexity of an organism by the size (in kilo base-pairs) of the genome and not by the organism's physical size. The aim of the present section is to highlight some of the key scales and units that reveal themselves at the interface between biology and nanotechnology. Indeed, we go further and assert that as yet we are still in the process of searching for the most suitable units to characterize the biology–nanotechnology interface. Although our attempt to determine such units and scales might involve seemingly complicated interconversions, such as measuring distances in terms of time (via diffusion), or measuring concentrations in terms of distances, we hold that the approximate *numerical* characterization of the scales of interest is of crucial importance to the endeavor of considering nanomechanics at the biology–nanotechnology interface. In particular, the right choice of units can assist us in building intuition about these systems. Our goal in this section is to emphasize the scales in length, time, force, energy and power that are relevant when contemplating the nanomechanics of biological systems.

39.2.1 Spatial Scales and Structures

We begin with a discussion of the length scales that arise when contemplating nanomechanics at the biology–nanotechnology interface. In this case, the prefix *nano* leads justifiably to a consideration of the nanometer as one possible choice for the fundamental unit of length. However, to prepare ourselves for the question of how best to describe the dimensions of the spatial structures of interest here, it is important to consider the *hierarchy* of length scales that arise in the nano–bio arena. After examining this hierarchy of structures, we reformulate these length scales in terms of the volumes of these structures measured in units of the volume of a typical bacterial cell, and conclude the present section with a discussion of the way that chemical concentrations can also be interpreted as determining a length scale.

As noted above, a first step when developing intuition about the spatial scales found at the biology–nanotechnology interface is through reference to the hierarchy of scales and structures that arise in this arena. The shortest distance in this hierarchy of scales that will interest us is that associated with the sizes of individual atoms. We recall that the size of a hydrogen atom is roughly 0.1 nm. The scale characterizing the linkage between atoms is that of typical bond lengths that range from roughly 0.1 to 0.3 nm. A step further in this hierarchy brings us to the basic building blocks of the biological world, such as amino acids, nucleotides, individual sugars and lipid molecules. A typical length scale that characterizes these building blocks is the nanometer itself. For example, we have already made reference to the importance of lipid bilayer membranes in bounding different regions of the cell. Phosphatidylcholine is one of the molecular building blocks of many such membranes. It has a polar head group and a hydrocarbon tail with an overall dimension of 2–3 nm. Similarly, the dimensions of single amino acids and nucleotides are on the order of 1 nm too.

As is well-known, individual nucleotides are assembled to form nucleic acids such as DNA, amino acids are assembled to form proteins, sugars combine to form polysaccharides, and lipids self-assemble to form membranes. One way of estimating the size of the resulting molecules is by taking the scale of the individual units and scaling up with the number of such units. The various molecular actors of relevance to the present discussion can also be characterized by a length scale known as the persistence length, which gives a rough description of the length over which the molecule behaves as a stiff rod. For example, the persistence length of DNA is ≈ 50 nm, while that of the cytoskeletal filaments is in the range of 15 μm for actin and 6 mm for microtubules [39.11]. We note that in our later discussions of viruses as a profound example of biological nanotechnology, the ratio of the persistence length of DNA to the size of the viral capsid (the container within which the DNA is packaged) will serve as a measure of the energetic cost of packing the DNA within the capsid and will signal the need for molecular motors to take active part in the packaging process.

The next scale above that of the various macromolecular building blocks are assemblies of such molecules in the form of various molecular machines which are some of the most compelling examples of nature's nanotechnology. The ability to begin formulating mechanistic models of such machines is founded upon key advances in both X-ray crystallography and cryoelectron microscopy [39.33, 34]. Examples of such machines and their associated dimensions include: the machine responsible for making the message carried by DNA readable by the protein synthesis machinery, namely RNA polymerase (≈ 15 nm) [39.35], the machine that produces ATP, the energy currency of the cell, namely ATP synthase (≈ 10 nm) [39.9], and the machine that carries out protein synthesis, namely the ribosome (≈ 25 nm) [39.36]. A second class of assemblies of particular relevance to the present article are viruses, representative examples of which include lambda phage (≈ 27 nm), tobacco mosaic virus (≈ 250 nm in length), and the HIV virus (≈ 110–125 nm) [39.37].

From the standpoint of cellular function, the next level of structural organization is associated with the various organelles within the cell, structures such as the cell nucleus (3–10 μm), the mitochondria that serve as the power plant of the cell (≈ 1000 nm), the Golgi apparatus wherein modifications are made to newly synthesized macromolecular components (≈ 1000 nm), and so on [39.1]. These organelles should be thought of as factories in which many molecular motors of the type described earlier do their job simultaneously. At larger scales still, and constituting a higher level of overall organization, life as nanotechnology is revealed as cells themselves, the fundamental unit of life that is self-replicating and self-sustaining. We will make special reference to one particular bacterial cell, namely *E. coli*, with typical dimensions of 1 μm. This should be contrasted with a typical eukaryotic cell such as yeast, which has linear dimensions roughly a factor of ten larger than the *E. coli* cell.

We note that the various biological structures described above should be seen with reference to the sorts of man-made structures to which they are interfaced. For example, earlier in the chapter we mentioned the importance of optical tweezers as a means of communicating forces to macromolecules and their assemblies. Typical dimensions for the optical beads used in optical tweezers are on the scale of 500–1000 nm. We note that though such dimensions are characteristic of organelles, they are much larger than the individual molecules they are used to study. A second way to communicate with individual macromolecules is through small tips such as those found on an AFM. In this case, the size of the tip can be understood through reference to its radius of curvature, which is typically of the order of 50 nm.

Our discussion thus far has centered on the use of a single characteristic length to describe the spatial extent of biological structures. In our quest to develop intuition about the typical spatial scales found at the biology–nanotechnology interface, we note that a second way to evaluate such scales is through reference to the typical *volumes* of the various structures of interest. As noted in the introduction to this section, we claim that a key way to develop intuition is by making sure that appropriate and revealing units are used. For the present purposes, we argue that one useful unit of volume is that of an *E. coli* bacterium, which, if idealized as a cylinder with diameter 1 μm and height 2 μm, has a volume of $\approx 1.5 \times 10^9$ nm^3. In particular, we will measure the sizes of the various entities described above in terms of how many of them can fit into a single *E. coli* bacterium. Our estimates of the comparative sizes of various structures of interest are given in Table 39.1. Note that we do not claim that this is literally how many of each of these entities are found in an *E. coli* bacterium, but rather we seek to give an impression of the relative volumes of some of the different structures that arise when contemplating biological nanotechnology.

The idea of counting the number of molecules that fit into a given volume is actually quite standard. This is exactly what chemists do when they invoke the notion of concentration: measuring the strength of an acid or base using pH amounts to expressing the concentration of H$^+$ ions in solution. Similarly, when we refer to the molarity of a given solution, it is a statement of how many copies of a given molecule will be found per liter of solution. These ideas are pertinent for our examination of the scales that arise when contemplating biological nanotechnology. As an example, we consider the action of molecular pumps such as that which maintains the concentration gradient in Ca^{2+} ions between

Table 39.1 Sizes of entities in comparison to the size of *E. coli*

Entity	Size (nm)	*f* itting inside an *E. coli*
Amino acid	1	5.0×10^9
Nucleotide	1	5.0×10^9
Monosaccharide	1	5.0×10^9
Phosphatidyl choline	2–3	2.0×10^8
Proteins	5–6	3.0×10^7
Ribosome	20	1.0×10^6
Mitochondria	500–1000	3–4
T4 phage	100	5000

the cellular interior and the extracellular medium. Note that this is described in detail in chapter 12 of *Alberts* et al. [39.1]. The Ca^{2+} pump is responsible for insuring that the intercellular concentration of Ca^{2+} ions is 10^{-7} smaller than that in the cellular exterior. One explanation for this low concentration of Ca^{2+} in the cell is the role played by these ions in signaling other activities. The low concentration might serve as a scheme for increasing the signal-to-noise ratio. The reported intracellular concentration of such ions is roughly 10^{-7} mM. It is interesting to ask how many Ca^{2+} ions this corresponds to in a eukaryotic cell. The volume of a eukaryotic cell measuring 20 μm in diameter is about 4×10^{-12} liters, which translates into ≈ 250 Ca^{2+} ions in the cytoplasm. For the case of *E. coli*, which has a volume that is approximately 1000 smaller, this concentration would correspond to one ion for every four cells. This number gives us far greater insight than the standard molar (M = mol/liter) method of expressing concentrations. In a similar vein, we argue that distances between individual ions/molecules in solution are perhaps a better way of thinking about pH and molarity when using them in the context of biological nanosystems. To this end, Fig. 39.5 shows a plot of pH and molarity versus average distance d in nm between individual ions/molecules. Note that, in addition, we have also translated these concentrations into the number of copies per *E. coli* cell and the time for an ion to diffuse over the average separation distance. For the time estimate we make use of the typical diffusion constant for small ions in water at 25 °C, $D = 2000 \, \mu m^2/s$.

We have seen that the world of biological nanotechnology utilizes length scales that span a rather wide range, from fractions of a nanometer to tens of microns. This provides a challenge to modeling, whereby methods with atomic-scale resolution need to be combined in

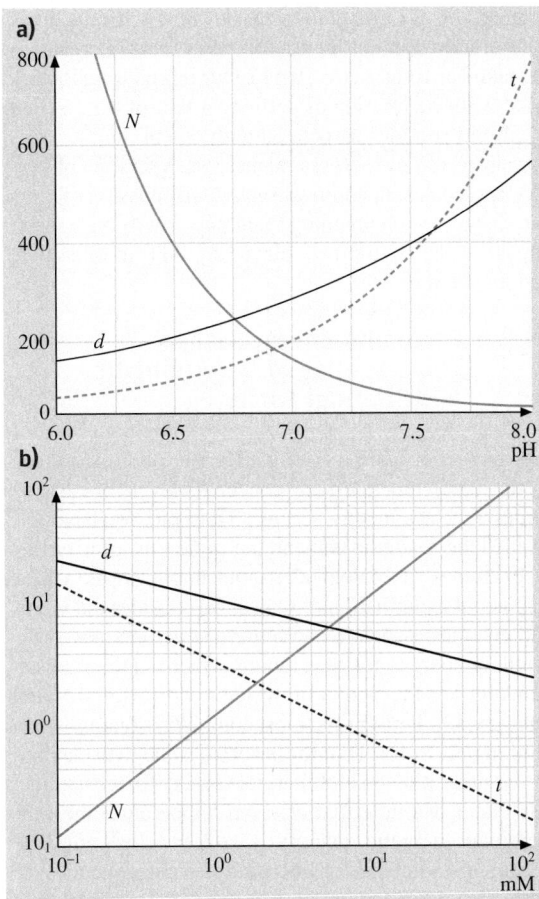

Fig. 39.5 (a) Representations of pH in terms of average distance between ions (d in nm), the number of ions per one *E. coli* cell volume (N), and the typical time for an ion to diffuse over a distance equal to their average separation (t in 100 ns). **(b)** Concentration unit millimoles (mM) represented in terms of d (nm), N (millions of molecules), and t (10 ns)

a consistent and seamless way with coarse-grained continuum descriptions to provide a complete picture. An example of such modeling will be provided in later sections when we take up the mechanical aspect of the viral life cycle.

39.2.2 Temporal Scales and Processes

We have seen that a study of spatial scales is a first step in our quest to understand the units that are suited to characterizing the biology–nanotechnology interface. We note that our understanding of spatial structures both in the

living world, as well as those used in intriguing manmade technologies such as microelectronics have been built around important advances such as electron microscopy, X-ray crystallography and nuclear magnetic resonance that have revolutionized structural biology and materials science alike. On the other hand, one of the key challenges that remain for really appreciating the structure–properties paradigm, which is as central to biology as it is to materials science, is the need to acknowledge the dynamical evolution of these structures.

We note that just as there is a hierarchy of spatial scales that are important to consider when describing biological nanotechnology, there is also a hierarchy of *temporal processes* that also demand a careful consideration of relevant units. An impressive representation of the temporal hierarchy that must be faced when contemplating biological systems is given by *Chan* and *Dill* [39.38] in their Fig. 5. They organize their temporal hierarchy according to factors of 1000, starting at the femtosecond time scale and ending with time scales representative of the cell cycle itself. One of the elementary dynamical processes undergone by all of the molecular actors of the living world is thermal vibration. Vibrations of atoms occur at time scales of 10^{-15} to 10^{-12} s. As will be noted later, from a modeling perspective, this depressing fact manifests itself in the necessity to use time steps of the same order when integrating the equations of motion describing molecular dynamics. The reason that this is unfortunate is that almost everything of dynamical interest occurs at time scales much longer than the femtosecond time scale characteristic of molecular jiggling, thus making molecular dynamics investigations computationally very expensive.

We follow *Chan* and *Dill* [39.38] in examining how successive thousandfold increases in our temporal resolution smears out features such as vibrational motion and brings into focus more interesting processes characteristic of macromolecular function. Indeed, the side chains of amino acids rotate with a characteristic period on the order of 10^{-9} s. Another thousandfold increase in time scale begins to bring key biological processes such as polymerization into view. Just as a length scale on the order of $10-100$ nm is perhaps most characteristic of the length scales of biological nanotechnology, the time scale that is most relevant when considering the processes associated with biological nanotechnology is something between a microsecond and a millisecond. To drive home this point, we reconsider the machines described in the previous section from the length scale perspective, but now with an eye toward how fast they perform their function. Several of the molecular machines considered in the previous section mediate polymerization reactions. For example, RNA polymerase produces messenger RNA molecules by the repeated addition of nucleotides to a chain of ever increasing length. Such messenger RNA molecules serve in turn as the template for synthesis of new proteins by the ribosome, which reads the message contained in the RNA and adds individual amino acids onto the protein. From the temporal perspective of the present section, it is of interest to examine the rate at which new monomers are added to these polymers – roughly ten per second in the case of RNA polymerase [39.39], and two per second in the case of the ribosome in eukaryotic cells [39.1] (bacterial ribosomes are ten times faster). Once proteins are synthesized, in order to assume their full cellular responsibilities they must fold into their native state, a process that occurs on time scales on the order of a millisecond.

In addition to translational machines like RNA polymerase and the ribosome, there are a host of intriguing rotary machines such as ATP synthase and the flagellar rotary motor. The flagellar motor, which is the means of locomotion for several bacteria, rotates at about 100 rpm [39.6]. ATP synthase makes use of a proton gradient across a lipid bilayer to provide rotation at a rate of roughly 6000 rpm [39.41], approaching that of turbojet engines.

Thus far, our discussion has emphasized the rate of a variety of active processes of importance to biological nanotechnology. It is of interest to contrast these scales with those pertaining to diffusion. We note that in many instances in the biological setting the time scale of interest is that determined by the time it takes for diffusive communication of two spatially separated regions. In particular, as noted earlier in the chapter, the operative time scale is given by $t_{\text{diffusive}} = x^2/2D$, where D is the diffusion constant and x is the distance over which the diffusion has taken place.

To get a sense for all of the time scales described in this section, Table 39.2 presents a variety of results con-

sidered above in tabular form, but now with all times measured in units of the time it takes for ATP synthase to make one rotation ($\approx 10\,\text{ms}$). Our reason for adopting ATP synthase rotation as defining a unit of time is that it exhibits motions associated with one of the most important of life's processes, namely the synthesis of new ATP molecules. In addition, our later discussion of units associated with both energy and power will once again appeal to the central role played by ATP in biological nanotechnology.

39.2.3 Force and Energy Scales: The Interplay of Deterministic and Thermal Forces

Another set of scales of great relevance and importance to the nano–bio interface concerns the nature of the forces that act in this setting. Thus far, we have examined the spatial extent and the cycle time of a variety of examples of biological nanotechnology. As a next step in our examination of scales, we consider the forces and energies associated with these structures. Two outstanding sources for developing a feel for the relevant numbers are the books of *Wrigglesworth* [39.42] and *Smil* [39.43]. Perhaps the most compelling feature when thinking about forces at these scales is the interplay between deterministic and thermal forces. To substantiate this claim, we note that at room temperature the fundamental energetic quantity is $k_{\text{B}}T \approx 4.1\,\text{pN nm}$. The reason this number is of interest to the current endeavor is the realization that many of the molecular motors that have thus far been investigated act with forces on the piconewton scale over distances of nanometers.

The fact that the piconewton is the relevant unit of force can be gleaned from a simple estimate. Namely, consider a typical skeletal muscle in the human arm. The cross-sectional size of the muscle is of order 3 cm. The muscle consists of cylindrical rods of protein called myofibrils, which are roughly 2 µm in diameter, while the myofibrils themselves are made of strands

Table 39.2 Time scales of various events in biological nanotechnology measured in units of the average time taken for ATP synthase to make a single rotation

Process	Time scale (units of ATPase rotation)	Ref.
1/(frequency of amino acid addition) – eukaryotic ribosome	50	[39.1]
1/(frequency of monomer addition) – RNA polymerase	10	[39.39]
Time between motion reset of *E. coli*	10	[39.6]
Kinesin step	1	[39.26]
Diffusion time for protein (ion) to cross an *E. coli*	0.5 (0.02)	[39.11]
Vibrational period of nanocantilever	$10^{-3} - 10^{-5}$	[39.40]

of actin and myosin filaments, which total some 60 nm in diameter [39.1]. This gives 10^{12} myosin filaments per cross-sectional area of the muscle. As each myosin filament over the length of a single sarcomere (the contractile unit of myofibrils, some 2.5 μm long) contains some 300 myosin heads, lifting a 30 kg load corresponds to a force of 1 pN per myosin head. This is certainly an underestimate, since not all myosin heads are attached to the actin filament at the same time. Since our estimate leaves out many details, we might wonder how it compares to the measured forces exerted by molecular motors. Sophisticated experiments with optical tweezers have revealed that actin–myosin motors stall at a force of around 5 pN [39.44]; RNA polymerase stalls at a force of about 20 pN [39.39]; the portal motor of the ϕ-29 bacteriophage exerts forces of up to 50 pN [39.29].

As noted earlier, RNA polymerase is a molecular motor that moves along DNA while transcribing genes into messenger RNA. The DNA itself is an elastic object that is deformed by forces exerted on it by various proteins. At forces less than 0.1 pN, its response is that of an entropic spring with a stretch modulus of 0.1 pN, while at forces exceeding 10 pN its stretch modulus is determined by hydrogen bonding of the base pairs and is roughly 1000 pN [39.45]. All of the abovementioned data reinforces the argument that the piconewton is the relevant unit of force in the nanomechanical world of the cell.

The concept of "stress" is closely related to that of force. Stress is a continuum mechanical concept of force per unit area, and it has been used with great success in solid and fluid mechanics. We extend the idea of stress to the nanomechanical level to see what numbers we arrive at. From the data above we can deduce that a single myosin fiber sustains a stress of about 10^{-2} pN/nm^2, which is the same as 10^{-2} MPa. Migratory animal cells such as fibroblasts, which are responsible for scavenging and destroying undesirable products in tissues, can generate a maximum stress on their substrate on the order of 3.0×10^{-2} MPa [39.46]. DNA can sustain stresses in excess of 20 pN/nm^2, or 20 MPa, at which point an interesting structural transition accompanied by an overall increase in contour length is observed [39.45]. Engineering materials such as steel and aluminium, on the other hand, can sustain stresses of about 100 MPa. This goes to show that nanotechnology in the context of biological systems is built from rather soft materials.

It is of interest to translate our intuition concerning piconewton forces into corresponding energetic terms. The kinesin motor advances 8 nm in each step against forces as high as 5 pN [39.25]. This translates to a work done on the order of 40 pN nm. A myosin motor suffers displacements of about 15 nm with forces in the piconewton range. ATP hydrolysis (to ADP) releases energy (at pH 7 and room temperature) of about 50 pN nm [39.1]. When a titin molecule is pulled, it unfolds under forces of 30–300 pN, causing discrete expansions of 10–30 nm [39.19], implying energies in the range of 300–9000 pN nm. Experimental data provides a compelling argument in favor of thinking of the pN nm as a unit of energy for nanomechanics. However, the observation that $k_B T = 4.1$ pN nm at room temperature gives important insight, since it reveals that thermal forces and entropic effects play a competing role in biological nanomechanics. This provides nature with unique design challenges, whereby molecular motors that can perform useful work must do so in the presence of strong thermal fluctuations for the normal functioning of the cell. Operation of motors in such a noisy environment is governed by laws that are probabilistic in nature. This is apparent in single molecule experiments that observe motors stalling and sometimes reversing direction.

Energy conversion is crucial to any developmental or evolutionary process. The steam engine powered the industrial revolution. However, long before thinking beings with man-made machines founded the industrial revolution, power generation had already become a central part of life's nanotechnology. Indeed, the development of ATP synthase is one of the cornerstones of the evolution of higher life-forms. An inevitable concomitant of evolution is the necessity for faster and more efficient operations. The industrial revolution led to the emergence of bigger and faster modes of transport; biological evolution led to the emergence of complex and intelligent organisms. Invariably, a machine or an organism is limited in its abilities by the speed at which it can convert one form of energy (usually chemical) to other forms of energy (usually mechanical). This is why studying their "power plants" becomes important.

Power plants (or engines) are usually characterized by their force–velocity curves and compared using their power-to-weight (P/W) ratios. For example, the myosin motor has a P/W ratio of 2×10^4 W/kg, the bacterial flagellar motor stands at 100 W/kg, an internal combustion engine is at about 300 W/kg, and a turbojet engine stands at 3000 W/kg [39.47]. These figures tell us that linear motors like myosin and kinesin are extremely powerful machines.

39.3 Modeling at the Nano–Bio Interface

We have already provided a number of different views of the biology–nanotechnology interface, all of which reveal the insights that can emerge from model building. Indeed, one of the key thrusts of this entire chapter is the view that as the type of data that emerges concerning biological systems becomes increasingly quantitative, it must be responded to with models that are also quantitative [39.30]. The plan of this section is to show how atomistic and continuum analyses each offer insights into problems of nanotechnological significance, but under some circumstances both are found wanting, and it is only through a synthesis of both types of models that certain problems will surrender. The plan of this section is to examine the advantages and difficulties associated with adopting both atomistic and continuum perspectives and then to hint at the benefits of seeking mixed representations.

39.3.1 Tension Between Universality and Specificity

One of the key insights concerning model building in nanomechanics, whether we are talking about the nanoscale tribological questions pertinent to magnetic recording or the operations of molecular motors, is that such questions live in the no-man's land between traditional continuum analysis at one extreme and all-atom approaches such as molecular dynamics on the other. Indeed, there is much discussion about the breakdown of continuum mechanics in modeling the mechanics of systems at the nanoscale. This dichotomy between continuum theories, which treat matter as continuously distributed, and atomic-level models, which explicitly acknowledge the graininess of matter, can be restated in a different (and perhaps more enlightening) way. In particular, it is possible to see atomistic and continuum theories as offering complementary views of the same underlying physics. Continuum models are suitable for characterizing those features of a system that can be thought of as averages over the underlying microscopic fluctuations. By way of contrast, atomistic models reveal the details that a continuum model will never capture, and in particular, they shed light on the specificity of the problem at hand.

The perspective adopted here is that continuum models and atomistic models each reveal important features of a given problem. For example, in contemplating the competition between fracture and plasticity at crack tips, a continuum analysis provides critical insights into the nature of the elastic fields surrounding a defect such as a crack. These fields adopt a fundamental and universal form at large scales with all detailed material features buried in simple material parameters. By way of contrast, the precise details of the dissipative processes occurring at a crack tip (in particular, the competition between bond breaking with the creation of new free surface and dislocation nucleation) require detailed atomic-level descriptions of the energetics of bond stretching and breaking. In the biological setting, similar remarks can be made. In certain instances, the description of biological polymers as random coils suffices and yields insights into features such as the mean size of the polymer chain as a function of its length. On the other hand, if our objective is mechanistic understanding of processes such as how phosphorylation of a particular protein induces conformational change, this is an intrinsically atomic-level question. The language we invoke to describe this dichotomy is the use of the terms *universality* and *specificity*, where, as described above, insights of a universal character refer to those features of systems that are generic, while specificity refers to the features of systems that depend upon precise details such as whether or not a particular molecule is bound at a particular site. This fundamental tension between atomistic and continuum perspectives is elaborated in *Phillips* et al. [39.48].

39.3.2 Atomic-Level Analysis of Biological Systems

As already described in the Introduction, one of the intriguing roles of nanotechnology in the biological setting is that it has brought the Instron technology of traditional solid mechanics to the nanoscale and has permitted the investigation of the force–extension characteristics of nanoscale systems (macromolecules and their assemblies in particular). As noted earlier, mechanical force spectroscopy [39.21] is emerging as a profound tool for exploring the connection between structure, force, and chemistry, in much the same way that conventional stress–strain tests provided insights into the connection between structure and properties of conventional materials. Figure 39.3 gives one such example in the case of the muscle protein titin. Similar insights have been obtained through systematic examination of the force–extension properties of DNA [39.49].

The objective of this section is to call the reader's attention to the types of modeling that can be done from an all-atom perspective. What exactly is meant by a model in this setting? We begin by noting that for the purposes of the general discussion given here, the same basic ideas are present whether one is modeling tribological processes such as the sliding of adjacent surfaces, or attempting to examine the operation of a protein machine such as ATP synthase. The set of degrees of freedom considered by the atomic-scale modeler is the full set of atomic positions $(R_1, R_2, \cdots R_N)$, which we also refer to as $\{R_i\}$. For most purposes, one proceeds through reference to a *classical* potential energy function, $E_{tot}(\{R_i\})$, which is a rule that assigns an energy for every configuration $\{R_i\}$. While it would be most appealing to be able to perform a full quantum mechanical analysis, such calculations are computationally prohibitive. Given the potential energy, the forces on each and every atom can be computed where, for example, the force in the αth Cartesian direction on the ith atom is given by

$$F_{i,\alpha} = -\frac{\partial E_{tot}}{\partial R_{i,\alpha}}. \qquad (39.8)$$

For those interested in finding the energy minimizers, such forces can be used in conjunction with methods such as the conjugate gradient method or the Newton–Raphson method. Alternatively, many questions are of a dynamic character, and in these cases Newton's equations of motion are integrated, thus permitting an investigation of the temporal evolution of the system of interest. We note that we have neglected to discuss subtleties of how one maintains the system at constant temperature, and we leave such subtleties to the curious reader, who can learn more about them in *Frenkel* and *Smit* [39.50].

To give a flavor of where such calculations can lead we note that the same force–extension characteristics already shown in Fig. 39.3 have been computed in a molecular dynamics simulation [39.51]. One of the insights to emerge from these calculations was the particular dynamical pathway, namely the breaking of a particular collection of hydrogen bonds during the rupture process of each of the immunoglobulin domains. We further note that in the case of titin there has been an especially pleasing synergy between the atomic-scale calculations and the corresponding force–extension measurements. In particular, in response to the suggestion that it was a particular set of hydrogen bonds that were impugned in the rupture process, mutated versions of the titin protein were created in which the number of such hydrogen bonds was changed with the result that the rupture force was changed according to expectation [39.52]. This can be seen as a primitive example of the ultimate goal of tailoring new materials (both biological and otherwise) through appropriate computer modeling.

39.3.3 Continuum Analysis of Biological Systems

We have already noted that there are many appealing features to strictly continuum analyses of material systems. In particular, models based upon continuum mechanics result in a mathematical formulation that permits us to uncork the traditional tools associated with partial differential equations and functional analysis. In keeping with our argument that it is the role of models to serve our intuition, the ability to write down continuum models raises the possibility of obtaining analytic solutions to problems of interest, an eventuality that is nearly impossible once the all-atom framework has been adopted.

Macromolecules as Elastic Rods

Whether we contemplate the information carrying nucleic acids, the workhorse proteins, or energy storing sugars, ultimately the molecular business of the living world is dominated by long-chain molecules. For the model builder, such molecules suggest two complementary perspectives, each of which contains a part of the whole truth. On the one hand, polymer physicists have gained huge insights [39.53] by thinking of long-chain molecules as random walks. Stated simply, the key virtue of the random walk description of long-chain molecules is that it reflects the overwhelming importance of entropy in governing the geometric conformations that can be adopted by macromolecules. The other side of the same coin considers long-chain molecules as elastic rods with a stiffness that governs their propensity for bending. We will take up this perspective in great detail in the final section of this chapter when we consider the energetics of DNA packaging in viruses.

For our present purposes, we examine one case study in treating macromolecules as elastic rods. The example of interest here is chosen in part because it reflects another fascinating aspect of biological nanotechnology, namely, regulation and control. It is well-known that genes are switched on and off as they are needed. These topics are described beautifully in the work of *Ptashne* [39.54]. The basic idea can be elucidated through the example of a particular set of genes in *E. coli*: the *lac* operon. There are a set of enzymes exploited by this bacterial cell when it needs to digest the sugar lactose. The gene that codes for these enzymes is only turned on when lactose is present and certain other

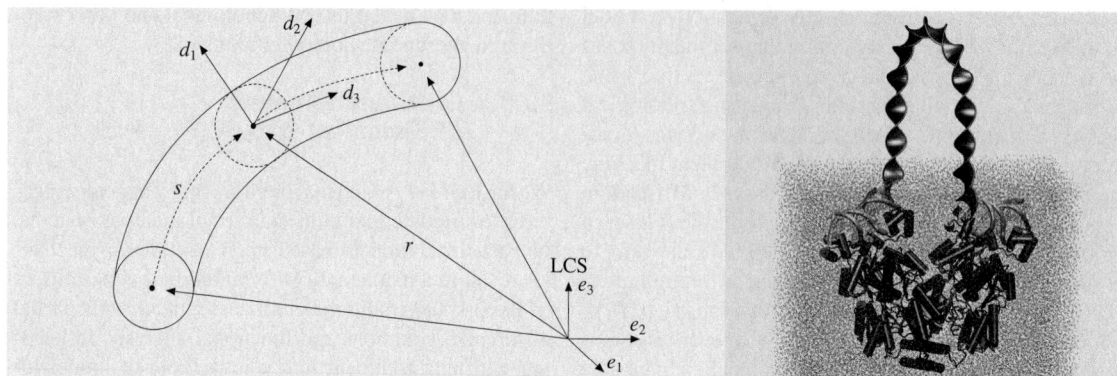

Fig. 39.6 Illustration of the way that the properties of a particular DNA–protein binding interaction were modeled by virtue of an elastic treatment of the looped region of the DNA [39.48]

sugars are absent. The nanotechnological solution to the problem of regulating this gene has been solved by nature through the presence of a molecule known as the *lac* repressor, which binds onto the DNA in the region of this *lac* gene and prevents the gene from being expressed. More specifically, the *lac* repressor binds onto two different regions of the DNA molecule simultaneously, forming a loop in the intervening region, and thus rendering that region inoperative for transcription.

We have belabored the difficulties that attend the all-atom simulation of most problems of nanotechnological interest – system sizes are too small and simulated times are too short. On the other hand, it is clearly of great interest to probe the atomic-level dynamics of the way that molecules such as *lac* repressor interact with DNA and thereby serve as gene regulators. A compromise position adopted by *Balaeff* et al. [39.55] was to use elasticity theory to model the structure and energetics of DNA, which then served effectively as a boundary condition for their all-atom calculation of the properties of the *lac* repressor/DNA complex. The simulation cell is shown in Fig. 39.6, which also shows the looped region of DNA that serves as a boundary condition for the atomic-level model of the complex.

Membranes as Elastic Media

One way to classify biological structures is along the lines of their dimensionality. In the previous discussion, we examined the sense in which many of the key macromolecules of the living world can be thought of as one-dimensional rods. The next level in this dimen-

sional hierarchy is to consider the various membranes that compartmentalize the cell and which can be examined from the perspective of two-dimensional elasticity. Just as there are a huge number of modeling questions to be posed concerning the structure and function of long-chain molecules such as nucleic acids and proteins, there is a similar list of questions that attend the presence of a host of different membranes throughout the cell. To name a few, we first remind the reader that such membranes are full of various proteins that serve in a variety of different capacities, some of which depend upon the mechanical state of the membrane itself. In a different vein, there has been great interest in examining the factors that give rise to the equilibrium shapes of cells [39.56, 57]. In both instances, the logic associated with building models of these phenomena centers on the construction of an elastic Hamiltonian that captures the energetics of deformations expressed in terms of surface area, curvature, and variations in thickness. Also, just as there has been great progress in measuring the properties of single proteins and nucleic acids, there has been considerable progress in examining the force response of membranes as well [39.58].

The prospect of bringing the tools of traditional continuum theory to bear on problems of biological significance is indeed a daunting one. As noted earlier, to our way of thinking one of the biggest challenges posed by biological problems is the dependence of these systems on detailed molecular structures and dynamics – what we have earlier characterized as "specificity", one of the hallmarks of biological action.

39.4 Nature's Nanotechnology Revealed: Viruses as a Case Study

Over the course of this chapter, we have presented different facets of the relationship between biology and nanotechnology and the modeling challenges they encompass. For example, we have noted that two key aspects of nature's nanotechnologies are the exploitation of self-assembly processes for the construction of molecular machines and the role of active processes mediated by molecular motors. We have also argued for the synergistic role of nanotechnology in producing new methods for the experimental analysis of biological systems with examples ranging from new molecular dyes to the use of optical tweezers. Finally, we have described some of the modeling challenges posed by contemplating the biology–nanotechnology interface and the sorts of coarse-grained models that have arisen to meet these challenges. In this final section, we present a discussion of viruses as a case study at the confluence of these different themes.

Though the importance of viruses from a health perspective are well-known even to the casual observer, they are similarly important from both the technological perspective (as will be explained below) and as a compelling and profound example of nature's nanotechnology par excellence. To appreciate the sense in which viruses serve as a compelling example of biological nanotechnology, we begin by reviewing the nature of the viral life cycle with special reference to the case of bacterial viruses (viruses that infect bacteria), known as bacteriophages. For concreteness, we consider the life cycle of a bacteriophage such as the famed lambda phage that infects *E. coli*. The life cycle of such a virus is shown schematically in Fig. 39.7. Upon an encounter with the *E. coli* host, the virus attaches to a receptor (protein) embedded in the bacterial membrane and ejects its DNA into the host cell, leaving an empty capsid as refuse from the process. As an aside, it is worth noting that experiments such as the famed Hershey–Chase experiment used tagged DNA and tagged proteins on viruses to settle the question of whether proteins or nucleic acids are the carriers of genetic information. The outcome of these experiments was the conclusion that DNA is the genetic material. For the purposes of the present discussion, the other interesting outcome of the Hershey–Chase experiment is that it provides insights into the mechanistic process associated with delivery of the viral genome.

Once the viral genome has been delivered to the host cell (we now oversimplify), the replication machinery of *E. coli* is hijacked to do the virus' bidding. In particular, the genes coded for in the viral DNA are

expressed, and the proteins needed to make copies of the phage particle (the ingredients for an eventual self-assembly process) are created. Interestingly, part of the gene products associated with this process are the components of the molecular motor, which is responsible for packaging the replicated viral DNA into the new protein capsids that will eventually become the next generation of viruses. Indeed, once assembled, this motor takes the replicated viral DNA and packs it into the viral capsid. Recall our insistence that another of the important themes presented by biological examples of nanotechnology is the huge role played by active processes that reflect mechanochemical coupling. Once the packaging process is completed, and the remaining parts of the self-assembly process have been effected (such as the attachment of the viral sheath and legs whereby the phage attaches to host cells), enzymes are released that breakdown the cell wall of the infected cell with the ultimate result that what started as a single bacteriophage has in

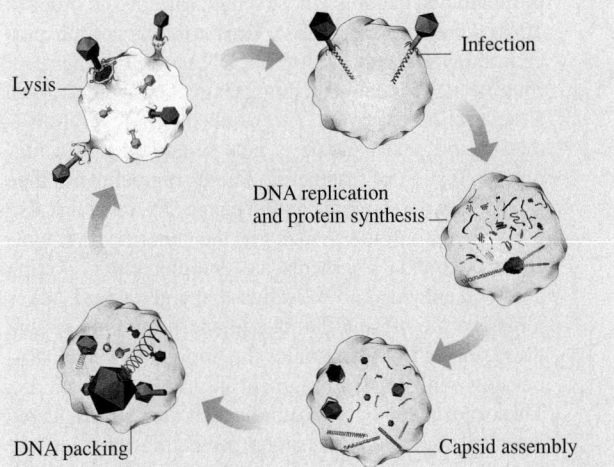

Fig. 39.7 Schematic representation of the viral life cycle illustrating the various nanotechnological actions in this cycle. During the infection stage, the viral genome is delivered to the victimized bacterial cell. During the process labeled "DNA replication and protein synthesis", the machinery of the host cell is enlisted to produce renegade DNA and proteins to be used in the assembly of new viral particles. The capsid assembly process involves the attachment of the various protein building blocks that make up the capsid. "DNA packing" refers to the active packaging of DNA into the newly formed capsids by a molecular motor. Finally, once the assembly process is complete, the infectious phage particles are released from the victimized cell to repeat their infectious act elsewhere

less than an hour become on the order of 100 new phage particles ready to infect new cells.

As noted above, a second sense in which viruses are deserving of case study status in the current chapter is the role viruses such as lambda phage play in biotechnology itself. For example, from the standpoint of both cloning and the construction of genomic libraries, the use of viruses is commonplace. From a more speculative perspective, viruses are also drawing increasing attention from the standpoint of gene therapy (as a way of delivering DNA to specific locations) and, more generally, as small-scale containers. To substantiate these assertions, we briefly consider the construction of genomic libraries and the role played by viruses in these manipulations. We pose the following question: Given that the length of the human genome is on the order of 10^9 base pairs, how can one organize and store all of this genetic information for the purposes of experiments such as the sequencing of the human genome? One answer to this challenge is the use of a bacteriophage to deliver phage DNA to *E. coli* cells, but with the subtlety that the delivered DNA fragments have ligated within them fragments of the human genome whose lengths are on the order of 10 kb (kb = kilo base-pairs). The particulars of this procedure involve first cutting the DNA into fragments of roughly 10 kb in length using a class of enzymes known as restriction enzymes. The result of this operation is that the original genome is now separated into a random collection of fragments. These fragments are then mixed with the original lambda phage DNA that has also been cut at a single site such that the genomic fragments and the lambda fragments are complementary. Using a second enzyme known as ligase, the genomic DNA is joined to the original lambda phage fragments so that the resulting DNA resembles the original lambda DNA, but now with a 10 kb fragment inserted in the middle. These cloned lambda DNA molecules are then packaged into the lambda phage using a packaging reaction (see the website www.stratagene.com for an example of these products). The resulting lambda phage, now fully packed with cloned DNA, are used to infect *E. coli* cells, and the cloned DNA, once in the bacterial cell, circularizes into a DNA fragment known as a plasmid. The plasmid is then passed from one generation of *E. coli* to the next. This is the so-called prophage pathway in which, unlike the lytic pathway shown in Fig. 39.7, the bacteriophage is latent and does not destroy the cell [39.1]. Note that different *E. coli* cells are infected with viruses containing different cloned fragments. As a result, the collection of all such *E. coli* cells constitute a library of the original genomic DNA.

We round off our introductory discussion on viruses with a discussion of the compelling recent experiments that have been performed to investigate the problem of viral packing and, similarly, how ideas like those described in this chapter can be used to model these processes. *Smith* et al. [39.29] used optical tweezers to measure the force applied by the packaging motor of the ϕ-29 bacteriophage during the DNA packaging process. In particular, they measured the force and rate of packaging as a function of the amount of DNA packed into the viral head with the result that as more DNA is packed, the resistive force due to the packed DNA increases, and the packing rate is reduced.

As noted above, the viral problem is interesting not only because it exemplifies many of the features of biological nanotechnology introduced throughout this chapter, but also because it illustrates the way in which model building has arisen in response to experimental insights. The various competing energies that are implicated in the DNA packing process have been described by *Riemer* and *Bloomfield* [39.59]. The energetics of viral packing is characterized by a number of different factors, including: (1) the entropic-spring effect that causes the DNA in solution to adopt a more spread out configuration than that in the viral capsid, (2) the energetics of elastic bending, which results from inducing curvature in the DNA on a scale that is considerably smaller than the persistence length of $\xi_p \approx 50$ nm, and (3) those factors related to the presence of charge both on the DNA itself and in the surrounding solution. As shown by *Riemer* and *Bloomfield* [39.59], the entropic contribution is smaller (by a factor of ten or more) than the bending energies and those mediated by the charges on the DNA and the surrounding solution, and hence we make no further reference to it. As a result, just as in earlier work [39.60, 61], we examine the interplay of elastic and interaction forces, although we neglect surface terms originating from DNA–capsid and DNA–solvent interactions.

We note that the viral packing process involves DNA segments with lengths on the order of 10 μm and that it takes place on a time scale of minutes. As a result, from a modeling perspective, it is clear that such problems are clearly beyond the reach of conventional molecular dynamics. As a result, we exploit a continuum description of the DNA packing process with the proviso that such models will ultimately need to be refined to account for the sequence dependence of the elasticity of DNA.

A mathematical description of the energetics of viral packing must account for two competing factors, namely the energy cost of bending the DNA and placing it in the capsid and the repulsive interaction between adjacent DNA segments that are too close together. The structural picture of the packaged DNA is inspired by experiments that indicate that the DNA is packed in concentric rings from large radii inwards [39.62–64]. The bending energy cost of accumulating hoops within the capsid is given by

$$E_{el} = \pi \xi_p k_B T \sum_i \frac{N(R_i)}{R_i} \,, \tag{39.9}$$

where $N(R_i)$ is the number of hoops that are packed at the radius R_i [39.59]. The basic idea of this expression is that we are adding up the elastic energy on a hoop-by-hoop basis, with each hoop penalized by the usual energy cost of bending an elastic beam into a circular arc. The presence of $N(R_i)$ reflects the fact that, due to the shape of the capsid, the DNA can pack higher up into the capsid as the radius gets smaller, thus increasing the number of allowed hoops. To make analytic progress with the expression for the stored elastic energy given above, we convert it into an integral of the form

$$E_{el} = \frac{\pi \xi_p k_B T}{\sqrt{3} d_s/2} \int_R^{R_{out}} \frac{N(R')}{R'} dR' \,. \tag{39.10}$$

The summation $\sum_i$ has been replaced by an integral $\int_R^{R_{out}} dR'/(\sqrt{3} d_s/2)$, where the integration bounds are the inner and outer radii of the inner spool and $\sqrt{3} d_s/2$ is the horizontal spacing between adjacent strands of the DNA. The geometrical factor $\sqrt{3}/2$ owes its presence to the hexagonal packing of the DNA strands. R is the radius of the innermost stack of hoops, R_{out} is the radius of the capsid, $N(R')$ is the number of hoops at radius R', d_s is the spacing between adjacent hoops, and ξ_p is the persistence length of DNA. We note that this expression is a manifestation of the first term in the energy functional in (39.3), used earlier in the context of the biofunctionalized cantilever and now specialized to the geometry of a partially filled capsid. Just as the energy functional in the cantilever context reflected a competition between different contributions to the total energy (in that case, the elastic bending energy and surface terms), the DNA packing problem reflects a similar competition, this time between the elastic bending and the interactions between nearby DNA segments by virtue of the charge along the DNA. In particular, the interaction energy between the DNA hoops packed in the viral capsid scales with the length and is given by

$$E_{int} = F_0 L(c^2 + c d_s) \exp(-d_s/c) \,, \tag{39.11}$$

where c and F_0 are constants for a given solvent and L is the length of the packed DNA. This form of the interaction energy was shown to be very robust for a variety of solvents containing monovalent and divalent cations in a series of experiments by *Rau* et al. [39.65, 66] and *Parsegian* et al. [39.67]. The forces arising from this kind of an energy are purely repulsive in character. Interestingly, in this setting, rather than using temperature as a control parameter, it is much more common to change the circumstances in such experiments by tuning concentrations of various chemical constituents. In particular, when trivalent or tetravalent cations are present in solution there is an effective attractive interaction between DNA segments. In that case, the forces are repulsive only when d_s is smaller than a certain critical separation d_0 (and attractive otherwise), and the form of the interaction energy changes to

$$E_{int} = F_0 L(c^2 + c d_s) \exp\left(\frac{d_0 - d_s}{c}\right) \,. \tag{39.12}$$

Once both the elastic and interaction contributions to the energy have been reckoned, we are in a position to compute the total energy. If we recognize that the length L of the packed DNA is given by

$$L = \frac{2}{\sqrt{3} d_s} \int_R^{R_{out}} 2\pi R' N(R') dR' \,, \tag{39.13}$$

then the total energy for the packed DNA, in terms of L and d_s, is given by

$$E(L, d_s) = \frac{\pi \xi_p k_B T}{\sqrt{3} d_s/2} \int_R^{R_{out}} \frac{N(R')}{R'} dR' \tag{39.14}$$

$$+ F_0(c^2 + c d_s) \exp(-d_s/c) \frac{2}{\sqrt{3} d_s} \int_R^{R_{out}} 2\pi R' N(R') dR' \,.$$

The spacing at a given length L is determined by requiring that the packed DNA be in a minimum energy configuration, which is equivalent to asking that $\partial E/\partial d_s = 0$. Hence, for a given geometry and length packed, one obtains d_s and thereby calculates the energy. Thus the energy is known as a function of the length packed, and to compute the force as a function of length of DNA packed one need only differentiate this energy with respect to the length packed. The result of this procedure is an expression for the force of the form

$$F(L) = F_0(c^2 + c d_s) \exp\left(-\frac{d_s}{c}\right) + \frac{\xi_p k_B T}{2R^2} \,. \tag{39.15}$$

This is a generic expression valid for a capsid of any shape. The effect of the geometry is captured in the variation of d_s as a function of the length packed, as well as through the inner packing radius R. The resulting force–packing curves are shown in Fig. 39.8. In particular, this figure shows the packing force as a function of a fraction of the genome packed for a number of different solutions, with the different curves revealing the large role played by positive counterions in dictating the overall energetics of these processes.

On the basis of calculations like those given above, *Kindt* et al. [39.61] estimate that the pressure inside the capsid of the phage is on the order of 35 atmospheres. A crude estimate by *Smith* et al. [39.29] gives a figure close to 60 atmospheres. These numbers are intriguing in their own right, but more importantly they demonstrate the promise that proteins and other biological materials hold as candidates for engineering materials. To further explore the structural integrity of viruses in their role as pressurized protein shells, we now turn to an examination of the rupture stress of viral capsids.

Throughout this chapter, one of our main arguments has been the idea that whether it is our ambition to model conventional materials or their biological counterparts, modeling at the nanoscale implies challenges that cannot be met either by purely continuum ideas or by traditional all-atom thinking. Indeed, one set of powerful methods is built around the attempt to borrow those features of continuum and atomistic thinking that are most robust, while rejecting those that are inapplicable at the atomic scale. As an example of this type of thinking, we examine another question drawn from the viruses-as-nanotechnology setting. In particular, since viruses have been considered a means of transporting material other than the genetic material of the virus itself, it is of interest to consider the maximum internal pressure viral capsids can sustain without rupture, as this will determine how much material can be safely packaged. To that end, we use continuum mechanics to estimate the stresses within the capsid walls. These stresses are then mapped onto atomic-level forces by appealing to the details of the protein structure of the monomers making up the capsid and a knowledge of the forces that link them. By relating the continuum and atomistic calculations, we then determine the maximum sustainable internal pressure.

We imagine the capsid to be a hollow sphere loaded by a pressure p_i from inside and a pressure p_o from outside. The inner and outer radii are R_i and R_o, respectively. As a representative example, bacteriophage GA is characterized geometrically by $R_i = 123\,\text{Å}$ and $R_o = 145\,\text{Å}$. Evaluation of a number of different capsid types suggests that treating capsids as though they have a mean thickness of roughly $15\,\text{Å}$ suffices for the level of modeling being considered here. For the purposes of computing the internal stresses within the capsid, we begin with a statement of equilibrium from continuum mechanics that requires that at every point in the capsid

$$\nabla \cdot \sigma = \mathbf{0}, \tag{39.16}$$

where σ is the stress tensor comprising three normal stresses and three shear stresses. For a problem with spherical symmetry, like that considered here, the stresses reduce to a radial stress σ_R and a circumferential stress σ_T. Solution of the equilibrium equations results in stresses of the form [39.32]

$$\sigma_R = \frac{C}{r^3} + D, \quad \sigma_T = -\frac{C}{2r^3} + D. \tag{39.17}$$

By using the boundary conditions $\sigma_R|_{r=R_i} = -p_i$ and $\sigma_R|_{r=R_o} = -p_o$, the constants C and D can be determined with the result

$$\sigma_R = \frac{p_o R_o^3 \left(r^3 - R_i^3\right)}{r^3 \left(R_i^3 - R_o^3\right)} + \frac{p_i R_i^3 \left(R_o^3 - r^3\right)}{r^3 \left(R_i^3 - R_o^3\right)}, \tag{39.18}$$

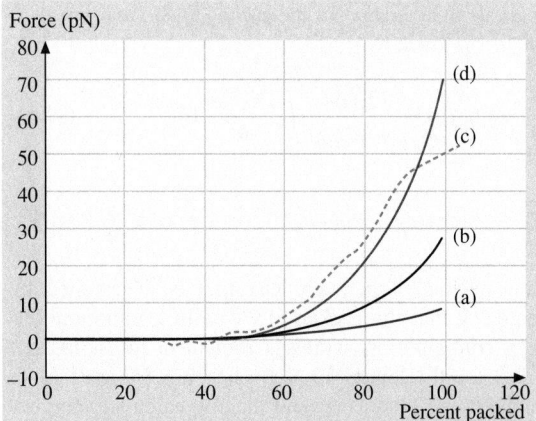

Fig. 39.8 Force as a function of percent of genome packed for a cylindrical capsid under purely repulsive solvent conditions. The dimensions of the capsid and the length of the genome packed were chosen to represent the $\phi 29$ phage: $R_{out} = 210\,\text{Å}$, $z = 540\,\text{Å}$, and $L = 6.6\,\mu\text{m}$. Curve (c) shows the experimental results of *Smith* et al. [39.29], while the theoretical curves (a), (b), and (d) are given by (39.15), with $F_0 = 10 \times 55\,000$, $40 \times 55\,000$, and $100 \times 55\,000\,\text{pN/nm}^2$, respectively

$$\sigma_T = \frac{p_o R_o^3 \left(2r^3 + R_i^3\right)}{2r^3 \left(R_i^3 - R_o^3\right)} - \frac{p_i R_i^3 \left(2r^3 + R_o^3\right)}{2r^3 \left(R_i^3 - R_o^3\right)} . \quad (39.19)$$

The stress σ_T is our primary concern, since it acts so as to tear the sphere apart. By looking at the expressions above, we can see that this stress is maximum at $r = R_i$ and the maximum value is given by

$$\sigma_T^{max} = \frac{3 p_o R_o^3 - p_i \left(2R_i^3 + R_o^3\right)}{2 \left(R_i^3 - R_o^3\right)} . \quad (39.20)$$

We note that elasticity theory itself is unable to comment on σ_T^{max}, since this is effectively a material parameter that characterizes the contacts between the various protein monomers that make up the capsid. As a result, we first examine how the rupture strength depends upon capsid dimensions in abstract terms and then turn to a concrete estimate of σ_T^{max} itself from several complementary perspectives. If (39.20) is rewritten using $p_i = p_o + \Delta p$ and $R_o = R_i + \Delta R$, and, further, it is realized that for typical capsid dimensions $\Delta R / R_i \ll 1$, it can be shown by rearranging (39.20) and by considering the case in which σ_T^{max} is much larger than p_o, that the maximum sustainable internal pressure is given by

$$p_i^{max} = \frac{2 \sigma_T^{max} \Delta R}{R_i} , \quad (39.21)$$

where p_i^{max} is really the quantity of interest, namely, the maximum sustainable internal pressure.

To make concrete progress to the point where we can actually estimate the rupture stress in atm, we need to consider capsids for which the structure is known and for which, at least approximately, the bonds between the monomers making up the capsid are understood. We note that in the language of fracture mechanics, what we seek is a cohesive surface model that provides a measure of the energy of interaction between two surfaces as a function of their separation [39.68]. There are a number of different ways to go about estimating the effective interaction between the monomers making up the capsid, one of which is by appealing to atomic-level calculations like those made by *Reddy* et al. [39.69]. The Viper website [39.70] has systematized such information for a number of capsids, and one of the avenues we take to estimate σ_T^{max} is to appeal directly to their calculations. To that end, we assume that the energy of interaction per unit area as a function of separation x between adjacent monomers making up the capsid can be written in the form

$$E(x) = V_0 \left[\frac{1}{4} \left(\frac{x^*}{x} \right)^{12} - \frac{1}{2} \left(\frac{x^*}{x} \right)^6 \right] . \quad (39.22)$$

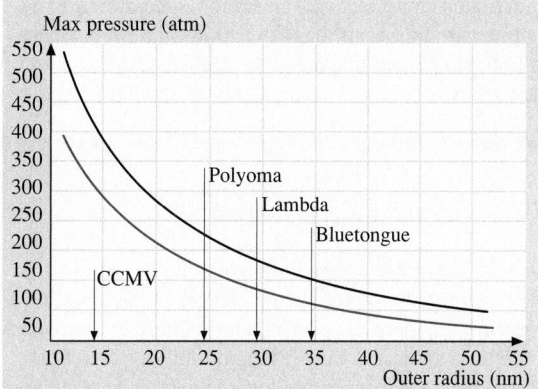

Fig. 39.9 Rupture pressure as a function of capsid radius for $x^* = 3$ Å (*upper curve*) and $x^* = 4$ Å (*lower curve*). The width of the capsid walls was set to $\Delta R = 1.5$ nm, while $V_0 = 180$ cal/molÅ^2

The motivation for this functional form is the idea that the energy of interaction between adjacent monomeric units making up the capsid is the result of van der Waals contacts, and hence our cohesive surface law has inherited the properties of the underlying atomic force fields. To proceed to an estimate of σ_T^{max} itself, we must determine the parameters in the cohesive model described above. To that end, we note that *Reddy* et al. [39.69] have computed the association energies of various inequivalent contacts throughout a number of different icosahedral capsids. Their calculations result in a roughly constant value of ≈ -45 cal/mole Å^2 as the association energy, which in the language of our cohesive potential results in $V_0 \approx 12.5$ pN/Å. This may be seen by noting that the association energy is given by $E(x^*) = -V_0/4$.

Once we have fixed x^*, which amounts to choosing a particular value for the equilibrium separation between two monomers, the material parameter σ_T^{max} is obtained by evaluating $\partial E(x)/\partial x$ at a value of x corresponding to the point of inflection ($\partial^2 E(x)/\partial x^2 = 0$) in the cohesive surface function. For the cohesive surface function used above, this results in a maximum stress of the form

$$\sigma_T^{max} = \frac{7^{7/6} \times 18}{13^{13/6}} \frac{V_0}{x^*} . \quad (39.23)$$

With σ_T^{max} in hand, the maximum sustainable pressure is obtained from (39.21), and the results are shown in Fig. 39.9. These estimates suggest that the pressures within capsids as a result of packed DNA, while large, are still smaller than our estimated rupture stresses. To more completely examine this question we have

undertaken finite element elasticity calculations to examine the stresses in capsids exhibiting irregularities
in both shape and thickness. In addition, further atomistic analysis of σ_T^{max} is needed with special reference
to its dependence on distance between protein units. It
would also be of interest to examine mutant versions of
the monomeric units making up the capsid to see the
implications of such mutations for σ_T^{max}.

39.5 Concluding Remarks

One of the most compelling areas to be touched by
nanotechnology is biological science. Indeed, we have
argued that there is a fascinating interplay between
these two subjects, with biology as a key beneficiary
of advances in nanotechnology as a result of a new
generation of single-molecule experiments that complement traditional assays involving statistical assemblages
of molecules. This interplay runs in both directions,
with nanotechnology continually receiving inspiration
from biology itself. The goal of this chapter has been to
highlight some representative examples of the interplay
between biology and nanotechnology and to illustrate
the role of nanomechanics in this field and how mechanical models have arisen in response to the emergence
of this new field. Primary attention has been given to
the particular example of the processes that attend the
life cycle of bacterial viruses. Viruses feature many of
the key lessons of biological nanotechnology, including
self-assembly, as evidenced in the spontaneous formation of the protein shell (capsid) within which the viral
genome is packaged and a motor-mediated biological
process, namely, the packaging of DNA in this capsid by
a molecular motor that pushes the DNA into the capsid.
We argue that these processes in viruses are a compelling
real-world example of nature's nanotechnology, and they
reveal the nanomechanical challenges that will continue to be encountered at the nanotechnology–biology
interface.

Though this chapter represents something of a departure from the rest of the chapters in this volume, it
advances the view that biological nanotechnology serves
as an inspiring vision of what nanotechnology can offer,
and it reflects the authors' views that biologically inspired nanotechnology will play an ever-increasing role
as Feynman's view that "there is plenty of room at the
bottom" continues to play out. Moreover, developing
simple models of nature's nanotechnology can provide important insights into viable strategies for making
machines at the nanoscale.

We have argued that the progress in single-molecule
technology presents both scientific and technological possibilities. Both classes of questions imply
significant modeling demands, just as earlier advances in other settings, such as in materials for
microelectronics applications, did too. We claim that
these modeling challenges are perhaps more acute
as a result of the vicious chemical and structural
nonhomogeneity of biological systems. As a result, we see an ever-increasing role for modeling
methods that aim to keep atomic-level specificity
where needed, while rejecting such resolution elsewhere.

References

39.1   B. Alberts, D. Bray, A. Johnson, J. Lewis, M. Raff,
        K. Roberts, P. Walter: *Essential Cell Biology* (Gar-
        land, New York 1997) Chap. 12
39.2   B. Alberts: The cell as a collection of protein ma-
        chines: Preparing the next generation of molecular
        biologists, Cell **92**, 291 (1998)
39.3   D. Bray: *Cell Movements From Molecules to Motility*
        (Garland, New York 2001)
39.4   S. J. Flint, L. W. Enquist, R. M. Krug, V. R. Racaniello,
        A. M. Skalka: *Principles of Virology* (ASM, Washing-
        ton, DC 2000)
39.5   J. C. M. van Hest, D. A. Tirrell: Protein-based ma-
        terials, toward a new level of structural control,
        Chem. Commun. **19**, 1807–1904 (2001)
39.6   H. Berg: Motile behavior of bacteria, Phys. Today
        53, 24 (2000)
39.7   J. E. Walker: ATP synthesis by rotary catalysis,
        Angew. Chem. Int. Ed. **37**, 2308 (1998)
39.8   P. D. Boyer: The ATP synthase – A splendid molecu-
        lar machine, Annu. Rev. Biochem. **66**, 717 (1997)
39.9   M. Yoshida, E. Muneyuki, T. Hisabori: ATP synthase
        – A marvelous rotary engine of the cell, Nature Rev.
        Mol. Cell Bio. **2**, 669 (2001)

39.10 R. Yasuda, H. Noji, K. Kinosita, M. Yoshida: F_1-ATPase is a highly efficient molecular motor that rotates with discrete 120° steps, Cell **93**, 1117 (1998)

39.11 J. Howard: *Mechanics of Motor Proteins and the Cytoskeleton* (Sinauer, Sunderland 2001)

39.12 B. Hille: *Ion Channels of Excitable Membranes* (Sinauer, Sunderland 2001)

39.13 S. Sukharev, S. R. Durell, H. R. Guy: Structural models of the MscL gating mechanism, Biophys. J. **81**, 917 (2001)

39.14 S. Sukharev, M. Betanzos, C.-S. Chiang, H. R. Guy: The gating mechanism of the large mechanosensitive channel MscL, Nature **409**, 720 (2001)

39.15 K. Kinosita, R. Yasuda, H. Noji: F_1-ATPase: A highly efficient rotary ATP machine, Essays Biochem. **35**, 3 (2000)

39.16 E. Racker, W. Stoeckenius: Reconstitution of purple membrane vesicles catalyzing light-driven proton uptake and adenosine triphosphate formation, J. Biol. Chem. **249**, 662 (1974)

39.17 G. Groth, J. E. Walker: ATP synthase from bovine heart mitochondria: reconstitution into unilamellar phospholipid vesicles of the pure enzyme in a functional state, Biochem. J. **318**, 351 (1996)

39.18 G. Wu, R. H. Datar, K. M. Hansen, T. Thundat, R. J. Cote, A. Majumdar: Bioassay of prostrate-specific antigen (PSA) using microcantilevers, Nature Biotech. **19**, 856 (2001)

39.19 M. Rief, M. Gautel, F. Oesterhelt, J. M. Fernandez, H. Gaub: Reversible unfolding of individual titin immunoglobulin domains by AFM, Science **276**, 1109 (1997)

39.20 R. P. Feynman: There's plenty of room at the bottom, Eng. Sci. **23**, 22–36 (1960), and www.zyvex.com/nanotech/feynman.html (1959)

39.21 E. Evans: Probing the relation between force-lifetime and chemistry in single molecular bonds, Annu. Rev. Biophys. Biomol. Struct. **30**, 105 (2001)

39.22 T. E. Fisher, P. E. Marszalek, J. M. Fernandez: Stretching single molecules into novel conformations using the atomic force microscope, Nature Struct. Bio. **7**, 719 (2000)

39.23 K. Svoboda, S. M. Block: Biological applications of optical forces, Annu. Rev. Biophys. Biomol. Struct. **23**, 247 (1994)

39.24 C. Bustamante, J. C. Macosko, G. J. L. Wuite: Grabbing the cat by the tail: Manipulating molecules one by one, Nature Rev. Mol. Cell Bio. **1**, 130 (2000)

39.25 K. Svoboda, S. M. Block: Force and velocity measured for single kinesin molecules, Cell **77**, 773 (1994)

39.26 M. J. Schnitzer, S. M. Block: Kinesin hydrolyses one ATP per 8-nm step, Nature **388**, 386 (1997)

39.27 B. Maier, T. R. Strick, V. Croquette, D. Bensimon: Study of DNA motors by single molecule micromanipulation, Single Mol. **1**, 145 (2000)

39.28 A. A. Simpson, Y. Tao, P. G. Leiman, M. O. Badasso, Y. He, P. J. Jardine, N. H. Olson, M. C. Morais, S. Grimes, D. L. Anderson, T. S. Baker, M. G. Rossmann: Structure of the bacteriophage ϕ-29 DNA packaging motor, Nature **408**, 745 (2000)

39.29 D. E. Smith, S. J. Tans, S. B. Smith, S. Grimes, D. L. Anderson, C. Bustamante: The bacteriophage ϕ-29 portal motor can package DNA against a large internal force, Nature **413**, 748 (2001)

39.30 D. Bray: Reasoning for results, Nature **412**, 863 (2001)

39.31 G. Wu, H. Ji, K. Hansen, T. Thundat, R. Datar, R. Cote, M. F. Hagan, A. K. Charkraborty, A. Majumdar: Origin of nanomechanical cantilever motion generated from biomolecular interactions, Proc. Natl. Acad. Sci. USA **98**, 1560 (2001)

39.32 L. D. Landau, E. M. Lifshitz: *Theory of Elasticity* (Pergamon, Oxford 1986)

39.33 D. H. Bamford, R. J. C. Gilbert, J. M. Grimes, D. I. Stuart: Macromolecular assemblies: greater than their parts, Curr. Opin. Struct. Biol. **11**, 107 (2001)

39.34 J. Frank: Single-particle imaging of macromolecules by cryo-electron microscopy, Annu. Rev. Biophys. Biomol. Struct. **31**, 303 (2002)

39.35 S. A. Darst: Bacterial RNA polymerase, Curr. Opin. Struct. Biol. **11**, 155 (2001)

39.36 N. Ban, P. Nissen, J. Hansen, P. B. Moore, T. A. Steitz: The complete atomic structure of the large ribosomal subunit at 2.4 Å resolution, Science **289**, 905 (2000)

39.37 T. S. Baker, N. H. Olson, S. D. Fuller: Adding the third dimension to virus life cycles: Three-dimensional reconstruction of icosahedral viruses from cryo-electron micrographs, Microbiol. Mol. Biol. Rev. **63**, 862 (1999)

39.38 H.-S. Chan, K. A. Dill: The protein folding problem, Phys. Today **46**, 24 (1993)

39.39 M. D. Wang, M. J. Schnitzer, H. Yin, R. Landick, J. Gelles, S. M. Block: Force and velocity measured for single molecules of RNA polymerase, Science **282**, 902 (1998)

39.40 M. L. Roukes: Nanoelectromechanical Systems, Technical Digest of the 2000 Solid-State Sensor and Actuator Workshop (2000)

39.41 K. Adachi, R. Yasuda, H. Noji, H. Itoh, Y. Harada, M. Yoshida, K. Kinosita: Stepping rotation of F_1-ATPase visualized through angle-resolved single-fluorophore imaging, Proc. Natl. Acad. Sci. USA **97**, 7243 (2000)

39.42 J. Wrigglesworth: *Energy and Life* (Taylor and Francis, London 1997)

39.43 V. Smil: *Energies* (MIT, Cambridge 1999)

39.44 J. T. Finer, R. M. Simmons, J. A. Spudich: Single myosin molecule mechanics: Piconewton forces and nanometre steps, Nature **368**, 113 (1994)

39.45 S. B. Smith, Y. Cui, C. Bustamante: Overstretching B-DNA: The elastic response of individual double-stranded and single-stranded DNA molecules, Science **271**, 795 (1996)

39.46   S. Munevar, Y. Wang, M. Dembo: Traction force microscopy of migrating normal and H-ras transformed 3T3 fibroblasts, Biophys. J. **80**, 1744 (2001)

39.47   L. Mahadevan, P. Matsudaira: Motility powered by supramolecular springs and ratchets, Science **288**, 95 (2000)

39.48   R. Phillips, M. Dittrich, K. Schulten: Quasicontinuum representations of atomic-scale mechanics: From proteins to dislocations, Ann. Rev. Mater. Sci. **32**, 219 (2002)

39.49   T. Strick, J.-F. Allemand, V. Croquette, D. Bensimon: Twisting and stretching single DNA molecules, Prog. Biophys. Mol. Bio. **74**, 115 (2000)

39.50   D. Frenkel, B. Smit: *Understanding Molecular Simulation* (Academic, San Diego 1996)

39.51   H. Lu, B. Isralewitz, A. Krammer, V. Vogel, K. Schulten: Unfolding of titin immunoglobulin domains by steered molecular dynamics simulation, Biophys. J. **75**, 662 (1998)

39.52   M. Carrion-Vazquez, A. F. Oberhauser, T. E. Fisher, P. E. Marszalek, H. Li, J. M. Fernandez: Mechanical design of proteins studies by single-molecule force spectroscopy and protein engineering, Prog. Biophys. Mol. Bio. **74**, 63 (2000)

39.53   A. Y. Grosberg, A. R. Khokhlov: *Giant Molecules* (Academic, San Diego 1997)

39.54   M. Ptashne, A. Gann: *Genes and Signals* (Cold Spring Harbor Laboratory Press, Cold Spring Harbor 2002)

39.55   A. Balaeff, L. Mahadevan, K. Schulten: Elastic rod model of a DNA loop in the Lac Operon, Phys. Rev. Lett. **83**, 4900 (1999)

39.56   U. Seifert: Configurations of fluid membranes and vesicles, Adv. Phys. **46**, 13 (1997)

39.57   D. Boal: *Mechanics of the Cell* (Cambridge Univ. Press, Cambridge 2002)

39.58   D. Leckband, J. Israelachvili: Intermolecular forces in biology, Q. Rev. Biophys. **34**, 105 (2001)

39.59   S. C. Riemer, V. A. Bloomfield: Packaging of DNA in bacteriophage heads: Some considerations on energetics, Biopolymers **17**, 785 (1978)

39.60   T. Odijk: Hexagonally packed DNA within bacteriophage T7 stabilized by curvature stress, Biophys. J. **75**, 1223 (1998)

39.61   J. Kindt, S. Tzlil, A. Ben-Shaul, W. Gelbart: DNA packaging and ejection forces in bacteriophage, Proc. Natl. Acad. Sci. USA **98**, 13671 (2001)

39.62   K. E. Richards, R. C. Williams, R. Calendar: Mode of DNA packing within bacteriophage heads, J. Mol. Bio. **78**, 255 (1973)

39.63   M. E. Cerritelli, N. Cheng, A. H. Rosenberg, C. E. McPherson, F. P. Booy, A. C. Steven: Encapsidated conformation of bacteriophage T7 DNA, Cell **91**, 271 (1997)

39.64   N. H. Olson, M. Gingery, F. A. Eiserling, T. S. Baker: The structure of isomeric capsids of bacteriophage T4, Virology **279**, 385 (2001)

39.65   D. C. Rau, B. Lee, V. A. Parsegian: Measurement of the repulsive force between polyelectrolyte molecules in ionic solution: Hydration forces between parallel DNA double helices, Proc. Natl. Acad. Sci. USA **81**, 2621 (1984)

39.66   D. C. Rau, V. A. Parsegian: Direct measurement of the intermolecular forces between counterion-condensed DNA double helices, Biophys. J. **61**, 246 (1992)

39.67   V. A. Parsegian, R. P. Rand, N. L. Fuller, D. C. Rau: Osmotic stress for the direct measurement of intermolecular forces, Meth. Enzymol. **127**, 400 (1986)

39.68   R. Phillips: *Crystals, Defects and Microstructures* (Cambridge Univ. Press, Cambridge 2001)

39.69   V. S. Reddy, H. A. Giesing, R. T. Morton, A. Kumar, C. B. Post, C. L. Brooks, J. E. Johnson: Energetics of quasiequivalence: Computational analysis of protein-protein interactions in icosahedral viruses, Biophys. J. **74** (1998) 546
The parameters can be found at http://www.scripps.edu/pub/olson-web/gmm/autodock/ad305/Using_AutoDock_305.a.html

39.70   V. S. Reddy, P. Natarajan, B. Okerberg, K. Li, K. V. Damodaran, R. T. Morton, C. L. Brooks, J. E. Johnson: Virus Particle Explorer (VIPER), a website for virus capsid structures and their computational analyses, J. Virol. **75** (2001) 11943
The website can be found at The Viper website can be found at http://mmtsb.scripps.edu/viper

40. Structural, Nanomechanical and Nanotribological Characterization of Human Hair Using Atomic Force Microscopy and Nanoindentation

Human hair is a nanocomposite biological fiber. Healthy, soft hair with good feel, shine, color and overall aesthetics is generally highly desirable. It is important to study hair care products such as shampoos and conditioners as well as damaging processes such as chemical dyeing and permanent wave treatments because they affect the maintenance and grooming process and therefore alter many hair properties. Nanoscale characterization of the cellular structure, the mechanical properties, as well as the morphological, frictional and adhesive properties (tribological properties) of hair is essential if we wish to evaluate and develop better cosmetic products, and crucial to advancing the understanding of biological and cosmetic science. The atomic/friction force microscope (AFM/FFM) and nanoindenter have recently become important tools for studying the micro/nanoscale properties of human hair. In this chapter, we present a comprehensive review of structural, mechanical, and tribological properties of various hair and skin as a function of ethnicity, damage, conditioning treatment, and various environments. Various cellular structures of human hair and fine sublamellar structures of the cuticle are identified and studied. Nanomechanical properties such as hardness, elastic modulus, creep and scratch resistance are discussed. Nanotribological properties such as roughness, friction, and adhesion are presented, as well as investigations of conditioner distribution, thickness and binding interactions.

Everybody wants beautiful, healthy hair and skin. Most people groom and maintain their hair and skin daily. The demand for products that improve the look and feel of these surfaces has created a huge industry for hair and skin care. Beauty care technology has advanced the cleaning, protection and restoration of desirable hair and skin properties by altering the hair surface. For many years (especially in the second half of the twentieth century) scientists have focused on studying the physical and chemical properties of hair in order to consistently develop products which alter the health, feel, shine, color, softness and overall aesthetics of the hair.

Table 40.1 Common hair care products/processes and their functions

Product/Process	Functions
Shampoos	Clean oils from the hair and skin
Conditioners	Repair hair damage and make the hair easier to comb; prevent flyaway; add feel, shine, softness
Combing/cutting/blowdrying	Style the hair
Chemical dyes/colorants/bleaches	Enhance or change the color and look of hair
Permanent wave treatments	Change the style and look of the hair

Table 40.1 displays common products and processes involved in hair care. Hair care products such as shampoos

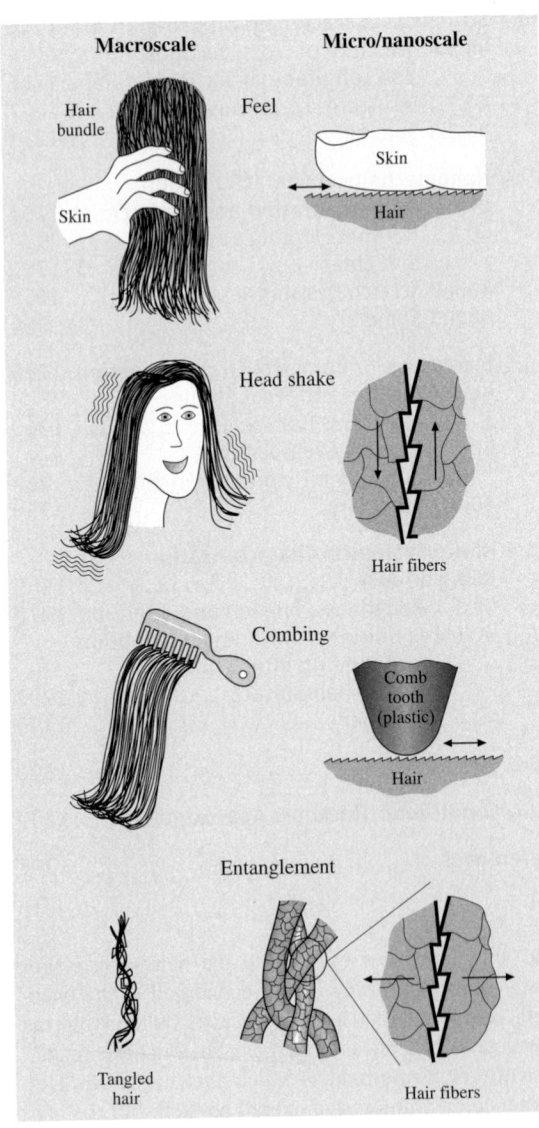

Fig. 40.1 Schematics illustrating various functions (with associated macroscale and micro/nanoscale mechanisms) of hair and skin friction during feel or touch, shaking and bouncing of the hair, combing, and entanglement (after [40.1])

and conditioners aid the maintenance and grooming process. Mechanical processes such as combing, cutting and blowdrying serve to style the hair. Chemical products and processes such as chemical dyes, colorants, bleaches and permanent wave treatments enhance the appearance and hue of the hair. Of particular interest is how all these common hair care items deposit onto and change hair properties, since these properties are closely tied to product performance. The fact that companies like Procter & Gamble (see http://www.pg.com), L'Oréal (see http://www.loreal.com) and Unilever (see http://unilever.com) have hair care product sales consistently measured in the billions of dollars suggests that understanding the science behind human hair can lead to more than just purely academic benefits, as well.

While products and processes such as combing, chemical dyeing and permanent wave treatment (Table 40.1) are used to enhance the appearance and style of the hair, they also contribute a large amount of chemical and mechanical damage to the fibers, which leads to degradation of structure and mechanical properties. As a result, the fibers become weak and more susceptible to breakage after time, which is undesirable for healthy hair. Shampoos and conditioners, which typically clean and repair the hair surface, respectively, have a distinct effect on its mechanical properties as well.

The tribology of the hair also changes as a function of the various hair care products and processes. Figure 40.1 illustrates schematically various functions, along with the macro- and micro/nanoscale mechanisms behind these interactions, that make surface roughness, friction and adhesion very important to hair and skin [40.1]. Desired features and corresponding tribological attributes of conditioners are listed in Table 40.2 [40.2]. For a smooth wet and dry feel, friction between hair and

Table 40.2 Desired features and corresponding tribological attributes of conditioners

Desired hair feature	Tribological attributes
Smooth feel in wet and dry environments	Low friction between hair and skin in respective environments
Shaking and bouncing during daily activities	Low friction between hair fibers and groups of hair
Easy combing and styling	Low friction between hair and comb (plastic) and low adhesion Note: More complex styles may require higher adhesion between fibers

skin should be minimized in wet and dry environments, respectively. For a good feel with respect to bouncing and shaking of the hair during walking or running, friction between hair fibers and groups of hair fibers should be low. The friction one feels during combing is a result of interactions between hair and the comb material (generally a plastic), and this too needs to be low in order to easily maintain, sculpt and comb the hair. To minimize entanglement, adhesive force (the force required to separate the hair fibers) needs to be low. In other cases, a certain level of adhesion may be acceptable and is often a function of the hair style. For individuals seeking "hair alignment", where hair fibers lay flat and parallel to each other, a small amount of adhesive force between fibers may be desired. For more complex and curly styles, even higher adhesion between fibers may be optimal.

Early research into human hair was done primarily on the chemical and physical properties of the hair fiber itself. Key topics dealt with analysis of chemical composition in the fiber, microstructure, and hair growth, to name a few. Mechanical properties were also of interest. Most of the mechanical property studies of human hair were on the macroscale and used conventional methods, such as tension, torsion and bending tests [40.3–8]. Efforts were also made to study the effects of environmental and chemical damage and treatment, such as dyeing, bleaching and polymer application; these topics have stayed a mainstream area of investigation due to the availability and formation of new chemicals and conditioning ingredients. Tribology has generally been studied via the macroscale friction force of hair. As a matter of fact, much of the tribological work performed by the hair care industry today still focuses on the measurement of macroscale friction, particularly between a skin replica and a hair swatch of interest [40.3]. The intrinsic differences between hair as a function of ethnicity eventually became a concern as well. For instance, research has shown that African-American hair has higher resistance to combing, higher static charge, and lower moisture content than Caucasian hair [40.9]. Because of differences like these, a growing number of hair care products specifically tailored to ethnic hair care have been developed and sold with commercial success.

Modern research since the late 1990s has been primarily concerned with using micro/nanoscale experimental methods such as atomic force/friction force microscopies (AFM/FFM) and nanoindentation to answer the complex questions surrounding the structure and behavior of the hair. Nanoscale characterization of the cellular structure, the mechanical properties and the tribological properties of hair are critical to evaluating and developing better cosmetic products and to advancing the understanding of composite biological systems, cosmetic science and dermatology.

AFM/FFM have been used to effectively study the structure of the hair surface and cross-section. AFM provides the potential to image the cellular structure and molecular assembly of hair, to determine various properties of hair, such as elastic modulus and viscoelastic properties, and to investigate the physical behavior of various cellular structures of hair in various environments [40.10, 11]. As a noninvasive technique, AFM has been used to evaluate the effect of hair treatment and can be operated in ambient conditions in order to study the effect of the environment on various physical properties. AFM has also been used to study the tribological properties of hair and skin and the effects of hair care products on hair [40.1, 2, 12]. Roughness parameters have been measured in order to compare changes due to damaging processes. Friction force has been measured in order to understand damage or conditioner distribution and its effect on hair tribology. Adhesive force mapping has been shown to be useful for observing conditioner distribution as well.

The nanoindenter has been used to characterize the nanomechanical behavior of the hair surface and cross-section using nanoindentation and nanoscratch techniques [40.13, 14]. These properties are important when evaluating cosmetic products by comparing the nanomechanical properties, such as hardness, elastic modulus and scratch resistance, of the hair surface before and after chemical damage or conditioner treatment. Since hair is a nanocomposite biological fiber with well-characterized structures, which will be described in detail in the next section, it is a good model for studying the roles of various structural and chemical components in providing mechanical strength for composite biological fibers [40.13]. Furthermore, a quantitative determination of the mechanical properties of human hair can also provide dermatologists with some useful markers for diagnosing hair disorders and for evaluating their response to therapeutic regimens [40.15, 16]. Combing results in physical damage such as scratching, and therefore it is useful to study scratch resistance, especially when conditioner is applied.

In this chapter, we present a comprehensive study of various hair and skin structural, mechanical and tri-bological properties as a function of ethnicity, damage, conditioning treatment and various environments. Various cellular structures (such as the cortex and the cuticle) of human hair and fine sublamellar structures of the cuticle, such as the A-layer, the exocuticle, the endocuticle and the cell membrane complex, are easily identified and studied. Nanomechanical properties such as hardness, elastic modulus, creep and scratch resistance are discussed. Nanotribological properties such as surface roughness, friction, adhesion and wear are presented, as well as investigations of conditioner localization and thickness.

40.1 Human Hair, Skin and Hair Care Products

40.1.1 Human Hair and Skin

Figure 40.2 shows a schematic of a human hair fiber with its various layers of cellular structure [40.3, 4, 17–19]. Hair fibers (about 50 to 100 μm in diameter) consist of the cuticle and cortex, and in some cases medulla in the central region. All are composed of dead cells, which are mainly filled with keratin protein. Table 40.3 displays a summary of the chemical species in hair [40.10].

Depending on its moisture content, human hair consists of approximately 65–95% proteins, which are condensation polymers of amino acids. The remaining constituents are water, lipids (structural and free), pigment and trace elements. Among numerous amino acids in human hair, cystine is one of the most important amino acids. The distinct cystine content of various cellular structure of human hair results in a significant effect on its physical properties. Cystine has the capacity to crosslink the protein by intermolecular disulfide linkages. A high cystine content corresponds to rich disulfide cross-links, leading to high mechanical properties. In addition to disulfide bonds, hair is also rich in peptide bonds and the abundant CO- and NH-groups present give rise to hydrogen bonds between groups of neighboring chain molecules. The species responsible for color in hair is the pigment melanin, which is located in the cortex of the hair in granular form.

An average head contains over 100 000 hair follicles, which are the cavities in the skin surface from which hair fibers grow. Each follicle grows about 20 new hair fibers in a lifetime. Each fiber grows for several years until it falls out and is replaced by a new fiber. Hair typically grows at a rate of around 10 mm/month.

The Cuticle

The cuticle consists of flat overlapping cells (scales). The cuticle cells are attached at the root end and they point forward towards the tip end of the hair fiber, like roof tiles. Each cuticle cell is approximately 0.3 to 0.5 μm thick and the visible length of each cuticle cell is approximately 5 to 10 μm. The cuticle in human hair is generally 5–10 scales thick. Each cuticle cell consists

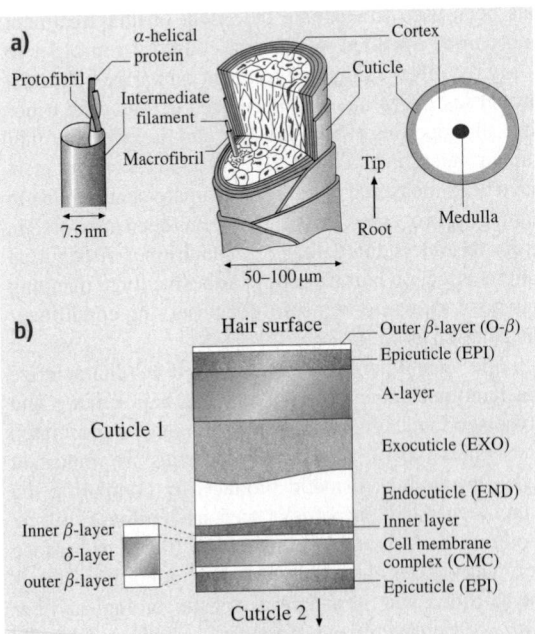

Fig. 40.2 Schematic of hair fiber structure (**a**) and cuticle sublamellar structure (**b**)

Table 40.3 Summary of chemical species present in human hair

Keratin protein	65–95%
(Amino acids)	$\overset{\oplus}{N}H_3-\underset{\underset{CO_2^{\ominus}}{\|}}{CH}-R$ (R: functional group)
Cystine	$\overset{\oplus}{N}H_3-\underset{\underset{CO_2^{\ominus}}{\|}}{CH}-CH_2-S-S-CH_2-\underset{\underset{CO_2^{\ominus}}{\|}}{CH}-\overset{\oplus}{N}H_3$
Lipids	Structural and free
18-methyl eicosanoic acid (18-MEA)	$H_3C\overset{(CH_2)_{16}}{\underset{CH_3}{\diagdown}}COOH$
Water	Up to 30%
Pigment and trace elements	Melanin

Table 40.4 Various layers of the cuticle and their details

Cuticle layer	Cystine component	Details
Epicuticle	≈ 12%	18-MEA lipid layer attached to outer epicuticle contributes to lubricity of the hair
A-layer	≈ 30%	Highly cross-linked
Exocuticle	≈ 15%	Mechanically tough
		Chemically resilient
Endocuticle	≈ 3%	
Inner layer	–	
Cell membrane complex (CMC)	≈ 2%	Lamellar structure consists of inner β-layer, δ-layer, and outer β-layer

of various sublamellar layers (the epicuticle, the A-layer, the exocuticle, the endocuticle and inner layer) and the cell membrane complex (Fig. 40.2). Table 40.4 displays the various layers of the cuticle, their respective cystine levels [40.3] and other details. The outer layer is the epicuticle, which is covered with a thin

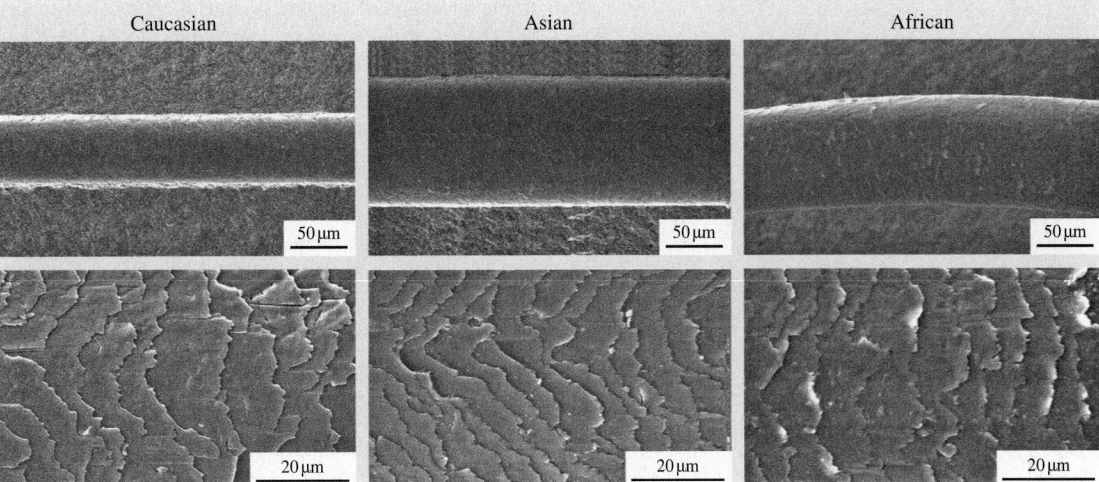

Fig. 40.3 SEM images of various hair (after [40.13])

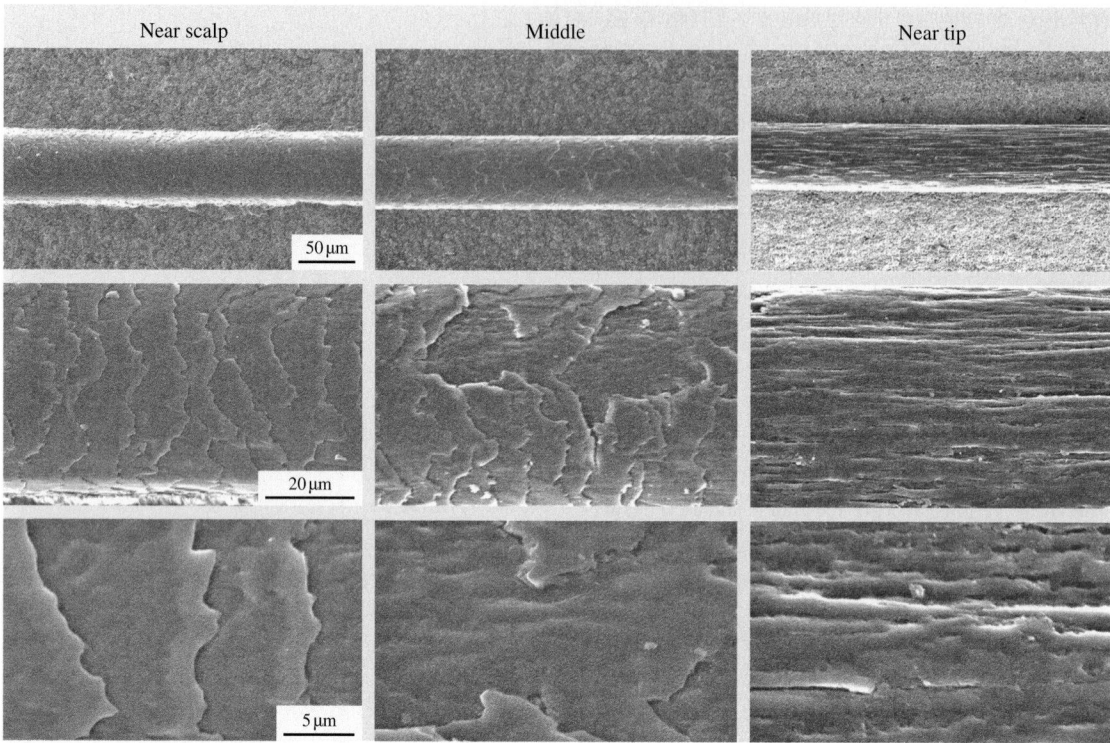

Fig. 40.4 SEM images of virgin Caucasian hair at three locations (after [40.13])

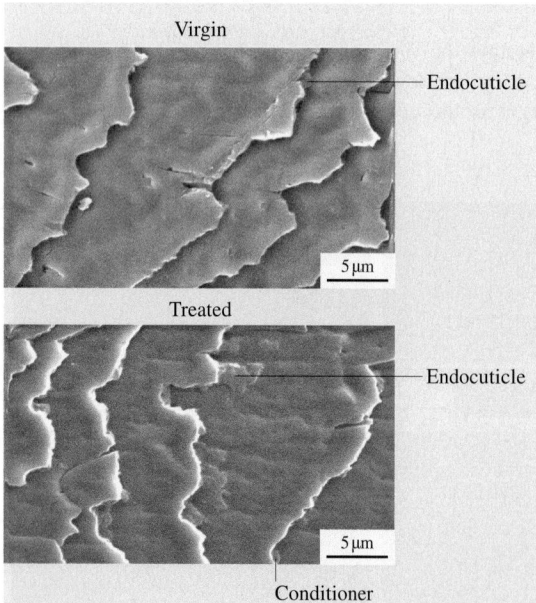

Fig. 40.5 SEM images of virgin and treated Caucasian hair (after [40.13])

layer of covalently attached lipid 18-methyl eicosanoic acid (18-MEA) (Table 40.3). The A-layer is a component of high cystine content ($\approx 30\%$) and is located on the outer-facing aspect of each cell. The A-layer is highly crosslinked, which gives this layer considerable mechanical toughness and chemical resilience, and the swelling in water is presumed to be minimal. The exocuticle, which is immediately adjacent to the A-layer, is also of high cystine content ($\approx 15\%$). On the inner facing aspect of each cuticle cell is a thin layer of material which is known as the inner layer. Between the exocuticle and inner layer is the endocuticle, which is low in cystine ($\approx 3\%$). The cell membrane complex itself is a lamellar structure, which consists of the inner β-layer, the δ-layer and the outer β-layer.

Figure 40.3 shows SEM images of virgin Caucasian, Asian and African hair [40.13]. It can be seen that the Asian hair is the thickest (about $100\,\mu m$), followed by African hair (about $80\,\mu m$) and Caucasian hair (about $50\,\mu m$). The visible cuticle cell is about $5–10\,\mu m$ long for the three hair. A list of various cross-section dimensional properties is presented in Table 40.5 [40.13].

Table 40.5 Cross-sectional dimensions of various ethnicities of human hair

	Shape	Maximum diameter (D_1) (μm)	Minimum diameter (D_2)(μm)	Ratio D_1/D_2	Number of cuticle scales	Cuticle scale thickness (μm)
Caucasian	Nearly oval	74	47	1.6	6–7	0.3–0.5
Asian	Nearly round	92	71	1.3	5–6	0.3–0.5
African	Oval–flat	89	44	2.0	6–7	0.3–0.5

Average length of visible cuticle scale: about 5 to 10 μm

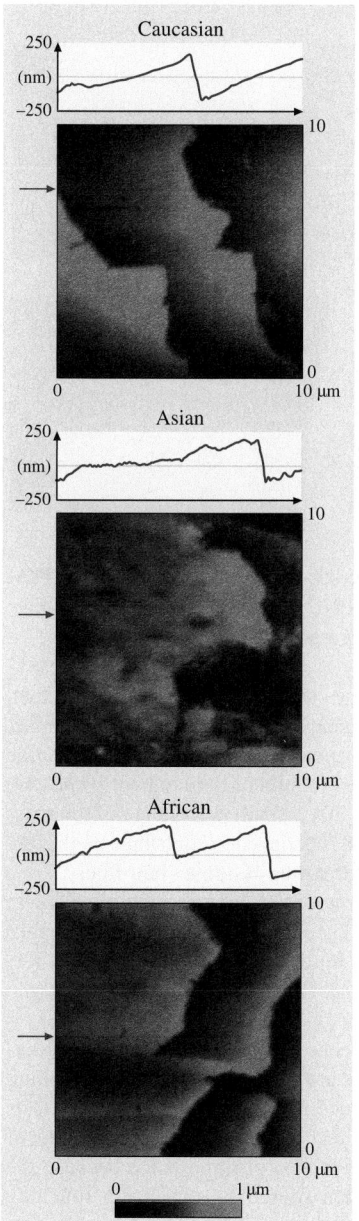

While Caucasian and Asian hair typically have similar cross-sectional shapes (Asian hair being the most cylindrical), African hair has a highly elliptical shape. African hair is much more curly and wavy along the hair fiber axis than Caucasian or Asian hair. Figure 40.4 shows the SEM images of virgin Caucasian hair at three locations: near the scalp, in the middle and near the tip. Three magnifications were used to show the main differences. The hair near the scalp had complete cuticles, while no cuticles were found on the hair near the tip. This may be because that the hair near the tip experiences more mechanical damage during its life than the hair near the scalp. The hair in the middle experiences intermediate damage, so one or more scales of the cuticles are worn away, but many cuticles stay complete. If some substructures of one cuticle scale, like the A-layer or the A-layer and exocuticle (Fig. 40.2) are removed, or (even worse) one or several cuticle scales are worn away, it is impossible to heal the hair biologically, because hair fibers are composed of dead cells. However, it is possible to physically "repair" the damaged hair using conditioner, one of the functions of which is to cover or fill the damaged area of the cuticle. Figure 40.5 shows high magnification SEM images of virgin and treated Caucasian hair. The endocuticles (pointed at by arrows) were found in both types of hair. In order for the conditioner to physically repair the hair, it is expected that it should cover the endocuticles. In the case of severely damaged hair – for example, an edge of one whole cuticle scale worn away – the conditioner may fill that damaged edge. In the SEM image of the treated hair in Fig. 40.5, the substance which stayed near the cuticle edge is probably the conditioner (pointed by an arrow).

Figure 40.6 shows the AFM images of various virgin hair, along with section plots [40.1]. The arrows point to the position that the section plots were taken from. Each cuticle cell is nearly parallel to the underlying cuticle cell, and they all have similar angles to the hair axis, forming a tile-like hair surface structure. The visible cuticle cell is approximately 0.3–0.5 μm thick and about 5–10 μm long for all three types of hair.

Fig. 40.6 AFM images of various virgin hair (after [40.1])

Part D | 40.1

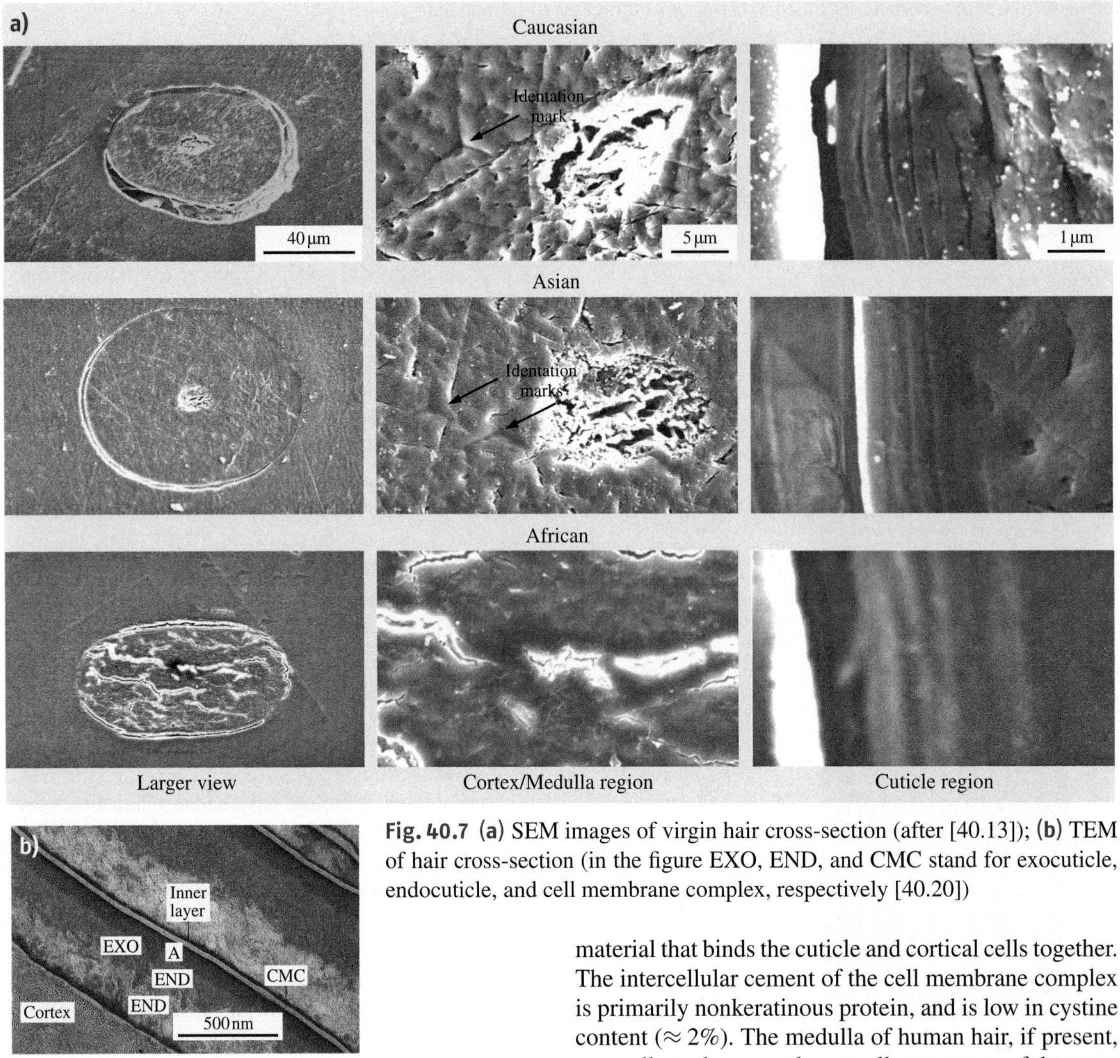

a)

Caucasian

Indentation mark

40 µm

5 µm

1 µm

Asian

Indentation marks

African

Larger view

Cortex/Medulla region

Cuticle region

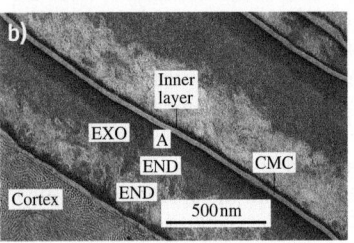

b)

Inner layer

EXO A
END   CMC
Cortex END

500 nm

Fig. 40.7 (a) SEM images of virgin hair cross-section (after [40.13]); **(b)** TEM of hair cross-section (in the figure EXO, END, and CMC stand for exocuticle, endocuticle, and cell membrane complex, respectively [40.20])

The Cortex and Medulla

The cortex contains cortical cells and the intercellular binding material, or the cell membrane complex. The cortical cells are generally 1–6 µm thick and 100 µm long, run longitudinally along the hair fiber axis and form the majority of the inner hair fiber content [40.21]. The macrofibrils (about 0.1–0.4 µm in diameter) comprise a major portion of the cortical cells. Each macrofibril consists of intermediate filaments (about 7.5 nm in diameter), previously called microfibrils, and the matrix. The intermediate filaments are low in cystine ($\approx 6\%$), and the matrix is rich in cystine ($\approx 21\%$). The cell membrane complex consists of cell membranes and adhesive

material that binds the cuticle and cortical cells together. The intercellular cement of the cell membrane complex is primarily nonkeratinous protein, and is low in cystine content ($\approx 2\%$). The medulla of human hair, if present, generally makes up only a small percentage of the mass of the whole hair, and is believed to contribute negligibly to the mechanical properties of human hair fibers.

Figure 40.7a shows SEM images of virgin hair in cross-section [40.13] and Fig. 40.7b shows TEM images of a cross-section of human hair [40.20].

Skin

Skin covers and protects our bodies. The skin at the forehead and scalp areas are of most interest when dealing with human hair, since most of the hair care products are developed specifically for head hair. The skin of the hand and fingers is also of importance because the "feel" of hair is often sensed by physically touching the fibers with these regions. In general, skin is composed of three main parts: the epidermis, dermis, and subcutaneous tissue (L'Oréal; Fig. 40.8).

The epidermis contains four distinct cellular layers: basal, spinous, granular and horny. In the basal layer, melanocytes deliver the pigment melanin to keratinocytes. Keratinocyte cells that have been cornified are referred to as corneocytes [40.22]. Hexagonally shaped corneocyte cells compose the horny layer, or stratum corneum. The stratum corneum is the outer layer of the skin; at about 15 μm thick, it acts as a mechanical, thermal and chemical barrier to environmental factors and contamination. The complex organization of corneocytes and intercellular matrix contributes to the success of the barrier [40.23]. In fact, *Wertz* et al. [40.24] developed a structural model which treats the matrix as a lamellar phase composition of various lipids which provide a glue-like system, providing the barrier effect.

The dermis structure is known for its ability to handle most of the physical stresses imposed on the skin, and takes up roughly 90% of the mass [40.22]. The dermis is divided into an outermost papillary layer and an underlying reticular layer.

40.1.2 Hair Care: Cleaning and Conditioning Treatments and Damaging Processes

Cleaning and Conditioning Treatments: Shampoos and Conditioners

Shampoos are used primarily to clean the hair and scalp of dirt and other greasy residues that can build up after time. Shampoos also have many secondary functions, including controlling dandruff, reducing irritation, and even conditioning. Conditioners, on the other hand, are used primarily to give the hair a soft, smooth feel, which results in easier hair combing. Secondary functions include preventing "flyaway" hair due to static electricity, giving the hair a shiny appearance, and protecting the

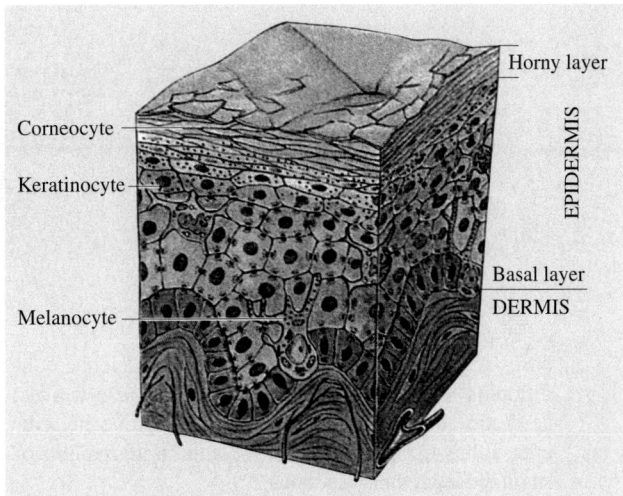

Fig. 40.8 Schematic image of human skin structure with different layers: dermis, epidermis and horny layer (L'Oréal)

hair from further damage by forming a thin coating over the fibers.

Further developments in marketing and aesthetic factors (brand name, fragrance, feel and color of the shampoos and conditioners) have created new market segments. In many instances, these factors have become primary reasons for use.

Shampoo: Constitution and Main Functions

The following discussion is based on *Gray* [40.25, 26]. As stated above, shampoos serve various cleaning functions for the hair and scalp. In the past, typical shampoos were mainly soap-based products. However, soaps did not have very good lathering capability, and often left a residual "scum" layer on the hair that was undesirable and could not be rinsed off.

In modern shampoos, advances in chemistry and technology have made it possible to replace the soap

Table 40.6 Components of common shampoos and their functions

Shampoo component	Functions
Cleansing agents	• Produce lather to trap greasy matter and prevent redeposition
	• Remove dirt and grease from hair and scalp
	• Stabilize the mixture and help keep the network of ingredients bound together
	• Thicken the shampoo to the desired viscosity
Conditioning agents	Condition the hair
Functional additives	Control the viscosity and pH level of the shampoo
Preservatives	Prevent decomposition and contamination of shampoo
Aesthetic additives	Enhance color, scent and luminescence of shampoo
Medically active ingredients	Aid treatment of dandruff or hair loss

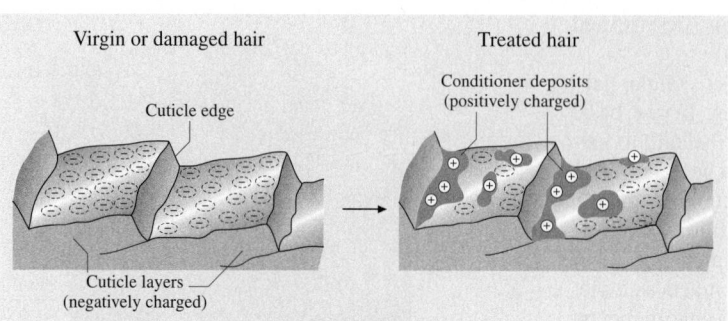

Fig. 40.9 Negatively charged hair and the deposition of positively charged conditioner on the cuticle surface (after [40.2])

bases with complex formulae of cleansing agents, conditioning agents, functional additives, preservatives, aesthetic additives and even medically active ingredients. Table 40.6 shows the most common ingredients of shampoos and their functions.

Cleansing Agents. In most modern shampoos, the primary cleansing agents are surfactants. Dirt and greasy residues are removed from the hair and scalp by these surfactants, making them the most important part of the shampoo. Surfactants have great lathering capabilities and rinse off very easily; see Table 40.6 for a full list of features.

Surfactant molecules have two different ends, one of which is negatively charged and soluble in water (unable to mix with greasy matter), while the other is soluble in greasy matter (unable to mix with water). In general, surfactants clean the hair by the following process. Surfactant molecules encircle the greasy matter on the hair surface. The molecule end that is soluble in greasy matter buries itself in the grease, which leaves the water-soluble end to face outward with a negative charge. Since the hair fibers are negatively charged as well, the two negative charges repel each other. Thus, the greasy matter is easily removed from the hair surface and rinsed off. Shampoos contain several surfactants, generally up to four, which clean the hair differently depending on the hair type of the individual. Mild cleansing systems which

do not damage or irritate the scalp, hair and eyes are now quite common.

Conditioning Agents. Many shampoos contain conditioning agents which serve many of the same roles as full conditioners. Conditioning agents are further described in the following subsection.

Functional Additives. Functional additives can aid in controlling the thickness and feel of the shampoo itself. Simply stated, the right blend is required so that the shampoo is not too thin and not too thick. Functional additives can also control the acidity of the shampoo by obtaining a goal pH level, typically of around 4.

Preservatives. Preservatives detract germs and prevent decomposition of the shampoos. They also prevent various other health risks that accompany contamination by germs and bacteria. Typical preservatives in shampoos are sodium benzoate, parabens, DMDM hydantoin and tetrasodium EDTA.

Aesthetic Additives. Shampoos contain many aesthetic additives that enhance the appearance, color and smell of the mixture. These additives typically give the shampoo the luminous shine and pleasant fragrance to which many consumers are accustomed.

Medically Active Ingredients. For people with dandruff and other more serious hair and scalp disorders, shampoos are available with active ingredients which aim to treat or control these conditions. In the treatment of dandruff, zinc pyrithione is a common shampoo additive. For hair loss issues, panthenol is commonly added to shampoos to aid in hair growth and moisture content.

Conditioner: Constitution and Main Functions
As stated earlier, many shampoos have certain levels of conditioning agents which mimic the functions of

Oil/water emulsion	Lamellar liquid crystal	Gel network

─● Surfactant
─► Fatty alcohol

Fig. 40.10 Conditioner formation, from emulsion to gel network

Table 40.7 Ingredients of conditioners and their benefits in terms of wet and dry feels

Gel network chassis for superior wet feel	
Key ingredients	Benefits
• Cationic surfactant	• Creamy texture
• Fatty alcohols	• Ease of spreading
• Water	• Slippery feel upon application
	• Soft rinsing feel
Combination of "conditioning actives" for superior dry feel	
Key ingredients	Benefits
• Silicones	• Moist
• Fatty alcohols	• Soft
• Cationic surfactant	• Makes dry combing easier

a full conditioner product. Conditioner molecules contain cationic surfactants which give a positive electrical charge to the conditioner. The negative charge of the hair is attracted to the positively charged conditioner molecules, which results in conditioner deposition on the hair (Fig. 40.9). This is especially true for damaged hair, since damaging processes result in hair fibers being even more negatively charged. The attraction of the conditioner to hair results in a reduction in static electricity on the fiber surface, and consequently a reduction in "fly-away" behavior. The conditioner layer also flattens the cuticle scales against each other, which improves shine and color. The smooth feel resulting from conditioner use results in easier combing and detangling in both wet and dry conditions (Table 40.7).

Conditioner consists of a gel network chassis (cationic surfactants, fatty alcohols and water) for a superior wet feel and a combination of conditioning actives (cationic surfactants, fatty alcohols and silicones) for a superior dry feel. Figure 40.10 shows the transformation of the cationic surfactants and fatty alcohol mixture into the resulting gel network, which is a frozen lamellar liquid crystal gel phase. The process starts as an emulsion of the surfactants and alcohols in water. The materials then go through a strictly controlled heating and cooling process: the application of heat causes the solid compounds to melt, and the solidification process enables the lamellar assembly molecules to be set in a fully extended conformation, creating a lamellar gel network. When this network interacts with the hair surface, the high concentration of fatty alcohols make it the most deposited ingredient group, followed by the silicones and cationic surfactants. Typical deposition levels for cationic surfactant, fatty alcohol and silicone are around 500–800 ppm, 1000–2000 ppm and 200 ppm,

Table 40.8 Chemical structures and functions of ingredients of conditioners

Ingredient		Chemical structure	Purpose/Function
Water			
Cationic surfactants	Stearamidopropyl dimethylamine		• Aids formation of lamellar gel network
	Behenyl amidopropyl dimethylamine glutamate (BAPDMA)		• Lubricates and static control agent
	Behentrimonium chloride (BTMAC) $CH_3(CH_2)_{21}N(Cl)(CH_3)_3$		
Fatty alcohols	Stearyl alcohol ($C_{18}OH$)		• Lubricates and moisturizes
	Cetyl alcohol ($C_{16}OH$)		• Aids formation of lamellar gel network along with cationic surfactant
Silicones	PDMS blend (Dimethicone)		• Primary source of lubrication • Gives hair a soft and smooth feel

Part D | 40.1

respectively. Typical concentrations are approximately 2–5 wt%, 5–10 wt% and 1–10 wt%, respectively [40.2].

The benefits of the conditioner are shown in Table 40.7 [40.2]. The wet feel benefits are creamy texture, ease of spreading, slippery feel while applying, and soft rinsing feel. The dry feel benefits are moistness, softness and easier dry combing. Each of the primary conditioner ingredients also has specific functions and roles that affect the performance of the entire product. Table 40.8 displays the functions of the major conditioner ingredients and their chemical structures [40.2]. Cationic surfactants are critical to the formation of the lamellar gel network in conditioner, and also act as a lubricant and static control agent, since their positive charge aids in counteracting the negative charge of the hair fibers. Fatty alcohols are used to lubricate and moisturize the hair surface, and to form the gel network. Finally, silicones are the main source of lubrication in the conditioner formulation.

Damaging Processes

In Sect. 40.1.2 we discussed some of the products which aid in "treating" the hair. There are other hair care products and processes which, while creating a desired look or style to the hair, also bring about significant damage to the fibers (Table 40.1). Most of these processes occur on some type of periodic schedule, whether it be daily (while combing the hair), or monthly (haircut and coloring at a salon). In general, hair fiber damage occurs most readily by mechanical or chemical means, or by a combination of both (chemomechanical).

Mechanical Damage. Mechanical damage occurs on a daily basis for many individuals. The damage results from large physical forces or temperatures which degrade and wear the outer cuticle layers. Common causes are:

- combing (generally with plastic objects, and often multiple times over the same area, leading to scratching and wearing of the cuticle layers)
- scratching (usually with fingernails around the scalp)
- cutting (affects the areas surrounding the fiber tips)
- blowdrying (high temperatures thermally degrade the surface of the hair fibers)

Permanent Wave Treatment. Permanent wave treatments saw many advances at the beginning of the

twentieth century, but have not changed much since the invention of the Cold Wave around the turn of that century. Generally speaking, the Cold Wave uses mercaptans (typically thioglycolic acid) to break down disulfide bridges and style the hair without much user interaction (at least in the period soon after perm application) [40.25]. The Cold Wave process does not need increased temperatures (so no thermal damage is caused to the hair), but generally consists of a reduction period (whereby molecular reorientation to the cuticle and cortex occurs via a disulfide–mercaptan interchange pathway [40.3]) followed by rinsing, setting of the hair to the desired style, and finally neutralization to decrease the mercaptan levels and stabilize the style. The chemical damage brought on by the permanent wave can increase dramatically unless performed with care.

Chemical Relaxation. Commonly used as a means of straightening hair (especially in highly curved, tightly curled African hair), this procedure uses an alkaline agent, an oil phase, and a water phase of a high viscosity emulsion to relax and reform bonds in extremely curly hair. A large part of the ability to sculpt the hair to a desired straightness comes from the breakage of disulfide bonds of the fibers.

Coloring and Dyeing. Hair coloring and dyeing have become extremely successful hair care procedures, due in part to "over-the-counter" style kits which allow home hair care without professional assistance. The most common dyes are para dyes, which contain paraphenylenediamine (PPD) solutions accompanied by conditioners and antioxidants. Hydrogen peroxide (H_2O_2) is combined with the para dyes to effectively create tinted, insoluble molecules which are contained within the cortex and are not small enough to pass through the cuticle layers, leaving a desired color to the hair. Due to the levels of hydrogen peroxide, severe chemical damage can ensue in the cuticle and cortex.

Bleaching. Like dyeing, bleaching consists of using hydrogen peroxide to tint the hair. However, bleaching can only lighten the shade of hair color, as the H_2O_2 releases oxygen to bind hair pigments [40.25]. Bleaching may also be applied to limited areas of the hair (such as in highlights) to create a desired look. The chemical damage brought on by bleaching leads to high porosity and severe wear of the cuticle layer.

40.2 Experimental Techniques

To date, most of the information about the detailed structure of human hair has been obtained from scanning electron microscope (SEM) and transmission electron microscope (TEM) observations [40.3, 13, 20, 29]. Table 40.9 shows a comparison between the various tools used to study hair on the micro/nanoscale. The SEM has long been the standard means of investigating the surface topography of human hair. The SEM uses an electron beam to give a 'photographic' image of the sample, but cannot provide quantitative data regarding the surface. SEM requires the hair sample to be covered with a very thin layer of a conductive material and needs to be operated under vacuum during both metallization and measurements. Surface metallization and vacuum exposure could potentially induce modifications to the surface details. Figures 40.7a and 40.7b show typical images of human hair obtained by SEM and TEM, respectively. TEM examinations provide fine detailed internal structures of human hair. However, thin sections of 50–100 nm thickness and heavy metal compound staining treatments are required for TEM examinations. The cutting of these thin sections with the aid of an ultramicrotome is not an easy task. Moreover, since neither the SEM nor the TEM techniques can measure the phys-

ical properties (mechanical, tribological, and so on) of various cellular structures in human hair of interest and they do not allow ambient imaging conditions, many outstanding issues remain to be answered. For example, how do the various cellular structures of hair behave physically in various environments (temperature, humidity, and so forth)? How do they swell in water? For conditioner-treated hair, how thick is the conditioner layer and how is the conditioner distributed on the hair surface?

AFM has been commonly used to characterize tribological and mechanical properties of surfaces [40.27, 30–33]. Since it is a noninvasive technique, AFM can be used to evaluate the effect of sample treatment since it can be used without requiring any specific treatment. Most importantly, it can be operated under ambient conditions in order to study the effect of the environment. AFM provides the potential to image the cellular structures and molecular assembly of hair, to determine various properties of hair, such as friction, adhesion, wear, elastic stiffness and viscoelastic properties, and to investigate the physical behavior of various cellular structure of hair in various environments. To date, most AFM studies on hair fibers have focused on surface to-

Table 40.9 Comparison of methods used to characterize hair on micro/nanoscale

Method	Type of information	Quantitative data	Normal load	Resolution (nm)	Limitations
Scanning electron microscopy (SEM)	Structural	Gross dimensions	None	0.2–1 (spatial)	• Requires thin conductive coating on sample • Requires vacuum environment • Expensive instrumentation [40.27] • Tedious [40.27]
Transmission electron microscopy (TEM)	Structural	Gross dimensions	None	0.2–1 (spatial)	• Requires thin sections (<100 nm) and heavy metal compound staining treatment • Requires vacuum environment • Expensive instrumentation [40.27] • Tedious [40.27]
Atomic force/friction force microscopy (AFM/FFM)	Structural Mechanical Tribological	• Surface roughness • Elastic modulus • Viscoelasticity • Friction • Adhesion • Conditioner thickness	< 0.1 nN −500 nN [40.27]	0.2–1 (spatial) [40.27] 0.02 (vertical) [40.27]	None
Nanoindenter	Mechanical	• Hardness • Elastic modulus • Creep • Scratch resistance	< 0.1 mN −350 mN	400 (spatial) 0.1 (vertical) [40.28]	None

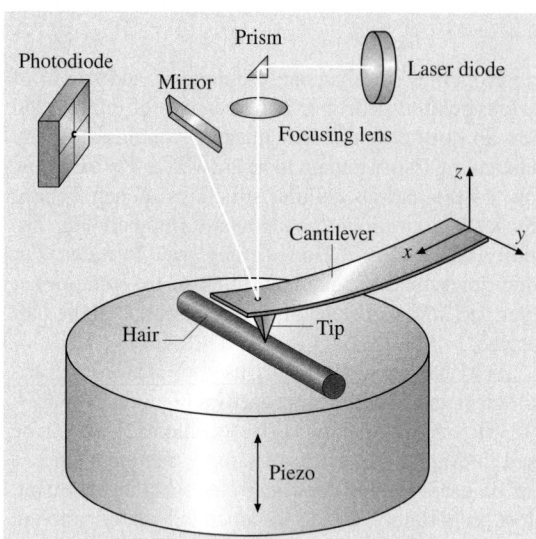

Fig. 40.11 Schematic diagram of operation of an AFM with a human hair sample

pographic imaging [40.1, 19] and friction, adhesion and wear properties [40.1, 2, 12].

A schematic of an AFM imaging a hair fiber is shown in Fig. 40.11. AFM/FFM uses a sharp tip with a radius of approximately 10–50 nm. This significant reduction in tip-to-sample interaction compared to the macroscale allows the simulation of single-asperity contact, giving detailed surface information. Contact mode allows simultaneous measurement of the surface roughness and the friction force. Different AFM operating modes, such as tapping mode and the recently introduced torsional resonance (TR) mode, can be used for measurements of material stiffness and to map viscoelastic properties using amplitude and phase angle imaging.

When skin comes into contact with hair, actual contact occurs over a large number of asperities. During relative motion, friction and adhesion are governed by the surface interactions that occur at these asperities. To date, much of the work in the industry has focused on the measurement of macroscale friction, particularly between a skin replica and a hair swatch of interest [40.3]. Figure 40.12 shows schematics of typical macro- and micro/nanoscale test apparatuses. However, there are many problems associated with these types of measurements. Factors such as topographical variations, lumping of the hair fibers, the large size of the synthetic skin, and traditional measurement system errors can all lead to uncertainty in the data.

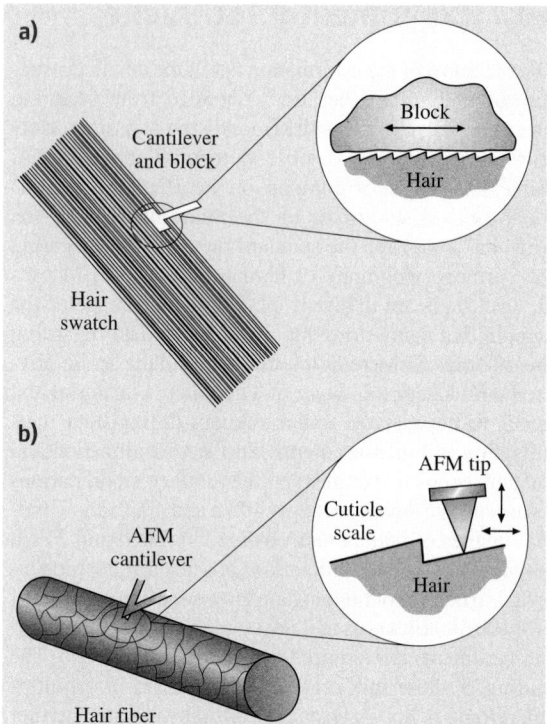

Fig. 40.12a,b Comparison of macroscale (**a**) and micro/nanoscale (**b**) friction test apparatuses

In the past, most of the mechanical property studies of human hair were on the macroscale and were based on conventional methods, such as tension, torsion and bending tests [40.3–8]. The mechanical properties obtained from these tests are the overall mechanical properties of the hair, not just the hair surface. Depth-sensing nanoindentation measurement techniques are commonly used to measure the nanomechanical properties of surface layers of bulk materials and ultrathin coatings [40.28, 31–33]. More recently, the nanoindentation technique has been used to investigate nanomechanical properties (such as hardness, Young's modulus, creep and scratch resistance) of various cellular layers of glass fibers, keratin fibers and hair surfaces/cross-sections [40.13, 14, 34, 35].

40.2.1 Experimental Procedure

Structural Characterization Using an AFM
An AFM (Multimode Nanoscope IIIa, DI-Veeco, Santa Barbara, CA), modified for the TR mode and with a nanoscope extender electronic circuit for the measure-

ment of the phase angle, were used in the study reported in this chapter [40.10, 11]. All measurements were conducted in ambient conditions ($22 \pm 1\,°C$, $50 \pm 5\%$ relative humidity). The probes used in this study were single beam etched Si probes (MikroMasch) with a fundamental flexural mode frequency of 75 kHz and a fundamental torsional resonance frequency of 835 kHz, with a quality factor of around 1000. The dimension of the cantilever was typically $230\,\mu m \times 40\,\mu m \times 3\,\mu m$ with a flexural spring constant of $1-5\,N/m$ and a torsional spring constant estimated to be $30-150\,N/m$. The radius of curvature of the tip was about 10 nm. The surface height images shown in this study were processed using the first-order planefit command available in the AFM software, which eliminates tilt in the image. Amplitude and phase angle images were processed using the zero-order flatten command, which only modifies the offset of the image.

A schematic diagram of the tip–cantilever assembly in an AFM is shown in Fig. 40.11. The cantilever can scan a sample with its tip in constant contact, intermittent contact, or without contact with the sample surface [40.27, 30–33]. The scanning is implemented by the motion of a cylindrical piezoelectric tube, which can act as the holder of either the cantilever or the sample. The deflection of the cantilever is generally measured using an optical lever method. A laser beam is projected on and reflected from a location on the upper surface of the cantilever close to the tip, and led by a mirror into a four-segment photodiode. The normal and lateral deflections of the cantilever at that location can then be obtained by calibrating the voltage output of the photo diode. AFM measurements can be performed with one of several modes: tapping mode I, torsional resonance modes (TR mode) [40.36], and contact mode, as shown in Fig. 40.13. The phase angle is also defined in Fig. 40.13 as a phase delay between the input/output strain and/or the stress profiles. Table 40.10 summarizes the characteristics of the tapping mode, the torsional resonance modes and the contact mode. The contact mode is a static mode

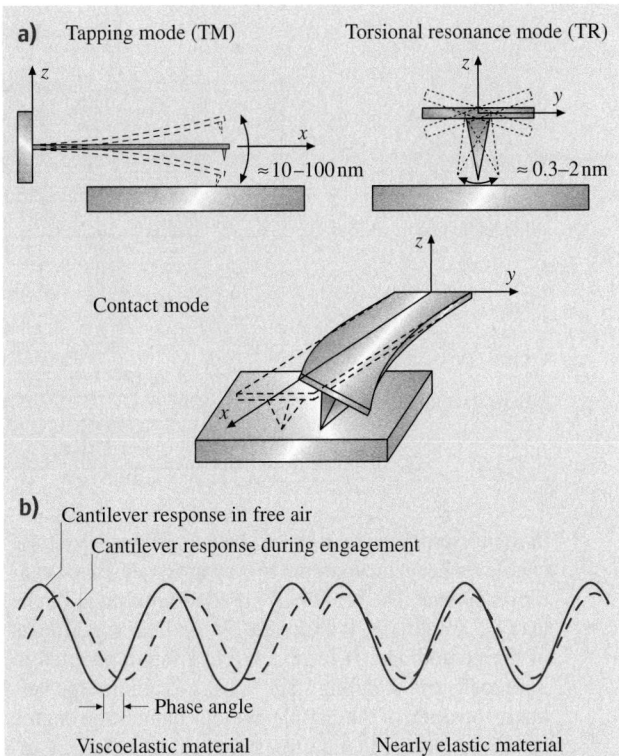

Fig. 40.13 (a) Three different AFM settings are compared at the *top*: tapping mode (TM), torsional resonance (TR) mode and contact mode. The TR mode is a dynamic approach with a laterally vibrating cantilever tip that can interact with the surface more intensively than tapping mode, so more detailed near-surface information is available using this technique. **(b)** The phase angle is defined, and two examples of phase angle response are shown at the *bottom*: one is for material exhibiting viscoelastic properties (*left*) and the other for nearly elastic material (*right*)

and uses a nonvibrating tip, so phase analysis is not available.

The TR modes measure surface height and phase angle (and amplitude) images as follows. The tip is vibrated

Table 40.10 Summary of various operating modes of AFM for surface imaging

Operating modes	Direction of cantilever vibration	Parameter controlled	Data obtained
Tapping	Vertical	Setpoint (constant amplitude)	Surface height, phase angle (normal viscoelasticity)
TR mode I	Torsional (lateral)	Setpoint (constant amplitude)	Surface height, phase angle (lateral viscoelasticity)
TR mode II	Torsional (lateral)	Constant load	Surface height, amplitude and phase angle (lateral stiffness and viscoelasticity)
Contact	n/a	Constant load	Surface height, friction

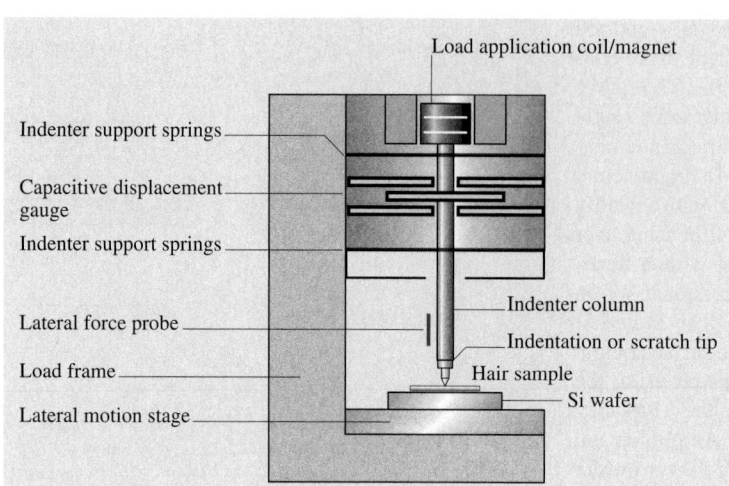

Fig. 40.14 Schematic of nanoindentation and nanoscratch tests on hair using nanoindenter

in the torsional mode at the resonance frequency of the cantilever beam in air driven by a specially designed cantilever holder. The torsional vibration amplitude of the tip (TR amplitude) is detected by the lateral segments of the photodiode. A feedback loop system coupled to a piezoelectric scanning stage is used to control the vertical z-position of the sample, which changes the degree of the in-plane (lateral) tip–sample interaction of interest. The z-displacement of the sample gives a surface height image of the sample. There are two possible operating modes depending on which parameter is controlled (Table 40.10):

1. TR mode I: constant TR amplitude;
2. TR mode II: constant normal cantilever deflection (constant load).

Both modes are operated at the resonance frequency of the cantilever in air, which is different from the TR friction mode used in previous studies [40.37], in which the tip is vibrated at the resonance frequency of the cantilever after engagement. During the measurement, the cantilever/tip assembly is first vibrated at its resonance at some amplitude before the tip engages the sample. Then, for TR mode I, the tip engages the sample at some setpoint which is reported as a ratio of the vibration amplitude after engagement to the vibration amplitude in free air before the engagement [40.36, 38, 39].

For TR mode II, instead of maintaining a constant setpoint, a constant normal load measured using vertical segments of the photodiode is applied. Under in-plane tip–sample interactions, the torsional resonance frequency, the amplitude and the phase of the cantilever all change from those observed when it is far away from

the sample surface and can be used to contrast and image the in-plane lateral surface properties.

Compared to TM and TR mode I, the AFM tip interacts with the surface more intensively in TR mode II, so more detailed in-plane surface information can be obtained using this mode [40.10]. When using TR mode II, TR amplitude and TR phase angle images give even greater contrast. Previous studies [40.36, 38, 39] indicated that the phase shift can be related to the energy dissipation through the viscoelastic deformation process between the tip and the sample. Recent theoretical analysis [40.40] has established a quantitative correlation between the lateral surface properties (lateral stiffness and viscoelasticity) of materials and the amplitude/phase angle shift in TR measurements. The contrast in the TR amplitude and phase angle images is due to the in-plane (lateral) heterogeneity of the surface. The lateral surface properties (lateral stiffness and viscoelasticity) of materials can be mapped using the TR amplitude and phase angle images. In the work presented in this chapter, TR mode II amplitude and phase images were obtained in order to characterize the cellular structure of human hair. For convenience, the TR amplitude is recorded in volts, and 1 V corresponds to a TR amplitude of about 0.5 nm.

Nanomechanical Characterization Using Nanoindentation

Nanoindentation. Figure 40.14 shows a schematic of nanoindentation and nanoscratch tests on the surface of human hair performed with a Nano Indenter II® (MTS Systems Corp.). This instrument monitors and records the dynamic load and displacement of the indenter during indentation with a force resolution of

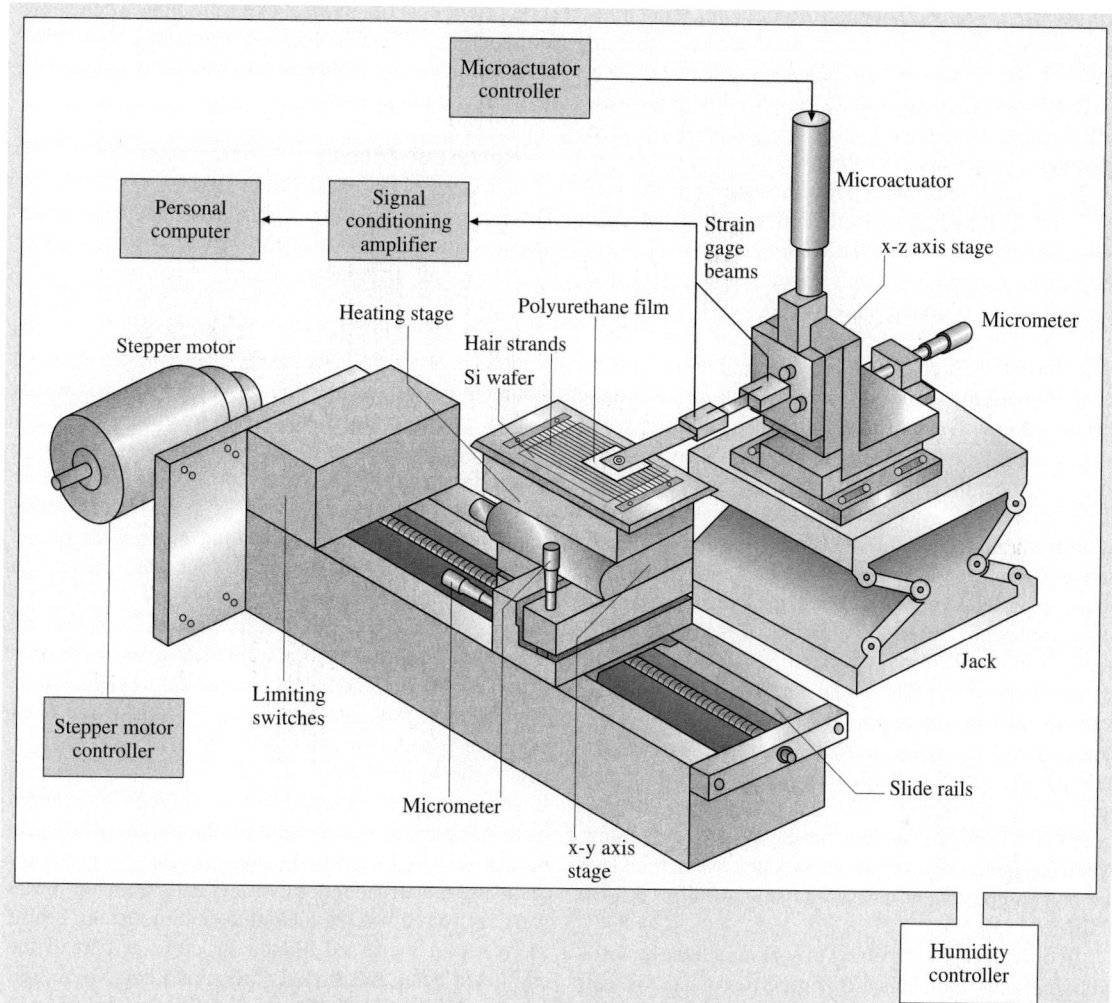

Fig. 40.15 Schematic of the reciprocating tribometer. Normal load is applied by lowering the x–z stage (mounted on a laboratory jack). Normal and friction forces are measured by semiconductor strain gauges mounted on a crossed-I-beam structure (after [40.41])

about 75 nN and a displacement resolution of about 0.1 nm [40.28, 31, 33].

A three-sided (triangular-based) pyramidal diamond Berkovich indenter tip (radius 100–200 nm) was used for hardness, Young's modulus and creep measurements [40.13, 14]. A wide load range (0.1–300 mN) was used during nanoindentation on all virgin hair surfaces except for African hair, in order to study the variation in mechanical properties with indentation depth. In the case of virgin African hair, a load of 300 mN was not used because the hair was too soft to get reasonable data at 300 mN. For the damaged, treated hair and the

hair near the scalp, in the middle and near the tip, normal loads of 0.1, 1.0, 10 and 100 mN were used. At each load, five indents were made, and the hardness and elastic modulus values were averaged from these and standard deviations were calculated. For nanoindentation on virgin hair cross-sectional samples, only one normal load (1.0 mN) was used to make indents at cuticle, cortex and medulla. A normal load was applied for the creep test, and then the tip was held for 600 s. The change in displacement during the holding time was recorded. The loads used were 0.1, 1.0 and 10 mN.

Nanoscratch. For coefficient of friction and scratch resistance measurements, a conical diamond tip with a tip radius of about 1 μm and an included angle of 60° was used. Before scratching, the hair sample holder was manually rotated so that the hair axis is parallel to the scratch direction.

The scratch tests were performed on both the single cuticle cell and on multiple cuticle cells from each hair sample by controlling the scratch length to be 5 μm and 50 μm, respectively. Each scratch test was repeated at least five times on the same hair to verify data reproducibility. For the 50 μm-long scratch test, two scratch directions were used to study the "directionality effect" of the cuticle. One scratch direction was along the cuticle and the other was against the cuticle. The coefficient of friction was monitored during scratching. In order to obtain scratch depths during scratching, the surface profile of the human hair was first obtained by translating the sample at a low load of about 0.01 mN, which is insufficient to damage the surface of the hair. The 5 μm- and 50 μm-long scratches were then made by translating the hair sample at a constant tip velocity of 0.5 μm/s while ramping the load from 0.01 to 1 mN, and from 0.01 to 10 mN, respectively. The actual depth during scratching was obtained by subtracting the initial profile from the scratch depth measured during scratching. In order to measure the residual depth after scratch, the scratched surface was profiled at a low load of 0.01 mN and was subtracted from the surface profile before scratching. After the nanoscratch tests, the scratch wear tracks of the hair samples were measured using a Philips XL-30 ESEM.

In order to study the effect of soaking on the nanotribological and nanomechanical properties of hair, the hair samples were soaked in deionized water for 5 min, which

is representative of a typical exposure time when showering/bathing. Then the hair was immediately mounted on Si wafer and 5 μm- and 50 μm-long scratch tests were performed on them.

Macroscale Tribological Characterization Using a Friction Test Apparatus

The macroscale tribological (friction and wear) characterizations of human hair were conducted using a flat-on-flat tribometer under reciprocating motion [40.27, 30, 41] (Fig. 40.15). A piece of square polyurethane film (film area 25–400 mm^2) was fixed at the end of a cantilever beam. The hair strands were mounted on the Si wafer in such a way that all strands of hair were separated and parallel to each other. For high-temperature studies, the Si wafer was placed on a heating stage, which could increase the temperature of the hair sample up to 100°C [40.41]. The heat-generating elements of the heating stage were ohmic resistors encapsulated in a steel holder that were kept in good thermal contact using thermal paste. J-type thermocouples were used to measure the sample temperature. A thermal controller and a solid state relay were used to control the temperature by adjusting the on/off time. A glass plate was attached at the bottom of the heating stage to isolate it thermally from the x–y axis stage, where the heating stage was mounted. The x–y axis stage was a motor-driven lead-screw-type stage driven by a stepper motor. The load on the polyurethane film was applied by lowering the beam against the hair sample using a microactuator. Normal and frictional forces were measured with semiconductor strain gauge beams mounted on a crossed-I-beam structure as part of the cantilever beam mentioned earlier, and data were digitized and collected on a personal computer. The effect of relative humidity on hair friction was studied in an environmentally controlled chamber, in which the humidity was controlled by introducing a combination of dry air and moist air.

In order to select the relevant load, velocity, and film area, one needs to be guided by the application. For a finger feeling hair, the normal load applied by the finger was estimated as 50–100 mN, measured by pressing the finger on a microbalance. The estimated apparent contact size and stroke length were 10–100 mm^2 and 5–20 mm, respectively. The sliding velocity was estimated as 5–20 mm/s. To perform a parametric study, tests were performed at a range of operating conditions in the range of interest for the application at a temperature of 22± 1°C and 50±5% RH. The following test conditions were used: stroke length, 3 mm; slid-

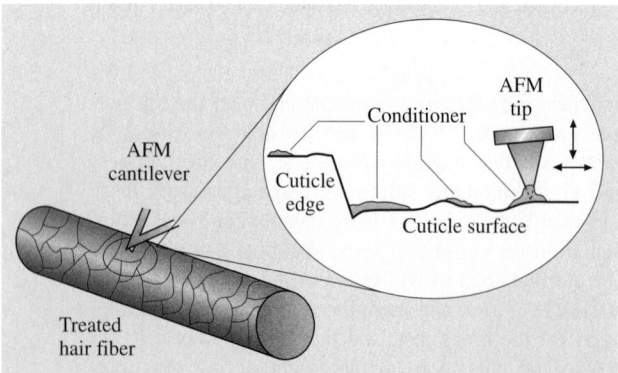

Fig. 40.16 Interactions between the AFM tip and conditioner on the cuticle surface for treated hair (after [40.12])

ing velocity, 0.4–4.5 mm/s; normal load, 50–100 mN; film size, 25–400 mm². For the friction studies of polyurethane film vs. various hair samples, hair vs. hair, and polyurethane film vs. virgin and treated hair samples at dry and wet conditions, the following nominal test conditions were chosen: sliding velocity, 1.4 mm/s; normal load, 50 mN; and film area, 100 mm². To simulate the wet conditions, plumber's putty was placed around the stage and the hair region was filled with water. For the wear measurements, the polyurethane film was rubbed against the virgin and treated Caucasian hair for 24 h at the above nominal conditions and the coefficient of friction was measured. The hair surface was studied using an optical microscope prior to and after wear tests. To study the effect of temperature and humidity on hair friction, the tests were conducted at the nominal test conditions and the following parameters were used: temperature, 22–80°C; relative humidity, 35–85 %.

Nanotribological Characterization Using an AFM

Specimen Mounting. Hair specimens were mounted onto AFM sample pucks using Liquid Paper (Sanford Corp., Oak Brook, IL, USA) correction fluid. A thin layer of the fluid was brushed onto the puck, and when the fluid hardened into a tacky state, the hair sample was carefully placed on it. The Liquid Paper dries quickly to keep the hair firmly in place. An optical microscope was used to preliminarily image the specimen to ensure that none of the Liquid Paper was deposited on the hair surface.

Synthetic materials were attached to AFM sample pucks using double-sided adhesive tape.

Surface Roughness, Friction Force and Adhesive Force Measurements. Surface roughness and friction force measurements were performed using a commercial AFM system (MultiMode Nanoscope IIIa, Digital Instruments, Santa Barbara, CA) under ambient conditions (22± 1°C, 50±5% relative humidity) [40.1, 2, 12]. Square pyramidal Si_3N_4 tips of nominal 30–50 nm radius attached to the end of a soft Si_3N_4 cantilever beam (spring constant of 0.06 N/m) were used for surface roughness and friction force measurements. A softer cantilever was used to minimize damage to the hair. After engagement of the tip with the cuticle surface, the tip was scanned perpendicular to the longitudinal axis of the fiber. The tip was centered over the cross-section in order to be at the very top of the fiber, so as to negate effects caused by the AFM tip hitting the sides of the hair and adding error to the measure-

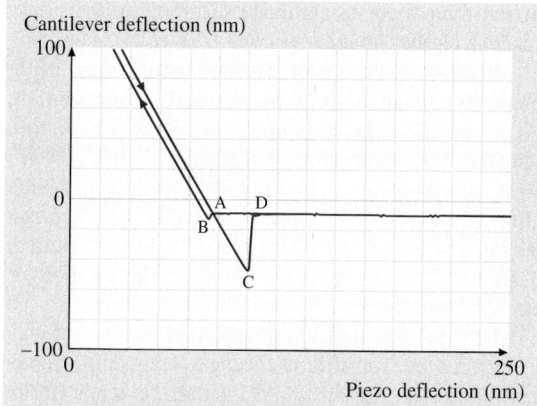

Fig. 40.17 Typical force calibration plot for Caucasian virgin hair. Contact between the tip and hair occurs at point B. At point C, the elastic force of the cantilever becomes equivalent to the adhesive force, causing the cantilever to snap back to point D

ments. In order to minimize scanning artifacts, a scan rate of 1 Hz was used for all measurements. Topographical images characterizing the shape and structure of the various hair were taken at scan sizes of 5×5, 10×10, and $20 \times 20 \mu m^2$ at a normal load of 5 nN. These scan sizes were suitable for capturing the surface features of multiple scales and scale edges of the cuticle. To characterize roughness, $2 \times 2 \mu m^2$ scans of the cuticle surface (without edges) were used. Friction force mapping of the scan area was performed at the same time as the roughness mapping. Figure 40.12 (bottom cartoon) and Fig. 40.16 show the AFM tip scanning over the hair surface for untreated and conditioner-treated hair, respectively. The effects of the conditioner can be examined by comparing the friction information. A quantitative measure of friction force was calculated by first calibrating the force based on a method by *Bhushan* [40.27, 30]. The normal load was varied from 5 nN to 45 nN in roughly 5 nN increments, and a friction force measurement was taken at each increment. By plotting the friction force as a function of normal force, the average coefficient of friction was determined from the slope of the least squares fit line of the data.

The surface roughness images shown in this study were processed using a first-order "planefit" command available in the AFM software, which eliminates tilt in the image. Roughness data as well as friction force data were taken after the "planefit" command was employed. A first-order "flatten" command was also applied to the friction force maps to eliminate scanning artifacts and present a cleaner image.

Adhesive force measurements were made with square pyramidal Si_3N_4 tips attached to the end of a Si_3N_4 cantilever beam (spring constant of 0.58 N/m), using the "force calibration plot technique". In this technique, the AFM tip is brought into contact with the sample by extending the piezo vertically and then retracting the piezo and calculating the force required to separate the tip from the sample. The method is described in detail by *Bhushan* [40.27, 30–32].

The cantilever deflection is plotted on the vertical axis against the z-position of the piezo scanner in a force calibration plot, Fig. 40.17. As the piezo extends (from right to left in the graph), it approaches the tip and does not show any deflection while in free air. The tip then approaches within a few nanometers (point A) and becomes attached to the sample via attractive forces at very close range (point B). After initial contact, any extension of the piezo results in a further deflection of the tip. Once the piezo reaches the end of its designated ramp size, it retracts to its starting position (from left to right). The tip goes beyond zero deflection and enters the adhesive regime. At point C, the elastic force of the cantilever becomes equivalent to the adhesive force, causing the cantilever to snap back to point D. The horizontal distance between A and D multiplied by the stiffness of the cantilever results in a quantitative measure of adhesive force [40.27, 30, 31, 33].

The force calibration plot allows us to calculate the adhesive force at a distinct point on the sample surface. Consequently, by constructing a force calibration plot at discrete sampling intervals over an entire scan area, adhesive force mapping (also known as force–volume mapping) can be performed in order to study the variation in adhesive force over the surface [40.42]. In this work, plots were taken at 64×64 distinct points over a scan area of $2 \times 2 \, \mu m^2$ for all hair types and ethnicities. Current Digital Instruments software for the Multimode AFM does not allow direct calculation of the adhesive force from the force–volume maps. Thus, a custom program coded in Matlab was used to display the force–volume maps. The adhesive force for each force calibration plot was obtained by multiplying the spring constant by the horizontal distance (in the retract mode) traveled by the piezo from the point of zero applied load to the point where the tip snaps off.

Relative Humidity, Temperature, Soaking and Durability Measurements. A humidity–temperature detector was used to monitor the humidity inside a Plexiglas test

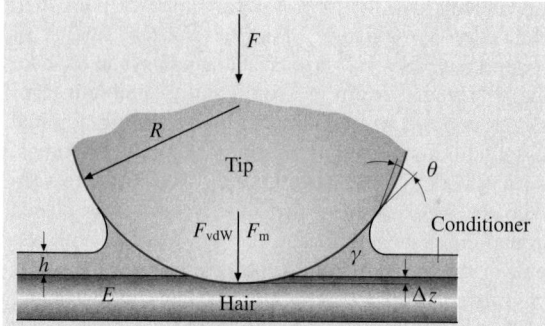

Fig. 40.18 An enlarged cross-sectional view of the region around the AFM tip, conditioner and hair surface. The parameters used in the text are defined. Here, R is the tip radius, h is the conditioner film thickness, γ is the surface tension of the conditioner, θ is the contact angle between the tip and conditioner, F is the load applied, F_a is the adhesive force acting on the tip, which is the sum of van der Waals force F_{vdW} and the meniscus force F_m, Δz is the indention on the hair surface under the applied load F and the adhesive force F_a, and E is the Young's modulus of the sample

chamber enclosing the AFM system. An appropriate experimental set-up was used to control the humidity inside the chamber. Measurements were taken at nominal relative humidities of 5, 50 and 80%. Hair fibers were exposed to each humidity level for approximately 2 h prior to measurements.

A home-made thermal stage was used to study the effect of varying temperature between 20, 37, 50, and 80°C. Hair fibers were exposed to each temperature condition for approximately 30 min prior to testing.

Soak tests were performed as follows. A dry hair fiber was taken from a swatch, and a sample was cut from the fiber (approximately 7 mm long) for coefficient of friction measurements. An adjacent sample was also taken from the fiber and placed in a small beaker filled with deionized water. The sample was subjected to the aqueous environment for 5 min, which is representative of a typical exposure time when showering/bathing, and then immediately analyzed with AFM. It should be noted that hair becomes saturated when wet in about 1 min and remains saturated if kept in a humid environment. It was determined from unpublished results that if the wet hair was exposed to the ambient environment for more than 20 min while in the AFM, the hair became dry and the coefficient of friction became similar to that of dry hair. Thus, the coefficient of friction measurements were made within a 20-min timeframe for each sample.

In order to simulate the scratching that can occur on the surface of hair and its effect on the friction force on the cuticle surface, a durability test was conducted using a stiff silicon tip (spring constant of 40 N/m). A load of 10 μN was used on a 2 μm section of the cuticle. A total of 1000 cycles were performed at 2 Hz. Measurements were conducted using an AFM. The friction force signal was recorded with respect to cycling time.

Conditioner Thickness. A commercial AFM system (MultiMode Nanoscope IIIa, Veeco, Santa Barbara, CA) was used in ambient conditions (22± 1°C, 50±5% relative humidity) [40.11]. The measurements were made with FESP (force modulation etched Si probe) tips (100 nm radius, spring constant of 5 N/m), using the force calibration plot technique ("force–volume mode" of the nanoscope software). The typical FESP tip radius was 5–20 nm, but blunt tips were preferred in the study so that when the tip compressed on the surface, the surface tended to deform elastically instead of being indented (plastically deformation). Then Hertz analysis can be applied in order to calculate the effective Young's modulus. Figure 40.11 shows a schematic diagram of the AFM, and an enlarged cross-sectional view of the AFM tip, conditioner and hair surface is shown in Fig. 40.18.

The force calibration plot allows us to obtain the deflection of the cantilever as a function of the piezo position at a distinct point on the sample surface. Figure 40.19a shows a typical force calibration plot curve for commercial conditioner-treated hair. The conditioner thickness and adhesive force can be extracted from this plot. The snap-in distance H_s, which is proportional to the real film thickness, is the horizontal distance between point B and point D. The adhesive force F_a (on the retract curve), which is the force needed to pull the sample away from the tip and is the sum of the van der Waals force F_{vdW} mediated by the adsorbed water or conditioner layer and the meniscus force F_m due to the Laplace pressure ($F_a = F_{vdW} + F_m$), can be calculated from the force calibration plot by multiplying the spring constant by the vertical distance between point B and point F. The meniscus force F_m is given by

$$F_m = 2\pi R\gamma(1 + \cos\theta) , \qquad (40.1)$$

where R is the tip radius, γ is the surface tension of the conditioner, and θ is the contact angle between the tip and conditioner.

In addition, the Young's modulus of the sample can be determined using Hertz analysis,

$$F + F_a = \frac{4}{3}\sqrt{R}E\Delta z^{3/2} , \qquad (40.2)$$

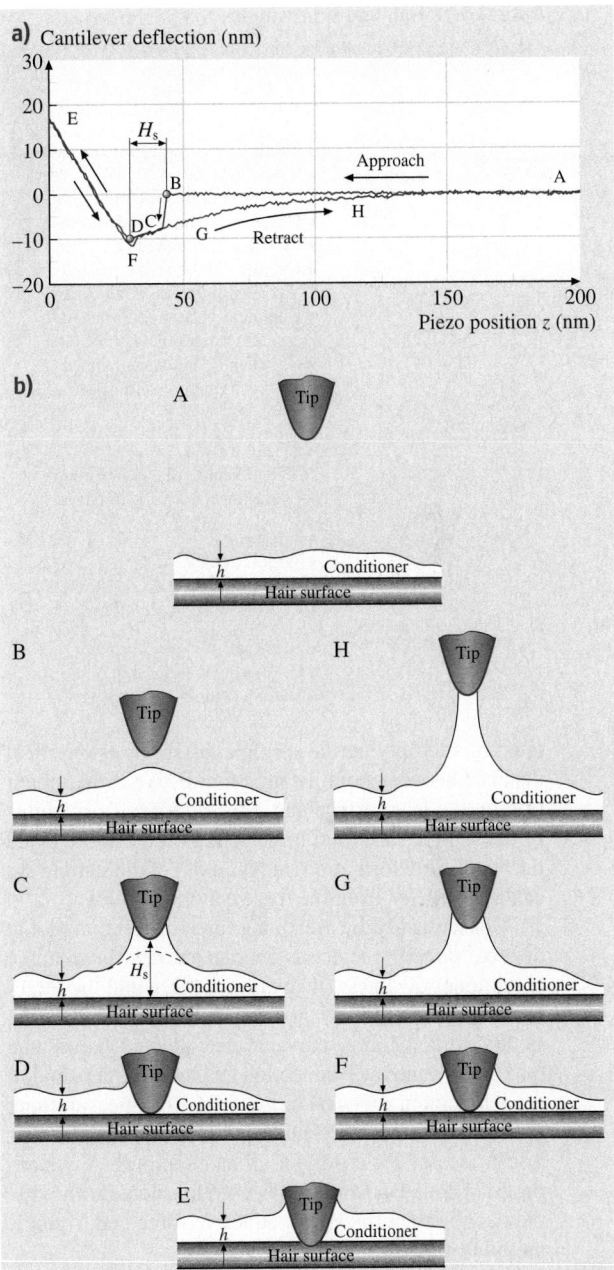

Fig. 40.19 (a) A typical force calibration plot for commercial conditioner-treated (three cycles) hair; (b) schematic diagrams of the AFM tip, conditioner and hair surface at different tip–sample separations, as labeled in (a) (after [40.11])

where R is the tip radius, $F + F_a$ is the total force acting on the surface on the approach curve, which is calcu-

Table 40.11 Hair and skin samples

Hair and skin samples	
Sample	**Type**
Caucasian hair	• Virgin • Virgin, treated (1 cycle commercial conditioner) • Chemomechanically damaged • Chemically damaged • Chemically damaged, treated (1 cycle commercial conditioner) • Chemically damaged, treated (3 cycles commercial conditioner) • Chemically damaged, treated (various ingredients, Table 40.12)
Asian hair	• Virgin • Virgin, treated (1 cycle commercial conditioner) • Chemomechanically damaged
African hair	• Virgin • Virgin, treated (1 cycle commercial conditioner) • Chemomechanically damaged
Synthetic materials	• Artificial collagen film (hair) • Polyurethane film (skin) • Human skin (putty replica)

lated by multiplying the spring constant by the vertical distance between point D and point E, Δz is the indention on the hair surface and E is the Young's modulus of the sample. The total force acting on the surface and the resulting deformation (indentation) of the sample Δz can be extracted from the force calibration plot.

Consequently, by creating a force calibration plot at discrete sampling intervals over an entire scan area, the conditioner thickness, the adhesive force and the effective Young's modulus can be mapped over the surface. In this work, the force curves were collected at the same maximum cantilever deflection (relative trigger mode), at each point in a 64×64 array (total 4096 measurement points), with each force curve sampled at 64 points over a scan area of $2 \times 2 \,\mu m^2$ for all hair samples. A custom program coded in Matlab was used to calculate and display conditioner thickness, adhesive force and Young's modulus maps.

40.2.2 Hair and Skin Samples

For the research reported in this chapter, all hair samples were received from Procter & Gamble (Cincinnati, OH) and Procter & Gamble Far East (Kobe, Japan), and the shampoo and conditioner treatments used were as follows.

Shampoo treatments consisted of applying a commercial shampoo evenly down a hair swatch with a syringe. Hair was lathered for 30 s, rinsed with tap water for 30 s, and then this was repeated. The amount of shampoo used for each hair swatch was $0.1 \,cm^3$ shampoo per gram of hair. The swatches were then hung up to dry in an environmentally controlled laboratory, and finally wrapped in aluminium foil.

When a commercial conditioner was applied, $0.1 \,cm^3$ of conditioner was used per gram of hair. The conditioner was applied in a downward direction (scalp to tip) thoroughly throughout the hair swatch for 30 s, and then allowed to sit on the hair for another 30 s. The swatch was then rinsed thoroughly for 30 s. The swatches were then hung up to dry in an environmentally controlled laboratory, and finally wrapped in aluminium foil.

Samples arrived as hair swatches that were approximately 0.3 m long. Although the exact location from the root is unknown, it is estimated that hair samples used for testing were between 0.1 to 0.2 m from the scalp. Table 40.11 presents a list of all samples used. The main hair categories of interest are: virgin (untreated), virgin (treated with one cycle of commercial conditioner), chemomechanically damaged (untreated), chemically damaged (untreated), and chemically damaged (treated with one or three cycles of commercial conditioner or matrix of conditioner ingredients). Virgin samples are considered to be baseline specimens and are defined as having an intact cuticle and an absence of chemical damage. Chemomechanically damaged hair fibers have been exposed to one or more cycles of coloring and permanent wave treatment, washing and drying, as well as combing (to contribute mechanical damage), which are representative of common hair management and alteration. In the case of African damaged hair samples, chemical damage occurred only by chemical straightening. Chemically damaged fibers were not exposed to the combing sequence in their preparation. All treated hair samples were treated with either one or three rinse/wash cycles of a conditioner, similar to a Procter & Gamble commercial product, or were treated with various combinations of surfactant, fatty alcohol, and silicone types and deposition levels, presented in the matrix of Table 40.12 [40.2]. Two different types of cationic surfactants were used: behentrimonium chloride (BTMAC) and behenyl amidopropyl dimethylamine (BAPDMA). Only one group of fatty alcohols were used for all samples. In the last set of samples, two different silicones were used: a PDMS (a blend with high MW) silicone and an amino silicone.

Table 40.12 Matrix of hair samples treated with various combinations of ingredients

	Ingredient	Damaged and treated (commercial conditioner)	BTMAC surfactant, no silicone	PDMS-blend silicone, low deposition	PDMS-blend silicone, high deposition	Amino silicone, low deposition	Amino silicone, high deposition	BAPDMA surfactant, no silicone
Cationic surfactants (wt%)	Behentrimonium chloride (BTMAC)	−	×	×	×	×	×	−
	Behenyl amidopropyl dimethylamine (BAPDMA)	−	−	−	−	−	−	×
	Stearamidopropyl dimethylamine (SAPDMA)	×	−	−	−	−	−	−
Fatty alcohols (wt%)	Fatty alcohols	×	×	×	×	×	×	×
Silicones (wt%)	PDMS-blend (dimethicone)	×	−	−	−	−	−	−
	PDMS-blend (with blend of high MW)	−	−	Low	High	−	−	−
	Amino silicone	−	−	−	−	Low	High	−
Deposition levels for conditioner ingredients on hair								
Cationic surfactant deposition level	Similar to commercial conditioner	×	×	×	×	×	×	
Fatty alcohol deposition level	Similar to commercial conditioner	×	×	×	×	×	×	
Silicone deposition level	Similar to commercial conditioner	None	Low	High	Low	High	None	

Table 40.13 Contact angles and surface energies of materials associated with nanotribological characterization of hair

		Contact angle (°)	Surface energy (N/m)
Human hair	• untreated	103 [40.43]	0.024 [40.43]
		100 [40.2]	0.028 [40.2]
	• damaged		
	brown color	60 [40.2]	0.038 [40.2]
	blonde color	55 [40.2]	0.047 [40.2]
PDMS (bulk)		105 [40.44]	0.020 [40.45]
Human skin	• forehead	55 [40.46]	0.043 [40.46]
	• forearm	88 [40.46]	0.038 [40.46]
		84 [40.47]	0.029 [40.47]
	• finger	74 [40.47]	0.024 [40.47]
		58 [40.48] (before soap-washing)	0.027 [40.48]
		104 [40.48] (after soap-washing)	
Si_3N_4 film		35 [40.2]	0.047 [40.49]

Collagen film is typically used as a synthetic hair material for testing purposes. Polyurethane films represent synthetic human skin. They are cast from human skin and have similar surface energy, which also makes them suitable test specimens when real skin cannot be used. To characterize the surface roughness of human skin, it was also replicated using a two-part silicone elastomer putty (DMR-503 Replication Putty, Dynamold, Inc., Fortworth, TX). The thickness of the film was approximately 3 mm.

In order to simulate hair conditioner–skin contact in AFM experiments, it is important to have the contact angle and surface energy of the AFM tip close to that of skin. Table 40.13 shows the contact angles and surface energies of materials important to the nanocharacterization of the hair samples: untreated human hair; PDMS, which is used in conditioners; skin, which comes into contact with hair; and Si_3N_4 film, which is used for nanotribological measurements in the form of an AFM tip [40.2]. The static contact angle of the Si_3N_4 film with high-purity deionized water was measured in air using a sessile-drop method with a contact angle goniometer (Model 100, Rame-Hart Inc., Mountain Lakes, NJ, USA). 5 μl of deionized water was applied to two samples using a micropipette, and three contact angle measurements were taken on each sample and averaged.

40.3 Structural Characterization Using an AFM

The schematic diagram in Fig. 40.2 provides an overview of various cellular structures in human hair [40.3, 20]. Human hair is a complex tissue consisting of several morphological components (Fig. 40.2), and each component consists of several different chemical species. Table 40.3 summarizes the main chemical species present in human hair.

Traditionally, most characterizations of the cellular structure of human hair or wool fiber have been done using SEM and TEM. More recently, the cellular structure of human hair have been characterized using AFM and TR mode II (described earlier) [40.10, 11]. The structure of hair in cross-section and in longitudinal section is presented in Sect. 40.3.1, while the structures of various cuticle layers present in human hair are presented in Sect. 40.3.2.

40.3.1 Structure of Cross-Sections and Longitudinal Sections of Hair

Cross-Section of Hair
Figure 40.20a shows AFM images of Caucasian hair in cross-section. The hair fiber is seen embedded in epoxy resin. The cortex region, the cuticle region and the epoxy resin region are easily identified from the TR amplitude and TR phase images. Five layers of cuticle cells are seen

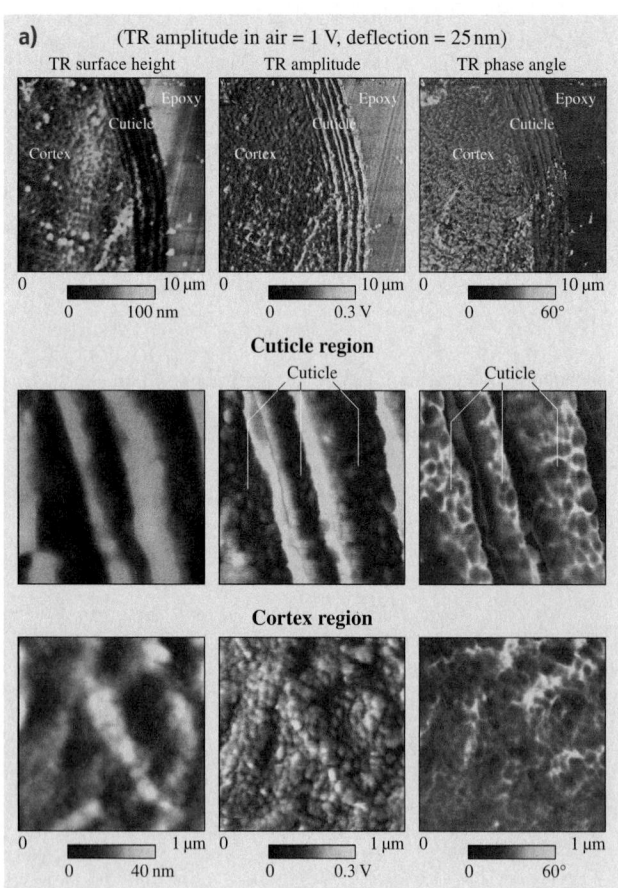

Fig. 40.20 (a) Cross-section images of virgin Caucasian hair (*upper*) and fine detail images of the cuticle region (*middle*) and cortex region (*lower*); (**b**) longitudinal section images of virgin Caucasian hair (*upper*) and fine detail images of the cuticle region (*middle*) and cortex region (*lower*). Note that the longitudinal section is not perfectly parallel to the long axis of the hair fiber but is skewed from it by a small angle, so the thicknesses found for the sublamellar layers of the cuticle are not the real thicknesses (after [40.10]) ◄►

in the cuticle region, and the total thickness of the cuticle region is about $2\,\mu m$ for this sample. In the detailed images of the cuticle region, three layers of cuticle cells are shown, and various sublamellar structures of the cuticle can be seen. The thickness of the cuticle cell varies from 200 nm to 500 nm. The cortex region shows very a fine circular structure about 50 nm in size, which represents the transverse face of the macrofibril and matrix. No intermediate filament structure can be seen at this scale.

Longitudinal Section of Hair

Figure 40.20b shows the AFM images of a longitudinal section of virgin Caucasian hair. Different regions (the cortex region, the cuticle region and the embedding epoxy resin) can be seen. As shown in the detailed image of the cuticle region, various sublamellar structures of the cuticle – the A-layer, the exocuticle, the endocuticle and the cell membrane complex – which are not easy to make out in TR surface height images, are clearly resolved in the TR amplitude and TR phase images due to the different contrast. Most of the sublamellar structural features of the cuticle shown in the TEM image of Fig. 40.7b can be identified in the TR amplitude and phase angle images. Previously, these sublamellar structures could only be distinguished by TEM [40.20]. The various cellular sublamellar structures in the cuticle have very different chemical compositions [40.3, 20]: the A-layer is rich in disulfide crosslinks due to a very high cystine content of up to 35%; the exocuticle is also rich in disulfide crosslinks (15% cystine); however, the endocuticle is relatively lightly crosslinked, since it contains only about 3% cystine. Consequently, these layers exhibit distinctly different stiffness and viscoelastic properties, and TR mode II imaging technique (TR amplitude and phase angle images) can easily detect these differences. Note that this longitudinal section is not perfectly parallel to the long axis of the hair fiber but skews away from it by a small angle; therefore, the thicknesses of various sublamellar layers of the cuticle do not represent the actual layer thicknesses. Two different morphological regions can be seen in the cortex region: the macrofibril and the matrix. The macrofibril is a bundle of intermediate filaments aligned parallel to each other, and it looks like a tree trunk; the matrix surrounds the macrofibril region. The matrix region has a high cystine content compared to that of the macrofibril region. This difference in chemical composition between the macrofibril and the matrix make it possible to reveal the fine internal cellular structure of hair using the AFM TR mode II technique.

40.3.2 Structures of Various Cuticle Layers

Virgin Hair

Figure 40.21 shows AFM images of the surface of virgin Caucasian hair. Two typical sample positions are shown: position 1 is near the root end of the hair, and position 2 is near the tip end of the hair. One cuticle edge is shown in position 1. (This is also seen in the TR amplitude and phase images as black strips due to topographic effects near the cuticle edge. The topographic effects tend to be significant only when there is a large change in local geometry.) The cuticle edge shows little natural weathering damage and is still intact with a step height of about 500 nm, and the general cuticle surface, which is covered with a lipid layer (the outer β-layer), is relatively uniform at large scales. In contrast, the surface near the tip of the hair (position 2) shows lots of damage, which may be simply due to natural weathering and mechanical damage from the effects of normal

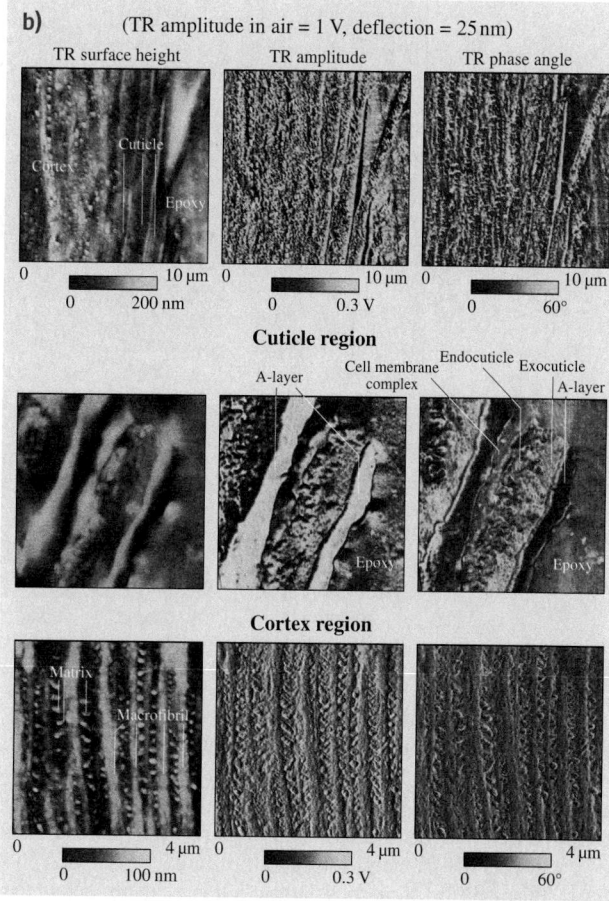

Part D | 40.3

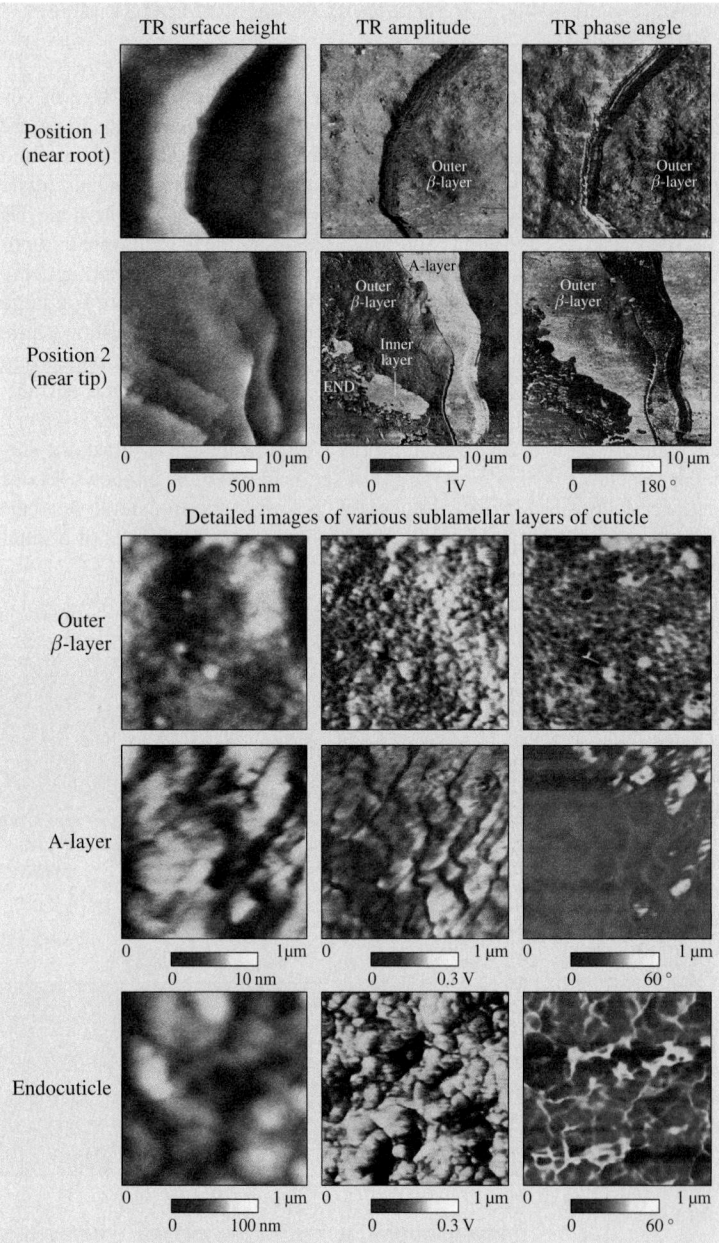

Fig. 40.21 Images of the surface of a virgin Caucasian hair. Two typical samples are shown: near the root end, where intact cuticle edges can be seen, and near the tip end, where damage has occurred – part of the cuticle top layer has been removed and the sublamellar layers beneath are exposed. *Detailed images* of the outer β-layer, the A-layer and the endocuticle are shown at the *bottom* (after [40.10])

grooming actions, such as combing, brushing and shampooing. Part of the cuticle's outer sublamellar layers was removed and the layers underneath (the A-layer, the endocuticle and the inner layer) are exposed. Because the different chemical contents of the various sublamellar layers of the cuticle results in different stiffnesses and viscoelastic properties, large contrasts can be seen in the TR amplitude and TR phase angle images. Note that the surface height within each individual sublamellar layer (the A-layer, the outer β-layer and the inner layer) is relatively uniform, so the topographic effects on the TR amplitude and phase are minor. Detailed images of the outer β-layer, the A-layer and the endocuticle are shown at the bottom of Fig. 40.21. All of these layers

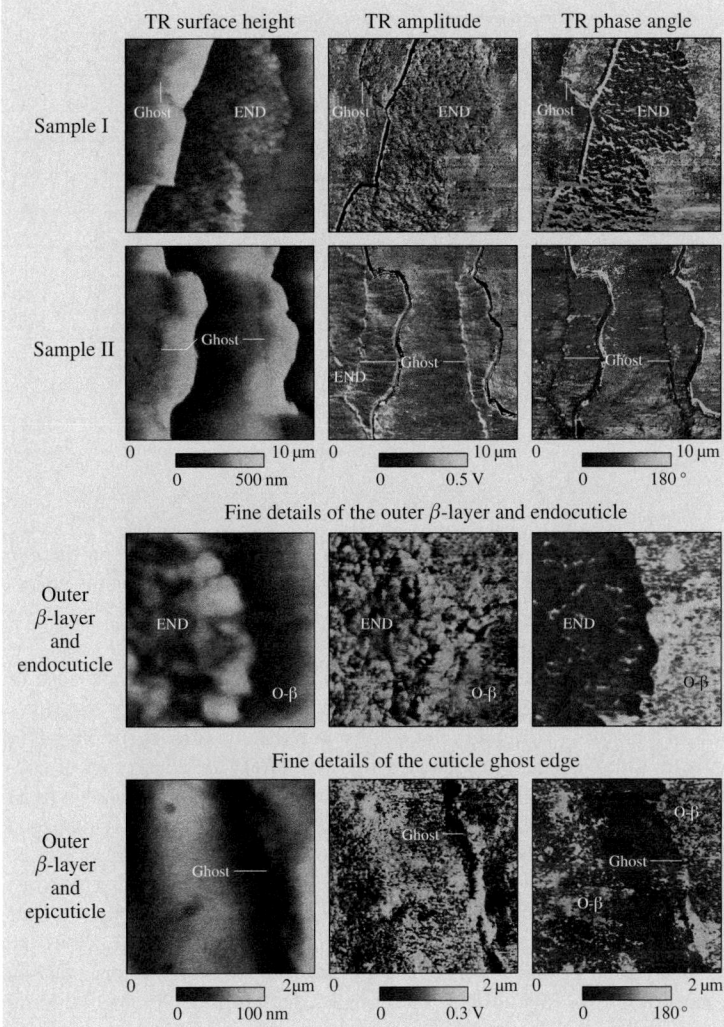

Fig. 40.22 Images of the surface of a chemically damaged Caucasian hair. Two typical images are shown: sample I in which large areas of the cuticle's top sublamellar layers are removed and the much rougher endocuticle layer is exposed; and sample II in which entire pieces of the cuticle are removed and only the cuticle ghost edges are left. *Detailed images* of the outer β-layer, the endocuticle and the epicuticle are shown at the *bottom* (after [40.10])

show distinct morphologies which can be clearly seen in the TR amplitude and phase angle images: the outer β-layer shows very fine granular structure; the A-layer shows few discernable features; while the endocuticle has a much rougher granular structure [40.20, 29] than that of the outer β-layer. Previous friction force microscopy (FFM) studies [40.19] on keratin fibers have indicated that no image contrast can be observed on the outer facing surfaces of the scales of untreated (virgin) fibers. These results indicate that the outer lipid layer may form fine domains, which cause the fine granular structure shown in the TR amplitude and phase images (Fig. 40.21). The TR mode II technique has higher sensitivity than the FFM technique, so the fine chemical distribution, which cannot normally be detected by FFM or other techniques, is readily revealed by TR mode II.

Chemically Damaged Hair

Figure 40.22 shows AFM images of surfaces of chemically damaged Caucasian hair. Two typical samples are shown. More damage can be seen than for the surface of virgin hair. More cuticle edges are removed and larger areas of rough granular endocuticle layers are often exposed (Sample I). Of the components within each cuticle cell (the A-layer, the exocuticle, the endocuticle, etc.), the endocuticle is the least crosslinked [40.3]. Under wet conditions it will swell preferentially and is the preferred plane for lamellar fracture under mechanical stress. In-

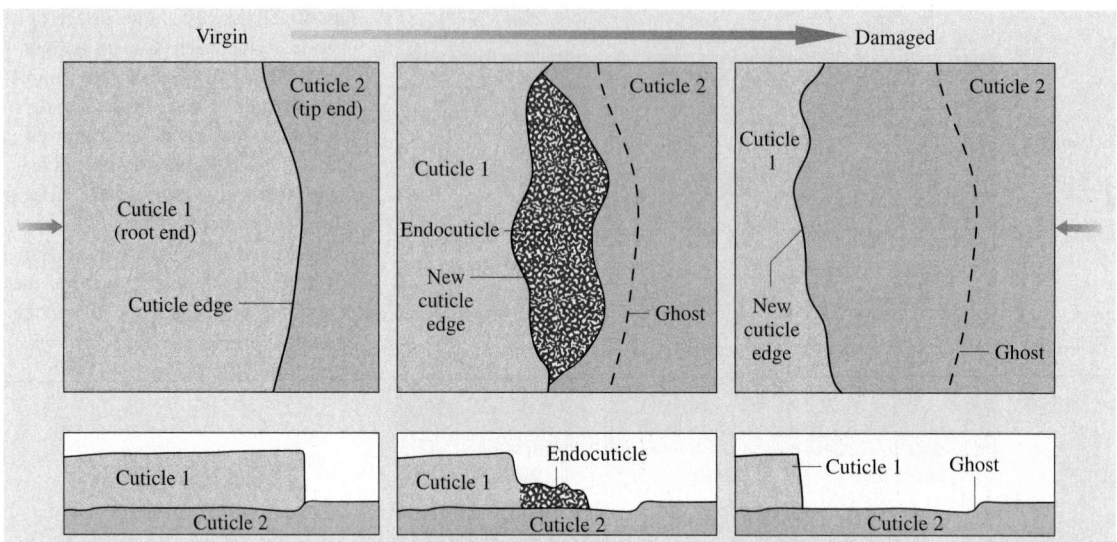

Fig. 40.23 Schematic of the progress of hair damage. Images of the hair surface at each stage are shown at the *top*. Cross-sectional views are shown at the *bottom*, which were taken at the locations denoted by the corresponding arrows (after [40.10])

deed, many examples have been observed where the cuticle has come off, leaving this granular endocuticle layer that is approximately half the thickness of the original scale and is located at the scale margins. In sample II, the endocuticle layers are eliminated, entire pieces of cuticles are removed and some fine lines on the cuticle surface which delineate the original boundaries of the cuticle edges are clearly apparent in TR amplitude images. These lines are referred to as the cuticle edge "ghost" in the literature [40.19, 29]. The actual fracture occurs at the interface between the outer β-layer and the δ-layer (Fig. 40.2). The outer β-layer was originally present, but because of its location (under the original overlying cuticle but close to the scale edge), it may have undergone oxidative loss through environmental or chemical exposure.

For easy visualization, Fig. 40.23 illustrates increasing hair damage. Virgin hair has an intact smooth cuticle edge; as damage occurs (natural weathering or mechanical damage), some of the cuticle outer sublamellar layers wear off and the layers beneath (such as the endocuticle) are exposed. Further damage will cause entire portions of cuticle to break off and the ghost which delineates the original boundary of the cuticle edge is seen.

Conditioner–Treated Hair

Various sublamellar layers of cuticle can be exposed at the surfaces of virgin hair and chemically damaged hair depending on the degree of damage. Because of their distinct chemical natures, these layers may have different interactions with the conditioner (or other hair care products) which will affect the adsorption of conditioner onto the hair surface. Figure 40.24 shows AFM surface images of two samples of chemically damaged treated hair. In sample I, intact cuticle edges can be seen. In the TR phase angle image, a higher contrast can be seen near the cuticle edges. We will see later that the conditioner is unevenly distributed across the hair surface, and a thicker layer of conditioner can be found near the cuticle edges [40.1, 12]. This build-up of conditioner might simply be caused by physical entrapment at the steps along the cuticle edges. Uneven thicknesses of conditioner cause the contrast seen in TR phase angle images. For sample II, a sharp cuticle edge and a cuticle ghost edge can be readily seen in the TR amplitude image. No endocuticle or other sublamellar layers can be seen because further treatments removed these layers. As shown in the TR phase angle image of sample II, the region between the cuticle edge and the cuticle ghost edge shows a different contrast to the other parts of the surface of the hair. This newly formed region of the surface may have an exposed epicuticle layer, while the other parts of the hair are still covered by the outer β-layer. The outer β-layer is basically a covalently attached lipid-covered layer that is hydrophobic. The interaction between the conditioner

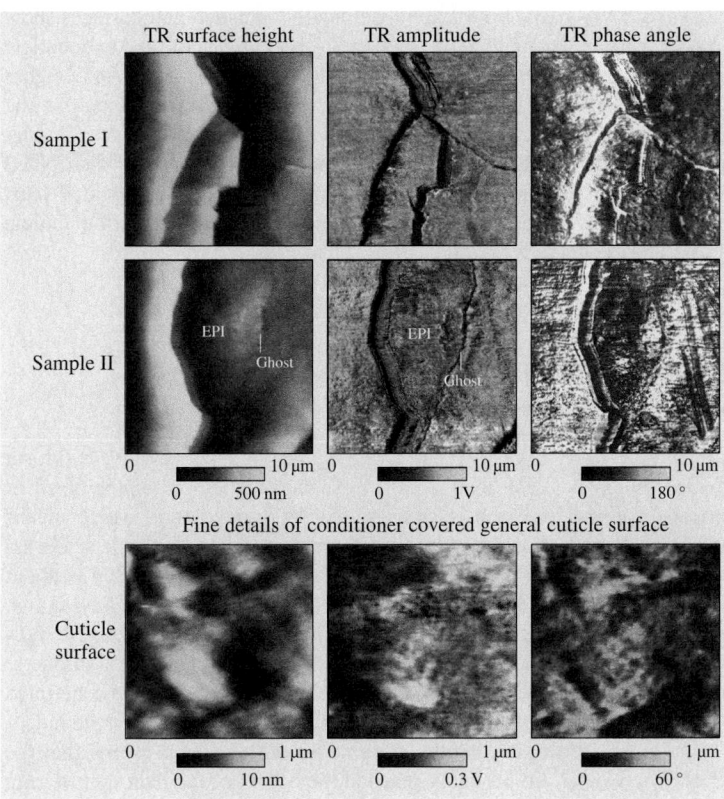

Fig. 40.24 Images of the surface of chemically damaged and treated hair (three cycles of conditioner treatment). Conditioner is unevenly distributed on the hair surface. A thicker layer of conditioner can be seen near the cuticle edges. *Fine details* of the conditioner-covered hair surface are shown at the *bottom*. No features are discernable (after [40.10])

and the outer β-layer or other sublamellar layers of the cuticle (such as the epicuticle) can be very different, so the adsorption of conditioner onto these surfaces vary too.

Fine details of the chemically damaged and treated cuticle surface (the outer β-layer) are shown at the bottom of Fig. 40.24. Unlike the fine granular structure of the outer β-layer shown in Fig. 40.21, no features can be discriminated now since the entire surface is covered with a layer of conditioner.

In summary, although the morphology of the fine cellular structure of human hair has traditionally been investigated using SEM and TEM, these techniques have limited ability to study the effect of the environment on the physical behavior of hair in situ. The AFM TR mode can be used to characterize the fine cellular structure of human hair, and many features previously only seen with SEM and TEM can be identified. AFM provides the ability to enhance studies of the effects of the environment (temperature, humidity, etc.) and hair-care product treatment on the physical behavior of human hair in situ.

40.3.3 Summary

SEM studies of hair in cross-section and AFM studies of the hair surface show that the cuticle is about 5–7 scales thick, and that each cuticle cell is about 0.3 to 0.5 μm thick. The visible length of each cuticle cell is about 5–10 μm. It appears that the morphology of hair is different from root to tip. That is, the hair near the scalp has complete cuticles, the hair in the middle has worn cuticles, and the hair near the tip seldom has cuticles. The size and shape of Caucasian, Asian and African hair have been measured from the hair surface and from cross-sections of hair. Asian hair seems to be the thickest (nearly round), followed by African hair (oval–flat) and Caucasian hair (nearly oval).

Cross-sections and longitudinal sections of virgin Caucasian human hair were investigated using the TR mode II. The cortex and the cuticle, the macrofibril and the matrix of human Caucasian hair were readily revealed. Various sublamellar cellular structures of cuticle, such as the A-layer, the exocuticle, the endocuticle and the cell membrane complex, are easily observed because of their distinct stiffnesses and viscoelastic properties. The surface features of various Caucasian human hair (virgin, chemically damaged, and chemically damaged and treated) were readily revealed. Sublamellar layers show distinct contrast in TR amplitude and phase angle images. The fine granular structure of the outer β-layer, which had not previously been seen using SEM, TEM and other AFM studies, is a result of the fine domain formation of

the lipid layer. Chemically damaged hair surfaces show much more damage; larger areas of the endocuticle are exposed. The endocuticle has a much rougher structure than the general cuticle surface (such as the outer β-layer), which could be part of the reason why damaged hair loses shine. Conditioner is distributed unevenly on the surface of damaged andtreated hair; thicker conditioner films are found near the cuticle edges.

40.4 Nanomechanical Characterization Using Nanoindentation and Nanoscratch

Nanomechanical characterization of human hair using nanoindentation and nanoscratch provides valuable information on the hair fiber itself, and also shows how damage and treatment affect important mechanical properties of the fiber [40.13, 14]. In Sect. 40.4.1, the hardness, Young's modulus and creep results for both the hair surface and cross-sections of hair are discussed. In Sect. 40.4.2, the coefficient of friction and scratch resistance of the hair surface are presented for unsoaked and soaked hair.

40.4.1 Hardness, Young's Modulus and Creep

Hair Surface

Figure 40.25 shows an optical micrograph of three indents on virgin Asian hair made at a normal load of 100 mN. The indentation depths and residual depths were about 5 μm and 3 μm, respectively. These indents are about 15 to 20 μm in diameter. This image clearly shows that the Nano Indenter II system can successfully make indents on the surface of human hair.

Figure 40.26a shows the load–displacement curves for virgin, chemomechanically damaged and virgin treated Caucasian hair obtained at four loads: 0.1, 1.0, 10 and 100 mN. The hardness and elastic modulus values corresponding to each load–displacement curve are listed in the figure boxes. As mentioned in the Experimental section, five indents were made at each load.

Figure 40.26a presents just one representative result for each load. At 0.1 mN, the indentation depths of all of these hair types were less than 150 nm, which means that the indents were made within one cuticle scale, assuming that the thickness of one cuticle scale is about 0.3 to 0.5 μm for all these hair samples. At 1.0 mN, the indentation depths were about 0.4 to 0.6 μm, indicating that indents were probably made through one to two cuticle scales. At 10 mN, the indenter tip penetrated about three to five cuticle scales. At 100 mN, the indentation depths were about 5 μm, which means that the tip probably reached the cortex of the hair, considering that a hair fiber generally has about 5–10 cuticle scales. It is interesting to observe that the loading curves of chemomechanically damaged hair obtained at 0.1 mN and 1.0 mN are not as smooth as those for virgin and virgin treated hair, especially at the beginning, indicating that the chemomechanically damaged hair was very soft over the first 30 to 50 nm or so, probably due to the changes to the exposed surface by chemical damage. At 0.1 and 1.0 mN, the hardnesses and elastic moduli of virgin treated and chemomechanically damaged hair are lower than those of virgin hair, indicating that chemical damage and treatment with conditioner led to the softness of the hair surface. If we consider human hair as a polymeric cylinder, the absorption of chemicals used in hair coloring and conditioner ingredients in the first micron or so might plasticize the polymer and hence reduce its mechanical properties. At 10 mN and 100 mN, the hardnesses and elastic moduli of all three hair types look similar. This result indicates that the effective depth of influences of the chemicals/conditioner is probably less than 1.5 μm; in other words, the first 3–4 scales of the cuticles probably interact with the chemicals and conditioner ingredients more effectively than the rest of the scales.

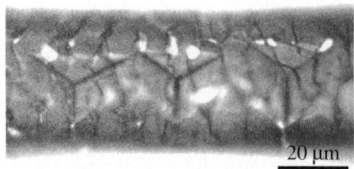

Fig. 40.25 Image of indents on hair surface (virgin Asian hair) (after [40.13])

20 μm

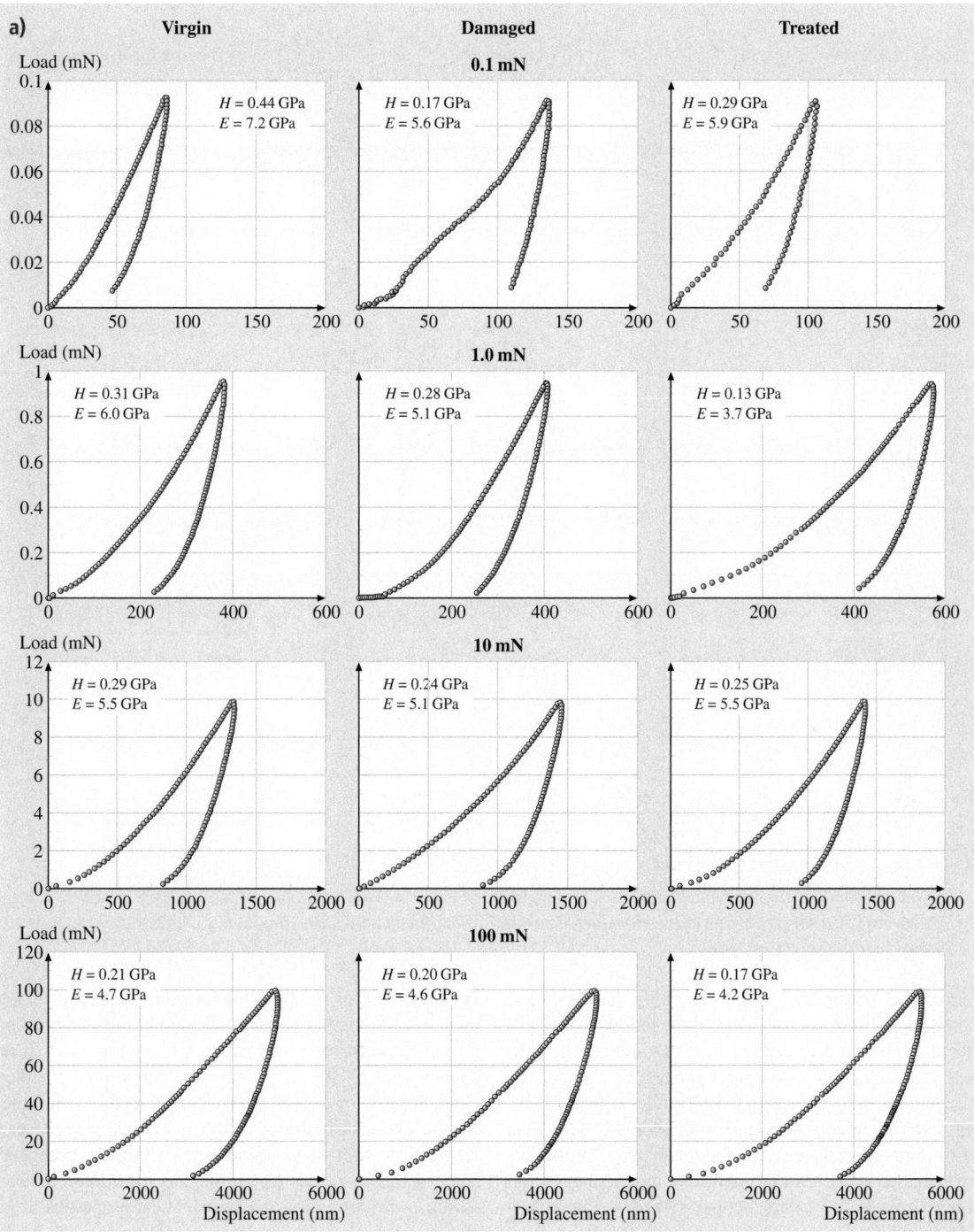

Figure 40.26b shows the hardnesses and moduli vs. indentation depth for various hair types. Each data point (averaged hardness and elastic modulus value) and each error bar in Fig. 40.26b was calculated from five in-

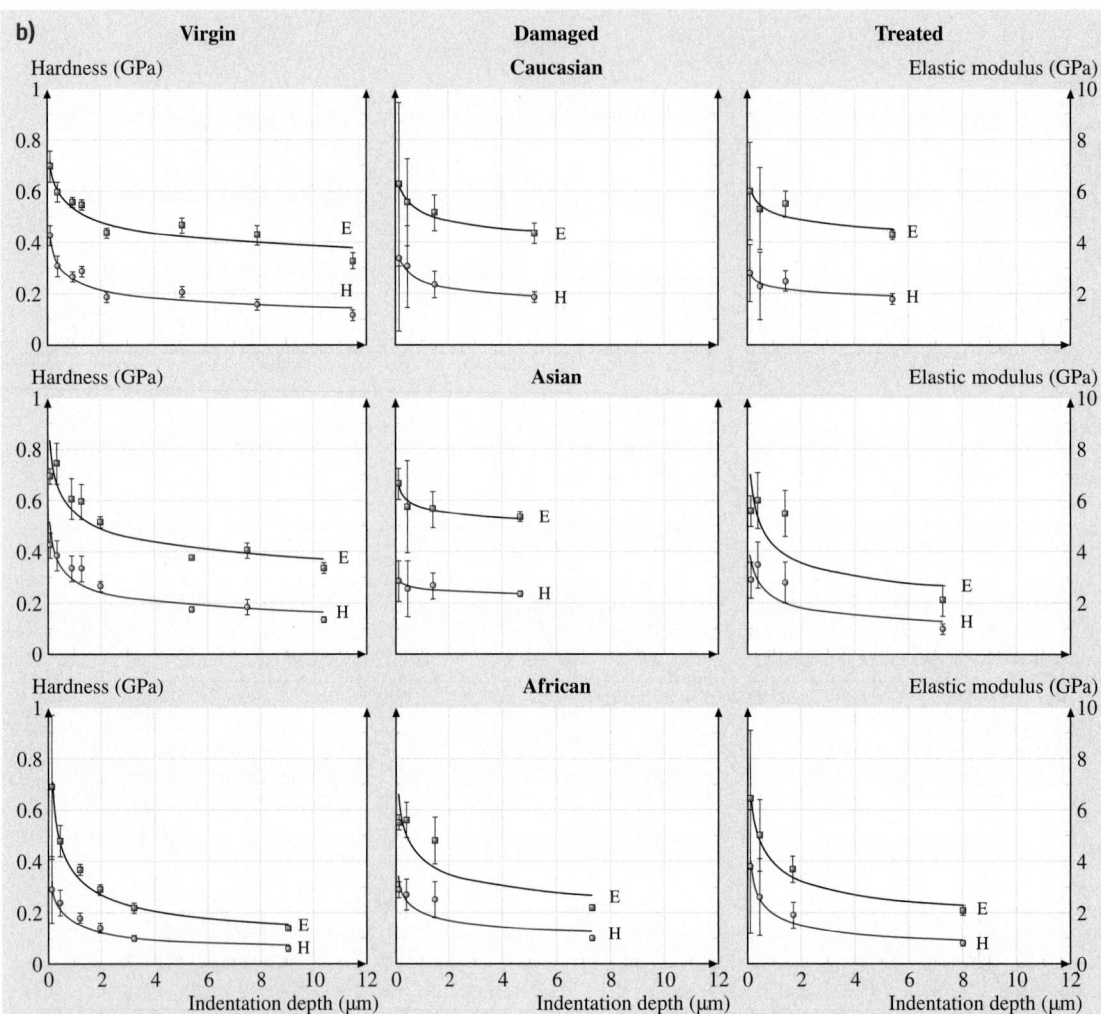

Fig. 40.26 (a) Load–displacement curves for Caucasian hair; (b) hardness and elastic modulus vs. indentation depth plots for various hair (after [40.13])◄ ▲

dentations. According to Fig. 40.26b, the hardness and elastic modulus of hair decrease as the indentation depth increases. In order to explain this, the indentation process is divided into two stages. In the first stage, the indenter tip only penetrated the cuticle scales, meaning that the indentation depth was probably less than 5 μm. For one cuticle scale, the mechanical properties are expected to decrease from the top layer to the bottom layer, because the cystine content (and so the disulfide crosslink density) decreases from the A-layer to the exocuticle to the endocuticle (Fig. 40.2, Table 40.4). The cuticle scales are bound together by the cell membrane complex, which is one of the weakest parts of

the hair fiber in terms of mechanical properties. The intercellular cement of the cell membrane complex is primarily nonkeratinous protein, and is low in cystine content (≈2%). As the indenter tip penetrates the cuticle scales one by one, the number of the cell membrane complex layers penetrated by the tip gradually increases. Joining together these weak cell membrane complex layers might lead to a deeper displacement upon indentation, contributing to lower mechanical properties. It is also possible that the outer scales of the cuticles have higher mechanical properties than the inner scales. In the second stage, the tip began to penetrate the cortex. In general, the cuticles are richer in disulfide crosslinks

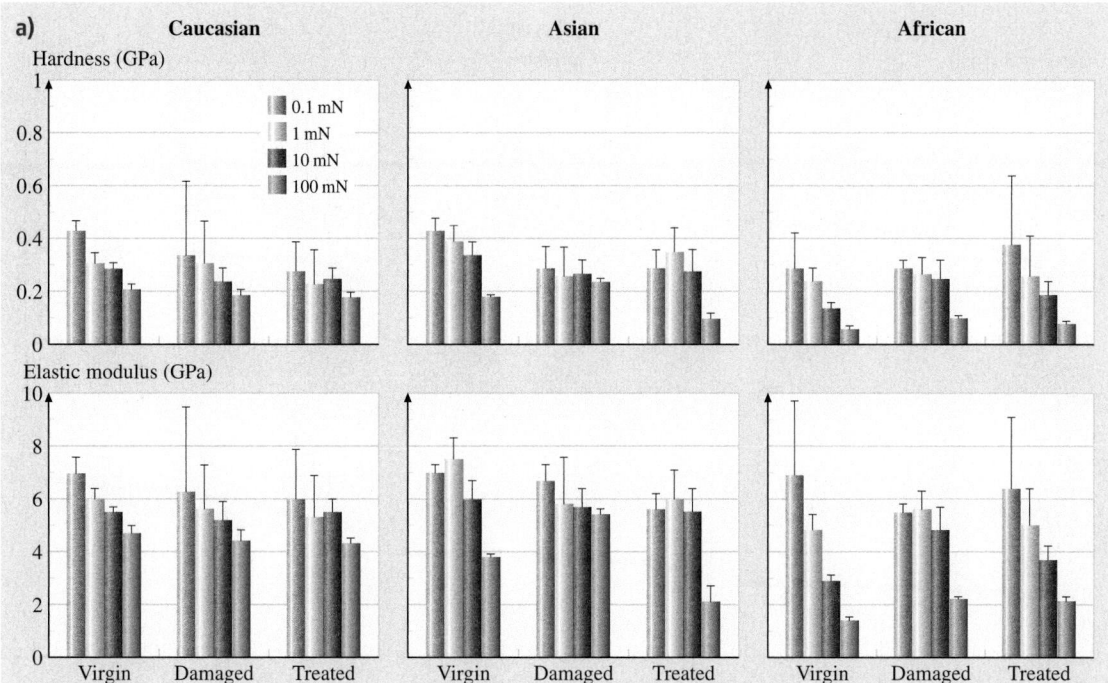

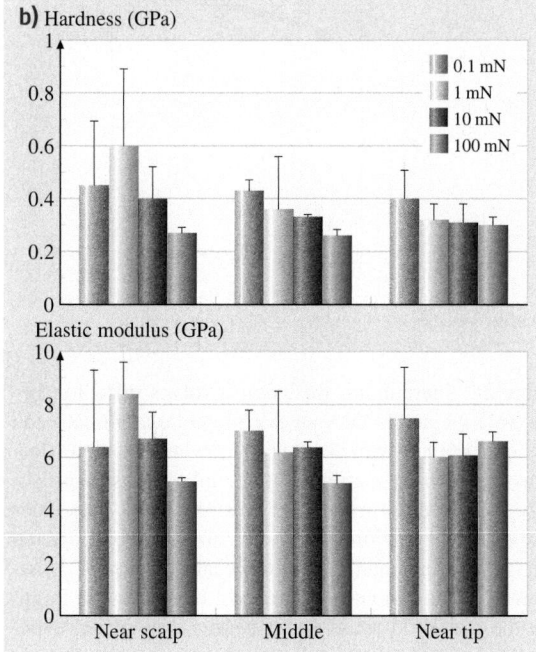

Fig. 40.27 (**a**) Summary of hardness and elastic modulus for various hair; (**b**) summary of hardness and elastic modulus for virgin Caucasian hair at three locations (after [40.13])

of the cuticle. Putting the two indentation processes together, the hardness and elastic modulus of hair will decrease as a function of indentation depth.

Figure 40.26b also indicates that at normal loads of 0.1, 1.0 and 10 mN, corresponding to an indentation depth of less than 1.5 μm, the chemomechanically damaged and virgin treated hair generally had lower hardnesses and elastic moduli, but larger error bars (larger deviations in the data), than the virgin hair for each ethnicity. This result means that the effective interaction depths were probably less than 1.5 μm for all three ethnicities of hair, and that the effects and distributions of the conditioner on the hair surfaces were not uniform. It is believed that most of the important interactions between shampoo/conditioner and hair occur at or near to the hair surface (the first few micrometers of the fiber periphery). Nanomechanical characterization of the hair surface shows that the effective interaction depth (< 1.5 μm) may be shallower than what was thought before. In general, two types of interaction occur between chemical/conditioner ingre-

than the cortex [40.3,4]. Therefore, the mechanical properties of the cortex are expected to be lower than those

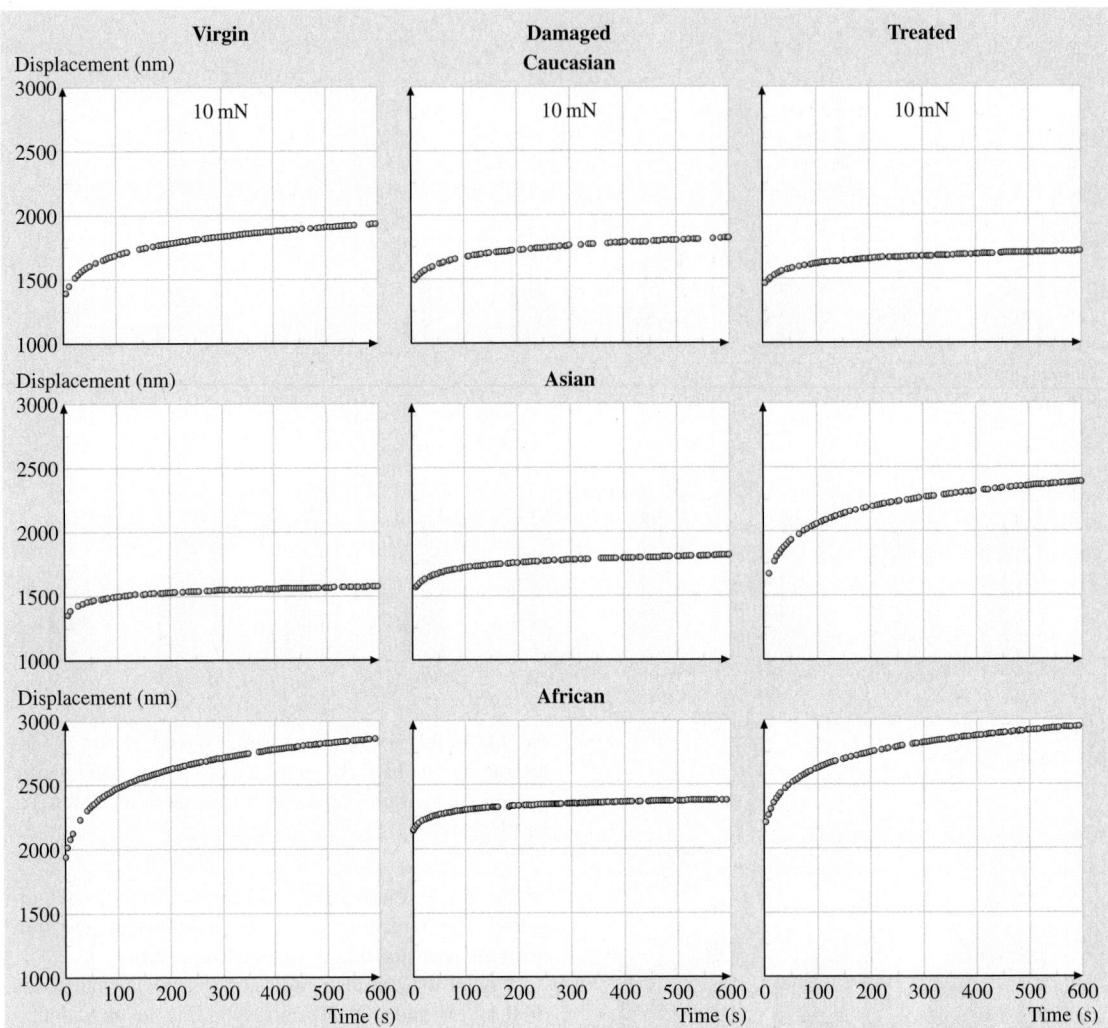

Fig. 40.28 Creep displacement vs. time curves for various hair (after [40.13])

dients and hair: adsorption and absorption. It has been suggested that for conditioning ingredients in hair conditioners, adsorption is more critical than absorption, because the conditioning ingredients are relatively large species [40.3]. If this is the case, then the data variation was probably caused by the nonuniform adsorption of chemical molecules and the conditioning ingredients to the hair surface. Because the interaction affected the hair up to a depth of 1.5 μm, absorption should also play an important role here. Transcellular and intercellular diffusion are the two theoretical pathways for absorption to occur. The transcellular route involves diffusion across cuticle cells through both high and low crosslinked

proteins. Intercellular diffusion involves diffusion between cuticlecells through the intercellular cement and the endocuticle, which are low in cystine content (low crosslink density regions). Intercellular diffusion is usually the preferred method of entry for most molecules (especially large ones such as surfactants, and even for species as small as sulfite near neutral pH). However, for small molecules, transcellular diffusion might be the preferred route under certain conditions, especially if the highly crosslinked A-layer and exocuticle are chemomechanically damaged [40.3]. Depending on the molecular size and the condition of the hair, the diffusion pathway and diffusion rate might vary from site to site on the hair surface, and so the distribution of conditioner might not be uniform. To sum up, for chemomechanically damaged and virgin treated hair, since the adsorption and absorption of chemicals and conditioner ingredients were probably not uniform on the hair surface, the nanomechanical properties of the hair surface (depth $< 1.5\,\mu m$) were not affected (they generally decreased) uniformly, leading to larger variations in the data compared to that for virgin hair. This implies that the nanoindentation technique can be used to quantitatively evaluate the effective depth of the conditioned hair and the distribution of conditioner by measuring the hardness and elastic modulus of the hair surface before and after conditioner treatment as a function of depth and location.

Figure 40.27a summarizes the hardnesses and elastic moduli of various hair types. In general, the chemomechanically damaged and virgin treated hair had lower nanomechanical properties and larger error bars than the corresponding virgin hair, as discussed above. The data for African hair was a little strange. For example, the virgin treated African hair seemed to have higher hardness than virgin African hair. It should be noted that African hair is naturally curly and highly elliptical, and so it was very difficult to mount it and make indentations on its surface. The curly and highly elliptical surface of African hair could cause the indentation results to vary somewhat from the actual values. If the hardness and elastic modulus measured at 1.0 mN is taken as the hair surface hardness and elastic modulus, then upon comparing the virgin Caucasian, Asian and African hair, it is seen that the Asian hair has the highest hardness (0.39 ± 0.06 GPa) and elastic modulus (7.5 ± 0.8 GPa), followed by Caucasian hair with a hardness of 0.31 ± 0.04 GPa and an elastic modulus of 6.0 ± 0.4 GPa. The African hair seems to have the lowest mechanical properties; its hardness is 0.24 ± 0.05 GPa and its elastic modulus is 4.8 ± 0.6 GPa. Note that all of these mechanical properties were measured in the middle part of the hair.

Figure 40.27b summarizes the hardness and elastic modulus for virgin Caucasian hair at three locations: near the scalp, in the middle and near the tip. As expected, the hardness and elastic modulus of the hair surface decreases from root to tip because of the cuticle damage. Considering that the hair near the scalp has complete cuticles, while the hair near the tip has only exposed cortex, comparing these areas provides a good way to compare the nanomechanical properties of hair cuticle and cortex in the lateral direction. At 1.0 mN, the cuticle (hair near the scalp) has a higher hardness ($0.6 \pm$

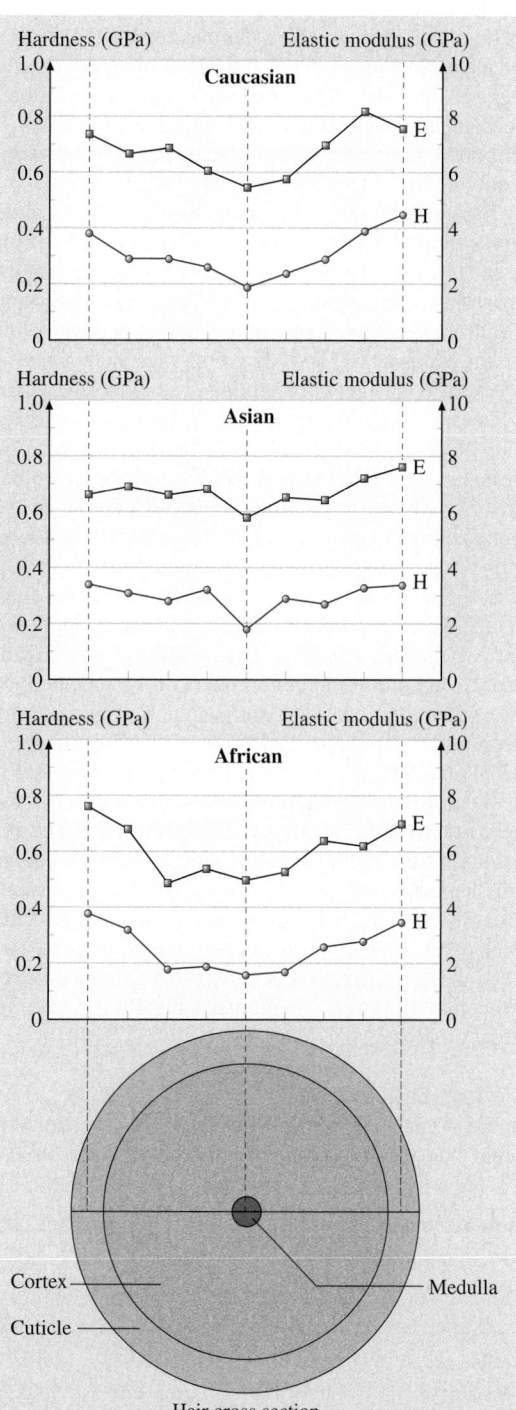

Fig. 40.29 Hardness and elastic modulus plots across various hair (after [40.13])

0.29 GPa) and elastic modulus (8.4 ± 1.2 GPa) than cortex (hair near the tip), whose hardness is 0.3 ± 0.06 GPa and elastic modulus is 6.0 ± 0.6 GPa. This result clearly suggests that the cuticles contribute more to the lateral mechanical properties of the hair than the cortex, which is in good agreement with theoretical models for wool fibers [40.50].

Figure 40.28 shows the creep displacement vs. time curves for various hair types. The normal load used for creep tests was 10 mN. In all cases, the displacement increased as time passed. The creep behavior of hair may arise from several sources. Hair is rich in peptide bonds and the abundant CO and NH groups present give rise to hydrogen bonds between groups of neighboring chain molecules. Other linkages, such as side-chain interactions of disulfide type and chain folding may also be present in hair. When hair is compressed, the creep behavior is a result of deformation and relaxation of the chemical bonds, the polypeptide chains and the noncrystalline regions [40.7]. It should be noted that the creep behavior was not obvious at normal loads lower than 10 mN. Assuming that the diameter of Caucasian, Asian and African hair was about 50 µm, 100 µm and 80 µm, respectively, the compression ratios of the indented areas of these hair types at 10 mN at the beginning of the creep tests were about: Caucasian ($\approx 2.6\%$), Asian ($\approx 1.3\%$) and African ($\approx 2.5\%$). This may suggest that if the local compression ratio was less than the values for the corresponding hair types, the deformation and relaxation of the chemical bonds, the polypeptide chains and the noncrystalline regions may have been too small to cause the creep behavior to occur. According to the creep displacement vs. time curves, it is difficult to correlate the creep behavior of each hair with its ethnicity and condition (virgin, chemomechanically damaged or virgin treated).

Cross-Sections

Figure 40.7a shows SEM images of cross-sections of virgin hair. These SEM images represent the typical shapes of Caucasian (nearly oval), Asian (nearly round) and African (oval-flat) hair. The Asian hair seems to have the greatest diameter, followed by African and Caucasian hair, which is in good agreement with SEM studies of the hair surface. The central column of Fig. 40.7a shows the cortex and medulla of each hair. The arrows point to the indents made at the cortex. The medulla of African hair is not so obvious, and it is believed that not all hair has medulla [40.3]. The right-hand column shows images of the cuticles. The top-right image clearly shows that the cuticle of Caucasian hair is about 6–7 scales thick, and that each cuticle cell is about 0.3 to 0.5 µm thick. Note that the cuticle scales were separated due to polishing, implying that the binding strength of the cell membrane complex between the cuticle scales might not be very strong.

The hardnesses and elastic moduli of hair cuticle, cortex and medulla were measured from the cross-sectional samples, and Fig. 40.29 shows hardness and elastic modulus plots for various virgin hair. As expected, the cuticles have the highest mechanical properties, followed by the cortex and medulla. Table 40.14 summarizes the hardness and elastic modulus of various hair types [40.13]. The hardness of the cuticles was found from the hair surface measurements (Fig. 40.27a). By comparing the mechanical properties of Caucasian, Asian and African hair cortex, it can be seen that the Asian cortex appears to have the highest properties, followed by Caucasian and African hair. This trend is in agreement with the trends seen in hair surface measurement results (Fig. 40.27a). Table 40.14 shows that the hardness of the cuticle is greater than that of the cortex, but the elastic modulus of the cortex is comparable to that of the cuticle. Comparing the hardness (0.3 ± 0.06 GPa) and elastic modulus (6.0 ± 0.6 GPa) of Caucasian cortex in the lateral direction (Fig. 40.27b) with its hardness (0.27 ± 0.02 GPa) and elastic modulus (6.5 ± 0.5 GPa) in the longitudinal direction, it can be seen that the hardness and elastic modulus of the hair cor-

Table 40.14 Summary of hardness and elastic modulus of different parts of different ethnicities of human hair. Mean and $\pm 1\sigma$ values are presented

	Hardness (GPa)			Elastic modulus (GPa)		
	Cuticle[a]	Cortex[b]	Medulla[b]	Cuticle[a]	Cortex[b]	Medulla[b]
Caucasian	0.32 ± 0.04	0.27 ± 0.02	≈ 0.19	6.0 ± 0.4	6.5 ± 0.5	≈ 5.5
Asian	0.39 ± 0.06	0.30 ± 0.02	≈ 0.18	7.5 ± 0.8	6.7 ± 0.3	≈ 5.8
African	0.24 ± 0.05	0.23 ± 0.06	≈ 0.16	4.8 ± 0.6	5.8 ± 0.7	≈ 5.0

[a] Obtained from the hair surface at normal load of 1.0 mN   [b] Obtained from the hair cross-section at normal load of 1.0 mN

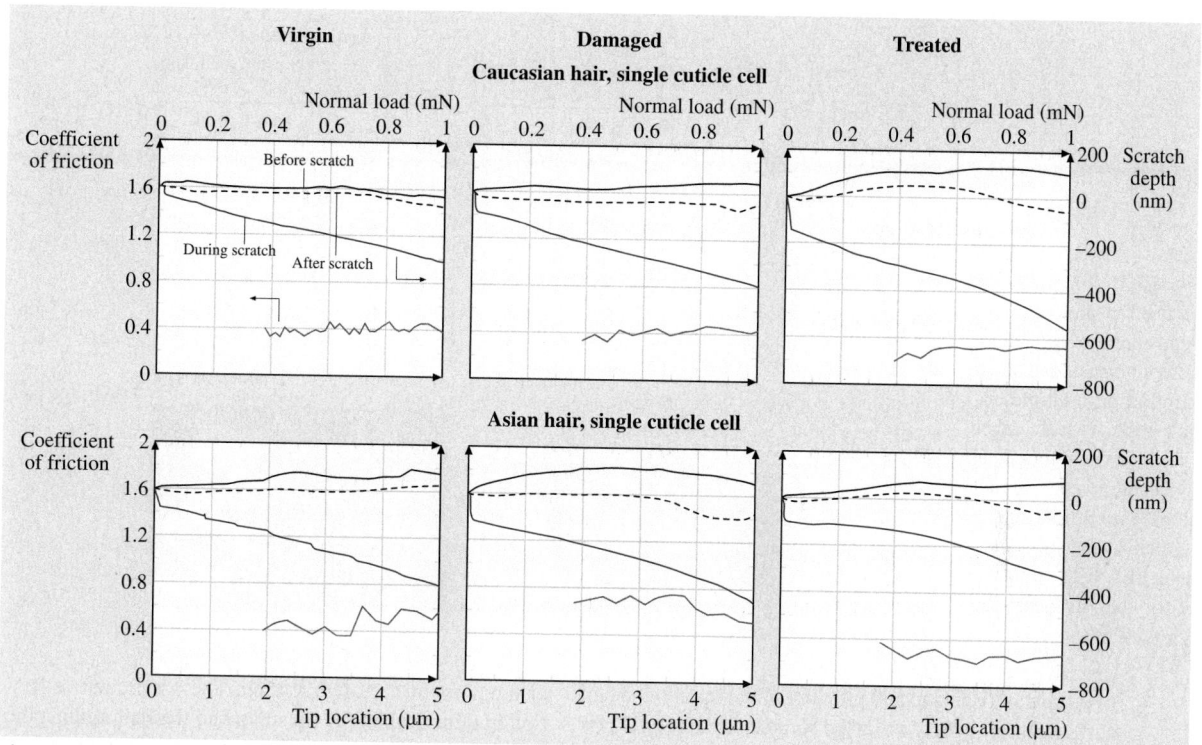

Fig. 40.30 Coefficient of friction and scratch depth profiles as a function of normal load and tip location on a single cuticle cell from Caucasian and Asian hair (virgin, damaged and treated) (after [40.14])

tex in the longitudinal direction are lower and higher, respectively, than those in the lateral direction.

40.4.2 Scratch Resistance

The nanoscratch technique is capable of simulating scratch phenomena on the surface of hair on the nanoscale by scratching the hair surface using a conical diamond tip (radius about 1 μm) and recording the coefficient of friction, the in situ scratch depth and residual depth.

Nanoscratch on a Single Cuticle Cell
Figure 40.30 shows the coefficient of friction and scratch depth profiles as a function of normal load and tip location on a single cuticle cell of Caucasian and Asian hair (virgin, chemomechanically damaged and virgin treated). The scratch direction is from left to right. The scratch length was 5 μm and the normal load was increased from 0.01 to 1 mN during scratching. The coefficients of friction for all of the hair samples ranged from 0.3 to 0.6 [40.14]. The coefficient of friction for

virgin treated Caucasian hair (≈ 0.3) is lower than that for virgin Caucasian hair (≈ 0.4), and the coefficient of friction of virgin treated Asian hair (≈ 0.3) is also lower than that for virgin Asian hair (≈ 0.5). The conditioner acts as a thin layer of lubricant on the hair surface and it reduces the coefficient of friction of hair during scratch. The coefficient of friction of chemomechanically damaged hair depends on the type and extent of damage. If the chemical damage softens the hair surface, then the tip plows into the hair easily during scratching, leading to a higher coefficient of friction. If the damage hardens the hair surface or does not change the mechanical properties of the hair surface, then the coefficient of friction will probably decrease or stay the same during scratch.

Scratch depth profiles include the profiles obtained before (pre-scratch), during (in situ scratch) and after (post-scratch) scratching, as indicated in Fig. 40.30. A reduction in scratch depth is observed after scratching compared to that observed during scratching. This reduction in scratch depth is attributed to elastic recovery after removal of the normal load. The post-scratch depth indicates the final depth, which reflects the extent

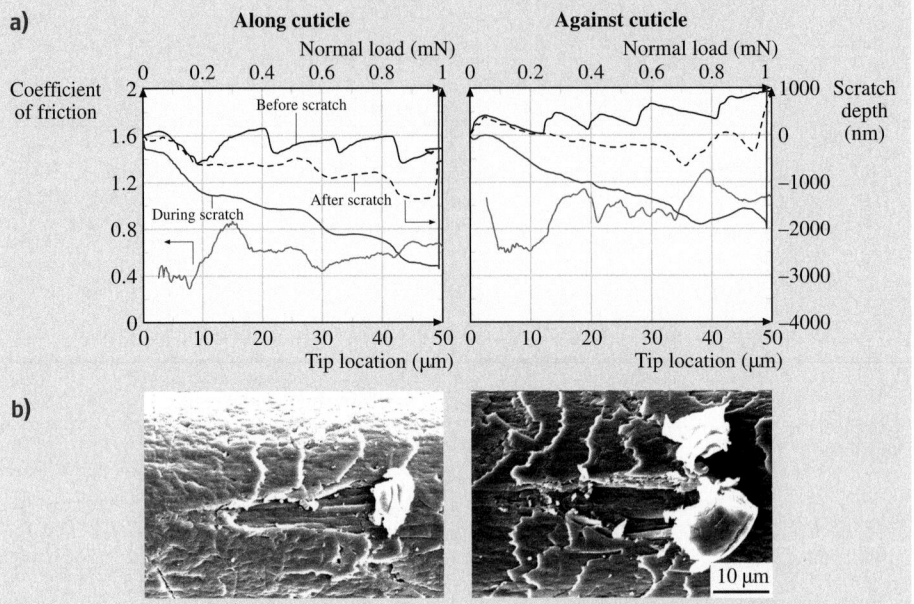

Fig. 40.31 (a) Coefficient of friction and scratch depth profiles as a function of normal load and tip location on multiple cuticle cells from damaged Caucasian hair obtained in two scratch tests: scratch along the cuticle and scratch against the cuticle; (b) SEM images of the hair surface after scratch (after [40.14])

of permanent damage and plowing of the tip into the hair surface. Interestingly, from the scratch depth profiles for all of the hair samples, we can see that the in situ displacement (30–200 nm) increased greatly at very low load at the very beginning of the scratch. After that, the in situ displacement increased gradually. This observation indicates that the first ≈ 200 nm of the hair surface may be softer than the underlying layer. The scratch depth profile also shows that the reference surface profile before scratch is not very flat, indicating that human hair has a rough surface. AFM studies have shown that the RMS roughnesses of Caucasian and Asian hair surfaces range from 7 nm to 48 nm [40.1]. At 1 mN, the in situ scratch depths of all of the hair samples range from 300 nm to 600 nm, and the residual depths range from 50 nm to 200 nm. Since the thickness of one cuticle cell is about 300 to 500 nm, the scratch tip might only penetrate one cuticle cell layer during the 1 mN nanoscratch test.

Nanoscratch on Multiple Cuticle Cells

When we comb our hair, the comb is usually scratching multiple cuticle cells. Figure 40.31a shows the coefficient of friction and scratch depth profiles as a function of normal load and tip location on multiple cuticle cells of chemomechanically damaged Caucasian hair obtained in two scratch tests: scratch along the cuticle and scratch against the cuticle, and Fig. 40.31b shows SEM images

of the hair surface after scratch. The coefficient of friction obtained when the tip scratched the hair against the cuticle is significantly higher than the coefficient of friction obtained when the tip scratched the hair along the cuticle, which is known as the "directionality effect". This is understandable, because when the tip scratches the hair surface against the cuticle, the 300–500 nm-high cuticle "wall" resists tip movement, leading to a higher coefficient of friction [40.14].

The surface profiles (before scratch) in Fig. 40.31a clearly show the shape (height and visible length) of each cuticle cell: the height is about 300–500 nm and the visible length is about 5–10 μm, which is in good agreement with SEM and AFM data. The scratch tip acts as a surface profiler before scratching. During scratching, the in situ displacement increased up to about 3 μm at 10 mN, while the residual depth was about 1.5 μm. Considering that the thickness of the cuticle is about 1.5 to 5 μm, it is likely that the tip reached the cortex during the 10 mN scratch test. The SEM images (Fig. 40.31b) clearly show that the cuticle cells were worn away in both the along-cuticle and against-cuticle cases. The topography of the exposed surface is totally different from the cuticle topography and it is believed that the exposed surface is the cortex. It can also be seen from Fig. 40.31b that scratching against the cuticle caused much more damage to the hair than scratching along the cuticle.

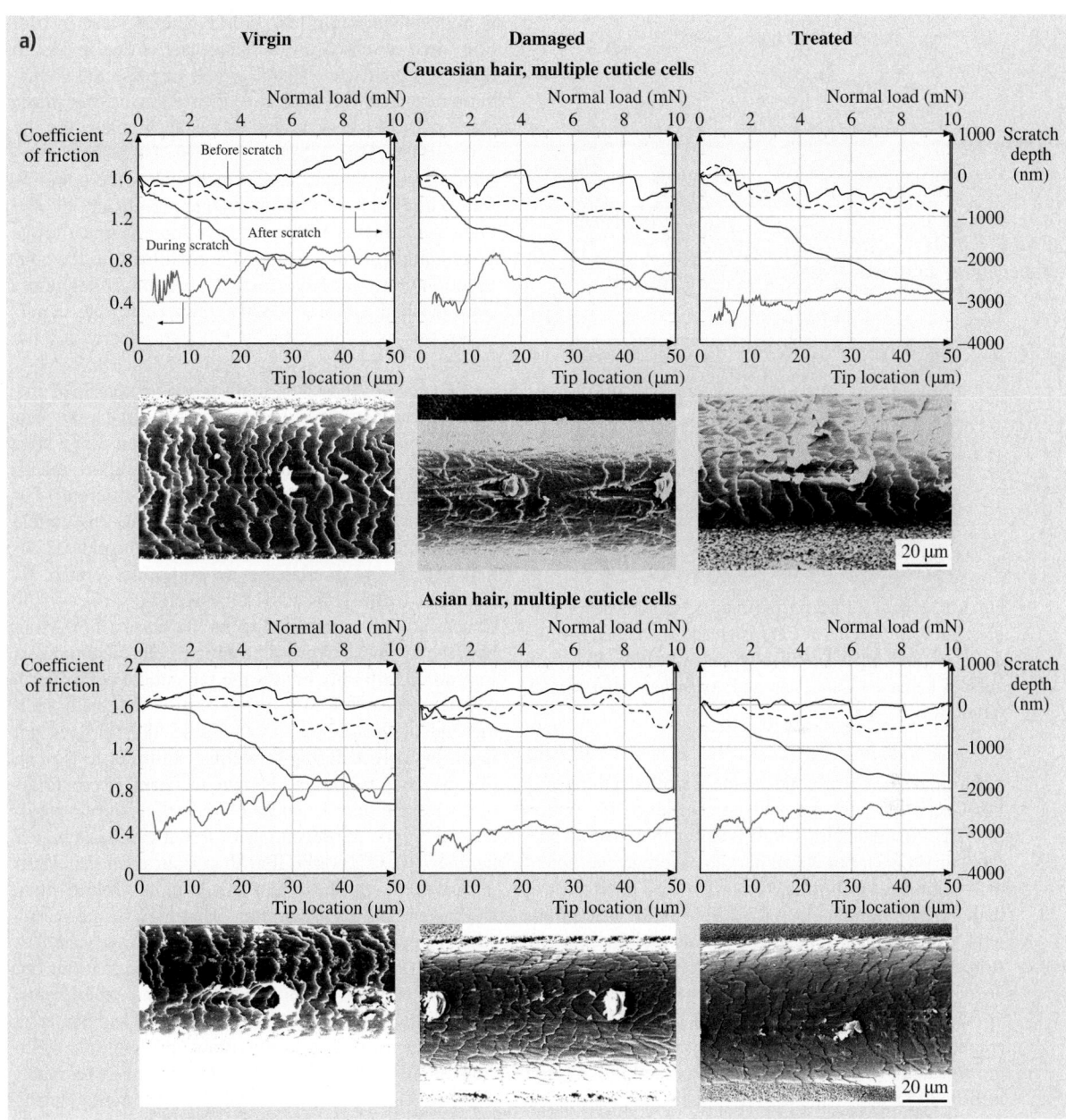

Given the fact that the "directionality effect" is universal for human hair from all races, and that scratching along the cuticle is more relevant to our daily life, we now focus on the scratch tests along the cuticle. Figure 40.32a shows the coefficient of friction and scratch depth profiles as a function of normal load and tip location on multiple cuticle cells of Caucasian and Asian hair (virgin, chemomechanically damaged and virgin treated), and the SEM images of the scratch wear tracks. Note that since at least five scratches were made on the same hair, some SEM images may show more than one scratch wear track. For example, the SEM image of chemomechanically damaged Caucasian hair shows two scratches, and the scratch wear track on the

b)

Damaged Caucasian hair

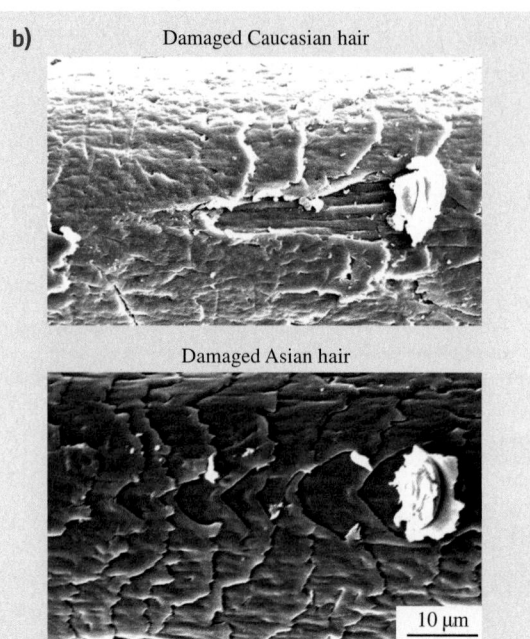

Damaged Asian hair

10 μm

Fig. 40.32 (a) Coefficient of friction, scratch depth profiles and SEM images of Caucasian and Asian hair (virgin, damaged and treated); **(b)** high-magnification SEM images of damaged Caucasian and Asian hair after scratch (after [40.14]) ◄ ▲

right-hand side corresponds to the scratch depth profile. For Caucasian hair, the average coefficient of friction for virgin treated hair (≈ 0.4) is lower than that for virgin hair (≈ 0.7). For Asian hair, the average coefficient of friction for virgin treated hair (≈ 0.5) is also lower than that for virgin hair (≈ 0.8). This trend corresponds well with the nanoscratch results on a single cuticle. It is therefore clear that conditioner treatment does indeed reduce the coefficient of friction of the hair surface upon scratching. Note that the coefficient of friction of chemo-mechanically damaged hair varies depending on the type and extent of damage (as discussed above), so it is difficult to make comparisons with the virgin or virgin treated hair.

It is worth mentioning that the coefficient of friction of human hair measured by the nanoscratch technique is on the microscale property, not a nanoscale property, since the tip radius is 1 μm and the normal load range is 1 to 10 mN. It will be shown later that *LaTorre* and *Bhushan* measured the coefficient of friction of human hair using an AFM tip (radius 30–50 nm) and found that the coefficient of friction of conditioner-treated hair

is higher than virgin hair [40.1]. The increase in friction force observed on the nanoscale is due in part to increased meniscus effects, which increase the contribution from adhesive forces to the friction at sites where the conditioner is deposited or has accumulated on the hair surface. This adhesive force is of the same magnitude as the normal load, which makes the adhesive force contribution to friction rather significant. On the microscale, however, the adhesive force is much lower in magnitude than the applied normal load, so the contribution from adhesive force to friction is negligible over the hair surface. On the microscale, the conditioner acts as a thin layer of lubricant, decreasing the friction.

Figure 40.32b shows high-magnification SEM images of chemomechanically damaged Caucasian hair and chemomechanically damaged Asian hair after scratch. It is interesting to note that the failure mechanisms for these two hair types are different. For Caucasian hair, it seems that the tip plows the cuticle cells continuously during scratching and the plowed cuticle cells are accumulated at the end of the scratch. As discussed before (Fig. 40.31), the exposed surface of the Caucasian hair is believed to be the cortex. For Asian hair, the tip does not plow the cuticle cells continuously. Instead, the tip only breaks the top cuticle cell of each cuticle and carries away the broken cuticle cells until the end of the scratch. In this case, the tip does not reach the cortex during scratching. In order to explain this, we need to look at the nanomechanical properties of Caucasian and Asian hair. According to our previous study [40.13], the cuticle of Asian hair has a higher hardness (0.39 ± 0.06 GPa) than Caucasian hair (0.24 ± 0.05 GPa). So the Asian hair is more "brittle" than Caucasian hair during scratch. That may be the reason why Asian hair is easier to fracture than Caucasian hair during scratch. However, it must be noted that our observation is based on limited samples and experiments. Since human hair varies from one hair to another, even in the same race, it is hard to draw any generic conclusions about hair failure mechanisms in terms of hair race. What we can say is that the hair fails differently during scratching depending on the nanomechanical properties of the cuticle of the hair.

Figure 40.33 shows a schematic of the various failure mechanisms for nanoscratch on hair. The top and middle figures show scratching along the cuticle, and the bottom diagram shows scratching against the cuticle. In the case of scratching along the cuticle, if the hair cuticle is soft (top figure), then the scratch tip will plow the cuticle and carry away the worn cuticle cells (scales). If the

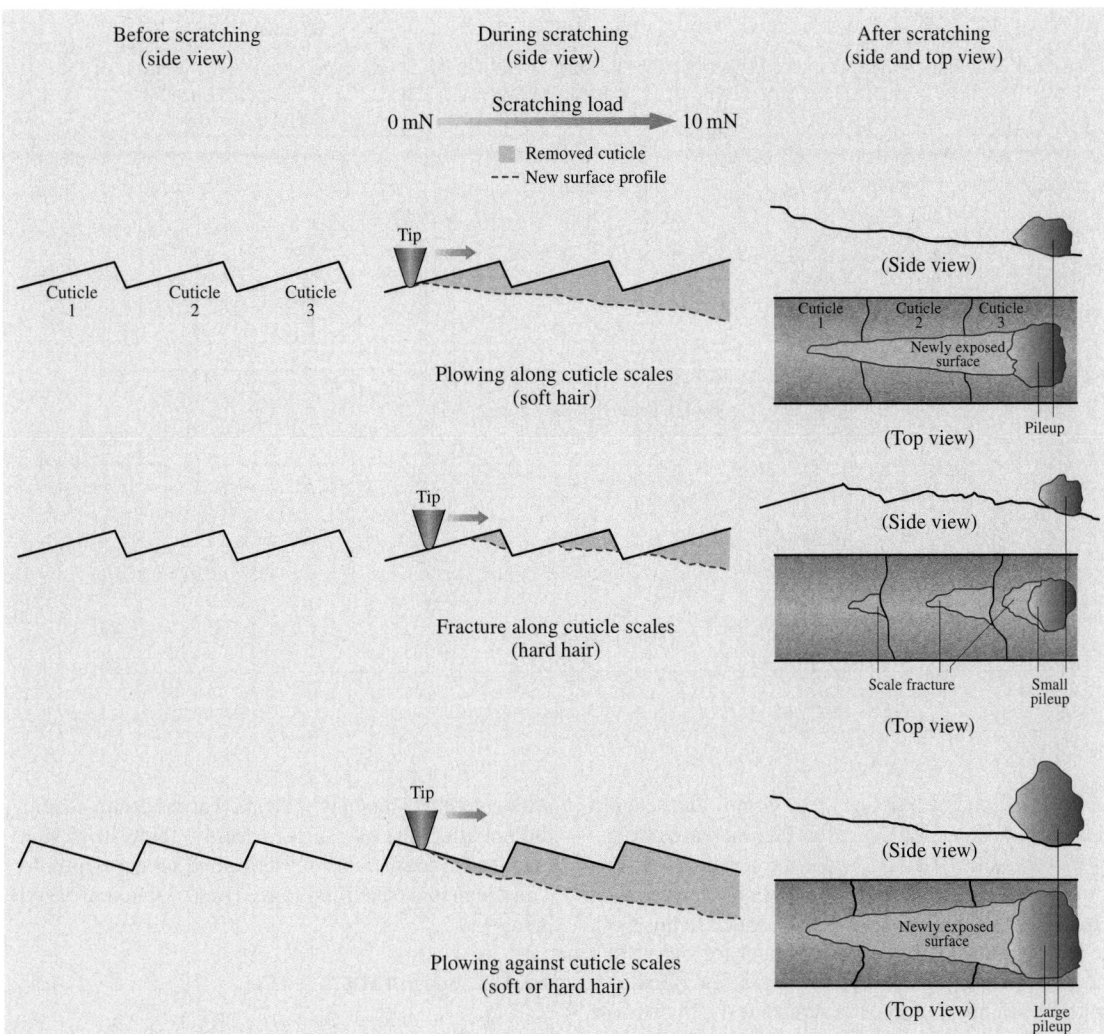

Fig. 40.33 Schematic of the various failure mechanisms for nanoscratching on hair (after [40.14])

load is high enough, then the tip can reach the cortex during scratching. After scratching, a newly exposed surface will be created and some pile-up is formed at the end of the scratch wear track. If the hair cuticle is hard (middle figure), then the scratch tip will fracture the cuticle cells instead of plowing deep into them. After scratching, part of each cuticle cell that has undergone scratching is carried away by the tip, resulting in the formation of a small amount of pile-up at the end of the scratch wear track, and a series of incomplete cuticle cells are left behind. In the case of scratching against the cuticle (bottom figure), the tip will plow the cuticle cells (whether soft or hard) and create a newly exposed surface with large wedge formation and pile-up at the end of the scratch wear track.

Nanomechanical measurements were also performed on African hair in previous studies [40.13]. In this study, the curly shape and structure of this hair made it very difficult to perform nanoscratching on its surface.

Effect of Soaking

Figures 40.34a and 40.34b compare the coefficients of friction and scratch depth profiles of unsoaked and soaked Caucasian hair, obtained at 1 mN and 10 mN scratch loads, respectively. At 1 mN, (Fig. 40.34a),

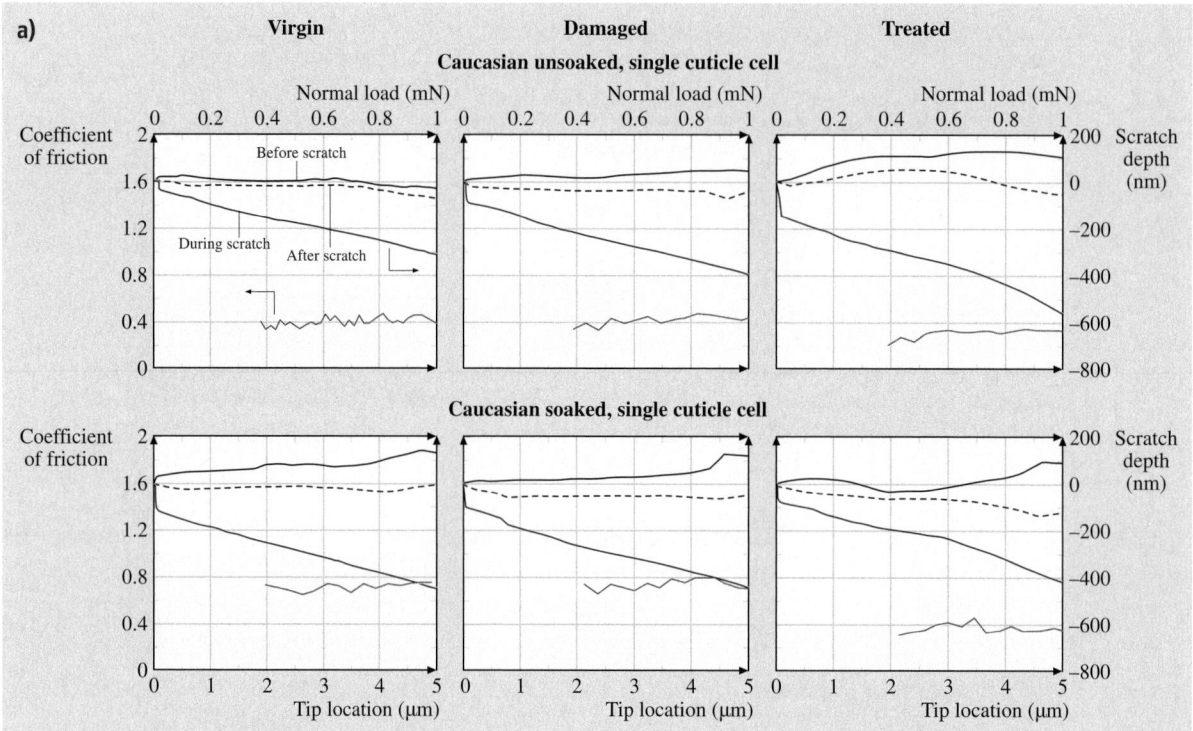

Part D | 40.4

the coefficients of friction for virgin and chemo-mechanically damaged Caucasian hair increased from ≈ 0.4 to ≈ 0.7 after soaking, while the coefficient of friction of virgin treated hair (≈ 0.3) did not change much. It is known that human hair swells in water. In this work, the hair was soaked in deionized water for only 5 min. After the sample was soaked, it took a few minutes to mount the sample and run the scratch tests. In this case, it is possible that only a few hundred nanometers of the hair surface contained considerable amounts of water and were softened. During scratching, it is easier for the tip to plow the softer hair surface, leading to a higher coefficient of friction. This may be the reason that for 1 mN scratches, where the maximum in situ scratch depths were less than 600 nm, the coefficients of friction for virgin and chemomechanically damaged hair increased. For virgin treated hair in the 1 mN scratch test, however, some of the conditioner molecules could have occupied the pathways of water molecules, so the swelling of virgin treated hair was not as significant as virgin and chemomechanically damaged hair. Therefore, the virgin treated hair shows little change in the coefficient of friction after soaking. At 10 mN, (Fig. 40.34b), the coefficients of friction of all three hair types hardly change

after soaking. This may indicate that the 5 min soaking did not affect the hair surface deeply. Table 40.15 summarizes the coefficients of friction and scratch depths for Caucasian (unsoaked and soaked) and Asian (unsoaked) hair.

40.4.3 Summary

A nanoindenter has been used to perform nanomechanical studies on human hair. Chemical damage and conditioning treatment caused the hardness and elastic modulus of the hair surface to decrease within a depth of less than 1.5 μm. That is, the first 3–4 cuticle scales may interact with the chemicals and ingredients of the conditioner more effectively than the rest of the scales. It is found that the hair cuticle has a higher hardness and elastic modulus than the cortex in the lateral direction. The hardness and elastic modulus of the hair decreased as the indentation depth increased. The differences in cystine content between cuticle substructures (A-layer, exocuticle, endocuticle, cell membrane complex) and the cortex are believed to be responsible for this result.

Nanoscratch tests were performed on single and multiple cuticles of various hair types, both unsoaked and

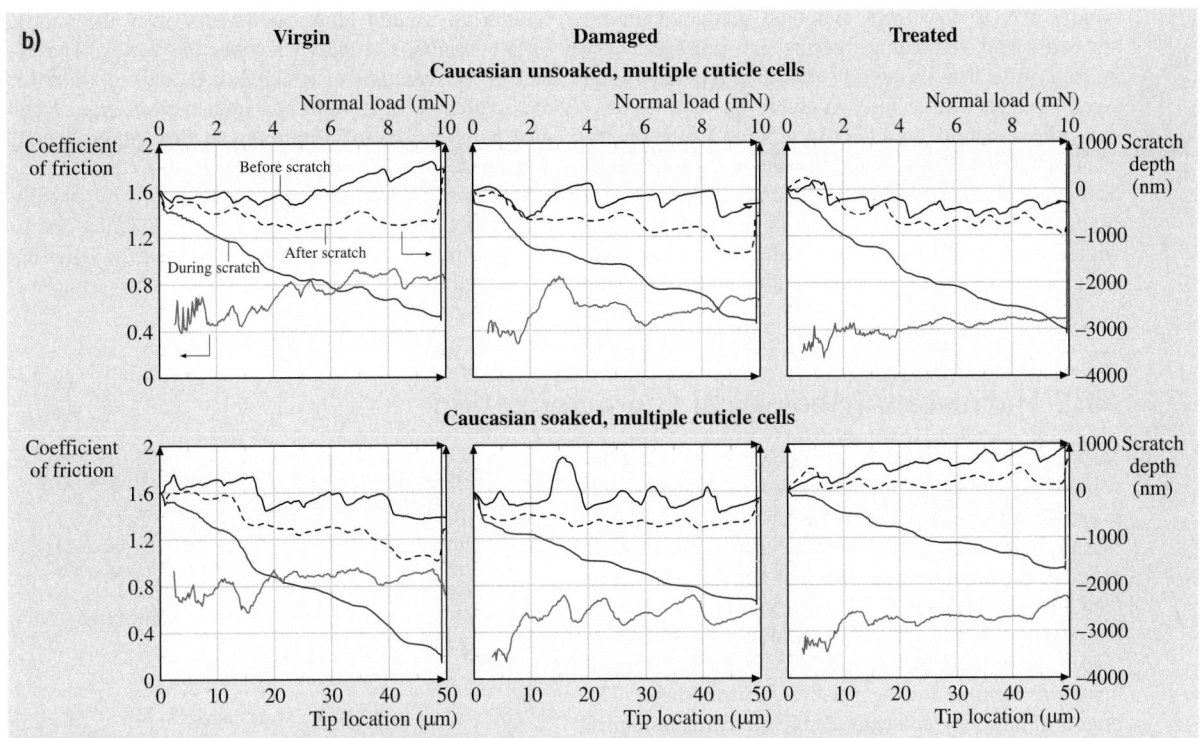

Fig. 40.34 Comparison of nanoscratch results on Caucasian unsoaked and soaked hair samples: (**a**) scratch on a single cuticle cell; (**b**) scratch on multiple cuticle cells (after [40.14]) ◄ ▲

Table 40.15 Summary of coefficient of friction and scratch depths for Caucasian (unsoaked and soaked) and Asian (unsoaked) hair

Max. normal load/No. of cuticle cells	1 mN/single cuticle cell			10 mN/multiple cuticle cells		
	Unsoaked Caucasian					
Hair condition	Virgin	Damaged	Treated	Virgin	Damaged	Treated
Average coefficient of friction	0.4	0.4	0.3	0.7	0.6	0.4
Max. in situ depth (nm)	280	440	650	3250	2500	2750
Max. residual depth (nm)	25	100	150	1100	1000	750
	Soaked Caucasian					
Hair condition	Virgin	Damaged	Treated	Virgin	Damaged	Treated
Average coefficient of friction	0.7	0.7	0.3	0.8	0.6	0.5
Max. in situ depth (nm)	570	540	500	3000	2100	2600
Max. residual depth (nm)	130	170	210	1000	500	800
	Unsoaked Asian					
Hair condition	Virgin	Damaged	Treated	Virgin	Damaged	Treated
Average coefficient of friction	0.5	0.6	0.3	0.8	0.4	0.5
Max. in situ depth (nm)	480	500	400	2500	2400	1800
Max. residual depth (nm)	60	150	130	700	900	480

soaked. For Caucasian and Asian hair, the coefficients of friction for virgin treated hair were lower than those for virgin hair in single cuticle scratch and multiple cu-ticle scratch tests. The thin layer of conditioner acts as a lubricant, reducing the coefficient of friction during scratching. In situ displacement (30–200 nm) increased rapidly at very low loads and then increased gradually with load, indicating that the first approximately 200 nm of the hair surface is softer than the underlying layer. The nanoscratch tests on multiple cuticles clearly show the directionality effect in relation to the coefficient of friction. It was found that the hair surface fails differently during scratching depending on the nanomechanical properties of the cuticle of the hair. For hair with a hard cuticle, the cuticle cells tend to be fractured during scratching. For hair with a soft cuticle, the

scratch tip usually plows and wears away the cuticle cells continuously until it reaches the cortex. The effect of five minutes of soaking in deionized water on the coefficient of friction and scratch resistance of human hair was limited within the shallow region (about 600 nm deep) of the hair surface. In this case, the coefficients of friction of virgin and chemo-mechanically damaged Caucasian hair increase after soaking due to the swelling of the water, which softens the hair surface.

40.5 Macroscale Tribological Characterization

Tribology is very important to hair care and product development. While the current state-of-the-art approach is to use AFM to measure nanoscale tribological properties of hair, macroscale tribological measurements provide an excellent simulation of skin–hair and hair–skin contacts [40.41]. In Sect. 40.5.1, friction and wear studies on various hair are presented, including the effects of load, velocity and skin size. The friction and wear of hair were measured using a flat-on-flat tribometer. The effects of humidity and temperature on hair tribological properties are discussed in Sect. 40.5.2.

40.5.1 Friction and Wear Studies of Various Hair Types

Figure 40.35a shows raw friction force data and coefficients of friction measured from hair strands [40.41]. The data shows that the coefficient of friction for virgin Caucasian hair is about 0.14 along the cuticle and about 0.23 against the cuticle. As with most animal fibers, human hair shows a directionality friction effect; that is, it is easier to move a surface over the hair in the root-to-tip direction than in the tip-to-root direction because of the anisotropic orientation of the hair cuticles [40.3,27,30,51]. The data shows that the flat-on-flat tribometer can measure the directionality dependence of friction. Note that single hair strands were used to obtain the data shown in Fig. 40.35a,, so there was no interaction between hairs during the friction tests. The output signals for normal force and friction force are smooth and the coefficient of friction only shows small variations.

In industry, many friction tests are performed on bundles of hair, in which case some hairs overlap. This means that hair–hair interactions occur during the

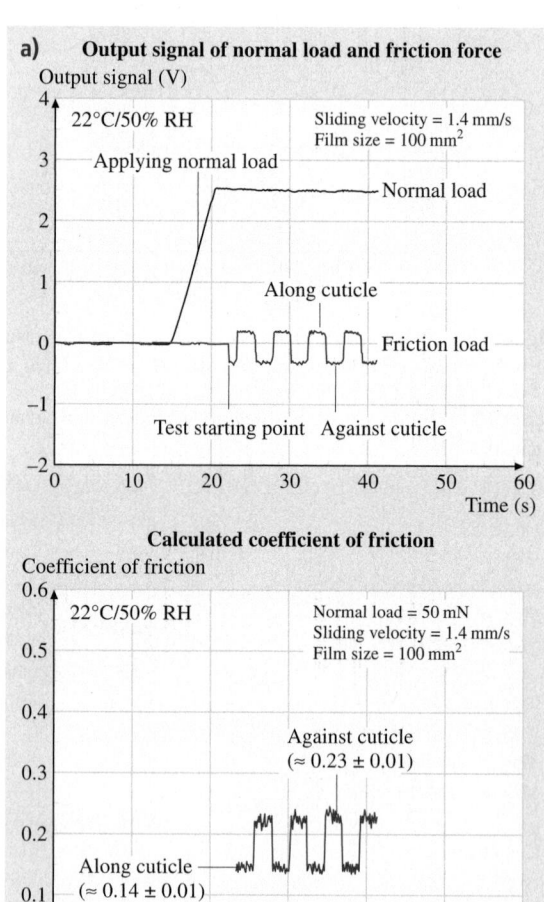

a) **Output signal of normal load and friction force**

Output signal (V)

22°C/50% RH

Sliding velocity = 1.4 mm/s
Film size = 100 mm²

Applying normal load

Normal load

Along cuticle

Friction load

Test starting point   Against cuticle

Time (s)

Calculated coefficient of friction

Coefficient of friction

22°C/50% RH

Normal load = 50 mN
Sliding velocity = 1.4 mm/s
Film size = 100 mm²

Against cuticle
($\approx 0.23 \pm 0.01$)

Along cuticle
($\approx 0.14 \pm 0.01$)

Time (s)

friction tests and large variations occur in the data. Figure 40.35b shows the output signal and coefficient of friction obtained from a bunch of hairs. It can be seen that the output signal for the normal force fluctuates a lot, and the output signal for the friction force is not smooth, leading to big variations in the coefficient of friction. Both the coefficient of friction along the cuticle and that against the cuticle are greater for bundles of hair compared to hair strands due to the hair–hair interactions during friction tests. It has been observed that if a bunch of hairs are used for friction tests, the data shows much poorer reproducibility than if hair strands are used. This

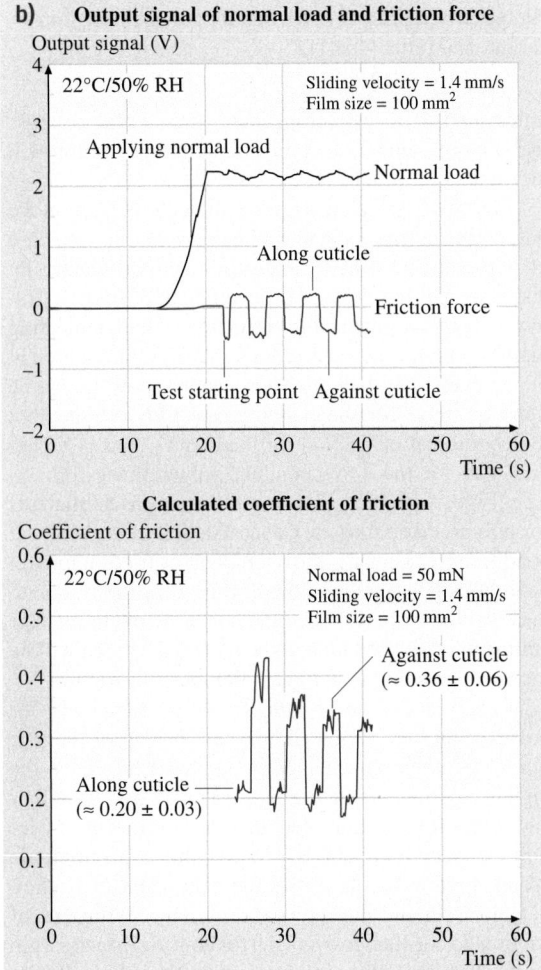

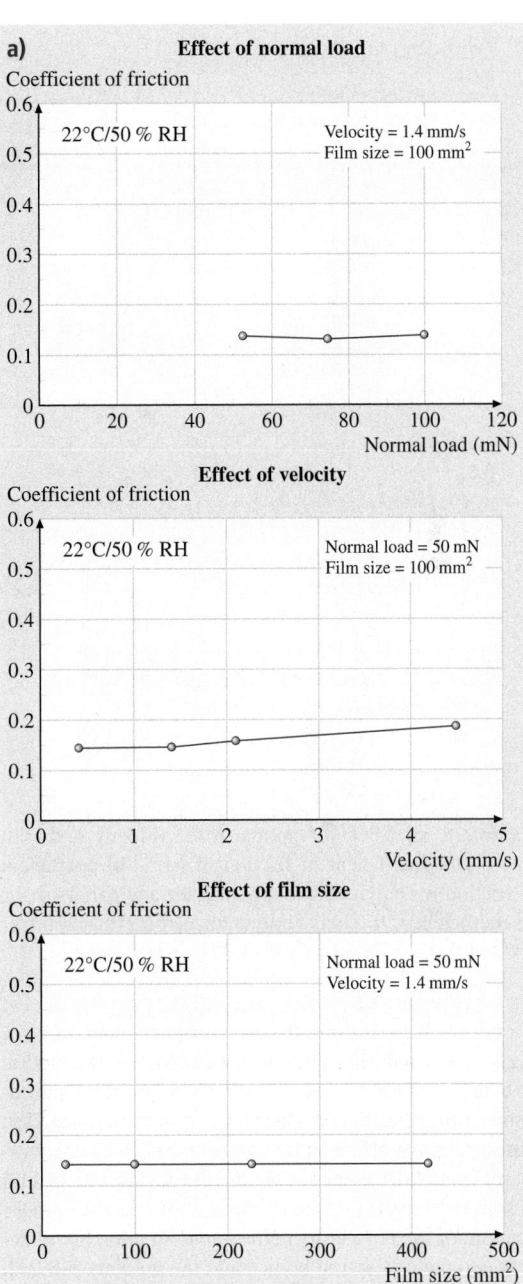

Part D | 40.5

Fig. 40.35 (a) Raw force data and coefficient of friction measured from hair strands; **(b)** raw force data and coefficient of friction measured from a bundle of hair (after [40.41]) ◀ ▲

may be because the hair is placed randomly in a bundle and so the positions of the hairs may not remain the same between tests. For hair strands, the positions of the hairs are easy to control since they are separated and parallel to each other. Therefore, in the work described here, hair strands were used to make further measurements.

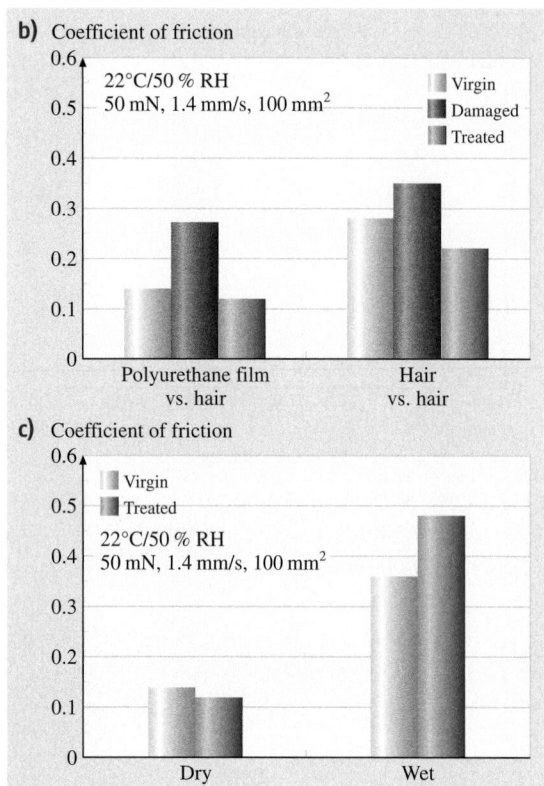

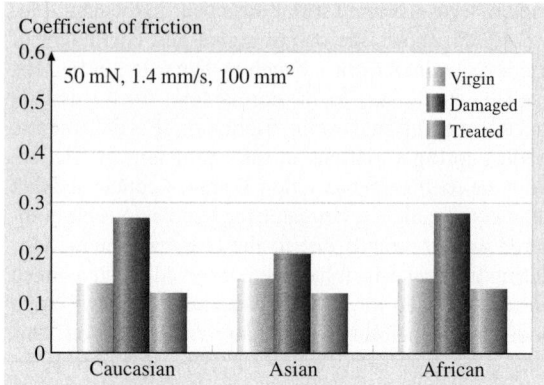

Fig. 40.37 Coefficient of friction for polyurethane film vs. various hair (after [40.41])

Fig. 40.36 (a) Effect of normal load, velocity and film size on the coefficient of friction of hair; **(b)** coefficient of friction for polyurethane film vs. hair and hair vs. hair; **(c)** coefficient of friction of polyurethane film vs. hair under dry and wet conditions (after [40.41]) ◀ ▲

Figure 40.36a shows the effects of normal load, sliding velocity and film area on the coefficient of friction of hair. All of the coefficient of friction values were obtained along the cuticle. This shows that the load and the film area has no effect on the hair friction, but a higher velocity leads to a higher coefficient of friction. According to some observations, the coefficient of friction is independent of the normal load, the apparent area of contact between the contacting bodies and the velocity [40.27]. Macroscale hair friction appears to obey the first two observations. The third rule of friction, which states that friction is independent of velocity, is not valid in the case of hair friction. Changes in the sliding velocity may result in a change in the shear rate, which can influence the mechanical properties of the hair and polyurethane film (shear strength, elastic modulus, yield strength and hardness) [40.52]. If the changes in mechanical prop-

erties lead to a less strong hair surface and a larger real area of contact, then the coefficient of friction will increase.

Figure 40.36b shows the coefficients of friction for polyurethane film vs. hair and hair vs. hair. In the case of polyurethane film vs. hair, the chemomechanically damaged hair has the highest coefficient of friction, followed by the virgin and the virgin treated hair, indicating that the conditioner can reduce the friction coefficient of the hair. The hair vs. hair results show the same trend as those for the polyurethane film vs. hair, but the coefficient of friction for the hair vs. hair is higher than that for the corresponding polyurethane film vs. hair. Figure 40.36c compares the friction coefficients for polyurethane film vs. Caucasian hair at dry and wet conditions. Obviously, the coefficient of friction for wet conditions is higher than that for dry conditions, due to the swelling of hair. When the hair is swollen, the hair cuticle will be lifted up and the real contact area will be increased, leading to a higher coefficient of friction.

Figure 40.37 shows the friction coefficients for polyurethane vs. various hair types. For all of the hair types, the chemomechanically damaged hair gave the highest coefficients of friction, followed by virgin hair and virgin treated hair. Note that the coefficient of friction can vary by about 10–15% within a given ethnic group. Figure 40.38a shows the coefficient of friction of Caucasian hair during wear tests. The coefficients of friction do not change over 24 h for both virgin and virgin treated hair. From the optical micrographs in Fig. 40.38b, it can be seen that some cuticles were damaged after wear tests. Polyurethane film is soft and does not damage the hair much. Since the observed damaged area was small, it may not affect the overall coefficient of friction.

Fig. 40.38 (a) Coefficient of friction for virgin and virgin treated Caucasian hair during wear tests; **(b)** optical micrographs of Caucasian hair before and after a wear test (after [40.41])

40.5.2 Effects of Temperature and Humidity on Hair Friction

Figure 40.39 shows the effects of temperature and humidity on hair friction. The coefficient of friction of hair is a strong function of the humidity, and it increases as the relative humidity increases. In addition, the differential friction effect also increases with increasing relative humidity, as shown in Fig. 40.39. It is interesting to note that the differential friction effect of virgin treated hair is less dependent on relative humidity than that for virgin hair. This may be because some of the conditioner molecules occupy the pathways of water molecules, so that the swelling of this hair was not as significant as in virgin hair.

For virgin hair, temperature has no effect on the coefficient of friction. This is in agreement with *Scott* and *Robbins's* work [40.53]. For virgin treated hair, it is found that the coefficient of friction increases as the temperature increases. After the hair was treated with conditioner, the hair surface properties can change, which may lead to them being affected by temperature. For instance, a higher temperature might lead to a softer treated hair surface, leading to a higher coefficient of friction.

40.5.3 Summary

A flat-on-flat tribometer has been used to measure macroscale friction and wear of polyurethane film (synthetic skin) vs. hair and hair vs. hair. In the case

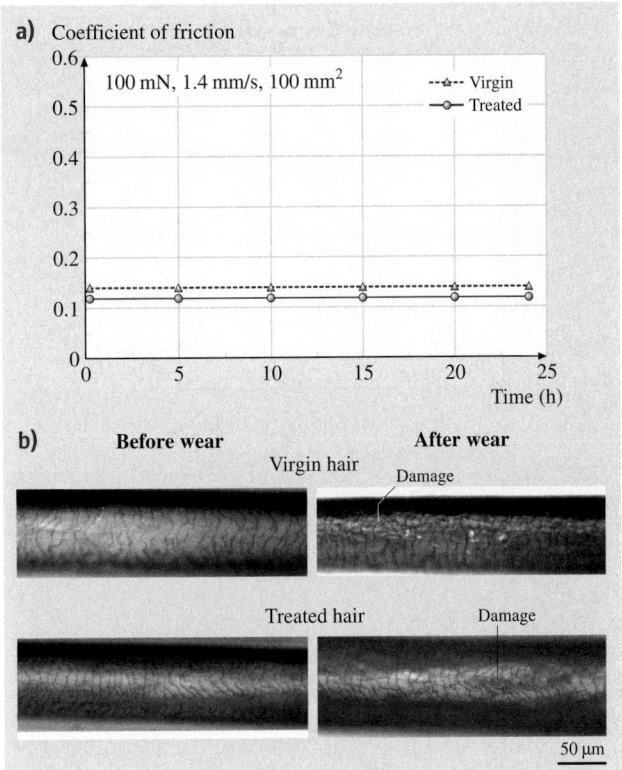

of polyurethane film vs. hair, the chemomechanically damaged hair shows the highest coefficient of friction, followed by the virgin and the virgin treated hair types. The coefficient of friction obtained in the case of hair vs. hair is greater than that of polyurethane film vs. hair. In the skin vs. hair wear test, the coefficient of friction did not change after 24 h, while some of the cuticles were damaged.

40.6 Nanotribological Characterization Using an AFM

Nanoscale tribological characterization is essential when studying hair and evaluating/developing better cosmetic products. It becomes especially important when studying the effects of damage and conditioner treatment. How common hair care products, such as conditioner, deposit onto and change hair roughness, friction, and adhesion are of interest, since these properties are closely tied to product performance. Before the advent of nanoscale tribological characterization, cosmetic scientists had also previously struggled to

address other important issues, such as the thickness distribution of conditioner on the hair surface, which is important when determining the proper functions of the conditioner.

The roughnesses, friction and adhesion of various hair types are studied in Sect. 40.6.1 [40.1, 2, 12]. In Sect. 40.6.2, a new method based on the AFM technique for determining thin liquid film thickness is developed, and the thickness distributions of conditioner, the adhesive forces and the effective Young's

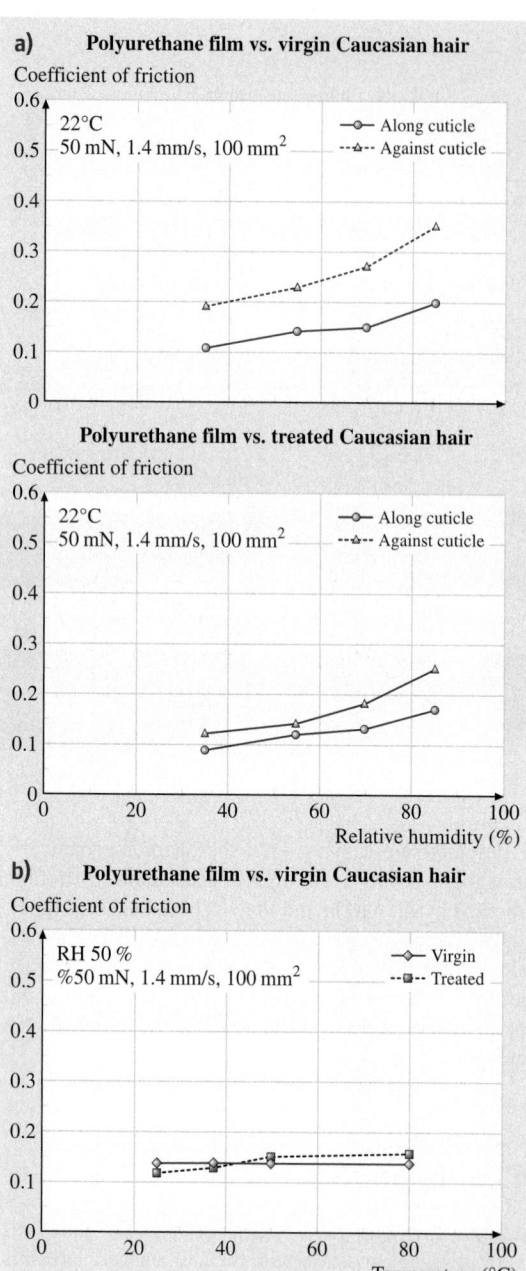

a) Polyurethane film vs. virgin Caucasian hair

Polyurethane film vs. treated Caucasian hair

b) Polyurethane film vs. virgin Caucasian hair

Fig. 40.39 Effect of humidity (**a**) and temperature (**b**) on virgin Caucasian hair friction (after [40.41])

moduli are presented for various hair samples [40.10, 11]. The binding interactions between the conditioner molecules and the hair surface are discussed too.

40.6.1 Roughness, Friction and Adhesion

Various Ethnicities

Topographical images of Caucasian, Asian, and African hair were taken up to scan sizes of $20 \times 20\,\mu m^2$, as shown in Fig. 40.40. Lighter areas of the images correspond to higher topography, and darker areas correspond to lower topography. Only virgin and chemomechanically damaged hair types are shown in Fig. 40.40 because virgin treated samples closely resemble virgin hair samples. One can see the variation in cuticle structure even in virgin hair. Cracking and miscellaneous damage at the cuticle edges is evident for both virgin and chemomechanically damaged conditions. In virgin hair, the damage is likely to be caused by mechanical damage resulting from daily activities such as washing, drying and combing. Most of the virgin cuticle scales that were observed, however, were relatively intact. Long striations similar to scratches and "scale edge ghosts" (outlines of a former overlying cuticle scale edge left on the underlying scale before it was broken away) were found on the surface. In some instances, the areas surrounding the cuticle edges appeared to show residue or debris on the surface, which are most likely the remnants of a previous cuticle or the undersides of cuticle edges that are now exposed (such as the endocuticle). Caucasian and Asian virgin hair displayed similar surface structures, while the African hair samples showed more signs of endocuticular remains along the scale edges. One can also see more curvature in the cuticle scales of African hair, which is attributed to its elliptical cross-sectional shape and curliness, which can partially uplift the scales in different places. Several regions seem to exist in chemomechanically damaged hair samples, ranging from intact cuticle scales to high levels of wear on the surface. In many cases these regions occur side-by-side. This wide variation in chemomechanically damaged cuticle structure results in a wider range of tribological properties on the micro/nanoscale for these fibers. Caucasian and Asian chemomechanically damaged hair showed more worn away cuticle scales than chemomechanically damaged African hair, which showed mostly endocuticle remnants. This is most likely due to the different effects that chemical straightening has on the hair versus multiple cycles of perming the hair.

A more focused look into roughness and friction on the cuticle surface can be achieved by comparing Caucasian, Asian and African virgin and chemomechanically damaged hair, Figs. 40.41 a and 40.41b, and virgin and virgin treated hair, Figs. 40.42a and 40.42b. Virgin hair was used as the baseline to compare vari-

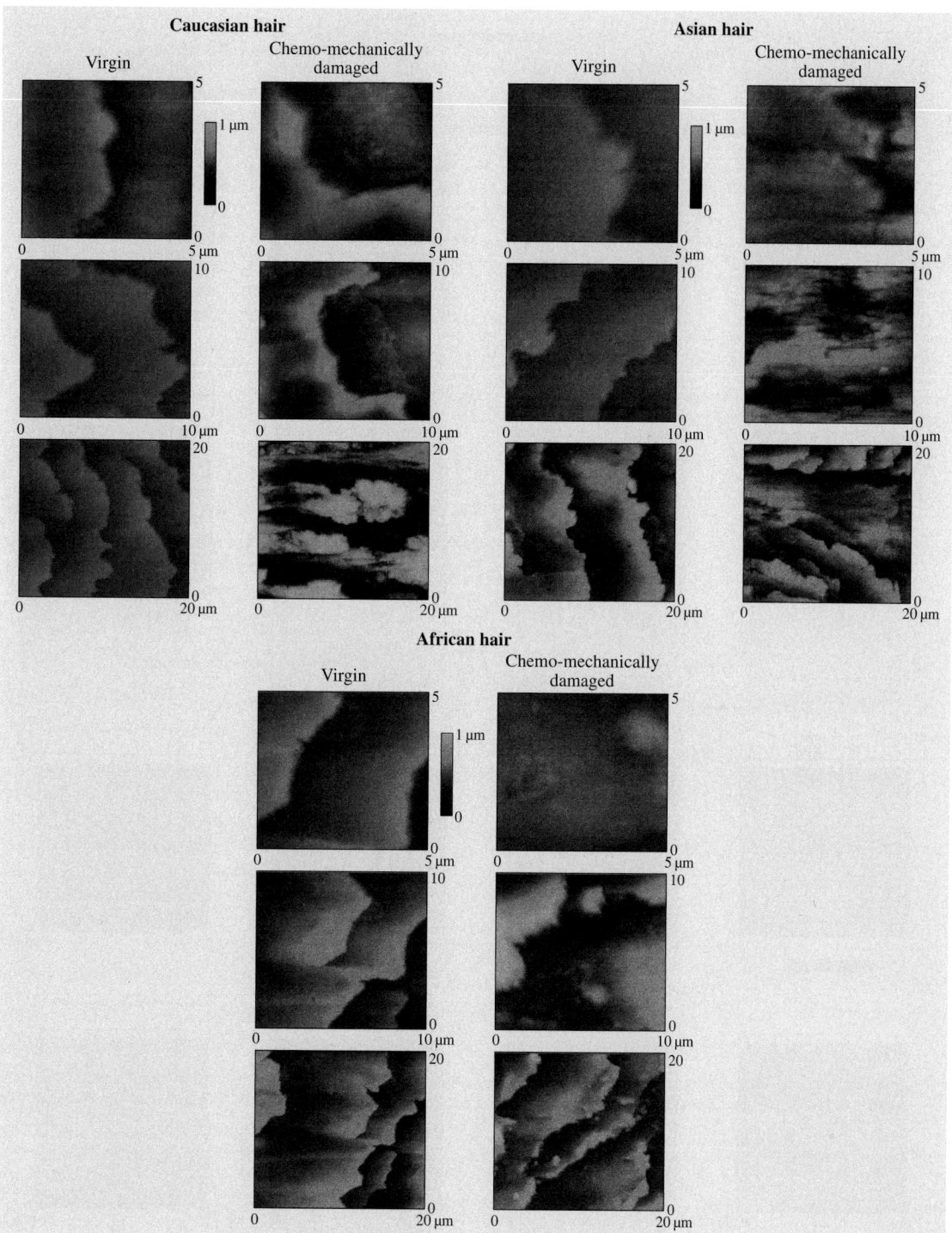

Fig. 40.40 Surface roughnesses of virgin and chemomechanically damaged Caucasian, Asian and African hair at scan sizes of 5, 10 and 20 μm (after [40.1])

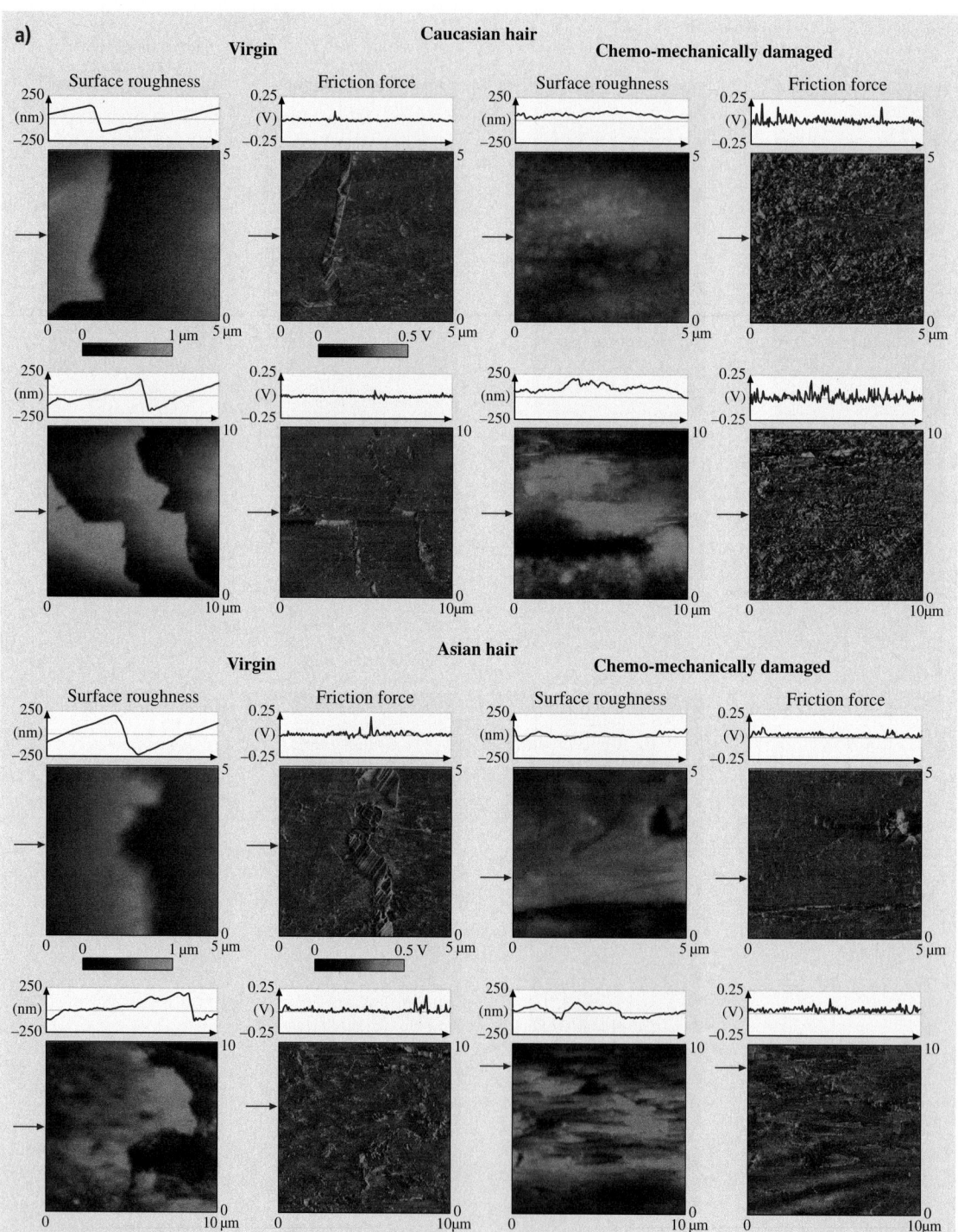

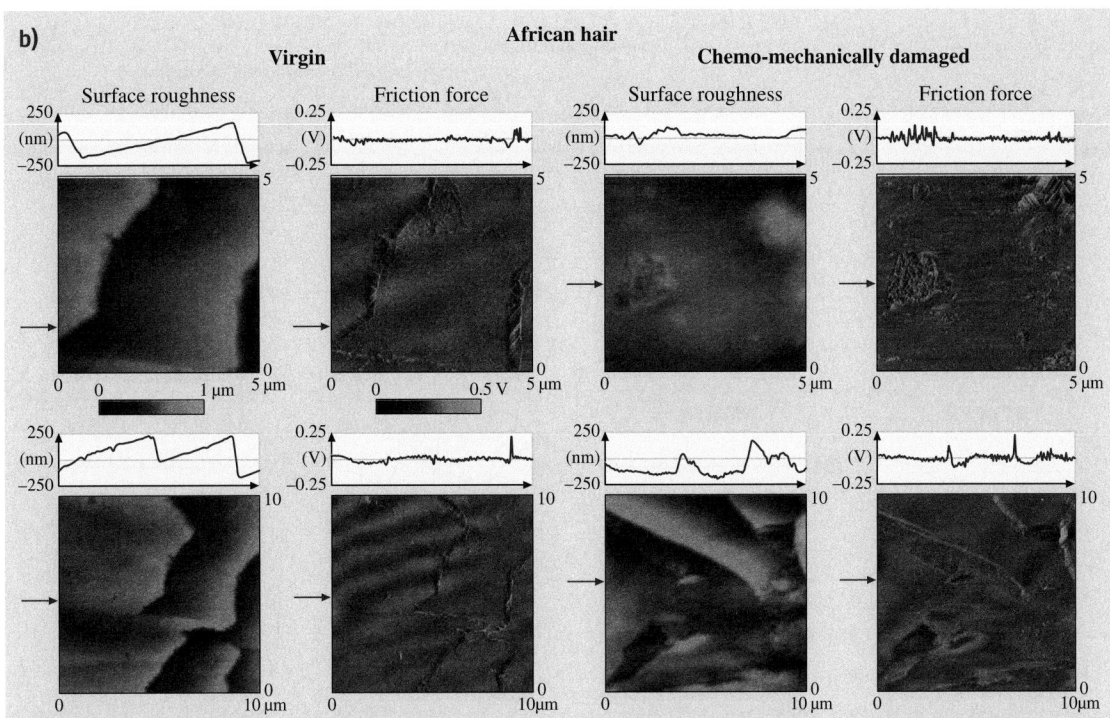

Fig. 40.41 (a) Surface roughness and friction images for virgin and chemomechanically damaged Caucasian and Asian hair at scan sizes of 5 and 10 μm. Shown above each image is a cross-section taken at the position denoted by the corresponding *arrows* showing roughness and friction force information; **(b)** surface roughness and friction force images for virgin and chemomechanically damaged African hair at scan sizes of 5 and 10 μm. Shown above each image is a cross-section taken at the position denoted by the corresponding *arrows* showing roughness and friction force information (after [40.1]) ◀ ▲

ations in roughness and friction force against modified hair (chemomechanically damaged or virgin treated). Scan sizes of $5 \times 5 \, \mu m^2$ and $10 \times 10 \, \mu m^2$ are displayed. Above each AFM and FFM image are cross-sectional plots of the surface (taken at the accompanying arrows) corresponding to surface roughness or friction force, respectively. In the surface roughness images, step heights of one or more cuticle edges can be clearly seen. Step heights range from approximately 0.3 to 0.5 μm.

If the surface is assumed to have Gaussian height distribution and an exponential autocorrelation function, then the surface can be statistically characterized by just two parameters: a vertical descriptor, the height standard deviation σ, and a spatial descriptor, the correlation distance β^* [40.27,30,31]. The standard deviation σ is the square root of the arithmetic mean of the square of the vertical deviation from the mean line. The correlation length can be referred to as the length at which two data points on a surface profile can be

regarded as being independent, thus serving as a measure of randomness [40.27]. Table 40.16 displays these roughness parameters for each ethnicity as a function of hair type (virgin, chemomechanically damaged, and virgin treated) [40.1]. Virgin hair was shown to have the lowest roughness values in general, with virgin treated hair closely resembling virgin hair. Chemomechanically damaged hair showed a significantly higher standard deviation of surface height. This variation is expected because of the nonuniformity of the mechanical and chemical damage that occurs throughout a whole head of hair as well as each individual fiber. This is in agreement with the images of chemomechanically damaged hair shown previously, where regions of intact cuticle and severe degradation of the surface are intermingled. The trends observed for standard deviations were not as evident for the correlation length β^*. For each ethnicity, chemomechanically damaged and virgin treated hair showed similar β^* values.

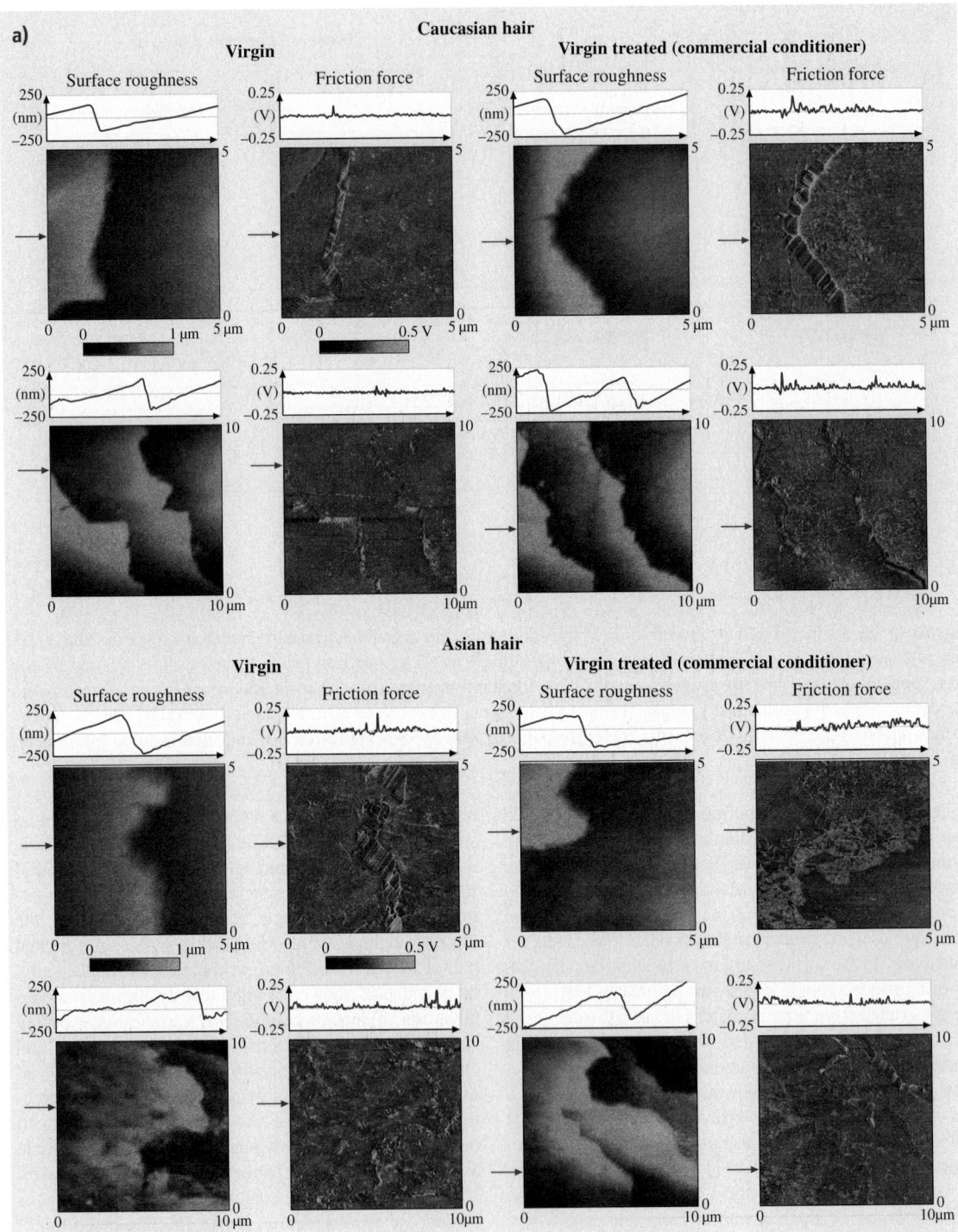

a) Caucasian hair

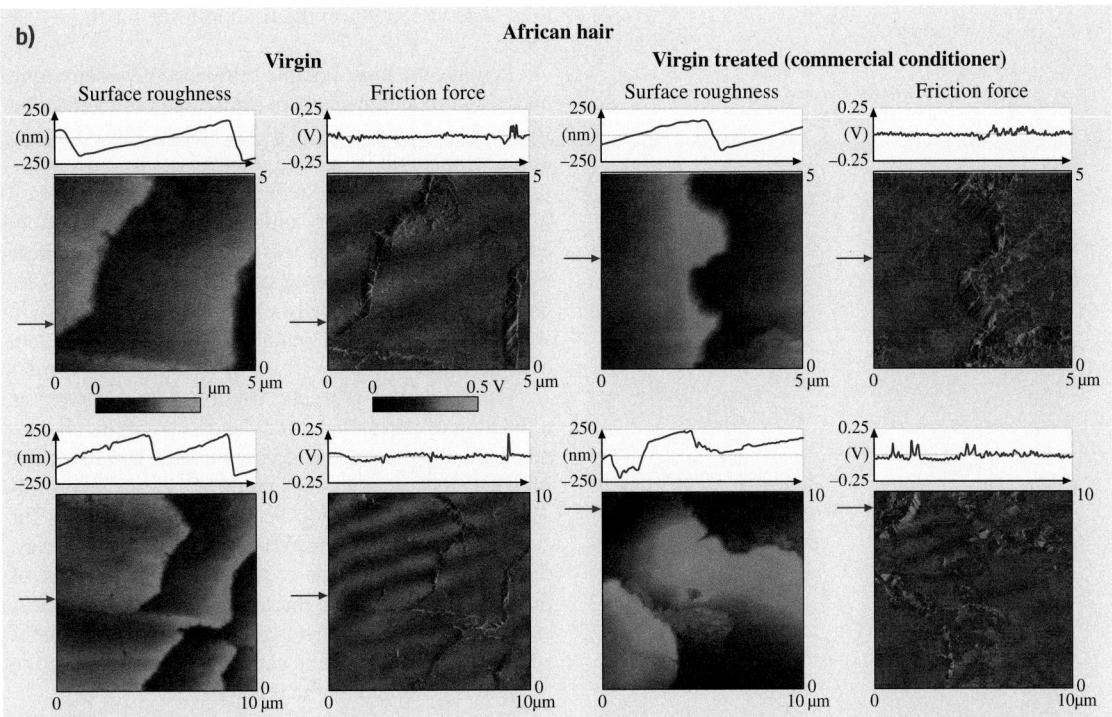

Fig. 40.42 (a) Surface roughness and friction images for virgin and virgin treated Caucasian and Asian hair at scan sizes of 5 and 10 μm^2. Shown above each image is a cross-section taken at the position denoted by the corresponding *arrows* showing roughness and friction force information; **(b)** surface roughness and friction force images for virgin and virgin treated African hair at scan sizes of 5 and 10 μm^2. Shown above each image is a cross-section taken at the position denoted by the corresponding *arrows* showing roughness and friction force information (after [40.1]) ◄ ▲

From Figs. 40.41a and 40.41b, friction forces are generally seen to be higher on chemomechanically dam-aged hair than on virgin hair. Although friction forces were similar in magnitude, it was observed that the

Table 40.16 Surface roughness, coefficient of friction, and adhesive force values for virgin, chemomechanically damaged and virgin treated (with one cycle of commercial conditioner) hair of various ethnicities

	Virgin hair		Chemomechanically damaged hair		Virgin treated hair (commercial conditioner)	
	Surface roughness parameters [σ(nm), β^*(μm)]					
	σ(nm)	β^*(μm)	σ(nm)	β^*(μm)	σ(nm)	β^*(μm)
Caucasian	12±8	0.61±0.3	17±10	1.0±0.3	12±4	0.90±0.3
Asian	9.7±4	0.73±0.3	33±15	0.94±0.3	7.1±0.1	0.97±0.3
African	12±5	0.92±0.3	21±16	0.78±0.3	11±4	0.89±0.2
	Average coefficient of friction μ					
Caucasian	0.02±0.01		0.13±0.05		0.03±0.01	
Asian	0.03±0.01		0.13±0.04		0.06±0.04	
African	0.04±0.02		0.14±0.08		0.05±0.01	
	Adhesive force F_m (nN)					
Caucasian	25		16		32	
Asian	31		18		79	
African	35		38		63	

Part D | 40.6

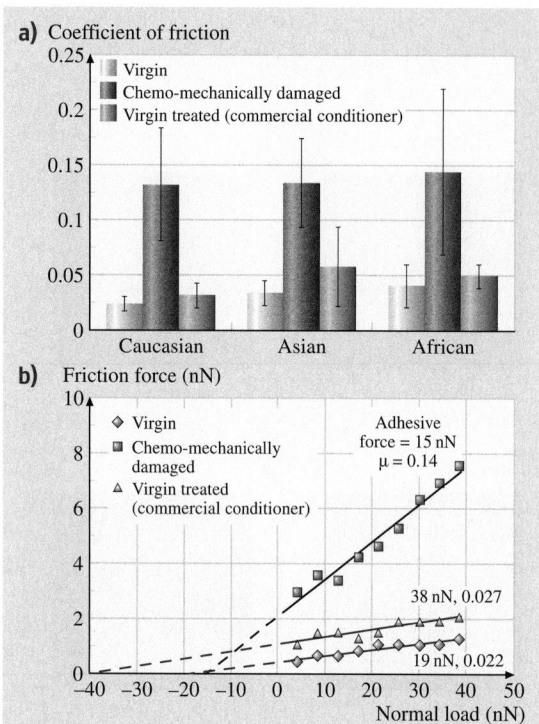

Fig. 40.43 Average coefficient of friction values for virgin, damaged and treated hair of each ethnicity (**a**). *Error bars* represent $\pm 1\sigma$ on the average coefficient value. The *bottom plot* (**b**) shows friction force vs. normal load curves with typical values for virgin, chemomechanically damaged and virgin treated Caucasian hair. While the coefficients of friction are similar for virgin and treated hair, the adhesive force of treated hair is higher than that of virgin hair (after [40.1])

friction force on the cuticle surface of chemomechanically damaged hair showed a much larger variance, which contributed to the higher values of friction. Another contribution to the higher friction could be that the tiny peaks developed after damage also create a ratchet mechanism on the nanoscale, which affects the friction between the AFM tip and the surface. These peaks could then add to the friction signal. The chemical and mechanical damage caused to the hair have shown high reproducibility in the lab in terms of structural alteration, which explains the similar friction properties across ethnicities for chemomechanically damaged hair. With virgin and virgin treated hair, however, the prior mechanical damage and level of exposure of the fibers to the sun are not known, and they depend largely on the individuals. Thus, across ethnicities

there was variability in the friction force for those hair types.

Perhaps the most notable difference between virgin and virgin treated hair fibers can be seen in the friction force maps shown in Figs. 40.42a and 40.42b. Although quite comparable in surface roughness, close examination of the virgin treated hair surface shows an increase in friction force, usually only surrounding the bottom edge of the cuticle. This was unlike virgin hair, where friction generally remained constant along the surface, unlike chemomechanically damaged hair, where there was a large amount of variability which was random over the entire surface.

Figure 40.43 presents friction force curves as a function of normal load for Caucasian virgin, chemomechanically damaged and virgin treated hair to further illustrate the previous discussion. One can see a relatively linear relationship between the data points for each type of hair sample. When plotted in such a way, the coefficient of friction is determined by the slope of the least squares fit line through the data. If this line is extended to intercept the horizontal axis, then a value for adhesive force can also be calculated, since the friction force F is governed by the relationship

$$F = \mu(W + F_m), \tag{40.3}$$

where μ is the coefficient of friction, W is the applied normal load, and F_m is the adhesive force [40.27, 30].

One explanation for the increase in friction force of virgin treated hair on the micro/nanoscale is that meniscus forces between the tip and the conditioner/cuticle become large during tip contact as the tip rasters over the surface, causing an increase in the adhesive force. This adhesive force is of the same magnitude as the normal load, which makes the contribution of adhesive force to friction rather significant. Thus, at sites where conditioner is accumulated on the surface (namely around the cuticle scale edges), friction force actually increases. On the macroscale, however, the adhesive force is much lower in magnitude than the applied normal load, so the contribution of adhesive force to friction is negligible over the hair swatch. As a result, virgin treated hair shows a decrease in friction force on the macroscale, which is the opposite to the micro/nanoscale trend. The friction and adhesion data taken on the micro/nanoscale are useful, however, because they relates to the presence of conditioner on the cuticle surface and allow us to obtain the estimated conditioner distribution.

It is also observed from Fig. 40.43 that while chemomechanically damaged hair displays a higher friction force upon the application of normal load, and

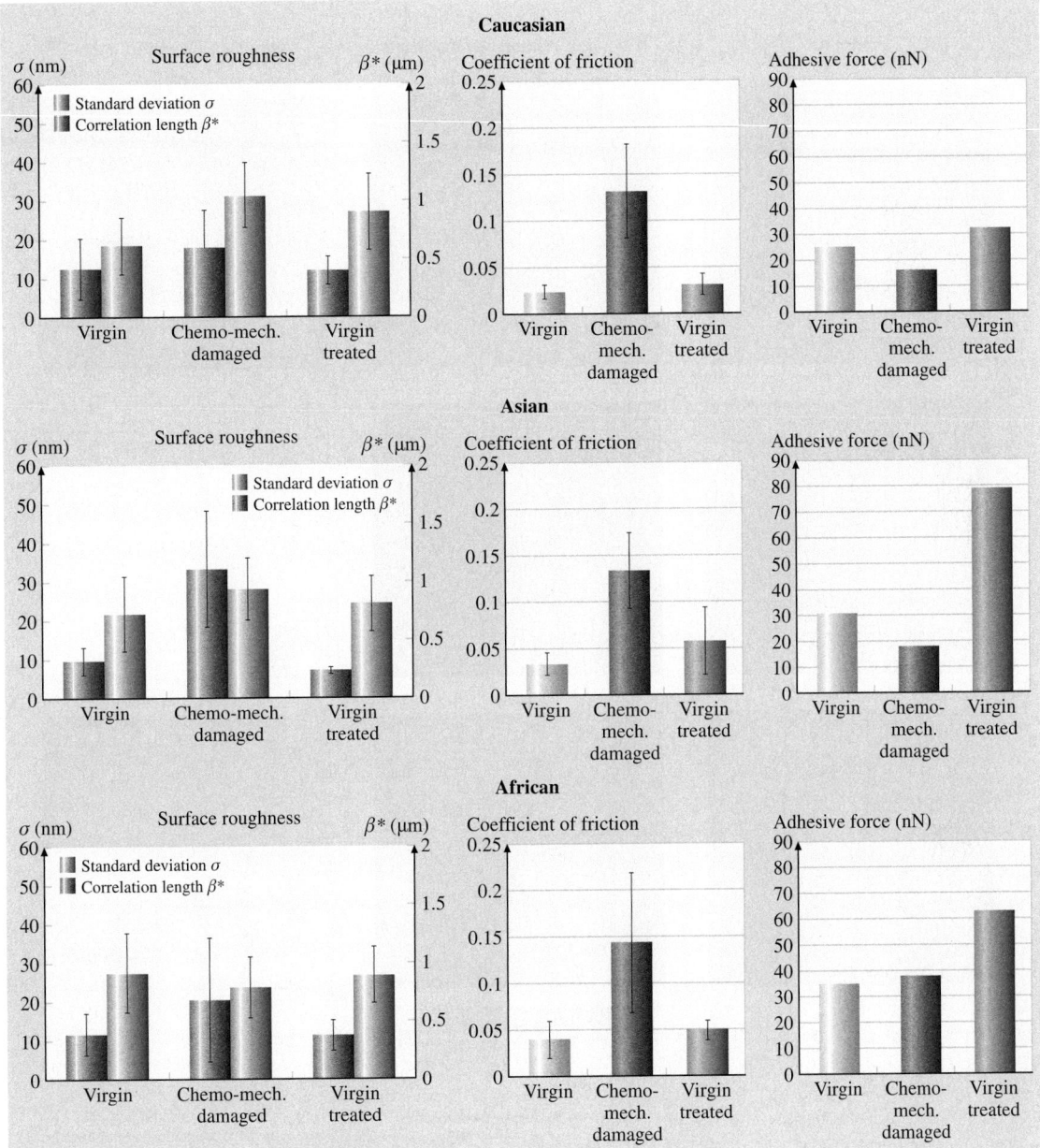

Fig. 40.44 Surface roughnesses, coefficients of friction, and adhesive force data for virgin, chemomechanically damaged and virgin treated hair of each ethnicity (after [40.1])

consequently a higher coefficient of friction, chemomechanically damaged hair friction is not as strongly dependent on the adhesive force contribution as it is for the virgin and virgin treated hair types. Average values for μ were calculated and are compiled in Fig. 40.43, Fig. 40.44 and Table 40.16 for all hair ethnicities and types [40.1]. Error bars represent $\pm 1\sigma$ on the average coefficient value. The coefficients of friction for virgin, chemomechanically damaged and virgin treated Caucasian hair are 0.02, 0.13 and 0.03, respectively. For

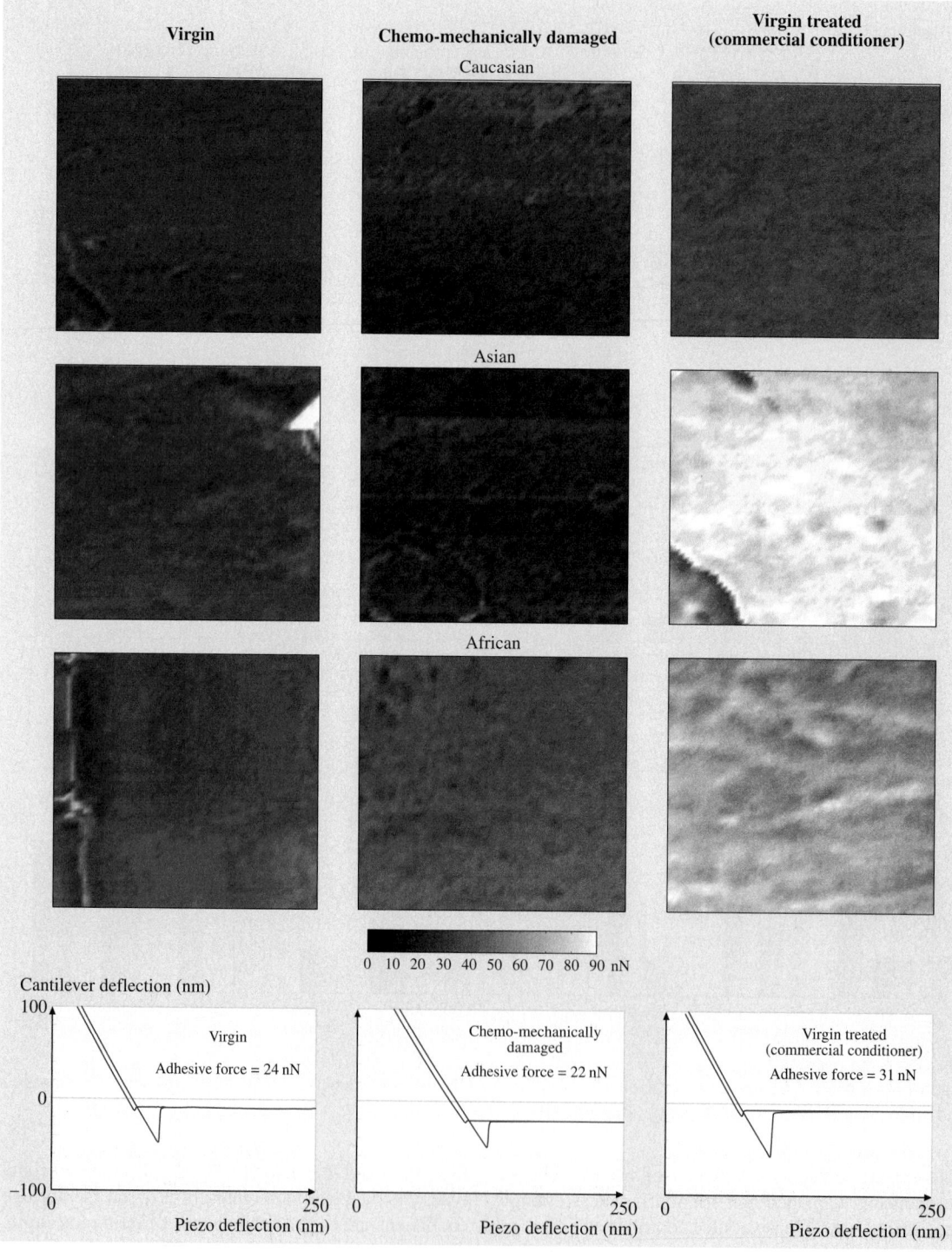

Fig. 40.45 Force–volume maps of virgin, chemomechanically damaged and virgin treated hair of each ethnicity. Examples of the individual force calibration plots (*bottom*) used to construct the FV maps are presented for Caucasian hair of each type (after [40.1]) ◄

virgin, chemomechanically damaged and virgin treated Asian hair, the coefficients of friction are 0.03, 0.13 and 0.06, respectively. Finally, the coefficients of friction for virgin, chemomechanically damaged and virgin treated African hair are 0.04, 0.14 and 0.05, respectively. Chemomechanically damaged hair presents the highest coefficient of friction, but also displays the largest standard deviation, due to the large variations in chemical

and mechanical damage that each hair or hair bundle experiences. The coefficient of friction of virgin treated hair is slightly larger than that of virgin hair for all ethnicities. While the coefficients of friction are similar in virgin and virgin treated hair, the contribution of the adhesive force to the friction for Caucasian virgin treated hair is higher than it is for Caucasian virgin hair, when calculated according to the method described above. However, this was not always the trend for Asian and African virgin treated hair samples. It should be noted that since the tip moves laterally over the surface in friction force measurement, the conditioner layer could become smeared out, which would account for the inconsistent trend. In the adhesive force mapping described

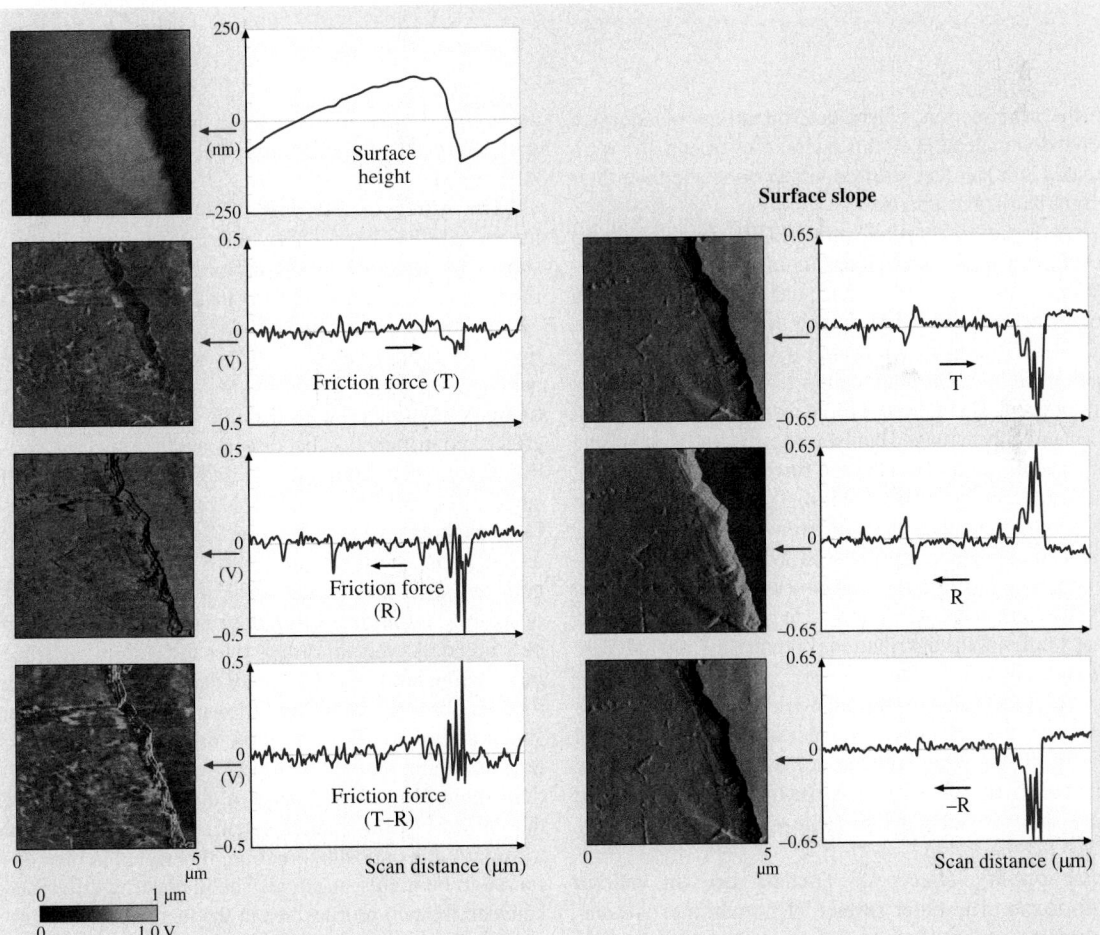

Fig. 40.46 Surface roughness, friction force and slope across a cuticle scale edge showing the directionality dependence of friction. *Left*: Surface roughness and friction force maps with accompanying 2-D profiles. *Right*: Surface slope maps with accompanying 2-D profiles (after [40.1])

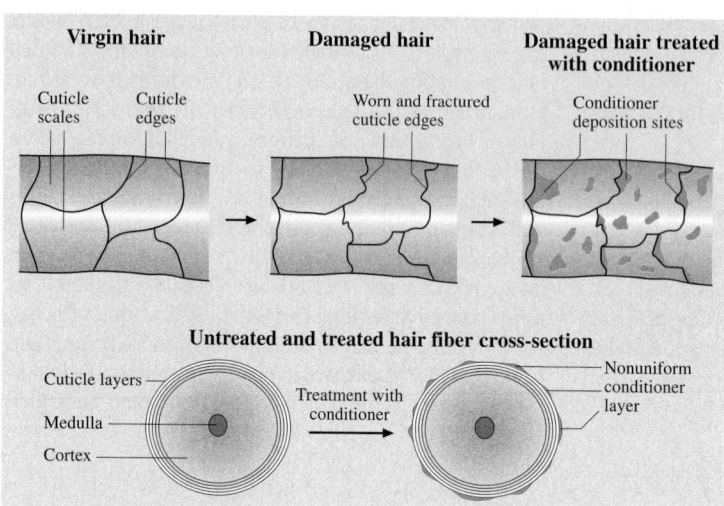

Fig. 40.47 The effect of damage to the cuticle scales and the deposition of conditioner on the cuticle surface. The cross-sections of the hair with and without conditioner are shown below (after [40.2])

in the next section, where determinations of adhesive force do not depend on this lateral movement, all virgin treated hair samples showed higher adhesion than their virgin hair counterparts.

A force calibration plot (FCP) technique and resulting adhesive force maps (commonly called force–volume maps) can be used to understand the adhesive forces between the AFM tip and the sample [40.31, 33, 42, 54]. FV maps are shown in Fig. 40.45, as is a sample individual force calibration plots from which these maps are created. The adhesive force distribution for chemo-mechanically damaged hair was shown to be comparable to that for virgin hair adhesive force, but slightly lower. A significant increase in adhesive force over the entire map was found in all cases of virgin treated hair as compared to virgin hair, especially in Asian and African hair. Conditioner distributions can be seen from these images. This technique has the potential to be very useful in further studies of the distributions of materials and hair care products on hair surfaces.

A typical value for the adhesive force shown in each FV map was calculated. Values are shown in the plot of Fig. 40.44, along with surface roughness and coefficient of friction data for all hair samples. Adhesive force values are also tabulated in Table 40.16.

Directionality Effects of Friction on the Micro/Nanoscale. The outer surface of human hair is composed of numerous cuticle scales running along the fiber axis, generally stacked on top of each other. As previously discussed, the heights of these step changes are approximately 300 nm. These large changes in

topography make the cuticle an ideal surface for investigating the directionality effects of friction using AFM/FFM.

The directionality effect of friction on the macroscale has been well studied in the past. It was shown by *Robbins* [40.3] and *Bhushan* [40.41] that rubbing the hair from the tip to the root (against the cuticle steps) results in a higher coefficient of friction than rubbing the hair from root to tip (with the cuticles). On the micro/nanoscale, it is important to distinguish how material effects and topography-induced effects contribute to the directionality effect of friction force when scanning over and back across a small surface region [40.27, 31, 33]. Figure 40.46 shows surface roughness, friction force and surface slope maps of a Caucasian virgin hair fiber, each coupled to their accompanying 2-D cross-sectional profiles. The scan size of 5 μm × 5 μm means that one cuticle step height can be studied. As the tip rasters over the step in the trace mode (from left to right), a small decrease in the friction force is observed as the tip follows the step downward. When the tip comes back in the retrace mode (right to left), climbing up the sharp peak results in a high friction signal. However, because of the sign convention of the AFM/FFM that causes a reversal in the sign when traversing the opposite direction, this signal is now observed to be highly negative. The interesting difference between the two profiles lies in the fact that the magnitude of the decrease in friction when going up the step is much larger than the magnitude of the friction when the tip is going down the step, yet both of the signals are in the same direction. The important point is that, even

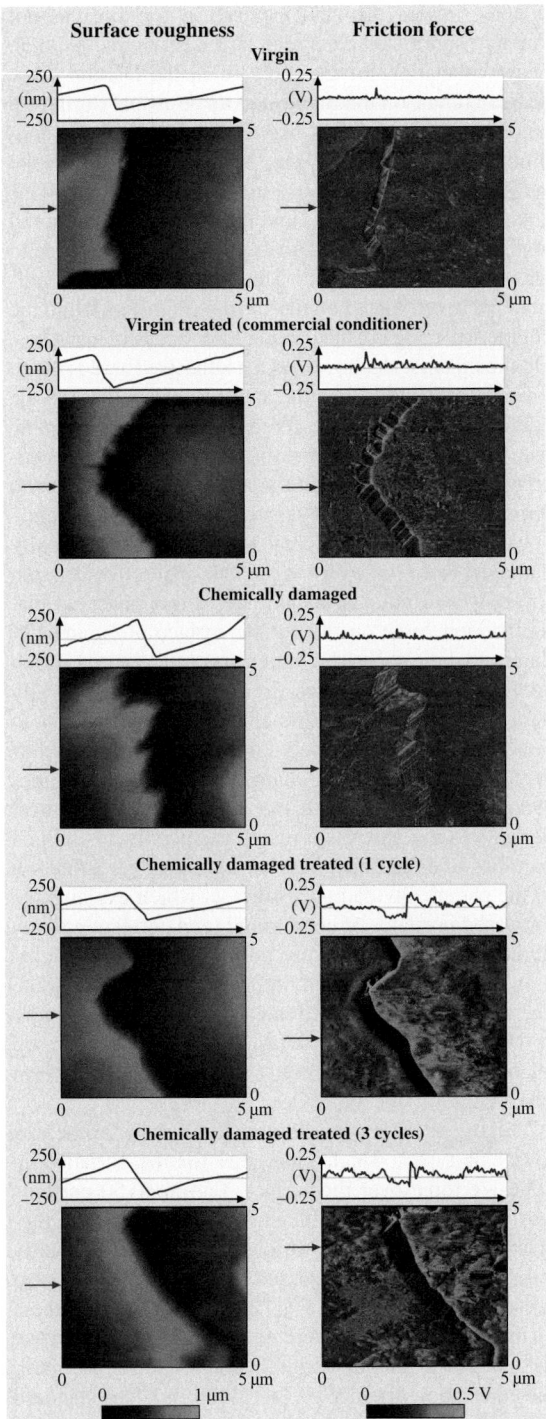

when the difference between the two signals is taken (T−R), there is still a gross variation in the image due to topography effects. These topography effects yield variations in friction in the same direction, whereas material effects show up in opposite directions. It can be shown that the cuticle edge provides a local ratchet and collision mechanism that increases the friction signal at that point. It was concluded that surface slope variations always correlate with friction force variations with respect to topography effects, and the data presented in Fig. 40.46 show the same trend when comparing trace and negative retrace slope profiles.

Virgin and Chemically Damaged Caucasian Hair (With and Without Commercial Conditioner Treatment)

The hair surface is negatively charged and can be damaged by a variety of chemical (permanent hair waving, chemical relaxation, coloring, bleaching) and mechanical (combing, blow-drying) factors [40.3, 25, 55]. Figure 40.47 shows the transformation and wear of cuticle scales before and after damage. Chemical damage causes some of the scales to fracture and reveal underlying cuticle remnants. Application of conditioner provides a protective coating to the hair surface, preventing future damage.

Shown in Fig. 40.48 are surface roughness and friction force plots for virgin, virgin treated, chemically damaged, chemically damaged and treated (one cycle of conditioner), and chemically damaged treated (three cycles of conditioner) hair. Above each AFM and FFM image are cross-sectional plots of the surface (taken at the accompanying arrows) corresponding to the surface roughness and friction force, respectively. Although virgin and virgin treated hair are quite comparable in surface roughness maps, examination of the treated hair surface shows an increase in friction force, especially in the area surrounding the bottom level of the scale edge. These frictional patterns observed in treated hair were not like anything observed in the virgin or chemically damaged cases. Images of all hair types have shown friction variation due to edge contributions and cuticle mechanical damage that has left only remnants of

Fig. 40.48 Surface roughness and friction force images for virgin, virgin treated, damaged, damaged treated (one cycle of conditioner), and damaged treated (three cycles of conditioner) hair at scan sizes of 5 μm. Shown *above each image* is a cross-section taken at the positions denoted by the corresponding *arrows* showing roughness and friction force information (after [40.12])

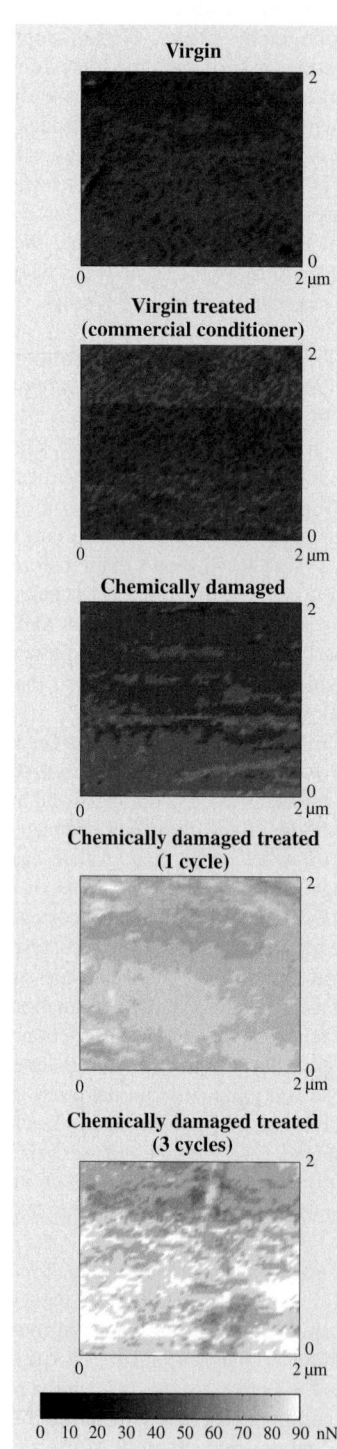

Fig. 40.49 Adhesive force maps displaying variations in adhesive force on the cuticle surface of various hair. Treated hair is seen to have higher adhesive force due to meniscus effects (after [40.12])

cuticle sublayers (such as the endocuticle). Further investigation of the corresponding treated hair roughness images showed that this increase in friction was not due to a significant change in surface roughness either. One explanation for the increase in the friction force of treated hair on the micro/nanoscale is that during tip contact meniscus forces between the tip and the conditioner/cuticle become large as the tip rasters over the surface, causing an increase in the adhesive force. This adhesive force is of the same magnitude as the normal load, which makes the contribution of adhesive force to friction rather significant. Thus, at sites where conditioner is accumulated on the surface (namely around the cuticle scale edges), the friction force actually increases. On the macroscale, however, the adhesive force is much lower in magnitude than the applied normal load, so the contribution of adhesive force to friction is negligible over the hair swatch. As a result, treated hair shows a decrease in friction force on the macroscale, which is the opposite trend to what is seen on the micro/nanoscale.

In general, friction forces are higher on chemically damaged hair than on virgin hair. Although friction forces were similar in magnitude, it was observed that the friction force on the cuticle surface of chemically damaged hair showed a much larger variance, which contributed to the higher friction values. Chemically damaged treated hair shows a much stronger affinity to the conditioner. It is widely known that the cuticle surface of hair is negatively charged. This charge becomes even more negative with the application of chemical damage to the hair. As a result, the positively charged particles of conditioner have an even stronger attraction to the chemically damaged surface, which explains the increased presence of conditioner (and correspondingly higher friction forces) when compared to virgin treated hair. After the application of three conditioner cycles to chemically damaged and treated hair, the friction force is still higher near the cuticle edge, but it is also increased all over the cuticle surface, showing a more uniform placement of the conditioner.

Figure 40.49 shows adhesive force maps for the various hair types, which gives feel for the variation in adhesive force over the surface. Figure 40.50 presents surface roughness, coefficient of friction, and adhesive force plots for the various virgin and chemically damaged hair types discussed above. The data is also presented in Table 40.17 [40.12]. The surface roughness of human hair is characterized by a vertical descriptor, the height standard deviation σ, and a spatial descriptor, the correlation distance β^* [40.27, 30, 31]. The standard deviation σ is the square root of the arithmetic mean of the square of the vertical deviation from the mean line. The correlation length can be referred to as the length at which two data points on a surface profile can be regarded as being independent, thus serving as a measure of randomness. These two parameters are all that is needed if the surface is assumed to have a Gaussian height distribution and an exponential autocorrelation function [40.27, 30, 31]. Virgin and virgin treated hair types showed similar σ values, while β^* was higher in virgin treated hair. Chemically damaged hair and both types of chemically damaged and treated hair showed similar roughness values, although σ was higher for the treated cases. The chemically damaged hair presented in this work is different from the chemomechanically damaged hair studied in [40.1]. It seems that chemically damaging the surface does not lead to as much of an increase in wear and surface roughness as a combination of both chemical and mechanical damage. Thus, it should be noted (and it is understandable) that there are differences between chemomechanically and chemically damaged hair.

The coefficients of friction for virgin and virgin treated hair types are similar, but slightly higher for the treated cases. Chemically damaged hair shows a much higher coefficient of friction, and also more data variation, since the chemical damage varies throughout each individual fiber. An interesting finding was that, contrary to the results for virgin and virgin treated hair types, the coefficient of friction for chemically damaged hair decreased with the application of conditioner (both one and three cycles). One possible explanation for this is that the stronger negative charge on chemically damaged hair results in more attraction to the conditioner, which leads to greater adhesive forces, as well as (more importantly) a lower shear strength on the surface. This creates an overall effect of lubrication and ultimately lowers the coefficient of friction.

Average adhesive force values were taken from the adhesive force maps described previously. Virgin treated hair shows a higher adhesive force than virgin hair due to the meniscus effects that arise from the interaction of the AFM tip with the conditioner on the cuticle surface [40.27, 30, 31]. The same trend is true and even more evident for chemically damaged treated hair compared to chemically damaged hair. A possible reason that one cycle of conditioner on chemically damaged hair gave higher average adhesive force than three cycles could be that three cycles generally places the conditioner more uniformly on the surface rather than accumulating it most near the bottom surface (near the cuticle edge), which is where

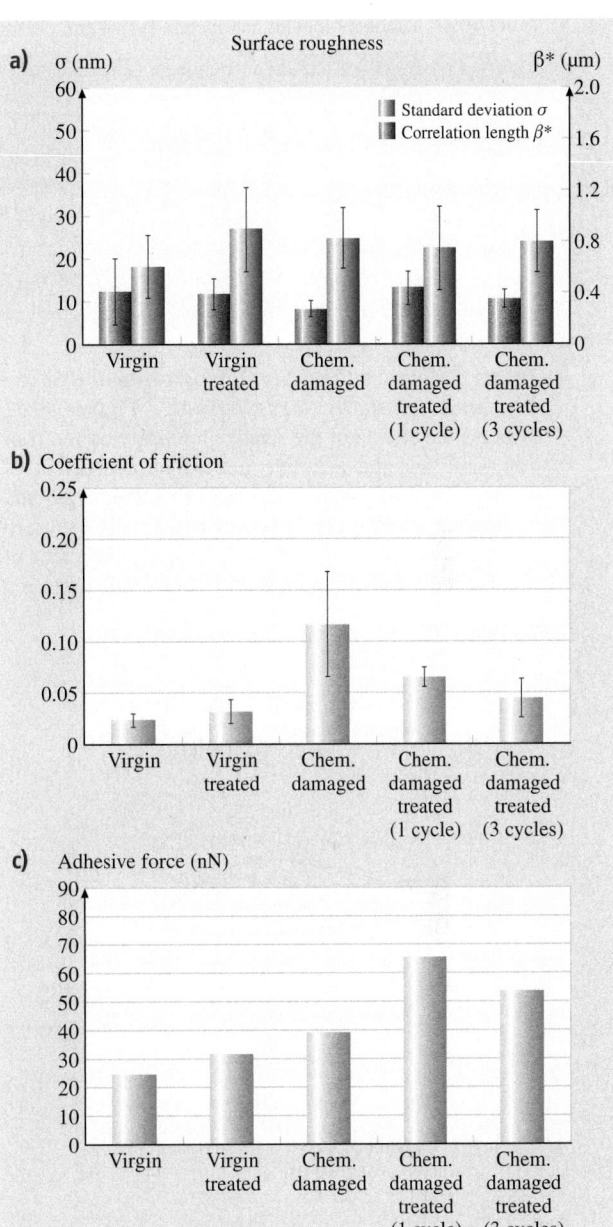

Fig. 40.50 Surface roughness (**a**), coefficient of friction (**b**), and adhesive force (**c**) plots for various hair (after [40.12])

the adhesive force maps were generally taken. Nevertheless, the increased adhesive force shown in the plots is a clear indication of the presence of conditioner on the hair surface, and its localization can be observed.

Table 40.17 Nanotribological parameters for virgin, chemically damaged, and treated hair

Surface roughness, coefficient of friction and adhesive force for virgin and chemically damaged hair, with and without conditioner treatment				
Hair type	**Surface roughness**		**Coefficient of friction**	**Adhesive force (nN)**
	σ (nm)	β^* (μm)		
Virgin	12±8	0.61±0.2	0.02±0.01	25±5
Virgin, treated	12±4	0.90±0.3	0.03±0.01	32±5
Chemically damaged	8.4±2	0.83±0.2	0.13±0.06	39±0.5
Damaged, treated (1 cycle)	13±4	0.75±0.3	0.05±0.02	66±0.7
Damaged, treated (3 cycles)	11±2	0.80±0.2	0.04±0.02	54±33

Effects of Relative Humidity, Temperature and Soaking, and Durability Measurements. Figure 40.51 displays the effects of the relative humidity on the friction force and the adhesive force. The coefficient of friction remained relatively constant for the virgin and virgin treated hair types. However, chemically damaged hair experienced a large increase in the coefficient of friction at the high humidity, while chemically damaged

and treated hair experienced the opposite trend. This clearly shows that heavy moisture in the air plays a role in the frictional properties of chemically damaged hair. When combined with conditioner, a lubricating effect once again dominates as the water helps form a liquid layer which is more easily sheared. In terms of the adhesive force, most of the samples showed a decrease

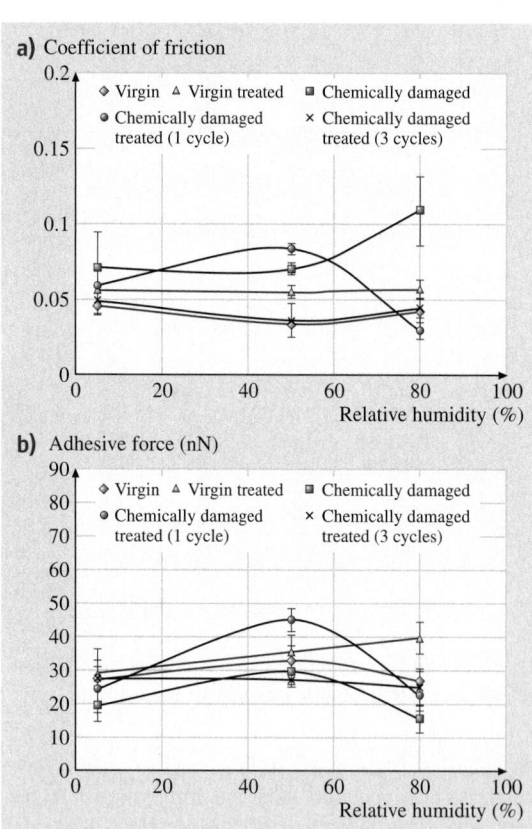

Fig. 40.51 Effect of relative humidity on the nanotribological properties of various hair (after [40.12])

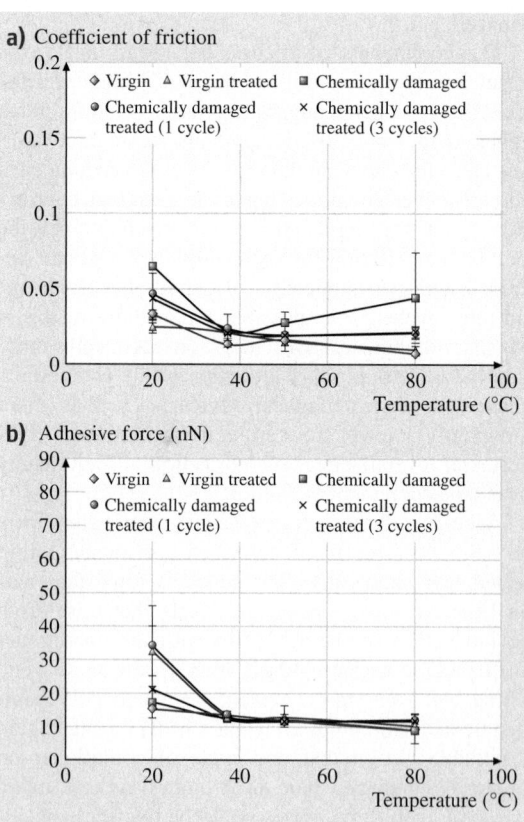

Fig. 40.52 Effect of temperature on the nanotribological properties of various hair; (**a**) coefficient of friction, (**b**) adhesive force (after [40.12])

in adhesive force with high humidity. It is expected that meniscus effects diminish as the water builds up on a surface, and as a result they do not readily contribute to adhesive force. Thus, the adhesive force is expected to drop at very high humidity.

Figure 40.52 displays the effects of the temperature on the friction force and the adhesive force. The coef-

ficient of friction generally decreased with increasing temperature. As the hair fiber heats up, any conditioner present on the surface decreases in viscosity, causing the film to thin and the friction force to drop. The decrease in friction force ultimately leads to lower coefficient of friction values. Adhesive force was shown to decrease with increasing temperature as well. This

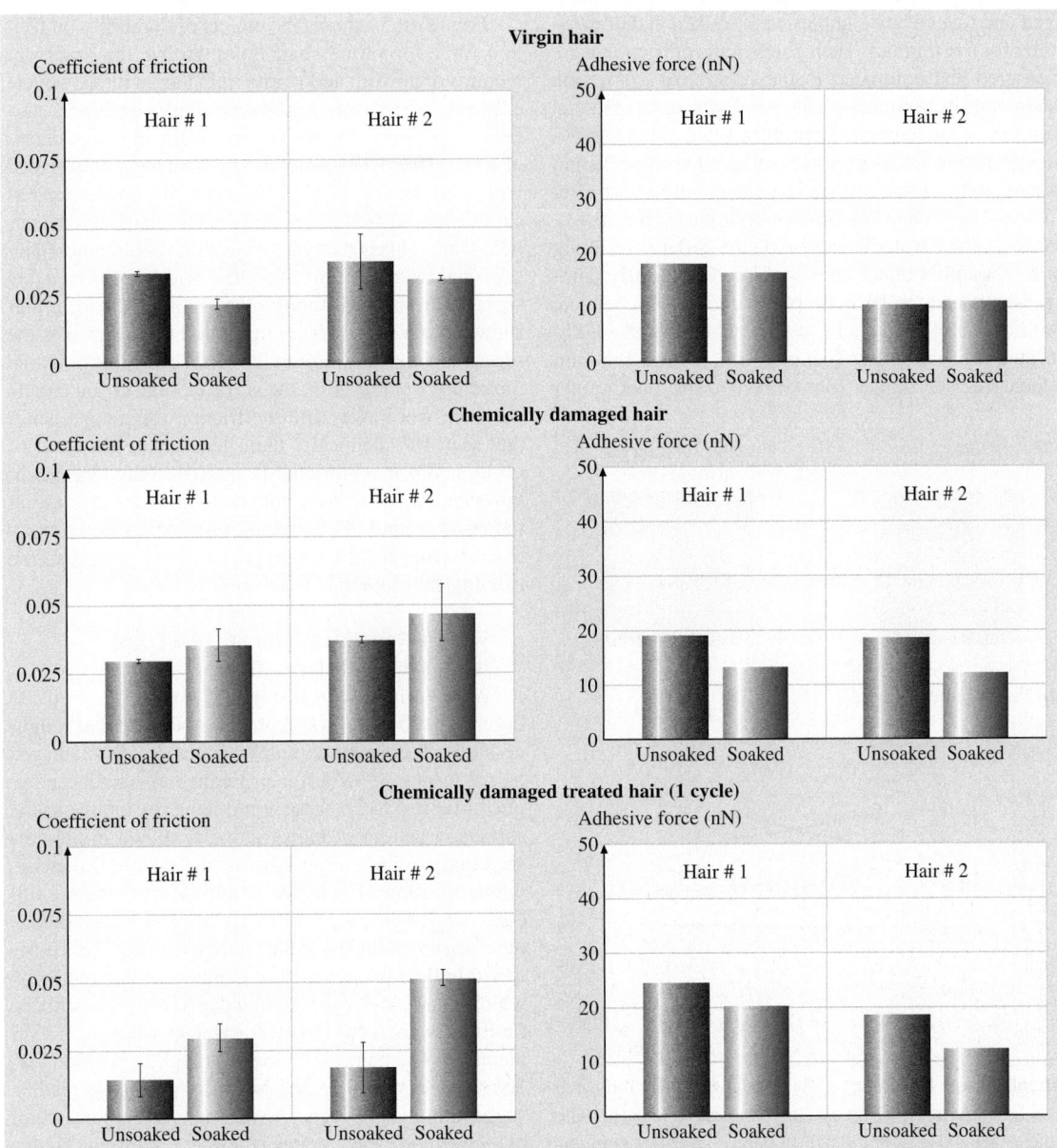

Fig. 40.53 Effect of soaking hair in deionized water on the coefficient of friction and adhesive force for virgin, chemically damaged, and chemically damaged treated (one cycle of commercial conditioner) hair (after [40.2])

was especially evident for treated hair fibers, whereas the large adhesive force seen at room temperature decreased rapidly to adhesion values similar to untreated fibers. It is most likely that the thinning conditioner layer causes a reduced surface tension at higher temperatures, which directly relates to the drop in adhesive force.

Virgin, chemically damaged and chemically damaged and treated hair samples were soaked in deionized water for five minutes. Their coefficients of friction were measured and compared to the coefficient of friction values for dry samples which were adjacent to the wet samples on the respective hair fiber. Figure 40.53 shows the results for two hair samples of each hair type. Virgin hair exhibits a decrease in the coefficient of friction after soaking. Virgin hair is more hydrophobic (Table 40.13), so more of the water is present on the surface, resulting in a lubrication effect after soaking. Chemically damaged hair tends to be hydrophilic due to the chemical degradation of the cuticle surface, which results in absorption of water after soaking. This softens the hair, which leads to higher friction, even with conditioner

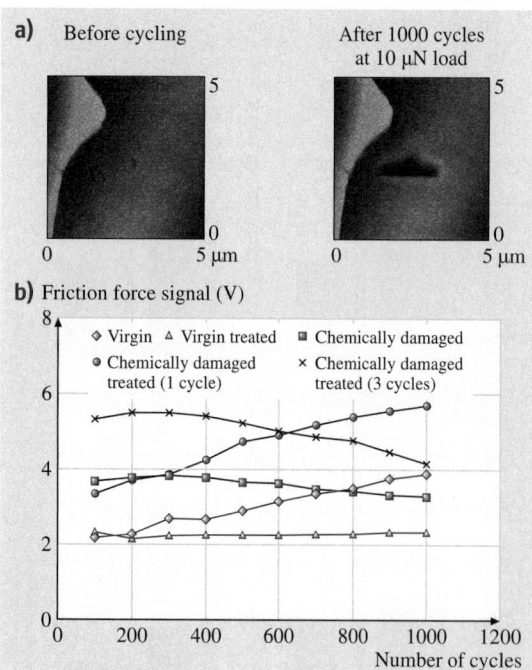

Fig. 40.54a,b Durability study of change in friction force as a function of AFM tip cycling for various hair (**b**). The images (**a**) above the plot signify before and after comparisons of a cuticle surface subjected to cycling at a load of 10 μN (after [40.12])

treatment. This is yet another indication that the virgin and chemically damaged hair types have significantly different surface properties, which in many cases results in opposite trends to their nanoscale tribological properties. Adhesive force for virgin hair remained approximately the same before and after soaking, while it decreased for chemically damaged and chemically damaged and treated hair after soaking.

Figure 40.54 shows the effects of durability on friction force for various hair types. Above the graph are pictures of unworn and worn virgin hair, with the cuticle edge serving as a reference point. Before testing the surface is relatively smooth and free from any large debris or wear. After 1000 cycles at approximately 10 μN load with a stiff silicon AFM tip, however, the interaction has caused degradation and wear (scratch) marks on the cuticle scale. This is the type of wear one could potentially see if hair were to come in contact with sand from a day spent at the beach, among other activities. Virgin hair shows an obvious increase in friction force signal as the scratching digs further into the surface. By this point the lubricious lipid layer on the surface of the virgin cuticle has been worn away and the friction force approaches that seen for chemically damaged hair at the onset of cycling. When conditioner is applied to the virgin hair, however, the wear does not show up as an increase in the friction force. Thus, conditioner serves as a protective covering for the virgin hair, helping to protect the tribological properties when wear occurs.

Surface Roughness and Friction Force for Various Hair Types Treated with Various Conditioner Matrices

Figure 40.55 displays representative surface roughness and friction force maps for chemically damaged hair to which seven different treatments have been applied (Table 40.12). When conditioner is applied to the surface, a pattern of high friction is shown in the area surrounding the bottom edge of the cuticle. Likewise, the application of a BTMAC surfactant (with no silicone deposition) or a BAPDMA surfactant (no silicone deposition) results in similar friction features. This is believed to be due to an area of conditioner accumulation which causes increased friction due to meniscus effects. Friction maps for BTMAC surfactant where a PDMS blend of silicone (at both low and high deposition levels) has been added do not show this increase as readily, suggesting that this type of silicone is not a contributor to high friction force on the nanoscale. This may be due to the fact that a PDMS-type silicone is fairly mobile on the surface and so does not cause the same meniscus ef-

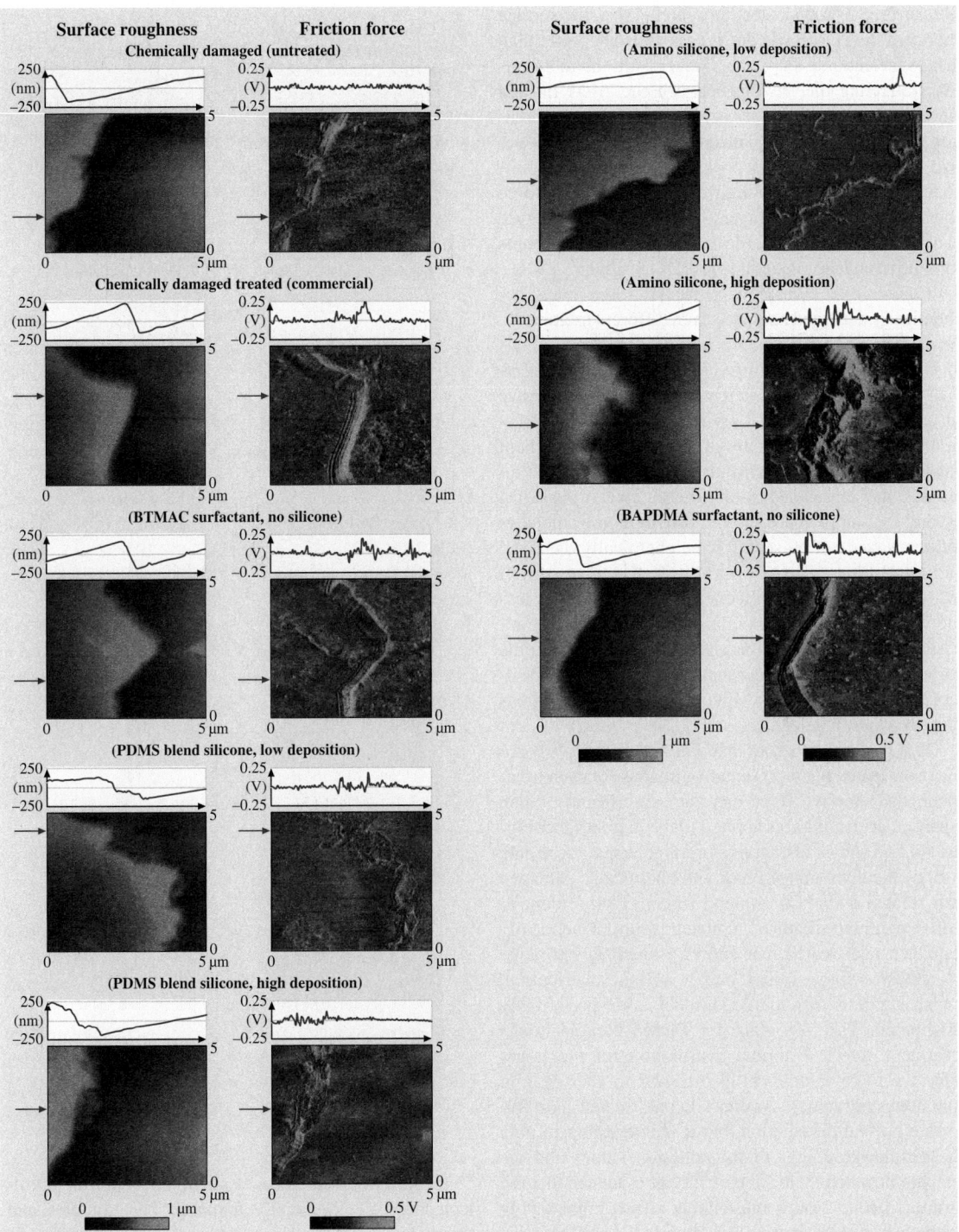

Fig. 40.55 Surface roughness and friction force maps for Caucasian chemically damaged hair after various treatments (after [40.2])

fects as the AFM tip rasters through it. Also, the surface energy of the PDMS silicone is believed to be lower than that of the cationic surfactant. As a result, the meniscus force affecting the friction between the AFM tip and silicones in the conditioner is lower than that of conditioner without silicones. Thus, the overall friction force will be lower. For high deposition levels of amino silicone, however, there is a high variation in friction force and a more distributed layer. The amino group is typically less mobile and harder to move, which accounts for a different slip plane flow to PDMS silicone.

Figure 40.56a displays adhesive force maps for chemically damaged hair and seven different treatments. As shown in the legend, a lighter area corresponds to a higher tip pull-off force (adhesive force). Chemically damaged untreated hair has relatively low adhesive force, and is more or less consistent over the hair surface. In almost all cases, addition of a conditioner treatment caused an increase in meniscus force, which in turn increased the adhesive pull-off force between the AFM tip and the sample. The uneven distribution of the conditioner layer can be seen for the chemically damaged and treated hair. This uneven distribution is also most evident for the amino silicone images, where the less mobile silicone brings about a significant change in adhesion over the surface. It is clear that adhesion over the surface is much more consistent for the PDMS blend silicone than the for amino silicone, where various areas of high adhesion occur.

It is important to note that while the adhesive force maps presented are representative images for each treatment, the adhesive force can vary significantly when a particular treatment is applied to the hair surface. Figure 40.56b shows histograms of all adhesive force data for chemically damaged hair and chemically damaged hair treated with PDMS blend silicones (high deposition) and amino silicones (high deposition). Chemically damaged and treated hair shows a much larger range of adhesive force values and a normal distribution, which suggests that the conditioner layer is normally distributed. The histogram for PDMS-blend silicone treatment shows a normal distribution for the larger adhesive force values, but also shows another peak at low adhesion values. Amino silicone-treated hair follows a normal distribution, but it is interesting to note the distinct groupings of the adhesion values and the spacing between them. This is further evidence that the amino silicone groups most likely attach immediately to the hair surface and are less mobile than PDMS silicone, causing distinct regions of high and low adhesion values over the cuticle surface.

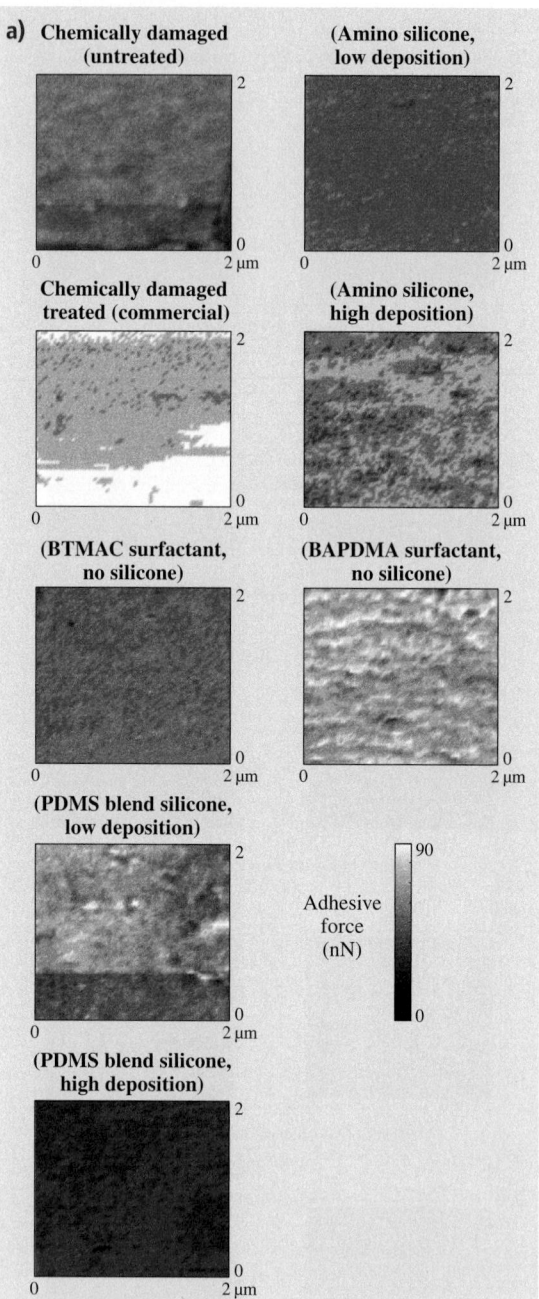

Figure 40.57 displays a summary of the data collected for all chemically damaged hair samples and their treatments. The figure also includes the macroscale coefficient of friction data obtained using a technique similar to the macroscale measurement technique de-

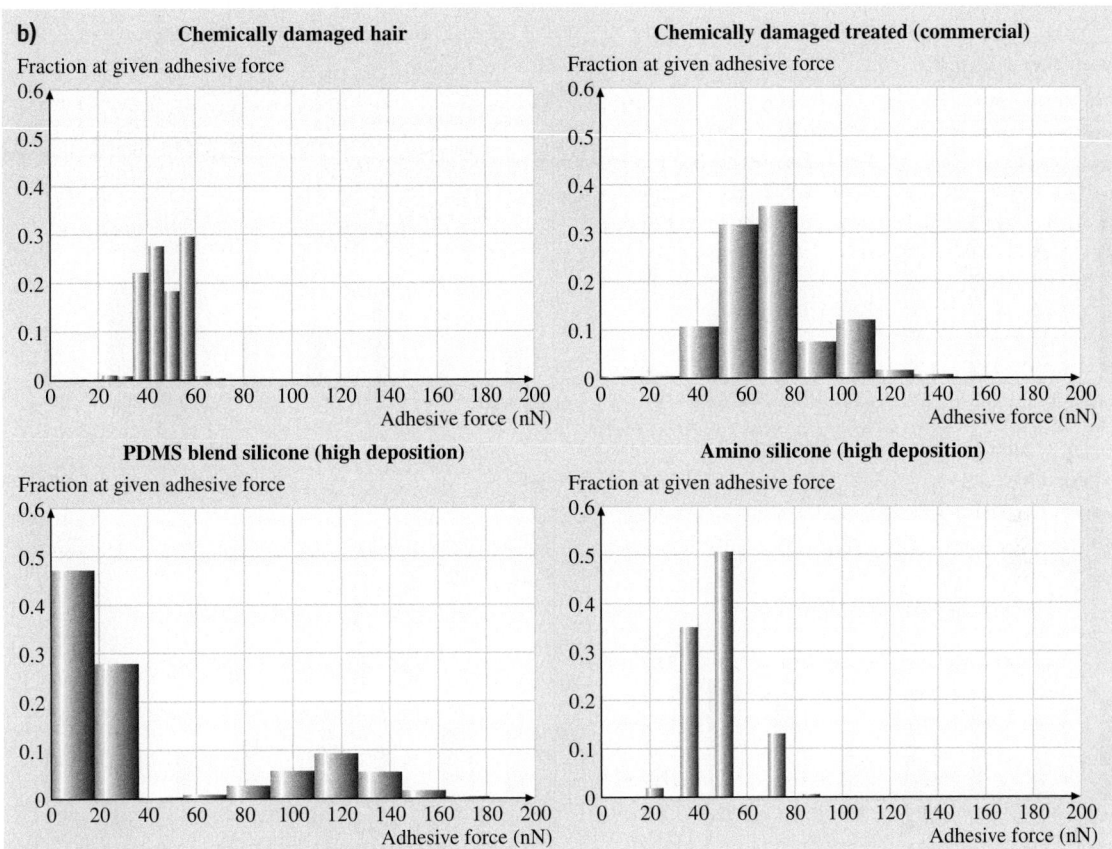

Fig. 40.56 (**a**) Adhesive force maps for damaged hair after various treatments; (**b**) adhesive force histograms for various hair (after [40.2]) ◄ ▲

Part D | 40.6

scribed earlier (Fig. 40.15). Table 40.18 reviews some of the observations and corresponding mechanisms, which help to explain the trends found [40.2]. The application of the commercial conditioner to the chemically damaged hair caused a decreased coefficient of friction and a large increase in adhesive force. The decreased coefficient of friction may be explained by the fact that the chemically damaged hair accumulates much of the positively charged conditioner on the surface due to its highly negative charge, which in turn makes it easier to shear the liquid on the surface, causing a lower coefficient of friction. However, the nanoscale pull-off force (adhesive force) is much larger than on the untreated hair because of meniscus effects. In general, the adhesive force varied widely but typically showed a significant increase with the presence of conditioner. As discussed previously, this is a clear sign that meniscus effects are influencing the pull-off force between the tip and the sample.

In most cases, the macroscale and microscale coefficient of friction followed the same trend, in which a decrease was observed with the addition of the PDMS-blend or amino silicones to the BTMAC surfactant. Silicones are typically used for lubrication and thus give the conditioner more mobility on the hair surface compared to just surfactants and fatty alcohols. However, an inverse trend was seen for the amino silicone group.

The amino silicones have a strong electrostatic attraction to the negatively charged hair surface, which in turn creates higher binding forces and less mobility. The dampened mobility of the amino silicone at high deposition levels with respect to the hair surface and tip may account for this wide variation in the coefficient of friction and the large adhesive force values. In terms of adhesive force, it was previously observed in Fig. 40.56a that the amino silicone treatments showed much more distinct regions of higher and lower adhesion

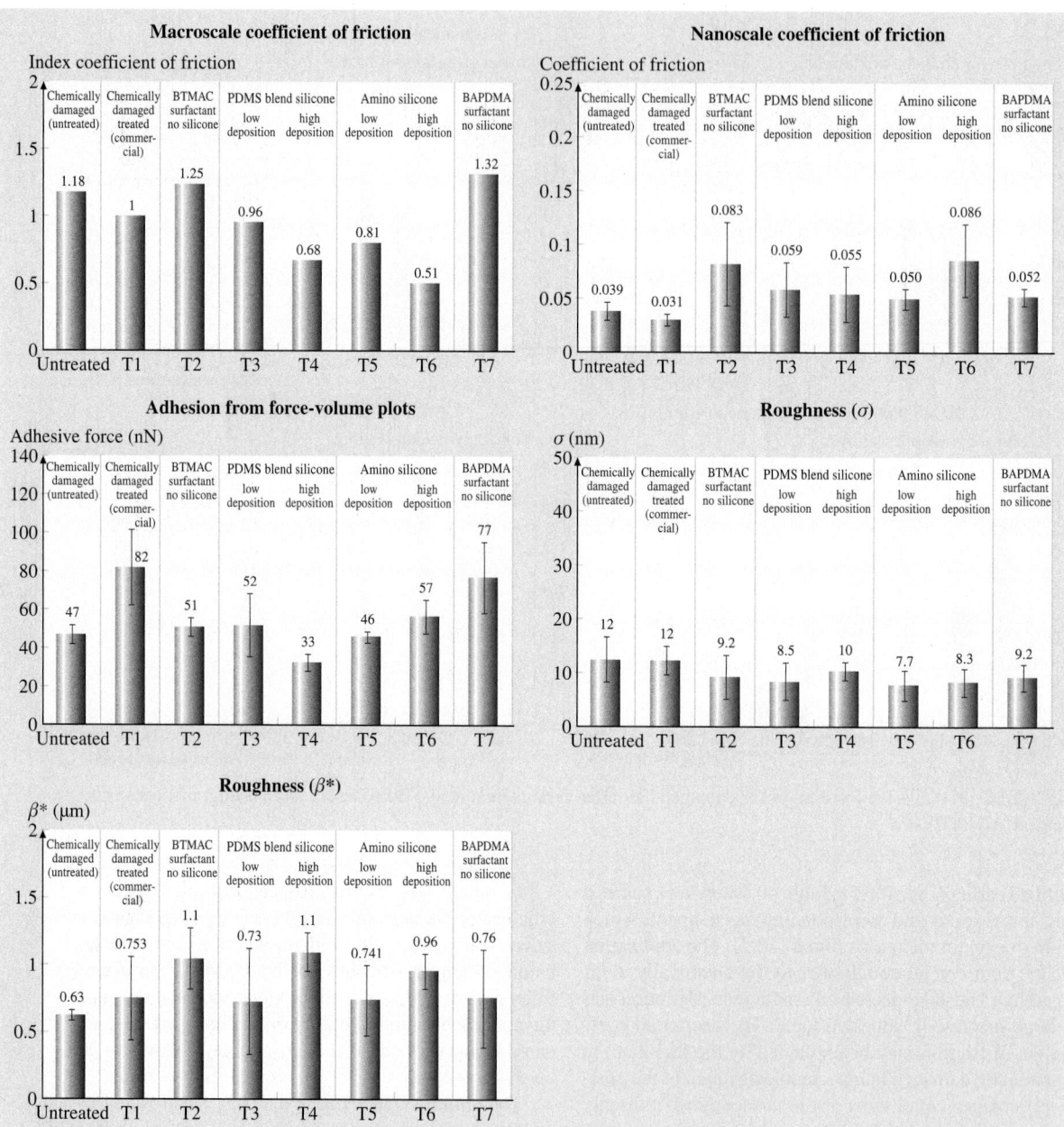

Fig. 40.57 Coefficient of friction, adhesive force and surface roughness plots for damaged hair after various treatments (after [40.2])

compared to PDMS-blend silicones. This nonuniform amino silicone thickness distribution on hair is also believed to be caused by the inhibited mobility, as the molecules immediately attach to the hair at contact and do not redistribute as a uniform coating. The increased polarity of the amino silicones compared to the PDMS-

blend may also be a major contributor to the higher friction and adhesion at high deposition levels.

If we consider hair coated only with the BTMAC surfactant and fatty alcohols, and then add low deposition levels of PDMS-blend (with high MW weight) or amino silicone, it is seen that the coefficient of friction

Table 40.18 Observations and corresponding mechanisms regarding coefficient of friction and adhesion for various hair treatments

Observation	Mechanism
Damaged vs. damaged treated hair	
Damaged hair shows a decrease in the coefficient of friction but an increase in adhesion upon the application of commercial conditioner.	The layer of conditioner deposited on the surface of the damaged hair results in a lower shear strength, which in turn lowers the coefficient of friction, while meniscus effects increase the pull-off (adhesive) force between the tip and hair sample.
PDMS-blend vs. amino silicone (at high deposition)	
Amino silicones interact strongly with the negatively charged hair surface, especially at high deposition levels.	A stronger electrostatic attraction exists, which results in stronger binding forces (leading to higher adhesion) for high-deposition amino silicone.
The thickness distribution of amino silicone on hair is less uniform than that for the PDMS blend.	Amino silicone is less mobile, so molecules attach to hair upon contact and do not redistribute easily.
Surfactant vs. surfactant plus addition of silicone	
Adhesion remains approximately the same while the coefficient of friction decreases with the addition of PDMS-blend silicone and amino silicone (low deposition). Coefficient of friction remains about same for high-deposition amino silicone.	Mobility increases with the addition of PDMS-blend silicone, which leads to a lower coefficient of friction. For a large deposition of amino silicone, mobility ceases and the coefficient of friction becomes high again.
BTMAC surfactant vs. BAPDMA surfactant	
BTMAC surfactant has lower adhesion but a higher coefficient of friction than the BAPDMA surfactant.	Interaction of surfactants with damaged hair surface causes inherent differences in coefficients of friction and adhesion. BAPDMA has both amino and amine groups, which increase its polarity.

decreases, while the adhesion remains approximately the same. If the deposition level is increased, it is observed that the coefficient of friction for the PDMS blend is still lower, but now the coefficient for the amino silicone is about the same as that for the BTMAC only samples. The mobility of the conditioner layer is again a major issue, since this easier mobility accounts for a lower coefficient of friction. However, when there is a large amount of amino silicone on the hair surface, the mobility ceases and the coefficient of friction becomes high again.

The BAPDMA surfactant typically showed a higher adhesive force than the BTMAC surfactant. However, there may be an increased accumulation of the BAPDMA surfactant such that the layer of conditioner is more easily sheared by the tip, as is the case for chemically damaged and treated hair. The inherent differences in surfactant composition and how they interact with the

Table 40.19 Surface roughness parameters σ, β^* for collagen and polyurethane films and human skin

Surface roughness parameters σ, β^* for collagen and polyurethane films and human skin		
	σ (nm)	β^* (μm)
Collagen (synthetic hair)	36 ± 11	0.50 ± 0.1
Polyurethane (synthetic skin)	33 ± 6	0.71 ± 0.1
Human skin	80 ± 28	0.59 ± 0.2

chemically damaged hair surface most likely account for the differences in the coefficients of friction.

With respect to the roughness, the vertical standard deviation decreased slightly with most treatments, although the standard deviations were similar. The spatial parameter increased slightly with treatment, but the variation also became extremely high.

Skin

Synthetic materials were also studied for surface roughness and friction force information, as shown in Fig. 40.58a. While macroscale dimples could be seen on the surface of collagen film, it was interesting to find similar pits and dimples on the micro/nanoscale, with a large variation in dimple size and depth. Polyurethane films are shown to have quite different topographies and friction forces, while their coefficients of friction are very similar to collagen. Human skin shows a rougher texture, with higher peaks, Fig. 40.58b. The roughness parameters for collagen and polyurethane films as well as for human skin are presented in Table 40.19 [40.1]. The standard deviation in the vertical height σ was approximately three times larger than that of virgin hair for both synthetic materials. However, the correlation length β^* was lower than that typically observed in hair. The average coefficients of friction for these synthetic materials are shown in Fig. 40.59 plotted next to virgin Caucasian hair as a reference. These values were calculated using

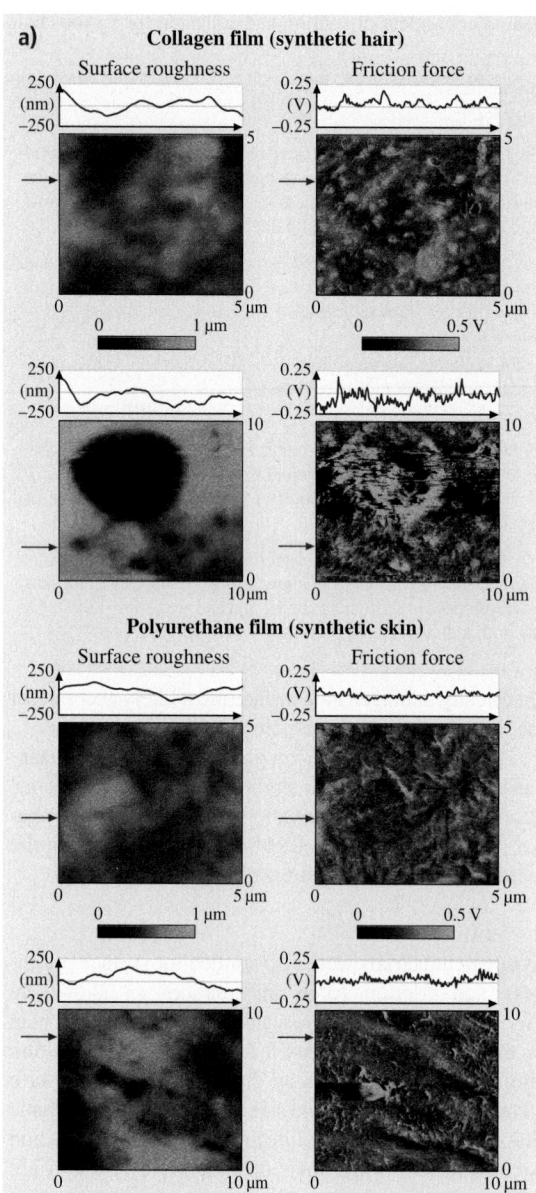

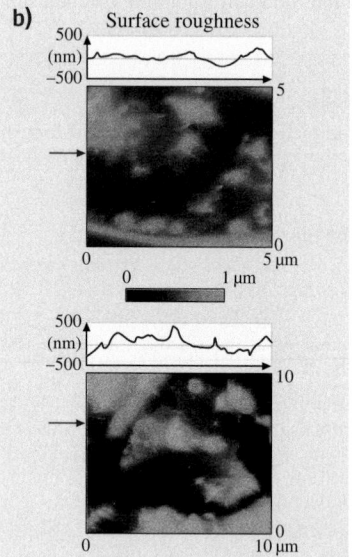

Fig. 40.58 (**a**) Surface roughness and friction force images for collagen and polyurethane films at scan sizes of 5 and 10 μm. Shown *above each image* is a cross-section taken at the position denoted by the corresponding *arrows* showing roughness and friction force information; (**b**) surface roughness for human skin at scan sizes of 5 and 10 μm. Note that the *vertical scales* in the 2-D section profiles have been doubled (after [40.1])

40.6.2 Conditioner Thickness Distribution and Binding Interactions on the Hair Surface

How common hair care products, such as conditioner, deposit onto and change hair properties is of great interest to those working in beauty care science, since these properties are closely tied to product performance. Conditioner is one of the hair care products that most people use on a daily basis. Conditioner coats the hair with a thin layer that can cause drastic changes to the surface properties of hair. Silicones are the main source of lubrication in the conditioner formulation. Silicone molecules remain as droplets surrounded by water, and their high molecular weight causes them to remain liquid and drain off the hair surface gradually, which creates a long-lasting, soft and smooth feel for conditioner-treated hair. The binding interaction between these molecules and the hair surface is one of the main factors that determines the condi-

the slopes of the friction force curves, as described previously. Both the collagen and the polyurethane films displayed similar coefficient of friction values of 0.22 and 0.24, respectively. Virgin hair displays a much lower coefficient of friction than both materials, approximately eight times lower.

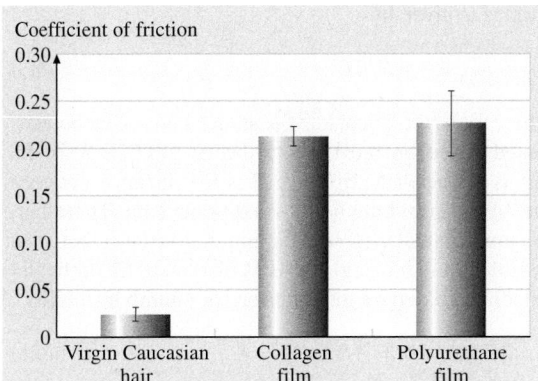

Fig. 40.59 Average coefficient of friction values for collagen and polyurethane films. Error bars represent $\pm 1\sigma$ on the average coefficient value (after [40.1])

tioner thickness distribution and consequently whether the conditioner works properly.

The morphology of the cellular structure of human hair has traditionally been investigated using SEM and TEM [40.3]. However, these techniques have limited ability to perform measurements in ambient conditions, and they cannot measure the physical properties of the hair. Cosmetic scientists have struggled to answer some important issues, such as the thickness and distribution of conditioner on the hair surface, for decades. In order to determine the thickness and distribution of thin liquid film on a substrate, several possible techniques can be applied: Fourier transform infrared spectroscopy (FTIR), ellipsometry, angle-resolved X-ray photon spectroscopy (XPS) and AFM [40.31, 33]. Ellipsometry and angle-resolved XPS have excellent vertical resolutions, on the order of 0.1 nm, but their lateral resolutions are on the order of 1 and 0.2 mm, respectively. Therefore, these techniques cannot be used for hair since the diameter of a hair fiber is only 50–100 μm. AFM on the other hand is a versatile tool and has a lateral resolution on the order of the tip radius (about 100 nm), which is difficult to achieve by other techniques. In the study reported here, force calibration plot measurements are conducted using an AFM in order to obtain the local conditioner thickness distributions on various hair surfaces. The conditioner thickness is extracted by measuring the force on the AFM tip as it approaches, contacts with and pushes through the conditioner layer. The conditioner thickness distribution on hair is effectively measured using an AFM. The effective Young's moduli of various hair surfaces are also calculated from the force–distance curves using Hertz analysis. The binding interactions of different silicones on the hair surface, as well as the effect of the hair on their effective Young's moduli, are also discussed.

Conditioner Thickness and Adhesive Force Mapping

Figure 40.19a shows a typical force calibration plot for commercial conditioner-treated hair, and the detailed interactions of the AFM tip, conditioner and hair surface at different tip–sample separations are illustrated in Fig. 40.19b as the sample first approaches the tip, makes contact, and then recedes from the tip. The measurement starts at large separation (point A) for each approach–retract cycle. There is no interaction between the tip and conditioner and no deflection of the cantilever; as the piezo (sample) moves toward the tip, because of the van der Waals attraction between the tip and the conditioner, the conditioner layer may deform a little and bulge toward the tip (point B). Note that this attractive force is very small and there is still no deflection of the cantilever. Then a sudden mechanical instability occurs (point C), and the conditioner jumps into contact with the tip and wicks up around the tip to form a meniscus. The cantilever now bends downward because of the meniscus (attractive) force acting on the tip. As the sample further approaches the tip, the deflection of the cantilever gradually increases while the tip is in the conditioner film on the hair surface. At point D, the tip contacts the underlying hair surface, the cantilever starts to bend the other way, and the surface starts to deform elastically as the tip is pushed against the hair surface. When the maximum load (relative trigger) is reached at point E, the retract cycle begins, the sample is pulled away from the tip, and a slightly larger force is needed to break the contact between the tip and the surface (point F). As the sample is retracted further from the tip, a long meniscus bridge is formed and stretched, and the force (the deflection of the cantilever) on the tip slowly decreases (point G). The meniscus bridge is stretched further (point H) until it eventually breaks at large separation. One approach–retract cycle has then been performed, and the system recovers to the initial state (point A).

Conceptually, the distance that the sample moves after the tip snaps in until it contacts the hair surface (the horizontal distance between point B and point D in Fig. 40.19a) should be a measure of the thickness of the conditioner film. Here we define it as snap-in distance H_s. As shown in Fig. 40.19a, the snap-in distance H_s is about 15 nm for this particular measurement. This snap-in distance H_s is not the real conditioner thickness h, and it tends to be thicker than the actual film thickness. In previous studies determining lubricant film thicknesses on a particulate disk surface by AFM [40.56], it was realized that as the ellipsometry thickness of the lubricant film increases, the thickness as measured by AFM also increases; however, AFM tends to give a thicker film than measured by ellipsometry. There is an offset of a few nanometers between the snap-in distance H_s determined by AFM and the film thickness h determined by ellipsometry. This offset was attributed to a thin coating of the lubricant on the surface of the AFM tip, which results from the previous contact of the tip with the lubricant film. Although this may account for part of the offset, the main reason for this offset actually likely to be a deformation of the liquid film due to its interaction with the AFM tip as the liquid film approaches to the tip. The liquid film can only approach the AFM tip to a finite minimal distance (point B as shown in Fig. 40.19b), below which the liquid surface is no longer stable due to the van der Waals attractive force between the liquid film and the AFM tip. For smaller distances, surface tension and adhesion to the substrate cannot keep the liquid surface from bulging and jumping into contact with the tip. *Forcada* et al. [40.57] theoretically analyzed the hydrostatics of the liquid film in the force fields originated by the tip and solid substrate, and indicated that the offset between the snap-in distance and the film thickness measured by ellipsometry arises from the bulging and posterior instability of the liquid film. A number of theoretical publications [40.58, 59] have addressed the issue of liquid coalescence in terms of the 'effective stiffness' of a liquid surface or interface, and have concluded that a liquid surface behaves like a Hookian spring with an effective spring constant K_{eff} equal to its surface or interfacial tension γ, viz:

$$K_{eff} \approx \gamma . \tag{40.4}$$

Therefore, an effective spring constant of only 20 mN/m is expected in air for one of the main components of conditioner, PDMS silicone, as the surface tension of PDMS is about 20 mN/m [40.60] and a film of water has an effective spring constant of 72 mN/m. These are extremely low values compared to the cantilever spring constant of 5 N/m. This suggests that even very weak forces can deform liquid surfaces. When the hair surface approaches the AFM tip, the weak van der Waals attractive force between the liquid conditioner film and the tip will cause the liquid film to deform and jump-in at the tip at a finite distance. The theoretically expected van der Waals force F_{vdW} between a sphere and a flat surface is given by

$$F_{vdW} = -AR/6D^2 , \tag{40.5}$$

where A is the Hamaker constant, which can be estimated based on the Lifshitz theory [40.61], R is the radius of the AFM tip, and D is the distance between the AFM tip and the liquid conditioner film. Therefore, the jump-in distance D_J (the minimal distance between a stable liquid film and the AFM tip) can be theoretically calculated based on the criterion for a jump instability:

$$dF_{vdW}/dD = AR/3D_J^3 = K , \tag{40.6}$$

where K is the spring constant. For a conditioner-treated hair surface, K is the effective spring constant of the liquid film K_{eff} since it is much weaker than that of the cantilever. From (40.6), we obtain

$$D_J = (AR/3K_{eff})^{1/3} . \tag{40.7}$$

For a silicon tip covered by SiO_2 (the sphere) interacting with a liquid conditioner film (the flat surface) in air, the Hamaker constant A is estimated to be on the order of 10^{-20} J, R is about 100 nm, and K_{eff} is in the range of 20–72 mN/m (the surface tensions of PDMS and water are 20 mN/m and 72 mN/m, respectively). Then a jump-in distance D_J of about 2 nm can be obtained based on (40.7). This jump-in distance is basically the offset between the snap-in distance H_s and the actual film thickness h. This 2 nm offset (jump-in distance) is surprisingly close to previous measurements on the localized lubricant-film thickness on a particulate type magnetic rigid disk, which indicates that the thickness measured using an AFM is about 2 nm larger than the actual thickness based on the ellipsometry measurements [40.56]. Note that the offset (jump-in distance D_J) only depends on the radius of the tip and the intrinsic properties of the film (surface tension and Hamaker constant); it is independent of the film thickness. Recent surface force apparatus (SFA) experiments indicate that for two 25 nm-thick liquid PDMS films interacting in air in a quasi-equilibrium state (very slow approach rate of about 0.3 nm/s), the jump-in distance is about 200 nm [40.62]. SFA measurements give a much larger jump-in distance due to the much larger value of the radius (in SFA experiments, the radius R is about 2 cm). Therefore, although the snap-in distance H_s in the force calibration plot overestimates the actual film thickness (by approximately 2 nm), it still provides a very good measurement of the actual thickness of thin conditioner film.

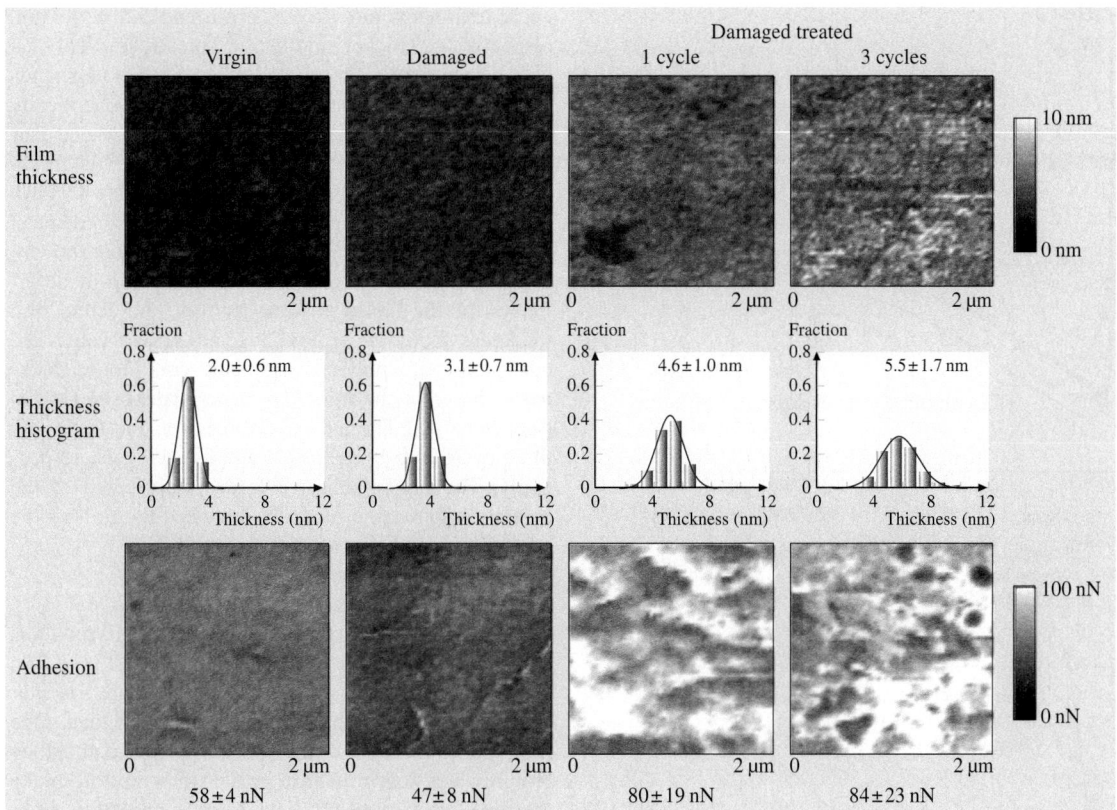

Fig. 40.60 Film thickness maps, histograms and adhesive force maps for various hair samples. The *dotted lines* in the histograms are the Gaussian fits for film thickness (after [40.11])

Figure 40.60 shows typical film thickness (snap-in distance H_s) and adhesive force maps of virgin, chemically damaged and chemically damaged and treated (one cycle and three cycles) hair types. The snap-in distance H_s of virgin hair is 2.0 ± 0.6 nm, with a very narrow distribution, indicating that the virgin hair surface is relative uniform and undamaged. The virgin hair surface is covered with a layer of a saturated fatty acid lipid called 18-methyleicosanoic acid (18-MEA) [40.3], which makes the hair surface hydrophobic (Table 40.13 for contact angle data). Therefore, a continuous water film is absent from the virgin hair surface. Instead of deforming the liquid film, the AFM tip will jump into contact with the hair surface at a finite tip–sample distance because of the van der Waals attractive force.

The snap-in distance of chemically damaged hair is larger than that of virgin hair, and it increases to 3.1 ± 0.7 nm. The outermost surface of the hair is the cuticle, which consists of a large amount of cystine. Chemical treatment will partially remove the fatty acid lipid layer

(18-MEA) covering the hair surface and break the disulfide bonds in cystine to form new ionic groups (such as cysteic acid residues by the oxidation of cystine acids). The hair surface becomes hydrophilic and the contact angle with water decreases [40.43], so the amount of water adsorbed on the hair surface increases. As the sample approaches the tip, the AFM tip will jump into contact with the hair surface at a finite distance because of van der Waals attractive force, and a small meniscus bridge will form between the tip and adsorbed water layer. The adhesive force of the chemically damaged hair surface is 47 ± 8 nN, less than that of virgin hair (58 ± 4 nN). The existence of a thin adsorbed water layer tends to decrease the van der Waals (attractive) force, which is now mediated by the adsorbed water layer, but to increase the meniscus force. Note that the tip and the hair surface are not as smooth as shown in Fig. 40.18 (rough on the molecular scale), and the adsorbed water layer is very thin, so many small menisci (not a single large meniscus) form between the tip and the hair surface. The total

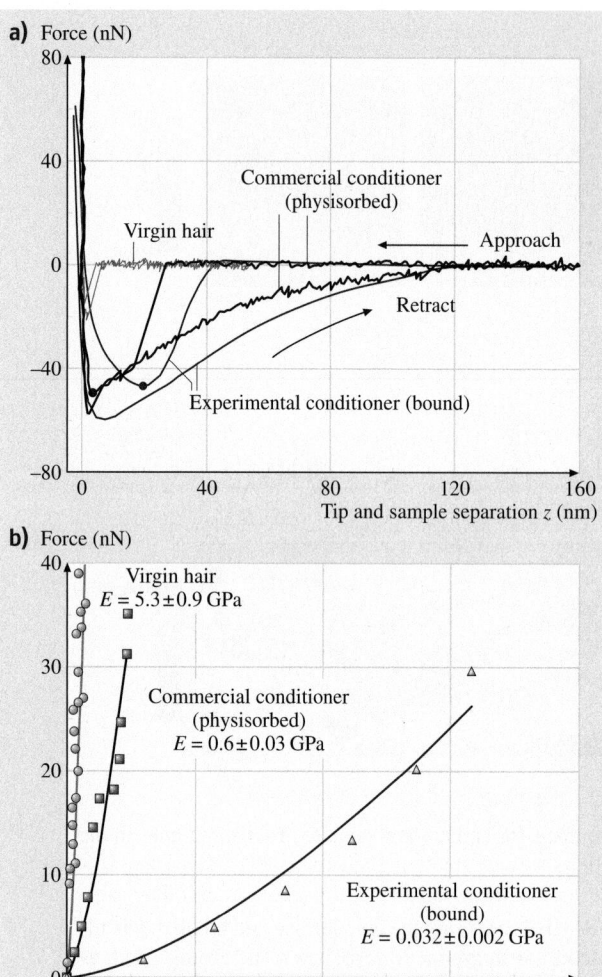

Fig. 40.61 (a) The force acting on the tip as a function of tip–sample separation for various hair samples. The sample is moved with a velocity of 200–400 nm/s, and the zero tip sample separation is defined to be the position where the force on the tip is zero when in contact with the sample. The *arrows* show the direction of motion of the sample relative to the tip. A negative force indicates an attractive force; **(b)** surface effective Young's modulus extracted from the force–distance curves. The *solid lines* show best fits for the data points for various hair samples (after [40.11])

attractive force of chemically damaged hair therefore is smaller than that of virgin hair.

After conditioner treatment, the snap-in distance H_s increases with the number of cycles of treatment. The snap-in distances H_s for one cycle treatment and three

cycle treatments are 4.6 ± 1.0 nm and 5.5 ± 1.7 nm, respectively. Thicker liquid film spreads unevenly on conditioner-treated hair surfaces, and the thickness tends to have a broader distribution than that of chemically damaged hair. This is more obvious for damaged hair treated with three conditioning cycles, where the thickness distribution is much broader and tends to have a long tail to larger thicknesses. The measured mean thickness varies from 5 nm up to 25 nm (data not shown). Excluding the 2 nm offset due to the jump-in deformation of the liquid conditioner film, the actual film thickness should be around 3–23 nm. These values are consistent with previous estimated conditioner thicknesses based on the amount of material deposited [40.2] (Appendix A). The adhesive forces for one cycle and three cycles are 80 ± 19 nN and 84 ± 23 nN, respectively. The larger adhesive force is attributed to the formation of a large meniscus between the tip and the conditioner layer on the hair surface.

Effective Young's Modulus Mapping

One method of measuring the effective Young's modulus is to use the nanoindentation technique [40.13, 31, 33], which involves pressing a sharp diamond indenter against a sample with a steadily increasing load. One limitation with these types of measurements is that they require static loads that are greater than 1 μN to make accurate measurements. The force modulation technique [40.63] has been applied to obtain quantitative measurements of the elasticities of compliant samples with effective Young's moduli as high as a few GPa. A knowledge of the adhesive force is necessary to accurately measure the effective Young's modulus. In these studies, the applied load was much greater than the measured adhesive force, so uncertainties in the adhesive force did not significantly affect the measurements.

In this study, the adhesive force as well as the total force acting on the tip at each measurement point can be accurately measured from the force calibration plot. If zero tip–sample separation is defined to be the position where the force on the tip is zero when in contact with the sample, the force calibration plot (cantilever deflection vs. piezo position plot as shown in Fig. 40.19a) can be converted to a force vs. tip–sample separation curve. Figure 40.61a shows the forces acting on the tip as a function of tip–sample separation for virgin, commercial conditioner-treated and experimental conditioner-treated hair. The lowest point on the approach curve, which corresponds to point D in Fig. 40.19a, is assumed to be the point that the tip contacts the hair surface. After this, the hair surface deforms elastically under the load, which can be used to extract the deformation (indentation Δz) of the surface. Plotting the deformation obtained (indentation Δz) against the total force (on the approach curve) acting on the surface gives the force vs. indentation (deformation) curve. Figure 40.61b shows the force vs. indentation curves for virgin, commercial conditioner-treated and experimental conditioner-treated hair surfaces, which are extracted from the force vs. tip–sample separation curves shown in Fig. 40.61a, and the effective Young's moduli of various hair surfaces can be determined from these curves by fitting them to (40.2). The effective Young's moduli of virgin, commercial conditioner-treated and experimental conditioner-treated hair surfaces are 5.3 $\pm$ 0.9 GPa, 0.60 $\pm$ 0.03 GPa and 0.032 $\pm$ 0.002 GPa, respectively, for these three specific curves. For these calculations, the radius of the tip was measured to be approximately 100 nm. The effective Young's modulus of a conditioner-treated hair surface can be one to two orders of magnitude less than that of virgin hair.

Repeating these calculations over the whole surface on various hair samples yields maps of the effective Young's modulus for various hair surface, as shown in Fig. 40.62. The virgin hair surface has an effective Young's modulus of about 7.1 $\pm$ 2.9 GPa, which is the most stiff amongst all of the samples and is consistent with previous nanoindentation measurement results [40.13]. The high density of disulfide crosslinks in the outermost layer of the hair surface accounts for this stiffness; the effective Young's modulus of a chemically damaged hair surface (6.6 $\pm$ 3.3 GPa) is slightly smaller than that of a virgin hair surface. Chemical treatment partially breaks the disulfide crosslinks in the outermost layer of the hair surface, and weakens the hair surface. A commercial conditioner-treated hair surface tends to have much smaller effective Young's modulus than that of a chemically damaged hair surface (5.5 $\pm$ 4.6 GPa for one cycle of treatment, and 3.3 $\pm$ 3.2 GPa for three cycles of treatment). The effective Young's modulus decreases as the number of cycles of treatment increases. Large deviations in the

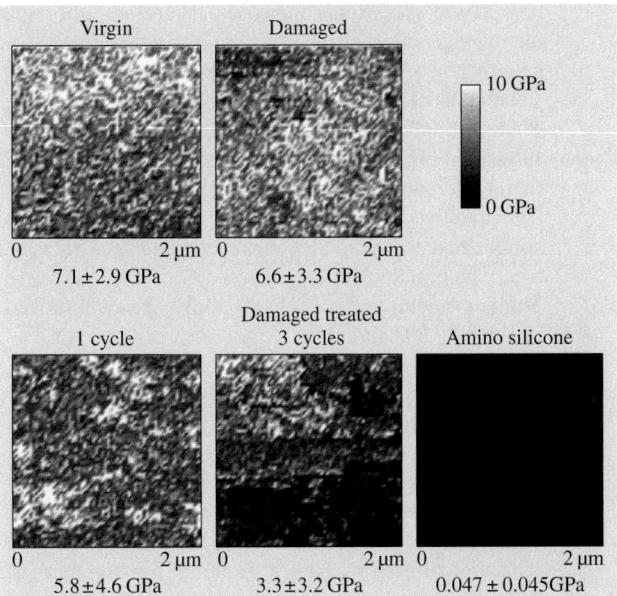

Fig. 40.62 Effective Young's modulus maps for various hair samples (same scanning area as shown in Fig. 40.60). The average value and standard deviation for each map are listed. Large deviations compared to average values, especially for conditioner-treated hair surface, indicate uneven distribution and adsorption of conditioner layers (after [40.11])

effective Young's moduli of conditioner-treated hair surfaces are observed, indicating uneven distributions and adsorption of the conditioner on the hair surface. This large deviation in the effective Young's moduli of conditioner-treated hair surfaces is due to a large area of the hair surface having a low effective Young's modulus (dark region). Comparing the conditioner thickness map (Fig. 40.60) and the effective Young's modulus map (Fig. 40.62) for the sample treated with three conditioning cycles, it is clear that the region with a low effective Young's modulus (dark region) is closely correlated to the region with a thick conditioner layer (bright region) on the hair surface. The effective Young's modulus decreases as the thick-

Table 40.20 Summary of film thicknesses, adhesive forces and effective Young's moduli for various hair samples

Samples		Film thickness (nm)	Adhesion (nN)	Young's modulus (GPa)
Virgin		2.0 $\pm$ 0.6	58 $\pm$ 4	7.1 $\pm$ 2.9
Chemically damaged		3.1 $\pm$ 0.7	47 $\pm$ 8	6.6 $\pm$ 3.1
Chemically damaged treated	1 cycle	4.9 $\pm$ 1.0	80 $\pm$ 19	5.8 $\pm$ 4.6
	3 cycles	5.5 $\pm$ 1.7	84 $\pm$ 23	3.3 $\pm$ 3.2
	Amino silicone	n/a	n/a	0.047 $\pm$ 0.046

ness of the conditioner increases. Therefore, although the average value of the effective Young's modulus of hair treated with three conditioning cycles is 3.3 GPa, some areas of the surface could have very low effective Young's modulus values (as low as the 0.60 GPa value shown in Fig. 40.61b). The experimental conditioner (amino silicone)-treated hair surface has an extremely low effective Young's modulus (0.047 ± 0.045 GPa) compared to the value for a chemically damaged hair surface. The film thicknesses, adhesive forces and effective Young's moduli of various hair samples are summarized in Table 40.20 [40.11].

Binding Interactions Between the Conditioner and the Hair Surface

Small molecules present in the conditioner, such as water and surfactant, may diffuse into the outermost layers of the hair surface and swell the disulfide crosslink network within these layers, which weakens these layers, yielding a smaller effective Young's modulus for conditioner-treated hair surfaces. Even more significantly, although the silicones in the conditioner cannot diffuse into the hair surface layers because of the size of the molecules, they can physically adsorb via van der Waals interactions or bind (chemically or electrostatically) on the hair surface, which will significantly affect the physical properties of the hair surface. Although the AFM cannot give an absolute distance, unlike the surface forces apparatus (SFA) [40.61], the force–distance curves obtained using an AFM can still show the interactions between the surface and the tip, and so reveal the nature of various surfaces. As shown in Fig. 40.61a, virgin hair and conditioner-treated hair behave differently as the tip approaches, compresses and retracts from the surface. Virgin hair behaves like a stiff elastic solid surface. As the virgin hair surface approaches the tip, the tip jumps into contact with the hair surface at a small separation of around 4 nm because of the van der Waals attractive interaction between the hair surface and the tip. Then the force quickly turns repulsive and the tip reaches the hard wall contact (tip–hair solid–solid contact) with very small deformation. When the virgin hair surface is withdrawn, the tip will simply jump out and no large liquid deformations occur because of the lack of a liquid layer.

Conditioner-treated hair surfaces show a much greater range of interactions than virgin hair because of the presence of the liquid conditioner layer. The tip will snap in at a large separation because of the van der Waals forces as well as the meniscus formation between the tip and the layer of conditioner. When the sample is pulled away from the tip, the forces on the tip will slowly decrease to zero as a long meniscus bridge of liquid is drawn out from the surface and eventually breaks at a large separation. Another important feature is that, before the tip reaches the tip–hair solid–solid contact, the commercial conditioner-treated hair surface shows more deformation than the virgin hair surface, indicating the presents of a thin physically adsorbed layer on the hair surface. Experimental conditioner-treated hair shows very large deformation and no tip–hair solid–solid contact is reached under the given experimental conditions, indicating that there is a strongly bound layer of conditioner. The large hysteresis between the approaching and retracting curves indicates that this bound layer deforms viscoelastically under the load. The adhesive force (on the approach curve) of the experimental conditioner-treated hair surface is smaller than that of the commercial conditioner-treated hair surface. The meniscus forces in both cases should be similar, and the extra repulsive force on the tip for the experimental conditioner-treated surface comes from the compression of the molecules beneath the tip. The tip has to overcome the intermolecular polar or hydrogen bonding and the strong binding between amino silicones and the hair surface to squeeze out silicone molecules from between the tip and the hair surface.

Figure 40.63 illustrates the difference between physisorbed conditioner and bound conditioner. Chemical treatments break the disulfide bonds of the cystine crosslinked network in the cuticle and form new ionic groups on the hair surface, so the surface of chemically damaged hair is negatively charged. Silicones in commercial conditioners are nonpolar and do not contain any

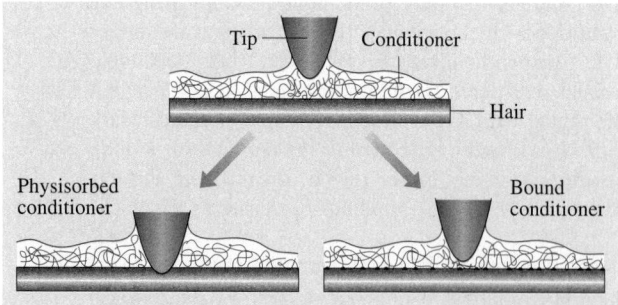

Fig. 40.63 Comparison between physisorbed conditioner and bound conditioner. Physisorbed conditioner can be easily squeezed out from the contact region under the load, while bound conditioner cannot escape from the contact region under the load because of the strong electrostatic binding between the silicone molecules and the hair surface (*black dots* are the binding sites) (after [40.11])

functional groups, so most of the silicone molecules are free and mobile except that the last layer of molecules adjacent to the hair surface may be adsorbed physically via van der Waals interactions. As the tip starts to touch and compress the conditioner film, free conditioner molecules can easily escape from the contact region. The last layer of adsorbed molecules may sustain some load before they are eventually squeezed out at higher loads. Then the tip penetrates the layer of conditioner and reaches tip–hair solid–solid contact and compresses directly on the hair surface. The last layer of physisorbed conditioner molecules is very thin but effectively lowers the Young's modulus of the hair surface by one order of magnitude (Figs. 40.61b and 40.62).

On the other hand, experimental conditioner consists of amino silicones, which are positively charged in aqueous conditions and can strongly bind with the anionic groups present on chemically damaged hair surface. The bonds between amino silicone molecules and the hair surface, which are much stronger than van der Waals attractions in the commercial conditioner-treated hair surface, make it difficult for bound conditioner molecules to escape from the contact region, even at high load. Therefore, after the tip touches the conditioner, instead of penetrating the conditioner layer and compressing directly on the hair surface, it compresses on a soft bound conditioner layer. The effective Young's modulus of this layer (silicones) is about 0.047 ± 0.045 GPa. Although it is much softer than hair, it is still much stiffer than bulk PDMS, whose Young's modulus is only 680 kPa [40.64]. Two reasons for this increase in effective Young's modulus of silicones (PDMS) have been proposed: the bound silicone layer is very thin (a few nanometers), and it is in a confined geometry during the measurements. It is well known that thin films can exhibit very different physical properties to bulk materials. Thin liquid films can behave like a solid under confinement [40.61]. On the other hand, underneath this thin bound silicone layer is the stiff hair surface, which has a much higher Young's modulus (chemically damaged hair surface 6.6 ± 3.3 GPa, Fig. 40.62). Since the bound layer is very thin, the measured Young's modulus contains contributions from both the soft bound silicone layer and the stiff hair surface.

Amino silicones act as a cushion, protecting the hair surface. Although their strong affinity to the hair surface can dramatically increase the load or contact pressure that a conditioner film can support before solid–solid contact, a recent AFM study on hair [40.2] indicated that hair treated with excess amino silicone has a higher coefficient of friction than hair treated with commercial (physisorbed) conditioner . It is well known that, in general, lubricant film with mobile and immobile fractions (partially bonded lubricant film) is desirable for low friction and high durability [40.30]. Biolubrication studies using SFA [40.65] also indicate that a low coefficient of friction is not necessarily a good measure of wear protection. A conditioner layer strongly bound to the hair surface is good for preventing hair from further damage, but it may be not good from a lubrication point of view. A balance between good lubrication and good protection must be reached for a good conditioning product.

40.6.3 Summary

Contact mode AFM has been used to perform nanotribological studies on various hair and skin. Friction force and the resulting coefficient of friction were found to be higher on chemomechanically damaged hair than on virgin hair, due to the increase in surface roughness and the change in surface properties that results from exposure to chemical damage. Generally speaking, the average coefficient of friction is similar on virgin and virgin treated hair for each ethnicity. However, there is an increase in friction force for virgin treated hair around the cuticle edges and surrounding area. It is currently believed that this increase in friction force is due in part to increased meniscus effects, which increase the contribution of the adhesive force to friction at sites where conditioner is deposited or accumulated on the surface, namely around the cuticle scale edges. This adhesive force is of the same magnitude as the normal load, which makes this contribution to friction from the adhesive force rather significant. On the macroscale, however, the adhesive force is much lower in magnitude than the applied normal load, so the contribution from adhesive force to friction is negligible over the hair swatch. As a result, treated hair shows a decrease in friction force on the macroscale, which is the opposite trend to that seen on the micro/nanoscale. Friction and adhesion data on the micro/nanoscale are useful, however, because they relate to the presence of conditioner on the cuticle surface and allow us to obtain an estimate for the conditioner distribution. Studies using the force calibration plot technique show a decrease in adhesive force with damaged hair, and significantly higher adhesive force for treated hair. This increase on the micro/nanoscale is most likely due to meniscus force contributions from the accumulation and localization of a layer of conditioner on the hair surface. Thus, the presence of conditioner can be detected using this increase in adhesive force. The directionality dependence of friction is evident when the cuticle edge is examined using FFM.

Chemically damaged and treated hair shows a much stronger affinity to conditioner than virgin hair. The negative charge of the hair fiber becomes even more negative after chemical damage. As a result, the positively charged particles of the conditioner have an even stronger attraction to the chemically damaged surface, and this results in the presence of an increased amount of conditioner (and correspondingly higher friction forces) when compared to virgin treated hair. When three conditioning cycles are applied to chemically damaged hair, friction force increases all over the cuticle surface, showing a more uniform placement of the conditioner. Contrary to results for the virgin and virgin treated hair, the coefficient of friction for chemically damaged hair decreases with the application of commercial conditioner (for both one and three cycles of conditioning). One possible explanation for this is that the stronger negative charge on damaged hair results in a stronger attraction to the conditioner, which leads to higher adhesive forces but also, more importantly, lower surface shear strength.

Environmental effects were also studied for various hair types. The coefficient of friction generally decreased with increasing temperature. After soaking hair in deionized water, virgin hair exhibits a decrease in the coefficient of friction after soaking. Virgin hair is more hydrophobic (based on contact angle data), so more of the water is present on the surface, resulting in a lubrication effect after soaking. Chemically damaged hair tends to be hydrophilic due to the chemical degradation of the cuticle surface, resulting in absorption of water after soaking. This softens the hair, which leads to higher friction, even with conditioning treatment. Durability tests show that once conditioner is applied to virgin hair, wear does not exhibit itself as an increase in friction force. Thus, conditioner serves as a protective coating for virgin hair and helps protect its tribological properties when wear occurs.

In most cases, the coefficient of friction of chemically damaged hair decreased upon the addition of PDMS-blend or amino silicones to the BTMAC surfactant. Silicones are typically used to lubricate and thus give the conditioner more mobility on the hair surface compared to just surfactants and fatty alcohols. However, the opposite trend was seen for large depositions of amino silicone. The dampened mobility of the amino silicone at high deposition levels with respect to the hair surface and tip may account for this wide variation in the coefficient of friction. Adhesive forces varied widely, but typically showed a significant increase with the presence of conditioning ingredients. This is a clear sign that meniscus effects influence the pull-off force between the tip and the sample. At high deposition levels, the amino silicones showed much more distinct regions of high and low friction and adhesion, which shows that these molecules are less mobile and are redistributed to a much smaller degree as they coat the hair. Force calibration plots indicate that commercial conditioner containing only nonpolar silicones and experimental conditioner containing polar amino silicones exhibit different affinities for chemically damaged hair surfaces. Commercial conditioner, which is physically adsorbed on hair surface via van der Waals attractions, can be easily squeezed out from the contact region when it is under load; on the other hand, the amino silicones in the experimental conditioner lock on to the hair surface and cannot escape from the contact region under load due to the strong electrostatic binding between the polar amino groups of the silicone molecules and the hair surface. Amino silicones provide a much better load-bearing capacity than nonpolar silicones, and act as a cushion that prevents the hair surface from being damaged further.

The snap-in distance H_s in the force calibration measurement provides a good estimate for the thickness of the film of conditioner on the hair. The conditioner is unevenly distributed over the hair surface. Conditioner thickness increases with the number of the cycles of conditioning treatment. The mean conditioner thickness varies from 5 nm up to 25 nm.

40.7 Closure

A comprehensive nanoscale characterization of human hair and skin has been performed. SEM studies of cross-sections of hair and AFM studies of hair surfaces show that the cuticle is about 5–10 scales thick, and that each cuticle cell is about 0.3 to 0.5 μm thick. The visible length of each cuticle cell is about 5–10 μm. Asian hair seems to be the thickest (nearly round), followed by African hair (oval–flat) and Caucasian hair (nearly oval).

Cross-sections and longitudinal sections of virgin Caucasian human hair have been investigated using TR mode II. The cortex and the cuticle, the macrofibril and the matrix of human Caucasian hair were readily revealed. The fine granular structure of the outer β-layer, which has not previously been seen in SEM, TEM and other AFM studies, is a result of the formation of fine domains by the lipid layer. Chemically damaged hair surfaces show much more damage; larger areas of the endocuticle were exposed. Conditioner unevenly distributes over damaged hair surfaces; thicker conditioner films are found near the cuticle edges.

A nanoindenter has been used to perform nanomechanical studies on human hair. Chemical damage and conditioning treatment caused the hardness and elastic modulus of the hair surface to decrease up to a depth of less than $1.5\,\mu m$. That is, the first 3–4 cuticle scales may interact with the chemicals and ingredients of the conditioner more effectively than the remaining scales. Nanoscratch tests show that, for Caucasian and Asian hair, the coefficient of friction for virgin treated hair is lower than that for virgin hair, as seen in both single cuticle and multiple cuticle scratch tests. This thin layer of conditioner acts as a layer of lubricant, reducing the coefficient of friction during scratching. In situ displacement (30–200 nm) increased greatly at very low initial loads and then increased gradually with load, indicating that the first approximately 200 nm of the hair surface is softer than the layers below. The cuticle cells tend to be fractured during scratching in hair with a hard cuticle, while the scratch tip usually plows and wears away the cuticle cells continuously until it reaches the cortex in hair with soft cuticle. The friction coefficients for virgin and chemomechanically damaged Caucasian hair increase after soaking due to the swelling of the water, which softens the hair surface.

A flat-on-flat tribometer has been used to measure macroscale friction and wear of polyurethane film (synthetic skin) vs. hair and hair vs. hair. In the case of polyurethane film vs. hair, chemomechanically damaged hair shows the highest coefficient of friction, followed by virgin and virgin treated hair. The coefficient of friction did not change after testing skin vs. hair for 24 h, but some of the cuticles were damaged.

Contact mode AFM has been used to perform nanotribological studies on various hair and skin samples. Friction force and the resulting coefficient of friction are seen to be higher on chemomechanically damaged hair than on virgin hair. In virgin treated hair there is a slight increase in friction force around the cuticle edges and surrounding area. It is currently believed that this increase in friction force is due in part to increased meniscus effects, which increase the contribution from adhesive force to the friction at sites where conditioner is deposited or accumulated on the surface, namely around the cuticle scale edges. Adhesive force studies showed a decrease in adhesive force for chemomechanically damaged hair and significantly higher adhesive force for virgin treated hair, most likely due to meniscus force contributions from the conditioner layer.

Chemically damaged hair shows a much stronger affinity to conditioner than virgin hair. After the application of three conditioning cycles to chemically damaged hair, the friction force increases all over the cuticle surface, showing a more uniform distribution of the conditioner. Unlike the results for virgin and virgin treated hair, the coefficient of friction for chemically damaged hair decreased with the application of commercial conditioner (both one and three cycles), possibly because the stronger negative charge on damaged hair results in it being more attractive to conditioner (and a lower shear strength), which leads to an overall lubrication effect and ultimately to a lower coefficient of friction.

Environmental studies show that the coefficient of friction generally decreases with increasing temperature. After soaking hair in deionized water, virgin hair exhibits a decrease in the coefficient of friction after soaking while chemically damaged hair exhibits an increase. Virgin hair is more hydrophobic, so more of the water is present on the surface which results in a lubrication effect after soaking. Chemically damaged hair tends to be hydrophilic due to the chemical degradation of the cuticle surface, and this results in the absorption of water after soaking, which softens the hair, leading to high nanoscale friction even with conditioner treatment. Durability tests show that once conditioner is applied to virgin hair, wear does not show up as an increase in friction force. Thus, conditioner serves as a protective covering for virgin hair that helps maintain its tribological properties as wear kicks in.

Usually a decrease in the coefficient of friction was observed for chemically damaged hair upon the addition of PDMS-blend or amino silicones to BTMAC surfactant. Silicones are typically used for lubrication in order to give the conditioner more mobility on the hair surface compared to just surfactants and fatty alcohols. However, at high deposition, the inverse trend was seen for the amino silicone, due to the dampened mobility of the amino silicone at high deposition levels. Commercial conditioner physically adsorbed on hair surface via van der Waals attractions was easily

40.A Conditioner Thickness Approximation

squeezed out from the contact region under load; on the other hand, the amino silicone in experimental conditioner binds to the hair surface and is not displaced from the contact region under load due to the strong electrostatic binding between the polar amino groups of the silicone and the hair surface. Amino silicone gives enhanced load-bearing capacity over nonpolar silicones and cushions the surface of the hair from further damage.

The snap-in distance H_s in the force calibration measurement provides a good estimate for the thickness of a film of conditioner on hair. The mean conditioner thickness varies from 5 nm up to 25 nm.

A cylindrical hair fiber of diameter $D = 50\,\mu m$ (radius $R = 25\,\mu m$) is considered. For conditioner thickness calculations, the following assumptions are made: (1) hair and the material being added have the same density, (2) coating of material is uniform on the hair surface, (3) the cross-sectional area of a hair fiber remains constant along the longitudinal axis of the fiber (i.e. from root to tip); the hair fiber is perfectly cylindrical (circular cross-section), and (4) the deposited conditioner remains bonded to the cuticle surface (no absorption into the cuticle layer).

The cross-sectional area of an untreated hair fiber is initially calculated. By adding a specified amount of conditioner, this area will increase and cause the radius of the hair treated fiber to increase. This increase in the radius of the treated hair will be equivalent to the

thickness of the conditioner layer. The original cross-sectional area A_c of hair fiber is

$$A_c = \pi R^2 = \pi\,(25\,\mu m)^2 = 1963.4954\,\mu m^2 \quad (40.A1)$$

Adding 200 ppm material to the surface (which is comparable to the amount that commercial conditioners typically deposit) will cause an increase in volume (for a unit fiber length) by 200 ppm, or by 0.0002. Thus, the cross-sectional area $A_{c,\,conditioner}$ of the treated hair will increase by the same amount to

$$A_{c,\,conditioner} = 1.0002 A_c = 1963.888\,\mu m^2 \quad (40.A2)$$

which results in a new radius $R_{conditioner}$,

$$R_{conditioner} = \sqrt{\frac{A_{c,\,conditioner}}{\pi}} = 25.0025\,\mu m \quad (40.A3)$$

Therefore, subtracting the original radius from the radius after treatment increases the thickness of the hair by $0.0025\,\mu m$, or 2.5 nm.

It is important to note that the approximation of the conditioner thickness as 2.5 nm was determined for a particular hair diameter and material deposition amount (with the hair and material having equal densities). Although these are generally realistic approximations, hair diameter often varies by a factor of 2 and the deposition level can vary up to an order of magnitude. The conditioner layer has been shown in previous work to be nonuniform as well. Thus, actual conditioner thickness can deviate significantly from this number.

References

40.1   C. LaTorre, B. Bhushan: Nanotribological characterization of human hair and skin using atomic force microscopy, Ultramicroscopy **105**, 155–175 (2005)

40.2   C. LaTorre, B. Bhushan, J. Z. Yang, P. M. Torgerson: Nanotribological effects of silicone type, silicone deposition level, and surfactant type on human hair using atomic force microscopy, J. Cosmetic Sci. **57**, 37–56 (2006)

40.3   C. Robbins: *Chemical and Physical Behavior of Human Hair*, 3rd edn. (Springer, Berlin, Heidelberg 1994)

40.4   A. Feughelman: *Mechanical Properties and Structure of Alpha-Keratin Fibres: Wool, Human Hair and Related Fibres* (Univ. South Wales Press, Sydney 1997)

40.5   J. A. Swift: The mechanics of fracture of human hair, Int. J. Cosmet. Sci. **21**, 227–239 (1999)

40.6   J. A. Swift: The cuticle controls bending stiffness of hair, J. Cosmet. Sci. **51**, 37–38 (2000)

40.7   H. A. Barnes, G. P. Roberts: The non-linear viscoelastic behaviour of human hair at moderate extensions, Int. J. Cosmet. Sci. **22**, 259–264 (2000)

40.8   J. Jachowicz, R. McMullen: Mechanical analysis of elasticity and flexibility of virgin and polymer-treated hair fiber assemblies, J. Cosmet. Sci. **53**, 345–361 (2002)

40.9   A. N. Syed, A. Kuhajda, H. Ayoub, K. Ahmad, E. M. Frank: African-American hair: Its physical properties and differences relative to Caucasian hair. In: *Hair Care*, Cosmet. Toil. Appl. Res. Ser. (Allured, Carol Stream 1996)

40.10   N. H. Chen, B. Bhushan: Morphological, nanome-
        chanical and cellular structural characterization of
        human hair and conditioner distribution using tor-
        sional resonance mode with an AFM, J. Microscopy
        220, 96–112 (2005)

40.11   N. H. Chen, B. Bhushan: Atomic force microscopy
        studies of conditioner thickness distribution and
        binding interactions on the hair surface, J. Mi-
        croscopy **221**, 203–215 (2006)

40.12   C. LaTorre, B. Bhushan: Nanotribological effects of
        hair care products and environment on human hair
        using atomic force microscopy, J. Vac. Sci. Technol.,
        A **23**, 1034–1045 (2005)

40.13   G. Wei, B. Bhushan, P. M. Torgerson: Nanome-
        chanical characterization of human hair using
        nanoindentation and SEM, Ultramicroscopy **105**,
        248–266 (2005)

40.14   G. Wei, B. Bhushan: Nanotribological and
        nanomechanical characterization of human hair
        using a nanoscratch technique, Ultramicroscopy
        106, 742–754 (2006)

40.15   G. Swanbeck, J. Nyren, L. Juhlin: Mechanical prop-
        erties of hair from patients with different types
        of hair diseases, J. Invest. Dermatol. **54**, 248–251
        (1970)

40.16   G. Nikiforidis, C. Balas, D. Tsambaos: Mechanical
        parameters of human hair: possible applications
        in the diagnosis and follow-up of hair disorders,
        Clin. Phys. Physiol. Meas. **13**, 281–290 (1992)

40.17   C. Zviak (Ed.): *The Science of Hair Care* (Marcel
        Dekker, New York 1986)

40.18   P. Jolles, H. Zahn, H. Höcker (Eds.): *Formation and
        Structure of Human Hair* (Birkhäuser, Basel 1997)

40.19   J. R. Smith, J. A. Swift: Lamellar subcomponents
        of the cuticular cell membrane complex of mam-
        malian keratin fibres show friction and hardness
        contrast by AFM, J. Microsc. **206**, 182–193 (2002)

40.20   J. A. Swift: Morphology and histochemistry of
        human hair. In: *Formation and Structure of Hu-
        man Hair*, ed. by P. Jolles, H. Zahn, H. Höcker
        (Birkhäuser, Basel 1997) pp. 149–175

40.21   R. J. Randebrook: J. Soc. Cosmet. Chem. **15**, 691
        (1964)

40.22   P. T. Pugliese: *Physiology of the Skin* (Allured, Carol
        Stream 1996)

40.23   P. W. Wertz, D. T. Downing: Stratum corneum:
        biological and biochemical considerations. In:
        Transdermal Drug Delivery, ed. by J. Swarbrick,
        R. H. Guy (Marcel Dekker, New York 1989)

40.24   P. W. Wertz, K. C. Madison, D. T. Downing: Cova-
        lently bound lipids of human stratum corneum, J.
        Invest. Dermatol. **92**, 109 (1989)

40.25   J. Gray: Hair care and hair care products, Clin.
        Dermatol. **19**, 227–236 (2001)

40.26   J. Gray: *The World of Hair* (Procter & Gamble,
        Cincinnati 2003)

40.27   B. Bhushan: *Introduction to Tribology* (Wiley, New
        York 2002)

40.28   B. Bhushan, X. Li: Nanomechanical properties of
        solid surfaces and thin films, Int. Mater. Rev. **48**,
        125–164 (2003)

40.29   J. A. Swift: Fine details on the surface of human
        hair, Int. J. Cosmet. Sci. **13**, 143–159 (1991)

40.30   B. Bhushan: *Principles and Applications of Tribol-
        ogy* (Wiley, New York 1999)

40.31   B. Bhushan: *Handbook of Micro/Nanotribology*,
        2nd edn. (CRC, Boca Raton 1999)

40.32   B. Bhushan (Ed.): *Springer Handbook of Nanotech-
        nology* (Springer, Berlin, Heidelberg 2004)

40.33   B. Bhushan: *Nanotribology and Nanomechanics –
        An Introduction* (Springer, Berlin, Heidelberg 2005)

40.34   X. Li, B. Bhushan, P. B. McGinnis: Nanoscale me-
        chanical characterization of glass fibers, Mater.
        Lett. **29**, 215–220 (1996)

40.35   A. N. Parbhu, W. G. Bryson, R. Lal: Disulfide bonds
        in the outer layer of keratin fibers confer higher
        mechanical rigidity: correlative nano-indentation
        and elasticity measurement with an AFM, Bio-
        chemistry **38**, 11755–11761 (1999)

40.36   T. Kasai, B. Bhushan, L. Huang, C. M. Su: To-
        pography and phase imaging using the torsional
        resonance mode, Nanotechnology **15**, 731–742
        (2004)

40.37   B. Bhushan, T. Kasai: A surface topography-
        independent friction measurement technique
        using torsional resonance mode in an AFM, Nano-
        technology **15**, 923–935 (2004)

40.38   B. Bhushan, J. Qi: Phase contrast imaging of
        nanocomposites and molecularly thick lubricant
        films in magnetic media, Nanotechnology **14**, 886–
        895 (2003)

40.39   W. W. Scott, B. Bhushan: Use of phase imaging in
        atomic force microscopy for measurement of vis-
        coelastic contrast in polymer nanocomposites and
        molecularly thick lubricant films, Ultramicroscopy
        97, 151–169 (2003)

40.40   Y. Song, B. Bhushan: Quantitiative extraction of in-
        plane surface properties using torsional resonance
        mode of atomic force microscopy, J. Appl. Phys. **97**,
        083533 (2005)

40.41   B. Bhushan, G. Wei, P. Haddad: Friction and wear
        studies of human hair and skin, Wear **259**, 1012–
        1021 (2005)

40.42   B. Bhushan, C. Dandavate: Thin-film friction and
        adhesion studies using atomic force microscopy, J.
        Appl. Phys. **87**, 1201–1210 (2000)

40.43   R. Molina, F. Comelles, M. R. Julia, P. Erra: Chemical
        modifications on human hair studied by means of
        contact angle determination, J. Colloid Interf. Sci.
        237, 40–46 (2001)

40.44   B. Bhushan, Z. Burton: Adhesion and friction
        properties of polymers in microfluidic devices,
        Nanotechnology **16**, 467–478 (2005)

40.45   C. Jalbert, J. T. Koberstein, I. Yilgor, P. Gal-
        lagher, V. Krukonis: Molecular weight dependence
        and end-group effects on the surface tension of

poly(dimethylsiloxane), Macromolecules **26**, 3069–3074 (1993)

40.46   G. Lerebour, S. Cupferman, C. Cohen, M. N. Bellon-Fontaine: Comparison of surface free energy between reconstructed human epidermis and in situ human skin, Skin Res. Technol. **6**, 245–249 (2000)

40.47   H. Schott: Contact angles and wettability of human skin, J. Pharm. Sci. **60**, 1893–1895 (1971)

40.48   M. E. Ginn, C. M. Noyes, E. Jungermann: The contact angle of water on viable human skin, J. Colloid Interf. Sci. **26**, 146–151 (1968)

40.49   H. Yanazawa: Adhesion model and experimental-verification for polymer SiO_2 system, Colloids Surface **9**, 133–145 (1984)

40.50   F. J. Wortmann, H. Zahn: The stress/strain curve of α-keratin fibers and the structure of the intermediate filament, Text. Res. J. **64**, 737–743 (1994)

40.51   B. J. Briscoe, A. Winkler, M. J. Adams: A statistical analysis of the frictional forces generated between monofilaments during intermittent sliding, J. Phys. D **18**, 2143–2167 (1985)

40.52   B. Bhushan: Influence of test parameters in the measurement of the coefficient of friction of magnetic tapes, Wear **93**, 81–89 (1984)

40.53   G. V. Scott, C. R. Robbins: Effects of surfactant solutions on hair fibre friction, J. Soc. Cosmet. Chem. **31**, 179–200 (1980)

40.54   H. Liu, B. Bhushan: Nanotribological characterization of molecularly thick lubricant films for applications to MEMS/NEMS by AFM, Ultramicroscopy **97**, 321–340 (2003)

40.55   C. Bolduc, J. Shapiro: Hair care products: waving, straightening, conditioning, and coloring, Clin. Dermatol. **19**, 431–436 (2001)

40.56   B. Bhushan, G. S. Blackman: Atomic force microscopy of magnetic rigid disks and sliders and its applications to tribology, J. Tribol. Trans. ASME **113**, 452–457 (1991)

40.57   M. L. Forcada, M. M. Jakas, A. Grasmarti: On liquid-film thickness measurements with the atomic-force microscope, J. Chem. Phys. **95**, 706–708 (1991)

40.58   P. Attard, S. J. Miklavcic: Effective spring constant of bubbles and droplets, Langmuir **17**, 8217–8223 (2001)

40.59   D. Bhatt, J. Newman, C. J. Radke: Equilibrium force isotherms of a deformable buble/drop interacting with a solid particle across a thin liquid film, Langmuir **17**, 116–130 (2001)

40.60   J. Brandup, E. H. Immergut, E. A. Grulke (Eds.): *Polymer Handbook*, 4th edn. (Wiley, New York 1999)

40.61   J. N. Israelachvili: *Intermolecular and Surface Forces*, 2nd edn. (Academic, London 1991)

40.62   N. H. Chen, T. Kuhl, R. Tadmor, Q. Lin, J. N. Israelachvili: Large deformations during the coalescence of fluid interfaces, Phys. Rev. Lett. **92**, Art.–No. 024501 (2004)

40.63   D. DeVecchio, B. Bhushan: Localized surface elasticity measurements using an atomic force microscope, Rev. Sci. Instrum. **68**, 4498–4505 (1997)

40.64   J. E. Mark: *Polymer Data Handbook* (Oxford Univ. Press, Oxford 1999)

40.65   M. Benz, N. H. Chen, J. Israelachvili: Lubrication and wear properties of grafted polyelectrolytes, hyaluronan and hylan, measured in the surface forces apparatus, J. Biomed. Mater. Res. A **71A**, 6–15 (2004)

41. Mechanical Properties of Nanostructures

Structural integrity is of paramount importance in all devices. Load applied during the use of devices can result in component failure. Cracks can develop and propagate under tensile stresses, leading to failure. Knowledge of the mechanical properties of nanostructures is necessary for designing realistic MEMS/NEMS and BioMEMS/BioNEMS devices. Elastic and inelastic properties are needed to predict deformation from an applied load in the elastic and inelastic regimes, respectively. The strength property is needed to predict the allowable operating limit. Some of the properties of interest are hardness, elastic modulus, bending strength, fracture toughness and fatigue strength. Many of the mechanical properties are scale dependent therefore these should be measured at relevant scales. Atomic force microscopy and nanoindenters can be used satisfactorily to evaluate the mechanical properties of micro/nanoscale structures.

Commonly used materials in MEMS/NEMS are single-crystal silicon and silicon-based materials, e.g., SiO_2 and polysilicon films deposited by low-pressure chemical vapor deposition. An early study showed silicon to be a mechanically resilient material in addition to its favorable electronic properties. Single-crystal SiC deposited on large-area silicon substrates is used for high-temperature micro/nanosensors and actuators. Amorphous alloys can be formed on both metal and silicon substrates by sputtering and plating techniques, providing more flexibility in surface-integration. Electroless deposited Ni-P amorphous thin films have been used to construct microdevices, especially using the so-called LIGA techniques. Micro/nanodevices need conductors to provide power, as well as electrical/magnetic signals to make them functional. Electroplated gold films have found wide applications in electronic devices because of their ability to make thin films and process simply.

Polymers, such as poly (methyl methacrylate) (PMMA) and poly (dimethylsiloxane) (PDMS) are commonly used in BioMEMS/BioNEMS, such as micro/nanofluidic devices, because of ease of manufacturing and reduced cost. The use of poly-

Part D | 41

mers also offers a wide range of material properties to allow tailoring of biological interactions for improved biocompatibility.

This chapter presents a review of mechanical property measurements on the micro/nanoscale of various materials of interest and stress and deformation analyses of nanostructures.

Microelectromechanical systems (MEMS) refer to microscopic devices that have a characteristic length of less than 1 mm but more than 100 nm (1 μm), and nanoelectromechanical systems (NEMS) refer to nanoscopic devices that have a characteristic length of less than 100 nm (or 1 μm). These are referred to as an intelligent miniaturized system comprising sensing, processing, and/or actuating functions and combine electrical and mechanical components. The acronym MEMS originated in the U.S.A. The term commonly used in Europe is microsystem technology (MST) and in Japan is micromachines. Another term generally used is micro/nanodevices. MEMS/NEMS terms are also now used in a broad sense and include electrical, mechanical, optical, biological, and/or fluidic functions. To put the dimensions in perspective, individual atoms are typically fraction of a nanometer in diameter, DNA molecules are about 2.5 nm wide, biological cells are in the range of thousands of nm in diameter, and human hair is about 75 μm in diameter. The mass of a micromachined silicon structure can be as low as 1 nN and NEMS can be built with mass as low as 10^{-20} N with cross sections of about 10 nm. In comparison, the mass of a drop of water is about 10 μN and the mass of an eyelash is about 100 nN.

A wide variety of MEMS, including Si-based devices, chemical and biological sensors and actuators, and miniature non-silicon structures (e.g., devices made from plastics or ceramics) have been fabricated with dimensions in the range of a couple to a few thousand microns (see e.g., [41.1–13]). A variety of NEMS have also been produced (see e.g., [41.14–19]). MEMS/NEMS technology and fabrication processes have found a variety of applications in biology and biomedicine, leading to the establishment of an entirely new field known as BioMEMS/BioNEMS [41.20–28]. The ability to use micro/nanofabrication processes to develop precision devices that can interface with biological environments at the cellular and molecular level has led to advances in the fields of biosensor technology [41.27,29–32], drug delivery [41.33–35], and tissue engineering [41.36–38]. The miniaturization of fluidic systems using micro/nanofabrication techniques has led to new and more efficient devices for medical diagnostics and biochemical analysis [41.39]. Largest "killer" industrial applications of MEMS include accelerometers (over a billion dollars a year in 2004), pressure sensors for manifold absolute pressure sensing for engines (more than 30 million units in 2004), and tire pressure measurements, inkjet printer heads (more than 500 million units in 2004), and digital micromirror devices (about \$700 million revenues in 2004). BIOMEMS and BIONEMS are increasingly used in commercial applications. Largest applications of BIOMEMS include silicon-based disposable blood pressure sensor chips for blood pressure monitoring (more than 25 million units in 2004), and a variety of biosensors.

Structural integrity is of paramount importance in all devices. Load applied during the use of devices can result into component failure. Cracks can develop and propagate under tensile stresses leading to failure [41.31, 40]. Friction/stiction and wear limit the lifetimes and compromise the performance and reliability of the devices involving relative motion [41.4, 5, 41]. Most MEMS/NEMS applications demand extreme reliability. Stress and deformation analyses are carried out for an optimal design. MEMS/NEMS designers require mechanical properties on the nanoscale. Mechanical properties include elastic, inelastic (plastic, fracture or viscoelastic), and strength. Elastic and inelastic properties are needed to predict deformation from an applied load in the elastic and inelastic regimes, respectively. The strength property is needed to predict the allowable operating limit. Some of the properties of interest are hardness, elastic modulus, creep, bending strength (fracture stress), fracture toughness, and fatigue strength. Micro/nanostructures have some surface topography and local scratches dependent upon the manufacturing process. Surface roughness and local scratches may compromise the reliability of the devices and their effect needs to be studied.

Most mechanical properties are scale dependent [41.5, 40, 42]. Several researchers have measured mechanical properties of silicon and silicon-based milli- to microscale structures including tensile tests and bending tests [41.43–52], resonant structure tests for measurement of elastic properties [41.53], fracture toughness tests [41.44, 46, 54–58], and fatigue tests [41.56, 59, 60]. Most recently, few researchers have measured mechanical properties of nanoscale structures using atomic force microscopy (AFM) [41.61, 62] and nanoindentation [41.63–65]. For stress and deformation analyses of simple geometries and boundary condition, analytical models can be used. For analysis of complex geometries and boundary condition, numerical models are needed. Conventional finite element method (FEM) can be used down to few tens of nanometer scale although its applicability is questionable at nanoscale. FEM has been used for simulation and prediction of residual stresses and strains induced in MEMS devices during fabrication [41.66], to perform fault analysis in order to study MEMS faulty behavior [41.67], to compute mechanical strain resulting from doping of silicon [41.68], analyze micromechanical experimental data [41.46, 69, 70], and nanomechanical experimental data [41.62]. FEM analysis of nanostructures has been carried out to analyze the effect of types of surface roughness and scratches on stresses in nanostructures [41.71, 72].

Commonly used materials for MEMS/NEMS are single-crystal silicon and silicon-based materials (e.g., SiO_2 and polysilicon films deposited by low pressure chemical vapor deposition (LPCVD) process) [41.5]. An early study showed silicon to be a mechanically resilient material in addition to its favorable electronic properties [41.73]. Single-crystal 3C-SiC (cubic or β-SiC) films, deposited by atmospheric pressure chemical vapor deposition (APCVD) process on large-area silicon substrates are produced for high-temperature micro/nanosensor and actuator applications [41.74–76]. Amorphous alloys can be formed on both metal and silicon substrates by sputtering and plating techniques, providing more flexibilities in surface-integration. Electroless deposited Ni-P amorphous thin films have been used to construct microdevices, especially using the so-called LIGA techniques [41.5, 64]. Micro/nanodevices need conductors to provide power as well as electrical/magnetic signals to make them functional. Electroplated gold films have found wide applications in electronic devices because of their ability to make thin films and process simplicity [41.64].

As the field of MEMS/NEMS has progressed, alternative materials, especially polymers have established an important role in the advancement of the technology. This trend has been driven by the reduced cost associated with polymer materials. Polymer microfabrication processes, including the micromolding and hot embossing techniques [41.77], can be orders of magnitude less expensive than traditional silicon photolithography processes. The use of polymers in the BioMEMS/BioNEMS field has additional functional advantages, as polymers offer a wide range of material properties to allow tailoring of biological interactions for improved biocompatibility. Polymer BioMEMS structures involving microbeams have been designed to measure cellular forces [41.65]. Polymer materials most commonly used for biomedical applications include poly(methyl methacrylate) (PMMA), and poly(dimethylsiloxane) (PDMS) [41.77, 78]. Another material of interest due to ease of fabrication is poly(propyl methacrylate) (PPMA), which has lower glass transition temperature (T_g) (35–43 °C) [41.79] than PMMA (104–106 °C) [41.80, 81], which permits low temperature thermal processing.

This chapter presents a review of mechanical property measurements on nanoscale of various materials of interest and stress and deformation analyses of nanostructures.

41.1 Experimental Techniques for Measurement of Mechanical Properties of Nanostructures

41.1.1 Indentation and Scratch Tests Using Micro/Nanoindenters

A nanoindenter is commonly used to measure hardness, elastic modulus, and fracture toughness, and to perform micro/nanoscratch studies to get a measure of scratch/wear resistance of materials [41.5, 82].

Hardness and Elastic Modulus

The nanoindenter monitors and records the dynamic load and displacement of the three-sided pyramidal diamond (Berkovich) indenter during indentation with a force resolution of about 75 nN and displacement resolution of about 0.1 nm. Hardness and elastic modulus are calculated from the load-displacement data [41.5, 82]. The peak indentation load depends on mechanical properties of the specimen; a harder material requires higher load for a reasonable indentation depth.

Fracture Toughness

The indentation technique for fracture toughness measurement of brittle samples, on the microscale, is based on the measurement of the lengths of median-radial cracks produced by indentation. A Vickers indenter (a four-sided diamond pyramid) is used in a microhardness tester. A load on the order of 0.5 N is typically used in making the Vickers indentations. The indentation impressions are examined using an optical microscope with Nomarski interference contrast to measure the length of median-radial cracks, c. The fracture toughness (K_{IC}) is calculated by the following relation [41.83],

$$K_{IC} = \alpha \left(\frac{E}{H} \right)^{\frac{1}{2}} \left(\frac{P}{c^{\frac{3}{2}}} \right) \qquad (41.1)$$

where α is an empirical constant depending on the geometry of the indenter, E and H are hardness and elastic moduli, and P is the peak indentation load. For Vickers

indenters, α has been empirically found based on experimental data and is equal to 0.016 [41.5]. Both E and H
values are obtained from the nanoindentation data. The
crack length is measured from the center of the indent
to the end of crack using an optical microscope. For one
indent, all crack lengths are measured. The crack length
c is obtained from the average values of several indents.

Indentation Creep

For indentation creep test of polymer samples, the test
is performed using a continuous stiffness measurements (CSM) technique [41.82]. In a study by Wei
et al. [41.65], the indentation load was typically $30\,\mu N$
and the loading rate was $3\,\mu N/s$. The tip was held typically for $600\,s$ after the indentation load reached $30\,\mu N$.
To measure the mean stress and contact stiffness, during
the hold segment the indenter was oscillated at a peak-to-
peak load amplitude of $1.2\,\mu N$ and a frequency of $45\,Hz$.

Scratch Resistance

In micro/nanoscratch studies, in a nanoindenter, a conical diamond indenter having a tip radius of about $1\,\mu m$
and an included angle of $60°$, is drawn over the sample surface, and the load is ramped up until substantial
damage occurs [41.5, 82]. The coefficient of friction is
monitored during scratching. In order to obtain scratch
depths during scratching, the surface profile of the sample surface is first obtained by translating the sample at
a low load of about $0.2\,mN$, which is insufficient to damage a hard sample surface. The $500\,\mu m$ long scratches
are made by translating the sample while ramping the
loads on the conical tip over different loads dependent
upon the material hardness. The actual depth during
scratching is obtained by subtracting the initial profile
from the scratch depth measured during scratching. In
order to measure the scratch depth after the scratch, the
scratched surface is profiled at a low load of $0.2\,mN$
and is subtracted from the actual surface profile before
scratching.

41.1.2 Bending Tests of Nanostructures Using an AFM

Quasi-static bending tests of fixed nanobeam arrays are
carried out using an AFM [41.62, 84]. A three-sided
pyramidal diamond tip (with a radius of about $200\,nm$)
mounted on a rectangular stainless steel cantilever is
used for the bending tests. The beam stiffness is selected
based on the desired load range. The stiffness of the
cantilever beams for application of normal load up to
$100\,\mu N$ is about $150–200\,N/m$.

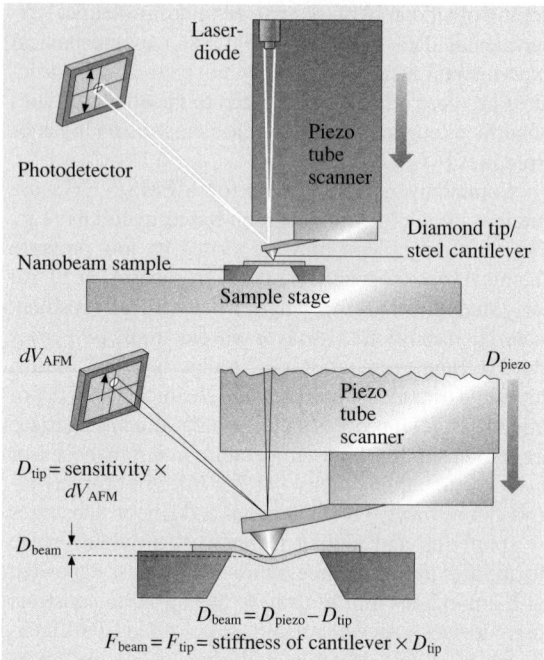

Fig. 41.1 Schematic showing the details of a nanoscale
bending test using an AFM. The AFM tip is brought to
the center of the nanobeam and the piezo is extended
over a known distance. By measuring the tip displacement, a load displacement curve of the nanobeam can be
obtained [41.62]

The wafer with nanobeam array is fixed onto a flat
sample chuck using double-stick tape [41.62]. For the
bending test, the tip is brought over the nanobeam array
with the help of the sample stage of the AFM and a built-
in high magnification optical microscope Fig. 41.1. For
the fine positioning of the tip over a chosen beam, the
array is scanned in contact mode at a contact load of
about 2 to $4\,\mu N$, which results in a negligible damage
to the sample. After scanning, the tip is located at one
end of a chosen beam. To position the tip at the center of
the beam span, the tip is moved to the other end of the
beam by giving the X-piezo an offset voltage. The value
of this offset is determined after several such attempts
have been made in order to minimize effects of piezo
drift. Half of this offset is then applied to the X-piezo
after the tip is positioned at one end of the beam, which
usually results in the tip being moved to the center of
the span. Once the tip is positioned over the center of
the beam span, the tip is held stationary without scanning and the Z-piezo is extended by a known distance,
typically about $2.5\,\mu m$, at a rate of $10\,nm/s$, as shown

in Fig. 41.1. During this time, the vertical deflection signal (dV_{AFM}), which is proportional to the deflection of the cantilever (D_{tip}), is monitored. The displacement of the piezo is equal to the sum of the displacements of the cantilever and the nanobeam. Hence the displacement of the nanobeam (D_{beam}) under the point of load can be determined as

$$D_{beam} = D_{piezo} - D_{tip} \qquad (41.2)$$

The load (F_{beam}) on the nanobeam is the same as the load on the tip/cantilever (F_{tip}) and is given by

$$F_{beam} = F_{tip} = D_{tip} \times k \qquad (41.3)$$

where k is the stiffness of the tip/cantilever. In this manner, a load displacement curve for each nanobeam can be obtained.

The photodetector sensitivity of the cantilever needs to be calibrated to obtain D_{tip} in nm. For this calibration, the tip is pushed against a smooth diamond sample by moving the Z-piezo over a known distance. For the hard diamond material, the actual deflection of the tip can be assumed to be the same as the Z-piezo travel (D_{piezo}), and the photodetector sensitivity (S) for the cantilever setup is determined as

$$S = D_{piezo}/dV_{AFM} \, \text{nm/V} \qquad (41.4)$$

In the measurements, D_{tip} is given as $dV_{AFM} \times S$.

Since a sharp tip would result in an undesirable large local indentation, *Sundararajan* and *Bhushan* [41.62] used a diamond tip which was a worn (blunt) diamond tip. Indentation experiments using this tip on a silicon substrate yielded a residual depth of less than 8 nm at a maximum load of 120 μN, which is negligible compared to displacements of the beams (several hundred nm). Hence we can assume that negligible local indentation or damage is created during the bending process of the beams and that the displacement calculated from (41.2) is from the beam structure.

Elastic Modulus and Bending Strength

Elastic modulus and bending strength (fracture stress) of the beams can be estimated by equations based on the assumption that the beams follow linear elastic theory of an isotropic material. This is probably valid since the beams have high length-to-width ℓ/w and length-to-thickness ℓ/t ratios and also since the length direction is along the principal stress direction during the test. For a fixed elastic beam loaded at the center of the span, the elastic modulus is expressed as

$$E = \frac{\ell^3}{192I} m \qquad (41.5)$$

where ℓ is the beam length, I is the area moment of inertia for the beam cross-section and m is the slope of the load-displacement curve during bending [41.85]. The area moment of inertia for a beam with trapezoidal cross section, is calculated from the following equation

$$I = \frac{w_1^2 4 w_1 w_2 + w_2^2}{36 (w_1 + w_2)} t^3 \qquad (41.6)$$

where w_1 and w_2 are the upper and lower widths, respectively, and t is the thickness of the beam. According to linear elastic theory, for a centrally loaded beam, the moment diagram is shown in Fig. 41.2. The maximum moments are generated at the ends (negative moment) and under the loading point (positive moment) as shown in Fig. 41.2. The bending stresses generated in the beam are proportional to the moments and are compressive or tensile about the neutral axis (line of zero stress). The maximum tensile stress (σ_b, which is the bending strength or fracture stress) is produced on the top surface at both the ends and is given by [41.85]

$$\sigma_b = \frac{F_{max} \ell e_1}{8I} \qquad (41.7)$$

where F_{max} is the applied load at failure and e_1 is the distance of the top surface from the neutral plane of the beam cross-section and is given by [41.85]

$$e_1 = \frac{t (w_1 + 2w_2)}{3 (w_1 + w_2)} \qquad (41.8)$$

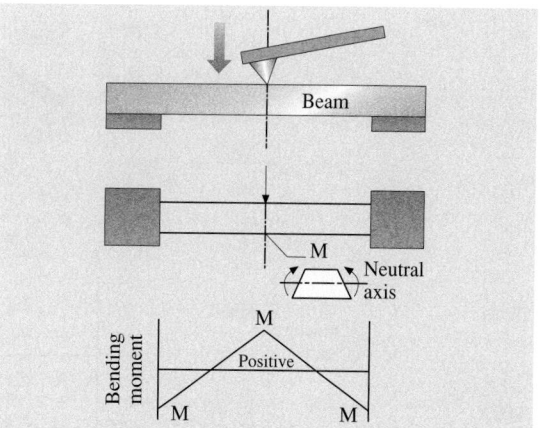

Fig. 41.2 A schematic of the bending moments generated in the beam during a quasi-static bending experiment, with the load at the center of the span. The maximum moments occur under the load and at the fixed ends. Due to the trapezoidal cross section, the maximum tensile bending stresses occur at the top surfaces at the fixed ends

Although the moment value at the center of the beam is the same as at the ends, the tensile stresses at the center (generated on the bottom surface) are less than those generated at the ends (per (41.7)) because the distance from the neutral axis to the bottom surface is less than e_1. This is because of the trapezoidal cross section of the beam, which results in the neutral axis being closer to the bottom surface than the top (Fig. 41.2).

In the preceding analysis, the beams were assumed to have fixed ends. However, in the nanobeams used by *Sundararajan* and *Bhushan* [41.62], the underside of the beams was pinned over some distance on either side of the span. Hence a finite element model of the beams was created to see if the difference in the boundary conditions affected the stresses and displacements of the beams. It was found that the difference in the stresses was less than 1%. This indicates that the boundary conditions near the ends of the actual beams are not that different from that of fixed ends. Therefore the bending strength values can be calculated from (41.7).

Fracture Toughness

Fracture toughness is another important parameter for brittle materials such as silicon. In the case of the nanobeam arrays, these are not best suited for fracture toughness measurements because they do not possess regions of uniform stress during bending. *Sundararajan* and *Bhushan* [41.62] developed a methodology and its steps are outlined schematically in Fig. 41.3a. First,

a crack of known geometry is introduced in the region of maximum tensile bending stress, i. e. on the top surface near the ends of the beam. This is achieved by generating a scratch at high normal load across the width (w_1) of the beam using a sharp diamond tip (radius < 100 nm). A typical scratch thus generated is shown in Fig. 41.3b. By bending the beam as shown, a stress concentration will be formed under the scratch. This will lead to failure of the beam under the scratch once a critical load (fracture load) is attained. The fracture load and relevant dimensions of the scratch are input into the FEM model, which is used to generate the fracture stress plots. Figure 41.3c shows an FEM simulation of one such experiment, which reveals that the maximum stress does occur under the scratch.

If we assume that the scratch tip acts as a crack tip, a bending stress will tend to open the crack in Mode I. In this case, the stress field around the crack tip can be described by the stress intensity parameter K_I (for Mode *I*) for linear elastic materials [41.86]. In particular the stresses corresponding to the bending stresses are described by

$$\sigma = \frac{K_I}{\sqrt{2\pi r}} \cos\left(\frac{\theta}{2}\right) \left[1 + \sin\left(\frac{\theta}{2}\right) \sin\left(\frac{3\theta}{2}\right)\right]$$
(41.9)

for every point $p(r, \theta)$ around the crack tip as shown in Fig. 41.4. If we substitute the fracture stress (σ_f) into the left hand side of (41.9), then the K_I value can be substi-

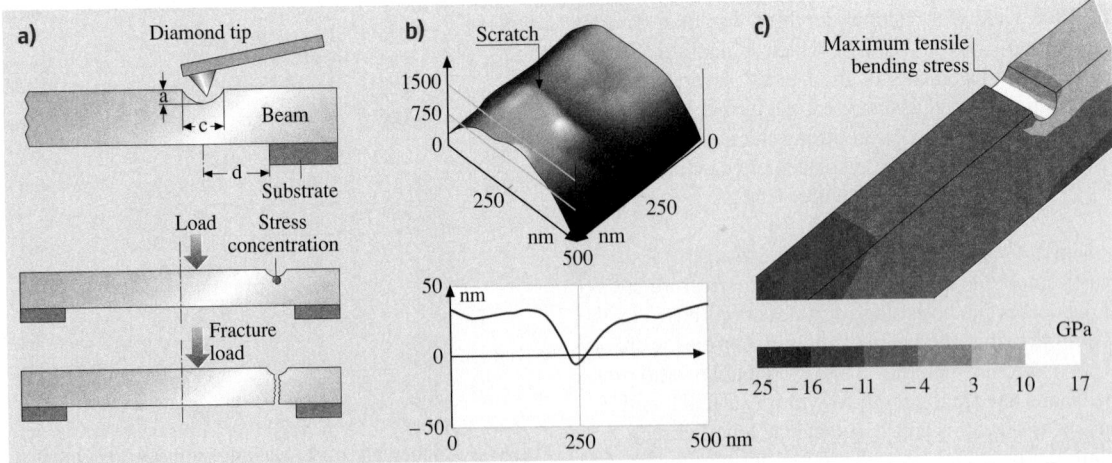

Fig. 41.3 (a) Schematic of technique to generate a defect (crack) of known dimensions in order to estimate fracture toughness. A diamond tip is used to generate a scratch across the width of the beam. When the beam is loaded as shown, a stress concentration is formed at the bottom of the scratch. The fracture load is then used to evaluate the stresses using FEM. **(b)** AFM 3-D image and 2-D profile of a typical scratch. **(c)** Finite element model results verifying that the maximum bending stress occurs at the bottom of the scratch [41.62]

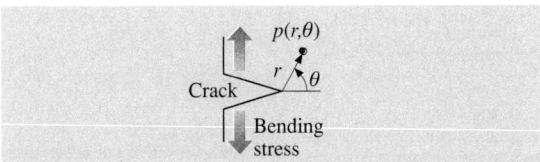

Fig. 41.4 Schematic of crack tip and coordinate systems used in (41.9) to describe a stress field around the crack tip in terms of the stress intensity parameter, K_I [41.62]

tuted by its critical value, which is the fracture toughness K_{IC}. Now, the fracture stress can be determined for the point $(r = 0, \theta = 0)$, i. e. right under the crack tip as explained above. However, we cannot substitute $r = 0$ in (41.9). The alternative is to substitute a value for r, which is as close to zero as possible. For silicon, a reasonable number is the distance between neighboring atoms in the (111) plane, the plane along which silicon exhibits the lowest fracture energy. This value was calculated from silicon unit cell dimensions of 0.5431 nm [41.87] to be 0.4 nm (half of the face diagonal). This assumes that Si displays no plastic zone around the crack tip, which is reasonable since in tension, silicon is not known to display much plastic deformation at room temperature. *Sundararajan* and *Bhushan* [41.62] used values $r = 0.4$ to 1.6 nm (i. e. distances up to 4 times the distance between the nearest neighboring atoms) to estimate the fracture toughness for both Si and SiO$_2$ according to the following equation

$$K_{IC} = \sigma_f \sqrt{2\pi r} \quad r = 0.4 \text{ to } 1.6 \text{ nm} \tag{41.10}$$

Fatigue Strength
In addition to the properties mentioned so far that can be evaluated from quasi-static bending tests, the fatigue properties of nanostructures are also of interest. This is especially true for MEMS/NEMS involving vibrating structures such as oscillators and comb drives [41.88] and hinges in digital micromirror devices [41.89]. To study the fatigue properties of the nanobeams, *Sundararajan* and *Bhushan* [41.62] applied monotonic cyclic stresses using an AFM, Fig. 41.5a. Similar to the bending test, the diamond tip is first positioned at the center of the beam span. In order to ensure that the tip is always in contact with the beam (as opposed to impacting it), the piezo is first extended by a distance D_1, which ensures a minimum stress on the beam. After this extension, a cyclic displacement of amplitude D_2 is applied continuously until failure of the beam occurs. This results in the application of a cyclic load to the beam. The maximum frequency of the cyclic load that

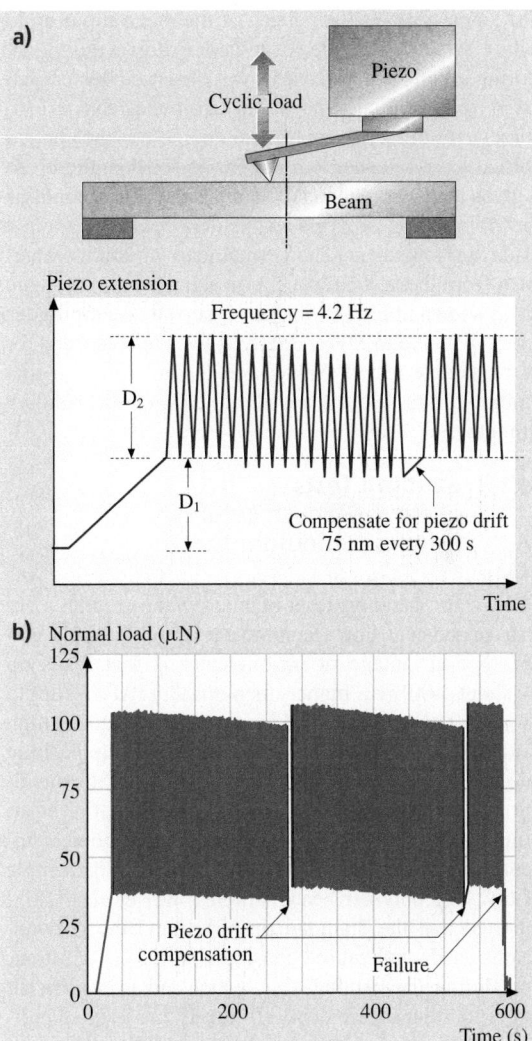

Fig. 41.5 (a) Schematic showing the details of the technique to study fatigue behavior of the nanobeams. The diamond tip is located at the middle of the span and a cyclic load at 4.2 Hz is applied to the beam by forcing the piezo to move in the pattern shown. An extension is made every 300 s to compensate for the piezo drift to ensure that the load on the beam is kept fairly constant. **(b)** Data from a fatigue experiment on a nanobeam until failure. The normal load is computed from the raw vertical deflection signal. The compensations for piezo drift keep the load fairly constant [41.62]

could be attained using the AFM by *Sundararajan* and *Bhushan* [41.62] was 4.2 Hz. The vertical deflection signal of the tip is monitored throughout the experiment.

The signal follows the pattern of the piezo input up to failure, which is indicated by a sudden drop in the signal. During initial runs, piezo drift was observed that caused the piezo to gradually move away from the beam (i. e. to retract), resulting in a continuous decrease in the applied normal load. In order to compensate for this, the piezo is given a finite extension of 75 nm every 300 seconds as shown in Fig. 41.5a. This results in keeping the applied loads fairly constant. The normal load variation (calculated from the vertical deflection signal) from a fatigue test is shown in Fig. 41.5b. The cyclic stress amplitudes (corresponding to D_2) and fatigue lives are recorded for every sample tested. Values for D_1 are set such that minimum stress levels are about 20% of the bending strengths.

41.1.3 Bending Tests of Micro/Nanostructures Using a Nanoindenter

Quasi-static bending tests of micro/nanostructures are also carried out using a nanoindenter (Fig. 41.6) [41.63–65]. The advantage of nanoindenter is that loads up to about 400 mN, higher than that in AFM (up to about 100 μN), can be used for structures requiring high loads for experiments. To avoid the indenter tip pushing into the specimen, a blunt tip is used in the bending and fatigue tests. For ceramic and metallic beam samples, *Li* et al. [41.64] used a diamond conical indenter with a radius of 1 μm and an included angle of 60°. For polymer beam samples, *Wei* et al. [41.65] reported that the diamond tip penetrated the polymer beams easily and caused considerable plastic deformation during the bending test, which led to significant errors in the measurements. To avoid this issue, the diamond tip was dip-coated with PMMA (about 1–2 μm thick) by dipping the tip in the 2% PMMA (wt/wt) solution for about 5 seconds. Load position used was

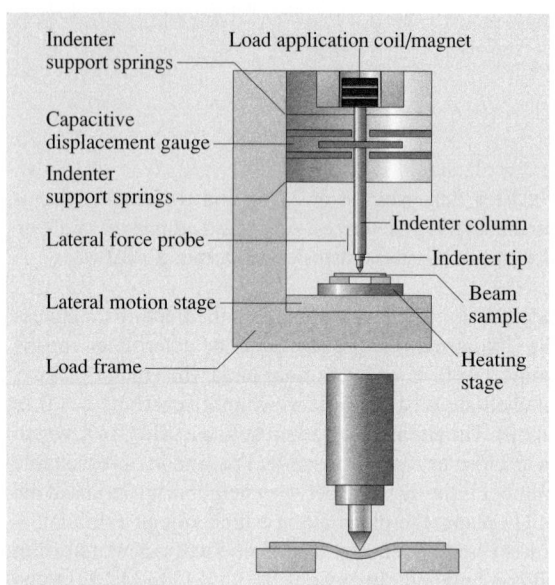

Fig. 41.6 Schematic of micro/nanoscale bending test using a nanoindenter

at the center of the span for the bridge beams and at 10 μm off from the free end of cantilever beams. An optical microscope with a magnification of 1500 × or an in-situ AFM is used to locate the loading position. Then the specimen is moved under the indenter location with a resolution of about 200 nm in longitudinal direction and less than 100 nm in lateral direction.

Using the analysis presented earlier, elastic modulus and bending strength of the beams can be obtained from the load displacement curves [41.64, 65]. For fatigue tests, an oscillating load is applied and contact stiffness is measured during the tests. A significant drop in the contact stiffness during the test is a measure of number of cycles to failure [41.63].

41.2 Experimental Results and Discussion

41.2.1 Indentation and Scratch Tests of Various Ceramic and Metals Using a Micro/Nanoindenter

Studies have been conducted on five different materials: undoped single-crystal Si(100), undoped polysilicon film, SiO_2 film, SiC film, electroless deposited Ni − 11.5 wt % P amorphous film, and electroplated Au film [41.64, 75, 76]. A 3 μm thick polysilicon film was

deposited by a low pressure chemical vapor deposition (LPCVD) process on an Si(100) substrate. The 1 μm thick SiO_2 film was deposited by a plasma enhanced chemical vapor deposition (PECVD) process on an Si(111) substrate. A 3 μm thick 3C-SiC film was epitaxially grown using an atmospheric pressure chemical vapor deposition (APCVD) process on Si(100) substrate. A 12 μm thick Ni-P film was electroless plated on a 0.8 mm thick Al − 4.5 wt % Mg alloy substrate.

A 3 µm thick Au film was electroplated on an Si(100) substrate.

Hardness and Elastic Modulus

Hardness and elastic modulus measurements are made using a nanoindenter [41.64]. The hardness and elastic modulus values of various materials at a peak indentation depth of 50 nm are summarized in Fig. 41.7 and Table 41.1. The SiC film exhibits the highest hardness of about 25 GPa an elastic modulus of about 395 GPa among the samples examined, followed by the undoped Si(100), undoped polysilicon film, SiO$_2$ film, Ni-P film, and Au film. The hardness and elastic modulus data of the undoped Si(100) and undoped polysilicon film are comparable. For the metal alloy films, the Ni-P film exhibits higher hardness and elastic modulus than the Au film.

Fracture Toughness

The optical images of Vickers indentations made using a microindenter, at a normal load of 0.5 N held for 15 s on the undoped Si(100), undoped polysilicon film, and SiC film are shown in Fig. 41.8 [41.76]. The SiC film exhibits the smallest indentation mark, followed by the undoped polysilicon film and undoped Si(100). These Vickers indentation depths are smaller than one-third of the film thickness. Thus, the influence of substrate on the fracture toughness of the films can be ignored. In addition to the indentation marks, radial cracks are observed, emanating from the indentation corners. The SiC film shows the longest radial crack length, followed by the undoped Si(100) and undoped polysilicon film. The radial cracks for the undoped Si(100) are straight

Fig. 41.7 Bar chart summarizing the hardness, elastic modulus, fracture toughness, and critical load (from scratch tests) results of the bulk undoped single-crystal Si(100) and thin films of undoped polysilicon, SiO$_2$, SiC, Ni-P, and Au [41.64]

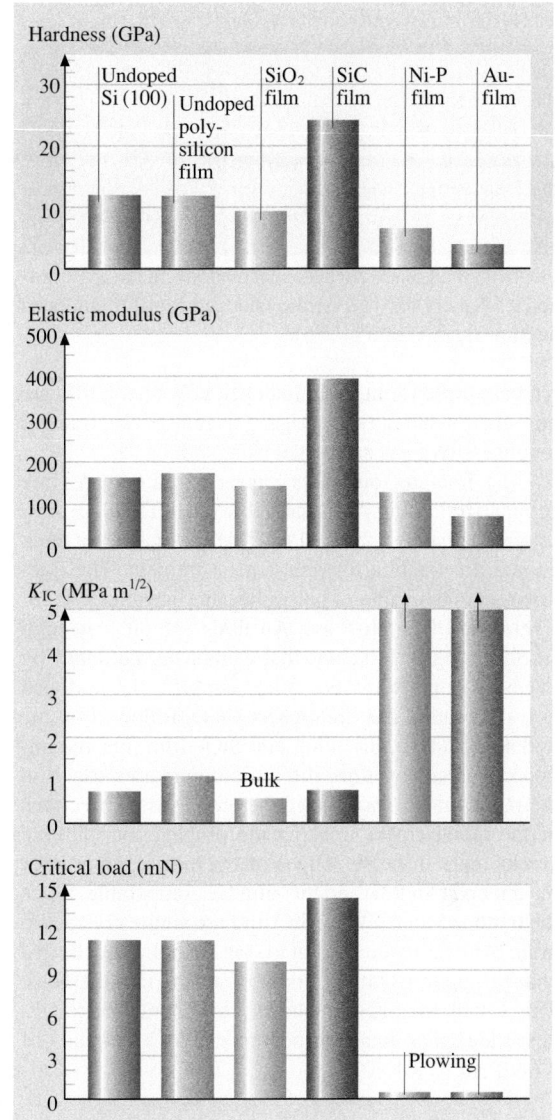

Table 41.1 Hardness, elastic modulus, fracture toughness and critical load results of the bulk single-crystal Si(100) and thin films of undoped polysilicon, SiO$_2$, SiC, Ni−P and Au

Samples	Hardness (GPa)	Elastic modulus (GPa)	Fracture toughness (MPa m$^{1/2}$)	Critical load (mN)
Undoped Si(100)	12	165	0.75	11
Undoped polysilicon film	12	167	1.11	11
SiO$_2$ film	9.5	144	0.58 (Bulk)	9.5
SiC film	24.5	395	0.78	14
Ni−P film	6.5	130		0.4 (Plowing)
Au film	4	72		0.4 (Plowing)

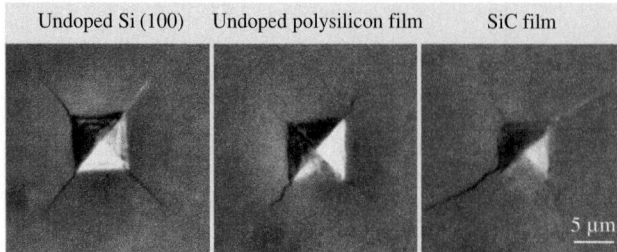

Undoped Si (100)	Undoped polysilicon film	SiC film

5 μm

Fig. 41.8 Optical images of Vickers indentations made at a normal load of 0.5 N held for 15 s on the undoped Si(100), undoped polysilicon film, and SiC film [41.76]

whereas those for the SiC, undoped polysilicon film are not straight but go in a zigzag manner. The fracture toughness (K_{IC}) is calculated using (41.1).

The fracture toughness values of all samples are summarized in Fig. 41.7 and Table 41.1. The SiO_2 film used in this study is about 1 μm thick, which is not thick enough for fracture toughness measurement. The fracture toughness value of bulk silica are listed instead for a reference. The Ni-P and Au films exhibit very high fracture toughness values that cannot be measured by indentation methods. For other samples, the undoped polysilicon film has the highest value, followed by the undoped Si(100), SiC film, and SiO_2 film. For the undoped polysilicon film, the grain boundaries can stop the radial cracks and change the propagation directions of the radial cracks, making the propagation of these cracks more difficult. Values of fracture toughness for the undoped Si(100) and SiC film are comparable. Since the undoped Si(100) and SiC film are single crystal, no grain boundaries are present to stop the radial cracks and change the propagation directions of the radial cracks. This is why the SiC film shows a lower fracture toughness value than the bulk polycrystal SiC materials of 3.6 MPam$^{1/2}$ [41.90].

Scratch Resistance

Scratch resistance of various materials have been studied using a nanoindenter by *Li* et al. [41.64]. Figure 41.9 compares the coefficient of friction and scratch depth profiles as a function of increasing normal load and optical images of three regions over scratches: at the beginning of the scratch (indicated by A on the friction profile), at the point of initiation of damage at which the coefficient of friction increases to a high value or increase abruptly (indicated by B on the friction profile), and towards the end of the scratch (indicated by C on the friction profile) for all samples. Note that the ramp loads for the Ni-P and Au range are from 0.2 to 5 mN

Fig. 41.9 Coefficient of friction and scratch depth profiles as a function of increasing normal load and optical images of three regions over scratches: at the beginning of the scratch (indicated by A on the friction profile), at the point of initiation of damage at which the coefficient of friction increases to a high value or increase abruptly (indicated by B on the friction profile), and towards the end of the scratch (indicated by C on the friction profile) for all samples [41.64] ▶

whereas the ramp loads for other samples are from 0.2 to 20 mN. All samples exhibit a continuous increase in the coefficient of friction with increasing normal load from the beginning of the scratch. The continuous increase in the coefficient of friction during scratching is attributed to the increasing plowing of the sample by the tip with increasing normal load, as shown in the SEM images in Fig. 41.9. The abrupt increase in the coefficient of friction is associated with catastrophic failure as well as significant plowing of the tip into the sample. Before the critical load, the coefficient of friction of the undoped polysilicon, SiC and SiO_2 films increased at a slower rate, and was smoother than that of the other samples. The undoped Si(100) exhibits some bursts in the friction profiles before the critical load. At the critical load, the SiC and undoped polysilicon films exhibit a small increase in the coefficient of friction whereas the undoped Si(100) and undoped polysilicon film exhibit a sudden increase in the coefficient of friction. The Ni-P and Au films show a continuous increase in the coefficient of friction, indicating the behavior of a ductile metal. The bursts in the friction profile might result from the plastic deformation and material pile-up in front of the scratch tip. The Au film exhibits a higher coefficient of friction than the Ni-P film. This is because the Au film has lower hardness and elastic modulus values than the Ni-P film.

The SEM images show that below the critical loads the undoped Si(100) and undoped polysilicon film were damaged by plowing, associated with the plastic flow of the material and formation of debris on the sides of the scratch. For the SiC and SiO_2 films, in region A, a plowing scratch track was found without any debris on the side of the scratch, which is probably responsible for the smoother curve and slower increase in the coefficient of friction before the critical load. After the critical load, for the SiO_2 film, delamination of the film from the substrate occurred, followed by cracking along the scratch track. For the SiC film, only several small debris particles were found without any cracks on the side of the scratch, which is responsible for the small in-

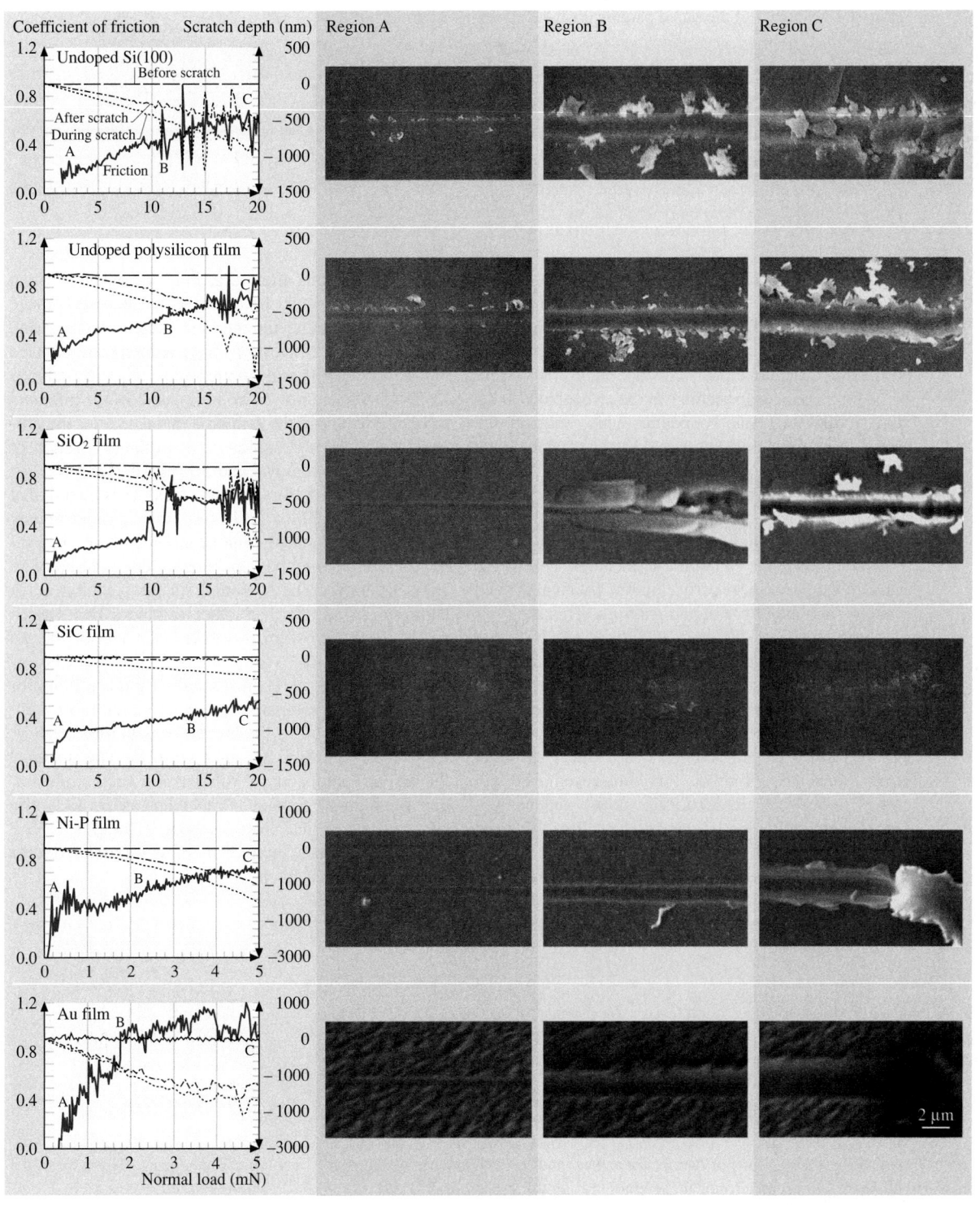

Table 41.2 Summary of measured parameters from quasi-static bending tests

Sample	Elastic modulus E (GPa)		Bending strength σ_b (GPa)		Fracture toughness K_{IC} (MPa $\sqrt{m}$)		
	Measured	Bulk value	Measured	Reported (microscale)	Estimated	Reported (microscale)	Bulk value
Si	182 ± 11	169^a	18 ± 3	$< 10^c$	1.67 ± 0.4	$0.6-1.65^e$	0.9^f
SiO$_2$	85 ± 13	73^b	7.6 ± 2	$< 2^d$	0.60 ± 0.2	$0.5-0.9^4$	–

[a]Si[110] [41.91]   [b][41.92]   [c][41.43, 44, 46–49, 70, 93, 94]   [d][41.58]   [e][41.54–57]   [f][41.87]

crease in the coefficient of friction at the critical load. For the undoped Si(100), cracks were found on the side of the scratch right from the critical load and up, which is probably responsible for the big bursts in the friction profile. For the undoped polysilicon film, no cracks were found on the side of the scratch at the critical load. This might result from grain boundaries which can stop the propagation of cracks. At the end of the scratch, some of the surface material was torn away and cracks were found on the side of the scratch in the undoped Si(100). A couple of small cracks were found in the undoped polysilicon and SiO$_2$ films. No crack was found in the SiC film. Even at the end of the scratch, less debris was found in the SiC film. A curly chip was found at the end of the scratch in both Ni-P and Au films. This is a typical characteristic of ductile metal alloys. The Ni-P and Au films were damaged by plowing right from the beginning of the scratch with material pile-up at the side of the scratch.

The scratch depth profiles obtained during and after the scratch on all samples with respect to initial profile, after the cylindrical curvature is removed, are plotted in Fig. 41.9. Reduction in scratch depth is observed after scratching as compared to that of during scratching. This reduction in scratch depth is attributed to an elastic recovery after removal of the normal load.

The scratch depth after scratching indicates the final depth which reflects the extent of permanent damage and plowing of the tip into the sample surface, and is probably more relevant for visualizing the damage that can occur in real applications. For the undoped Si(100), undoped polysilicon film, and SiO$_2$ film, there is a large scatter in the scratch depth data after the critical loads, which is associated with the generation of cracks, material removal and debris. The scratch depth profile is smooth for the SiC film. It is noted that the SiC film exhibits the lowest scratch depth among the samples examined. The scratch depths of the undoped Si(100), undoped polysilicon film and SiO$_2$ film are comparable. The Ni-P and Au films exhibit much lager scratch depth than other samples. The scratch depth of the Ni-P film is smaller than that of the Au film.

The critical loads estimated from friction profiles for all samples are compared in Fig. 41.7 and Table 41.1. The SiC film exhibits the highest critical load of about 14 mN, as compared to other samples. The undoped Si(100) and undoped polysilicon film show comparable critical load of about 11 mN whereas the SiO$_2$ film shows a low critical load of about 9.5 mN. The Ni-P and Au films were damaged by plowing right from the beginning of the scratch.

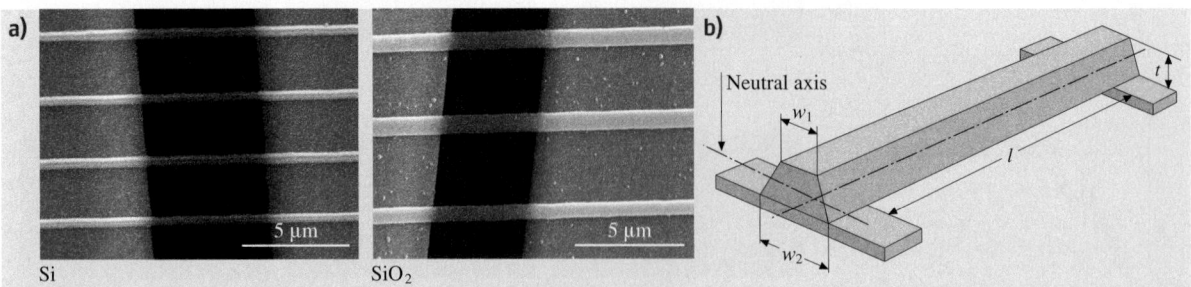

Fig. 41.10 (a) SEM micrographs of nanobeam arrays, and (b) a schematic of the shape of a typical nanobeam. The trapezoidal cross-section is due to the anisotropic wet etching during the fabrication [41.84]

41.2.2 Bending Tests of Ceramic Nanobeams Using an AFM

Bending tests have been performed on Si and SiO_2 nanobeam arrays [41.62, 84]. The single-crystal silicon bridge nanobeams were fabricated by bulk micromachining incorporating enhanced-field anodization using an AFM [41.61]. The Si nanobeams are oriented along the [110] direction in the (001) plane. Subsequent thermal oxidation of the beams results in formation of SiO_2 beams. The cross section of the nanobeams is trapezoidal owing to the anisotropic wet etching process. SEM micrographs of Si and SiO_2 nanobeam arrays and a schematic of the shape of a typical nanobeam are shown in Fig. 41.10. The actual widths and thicknesses of nanobeams were measured using an AFM in tapping mode prior to tests using a standard Si tapping mode tip (tip radius < 10 nm). Surface roughness measurements of the nanobeam surfaces in tapping mode yielded a σ of 0.7 ± 0.2 nm and peak-to-valley (P–V) distance of 4 ± 1.2 nm for Si and a σ of 0.8 ± 0.3 nm and a P–V distance of 3.1 ± 0.8 nm for SiO_2. Prior to testing, the Si nanobeams were cleaned by immersing them in a "piranha etch" solution (3:1 solution by volume of 98% sulphuric acid and 30% hydrogen peroxide) for 10 minutes to remove any organic contaminants.

Bending Strength

Figure 41.11 shows typical load displacement curves for Si and SiO_2 beams that were bent to failure [41.62, 84]. The upper width (w_1) of the beams is indicated in the figure. Also indicated in the figure are the elastic modulus values obtained from the slope of the load displacement curve (41.5). All the beams tested showed linear elastic behavior followed by abrupt failure, which is suggestive of brittle fracture. Figure 41.12 shows the scatter in the values of elastic modulus obtained for both Si and SiO_2 along with the average values ($\pm$ standard deviation). The scatter in the values may be due to differences in orientation of the beams with respect to the trench and the loading point being a little off-center with respect to the beam span. The average values are a little higher than the bulk values (169 GPa for Si[110] and 73 GPa for SiO_2 in Table 41.2). However the values of E obtained from (41.5) have an error of about 20% due to the uncertainties in beam dimensions and spring constant of the tip/cantilever (which affects the measured load). Hence the elastic modulus values on the nanoscale can be considered to be comparable to bulk values.

Most of the beams when loaded quasi-statically at the center of the span broke at the ends as shown

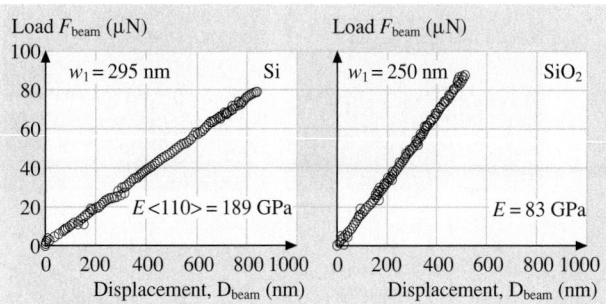

Fig. 41.11 (a) Typical load displacement curves of silicon and SiO_2 nanobeams. The curves are linear until sudden failure, indicative of brittle fracture of the beams. The elastic modulus (E) values calculated from the curves are shown. The dimensions of the Si beam were $w_1 = 295$ nm, $w_2 = 484$ nm and $t = 255$ nm, while those of the SiO_2 beam were $w_1 = 250$ nm, $w_2 = 560$ nm and $t = 425$ nm [41.84]

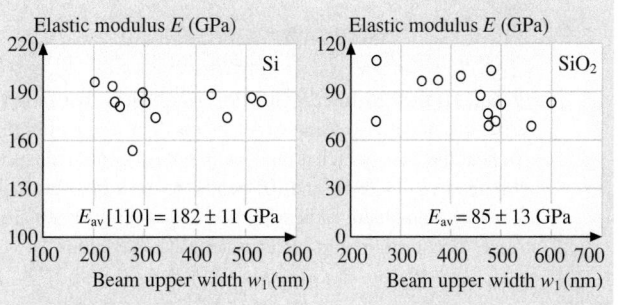

Fig. 41.12 Elastic modulus values measured for Si and SiO_2. The average values are shown. These are comparable to bulk values, which shows that elastic modulus shows no specimen size dependence [41.62]

in Fig. 41.13a, which is consistent with the fact that maximum tensile stresses occur on the top surfaces near the ends. (See FEM stress distribution results in Fig. 41.13b.) Figure 41.14 shows the values of bending strength obtained for different beams. There appears to be no trend in bending strength with the upper width (w_1) of the beams. The large scatter is expected for the strength of brittle materials, since they are dependent on pre-existing flaw population in the material and hence are statistical in nature. The Weibull distribution, a statistical analysis, can be used to describe the scatter in the bending strength values. The means of the Weibull distributions were found to be 17.9 GPa and 7.6 GPa for Si and SiO_2, respectively. Previously reported numbers of strengths range from 1–6 GPa for silicon [41.44, 46–49, 51, 54, 70, 93, 94] and about

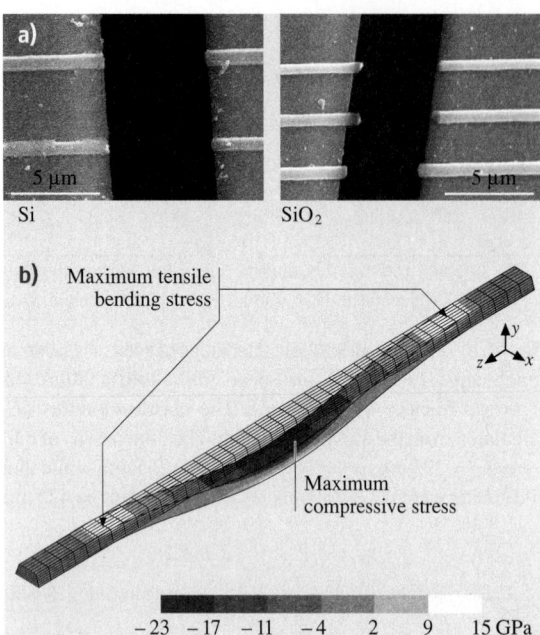

Fig. 41.13 (a) SEM micrographs of nanobeams that failed during quasi-static bending experiments. The beams failed at or near the ends, which is the location of maximum tensile bending stress [41.84], and **(b)** bending stress distribution for silicon nanobeam indicating that the maximum tensile stresses occur on the top surfaces near the fixed ends

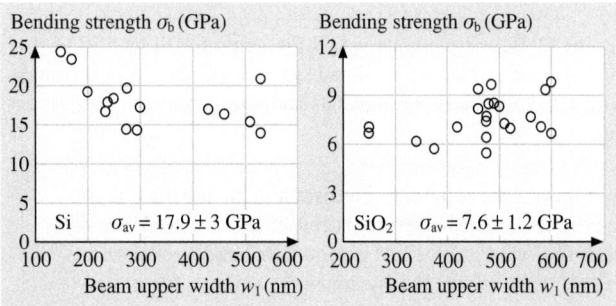

Fig. 41.14 Bending strength values obtained from bending experiments. Average values are indicated. These values are much higher than values reported for microscale specimens, indicating that bending strength shows a specimen size effect [41.62]

1 GPa for SiO_2 [41.58] microscale specimens. This clearly indicates that bending strength shows a specimen size dependence. Strength of brittle materials is dependent on pre-existing flaws in the material. Since for nanoscale specimens, the volume is smaller than for micro and macroscale specimens, the flaw population

will be smaller as well, resulting in higher values of strength.

Fracture Toughness
Estimates of fracture toughness calculated using (41.10) for Si and SiO_2 are shown in Fig. 41.15 [41.62]. The results show that the K_{IC} estimate for Si is about $1-2\,MPa\sqrt{m}$ whereas for SiO_2 the estimate is about $0.5-0.9\,MPa\sqrt{m}$. These values are comparable to values reported by others on larger specimens for Si [41.54–57] and SiO_2 [41.58]. The high values obtained for Si could be due to the fact that the scratches, despite being quite sharp, still have a finite radius of about 100 nm. The bulk value for silicon is about $0.9\,MPa\sqrt{m}$. Fracture toughness is considered to be a material property and is believed to be independent of specimen size. The values obtained in this study, given its limitations, appear to show that fracture toughness is comparable, if not a little higher on the nanoscale.

Fatigue Strength
Fatigue strength measurements of Si nanobeams have been carried out by *Sundararajan* and *Bhushan* [41.62] using an AFM and *Li* and *Bhushan* [41.63] using a nanoindenter. Various stress levels were applied to nanobeams by *Sundararajan* and *Bhushan* [41.62]. The minimum stress was 3.5 GPa for Si beams and 2.2 GPa for SiO_2 beams. The frequency of applied load was 4.2 Hz. In general, the fatigue life decreased with increasing mean stress as well as increasing stress amplitude. When the stress amplitude was less than 15% of the bending strength, the fatigue life was greater than 30 000 cycles for both Si and SiO_2. However, the mean

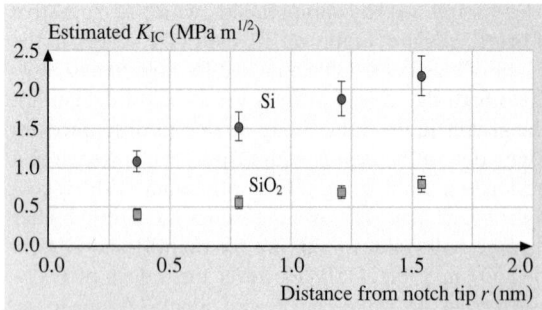

Fig. 41.15 Fracture toughness (K_{IC}) values of for increasing values of r corresponding to distance between neighboring atoms in {111} planes of silicon (0.4 nm). Hence r values between 0.4 and 1.6 nm are chosen. The K_{IC} values thus estimated are comparable to values reported by others for both Si and SiO_2 [41.62]

stress had to be less than 30% of the bending strength for a life of greater than 30 000 for Si whereas even at a mean stress of 43% of the bending strength, SiO$_2$ beams showed a life greater than 30 000. During fatigue, the beams broke under the loading point or at the ends, when loaded at the center of the span. This was different from the quasi-static bending tests, where the beams broke at the ends almost every time. This could be due to the fact that the stress levels under the load and at the ends are not that different and fatigue crack propagation could occur at either location. Figure 41.16 shows a nanoscale S–N curve, with bending stress (S) as a function of fatigue in cycles (N) with an apparent endurance life at lower stress. This study clearly demonstrates that fatigue properties of nanoscale specimens can be studied.

SEM Observations of Fracture Surfaces

Figure 41.17 shows SEM images of the fracture surfaces of nanobeams broken during quasi-static bending as well as fatigue [41.62]. In the quasi-static cases, the maximum tensile stresses occur on the top surface, so it is reasonable to assume that fracture initiated at or near the top surface and propagated downward. The fracture surfaces of the beams suggest a cleavage type of fracture. Silicon beam surfaces show various ledges or facets, which is typical for crystalline brittle materials. Silicon usually fractures along the (111) plane due to this plane having the lowest surface energy to overcome by a propagating crack. However, failure has also been known to occur along the (110) planes in microscale specimens, despite the higher energy required as compared to the (111) planes [41.70]. The plane normal to the beam direction in these samples is the (110) plane while (111) planes will be oriented at 35° from the (110) plane. The presence of facets and irregularities on the silicon surface in Fig. 41.17a suggest that it is a combination of these two types of fractures that has occurred. Since the stress levels are very high for these specimens, it is reasonable to assume that crack propagating forces will be high enough to result in (110) type failures.

In contrast, the silicon fracture surfaces under fatigue, shown in Fig. 41.17b, appear very smooth without facets or irregularities. This is suggestive of low energy fracture, i.e. of (111) type fracture. We do not see evidence of fatigue crack propagation in the form of steps or striations on the fracture surface. We believe that for the stress levels applied in these fatigue experiments, failure in silicon occurred via cleavage associated with 'static fatigue' type of failures.

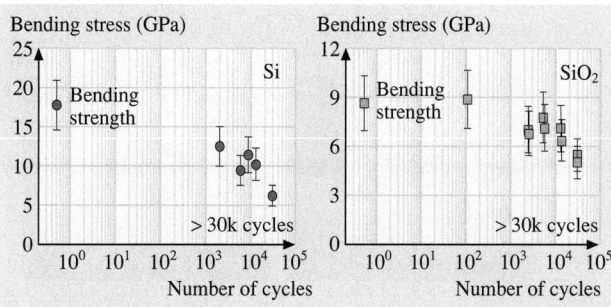

Fig. 41.16 Fatigue test data showing applied bending stress as a function of number of cycles. A single load–unload sequence is considered as 1 cycle. The bending strength data points are therefore associated with 1/2 cycle, since failure occurs upon loading [41.62]

SiO$_2$ shows very smooth fracture surfaces for both quasi-static bending and fatigue. This is in contrast to the hackled surface one might expect for the brittle failure of an amorphous material on the macroscale. However, in larger scale fracture surfaces for such materials, the region near the crack initiation usually appears smooth or mirror-like. Since the fracture surface here is so small and very near the crack initiation site, it is not unreasonable to see such a smooth surface for SiO$_2$ on this scale. There appears to be no difference between the fracture surfaces obtained by quasi-static bending and fatigue for SiO$_2$.

Summary of Mechanical Properties Measured Using Quasi-Static Bending Tests

Table 41.2 summarizes the various properties measured via quasi-static bending in this study [41.62]. Also shown are bulk values of the parameters along with values reported on larger scale specimens by other researchers. Elastic modulus and fracture toughness values appear to be comparable to bulk values and show no dependence on specimen size. However bending strength shows a clear specimen size dependence with nanoscale numbers being twice as large as numbers reported for larger scale specimens.

41.2.3 Bending Tests of Metallic Microbeams Using a Nanoindenter

Bending tests have been performed on Ni-P and Au microbeams [41.64]. The Ni-P cantilever microbeams were fabricated by focused ion beam machining technique. The dimensions were $10 \times 12 \times 50\,\mu m^3$. Notches with a depth of 3 µm and a tip radius of 0.25 µm were introduced in the microbeams to facilitate failure at a lower

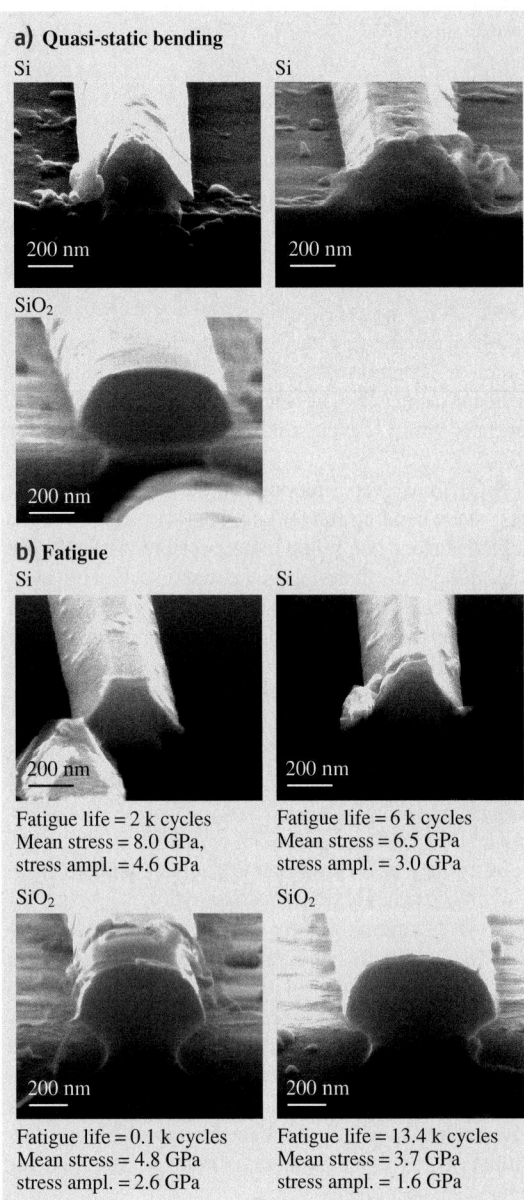

a) Quasi-static bending

Si   Si

200 nm   200 nm

SiO$_2$

200 nm

b) Fatigue

Si   Si

200 nm   200 nm

Fatigue life = 2 k cycles   Fatigue life = 6 k cycles
Mean stress = 8.0 GPa,   Mean stress = 6.5 GPa
stress ampl. = 4.6 GPa   stress ampl. = 3.0 GPa

SiO$_2$ SiO$_2$

200 nm   200 nm

Fatigue life = 0.1 k cycles Fatigue life = 13.4 k cycles
Mean stress = 4.8 GPa   Mean stress = 3.7 GPa
stress ampl. = 2.6 GPa   stress ampl. = 1.6 GPa

Fig. 41.17a,b SEM micrographs of fracture surfaces of silicon and SiO$_2$ beams subjected to (**a**) quasi-static bending and (**b**) fatigue [41.62]

load in the bending tests. The Au bridge microbeams were fabricated by electroplating technique.

Figure 41.18 shows the SEM images, load displacement curve and FEM stress contour for the notched Ni-P cantilever microbeam that was bent to failure [41.64]. The distance between the loading position and the fixed end is 40 µm. The 3 µm deep notch is 10 µm from the fixed end. The notched beam showed linear behavior followed by abrupt failure. The FEM stress contour shows that there is higher stress concentration at the notch tip. The maximum tensile stress σ_m at the notch tip can be analyzed by using Griffith fracture theory as follows [41.83],

$$\sigma_m \approx 2\sigma_0 \left(\frac{c}{\rho}\right)^{\frac{1}{2}} \qquad (41.11)$$

where σ_0 is the average applied tensile stress on the beam, c is the crack length, and ρ is the crack tip radius. Therefore, elastic–plastic deformation will first occur locally at the end of the notch tip, followed by abrupt fracture failure after the σ_m reaches the ultimate tensile strength of Ni-P, even though the rest of the beam is still in the elastic regime. The SEM image of the fracture surface shows that the fracture started right from the notch tip with plastic deformation characteristic. This indicates that although local plastic deformation occurred at the notch tip area, the whole beam failed catastrophically. The present study shows that FEM simulation can predict well the stress concentration, and helps in understanding the failure mechanism of the notched beams.

Figure 41.19 shows the SEM images, load displacement curve and FEM stress contour for the Au bridge microbeam that was deformed by the indenter [41.64]. The recession gap between the beam and substrate is about 7 µm, which is not large enough to break the beam at the load applied. From the load–displacement curve, we note that the beam experienced elastic–plastic deformation. The FEM stress contour shows that the maximum tensile stress is located at the fixed ends whereas the minimum compressive stress is located around the center of the beam. The SEM image shows that the beam has been permanently deformed. No crack was found on the beam surface. The present study shows a possibility for mechanically forming the Au film into the shape as needed. This may help in designing/fabricating functionally complex smart micro/nanodevices which need conductors for power supply and input/output signals.

41.2.4 Indentation and Scratch Tests of Polymeric Microbeams Using a Nanoindenter

Studies have been conducted on two polymer microbeams made of PMMA and PPMA with thickness of about 2 to 5 µm [41.65]. PPMA was chosen due to its relatively low glass transition temperature, allowing

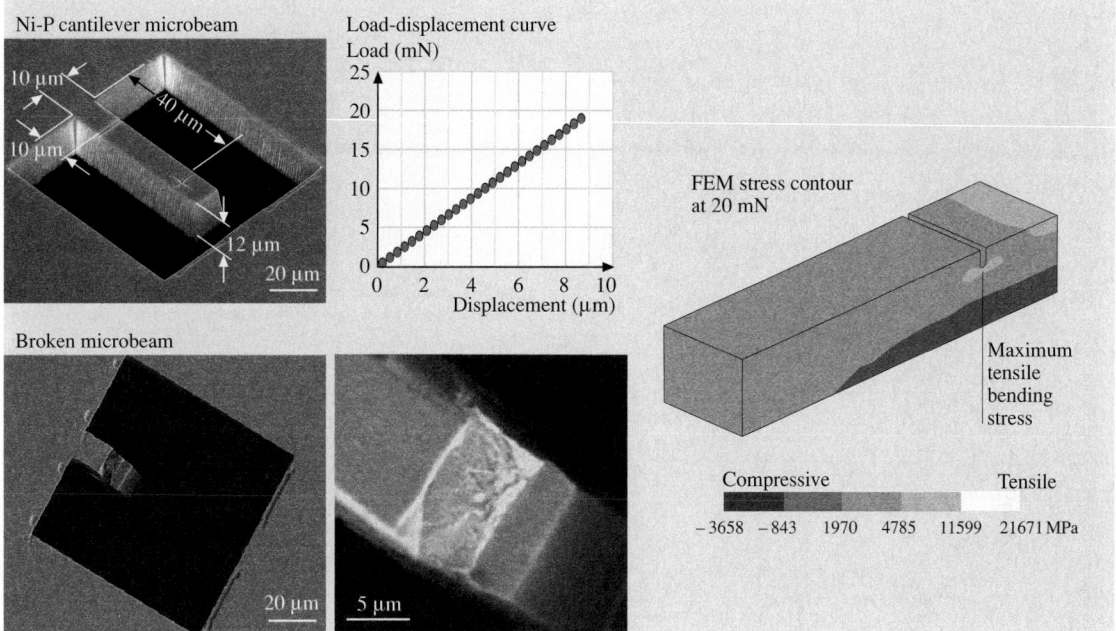

Fig. 41.18 SEM micrographs of the new and broken beams, load displacement curve and FEM stress contour for the notched Ni-P cantilever microbeam [41.64]

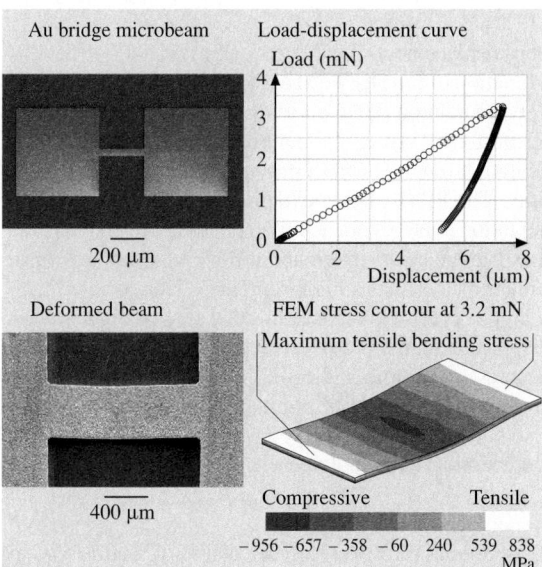

Fig. 41.19 SEM micrographs of the new and deformed beams, load displacement curve and FEM stress contour for the Au bridge microbeam [41.64]

easy thermal processing of the material during the microfabrication process. PMMA was chosen due to its wide

use in commercial biomedical applications. Table 41.3 summarizes published data on these materials.

Hardness and Elastic Modulus

The hardness (H), elastic modulus (E) and creep of PPMA and PMMA beams were measured at the supported region of the beams [41.65]. In Fig. 41.20, the indentation location, where the H, E and creep were measured, is indicated by an arrow. Figure 41.21a shows the H and E of 2.9 μm PPMA and 3.4 μm PMMA beams as a function of contact depth [41.65]. The H and E were calculated by averaging the H and E values obtained at contact depth of 100 nm from five indents. The H (0.41 GPa) and E (5.0 GPa) of the PMMA beams are both higher than H (0.19 GPa) and E (3.7 GPa) of the PPMA beams. Table 41.3 compares the present data with the published data. We note that the E of PMMA is comparable to the published data. It should be noted that the H and E were also measured on 3 μm PPMA and 3 μm PMMA thin films/SU8-25, and the results were very close to the values obtained from the beams.

Indentation Creep

In the creep tests, the changes in indentation displacement, mean stress (hardness) and contact stiffness were monitored [41.65]. Figure 41.21b shows the indenta-

Part D | 41.2

Table 41.3 Summary of physical properties of PPMA and PMMA materials

Structure	T_g (°C)	Water absorption after 24 h (%)	Thermal expansion coefficient (K^{-1})	Thermal Conduc. (0–50 °C) (W/mK)	Elastic modulus (E) (GPa) Bulk (literature)	Nano-indentation (literature)	Nano-indentation (measured)	Beam bending	Hardness (H) (measured) (MPa)
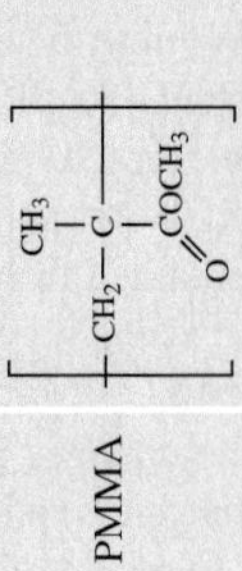 PMMA CH$_3$–C(CH$_2$–)COCH$_3$	104–106[1]	0.3–0.4[3]	2–3×10^{-4} (<T_g)[5]	0.193[6]	3.1–3.3[1]	4.43[7]	5.0	2.0	410
PPMA CH$_2$C(CH$_3$)COCH$_2$CH$_2$CH$_3$	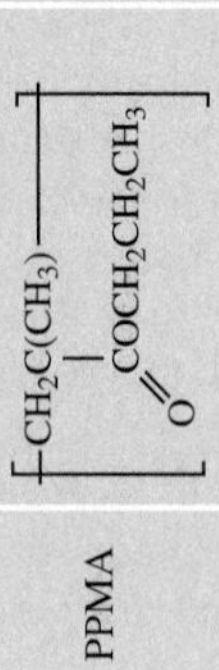 35–43[2]	0.1–0.3[4]	6×10^{-4} (>T_g)[4]	–	–	–	3.7	0.7	190

[1] [41.80,81] [2] [41.79] [3] [41.95] [4] [41.96] [5] [41.97] [6] [41.98,99] [7] [41.100]

Fig. 41.20 SEM images of PPMA beams at low and high magnifications

Bending location

Indentation location (H, E, creep)

tion displacement, mean stress and contact stiffness as a function of time for PPMA and PMMA beams [41.65]. The indentation depth corresponding to 30 μN is about 60 nm for PPMA and about 30 nm for PMMA. The indentation depths of both polymer beams increase with time. The PPMA beam exhibits a faster increase in indentation depth (from about 60 nm to about 90 nm) than the PMMA beam (from about 30 nm to about 50 nm). This indicates that a higher hardness is associated with a higher creep resistance. In contrast with indentation displacement, the mean stresses of both polymer beams decrease with time, indicating that stress relaxation occurred during the hold segment. The creep tests were also conducted on the polymer thin films and the results were similar to the polymer beams.

Scratch Resistance

The polymer films were annealed at 95 °C and 175 °C to PPMA and PMMA respectively to simulate the thermal processing used in the micromolding process for polymer beam fabrication. The annealing treatment seemed to effectively simulate the beam fabrication process, because the H, E and creep measurements show that the PPMA and PMMA thin films have similar mechanical properties with PPMA and PMMA beams. In order to

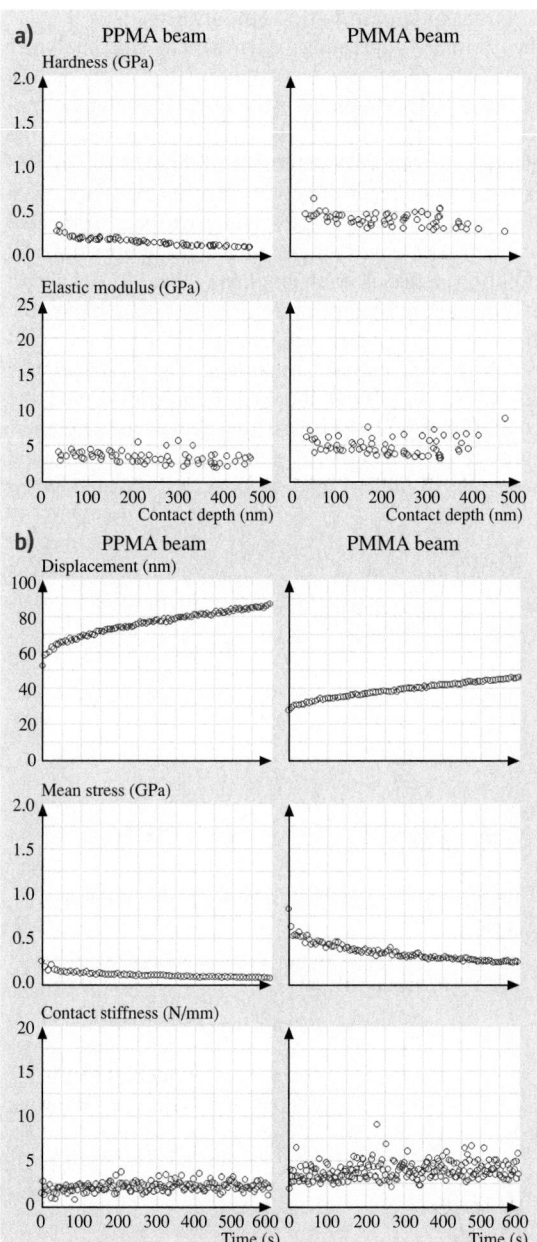

Fig. 41.21a–c (**a**) Hardness and elastic modulus of 2.9 μm thick PPMA and 3.4 μm thick PMMA beams as a function of contact depth, (**b**) indentation displacement, mean stress and contact stiffness as a function of time for 2.9 μm thick PPMA and 3.4 μm thick PMMA beams

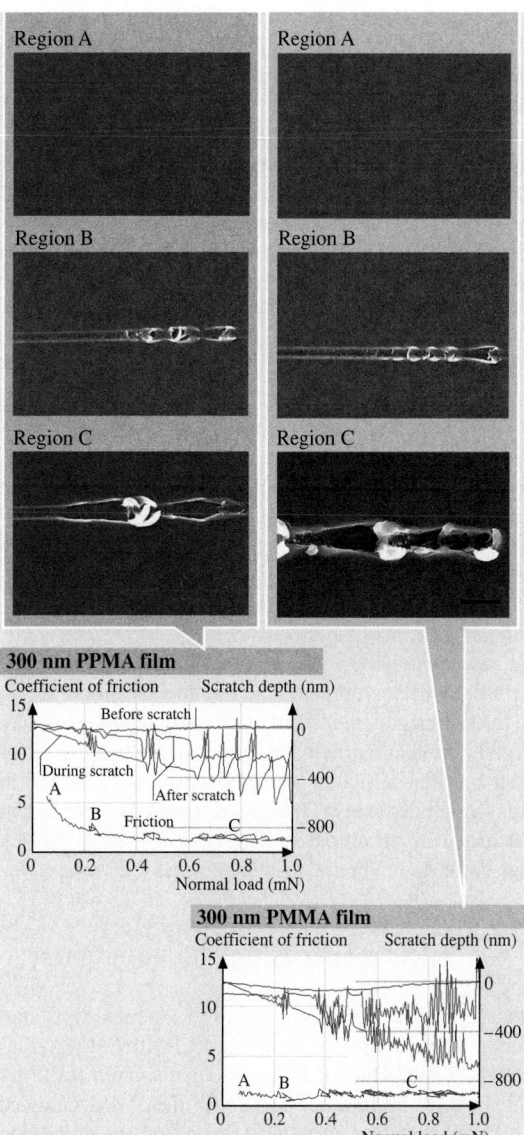

Fig. 41.21 (**c**) scratch depth profiles and coefficient of friction as a function of increasing normal load and SEM images of three regions over scratches: at the beginning of the scratch (indicated by A on the scratch depth profile), and at the point of initiation of damage (indicated by B on the scratch depth profile) and at the end of scratch (indicated by C on the scratch depth profile) for PPMA and PMMA thin films (after [41.65])

evaluate the scratch resistance and adhesion of PPMA and PPMA beams/SU8-25 substrate, nanoscratch meas-

urements were needed. Since polymer beams were too narrow (width 5 μm) to perform nanoscratch tests, the nanoscratch experiments were conducted on 300 nm PPMA and PMMA thin films/SU8-25, assuming that they have similar scratch resistance and adhesion with PPMA and PMMA beams.

Figure 41.21c shows that the scratch depth profiles, coefficient of friction, and SEM images of three regions over scratches: at the beginning of the scratch (indicated by A on the scratch depth profile), at the point of initiation of damage (indicated by B on the scratch depth profile), and towards the end of the scratch (indicated by C on the scratch depth profile) for PPMA and PMMA thin films [41.65]. The scratch results of polymer thin films are very different from metallic and ceramic thin films [41.82]. Firstly, the coefficient of friction value is very high, even greater than one, and the coefficient of friction profile jumps up and down frequently. Secondly, after the polymer thin films were damaged, the scratch depth profile also jumps up and down considerably, indicating the scratch tip moved up and down during scratching. The SEM images of PPMA show that during scratching, the materials were plowed and accumulated in front of the tip, instead of being pushed aside, because the polymers were so soft. When the plowed materials reached certain amount, the tip was almost stuck, so the tip jumped up instead of ramping down. This may explain the oscillation in the scratch depth profile. At region B., the coefficient of friction increased and the in situ scratch depth also changed abruptly, indicating that the film was delaminated. The critical load of PPMA and PMMA are about the same, around 0.22 mN.

41.2.5 Bending Tests of Polymeric Microbeams Using a Nanoindenter

As mentioned earlier, to avoid tip penetration into polymer beams, the tip was dip-coated with PMMA. Figure 41.22 shows the load–displacement curves of PPMA and PMMA beams measured with the PMMA coated tip [41.65] loading curve was linear and the unloading curve does not exhibit much plastic deformation, which means that the bending tests were performed mostly in the elastic region of the samples and the elastic modulus can be calculated from the loading curve. Any slight degree of deformation of the tip coating during the bending tests was assumed negligible compared to beam bending. Based on Fig. 41.22, the calculated elastic modulus for PPMA and PMMA beams are 0.7 ± 0.1 GPa and 2.0 ± 0.1 GPa, respectively. The standard deviation was calculated based on five measurements. Calculated elastic moduli of PPMA and PMMA beams from bending tests are lower than the values obtained from indentation.

Effect of Soaking and Temperature

The primary operating environment for polymer BioMEMS is an aqueous solution (e.g., cell culture medium). Absorption of water by the polymer matrix has the potential to change the properties of the material. In order to determine if immersion in water has a significant effect on material behavior, tests need to be run on samples that have been subjected to an aqueous environment for a considerable time. In addition, in the case of employing BioMEMS for in vivo or elevated temperature in vitro applications, the effect of human body temperature (37.5 °C) on polymer properties may be important. The T_g of PPMA (35–43 °C) is around the body temperature and the T_g of PMMA (104–109 °C) is above body temperature. It is necessary to study how the mechanical properties of PPMA and PMMA are affected when

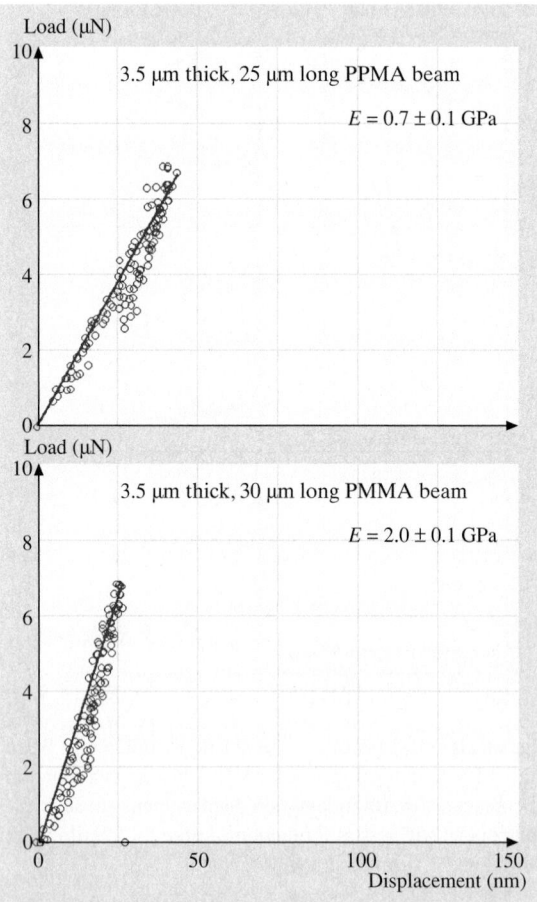

Fig. 41.22 Bending results on PPMA and PMMA beams with PMMA coated diamond conical tip (after [41.65])

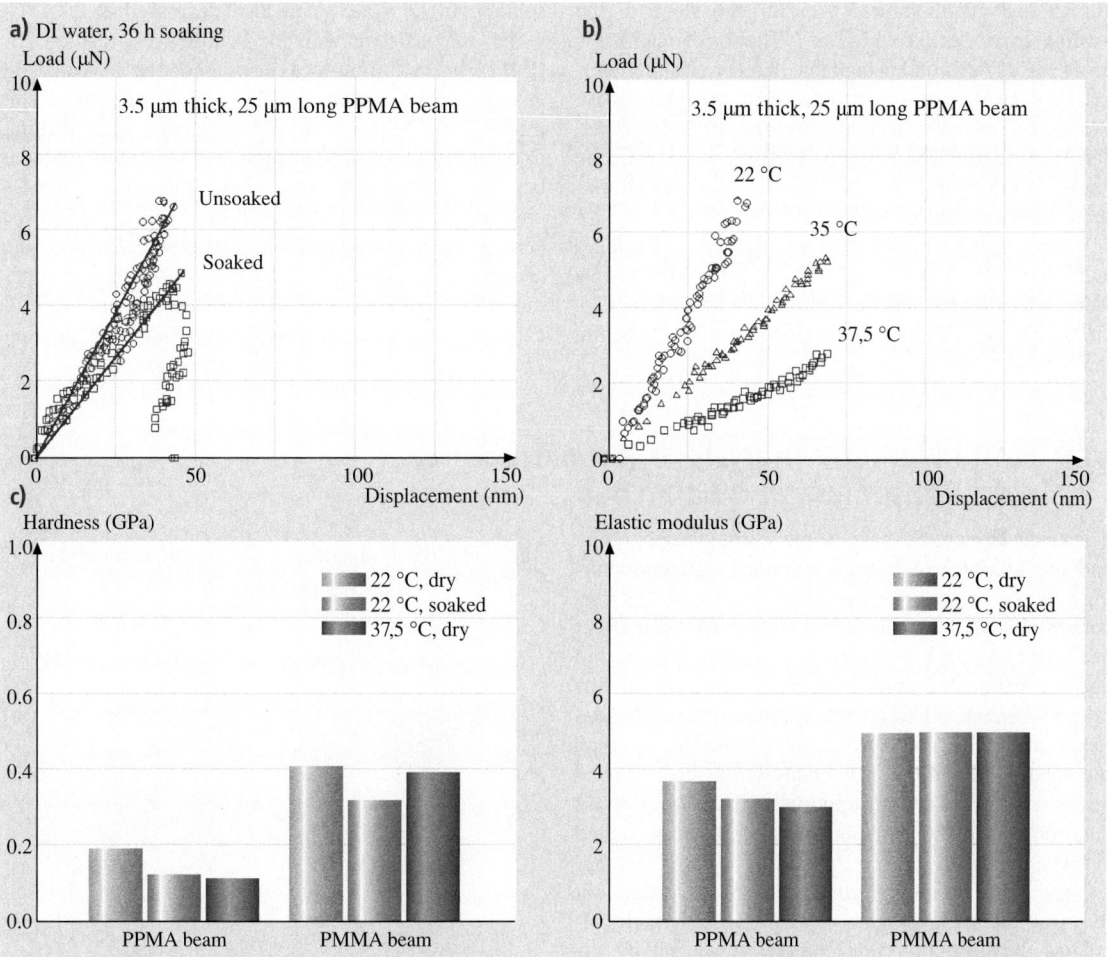

Fig. 41.23 (a) Soaking effect on bending of PMMA beam, **(b)** temperature effect on bending of PPMA beam, and **(c)** soaking and temperature effects on H and E of PPMA and PMMA beams [41.65]

operating at body temperature as compared to room temperature.

To simulate the aqueous working environment of the polymer BioMEMS, the bending tests were also performed on soaked samples in addition to the dry samples and the results were compared. To make the soaked samples, the PMMA and PPMA beams were soaked in DI water for 36 hours before the bending tests. In addition, the nanoindentation tests were conducted in a range of temperature (22–37.5 °C) to study the effect of temperature on nanomechanical properties of polymer beams. In order to heat the sample, the polymer beam sample was placed on a heating stage, which can increase the temperature up to 200 °C.

Figure 41.23a shows the effect of soaking on elastic modulus of PPMA beams. After soaking, the calculated elastic modulus (0.5 ± 0.1 GPa) of PPMA was lower than the elastic modulus (0.7 ± 0.1 GPa) obtained at dry conditions [41.65]. But for PMMA, after soaking, the calculated elastic modulus does not change (figure not shown). The more open structure of PPMA versus PMMA might allow more water to penetrate into the matrix, thus having a greater effect on the modulus than with PMMA (closed, hydrophobic structure). The open structure of PPMA is due to the larger side chain compared to the side chain of PMMA.

Figure 41.23b shows the effect of temperature on load–displacement curves of PPMA beams. As the temperature increased from room temperature (22 °C) to human body temperature (37.5 °C), the slope of the loading curve decreased [41.65]. The T_g of PMMA is 35–43 °C, so it is understandable that the elastic modulus, thus the slope of the loading curve, would decrease when the temperature increased beyond 35 °C. However, since the T_g of PMMA is about 100 °C, at 37.5 °C, the load–displacement curve of PMMA beam was similar to 22 °C (figure not shown).

The soaking and temperature effects on hardness and elastic modulus of PPMA and PMMA beams were also studied and the results are shown in Figure 41.23c [41.65]. It can be seen that for PPMA beam, the hardness and elastic modulus decreased after 36 h

soaking in DI water, and also decreased at 37.5 °C, which is consistent with the bending test results. For PMMA beam, the hardness decreased after soaking, and also decreased slightly at 37.5 °C. However, the elastic modulus of PMMA beam did not change after soaking or heating up to 37.5 °C, which is in agreement with the bending results. The lower elastic modulus of PPMA compared to PMMA makes it an attractive option for certain applications, such as cantilevers. However, the changes in the properties of PPMA as a function of aqueous environment and temperature would require careful calibration at operating conditions when implementing PPMA microstructures in a biological setting.

41.3 Finite Element Analysis of Nanostructures with Roughness and Scratches

Micro/nanostructures have some surface topography and local scratches dependent upon the manufacturing process. Surface roughness and local scratches may compromise the reliability of the devices and their effect needs to be studied. Finite element modeling is used to perform parametric analysis to study the effect of surface roughness and scratches in different well defined forms on tensile stresses which are responsible for crack propagation [41.71, 72]. The analysis has been carried out on trapezoidal beams supported at the bottom whose data (on Si and SiO_2 nanobeams) have been presented earlier.

The finite element analysis has been carried out by using the static analysis of ANSYS 5.7 which calculates the deflections and stresses produced by the applied loading. The type of element selected for the present study was SOLID95 type which allows the use of different shapes without much loss of accuracy. This element is 3-D with 20 nodes, each node having three degrees of freedom which implies translation in the x, y and z directions. The nanobeam cross section is divided into six elements each along the width and the thickness and forty elements along the length. SOLID95 has plasticity, creep, stress stiffening, large

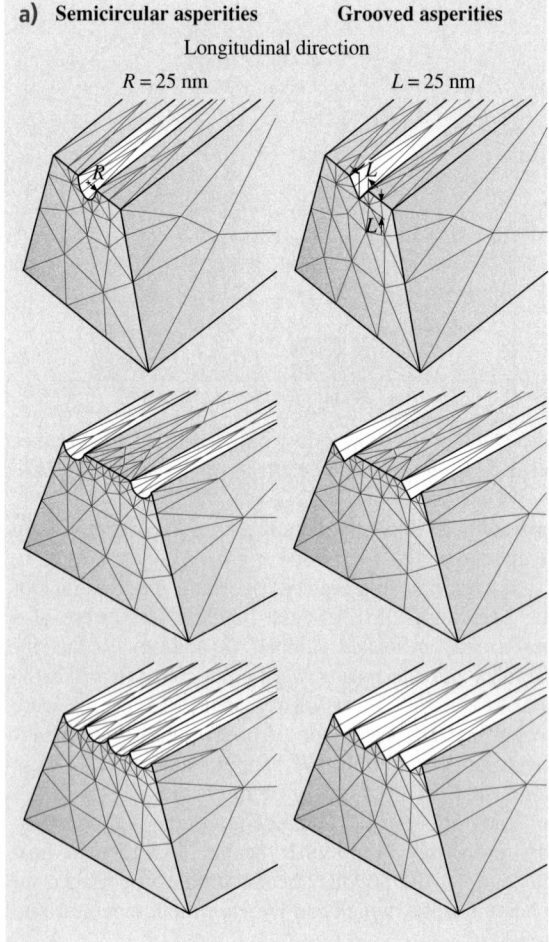

Fig. 41.24 (a) Plots showing the geometries of modeled roughness–semicircular and grooved asperities along the nanobeam length with defined geometrical parameters, (b) Schematic showing semicircular asperities and scratches in the transverse direction followed by the illustration of the mesh created on the beam with fine mesh near the asperities and the scratches. Also shown are the semicircular asperities and scratches at different pitch values

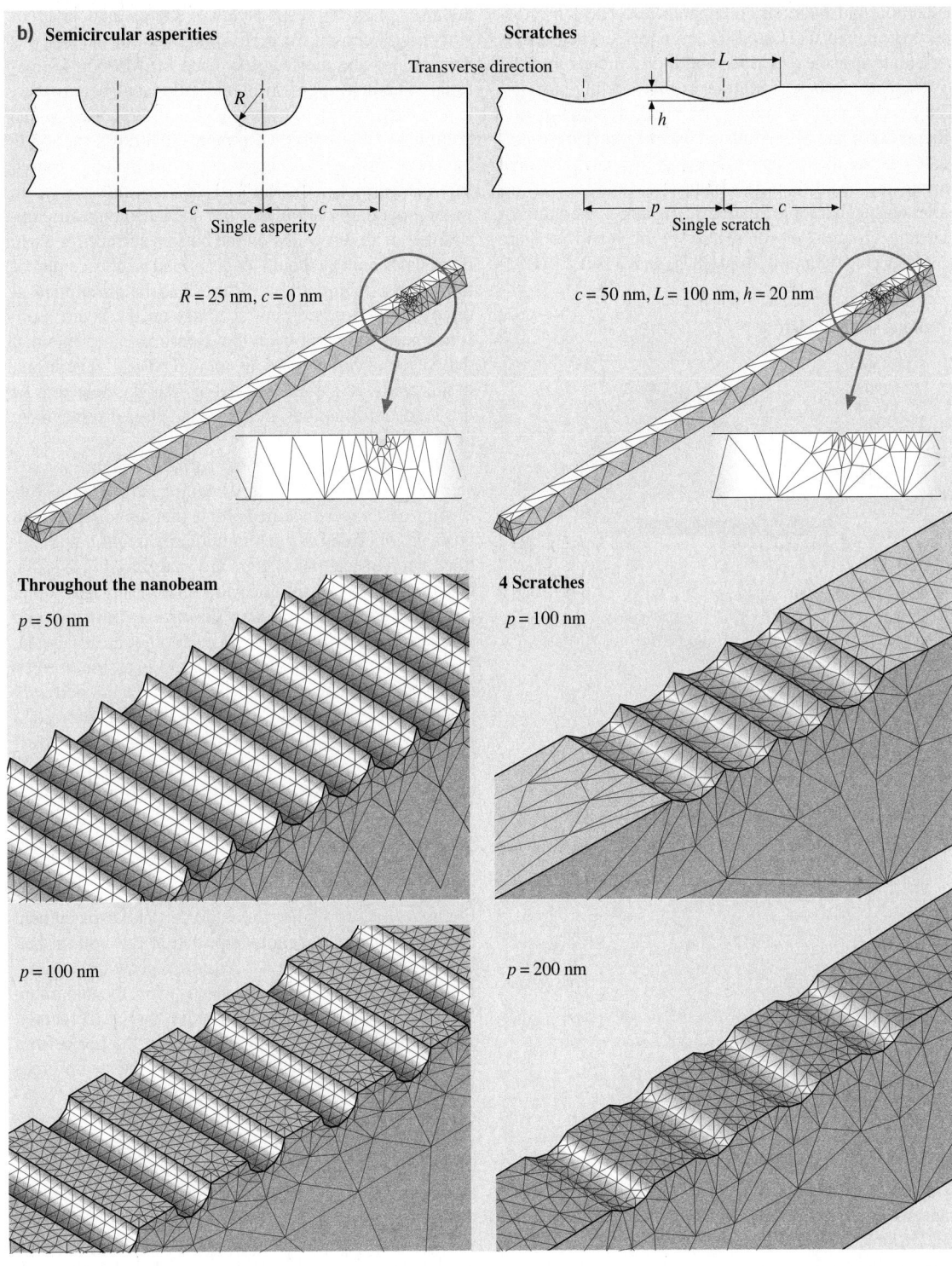

deflection and large strain capabilities. The large displacement analysis is used for large loads. The mesh is kept finer near the asperities and the scratches in order to take into account variation in the bending stresses. The beam materials studied are made of single-crystal silicon (110) and SiO$_2$ films whose data have been presented earlier. Based on bending experiments presented earlier, the beam materials can be assumed to be linearly elastic isotropic materials. Young's modulus of elasticity (E) and Poisson's ratio (ν) for Si and SiO$_2$ are 169 GPa [41.91] and 0.28 [41.87], and 73 GPa [41.92]

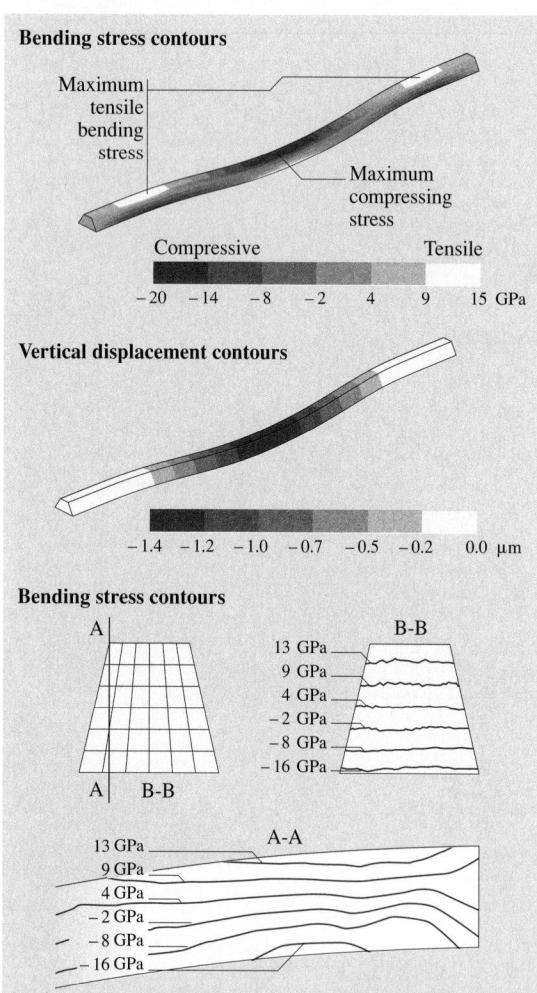

Fig. 41.25 Bending stress contours, vertical displacement contours, and bending stress contours after loading trapezoidal Si nanobeam ($w_1 = 200$ nm, $w_2 = 370$ nm, $t = 255$ nm, $\ell = 6$ μm, $E = 169$ GPa, $\nu = 0.28$) at 70 μN load [41.72]

and 0.17 [41.101], respectively. A sample nanobeam of silicon was chosen for performing most of the analysis as silicon is the most widely used MEMS/NEMS material. The cross section of the fabricated beams used in the experiment is trapezoidal and supported at the bottom so nanobeams with trapezoidal cross section is modeled, Fig. 41.10. The following dimensions are used $w_1 = 200$ nm, $w_2 = 370$ nm, $t = 255$ nm, and $P = 6$ μm. In the boundary conditions, the displacements are constrained in all directions on the bottom surface for 1 μm from each end. A point load applied at the center of the beam is simulated with the load being applied at three closely located central nodes on the beam used. It has been observed from the experimental results that the Si nanobeam breaks at around 80 μN. Therefore, in this analysis, a nominal load of 70 μN is selected. At this load, deformations are large, and large displacement option is used.

To study the effect of surface roughness and scratches on the maximum bending stresses the following cases were studied. First the semicircular and grooved asperities in the longitudinal direction with defined geometrical parameters are analyzed, Fig. 41.24a. Next semicircular asperities and scratches placed along the transverse direction at a distance c from the end and separated by pitch p from each other are analyzed, Fig. 41.24b. Lastly the beam material is assumed to be either purely elastic, elastic–plastic, or elastic–perfectly plastic. In the following, we begin with the stress distribution in smooth nanobeams followed by the effect of surface roughness in the longitudinal and transverse directions and scratches in the transverse direction.

41.3.1 Stress Distribution in a Smooth Nanobeam

Figure 41.25 shows the stress and vertical displacement contours for a nanobeam supported at the bottom and loaded at the center, [41.71, 72]. As expected, the maximum tensile stress occurs at the ends while the maximum compressive stress occurs under the load at the center. Stress contours obtained at a section of the beam from the front and side are also shown. In the beam cross section, the stresses remain constant at a given vertical distance from one side to another and change with a change in vertical location. This can be explained due to the fact that the bending moment is constant at a particular cross section so the stress is only dependent on the distance from the neutral axis. However, in cross section A-A the high tensile and compressive stresses are localized near the end of the beam at top and bottom, respectively, whereas the lower values are spread out away from the ends. High value of tensile stresses occurs near the ends because of high bending moment.

41.3.2 Effect of Roughness in the Longitudinal Direction

The roughness in the form of semicircular and grooved asperities in the longitudinal direction on the maximum bending stresses are analyzed [41.72]. The radius R and depth L are kept fixed at 25 nm while the number of asperities is varied and their effect is observed on the maximum bending stresses. Figure 41.26 shows the variation of maximum bending stresses as a function of asperity shape and the number of asperities. The maximum bending stresses increase as the asperity number increases for both semicircular and grooved asperities. This can be attributed to the fact that as asperity number increases, the moment of inertia decreases for that cross section. Also the distance from the neutral axis increases because neutral axis shifts downwards. Both these factors lead to the increase in the maximum bending stresses and this effect is more pronounced in the case of semicircular asperity as it exhibits a higher value of maximum bending stress than that in grooved asperity. Figure 41.26 shows the stress contours obtained at a section of the beam from front and from the side for both cases when we have a single semicircular asperity and when four adjacent semicircular asperities are present. Trends are similar to that observed earlier for a smooth nanobeam (Fig. 41.25).

41.3.3 Effect of Roughness in the Transverse Direction and Scratches

We analyze semicircular asperities when placed along the transverse direction followed by the effect of scratches on the maximum bending stresses in varying numbers and different pitch [41.72]. In the analysis of semicircular transverse asperities three cases were considered which included a single asperity, asperities throughout the nanobeam surface separated by pitch equal to 50 nm and pitch equal to 100 nm. In all of these cases, c value was kept equal to 0 nm. Figure 41.27 shows that the value of maximum tensile stress is 42 GPa which is much larger than the maximum tensile stress value with no asperity of 16 GPa or when the semicircular asperity is present in the longitudinal direction. It is also observed that the maximum tensile stress does not vary with the number of asperities or the pitch while the

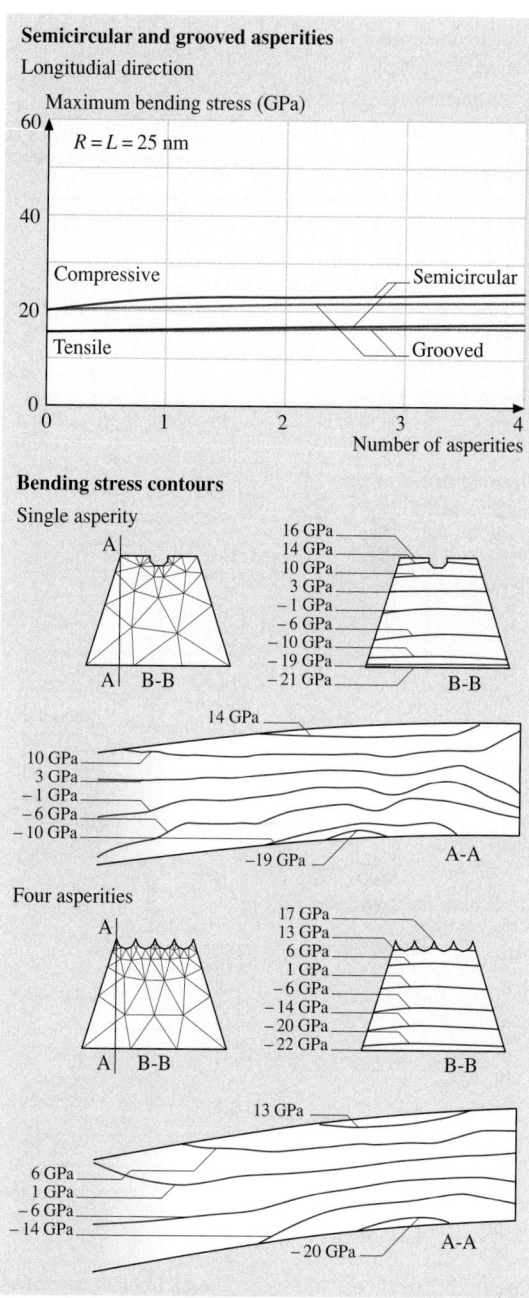

Fig. 41.26 Effect of longitudinal semicircular and grooved asperities in different numbers on maximum bending stresses after loading trapezoidal Si nanobeams ($w_1 = 200$ nm, $w_2 = 370$ nm, $t = 255$ nm, $\ell = 6\,\mu$m, $E = 169$ GPa, $\nu = 0.28$, Load $= 70\,\mu$N). Bending stress contours obtained in the beam with semicircular single asperity and four adjacent asperities of $R = 25$ nm [41.72]

Part D | 41.3

Part D | 41.3

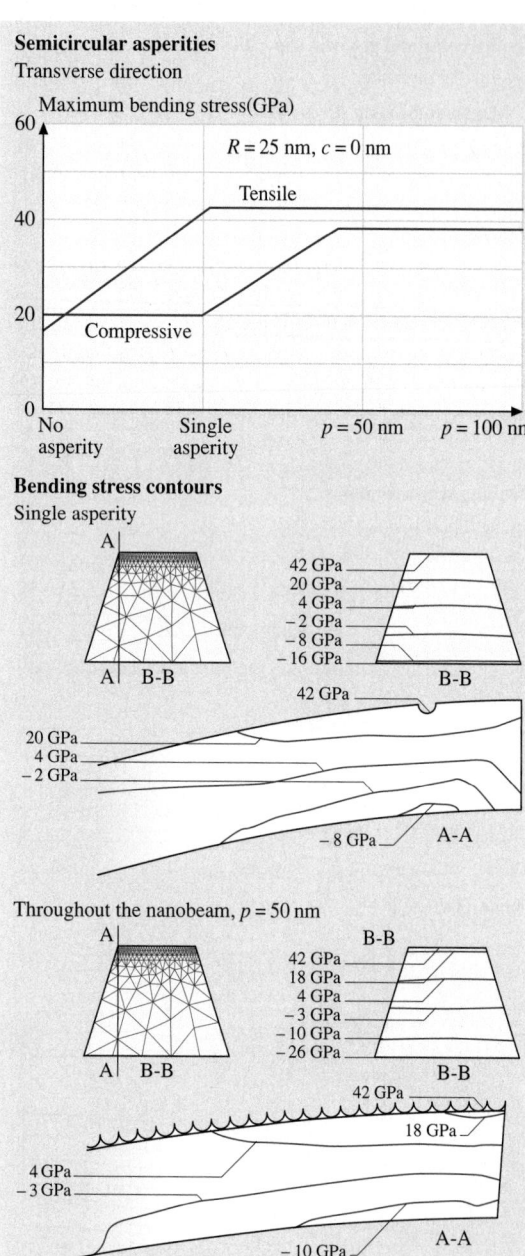

Fig. 41.27 Effect of transverse semicircular asperities located at different pitch values on the maximum bending stresses after loading trapezoidal Si nanobeams ($w_1 = 200$ nm, $w_2 = 370$ nm, $t = 255$ nm, $\ell = 6\,\mu$m, $E = 169$ GPa, $v = 0.28$, Load $= 70\,\mu$N). Bending stress contours obtained in the beam with semicircular single asperity and semicircular asperities throughout the nanobeam surface at $p = 50$ nm [41.72]

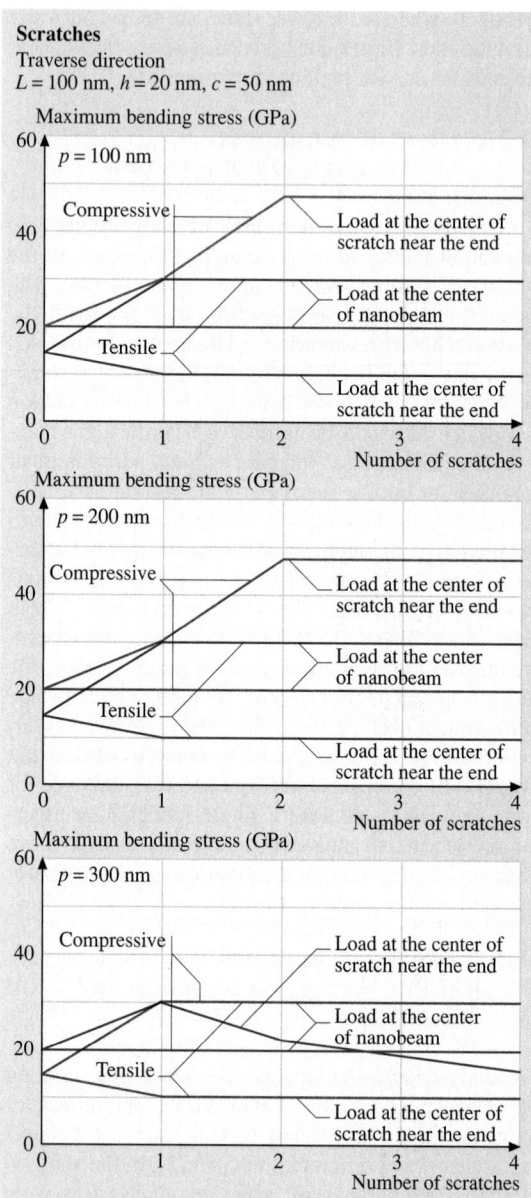

Fig. 41.28 Effect of number of scratches along with the variation in the pitch on the maximum bending stresses after loading trapezoidal Si nanobeams ($w_1 = 200$ nm, $w_2 = 370\,v\mu$, $\tau = 255\,v\mu$, $\ell = 6\,\mu$m, $E = 169$ GPa, $v = 0.28$, Load $= 70\,\mu$N.) Also shown is the effect of load when applied at the center of the beam and at the center of the scratch near the end [41.72]

maximum compressive stress does increase dramatically for the asperities present throughout the beam surface from its value when a single asperity is present. Maximum tensile stress occurs at the ends and an increase in p does not add any asperities at the ends whereas asperities are added in the central region where compressive stresses are maximum. The semicircular asperities present at the center cause the local perturbation in the stress distribution at the center of the asperity where load is being applied leading to a high value of maximum compressive stress [41.101]. Figure 41.27 also shows the stress contours obtained at a section of the beam from front and from the side for both cases when there is a single semicircular asperity and when asperities are present throughout the beam surface at a pitch equal to 50 nm. Trends are similar to that observed earlier for a smooth nanobeam (Fig. 41.25).

In the study pertaining to scratches, the number of scratches are varied along with the variation in the pitch as well. Furthermore the load is applied at the center of the beam and at the center of the scratch near the end as well for all the cases. In all of these c value was kept equal to 50 nm and L value was equal to 100 nm with h value being 20 nm. Figure 41.28 shows that the value of maximum tensile stress remains almost the same with the number of scratches for both types of loading that is when load is applied at the center of the beam and at the center of the scratch near the end. This is because that the maximum tensile stress occurs at the beam ends no matter where the load gets applied. But the presence of scratch does increase the maximum tensile stress as compared to its value for a smooth nanobeam although the number of scratches no longer matter as the maximum tensile stress occurring at the nanobeam end is unaffected by the presence of more scratches beyond the first scratch in the direction towards the center. The value of the tensile stress is much lower when the load is applied at the center of scratch and it can be explained as follows. The negative bending moment at the end near the load applied decreases with load offset after two-thirds of the length of the beam [41.102]. Since this negative bending moment is responsible for tensile stresses, their behavior with the load offset is the same as the negative bending moment. Also the value of maximum compressive stress when load is applied at the center of the nanobeam remains almost the same as the center geometry is unchanged due to the number of scratches and hence the maximum compressive stress occurring below the load at the center is same. On the other hand when the load is applied at the center of the scratch we observe that the maximum compressive stress increases dramatically because the local perturbation in the stress distribution at the center of the scratch where load is being applied leads

to a high value of maximum compressive stress [41.101]. It increases further with the number of scratches and then levels off. This can be attributed to the fact that when there is another scratch present close to the scratch near the end, the stress concentration is more as the effect of local perturbation in the stress distribution is more significant. However, this effect is insignificant when more than two scratches are present.

Now we address the effect of pitch on the maximum compressive stress when the load is applied at the center of the scratch near the end. When the pitch is up to a value of 200 nm the maximum compressive stress increases with the number of scratches as discussed earlier. On the other hand, when the pitch value goes beyond 225 nm this effect is reversed. This is because the presence of another scratch no longer affects the local perturbation in the stress distribution at the scratch near the end. Instead more scratches at a fair distance distribute the

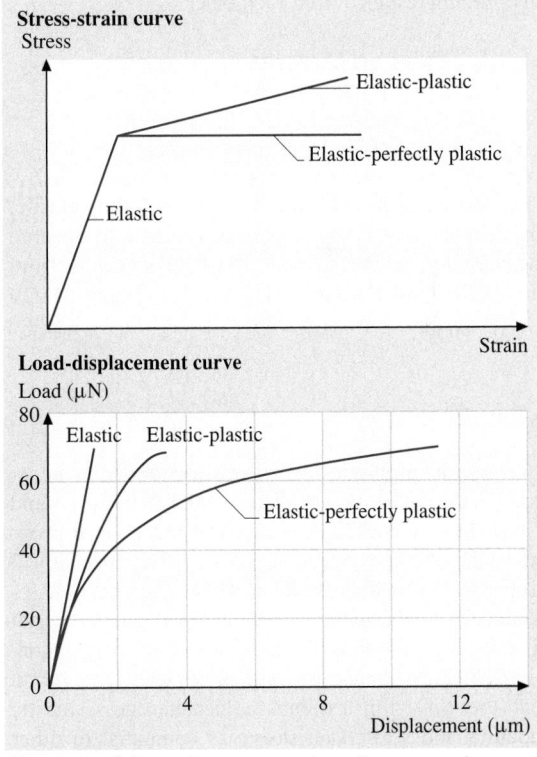

Fig. 41.29 Schematic representation of stress–strain curves and load-displacement curves for material when it is elastic, elastic-plastic, or elastic-perfectly plastic for a Si nanobeam ($w_1 = 200$ nm, $w_2 = 370$ nm, $t = 255$ nm, $\ell = 6$ μm, $E = 169$ GPa, tangent modulus in plastic range $= 0.5E$, $\nu = 0.28$) [41.72]

Table 41.4 Stresses and displacements for materials that are elastic, elastic–plastic or elastic–perfectly plastic (load $= 70\,\mu\mathrm{N}$, $w_1 = 200\,\mathrm{nm}$, $w_2 = 370\,\mathrm{nm}$, $t = 255\,\mathrm{nm}$, $\ell = 6\,\mu\mathrm{m}$, $R = 25\,\mathrm{nm}$, $E = 169\,\mathrm{GPa}$, tangential modulus in plastic range $= 0.5E$, $\nu = 0.28$)

	Elastic		Elastic–plastic		Elastic–perfectly plastic	
	Smooth nanobeam	Single semicircular longitudinal asperity	Smooth nanobeam	Single semicircular longitudinal asperity	Smooth nanobeam	Single semicircular longitudinal asperity
Max. von Mises stress (GPa)	18.2	19.3	13.5	15.2	7.8	9.1
Max. displacement (μm)	1.34	1.40	3.35	3.65	11.5	12.3

maximum compressive stress at the scratch near the end and the stress starts going down. Such observations of maximum bending stresses can help in identifying the number of asperities and scratches allowed separated by an optimum distance from each other.

41.3.4 Effect on Stresses and Displacements for Materials Which are Elastic, Elastic–Plastic or Elastic– Perfectly Plastic

This section deals with the beam modeled as elastic, elastic–plastic and elastic–perfectly plastic to observe the variation in the stresses and displacements from an elastic model used so far [41.72]. Figure 41.29 shows the typical stress-strain curves for the three types of deformation regimes and their corresponding load–displacement curves obtained from the model of an Si nanobeam which are found to exhibit same trends.

Table 41.4 shows the comparison of maximum von Mises stress and maximum displacements for both smooth nanobeam and nanobeam with a defined roughness which is single semicircular longitudinal asperity of R value equal to 25 nm for the three different models. It is observed that the maximum value of stress is obtained at a given load for elastic material whereas the displacement is maximum for elastic–perfectly plastic material. Also the pattern that the maximum bending stress value increases for a rough nanobeam still holds true in the other models as well.

41.4 Closure

Mechanical properties of nanostructures are necessary in designing realistic MEMS/NEMS and BioMEMS/BioNEMS devices. Most mechanical properties are scale dependent. Micro/Nanomechanical properties, hardness, elastic modulus, and scratch resistance of bulk materials of undoped single-crystal silicon (Si) and thin films of undoped polysilicon, SiO_2, SiC, Ni-P, and Au are presented. It is found that the SiC film exhibits higher hardness, elastic modulus and scratch resistance as compared to other materials.

Bending tests have been performed on the Si and SiO_2 nanobeams and Ni-P, Au, PPMA and PMMA microbeams using an AFM and a depth-sensing nanoindenter, respectively. The bending tests were used to evaluate elastic modulus, bending strength (fracture stress), fracture toughness (K_{IC}), and fatigue strength of the beam materials. The Si and SiO_2 nanobeams exhibited elastic linear response with sudden brittle fracture. The notched Ni-P beam showed linear deformation behavior followed by abrupt failure. The Au beam showed elastic–plastic deformation behavior. Elastic modulus values of 182 ± 11 GPa for Si(110) and 85 ± 3 GPa for SiO_2 were obtained, which are comparable to bulk values. Bending strength values of 18 ± 3 GPa for Si and 7.6 ± 2 GPa for SiO_2 were obtained, which are twice as large as values reported on larger scale specimens. This indicates that bending strength shows a specimen size dependence. Fracture toughness value estimates obtained were 1.67 ± 0.4 MPa$\sqrt{m}$ for Si and $0.6. \pm 0.2$ MPa$\sqrt{m}$ for SiO_2, which are also comparable to values obtained on larger specimens. At stress amplitudes less than 15% of their bending strength and at mean stresses of less than 30% of the bending strength, Si and SiO_2 displayed an apparent endurance life of greater than 30 000 cycles. SEM observations of the fracture surfaces revealed a cleavage type of fracture for both materials when subjected to bending as well as fatigue.

The hardness, elastic modulus and creep behavior of PPMA and PMMA polymeric beams are also presented. The hardness (0.41 GPa), elastic modulus (5.0 GPa), and creep resistance of PMMA beams are higher than the hardness (0.19 GPa, elastic modulus (3.7 GPa), and creep resistance of PPMA beams. The scratch behavior of PPMA and PMMA films is different from metallic and ceramic films in that the in situ displacement oscillated and coefficient of friction are greater than one during the scratch test. This might be because the polymer is so soft and easily plowed. After 36 hours soaking in DI water, the hardness and elastic modulus of PPMA beam were decreased. The soaking seems to have had no effect on elastic modulus of PMMA beam. At 37.5 °C, the hardness and elastic modulus of PPMA beam were decreased as compared to room temperature, while this temperature appears to have no effect on elastic modulus of PMMA beam.

The AFM and nanoindenters used in this study can be satisfactorily used to evaluate the mechanical properties of micro/nanoscale structures for use in MEMS/NEMS.

FEM simulations are used to predict the stress and deformation in nanostructures. The FEM has been used to analyze the effect of type of surface roughness and scratches on stresses and deformation of nanostructures. We find that roughness affects the maximum bending stresses. The maximum bending stresses increase as the asperity number increases for both semicircular and grooved asperities in longitudinal direction. When the semicircular asperity is present in the transverse direction the maximum tensile stress is much larger than the maximum tensile stress value with no asperity or when the semicircular asperity is present in the longitudinal direction. This observation suggests that the asperity in the transverse direction is more detrimental. The presence of scratches increases the maximum tensile stress. The maximum tensile stress remains almost the same with the number of scratches for two types of loading, that is, when load is applied at the center of the beam or at the center of the scratch near the end, although the value of the tensile stress is much lower when the load is applied at the center of the scratch. This means that the load applied at the ends is less damaging. This analysis shows that FEM simulations can be useful to designers to develop the most suitable geometry for nanostructures.

References

41.1   R. S. Muller, R. T. Howe, S. D. Senturia, R. L. Smith, R. M. White: *Microsensors* (IEEE, New York 1990)

41.2   I. Fujimasa: *Micromachines: A New Era in Mechanical Engineering* (Oxford Univ. Press, Oxford 1996)

41.3   W. S. Trimmer (ed.): *Micromachines and MEMS, Classic and Seminal Papers to 1990* (IEEE, New York 1997)

41.4   B. Bhushan: *Tribology Issues and Opportunities in MEMS* (Kluwer, Dordrecht 1998)

41.5   B. Bhushan: *Handbook of Micro/Nanotribology*, 2nd edn. (CRC, Boca Raton 1999)

41.6   B. Bhushan: *Nanotribology and Nanomechanics – An Introduction* (Springer, Heidelberg 2005)

41.7   G. T. A. Kovacs: *Micromachined Transducers Sourcebook* (WCB McGraw-Hill, Boston 1998)

41.8   S. D. Senturia: *Microsystem Design* (Kluwer, Boston 2001)

41.9   M. Elwenspoek, R. Wiegerink: *Mechanical Microsensors* (Springer, Berlin 2001)

41.10 M. Gad-el-Hak: *The MEMS Handbook* (CRC, Boca Raton 2002)

41.11 T. R. Hsu: *MEMS and Microsystems: Design and Manufacture* (McGraw-Hill, Boston 2002)

41.12 M. Madou: *Fundamentals of Microfabrication: The Science of Miniaturization*, 2nd edn. (CRC, Boca Raton 2002)

41.13 A. Hierlemann: *Integrated Chemical Microsensor Systems in CMOS Technology* (Springer, Berlin 2005)

41.14 K. E. Drexler: *Nanosystems: Molecular Machinery, Manufacturing and Computation* (Wiley, New York 1992)

41.15 G. Timp (ed.): *Nanotechnology* (Springer, Berlin, Heidelberg 1999)

41.16 M. S. Dresselhaus, G. Dresselhaus, Ph. Avouris: *Carbon Nanotubes – Synthesis, Structure, Properties and Applications* (Springer, Berlin 2001)

41.17 E. A. Rietman: *Molecular Engineering of Nanosystems* (Springer, Berlin, Heidelberg 2001)

41.18 H. S. Nalwa (ed.): *Nanostructured Materials and Nanotechnology* (Academic, San Diego 2002)

41.19 W. A. Goddard, D. W. Brenner, S. E. Lyshevski, G. J. Iafrate: *Handbook of Nanoscience, Engineering, and Technology* (CRC, Boca Raton 2003)

41.20   A. Manz, H. Becker (eds.): *Microsystem Technology in Chemistry and Life Sciences, Topics in Current Chemistry 194* (Springer, Heidelberg 1998)

41.21   J. Cheng, L.J. Kricka (eds.): *Biochip Technology* (Harwood Academic Publishers, Philadelphia 2001)

41.22   M.J. Heller, A. Guttman (eds.): *Integrated Microfabricated Biodevices* (Dekker, New York, N.Y. 2001)

41.23   C. Lai Poh San, E.P.H. Yap (eds.): *Frontiers in Human Genetics* (World Scientific, Singapore 2001)

41.24   C.H. Mastrangelo, H. Becker (eds.): *Microfluidics and BioMEMS, Proc. of SPIE*, Vol. 4560 (SPIE, Bellingham, Washington 2001)

41.25   H: Becker, L.E. Lacascio: Polymer Microfluidic Devices, Talanta **56**, 267–287 (2002)

41.26   D.J. Beebe, G.A. Mensing, G.M. Walker: Physics and applications of microfluidics in biology, Annu. Rev. Biomed. Eng. **4**, 261–286 (2002)

41.27   A. van der Berg (ed.): *Lab-on-a-Chip: Chemistry in Miniaturized Synthesis and Analysis Systems* (Elsevier, Amsterdam, Netherlands 2003)

41.28   C.P. Poole, F.J. Owens: *Introduction to Nanotechnology* (Wiley, Hoboken, New Jersey 2003)

41.29   J.V. Zoval, M.J. Madou: Centrifuge-based fluidic platforms, Proc. IEEE **92**, 140–153 (2000)

41.30   R. Raiteri, M. Grattarola, H. Butt, P. Skladal: Micromechanical cantilever-based biosensors, Sensors and Actuators B: Chemical **79**, 115–126 (2001)

41.31   W.C. Tang, A.P. Lee: Defense applications of MEMS, MRS Bulletin **26**, 318–319 (2001)

41.32   M.R. Taylor, P. Nguyen, J. Ching, K.E. Peterson: Simulation of microfluidic pumping in a genomic DNA blood-processing cassette, J. Micromech. Microeng. **13**, 201–208 (2003)

41.33   K. Park (ed.): *Controlled Drug Delivery: Challenges and Strategies* (American Chemical Society, Washington, DC 1997)

41.34   R.S. Shawgo, A.C.R. Grayson, Y. Li, M.J. Cima: BioMEMS for drug delivery, Current Opinion in Solid State & Mat. Sci. **6**, 329–334 (2002)

41.35   P.A. Oeberg, T. Togawa, F.A. Spelman: *Sensors in Medicine and Health Care* (Wiley, New York 2004)

41.36   S.N. Bhatia, C.S. Chen: Tissue engineering at the micro-scale, Biomedical Microdevices **2**, 131–144 (1999)

41.37   R.P. Lanza, R. Langer, J. Vacanti (eds.): *Principles of Tissue Engineering* (Academic Press, San Diego 2000)

41.38   E. Leclerc, K.S. Furukawa, F. Miyata, T. Sakai, T. Ushida, T. Fujii: Fabrication of microstructures in photosensitive biodegradable polymers for tissue engineering applications, Biomaterials **25**, 4683–4690 (2004)

41.39   T.H. Schulte, R.L. Bardell, B.H. Weigl: Microfluidic technologies in clinical diagnostics, Clinica Chimica Acta **321**, 1–10 (2002)

41.40   B. Bhushan: *Introduction to Tribology* (Wiley, New York 2002)

41.41   B. Bhushan: Macro- and microtribology of MEMS materials. In: *Modern Tribology Handbook*, ed. by B. Bhushan (CRC, Boca Raton 2001) pp. 1515–1548

41.42   B. Bhushan: *Principles and Applications of Tribology* (Wiley, New York 1999)

41.43   S. Johansson, J.A. Schweitz, L. Tenerz, J. Tiren: Fracture testing of silicon microelements in-situ in a scanning electron microscope, J. Appl. Phys. **63**, 4799–4803 (1988)

41.44   F. Ericson, J.A. Schweitz: Micromechanical fracture strength of silicon, J. Appl. Phys. **68**, 5840–5844 (1990)

41.45   E. Obermeier: Mechanical and thermophysical properties of thin film materials for MEMS: Techniques and devices, Micromechan. Struct. Mater. Res. Symp. Proc. **444**, 39–57 (1996)

41.46   C.J. Wilson, A. Ormeggi, M. Narbutovskih: Fracture testing of silicon microcantilever beams, J. Appl. Phys. **79**, 2386–2393 (1995)

41.47   W.N. Sharpe, Jr., B. Yuan, R.L. Edwards: A new technique for measuring the mechanical properties of thin films, J. Microelectromech. Syst. **6**, 193–199 (1997)

41.48   K. Sato, T. Yoshioka, T. Anso, M. Shikida, T. Kawabata: Tensile testing of silicon film having different crystallographic orientations carried out on a silicon chip, Sens. Actuators A **70**, 148–152 (1998)

41.49   S. Greek, F. Ericson, S. Johansson, M. Furtsch, A. Rump: Mechanical characterization of thick polysilicon films: Young's modulus and fracture strength evaluated with microstructures, J. Micromech. Microeng. **9**, 245–251 (1999)

41.50   D.A. LaVan, T.E. Buchheit: Strength of polysilicon for MEMS devices, Proc. SPIE **3880**, 40–44 (1999)

41.51   E. Mazza, J. Dual: Mechanical behavior of a μm-sized single crystal silicon structure with sharp notches, J. Mech. Phys. Solids **47**, 1795–1821 (1999)

41.52   T. Yi, C.J. Kim: Measurement of mechanical properties for MEMS materials, Meas. Sci. Technol. **10**, 706–716 (1999)

41.53   H. Kahn, M.A. Huff, A.H. Heuer: Heating effects on the Young's modulus of films sputtered onto micromachined resonators, Microelectromech. Struct. Mater. Res. Symp. Proc. **518**, 33–38 (1998)

41.54   S. Johansson, F. Ericson, J.A. Schweitz: Influence of surface-coatings on elasticity, residual-stresses, and fracture properties of silicon microelements, J. Appl. Phys. **65**, 122–128 (1989)

41.55   R. Ballarini, R.L. Mullen, Y. Yin, H. Kahn, S. Stemmer, A.H. Heuer: The fracture toughness of polysilicon microdevices: A first report, J. Mater. Res. **12**, 915–922 (1997)

41.56   H. Kahn, R. Ballarini, R.L. Mullen, A.H. Heuer: Electrostatically actuated failure of microfabricated polysilicon fracture mechanics specimens, Proc. R. Soc. London A **455**, 3807–3823 (1999)

41.57 A. M. Fitzgerald, R. H. Dauskardt, T. W. Kenny: Fracture toughness and crack growth phenomena of plasma-etched single crystal silicon, Sensor. Actuat. A **83**, 194–199 (2000)

41.58 T. Tsuchiya, A. Inoue, J. Sakata: Tensile testing of insulating thin films: Humidity effect on tensile strength of SiO_2 films, Sens. Actuators A **82**, 286–290 (2000)

41.59 J. A. Connally, S. B. Brown: Micromechanical fatigue testing, Exp. Mech. **33**, 81–90 (1993)

41.60 K. Komai, K. Minoshima, S. Inoue: Fracture and fatigue behavior of single-crystal silicon micro-elements and nanoscopic AFM damage evaluation, Microsyst. Technol. **5**, 30–37 (1998)

41.61 T. Namazu, Y. Isono, T. Tanaka: Evaluation of size effect on mechanical properties of single-crystal silicon by nanoscale bending test using AFM, J. Microelectromech. Syst. **9**, 450–459 (2000)

41.62 S. Sundararajan, B. Bhushan: Development of AFM-based techniques to measure mechanical properties of nanoscale structures, Sens. Actuators A **101**, 338–351 (2002)

41.63 X. Li, B. Bhushan: Fatigue studies of nanoscale structures for MEMS/NEMS applications using nanoindentation techniques, Surf. Coat. Technol. **163-164**, 521–526 (2003)

41.64 X. Li, B. Bhushan, K. Takashima, C. W. Baek, Y. K. Kim: Mechanical characterization of micro/nanoscale structures for MEMS/NEMS applications using nanoindentation techniques, Ultramicroscopy **97**, 481–494 (2003)

41.65 G. Wei, B. Bhushan, N. Ferrell, D. Hansford: Microfabrication and nanomechanical characterization of polymer MEMS for biological applications, J. Vac. Sci. Technol. A **23**, 811–819 (2005)

41.66 T. Hsu, N. Sun: Residual stresses/strains analysis of MEMS. In: *Proc. Int. Conf. on Modeling and Simulation of Microsystems, Semiconductors, Sensors and Actuators*, ed. by M. Laudon, B. Romanowicz (Computational Publications, Cambridge 1998) pp. 82–87

41.67 A. Kolpekwar, C. Kellen, R. D. (Shawn) Blanton: Fault model generation for MEMS. In: *Proc. Int. Conf. on Modeling and Simulation of Microsystems, Semiconductors, Sensors and Actuators*, ed. by M. Laudon, B. Romanowicz (Computational Publications, Cambridge 1998) pp. 111–116

41.68 H. A. Rueda, M. E. Law: Modeling of strain in boron-doped silicon cantilevers. In: *Proc. Int. Conf. on Modeling and Simulation of Microsystems, Semiconductors, Sensors and Actuators*, ed. by M. Laudon, B. Romanowicz (Computational Publications, Cambridge 1998) pp. 94–99

41.69 M. Heinzelmann, M. Petzold: FEM analysis of microbeam bending experiments using ultra-micro indentation, Comput. Mater. Sci. **3**, 169–176 (1994)

41.70 C. J. Wilson, P. A. Beck: Fracture testing of bulk silicon microcantilever beams subjected to a side load, J. Microelectromech. Syst. **5**, 142–150 (1996)

41.71 B. Bhushan, G. B. Agrawal: Stress analysis of nanostructures using a finite element method, Nanotechnology **13**, 515–523 (2002)

41.72 B. Bhushan, G. B. Agrawal: Finite element analysis of nanostructures with roughness and scratches, Ultramicroscopy **97**, 495–501 (2003)

41.73 K. E. Petersen: Silicon as a mechanical material, Proc. IEEE **70**, 420–457 (1982)

41.74 B. Bhushan, S. Sundararajan, X. Li, C. A. Zorman, M. Mehregany: Micro/nanotribological studies of single-crystal silicon and polysilicon and SiC films for use in MEMS devices. In: *Tribology Issues and Opportunities in MEMS*, ed. by B. Bhushan (Kluwer, Dordrecht 1998) pp. 407–430

41.75 S. Sundararajan, B. Bhushan: Micro/nano-tribological studies of polysilicon and SiC films for MEMS applications, Wear **217**, 251–261 (1998)

41.76 X. Li, B. Bhushan: Micro/nanomechanical characterization of ceramic films for microdevices, Thin Solid Films **340**, 210–217 (1999)

41.77 H: Becker, C. Gaertner: Polymer Microfabrication Methods for Microfluidic Analytical Applications, Electrophoresis **21**, 12–26 (2000)

41.78 J. C. McDonald, D. C. Duffy, J. R. Anderson, D. T. Chiu, H. Wu, O. J. A. Schueller, G. M Whitesides: Fabrication of microfluidic systems in poly(dimethylsiloxane), Electrophoresis **21**, 27–40 (2000)

41.79 B. Ellis: *Polymers: A Property Database, Available on compact disk* (CRC Press, Boca Raton 2000) also see www.polymersdatabase.com

41.80 J. Brandrup, E. H. Immergut: *Polymer Handbook* (Wiley, New York 1989)

41.81 J. E. Mark: *Polymers Data Handbook* (Oxford Univ. Press, Oxford 1999)

41.82 B. Bhushan, X. Li: Nanomechanical characterization of solid surfaces and thin films, Int. Mater. Rev. **48**, 125–164 (2003)

41.83 B. R. Lawn, A. G. Evans, D. B. Marshall: Elastic/plastic indentation damage in ceramics: The median/radial system, J. Am. Ceram. Soc. **63**, 574 (1980)

41.84 S. Sundararajan, B. Bhushan, T. Namazu, Y. Isono: Mechanical property measurements of nanoscale structures using an atomic force microscope, Ultramicroscopy **91**, 111–118 (2002)

41.85 R. J. Roark: *Formulas for Stress and Strain*, 6th edn. (McGraw-Hill, New York 1989)

41.86 R. W. Hertzberg: *Deformation and Fracture Mechanics of Engineering Materials*, 3rd edn. (Wiley, New York 1989) pp. 277–278

41.87 T. K. Ning (ed): *Properties of Silicon*, EMIS Datareviews Series No. 4 (INSPEC, Institution of Electrical Engineers, London 1988)

Part D | 41

41.88   C. T. C. Nguyen, R. T. Howe: An integrated CMOS micromechanical resonator high-Q oscillator, IEEE J. Solid-State Circ. **34**, 440–455 (1999)

41.89   L. J. Hornbeck: A digital light processing ™ update – status and future applications, Proc. Soc. Photo-Opt. Eng. **3634**, 158–170 (1999)

41.90   M. Tanaka: Fracture toughness and crack morphology in indentation fracture of brittle materials, J. Mater. Sci. **31**, 749 (1996)

41.91   B. Bhushan, S. Venkatesan: Mechanical and tribological properties of silicon for micromechanical applications: A review, Adv. Info. Stor. Syst. **5**, 211–239 (1993)

41.92   B. Bhushan, B. K. Gupta: *Handbook of Tribology: Materials, Coatings, and Surface Treatments* (McGraw-Hill, New York 1997)

41.93   T. Tsuchiya, O. Tabata, J. Sakata, Y. Taga: Specimen size effect on tensile strength of surface-micromachined polycrystalline silicon thin films, J. Microelectromech. Syst. **7**, 106–113 (1998)

41.94   T. Yi, L. Li, C. J. Kim: Microscale material testing of single crystalline silicon: Process effects on surface morphology and tensile strength, Sens. Actuators A **83**, 172–178 (2000)

41.95   F. W. J. Billimeyer: *Textbook of Polymer Science* (Wiley, New York 1984)

41.96   Anonymous: Rohm and Haas General Information on PMMA

41.97   W. Wunderlich (ed): *Physical Constants of Poly(methyl methacrylate)*, 2nd edn. (Wiley, New York 1975)

41.98   E. Calvet, J. P. Bros, H. Pinelle: Compt. Rend. **260**, 1164 (1965)

41.99   K. Eiermann: Kolloid-Z. **198**, 5 (1965)

41.100 G. Hochstetter, A. Jimenez, J. P. Cano, E. Felder: An attempt to determine the true stress–strain curves of amorphous polymer by nanoindentation, Tribol. Inter. **36**, 973–985 (2003)

41.101 S. P. Timoshenko, J. N. Goodier: *Theory of Elasticity*, 3rd edn. (McGraw-Hill, New York 1970)

41.102 J. E. Shigley, L. D. Mitchell: *Mechanical Engineering Design*, 4th edn. (McGraw-Hill, New York 1993)

Part E
Molecular

Part E Molecularly Thick Films for Lubrication

42. Nanotribology of Ultrathin and Hard Amorphous Carbon Films

One of the best materials to use in applications that require very low wear and reduced friction is diamond, especially in the form of a diamond coating. Unfortunately, true diamond coatings can only be deposited at high temperatures and on selected substrates, and they require surface finishing. However, hard amorphous carbon – commonly known as diamond-like carbon or a DLC coating – has similar mechanical, thermal and optical properties to those of diamond. It can also be deposited at a wide range of thicknesses using a variety of deposition processes on various substrates at or near room temperature. The coatings reproduce the topography of the substrate, removing the need for finishing. The friction and wear properties of some DLC coatings make them very attractive for some tribological applications. The most significant current industrial application of DLC coatings is in magnetic storage devices.

In this chapter, the state-of-the-art in the chemical, mechanical and tribological characterization of ultrathin amorphous carbon coatings is presented.

EELS and Raman spectroscopies can be used to characterize amorphous carbon coatings chemically. The prevailing atomic arrangement in the DLC coatings is amorphous or quasi-amorphous, with small diamond (sp^3), graphite (sp^2) and other unidentifiable micro- or nanocrystallites. Most DLC coatings, except for those produced using a filtered cathodic arc, contain from a few to about 50 at. % hydrogen. Sometimes hydrogen is deliberately incorporated into the sputtered and ion-plated coatings in order to tailor their properties.

Amorphous carbon coatings deposited by different techniques exhibit different mechanical and tribological properties. Thin coatings deposited by filtered cathodic arc, ion beam and ECR-CVD hold much promise for tribological applications. Coatings of 5 nm or even less provide wear protection. A nanoindenter can be used to measure DLC coating hardness, elastic modulus, fracture toughness and fatigue life. Microscratch and microwear tests can be performed on the coatings using either a nanoindenter or an AFM, and along with accelerated wear testing, can be used to screen potential industrial coatings. For the examples shown in this chapter, the trends observed in such tests were similar to those found in functional tests.

Carbon exists in both crystalline and amorphous forms and exhibits both metallic and nonmetallic characteristics [42.1–3]. Forms of crystalline carbon include graphite, diamond, and a family of fullerenes, Fig. 42.1. The graphite and diamond are infinite periodic network solids with a planar structure, whereas the fullerenes are a molecular form of pure carbon with a finite network and a nonplanar structure. Graphite has a hexagonal, layered structure with weak interlayer bonding forces and it exhibits excellent lubrication properties. The graphite crystal may be visualized as infinite parallel layers of hexagons stacked 0.34 nm apart with an interatomic distance of 0.1415 nm between the carbon atoms in the basal plane. Each atom lying in the basal planes is trigonally coordinated and closely packed with strong σ (covalent) bonds to its three carbon neighbors via hybrid sp^2 orbitals. The fourth electron lies in a p_z orbital lying normal to the σ bonding plane and forms a weak π bond by overlapping side-to-side with a p_z orbital of an adjacent atom to which the carbon is attached by a σ

bond. The layers (basal planes) themselves are relatively far apart and the forces that bond them are weak van der Waals forces. These layers can align themselves parallel to the direction of the relative motion and slide over one another with relative ease, meaning low friction. Strong interatomic bonding and packing in each layer is thought to help reduce wear. The operating environment has a significant influence on the lubrication – low friction and low wear – properties of graphite. It lubricates better in a humid environment than a dry one, due to the adsorption of water vapor and other gases from the environment, which further weakens the interlayer bonding forces and results in easy shear and transfer of the crystallite platelets to the mating surface. Thus, transfer plays an important role in controlling friction and wear. Graphite oxidizes at high operating temperatures and can be used up to about 430 °C.

One of the most well-known fullerene molecules is C_{60}, commonly known as buckyballs. Since these C_{60} molecules are very stable and do not require additional atoms to satisfy chemical bonding requirements, they are expected to have low adhesion to the mating surface and low surface energy. Since the C_{60} molecule, which has a perfect spherical symmetry, bonds only weakly to other molecules, C_{60} clusters readily become detached, similar to other layered lattice structures, and either get transferred to the mating surface by mechanical compaction or are present as loose wear particles that may roll like tiny ball bearings in a sliding contact, resulting in low friction and wear. The wear particles are expected to be harder than as-deposited C_{60} molecules, because of their phase transformation at the high-asperity contact pressures present in a sliding interface. The low surface energy, the spherical shapes of C_{60} molecules, the weak intermolecular bonding, and the high load bearing capacity offer vast potential for various mechanical and tribological applications. Sublimed C_{60} coatings and fullerene particles used as an additive to mineral oils and greases have been reported to be good solid lubricants comparable to graphite and MoS_2 [42.4–6].

Diamond crystallizes in a modified face-centered cubic (fcc) structure with an interatomic distance of 0.154 nm. The diamond cubic lattice consists of two interpenetrating fcc lattices displaced by a quarter of the cube diagonal. Each carbon atom is tetrahedrally coordinated, making strong σ (covalent) bonds to its four carbon neighbors using the hybrid sp^3 atomic orbitals, which accounts for it having the highest hardness (80–104 GPa) and thermal conductivity (900–2100 W/mK, on the order of five times that of copper) of any known solid, as well as high electrical re-

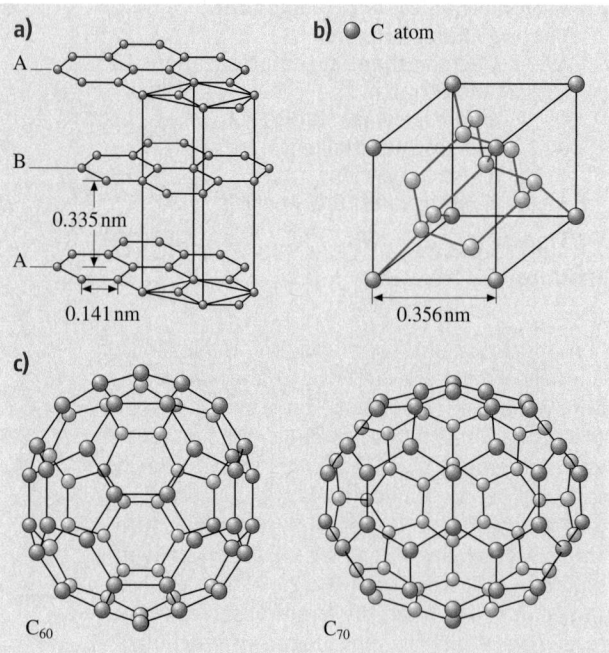

Fig. 42.1a–c The structures of the three known forms of crystalline carbon: (**a**) hexagonal structure of graphite, (**b**) modified face-centered cubic (fcc) structure (two interpenetrating fcc lattices displaced by a quarter of the cube diagonal) of diamond (each atom is bonded to four others that form the corners of a tetrahedron), and (**c**) the structures of the two most common fullerenes: a soccer ball C_{60} and a rugby ball C_{70} molecules

sistivity and optical transmission and a large optical band gap. It is relatively chemically inert, and it exhibits poor adhesion with other solids, enhancing its low friction and wear properties. Its high thermal conductivity permits the dissipation of frictional heat during sliding and it protects the interface, and the dangling carbon bonds on the surface react with the environment to form hydrocarbons that act as good lubrication films. These are some of the reasons for the low friction and wear of the diamond. Diamond and its coatings find many industrial applications: tribological applications (low friction and wear), optical applications (exceptional optical transmission, high abrasion resistance), and thermal management or heat sink applications (high thermal conductivity). Diamond can be used at high temperatures; it starts to graphitize at about 1000 °C in ambient air and at about 1400 °C in vacuum. Diamond is an attractive material for cutting tools, abrasives for grinding wheels and lapping compounds, and other extreme wear applications.

Natural diamond – particularly in large quantities – is very expensive, and so diamond coatings – a low-cost alternative – are attractive. True diamond coatings are deposited by chemical vapor deposition (CVD) processes at high substrate temperatures (on the order of 800 °C). They adhere best on silicon substrate and require an interlayer for other substrates. One major hin-

drance to the widespread use of true diamond films in tribological, optical and thermal management applications is their surface roughness. Growth of the diamond phase on a non-diamond substrate is initiated by nucleation at either randomly seeded sites or at thermally favored sites, due to statistical thermal fluctuations at the substrate surface. Depending on the growth temperature and pressure conditions, certain favored crystal orientations dominate the competitive growth process. As a result, the films grown are polycrystalline in nature with a relatively large grain size (> 1 μm) and they terminate in very rough surfaces, with RMS roughnesses ranging from a few tenths of a micron to tens of microns. Techniques for polishing these films have been developed. It has been reported that laser polished films exhibit friction and wear properties almost comparable to those of bulk polished diamond [42.7, 8].

Amorphous carbon has no long-range order, and the short-range order of the carbon atoms in it can have one or more of three bonding configurations: sp^3 (diamond), sp^2 (graphite), or sp^1 (with two electrons forming strong σ bonds, and the remaining two electrons left in orthogonal p_y and p_z orbitals, that form weak π bonds). Short-range order controls the properties of amorphous materials and coatings. Hard amorphous carbon (a-C) coatings, commonly known as diamond-like carbon or

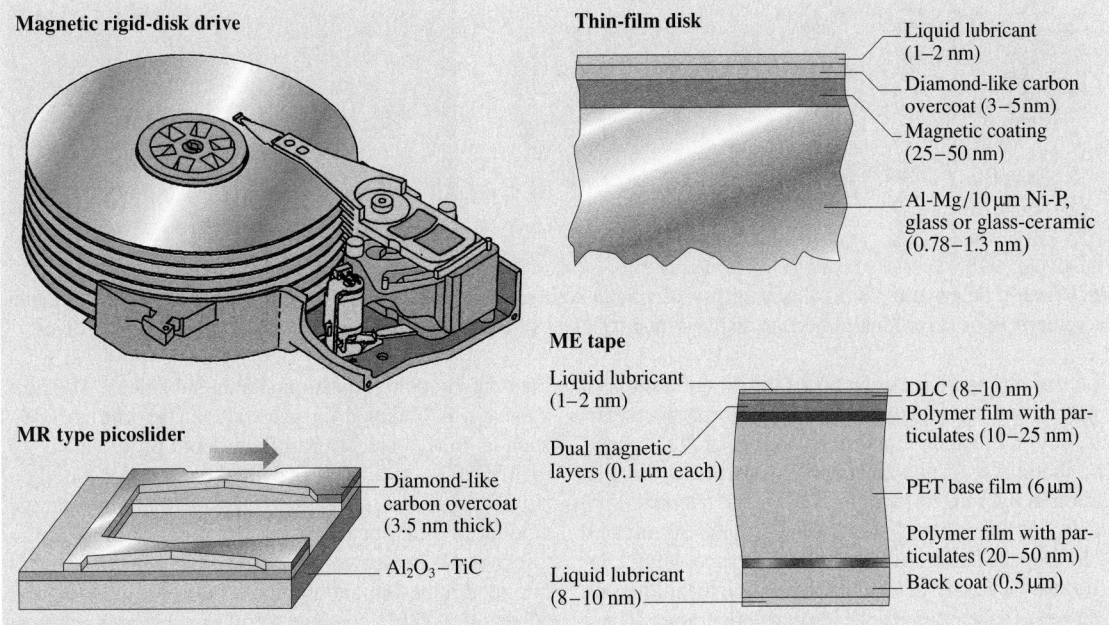

Fig. 42.2 Schematic of a magnetic rigid-disk drive and MR type picoslider, and cross-sectional schematics of a magnetic thin film rigid disk and a metal evaporated (ME) tape

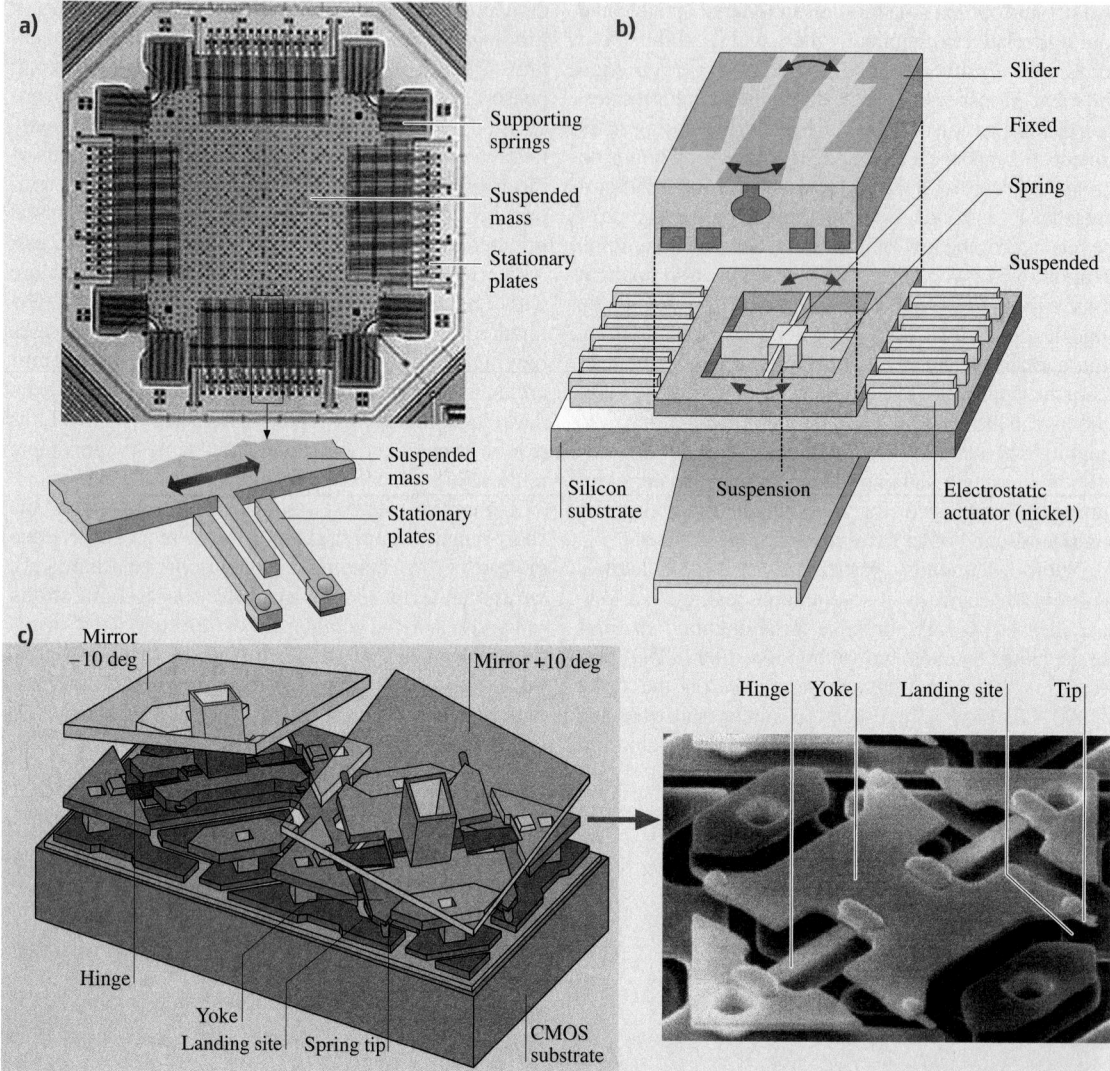

Fig. 42.3a–c Schematics of (**a**) a capacitive-type silicon accelerometer for automotive sensory applications, (**b**) digital micrometer devices for high-projection displays, and (**c**) a polysilicon rotary microactuator for a magnetic disk drives

DLC (implying high hardness) coatings, are a class of coatings that are mostly metastable amorphous materials, but that include a micro- or nanocrystalline phase. The coatings are random networks of covalently bonded carbon in hybridized tetragonal (sp^3) and trigonal (sp^2) local coordination with some of the bonds terminated by hydrogen. These coatings have been successfully deposited by a variety of vacuum deposition techniques on a variety of substrates at or near room temperature. These coatings generally reproduce substrate topography and do not require any post-finishing. However, these coat-

ings mostly adhere best on silicon substrates. The best adhesion is obtained on substrates that form carbides, such as Si, Fe and Ti. Based on depth profile analyses (using Auger and XPS) of DLC coatings deposited on silicon substrates, it has been reported that a substantial amount of silicon carbide (on the order of 5–10 nm in thickness) is present at the carbon–silicon interface, giving good adhesion and hardness (see [42.9]). For good adhesion of DLC coatings to other substrates, in most cases, an interlayer of silicon is required in most cases, except for coatings deposited by a cathodic arc.

There is significant interest in DLC coatings due to their unique combination of desirable properties. These properties include high hardness and wear resistance, chemical inertness to both acids and alkalis, lack of magnetic response, and an optical band gap ranging from zero to a few eV, depending upon the deposition conditions. These are used in a wide range of applications, including tribological, optical, electronic and biomedical applications [42.1, 10, 11]. The high hardness, good friction and wear properties, versatility in deposition and substrates, and no requirement for post-deposition finishing make them very attractive for tribological applications. Two primary examples include overcoats for magnetic media (thin film disks and ME tapes) and MR-type magnetic heads for magnetic storage devices, Fig. 42.2 [42.12–20], and the emerging field of microelectromechanical systems, Fig. 42.3 [42.21–24]. The largest industrial application of the family of amorphous carbon coatings, typicallydeposited by DC/RF magnetron sputtering, plasma-enhanced chemical vapor deposition or ion beam deposition techniques, is in magnetic storage devices. These are employed to protect magnetic coatings on thin film rigid disks and metal evaporated tapes and the thin film head structure of a read/write disk head against wear and corrosion (Fig. 42.2). Thicknesses ranging from 3 to 10 nm are employed to maintain low physical spacing between the magnetic element of a read/write head and the magnetic layer of the storage media. Mechanical properties affect friction wear and therefore need to be optimized. In 1998, Gillette introduced Mach 3 razor blades with ultrathin DLC coatings, which could potentially become a very large industrial application. DLC coatings are also used in other commercial applications such as the glass windows of supermarket laser barcode scanners and sunglasses. These coatings are actively pursued in microelectromechanical systems (MEMS) components [42.23].

In this chapter, a state-of-the-art review of recent developments in the field of chemical, mechanical, and tribological characterization of ultrathin amorphous carbon coatings is presented. An overview of the most commonly used deposition techniques is provided, followed by typical chemical and mechanical characterization data and typical tribological data from both coupon-level testing and functional testing.

42.1 Description of Common Deposition Techniques

The first hard amorphous carbon coatings were deposited by a beam of carbon ions produced in an argon plasma on room temperature substrates, as reported by *Aisenberg* and *Chabot* [42.25]. Subsequent confirmation by *Spencer* et al. [42.26] led to the explosive growth of this field. Following this first work, several alternative techniques were developed. Amorphous carbon coatings have been prepared by a variety of deposition techniques and precursors, including evaporation, DC, RF or ion beam sputtering, RF or DC plasma-enhanced chemical vapor deposition (PECVD), electron cyclotron resonance chemical vapor deposition (ECR-CVD), direct ion beam deposition, pulsed laser vaporization and vacuum arc, from a variety of carbon-bearing solids or gaseous source materials [42.1, 27]. Coatings with both graphitic and diamond-like properties have been produced. Evaporation and ion plating techniques have been used to produce coatings with graphitic properties (low hardness, high electrical conductivity, very low friction, and so on, and all of the techniques have been used to produce coatings with diamond-like properties.

The structure and properties of a coating are dependent upon the deposition technique and parameters.

High-energy surface bombardment has been used to produce harder and denser coatings. It is reported that the sp^3/sp^2 fraction decreases in the order: cathodic arc deposition, pulsed laser vaporization, direct ion beam deposition, plasma-enhanced chemical vapor deposition, ion beam sputtering, DC/RF sputtering [42.12, 28, 29]. A common feature of these techniques is that the deposition is energetic; in other words the carbon species arrive with an energy significantly greater than that represented by the substrate temperature. The resultant coatings are amorphous in structure, with hydrogen contents of up to 50%, and display a high degree of sp^3 character. From the results of previous investigations, it has been proposed that deposition of sp^3-bonded carbon requires that the depositing species have kinetic energies on the order of 100 eV or higher, well above those obtained in thermal processes like evaporation (0–0.1 eV). The species must then be quenched into the metastable configuration via rapid energy removal. Excess energy, such as that provided by substrate heating, is detrimental to the achievement of a high sp^3 fraction. In general, the higher the fraction of sp^3-bonded carbon atoms in the amorphous network, the greater the

Table 42.1 Summary of common deposition techniques, the kinetic energies of the depositing species and deposition rates

Deposition technique	Process	Kinetic energy (eV)	Deposition rate (nm/s)
Filtered cathodic arc (FCA)	Energetic carbon ions produced by a vacuum arc discharge between a graphite cathode and a grounded anode	100–2500	0.1–1
Direct ion beam (IB)	Carbon ions produced from methane gas in an ion source and accelerated toward a substrate	50–500	0.1–1
Plasma-enhanced chemical vapor deposition (PECVD)	Hydrocarbon species produced by plasma decomposition of a hydrocarbon gas (such as acetylene) are accelerated toward a DC-biased substrate	1–30	1–10
Electron cyclotron resonance plasma chemical vapor deposition (ECR-CVD)	Hydrocarbon ions produced by the plasma decomposition of ethylene gas in the presence of a plasma at the electron cyclotron resonance condition are accelerated toward a RF-biased substrate	1–50	1–10
DC/RF sputtering	Sputtering of graphite target by argon ion plasma	1–10	1–10

hardness [42.29–36]. The mechanical and tribological properties of a carbon coating depend on the sp^3/sp^2-bonded carbon ratio, the amount of hydrogen in the coating, and the adhesion of the coating to the substrate, which are influenced by the precursor material, the kinetic energy of the carbon species prior to deposition, the deposition rate, the substrate temperature, the substrate biasing, and the substrate itself [42.29, 33, 35, 37–46]. The kinetic energies and deposition rates involved in selected deposition processes used in the deposition of DLC coatings are compared in Table 42.1 [42.1, 28].

In the studies by *Gupta* and *Bhushan* [42.12, 47], *Li* and *Bhushan* [42.48, 49], and *Sundararajan* and *Bhushan* [42.50], DLC coatings typically ranging in thickness from 3.5 nm to 20 nm were deposited on single-crystal silicon, magnetic Ni-Zn ferrite, and Al_2O_3-TiC substrates (surface roughness ≈ 1–3 nm

RMS) by filtered cathodic arc (FCA) deposition, (direct) ion beam deposition (IBD), electron cyclotron resonance chemical vapor deposition (ECR-CVD), plasma-enhanced chemical vapor deposition (PECVD), and DC/RF planar magnetron sputtering (SP) deposition techniques [42.51]. In this chapter, we will limit the presentation of data to coatings deposited by FCA, IBD, ECR-CVD and SP deposition techniques.

42.1.1 Filtered Cathodic Arc Deposition

When the filtered cathodic arc deposition technique is used to create carbon coatings [42.29, 52–59], a vacuum arc plasma source is used to form the carbon film. In the FCA technique used by *Bhushan* et al. (see [42.12]), energetic carbon ions are produced by a vacuum arc discharge between a planar graphite cathode and a

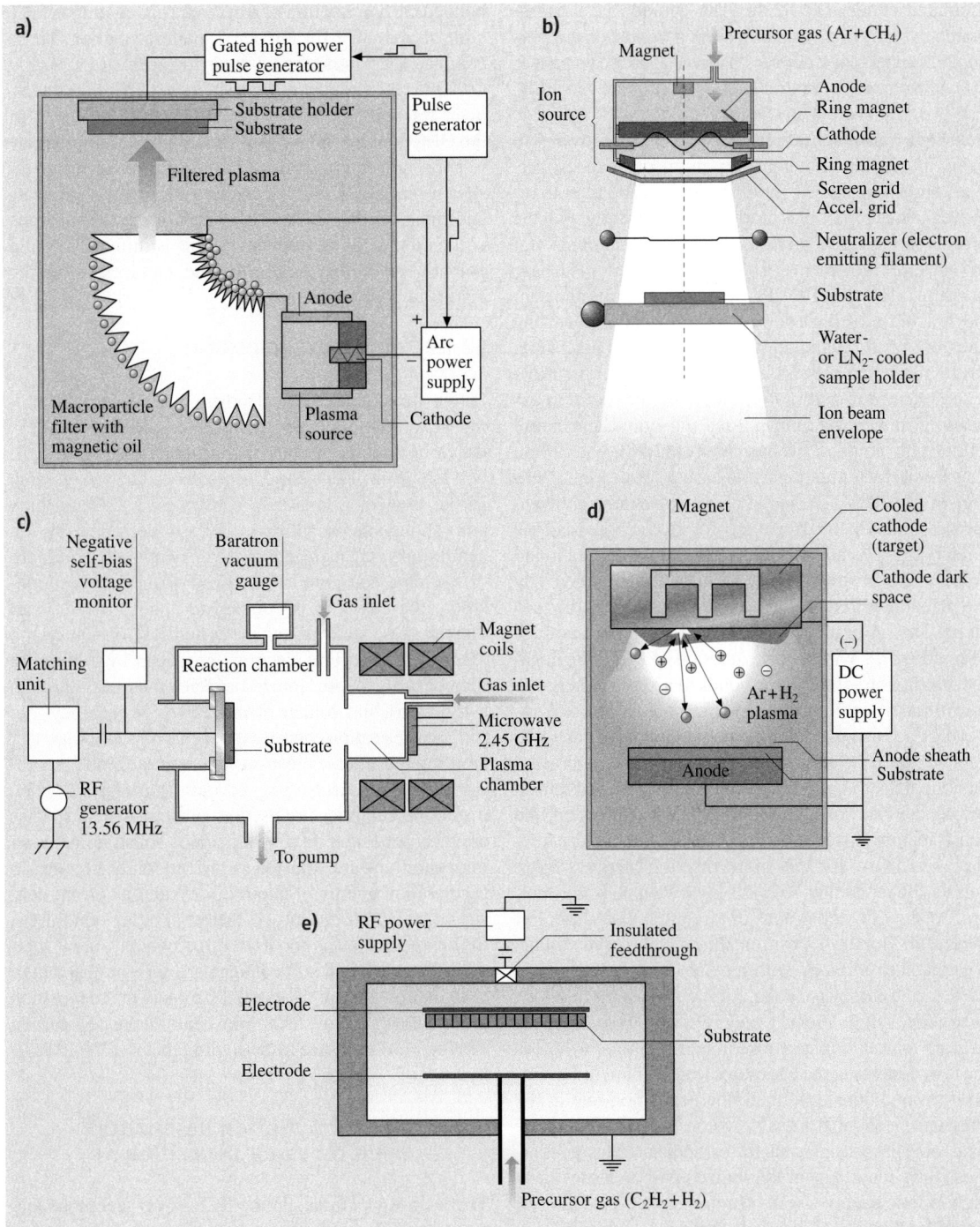

Fig. 42.4a–e Schematic diagrams of deposition by (**a**) filtered cathodic arc deposition, (**b**) ion beam deposition, (**c**) electron cyclotron resonance chemical vapor deposition (ECR-CVD), (**d**) DC planar magnetron sputtering, and (**e**) plasma-enhanced chemical vapor deposition (PECVD)

grounded anode, Fig. 42.4a. The cathode is a 6 mm-diameter high-density graphite disk mounted on a water-cooled copper block. The arc is driven at an arc current of 200 A, with an arc duration of 5 ms and an arc repetition rate of 1 Hz. The plasma beam is guided by a magnetic field that transports current between the electrodes to form tiny, rapidly moving spots on the cathode surface. The source is coupled to a 90° bent magnetic filter to remove the macroparticles produced concurrently with the plasma in the cathode spots. The ion current density at the substrate is in the range of $10–50 \, mA/cm^2$. The base pressure is less than 10^{-4} Pa. A much higher plasma density is achieved using a powerful arc discharge than using electron beam evaporation with auxiliary discharge. In the discharge process, the cathodic material suffers a complicated transition from the solid phase to an expanding, nonequilibrium plasma via liquid and dense equilibrium nonideal plasma phases [42.58]. The carbon ions in the vacuum arc plasma have a direct kinetic energy of 20–30 eV. The high voltage pulses are applied to the substrate which is mounted on a water-cooled sample holder, and ions are accelerated through the sheath and arrive at the substrate with an additional energy given by the potential difference between the plasma and the substrate. The substrate holder is pulsed-biased to a voltage of up to −2 kV with a pulse duration of 1 μs. The negative biasing of −2 kV corresponds to a kinetic energy of 2 keV for the carbon ions. The use of a pulsed bias instead of a DC bias enables much higher voltages to be applied and it permits a surface potential to be created on nonconducting films. The ion energy is varied during the deposition. For the first 10% of the deposition, the substrates are pulsed-biased to −2 keV with a pulse duty cycle of 25%, so for 25% of the time the energy is 2 keV and for the remaining 75% it is 20 eV, which is the "natural" energy of carbon ions in a vacuum discharge. For the last 90% of the deposition, the pulsed-biased voltage is reduced to −200 eV with a pulsed bias duty cycle of 25%, so the energy is 200 eV for 25% and 20 eV for 75% of the deposition. The high energy at the beginning leads to good intermixing and adhesion of the films, whereas the lower energy at the later stage leads to hard films. Under the conditions described, the deposition rate at the substrate is about 0.1 nm/s, which is slow. Compared with most gaseous plasma, the cathodic arc plasma is almost fully ionized, and the ionized carbon atoms have high kinetic energies which promotes the formation of a high fraction of sp^3-bonded carbon ions, which in turn results in high hardness and higher interfacial adhesion. *Cuomo* et al. [42.42] have reported, based on electron energy loss spectroscopy (EELS) analysis, that the sp^3-bonded carbon fraction of a cathodic arc coating is 83% compared to 38% for ion beam sputtered carbon. These coatings are reported to be *nonhydrogenated*.

This technique does not require an adhesion underlayer for non-silicon substrates. However, adhesion of the DLC coatings on the electrically insulating substrate is poor, as negative pulsed biasing forms an electrical sheath that accelerates depositing ions to the substrate and enhances the adhesion of the coating to the substrate with associated ion implantation. It is difficult to build potential on an insulating substrate, and lack of biasing results in poor adhesion.

42.1.2 Ion Beam Deposition

In the direct ion beam deposition of a carbon coating [42.60–64], as used by *Bhushan* et al. (see [42.12]), the carbon coating is deposited from an accelerated carbon ion beam. The sample is precleaned by ion etching. In the case of non-silicon substrates, a 2–3 nm-thick amorphous silicon adhesion layer is deposited by ion beam sputtering using an ion beam containing a mixture of methane and argon at 200 V. For the carbon deposition, the chamber is pumped to about 10^{-4} Pa, and methane gas is fed through the cylindrical ion source and ionized by energetic electrons produced by a hot-wire filament, Fig. 42.4b. Ionized species then pass through a grid with a bias voltage of about 50 eV, where they gain a high acceleration energy and reach a hot-wire filament, emitting thermionic electrons that neutralize the incoming ions. The discharging of ions is important when insulating ceramics are used as substrates. The species are then deposited on a water-cooled substrate. Operating conditions are adjusted to give an ion beam with an acceleration energy of about 200 eV and a current density of about $1 \, mA/cm^2$. At these operating conditions, the deposition rate is about 0.1 nm/s, which is slow. Incidentally, tough and soft coatings are deposited at a high acceleration energy of about 400 eV and at a deposition rate of about 1 nm/s. The ion beam-deposited carbon coatings are reported to be hydrogenated (30–40 at. % hydrogen).

42.1.3 Electron Cyclotron Resonance Chemical Vapor Deposition

The lack of electrodes in the ECR-CVD technique and its ability to create high densities of charged and excited species at low pressures ($\leq 10^{-4}$ Torr) make it attractive for coating deposition [42.65]. In the ECR-CVD carbon deposition process described by *Suzuki* and

Okada [42.66] and used by *Li* and *Bhushan* [42.48, 49] and *Sundararajan* and *Bhushan* [42.50], microwave power is generated by a magnetron operating in continuous mode at a frequency of 2.45 GHz, Fig. 42.4c. The plasma chamber functions as a microwave cavity resonator. The magnetic coils arranged around the plasma chamber generate a magnetic field of 875 G, necessary for electron cyclotron resonance. The substrate is placed on a stage that is connected capacitively to a 13.56 MHz RF generator. The process gas is introduced into the plasma chamber and the hydrocarbon ions generated are accelerated by a negative self-bias voltage, which is generated by applying RF power to the substrate. Both the substrate stage and the plasma chamber are water-cooled. The process gas used is 100% ethylene and its flow rate is held constant at 100 sccm. The microwave power is 100–900 W. The RF power is 30–120 W. The pressure during deposition is kept close to the optimum value of 5.5×10^{-3} Torr. Before the deposition, the substrates are cleaned using Ar ions generated in the ECR plasma chamber.

42.1.4 Sputtering Deposition

In DC planar magnetron carbon sputtering [42.13, 33, 37, 40, 67–71], the carbon coating is deposited by the sputtering of a graphite target with Ar ion plasma. In the glow discharge, positive ions from the plasma strike the target with sufficient energy to dislodge the atoms by momentum transfer, which are intercepted by the substrate. An $\sim$5 nm-thick amorphous silicon adhesion layer is initially deposited by sputtering if the deposition is to be carried out on a non-silicon surface. In the process used by *Bhushan* et al. (see [42.12]), the coating is deposited by the sputtering of a 200 mm-diameter graphite target with Ar ion plasma at 300 W power and a pressure of about 0.5 Pa (6 mTorr), Fig. 42.4d. Plasma is generated by applying a DC potential between the substrate and a target. *Bhushan* et al. [42.35] reported that the sputtered carbon coating contains about 35 at. % hydrogen. The hydrogen comes from the hydrocarbon contaminants present in the deposition chamber. In order to produce a hydrogenated carbon coating with a larger concentration of hydrogen, the deposition is carried out in Ar and hydrogen plasma.

42.1.5 Plasma–Enhanced Chemical Vapor Deposition

In the RF-PECVD deposition of carbon, as used by *Bhushan* et al. (see [42.12]), the carbon coating is deposited by adsorbing hydrocarbon free radicals onto the substrate and then via chemical bonding to other atoms on the surface. The hydrocarbon species are produced by the RF plasma decomposition of hydrocarbon precursors such as acetylene (C_2H_2), Fig. 42.4e [42.27, 69, 72–75]. Instead of requiring thermal energy, as in thermal CVD, the energetic electrons in the plasma (at a pressure of $1–5 \times 10^2$ Pa, and typically less than 10 Pa) can activate almost any reaction among the gases in the glow discharge at relatively a low substrate temperature of 100 to 600 °C (typically less than 300 °C). To deposit the coating on non-silicon substrates, an amorphous silicon adhesion layer about 4 nm thick

is first deposited under similar conditions from a gas mixture of 1% silane in argon in order to improve adhesion [42.76]. In the process used by Bhushan and coworkers [42.12], the plasma is sustained in a parallel-plate geometry by a capacitive discharge at 13.56 MHz, at a surface power density of around 100 mW/cm^2. The deposition is performed at a flow rate on the order of 6 sccm and a pressure on the order of 4 Pa (30 mTorr) on a cathode-mounted substrate maintained at a substrate temperature of 180 °C. The cathode bias is held fixed at about -120 V with an external DC power supply attached to the substrate (powered electrode). The carbon coatings deposited by PECVD usually contain hydrogen at levels of up to 50% [42.35, 77].

42.2 Chemical and Physical Coating Characterization

The chemical structures and properties of amorphous carbon coatings depend on the deposition conditions employed when they are formed. It is important to understand the relationship between the chemical structure of a coating and its properties since it allows useful deposition parameters to be defined. Amorphous carbon films are metastable phases formed when carbon particles are condensed on a substrate. The prevailing atomic arrangement in the DLC coatings is amorphous or quasi-amorphous, with small diamond (sp^3), graphite (sp^2) and other unidentifiable micro- or nanocrystallites. The coating is dependent upon the deposition process and the deposition conditions used because these influence the sp^3/sp^2 ratio and the proportion of hydrogen in the coating. The sp^3/sp^2 ratios of DLC coatings typically range from 50% to close to 100%, and hardness increases with the sp^3/sp^2 ratio. Most DLC coatings, except those produced by a filtered cathodic arc, contain from a few to about 50 at. % hydrogen. Sometimes hydrogen and nitrogen are deliberately added to produce hydrogenated (a-C:H) and nitrogenated amorphous carbon (a-C:N) coatings, respectively. Hydrogen helps to stabilize sp^3 sites (most of the carbon atoms attached to hydrogen have a tetrahedral structure), so the sp^3/sp^2 ratio for hydrogenated carbon is higher [42.30]. The optimum sp^3/sp^2 ratio for a random covalent network composed of sp^3 and sp^2 carbon sites (N_{sp^2} and N_{sp^3}) and hydrogen is [42.30]:

$$\frac{N_{sp^3}}{N_{sp^2}} = \frac{6X_H - 1}{8 - 13X_H} , \tag{42.1}$$

where X_H is the atomic fraction of hydrogen. The hydrogenated carbon has a larger optical band gap, higher electrical resistivity (semiconductor), and a lower optical absorption or high optical transmission. Hydrogenated coatings have lower densities, probably because of the reduced cross-linking due to hydrogen incorporation. However, the hardness decreases with increasing hydrogen, even though the proportion of sp^3 sites increases (that is, as the local bonding environment becomes more diamond-like) [42.78, 79]. It is speculated that the high hydrogen content introduces frequent terminations in the otherwise strong 3-D network, and hydrogen increases the soft polymeric component of the structure more than it enhances the cross-linking sp^3 fraction.

A number of investigations have been performed to identify the microstructure of amorphous carbon films using a variety of techniques, such as Raman spectroscopy, EELS, nuclear magnetic resonance, optical measurements, transmission electron microscopy, and X-ray photoelectron spectroscopy [42.33]. The structure of diamond-like amorphous carbon is amorphous or quasi-amorphous, with small graphitic (sp^2), tetrahedrally coordinated (sp^3) and other types of nanocrystallites (typically on the order of a couple of nm in size, randomly oriented) [42.33, 80, 81]. These studies indicate that the chemical structure and physical properties of the coatings are quite variable, depending on the deposition techniques and film growth conditions. It is clear that both sp^2- and sp^3-bonded atomic sites are incorporated in diamond-like amorphous carbon coatings and that the physical and chemical properties of the coatings depend strongly on their chemical bonding and microstructures. Systematic studies have been conducted to carry out chemical characterization and to investigate how the physical and chemical properties of amorphous carbon coatings vary as a function

of deposition conditions (see [42.33, 35, 40]). EELS and Raman spectroscopy are commonly used to characterize the chemical bonding and microstructure. The hydrogen concentration in the coating is obtained via forward recoil spectrometry (FRS). A variety of physical properties relevant to tribological performance are measured.

In order to give the reader a feel for typical data obtained when characterizing amorphous carbon coatings and their relationships to physical properties, we present data on several sputtered coatings, RF-PECVD amorphous carbon and microwave-PECVD (MPECVD) diamond coatings [42.33, 35, 40]. The sputtered coatings were DC magnetron sputtered at a chamber pressure of 10 mTorr under sputtering power densities of 0.1 and 2.1 W/cm^2 (labeled as coatings W1 and W2, respectively) in a pure Ar plasma. These coatings were prepared at a power density of 2.1 W/cm^2 with various hydrogen fractions (0.5, 1, 3, 5, 7 and 10%) of Ar/H; the gas mixtures were labeled as H1, H2, H3, H4, H5, and H6, respectively.

42.2.1 EELS and Raman Spectroscopy

EELS and Raman spectra of four sputtered (W1, W2, H1, and H3) carbon samples and one PECVD carbon sample were obtained. Figure 42.5 shows the EELS spectra of these carbon coatings. EELS spectra (up to 50 eV) for bulk diamond and polycrystalline graphite are also shown in Fig. 42.5. One prominent peak is seen at 35 eV in diamond, while two peaks are seen at 27 eV and 6.5 eV in graphite, which are called the $(\pi + \sigma)$ and (π) peaks, respectively. These peaks are produced by the loss of transmitted electron energy to plasmon oscillations of the valence electrons. The $\pi + \sigma$ peak in each coating is positioned at a lower energy region than that of graphite. The π peaks in the W series and PECVD samples also occur at a lower energy region than that of the graphite. However, the π peaks in the H-series are comparable to or higher than those of graphite (see Table 42.2). The plasmon oscillation frequency is proportional to the square root of the corresponding electron density to a first approximation. Therefore, the samples in the H-series most likely have a higher density of π electrons than the other samples.

Amorphous carbon coatings contain (mainly) a mixture of sp^2- and sp^3-bonds, even though there is some evidence for the presence of sp-bonds as well [42.82]. The PECVD coatings and the H-series coatings in this study have almost the same mass density (as seen in Table 42.4, discussed in more detail later), but the former have a lower concentration of hydrogen (18.1%) than the H-series (35–39%) (as seen in Table 42.3, also discussed in more detail later). The relatively low-energy positions of the π peaks of the PECVD coatings compared to those of the H-series indicates that the PECVD coatings contain a higher fraction of sp^3-bonds than the sputtered hydrogenated carbon coatings (H-series).

Figure 42.5b shows EELS spectra associated with the inner-shell (K-shell) ionization. Again, the spectra for diamond and polycrystalline graphite are included for comparison. Sharp peaks are observed at 285.5 eV and 292.5 eV in graphite, while no peak is seen at 285.5 eV in diamond. The general features of the K-shell EELS spectra for the sputtered and PECVD carbon samples resemble those of graphite, but with the higher energy features smeared. The peak at 285.5 eV in the sputtered and PECVD coatings also indicates the presence of sp^2-bonded atomic sites in the coatings. All of these spectra peak at 292.5 eV, similar to the spectra of graphite, but the peak in graphite is sharper.

Raman spectra from samples W1, W2, H1 and PECVD are shown in Fig. 42.6. Raman spectra could not be observed in specimens H2 and H3 due to high fluorescence signals. The Raman spectra of single-crystal diamond and polycrystalline graphite are also shown for comparison in Fig. 42.6. The results from the spectral fits are summarized in Table 42.2. We will focus

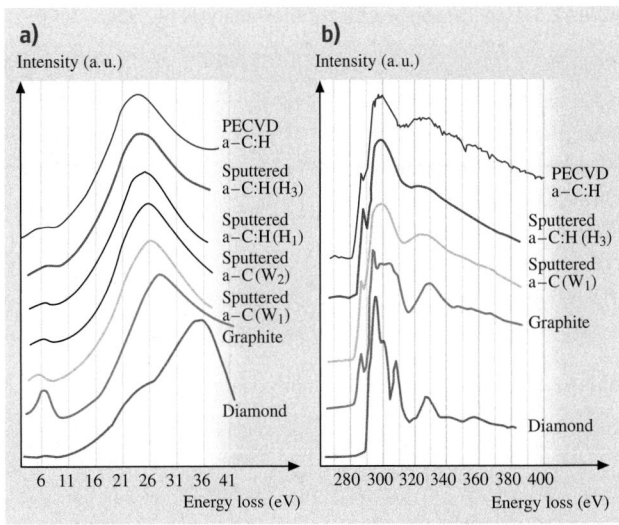

Fig. 42.5 (a) Low-energy and **(b)** high-energy EELS of DLC coatings produced by the DC magnetron sputtering and RF-PECVD techniques. Data for bulk diamond and polycrystalline graphite are included for comparison [42.35]

on the position of the G-band, which has been shown to be related to the fraction of sp^3-bonded sites. Increasing the power density in the amorphous carbon coatings (W1 and W2) results in a higher G-band frequency, implying a smaller fraction of sp^3-bonding in

Table 42.2 Experimental results from EELS and Raman spectroscopy [42.35]

Sample	EELS peak position		Raman peak position		Raman FWHM[a]		I_D/I_G^d
	π (eV)	$\pi + \sigma$ (eV)	G-band[b] (cm^{-1})	D-band[c] (cm^{-1})	G-band (cm^{-1})	D-band (cm^{-1})	
Sputtered a-C coating (W1)	5.0	24.6	1541	1368	105	254	2.0
Sputtered a-C coating (W2)	6.1	24.7	1560	1379	147	394	5.3
Sputtered a-C:H coating (H1)	6.3	23.3	1542	1334	95	187	1.6
Sputtered a-C:H coating (H3)	6.7	22.4	e	e	e	e	e
PECVD a-C:H coating	5.8	24.0	1533	1341	157	427	1.5
Diamond coating	...	...	1525[f]	1333[g]	...	8[g]	...
Graphite (for reference)	6.4	27.0	1580	1358	37	47	0.7
Diamond (for reference)	...	37.0	...	1332[g]	...	2[g]	...

[a] Full width at half maximum
[b] Peak associated with sp^2 "graphite" carbon
[c] Peak associated with sp^2 "disordered" carbon (not sp^3-bonded carbon)
[d] Intensity ratio of the D-band to the G-band
[e] Fluorescence
[f] Includes D and G band, signal too weak to analyze
[g] Peak position and width for diamond phonon

Table 42.3 Experimental results of FRS analysis [42.35]

Sample	Ar/H ratio	C (at.% ± 0.5)	H (at.% ± 0.5)	Ar (at.% ± 0.5)	O (at.% ± 0.5)
Sputtered a-C coating (W2)	100/0	90.5	9.3	0.2	...
Sputtered a-C:H coating (H2)	99/1	63.9	35.5	0.6	...
Sputtered a-C:H coating (H3)	97/3	56.1	36.5	...	7.4
Sputtered a-C:H coating (H4)	95/5	53.4	39.4	...	7.2
Sputtered a-C:H coating (H5)	93/7	58.2	35.4	0.2	6.2
Sputtered a-C:H coating (H6)	90/10	57.3	35.5	...	7.2
PECVD a-C:H coating	99.5% CH$_4$	81.9	18.1	...	...
Diamond coating	H$_2$-1 mole % CH$_4$	94.0	6.0	...	...

Table 42.4 Experimental results of physical properties [42.35]

Sample	Mass density (g/cm^3)	Nano-hardness (GPa)	Elastic modulus (GPa)	Electrical resistivity (Ohm-cm)	Compressive residual stress (GPa)
Sputtered a-C coating (W1)	2.1	15	141	1300	0.55
Sputtered a-C:H coating (W2)	1.8	14	136	0.61	0.57
Sputtered a-C:H coating (H1)	...	14	96	...	> 2
Sputtered a-C:H coating (H3)	1.7	7	35	> 10^6	0.3
PECVD a-C:H coating	1.6–1.8	33–35	~ 200	> 10^6	1.5–3.0
Diamond coating	...	40–75	370–430	...	...
Graphite (for reference)	2.267	Soft	9–15	5×10^{-5a}, 4×10^{-3b}	0
Diamond (for reference)	3.515	70–102	900–1050	10^7–10^{20}	0

[a] Parallel to layer planes
[b] Perpendicular to layer planes

W2 than in W1. This is consistent with the higher density of W1. H1 and PEVCD have even lower G-band positions than W1, implying an even higher fraction of sp^3-bonding, which is presumably caused by the incorporation of H atoms into the lattice. The high hardness of H3 might be attributed to efficient sp^3 cross-linking of small sp^2-ordered domains.

The Raman spectrum of a MPECVD diamond coating is shown in Fig. 42.6. The Raman peak of diamond is at 1333 cm^{-1}, with a line width of 7.9 cm^{-1}. There is a small broad peak at around 1525 cm^{-1}, which is attributed to a small amount of a-C:H. This impurity peak is not intense enough to be able to separate the G- and D-bands. The diamond peak frequency is very close to that of natural diamond (1332.5 cm^{-1}, see Fig. 42.6), indicating that the coating is not under stress [42.83]. The large line width compared to that of natural diamond (2 cm^{-1}) indicates that the microcrystallites probably have a high concentration of defects [42.84].

42.2.2 Hydrogen Concentrations

A FRS analysis of six sputtered (W2, H2, H3, H4, H5, and H6) coatings, one PECVD coating, and one diamond coating was performed. Figure 42.7 shows an overlay of the spectra from the six sputtered samples. Similar spectra were obtained from the PECVD and the diamond films. Table 42.3 shows the H and C fractions as well as the amount of impurities (Ar and O) in the films in atomic %. Most apparent is the large fraction of H in the sputtered films. Regardless of how much H$_2$ is in the Ar sputtering gas, the H content of the coatings is about the same, ≈ 35 at. %. Interestingly, there is still $\approx 10\%$ H present in the coating sputtered in pure Ar (W2). It is interesting to note that Ar is present only in coatings grown using Ar carrier gas with a low (< 1%) H content. The presence of O in the coatings combined with the fact that the coatings were prepared approximately nine months before the FRS analysis caused suspicion that they had absorbed

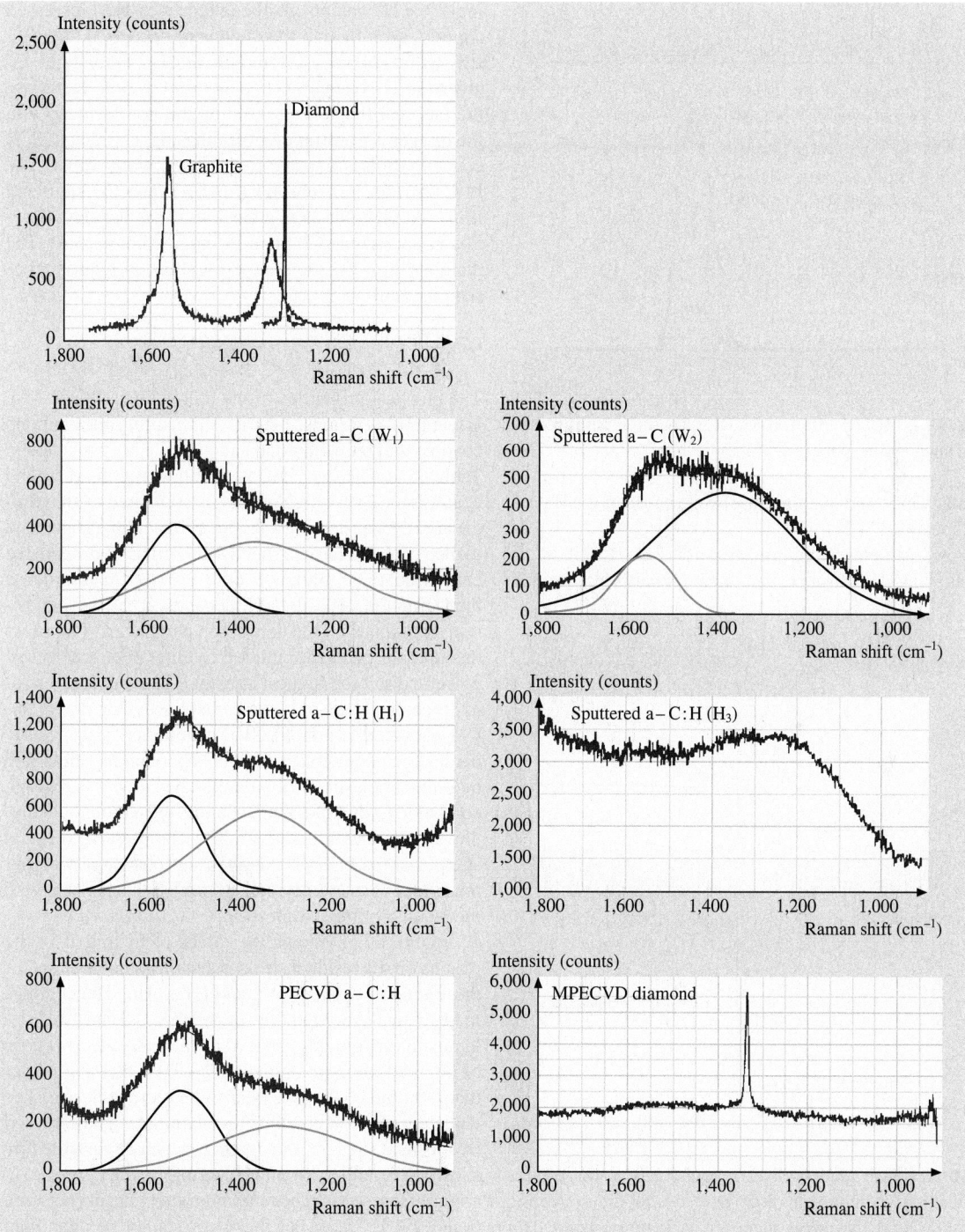

Fig. 42.6 Raman spectra of the DLC coatings produced by DC magnetron sputtering and RF-PECVD techniques and a diamond film produced by the MPE-CVD technique. Data for bulk diamond and microcrystalline graphite are included for comparison [42.35]

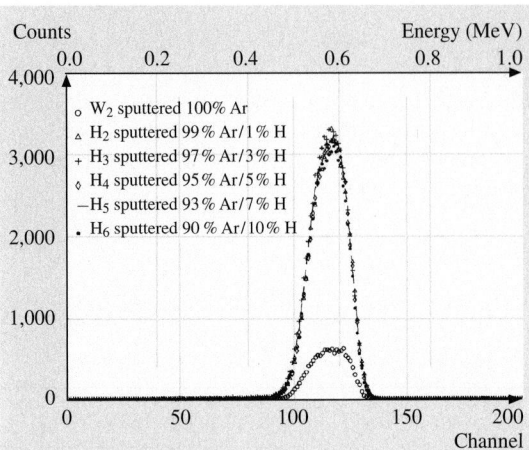

Fig. 42.7 FRS spectra for six DLC coatings produced by DC magnetron sputtering [42.35]

water vapor, and that this may be the cause of the H peak in specimen W2.

All samples were annealed for 24 h at 250 °C in a flowing He furnace and then reanalyzed. Surpris-

ingly, the H contents of all coatings measured increased slightly, even though the O content decreased, and W2 still had a substantial amount of H_2. This slight increase in H concentration is not understood. However, the fact that the H concentration did not decrease with the oxygen as a result of annealing suggests that high H concentration is not due to adsorbed water vapor. The PECVD film has more H ($\approx 18\%$) than the sputtered films initially, but after annealing it has the same fraction as specimen W2, the film sputtered in pure Ar. The diamond film has the smallest amount of hydrogen, as seen in Table 42.3.

42.2.3 Physical Properties

The physical properties of the four sputtered (W1, W2, H1, and H3) coatings, one PECVD coating, one diamond coating, and bulk diamond and graphite are presented in Table 42.4. The hydrogenated carbon and the diamond coatings have very high resistivity compared to unhydrogenated carbon coatings. It appears that unhydrogenated carbon coatings have higher densities than hydrogenated carbon coatings, although both groups are less dense than graphite. The density depends upon the deposition technique and the deposition parameters. It appears that unhydrogenated sputtered coatings deposited at low power exhibit the highest density. The nanohardness of hydrogenated carbon is somewhat lower than that of unhydrogenated carbon. PECVD coatings are significantly harder than sputtered coatings. The nanohardness and modulus of elasticity of a diamond coating is very high compared to that of a DLC coating, even though the hydrogen contents are similar. The compressive residual stresses of the PECVD coatings are substantially higher than those of sputtered coatings, which is consistent with the results for the hardness.

Figure 42.8a shows the effect of hydrogen in the plasma on the residual stresses and the nanohardness for the sputtered coatings W2 and H1 to H6. The coatings made with a hydrogen flow of between 0.5 and 1.0% delaminate very quickly, even when they are only a few tens of nm thick. In pure Ar and at H_2 flows that are greater than 1%, the coatings appear to be more adhesive. The tendency of some coatings to delaminate can be caused by intrinsic stress in the coating, which is measured by substrate bending. All of the coatings in the figure are in compressive stress. The maximum stress occurs between 0 and 1% H_2 flow, but the stress cannot be quantified in this range because the coatings instantly delaminate upon exposure to air. At higher hydrogen concentrations the stress gradually diminishes. A generally decreasing

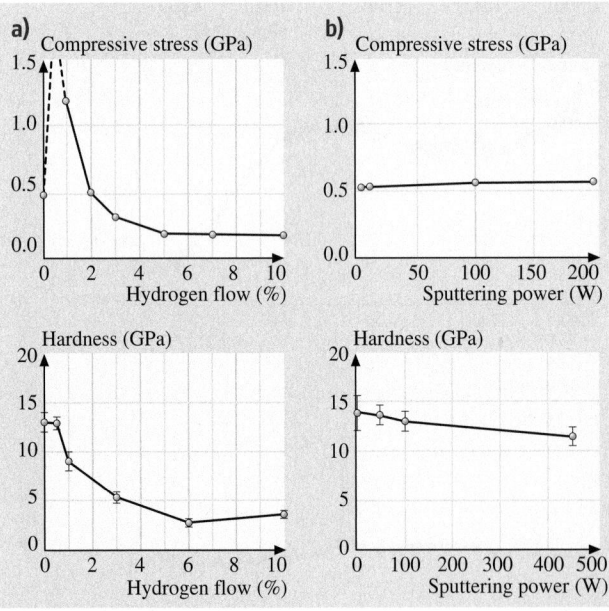

Fig. 42.8a,b Residual compressive stresses and nanohardness (**a**) as a function of hydrogen flow rate, where the sputtering power is 100 W and the target diameter is 75 mm (power density $= 2.1\,\mathrm{W/cm^2}$), and (**b**) as a function of sputtering power over a 75 mm diameter target with no hydrogen added to the plasma [42.40]

trend is observed for the hardness of the coatings as the hydrogen content is increased. The hardness decreases slightly, going from 0% H$_2$ to 0.5% H$_2$, and then decreases sharply. These results are probably lower than the true values because of local delamination around the indentation point. This is especially likely for the 0.5% and 1.0% coatings, where delamination is visually apparent, but may also be true to a lesser extent for the other coatings. Such an adjustment would bring the hardness profile into closer correlation with the stress profile. *Weissmantel* et al. [42.68] and *Scharff* et al. [42.85] observed a downturn in hardness for high bias and a low hydrocarbon gas pressure for ion-plated carbon coating, and, therefore, presumably low hydrogen content in support of the above contention.

Figure 42.8b shows the effect of sputtering power (with no hydrogen added to the plasma) on the residual stresses and nanohardness for various sputtered coatings. As the power decreases, the compressive stress does not seem to change while the nanohardness slowly increases. The rate of change becomes more rapid at very low power levels.

The addition of H$_2$ during sputtering of the carbon coatings increases the H concentration in the coating. Hydrogen causes the character of the C–C bonds to shift from sp^2 to sp^3, and a rise in the number of C–H bonds, which ultimately relieves stress and produces a softer "polymer-like" material. Low power deposition, like the presence of hydrogen, appears to stabilize the formation of sp^3 C–C bonds, increasing hardness. These coatings relieve stress and lead to better adhesion. Increasing the temperature during deposition at high power density results in graphitization of the coating material, producing a decrease in hardness with an increase in power density. Unfortunately, low power also means impractically low deposition rates.

42.2.4 Summary

Based on analyses of EELS and Raman data, it is clear that all DLC coatings have both sp^2 and sp^3 bonding characteristics. The sp^2/sp^3 bonding ratio depends upon the deposition technique and parameters used. DLC coatings deposited by sputtering and PECVD contain significant concentrations of hydrogen, while diamond coatings contain only small amounts of hydrogen impurities. Sputtered coatings with no deliberate addition of hydrogen in the plasma contain a significant amount of hydrogen. Regardless of how much hydrogen is in the Ar sputtering gas, the hydrogen content of the coatings increases initially but then does not increase further.

Hydrogen flow and sputtering power density affect the mechanical properties of these coatings. Maximum compressive residual stress and hardness occur between 0 and 1% hydrogen flow, resulting in rapid delamination. Low sputtering power moderately increases hardness and also relieves residual stress.

42.3 Micromechanical and Tribological Coating Characterization

42.3.1 Micromechanical Characterization

Common mechanical characterizations include measurements of hardness and elastic modulus, fracture toughness, fatigue life, and scratch and wear testing. Nanoindentation and atomic force microscopy (AFM) are used for the mechanical characterization of ultrathin films.

Hardness and elastic modulus are calculated from the load displacement data obtained by nanoindentation at loads of typically 0.2 to 10 mN using a commercially available nanoindenter [42.23,86]. This instrument monitors and records the dynamic load and displacement of the three-sided pyramidal diamond (Berkovich) indenter during indentation. For fracture toughness measurements of ultrathin films 100 nm to a few μm thick, a nanoindentation-based technique is used in which through-thickness cracking in the coating is detected from a discontinuity observed in the load–displacement curve, and the energy released during the cracking is obtained from the curve [42.87–89]. Based on the energy released, fracture mechanics analysis is then used to calculate the fracture toughness. An indenter with a cube-corner tip geometry is preferred because the through-thickness cracking of hard films can be accomplished at lower loads. In fatigue measurement, a conical diamond indenter with a tip radius of about one micron is used and load cycles with sinusoidal shapes are applied [42.90, 91]. The fatigue behavior of a coating is studied by monitoring the change in contact stiffness, which is sensitive to damage formation.

Hardness and Elastic Modulus

For materials that undergo plastic deformation, high hardness and elastic modulus are generally needed for low friction and wear, whereas for brittle materials, high fracture toughness is needed [42.2,3,21]. The DLC coatings used for many applications are hard and brittle, and values of hardness and fracture toughness need to be optimized.

Representative load–displacement plots of indentations made at 0.2 mN peak indentation load on 100 nm-thick DLC coatings deposited by the four deposition techniques on a single-crystal silicon substrate are compared in Fig. 42.9. The indentation depths at the peak load range from about 18 to 26 nm, smaller than that of the coating thickness. Many of the coatings exhibit a discontinuity or pop-in marks in the loading curve, which indicate a sudden penetration of the tip into the sample. A nonuniform penetration of the tip into a thin coating possibly results from formation of cracks in the coating, formation of cracks at the coating–substrate interface, or

debonding or delamination of the coating from the substrate.

The hardness and elastic modulus values for a peak load of 0.2 mN on the various coatings and single-crystal silicon substrate are summarized in Table 42.5 and Fig. 42.10 [42.47,49,89,90]. Typical values for the peak and residual indentation depths range from 18 to 26 nm and 6 to 12 nm, respectively. The FCA coating exhibits the greatest hardness of 24 GPa and the highest elastic modulus of 280 GPa of the various coatings, followed by the ECR-CVD, IB and SP coatings. Hardness and elastic modulus have been known to vary over a wide range with the sp^3-to-sp^2 bonding ratio, which depends on the kinetic energy of the carbon species and the amount of hydrogen [42.6, 30, 47, 92, 93]. The high hardness and elastic modulus of the FCA coatings are attributed to the high kinetic energy of the carbon species involved in the FCA deposition [42.12, 47]. *Anders* et al. [42.57] also reported a high hardness, measured by nanoindentation, of about 45 GPa for cathodic arc carbon coatings. They observed a change in hardness from 25 to 45 GPa with a pulsed bias voltage and bias duty cycle. The high hardness of cathodic arc carbon was attributed to the high percentage (more than 50%) of sp^3 bonding. *Savvides* and *Bell* [42.94] reported an increase in hardness from 12 to 30 GPa and an increase in elastic modulus from 62 to 213 GPa with an increase in the sp^3-to-sp^2 bonding ratio, from 3 to 6, for a C:H coating deposited by low-energy ion-assisted unbalanced magnetron sputtering of a graphite target in an Ar-H$_2$ mixture.

Bhushan et al. [42.35] reported hardnesses of about 15 and 35 GPa and elastic moduli of about 140 and 200 GPa, measured by nanoindentation, for a-C:H coatings deposited by DC magnetron sputtering and RF-plasma-enhanced chemical vapor deposition techniques, respectively. The high hardness of RF-PECVD a-C:H coatings is attributed to a higher concentration of sp^3 bonding than in a sputtered hydrogenated a-C:H coating. Hydrogen is believed to play a crucial role in the bonding configuration of carbon atoms by helping to stabilize the tetrahedral coordination (sp^3 bonding) of carbon species. *Jansen* et al. [42.78] suggested that the incorporation of hydrogen efficiently passivates the dangling bonds and saturates the graphitic bonding to some extent. However, a large concentration of hydrogen in the plasma in sputter deposition is undesirable. *Cho* et al. [42.33] and *Rubin* et al. [42.40] observed that the hardness decreased from 15 to 3 GPa with increased hydrogen content. *Bhushan* and *Doerner* [42.95] reported a hardness of about 10–20 GPa and an elastic modulus of about 170 GPa, measured by nanoindenta-

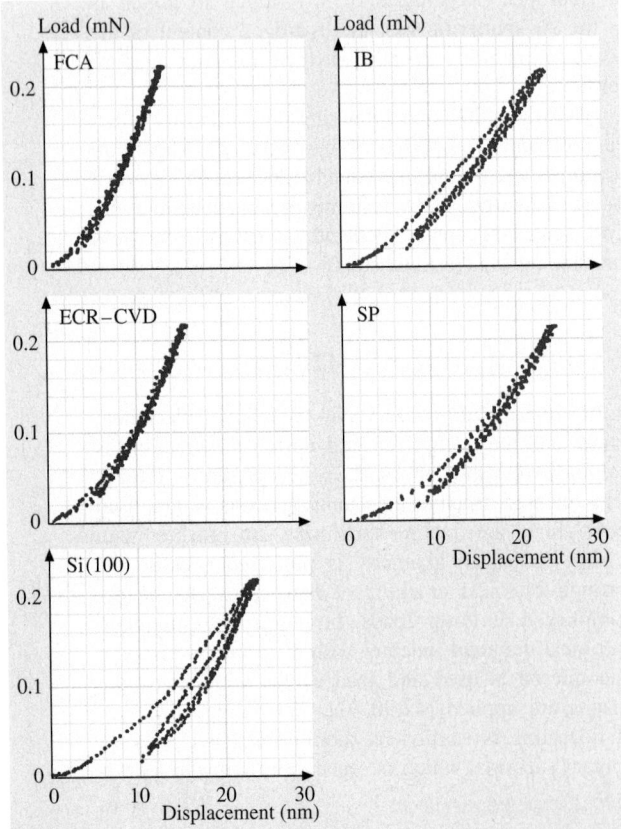

Fig. 42.9 Load versus displacement plots for various 100 nm-thick amorphous carbon coatings on single-crystal silicon substrate and bare substrate

Table 42.5 Hardness, elastic modulus, fracture toughness, fatigue life, critical load during scratch, coefficient of friction during accelerated wear testing and residual stress for various DLC coatings on single-crystal silicon substrate

Coating	Hardness[a] [42.48] (GPa)	Elastic modulus[a] [42.48] (GPa)	Fracture toughness[a] [42.89] (MPa m$^{1/2}$)	Fatigue life[b], N_f[d] [42.90] ×10^4	Critical load during scratch[b] [42.48] (mN)	Coefficient of friction during accelerated wear testing[b] [42.48]	Compressive residual stress[c] [42.47] (GPa)
Cathodic arc carbon coating (a-C)	24	280	11.8	2.0	3.8	0.18	12.5
Ion beam carbon coating (a-C:H)	19	140	4.3	0.8	2.3	0.18	1.5
ECR-CVD carbon coating (a-C:H)	22	180	6.4	1.2	5.3	0.22	0.6
DC sputtered carbon coating (a-C:H)	15	140	2.8	0.2	1.1	0.32	2.0
Bulk graphite (for comparison)	Very soft	9–15	–	–	–	–	–
Diamond (for comparison)	80–104	900–1050	–	–	–	–	–
Si(100) substrate	11	220	0.75	–	0.6	0.55	0.02

[a] Measured on 100 nm-thick coatings
[b] Measured on 20 nm-thick coatings
[c] Measured on 400 nm-thick coatings
[d] N_f was obtained for a mean load of 10 μN and a load amplitude of 8 μN

tion, for 100 nm-thick DC magnetron sputtered a-C:H on the silicon substrate.

Residual stresses measured using a well-known curvature measurement technique are also presented in Table 42.5. The DLC coatings are under significant compressive internal stresses. Very high compressive stresses in FCA coatings are believed to be partly responsible for their high hardness. However, high stresses result in coating delamination and buckling. For this reason, the coatings that are thicker than about 1 μm have a tendency to delaminate from the substrates.

Fracture Toughness

Representative load–displacement curves of indentations on 400 nm-thick cathodic arc carbon coating on silicon for various peak loads are shown in Fig. 42.11. Steps are found in all of the curves, as shown by arrows in Fig. 42.11a. In the 30 mN SEM micrograph, in addi-

tion to several radial cracks, ring-like through-thickness cracking is observed with small lips of material overhanging the edge of indentation. The steps at about 23 mN in the loading curves of indentations made with 30 and 100 mN peak indentation loads result from the ring-like through-thickness cracking. The step at 175 mN in the loading curve of the indentation made with 200 mN peak indentation load is caused by spalling and second ring-like through-thickness cracking.

According to Li et al. [42.87], the fracture process progresses in three stages: (1) ring-like through-thickness cracks form around the indenter due to high stresses in the contact area, (2) delamination and buckling occur around the contact area at the coating/substrate interface due to high lateral pressure, and (3) second ring-like through-thickness cracks and spalling are generated by high bending stresses at the edges of the buckled coating, see Fig. 42.12a. In the

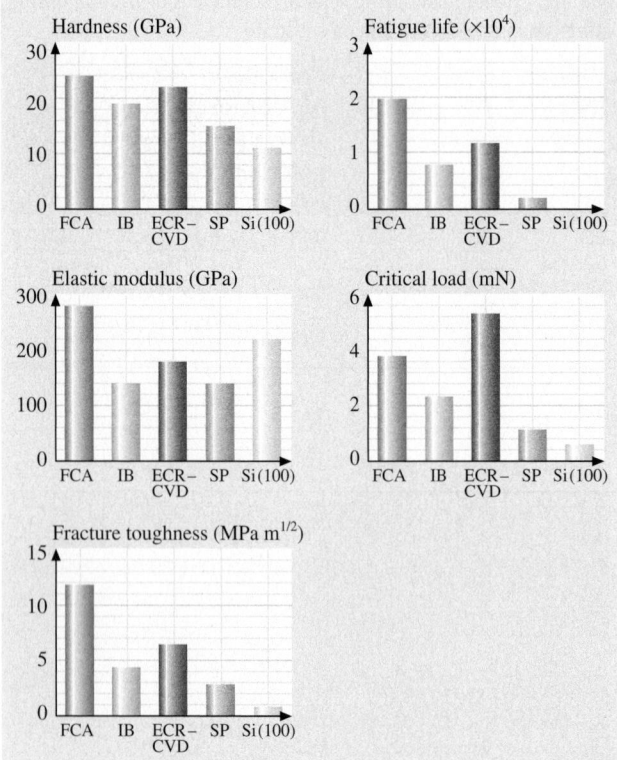

Fig. 42.10 Bar charts summarizing data for various coatings and single-crystal silicon substrate. Hardnesses, elastic moduli, and fracture toughnesses were measured on 100 nm-thick coatings, and fatigue lifetimes and critical loads during scratch were measured on 20 nm-thick coatings

first stage, if the coating under the indenter is separated from the bulk coating via the first ring-like through-thickness cracking, a corresponding step will be present in the loading curve. If discontinuous cracks form and the coating under the indenter is not separated from the remaining coating, no step appears in the loading curve, because the coating still supports the indenter and the indenter cannot suddenly advance into the material. In the second stage, for the coating used in the present study, the advances of the indenter during the radial cracking delamination, and buckling are not big enough to form steps in the loading curve, because the coating around the indenter still supports the indenter, but they generate discontinuities that change the slope of the loading curve with increasing indentation load. In the third stage, the stress concentration at the end of the interfacial crack cannot be relaxed by the propagation

of the interfacial crack. With an increase in indentation depth, the height of the bulged coating increases. When the height reaches a critical value, the bending stresses caused by the bulged coating around the indenter will result in second ring-like through-thickness crack formation and spalling at the edge of the buckled coating, as shown in Fig. 42.12a, which leads to a step in the loading curve. This is a single event and it results in the separation of the part of the coating around the indenter from the bulk coating via cracking through coatings. The step in the loading curve results (completely) from the coating cracking and not from the interfacial cracking or the substrate cracking.

The area under the load–displacement curve is the work performed by the indenter during the elastic–plastic deformation of the coating/substrate system. The strain energy release in the first/second ring-like cracking and spalling can be calculated from the corresponding steps in the loading curve. Figure 42.12b shows a modeled load–displacement curve. OACD is the loading curve and DE is the unloading curve. The first ring-like through-thickness crack should be considered. It should be emphasized that the edge of the buckled coating is far from the indenter and, therefore, it does not matter if the indentation depth exceeds the coating thickness, or if deformation of the substrate occurs around the indenter when we measure the fracture toughness of the coating from the energy released during the second ring-like through-thickness cracking (spalling). Suppose that the second ring-like through-thickness cracking occurs at AC. Now, let us consider the loading curve OAC. If the second ring-like through-thickness crack does not occur, OA will extend to OB to reach the same displacement as OC. This means that crack formation changes the loading curve OAB into OAC. For point B, the elastic–plastic energy stored in the coating/substrate system should be OBF. For point C, the elastic–plastic energy stored in the coating/substrate system should be OACF. Therefore, the energy difference before and after the crack generation is the area of ABC, so this energy stored in ABC will be released as strain energy, creating the ring-like through-thickness crack. According to the theoretical analysis by *Li* et al. [42.87], the fracture toughness of a thin film can be written as

$$K_{Ic} = \left[\left(\frac{E}{(1-\nu^2)\, 2\pi C_R} \right) \left(\frac{U}{t} \right) \right]^{1/2} , \qquad (42.2)$$

where E is the elastic modulus, ν is the Poisson ratio, $2\pi C_R$ is the crack length in the coating plane, t is the coating thickness, and U is the strain energy difference before and after cracking.

The fracture toughness of the coatings can be calculated using (42.2). The loading curve is extrapolated along the tangential direction of the loading curve from the starting point of the step up to reach the same displacement as the step. The area between the extrapolated line and the step is the estimated strain energy difference before and after cracking. C_R is measured from SEM micrographs or AFM images of indentations. The second ring-like crack is where the spalling occurs. For example, for the 400 nm-thick cathodic arc carbon coating data presented in Fig. 42.11, the U value of 7.1 nNm is assessed from the steps in Fig. 42.11a at peak indentation loads of 200 mN. For a C_R value of 7.0 μm, from Fig. 42.11b, with $E = 300$ GPa (measured using a nanoindenter and an assumed value of 0.25 for v), fracture toughness values are calculated as 10.9 MPa$\sqrt{m}$ [42.87, 88]. The fracture toughness and related data for various 100 nm-thick DLC coatings are presented in Fig. 42.10 and Table 42.5.

Nanofatigue

Delayed fracture resulting from extended service is called fatigue [42.96]. Fatigue fracturing progresses through a material via changes within the material at the tip of a crack, where there is a high stress intensity. There are several situations: cyclic fatigue, stress corrosion and static fatigue. Cyclic fatigue results from cyclic loading of machine components. In a low-flying slider in a magnetic head-disk interface, isolated asperity contacts occur during use and the fatigue failure occurs in the multilayered thin film structure of the magnetic disk [42.13]. Impact occurs in many MEMS components and the failure mode is cyclic fatigue. Asperity contacts can be simulated using a sharp diamond tip in oscillating contact with the component.

Figure 42.13 shows the schematic of a fatigue test on a coating/substrate system using a continuous stiffness measurement (CSM) technique. Load cycles are applied to the coating, resulting in cyclic stress. P is the cyclic load, P_{mean} is the mean load, P_0 is the oscillation load amplitude, and ω is the oscillation frequency. The following results can be obtained: (1) endurance limit (the maximum load below which there is no coating failure for a preset number of cycles); (2) number of cycles at which the coating failure occurs; and (3) changes in contact stiffness (measured using the unloading slope of each cycle), which can be used to monitor the propagation of interfacial cracks during a cyclic fatigue process.

Figure 42.14a shows the contact stiffness as a function of the number of cycles for 20 nm-thick FCA

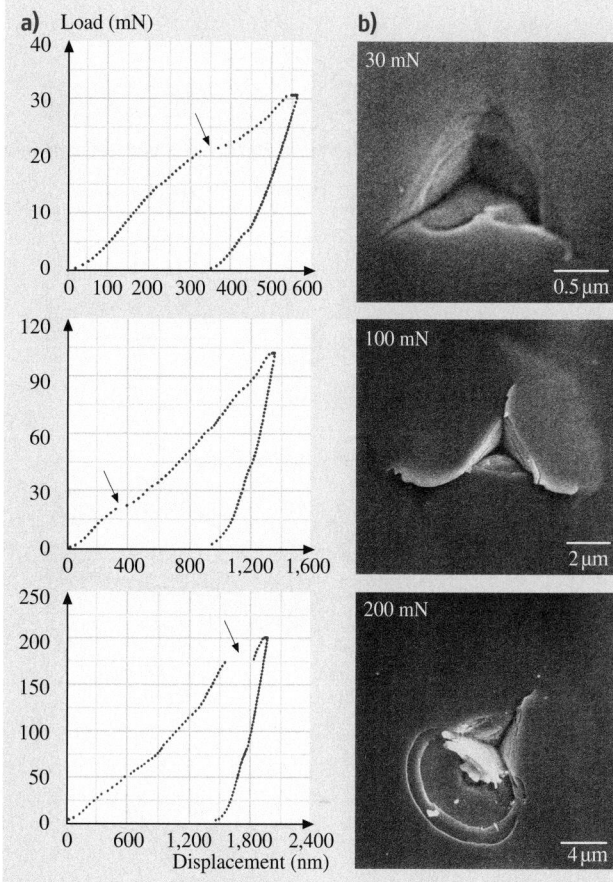

Fig. 42.11 (a) Load–displacement curves of indentations made with 30, 100, and 200 mN peak indentation loads using the cube corner indenter, and (b) SEM micrographs of indentations on a 400 nm-thick cathodic arc carbon coating on silicon. *Arrows* indicate steps during the loading portion of the load–displacement curve [42.87]

coatings cyclically deformed by various oscillation load amplitudes with a mean load of 10 μN at a frequency of 45 Hz. At 4 μN load amplitude, no change in contact stiffness was found for all coatings. This indicates that 4 μN load amplitude is not high enough to damage the coatings. At 6 μN load amplitude, an abrupt decrease in contact stiffness was found after a certain number of cycles for each coating, indicating that fatigue damage had occurred. With increasing load amplitude, the number of cycles to failure, N_f, decreases for all coatings. Load amplitude versus N_f, a so-called S–N curve, is plotted in Fig. 42.14b. The critical load amplitude below which no fatigue damage occurs (an endurance limit), was identified for each coating. This critical load amplitude,

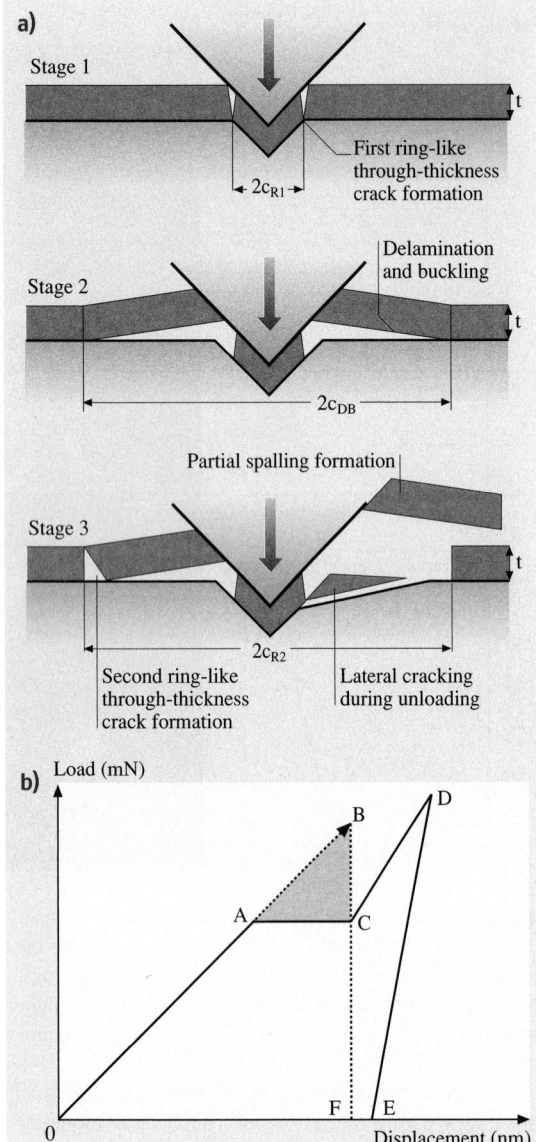

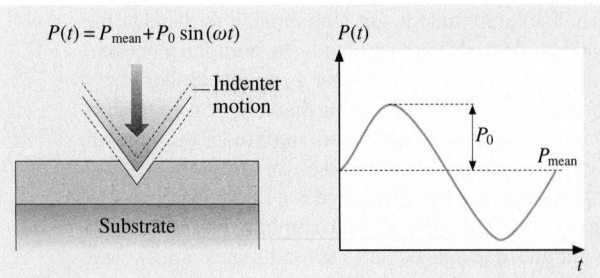

Fig. 42.12 (a) Schematic of various stages in nanoindentation fracture for the coating/substrate system, and (b) schematic of a load–displacement curve showing a step during the loading cycle and the associated energy release

together with the mean load, are of critical importance to the design of head-disk interfaces or MEMS/NEMS device interfaces.

To compare the fatigue lives of the different coatings studied, the contact stiffness is shown as a function of the number of cycles for 20 nm-thick FCA, IB, ECR-CVD and SP coatings cyclically deformed by an oscillation load amplitude of 8 μN with a mean load of 10 μN at a frequency of 45 Hz in Fig. 42.14c. The FCA coating has the largest N_f, followed by the ECR-CVD, IB and SP coatings. In addition, after N_f, the contact stiffness of the FCA coating shows a slower decrease than the other coatings. This indicates that the FCA coating was less damaged than the others after N_f. The fatigue behaviors of FCA and ECR-CVD coatings of different thicknesses are compared in Fig. 42.14d. For both coatings, N_f decreases with decreasing coating thickness. At 10 nm, FCA and ECR-CVD have almost the same fatigue life. At 5 nm, the ECR-CVD coating shows a slightly longer fatigue life than the FCA coating. This indicates that the microstructure and residual stresses are not uniform across the thickness direction, even for nanometer-thick DLC coatings. Thinner coatings are more influenced by interfacial stresses than thicker coatings.

Figure 42.15a shows high-magnification SEM images of 20 nm-thick FCA coatings before, at, and after N_f. In the SEM images, the net-like structure is the gold film coated on the DLC coating, which should be ignored when analyzing the indentation fatigue damage. Before N_f, no delamination or buckling was found except for the residual indentation mark at magnifications of up to 1 200 000× using SEM. This suggests that only plastic deformation occurred before N_f. At N_f, the coating around the indenter bulged upwards, indicating delamination and buckling. Therefore, it is believed that the decrease in contact stiffness at N_f results from delamination and buckling of the coating from the substrate. After N_f, the buckled coating was broken down around the edge of the buckled area, forming a ring-like crack. The remaining coating overhung at the edge of the buckled area. It is noted that the in-

Fig. 42.13 Schematic of a fatigue test on a coating/substrate system using the continuous stiffness measurement technique

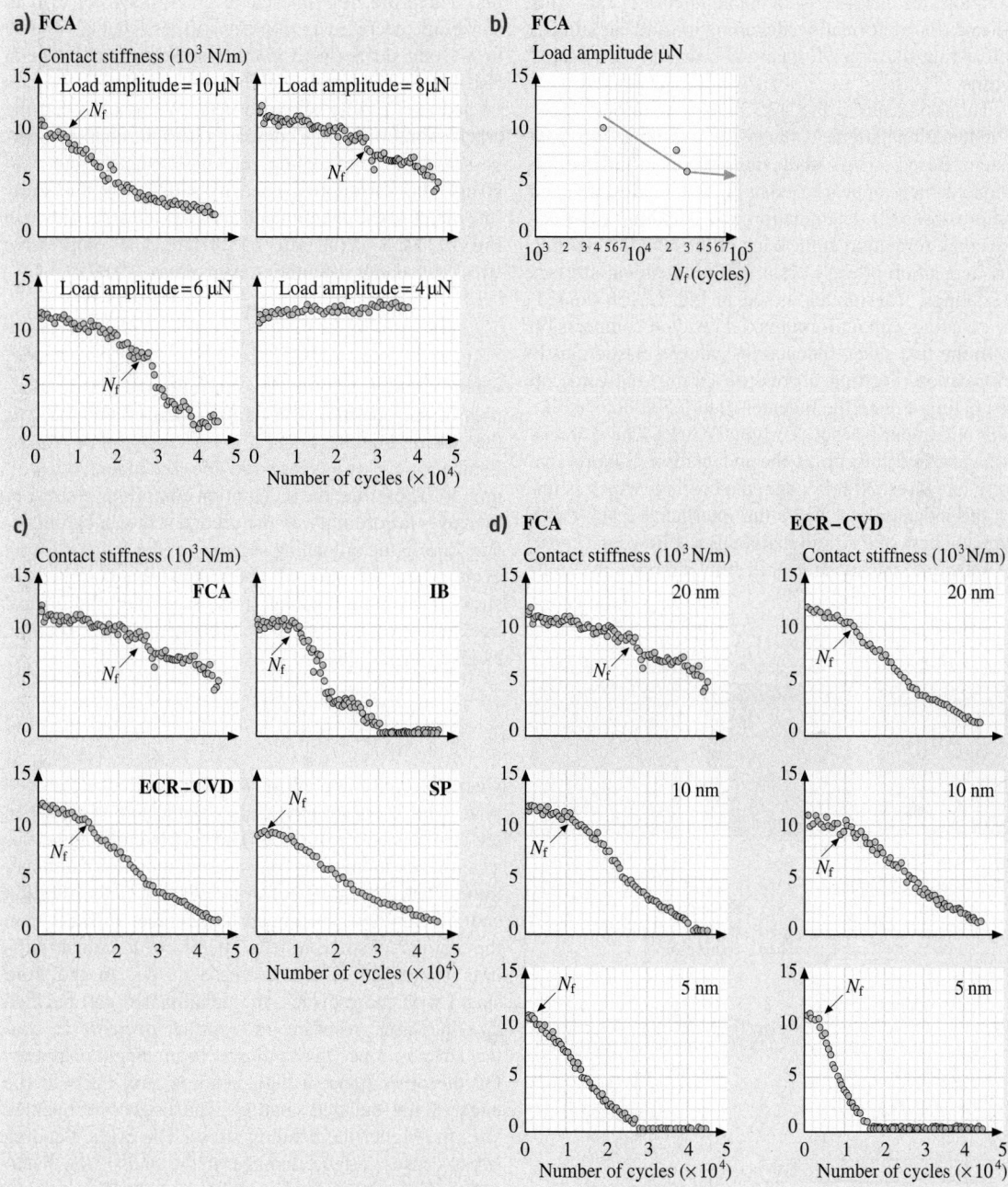

Fig. 42.14 (a) Contact stiffness as a function of the number of cycles for 20 nm-thick FCA coatings cyclically deformed by various oscillation load amplitudes with a mean load of 10 μN at a frequency of 45 Hz; (b) plot of load amplitude versus N_f; (c) contact stiffness as a function of the number of cycles for four different 20 nm-thick coatings with a mean load of 10 μN and a load amplitude of 8 μN at a frequency of 45 Hz; and (d) contact stiffness as a function of the number of cycles for two coatings of different thicknesses at a mean load of 10 μN and a load amplitude of 8 μN at a frequency of 45 Hz

dentation size increases with the number of cycles. This indicates that deformation, delamination and buckling as well as ring-like crack formation occurred over a period of time.

The schematic in Fig. 42.15b shows various stages of indentation fatigue damage for a coating/substrate system. Based on this study, three stages of indentation fatigue damage appear to exist: (1) indentation-induced compression; (2) delamination and buckling; (3) ring-like crack formation at the edge of the buckled coating. The deposition process often induces residual stresses in coatings. The model shown in Fig. 42.15b considers a coating with uniform biaxial residual compression σ_r. In the first stage, indentation induces elastic/plastic deformation, exerting a pressure (acting outward) on the coating around the indenter. Interfacial defects like voids and impurities act as original cracks. These cracks propagate and link up as the indentation compressive stress increases. At this stage, the coating, which is under the indenter and above the interfacial crack (with a crack length of $2a$), still maintains a solid contact with the substrate; the substrate still fully supports the coat-

ing. Therefore, this interfacial crack does not lead to an abrupt decrease in contact stiffness, but gives rise to a slight decrease in contact stiffness, as shown in Fig. 42.14. The coating above the interfacial crack is treated as a rigidly clamped disk. We assume that the crack radius, a, is large compared with the coating thickness t. Since the coating thickness ranges from 5 to 20 nm, this assumption is easily satisfied in this study (the radius of the delaminated and buckled area, shown in Fig. 42.15a, is on the order of 100 nm). The compressive stress caused by indentation is given as [42.97]:

$$\sigma_i = \frac{E}{(1-v)}\varepsilon_i = \frac{EV_i}{2\pi t a^2 (1-v)} \, , \qquad (42.3)$$

where v and E are the Poisson ratio and elastic modulus of the coating, V_i is the indentation volume, t is the coating thickness, and a is the crack radius. As the number of cycles increases, so does the indentation volume V_i. Therefore, the indentation compressive stress σ_i increases accordingly. In the second stage, buckling occurs during the unloading segment of the fatigue testing cycle when the sum of the indentation compressive stress σ_i and the residual stress σ_r exceed the critical buckling stress σ_b for the delaminated circular section, as given by [42.98]

$$\sigma_b = \frac{\mu^2 E}{12\left(1-v^2\right)}\left(\frac{t}{a}\right)^2 \, , \qquad (42.4)$$

where the constant μ equals 42.67 for a circular clamped plate with a constrained center point and 14.68 when the center is unconstrained. The buckled coating acts as a cantilever. In this case, the indenter indents a cantilever rather than a coating/substrate system. This ultrathin coating cantilever has much less contact stiffness than the coating/substrate system. Therefore, the contact stiffness shows an abrupt decrease at N_f. In the third stage, with more cycles, the delaminated and buckled size increases, resulting in a further decrease in contact stiffness since the cantilever beam length increases. On the other hand, a high bending stress acts at the edge of the buckled coating. The larger the buckled size, the higher the bending stress. The cyclic bending stress causes fatigue damage at the end of the buckled coating, forming a ring-like crack. The coating under the indenter is separated from the bulk coating (caused by the ring-like crack at the edge of the buckled coating) and the substrate (caused by the delamination and buckling in the second stage). Therefore, the coating under the indenter is not constrained; it is free to move with the indenter during fatigue test-

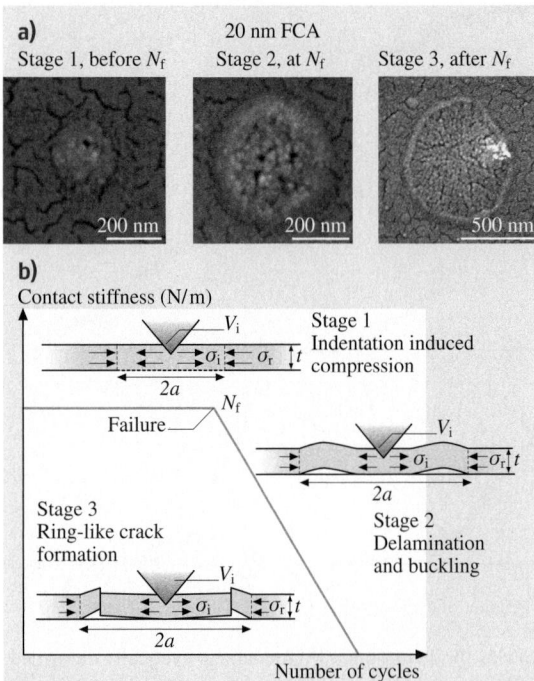

Fig. 42.15 (a) High-magnification SEM images of a coating before, at, and after N_f, and (b) schematic of various stages of indentation fatigue damage for a coating/substrate system [42.90]

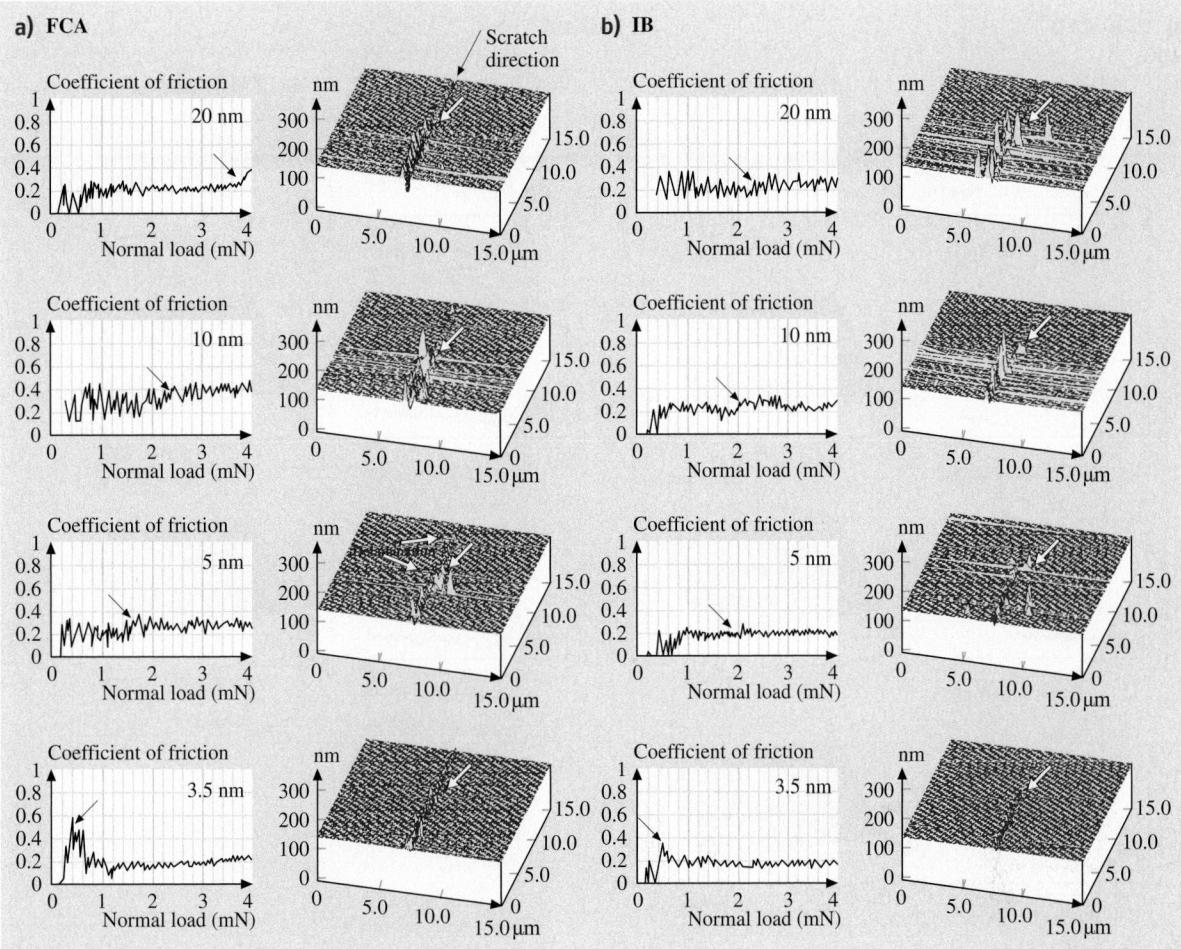

Fig. 42.16a,b

ing. At this point, the sharp nature of the indenter is lost, because the coating under the indenter gets stuck on the indenter. The indentation fatigue experiment results in the contact of a (relatively) huge blunt tip with the substrate. This results in a low contact stiffness value.

Compressive residual stresses result in delamination and buckling. A coating with a higher adhesion strength and a lower compressive residual stress is required for a higher fatigue life. Interfacial defects should be avoided in the coating deposition process. We know that ring-like crack formation occurs in the coating. Formation of fatigue cracks in the coating depends upon the hardness and the fracture toughness. Cracks are more difficult to form and propagate in the coating with higher strength and fracture toughness.

It is now accepted that long fatigue life in a coating/substrate almost always involves "living with a crack", that the threshold or limit condition is associated with the nonpropagation of existing cracks or defects, even though these cracks may be undetectable [42.96]. For all of the coatings studied at $4\,\mu N$, the contact stiffness does not change much. This indicates that delamination and buckling did not occur within the number of cycles tested in this study. This is probably because the indentation-induced compressive stress was not high enough to allow the cracks to propagate and link up under the indenter, or the sum of the indentation compressive stress σ_i and the residual stress σ_r did not exceed the critical buckling stress σ_b.

Figure 42.10 and Table 42.5 summarize the hardnesses, elastic moduli, fracture toughnesses, and fatigue

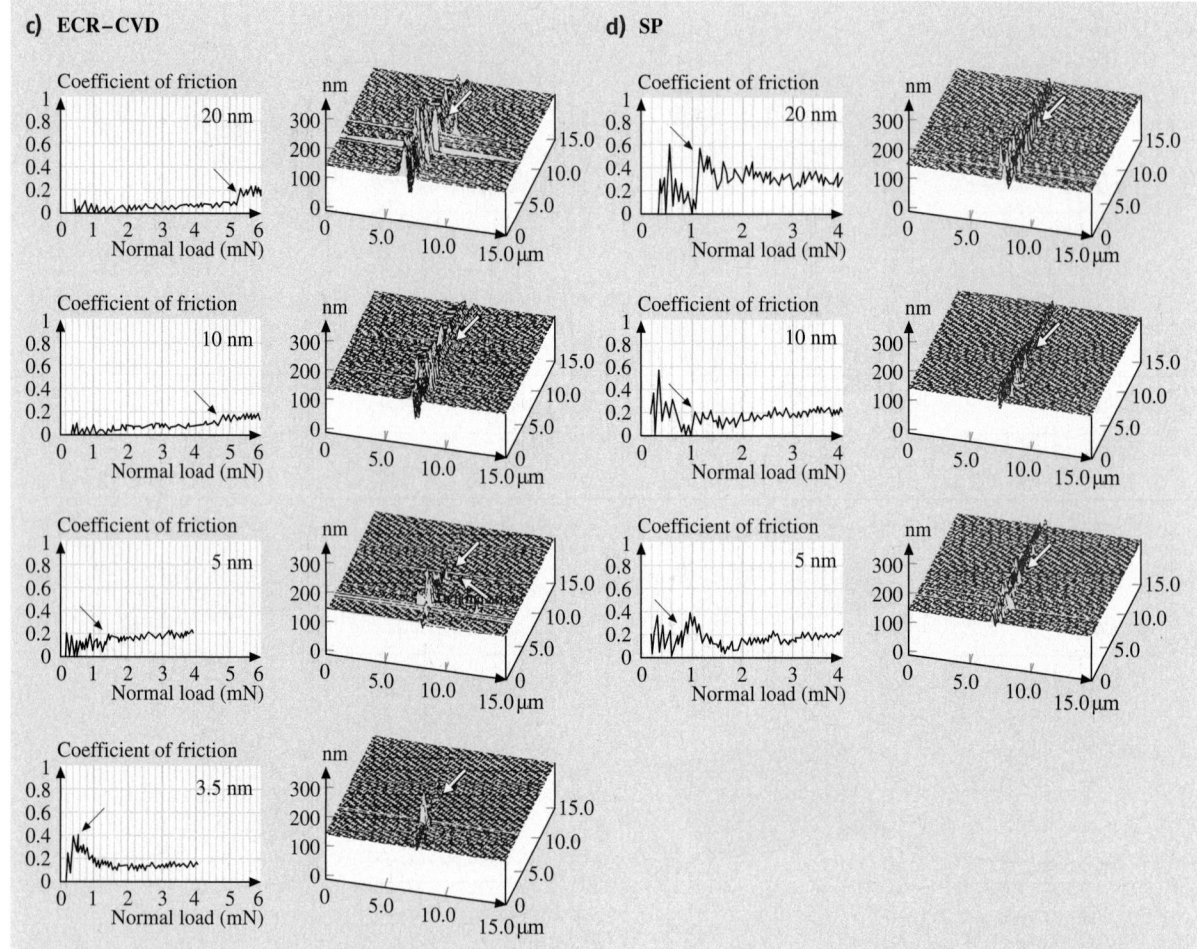

Fig. 42.16a–d Coefficient of friction profiles as a function of normal load, as well as corresponding AFM surface height maps of regions over scratches at the respective critical loads (indicated by the *arrows* in the friction profiles and AFM images), for coatings of different thicknesses deposited by various deposition techniques: (**a**) FCA, (**b**) IB, (**c**) ECR-CVD, (**d**) SP

lifetimes of all of the coatings studied. A good correlation exists between the fatigue life and other mechanical properties. Higher mechanical properties result in a longer fatigue life. The mechanical properties of DLC coatings are controlled by the sp^3-to-sp^2 ratio. An sp^3-bonded carbon exhibits the outstanding properties of diamond [42.51]. Higher kinetic energy during deposition will result in a larger fraction of sp^3-bonded carbon in an amorphous network. Thus, the higher kinetic energy for the FCA could be responsible for its enhanced carbon structure and mechanical properties [42.48–50, 99]. Higher adhesion strength between the FCA coating and substrate makes the FCA coating more difficult to delaminate from the substrate.

42.3.2 Microscratch and Microwear Studies

For microscratch studies, a conical diamond indenter (that has a tip radius of about one micron and a cone angle of 60° for example) is drawn over the sample surface, and the load is ramped up (typically from 2 mN to 25 mN) until substantial damage occurs. The coefficient of friction is monitored during scratching. Scratch-induced coating damage, specifically fracture or delamination, can be monitored by in situ friction force measurements and using optical and SEM imaging of the scratches after tests. A gradual increase in friction is associated with plowing, and an abrupt increase in friction is associated with fracture or catastrophic failure [42.100]. The load corresponding to an abrupt increase in friction or an increase in friction above a certain value (typically $2\times$ the initial value) provides a measure of the scratch resistance or the adhesive strength of a coating, and is called the "critical load". The depths of scratches are measured with increasing scratch length or normal load using an AFM, typically with an area of $10\times10\,\mu m$ [42.48, 49, 101].

Microscratch and microwear studies are also conducted using an AFM [42.23, 50, 99, 102, 103]. A square pyramidal diamond tip (tip radius $\approx 100\,nm$) or a three-sided pyramidal diamond (Berkovich) tip with an apex angle of $60°$ and a tip radius of about $100\,nm$, mounted on a platinum-coated, rectangular stainless steel cantilever of stiffness of about $40\,N/m$, is scanned orthogonal to the long axis of the cantilever to generate scratch and wear marks. During the scratch test, the normal load is either kept constant or is increased (typically from 0 to $100\,\mu N$) until damage occurs. Topographical images of the scratch are obtained in situ with the AFM at a low load. By scanning the sample during scratching, wear experiments can be conducted. Wear is monitored as a function of the number of cycles at a constant load. Normal loads ($10-80\,\mu N$) are typically used.

Microscratch

Scratch tests conducted with a sharp diamond tip simulate a sharp asperity contact. In a scratch test, the cracking or delamination of a hard coating is signaled by a sudden increase in the coefficient of friction [42.23]. The load associated with this event is called the "critical load".

Wu [42.104], *Bhushan* et al. [42.70], *Gupta* and *Bhushan* [42.12, 47], and *Li* and *Bhushan* [42.48, 49, 101] have used a nanoindenter to perform microscratch (mechanical durability) studies of various carbon coatings. The coefficient of friction profiles as a function of increasing normal load as well as AFM surface height maps of regions over scratches at the respective critical loads (indicated by the arrows in the friction profiles and AFM images) observed for coatings with different thicknesses and single-crystal silicon substrate using a conical tip are compared in Figs. 42.16 and 42.17. *Bhushan* and *Koinkar* [42.102], *Koinkar* and *Bhushan* [42.103], *Bhushan* [42.23], and *Sundararajan* and *Bhushan* [42.50, 99] used an AFM to perform microscratch studies. Data obtained for coatings with different thicknesses and silicon substrate using a Berkovich tip are compared in Figs. 42.18 and 42.19. Critical loads for various coatings tested using a nanoindenter and AFM are summarized in Fig. 42.20. Selected data for

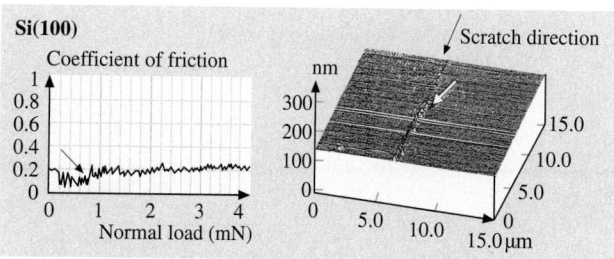

Fig. 42.17 Coefficient of friction profiles as a function of normal load as well as corresponding AFM surface height maps of regions over scratches at the respective critical loads (indicated by the *arrows* in the friction profiles and AFM images) for Si(100)

20 nm-thick coatings obtained using nanoindenter are also presented in Fig. 42.10 and Table 42.5.

It is clear that a well-defined critical load exists for each coating. The AFM images clearly show that below the critical loads the coatings were plowed by the scratch tip, associated with the plastic flow of materials. At and after the critical loads, debris (chips) or buckling was observed on the sides of the scratches. Delamination or buckling can be observed around or after the critical loads, which suggests that the damage starts from delamination and buckling. For the 3.5- and 5 nm-thick FCA coatings, before the critical loads small debris is observed on the sides of the scratches. This suggests that the thinner FCA coatings are not so durable. It is obvious that, for a given deposition method, the critical loads increase with increasing coating thickness. This indicates that the critical load is determined not only by the strength of adhesion to the substrate, but also by the coating thickness. We note that more debris generated on the thicker coatings than thinner coatings. A thicker coating is more difficult to break; the broken coating chips (debris) seen for a thicker coating are larger than those for the thinner coatings. The different residual stresses of coatings of different thicknesses may also affect the size of the debris. The AFM image shows that the silicon substrate was damaged by plowing, associated with the plastic flow of material. At and after the critical load, a small amount of uniform debris is observed and the amount of debris increases with increasing normal load.

Since the damage mechanism at the critical load appears to be the onset of plowing, harder coatings with more fracture toughness will therefore require a higher load for deformation and hence a higher critical load. Figure 42.21 gives critical loads of various coatings, obtained with AFM tests, as a function of the coating hardness and fracture toughness (from Table 42.5). It can be seen that, in general, increasing coating hard-

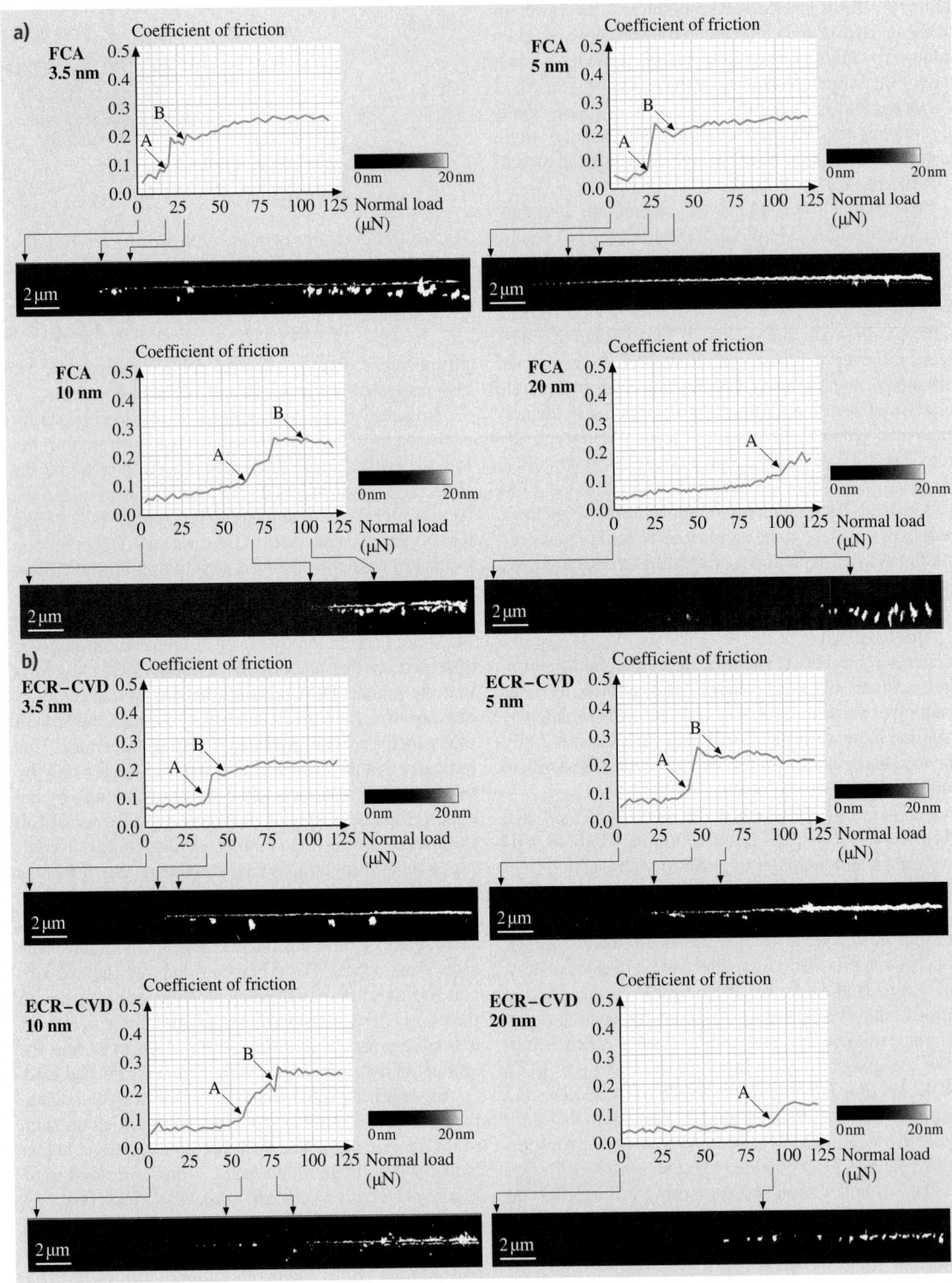

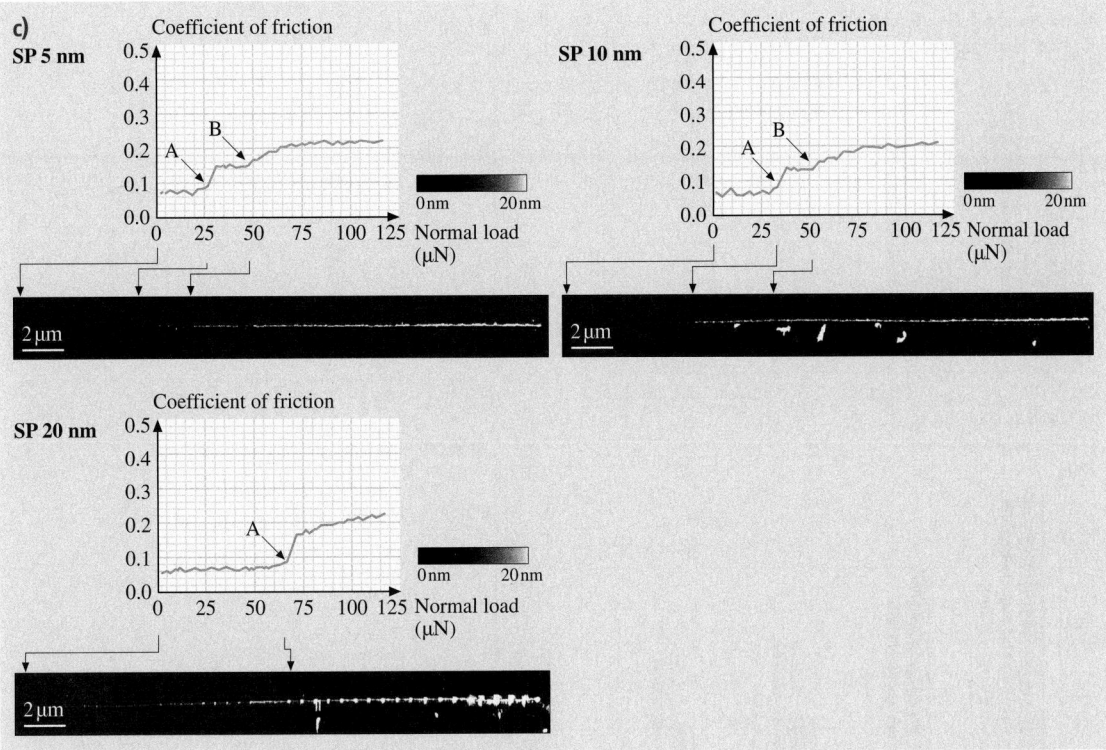

Fig. 42.18a–c Coefficient of friction profiles during scratch as a function of normal load and corresponding AFM surface height maps for (**a**) FCA, (**b**) ECR-CVD, and (**c**) SP coatings [42.99] ◀ ▲

ness and fracture toughness results in a higher critical load. The only exceptions are the FCA coatings at 5 and 3.5 nm thickness, which show the lowest critical loads despite their high hardness and fracture toughness. The brittleness of the thinner FCA coatings may be one

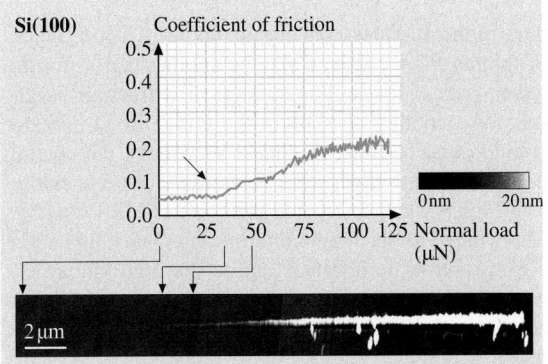

Fig. 42.19 Coefficient of friction profiles during scratch as a function of normal load and corresponding AFM surface height maps for Si(100) [42.99]

reason for their low critical loads. The mechanical properties of coatings that are less than 10 nm thick are not known. The FCA process may result in coatings with low hardness at low thickness due to differences in coating stoichiometry and structure compared to coatings of higher thickness. Also, at these thicknesses stresses at the coating–substrate interface may affect coating adhesion and load-carrying capacity.

Based on the experimental results, a schematic of the scratch damage mechanisms encountered for the DLC coatings used in this study is shown in Fig. 42.22. Below the critical load, if a coating has a good combination of strength and fracture toughness, plowing associated with the plastic flow of materials is responsible for the coating damage (Fig. 42.22a). However, if the coating has a low fracture toughness, cracking could occur during plowing, resulting in the formation of small amounts of debris (Fig. 42.22b). When the normal load is increased to the critical load, delamination or buckling will occur at the coating/substrate interface (Fig. 42.22c). A further increase in normal load will result in coating breakdown via through-coating thickness

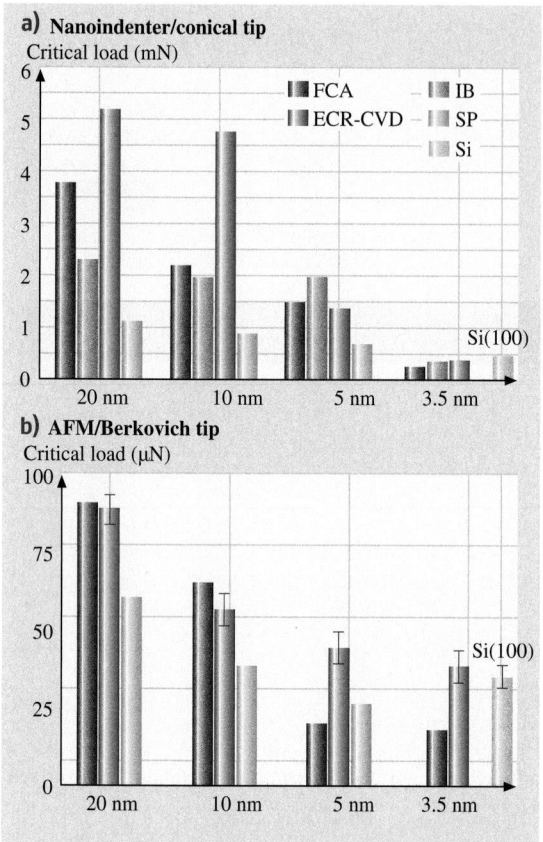

Fig. 42.20a,b Critical loads estimated from the coefficient of friction profiles from (**a**) nanoindenter and (**b**) AFM tests for various coatings of different thicknesses and Si(100) substrate

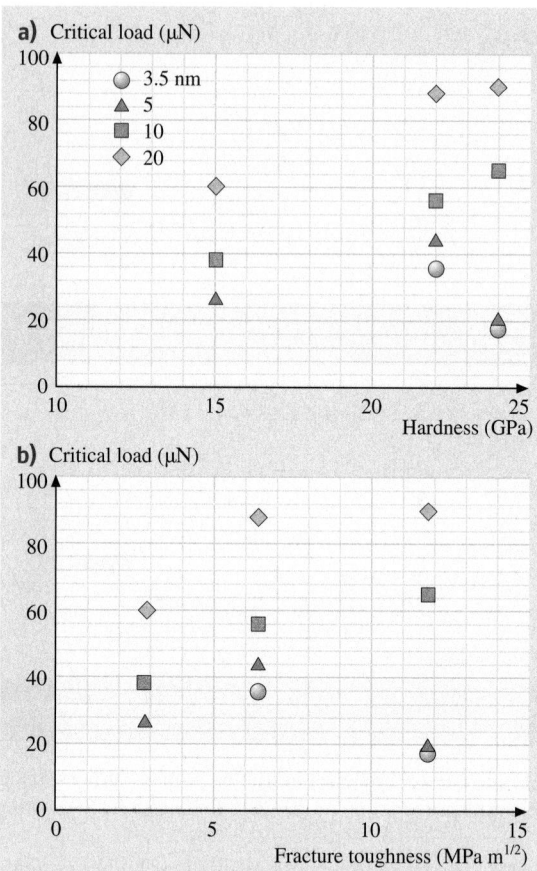

Fig. 42.21a,b Measured critical loads estimated from the coefficient of friction profiles from AFM tests as a function of (**a**) coating hardness and (**b**) fracture toughness. Coating hardness and fracture toughness values were obtained using a nanoindenter on 100 nm-thick coatings (Table 42.5)

cracking, as shown in Fig. 42.22d. Therefore, adhesion strength plays a crucial role in the determination of critical load. If a coating adheres strongly to the substrate, the coating is more difficult to delaminate, which will result in a higher critical load. The interfacial and residual stresses of a coating may also greatly affect the delamination and buckling [42.1]. A coating with higher interfacial and residual stresses is more easily delaminated and buckled, which will result in a low critical load. It was reported earlier that FCA coatings have higher residual stresses than other coatings [42.47]. Interfacial stresses play an increasingly important role as the coating gets thinner. A large mismatch in elastic modulus between the FCA coating and the silicon substrate may cause large interfacial stresses. This may be why thinner FCA coatings show lower critical loads than thicker FCA coatings, even though the FCA coatings

have higher hardness and elastic moduli. The brittleness of thinner FCA coatings may be another reason for the lower critical loads. The strength and fracture toughness of a coating also affect the critical load. Greater strength and fracture toughness will make the coating more difficult to break after delamination and buckling. The high scratch resistance/adhesion of FCA coatings is attributed to the atomic intermixing that occurs at the coating–substrate interface due to the high kinetic energy (2 keV) of the plasma formed during the cathodic arc deposition process [42.57]. This atomic intermixing provides a graded compositional transition between the coating and the substrate material. In all other coatings used in this study, the kinetic energy of the plasma is insufficient for atomic intermixing.

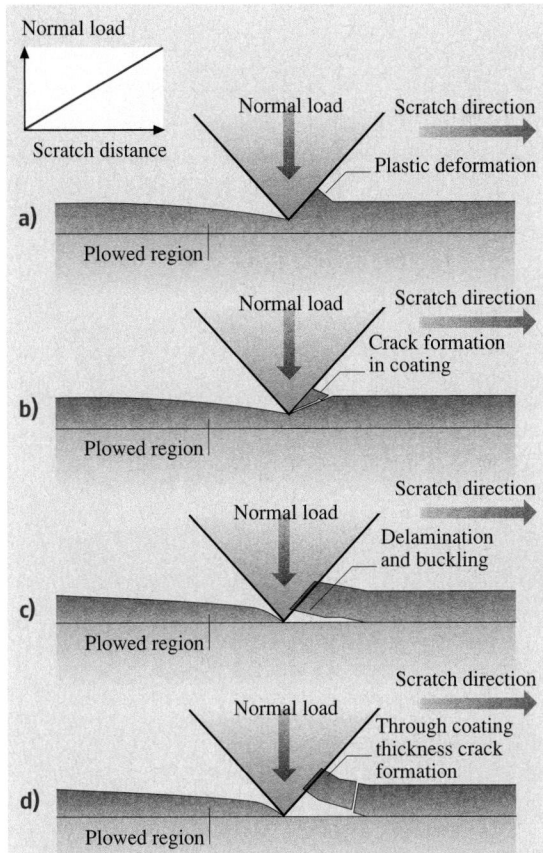

Fig. 42.22a–d Schematic of scratch damage mechanisms for DLC coatings: (**a**) plowing associated with the plastic flow of materials, (**b**) plowing associated with the formation of small debris, (**c**) delamination and buckling at the critical load, and (**d**) breakdown via through-coating thickness cracking at and after the critical load [42.48]

Gupta and *Bhushan* [42.12, 47] and *Li* and *Bhushan* [42.48, 49] measured the scratch resistances of DLC coatings deposited on Al_2O_3-TiC, Ni-Zn ferrite and single-crystal silicon substrates. An interlayer of silicon is required to adhere the DLC coating to other substrates, except in the case of cathodic arc-deposited coatings. The best adhesion with cathodic arc carbon coating is obtained on electrically conducting substrates such as Al_2O_3-TiC and silicon rather than Ni-Zn ferrite.

Microwear

Microwear studies can be conducted using an AFM [42.23]. For microwear studies, a three-sided pyramidal single-crystal natural diamond tip with an apex angle of about 80° and a tip radius of about 100 nm is used at relatively high loads of 1–150 μN. The diamond tip is mounted on a stainless steel cantilever beam with a normal stiffness of about 30 N/m. The sample is generally scanned in a direction orthogonal to the long axis of the cantilever beam (typically at a rate of 0.5 Hz). The tip is mounted on the beam such that one of its edges is orthogonal to the beam axis. In wear studies, an area of 2 μm × 2 μm is typically scanned for a selected number of cycles.

Microwear studies of various types of DLC coatings have been conducted [42.50, 102, 103]. Figure 42.23a shows a wear mark on uncoated Si(100). Wear occurs uniformly and material is removed layer-by-layer via plowing from the first cycle, resulting in constant friction force during the wear (Fig. 42.24a). Figure 42.23b shows AFM images of the wear marks on 10 nm-thick coatings. It is clear that the coatings wear nonuniformly. Coating failure is sudden and accompanied by a sudden rise in the friction force (Fig. 42.24b). Figure 42.24 shows the wear depth of Si(100) substrate and various DLC coatings at two different loads. FCA- and ECR-CVD-deposited 20 nm-thick coatings show excellent wear resistance up to 80 μN, the load that is required for the IB 20 nm coating to fail. In these tests, "failure" of a coating occurs when the wear depth exceeds the quoted coating thickness. The SP 20 nm coating fails at a much lower load of 35 μN. At 60 μN, the coating hardly provides any protection. Moving on to the 10 nm coatings, the ECR-CVD coating requires about 45 cycles at 60 μN to fail, whereas the IB and FCA, coatings fail at 45 μN. The FCA coating exhibits slight roughening in the wear track after the first few cycles, which leads to an increase in the friction force. The SP coating continues to exhibit poor resistance, failing at 20 μN. For the 5 nm coatings, the load required to make the coatings fail continues to decrease, but IB and ECR-CVD still provide adequate protection compared to bare Si(100) in that order, with the silicon failing at 35 μN, the FCA coating at 25 μN and the SP coating at 20 μN. Almost all of the 20, 10, and 5 nm coatings provide better wear resistance than bare silicon. At 3.5 nm, FCA coating provides no wear resistance, failing almost instantly at 20 μN. The IB and ECR-CVD coatings show good wear resistance at 20 μN compared to bare Si(100). However, IB only lasts about ten cycles and ECR-CVD about three cycles at 25 μN.

The wear tests highlight the differences between the coatings more vividly than the scratch tests. At higher thicknesses (10 and 20 nm), the ECR-CVD and FCA coatings appear to show the best wear resistance. This is probably due to the greater hardness of the coat-

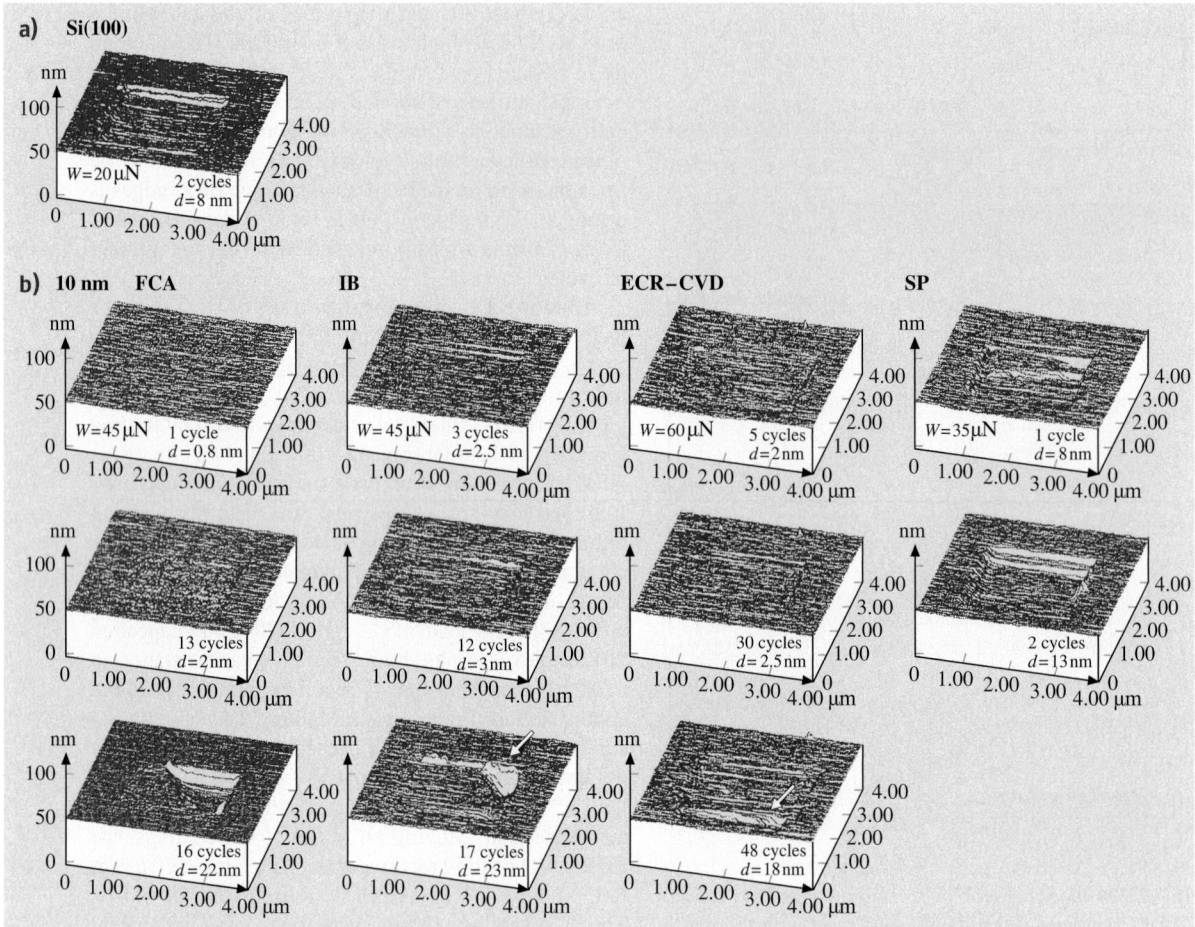

Fig. 42.23a,b AFM images of wear marks on (**a**) bare Si(100), and (**b**) various 10 nm thick DLC coatings [42.50]

ings (see Table 42.5). At 5 nm, the IB coating appears to be the best. FCA coatings show poorer wear resistance with decreasing coating thickness. This suggests that the trends in hardness seen in Table 42.5 no longer hold at low thicknesses. SP coatings show consistently poor wear resistance at all thicknesses. The 3.5 nm IB coating does provide reasonable wear protection at low loads.

42.3.3 Macroscale Tribological Characterization

So far, we have presented data on mechanical characterization and microscratch and microwear studies using a nanoindenter and an AFM. Mechanical properties affect the tribological performance of a coating, and microwear studies simulate a single asperity con-

tact, which helps us to understand the wear process. These studies are useful when screening various candidate coatings, and also aid our understanding of the relationships between deposition conditions and properties of the samples. In the next step, macroscale friction and wear tests need to be conducted to measure the tribological performance of the coating.

Macroscale accelerated friction and wear tests have been conducted to screen a large number of candidates, as have functional tests on selected candidates. An accelerated test is designed to accelerate the wear process such that it does not change the failure mechanism. The accelerated friction and wear tests are generally conducted using a ball-on-flat tribometer under reciprocating motion [42.70]. Typically, a diamond tip with a tip radius of 20 μm or a sapphire ball with a diameter of 3 mm and a surface finish of about 2 nm RMS is

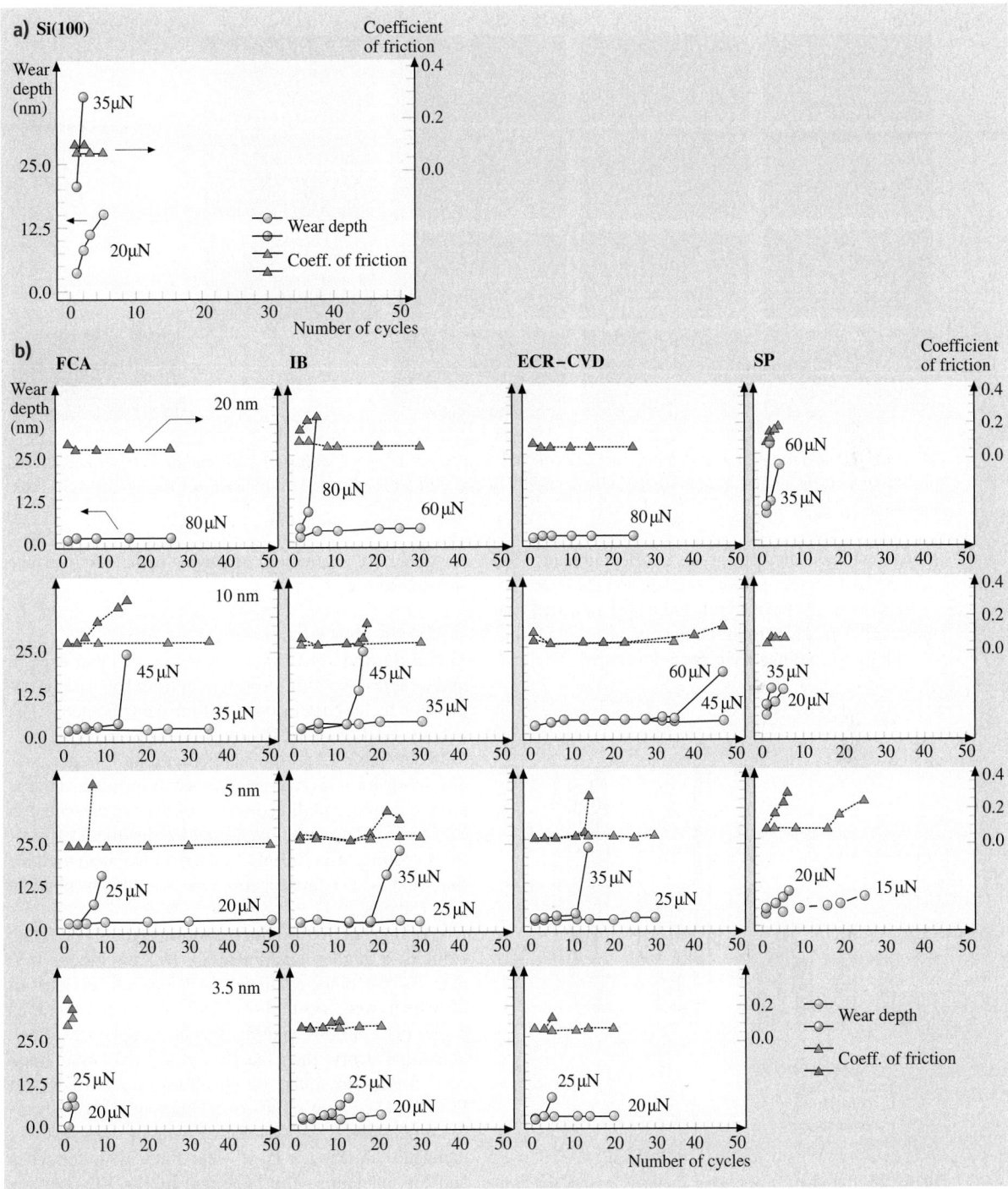

Fig. 42.24a,b Wear data for (**a**) bare Si(100) and (**b**) various DLC coatings. Coating thickness is constant along each row in (**b**). Both the wear depth and the coefficient of friction during wear are plotted for a given cycle [42.50]

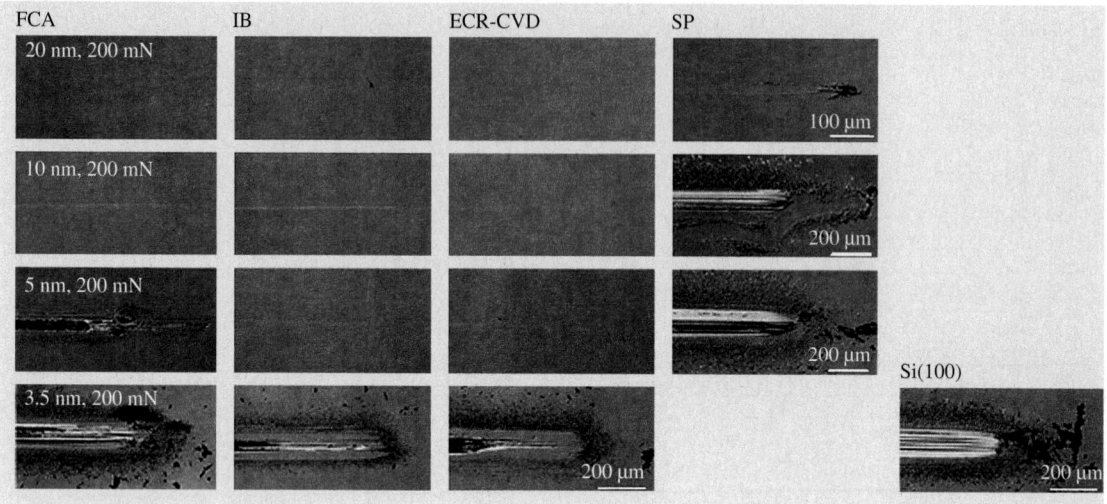

Fig. 42.25 Optical micrographs of wear tracks and debris formed on various coatings of different thicknesses and silicon substrate when slid against a sapphire ball after a sliding distance of 5 nm. The end of the wear track is on the right-hand side of the image

slid against the coated substrates at selected loads. The coefficient of friction is monitored during the tests.

Functional tests are conducted using an actual machine running at close to the actual operating conditions for which the coatings have been developed. The tests are generally accelerated somewhat to fail the interface in a short time.

Accelerated Friction and Wear Tests

Li and *Bhushan* [42.48] conducted accelerated friction and wear tests on DLC coatings deposited by various deposition techniques using a ball-on-flat tribometer. The average coefficient of friction values observed are presented in Table 42.5. Optical micrographs of wear tracks and debris formed on all samples when slid against a sapphire ball after a sliding distance of 5 m are presented in Fig. 42.25. The normal load used for the 20 and 10 nm-thick coatings was 200 mN, and the normal load used for the 5 and 3.5 nm-thick coatings and the silicon substrate was 150 mN.

Among the 20 nm-thick coatings, the SP coating exhibits a higher coefficient of friction (about 0.3) than for the other coatings coefficient of friction (all of which were about 0.2). The optical micrographs show that the SP coating has a larger wear track and more debris than the IB coating. No wear track or debris were found on the 20 nm-thick FCA and ECR-CVD coatings. The optical micrographs of 10 nm-thick coatings show that the SP coating was severely damaged, showing a large wear track with scratches and lots of debris. The FCA and ECR-CVD coatings show smaller wear tracks and less debris than the IB coatings.

Fig. 42.26 Bar chart of the wear damage indices for various coatings of different thicknesses and Si(100) substrate based on optical examination of the wear tracks and debris

Damage index

0 - No apparent damage
1 - Small damage
2 - Medium damage
3 - Large damage
4 - Heavy damage

FCA   IB
ECR-CVD   SP

For the 5 nm-thick coatings, the wear tracks and debris of the IB and ECR-CVD coatings are comparable. The bad wear resistance of the 5 nm-thick FCA coating is in good agreement with the low critical scratch load, which may be due to the higher interfacial and residual stresses as well as the brittleness of the coating.

At 3.5 nm, all of the coatings exhibit wear. The FCA coating provides no wear resistance, failing instantly like the silicon substrate. Large block-like debris is observed on the sides of the wear track of the FCA coating. This indicates that large region delamination and buckling occurs during sliding, resulting in large block-like debris. This block-like debris, in turn, scratches the coating, making the damage to the coating even more severe. The IB and ECR-CVD coatings are able to provide some protection against wear at 3.5 nm.

In order to better evaluate the wear resistance of various coatings, based on an optical examination of the wear tracks and debris after tests, a bar chart of the wear damage index for various coatings of different thicknesses and an uncoated silicon substrate is presented in Fig. 42.26. Among the 20 and 10 nm-thick coatings, the SP coatings show the worst damage, followed by FCA/ECR-CVD. At 5 nm, the FCA and SP coatings show the worst damage, followed by the IB and ECR-CVD coatings. All of the 3.5 nm-thick coatings show the same heavy damage as the uncoated silicon substrate.

The wear damage mechanisms of the thick and thin DLC coatings studied are believed to be as illustrated in Fig. 42.27. In the early stages of sliding, deformation zone, Hertzian and wear fatigue cracks that have formed beneath the surface extend within the coating upon subsequent sliding [42.1]. Formation of fatigue cracks depends on the hardness and subsequent cycles. These are controlled by the sp^3-to-sp^2 ratio. For thicker coatings, the cracks generally do not penetrate the coating. For a thinner coating, the cracks easily propagate down to the interface aided by the interfacial stresses and get diverted along the interface just enough to cause local delamination of the coating. When this happens, the coating experiences excessive plowing. At this point, the coating fails catastrophically, resulting in a sudden rise in the coefficient of friction. All 3.5 nm-thick coatings failed much quicker than the thicker coatings. It appears that these thin coatings have very low load-carrying capacities and so the substrate undergoes deformation almost immediately. This generates stresses at the interface that weaken the coating adhesion and lead to delamination of the coating. Another reason may be that the thickness is insufficient to produce a coating that has

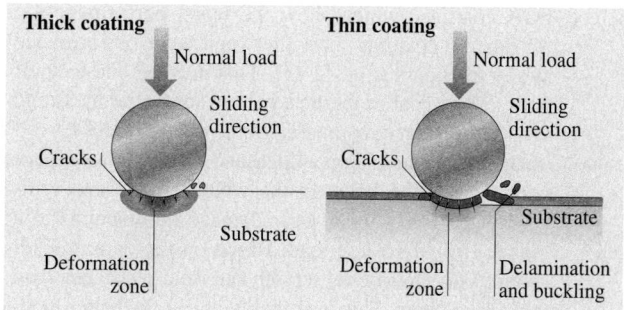

Fig. 42.27 Schematic of wear damage mechanisms for thick and thin DLC coatings [42.48]

the DLC structure. Instead, the bulk may be made up of a matrix characteristic of the interface region where atomic mixing occurs with the substrate and/or any interlayer used. This would also result in poor wear resistance and silicon-like behavior of the coating, especially for

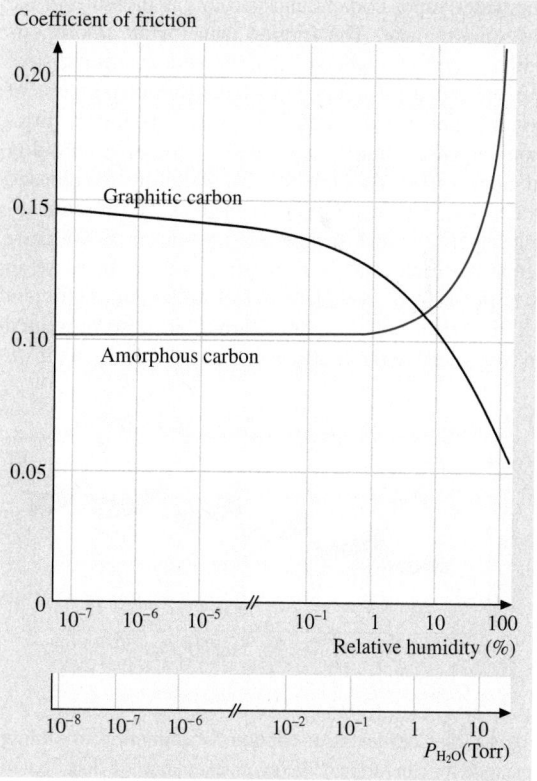

Fig. 42.28 Coefficient of friction as a function of relative humidity and water vapor partial pressure for a RF-plasma deposited amorphous carbon coating and a bulk graphitic carbon coating sliding against a steel ball

FCA coatings, which show the worst performance at 3.5 nm. SP coatings show the worst wear performance at any thickness (Fig. 42.25). This may be due to their poor mechanical properties, such as lower hardness and scratch resistance, compared to the other coatings.

Comparison of Figs. 42.20 and 42.26 shows a very good correlation between the wear damage and critical scratch loads. Less wear damage corresponds to a higher critical scratch load. Based on the data, thicker coatings do show better scratch and wear resistance than thinner coatings. This is probably due to the better load-carrying capacities of the thick coatings compared to the thinner ones. For a given coating thickness, increased hardness and fracture toughness and better adhesion strength are believed to be responsible for the superior wear performance.

Effect of Environment

The friction and wear performance of an amorphous carbon coating is known to be strongly dependent on the water vapor content and partial gas pressure in the test environment. The friction data for an amorphous carbon film on a silicon substrate sliding against steel are presented as a function of the partial pressure of water vapor in Fig. 42.28 [42.1, 13, 69, 105, 106]. Friction increases dramatically above a relative humidity of about 40%. At high relative humidity, condensed water vapor forms meniscus bridges at the contacting asperities, and the menisci result in an intrinsic attractive force that is responsible for an increase in the friction. For completeness, data on the coefficient of friction of bulk graphitic carbon are also presented in Fig. 42.28. Note that the friction decreases with in-

creased relative humidity [42.107]. Graphitic carbon has a layered crystal lattice structure. Graphite absorbs polar gases (such as H_2O, O_2, CO_2, NH_3) at the edges of the crystallites, which weakens the interlayer bonding forces facilitating interlayer slip and results in lower friction [42.1].

A number of tests have been conducted in controlled environments in order to better study the effects of environmental factors on carbon-coated magnetic disks. *Marchon* et al. [42.108] conducted tests in alternating environments of oxygen and nitrogen gases, Fig. 42.29. The coefficient of friction increases as soon as oxygen is added to the test environment, whereas in a nitrogen environment the coefficient of friction reduces slightly. Tribochemical oxidation of the DLC coating in the oxidizing environment is responsible for an increase in the coefficient of friction, implying wear. *Dugger* et al. [42.109], *Strom* et al. [42.110], *Bhushan* and *Ruan* [42.111] and *Bhushan* et al. [42.71] conducted tests with DLC-coated magnetic disks (with about 2 nm-thick perfluoropolyether lubricant film) in contact with Al_2O_3-TiC sliders in different gaseous environments, including a high vacuum of 2×10^{-7} Torr, Fig. 42.30. The wear lives are the shortest in high vacuum and the longest in atmospheres of mostly nitrogen and argon with the following order (from best to worst): argon or nitrogen, Ar + H_2O, ambient, Ar + O_2, Ar + H_2O, vacuum. From this sequence of wear per-

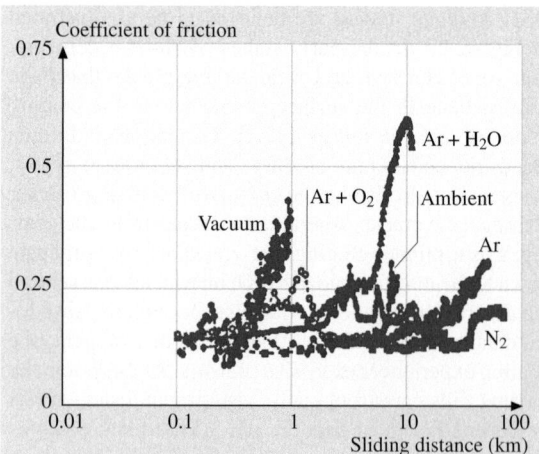

Fig. 42.30 Durability, measured by sliding a Al_2O_3-TiC magnetic slider against a magnetic disk coated with a 20 nm-thick DC sputtered amorphous carbon coating and 2 nm-thick perfluoropolyether film, measured at a speed of 0.75 m/s and for a load of 10 g. Vacuum refers to 2×10^{-7} Torr [42.71]

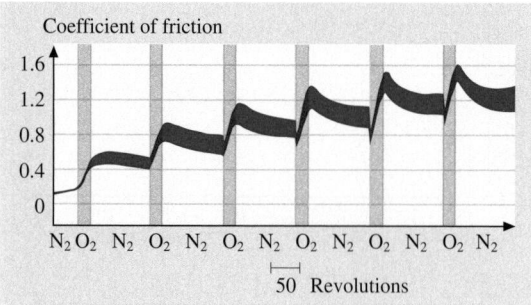

Fig. 42.29 Coefficient of friction as a function of sliding distance for a ceramic slider against a magnetic disk coated with a 20 nm-thick DC magnetron sputtered DLC coating, measured at a speed of 0.06 m/s for a load of 10 g. The environment was alternated between oxygen and nitrogen gases [42.108]

formance, we can see that having oxygen and water in an operating environment worsens the wear performance of the coating, but having a vacuum is even worse. Indeed, failure mechanisms differ in different environments. In high vacuum, intimate contact between the disk and the slider surface results in significant wear. In ambient air, $Ar + O_2$ and $Ar + H_2O$, tribochemical oxidation of the carbon overcoat is responsible for interface failure. For experiments performed in pure argon and nitrogen, mechanical shearing of the asperities causes the formation of debris, which is responsible for the formation of scratch marks on the carbon surface, which were observed with an optical microscope [42.71].

Functional Tests

Magnetic thin film heads made with Al_2O_3-TiC substrate are used in magnetic storage applications [42.13]. A multilayered thin film pole-tip structure present on the head surface wears more rapidly than the much harder Al_2O_3-TiC substrate. Pole-tip recession (PTR) is a serious concern in magnetic storage [42.15–19, 112]. Two of the diamond-like carbon coatings with superior mechanical properties – ion beam and cathodic arc carbon coatings – were deposited on the air-bearing surfaces of Al_2O_3-TiC head sliders [42.15]. Functional tests were conducted by running a metal particle (MP) tape in a computer tape drive. The average PTR as a function of sliding distance is presented in Fig. 42.31. We note that the PTR increases for the uncoated head, whereas there is a slight increase in PTR for the coated heads during early sliding followed by little change. Thus, the coatings provide protection.

The micromechanical as well as the accelerated and functional tribological data presented here clearly suggest that there is a good correlation between the scratch resistance and wear resistance measured using accelerated tests and functional tests. Thus, scratch tests can be successfully used to screen coatings for wear applications.

42.3.4 Coating Continuity Analysis

Ultrathin (less than 10 nm) coatings may not uniformly coat the sample surface. In other words, the coating may be discontinuous and deposited in the form of islands on the microscale. Therefore, one possible reason for poor wear protection and the nonuniform failure of thin coatings may be poor coverage of the substrate. Coating continuity can be studied using surface analytical techniques such as Auger and/or XPS analyses.

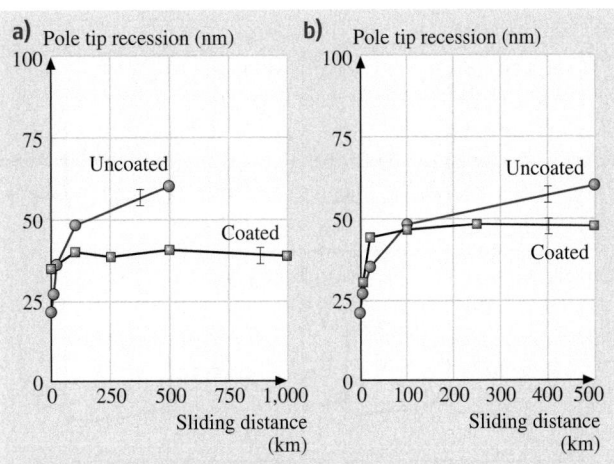

Fig. 42.31a,b Pole-tip recession as a function of sliding distance, measured with an AFM, for (**a**) uncoated and 20 nm-thick ion beam carbon coated, and (**b**) uncoated and 20 nm-thick cathodic arc carbon coated Al_2O_3-TiC heads run against MP tapes [42.15]

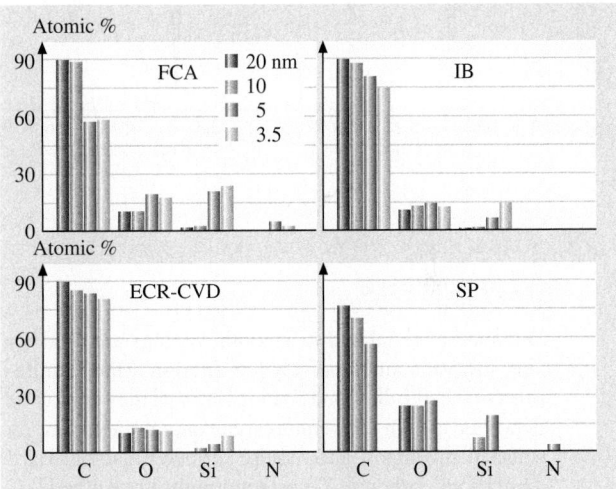

Fig. 42.32 Quantified XPS data for various DLC coatings on Si(100) substrate [42.50]. Atomic concentrations are shown

Any discontinuity in coating thickness that is less than the sampling depth of the instrument result in the local detection of the substrate species [42.49, 50, 102].

The results from an XPS analysis of 1.3 mm² regions (single point measurement with spot diameter of 1300 μm) on various coatings deposited on Si(100) substrates are shown in Fig. 42.32. The sampling depth is about 2–3 nm. The poor SP coatings and the poor 5 nm and 3.5 nm FCA coatings () show much lower carbon contents (atomic concentrations of $< 75\%$ and $< 60\%$

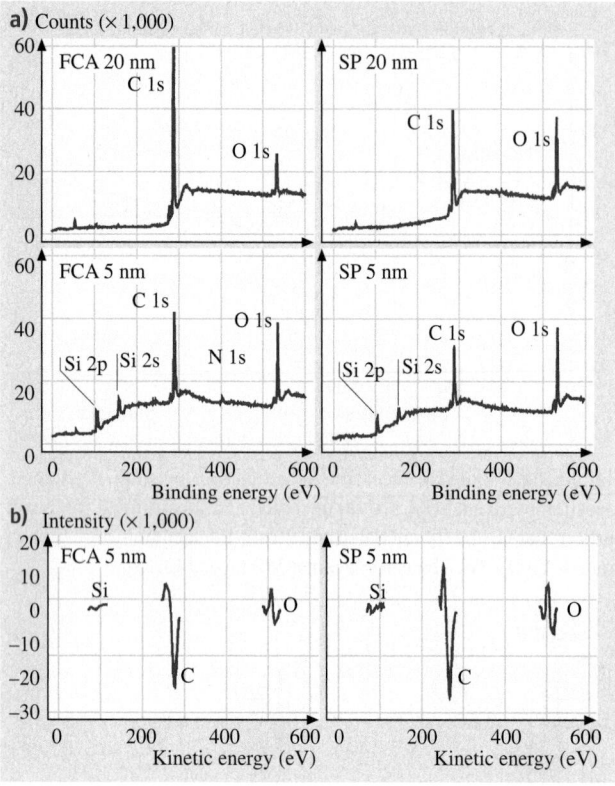

a) Counts (× 1,000)

FCA 20 nm
C 1s
O 1s

SP 20 nm
C 1s
O 1s

FCA 5 nm
C 1s
Si 2p Si 2s O 1s N 1s

SP 5 nm
Si 2p Si 2s C 1s O 1s

Binding energy (eV)

b) Intensity (× 1,000)

FCA 5 nm
Si O
C

SP 5 nm
Si O
C

Kinetic energy (eV)

Fig. 42.33 (**a**) XPS spectra for 5 nm- and 20 nm-thick FCA and SP coatings on Si(100) substrate, and (**b**) AES spectra for FCA and SP coatings of 5 nm thickness on Si(100) substrate [42.50]

respectively) than the IB and ECR-CVD coatings. Silicon is detected in all of the 5 nm coatings. From the data it is hard to infer whether the Si is from the substrate or from exposed regions due to discontinuous coating. Based on the sampling depth, any Si detected in 3.5 nm coatings would likely be from the substrate. The other interesting observation is that all poor coatings (all SP and FCA 5 and 3.5 nm coatings) have almost twice the oxygen content of the other coatings. Any oxygen present may be due to leaks in the deposition chamber, and it is present as silicon oxides.

AES measurements averaged over a scan area of 900 μm^2 were conducted on FCA and SP 5 nm coatings at six different regions on each sample. Very little silicon was detected on this scale, and the detected peaks were characteristic of oxides. The oxygen levels were comparable to those seen for good coatings via XPS. These results contrast with the XPS measurements performed at a larger scale, suggesting that the coatings only possess discontinuities at isolated areas and that the 5 nm coatings are generally continuous on the microscale. Figure 42.33 shows representative XPS and AES spectra of selected samples.

42.4 Closure

Diamond material and its smooth coatings are used for very low wear and relatively low friction. Major limitations of the true diamond coatings are that they need to be deposited at high temperatures, can only be deposited on selected substrates, and require surface finishing. Hard amorphous carbon (a-C) or commonly known as DLC coatings exhibit mechanical, thermal and optical properties close to that of diamond. These can be deposited with a large range of thicknesses by using a variety of deposition processes, on variety of substrates at or near room temperature. The coatings reproduce substrate topography avoiding the need of post finishing. Friction and wear properties of some DLC coatings can be very attractive for tribological applications. The largest industrial application of these coatings is in magnetic storage devices. They are expected to be used in MEMS/NEMS.

EELS and Raman spectroscopies can be successfully used for chemical characterization of amorphous carbon coatings. The prevailing atomic arrangement in the DLC coatings is amorphous or quasi-amorphous with small diamond (sp^3), graphite (sp^2) and other unidentifiable micro- or nanocrystallites. Most DLC coatings except those produced by filtered cathodic arc contain from a few to about 50 at% hydrogen. Sometimes hydrogen is deliberately incorporated in the sputtered and ion plated coatings to tailor their properties.

Amorphous carbon coatings deposited by various techniques exhibit different mechanical and tribological properties. The nanoindenter can be successfully used for measurement of hardness, elastic modulus, fracture toughness, and fatigue life. Microscratch and microwear experiments can be performed using either a nanoindenter or an AFM. Thin coatings deposited by filtered cathodic arc, ion beam and ECR-CVD hold a promise for tribological applications. Coatings as thin as 5 nm or even thinner in thickness provide wear protection. Microscratch, microwear, and accelerated wear testing, if simulated properly can be successfully used to screen coating candidates for industrial applications. In the examples shown in this chapter, trends observed in the microscratch, microwear, and accelerated macrofriction wear tests are similar to that found in functional tests.

References

42.1   B. Bhushan, B. K. Gupta: *Handbook of Tribology: Materials, Coatings, and Surface Treatments*, reprint edn. (Krieger, Malabar 1997)

42.2   B. Bhushan: *Principles and Applications of Tribology* (Wiley, New York 1999)

42.3   B. Bhushan: *Introduction to Tribology* (Wiley, New York 2002)

42.4   B. Bhushan, B. K. Gupta, G. W. VanCleef, C. Capp, J. V. Coe: Fullerene (C_{60}) films for solid lubrication, Tribol. Trans. **36**, 573–580 (1993)

42.5   B. K. Gupta, B. Bhushan, C. Capp, J. V. Coe: Material characterization and effect of purity and ion implantation on the friction and wear of sublimed fullerene films, J. Mater. Res. **9**, 2823–2838 (1994)

42.6   B. K. Gupta, B. Bhushan: Fullerene particles as an additive to liquid lubricants and greases for low friction and wear, Lubr. Eng. **50**, 524–528 (1994)

42.7   B. Bhushan, V. V. Subramaniam, A. Malshe, B. K. Gupta, J. Ruan: Tribological properties of polished diamond films, J. Appl. Phys. **74**, 4174–4180 (1993)

42.8   B. Bhushan, B. K. Gupta, V. V. Subramaniam: Polishing of diamond films, Diam. Films Technol. **4**, 71–97 (1994)

42.9   P. Sander, U. Kaiser, M. Altebockwinkel, L. Wiedmann, A. Benninghoven, R. E. Sah, P. Koidl: Depth profile analysis of hydrogenated carbon layers on silicon by X-ray photoelectron spectroscopy, auger electron spectroscopy, electron energy-loss spectroscopy, and secondary ion mass spectrometry, J. Vacuum Sci. Technol. A **5**, 1470–1473 (1987)

42.10   A. Matthews, S. S. Eskildsen: Engineering applications for diamond-like carbon, Diam. Relat. Mater. **3**, 902–911 (1994)

42.11   A. H. Lettington: Applications of diamond-like carbon thin films, Carbon **36**, 555–560 (1998)

42.12   B. K. Gupta, B. Bhushan: Mechanical and tribological properties of hard carbon coatings for magnetic recording heads, Wear **190**, 110–122 (1995)

42.13   B. Bhushan: *Tribology and Mechanics of Magnetic Storage Devices*, 2nd edn. (Springer, Berlin, Heidelberg 1996)

42.14   B. Bhushan: *Mechanics and Reliability of Flexible Magnetic Media*, 2nd edn. (Springer, Berlin, Heidelberg 2000)

42.15   B. Bhushan, S. T. Patton, R. Sundaram, S. Dey: Pole tip recession studies of hard carbon-coated thin-film tape heads, J. Appl. Phys. **79**, 5916–5918 (1996)

42.16   J. Xu, B. Bhushan: Pole tip recession studies of thin-film rigid disk head sliders II: Effects of air bearing surface and pole tip region designs and carbon coating, Wear **219**, 30–41 (1998)

42.17   W. W. Scott, B. Bhushan: Corrosion and wear studies of uncoated and ultra-thin DLC coated magnetic tape-write heads and magnetic tapes, Wear **243**, 31–42 (2000)

42.18   W. W. Scott, B. Bhushan: Loose debris and head stain generation and pole tip recession in modern tape drives, J. Info. Stor. Proc. Syst. **2**, 221–254 (2000)

42.19   W. W. Scott, B. Bhushan, A. V. Lakshmikumaran: Ultrathin diamond-like carbon coatings used for reduction of pole tip recession in magnetic tape heads, J. Appl Phys **87**, 6182–6184 (2000)

42.20   B. Bhushan: Macro- and microtribology of magnetic storage devices. In: *Modern Tribology Handbook*, ed. by B. Bhushan (CRC, Boca Raton 2001) pp. 1413–1513

42.21   B. Bhushan: *Nanotribology and nanomechanics of MEMS devices*, Proc. Ninth Annual Workshop on Micro Electro Mechanical Systems (IEEE, New York 1996) pp. 91–98

42.22   B. Bhushan (ed): *Tribology Issues and Opportunities in MEMS* (Kluwer, Dordrecht 1998)

42.23   B. Bhushan: *Handbook of Micro/Nanotribology*, 2nd edn. (CRC, Boca Raton 1999)

42.24   B. Bhushan: Macro- and microtribology of MEMS materials. In: *Modern Tribology Handbook*, ed. by B. Bhushan (CRC, Boca Raton 2001) pp. 1515–1548

42.25   S. Aisenberg, R. Chabot: Ion beam deposition of thin films of diamond-like carbon, J. Appl. Phys. **49**, 2953–2958 (1971)

42.26   E. G. Spencer, P. H. Schmidt, D. C. Joy, F. J. Sansalone: Ion beam deposited polycrystalline diamond-like films, Appl. Phys. Lett. **29**, 118–120 (1976)

42.27   A. Grill, B. S. Meyerson: Development and status of diamondlike carbon. In: *Synthetic Diamond: Emerging CVD Science and Technology*, ed. by K. E. Spear, J. P. Dismukes (Wiley, New York 1994) pp. 91–141

42.28   Y. Catherine: Preparation techniques for diamond-like carbon. In: *Diamond and Diamond-Like Films and Coatings*, ed. by R. E. Clausing, L. L. Horton, J. C. Angus, P. Koidl (Plenum, New York 1991) pp. 193–227

42.29   J. J. Cuomo, D. L. Pappas, J. Bruley, J. P. Doyle, K. L. Seagner: Vapor deposition processes for amorphous carbon films with sp³ fractions approaching diamond, J. Appl. Phys. **70**, 1706–1711 (1991)

42.30   J. C. Angus, C. C. Hayman: Low pressure metastable growth of diamond and diamondlike phase, Science **241**, 913–921 (1988)

42.31   J. C. Angus, F. Jensen: Dense diamondlike hydrocarbons as random covalent networks, J. Vacuum Sci. Technol. A **6**, 1778–1782 (1988)

42.32   D. C. Green, D. R. McKenzie, P. B. Lukins: The microstructure of carbon thin films, Mater. Sci. Forum **52–53**, 103–124 (1989)

42.33   N. H. Cho, K. M. Krishnan, D. K. Veirs, M. D. Rubin, C. B. Hopper, B. Bhushan, D. B. Bogy: Chemical structure and physical properties of diamond-like amorphous carbon films prepared by magnetron sputtering, J. Mater. Res. **5**, 2543–2554 (1990)

42.34   J. C. Angus: Diamond and diamondlike films, Thin Solid Films **216**, 126–133 (1992)

42.35   B. Bhushan, A. J. Kellock, N. H. Cho, J. W. Ager III: Characterization of chemical bonding and physical characteristics of diamond-like amorphous carbon and diamond films, J. Mater. Res. **7**, 404–410 (1992)

42.36   J. Robertson: Properties of diamond-like carbon, Surf. Coat. Technol. **50**, 185–203 (1992)

42.37   N. Savvides, B. Window: Diamondlike amorphous carbon films prepared by magnetron sputtering of graphite, J. Vacuum Sci. Technol. A **3**, 2386–2390 (1985)

42.38   J. C. Angus, P. Koidl, S. Domitz: Carbon thin films. In: *Plasma Deposited Thin Films*, ed. by J. Mort, F. Jensen (CRC, Boca Raton 1986) pp. 89–127

42.39   J. Robertson: Amorphous carbon, Adv. Phys. **35**, 317–374 (1986)

42.40   M. Rubin, C. B. Hooper, N. H. Cho, B. Bhushan: Optical and mechanical properties of DC sputtered carbon films, J. Mater. Res. **5**, 2538–2542 (1990)

42.41   G. J. Vandentop, M. Kawasaki, R. M. Nix, I. G. Brown, M. Salmeron, G. A. Somorjai: Formation of hydrogenated amorphous carbon films of controlled hardness from a methane plasma, Phys. Rev. B **41**, 3200–3210 (1990)

42.42   J. J. Cuomo, D. L. Pappas, R. Lossy, J. P. Doyle, J. Bruley, G. W. Di Bello, W. Krakow: Energetic carbon deposition at oblique angles, J. Vacuum Sci. Technol. A **10**, 3414–3418 (1992)

42.43   D. L. Pappas, K. L. Saegner, J. Bruley, W. Krakow, J. J. Cuomo: Pulsed laser deposition of diamondlike carbon films, J. Appl. Phys. **71**, 5675–5684 (1992)

42.44   H. J. Scheibe, B. Schultrich: DLC film deposition by laser-arc and study of properties, Thin Solid Films **246**, 92–102 (1994)

42.45   C. Donnet, A. Grill: Friction control of diamond-like carbon coatings, Surf. Coat. Technol. **94–95**, 456 (1997)

42.46   A. Grill: Tribological properties of diamondlike carbon and related materials, Surf. Coat. Technol. **94–95**, 507 (1997)

42.47   B. K. Gupta, B. Bhushan: Micromechanical properties of amorphous carbon coatings deposited by different deposition techniques, Thin Solid Films **270**, 391–398 (1995)

42.48   X. Li, B. Bhushan: Micro/nanomechanical and tribological characterization of ultra-thin amorphous carbon coatings, J. Mater. Res. **14**, 2328–2337 (1999)

42.49   X. Li, B. Bhushan: Mechanical and tribological studies of ultra-thin hard carbon overcoats for magnetic recording heads, Z. Metallkd. **90**, 820–830 (1999)

42.50   S. Sundararajan, B. Bhushan: Micro/nanotribology of ultra-thin hard amorphous carbon coatings using atomic force/friction force microscopy, Wear **225–229**, 678–689 (1999)

42.51   B. Bhushan: Chemical, mechanical, and tribological characterization of ultra-thin and hard amorphous carbon coatings as thin as 3.5 nm: Recent developments, Diam. Relat. Mater. **8**, 1985–2015 (1999)

42.52   I. I. Aksenov, V. E. Strel'Nitskii: Wear resistance of diamond-like carbon coatings, Surf. Coat. Technol. **47**, 252–256 (1991)

42.53   D. R. McKenzie, D. Muller, B. A. Pailthorpe, Z. H. Wang, E. Kravtchinskaia, D. Segal, P. B. Lukins, P. J. Martin, G. Amaratunga, P. H. Gaskell, A. Saeed: Properties of tetrahedral amorphous carbon prepared by vacuum arc deposition, Diam. Relat. Mater. **1**, 51–59 (1991)

42.54   R. Lossy, D. L. Pappas, R. A. Roy, J. J. Cuomo: Filtered arc deposition of amorphous diamond, Appl. Phys. Lett. **61**, 171–173 (1992)

42.55   I. G. Brown, A. Anders, S. Anders, M. R. Dickinson, I. C. Ivanov, R. A. MacGill, X. Y. Yao, K. M. Yu: Plasma synthesis of metallic and composite thin films with atomically mixed substrate bonding, Nucl. Instrum. Meth. B **80–81**, 1281–1287 (1993)

42.56   P. J. Fallon, V. S. Veerasamy, C. A. Davis, J. Robertson, G. A. J. Amaratunga, W. I. Milne, J. Koskinen: Properties of filtered-ion-beam-deposited diamond-like carbon as a function of ion energy, Phys. Rev. B **48**, 4777–4782 (1993)

42.57   S. Anders, A. Anders, I. G. Brown, B. Wei, K. Komvopoulos, J. W. Ager III, K. M. Yu: Effect of vacuum arc deposition parameters on the properties of amorphous carbon thin films, Surf. Coat. Technol. **68–69**, 388–393 (1994)

42.58   S. Anders, A. Anders, I. G. Brown, M. R. Dickinson, R. A. MacGill: Metal plasma immersion ion implantation and deposition using arc plasma sources, J. Vacuum Sci. Technol. B **12**, 815–820 (1994)

42.59   S. Anders, A. Anders, I. G. Brown: Transport of vacuum arc plasma through magnetic macroparticle filters, Plasma Sources Sci. **4**, 1–12 (1995)

42.60   D. M. Swec, M. J. Mirtich, B. A. Banks: *Ion Beam and Plasma Methods of Producing Diamondlike Carbon Films*, Report No. NASA TM102301 (NASA, Cleveland 1989)

42.61   A. Erdemir, M. Switala, R. Wei, P. Wilbur: A tribological investigation of the graphite-to-diamond-like behavior of amorphous carbon films ion beam deposited on ceramic substrates, Surf. Coat. Technol. **50**, 17–23 (1991)

42.62   A. Erdemir, F. A. Nicols, X. Z. Pan, R. Wei, P. J. Wilbur: Friction and wear performance of ion-beam deposited diamond–like carbon films on steel substrates, Diam. Relat. Mater. **3**, 119–125 (1993)

42.63   R. Wei, P. J. Wilbur, M. J. Liston: Effects of diamond-like hydrocarbon films on rolling contact fatigue of bearing steels, Diam. Relat. Mater. **2**, 898–903 (1993)

42.64   A. Erdemir, C. Donnet: Tribology of diamond, diamond-like carbon, and related films. In: *Modern Tribology Handbook*, ed. by B. Bhushan (CRC, Boca Raton 2001) pp. 871–908

42.65   J. Asmussen: Electron cyclotron resonance microwave discharges for etching and thin-film deposition, J. Vacuum Sci. Technol. A **7**, 883–893 (1989)

42.66   J. Suzuki, S. Okada: Deposition of diamondlike carbon films using electron cyclotron resonance plasma chemical vapor deposition from ethylene gas, Jpn. J. Appl. Phys. **34**, L1218–L1220 (1995)

42.67   B. A. Banks, S. K. Rutledge: Ion beam sputter deposited diamond like films, J. Vacuum Sci. Technol. **21**, 807–814 (1982)

42.68   C. Weissmantel, K. Bewilogua, K. Breuer, D. Dietrich, U. Ebersbach, H. J. Erler, B. Rau, G. Reisse: Preparation and properties of hard i-C and i-BN coatings, Thin Solid Films **96**, 31–44 (1982)

42.69   H. Dimigen, H. Hubsch: Applying low-friction wear-resistant thin solid films by physical vapor deposition, Philips Tech. Rev. **41**, 186–197 (1983/84)

42.70   B. Bhushan, B. K. Gupta, M. H. Azarian: Nanoindentation, microscratch, friction and wear studies for contact recording applications, Wear **181-183**, 743–758 (1995)

42.71   B. Bhushan, L. Yang, C. Gao, S. Suri, R. A. Miller, B. Marchon: Friction and wear studies of magnetic thin-film rigid disks with glass-ceramic, glass and aluminum-magnesium substrates, Wear **190**, 44–59 (1995)

42.72   L. Holland, S. M. Ojha: Deposition of hard and insulating carbonaceous films of an RF target in butane plasma, Thin Solid Films **38**, L17–L19 (1976)

42.73   L. P. Andersson: A review of recent work on hard i-C films, Thin Solid Films **86**, 193–200 (1981)

42.74   A. Bubenzer, B. Dischler, B. Brandt, P. Koidl: R.F. plasma deposited amorphous hydrogenated hard carbon thin films, preparation, properties and applications, J. Appl. Phys. **54**, 4590–4594 (1983)

42.75   A. Grill, B. S. Meyerson, V. V. Patel: Diamond-like carbon films by RF plasma-assisted chemical vapor deposition from acetylene, IBM J. Res. Dev. **34**, 849–857 (1990)

42.76   A. Grill, B. S. Meyerson, V. V. Patel: Interface modification for improving the adhesion of a-C:H to metals, J. Mater. Res. **3**, 214 (1988)

42.77   A. Grill, V. V. Patel, B. S. Meyerson: Optical and tribological properties of heat-treated diamond-like carbon, J. Mater. Res. **5**, 2531–2537 (1990)

42.78   F. Jansen, M. Machonkin, S. Kaplan, S. Hark: The effect of hydrogenation on the properties of ion beam sputter deposited amorphous carbon, J. Vacuum Sci. Technol. A **3**, 605–609 (1985)

42.79   S. Kaplan, F. Jansen, M. Machonkin: Characterization of amorphous carbon-hydrogen films by solid-state nuclear magnetic resonance, Appl. Phys. Lett. **47**, 750–753 (1985)

42.80   H. C. Tsai, D. B. Bogy, M. K. Kundmann, D. K. Veirs, M. R. Hilton, S. T. Mayer: Structure and properties of sputtered carbon overcoats on rigid magnetic media disks, J. Vacuum Sci. Technol. A **6**, 2307–2315 (1988)

42.81   B. Marchon, M. Salmeron, W. Siekhaus: Observation of graphitic and amorphous structures on the surface of hard carbon films by scanning tunneling microscopy, Phys. Rev. B **39**, 12907–12910 (1989)

42.82   B. Dischler, A. Bubenzer, P. Koidl: Hard carbon coatings with low optical absorption, Appl. Phys. Lett. **42**, 636–638 (1983)

42.83   D. S. Knight, W. B. White: Characterization of diamond films by Raman spectroscopy, J. Mater. Res. **4**, 385–393 (1989)

42.84   J. W. Ager, D. K. Veirs, C. M. Rosenblatt: Spatially resolved Raman studies of diamond films grown by chemical vapor deposition, Phys. Rev. B **43**, 6491–6499 (1991)

42.85   W. Scharff, K. Hammer, O. Stenzel, J. Ullman, M. Vogel, T. Frauenheim, B. Eibisch, S. Roth, S. Schulze, I. Muhling: Preparation of amorphous i-C films by ion-assisted methods, Thin Solid Films **171**, 157–169 (1989)

42.86   B. Bhushan, X. Li: Nanomechanical characterization of solid surfaces and thin films, Int. Mater. Rev. **48**, 125–164 (2003)

42.87   X. Li, D. Diao, B. Bhushan: Fracture mechanisms of thin amorphous carbon films in nanoindentation, Acta Mater. **45**, 4453–4461 (1997)

42.88   X. Li, B. Bhushan: Measurement of fracture toughness of ultra-thin amorphous carbon films, Thin Solid Films **315**, 214–221 (1998)

42.89   X. Li, B. Bhushan: Evaluation of fracture toughness of ultra-thin amorphous carbon coatings deposited by different deposition techniques, Thin Solid Films **355-356**, 330–336 (1999)

42.90   X. Li, B. Bhushan: Development of a nanoscale fatigue measurement technique and its application to ultrathin amorphous carbon coatings, Scripta Mater. **47**, 473–479 (2002)

42.91   X. Li, B. Bhushan: Nanofatigue studies of ultrathin hard carbon overcoats used in magnetic storage devices, J. Appl. Phys. **91**, 8334–8336 (2002)

42.92   J. Robertson: Deposition of diamond-like carbon, Philos. Trans. R. Soc. Lond. A **342**, 277–286 (1993)

42.93   S. J. Bull: Tribology of carbon coatings: DLC, diamond and beyond, Diam. Relat. Mater. **4**, 827–836 (1995)

42.94   N. Savvides, T. J. Bell: Microhardness and Young's modulus of diamond and diamondlike carbon films, J. Appl. Phys. **72**, 2791–2796 (1992)

42.95   B. Bhushan, M. F. Doerner: Role of mechanical properties and surface texture in the real area of contact of magnetic rigid disks, ASME J. Tribol. **111**, 452–458 (1989)

42.96   S. Suresh: *Fatigue of Materials* (Cambridge Univ. Press, Cambridge 1991)

42.97   D. B. Marshall, A. G. Evans: Measurement of adherence of residual stresses in thin films by indentation. I. Mechanics of interface delamination, J. Appl. Phys. **15**, 2632–2638 (1984)

42.98   A. G. Evans, J. W. Hutchinson: On the mechanics of delamination and spalling in compressed films, Int. J. Solids Struct. **20**, 455–466 (1984)

42.99   S. Sundararajan, B. Bhushan: Development of a continuous microscratch technique in an atomic force microscope and its application to study scratch resistance of ultrathin hard amorphous carbon coatings, J. Mater. Res. **16**, 437–445 (2001)

42.100 B. Bhushan, B. K. Gupta: Micromechanical characterization of Ni-P coated aluminum-magnesium, glass and glass-ceramic substrates and finished magnetic thin-film rigid disks, Adv. Info. Stor. Syst. **6**, 193–208 (1995)

42.101 X. Li, B. Bhushan: Micromechanical and tribological characterization of hard amorphous carbon coatings as thin as 5 nm for magnetic recording heads, Wear **220**, 51–58 (1998)

42.102 B. Bhushan, V. N. Koinkar: Microscale mechanical and tribological characterization of hard amorphous coatings as thin as 5 nm for magnetic disks, Surf. Coat. Technol. **76-77**, 655–669 (1995)

42.103 V. N. Koinkar, B. Bhushan: Microtribological properties of hard amorphous carbon protective coatings for thin-film magnetic disks and heads, Proc. Inst. Mech. Eng. Part J **211**, 365–372 (1997)

42.104 T. W. Wu: Microscratch and load relaxation tests for ultra-thin films, J. Mater. Res. **6**, 407–426 (1991)

42.105 R. Memming, H. J. Tolle, P. E. Wierenga: Properties of polymeric layers of hydrogenated amorphous carbon produced by plasma-activated chemical vapor deposition: tribological and mechanical properties, Thin Solid Films **143**, 31–41 (1986)

42.106 C. Donnet, T. Le Mogne, L. Ponsonnet, M. Belin, A. Grill, V. Patel: The respective role of oxygen and water vapor on the tribology of hydrogenated diamond-like carbon coatings, Tribol. Lett. **4**, 259 (1998)

42.107 F. P. Bowden, J. E. Young: Friction of diamond, graphite and carbon and the influence of surface films, Proc. R. Soc. Lond. **208**, 444–455 (1951)

42.108 B. Marchon, N. Heiman, M. R. Khan: Evidence for tribochemical wear on amorphous carbon thin films, IEEE Trans. Magn. **26**, 168–170 (1990)

42.109 M. T. Dugger, Y. W. Chung, B. Bhushan, W. Rothschild: Friction, wear, and interfacial chemistry in thin film magnetic rigid disk files, ASME J. Tribol. **112**, 238–245 (1990)

42.110 B. D. Strom, D. B. Bogy, C. S. Bhatia, B. Bhushan: Tribochemical effects of various gases and water vapor on thin film magnetic disks with carbon overcoats, ASME J. Tribol. **113**, 689–693 (1991)

42.111 B. Bhushan, J. Ruan: Tribological performance of thin film amorphous carbon overcoats for magnetic recording rigid disks in various environments, Surf. Coat. Technol. **68/69**, 644–650 (1994)

42.112 B. Bhushan, G. S. A. M. Theunissen, X. Li: Tribological studies of chromium oxide films for magnetic recording applications, Thin Solid Films **311**, 67–80 (1997)

43. Self–Assembled Monolayers (SAMs) for Controlling Adhesion, Friction, and Wear

Making micro- and nanodevices as well as magnetic storage devices reliable necessitates the use of protective hydrophobic lubricating films that can minimize the adhesion, friction, and wear of sliding surfaces. Because of the small clearances associated with these devices, such films need to be very thin (on the order of a few molecules thick). Chemically–bonded low surface tension liquid films are suitable for this purpose, as are a select number of hydrophobic solid films. Highly hydrophobic ordered molecular assemblies can also be used; these are engineered by chemically grafting various polymer molecules with suitable functional head groups, spacer chains and nonpolar surface terminal groups to the surface involved.

In this chapter, we focus on the use of self-assembled monolayers (SAMs) for high hydrophobicity and/or low adhesion, friction and wear applications. SAMs are produced by various organic precursors, so the chapter starts with a primer for the organic chemistry associated with this field. This is followed by an overview of selected SAMs with various spacer chains and terminal groups in their molecular chains on a variety of substrates, and a summary of the tribological properties of SAMs. The adhesion, friction and wear properties of SAMs with various spacer chains and surface terminal and head groups (hexadecane thiol,

iphenyl thiol, alkylsilane, perfluoroalkylsilane and alkylphosphonate) on various substrates (Au, Si, and Al) are then surveyed. Degradation mechanisms and environmental effects are studied. Nanotribological studies of various SAM films by atomic force microscopy (AFM), show that perfluoroalkylsilane SAMs in particular exhibit attractive hydrophobic and tribological properties.

Maximizing reliability is an important issue in the field of micro/nanodevices, commonly referred to as micro/nanoelectromechanical systems (MEMS/NEMS) and BioMEMS/BioNEMS, as well as for magnetic storage devices (which include magnetic rigid disk drives, flexible disk drives and tape drives). It often necessitates the application of molecular films to sliding surfaces in order to lubricate and protect them [43.1–8]. A solid or liquid film is generally required in order to obtain acceptable tribological properties for sliding interfaces. However, the presence of a small quantity of liquid be-

tween smooth surfaces can substantially increase the adhesion, friction and wear due to the formation of menisci or adhesive bridges [43.9, 10]. This becomes a major concern in micro/nanodevices operating at ultralow loads, as the magnitude of the liquid-mediated adhesive force may be similar to that of the external load.

The liquid film may be pre-existing and/or it can be a capillary condensate of water vapor from the environment. If the liquid wets the surface ($0 \leq \theta < 90°$, where θ is the contact angle between the liquid–vapor inter-

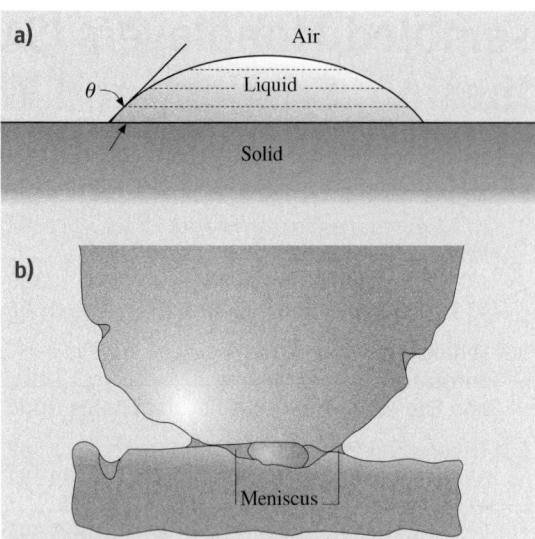

Fig. 43.1 (a) Schematic of a sessile drop on a solid surface and the definition of contact angle, and **(b)** the formation of meniscus bridges due to the presence of liquid at an interface

face and the liquid–solid interface for a liquid droplet sitting on a solid surface, Fig. 43.1a), the surface of the liquid is constrained to lie parallel with the solid surface [43.11–13], and so the surface of the liquid must be concave, Fig. 43.1b. The direct measurement of the contact angle θ is often achieved using sessile drops. The angle is generally measured by finding the tangent to the drop profile at the point of contact with the solid surface using a telescope equipped with a goniometer eyepiece.

Surface tension results in a pressure difference across any meniscus surface, referred to as the capillary pressure or the Laplace pressure, and this is negative for a concave meniscus [43.9, 10]. The negative Laplace pressure results in an intrinsic attractive (adhesive) force that depends on the roughness (the local geometries of interacting asperities and the number of asperities) of the interface, the surface tension and the contact angle. During sliding, frictional effects must be overcome, not only because of external load but also because of intrinsic adhesive force. A high static friction force deriving largely from liquid-mediated adhesion (contribution of the meniscus) is generally referred to as "stiction". There are three basic ways to minimize the effect of liquid-mediated adhesion: to increase the surface roughness, to use hydrophobic (water-fearing) rather than hydrophilic (water-loving) surfaces, and/or to use a liquid with low surface tension [43.9, 10, 14–16].

As an example, bulk silicon and polysilicon films used in the construction of micro/nanodevices can be dipped in hydrofluoric (HF) acid to make them hydrophobic [43.17]. In the etching of silicon with HF, hydrogen passivates the silicon surface by saturating the dangling bonds, resulting in a hydrogen-terminated silicon surface that adsorbs less water. However, after the treated surface has been exposed to the environment it reoxidizes, which means that it can adsorb water and so the surface again becomes hydrophilic.

The surfaces can also be treated or coated with a liquid with relatively low surface tension or with certain types of solid film to make them hydrophobic and/or to control adhesion, friction and wear. The liquid lubricant film should be thin (about half the composite roughness of the interface) to minimize the liquid-mediated adhesion [43.9, 10, 14, 15, 18, 19]. Thus, for ultra-smooth surfaces with RMS roughnesses of a few nm, molecularly thick liquid films are required for liquid lubrication. The classical approach to lubrication uses freely supported multimolecular layers of liquid lubricants [43.2, 4, 9, 10, 19–23]. Boundary lubricant films are formed by either physisorption, chemisorption or chemical reaction. The thickness of the physisorbed film can be either monomolecular or polymolecular. The chemisorbed films are monomolecular, but stoichiometric films formed by chemical reaction can be multilayers. In general, the stabilities and durabilities of surface films decrease in the following order: chemically reacted films, chemisorbed films and physisorbed films. A good boundary lubricant should have a high degree of interaction between its molecules and the sliding surface. As a general rule, liquids will perform better when they are polar and thus able to grip onto solid surfaces (or when they can be adsorbed). Polar lubricants contain reactive functional end groups. Boundary lubrication properties are also dependent upon the molecular conformation and lubricant spreading. It should be noted that a liquid film with a thickness of a few nanometers may be discontinuous and may deposit in "islands" where the layer has nonuniform thickness with a lateral resolution on the nanometer scale.

Solid films are also commonly used to control hydrophobicity and/or adhesion, friction and wear. Hydrophobic films have nonpolar surface terminal groups (described later) that repel water. These films have low surface energies (15 to 30 dyn/cm) and high contact angles ($\theta \geq 90°$) which minimize wetting [43.21, 24, 25]. Multimolecularly thick (a few tenths of a nanometer in thickness) films of conventional solid lubricants have been studied. *Hansma* et al. [43.26] reported the deposition of multimolecularly thick, highly oriented PTFE films from the melt, vapor phase or from solution by a mechanical deposition technique achieved by dragging the polymer at controlled temperature, pressure and speed against a smooth glass substrate. *Scandella* et al. [43.27] reported that the coefficient of nanoscale friction for MoS_2 platelets on mica, obtained by exfoliation of lithium-intercalated MoS_2 in water, was a factor of 1.4 less than that of mica itself. However, MoS_2 reacts with water and its friction and wear properties degrade with increasing humidity [43.9, 10]. Amorphous diamondlike carbon (DLC) coatings can be produced with extremely high hardnesses and are used commercially as wear-resistant coatings [43.28, 29]. They are widely used in magnetic storage devices [43.2]. Doping the DLC matrix with elements like hydrogen, nitrogen, oxygen, silicon and fluorine influences its hydrophobicity and tribological properties [43.28, 30, 31]. Nitrogen and oxygen reduce the contact angle (or increase the surface energy) due to the strong polarity induced when these elements are bonded to carbon. On the other hand, silicon and fluorine increase the contact angle to 70–100° (or reduce the surface energy to 20–40 dyn/cm) making them hydrophobic [43.32, 33]. Nanocomposite coatings with diamondlike carbon (a-C:H) networks and glasslike a-Si:O networks are generally deposited using a PECVD (plasma-enhanced chemical vapor deposition) technique in which plasma is formed from a siloxane precursor using a hot filament. For a fluorinated DLC, CF_4 is added to acetylene plasma to provide the fluorocarbon source. In addition, fluorination of DLC can be achieved by the post-deposition treatment of DLC coatings in a CF_4 plasma. Silicon- and fluorine-containing DLC coatings usually have reduced polarity due to the loss of sp^2-bonded carbon (resulting in reduced polarization potential from π electrons) and the dangling bonds of the DLC network. As silicon and fluorine are unable to form double bonds, they force carbon into an sp^3 bonding state [43.33]. The friction and wear properties of both silicon-containing and fluorinated DLC coatings have been reported to be superior to those of conventional DLC coatings [43.34, 35]. However, DLC coatings require a line-of-sight deposition process, which prevents deposition on complex geometries. Furthermore, it has been reported that some self-assembled monlayers (SAMs) are superior to DLC coatings in terms of hydrophobicity and tribological performance [43.36, 37].

Organized and dense molecular scale layers of, preferably, long-chain organic molecules are known to be better lubricants on macro-, micro- and nanoscales than freely supported multimolecular layers [43.4, 5, 38–48]. Common techniques used to produce molecular scale organized layers include Langmuir–Blodgett (LB) deposition and chemical grafting of organic molecules to the surface to realize SAMs [43.24, 25]. In the LB technique, organic molecules from suitable amphiphilic molecules are first organized at the air–water interface and then physisorbed onto a solid surface to form mono- or multimolecular layers [43.49]. On the other hand, SAMs are produced when functional groups of molecules chemisorb on a solid surface, which results in the spontaneous formation of robust, highly ordered, oriented and dense monolayers [43.25]. In both cases, the organic molecules used have well distinguished amphiphilic properties (a hydrophilic functional head and a hydrophobic aliphatic tail) so that adsorption of the molecules on an active inorganic substrate leads to their firm attachment to the surface. Direct organization of SAMs on the solid surfaces allows tight areas such as the bearing and journal surfaces in an assembled bearing to be coated. The weak adhesion of classical LB films to the substrate surface restricts their lifetimes during sliding, whereas certain SAMs can be very durable. As a result, SAMs are of great interest for tribological applications.

Much research into the application of SAMs has been associated with the so-called soft lithographic technique [43.50, 51]. This is a nonphotolithographic technique. Photolithography is based on a projection printing system used to project an image from a mask to a thin film photoresist, and its resolution is limited by optical diffraction limits. In soft lithography, an elastomeric stamp or mold is used to generate micropatterns of SAMs using either contact printing (known as microcontact printing or μCP [43.52]), embossing (nanoimprint lithography) [43.53], or replica molding [43.54], which all circumvent the diffraction limits of photolithography. The stamps are generally cast from photolithographically generated patterned masters, and the stamp material is generally polydimethylsiloxane (PDMS). In μCP, the ink is a SAM precursor, and nanometer-thick resists with lines thinner than 100 nm are produced. Soft lithography requires little capitol investment. μCP and embossing techniques can be used to produce microdevices that are substantially cheaper and more flexible in terms of construction materials than conventional photolithography (examples include SAMs and non-SAM entities for μCP, and elastomers for embossing).

The largest industrial application of SAMs is in digital micromirror devices (DMD) used in optical projection displays [43.55, 56]. The chip set of a DMD consists of half a million to more than two million independently controlled reflective aluminium alloy micromirrors each about $12\,\mu$m square. These micromirrors switch forward and backward at a frequency of around $5-7$ kHz with a rotation of $\pm 12°$ with respect to the horizontal plane; the movement is limited by a mechanical stop. Mechanical contact leads to stiction and wear of the contacting surfaces. A SAM of vapor-deposited perfluorinated n-alkanoic acid ($C_n F_{2n-1} O_2 H$) (such as perfluorodecanoic acid, or PFDA, $CF_3(CF_2)_8 COOH$), is used to coat the contacting surfaces to make them hydrophobic in order to minimize meniscus formation. Furthermore, the entire DMD chip set is hermetically sealed in order to prevent particulate contamination and excessive condensation of water at the contacting surfaces. A so-called "getter" strip of PFDA is included inside the hermetically sealed enclosure containing the chip, which acts as a reservoir that maintains PFDA vapor within the package. Degradation mechanisms of SAMs leading to stiction have been studied by *Liu* and *Bhushan* [43.57, 58].

Other industrial applications of SAMs are in the areas of bio/chemical and optical sensors, devices for use as drug delivery vehicles, and in the construction of electronic components [43.59–62]. Bio/chemical sensors require highly sensitive organic layers with tailored biological properties that can be incorporated into electronic, optical or electrochemical devices. Self-assembled microscopic vesicles are being developed to ferry potentially life-saving drugs to cancer patients. By assembling organic, metal and phosphonate molecules (complexes of phosphorous and oxygen atoms) into conductive materials, these can be produced as sandwiches for use as electronic components. Several applications have been proposed based on silicon, glass or polymer nanochannels, including cell immunoisolation chambers, protective DNA separation devices, and biocapsules for drug delivery. Using hydrophobic surfaces in gas-based separations based on nanochannels provides several advantages, including low fouling and high gas transport rates.

An overview of molecularly thick layers of liquid lubricants and conventional solid lubricants can be found in various references, such as *Bhushan* [43.2, 4, 9, 10, 18, 28], *Bhushan* and *Zhao* [43.19], and *Liu* and *Bhushan* [43.23]. In this chapter, we focus on the use of SAMs for high hydrophobicity, low adhesion, friction and wear applications. SAMs are produced by various organic precursors, and so we first present a primer to the organic chemistry associated with this topic. This is followed by an overview of suitable substrates, spacer chains and molecular chain end groups, a summary of the tribological properties of various SAMs, and finally some concluding remarks.

43.1 A Brief Organic Chemistry Primer

All organic compounds contain the carbon (C) atom. The fact that carbon can be combined with hydrogen, oxygen, nitrogen, sulfur and phosphorus results in a vast number of potential organic compounds. The atomic number of carbon is six, and its electron structure is $1s^2\, 2s^2\, 2p^2$. Two stable isotopes of carbon, ^{12}C and ^{13}C, exist. With four electrons in its outer shell, carbon forms four covalent bonds, with each bond resulting from two atoms sharing a pair of electrons. The number of electron pairs that two atoms share determines whether or not the bond is single or multiple. In a single bond, only one pair of electrons is shared by the atoms. Carbon can also form multiple bonds by sharing two or three pairs of electrons between the atoms. For example, the double bond formed by sharing two electron pairs is stronger than a single bond and it is shorter than a single bond. An organic compound is classified as saturated if it contains only a single bond and unsaturated if the molecules possess one or more multiple carbon–carbon bonds.

43.1.1 Electronegativity/Polarity

When two different kinds of atoms share a pair of electrons a bond is formed in which electrons are shared unequally; one atom assumes a partial positive charge and the other a negative charge with respect to each other. This difference in charge occurs because the two atoms exert different levels of attraction on the pair of shared electrons. The attractive force that an atom of a particular element has for shared electrons in a molecule or polyatomic ion is known as its electronegativity. Elements differ in their electronegativities. A scale of relative electronegativities, in which the most electronegative el-

Self-Assembled Monolayers (SAMs) for Controlling Adhesion, Friction, and Wear | 43.1 A Brief Organic Chemistry Primer   1383

Part E | 43.1

Table 43.1 Relative electronegativities of selected elements

Element	Relative electronegativity
F	4.0
O	3.5
N	3.0
Cl	3.0
C	2.5
S	2.5
P	2.1
H	2.1

ement, fluorine, is assigned a value of 4.0, was developed by Pauling. Relative electronegativities of the elements in the periodic table can be found in most undergraduate chemistry textbooks (see [43.63] for example). The relative electronegativities of nonmetals are high compared to those of metals. The relative electronegativities of selected elements of interest (those with high electronegativities) are presented in Table 43.1.

The polarity of a bond is determined by the difference in electronegativity between the atoms forming the bond. If the electronegativities are the same, the bond is completely nonpolar and the electrons are shared equally. In this type of bond there is no separation of positive and negative charge between atoms. If the atoms have vastly different electronegativities the bond is very polar. A dipole is a molecule that is electrically asymmetrical, causing it to be oppositely charged at two points. As an example, in hydrogen chloride (HCl), both hydrogen and chlorine need one electron to form stable electron configurations. They share a pair of electrons. Chlorine is more electronegative and there-

fore has a greater attraction for the shared electrons than hydrogen does. As a result, the pair of electrons is displaced towards the chlorine atom, giving it a partial negative charge and leaving the hydrogen atom with a partial positive charge, (Fig. 43.2). However, the entire molecule, HCl, is electrically neutral. The hydrogen atom with its partial positive charge (exposed proton on one end) can be easily attracted to the negative charge of another molecule, and this is responsible for the polarity of the molecule. A partial charge is usually indicated by δ, and the electronic structure of HCl is given as:

$$\overset{\delta +}{H} \overset{\delta -}{:\overset{..}{Cl}:} \ .$$

Similar to the HCl molecule, HF is polar and behaves as a small dipole. On the other hand, methane (CH_4), carbon tetrachloride (CCl_4) and carbon dioxide (CO_2) are nonpolar. In CH_4 and CCl_4, the four C–H and C–Cl polar bonds are identical, and because these bonds emanate from the center to the corners of a tetrahedron in the molecule, their polarities cancel when considering the whole molecule. $CO_2(O=C=O)$ is nonpolar because the carbon–oxygen dipoles cancel each other by acting in opposite directions. Water (H–O–H) is a polar molecule. If the atoms in water were linear as in CO_2, the two O–H dipoles would cancel each other, and the molecule would be nonpolar. However, water has a bent structure, with an angle of 105° between the two bonds, which causes water to be a polar molecule.

43.1.2 Classification and Structures of Organic Compounds

Table 43.2 presents selected organic compounds grouped into classes.

Hydrocarbons

Hydrocarbons are compounds that are composed entirely of carbon and hydrogen atoms bonded to each other by covalent bonds. Saturated hydrocarbons (alkanes) contain single bonds. Unsaturated hydrocarbons that contain carbon–carbon double bonds are called alkenes, and ones with triple bonds are called alkynes. Unsaturated hydrocarbons that contain aromatic rings, including benzene rings, are called aromatic hydrocarbons.

Saturated Hydrocarbons: Alkanes. The alkanes, also known as paraffins, are saturated hydrocarbons: straight- or branched-chain hydrocarbons with only single co-

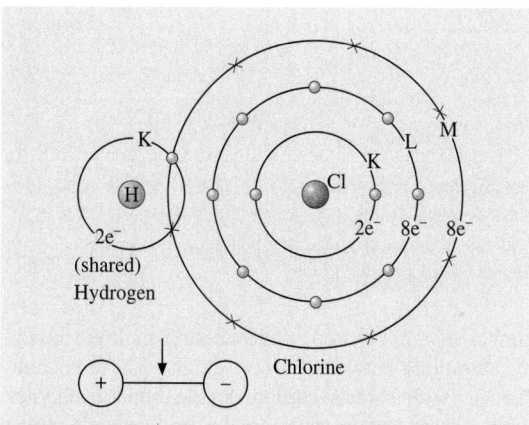

Fig. 43.2 Schematic representation of the formation of a polar HCl molecule

Table 43.2 Names and formulae of selected hydrocarbons

Name	Formula
Saturated hydrocarbons	
Straight-chain alkanes	C_nH_{2n+2}
e.g., methane	CH_4
ethane	C_2H_6 or CH_3CH_3
Alkyl groups	C_nH_{2n+1}
e.g., methyl	$-CH_3$
ethyl	$-CH_2CH_3$
Unsaturated hydrocarbons	
Alkenes	$(CH_2)_n$
e.g., ethene	C_2H_4 or $CH_2 = CH_2$
propene	C_3H_6 or $CH_3CH = CH_2$
Alkynes	
e.g., acetylene	$HC \equiv CH$
Aromatic hydrocarbons	
e.g., benzene	C_6H_5OH or

Table 43.3 Names and formulae of selected alcohols, ethers, phenols and thiols

Name	Formula
Alcohols	$R-OH$
e.g., methanol	CH_3OH
ethanol	CH_3CH_2OH
Ethers	$R-O-R'$
e.g., dimethyl ether	CH_3-O-CH_3
diethyl ether	$CH_3CH_2-O-CH_2CH_3$
Phenols	C_6H_5OH or
Thiols	$-SH$
e.g., methanethiol	CH_3SH

The letters R– and R'– represent an alkyl group. The R– groups in ethers can be the same or different and can be alkyl or aromatic (Ar) groups

Table 43.4 Names and formulae of selected aldehydes and ketones

Name	Formula
Aldehydes	RCHO or
	ArCHO or
e.g., methanal or formaldehyde	$HCHO$
ethanal or acetaldehyde	CH_3CHO
Ketones	RCOR' or
	RCOAr or
	ArCOAr or
e.g., butanone or methyl ethyl ketone	$CH_3COCH_2CH_3$

The letters R and R' represent an alkyl group and Ar represents an aromatic group

valent (saturated) bonds between the carbon atoms. The general molecular formula for alkanes is C_nH_{2n+2}, where n is the number of carbon atoms in the molecule. Each carbon atom is connected to four other atoms by four single covalent bonds. These bonds are separated by angles of $109.5°$ (the angle subtended by lines drawn from the center of a regular tetrahedron to its corners). Alkane molecules contain only carbon–carbon and carbon–hydrogen bonds, which are symmetrically directed towards the corners of a tetrahedron. Alkane molecules are therefore essentially nonpolar.

Common alkyl groups have the general formula C_nH_{2n+1} (one hydrogen atom less than the corresponding alkane). The missing H atom can be detached from any carbon in the alkane. The name of the group is formed from the name of the corresponding alkane by replacing -ane with -yl. Some examples are shown in Table 43.2.

Unsaturated Hydrocarbons. Unsaturated hydrocarbons consist of three families of compounds that contain fewer hydrogen atoms than the alkane with the corresponding number of carbon atoms, and contain multiple bonds (unsaturation) between carbon atoms. These include alkenes (with carbon–carbon double bonds), alkynes (with carbon–carbon triple bonds), and aromatic compounds (with benzene rings, which are arranged in a six-membered ring with one hydrogen atom bonded

Table 43.5 Names and formulae of selected carboxylic acids and esters

Name	Formula
Carboxylic acid [a]	$\underset{\text{ }}{\text{RCOOH or R}}\overset{\displaystyle O}{\overset{\|}{\underset{\text{ }}{—C—}}}\text{OH}$
	$\text{ArCOOH or Ar}\overset{\displaystyle O}{\overset{\|}{—C—}}\text{OH}$
e.g., methanoic acid (formic acid)	HCOOH
ethanoic acid (acetic acid)	CH_3COOH
octadecanoic acid (stearic acid)	$CH_3(CH_2)_{16}COOH$
Esters [b]	$\text{RCOOR' or } \underset{\text{acid}}{\text{R}}\overset{\displaystyle O}{\overset{\|}{—C—}}\underset{\text{alcohol}}{\text{O—R'}}$
e.g., methyl propanoate	$CH_3CH_2COOCH_3$

[a] The letter R represents an alkyl group and Ar represents an aromatic group
[b] The letter R represents a hydrogen, an alkyl group or an aromatic group and R' represents an alkyl group or an aromatic group

Table 43.6 Names and formulae of selected organic nitrogen compounds (amides and amines)

Name	Formula
Amides	$\text{RCONH}_2 \text{ or } \text{R}\overset{\displaystyle O}{\overset{\|}{—C—}}\text{NH}_2$
e.g., methanamide (formamide)	$HCONH_2$
ethanamide (acetamide)	CH_3CONH_2
Amines	$\text{RNH}_2 \text{ or } \text{R—N}\begin{smallmatrix}H \\ \\ H\end{smallmatrix}$
	R_2NH
	R_3N
e.g., methylamine	CH_3NH_2
ethylamine	$CH_3CH_2NH_2$

The letter R represents an alkyl group or aromatic group

Table 43.7 Some examples of polar (hydrophilic) and non-polar (hydrophobic) groups

Name	Formula
Polar	
Alcohol (hydroxyl)	−OH
Carboxylic acid	−COOH
Aldehyde	−COH
Ketone	$\text{R}\overset{\displaystyle O}{\overset{\|}{—C—}}\text{R}$
Ester	−COO−
Carbonyl	>C=O
Ether	R−O−R
Amine	$−NH_2$
Amide	$\overset{\displaystyle O}{\overset{\|}{—C—}}NH_2$
Phenol	benzene ring−OH
Thiol	−SH
Trichlorosilane	$SiCl_3$
Nonpolar	
Methyl	$−CH_3$
Trifluoromethyl	$−CF_3$
Aryl (benzene ring)	benzene ring

The letter R represents an alkyl group

Alcohols, Ethers, Phenols and Thiols

Organic molecules with certain functional groups are synthesized because they have desirable properties. Alcohols, ethers and phenols are derived from the structure of water by replacing the hydrogen atoms of water with hydroxy groups (OH), alkyl groups (R) or aromatic rings (Ar), respectively. For example, phenol is a class of compounds that has a hydroxy group attached to an aromatic ring (benzene ring). Organic compounds that contain the −SH group are analogs of alcohols, and are known as thiols. Some examples are shown in Table 43.3.

to each carbon atom). Some examples are shown in Table 43.2.

Aldehydes and Ketones

Both aldehydes and ketones contain the carbonyl group (>C=O), a carbon–oxygen double bond (C=O). Aldehydes have at least one hydrogen atom bonded to the carbon of the carbonyl group, whereas ketones have only alkyl or aromatic groups bonded to the carbon of the carbonyl group. The general formula for the saturated homologous series of aldehydes and ketones is $C_nH_{2n}O$. Some examples are shown in Table 43.4.

Carboxylic Acids and Esters

The functional group of the carboxylic acids is known as the carboxyl group. This group is essentially a carbonyl group where a hydroxy group is bonded to the carbon (−COOH). Carboxylic acids can be either aliphatic (RCOOH) or aromatic (ArCOOH). Carboxylic acids with even numbers of carbon atoms, n (ranging from 4 to about 20), are called fatty acids (for example, $n = 10, 12, 14, 16$ and 18 are called capric acid, lauric acid, myristic acid, palmitic acid, and stearic acid, respectively).

Table 43.8 Organic groups listed in increasing order of polarity

Alkanes
Alkenes
Aromatic hydrocarbons
Ethers
Trichlorosilanes
Aldehydes, ketones, esters, carbonyls
Thiols
Amines
Alcohols, phenols
Amides
Carboxylic acids

Esters are alcohol derivatives of carboxylic acids, where the hydrogen of the hydroxyl group is replaced with an alkyl or aryl group. Therefore, their general formula is RCOOR′, where R can be a hydrogen, alkyl group or aromatic group, and R′ may be an alkyl group or aromatic group but not a hydrogen. Esters are found in fats and oils. Some examples are shown in Table 43.5.

Amides and Amines

Amides and amines are organic compounds containing nitrogen. Amides are nitrogen derivatives of carboxylic acids where the carbon atom of the carbonyl group is bonded directly to the nitrogen atom of an −NH₂, −NHR, or −NR₂ group instead of the oxygen of a hydroxyl group. The characteristic structure of an amide is RCONH₂.

An amine is a substituted ammonia molecule where at least one of the hydrogens attached to nitrogen is replaced with an alkyl or aryl group. It therefore has a general structure of RNH₂, R₂NH or R₃N, where R is an alkyl or aromatic group. Some examples are shown in Table 43.6.

43.1.3 Polar and Nonpolar Groups

Table 43.7 summarizes the polar and nonpolar groups commonly used to construct hydrophobic and hydrophilic molecules. Table 43.8 lists the relative polarities of selected polar groups [43.64]. Thiol, silane, carboxylic acid, and alcohol (hydroxyl) groups are the polar anchor groups most commonly used to attach to surfaces. Silane anchor groups are commonly used for Si or SiO_2 surfaces because −Si−O− bonds are strong. Methyl and trifluoromethyl are commonly used as end groups for hydrophobic film surfaces.

43.2 Self-Assembled Monolayers: Substrates, Spacer Chains; and End Groups in the Molecular Chains

SAMs are formed as a result of the spontaneous self-organization of functionalized organic molecules into stable, well defined structures on substrate surfaces, Fig. 43.3. The final structure is close to or at thermodynamic equilibrium and so it tends to form spontaneously and it rejects defects. SAMs consist of three building groups: a head group that binds strongly to a substrate, a surface terminal (tail or end) group that constitutes the outer surface of the film, and a spacer chain (backbone chain) that connects the head and surface terminal groups. The SAMs are named according to the surface terminal group (as opposed to the spacer chain and the head group), or the type of compound formed at the surface. In order to control hydrophobicity, adhesion, friction, and wear, it should strongly adhere to the substrate and the surface terminal group of the organic molecular chain should be nonpolar. To obtain strong attachment of the organic molecules to the substrate, the

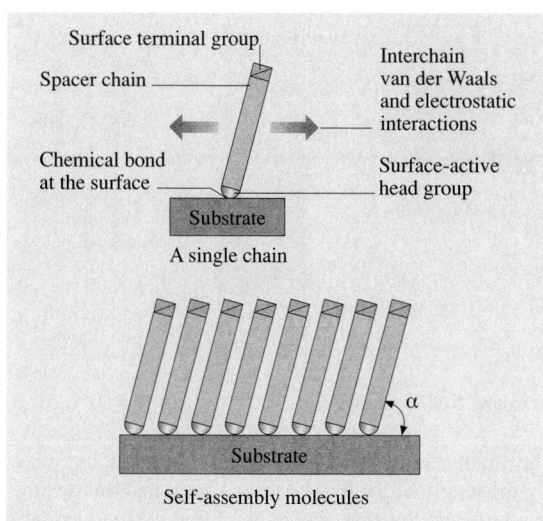

Fig. 43.3 Schematic of a self-assembled monolayer on a surface and associated forces

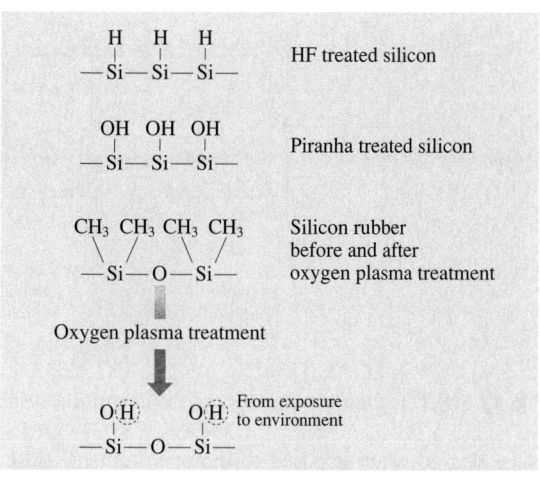

Fig. 43.4 Schematic showing HF-treated silica and the hydroxylation process that occurs on silica and elastomeric surfaces treated with Piranha solution and oxygen plasma, respectively

head group of the molecular chain should contain a polar terminal group that associates with the surface in an exothermic process (energies on the order of tens of kcal/mol); in other words it results in an apparent pinning of the head group to a specific site on the surface through a chemical bond. Furthermore, their molecular structures and the presence of crosslinking both have a significant effect on their friction and wear performance. The substrate surface should have a high surface energy (hydrophilic), so that there is a strong tendency for molecules to adsorb on the surface. The surface should be highly functional, with polar groups and dangling bonds (generally unpaired electrons), so that it can react with organic molecules and provide a strong bond. Because of the exothermic head group–substrate interactions, molecules try to occupy every available binding site on the surface, and during this process they generally push together molecules that have already been adsorbed. This process results into the formation of ordered molecular assemblies. The interactions between molecular chains are van der Waals or electrostatic in nature, with energies on the order of a few (< 10) kcal/mol (exothermic). The molecular chains in SAMs are not perpendicular to the surface; the tilt angle depends on the anchor group as well as on the substrate and the spacer group. For example, the tilt angle for alkanethiolate on Au is typically about $30-35°$ with respect to the substrate normal.

Table 43.9 lists selected systems that have been used to form SAMs [43.51]. The spacer chain of the SAM is often an alkyl chain (($-CH_2)_n$) or a derivatized alkyl group. By attaching different terminal groups to the surface, the film surface can be made to attract or repel water. Commonly used surface terminal groups in hydrophobic films with low surface energy, in the case of a single alkyl chain, include nonpolar methyl ($-CH_3$) or trifluoromethyl ($-CF_3$) groups. For hydrophilic films, commonly used surface terminal groups include alcohol ($-OH$) or carboxylic acid ($-COOH$) groups. Commonly used surface-active head groups include thiol ($-SH$), silane (such as trichlorosilane or $-SiCl_3$) and carboxyl ($-COOH$) groups. The substrates most commonly used are gold, silver, platinum, copper, hydroxylated (activated) surfaces of SiO_2 on Si, Al_2O_3 on Al, and glass.

The substrate surface should be clean before deposition. For silicon substrates, a concentrated HF solution (typically 49% HF) is commonly used to remove the oxide layer, and then the surface is rinsed with deionized water [43.45, 46]. Hydrogen passivates the surface by saturating the dangling bonds resulting in a hydrogen-terminated silicon surface with hydrophobic properties. In the deposition of multimolecularly thick polymer films with nonpolar ends, hydrophobic substrates can lead to a coated surface with a high contact angle, and are therefore preferred. The substrate should be hydrophilic for SAM deposition in order to ensure strong interfacial bonds with head groups. Hydroxylation of oxide surfaces is carried out to make them hydrophilic. Silicon and other metals get oxidized and hydroxylated to

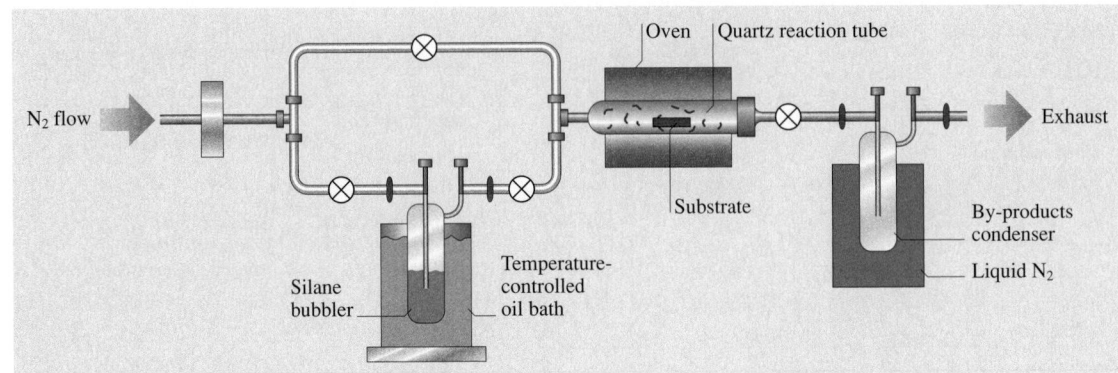

Fig. 43.5 Schematic showing vapor phase deposition system for silane SAMs [43.43]

some degree when exposed to the environment. Bulk silicon, polysilicon film or SiO_2 film surfaces are commonly treated by immersing them in Piranha solution (a mixture of typically 3:1 v/v 98% H_2SO_4 : 30% H_2O_2) at temperatures of about 90 °C for about 30 min, followed by rinsing in deionized water, to produce a hydroxylated silica surface [43.23,45,46]. Piranha solution also removes any organic and metallic contaminants, whereas HF would not necessarily remove organics. Oxygen plasma is another technique used to hydroxylate SiO_2 as well as polymer surfaces [43.43,44]. For complex silicon geometries, oxygen plasma may be preferable. Figure 43.4 shows schematics of surfaces after various surface treatments. Surfaces after piranha or oxygen plasma treatment remain hydrophilic for a few hours to about a day and become hydrophobic when they come into contact with carbon. They can retain hydrophilicity longer in dry nitrogen. To retain their hydrophobicity, the polymers are generally stored in DI water. Surfaces treated with HF remain hydrophobic for about 2–3 h and retain their hydrophobicity longer in dry nitrogen.

Epitaxial Au film on glass, mica or single-crystal silicon (produced by e-beam evaporation) substrates are commonly used because they can be deposited on smooth surfaces as an atomically flat and defect-free film. To get the organic molecules to pack together and provide a better ordering, the substrate should be chosen such that the cross-sectional diameter of the spacer chain is equal to or smaller than the distance between the anchor groups attached to the substrate. In the case of alkanethiol film, the advantage of Au substrate over SiO_2 substrate is that it results in better ordering because the cross-sectional diameter of the alkane molecule is slightly smaller than the distance between the sulfur atoms attached to the Au substrate (≈ 0.53 nm).

The thickness of the film can be controlled by varying the length of the hydrocarbon chain, and the surface properties of the film can be modified by the terminal group.

SAMs are usually produced by immersing a substrate in the solution containing a precursor (ligand) that reacts with the substrate surface or by exposing the substrate to the vapor of the reactive chemical precursor [43.24]. A schematic of a vapor deposition system is shown in Fig. 43.5. Samples are placed in a quartz reaction tube. A silane bubbler is used to introduce gaseous silane into the quartz reaction tube, which is placed in an oven at a controlled temperature. An inert gas (N_2) is used as the carrier gas. A by-product condenser is used to trap the by-products and/or unreacted silane.

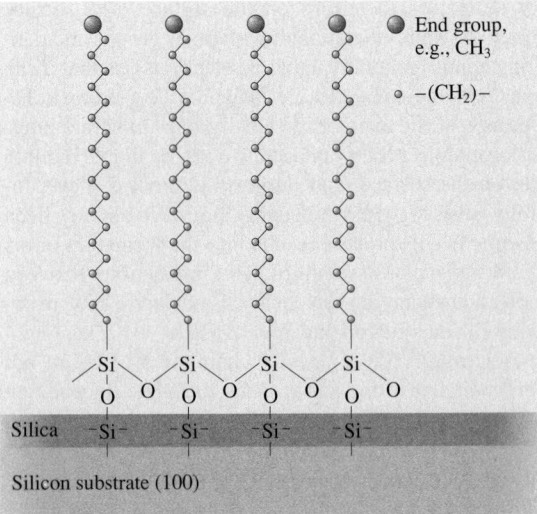

Fig. 43.6 Schematics of a methyl-terminated, n-alkylsiloxane monolayer on Si/SiO_2

Some SAMs have been widely reported. SAMs of long-chain fatty acids $C_nH_{2n+1}COOH$ or $(CH_3)(CH_2)_nCOOH$ ($n = 10$, 12, 14 or 16) on glass or alumina substrates have been widely studied since the 1950s [43.12, 20, 21, 24, 25]. Probably the most studied SAMs to date are n-alkanethiolate (n-alkyl and n-alkane are used interchangeably) monolayers $CH_3(CH_2)_nS-$ prepared by the adsorption of alkanethiol $-(CH_2)_nSH$ solution onto Au film [43.39–41, 51], and n-alkylsiloxane monolayers produced by the adsorption of n-alkyltrichlorosilane ($-CH_2)_nSiCl_3$ onto hydroxylated Si/SiO_2 substrate with siloxane ($Si-O-Si$) binding; see Fig. 43.6 [43.43, 44, 65]. (Note that "siloxane" ($Si-O-Si$) refers to the bond, while "silane" (Si_nX_{2n+2}), which describes covalently bonded compounds containing the elements Si and other atoms or groups such as H and Cl to form SiH_4

and $SiCl_4$, refers to the head group of the precursor. These terms are used interchangeably.) *Jung* et al. [43.66] have produced organosulfur monolayers – decanethiol $(CH_3)(CH_2)_9SH$ and didecyl sulfide $CH_3(CH_2)_9-S-(CH_2)_9CH_3$ – on Au films. *Geyer* et al. [43.67], *Bhushan* and *Liu* [43.39], *Liu* et al. [43.40], and *Liu* and *Bhushan* [43.41] have produced monolayers of 1, 1′-biphenyl-4-thiol (BPT) on Au surfaces, where the spacer chain of the film consists of two phenyl rings with hydrogen end groups. *Bhushan* and *Liu* [43.39] and *Liu* and *Bhushan* [43.41] have also reported monolayers of 4, 4′-dihydroxybiphenyl on Si surfaces. *Bhushan* et al. [43.43, 46], *Kasai* et al. [43.44], *Lee* et al. [43.45] and *Tao* and *Bhushan* [43.48] have produced perfluoroalkylsilane on Si surfaces and *Tambe* and *Bhushan* [43.47] have produced alkylphosphonate on Al surfaces.

43.3 Tribological Properties of SAMs

The basis for the molecular design and tailoring of SAMs should be complete knowledge of the interrelationships between the molecular structures and tribological properties of SAMs, as well as a deep understanding of the adhesion, friction and wear mechanisms of SAMs at the molecular level. Friction and wear properties of SAMs have been studied on the macro- and nanoscale. Macroscale tests are conducted using a so-called pin-on-disk tribotester apparatus, in which a ball specimen slides against a lubricated flat specimen [43.9, 10]. Nanoscale tests are conducted using an atomic force/friction force microscope (AFM/FFM) [43.4, 5, 9, 10]. In the AFM/FFM experiments, a sharp tip of radius 5–50 nm slides against a SAM specimen. A Si_3N_4 tip is commonly used for friction studies and a Si or natural diamond tip is commonly used for scratch, wear and indentation studies.

In early studies, the effect of the chain length of the fatty acid monolayers on the coefficient of friction and wear on the macroscale was studied by *Bowden* and *Tabor* [43.20] and *Zisman* [43.21]. *Zisman* [43.21] reported that there is a steady decrease in friction with increasing chain length for monolayers deposited on a glass surface sliding against a stainless steel surface. At a significantly long chain length, the coefficient of friction reaches a lower limit, Fig. 43.7a. He also reported that monolayers with a chain length of below 12 carbon atoms behave as liquids (poor durability), those with chain lengths of 12–15 carbon atoms behave like

plastic solids (medium durability), whereas those with chain lengths of above 15 carbon atoms behave like crystalline solids (high durability). Investigations by *Ruhe* et al. [43.68] indicated that the lifetime of the alkylsilane monolayer coating on a silicon surface increases greatly with increasing chain length of the alkyl substituent. *De-Palma* and *Tillman* [43.69] showed that a monolayer of n-octadecyltrichlorosilane (n-$C_{18}H_{37}SiCl_3$, OTS) is an effective lubricant on silicon.

With the development of AFM techniques, researchers have successfully characterized the nanotribological properties of self-assembled monolayers [43.1, 4, 5, 38–44, 47]. Studies by *Bhushan* et al. [43.38] showed that C_{18} alkylsiloxane films exhibit the lowest coefficient of friction and can withstand much higher normal loads during sliding than LB films, soft Au films and hard SiO_2 coatings. *McDermott* et al. [43.70] studied the effect of alkyl chain length on the frictional properties of methyl-terminated n-alkylthiolate $CH_3(CH_2)_nS-$ films chemisorbed on Au(111) using an AFM. They reported that the longer chain monolayers exhibit markedly lower friction and reduced propensity to wear than shorter chain monolayers, Fig. 43.7b. These results are in good agreement with the macroscale results published by *Zisman* [43.21]. They also conducted infrared reflection spectroscopy in order to measure the bandwidth of the methylene stretching mode [$\nu_a(CH_2)$], which exhibits a qualitative correlation with the packing density of the chains. It was found that the chain structures of monolay-

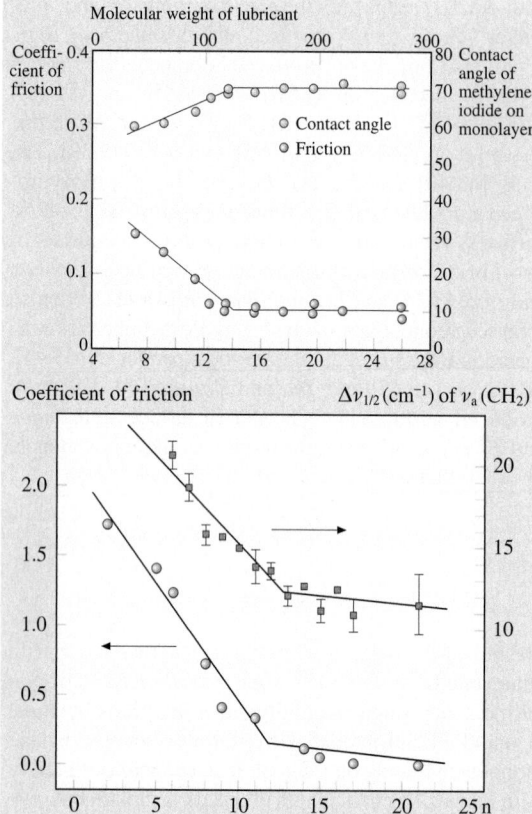

Fig. 43.7 (a) Effect of chain length (or molecular weight) on coefficient of macroscale friction of stainless steel sliding on glass lubricated with a monolayer of fatty acid, and contact angle of methyl iodide on condensed monolayers of fatty acid on glass [43.21]. **(b)** Effect of chain length of methyl-terminated, n-alkanethiolate over Au film $AuS(CH_2)_n CH_3$ on the coefficient of microscale friction and peak bandwidth at half-maximum ($\Delta\nu_{1/2}$) for the bandwidth of the methylene stretching mode [$\nu_a (CH_2)$] [43.70]

ers prepared with longer chain lengths are more ordered and more densely packed than those of monolayers prepared with shorter chain lengths. They further reported that the ability of the longer chain monolayers to retain molecular scale order during shear leads to a decrease in friction. Monolayers with a chain length of more than 12 carbon atoms – preferably 18 or more – are desirable for tribological applications. (Incidentally, monolayers with 18-carbon- atom octadecanethiol films are often studied.)

Xiao et al. [43.71] and *Lio* et al. [43.72] also studied the effect of the length of the alkyl chain on the fric-

tional properties of n-alkanethiolate films on gold and n-alkylsilane films on mica. Friction was found to be particularly high for short chains of less than eight carbon atoms. Thiols and silanes exhibit similar friction force for the same n when $n > 11$; for $n < 11$, silanes exhibit higher friction, about three times larger than that for thiols for $n = 6$. The increase in friction was attributed to the large number of dissipative modes in the less ordered chains that occur when going from a thiol to a silane anchor or when decreasing n. Longer chains ($n > 11$), stabilized by van der Waals attractions, form more compact and rigid layers and are better lubricants. *Schonherr* and *Vancso* [43.73] also correlated the magnitude of the friction with the order among the alkane chains. The disorder in short-chain hydrocarbon disulfide SAMs was found to result in a significant increase in the magnitude of the friction.

Tsukruk and *Bliznyuk* [43.74] studied the adhesion and friction between a Si sample and a Si_3N_4 tip, in which both surfaces were modified by $-CH_3$-, $-NH_2$- and $-SO_3H$-terminated silane-based SAMs. Various polymer molecules were used for the backbone. They reported a very broad maximum adhesive force at pH 4–8, with minimum adhesion at pH > 9 and pH < 3, for all of the studied mating surfaces. This observation can be understood by considering a balance of electrostatic and van der Waals interactions between composite surfaces with multiple isoelectric points. The friction coefficients of NH_2/NH_2- and SO_3H/SO_3H-mating SAMs are very high in aqueous solution. Capping NH_2-modified surfaces (3-aminopropyltriethoxysilane) with rigid and soft polymer layers resulted in a significant reduction in adhesion, to a level lower than that of the untreated surface [43.75]. *Fujihira* et al. [43.76] studied the influence of surface terminal groups of SAMs and functional tips on adhesive force. It was found that the adhesive force measured in air increases in the order CH_3/CH_3, $CH_3/COOH$, $COOH/COOH$.

Bhushan and *Liu* [43.39], *Liu* et al. [43.40], and *Liu* and *Bhushan* [43.41, 42] studied adhesion, friction and wear properties of alkylthiol and biphenylthiol SAMs. They explained the friction mechanisms using a molecular spring model in which local stiffness and intermolecular forces govern friction properties. They studied the influence of relative humidity, temperature and velocity on adhesion and friction. They also investigated the wear mechanisms of SAMs using a continuous microscratch AFM technique.

Fluorinated carbon (fluorocarbon) molecules are known to have low surface energy and are commonly used for lubrication [43.2, 9, 10]. *Bhushan* et al. [43.43,

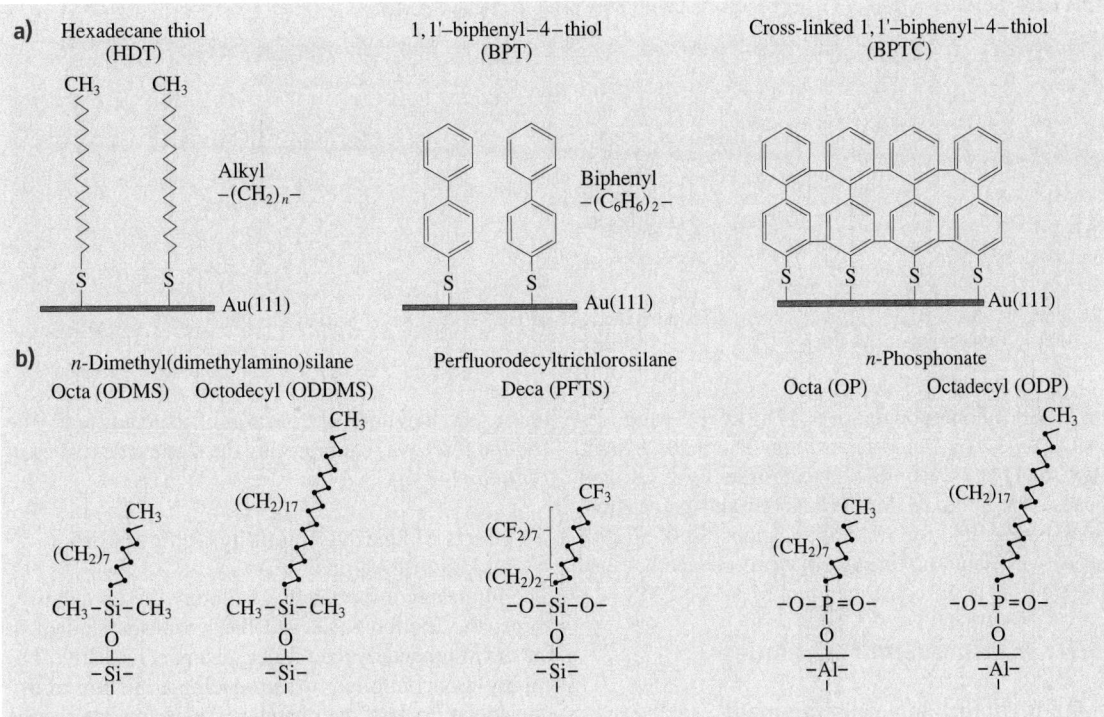

Fig. 43.8a,b Schematics of the structures of (**a**) hexadecane and biphenyl thiol SAMs on Au(111) substrates, and (**b**) perfluoroalkylsilane and alkylsilane SAMs on Si with native oxide substrates, as well as alkylphosphonate SAMs on Al with native oxide

46], *Kasai* et al. [43.44] and *Lee* et al. [43.45] studied the friction and wear properties of methyl- and perfluoro-terminated alkylsilanes on silicon. *Bhushan* et al. [43.5] and *Kasai* et al. [43.44] reported that perfluoroalkylsilane SAMs exhibit lower surface energies, higher contact angles, lower adhesive forces, and lower wear than those of alkylsilanes. *Kasai* et al. [43.44] also reported the influence of relative humidity, temperature and velocity on adhesion and friction. *Tambe* and *Bhushan* [43.47] studied the nanotribological properties of methyl-terminated alkylphosphonate on aluminium, which is of industrial interest. They found that these SAMs perform as well on aluminium as on silicon. *Tao* and *Bhushan* [43.48] studied degradation mechanisms of SAMs. They reported that oxygen from the air causes thermal oxidation of SAMs.

To date, the nanotribological properties of alkanethiol, biphenylthiol, alkylsilane and perfluoroalkylsilane SAMs have been widely studied. In this chapter, we review, in some detail, the nanotribological properties of various SAMs that have alkyl and biphenyl spacer chains with different surface terminal groups

($-CH_3$, $-CF_3$) and head groups ($-S-H$, $-Si-O-$, $-OH$, and $P-O-$), which were investigated by AFM at various operating conditions, Fig. 43.8a,b [43.5, 39–42, 44, 47, 48]. Hexadecane thiol (HDT), 1,1'-biphenyl-4-thiol (BPT) and crosslinked BPT (BPTC) were deposited on Au(111) substrates by immersing the substrate in a solution containing a precursor (ligand) that reacts with the substrate surface. Crosslinked BPTC was produced by irradiating BPT monolayers with low-energy electrons. Perfluorodecyltricholorosilane (PFTS), $CF_3-(CF_2)_7-(CH_2)_2-SiCl_3$, n-octyldimethyl (dimethylamino) silane (ODMS), $CH_3-(CH_2)_n-Si(CH_3)_2-N(CH_3)_2$ ($n = 7$), and n-octadecylmethyl (dimethylamino) silane ($n = 17$) (ODDMS) were deposited on Si by exposing the substrate to the vapor of the reactive chemical precursor. Octylphosphonate (OP)

$$CH_3-(CH_2)_n-\overset{\displaystyle O}{\underset{\displaystyle O}{\overset{|}{\underset{||}{P}}}}-OH (n = 7)$$

Table 43.9 Selected substrates and precursors commonly used to create SAMs

Substrate	Precursor	Binding with Substrate
Au	R SH (thiol)	RS−Au
Au	Ar SH (thiol)	ArS−Au
Au	RSSR'(disulfide)	RS−Au
Au	RSR' (sulfide)	
Si/SiO$_2$, glass	RSiCl$_3$ (trichlorosilane)	Si−O−Si (siloxane)
Si/Si−H	RCOOH (carboxyl)	R−Si
Metal oxides	RCOOH (carboxyl)	RCOO−...MO$_n$
(e.g., Al$_2$O$_3$, SnO$_2$, TiO$_2$)		

R represents alkane (C$_n$H$_{2n+2}$) and Ar represents aromatic hydrocarbon. These consist of various surface-active headgroups, usually with a methyl terminal group

and octadecylphosponate $(n = 17)$ (ODP) were deposited on Al by liquid deposition. Thermally evaporated Au(111) films on Si(111) substrate were selected because the epitaxial film is smooth and defect-free, which is desirable for SAM applications. Bulk Si(100) and Al with natural oxide layers were selected because they are used in the construction of MEMS/NEMS.

43.3.1 Measurement Techniques

Static Contact Angle Measurement Using DI Water

The static contact angle, a measure of how water repellent a material is, was measured using a Rame–Hart model 100 contact angle goniometer (Mountain Lakes, NJ, USA) [43.77,78]. Ten microliter droplets of DI water were typically used for the contact angle measurements. At least two measurements of the contact angle were taken. The contact angles were reproducible within ±2°.

AFM Adhesion and Friction Measurements

Adhesion and friction tests were conducted using a commercial AFM system (Dimension 3000, Nanoscope IIIa controller, DI, Santa Barbara, USA). Square-pyramidal Si$_3$N$_4$ tips with a 30–50 nm tip radius were used on a gold back-coated triangular Si$_3$N$_4$ cantilever with a typical spring constant of 0.58 N/m. The adhesion can be calculated using either force calibration plots or from the negative intercepts on friction force versus normal load plots. Both methods generally yield similar results. The force calibration plot technique was used in this study. The coefficient of friction was obtained from the slope of a plot of the friction force versus the normal load. Normal loads typically ranged from 5 to 100 nN. Friction force measurements were performed at a scan rate of 1 Hz along the fast scan axis and over a scan size of 2 μm. The fast scan axis was perpendic-

ular to the longitudinal direction of the cantilever. The friction force was calibrated by the method described in *Bhushan* [43.4].

Effects of Relative Humidity, Temperature and Sliding Velocity

The influence of the relative humidity on the adhesive force, the friction force and the wear was studied in an environmentally controlled chamber. Relative humidity was controlled by introducing a mixture of dry and moist air into the chamber. The temperature was maintained at 22 ± 1 °C. The sample was kept in the environmental chamber at the desired humidity for at least 2 h prior to the tests so that the system could reach equilibrium.

In order to study the effect of temperature on adhesion and friction force, the samples were placed on a thermal stage during the measurements. A glass plate was placed under the thermal stage to prevent the heat from being transported away. The temperature range studied was from 20 to 110 °C. The relative humidity was maintained at 50 ± 5% during the measurements.

The effect of sliding velocity on friction force was monitored in ambient conditions using a high-velocity piezo stage designed to achieve high relative sliding velocities on a commercial AFM set-up [43.79]. The traveling distance of the sample (the scan size) was set at 25 μm, while the scan frequency was varied between 0.1 Hz (5 μm/s) and 100 Hz (5000 μm/s).

AFM Wear Measurements

Wear tests were conducted using a diamond tip with a nominal radius of 50 nm and a cantilever with a nominal stiffness of 10 N/m. Wear tests were performed on a scan area of 1 μm × 1 μm at the desired normal load and at a scan rate of 1 Hz. After each wear test, an area of 3 μm × 3 μm was imaged and the average wear depth was calculated.

Degradation and Environmental Studies

The lubricant degradation experiments were carried out in a high-vacuum tribotest apparatus [43.82, 83]. The system was equipped with a mass spectrometer so that gaseous emissions from the interface could be monitored in situ during the sliding in high vacuum. The normal loads and friction forces at the contacting interface were measured using resistive-type strain-gauge transducers. Sliding tests were conducted by rubbing the sample against a flat sample of Si(100) at a vacuum pressure of 2×10^{-7} Torr at a sliding speed of 0.3 m/s. The environmental effects were investigated in high vacuum (2×10^{-7} Torr), argon, dry air (less than 2% RH), ambient air (30% RH) and in high humidity air (70% RH).

43.3.2 Hexadecane Thiol and Biphenyl Thiol SAMs on Au(111)

Hexadecane thiol on Au(111) was selected as it is a widely studied film. Biphenyl thiol was selected to study the effect of rigidity on nanotribological performance. The biphenyl thiol film was crosslinked to further increase its stiffness.

Surface Roughness, Adhesion and Friction

Surface height and friction force images of SAMs were recorded simultaneously on an area of 1 μm × 1 μm by an AFM, and adhesive forces were measured using an AFM in force calibration mode [43.39].

Detailed analysis is presented later in this chapter, but the measured roughnesses, thickness, tilt angles and spacer chain lengths of Si(111), Au(111) and various

SAMs are listed in Table 43.10 [43.39]. The roughness of BPT is very close to that of Au(111), but the roughness of BPTC is lower than that of Au(111) and BPT; this is caused by electron irradiation. Table 43.10 indicates that the roughness of HDT is much higher than the substrate roughness of Au(111). This is caused by local aggregation of organic compounds on the substrates during SAM deposition. Table 43.5 also indicates that the thicknesses of biphenyl thiol SAMs are generally smaller than those of the alkylthiol, because of the shorter spacer chain in biphenyl thiol.

Average values and standard deviations for the adhesive force and coefficient of friction are presented in Fig. 43.9 [43.39]. Based on the data, the adhesive forces and friction coefficients of SAMs are less than those of their corresponding substrates. Among the various films, HDT exhibits the lowest values. The adhesive force F_a is ranked as follows: $F_{a-Au} > F_{a-BPT} > F_{a-BPTC} > F_{a-HDT}$. The rankings for the friction coefficients μ are: $\mu_{Au} > \mu_{BPTC} > \mu_{BPT} > \mu_{HDT}$. Note that many SAMs have similar rankings for both adhesive force and coefficient of friction. This suggests that alkylthiol and biphenyl SAMs would both make effective molecular lubricants for micro/nanodevices.

Liquid capillary condensation is one source of adhesion and friction in micro/nanoscale contact. For a sphere in contact with a flat surface, the attractive Laplace force caused by a water capillary is

$$F_L = 2\pi R \gamma_{la}(\cos\theta_1 + \cos\theta_2) , \qquad (43.1)$$

where R is the radius of the sphere, γ_{la} is the surface tension of the liquid against air, and θ_1 and θ_2 are the contact angles between the liquid and flat and spherical surfaces, respectively [43.9, 10]. In an AFM adhesive study, the tip–flat sample contact is just like a sphere in contact with a flat surface, and the liquid is water. Since

Table 43.10 R_a roughnesses, thicknesses, tilt angles and spacer chain lengths of SAMs

Samples	R_a roughness [a] (nm)	Thickness [b] (nm)	Tilt angle [b] (degrees)	Spacer length [c] (nm)
Si(111)	0.07			
Au(111)	0.37			
HDT	0.92	1.89	30	1.91
BPT	0.36	1.25	15	0.89
BPTC	0.14	1.14	25	0.89

[a] Measured by an AFM with a 1 μm × 1 μm scan size, using a Si_3N_4 tip under 3.3 nN normal load
[b] The thickness and tilt angles of BPT and BPTC are reported by *Geyer* et al. [43.67]. The thickness and tilt angles of HDT are reported by *Ulman* [43.25]
[c] The spacer chain lengths of alkylthiols were calculated by the method reported by *Miura* et al. [43.80]. The spacer chain lengths of biphenyl thiols were calculated from the data reported by *Ratajczak-Sitarz* et al. [43.81]

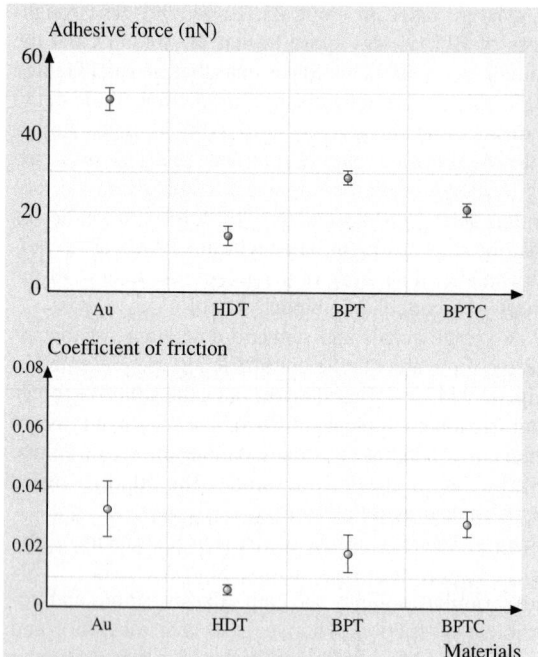

Fig. 43.9 Adhesive forces and coefficients of friction for Au(111) and various SAMs

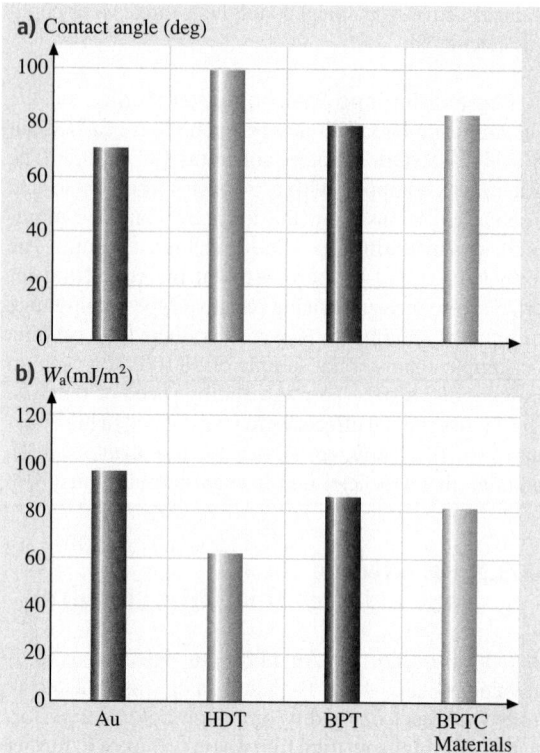

Fig. 43.10 (a) The static advancing contact angle, and (b) the work of adhesion for Au(111) and various SAMs. Each point in this figure represents the mean value of six measurements. The uncertainty associated with the average contact angle is $\pm 2°$

a single tip is used in the adhesion measurements, $\cos \theta_2$ can be treated as a constant. Therefore,

$$F_L = 2\pi R \gamma_{la}(1 + \cos\theta_1) - 2\pi R \gamma_{la}(1 - \cos\theta_2)$$
$$= 2\pi R \gamma_{la}(1 + \cos\theta_1) - C , \qquad (43.2)$$

where C is a constant.

Based on the following Young–Dupre equation, the work of adhesion W_a (the work required to pull apart a unit area of the solid–liquid interface) can be written as [43.84]

$$W_a = \gamma_{la}(1 + \cos\theta_1) . \qquad (43.3)$$

This indicates that W_a is determined by the SAM contact angle; in other words it is influenced by the surface chemistry properties (polarization and hydrophobicity) of the SAM. By substituting (43.3) into (43.2), F_L can be expressed as

$$F_L = 2\pi R W_a - C \qquad (43.4)$$

When the influence of other factors, such as van der Waals force, on the adhesive force is very small, then the adhesive force $F_a \approx F_L$. Thus the adhesive force F_a should be proportional to the work of adhesion W_a.

The contact angle is a measure of the wettability of a solid by a liquid, and it determines the

W_a value [43.77, 78]. The contact angles for distilled water on Au(111) and SAMs have been measured, and are summarized in Fig. 43.10a [43.39]. For water, $\gamma_{la} = 72.6 \, \text{mJ/m}^2$ at $22°C$. Therefore, using this value and (43.3), it is possible to obtain W_a data, and these are presented in Fig. 43.10b. The W_a values can be ranked in the following order: $W_{a-Au}(97.1) > W_{a-BPT}(86.8) > W_{a-BPTC}(82.1) > W_{a-HDT}(61.4)$. Except for W_{a-Au}, this order mimics the order of adhesion force in Fig. 43.9. The relationship between F_a and W_a is summarized in Fig. 43.11 [43.39]. It indicates that the adhesive force F_a (nN) increases with the work of adhesion W_a (mJ/m^2) as follows:

$$F_a = 0.57 W_a - 22 . \qquad (43.5)$$

These experimental results agree well with the modeling prediction presented earlier in (43.4). It proves that, on the nanoscale, and under ambient conditions, the adhesive forces of SAMs are mainly influenced by the

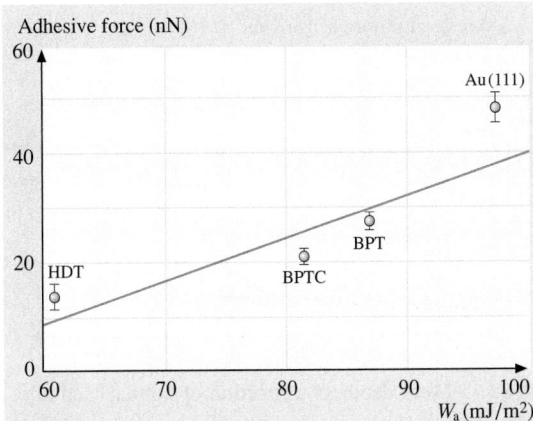

Fig. 43.11 Relationship between the adhesive force and the work of adhesion for different specimens

water capillary force. Though neither HDT nor BPT has polar surface groups, the surface terminal group of HDT has a symmetrical structure, which causes a smaller electrostatic attractive force and yields a smaller adhesive force than BPT. It is believed that the easy attachment of Au to the tip is one of the reasons for the unexpectedly large adhesive force, which means that it does not fit the linear relationship described by (43.5).

Stiffness, Molecular Spring Model and Micropatterned SAMs

The friction mechanisms of SAMs were also examined. Monte Carlo simulation of the mechanical relaxation of a $CH_3(CH_2)_{15}SH$ self-assembled monolayer, as performed by *Siepman* and *McDonald* [43.85], indicated that SAMs compress and respond nearly elastically to microindentation by an AFM tip when the load is below a critical normal load. Compression can lead to major changes in the mean molecular tilt (the orientation), but the original structure is recovered as the normal load is removed.

Stiffness properties were measured by an AFM in force modulation mode [43.4, 5, 41]. They reported that BPT was stiffer than HDT. Since BPT has rigid benzene structure, it is more difficult to compress than HDT. Figure 43.12 shows the variation in the displacement with normal load during indentation mode. It clearly indicates that SAMs can be compressed. At a given normal load, SAMs with long carbon chain structures such as HDT are easy to compress compared to SAMs with rigid benzene ring structures, such as BPT. *Garcia-Parajo* et al. [43.86] have also reported on the compression and relaxation of octadecyltrichlorosilane (OTS) film obtained from loading and unloading tests.

A molecular spring model is presented in Fig. 43.13 to explain the difference in friction between the SAMs observed in the friction and stiffness measurements by AFM and the Monte Carlo simulation. It is believed that the self-assembled molecules on a substrate act just like assembled molecular springs anchored to the substrate [43.39]. A Si_3N_4 tip sliding on the surface of a SAM is like a tip sliding on the top of molecular springs or a brush. The molecular spring assembly has compliant features and can experience compression and orientation under normal load. The orientation of the molecular springs or brush reduces the shearing force

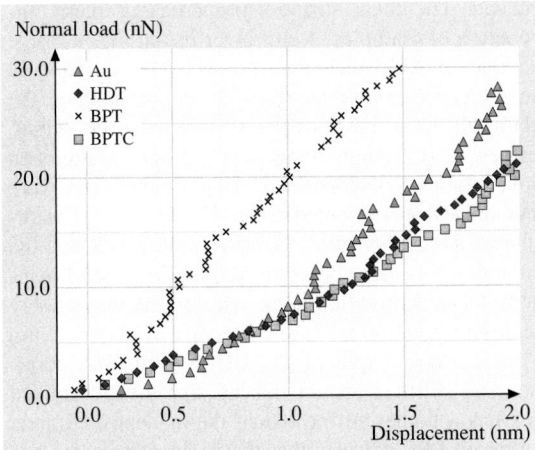

Fig. 43.12 Normal load versus displacement curves for Au(111) and various SAMs

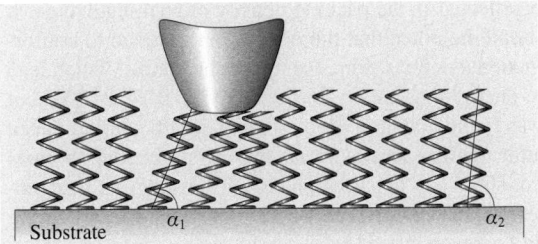

Fig. 43.13 Molecular spring model for SAMs. In this figure, $\alpha_1 < \alpha_2$, which is caused by reorientation under the normal load applied by the AFM tip. The reorientation of the molecular springs reduces the shearing force at the interface, which in turn reduces the friction force. The molecular spring constant and the intermolecular forces determine the magnitude of the coefficient of friction. In this figure, the size of the tip and the molecular springs are not drawn to the same scale [43.39]

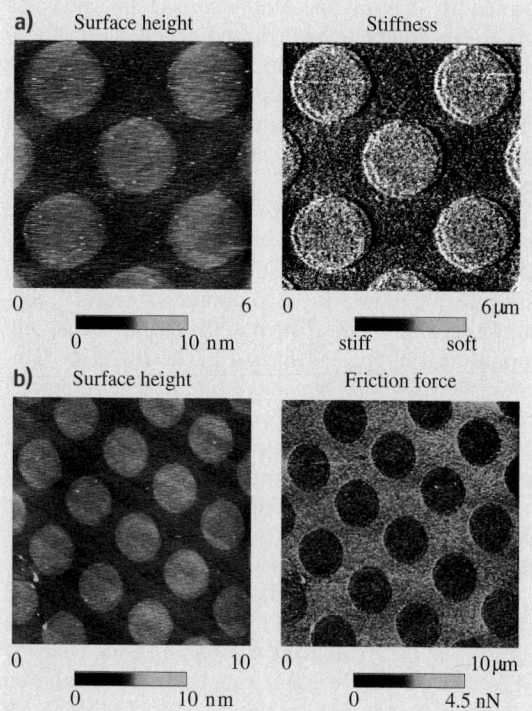

a)   Surface height   Stiffness

0   6   0   6 μm
    0   10 nm   stiff   soft

b)   Surface height   Friction force

0   10   0   10 μm
    0   10 nm   0   4.5 nN

Fig. 43.14 (a) AFM grayscale surface height and stiffness images, and (b) AFM grayscale surface height and friction force images of micropatterned BDCS [43.41]

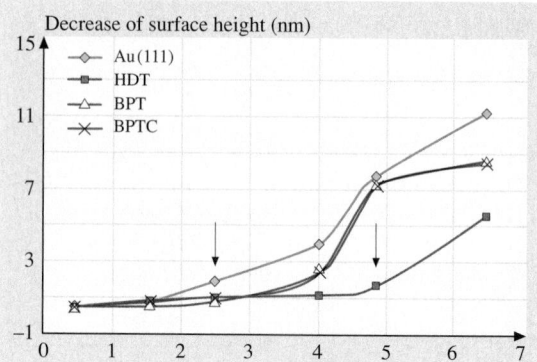

Decrease of surface height (nm)

- Au(111)
- HDT
- BPT
- BPTC

Fig. 43.15 Wear depth as a function of normal load after one scan cycle

at the interface, which in turn reduces the friction force. The possibility of orientation is determined by the spring constant of a single molecule (local stiffness), as well as the interactions between neighboring molecules, which is reflected in the packing density or packing energy. It should be noted that the orientation can lead to conformational defects along the molecular chains, which lead to energy dissipation. In the study of BPT by AFM, it was found that the friction force is significantly reduced after the first several scans, but the surface height does not show any apparent change. This suggests that molecular orientation can be facilitated by initial sliding and is reversible [43.42].

Based on the stiffness measurement results presented in Fig. 43.12 and the view of molecular structures given in Fig. 43.13, biphenyl is a more rigid structure due to the contribution of the two rigid benzene rings. Therefore, the spring constant of BPT is larger than that of HDT. The hydrogen (H^+) in the biphenyl chain has an electrostatic attraction to the π electrons in the neighboring benzene ring. Thus the intermolecular forces between biphenyl chains are stronger than those between alkyl

chains. The larger spring constant of BPT and stronger intermolecular forces mean that it requires a larger external force to allow it to orient, thus causing a higher coefficient of friction. The crosslinking of BPT leads to a larger packing energy for BPTC. Therefore BPTC orientation requires a larger external force: the coefficient of BPTC is higher than BPT.

An elegant way to demonstrate the influence of molecular stiffness on friction is to investigate SAMs with different structures on the same wafer. A micropatterned SAM was prepared for this purpose. First biphenyldimethylchlorosilane (BDCS) was deposited on the silicon using a typical self-assembly method [43.41]. Then the film was partially crosslinked using a mask technique by low-energy electron irradiation. Finally, micropatterned BDCS films that had the different coating regions on the same wafer were realized. The local stiffness properties of these micropatterned samples were investigated by a force modulation AFM technique [43.87]. The variation in the deflection amplitude provides a measure of the relative local stiffness of the surface. Surface height, stiffness and friction images of the micropatterened biphenyldimethylchlorosilane (BDCS) specimen were obtained and are presented in Fig. 43.14 [43.41]. The circular areas correspond to the as-deposited film, and the remaining area to the crosslinked film. Figure 43.14a indicates that crosslinking caused by the low-energy electron irradiation leads to a decrease of about 0.5 nm in the surface height of the BDCS film. The corresponding stiffness images indicate that the crosslinked area has a higher stiffness than the as-deposited area. Figure 43.14b indicates that the as-deposited area has lower friction force. Obviously, these data for the micropatterned sample prove that the local stiffness of the

SAM influences its friction performance. Higher stiffness leads to larger friction force. These results correlate well with the suggested molecular spring model.

In summary, it was found that SAMs exhibit compliance and can experience compression and orientation under normal load. SAM orientation reduces the shear stress at the interface, so SAMs provide good lubricants. The molecular spring constant (local stiffness) and intermolecular forces can both influence the friction coefficient of a SAM.

Wear and Scratch Resistance

Wear resistance was studied on an area of 1 μm × 1 μm. The variation of wear depth with normal load is presented in Fig. 43.15 [43.39]. HDT exhibits the best wear resistance. A critical normal load (marked by arrows in Fig. 43.15) appears in the curves for all of the SAMs tested. When the normal load is smaller than the critical normal load, the monolayer only changes in height very slightly in the scan area. When the normal load is higher than the critical value, the change in height of the SAM increases dramatically. Relocation and accumulation of BPT molecules is observed during the first few scans, which leads to the formation of a larger terrace. Wear studies of a single BPT terrace indicate that the wear life of BPT increases exponentially with terrace size [43.40, 41].

The scratch resistances of Au(111) and SAMs were studied using a continuous AFM microscratch technique. Figure 43.16a shows coefficient of friction profiles as a function of increasing normal load, and corresponding tapping mode AFM surface height images of the scratches captured on Au(111) and SAMs [43.41]. Figure 43.16a indicates that there is an abrupt increase in the coefficient of friction for all of the tested samples. The normal load associated with this event is termed the critical load (indicated by the arrows labeled "A"). Initially during scratching, all of the samples exhibit a low coefficient of friction, indicating that the friction force is dominated by the shear component. This is in agreement with AFM image analysis, which shows negligible damage on the surfaces prior to the critical load. At the critical load, a clear groove is formed, which is accompanied by material piling up at the sides of the scratch. This suggests that the initial damage that occurs at the critical load is due to plowing associated with plastic deformation, and this causes the sharp rise in the coefficient of friction. Beyond the critical load, debris can be seen as well as material pile-up at the sides of the scratch. Figure 43.16b summarizes the critical loads for the various samples

obtained in this study. It clearly indicates that all of the SAMs increase the critical load of the corresponding substrate.

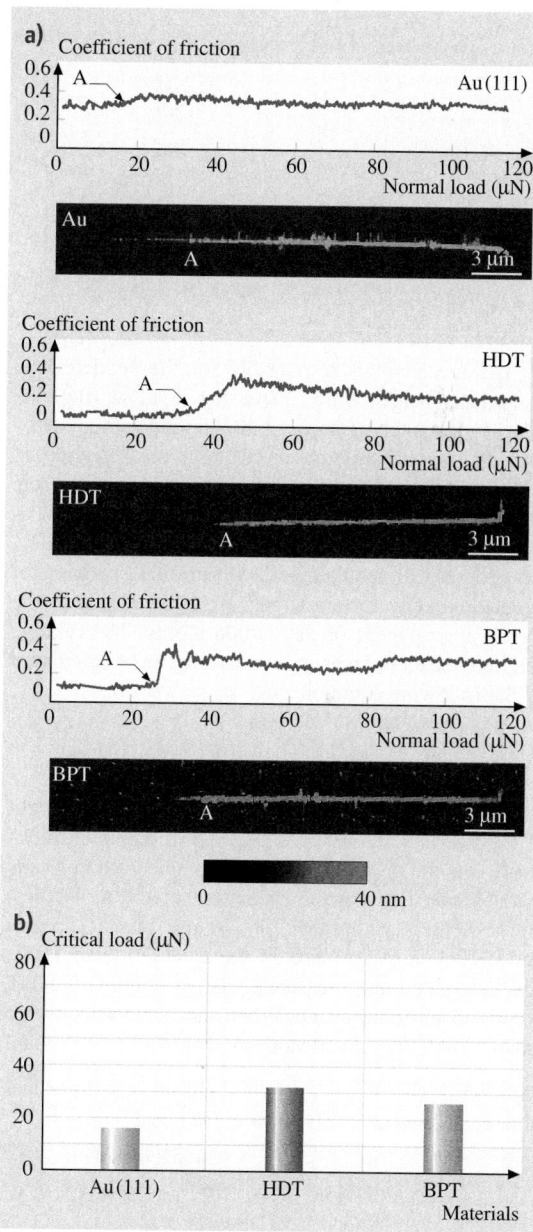

Fig. 43.16 (a) Coefficient of friction profiles during scratch as a function of normal load, and corresponding AFM surface height images as well as (b) critical loads estimated from coefficient of friction profiles and AFM images for Au(111), HDT/Au(111) and BPT/Au(111) [43.41]

Table 43.11 Calculated $\left[1+\left(\frac{na}{d}\right)^2\right]^{-\frac{1}{2}}$ and measured $\left(\frac{h}{L}\right)$ relative heights of an HDT self-assembled monolayer [43.39]

Steps (n)	Calculated[a] $\left[1+\left(\frac{na}{d}\right)^2\right]^{-\frac{1}{2}}$	Measured $\left(\frac{h}{L}\right)$
1	0.883	
2	0.685	0.674[b]
3	0.531	0.532[c]
4	0.425	0.416[d]
5	0.352	0.354[e]
6	0.299	

[a] Calculations are based on the assumption that the molecules tilt in discrete steps (n) upon compression with a diamond AFM tip [43.88]
[b,c,d,e] These measured values correspond to normal loads of 0.50 μN, 1.57 μN, 2.53 μN and 4.03 μN, respectively

The mechanisms responsible for the sudden drop in surface height with increasing load during the wear and scratch tests need to be understood. *Barrena* et al. [43.88] observed that the heights of self-assembled alkylsilane layers decrease in discrete amounts with normal load. This step-like behavior is due to discrete molecular tilts, which are dictated by the geometrical requirements of the close packing of molecules. Only certain angles are allowed due to the zigzag arrangement of the carbon atoms. The relative height of the monolayer under pressure can be calculated by the following equation

$$\left(\frac{h}{L}\right)=\left[1+\left(\frac{na}{d}\right)^2\right]^{-1/2}, \tag{43.6}$$

where L is the total length of the molecule, h is the height of the SAMs in the tilt configuration (monolayer thickness), a is the distance between alternate carbon atoms in the molecule, d is the separation of the molecules, and n is the step number. The values of a (0.25 nm) and d (0.47 nm) are used in the calculation of HDT. The calculated and measured relative heights of HDT are listed in Table 43.11. When the normal loads are smaller than the critical values in Fig. 43.15, the meas-

Table 43.12 Calculated L_0 values and measured residual film thicknesses of SAMs under critical load

	L_0[a] (nm)	Residual thickness[b] (nm)
HDT	0.24	0.25
BPT	0.39	0.42
BPTC	0.33	0.38

[a] Calculated using the equation $h = b\cos(\alpha)n + L_0$ [43.80]
[b] Measured by AFM using a diamond tip under critical normal load. All of the data are the mean values of three tests

ured relative height values of HDT are very close to the calculated values. This means that HDT underwent step tilting below the critical normal load.

The residual SAM thickness after wear under critical normal load was measured by profiling the worn film using AFM. The results are listed in Table 43.12. For an alkanethiol monolayer, the relationship between the monolayer thickness h and the intercept length L_0 can be expressed as (Fig. 43.17)

$$h = b\cos(\alpha)n + L_0, \tag{43.7}$$

where b is the length of the projection of the C−C bond onto the main chain axis ($b = 0.127$ nm for alkanethiol), n is the chain length defined by $CH_3(CH_2)_n SH$, and α is the tilt angle [43.80]. The L_0 values have also been calculated for BPT and BPTC, based on the same principle, and using the bond lengths reported in reference [43.81], Table 43.7. This indicates that the measured residual thickness values of SAMs under critical load are very close to the calculated intercept length (L_0) values. This means that the Si_3N_4 tip approaches the interface and SAMs are severely worn away from the substrate under the critical normal load. This is because chemical adsorption bond strength at the interface (S−Au) is gen-

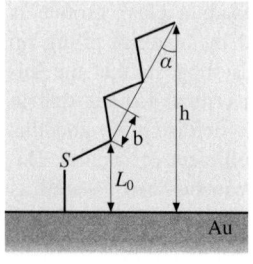

Fig. 43.17 Illustration of the relationship between the components of the equation $h = b\cos(\alpha)n + L_0$ [43.39]

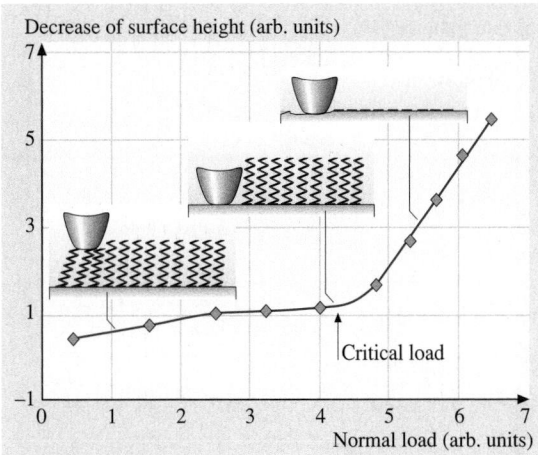

Fig. 43.18 Illustration of the wear mechanisms of SAMs with increasing normal load [43.41]

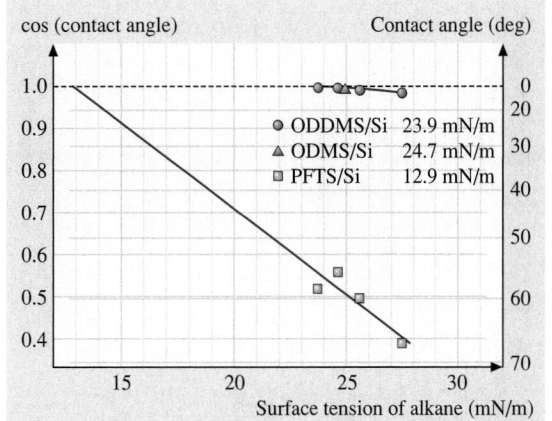

Fig. 43.19 Zisman plots for PFTS/Si, ODMS/Si and ODDMS/Si used to calculate the critical surface tension, a measure of surface energy, which is given by the x-intercept [cos(contact angle) $= 1$] of the line fitted to the data

erally smaller than other chemical bond strengths in SAM spacer chains (Table 43.8).

The change in the wear mechanism of a SAM with increasing normal load is therefore illustrated in Fig. 43.18. Below the critical normal load SAMs undergo step orientation; at the critical load SAMs wear away from the substrate due to the weak interface bond strengths; while above the critical normal load severe wear takes place on the substrate. To improve the wear resistance, the interface bond must be enhanced; a rigid spacer chain and a hard substrate are also preferable in this regard.

43.3.3 Alkylsilane and Perfluoroalkylsilane SAMs on Si(100) and Alkylphosphonate SAMS on Al

The nanotribological parameters of perfluorodecyltricholorosilane (PFTS), $CF_3-(CF_2)_7-(CH_2)_2-SiCl_3$, n-octyldimethyl (dimethylamino)silane (ODMS), $CH_3-(CH_2)_n-Si(CH_3)_2-N(CH_3)_2$ ($n = 7$), and n-octadecyl-methyl(dimethylamino)silane ($n = 17$) (ODDMS) vapor deposited on Si substrate were investigated, as were those of octylphosphonate (OP),

$$CH_3 - (CH_2)_n - \overset{\overset{\displaystyle O}{|}}{\underset{\underset{\displaystyle O}{||}}{P}} -OH$$

($n = 7$) and octadecylphosponate (ODP) ($n = 17$) on Al substrate. The perfluoroalkylsilane SAM was selected because fluorinated films are known to have low surface energy. Alkylsilanes with two different chain lengths

(with 8 and 18 carbon atoms) were selected in order to compare their nanotribological performance with that of PFTS as well as to study the influence of chain length. The alkylphosphonate SAMs (with 8 and 18 carbon atoms) on Al were selected due to their industrial use in applications such as digital projection displays.

Surface Free Energy and Contact Angle Measurement

As stated earlier, adhesive force arises from the presence of a thin liquid film such as a mobile lubricant or an adsorbed water layer that causes meniscus bridges to build up around asperities due to surface energy effects. The intrinsic attractive force arising from meniscus contributions depends on the surface tension of the film and the contact angle and may result in high friction and wear. Surface energies and contact angles were measured in order to evaluate the hydrophobicity. Figure 43.19 shows a Zisman plot for the SAMs deposited on Si and their surface energies obtained using various alkane liquids [43.43]. Zisman analysis data was not available for the Si substrate because the alkane liquids used for the measurement spreads instantly across such surfaces. A significantly lower critical surface tension or surface energy was observed for PFTS (12.9 mN/m for PFTS/Si) than for ODMS (24.7 mN/m for ODMS/Si) or ODDMS (23.9 mN/m for ODDMS/Si). The surface energies for ODMS and ODDMS were comparable. This suggests that the surface was covered by these SAMs to a comparable degree without bare substrate appearing.

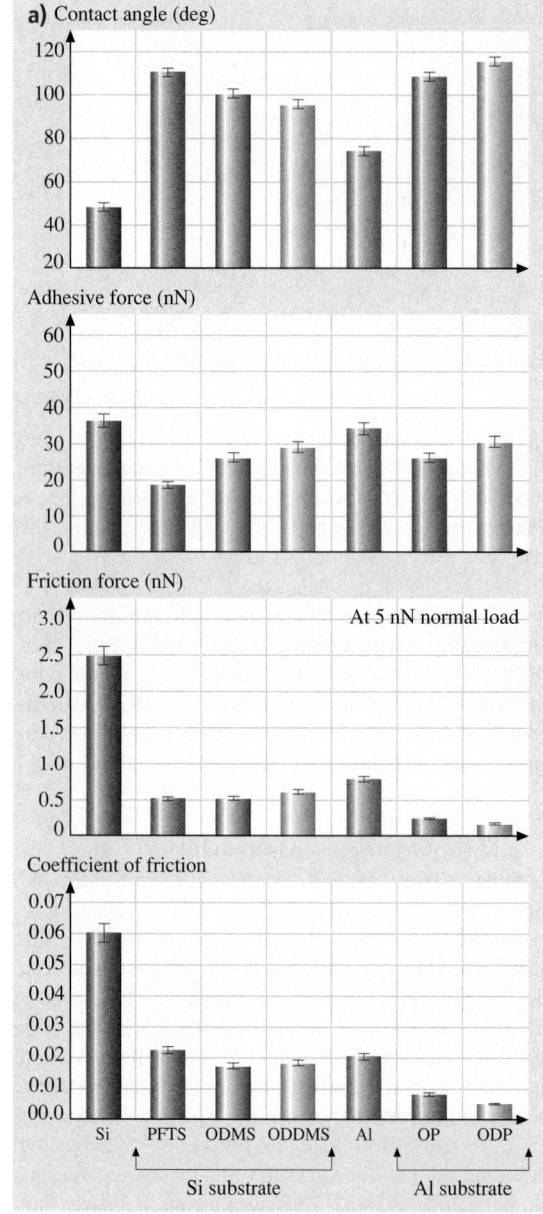

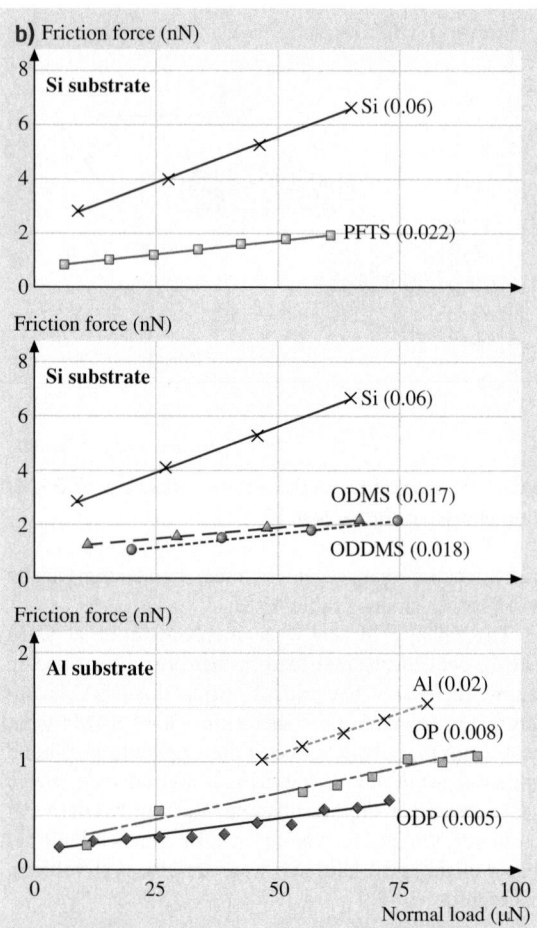

Fig. 43.20 (a) The static contact angle, adhesive force, friction force and coefficient of friction measured using an AFM for various SAMs on Si and Al substrates, and (b) friction force vs. normal load plots for various SAMs on Si and Al substrates

Contact angles were measured for various SAMs on Si and Al substrates so that they could be compared. The measured values for the samples are shown in Fig. 43.20a [43.43, 47]. A summary of RMS roughness measured using an AFM, contact angles and film thickness measured using an ellipsometer are summarized in Table 43.13. Significantly better water repellency was observed for PFTS compared to bare Si with natu-

ral oxide. The contact angle for PFTS/Si was $\approx 110°$ whereas it was $\approx 50°$ for Si. The contact angle generally increases with decreasing surface energy [43.89], which is consistent with the data obtained. The water contact angles of ODMS and ODDMS were also large ($\approx 100°$), implying high surface hydrophobicity. These contact angles can be influenced by the packing density as well as the sample roughness [43.90], which probably accounts for the slightly higher contact angles for the SAMs deposited on Al substrates compared to those on Si substrate. The $-CH_3$ groups in ODMS and ODDMS are nonpolar and are known to contribute to the water repellency; however, OP and ODP (which have different

Table 43.13 A summary of RMS roughness, contact angle and film thicknesses of various SAMs

SAM/substrate	Acronym	RMS roughness (nm)	Contact angle (deg.)	Film thickness (nm)
Silicon(111)	Si	0.07	48	–
Perfluorodecyltricholorosilane/Si	PFTS/Si	0.09	112	1.8 [a]
Octyldimethyl(dimethylamino)silane/Si	ODMS/Si	0.10	99	1.9 [a]
Octadecyldimethyl(dimethylamino)silane/Si	ODDMS/Si	0.10	92	2.1 [a]
Aluminium	Al	1.73	74	–
Octylphosphonate/Al	OP/Al	–	108	≈ 1.9 [b]
Octadecylphosphonate/Al	ODP/Al	–	115	≈ 2.1 [b]

[a] *Kasai* et al. [43.44]
[b] *Kulik* (personal communication)

head groups to ODMS and ODDMS) exhibited greater surface hydrophobicity. Among the SAMs, PFTS and ODP exhibited the highest contact angle.

AFM Adhesion and Friction Measurements under Ambient Conditions

Figure 43.20a gives the adhesive force, friction force and the coefficient of friction measured under ambient conditions using AFM for various SAMs deposited onto Si and Al substrates, while Fig. 43.20 shows friction force–normal load plots for these systems [43.43, 47]. Figure 43.21 shows surface height and friction force maps for Si and PFTS, ODMS and ODDMS on Si [43.43].

The bare substrates gave much higher adhesive force than the SAM coatings. ODMS and ODDMS show adhesive forces comparable to OP and ODP despite their lower water contact angles. These SAMs have the same tail groups, and during AFM measurements the AFM tip only interacts with the tail groups, whereas the contact angles can also be influenced by the head groups in these SAMs. This is probably why the adhesive forces for these SAMs are comparable. PFTS, which has one of the highest contact angles, showed the lowest adhesion.

Friction force images of SAMs are more uniform than those of Si. The coefficient of friction was higher for the bare substrates than for the corresponding SAMs deposited on them. The SAMs deposited on the Si substrate showed higher coefficients of friction than those deposited on the Al substrates. The primary reason for this is believed to be the greater roughness of the Al substrates and possibly higher packing densities of SAMs on Al as mentioned earlier. For the SAMs deposited on Si substrates, the ones with fluorocarbon backbones were found to have higher coefficients of friction than those

with hydrocarbon backbones. This might be due to the higher stiffness of the fluorocarbon backbone [43.44]. It is harder to rotate a fluorocarbon backbone because the F atom is larger than the H atom [43.91]. The C−C bonds of hydrocarbon chains, on the other hand, have more freedom to rotate. We presented a molecular spring or brush model earlier that explained why less compliant SAMs exhibit more friction. SAMs with higher spring constants or stiffer backbones may need more energy to be elastically deformed during sliding, so these SAMS show more friction. In terms of the influence of the chain length, it has been reported that the friction coefficients for SAM surfaces decrease with the carbon backbone chain length (n) up to 12 carbon atoms ($n \approx 12$) [43.70]. However, this effect of chain length on the coefficient of friction was not apparent in these data.

Effect of Relative Humidity, Temperature and Sliding Velocity on AFM Adhesion and Friction

The influence of relative humidity was studied for various SAMs. Its effects on adhesive force, friction force at a normal load of 5 nN, the coefficient of friction and on microwear are shown in Fig. 43.22 [43.44, 47]. The adhesive force of silicon showed an increase with relative humidity, Fig. 43.22a. This is expected since the surface of silicon is hydrophilic, as shown in Fig. 43.20a. Greater condensation of water at the tip–sample interface increases the adhesive force due to the capillary effect. On the other hand, the adhesive forces of the SAMs showed very weak dependencies on humidity. This may be because the surfaces of the SAMs are hydrophobic. The adhesive forces of ODMS/Si and ODDMS/Si showed slight increases from 75 to 90% RH. This increase was absent for PFTS/Si. This may result from the more hydrophobic surface properties of PFTS/Si. The

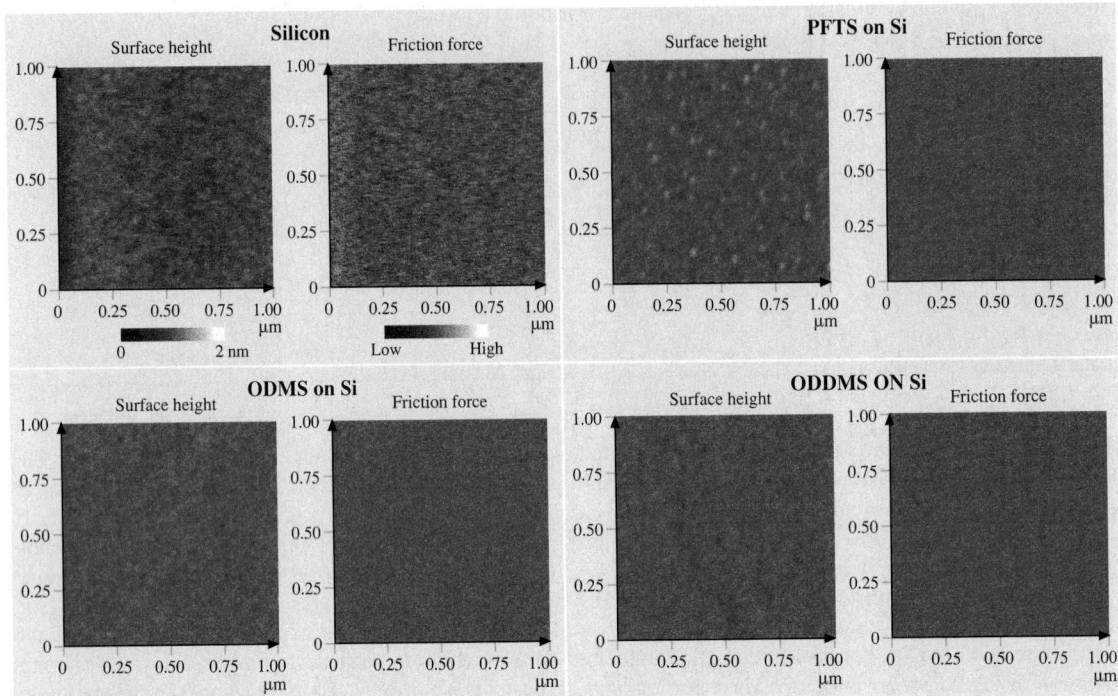

Fig. 43.21 Surface height (*left*) and friction force (*right*) maps for (a)Si, PFTS/Si, ODMS/Si and ODDMS/Si [43.43]

Al substrate is partially hydrophobic and hence does not show much of a dependence on humidity, Fig. 43.22b. The OP and ODP SAMs deposited on Al substrates showed almost no change in adhesive force with humidity. The highly hydrophobic nature of these monolayers means that the contribution of the water menisci to the overall adhesive force is negligible at all humidities.

The friction force of silicon showed an increase with relative humidity up to about 75% RH and then a slight decrease beyond this point, see Fig. 43.22a. The initial increase could result from the increase in adhesive force. The decrease in friction force at higher humidities could be attributed to the lubricant effects of the water layer. This effect is more pronounced in the coefficient of friction. Since the adhesive force increased and the coefficient of friction decreased in this range, those effects cancelled each other out and the resulting friction force showed only slight changes. On the other hand, the friction forces and coefficients of friction of the SAMs showed very small changes with relative humidity; similar behavior to that observed for adhesive force. This suggests that the adsorbed water layer on the surface maintained a similar thickness through-

out the relative humidity range tested. The differences among the SAM types were small within measurement errors, but a closer look at the coefficient of friction for ODMS/Si showed a slight increase from 75 to 90% RH compared to PFTS/Si, possibly due to the same reasons as for the increase in adhesive force. The inherent hydrophobicity of the SAMs means that they do not show much relative humidity dependence.

Figure 43.23 shows the effect of temperature on the adhesive force, the friction force at a normal load of 5 nN and the coefficient of friction [43.44, 47]. The adhesive force for silicon increased with the temperature from room temperature (RT) to 55 °C, and then decreased from 55 to 75 °C before eventually leveling off from 75 to 110 °C, see Fig. 43.23a. The adhesive forces of the SAMs showed a similar tendency except that the initial increase was not pronounced. The initial increase in adhesive force for silicon at low temperatures is not understood. The decrease observed for silicon could be attributed to the desorption of water molecules on the surface. After the water layer has been almost completely depleted, the adhesive force may stay constant. The SAMs with hydrocarbon backbone chains

(OP, ODP, ODMS and ODDMS) showed similar behavior to the Al substrate but the initial increase in the adhesive force with temperature was much smaller. The SAMs with fluorocarbon backbone chains showed almost no temperature dependence. The adhesive force shows some temperature dependence for the SAMs with hydrocarbon backbone chains. This increase in adhesive force is believed to be caused by the melting of the SAM film. The typical melting point for a linear carbon chain molecule such as $CH_3(CH_2)_{14}CH_2OH$ is 50 °C [43.92]. As the temperature increases, the SAM film softens, thereby increasing the real area of contact and consequently the adhesive force. Once the temperature is higher than the melting point, the lubrication regime is changed from boundary lubrication in the solid SAM to liquid lubrication in the melted SAM [43.41].

The friction force of silicon increased with temperature and then steadily decreased. The friction force is highly affected by changes in adhesion. The decrease in friction could therefore result from the depletion of the water layer. The coefficient of friction for silicon remained constant and then decreased, starting at about 80 °C. For the SAMs, the coefficient of friction exhibited a monotonic decrease with temperature. The decrease in the friction and the coefficient of friction for the SAMs possibly results from the decrease in stiffness. As mentioned before, the spring model suggests lower friction for more compliant SAMs [43.41]. The different types of SAM types behaved reasonably similarly. PFTS kept its stiffness more than ODMS and ODDMS when the temperature was increased [43.93], but this was not a particularly pronounced effect in the results.

Figure 43.23b shows the effect of temperature on the adhesive force for SAMs on Al. The adhesive force increased for Al substrate up to 50 °C and then decreased to a stable value for higher temperatures. This initial increase in the adhesive force with temperature for Al is not understood. The inherently high hydrophobicity of the SAMs over the corresponding substrates meant that they did not show much of a relative humidity or temperature dependence.

Figure 43.24 shows the effect of sliding velocity on the adhesive force, the friction force and the coefficient of friction [43.44, 47]. The adhesive force for silicon

Fig. 43.22a,b Relative effect of humidity on the adhesive force, the friction force, the coefficient of friction and the microwear for various SAMs on (**a**) Si substrates [43.44], and (**b**) Al substrates

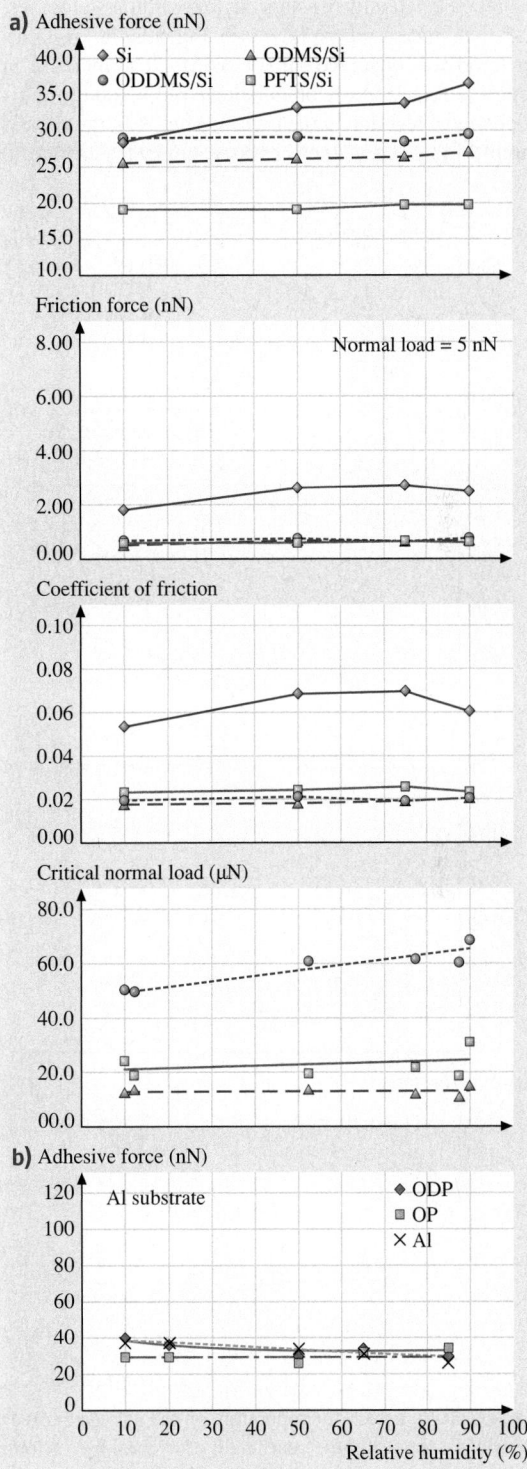

Part E | 43.3

remained relatively constant at low sliding velocities, and then increased rapidly: see Fig. 43.24a. A similar trend was observed for the SAMs. The increase in adhesive force for silicon is believed to be due to a tribochemical reaction at the interface between the tip and sample [43.23] and the increased contact area due to mechanical plowing. For the SAMs, the increased adhesive force at high velocities may result from viscous drag of SAM molecules [43.94]. SAMs can be detached from the surface and attached to an AFM tip. In addition, the increased contact area may be caused by greater penetration of the AFM tip into the SAM. The rate of increase was larger for ODMS than for PFTS, presumably because of the higher stiffness and more dense structure of PFTS.

The coefficient of friction increased with sliding velocity and then reached a plateau for Si, ODMS/Si and ODDMS/Si. As the sliding speed is increased, extra work is needed for SAM reorientation, which may lead to increased friction. For PFTS, the coefficient of friction decreased at large sliding velocities, resulting in a peak. This peak structure may result from the viscoelastic properties of SAMs [43.95].

Figure 43.24b shows the effect of sliding velocity on the friction force. Friction force was found to remain constant over a range of sliding velocities for Al substrate as well as the OP and ODP SAMs deposited on Al substrates. The increase in friction force at high velocities (> 1 mm/s) is the result of asperity impacts and correspondingly high frictional energy dissipation at the sliding interface for Al [43.96]. For the OP and ODP SAMs, the increase in friction force is believed to result from the SAM molecules reorienting under the tip load and during the tip motion. The SAM reorientation can act as an additional hindrance to tip motion when the AFM tip reverses during scanning, resulting in higher friction. The molecules can get entangled and/or get detached from the substrate and attach to the AFM tip. *Tambe* and *Bhushan* [43.94] extended the molecular spring model presented by *Bhushan* and *Liu* [43.39] to explain this velocity-dependent increase in friction force for compliant SAM molecules.

Overall, the SAMs deposited on the Al substrate showed much lower friction at all sliding velocities than those deposited on the Si substrates. As discussed, the primary reason for this is believed to be the ease with which the molecules on Al substrate can rotate due to the absence of either the $-CH_3$ groups or any crosslinking at the head groups. The higher stiffness of the fluorocarbon backbone chains than the hydrocarbon backbone chains [43.44] is also believed to result in the higher friction for PFTS than ODMS and ODDMS.

AFM Wear Measurements

Figure 43.25a shows the relationships between surface height and normal load found for various SAMs during wear tests [43.44, 47]. As shown in the figure, the

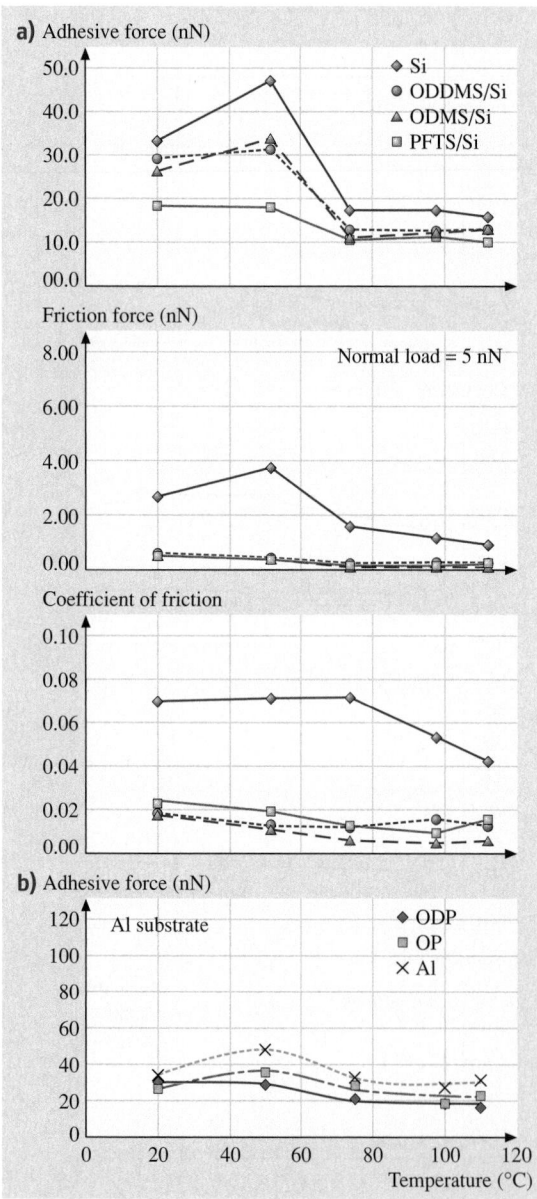

Fig. 43.23a,b Effect of temperature on the adhesive force, the friction force and the coefficient of friction for various SAMs on (**a**) Si substrates [43.44], and (**b**) Al substrates

SAMs exhibit a critical normal load beyond which the surface height decreases drastically. Figure 43.25a also shows the wear behavior of the Al and Si substrates. Unlike the SAMs, the substrates show a monotonic decrease in surface height with increasing normal load,

with wear initiating from the very beginning, even for low normal loads. Si (Young's modulus of elasticity, $E = 130\,\text{GPa}$ [43.97], hardness, $H = 11\,\text{GPa}$ [43.28]) is relatively hard in comparison to Al ($E = 77\,\text{GPa}$,

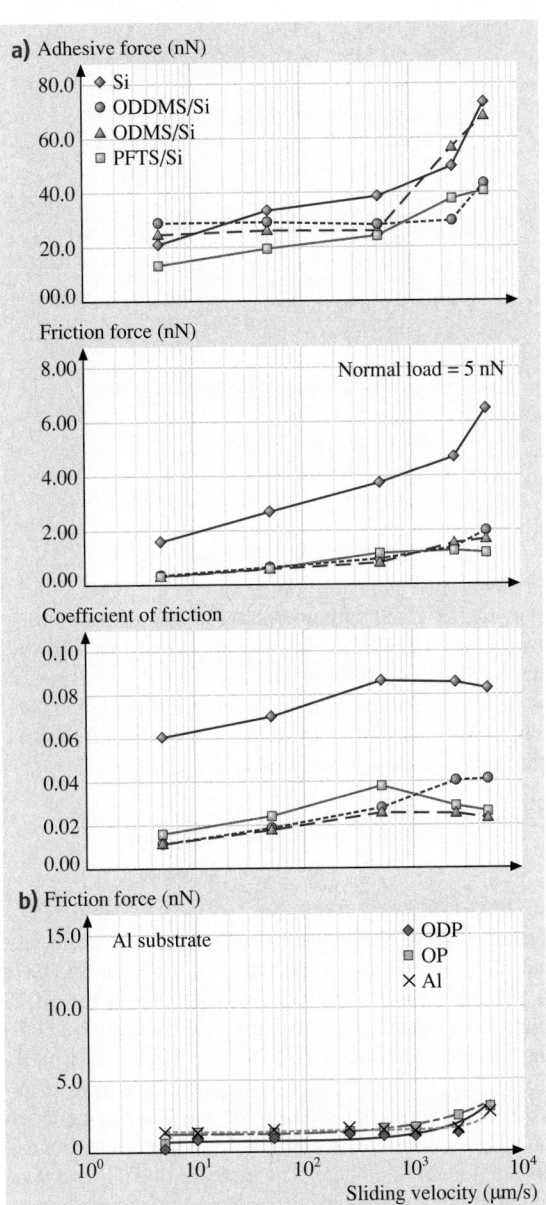

Fig. 43.24a,b Effect of sliding velocity on the adhesive force, the friction force and the coefficient of friction for various SAMs on (**a**) Si substrates [43.44], and (**b**) Al substrates (22 °C, 50% RH, 70 nN normal load)

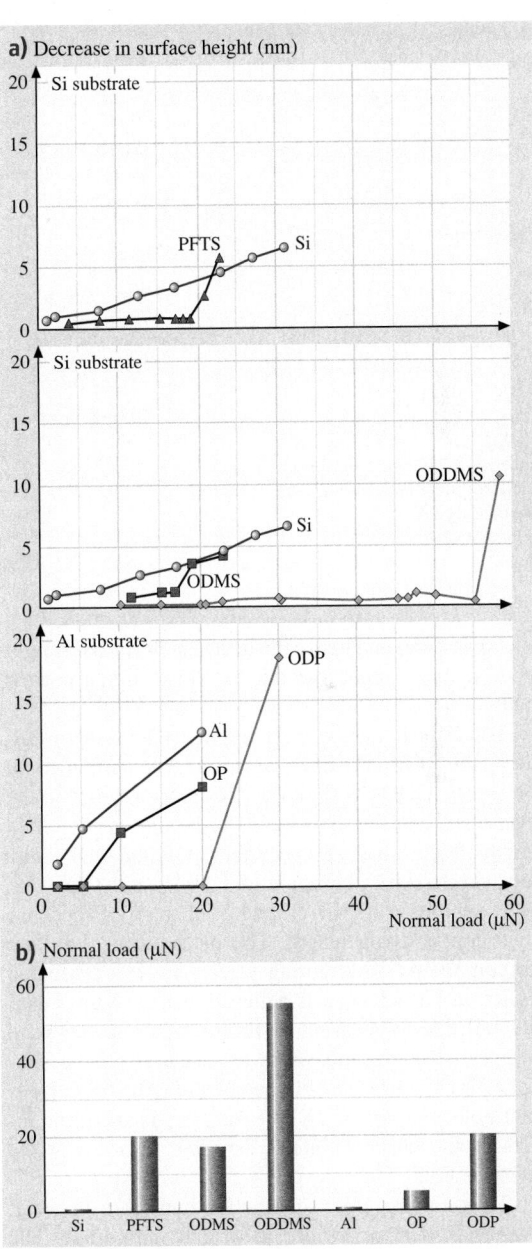

Fig. 43.25 (**a**) Decrease in surface height as a function of normal load after one scan cycle for various SAMs on Si and Al substrates, and (**b**) comparison of critical loads for failure during wear tests for various SAMs

Table 43.14 Typical bond strengths[a] in SAMs

Bond	Hexadecanethiol (HDT) (kJ/mol)	Biphenylthiol (BPT) (kJ/mol)	Bond	Perfluoroalkylsilane (PFTS) (kJ/mol)	Alkylsilane (ODMS or ODDMS)	Akylphosphonate (OP and ODP)
Interfacial bonds						
S–Au	184[b]	184[b]	Si–O	242[c]	242[c]	242[c]
S–C	286[a]			800[d]	800[d]	800[d]
C_6H_5–S		362[a]	Si–C	414[a]	414[a]	
			Al–O			511[a]
			P–C			513[a]
			P–O			599[a]
Bonds in backbone						
C–C			C–C			
CH_2–CH_2	326[e]		CH_2–CH_2	326[e]	326[e]	
CH_3–CH_2	≈ 305[a]		CF_2–CF_2	≈ 326[f]		
C_6H_5		strong	CF_2–CH_2	≈ 326[f]		
			CF_3–CF_2	≈ 326[f]		
			CH_3–CH_2		≈ 305[a]	

[a] *Lide* [43.92]
[b] Chemical adsorption bond from *Lio* et al. [43.72]
[c] Chemical adsorption bond from *Hoshino* [43.98]
[d] In diatomic molecules
[e] *Cottrell* [43.99]
[f] Because of the C–C bond it is expected to be close to that of CH_2–CH_2

$H = 0.41$ GPa) and hence the decrease in surface height for Al is much larger than that for Si for similar normal loads.

The critical loads corresponding to the sudden failure of the SAM are shown in Fig. 43.25b. Amongst all the SAMs, ODDMS shows the best performance in the wear tests, and this is believed to be due to the chain length effect (it has a longer chain). OP and ODP show very similar wear behavior to ODMS and ODDMS. ODP exhibits a higher critical load than OP because of its longer chain length. The mechanism of failure for compliant SAMs during wear tests was presented earlier, in Fig. 43.18. It is believed that the SAMs usually fail due to shearing of the molecule at the head group; that is, the molecules are sheared off the substrate. Table 43.14 gives the bond strengths for various intermolecular bonds. The weakest bonds are at the interface, and hence failure is expected to occur at the interface first.

To study the effect of relative humidity on wear, wear tests were performed at various humidities. The bottom of Fig. 43.22a shows the critical normal load as a function of relative humidity. The critical normal load shows a weak dependency on the relative humidity for ODMS/Si and PFTS/Si, and was larger for ODMS/Si

than for PFTS/Si throughout the range of humidities used. For ODDMS/Si, the critical normal load showed an increase with relative humidity. This suggests that water molecules can penetrate into ODDMS, which then might work as lubricant [43.41, 100]. This effect was absent for PFTS/Si and ODMS/Si.

43.3.4 Degradation and Environmental Studies

Degradation Studies

The coefficient of friction and the gaseous products detected for HDT/Au are shown in Fig. 43.26a [43.48]. A normal pressure of 50 kPa was applied to the HDT films. The coefficient of friction increased after a sliding distance of about 10 m. During sliding, $(CH_2)_{15}S$, C_2H_3, CH_3, CH_2 and H_2 were detected by a mass spectrometer. The partial pressure of the HS fragments is of interest since it corresponds to the interface bonds, and so it is reported here. The increase in $(CH_2)_{15}S$ was much more than that of other species, due to the breaking of the S–Au bond. The partial pressures of C_2H_3, CH_3, CH_2, and H_2 were also found to increase during the sliding. There was no noticeable change in the partial pressure of HS.

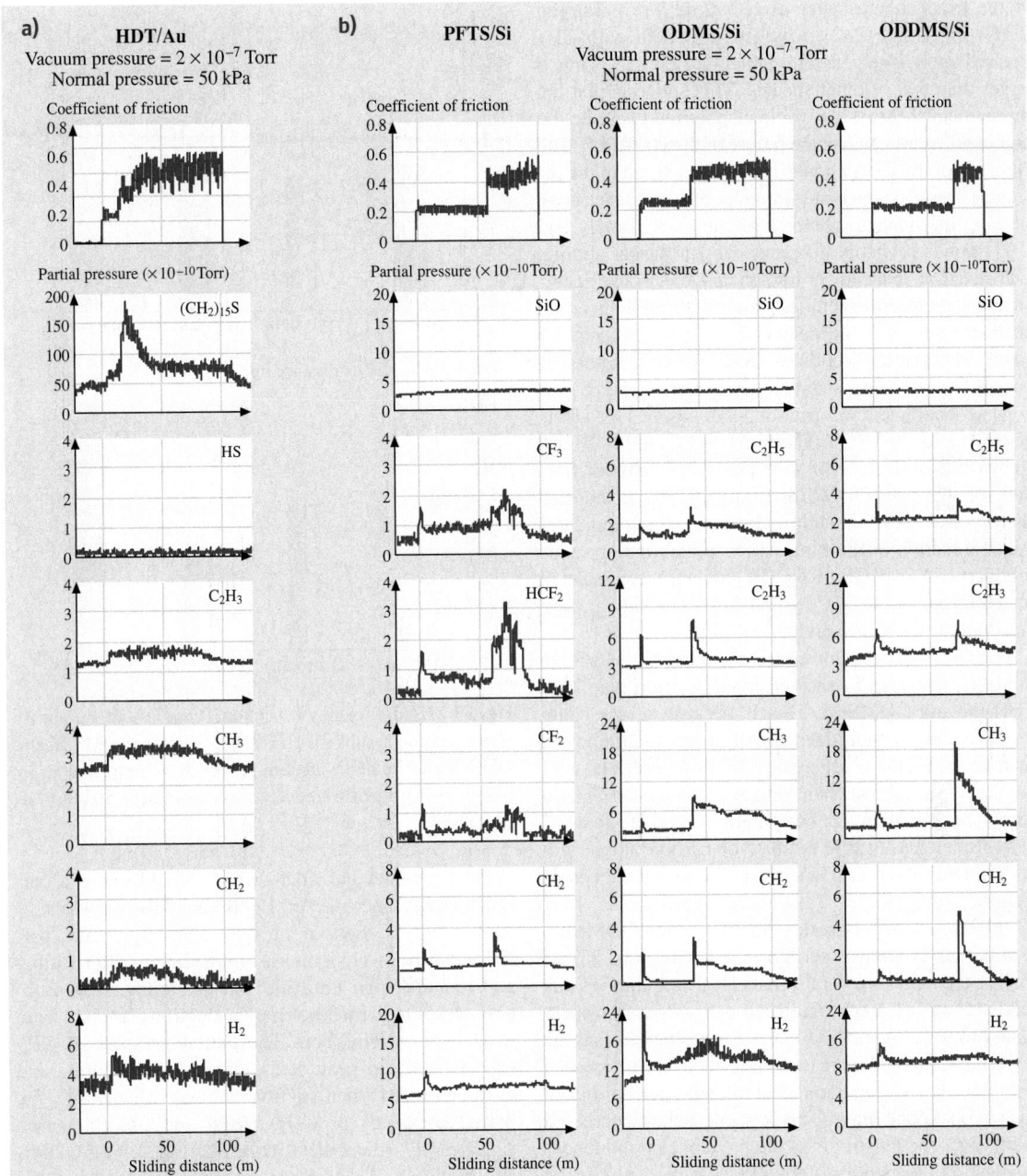

Fig. 43.26a,b Coefficients of friction and mass spectral data on (**a**) HDT/Au (1.9 nm), (**b**) PFTS/Si (1.8 nm), ODMS/Si ($\approx$ 1.9 nm) and ODDMS/Si ($\approx$ 2.1 nm) in high vacuum [43.48]

The HDT film was deposited on an Au(111) layer. The bond strength of S−Au is 184 kJ/mol (Table 43.14), which is lower than those of the C−C bonds (425 kJ/mol), C−H bonds (422 kJ/mol), and C−S bonds (286 kJ/mol) in the alkyl chains. Since the S−Au bond is the weakest bond in the alkanethiol chain, the whole chain should be sheared away from the substrate. Because the upper atomic mass unit (amu) limit of the mass spectrometer used is 250, we monitored $(CH_2)_{15}S$ (amu = 242), which is the chain with CH_3 sheared away. The generation rate of $(CH_2)_{15}S$ is much larger than that of other species. This suggests that the mechanical shear of the whole alkanethiol chain be the dominant factor causing the failure of the HDT film. The cleavage of the S−Au bonds has been reported in the literature. Based on the bond strengths, as well as the above studies, mechanical shearing of the C−C bonds and C−H bonds probably does not happen during sliding. The reaction induced by low-energy electrons, generated by triboelectrical emission during the sliding, could be responsible for the degradation of the alkanethiol chain. Thermal desorption of HDT from Au is another possibility for the degradation mechanism of HDT.

The coefficient of friction and gaseous products generated for PFTS/Si, ODMS/Si and ODDMS/SI are shown in Fig. 43.26b [43.48]. The coefficients of friction for PFTS/Si, ODMS/Si, and ODDMS/Si increase sharply after a certain sliding distance, which indicates the degradation of the film. At the same time, gaseous products of CF_3, HCF_2, CF_2, CH_2 and H_2 were detected for PFTS/Si, and C_2H_5, C_2H_3, CH_3, CH_2 and H_2 were detected for ODMS/Si and ODDMS/Si.

PFTS/Si showed lower friction than ODMS/Si in the tests. ODDMS/Si showed lower friction than both PFTS/Si and ODMS/Si. This is because of the chain length effect; as mentioned earlier, it has been reported that for SAMs the coefficient of friction decreases with the carbon backbone chain length (n) when the carbon atoms are less than 12. For chains with more than 12 carbons, increasing the number of carbon atoms will not influence the coefficient of friction to any noticeable extent.

PFTS/Si showed greater durability than ODMS/Si. It is harder to rotate a perfluorinated carbon backbone (due to the larger size of F versus H) which implies that this structure is more rigid than a hydrocarbon backbone [43.91]. *Chambers* [43.101] has reported that the C−C bond strength increases when hydrogen is replaced with fluorine. This suggests that the rigid perfluorinated carbon backbone may be responsible for the increased durability. The length of the alkyl chain also influences the desorption energies of alkanes. Based on studies of the adsorption of alkanes on Cu(100), Au(111), Pt(110) and others, the physisorption energy increases with the alkyl chain length [43.102–104]. Therefore, ODDMS are more durable than ODMS.

During sliding on PFTS films, gaseous products of CF_3, HCF_2, CF_2, CH_2 and H_2 were detected. From the structure of perfluoroalkylsilane, the only source of

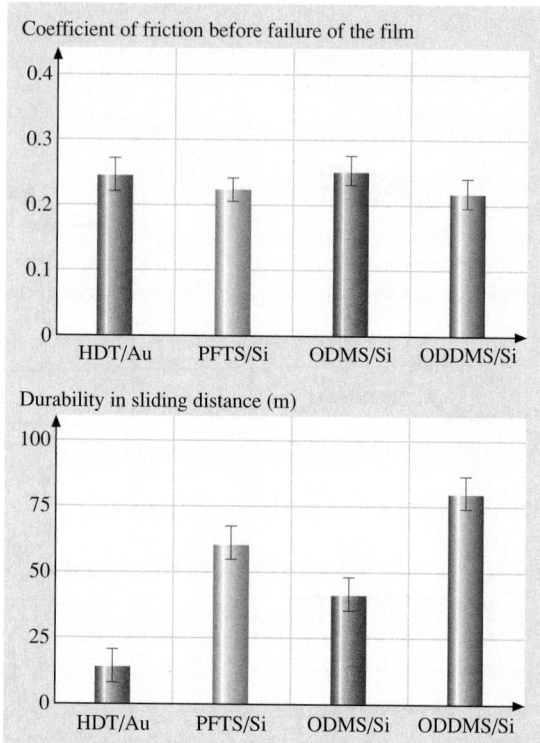

Fig. 43.27 Coefficient of friction (*upper*) and durability (*lower*) comparisons for HDT/Au, PFTS/Si, ODMS/Si and ODDMS/Si in high vacuum. *Error bars* represent $\pm 3\sigma$ based on five measurements (normal pressure 50 kPa) (after [43.48])

H on the molecular chain which would cause a partial pressure increase of H_2 is the $(CH_2)_2$, which is located at the bottom of the chain. Since the partial pressure of H_2 increases immediately after sliding and remains high until the end of sliding, it is probably generated by low-energy electrons arising from triboelectrical emission. The partial pressure of CH_2 exhibits a sharp peak at the beginning of sliding and at the moment when friction changes. Meanwhile, the partial pressures of CH_3, HSF_2 and CF_2 increased significantly when the friction increased. For ODMS and ODDMS, C_2H_5, C_2H_3, CH_3, CH_2 and H_2 were detected during sliding. The partial pressures of the carbon-related products increase considerably when the friction is increased. SiO, which is associated with interface bonds, shows no noticeable change during sliding.

Perfluoroalkylsilanes and alkylsilanes are attached to the naturally oxidized silicon by Si−O bonds. The Si−O

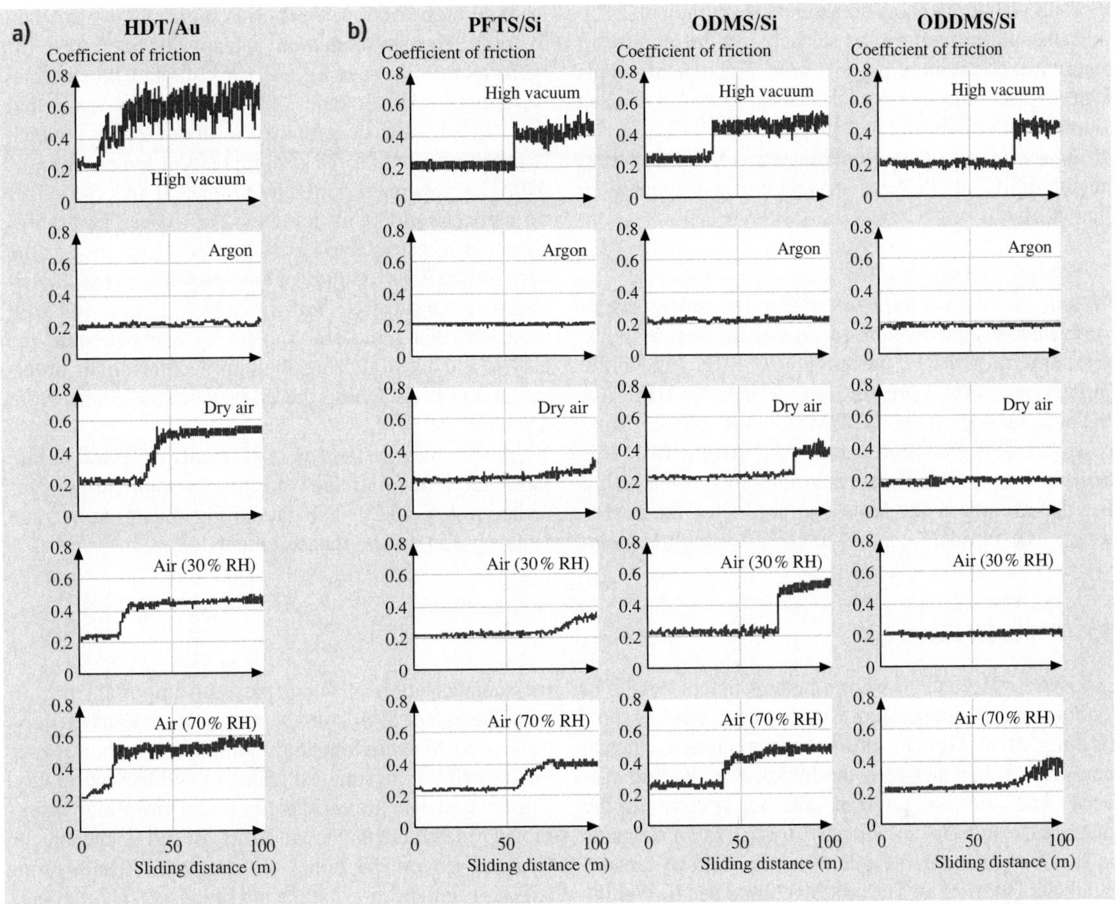

a) HDT/Au   **b)** PFTS/Si   ODMS/Si   ODDMS/Si

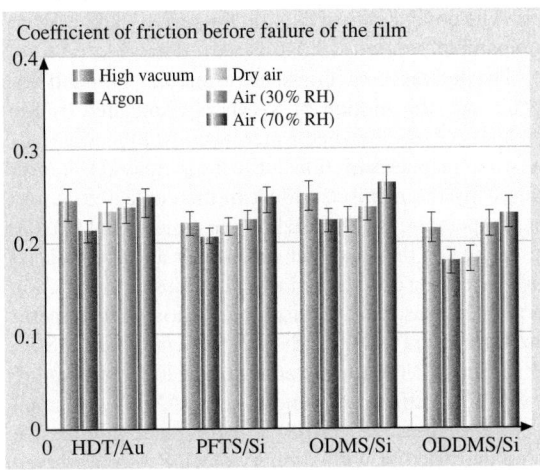

Fig. 43.28a,b Coefficient of friction data taken in high vacuum, argon and air with different humidity levels for (**a**) HDT/Au (1.9 nm), (**b**) PFTS/Si (1.8 nm), ODMS/Si (≈ 1.9 nm) and ODDMS/Si (≈ 2.1 nm) (RH: relative humidity, normal pressure 50 kPa) (after [43.48])

C$-$Si bond strength (414 kJ/mol) is slightly lower than the C$-$C bond strength. Based on Table 43.14, the interfacial bonds (Si$-$O) are weaker than the C$-$C bonds in the backbone. Therefore, it is believed that film cleavage occurs at the interface. We have previously reported evidence of the cleavage of interfacial bonds using an AFM. To explain the hydrogen, C_1 and C_2 hydrocarbon (in the tests for PFTS/Si, ODMS/Si and ODDMS/Si) or fluorocarbon (in the tests for PFTS/Si) products, *Kluth* et al. [43.105] suggested that the alkylsilane (or per-

Fig. 43.29 Coefficient of friction comparison for HDT/Au, PFTS/Si, ODMS/Si and ODDMS/Si in high vacuum, argon and air with different humidity levels. *Error bars* represent $\pm 3\sigma$ based on five measurements (normal pressure 50 kPa) (after [43.48])

fluoroalkylsilane) chains break and create radicals. The radical could remain on the surface and decompose to generate a short radical and an alkene. The radical can repeatedly decompose to ever shorter radicals and alkenes so long as it remains on the surface.

A summary of the coefficients of friction and the durability of all films in vacuum is presented in Fig. 43.27.

Environmental Studies

To study the effect of the environment, friction tests were conducted in high vacuum, argon, dry air (less than 2% RH), air with 30% RH, and air with 70% RH; Fig. 43.28. The applied normal pressure was 50 kPa for HDT/Au, PFTS/Si, ODMS/Si and ODDMS/Si, which is the same as used in the degradation tests. By comparing the coefficients of friction in different environments, it was found that the friction is the lowest in argon for the SAMs tested. The intimate contact induced by high vacuum

leads to high friction. Friction is higher in dry air than in argon. This shows that oxygen appears to effect SAM performance. *Kim* et al. [43.106] studied the thermal stability of alkylsiloxane SAMs in air. They found that the alkylsiloxane decomposes at about 200 °C, which is much lower than the decomposition temperature of 470 °C in vacuum reported by *Kluth* et al. [43.105]. This difference could be attributed to the oxygen in air. Water in the air is found to have significant influence on the friction of SAMs. A study of humidity effects for alkylsilane on mica substrate performed by *Tian* et al. [43.100] indicated that the water molecules can penetrate the alkylsilane film, altering their molecular chain ordering and also detaching the alkylsilane molecules from the substrate.

A summary of friction coefficients observed before lubricant film failure in various environments is presented in Fig. 43.29. The data in Fig. 43.29 are average values based on five measurements.

43.4 Closure

Exposure of devices to a humid environment results in condensates of water vapor. Condensed water or a pre-existing film of liquid results in the formation of concave meniscus bridges between the hydrophilic mating surfaces. The negative Laplace pressure present in the meniscus results in an adhesive force which depends on the roughness of the interface, the surface tension and the contact angle. The adhesive force can be significant in an interface with ultrasmooth surfaces, and it can be on the same order as the external load if the latter is small, such as in micro/nanodevices. Surfaces with high hydrophobicity can be produced by surface treatment. In many applications, hydrophobic films are expected to provide low adhesion, friction and wear. Because of the small clearances inherent to micro/nanodevices, these films should be molecularly thick. Liquid films with low surface tension or hydrophobic solid films can be used. Ordered molecular assemblies with high hydrophobicity can be engineered by chemically grafting various polymer molecules with suitable functional head groups and nonpolar surface terminal groups.

The adhesion, friction and wear properties of SAMs with alkyl, biphenyl and perfluoroalkyl spacer chains and different surface terminal ($-CH_3$ and $-CF_3$) and head groups ($-SH$, $-Si-O-$, $-OH$, and $P-O-$), studied using an AFM, are reported in this chapter. It was found that the adhesive force varies linearly with the W_a value of the SAM, which indicates that capil-

lary condensation of water plays an important role in the adhesion of SAMs on the nanoscale at ambient conditions. SAMs with long high-compliance carbon spacer chains exhibit the smallest adhesive and friction forces. The friction data are explained using a molecular spring model, in which the local stiffness and intermolecular force govern frictional performance. Results from stiffness and friction characterizations of a micropatterned sample with different structures on it support this model. Perfluoroalkylsilane SAMs exhibit lower surface energies, higher contact angles and lower adhesive forces than alkylsilane SAMs. The substrate had little effect. The coefficients of friction for various SAMs were comparable.

The influence of the relative humidity on the adhesion and the friction of SAMs is dominated by the thickness of the adsorbed water layer. At higher humidity, water increases friction due to the increased adhesion caused by the meniscus effect in the contact zone. As the temperature is increased, in the case of Si(111), the desorption of the adsorbed water layer and the reduced surface tension of water reduces the adhesive and friction forces. Decreases in adhesion and friction with temperature were found for PFTS/Si, ODMS/Si and ODDMS/Si. PFTS showed the lowest adhesion at low temperature (≈ 70 °C). Differences among the SAMs were small at high temperature (≈ 100 °C). Increases in adhesive force and friction with sliding velocity were observed

Table 43.15 Summary of nanotribological characterization studies of SAMs on Si and Al substrates

SAM property		Friction force	Adhesive force	Wear
Substrate	Hard	High	Low	Low
	Soft	Low	Low	High
Chemical structure	Linear chain molecule	High	Low	High
	Ring molecule	High	High	Low
Backbone	Fluorocarbon backbone	Low	Low	Low
	Hydrocarbon backbone	Low	High	High
Chain length	Long backbone chain		High	High
	Short backbone chain		Low	Low

for PFTS/Si, ODMS/Si and ODDMS/Si. A peak in the coefficient of friction appeared for PFTS.

PFTS/Si showed better wear resistance than ODMS/Si. ODDMS/Si showed better wear resistance than ODMS/Si due to the chain length effect. SAM wear behavior is mostly determined by the molecule–substrate bond strength. The long chain molecules of ODP/Al and ODDMS/Si showed higher critical loads for failure.

The results from nanotribological characterization studies of SAMs deposited on Si and Al substrates are summarized in Table 43.15 [43.47]. SAMs deposited on Si and Al substrates show low friction and low adhesion, both of which are desirable for MEMS/NEMS applications.

Based on studies in high vacuum (2×10^{-7} Torr), the friction coefficients for SAMs have been found to follow the order (from low to high): ODDMS/Si, PFTS/Si, HDT/Au, ODMS/Si. HDT on Au is less

durable than perfluoroalkylsilane and alkylsilane on Si due to its weak interfacial bonding. PFTS/Si is more durable than ODMS/Si. This indicates that fluorinating alkylsilane can improve durability. ODDMS/Si is more durable than ODMS/Si and PFTS/Si because of the chain length effect. SAM friction is higher in high vacuum than in argon because of the intimate contact involved. Based on studies in argon and air with various relative humidities, oxygen can increase the friction and durability of SAMs. The water molecule can detach SAM molecules from the substrate, resulting in high friction and low durability.

In summary, nanotribological studies of SAM films using AFM demonstrate that they exhibit attractive hydrophobic and tribological properties. Fluorinated SAMs appear to be the most performant. SAM films should find application in many fields, including micro-/nanodevices.

References

43.1 B. Bhushan, J.N. Israelachvili, U. Landman: Nanotribology: friction, wear and lubrication at the atomic scale, Nature **374**, 607–616 (1995)

43.2 B. Bhushan: *Tribology and Mechanics of Magnetic Storage Devices*, 2nd edn. (Springer, Berlin, Heidelberg 1996)

43.3 B. Bhushan (Ed.): *Tribology Issues and Opportunities in MEMS* (Kluwer Academic, Dordrecht 1998)

43.4 B. Bhushan (Ed.): *Handbook of Micro/Nanotribology*, 2nd edn. (CRC, Boca Raton 1999)

43.5 B. Bhushan: *Nanotribology and Nanomechanics – An Introduction* (Springer, Berlin, Heidelberg 2005)

43.6 K.F. Man, B.H. Stark, R. Ramesham: *A Resource Handbook for MEMS Reliability* (JPL Press, Jet Propulsion Laboratory, California Institute of Technology, Pasadena 1998)

43.7 K.F. Man: MEMS reliability for space applications by elimination of potential failure modes through testing and analysis, http://www.rel.jpl.nasa.gov

43.8 /Org/5053/atpo/products/Prod-map.html, (2002)

D.M. Tanner, N.F. Smith, L.W. Irwin et al.: *MEMS Reliability: Infrastructure, Test Structure, Experiments, and Failure Modes*, SAND2000-0091 (Sandia National Laboratories, Albuquerque 2000)

43.9 B. Bhushan: *Principles and Applications of Tribology* (Wiley, New York 1999)

43.10 B. Bhushan: *Introduction to Tribology* (Wiley, New York 2002)

43.11 M.E. Schrader, G.I. Loeb (Eds.): *Modern Approaches to Wettability* (Plenum, New York 1992)

43.12 A. Ulman (Eds.): *Characterization of Organic Thin Films* (Butterworth–Heinemann, Boston 1995)

43.13 A.W. Neumann, J.K. Spelt (Eds.): *Applied Surface Thermodynamics* (Marcel Dekker, New York 1996)

43.14 B. Bhushan: Contact mechanics of rough surfaces in tribology: multiple asperity contact, Tribol. Lett. **4**, 1–35 (1998)

43.15   B. Bhushan, W. Peng: Contact mechanics of multilayered rough surfaces, Appl. Mech. Rev. **55**, 435–480 (2002)

43.16   M. Nosonovsky, B. Bhushan: Roughness optimization for biomimetic superhydrophobic surfaces, Microsyst. Technol. **11**, 535–549 (2005)

43.17   R. Maboudian: Surface processes in MEMS technology, Surf. Sci. Rep. **30**, 209–269 (1998)

43.18   B. Bhushan (Ed.): *Modern Tribology Handbook, Vol. 1 – Principles of Tribology; Vol. 2 – Materials, Coatings, and Industrial Applications* (CRC, Boca Raton 2001)

43.19   B. Bhushan, Z. Zhao: Macro- and microscale tribological studies of molecularly-thick boundary layers of perfluoropolyether lubricants for magnetic thin-film rigid disks, J. Info. Stor. Proc. Syst. **1**, 1–21 (1999)

43.20   F. P. Bowden, D. Tabor: *The Friction and Lubrication of Solids, Part I* (Clarendon, Oxford 1950)

43.21   W. A. Zisman: Friction, durability and wettability properties of monomolecular films on solids. In: *Friction and Wear*, ed. by R. Davies (Elsevier, Amsterdam 1959) pp. 110–148

43.22   V. N. Koinkar, B. Bhushan: Microtribological studies of unlubricated and lubricated surfaces using atomic force/friction force microscopy, J. Vac. Sci. Technol. A **14**, 2378–2391 (1996)

43.23   H. Liu, B. Bhushan: Nanotribological characterization of molecularly-thick lubricant films for applications to MEMS/NEMS by AFM, Ultramicroscopy **97**, 321–340 (2003)

43.24   A. Ulman: *An Introduction to Ultrathin Organic Films: From Langmuir-Blodgett to Self-Assembly* (Academic, San Diego 1991)

43.25   A. Ulman: Formation and structure of self-assembled monolayers, Chem. Rev **96**, 1533–1554 (1996)

43.26   H. Hansma, F. Motamedi, P. Smith, P. Hansma, J. C. Wittman: Molecular resolution of thin, highly oriented poly(tetrafluoroethylene) films with the atomic force microscope, Polym. Commun. **33**, 647–649 (1992)

43.27   L. Scandella, A. Schumacher, N. Kruse, R. Prins, E. Meyer, R. Luethi, L. Howald, M. Scherge, J. A. Schaefer: Surface modification and mechanical properties of bulk silicon. In: *Tribology Issues and Opportunities in MEMS*, ed. by B. Bhushan (Kluwer, Dordrecht 1998) pp. 529–537

43.28   B. Bhushan: Chemical, mechanical and tribological characterization of ultra-thin and hard amorphous carbon coatings as thin as 3.5 nm: recent developments, Diamond Relat. Mater. **8**, 1985–2015 (1999)

43.29   A. Erdemir, C. Donnet: Tribology of diamond, diamond-like carbon, and related films. In: *Modern Tribology Handbook*, Vol. 2: Materials, Coatings, and Industrial Applications, ed. by B. Bhushan (CRC, Boca Raton 2001) pp. 871–908

43.30   V. F. Dorfman: Diamond-like nanocomposites (DLN), Thin Solid Films **212**, 267–273 (1992)

43.31   M. Grischke, K. Bewilogua, K. Trojan, H. Dimigan: Application-oriented modification of deposition process for diamond-like carbon based coatings, Surf. Coat. Technol. **74–75**, 739–745 (1995)

43.32   R. S. Butter, D. R. Waterman, A. H. Lettington, R. T. Ramos, E. J. Fordham: Production and wetting properties of fluorinated diamond-like carbon coatings, Thin Solid Films **311**, 107–113 (1997)

43.33   M. Grischke, A. Hieke, F. Morgenweck, H. Dimigan: Variation of the wettability of DLC coatings by network modification using silicon and oxygen, Diamond Relat. Mater. **7**, 454–458 (1998)

43.34   C. Donnet, J. Fontaine, A. Grill, V. Patel, C. Jahnes, M. Belin: Wear-resistant fluorinated diamondlike carbon films, Surf. Coat. Technol. **94–95**, 531–536 (1997)

43.35   D. J. Kester, C. L. Brodbeck, I. L. Singer, A. Kyriakopoulos: Sliding wear behavior of diamond-like nanocomposite coatings, Surf. Coat. Technol. **113**, 268–273 (1999)

43.36   H. Liu, B. Bhushan: Adhesion and friction studies of microelectromechanical systems/nanoelectromechanical systems materials using a novel microtriboapparatus, J. Vac. Sci. Technol. A **21**, 1528–1538 (2003)

43.37   B. Bhushan, H. Liu, S. M. Hsu: Adhesion and friction studies of silicon and hydrophobic and low friction films and investigation of scale effects, ASME J. Tribol. **126**, 583–590 (2004)

43.38   B. Bhushan, A. V. Kulkarni, V. N. Koinkar, M. Boehm, L. Odoni, C. Martelet, M. Belin: Microtribological characterization of self-assembled and Langmuir–Blodgett monolayers by atomic and friction force microscopy, Langmuir **11**, 3189–3198 (1995)

43.39   B. Bhushan, H. Liu: Nanotribological properties and mechanisms of alkylthiol and biphenyl thiol self-assembled monolayers studied by atomic force microscopy, Phys. Rev. B **63**, 245412-1–11 (2001)

43.40   H. Liu, B. Bhushan, W. Eck, V. Stadler: Investigation of the adhesion, friction, and wear properties of biphenyl thiol self-assembled monolayers by atomic force microscopy, J. Vac. Sci. Technol. A **19**, 1234–1240 (2001)

43.41   H. Liu, B. Bhushan: Investigation of nanotribological properties of alkylthiol and biphenyl thiol self-assembled monolayers, Ultramicroscopy **91**, 185–202 (2002)

43.42   H. Liu, B. Bhushan: Orientation and relocation of biphenyl thiol self-assembled monolayers, Ultramicroscopy **91**, 177–183 (2002)

43.43   B. Bhushan, T. Kasai, G. Kulik, L. Barbieri, P. Hoffmann: AFM study of perfluorosilane and alkylsilane self-assembled monolayers for anti-stiction in MEMS/NEMS, Ultramicroscopy **105**, 176–188 (2005)

43.44 T. Kasai, B. Bhushan, G. Kulik, L. Barbieri, P. Hoffmann: Nanotribological study of perfluorosilane SAMs for anti-stiction and low wear, J. Vac. Sci. Technol. B **23**, 995–1003 (2005)

43.45 K. K. Lee, B. Bhushan, D. Hansford: Nanotribological characterization of perfluoropolymer thin films for BioMEMS applications, J. Vac. Sci. Technol. A **23**, 804–810 (2005)

43.46 B. Bhushan, D. Hansford, K. K. Lee: Surface modification of silicon surfaces with vapor phase deposited ultrathin fluorosilane films for biomedical devices, J. Vac. Sci. Technol. A **24**, 1197–1202 (2006)

43.47 N. S. Tambe, B. Bhushan: Nanotribological characterization of self assembled monolayers deposited on silicon and aluminum substrates, Nanotechnology **16**, 1549–1558 (2005)

43.48 Z. Tao, B. Bhushan: Degradation mechanisms and environmental effects on perfluoropolyether self assembled monolayers and diamondlike carbon films, Langmuir **21**, 2391–2399 (2005)

43.49 J. A. Zasadzinski, R. Viswanathan, L. Madsen, J. Garnaes, D. K. Schwartz: Langmuir–Blodgett films, Science **263**, 1726–1733 (1994)

43.50 J. Tian, Y. Xia, G. M. Whitesides: Microcontact printing of SAMs. In: *Thin Films − Self-Assembled Monolayers of Thiols*, Vol. 24, ed. by A. Ulman (Academic, San Diego 1998) pp. 227–254

43.51 Y. Xia, G. M. Whitesides: Soft lithography, Angew. Chem. Int. Ed. **37**, 550–575 (1998)

43.52 A. Kumar, G. M. Whitesides: Features of gold having micrometer to centimeter dimensions can be formed through a combination of stamping with an elastomeric stamp and an alkanethiol ink followed by chemical etching, Appl. Phys. Lett. **63**, 2002–2004 (1993)

43.53 S. Y. Chou, P. R. Krauss, P. J. Renstrom: Imprint lithography with 25-nanometer resolution, Science **272**, 85–87 (1996)

43.54 Y. Xia, E. Kim, X. M. Zhao, J. A. Rogers, M. Prentiss, G. M. Whitesides: Complex optical surfaces formed by replica molding against elastomeric masters, Science **273**, 347–349 (1996)

43.55 L. J. Hornbeck: The DMD™ projection display chip: a MEMS-based technology, MRS Bull. **26**, 325–328 (2001)

43.56 M. R. Douglass: Lifetime estimates and unique failure mechanisms of the digital micromirror device (DMD). In: *1998 International Reliability Physics Proceedings*, IEEE Catalog No. 98 CH 36173 (, 1998) pp. 9–16 Presented at the 36th Annual International Reliability Physics Symposium, Reno

43.57 H. Liu, B. Bhushan: Nanotribological characterization of digital micromirror devices using an atomic force microscope, Ultramicroscopy **100**, 391–412 (2004)

43.58 H. Liu, B. Bhushan: Investigation of nanotribological and nanomechanical properties of the digital micromirror device by atomic force microscope, J. Vac. Sci. Technol. A **22**, , 1388–1396 (2004)

43.59 A. Manz, H. Becker (Eds.): *Microsystem Technology in Chemistry and Life Sciences* (Springer, Berlin, Heidelberg 1998)

43.60 J. Cheng, L. J. Krica (Eds.): *Biochip Technology* (Harwood, New York 2001)

43.61 M. J. Heller, A. Guttman (Eds.): *Integrated Microfabricated Biodevices* (Marcel Dekker, New York 2001)

43.62 A. van der Berg (Ed.): *Lab-on-a-Chip: Chemistry in Miniaturized Synthesis and Analysis Systems* (Elsevier, Amsterdam 2003)

43.63 M. Hein, L. R. Best, S. Pattison, S. Arena: *Introduction to General, Organic, and Biochemistry*, 6th edn. (Brooks/Cole, Pacific Grove 1997)

43.64 J. R. Mohrig, C. N. Hammond, T. C. Morrill, D. C. Neckers: *Experimental Organic Chemistry* (W. H. Freeman, New York 1998)

43.65 S. R. Wasserman, Y. T. Tao, G. M. Whitesides: Structure and reactivity of alkylsiloxane monolayers formed by reaction of alkylchlorosilanes on silicon substrates, Langmuir **5**, 1074–1089 (1989)

43.66 C. Jung, O. Dannenberger, Y. Xu, M. Buck, M. Grunze: Self-assembled monolayers from organosulfur compounds: a comparison between sulfides, disulfides, and thiols, Langmuir **14**, 1103–1107 (1998)

43.67 W. Geyer, V. Stadler, W. Eck, M. Zharnikov, A. Golzhauser, M. Grunze: Electron-induced crosslinking of aromatic self-assembled monolayers: negative resists for nanolithography, Appl. Phys. Lett. **75**, 2401–2403 (1999)

43.68 J. Ruhe, V. J. Novotny, K. K. Kanazawa, T. Clarke, G. B. Street: Structure and tribological properties of ultrathin alkylsilane films chemisorbed to solid surfaces, Langmuir **9**, 2383–2388 (1993)

43.69 V. DePalma, N. Tillman: Friction and wear of self-assembled tricholosilane monolayer films on silicon, Langmuir **5**, 868–872 (1989)

43.70 M. T. McDermott, J. B. D. Green, M. D. Porter: Scanning force microscopic exploration of the lubrication capabilities of *n*-alkanethiolate monolayers chemisorbed at gold: structural basis of microscopic friction and wear, Langmuir **13**, 2504–2510 (1997)

43.71 X. Xiao, J. Hu, D. H. Charych, M. Salmeron: Chain length dependence of the frictional properties of alkylsilane molecules self-assembled on mica studied by atomic force microscopy, Langmuir **12**, 235–237 (1996)

43.72 A. Lio, D. H. Charych, M. Salmeron: Comparative atomic force microscopy study of the chain length dependence of frictional properties of alkanethiol on gold and alkylsilanes on mica, J. Phys. Chem. B **101**, 3800–3805 (1997)

43.73 H. Schonherr, G. J. Vancso: Tribological properties of self-assembled monolayers of fluorocarbon and hydrocarbon thiols and disulfides on Au(111) studied by scanning force microscopy, Mater. Sci. Eng. C **8–9**, 243–249 (1999)

43.74   V. V. Tsukruk, V. N. Bliznyuk: Adhesive and friction forces between chemically modified silicon and silicon nitride surfaces, Langmuir **14**, 446–455 (1998)

43.75   V. V. Tsukruk, T. Nguyen, M. Lemieux, J. Hazel, W. H. Weber, V. V. Shevchenko, N. Klimenko, E. Sheludko: Tribological properties of modified MEMS surfaces. In: *Tribology Issues and Opportunities in MEMS*, ed. by B. Bhushan (Kluwer, Dordrecht 1998) pp. 607–614

43.76   M. Fujihira, Y. Tani, M. Furugori, U. Akiba, Y. Okabe: Chemical force microscopy of self-assembled monolayers on sputtered gold films patterned by phase separation, Ultramicroscopy **86**, 63–73 (2001)

43.77   R. J. Good, C. J. V. Oss: *Modern Approaches to Wettability – Theory and Applications* (Plenum, New York 1992)

43.78   M. H. V. C. Adao, B. J. V. Saramago, A. C. Fernandes: Estimation of the surface properties of styrene-acrylonitrile random copolymers from contact angle measurements, J. Colloid Interf. Sci. **217**, 94–106 (1999)

43.79   N. S. Tambe, B. Bhushan: A new atomic force microscopy based technique for studying nanoscale friction at high sliding velocities, J. Phys. D **38**, 764–773 (2005)

43.80   Y. F. Miura, M. Takenga, T. Koini, M. Graupe, N. Garg, R. L. Graham, T. R. Lee: Wettability of self-assembled monolayers generated from CF_3-terminated alkanethiols on gold, Langmuir **14**, 5821–5825 (1998)

43.81   M. Ratajczak-Sitarz, A. Katrusiak, Z. Kaluski, J. Garbarczyk: 4,4'-biphenyldithiol, Acta Crystallogr. **C.43**, 2389–2391 (1987)

43.82   B. Bhushan, J. Ruan: Tribological performance of thin film amorphous carbon overcoats for magnetic recording disks in various environments, Surf. Coat. Technol. **68/69**, 644–650 (1994)

43.83   B. Bhushan, L. Yang, C. Gao, S. Suri, R. A. Miller, B. Marchon: Friction and wear studies of magnetic thin film rigid disks with glass-ceramic, glass and aluminum-magnesium substrates, Wear **190**, 44–59 (1995)

43.84   J. N. Israelachvili: *Intermolecular and Surface Forces*, 2nd edn. (Academic, London 1992)

43.85   J. I. Siepman, I. R. McDonald: Monte Carlo simulation of the mechanical relaxation of a self-assembled monolayer, Phys. Rev. Lett. **70**, 453–456 (1993)

43.86   M. Garcia-Parajo, C. Longo, J. Servat, P. Gorostiza, F. Sanz: Nanotribological properties of octadecyltrichlorosilane self-assembled ultrathin films studied by atomic force microscopy: contact and tapping modes, Langmuir **13**, 2333–2339 (1997)

43.87   D. DeVecchio, B. Bhushan: Localized surface elasticity measurements using an atomic force microscope, Rev. Sci. Instrum. **68**, 4498–4505 (1997)

43.88   E. Barrena, S. Kopta, D. F. Ogletree, D. H. Charych, M. Salmeron: Relationship between friction and molecular structure: alkysilane lubricant films under pressure, Phys. Rev. Lett. **82**, 2880–2883 (1999)

43.89   N. Eustathopoulos, M. Nicholas, B. Drevet: *Wettability at High Temperature* (Pergamon, Amsterdam 1999)

43.90   S. Ren, S. Yang, Y. Zhao, T. Yu, X. Xiao: Preparation and characterization of ultrahydrophobic surface based on a stearic acid self-assembled monolayer over polyethyleneimine thin films, Surf. Sci. **546**, 64–74 (2003)

43.91   E. S. Clark: The molecular conformations of polytetrafluoroethylene: forms II and IV, Polymer **40**, 4659–4665 (1999)

43.92   D. R. Lide: *CRC Handbook of Chemistry and Physics*, 85th edn. (CRC, Boca Raton 2004)

43.93   W. D. Callister: *Mater. Sci. Eng.*, 4th edn. (Wiley, New York 1997) p. 4

43.94   N. S. Tambe, B. Bhushan: Friction model for velocity dependence of nanoscale friction, Nanotechnology **16**, 2309–2324 (2005)

43.95   S. C. Clear, P. F. Nealey: The effect of chain density on the frictional behavior of surfaces modified with alkylsilanes and immersed in *n*-Alcohols, J. Chem. Phys. **114**, 2802–2811 (2001)

43.96   N. S. Tambe, B. Bhushan: Durability studies of micro/nanoelectromechanical systems materials, coatings, and lubricants at high sliding velocities (up to 10 mm/s) using a modified atomic force microscope, J. Vac. Sci. Technol. A **23**, 830–835 (2005)

43.97   INSPEC: *Properties of Silicon*, EMIS Data Reviews Series No. 4 (INSPEC, Institution of Electrical Engineers, London 1988)

43.98   T. Hoshino: Adsorption of atomic and molecular oxygen and desorption of silicon monoxide on Si(111) surfaces, Phys. Rev. B **59**, 2332–2340 (1999)

43.99   T. L. Cottrell: *The Strength of Chemical Bonds*, 2nd edn. (Butterworths, London 1958)

43.100   F. Tian, X. Xiao, M. M. T. Loy, C. Wang, C. Bai: Humidity and temperature effect on frictional properties of mica and alkylsilane monolayer self-assembled on mica, Langmuir **15**, 244–249 (1999)

43.101   R. D. Chambers: *Fluorine in Organic Chemistry* (Wiley, New York 1973)

43.102   B. A. Sexton, A. E. Hughes: A comparison of weak molecular adsorption of organic-molecules on clean copper and platinum surfaces, Surf. Sci. **140**, 227–248 (1984)

43.103   L. H. Dubois, B. R. Zegarski, R. G. Nuzzo: Fundamental studies of microscopic wetting on organics surfaces 2. Interaction of secondary adsorbates with chemically textured organic monolayers, J. Am. Chem. Soc. **112**, 570–579 (1990)

43.104 M.C. McMaster, S.L.M. Schroeder, R.J. Madix: Molecular propane adsorption dynamics on Pt(110)–(1×2), Surf. Sci. **297**, 253–271 (1993)

43.105 G.J. Kluth, M. Sander, M.M. Sung, R. Maboudian: Study of the desorption mechanism of alkylsiloxane self-assembled monolayers through isotopic labeling and high resolution electron energy-loss spectroscopy experiments, J. Vac. Sci. Technol. A **16**, 932–936 (1998)

43.106 H.K. Kim, J.P. Lee, C.R. Park, H.T. Kwak, M.M. Sung: Thermal decomposition of alkylsiloxane self-assembled monolayers in air, J. Phys. Chem. B **107**, 4348–4351 (2003)

44. Nanoscale Boundary Lubrication Studies

Boundary films are formed by physisorption, chemisorption, and chemical reaction. With physisorption, no exchange of electrons takes place between the molecules of the adsorbate and those of the adsorbant. The physisorption process typically involves van der Waals forces, which are relatively weak. In chemisorption, there is an actual sharing of electrons or electron interchange between the chemisorbed species and the solid surface. The solid surfaces bond very strongly to the adsorption species through covalent bonds. Chemically reacted films are formed by the chemical reaction of a solid surface with the environment. The physisorbed film can be either monomolecularly or polymolecularly thick. The chemisorbed films are monomolecular, but stoichiometric films formed by chemical reaction can have a large film thickness. In general, the stability and durability of surface films decrease in the following order: chemically reacted films, chemisorbed films, and physisorbed films. A good boundary lubricant should have a high degree of interaction between its molecules and the sliding surface. As a general rule, liquids are good lubricants when they are polar and, thus, able to grip solid surfaces (or be adsorbed). In this chapter, we focus on perfluoropolyethers (PFPEs). We first introduce details of the commonly used PFPE lubricants; then present a summary of nanodeformation, molecular conformation, and lubricant spreading studies; followed by an overview of nanotribological properties of polar and nonpolar PFPEs studied by atomic force microscopy (AFM) and some concluding remarks.

Boundary films are formed by physisorption, chemisorption, and chemical reaction. With physisorption, no exchange of electrons takes place between the molecules of the adsorbate and those of the adsorbant. The physisorption process typically involves van der Waals forces, which are relatively weak. In chemisorption, there is an actual sharing of electrons or electron interchange between the chemisorbed species and the solid surface. The solid surfaces bond very strongly to the adsorption species through covalent bonds. Chemically reacted films are formed by the chemical reaction of a solid surface with the environment. The physisorbed film can be either monomolecularly or polymolecularly thick. The chemisorbed films are monomolecular, but stoichiometric films formed by chemical reaction can have a large film thickness. In general, the stability and durability of surface films decrease in the following order: chemically reacted films, chemisorbed films, and physisorbed films. A good boundary lubricant should have a high degree of interaction between its molecules and the sliding surface. As a general rule, liquids are good lubricants when they are polar and, thus, able to grip solid surfaces (or be adsorbed). Polar lubricants contain reactive functional groups with low ionization potential, or groups having high polarizability [44.1–3]. Boundary lubrication properties of lubricants are also dependent upon the molecular conformation and lubricant spreading [44.4–7].

Self-assembled monolayers (SAMs), Langmuir–Blodgett (LB) films, and perfluoropolyether (PFPE) films can be used as boundary lubricants [44.2,3,8–10].

Part E | 44

PFPE films are commonly used for lubrication of magnetic rigid disks and metal evaporated magnetic tapes to reduce friction and wear of a head–medium interface [44.10]. PFPEs are well suited for this application because of the following properties: low surface tension and low contact angle, which allow easy spreading on surfaces and provide a hydrophobic property; chemical and thermal stability, which minimizes degradation under use; low vapor pressure,which provides low outgassing; high adhesion to substrate via organofunctional bonds; and good lubricity, which reduces the interfacial friction and wear [44.10–12]. While the structure of the lubricants employed at the head–medium interface has not changed substantially over the past decade, the thickness of the PFPE film used to lubricate the disk has steadily decreased from multilayer thicknesses to the sub-monolayer thickness regime [44.11, 13]. Molecularly thick PFPE films are also being considered for lubrication purposes of the evolving microelectromechanical systems (MEMS) industry [44.14]. It is well known that the properties of molecularly thick liquid films confined to solid surfaces can be dramatically different from those of the corresponding bulk liquid. In order to efficiently develop lubrication systems that meet the requirements of the advanced rigid disk drive and MEMS industries, the impact of thinning the PFPE lubricants on the resulting nanotribology should be fully understood [44.15, 16]. It is also important to understand lubricant–substrate interfacial interactions and the influence of the operating environment on the nanotribological performance of molecularly thick PFPEs.

An overview of nanotribological properties of SAMs and LB films can be found in many references, such as [44.17]. In this chapter, we focus on PFPEs. We first introduce details of the commonly used PFPE lubricants; then present a summary of nanodeformation, molecular conformation, and lubricant spreading studies; followed by an overview of nanotribological properties of polar and nonpolar PFPEs studied by atomic force microscopy (AFM) and some concluding remarks.

44.1 Lubricants Details

Properties of two commonly used PFPE lubricants (Z-15 and Z-DOL) are reviewed here. Their molecular structures are shown schematically in Fig. 44.1. Z-15 has nonpolar $-CF_3$ end groups, whereas Z-DOL is a polar lubricant with hydroxyl ($-OH$) end groups. Their typical properties are summarized in Table 44.1; it shows that Z-15 and Z-DOL have almost the same density and surface tension. But Z-15 has larger molecular weight and higher viscosity. Both of them have low surface tension, low vapor pressure, low evaporation weight loss, and good oxidative stability [44.10, 12]. Generally, a single-crystal Si(100) wafer with a native oxide layer was used as a substrate for deposition of molecularly thick lubricant films for nanotribological characterization. Z-15 and Z-DOL films can be deposited directly

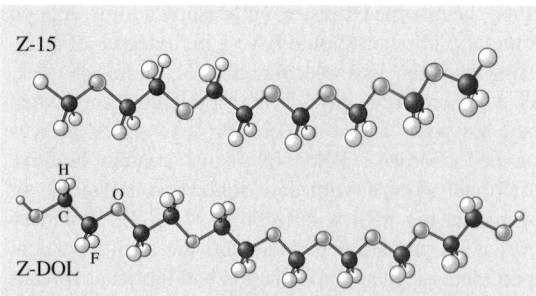

Fig. 44.1 Schematics of the molecular structures of Z-15 and Z-DOL. In this figure the m/n value, shown in Table 44.1, equals 2/3

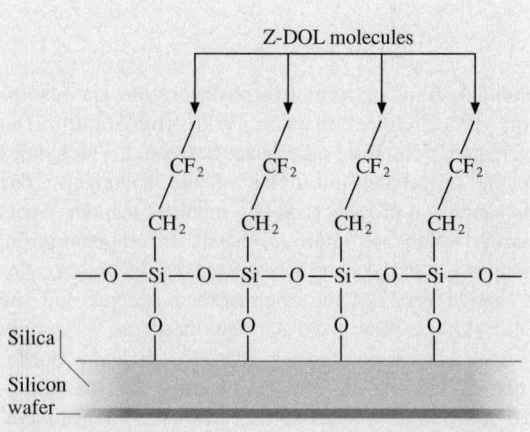

Fig. 44.2 Schematic of Z-DOL molecules that are chemically bonded on Si(100) substrate surface (which has native oxide) after thermal treatment at 150 °C for 30 min

Table 44.1 Typical properties of Z-15 and Z-DOL (data from Montefluous S.P.A., Milan, Italy)

	Z-15	Z-DOL (2000)
Formula	$CF_3-O-(CF_2-CF_2-O)_m-$ $(CF_2-O)_n-CF_3{}^*$	$HO-CH_2-CF_2-O-(CF_2-CF_2-O)_m-$ $(CF_2-O)_n-CF_2-CH_2-OH^*$
Molecular weight (Daltons)	9100	2000
Density (ASTM D891) 20 °C (g/cm^3)	1.84	1.81
Kinematic viscosity (ASTM D445) (cSt)		
20 °C	148	85
38 °C	90	34
99 °C	25	–
Viscosity index (ASTM D2270)	320	–
Surface tension (ASTM D1331) (dyn/cm) 20 °C	24	24
Vapor pressure (torr)		
20 °C	1.6×10^{-6}	2×10^{-5}
100 °C	1.7×10^{-5}	6×10^{-4}
Pour point (ASTM D972)		
°C	-80	–
Evaporation weight loss (ASTM D972)		
149 °C, 22 h (%)	0.7	–
Oxidative stability (°C)	–	320
Specific heat (cal/g °C)		
38 °C	0.21	–

* $m/n \sim 2/3$

onto the Si(100) wafer by the dip-coating technique. The clean silicon wafer is vertically submerged into a dilute solution of lubricant in hydrocarbon solvent (HT-70) for a certain time. The silicon wafers are pulled up vertically from the solution with a motorized stage at a constant speed for deposition of the desired thicknesses of Z-15 and Z-DOL lubricants. The lubricant film thickness obtained in dip coating is a function of the concentration and pull-up speed, among other factors. The Z-DOL film is bonded to the silicon substrate by heating the as-deposited Z-DOL samples in an oven at 150 °C for about 30 minutes. The native oxide layer of Si(100) wafer reacts with the −OH groups of the lubricants during thermal treatment [44.18–21]. Subsequently, fluorocarbon solvent (FC-72) washing of the thermally treated specimen removes loosely absorbed species, leaving the chemically bonded phase on the substrate. The chemical bonding between Z-DOL molecules and silicon substrate is illustrated in Fig. 44.2. The bonded and washed Z-DOL film is referred to as Z-DOL(BW) in this chapter. The as-deposited Z-15 and Z-DOL films are mobile-phase lubricants (i.e., liquid-like lubricants), whereas the Z-DOL(BW) films are fully bonded soft solid phase (i.e., solid-like) lubricants. This will be further discussed in the next section.

44.2 Nanodeformation, Molecular Conformation, and Lubricant Spreading

Nanodeformation behavior of Z-DOL lubricants was studied using an AFM by *Blackman* et al. [44.22, 23]. Before bringing a tungsten tip into contact with a molecular overlayer, it was brought into contact with a bare clean-silicon surface, Fig. 44.3. As the sample approaches the tip, the force initially is zero, but at point A the force suddenly becomes attractive (top curve), which increases until point B, where the sample and tip come into intimate contact and the force becomes repulsive. As the sample is retracted, a pull-off force of 5×10^{-8} N (point D) is required to overcome adhesion between the tungsten tip and the silicon surface. When an AFM tip is brought into contact with an unbonded Z-DOL film, a sudden jump into adhesive contact is also observed. A much larger pull-off force is required to overcome the adhesion. The adhesion is initiated by the formation of a lubricant meniscus surrounding the tip. This suggests that the unbonded Z-DOL lubricant shows liquid-like behavior. However, when the tip was brought into contact with a lubricant film, which was firmly bonded to the surface, the liquid-like behavior disappears. The initial attractive force (point A) is no longer sudden, as with the liquid film, but, rather, gradually increases as the tip penetrates the film.

According to *Blackman* et al. [44.22, 23], if the substrate and tip were infinitely hard with no compliance and/or deformation in the tip and sample supports, the line for B to C would be vertical with an infinite slope. The tangent to the force–distance curve at a given point is referred to as the stiffness at that point and was determined by fitting a least-squares line through the nearby data points. For silicon, the deformation is reversible (elastic), since the retracting (outgoing) portion of the curve (C to D) follows the extending (ingoing) portion (B to C). For bonded lubricant film, at the point where the slope of the force changes gradually from attractive to repulsive, the stiffness changes gradually, indicating compression of the molecular film. As the load is increased, the slope of the repulsive force eventually approaches that of the bare surface. The bonded film was found to respond elastically up to the highest loads of 5 μN that could be applied. Thus, bonded lubricant behaves as a soft polymer solid.

Figure 44.4 illustrates two extremes for the conformation on a surface of a linear liquid polymer without any reactive end groups and at submonolayer coverages [44.4, 6]. At one extreme, the molecules lie flat on the surface, reaching no more than their chain diameter δ above the surface. This would be the case if a strong attractive interaction exists between the molecules and the solid. On the other extreme, when a weak attraction exists between polymer segments and the solid, the molecules adopt a conformation close to that of the molecules in the bulk, with the microscopic thickness equal to about the radius of gyration R_g. *Mate* and *Novotny* [44.6] used AFM to study conformation of 0.5–1.3 nm-thick Z-15 molecules on clean Si(100) surfaces. They found that the thickness measured by AFM is thicker than that measured by ellipsometry, with the offset ranging from 3–5 nm. They found that the offset was the same for very thin submonolayer coverages. If the coverage is submonolayer and inadequate to make a liquid film, the relevant thickness is then the height (h_e) of the molecules extended above the solid surface. The offset should be equal to $2h_e$,

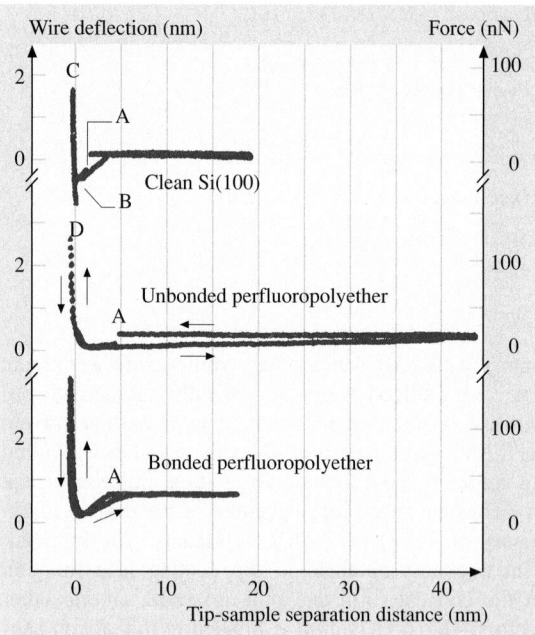

Fig. 44.3 Deflection (normal load) as a function of tip–sample separation-distance curves comparing the behavior of clean Si(100) surface to a surface lubricated with free and unbonded PFPE lubricant, and a surface where the PFPE lubricant film was thermally bonded to the surface [44.22]

assuming that the molecules extend the same height above both the tip and silicon surfaces. They therefore concluded that the molecules do not extend more than 1.5–2.5 nm above a solid or liquid surface, much smaller than the radius of gyration of the lubricants, which ranges between 3.2 and 7.3 nm, and to the approximate cross-sectional diameter of 0.6–0.7 nm for the linear polymer chain. Consequently, the height that the molecules extend above the surface is considerably less than the diameter of gyration of the molecules and only a few molecular diameters in height, implying that the physisorbed molecules on a solid surface have an extended, flat conformation. They also determined the disjoining pressure of these liquid films from AFM measurements of the distance needed to break the liquid meniscus that forms between the solid surface and the AFM tip. (Also see [44.7].) For a monolayer thickness of about 0.7 nm, the disjoining pressure is about 5 MPa, indicating strong attractive interaction between the liquid molecules and the solid surface. The disjoining pressure decreases with increasing film thickness in a manner consistent with a strong attractive van der Waals interaction between the liquid molecules and the solid surface.

Rheological characterization shows that the flow activation energy of PFPE lubricants is weakly dependent on chain length and is strongly dependent on the functional end groups [44.25]. PFPE lubricant films that contain polar end groups have lower mobility than those with nonpolar end groups of similar chain length [44.26]. The mobility of PFPE also depends on the surface chemical properties of the substrate. The spreading of Z-DOL on amorphous carbon surface has been studied as a function of hydrogen or nitrogen content in the carbon film, using scanning microellipsometry [44.24]. The diffusion coefficient data presented in Fig. 44.5 is thickness-dependent. It shows that the surface mobility of Z-DOL increased as the hydrogen content increased, but decreased as nitrogen content increased. The enhancement of Z-DOL surface mobility by hydrogenation may be understood from the fact that the interactions between Z-DOL molecules and the carbon surface can be significantly weakened, due to a reduction of the number of high-energy binding sites on the carbon surface. The stronger interactions between the Z-DOL molecules and carbon surface, as the nitrogen content in the carbon coating increases, leads to the lowering of Z-DOL surface mobility.

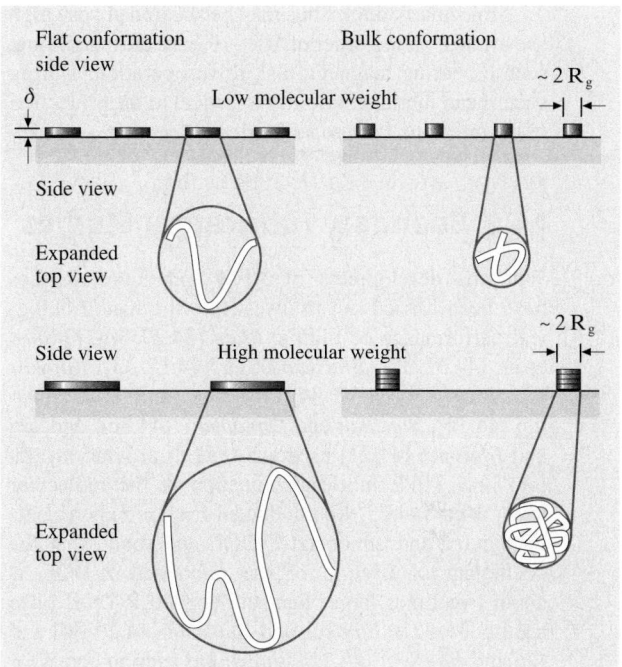

Fig. 44.4 Schematic representation of two extreme liquid conformations at the surface of the solid for low and high molecular weights at low surface coverage. δ is the cross-sectional diameter of the liquid chain, and R_g is the radius of gyration of the molecules in the bulk [44.6]

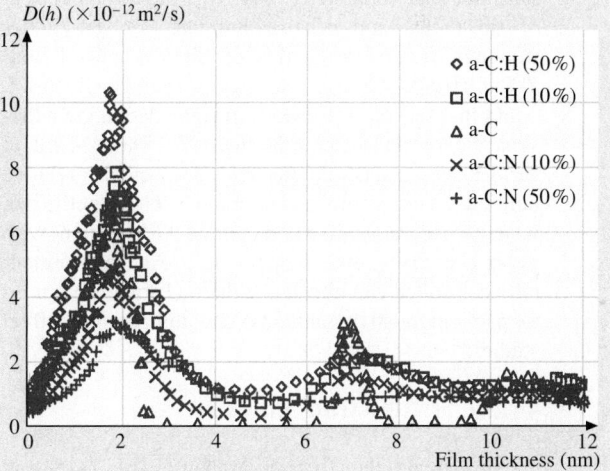

Fig. 44.5 Diffusion coefficient $D(h)$ as a function of lubricant film thickness for Z-DOL on different carbon films [44.24]

Part E | 44.2

Molecularly thick films may be sheared at very high shear rates, on the order of 10^8–$10^9\,\mathrm{s}^{-1}$ during sliding, such as during magnetic disk drive operation. During such shear, lubricant stability is critical to the protection of the interface. For proper lubricant selection, viscosity at high shear rates and associated shear thinning need to be understood. Viscosity measurements of eight different types of PFPE films show that all eight lubricants display Newtonian behavior and their viscosity remains constant at shear rates up to $10^7\,\mathrm{s}^{-1}$ [44.27, 28].

44.3 Boundary Lubrication Studies

With the development of AFM techniques, studies have been carried out to investigate the nanotribological performance of PFPEs. *Mate* [44.29, 30], *O'Shea et al.* [44.31, 32], *Bhushan et al.* [44.15, 33], *Koinkar and Bhushan* [44.20, 34], *Bhushan and Sundararajan* [44.35], *Bhushan and Dandavate* [44.36], and *Liu and Bhushan* [44.21] used an AFM to provide insight into how PFPE lubricants function at the molecular level. *Mate* [44.29, 30] conducted friction experiments on bonded and unbonded Z-DOL and found that the coefficient of friction of the unbonded Z-DOL is about two times larger than the bonded Z-DOL (also see [44.31, 32]). *Koinkar and Bhushan* [44.20, 34] and *Liu and Bhushan* [44.21] studied the friction and wear performance of a Si(100) sample lubricated with Z-15, Z-DOL, and Z-DOL(BW) lubricants. They found that using Z-DOL(BW) could significantly improve the adhesion, friction, and wear performance of Si(100). They also discussed the lubrication mechanisms on the molecular level. *Bhushan and Sundararajan* [44.35] and *Bhushan and Dandavate* [44.36] studied the effect of tip radius and relative humidity on the adhesion and friction properties of Si(100) coated with Z-DOL(BW).

In this section, we review, in some detail, the adhesion, friction, and wear properties of Z-15 and Z-DOL at various operating conditions (rest time, velocity, relative humidity, temperature, and tip radius). The experiments were carried out using a commercial AFM system with pyramidal Si_3N_4 and diamond tips. An environmentally controlled chamber and a thermal stage were used to perform relative humidity and temperature effect studies.

44.3.1 Friction and Adhesion

To investigate the friction properties of Z-15 and Z-DOL(BW) films on Si(100), the curves of friction force versus normal load were measured by making friction measurements at increasing normal loads [44.21]. The representative results of Si(100), Z-15, and Z-DOL(BW) are shown in Fig. 44.6. An approximately linear response of all three samples is observed in the load range of 5–130 nN. The friction force of solid-like Z-DOL(BW) is consistently smaller than that for Si(100), but the friction force of liquid-like Z-15 lubricant is higher than that of Si(100). *Sundararajan and Bhushan* [44.37] have studied the static friction force of silicon micromotors lubricated with Z-DOL by AFM. They also found that liquid-like lubricants of Z-DOL significantly increase the static friction force, whereas solid-like Z-DOL(BW) coatings can dramatically reduce the static friction force. This is in good agreement with the results of *Liu* and *Bhushan* [44.21]. In Fig. 44.6, the nonzero value of the friction force signal at zero external load is due to the adhesive forces. It is well known that the following relationship exists between the friction force F and external normal load W [44.2, 3]

$$F = \mu(W + W_a), \tag{44.1}$$

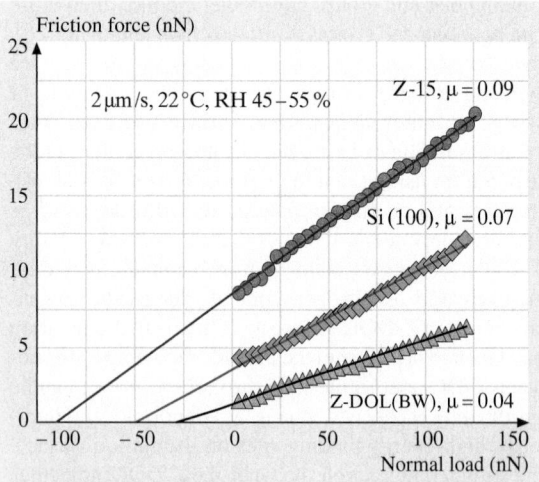

Fig. 44.6 Curves of friction force versus normal load for Si(100), 2.8-nm-thick Z-15 film, and 2.3-nm-thick Z-DOL(BW) film at 2 μm/s, and in ambient air sliding against a Si_3N_4 tip. Based on these curves, the coefficient of friction μ and adhesion force of W_a can be calculated [44.21]

where μ is the coefficient of friction and W_a is the adhesive force. Based on this equation and the data in Fig. 44.6, we can calculate the μ and W_a values. The coefficients of friction of Si(100), Z-15, and Z-DOL are 0.07, 0.09, and 0.04, respectively. Based on (44.1), the adhesive force values are obtained from the horizontal intercepts of the curves of friction force versus normal load at a zero value of friction force. Adhesive force values of Si(100), Z-15, and Z-DOL are 52 nN, 91 nN, and 34 nN, respectively.

The adhesive forces of these samples were also measured using a force calibration plot (FCP) technique. In this technique, the tip is brought into contact with the sample and the maximum force needed to pull the tip and sample apart is measured as the adhesive force. Figure 44.7 shows the typical FCP curves of Si(100), Z-15, and Z-DOL(BW) [44.21]. As the tip approaches the sample within a few nanometers (point A), an attractive force exists between the tip and the sample surfaces. The tip is pulled toward the sample, and contact occurs at point B on the graph. The adsorption of water molecules and/or presence of liquid lubricant molecules on the sample surface can also accelerate this so-called snap-in, due to the formation of meniscus of the water and/or liquid lubricant around the tip. From this point on, the tip is in contact with surface, and as the piezo extends further, the cantilever is further deflected. This is represented by the slope portion of the curve. As the piezo retracts, at point C the tip goes beyond the zero deflection (flat) line, because of the attractive forces, into the adhesive force regime. At point D, the tip snaps free of the adhesive forces and is again in free air. The adhesive force (pull-off force) is determined by multiplying the cantilever spring constant (0.58 N/m) by the horizontal distance between points C and D, which corresponds to the maximum cantilever deflection toward the samples before the tip is disengaged. Incidentally, the horizontal shift between the loading and unloading curves results from the hysteresis of the PZT tube.

The adhesive forces of Si(100), Z-15, and Z-DOL(BW) measured by FCP and plots of friction force versus normal load are summarized in Fig. 44.8 [44.21]. The results measured by these two methods are in good agreement. Figure 44.8 shows that the presence of mobile Z-15 lubricant film increases the adhesive force compared to that of Si(100). In contrast, the presence of solid-phase Z-DOL(BW) film reduces the adhesive force compared to that of Si(100). This result is in good agreement with the results of *Blackman* et al. [44.22] and *Bhushan* and *Ruan* [44.38]. Sources of adhesive forces between the tip and the sample surfaces

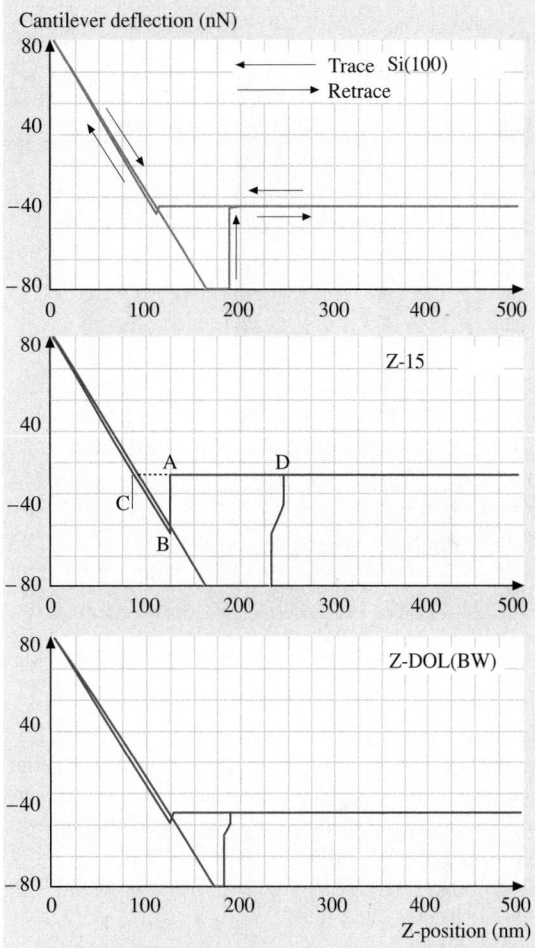

Fig. 44.7 Typical force calibration plots of Si(100), 2.8-nm-thick Z-15 film, and 2.3-nm-thick Z-DOL(BW) film in ambient air. The adhesive forces can be calculated from the horizontal distance between points C and D, and the cantilever spring constant of 0.58 N/m [44.21]

are van der Waals attraction and long-range meniscus forces [44.2,3,16]. The relative magnitudes of the forces from the two sources are dependent on various factors, including the distance between the tip and the sample surface, their surface roughness, their hydrophobicity, and relative humidity [44.39]. For most surfaces with some roughness, meniscus contribution dominates at moderate to high humidities.

The schematic (bottom) in Fig. 44.8 shows relative size and sources of meniscus. The native oxide layer (SiO$_2$) on the top of Si(100) wafer exhibits hydrophilic properties, and some water molecules can be adsorbed

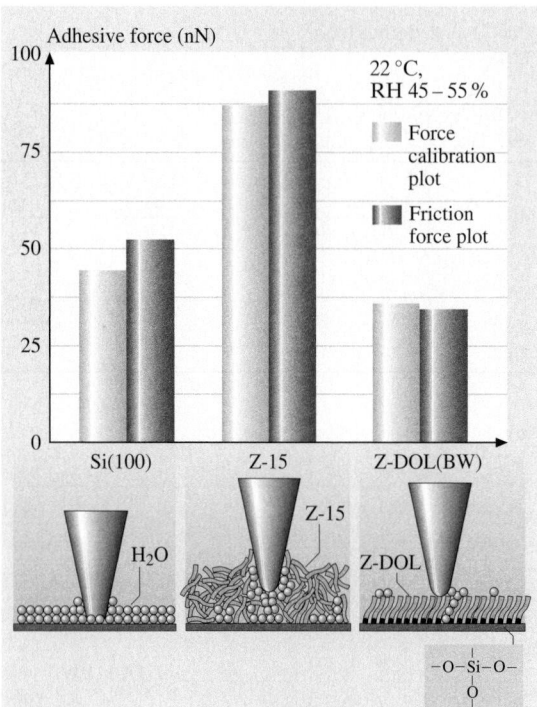

Fig. 44.8 Summary of the adhesive forces of Si(100), 2.8-nm-thick Z-15 film, and 2.3-nm-thick Z-DOL(BW) film measured by force calibration plots and friction force versus normal load in ambient air. The schematic (*bottom*) shows the effect of meniscus formed between the AFM tip and the sample surface on the adhesive and friction forces [44.21]

on this surface. The condensed water will form a meniscus as the tip approaches the sample surface. In the case of a sphere (such as a single-asperity AFM tip) in contact with a flat surface, the attractive Laplace force (F_L) caused by capillary is:

$$F_L = 2\pi R \gamma_{la}(\cos\theta_1 + \cos\theta_2), \qquad (44.2)$$

where R is the radius of the sphere, γ_{la} is the surface tension of the liquid against air, θ_1 and θ_2 are the contact angles between liquid and flat and spherical surfaces, respectively [44.2, 3, 40]. As the surface tension value of Z-15 (24 dyn/cm) is smaller than that of water (72 dyn/cm), the larger adhesive force in Z-15 cannot only be caused by the Z-15 meniscus. The non-polarized Z-15 liquid does not have complete coverage and strong bonding with Si(100). In the ambient environment, the condensed water molecules will permeate through the liquid Z-15 lubricant film and compete with

the lubricant molecules present on the substrate. The interaction of the liquid lubricant with the substrate is weakened, and a boundary layer of the liquid lubricant forms puddles [44.20, 34]. This dewetting allows water molecules to be adsorbed on the Si(100) surface as aggregates along with Z-15 molecules. And both of them can form meniscus while the tip approaches the surface. In addition, as the Z-15 film is pretty soft compared to the solid Si(100) surface, penetration of the tip in the film occurs while pushing the tip down. This leads to a large area of the tip involved to form the meniscus at the tip–liquid (water aggregates along with Z-15) interface. These two factors of the liquid-like Z-15 film result in higher adhesive force. It should also be noted that Z-15 has a higher viscosity compared to that of water. Therefore, Z-15 film provides higher resistance to sliding motion and results in a larger coefficient of friction. In the case of Z-DOL(BW) film, both of the active groups of Z-DOL molecules are strongly bonded on Si(100) substrate through the thermal and washing treatment. Thus, the Z-DOL(BW) film has relatively low free surface energy and cannot be displaced readily by water molecules or readily adsorb water molecules. Thus, the use of Z-DOL(BW) can reduce the adhesive force. We further believe that the bonded Z-DOL molecules can be orientated under stress (behaving as a soft polymer solid), which facilitates sliding and reduces coefficient of friction.

These studies suggest that, if the lubricant films exist as liquid-like, such as Z-15 films, they easily form meniscus (by themselves and the adsorbed water molecules), and thus have higher adhesive force and higher friction force. Whereas, if the lubricant film exists in solid-like

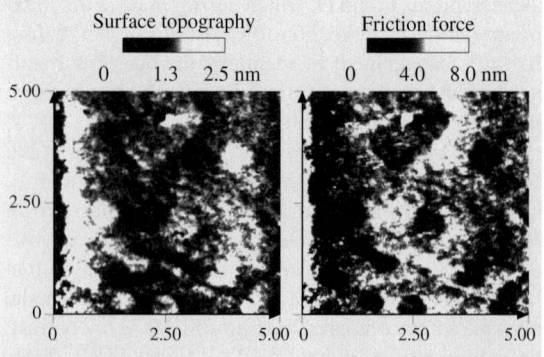

Fig. 44.9 Gray scale plots of the surface topography and friction force obtained simultaneously for unbonded 2.3-nm-thick Demnum-type PFPE lubricant film on silicon [44.20]

Fig. 44.10 Gray scale plots of the adhesive force distribution of a uniformly coated, 3.5-nm-thick unbonded Z-DOL film on silicon and 3–10 nm-thick unbonded Z-DOL film on silicon that was deliberately coated nonuniformly by vibrating the sample during the coating process [44.36]

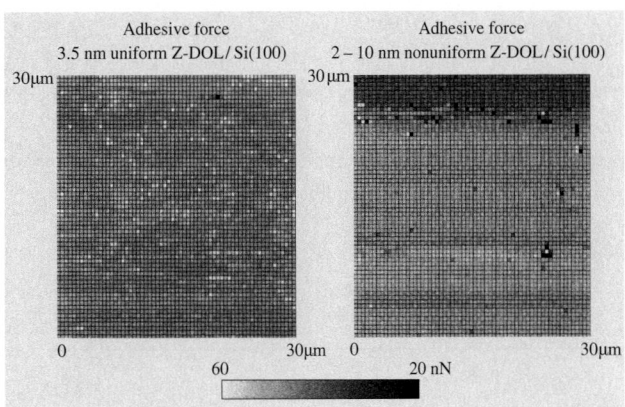

phase, such as Z-DOL(BW) films, they are hydrophobic with low adhesion and friction.

In order to study the uniformity of lubricant film and its influence on friction and adhesion, friction force mapping and adhesive force mapping of PFPE have been carried out by *Koinkar* and *Bhushan* [44.34] and *Bhushan* and *Dandavate* [44.36], respectively. Figure 44.9 shows gray scale plots of surface topography and friction force images obtained simultaneously for unbonded Demnum-type PFPE lubricant film on silicon [44.34]. The friction force plot shows well-distinguished low- and high-friction regions corresponding roughly to high- and low-surface-height regions in the topography image (thick- and thin-lubricant regions). A uniformly lubricated sample does not show such a variation in friction. Figure 44.10 shows the gray scale plots of the adhesive force distribution for silicon samples coated uniformly and nonuniformly with Z-DOL lubricant. It can be clearly seen that there exists a region that has an adhesive force distinctly different from the other region for the nonuniformly coated sample. This implies that the liquid film thickness is nonuniform, giving rise to a difference in the meniscus forces.

44.3.2 Rest Time Effect

It is well known that, in the computer rigid disk drive, the stiction force increases rapidly with an increase in rest time between the head and magnetic-medium disk [44.10, 11]. Considering that the stiction and friction are two of the major issues that lead to the failure of computer rigid disk drives and MEMS, it is very important to find out if the rest time effect also exists on the nanoscale. First, the rest time effect on the friction force, adhesive force, and coefficient of Si(100) sliding against Si$_3$N$_4$ tip was studied, Fig. 44.11a [44.21]. It was found

Fig. 44.11 (a) Rest time effect on friction force, adhesive force, and coefficient of friction of Si(100). **(b)** (*see next page*) Summary of the rest time effect on friction force, adhesive force, and coefficient of friction of Si(100), 2.8-nm-thick Z-15 film, and 2.3-nm-thick Z-DOL(BW) film. All of the measurements were carried out at 70 nN, 2 μm/s, and in ambient air [44.21]

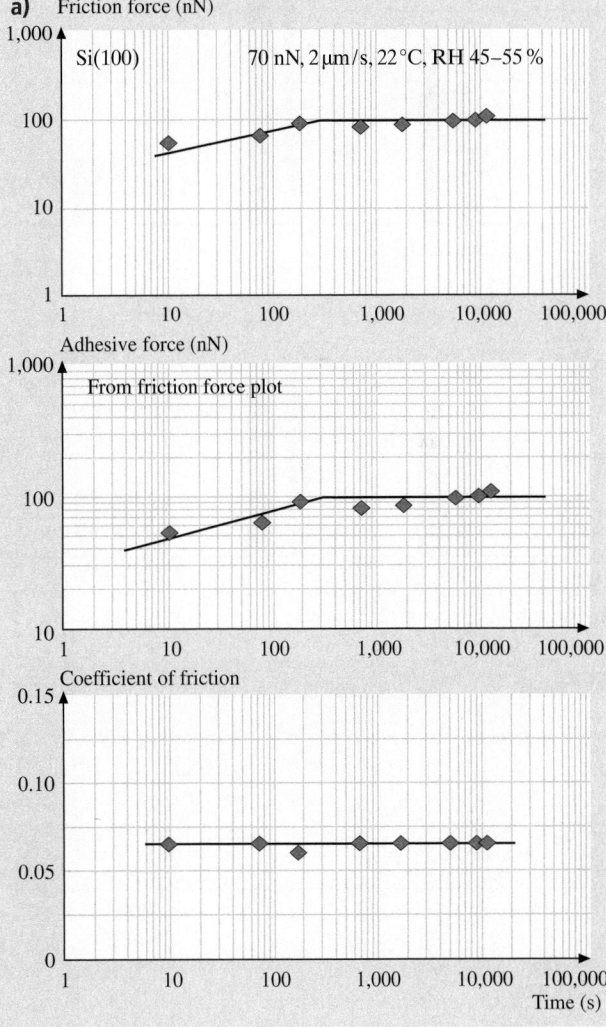

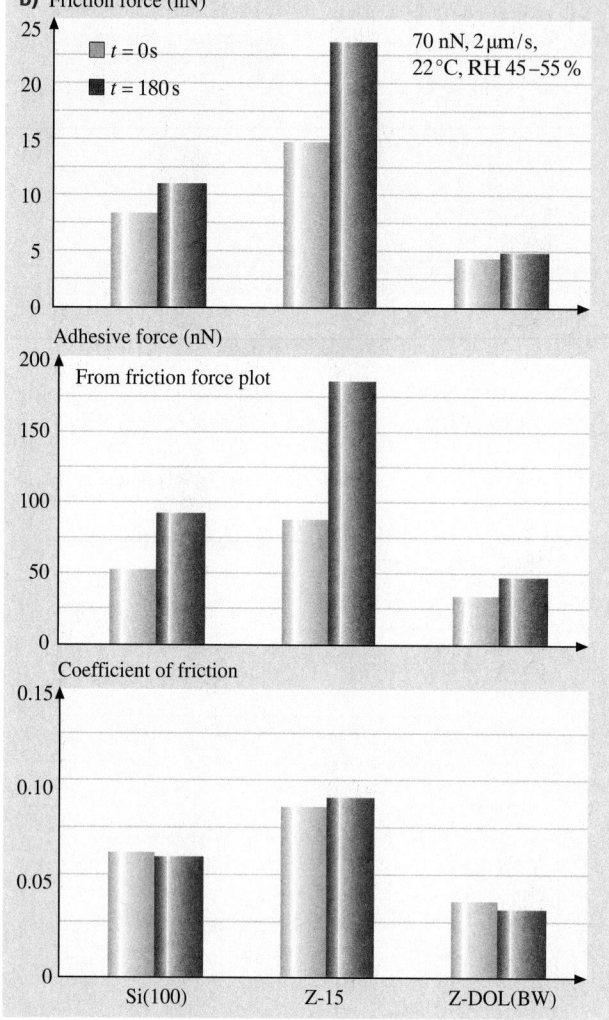

Fig. 44.11b

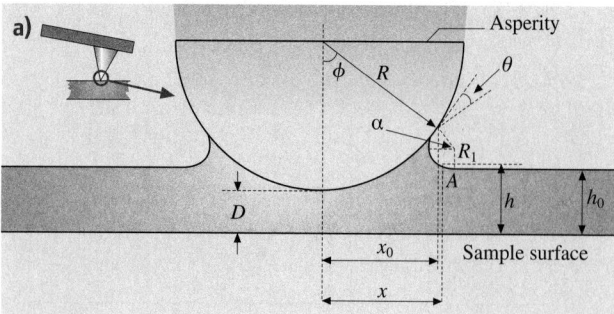

Fig. 44.12a

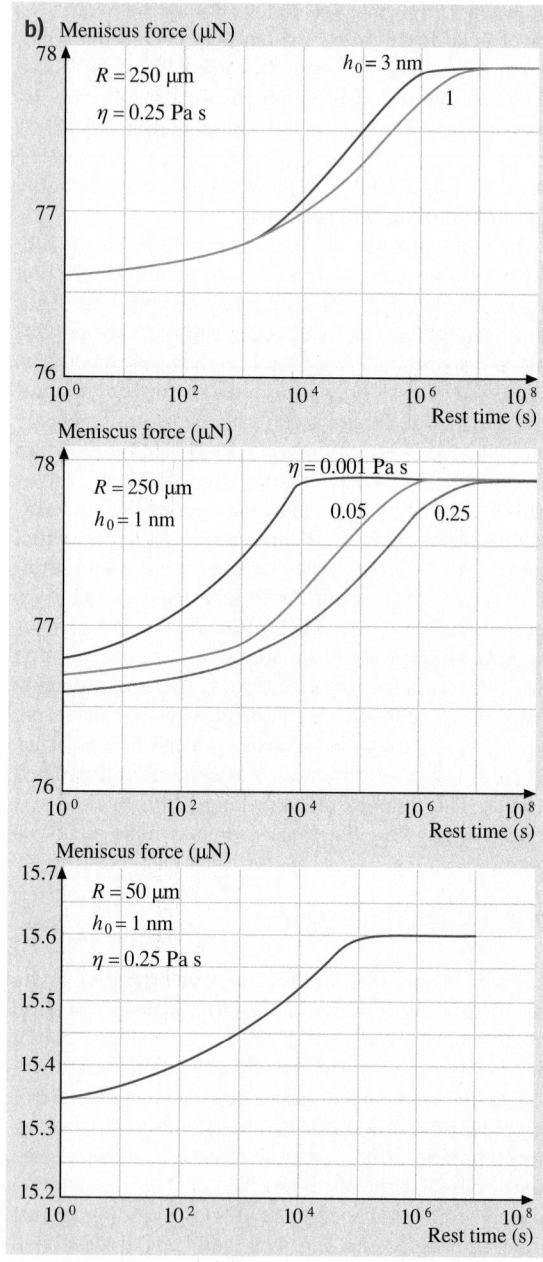

Fig. 44.12 (a) Schematic of a single asperity in contact with a smooth flat surface in the presence of a continuous liquid film when ϕ is large. (b) Results of the single-asperity model. Effect of viscosity of the liquid, radius of the asperity, and film thickness is studied with respect to the time-dependent meniscus force [44.41]

that the friction and adhesive forces logarithmically increase up to a certain equilibrium time after which they remain constant. Figure 44.11a also shows that the rest time does not affect the coefficient of friction. These results suggest that the rest time can result in the growth of the meniscus, which causes a higher adhesive force, and in turn, a higher friction force. But in the whole testing range the friction mechanisms do not change with the rest time. Similar studies were also performed on Z-15 and Z-DOL(BW) films. The results are summarized in Fig. 44.11b [44.21]. It is seen that a similar time effect has been observed on Z-15 film, but not on Z-DOL(BW) film.

An AFM tip in contact with a flat sample surface can be treated as a single-asperity contact. Therefore, a Si_3N_4 tip in contact with Si(100) or Z-15/Si(100) can be modeled as a sphere in contact with a flat surface covered by a layer of liquid (adsorbed water and/or liquid lubricant), Fig. 44.12a. Meniscus forms around the contacting asperity and grows with time until equilibrium occurs [44.41]. The meniscus force, which is the product of meniscus pressure and meniscus area, depends on the flow of liquid phase toward the contact zone. The flow of the liquid toward the contact zone is governed by the capillary pressure P_c, which draws liquid into the meniscus, and the disjoining pressure Π, which tends to draw the liquid away from the meniscus. Based on the Young and Laplace equation, the capillary pressure, P_c, is:

$$P_c = 2K\gamma , \qquad (44.3)$$

where $2K$ is the mean meniscus curvature ($= K_1 + K_2$, where K_1 and K_2 are the curvatures of the meniscus in the contact plane and perpendicular to the contact plane) and γ is the surface tension of the liquid. Mate and Novotny [44.6] have shown that the disjoining pressure decreases rapidly with increasing liquid film thickness in a manner consistent with a strong van der Waals attraction. The disjoining pressure, Π, for these liquid films can be expressed as:

$$\Pi = \frac{A}{6\pi h^3} , \qquad (44.4)$$

where A is the Hamaker constant and h is the liquid film thickness. The driving forces that cause the lubricant flow that results in an increase in the meniscus force are the disjoining pressure gradient, due to a gradient in film thickness, and the capillary pressure gradient, due to the curved liquid–air interface. The driving pressure, P, can then be written as:

$$P = -2K\gamma - \Pi . \qquad (44.5)$$

Based on these three basic relationships, the following differential equation has been derived by *Chilamakuri* and *Bhushan* [44.41], which can describe the meniscus at time t:

$$2\pi x_0 \left(D + \frac{x_0^2}{2R} - h_0 \right) \frac{dx_0}{dt}$$
$$= \frac{2\pi h_0^3 \gamma}{3\eta} \frac{(1+\cos\theta)}{D+a-h_0} - \frac{A x_0}{3\eta h} \cot\alpha , \qquad (44.6)$$

where η is the viscosity of the liquid and a is given as

$$a = R(1-\cos\phi) \sim \frac{R\phi^2}{2} \sim \frac{x_0^2}{2R} . \qquad (44.7)$$

The differential equation (44.4) was solved numerically using Newton's iteration method. The meniscus force at any time t less than the equilibrium time is proportional to the meniscus area and meniscus pressure ($2K\gamma$), and it is given by

$$f_m(t) = 2\pi R\gamma(1+\cos\theta) \left(\frac{x_0}{(x_0)_{eq}} \right)^2 \left(\frac{K}{K_{eq}} \right) , \qquad (44.8)$$

where $(x_0)_{eq}$ is the value of x_0 at the equilibrium time

$$[(x_0)_{eq}]^2 = 2R \left[\frac{-6\pi h_0^3 \gamma(1+\cos\theta)}{A} + (h_0 - D) \right] . \qquad (44.9)$$

This modeling work (at the microscale) showed that the meniscus force initially increases logarithmically with the rest time up to a certain equilibrium time, after which it remains constant. Equilibrium time decreases with an increase in liquid film thickness, a decrease in viscosity, and a decrease in the tip radius, Fig. 44.12b. This early numerical modeling work and the data at the nanoscale in Fig. 44.11a are in good agreement.

44.3.3 Velocity Effect

To investigate the velocity effect on friction and adhesion, the relationships between friction force and normal load for Si(100), Z-15, and Z-DOL(BW) at different velocities were measured, Fig. 44.13 [44.21]. Based on these data, the adhesive force and coefficient of friction values can be calculated by (44.1). The variation of friction force, adhesive force, and coefficient of friction of Si(100), Z-15, and Z-DOL(BW) as a function of velocity are summarized in Fig. 44.14. This indicates that, for silicon wafer, the friction force decreases logarithmically with increasing velocity. For Z-15, the

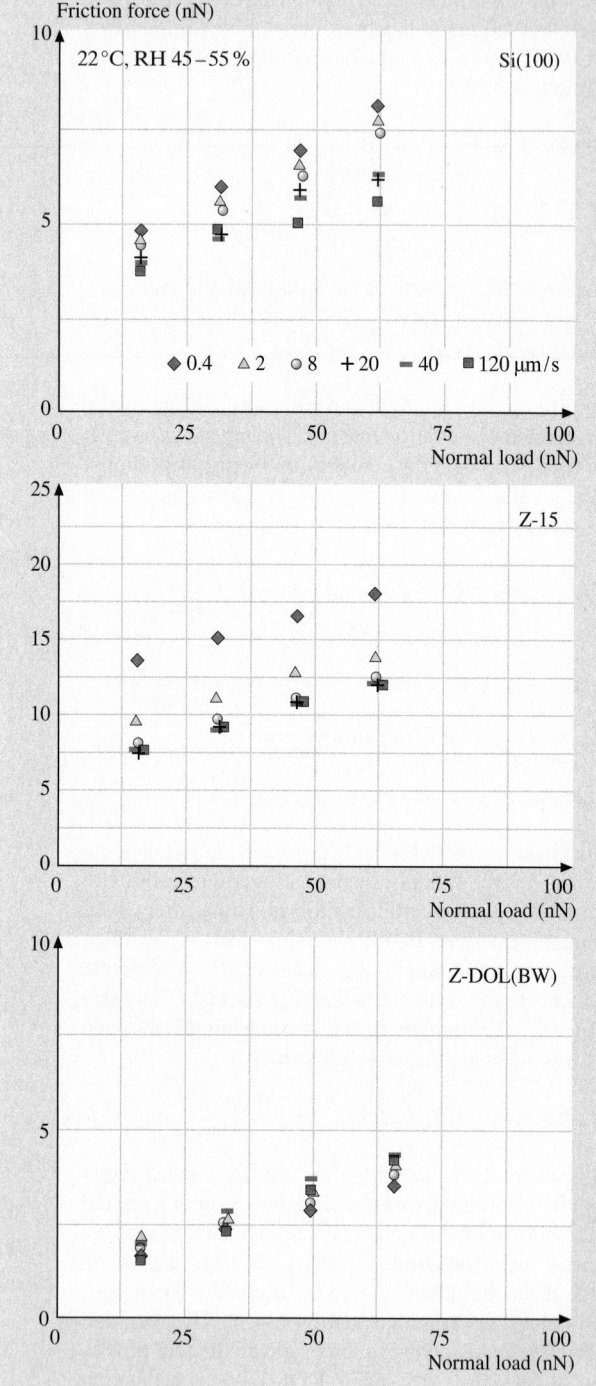

Fig. 44.13 Friction forces versus normal load data of Si(100), 2.8-nm-thick Z-15 film, and 2.3-nm-thick Z-DOL(BW) film at various velocities in ambient air [44.21]

friction force decreases with increasing velocity up to 10 μm/s, after which it remains almost constant. The velocity has a much smaller effect on the friction force of Z-DOL(BW); it reduced slightly only at very high velocity. Figure 44.14 also indicates that the adhesive force of Si(100) is increased when the velocity is higher than 10 μm/s. The adhesive force of Z-15 is reduced dramatically with a velocity increase up to 20 μm/s, after which it is reduced slightly; the adhesive force of Z-DOL(BW) also decreases at high velocity. In the testing range of velocity, only the coefficient of friction of Si(100) decreases with velocity, but the coefficients of friction of Z-15 and Z-DOL(BW) almost remain constant. This implies that the friction mechanisms of Z-15 and Z-DOL(BW) do not change with the variation of velocity.

The mechanisms of the effect of velocity on the adhesion and friction are explained based on the schematics shown in Fig. 44.14 (right). For Si(100), tribochemical reaction plays a major role. Although at high velocity the meniscus is broken and does not have enough time to rebuild, the contact stresses and high velocity lead to tribochemical reactions of the Si(100) wafer and Si_3N_4 tip, which have native oxide (SiO_2) layers with water molecules. The following reactions occur:

$$SiO_2 + 2H_2O \rightarrow Si(OH)_4 \qquad (44.10)$$

$$Si_3N_4 + 16H_2O \rightarrow 3Si(OH)_4 + 4NH_4OH . \quad (44.11)$$

The $Si(OH)_4$ is removed and continuously replenished during sliding. The $Si(OH)_4$ layer between the tip and Si(100) surface is known to be of low shear strength and causes a decrease in friction force and coefficient of friction in the lateral direction [44.42–46]. The chemical bonds of Si−OH between the tip and Si(100) surface induce large adhesive force in the normal direction. For Z-15 film, at high velocity the meniscus formed by condensed water and Z-15 molecules is broken and does not have enough time to rebuild. Therefore, the adhesive force and, consequently, friction force is reduced. For Z-DOL(BW) film, the surface can adsorb few water molecules in ambient condition, and at high velocity these molecules are displaced, which is responsible for a slight decrease in friction force and adhesive force. Even in the high-velocity range, the friction mechanisms for Z-15 and Z-DOL(BW) films are still shearing of the viscous liquid and molecular orientation, respectively. Thus the coefficients of friction of Z-15 and Z-DOL(BW) do not change with velocity.

Koinkar and *Bhushan* [44.20, 34] have suggested that, in the case of samples with mobile films, such as condensed water and Z-15 films, alignment of liquid

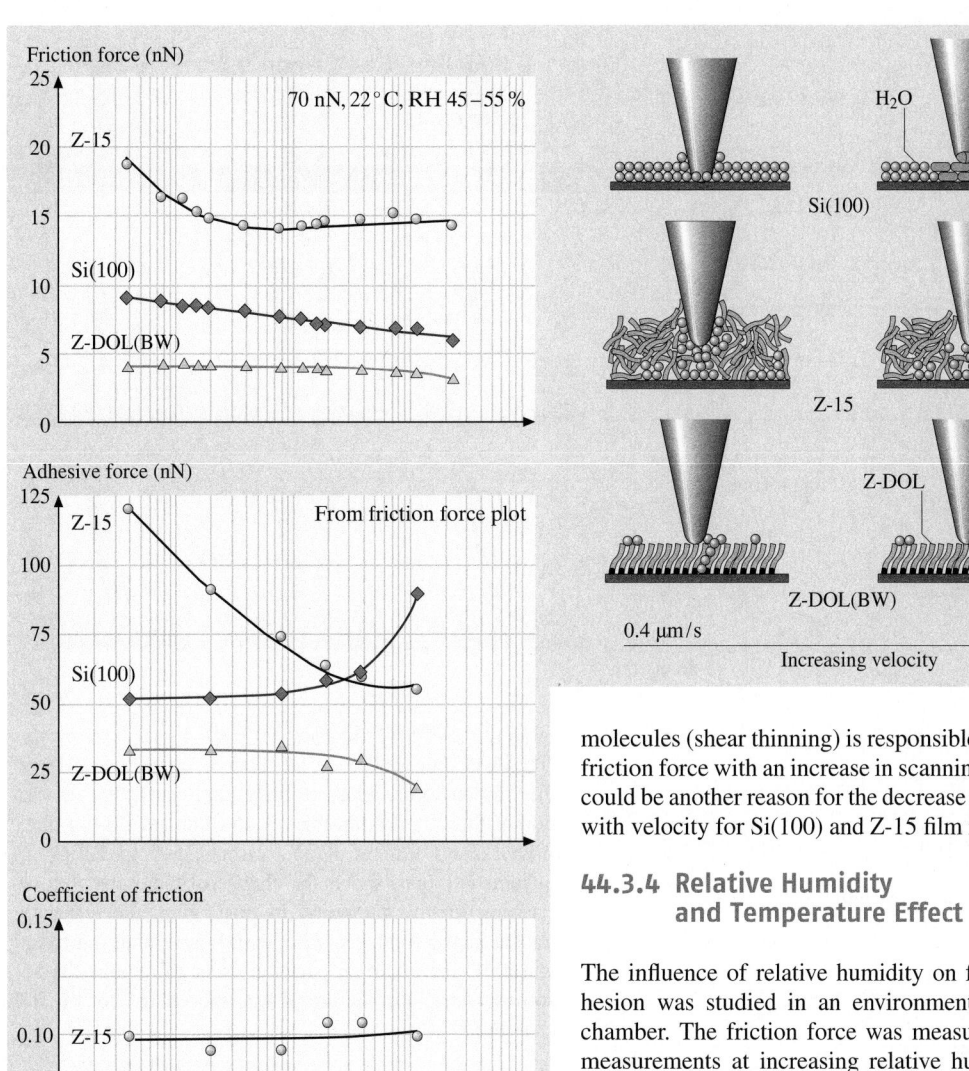

Fig. 44.14 The influence of velocity on the friction force, adhesive force, and coefficient of friction of Si(100), 2.8-nm-thick Z-15 film, and 2.3-nm-thick Z-DOL(BW) film at 70 nN, in ambient air. The schematic (*right*) shows the change of surface composition (by tribochemical reaction) and change of meniscus with increasing velocity [44.21]

molecules (shear thinning) is responsible for the drop in friction force with an increase in scanning velocity. This could be another reason for the decrease in friction force with velocity for Si(100) and Z-15 film in this study.

44.3.4 Relative Humidity and Temperature Effect

The influence of relative humidity on friction and adhesion was studied in an environmentally controlled chamber. The friction force was measured by making measurements at increasing relative humidity, the results are presented in Fig. 44.15 [44.21]. These shows that, for Si(100) and Z-15 film, the friction force increases with a relative humidity increase up to RH 45%, and then it shows a slight decrease with a further increase in relative humidity. Z-DOL(BW) has a smaller friction force than Si(100) and Z-15 in the whole testing range, and its friction force shows a relatively apparent increase when the relative humidity is higher than RH 45%. For Si(100), Z-15, and Z-DOL(BW), adhesive forces increase with relative humidity. And their coefficients of friction increase with relative humidity up to RH 45%, after which they decrease with further increases of the relative humidity. It is also observed that the humidity effect on Si(100) really depends on the history of the Si(100) sample. As the surface of Si(100)

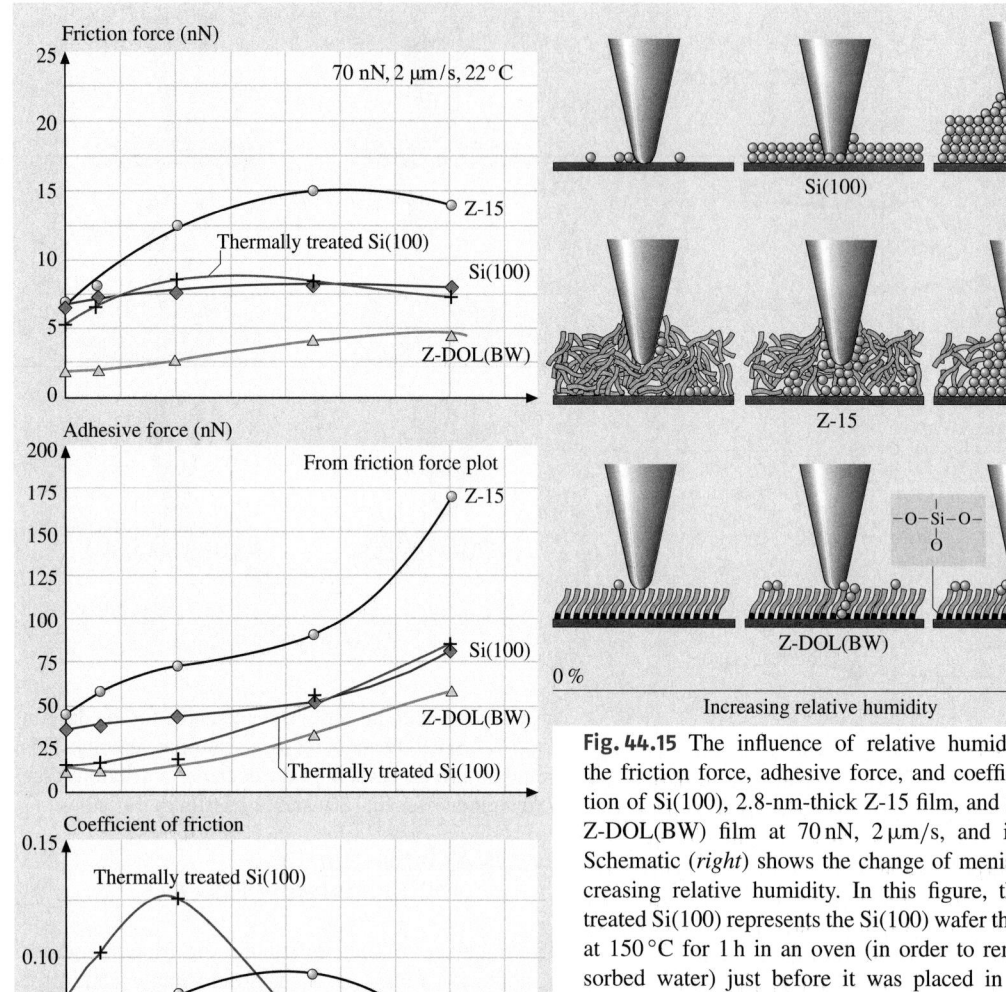

Fig. 44.15 The influence of relative humidity (RH) on the friction force, adhesive force, and coefficient of friction of Si(100), 2.8-nm-thick Z-15 film, and 2.3-nm-thick Z-DOL(BW) film at 70 nN, 2 μm/s, and in 22 °C air. Schematic (*right*) shows the change of meniscus with increasing relative humidity. In this figure, the thermally treated Si(100) represents the Si(100) wafer that was baked at 150 °C for 1 h in an oven (in order to remove the adsorbed water) just before it was placed in the 0% RH chamber [44.21]

The schematic (right) in Fig. 44.15 shows that, because of its high free surface energy, Si(100) can adsorb more water molecules with increasing relative humidity. As discussed earlier, for Z-15 film in a humid environment, condensed water competes with the lubricant film present on the sample surface. Obviously, more water molecules can also be adsorbed on a Z-15 surface with increasing relative humidity. Thermal adsorbed water molecules in the case of Si(100), along with lubricant molecules in Z-15 film, form a bigger water meniscus, which leads to an increase of the friction force, adhesive force, and coefficient of friction of Si(100) and Z-15 with humidity. But at the very high humidity of RH 70%, large quantities of adsorbed water can form a continuous water layer that separates the

wafer readily adsorbs water in air, without any pretreatment the Si(100) used in our study almost reaches its saturated stage of adsorbing water and is responsible for a smaller effect with increasing relative humidity. However, once the Si(100) wafer was thermally treated by baking at 150 °C for 1 h, a bigger effect was observed.

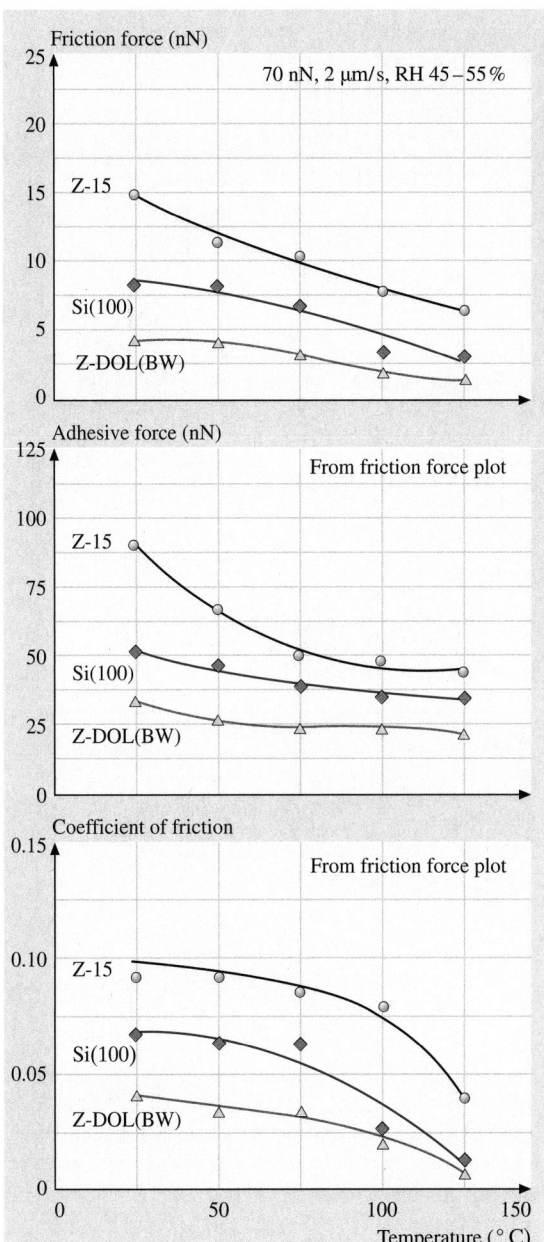

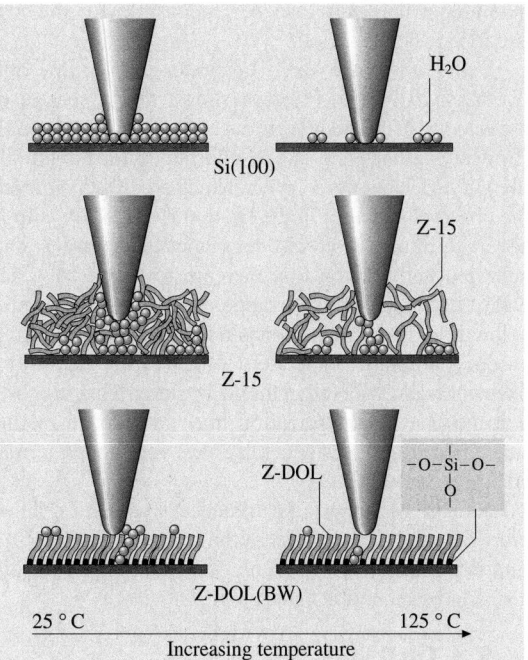

Fig. 44.16 The influence of temperature on the friction force, adhesive force, and coefficient of friction of Si(100), 2.8-nm-thick Z-15 film, and 2.3-nm-thick Z-DOL(BW) film at 70 nN, at 2 μm/s, and in RH 40–50% air. The schematic (*right*) shows that, at high temperature, desorption of water decreases the adhesive forces. And the reduced viscosity of Z-15 leads to the decrease of coefficient of friction. High temperature facilitates orientation of molecules in Z-DOL(BW) film, which results in a lower coefficient of friction [44.21]

The effect of temperature on friction and adhesion was studied using a thermal stage attached to the AFM. The friction force was measured at increasing temperature from 22–125 °C. The results are presented in Fig. 44.16 [44.21]. It shows that the increasing temperature causes a decrease of friction force, adhesive force, and coefficient of friction of Si(100), Z-15, and Z-DOL(BW). The schematic (right) in Fig. 44.16 indicates that, at high temperature, desorption of water leads to a decrease in the friction force, adhesive force, and coefficient of friction for all of the samples. Besides that, the reduction of surface tension of water also contributes to the decrease of friction and adhesion. For Z-15 film, the reduction of viscosity at high temperature makes an additional contribution to the decrease of friction. In the case of Z-DOL(BW) film, molecules are more easily oriented at high tem-

tip and sample surface, and acts as a kind of lubricant, which causes a decrease in the friction force and coefficient of friction. For the Z-DOL(BW) film, because of its hydrophobic surface properties, water molecules can only be adsorbed at high humidity (≥ RH 45%), which causes an increase in the adhesive force and friction force.

perature, which may also be responsible for the low friction.

Using a surface force apparatus, *Yoshizawa* and *Israelachvili* [44.47, 48] have shown that a change in the velocity or temperature induces phase transformation (from crystalline solid-like to amorphous, then to liquid-like) in surfactant monolayers, which are responsible for the observed changes in the friction force. Stick–slip is observed in a low-velocity regime of a few μm/s, and adhesion and friction first increase followed by a decrease in the temperature range of 0–50 °C. Stick–slip at low velocity and adhesion and friction curves peaking at some particular temperature (observed in their study), have not been observed in the AFM study. This suggests that the phase transformation may not happen in this study, because PFPEs generally have very good thermal stability [44.10, 12].

As a brief summary, the influence of velocity, relative humidity, and temperature on the friction force of Z-15 film is presented in Fig. 44.17. The changing trends are also addressed in this figure.

44.3.5 Tip Radius Effect

The tip radius and relative humidity affect adhesion and friction for dry and lubricated surfaces [44.35, 36]. Figure 44.18a shows the variation of single-point adhesive force measurements as a function of tip radius on a Si(100) sample for several humidities. The adhesive force data are also plotted as a function of relative humidity for various tip radii. Figure 44.18a indicates that the tip radius has little effect on the adhesive forces at low humidities, but the adhesive force increases with tip radius at high humidity. Adhesive force also increases with increasing humidity for all tips. The trend in adhesive forces as a function of tip radii and relative humidity, in Fig. 44.18a, can be explained by the presence of meniscus forces, which arise from capillary condensation of water vapor from the environment. If enough liquid is present to form a meniscus bridge, the meniscus force should increase with an increase in tip radius, based on (44.2). This observation suggests that the thickness of the liquid film at low humidities is insufficient to form continuous meniscus bridges and to affect adhesive forces in the case of all tips.

Figure 44.18a also shows the variation in coefficient of friction as a function of tip radius at a given humidity and as a function of relative humidity for a given tip radius on the Si(100) sample. It can be observed that, for RH 0%, the coefficient of friction is about the same

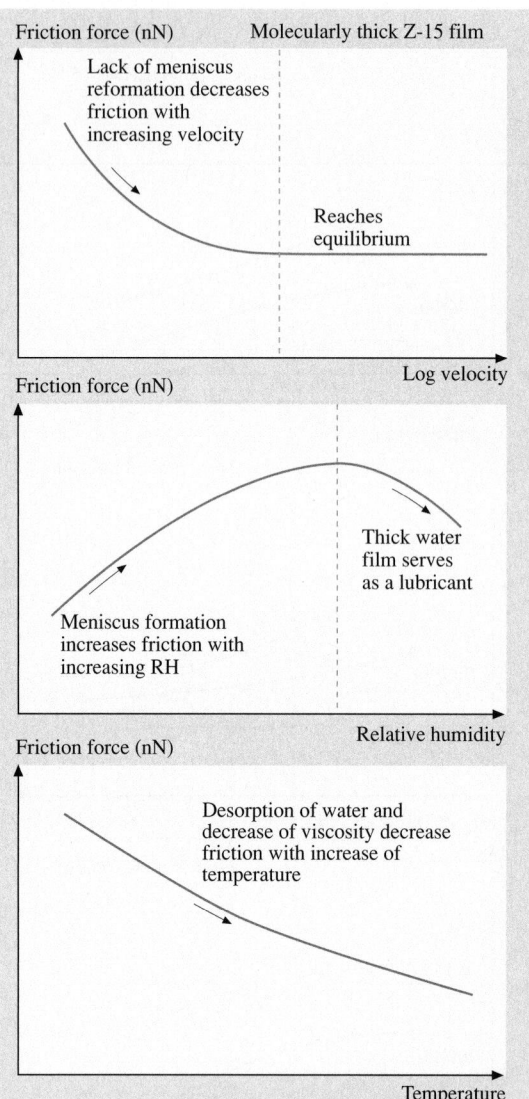

Fig. 44.17 Schematic showing the change of friction force of molecularly thick Z-15 films with log velocity, relative humidity, and temperature [44.21]

for the tip radii investigated except for the largest tip, which shows a higher value. At all other humidities, the trend consistently shows that the coefficient of friction increases with tip radius. An increase in friction with tip radius at low to moderate humidities arises from increased contact area (i. e., higher van der Waals forces) and the higher values of shear forces required for a larger contact area. At high humidities, similarly to adhesive force data, an increase with tip radius occurs due to

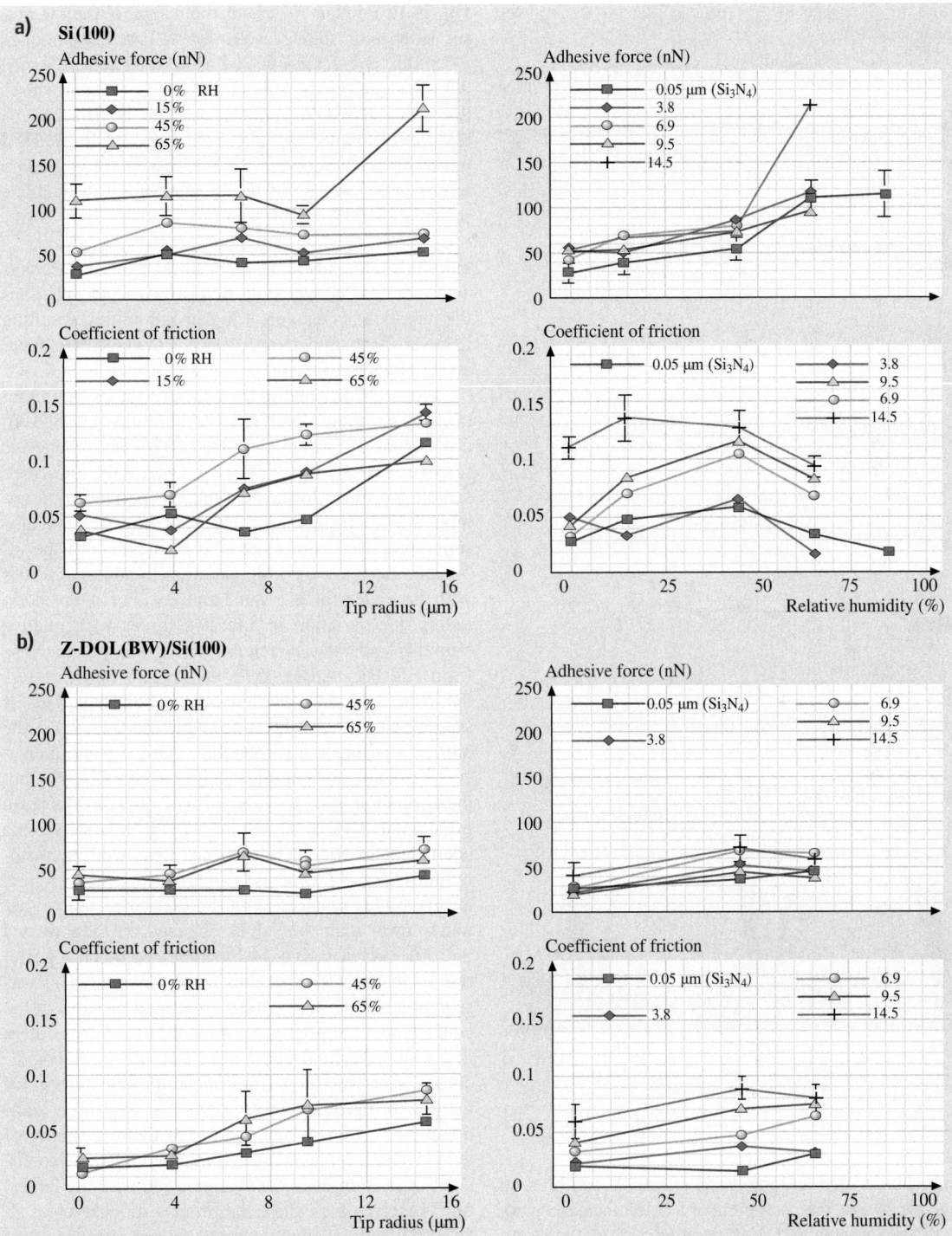

Fig. 44.18a,b Adhesive force and coefficient of friction as a function of tip radius at several humidities and as a function of relative humidity at several tip radii on (**a**) Si(100) and (**b**) 0.5-nm Z-DOL(BW) films [44.35]

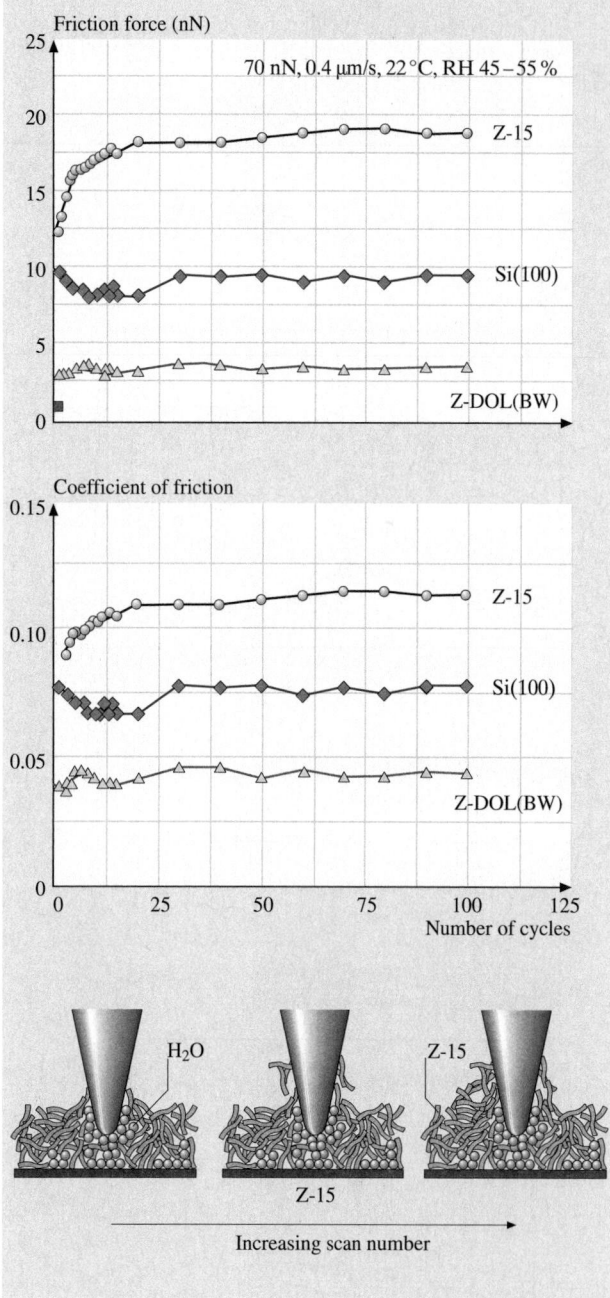

Fig. 44.19 Friction force and coefficient of friction versus number of sliding cycles for Si(100), 2.8-nm-thick Z-15 film, and 2.3-nm-thick Z-DOL(BW) film at 70 nN, 0.8 μm/s, and in ambient air. Schematic (*bottom*) shows that some liquid Z-15 molecules can be attached on the tip. The molecular interaction between the attached molecules on the tip with the Z-15 molecules in the film results in an increase of the friction force with multiple scanning [44.21]

humidities, the adsorbed water film on the surface acts as a lubricant between the two surfaces [44.21]. Thus the interface is changed at higher humidities, resulting in lower shear strength and, hence, a lower friction force and coefficient of friction.

Figure 44.18b shows the adhesive forces as a function of tip radius and relative humidity on Si(100) coated with 0.5 nm-thick Z-DOL(BW) film. Adhesive forces for all the tips with the Z-DOL(BW)-lubricated sample are much lower than those experienced on unlubricated Si(100), shown in Fig. 44.18a. The data also show that, even with a monolayer thickness of the lubricant, there is very little variation in adhesive forces with tip radius at a given humidity. For a given tip radius, the variation in adhesive forces with relative humidity indicates that these forces slightly increase from RH 0% to RH 45%, but remain more or less the same with further increases in humidity. This is seen even with the largest tip, which indicates that the lubricant is indeed hydrophobic; there is some meniscus formation at humidities higher than RH 0%, but the formation is very minimal and does not increase even up to RH 65%. Figure 44.18b also shows the coefficient of friction for various tips at different humidities for the Z-DOL(BW)-lubricated sample. Again, all the values obtained with the lubricated sample are much lower than the values obtained on unlubricated Si(100), shown in Fig. 44.18a. The coefficient of friction increases with tip radius for all humidities, as was seen on unlubricated Si(100), due to an increase in the contact area. Similarly to the adhesive forces, there is an increase in friction from RH 0% to RH 45%, due to a contribution from an increasing number of menisci bridges. However, thereafter very little additional water film forms, due to the hydrophobicity of the Z-DOL(BW) layer, and, consequentially, the coefficient of friction does not change, even with the largest tip. These findings show that even a monolayer of Z-DOL(BW) offers good hydrophobic performance of the surface.

both contact area and meniscus effects. It can be seen that, for all tips, the coefficient of friction increases with humidity to about RH 45%, beyond which it starts to decrease. This is attributed to the fact that, at higher

44.3.6 Wear Study

To study the durability of lubricant films at the nanoscale, the friction of Si(100), Z-15, and Z-DOL(BW) as a function of the number of scanning cycles was measured, Fig. 44.19 [44.21]. As observed earlier, friction force and coefficient of friction of Z-15 is higher than that of Si(100), and Z-DOL(BW) has the lowest values. During cycling, friction force and coefficient of friction of Si(100) show a slight variation during the initial few cycles then remain constant. This is related to the removal of the top adsorbed layer. In the case of Z-15 film, the friction force and coefficient of friction show an increase during the initial few cycles and then approach higher and stable values. This is believed to be caused by the attachment of the Z-15 molecules to the tip. The molecular interaction between these attached molecules on the tip and molecules on the film surface is responsible for an increase in the friction. However, after several scans, this molecular interaction reaches equilibrium, and after that, the friction force and coefficient of friction remain constant. In the case of Z-DOL(BW) film, the friction force and coefficient of friction start out low and remain low during the entire test for 100 cycles. This suggests that Z-DOL(BW) molecules do not become attached or displaced as readily as Z-15.

Koinkar and *Bhushan* [44.20, 34] conducted wear studies using a diamond tip at high loads. Figure 44.20 shows the plots of wear depth as a function of normal force, and Fig. 44.21 shows the wear profiles of the worn samples at 40 μN normal load. The 2.3 nm-thick Z-DOL(BW)-lubricated sample exhibits better wear resistance than unlubricated and 2.9 nm-thick

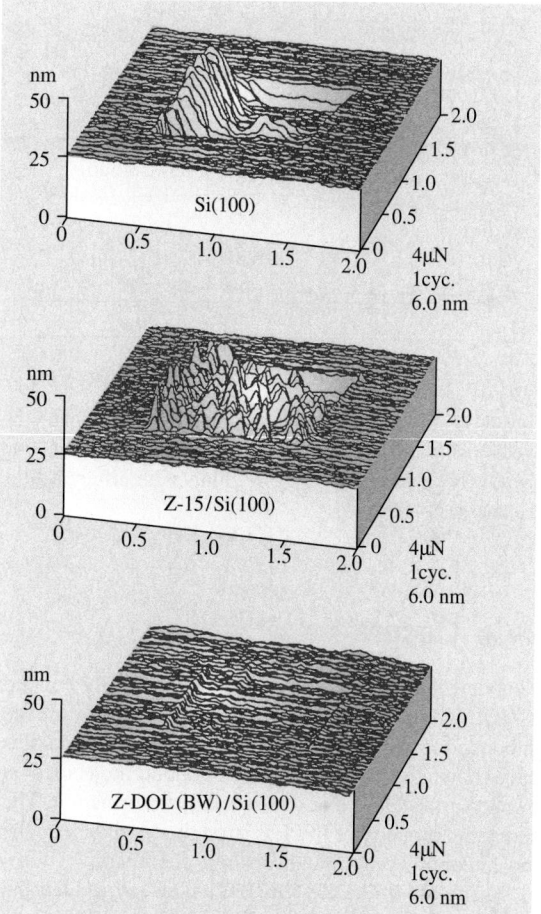

Fig. 44.21 Wear profiles for Si(100), 2.9-nm-thick Z-15 film, and 2.3-nm-thick Z-DOL(BW) film after wear studies using a diamond tip. The normal force used and wear depths are listed in the figure [44.20]

Z-15-lubricated silicon samples. Wear resistance of a Z-15-lubricated sample is little better than that of the unlubricated sample. The Z-15-lubricated sample shows debris inside the wear track. Since Z-15 is a liquid lubricant, debris generated is held by the lubricant and they become sticky. The debris moves inside the wear track and does damage, Fig. 44.20. These results suggest that using Z-DOL(BW) can improve the wear resistance of the substrate.

To study the effect of the degree of chemical bonding, the durability tests were conducted on both fully bonded and partially bonded Z-DOL films. Durability results for Z-DOL(BW) and Z-DOL bonded and unwashed (Z-DOL(BUW), a partially

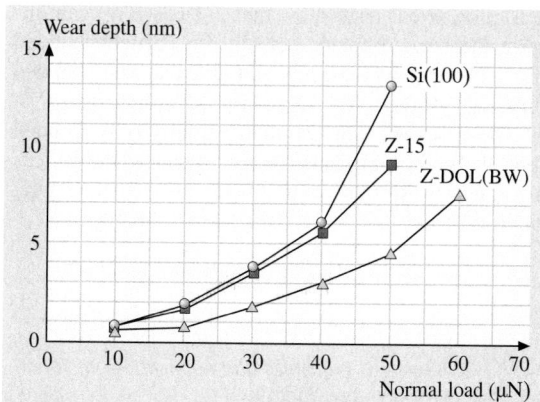

Fig. 44.20 Wear depth as a function of normal load using a diamond tip for Si(100), 2.9-nm-thick Z-15 film, and 2.3-nm-thick Z-DOL(BW) after one cycle [44.20]

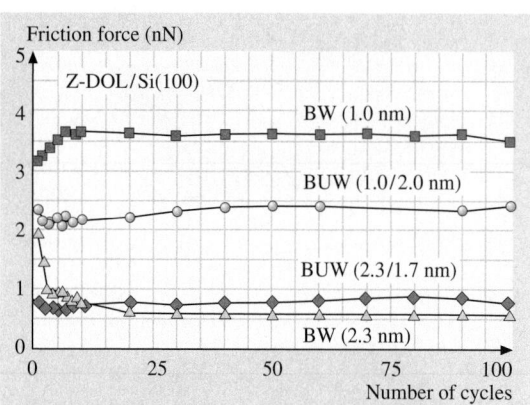

Fig. 44.22 Friction force as a function of number of cycles using a Si$_3$N$_4$ tip at a normal load of 300 nm for Z-DOL(BW) and Z-DOL(BUW) films with different film thicknesses [44.34]

bonded film that contains both bonded and mobile phase lubricants) with different film thicknesses are shown in Fig. 44.22 [44.34]. Thicker films, such as Z-DOL(BUW), with a thickness of 4.0 nm (bonded/mobile = 2.3 nm/1.7 nm) exhibit behavior similar to 2.3 nm-thick Z-DOL(BW) film. Figure 44.22 also indicates that Z-DOL(BW) and Z-DOL(BUW) films with a thinner film thickness exhibit a higher friction value. Comparing 1.0 nm-thick Z-DOL(BW) with 3.0 nm-thick Z-DOL(BUW) (bonded/mobile = 1.0 nm/2.0 nm), the Z-DOL(BUW) film exhibits a lower and stable friction value. This is because the mobile phase on the surface acts as a source of lubricant replenishment. A similar conclusion has also been reported by *Ruhe* et al. [44.19], *Bhushan and Zhao* [44.13], and *Eapen* et al. [44.49]. All of them indicate that using partially bonded Z-DOL films can dramatically reduce the friction and improve the wear life.

44.4 Closure

Nanodeformation studies have shown that fully bonded Z-DOL lubricants behave as polymer coatings, while the unbonded lubricants behave like liquids. AFM studies have shown that the physisorbed nonpolar molecules on a solid surface have an extended, flat conformation. The spreading property of PFPE is strongly dependent on the molecular end groups and substrate chemistry.

Using solid-like Z-DOL(BW) film can reduce the friction and adhesion of Si(100), while using mobile lubricant of Z-15 shows a negative effect. Si(100) and Z-15 film show apparent time effects. The friction and adhesion forces increase as a result of growth of meniscus up to an equilibrium time, after which they remain constant. Using Z-DOL(BW) film can prevent time effects. High velocity leads to the rupture of meniscus and prevents its reformation, which leads to a decrease of friction and adhesive forces of Z-15 and Z-DOL(BW). The influence of relative humidity on the friction and adhesion is dominated by the amount of the adsorbed water

molecules. Increasing humidity can either increase friction through increased adhesion by the water meniscus, or reduce friction through an enhanced water-lubricating effect. Increasing temperature leads to desorption of the water layer, decrease of the water surface tension, decrease of viscosity, and easier orientation of the Z-DOL(BW) molecules. These changes cause a decrease of the friction force and adhesion at high temperature. During cycling tests, the molecular interaction between the Z-15 molecules attached to the tip and the Z-15 molecules on the film surface causes the initial rise of friction. Wear tests show that Z-DOL(BW) can improve the wear resistance of silicon. Partially bonded PFPE film appears to be more durable than fully bonded films.

These results suggest that partially/fully bonded films are good lubricants for micro/nanoscale devices operating in different environments and under varying conditions.

References

44.1   B. Bhushan: Magnetic Recording Surfaces. In: *Characterization of Tribological Materials*, ed. by W. A. Glaeser (Butterworth−Heinemann, Boston 1993) pp. 116−133

44.2   B. Bhushan: *Principles and Applications of Tribology* (Wiley, New York 1999)

44.3   B. Bhushan: *Introduction to Tribology* (Wiley, New York 2002)

44.4 V. J. Novotny, I. Hussla, J. M. Turlet, M. R. Philpott: Liquid polymer conformation on solid surfaces, J. Chem. Phys. **90**, 5861–5868 (1989)

44.5 V. J. Novotny: Migration of liquid polymers on solid surfaces, J. Chem. Phys. **92**, 3189–3196 (1990)

44.6 C. M. Mate, V. J. Novotny: Molecular conformation and disjoining pressures of polymeric liquid films, J. Chem. Phys. **94**, 8420–8427 (1991)

44.7 C. M. Mate: Application of disjoining and capillary pressure to liquid lubricant films in magnetic recording, J. Appl. Phys. **72**, 3084–3090 (1992)

44.8 G. G. Roberts: *Langmuir–Blodgett Films* (Plenum, New York 1990)

44.9 A. Ulman: *An Introduction to Ultrathin Organic Films* (Academic, Boston 1991)

44.10 B. Bhushan: *Tribology and Mechanics of Magnetic Storage Devices*, 2nd edn. (Springer, New York 1996)

44.11 B. Bhushan: Macro- and microtribology of magnetic storage devices. In: *Modern Tribology Handbook Vol. 2: Materials, Coatings, and Industrial Applications*, ed. by B. Bhushan (CRC, Boca Raton 2001) pp. 1413–1513

44.12 Anonymous: Fomblin Z Perfluoropolyethers, Data sheet Montedism Group, Milan (2002)

44.13 B. Bhushan, Z. Zhao: Macroscale and microscale tribological studies of molecularly thick boundary layers of perfluoropolyether lubricants for magnetic thin-film rigid disks, J. Info. Storage Proc. Syst. **1**, 1–21 (1999)

44.14 B. Bhushan: *Tribology Issues and Opportunities in MEMS* (Kluwer, Dordrecht 1998)

44.15 B. Bhushan, J. N. Israelachvili, U. Landman: Nanotribology: Friction, wear and lubrication at the atomic scale, Nature **374**, 607–616 (1995)

44.16 B. Bhushan: *Handbook of Micro/Nanotribology*, 2nd edn. (CRC, Boca Raton 1999)

44.17 B. Bhushan: Self-assembled monolayers for controlling hydrophobicity and/or friction and wear. In: *Modern Tribology Handbook Vol. 2: Materials, Coatings, and Industrial Applications*, ed. by B. Bhushan (CRC, Boca Raton 2001) pp. 909–929

44.18 J. Ruhe, G. Blackman, V. J. Novotny, T. Clarke, G. B. Street, S. Kuan: Thermal attachment of perfluorinated polymers to solid surfaces, J. Appl. Polym. Sci. **53**, 825–836 (1994)

44.19 J. Ruhe, V. Novotny, T. Clarke, G. B. Street: Ultrathin perfluoropolyether films—influence of anchoring and mobility of polymers on the tribological properties, ASME J. Tribol. **118**, 663–668 (1996)

44.20 V. N. Koinkar, B. Bhushan: Microtribological studies of unlubricated and lubricated surfaces using atomic force/friction force microscopy, J. Vac. Sci. Technol. A **14**, 2378–2391 (1996)

44.21 H. Liu, B. Bhushan: Nanotribological characterization of molecularly-thick lubricant films for applications to MEMS/NEMS by AFM, Ultramicroscopy **97**, 321–340 (2003)

44.22 G. S. Blackman, C. M. Mate, M. R. Philpott: Interaction forces of a sharp tungsten tip with molecular films on silicon surface, Phys. Rev. Lett. **65**, 2270–2273 (1990)

44.23 G. S. Blackman, C. M. Mate, M. R. Philpott: Atomic force microscope studies of lubricant films on solid surfaces, Vacuum **41**, 1283–1286 (1990)

44.24 X. Ma, J. Gui, K. J. Grannen, L. A. Smoliar, B. Marchon, M. S. Jhon, C. L. Bauer: Spreading of PFPE lubricants on carbon surfaces: Effect of hydrogen and nitrogen content, Tribol. Lett. **6**, 9–14 (1999)

44.25 C. A. Kim, H. J. Choi, R. N. Kono, M. S. Jhon: Rheological characterization of perfluoropolyether lubricant, Polym. Prepr. **40**, 647–649 (1999)

44.26 M. Ruths, S. Granick: Rate-dependent adhesion between opposed perfluoropoly(alkylether) layers: Dependence on chain-end functionality and chain length, J. Phys. Chem. B **102**, 6056–6063 (1998)

44.27 U. Jonsson, B. Bhushan: Measurement of rheological properties of ultrathin lubricant films at very high shear rates and near-ambient pressure, J. Appl. Phys. **78**, 3107–3109 (1995)

44.28 C. Hahm, B. Bhushan: High shear rate viscosity measurement of perfluoropolyether lubricants for magnetic thin-film rigid disks, J. Appl. Phys. **81**, 5384–5386 (1997)

44.29 C. M. Mate: Atomic-force-microscope study of polymer lubricants on silicon surface, Phys. Rev. Lett. **68**, 3323–3326 (1992)

44.30 C. M. Mate: Nanotribology of lubricated and unlubricated carbon overcoats on magnetic disks studied by friction force microscopy, Surf. Coat. Technol. **62**, 373–379 (1993)

44.31 S. J. O'Shea, M. E. Welland, T. Rayment: Atomic force microscope study of boundary layer lubrication, Appl. Phys. Lett. **61**, 2240–2242 (1992)

44.32 S. J. O'Shea, M. E. Welland, J. B. Pethica: Atomic force microscopy of local compliance at solid-liquid interface, Chem. Phys. Lett. **223**, 336–340 (1994)

44.33 B. Bhushan, T. Miyamoto, V. N. Koinkar: Microscopic friction between a sharp diamond tip and thin-film magnetic rigid disks by friction force microscopy, Adv. Info. Storage Syst. **6**, 151–161 (1995)

44.34 V. N. Koinkar, B. Bhushan: Micro/nanoscale studies of boundary layers of liquid lubricants for magnetic disks, J. Appl. Phys. **79**, 8071–8075 (1996)

44.35 B. Bhushan, S. Sundararajan: Micro/nanoscale friction and wear mechanisms of thin films using atomic force and friction force microscopy, Acta Mater. **46**, 3793–3804 (1998)

44.36 B. Bhushan, C. Dandavate: Thin-film friction and adhesion studies using atomic force microscopy, J. Appl. Phys. **87**, 1201–1210 (2000)

44.37   S. Sundararajan, B. Bhushan: Static friction and surface roughness studies of surface micromachined electrostatic micromotors using an atomic force/friction force microscope, J. Vac. Sci. Technol. A **19**, 1777–1785 (2001)

44.38   B. Bhushan, J. Ruan: Atomic-scale friction measurements using friction force microscopy: Part II— application to magnetic media, ASME J. Tribol. **116**, 389–396 (1994)

44.39   T. Stifter, O. Marti, B. Bhushan: Theoretical investigation of the distance dependence of capillary and van der Waals forces in scanning probe microscopy, Phys. Rev. B **62**, 13667–13673 (2000)

44.40   J. N. Israelachvili: *Intermolecular and Surface Forces*, 2nd edn. (Academic, London 1992)

44.41   S. K. Chilamakuri, B. Bhushan: A comprehensive kinetic meniscus model for prediction of long-term static friction, J. Appl. Phys. **15**, 4649–4656 (1999)

44.42   H. Ishigaki, I. Kawaguchi, M. Iwasa, Y. Toibana: Friction and wear of hot pressed silicon nitride and other ceramics, ASME J. Tribol. **108**, 514–521 (1986)

44.43   T. E. Fischer: Tribochemistry, Annu. Rev. Mater. Sci. **18**, 303–323 (1988)

44.44   K. Mizuhara, S. M. Hsu: Tribochemical reaction of oxygen and water on silicon surfaces. In: *Wear Particles*, ed. by D. Dowson (Elsevier, New York 1992) pp. 323–328

44.45   S. Danyluk, M. McNallan, D. S. Park: Friction and wear of silicon nitride exposed to moisture at high temperatures. In: *Friction and Wear of Ceramics*, ed. by S. Jahanmir (Dekker, New York 1994) pp. 61–79

44.46   V. A. Muratov, T. E. Fischer: Tribochemical polishing, Annu. Rev. Mater. Sci. **30**, 27–51 (2000)

44.47   H. Yoshizawa, Y. L. Chen, J. N. Israelachvili: Fundamental mechanisms of interfacial friction I: Relationship between adhesion and friction, J. Phys. Chem. **97**, 4128–4140 (1993)

44.48   H. Yoshizawa, J. N. Israelachvili: Fundamental mechanisms of interfacial friction II: Stick slip friction of spherical and chain molecules, J. Phys. Chem. **97**, 11300–11313 (1993)

44.49   K. C. Eapen, S. T. Patton, J. S. Zabinski: Lubrication of microelectromechanical systems (MEMS) using bound and mobile phase of Fomblin Z-DOL, Tibol. Lett. **12**, 35–41 (2002)

45. Kinetics and Energetics in Nanolubrication

In the 19th century, lubrication, one of humankind's oldest engineering disciplines, gained a theoretical base from Reynolds's classical hydrodynamic description that was unmatched by most of the theories developed in tribology to date. In the 20th century, however, increasing demands on lubricants shifted attention from bulk films to ultra-thin film lubrication. Finite-size limitations imposed constraints on the lubrication process that were not considered in the bulk phenomenological treatments introduced by Reynolds. At this point, as is common in many engineering applications, empiricism took over. Functional relationships derived from the classical theories were tweaked to accommodate the new situation of reduced scales by introducing *effective* or *apparent* properties.

With the inception of nanorheological tools of complementary nature in the later decades of the 20th century (e.g., the surface forces apparatus and scanning force microscopy), tribology entered the realm of nanoscience. Through an increasing confidence in experimental findings on the nanoscale, kinetic and energetic theories incorporated interfacial and molecular constraints.

The very fundamentals have been challenged in recent years. Researchers have realized that bulk perceptions, such as *solid* and *liquid* are defied on the nanoscale. The reduction in dimensionality of the nanoscale imposes constraints that bring into question the use of classical statistical mechanics of decoupled events. The diffusive description of lubrication is failing in a system that is thermodynamically not well-equilibrated.

Part E | 45

The challenge any nanotechnological endeavor encounters is the development of a theoretical framework based on an appropriate statistics. In tribology this is met with spectral descriptions of the dynamic sliding process. Statistical kernels are being developed for probability density functions to explain anomalous transport processes that involve long-range spatial or temporal correlations. With such theoretical developments founded in nanorheological experiments, a more realistic foundation will be laid to describe the behavior of lubricants in the confined geometries of the nanometer length scale.

What is inaccessible today may become accessible tomorrow as has happened by the invention of the microscope. ... Coherent assumptions on what is still invisible may increase our understanding of the visible. ... Strong reasons have come to support a growing *probability, and it can finally be said the certainty, in favor of the hypothesis of the atomists.* (Jean Baptiste Perrin – Nobel Lecture, December 11, 1926)

Since technology is driving lubricant films to molecular thickness, kinetic friction and its dependence on the sliding parameters – especially the sliding velocity – have become of great interest. The complexity of the frictional resistance in lubricated sliding is illustrated in Fig. 45.1 with the *Stribeck curve* that identifies various regimes of lubrication in respect to the lubrication impact of the hydrodynamic pressure. In the ultra-low-speed regime, called the *boundary lubrication regime*, no hydrodynamic pressure is built up in the lubricant. Consequently the load is carried by contact asperities coated with adsorbed lubricant molecules. If the speed is raised, hydrodynamic pressure builds that leads to *mixed lubrication*, in which the load is carried by both asperities and hydrodynamic pressure. At even higher speeds, elastic contributions of the solid surfaces have to be considered to be paired with hydrodynamic pressure effects (elasto-hydrodynamic lubrication), until only *hydrodynamic lubrication* matters. Hence, the Stribeck curve combines various aspects of lubrication. The curve cannot be discussed without considering the lubricant thickness and the different models of asperity contact sliding.

In one of the first comprehensive physical models of *dry* friction, Bowden and Tabor introduced a plastic asperity model, in which the material's yield stress and adhesive properties play an important role [45.1]. Considering this model, which depends on surface energies and mechanical yield properties paired with all the properties that accompany a surface adsorbent lubricant, one can hardly grasp the difficulty level involved in describing the frictional kinetics in lubrication.

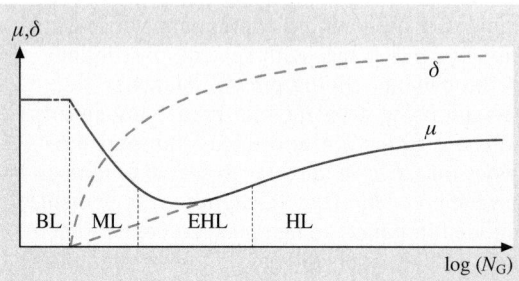

Fig. 45.1 Stribeck curve (schematic) relates the fluid lubricant thickness, δ, and the friction coefficient μ to the Gumbel number $N_{\mathrm{G}} = \eta\omega P^{-1}$; i.e., the product of the liquid bulk viscosity, η, the sliding speed (or more precisely the shaft frequency), ω, and the inverse of the normal pressure, P. BL: boundary lubrication, ML: mixed lubrication, EHL: elasto-hydrodynamic lubrication, HL: hydrodynamic lubrication

Past and current engineering challenges in lubrication have been met with great and complex empiricism. The theoretical modeling of lubrication junctions generally involved only bulk property considerations with inadequately known adsorption mechanisms. The complexity of today's lubricants, most of them, such as motor oil, a product of empirical design over many years, increased exponentially, making it very difficult to meet future challenges. The problem of empiricism is that conventional laws and perceptions are unchallenged. *Effective* quantities are invented (e.g., effective viscosity), exponential fitting parameters are introduced (e.g., the Kohlrausch relaxation parameter), and terminologies such as *solid* and *liquid* are taken for granted. Progress based on empiricism is only incremental and rarely revolutionary.

One of the reasons for empiricism is a lack of access to a system with fewer and better-controlled parameters. In lubrication sliding that challenge has been addressed over the last two decades with the inception of the surface forces apparatus (SFA) by *Tabor* et al. [45.2] and scanning force microscopy (SFM) by *Binnig* et al. [45.3]. These two instrumental methods allow lubrication studies where roughness effects can be neglected, surface energies controlled, and wear from wearless friction distinguished. Lubricant properties can be studied at nearly mathematically described boundaries, atomistic friction events can be recorded, and fundamental models that have been considered to be mere Gedanken experiments, such as the *Tomlinson model of friction*, can be verified. In the wake of these nanoscopic tools, exciting new theoretical lubrication and friction models have appeared.

This chapter considers these recent experimental and theoretical developments with a particular focus on sliding speed and real or apparent changes in the lubricant material properties. We will discuss kinetics and energetics in the simplified world of nanolubrication, in which our conventional perception is challenged. After a brief review of some of the classical lubrication concepts (hydrodynamic lubrication and boundary lubrication), we will turn our attention to a thermal activation model of friction, functional behavior of lubricated friction with velocity, and models based on small nonconforming contacts. We will critically discuss the limitation of the underlying Gaussian statistics, introduce fractal dynamics in lubrication, and will end our discussion with metastable lubricant systems.

45.1 Background: From Bulk to Molecular Lubrication

45.1.1 Hydrodynamic Lubrication and Relaxation

In the classical theories of tribology by da Vinci, Amonton, and Coulomb, not much attention was given to the dependence of kinetic friction on the sliding velocity. This clearly changed in the 19th century, during the first industrial revolution, at which time lubricants became increasingly important, for instance, in ball and journal bearings. It was *Petrov* [45.4–7], *Tower* [45.8], and *Reynolds* [45.9] who established that the liquid viscous-shear properties determine the frictional kinetics. *Reynolds* [45.9] combined the pressure-gradient-determined *Poisseuille flow* with the bearing-surface-induced *Couette flow* assuming, based on Petrov's law [45.4–7], a no-slip condition at the interface between lubricant and solid. This led to the widely used linear relationship between friction and velocity. Reynolds's hydrodynamic theory of lubrication can be applied to steady-state sliding at constant relative velocity and to transient decay sliding (sliding is stopped from an initial velocity v and a corresponding shear stress τ_0), which leads to the classical Debye exponential relaxation behavior, i. e.:

$$\tau = \tau_0 \exp\left(\frac{-D}{A\eta}t\right) ; \quad \tau_0 \propto \frac{v\eta}{D} . \tag{45.1}$$

D is the lubricant thickness, A the area of the slider, and η the viscosity of the fluid. We will later see that this classical exponential relaxation behavior, obtained in a thermodynamically well-mixed three-dimensional medium, is distorted when the liquid film thickness is reduced to molecular dimensions.

45.1.2 Boundary Lubrication

Reynolds's hydrodynamic description of lubrication was found to work well for thick lubricant films but to break down for thinner films. One manifestation is that for films on the order of ten molecular diameters the stress in the film does not allow the tension to return to zero. It was also found that the motion in the steady-state sliding regime was disrupted, exhibiting a stick–slip-like slider motion [45.10]). Consequently, this non-Newtonian behavior was treated with a modified viscosity parameter (effective viscosity), which was composed of the pressure, temperature, and rate of shear.

The term *boundary lubrication* is used to describe a lubricant that is reduced in thickness to molecular dimension and effectively reduces the friction between two opposing solid surfaces. *Hardy* et al. [45.11] recognized that molecular properties such as molecular weight and molecular arrangement govern the frictional force. This confined concept of lubrication, often visualized by two highly ordered opposing films with shear taking place somewhere in between the two layers, contains many of the rate-dependent manifestations of frictional sliding; e.g., stick–slip, ultra-low friction, transitions from high to low friction, phase transitions, dissipation due to dislocations (e.g., gauche and cis-transformations), and memory effects.

Boundary lubrication was found to be in many respects unique [45.12]. In macroscopic experiments, which involved rough surfaces, friction–velocity plots resembled logarithmic functions at moderate speeds. No static-stiction force peaks were observed in boundary lubricants close to zero speed. On the contrary, retractive slips could be observed upon halting, constituting a static friction coefficient exceeded by the dynamic friction coefficient [45.12]. These unique manifestations of boundary lubrication were discussed in terms of a lubricated asperity–junction mechanism, which associated "an increase in the coefficient of friction with a decrease in the adsorptive coverage of the rubbing surfaces by the lubricant substance" [45.12]. It was argued that, in the course of the sliding process of a macroscopic slider, more adsorbed lubricant is expected to exist within the interfacial area than outside the contact zone. This would lead upon halting to a relaxation process of the elastic restraints on the slider, causing the slider to retractively slip.

45.1.3 Stick–Slip and Collective Phenomena

Based on numerous friction experiments at the initiation of sliding with rough macroscopic contact, it was argued that the distinction between static and kinetic friction is not categorical but rather a manifestation of the apparatus [45.12]. This was a widely held opinion prior to *Briscoe* et al.'s [45.13] molecularly smooth monolayer SFA experiments of aliphatic carboxylic acids and their soaps. Briscoe found that the character of sliding motion (continuous vs. discontinuous), depends not only on the apparatus but also on the properties (chemistry) of the monolayer. As in the rough boundary layer experiments discussed above, Briscoe's molecularly smooth monolayer experiments exhibited logarithmic-like friction–velocity behaviors.

It was *Israelachvili* et al. [45.10] who, based on SFA experiments and computer simulations, provided a molecular picture of the stick–slip behavior caused by the lubricant material. The major achievement of this work was to draw our attention to the molecular structure properties of the lubricant that are different under nanoconstraints and transient during sliding motion. Nanolubrication involving in-plane structuring and freezing-melting transitions was found to be inadequately described with bulk rheology [45.10].

The interpretation involving a "freezing-melting transition" is motivated by the stick-slip motion observed at low velocity sliding: While the "stick regime" is interpreted as a solid-like (Hookian) response to stress, the "slip regime" is explained in terms of a stress-strain rate response; i.e., a Newtonian liquid-like effect. In most cases, this simplified binary rheological explanation of the shear induced stick-slip motion may however not be confused with a true freezing-melting transition. Any deformation of a solid can be both, coordinated or uncoordinated, and thus can exhibit both solid-like and liquid-like behavior. For instance, most of the plastic yielding processes are uncoordinated. On the other hand, slipping within a solid, along a crystal plane in a thermally activated strain-release process for instance, is a highly coordinated molecular process [45.15].

Similar arguments can be made for a liquid. For example, stick–slip behaviors were observed in more complex fluidic systems by *Reiter* et al. [45.16], who compared a molecularly "wet" lubricant film with a "dry" self-assembled monolayer lubricant. They concluded that sliding in liquid films is the result of slippage along an interface. In other words, the degree of molecular cooperation determined the frictional resistance.

The concept of local versus cooperative yield to shear is briefly illustrated here with a frictional-load study of a molecularly entangled polymer melt obtained in a SFM study of *Buenviaje* et al. [45.14]. Each of the curves presented in Fig. 45.2a represents a polymer film of polyethylene co-propylene of distinctly different degree of entanglement. Films of thickness above 230 nm exhibit the strongest entanglement strength. Films of 20-nm thickness or thinner are fully disentangled. The reason for the film-thickness-dependent entanglement strength is given by the substrate-distance-dependent shear strength during the spin-coating process of the thin films. For entangled films SFM friction studies exhibit a critical applied load (identified by P_t, and the thickness t) that separates two friction regimes: one

identified by a high friction coefficient and the other by a low friction coefficient. At loads below P_t the friction coefficients are high, indicating plastic yielding during sliding. In these plastic regimes of sliding, molecular cooperation is low, leading to high local shear stresses compensated by local yielding of the material. Above the critical load, the friction coefficient drops, independent of the film thickness, to a low value of 3.0, corresponding to the value obtained from the fully disentangled film. Note that the polymer molecules in the 20-nm-thick film experience high substrate tangential stresses during the spin-coating process. Hence the disentangled polymer molecules can be considered to be aligned preferentially along the substrate surface,

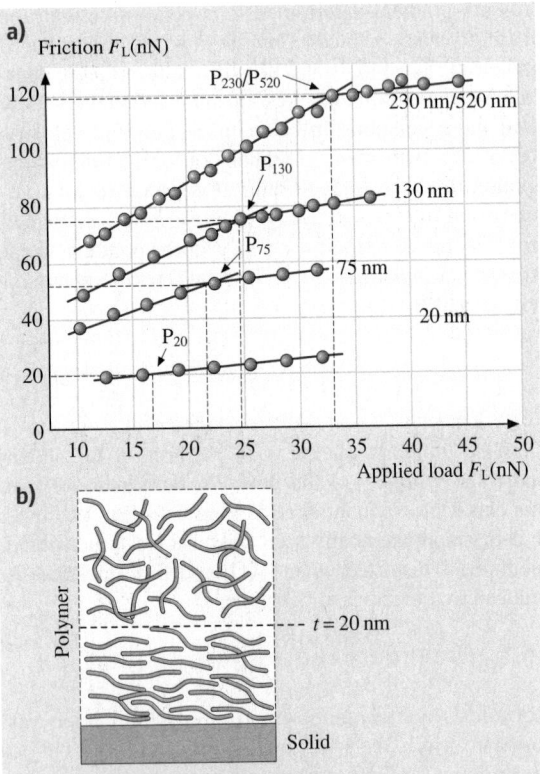

Fig. 45.2 (a) SFM friction measurements at a speed of 1 μm/s: cooperative molecular response of polyethylene co-propylene to frictional shear forces as a function of the applied load. P_t (t corresponds to the thickness of the polymer film) represents the critical activation load at which collective sliding is energetically more favorable than local plastic yielding. Adapted from [45.14]. (b) Sketch of the degree of disentanglement in the vicinity of the solid substrate surface

as sketched in Fig. 45.2b. This leads to a decrease of the structural entropy the closer the material is to the solid substrate surface. Considering the matching friction coefficient of 0.3 above P_t for thicker films, we can assume that any entangled film above a critical load exhibits a similar molecular collective response toward shear as the 20-nm film during spin-coating. The critical load and its related pressure represent a barrier that has to be overcome before a collective phenomenon is activated.

45.2 Thermal Activation Model of Lubricated Friction

With the discussion of shear in entangled polymer systems we have introduced structural entropy as one of the key players that affect frictional resistance in lubricants. We found that the structural entropy was affected by the load of the slider, which introduces an activation barrier in the form of a critical pressure. The terminology used here resembles that of the Eyring theory of molecular liquid transport [45.17].

Eyring discussed a pure liquid at rest in terms of a thermal activation model. The individual liquid molecules experience a *cage-like* barrier that hinders molecular free motion, because of the close-packing in liquids. To escape from the cage an activation barrier needs to be surmounted. In Eyring's model, two processes are considered in order to overcome the potential barrier: (i) shear stresses and (ii) thermal fluctuations. The potential barrier in the thermal activation model is depicted in Fig. 45.3 indicating the barrier modification by the applied pressure force P, and shear stress τ. *Briscoe* et al. [45.13] picked up on this idea to interpret the frictional behavior observed on molecularly smooth monolayer systems. Starting from the overall barrier height $E = Q + P\Omega - \tau\phi$ that is repeatedly overcome during a discontinuous sliding motion, using a Boltzmann distribution to determine the average time for single molecular barrier-hopping, and assuming a regular series of barriers and a high stress limit ($\tau\Phi/kT > 1$), the following shear strength versus velocity v relationship was derived [45.13]:

$$\tau = \frac{k_B T}{\phi} \ln\left(\frac{v}{v_0}\right) + \frac{1}{\phi}(Q + P\Omega) . \tag{45.2}$$

The barrier height, E, is composed of the process activation energy Q, the compression energy $P\Omega$, where P is the pressure acting on the volume of the junction Ω, and the shear energy $\tau\phi$, where τ is the shear strength acting on the stress activation volume ϕ. T represents the absolute temperature. The stress activation volume ϕ can be conceived as a process coherence volume and interpreted as the size of the moving segment in the unit shear process, whether it is a part of a molecule or a dislocation line. The most critical parameter in (45.2), v_0, is a characteristic velocity related to the frequency of the process and to a jump distance (discussed further below).

From (45.2) the following iso-relationships can be directly deduced [45.13]:

$$\tau = \tau_0 + \alpha P ; \quad \tau_0 = \frac{1}{\phi}\left[k_B T \ln\left(\frac{v}{v_0}\right) + Q\right] ;$$

$$\alpha = \frac{\Omega}{\phi} ; \text{ at constant } v, T , \tag{45.3a}$$

$$\tau = \tau_1 - \beta T ; \quad \tau_1 = \frac{1}{\phi}(Q + P\Omega) ;$$

$$\beta = -\frac{k}{\phi} \ln\left(\frac{v}{v_0}\right) , \quad \text{at constant } P, v , \tag{45.3b}$$

$$\tau = \tau_2 + \theta \ln v ; \quad \tau_2 = \frac{1}{\phi}(Q + P\Omega - kT \ln v_0) ;$$

$$\theta = \frac{kt}{\phi} , \quad \text{at constant } P, T . \tag{45.3c}$$

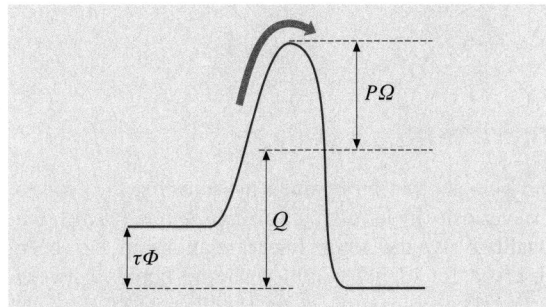

Fig. 45.3 Potential barrier in a lubricant based on Eyring's thermodynamic *cage model*. The normal pressure P and the shear stress τ are modifying the barrier height Q. Modified from [45.13]

Thus Eyring's model predicts a linear relationship of friction (the product of the shear strength and the active process area) in pressure and temperature and a logarithmic relationship in velocity.

Fig. 45.4 (a) Logarithmic $F_F(v)$ plots. $F_F(v) = F_0 + \alpha \ln(v[\mu m/s])$: ● "dry" contact (18% relative humidity) with $F_0 = 16.4$ nN and $\alpha = 0.91$ nN, △ OMCTS lubricated with $F_0 = 11.3$ nN and $\alpha = 3.4$ nN, and ■ n-hexadecane (n-C$_{16}$H$_{34}$) lubricated with $F_0 = 7.1$ nN and $\alpha = 2.5$ nN. The measurements were obtained with rectangular SFM cantilevers (0.4−0.8 N/m) at 100-nN load and 21 °C, both feedback-controlled. **(b)** Stress activation length, ϕ/A. (A area of contact) for OMCTS, n-hexadecane and dry contact. The *inset* provides a linear relationship between friction and temperature at a velocity of 1 μm/s and a normal load of 100 nN. Adapted from [45.18]

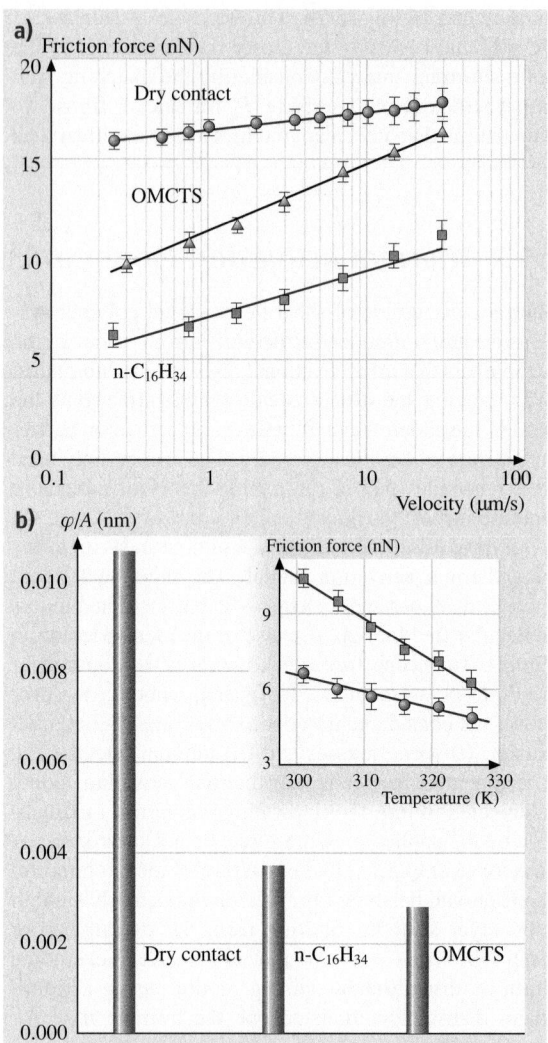

Eyring's model has been verified in lubrication experiments of solid (soap-like) lubricants by Briscoe and liquid lubricants by *He* et al. [45.18] within three logarithmic decades of velocities. While *Briscoe* et al. [45.13] employed an SFA that confines and pressurizes the film over several square microns, He et al. used a SFM system in which the contact is on the order of the lubricant molecular dimension.

He et al. determined the degree of interfacial structuring and its effect on lubrication of n-hexadecane and octamethylcyclotetrasiloxane (OMCTS). For spherically shaped OMCTS molecules, only an interfacial *monolayer* was found; in contrast, a 2-nm-thick entropically cooled layer was detected for n-hexadecane in the boundary regime to an ultra-smooth silicon wafer. SFM measurements of the two lubricants (with similar chemical affinity to silicon) identified the molecular shape of n-hexadecane as being responsible for augmented interfacial structuring. Consequently, interfacial liquid structuring was found to reduce lubricated friction, Fig. 45.4. Again as reasoned above, these results can be discussed in terms of a collective phenomenon, i.e., in terms of increased molecular coordination in n-hexadecane versus OMCTS.

45.3 Functional Behavior of Lubricated Friction

Friction-rate experiments are well suited to evaluate the rheological nature of interfacial liquids. In classical theories, such as the Reynolds's hydrodynamic theory discussed above, drag forces in lubricated sliding over thick liquid films were found to depend linearly on the rate of sliding and on the viscosity of the bulk fluid. In high-pressure lubrication, described by the elasto-hydrodynamic lubrication theory [45.19, 20], it was found that the linear relationship between friction

and velocity can be retained by adjusting the (apparent) viscosity by introducing an *apparent viscosity term*. Qualitatively, the same linear relationship has been observed for highly confined simple liquids between ultra-smooth mica surfaces such as alkanes [45.21]. Note that the lubricated contact area in SFA experiments is on the micron-scale. It significantly exceeds the size of the confined molecules. For small and unbranched molecules, such as simple alkanes, it is possible that the confined material undergoes a pressure-induced phase reconstruction, which leads to material properties that deviate significantly from the bulk. Larger and more complex (branched) molecules are less likely to exhibit pressure-induced phase reconstruction due to internal constraints and poor mixing within the contact area. This was shown by *Drummond* et al. [45.22] in SFA shear experiments. They found that the linear friction–velocity dependence does not apply for branched hydrocarbon lubricants. Also Drummond discussed "molecular lubrication" in terms of a logarithmic friction–velocity relationship, which is in accordance with the aforementioned thermal activation model, the solid lubricant SFA study by *Briscoe* et al. [45.13], and the liquid lubricant SFM study by *He* et al. [45.18].

Common to the three studies by *Briscoe*, *He*, and *Drummond* is that they operate on a single material phase that is disrupted or relaxed over a very specific lateral length scale. In the Eyring model, the length scale is deduced by assuming a regular series of barriers, separated by a *virtual jump distance*. The distance is embedded in v_0, the characteristic velocity, which is the product of the jump distance and the frequency of the process. *Briscoe* et al. [45.13] used the lattice constant of the highly oriented monolayers as the virtual jump distance. It was assumed that the process frequency was related to the vibrational frequency of the molecules ($10^{11}\,\mathrm{s}^{-1}$), neglecting sliding velocity, temperature, and pressure effects. *He* et al. [45.18] assumed a jump distance of $0.2\,\mathrm{nm}$ and considered frequencies between a perfectly structured alkane layer ($10^{11}\,\mathrm{Hz}$) and the bulk fluid (10^{13}–$10^{15}\,\mathrm{Hz}$, estimated from infrared absorption data for typical covalent bonds). With these assumptions *He* determined total *jump energies* of 4–$8 \times 10^{-20}\,\mathrm{J}$. *Briscoe* and *He* pointed out that a friction–velocity study alone provides only a qualitative measure of the microscopic origin of friction. Additional measurements have to be conducted that quantitatively address jump distances and frequencies.

The issue of the jump distance has been addressed by *Overney* et al. [45.23] in a SFM study on a highly ordered lubricant model system. This study avoided two levels of difficulties *Briscoe* et al. [45.13] and *He* et al. [45.18] encountered: (a) large contact areas of SFA studies, and (b) complex rheology with unknown structure parameters, as in liquid lubricant studies. It involved contact dimensions on the order of $1\,\mathrm{nm}^2$, and the crystalline form a bilayer model-lipid lubricant with in-plane lattice spacings of 0.6 and 1.1 nm. The study mainly focused on the effect of the depth of the corrugation potential (barrier height) on the static and

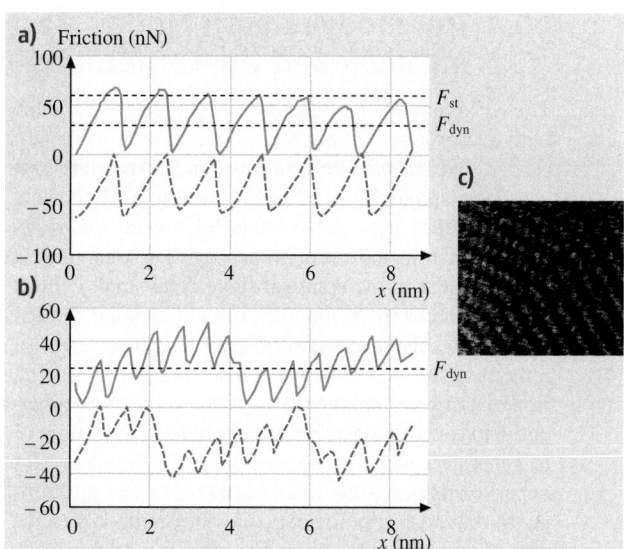

Fig. 45.5 SFM molecular stick–slip measurements of a bilayer lipid system (5-(4'-N,N-dihexadecylamino)benzylidene barbituric acid). (**a**) High-amplitude frictional stick–slip behavior is observed for scans perpendicular to molecular rows as imaged in (**c**). F_{st}, static friction, is assigned to the maximum force occurrence. The average value corresponds to the dynamic friction value, F_{dyn}, determined on large-scale micrometer scans. (**b**) A scan at 30° to the row direction leads to decreased frictional stick–slip behavior due to smaller molecular corrugations. (**c**) $12 \times 12\,\mathrm{nm}^2$ SFM lateral force image of a highly structured lipid bilayer. Two crystalline domains with a boundary are imaged. The anisotropic row-like structure is responsible for direction-dependent friction forces. The molecular corrugation between the rows is larger than the molecular corrugation in between a single row. (**a**) and (**b**) are adapted from [45.23] and (**c**) from [45.24].

dynamic friction force. This is illustrated in Fig. 45.5 in the form of the stick–slip amplitude plotted as a function of the drag direction (i.e., sliding with respect to the anisotropic row-like film structure). Relevant to our discussion about jump distance in lubrication events is Overney's discussion about the sliding speed and its effect on the slip distance. They demonstrated that within sliding speeds of 36–100 nm/s, the jump distance corresponded to the lattice spacing. At higher velocities, however, they could observe jumps over multiples of lattice distances and found the jump-length distribution to become increasingly stochastic. They proposed molecular (or atomistic) friction as a white-noise-driven system, which obeys a Gaussian fluctuation–dissipation relation. Hence, based on this finding one should consider discussing kinetic friction in terms of a statistical

45.4 Thermodynamical Models Based on Small and Nonconforming Contacts

The SFM approach simulates a single-asperity contact with a very high compliance, provided by a microfabricated and etched ultra-sharp tip and a typically soft cantilever spring. From a realistic, tribological perspective, the SFM approach is targeted toward the study of the intrinsic lubricant properties of a thin film in close vicinity to the solid substrate. The small contact area, on the order of the lubricant's molecular dimension, allows the discussion of SFM results in terms of a thermodynamic equilibrium. The area is insufficient, in terms of reorganizing the lubricant molecules coherently, to cause an apparent material phase transition, or to generate a metastable situation, as observed in SFA experiments (see below). SFM is therefore not appropriate to reflect on tribological issues involving large-area confinement effects.

In our prior molecular discussion of friction above, we introduced a thermal activation model for solid and liquid lubrication, the *Eyring model*, which employed a regular series of potential barriers. Note that the concept applies for a solid lubricant of an inherently highly ordered structure (e.g., [45.13]), but also for a liquid system in which the series of potentials is built up and overcome in the course of the shear process (e.g., [45.18]).

Gnecco et al. [45.25] showed in a ultra-high-vacuum study on sodium chloride that the concept of the Eyring model also applies for dry SFM friction studies. Thus a molecular theory of lubricated friction involving a molecular contact could be derived from a very simplistic model of an apparent sinusoidally corrugated surface over which a cantilever tip is pulled. In a first attempt one could assume that the corresponding wavelength of the shear process corresponds to the apparent lattice spacing of the corrugated surface. With such a simple attempt it is, however, assumed that no noise, such as thermal noise, exists in the system, and thus the driven tip leaves the total potential well when the barrier vanishes at the instability point. In the presence of noise, the transition to sliding can be expected to occur before the top of the barrier is reached. Such barrier-hopping fluctuations have been discussed theoretically by *Sang* et al. [45.26] and *Dudko* et al. [45.27].

The relationship between thermal fluctuations and velocity must be handled thoughtfully. *Sang* et al. [45.26] pointed out that, in previous considerations of thermal fluctuations by *Heslot* et al. [45.28], the fluctuations were proportionally related to the veloc-

ity, which led to a friction force that is logarithmically dependent on the velocity. In Heslot's *linear creep model*, the barrier height is proportional to the frictional force. *Sang* argued that, if one considered an absorbing boundary condition (i. e., an elastic deformation of the overall potential which is accomplished by shifting the *x*-axis), the barrier height becomes proportional to a 3/2-power law in the friction force. *Sang*'s extended linear creep model resembles a *ramped creep model* and leads analytically to a logarithmic distorted dynamic friction–velocity relationship; i. e.,

$$F = F_c - \Delta F \left| \ln v^* \right|^{2/3} . \tag{45.4}$$

In (45.4) v^* represents a dimensionless velocity, $\Delta F \propto T^{2/3}$, and F_c is an experimentally determined constant (found by plotting F versus $T^{2/3}$ for a fixed ratio $T/v = 1 \, \mathrm{K/(nm/sec)}$) that contains the critical position of the cantilever support. The same relationship between friction and velocity was also derived for the maximum spring force by *Dudko* et al. [45.27]. ΔF and v^* in (45.4) were derived as follows by *Sang*:

$$v^* = 2 \left(\frac{v \beta \omega_0^2 U_0}{k_B T \lambda} \right) \frac{\Omega_k^2}{\left(1 - \Omega_k^4 \right)^{1/2}} ;$$

$$\Omega_k = \frac{\omega_0}{2\pi \omega_k} ;$$

$$\omega_0 = \sqrt{\frac{M \lambda^2}{U_0}} ; \quad \omega_k = \sqrt{\frac{M}{k}} , \tag{45.5a}$$

$$\Delta F = \frac{\pi U_0}{\lambda} \left(\frac{3}{2} \frac{k_B T}{U_0} \right)^{\frac{2}{3}} \left(\frac{\left(1 - \Omega_k^4 \right)^{\frac{1}{6}}}{1 + \Omega_k^2} \right) . \tag{45.5b}$$

In (45.5a,b) v is the velocity of the cantilever stage, β is the microscopic friction coefficient or dissipation (damping) factor, ω_0 is the frequency of the small oscillations of the tip in the minima of the periodic potential, λ is the lattice constant, U_0 is the surface barrier potential height, M and k represent the mass and the spring constant of the cantilever, respectively, and $2\pi\Omega_k$ represents the ratio of ω_0 to the intrinsic cantilever resonance frequency ω_k.

Sang and *Dudko*'s model was experimentally confirmed by *Sills* and *Overney* [45.29] on an unstructured amorphous surface of atactic polystyrene, Fig. 45.6. *Dudko* determined that the weak spring constants typ-

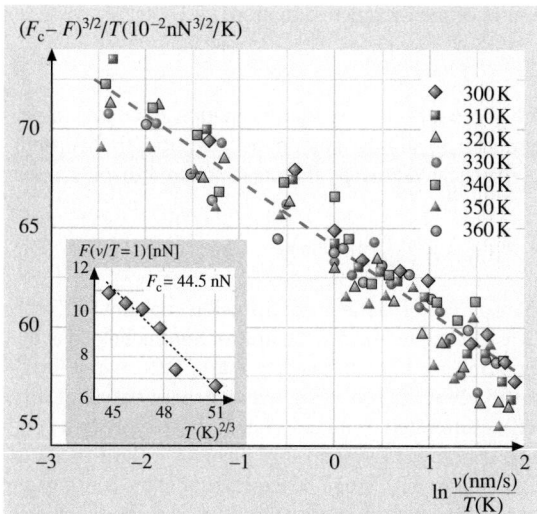

Fig. 45.6 Collapse of SFM friction data obtained on atactic polystyrene using the ramped creep scaling model. The regression parameters from the linear fit (*dashed line*) are -2.158×10^{-2} N$^{3/2}$ s/m and 40.186×10^{-2} nN$^{-3/2}$/K ($R^2 = 0.9124$). Lower *inset*: the constant F_c is determined from the intercept of the friction force F versus $T^{2/3}$ for a fixed ratio $T/v = 1$ K/(nm/s): $F_c = 44.5$ nN. (Adapted from *Sills* and *Overney* [45.29])

ically used in SFM measurements are responsible for the more pronounced logarithmic behavior of friction in velocity as found by *Gnecco* et al. [45.25] and *He* et al. [45.18]. The ramped creep model is also supported by numerical solutions of the Langevin equation.

The Langevin equation combines the equation of motion (including the sinusoidal potential and perfect cantilever oscillator in the total potential energy E) with the thermal noise in the form of the random force, $\xi(t)$, i.e.,

$$M\ddot{x} + M\beta\dot{x} + \frac{\partial E(x, t)}{\partial x} = \xi(t) , \qquad (45.6)$$

where $E(x, t) = \frac{k}{2}(R(t) - x)^2 - U_0\left(\frac{2\pi x}{\lambda}\right) .$

Equation (45.6) was solved numerically by both *Sang* et al. [45.26] and *Dudko* et al. [45.27] independently, assuming a Gaussian fluctuation–dissipation relation, $\langle \xi(t)\xi(t') \rangle = 2M\beta k_b T\delta(t - t')$ to express the random force. *Sang* confirmed the *ramped creep model*, and *Dudko* showed that a force reconstruction approach from the density of states (accumulated from the corresponding Fokker–Planck equation) is equivalent to the Langevin equation. From the dynamic spectral analysis, it could be concluded that the locked states (states within the potential wells) contribute mostly to the potential component of the friction force that dominates at low driving velocities, and sliding states contribute to viscous friction dominating at high driving velocities [45.27].

It should be noted that all of the above results considered an overdamped SFM system with respect to the driven spring (i.e., $\beta^2 > 4k$M), and an underdamped system with respect to the periodic potential (i.e., $\beta^2 < 4(M\omega)^2$). This aspect will be further addressed below in our discussion of metastable lubricant systems in large conforming contacts.

45.5 Limitation of the Gaussian Statistics – The Fractal Space

The spectral description of dynamic processes involving probability density functions has recently been the focus of numerous theoretical papers that treat various statistical kernels [45.30–33]. *Dudko*'s et al. [45.27] Fokker–Planck discussion of kinetic friction and *Luedtke*'s et al. [45.30] Lévy flight model of slip diffusion of adsorbed nanoclusters are two examples in which statistical methods are applied to describe diffusive properties relevant to the kinetics in tribology.

Currently most models used to describe tribological processes assume Gaussian statistics (e.g., [45.26], and [45.27]). One of the limitations of Gaussian statistics is that there are no correlations between statistical incidences. In other words, the Gaussian dynamic system is

without memory. This is important to remember as the Eyring model discussed above, with its equally spaced potential barriers, used Gaussian statistics. Simple *inert* lubricants, such as short-chain alkanes embedded between silicon wavers, are described satisfactorily with such statistics; however, confined complex liquids are not, including branched molecules, polymers, and generally chemically interactive and entropically confined systems (e.g., perfluoropolyether lubricants as discussed below).

Confined complex liquids, for instance, easily exhibit strongly interacting glass-like behavior. The dynamic and stress-relaxation behaviors in glasses, frequently discussed only as a weakly interacting system with Arrhenius laws (Gaussian statistics), are often distorted from processes described by independently occurring microscopic processes. For instance, deviations from the Debye exponential relaxation as introduced in (45.1) are expressed in the form of an *extended exponential* Kohlrausch relaxation function over time t; i.e.,

$$F(t) \equiv \left[\frac{X(t) - X(\infty)}{X(0) - X(\infty)} \right] = \mathrm{e}^{-\left(\frac{t}{\tau}\right)^b}; \quad 0 < b < 1 \,,$$
(45.7)

where X is the property that is relaxed. The exponent b, the Kohlrausch exponent, can theoretically be determined if one assumes that the process occurs in series, representing a well-determined microscopic origin that correlates the various degrees of freedom [45.34]. This approach is borrowed from magnetic spin models such as the *Ising spin model* [45.35]. The idea is that a given molecular motion is dependent on the availability of other degrees of freedom of mobile neighboring structural units. Finite relaxation times, $t_{\max}$, are gradually obtained with increasing spin levels (the ergodic limit).

As mentioned above, the models by *Dudko* et al. [45.27] and *Sang* et al. [45.26] assumed Gaussian statistics. To illustrate how a diffusion process can deviate from Gaussian statistics, we introduce a simplified version of the Langevin equation, i.e.,

$$\ddot{x} = -\eta \dot{x} + \varsigma(t)$$
(45.8)

with the coordinate x, the dissipation (or dampening) parameter η, and the random acceleration $\zeta(t)$. Assuming Gaussian statistics, the mean squared displacement is

$$\left\langle x^2(t) \right\rangle = 2k_{\mathrm{B}} T \eta t = 2Dt \,,$$
(45.9)

where $D = \eta k T$ defines the diffusion constant [45.36]). It was already realized at the time of Smoluchovski at the beginning of the 20th century that a diffusive description of a dynamic process demands a thermodynamically well-equilibrated or mixed system. Especially in a confined tribological system that involves a third medium (e.g., a lubricant), it can be expected that the Markovian nature of the underlying stochastic process could be disturbed. Consequently, for a monolayer lubricant that is chemically interacting, a nonlinear relationship of the mean squared displacement in time can be expected. Manifestations of anomalous transport are long-range spatial or temporal correlations. Two extreme limits can be distinguished: (a) processes with strong temporal relations (*fractal time*) [45.32], and (b) systems that exhibit long jumps (*Lévy flights*) [45.33].

45.6 Fractal Mobility in Reactive Lubrication

The importance of the underlying kinetics is illustrated by ultra-thin wetting lubricants. The spreading of *completely wetting* polymer liquids on solid surfaces has revealed unexpected spatial and temporal features when examined at the molecular level. The spreading profile is typically characterized by the appearance of a precursor film of monomolecular thickness extending over macroscopic distances and, in many cases, a terracing (also on the order of molecular dimensions) of the fluid remaining in the reservoir [45.37]. These spatial features have been shown to be consistent with a Poiseuille-like flow in which the disjoining pressure gradients with film thickness drive the spreading process [45.38]. The temporal evolution of the spreading profile in this film thickness regime is, however, found to scale universally as $t^{1/2}$ even at short times [45.37]. That the spreading dynamics are reflective of a diffusive transport mechanism and not of a pressure driven *liquid* flow suggests that interfacial confinement substantially alters the mobility of molecularly thin polymer fluids [45.39].

The molecular mobility is of fundamental importance for monolayer lubrication purposes, such as in magnetic storage devices. It has, for instance, been shown that, for low-surface-energy hydroxyl-terminated perfluoropolyether (PFPE-OH) films, the lubricant exhibits spatially terraced flow profiles indicative of film layering [45.38] and spreading dynamics that are diffusive in nature [45.40, 41]. In magnetic storage devices the hydroxylated chain ends of molecularly thin PFPE-OH films interact with the solid surface, an amorphous carbon surface, via the formation of hydrogen bonds with the polar, carbon–oxygen functionalities located on the carbon surface. The bonding of the PFPE-OH polymer to carbon is predicated on the ability of the PFPE backbone to deliver the hydroxyl end-group spatially to within a sufficiently close distance to the surface-active sites. Kinetic measurements probing the bonding of the PFPE-OH polymer to the carbon reveal two distinctive kinetic behaviors, as illustrated in Fig. 45.7 at two representative temperatures: 50 °C and 90 °C for the

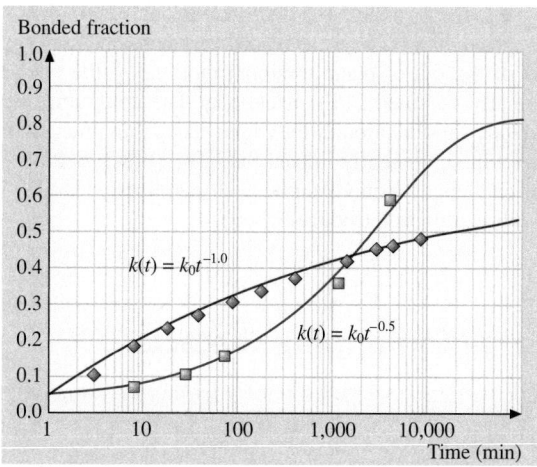

Fig. 45.7 Representative kinetic data for the bonding of PFPE-OH (tradename: Fomblin ZDOL) to amorphous carbon at $T = 50\,°C$ and $T = 90\,°C$. *Solid lines* represent fits using a rate coefficient of the form, $k(t) = k_0 t^{-\alpha}$ with $\alpha = 0.5$ for $T = 50\,°C$ and $\alpha = 1.0$ for $T = 90\,°C$

two temperature regimes below 56 °C and above 85 °C. Below 56 °C the kinetics are described with a time-dependent (fractal-time-dependent) rate coefficient of the form

$$k(t) = k_0 t^{-1/2} \tag{45.10}$$

and at temperatures above 85 °C with the form

$$k(t) = k_0 t^{-1.0} . \tag{45.11}$$

The initial bonding rate constants, k_0, increased abruptly as the temperature rose above 50 °C.

The bonding kinetics in the low-temperature regime is characteristic of a diffusion-limited reaction occurring from a glass-like state of the molecularly thin PFPE-OH film [45.42]. The mobility of the PFPE chain in the glass-like state is limited by the propagation of holes or packets of free volume, which facilitate configurational rearrangements of the chain. The onset of changes in the bonding kinetics at nominally $T > 56°$ signifies a fundamental change in the mobility of the molecularly thin PFPE-OH film. Specifically, the transition in the fractal time dependence suggests that delivery of the hydroxyl moiety to the surface is no longer limited by hole diffusion, and the increase in the initial rate constant indicates an enhancement in the backbone flexibility. These results are consistent with a transition in the film from a glass-like to a liquid-like state in which the enhanced PFPE-OH segmental mobility results from rotations about the ether oxygen linkages in

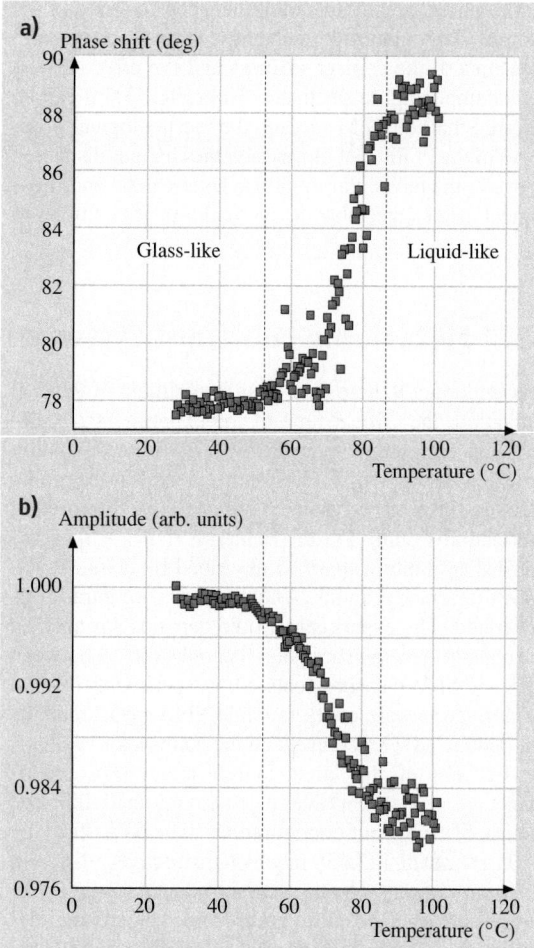

Fig. 45.8a,b Shear-modulated SFM experiments performed on a $10.7 \pm 0.5\,\text{Å}$ Fomblin Zdol film: (**a**) phase-shift response between disturbance and response, and (**b**) contact stiffness response vs. temperature

the chain, which become increasingly facile. The time dependence observed in the high-temperature rate coefficients, $k(t) = k_0 t^{-1.0}$ is characteristic of a process occurring from a confined liquid-like state in which the activation energy increases as the extent of the reaction increases.

The impact of this transition in the molecular mobility on tribology can be illustrated with sinusoidally modulated shear force experiments [45.43, 44]. In brief, a molecularly thin ($10.7 \pm 0.5\,\text{Å}$) PFPE-OH film is subjected to a local shear stress by means of a sinusoidal force applied *laterally* to a SFM probe (at constant load) where the modulation amplitude is initially set below that required to initiate sliding between the tip and the sample. The amplitude and phase-shift responses are measures of the contact stiffness and the effective viscous dampening, respectively. From Fig. 45.8 it can be inferred that the SFM-measured nanorheological properties of the PFPE-OH film exhibit the changes discussed above in the molecular mobility. Thus kinetic and rheological data suggest that the thermal transition observed

is due to the formation of a two-dimensional (2-D) glass. The *glass transition* results from the preferential *freezing out* of the out-of-plane torsional motions of the energetically confined PFPE backbone. The confinement-induced solidification in the molecularly thin precursor film will significantly impact the lubrication properties and challenge thermodynamic, well-equilibrated models of lubrication as introduced above.

45.7 Metastable Lubricant Systems in Large Conforming Contacts

It is important to note that, in an experiment of thermal activation, the critical time of the experiment t_{exp} determines the system response with its finite relaxation time t_{max}. If $t_{exp} > t_{max}$, the system behaves in an ergodic manner, and thermodynamic laws apply for interpreting lubrication results. On the contrary, if $t_{exp} < t_{max}$ the thermal evolution cannot be described by classical statistical thermodynamics. A metastable configuration is generated. The experimental time depends strongly on the contact area, the parameter that most differs between SFA and SFM measurements. SFA experiments involve large micron-scale contacts while SFM measurements are conducted with contacts on the nanoscale.

Because of the large contact area, SFA experiments are very susceptible to generating unequilibrated metastable lubricant configurations. The SFA study by *Yoshizawa* et al. [45.45], in which distinctively different *dynamic states of friction* were introduced, could be interpreted as such. To date there are three velocity regimes used to describe the dynamic state of friction for a system that is "underdamped" [45.46]. In an underdamped system, realized by a stiff spring compared to the friction constant, the characteristic slip time is comparable or smaller than the response time of the mechanical system. One distinguishes three velocity regimes, as depicted in Fig. 45.9. The three regimes distinguish themselves by a single discriminator, v_c, a material, pressure and temperature dependent critical velocity. The regimes are described as

1. highly regular with high-amplitude stick–slip for $v \ll v_c$,
2. intermittent stochastic stick–slip for $v < v_c$, and
3. smooth low-friction sliding for $v > v_c$.

Various models have been suggested to describe the different dynamic states, including a melting–freezing transition [45.47], chain adsorption on the substrate [45.48], and embedded particles [45.46]. Until

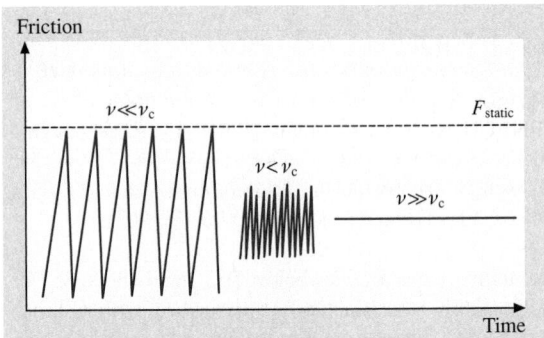

Fig. 45.9 Dynamic states of friction for an underdamped spring system

the embedded-particle model by Rozman was introduced, the SFA approach seemed to be the tool of choice to investigate metastable lubricant configurations. Rozman's single-particle model alerts us, however, to drawing unambiguous conclusions on the dynamical structure of a molecular system embedded between two plates and driven by an underdamped system. In Rozman's simple theoretical model, sketched in Fig. 45.10, a single particle is embedded between two corrugated surfaces (*two-wave potential*). The top plate is pulled at constant velocity by a linear spring, and the plate motion is monitored. Interestingly, the plate exhibits exactly the same dynamic state of frictional motion as introduced above with the three regimes, which were at-

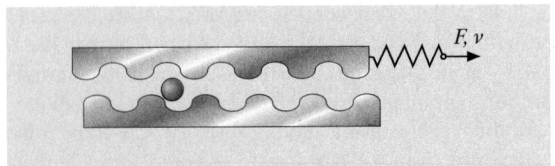

Fig. 45.10 Model of a single particle embedded in a two-wave potential driven at constant force, F, and velocity v

tributed to a rate-dependent configurational change of the lubricant.

Finally it shall be noted that the slip relaxation pattern depends under which conditions the stick–slip motion is being studied. In an *overdamped* system, i. e.,

a system in which the spring constant is weak compared to the friction constant, *Rozman* et al. [45.49] found that one can control the experimentally observed relaxation pattern of the slip by controlling the spring constant.

45.8 Conclusion

Starting from a classical tribological master curve, the Stribeck curve, with its complex description of lubrication for thin lubricants, we launched a discussion of molecular lubrication, pulling together disparate approaches to studying friction. We found that phenomenological descriptions of lubrication, such as the Reynolds's theory, were kept alive for ultra-thin lubricants by *adjusting* material properties, such as the viscosity. This is a common engineering approach to introduce *effective* or *apparent* properties, if tested models fail to describe new situations.

In the case of molecular lubrication, significant progress has been made in the last ten years. Two instrumental techniques have in particular contributed to this progress: the SFA and the SFM. The contributions of these two techniques have been complementary. While the SFA has tested lubricants under pressure constraints

with large contacts with respect to the size of the trapped molecules, the SFM has probed the degree of collective mobility with disturbances in the size of the molecules themselves.

One feature common to the interpretation of lubrication results, and which was discussed here in detail, is the problem of finding the appropriate underlying statistics to describe the lubrication process. Most current molecular models that have been used to describe molecular friction and lubrication assumed Gaussian statistics. Only recently has it been recognized as important to consider statistics that embrace long-range spatial or temporal correlations. In the near future, it can be expected that the next leap in an improved fundamental understanding of kinetics and energetics in nanolubrication will come from interpretations that are challenging the Markovian nature of the underlying stochastic process.

Part E | 45

References

45.1   F. P. Bowden, D. Tabor: *The Friction and Lubrication of Solids* (Clarendon, Oxford 1951)

45.2   D. Tabor, R. H. S. Winterton: The direct measurement of normal and retarded van der Waals forces, Proc. R. Soc. Lond. A **312**, 435–450 (1969)

45.3   G. Binnig, C. F. Quate, C. Gerber: Atomic force microscope, Phys. Rev. Lett. **56**, 930–933 (1986)

45.4   N. P. Petrov: *Friction in Machines and the Effect of the Lubricant*, Vol. 1 (Inzh. Zh. St. Petersburgo, St. Petersburg 1883) pp. 71–140

45.5   N. P. Petrov: *Friction in Machines and the Effect of the Lubricant*, Vol. 2 (Inzh. Zh. St. Petersburgo, St. Petersburg 1883) pp. 227–279

45.6   N. P. Petrov: *Friction in Machines and the Effect of the Lubricant*, Vol. 3 (Inzh. Zh. St. Petersburgo, St. Petersburg 1883) pp. 377–463

45.7   N. P. Petrov: *Friction in Machines and the Effect of the Lubricant*, Vol. 4 (Inzh. Zh. St. Petersburgo, St. Petersburg 1883) pp. 535–564

45.8   B. Tower: First report on friction experiments (friction of lubricated bearings), Proc. Inst. Mech. Eng., 632–659 (November 1883)

45.9   O. Reynolds: On the theory of lubrication and its application to Mr. Beauchamp Tower's experiments, including an experimental determination of the viscosity of olive oil, Philos. Trans. R. Soc. Lond. **177**, 157–234 (1886)

45.10   J. P. Israelachvili, G. M. McGuiggan, M. Gee, A. Homola, M. Robbins, P. Thompson: Liquid dynamics in molecularly thin films, J. Phys. **2**, 89–98 (1990)

45.11   W. B. Hardy, I. Doubleday: Boundary lubrication—The paraffin series, Proc. R. Soc. Lond. A **100**, 550–574 (1922)

45.12   A. Dorinson, K. C. Ludema: *Mechanics and Chemistry in Lubrication* (Elsevier, Amsterdam 1985)

45.13   B. J. Briscoe, D. C. B. Evans: The shear properties of Langmuir–Blodgett layers, Proc. R. Soc. Lond. A **380**, 389–407 (1982)

45.14 C. Buenviaje, S. Ge, M. Rafailovich, J. Sokolov, J. M. Drake, R. M. Overney: Confined flow in polymer films at interfaces, Langmuir **19**, 6446–6450 (1999)

45.15 S. Blunier, H. Zogg, A. N. Tiwari, R. M. Overney, H. Haefke, P. Buffat, G. Kostorz: Lattice and thermal misfit dislocations in epitaxial $CaF_2/Si(111)$ and $BaF_2/CaF_2/Si(111)$ structures, Phys. Rev. Lett. **68**, 3599–3602 (1992)

45.16 G. Reiter, A. L. Demirel, J. Peanasky, L. L. Cai, S. Granick: Stick to slip transition and adhesion of lubricated surfaces in moving contact, J. Chem. Phys. **101**, 2606–2615 (1994)

45.17 S. Glasstone, K. J. Laidler, H. Eyring: *Theory of Rate Processes* (McGraw-Hill, New York 1941)

45.18 M. He, A. Szuchmacher Blum, G. Overney, R. M. Overney: Effect of interfacial liquid structuring on the coherence length in nanolubrication, Phys. Rev. Lett. **88**(15), 154302/1–4 (2002)

45.19 K. L. Johnson: *Contact Mechanics* (Cambridge Univ. Press, Cambridge 1987)

45.20 E. Meyer, R. M. Overney, K. Dransfeld, T. Gyalog: *Nanoscience: Friction and Rheology on the Nanometer Scale* (World Scientific, Singapore 1998)

45.21 H. K. Christenson, D. W. R. Gruen, R. G. Horn, J. N. Israelachvili: Structuring in liquid alkanes between solid-surfaces—Force measurements and mean-field theory, J. Chem. Phys. **87**(3), 1834–1841 (1987)

45.22 C. Drummond, J. Israelachvili: Dynamic behavior of confined branched hydrocarbon lubricant fluids under shear, Macromolecules **33**(13), 4910–4920 (2000)

45.23 R. M. Overney, H. Takano, M. Fujihira, W. Paulus, H. Ringsdorf: Anisotropy in friction and molecular stick–slip motion, Phys. Rev. Lett. **72**, 3546–3549 (1994)

45.24 R. M. Overney, H. Takano, M. Fujihira: Elastic compliances measured by atomic force microscopy, Europhys. Lett. **26**(6), 443–447 (1994)

45.25 E. Gnecco, R. Bennewitz, T. Gyalog, C. Loppacher, M. Bammerlin, E. Meyer, H.-J. Güntherodt: Velocity dependence of atomic friction, Phys. Rev. Lett. **84**(6), 1172–1175 (2000)

45.26 Y. Sang, M. Dube, M. Grant: Thermal effects on atomic friction, Phys. Rev. Lett. **87**(17), 174301/1–4 (2001)

45.27 O. K. Dudko, A. E. Filippov, J. Klafter, M. Urbakh: Dynamic force spectroscopy: A Fokker–Planck approach, Chem. Phys. Lett. **352**, 499–504 (2002)

45.28 F. Heslot, T. Baumberger, B. Perrin, B. Caroli, C. Caroli: Creep, stick–slip, and dry-friction dynamics: Experiments and a heuristic model, Phys. Rev. E **49**, 4973–4988 (1994)

45.29 S. Sills, R. M. Overney: Creeping friction dynamics and molecular dissipation mechanisms in glassy polymers, Phys. Rev. Lett. **91**, 095501(1–4) (2003)

45.30 W. D. Luedtke, U. Landman: Slip diffusion and Levy flights of an adsorbed gold nanocluster, Phys. Rev. Lett. **82**, 3835–3838 (1999)

45.31 I. M. Sokolov: Levy flights from a continuous-time process, Phys. Rev. E **63**, 011104/1–10 (2000)

45.32 R. Metzler, J. Klafter: The random walks guide to anomalous diffusion: A fractional dynamics approach, Phys. Rep. **339**, 1–77 (2000)

45.33 R. Metzler, J. Klafter: Levy meets Boltzmann: Strange initial conditions for Brownian and fractional Fokker–Planck equations, Physica A **302**, 290–296 (2001)

45.34 R. G. Palmer, D. L. Stein, E. Abrahams: Models of hierarchically constrained dynamics for glass relaxation, Phys. Rev. Lett. **53**, 958–961 (1984)

45.35 N. W. Ashcroft, N. D. Mermin: *Solid State Physics* (CBS Asia, Philadelphia 1976)

45.36 R. Becker: *Theorie der Wärme* (Springer, Berlin 1985)

45.37 F. Heslot, N. Fraysse, A. M. Cazabat: Molecular layering in the spreading of wetting liquid drops, Nature **338**, 640–642 (1989)

45.38 T. E. Karis, G. W. Tyndall: Calculation of spreading profiles for molecularly-thin films from surface energy gradients, J. Non-Newtonian Fluid Mech. **82**, 287–302 (1999)

45.39 S. F. Burlatsky, G. Oshanin, A. M. Cazabat, M. Moreau: Microscopic model of upward creep of an ultrathin wetting film, Phys. Rev. Lett. **76**, 86–89 (1996)

45.40 T. M. O'Connor, Y. R. Back, M. S. Jhon, B. G. Min, D. Y. Yoon, T. E. Karis: Surface diffusion of thin perfluoropolyalkylether films, J. Appl. Phys. **79**, 5788–5790 (1996)

45.41 X. Ma, J. Gui, L. Smoliar, K. Grannen, B. Marchon, C. L. Bauer, M. S. Jhon: Complex terraced spreading of perfluoropolyalkylether films on carbon surfaces, Phys. Rev. E **59**, 722–727 (1999)

45.42 A. Plonka, J. Bednarek, K. Pietrucha: Reaction dynamics in glass transition region: propagating radicals in ultraviolet-irradiated poly(methyl methacrylate), J. Chem. Phys. **104**, 5279–5283 (1996)

45.43 R. M. Overney, C. Buenviaje, R. Luginbuehl, F. Dinelli: Glass and structural transitions measured at polymer surfaces on the nanoscale, J. Therm. Anal. Calorimetry **59**, 205–225 (2000)

45.44 S. Ge, Y. Pu, W. Zhang, M. Rafailovich, J. Sokolov, C. Buenviaje, R. Buckmaster, R. M. Overney: Shear modulation force microscopy study of near surface glass transition temperature, Phys. Rev. Lett. **85**(11), 2340–2343 (2000)

45.45 H. Yoshizawa, P. McGuiggan, J. N. Israelachvili: Identification of a second dynamic state during stick–slip motion, Science **259**, 1305–1308 (1993)

45.46 M. G. Rozman, M. Urbakh, J. Klafter: Stick–slip motion and force fluctuations in a driven two-wave potential, Phys. Rev. Lett. **77**, 683–686 (1996)

45.47 J. M. Carlson, A. A. Batista: Constitutive relation for the friction between lubricated surfaces, Phys. Rev. E **53**, 4153–4164 (1996)

45.48 Y. Braiman, F. Family, H. G. E. Hentschel: Array-enhanced friction in the periodic stick–slip motion of nonlinear oscillators, Phys. Rev. E **53**, R3005–R3008 (1996)

45.49 M. G. Rozman, M. Urbakh, J. Klafter: Controlling chaotic frictional forces, Phys. Rev. E **57**, 7340–7343 (1998)

Part F Industrial

Industrial

Part F Industrial Applications

46. The "Millipede" – A Nanotechnology-Based AFM Data-Storage System

The "millipede" concept presented here is a new approach for storing data at high speed and ultra-high density. The interesting part is that millipede stores digital information in a completely different way from magnetic hard disks, optical disks, and transistor-based memory chips. The ultimate locality is provided by a tip, and high data rates are a result of massive parallel operation of such tips. As storage medium, polymer films are being considered, although the use of other media, in particular magnetic materials, has not been ruled out. The current effort is focused on demonstrating the millipede concept with areal densities of up to 0.5−1Tb/in^2 and parallel operation of very large 2-D (up to 64×64) AFM cantilever arrays with integrated tips and write/read/erase functionality. The fabrication and integration of such a large number of mechanical devices (cantilever beams) will lead to what we envision as the VLSI age of micro- and nanomechanics.

In this chapter, the millipede concept for a MEMS-based storage device is described in detail. In particular, various aspects pertaining to AFM thermomechanical read/write/erase functions, 2-D array fabrication and characteristics, $x/y/z$ microscanner design, polymer media properties, read channel modeling, servo control and synchronization, as well as modulation coding techniques suitable for probe-based data-storage devices are discussed.

In the twenty-first century, the nanometer will probably play a role similar to the one played by the micrometer in the twentieth century. The nanometer scale will presumably also pervade the field of data storage, although there is so far no obvious way in conventional magnetic, optical, or transistor-based storage to achieve the nanometer scale in all three dimensions. After decades of spectacular progress, those mature technologies have entered the homestretch; imposing physical limitations loom before them.

One promising method involves the use of patterned magnetic media for which the ideal write/read concept still needs to be demonstrated. The biggest challenge, however, is to pattern the magnetic disk in a cost-effective way. If such approaches are successful, the basis for large-capacity storage in the twenty-first century might still be magnetism.

Other proposals call for totally different media and techniques such as local probes, near-field optics, magnetic super-resolution or holographic methods. In general, when an existing technology is about to reach its limits in the course of its evolution and alternatives are emerging in parallel, two things usually happen: First, the existing and well-established technology is explored further and every effort is made to push its limits to the utmost in order to get the maximum return on the considerable investments made. Then, when all possibilities for improvement have been exhausted, the technology may still survive for certain niche applications, but the emerging technology will take over, opening new perspectives and new directions.

In many fields today we are witnessing the transition of structures from the micrometer to the nanometer scale, a dimension that nature has long been using to build the finest devices with a high degree of local functionality. Many of the techniques we use today are not suitable for the nanometer age; some will require minor or major modifications, others will be partially or entirely replaced. It is certainly difficult to predict which techniques will fall into which category. In key information technology areas, such as nano-electronics and data storage, it is not yet obvious which technologies and materials will be used in the future.

In any case, for an emerging technology to be seriously considered as a candidate to replace an existing technology that is approaching its inherent limits, it must provide a long-term perspective. For instance, the silicon microelectronics and storage industries are huge and have exacted correspondingly enormous investments, which makes them long-term oriented by nature. The consequence for storage is that any novel technology with higher areal storage density than today's magnetic recording [46.1, 2] should have long-term potential for further scaling, preferably down to the nanometer or even atomic scale.

The only available tool known today that is simple and yet provides this very long-term perspective is the nanometer-sharp tip. The simple tip is a very reliable tool that concentrates on one functionality: the ultimate local confinement of interaction. Techniques that use nanometer-sharp tips for imaging and investigating the structure of materials down to the atomic scale, such as the atomic force microscope (AFM) and the scanning tunneling microscope (STM) [46.3, 4], are suitable for the development of ultrahigh-density storage devices.

In the early 1990s, Mamin and Rugar at the IBM Almaden Research Center pioneered the capability of using an AFM tip for writing and readback of topographic features for data storage. In one of their schemes [46.5], writing and reading were demonstrated with a single AFM tip in contact with a rotating polycarbonate substrate. The writing was performed thermomechanically by heating the tip. In this way, storage densities of up to $30\,\text{Gb/in}^2$ were achieved, constituting a significant advance over the densities of that time. Later refinements included increasing readback speeds to attain a data rate of $10\,\text{Mb/s}$ [46.6] and the implementation of track servoing [46.7].

When using single tips in AFM or STM operation for storage, one has to deal with their fundamental data-rate limitations. At present, the mechanical resonance frequencies of the AFM cantilevers limit the data rates of a single cantilever to a few Mb/s for AFM data storage [46.8, 9]. The feedback speed and low tunneling currents limit STM-based storage approaches to even lower data rates.

Currently a single AFM operates at best on the microsecond time scale. Conventional magnetic storage, however, operates at best on the nanosecond time scale, making it clear that AFM data rates have to be improved by at least three orders of magnitude to be competitive with current and future magnetic recording. One solution for substantially increasing the data rates achievable by tip-based storage devices is to employ micro-electro-mechanical system (MEMS) arrays of cantilevers operating in parallel, with each cantilever performing write/read/erase operations in an individual storage field. It is our conviction that very large-scale integrated (VLSI) micro/nanomechanics will greatly complement future micro- and nanoelectronics (integrated or hybrid) and may generate hitherto inconceivable applications of VLSI-MEMS.

Various efforts are under way to develop MEMS-based storage devices. For example, a MEMS-actuated magnetic-probe-based storage device capable of storing $2\,\text{GB}$ of data on $2\,\text{cm}^2$ of die area and whose fabrication is compatible with a standard integrated circuit manufacturing process is described by *Carley* et al. [46.10]. In their device, a magnetic storage medium is positioned in the x/y plane, and writing is achieved magnetically by using an array of probe tips, each tip being actuated in the z-direction. Another concept is the atomic resolution storage described by *Gibson* et al. [46.11], who employ electron field emitters to change the state of a phase-change medium in a bit-wise fashion from polycrystalline to amorphous or vice versa. Reading is done with lower currents by detecting either back-scattered electrons or changes in the semiconductor properties in the media.

The "millipede" concept presented here is a new approach for storing data at high speed and ultrahigh density. The interesting part is that millipede stores digital information in a completely different way from magnetic hard disks, optical disks, and transistor-based memory chips. The ultimate locality is provided by a tip, and high data rates are a result of massive parallel operation of such tips. As storage medium, polymer films are being considered, although the use of other media, in particular magnetic materials, has not been ruled out. Our current effort is focused on demonstrating the millipede concept with areal densities of up to $0.5–1\,\mathrm{Tb/in^2}$ and parallel operation of very large 2-D (up to 64×64) AFM cantilever arrays with integrated tips and write/read/erase functionality. The fabrication and integration of such a large number of mechanical devices (cantilever beams) will lead to what we envision as the VLSI age of micro- and nanomechanics.

In this chapter, the millipede concept for a MEMS-based storage device is described in detail. In particular, various aspects pertaining to AFM thermomechanical read/write/erase functions, 2-D array fabrication and characteristics, $x/y/z$ microscanner design, polymer media properties, read channel modeling, servo control and synchronization, as well as modulation coding techniques suitable for probe-based data-storage devices are discussed.

46.1 The Millipede Concept

A 2-D AFM cantilever array storage technique [46.12–15], internally called the millipede, is illustrated in Fig. 46.1. Information is stored as sequences of indentations and no indentations written on nanometer-thick polymer films using the array of AFM cantilevers. The presence and absence of indentations will also be referred to as logical marks. Each cantilever performs write/read/erase operations over an individual storage field with an area on the order of $100 \times 100\,\mu\mathrm{m}^2$. Write/read operations depend on a mechanical parallel x/y scanning of either the entire cantilever array chip or the storage medium. The tip-medium contact is maintained and controlled globally, i.e., not on an individual cantilever basis, by using a feedback control in the z-direction for the entire chip, which greatly simplifies the system. This basic concept of the entire chip approach/leveling has been tested and demonstrated for the first time by parallel imaging with a 5×5 array chip [46.16, 17]. These parallel imaging results have shown that all 25 cantilever tips have approached the substrate within less than $1\,\mu\mathrm{m}$ of z-actuation, which indicates that overall chip tip-apex height control to within 500 nm is feasible. The stringent requirement for tip-apex uniformity over the entire chip is determined by the uniform force required to reduce tip and medium wear due to large force variations resulting from large tip-height nonuniformities [46.7]. Moreover, as the entire array is tracked without individual lateral cantilever positioning, thermal expansion of the array chip has to be small or well controlled. Thermal expansion considerations are a strong argument for a 2-D instead of a 1-D array arrangement.

Efficient parallel operations of large 2-D arrays can be achieved by a row/column time-multiplexed addressing scheme similar to that implemented in DRAMs. In the case of the millipede, the multiplexing scheme is used to address the array column by column with full parallel write/read operation within one column [46.18]. In particular, readback signal samples are obtained by applying an electrical read pulse to the cantilevers in a column of the array, low-pass filtering the cantilever response signals, and finally sampling the filter output signals. This process is repeated sequentially until all columns of the array have been addressed, and then restarted from the first column. The time between two pulses applied to the cantilevers of the same column corresponds to the time it takes for a cantilever to move from one logical mark position to the next. An alternative approach is to access all or a subset of the cantilevers simultaneously without resorting to the row/column multiplexing scheme. Clearly, the latter scheme yields higher data rates, whereas the former leads to lower implementation complexity of the channel electronics.

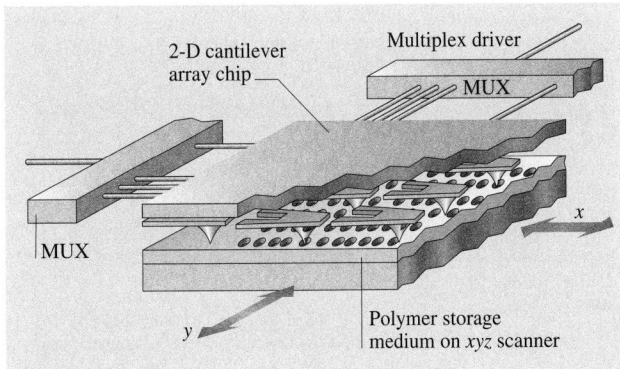

Fig. 46.1 The millipede concept (after [46.12])

46.2 Thermomechanical AFM Data Storage

In recent years, AFM thermomechanical recording in polymer storage media has undergone extensive modifications mainly with respect to the integration of sensors and heaters designed to enhance simplicity and to increase data rate and storage density. Thermomechanical writing in polycarbonate films and optical readback was first investigated and demonstrated by Mamin and Rugar (1992). Using heater cantilevers, thermomechanical recording at $30\,\mathrm{Gb/in^2}$ storage density and data rates of a few Mb/s for reading and $100\,\mathrm{kb/s}$ for writing have been demonstrated [46.5, 6, 19]. Although the storage density of $30\,\mathrm{Gb/in^2}$ obtained originally was not overwhelming, the results were encouraging enough to consider using polymer films to achieve density improvements. The current millipede storage approach is based on a new thermomechanical write/read process in nanometer-thick polymer films.

Thermomechanical writing is achieved by applying a local force through the cantilever/tip to the polymer layer and simultaneously softening the polymer layer by local heating. Initially, the heat transfer from the tip to the polymer through the small contact area is very poor, but it improves as the contact area increases. This means that the tip must be heated to a relatively high temperature of about 400 °C to initiate softening. Once softening has been initiated, the tip is pressed into the polymer, and hence the indentation size is increased. Rough estimates [46.20, 21] indicate that at the beginning of the writing process only about 0.2% of the heating power is used in the very small contact zone ($10–40\,\mathrm{nm^2}$) to soften the polymer locally, whereas about 80% is lost through the cantilever legs to the chip body and about 20% is radiated from the heater platform through the air gap to the medium/substrate. After softening has started and the contact area has increased, the heating power available for generating the indentations increases at least ten times to reach 2% or more of the total heating power.

With this highly nonlinear heat-transfer mechanism it is very difficult to achieve small tip penetration and hence small bit sizes, as well as to control and reproduce the thermomechanical writing process. This situation can be improved if the thermal conductivity of the substrate is increased and if the depth of tip penetration is limited. We have explored the use of very thin polymer layers deposited on Si substrates to improve these characteristics [46.22, 23], as illustrated in Fig. 46.2. The hard Si substrate prevents the tip from penetrating farther than the film thickness, and it enables more rapid transport of heat away from the heated region, as Si is a much better conductor of heat than the polymer. Using coated Si substrates with a 50 nm film of polymethylmethacrylate (PMMA), we have achieved indentation sizes of between 10 and 50 nm. However, increased tip wear has occurred, probably caused by contact between the Si tip and the Si substrate during writing. Therefore, a 70 nm layer of crosslinked photoresist (SU-8) was introduced between the Si substrate and the PMMA film to act as a softer penetration stop that avoids tip wear, but remains thermally stable.

Using this layered storage medium, indentations 40 nm in diameter have been written as shown in Fig. 46.3. These experiments were performed

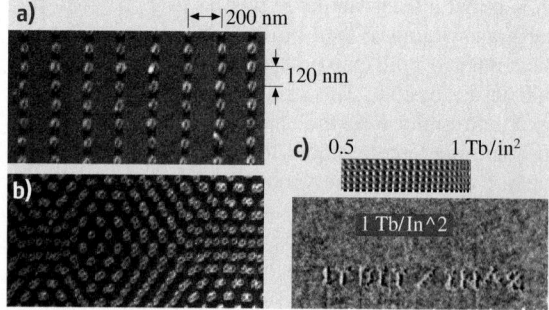

Fig. 46.3a–c Series of 40 nm indentations formed in a uniform array with (**a**) 120 nm pitch and (**b**) variable pitch (≥ 40 nm), resulting in areal densities of up to $400\,\mathrm{Gb/in^2}$. Images obtained with a thermal readback technique. (**c**) Ultra-high-density bit writing with areal densities approaching $1\,\mathrm{Tb/in^2}$. The scale is the same for all three images (from [46.15] © 2002 IEEE)

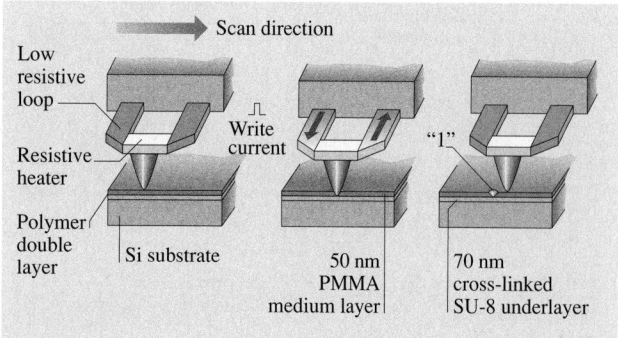

Fig. 46.2 New storage medium used for writing small bits. A thin, writable PMMA layer is deposited on a Si substrate separated by a cross-linked film of epoxy photoresist (after [46.15])

using a $1\,\mu m$ thick, $70\,\mu m$long, two-legged Si cantilever [46.19]. The cantilever legs are made highly conducting by high-dose ion implantation, whereas the heater region remains low-doped. Electrical pulses $2\,\mu s$ in duration were applied to the cantilever with a period of $50\,\mu s$. Figure 46.3a demonstrates that 40 nm bits can be written with 120 nm pitch or, as shown in Fig. 46.3b, very close to each other without merging, implying a potential areal density of $400\,Gb/in^2$. Figure 46.3c shows results from a single-lever experiment, where indentations are spaced as closely as 25 nm apart, resulting in areal densities of up to $1\,Tb/in^2$, although with a somewhat degraded write/read quality.

Imaging and reading are performed by a new thermomechanical sensing concept [46.24]. To read the written information, the heater cantilever originally used for writing is given the additional function of a thermal readback sensor by exploiting its temperature-dependent resistance. In general, the resistance increases nonlinearly with heating power/temperature from room temperature to a peak value at $500-700\,°C$. The peak temperature is determined by the doping concentration of the heater platform, which ranges from 1×10^{17} to $2 \times 10^{18}\,at/cm^3$. Above the peak temperature, the resistance drops as the number of intrinsic carriers increases due to thermal excitation [46.25]. For sensing, the resistor is operated at about $350\,°C$, a temperature that is not high enough to soften the polymer as in the case of writing. The principle of thermal sensing is based on the fact that the thermal conductance between the heater platform and the storage substrate changes according to the distance between them. The medium between the heater platform and the storage substrate, in our case air, transports heat from the cantilever to the substrate. When the distance between cantilever and storage substrate is reduced as the tip moves into an indentation, the heat transport through the air becomes more efficient. As a result, the evolution of the heater temperature differs in response to a pulse applied to the cantilever. In particular, the maximum value achieved by the heater temperature is higher in the absence of an indentation. As the value of the variable resistance depends on the temperature of the cantilever, the maximum value achieved by the resistance will be lower as the tip moves into an indentation. Therefore, during the read process, the cantilever resistance reaches different values depending on whether the tip moves into an indentation (logical bit "1") or over a region without an indentation (logical bit "0"). Figure 46.4 illustrates this concept.

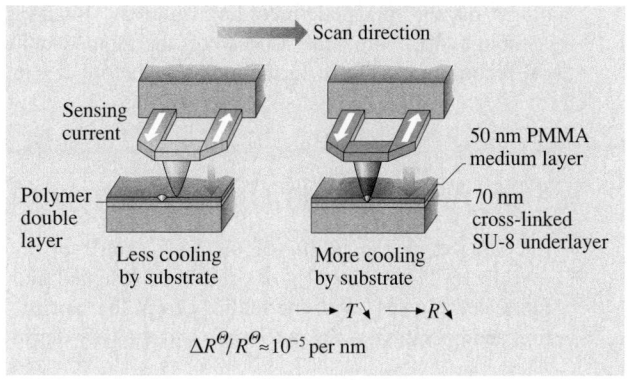

Fig. 46.4 Principle of AFM thermal sensing. The tip of the heater cantilever is continuously heated by a dc power supply while the cantilever is being scanned and the heater resistivity measured (after [46.15])

Under typical operating conditions, the sensitivity of thermomechanical sensing is even greater than that of piezoresistive-strain sensing, which is not surprising because thermal effects in semiconductors are stronger than strain effects. The good sensitivity is demonstrated by the images of the 40 nm sized indentations in Fig. 46.3, which were obtained using the described thermal-sensing technique.

The thermomechanical cantilever sensor, which transforms temperature into an electrical signal that carries information, is the electrical equivalent, to a first degree of approximation, of a variable resistance. A detection circuit must, therefore, sense a voltage that depends on the value of the cantilever resistance to decide whether a "1" or a "0" is written. The relative variation of thermal resistance is on the order of $10^{-5}/nm$. Hence a written "1" typically produces a relative change of the cantilever thermal resistance $\Delta R^{\Theta}/R^{\Theta}$ of about 10^{-4} to 5×10^{-4}. Note that the relative change of the cantilever electrical resistance is of the same order of magnitude. Thus, one of the most critical issues in detecting the presence or absence of an indentation is the high resolution required to extract the signal that contains the information about the logical bit being "1" or "0". The signal carrying the information can be regarded as a small signal superimposed on a very large offset signal. The large offset problem can be mitigated by subtracting a suitable reference signal [46.12, 26, 27].

More recently, single-probe experimental results have been obtained, in which large data sets were recorded at $641\,Gbit/in^2$ and read back with raw bit-error rates better than 10^{-4}, measured using the methodology of the magnetic-recording industry [46.28]. Although there are still aspects of thermomechanical recording that need further scrutiny before it can be employed in useful storage devices, it is the first scanning-probe-based method for which a thorough ultrahigh areal density demonstration has been performed.

46.3 Array Design, Technology, and Fabrication

Encouraged by the results of the 5×5 cantilever array [46.16, 17], a 32×32 array chip was designed and fabricated [46.18]. With the findings from the fabrication and operation of the 5×5 array and the very dense

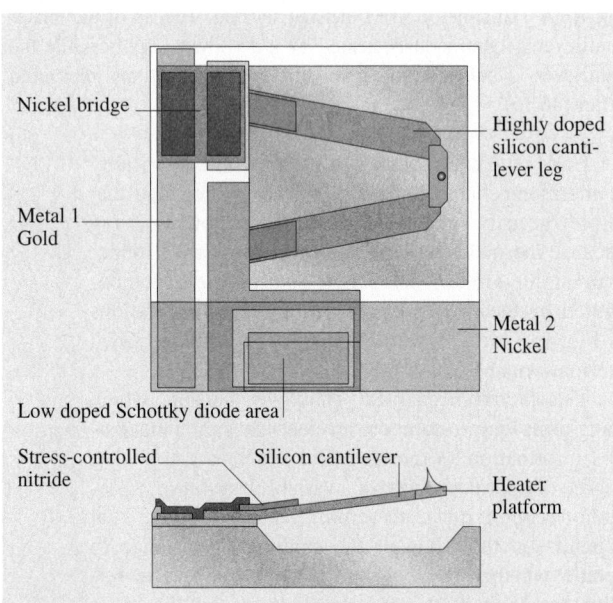

Fig. 46.5 Layout and cross section of one cantilever cell (after [46.29])

Fig. 46.6 Photograph of fabricated chip (14×7 mm^2). The 32×32 cantilever array is located at the center with bond pads distributed on either side (from [46.29] © 1999 IEEE)

thermomechanical writing/reading in thin polymers with single cantilevers, some important changes of the chip functionality and fabrication processes were made. The major differences are (1) surface micromachining to form cantilevers at the wafer surface, (2) all-silicon cantilevers, (3) thermal instead of piezoresistive sensing, and (4) first- and second-level wiring with an insulating layer for a multiplexed row/column-addressing scheme.

As the heater platform functions as a read/write element and no individual cantilever actuation is required, the basic array cantilever cell becomes a simple two-terminal device addressed by a multiplexed x/y wiring, as shown in Fig. 46.5. The cell area and x/y cantilever pitch are 92×92 μm^2, which results in a total array size of less than 3×3 mm^2 for the 1024 cantilevers. The cantilevers are fabricated entirely of silicon for good thermal and mechanical stability. They consist of a heater platform with the tip on top, legs acting as soft mechanical springs, and electrical connections to the heater. They are highly doped to minimize interconnect resistance and to replace the metal wiring on the cantilever in order to eliminate electromigration and parasitic z-actuation of the cantilever due to a bimorph effect. The resistive ratio between the heater and the silicon interconnect sections should be as high as possible; currently the resistance of the highly doped interconnections is $\approx 400\,\Omega$ and that of the heater platform is $5\,k\Omega$ (at 3 V reading bias).

The cantilever mass has to be minimized to obtain soft, high-resonance-frequency cantilevers. Soft cantilevers are required for a low loading force in order to eliminate or reduce tip and medium wear, whereas a high resonance frequency allows high-speed scanning. In addition, sufficiently wide cantilever legs are required for a small thermal time constant, which is partly determined by cooling via the cantilever legs [46.19]. These design considerations led to an array cantilever with 50 μm long, 10 μm wide and 0.5 μm thick legs, and a 5 μm wide, 10 μm long and 0.5 μm thick heater platform. Such a cantilever has a stiffness of $\approx 1\,N/m$ and a resonance frequency of $\approx 200\,kHz$. The heater time constant is a few microseconds, which should allow a multiplexing rate of up to 100 kHz.

The tip height should be as small as possible because the heater platform sensitivity depends strongly on the platform-to-medium distance. This contradicts the requirement of a large gap between the chip surface and the storage medium to ensure that only the tips, and not the chip surface, make contact with the medium. Instead of making the tips longer, we purposely bent the cantilevers a few micrometers out of the chip plane by depositing a stress-controlled plasma-enhanced chemical vapor deposition (PECVD) silicon-nitride layer at the base of the cantilever (see Fig. 46.5). This bending as well as the tip height must be well controlled in order to maintain an equal loading force for all cantilevers of an array.

Cantilevers are released from the crystalline Si substrate by surface micromachining using either plasma or wet chemical etching to form a cavity underneath the cantilever. Compared to a bulk-micromachined through-wafer cantilever-release process, as was done for our 5×5 array [46.16, 17], the surface micromachining technique allows an even higher array density and yields better mechanical chip stability and heat sinking. As mentioned above, the entire array is tracked without individual lateral cantilever positioning, therefore thermal expansion of the array chip has to be small or well controlled. For a $3 \times 3\,mm^2$ silicon array area and 10 nm tip-position accuracy, the temperature difference between array chip and medium substrate has to be controlled to about 1 °C. This is ensured by four temperature sensors in the corners of the array and heater elements on each side of the array.

The photograph in Fig. 46.6 shows a fabricated chip with the 32×32 array located in the center ($3 \times 3\,mm^2$) and the electrical wiring interconnecting the array with the bonding pads at the chip periphery. Figure 46.7

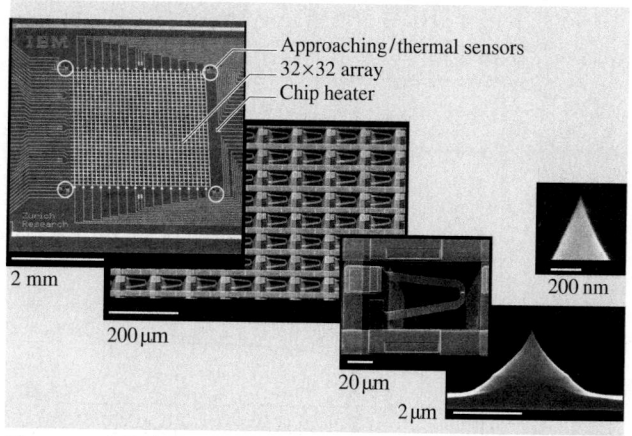

Fig. 46.7 SEM images of the cantilever array section with approaching and thermal sensors in the corners, array and single cantilever details, and tip apex (© 2000 International Business Machines Corporation; after [46.18])

shows the 32×32 array section of the chip with the independent approach/heat sensors in the four corners and the heaters on each side of the array, as well as zoomed scanning electron micrographs (SEMs) of an array section, a single cantilever, and a tip apex. The tip height is 1.7 μm and the apex radius is smaller than 20 nm, which is achieved by oxidation sharpening [46.30]. The cantilevers are connected to the column and row address lines using integrated Schottky diodes in series with the cantilevers. The diode is operated in reverse bias (high resistance) if the cantilever is not addressed, thereby greatly reducing cross talk between cantilevers. More details about the array fabrication are given in [46.29, 31].

46.4 Array Characterization

The array's independent cantilevers, which are located in the four corners of the array and used for approaching and leveling the chip and storage medium, serve to initially characterize the interconnected array cantilevers. Additional cantilever test structures are distributed over the wafer; they are equivalent to but independent of the array cantilevers. Figure 46.8 shows an I/V curve of such a cantilever; note the nonlinearity of the resistance. In the low-power part of the curve, the resistance increases as a function of heating power, whereas in the high-power regime, it decreases.

In the low-power, low-temperature regime, silicon mobility is affected by phonon scattering, which depends

on temperature, whereas at higher power, the intrinsic temperature of the semiconductor is reached, which results in a resistivity drop owing to the increasing number of carriers [46.25]. Depending on the heater-platform doping concentration of 1×10^{17} to 2×10^{18} at/cm^3, our calculations estimate a resistance maximum at a temperature of 500 to 700 °C, respectively.

The cantilevers within the array are electrically isolated from one another by integrated Schottky diodes. As every parasitic path in the array to the cantilever addressed contains a reverse-biased diode, the cross talk current is drastically reduced, as shown in Fig. 46.9. Thus, the current response of an addressed cantilever in

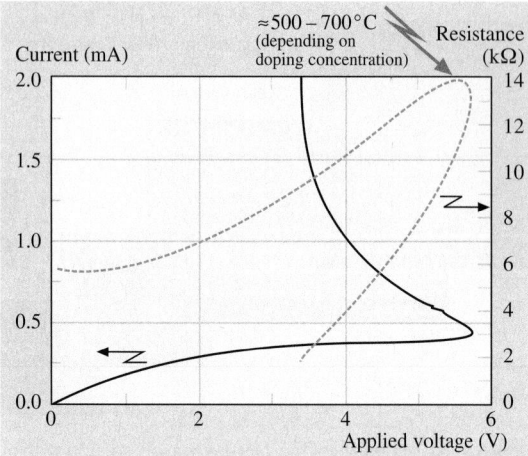

Fig. 46.8 *I/V* curve of one cantilever. The curve is nonlinear owing to the heating of the platform as the power and temperature are increased. For doping concentrations between 1×10^{17} and 2×10^{18} at/cm^3, the maximum temperature varies between 500 and 700 °C (after [46.31])

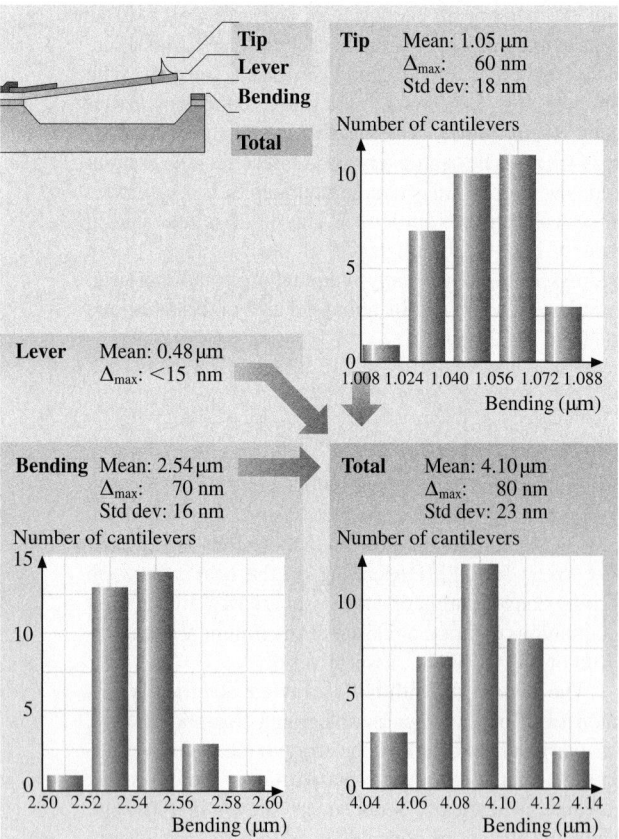

Fig. 46.10 Tip-apex height uniformity across one cantilever row of the array with individual contributions from the tip height and cantilever bending. (© 2000 International Business Machines Corporation; after [46.18])

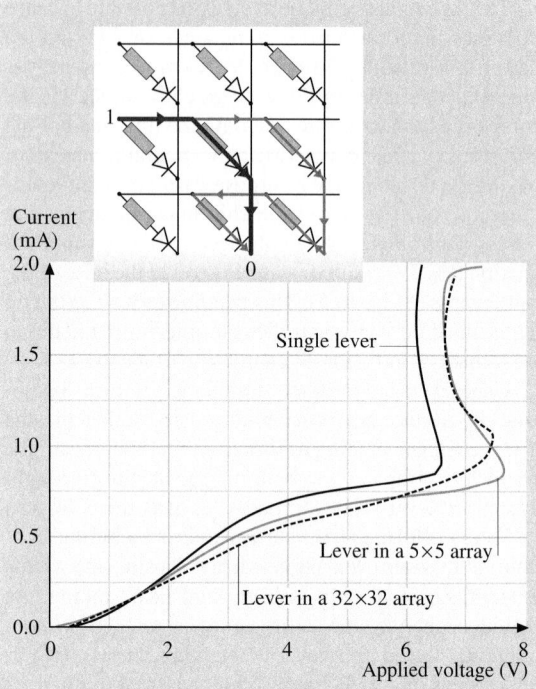

Fig. 46.9 Comparison of the *I/V* curve of an independent cantilever (*solid black line*) with the current response when addressing a cantilever in a 5×5 (*solid brown line*) or a 32×32 (*dashed line*) array with a Schottky diode serially connected to the cantilever. Little change is observed in the *I/V* curve between the different cases. Inset: sketch of the direct path (*bold line*) and a parasitic path (*thin line*) in a cantilever-diode array. In the parasitic path there is always one diode in reverse bias that reduces the parasitic current (after [46.31])

an array is nearly independent of the size of the array, as demonstrated by the *I/V* curves in Fig. 46.9. Hence, the power applied to address a cantilever is not shunted by other cantilevers, and the reading sensitivity is not degraded – not even for very large arrays (32×32). The introduction of the electrical isolation using integrated Schottky diodes turned out to be crucial for the successful operation of interconnected cantilever arrays with a simple time-multiplexed addressing scheme.

The tip-apex height uniformity within an array is very important because it determines the force of each cantilever while in contact with the medium and hence influences write/read performance, as well as medium and tip wear. Wear investigations suggest that a tip-apex height uniformity across the chip of less than 500 nm is required [46.7], with the exact number depending on the spring constant of the cantilever. In the case of the millipede, the tip-apex height is determined by the tip height and the cantilever bending. Figure 46.10 shows the tip-apex height uniformity of one row of the array (32 tips) due to tip height and cantilever bending. It demonstrates that our uniformity is on the order of 100 nm, thus meeting the requirements.

46.5 *x/y/z* Medium Microscanner

A key issue for the millipede concept is the need for a low-cost, miniaturized scanner with *x/y/z* motion capabilities and a lateral scanning range on the order of 100 μm. Multiple-probe systems arranged as 1-D or 2-D arrays [46.18] must also be able to control, by means of tilt capabilities, the parallelism between the probe array and the sample [46.18, 33].

We have developed a microscanner with these properties based on electromagnetic actuation. It consists of a mobile platform supported by springs and containing integrated planar coils positioned over a set of miniature permanent magnets [46.34]. A suitable arrangement of the coils and magnets allows us, by electrically addressing the various coils, to apply magnetically induced forces to the platform and drive it in the *x*, *y*, *z*, and tilt directions. Our first silicon/copper-based version of this device has proved the validity of the concept [46.35], and variations of it have since been used elsewhere [46.36]. However, the undamped copper spring system gave rise to excessive cross talk and ringing when driven in an open loop, and its layout limited the compactness of the overall device.

We investigate a modified microscanner that uses flexible rubber posts as a spring system and a copper-epoxy-based mobile platform, Fig. 46.11. The platform is made of a thick, epoxy-based SU-8 resist [46.37], in which the copper coils are embedded. The posts are made of polydimethylsiloxane (PDMS, Sylgard 184 silicon elastomer, Dow Corning, Midland, MI) and are fastened at the corners of the platform and at the ground plate, providing an optimally compact device by sharing the space below the platform with the magnets. The shape of the posts allows their lateral and longitudinal stiffness to be adjusted, and the dissipative rubber-like properties of PDMS provide damping to avoid platform ringing and to suppress nonlinearities.

Figure 46.12 shows the layout of the platform, which is scaled laterally, so that the long segments of the "racetrack" coils used for in-plane actuation coincide with commercially available 24 mm^2 SmCo magnets.

The thickness of the device is determined by that of the magnets (1 mm), the clearance between magnet and platform (500 μm), and the thickness of the platform itself, which is 250 μm and determined mainly by the aspect

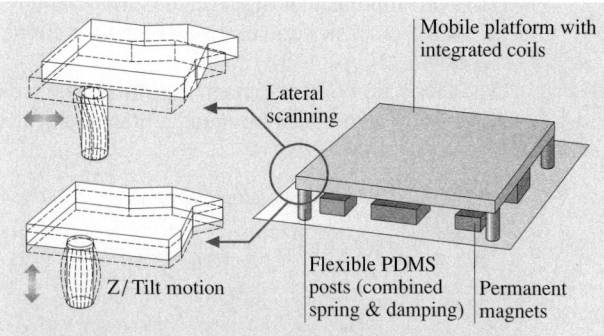

Fig. 46.11 Microscanner concept using a mobile platform and flexible posts (after [46.32])

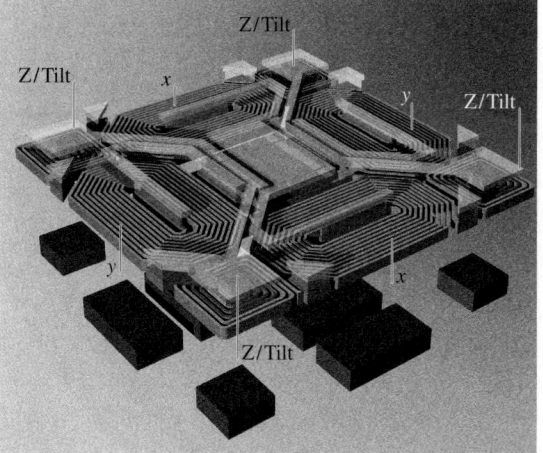

Fig. 46.12 Arrangement of the coils, the interconnects, and the permanent magnets, as well as the various motions addressed by the corresponding coils (after [46.15] © 2002 IEEE)

ratio achievable in SU-8 resist during the exposure of the coil plating mold. The resulting device volume is approximately $15 \times 15 \times 1.6 \, \text{mm}^3$.

The SmCo magnets produce a measured magnetic field intensity of $\approx 0.14 \, \text{T}$ at the mid-thickness of the coils. The effective coil length is 320 mm, yielding an expected force $F_{x,y}$ of 45 μN per mA of drive current.

The principal design issue of the spring system is the ratio of its stiffnesses for in-plane and out-of-plane motion. Whereas for many scanning probe applications the required z-axis range need not be much larger than a few microns, it is necessary to ensure that the z-axis retraction of the platform due to the shortening of the posts as they take on an S-shape at large in-plane deflections can be compensated for at acceptable z-coil current levels. Various PDMS post shapes have been investigated to optimize and trade off the various requirements. Satisfactory performance was found for simple O-shapes [46.32].

The fabrication of the scanner, Fig. 46.13, starts on a silicon wafer with a seed layer and a lithographically patterned 200 μm thick SU-8 layer, in which copper is electroplated to form the coils (Fig. 46.13a). The coils typically have 20 turns, with a pitch of 100 μm and a spacing of 20 μm. Special care was taken in the resist processing and platform design to achieve the necessary aspect ratio and to overcome adhesion and stress problems of SU-8. A second SU-8 layer, which serves as an insulator, is patterned with via holes, and another seed layer is then deposited (Fig. 46.13b). Next, an interconnect level is formed using a Novolac-type resist mask and a second copper-electroplating step (Fig. 46.13c). After stripping the resist, the silicon wafer is dissolved by a sequence of wet and dry etching, and the exposed seed layers are sputtered away to prevent shorts (Fig. 46.13d).

The motion of the scanner was characterized using a microvision strobe technique [46.38]. The results presented below are based on O-type PDMS ports. Frequency response curves for in-plane motion (Fig. 46.14) show broad peaks (characteristic of a large degree of damping) at frequency values that are consistent with expectations based on the measured mass of the platform (0.253 mg). The amplitude response (Fig. 46.15) displays the excellent linearity of the spring system for displacement amplitudes up to 80 μm (160 μm displacement range). Based on these near-DC (10 Hz) responses ($\approx 1.4 \, \text{μm/mA}$) and a measured circuit resistance of 1.9 Ω, the power necessary for a 50 μm displacement amplitude is approximately equal to 2.5 mW.

Owing to limitations of the measurement technique, it was not possible to measure out-of-plane

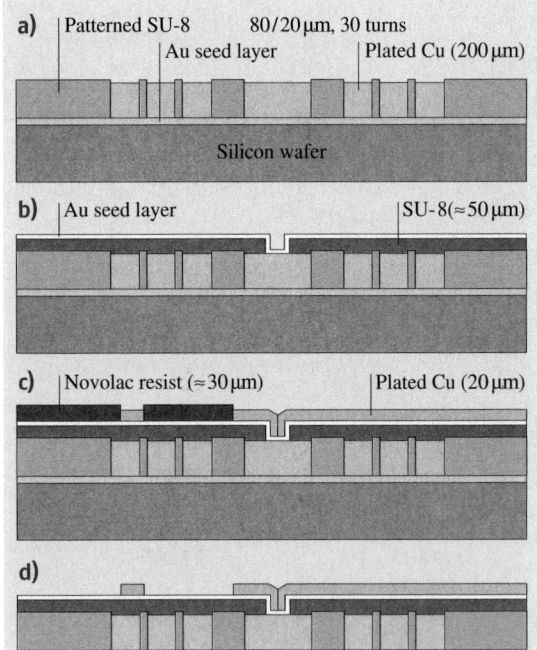

Fig. 46.13a–d Cross section of the platform fabrication process. (**a**) Coils are electroplated through an SU-8 resist mask, which is retained as the body of the platform; (**b**) an insulator layer is deposited; (**c**) interconnects are electroplated; (**d**) the platform is released from the silicon substrate (after [46.32])

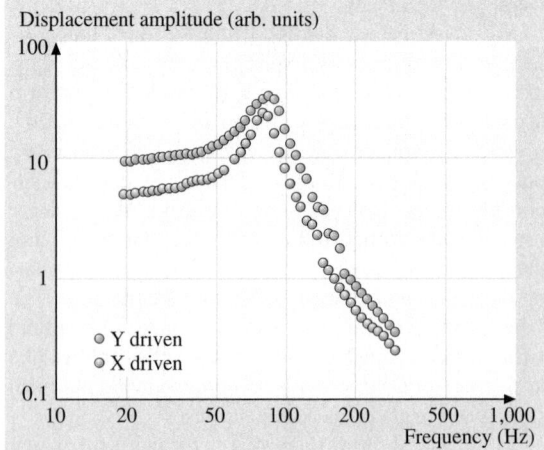

Fig. 46.14 Frequency response for in-plane x- and y-axis motion. The mechanical quality factors measured are between 3.3 and 4.6 (after [46.15])

The "Millipede" – A Nanotechnology-Based AFM Data-Storage System | 46.6 First Write/Read Results with the 32×32 Array Chip   1467

Part F | 46.6

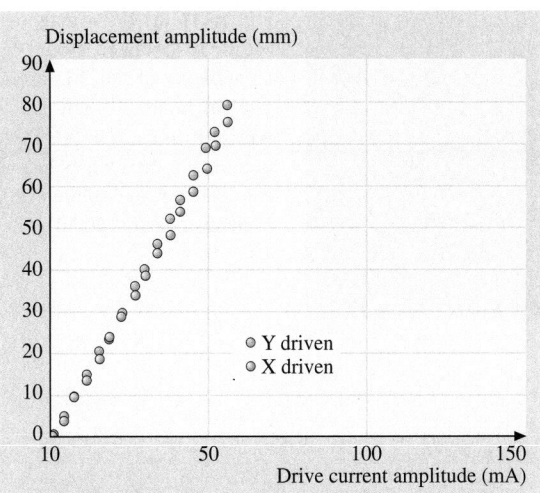

Fig. 46.15 In-plane displacement amplitude response for an ac drive current at 10 Hz (off resonance) (after [46.15])

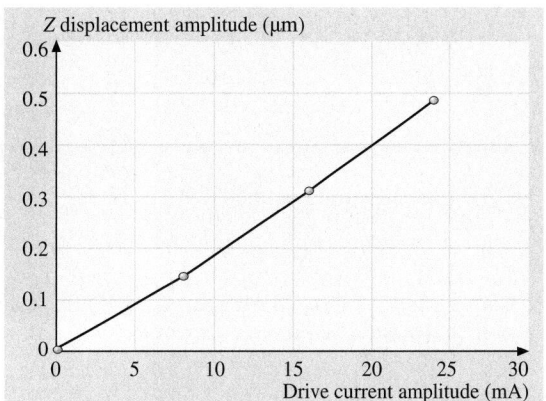

Fig. 46.16 Out-of-plane amplitude response for an ac drive current at 3 Hz. The drive current is the total for all four corner coils, which are driven in phase (after [46.32])

displacements greater than 0.5 μm. However, the small-amplitude response for z-motion when all four corner coils are driven in-phase also displays good linearity over the range that can be measured (Fig. 46.16).

The electromagnetic scanner performs reliably and as predicted in terms of the scan range, device volume, and power requirements, achieving overall displacement ranges of 100 μm with approximately 3 mW of power. The potential access time in the 100 μm storage field is on the order of a few milliseconds. By being potentially cheap to manufacture, the integrated scanner presents a good alternative actuation system for many scanning probe applications.

An alternative design that has been realized for the displacement of the storage medium relative to the array of cantilevers is based on using a silicon microscanner with x/y-displacement capabilities of about 120 μm, i. e., approximately 20% larger than the pitch between adjacent cantilevers in the array. The scanner consists of a 6.8 mm × 6.8 mm scan table and a pair of voice-coil-type actuators, all of which are supported by springs. The mechanical components of the scanner are fabricated from a 400 μm thick silicon wafer using a deep-trench-etching process. This scanner chip is then mounted on a silicon base plate that acts as the mechanical ground of the system. The base plate has been designed to provide a clearance of about 20 μm between its top surface and the bottom surface of the moving parts of the scanner. The scan table, which carries the polymer storage medium, can be displaced in two orthogonal directions (x and y) in the plane of the silicon wafer. Each voice-coil actuator consists of a pair of permanent magnets glued into a silicon frame, with a miniature 8.4 Ω coil mounted between them on the base plate. Actuation is achieved by applying a current to the coil, which generates a force on the magnets and induces a displacement of the actuator. This motion is coupled to the scan table by means of a mass-balancing scheme that provides a 1:1 translation of the motion while making the scanner robust against external shock and vibrations [46.39].

46.6 First Write/Read Results with the 32×32 Array Chip

We have built a prototype that includes all the basic building blocks of the millipede concept (see Fig. 46.1) [46.40]. A 3 × 3-mm² silicon substrate is spin-coated with the SU-8/PMMA polymer medium, as described in Sect. 46.3. This storage medium is attached to the $x/y/z$ microscanner and approaching device. The magnetic z-approaching actuators bring the medium into contact with the tips of the array chip. The z-distance between the medium and the millipede chip is controlled by the approaching sensors in the corners of the array. The signals from these cantilevers are used to determine the forces on the z-actuators and, hence, the forces of the cantilever while it is in contact with the medium. This sensing/actuation feedback loop continues to operate

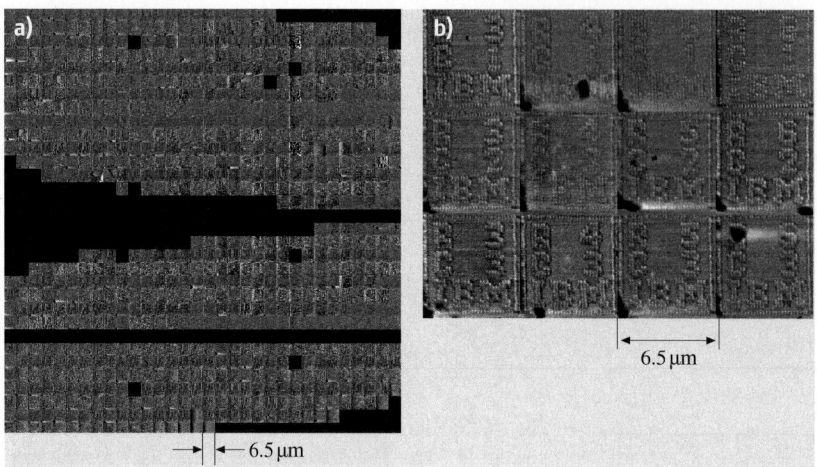

6.5 μm

6.5 μm

Fig. 46.17 (a) 1024 images, each one obtained from a cantilever of the 2-D array. **(b)** Enlarged view of typical images from **(a)**. The numbers in the images indicate the row and column of each lever (from [46.40])

during x/y scanning of the medium. The PC-controlled write/read scheme addresses the 32 cantilevers of one row in parallel. Writing is performed by connecting the addressed row for 20 μs to a high negative voltage and simultaneously applying data inputs ("0" or "1") to the 32 column lines. The data input is a high positive voltage for a "1" and ground for a "0". This row-enabling and column-addressing scheme supplies a heater current to all cantilevers, but only those cantilevers with high positive voltage generate an indentation ("1"). Those grounded are not hot enough to form an indentation and thus write a "0". When the scan stage has moved to the next logical mark position, the process is repeated, and this is continued until the line scan is finished. In the read process, the selected row line is connected to a moderate negative voltage, and the column lines are grounded via a protection resistor of about 10 kΩ, which keeps the cantilevers warm. During scanning, the voltages across the resistors are measured. Depending on the topography of the recording surface, the degree of cooling of each cantilever varies, thus changing the resistance and voltage across the series resistor and allowing written data to be read back.

The results of writing and reading in this fashion can be seen in Fig. 46.17, which shows 1024 images corresponding to the 1024 storage fields and associated cantilevers. Of the 1024 levers, 834 were able to write and read data back, i.e., a success rate of more than 80%. The sequence is as follows. First, a bit pattern is written simultaneously by each of the levers in row 1, then read back simultaneously, followed by

row 2, etc. through row 32. The data sent to the levers is different, each lever writing its own row and column number in the array. The bit pattern consists of 64×64 bits, where odd-numbered bits are always set to 0. In this case, the area used is 6.5×6.5 μm^2. The readback image is a grey-scale bit map of 128×128 pixels. The distance between levers is 92 μm, so the images in Fig. 46.17 are also 92 μm apart. The areal density of the written information shown in Fig. 46.17 corresponds to 15–30 Gb/in^2, depending on the coding scheme adopted. More recently, an areal density of 150–200 Gb/in^2, at an array yield of about 60%, has been demonstrated.

Those levers that did not read back failed for one of four reasons: (i) a defective chip connector rendered an entire column unusable, (ii) a point defect occurred, meaning that a single lever or tip was broken, (iii) there was a nonuniformity of the tip contact due to tip/lever variability or storage substrate bowing due to mounting, (iv) there were thermal drifts. The latter two reasons were the most likely and major failure sources. At present, there is clearly a tradeoff between the number of working levers and the density, which will most likely be resolved by a better substrate/chip mounting technique and lower thermal drifts.

The writing and readback rates achieved with this system are 1 kb/s per lever, thus, the total data rate is about 32 kb/s. This rate is limited by the rate at which data can be transferred over the PC ISA bus, not by a fundamental time limitation of the read/write process.

46.7 Polymer Medium

The polymer storage medium plays a crucial role in millipede-like thermomechanical storage systems. The thin-film multilayer structure with PMMA as active layer (see Fig. 46.2) is not the only possible choice, considering the almost unlimited range of polymer materials available. The ideal medium should be easily deformable for writing, yet indentations should be stable against tip wear and thermal degradation. Finally, one would also like to be able to erase and rewrite data repeatedly. In order to be able to address all important aspects properly, some understanding of the basic physical mechanism of thermomechanical writing and erasing is required.

46.7.1 Writing Mechanism

In a *gedanken* experiment we visualize writing of an indentation as the motion of a rigid body (the tip) in a viscous medium (the polymer melt). Let us initially assume that the polymer, i.e., PMMA, behaves like a simple liquid after it has been heated above the glass-transition temperature in a small volume around the tip. As viscous drag forces must not exceed the loading force applied to the tip during indentation, we can estimate an upper bound for the viscosity ζ of the polymer melt using Stokes's equation:

$$F = 6\pi\zeta\varrho v \,. \tag{46.1}$$

In actual indentation formation, the tip loading force is on the order of $F = 50\,\text{nN}$ and the radius of curvature at the apex of the tip is typically $\varrho = 20\,\text{nm}$. Assuming a depth of the indentation of, say, $h = 50\,\text{nm}$ and a heat pulse of $\tau_h = 10\,\mu\text{s}$ duration, the mean velocity during indentation formation is on the order of $v = h/\tau_h = 5\,\text{mm/s}$. Note that thermal relaxation times are on the order of microseconds [46.20, 21] and, hence, the heating time can be equated to the time it takes to form an indentation. With these parameters we obtain $\zeta < 25\,\text{Pa\,s}$, whereas typical values for the shear viscosity of PMMA are at least seven orders of magnitude larger even at temperatures well above the glass-transition point [46.41].

This apparent contradiction can be resolved by considering that polymer properties are strongly dependent on the time scale of observation. At time scales on the order of 1 ms and below, entanglement motion is in effect frozen in and the PMMA molecules form a relatively static network. Deformation of the PMMA now proceeds by means of uncorrelated deformations of short molecular segments, rather than by a flow mechanism involving the coordinated motion of entire molecular chains. The price one has to pay is that elastic stress builds up in the molecular network as a result of the deformation (the polymer is in a so-called rubbery state). On the other hand, corresponding relaxation times are orders of magnitude smaller, giving rise to an effective viscosity at millipede time scales on the order of 10 Pa s [46.41], as required by our simple argument (see (46.1)). Note that, unlike normal viscosity, this high-frequency viscosity is basically independent of the detailed molecular structure of the PMMA, i.e., chain length, tacticity, polydispersity, etc. In fact, we can even expect that similar high-frequency viscous properties can be found in a large class of other polymer materials, which makes thermomechanical writing a rather robust process in terms of material selection.

We have argued above that elastic stress builds up in the polymer film during the formation of an indentation, creating a corresponding reaction force on the tip on the order of $F_r \sim 2\pi G\varrho^2$, where G denotes the elastic shear modulus of the polymer [46.42]. An important property for millipede operation is that the shear modulus drops by several orders of magnitude in the glass-transition regime, i.e., for PMMA from $\approx 1\,\text{GPa}$ below Θ_g to ≈ 0.5–$1\,\text{MPa}$ above Θ_g, where Θ_g denotes the glass-transition temperature [46.41]. The bulk modulus, on the other hand, retains its low-temperature value of several GPa. Hence, in this elastic regime, formation of an indentation above Θ_g constitutes a volume-preserving deformation. For proper indentation formation, the tip load must be balanced between the extremes of the elastic reaction force F_r for temperatures below and above Θ_g, i.e., $F \ll 2.5\,\mu\text{N}$ for PMMA to prevent indentation of the polymer in the cold state and $F \gg 2.5\,\text{nN}$ to overcome the elastic reaction force in the hot state. Unlike the deformation of a simple liquid, the indentation represents a metastable state of the entire deformed volume, which is under elastic tension. Recovery of the unstressed initial state is prevented by rapid quenching of the indentation below the glass temperature with the tip in place. As a result, the deformation is frozen in, because below Θ_g motion of molecular-chain segments is, in effect, inhibited (see Fig. 46.18).

This mechanism also allows indentations to be erased locally – it suffices to heat the deformed volume locally above Θ_g, whereupon the indented volume reverts to its unstressed flat state driven by internal elastic stress. In addition, erasing is promoted by surface tension forces, which give rise to a restoring surface

a)

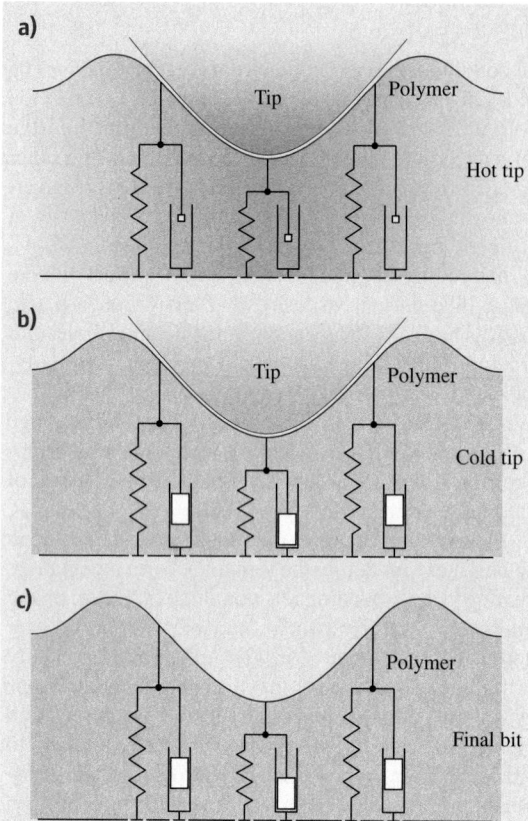

b)

c)

Fig. 46.18a–c Viscoelastic model of indentation writing. (**a**) The hot tip heats a small volume of polymer material to more than Θ_g. The shear modulus of the polymer drops drastically from GPa to MPa, which in turn allows the tip to indent the polymer. In response, elastic stress (represented as compression springs) builds up in the polymer. In addition, viscous forces (represented as pistons) associated with the relaxation time for the local deformation of molecular segments limit the indentation speed. (**b**) At the end of the writing process, the temperature is quenched on a microsecond time scale to room temperature: The stressed configuration of the polymer is frozen-in (represented by the locked pistons). (**c**) The final indentation corresponds to a metastable configuration. The original unstressed flat state of the polymer can be recovered by heating the indentation volume to more than Θ_g, which unlocks the compressed springs (after [46.15])

pressure on the order of $\gamma(\pi/\varrho)^2 h \approx 25$ MPa, where $\gamma \approx 0.02$ N/m denotes the polymer-air surface tension.

One question immediately arises from these speculations: If the polymer behavior can be determined from

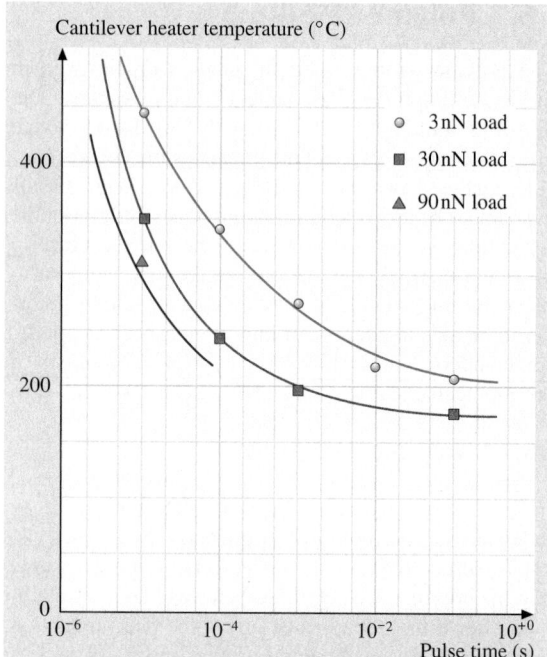

Fig. 46.19 Indentation-writing threshold measurements. The load was controlled by pushing the cantilever/tip into the sample with a controlled displacement and a known spring constant of the cantilever. When a certain threshold is reached, the indentations become visible in subsequent imaging scans (see also Fig. 46.21). The *solid lines* are guides to the eye. Curves of similar shape would be expected from the time-temperature superposition principle (after [46.15])

the macroscopic characteristics of the shear modulus as a function of time, temperature, and pressure, can the time-temperature superposition principle also be applied in this case? The time-temperature superposition principle is a powerful concept of polymer physics [46.43]. It basically states that the time scale and the temperature are interdependent variables that determine the polymer behavior such as the shear modulus. A simple transformation can be used to translate time-dependent into temperature-dependent data and vice versa. It is not clear, however, whether this principle can be applied in our case, i.e., under such extreme conditions (high pressures, short time scales, and nanometer-sized volumes, which are clearly below the radius of gyration of individual polymer molecules).

To test this, we varied the heating time, the heating temperature, and the loading force in indentation-writing experiments on a standard PMMA sample. The results are summarized in Fig. 46.19. The minimum heater temperature at which the formation of an indentation starts for a given heating-pulse length and loading force was determined. This so-called threshold temperature is plotted against the heating-pulse length. A careful calibration of the heater temperature has to be done to allow a comparison of the data. The heater temperature was determined by assuming proportionality between temperature and electrical power dissipated in the heater resistor at the end of the heating pulse when the tip has reached its maximum temperature. An absolute temperature scale is established using two well-defined reference points. One is room temperature, corresponding to zero electrical power. The other is provided by the point of turnover from positive to negative differential resistance (see Fig. 46.8), which corresponds to a heater temperature of approximately 550 °C. The general shape of the measured threshold temperature versus heating time curves indeed shows the characteristics of time-temperature superposition. In particular, the curves are identical up to a load-dependent shift with respect to the time axis. Moreover, we observe that the time it takes to form an indentation at constant heater temperature is inversely proportional to the tip load. This property is exactly what one would expect if internal friction (owing to the high-frequency viscosity) is the rate-limiting step in forming an indentation (46.1).

The time it takes to heat the indentation volume of polymer material higher than the glass-transition temperature is another potentially rate-limiting step. Here, the spreading resistance of the heat flow in the polymer and the thermal contact resistance are the most critical parameters. Simulations suggest [46.20, 21] that equilibration of temperature in the polymer occurs within less than 1 μs. Very little is known, however, about the thermal coupling efficiency across the tip-polymer interface. We have several indications that the heat transfer between tip and sample plays a crucial role, one of them being the asymptotic heater temperature for long writing times, which according to the graph in Fig. 46.19 is approximately 200 °C. The exact temperature of the polymer is unknown. However, the polymer temperature should approach the glass-transition temperature (around 120 °C for PMMA) asymptotically. Hence, the temperature drop between heater and polymer medium is substantial. Part of the temperature difference is due to a temperature drop along the tip, which according to heat-flow simulations [46.20, 21] is expected to be on the order of 30 °C at most. Therefore, a significant temperature gradient must exist in the tip-polymer contact zone. Further experiments on the heat

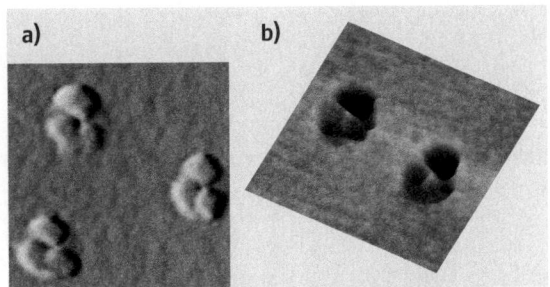

Fig. 46.20a,b Topographic image of individual indentations. (**a**) The region around the actual indentations clearly shows the threefold symmetry of the tip, here a three-sided pyramid. (**b**) The indentations themselves exhibit sharp edges, as can be seen from the inverted 3-D image. Image size is $2 \times 2\,\mu m^2$ (from [46.15] © 2002 IEEE)

transfer from tip to surface are needed to clarify this point.

We also find that the heat transfer for a nonspherical tip is anisotropic. As shown in Fig. 46.20, in the case of a pyramid-shaped tip, the indentation not only exhibits sharp edges, but also the region around the indentation, where polymer material is piled up, is anisotropic. The pile-up characteristics will be discussed in detail below. At this point we take it as an indication of the relevance of the heat transfer to the measurements.

One of the most striking conclusions of our model of the indentation-writing process is that it should in principle work for most polymer materials. The general behavior of the mechanical properties as a function of temperature and frequency is similar for all polymers [46.43]. The glass-transition temperature Θ_g would then be one of the main parameters that determines the threshold writing temperature.

A verification of this was found experimentally by comparing various polymer films. The samples were prepared in the same way as the PMMA samples discussed above [46.18]: by spin-casting thin films (10–30 nm) onto a silicon wafer with a photo-resist buffer. Then, threshold measurements were done by applying heat pulses with increasing current (or temperature) to the tip while the load and the heating time were held constant (load about 10 nN and heating time 10 μs). Examples of such measurements are shown in Fig. 46.21, where the increasing size and depth of indentations can be seen for different heater temperatures. A threshold can be defined based on such data and compared with the glass-transition temperature of these materials. The results show a clear correlation between the threshold

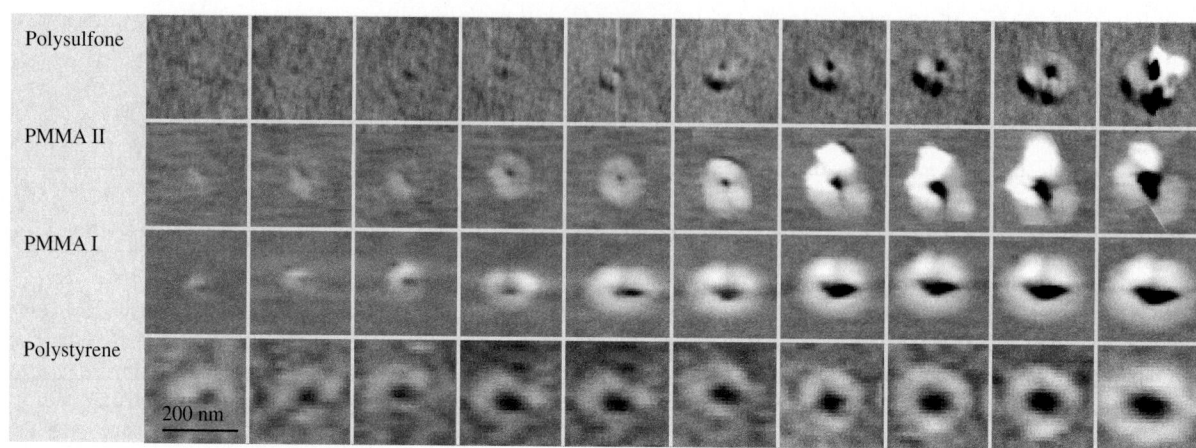

Fig. 46.21 Written indentations for different polymer materials. The heating pulse length was 10 μs, the load about 10 nN. The grey scale is the same for all images. The heater temperatures for the indentation on the left-hand side are 445, 400, 365, and 275 °C for the polymers Polysulfone, PMMA II (anionically polymerized PMMA, $M \approx 26$ k), PMMA I (Polymer Standard Service, Germany, $M \approx 500$ k), and Polystyrene, respectively. The temperature increase between events on the horizontal axis is 14, 22, 20, and 9 °C, respectively (from [46.15] © 2002 IEEE)

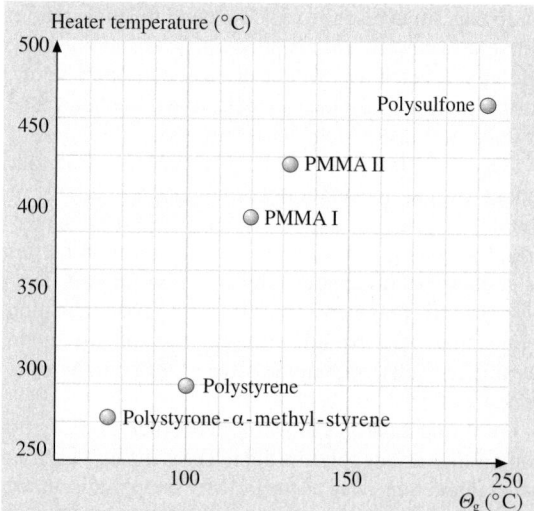

Fig. 46.22 The heater temperature threshold for writing indentations with the same parameters as in Fig. 46.21 is plotted against the glass-transition temperature for these polymers, including poly-α-methyl-styrene (after [46.15])

heater temperature and the glass-transition temperature (see Fig. 46.22).

With our simple viscoelastic model of writing we are able to formulate a set of requirements that potential candidate materials for millipede data storage have to fulfill. First, the material should ideally exhibit a well-defined glass-transition point with a large drop of the shear modulus at Θ_g. Second, a rather high value of Θ_g (on the order of 150 °C) is preferred to facilitate thermal read back of the data without destroying the information. We have investigated a number of materials to explore the Θ_g parameter space. The fact that all polymer types tested are suitable for forming small indentations leaves us free to choose which polymer type to optimize in terms of the technical requirements for a device such as lifetime of indentations, polymer endurance of the read and write process, power consumption, etc. These are subjects of ongoing research.

46.7.2 Erasing Mechanism

It is worthwhile to look at the detailed shapes of the written indentations. The polymer material around an indentation appears piled up as can be seen, for example, in Fig. 46.20. This is not only material that was pushed aside during indentation formation as a result of volume conservation. Rather, the flash heating by the tip and subsequent rapid cooling result in an increase of the specific volume of the polymer. This phenomenon, that the specific volume of a polymer can be increased by rapidly cooling a sample through the glass transition [46.43], is well-known. Our system allows a cooling time on the order of microseconds, which is much faster than the highest rates that can be achieved with standard polymer-analysis tools. However, a quantitative measurement of the specific volume change cannot be easily performed with our type of experiments. On the other hand, the pile-up effect serves as a convenient threshold thermometer. The outer perimeter of the pile-ups surrounding the indentations corresponds to the Θ_g isotherm, and the temperature in the enclosed area has certainly reached values greater than Θ_g during the indentation process. Based on our viscoelastic model, one would thus conclude that previously written indentations that overlap with the piled up region of a subsequently written indentation should be erased.

That this pile-up effect actually works against the formation of an indentation can clearly be seen in the line scans of a series of indentations written in Polysulfone (Fig. 46.23). Here, the heating of the tip was accompanied by a rather high normal force. The force was high enough to create a small indentation, even if the tip was too cold to modify the polymer (Fig. 46.23a). Then, with increasing tip heating, the indentations initially fill up in the piled up region (Fig. 46.23b) before they finally become deeper (Fig. 46.23c).

The pile-up phenomenon turns out to be particularly beneficial for data-storage applications. The following example demonstrates the effect. If we look at the sequence of images in Fig. 46.24, taken on a standard PMMA sample, we find that the piled up regions can overlap each other without disturbing the indentations. However, if the piled up region of an individual writing event extends over the indented area of a previously written "1", the depth of the corresponding indentation decreases markedly (Fig. 46.24d). This can be used for erasing written data. On the other hand, if the pitch between two successive indentations is decreased even further, this erasing process will no longer work. Instead, a broader indentation is formed, as shown in Fig. 46.24e. Hence, to exclude mutual interference, the minimum pitch between successive indentations, which we denote by minimum-indentation pitch (BP_{min}), must be larger than the radius of the piled up area around an indentation.

In the example shown in Fig. 46.24, the temperature chosen was so high that the ring around the indentations was very large, whereas the depth of the indentation was limited by the stop layer underneath the PMMA material. Clearly, the temperature was too high here to form small indentations, the minimum pitch of which is around 250 nm. However, by carefully optimizing all parameters it is possible to achieve areal densities of up to 1 Tb/in^2, as demonstrated in Fig. 46.3c.

The new erasing scheme based on this volume effect switches from writing to erasing merely by de-

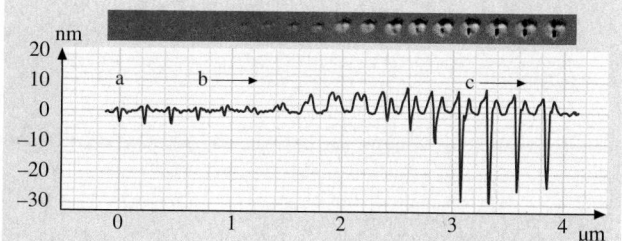

Fig. 46.23 Section through a series of indentations similar to Fig. 46.21. Here, a load of about 200 nN was applied before a heating pulse of 10 µs length was fired. The temperature of the heater at the end of the pulse has been increased from 430 to 610 °C in steps of about 10.6 °C. (a) The load was sufficient to form a plastic indentation even if the polymer was not heated enough to come near the glass transition. (b) With increasing heater temperature, the polymer swells. This eliminates the indentation, thus erasing previously written "cold" marks. (c) As this process continues, the thermomechanical formation of indentations begins to dominate until, finally, normal thermomechanical indentation writing occurs (after [46.15])

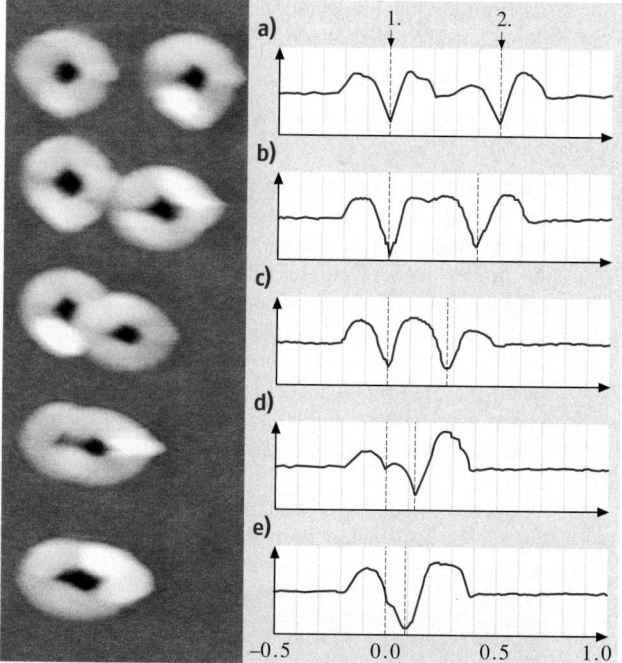

Fig. 46.24a–e Indentations in a PMMA film at several distances. The depth of the indentations is ≈ 15 nm, roughly the same as the thickness of the PMMA layer. The indentations on the left-hand side were written first, then a second series of indentations was made with decreasing distance from the first series going from (a) to (e) (after [46.15])

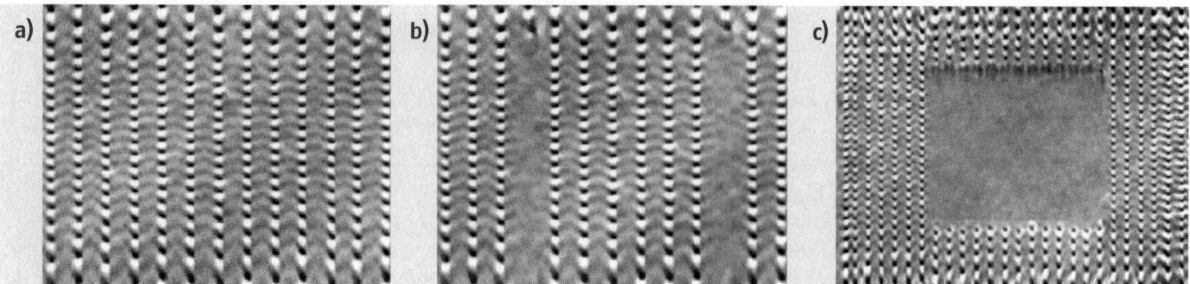

Fig. 46.25a–c Demonstration of the new erasing scheme: (**a**) A bit pattern recorded with variable pitch in the vertical axis (fast scan axis) and constant pitch in the horizontal direction (slow scan axis) was prepared. (**b**) Then two of the lines were erased by decreasing the pitch in the vertical direction by a factor of three, showing that the erasing scheme works for individual lines. One can also erase entire fields of indentations without destroying indentations at the edges of the fields. This is demonstrated in (**c**), where a field has been erased from an indentation field similar to the one shown in (**a**). The distance between the lines is 70 nm (from [46.15] © 2002 IEEE)

creasing the pitch of writing indentations. This can be done in a very controlled fashion, as shown in Fig. 46.25, where individual lines or predefined subareas are erased. Hence, this new erasing scheme can be made to work in a way that is controlled on the scale of individual indentations. Compared with earlier global erasing schemes [46.23], this simplifies erasing significantly.

46.7.3 Overwriting Mechanism

Overwriting data on some part of the storage medium can be achieved by first erasing the entire area and then writing the desired data on the erased surface. Although this process works well, it is time-consuming and dissipates a significant amount of power. In a millipede-based storage device, where data rate and power consumption are at a premium, such a two-step overwriting mechanism may be impractical. Instead, a one-step, direct overwriting process similar to those applied in magnetic hard-disk drives and rewritable optical drives is desired.

As discussed above, switching from writing to erasing may be achieved by decreasing the pitch of writing.

It has been found experimentally that erasing can be performed effectively by halving the pitch of writing successive indentations, which is denoted as BP, provided the condition $BP \approx BP_{min}$ is satisfied. This suggests that the basic distance unit for combined write-erase operations should be $BP/2$. Written indentations are spaced n units apart, where $n \geq 2$. Let us recall that the presence of an indentation corresponds to a logical bit "1" and the absence of an indentation to a logical bit "0". Logical bits are then stored in the medium at the points of a regular lattice with minimum distance between points equal to $BP/2$ in the on-track direction, with successive "1"s separated by at least one "0". This condition is necessary in order to avoid mutual interference between successive "1"s. It is also the basis for an important category of codes known as (d, k) codes, which are described in Sect. 46.9.3. Coding can thus be used to enable direct data overwriting in an elegant way. Direct overwriting requires the simultaneous realization of two conditions: If previously written "1"s exist where "0"s are to be written, then these "1"s have to be *erased*. On the other hand, if "0"s exist where "1"s are to be written, then these "1"s have to be *written*. Writing an indentation is performed thermomechanically as

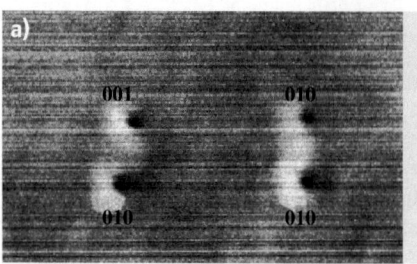
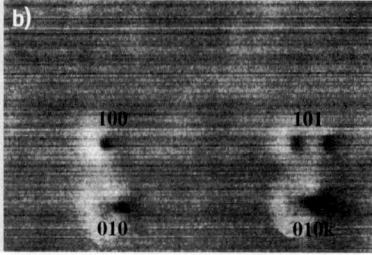

Fig. 46.26a,b Bit strings (**a**) 001 and 010, and (**b**) 100 and 101, overwritten to 010

described above. Erasing an existing indentation is done by writing another indentation next to it, at a distance of BP/2 units. However, as this operation creates a new indentation shifted by BP/2 with respect to the one erased, the erasing process must be performed repeatedly until the newly created indentation lies at a position corresponding to a "1" in the new data pattern. The basic principle of erasing is illustrated in Fig. 46.26a and b. The figures show how the four bit strings 001, 010, 100, and 101 are modified into the string 010. Figure 46.27 depicts the results of a rewriting experiment; the top track shows a prestored sequence, which is to be overwritten by another sequence, shown on the bottom track for comparison. The result of direct overwriting of the prestored sequence is shown on the middle track.

Comparison with the sequence on the bottom track, which is written on a clean surface, illustrates the effectiveness of the proposed procedure. Although the write/read quality of overwritten data is somewhat in-

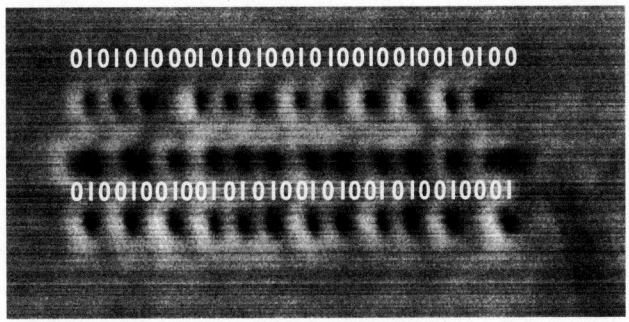

Fig. 46.27 Experimental result of overwriting a bit sequence

ferior to that of data written on a clean storage surface, detection of the newly written sequence is not affected. However, repeated overwriting may further degrade the quality of stored data. As the extent and rate of degradation are important characteristics of a storage system, this remains an area of ongoing investigation.

46.8 Read Channel Model

Let us now consider the readback channel for a single cantilever that is scanning a storage field where bits are written as indentations or absence of indentations in the storage medium. As discussed above, a cantilever is modeled as a variable resistor that depends on the temperature at the cantilever tip. The model of the read channel, used for the design and analysis of the detection system, is illustrated in Fig. 46.28 [46.12, 26, 27].

To evaluate the evolution of the temperature of a heated cantilever during the read process, we resort to a simple RC-equivalent thermal circuit, illustrated in Fig. 46.29, where $(1 + \eta_x) R^\Theta$ and C^Θ denote the thermal resistance and capacitance, respectively. The parameter $\eta_x = \Delta R^\Theta(x)/R^\Theta$ indicates the relative variation of thermal resistance that results from the small change in air-gap width between the cantilever

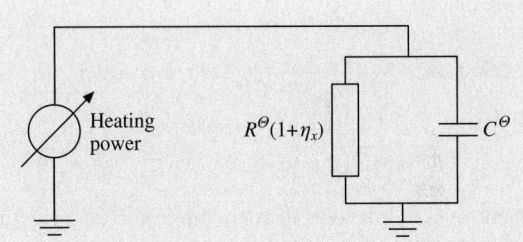

Fig. 46.29 RC-equivalent thermal model of the heat transfer process (after [46.12])

and the storage medium. The subscript x indicates the distance in the direction of scanning from the initial point. Therefore, the parameter η_x will assume the largest absolute value when the tip of the cantilever

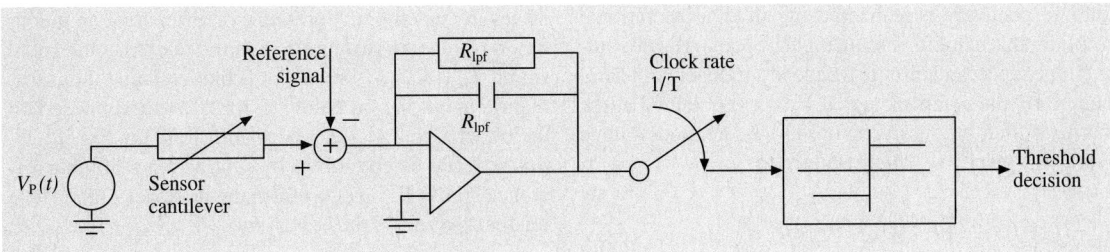

Fig. 46.28 Block diagram of the detection circuit

is located at the center of an indentation. The heating power that is dissipated in the cantilever heater region is expressed as

$$P^e[t, \Theta(t, x)] = \frac{V_C^2(t)}{R^e[\Theta(t, x)]}, \qquad (46.2)$$

where $V_C(t)$ is the voltage across the cantilever, $\Theta(t, x)$ is the cantilever temperature, and $R^e[\Theta(t, x)]$ is the temperature-dependent cantilever resistance.

As the heat-transfer process depends on the value of thermal resistance and on the read pulse waveform, the cantilever temperature $\Theta(t, x)$ depends on time t and distance x. However, because the time it takes for the cantilever to move from the center of one logical mark to the next is greater than the duration of a read pulse, we assume that $\Theta(t, x)$ does not vary significantly as a function of x while a read pulse is being applied, and that it decays to the ambient temperature Θ_0 before the next pulse is applied. Therefore, the evolution of the cantilever temperature in response to a pulse applied at time $t_0 = x_0/v$ at a certain distance x_0 from the initial point of scanning and for a certain constant velocity v of the scanner obeys a differential equation expressed as

$$\Theta'(t, x_0) + \frac{1}{(1 + \eta_{x_0})R^\Theta C^\Theta}[\Theta(t, x_0) - \Theta_0]$$
$$= \frac{1}{C^\Theta} \frac{V_C^2(t)}{R^e[\Theta(t, x_0)]}, \qquad (46.3)$$

where $\Theta'(t, x_0)$ denotes the derivative of $\Theta(t, x_0)$ with respect to time.

With reference to the block diagram of the read channel illustrated in Fig. 46.28, the source generates the read pulse $V_P(t)$ applied to the cantilever. Clearly, because of the virtual ground at the operational amplifier input, the voltage $V_C(t)$ across the cantilever variable resistance is equal to $V_P(t)$. Furthermore, the active low-pass RC detector filter, where R_{lpf} and C_{lpf} denote the resistance and capacitance of the low-pass filter, respectively, is realized using an ideal operational amplifier that exhibits infinite input impedance, zero output impedance, and infinite frequency-independent gain. Therefore, the readback signal $V_o(t, x_0)$ obtained at the low-pass filter output in response to the applied voltage $V_P(t) = A \, \mathrm{rect}[(t - T_0)/\tau]$, where

$$\mathrm{rect}\left(\frac{t}{\tau}\right) = \begin{cases} 1 & \text{if } 0 \le t \le \tau \\ 0 & \text{otherwise} \end{cases}, \qquad (46.4)$$

and A denotes the pulse amplitude, obeys the differential equation

$$V_o'(t, x_0) = \frac{1}{R_{lpf}C_{lpf}}\left[-V_o(t, x_0) + \frac{R_{lpf}}{R^e[\Theta(t, x_0)]}V_P(t)\right]. \qquad (46.5)$$

As the voltage at the output of the low-pass filter depends on the value of the variable resistance $R^e[\Theta(t, x_0)]$, the readback signal is determined by solving jointly the differential equations (46.3) and (46.5), with initial conditions $\Theta(t_0, x_0) = \Theta_0$ and $V_o(t_0, x_0) = 0$. For example, a comparison between experimental and synthetic readback signals is shown in Figs. 46.30 and 46.31 for a time constant of the low-pass filter $\tau_{lpf} = 1.18\,\mu s$ and two values of the duration of the applied rectangular pulse. For a given cantilever design the function $R^e(\Theta)$ is determined experimentally. Finally, the parameters R^Θ and C^Θ used in the simple readback channel model are obtained via simulated annealing, where the cost function is given by the mean-square error between experimental and synthetic signals at the low-pass filter output.

Assuming that ideal control of the scanner is performed such that the time of application of a read pulse corresponds either to the cantilever located at the center of an indentation for detecting a "1" bit, or away from an indentation for detecting a "0" bit, two possible responses are obtained at the output of the low-pass filter as solutions of (46.3) and (46.5), which we denote as $V_{o,1}(t, x_0)$ and $V_{o,0}(t, x_0)$, respectively. By sampling the readback signal at the instant $t_s = t_0 + \tau$, simple threshold detection may in principle be applied to detect a written bit, where the value of the threshold is given by

$$V_{Th} = \frac{1}{2}\left[V_{o,1}(t_s, x_0) + V_{o,0}(t_s, x_0)\right]. \qquad (46.6)$$

As mentioned in Sect. 46.2, one of the most critical issues in detecting the presence or absence of an indentation is the high resolution required to extract the small signal $V_{o,1}(t_s, x_0) - V_{o,0}(t_s, x_0)$ that contains the information about the bit being "1" or "0" superimposed on the offset signal $V_{o,0}(t, x_0)$. As illustrated in Fig. 46.28, this problem can be solved by subtracting a suitable reference signal $V_{o,ref}(t, x_0)$ from the readback signal. The readback signal is thus given by

$$\tilde{V}_o(t, x_0) = V_o(t, x_0) - V_{o,ref}(t, x_0), \qquad (46.7)$$

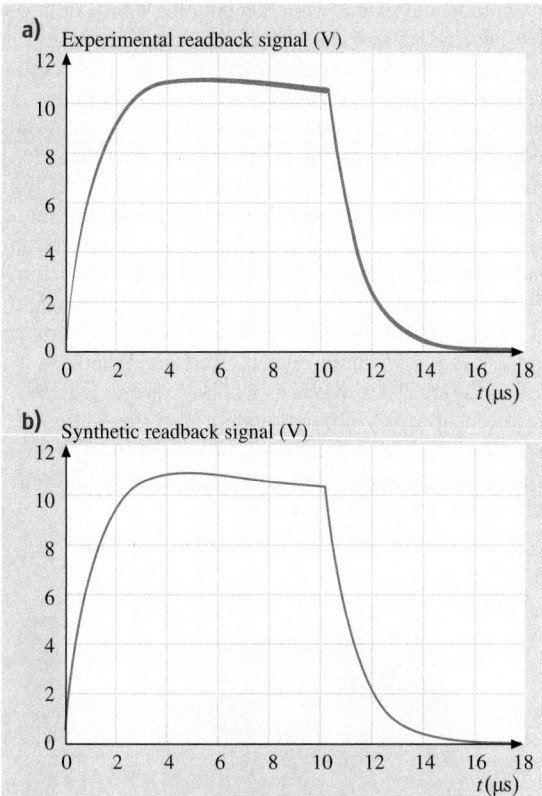

Fig. 46.30 (a) Experimental and **(b)** synthetic readback signal for $\tau = 10.25\,\mu s$ (after [46.12])

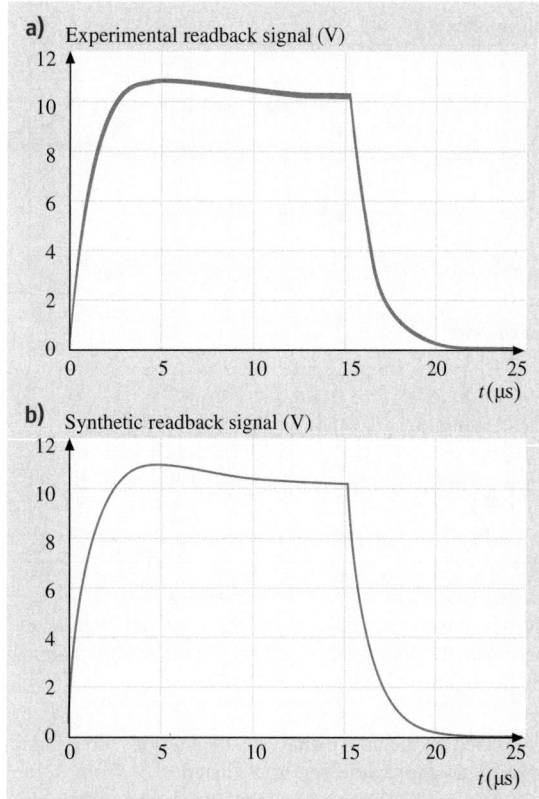

Fig. 46.31 (a) Experimental and **(b)** synthetic readback signal for $\tau = 15.25\,\mu s$ (after [46.12])

and the threshold is set at $\tilde{V}_{\text{Th}} = \frac{1}{2}\left[V_{o,1}(t_s, x_0) - V_{o,0}(t_s, x_0)\right]$. A VLSI implementation of the detection scheme analyzed here is presented in [46.44].

Now consider read pulses of duration τ that are periodically applied at instants $t_n = nT$, where $1/T$ denotes the symbol rate. Assuming that the response of the previous pulse has vanished as a new pulse is applied and that the temperature of the cantilever has approached the ambient temperature, i.e., $V_o(t_n, x_n) = 0$ and $\Theta(t_n, x_n) = \Theta_0$, then the analysis presented above still holds. In particular, the readback signal samples obtained in response to N pulses applied to the cantilever for detecting a sequence of N binary written symbols are expressed as

$$s(t_{s,n}) = \tilde{V}_o(t_{s,n}, x_n), \quad t_{s,n} = nT + \tau,$$
$$n = 0, \dots, N-1, \tag{46.8}$$

where $\tilde{V}_o(t, x_n)$ is given by (46.7) for pulses applied at time t_n and at distance $x_n = nTv$, $n = 0, \dots, N-1$

from the initial point of scanning. Note that the functions $V_o(t, x_n)$ and $V_{o,\text{ref}}(t, x_n)$ in (46.7) are ideally given by the solution of the differential equations (46.3) and (46.5) for $\eta_x = \Delta R^\Theta(x_n)/R^\Theta$ and $\eta_x = 0$, respectively.

The readback signal (46.7) at the output of the low-pass filter is observed in the presence of additive noise. Therefore, the readback signal for the detection of the n-th binary symbol is given by

$$r(t_{s,n}) = s(t_{s,n}) + w(t_{s,n}), \tag{46.9}$$

where $w(t)$ denotes the noise signal. The components of the noise signal that must be taken into account are thermal noise (Johnson's noise) from the sensor, which reaches a temperature of about $\Theta_1 = 350\,°C$ during the read process, as well as from front-end analog circuitry. However, note that besides thermal noise, medium-related noise also affects the overall system performance.

a) **Experimental data**

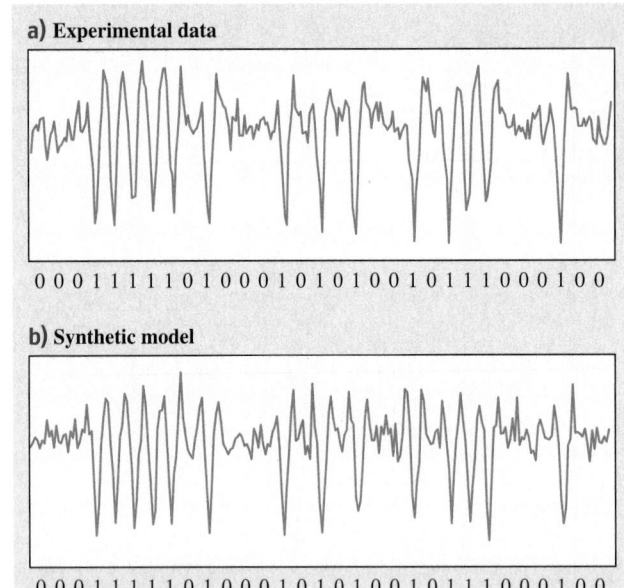

0 0 0 1 1 1 1 1 0 1 0 0 0 1 0 1 0 1 0 0 1 0 1 1 1 0 0 0 1 0 0

b) **Synthetic model**

0 0 0 1 1 1 1 1 0 1 0 0 0 1 0 1 0 1 0 0 1 0 1 1 1 0 0 0 1 0 0

Fig. 46.32a,b Comparison between (**a**) the readback signal obtained experimentally along a data track and (**b**) the readback signal obtained by the synthetic model (after [46.12])

Based on the above analysis, the response to a pulse applied to the cantilever at a distance x from some initial point can be calculated given the parameter $\eta_x = \Delta R^{\Theta}(x)/R^{\Theta}$. Recall that the value of η_x is proportional to the distance of the cantilever from the storage medium at the current location of the tip. Therefore, during tip displacement due to scanner motion, η_x is modulated from the topographical features of the storage surface such as written indentations, rings, and dust particles. This indicates that the modeling of the readback signal is a two-step process. First, a model for the storage surface topography is developed, which directly determines η_x, and then the above procedure is used to calculate the readback signal samples in response to pulses applied at selected points in the particular storage area.

In the absence of any imperfections during the manufacturing and the writing process, the storage surface would consist of completely flat regions interrupted by uniformly shaped indentations, possibly surrounded by polymer rings. A 1-D cross section of an indentation along the scanning direction is modeled by a function with one main lobe and two side lobes, one on each side of the main lobe, the magnitude and the extent of which can be varied independently. The side lobes are of opposite sign than the main lobe and simulate the polymer rings around written indentations. By varying their

magnitude and extent while keeping the total extent of the pulse fixed, one can simulate indentations/rings of varying width and asymmetric ring formation, phenomena that are caused by different recording conditions. In practice, however, no polymer surface is entirely flat and indentation shapes are far from uniform. The deviation of indentations from uniformity is simulated by scaling the amplitude of each pulse shape by a random number drawn from a Gaussian distribution with unit mean and adjustable variance. The surface roughness is in turn simulated by adding white Gaussian noise to the height of every point in the area of interest. Note that surface roughness is a medium-related effect and manifests itself in the readback signal as a noise process, which is, however, of a very different nature than thermal noise. The advantage of the adopted two-step model is that it naturally decouples these unrelated noise sources, as well as the write and the read processes.

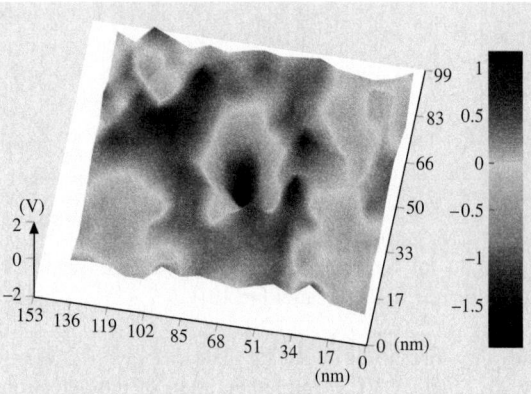

Fig. 46.33 Three-dimensional view of an isolated indentation obtained experimentally

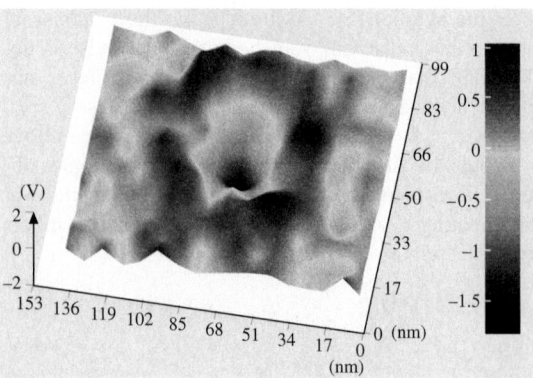

Fig. 46.34 Three-dimensional view of an isolated indentation obtained by the synthetic model

Figure 46.32 illustrates the experimental and synthetic readback signals obtained along a data track. The waveforms shown in Fig. 46.32 have been obtained by applying pulses at the oversampling rate of q/T, where q denotes the oversampling factor. For a more detailed comparison between model and actual signals, Figs. 46.33 and 46.34 illustrate 3-D views of isolated indentations from experimental and synthetic readback signals, respectively. The dark regions in the center of both figures correspond to the indentation centers, whereas dark regions around them are due to rings. Note also the irregular height of the surrounding surface, which is attributed to the roughness of the storage medium.

46.9 System Aspects

In this section, we describe various aspects of a storage system based on our millipede concept. Each cantilever can write data to and read data from a dedicated area of the polymer surface, called a *storage field*. As mentioned above, in each storage field the presence (absence) of an indentation corresponds to a logical "1" ("0"). All indentations are nominally of equal depth and size. The logical marks are placed at a fixed horizontal distance from each other along a data track. We refer to this distance, measured from one logical mark center to the next, as the *bit pitch* (BP). The vertical (cross-track) distance between logical mark centers, the *track pitch* (TP), is also fixed. To read and write data the polymer medium is moved under the (stationary) cantilever array at a constant velocity.

A robust way to achieve synchronization and servo control in an x/y-actuated large 2-D array is by reserving a small number of storage fields exclusively for timing recovery and servo-control purposes, as illustrated in Fig. 46.35. Because of the large number of levers in the millipede, this solution is advantageous in terms of overhead compared with the alternative of timing and servo information being embedded in all data fields.

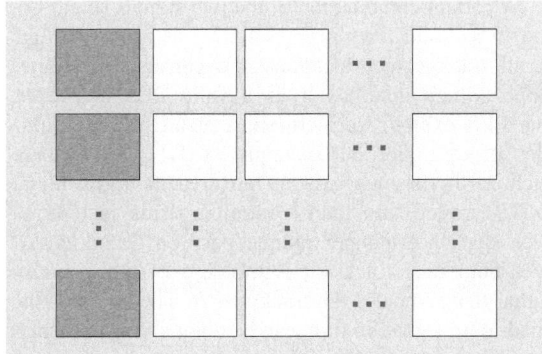

Fig. 46.35 Layout of data and servo/timing fields. *Dark boxes* represent dedicated servo/timing fields, *white boxes* represent data fields

46.9.1 PES Generation for the Servo Loop

With logical marks as densely spaced as in the millipede, accurate track following becomes a critical issue. Track following means controlling the position of each tip such that the tip is always positioned over the center of a desired track during reading. During writing, the tip position should be such that the written marks are aligned in a predefined way. In electromechanical systems, track following is performed in a servo loop, which is driven by an appropriate error signal called position-error signal (PES). Ideally, its magnitude is a direct estimate of the vertical (cross-track) distance of the tip from the closest track centerline, and its polarity indicates the direction of this offset.

Several approaches exist to generate a PES for AFM-based storage devices [46.9]. However, based on the results reported, none of these methods can achieve the track-following accuracy required for the millipede system. The quality of the PES directly affects the stability and robustness of the associated tracking servo loop [46.45].

We describe a method for generating a uniquely decodable PES for the millipede system [46.12, 27]. The method is based on the concept of *bursts* that are vertically displaced with respect to each other, arranged in such a way as to produce two signals in quadrature, which can be combined to provide a robust PES. This concept is borrowed from magnetic recording [46.45]. However, servo marks, as opposed to magnetic transitions, are placed in bursts labeled A and B for the in-phase signal and C and D for the quadrature signal. The centers of servo marks in burst B are vertically offset from mark centers in burst A by d' units of length. This amount of vertical spacing is related to the diameter of the written marks. The same principle applies to marks in the quadrature bursts C and D, with the additional condition that mark centers in burst C are offset by $d'/2$ units from mark centers in A in the cross-track direction. The latter condition is required in order to generate

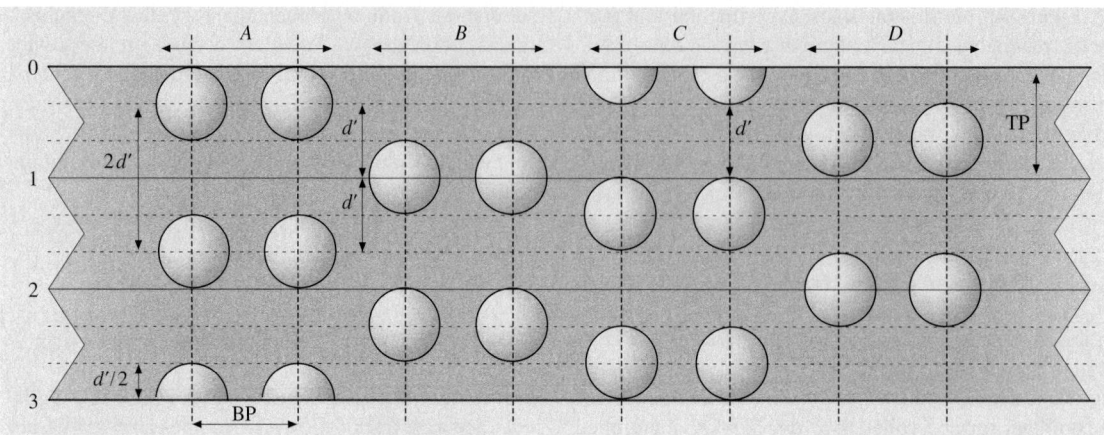

Fig. 46.36 Servo burst configuration (after [46.12])

a quadrature signal. The configuration of servo bursts is illustrated in Fig. 46.36 for a case where TP = $3d'/2$. Although each burst typically consists of many marks to enable averaging of the corresponding readout signals, only two marks per burst are shown here to simplify the presentation. The solid horizontal lines depict track centerlines, and circles represent written marks, which are modeled here as perfect conical indentations on the polymer storage surface.

To illustrate the principle of PES generation, let us assume that marks in all bursts are spaced BP units apart in the longitudinal direction, and that sampling occurs exactly at mark centers, so that timing is perfect. The assumption of perfect timing is made only for the purpose of illustration. In actual operation, sampling is performed with the aid of a timing recovery loop, as described in Sect. 46.9.2. Referring to Fig. 46.36, let us further assume that the cantilever/tip is located on the line labeled "0" and moves vertically toward line "3" in a line crossing the centers of the left-most marks in burst A (shown as a brown dashed line). The tip moves from the edge of the top mark toward its center, then toward its bottom edge, then to a blank space, again to a mark, and so on. The readout signal magnitude decreases linearly with the distance from the mark center and reaches a constant, background level value at a dis-

tance greater than the mark radius from the mark center, according to the adopted (conical mark) model. To synthesize the in-phase signal, the readout signal is also captured as the tip (conceptually) moves in a vertical line crossing the mark centers of burst B (brown dashed line in Fig. 46.36). The in-phase signal is then formed as the difference $\bar{A} - \bar{B}$, where $\bar{A}$ and $\bar{B}$ stand for the measured signal amplitudes in bursts A and B, respectively. This signal is represented by the line labeled "I" in Fig. 46.37. It has zero-crossings at integer multiples of d', which do not generally correspond to track centers because we set TP = $3d'/2$ in this example. Therefore, the I-signal is not a valid PES in itself. This is why the quadrature (Q) signal becomes necessary in this case. The Q-signal is generated from the servo readback signals of bursts C and D as $\bar{C} - \bar{D}$ and is also shown in Fig. 46.37 (Q-curve). Note that it exhibits zero-crossings at points where the I-signal has local extrema.

A certain combination of the two signals (I and Q), shown as solid lines in Fig. 46.37, has zero-crossings at all track center locations and constant (absolute) slope, which qualifies it as a valid PES. However, this PES exhibits zero-crossings at all integer multiples of $d'/2$. For our example of TP = $3d'/2$, three such zero-crossings exist in an area of width equal to TP around any track centerline. This fact, however, does not hamper unique position decoding. At even-numbered tracks, it is the zero of the *in-phase* signal that indicates the track center. The zeros of the quadrature signal, in turn, can be uniquely mapped into a position estimate by examining the polarity of the in-phase signal at the corresponding positions. This holds for any value of the combined PES within an area of width equal to TP around each current track cen-

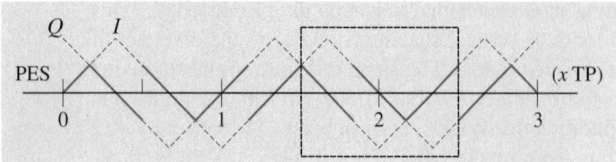

Fig. 46.37 Ideal position-error signal (after [46.12])

terline. The signals exchange roles for odd-numbered tracks. The current track number, which is known a priori from the seek operation, is used to determine the mode of operation for the position demodulation procedure.

The principle of PES generation based on servo marks has been verified experimentally. For this purpose, A, B, C, and D bursts were written by an AFM cantilever/tip on an appropriate polymer medium consisting of a polymer coating on a silicon substrate. The bit pitch was set to 42 nm, and the track pitch was taken to be approximately equal to d', the cross-track distance between A (C) and B (D) bursts. An image created by reading the written pattern with the same cantilever is shown in Fig. 46.38. Shaded areas indicate indentations. The readout signal from the cantilever was also used for servo demodulation, as described above. The resulting in-phase and quadrature signals are shown in Fig. 46.39. The track centerlines are indicated by vertical dotted lines in the graph.

It can be observed that the zero-crossings of the in-phase signal are closely aligned with the track centerlines, as well as with the minima and maxima of the quadrature signal, as required for unique position decoding across all possible cross-track positions, at least in cases where TP $\neq d'$. Moreover, the PES slope is nearly linear along a cross-track width of one track pitch around each track center, as TP $\approx d'$ in this case, although deviations from the ideal signal shape exist. These deviations occur mainly because written indentations do not have perfect conical shapes, and also because of medium noise due to the roughness of the recording medium. Nevertheless, the experimentally generated error signals indicate that the proposed concept is valid and promising. Specifically, the results indicate that servo self-writing is feasible, that servo demodulation is almost identical to data readout and can be performed by any cantilever without special provisions, and that the PES generated closely approximates the desirable features described above.

46.9.2 Timing Recovery

Similar to obtaining servo information based on using dedicated servo fields, we employ separate dedicated clock fields for recovery of timing information [46.12, 26, 27]. The concept is to have continuous access to a pilot signal for synchronization after initial phase acquisition and gain estimation. The recovered clock is then distributed to all remaining storage fields to allow reliable detection of random data. Initial phase

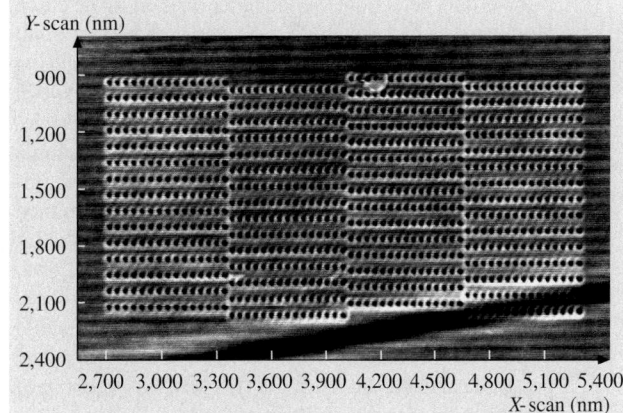

Fig. 46.38 Experimental A, B, C, and D servo bursts (BP = 42 nm) (after [46.12] © 2003 IEEE)

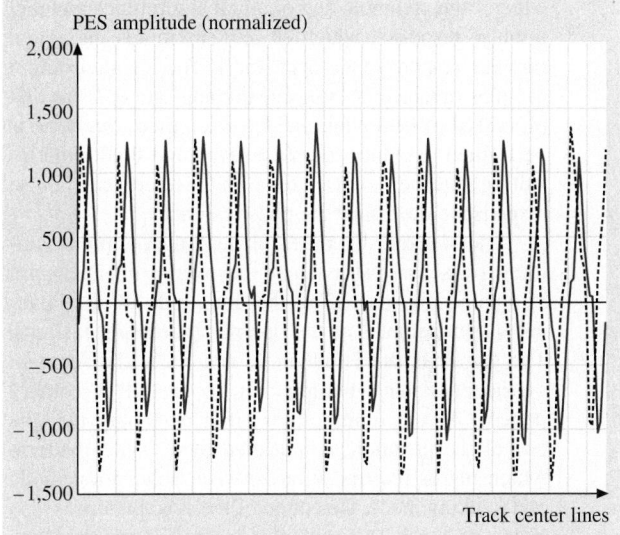

Fig. 46.39 Demodulated in-phase (*solid line*) and quadrature (*dashed line*) PES based on the servo burst of Fig. 46.38 (after [46.12])

acquisition is obtained by a robust correlation algorithm, gain estimation is based on averaging of the readback signal obtained from a predefined stored pattern, and tracking of the optimum sampling phase is achieved by a second-order digital loop.

At the beginning of the read process, several signal parameters have to be estimated prior to data detection. Besides the clock phase and frequency, it is necessary to estimate the gain of the overall read channel. To solve the problem of initial estimation of signal parameters prior to data detection, the sequence written in the clock field consists of a preamble, followed by a pattern of all "1"s for tracking the optimum sampling phase during the detection of random data. The transition between the preamble and the pattern of all "1"s must be detected reliably, as it indicates the start of data records to the remaining storage fields. Assuming that the initial frequency offset is within a small, predetermined range, usually 1000 parts-per-million (PPM), we distinguish the tasks needed for timing recovery as follows: (i) acquisition of the optimum sampling phase; (ii) estimation of the overall channel gain needed for threshold detection; (iii) detection of the transition between the preamble and the pattern of all "1"s; and (iv) tracking of the optimum sampling phase.

At the beginning of the acquisition process, an estimate of the optimum sampling phase is obtained by resorting to a correlation method. We rely on the knowledge of the preamble and of an ideal reference-channel impulse response, which closely resembles the actual impulse response (see Sect. 46.9). The channel output samples obtained at the oversampling rate q/T are first processed by removing the dc-offset, then averaging is performed to reduce the noise level, and finally the resulting sequence is correlated with the reference impulse response to determine the phase estimate.

After determining the estimate of the optimum sampling phase, an estimate of the overall channel gain is obtained by averaging the amplitude of the channel output samples at the optimum sampling instants. The gain estimate is obtained from an initial segment of the preamble corresponding to an "all 1" binary pattern. As mentioned above, it is necessary that the end of the preamble is indicated by a "sync" pattern, which marks the transition between acquisition mode and tracking mode. Detection of the sync pattern is also based on a robust correlation method. After the sync pattern, an "all 1" pattern, as in the case of robust phase acquisition and gain estimation, is employed for tracking. The "all 1" pattern corresponds to regularly spaced indentations, which convey reliable timing information.

Tracking of the optimum sampling phase is achieved by the second-order loop configuration shown in Fig. 46.40. Assuming data detection is performed at instants that correspond to integer multiples of the oversampling factor q, the deviation of the sampling phase from the optimum sampling phase is estimated as

$$\Delta \tau_n = r(t_{s,nq+1}) - r(t_{s,nq-1}). \tag{46.10}$$

This estimate of the phase deviation is input to a second-order loop filter, which provides an output

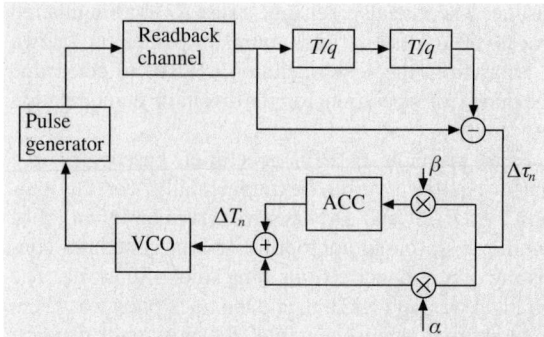

Fig. 46.40 Second-order loop for tracking the optimum sampling phase (after [46.12])

given by

$$\Delta T_n = u_n + \alpha \, \Delta \tau_n , \tag{46.11}$$

where the discrete-time integrator is recursively updated as

$$u_{n+1} = u_n + \beta \, \Delta \tau_n . \tag{46.12}$$

The loop-filter output then determines the control signal for a voltage-controlled oscillator (VCO).

Note that a similar concept for timing recovery can also be applied if no separate clock field is available. In this case, the timing information is extracted from the random user data on each storage field.

46.9.3 Considerations on Capacity and Data Rate

The ultimate locality provided by nanometer-sharp tips represents the pathway to the high areal densities that will be needed in the foreseeable future. The intrinsic nonlinear interactions between closely spaced indentations, however, may limit the minimum distance between successive indentations and, hence, the areal density. The storage capacity of a millipede-based storage device can be further increased by applying modulation or constrained codes [46.12].

With modulation coding, a desired constraint is imposed on the data-input sequence, so that the encoded data stream satisfies certain properties in the time or frequency domain. These codes are very important in digital recording devices and have become ubiquitous in all data-storage applications. The particular class of codes that imposes restrictions on the number of consecutive "1"s and "0"s in the encoded data sequence, generally known as run-length-limited (RLL)

Fig. 46.41 Areal density versus indentation spacing. Curve 1: $d = 0$; curve 2: $d = 1$; and curve 3: $d = 2$ ▶

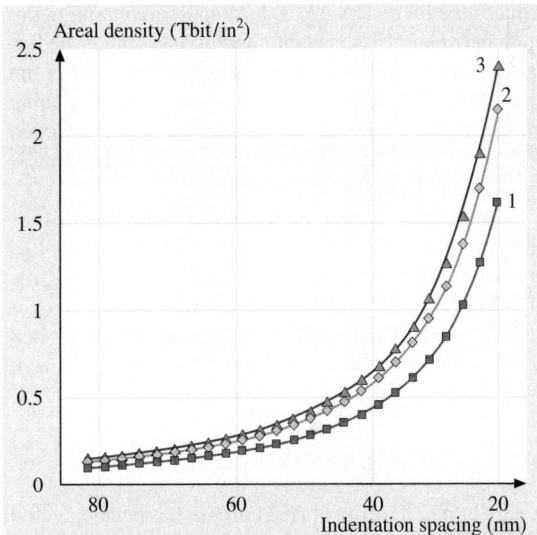

(d, k) codes [46.46], can be used to facilitate overwriting and also increase the effective areal density of a millipede-based storage device. The code parameters d and k are nonnegative integers with $k > d$, where d and k indicate the minimum and maximum number of "0"s between two successive "1"s, respectively. In the past, the precoded (RLL) (d, k) codes were mainly used for spreading the magnetic transitions further apart via the d-constraint, thereby minimizing intersymbol interference and nonlinear distortion, and for preventing loss of clock synchronization via the k-constraint. In optical recording, precoded RLL codes are primarily used for increasing the shortest pit length in order to improve the reliability of bit detection, as well as for limiting the number of identical symbols, so that useful timing information can be extracted from the readback signal.

For the millipede application, where dedicated clock fields are used, the k-constraint does not really play an important role and, therefore, can, in principle, be set to infinity, thereby facilitating the code design process. In a precoderless RLL code design, where the presence or absence of an indentation represents a "1" or "0", respectively, the d-constraint is instrumental in limiting the interference between successive indentations, as well as in increasing the effective areal density of the storage device. In particular, the quantity $(d + 1)R$, where R denotes the rate of the (d, k) code, is a direct measure of the increase in linear recording density. Clearly, the packing density can be made larger by increasing d. On the other hand, large values of d lead to codes with very low rate, which implies high recording symbol rates, thus rendering these codes impractical for storage systems that are limited by the clock speed. The choice of $d = 1$ and $k \geq 6$ guarantees the existence of a code with rate $R = 2/3$. Use of $(d = 1, k \geq 6)$ modulation coding reduces the bit distance by half while maintaining the pitch between "1"s constant, thereby increasing the linear density by a factor of $4/3$. Similarly, the choice of $d = 2$ and $k > 6$ guarantees the existence of a code with rate $R = 1/2$. Use of $(d = 2, k > 6)$ modulation coding

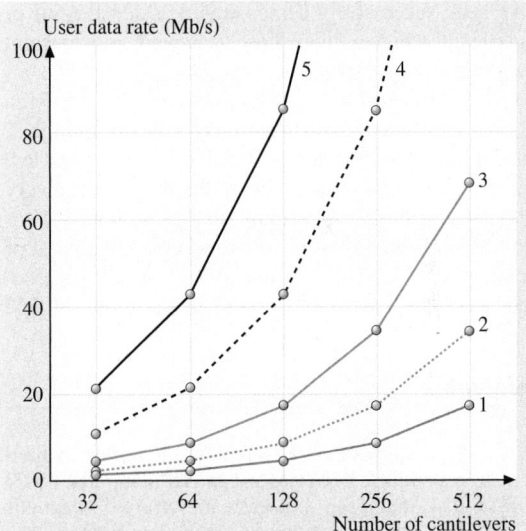

Fig. 46.42 User data rate versus number of active cantilevers for the $(d = 1, k \geq 6)$ coding scheme. Curve 1: $T = 20\,\mu s$; curve 2: $T = 10\,\mu s$; curve 3: $T = 5\,\mu s$; curve 4: $T = 2\,\mu s$; and curve 5: $T = 1\,\mu s$ (after [46.12])

Table 46.1 Areal density and storage capacity (after [46.12] with permission © 2003 IEEE)

Coding	Linear density (kb/in)	Track density (kt/in)	Areal density (Gb/in²)	Capacity (Gb)
Uncoded	847	847	717	1.21
$(d = 1, k \geq 6)$	1129	847	956	1.61
$(d = 2, k > 6)$	1269	847	1075	1.81

reduces the bit distance to a third while maintaining the pitch between "1"s constant, thereby increasing the linear density by a factor of 3/2. Figure 46.41 shows the areal density as a function of the indentation spacing for an uncoded system, as well as for systems coded with $(d = 1, k \geq 6)$ and $(d = 2, k > 6)$, where coding is applied only in the on-track direction.

Table 46.1 shows the achievable areal densities and storage capacities for a 32×32 cantilever array with 1024 storage fields, each having an area of $100 \times 100 \, \mu m^2$, resulting in a total storage area of $3.2 \times 3.2 \, mm^2$. The indentation pitch and the track pitch are set equal to 30 nm. Finally, for the computation of the storage capacity an overall efficiency of 85% has been assumed, taking into account the redundancy of the outer error-correction coding, as well as the presence of dedicated servo and clock fields.

Figure 46.42 shows the user data rate as a function of the total number of cantilevers accessed simultaneously, for various symbol rates and a $(d = 1, k \geq 6)$ modulation coding scheme. For example, for a 32×32 cantilever array, a system designed to access a maximum of 256 cantilevers every $T = 5 \, \mu s$ provides a user data rate of 34.1 Mb/s. Alternatively, by resorting to the row/column multiplexing scheme with $T = 80 \, \mu s$ a data rate of 8.5 Mb/s is achieved.

46.10 Conclusions

A very large 2-D array of AFM probes has been operated for the first time in a multiplexed/parallel fashion, and write/read/erase operations in a thin polymer medium have been successfully demonstrated at densities of or significantly higher than those achieved with current magnetic storage systems.

The millipede has the potential to achieve ultrahigh storage areal densities on the order of 1 Tb/in^2 or higher. The high areal storage density, small form factor, and low power consumption render the millipede concept a very attractive candidate as a future storage technology for mobile applications, as it offers several Gigabytes of capacity at data rates of several Megabytes per second. Dedicated servo and timing fields allow reliable system operation with a very small overhead. The read channel model provides the methodology for analyzing system performance and assessing various aspects of the detection and servo/timing algorithms that are key to achieving the system reliability required by the applications envisaged.

Although the first high-density storage operations with the largest 2-D AFM array chip ever built have been demonstrated, there are a number of issues that need further investigation such as overall system reliability, including long-term stability of written indentations, tip and medium wear, limits of data rates, array and cantilever size, as well as trade-offs between data rate and power consumption.

References

46.1   E. Grochowski, R. F. Hoyt: Future trends in hard disk drives, IEEE Trans. Magn. **32**, 1850–1854 (1996)

46.2   D. A. Thompson, J. S. Best: The future of magnetic data storage technology, IBM J. Res. Dev. **44**, 311–322 (2000)

46.3   G. Binnig, H. Rohrer, C. Gerber, E. Weibel: 7×7 reconstruction on Si(111) resolved in real space, Phys. Rev. Lett. **50**, 120–123 (1983)

46.4   G. Binnig, C. F. Quate, C. Gerber: Atomic force microscope, Phys. Rev. Lett. **56**, 930–933 (1986)

46.5   H. J. Mamin, D. Rugar: Thermomechanical writing with an atomic force microscope tip, Appl. Phys. Lett. **61**, 1003–1005 (1992)

46.6   R. P. Ried, H. J. Mamin, B. D. Terris, L. S. Fan, D. Rugar: 6-MHz 2-N/m piezoresistive atomic-force-microscope cantilevers with incisive tips, J. Microelectromech. Syst. **6**, 294–302 (1997)

46.7   B. D. Terris, S. A. Rishton, H. J. Mamin, R. P. Ried, D. Rugar: Atomic force microscope-based data storage: Track servo and wear study, Appl. Phys. A **66**, S809–S813 (1998)

46.8   H. J. Mamin, B. D. Terris, L. S. Fan, S. Hoen, R. C. Barrett, D. Rugar: High-density data storage using proximal probe techniques, IBM J. Res. Dev. **39**, 681–699 (1995)

46.9   H. J. Mamin, R. P. Ried, B. D. Terris, D. Rugar: High-density data storage based on the atomic force microscope, Proc. IEEE **87**, 1014–1027 (1999)

46.10   L. R. Carley, J. A. Bain, G. K. Fedder, D. W. Greve, D. F. Guillou, M. S. C. Lu, T. Mukherjee, S. Santhanam, L. Abelmann, S. Min: Single-chip computers with microelectromechanical systems-based magnetic memory, J. Appl. Phys. **87**, 6680–6685 (2000)

46.11 G. Gibson, T. I. Kamins, M. S. Keshner, S. L. Neber-huis, C. M. Perlov, C. C. Yang: Ultra-high density storage device, (1996)US Patent 5,557,596

46.12 E. Eleftheriou, T. Antonakopoulos, G. K. Binnig, G. Cherubini, M. Despont, A. Dholakia, U. Dürig, M. A. Lantz, H. Pozidis, H. E. Rothuizen, P. Vettiger: millipede – A MEMS-based scanning-probe data-storage system, IEEE Trans. Magn. **39**, 938–945 (2003)

46.13 G. K. Binnig, H. Rohrer, P. Vettiger: Mass-storage applications of local probe arrays, (1998)US Patent 5,835,477

46.14 P. Vettiger, J. Brugger, M. Despont, U. Drechsler, U. Dürig, W. Häberle, M. Lutwyche, H. Rothuizen, R. Stutz, R. Widmer, G. Binnig: Ultrahigh density, high-data-rate NEMS-based AFM data storage system, J. Microelectron. Eng. **46**, 11–17 (1999)

46.15 P. Vettiger, G. Cross, M. Despont, U. Drechsler, U. Dürig, B. Gotsmann, W. Häberle, M. A. Lantz, H. E. Rothuizen, R. Stutz, G. K. Binnig: The "millipede" – Nanotechnology entering data storage, IEEE Trans. Nanotechnol. **1**, 39–55 (2002)

46.16 M. Lutwyche, C. Andreoli, G. Binnig, J. Brugger, U. Drechsler, W. Häberle, H. Rohrer, H. Rothuizen, P. Vettiger: Microfabrication and parallel operation of 5 × 5 2D AFM cantilever array for data storage and imaging, Proc. IEEE 11th Int. Workshop MEMS, Heidelberg 1998 (IEEE, Piscataway 1998) 8–11

46.17 M. Lutwyche, C. Andreoli, G. Binnig, J. Brugger, U. Drechsler, W. Häberle, H. Rohrer, H. Rothuizen, P. Vettiger, G. Yaralioglu, C. Quate: 5 × 5 2D AFM cantilever arrays: A first step towards a ter-abit storage device, Sens. Actuators A **73**, 89–94 (1999)

46.18 P. Vettiger, M. Despont, U. Drechsler, U. Dürig, W. Häberle, M. I. Lutwyche, H. E. Rothuizen, R. Stutz, R. Widmer, G. K. Binnig: The "millipede" – More than one thousand tips for future AFM data storage, IBM J. Res. Dev. **44**, 323–340 (2000)

46.19 B. W. Chui, H. J. Mamin, B. D. Terris, D. Rugar, K. E. Goodson, T. W. Kenny: Micromachined heaters with 1-μs thermal time constants for AFM thermo-mechanical data storage, Proc. IEEE Transducers, Chicago 1997 (IEEE, Piscataway 1997) 1085–1088

46.20 W. P. King, J. G. Santiago, T. W. Kenny, K. E. Good-son: Modelling and prediction of sub-micrometer heat transfer during thermomechanical data stor-age, 1999 Microelectromechanical Systems (MEMS). Proc. ASME Intl. Mechanical Engineering Congress and Exposition, ed. by A. P. Lee, L. Lin, F. K. Forster, Y. C. Young, K. Goodson, R. S. Keynton (ASME, New York 1999) 583–588

46.21 W. P. King, T. W. Kenny, K. E. Goodson, G. L. W. Cross, M. Despont, U. Dürig, H. Rothuizen, G. Binnig, P. Vettiger: Design of atomic force microscope can-tilevers for combined thermomechanical writing and thermal reading in array operation, J. Micro-electromech. Syst. **11**, 765–774 (2002)

46.22 G. K. Binnig, M. Despont, W. Häberle, P. Vettiger: Method of forming ultrasmall structures and ap-paratus therefore, (March 1999)US Patent Office, Application No. 147865

46.23 G. Binnig, M. Despont, U. Drechsler, W. Häberle, M. Lutwyche, P. Vettiger, H. J. Mamin, B. W. Chui, T. W. Kenny: Ultra high-density AFM data storage with erase capability, Appl. Phys. Lett. **74**, 1329–1331 (1999)

46.24 G. K. Binnig, J. Brugger, W. Häberle, P. Vettiger: Investigation and/or manipulation device, (March 1999)US Patent Office, Application No. 147867

46.25 S. M. Sze: *Physics of Semiconductors Devices* (Wiley, New York 1981)

46.26 G. Cherubini, T. Antonakopoulos, P. Bächtold, G. K. Binnig, M. Despont, U. Drechsler, A. Dho-lakia, U. Dürig, E. Eleftheriou, B. Gotsmann, W. Häberle, M. A. Lantz, T. Loeliger, H. Pozidis, H. E. Rothuizen, R. Stutz, P. Vettiger: The milli-pede, a very dense, highly parallel scanning-probe data-storage system, ESSCIRC – Proceedings 28th European Solid-State Circuits Conference, ed. by A. Baschirotto, P. Malcovati (Univ. Bologna, Bologna 2002) 121–125

46.27 E. Eleftheriou, T. Antonakopoulos, G. K. Binnig, G. Cherubini, M. Despont, A. Dholakia, U. Dürig, M. A. Lantz, H. Pozidis, H. E. Rothuizen, P. Vettiger: "millipede": A MEMS-based scanning-probe data-storage system, Digest of the Asia-Pacific Magnetic Recording Conference 2002, APMRC '02 (IEEE, Pis-cataway 2002) CE–2–1–CE2–2

46.28 H. Pozidis, W. Häberle, D. Wiesmann, U. Drech-sler, M. Despont, T. R. Albrecht, E. Eleftheriou: Demonstration of thermomechanical recording at 641 Gbit/sq.in., IEEE Trans. Magn. **40**, 2531–2536 (2004)

46.29 M. Despont, J. Brugger, U. Drechsler, U. Dürig, W. Häberle, M. Lutwyche, H. Rothuizen, R. Stutz, R. Widmer, G. Binnig, H. Rohrer, P. Vettiger: VLSI-NEMS chip for AFM data storage, Technical Digest 12th IEEE Int. Micro Electro Mechanical Systems Conf. "MEMS '99", Orlando 1999 (IEEE, Piscataway 1999) 564–569

46.30 T. S. Ravi, R. B. Marcus: Oxidation sharpening of silicon tips, J. Vac. Sci. Technol. B **9**, 2733–2737 (1991)

46.31 M. Despont, J. Brugger, U. Drechsler, U. Dürig, W. Häberle, M. Lutwyche, H. Rothuizen, R. Stutz, R. Widmer, G. Binnig, H. Rohrer, P. Vettiger: VLSI-NEMS chip for parallel AFM data storage, Sens. Actuators A **80**, 100–107 (2000)

46.32 H. Rothuizen, M. Despont, U. Drechsler, G. Genolet, W. Häberle, M. Lutwyche, R. Stutz, P. Vettiger: Com-pact copper/epoxy-based micromachined electro-magnetic scanner for scanning probe applications, Technical Digest, 15th IEEE Int. Conf. on Micro Electro Mechanical Systems "MEMS 2002" (IEEE, Pis-cataway 2002) 582–585

46.33   S. C. Minne, G. Yaralioglu, S. R. Manalis, J. D. Adams, A. Atalar, C. F. Quate: Automated parallel high-speed atomic force microscopy, Appl. Phys. Lett. **72**, 2340–2342 (1998)

46.34   M. Lutwyche, U. Drechsler, W. Häberle, R. Widmer, H. Rothuizen, P. Vettiger, J. Thaysen: Planar micromagnetic x/y/z scanner with five degrees of freedom. In: *Magnetic Materials, Processes, and Devices: Applications to Storage and Micromechanical Systems (MEMS)*, Vol. 98-20, ed. by L. Romankiw, S. Krongelb, C. H. Ahn (Electrochemical Society, Pennington 1999) pp. 423–433

46.35   H. Rothuizen, U. Drechsler, G. Genolet, W. Häberle, M. Lutwyche, R. Stutz, R. Widmer, P. Vettiger: Fabrication of a micromachined magnetic x/y/z scanner for parallel scanning probe applications, Microelectron. Eng. **53**, 509–512 (2000)

46.36   J.-J. Choi, H. Park, K. Y. Kim, J. U. Jeon: Electromagnetic micro x-y stage for probe-based data storage, J. Semicond. Technol. Sci. **1**, 84–93 (2001)

46.37   H. Lorenz, M. Despont, N. Fahrni, J. Brugger, P. Vettiger, P. Renaud: High-aspect-ratio, ultrathick, negative-tone near-UV photoresist and its applications for MEMS, Sens. Actuators A **64**, 33–39 (1998)

46.38   C. Q. Davis, D. Freeman: Using a light microscope to measure motions with nanometer accuracy, Opt. Eng. **37**, 1299–1304 (1998)

46.39   A. Pantazi, A. Lantz, G. Cherubini, H. Pozidis, E. Eleftheriou: A servomechanism for a micro-electro-mechanical-system-based scanning-probe data storage device, Nanotechn. **15**, S612–S621 (2004)

46.40   M. I. Lutwyche, M. Despont, U. Drechsler, U. Dürig, W. Häberle, H. Rothuizen, R. Stutz, R. Widmer, G. K. Binnig, P. Vettiger: Highly parallel data storage system based on scanning probe arrays, Appl. Phys. Lett. **77**, 3299–3301 (2000)

46.41   K. Fuchs, C. Friedrich, J. Weese: Viscoelastic properties of narrow-distribution poly(methyl metacrylates), Macromolecules **29**, 5893–5901 (1996)

46.42   U. Dürig, B. Gotsman: This estimate is based on a fluid dynamic deformation model of a thin film, private communication

46.43   J. D. Ferry: *Viscoelastic Properties of Polymers, 3rd edition* (Wiley, New York 1980)

46.44   T. Loeliger, P. Bächtold, G. K. Binnig, G. Cherubini, U. Dürig, E. Eleftheriou, P. Vettiger, M. Uster, H. Jäckel: CMOS sensor array with cell-level analog-to-digital conversion for local probe data storage, ESSCIRC – Proceedings 28th European Solid-State Circuits Conference, ed. by A. Baschirotto, P. Malcovati (Univ. Bologna, Bologna 2002) 623–626

46.45   A. H. Sacks: Position signal generation in magnetic disk drives. Ph.D. Thesis (Carnegie Mellon University, Pittsburgh 1995)

46.46   K. A. S. Immink: *Coding Techniques for Digital Recorders* (Prentice Hall, Hemel 1991)

47. Nanotechnology for Data Storage Applications

This chapter considers atomic force microscopy (AFM) as an enabling technology for data storage applications, considering already existing technologies such as hard disk drives (HDD), optical disk drives (ODD) and flash memories that currently dominate the nonvolatile data storage market, together with future devices based on magnetoresistive and phase change effects. The issue at hand is the question of whether the novel AFM-based storage, dubbed *probe storage*, can offer a competing approach to the currently available technologies by playing the role of a disruptive technology. Probe storage will be contrasted to HDD and ODD, which are purely mechanical as they are based on a rotating disk that uses just a single probe to address billions of bits of data, and nonvolatile random-access memory (RAM) that has no moving parts yet requires billions of interconnects. In particular, capacity, areal density, transfer rate, form factor and the cost of various data storage devices will be discussed and the unique opportunity offered by probe storage in employing massive parallelism will be outlined. It will be shown that probe storage bridges the gap between HDD, ODD and other nonvolatile RAM, drawing from the strength of each one of these and adding a significant attribute neither of these

has; namely, the possibility of addressing a very large number of nanoscale bits of data in parallel. This chapter differs from the other chapters in this book in that it addresses the important issue of whether a given scientific effort, namely, probe storage, is mature enough to evolve into a commercially viable technology. The answer seems to indicate that there is indeed a huge niche in the data storage arena that such a technology is uniquely qualified to fill, which is large enough to justify a major investment in research and development. Indeed, as other chapters indicate, such an effort is developing at a rapid pace, with hopes of having a viable product within a few years.

This chapter will differ from the other chapters in this handbook in that it addresses the important issue of whether a given scientific effort is mature enough to evolve into a commercially viable technology. Specifically, we consider scanning probe microscopy (SPM), consisting of scanning tunneling microscopy (STM) [47.1–3] and atomic force microscopy (AFM) with all their variants, as a means for storing and retrieving nanoscale bits of data to and from a substrate. Indeed, many studies have already been published in which researchers demonstrated the feasibility of using SPM techniques that hold promise for realistic applications [47.4]. However, in approaching a topic from a practical point of view, that is, whether it has the

ingredients that will spawn a commercial product, it is imperative to obtain a clear view of the status of competing technologies. The commercially available technologies that will concern us here entail two classes of devices, one consisting of hard disk drives (HDD) and optical disk drives (ODD), and the other consisting of nonvolatile random-access memories (NVRAM) based on charge trapping (flash memory) that currently dominate the data storage market. The latter may soon be supplemented by magnetoresistive RAM (MRAM) and phase change RAM (PC-RAM). For brevity, we will refer to the first class as HDD and the second as NVRAM.

It is commonly thought that nanotechnology is a breakthrough technology that is a quantum leap beyond existing technologies in its capabilities. In reality, however, nanotechnology is in many cases just a limiting case of the already commercially available technologies. The latter may progress at such a high rate that the distinction between what is considered *conventional* and *nano* becomes blurred. A case in point is data storage, in which distances and times are already specified by nanometers and nanoseconds. For example, the head of a HDD in a PC's hard disk flies across the storage medium (platter) at a speed of $10 \, \text{m/s}$ (≈ 22 miles per hour) at a height of $\approx 10 \, \text{nm}$, reading and writing bits of data on a nanosecond time scale. Also, both giant magnetoresistance (GMR) devices that read magnetic data in a HDD and transistors belonging to NVRAM devices use nanoscale structures, one employing spintronic effects and the other tunneling junctions between pairs of transistors, where the quantum laws of physics have to be used to describe their operation.

The issue at hand is, therefore, whether novel nanotechnologies can offer a competing approach to currently available technologies by playing the role of a *disruptive technology*. This term has been coined in the high-technology industry to describe a situation where, figuratively speaking, a discovery developed in a basement can lead to a development of a new device that has the potential of becoming a commercial alternative to a commonly used one. Such a disruption of the success of a commercially established technology by a product developed by a new start-up, for example, has been witnessed more and more frequently in recent years. One would even venture to assume that the support that companies give to research groups at universities can be attributed to their concern about a possible disruption of their own technology.

The important issue here is the question of what nanotechnology can offer to an already mature field such as data storage. As will be discussed later, HDD and NVRAM are worlds apart in technology. HDD is purely mechanical, as it is based on a rotating disk that uses just a single probe (head) to address billions of bits of data. NVRAM, on the other hand, has no moving parts yet requires billions of probes, i. e., interconnects, positioned at the intersection of horizontal and vertical interconnects, with one probe for every bit of data. Another distinction between HDD and NVRAM is the time it takes to access a random bit of data. For a HDD the random-access time is $10 \, \text{ms}$ ($75 \, \text{ms}$ for ODD), while for NVRAM it is less than one μs. Clearly, NVRAM beats HDD (and ODD) in the random-access arena, if that is the criteria for performance. On the other hand, the rate at which a HDD can read data sequentially is

rather impressive. The current performance of a HDD stands at $436 \, \text{Mb/s}$ and $22 \, \text{Mb/s}$ for an ODD, comparable to what NVRAM can offer, being of the order of $10 \, \text{MByte/s}$. Another playing field in the battle between HDD and NVRAM performance concerns capacity and the areal density of bits of data. Here, capacity measures the total number of bytes of data, while areal density measures the number of bits per square inch. The first addresses the ever growing need for storing larger and larger amounts of data, say, for high-definition television applications, while the latter is related to what is commonly called the form factor, namely, the physical size of a storage device. The form factor is intimately connected to portability, as more and more hand-held devices have restrictions on the physical size of each of its components. Note also that a major consideration in comparing HDD and NVRAM is cost, a topic that will be addressed later.

As discussed before, both HDD and ODD use just a single probe that addresses all the bits of data on a platter where data is stored. NVRAM, on the other hand, has to resort to a grid of interconnects that contact each bit of information individually. Such a grid, in which each bit is interfaced with one *word line* and one *bit line*, occupies expensive real estate across a storage device, and, even more importantly, it requires that the size of the interconnects be on the order of the bits of data they address. As long as these bits have dimensions on the order of, say, $1 \, \mu\text{m}$, one is able to use conventional lithography for the fabrication of the interconnects. However, when the bits of data shrink to, say, $25 \, \text{nm}$ in size, interconnects are no longer a viable option due to lithographic limitations. HDD and ODD, in contrast, require no interconnects, so this limitation on the size of bits is not a concern. However, there are many other problems that arise as one tries to push the size of the stored bits to nanoscale dimensions, as will be discussed later.

By now we have touched on the heart of the challenge that nanotechnology-based storage faces if it is to compete with HDD and NVRAM in both performance and cost. Practically, one should restate this issue and ask a more modest question, namely, what niche nanotechnology will fill in consumer data storage applications. To address this question, one notes that nanotechnology is concerned with the ability to miniaturize and characterize nanometer-sized structures. For data storage, the question is how to address nanoscale data bits rapidly without having to resort to the use of too many interconnects.

A possible solution to this challenge emerged from a brilliant idea developed at IBM and Stanford laboratories, as part of their scanning probe microscopy research [47.5–9].

A conventional AFM employs a single cantilever with a sharp tip at its end to raster-scan a sample and its minute deflections, due to protrusion on the surface, and so generate a map of the surface topography. Under ideal conditions one can even obtain atomically resolved images. The IBM–Stanford group demonstrated that it is possible to employ a large number of such cantilevers and operate them in parallel. Such an operation provides a faster means of obtaining images of different parts of a sample simultaneously. These cantilever arrays are often referred to as types of microelectromechanical systems (MEMS), or if they are small enough, nanoelectromechanical systems (NEMS).

One can view this breakthrough concept as an enabling technology for data storage applications in that it bridges the gap between HDD and NVRAM technologies, drawing from the strength of each. To appreciate this concept, consider a medium whose area is sectioned into 1000 squares, each containing 1 000 000 bits of data. Each square is addressed by its own interconnect whose width can now be as wide as 1000 bits of data. Consider now a square structure consisting of 1000 cantilevers whose tips are aligned such that each one addresses its own square. Such an arrangement, known as probe storage, and dubbed *millipede* by IBM, has some attributes belonging to an HDD, some attributes belonging to NVRAM, and potentially one significant attribute neither of these has: the possibility of addressing nanoscale bits of data. While random access is faster than that of a HDD, it is slower than that of NVRAM.

In this chapter, we will address the issues mentioned in the introduction with an emphasis on a comparison between existing and future development of conventional technologies and nanotechnologies. We will not address more futuristic approaches such as DNA and molecular electronics for data storage applications—fascinating topics on their own merit, yet deemed too far in the future for our more technological approach.

47.1 Current Status of Commercial Data Storage Devices

Before describing the current status of commercial storage devices, it is worthwhile to reiterate definitions and cover some new key features associated with HDD [47.10]. A schematic of a hard disk drive, shown in Fig. 47.1, is configured into sectors, each having radial tracks; there are typically 58 000 tracks per inch for HDD and 34 000 tracks per inch for ODD, 562 kbits per inch for HDD, and 100 kbits per inch for ODD. In 1993, a HDD head was flying above the storage media at a typical height of 100 nm, which dropped to 10 nm in 2002, and is expected to drop to 4 nm in 2004.

Hard Disk Drives and Optical Disk Drives

The time it takes the read–write head to get to a random bit along a track is called latency, while the time it takes the head to get to a random sector is called the seek time. The time to get to any random bit is called the access time. From 1990 to 2002 this time decreased linearly from 20 ms to 8 ms, and is expected to decrease to 4 ms by 2004.

The transfer time is the time it takes to transfer a given number of bits from the hard drive to a given destination in a computer. The total number of bits per unit area, usually given in bits per square inch (bpsi), is called the areal density. The total number of bytes in a storage device is called he capacity.

The physical size of a storage device and its price are paramount factors in its commercial viability. Disk drives are usually divided into magnetic and optical, although a combination of both is currently advanced by several companies and research groups around the world. A magnetic hard drive (HDD) has a higher capacity, lower cost, and higher performance than an optical hard drive (ODD), but its disk is not removable. The ODD, on the other hand, has a removable disk that can be mass-produced by parallel replication (stamping), making it cheaper for large-scale distribution. Also, an ODD is configured to operate in such a way that the head reads data sequentially, while a HDD is optimized for random-access operation. Note that units of bits are usually denoted by a lower case "b" and used for areal density and transfer rate, while units of bytes are usually denoted by an upper case "B" and used for capacity. A note in passing: areas are, oddly enough, expressed in square inches, and the holy grail for areal density is denoted by 1 tbpsi (one terabit per square inch).

Now, a crucial question when assessing the viability of a promised technology concerns the demand for the commercial product in question. Figure 47.2 estimates that a grand total of 2.12 million terabytes of information were produced in 1999 [47.11]. The breakdown is 83 terabytes for optical, 240 terabytes for paper,

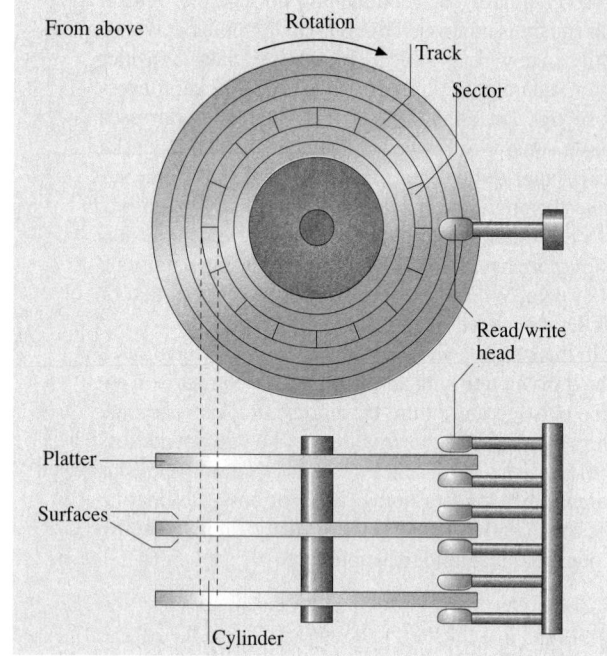

Fig. 47.1 A schematic diagram of a hard disk drive illustrating the concepts of cylinder, track, and sector

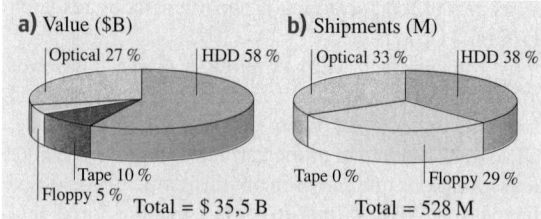

Fig. 47.3a,b Storage drive industry shipments in [47.10]. (**a**) Categorized by value and (**b**) units shipped

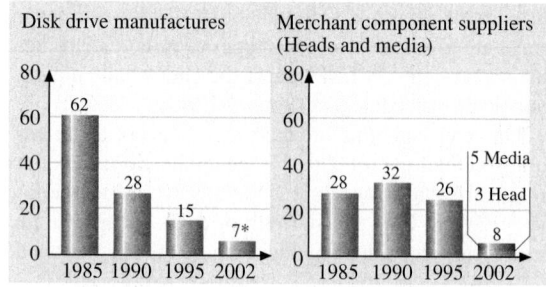

Fig. 47.4 Disk drive manufacturers and merchant component suppliers [47.12]. The number of disk drive manufacturers decreased from 62 in 1985 to a mere seven by 2002

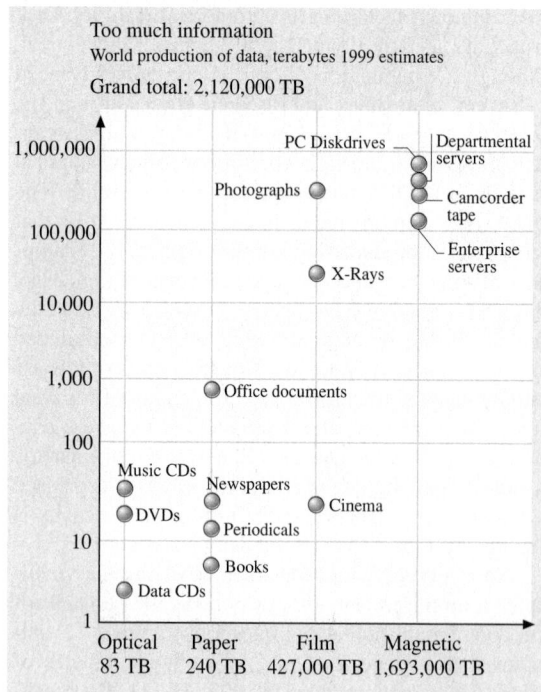

Fig. 47.2 World production of data in 1999 estimates

427 000 terabytes for film, and 1.6 million terabytes for magnetic data storage. PC disk drives take the first place in demand: 700 000 terabytes of data.

To appreciate the economic impact of the HDD storage drive industry (NVRAM will be discussed later), consider Fig. 47.3, which is current to 2002 [47.10]. Here, one observes that the industry shipped 528 million units with a value of $ 35.5 billion, 58% of the value consisting of HDD and 27% of ODD, while the numbers shipped are almost evenly divided between HDD, ODD, and floppy disks. Based on Fig. 47.3, one would think that such a huge market would involve a large number of companies competing with each other on price and performance.

However, as shown in Fig. 47.4, the cut-throat competition among various companies gave rise to the disappearance of many of them and mergers between the others [47.12]. Note that the number of disk drive manufacturers dropped from 62 in 1985 to only seven in 2002, while merchant components suppliers dropped from 28 to eight during that period. For example, Maxtor and Quantum merged, IBM sold its disk drive division to Hitachi, and Fujitsu exited desktop HDD altogether. In 2002, Seagate shipped 1 959 000 terabytes of storage, 10.2 million enterprise drives, 1.2 million 15 000-rpm drives, and 44.8 million personal storage devices; they expect that there will be migration to smaller form factors and increased areal density in all segments of the industry and that mobility will be become increasingly important.

A projection of HDD areal density on a logarithmic scale is shown in Fig. 47.5 [47.13]. Whereas in 1990 the areal density was only 1 gigabit per square inch, it reached 100 gigabits per square inch in 2002 and is projected to reach 1 Tbpsi in 2006. One wonders whether it will take nanotechnology to get to 50 Tbpsi in 2017, as the figure suggests. As quoted from a Seagate presentation, "in traditional applications, areal density has grown faster than consumption, moving to a single platter society in PC's that enables smaller form factors. In non-traditional markets, new applications drive capacity growth, especially with video content." [47.12].

Table 47.1 [47.10] presents several key parameters characterizing HDD and ODD that should serve as a reality check for the potential viability of any nanotechnology approach that aims to act as a disruptive technology. Thus any new concept should be judged against this current status of HDD and ODD and, as described later, NVRAM. Note that, in contrast to these impressive parameters associated with HDD and ODD, seek time is limited to 10 ms and 75 ms, respectively.

A comparison of the perceived limits to HDD and ODD performance is presented in Table 47.2 [47.10]. For magnetic disk drives, the limit to areal density is the superparamagnetic effect in which too small a magnetic domain can be thermally excited into an opposite di-

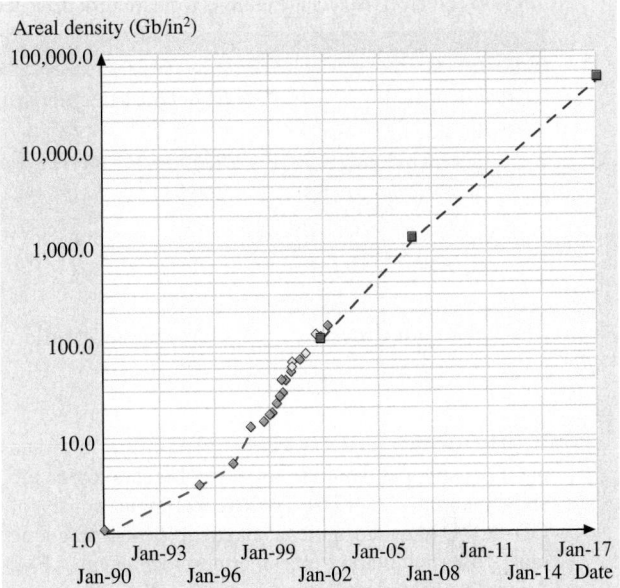

Fig. 47.5 Projected areal density for hard disk drives [47.13]

rection, losing the reliability of the stored data. Ways and means for approaching this limit and maintaining reliability will be discussed later. For optical data storage, the conventional limit is clearly the spot size of the focused laser beam. However, near-field methods are currently being developed to circumvent this limit. As far as speed goes, switching times for magnetic domains is around 1 ns, while for laser-induced thermal effects it is ten times slower.

Table 47.1 A comparison of the parameters of HDD and ODD

Product comparison		
Parameter	HDD (80 Gb desktop)	ODD (4.7 Gb DVD-RAM)
Areal density	32.6 Gb/in^2	3.3 Gb/in^2
	Best = 50.0 Gb/in^2 (2.5″)	Best = 4.6 Gb/in^2 (50 mm)
TPI/BPI	58 K/562 K	34 K/100 K
Capacity/platter	40 Gb (2 side, 95 mm)	9.4 Gb (2 side, 120 mm)
Unit price (drive)	$80–$120	$365
Price/Gb (drive)	$1–$1.5	$39
Price/Gb (media)	–	~ $1.5–$2
Seek time	10 ms	75 ms
Transfer rate (writing)	436 Mb/s	22 Mb/s
Research laboratory comparison		
Parameter	HDD (longitudinal)	ODD (blue near-field)
Areal density	130 Gb/in^2	45 Gb/in^2
TPI/BPI	213 K/610 K	141 K/319 K

Table 47.2 Perceived technical limits for hard disk drive technology [47.10]

	Magnetic	Optical

Areal density (Gb/in^2)	**Superparamagnetism**
	1994: 35 (scaling)
	1997: 100 (lower bit aspect ratio)
	2002: 500 (perpendicular)
	         1000 (patterned)
	         > 1000 (HAMR)
		Focused spot size
		1994: 0.7 ($\lambda = 780$ nm)
		1997: 3.3 ($\lambda = 650$ nm)
		2002: 19 (Blu-Ray)
		         2× (ML)
		         2–3× (NF)
		         ? × (Volume)

Switching time (ns)	**Gyromagnetic ratio** ≈ 1	**Media crystallization** ≈ 10

The current and projected relationship between areal density and data rate for several technologies are shown in Fig. 47.6 [47.5]. These technologies consist of CD-ROM, magneto-optical, magnetic disks, magnetic tape, SIL, holography, and thermo-mechanical AFM. According to this projection, magnetic disks are in the forefront of technology and are expected to have the same areal density as that of the probe storage concept, yet have a much faster data rate. The latter technology, as predicted in this figure, will grow from a data rate of only 1 Mb/s to more than 10 Mb/s, accompanied by an increase in areal density from 10 gigabits per square inch to hundreds of gigabits per square inch. The probe storage concept will be further discussed later in this chapter.

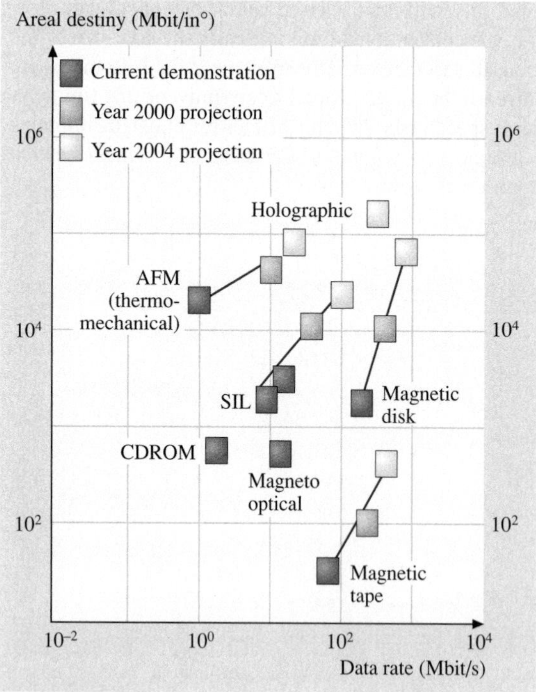

47.1.1 Nonvolatile Random–Access Memory

Demand for flash memory has been growing rapidly over the last few years, driven by the rapid expansion of several markets, including mobile communications and portable media devices (notably cameras and MP3 players). Flash memory is by now the fastest growing segment of the memory market. Figure 47.7 [47.14] shows the total value of the flash memory market year by year. It can be seen that the market is increasing by nearly a factor of ten every five years and up to 100%

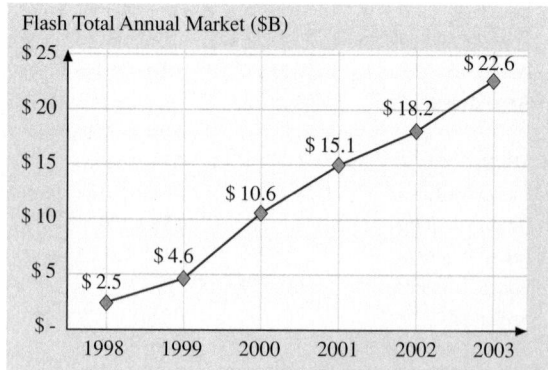

Fig. 47.6 Projected areal density versus data rate for several technologies

Fig. 47.7 Total value of the flash memory market, 1998–2003. The market has been growing rapidly over the last few years, and is now the fastest growing sector of the memory market

Fig. 47.8 Market segments that are driving the demand for flash memory, where cellular phones are leading the demand

every year. The market is projected to reach $23 billion by 2003, up from $2.5 billion in 1998. The reasons for this dramatic growth are depicted in Fig. 47.8 [47.14], which shows the demand by industry sector and the growth in these areas. By far the biggest driver of demand for flash memory is consumer electronics and, in particular, cellular phones, which account for nearly half the market and are growing at a compounded annual growth rate of 120%.

Flash Memory

Flash memory, the most common nonvolatile, programmable memory device, uses rows and columns of interconnects that address each data cell. At the intersection of these are two transistors, which comprise the flash memory cell, as shown in Fig. 47.9. Each cell is addressed by a network of interconnects, where the common connection to the transistor gates is called the word line and the common connection to the drain is called the bit line. Every cell in the memory array can be accessed by activating a unique combination of these lines. Here, the floating gate is linked to a word line via a second gate that serves as a control gate. The floating gate potential is altered by Fowler–Nordheim tunneling of electrons, an effect that takes place when the applied bias needed for switching is larger than the work function of the relevant electrode. The electrons arrive from the column, or bit line, and enter the floating gate from which they drain to the ground. A cell sensor monitors the flow of electrons through the gate. A value of 1 is assigned when the current is greater than a given threshold value, and 0 if it is below that value. Approximately 30 000 electrons are stored in the gate to make a 1 and 5000 for a 0. The memory of the device derives from the very long storage time (tens of years) of the tunneling electrons in the gate capacitor. For the data to have a lifetime of ten years, the electrons can leak at a rate of no more than five a day. Flash memory can only be rewritten a limited number of times (10^{5-6}), as electrons get permanently trapped in the gate over time, impairing device efficiency. Memory devices using this technology are SmartMedia, CompactFlash cards, and Memory Sticks, for example, that read and write memory in 256- or 512-byte increments, enhancing their speed of operation. A flash memory device, in contrast to a hard disk drive, is noiseless, has no moving parts, is lighter and smaller, and, above all, has access times orders of magnitude faster than HDD.

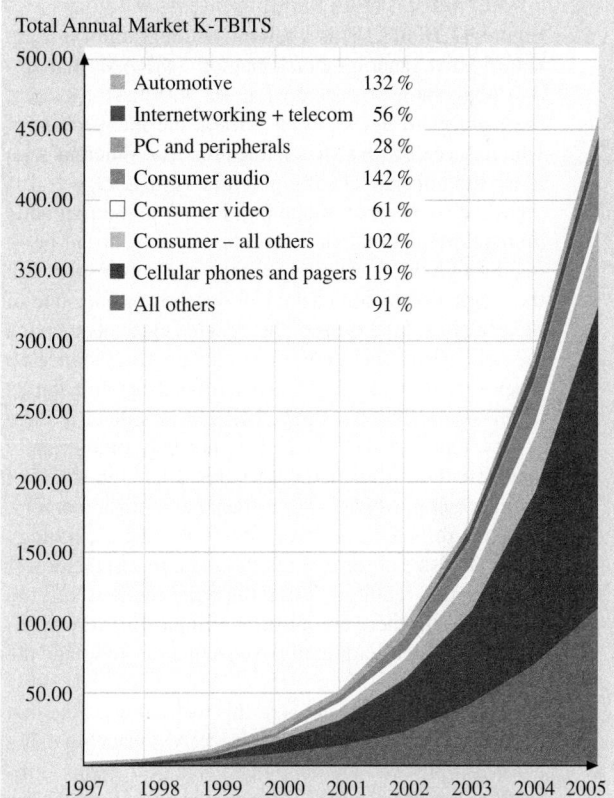

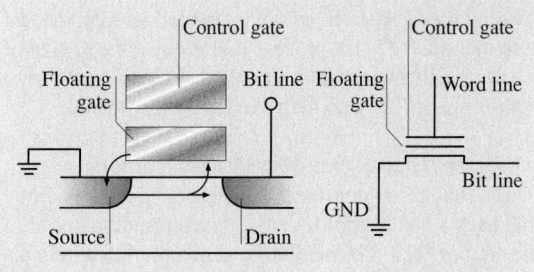

Fig. 47.9 Schematics diagram of a typical flash memory cell

SmartMedia cards, which serve as solid-state floppy-disk cards, were originally developed by Toshiba. They are currently available in a capacity of 128 Mbytes and higher and measure roughly 45 mm × 37 mm × 1 mm. In comparison to a HDD, however, SmartMedia cards are much more expensive. Note that a single storage cell occupies roughly $0.1\,\mu m^2$, a size limited by the complexity of its physical structure and width of the addressing interconnects.

Part F | 47.1

Magnetoresistive Random–Access Memory

Figure 47.10 [47.15] is a schematic diagram of a magnetoresistive random-access memory (MRAM) that acts as a nonvolatile storage device. As with flash memory, there are word lines and bit lines at the intersection of which is a cell with a switchable property. Whereas with Flash memory the electric potential of a storage cell is modified by electron tunneling, here it is the magnetic domain orientation that is switched by the current flowing through the electrodes. Passage of electrons through the magnetic domain of the cell depends on the state of polarization of the spins of the injected electrodes, much the same as the operation of giant magnetoresistance elements used to read data in a hard disk drive. In contrast to HDD and flash memory, MRAM is still in a stage of development and, as such, is not yet commercially available. One of its main attributes is its high speed of programming, which is faster than that of flash memory. The main challenges confronting MRAM technology are: (i) the size of the cell is still too large, (ii) the power consumption required during the write mode is still too high, and (iii) there are manufacturing difficulties associated with high-temperature steps that may damage the magnetic layers.

To exemplify the status of this technology, note that Motorola demonstrated a 1 Mbit MRAM test chip using a 0.6-μm process, yielding a cell size of 7.2 μm^2. Projections are that using a 0.18-μm process will shrink the cell size to about 0.7 μm^2, and thus be competitive with flash memory. Also, Sony demonstrated an MRAM test chip using a 0.35-μm process that yielded a cell size of 5.8 μm^2. Other companies like IBM, which is partnering with Infineon, are also active in MRAM research.

Phase–Change Random–Access Memory

Phase-change random-access memory is a technology that uses a chalcogenide as the data storage material in the memory cell. The medium is similar to the switchable material used in rewritable compact disks (CDs). It is being pioneered by Ovonyx, in association with Intel. A schematic diagram of an Ovonyx unified memory (OUM) is shown in Fig. 47.11 [47.16]. The OUM is an example of a phase-change random-access memory that operates much as the flash memory and MRAM, as far as the electrodes are concerned, except that the memory bit cell consists of a chalcogenide material that undergoes a phase change. Upon passing current from the word line

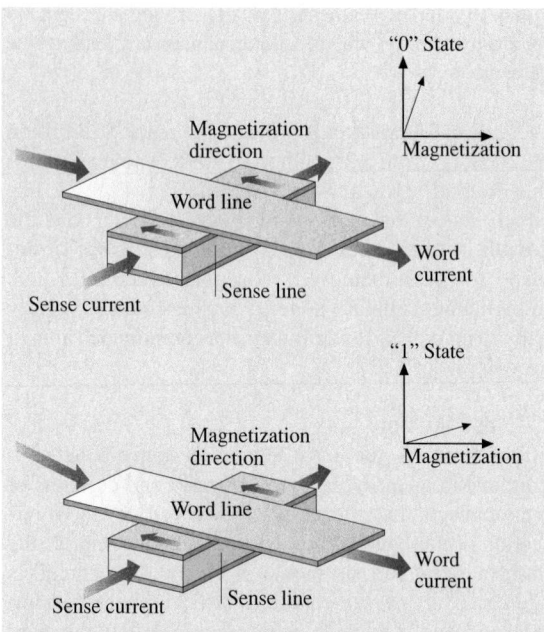

Fig. 47.10 A diagram of a MRAM cell

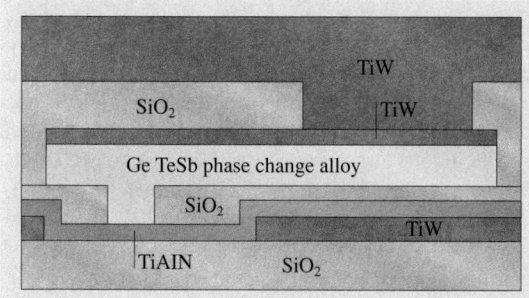

Fig. 47.11 A cross section of an OUM cell

to the bit line, the chalcogenide, a GeTeSb alloy, changes its phase reversibly between amorphous and crystalline states, depending on the temperature it is heated to and the rate of cooling. This change in phase is accompanied by a several orders of magnitude change in conductivity, which makes it possible to obtain a high reading signal-to-noise ratio. Similar to flash memory and MRAM, the OUM requires a protective shield from the environment for reliable operation over a long period. We shall come back to this issue later.

47.2 Opportunities Offered by Nanotechnology for Data Storage

We have summarized the two mainstream commercial data storage devices, HDD and NVRAM, and the two near-term candidates for future commercial success, the MRAM and PC-RAM devices. As stated before, the technology utilized by these devices should be used as a measure against which novel nanotechnology approaches are judged. Note that, even if a promising novel nanotechnology-based device comes into being, it will take time to bring it into maturity, and during that period, the state-of-the-art of HDD, flash memory, MRAM, and PC-RAM will steadily improve. One could state that we have here a competition on the run that will be decided based on the following four criteria: (i) capacity, (i) cell size, which translates to areal density or form factor, (iii) access time, and (iv) price, where each technology is looking for its own business niche.

Nanotechnology offers two unique opportunities to solve the problem of how to miniaturize storage devices. The first opportunity entails the use of a variety of chemical and physical processes to fabricate memory media capable of supporting nanoscale features that have reversible properties, be it magnetic, electric, or phase change. Such media should satisfy the first and second criteria, namely, a large data storage capacity and a small form factor. The second opportunity derives from the ability of scanning probe microscopy, the champion of nanotechnology, to characterize and modify structures down to atomic dimensions. Here, the use of a large number of probes operating in parallel should satisfy the third criterion, namely, fast access times. Also, because the fabrication of multiple cantilevers lends itself rather readily to mass production using currently available photolithographic processes, it should satisfy the fourth criterion: low cost.

In the next section, we describe how probe storage can be realized as a data storage technology that may find a commercially viable niche. It will be useful to divide this section into three parts, each one describing different aspects of what is known as probe storage. The first part, titled "Motors", will deal with the question of how one can access a large number of nanoscale memory cells using parallel and sequential processes. The second part, denoted by "Heads", will describe methods for reading and writing each individual memory cell using different SPM techniques. The third part, called "Media and Experimental Results", will describe the properties of different media applicable to data storage that have or can be used for probe storage.

47.2.1 Motors

This section considers two possible geometric configurations of data storage devices that operate in parallel. One consists of a rotating cylinder and an array of cantilevers positioned in an axial direction, as shown in Fig. 47.12, and the other is a 2-D, all-MEMS configuration, as shown in Fig. 47.13 [47.17]. Very little work has been done in the first, *mixed* configuration, and we will, therefore, present only a short survey of this work. Most of the section will be dedicated to the all-MEMS configuration, on which massive work has been proposed, tested, and published.

Cylinder

A cylindrical ROM media for optical data storage is currently being explored as an alternative to digital versatile disk (DVD) devices, using an adhesive tape film as the recordable medium [47.18]. The purpose of the study is to find out whether this geometry will result in a smaller form factor and reduced vibrations relative to

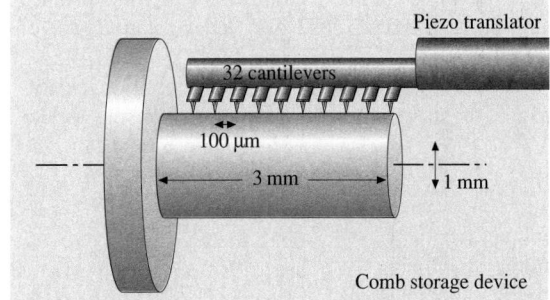

Fig. 47.12 A schematic diagram of a comb storage device

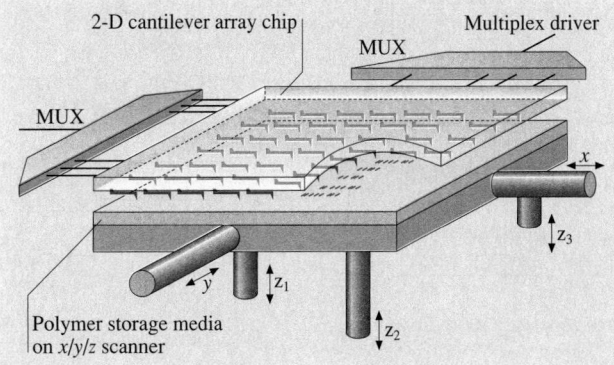

Fig. 47.13 A schematic diagram of a MEMS storage device

a commercial DVD. These authors find that this geometry offers a larger writing area than that of a disk. For example, consider a disk with inner and outer radii of r and R, respectively, and a cylinder with the same radius and a height of h. For equal areas, the height of the cylinder is given by $h = (R^2 - r^2)/2R$. Therefore, for $r = 2.5$ mm and $R = 15$ mm, one obtains $h = 7.3$ mm. As a result, a cylinder with $R = h = 15$ mm offers more than twice the area of a disk. Such a device presents an interesting opportunity that the authors are currently pursuing. Figure 47.12 is an artist's concept of a possible geometry where a linear array of 32 cantilevers with their tips is mounted along the axis of a rotating cylinder. This array, which resembles a comb, is mounted on a piezoelectric actuator that can move the minute array very fast across a distance that equals the gap between adjacent cantilever tips. Let the cylinder circumference be divided into 32 sectors, each containing data bits arranged as circular tracks within each sector. As the cylinder rotates, each of the 32 tips reads independently a single track. Using the piezoelectric actuator, one can now move all the tips such that they read the next 32 tracks. Consider the particular geometry presented in Fig. 47.12, where the diameter of the cylinder is 1 mm, its length 3 mm, the distance between tips 100 μm, and a cylinder rotation rate of 10 020 rpm.

A crude assessment of the possible performance of such a device indicates promising values for the capacity, latency, access time, and transfer rate [47.19].

Note that, while the rotational rates of these devices are expected to be similar to those of a HDD, one can conceive of MEMS-derived motors increasing the rotation rate. For example, a micro-engine with a diameter of 64 μm driven electrostatically at 300 000 rpm has recently been investigated [47.20]. The purpose here was to study the out-of-plane displacement and tilts about the rotating hub. Optical methods have been developed to characterize the operation of this device, together with a model that takes into account the various parameters relating to its operation. One wonders whether a combination of this MEMS technology with a disk or cylindrical recordable medium is a viable option for reducing latency and decreasing access time for data storage applications.

Probe Storage

The next section provides a detailed description of a probe storage device, so here we will only outline briefly the main principles of its operation [47.8, 9, 21–26]. Consider a typical schematic of a MEMS-based storage presented in Fig. 47.13 [47.17], where the cantilevers are laid out on a *sled* in both the x and y directions. The sled is translated, typically by electrostatic actuators, over the medium surface. This means that, unlike in conventional hard disks, the sectors are laid out adjacent to each other.

There are two possible architectures to this arrangement, one in which the media is moved relative to the tips and one in which the media is fixed and the tips are moved. The moving-tips model is very space inefficient, allowing only about 1% of the media area to be used for data storage, and, therefore, making it unviable in comparison to other storage technologies. In this configuration, the media is suspended by springs over the tips with actuators that control their positioning. In contrast, however, a typical design for a moving media model could have 10 000 tips in a 1 cm^2 area with a bit size of 50 nm^2. This geometry allows 30–50% of the area of the media to be utilized. Allowing for overheads such as error correction, this gives a storage capacity of 4 Gb/cm^2.

The disadvantage of this scheme is the greater mass that needs to be moved, resulting in long seek times. There is a settling time associated with each movement of the system, as the springs cause the sled to oscillate around its final position before coming to rest.

As shown in Fig. 47.14, the media is divided into $M \times N$ rectangular regions, each of which is accessible only

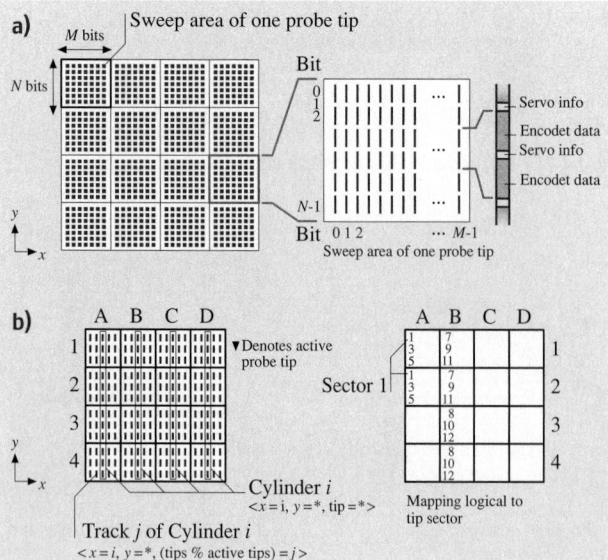

Fig. 47.14 A schematic diagram of probe storage data organization

Table 47.3 A table of probe storage operating parameters

Sled mobility in x and y	100 μm
Bit cell width (area)	50 nm (0.0025 μm^2)
Number of tips	6400
Simultaneously active tips	1280
Tip sector length	80 bits (8 bytes)
Servo overhead	10 bits per tip sector
Device capacity (per sled)	2.1 GByte
Sled acceleration	114.8 m/s^2
Per-tip data rate	400 kbit/s
Settling time constants	1
Sled resonant frequency	220 Hz
Spring factor	75%

by a single tip. The major advantage of this is that the data can be *striped* across multiple regions, dramatically increasing read times, as one seek operation is sufficient to read many bits.

The notation to address each bit is of the form $\langle x, y, tip \rangle$, where x and y are the horizontal and vertical coordinates of the bit within a rectangular region accessible by tip. A tip track is the data for which $\langle x, tip \rangle$ is the same, i. e., a vertical column within a rectangular region. By analogy with hard-disk architecture, all data with identical x form a cylinder, accessible without moving the sled in the x direction. Cylinders are divided into tracks, which are accessible by the currently active tips, as not all tips can be active simultaneously, due to power considerations.

As mentioned, a benefit of this architecture is that data tracks are striped across tip tracks. This is depicted in Fig. 47.14 [47.23], where each data sector is mapped across several tip tracks. Sectors 1 and 2 can be read in parallel by activating tips 1–4 and moving them vertically downwards.

Table 47.4 Probe storage projected performance

Average service time	1.49 ms (0.25)
Maximum service time	4.51 ms
Average seek time	1.27 ms (0.25)
Maximum seek time	1.66 ms
Average x seek time	1.24 ms (0.21)
Maximum x seek time	1.66 ms
Average y seek time	0.90 ms (0.31)
Maximum y seek time	1.62 ms
Settling time	0.72 ms
Average per-request turnaround time	0.20 ms (0.20)
Maximum per-request turnaround time	1.34 ms

Simulated device performance for a prototype sled with the parameters given in Table 47.3 is shown in Table 47.4. Note that an important breakthrough in MEMS technology has recently been reported in which over one million cantilevers per square cm have been fabricated by anisotropic etching of silicon [47.27]. The cantilevers, which had tips at their end, measured a few micrometers in length, had spring constants of a few N/m, resonance frequencies around 10 MHz, and Q factors of 5 in air. Significantly, the resonance frequencies within the same row deviated only by 0.01%. Cantilevers as short as 100 nm and as thin as 20 nm were also fabricated, demonstrating the latitude that this technology offers. Also reported was the fabrication of a large number of cantilevers operating at frequencies up to 1 GHz with Q factors up to 8000 [47.28]. These results will no doubt play a pivotal role in MEMS-related data storage technology for which mass parallelism combined with superfast sensors offer an important advantage.

Table 47.4 [47.22] takes into account the physical performance of the device, including the acceleration of the actuators, settling time, etc. One observes that the results of the model compare favorably to current hard disk technology, having access times that are approximately six times faster. Note that recent results by the IBM group imply a potential bit areal densities of 400 Gbpsi [47.29].

47.2.2 Sensors

As stated before, the key problem associated with densely packed nanoscale data bits is how to address them, realizing that no interconnects of this small size exist. Clearly, the only known method is to utilize MEMS technology where sharp tips are used to read, write, and rewrite the data bits. Although such a technology is attractive, one has to realize that the number of physical phenomena available for effecting such an interaction is quite limited.

There are, in principle, three modes for interfacing, or sensing nanoscale data bits. In the first mode, denoted as contact mode, a sharp tip is constantly in contact with the medium on which the bits are written. Here, the apex of the tip has to be comparable in size to that of the bit. The second mode involves such a tip in close proximity to the bit without actually touching it. The third mode, dubbed the noncontact mode, is a combination of the other two modes. Here, the tip vibrates in close proximity to the media in such a way that on each cycle it touches or taps its surface for a short period. This mode is referred to as the tapping mode [47.30], which turns out to have several advantages relative to the other two modes. In either mode one can use an electric current within the tip to heat up a data bit, inject a current into it, or impose an electric field across it.

For the noncontact mode, one can use field emission from a sharp tip that is positioned several nanometers away from a sample, or use a tunneling current when the tip is within a fraction of a nanometer of from the sample. The current can then be used to heat a data bit and probe its conductivity as a means for data storage and retrieval. Yet both of these are accompanied with electronic noise when operating in ambient, thus requiring vacuum for a more steady operation. Note that a tunneling current is so sensitive to the tip–sample gap that a change of 0.1 nm in gap size will change the tunneling current by one order of magnitude. One can, therefore, rule out such a technique, as its restrictions on media flatness are too strenuous. In spite of some inherent difficulties, field emission maintained its position as a viable technology for data storage applications. A schematic diagram of the operation of a device based on this technology is shown in Fig. 47.15 [47.31].

Here, a multiplicity of stationary tips is positioned in close proximity to a conducting medium, such as a chalcogenide, and a controlled current emission for each tip is used to transform its phase reversibly from crystalline to amorphous. Either the tips or the media are raster-scanned across each other such that, as with the probe storage device, one can address a large number of data bits. Both tips and media are placed under a mild vacuum that has to be maintained at all times. Although writing bits may be utilized efficiently, reading them back, namely, measuring their conductivity, is a nontrivial task that requires further investigation. Along these lines, one may employ a MEMS-based ap-

proach and use conducting tips that are in direct contact with a conducting medium.

A different approach to locally heating a sample that does not require direct tip–sample contact, vacuum conditions, or conducting media, is to have a heating element embedded inside each tip. In spite of the fact that each tip has a complex structure, as shown in Fig. 47.16b, it was found possible to fabricate a 64×64 2-D array that raster-scans a polymer medium. Heating a data bit and pressing a sharp tip into it generates a nanoscale pit that is estimated to persist for years. Sensing the pit for readout is accomplished by monitoring the heat dissipation of the heated tip as a function of the media topography. When the tip is across a smooth surface, it is exposed mostly to air. When the tip is sunk into a pit, it is surrounded partially by the polymer material. Thus, the difference in heat conductivity between these two

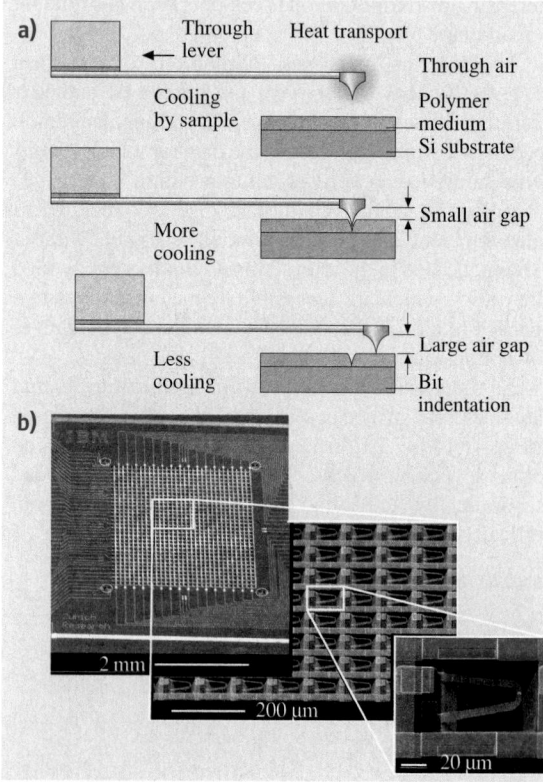

Fig. 47.16 (a) A schematic diagram of a cantilever with a heating and heat-sensing tip. Indentations are inscribed into a polymeric medium by means of a heated tip. The bits are read by sensing the increased heat loss in an indentation due to the greater surface area in contact. (b) Details of the tip structure

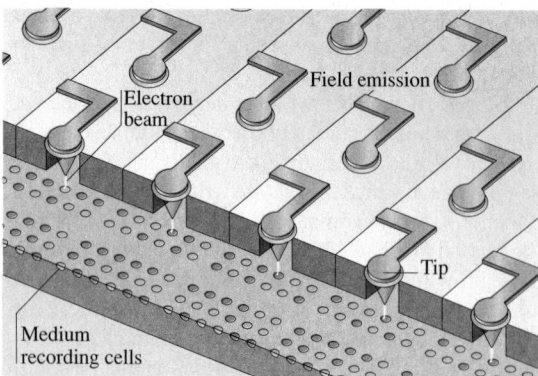

Fig. 47.15 A schematic diagram of field-emission tips

cases serves as a measure of the state of the polymer, whether it is smooth or dimpled. A key difficulty with this technology has been the erase mechanism, which is currently being addressed [47.29].

In the noncontact mode, the tip is vibrated at a distance such that the tip never touches the surface directly. However, the tip senses the surface of the sample indirectly through force gradients that modify the resonance frequency of the cantilever. This mode can be useful in magnetic recording if one attaches a small magnet to the tip and uses it to both modify and sense magnetic domains. This mode, however, is slow, as the tip has to vibrate many times during the sensing of a single data bit and, therefore, can be ruled out as a candidate for data storage.

We shall now consider the tapping mode, which has three main advantages over the contact mode. First, the tip–sample contact time is reduced by almost two orders of magnitude, thus minimizing tip–media wear. The second advantage is that it touches the media in a much gentler way and, therefore, does not twist or bend. The third advantage is that the tip jumps between points of contacts, which constitute the data bits, so it is less prone to crash due to media roughness. If one uses the tapping mode with a conducting tip and taps on a conducting substrate, it is possible to follow the motion of the vibrating cantilever by monitoring the time the tip makes contact with the sample. This method makes it easy to control a feedback system that keeps the motion of the cantilever under control, provided the contact is good enough. The problem one encounters using this mode, however, is that such an operation is usually chaotic and, therefore, difficult to control. If one were to overcome this problem, this mode would be fast enough relative to the noncontact mode to be considered a viable method for data storage.

We will now consider the theory and experimental results of the tapping mode by first treating it as a grazing impact oscillator (GIO) for which one can obtain closed-form expressions connecting all of its parameters [47.32]. The GIO consists of a cantilever driven at a constant frequency and an amplitude by an external force. The vibrating cantilever is then brought into close proximity to the surface of a sample until its tip starts impacting the surface. At this point the amplitude of vibration of the GIO changes, together with the relative phase between the driver and the tip. Let the amplitude of vibration of the driver be denoted by a and the spring constant and quality factor of the cantilever be denoted by k_0 and Q, respectively. On impact,

the GIO acquires amplitudes of vibration A and effective spring constant k_{eff}, establishing a new value for the input power and phase:

$$A = \frac{aQ}{\sqrt{Q^2\left(1 - \frac{k}{k_0}\right)^2 + \frac{k}{k_0}}}, \tag{47.1a}$$

$$\theta = \tan^{-1}\left(\frac{\sqrt{k/k_0}}{Q(1 - k/k_0)}\right), \tag{47.1b}$$

$$k = k_0\left[\sqrt{\left(\frac{a}{AA}\right)^2 + \frac{1}{4}\left(\frac{1}{Q^4} - \frac{4}{Q^2}\right)}\right.$$
$$\left. - \frac{1}{2Q^2} + 1\right]. \tag{47.1c}$$

Equation (47.1) shows the relationship among the various parameters governing the operation of the GIO in a closed form, making it possible to model the effect that impacting has on the tapping cantilever. Figure 47.17a shows that, on bringing the vibrating cantilever closer to the surface, its amplitude of vibration decreases and its effective spring constant increases. Figure 47.17b shows the phase of the vibrating cantilever as a function of the amplitude of vibration. Figure 47.17c and d depict the input power as a function of the amplitude of vibration of the cantilever and of its phase, respectively. The double-valued functions express the fact that one can operate either below or above the original resonance frequency of the cantilever.

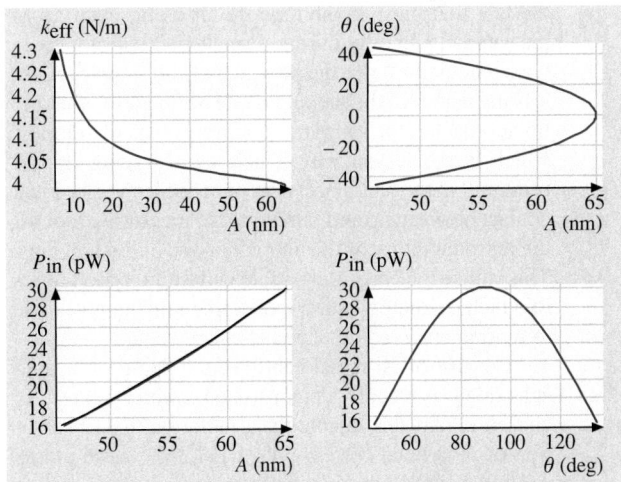

Fig. 47.17 Calculated tapping behavior. Modeling of the relationships among the amplitude A, effective spring constant k_{eff}, input power P_{in}, and phase ϕ

This simplified model of the tapping mode neglects many effects, such as indentation of the surface on impact and adhesion and other interface tip–sample forces. For our purpose, however, we will neglect these effects and explore the feasibility of using the tapping mode for data storage applications. The idea here is to have the tip jump from one pixel, or bit of stored data on the surface of a sample to another pixel without touching the surface in between. And while briefly touching the surface, the tip, which is conducting, will inject a pulse of current into the pixel and modify its conductivity [47.33]. Upon returning to the same pixel, the tip could monitor the conductivity of that particular pixel.

Considering the fact that the tapping mode involves a driven, damped oscillator, it is apparent that the motion of the cantilever traces a chaotic trajectory, necessitating an averaging process. For most imaging or nanofabrication purposes, such an averaging, combined with an appropriate image processing procedure, can indeed yield images with nanometer resolution. For data storage applications, however, it is desirable to have non-chaotic, reproducible interactions for every tap because one could not afford an averaging procedure.

There are two key issues regarding the requirement that each tap produce a controlled interaction. The first concerns the identification of conditions under which (i) chaotic motion of the cantilever is minimized and (ii) good tip–sample contact is established. The second issue concerns the means by which one can observe these individual interactions. The difficulty in measuring tip–sample interactions during an individual tap stems from the fact that most of the time the tip is far away from the sample, and only during a small fraction of a cycle does it actually sense the presence of the sample. Also, it is the nature of the tapping mode of operation that the tip approaches the sample at a relatively small velocity, which is the reason why it acts as a grazing impact oscillator in the first place. One can choose the electrical contact between tip and sample as the measuring tool for the reproducibility of the interaction for individual taps. This choice is clearly compatible with storage modalities in which current injection into storage media gives rise to writing and reading of phase-change data.

The use of all-metal cantilevers tapping on a gold substrate was recently demonstrated, whereby carefully chosen parameters enable the generation and observation of individual current pulses [47.34]. These pulses were generated by applying a short bias pulse during the tip–sample contact time. Each current pulse, lasting up to $20\,\mu s$, could produce a current of $\approx 10\,\mu A$ in a nm^2 region, yielding a current density of $10^{13}\,A/m^2$.

This density is large enough to be applicable for phase change. The cantilever was fabricated from a rectangular PtIr wire whose length, width, and thickness were $1200\,\mu m$, $100\,\mu m$, and $50\,\mu m$, respectively. One side of the cantilever was attached to a driving bimorph, and the other side etched to produce a tip whose apex radius was estimated to be $70\,nm$. The spring constant, resonance frequency, amplitude of vibration, and quality factor of the cantilever were $300\,N/m$, $13\,kHz$, $100\,nm$, and 100, respectively. The all-metal cantilever ensured good conductivity and could withstand a large number of tapping impacts without degrading and had a shiny surface that acted as a good mirror for the optical deflection technique.

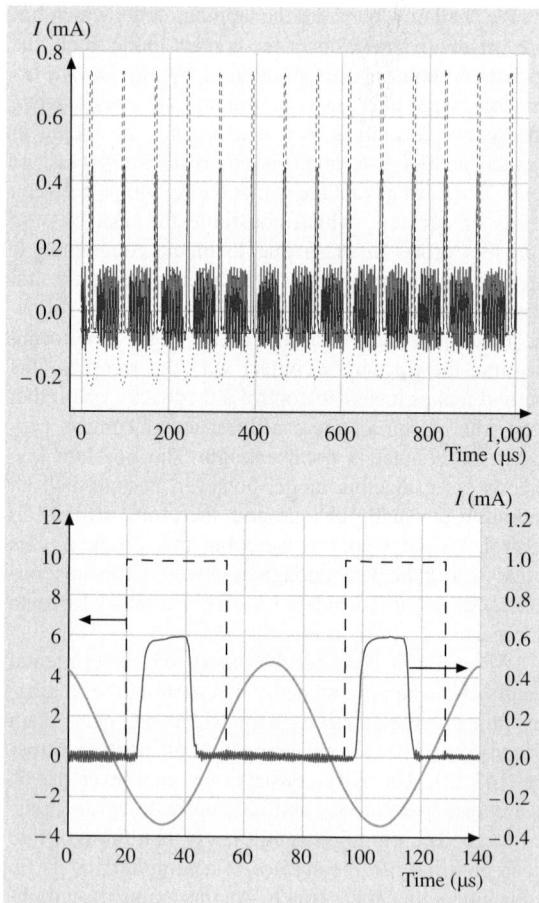

Fig. 47.18 (a) A typical current pulse train (*solid line*) obtained with pulsed voltage bias in phase with the cantilever's lowest position and the associated cantilever motion (*dotted line*), (b) two current pulses (*light line*), bias voltage pulses (*dashed line*), and cantilever motion (*dark line*)

The experimental results yielded long trains of individually injected current pulses. Figure 47.18 (left-hand side) shows 14 non-averaged experimental current pulses (solid line) and their associated cantilever motion (dashed line) with a background of instrumental noise [47.34]. The current pulses demonstrate the reproducibility of the conductance established during each tap. Figure 47.18 [47.34] (right-hand side) shows two averaged current pulses (solid line) with their input bias pulses (dashed line) and the associated cantilever motion (dotted line). Note that the tip–sample contact time is revealed by the duration of the current pulse, which is shorter than the duration of the applied bias pulse. Here, the amplitude of the bias pulse was 10 mV, its duration was 40 μs, and the timing of its application relative to the motion of the cantilever was provided by the output of the atomic force microscope photodiode. The current pulses, with a magnitude of 0.5 mA, are much larger than the tunneling current occurring while the tip is within a fraction of a nanometer from the surface. Also, the displacement current, due to the motion of the cantilever and the time varying nature of the applied bias, is expected to be much smaller than the observed current pulses. The contact conductance is shown in Fig. 47.19, where a linear relation between voltage and current implies a good ohmic contact. These results demonstrate that it is possible to minimize the chaotic behavior

of a tapping cantilever and obtain reproducible conductance on each individual tap. The next step is to use the tapping mode in conjunction with current injection to write and read data bits on conducting media.

47.2.3 Media and Experimental Results

This section summarizes several typical experimental results using different data storage media, electric charging, electric domain orientation, magnetic domain orientation, chemical modification, and phase change in chalcogenides and organic films.

Electric Charging
Quate's group at Stanford has studied charge trapping in a silicon nitride/silicon dioxide/silicon system using a scanning capacitance microscope [47.6]. The charge trapping in this system has been used for many years for nonvolatilesemiconductor memory. Using chemical vapor deposition, 500 Å of silicon nitride was deposited onto a silicon wafer with a 50-Å oxide coating. By selectively storing charge into the substrate, they were able to write bits as small as $0.02\,\mu m^2$. Using this method, they were able to write the founding document of the Swiss Confederation onto a $120\,\mu m^2$ area, corresponding to 256 kilobits of information, which yields a data density of 10 Gbpsi.

Electric Domains
Hidaka et al. have demonstrated the possibility of ultra-high-density charge storage in a ferroelectric film, $PbZr_{1-x}Ti_xO_3$ [47.35]. They show that a change in the topography of the film can be introduced by domain reorientation induced by the application of a 10-V pulse with a duration of 100 μs. This is due to the piezoelectric effect caused by domain structures. Data bits as small as 40 nm were written this way. This result implies a data density of approximately 300 Gbpsi, suggesting that this medium is a good candidate for high-density data storage.

Thin Oxide Films
IBM Research in Switzerland has demonstrated reproducible switching in thin oxide perovskite films [47.36]. By constructing metal–insulator–metal capacitor-like devices, with $SrZrO_3$ as the insulator, they were able to observe an order of magnitude increase in the leakage current of the device at a threshold voltage of $-0.5\,V$. This process is perfectly reversible, and the leakage current can be restored to its initial state upon appli-

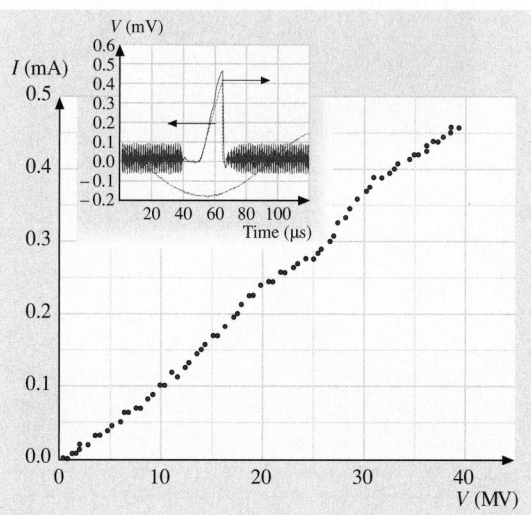

Fig. 47.19 An *I*–*V* curve taken with an individual ramped input voltage pulse demonstrating an ohmic contact. *Insert:* The current pulse (*solid line*), ramped input bias voltage (*dashed line*), and cantilever motion (*dotted line*)

cation of a positive voltage. The switching occurs very rapidly, taking less than 100 ns to change from one conductivity state to another. Interestingly, the magnitude of the transition depends on the voltage applied, suggesting the possibility of storing multiple bits in one pixel.

Magnetic Domains

Sun et al. have demonstrated a novel way to overcome the limitations of media currently used for magnetic data storage [47.37]. By reducing iron and platinum-containing compounds, they demonstrated the formation of large-scale self-ordered arrays of magnetic particles. Initial studies suggest that an assembly of magnetic particles as small as 4 nm can support stable magnetization transitions at room temperature. By using the particles in these arrays as bits, the possibility of terabit-per-square-inch magnetic data storage could be realized.

Figure 47.20 shows transmission electron microscopy of these arrays that indicate a remarkable uniformity of particle size and large-scale periodicity. The figure also shows assemblies of these particles deposited onto a silicon oxide substrate under varying experimental conditions.

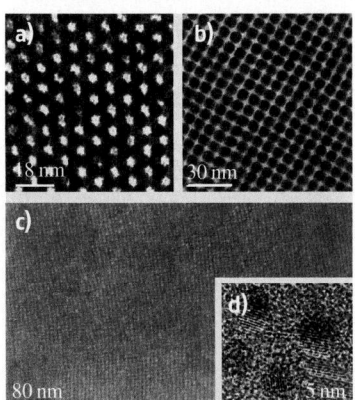

Fig. 47.20 (a) TEM micrograph of a 3-D assembly of 6-nm as-synthesized $Fe_{50}Pt_{50}$ particles deposited from a hexane/octane (v/v 1/1) dispersion onto a SiO-coated copper grid. (b) TEM micrograph of a 3-D assembly of a 6-nm $Fe_{50}Pt_{50}$ sample after replacing oleic acid/oleyl amine with hexanoic acid/hexylamine. (c) HRSEM image of a $\approx$ 180-nm-thick, 4-nm $F_{52}Pt_{48}$ nanocrystal assembly annealed at 560 °C for 30 min under 1 atm of N_2 gas. (d) High-resolution TEM image of 4-nm $Fe_{52}Pt_{48}$ nanocrystals annealed at 560 °C for 30 min on a SiO-coated copper grid reference

Chemical Modification

We have discussed the fact that the tapping mode is considerably less damaging to the tip and sample, as the tip spends less time contacting the sample and does not drag across it while raster-scanning. For this purpose, diamond-coated cantilevers have been found useful for experiments in which wear is a problem. For sample oxidation purposes, the tip is first oscillated away from the sample with peak-to-peak amplitude of 10–15 nm, and then a feedback system reduces the amplitude to 7–10 nm by pressing the tip further into the sample. A 50 ms long pulse with a 10–12 V magnitude is then applied to oxidize a thin Ti film. It was found that this technique produces much finer structures. Figure 47.21 [47.38] is a tapping-mode AFM image of four dots created in such a way. The dots in this image have a diameter of less than 10 nm and a height of 3.5 nm above the surrounding Ti surface with a center separation of 20 nm. Around the dots there is what appears to

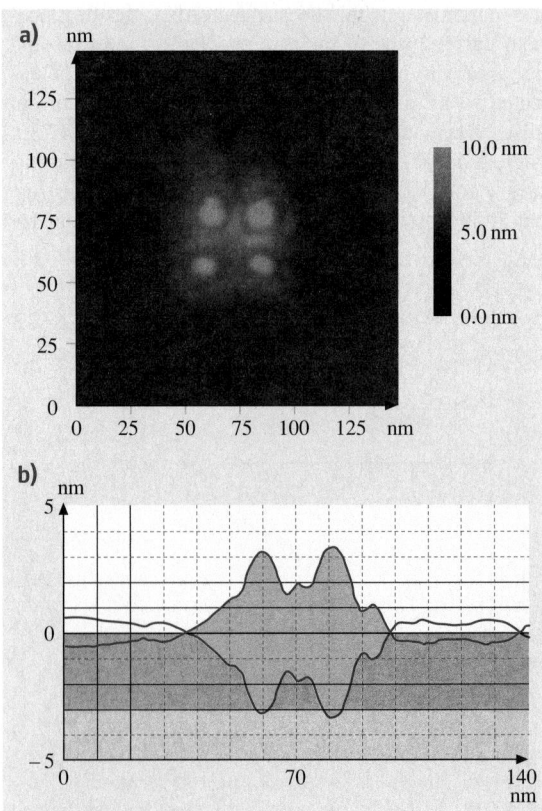

Fig. 47.21a,b Experimental results of local chemical modification using AFM tip: (a) 20-nm dots written on Ti film, (b) cross section of written bit

be a lower ring of material with a channel in between. This experiment demonstrates that one can use the tapping mode to write fine pixels, although not to erase them.

Phase Change in Chalcogenides

Chalcogenides, the active material in rewritable CDs, have the advantage of being a well-characterized and well-understood bistable phase-change material that has already had successful commercial data storage applications, i. e., read–writable CDs (CDRWs). Currently, NVRAM using chalcogenides as the active material is being pursued by an alliance of Ovonyx and Intel. Chalcogenides are materials that have an easily reversible stable change between amorphous and crystalline phases. To switch from the crystalline to the amorphous phase, the material is heated sufficiently, so that the material subsequently cools rapidly into the amorphous phase. The reverse change is achieved by heating the material above the glass-transition temperature, so that it recrystallizes upon cooling. For commercial CDRWs, writing is done by means of laser-light heating and reading is done by measuring the difference in reflectivity of the laser light from each of these two phases.

Scanning probe microscopy can be used to switch the phase of a chalcogenide material by the application of a pulse of current that heats the material as it passes through it [47.39]. The state of the material can be read typically by measuring the resistance of the sample. Figure 47.22 shows the two different states, amorphous and crystalline of the chalcogenides. The reversibility of this transition is remarkable, and has been shown to be stable over 10^{12} repeated read–write cycles. *Gotoh* et al. demonstrated this effect using conducting AFM tips, where a transition from the high-resistance amorphous state to the lower-resistance crystalline state was observed both the in topography and the conductivity of the sample. The change from amorphous to crystalline phases was confirmed by X-ray diffraction. Figure 47.23a and a′ show a conductance AFM image and a profile of the surface after the application of varying voltage pulses. The topography of the substrate is shown in Fig. 47.23b. Figure 47.23c and c′ show the equivalent conductance AFM images after the substrate has been scanned at a negative voltage, erasing the lower power marks on the substrate.

These materials have been further characterized by *Gidon* et al. [47.26], who showed that the resistivity of the film in its amorphous phase was $100\,\Omega\,cm^{-1}$, while that of the crystalline state was $0.1\,\Omega\,cm^{-1}$. They also show that a data bit requires a 200-ns pulse with a power of $50\,\mu W$, which translates to a total energy of 10 pJ.

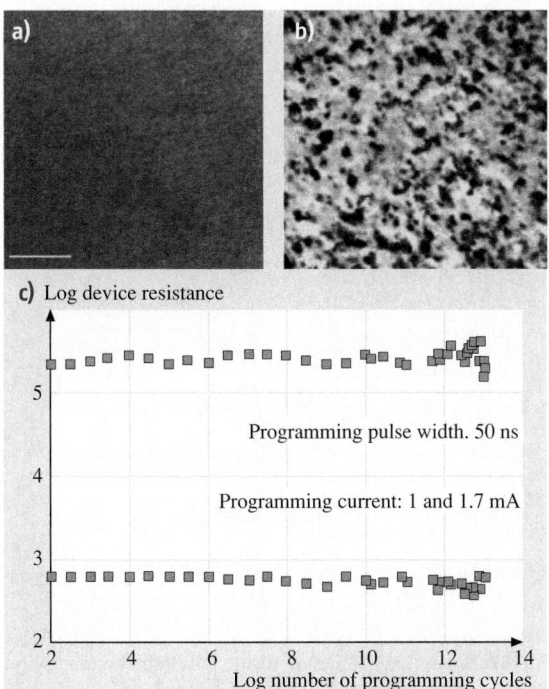

Fig. 47.22a–c Experimental results of phase change using interconnects. (**a**) Crystalline area, (**b**) amorphous area, (**c**) stability of bits over 1012 rewrites (Ovonyx)

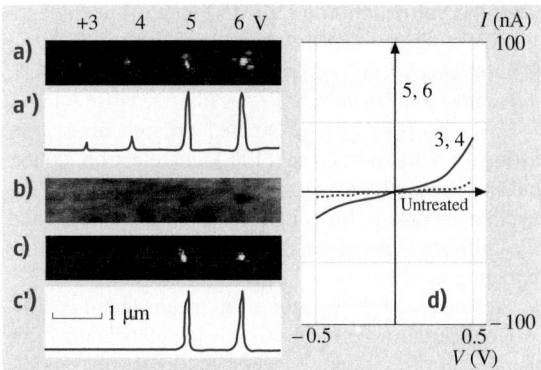

Fig. 47.23 Experimental results of phase change using an AFM tip (**a**), (**a′**), conducting AFM images after applying 100-ns pulses to the substrate, (**b**) topography AFM image of the substrate, (**c**), (**c′**) conducting AFM images showing erasure after scanning the substrate with an applied bias of −1 V, (**d**) *I*–*V* curves of untreated (*dotted*) and treated (*solid*) regions

Phase Change in Organics

Organics have proved important in a variety of electronic and optoelectronic applications such as light-emitting diodes, thin-film displays, transistors, photovoltaics, and lasers. There are several advantages to organics for economic and manufacturing reasons, and a large body of research already exists in this field. As such, there is significant interest and motivation in finding data storage applications based on organic technology.

There has been much research recently into organic materials and systems capable of reversible switching and thus capable of being exploited for data storage applications. There are two broad categories of organics data storage; one where the active material (which can be homogenous or heterogeneous) has inherent bistability, and one where the bistability is a property of the device structure.

There are many different materials that have been investigated for data storage. The fundamental property is that it should have two states (electronic, morphological, or some other easily measured physical characteristic) that the material can exist in, and it should be stable indefinitely in either state. Furthermore, the material should be able to be easily and reversibly switched between these states, and this should be reproducible over many millions of cycles for it to be useful as a data storage medium.

Several groups have been working on reversible transitions in organic complexes suitable for switching by scanning probe techniques. Many complexes have been demonstrated to exhibit reversible switching, e.g., 3-nitrobenzal malonitrile (NBMN) and 1,4-phenylenediamine (pDA) [47.40], tetrathiofulvane (TTF) and m-nitrobenzylidene propanedinitrile (m-NBP) [47.41], and many other variants.

Li et al. [47.41] have studied organic films consisting of tetrathiofulvane (TTF), an electron donor, and m-nitrobenzylidene propanedinitrile (m-NBP), an electron acceptor. This complex demonstrates electrical bistability, which can be written and read by scanning tunneling microscopy.

The nature of the measurement means that it is difficult to ascertain whether this is due to solely conductive changes in the sample, or whether topographic modifications might also contribute. By applying pulses of 1-ms duration, bits of 1.2-nm diameter could be written. This implies an areal density of $\approx 10^{18}$ bits/in^2. This change was characterized electrically, and the resistance was measured as changing from $10^8 \, \Omega$ to $10^3 \, \Omega$ with a threshold switching voltage of 3.1 V. The proposed switching mechanism is charge transfer between

the two film constituents. This change was stable over time and is reversible. This change is reversible, but only by heating for prolonged periods, making it as yet unsuitable for applications.

Gao et al. have demonstrated stable, local, reversible nanoscale switching [47.40] using a charge-transfer complex of 3-nitrobenzal malonitrile (NBMN) and 1,4-phenylenediamine (pDA). A 200-nm-thick film of this complex was deposited onto highly ordered pyrolitic graphite. Macroscopic electrical measurements demonstrate the bistability and switching of the material. Figure 47.24a shows the current–voltage characteristics of the sample in these two conductivity states. Initially, the material is in a low-conductivity state, but upon reaching a certain threshold voltage, in this case 3.2 V, the material abruptly changes to a conductive state. The transition time is very abrupt, 80 ns, as shown in Fig. 47.24b.

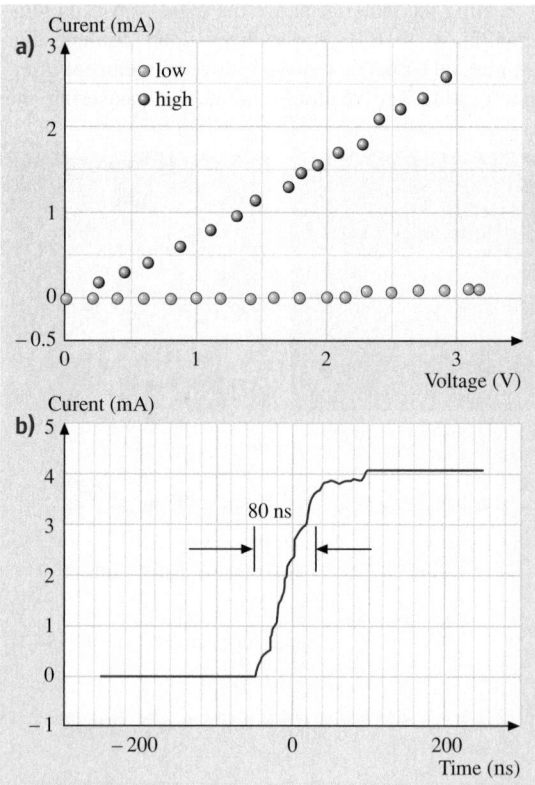

Fig. 47.24a,b Experimental results of bulk conductivity changes of organic films. (**a**) *I–V* curves of a 200-nm-thick film showing the two conductivity states. The switching voltage is 3.2 V. (**b**) Change in conductivity against time, showing a transition time of 80 ns

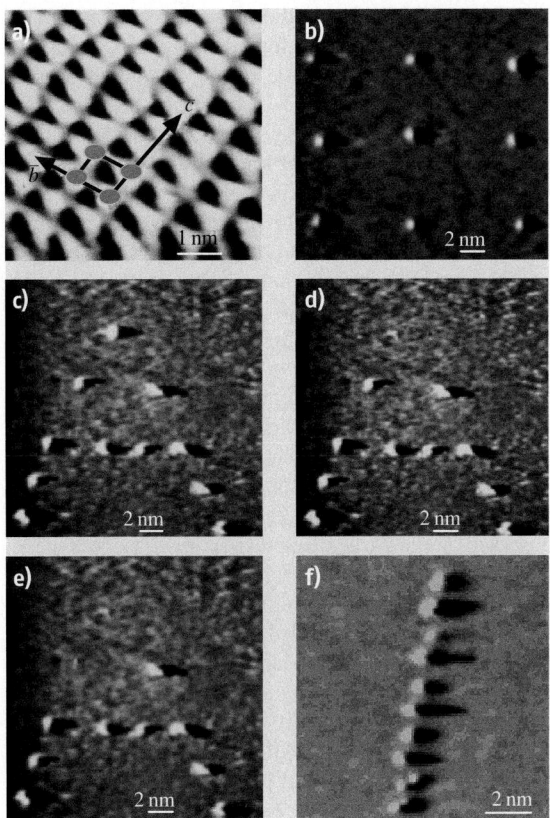

Fig. 47.25a–f STM of the NBMN–pDA film on highly ordered pyrolytic graphite (HOPG). (**a**) 6 nm^2 image showing the crystalline order of the film; (**b**) a 3×3 array of bits (4 V, 1 μs); (**c**) an "A" pattern (3.5 V, 2 μs); (**d**), (**e**), (**f**) STM images after erasing marks with reverse-biased pulses (−4.5 V, 50 μs). The distance between neighboring marks is 1.7 nm

Of further interest for data storage applications is that, having demonstrated reversible switching in macroscopic films, Gao et al. also showed that it was possible to switch these materials on the nanoscale using an STM to apply pulses, i.e., to write bits, and also to measure conductivity, i.e., to read bits.

Figure 47.25a shows the local nanoscale film morphology of the sample. Figure 47.25b and c show patterns written on a sample by application of voltage pulses from the STM tip. Figure 47.25d demonstrates the erasure of a single bit from the pattern by application of a reverse-biased voltage pulse. Similar to the work of Li et al., the entire substrate could be erased by prolonged heating. Of much greater interest is the demonstration of individual bit erasure, achieved by ap-

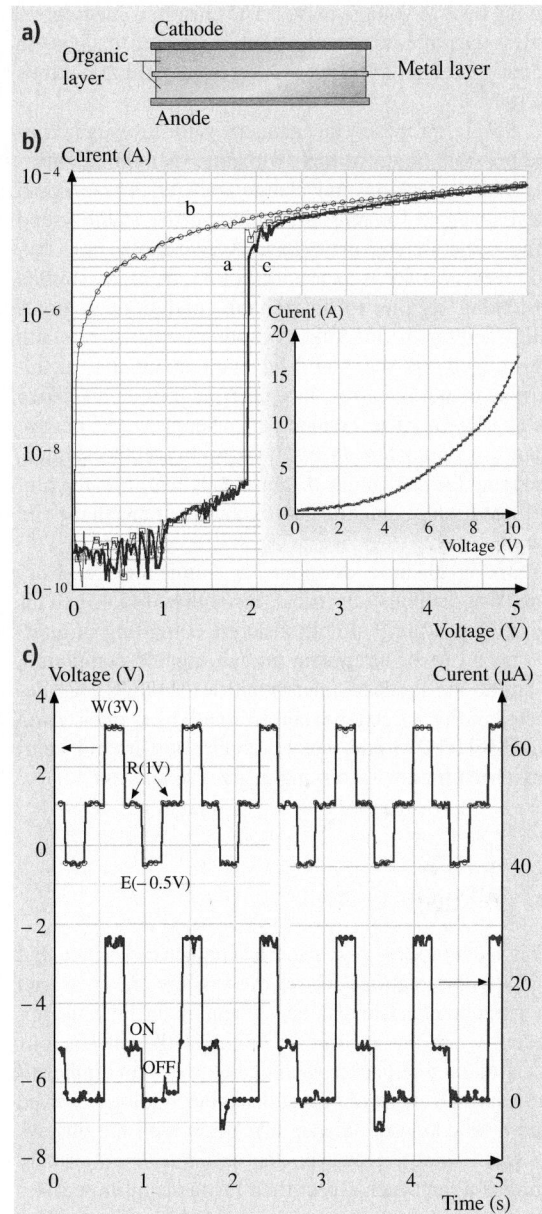

Fig. 47.26a–c Experimental results of conductivity changes of organic device structures: (**a**) the structure of the bistable device, (**b**) *I*–*V* characteristics. Voltage was scanned in steps of 0.1 V from 0 to 5 V. Curves a and b represent the *I*–*V* characteristics of the first and second bias scan, respectively. Curve c is the *I*–*V* curve of the third bias scan after the application of a reverse voltage pulse (−3 V). (**c**) Typical current responses of the device being written, read, erased, read over time

plying reverse voltage pulses. The erasure is incomplete unless the pulse is reverse-biased, indicating that it is the combined effect of heating and electric field that erases the bits.

Several proposed mechanisms were investigated as the possible origin of this switching, including conformational switching and charge transfer. These could be discounted for reasons, including the stability and time scale of the switching. A more plausible hypothesis is a local phase transition from crystalline insulating regions to amorphous conducting regions. TEM and diffraction studies of the samples agree with this hypothesis. It should be pointed out that an alternative interpretation has been suggested by *Zhao* et al. [47.42]. The conductivity change could be due mainly to a change in the topography of the sample, reducing the tip–sample distance and, therefore, the tunneling distance, manifested as an apparent change in conductivity.

Ma et al. have developed a simple device structure that exhibits electrical bistability [47.43]. This device has a fairly simple structure consisting of alternating layers of aluminum and an organic compound, 2-amino-4,5-imidazoledicarbonitrile (AIDCN), as depicted in Fig. 47.26a. Several alternative organics (Alq3 and MEH–PPV) and metals (copper, silver, and gold) have been tried with varying degrees of success.

Figure 47.26b shows the current–voltage bistability of a device with structure Al/AIDCN (50 nm)/Al (20 nm)/AIDCN (50 nm)/Al. A first run shows an initially low current, followed by an abrupt and dramatic increase in the current of four orders of magnitude at 2 V. Subsequent repeated voltage runs (b) show that this transition is stable, being orders of magnitude higher than the initial measurement. The devices proved to be stable in this state over periods ranging from days to weeks.

Of further interest is that the device proved to be erasable by purely electrical means, the application of a $-0.5-$ V pulse bringing the device back to its original state. This is denoted by c in Fig. 47.26b. Furthermore, the devices showed excellent stability over time, with over 1 000 000 read–write–erase cycles being performed on the devices with excellent repeatability, as depicted in Fig. 47.26c. The origin of this bistability is unclear. It is clear that the central electrode is critical to the observed effect, as the bistability is only observed when the central electrode thickness is above a threshold value of 10 nm. Organics with high dielectric constants and low conductivity are believed to be important for device performance. The exact mechanism, however, is as yet unclear. Preliminary results on novel materials [47.44] have suggested the possibility of an all-organic variation of this concept, with an AFM tip switching between bistable states of the system.

47.3 Conclusion

Nanotechnology is perceived as a branch of science that deals with structures whose dimensions are measured in nanometers. Clearly, new phenomena become apparent when the number of atoms at the interface of such a structure becomes comparable to the number of atoms in its volume. As a result, nanotechnology evolved into a new kind of science that deals with the physics of small things. However, the suffix *technology* here implies a far wider aspect than just a scientific curiosity; this branch of science should also be of some use. This poses a dilemma for researchers whose main interest is in understanding new phenomena, rather than in finding a use for them. Consider scanning probe microcopy, which is celebrating its 20th birthday now with uncountable contributions to the science of measurement of surfaces and near-surface phenomena. Despite all the contributions scanning probe microscopy has made, it has always served as a tool for investigation, rather than as a practical device that can become a com-

mercial commodity. And yet this is exactly what this chapter deals with. The challenge of converting scanning probe microscopy from a research tool into a probe storage device is daunting indeed and will require new ideas to make it happen. The danger associated with this challenge is that novel ideas developed by research groups are sometimes too immature to be measured against any existing technology, in general, and against data storage, in particular. One has to keep pursuing research opportunities whose outcomes are unknown and whose applications are unexpected with the hope that they will bring about good science and hopefully useful applications.

However, based on the technical breakthroughs delineated in this and the following chapter and judging from the history of data storage evolution, it seems plausible that probe storage will serve as a disruptive technology that replaces data storage devices in a particular market niche. These new devices should be nonvolatile, have terabyte capacity with microsecond access times and gigabyte transfer rates, all for a cost that is highly competitive with current commercially available data storage devices.

References

47.1 D. Sarid: *Scanning Force Microscopy, with Applications to Electric, Magnetic, and Atomic Forces* (Oxford Univ. Press, Oxford 1991)

47.2 D. Sarid: *Scanning Force Microscopy, with Applications to Electric, Magnetic, and Atomic Forces*, revised edn. (Oxford Univ. Press, Oxford 1994)

47.3 D. Sarid: *Exploring Scanning Probe Microscopy with Mathematica* (Wiley, New York 1997)

47.4 B. G. Levi: Do atomic force microscope arrays have the write stuff?, Phys. Today **55**(10), 14–17 (2002)

47.5 H. J. Mamin, B. D. Terris, L. S. Fan, S. Hoen, R. C. Barrett, D. Rugar: High–density data storage using proximal probe techniques, IBM J. Res. Dev. **39**(6), 681–699 (1995)

47.6 R. C. Barrett, C. F. Quate: Large-scale charge storage by scanning capacitance microscopy, Ultramicroscopy **42–44**, 262–267 (1992)

47.7 H. J. Mamin, R. P. Ried, B. D. Terris, D. Rugar: High-density data storage based on the atomic force microscope, Proc. IEEE **87**(6), 1014–1027 (1999)

47.8 P. Vettiger, J. Brugger, M. Despont, U. Drechsler, U. Durig, W. Haberle, H. Lutwyche, H. Rothuizen, R. Stutz, R. Widmer, G. Binnig: Ultrahigh density, high-data-rate NEMS-based AFM data storage system, Microelectron. Eng. **46**, 11–17 (1999)

47.9 P. Vettiger, M. Despont, U. Drechsler, U. Durig, W. Haberle, M. I. Lutwyche, H. E. Rothuizen, R. Stutz, R. Widmer, G. K. Binnig: The millipede—more than one thousand tips for future AFM data storage, IBM J. Res. Dev. **44**(3), 323–340 (2000)

47.10 B. Schechtman: The future of optical disk storage, NSIC Annual Meeting Symposium, Monterey 2002, 1–14

47.11 P. Lyman, H. L. Varian: How much information (2000) retrieved from http://www.sims.berkeley.edu/how–much–info

47.12 W. Perdue: Private Communication (September, 2002)

47.13 P. Frank: private communication, (2002)

47.14 H. De J. Ruiz, S. D. Morgan: , Witter Semiconductor and Systems Conference, Laguna Michael 2001

47.15 J. Daughton, NVE Inc., private communication

47.16 M. Gill, T. Lowrey, J. Park: Ovonics unified memory—A high performance nonvolatile memory technology for stand alone memory and embedded applications, 2002 IEEE Int. Solid State Circuits Conf., San Francisco 2002 (IEEE, Piscataway 2002)

47.17 M. Despont, J. Brugger, U. Drechsler, U. Dürig, W. Häberle, M. Lutwyche, H. Rothuizen, R. Stutz, R. Widmer, G. Binnig, H. Rohrer, P. Vettiger: VLSI–NEMS chip for AFM data storage, Technical Digest 12th IEEE Int. Conf. on Micro Electro Mechanical Systems (MEMS'99) (IEEE, Piscataway 1999) 564–569

47.18 K. Schulte-Wieking, S. Noehte, C. Dietrich, M. Mayer: A new cylindrically shaped optical data storage medium based on adhesive tape—concept and first results, IEEE Proc. Technical Digest of ISOM/ODS 2002, Joint Int. Symp. on Optical Memory and Optical Data Storage 2002 (IEEE, Piscataway 2002) 117–120

47.19 D. Sarid: unpublished results, (2002)

47.20 L. A. Romero, F. M. Dickey: A method for achieving constant rotation rates in a microorthogonal linkage system, J. Microelectromech. Syst. **9**(2), 236–244 (2000)

47.21 L. R. Carley, G. R. Ganger, D. F. Nagle: MEMS-based integrated-circuit mass-storage systems, Commun. ACM **43**(11), 71–80 (2000)

47.22 J. L. Griffin, S. W. Schlosser, G. R. Ganger, D. F. Nagle: Modeling and performance of MEMS-based storage devices, Int. Conf. on Measurement and Modeling of Computer Systems, Proc. of ACM Sigmetrics, Santa Clara 2000, **28**(1), 56–65 (2000), published as Performance Evaluation Review

47.23 S. W. Schlosser, J. L. Griffin, D. F. Nagle, G. R. Ganger: Designing computer systems with MEMS-based storage, Proc. of the 9th International Conf. on Architectural Support for Programming Languages and Operating Systems, Boston 2000, 1–12

47.24 L. R. Carley, G. Ganger, D. F. Guillou, D. Nagle: System design considerations for MEMS-actuated magnetic-probe-based mass storage, IEEE Trans. Magn. **37**(2), 657–662 (2001)

47.25 R. T. El-Sayed, L. R. Carley: Performance analysis of beyond 100 Gb/in^2 MFM-based MEMS-actuated mass storage devices, IEEE Trans. Magn. **38**(5), 1–3 (2002)

47.26 O. Bichet, S. Gidon, Y. Samson: Scanning probe based storage on phase change media, OSA Trends in Optics and Photonics Vol. 88 Optical Data Storage (OSA, Washington D.C. 2003)

47.27 H. Kawakatsu, D. Saya, A. Kato, K. Fukushima, H. Toshiyoshi, H. Fujita: Millions of cantilevers for atomic force microscopy, Rev. Sci. Instrum. **73**(3), 1188–1192 (2002)

47.28 H. Kawakatsu, S. Kawai, D. Saya, M. Nagashio, D. Kobayashi, H. Toshiyoshi, H. Fujita: Towards atomic force microscopy up to 100 MHz, Rev. Sci. Instrum. **73**(6), 2317–2320 (2002)

47.29   P. Vettiger, G. Cross, M. Despont, U. Drechsler, U. Dürig, B. Gotsman, W. Häberle, M. A. Lantz, H. E. Rothuizen, R. Stutz, G. K. Binnig: The 'millipede'—Nanotechnology entering data storage, IEEE Trans. Nanotechnol. **1**, 39–55 (2002)

47.30   Q. Zhong, D. Inniss, K. Kjoller, V. B. Elings: Fractured polymer/silica fiber surface studied by tapping mode atomic force microscopy, Surf. Sci. **290**(1–2), L688–L692 (1993)

47.31   A. M. Hayashi: Punch cards of the future, Sci. Am. (May 2000)

47.32   J. P. Hunt, D. Sarid: Kinetics of lossy grazing impact oscillators, Appl. Phys. Lett. **72**(23), 2969–2971 (1998)

47.33   D. Sarid: Tapping-mode scanning force microscopy: Metallic tips and samples, Comp. Mater. Sci. **5**, 291–297 (1996)

47.34   A. Fein, Y. Zhao, C. A. Peterson, G. E. Jabbour, D. Sarid: Individually injected current pulses with conducting-tip, tapping-mode atomic force microscopy, Appl. Phys. Lett. **79**(24), 3935–3937 (2001)

47.35   T. Hidaka, T. Maruyama, M. Saitoh, N. Mikoshiba, M. Shimizu, T. Shiosaki, L. A. Willis, R. Hiskes, S. A. Dicarolis, J. Amano: Formation and observation of 50 nm polarized domains in $PbZr_{1-x}Ti_xO_3$ thin film using scanning probe microscope, Appl. Phys. Lett. **68**(17), 2358–2359 (1996)

47.36   A. Beck, J. G. Bednorz, Ch. Gerber, C. Rossel, D. Widmer: Reproducible switching effect in thin oxide films for memory applications, Appl. Phys. Lett. **77**(1), 139–141 (2000)

47.37   S. Sun, C. B. Murray, D. Weller, L. Folks, A. Moser: Monodisperse FePt nanoparticles and ferromagnetic FePt nanocrystal superlattices, Science **287**, 1989–1992 (2000)

47.38   C. Peterson: Characterization of semiconductor devices through scanned probe microscopy. Ph.D. Thesis (University of Arizona, Tucson 2002)

47.39   T. Gotoh, K. Sugawara, K. Tanaka: Nanoscale electrical phase-change in $GeSb_2Te_4$ films with scanning probe microscopes, J. Non-Cryst. Solids **299-302**, 968–972 (2002)

47.40   H. J. Gao, K. Sohlberg, Z. Q. Xue, H. Y. Chen, S. M. Hou, L. P. Ma, X. W. Fang, S. J. Pang, S. J. Pennycook: Reversible, nanometer-scale conductance transitions in an organic complex, Phys. Rev. Lett. **84**(8), 1780–1783 (2000)

47.41   J. C. Li, Z. Q. Xue, W. M. Liu, S. M. Hou, X. L. Li, X. Y. Zhao: Study on a new organic-complex thin film with electrical bistable properties using a scanning tunneling microscope, Phys. Lett. A **266**, 441–445 (2000)

47.42   Y. Zhao, A. Fein, C. A. Peterson, D. Sarid: Comment on reversible, nanometer-scale conductance transitions in an organic complex, Phys. Rev. Lett. **87**(17), 179706-1 (2001)

47.43   L. P. Ma, J. Liu, Y. Yang: Organic electrical bistable devices and rewritable memory cells, Appl. Phys. Lett. **80**(16), 2997–2999 (2002)

47.44   B. McCarthy, K. Yamnitskiy, G. E. Jabbour, D. Sarid: unpublished results, (2002)

48. Microactuators for Dual-Stage Servo Systems in Magnetic Disk Files

This chapter discusses the design and fabrication of electrostatic MEMS microactuators and the design of dual-stage servo systems in disk drives. It introduces fundamental requirements of disk drive servo systems, along with challenges posed by storage densities increases. It describes three potential dual-stage configurations and focuses on actuated slider assemblies using electrostatic MEMS microactuators. The authors discuss major electrostatic actuator design issues, such as linear versus rotary motion, electrostatic array configuration, and differential operation. Capacitive and piezoresistive elements may be used to sense relative slider position, while integrated gimbals and structural isolation may prove useful in improving performance. A detailed design example based on a translational, micromolded actuator illustrates several of these concepts and is accompanied by theoretical and experimental results.

The chapter continues to discuss MEMS microactuator fabrication. It describes several processes for obtaining appropriate electrostatic devices including micromolding, deep reactive-ion etching, and electroplating. The primary goal of these processes is to obtain very-high-aspect ratio structures, which improve both actuation force and structural robustness. Other fabrication issues, such as electrical interconnect formation, material selection, and processing cost, are also considered. Actuated slider fabrication is compared to that of actuated suspension and actuated head assemblies; this includes a description of an instrumented suspension, in which the suspension is instrumented with strain sensors to improve vibration detection.

The section on controller design reviews dual-stage servo control design architectures and methodologies. The considerations for controller design of MEMS microactuator dual-stage servo systems are discussed. The details of control designs using a decoupled single-input single-output (SISO) design method and the robust mixed H_2/H_∞ multivariable design method of µ-synthesis are also presented. In the decoupled design, a self-tuning control algorithm has been developed to compensate for the variations in the microactuator's resonance mode. In the mixed H_2/H_∞ design method, controllers can use additive or parametric uncertainties to characterize unmodeled dynamics of the VCM and variations in the microactuator's resonant mode, while providing robust performance. A comprehensive robust controller design example using multi-rate control follows. Finally, the chapter also introduces a dual-stage short-span seek control scheme based on decoupled feed-forward reference-trajectory generation.

This chapter discusses the design and fabrication of electrostatic microelectromechanical systems (MEMS) microactuators, and the design of dual-stage servo systems, in disk drives. The focus of the chapter is an actuated slider assembly using an electrostatic MEMS microactuator. We discuss major design issues, including linear versus rotary actuation, electrostatic array configuration, and integrated sensing capability. We describe several fabrication processes for obtaining the necessary devices, such as micromolding, deep reactive-ion etching, and electroplating. Dual-stage servo control design architectures and methodologies are then reviewed. We present in detail track-following controller designs based on a sensitivity-function-decoupling single-input single-output design methodology and the robust µ-synthesis design methodology. Finally, we introduce a two-degree-of-freedom (2-DOF) short-span seek control design using a dual-stage actuator.

Since the first hard disk drive (HDD) was invented in the 1950s by IBM, disk drive storage density has been following Moore's law, doubling roughly every 18 months. The current storage density is 10 million times larger than that of the first HDD [48.1]. Historically, increases in storage density have been achieved by almost equal increases in track density, the number of tracks encircling the disk, and bit density, the number of bits in each track. However, because of limitations to superparamagnetism, future areal storage density increases in HDDs are predicted to be achieved mainly through an increase in track density [48.2].

Research in the HDD industry is now targeting an areal density of one terabit per square inch. For a predicted bit aspect ratio of 4 : 1, this translates to a linear bit density of 2 Mbits per inch (BPI) and a radial track density of 500 k tracks per inch (TPI), which in turn implies a track width of 50 nm. A simple rule of thumb for servo design in HDDs is that three times the statistical standard deviation of the position error between the head and the center of the data track should be less than 1/10 of the track width. Thus, to achieve such a storage density, nanometer-level precision of the servo system will be required.

A disk drive stores data as magnetic patterns, forming bits on one or more disks. The polarity of each bit is detected (read) or set (written) by an electromagnetic device known as the read/write head. The job of a disk drive's servo system is to position the read/write head over the bits to be read or written as they spin by on the disk. In a conventional disk drive, this is done by sweeping over the disk or disks a long arm consisting of a voice coil motor (VCM), an E-block, suspensions, and sliders, as shown in Figs. 48.1 and 48.2. A read/write head is fabricated on the edge of each slider (one for each disk surface). Each slider is supported by a suspension and flies over the surface of a disk on an air bearing. The VCM actuates the suspensions and sliders about a pivot in the center of the E-block. We describe this operation in more detail in the following section.

The key to increasing HDD servo precision is to increase the servo control bandwidth. However, the bandwidth of a traditional single-stage servo system, as shown in Fig. 48.2, is limited by the multiple mechanical resonance modes of the pivot, the E-block, and the suspension between the VCM and the head. Nonlinear friction of the pivot bearing also limits achievable servo precision. Dual-stage actuation, with a second stage actuator placed between the VCM and the head, has been proposed as a solution that would increase servo bandwidth and precision.

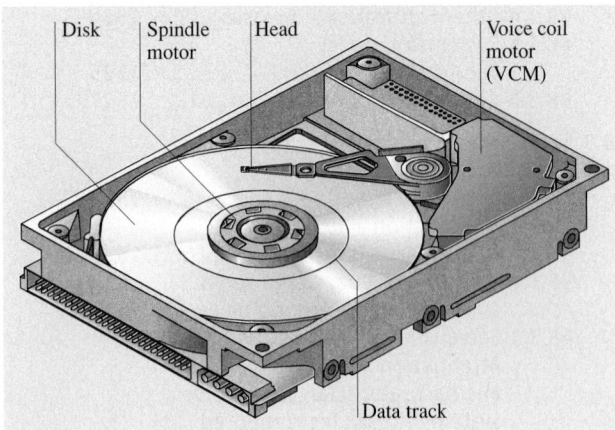

Fig. 48.1 A schematic diagram of a HDD

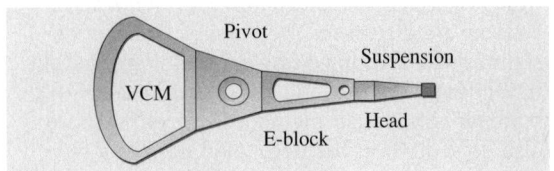

Fig. 48.2 VCM actuator in a HDD

Several different secondary actuation forces and configurations have been proposed, each having strengths and weaknesses given the requirements of HDDs. The dual-stage configurations can be categorized into three groups: actuated suspension, actuated slider, and actuated head. Within these, actuation forces include piezoelectric, electrostatic, and electromagnetic. In this chapter, we discuss design, fabrication and control of an electrostatic MEMS microactuator (MA) for actuated slider dual-stage positioning.

48.1 Design of the Electrostatic Microactuator

The servo system of a hard disk drive is the mechatronic device that locates and reads data on the disk. In essence, it is a large arm that sweeps across the surface of the disk. At the end of the arm is the read/write head, containing the magnetic reading and writing elements that transfer information to and from the disk.

48.1.1 Disk Drive Structural Requirements

The read/write head is contained in a box-like structure known as a slider. The slider has a contoured lower surface that acts as an air bearing between the head and the disk. The high-velocity airflow generated by the spinning disk pushes up on the air-bearing surface, maintaining the slider and read/write head at a constant distance from the disk, despite unevenness of the disk, permitting reliable data reading and writing.

The arm over a HDD's disk has three primary stages: the voice-coil motor (VCM), the E-block, and the suspension. In a conventional disk drive, the VCM performs all positioning of the head, swinging it back and forth across the disk. The E-block lies between the VCM and the suspension and contains the pivot point. The suspension projects from the E-block over the disk as a thin flexible structure, generally narrowing to a point at the location of the slider.

For a disk drive servo to operate effectively, it must maintain the read/write head at a precise height above the disk surface and within a narrow range between the disk tracks, which are arranged in concentric circles around the disk. It must also be able to seek from one track on the disk to another. Information about the track that the head is following is encoded in sectors radiating from the center of the disk, allowing the head to identify its position and distance from the center of the track. To maintain the correct flying height, the suspension must be designed with an appropriate stiffness in the vertical direction to balance an air-bearing force corresponding to the slider design in use in the drive. Meanwhile, the suspension must be flexible to roll and pitch at the slider location to permit adaptation to unevenness of the disk

surface. This is accomplished using a gimbal structure. The suspension as a whole, however, should not bend or twist during the operation, as this would misdirect the head away from the track it is following.

As data densities in HDDs increase and track widths diminish, single-stage, conventional servo systems become less able to position the head precisely. Because the VCM/E-block/suspension assembly is large and massive as a unit, the speed at which the head can be controlled is limited. Furthermore, the assembly tends to have a low natural frequency, which can accentuate vibration in the disk drive and cause off-track errors. At track densities in the future approaching one terabit per square inch, the vibration induced in a disk drive by airflow alone is enough to force the head off-track.

A solution to these problems is to complement the VCM with a smaller, secondary actuator to form a dual-stage servo system. The VCM continues to provide rough positioning, while the second stage actuator does fine positioning and damps out vibration and other disturbances. The smaller second-stage actuator can typically be designed to have a much higher natural frequency and be less susceptibility to vibration than the VCM. Any actuator used in a dual-stage system should be inexpensive to build, require little power to operate, and preserve the stiffness properties described above that are necessary to preserve the flying height.

48.1.2 Dual–Stage Servo Configurations

In the past six years, much research work has been dedicated to the exploration of suitable secondary actuators for constructing dual-stage servo systems for HDDs. These dual-stage configurations can be categorized into three groups: actuated suspension, actuated slider, and actuated head.

Actuated Suspension
In this approach, the suspension is redesigned to accommodate an active component, typically a piezoelectric material. This piezoelectric material stretches or flexes

the suspension to position the slider and magnetic head. Piezoelectric material is an active actuation element that produces a large actuation force but small actuation stroke. In the actuated suspension configuration, therefore, the piezoelectric actuators are usually implemented in a leverage mechanism that can convert small actuation displacements into large head displacements. Typically this is done by placing the piezoelectric actuators away from the magnetic head (between the E-block and suspension) so that they can have a long leverage arm to gain mechanical amplification and produce a sufficient magnetic-head motion. The advantage of this approach is that the suspension can be fabricated by a conventional suspension-making process, and its dual-stage servo configuration is effective in attaining low frequency runout attenuation in the positioning servo loop. The major drawback of this approach is that the system is still susceptible to instabilities due to the excitation of suspension resonance modes. Thus, track-per-inch (TPI) servo performance can be increased but remains limited when compared to the alternative approaches. The actuated suspension approach, nonetheless is expected to be the first deployed in commercial HDDs [48.3, 4].

Actuated Slider

In this approach, a microactuator is placed between the slider and gimbal to position the slider/magnetic heads. The resulting servo bandwidth can be higher than the previous approach because the secondary actuation bypasses the mechanical resonances of the suspension. This approach uses existing sliders and microactuators that can be batch fabricated, and thus could be cost effective. However, the size and mass of the microactuator are significant relative to those of current sliders and may interfere with the slider flying stability. Current suspensions, therefore, need to be redesigned to adopt this secondary actuator. Suitable driving forces in this approach include electrostatic, electromagnetic, and piezoelectric [48.5–8]. To further reduce the assembly task of placing the microactuator in between the gimbal structure and slider, some researchers have proposed microactuators that are either integrated with the gimbal structure [48.6] or the slider [48.9].

Actuated Head

In this approach, the slider is redesigned so that the microactuators can be placed inside the slider block and actuate the magnetic heads with respect to the rest of the slider body. As these microactuators are very small, they only slightly increase the slider weight and are thus capable of working with the current suspension as-

sembly. Researchers have successfully demonstrated the integrated fabrication process for fabricating the electrostatic microactuators and magnetic heads within one piece of ceramic block (slider). The embedded electrostatic microactuator has its resonance close to 30 kHz and was able to position the magnetic heads relative to the rest of the slider body by 0.5 μm [48.10, 11]. Fully fledged integration of slider, actuator, and read/write head remains a challenge.

In this chapter, we focus on actuated slider configurations, as they involve a great deal of interesting microscale engineering. In particular, we will discuss electrostatic actuation, probably the most common method of implementing microactuation in microelectromechanical devices. Any such microactuator will exhibit certain features:

- a fixed base, which attaches to the suspension,
- a movable platform, upon which the slider rests,
- springs between the base and platform, flexible in the direction of desired motion, and stiff in all other directions, and
- an electrostatic actuation array that generates the force used to move the platform and slider.

Microactuators must also include a wiring scheme for transferring signals to and from the slider and often incorporate a structure for sensing the motion of the slider relative to the suspension. Electrostatic microactuators to be discussed in this chapter include HexSil and deep reactive-ion etching (DRIE) fabricated actuators from the University of California, Berkeley, and electroplating-formed actuators by IBM and the University of Tokyo.

Electromagnetic or piezoelectric force are alternatives to electrostatic actuation in the actuated slider configuration. Electromagnetic microactuators use ferromagnetic films to produce a force perpendicular to an applied electric field. This type of actuation potentially has low voltage requirements but requires special fabrication techniques to integrate the magnetic components into the assembly. Piezoelectric microactuators use a piezoelectric material, which expands or contracts in response to applied voltage, to move the slider. These actuators have simple fabrication, the patterning of a piece of piezoelectric material to sit between the suspension and slider, but it is difficult to obtain an adequate range of motion. A short stroke from the piezoelectric piece must be leveraged into a much larger motion at the read/write head. A piezoelectric microactuator has been produced by the TDK corporation with a 0.5 μm stroke length at 10 V, with a 10 V bias [48.12].

48.1.3 Electrostatic Microactuators: Comb Drives versus Parallel Plates

Electrostatic microactuators have been studied as the secondary actuators in HDDs for their relative ease of fabrication, particularly in the actuated slider and actuated head configurations, since the structural material only needs to be conductive rather than ferromagnetic or piezoelectric. The electrostatic force is generated by applying a voltage difference between the moving shuttle and a fixed stator element. Depending on the designated motion for the shuttle, electrostatic actuators are often categorized into two groups: comb drives and parallel plates, as illustrated in Fig. 48.3.

The magnitude of the electrostatic force generated equals the rate of change of energy that is retained within the finger-like structure and varied by shuttle motion. Therefore, the electrostatic force for comb-drive actuators, in which the designated shuttle motion moves along the x-direction, as shown in Fig. 48.3, equals

$$F_{\text{comb}} = \frac{\partial E}{\partial x} = \frac{\epsilon h}{2d} V^2 , \qquad (48.1)$$

where ϵ is the permittivity of air, x is the overlap between two adjacent plates, h is plate thickness, and d is the gap between two parallel plates. Similarly, the electrostatic force for parallel-plate actuators is

$$F_{\text{parallel}} = \frac{\partial E}{\partial y} = \frac{\epsilon x h}{2d^2} V^2 , \qquad (48.2)$$

where x is the finger overlap.

As indicated in (48.1) and (48.2), the electrostatic force for comb-drive actuators does not depend on the displacement of the moving shuttle and thus allows a long stroke while maintaining a constant electrostatic force. The electrostatic force for parallel-plate actuators, in contrast, is a nonlinear function of its shuttle motion ($\propto 1/d^2$), and the maximum stroke is limited by the nominal gap between shuttle and stator. A longer stroke is achieved with a larger gap, at the expense of lower electrostatic force. For applications that require small stroke but large force output, parallel-plate actuators are preferred since the output force from parallel plates can be x/d times larger than the force from comb drives. The following equation (48.3) is easily derived from (48.1) and (48.2).

$$\frac{F_{\text{parallel}}}{F_{\text{comb}}} = \frac{x}{d} . \qquad (48.3)$$

A simplified second-order differential equation is often utilized to describe the dynamic response of an

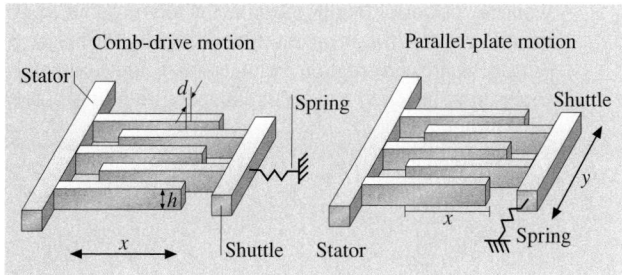

Fig. 48.3 Electrostatic microactuators: comb drives versus parallel plates

electrostatic microactuator

$$m\ddot{x}(t) + b\dot{x}(t) + K_{\text{m}}x(t) = F[V, x(t)] , \qquad (48.4)$$

where m is the mass of the moving shuttle, b is the damping coefficient of the microactuator, K_{m} is the spring constant of the mechanical spring that connects the moving shuttle to an anchor point, and F is the electrostatic force that can be obtained from (48.1) and (48.2).

Differential Drives

Because electrostatic force is always attractive, electrostatic microactuators need other features to actively control the direction of shuttle motion in servo applications, as opposed to relying on the restoring force from a mechanical spring. For this reason, the differential-drive approach, as shown in Fig. 48.4, is frequently adopted in electrostatic microactuator designs. Based on the differential-drive configuration, the simplified second-order differential equation (48.4) is rewritten as

$$m\ddot{x}(t) + b\dot{x}(t) + K_{\text{m}}x(t) = F[V_{\text{bias}} + V_{\text{dr}}, x_0 - x(t)] - $$
$$F[V_{\text{bias}} - V_{\text{dr}}, x_0 + x(t)] , \qquad (48.5)$$

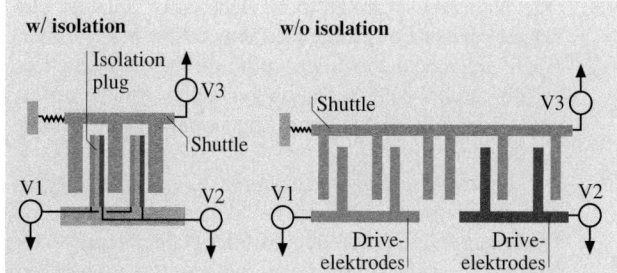

Fig. 48.4 Differential parallel-plate actuators and electrical-isolation features

where x_0 is the nominal position of the moving shuttle. If the differential drive is operated at the bias voltage (V_{bias}) with a small perturbation voltage (V_{dr}), the nonlinear force input in (48.5) can be linearized with a first-order approximation.

$$m\ddot{x}(t) + b\dot{x}(t) + (K_m - K_e)x(t) = K_v V_{dr},$$
$$K_e = 2\frac{\partial F}{\partial x}\big|_{V_{bias}},$$
$$K_v = 2\frac{\partial F}{\partial V_{dr}}\big|_{V_{bias}}. \qquad (48.6)$$

Here K_e represents a softening electrostatic spring constant, and K_v represents the voltage-to-force gain. The electrostatic spring constant acts as a negative spring during the electrostatic microactuator operation, and its value varies with the bias voltage, V_{bias}. When the electrostatic spring constant K_e exceeds the spring constant K_m of the mechanical spring, the microactuator becomes unstable; this is often described as pull-in instability. As shown in (48.5) and (48.6), the differential configuration cancels the even-order harmonics in voltage and thus linearizes the voltage–force relation to some extent. Furthermore, in parallel-plate actuators, the differential configuration reduces the nonlinearity in actuation voltage as well as in shuttle displacement.

Electrical Isolation

Because electrostatic actuation requires multiple voltage levels for actuation force and position sensing (as discussed in the following section), electrical isolation is another challenge for designing an electrostatic microactuator. Generally speaking, when multiple voltage levels are needed in MEMS devices, electrical isolation is achieved by breaking up the parts that need to be on different voltage level and anchoring them separately to a nonconductive substrate. This approach has many drawbacks, not only because it requires a substrate in a device but also because structures have to be mechanically separated to be electrically isolated. The electrical isolation problem is far more severe in parallel-plate microactuators than comb-drive microactuators since parallel-plate actuation generally requires different voltage levels for stator fingers pulling in opposite directions.

Figure 48.4 shows an example of how an electrical-isolation feature can be utilized to increase the actuation force output in a differential parallel-plate microactuator design. As shown in the figure, without the proper electrical isolation, drive electrodes with different voltage potentials have to be placed in separate groups and result

in the same voltage difference between drive electrodes and shuttle on both sides of each shuttle finger [48.5, 6]. Since electrostatic forces are always attractive, gaps on two sides of the interlaced structure cannot be made equal, otherwise the forces on two sides of a shuttle finger will be equal and the shuttle's movement direction will be uncontrollable. With such electrical-isolation features as the isolation plug, shown on the left in Fig. 48.4, gaps in the interlaced structure can be the same width, since different voltages can be applied on the two sides of the shuttle fingers. As shown in Fig. 48.4, the design with integrated electrical-isolation features is more compact than without isolation features. Consequently, more finger structures can fit into the same amount of space, and the actuation voltage can be reduced.

48.1.4 Position Sensing

Most proposed HDD dual-stage servo controllers utilize only the position of the magnetic head relative to the center of data track, known in the industry as the position error signal, or PES, for closed-loop track-following control. These systems have a single-input multi-output (SIMO) control architecture. In some instances, however, it is also possible to measure the relative position error signal (RPES) of the magnetic head relative to the VCM. In this case, the control architecture is multi-input multi-output (MIMO). As shown in [48.13], RPES can be used in a MIMO controller to damp out the second-stage actuator's resonance mode and enhance the overall robustness of the servo system.

Capacitive position sensing and piezoresistive position sensing are two popular sensing mechanisms among electrostatic microactuator designs. Each of these sensing mechanisms is discussed in more detail in the following sections.

Capacitive Position Sensing

Capacitive position sensing is based on shuttle movement causing a capacitance change between the moving shuttle and fixed stators. By measuring the change in capacitance, it is possible to determine the shuttle location relative to the fixed stator. The output voltage (V_o) for both differential drives in Fig. 48.5 equals $2\delta C/C_i V_s$. The capacitance change due to shuttle movement (dC/dx) can be derived and the output voltage (V_o) for the comb drives and parallel plates can be formulated as a function of shuttle displacement.

$$V_{comb} = 2\frac{C_s}{C_i}\frac{\delta x}{x_0}V_s,$$

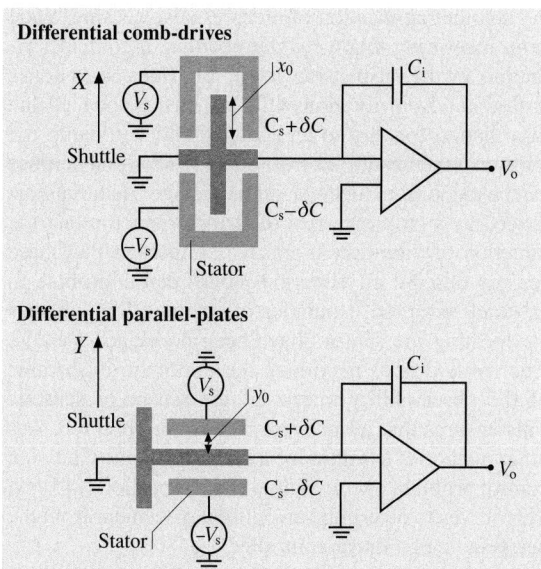

Fig. 48.5 Capacitive position sensing. Comb-drive versus parallel-plate motion

$$V_{\text{parallel}} = 2\frac{C_s}{C_i}\frac{y_0\delta y}{y_0^2 - \delta y^2}V_s ,$$

$$\approx 2\frac{C_s}{C_i}\frac{\delta y}{y_0}V_s , \qquad (48.7)$$

where x_0 is the nominal overlap for interlaced fingers and y_0 is the nominal gap between overlapped fingers, as shown in Fig. 48.5.

As indicated by (48.7), the voltage output for the comb-drive sensing structure, V_{comb}, is linear with shuttle displacement. On the other hand, the sensing configuration that makes use of the parallel-plate motion has better sensitivity for detecting shuttle motion since y_0 is usually smaller than x_0. Although the nonlinearity in parallel-plate sensing can be linearized by the differential drive configuration to some extent, in a design example of a 4 μm gap with a 1 μm stroke, the linear model shown in (48.7) still produces 6% deviation from the nonlinear model.

Among electrostatic microactuators that use capacitive position sensing, the capacitance variation due to shuttle motion (dC/dx) is typically at the level of 100 fF/μm. In order to obtain 10 nm position sensing resolution, the capacitance-sensing circuit must be able to detect capacitance variation of 1 fF in the presence of parasitic capacitance and offset/mismatches from op-amps, which can easily result in an output voltage orders of magnitude larger than the output voltage from

the designated capacitance variation. In most capacitive position sensing, the limiting factor for the sensing resolution is not the thermal noise but the sensing circuit's design.

Here we introduce two basic capacitance-sensing circuits suitable for high-resolution position sensing [48.14]. The concept of the synchronous scheme, as shown in Fig. 48.6 I, is to reduce the impedance of sense capacitors as well as the offset and $1/f$ noise from op-amps by applying modulation techniques to the sense voltage (V_s). The R_{dc} resistor on the feedback loop sets the DC voltage level at the input nodes of the charge integrator. The effect of the presence of parasitic capacitance (C_p) is nullified by the virtual ground condition from the op-amps. The major drawback of the synchronous scheme is that the DC-setting resistor (R_{dc}) has to be large to ensure the proper gain for the capacitance sensing [48.14], which introduces excessive thermal noise into the sensing circuit. In addition, a large resistor usually consumes a large die space in implementation.

A switched-capacitance scheme, as shown in Fig. 48.6 II, is one alternative that avoids the use of the

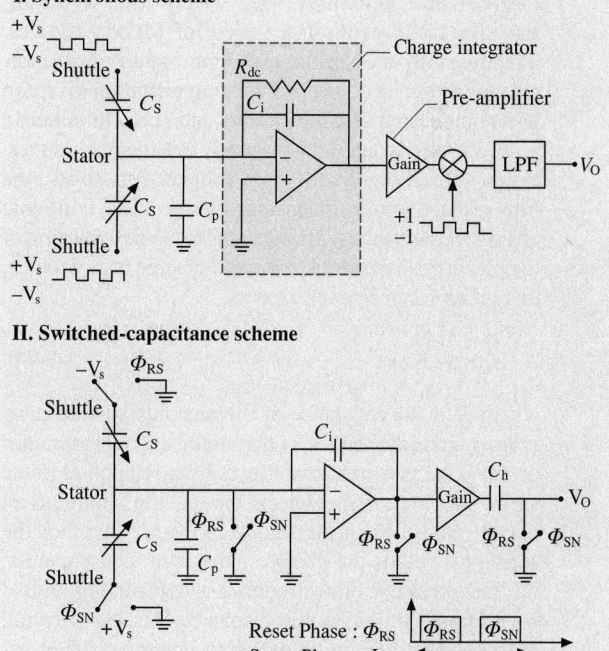

Fig. 48.6 Capacitive position-sensing circuits. The synchronous scheme versus the switched-capacitance scheme

DC-setting resistor. The capacitance-sensing period is broken into two phases: the reset phase and sense phase. During the reset phase, the input/output nodes of the capacitors and the input voltage to op-amps are set to the ground or reference level to ensure proper DC voltage for the charge integrator. During the sense phase, the sense voltage $\pm V_s$ is applied to the sense capacitors, and the amount of charge proportional to the mismatch in the sense capacitors is integrated on the capacitor C_i, thus producing an output voltage proportional to the capacitance mismatch from the sense capacitors. This approach replaces the large DC-setting resistor by capacitors and switches and results in a much smaller die compared to the synchronous scheme. Furthermore, the switching technique allows more design flexibility for system integration and performance improvement because of the ability to allocate separate phases for various operations. The major drawback of this design is that it draws noise into the sensing circuits from switches and the sampling capacitor C_h. However, these noises can be compensated by dividing the sense phase into $2 \approx 3$ sub-sense phases [48.14], at the expense of a complicated circuit design.

Piezoresistive Sensing

Piezoresistive films have been widely used as strain-sensitive components in a variety of MEMS devices, including pressure sensors and vibration sensors. Generally speaking, piezoresistive sensing techniques require less-complicated sensing circuits and perform better in a severe environment than other sensing techniques. When a piezoresistive film is subjected to stress, the film resistivity and dimensions change. The fractional change of resistance is proportional to the deformation of the piezoresistive film. For a small change of resistance, this relation can be expressed as,

$$\frac{\Delta R}{R} = K \cdot \epsilon , \tag{48.8}$$

where R is the resistance of the piezoresistive film, K is its gage factor and ϵ is the strain. In microactuator designs, the piezoresistive film is usually applied to the spring structure that connects the moving shuttle to an anchor point. When the shuttle moves, it stretches the spring as well as the piezoresistive film; consequently, the piezoresistive film produces a deformation signal proportional to the shuttle displacements. As a result, piezoresistive sensing is easier to implement than capacitive position sensing, but the sensing resolution is usually lower due to the larger thermal noise introduced by the resistance of the piezoresistive film.

Another application of piezoresistive sensing, aside from measuring relative slider position, is to detect vibration in the suspension itself. The idea is to sense airflow-induced vibration of the suspension and feed that information forward to an actuated slider to damp out motion at the head. Piezoresistors used for this purpose can be made from metal or semiconductor materials, arranged as a strip or series of strips oriented along the direction of vibration strain. It is important that these sensors observe all vibration modes that contribute to off-track error, so a number of optimization schemes for locating the sensors have been developed [48.15]. One method is to maximize the minimum eigenvalue of the observability matrix of the sensor or sensors. This ensures that all relevant modes are observed. Another method is to minimize a linear quadratic Gaussian control problem over potential sensor locations [48.16]. This serves to determine an optimal placement from the perspective of a linear controller.

48.1.5 Piezoelectric Sensing

Piezoelectric materials are also widely used to build sensors such as surface-acoustic-wave devices, pressure sensors and accelerometers. Devices fabricated from piezoelectric materials are energy convertors that convert mechanical to electrical energy and vice versa. Depending on how the devices are used, they can be utilized as sensors or as actuators. When acting as sensors, the external stress causes electrical asymmetry inside the material structure. As a consequence, charges are generated to compensate the asymmetry. The amount of generated charge depends both on the applied force and the crystal structure of the material. This phenomenon, called the direct piezoelectric effect, is usually expressed in an orthogonal coordinate system as

$$D_i = d_{ij} T_j + \varepsilon_{ik}^T E_k , \tag{48.9}$$

where D_i, d_{ij}, T_j, ε_{ik}^T and E_k are the electric displacement, piezoelectric constant, stress, dielectric constant and electric field, respectively. The superscript T, denoting the dielectric constant ϵ_{ik}, is measured at constant stress. The subscripts i and k take the values 1, 2, 3 while j takes the values 1, 2, 3, 4, 5, 6 and the summation convention is used for repeated indices. The piezoelectric constant d_{ij} represents how electrical displacement, in charge per unit area, is related to the external force. For example, d_{33} is the proportional constant that describes the amount charge generated on the surfaces perpendicular to the direction of applied stress.

Piezoelectric sensors have very different characteristics to piezoresistive materials. They have less intrinsic thermal noise, which enables them to achieve better resolution and larger dynamic range. While piezoresistors are especially good at static measurements, piezoelectric sensors are capable of resolving vibration signals over a wide frequency range. Piezoelectric materials can also be made into modal sensors [48.17, 18], which are used to detect vibrations at specific frequencies. These characteristics of piezoelectric materials make them ideal candidates for vibration sensors at the MEMS scale. The chief disadvantages to piezoelectric sensing are complex processing requirements and the need for precise circuitry to condition and amplify the sensing signals if ultra-high resolution is desirable, as piezoelectric sensors are very sensitive to parasitic capacitances.

48.1.6 Electrostatic Microactuator Designs for Disk Drives

Various electrostatic microactuators have been designed for secondary actuation in HDDs. To incorporate an electrostatic microactuator into a HDD without altering much of the current suspension configuration, many design constraints are imposed. In this section, we will first discuss some design issues and then present one specific design example.

Translational Microactuators versus Rotary Microactuators

Depending on the motion of the magnetic head actuated, microactuator designs are categorized into two groups: translational actuators and rotary actuators. Either type can be implemented by comb-drive [48.8, 19] or parallel-plate [48.5, 6, 20] actuation.

When employing a translational microactuator in a dual-stage HDD servo, previous research [48.13] has shown that a force coupling between the suspension and the translational microactuator exists, consisting of transmitted actuation force from the VCM and suspension vibration induced by windage. The force coupling from the VCM not only complicates the dual-stage servo controller but also imposes a design constraint on a translational microactuator design, in that the translational microactuator has to provide a large force output to counterbalance the coupling force. When the VCM makes a large movement, as in seeking a new data track, the microactuator may be overpowered. One solution is to pull the actuator to one side and lock it momentarily in place. Even then, the use of the two actuation-stages must be carefully coordinated to mod-

erate the influence of the VCM on the microactuator. On the other hand, the linear springs in the translational microactuator can also aid in damping out motion of the suspension. The portion of suspension vibration induced by windage mostly consists of high-frequency excitation, so the resulting magnetic head's position error can be passively attenuated by low-resonant-frequency translational microactuators. This suppression makes translational microactuators preferable during track-following control.

Generally speaking, rotary actuators are more difficult to design/analyze than translational actuators because of their nonuniform gap between the shuttle and stator. Still, their different operating properties have both strengths and weaknesses. Unlike a translational actuator, no obvious force coupling is transmitted from the suspension to a rotary actuator, as the microactuator is nearly always attached to the end of suspension at the microactuator's center of rotation, which acts as a pivot point. With no mechanical coupling, the dual-stage servo system using a rotary microactuator does not suffer from the force coupling between VCM and microactuator seen in translational designs. However, the rotary microactuator has to compensate for the magnetic head's position error induced by suspension vibration without any passive attenuation of the vibration. Overall, a rotary actuator is likely to behave better than a translational microactuator during track seeking and worse during track following.

Gimballed Microactuator Design

Proper flying height and orientation of a slider and read/write head over a hard disk is maintained by the interaction of the suspension, the air bearing of the slider, and a gimbal structure. The gimbal structure is located at the tip of the suspension and holds the slider/microactuator in its center coupon. The dynamic characteristic requirements of gimbal structures are that they be flexible in pitch and roll motion but stiff to in-plane and out-of-plane bending motion. To meet all these requirements using only one piece of metal is a highly challenging task. For this reason, most commercially available suspension/gimbal designs consist of two to four pieces of steel, each with different thickness.

The goal of a gimballed microactuator design is to integrate an actuator and gimbal seamlessly into a one-piece structure, as shown in Fig. 48.7, to simplify both suspension design and HDD assembly. A fully integrated suspension that includes suspension, gimbal, and microactuators in one part has also been pro-

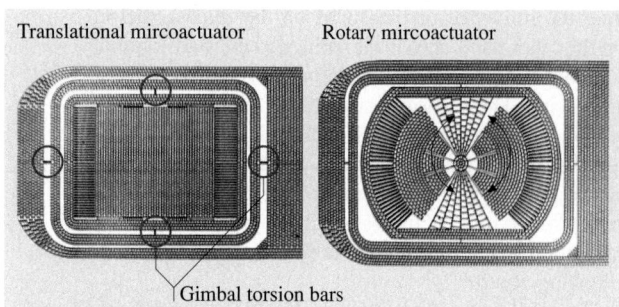

Fig. 48.7 Schematics of translational microactuator versus rotary microactuator. Courtesy of *Lilac Muller* [48.6]

posed [48.21]. The dimple structure, existing in most current gimbal structures, is excluded from the gim-balled microactuator design, and electrical interconnects are in situ fabricated on the gimballed microactuator, re-

Fig. 48.8 Pico-slider mounted on a translational microactuator. Courtesy of *Horsley* 1998 [48.5]

placing the flexible cable in the current HDD suspension assembly.

The dimple structure in current suspension assemblies provides out-of-plane stiffness while preserving the necessary torsional compliance in the gimbal structure. Without a dimple structure in the suspension assembly, the gimbal itself must provide high out-of-plane bending stiffness. Otherwise, it would unbalance the suspension preload, which is an overbend of the suspension that balances the upward air-bearing force on the suspension during operation. The electrical interconnects are implemented to transmit data between magnetic heads, located at the center coupon of the gimbal structure, and a flex cable leading to interface circuits in the base of the disk drive. The electrical interconnects fabricated in situ are inevitably passed through torsion bars of the gimbal and thus set a design constraint for the minimum width of the torsional bars. Furthermore, the gimbal structure and microactuator should be the same thickness to simplify the MEMS fabrication process.

To summarize the design constrains discussed above, the integrated gimbal structure has to meet performance requirements with a single, uniform piece of material that would previously have been achieved by two to four metal pieces with different thicknesses, while the minimum width of any torsion bars used is predetermined. To solve this problem, *Muller* [48.6] proposed a T-shaped structure (a beam structure with an overhanging surface sheet) for the torsion bars, and *Chen* [48.21] proposed double-flexured torsion bars. Additionally, many suspension manufacturers have developed new gimbal structures for their suspensions designed specifically for use with MEMS microactuators, moving

Table 48.1 Parameters of the electrostatic microactuator design by *Horsley* 1998 [48.5]

Parameters	Source*	Value
Nominal gap	D	$10\,\mu m$
Structure thickness	D	$45\,\mu m$
Rotor mass, m	I	$44\,\mu g$
dC/dx	C	$68\,fF/\mu m$
Actuation voltage, bias voltage	D	$40\,V$
Actuation voltage, maximum driving voltage	D	$\pm 40\,V$
Voltage-to-force gain, K_v	I	$50\,nN/V$
Mechanical spring constant, K_m	I	$29\,N/m$
Electrostatic spring constant, K_e	I	$9.6\,N/m$
Damping coefficient, b	I	$1.03 \times 10^{-4}\,N/(m/s)$
Voltage-to-position DC gain	M	$0.05\,\mu m/V$
Resonance frequency, w_r	M	$550\,Hz$

* D = design value, C = calculation, M = measurement, I = inferred from measurements

the gimbal location back to the suspension from the microactuator.

An Electrostatic Microactuator Design Example

Figure 48.8 shows a translational microactuator design suitable for the HDD dual-stage actuation by *Horsley* in 1998 [48.5]. The translational electrostatic microactuator dimensions are $2.2 \text{ mm} \times 2.0 \text{ mm} \times 0.045 \text{ mm}$ and the weight is $67 \mu\text{g}$. The dimensions of the pico-slider on the top are $1.2 \text{ mm} \times 1.0 \text{ mm} \times 0.3 \text{ mm}$ and the weight is 1.6 mg. This microactuator design does not include electrical-isolation features, and thus the electrical isolation and electrical interconnects were fabricated on a separate substrate and subsequently bonded to the microactuator. This microactuator uses parallel plates for the actuation force but does not have dedicated position-sensing structures due to fabrication process limitations. Table 48.1 summarizes the key parameters of this microactuator design.

Based on these parameters, the characteristics of this electrostatic microactuator can be estimated by the linear differential equation shown in (48.6).

Figure 48.9 shows the schematics of a circuit design by *Wongkomet* in 1998 [48.14], in which the actuation driving voltage and capacitive position sensing were implemented for the electrostatic microactuator designed by Horsley. As mentioned before, the electrostatic microactuator design does not have a dedicated structure for a position sensing;

as a consequence, the input nodes for actuation and output nodes for capacitive position sensing have to share the same electrodes. The capacitors C_c and C_{c0}, shown in Fig. 48.9, are carefully designed to shield the high voltage presented in the actuation circuit from the sensing circuit, which is mostly low voltage, and thus enable driving/sensing circuit integration.

The driving circuit, shown in the left in Fig. 48.9, demonstrates how to generate the bias voltage (V_{bias}, $\pm40 \text{ V}$) and drive voltage (V_{dr}, $-40-+40 \text{ V}$) from $0-5 \text{ V}$ complementary metal oxide semiconductor (CMOS)-compatible circuits. The switches at the output of the charge pumps were synchronized with the switching period ϕ_{RS}. During the sense phases ϕ_{SN1} and ϕ_{SN2}, therefore, the switches are left open and thus no voltage fluctuation is seen by the sensing circuits. This arrangement was utilized to reduce feed-through from the driving circuit to the position-sensing circuit.

The design target for the capacitive position-sensing circuit was to achieve a position-sensing resolution of 10 nm, and this goal was approached by two main techniques implemented in the circuit: differential sensing and correlated double sampling (CDS). The main benefits of the differential sensing scheme are reduced noise coupling and feed-through, elimination of even-order harmonics, and improvement of dynamic range by doubling the output swing. To adapt this differential sensing scheme in a differential parallel-plate electrostatic microactuator, a bias voltage (V_{bias}) was applied

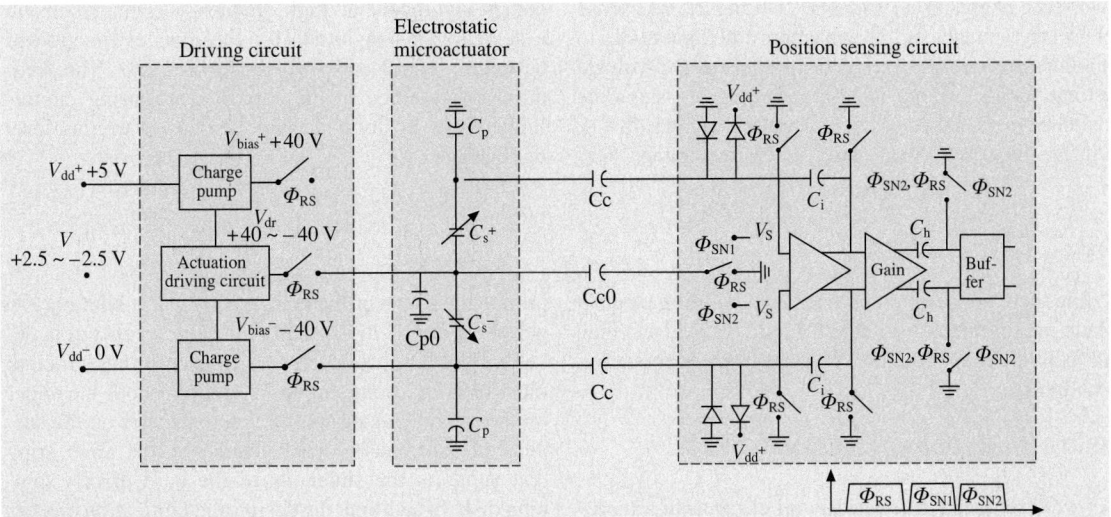

Fig. 48.9 A simplified schematic including driving and sensing circuits for the electrostatic microactuator. Courtesy of *Wongkomet* 1998 [48.14]

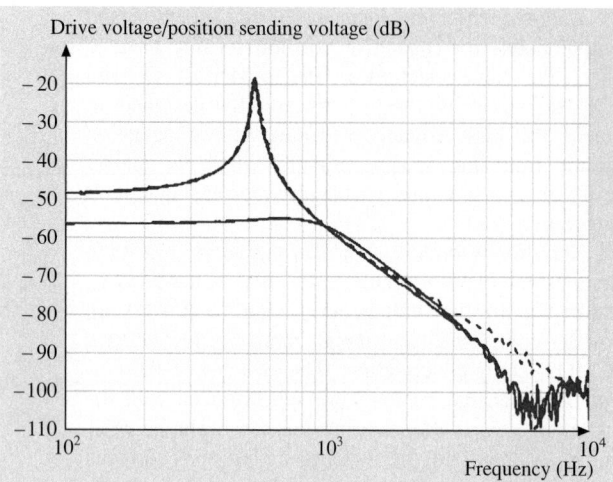

Fig. 48.10 Open-loop and closed-loop frequency response of the prototype microactuator with a pico-slider. The capacitive position measurement (*solid line*) is compared to the measurements from LDV (*dashed line*). Courtesy of *Wongkomet* [48.14]

to the stator and the drive voltage (V_{dr}) was applied to the shuttle. The CDS technique, a modified capacitance-sensing technique based on the switched-capacitance scheme, was implemented along with the differential sensing scheme to compensate for sensing noises, including $1/f$, thermal noise, switch charge injection and offset from the op-amps [48.14]. The concepts of CDS can be briefly described as follows. The sense period is broken into three phases: one reset phase, ϕ_{RS}, and two sense phases, ϕ_{SN1} and ϕ_{SN2}. During the ϕ_{RS} sense phase, the voltages for the capacitors and input nodes to op-amps are set to the reference level of the DC-voltage setting for the charge integrator, the same as for the switched-capacitance scheme discussed in Sect. 48.1.4. During the ϕ_{SN1} sense phase, a sensing voltage $-V_s$

is applied to the shuttle and results in a voltage difference, $\alpha(-V_s) + V_{error}$, across the sampling capacitor C_h, where α is the transfer function from the sense voltage to the output voltage at the preamplifier, shown as *Gain* in the plot, and V_{error} is the voltage at the output node of the preamplifier resulting from noise, leakage charge, and offset of op-amps. Lastly, during the ϕ_{SN2} sense phase, the sensing voltage is switched from $-V_s$ to V_s and the switch next to C_h is switched open. This results in a voltage output $\alpha(V_s) + V_{error}$ at the output node of the preamplifier and a voltage, $2\alpha(V_s)$, at the input node of the buffer. As a consequence, the voltage resulting from the sensing error (V_{error}), which appears at the output node of the preamplifier, disappears from the voltage-input node to the buffer. The switches utilized in the circuit and their corresponding timings are shown in Fig. 48.9. Be aware that, for simplicity of presentation, components in the preamplifier and buffer are not shown in detail in Fig. 48.9.

The frequency response, both open-loop and closed-loop with a proportional-derivative (PD) controller, of the electrostatic microactuator measured by a capacitive position sensing circuit is shown in Fig. 48.10. The position measurements from an laser doppler velocimeter (LDV) are also shown in the same plot for comparison. The deviation between the measurements from the different position-sensing devices appears in the high-frequency region of the plot and has been identified as the feed-through from the capacitance-sensing circuit. The effect of feed-through was negligible in the low-frequency region but becomes significant at high frequencies and results in a deviation in magnitude for the transfer function of between -80 dB and -90 dB above 2 kHz. The feed-through presented in the capacitance-sensing circuits limited the position-sensing resolution to the level of 10 nm.

48.2 Fabrication

While there are several approaches to building electrostatic microactuators suitable for hard disk drives, they all exhibit certain common features from a fabrication standpoint.

48.2.1 Basic Requirements

As we discussed earlier, nearly all electrostatic microactuators rely on a system of interlaced fingers or plates to provide actuation force. As a result, a method for

producing arrays of these fingers or plates with narrow gaps between them is usually the central concern in developing a fabrication process. The resulting structure must then be strong enough to support both the slider on the microactuator and the microactuator on the suspension, particularly when loaded by the air bearing that supports the slider above the hard drive's spinning disk. In addition, the design and fabrication process must include a way to perform electrical interconnection on the microactuator. This involves transferring signals to and from the actuator and slider and isolating the parts of the microactuator that require different voltage levels.

Meanwhile, the microfabrication process is subject to certain basic constraints. The materials used in its fabrication must either be thermally and chemically compatible with any processing steps that take place after their deposition or must somehow be protected during steps that would damage them. This often constrains the choice of materials, deposition techniques, and processing order for microdevices. Another concern is that the surface of the structure be planar within photoresist spinning capabilities and lithography depth-of-focus limits if patterning is to be performed. This can be a major challenge for disk drive microactuators, which are large in size and feature high-aspect-ratio trenches compared to other MEMS devices.

48.2.2 Electrostatic Microactuator Fabrication Example 1

Section 48.1.5 described the design and operation of a translational electrostatic microactuator. This section examines the fabrication process by which that microactuator was built [48.5]. The process is a variation on a micromolding process known as HexSil [48.22]. In a micromolding process, a mold wafer defines the structure of the microactuator and may be reused many times like dies in macro-scale molding; in the HexSil version of the procedure, the mechanical parts of the microactuator are formed in the mold by polysilicon. A second wafer is used to create metalized, patterned target dies. Upon extraction from the mold, the HexSil structure is bonded to this target substrate, which is patterned to determine which sections of the HexSil structure are electrically connected.

Naturally, the mold wafer is the first item to be processed and is quite simple. It is a negative image of the desired structure, etched down into the wafer's surface, as shown in Fig. 48.11a. A deep but very straight etch is critical for successful fabrication and subsequent operation of the devices. If the trench is too badly bowed or is undercut, the finished devices will become stuck in the mold and difficult to release. On the other hand, a tall device will have a larger electrostatic array area, will generate more force, and be better able to support a slider while in a disk drive.

Fabrication of the HexSil structure forms the majority of the processing sequence Fig. 48.11b–g. HexSil fabrication begins by coating the surface of the mold wafer with a sacrificial silicon oxide. This layer, de-

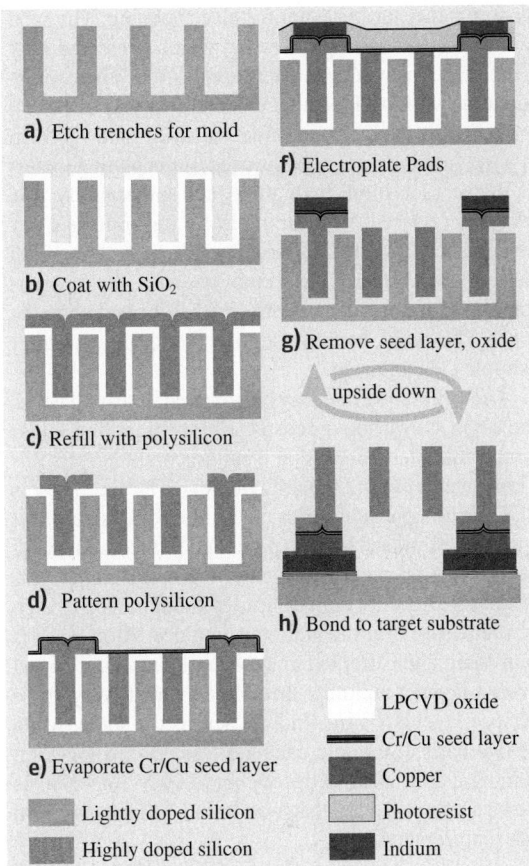

Fig. 48.11 Fabrication example 1: HexSil

posited by low-pressure chemical vapor deposition (LPCVD), must coat all surfaces of the mold, so that removal of the oxide at the end of the process will leave the device completely free of the mold. This microactuator uses a 3–4 μm-thick oxide layer to ensure clearance between the polysilicon structure and the trench walls during release. The mold is then refilled completely with LPCVD polysilicon, which will form the desired mechanical structure. Polysilicon is chosen for its conductivity and good conformality in refilling trenches. This also leaves a planar surface ready for photolithography to cover the bonding locations. After the lithography step, the polysilicon deposited on top of the wafer is removed except where the bonding points were defined.

The structure is then prepared for bonding by forming a soldering surface. First a chrome/copper seed layer is evaporated onto the surface of the wafer. The chrome promotes copper adhesion to the silicon, while the copper forms the starting point for electroplating. The seed layer is covered with photoresist, which is cleared only in the locations where contacts are desired. There, electroplating will produce a metal film, forming soldering points for the device to the target substrate that will form its base. In this case, copper is used as the plating material, thanks to its high conductivity and solderability. The device is released from the mold by ion milling away the thin seed layer, then dissolving the oxide lining with a hydrofluoric (HF) acid wet etch. The free-standing portion of the microactuator is thus ready for bonding to the target. The mold, meanwhile, may be used again after a simple cleaning step.

The target substrate may be prepared in several ways. The target shown above consists simply of indium solder bumps and interconnects on a patterned seed layer; this is known as a *plating bus* arrangement. The seed layer is used for the same reasons as on the HexSil structure, but in this case the seed layer is patterned immediately, with the unwanted portions removed by sputter etching. This provides isolation where desired without having to etch the seed layer after the indium is in place. Photoresist is spun again and patterned to uncover the remaining seed layer. A layer of indium is then electroplated everywhere the seed layer is visible. Indium acts as the solder when the two parts of the microactuator are pressed together, forming a cold weld to the copper at 200–300 MPa, as shown in Fig. 48.11h, that was found suitable for hard disk requirements.

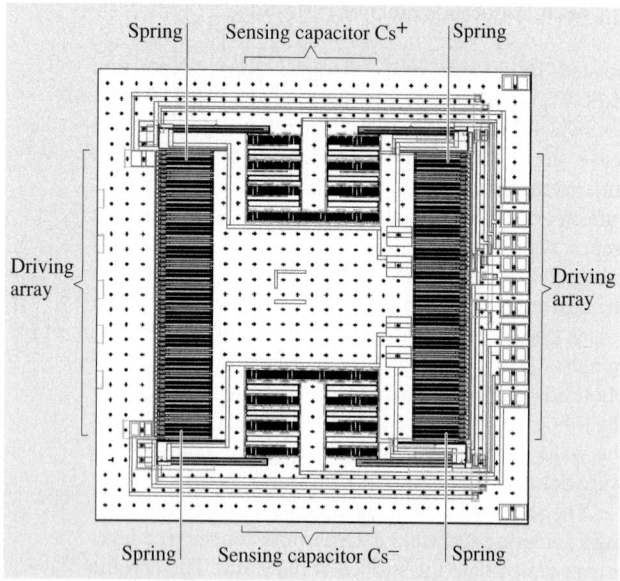

Fig. 48.12 Translational high-aspect-ratio microactuator

48.2.3 Electrostatic Microactuator Fabrication Example 2

The high-aspect ratio microactuator described in this section is in development at the University of California, Berkeley, and demonstrates another way to create the features required of an actuated slider. The microactuator is translational, with parallel-plate actuation, and includes two of the design options described previously: integrated isolation plugs, as described in Sect. 48.1.3, and a capacitive sensing array, as described in Sect. 48.1.4. A picture of the basic design is shown in Fig. 48.12; a central shuttle holding the slider is supported by four folded-flexure springs and driven by parallel-plate arrays.

Basic Process: Silicon-on-Insulator

Successful processing of the design is centered around the ability to etch very straight, narrow trenches using deep reactive-ion etching (DRIE). DRIE is a special plasma etching sequence that uses polymerization of the sidewalls of a trench to keep the walls straight even at extreme aspect ratios. Deep, narrow trenches make possible a microactuator with both closely packed electrostatic arrays and good mechanical strength. Conventional surface micromaching is used to produce the majority of the electrical interconnects.

The basic fabrication process uses a silicon-on-insulator (SOI) wafer to control the DRIE trench depth. SOI wafers are very useful in microfabrication as they provide a layer of single-crystal silicon (the device layer) separated from the bulk of the wafer (the handle layer) by a thin layer of buried silicon oxide. This gives a very well-controlled thickness to the finished device but is very expensive. The microactuators described here are fabricated from an SOI wafer with a $100\,\mu\text{m}$ device layer. A variation on the processing sequence and layout eliminates the need for SOI wafers and will be discussed in the following section.

The first stage of fabrication is the creation of deep isolation trenches (see Fig. 48.13a–c). This procedure has been adapted to MEMS from integrated circuit processing for isolating thick MEMS structures [48.23]. The isolation pattern is formed by photolithography and etched by DRIE down to the buried oxide. With trenches only $2\,\mu\text{m}$ wide, this corresponds to an aspect ratio of $50:1$. The wafer surface and trenches are then coated with LPCVD silicon nitride, which acts as the electrical insulator. The trench is then refilled in its entirety with LPCVD polysilicon, which is a more conformal material that better fills the trench than silicon nitride would alone; polysilicon also has much lower residual stresses than silicon nitride. The refill leaves a layer of polysilicon on the surface of the substrate, which must be etched or polished back to the silicon nitride. After etch-back, a second layer of LPCVD nitride completes the isolation between regions.

Next, electrical interconnects are formed, as shown in Fig. 48.13d–f. Contact holes to the substrate are defined by photolithography and etched by reactive-ion etching of the silicon nitride. Metal lines may then be patterned directly by photolithography or formed by lift-off. In a lift-off process, photoresist is deposited and patterned before the metal. The metal is then deposited vertically, so that the sidewalls of the resist are uncovered. Dissolving the resist allows the metal on top to float away, leaving behind the interconnect pattern. In the figures shown here, the interconnects are formed by a metal/polysilicon stack. A highly doped

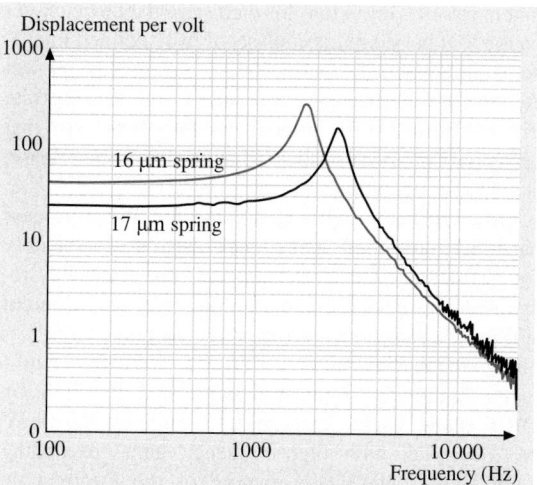

Fig. 48.14 Dynamic response of microactuators from SOI wafer with 16 μm and 17 μm spring widths

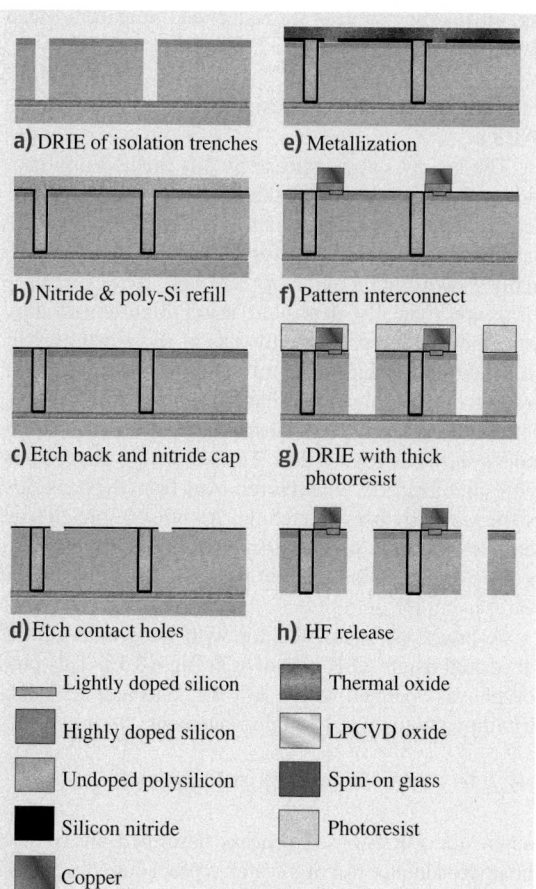

a) DRIE of isolation trenches

b) Nitride & poly-Si refill

c) Etch back and nitride cap

d) Etch contact holes

e) Metallization

f) Pattern interconnect

g) DRIE with thick photoresist

h) HF release

Lightly doped silicon

Highly doped silicon

Undoped polysilicon

Silicon nitride

Copper

Thermal oxide

LPCVD oxide

Spin-on glass

Photoresist

Fig. 48.13 Basic silicon-on-insulator fabrication process

Fig. 48.15 Electrostatic driving array of microactuator from SOI

polysilicon layer provides a robust base layer for the interconnects, while the overlaying metal keeps resistivity low. The ability to isolate portions of the substrate is useful for simplifying the interconnect layout, as one interconnect can cross another by passing underneath it through an isolated portion of the substrate.

Finally, structural lithography is performed and a second etch using DRIE is done, as shown in Fig. 48.13g. These trenches will define the shape of the microactuator, and are larger, 4 μm wide, to allow sufficient rotor travel. After this etch step, the device layer on the SOI buried oxide consists of fully defined microactuators. When the buried oxide is removed by wet etching, the microactuator is released from the substrate Fig. 48.13h and, after removal of the photoresist that protected the device surface from the HF during release, is ready for operation.

Microactuators fabricated using the above process have been installed into a disk drive test assembly for dual-stage controller evaluation. The frequency responses of two such microactuators, with two different spring sizes, are shown in Fig. 48.14. Microactuator resonant frequencies range from 1.7 to 2.6 kHz. Thanks to the high-aspect-ratio processing made possible by DRIE, static displacement gains as large as 200 nm/V are observed, with high-frequency gains exceeding 1 nm/V past 10 kHz. A picture of the electrostatic structure used to drive this motion, with rotor fingers on the right and stator fingers on the left, divided down the center by isolation trenches, is shown in Fig. 48.15.

An alternative fabrication approach is to build the microactuator from a regular silicon wafer ($\approx 550\,\mu$m thick), but then, since the microactuator should not be more than $100\,\mu$m thick to fit in a HDD, etch away the back side of the wafer to obtain devices of that size. The main reason for using a silicon-only wafer is to reduce the need for expensive SOI wafers, which cost approximately ten times as much as single-material wafers. Another benefit is the elimination of the need for etch holes to reach the buried oxide layer, which increase the vulnerability of microactuators to particle or moisture contamination. However, the thickness of the devices cannot be controlled as accurately as with the SOI process, and the processing sequence is more complicated.

Control of back-side etching is difficult, making it hard to obtain devices with uniform thickness across a wafer. A proposed solution is to use an anisotropic wet etchant during the back-side etch step. Anisotropic etchants work very slowly on certain crystal planes of a single-crystal silicon wafer; a coating on trench sidewalls can protect fast-etching planes when the etchant reaches the trenches from the back, causing slow-etching crystal planes to begin to come together, which slows the etch. The result is a nearly self-stopping etch with a V-shaped profile that prevents overetching even in the presence of nonuniformities in etch rate and trench depth across the wafer. The most common anisotropic etchants are potassium hydroxide (KOH) and tetramethyl ammonium hydroxide (TMAH). This release process is

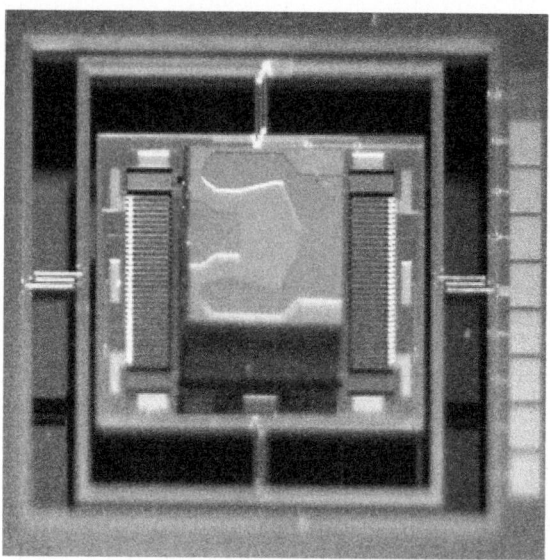

Fig. 48.16 Microactuator with integrated gimbal fabricated by ABR

referred to here as an anisotropic back-side release (ABR).

The key to a good release by this method it to provide reliable protection of the top of the wafer and the deep trenches from the etchant. The most effective etch barriers are silicon nitride (for KOH or TMAH) or thermally grown silicon oxide (for TMAH). However, these films are typically deposited using high-temperature processes, which preclude the use of most metals. Microactuators produced by this method have a similar process flow to those described in the previous section, but the metal layer of the surface interconnects is omitted, as shown in Fig. 48.14a. The wafers are then coated with silicon nitride, which is removed from the back side of the wafer by plasma etching. An anisotropic etchant removes the bulk of the silicon wafer, and finally the protective nitride is removed from the front side of the wafer using HF acid.

A prototype microactuator with integrated gimbal fabricated using ABR is shown in Fig. 48.16. This prototype was operated with a 30 V DC bias and ± 8 V AC driving voltage for $\pm 1\,\mu$m displacement.

48.2.4 Other Fabrication Processes

Where the DRIE-based process described above *digs* the microactuator out of a wafer, a procedure developed by IBM *grows* a microactuator with high-aspect-ratio trenches on top of a wafer [48.24].

IBM Electroplated Microactuator

The microactuator is fabricated by a clever sequence of electroplating steps and sacrificial depositions. A sacrificial oxide film is first patterned then covered with a metal seed layer. When later removed, the sacrificial film will have left a gap between the microactuator and substrate portions of the microactuator needed to move freely. A polymer, 40 μm thick, is spun onto the wafer and patterned with a reverse image of the main structural layer. Trenches are etched down to the sacrificial layer by plasma etching then refilled by electroplating, much like micromolding. Another sacrificial polymer, a photoresist this time, separates the top of the fixed portions of the structure from the platform upon which the slider will sit, while conveniently planarizing the surface for further lithography. The platform is created by two more electroplating steps on top of the parts of the structure that will move freely. Removal of the sacrificial layers (photoresist, polymer, and oxide) releases the microactuator for use.

This process has been used to produce both rotary and linear microactuators and is comparatively advanced from a commercial standpoint. The resulting devices are thinner than those described in the previous example, but a similar aspect ratio (20 : 1) can be achieved. This procedure also has the benefits of being a low-temperature process, which helps decrease processing cost and is compatible with the thin-film magnetic-head manufacturing process. Moreover, it includes an upper structure that covers the fingers, which improves device reliability by shielding the fingers from particles.

Additional fabrication details and dynamic testing results for the IBM design may be found in references [48.24] and [48.25]. A microactuator fabricated by this process has demonstrated an open-loop bandwidth of 8 kHz after installation into a hard disk drive [48.26]. Certain closed-loop control results for such a microactuator are described in Sect. 48.3.2.

Electroplated Microactuators with Fine Gaps

Finally, fabrication processes developed at the University of Tokyo demonstrate additional techniques for obtaining very small gaps between electrostatic fingers [48.27]. Similar to the IBM process, electrostatic fingers are electroplated, in this case by nickel, within a polymer pattern. The gap between closely spaced fingers, however, is then formed by a sacrificial metal. Photoresist is used to cover the fingers except for the surfaces where a small gap is desired; these are electroplated with sacrificial copper, making narrow, well-defined gaps possible, in perhaps the most reliable of the pro-

cesses described here. Continuing with a second nickel electroplating step creates the interlaced fingers, which will also support the slider; the wafer is polished down to the level of the structure. Finally the copper, photoresist, and seed layer are etched away, leaving the free-standing structure.

In another version of process, photoresist alone separates the stator and rotor fingers. In this case, a thin layer of photoresist is left where small gaps are desired, instead of growing a layer of copper. This method is dependent on excellent alignment but can eliminate the need for polishing after the second electroplating step.

48.2.5 Suspension-Level Fabrication Processes

The long, highly integrated fabrication processes required by actuated slider microactuators contrast greatly with the fabrication of actuated suspension microactuators and instrumented suspension, though the material requirements and environments of the suspension add interesting challenges to the use of such devices.

PZT Actuated Suspensions

In most cases, actuated suspensions use piezoelectric (lead zirconate titanate, PZT) drivers cut or etched as single, homogeneous pieces and attached to the steel suspension. Nevertheless, microfabrication techniques can still be useful in suspension processing. For example, some recent suspensions have used thin-film PZT deposited on a substrate by sputtering or spinning on of a sol–gel for incorporation in a suspension. This is intended to produce higher-quality PZT films and begins to introduce MEMS-style processing techniques even into suspension-scale manufacturing [48.28, 29].

Integrated Silicon Suspension

Another area where silicon processing has been suggested for use on a suspension is the gimbal region, or even the entire suspension, as described in [48.21]. An integrated silicon suspension would incorporate aspects of both actuated suspension and actuated slider fabrication. Force is provided by four pieces of bulk PZT cut down to size, as in other actuated suspensions, but the rest of the suspension is fabricated from single-crystal silicon by bulk micromachining.

The fabrication process for the integrated suspension is the precursor to the ABR process used to fabricate the microactuator in Sect. 48.2.3. Layers of LPCVD silicon nitride and polysilicon are deposited on bare silicon wafers. The polysilicon is doped by ion implantation

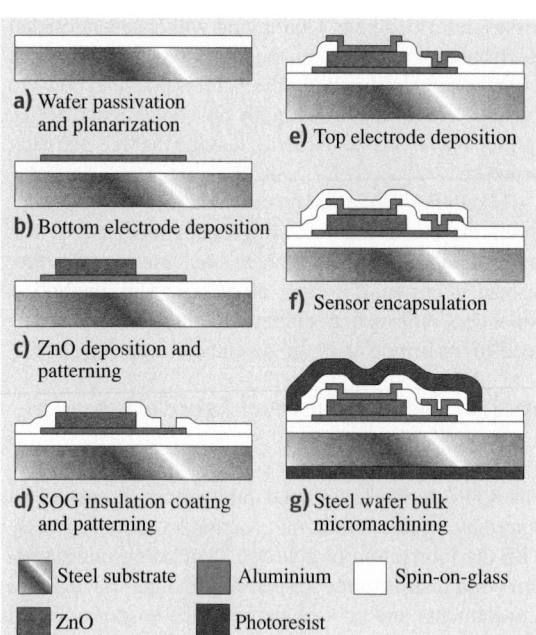

a) Wafer passivation and planarization

b) Bottom electrode deposition

c) ZnO deposition and patterning

d) SOG insulation coating and patterning

e) Top electrode deposition

f) Sensor encapsulation

g) Steel wafer bulk micromachining

█ Steel substrate ▓ Aluminium ☐ Spin-on-glass
■ ZnO ▒ Photoresist

Fig. 48.17 Fabrication process for a zinc oxide sensor on a steel substrate

and covered with a second layer of nitride. The outline of the suspension and any spaces within it are then lithographically patterned. After a reactive-ion nitride etch to reach the substrate, trenches are etched by DRIE. The trenches are then plugged with spin-on polymer for further lithography. First, the top layer of polysilicon is patterned into piezoresistive strips for sensing vibration. Second, a layer of photoresist is patterned for copper lift-off, to create metal interconnects. Last, the planarizing polymer is removed, and the entire surface is coated with silicon nitride. This final coating protects the top surface during an anisotropic back-side wet etch, in this case by potassium hydroxide.

Instrumented Suspensions

Finally, microfabrication techniques may be useful for installing sensors on conventional steel suspensions, in order to help a microactuator suppress suspension vibration. Strain gages located at key points on the suspension can detect structural vibrations of the suspension. These measurements may be fed forward to a microactuator at very high sampling rates, and used to cancel out the vibration at the read/write head. For effective operation, the strain gages must detect strain levels on the order of 10 s of nanostrain, requiring highly sensitive sensor materials, such as those used in microfabrica-

tion on silicon wafers. Because the sensors are built on steel substrates to be assembled in suspensions, however, only low-temperature, low-stress microfabrication processes are permissible. This complicates sensor fabrication, as does the comparatively rough surface of steel wafers. In addition, the performance of sensors on a disk drive suspension is very sensitive to sensor placement, as discussed in [48.30–32].

Two types of strain sensors are currently under development: piezoresistive strain sensors and piezoelectric strain sensors. The critical step in sensor fabrication is the addition of a strain-sensitive material under the temperature and stress constraints imposed by fabrication on steel. In either case, processing begins with the deposition of an insulating film to isolate the sensors electrically from the steel substrate. Possible materials include spin-on glasses, polymer films, or metal oxides. Sensors must also include metal interconnects to provide electrical connections to an external controller, typically implemented through metal evaporation and patterning or lift-off.

The function of piezoresistive strain gages is relatively simple, and was discussed in Sect. 48.1.4. To build such sensors on a steel substrate, the simplest procedure is to deposit a piezoresistive film directly onto the wafer. For instance, doped polysilicon may be sputtered onto steel at low temperatures with low residual stress. Unfortunately, such a film has a very high resistivity, resulting in a very large source of thermal noise. Alternatively, high-quality sensors may be built separately from any piezoresistive material, such as single-crystal silicon, then installed on the steel substrate using metal-on-metal bonding, but this process is very expensive.

A very promising alternative to piezoresistive strain gages is piezoelectric vibration sensors. As described in Sect. 48.1.4, piezoelectric materials are very good at detecting extremely small dynamic signals. In order to fabricate piezoelectric sensors directly on steel wafers, however, it is important to choose a material that has good microfabrication compatibility in addition to good piezoelectric properties. Thermal-coefficient mismatch between steel substrates and all nonmetal materials is especially constraining for piezoelectric materials, as it limits the maximum permissible temperature over the course of the fabrication process. A survey shows that the most well-known piezoelectric material lead zirconate titanate (PZT) does not provide sufficient process compatibility. Not only iw it difficult to deposit, but it also requires high temperatures for curing and polarization. Aluminum nitride, a material commonly used to build MEMS devices, has much better fabri-

Fig. 48.18 Piezoelectric strain sensor on steel substrate

cation compatibility but still requires relatively high temperatures for deposition. For instrumented suspensions, zinc oxide (ZnO) has been identified as having good process compatibility and decent piezoelectric properties.

Zinc oxide sensor fabrication, shown in Fig. 48.17, starts with putting a spin-on glass (SOG) layer onto a bare steel wafer. The SOG layer serves as an electrical-insulation layer between the steel substrate and the sensors, while also planarizing the rough steel surface and preventing steel from oxidizing during processing. Next, an aluminum layer is evaporated and patterned to form the bottom electrode of the sensor. Then, a ZnO layer is sputtered and patterned onto the electrode, followed by the second SOG coating as another insulation layer. Using the bottom electrode and ZnO as etch stoppers, the contact holes are opened through the SOG layer using reactive-ion etching (RIE). Another metal layer is then deposited to form the top electrode and leads. To prevent leads from shorting out accidentally during operation, the last SOG layer is coated on the wafer to encapsulate the sensor and leads. After the fabrication of the sensors, two more etching steps are needed to transform steel wafers into suspension parts. First, the residual SOG has to be cleared out so that steel is exposed for welding during suspension assembly. Secondly, the wafer is patterned into suspension parts and bulk micromachining is performed to etch through the wafer, leaving suspension geometries at the end. The cross section of a sensor built by this method is shown in Fig. 48.17. A picture of a ZnO piezoelectric sensor, 200 μm by 200 μm in area and 0.8 μm thick, is shown in Fig. 48.18. Slightly larger sensors, 300 μm by

300 μm, have demonstrated strain resolutions down to 50 nanostrain.

48.2.6 Actuated Head Fabrication

Several interesting fabrication processes have been proposed for another approach to creating a dual-stage disk drive servo: the actuated head. In an actuated head, the actuator is built into the slider and moves just the read/write head. This could potentially eliminate many of the mechanical limits of actuated suspensions or actuated sliders. Actuation techniques are typically similar to those of actuated sliders, with the key complication being the need to integrate an actuator fabrication process with a slider fabrication process. We discuss ways in which this might be done below.

One fabrication approach to actuating a read/write head is to enclose an electrostatic driving array inside a slider. The slider is built on a glass or silicon substrate, beginning with the air-bearing surface (ABS), followed by a microactuator, and then surrounded by the remainder of the slider. The ABS is formed by silicon oxide deposited over aluminum and tapered photoresist layers that give it a contoured shape. The majority of the ABS is then covered with photoresist, while a block in the center is plated with nickel and etched to form the electrostatic actuator. The electrostatic microactuator, in turn, is covered while the remainder of the slider body is electroplated. The slider may then be bonded to a suspension and the ABS surface released from the substrate by etching away the original aluminum and photoresist surface [48.9].

The sacrificial metal technique described in Sect. 48.2.4 is another possibility for building an electrostatic array onto a slider. It can be used to form a small array with very narrow gaps, and the process is compatible with slider and head materials [48.33].

The third actuated-head process makes use of an SOI wafer and silicon-based materials. For this structure, a 20 μm device layer is coated with silicon nitride. The nitride is patterned with contact holes for connections to the substrate and used as insulation over the rest of the device. A layer of molybdenum is sputter-deposited to perform interconnections to both the head and the electrostatic array. This array is formed by DRIE to the buried oxide, using a lithographically patterned hard mask of tetraethylorthosilicate. The result is a simple, tiny, actuator that must be bonded to the edge of a slider after release from the SOI wafer [48.34].

48.3 Servo Control Design of MEMS Microactuator Dual-Stage Servo Systems

The objective of disk drive servo control is to move the read/write head to the desired track as quickly as possible, referred to as track-seek control, and, once on-track, position the head on the center of the track as precisely as possible, referred to as track-following control, so that data can be read/written quickly and reliably. The implementation of the servo controller relies on the position error signal (PES), which is obtained by reading the position information encoded on the disk's data tracks.

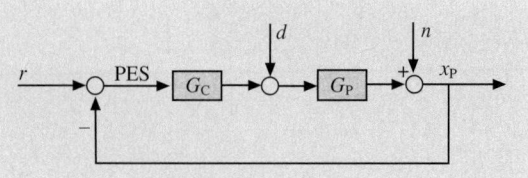

Fig. 48.19 Disk-drive servo control

48.3.1 Introduction to Disk Drive Servo Control

The position error is also called track misregistration (TMR) in the disk drive industry. Major TMR sources in the track-following mode include spindle runout, disk fluttering, bias force, external vibration/shock disturbance, arm and suspension vibrations due to air turbulence, PES noise, written-in repeatable runout, and residual vibration due to seek/settling [48.35]. These sources can be categorized as runout, r, input disturbance, d, and measurement noise, n, by the locations they are injected into the control system, as shown in Fig. 48.19. In Fig. 48.19, $G_P(s)$ and $G_C(s)$ represent the disk drive actuator and the controller, respectively; r and x_p represent track runout and head position, respectively.

From Fig. 48.19, the PES can be written as

$$\text{PES} = S(s)r + S(s)G_P(s)d - S(s)n , \qquad (48.10)$$

where $S(s)$ is the closed-loop sensitivity function defined by

$$S(s) = \frac{1}{1 + G_C(s)G_P(s)} . \qquad (48.11)$$

The higher the bandwidth of the control system, the higher the attenuation of the sensitivity function $S(s)$ below the bandwidth. Thus, one of the most effective methods to reduce PES and increase servo precision is to increase the control system bandwidth.

Traditional disk drive servo systems utilize a single voice coil motor (VCM) to move the head. Multiple structural resonance modes of the E-block arm and the suspension located between the pivot and the head impose a major limitation on the achievable control bandwidth. A dual-stage servo system using a MEMS actuated-slider microactuator can achieve high bandwidth because the microactuator is located between the

suspension and the slider, so actuation of the microactuator does not suffer from the resonance of the pivot, E-block, and suspension. Cooperation of a VCM and a MEMS microactuator thus provides the possibility of achieving higher servo bandwidth.

When the MEMS microactuator is translational, it even has the capability of filtering out those structural resonance modes. This feature is preferable and even crucial when the PES has been reduced down to the order of a few nanometers $(1 - \sigma)$ and airflow-excited suspension vibrations have become a significant component in TMR. As a solution, an instrumented suspension can be employed, and vibration information from the suspension can be fed to the VCM for feedback damping, or to the microactuator for feed-forward compensation.

One basic task of dual-stage servo control design is to increase the control bandwidth using the second-stage actuator. Many design methodologies have been developed to accomplish this objective. In this section, we first review major dual-stage servo control design methodologies, and then we discuss design considerations for controller design of a MEMS microactuator dual-stage servo system. As design examples, the details of track-following control designs using a sensitivity-function-decoupling design method and the mixed H_2/H_∞ design method are presented in Sect. 48.3.3. A comprehensive approach to robust control using a multi-rate controller is described in Sect. 48.3.4. In Sect. 48.3.5, we introduce a short-span seek control scheme using a dual-stage actuator based on decoupled feed-forward reference-trajectory generation.

48.3.2 Overview of Dual-Stage Servo Control Design Methodologies

Various control design architectures and methodologies have been developed for dual-stage servo control design. They can be largely classified into two categories: those based on decoupled or sequential single-input single-output (SISO) designs, and those based on modern optimal design methodologies, such as linear quadratic Gaussian (LQG), LQG/loop-transfer recovery (LTR), H_∞, μ-synthesis, and mixed H_2/H_∞ in which the dual-stage controllers are obtained simultaneously.

Two constraints must be considered in dual-stage servo control design. First, the contribution from each actuator must be properly allocated. Usually the first-stage actuator, or the coarse actuator, has a large moving range but a low bandwidth while the second-stage actuator, or the fine actuator, has a high bandwidth but small moving range. Second, the destructive effect, in which the two actuators fight each other by moving in opposite directions, must be avoided.

Classical SISO Design Methodologies

Several architectures and design methodologies have been proposed to transform the dual-stage control design problem into decoupled or sequential multiple SISO compensator design problems, for example, master–slave design, decoupled design [48.36], *PQ* method [48.37], and direct parallel design [48.38]. Figures 48.20–48.23 show the block diagrams of dual-stage controller designs using these methods. In these figures, G_1 and G_2 represent the coarse actuator (VCM) and fine actuator (microactuator) respectively; x_1 is the position of the coarse actuator; x_r is the position of the fine actuator relative to the coarse actuator; x_p is the total position output; r is the reference input (runout), and PES is the position error, $e = r - x_p$.

Master–Slave Design. In a traditional master–slave structure, the absolute position error is fed to the fine

actuator, and the output of the fine actuator is fed to the coarse actuator, as shown in Fig. 48.20. The position error will be compensated by the high-bandwidth fine actuator. The coarse actuator will follow the fine actuator to prevent its saturation.

Decoupled Design. Figure 48.21 shows the decoupled design structure [48.36], which is similar to the master–slave structure. A summation of the fine actuator output and the position of the coarse actuator is fed the coarse actuator. A very nice feature of this structure is that the control system is decoupled into two independent control loops, and the total sensitivity function is the product of the sensitivity functions of each of the control loops [48.13]. Thus the two compensators C_1 and C_2 can be designed independently. Decoupled design is also referred to as decoupled master–slave design, or sensitivity-function-decoupling design. In Sect. 48.3.3, details of a track-following controller designed using this method will be described.

Both the master–slave and the decoupled designs use the relative position of the fine actuator, x_r. If a relative position sensor is unavailable, x_r can be estimated using a model of the fine actuator.

PQ Design Method. The *PQ* method is another innovative design technique for control design of dual-input single-output systems [48.37]. A block diagram of a dual-stage control design using this method is shown in Fig. 48.22.

In *PQ* design, *P* is defined by

$$P = \frac{G_1}{G_2}, \qquad (48.12)$$

and a dual-stage controller can be designed in two steps. The first involves the design of an auxiliary compensator *Q* for plant *P*, which is defined by

$$Q = \frac{C_1}{C_2}. \qquad (48.13)$$

Q is designed to parameterize the relative contribution of the coarse and fine actuators. The 0 dB crossover

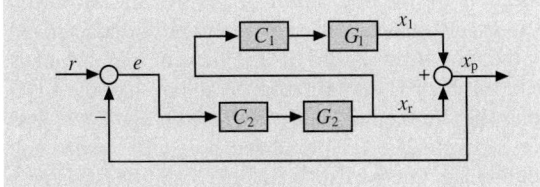

Fig. 48.20 Master–slave design structure

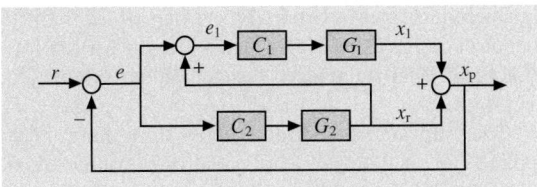

Fig. 48.21 Decoupled control design structure

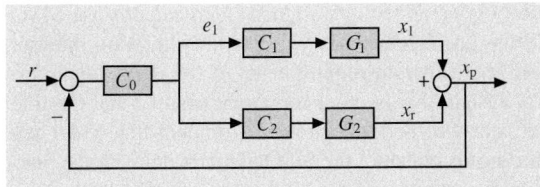

Fig. 48.22 *PQ* control design structure

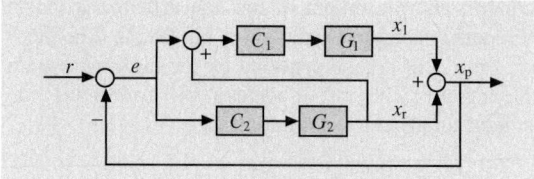

Fig. 48.23 Parallel control design structure

frequency and phase margin of the open-loop transfer function PQ are the design parameters in the design of Q. At frequencies below the 0 dB crossover frequency of PQ, the output is dominated by the coarse actuator, while at frequencies above the 0 dB crossover frequency, the output is dominated by the fine actuator. At the 0 dB crossover frequency, the contributions from the two actuators are equal. A large phase margin of PQ will ensure that the two actuators will not fight each other when their outputs are close in magnitude, thus avoiding any destructive effects.

The second step in the PQ design methodology is to design a compensator C_0 for the SISO plant PQ such that the bandwidth (crossover frequency), gain margin, phase margin, and error rejection requirements of the overall control system are satisfied.

Direct Parallel Design. It is also possible to design the dual-stage controller directly using a parallel structure, as shown in Fig. 48.23, by imposing some design constraints and by sequential loop closing [48.38].

The two constraints for parallel design in terms of the PES open-loop transfer functions are [48.38]:

$$C_1(s)G_1(s) + C_2(s)G_2(s) \rightarrow C_2(s)G_2(s) , \quad (48.14)$$

at high frequencies and

$$|C_1(s)G_1(s) + C_2(s)G_2(s)| \gg |C_2(s)G_2(s)| , \quad (48.15)$$

at low frequencies. The first constraint implies that the open-loop frequency response of the dual-stage control system at high frequencies approximately equals that of the fine-actuator control loop. Thus the compensator $C_2(s)$ can be first designed independently as a SISO design problem to satisfy the bandwidth, gain margin, and phase margin requirements of the dual-stage control system. The second constraint ensures that the fine actuator will not be saturated. Compensator $C_1(s)$ can then be designed for the SISO plant model with the fine-actuator control loop closed, such that the low-frequency constraint and overall stability requirement are satisfied.

This model is defined by:

$$G(s) = \frac{G_1(s)}{1 + C_2(s)G_2(s)} . \quad (48.16)$$

A dual-stage controller for a MEMS dual-stage actuator servo system has been designed using this method and implemented at IBM. An open-loop gain-crossover frequency of 2.39 kHz with a gain margin of 5.6 dB and a phase margin of 33° was obtained. In experimental testing, a 1-σ TMR of 0.024 μm has been achieved [48.38].

Modern MIMO Design Methodologies

Since the dual-stage actuator servo system is a MIMO system, it is natural to utilize modern MIMO optimal design methodologies, such as LQG, LQG/LTR, H_∞, and μ-synthesis, to design the dual-stage controller. Usually MIMO optimal designs are based on the parallel structure shown in Fig. 48.23, augmented with noise/disturbances models and other weighting functions to specify the control design performance objectives.

Linear quadratic Gaussian (LQG) control combines a Kalman filter and optimal state feedback control based on the separation principle. However, the Kalman filter weakens the desirable robustness properties of the optimal state feedback control. Linear quadratic Gaussian/loop-transfer recovery (LQG/LTR) control recovers robustness by a Kalman filter redesign process. Examples of dual-stage control designs using LQG and LQG/LTR have been reported in [48.39–41].

μ-synthesis design methodology is based on H_∞ design and further accounts for plant uncertainties during the controller synthesis process with guaranteed robustness. In μ-synthesis, system performance is characterized by the H_∞ norm of the system. However, system performance can be more naturally characterized by the root-mean-square (RMS) value of PES, which is reduced to an H_2 optimization problem, like the LQG controller. In the mixed H_2/H_∞ design approach, system performance is formulated as an H_2 norm, and bounds are imposed to guarantee stability robustness to plant uncertainties. Example of dual-stage control designs using H_∞, μ-synthesis and mixed H_2/H_∞ design methodologies have been reported in [48.39,40,42]. A control design using mixed H_2/H_∞ for the MEMS microactuator dual-stage servo system will be presented in Sect. 48.3.3.

Other advanced control theories also have been applied to dual-stage servo control designs, such as sliding-mode control [48.43] and neural networks [48.44].

48.3.3 Servo Control and Vibration Attenuation of a MEMS Microactuator Dual-Stage Servo System with an Instrumented Suspension

System Modeling

MEMS microactuators for dual-stage servo applications are usually designed to have a single flexure resonance mode between approximately 1–3 kHz [48.45, 46]. This resonance mode is usually lightly damped with a damping ratio of about 0.1, and hence the microactuator is susceptible to airflow turbulence and external disturbances. Capacitive sensing can be imbedded in MEMS microactuators to measure the position of the microactuator relative to the suspension tip [48.14, 45]. This relative position signal can be utilized to damp the microactuator's resonance mode in order to make a well-behaved microactuator and ease the control design that follows [48.47].

In addition to the use of dual-stage actuators for achieving higher servo bandwidth, instrumented suspensions have also been proposed as a means of providing information on airflow-excited suspension vibrations. The sensor output can be used for VCM–suspension mode damping to increase its bandwidth [48.30], or be used for feed-forward vibration compensation by the microactuator [48.48], or both [48.42].

The block diagram of a dual-stage system with a MEMS microactuator and an instrumented suspension is illustrated in Fig. 48.24. In the figure, G_V and G_M represents the VCM and microactuator dynamics respectively, w_v and w_m are airflow disturbances acting on the VCM and microactuator respectively, u_v and u_m

are the control inputs. x_p, x_s, x_m, and x_v are respectively the read/write head position, the strain sensor output from the instrumented suspension, the relative motion output from the capacitive sensor embedded in the microactuator, and the suspension tip position. G_M can be modeled as a second-order system with the following transfer function

$$G_M(s) = \frac{A_m}{s^2 + 2\zeta_m\omega_m s + \omega_m^2}, \qquad (48.17)$$

where A_m, ω_m and ζ_m are the modal constant, natural frequency and damping ratio, respectively. G_C represents the coupling dynamics from x_v to x_p. If the microactuator is rotary, G_C is a constant of value 1, and x_m is solely the output of the microactuator driven by u_m, implying no coupling between the two actuators [32.13]. If the microactuator is translational, then G_C can be derived from the microactuator to be

$$G_C(s) = \frac{2\zeta_m\omega_m s + \omega_m^2}{s^2 + 2\zeta_m\omega_m s + \omega_m^2}. \qquad (48.18)$$

This coupling effect implies that actuation of the VCM will excite the microactuator dynamics, and x_m then becomes the combined output of G_V and G_M. In both cases, rotary and translational, actuation of the microactuator can be assumed to have little effect on the VCM dynamics, given the fact that the inertia of the microactuator is very small compared to that of the VCM. This assumption implies that the transfer function from u_m to x_v can be neglected.

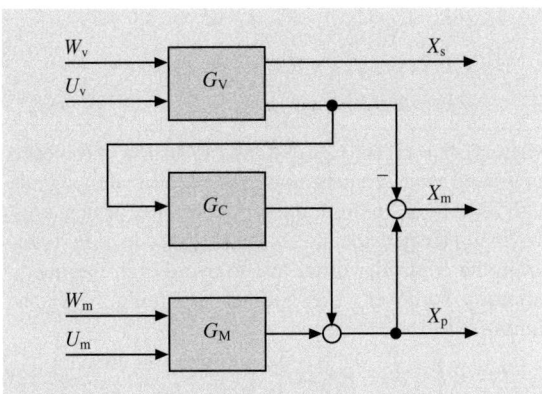

Fig. 48.24 General plant for the dual-stage actuator with an instrumented suspension

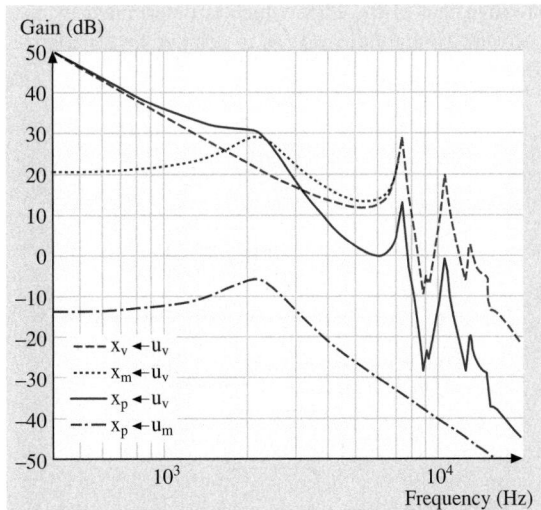

Fig. 48.25 Frequency responses of the dual-stage plant

Figure 48.25 shows frequency responses of the dual-stage plant with a translational microactuator of the type described in Section 48.2.3. The VCM–suspension assembly has several suspension modes in the high-frequency range. The microactuator resonance mode is located at about 2.2 kHz. Due to the coupling effect G_C, the responses of x_p and x_v from u_v are different: x_v is filtered by G_C in the frequency range above the microactuator resonance mode. In the rotary microactuator case, the two responses are the same, and the response from u_v to x_m is always zero. This coupling/decoupling effect makes a difference between the rotary and translational microactuators. From the viewpoint of suspension vibration attenuation, we would like to have a soft spring and a light damping effect, such that most high-frequency suspension vibrations can be attenuated when passing through the microactuator. On the other hand, we want a stiff spring and heavy damping so that the response from u_v to x_p has high gains in a wide frequency range for effective track following by the VCM. A tradeoff is therefore necessary between high-frequency vibration attenuation and low-frequency track following, depending on the relative magnitude of runout, which is mainly in the low frequency range, versus suspension vibration, which is mainly in the high frequency range.

Minor-Loop Vibration Damping Design

When x_m and x_s are assumed to be available as auxiliary information, it is possible to first design minor-loop vibration damping controllers before designing an outer-loop tracking controller for the dual-stage servo system. In addition, the two signals can be sampled at a higher rate than that of the PES, which is determined by the hardware configuration, so as to achieve better control effects.

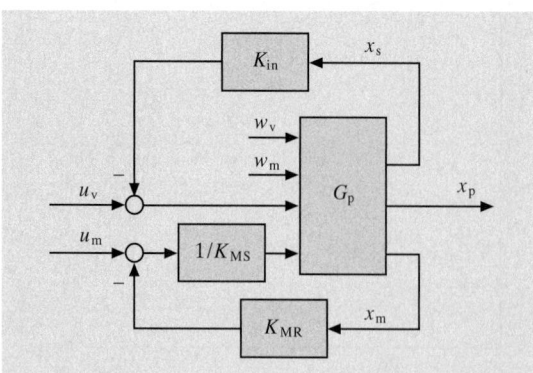

Fig. 48.26 Minor-loop vibration damping and compensation

Microactuator Resonance Mode Damping. The basic use of the relative motion signal, x_m, is to actively damp the microactuator resonance mode to make for a well-behaved microactuator and to simplify the control design that follows. This can be implemented as a minor loop around the microactuator, as shown in the lower part of Fig. 48.26. Consider the discrete-time open-loop plant

$$G_M(q^{-1}) = \frac{q^{-1}B_0(q^{-1})}{A_0(q^{-1})} , \qquad (48.19)$$

where q^{-1} is the one-step delay operator. $B_0(q^{-1})$ is the plant zero polynomial, and $A_0(q^{-1}) = 1 + a_1 q^{-1} + a_2 q^{-1}$ is the plant pole polynomial. The desired damped microactuator can be expressed as

$$G_{MD}(q^{-1}) = \frac{q^{-1}B_0(q^{-1})}{A_D(q^{-1})} . \qquad (48.20)$$

With the controller structure shown in Fig. 48.26, G_{MD} can be achieved by solving the following Diophantine equation

$$\begin{aligned} A_D(q^{-1}) = \; & A_0(q^{-1})K_{MS}(q^{-1}) \\ & + q^{-1}B_0(q^{-1})K_{MR}(q^{-1}) . \end{aligned} \qquad (48.21)$$

The closed-loop polynomial $A_D(q^{-1})$ is chosen by the designer. Usually the damping coefficient for G_{MD} is set to one, and the natural frequency ω_{MR} can be replaced to achieve adequate gains in the low frequency range.

Suspension Resonance Modes Damping. After the minor loop around the microactuator is closed, a vibration controller K_{in} is designed using x_s to provide more damping of the suspension resonance modes. The design of K_{in} is formulated as a standard LQG problem. Consider the discrete-time representation of G_V as shown in Fig. 48.26:

$$\begin{aligned} x(k+1) &= Ax(k) + Bu_v(k) + B_w w_v(k) , \\ y(k) &= Cx(k) + n(k) , \end{aligned} \qquad (48.22)$$

where $y(k) = [x_v(k)\ x_s(k)]^T$, and the airflow turbulence $w_v(k)$ and measurement noise $n(k)$ are random signals with zero mean. In this model for controller design, only two major suspension modes are included in order to restrain the controller order and to avoid high-frequency vibration spillover. The goal of designing K_{in} is to minimize the cost function

$$J = E\left\{x_v^2(k) + Ru_v^2(k)\right\} , \qquad (48.23)$$

where $E\{\cdot\}$ is the expectation operator and R is a control weight.

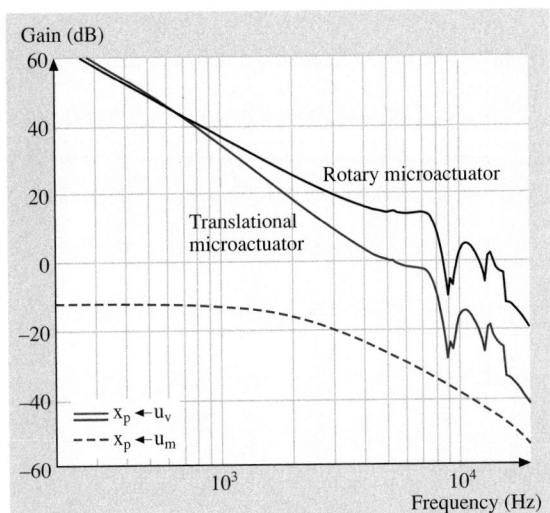

Fig. 48.27 Frequency responses of the damped plant

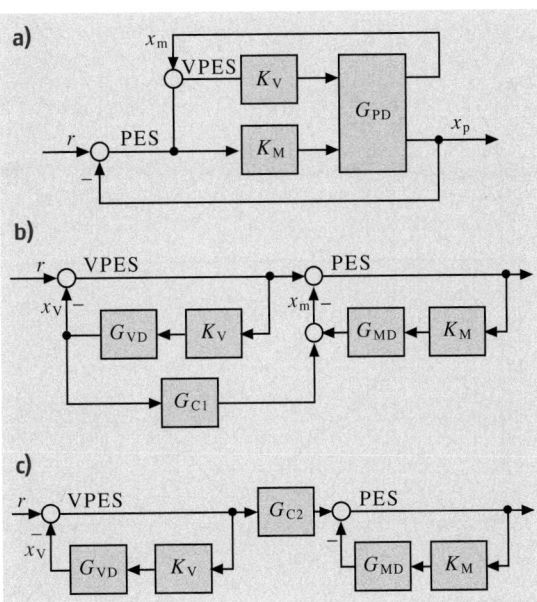

Fig. 48.28 Sensitivity-decoupling design

Figure 48.27 shows the frequency responses of the two damped subsystems. Two major resonance modes of the VCM–suspension assembly and the microactuator mode have been damped using x_s and x_m, respectively. As can be seen, the frequency response from u_v to x_p for the translational microactuator case has stronger high-frequency attenuation than that for the rotary microactuator case due to the coupling effect. This implies that the translational microactuator will behave better in suspension vibration attenuation, which happens in the high frequency range.

Track-Following Control Design Using the Sensitivity-Decoupling Method

Sensitivity-Decoupling Design Methodology. The sensitivity-decoupling design approach, originally proposed in [48.36] and explored in detail in [48.13], is popularly applied in the design of track-following controllers for dual-stage systems. This approach utilizes the PES and y_m to estimate the position error of the suspension tip relative to the data track center, which is also called the VPES,

$$\text{VPES} = \text{PES} + x_m = r - x_v , \qquad (48.24)$$

where r represents track runout. This configuration is illustrated in Fig. 48.28a, where G_{PD} is the damped plant, as shown in Fig. 48.26. It can be shown that, for the dual-stage system shown in Fig. 48.24, with a either rotary or translational microactuator, it is equivalent to the sensitivity block diagrams as shown in Fig. 48.28b and c, where G_{VD} and G_{MD} are the damped versions of G_V

and G_M respectively, and G_{C1} and G_{C2} are derived to be

$$G_{C1} = \frac{K_{MS}}{K_{MS} + G_M K_{MR}}(G_C - 1) . \qquad (48.25)$$

$$G_{C2} = 1 - G_{C1} G_V K_V . \qquad (48.26)$$

The total closed-loop sensitivity from r to the PES can then be expressed as

$$S_T = S_V G_{C2} S_M , \qquad (48.27)$$

where

$$S_V = \frac{1}{1 + K_V G_{VD}} , \qquad S_M = \frac{1}{1 + K_M G_{MD}} . \qquad (48.28)$$

With the configuration shown in Fig. 48.28c, decoupling the designs of K_V and K_M is made possible: the VCM loop and the microactuator loop can be designed sequentially using conventional SISO design techniques.

Coupling Compensation for Translational Microactuators. It is worthwhile to pay special attention to the coupling effect in this design process. The blocks, G_{C1} and G_{C2}, in Fig. 48.28 are derived from the coupling effect G_C shown in Fig. 48.24 and (48.18). As can be seen, for a rotary microactuator, $G_C = 1$, then $G_{C1} = 0$ and $G_{C2} = 1$, implying exact decoupling between the

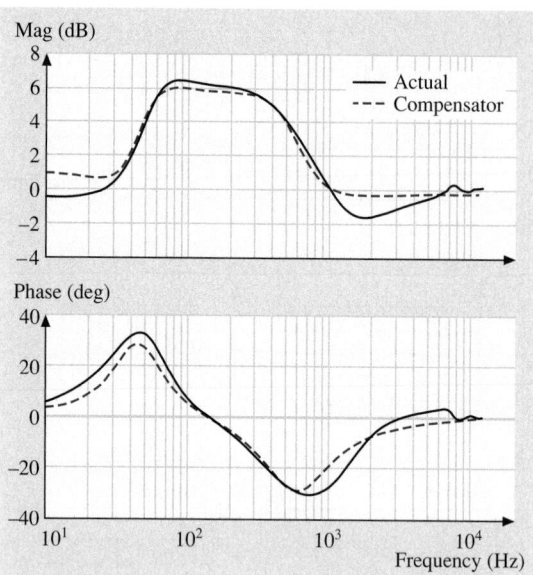

Fig. 48.29 Frequency responses of G_{C2} and the compensator K_C

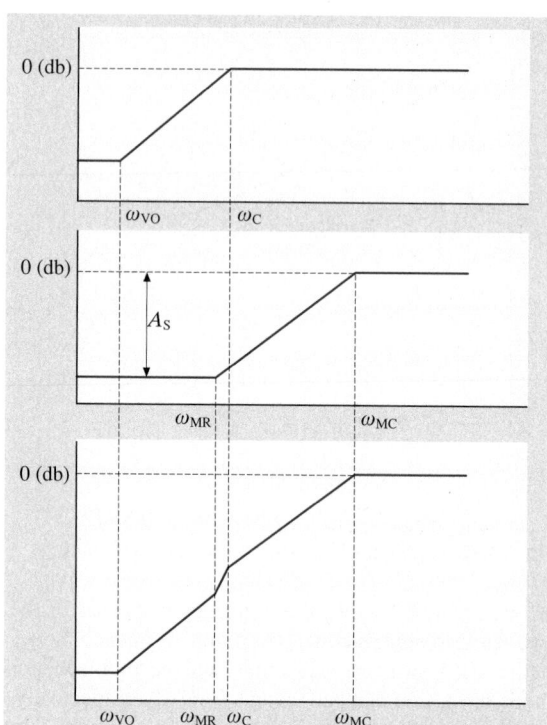

Fig. 48.30 Illustration of dual-stage sensitivity S_T design

two minor loops in Fig. 48.28c. In the translational microactuator case, G_{C2} involves all plant dynamics and designed controllers except for K_M. The frequency response of G_{C2} is shown in Fig. 48.29. It is seen that the dynamics of G_{C2} are fairly mild: it is about 6 dB in the region of 60–500 Hz, and close to 0 dB beyond 1 kHz. Its phase property is mild as well: 30° around 60 Hz and −30° around 600 Hz. This implies that, based on the K_M designed for the rotary microactuator case, a modifier, which has similar dynamics to G_{C2}, can be designed to compensate for the coupling effect, yielding a sensitivity response similar to that of the rotary microactuator case:

$$K_{MC} = K_M K_C \approx K_M G_{C2} \, , \qquad (48.29)$$

where K_M is the controller designed for the rotary microactuator case, K_C is a modifier to compensate for G_{C2}, and its dynamics are also shown in Fig. 48.29 in dashed lines.

Dual-Stage Sensitivity-Function Design. In the rotary microactuator case, the design of K_V and K_M is to use pole placement. The VCM loop sensitivity is designed to achieve enough attenuation in the low frequency range, to deal with large-amplitude disturbance components. The microactuator loop sensitivity is designed to expand the bandwidth and to achieve further attenuation of low-

frequency runout and disturbances. The compensators in the dual-stage servo system depicted in Fig. 48.28 can be designed by a two-step design process, as illustrated in Fig. 48.30.

First, the VCM loop compensator, K_V, is designed to attain a desired VCM closed-loop sensitivity S_V, as shown in the top part of Fig. 48.30. Its bandwidth, ω_{VC} in Fig. 48.30, is generally limited by the E-block and suspension resonance modes. The design of this compensator can be accomplished using conventional SISO frequency-shaping techniques.

The second step of the design process involves the design of the microactuator loop compensators to attain additional attenuation S_M, as shown in the middle part of Fig. 48.30. As before, the minor loop compensator, K_{MR}/K_{MS}, is determined in order to damp the microactuator resonance mode and place the poles of G_{MD}, or equivalently ω_{MR} in Fig. 48.30, at a desired location. The poles of G_{MD} will become the zeros of S_M.

Finally, the microactuator loop compensator, K_M, is designed to place the poles, or equivalently ω_{MC} in Fig. 48.30, of the microactuator loop sensitivity S_M. ω_{MC} is limited by the PES sampling frequency and computational time delay [48.49].

The total dual-stage sensitivity is shown in the bottom part of Fig. 48.30. For a given ω_{MC}, the additional attenuation A_S, provided by the microactuator loop, will be determined by ω_{MR}. In our proposed procedure, the initial value of ω_{MR} can be chosen to be the same as ω_{VC}. It is then adjusted so that the desired attenuation and phase-margin requirements of the overall dual-stage system are satisfied. Decreasing ω_{MR} increases the low-frequency attenuation of the closed-loop sensitivity function S_T, but also generally reduces the phase margin of open-loop transfer function G_T.

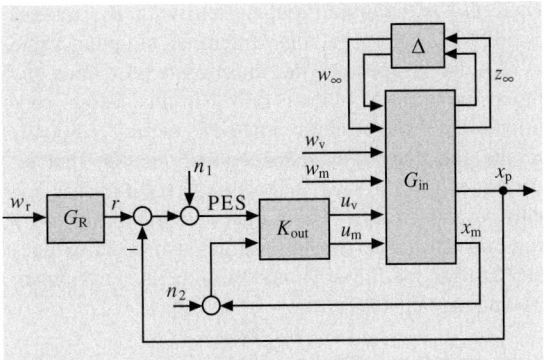

Fig. 48.31 Multi-objective optimization

Servo Design Using the Mixed H_2/H_∞ Design Approach. In this section, a multi-objective optimization method, also called the mixed H_2/H_∞ approach, is applied to the design of tracking controllers [48.42]. Unlike the sensitivity-decoupling approach, which proceeds sequentially by using SISO design techniques, the mixed H_2/H_∞ approach is a MIMO design approach. It formulates the optimization of tracking performance as a H_2 control problem, and stability robustness is explicitly considered by posing some H_∞ norm bounds. Figure 48.31 shows the schematic for this method.

In Fig. 48.31, G_{RO} is the frequency-shaping function for generating track runout from the normalized noise w_r. n_1 and n_2 are measurement noises. G_{in} is the damped plant model with the microactuator damping and K_{in} closed. $\Delta := \text{diag}(\Delta_V, \Delta_M)$ represents multiplicative uncertainties of the nominal actuator model and will be explained later.

Tracking-Following Error Minimization. The main objective of the dual-stage servo system is to make the PES as small as possible, in order to achieve high areal density and low readout error rates. Since the whole system has been adequately modeled as a stochastic system, that is, all disturbances are modeled as random signals with Gaussian distributions, tracking performance is normally characterized by the 3-σ value of the PES. When all those disturbances are normalized using corresponding weighting functions, minimizing the RMS value is then equivalent to minimizing the H_2 norm of the following transfer function

$$\min \text{RMS (PES)} \Leftrightarrow \min \|G_{z_2 w_2}\|_2 , \qquad (48.30)$$

where $z_2 = [\text{PES } u_v \ u_m]^T$ and $w_2 = [w_r \ w_v \ w_m \ n_1 \ n_2]^T$. This is just an H_2 optimization problem.

Stability Robustness Consideration. Stability robustness is an important issue for practical implementation of hard-disk servo controllers, since there always exist uncertainties and variations in disk drives' plant dynamics. It is infeasible to fine-tune controller parameters for each individual disk drive and actual working situations. Therefore, the designed controller should retain stability over a large batch of drives. In other words, it should exhibit stability robustness to plant uncertainties and variations. To this end, both qualitative and quantitative information about plant uncertainties need to be obtained to some extent and be taken into consideration during the design process. In this design example, the plant uncertainty is modeled to be multiplicative uncertainty in the following form:

$$G_V(s) = G_{Vnom}(s)[1 + \Delta_V(s)W_V(s)] ,$$
$$G_M(s) = G_{Mnom}(s)[1 + \Delta_M(s)W_M(s)] . \qquad (48.31)$$

Here, G_{Vnom} and G_{Mnom} are the nominal dynamics of the VCM and microactuator, respectively. Δ_V and Δ_M have been scaled by the gain bounding functions W_V and W_M respectively, such that $\|\Delta_V\|_\infty \le 1, \|\Delta_M\|_\infty \le 1$. Usually the bounding functions, W_V and W_M, have large magnitudes in the high frequency range, implying more uncertainty due to unmodeled dynamics and plant variation

$$W_V(s) = \frac{3s^2 + 3 \times 10^4 s + 1.4 \times 10^8}{s^2 + 4.6 \times 10^4 s + 1.4 \times 10^9} ,$$
$$W_M(s) = \frac{s + 6.28 \times 10^3}{s + 6.28 \times 10^4} . \qquad (48.32)$$

To guarantee stability robustness of the closed-loop system, the following conditions should be satisfied [48.50],

$$\sup_\omega \mu_\Delta [G_{cl}(j\omega)] < 1 , \qquad (48.33)$$

where G_{cl} is the closed-loop system with K_{out} closed around G_{in}, and μ is the structured singular value of G_{cl} with respect to the uncertainty Δ. Since the dimension of the Δ block is only 2 in this design, condition (48.33) can be approximated by setting μ-bounds on the two uncertain channels individually, that is, $\sup \mu_{\Delta_V}[G_{cl}(j\omega)] < \gamma_V$ and $\sup \mu_{\Delta_M}[G_{cl}(j\omega)] < \gamma_M$, with γ_V and γ_M strictly less than 1 and determined by trial and error. The two inequalities can be further reduced to $\|G_{cl,\Delta_V}\|_\infty < \gamma_V$ and $\|G_{cl,\Delta_M}\|_\infty < \gamma_M$ when Δ_V and Δ_M are scalars.

Multi-objective Optimization through Linear Matrix Inequalities (LMIs). From the above discussions, we have shown that the multi-objective optimization problem can be cast as an H_2 minimization problem with some H_∞ bounds, i.e., to design an output dynamic feedback controller K_{out} such that

$$K_{out} = \arg \min_{K_{out}} \gamma_2 , \qquad (48.34)$$

with

$$\|G_{z_2 w_2}\|_2 < \gamma_2 , \qquad (48.35)$$

$$\|G_{z_\infty w_\infty, \Delta_V}\|_\infty < \gamma_V , \qquad (48.36)$$

and

$$\|G_{z_\infty w_\infty, \Delta_M}\|_\infty < \gamma_M . \qquad (48.37)$$

This minimization problem with inequalities can be transformed to a set of LMIs and be solved with a convex optimization algorithm, such as SeDuMi [48.51].

The designed controller, K_{out}, is a MIMO controller. Figure 48.31 shows the frequency response of K_{out} for the translational microactuator case, from the PES to u_v. We can see that there is a large peak for the H_2 design, which implies the potential for poor stability robustness due to plant variation. When the H_∞ bounds are imposed, that peak is suppressed significantly in order to achieve more stability robustness.

Adaptive Feed-Forward Compensation for Suspension Vibration

As mentioned before, some suspension resonance modes can be damped by using the strain signal x_s measured from the suspension. Actually this signal can further be exploited to drive the microactuator to compensate for airflow-excited suspension vibrations appearing at x_p [48.13].

Define G_{ws} and G_{wp} to be the transfer functions from w_v to x_s and x_p respectively. We want the feed-forward compensator, K_{MF}, to minimize airflow-excited vibrations at the head, i.e., to minimize

$$e_a = G_M K_{MF} G_{ws} w_v + G_{wp} w_v . \qquad (48.38)$$

This scheme is different from feedback damping of the VCM in that the motion generated by the microactuator cannot directly affect the suspension outputs, x_v and x_s. Instead, what it can do is to compensate for those vibrations at the head that result from suspension vibration. It is also desirable to tune the coefficients of K_{MF} in real time in order to take into account the following factors that are slowly time-varying: suspension dynamics variation from drive to drive, and the dependence of strain-sensor properties on ambient temperature and time.

In this design, K_{MF} assumes the form of a finite impulse response (FIR) for stability consideration:

$$K_{MF}(\theta, q^{-1}) = h_0 + h_1 q^{-1} + \cdots + h_n q^{-n} , \qquad (48.39)$$

where θ is the filter coefficient vector $\theta = [h_0 \, h_1 \cdots h_n]^T$ and n is the order of K_{MF}. The feed-forward compensation motion can be expressed as

$$\begin{aligned} x_{MF}(k) &= G_M(q^{-1}) K_{MF}(q^{-1}) x_s(k) \\ &= K_{MF}(q^{-1}) G_M(q^{-1}) x_s(k) \\ &= K_{MF}(q^{-1}) x_f(k) \\ &= \theta^T \phi(k-1) , \end{aligned} \qquad (48.40)$$

where $x_f(k) = G_M(q^{-1}) x_s(k)$, $\phi(k) = [x_f(k) \, x_f(k-1) \cdots x_f(k-n)]^T$. Since $x_f(k)$ is not directly measurable, it is estimated by passing $x_s(k)$ through the model of the microactuator $\hat{G}_M$:

$$x_f(k) = \hat{G}_M(q^{-1}) x_s(k) . \qquad (48.41)$$

The coefficients of θ is tuned in such a way that $E\{e_a^2(k)\}$ is minimized. However, e_a is not directly measurable. What we have is the PES, and it can be expressed as

$$PES(k) = e_a(k) + e_r(k) , \qquad (48.42)$$

where e_r represents the tracking error resulting from all other disturbance sources except for the airflow turbulence acting on the suspension. It is roughly valid to assume that w_v and r are uncorrelated, then we have

$$E\{PES^2(k)\} = E\left\{e_a^2(k)\right\} + E\left\{e_r^2(k)\right\} . \qquad (48.43)$$

Thus, minimizing $E\{e_a^2(k)\}$ is equivalent to minimizing $E\{PES^2(k)\}$, and we can utilize the PES as a corrupted error signal to perform the adaptation. With e_a corrupted by e_r, there will be some degradation in feed-forward compensation performance, and a little longer time is needed for the adaptation process to converge.

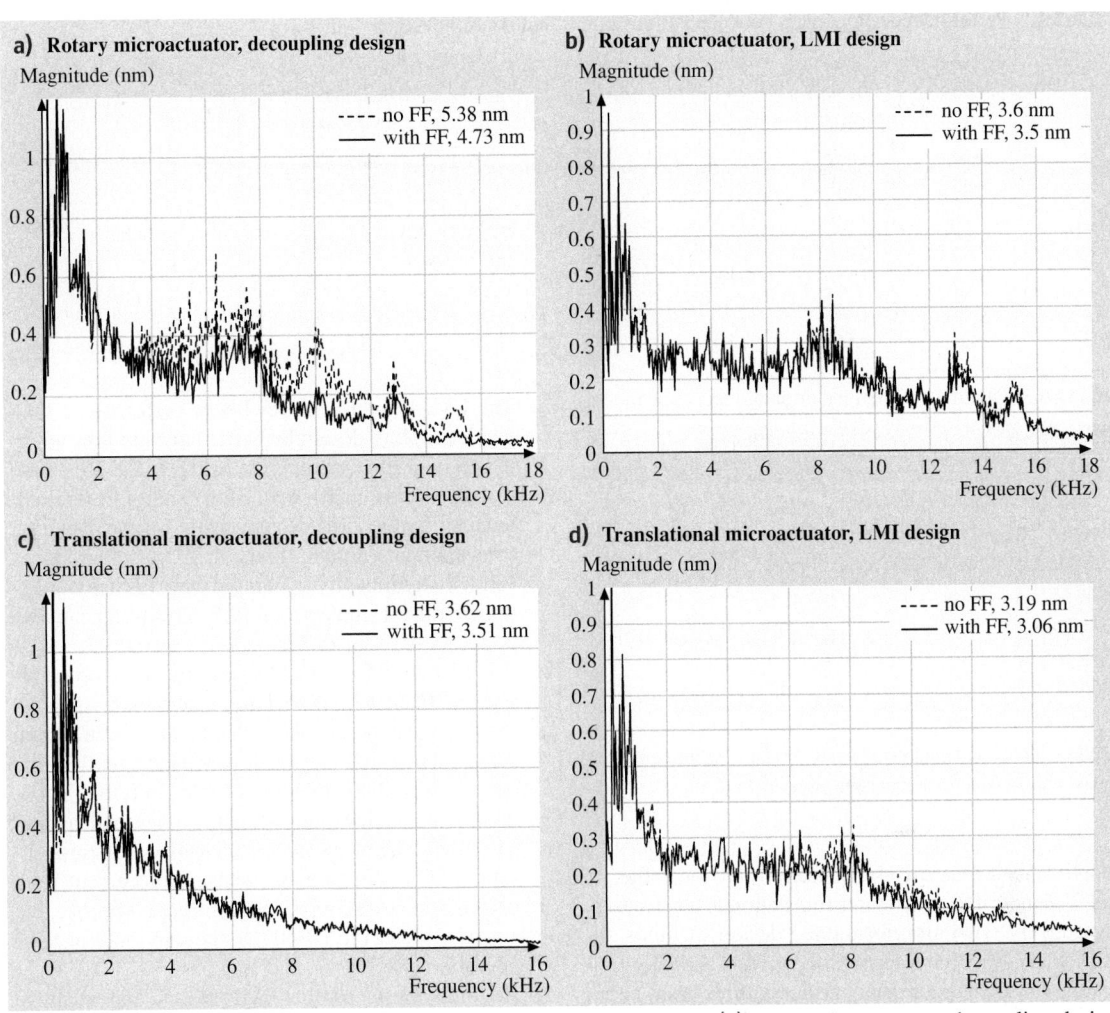

Fig. 48.32a–d Power spectra of the PES for various system configurations (**a**)Rotary microactuator, decoupling design (**b**)Rotary microactuator, LMI design (**c**)Translational microactuator, decoupling design (**d**)Translational microactuator, LMI design

Simulation Results

Simulations are conducted to investigate the performance of systems with different structures and controllers. In the simulation, the reference signal r is generated from a combination of various sources, such as repeatable track runout, disk fluttering, low-frequency torque disturbances to the VCM, etc. Measurement noises are injected into the system at proper locations.

Figure 48.32 shows the simulation results for various configurations, in which FF means adaptive feedforward compensation using x_s. Comparing (a) and (c), we can see significant differences between the rotary and translational microactuators, especially in the high frequency range. There are almost no resonance modes showing up in (c) due to the coupling effect. Comparing (a) and (b), we see that the LMI design achieves better performance than the decoupling design by optimally shaping the sensitivity function and making a better balance between the attenuation of r and w_v. From (c) and (d), we can see more clearly the improvement achieved by optimal frequency shaping.

Table 48.2 summarizes the simulation results in more detail. In the table, the total PES has been decomposed into two parts: $PES(w_v)$ and $PES(r)$, where $PES(w_v) \equiv e_a$ and $PES(r) \equiv e_r$. After this decomposition, we can more clearly see the improvement by

Table 48.2 Performance comparison between various plants and control designs

Plant type	Parameter variation	Tracking controller	PES(r) (nm)		PES(w_v) (nm)		Total PES (nm)	
			no FF	with FF	no FF	with FF	no FF	with FF
R	Nom	Decp	3.62	3.57	3.98	3.10 (22%)	5.38	4.73 (12%)
T	Nom.	Decp	3.49	3.51	0.96	0.65 (32%)	3.62	3.57 (2%)
R	Min.	LMI	2.75	2.73	2.75	2.53 (8%)	3.89	3.72 (5%)
R	Nom.	LMI	2.77	2.74	2.32	2.18 (6%)	3.61	3.50 (3%)
R	Max.	LMI	2.99	2.93	2.46	2.09 (15%)	3.87	3.60 (7%)
T	Min.	LMI	2.81	2.82	1.90	1.40 (26%)	3.39	3.15 (7%)
T	Nom.	LMI	2.78	2.78	1.56	1.28 (18%)	3.19	3.06 (4%)
T	Max.	LMI	2.81	2.79	1.67	1.35 (19%)	3.27	3.10 (5%)

Table 48.3 Parameter variations of plant resonance modes

Parameters	Variations		
	Min.[a]	Nom.[c]	Max.[b]
Suspension natural frequencies (ω_i)	−10%	0	+10%
Microactuator natural frequency (ω_m)	−15%	0	+15%
Damping coefficient (ζ_i, ζ_m)	−20%	0	+20%
Modal constants (A_i, A_m)	−5%	0	+5%

[a] Plant variables at minimum bound, [b] plant variables at maximum bound, [c] nominal plant variables

adaptive feed-forward compensation. In the table, R and T denote the rotary and translational plant cases, respectively. Two aforementioned design methods are employed: sensitivity-decoupling design and the LMI design. For plant parameter variation, three plant situations are considered: minimum, nominal and maximum. They indicate how much the plant's modal parameters vary, as specified in Table 48.3. In Table 48.2, those percentage numbers indicate how much improvement that is achieved by adaptive feed-forward compensation compared to the cases without FF as shown in the left columns.

Several conclusions can be drawn from Table 48.2.

1. The translational microactuator case always performs better than the rotary microactuator case in the attenuation of airflow-excited suspension vibrations, no matter what kind of design approach is used. We also see that the improvement is mainly achieved by reducing PES(w_v). Obviously this is due to the coupling/filtering effect of the translational microactuator on suspension vibration.

2. The LMI design achieves better performance than the decoupling design by better balancing the attenuation of PES(r) and PES(w_v). For the R-N case (rotational, nominal), both PES(r) and PES(w_v) are reduced by the LMI design, while for the T-N case (translational, nominal), the LMI design reduces PES(r) significantly by amplifying PES(w_v) a little, yielding a smaller total PES. This trend can also be seen from Fig. 48.32.

3. Adaptive feed-forward compensation can further attenuate PES(w_v), especially when there is plant variation. In practical implementation, the strain sensor may pick up some resonance modes that do not contribute to head offtrack motion. These modes are called non-offtrack modes and appear as measurement noise in feed-forward compensation. Optimization in sensor placement and orientation is therefore necessary to improve the signal-to-noise ratio (SNR) for better vibration compensation [48.32].

4. A combination of the LMI design and adaptive feed-forward compensation can achieve the best performance with fairly good performance robustness. For the rotary microactuator case, performance degradation under plant variation is less than 6%, while the degradation is less than 3% for the translational microactuator case. Again, the translational microactuator case achieves better performance robustness partially due to the coupling/filtering effect, which is very robust to parameter variations.

48.3.4 Multi-Rate Robust Track-Following Control: A Direct Approach

The controller design procedure in the previous section consists of three steps: fast-rate minor-loop damping controller design, down-sampling, and slow-rate outer-loop track-following controller design. In other words, controllers with different sampling rates have been designed separately. In contrast, in this section, we will present a robust control technique which directly designs a multi-rate controller. The model under consideration in this section is the same as the one explained in Sect. 48.3.3.

Design Procedure for Multi–Rate Controllers

A multi-rate robust track-following problem is formulated as follows. Design a discrete-time controller K which has two outputs (u_v, u_m) and three inputs (PES, y_s, y_m), where $y_s := x_s + n_3$ and $y_m := x_m + n_2$ are noise-corrupted signals, such that the PES has a small RMS value, and that the closed-loop system is robustly stable for uncertainties Δ_V and Δ_M, as well as parametric uncertainties in G_V and G_M. Next, we shall provide a procedure to design a controller which solves this problem.

A Generalized Plant

First, we construct a continuous-time generalized plant, which reflects the robust track-following problem, as depicted in Fig. 48.33, by extracting the uncertainties and by connecting a discrete-time controller K with a multi-rate sampler S and a zero-order hold H. We discretize the generalized plant by using the zeroth-order hold with a sampling rate of 50 kHz.

Although our main goal is to obtain a small PES, we have included (u_v, u_m) in the controlled output channel, to impose constraints on input magnitudes.

Derivation
of a Periodically Time-Varying System

Next, following the technique used in [48.52], the multi-rate sampler S can be written mathematically as $S : \tilde{y}(k) = \Gamma(k)y(k), k = 0, 1, 2, \ldots$, where $\{\Gamma(k)\}_{k \geq 0}$ is a sequence of matrices defined by

$$\Gamma(k) := \begin{cases} I_3, & \text{if } k \text{ is even} \\ \begin{pmatrix} 0 & 0 \\ 0 & I_2 \end{pmatrix}, & \text{if } k \text{ is odd} \end{cases}_{k \geq 0} \tag{48.44}$$

The sequence $\{\Gamma(k)\}_{k \geq 0}$ captures the multi-rate nature of our problem; at even times, we sample all the measurements, while at odd times, we cannot measure PES because of its slow sampling rate. Similarly, we can express the hold H as $H : u(k) = \tilde{u}(k)$, for $k = 0, 1, 2, \ldots$. This means that we send both control signals at the fast sampling rate of 50 kHz, in order to achieve high track-following performance of the closed-loop system.

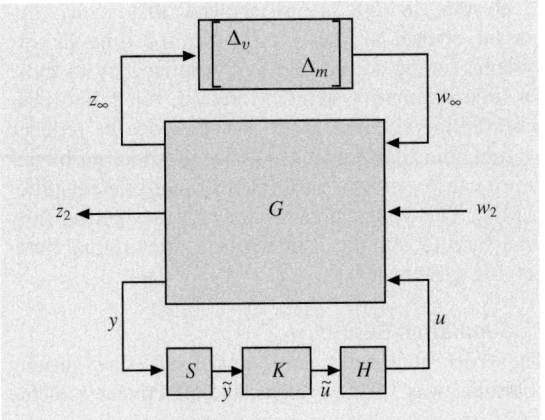

Fig. 48.33 A generalized plant G with an uncertainty block and a multi-rate controller

Combining the sampler S and the hold H with the discretized generalized plant results in the following periodically time-varying generalized plant. In our case, this plant is periodically time-varying with period 2.

A Mixed H_2/H_∞ Control Problem
for the Periodically Time-Varying System

Using the generalized plant G, we formulate an optimization problem which reflects our control problem as follows:

$$\min_{K} \gamma_0 \text{ subject to } \begin{cases} K \text{ stabilizes the nominal system,} \\ |T_{z_2 w_2}|_2 < \gamma_0 \\ |e_i^T T_{z_\infty w_\infty} e_i| < \gamma_i, \quad i = 1, 2, \end{cases} \tag{48.45}$$

where T_{qp} denotes the map from the channel p to the channel q, e_i is the i-th unit vector, and γ_i is a positive scalar to be tuned by the designer.

We remark that the norms used in (48.45) are for time-varying systems. The norm $|\cdot|_2$ is a generalization of the standard H_2 norm to a norm for time-varying systems, and is defined in [48.53]. On the other hand, the norm $|\cdot|$ in (48.45) is used with an abuse of notation; it means the ℓ_2-induced norm. We use this notation to emphasize that this optimization is a generalization of the standard mixed H_2/H_∞ optimization for time-invariant cases to that for time-varying cases. Roughly speaking, the minimization of $|T_{z_2 w_2}|_2$ corresponds to the minimization of the RMS value of the PES, while the norm constraints on $|e_i^T T_{z_\infty w_\infty} e_i|_\infty$ lead to robust stability against plant uncertainties Δ_V and Δ_M.

Results in [48.52–54] showed that many important control synthesis problems for time-varying systems can be solved in a very similar way to those for time-invariant systems. Moreover, for periodically time-varying systems, these problems can be reduced to finite-dimensional convex optimization problems, which can be solved by using linear matrix inequalities (LMIs). The designed controller becomes periodically time-varying with the same period as the original time-varying system.

Simulation Results

The order of the designed periodically time-varying controller was 13. To evaluate the effectiveness of the multi-rate controller, we have also designed a single-rate controller, in which all output signals are sampled with the same 25 kHz rate as the PES. For comparison, we have set the same robust stability constraints, γ_1, γ_2 in (48.45), during both the multi-rate and single-rate designs. For the single-rate case, the controller is time-invariant, and the order of the controller is also 13.

First, time domain simulations have been done for both controllers. The PES signals are shown in Fig. 48.34, where we can notice that the multi-rate controller achieves a smaller RMS value than its single-rate counterpart. This is also confirmed by the PES RMS values shown in Table 48.4. As can be seen in both the figures and the table, the track-following property can be greatly (about 30%) improved by using the multi-rate controller, with the same guaranteed robust stability margin. This illustrates the efficiency of the multi-rate controller for track-following.

Table 48.4 Comparison of RMS values of the PES for the multi-rate and the single-rate controllers

		RMS of PES (nm)
Mixed H_2/H_∞	Multi-rate	4.414
	Single-rate	6.274
LQG	Multi-rate	1.3659
	Single-rate	2.1958

Table 48.5 Robustness test

Controller	**Number of stabilized cases**
Mixed H_2/H_∞	956/1000
LQG	648/1000

Next, we test the robustness of the controllers. We have randomly generated, 1000 times, parameter variations within the ranges in Table 48.4, and checked how many times the multi-rate mixed H_2/H_∞ controller and the multi-rate H_2 (LQG) controller stabilize the closed-loop system. A simulation result is shown in Table 48.5, where one can see that the multi-rate mixed H_2/H_∞ controller is significantly more robust than the multi-rate LQG controller.

48.3.5 Dual–Stage Seek Control Design

Because the inertia of a microactuator is much smaller than that of the VCM, it can produce a larger acceleration and move faster than the VCM. The motion range of a microactuator is usually limited to a few micrometers, however. Thus the performance improvement in seek control madepossible by using

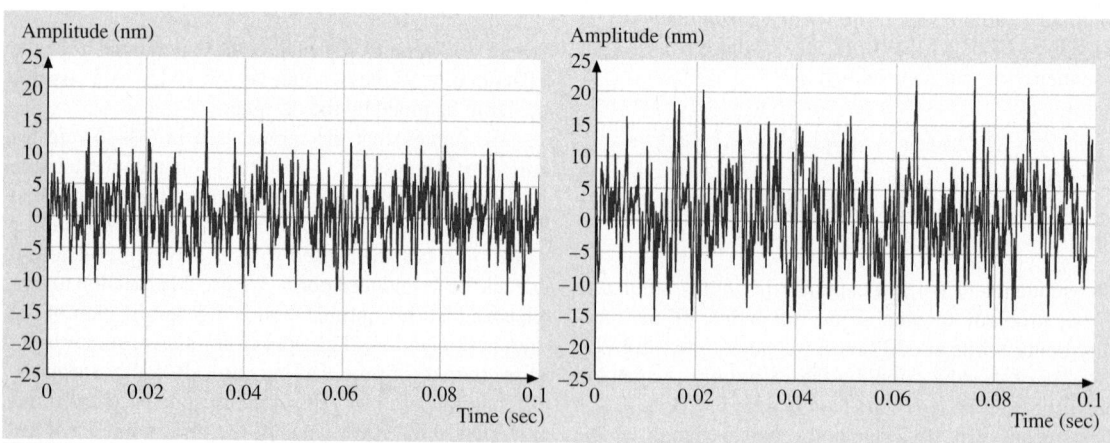

Fig. 48.34 The position error signals for the multi-rate controller (*left figure*) and for the single-rate controller (*right figure*)

dual-stage actuation will mainly be in short-distance seeks.

Two-degree-of-freedom (2-DOF) position control has been a very popular control technique in short-distance seeks. A 2-DOF control technique utilizing decoupled feed-forward reference trajectories has been developed for short-span seek control using a PZT-actuated suspension dual-stage servo system [48.55]. The same technique can also be applied to seek control of MEMS microactuator dual-stage servo systems.

Figure 48.35 shows a block diagram for dual-stage short-span seek control design using this method. In the figure, x_V^d and x_R^d are the desired seek trajectories of the VCM and the microactuator respectively. K_{VF} and K_{RF} are zero-phase-error tracking feed-forward controllers (ZPETFFC) generated using the VCM model G_V and the damped microactuator model G_R [48.56].

To minimize the residual vibration after a seek operation, minimum-jerk seek trajectories were applied to both the VCM and the microactuator. These can be generated by [48.57]

$$x_V^d(t) = 60\, d_s \left[\frac{1}{10}\left(\frac{t}{T_V}\right)^5 - \frac{1}{4}\left(\frac{t}{T_V}\right)^4 + \frac{1}{6}\left(\frac{t}{T_V}\right)^3 \right] ,$$

(48.46)

$$x_R^d = \begin{cases} 60\left(d_s - x_V^d(T)\right) \\ \times \left[\frac{1}{10}\left(\frac{t}{T}\right)^5 - \frac{1}{4}\left(\frac{t}{T}\right)^4 + \frac{1}{6}\left(\frac{t}{T}\right)^3 \right] & t \le T , \\ d_s - x_V^d(t) & t > T \end{cases}$$

(48.47)

where d_s is the distance of the head from the target track, T is the time when the head reaches the target track if dual-stage actuator is used, while T_V is the time when the head reaches the target track if only VCM is

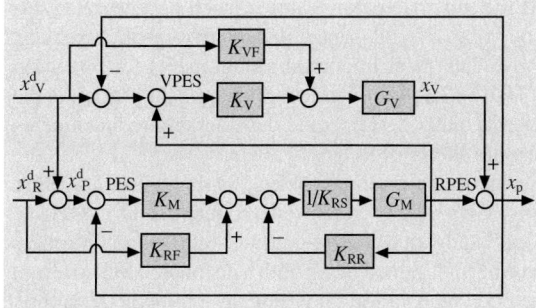

Fig. 48.35 Dual-stage seek control design

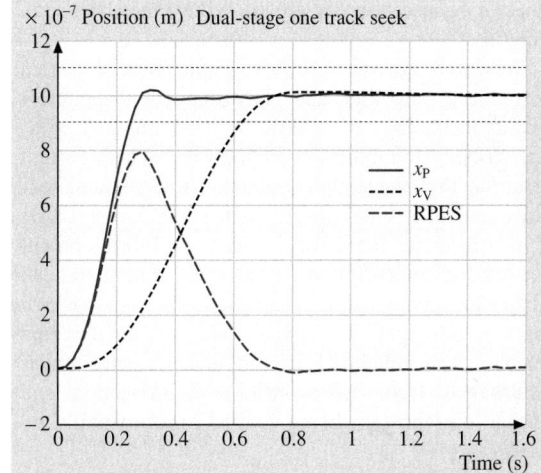

Fig. 48.36 Dual-stage short-span seek response

used. T_V and T can be chosen based on the control force saturation and seek performance requirements.

Figure 48.36 shows the 1 μm seek responses of the dual-stage actuator for the MIMO design. Overshoot was eliminated, and no obvious residual vibrations occurred. The seek time using the dual-stage actuator is about 0.25 ms, compared to a seek time of about 0.7 ms if only the VCM is used.

48.4 Conclusions and Outlook

In order to increase hard-disk-drive storage density, high-bandwidth dual-stage servo systems are necessary to suppress disturbances and increase servo precision. Various prototype MEMS microactuators have been designed and fabricated to provide a dual-stage actuation system. The most common approach for these proto-

types is to use electrostatic driving arrays produced by MEMS fabrication methods.

MEMS microactuators provide a potentially high-performance and low-cost solution for achieving the servo requirements for extremely high HDD storage density. Simulations and experiments show that many

MEMS microactuator features, such as integrated silicon gimbals and capacitive relative-position-sensing arrays, can meet important requirements for operation of HDDs. The use of structural electrical isolation in the microactuator design could considerably reduce microactuator driving voltages, while processes that avoid the use of silicon-on-insulator wafers or high-temperature processing steps could reduce manufacturing costs. High-bandwidth dual-stage track-following controllers have been designed using both a decoupled SISO design method and a variety of robust design methods, among others. In addition, short-span seeking control using a 2-DOF control structure with decoupled feed-forward trajectory generation can greatly reduce the short-span seek time.

Research remains to be done in the areas of system integration, reliability, and performance enhancement for this technology to be utilized in commercial products. Cost and reliability are probably the most important obstacles to commercial implementation of actuated-slider dual-stage servo systems at this time. Streamlining of microactuator fabrication processes or development of a process compatible with that of the slider and head fabrication could reduce manufacturing costs, which are not quite yet economical. Meanwhile, dynamic behavior and reliability of the microactuator under disturbances from airflow, head–disk interaction, and particle presence are being studied experimentally by industry.

Further in the future, to achieve nanometer servo precision, third-generation dual-stage servo systems will have to employ an actuated head approach. Research in this area has just started, with the key problem being how to combine microactuator fabrication with read/write head fabrication. Another area of research is the use of MEMS technology to incorporate additional sensors, such as accelerometers and strain-gage vibration sensors, to suppress the TMR due to structural vibrations excited by airflow and external disturbances. New robust, adaptive MIMO control architectures and algorithms must be developed for such a multisensing multiactuation servo system.

Besides the application of MEMS microactuators in dual-stage servo control of traditional rotating media magnetic HDDs, recent advances in MEMS and nanotechnology have made possible the development of a new class of miniaturized ultra-high-density storage devices that use probe arrays for recording and reading data. This new architecture abandons the traditional rotating media and flying-slider paradigm in favor of parallel x–y scanning of entire probe arrays over a storage medium. The probe arrays and/or the media are moved with electrostatic or electromagnetic microactuators that generate x–y in-plane relative motion, as well as z-axis out-of-plane relative motion. One example of this new generation of storage devices is IBM's *millipede*, which utilizes atomic force microscopy (AFM) recording technology [48.58]. Data is recorded on a polymer media by tiny depressions melted by a thermomechanical process by the AFM tips. $400\,\mathrm{Gb/in^2}$ bit patterns have been demonstrated using thermal probes.

By any of the paths above, MEMS and nanotechnology will become a basic enabling technology for the development of future data storage devices, offering the capabilities of small size, low power consumption, ultra-high densities, low cost, and high performance.

References

48.1   T. Yamaguchi: Modelling and control of a disk file head-positioning system, Proc. Inst. Mech. Eng. I (J. Syst. Cont. Eng.) **215**, 549–567 (2001)

48.2   T. Howell, R. Ehrlich, M. Lippman: TPI growth is key to delaying superparamagnetism's arrival, Data Stor., 21–30 (1999)

48.3   R. B. Evans, J. S. Griesbach, W. C. Messner: Piezoelectric microactuator for dual-stage control, IEEE Trans. Magn. **35**, 977–981 (1999)

48.4   I. Naniwa, S. Nakamura, S. Saegusa, K. Sato: Low voltage driven piggy-back actuator of hard disk drives, IEEE International MEMS 99 Conference, Orlando 1999 (IEEE, Piscataway 1999) 49–52

48.5   D. Horsley: Microfabricated Electrostatic Actuators for Magnetic Disk Drives. Ph.D. Thesis (University of California, Berkeley 1998)

48.6   L. Muller: Gimballed Electrostatic Microactuators with Embedded Iterconnects. Ph.D. Thesis (University of California, Berkeley 2000)

48.7   S. Koganezawa, K. Takaishi, Y. Mizoshita, Y. Uematsu, T. Yamada, S. Hasegawa, T. Ueno: A flexural piggyback milli-actuator for over 5 Gbit/in.2 density magnetic recording, IEEE Trans. Magnetics **32**(5), 3908–3910 (1996)

48.8   L.-S. Fan, T. Hirano, J. Hong, P. R. Webb, W. H. Juan, W. Y. Lee, S. Chan, T. Semba, W. Imaino, T. S. Pan, S. Pattanaik, F. C. Lee, I. McFadyen,

S. Arya, R. Wood: Electrostatic microactuator and design considerations for HDD application, IEEE Trans. Magn. **35**, 1000–1005 (1999)

48.9 T. Imamura, M. Katayama, Y. Ikegawa, T. Ohwe, R. Koishi, T. Koshikawa: MEMS-based integrated head/actuator/slider for hard disk drives, IEEE/ASME Trans. Mechatron. **3**, 166–174 (1998)

48.10 T. Imamura, T. Koshikawa, M. Katayama: Transverse mode electrostatic microactuator for MEM-based HDD slider, Proc. IEEE MEMS Workshop, San Diego 1996 (IEEE, New York 1996) 216–221

48.11 S. Nakamura, K. Suzuki, M. Ataka, H. Fujita: *An electrostatic micro actuator for a magnetic head tracking system of hard disk drives*, Transducers '97 (IEEE, Piscataway 1997) pp. 1081–1084

48.12 Y. Soeno, S. Ichikawa, T. Tsuna, Y. Sato, I. Sato: Piezoelectric piggy-back microactuator for hard disk drive, IEEE Trans. Magn. **35**, 983–987 (1999)

48.13 Y. Li, R. Horowitz: Mechatronics of electrostatic microactuator for computer disk drive dual-stage servo systems, IEEE/ASME Trans. Mechatron. **6**, 111–121 (2001)

48.14 N. Wongkomet: Position Sensing for Electrostatic Micropositioners. Ph.D. Thesis (University of California, Berkeley 1998)

48.15 A. Hac, L. Liu: Sensor and actuator location in motion control of flexible structures, J. Sound Vib. **167**, 239–261 (1993)

48.16 K. Hiramoto, H. Doki, G. Obinata: Optimal sensor/actuator placement for active vibration control using explicit solution of algebraic Riccati equation, J. Sound Vib. **229**, 1057–1075 (2000)

48.17 C. K. Lee: Theory of laminated piezoeleictric plates for the design of distributed sensors/actuators. Part I: Governing equations and reciprocal relationships, J. Acoustic Soc. Am. **87**, 1144–1158 (1990)

48.18 C. K. Lee, F. C. Moon: Modal sensors/actuators, ASME, J. Appl. Mech. **57**, 434–441 (1990)

48.19 P. Cheung, R. Horowitz, R. Howe: Design, fabrication and control of an electrostatically driven polysilicon microactuator, IEEE Trans. Magn. **32**, 122–128 (1996)

48.20 T. Chen, Y. Li, K. Oldham, R. Horowitz: MEMS application in computer disk drive drive dual-stage servo systems, J. Soc. Instrum. Control Eng. **41**, 412–420 (2002)

48.21 T.-L. Chen: Design and Fabrication of PZT-Actuated Silicon Suspensions for Hard Disk Drives. Ph.D. Thesis (University of California, Berkeley 2001)

48.22 C. G. Keller, R. T. Howe: HexSil tweezers for teleoperated micro-assembly, 10th Intl Workshop on Micro Electro Mechaninical Systems (MEMS'97), Nagoya 1997 (IEEE, New York 1997) 72–77

48.23 T. J. Brosnihan, J. M. Bustillo, A. P. Pisano, R. T. Howe: Embedded interconnect and electrical isolation for high-aspect-ratio, SOI inertial instruments, International Conference on Solid-State Sensors and Actuators, New York 1997 (IEEE, New York 1997) 637–640

48.24 T. Hirano, L.-S. Fan, T. Semba, W. Y. Lee, J. Hong, S. Pattanaik, P. Webb, W.-H. Juan, S. Chan: Microactuator for tera-storage, IEEE Intl MEMS 1999 Conference, Orlando 1999 (IEEE, Piscataway 1999) 6441–6446

48.25 T. Hirano, L.-S. Fan, T. Semba, W. Lee, J. Hong, S. Pattanaik, P. Webb, W.-H. Juan, S. Chan: High-bandwidth HDD tracking servo by a moving-slider micro-actuator, IEEE Trans. Magn. **35**, 3670–3672 (1999)

48.26 M. White, T. Hirano, H. Yang, K. Scott, S. Pattanaik, F. Y. Huang: *High-bandwidth hard disk drive tracking using a moving-slider MEMS microactuator*, 8th IEEE Int. Wkshop. Motion Control (IEEE, Piscataway 2004) pp. 299–304

48.27 T. Iizuka, T. Oba, H. Fugita: *Electrostatic micro actuators with high-aspect-ratio driving gap for hard disk drive applications*, Intl Symposium on micromechatronics and human science (IEEE, Piscataway 2000) pp. 229–236

48.28 H. Kuwajima, K. Matsuoka: *Thin film piezoelectric dual-stage actuator for HDD*, InterMag Europe, Session BS04 (IEEE, Piscaraway 2000) p. BS4

48.29 Y. Lou, P. Gao, B. Qin, G. Guo, E.-H. Ong, A. Takada, K. Okada: Dual-stage servo with on-slider PZT microactuator for hard disk drives, InterMag Europe, Session BS03, Amsterdam 2002 (IEEE, Piscataway 2002)

48.30 Y. Huang, M. Banthur, P. Mathur, W. Messner: Design and analysis of a high bandwidth disk drive servo systemusing an instrumented suspension, IEEE/ASME Trans. Mech. **4**, 196–206 (1999)

48.31 F. Y. Huang, T. Semba, W. Imaino, F. Lee: Active damping in HDD actuator, IEEE Trans. Mag. **37**, 847–849 (2001)

48.32 K. Oldham, S. Kon, R. Horowitz: *Fabrication and optimal strain sensor placement on an instrumented disk drive suspension for vibration suppression*, Proc. 2004 Am. Control Conf. (IEEE, Piscataway 2004) pp. 1855–1861

48.33 H. Fujita, K. Suzuki, M. Ataka, S. Nakamura: A microactuator for head positioning system of hard disk drives, IEEE Trans. Magn. **35**, 1006–1010 (1999)

48.34 B.-H. Kim, K. Chun: Fabrication of an electrostatic track-following micro actuator for hard disk drives using SOI wafer, J. Micromech. Microeng. **11**, 1–6 (2001)

48.35 R. Ehrlich, D. Curran: Major HDD TMR sources, and projected scaling with TPI, IEEE Trans. Magn. **35**, 885–891 (1999)

48.36 K. Mori, T. Munemoto, H. Otsuki, Y. Yamaguchi, K. Akagi: A dual-stage magnetic disk drive actuator using a piezoelectric device for a high track density, IEEE Trans. Magn. **27**, 5298–5300 (1991)

48.37   S. J. Schroeck, W. C. Messner, R. J. McNab: On compensator design for linear time-invariant dual-input single-output systems, IEEE/ASME Trans. Mechatron. **6**, 50–57 (2001)

48.38   T. Semba, T. Hirano, L.-S. Fan: Dual-stage servo controller for HDD using MEMS actuator, IEEE Trans. Magn. **35**, 2271–2273 (1999)

48.39   T. Suzuki, T. Usui, M. Sasaki, F. Fujisawa, T. Yoshida, H. Hirai: *Comparison of robust track-following control systems for a dual stage hard disk drive*, Proc. of International Conference on Micromechatronics for Information and Precision Equipment, ed. by B. Bhushan, K. Ohno (Word Scientific, Singapore 1997) pp. 101–118

48.40   X. Hu, W. Guo, T. Huang, B. M. Chen: *Discrete time LQG/LTR dual-stage controller design and implementation for high track density HDDs*, Proc. of American Automatic Control Conference (IEEE, Piscataway 1999) pp. 4111–4115

48.41   S.-M. Suh, C. C. Chung, S.-H. Lee: Design and analysis of dual-stage servo system for high track density HDDs, Microsyst. Technol. **8**, 161–168 (2002)

48.42   X. Huang, R. Nagamune, R. Horowitz, Y. Li: *Design and analysis of a dual-stage disk drive servo system using an instrumented suspension*, Proc. Am. Control Conf. (IEEE, Piscataway 2004) pp. 535–540

48.43   S.-H. Lee, S.-E. Baek, Y.-H. Kim: *Design of a dual-stage actuator control system with discrete-time sliding mode for hard disk drives*, Proc. of the 39th IEEE Conference on Decision and Control (IEEE, Piscataway 2000) pp. 3120–3125

48.44   M. Sasaki, T. Suzuki, E. Ida, F. Fujisawa, M. Kobayashi, H. Hirai: Track-following control of a dual-stage hard disk drive using a neuro-control system, Eng. Appl. Artif. Intell. **11**, 707–716 (1998)

48.45   D. Horsley, N. Wongkomet, R. Horowitz, A. Pisano: Precision positioning using a microfabricated electrostatic actuator, IEEE Trans. Magn. **35**, 993–999 (1999)

48.46   L. S. Fan, T. Hirano, J. Hong, P. R. Webb, W. H. Juan, W. Y. Lee, S. Chan, T. Semba, W. Imaino, T. S. Pan, S. Pattanaik, F. C. Lee, I. McFadyen, S. Arya, R. Wood: Electrostatic microactuator and design considerations for hdd application, IEEE Tran. Mag. **35**, 1000–1005 (1999)

48.47   M. T. White, T. Hirano: *Use of relative position signal for microactuators in hard disk drives*, Proc. Am. Control Conf. (IEEE, Piscataway 2003) pp. 2335–2540

48.48   Y. Li, F. Marcassa, R. Horowitz, R. Oboe, R. Evans: *Track-following control with active vibration damping of a PZT-actuated suspension dual-stage servo system*, Proc. Am. Control Conf. (IEEE, Piscataway 2004) pp. 2553–2559

48.49   M. T. White, W. M. Lu: *Hard disk drive bandwidth limitations due to sampling frequency and computational delay*, Proc. 1999 IEEE/ASME Int. Conf. Intelligent Mechatronics (IEEE, Piscataway 1999) pp. 120–125

48.50   G. J. Balas, J. C. Doyle, K. Glover, A. Packard, R. Smith: *μ-Analysis and Synthesis ToolBox* (MUSYN Inc. and The MathWorks, Natick 1995)

48.51   J. F. Sturm: Using Sedumi 1.05, A MATLAB Toolbox for Optimization over Symmetric Cones, http://fewcal.kub.nl/sturm/sorftware/sedumi.html (2001)

48.52   G. E. Dullerud, S. Lall: A new approach for analysis and synthesis of time-varying systems, IEEE Trans. Automatic Control **44**, 1486–1497 (1999)

48.53   M. A. Peters, P. A. Iglesias: *Minimum Entropy Control for Time-Varying Systems* (Birkhäuser, Basel 1997)

48.54   S. Lall, G. Dullerud: An LMI solution to the robust synthesis problem for multi-rate sampled-data systems, Automatica **37**, 1909–1922 (2001)

48.55   M. Kobayashi, R. Horowitz: Track seek control for hard disk dual-stage servo systems, IEEE Trans. Magn. **37**, 949–954 (2001)

48.56   M. Tomizuka: Zero phase error tracking algorithm for digital control, Trans. ASME J. Dynam. Syst. Meas. Control **109**, 65–68 (1987)

48.57   Y. Mizoshita, S. Hasegawa, K. Takaishi: Vibration minimized access control for disk drive, IEEE Trans. Magn. **32**, 1793–1798 (1996)

48.58   P. Vettiger, G. Cross, M. Despont, U. Drechsler, U. Durig, B. Gotsmann, W. Haberle, M. A. Lantz, H. W. Rothuizen, R. Stutz, G. K. Binnig: The "millipede"—nanotechnology entering data storage, IEEE Trans. Nanotechnol. **1**, 39–55 (2002)

49. Nanorobotics

Nanorobotics is the study of robotics at the nanometer scale, and includes robots that are nanoscale in size and large robots capable of manipulating objects that have dimensions in the nanoscale range with nanometer resolution. With the ability to position and orient nanometer-scale objects, nanorobotic manipulation is a promising way to enable the assembly of nanosystems including nanorobots.

This chapter overviews the state of the art of nanorobotics, outlines nanoactuation, and focuses on nanorobotic manipulation systems and their application in nanoassembly, biotechnology and the construction and characterization of nanoelectromechanical systems (NEMS) through a hybrid approach.

Because of their exceptional properties and unique structures, carbon nanotubes (CNTs) and SiGe/Si nanocoils are used to show basic processes of nanorobotic manipulation, structuring and assembly, and for the fabrication of NEMS including nano tools, sensors and actuators.

A series of processes of nanorobotic manipulation, structuring and assembly has been demonstrated experimentally. Manipulation of individual CNTs in 3-D free space has been shown by grasping using dielectrophoresis and placing with both position and orientation control for mechanical and electrical property characterization and assembly of nanostructures and devices. A variety of material property investigations can be performed, including bending, buckling, and pulling to investigate elasticity as well as strength and tribological characterization. Structuring of CNTs can be performed including shape modification, the exposure of nested cores, and connecting CNTs by van der Waals forces, electron–beam–induced deposition and mechanochemical bonding.

Nanorobotics provides novel techniques for exploring the bio-domain by manipulation and characterization of nanoscale objects such as cellular membranes, DNA and other biomolecules. Nano tools, sensors and actuators can provide measurements and/or movements that are calculated in nanometers, gigahertz, piconewtons, femtograms, etc., and are promising for molecular machines and bio- and nanorobotics applications. Efforts are focused on developing enabling technologies for nanotubes and other nanomaterials and structures for NEMS and nanorobotics. By combining bottom-up nanorobotic manipulation and top-down nanofabrication processes, a hybrid approach is demonstrated for creating complex 3-D nanodevices. Nanomaterial science, bionanotechnology, and nanoelectronics will benefit from advances in nanorobotics.

Part F | 49

49.1 Overview of Nanorobotics

Progress in robotics over the past years has dramatically extended our ability to explore the world from perception, cognition and manipulation perspectives at a variety of scales extending from the edges of the solar system down to individual atoms (Fig. 49.1). At the bottom of this scale, technology has been moving toward greater control of the structure of matter, suggesting the feasibility of achieving thorough control of the molecular structure of matter atom by atom as Richard Feynman first proposed in 1959 in his prophetic article on miniaturization [49.1]: "What I want to talk about is the problem of manipulating and controlling things on a small scale ... I am not afraid to consider the final question as to whether, ultimately – in the great future – we can arrange the atoms the way we want: the very atoms, all the way down!" He asserted that "At the atomic level, we have new kinds of forces and new kinds of possibilities, new kinds of effects. The problems of manufacture and reproduction of materials will be quite different. The principles of physics, as far as I can see, do not speak against the possibility of maneuvering things atom by atom." This technology is now labeled *nanotechnology*.

The "great future" of Feynman began to be realized in the 1980s. Some of the capabilities he dreamed of have been demonstrated, while others are being developed. Nanorobotics represents the next stage in miniaturization for maneuvering nanoscale objects. Nanorobotics is the study of robotics at the nanometer scale, and includes robots that are nanoscale in size,

i. e., nanorobots, and large robots capable of manipulating objects that have dimensions in the nanoscale range with nanometer resolution, i. e., nanorobotic manipulators. The field of nanorobotics brings together several disciplines, including nanofabrication processes used for producing nanoscale robots, nanoactuators, nanosensors, and physical modeling at nanoscales. Nanorobotic manipulation technologies, including the assembly of nanometer-sized parts, the manipulation of biological cells or molecules, and the types of robots used to perform these types of tasks also form a component of nanorobotics.

As the 21st century unfolds, the impact of nanotechnology on the health, wealth, and security of humankind is expected to be at least as significant as the combined influences in the 20th century of antibiotics, the integrated circuit, and human-made polymers. For example, *Lane* stated in 1998, "If I were asked for an area of science and engineering that will most likely produce the breakthroughs of tomorrow, I would point to nanoscale science and engineering." [49.2] The great scientific and technological opportunities nanotechnology presents have stimulated extensive exploration of the nano world and initiated an exciting worldwide competition, which has been accelerated by the publication of the *National Nanotechnology Initiative* by the U.S. government in 2000 [49.3]. Nanorobotics will play a significant role as an enabling nanotechnology and could ultimately be a core part of nanotechnology if Drexler's machine-phase nanosystems based on self-

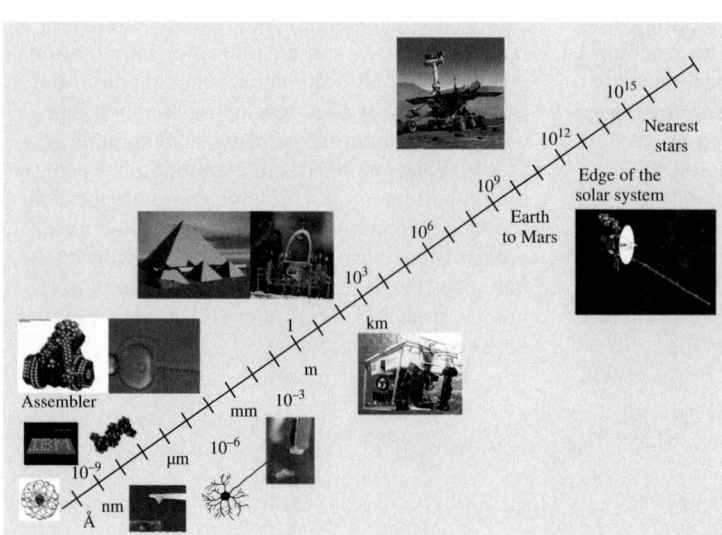

Fig. 49.1 Robotic exploration

replicative molecular assemblers via mechanosynthesis can be realized [49.4].

By the early 1980s, scanning tunneling microscopes (STMs) [49.5] radically changed the ways in which we interacted with and even regarded single atoms and molecules. The very nature of proximal probe methods encourages exploration of the nanoworld beyond conventional microscopic imaging. Scanned probes now allow us to perform engineering operations on single molecules, atoms, and bonds, thereby providing a tool that operates at the ultimate limits of fabrication. They have also enabled exploration of molecular properties on an individual nonstatistical basis.

STMs and other nanomanipulators are nonmolecular machines but use bottom-up strategies. Although performing only one molecular reaction at a time is obviously impractical for making large amounts of a product, it is a promising way to provide the next generation of nanomanipulators. Most importantly, it is possible to realize the directed assembly of molecules or supermolecules to build larger nanostructures through nanomanipulation. The products produced by nanomanipulation could be the first step of a bottom-up strategy in which these assembled products are used to self-assemble into nanomachines.

One of the most important applications of nanorobotic manipulation will be nanorobotic assembly. However, it appears that until assemblers capable of replication can be built, the parallelism of chemical synthesis and self-assembly are necessary when starting from atoms; groups of molecules can self-assemble quickly due to their thermal motion, enabling them to *explore* their environments and find (and bind to) complementary molecules. Given their key role in natural molecular machines, proteins are obvious candidates for early work in self-assembling artificial molecular systems. *Degrado* [49.6] demonstrated the feasibility of designing protein chains that predictably fold into solid molecular objects. Progress is also being made in artificial enzymes and other relatively small molecules that perform functions like those of natural proteins; the 1987 Nobel prize for chemistry went to *Cram* and *Lehn* for such work on supramolecular chemistry [49.7]. Several bottom-up strategies using self-assembly appear feasible [49.8]. *Fujita*'s pioneering work has shown that self-assembly can be directed by adroitly exploiting the chemical and electrical bonds that hold natural molecules together, and hence get molecules to form desired nanometer-scale structures [49.9]. Chemical synthesis, self assembly, and supramolecular chemistry make it possible to provide building blocks at relatively large sizes beginning from the nanometer scale. Nanorobotic manipulation serves as the base for a hybrid approach to construct nanodevices by structuring these materials to obtain building blocks and assembling them into more complex systems.

This chapter focuses on nanorobotics including actuation, manipulation and assembly at the nanoscale. The main goal of nanorobotics is to provide an effective technology for the experimental exploration of the nanoworld, and to push the boundaries of this exploration from a robotics research perspective.

49.2 Actuation at Nanoscales

The positioning of nanorobots and nanorobotic manipulators depends largely on nanoactuators. While nano-sized actuators for nanorobots are still under exploration and relatively far from implementation, microelectromechanical system (MEMS)-based efforts are focused on shrinking their sizes [49.16]. Nanometer-

Table 49.1 Actuation with MEMS

Actuation principle	Type of motion	Volume (mm³)	Speed (s⁻¹)	Force (N)	Stroke (m)	Resolution (m)	Power density (W/m³)	Ref.
Electrostatic	Linear	400	5000	1×10^{-7}	6×10^{-6}	NA	200	[49.10]
Magnetic	Linear	$0.4\times0.4\times0.5$	1000	2.6×10^{-6}	1×10^{-4}	NA	3000	[49.11]
Piezoelectric	Linear	$25.4\times12.7\times1.6$	4000	350	1×10^{-3}	7×10^{-8}	NA	[49.12]
Actuation principle	**Type of motion**	**Volume (mm³)**	**Speed (rad/s)**	**Torque (Nm)**	**Stroke (rad)**	**Resolution (rad)**	**Power density (W/m³)**	**Ref.**
Electrostatic	Rotational	$\pi/4\times0.5^2\times3$	40	2×10^{-7}	2π	NA	900	[49.13]
Magnetic	Rotational	$2\times3.7\times0.5$	150	1×10^{-6}	2π	$5/36\pi$	3000	[49.14]
Piezoelectric	Rotational	$\pi/4\times1.5^2\times0.5$	30	2×10^{-11}	0.7	NA	NA	[49.15]

resolution motion has been extensively investigated and can be generated using various actuation principles. Electrostatics, electromagnetics, and piezoelectrics are the most common ways to realize actuation at nanoscales. For nanorobotic manipulation, besides nanoresolution and compact sizes, actuators generating large strokes and high forces are best suited for such applications. The speed criteria are of less importance as long as the actuation speed is in the range of a couple of hertz and above. Table 49.1 provides a small selection of early works on actuators [49.10–14, 17] suitable in actuation principle actuators suitable for nanorobotic applications (partially adapted from [49.16]).

Several extensive reviews on various actuation principles have been published [49.15, 18–21]. During the design of an actuator, the tradeoffs among range of motion, force, speed (actuation frequency), power consumption, control accuracy, system reliability, robustness, load capacity, etc. must be taken into consideration. This section reviews basic actuation technologies and potential applications at nanometer scales.

49.2.1 Electrostatics

Electrostatic charge arises from a build up or deficit of free electrons in a material, which can exert an attractive force on oppositely charged objects, or a repulsive force on similarly charged objects. Since electrostatic fields arise and disappear rapidly, such devices will likewise demonstrate very fast operation speeds and be little affected by ambient temperatures.

Recent investigations have produced many examples of miniature devices using electrostatic force for actuation including silicon micro motors [49.22, 23], micro valves [49.24], and micro tweezers [49.25]. This type of actuation is important for achieving nano-sized actuation.

Electrostatic fields can exert great forces, but generally across very short distances. When the electric field must act over larger distances, a higher voltage will be required to maintain a given force. The extremely low-current consumption associated with electrostatic devices makes for highly efficient actuation.

49.2.2 Electromagnetics

Electromagnetism arises from electric current moving through a conducting material. Attractive or repulsive forces are generated adjacent to the conductor and pro-

portional to the current flow. Structures can be built which gather and focus electromagnetic forces, and harness these forces to create motion.

Electromagnetic fields arise and disappear rapidly, thus permitting devices with very fast operation speeds. Since electromagnetic fields can exist over a wide range of temperatures, performance is primarily limited by the properties of the materials used in constructing the actuator.

One example of a microfabricated electromagnetic actuator is a micro valve which uses a small electromagnetic coil wrapped around a silicon micromachined valve structure [49.26]. However, the downward scalability of electromagnetic actuators into the micro and nano realm may be limited by the difficulty of fabricating small electromagnetic coils. Furthermore, most electromagnetic devices require perpendicularity between the current conductor and the moving element, presenting a difficulty for planar fabrication techniques commonly used to make silicon devices.

An important advantage of electromagnetic devices is their high efficiency in converting electrical energy into mechanical work. This translates into less current consumption from the power source.

49.2.3 Piezoelectrics

Piezoelectric motion arises from the dimensional changes generated in certain crystalline materials when subjected to an electric field or to an electric charge. Structures can be built which gather and focus the force of the dimensional changes, and harness them to create motion. Typical piezoelectric materials include quartz (SiO_2), lead zirconate titanate (PZT), lithium niobate, and polymers such as polyvinyledene fluoride (PVDF).

Piezoelectric materials respond very quickly to changes in voltages and with great repeatability. They can be used to generate precise motions with repeatable oscillations, as in quartz timing crystals used in many electronic devices. Piezo materials can also act as sensors, converting tension or compression strains to voltages.

On the micro scale, piezoelectric materials have been used in linear inchworm drive devices [49.27], and micro pumps [49.28]. STMs and most nanomanipulators use piezoelectric actuators.

Piezo materials operate with high force and speed, and return to a neutral position when unpowered. They exhibit very small strokes (under 1 percent). Alternating electric currents produce oscillations in

Table 49.2 Comparison of nanoactuators

Method	Efficiency	Speed	Power density
Electrostatic	Very high	Fast	Low
Electromagnetic	High	Fast	High
Piezoelectric	Very high	Fast	High
Thermomechanical	Very high	Medium	Medium
Phase change	Very high	Medium	High
Shape memory	Low	Medium	Very high
Magnetostrictive	Medium	Fast	Very high
Electrorheological	Medium	Medium	Medium
Electrohydrodynamic	Medium	Medium	Low
Diamagnetism	High	Fast	High

the piezo material, and operation at the sample's fundamental resonant frequency produces the largest elongation and highest power efficiency [49.29]. Piezo actuators working in the stick–slip mode can provide millimeter to centimeter strokes. Most commercially available nanomanipulators adopt this type of actuators, such as Picomotors^{TM} from New Focus and Nanomotors^{TM} from Klock.

49.2.4 Other Techniques

Other techniques include thermomechanical, phase change, shape memory, magnetostrictive, electrorheological, electrohydrodynamic, diamagnetism, magnetohydrodynamic, shape changing, polymers, biological methods (living tissues, muscle cells, etc.) and so on. Table 49.2 lists a comparison of these.

49.3 Nanorobotic Manipulation Systems

49.3.1 Overview

Nanomanipulation, or positional and/or force control at the nanometer scale, is a key enabling technology for nanotechnology by filling the gap between top-down and bottom-up strategies, and may lead to the appearance of replication-based molecular assemblers [49.4]. These types of assemblers have been proposed as general-purpose manufacturing devices for building a wide range of useful products as well as copies of themselves (self-replication).

Presently, nanomanipulation can be applied to the scientific exploration of mesoscopic physical phenomena, biology and the construction of prototype nanodevices. It is a fundamental technology for property characterization of nanomaterials, structures and mechanisms, for the preparation of nano building blocks, and for the assembly of nanodevices such as nanoelectromechanical systems (NEMS).

Nanomanipulation was enabled by the inventions of the STM [49.5], atomic force microscope (AFM) [49.30], and other types of scanning probe microscope (SPM). Besides these, optical tweezers (laser trapping) [49.31] and magnetic tweezers [49.32] are also

possible nanomanipulators. Nanorobotic manipulators (NRMs) [49.33, 34] are characterized by the capability of 3-D positioning, orientation control, independently actuated multiple end-effectors, and independent real-

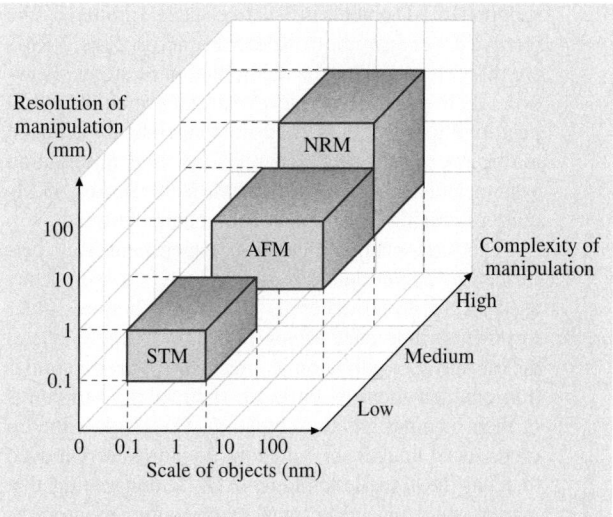

Fig. 49.2 Comparison of nanomanipulators

time observation systems, and can be integrated with scanning probe microscopes. NRMs largely extend the complexity of nanomanipulation.

A concise comparison of STM, AFM, and NRM technology is shown in Fig. 49.2. With its incomparable imaging resolution, an STM can be applied to particles as small as atoms with atomic resolution. However, limited by its 2-D positioning and available strategies for manipulations, standard STMs are ill-suited for complex manipulation and cannot be used in 3-D space. An AFM is another important type of nanomanipulator. There are three imaging modes for AFMs, i. e., contact mode, tapping mode (periodic contact mode), and non-contact mode. The later two are also called dynamic modes and can attain higher imaging resolution than the contact mode. Atomic resolution is obtainable with non-contact mode. Manipulation with an AFM can be done in either contact or dynamic mode. Generally, manipulation with an AFM involves moving an object by touching it with a tip. A typical manipulation is like this: image a particle first in non-contact mode, then remove the tip oscillation voltage and sweep the tip across the particle in contact with the surface and with the feedback disabled. Mechanical pushing can exert larger forces on objects and, hence, can be applied for the manipulation of relatively larger objects. 1-D to 3-D objects can be manipulated on a 2-D substrate. However, the manipulation of individual atoms with an AFM remains a challenge. By separating the imaging and manipulation functions, nanorobotic manipulators can have many more degrees of freedom including rotation for orientation control, and, hence, can be used for the manipulation of 0-D (symmetric spheres) to 3-D objects in 3-D free space. Limited by the relative lower resolution of electron microscopes, NRMs are difficult to use for the manipulation of atoms. However, their general robotic capabilities including 3-D positioning, orientation control, independently actuated multiple end-effectors, separate real-time observation system, and integrations with SPMs inside makes NRMs quite promising for complex nanomanipulation.

The first nanomanipulation experiment was performed by *Eigler* and *Schweizer* in 1990 [49.35]. They used an STM and materials at low temperatures (4 K) to position individual xenon atoms on a single-crystal nickel surface with atomic precision. The manipulation enabled them to fabricate rudimentary structures of their own design, atom by atom. The result is the famous set of images showing how 35 atoms were moved to form the three-letter logo *IBM*, demonstrating that matter could indeed be maneuvered atom by atom as *Feynman* suggested [49.1].

A nanomanipulation system generally includes nanomanipulators as the positioning device, microscopes as *eyes*, various end-effectors including probes and tweezers among others as its *fingers*, and types of sensors (force, displacement, tactile, strain, etc.) to facilitate the manipulation and/or to determine the properties of the objects. Key technologies for nanomanipulation include observation, actuation, measurement, system design and fabrication, calibration and control, communication, and human–machine interface.

Strategies for nanomanipulation are basically determined by the environment – air, liquid or vacuum – which is further decided by the properties and size of the objects and observation methods. Figure 49.3 depicts the microscopes, environments and strategies of nanomanipulation. In order to observe manipulated objects, STMs can provide sub-angstrom imaging resolution, whereas AFMs can provide atomic resolution. Both can obtain 3-D surface topology. Because AFMs can be used in an ambient environment, they provide a powerful tool for biomanipulation that may require a liquid environment. The resolution of scanning electron microscopes (SEM) is limited to about 1 nm, whereas field-emission SEMs (FESEM)

can achieve higher resolutions. SEMs/FESEM can be used for 2-D real-time observation for both the objects and end-effectors of manipulators, and large ultra-high-

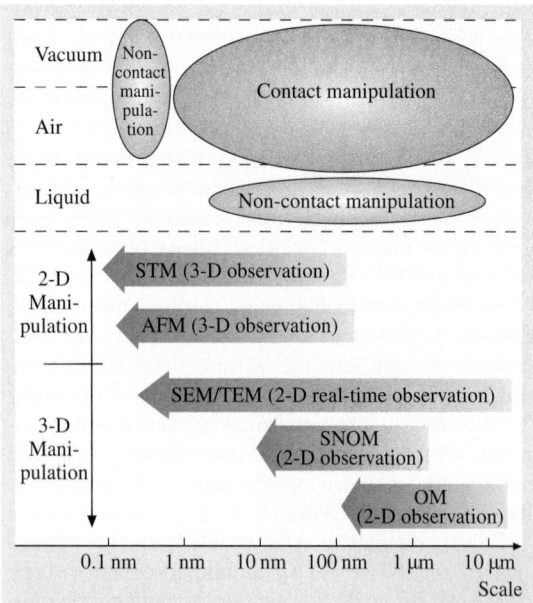

Fig. 49.3 Microscopes, environments and strategies of nanomanipulation

vacuum (UHV) sample chambers provide enough space to contain an NRM with many degrees of freedom (DOFs) for 3-D nanomanipulation. However, the 2-D nature of the observation makes positioning along the electron-beam direction difficult. High-resolution transmission electron microscopes (HRTEM) can provide atomic resolution. However, the narrow UHV specimen chamber makes it difficult to incorporate large manipulators. In principle, optical microscopes (OMs) cannot be used for nanometer-scale (smaller than the wavelength of visible lights) observation because of diffraction limits. Scanning near-field OMs (SNOMs) break this limitation and are promising as a real-time observation device for nanomanipulation, especially for ambient environments. SNOMs can be combined with AFMs, and potentially with NRMs for nanoscale biomanipulation.

Nanomanipulation processes can be broadly classified into three types: (1) lateral non-contact, (2) lateral contact, and (3) vertical manipulation. Generally, lateral non-contact nanomanipulation is mainly applied for atoms and molecules in UHV with an STM or bio-object in liquid using optical or magnetic tweezers. Contact nanomanipulation can be used in almost any environment, generally with an AFM, but is difficult for atomic manipulation. Vertical manipulation can be performed by NRMs. Figure 49.4 shows the processes of the three basic strategies.

Motion of the lateral non-contact manipulation processes are shown in Fig. 49.4a. Applicable effects [49.36] able to cause the motion include long-range van der Waals (vdW) forces (attractive) generated by the proximity of the tip to the sample [49.37], electric-field-induced fields caused by the voltage bias between the tip and the sample [49.38, 39], tunneling current local heating or inelastic tunneling vibration [49.40, 41]. With these methods, some nanodevices and molecules have been assembled [49.42, 43]. Laser trapping (optical tweezers) and magnetic tweezers are possible for non-contact manipulation of nano-order biosamples, e.g. DNA [49.44, 45].

Non-contact manipulation combined with STMs has revealed many possible strategies for manipulating atoms and molecules. However, for the manipulation of CNTs no examples have been demonstrated.

Pushing or pulling nanometer objects on a surface with an AFM is a typical manipulation using this method as shown in Fig. 49.4b. Early work showed the effectiveness of this method for the manipulation of nanoparticles [49.46–49]. This method has also been shown in nanoconstruction [49.50] and biomanipulation [49.51]. A virtual-reality interface facilitates such

manipulation [49.52, 53] and may create an opportunity for other types of manipulation. This technique has been used in the manipulation of nanotubes on a surface, and some examples will be introduced later in this chapter.

The pick-and-place task as shown in Fig. 49.4c is especially significant for 3-D nanomanipulation since its main purpose is to assemble prefabricated building blocks into devices. The main difficulty is in achieving sufficient control of the interaction between the tool and object and between the object and the substrate. Two strategies have been presented for micromanipulation [49.54] and have also proven to be effective for nanomanipulation [49.34, 55]. One strategy is to apply

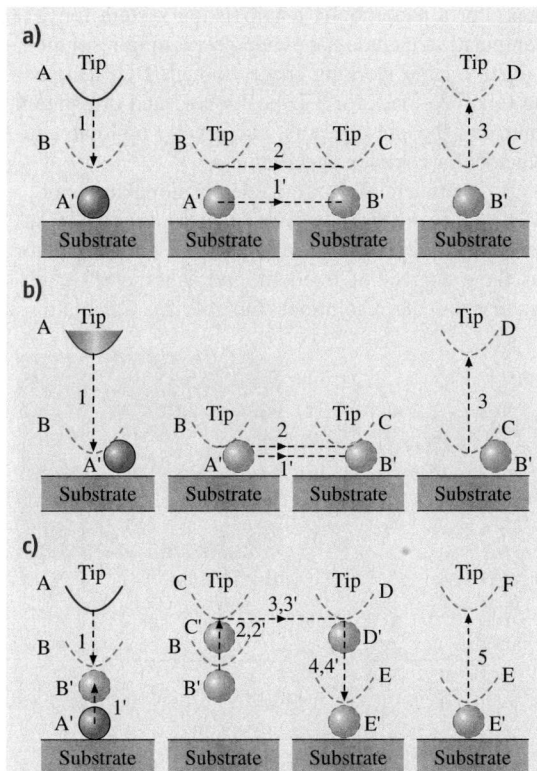

Fig. 49.4a–c Basic strategies of nanomanipulation. In the figure, A, B, C, ... represent the positions of end-effector (e.g., a tip), A, B, C, ... the positions of objects, 1, 2, 3, ... the motions of end-effector, and 1, 2, 3, ... the motions of objects. Tweezers can be used in pick-and-place to facilitate the picking-up, but are generally not necessarily helpful for placing. (**a**) Lateral non-contact nanomanipulation (sliding). (**b**) Lateral contact nanomanipulation (pushing/pulling). (**c**) Vertical nanomanipulation (picking and placing)

a dielectrophoretic force between a tool and an object as a controllable additional external force by applying a bias between the tool and the substrate on which the object is placed. Another strategy is to modify the van der Waals and other intermolecular and surface forces between the object and the substrate. For the former, an AFM cantilever is ideal as one electrode to generate a nonuniform electrical field between the cantilever and the substrate.

49.3.2 Nanorobotic Manipulation Systems

Nanorobotic manipulators are the core components of nanorobotic manipulation systems. The basic requirements for a nanorobotic manipulation system for 3-D manipulation include nanoscale positioning resolution, a relative large working space, enough DOFs including rotational ones for 3-D positioning and orientation control of the end-effectors, and usually multiple end-effectors for complex operations.

A commercially available nanomanipulator (MM3A^{TM} from Kleindiek) installed inside a SEM (Carl Zeiss DSM962) is shown in Fig. 49.5. The manipulator has three degrees of freedom, and nanometer to sub-nanometer-scale resolution (Table 49.3). Calculations

Table 49.3 Specifications of MM3A

Item	Specification
Operating range q_1 and q_2	240°
Operating range Z	12 mm
Resolution A (horiz.)	10^{-7} rad (5 nm)
Resolution B (vert.)	10^{-7} rad (3.5 nm)
Resolution C (linear)	0.25 nm
Fine (scan) range A	20 μm
Fine (scan) range B	15 μm
Fine (scan) range C	1 μm
Speed A, B	10 mm/s
Speed C	2 mm/s

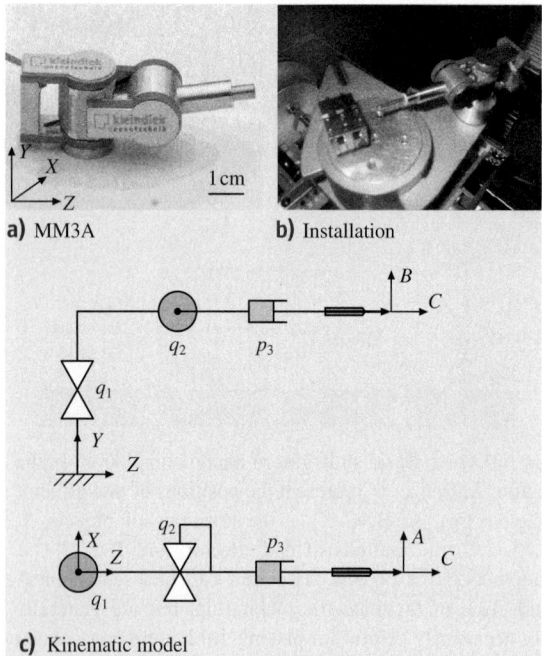

a) MM3A   **b)** Installation

c) Kinematic model

Fig. 49.5 Nanomanipulator (MM3A^{TM} from Kleindiek) inside an SEM

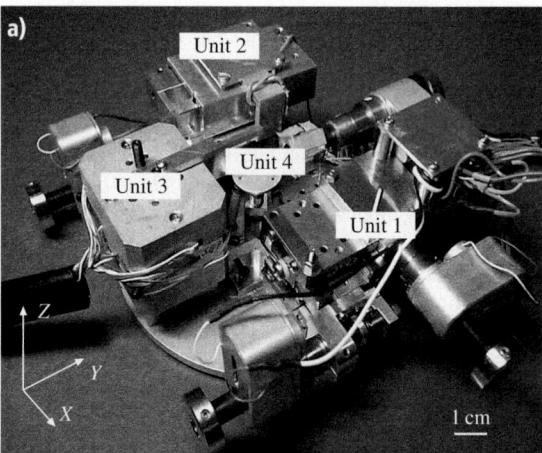

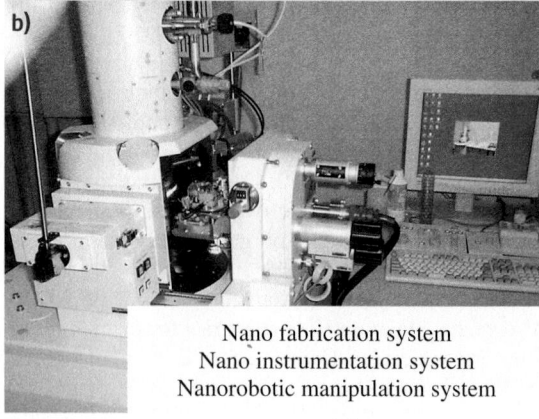

Nano fabrication system
Nano instrumentation system
Nanorobotic manipulation system

Fig. 49.6a,b Nanorobotic system. (**a**) Nanorobotic manipulators. (**b**) System set-up

show that, when moving/scanning in A/B direction by joint q_1/q_2, the additional linear motion in C is very small. For example, when the arm length is 50 mm, the

Table 49.4 Specifications of a nanorobotic manipulation system

Item	Specification
Nanorobotic manipulation system	
DOFs	Total: 16 DOFs
	Unit 1: 3 DOFs (x, y and β; coarse)
	Unit 2: 1 DOF (z; coarse), 3-DOF (x, y and z; fine)
	Unit 3: 6 DOFs (x, y, z, α, β, γ; ultrafine)
	Unit 4: 3 DOFs (z, α, β; fine)
Actuators	4 Picomotors^{TM}(Units 1&2)
	9 PZTs (Units 2&3)
	7 Nanomotors^{TM}(Units 2&4)
End-effectors	3 AFM cantilevers + 1 substrate or
	4 AFM cantilevers
Working space	18 mm × 18 mm × 12 mm × 360° (coarse, fine),
	26 μm × 22 μm × 35 μm (ultrafine)
Positioning resolution	30 nm (coarse), 2 mrad (coarse), 2 nm (fine), sub-nm (ultrafine)
Sensing system	FESEM (imaging resolution: nm) and AFM cantilevers
Nanoinstrumentation system	
FESEM	Imaging resolution: 1.5 nm
AFM cantilever	Stiffness constant: 0.03 nN/nm
Nanofabrication system	
EBID	FESEM emitter: T-FE
	CNT emitter

additional motion in the C direction is only 0.25–1 nm when moving in the A direction for 5–10 μm; these errors can be ignored or compensated with an additional motion of the prismatic joint p_3, which has a 0.25-nm resolution.

Figure 49.6a shows a nanorobotic manipulation system that has 16 DOFs in total and can be equipped with three to four AFM cantilevers as end-effectors for both manipulation and measurement. Table 49.4 lists the specifications of the system. Table 49.5 shows the functions of the nanorobotic manipulation system for nanomanipulation, nanoinstrumentation, nanofabrication and nanoassembly. The positioning resolution is sub-nanometer order and strokes are centimeter scale. The manipulation system is not only for nanomanipulation, but also for nanoassembly, nanoinstrumentation and nanofabrication. Four-probe semiconductor mea-surements are perhaps the most complex manipulation this system can perform, because it is necessary to actuate four probes independently by four manipulators. Theoretically, 24 DOFs are needed for four manipulators for general-purpose manipulations, i.e., six DOFs for each manipulator for complete control of three linear DOFs and three rotation DOFs. However, 16 DOFs are sufficient for this specific purpose. In general, two manipulators are sufficient for most tasks. More probes provide for more potential applications. For example, three manipulators can be used to assemble a nanotube transistor, a third probe can be applied to cut a tube supported on the other two probes, four probes can be used for four-terminal measurements to characterize the electric properties of a nanotube or a nanotube cross-junction. There are many potential applications for the manipulators if all four probes are used together. With

Table 49.5 Functions of a nanorobotic manipulation system

Functions	Manipulations involved
Nanomanipulation	Picking up nanotubes by controlling intermolecular and surface forces, and positioning them together in 3-D space
Nanoinstrumentation	Mechanical properties: buckling or stretching Electrical properties: placing between two probes (electrodes)
Nanofabrication	EBID with a CNT emitter and parallel EBID Destructive fabrication: breaking Shape modification: deforming by bending and buckling, and fixing with EBID
Nanoassembly	Connecting with van der Waals Soldering with EBID Bonding through mechanochemical synthesis

the advancement of nanotechnology, one could shrink the size of nanomanipulators and insert more DOFs inside the limited vacuum chamber of a microscope, and, perhaps, the molecular version of manipulators such as that dreamed of by *Drexler* could be realized [49.4].

For the construction of multiwalled carbon nanotubes (MWNT)-based nanostructures, manipulators position and orient nanotubes for the fabrication of nanotube probes and emitters, for performing nanosoldering with electron-beam-induced deposition (EBID) [49.56], for the property characterization of single nanotubes for selection purposes and for characterizing junctions to test connection strength.

A nano laboratory is shown in Fig. 49.6b, and its specifications are listed in Table 49.4. The nano laboratory integrates a nanorobotic manipulation system with a nano analytical system and a nanofabrication system, and can be applied for manipulating nano materials, fabricating nano building blocks, assembling nanodevices, and for in situ analysis of the properties of such materials, building blocks and devices. Nanorobotic manipulation within the nanolaboratory has opened a new path for constructing nanosystems in 3-D space, and will create opportunities for new nanoinstrumentation and nanofabrication processes.

Table 49.6 Properties of carbon nanotubes

Property	Item	Data
Geometrical	Layers	Single/multiple
	Aspect ratio	10–1000
	Diameter	≈ 0.4 nm to >3 nm (SWNTs)
		≈ 1.1 to > 100 nm (MWNTs)
	Length	Several μm (rope up to cm)
Mechanical	Young's modulus	≈ 1 TPa (steel: 0.2 TPa)
	Tensile strength	45 GPa (steel: 2 GPa)
	Density	$1.33 \approx 1.4$ g/cm^3 (Al: 2.7 g/cm^3)
Electronic	Conductivity	Metallic/semiconductivity
	Current carrying	
	Capacity	≈ 1 TA/cm^3 (Cu: 1 GA/cm^3)
	Field emission	Activate phosphorus at $1-3$ V
Thermal	Heat transmission	> 3 kW/mK (diamond: 2 kW/mK)

49.4 Nanorobotic Assembly

49.4.1 Overview

Nanomanipulation is a promising strategy for nano-assembly. Key techniques for nanoassembly include the structuring and characterization of nano building blocks, the positioning and orientation control of the building blocks with nanometer-scale resolution, and effective connection techniques. Nanorobotic manipulation, which is characterized by multiple DOFs with both position and orientation controls, independently actuated multi-probes, and a real-time observation system, have been shown to be effective for assembling nanotube-based devices in 3-D space.

The well-defined geometries, exceptional mechanical properties, and extraordinary electric characteristics, among other outstanding physical properties (as listed in Table 49.6), of CNTs [49.57] qualify them for many potential applications (as concisely listed in Table 49.7), especially in nanoelectronics [49.58–60], NEMS, and other nanodevices [49.61]. For NEMS, some of the most important characteristics of nanotubes include their nanometer diameter [49.62], large aspect ratio (10–1000) [49.63, 64], TPa-scale Young's modulus [49.65–71], excellent elasticity [49.33, 72], ultra-small interlayer friction, excellent capability for field emission, various electric conductivities [49.73–75], high thermal conductivity [49.76], high current-carrying capability with essentially no heating [49.77, 78], sensitivity of conductance to various physical or chemical changes, and charge-induced bond-length change.

Helical 3-D nanostructures, or nanocoils, have been synthesized from various materials, including helical carbon nanotubes [49.79] and zinc oxide nanobelts [49.80]. A new method of creating structures with nanometer-scale dimensions has recently been presented [49.81] and can be fabricated in a controllable way [49.82]. The structures are created through a top-down fabrication process in which a strained nanometer-thick heteroepitaxial bilayer curls up to form 3-D structures with nanoscale features. Helical geometries and tubes with diameters between 10 nm and 10 μm have been achieved. Because of their interesting morphology, mechanical, electrical, and electromagnetic properties, potential applications of these nanostructures in NEMS include nanosprings [49.83], electromechanical sensors [49.84], magnetic-field detectors, chemical or biological sensors, generators of magnetic beams, inductors, actuators, and high-performance electromagnetic-wave absorbers.

NEMS based on individual single or multiwalled carbon nanotubes (SWNTs [49.85, 86] or MWNTs [49.57]) and Nanocoils are of increasing interest, indicating that capabilities for incorporating these individual building blocks at specific locations on a device must be developed. Random spreading [49.87], direct growth [49.88], self-assembly [49.89], dielectrophoretic assembly [49.90] and nanomanipulation [49.91] have been demonstrated for positioning as-grown nanotubes on electrodes for the construction of these devices. However, for nanotube-based structures, nanorobotic

Table 49.7 Applications of carbon nanotubes

State	Device	Main properties applied
Bulk/array	Composite	High strength, conductivity, etc.
	Field emission devices: flat display, lamp, gas discharge tube, X-ray source, microwave generator, etc.	Field emission: stable emission, long lifetimes, and low emission threshold potentials, high current densities
	Electrochemical devices: supercapacitor, battery cathode, electromechanical actuator, etc.	Large surface area conductivity, high strength, high reversible component of storage capacity
	Fuel cell, hydrogen storage, etc.	Large surface area
Individual	Nanoelectronics: wire, diode, transistor, switch memory, etc.	Small sizes, semiconducting/metallic
	NEMS: probe, tweezers, scissors, sensor, actuator, bearing, gear, etc.	Well-defined geometrics, exceptional mechanical and electronic properties

assembly is still the only technique capable of in situ structuring, characterization and assembly. Because the as-fabricated nanocoils are not free-standing from their substrate, nanorobotic assembly is virtually the only way to incorporate them into devices at present.

49.4.2 Carbon Nanotubes

Nanotube manipulation in two dimensions on a surface was first performed with an AFM by contact pushing on a substrate. Figure 49.7 shows the typical methods for 2-D pushing. Although similar to that shown in Fig. 49.4b, the same manipulation caused various results because nanotubes cannot be regarded as 0-D points. The first demonstration was given by *Lieber* and coworkers for measuring the mechanical properties of a nanotube [49.66]. They used the method shown in Fig. 49.7b, i. e., bending a nanotube by pushing on one end of it and fixing the other end. The same strategy was used for the investigation of the behaviors of nanotubes under large strain [49.92]. *Dekker* and coworkers applied the strategies shown in Figs. 49.7c and 49.7d to get a kinked junction and crossed nanotubes [49.93,94]. *Avouris* and

coworkers combined this technique with an inverse process, namely straightening by pushing along a bent tube, and realized the translation of the tube to another location [49.95] and between two electrodes to measure the conductivity [49.96], and to form a field-effect transistor (FET) [49.97]. This technique was also used to place a tube on another tube to form a single electron transistor (SET) with cross-junction of nanotubes [49.98]. Pushing-induced breaking (Fig. 49.7d) has also been demonstrated for an adsorbed nanotube [49.95] and a freely suspended SWNT rope [49.72]. The simple assembly of two bent tubes and a straight one formed the Greek letter θ. To investigate the dynamics of rolling at the atomic level, rolling and sliding of a nanotube (as shown in Figs. 49.7e and 49.7f) are performed on graphite surfaces using an AFM [49.99, 100].

Manipulation of CNTs in 3-D space is important for assembling CNTs into structures and devices. The basic techniques for the nanorobotic manipulation of carbon nanotubes are shown in Fig. 49.8 [49.101]. These serve as the basis for handling, structuring, characterizing and assembling NEMS.

The basic procedure is to pick up a single tube from nanotube soot, Fig. 49.8a. This has been shown first by using dielectrophoresis [49.34] through nanorobotic manipulation (Fig. 49.8b). By applying a bias between a sharp tip and a plane substrate, a nonuniform electric field can be generated between the tip and the substrate with the strongest field near the tip. This field can cause a tube to orient along the field or further *jump* to the tip by electrophoresis or dielectrophoresis (determined by the conductivity of the objective tubes). Removing the bias, the tube can be placed at other locations at will. This method can be used for free-standing tubes on nanotube soot or on a rough surface on which surface van der Waals forces are generally weak. A tube strongly rooted in CNT soot or lying on a flat surface cannot be picked up in this way. The interaction between a tube and the atomically flat surface of the AFM cantilever tip has been shown to be strong enough to pick up a tube with the tip [49.102] (Fig. 49.8c). By using EBID, it is possible to pick up and fix a nanotube onto a probe [49.103] (Fig. 49.8d). For handling a tube, a weak connection between the tube and the probe is desired.

Bending and buckling a CNT as shown in Figs. 49.8e and 49.8f are important for in situ property characterization of a nanotube [49.104, 105], which is a simple way to get the Young's modulus of a nanotube without damaging the tube (if performed within its elastic range) and, hence, can be used for the se-

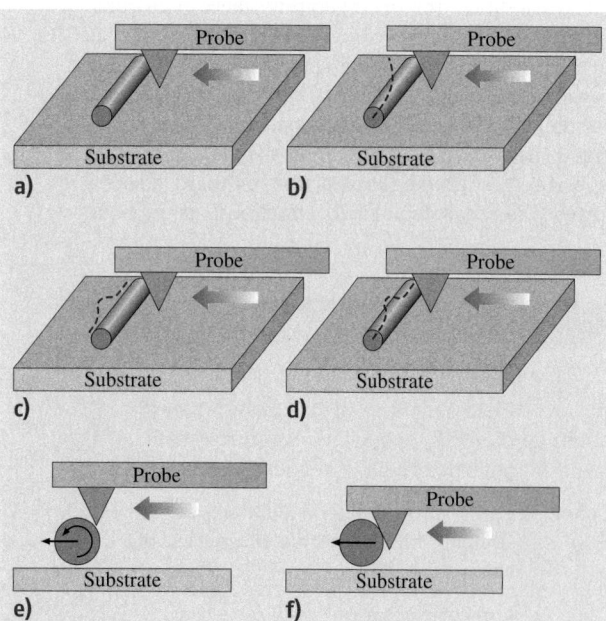

Fig. 49.7a–f 2-D manipulation of CNTs. Starting from the original state shown in (**a**), pushing the tube at a different site with a different force may cause the tube to deform as in (**b**) and (**c**), to break as in (**d**), or to move as in (**e**) and (**f**). (**a**) Original state. (**b**) Bending. (**c**) Kinking. (**d**) Breaking. (**e**) Rolling. (**f**) Sliding

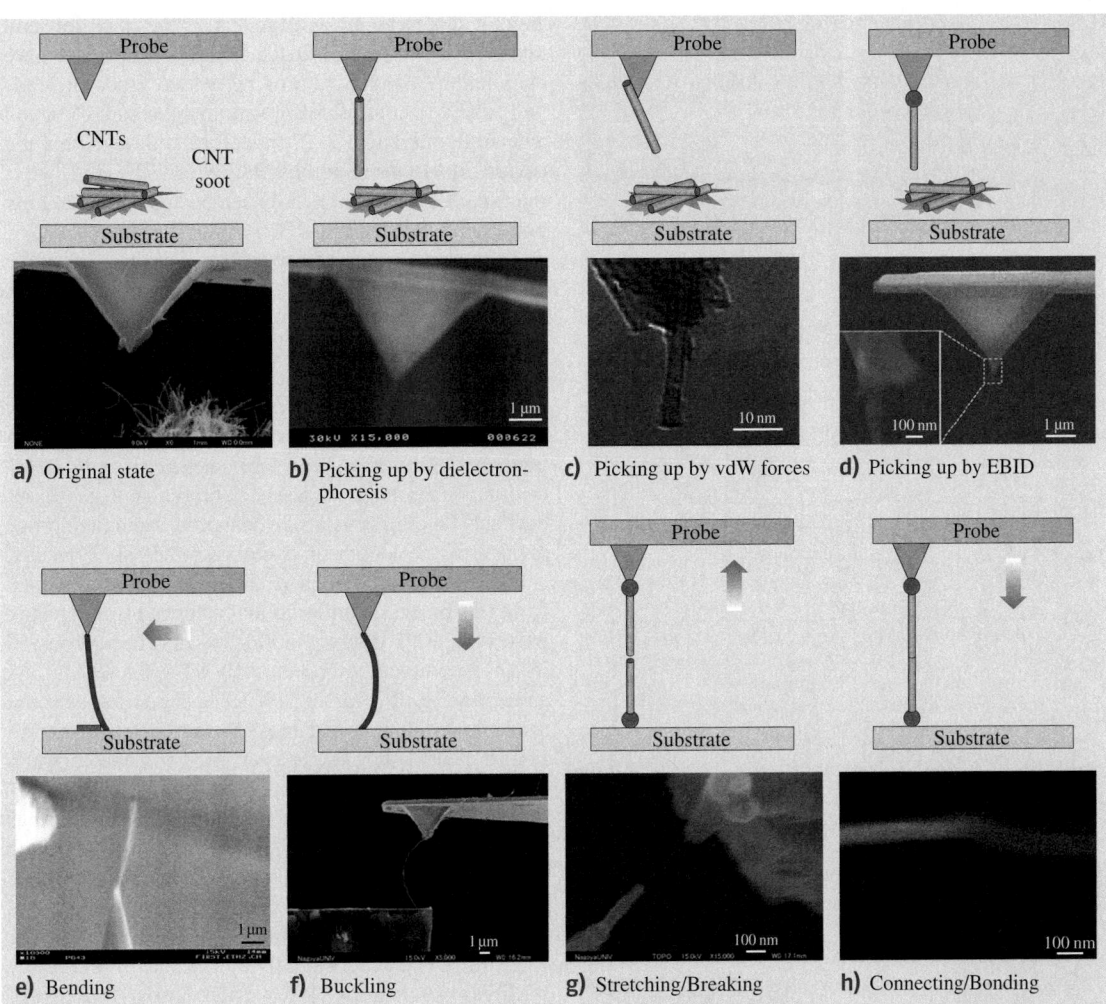

Fig. 49.8a–h Nanorobotic manipulation of CNTs. The basic technique is to pick up an individual tube from CNT soot (**a**) or from an oriented array; (**b**) shows a free-standing nanotube picked up by dielectrophoresis generated by a nonuniform electric field between the probe and substrate, (**c**) (after [49.102]) and (**d**) show the same manipulation by contacting a tube with the probe surface or fixing (e.g. with EBID) a tube to the tip (*inset* shows the EBID deposit). Vertical manipulation of nanotubes includes bending (**e**), buckling (**f**), stretching/breaking (**g**), and connecting/bonding (**h**). All examples with the exception of (c) are from the authors' work

lection of a tube with desired properties. The process is shown in Fig. 49.9a. The left figure shows an individual MWNT, whereas the right four show a bundle of MWNTs being buckled. Figure 49.9b depicts the property curve of the elastic and plastic deformations of the MWNT (ϕ133 nm × 6.055 μm), where d and F are the axial deformation and buckling force, as shown in Fig. 49.9a (left). By using the model and analysis method presented in [49.55], the flexural rigidity of the

MWNT bundle shown in Fig. 49.9a (right) is found to be $EI = 2.086 \times 10^{-19}$ [Nm^2]. This result suggests that the diameter of the nanotube is 46.4 nm if the theoretical value of the Young's modulus of the nanotube $E = 1.26$ TPa is used. The SEM image shows that the bundle of nanotubes includes at least three single ones with diameters of 31 nm, 34 nm, and 41 nm. Hence, it can be determined that there must be damaged parts in the bundle because the stiffness is too low, and it is nec-

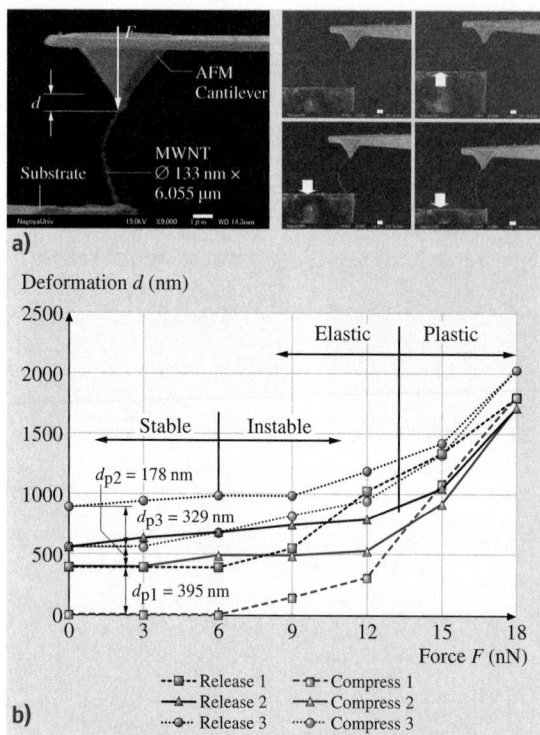

a)

b)

Fig. 49.9a,b In situ mechanical property characterization of a nanotube by buckling it (scale bars: 1 μm). (**a**) Process of buckling. (**b**) Elastic and plastic properties of an MWNT

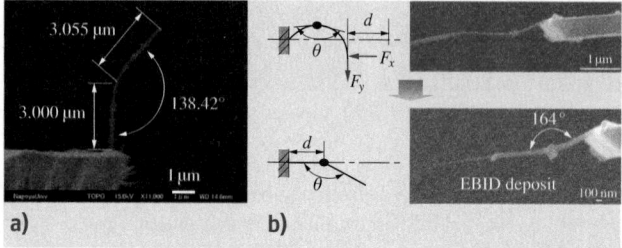

a)   **b)**

Fig. 49.10a,b Shape modification. (**a**) Kinked structure of a MWNT through plastic deformation, (**b**) Shape modifications of a MWNT by elastic bending and buckling deformation and shape fixing through EBID

essary to select another one without defects. By buckling an MWNT over its elastic limit, a kinked structure can be obtained. After three loading/releasing rounds, as shown in Fig. 49.9b, a kinked structure is obtained, as shown in Fig. 49.10a [49.106]. To obtain any desired angle for a kinked junction it is possible to fix the shape of a buckled nanotube within its elastic limit by using EBID (Fig. 49.10b) [49.107]. For a CNT, the maximum

angular displacement will appear at the fixed left end under pure bending or at the middle point under pure buckling. A combination of these two kinds of loads will achieve a controllable position of the kink point and a desired kink angle θ. If the deformation is within the elastic limit of the nanotube, it will recover as the load is released. To avoid this, EBID can be applied at the kink point to fix the shape.

Stretching a nanotube between two probes or a probe and a substrate has generated several interesting results (Fig. 49.8g). The first demonstration of 3-D nanomanipulation of nanotubes took this as an example to show the breaking mechanism of an MWNT [49.33], and to measure the tensile strength of CNTs [49.71]. By breaking an MWNT in a controlled manner, interesting nanodevices have been fabricated. This technique – destructive fabrication – has been presented to get sharpened and layered structures of nanotubes and to improve control of the length of nantubes [49.108]. Typically, a layered and a sharpened structure can be obtained from this process, similar to that achieved from electric pulses [49.109]. Bearing motion has also been observed in an incompletely broken MWNT (Fig. 49.11). As shown in Fig. 49.11a, an MWNT is supported between a substrate (left end) and an AFM cantilever (right end). Figure 49.11b shows a zoomed image of the centrally blocked part of Fig. 49.11a, and the inset shows its structure schematically. It can be found that the nanotube has a thinner neck (part B in Fig. 49.11b) that was formed by destructive fabrication, i. e., by moving the cantilever to the right. To move it further in the same direction, a motion like a linear bearing is observed, as shown in Fig. 49.11c and schematically by the inset. By comparing Figs. 49.11b and 49.11c, we find that part B remained unchanged in its length and diameter, while its two ends brought out two new parts I and II from parts A and B, respectively. Part II has uniform diameter (ϕ 22 nm), while part I is a tapered structure with a smallest diameter of ϕ 25 nm. The interlayer friction has been predicted to be very small [49.110], but direct measurement of the friction remains a challenging problem.

The reverse process, namely the connection of broken tubes (Fig. 49.8h), has been demonstrated recently, and the mechanism is revealed as rebonding of unclosed dangling bonds at the ends of broken tubes [49.111]. Based on this interesting phenomenon, mechanochemical nanorobotic assembly has been performed [49.112].

Assembly of nanotubes is a fundamental technology for enabling nanodevices. The most important tasks include the connection of nanotubes and placing of nanotubes onto electrodes. Pure nanotube circuits [49.113–115] created by interconnecting nanotubes of different diameters and chirality could lead to further size reductions in devices. Nanotube intermolecular and intramolecular junctions are basic elements for such systems. An intramolecular kink junction behaving like a rectifying diode has been reported [49.116]. Room-temperature (RT) SETs [49.117] have been shown with a short ($\approx$ 20-nm) nanotube section that is created by inducing local barriers into the tube with an AFM, and Coulomb charging has been observed. With a cross-junction of two SWNTs (semiconducting/metallic), three- and four-terminal electronic devices have been made [49.118]. A suspended cross-junction can function as electromechanical nonvolatile memory [49.119].

Although some kinds of junctions have been synthesized with chemical methods, there is no evidence yet showing that a self-assembly-based approach can provide more complex structures. SPMs were also used to fabricate junctions, but they are limited to a 2-D plane. We have presented 3-D nanorobotic manipulation-based nanoassembly, which is a promising strategy, both for the fabrication of nanotube junctions and for the construction of more complex nanodevices with such junctions as well.

Nanotube junctions can be classified into different types by: the kind of components – SWNTs or MWNTs; geometric configuration – V (kink), I, X-(cross), T-, Y- (branch), and 3-D junctions; conductivity – metallic or semiconducting; and connection methods – intermolecular (connected with van der Waals force, EBID, etc.) or intramolecular (connected with chemical bonds) junctions. Here we show the fabrication of several kinds of MWNT junctions by emphasizing the connection methods. These methods will also be effective for SWNT junctions. Figure 49.12 shows CNT junctions constructed by connecting with van der Waals forces (a), joining by electron-beam-induced deposition (b), and bonding through mechanochemistry (c).

MWNT junctions connected with van der Waals forces are the basic forms of junctions. To fabricate such junctions, the main process is to position two or more nanotubes together with nanometer resolution; they will then be connected naturally by intermolecular van der Waals forces. Such junctions are mainly for structures where contact rather than strength is emphasized. Placing them onto a surface can make them more stable. In some cases, when lateral movement along the surface of nanotubes is desired while keeping them in contact, van der Waals-type connections are the only ones that are suitable.

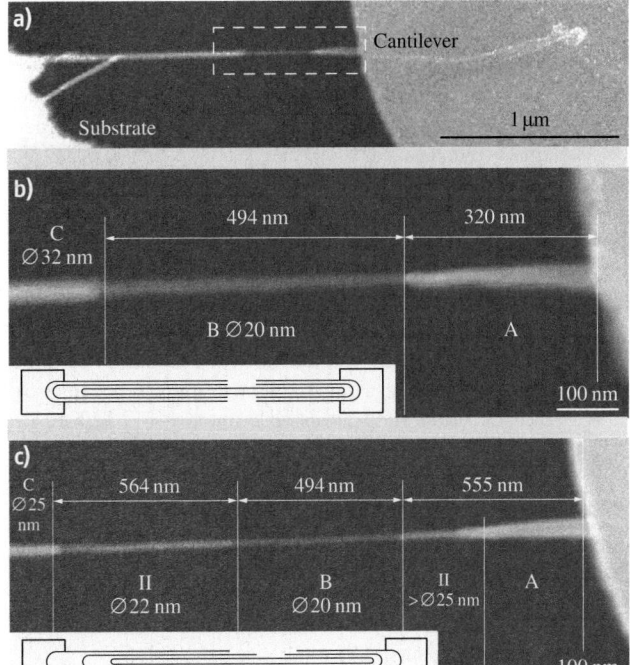

Fig. 49.11 Destructive fabrication of a MWNT and its bearing-like motion

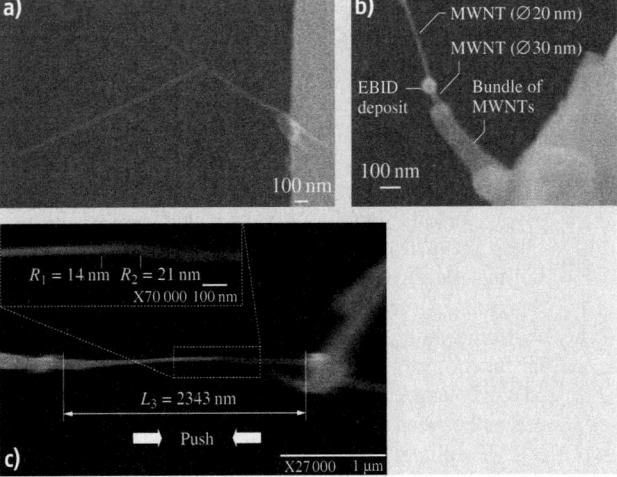

Fig. 49.12a–c MWNT junctions. (a) MWNTs connected with van der Waals force. (b) MWNTs joined with EBID. (c) MWNTs bonded with a mechanochemical reaction

Figure 49.12a shows a T-junction connected with van der Waals forces, which is fabricated by positioning the tip of an MWNT onto another MWNT until they

Part F | 49.4

form a bond. The contact is checked by measuring the shear connection force.

EBID provides a soldering method to obtain stronger nanotube junctions than those connected through van der Waals forces. Hence, if the strength of nanostructures is emphasized, EBID can be applied. Figure 49.12b shows an MWNT junction connected through EBID, in which the upper MWNT is a single one with a diameter of 20 nm and the lower one is a bundle of MWNTs with an extruded single CNT with ϕ 30 nm. The development of conventional EBID has been limited by the expensive electron filament used and low productivity. We have presented a parallel EBID system by using CNTs as emitters because of their excellent field-emission properties [49.120, 121]. The feasibility of parallel EBID is presented. It is a promising strategy for large-scale fabrications of nanotube junctions. Similar to its macro counterpart, welding, EBID works by adding material to obtain stronger connections, but in some cases, added material might influence normal functions for nanosystems. So, EBID is mainly applied to nanostructures rather than nanomechanisms.

To construct stronger junctions without adding additional material, mechanochemical nanorobotic assembly is an important strategy. Mechanochemical nanorobotic assembly is based on solid-phase chemical reactions, or mechanosynthesis, which is defined as chemical synthesis controlled by mechanical systems operating with atomic-scale precision, enabling direct positional selection of reaction sites [49.4]. By picking up atoms with dangling bonds rather than natural atoms only, it is easier to form primary bonds, which provides a simple but strong connection. Destructive fabrication provides a way to form dangling bonds at the ends of broken tubes. Some of the dangling bonds may close with neighboring atoms, but generally a few bonds will remain dangling. A nanotube with dangling bonds at its end will bind more easily to another to form intramolecular junctions. Figure 49.12c shows such a junction. An MWNT (length $L_1 = 1329$ nm, diameter $D_1 = 42$ nm) is placed between a substrate and an AFM cantilever with a CNT tip, and the two ends are fixed. By pulling the two ends of the MWNT, it is broken into two parts. By pushing the two nanotubes head to head close enough, a new one is formed. To test the strength of this nanotube, it was broken again. The fact that the nanotube breaks at a different site suggests that the tensile strength of the connected nanotubes is not weaker than that of the original nanotube itself. We have determined that no type of connection based on van der Waals interactions

can provide such a strong connection strength [49.112]. Also, we have shown that, from the measured tensile strength (1.3 TPa), chemical bonds must have been formed when the junction formed, and that these are most likely to be covalent bonds (sp^2-hybrid type, as in a nanotube).

3-D nanorobotic manipulation has opened a new route for structuring and assembly nanotubes into nanodevices. However, at present nanomanipulation is still performed in a serial manner with master–slave control, which is not a large-scale production-oriented technique. Nevertheless, with advances in the exploration of mesoscopic physics, better control of the synthesis of nanotubes, more accurate actuators, and effective tools for manipulation, high-speed and automatic nanoassembly will be possible. Another approach might be parallel assembly by positioning building blocks with an array of probes [49.122] and joining them together simultaneously, e.g., with the parallel EBID [49.103] approach we presented. Further steps might progress towards exponential assembly [49.123], and in the far future to self-replicating assembly [49.4].

49.4.3 Nanocoils

The construction of nanocoil-based NEMS involves the assembly of as-grown or as-fabricated nanocoils, which is a significant challenge from a fabrication standpoint. Focusing on the unique aspects of manipulating nanocoils due to their helical geometry, high elasticity,

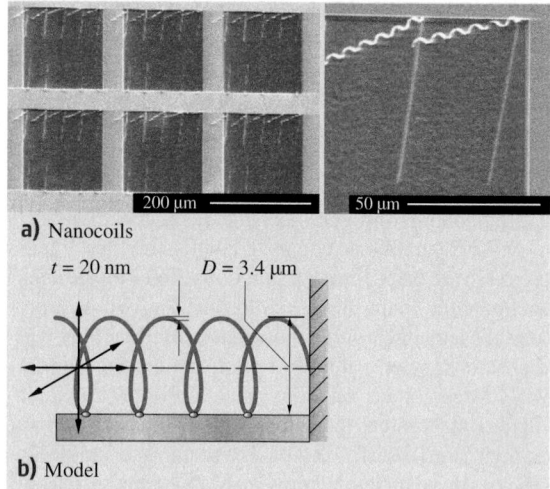

a) Nanocoils

$t = 20$ nm $D = 3.4$ μm

b) Model

Fig. 49.13 As-fabricated nanocoils with a thickness of $t = 20$ nm (without Cr layer) or 41 nm (with Cr layer). Diameter: $D = 3.4$ μm

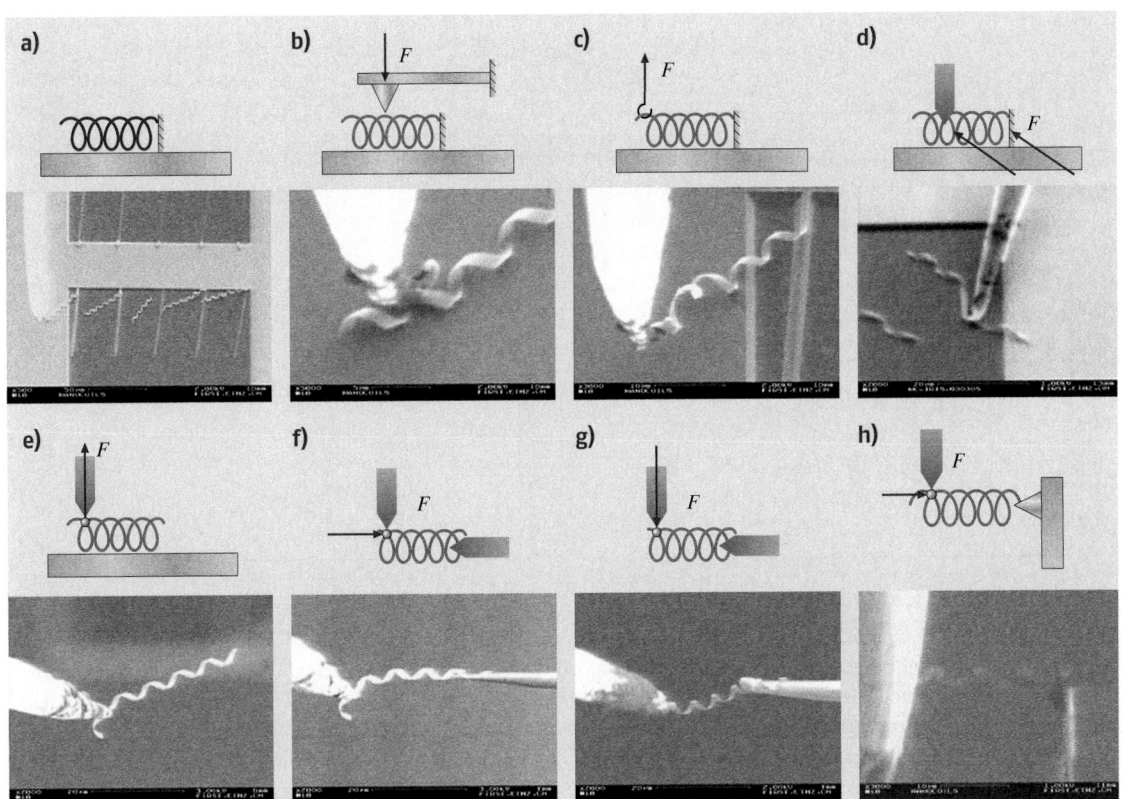

Fig. 49.14a–h Nanorobotic manipulation of nanocoils (**a**) original state, (**b**) compressing/releasing, (**c**) hooking, (**d**) lateral pushing/breaking, (**e**) picking, (**f**) placing/inserting, (**g**) bending, and (**h**) pushing and pulling

single end fixation, and strong adhesion to the substrate from wet etching, a series of new processes is presented using a manipulator (MM3A, Kleindiek) installed in an SEM (Zeiss DSM962). As-fabricated SiGe/Si bilayer nanocoils are shown in Fig. 49.13. Special tools have been fabricated including a nanohook prepared by controlled *tip-crashing* of a commercially available tungsten sharp probe (Picoprobe T-4-10-1 mm and T-4-10) onto a substrate, and a *sticky* probe prepared by tip dipping into a double-sided SEM silver conductive tape (Ted Pella, Inc.). As shown in Fig. 49.14, experiments demonstrate that nanocoils can be released from a chip by lateral pushing, picked up with a nanohook or a *sticky* probe, and placed between the probe/hook and another probe or an AFM cantilever (Nano-probe, NP-S). Axial pulling/pushing, radial compressing/releasing, and bending/buckling have also been demonstrated. These processes have shown the effectiveness of manipulation for the characterization of coil-shaped nanostructures and their assembly for NEMS, which have been otherwise unavailable.

Configurations of nanodevices based on individual nanocoils are shown in Fig. 49.15. Cantilevered nanocoils as shown in Fig. 49.15a can serve as nanosprings. Nanoelectromagnets, chemical sensors and nanoinductors involve nanocoils bridged between two electrodes, as shown in Fig. 49.15b. Electromechanical sensors can use a similar configuration but with one end connected to a moveable electrode, as shown in Fig. 49.15c. Mechanical stiffness and electric conductivity are fundamental properties for these devices that must be further investigated.

As shown in Fig. 49.14h, axial pulling is used to measure the stiffness of a nanocoil. A series of SEM images are analyzed to extract the AFM tip displacement and the nanospring deformation, i.e. the relative displacement of the probe from the AFM tip. From this displacement data and the known stiffness of the AFM cantilever, the tensile force acting on the nanospring versus the nanospring deformation was plotted. The deformation of the nanospring was measured relative to the first measurement point. This was necessary because

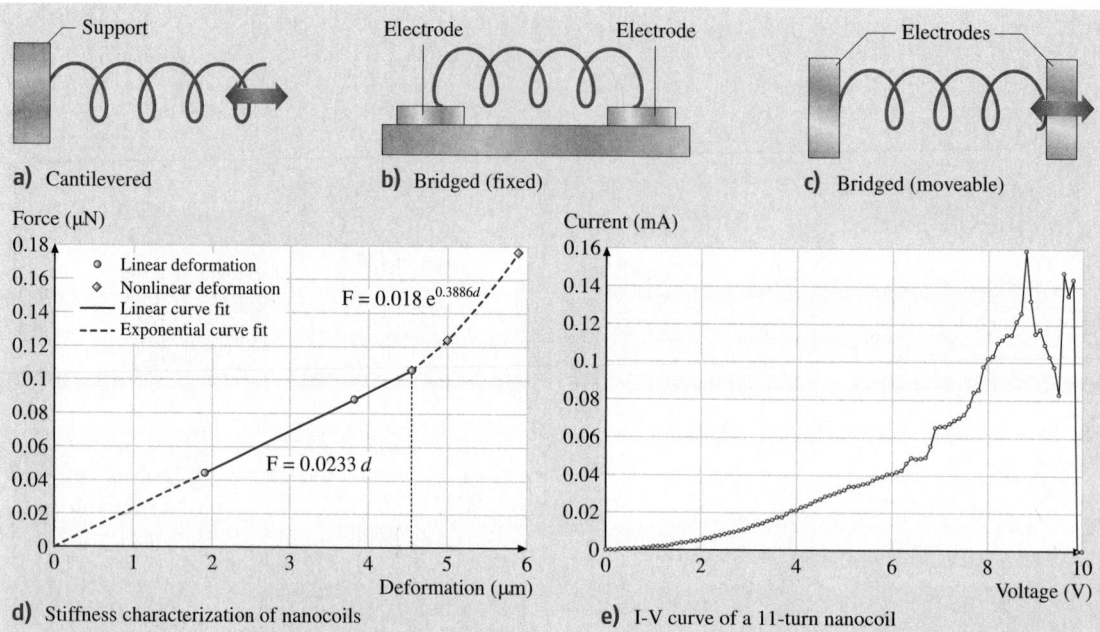

Fig. 49.15a–e Nanocoil-based devices. Cantilevered nanocoils (**a**) can serve as nanosprings. Nanoelectromagnets, chemical sensors, and nanoinductors involve nanocoils bridged between two electrodes (**b**). Electromechanical sensors can use a similar configuration but with one end connected to a moveable electrode (**c**). Mechanical stiffness (**d**) and electric conductivity (**e**) are basic properties of interest for these devices

proper attachment of the nanospring to the AFM cantilever must be verified. Afterwards, it was not possible to return to the point of zero deformation. Instead, the experimental data, as presented in Fig. 49.15d, has been shifted such that, with the calculated linear elastic spring stiffness, the line begins at zero force and zero deformation. From Fig. 49.15d, the stiffness of the spring was estimated to be 0.0233 N/m. The linear elastic region of the nanospring extends to a deformation of 4.5 μm. An exponential approximation was fitted to the nonlinear region. When the applied force reached 0.176 μN, the attachment between the nanospring and the AFM cantilever broke. Finite element simulation (ANSYS 9.0) was used to validate the experimental data [49.84]. Since the exact region of attachment cannot be identified according to the SEM images, simulations were conducted for 4, 4.5, and 5 turns to get an estimate of the possible range, given that the apparent number of turns of the nanospring is between 4 and 5. The nanosprings in the simulations were fixed on one end and had an axial load of 0.106 μN applied to the other end. The simulation results for the spring with 4 turns yield a stiffness of 0.0302 N/m; for the nanospring with 5 turns it is 0.0191 N/m. The measured stiffness falls into this range

at 22.0% above the minimum value and 22.8% below the maximum value, and very close to the stiffness of a 4.5-turn nanospring, which has a stiffness of 0.0230 N/m according to the simulation.

Figure 49.15e shows the results from electrical characterization experiments on a nanospring with 11 turns using the configuration as shown in Fig. 49.14g. The I–V curve is nonlinear, which may be caused by the resistance change of the semiconductive bilayer due to ohmic heating. Another possible reason is the decrease in contact resistance caused by thermal stress. The maximum current was found to be 0.159 mA under an 8.8-V bias. Higher voltage causes the nanospring to *blow off*. From the fast scanning screen of the SEM, the extension of the nanospring on the probes was observed around the peak current so that the current does not drop abruptly. At 9.4 V, the extended nanospring is broken down, causing an abrupt drop in the I–V curve.

From fabrication and characterization results, the helical nanostructures appear to be suitable to function as inductors. They would allow further miniaturization compared to state-of-the-art micro inductors. For this purpose, higher doping of the bilayer and an additional metal layer would result in the required conductance.

Conductance, inductance, and quality factor can be further improved if, after curling up, additional metal is electroplated onto the helical structures. Moreover, a semiconductive helical structure, when functionalized with binding molecules, can be used for chemical

sensing using the same principle as demonstrated with other types of nanostructures [49.124]. With bilayers in the range of a few monolayers, the resulting structures would exhibit a very high surface-to-volume ratio with the whole surface exposed to an incoming analyst.

49.5 Applications

Material science, biotechnology, electronics, and mechanical sensing and actuation will benefit from advances in nanorobotics. Research topics in bio-nanorobotics include the autonomous manipulation of single cells or molecules, the characterization of biomembrane mechanical properties using nanorobotic systems with integrated vision and force-sensing modules, and more. The objective is to obtain a fundamental understanding of single-cell biological systems and provide characterized mechanical models of biomembranes for deformable cell tracking during biomanipulation and cell injury studies. Robotic manipulation at nanometer scales is a promising technology for structuring, characterizing and assembling nano building blocks into NEMS. Combined with recently developed nanofabrication processes, a hybrid approach is realized to build NEMS and other nanorobotic devices from individual carbon nanotubes and SiGe/Si nanocoils.

49.5.1 Robotic Biomanipulation

Biomanipulation –
Autonomous Robotic Pronuclei DNA Injection
To improve the low success rate of manual operation, and to eliminate contamination, an autonomous robotic system (shown in Fig. 49.16) has been developed to deposit DNA into one of the two nuclei of a mouse embryo without inducing cell lysis [49.125, 126]. The laboratory's experimental results show that the success rate for the autonomous embryo pronuclei DNA injection is dramatically improved over manual conventional injection methods. The autonomous robotic system features a hybrid controller that combines visual servoing and precision position control, pattern recognition for detecting nuclei, and a precise autofocusing scheme. Figure 49.17 illustrates the injection process.

To realize large-scale injection operations, a MEMS cell holder was fabricated using anodic wafer-bonding techniques. Arrays of holes are aligned on the cell holder, which are used to contain and fix individual cells for injection. When well-calibrated, the system with the cell

Cell holding unit   Inverted microscope   3-DOF-microrobot

Readout circuit board with a wire bonded force sensor

Fig. 49.16 Robotic biomanipulation system with vision and force feedback

holder makes it possible to inject large numbers of cells using position control. The cell-injection operation can be conducted in a move–inject–move manner.

Successful injection is determined greatly by injection speed and trajectory, and the forces applied to cells (Figure 49.17). To further improve the robotic system's performance, a multi-axial MEMS-based capacitive cellular force sensor is being designed and fabricated to provide real-time force feedback to the robotic system. The MEMS cellular force sensor also aids our research in biomembrane mechanical property characterization.

MEMS-Based Multi–Axis
Capacitive Cellular Force Sensor
The MEMS-based two-axis cellular force sensor [49.127] shown in Fig. 49.18 is capable of resolving normal forces applied to a cell as well as tangential forces generated by improperly aligned cell probes. A high-yield microfabrication process was developed to form the 3-D high-aspect-ratio structure using deep reactive

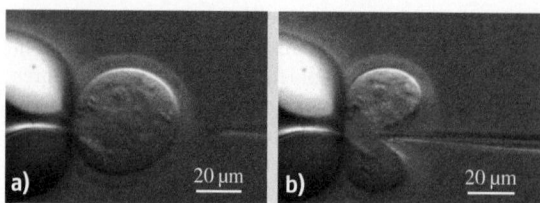

Fig. 49.17 Cell injection process. (a,b) Cell injection of a mouse oocyte

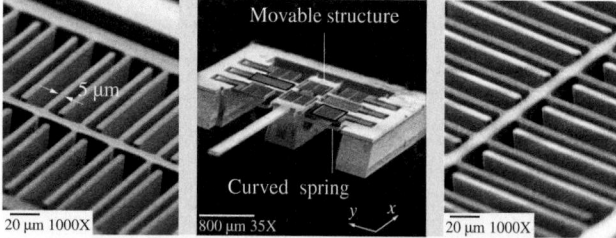

Fig. 49.18 A cellular force sensor with orthogonal comb drives detailed

ion etching (DRIE) on silicon-on-insulator (SOI) wafers. The constrained outer frame and the inner movable structure are connected by four curved springs. A load applied to the probe causes the inner structure to move, changing the gap between each pair of interdigitated comb capacitors. Consequently, the total capacitance change resolves the applied force. The interdigitated capacitors are orthogonally configured to make the force sensor capable of resolving forces in both the x- and y-directions. The cellular force sensors used in the experiments are capable of resolving forces up to 25 μN with a resolution of 0.01 μN.

Tip geometry affects the quantitative force measurement results. A standard injection pipette (Cook K-MPIP-1000-5) tip section with a tip diameter of 5 μm is attached to the probe of the cellular force sensors.

The robotic system and high-sensitivity cellular force sensor are also applied to biomembrane mechanical property studies [49.128]. The goal is to obtain

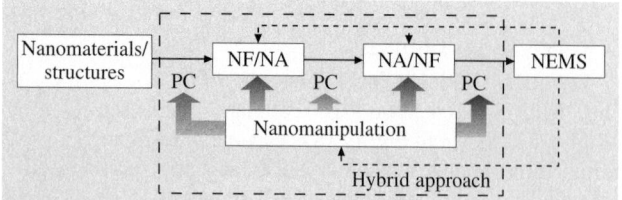

Fig. 49.19 Hybrid approach to NEMS (PC: property characterization, NF: nanofabrication, NA: nanoassembly)

a general parameterized model describing cell membrane deformation behavior when an external load is applied. This parameterized model serves two chief purposes. First, in robotic biomanipulation, it allows online parameter recognition so that cell membrane deformation behavior can be predicted. Second, for a thermodynamic model of membrane damage in cell injury and recovery studies, it is important to appreciate the mechanical behavior of the membranes. This allows the interpretation of such reported phenomena as mechanical resistance to cellular volume reduction during dehydration, and its relationship to injury. The establishment of such a biomembrane model will greatly facilitate cell injury studies.

Experiments demonstrate that robotics and MEMS technology can play important roles in biological studies such as automating biomanipulation tasks. Aided by robotics, the integration of vision and force-sensing modules, and MEMS design and fabrication techniques, investigations are being conducted in biomembrane mechanical property modeling, deformable cell tracking, and single-cell and biomolecule manipulation.

49.5.2 Nanorobotic Devices

Nanorobotic devices involve tools, sensors, and actuators at the nanometer scale. Shrinking device size makes it possible to manipulate nanosized objects with nanosized tools, measure mass in femtogram ranges, sense force at piconewton scales, and induce GHz motion, among other amazing advancements.

Top-down and bottom-up strategies for manufacturing such nanodevices have been independently investigated by a variety of researchers. Top-down strategies are based on nanofabrication and include technologies such as nanolithography, nanoimprinting, and chemical etching. Presently, these are 2-D fabrication processes with relatively low resolution. Bottom-up strategies are assembly-based techniques. At present, these strategies include such techniques as self-assembly, dip-pen lithography, and directed self-assembly. These techniques can generate regular nano patterns at large scales. With the ability to position and orient nanometer-scale objects, nanorobotic manipulation is an enabling technology for structuring, characterizing and assembling many types of nanosystems [49.101]. By combining bottom-up and top-down processes, a hybrid nanorobotic approach (as shown in Fig. 49.19) based on nanorobotic manipulation provides a third way to fabricate NEMS by structuring as-grown nanomaterials or nanostructures. This new nanomanu-

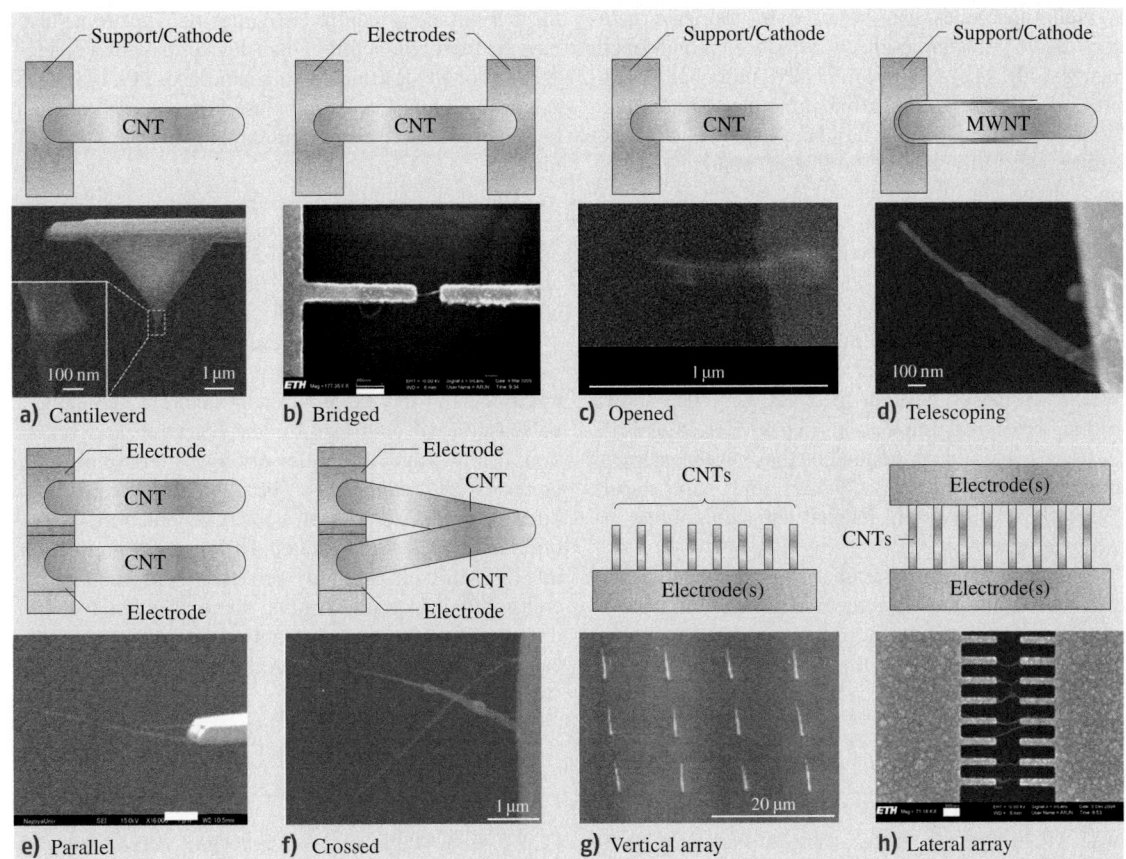

Fig. 49.20a–h Configurations of individual nanotube-based NEMS. Scale bars: (**a**) 1 μm (*inset*: 100 nm), (**b**) 200 nm, (**c**) 1 μm, (**d**) 100 nm, (**e**) and (**f**) 1 μm, (**g**) 20 μm, and (**h**) 300 nm. All examples are from the authors' work

facturing technique can be used to create complex 3-D nanodevices with such building blocks. Nanomaterial science, bionanotechnology, and nanoelectronics will also benefit from advances in nanorobotic assembly.

The configurations of nano tools, sensors, and actuators based on individual nanotubes that have been experimentally demonstrated are summarized as shown in Fig. 49.20.

For detecting deep and narrow features on a surface, cantilevered nanotubes (Fig. 49.20a, [49.103]) have been demonstrated as probe tips for an AFM [49.129], an STM and other types of SPM. Nanotubes provide ultra-small diameters, ultra-large aspect ratios, and excellent mechanical properties. Manual assembly, direct growth [49.130] and nanoassembly [49.131] have proven effective for their construction. Cantilevered nanotubes have also been demonstrated as probes for the measurement of ultra-small physical quantities, such as

femtogram masses [49.132], mass flow sensors [49.133], and piconewton-order force sensors [49.134] on the basis of their static deflections or change of resonant frequencies detected within an electron microscope. Deflections cannot be measured from micrographs in real time, which limits the application of this kind of sensor. Inter-electrode distance changes cause emission current variation of a nanotube emitter and may serve as a candidate to replace microscope images.

Bridged individual nanotubes (Fig. 49.20b, [49.135]) have been the basis for electric characterization. A nanotube-based gas sensor adopted this configuration [49.136].

Opened nanotubes (Fig. 49.20c, [49.137]) can serve as an atomic or molecular container. A thermometer based on this structure has been shown by monitoring the height of the gallium inside the nanotube using TEM [49.138].

Part F | 49

Bulk nanotubes can be used to fabricate actuators based on charge-injection-induced bond-length change [49.139], and, theoretically, individual nanotubes also work on the same principle. Electrostatic deflection of a nanotube has been used to construct a relay [49.140]. A new family of nanotube actuators can be constructed by taking advantage of the ultra-low inter-layer friction of a multiwalled nanotube. Linear bearings based on telescoping nanotubes have been demonstrated [49.110]. Recently, a micro actuator with a nanotube as a rotation bearing has been demonstrated [49.141]. A preliminary experiment on a promising nanotube linear motor with field emission current serving as position feedback has been shown with nanorobotic manipulation (Fig. 49.20d, [49.137]).

Cantilevered dual nanotubes have been demonstrated as nanotweezers [49.142] and nanoscissors (Fig. 49.20e, [49.91]) by manual and nanorobotic assembly, respectively.

Based on electric resistance change under different temperatures, nanotube thermal probes (Fig. 49.20f) have been demonstrated for measuring the temperature at precise locations. These thermal probes are more advantageous than nanotube-based thermometers because the thermometers require TEM imaging. The probes also have better reproducibility than devices based on dielectrophoretically assembled bulk nanotubes [49.143]. Gas sensors and hot-wire-based mass-flow sensors can also be constructed in this configuration rather than a bridged one.

The integration of these devices can be realized using the configurations shown in Figs. 49.20g [49.144] and 49.20h [49.90]. Arrays of individual nanotubes can also be used to fabricate nanosensors, such as position encoders [49.145].

Nanotube-based NEMS remains a rich research field with a large number of open problems. New materials and effects at the nanoscale will enable a new family of sensors and actuators for the detection and actuation of ultra-small quantities or objects with ultra-high precision and frequencies. Through random spreading, direct growth, and nanorobotic manipulation, prototypes have been demonstrated. However, for integration into NEMS, self-assembly processes will become increasingly important. Among them, we believe that dielectrophoretic nanoassembly will play a significant role for large-scale production of 2-D regular structures.

References

49.1   R. P. Feynman: There's plenty of room at the bottom, Caltech's Eng Sci **23**, 22–36 (1960)

49.2   M. C. Roco, R. S. Williams, P. Alivisatos: *Nanotechnology Research Directions: Interagency Working Group on Nanoscience, Engineering and Technology (IWGN) Workshop Report (Vision for Nanotechnology R&D in the Next Decade)* (Kluwer Academic, Dordrecht 2000)

49.3   Committee on Technology, M. L. Downey, D. T. Moore, G. R. Bachula, D. M. Etter, E. F. Carey, L. A. Perine et al.: *National Nanotechnology Initiative: Leading to the Next Industrial Revolution, A Report by the Interagency Working Group on Nanoscience, Engineering and Technology* (National Science and Technology Council, Washington D.C. 2000)

49.4   K. Drexler: *Nanosystems: Molecular Machinery, Manufacturing and Computation* (Wiley Interscience, New York 1992)

49.5   G. Binnig, H. Rohrer, Ch. Gerber, E. Weibel: Surface studies by scanning tunneling microscopy, Phys. Rev. Lett. **49**, 57–61 (1982)

49.6   W. F. Degrado: Design of peptides and proteins, Adv. Protein Chem. **39**, 51–124 (1998)

49.7   J.-M. Lehn: *Supramolecular Chemistry: Concepts and Perspectives* (VCH, Weinheim 1995)

49.8   G. M. Whitesides, B. Grzybowski: Self-assembly at all scales, Science **295**, 2418–2421 (2002)

49.9   M. Fujita, N. Fujita, K. Ogura, K. Yamaguchi: Spontaneous assembling of ten small components into a three-dimensionally interlocked compound consisting of the same two cage frameworks, Nature **400**, 52–55 (1999)

49.10   C. Liu, T. Tsao, Y.-C. Tai, C.-M. Ho: *Surface micromachined magnetic actuators*, Proc. 7th IEEE Int. Conf. Micro Electro Mechanical Systems (IEEE, Piscataway, Oiso 1994) pp. 57–62

49.11   J. Judy, D. L. Polla, W. P. Robbins: A linear piezoelectric stepper motor with submicron displacement and centimeter travel, IEEE Trans. Ultrason. Ferroelectrics Freq. Control **37**, 428–437 (1990)

49.12   K. Nakamura, H. Ogura, S. Maeda, U. Sangawa, S. Aoki, T. Sato: *Evaluation of the micro wobbler motor fabricated by concentric buildup process*, Proc. 8th IEEE Int. Conf. Micro Electro Mechanical Systems (IEEE, Piscataway, Amsterdam 1995) pp. 374–379

49.13   A. Teshigahara, M. Watanabe, N. Kawahara, I. Ohtsuka, T. Hattori: Performance of a 7-mm microfabricated car, IEEE/ASME J. MEMS **4**, 76–80 (1995)

49.14 K. R. Udayakumar, S. F. Bart, A. M. Flynn, J. Chen, L. S. Tavrow, L. E. Cross, R. A. Brooks, D. J. Ehrlich: *Ferroelectric thin-film ultrasonic micromotors*, Proc. 4th IEEE Int. Conf. Micro Electro Mechanical Systems (IEEE, Piscataway, Nara 1991) pp. 109–113

49.15 W. Trimmer, R. Jebens: *Actuators for micro robots*, Proc. of the 1989 IEEE Int. Conf. on Robotics and Automation (IEEE, Piscataway, Scottsdale 1989) pp. 1547–1552

49.16 T. Ebefors, G. Stemme: Microrobotics. In: *The MEMS Handbook*, ed. by M. Gad-el-Hak (CRC, Boca Raton 2002)

49.17 C.-J. Kim, A. P. Pisano, R. S. Muller: Silicon-processed overhanging microgripper, IEEE/ASME J. MEMS **1**, 31–36 (1992)

49.18 P. Dario, R. Valleggi, M. C. Carrozza, M. C. Montesi, M. Cocco: Review—Microactuators for microrobots: A critical survey, J. Micromech. Microeng. **2**, 141–157 (1992)

49.19 I. Shimoyama: *Scaling in microrobots*, Proc. IEEE/RSJ Intelligent Robots and Systems (IEEE, Piscataway, Pittsburgh 1995) pp. 208–211

49.20 R. S. Fearing: Powering 3-dimensional micro-robots: power density limitations. In: *Tutorial on Micro Mechatronics and Micro Robotics*, Proc. IEEE Int. Conf. on Robotics and Automation, ed. by G. Girait, P. Dario (IEEE, Piscataway, Leuven 1998)

49.21 R. G. Gilbertson, J. D. Busch: A survey of micro-actuator technologies for future spacecraft missions, J. Br. Interplanet. Soc. **49**, 129–138 (1996)

49.22 M. Mehregany, P. Nagarkar, S. D. Senturia, J. H. Lang: *Operation of microfabricated harmonic and ordinary side-drive motors*, Proc. 3rd IEEE Int. Conf. Micro Electro Mechanical Systems (IEEE, Piscataway, Napa Valley 1990) pp. 1–8

49.23 Y. C. Tai, L. S. Fan, R. S. Muller: *IC-processed micromotors: design, technology, and testing*, Proc. 2nd IEEE Int. Conf. Micro Electro Mechanical Systems (IEEE, Piscataway, Salt Lake City 1989) pp. 1–6

49.24 T. Ohnstein, T. Fukiura, J. Ridley, U. Bonne: *Micromachined silicon microvalve*, Proc. 3rd IEEE Int. Conf. Micro Electro Mechanical Systems (IEEE, Piscataway, Napa Valley 1990) pp. 95–99

49.25 L. Y. Chen, S. L. Zhang, J. J. Yao, D. C. Thomas, N. C. MacDonald: *Selective chemical vapor deposition of tungsten for microdynamic structures*, Proc. 2nd IEEE Int. Conf. Micro Electro Mechanical Systems (IEEE, Piscataway, Salt Lake City 1989) pp. 82–87

49.26 K. Yanagisawa, H. Kuwano, A. Tago: *An electro-magnetically driven microvalve*, Proc. of the 7th International Conference on Solid-State Sensors and Actuators (IEEE, Piscataway, Yokohama 1993) pp. 102–105

49.27 S. Brand: *New applications of piezo-electric actuators*, Proc. 3rd Int. Conf. on New Actuators (Messe Bremen GmbH, Bremen 1992) p. 59

49.28 M. Esashi, S. Shoji, A. Nakano: *Normally close microvalve and micropump fabricated on a silicon wafer*, Proc. 4th IEEE Int. Conf. Micro Electro Mechanical Systems (IEEE, Piscataway, Salt Lake City 1989) pp. 29–34

49.29 R. Petrucci, K. Simmons: *An introduction to piezoelectric crystals*, Sensors Magazine (Helmers, Peterborough 1994) p. 26

49.30 G. Binnig, C. F. Quate, Ch. Gerber: Atomic force microscope, Phys. Rev. Lett. **56**, 93–96 (1986)

49.31 A. Ashkin, J. M. Dziedzic: Optical trapping and manipulation of viruses and bacteria, Science **235**, 1517–1520 (1987)

49.32 F. H. C. Crick, A. F. W. Hughes: The physical properties of cytoplasm: A study by means of the magnetic particle method, Part I: Experimental, Exp. Cell Res. **1**, 37–80 (1950)

49.33 M. F. Yu, M. J. Dyer, G. D. Skidmore, H. W. Rohrs, X. K. Lu, K. D. Ausman, J. R. Von Ehr, R. S. Ruoff: Three-dimensional manipulation of carbon nanotubes under a scanning electron microscope, Nanotechnology **10**, 244–252 (1999)

49.34 L. X. Dong, F. Arai, T. Fukuda: *3-D nanorobotic manipulation of nano-order objects inside SEM*, Proc. of the 2000 Int. Symp. on Micromechatronics and Human Science (IEEE, Piscataway, Nagoya 2000) pp. 151–156

49.35 D. M. Eigler, E. K. Schweizer: Positioning single atoms with a scanning tunneling microscope, Nature **344**, 524–526 (1990)

49.36 Ph. Avouris: Manipulation of matter at the atomic and molecular levels, Acc. Chem. Res. **28**, 95–102 (1995)

49.37 M. F. Crommie, C. P. Lutz, D. M. Eigler: Confinement of electrons to quantum corrals on a metal surface, Science **262**, 218–220 (1993)

49.38 L. J. Whitman, J. A. Stroscio, R. A. Dragoset, R. J. Celotta: Manipulation of adsorbed atoms and creation of new structures on room-temperature surfaces with a scanning tunneling microscope, Science **251**, 1206–1210 (1991)

49.39 I.-W. Lyo, Ph. Avouris: Field-induced nanometer-scale to atomic-scale manipulation of silicon surfaces with the STM, Science **253**, 173–176 (1991)

49.40 G. Dujardin, R. E. Walkup, P. Avouris: Dissociation of individual molecules with electrons from the tip of a scanning tunneling microscope, Science **255**, 1232–1235 (1992)

49.41 T.-C. Shen, C. Wang, G. C. Abeln, J. R. Tucker, J. W. Lyding, Ph. Avouris, R. E. Walkup: Atomic-scale desorption through electronic and vibrational-excitation mechanisms, Science **268**, 1590–1592 (1995)

49.42 M. T. Cuberes, R. R. Schittler, J. K. Gimzewski: Room-temperature repositioning of individual C_{60} molecules at Cu steps: Operation of a molecular counting device, Appl. Phys. Lett. **69**, 3016–3018 (1996)

Part F | 49

49.43   H. J. Lee, W. Ho: Single-bond formation and characterization with a scanning tunneling microscope, Science **286**, 1719–1722 (1999)

49.44   T. Yamamoto, O. Kurosawa, H. Kabata, N. Shimamoto, M. Washizu: Molecular surgery of DNA based on electrostatic micromanipulation, IEEE Trans. IA **36**, 1010–1017 (2000)

49.45   C. Haber, D. Wirtz: Magnetic tweezers for DNA micromanipulation, Rev. Sci. Instrum. **71**, 4561–4570 (2000)

49.46   D. M. Schaefer, R. Reifenberger, A. Patil, R. P. Andres: Fabrication of two-dimensional arrays of nanometer-size clusters with the atomic force microscope, Appl. Phys. Lett. **66**, 1012–1014 (1995)

49.47   T. Junno, K. Deppert, L. Montelius, L. Samuelson: Controlled manipulation of nanoparticles with an atomic force microscope, Appl. Phys. Lett. **66**, 3627–3629 (1995)

49.48   P. E. Sheehan, C. M. Lieber: Nanomachining, manipulation and fabrication by force microscopy, Nanotechnology **7**, 236–240 (1996)

49.49   C. Baur, B. C. Gazen, B. Koel, T. R. Ramachandran, A. A. G. Requicha, L. Zini: Robotic nanomanipulation with a scanning probe microscope in a networked computing environment, J. Vac. Sci. Technol. B **15**, 1577–1580 (1997)

49.50   R. Resch, C. Baur, A. Bugacov, B. E. Koel, A. Madhukar, A. A. G. Requicha, P. Will: Building and manipulating 3-D and linked 2-D structures of nanoparticles using scanning force microscopy, Langmuir **14**, 6613–6616 (1998)

49.51   J. Hu, Z.-H. Zhang, Z.-Q. Ouyang, S.-F. Chen, M.-Q. Li, F.-J. Yang: Stretch and align virus in nanometer scale on an atomically flat surface, J. Vac. Sci. Technol. B **16**, 2841–2843 (1998)

49.52   M. Sitti, S. Horiguchi, H. Hashimoto: Controlled pushing of nanoparticles: Modeling and experiments, IEEE/ASME Trans. Mechatron. **5**, 199–211 (2000)

49.53   M. Guthold, M. R. Falvo, W. G. Matthews, S. Paulson, S. Washburn, D. A. Erie, R. Superfine, F. P. Brooks Jr., R. M. Taylor II: Controlled manipulation of molecular samples with the nanoManipulator, IEEE/ASME Trans. Mechatron. **5**, 189–198 (2000)

49.54   F. Arai, D. Andou, T. Fukuda: *Micro manipulation based on micro physics – strategy based on attractive force reduction and stress measurement*, Proc. of IEEE/RSJ Int. Conf. on Intelligent Robotics and Systems (IEEE, Piscataway, Pittsburgh 1995) pp. 236–241

49.55   L. X. Dong, F. Arai, T. Fukuda: 3-D nanorobotic manipulations of nanometer scale objects, J. Robotics Mechatron. **13**, 146–153 (2001)

49.56   H. W. P. Koops, J. Kretz, M. Rudolph, M. Weber, G. Dahm, K. L. Lee: Characterization and application of materials grown by electron-

49.57   beam-induced deposition, Jpn. J. Appl. Phys. **33**, 7099–7107 (1994)

49.57   S. Iijima: Helical microtubules of graphitic carbon, Nature **354**, 56–58 (1991)

49.58   S. J. Tans, A. R. M. Verchueren, C. Dekker: Room-temperature transistor based on a single carbon nanotube, Nature **393**, 49–52 (1998)

49.59   Y. Huang, X. F. Duan, Y. Cui, L. J. Lauhon, K.-H. Kim, C. M. Lieber: Logic gates and computation from assembled nanowire building blocks, Science **294**, 1313–1317 (2001)

49.60   A. Bachtold, P. Hadley, T. Nakanishi, C. Dekker: Logic circuits with carbon nanotube transistors, Science **294**, 1317–1320 (2001)

49.61   R. H. Baughman, A. A. Zakhidov, W. A. de Heer: Carbon nanotubes—the route toward applications, Science **297**, 787–792 (2002)

49.62   N. Wang, Z. K. Tang, G. D. Li, J. S. Chen: Single-walled 4Å carbon nanotube arrays, Nature **408**, 50–51 (2000)

49.63   Z. W. Pan, S. S. Xie, B. H. Chang, C. Y. Wang, L. Lu, W. Liu, W. Y. Zhou, W. Z. Li, L. X. Qian: Very long carbon nanotubes, Nature **394**, 631–632 (1998)

49.64   H. W. Zhu, C. L. Xu, D. H. Wu, B. Q. Wei, R. Vajtai, P. M. Ajayan: Direct synthesis of long single-walled carbon nanotube strands, Science **296**, 884–886 (2002)

49.65   M. J. Treacy, T. W. Ebbesen, J. M. Gibson: Exceptionally high Young's modulus observed for individual carbon nanotubes, Nature **381**, 678–680 (1996)

49.66   E. W. Wong, P. E. Sheehan, C. M. Lieber: Nanobeam mechanics: Elasticity, strength, and toughness of nanorods and nanotubes, Science **277**, 1971–1975 (1997)

49.67   P. Poncharal, Z. L. Wang, D. Ugarte, W. A. de Heer: Electrostatic deflections and electromechanical resonances of carbon nanotubes, Science **283**, 1513–1516 (1999)

49.68   M. F. Yu, O. Lourie, M. J. Dyer, K. Moloni, T. F. Kelley, R. S. Ruoff: Strength and breaking mechanism of multiwalled carbon nanotubes under tensile load, Science **287**, 637–640 (2000)

49.69   A. Krishnan, E. Dujardin, T. W. Ebbesen, P. N. Yianilos, M. M. J. Treacy: Young's modulus of single-walled nanotubes, Phys. Rev. B **58**, 14 013–14 019 (1998)

49.70   J. P. Salvetat, G. A. D. Briggs, J.-M. Bonard, R. R. Bacsa, A. J. Kulik, T. Stockli, N. A. Burnham, L. Forro: Elastic and shear moduli of single-walled carbon nanotube ropes, Phys. Rev. Lett. **82**, 944–947 (1999)

49.71   M. F. Yu, B. S. Files, S. Arepalli, R. S. Ruoff: Tensile loading of ropes of single wall carbon nanotubes and their mechanical properties, Phys. Rev. Lett. **84**, 5552–5555 (2000)

49.72   D. A. Walters, L. M. Ericson, M. J. Casavant, J. Liu, D. T. Colbert, K. A. Smith, R. E. Smalley: Elastic strain of freely suspended single-wall carbon nanotube ropes, Appl. Phys. Lett. **74**, 3803–3805 (1999)

49.73 R. Saito, M. Fujita, G. Dresselhaus, M.S. Dresselhaus: Electronic structure of graphene tubules based on C_{60}, Phys. Rev. B. **46**, 1804–1811 (1992)

49.74 T.W. Ebbesen, H.J. Lezec, H. Hiura, J.W. Bennett, H.F. Ghaemi, T. Thio: Electrical conductivity of individual carbon nanotubes, Nature **382**, 54–56 (1996)

49.75 H.J. Dai, E.W. Wong, C.M. Lieber: Probing electrical transport in nanomaterials: conductivity of individual carbon nanotubes, Science **272**, 523–526 (1996)

49.76 P. Kim, L. Shi, A. Majumdar, P.L. McEuen: Thermal transport measurements of individual multiwalled nanotubes, Phys. Rev. Lett. **87**, 215502 (2001)

49.77 W.J. Liang, M. Bockrath, D. Bozovic, J.H. Hafner, M. Tinkham, H. Park: Fabry–Perot interference in a nanotube electron waveguide, Nature **411**, 665–669 (2001)

49.78 S. Frank, P. Poncharal, Z.L. Wang, W.A. de Heer: Carbon nanotube quantum resistors, Science **280**, 1744–1746 (1998)

49.79 X.B. Zhang, D. Bernaerts, G. Van Tendeloo, S. Amelincks, J. Van Landuyt, V. Ivanov, J.B. Nagy, Ph. Lambin, A.A. Lucas: The texture of catalytically grown coil-shaped carbon nanotubules, Europhys. Lett. **27**, 141–146 (1994)

49.80 X.Y. Kong, Z.L. Wang: Spontaneous polarization-induced nanohelixes, nanosprings, and nanorings of piezoelectric nanobelts, Nano. Lett. **3**, 1625–1631 (2003)

49.81 S.V. Golod, V.Ya. Prinz, V.I. Mashanov, A.K. Gutakovsky: Fabrication of conducting GeSi/Si micro- and nanotubes and helical microcoils, Semicond. Sci. Technol. **16**, 181–185 (2001)

49.82 L. Zhang, E. Deckhardt, A. Weber, C. Schönenberger, D. Grützmacher: Controllable fabrication of SiGe/Si and SiGe/Si/Cr helical nanobelts, Nanotechnology **16**, 655–663 (2005)

49.83 D.J. Bell, L.X. Dong, Y. Sun, L. Zhang, B.J. Nelson, D. Grützmacher: *Manipulation of nanocoils for nanoelectromagnets*, Proc. 5th IEEE Conf. Nanotechnol. (IEEE, Piscataway, Nagoya 2005) pp.149–152

49.84 D.J. Bell, Y. Sun, L. Zhang, L.X. Dong, B.J. Nelson, D. Grützmacher: *Three-dimensional nanosprings for electromechanical sensors*, Proc. 13th Int. Conf. Solid-State Sensors, Actuators and Microsystems (IEEE, Piscataway, Seoul 2005) pp.15–18

49.85 S. Iijima, T. Ichihashi: Single-shell carbon nanotubes of 1-nm diameter, Nature **363**, 603–605 (1993)

49.86 D.S. Bethune, C.H. Kiang, M.S. de Vries, G. Gorman, R. Savoy, J. Vazquez, R. Beyers: Cobalt-catalysed growth of carbon nanotubes with single-atomic-layer walls, Nature **363**, 605–607 (1993)

49.87 R. Martel, T. Schmidt, H.R. Shea, T. Hertel, Ph. Avouris: Single- and multi-wall carbon nanotube field-effect transistors, Appl. Phys. Lett. **73**, 2447–2449 (1998)

49.88 N.R. Franklin, Y.M. Li, R.J. Chen, A. Javey, H.J. Dai: Patterned growth of single-walled carbon nanotubes on full 4-inch wafers, Appl. Phys. Lett. **79**, 4571–4573 (2001)

49.89 T. Rueckes, K. Kim, E. Joselevich, G.Y. Tseng, C.-Li Cheung, C.M. Lieber: Carbon nanotube-based non-volatile random access memory for molecular computing, Science **289**, 94–97 (2000)

49.90 A. Subramanian, B. Vikramaditya, L.X. Dong, D. Bell, B.J. Nelson: *Micro and Nanorobotic Assembly Using Dielectrophoresis. Robotics: Science and Systems I*, ed. by S. Thrun, G.S. Sukhatme, S. Schaal, O. Brock (MIT Press, Cambridge 2005) pp.327–334

49.91 T. Fukuda, F. Arai, L.X. Dong: Assembly of nanodevices with carbon nanotubes through nanorobotic manipulations, Proc. IEEE **91**, 1803–1818 (2003)

49.92 M.R. Falvo, G.J. Clary, R.M. Taylor, V. Chi, F.P. Brooks, S. Washburn, R. Superfine: Bending and buckling of carbon nanotubes under large strain, Nature **389**, 582–584 (1997)

49.93 H.W.C. Postma, A. Sellmeijer, C. Dekker: Manipulation and imaging of individual single-walled carbon nanotubes with an atomic force microscope, Adv. Mater. **12**, 1299–1302 (2000)

49.94 H.W.Ch. Postma, M. de Jonge, Z. Yao, C. Dekker: Electrical transport through carbon nanotube junctions created by mechanical manipulation, Phys. Rev. B **62**, R10653–R10656 (2000)

49.95 T. Hertel, R. Martel, Ph. Avouris: Manipulation of individual carbon nanotubes and their interaction with surfaces, J. Phys. Chem. B **102**, 910–915 (1998)

49.96 Ph. Avouris, T. Hertel, R. Martel, T. Schmidt, H.R. Shea, R.E. Walkup: Carbon nanotubes: nanomechanics, manipulation, and electronic devices, Appl. Surf. Sci. **141**, 201–209 (1999)

49.97 L. Roschier, J. Penttila, M. Martin, P. Hakonen, M. Paalanen, U. Tapper, E.I. Kauppinen, C. Journet, P. Bernier: Single-electron transistor made of multiwalled carbon nanotube using scanning probe manipulation, Appl. Phys. Lett. **75**, 728–730 (1999)

49.98 M. Ahlskog, R. Tarkiainen, L. Roschier, P. Hakonen: Single-electron transistor made of two crossing multiwalled carbon nanotubes and its noise properties, Appl. Phys. Lett. **77**, 4037–4039 (2000)

49.99 M.R. Falvo, R.M. Taylor II, A. Helser, V. Chi, F.P. Brooks Jr, S. Washburn, R. Superfine: Nanometre-scale rolling and sliding of carbon nanotubes, Nature **397**, 236–238 (1999)

49.100 M.R. Falvo, J. Steele, R.M. Taylor II, R. Superfine: Gearlike rolling motion mediated by commensurate contact: Carbon nanotubes on HOPG, Phys. Rev. B **62**, R10665–R10667 (2000)

49.101 L.X. Dong: *Nanorobotic manipulations of carbon nanotubes* (Nagoya University, Dissertation 2003)

49.102 J.H. Hafner, C.-L. Cheung, T.H. Oosterkamp, C.M. Lieber: High-yield assembly of individual single-walled carbon nanotube tips for scanning probe microscopies, J. Phys. Chem. B **105**, 743–746 (2001)

49.103 L.X. Dong, F. Arai, T. Fukuda: Electron-beam-induced deposition with carbon nanotube emitters, Appl. Phys. Lett. **81**, 1919–1921 (2002)

49.104 L.X. Dong, F. Arai, T. Fukuda: *3-D nanorobotic manipulations of multi-walled carbon nanotubes*, Proc. 2001 IEEE Int. Conf. Robotics and Automation (ICRA2001) (IEEE, Piscataway, Seoul 2001) pp. 632–637

49.105 L.X. Dong, F. Arai, T. Fukuda: Three-dimensional nanorobotic manipulations of carbon nanotubes, J. Robotics Mechatron JSME **14**, 245–252 (2002)

49.106 L.X. Dong, F. Arai, T. Fukuda: *Inter-process measurement of MWNT rigidity and fabrication of MWNT junctions through nanorobotic manipulations*, Conf. Proc. 590: Nanonetwork Materials: Fullerenes, Nanotubes, and Related Materials (AIP, Melville 2001) pp. 71–74

49.107 L.X. Dong, F. Arai, T. Fukuda: *Shape modification of carbon nanotubes and its applications in nanotube scissors*, Proc. IEEE Int. Conf. Nanotechnology (IEEE-NANO2002) (IEEE, Piscataway, Washington 2002) pp. 443–446

49.108 L.X. Dong, F. Arai, T. Fukuda: Destructive constructions of nanostructures with carbon nanotubes through nanorobotic manipulations, IEEE/ASME Trans. Mechatron. **9**, 350–357 (2004)

49.109 J. Cumings, P.G. Collins, A. Zettl: Peeling and sharpening multiwall nanotubes, Nature **406**, 58 (2000)

49.110 J. Cumings, A. Zettl: Low-friction nanoscale linear bearing realized from multiwall carbon nanotubes, Science **289**, 602–604 (2000)

49.111 L.X. Dong, F. Arai, T. Fukuda: *3-D nanoassembly of carbon nanotubes through nanorobotic manipulations*, Proc. 2002 IEEE Int. Conf. Robotics and Automation (ICRA2002) (IEEE, Piscataway, Washington 2002) pp. 1477–1482

49.112 L.X. Dong, F. Arai, T. Fukuda: Mechanochemical nanorobotic manipulations of carbon nanotubes, Jap. J. Appl. Phys. **42**, 295–298 (2003)

49.113 R. Saito, G. Dresselhaus, M.S. Dresselhaus: Tunneling conductance of connected carbon nanotubes, Phys. Rev. B **53**, 2044–2050 (1996)

49.114 L. Chico, V.H. Crespi, L.X. Benedict, S.G. Louie, M.L. Cohen: Pure carbon nanoscale devices: nanotube heterojunctions, Phys. Rev. Lett. **76**, 971–974 (1996)

49.115 M. Menon, D. Srivastava: Carbon nanotube 'T junctions': nanoscale metal-semiconductor-metal contact devices, Phys. Rev. Lett. **79**, 4453–4456 (1997)

49.116 Z. Yao, H.W.Ch. Postma, L. Balents, C. Dekker: Carbon nanotube intramolecular junctions, Nature **402**, 273–276 (1999)

49.117 H.W.Ch. Postma, T. Teepen, Z. Yao, M. Grifoni, C. Dekker: Carbon nanotube single-electron transistors at room temperature, Science **293**, 76–79 (2001)

49.118 M.S. Fuhrer, L. Shih, M. Forero, Y.-G. Yoon, M.S.C. Mazzoni, H.J. Choi, J. Ihm, S.G. Louie, A. Zettl, P.L. McEuen: Crossed nanotube junctions, Science **288**, 494–497 (2000)

49.119 T. Rueckes, K. Kim, E. Joselevich, G.Y. Treng, C.L. Cheung, C.M. Lieber: Carbon nanotube-based nonvolatile random access memory for molecular computing science, Science **289**, 94–97 (2000)

49.120 A.G. Rinzler, J.H. Hafner, P. Nikolaev, L. Lou, S.G. Kim, D. Tománek, P. Nordlander, D.T. Colbert, R.E. Smalley: Unraveling nanotubes: field emission from an atomic wire, Science **269**, 1550–1553 (1995)

49.121 Y. Saito, K. Hamaguchi, K. Hata, K. Uchida, Y. Tasaka, F. Ikazaki, M. Yumura, A. Kasaya, Y. Nishina: Conical beams from open nanotubes, Nature **389**, 554 (1997)

49.122 S.C. Minne, G. Yaralioglu, S.R. Manalis, J.D. Adams, J. Zesch, A. Atalar, C.F. Quate: Automated parallel high-speed atomic force microscopy, Appl. Phys. Lett. **72**, 2340–2342 (1998)

49.123 G.D. Skidmore, E. Parker, M. Ellis, N. Sarkar, R. Merkle: Exponential assembly, Nanotechnology **11**, 316–321 (2001)

49.124 Y. Cui, Q.Q. Wei, H.K. Park, C.M. Lieber: Nanowire nanosensors for highly sensitive and selective detection of biological and chemical species, Science **293**, 1289–1292 (2001)

49.125 Y. Sun, B.J. Nelson: *Microrobotic cell injection*, Proc. 2001 IEEE Int. Conf. Robotics and Automation (ICRA2001) (IEEE, Piscataway, Seoul 2001) pp. 620–625

49.126 Y. Sun, B.J. Nelson: *Autonomous injection of biological cells using visual servoing*, Int. Symp. on Experimental Robotics (ISER) (Lecture Notes Control Inform. Sci., Hawaii 2000) pp. 175–184

49.127 Y. Sun, B.J. Nelson, D.P. Potasek, E. Enikov: A bulk microfabricated multi-axis capacitive cellular force sensor using transverse comb drives, J. Micromech. Microeng. **12**, 832–840 (2002)

49.128 Y. Sun, K. Wan, K.P. Roberts, J.C. Bischof, B.J. Nelson: Mechanical property characterization of mouse zona pellucida, IEEE Trans. Nanobiosci. **2**, 279–286 (2003)

49.129 H.J. Dai, J.H. Hafner, A.G. Rinzler, D.T. Colbert, R.E. Smalley: Nanotubes as nanoprobes in scanning probe microscopy, Nature **384**, 147–150 (1996)

49.130 J.H. Hafner, C.L. Cheung, C.M. Lieber: Growth of nanotubes for probe microscopy tips, Nature **398**, 761–762 (1999)

49.131 H. Nishijima, S. Kamo, S. Akita, Y. Nakayama, K. I. Hohmura, S. H. Yoshimura, K. Takeyasu: Carbon-nanotube tips for scanning probe microscopy: preparation by a controlled process and observation of deoxyribonucleic acid, Appl. Phys. Lett. **74**, 4061–4063 (1999)

49.132 P. Poncharal, Z. L. Wang, D. Ugarte, W. A. de Heer: Electrostatic deflections and electromechanical resonances of carbon nanotubes, Science **283**, 1513–1516 (1999)

49.133 L. X. Dong, F. Arai, M. Nakajima, P. Liu, T. Fukuda: *Nanotube devices fabricated in a nano laboratory*, Proc. of the 2003 IEEE Int. Conf. Robotics and Automation (ICRA2003) (IEEE, Piscataway, Taipei 2003) pp. 3624–3629

49.134 M. Nakajima, F. Arai, L. X. Dong, T. Fukuda: Pico-Newton order force measurement using a calibrated carbon nanotube probe inside a scanning electron microscope, J. Robotics Mechatron. (JSME) **16**(2), 155–162 (2004)

49.135 A. Subramanian, B. J. Nelson, L. X. Dong, D. Bell: *Dielectrophoretic nanoassembly of nanotube-based NEMS with nanoelectrodes*, 6th IEEE Int. Symp. Assembly and Task Planning (IEEE, Piscataway, Montréal 2005) pp. 200–205

49.136 J. Kong, N. R. Franklin, C. W. Zhou, M. G. Chapline, S. Peng, K. J. Cho, H. J. Dai: Nanotube molecular wires as chemical sensors, Science **287**, 622–625 (2000)

49.137 L. X. Dong, B. J. Nelson, T. Fukuda, F. Arai: *Towards linear nano servomotors with integrated position sensing*, Proc. 2005 IEEE Int. Conf. Robotics and Automation (IEEE, Piscataway, Barcelona 2005) pp. 867–872

49.138 Y. H. Gao, Y. Bando: Carbon nanothermometer containing gallium, Nature **415**, 599 (2002)

49.139 R. H. Baughman, C. X. Cui, A. A. Zakhidov, Z. Iqbal, J. N. Barisci, G. M. Spinks, G. G. Wallace, A. Mazzoldi, D. De Rossi, A. G. Rinzler, O. Jaschinski, S. Roth, M. Kertesz: Carbon nanotube actuators, Science **284**, 1340–1344 (1999)

49.140 S. W. Lee, D. S. Lee, R. E. Morjan, S. H. Jhang, M. Sveningsson, O. A. Nerushev, Y. W. Park, E. E. B. Campbell: A three-terminal carbon nanorelay, Nano Lett. **4**, 2027–2030 (2004)

49.141 A. M. Fennimore, T. D. Yuzvinsky, W.-Q. Han, M. S. Fuhrer, J. Cumings, A. Zettl: Rotational actuators based on carbon nanotubes, Nature **424**, 408–410 (2003)

49.142 P. Kim, C. M. Lieber: Nanotube nanotweezers, Science **286**, 2148–2150 (1999)

49.143 C. K. M. Fung, V. T. S. Wong, R. H. M. Chan, W. J. Li: Dielectrophoretic batch fabrication of bundled carbon nanotube thermal sensors, IEEE Trans. Nanotech. **3**, 395–403 (2004)

49.144 A. Subramanian, L. X. Dong, B. J. Nelson: *Selective eradication of individual nanotubes from vertically aligned arrays*, Proc. 2005 IEEE/ASME Int. Conf. Advanced Intelligent Mechatronics (IEEE, Piscataway, Monterey 2005) pp. 105–110

49.145 L. X. Dong, A. Subramanian, B. J. Nelson: *Nano encoders based on arrays of single nanotube emitter*, Proc. 5th IEEE Conf. Nanotechnology (IEEE, Piscataway, Nagoya 2005) pp. 211–214

Part F | 49

Micro/Nan Part G

Part G Micro/Nanodevice Reliability

50. Nanotribology and Materials Characterization of MEMS/NEMS and BioMEMS/BioNEMS Materials and Devices

Micro/nanoelectromechanical systems (MEMS/NEMS) need to be designed to perform their expected functions with short duration, typically in millisecond to picosecond timescales. Expected life of the devices for high-speed contacts can vary from a few hundred thousand to many billions of cycles, e.g., over a hundred billion cycles for digital micromirror devices (DMDs), which puts serious requirements on materials. For BioMEMS/BioNEMS, adhesion between biological molecular layers and the substrate, and friction and wear of biological layers may be important. There is a need for the development of a fundamental understanding of adhesion, friction/stiction, wear, and the role of surface contamination, and environment. Most mechanical properties are known to be scale-dependent. Therefore, the properties of nanoscale structures need to be measured. MEMS/NEMS materials need to exhibit good mechanical and tribological properties on the micro/nanoscale. There is a need to develop lubricants and identify lubrication methods that are suitable for MEMS/NEMS. Methods need to be developed to enhance adhesion between biomolecules and the device substrate. Component-level studies are required to provide a better understanding of tribological phenomena occurring in MEMS/NEMS. The emergence of micro/nanotribology and techniques based on atomic force microscopy has provided researchers with a viable approach to address these problems. This chapter presents a review of micro/nanoscale adhesion, friction, and wear studies of materials and lubrication studies for MEMS/NEMS and BioMEMS/BioNEMS, and component-level studies of stiction phenomena in MEMS/NEMS devices.

50.1 Introduction

Microelectromechanical systems (MEMS) refers to microscopic devices that have a characteristic length of less than 1 mm but more than 100 nm and combine electrical and mechanical components. Nanoelectromechanical systems (NEMS) refer to nanoscopic devices that have a characteristic length of less than 100 nm and combine electrical and mechanical components. In mesoscale devices, if the functional components are on the micro- or nanoscale, they may be referred to as MEMS or NEMS, respectively. These are referred to as an intelligent miniaturized system consisting of sensing, processing, and/or actuating functions and combining electrical and mechanical components. The acronym MEMS originated in the U.S.A. The term commonly used in Europe is microsystem technology (MST) and in Japan is micromachines. Another term generally used is micro/nanodevices. The terms MEMS/NEMS are also now used in a broad sense and include electrical, mechanical, fluidic, optical, and/or biological functions. MEMS/NEMS for optical applications are referred to as micro/nano-optoelectromechanical systems (MOEMS/NOEMS). MEMS/NEMS for electronic applications are referred to as radio-frequency MEMS/NEMS (RF-MEMS/RF-NEMS). MEMS/NEMS for biological applications are referred to as BioMEMS/BioNEMS.

To put the dimensions of MEMS and NEMS in perspective, see Fig. 50.1 and Table 50.1. Individual atoms are typically a fraction of a nanometer in diameter, deoxyribonucleic acid (DNA) molecules are about 2.5 nm wide, biological cells are in the range of thousands of nm in diameter, and human hair is about 75 μm in diameter. NEMS shown in the figure have a size of 15–300 nm and MEMS have a scale of 12 000 nm. The mass of a micromachined silicon structure can be as low as 1 nN and NEMS can be built with mass as low as 10^{-20} N with cross sections of about 10 nm. In comparison, the mass of a drop of water is about 10 μN and the mass of an eyelash is about 100 nN.

Micro/nanofabrication techniques include top-down methods, in which one builds down from the large to the small, and bottom-up methods, in which one builds up from the small to the large. Top-down methods include micro/nanomachining methods and methods based on lithography as well as non-lithographic miniaturization, mostly for the fabrication of MEMS and a few NEMS. In bottom-up methods, also referred to as nanochemistry, the devices and systems are assembled from their elemental constituents for NEMS fabrication, much like the way nature uses proteins and other macromolecules to construct complex biological systems. The bottom-up approach has the potential to go far beyond the limits of top-down technology by producing nanoscale features through synthesis and subsequent assembly. Furthermore, the bottom-up approach offers the potential to produce structures with enhanced and/or completely new functions. It allows combination of materials with distinct chemical composition, structure, and morphology. For a brief overview of fabrication techniques, see Sect. 50.A.

Table 50.1 Dimensions and masses in perspective

(a) Dimensions in perspective	
NEMS characteristic length: < 100 nm	
MEMS characteristic length: < 1 mm and > 100 nm	
Molecular gear	≈ 10 nm
Vertical SWCNT transistor	≈ 15 nm
Quantum-dot transistor	300 nm
Digital micromirror	12 000 nm
Individual atoms	typically a fraction of a nm in diameter
DNA molecules	≈ 2.5 nm wide
Biological cells	in the range of thousands of nm in diameter
Human hair	≈ 75 000 nm in diameter
(b) Masses in perspective	
NEMS built with cross sections of about 10 nm – as low as 10^{-20} N	
Micromachines silicon structure – as low as 1 nN	
Water droplet	≈ 10 μN
Eyelash	≈ 100 nN

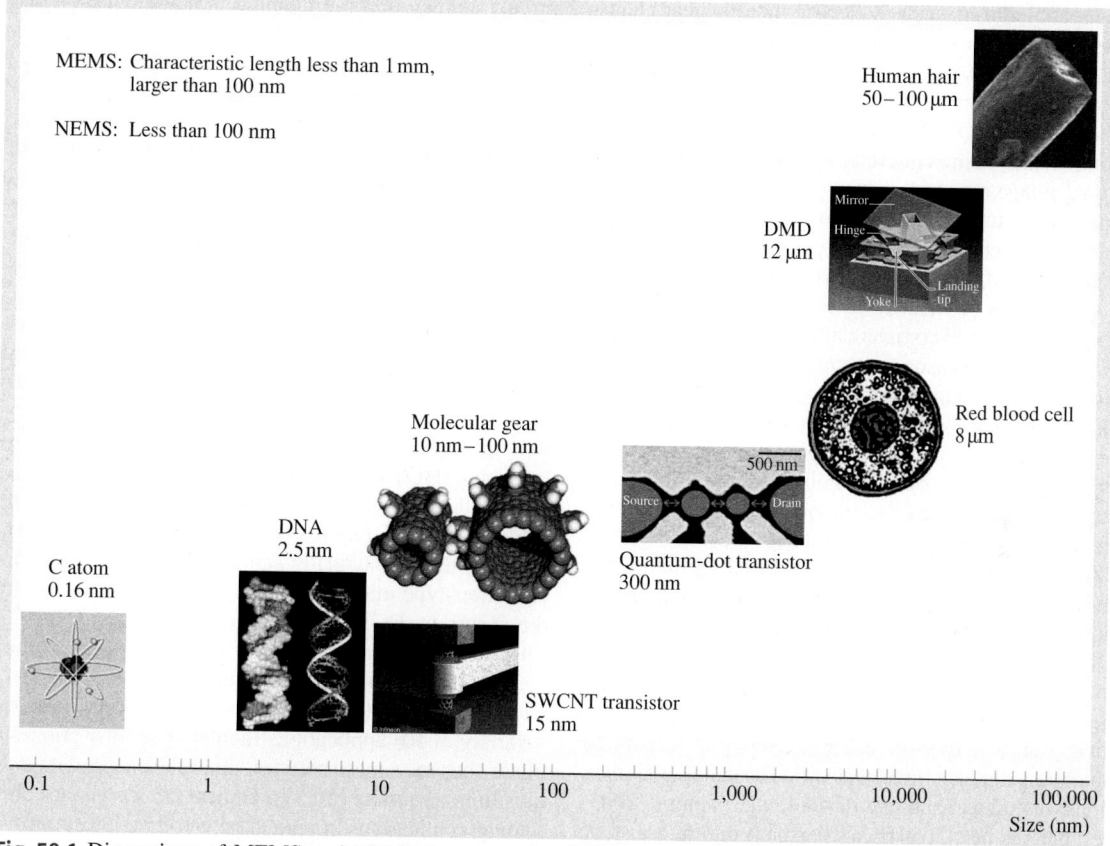

Fig. 50.1 Dimensions of MEMS and NEMS in perspective. MEMS/NEMS examples shown are of a vertical single-walled carbon nanotube (SWCNT) transistor (5 nm wide and 15 nm high) [50.7], of molecular dynamic simulations of a carbon-nanotube-based gear [50.8], quantum-dot transistor obtained from van der *Wiel* et al. [50.9], and DMD obtained from www.dlp.com

MEMS and emerging NEMS are expected to have a major impact on our lives, comparable to that of semiconductor technology, information technology, or cellular and molecular biology [50.1–4]. MEMS/NEMS are used in electromechanical, electronics, information/communication, chemical, and biological applications. The MEMS industry in 2004 was worth about $4.5 billion, with a projected annual growth rate of 17% [50.5]. Growth of Si-based MEMS may slow down and non-silicon MEMS may pick up during this decade. The NEMS industry was worth about $10 billion in 2004, mostly in nanomaterials [50.6]. It is expected to expand in this decade, in nanomaterials, biomedical applications as well as in nanoelectronics or molecular electronics. Due to the enabling nature of these systems and because of the significant impact they can have on both commercial and defense applications, industry as well as federal governments have taken special interest

in seeing growth nurtured in this field. MEMS/NEMS are the next logical step in the silicon revolution.

50.1.1 Introduction to MEMS

Advances in silicon photolithographic process technology since the 1960s led to the development of MEMS in the early 1980s. More recently, lithographic processes have also been developed to process non-silicon materials. The lithographic processes are being complemented with non-lithographic processes for fabrication of components or devices made from plastics or ceramics. Using these fabrication processes, researchers have fabricated a wide variety of devices with dimensions in the range of submicron to a few thousand microns (see e.g., [50.10–20]). MEMS for mechanical applications include acceleration, pressure, flow, and gas sensors, linear and rotary actuators, and other microstructures or microcomponents such as electric motors, gear chains, gas turbine engines, fluid pumps, fluid valves, switches, grippers, and tweezers. MEMS for chemical applications include chemical sensors and various analytical instruments. MOEMS are devices that include optical components, such as micromirror arrays for displays, infrared image sensors, spectrometers, bar-code readers, and optical switches. RF-MEMS include inductors, capacitors, antennas, and RF switches. High-aspect-ratio MEMS (HARMEMS) have also been introduced.

A variety of MEMS devices have been produced and some are used commercially [50.11, 13–16, 18–20]. A variety of sensors are used in industrial, consumer, defense, and biomedical applications. The largest *killer* industrial applications include accelerometers, pressure sensors, thermal and piezoelectric-type inkjet printheads, and digital micromirror devices. Integrated capacitive-type silicon accelerometers have been used in airbag deployment in automobiles since 1991 [50.21, 22]. Some 90 million units were installed in vehicles in 2004. Accelerometer technology was over a billion-dollar-a-year industry in 2004, dominated by Analog Devices followed by Freescale Semiconductor (formerly Motorola) and Bosch. These accelerometers are being used for many other applications such as vehicle stability, rollover control, and gyro sensors for automotive applications, and various consumer applications, including handheld devices, e.g., laptops (2003), cellular phones (2004), and personal digital assistants (PDAs) Silicon-based piezoresistive pressure sensors were launched in 1990 by GE NovaSensor for manifold absolute pressure (MAP) sensing for engines and for disposable blood-pressure sensors, and their annual sales were more than 30 million units and more than 25 million units, respectively, in 2004. MPA sensors measure the pressure in the intake manifold, which is fed to a computer that determines the optimum air/fuel mixture to maximize fuel economy. Most vehicles have these as part of the electronic engine-control system. Capacitive pressure sensors for tire-pressure measurements were launched by Freescale Semiconductor (formerly Motorola) in early 2000 and are also manufactured by Infineon/SensoNor and GE Novasensor (2003). Piezoresistive-type sensors are also used, manufactured by various companies such as EnTire Solutions (2003). The sensing module is located inside the rim of the wheel and relays the information via radio-frequency link to a central processing unit (CPU) in order to display it to the driver. In 2005, about 9.2 million vehicles were equipped with sensors, which translates to about 37 million units. These sales are expected to grow rapidly as they become required in automobiles. They will be required in the U.S. by 2008, which will affect 17 million vehicles (one in each tire) sold every year. Thermal inkjet printers were developed independently by HP and Cannon and commercialized in 1984 [50.23–26] and today are made by Canon, Epson, HP, Lexmark, Xerox, and others. They typically cost less initially than dry-toner laser printers and are the solution of choice for low-volume print runs. Annual sales of thermal inkjet printheads with microscale functional components were more than 500 million units in 2004.

Micromirror arrays are used for displays. Commercial digital light processing (DLP) equipments, using digital micromirror devices (DMD), were launched in 1996 by Texas Instruments for digital projection displays in computer projectors, high-definition television (HDTV) sets and movie projectors (DLP cinema) [50.27–29]. Several million projectors had been sold up to 2004 (about $700 million revenue for TI in 2004). Electrostatically actuated, membrane-type or cantilever-type microswitches have been developed for direct current (DC), RF, and optical applications [50.30]. There exists two basic forms of RF microswitches: the metal-to-metal contact microswitch (ohmic) and the capacitive microswitch. RF microswitches can be used in a variety of RF applications including cellular phones, phase shifters, smart antennas, multiplexers for data acquisition, and more [50.31]. Optical microswitches are finding applications in optical networking, telecommunications, and wireless technologies [50.30, 32].

Other applications of MEMS devices include chemical/biological and gas sensors [50.20, 33], microresonators, infrared detectors and focal-plane arrays for earth observation, space science and missile defense applications, pico-satellites for space applications, fuel cells, and many hydraulic, pneumatic, and other consumer products. MEMS devices are also being pursued in magnetic storage systems [50.34], where they are being developed for super-compact ultrahigh-recording-density magnetic disk drives. Several integrated head/suspension microdevices have been fabricated for contact recording applications [50.35]. High-bandwidth servo-controlled microactuators have been fabricated for ultrahigh-track-density applications which serve as the fine-position control element of a two-stage, coarse/fine servo system, coupled with a conventional actuator [50.36, 37].

Micro/nano-instruments and micro/nano-manipulators are used to move, position, probe, pattern, and characterize nanoscale objects and nanoscale features [50.38]. Miniaturized analytical equipments include gas chromatography and mass spectrometry. Other instruments include micro-STM, where STM stands for scanning tunneling microscope.

In some cases, MEMS devices are used primarily for their miniature size, while in others, as in the case of the air bags, because of their low-cost manufacturing techniques. This latter fact has been possible since semiconductor processing costs have reduced drastically over the last decade, allowing the use of MEMS in many fields.

50.1.2 Introduction to NEMS

NEMS are produced by nanomachining in a typical top-down approach (from large to small) and bottom-up approach (from small to large), largely relying on nanochemistry (see e.g., [50.39–45]). The NEMS field, in addition to fabrication of nanosystems, has provided impetus to the development of experimental and computation tools. Examples of NEMS include microcantilevers with integrated sharp nanotips for STM and atomic force microscopy (AFM) [50.46, 47], quantum corrals formed using STM by placing atoms one by one [50.48], AFM cantilever arrays (Millipede) for data storage [50.49], STM and AFM tips for nanolithography, dip-pen nanolithography for printing molecules, nanowires, carbon nanotubes, quantum wires (QWRs), quantum boxes (QBs), quantum transistors [50.9], nanotube-based sensors [50.7, 50], biological (DNA) motors, molecular gears made by attaching benzene molecules to the outer walls of carbon nanotubes [50.8], devices incorporating nm-thick films [e.g., in giant-magnetoresistive (GMR) read/write magnetic heads and magnetic media for magnetic rigid disk and magnetic tape drives], nanopatterned magnetic rigid disks, and nanoparticles (e.g., nanoparticles in magnetic tape substrates and nanomagnetic particles in magnetic tape coatings) [50.34, 51]. More than 2 billion read/write magnetic heads were shipped for magnetic disk and tape drives in 2004.

Nanoelectronics can be used to build computer memory using individual molecules or nanotubes to store bits of information [50.52], molecular switches, molecular or nanotube transistors, nanotube flat-panel displays, nanotube integrated circuits, fast logic gates, switches, nanoscopic lasers, and nanotubes as electrodes in fuel cells.

50.1.3 BioMEMS/BioNEMS

BioMEMS/BioNEMS are increasingly used in commercial and defense applications (see e.g., [50.53–60]).

They are used for chemical and biochemical analyses (biosensors) in medical diagnostics [e.g., DNA, ribonucleic acid (RNA), proteins, cells, blood pressure and assays, and toxin identification] [50.60, 61], tissue engineering [50.62–64], and implantable pharmaceutical drug delivery [50.65–67]. Biosensors, also referred to as biochips, deal with liquids and gases. There are two types. A large variety of biosensors are based on micro/nanofluidics [50.60, 68–70]. Micro/nanofluidic devices offer the ability to work with smaller reagent volumes and shorter reaction times, and perform multiple types of analysis at once. The second type of biosensors include micro/nanoarrays, which perform one type of analysis thousands of times [50.71–74].

A chip, called a lab-on-a-CD, with micro/nanofluidic technology embedded on the disk can test thousands of biological samples rapidly and automatically [50.68]). An entire laboratory can be integrated onto a single chip, called a lab-on-a-chip [50.60, 69, 70]. Silicon-based disposable blood-pressure sensor chips were introduced in the early 1990s by GE NovaSensor for blood-pressure monitoring (about 25 million units in 2004). A blood-sugar monitor, referred to as GlucoWatch, was introduced in 2002. It automatically checks blood sugar every 10 minutes by detecting glucose through the skin, without having to draw blood. If glucose is out of the acceptable range, it sounds an alarm so the diabetic can address the problem quickly. A variety of biosensors, many using plastic substrates, are manufactured by various companies including ACLARA, Agilent Technologies, Calipertech, and I-STAT.

The second type of biochips – micro/nanoarrays – are a tool used in biotechnology research to analyze DNA or proteins to diagnose diseases or discover new drugs. Also called DNA arrays, they can identify thousand of genes simultaneously [50.56, 71]. They include a microarray of silicon nanowires, roughly a few nm in size, to selectively bind and detect even a single biological molecule such as DNA or a protein by using nanoelectronics to detect the slight electrical charge caused by such binding, or a microarray of carbon nanotubes to detect glucose electrically.

After the tragedy of Sept. 11, 2001, concern over biological and chemical warfare has led to the development of handheld units with bio- and chemical sensors for the detection of biological germs, chemical or nerve agents, and mustard agents, and chemical precursors to protect subways, airports, water supplies, and the population at large [50.75].

BioMEMS/BioNEMS are also being developed for minimal-invasive surgery including endoscopic surgery, laser angioplasty, and microscopic surgery. Implantable artificial organs can also be produced. Other applications include implantable drug-delivery devices – e.g., micro/nanoparticles with drug molecules encapsulated in functionalized shells for a site-specific targeting applications, and silicon capsules with a nanoporous membrane filled with drugs for long-term delivery [50.65, 76–78] – nanodevices for sequencing single molecules of DNA in the Human Genome Project [50.60], cellular growth using carbon nanotubes for spinal-cord repair, nanotubes for nanostructured materials for various applications such as spinal fusion devices, organ growth, and the growth of artificial tissues using nanofibers.

50.1.4 Tribological Issues in MEMS/NEMS and BioMEMS/BioNEMS

Tribological issues are important in MEMS/NEMS and BioMEMS/BioNEMS requiring intended and/or unintended relative motion. In these devices, various forces associated with the device scale down with the size. When the length of the machine decreases from 1 mm to 1 μm, the area decreases by a factor of a million and the volume decreases by a factor of a billion. As a result, surface forces such as adhesion, friction, meniscus forces, viscous drag forces and surface tension that are proportional to area, become a thousand times larger than the forces proportional to the volume, such as inertial and electromagnetic forces. In addition to the consequences of a large surface-to-volume ratio, since these devices are designed for small tolerances, physical contact becomes more likely, which makes them particularly vulnerable to adhesion between adjacent components. Slight particulate or chemical contamination present at the interface can be detrimental. Since the start-up forces and torques involved in operation that are available to overcome retarding forces are small, the increase in resistive forces such as adhesion and friction become a serious tribological concern that limits the durability and reliability of MEMS/NEMS [50.13]. A large lateral force required to initiate relative motion between two surfaces, large static friction, is referred to as *stiction*, which has been studied extensively in tribology of magnetic storage systems [50.34, 46, 47, 79–82]. The source of stiction is generally liquid-mediated adhesion with a source of liquid being process fluid or capillary condensation of water vapor from the environment. Adhesion, friction/stiction (static friction), wear, and surface contamination affect MEMS/NEMS and BioMEMS/BioNEMS performance and, in some cases, can even prevent devices from working. Some examples of devices which experience tribological problems follow.

MEMS

Figure 50.2a shows examples of several microcomponents that can encounter the aforementioned tribological problems. The polysilicon electrostatic micromotor has 12 stators and a four-pole rotor and is produced by surface micromachining. The rotor diameter is 120 μm and the air gap between the rotor and stator is 2 μm [50.83]. It is capable of continuous rotation up to speeds of 100 000 revolutions per minute (RPM). The intermittent contacts at the rotor–stator interface and physical contact at the rotor–hub flange interface result in wear issues, and high stiction between the contacting surfaces limits the repeatability of operation or may even prevent operation altogether. Next, a bulk micromachined silicon stator/rotor pair is shown with bladed rotor and nozzle guide vanes on the stator with dimensions of less than a mm [50.84, 85]. These are being developed for high-temperature micro-gas turbine engine with rotor diameters of 4–6 mm and operating speed of up to 1*million* RPM (with a sliding velocity in excess of 500 m/s, comparable to velocities of large turbines operating at high velocities) to achieve high specific power, up to a total of about 10 W. Erosion of blades and vanes, and the design of microbearings required to operate at extremely high speeds used in the turbines are some of the concerns. Ultra-short high-speed micro hydrostatic gas journal bearings with a length-to-diameter (L/D) ratio of less than 0.1 are being developed for operation at surface speeds on the order of 500 m/s, which offer unique design challenges [50.86]

In Fig. 50.2a is an SEM micrograph of a surface-micromachined polysilicon six-gear chain from Sandia National Laboratory. (For more examples of an early version, see *Mehregany* et al., [50.87].) As an example of non-silicon components, a milligear system produced using the LIGA process (a German acronym for Lithographie Galvanoformung Abformung) for a DC brushless permanent-magnet millimotor (diameter = 1.9 mm, length = 5.5 mm with an integrated milligear box [50.88–90]) is also shown. The gears are made of metal (electroplated Ni–Fe) but can also be made from injected polymer materials [e.g., polyoxy-methylene (POM)] using the LIGA process. Even though the torque transmitted at the gear teeth is small, on the order of a fraction of a nN.m, because of the small dimensions of gear teeth, the bending stresses are large where the teeth mesh. Tooth breakage and wear at the contact of gear teeth is a concern.

Figure 50.2b shows a polysilicon, multiple-microgear speed-reduction unit and its components after laboratory wear tests conducted for 600 000 cycles at 1.8% RH [50.91]. These units have been developed for electrostatically driven microactuator (microengine) developed at Sandia National Laboratory for operation in the kHz frequency range [50.92]. Wear of various components is clearly observed in the figure. Humidity was shown to be a strong factor in the wear of rubbing surfaces. In order to improve the wear characteristics of rubbing surfaces, 20-nm-thick tungsten (W) coating deposited at 450 °C using the chemical vapor deposition (CVD) technique was used [50.93]. Tungsten-coated microengines tested for reliability showed improved wear characteristics with longer lifetimes than polysilicon microengines. However, these coatings have poor yield. Instead, vapor-deposited self-assembled monolayers of fluorinated (dimethylamino) silane are used [50.94]. They can be deposited with high yield, although durability is not as good.

Figure 50.3 shows a micromachined flow modulator; several micromachined flow channels are integrated in series with electrostatically actuated microvalves [50.95]. The flow channels lead to a central gas outlet hole drilled in the glass substrate. Gas enters the device through a bulk-micromachined gas inlet hole in the silicon cap. The gas, after passing through an open microvalve, flows parallel to the glass substrate through flow channels and exits the device through an outlet. The normally open valve structure consists of a freestanding double-end-clamped beam, which is positioned beneath the gas inlet orifice. When electrostatically deflected upwards, the beam seals against the inlet orifice and the valve is closed. In these microvalves used for flow control, the mating valve surfaces should be smooth enough to seal while maintaining a minimum roughness to ensure low adhesion [50.79–81, 96].

A second MEMS device, shown in Fig. 50.3, is an electrostatically driven rotary microactuator for a magnetic disk drive, surface-micromachined by a multilayer electroplating method [50.37]. This high-bandwidth servo-controlled microactuator, located between a slider and a suspension is being developed for ultrahigh-track-density applications, serves as the fine-position and high-bandwidth control element of a two-stage, coarse/fine servo system when coupled with a conventional actuator [50.36, 37]. A slider is placed on top of the central block of a microactuator, which gives rotational motion to the slider. The bottom of the silicon substrate is attached to the suspension. The radial flexure beams in the central block give the rotational freedom of

motion to the suspended mass (slider), and the electrostatic actuator drives the suspended mass. Actuation is accomplished via interdigitated, cantilevered electrode fingers, which are alternately attached to the central body of the moving part and to the stationary substrate

a)

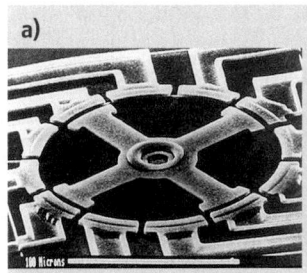

Electrostatic micromotor [83]

Microturbine bladed rotor and nozzle guide vanes on the stator [84]

Six gear chain (www.sandia.gov)

Ni-Fe Wolfrom-type gear system by LIGA [88]

b)

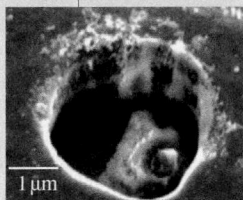

Multiple microgear speed reduction unit [91]

Hub

Gap

Clip

Pinhole

Fig. 50.2a,b Examples of MEMS devices and components that experience tribological problems: (**a**) several microcomponents, and (**b**) a polysilicon, multiple-microgear speed-reduction unit after laboratory wear testing for 600 000 cycles at 1.8% relative humidity.

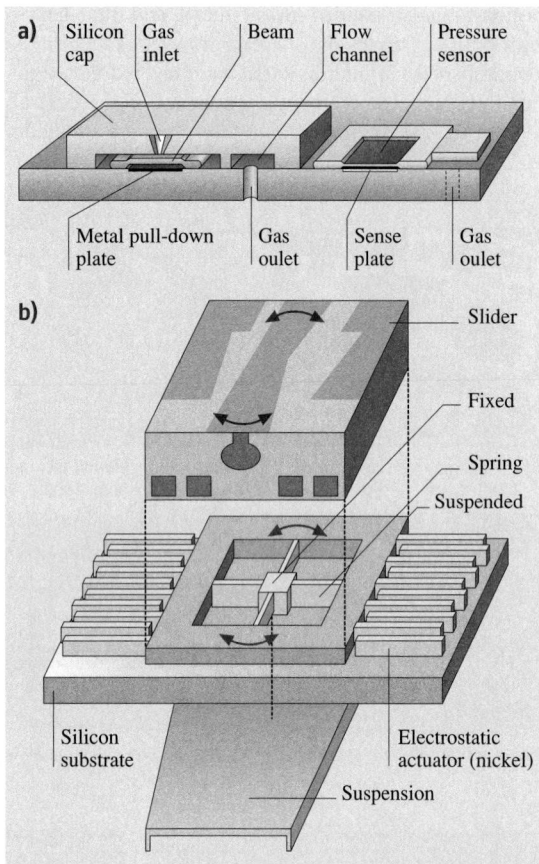

Fig. 50.3a,b Examples of MEMS devices that experience tribological problems (**a**) Low pressure flow modulator with electrostatically actuated microvalves [50.95] (**b**) Electroplated-nickel rotary microactuator for magnetic disc drives [50.37]

to form pairs. A voltage applied across these electrodes results in an electrostatic force, which rotates the central block. The inter-electrode gap width is about 2 μm. Any unintended contacts between the moving and stationary electroplated-nickel electrodes may result in wear and stiction.

Commercially available MEMS devices also exhibit tribological problems. Figure 50.4 shows an integrated capacitive-type silicon accelerometer fabricated using surface micromachining by Analog Devices, with dimensions of a couple of millimeters, which is used for deployment of airbags in automobiles, and more recently for various other consumer electronics markets [50.21, 97]. The central suspended beam mass (about 0.7 μg) is supported on the four corners by

spring structures. The central beam has interdigitated cantilevered electrode fingers (about 125 μm long and 3 μm thick) on all four sides that alternate with those of the stationary electrode fingers as shown, with a gap of about 1.3 μm. Lateral motion of the central beam causes a change in the capacitance between these electrodes, which is used to measure the acceleration. Here stiction between the adjacent electrodes as well as stiction of the beam structure with the underlying substrate are detrimental to the operation of the sensor [50.21, 97]. Wear during unintended contacts of these polysilicon fingers is also a problem. A molecularly thick diphenyl siloxane lubricant film with high resistance to temperature and oxidation applied by a vapor deposition process is used on the electrodes to reduce stiction and wear [50.98]. For deposition, a small amount of liquid is dispensed into each package before it is sealed. As the package is heated in the furnace, the liquid evaporates and coats the sensor surface. As sensors are required to sense low-g accelerations, they need to be more compliant and stiction becomes an even bigger concern.

Figure 50.4 also shows a cross-sectional view of a typical piezoresistive-type pressure sensor, which is used for various applications including manifold absolute pressure (MAP), and tire-pressure measurements, and disposable blood-pressure measurements. The sensing material is a diaphragm formed on a silicon substrate, which bends with applied pressure [50.99]. The deformation causes a change in the band structure of the piezoresistors that are placed on the diaphragm, leading to a change in the resistivity of the material. The MAP sensors are subjected to drastic conditions - extreme temperatures, vibrations, sensing fluid, and thermal shock. Fluid under extreme conditions could cause corrosive wear. Fluid cavitation could cause erosive wear. The protective gel encapsulent generally used can react with the sensing fluid and result in swelling or dissolution of the gel. Silicon cannot deform plastically, therefore any pressure spikes leading to deformation past its elastic limit will result in fracture and crack propagation. Pressure spikes could also cause the diaphragm to delaminate from the support substrate. Finally, cyclic loading of diaphragm during use can lead to fatigue and wear of silicon diaphragm or delamination.

The bottom schematic in Fig. 50.4 shows a cross-sectional view of a thermal printhead chip (on the order 10–50 cm³ in volume) used in inkjet printers [50.25]. They consist of a supply of ink and an array of elements with microscopic heating resistors on a substrate mated to a matching array of ink-injection orifices or nozzles (about 70 μm in diameter) [50.23, 24, 26]. In each ele-

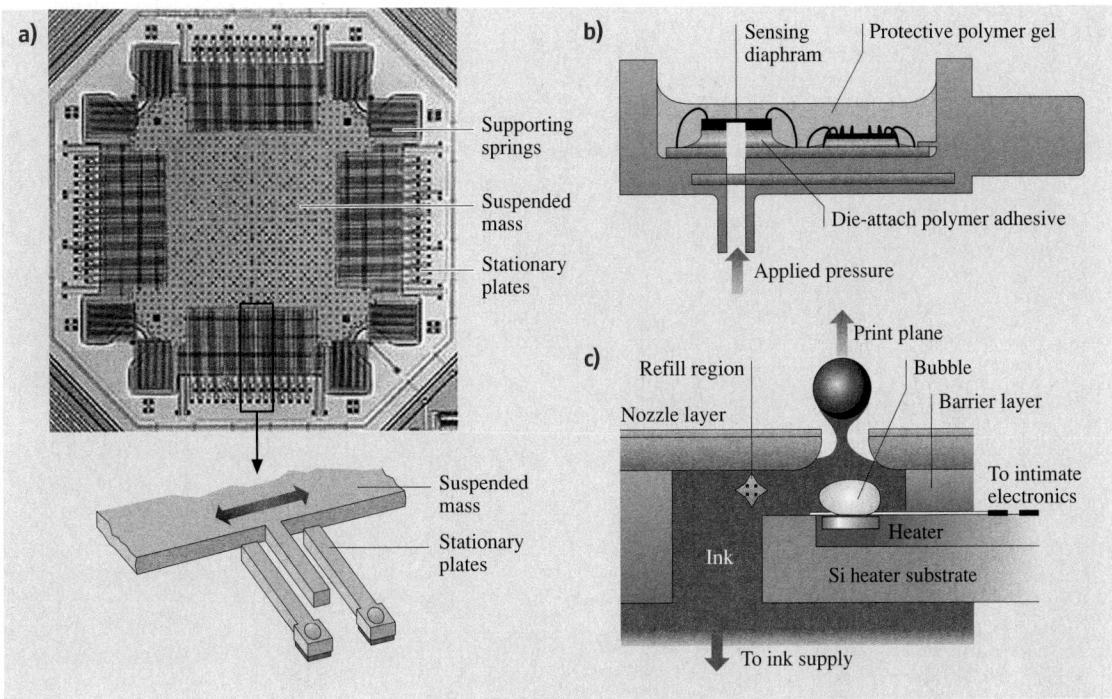

Fig. 50.4a–c Examples of MEMS devices having commercial use that experience tribological problems (**a**) Capacitive type silicon accelerometer for automotive sensory applications [50.97] (**b**) Piezoresistive type pressure sensor [50.101] (**c**) Thermal inkjet printhead [50.25]

ment, a small chamber is heated by the resistor, where a brief electrical impulse vaporizes part of the ink and creates a tiny bubble. The heaters operate at several kHz, and are therefore capable of high-speed printing. As the bubble expands, some of the ink is pushed out of the nozzle onto the paper. When the bubble pops, a vacuum is created and this causes more ink from the cartridge to move into the printhead. Clogged ink ports are the major failure mode. There are also various tribological concerns [50.23]. The surface of the printhead where the ink is shot out towards the paper can become scratched and damaged as a result of countless trips back and forth across the pages, which are somewhat rough. As a result of repeated heating and cooling, the heated resistors expand and contract. Over time, these elements will experience fatigue and may eventually fail. Bubble formation in the ink reservoir can lead to cavitation erosion of the chamber, which occurs when bubbles formed in the fluid become unstable and implode against the surface of the solid and apply impact energy on that surface. Fluid flow through nozzles may cause erosion and ink particles may also cause abrasive wear. Corrosion of the ink reservoir surfaces can also occur as

a result of exposure of the ink to high temperatures as well as due to ink pH. The substrate of the chip consists of silicon with a thermal barrier layer followed by thin film of resistive material and then conducting material. The conductor and resister layers are generally protected by an overcoat layer of a plasma-enhanced chemical vapor deposition (PECVD) α-SiC : H layer, which is 200–500 nm thick [50.100].

Figure 50.5 shows two digital micromirror device (DMD) pixels used in digital light processing (DLP) technology for digital projection displays in computer projectors, high-definition television (HDTV) sets, and movie projectors [50.27–29]. The entire array (chip set) consists of a large number of rotatable aluminium alloy micromirrors (digital light switches) which are fabricated on top of a complementary metal–oxide–semiconductor (CMOS) static random-access memory integrated circuit. The surface-micromachined array consists of half of a million to more than two million of these independently controlled reflective, micromirrors (mirror size on the order of 12 μm square and 13 μm pitch) which flip backward and forward at a frequency on the order of 5000–7000 times a second, as

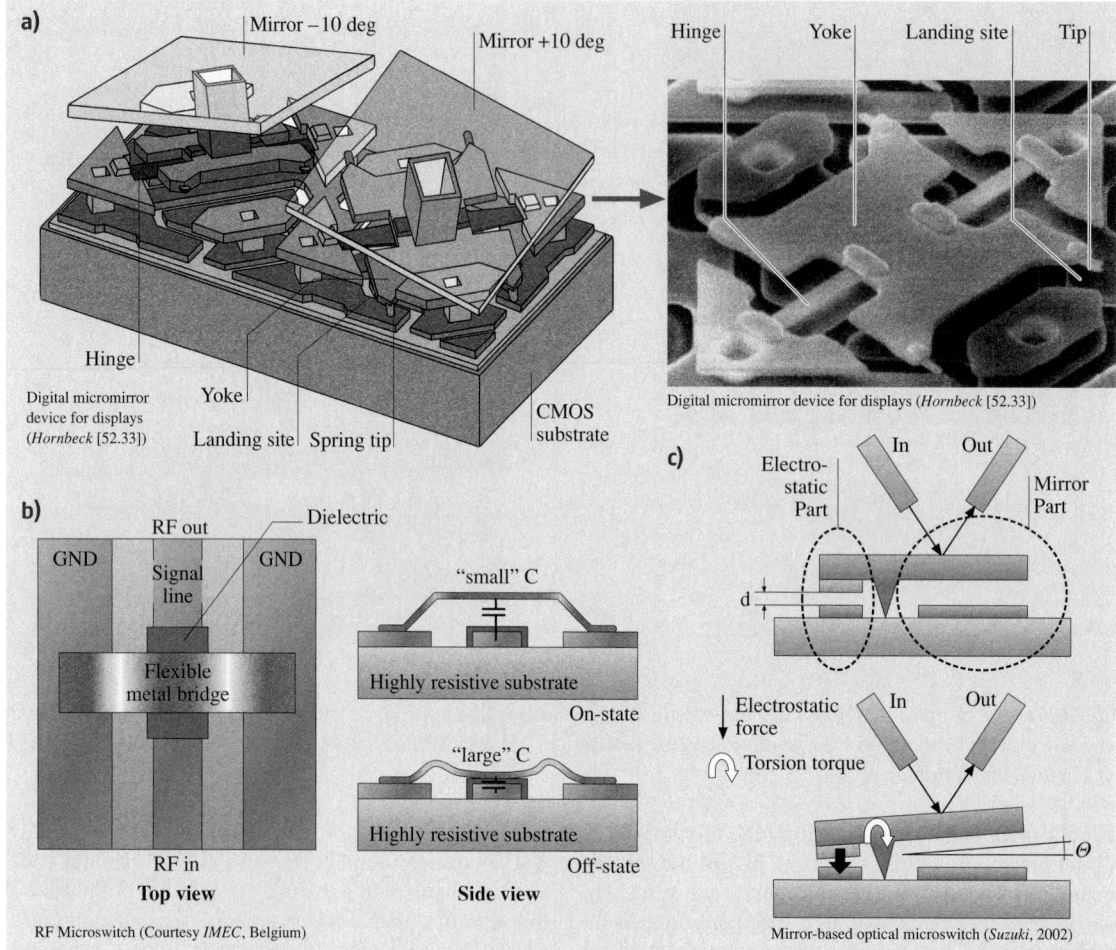

Fig. 50.5 Examples of MOEMS and RF-MEMS devices having commercial use that experience tribological problems

a result of electrostatic attraction between the micromirror structure and the underlying electrodes. For the binary operation, micromirror/yoke structure mounted on torsional hinges is rotated ±10° (with respect to the horizontal plane), and is limited by a mechanical stop. Contact between the cantilevered spring tips at the end of the yoke (four present on each yoke) with the underlying stationary landing sites is required for true digital (binary) operation. Stiction and wear during contact between the aluminium alloy spring tips and landing sites, hinge memory (metal creep at high operating temperatures), hinge fatigue, shock and vibration failure, and sensitivity to particles in the chip package and operating environment are some of the important issues affecting the reliable operation of a micromirror device [50.102–104]. Self-assembled monolayers of

a fatty acid – perfluorodecanoic acid (PFDA) – applied by a vapor deposition process is used on the surfaces of the tip and landing sites to reduce stiction and wear [50.105, 106]. However, these films are susceptible to moisture, and to keep moisture out and create a background pressure of PFDA, hermetic chip packages are used. The spring tip is used in order to use the stored spring energy to pop up the tip during pull-off. A lifetime estimate of over one hundred thousand operating hours with no degradation in image quality is the norm. At a mirror modulation frequency of 7 kHz, each micromirror element needs to switch about 2.5 trillion cycles.

Figure 50.5 also shows a schematic of an electrostatically actuated capacitive-type RF microswitch for switching of RF signals at microwave and low frequencies [50.107]. It is a membrane type and consists of a flexible metal (Al) bridge that spans the RF transmission line in the center of a coplanar waveguide. When the bridge is up, the capacitance between the bridge and RF transmission line is small and the RF signal passes without much loss. When a DC voltage is applied between the RF transmission line and the bridge, the latter is pulled down until it touches a dielectric isolation layer. The large capacitance thus created shorts the RF signal to the ground. The failure modes include creep in the metal bridge, fatigue of the bridge, charging and degradation of the dielectric insulator, and stiction of the bridge to the insulator [50.30, 107]. The stiction occurs due to capillary condensation of water vapor from the environment, van der Waals forces, and/or charging effects. If the restoring force in the bridge of the switch is not large enough to pull the bridge up again after the actuation voltage has been removed, the device fails due to stiction. Humidity-induced stiction can be avoided by hermetically sealing the microswitch. Some roughness of the surfaces reduces the probability of stiction. Selected actuation waveforms can be used to minimize charging effects.

The bottom schematic of Fig. 50.5 shows an electrostatically actuated mirror-based optical microswitch or attenuator [50.30]. It is a cantilever type and uses a hinged mirror which rotates to reflect light from one (in) fiber to another (out) fiber. It can either change light intensity or can be used as an on/off switch when the deflection angle of the reflective mirror is adjusted. A voltage is applied which creates an electrostatic force that pulls the mirror down until electrical contact is made. The primary source of failure arises from the rubbing present in the hinge, leading to wear issues. Failure in the hinge can occur either by mechanical drift due to wear or complete fracture. Stiction is a major concern after the mirror is kept in contact with the electrode underneath [50.30]. The dynamic impact of repeated contacts may lead to subsurface fatigue [50.108]. Bumps are placed on the mirror surface facing the electrode underneath in order to minimize the contact area and stiction. Lubricants are also used to reduce stiction.

NEMS

Figure 50.6 shows an AFM-probe-based nanoscale data-storage system for ultrahigh-density recording, which experiences tribological problems [50.49]. The system uses arrays of 1024 silicon microcantilevers ("Millipede") for thermomechanical recording and playback on a polymer [polymethyl methacrylate (PMMA)] medium about 40 nm thick with a harder Si substrate. The

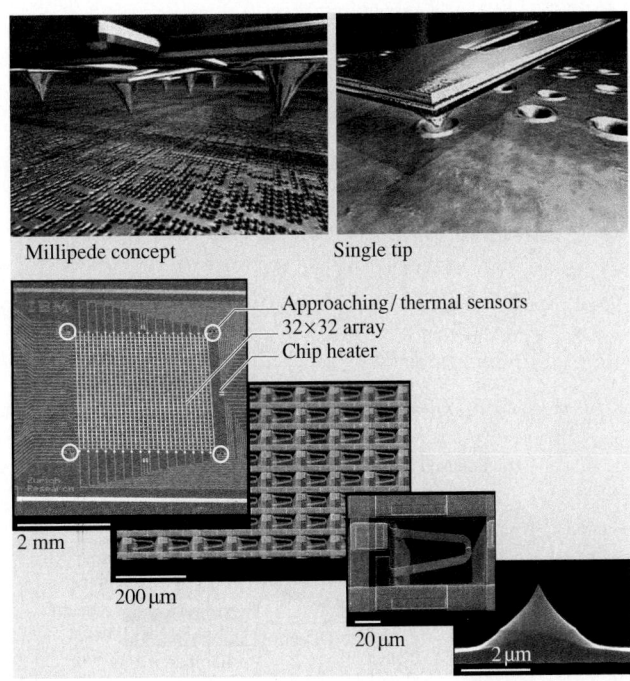

Fig. 50.6 Example of a NEMS device – AFM-probe-based nanoscale data storage system – that experience tribological problems

cantilevers consist of integrated tip heaters with tips of nanoscale dimensions. (Sharp tips themselves are also example of NEMS.) Thermomechanical recording is a combination of applying a local force to the polymer layer and softening it by local heating. The tip, heated to about $400\,°C$, is brought into contact with the polymer for recording. Reading is done using the heater cantilever, originally used for recording, as a thermal read-back sensor by exploiting its temperature-dependent resistance. The principle of thermal sensing is based on the fact that the thermal conductivity between the heater and the storage substrate changes according to the spacing between them. When the spacing between the heater and sample is reduced as the tip moves into a bit, the heater's temperature and hence its resistance will decrease. Thus, changes in temperature of the continuously heated resistor are monitored while the cantilever is scanned over data bits, providing a means of detecting the bits. Erasing for subsequent rewriting is carried out by thermal reflow of the storage field by heating the medium to $150\,°C$ for a few seconds. The smoothness of the reflown medium allows multiple rewriting of the same storage field. Bit sizes in the range 10–50 nm have been achieved by using a 32×32

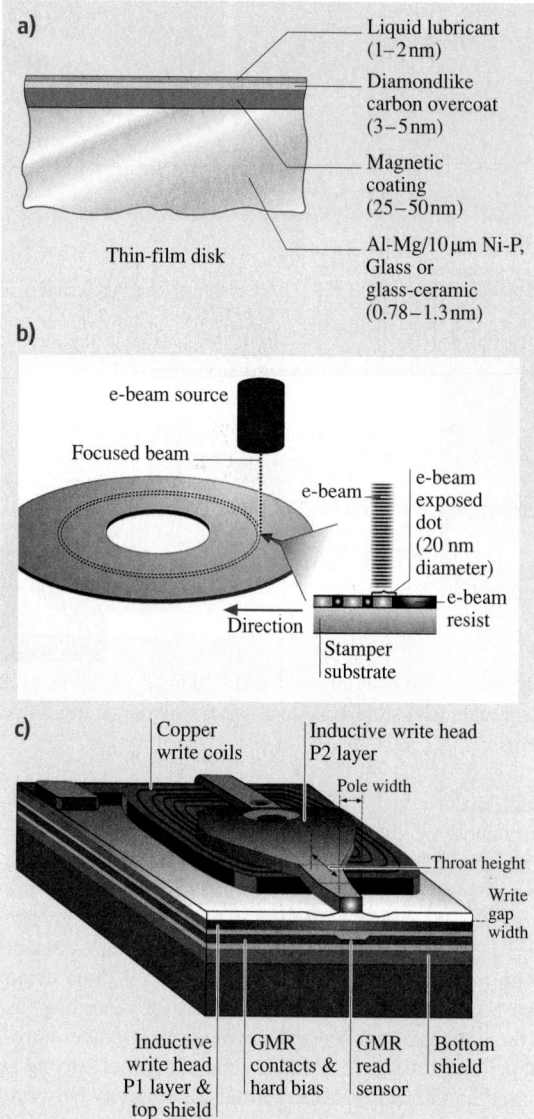

a)

Liquid lubricant (1–2 nm)

Diamondlike carbon overcoat (3–5 nm)

Magnetic coating (25–50 nm)

Al-Mg/10 μm Ni-P, Glass or glass-ceramic (0.78–1.3 nm)

Thin-film disk

b)

e-beam source

Focused beam

e-beam

e-beam exposed dot (20 nm diameter)

e-beam resist

Direction

Stamper substrate

c)

Copper write coils

Inductive write head P2 layer

Pole width

Throat height

Write gap width

Inductive write head P1 layer & top shield

GMR contacts & hard bias

GMR read sensor

Bottom shield

Fig. 50.7a–c Schematic of (**a**) sectional view of a conventional multigrain magnetic rigid disk, (**b**) nanopatterned magnetic rigid disk, and (**c**) an inductive-write GMR read magnetic head structure for magnetic data storage (Hitachi Global Storage Technologies)

head slider and a magnetic rigid disk [50.34]. Magnetic rigid disks and heads used today for magnetic data storage consist of one- to a few-nm-thick nanostructured films. Figure 50.7a shows the sectional view of a conventional multigrain magnetic rigid disk. The superparamagnetic effect poses a serious challenge for ever increasing areal density of disk drives. One of the promising methods to circumvent the density limitations imposed by this effect is the use of nanopatterned disks Fig. 50.7b. In conventional disks, the thin magnetic layer forms a random mosaic of nanometer-scale grains and each recorded bit consists of many tens of these random grains. In patterned disks, the magnetic layer is created as an ordered array of highly uniform islands, each island capable of storing an individual bit. These islands may be one or a few grains, rather than a collection of random decoupled grains. This increases the density by a couple of orders of magnitude. Figure 50.7c shows a schematic of an inductive-write GMR read head structure. These are constructed from a variety of materials: magnetic alloys, metal conductors, ceramic, and polymer insulators in a complex three-dimensional structure. The multilayered thin-film structures used to construct the sensor and individual films are only a few nm thick. The head slider surface, which flies over the disk surface, is coated with diamond-like carbon coatings that are about 3 nm thick to protect the thin-film structure from electrostatic discharge. Any isolated contacts between the disk and sensor and lubricant pickup pose tribological concerns [50.34].

BioMEMS

An example of a wristwatch-type biosensor based on microfluidics, referred to as a lab-on-a-chip system, is shown in Fig. 50.8a [50.60, 69]. These systems are designed to either detect a single or class of (bio)chemicals, or system-level analytical capabilities for a broad range of (bio)chemical species known as a micro total-analysis system (μTas), and have the advantage of incorporating sample handling, separation, detection, and data analysis onto one platform. The chip relies on microfluidics and involves manipulation of tiny amounts of fluids in microchannels using microvalves. The test fluid is injected into the chip, generally using an external pump or syringe, for analysis. Some chips have been designed with an integrated electrostatically actuated diaphragm-type micropump. The sample, which can have a volume measured in nanoliters, flows through microfluidic channels via an electric potential and capillary action using microvalves (having various designs including membrane type) for various analyses.

(1024) array write/read chip (3 mm × 3 mm). It has been reported that tip wear occurs due to contact between the tip and the Si substrate during writing. Tip wear is considered a major concern for device reliability.

In magnetic data storage, magnetic recording is accomplished by relative motion between the magnetic

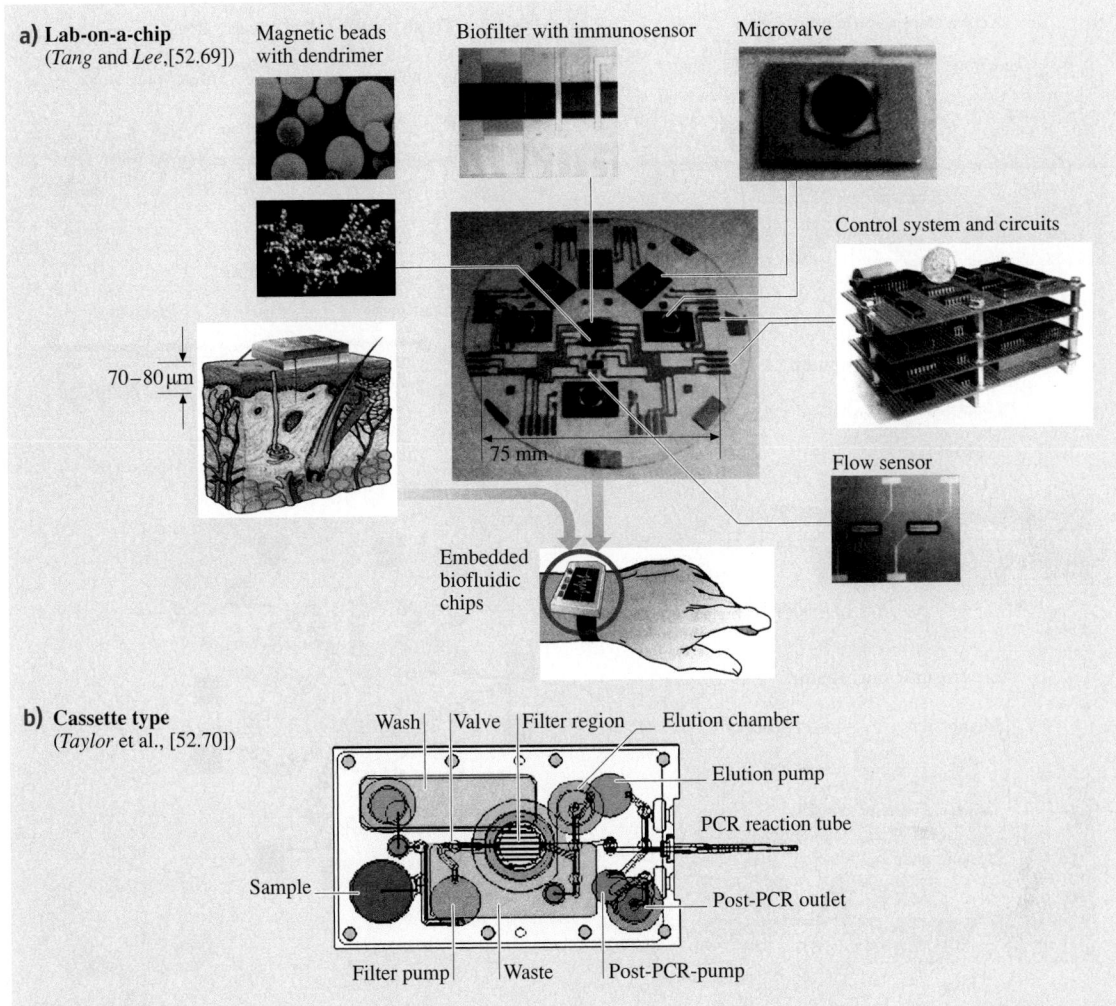

Fig. 50.8a–c (a) MEMS-based biofluidic chip, commonly known as a lab-on-a-chip, that can be worn like a wristwatch [50.69]. (b) Cassette-type biosensor used for human genomic DNA analysis [50.70].

The fluid is preprocessed and then analyzed using a biosensor. Another example of a biosensor is the cassette-type biosensor used for human genomic DNA analysis; integrated biological sample preparation is shown in Fig. 50.8b [50.70]. The implementation of micropumps and microvalves allows for fluid manipulation and multiple sample-processing steps in a single cassette. Blood or other aqueous solutions can be pumped into the system, where various processes are performed.

Microvalves, which are found in most microfluidic components of BioMEMS, can be classified into two categories: active microvalves (with an actuator) for flow regulation in microchannels and passive

microvalves integrated with micropumps. Active microvalves consist of a valve seat and a diaphragm actuated by an external actuator [50.61, 109]. Different types of actuators are based on piezoelectric, electrostatic, thermopneumatic, electromagnetic, and bimetallic materials, shape-memory alloys and solenoid plungers. An example of an electrostatic cantilever-type active microvalve is shown in Fig. 50.8c. Passive microvalves used in micropumps include mechanical check valves and a diffuser/nozzle [50.61, 109–112]. Check valves consist of a flap or membrane that is capable of opening and closing with changes in pressure; see Fig. 50.8c for schematics. A diffuser/nozzle

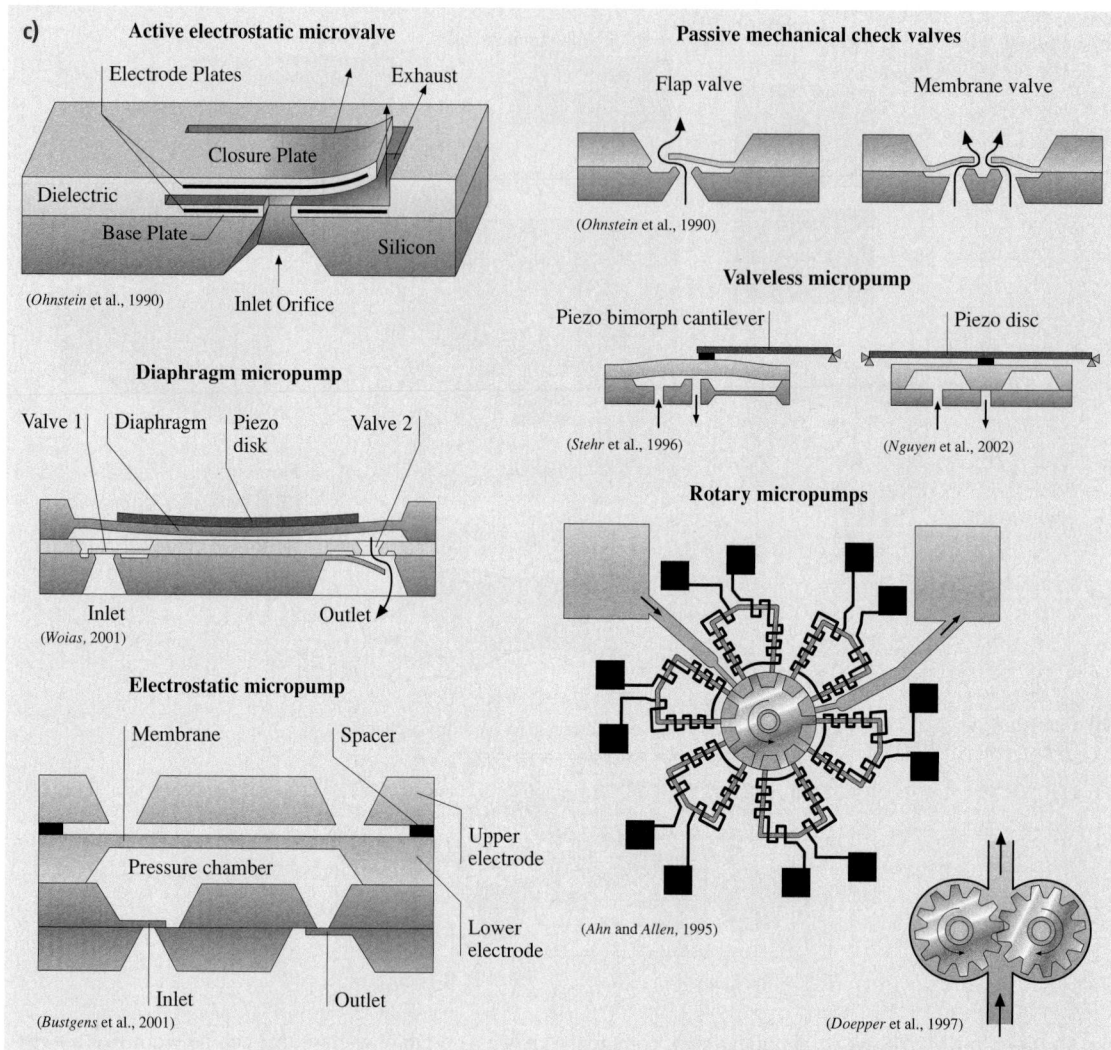

c) **Active electrostatic microvalve**

Electrode Plates

Exhaust

Closure Plate

Dielectric

Base Plate

Silicon

(*Ohnstein* et al., 1990)   Inlet Orifice

Diaphragm micropump

Valve 1 | Diaphragm | Piezo disk   Valve 2

Inlet   Outlet
(*Woias*, 2001)

Electrostatic micropump

Membrane   Spacer

Pressure chamber

Upper electrode

Lower electrode

Inlet   Outlet
(*Bustgens* et al., 2001)

Passive mechanical check valves

Flap valve   Membrane valve

(*Ohnstein* et al., 1990)

Valveless micropump

Piezo bimorph cantilever   Piezo disc

(*Stehr* et al., 1996)   (*Nguyen* et al., 2002)

Rotary micropumps

(*Ahn* and *Allen*, 1995)

(*Doepper* et al., 1997)

Fig. 50.8a–c (c) Multiple examples of valves and pumps found in BioMEMS devices. Mechanical check valves, diaphragm micropump, valveless micropump, and rotary micropump

uses an entirely different principle and only works with the presence of a reciprocating diaphragm. When one convergent channel works simultaneously with another convergent channel oriented in a specific direction, a change in pressure is possible.

There are four main types of mechanical micropumps, which include diaphragm micropumps that involves mechanical check valves, valveless rectification pumps that use diffuser/nozzle-type valves, valveless pumps without a diffuser/nozzle, electrostatic micropumps, and rotary micropumps [50.61, 109–112]. Diaphragm micropumps consist of a reciprocating di-

aphragm, which can be piezoelectrically driven, working in synchronization with two check valves Fig. 50.8c. Electrostatic micropumps have a diaphragm as well, but it is driven using two electrodes. Valveless micropumps also consist of a diaphragm that is piezoelectrically driven, but do not incorporate passive mechanical valves. Instead, these pumps use an elastic buffer mechanism or variable gap mechanisms. Finally, a rotary micropump has a rotating rotor that simply adds momentum to the fluid by the fast-moving action of the blades Fig. 50.8c. Rotary micropumps can be driven using an integrated electromagnetic motor or by the presence of an external electric field. All of these micropumps can be made of silicon or a polymer material.

During the operation of the microvalves and micropumps discussed above, adhesion and friction properties become important in which contacts occur due to relative motion. During operation, active mechanical microvalves have an externally actuated diaphragm which comes into contact with a valve seat to restrict the fluid flow. Adhesion between the diaphragm and valve seat will affect the operation of the microvalve. In diaphragm micropumps, two passive mechanical check valves are incorporated into the design. Passive mechanical check valves also exhibit adhesion when the flap or membrane comes into contact with the valve seat when fluid flow is removed. Adhesion also occurs during the operation of valveless micropumps when the diaphragm, which is piezoelectrically driven, comes into contact with the rigid outlet. Finally, adhesion and friction can also be seen during the operation of rotary micropumps when the gears rotate, come into contact and rub against one another.

If the adhesion between the microchannel surface and the biofluid is high, the biomolecules will stick to the microchannel surface and restrict flow. In order to facilitate flow, microchannel surfaces with low bioadhesion are required. Fluid flow in polymer channels can produce a triboelectric surface potential which may affect the flow. Polymers are known to generate surface potentials and the magnitude of the potential varies from one polymer to another [50.113–115]. Conductive surface layers on the polymer channels can be deposited to reduce triboelectric effects.

As mentioned, the microfluidic biosensor shown in Fig. 50.8a required the use of micropumps and microvalves. For example, a microdevice with 1000 channels requires 1000 micropumps and 2000 microvalves, which makes it bulky and poses reliability concerns. Two methods can be used to drive the flow of fluids in microchannels: pressure and electrokinetic drive. The electrokinetic flow is based on the movement of molecules in an electric field due to their charges. There are two components to electrokinetic flow: electrophoresis, which results from the accelerating force due to the charge of a molecule in an electric field, and electroosmosis, which uses electrically controlled surface tension to drive uniform liquid flow. Biosensors based on electrokinetic flow have also been developed. In so-called *digital-based microfluidics*, based on the electroosmosis process, electrically controlled surface tension is used to drive liquid droplets, thus eliminating the need for valves and pumps [50.116, 117]. These

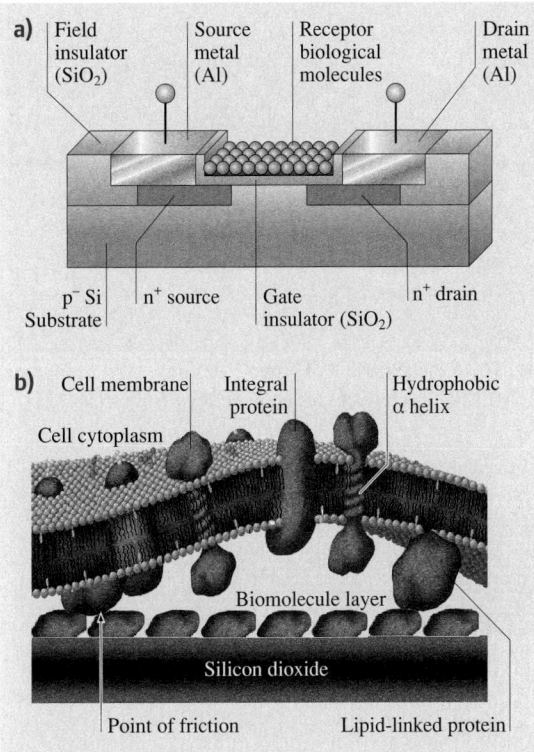

Fig. 50.9 (a) Schematic of MOSFET-based bioFET sensor [50.72], and **(b)** schematic showing the generation of friction and wear points due to interaction of implanted biomolecule layer on a biosensor with living tissue

microdevices consist of a rectangular grid of gold nanoelectrodes instead of micro/nanochannels. An externally applied electric field enables manipulation of samples of a few nanoliters through the capillary circuitry.

An example of a microarray-type biosensor under development in our laboratory is that based on a field-effect transistor (FET), which is shown in Fig. 50.9 [50.72, 118]. FETs are sensitive to the electrical field produced due to the charge at the surface of the gate insulator. In this sensor, the gate metal of a metal–oxide–semiconductor field-effect transistor (MOSFET) is removed and replaced with a protein (receptor layer) whose cognate is the analyte (e.g., virus or bacteria) that is meant to be sensed. The binding of the receptor layer with the analyte produces a change in the effective charge, which creates a change in the electrical field. This electrical field change may produce a measurable change in the current flow through the device. Adhesion between the protein and silica substrate affect the reliability of the biosensor. In the case

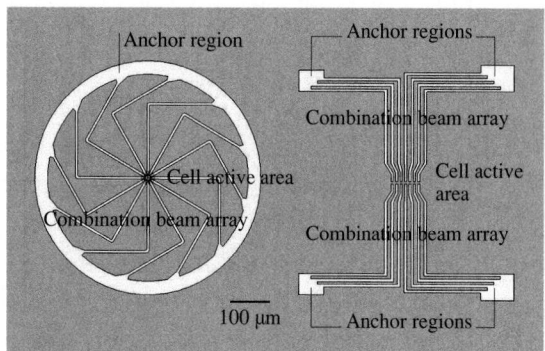

Fig. 50.10 Schematic of two designs for polymer bioMEMS structures to measure cellular forces [50.121]

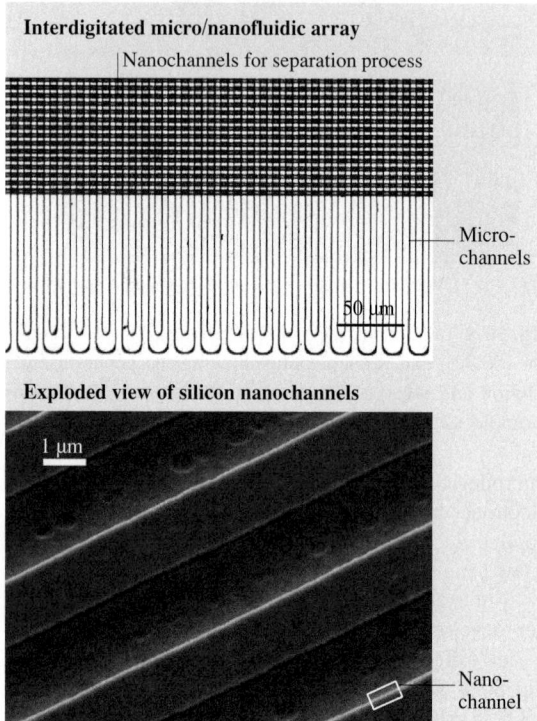

Fig. 50.11 Interdigitated micro/nanofluidic silicon array for the separation process [50.118]

of implanted biosensors, the biosensors come in contact with the exterior environment such as tissues and fluids, and any relative motion of the sensor surface with respect to exterior environment such as tissues or fluids may result in surface damage. A schematic of friction and points of wear generation when an implanted biosensor surface comes into contact with living tissue is shown in Fig. 50.9b. The, friction, wear, and adhe-

sion of the biosensor surface may be critical in these applications [50.72, 119, 120].

Polymer BioMEMS are designed to measure cellular surfaces. For two examples, see Fig. 50.10 [50.121]. The device on the left shows cantilevers anchored at the periphery of the circular structure, while the device on the right has cantilevers anchored at the two corners on the top and the bottom. The cell adheres to the center of the structure, and the contractile forces generated in the cells cytoskeleton cause the cantilever to deflect. The deflection of the compliant polymer cantilevers is measured optically and related to the magnitude of the forces generated by the cell. Adhesion between cells and polymer beam is desirable. In order to design the sensors, micro- and nanoscale mechanical properties of polymer structures are needed.

BioNEMS
Micro/nanofluidic devices provide a powerful platform for electrophoretic separations for a variety of biochemical and chemical analysis. Electrophoresis is a versatile analytical method which is used for separation of small ions, neutral molecules, and large biomolecules. Figure 50.11 shows an interdigitated micro/nanofluidic silicon array with nanochannels for the separation process. Figure 50.12a shows a schematic of an implantable, immuno-isolation submicroscopic biocapsules, aimed at drug delivery to treat significant medical condition such as type I diabetes [50.76, 77]. The purpose of the immuno-isolation biocapsules is to create an implantable device capable of supporting foreign living cells that can be transplanted into humans. It is a silicon capsule consisting of two nanofabricated membranes bonded together with the drug (e.g., encapsulated insulin-producing islet cells) contained within the cavities for long-term delivery. Pores or nanochannels in a semipermeable membrane as small as 6 nm are used as flux regulators for the long-term release of drugs. The nanomembrane also protects therapeutic substances from attack by the body's immune system. The pores are large enough to provide the flow of nutrients (e.g., glucose molecules) and drug (e.g., insulin), but small enough to block natural antibodies. Antibodies have the capability to pass through any orifice larger than 18 nm. The 50-nm pores in silicon were etched by using the sacrificial-layer lithography described in Sect. 50.A [50.77].

The main reliability concerns in the micro/nanofluidic silicon array and implantable biocapsules are biocompatibility and potential biofouling (undesirable accumulation of microorganisms) of the

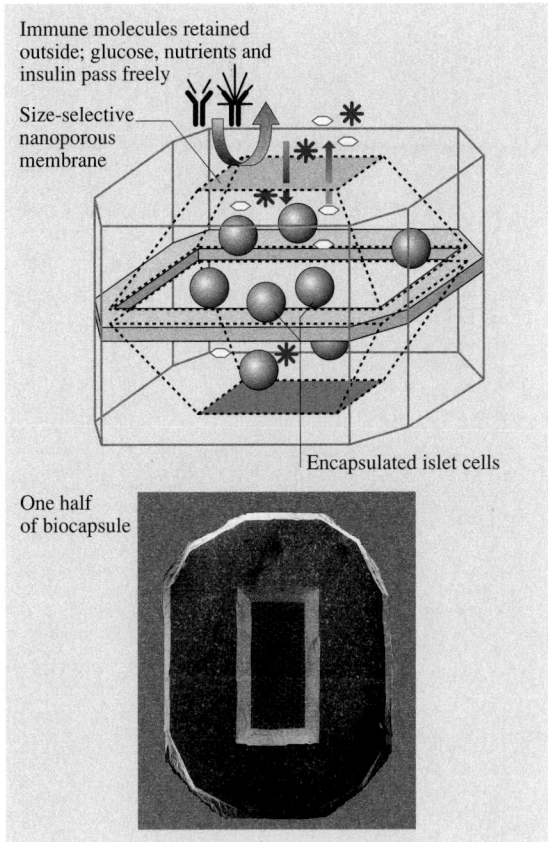

Immune molecules retained outside; glucose, nutrients and insulin pass freely

Size-selective nanoporous membrane

Encapsulated islet cells

One half of biocapsule

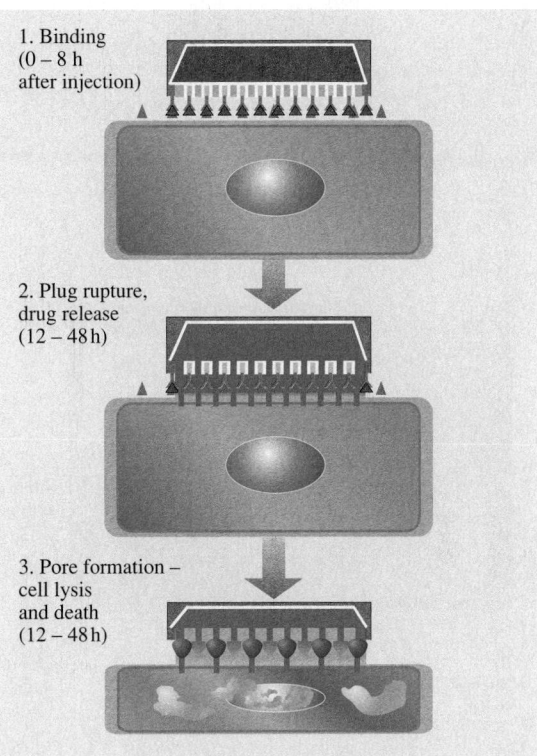

1. Binding (0 – 8 h after injection)

2. Plug rupture, drug release (12 – 48 h)

3. Pore formation – cell lysis and death (12 – 48 h)

Fig. 50.12a,b Schematics of (**a**) implantable, immunoisolation submicroscopic biocapsules (drug-delivery device) [50.77], and (**b**) intravascular nanoparticles to search for and destroy diseased blood cells [50.78]

channels/membrane by means of protein and cells adsorption from biological fluids. Biofouling can also result in the clogging of the nanochannels/nanopores, which potentially could render the device ineffective. The adhesion of proteins and cells to an implanted device can also cause detrimental results such as inflammation and excessive fibrosis. Deposition of the self-assembled monolayers of selected organic molecules on the channels implants, which makes them hydrophobic, presents an innovative solution to combat the adverse effects of the biological fluids [50.122–125].

Figure 50.12b shows a conceptual model of an intravascular drug-delivery device – nanoparticles used to search for and destroy disease cells [50.78]. With lateral dimensions of 1 μm or less, the particles are smaller than any blood cells. These particles can be injected into the blood stream and travel freely through the circulatory system. In order to direct these drug-delivery nanoparticles to cancer sites, their external surfaces are chemically modified to carry molecules that have lock-and-key binding specificity with molecules that support

a growing cancer mass. As soon as the particles dock onto the cells, a compound is released that forms a pore on the membrane of the cells, which leads to cell death and ultimately to that of the cancer mass that was being nourished by the blood vessel. Adhesion between nanoparticles and disease cells is required. Furthermore, the particles should travel close to the endothelium lining of vascular arteries. *Decuzzi* et al. [50.126] recently analyzed the margination of a particle circulating in the blood stream and calculated the speed and time for margination (motion of the particles towards the walls of the vessel) as a function of the density and diameter of particle, based on various forces present between the circulating particle and the endothelium lining. Human capillaries can have radii as small as 4–5 μm. They reported that the particles used for drug delivery should have a radius smaller than a critical value (of the order 100 nm).

In summary, adhesion, stiction/friction and wear clearly limit the lifetime and compromise the

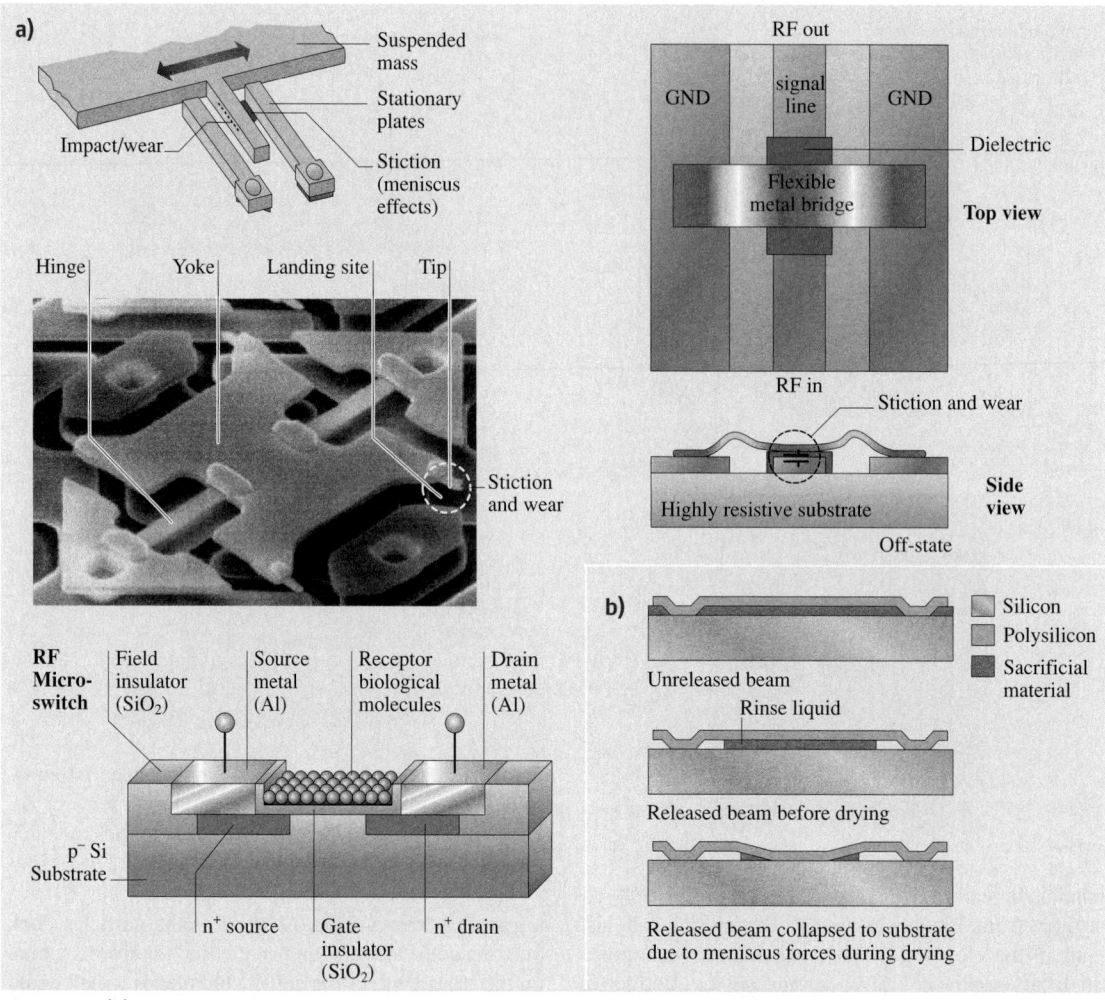

Fig. 50.13 (a) Summary of tribological issues in MEMS, MOEMS and RF-MEMS device operation, and **(b)** in microfabrication via surface micromachining

performance and reliability of MEMS/NEMS and BioMEMS/BioNEMS. Figure 50.13a summarizes tribological problems encountered in some of the MEMS, MOEMS, RF-MEMS, and BioMEMS devices discussed. In addition to in-use stiction, stiction issues are also present in some processes used for fabrication of MEMS/NEMS. For example, the last step in surface micromachining involves the removal of sacrificial layer(s), known as the release process, since the microstructures are released from the surrounding sacrificial layer(s). The release is accomplished by an aqueous chemical etch, rinsing and drying processes. Due to meniscus effects as a result of wet processes, the suspended structures can sometimes collapse and per-

manently adhere to the underlying substrate, as shown in Fig. 50.13b [50.127]. Adhesion is caused by water molecules adsorbed on the adhering surfaces and/or because of the formation of adhesive bonds by silica residues that remain on the surfaces after the water has evaporated. This so-called release stiction is overcome by using dry release methods (e.g., CO_2 critical-point drying or sublimation drying [50.128]).

Tribological Needs

MEMS/NEMS need to be designed to perform expected functions typically in the millisecond to picosecond range. Expected life of the devices for high-speed contacts can vary from a few hundred thousand to many billions of cycles, e.g., over a hundred billion cycles for DMDs, which puts serious requirements on materials [50.13, 91, 129–132]. Adhesion between a biological molecular layer and the substrate, referred to as *bioadhesion*, reduction of friction and wear of biological layers, biocompatibility, and biofouling for BioMEMS/BioNEMS are important. Most mechanical properties are known to be scale-dependent [50.133]. The properties of nanoscale structures need to be measured [50.134]. Tribology is an important factor affecting the performance and reliability of MEMS/NEMS and BioMEMS/BioNEMS [50.13, 46, 47, 80]. There is a need for the development of a fundamental understanding of adhesion, friction/stiction, wear, and the role of surface contamination, and the environment [50.13]. MEMS/NEMS materials need to exhibit good mechan-ical and tribological properties on the micro/nanoscale. There is a need to develop lubricants and identify lubrication methods that are suitable for MEMS/NEMS. Methods need to be developed to enhance adhesion between biomolecules and the device substrate, referred to as bioadhesion. Component-level studies are required to provide a better understanding of the tribological phenomena occurring in MEMS/NEMS. The emergence of micro/nanotribology and AFM-based techniques has provided researchers with a viable approach to address these problems [50.46, 47, 135]. This chapter presents an overview of micro/nanoscale tribological studies of materials and lubrication studies for MEMS/NEMS, bioadhesion, friction and wear of BioMEMS/BioNEMS, and component-level studies of stiction phenomena in MEMS/NEMS devices.

50.2 Tribological Studies of Silicon and Related Materials

The materials of most interest for planar fabrication processes using silicon as the structural material are undoped and boron-doped (p$^+$-type) single-crystal silicon for bulk micromachining, and phosphorus (n$^+$-type) doped and undoped low-pressure chemical vapor deposition (LPCVD) polysilicon films for surface micromachining. Silicon-based devices lack high-temperature capabilities with respect to both mechanical and electrical properties. SiC has been developed as a structural material for high-temperature microsensor and microactuator applications [50.138, 139]. SiC can also be desirable for high-frequency micromechanical resonators, in the GHz range, because of its high ratio of modulus of elasticity to density and consequently high resonance frequency. Table 50.2 compares selected bulk properties of SiC and Si(100). Researchers have found low-cost techniques of producing single-crystalline 3C–SiC (cubic or β-SiC) films via epitaxial growth on large-area silicon substrates for bulk micromachining [50.140] and polycrystalline 3C–SiC films on polysilicon and silicon dioxide layers for surface micromachining of SiC [50.141]. Single-crystalline 3C–SiC piezoresistive pressure sensors have been fabricated using bulk micromachining for high-temperature gas turbine applications [50.142]. Surface-micromachined polycrystalline SiC micromotors have been fabricated and satisfactory operation at high temperatures has been reported [50.143].

As will be shown, bare silicon exhibits inadequate tribological performance and needs to be coated with a solid and/or liquid overcoat or be surface treated (by, e.g., oxidation and ion implantation, commonly used in semiconductor manufacturing), which exhibits lower friction and wear. SiC films exhibit good tribological performance. Both macroscale and microscale tribological properties of virgin and treated/coated silicon, polysilicon films and SiC are presented next.

Table 50.2 Selected bulk properties of 3C (β- or cubic) SiC and Si(100)

Sample	Density^a (kg/m^3)	Hardness^a (GPa)	Elastic modulus^a (GPa)	Fracture toughness^a (MPam$^{1/2}$)	Thermal conductivity^b (W/mK)	Coeff. of thermal expansion^b ($\times 10^{-6}$ /$^\circ$C)	Melting point^a ($^\circ$C)	Band gap^a (eV)
β-SiC	3210	23.5–26.5	440	4.6	85–260	4.5–6	2830	2.3
Si(100)	2330	9–10	130	0.95	155	2–4.5	1410	1.1

^{a} Data from *Bhushan* and *Gupta* [50.136].
^{b} Data from *Shackelford* et al. [50.137].

50.2.1 Virgin and Treated/Coated Silicon Samples

Tribological Properties of Silicon and the Effect of Ion Implantation

Friction and wear of single-crystalline and polycrystalline silicon samples have been studied and the effect of ion implantation with various doses of C^+, B^+, N_2^+ and Ar^+ ion species at an energy of 200 keV to improve their friction and wear properties has been studied [50.144–146]. The coefficient of macroscale friction and wear factor of virgin single-crystal silicon and C^+-implanted silicon samples as a function of ion dose are presented in Fig. 50.14 [50.144]. The macroscale friction and wear tests were conducted using a ball-on-flat tribometer. Each data bar represents the average value of four to six measurements. The coefficient of friction and wear factor for bare silicon are very high and decrease drastically with ion dose. Silicon samples bombarded above an ion dose of 10^{17} C^+ cm^{-2} exhibit extremely low values of coefficients of friction (typically 0.03–0.06 in air) and the wear factor (reduced by as much as four orders of magnitude). *Gupta* et al. [50.144] reported that a decrease in the coefficient of friction and the wear factor of silicon as a result of C^+ ion bombardment occurred because of the formation of silicon carbide rather than amorphization of silicon. *Gupta* et al. [50.145] also reported an improvement in friction and wear with B^+ ion implantation.

Microscale friction measurements were performed using an atomic force/friction force microscope (AFM/FFM) [50.46, 47]. Table 50.3 shows values of surface roughness and coefficients of macroscale and microscale friction for virgin and doped silicon. There is a decrease in the coefficients of microscale and macroscale friction values as a result of ion implantation. When measured for the small contact areas and very low loads used in microscale studies, indentation hardness and elastic modulus are higher than those at the macroscale. This, added to the effect of the small apparent area of contact reducing the number of trapped particles on the interface, results in less plowing contribution and lower friction in the case of microscale friction measurements. Results of microscale wear resistance studies of ion-implanted silicon samples studied using a diamond tip in an AFM [50.147] are shown in Figs. 50.15a and 50.15b. For tests conducted at various loads on Si(111) and C^+-implanted Si(111), it is noted that wear resistance of implanted sample is slightly

Table 50.3 Surface roughness and micro- and macroscale coefficients of friction of selected samples

Material	RMS roughness (nm)	Coefficient of microscale friction[a]	Coefficient of macroscale friction[b]
Si(111)	0.11	0.03	0.33
C^+-implanted Si(111)	0.33	0.02	0.18

[a] Versus Si$_3$N$_4$ tip; tip radius of 50 nm in the load range 10–150 nN (2.5–6.1 GPa) at a scanning speed of 5 μm/s over a scan area of 1 μm × 1 μm in an AFM
[b] Versus Si$_3$N$_4$ ball, ball radius of 3 mm at a normal load of 0.1 N (0.3 GPa) at an average sliding speed of 0.8 mm/s using a tribometer

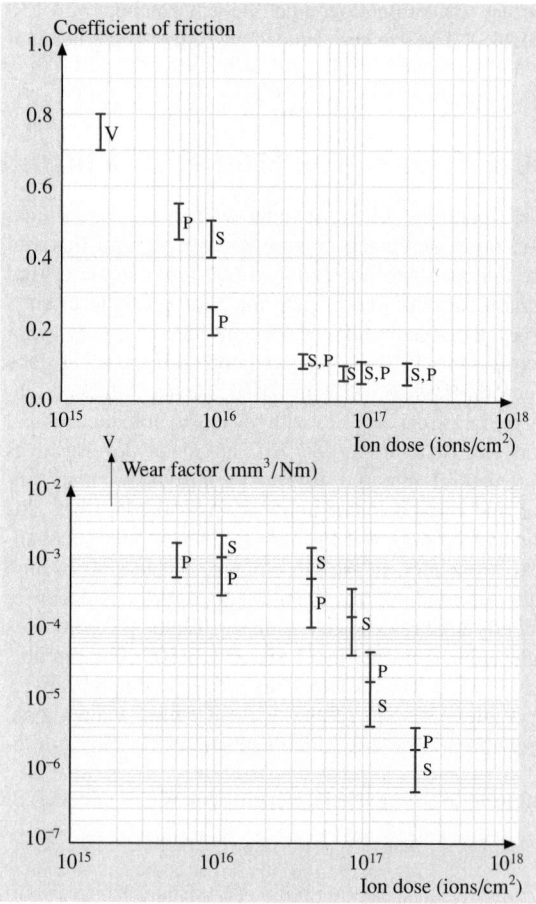

Fig. 50.14 Influence of ion doses on the coefficient of friction and wear factor on C^+-ion-bombarded single-crystal and polycrystalline silicon slid against alumina ball. V corresponds to virgin single-crystal silicon, while S and P denote tests that correspond to doped single-crystal and polycrystalline silicon, respectively [50.144]

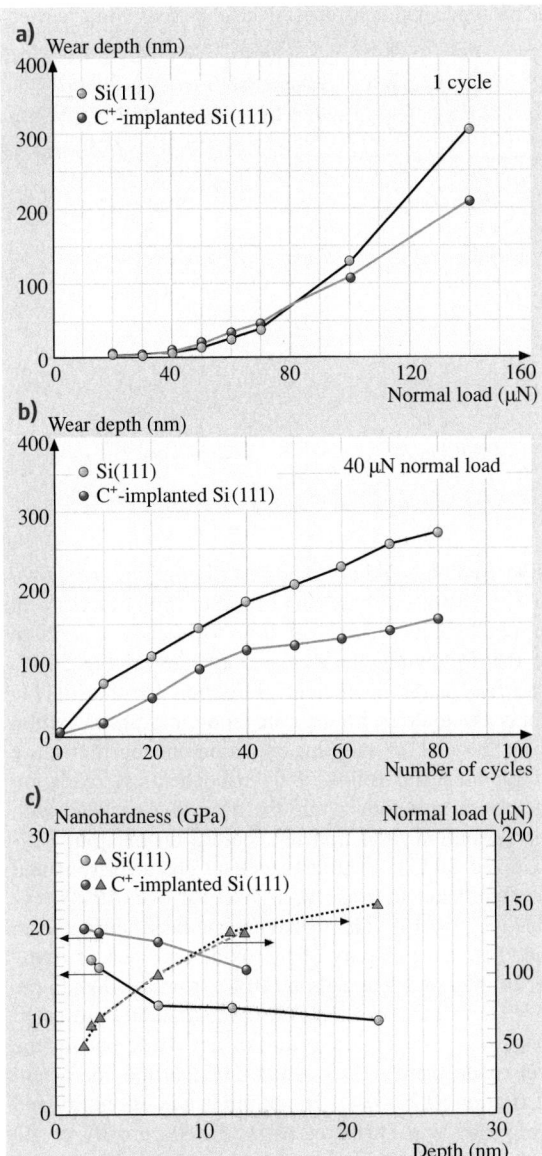

Fig. 50.15a–c Wear depth as a function of (**a**) load (after one cycle), and (**b**) cycles (normal load = 40 mN) for Si(111) and C^+-implanted Si(111). (**c**) Nanohardness and normal load as function of indentation depth for virgin and C^+-implanted Si(111) [50.147]

sistance than the unimplanted sample. Damage from the implantation in the top layer results in poorer wear resistance, however, the implanted zone at the subsurface is more wear-resistant than the virgin silicon.

Hardness values of virgin and C^+-implanted Si(111) at various indentation depths (normal loads) are presented in Fig. 50.15c [50.147]. The hardness at a small indentation depth of 2.5 nm is 16.6 GPa and it drops to a value of 11.7 GPa at a depth of 7 nm and a normal load of 100 μN. Higher hardness values obtained in low-load indentation may arise from the observed pressure-induced phase transformation during the nanoindentation [50.148, 149]. Additional increase in the hardness at an the even lower indentation depth of 2.5 nm reported here may arise from the contribution by complex chemical films (not from native oxide films) present on the silicon surface. At small volumes there is a lower probability of encountering material defects (dislocations, etc.). Furthermore, according to the strain-gradient plasticity theory advanced by *Fleck* et al. [50.150], large strain gradients inherent in small indentations lead to accumulation of geometrically necessary dislocations that cause enhanced hardening. These are some of the plausible explanations for an increase in hardness at smaller volumes. If the silicon material were to be used at very light loads such as in microsystems, the high hardness of surface films would protect the surface until it is worn.

From Fig. 50.15c, hardness values of C^+-implanted Si(111) at a normal load of 50 μN is 20.0 GPa with an indentation depth of about 2 nm, which is comparable to the hardness value of 19.5 GPa at 70 μN, whereas measured hardness value for virgin silicon at an indentation depth of about 7 nm (normal load of 100 μN) is only about 11.7 GPa. Thus, ion implantation with C^+ results in an increase in hardness in silicon. Note that the surface layer of the implanted zone is much harder compared with the subsurface, and may be brittle leading to higher wear on the surface. The subsurface of the implanted zone (SiC) is harder than the virgin silicon, resulting in higher wear resistance, which is also observed in the results of the macroscale tests conducted at high loads.

The Effect of Oxide Films on Tribological Properties of Silicon

Macroscale friction and wear experiments have been performed using a magnetic disk drive with bare, oxidized, and implanted pins sliding against amorphous-carbon-coated magnetic disks lubricated with a thin layer of perfluoropolyether lubricant [50.151–154].

poorer than that of virgin silicon up to about 80 μN. Above 80 μN, the wear resistance of implanted Si improves. As one continues to run tests at 40 μN for a larger number of cycles, the implanted sample, which forms hard and tough silicon carbide, exhibits higher wear re-

Table 50.4 RMS, microfriction, microscratching/microwear and nanoindentation hardness data for various virgin, coated and treated silicon samples.

Material	Rms roughness[a] (nm)	Coefficient of microscale friction[b]	Scratch depth[c] at 40μN (nm)	Wear depth[c] at 40μN (nm)	Nanohardness[c] at 100μN (GPa)
Si(111)	0.11	0.03	20	27	11.7
Si(110)	0.09	0.04	20		
Si(100)	0.12	0.03	25		
Polysilicon	1.07	0.04	18		
Polysilicon (lapped)	0.16	0.05	18	25	12.5
PECVD-oxide coated Si(111)	1.50	0.01	8	5	18.0
Dry-oxidized Si(111)	0.11	0.04	16	14	17.0
Wet-oxidized Si(111)	0.25	0.04	17	18	14.4
C[+]-implanted Si(111)	0.33	0.02	20	23	18.6

[a] Scan size of 500 nm × 500 nm using AFM
[b] Versus Si_3N_4 tip in AFM/FFM, radius 50 nm; at 1 μm × 1 μm scan size
[c] Measured using an AFM with a diamond tip of radius 100 nm

Representative profiles of the variation of the coefficient of friction with number of sliding cycles for Al_2O_3–TiC

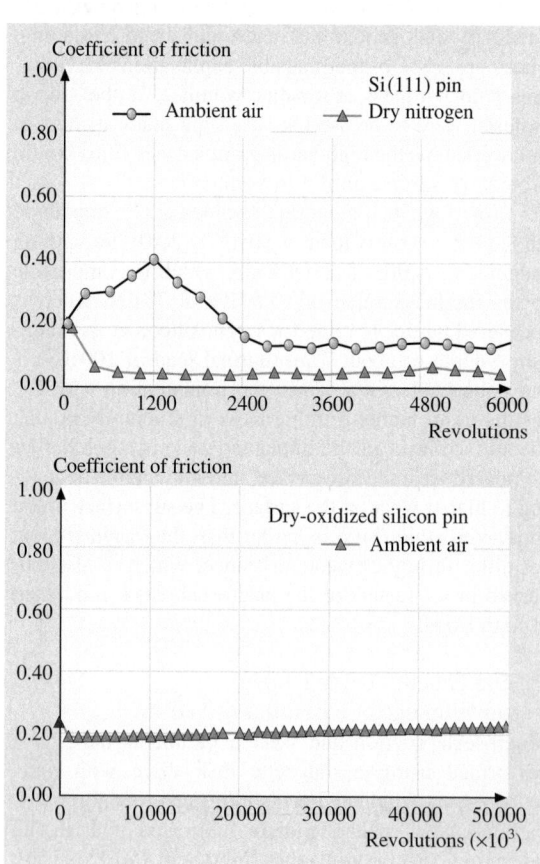

slider and bare and dry-oxidized silicon pins are shown in Fig. 50.16. For bare Si(111), after initial increase in the coefficient of friction, it drops to a steady-state value of 0.1 following the increase, as seen in Fig. 50.16. The rise in the coefficient of friction for the Si(111) pin is associated with the transfer of amorphous carbon from the disk to the pin, oxidation-enhanced fracture of pin material followed by tribochemical oxidation of the transfer film, while the drop is associated with the formation of a transfer coating on the pin. Dry-oxidized Si(111) exhibits excellent characteristics and no significant increase was observed over 50 000 cycles (Fig. 50.16). This behavior has been attributed to the chemical passivity of the oxide and lack of transfer of diamond-like carbon (DLC) from the disk to the pin. The behavior of PECVD oxide (data not presented here) was comparable to that of dry oxide but for the wet oxide there was some variation in the coefficient of friction (0.26–0.4). The difference between dry and wet oxide was attributed to increased porosity of the wet oxide [50.151]. Since tribochemical oxidation was determined to be a significant factor, experiments were conducted in dry nitrogen [50.152,153]. The variation of the coefficient of friction for a silicon pin sliding against a thin-film disk in dry nitrogen is shown in Fig. 50.16. It is seen that, in a dry nitrogen environment, the coefficient of friction of Si(111) sliding against a disk

Fig. 50.16 Coefficient of friction as a function of number of sliding revolutions in ambient air for a Si(111) pin in ambient air and dry nitrogen and a dry-oxidized silicon pin in ambient air [50.151]

decreased from an initial value of about 0.05–0.2 with continued sliding. Based on SEM and chemical analysis, this behavior has been attributed to the formation of a smooth amorphous-carbon/lubricant transfer patch and suppression of oxidation in a dry nitrogen environment. Based on macroscale tests using disk drives, it is found that the friction and wear performance of bare silicon is not adequate. With dry-oxidized or PECVD SiO_2-coated silicon, no significant friction increase or interfacial degradation was observed in ambient air.

Table 50.4 and Fig. 50.17 show surface roughness, microscale friction and scratch data and nanoindentation hardness for the various silicon samples [50.147]. Scratch experiments were performed using a diamond tip in an AFM. Results on polysilicon samples are also shown for comparison. Coefficients of microscale friction values for all the samples are about the same. These samples could be scratched at $10 \mu N$ load. Scratch depth increased with normal load. Crystalline orientation of silicon has little influence on scratch resistance because natural oxidation of silicon in ambient masks the expected effect of crystallographic orientation. PECVD-oxide samples showed the best scratch resistance, followed by dry-oxidized, wet-oxidized, and ion-implanted samples. Ion implantation with C^+ does not appear to improve scratch resistance.

Wear data on the silicon samples are also presented in Table 50.4 [50.147]. PECVD-oxide samples showed superior wear resistance followed by the dry-oxidized, wet-oxidized, and ion-implanted samples. This agrees with the trends seen in scratch resistance. In PECVD, ion bombardment during the deposition improves the

coating properties such as suppression of columnar growth, freedom from pinhole, decrease in crystalline size, and increase in density, hardness and substrate-coating adhesion. These effects may help in improving mechanical integrity of the sample surface. Coatings and treatments improved nanohardness of silicon. Note that dry-oxidized and PECVD films are harder than wet-oxidized films as these films may be porous. The high hardness of oxidized films may be responsible for the high measured scratch/wear resistance.

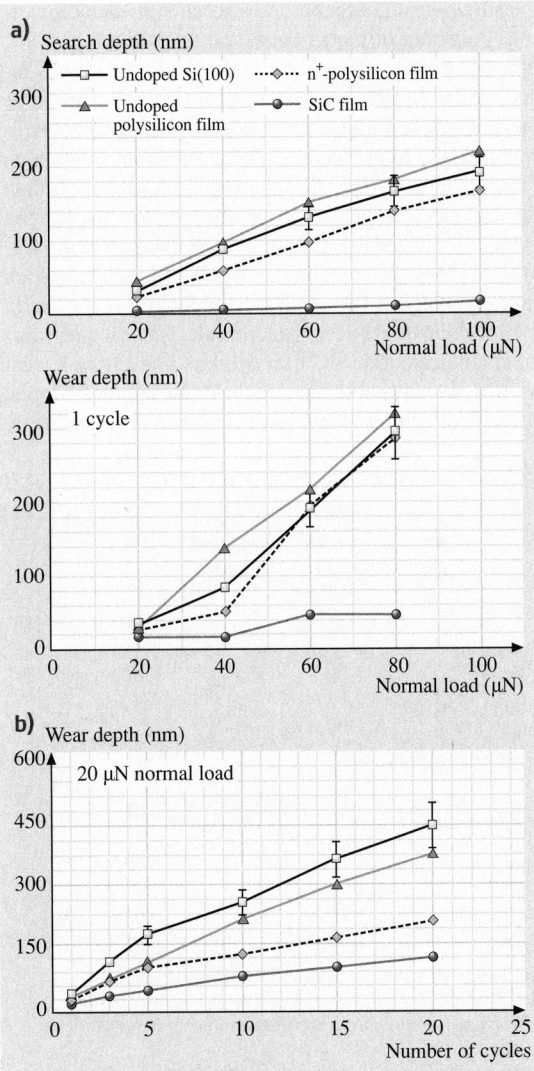

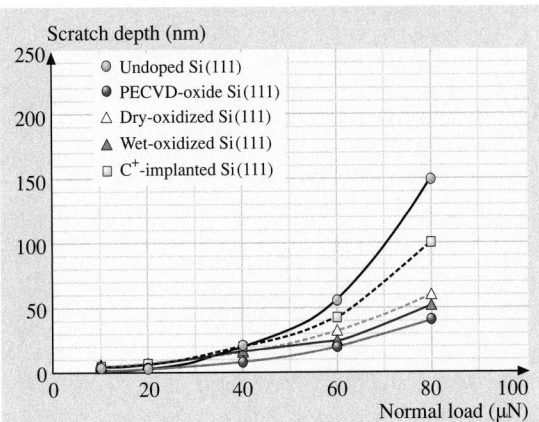

Fig. 50.17 Scratch depth as a function of normal load after 10 cycles for various silicon samples: virgin, treated, and coated [50.147]

Fig. 50.18 (a) Scratch depths for 10 cycles as a function of normal load and (b) wear depths as a function of normal load and of number of cycles for various samples [50.155]

50.2.2 Tribological Properties of Polysilicon Films and SiC Film

Studies have also been conducted on undoped polysilicon film, heavily doped (n$^+$-type) polysilicon film, heavily doped (p$^+$-type) single-crystal Si(100) and 3C–SiC (cubic or β-SiC) film [50.155–157]. The polysilicon films studied here are different from those discussed previously.

Table 50.5 presents a summary of the tribological studies conducted on polysilicon and SiC films. Values for single-crystal silicon are also shown for comparison. Polishing of the as-deposited polysilicon and SiC films drastically affect the roughness as the values reduce by two orders of magnitude. Si(100) appears to be the smoothest, followed by polished undoped polysilicon and SiC films, which have comparable roughness. The doped polysilicon film shows higher roughness than the undoped sample, which is attributed to the doping process. Polished SiC film shows the lowest friction followed by polished and undoped polysilicon film, which strongly supports the candidacy of SiC films for use in MEMS/NEMS devices. Macroscale friction measurements indicate that SiC film exhibits one of the lowest friction values as compared to the other samples. Doped polysilicon sample shows low friction on the macroscale

as compared to the undoped polysilicon sample, possibly due to the doping effect.

Figure 50.18a shows a plot of scratch depth versus normal load for various samples [50.155, 156]. Scratch depth increases with increasing normal load. Fig. 50.19 shows AFM three-dimensional (3D) maps and averaged two-dimensional (2D) profiles of the scratch marks on the various samples. It is observed that scratch depth increases almost linearly with the normal load. Si(100) and the doped and undoped polysilicon film show similar scratch resistance. From the data, it is clear that the SiC film is much more scratch-resistant than the other samples. Figure 50.18b shows results from microscale wear tests on the various films. For all the materials, wear depth increases almost linearly with increasing number of cycles. This suggests that the material is removed layer by layer in all the materials. Here also, SiC film exhibits lower wear depths than the other samples. Doped polysilicon film wears less than the undoped film. The higher fracture toughness and higher hardness of SiC compared to Si(100) is responsible for its lower wear. Also the higher thermal conductivity of SiC (see Table 50.2 compared to the other materials leads to lower interface temperatures, which generally results in less degradation of the surface [50.34, 46, 81]). Doping of the

Table 50.5 Summary of micro/nanotribological properties of the sample materials

Sample	Rms roughness[a] (nm)	P-V distance[a] (nm)	Coefficient of friction		Scratch depth[d] (nm)	Wear depth[e] (nm)	Nano-hardness[f] (GPa)	Young's modulus[f] (GPa)	Fracture toughness[g] K_{IC} (MPa m$^{1/2}$)
			Micro[b]	Macro[c]					
Undoped Si(100)	0.09	0.9	0.06	0.33	89	84	12	168	0.75
Undoped polysilicon film (as deposited)	46	340	0.05						
Undoped polysilicon film (polished)	0.86	6	0.04	0.46	99	140	12	175	1.11
n$^+$-type polysilicon film (as deposited)	12	91	0.07						
n$^+$-type polysilicon film (polished)	1.0	7	0.02	0.23	61	51	9	95	0.89
SiC film (as deposited)	25	150	0.03						
SiC film (polished)	0.89	6	0.02	0.20	6	16	25	395	0.78

[a] Measured using AFM over a scan size of 10 μm × 10 μm
[b] Measured using AFM/FFM over a scan size of 10 μm × 10 μm
[c] Obtained using a 3-mm-diameter sapphire ball in a reciprocating mode at a normal load of 10 mN and average sliding speed of 1 mm/s after 4 m sliding distance
[d] Measured using AFM at a normal load of 40 μN for 10 cycles; scan length of 5 μm
[e] Measured using AFM at normal load of 40 μN for 1 cycle, wear area of 2 μm × 2 μm
[f] Measured using nanoindenter at a peak indentation depth of 20 nm
[g] Measured using microindenter with Vickers indenter at a normal load of 0.5 N

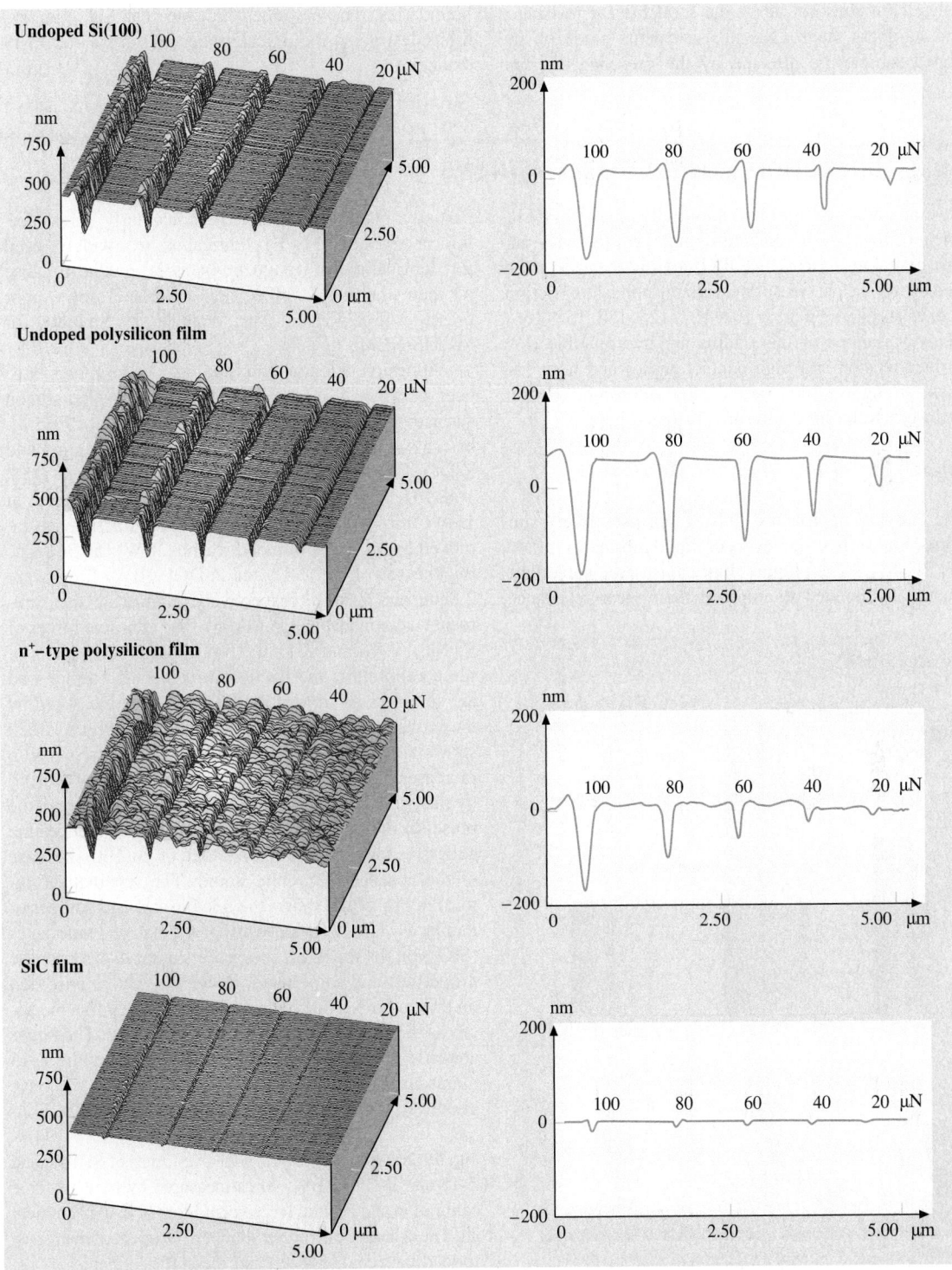

Fig. 50.19 AFM 3D maps and averaged 2D profiles of scratch marks on various samples [50.155]

polysilicon does not affect the scratch/wear resistance and hardness much. The measurements made on the doped sample are affected by the presence of grain boundaries. These studies indicate that SiC film exhibits desirable tribological properties for use in MEMS devices.

50.3 Lubrication Studies for MEMS/NEMS

Several studies of liquid perfluoropolyether (PFPE) lubricant films, self-assembled monolayers (SAMs), and hard diamond-like carbon (DLC) coatings have been carried out for the purpose of minimizing adhesion, friction, and wear [50.46, 47, 82, 102, 122–125, 154, 158–166]. Many variations of these films are hydrophobic (low surface tension and high contact angle) and have low shear strength, which provide low adhesion, friction, and wear. Relevant details are presented here.

50.3.1 Perfluoropolyether Lubricants

The classical approach to lubrication uses freely supported multimolecular layers of liquid lubricants [50.46, 47, 79, 81]. The liquid lubricants are sometimes chemically bonded to improve their wear resistance.

Partially chemically bonded, molecularly thick perfluoropolyether (PFPE) lubricants are widely used for lubrication of magnetic storage media because of their thermal stability and extremely low vapor pressure [50.34], and are found to be suitable for MEMS/NEMS devices.

Adhesion, friction, and durability experiments have been performed on virgin Si (100) surfaces and silicon surfaces lubricated with two commonly used PFPE lubricants – Z-15 (with $-CF_3$ nonpolar end groups) and Z-DOL (with $-OH$ polar end groups) [50.46, 47, 158, 160, 161, 165]. Z-DOL film was thermally bonded at 150 °C for 30 minutes and the unbonded fraction was removed by a solvent (bonded washed, BW) [50.34]. The thicknesses of the Z-15 and Z-DOL (BW) films were 2.8 nm and 2.3 nm, respectively. Nanoscale measurements were made using an AFM. The adhesive forces of Si(100), Z-15 and Z-DOL (BW) measured by plots of force calibration and friction force versus normal load are summarized in Fig. 50.20. The results measured by these two methods are in good agreement. Figure 50.20 shows that the presence of mobile Z-15 lubricant film increases the adhesive force compared to that of Si(100) by meniscus formation [50.79, 81, 167]. In contrast, the presence of solid phase Z-DOL (BW) film reduces the adhesive force as compared that of Si(100) because of the absence of mobile liquid. The schematic (bottom) in Fig. 50.20 shows the relative size and sources of meniscus. It is well known that the native oxide layer (SiO_2) on the top of Si(100) wafers exhibits hydrophilic properties, and some water molecules can be adsorbed on this surface. The condensed water will form a meniscus as the tip approaches the sample surface. The larger adhesive force in Z-15 is not only caused by the Z-15 meniscus, the nonpolarized Z-15 liquid does not have good wettability and strong bonding with Si(100). Con-

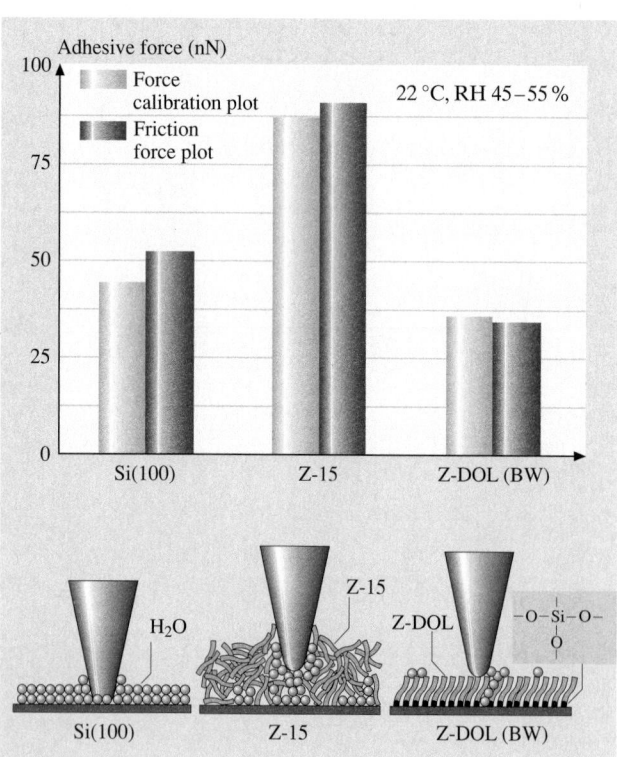

Fig. 50.20 Summary of the adhesive forces of Si(100) and Z-15 and Z-DOL (BW) films measured by plots of force calibration and friction force versus normal load in ambient air. The schematic (*bottom*) showing the effect of meniscus, formed between AFM tip and the surface sample, on the adhesive and friction forces [50.158]

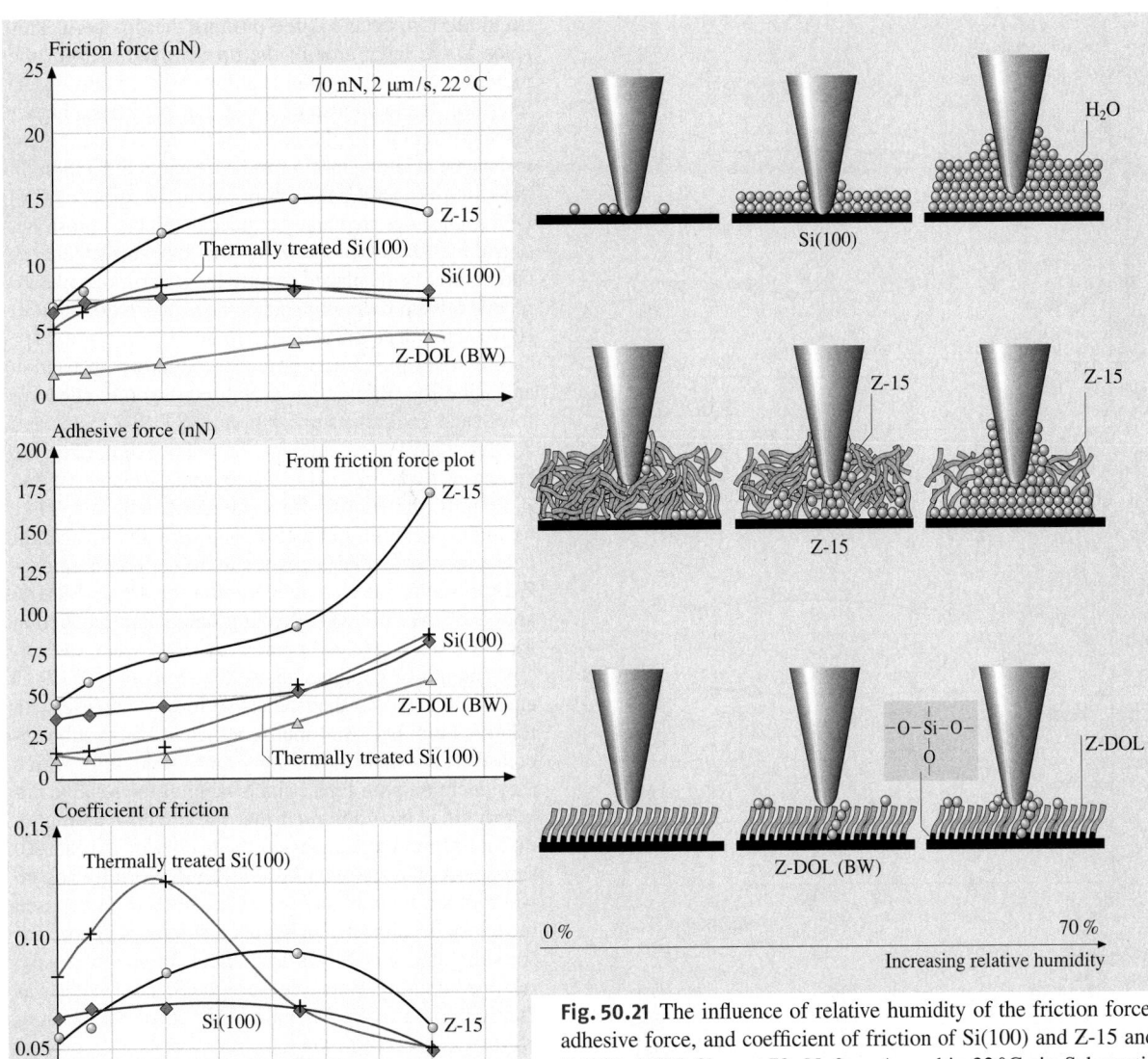

Fig. 50.21 The influence of relative humidity of the friction force, adhesive force, and coefficient of friction of Si(100) and Z-15 and Z-DOL (BW) films at 70 nN, 2 μm/s, and in 22 °C air. Schematic (*right* shows the change of meniscus while increasing the relative humidity. In this figure, the thermal treated Si(100) represents the Si(100) wafer that was baked at 150 °C for 1 h in an oven (in order to remove the adsorbed water) just before it was placed in the 0% RH chamber [50.158]

sequently, in the ambient environment, the condensed water molecules from the environment permeate through the liquid Z-15 lubricant film and compete with the lubricant molecules present on the substrate. The interaction of the liquid lubricant with the substrate is weakened, and a boundary layer of the liquid lubricant forms puddles [50.160, 161]. This dewetting allows wa-

ter molecules to be adsorbed on the Si(100) surface as aggregates along with Z-15 molecules. And both of them can form a meniscus while the tip approaches the surface. Thus, the dewetting of the liquid Z-15 film results in a higher adhesive force and poorer lubrication performance. In addition, as the Z-15 film is fairly soft compared to the solid Si(100) surface, penetration of the

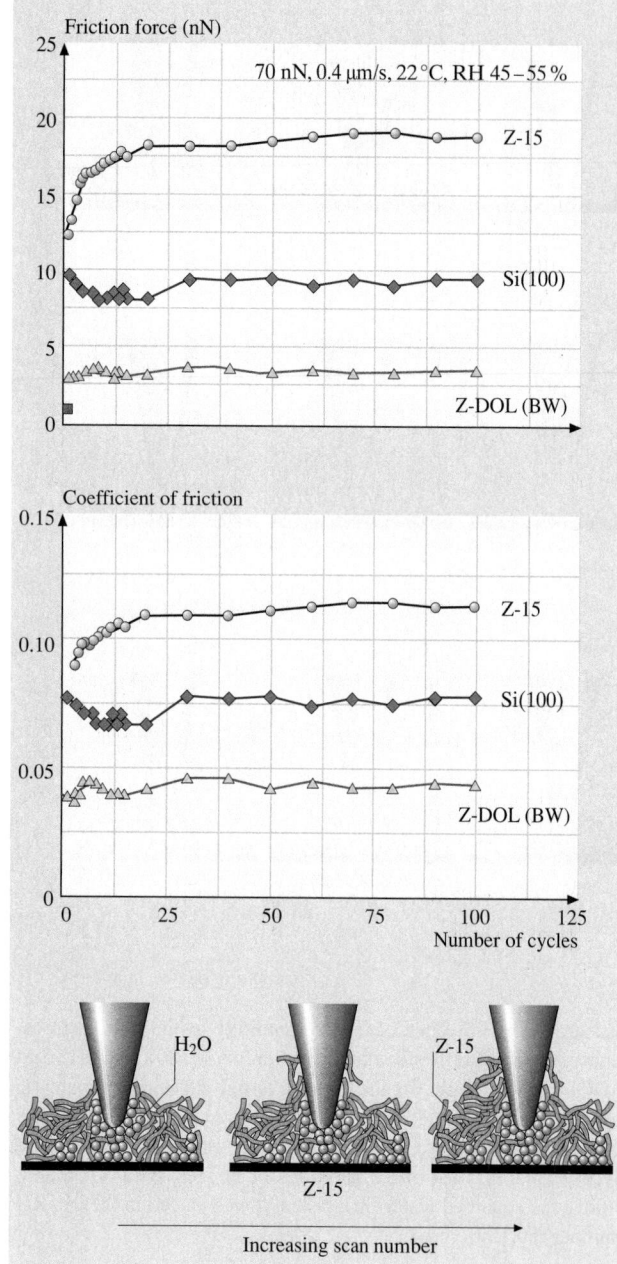

Fig. 50.22 Friction force and coefficient of friction versus number of sliding cycles for Si(100) and Z-15 and Z-DOL (BW) films at 70 nN, 0.4 μm/s, and in ambient air. Schematic (*bottom*) shows that some liquid Z-15 molecules can be attached onto the tip. The molecular interaction between the attached molecules onto the tip with the Z-15 molecules in the film results in an increase of the friction force with multiple scanning [50.158]

tip in the film occurs while pushing the tip down. This leads to the large area of the tip involved to form the meniscus at the tip–liquid (mixture of Z-15 and water) interface. It should also be noted that Z-15 has a higher viscosity than water, therefore Z-15 film provides higher resistance to lateral motion and coefficient of friction. In the case of Z-DOL (BW) film, the active groups of Z-DOL molecules are mostly bonded on Si(100) substrate, thus the Z-DOL (BW) film has low free surface energy and cannot be displaced readily by water molecules or readily adsorb water molecules. Thus, the use of Z-DOL (BW) can reduce the adhesive force.

To study the effect of relative humidity on friction and adhesion, the variation of friction force, adhesive force, and coefficient of friction of Si(100), Z-15, and Z-DOL (BW) as a function of relative humidity are shown in Fig. 50.21. This shows that, for Si(100) and Z-15 film, the friction force increases with a relative humidity increase up to 45%, and then shows a slight decrease with further increases in the relative humidity. Z-DOL (BW) has a smaller friction force than Si(100) and Z-15 over the whole testing range, and its friction force shows a relative apparent increase when the relative humidity is higher than 45%. For Si(100), Z-15 and Z-DOL (BW), their adhesive forces increase with relative humidity. And their coefficients of friction increase with a relative humidity up to 45%, after which they decrease with further increasing of the relative humidity. It is also observed that the effect of humidity on Si(100) really depends on the history of the Si(100) sample. As the surface of Si(100) wafer readily adsorb water in air, without any pretreatment the Si(100) used in our study almost reaches its saturated stage of adsorbed water, and shows less effect during increasing relative humidity. However, once the Si(100) wafer was thermally treated by baking at 150 °C for 1 h, a larger effect was observed.

The schematic (right) in Fig. 50.21 shows that Si(100), because of its high free surface energy, can adsorb more water molecules with increasing relative humidity. As discussed earlier, for the Z-15 film in the humid environment, the condensed water from the humid environment competes with the lubricant film present on the sample surface, and the interaction of the liquid lubricant film with the silicon substrate is weakened and a boundary layer of the liquid lubricant forms puddles. This dewetting allows water molecules to be adsorbed on the Si(100) substrate mixed with Z-15 molecules [50.160, 161]. Obviously, more water molecules can be adsorbed on the Z-15 surface with increasing relative humidity. The larger amount of adsorbed water in the case of Si(100), along with the lubricant molecules in the case of the Z-15 film, forms a larger water meniscus, which leads to an increase of the friction force, adhesive force, and coefficient of friction of Si(100) and Z-15 with humidity. However, at a very high humidity of 70%, large quantities of adsorbed water can form a continuous water layer that separates the tip and sample surface, acting as a kind of lubricant, which causes a decrease in the friction force and coefficient of friction. For Z-DOL (BW) film, because of their hydrophobic surface properties, water molecules can be adsorbed at humidity higher than 45%, and causes an increase in the adhesive force and friction force.

To study the durability of lubricant films at nanoscale, the friction of Si(100), Z-15, and Z-DOL (BW) as a function of the number of scanning cycles are shown in Fig. 50.22. As observed earlier, the friction force and coefficient of friction of Z-15 are higher than that of Si(100) with the lowest values for Z-DOL(BW). During cycling, the friction force and coefficient of friction of Si(100) show a slight decrease during the initial few cycles then remain constant. This is related to the removal of the top adsorbed layer. In the case of Z-15 film, the friction force and coefficient of friction show an increase during the initial few cycles and then approach higher stable values. This is believed to be caused by the attachment of the Z-15 molecules onto the tip. The molecular interaction between these attached molecules to the tip and molecules on the film surface is responsible for an increase in the friction. But after several scans, this molecular interaction reaches equilibrium and after that the friction force and coefficient of friction remain constant. In the case of Z-DOL (BW) film, the friction force and coefficient of friction start out low and remain low during the entire test for 100 cycles. This suggests that Z-DOL (BW) molecules do not become attached or displaced as readily as Z-15.

50.3.2 Self–Assembled Monolayers (SAMs)

For lubrication of MEMS/NEMS, another effective approach involves the deposition of organized and dense molecular layers of long-chain molecules. Two common methods to produce monolayers and thin films are Langmuir–Blodgett (LB) deposition and self-assembled monolayers (SAMs) by chemical grafting of molecules. LB films are physically bonded to the substrate by weak van der Waals attraction, while SAMs are chemically bonded via covalent bonds to the substrate. Because of the choice of chain length and terminal linking group that SAMs offer, they hold great promise for boundary

lubrication of MEMS/NEMS. A number of studies have been conducted to study the tribological properties of various SAMs [50.122–125, 159, 162–164, 166, 168].

Bhushan and *Liu* [50.162] studied the effect of film compliance on adhesion and friction. They used

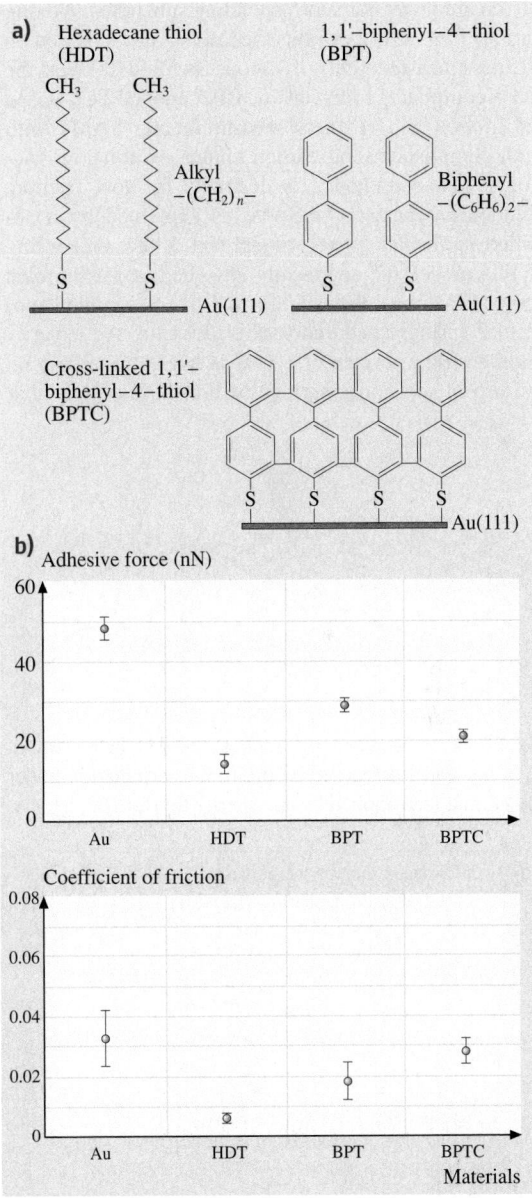

Fig. 50.23 (a) Schematics of structures of hexadecane thiol and biphenyl thiol SAMs on Au(111) substrates, and (b) adhesive force and coefficient of friction of Au(111) substrate and various SAMs

hexadecane thiol (HDT), 1,1,biphenyl-4-thiol (BPT), and crosslinked BPT (BPTC) solvent deposited on an Au(111) substrate, Fig. 50.23a. The average values and standard duration of the adhesive force and coefficient of friction are presented in Fig. 50.23b. Based on the data, the adhesive force and coefficient of friction of SAMs are lower than corresponding substrates. Among various films, HDT exhibits the lowest values. Based on stiffness measurements of various SAMs, HDT was the most compliant, followed by BPT and BPTC. Based on friction and stiffness measurements, SAMs with high-compliance long carbon chains exhibit low friction; chain compliance is desirable for low friction. Friction mechanism of SAMs is explained by a so-called *molecular spring* model Fig. 50.24. According to this model, the chemically adsorbed self-assembled molecules on a substrate are just like assembled molecular springs anchored to the substrate. An asperity sliding on the surface of SAMs is like a tip sliding on the top of molecular springs or brush. The molecular

spring assembly has compliant features and can experience orientation and compression under load. The orientation of the molecular springs or brush under normal load reduces the shearing force at the interface, which in turn reduces the friction force. The orientation is determined by the spring constant of a single molecule as well as the interaction between the neighboring molecules, which can be reflected by packing density or packing energy. It should be noted that the orientation can lead to conformational defects along the molecular chains, which lead to energy dissipation.

SAMs with high-compliance long carbon chains also exhibit the best wear resistance [50.162, 163]. In wear experiments, curves of wear depth as a function of nor-

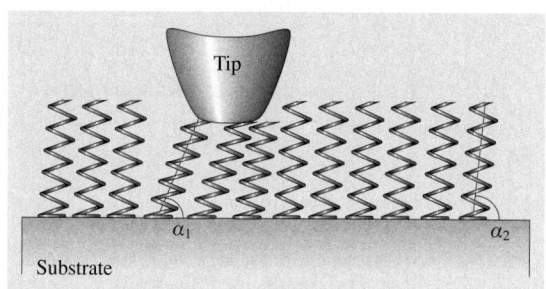

Fig. 50.24 Molecular spring model of SAMs. In this figure, $\alpha_1 < \alpha_2$, which is caused by the further orientation under the normal load applied by an asperity tip [50.162]

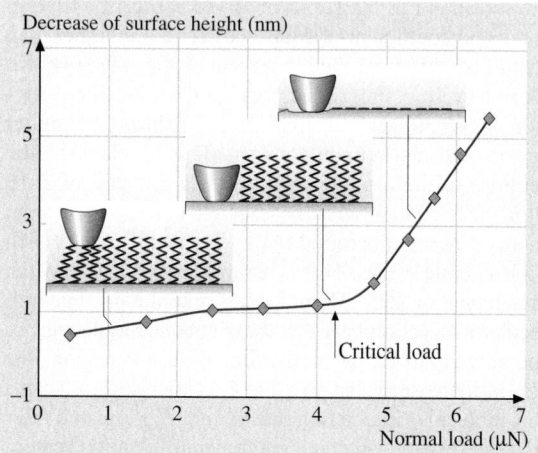

Fig. 50.25 Illustration of the wear mechanism of SAMs with increasing normal load [50.163]

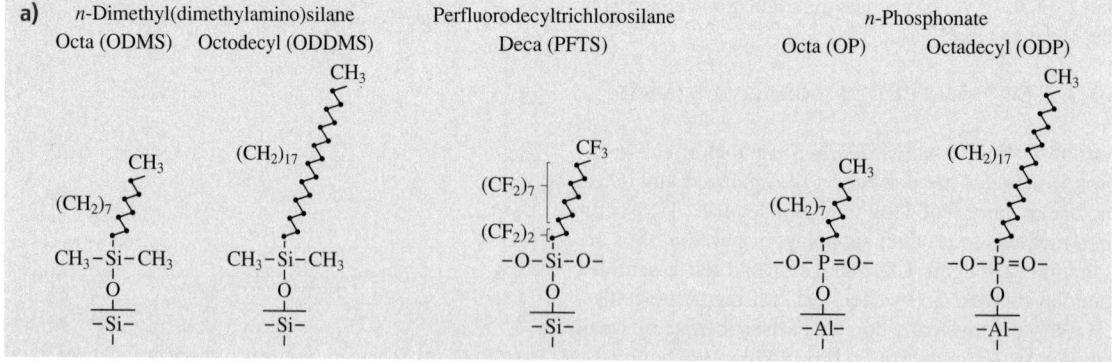

Fig. 50.26 **(a)** Schematics of structures of perfluoroalkylsilane and alkylsilane SAMs on Si with native oxide substrates, and alkylphosphonate SAMs on Al with native oxide, and **(b)** contact angle, adhesive force, friction force, and coefficient of friction of Si with native oxide and Al with native oxide substrates and with various SAMs ▶

Fig. 50.27 Decrease of surface height as a function of normal load after one scan cycle for various SAMs on Si and Al substrates, and comparison of critical loads for failure during wear tests for various SAMs▶

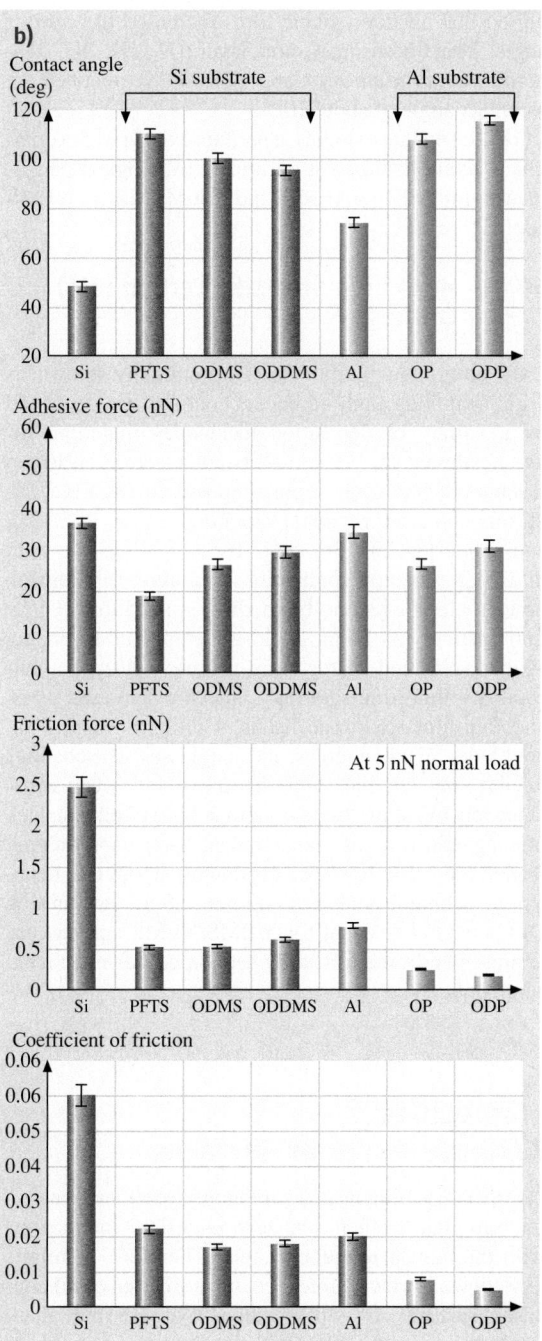

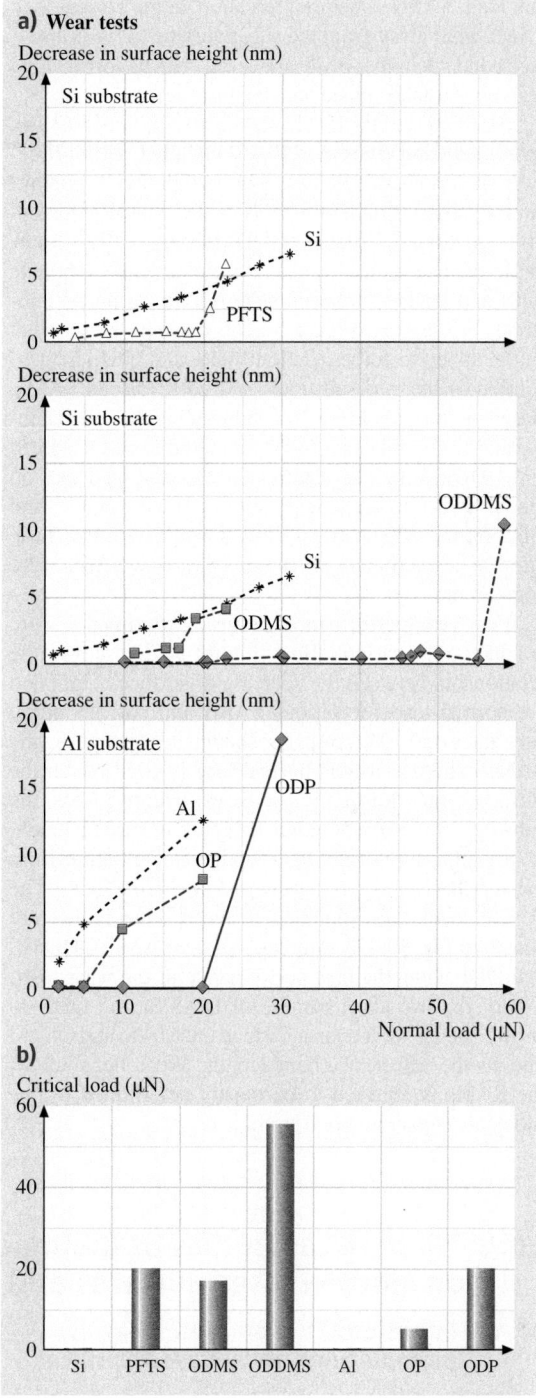

mal load show a critical normal load. A representative curve is shown in Fig. 50.25. Below the critical normal load, SAMs undergo orientation, at the critical load SAMs wear away from the substrate due to weak interface bond strengths, while above the critical normal load severe wear takes place on the substrate.

Bhushan et al. [50.122], *Kasai* et al. [50.124], and *Tambe* and *Bhushan* [50.164] studied perfluorodecyltricholorosilane (PFTS), *n*-octyldimethyl (dimethylamino) silane (ODMS) ($n = 7$), and *n*-octadecylmethyl (dimethylamino) silane (ODDMS) ($n = 17$) vapor-phase-deposited on a Si substrate, and octylphosphonate (OP) and octadecylphosphonate (ODP) on an Al substrate, Fig. 50.26a. Figure 50.26b presents the contact angle, adhesive force, friction force, and coefficient of friction of the two substrates and with various SAMs. Based on the data, PFTS/Si exhibits a higher contact angle and lower adhesive force compared to of ODMS/Si and ODDMS/Si. The data for ODMS and ODDMS on the Si substrate are comparable to those for OP and ODP on the Al substrate. Thus the substrate had little effect. The coefficient of friction of various SAMs were comparable.

For wear performance studies, experiments were conducted on various films. Figure 50.27a shows the relationship between the decrease of surface height and the normal load for various SAMs and corresponding substrates [50.124, 164]. As shown in the figure, the SAMs exhibit a critical normal load beyond which the surface height drastically decreases. Unlike SAMs, the substrates show a monotonic decrease in surface height with increasing normal load, with wear initiating from the very beginning, i. e., even for low normal loads. The critical loads corresponding to the sudden failure are shown in Fig. 50.27b. Amongst all the SAMs, ODDMS and ODP show the best performance in the wear tests. Out of the two alkyl SAMs, ODDMS/Si and ODP/Al showed better wear resistance than ODMS/Si and OD/Al due to the effect of chain length. Wear behavior of the SAMs is reported to be mostly determined by the molecule–substrate bond strengths.

Bhushan et al. [50.122] and *Lee* et al. [50.125] studied various fluoropolymer multilayers and fluorosilane monolayers on Si and a selected fluorosilane on PDMS surfaces. For nanoscale devices such as in nanochannels, monolayers are preferred. They reported that all fluorosilane films increased the contact angle. The fluorosilane monolayer 1H, 1H, 2H, 2H–perfluorodecyltriethoxysilane (PFDTES) resulted in a contact angle of about 100°.

Based on these studies, a perfluoro SAM with a compliant layer should have optimized tribological performance for MEMS/NEMS and BioMEMS/BioNEMS applications.

50.3.3 Hard Diamond–Like Carbon (DLC) Coatings

Hard amorphous carbon (a-C), commonly known as DLC (implying high hardness) coatings are deposited by a variety of deposition techniques including filtered cathodic arc (FCA), ion beam, electron cyclotron resonance chemical vapor deposition (ECR-CVD), plasma-enhanced chemical vapor deposition (PECVD), and sputtering [50.136, 154]. These coatings are used in a wide range on applications including tribological, optical, electronic, and biomedical applications. Ultra-thin coatings (3–10 nm thick) are employed to protect against wear and corrosion in magnetic storage applications – thin-film rigid disks, metal-evaporated tapes, and thin-film read/write heads –, Gillette Mach 3 razor blades, glass windows, and sunglasses. The coatings exhibit low friction, high hardness and wear resistance, chemical inertness to both acids and alkalis, lack of magnetic response, and optical band gaps ranging from zero to a few eV, depending upon the deposition technique and its conditions. Selected data on DLC coatings relevant for MEMS/NEMS applications is presented in the following section on adhesion measurements.

50.4 Tribological Studies of Biological Molecules on Silicon–Based Surfaces and of Coated Polymer Surfaces

50.4.1 Adhesion, Friction, and Wear of Biomolecules on Si–Based Surfaces

Proteins on silicon-based surfaces are of extreme importance in various applications including silicon microimplants, various bioMEMS such as biosensors, and therapeutics. Silicon is a commonly used substrate in microimplants, but it can have undesired interactions with the human immune system. Therefore, to mimic a biological surface, protein coatings are used on silicon-based surfaces as a passivation layer, so that these implants are compatible with the body and avoid rejection. Whether this surface treatment is applied to a large implant or a bioMEMS, the function of the protein passivation is obtained from the nanoscale 3D structural conformation of the protein. Proteins are also

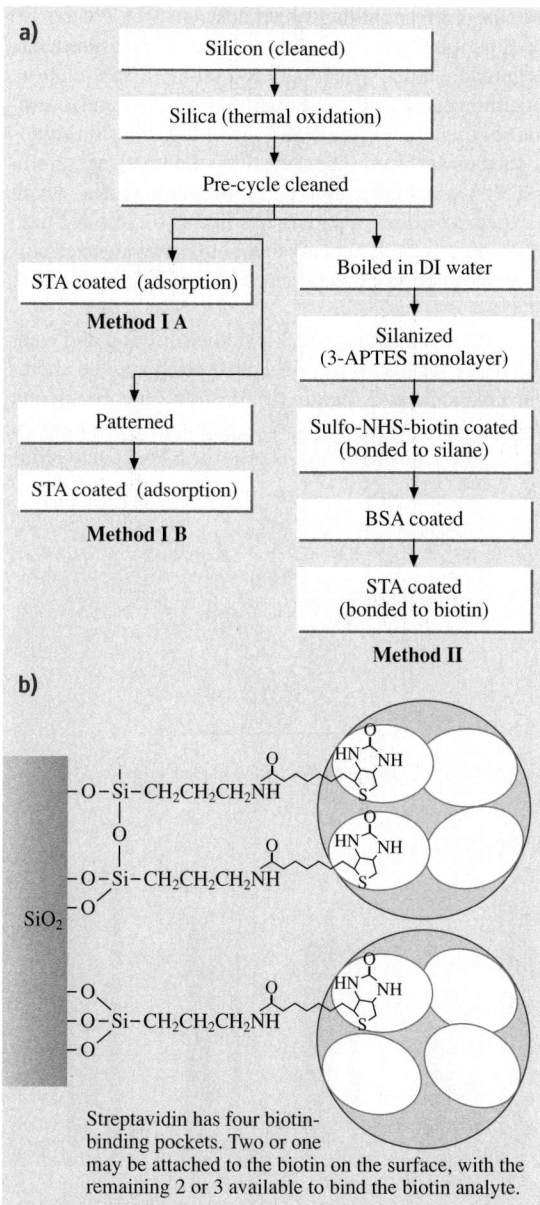

a)

Silicon (cleaned)

↓

Silica (thermal oxidation)

↓

Pre-cycle cleaned

STA coated (adsorption)

Method I A

Boiled in DI water

↓

Silanized
(3-APTES monolayer)

↓

Sulfo-NHS-biotin coated
(bonded to silane)

↓

BSA coated

↓

STA coated
(bonded to biotin)

Method II

Patterned

↓

STA coated (adsorption)

Method I B

b)

$-O-Si-CH_2CH_2CH_2NH$

$-O-Si-CH_2CH_2CH_2NH$

SiO_2

$-O-Si-CH_2CH_2CH_2NH$

Streptavidin has four biotin-binding pockets. Two or one may be attached to the biotin on the surface, with the remaining 2 or 3 available to bind the biotin analyte.

Fig. 50.28 (a) Flow chart showing the samples used and their preparation technique, and **(b)** a chemical structure showing streptavidin protein binding to the silica substrate by the chemical linker method

used in bioMEMS because of their function specificity. For biosensor applications, the extensive array of protein activities provides a rich supply of operations that may be performed at the nanoscale. Many antibodies (proteins) have an affinity to specific protein antigens. For example, pathogens (disease causing agents, e.g., virus or bacteria) trigger the production of antigens which can be detected when bound to a specific antibody on the biosensor. The specific binding behavior of proteins that has been applied to laboratory assays may also be redesigned for in vivo use as sensing elements of a bioMEMS. The epitope-specific binding properties of proteins to various antigens are useful in therapeutics. Adhesion between the protein and substrate affects the reliability of an application. Among other things, the morphology of the substrate affects the adhesion. Furthermore, for in vivo environments, the proteins on the biosensor surface should exhibit high wear resistance during direct contact with the tissue and circulatory blood flow without washing off.

Bhushan et al. [50.72] studied the step-by-step morphological changes and the adhesion of a model protein – streptavidin (STA) – on silicon-based surfaces. Figure 50.28a presents a flow chart showing the sequential modification of a silicon surface. In addition to physical adsorption, they also used nanopatterning and chemical linker methods to improve adhesion. Nanopatterned surfaces contain a large edge surface area, leading to high surface energy, which results in high adhesion. In the chemical linker method, sulfo-NHS-biotin was used as a cross linker because the bonds between the STA and the biotin molecule are some of the strongest noncovalent bonds known (Fig. 50.28b)). It was connected to the silica surface through a silane linker, 3-aminopropyltriethoxysilane (3-APTES). In order to make a bond between the silane linker and the silica surface, the silica surface was hydroxylated. Bovine serum albumin (BSA) was used before STA in order to block nonspecific binding sites of the STA protein with silica surface. Figure 50.29 shows the step-by-step morphological changes in the silica surface during the deposition process using the chemical linker method. There is an increase in roughness of the silica surface boiled in deionized (DI) water compared to the bare silica surface. After the silanization process, there are many free silane links on the surface which caused higher roughness. Once biotin was coated on the silanized surface, the surface became smoother. Finally, after the deposition of STA, surface shows large and small clumps. Presumably, the large clumps represent BSA and the smaller ones represent STA. To measure adhesion between STA

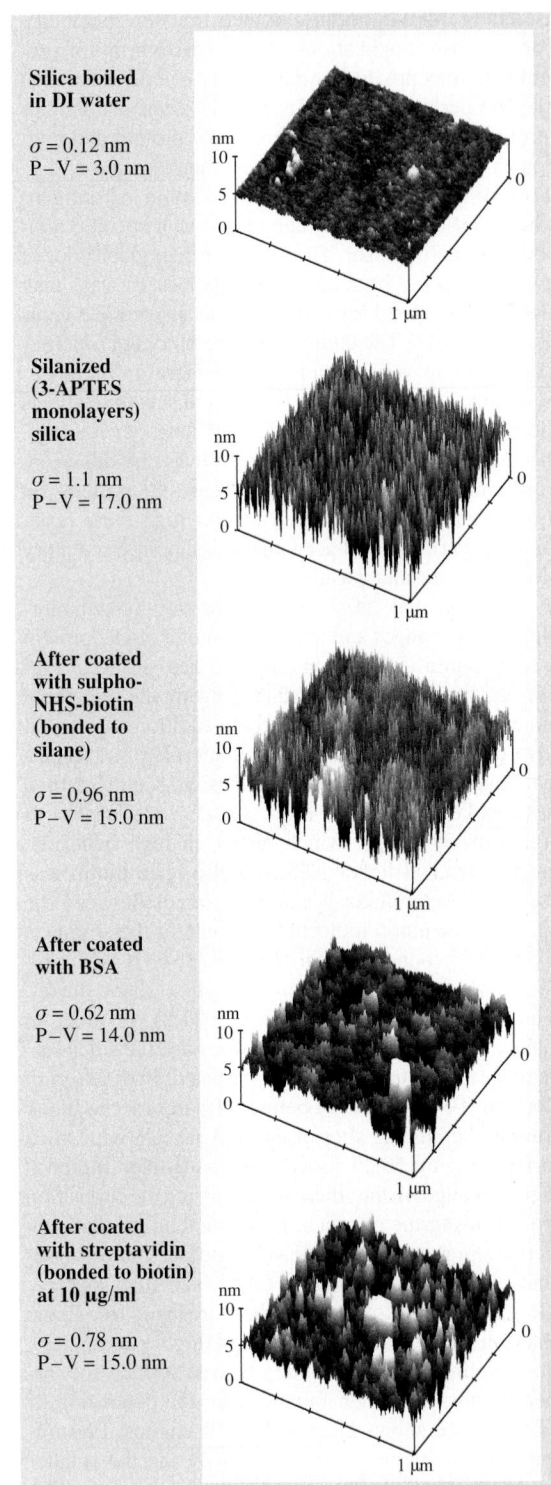

Silica boiled
in DI water

$\sigma = 0.12$ nm
$P-V = 3.0$ nm

Silanized
(3-APTES
monolayers)
silica

$\sigma = 1.1$ nm
$P-V = 17.0$ nm

After coated
with sulpho-
NHS-biotin
(bonded to
silane)

$\sigma = 0.96$ nm
$P-V = 15.0$ nm

After coated
with BSA

$\sigma = 0.62$ nm
$P-V = 14.0$ nm

After coated
with streptavidin
(bonded to biotin)
at 10 µg/ml

$\sigma = 0.78$ nm
$P-V = 15.0$ nm

Fig. 50.29 Morphological changes in silica surface during functionalization of silica surface by chemical linker imaged in PBS. Streptavidin is covalently bonded at a concentration of 10 µg/ml [50.72]

and the corresponding substrates, an STA-coated tip (or functionalized tip) was used and all measurements were made in phosphate buffered saline (PBS) solution, a medium commonly used in protein analysis and to simulate body fluid. Figure 50.30 shows the adhesion values of various surfaces. The adhesion value between biotin and STA was higher than that for other samples, which is expected. Edges of patterned silica also exhibited high adhesion values. It appears that both nanopatterned surfaces and chemical linker method increase adhesion with STA.

Tokachichu et al. [50.120] studied friction and wear of STA deposited by physical adsorption and the chemical linker method. Figure 50.31 shows the coefficient of friction between the Si_3N_4 tip and various samples. The coefficient of friction is less for STA-coated silica

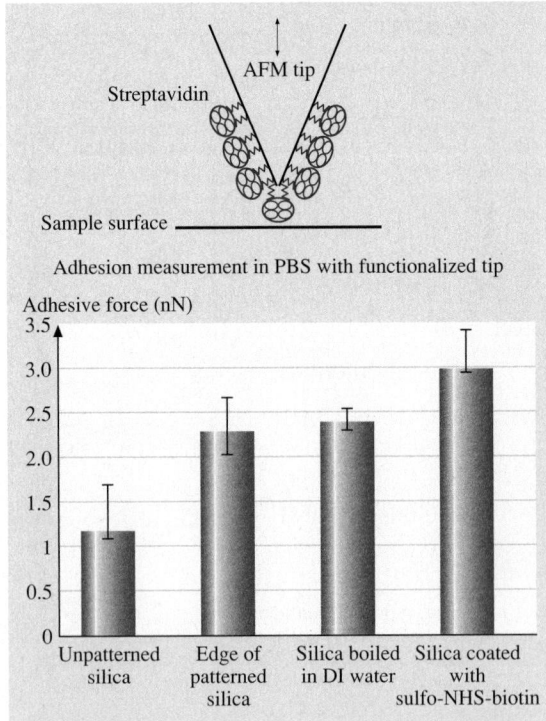

Fig. 50.30 Adhesion measurements of silica, patterned silicon, silica boiled in DI water, and sulfo-NHS-biotin using functionalized (with streptavidin) tips obtained from force–distance curves, captured in PBS

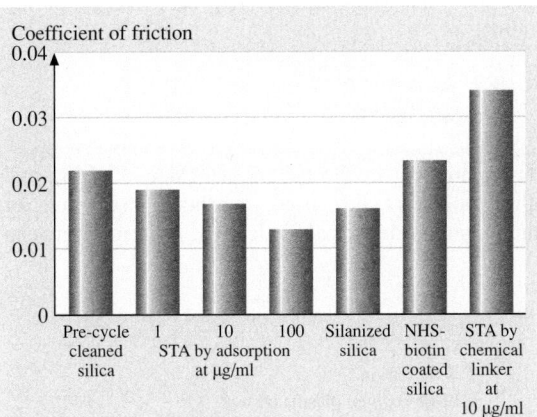

Fig. 50.31 Coefficient of friction for various surfaces with and without biomolecules

samples compared to uncoated sample. The streptavidin coating acts as a lubricant film. The coefficient of friction is found to be dependent upon the concentration of STA, and decreases with an increase in the concentration. *Bhushan* et al. [50.72] have reported that the density and distribution of the biomolecules vary with concentration. At higher concentration of the solution, the coated layer is more uniform and the silica substrate surface is highly covered with biomolecules than at lower concentration. This means that the surface forms a continuous lubricant film at higher concentration.

In the case of samples prepared by the chemical linker method, the coefficient of friction increases with an increase in the biomolecular chain length due to increased compliance. When normal load is applied on the surface, the surface becomes compressed, resulting in a larger contact area between the AFM tip and the

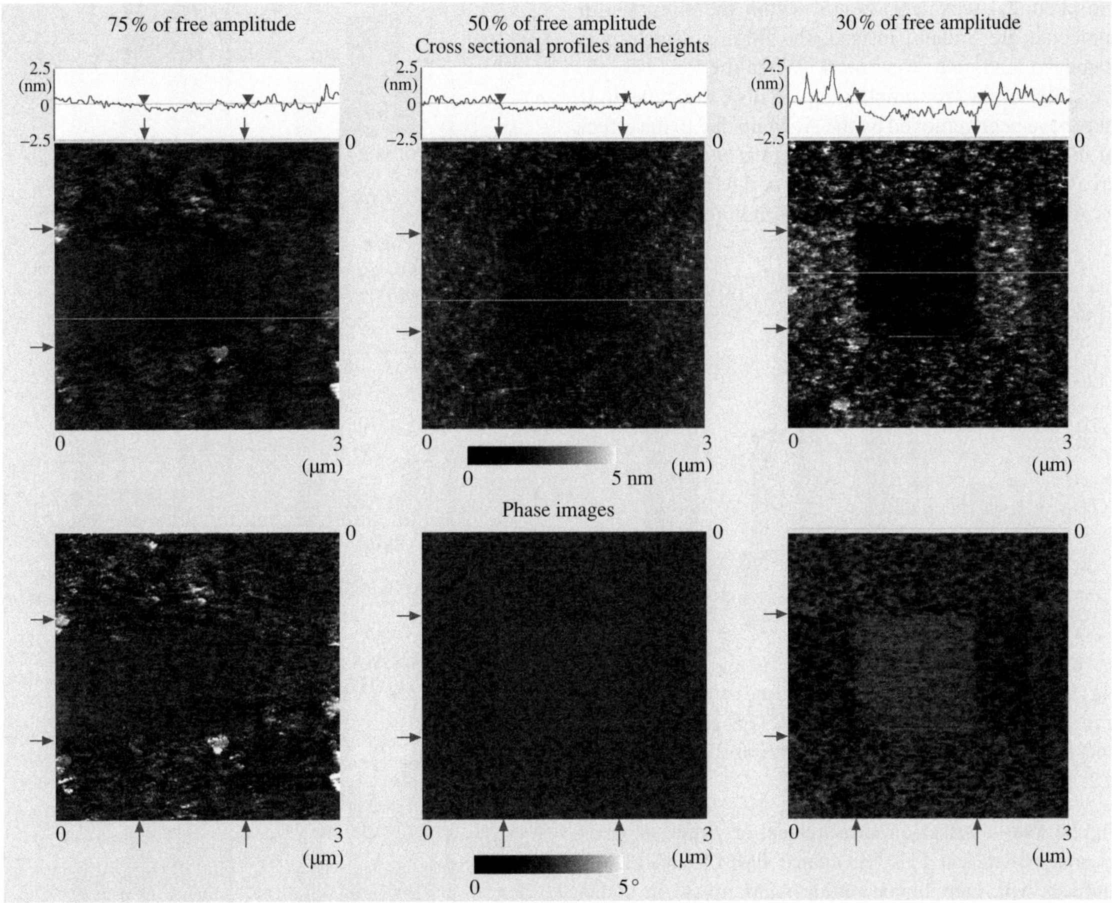

Fig. 50.32 Wear mark images and cross-sectional profiles of precycle cleaned silica coated with streptavidin at $10\,\mu g/ml$ by physical adsorption at three normal loads (increasing from *left to right*)

biomolecules. Besides that, the size of STA is much larger than that of APTES and biotin. This results in a tightly packed surface with the biomolecules, which results in very little lateral deflection of the linker in the case of STA-coated biotin. Due to this high contact area and low lateral deflection the friction force increases for the same applied normal load compared to directly adsorbed surface. These tests reveal that surfaces coated with biomolecules reduce the friction, but if the biomolecular coating of the surface is too thick or the surface has some cushioning effect, as seen in the chemical linker method, that increases the coefficient of friction.

Figure 50.32 shows the wear maps of STA deposited by physical adsorption at three normal loads. The wear depth increases with increasing normal load. An increase in normal load causes partial damage to the folding structure of the streptavidin molecules. It is unlikely that the chemical (covalent) bonds within the streptavidin molecule are broken; instead, the folding structure is damaged leading to wear mark. When the load is high, i.e., 30% of the free amplitude (≈ 8 nN), the molecules may have been removed by the AFM tip due to the effect of indentation/ Because of this, there is a significant increase in the wear depth from 50% of the free amplitude (≈ 6 nN) to 30% of the free amplitude (≈ 8 nN). The

data show that biomolecules will be damaged during sliding.

50.4.2 Adhesion of Coated Polymer Surfaces

As mentioned in Sect. 50.A, PMMA, PDMS, and other polymers are used in the construction of micro/nanofluidic-based biodevices. Adhesion between

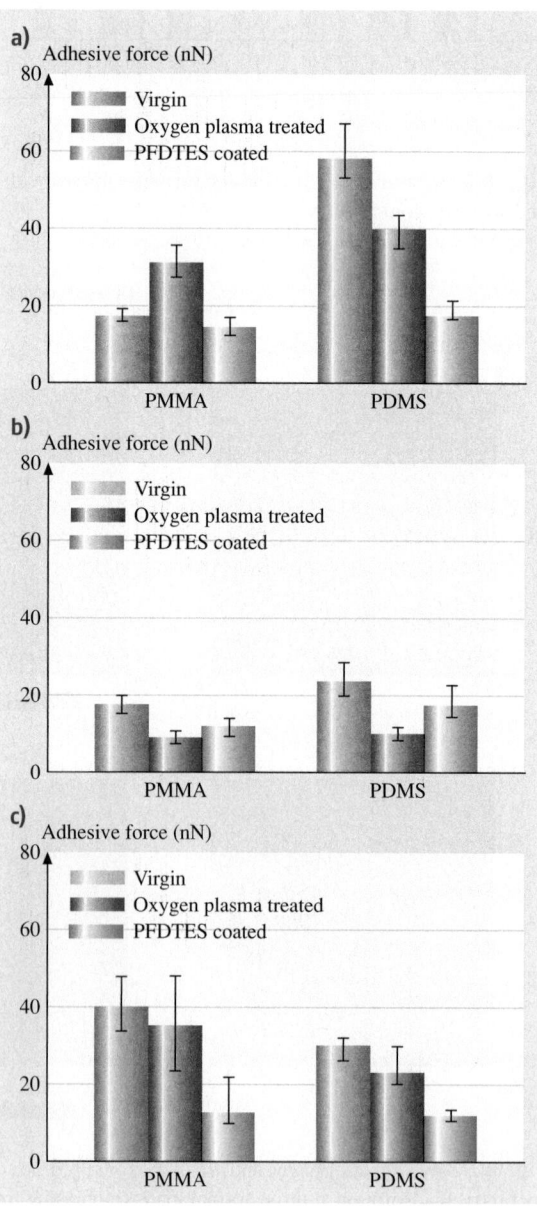

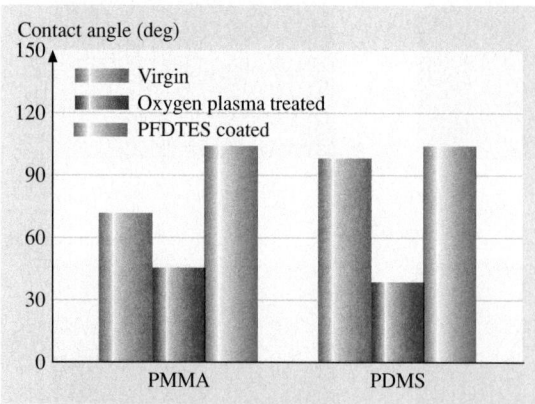

Fig. 50.33 Sessile drop contact-angle measurements of virgin, oxygen-plasma-treated and PFDTES-coated PMMA and PDMS surfaces. The maximum error in the data is $\pm 2°$ [50.120]

Fig. 50.34a–c Adhesion measurement of virgin, oxygen-plasma-treated and PFDTES-coated PMMA and PDMS surfaces with bare silicon nitride AFM tip (**a**) in ambient, and (**b**) in PBS environment, and (**c**) dip-coated tip with FBS in a PBS environment [50.120]

the moving parts needs to be minimized. Furthermore, if the adhesion between the microchannel surface and the biofluid is high, the biomolecules will stick to the microchannel surface and restrict flow. In order to facilitate flow, surfaces with low bioadhesion are required.

Tambe and *Bhushan* [50.169, 170] and *Bhushan* and *Burton* [50.171] have reported adhesive force data for PMMA and PDMS against an AFM Si_3N_4 tip and a silicon ball. *Tokachichu* and *Bhushan* [50.172] measured contact angle and adhesion of bare PMMA and PDMS and coated with a perfluoro SAM of perfluorodecyltriethoxysilane (PFDTES). Oxygen plasma treatment was used for hydroxylation of the surface to enhance chemical bonding of the SAM to the polymer surface. They made measurements in ambient and in PBS and fetal bovine serum (FBS); the latter is a blood component. Figs. 50.33 and 50.34 show the contact angle and adhesion data. SAM-coated surfaces have a high

contact angle Fig. 50.33, as expected. The adhesion value of PDMS in ambient is high because of electrostatic charge present on the surface. The adhesion values of PDMS are higher than PMMA because PDMS is softer than PMMA (elastic modulus $= 5\,$GPa and hardness $= 410\,$MPa [50.121]) and results in a higher contact area between the PDMS surface and the AFM tip, and PMMA does not develop electrostatic charge. When SAM is coated on PMMA and PDMS surfaces, the adhesion values are similar, which shows that electrostatic charge on virgin PDMS plays no role when the surface is coated. In the PBS solution, there is a decrease in adhesion values because of the lack of a meniscus contribution. The adhesion values in the FBS-coated tip in PBS are generally lower than for the uncoated tip in PBS. In summary, the adhesion values of SAM-coated surfaces are lower than bare surfaces in various environments.

50.5 Nanopatterned Surfaces

50.5.1 Analytical Model and Roughness Optimization

One of the crucial surface properties for surfaces in wet environments is nonwetting or hydrophobicity. It is usually desirable to reduce wetting in fluid flow applications and some conventional applications, such as glass windows and automotive windshields, in order for liquid to flow away along their surfaces. Reduction of wetting is also important in reducing meniscus formation, consequently reducing stiction, friction, and wear. Wetting is characterized by the contact angle, which is the angle between the solid and liquid surfaces. If the liquid wets the surface (referred to as a wetting liquid or a hydrophilic surface), the value of the contact angle is $0 \leq \theta \leq 90°$, whereas if the liquid does not wet the surface (referred to as a nonwetting liquid or a hydrophobic surface), the value of the contact angle is $90° < \theta \leq 180°$. A surface is considered superhydrophobic if θ is close to $180°$. Superhydrophobic surfaces should also have very low water contact angle hysteresis. One of the ways to increase the hydrophobic or hydrophilic properties of the surface is to increase surface roughness. It has been demonstrated experimentally that roughness changes contact angle. Some natural surfaces, including leaves of water-repellent plants such as lotus, are known to be superhydrophobic due to their high roughness and the presence of a wax coating Fig. 50.35. This phenomenon is called in the literature the *lotus effect* [50.173]).

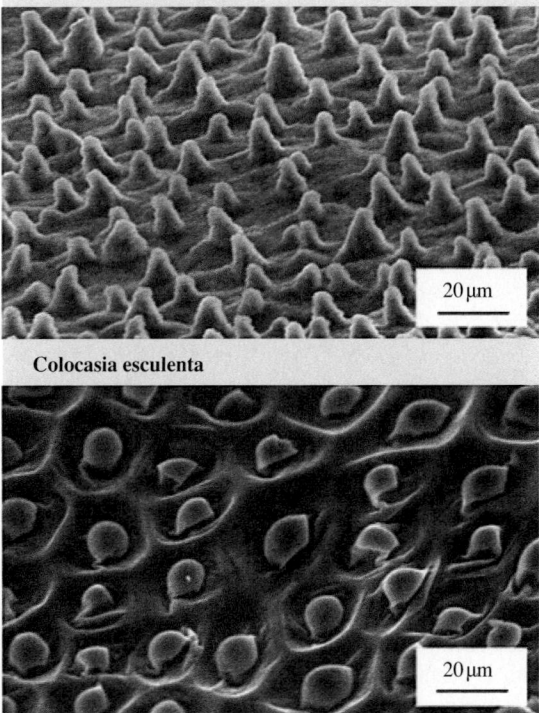

Fig. 50.35 SEM micrographs of two hydrophobic leaves, Nelumbo nucifera (lotus) and Colocasia esculenta

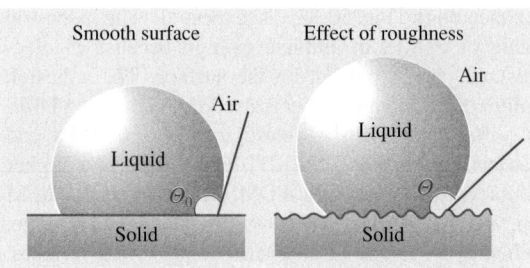

Fig. 50.36 Droplet of liquid in contact with a smooth solid surface (contact angle θ_0) and rough solid surface (contact angle θ) [50.173]

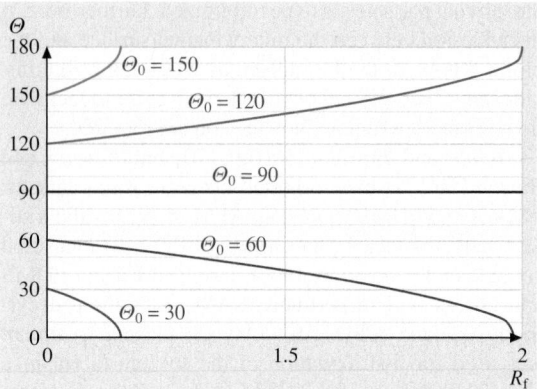

Fig. 50.37 Contact angle for rough surface (θ) as a function of the roughness factor (R_f) for various contact angles for a smooth surface (θ_0) [50.173]

If a droplet of liquid is placed on a smooth surface, the liquid and solid surfaces come together under equilibrium at a characteristic angle called the static contact angle θ_0; see Fig. 50.36. The contact angle can be determined from the condition of the total energy of the system being minimized. It can be shown that

$$\cos\theta_0 = dA_{LA}/dA_{SL} , \qquad (50.1)$$

where θ_0 is the contact angle for smooth surface, and A_{SL} and A_{LA} are the solid–liquid and liquid–air contact areas. Next, let us consider a rough solid surface with a typical size of roughness details smaller than the size of the droplet (on the order of a few hundred microns or larger), Fig. 50.36. For a rough surface, the roughness affects the contact angle due to the increased contact area A_{SL}. For a droplet in contact with a rough surface without air pockets, referred to as a homogeneous interface, based on the minimization of the total surface energy of the system, the contact angle is given as [50.174]

$$\begin{aligned}\cos\theta &= dA_{LA}/dA_F \\ &= \left(\frac{A_{SL}}{A_F}\right)(dA_{LA}/dA_{SL}) = R_f \cos\theta_0 , \end{aligned}$$
$$(50.2)$$

where A_F is the flat solid–liquid contact area (or a projection of the solid–liquid area A_{SL} onto the horizontal plane). R_f is a roughness factor defined as

$$R_f = \frac{A_{SL}}{A_F} . \qquad (50.3)$$

Equation (50.3) shows that, if the liquid wets a surface ($\cos\theta_0 > 0$), it will also wet the rough surface with a contact angle $\theta < \theta_0$, and for nonwetting liquids ($\cos\theta_0 < 0$), the contact angle with a rough surface will be greater than that with the flat surface, $\theta < \theta_0$. The dependence of the contact angle on the roughness factor

is resented in Fig. 50.37 for different values of θ_0, based on (50.2). It should be noted that (50.2) is valid only for moderate roughness, when $R_f \cos\theta_0 < 1$ [50.173].

For higher roughness, air pockets (composite solid–liquid–air interface) will be formed in the cavities of the surface [50.175]. In the case of partial contact, the contact angle is given by

$$\cos\theta = R_f f_{SL} \cos\theta_0 - f_{LA} , \qquad (50.4)$$

where f_{SL} and f_{LA} are fractional solid–liquid and liquid–air contact areas. The homogeneous and composite interfaces are two metastable states of the system. In reality, some cavities will be filled with liquid, and others with air, and the value of the contact angle is between the values predicted by Eqs. (50.2) and (50.4). If the distance is large between the asperities or if the slope changes slowly, the liquid–air interface can easily be destabilized due to imperfectness of the profile shape or due to dynamic effects, such as surface waves Fig. 50.38. *Nosonovsky* and *Bhushan* [50.176] proposed a stochastic model, which relates the contact angle to roughness and takes into account the possibility of destabilization of the composite interface due to imperfectness of the shape of the liquid–air interface, caused by effects such as capillary waves.

In addition to the surface roughness, sharp edges of asperities may affect wetting, because they result in pinning of the solid–liquid–air contact line and resist liquid flow. *Nosonovsky* and *Bhushan* [50.173] considered the effect of the surface roughness and sharp edges and found the optimum roughness distribution for nonwetting. They formulated five requirements for roughness-induced superhydrophobic surfaces. First, as-

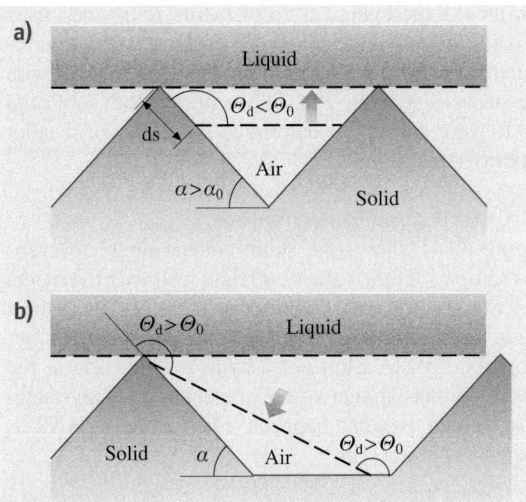

Fig. 50.38 (a) Formation of a composite solid–liquid–air interface for sawtooth and smooth profiles, and **(b)** destabilization of the composite interface for the sawtooth and smooth profiles due to dynamic effects. The dynamic contact angle $\theta_d > \theta_0$ corresponds to an advancing liquid–air interface, whereas $\theta_d < \theta_0$ corresponds to a receding interface [50.176]

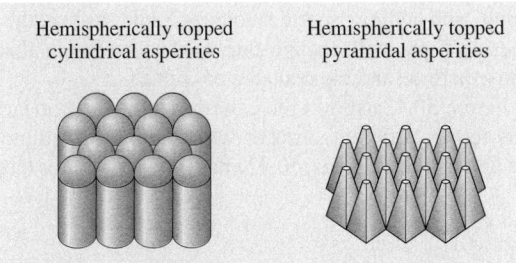

Fig. 50.39 Optimized roughness distribution – hemispherically topped cylindrical asperities and pyramidal asperities with square foundation and rounded tops. The square base gives a higher packing density but introduces undesirable sharp edges [50.173]

perities must have a high aspect ratio to provide a high surface area. Second, sharp edges should be avoided, to prevent pinning of the triple line. Third, asperities should be tightly packed to minimize the distance between them and avoid destabilization of the composite interface. Fourth, asperities should be small compared to typical droplet size (on the order of few hundred microns or larger). And fifth, in the case of hydrophilic surfaces, a hydrophobic film must be applied in order

to have initial $\theta > 90°$. These recommendations can be utilized for producing superhydrophobic surfaces. Remarkably, all these conditions are satisfied by biological water-repellent surfaces, such as some leaves: they have tightly packed hemispherically topped papillae with high (on the order of unity) aspect ratios and a wax coating [50.173]. Figure 50.39 shows two recommended geometries which use either hemispherically topped asperities with a hexagonal packing pattern or pyramidal asperities with a rounded top. These geometries can be used for producing superhydrophobic surfaces.

50.5.2 Experimental Validation

To validate the model of contact angle as a function of surface roughness, *Burton* and *Bhushan* [50.177] measured contact angles and adhesive forces on hydrophilic and hydrophobic polymer films with smooth surfaces

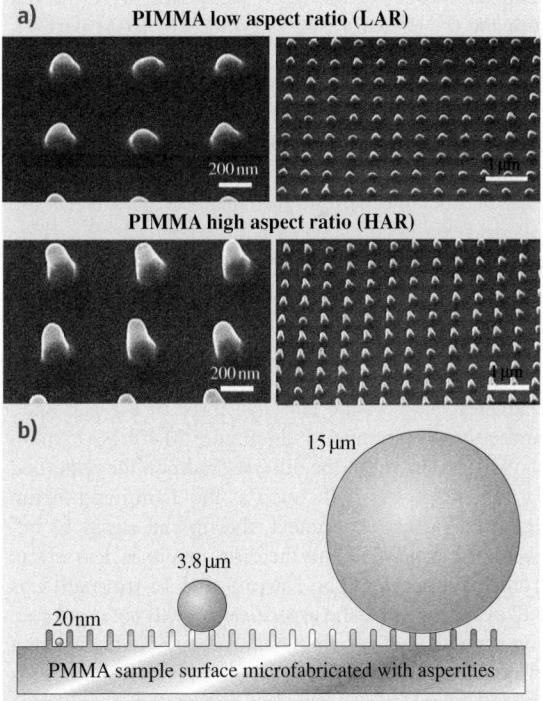

Fig. 50.40 (a) SEM micrographs of the patterned polymer surfaces. Both LAR and HAR are shown at two magnifications to see both the asperity shape and the asperity pattern on the surface. **(b)** Cartoon showing the effect of different radii on the patterned surface. Small radii can fit between asperities, while large radii rest on top of the asperities

and with discrete asperities. PMMA was chosen because it is a polymer often used in BioMEMS/BioNEMS devices. Three types of surface structures were measured: film, low-aspect-ratio asperities (LAR, 1:1 height-to-diameter ratio) and high-aspect-ratio asperities (HAR, 3:1 height-to-diameter ratio). Roughness (σ) and peak-to-valley distance (P–V) for PMMA film was measured using an AFM with values $\sigma = 0.98$ nm and $P - V = 7.3$ nm. The diameter of the asperities near the top is approximately 100 nm and the pitch of the asperities (distance between each asperity) is approximately 500 nm. Figure 50.40a shows SEM images of the two types of patterned structures, LAR and HAR, on a PMMA surface. According to the model presented earlier, by introducing roughness to a flat surface, the hydrophobicity will either increase or decrease depending on the initial contact angle on a flat surface. The material chosen was initially hydrophilic, so to obtain a sample that is hydrophobic, a self-assembled monolayer (SAM) was deposited on the sample surfaces. The samples chosen for the SAM deposition were the flat film and the HAR for each polymer. The SAM perfluorodecyltriethoxysilane (PFDTES) was deposited on the polymer surface using a vapor-phase deposition technique. PFDTES was chosen because of the hydrophobic nature of the surface. It should be noted that the bumps should be as close as possible to provide a high surface area. In order to benefit from an increase in contact angle and decrease in contact area with an increase in surface roughness, the pitch of the bumps should be smaller than the water droplet and the size of the contacting body.

To study the effect of scale dependence, the adhesive force between the four AFM tips with flat and patterned polymer films was examined. This allowed for complete characterization of the adhesive force of the patterned surfaces with varying tip radii. Figure 50.40b is a cartoon showing the effect of the different radii on the patterned surface. For small radii, such as the 20 nm and 50 nm tips used in this experiment, the tip can easily fit between the asperities and therefore, there is less effect from the asperities. The 3.8-μm- and 15-μm-radii tips will primarily sit on the asperities and will not come into contact with the flat polymer, thus reducing the real area of contact. Experiments in varying relative humidities show the dependence of hydrophobicity for a given surface roughness on adhesion and friction. Dry and wet friction and adhesion can vary dramatically because the dominant mechanism makes a transition from the real area of contact to meniscus forces. Therefore, the effect of relative humidity was also studied by performing measurements at 5, 50, and 80% relative humidity (RH).

For these experiments, a tip of radius 15 μm was used to measure both adhesion and coefficient of friction for both the film and patterned polymer surfaces along with the surfaces with the PFDTES coating. Both LAR and HAR were studied to determine the effect of a taller asperity on adhesion.

Contact Angle Measurements

Figure 50.41 shows the static contact angle for various samples. These values correlate well with the model describing roughness with hydrophobicity. The contact angle decreased with increased roughness for the hydrophilic PMMA film. For a hydrophobic surface, the model predicts an increase of contact angle with roughness, which is what happens when PMMA HAR is coated with PFDTES.

Adhesion Studies and Scale Dependence

Scale-dependent effects of adhesion are present because the tip–surface interface changes with size. Meniscus force will change by varying either the tip radius, the hydrophobicity of the sample or the number of contact and near-contacting points. Figure 50.42a shows the dependence of tip radius and hydrophobicity on adhesive force for PMMA and PFDTES coated on PMMA. By changing the radius of the tip, the contact angle of the sample, and adding asperities to the sample surface, the adhesive force will change due to the change in the meniscus force and the real area of contact.

Figure 50.42a shows the adhesive force on a linear scale for the different surfaces with varying tip radius. The first bar chart in Fig. 50.42a has PMMA film for the

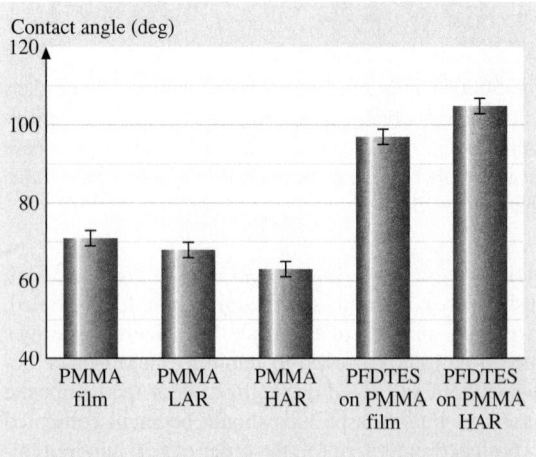

Fig. 50.41 Bar chart showing the contact angles for different materials and for different roughnesses [50.177]

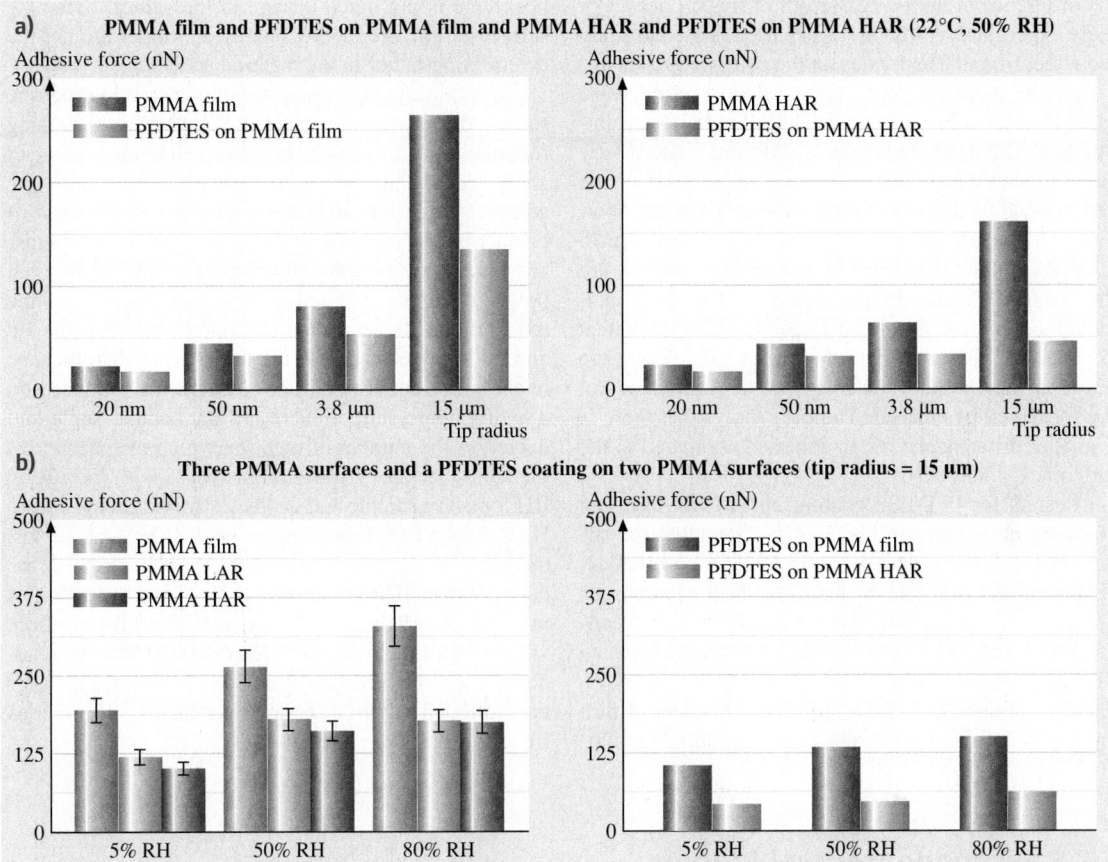

Fig. 50.42 (a) Scale-dependent adhesive force for PMMA film versus PFDTES on PMMA film and PMMA HAR versus PFDTES on PMMA HAR (*top*), and (b) the effect of relative humidity on the adhesive force for PMMA film, LAR and HAR and for PFDTES on PMMA film and HAR [50.177]

first bar while the second bar is for PFDTES coated on PMMA film and shows the effect of tip radius and hydrophobicity on the adhesive force. For an increase in radius, the adhesive force increases for each material, and decreases for PFDTES on PMMA film compared to PMMA film. With larger radius, the real area of contact and number of meniscus bridges increase and the adhesion is increased. The hydrophobicity of PFDTES on PMMA film reduces meniscus forces, which in turn reduces the adhesion on the PMMA film. The dominant mechanism for the hydrophobic film is the real area of contact and not the meniscus force, whereas with PMMA film there is a combination of real area of contact and meniscus forces.

The second bar chart in Fig. 50.42a shows the results for PMMA HAR and PFDTES coated on PMMA HAR. These samples show the same trends as the film samples,

but the increase in adhesion is not as dramatic. This is because of the decrease in real area of contact for each radius from a flat film. Again, meniscus forces do not play a large role in adhesion for PFDTES on PMMA HAR, and the increase in adhesion is due to the real area of contact. With PMMA HAR, a combination of both meniscus forces and real area of contact contribute to the adhesion.

Effect of Relative Humidity on Adhesive Force

The results from varying surface roughness, hydrophobicity and relative humidity are summarized in Fig. 50.42b. Experiments were also run on PMMA film and HAR with a PFDTES coating and are shown in Fig. 50.42b. Film HAR and LAR were used to see the change in adhesion for each type of surface structure. Only film and HAR were used for Fig. 50.42b to show the difference in the two extremes of the surfaces. For these experiments, only the 15-μm tip was used to study the effect from the asperities on the patterned surfaces.

For the adhesive force values, there is a decrease from PMMA film to LAR to HAR for the three humidities. The decrease between LAR and HAR is very small, which means that the contact area is about the same for a single-point measurement. There is, however, a large decrease in adhesive force from film to LAR and HAR. For a flat film, meniscus bridges are the dominant factor in the adhesion, but for a patterned sample the dominant factor is still the contact area and not the formation of menisci. At 5% RH the only factor in the adhesion is the area of contact and not the formation of menisci. The data shows that there is a smaller difference at 5% in adhesion compared to the difference at 80% RH.

Results for PFDTES coating on PMMA film and HAR are also shown in Fig. 50.42b. The adhesion values are much lower than those for PMMA and PMMA HAR and that is primarily due to the lack of meniscus bridge formation because of the hydrophobic contact angle of PFDTES. There is a decrease in adhesion between PFDTES film and PFDTES HAR, which is directly related to the area of contact difference between a film and HAR. Looking at the data across the three humidi-

ties, there is not much change in the values. Since the surfaces are hydrophobic, meniscus bridges are not the determining factor in the material adhesion.

In summary, increasing roughness on a hydrophilic surface decreases the contact angle, whereas increasing roughness on a hydrophobic surface increases contact angle. For a flat film, with increasing tip radius, the adhesive force increases due to increased real area of contact between the tip and the flat sample and meniscus force contributions. Introducing a pattern on a flat polymer surface will reduce adhesion because of the reduction of the real area of contact between the tip and the sample surface if the tip is larger than the size of the asperities. In addition, introducing a pattern on a hydrophobic surface increases the contact angle and decreases the number of menisci, which then decreases the adhesive force. Adhesion increases with increasing RH for every sample and decreases from film to LAR to HAR. When PFDTES is coated on the PMMA samples, the adhesion decreases but follows the same trend as the bare polymer. These trends are due to the formation of more menisci at higher relative humidities. In addition, with an increase in relative humidity, the increase in adhesive force for PMMA film is more dramatic than for the patterned samples due to larger menisci formations for a film sample.

50.6 Component-Level Studies

50.6.1 Surface Roughness Studies of Micromotor Components

Most of the friction forces resisting motion in the micromotor are concentrated near the rotor–hub interface where continuous physical contact occurs. The surface roughness of the surfaces usually has a strong influence

on the friction characteristics on the micro/nanoscale. A catalog of roughness measurements on various components of a MEMS device does not exist in the literature. Using an AFM, measurements on various component surfaces, for the first time, were made by *Sundararajan* and *Bhushan* [50.178]. Table 50.6 shows various surface roughness parameters obtained from

Table 50.6 Surface roughness parameters and microscale coefficient of friction for various micromotor component surfaces measured using an AFM. Mean and $\pm 1\sigma$ values are given

	RMS roughness[a] (nm)	Peak-to-valley distance[a] (nm)	Skewness[a] Sk	Kurtosis[a] K	Coefficient of microscale friction[b] (μ)
Rotor topside	21 ± 0.6	225 ± 23	1.4 ± 0.30	6.1 ± 1.7	0.07 ± 0.02
Rotor underside	14 ± 2.4	80 ± 11	-1.0 ± 0.22	3.5 ± 0.50	0.11 ± 0.03
Stator topside	19 ± 1	246 ± 21	1.4 ± 0.50	6.6 ± 1.5	0.08 ± 0.01

[a] Measured from a tapping-mode AFM scan of size $5\,\mu\text{m} \times 5\,\mu\text{m}$ using a standard Si tip scanning at $5\,\mu\text{m/s}$ in a direction orthogonal to the long axis of the cantilever.
[b] Measured using an AFM in contact mode at $5\,\mu\text{m} \times 5\,\mu\text{m}$ scan size using a standard Si_3N_4 tip scanning at $10\,\mu\text{m/s}$ in a direction parallel to the long axis of the cantilever.

$5 \times 5\,\mu m$ scans of the various component surfaces of several unlubricated micromotors using the AFM in tapping mode. A surface with a Gaussian height distribution should have a skewness of zero and kurtosis of three. Although the rotor and stator top surfaces exhibit comparable roughness parameters, the underside of the rotors exhibits lower root mean square (RMS) roughness and peak-to-valley distance values. More importantly, the rotor underside shows negative skewness and lower kurtosis than the topsides, both of which are conducive to high real area of contact and hence high friction [50.79, 81]. The rotor underside also exhibits a higher coefficient of microscale friction than the rotor topside and stator, as shown in Table 50.6. Figure 50.43 shows representative surface-height maps of the various surfaces of a micromotor measured using the AFM in tapping mode. The rotor underside exhibits a different topography from the outer edge to the middle and inner edge. At the outer edges, the topography shows smaller circular asperities, similar to the topside. The middle and inner regions show deep pits with fine edges that may have been created by the etchants used for etching of the sacrificial layer. It is known that etching can affect the roughness of surfaces in surface micromachining. The residence time of the etchant near the inner region is high, which is responsible for larger pits. Figure 50.44 shows the roughness of the surface directly beneath the rotors (the base polysilicon layer). There appears to be a difference in the roughness between the portion of this surface that was initially underneath the rotor (region B) during fabrication and the portion that was away from the rotor and hence always exposed (region A). The former region shows a lower roughness than the latter region. This suggests that the surfaces at the rotor–hub interface that come into contact at the end of the fabrication process exhibit large real areas of contact, which result in high friction.

50.6.2 Adhesion Measurements of Microstructures

Surface force apparatus (SFA) and AFMs are used to measure adhesion on micro- to nanoscales between two surfaces. In the SFA, adhesion of liquid films sandwiched between two curved and smooth surfaces is measured. In an AFM, as discussed earlier, adhesion between a sharp tip and the surface of interest is measured. The propensity for adhesion between two surfaces can be evaluated by studying the tendency of microstructures with well-defined contact areas, covering a wide of suspension compliances, to stick to the underlying sub-

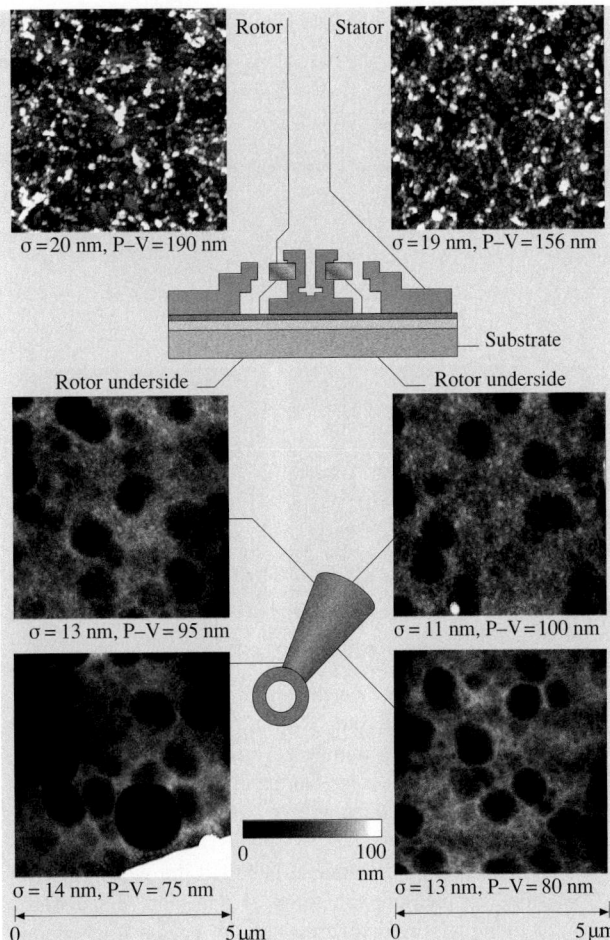

Fig. 50.43 Representative AFM surface-height images obtained in tapping mode ($5\,\mu m \times 5\,\mu m$ scan size) of various component surfaces of a micromotor. RMS roughness and peak-to-valley values of the surfaces are given. The underside of the rotor exhibits drastically different topography from the topside [50.178]

strate. The test structures which have been used include cantilever beam array (CBA) technique with different lengths [50.179–182] and stand-off multiple dimples mounted on microstructures with a range of compliances, standing above a substrate [50.183]. The CBA technique, more commonly used, utilizes an array of micromachined polysilicon beams (for Si MEMS applications), on the mesoscopic length scale, anchored to the substrate at one end and with different lengths parallel to the surface. It relies on peeling and detachment of cantilever beams. Changes in the free energy or reversible work done to separate unit areas of the

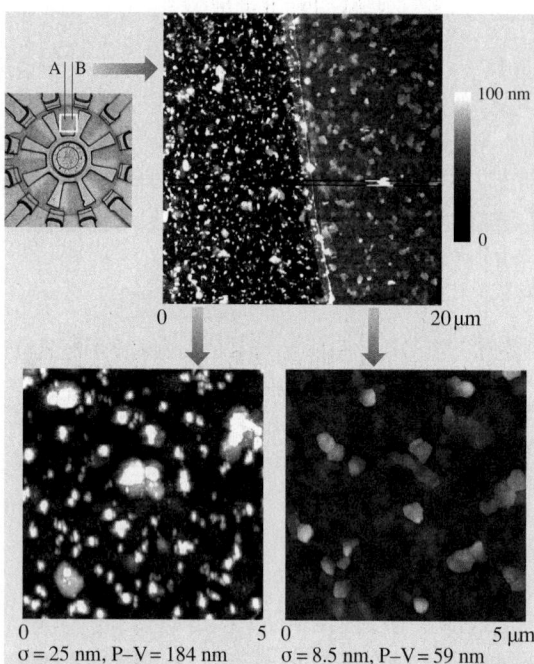

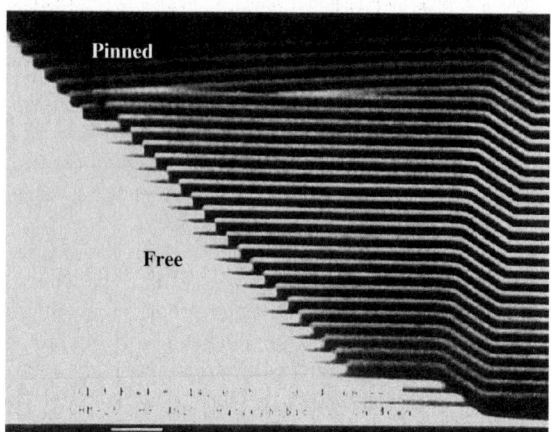

Fig. 50.45 SEM micrograph of micromachined array of polysilicon cantilever beams of increasing length. The micrograph shows the onset of pinning for beams longer than $34\,\mu\mathrm{m}$ [50.179]

Fig. 50.44 Surface-height images of polysilicon regions directly below the rotor. Region A is away from the rotor while region B was initially covered by the rotor prior to the release etch of the rotor. During this step, slight movement of the rotor caused region B to be exposed [50.178]

two surfaces from contact is called the work of adhesion. To measure the work of adhesion, electrostatic actuation is used to bring all the beams into contact with the substrate; see Fig. 50.45 [50.179, 181]. Once the actuation force is removed, the beams begin to peel themselves off the substrate, which can be observed with an optical interference microscope (e.g., a Wyko surface profiler). For beams shorter than a characteristic length, the so-called detachment length, their stiffness is sufficient to free them completely from the substrate underneath. Beams larger than the detachment length remain adhered. The beams at the transition region start to detach and remain attached to the substrate just at the tips. For this case, by equating the elastic energy stored within the beam and the beam–substrate interfacial energy, the work of adhesion, W_{ad}, can be calculated by the following equation [50.179]

$$W_{\mathrm{ad}} = \frac{3Ed^2t^3}{8\ell_d^4}\,, \tag{50.5}$$

where E is the Young's modulus of the beam, d is the spacing between the undeflected beam and the sub-

strate, t is the beam thickness, and P_d is the detachment length. The technique has been used to screen methods for adhesion reduction in polysilicon microstructures.

50.6.3 Microtriboapparatus for Adhesion, Friction, and Wear of Microcomponents

To measure adhesion, friction, and wear between two microcomponents, a microtriboapparatus has been used. Figure 50.46 shows a schematic of a microtriboapparatus, capable of adopting MEMS components [50.184]. In this apparatus, an upper specimen, mounted on a soft cantilever beam, comes into contact with a lower specimen mounted on a lower specimen holder. The apparatus consists of two piezos (x- and z-piezos), and four fiberoptic sensors (x- and z-displacement sensors, and x- and z-force sensors). For adhesion and friction studies, z- and x-piezos are used to bring the upper specimen and lower specimen in contact and to apply a relative motion in the lateral direction, respectively. The x- and z-displacement sensors are used to measure the lateral position of the lower specimen and vertical position of the upper specimen, respectively. The x- and z-force sensors are used to measure the friction force and normal load/adhesive force between these two specimens, respectively, by monitoring the deflection of the cantilever.

As most MEMS/NEMS devices are fabricated from silicon, study of silicon-on-silicon contacts is important. This contact was simulated by a flat single-crystal

Si(100) wafer (phosphorus-doped) specimen sliding against a single-crystal Si(100) ball (1 mm in diameter, 5×10^{17} atoms/cm^3 boron doped) mounted on a stainless-steel cantilever [50.184, 185]. Both of them have native oxide layer on their surfaces. The other materials studied were 10-nm-thick DLC deposited by filtered cathodic arc deposition on Si(100), 2.3-nm-thick chemically bonded PFPE (Z-DOL, BW) on Si(100), and hexadecane thiol (HDT) monolayer on evaporated Au(111) film to investigate their anti-adhesion performance.

It is well known that in computer rigid disk drives, the adhesive force increases rapidly with increasing rest time between a magnetic head and a magnetic disk [50.34]. Considering that adhesion and friction are the major issues that lead to the failure of MEMS/NEMS devices, the effect of rest time on the microscale on Si(100), DLC, PFPE, and HDT was studied, and the results are summarized in Fig. 50.47a. It is found that the adhesive force of Si(100) logarithmically increases with the rest time up to a certain equilibrium time ($t = 1000$ s), after which it remains constant. Figure 50.47a also shows that the adhesive force of DLC, PFPE, and HDT does not change with rest time. Single-asperity contact modeling of the dependence of meniscus force on the rest time has been carried out by *Chilamakuri* and *Bhushan* [50.186], and the modeling results Fig. 50.47b verify experimental observations. Due to the presence of thin-film adsorbed water on Si(100), the meniscus forms around the contacting asperities and grows with time until equilibrium occurs, which causes the effect of rest time on its adhesive force. The adhesive forces of DLC, PFPE, and HDT do not change with rest time, which suggests that the water meniscus is not present on their surfaces.

The measured adhesive forces of Si(100), DLC, PFPE, and HDT at a rest time of 1 s are summarized in Fig. 50.48, which shows that the presence of solid films of DLC, PFPE, and HDT greatly reduces the adhesive force of Si(100), whereas, the HDT film has the lowest adhesive force. It is well known that the native oxide layer (SiO$_2$) on the top of Si(100) wafers exhibits hydrophilic properties, and water molecules, produced by capillary condensation of water vapor from the environment, can easily be adsorbed onto this surface. The condensed water will form a meniscus as the upper specimen approaches to the lower specimen surface. The meniscus force is a major contributor to the adhesive force. In the case of DLC, PFPE, and HDT, the films are found to be hydrophobic based on contact angle measurements and the amount of condensed water vapor is

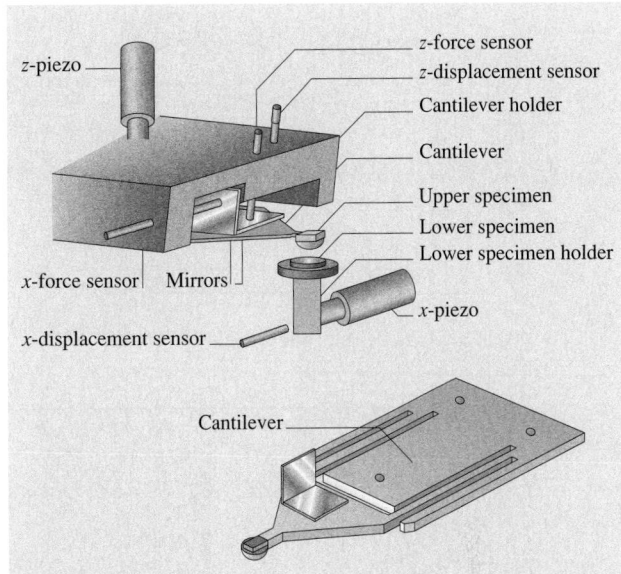

Fig. 50.46 Schematic of the microtriboapparatus including specially designed cantilever (with two perpendicular mirrors attached on the end), lower specimen holder, two piezos (*x*- and *z*-piezos), and four fiber-optic sensors (*x*- and *z*-displacement sensors and *x*- and *z*-force sensors) [50.184]

low compared to that on Si(100). It should be noted that the measured adhesive force is generally higher than that measured by AFM, because the larger radius of the Si(100) ball compared to that of an AFM tip induces a larger meniscus and van der Waals forces.

To investigate the effect of velocity on friction, the friction force as a function of velocity was measured and is summarized in Fig. 50.49a. It indicates that, for Si(100), the friction force initially decreases with increasing velocity until equilibrium occurs. Figure 50.49a also indicates that the velocity almost has no effect on the friction properties of DLC, PFPE, and HDT . This implies that the friction mechanisms on DLC, PFPE, and HDT do not change with the variation of velocity. For Si(100), at high velocity, the meniscus is broken and does not have enough time to rebuild. In addition, it is also believed that tribochemical reactions plays an important role. The high velocity leads to tribochemical reactions of Si(100) (which has a native oxide SiO$_2$) with water molecules to form a Si(OH)$_4$ film. This film is removed and continuously replenished during sliding. The Si(OH)$_4$ layer at the sliding surface is known to have low shear strength. The breaking of the water meniscus and the formation of a Si(OH)$_4$ layer results in a decrease in the friction force of Si(100). For DLC, PFPE,

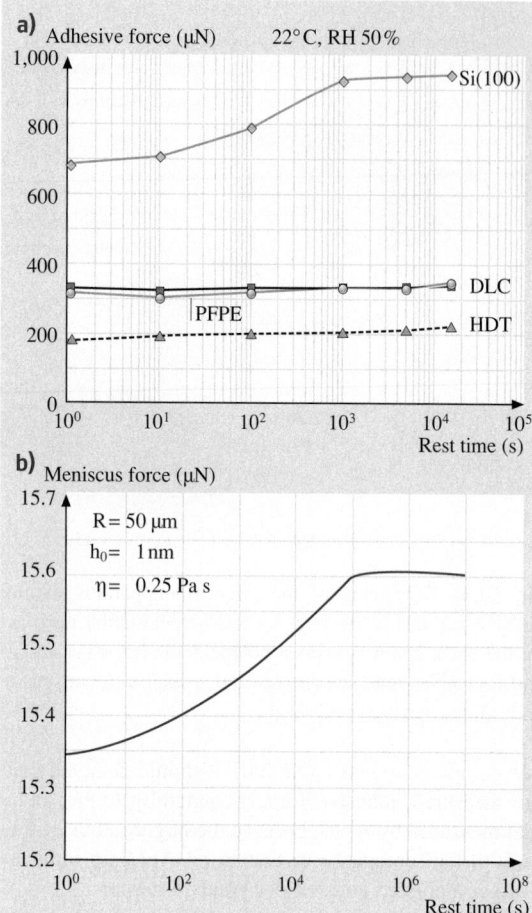

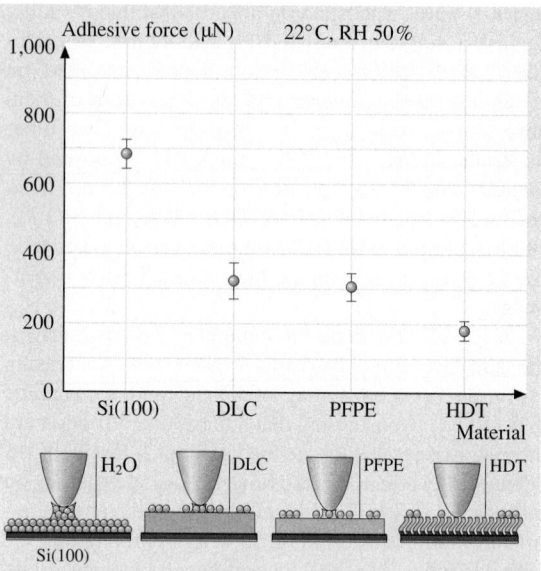

Fig. 50.48 Adhesive forces of Si(100), DLC, chemically bonded PFPE, and HDT at ambient condition and a schematic showing the relative size of the water meniscus on different specimens

Fig. 50.47 (**a**) The influence of rest time on the adhesive force of Si(100), DLC, chemically bonded PFPE, and HDT, and (**b**) Single-asperity contact modeling results of the effect of rest time on the meniscus force for an asperity of R in contact with a flat surface with a water film of thickness of h_0 and absolute viscosity of η_0 [50.186]

and HDT, their surfaces exhibit hydrophobic properties, and can only adsorb a few water molecules in ambient conditions. The aforementioned meniscus breaking and tribochemical reaction mechanisms do not exist for these films. Therefore, their friction force does not change with velocity.

The influence of relative humidity was studied in an environmentally controlled chamber. The adhesive force and friction force were measured by making measurements at increasing relative humidity, and the results are summarized in Fig. 50.49b, which shows that, for Si(100), the adhesive force increases with relative hu-

midity, but the adhesive forces for DLC and PFPE only show a slight increase when the humidity is higher than 45%, while the adhesive force of HDT does not change with humidity. Figure 50.49b also shows that, for Si(100), the friction force increases with relative humidity increases up to 45%, and then shows a slight decrease with further increases in the relative humidity. For PFPE, there is an increase in the friction force when the humidity is higher than 45%. In the whole testing range, the relative humidity does not have any apparent influence on the friction properties of DLC and HDT. In the case of Si(100), the initial increase in relative humidity up to 45% causes more adsorbed water molecules, and the formation of a larger water meniscus, which leads to an increase of the friction force. However, at very high humidity of 65%, large quantities of adsorbed water can form a continuous water layer that separates the tip and sample surfaces, and acts as a kind of lubricant, which causes a decrease in the friction force. For PFPE, dewetting of the lubricant film at humidities larger than 45% results in an increase in adhesive and friction forces. For DLC and HDT, their surfaces show hydrophobic properties, and increasing relative humidity does not play a large role in their friction force.

The influence of temperature was studied using a heated stage. The adhesive force and friction force

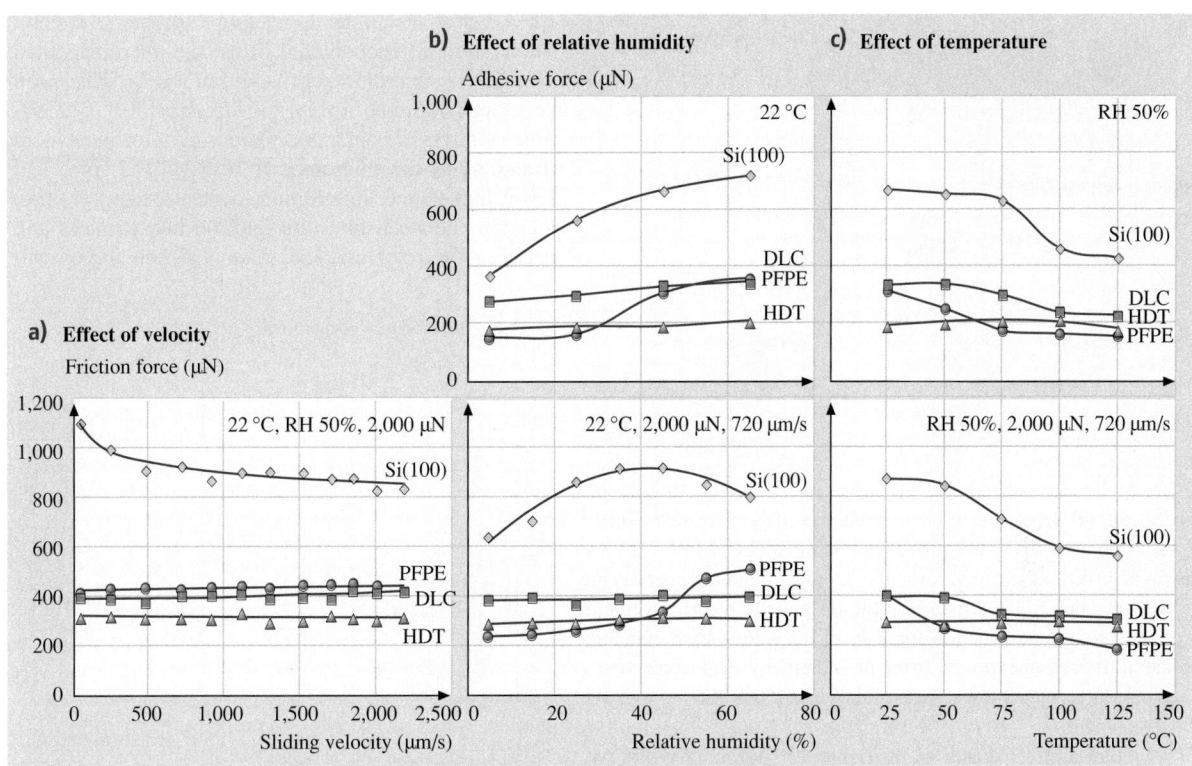

Fig. 50.49a–c The influence of (**a**) sliding velocity on the friction forces, (**b**) relative humidity on the adhesive and friction forces, and (**c**) temperature on the adhesive and friction forces of Si(100), DLC, chemically bonded PFPE, and HDT

were measured by making measurements at increasing temperatures of 22–125 °C. The results are presented in Fig. 50.49c, which shows that, once the temperature is higher than 50 °C, increasing temperature causes a significant decrease of adhesive and friction forces of Si(100) and a slight decrease in the case of DLC and PFPE. However, the adhesion and friction forces of HDT do not show any apparent change with test temperature. At high temperature, desorption of water, and the reduction of surface tension of water lead to the decrease of adhesive and friction forces of Si(100), DLC, and PFPE. However, in the case of HDT film, as only a few water molecules are adsorbed on the surface, the aforementioned mechanisms do not play a large role. Therefore, the adhesive and friction forces of HDT do not show any apparent change with temperature. Figure 50.49 shows that in the whole velocity, relative humidity, and temperature test range, the adhesive force and friction force of DLC, PFPE, and HDT are always smaller than that of Si(100), and that HDT has the smallest value.

To summarize, several methods can be used to reduce adhesion in microstructures. MEMS/NEMS surfaces can be coated with hydrophobic coatings such as PFPEs, SAMs, and passivated DLC coatings. It should be noted that other methods to reduce adhesion include the formation of dimples on the contact surfaces to reduce contact area [50.13, 79, 81, 177, 181]. Furthermore, an increase in hydrophobicity of the solid surfaces (high contact angle approaching 180°) can be achieved by using surfaces with suitable roughness, in addition to lowering their surface energy [50.173, 176]. The hydrophobicity of surfaces is dependent upon a subtle interplay between surface chemistry and mesoscopic topography. The self-cleaning mechanism or so-called lotus effect is closely related to the ultra-hydrophobic properties of the biological surfaces, which usually show microsculptures of specific dimensions.

50.6.4 Static Friction Force (Stiction) Measurements in MEMS

In MEMS devices involving parts in relative motion to each other, such as micromotors, large friction forces become the limiting factor for the successful opera-

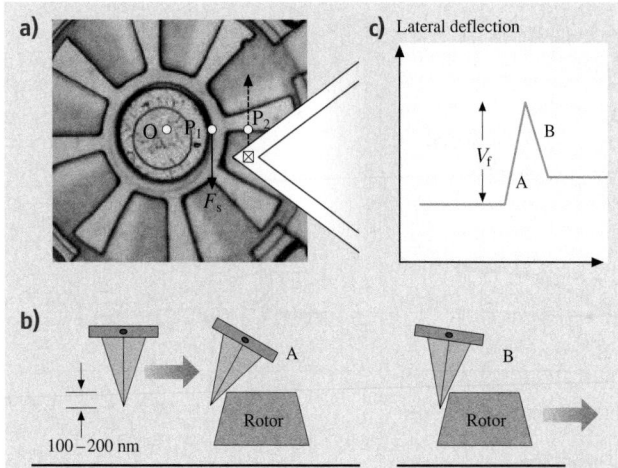

Fig. 50.50 (**a**) Schematic of the technique used to measure the force F_s required to initiate rotor movement using an AFM/FFM. (**b**) As the tip is pushed against the rotor, the lateral deflection experienced by the rotor due to the twisting of the tip prior to rotor movement is a measure of static friction force F_s of the rotors. (**c**) Schematic of lateral deflection expected from the aforementioned experiment. The peak V_f is related to the state of the rotor [50.178]

tion and reliability of the device. It is generally known that most micromotors cannot be rotated as manufactured and require some form of lubrication. It is therefore critical to determine the friction forces present in such MEMS devices. To measure in situ the static friction of a rotor–bearing interface in a micromotor, *Tai* and *Muller* [50.187] measured the starting torque (voltage) and pausing position for different starting positions under a constant bias voltage. A friction-torque model was used to obtain the coefficient of static friction. To measure the in situ kinetic friction of the turbine and gear structures, *Gabriel* et al. [50.188] used a laser-based measurement system to monitor the steady-state spins and decelerations. *Lim* et al. [50.189] designed and fabricated a polysilicon microstructure to measure in situ the static friction of various films. The microstructure consisted of a shuttle suspended above the underlying electrode by a folded beam suspension. A known normal force was applied and the lateral force was measured to obtain the coefficient of static friction. *Beerschwinger* et al. [50.190] developed a cantilever-deflection rig to measure friction of LIGA-processed micromotors. (Also see [50.191].) These techniques employ indirect methods to determine the friction forces or involve fabrication of complex structures.

A novel technique to measure the static friction force (stiction) encountered in surface-micromachined polysilicon micromotors using an AFM has been developed by *Sundararajan* and *Bhushan* [50.178]. Continuous physical contact occurs during rotor movement (rotation) in the micromotors between the rotor and lower hub flange. In addition, contact occurs at other locations between the rotor and the hub surfaces and between the rotor and the stator. Friction forces will

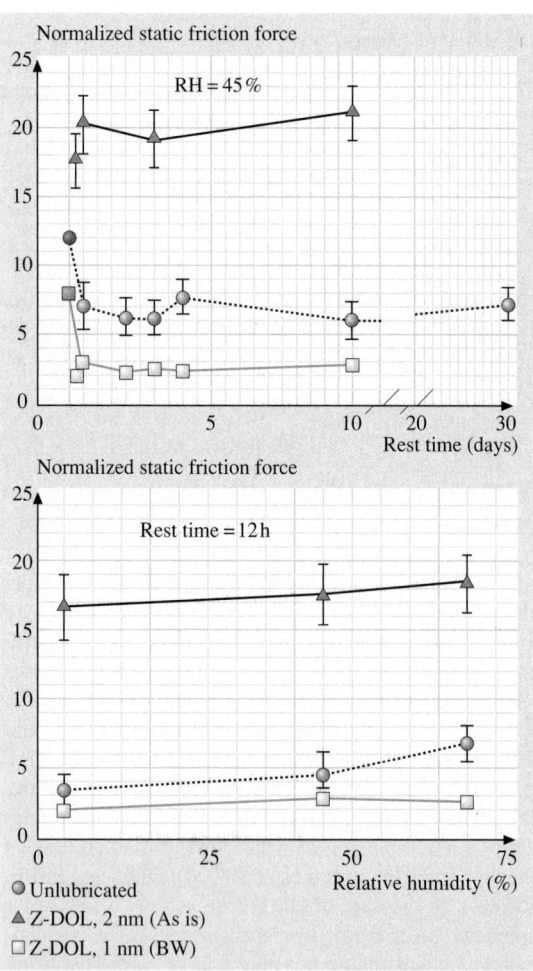

Fig. 50.51 Static friction force values of unlubricated motors and motors lubricated using PFPE lubricants, normalized over the rotor weight, as a function of rest time and relative humidity. Rest time is defined as the time elapsed between a given experiment and the first experiment in which motor movement was recorded (time 0). The motors were allowed to sit at a particular humidity for 12 h prior to measurement [50.178]

be present at these contact regions during motor operation. Although the actual distribution of these forces is not known, they can be expected to be concentrated near the hub where there is continuous contact. If we therefore represent the static friction force of the micromotor as a single force F_s acting at point P_1 (as shown in Fig. 50.50a), then the magnitude of the frictional torque about the center of the motor (O) that must be overcome before rotor movement can be initiated is

$$T_S = F_S \ell_1 \, , \tag{50.6}$$

where ℓ_1 is the distance OP_1, which is assumed to be the average distance from the center at which the friction force F_s occurs. Now consider an AFM tip moving against a rotor arm in a direction perpendicular to the long axis of the cantilever beam (the rotor-arm edge closest to the tip is parallel to the long axis of the cantilever beam), as shown in Fig. 50.50a. When the tip encounters the rotor at point P_2, the tip will twist, generating a lateral force between the tip and the rotor (event A in Fig. 50.50b). This reaction force will generate a torque about the center of the motor. Since the tip is trying to move further in the direction shown, the tip will continue to twist to a maximum value at which the lateral force between the tip and the rotor becomes high enough that the resultant torque T_f about the center of the motor equals the static friction torque T_s. At this point, the rotor will begin to rotate and the twist of the cantilever decreases sharply (event B in Fig. 50.50b). The twist of the cantilever is measured in the AFM as a change in the lateral deflection signal (in volts), which is the underlying concept of friction force microscopy (FFM). The change in the lateral deflection signal corresponding to the aforementioned events as the tip approaches the rotor is shown schematically in Fig. 50.50c. The value of the peak V_f is a measure of the force exerted on the rotor by the tip just before the static friction torque is matched and the rotor begins to rotate.

Using this technique, the viability of PFPE lubricants for micromotors has been investigated and the effect of humidity on the friction forces of unlubricated and lubricated devices was studied as well. Figure 50.51 shows static friction forces, normalized over the weight of the rotor, of unlubricated and lubricated micromotors as a function of rest time and relative humidity. Rest time here is defined as the time elapsed between the first experiment conducted on a given motor (solid symbol at time zero) and subsequent experiments (open symbols). Each open-symbol data point is an average of six measurements. It can be seen that, for the unlubricated motor and the motor lubricated with a bonded

layer of Z-DOL(BW), the static friction force is highest for the first experiment and then drops to an almost constant level. In the case of the motor with an as-is mobile layer of Z-DOL, the values remain very high up to 10 days after lubrication. In all cases, there is negligible difference in the static friction force at 0% and 45% RH. At 70% RH, the unlubricated motor exhibits a substantial increase in the static friction force, while the motor with bonded Z-DOL shows no increase in static friction force due to the hydrophobicity of the lubricant layer. The motor with an as-is mobile layer of the lubricant shows consistently high values of static friction force that vary little with humidity.

Figure 50.52 summarizes static friction force data for two motors, M1 and M2 along with schematics of the meniscus effects for the unlubricated and lubricated surfaces. Capillary condensation of water vapor from the environment results in the formation of meniscus bridges between contacting and near-contacting asper-

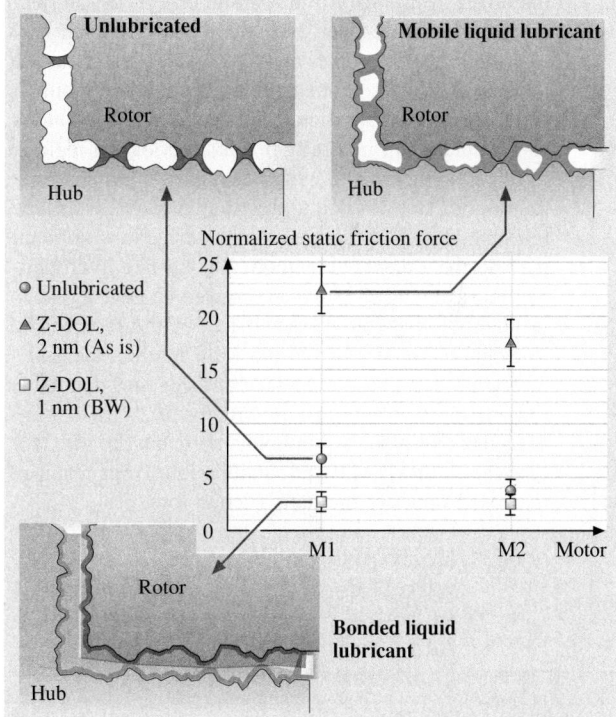

Fig. 50.52 Summary of the effect of liquid and solid lubricants on static friction force of micromotors. Despite the hydrophobicity of the lubricant used (Z-DOL), a mobile liquid lubricant (Z-DOL as is) leads to very high static friction force due to increased meniscus forces whereas a solid-like lubricant (bonded Z-DOL, BW) appears to provide some reduction in static friction force

ities of the two surfaces in close proximity to each other, as shown in Fig. 50.52. For unlubricated surfaces, more menisci are formed at higher humidity resulting in a higher friction force between the surfaces. The formation of meniscus bridges is supported by the fact that the static friction force for unlubricated motors increases at high humidity Fig. 50.52. Solid bridging may occur near the rotor–hub interface due to silica residues after the first etching process. In addition the drying process after the final etch can result in liquid bridging, formed by the drying liquid due to meniscus force at these areas [50.79, 81, 179, 180]. The initial static friction force therefore will be quite high, as evidenced by the solid data points in Fig. 50.52. Once the first movement of the rotor permanently breaks these solid and liquid bridges, the static friction force of the motors will drop (as seen in Fig. 50.52) to a value dictated predominantly by the adhesive energies of the rotor and hub surfaces, the real area of contact between these surfaces and meniscus forces due to water vapor in the air, at which point, the effect of lubricant films can be observed. Lubrication with a mobile layer, even a thin one, results in very high static friction forces due to meniscus effects of the lubricant liquid itself at and near the contact regions. It should be noted that a motor submerged in a liquid lubricant would result in a fully flooded lubrication regime. In this case there is no meniscus contribution and only the viscous contribution to the friction forces would be relevant. However, submerging the device in a lubricant may not be a practical method. A solid-like hydrophobic lubricant layer (such as bonded Z-DOL) results in favorable friction characteristics of the motor. The hydrophobic nature of the lubricant inhibits meniscus formation between the contact surfaces and maintains low friction even at high humidity Fig. 50.52. This suggests that solid-like hydrophobic lubricants are ideal for lubrication of MEMS while mobile lubricants result in increased values of the static friction force.

50.6.5 Mechanisms Associated with Observed Stiction Phenomena in Digital Micromirror Devices (DMD) and Nanomechanical Characterization

DMDs are used in digital projection displays, as described earlier. The DMD has a layered structure, consisting of an aluminium alloy micromirror layer, yoke and hinge layer, and metal layer on a CMOS memory array [50.27–29]. A blown-up view of the DMD and the corresponding AFM surface-height images are

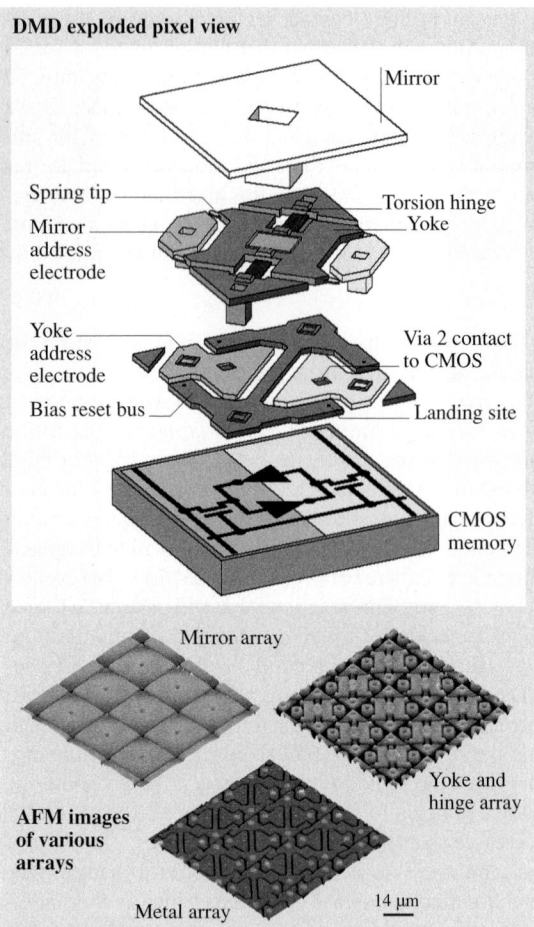

Fig. 50.53 Exploded view of a DMD pixel and AFM surface-height images of various arrays. The DMD layers were removed by an ultrasonic method [50.192]

presented in Fig. 50.53 [50.192]. Single-layered aluminium alloy films are used for the construction of micromirrors; these are also sometimes used for the construction of hinges, spring tips, and landing sites. The aluminium alloy films are overwhelmingly comprised of aluminium; trace elements (including Ti and Si) are present to suppress contact spiking and electromigration, which may occur if current densities become high during electrostatic operation. Multilayered sputtered $SiO_2 TiN/Al$ alloy films are now generally used for the landing site structure to minimize refraction throughout the visible region of the electromagnetic spectrum in order to increase the contrast ratio in projection display systems [50.193, 194]. These multilayered films are also generally used for hinges and spring tips. A low-

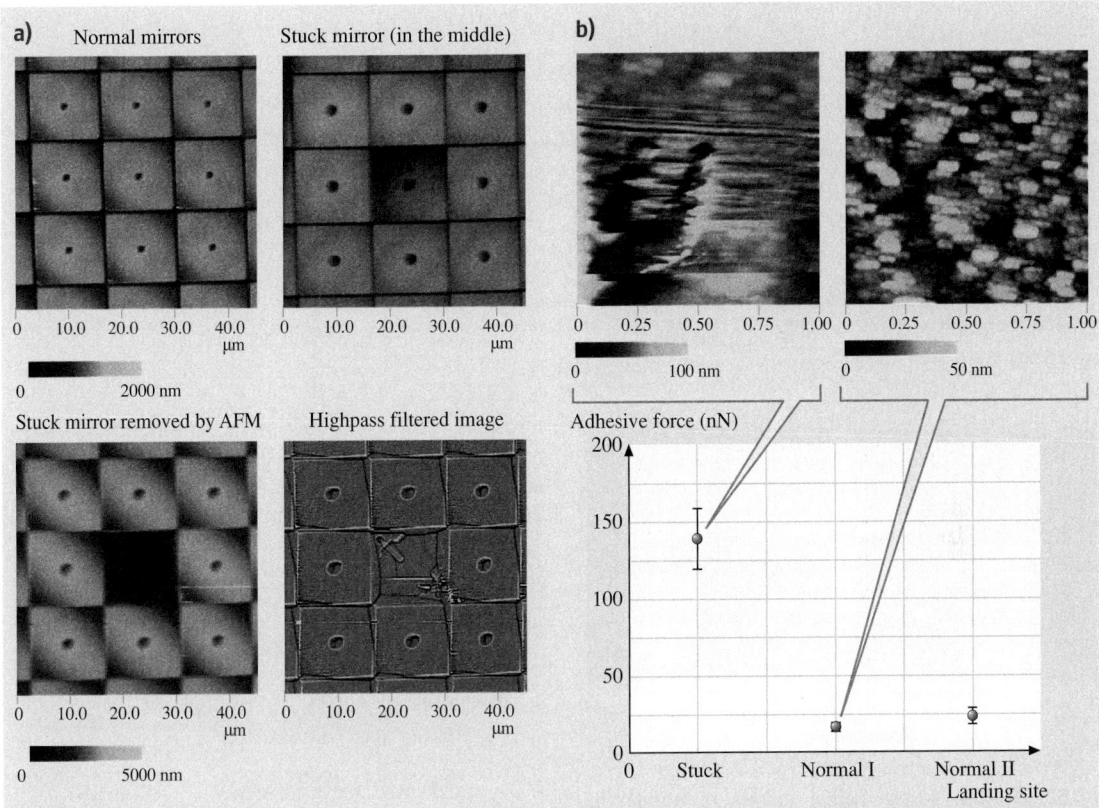

Fig. 50.54 (a) The *top row* shows AFM surface-height images of a stuck micromirror surrounded by eight normal micromirrors. The *left image* in the *bottom row* shows the stuck micromirror, which was removed by an AFM tip after repeated scanning at high normal load. The *right image* in the *bottom row* presents a high-pass-filtered image showing that the residual hinge that sits underneath the removed micromirror is clearly observed. (b) AFM surface-height images and adhesive forces of the landing sites underneath the two normal micromirrors and the stuck micromirror [50.192]

surface-energy SAM is maintained on the surfaces of the DMD, which is packaged in a hermetic environment to minimize stiction during contact between the spring tip and the landing site. An SAM of perfluorinated n-alkanoic acid ($C_nF_{2n-1}O_2H$) (e.g., perflurordecanoic acid or PFDA, $CF_3(CF_2)_8COOH$) applied by the vapor-phase deposition process is used. A getter strip of PFDA is included inside the hermetically sealed enclosure containing the chip, which acts as a reservoir in order to maintain a PFDA vapor within the package.

In order to identify a stuck mirror and characterize its nanotribological properties, the chip was scanned using an AFM [50.192]). It was found that it is hard to tilt the stuck micromirror back to its normal position by adding a normal load at the rotatable corner of the micromirror; thus, this is called a *hard* stuck micromirror. An example of a stuck micromirror is shown in Fig. 50.54a.

Once the stuck micromirror was found, the region was repeatedly scanned at a large normal load, up to 300 nN. After several scans, the stuck micromirror was removed. Once the stuck micromirror was removed, the surrounding micromirrors could also be removed by continuous scanning under a large normal load (Fig. 50.54a bottom row). The adhesive force of the landing site underneath the stuck micromirror and the normal micromirror are presented in Fig. 50.54b, which clearly indicates that the landing site underneath the stuck micromirror has much larger adhesion. A 1 µm × 1 µm view of the landing sites under stuck and normal micromirrors are also shown in Fig. 50.54b. The landing site under the stuck micromirror has an apparent U-shaped wear mark, which is surrounded by a smeared area.

Liu and *Bhushan* [50.192] calculated contact stresses to examine if the stresses were high enough to cause

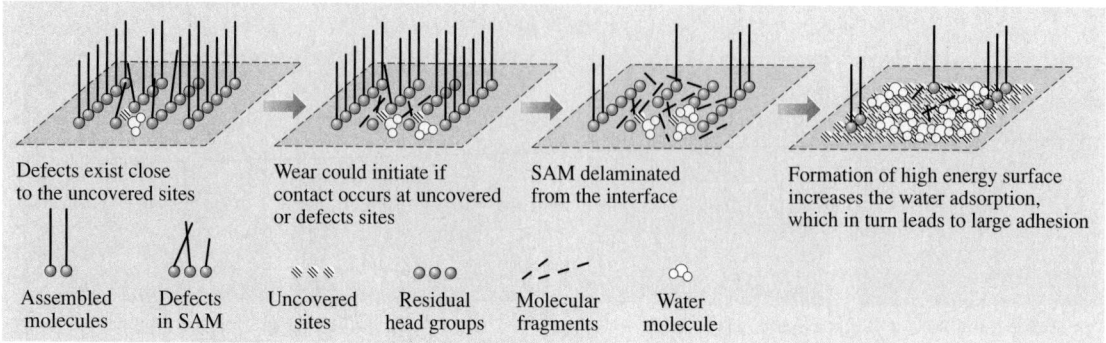

Fig. 50.55 Suggested mechanisms for wear and stiction [50.192]

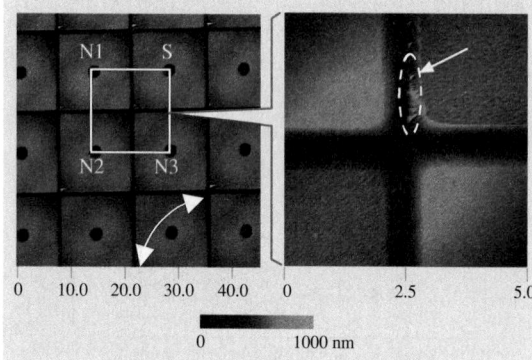

Fig. 50.56 AFM surface-height images of normal micromirrors and a soft-stuck micromirror. The soft-stuck micromirror was labeled S, and the normal micromirrors studied are labeled as N1, N2, and N3 [50.195]

wear at the spring tip–landing site interface. The calculated contact stress value was about 33 MPa which is substantially lower than the hardness, therefore much plastic deformation and consequently wear was not expected. The wear mark was only found on a very few landing sites on the DMD, which means that the SAM coating can generally endure such high contact stresses. Based on data reported in the literature, the coverage of vapor-deposited SAMs is expected to be about 97%. The bond strength of the molecules close to the boundary of the uncovered sites is expected to be weak. Thus, the uncovered sites and the adjacent molecules are referred to as defects in the SAM coating. Occasionally, if contact occurs at the defect sites, the large cyclic stress may be close to the critical load, and lead to the initial delamination of the SAM coating at the interface. The continuous contact leads to the formation of a high-surface-energy surface by exposure of the fresh substrate and the formation of SAM fragments. This eventually

leads to an increase in stiction by the formation of large menisci. Once this happens, the stress at the contact area is increased, which would accelerate the wear. Based on this hypothesis, suggested mechanisms for the wear and

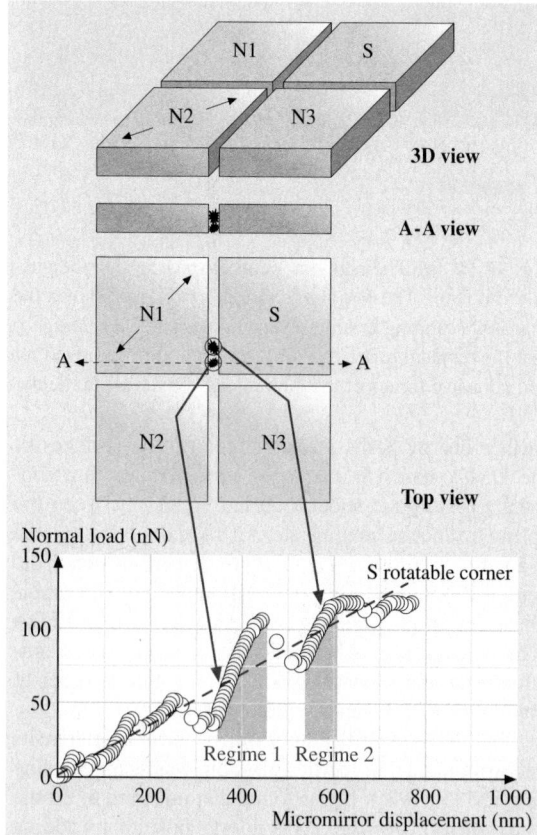

Fig. 50.57 Load–displacement curve obtained on the rotatable corner of the S micromirror and schematic to illustrate the suggested mechanism for the occurrence of soft stiction

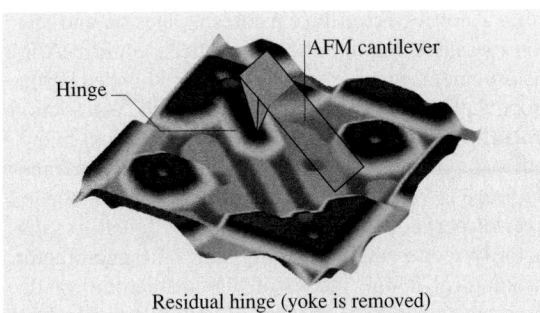

Fig. 50.58 AFM surface-height image of the residual hinge and schematic diagram of the relative position of the hinge and AFM tip during the nanoscale bending and fatigue tests. The tip is located at the free end of the hinge [50.196]

stiction of the landing site are summarized in Fig. 50.55. Wear initiates at the defect sites and consequent high stiction can result in high wear. Improving the coverage and wear resistance of SAM coatings could enhance the yield of DMD.

In some cases, the micromirrors are not fully stuck and can be moved by applying a load at the rotatable corner of the micromirror with a discontinuous motion, which is thus called *soft* stiction. Soft-stuck micromirrors studied by *Liu* and *Bhushan* [50.195] were identified in quality inspection. These micromirrors encountered slow transition from to end to the other end $(+1/-1)$. Figure 50.56 shows the AFM surface-height images of a location showing a stuck mirror (S) and surrounding normal micromirrors Ni ($i = 1, 2,$ and 3). Surprisingly, the images of the stuck and normal micromirror array are almost the same. On the micromirrors of interest, a tilting test was performed at the corner of the micromirrors, which are marked by white dots in Fig. 50.56. The rotatable direction of the microarray is indicated by an arrow bar in Fig. 50.56. The load–displacement curve for the stuck micromirror is presented in Fig. 50.57; it is not smooth and appears serrated. It is clearly indi-

cated that, although the S micromirror can be rotated, it rotates with hesitation. In regimes 1 and 2, as marked in Fig. 50.57, the slopes are much higher. In order to understand the mechanisms for the occurrence of stiction, stiction of landing sites of normal and stuck mirrors were measured. Unlike hard-stuck mirrors, the adhesive forces of soft-stuck and normal mirrors were comparable, which suggests that the SAM coating is intact with stuck mirror. It was found that a high normal load (about 900 nN) and on the order of a couple of hundred scans were required to remove the soft-stuck micromirrors by an AFM. However, only about 300 nN and about ten scans were required to remove a hard-stuck mirror. After careful examination of the AFM images of the micromirror sidewalls in Fig. 50.56 (bottom left), it is noted that there are contaminant particles attached to the sidewalls of the S mirror. It is, therefore, believed that, during the tilting test, the S micromirror (see schematic in Fig. 50.57) a sharper slope regime will occur in the displacement curve. Extra force is required to overcome the resistance that is induced by the sidewall contamination particles. This is believed to be the reason for the slow transition of the micromirror during quality inspection.

Finally, the nanomechanical characterization of various layers used in the construction of landing sites, hinge and micromirror materials have been measured by *Wei* et al. [50.193, 194]. Bending and fatigue studies of the hinge have been carried out by *Liu* and *Bhushan* [50.196] and *Bhushan* and *Liu* [50.197] to measure stiffness and fatigue properties. For these studies, the micromirror was removed. During removal, the micromirror/yoke structure was removed simultaneously, leaving the hinge mounted on one end of the array; see Fig. 50.58. The stiffness of the Al hinge was reported to be comparable to the stiffness of bulk Al. The Al hinge exhibited a higher modulus than the SiO$_2$ hinge. The fatigue properties depended upon the preparation of the hinge for testing.

50.7 Conclusion

The field of MEMS/NEMS has expanded considerably over the last decade. The length scale and large surface-to-volume ratio of the devices result in very high retarding forces such as adhesion and friction, which seriously undermine the performance and reliability of the devices. These tribological phenomena need to be studied and understood at the micro- to nanoscales.

In addition, materials for MEMS/NEMS must exhibit good microscale tribological properties. There is a need to develop lubricants and identify lubrication methods that are suitable for MEMS/NEMS. Using AFM-based techniques, researchers have conducted micro/nanotribological studies of materials and lubricants for use in MEMS/NEMS. In addition, component-level testing has also been carried out to aid understanding of the observed tribological phenomena in MEMS/NEMS.

Macroscale and microscale tribological studies of silicon and polysilicon films have been performed. The effect of doping and oxide films and environment on the tribological properties of these popular MEMS/NEMS materials have also been studied. SiC film is found to be a good tribological material for use in high-temperature MEMS/NEMS devices. Perfluoroalkyl self-assembled monolayers and bonded perfluoropolyether lubricants appear to be well suited for lubrication of microdevices under a range of environmental conditions. DLC coatings can also be used for low friction and wear. Adhesion of biomolecules on Si surfaces can be improved by nanopatterning and the chemical linker method. Roughness should be optimized for superhydrophobicity low adhesion and friction. Surface roughness measurements of micromachined polysilicon surfaces have been made using an AFM. The roughness distribution on surfaces is strongly dependent upon the fabrication process. Adhesion and friction of microstructures can be measured using a novel microtriboapparatus. Adhesion and friction measurements of silicon on silicon confirm AFM measurements that hexadecane thiol and bonded perfluoropolyether films exhibit superior adhesion and friction properties. Static friction force measurements of micromotors have been performed using an AFM. The forces are found to vary considerably with humidity. A bonded layer of perfluoropolyether lubricant is found to satisfactorily reduce the friction forces in the micromotor. Tribological failure modes of digital micromirror devices are either *hard* stiction or *soft* stiction. In hard stiction, the tip on the yoke remains stuck to the landing site underneath. The mechanism responsible for the hard stiction is localized damage to the SAM on the landing site. However, in soft stiction, the mirror–yoke assembly rotates with hesitation. The mechanism responsible for soft stiction is contaminant particles present on the mirror sidewalls.

AFM/FFM-based techniques show the capability to study and evaluate micro/nanoscale tribological phenomena related to MEMS/NEMS devices.

50.A Appendix Micro/Nanofabrication Methods

50.A.1 Top-Down Methods

The top-down fabrication methods used in the construction of MEMS include lithographic and non-lithographic techniques to produce micro- and nanostructures. The lithographic techniques fall into three basic categories: bulk micromachining, surface micromachining, and LIGA (a German acronym for Lithographie Galvanoformung Abformung), a German term for lithography, electroplating, and molding. The first two approaches, bulk and surface micromachining, mostly use planar photolithographic fabrication processes developed for semiconductor devices in producing two-dimensional (2D) structures [50.13, 19, 198, 199]. The various steps involved in these two fabrication processes are shown schematically in Fig. 50.59. Bulk micromachining employs anisotropic etching to remove sections through the thickness of a single-crystal silicon wafer, typically $250-500\,\mu m$ thick. Bulk micromachining is a proven high-volume production process and is routinely used to fabricate microstructures such as accelerometers, pressure sensors, and flow sensors. In surface micromachining, structural and sacrificial films are alternatively deposited, patterned, and etched to produce a free-standing structure. These films are typically made of low-pressure chemical vapor deposition (LPCVD) polysilicon film with a thickness of $2-20\,\mu m$. Surface micromachining is used to produce sensors, actuators, micromirror arrays, motors, gears, and grippers. The resolution in photolithography is dependent upon the wavelength of light. A commonly used light source is an argon fluoride excimer laser with a wavelength of 193 nm (ultraviolet or UV) used in patterning 90-nm lines and spaces. Deep-UV wavelengths, X-ray lithography, electron beam (e-beam) lithography, focused ion-beam lithography, maskless lithography, liquid-immersion lithography, and STM writing by removing material atom by atom are some of the recent developments for sub-100-nm patterning.

The fabrication of nanostructures such as nanochannels with sub-10-nm resolution can be accomplished through several routes: electron beam lithography and sacrificial-layer lithography (SLL). The process for e-beam lithographic technique is a finely focused electron beam that is exposed over a resist surface, where the exposure duration and location is controlled with the use of a computer [50.200, 201]. When the resist is exposed to the electron beam, the electrons either break or join the molecules in the resist, so the local characteristics are changed in such a way that further processes can either remove the exposed part (positive resist) or remove the unexposed part (negative resist). The resist material determines if the molecules will either break or join together and thus determines if a positive or negative image is produced. E-beam lithography can either be used to create photolithographic masks for replication or to create the devices directly. The masks that are created can be used for either optical or X-ray lithography. One

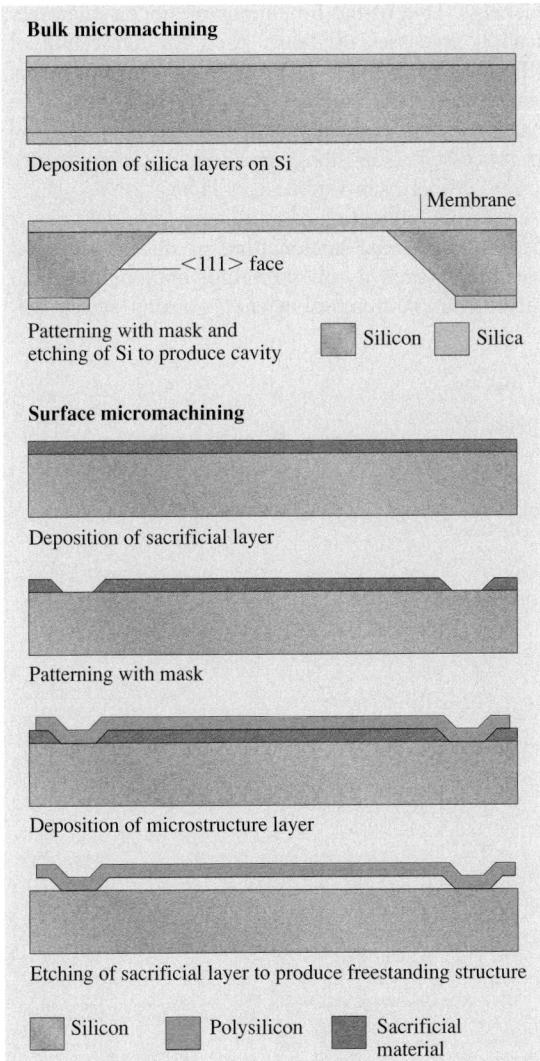

Bulk micromachining

Deposition of silica layers on Si

| Membrane

<111> face

Patterning with mask and
etching of Si to produce cavity   Silicon   Silica

Surface micromachining

Deposition of sacrificial layer

Patterning with mask

Deposition of microstructure layer

Etching of sacrificial layer to produce freestanding structure

Silicon   Polysilicon   Sacrificial
material

Fig. 50.59 Schematic of the process steps involved in bulk micromachining and surface-micromachining fabrication of MEMS

limitation of e-beam lithography is that throughput is drastically reduced since a single electron beam is used to create the entire exposure pattern on the resist. While this technique is slower than conventional lithographic techniques, it is ideal for prototype fabrication because no masks are required.

In SLL process, the use of a sacrificial layer allows the direct control of nanochannel dimensions as long as there exists a method for removing the sacrificial layer with absolute selectivity to the structural layers. A materials system with such selectivity is the silicon/silicon oxide system used widely in the microfabrication of MEMS devices. The use of sidewall deposition of the sacrificial layer and subsequent etching allows the fabrication of high-density nanochannels for biomedical applications. It is based on surface micromachining [50.77]. Figure 50.60 shows a schematic of the process steps in sacrificial-layer lithography based on *Hansford* et al.'s [50.77] work on fabrication of polysilicon membranes with nanochannels. As with all the membrane protocols, the first step in the fabrication is the etching of the support ridge structure into the bulk silicon substrate. A low-stress silicon nitride (LSN or simply nitride), which functions as an etch stop layer, is then deposited using LPCVD. The base structural polysilicon layer (base layer) is deposited on top of the etch stop layer. The plasma etching of holes in the base layer is what defines the shape of the pores. The buried nitride acts as an etch stop for the plasma etching of a polysilicon base layer. After the pore holes are etched through the base layer, the pore sacrificial thermal oxide layer is grown on the base layer. The basic requirement of the sacrificial layer is the ability to control the thickness with high precision across the entire wafer. Anchor points are defined in the sacrificial oxide layer to mechanically connect the base layer with the plug layer (necessary to maintain the pore spacing between layers). This is accomplished by using the same mask shifted from the pore holes. This produces anchors in one or two corners of each pore hole, which provide the desired connection between the structural layers while opening as much pore area as possible. After the anchor points are etched through the sacrificial oxide, the plug polysilicon layer is deposited (using LPCVD) to fill in the holes. To open the pores at the surface, the plug layer is planarized using chemical mechanical polishing (CMP) down to the base layer, leaving the final structure with the plug layer only in the pore hole openings. As the silicon wafer is ready for release, a protective nitride layer is deposited on the wafer (completely covering both sides of the wafer). The back-side

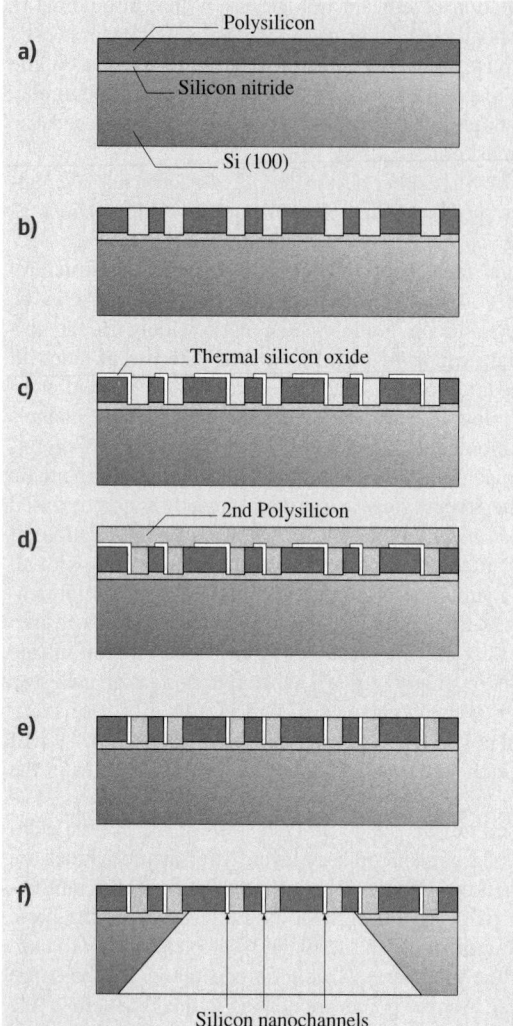

Fig. 50.60a–f Schematic of process steps involved in sacrificial-layer lithography: (**a**) Growth of silicon nitride layer (etch stop) and base polysilicon deposition, (**b**) hole definition in base, (**c**) growth of thin sacrificial oxide and patterning of anchor points, (**d**) deposition of plug polysilicon, (**e**) planarization of plug layer, and (**f**) deposition and patterning of the protective nitride layer through etch, followed by etching of protective, sacrificial and etch layers before final release of the structure in HF [50.77]

etch windows are etched in the protective layer, exposing the silicon wafer in the desired areas, and the wafer is placed in a KOH bath to etch. After the silicon wafer is completely removed up to the membrane (as evidenced by the smooth buried etch-stop layer), the protective,

sacrificial, and etch-stop layers are removed by etching in concentrated HF. Etching of sacrificial layer in polysilicon film defines nanochannels.

The LIGA process is based on the combined use of X-ray lithography, electroplating, and molding processes. The steps involved in the LIGA process are shown schematically in Fig. 50.61. LIGA is used to produce high-aspect-ratio MEMS (HARMEMS) devices that are up to 1 mm in height and only a few microns in width or length [50.202]. The LIGA process yields very sturdy 3D structures due to their increased thickness. One of the limitations of silicon microfabrication processes originally used for fabrication of MEMS devices is the lack of suitable materials that can be processed. With LIGA, a variety of non-silicon materials such as metals, ceramics and polymers can be processed. Non-lithographic micromachining processes, primarily in Europe and Japan, are also being used for the fabrication of millimeter-scale devices using direct material microcutting or micromechanical machining (such as microturning, micromilling, and microdrilling) or removal by energy beams (such as mi-

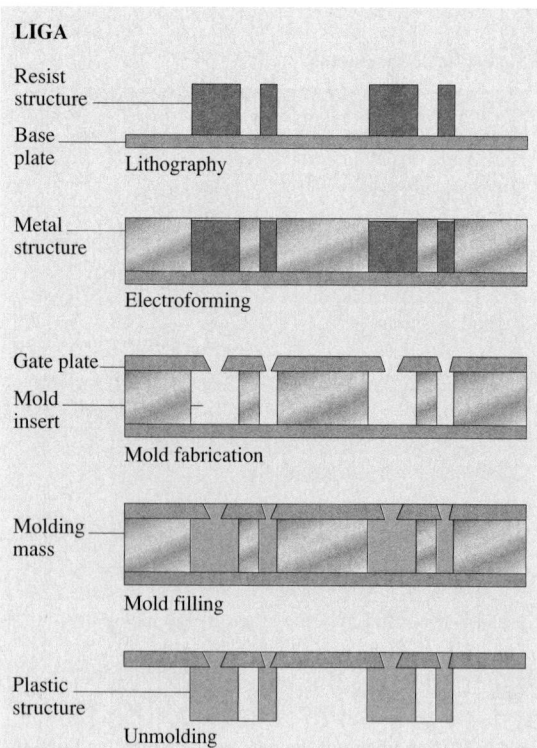

Fig. 50.61 Schematic of the process steps involved in LIGA fabrication of MEMS

crospark erosion, focused ion beam, laser ablation, and laser polymerization) [50.19, 203]. Hybrid technologies including LIGA and high-precision micromachining techniques have been used to produce miniaturized motors, gears, actuators, and connectors [50.88–90, 204]. These millimeter-scale devices may find more immediate applications.

A micro/nanofabrication technique, so-called *soft lithography*, is a non-lithographic technique [50.53, 59, 205, 206] in which a master or mold is used to generate patterns, defined by the relief on its surface, on polymers by replica molding [50.207], embossing (nanoimprint lithography) [50.208], or by contact printing (known as microcontact printing or μCP) [50.209]. Soft lithography is faster, less expensive, and more suitable for most biological applications than glass or silicon micromachining. Polymers have established an important role in BioMEMS/BioNEMS because of their reduced cost. The use of polymers also offers a wide range of material properties to allow tailoring of biological interactions for improved biocompatibility. Polymer fabrication is believed to be about an order of magnitude cheaper than silicon fabrication.

Replica molding is the transfer of a topographic pattern by curing or solidifying a liquid precursor against the original patterned mold. The mold or stamp is generally made of a two-part polymer (elastomer and curing agent), called poly(dimethylsiloxane) (PDMS) from photolithographically generated photoresist master. Solvent-based embossing, or imprinting, uses a solvent to restructure a polymer film. Hot embossing, also called nanoimprint lithography, usually refers to the transfer of pattern from a micromachined quartz or metal master to a pliable plastic sheet. Heat and high pressure allow the plastic sheet to become imprinted. These sheets can then be bonded to various plastics such as polymethyl methacrylate (PMMA). Nanoimprint lithography can produce patterns on a surface having 10-nm resolution. Contact printing uses a patterned stamp to transfer ink (mostly self-assembled monolayers) onto a surface in a pattern defined by the raised regions of a stamp. These techniques can be used to pattern line widths as small as 60 nm.

Replica molding is commonly used for mass-produced disposable plastic micro/nanocomponents, for example micro/nanofluidic chips, generally made of PDMS and PMMA [50.206, 210], and is also more flexible in choice of materials for construction than conventional photolithography.

To assemble microsystems, microrobots are used. Microrobotics include building blocks, such as steering links, microgrippers, conveyor system, and locomotive robots [50.17].

50.A.2 Bottom-Up Fabrication (Nanochemistry)

The bottom-up approach (from small to large) largely relies on nanochemistry [50.39, 40, 42–46]. The bottom-up approach includes chemical synthesis, the spontaneous *self-assembly* of molecular clusters (molecular self-assembly) from simple reagents in solution or biological molecules as building blocks to produce three-dimensional nanostructures as done by nature, quantum dots (nanocrystals) of arbitrary diameter (about $1 \times 10^1 - 1 \times 10^5$ atoms), molecular-beam epitaxy (MBE) and organometallic vapor-phase epitaxy (OMVPE) to create specialized crystals one atomic or molecular layer at a time, and manipulation of individual atoms by a scanning tunneling microscope or an atomic force microscope or atom optics. The self-assembly must be encoded, that is, one must be able to precisely assemble one object next to another to form a designed pattern. A variety of nonequilibrium plasma chemistry techniques are also used to produce layered nanocomposites, nanotubes, and nanoparticles. Nanostructures can also be fabricated using mechanosynthesis with proximal probes.

References

50.1   Anonymous: *Microelectromechanical Systems: Advanced Materials and Fabrication Methods*, NMAB–483 (National Academy, Washington, D.C. 1997)

50.2   M. Roukes: Nanoelectromechanical systems face the future, Physics World, 25–31 (Feb 2001)

50.3   Anonymous: *Small Tech 101 – An Introduction to Micro and Nanotechnology* (Small Times, 2003)

50.4   M. Schulenburg: *Nanotechnology – Innovation for Tomorrow's World* (European Commission, Research DG, Brussels 2004)

50.5   J. C. Eloy: *Status of the MEMS Industry 2005*, Report of Yole Developpement, France (SPIE Photonic West, San Jose 2005) presented at

50.6   S. Lawrence: Nanotech grows up, Technol. Rev. **108**(6), 31 (2005)

50.7   A. P. Graham, G. S. Duesberg, R. Seidel, M. Liebau, E. Unger, F. Kreupl, W. Hoenlein: Towards the integration of carbon nanotubes in microelectronics, Diam. Relat. Mater. **13**, 1296–1300 (2004)

50.8   D. Srivastava: Computational Nanotechnology of Carbon Nanotubes. In: *Carbon Nanotubes: Science and Applications*, ed. by M. Meyyappan (CRC, Boca Raton 2004) pp. 25–63

50.9   W. G. van der Wiel, S. De Franceschi, J. M. Elzerman, T. Fujisawa, S. Tarucha, L. P. Kouwenhoven: Electron transport through double quantum dots, Rev. Mod. Phys. **75**, 1–22 (2003)

50.10   R. S. Muller, R. T. Howe, S. D. Senturia, R. L. Smith, R. M. White: *Microsensors* (IEEE, New York 1990)

50.11   I. Fujimasa: *Micromachines: A New Era in Mechanical Engineering* (Oxford Univ. Press, Oxford 1996)

50.12   W. S. Trimmer (ed.): *Micromachines and MEMS, Classic and Seminal Papers to 1990* (IEEE, New York 1997)

50.13   B. Bhushan: *Tribology Issues and Opportunities in MEMS* (Kluwer Academic, Dordrecht 1998) Netherlands

50.14   G. T. A. Kovacs: *Micromachined Transducers Sourcebook* (WCB McGraw–Hill, Boston 1998)

50.15   S. D. Senturia: *Microsystem Design* (Kluwer Academic, Boston 2000)

50.16   M. Elwenspoek, R. Wiegerink: *Mechanical Microsensors* (Springer, Berlin, Heidelberg 2001)

50.17   M. Gad-el-Hak: *The MEMS Handbook* (CRC, Boca Raton 2002)

50.18   T. R. Hsu: *MEMS and Microsystems: Design and Manufacture* (McGraw–Hill, Boston 2002)

50.19   M. Madou: *Fundamentals of Microfabrication: The Science of Miniaturization*, 2nd edn. (CRC, Boca Raton 2002)

50.20   A. Hierlemann: *Integrated Chemical Microsensor Systems in CMOS Technology* (Springer, Berlin, Heidelberg 2005)

50.21   T. A. Core, W. K. Tsang, S. J. Sherman: Fabrication technology for an integrated surface-micromachined sensor, Solid State Technol. **36**, 39–47 (Oct 1993)

50.22   J. Bryzek, K. Peterson, W. McCulley: Micromachines on the march, IEEE Spectrum, 20–31 (May 1994)

50.23   J. S. Aden: The third-generation HP thermal inkjet printhead, HP J., 4–45 (Feb. 1994)

50.24   H. Le: Progress and trends in ink-jet printing technology, J. Imaging Sci. Technol. **42**, 49–62 (1998)

50.25   R. Baydo, A. Groscup: Getting to the heart of ink jet: Printheads, Beyond Recharger, 10–12 (2001)May 10, also visit

50.26   E. R. Lee: *Microdrop Generation* (CRC, Boca Raton 2003)

50.27   L. J. Hornbeck, W. E. Nelson: *Bistable Deformable Mirror Device*, Technical Digest Series: Spatial Light Modulators and Applications, Vol. 8 (OSA, Washington 1988) pp. 107–110

50.28   L. J. Hornbeck: A digital light processing™ update – Status and future applications, Proc. Soc. Photo-Opt. Eng. **3634**(Projection Displays V), 158–170 (1999)

50.29   L. J. Hornbeck: The DMD™ projection display chip: A MEMS-based technology, MRS Bull. **26**, 325–328 (2001)

50.30   K. Suzuki: Micro electro mechanical systems (MEMS) micro-switches for use in DC, RF, and optical applications, Jpn. J. Appl. Phys. **41**, 4335–4339 (2002)

50.31   V. M. Lubecke, J. C. Chiao: *MEMS Technologies for Enabling High Frequency Communication Cicuits*, Proc. IEEE 4th Int. Conf. on Telecom. In Modern Satellite, Cable and Broadcasting Services, Nis, Yugoslavia (IEEE, New York 1999) pp. 1–8

50.32   C. R. Giles, D. Bishop, V. Aksyuk: MEMS for lightwave networks, MRS Bull., 328–329 (April 2001)

50.33   A. Hierlemann, O. Brand, C. Hagleitner, H. Baltes: Microfabrication Techniques for Chemical/Biosensors. In: *Proc. of the IEEE, Chemical and Biological Microsensors*, Vol. Vol. 91, ed. by S. Casalnuovo, R. B. Brown (IEEE, New York 2003) pp. 839–863

50.34   B. Bhushan: *Tribology and Mechanics of Magnetic Storage Devices*, 2nd edn. (Springer, New York 1996)

50.35   H. Hamilton: Contact recording on perpendicular rigid media, J. Mag. Soc. Jpn. **15**((Suppl. S2)), 483–481 (1991)

50.36   D. A. Horsley, M. B. Cohn, A. Singh, R. Horowitz, A. P. Pisano: Design and fabrication of an angular microactuator for magnetic disk drives, J. Microelectromech. Syst. **7**, 141–148 (1998)

50.37   T. Hirano, L. S. Fan, D. Kercher, S. Pattanaik, T. S. Pan: HDD Tracking Microactuator and its Integration Issues. In: *Proc. ASME Int. Mech. Eng. Congress and Exp.*, Vol. MEMS- Vol. 2, ed. by A. P. Lee, J. Simon, F. K. Foster, R. S. Keynton (ASME, New York 2000) pp. 449–452

50.38   T. Fukuda, F. Arai, L. Dong: Assembly of nanodevices with carbon nanotubes through nanorobotic manipulations, Proc. IEEE **91**, 1803–1818 (2003)

50.39   K. E. Drexler: *Nanosystems: Molecular Machinery, Manufacturing and Computation* (Wiley, New York 1992)

50.40   G. Timp (ed.): *Nanotechnology* (Springer, New York 1999)

50.41   M. S. Dresselhaus, G. Dresselhaus, Ph. Avouris: *Carbon Nanotubes – Synthesis, Structure, Properties and Applications* (Springer, Berlin, Heidelberg 2001)

50.42   E. A. Rietman: *Molecular Engineering of Nanosystems* (Springer, New York 2001)

50.43   W. A. Goddard, D. W. Brenner, S. E. Lyshevski, G. J. Iafrate (eds): *Handbook of Nanoscience, Engineering, and Technology* (CRC, Boca Raton 2002)

50.44   H. S. Nalwa (ed.): *Nanostructured Materials and Nanotechnology* (Academic, San Diego 2002)

50.45   C. P. Poole, F. J. Owens: *Introduction to Nanotechnology* (Wiley, Hoboken 2003)

50.46   B. Bhushan: *Handbook of Micro/Nanotribology*, 2nd edn. (CRC, Boca Raton 1999)

50.47   B. Bhushan: *Nanotribology and Nanomechanics – An Introduction* (Springer, Berlin, Heidelberg 2005)

50.48   J. A. Stroscio, D. M. Eigler: Atomic and molecular manipulation with a scanning tunneling microscope, Science **254**, 1319 (1991)

50.49   P. Vettiger, J. Brugger, M. Despont, U. Drechsler, U. Duerig, W. Haeberle: Ultrahigh density, high data-rate NEMS based AFM data storage system, Microelec. Eng. **46**, 11–27 (1999)

50.50   C. Stampfer, A. Jungen, C. Hierold: *Fabrication of Discrete Carbon Nanotube Based Nanoscaled Force Sensor*, Proc. IEEE Sensors 2004, Vienna (IEEE, New York 2004) pp. 1056–1059

50.51   B. Bhushan: *Mechanics and Reliability of Flexible Magnetic Media*, 2nd edn. (Springer, New York 2000)

50.52   Anonymous: *International Technology Roadmap for Semiconductors* (2004), http://public.itrs.net/

50.53   A. Manz, H. Becker (eds): *Microsystem Technology in Chemistry and Life Sciences*, Topics in Current Chemistry 194 (Springer, Berlin, Heidelberg 1998)

50.54   J. Cheng, L. J. Kricka (eds): *Biochip Technology* (Harwood Academic, Philadelphia 2001)

50.55   M. J. Heller, A. Guttman (eds): *Integrated Microfabricated Biodevices* (Marcel Dekker, New York 2001)

50.56   C. Lai Poh San, E. P. H. Yap (eds.): *Frontiers in Human Genetics* (World Scientific, Singapore 2001)

50.57   C. H. Mastrangelo, H. Becker (eds): *Microfluidics and BioMEMS*, Proc. of SPIE Vol (SPIE, Bellingham 2001) p. 4560

50.58   H. Becker, L. E. Locascio: Polymer microfluidic devices, Talanta **56**, 267–287 (2002)

50.59   D. J. Beebe, G. A. Mensing, G. M. Walker: Physcis and applications of microfluidics in biology, Annu. Rev. Biomed. Eng. **4**, 261–286 (2002)

50.60   A. van der Berg (ed.): *Lab-on-a-Chip: Chemistry in Miniaturized Synthesis and Analysis Systems* (Elsevier, Amsterdam 2003)

50.61   P. Gravesen, J. Branebjerg, O. Jensen: Microfluidics – A Review, J. Micromech. Microeng. **3**, 168–182 (1993)

50.62   S. N. Bhatia, C. S. Chen: Tissue engineering at the micro-scale, Biomed. Microdevices **2**, 131–144 (1999)

50.63   R. P. Lanza, R. Langer, J. Vacanti (eds): *Principles of Tissue Engineering*, 2nd edn. (Academic, San Diego 2000)

50.64   E. Leclerc, K. S. Furukawa, F. Miyata, T. Sakai, T. Ushida, T. Fujii: Fabrication of microstructures in photosensitive biodegradable polymers for tissue engineering applications, Biomaterials **25**, 4683–4690 (2004)

50.65   K. Park (ed): *Controlled Drug Delivery: Challenges and Strategies* (American Chemical Society, Washington, D. C. 1997)

50.66   R. S. Shawgo, A. C. R. Grayson, Y. Li, M. J. Cima: BioMEMS for drug delivery, Curr. Opin. Solid State Mater. Sci. **6**, 329–334 (2002)

50.67   P. A. Oeberg, T. Togawa, F. A. Spelman: *Sensors in Medicine and Health Care* (Wiley, New York 2004)

50.68   J. V. Zoval, M. J. Madou: Centrifuge-based fluidic platforms, Proc. IEEE **92**, 140–153 (2000)

50.69   W. C. Tang, A. P. Lee: Defense applications of MEMS, MRS Bull. **26**, 318–319 (2001)Also see www.darpa.mil/mto/mems

50.70   M. R. Taylor, P. Nguyen, J. Ching, K. E. Peterson: Simulation of microfluidic pumping in a genomic DNA blood-processing cassette, J. Micromech. Microeng. **13**, 201–208 (2003)

50.71   R. Raiteri, M. Grattarola, M. Butt, P. Skladal: Micromechanical cantilever-based biosensor, Sens. Actuators B: Chem. **79**, 115–126 (2001)

50.72   B. Bhushan, D. R. Tokachichu, M. T. Keener, S. C. Lee: Morphology and adhesion of biomolecules on silicon based surfaces, Acta Biomateriala **1**, 327–341 (2005)

50.73   H. P. Lang, M. Hegner, C. Gerber: Cantilever array sensors, Mater. Today, 30–36 (April 2005)

50.74   F. Patolsky, C. Lieber: Nanowire nanosensors, Mater. Today, 20–28 (April 2005)

50.75   M. Scott: *MEMS and MOEMS for National Security ApplicationsReliability, Testing, and Characterization of MEMS/MOEMS II*, Proc. of SPIE, Vol. Vol. 4980 (SPIE, Bellingham 2003) pp. xxxvii–xliv

50.76   T. A. Desai, D. J. Hansford, L. Kulinsky, A. H. Nashat, G. Rasi, J. Tu, Y. Wang, M. Zhang, M. Ferrari: Nanopore technology for biomedical applications, Biomed. Devices **2**, 11–40 (1999)

50.77   D. Hansford, T. Desai, M. Ferrari: Nano-Scale Size-Based Biomolecular Separation Technology. In: *Biochip Technology*, ed. by J. Cheng, L. J. Kricka (Harwood Academic, New York 2001) pp. 341–361

50.78   F. J. Martin, C. Grove: Microfabricated drug delivery systems: concepts to improve clinical benefits, Biomed. Microdev. **3**, 97–108 (2001)

50.79   B. Bhushan: *Principles and Applications of Tribology* (Wiley, New York 1999)

50.80   B. Bhushan (ed.): *Modern Tribology Handbook* (CRC, Boca Raton 2001)

50.81   B. Bhushan: *Introduction to Tribology* (Wiley, New York 2002)

50.82   B. Bhushan: Adhesion and stiction: Mechanisms, measurement techniques, and methods for reduction, J. Vac. Sci. Technol. B **21**, 2262–2296 (2003)

50.83   Y. C. Tai, L. S. Fan, R. S. Muller: IC-Processed Micro-Motors: Design, Technology and Testing, Proc. IEEE Micro Electro Mechanical Systems, 1–6 (1989)

50.84   S. M. Spearing, K. S. Chen: Micro-gas turbine engine materials and structures, Ceramic Eng. Sci. Proc. **18**, 11–18 (2001)

50.85   L. G. Frechette, S. A. Jacobson, K. S. Breuer: High-speed microfabricated silicon turbomachinery and fluid film bearings, J. MEMS **14**, 141–152 (2005)

50.86 L. X. Liu, Z. S. Spakovszky: *Effect of Bearing Stiffness Anisotropy on Hydrostatic Micro Gas Journal Bearing Dynamic Behavior*, Proc. ASME Turbo Expo 2005, Paper No. GT-2005–68199 (Reno, Nevada 2005)

50.87 M. Mehregany, K. J. Gabriel, W. S. N. Trimmer: Integrated fabrication of polysilicon mechanisms, IEEE Trans. Electron. Dev. **35**, 719–723 (1988)

50.88 H. Lehr, S. Abel, J. Doppler, W. Ehrfeld, B. Hagemann, K. P. Kamper, F. Michel, Ch. Schulz, Ch. Thurigen: Microactuators as Driving Units for Microrobotic Systems. In: *Proc. Microrobotics: Components and Applications*, Vol. Vol. 2906, ed. by A. Sulzmann (SPIE, Bellingham 1996) pp. 202–210

50.89 H. Lehr, W. Ehrfeld, B. Hagemann, K. P. Kamper, F. Michel, Ch. Schulz, Ch. Thurigen: Development of micro-millimotors, Min. Invas. Ther. Allied Technol. **6**, 191–194 (1997)

50.90 F. Michel, W. Ehrfeld: Microfabrication Technologies for High Performance Microactuators. In: *Tribology Issues and Opportunities in MEMS*, ed. by B. Bhushan (Kluwer Academic, Dordrecht 1998) pp. 53–72

50.91 D. M. Tanner, N. F. Smith: *MEMS Reliability: Infrastructure, Test Structures, Experiments, and Failure Modes*, SAND2000-0091 (Sandia National Laboratories, Albuquerque 2000) Download from www.prod.sandia.gov

50.92 E. J. Garcia, J. J. Sniegowski: Surface micromachined microengine, Sens. Actuators A **48**, 203–214 (1995)

50.93 S. S. Mani, J. G. Fleming, J. A. Walraven, J. J. Sniegowski: *Effect of W Coating on Microengine Performance*, Proc. 38th Annual Inter. Reliability Phys. Symp. (IEEE, New York 2000) pp. 146–151

50.94 M. G. Hankins, P. J. Resnick, P. J. Clews, T. M. Mayer, D. R. Wheeler, D. M. Tanner, R. A. Plass: *Vapor Deposition of Amino-Functionalized Self-Assembled Monolayers on MEMS*, Proc. SPIE, Vol. 4980 (SPIE, Bellingham 2003) pp. 238–247

50.95 J. K. Robertson, K. D. Wise: An electrostatically actuated integrated microflow controller, Sens. Actuators A **71**, 98–106 (1998)

50.96 B. Bhushan: *Nanotribology and Nanomechanics of MEMS Devices*, Proc. Ninth Annual Workshop on Micro Electro Mechanical Systems (IEEE, New York 1996) pp. 91–98

50.97 R. E. Sulouff: MEMS Opportunities in Accelerometers and Gyros and the Microtribology Problems Limiting Commercialization. In: *Tribology Issues and Opportunities in MEMS*, ed. by B. Bhushan (Kluwer Academic, Dordrecht 1998) pp. 109–120

50.98 J. R. Martin, Y. Zhao: Micromachined Device Packaged to Reduce Stiction, US Patent 5,694,740 (1997) Dec. 9

50.99 G. Smith: The application of microtechnology to sensors for the automotive industry, Microelectron. J. **28**, 371–379 (1997)

50.100 L. S. Chang, P. L. Gendler, J. H. Jou: Thermal mechanical and chemical effects in the degradation of the plasma-deposited α-SC : H passivation layer in a multlayer thin-film device, J. Mater. Sci. **26**, 1882–1290 (1991)

50.101 M. Parsons: Design and manufacture of automotive pressure sensors, Sensors **18**(4), 32–46 (2001)

50.102 S. A. Henck: Lubrication of digital micromirror devices, Tribol. Lett. **3**, 239–247 (1997)

50.103 M. R. Douglass: *Lifetime Estimates and Unique Failure Mechanisms of the Digital Micromirror Devices (DMD)*, Proc. 36th Annual Inter. Reliability Phys. Symp. (IEEE, New York 1998) pp. 9–16

50.104 M. R. Douglass: *DMD Reliability: A MEMS Success Story*, Reliability, Testing, and Characterization of MEMS/MOEMS II, Proc. of SPIE Vol. 4980 (SPIE, Bellingham 2003) pp. 1–11

50.105 L. J. Hornbeck: Low Surface Energy Passivation Layer for Micromechanical Devices, US Patent 5,602,671 (1997) Feb. 11

50.106 R. A. Robbins, S. J. Jacobs: Lubricant Delivery for Micromechanical Devices, US Patent 6,300,294 B1 (2001) Oct. 9

50.107 I. DeWolf, W. M. van Spengen: Techniques to study the reliability of metal RF MEMS capacitive switches, Microelectron. Reliab. **42**, 1789–1794 (2002)

50.108 B. McCarthy, G. G. Adams, N. E. McGruer, D. Potter: A dynamic model, including contact bounce, of an electrostatically actuated microswitch, J. MEMS **11**, 276–283 (2002)

50.109 S. Shoji, M. Esashi: Microflow devices and systems, J. Micromech. Microeng. **4**, 157–171 (1994)

50.110 M. Stehr, S. Messner, H. Sandmaier, R. Zenergle: The VAMP – A new device for handing liquids or gases, Sens. Actuators A **57**, 153–157 (1996)

50.111 P. Woias: Micropumps – Summarizing the First Two Decades. In: *Proc. of SPIE – Microfluidics and BioMEMS*, Vol. Vol. 4560, ed. by C. H. Mastrangelo, H. Becker (SPIE, Bellingham 2001) pp. 39–52

50.112 N. T. Nguyen, X. Huang, T. K. Chuan: MEMS-micropumps: A review, ASME J. Fluids Eng. **124**, 384–392 (2002)

50.113 J. Henniker: Triboelectricity in polymers, Nature **196**, 474 (1962)

50.114 M. Sakaguchi, H. Kashiwabara: A generation mechanism of triboelectricity due to the reaction of mechaniradicals with mechanoions which are produced by mechanical fracture of solid polymer, Colloid Polym. Sci. **270**, 621–626 (1992)

50.115 G. R. Freeman, N. H. March: Triboelectricity and some associated phenomena, Mater. Sci. Eng. **15**, 1454–1458 (1999)

50.116 S. K. Cho, H. Moon, C. –J. Kim: Creating, transporting, cutting, and merging liquid droplets by electrowetting-based actuation for digital microfluidic circuits, J. MEMS **12**, 70–80 (2003)

50.117 A. R. Wheeler, H. Moon, C. A. Bird, R. R. O. Loo, C. –J. Kim, J. A. Loo, R. L. Garrell: Digital microfluidics with in-line sample purification for proteomics analysis with MALDI-MS, Anal. Chem. **77**, 534–540 (2005)

50.118 S. C. Lee, M. T. Keener, D. R. Tokachichu, B. Bhushan, P. D. Barnes, B. R. Cipriany, M. Gao, L. J. Brillson: Preparation of a thin protein surface on thermally grown silicon dioxide, J. Vac. Sci. Technol. B **23**, 1856–1865 (2005)

50.119 J. Black: *Biological Performance of Materials: Fundamentals of Biocompatibility* (Marcel Dekker, New York 1999)

50.120 D. R. Tokachichu, B. Bhushan: Bioadhesion of Polymers for BioMEMS, IEEE Trans. Nanotech. **5**, 228–231 (2006)

50.121 G. Wei, B. Bhushan, N. Ferrell, D. Hansford: Microfabrication and nanomechanical characterization of polymer microelectromechanical systems for biological applications, J. Vac. Sci. Technol. A **23**, 811–819 (2005)

50.122 B. Bhushan, T. Kasai, G. Kulik, L. Barbieri, P. Hoffman: AFM study of perfluorosilane and alkylsilane self-assembled monolayers for anti-stiction in MEMS/NEMS, Ultramicroscopy **105**, 176–188 (2005)

50.123 B. Bhushan, D. Hansford, K. K. Lee: Surface Modification of Silicon and PDMS Surfaces with Vapor Phase Deposited Ultrathin Fluorosilane Films for Biomedical Nanodevices, J. Vac. Sci. Technol. A **24**, 1197–1202 (2005)

50.124 T. Kasai, B. Bhushan, G. Kulik, L. Barbieri, P. Hoffman: Nanotribological study of perfluorosilane SAMs for anti-stiction and low wear, J. Vac. Sci. Technol. B **23**, 995–1003 (2005)

50.125 K. K. Lee, B. Bhushan, D. Hansford: Nanotribological characterization of perfluoropolymer thin films for biomems applications, J. Vac. Sci. Technol. A **23**, 804–810 (2005)

50.126 P. Decuzzi, S. Lee, B. Bhushan, M. Ferrari: A theoretical model for the margination of particles with blood vessels, Annals Biomed. Eng. **33**, 179–190 (2005)

50.127 H. Guckel, D. W. Burns: Fabrication of micromechanical devices from polysilicon films with smooth surfaces, Sens. Actuators **20**, 117–122 (1989)

50.128 G. T. Mulhern, D. S. Soane, R. T. Howe: *Supercritical Carbon Dioxide Drying of Microstructures*, Proc. Int. Conf. on Solid-State Sensors and Actuators (IEEE, New York 1993) pp. 296–299

50.129 K. F. Man, B. H. Stark, R. Ramesham: *A Resource Handbook for MEMS Reliability*, Rev. A. (JPL Press, Jet Propulsion Laboratory, California Institute of Technology, Pasadena 1998)

50.130 S. Kayali, R. Lawton, B. H. Stark: MEMS reliability assurance activities at JPL, EEE Links **5**, 10–13 (1999)

50.131 S. Arney: Designing for MEMS reliability, MRS Bull. **26**, 296–299 (2001)

50.132 K. F. Man: MEMS Reliability for Space Applications by Elimination of Potential Failure Modes Through Testing and Analysis, http://www.rel.jpl.nasa.gov/Org//atop/products/Prod-map.html (2001)

50.133 B. Bhushan, A. V. Kulkarni, W. Bonin, J. T. Wyrobek: Nano/picoindentation measurement using a capacitance transducer system in atomic force microscopy, Philos. Mag. **74**, 1117–1128 (1996)

50.134 S. Sundararajan, B. Bhushan: Development of AFM-based techniques to measure mechanical properties of nanoscale structures, Sens. Actuators A **101**, 338–351 (2002)

50.135 B. Bhushan, J. N. Israelachvili, U. Landman: Nanotribology: Friction, wear and lubrication at the atomic scale, Nature **374**, 607–616 (1995)

50.136 B. Bhushan, B. K. Gupta: *Handbook of Tribology: Materials, Coatings and Surface Treatments*, Reprint edition edn. (Krieger, Malabar 1997)

50.137 J. F. Shackelford, W. Alexander, J. S. Park (eds.): *CRC Material Science and Engineering Handbook*, 2nd edn. (CRC, Boca Raton 1994)

50.138 J. S. Shor, D. Goldstein, A. D. Kurtz: Characterization of n-type β-SiC as a piezoresistor, IEEE Trans. Electron. Dev. **40**, 1093–1099 (1993)

50.139 M. Mehregany, C. A. Zorman, N. Rajan, C. H. Wu: Silicon Carbide MEMS for Harsh Environments, Proc. IEEE **86**, 1594–1610 (1998)

50.140 C. A. Zorman, A. J. Fleischmann, A. S. Dewa, M. Mehregany, C. Jacob, S. Nishino, P. Pirouz: Epitaxial growth of 3C–SiC films on 4 in. diam Si(100) silicon wafers by atmospheric pressure chemical vapor deposition, J. Appl. Phys. **78**, 5136–5138 (1995)

50.141 C. A. Zorman, S. Roy, C. H. Wu, A. J. Fleischman, M. Mehregany: Characterization of polycrystalline silicon carbide films grown by atmospheric pressure chemical vapor deposition on polycrystalline silicon, J. Mater. Res. **13**, 406–412 (1998)

50.142 C. H. Wu, S. Stefanescu, H. I. Kuo, C. A. Zorman, M. Mehregany: *Fabrication and Testing of Single Crystalline 3C–SiC Piezoresistive Pressure Sensors*, Technical Digest – 11th Int. Conf. Solid State Sensors and Actuators – Eurosensors, Vol. XV (Munich 2001) pp. 514–517

50.143 A. A. Yasseen, C. H. Wu, C. A. Zorman, M. Mehregany: Fabrication and testing of surface micromachined polycrystalline sic micromotors, IEEE Electron. Dev. Lett. **21**, 164–166 (2000)

50.144 B. K. Gupta, J. Chevallier, B. Bhushan: Tribology of ion bombarded silicon for micromechanical applications, ASME J. Tribol. **115**, 392–399 (1993)

50.145 B. K. Gupta, B. Bhushan: Nanoindentation studies of ion implanted silicon, Surf. Coat. Technol. **68-69**, 564–570 (1994)

50.146 B. K. Gupta, B. Bhushan, J. Chevallier: Modification of tribological properties of silicon by boron ion implantation, Tribol. Trans. **37**, 601–607 (1994)

50.147 B. Bhushan, V. N. Koinkar: Tribological studies of silicon for magnetic recording applications, J. Appl. Phys. **75**, 5741–5746 (1994)

50.148 G. M. Pharr: The Anomalous Behavior of Silicon During Nanoindentation. In: *Thin Films: Stresses and Mechanical Properties III*, Vol. 239, ed. by W. D. Nix, J. C. Bravman, E. Arzt, L. B. Freund (Materials Research Soc., Pittsburgh 1991) pp. 301–312

50.149 D. L. Callahan, J. C. Morris: The extent of phase transformation in silicon hardness indentation, J. Mater. Res. **7**, 1612–1617 (1992)

50.150 N. A. Fleck, G. M. Muller, M. F. Ashby, J. W. Hutchinson: Strain gradient plasticity: theory and experiment, Acta Metall. Mater. **42**, 475–487 (1994)

50.151 B. Bhushan, S. Venkatesan: Friction and wear studies of silicon in sliding contact with thin-film magnetic rigid disks, J. Mater. Res. **8**, 1611–1628 (1993)

50.152 S. Venkatesan, B. Bhushan: The role of environment in the friction and wear of single-crystal silicon in sliding contact with thin-film magnetic rigid disks, Adv. Info Storage Syst. **5**, 241–257 (1993)

50.153 S. Venkatesan, B. Bhushan: The sliding friction and wear behavior of single-crystal, polycrystalline and oxidized silicon, Wear **171**, 25–32 (1994)

50.154 B. Bhushan: Chemical, mechanical and tribological characterization of ultra-thin and hard amorphous carbon coatings as thin as 3.5 nm: Recent developments, Diam. Relat. Mater. **8**, 1985–2015 (1999)

50.155 B. Bhushan, S. Sundararajan, X. Li, C. A. Zorman, M. Mehregany: Micro/Nanotribological Studies of Single-Crystal Silicon and Polysilicon and SiC Films for Use in MEMS Devices. In: *Tribology Issues and Opportunities in MEMS*, ed. by B. Bhushan (Kluwer Academic, Dordrecht 1998) pp. 407–430

50.156 S. Sundararajan, B. Bhushan: Micro/nanotribological studies of polysilicon and sic films for mems applications, Wear **217**, 251–261 (1998)

50.157 X. Li, B. Bhushan: Micro/nanomechanical characterization of ceramic films for microdevices, Thin Solid Films **340**, 210–217 (1999)

50.158 H. Liu, B. Bhushan: Nanotribological characterization of molecularly-thick lubricant films for applications to MEMS/NEMS by AFM, Ultramicroscopy **97**, 321–340 (2003)

50.159 B. Bhushan, A. V. Kulkarni, V. N. Koinkar, M. Boehm, L. Odoni, C. Martelet, M. Belin: Microtribological characterization of self-assembled and langmuir–blodgett monolayers by atomic force and friction force microscopy, Langmuir **11**, 3189–3198 (1995)

50.160 V. N. Koinkar, B. Bhushan: Micro/nanoscale studies of boundary layers of liquid lubricants for magnetic disks, J. Appl. Phys **79**, 8071–8075 (1996)

50.161 V. N. Koinkar, B. Bhushan: Microtribological studies of unlubricated and lubricated surfaces using atomic force/friction force microscopy, J. Vac. Sci. Technol. A **14**, 2378–2391 (1996)

50.162 B. Bhushan, H. Liu: Nanotribological properties and mechanisms of alkylthiol and biphenyl thiol self-assembled monolayers studied by AFM, Phys. Rev. B **63**, 245412:1–11 (2001)

50.163 H. Liu, B. Bhushan: Investigation of nanotribological properties of self-assembled monolayers with alkyl and biphenyl spacer chains, Ultramicroscopy **91**, 185–202 (2002)

50.164 N. S. Tambe, B. Bhushan: Nanotribological characterization of self assembled monolayers deposited on silicon and aluminum substrates, Nanotechnology **16**, 1549–1558 (2005)

50.165 Z. Tao, B. Bhushan: Bonding, degradation, and environmental effects on novel perfluoropolyether lubrications, Wear **259**, 1352–1361 (2005)

50.166 Z. Tao, B. Bhushan: Degradation mechanisms and environmental effects on perfluoropolyether, self assembled monolayers, and diamondlike carbon films, Langmuir **21**, 2391–2399 (2005)

50.167 T. Stifter, O. Marti, B. Bhushan: Theoretical investigation of the distance dependence of capillary and van der waals forces in scanning force microscopy, Phys. Rev. B **62**, 13667–13673 (2000)

50.168 H. Liu, B. Bhushan, W. Eck, V. Stadler: Investigation of the adhesion, friction, and wear properties of biphenyl thiol self-assembled monolayers by atomic force microscopy, J. Vac. Sci. Technol. A **19**, 1234–1240 (2001)

50.169 N. S. Tambe, B. Bhushan: Identifying materials with low friction and adhesion for nanotechnology applications, Appl. Phys. Lett. **86**, 061906–1 to –3 (2005)

50.170 N. S. Tambe, B. Bhushan: Micro/nanotribological characterization of PDMS and PMMA used for bioMEMS/NEMS applications, Ultramicroscopy **105**, 238–247 (2005)

50.171 B. Bhushan, Z. Burton: Adhesion and friction properties of polymers in microfluidic devices, Nanotechnology **16**, 467–478 (2005)

50.172 B. Bhushan, D. Tokachichu, M. T. Keener, S. C. Lee: Nanoscale adhesion, friction, and wear studies of biomolecules on silicon based surfaces, Acta Biomaterialia **2**, 39–49 (2005)

50.173 M. Nosonovsky, B. Bhushan: Roughness optimization for biomimetic superhydrophobic surfaces, Microsyst. Technol. **11**, 535–549 (2005)

50.174 R. N. Wenzel: Resistance of solid surfaces to wetting by water, Ind. Eng. Chem. **28**, 988–994 (1936)

50.175 A. Cassie, S. Baxter: Wetting of porous surfaces, Trans. Faraday Soc. **40**, 546–551 (1944)

50.176 M. Nosonovsky, B. Bhushan: Stochastic model for metastable wetting of roughness-induced superhydrophobic surfaces, Microsyst. Technol. **12**, 231–237 (2006)

50.177 Z. Burton, B. Bhushan: Hydrophobicity, adhesion and friction properties of nanopatterned polymers and scale dependence for micro- and nanoelectromechanical systems, Nano Lett. **20**, 83–90 (2005)

50.178 S. Sundararajan, B. Bhushan: Static friction and surface roughness studies of surface micromachined electrostatic micromotors using an atomic force/friction force microscope, J. Vac. Sci. Technol. A **19**, 1777–1785 (2001)

50.179 C. H. Mastrangelo, C. H. Hsu: Mechanical stability and adhesion of microstructures under capillary forces – Part II: Experiments, J. Microelectromech. Syst. **2**, 44–55 (1993)

50.180 R. Maboudian, R. T. Howe: Critical review: Adhesion in surface micromechanical structures, J. Vac. Sci. Technol. B **15**, 1–20 (1997)

50.181 C. H. Mastrangelo: Surface Force Induced Failures in Microelectromechanical Systems. In: *Tribology Issues and Opportunities in MEMS*, ed. by B. Bhushan (Kluwer Academic, Dordrecht 1998) pp. 367–395

50.182 M. P. De Boer, T. A. Michalske: Accurate method for determining adhesion of cantilever beams, J. Appl. Phys. **86**, 817 (1999)

50.183 R. L. Alley, G. J. Cuan, R. T. Howe, K. Komvopoulos: In: *Proc. Solid State Sensor and Actuator Workshop*, ed. by C. H. Mastrangelo, C. H. Hsu (IEEE, New York 1992) pp. 202–207

50.184 H. Liu, B. Bhushan: Adhesion and friction studies of microelectromechanical systems/nanoelectromechanical systems materials using a novel microtriboapparatus, J. Vac. Sci. Technol. A **21**, 1528–1538 (2003)

50.185 B. Bhushan, H. Liu, S. M. Hsu: Adhesion and friction studies of silicon and hydrophobic and low friction films and investigation of scale effects, ASME J. Tribol. **126**, 583–590 (2004)

50.186 S. K. Chilamakuri, B. Bhushan: A comprehensive kinetic meniscus model for prediction of long-term static friction, J. Appl. Phys. **15**, 4649–4656 (1999)

50.187 Y. C. Tai, R. S. Muller: Frictional study of IC processed micromotors, Sens. Actuators A **21–23**, 180–183 (1990)

50.188 K. J. Gabriel, F. Behi, R. Mahadevan, M. Mehregany: In situ friction and wear measurement in integrated polysilicon mechanisms, Sens. Actuators A. **21–23**, 184–188 (1990)

50.189 M. G. Lim, J. C. Chang, D. P. Schultz, R. T. Howe, R. M. White: *Polysilicon Microstructures to Characterize Static Friction*, Proc. IEEE Micro Electro Mechanical Systems (IEEE, New York 1990) pp. 82–88

50.190 U. Beerschwinger, S. J. Yang, R. L. Reuben, M. R. Taghizadeh, U. Wallrabe: Friction measurements on LIGA-processed microstructures, J. Micromech. Microeng. **4**, 14–24 (1994)

50.191 D. Matheison, U. Beerschwinger, S. J. Young, R. L. Rueben, M. Taghizadeh, S. Eckert, U. Wallrabe: Effect of progressive wear on the friction characteristics of nickel LIGA processed rotors, Wear **192**, 199–207 (1996)

50.192 H. Liu, B. Bhushan: Nanotribological characterization of digital micromirror devices using an atomic force microscope, Ultramicroscopy **100**, 391–412 (2004)

50.193 G. Wei, B. Bhushan, S. J. Jacobs: Nanomechanical characterization of digital multilayered thin film structures for digital micromirror devices, Ultramicroscopy **100**, 375–389 (2004)

50.194 G. Wei, B. Bhushan, S. J. Jacobs: Nanoscale indentation fatigue and fracture toughness measurements of multilayered thin film structures for digital micromirror devices, J. Vac. Sci. Technol. A **22**, 1397–1405 (2004)

50.195 H. Liu, B. Bhushan: Investigation of nanotribological and nanomechanical properties of the digital micromirror device by atomic force microscope, J. Vac. Sci. Technol. A **22**, 1388–1396 (2004)

50.196 H. Liu, B. Bhushan: Bending and fatigue study on a nanoscale hinge by an atomic force microscope, Nanotechnology **15**, 1246–1251 (2004)

50.197 B. Bhushan, H. Liu: Characterization of nanomechanical and nanotribological properties of digital micromirror devices, Nanotechnology **15**, 1785–1791 (2004)

50.198 R. C. Jaeger: *Introduction to Microelectronic Fabrication*, Vol. Vol. 5 (Addison–Wesley, Reading 1988)

50.199 J. W. Judy: Microelectromechanical systems (MEMS): Fabrication, design, and applications, Smart Mater. Struct. **10**, 1115–1134 (2001)

50.200 G. Brewer: *Electron-Bean Technology in Microelectronic Fabrication* (Academic, New York 1980)

50.201 K. Valiev: *The Physics of Submicron Lithography* (Plenum, New York 1992)

50.202 E. W. Becker, W. Ehrfeld, P. Hagmann, A. Maner, D. Munchmeyer: Fabrication of microstructures with high aspect ratios and great structural heights by synchrotron radiation lithography, galvanoforming, and plastic moulding (LIGA process), Microelectron. Eng. **4**, 35–56 (1986)

50.203 C. R. Friedrich, R. O. Warrington: Surface Characterization of Non-Lithographic Micromachining. In: *Tribology Issues and Opportunities in MEMS*, ed. by B. Bhushan (Kluwer Academic, Dordrecht 1998) pp. 73–84

50.204 M. Tanaka: Development of desktop machining microfactory, Riken Rev. **34**, 46–49 (April 2001)

50.205 Y. Xia, G. M. Whitesides: Soft lithography, Angew. Chem. Int. Ed. **37**, 550–575 (1998)

50.206 H. Becker, C. Gaertner: Polymer microfabrication methods for microfluidic analytical applications, Electrophoresis **21**, 12–26 (2000)

50.207 Y. Xia, E. Kim, X. M. Zhao, J. A. Rogers, M. Prentiss, G. M. Whitesides: Complex optical surfaces formed by replica molding against elastomeric masters, Science **273**, 347–349 (1996)

50.208 S. Y. Chou, P. R. Krauss, P. J. Renstrom: Imprint lithography with 25-nanometer resolution, Science **272**, 85–87 (1996)

50.209 A. Kumar, G. M. Whitesides: Features of gold hav-
        ing micrometer to centimeter dimensions can be
        formed through a combination of stamping with
        an elastomeric stamp and an alkanethiol ink fol-
        lowed by chemical etching, Appl. Phys. Lett. **63**,
        2002–2004 (1993)

50.210 J. C. McDonald, D. C. Duffy, J. R. Anderson,
        D. T. Chiu, H. Wu, O. J. A. Schueller, G. M. White-
        sides: Fabrication of microfluidic systems in
        poly(dimethylsiloxane), Electrophoresis **21**, 27–40
        (2000)

51. Experimental Characterization Techniques for Micro/Nanoscale Devices

This chapter introduces some commonly utilized techniques for the dynamic measurement and characterization of microelectromechanical systems (MEMS) and nanoelectromechanical systems (NEMS) devices. One important step in the design process is verifying that the devices perform as expected. In many MEMS applications, this involves quantifying the motion. In this chapter, a variety of techniques for characterizing MEMS and NEMS devices are presented. The techniques discussed here cover both dynamic and static/quasi-static characterization, as well as the extraction of model parameters. Techniques for external actuation and motion measurement are covered in depth. An example demonstrating methods for extracting linear and nonlinear device parameters from a microresonator is given. Error analysis techniques are also demonstrated. From this introduction, a user should be able to choose a suitable test technique for a variety of MEMS/NEMS devices.

51.1 Motivation

Since the development of the first micromachines, the need has existed for device test and characterization. Testing methods for macroscopic devices are often not applicable at the micro and nanoscale and this has led to significant problems in fully understanding their behavior. Difficulties include determining thin film material property variations and failing to capture essential dynamics. However, that has changed recently, as many researchers and vendors have developed an arsenal of MEMS test and characterization tools for many specialized measurements.

In this chapter, we outline many of the techniques currently available for the test and characterization of MEMS. These include on-chip and off-chip, optical, electrical, and mechanical methods for excitation and measurement. Due to the variation in micro- and nanoscale devices both in form and in function, the test techniques are often application-specific. An attempt is made to discuss these techniques in as broad

a context as possible and give guidance for choosing an appropriate technique to create and measure certain signals.

The chapter begins with a discussion of applications that strongly benefit from the ability to accurately characterize MEMS devices. Experimental techniques are then discussed. Optical methods are discussed first, including optical interferometry, optical stroboscopic image analysis, and laser Doppler vibrometry. External methods for quasi-static excitation and measurement of MEMS are covered next. Tools commonly used for quasi-static actuation include nanoindentation, AFM, as well as some of the optical techniques previously discussed. SEM-based testing is next and is followed by a discussion of actuation methods including a sampling of novel techniques for specific applications. Another class of characterization techniques are those developed to study the material properties of thin films and MEMS materials. These are beyond the scope of this chapter, but a good review is provided by *Saif* and *Haque* [51.1]. Finally, a through characterization example is presented

followed by some tips and tricks for achieving accurate measurements and designing test devices.

51.2 Applications Utilizing Dynamic MEMS/NEMS

The success of many MEMS-based applications ultimately depends on the ability to design, build, and package robust devices that perform their desired function. The verification of performance is known as testing/characterization. This important step in the design process can be accomplished in many different ways, and in micro/nanoscale systems, techniques can be highly application specific. The limitation of a particular test technique should not be a limiting factor in the device design. An adequate test technique will have appropriate resolution and bandwidth to capture the essential elements of the behavior. In the past, this was not often possible and led to a significantly reduced understanding of device motions and functionality, and caused much speculation as to actual device performance. By choosing an appropriate test technique, much can be learned about the device behavior during the design and qualification stages.

For example, in an optical switch, a flat platform (mirror) must be actuated, and in many cases rotates on a torsional spring support. For the switch to be successful, the device must be linear, the ring-down must be fast or nonexistent, the actuator must produce the expected power, and the mirror surface must be flat

and not deform under the applied load. Testing techniques are needed to measure and verify each of these features.

The optical switch test requirements are in sharp contrast to those for a micro-accelerometer or gyroscope (angular rate sensor). The inertial sensing applications require these devices to have adequate self-test mechanisms, appropriate bandwidth and dynamic range, functional sense transducers with appropriately functioning feedback, and calibration to a reference accelerometer standard.

These examples are just two of the many applications that require test and characterization for device development and qualification. We will present some test techniques that are widely applicable and some techniques developed for a specific application. Microsystem applications requiring dynamic test and characterization techniques include optical and radio frequency (RF) switches [51.2–4], mechanically based RF filters [51.5,6], accelerometers, gyroscopes [51.7,8], cantilever-based resonant sensors [51.9–11], disk drive head actuators [51.12], micro/nano positioners [51.13], nanowire resonant sensors [51.14, 15], and mass or chemical sensors [51.16].

51.3 Test/Characterization Techniques

51.3.1 Optical Methods for Dynamic Characterization

Optical test methods have many benefits, the most important being that they are usually noninvasive, and rarely require special modifications to the devices. They can be applied to many different device and experimental test set-ups. They often give high resolution, and can be extended to three-dimensional (3D) measurements. There are some limitations to optical techniques, including optical accessibility to a device, and in some cases, they are limited to periodic motion. In this section, techniques ranging from optical interferometry to video stroboscopy are covered. Each of the techniques

is introduced, and their benefits and limitations are discussed.

Stroboscopic Optical Techniques

Devices that move with a periodic, repeatable motion, such as a microstructure vibrating its natural frequency can be imaged with a stroboscopic method. The use of a strobed light source means that the minimum resolvable time interval is based on the limits of the light pulse rather than on the detector. Thus a slow detector, such as a video camera, can be used to image a microstructure with a much higher motional frequency than the refresh rate of the imager. Stroboscopic techniques can be used in conjunction with interferometry, or with video analysis to gain valuable information about the motion of the device.

Using an optical microscope with a charge-coupled device (CCD) camera and by replacing the light source with a light-emitting diode (LED) driven by a pulse generator, a computer microvision measurement system, shown in Fig. 51.1, has been demonstrated [51.18, 20]. The microstructure is driven at the same frequency as the LED, thus illuminating the device at one position in the motion. By adjusting the phase between the LED pulses and the microstructure drive signals, the entire motion of the device can be imaged. A computer is used to collect individual frames from the camera, as a function of phase difference, for post-processing. Image-recognition algorithms are used to perform a least-squares fit of the position of the device in adjacent frames in phase space, to determine the translational and rotational motion. In-plane motions have been shown with 2-nm resolution [51.20] and out-of-plane motions with 5-nm resolution [51.17, 18]. Resolution improvements are made by sampling sequences of phased images from different focal planes on the device.

The computer analysis technique is computationally intensive, limited to periodic motions, and requires the microstructure be a rigid body. Any deformation of the structure, as a function of position, would appear as motion of the device during analysis. The technique has been commercially released by at least two companies [51.21, 22] and has found good acceptance for non-contact in-plane device measurements.

Optical Stroboscopic Interferometry

A stroboscopic interferometry technique for measuring out of plane motions [51.19, 23]. This technique utilizes phase-shifting interferometry to resolve repeated dynamic motion has been developed. Figure 51.2 shows a schematic of this experimental set-up. A strobed laser diode is used to illuminate the device and a CCD camera is used to record the interferogram. By actuating the device with a pulse, the dynamic response can be determined by repeatedly applying the pulse and strobing the laser at different time delays (phase) from the excitation event. One point in the camera image can be analyzed to extract the dynamic motion of displacement as a function of time. The analysis can be repeated for each point in the CCD image and analyzed in the Fourier domain to resolve the natural frequencies of the device under test. The device can then be driven sinusoidally at one of its natural frequencies and a new set of interferograms collected by adjusting the phase of the strobe relative to the excitation as in the video method in the

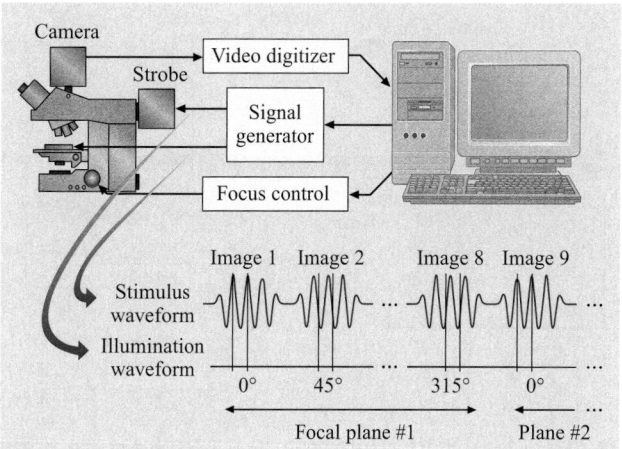

Fig. 51.1 Computer microvision measurement system. The computer controls the signal generator, which provides a stimulus waveform for the test structure, as well as an illumination waveform that determines the phase of motion illuminated by the stroboscopic light source. By controlling the focus of the microscope, the computer can take images at multiple planes of focus. (After [51.17, 18])

previous section. The resulting set of interferograms can be analyzed to extract the mode shape at the excitation frequency. The high brightness of the laser diode allows measurements to be made at a low illumination

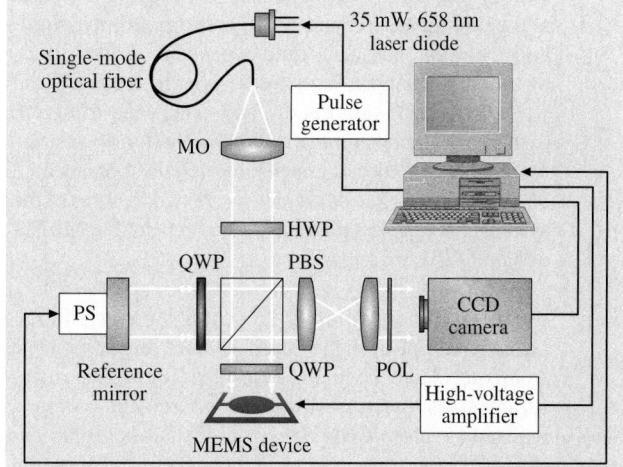

Fig. 51.2 Experimental arrangement for the computer-controlled stroboscopic interferometer. This interferometric measurement is capable of resolving motions as small as a few nm for out-of-plane deflections. (MO: microscope objective, HWP: half-wave plate, QWP: quarter-wave plate, PS: piezoelectric stage, PBS: polarizing beam-splitter cube, POL: linear polarizer). (After [51.19])

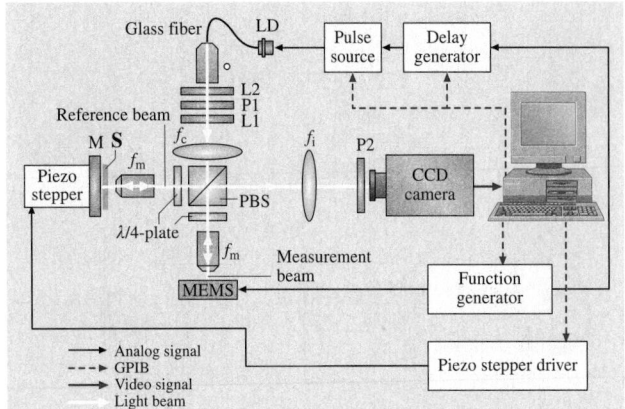

Fig. 51.3 Schematic for 3D stroboscopic interferometer and video analysis measurement. In-plane motions are resolved using stroboscopic video analysis, while out-of-plane motions are resolved with stroboscopic interferometry. Using software algorithms, three-dimensional motion information can be obtained. (After [51.24])

duty cycle. The authors have demonstrated interferometric stroboscopic imaging at duty cycles as small as 3×10^{-5} s using a standard 30-ms-exposure CCD. This system is capable of measurements from 10 Hz to 30 kHz range, and has amplitude repeatability better than 1.5 nm.

In addition, phase-measuring algorithms combined with spatial and temporal phase-unwrapping methods allow this measurement to analyze not only rigid-body motion but also time-dependent deformations of micromechanical structures, which are essential when characterizing micro- and nanoscale structure dynamics. The inherent two-dimensionality of interferometric optical measurements, as well as sophisticated computer-controlled data analysis are two features of the system that enable rapid analysis of repeatable MEMS dynamics [51.19].

Three Dimensional Stroboscopic Measurements

Using the out-of-plane stroboscopic interferometric technique together with image analysis for in-plane measurements, a three-dimensional measurement system is realized [51.24]. Again, based on the stroboscopic nature of the measurement technique, useful information will only be gained when the device motion is periodic or repeatable. A stroboscopic interferometry system can be used to obtain out-of-plane data, while a video image analysis system can be utilized for in-plane data. The experimental schematic is shown in Fig. 51.3. The two sets of data are combined using a Matlab routine to couple

the data and create a full three-dimensional motion analysis. Measurements from the in-plane data set are used to correct image shift in the out-of-plane data set resulting in greatly improved accuracy. It must be noted, however, that the in-plane motion will only interpret rigid-body motions. Deformations will be interpreted as rigid-body motion due to the nature of the video analysis. The three dimensional video analysis system [51.17, 18] was limited to rigid body motions. By adding the interferometric analysis this system can perfom three-dimensional measurements essential for devices which are integrated with stages, or any device which combines rigid-body motion with out-of-plane motion [51.24].

Optically Excited NEMS

As micro-oscillators become smaller (i.e. NEMS), it is often hard to fabricate in situ actuators to excite and measure an oscillation. In order to characterize the mechanical components of this motion, other techniques need to be used. *Craighead* and colleagues developed a technique for exciting MEMS and NEMS which utilizes an optical drive [51.25, 26]. In this work, disk-shaped micro-oscillators, shown in Fig. 51.4, were actuated by focusing a continuous-wave (CW) laser at the edge of the disk. Light interference in a Fabry–Perot-type resonator, created by the moving silicon disk and its stationary base, provides the mechanism for position-dependent absorption of light, leading to parametric amplification. In this case, the oscillators can be actuated merely by using a laser power which is over a threshold of $250 \, \mu W$ (for this particular device) [51.25]. The motion of the device was then monitored using HeNe and Ar^+ lasers focused at a 5-mm spot on the device surface. This optical technique can be very useful when devices have very small dimensions (i.e. NEMS) and/or when built in actuators are not easily integrated. This optical modulation technique has also been used to parametrically excite nanomechanical devices in air [51.26, 27]. Due to damping, nanomechanical devices often have very poor quality factors in air. This leads to very small responses that often cannot be measured. Using this optical excitation technique, cantilevers can be parametrically amplified, increasing the effective Q and the signal-to-noise ratio of the measurement signals.

In summary, optical actuation and amplification techniques can be useful when it is impractical or difficult to drive the structures in an on-chip manner. This will allow for mechanical characterization of the devices. Care must be taken when using this technique, however, as the laser power can heat the device, causing

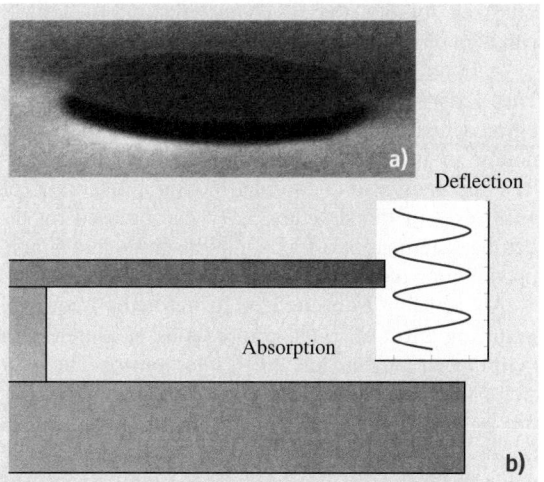

Fig. 51.4 (a) A scanning electron micrograph of a disk oscillator. The disk is 40 µm in diameter. (b) The position-dependent absorption caused by the moving disk and the substrate excites the device. (After [51.25])

mechanical stress gradients, as well as thermal variations in material properties.

Laser Doppler Vibrometry

Laser Doppler vibrometry (LDV) is a technique widely used to measure velocity and displacement of macroscopic systems [51.29, 30]. A coherent light beam is aimed at a moving object, Doppler shifting it for interference with a reference. Velocity information is obtained from the frequency modulation of the signals, while displacement information is obtained from the phase modulation. Operation range is from DC to 30 MHz with resolution from 0.5-µm/s velocity and 2-pm (2×10^{-12} m) displacement. The vibrometry technique measures the component of motion perpendicular to the incident laser beam.

Laser vibrometry has been applied to measurements of MEMS devices using a standard spot size of a few mm. Several groups have used laser vibrometers to characterize the motion of micromechanical membranes [51.31–35]. Membranes are generally large ($1-2$ mm^2) and move perpendicular to the wafer, making good vibrometer samples for measuring natural frequencies. A simple figure of voltage versus displacement for membrane actuators has been determined. Others have used vibrometry to analyze natural frequencies of narrow cantilever beams [51.22, 36] with laser spot sizes larger than the feature of interest, incurring a reduction in signal/noise ratio.

A fiber-optic-based vibrometer was integrated with a long-focal-length optical microscope and environmental chamber to study real-time device dynamics of MEMS [51.28, 37]. The microscope moves on a x–y stage and provides micrometer-scale spatial resolution for positioning the laser. The minimum diameter spot for the 633-nm vibrometer laser is 1.5 µm for a 50× ob-

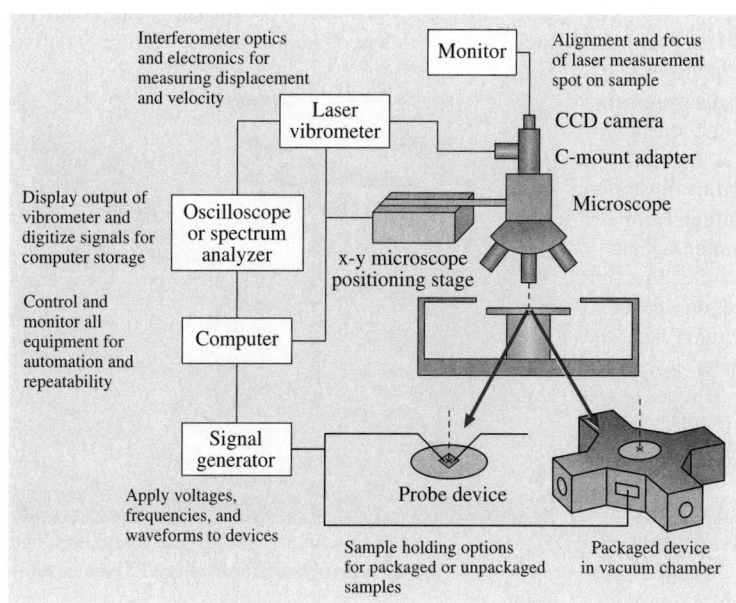

Fig. 51.5 Laser Doppler vibrometer with integrated microscope system for MEMS testing. (After [51.28])

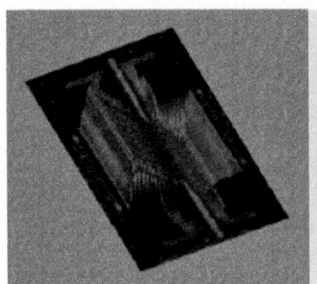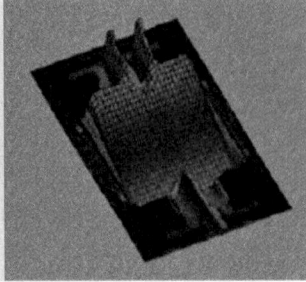

Fig. 51.6 Scanning laser Doppler vibrometer image of torsional mode shapes for a mirror device. (After [51.38])

jective lens. A schematic of the technique is shown in Fig. 51.5.

Mode Shape Measurements. The LDV system is a single-point measurement. With the addition of a two-degree-of-freedom mirror in the microscope adapter [51.39, 40] or a *x–y* stage to move the microscope [51.28], the measurement beam can be repositioned on the microdevice. Mode shapes and stress states can be measured by scanning the laser beam over the entire device and repeating the measurements at each point without changing the excitation on the device [51.28, 32]. Software can assemble the data and display the mode shape of the device, as shown in Fig. 51.6 for a torsional device tested at two different driving frequencies.

In-Plane Measurement. Laser vibrometry is limited to measuring motion parallel to the incident beam and thus is typically used only for characterizing the out-of-plane motions of MEMS. In-plane measurements can be made by fabricating an integrated micromirror adjacent to a MEMS device with a focused ion beam (FIB) system [51.28]. The mirror, illustrated in Fig. 51.8 reflects the incident laser light from the objective lens into the plane of the wafer, where it strikes the MEMS structure parallel to the direction of motion. Micromirrors are fabricated on the edge of bonding pads or other support structures near the moving devices. For a $\approx 1.5\,\mu$m-diameter laser spot size, the mirror, can be very small. Silicon has a 25% reflectivity for 633-nm photons [51.41] and is typically sufficient for good measurements without a coating.

The microscope LDV system has also been integrated with the computer microvision system discussed earlier as a product from Polytec [51.38]. This system combines the best of both techniques and provides

a turnkey solution for 3D characterization of MEMS with a lot of flexibility.

Cross–Axis Motion Measurement. When characterizing an actuator, measurements focus on looking at the motion in the primary direction(s) of motion. It is very important also to consider motion in the other (or cross) axes of the structure. LDV can be used for direct measurement of off-axis motions through a simple repositioning of the laser spot.

An actuator is excited at its resonant frequency producing $2\,\mu$m of in-plane motion as measured with a vibrometer and integrated FIB micromirror. The laser spot is then repositioned to the top surface of the structure as shown in Fig. 51.7 while the device is still in motion. A *z*-axis displacement of 17 nm at the resonant frequency was measured. This measurement was repeated at four positions, and produced readings from 10–22 nm, or approximately 1% of the *x*-axis displacement. The motion axes are well isolated from each other, as is typical of MEM devices with high-aspect-ratio springs.

Differential Measurements. Some models of fiber optic LDV systems have the reference beam of the interferometer on a second fiber optic cable to allow a differential measurement to be made. The fibers can both be used macroscopically or with one or both through the microscope.

A differential measurement allows an off-chip excitation method to be used. By focusing one beam on the device and a second on the test fixture, relative

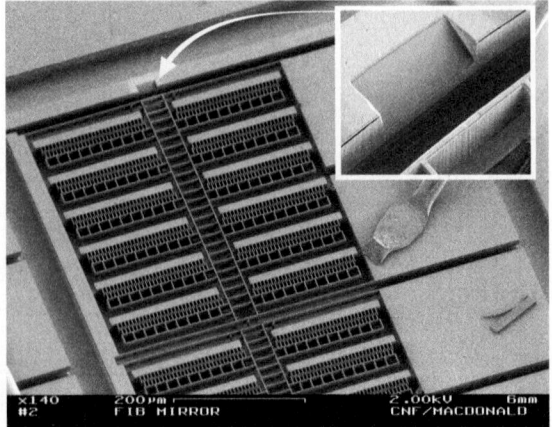

Fig. 51.7 A silicon micromachined in-plane actuator. The *inset* shows the 45° mirror used for in-plane LDV measurements

velocity and displacement can be measured to determine the response of the device to the excitation. This has been done successfully with a voice coil shaker excitation [51.13].

51.3.2 Static and Quasi-Static Measurements for Characterizing MEMS/NEMS

Size/Dimensional Measurements

One of the fundamental measurements necessary for MEMS or NEMS device characterization is determining the critical dimension of the finished structure. The mass and stiffness of the device can be determined with length, width, and thickness measurements and from models to describe the device and some assumptions on material properties such as density. Even with good device models, the accuracy of the calculations will only be as good as the measurements.

This section will cover some of the techniques used to characterize MEMS and NEMS devices dimensionally. Resolving the features of a microstructure requires high magnification, typically achieved in the form of optical or scanning electron microscopes. Access to the interior structure of a complex device necessitates cutting or cross-sectioning the device through the area of interest. Cross-sectioning can be accomplished in several ways by using a focused ion beam (FIB) system, diamond saw, polishing, or simply cleaving the device along a crystal plane if it is made from a single-crystal material such as silicon.

SEM Measurements and Issues. The scanning electron microscope (SEM) is one of the most commonly used tools for imaging micro- or nanoscale devices. Samples are mounted in a vacuum chamber on a stage, typically with five degrees of freedom—x, y, z, tilt, and rotation. The SEM has a large depth of focus and magnification that can be changed from the atomic scale to a field size of many millimeters. The positioning possibilities of the stage make the SEM ideal for imaging the 3D nature of most microstructures. Newer systems with low accelerating voltages and digital imaging electronics enhance the ability to image thin and nonconducting materials, like photoresist shown in Fig. 51.8, that have been difficult in the past. Avoiding coating a device with a conductive layer to enhance imaging is often necessary to preserve device functionality or prevent contamination if further processing is necessary.

A few key points must be reiterated when using the SEM on small devices. The scanning electron beam,

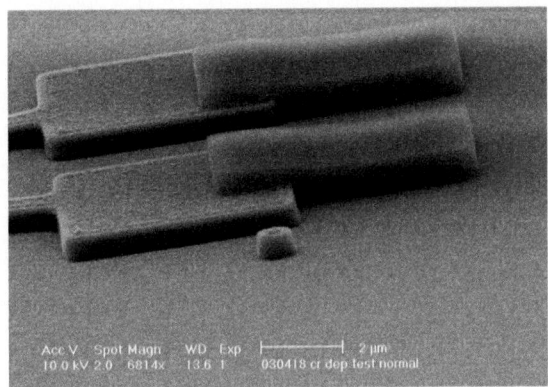

Fig. 51.8 Thick photoresist covering a step prior to metal etching. The micrograph shows the imaging capabilities of low-accelerating-voltage SEM imaging and the 3D nature enabled by good staging and depth of focus

which can have substantial accelerating voltage and beam current, can cause charging. This can result in actuation of the structures or even lead to the breakdown of dielectric films on devices. Care should be taken before imaging devices that could be damaged or to only analyze test structures with similar features to avoid damaging important devices.

The digital interface common to most modern SEMs includes a host of tools for measuring features in the image. When coupled with the powerful imaging capabilities of the SEM, it is easy to make quick measurements and trust that they are accurate. However, when samples are tilted and beam corrections such as scan rotation are employed the image can become rather distorted, greatly decreasing the accuracy of measurements. An excellent discussion of image distortions in scanning electron microscopy and correction methods is presented by *Kapur* [51.42]. Figure 51.9a shows a micrograph taken of a grid of squares with the sample at a 60° tilt. Most SEMs employ an automatic tilt-correction algorithm in the software to help correct the obvious distortion. The corrected image is shown in Fig. 51.9b. While *looking* greatly improved, there is still a distortion in the y-axis and a small rotation in-plane from the scan rotation present in Fig. 51.9a. Kapur developed an image-processing algorithm based on restoring the standard pattern for x/y and rotation and produced the image in Fig. 51.9c correctly showing the square pattern.

The differences between Figs. 51.9b and 51.9c may be subtle, but when SEM is used to measure microstructure features to verify models and extract device

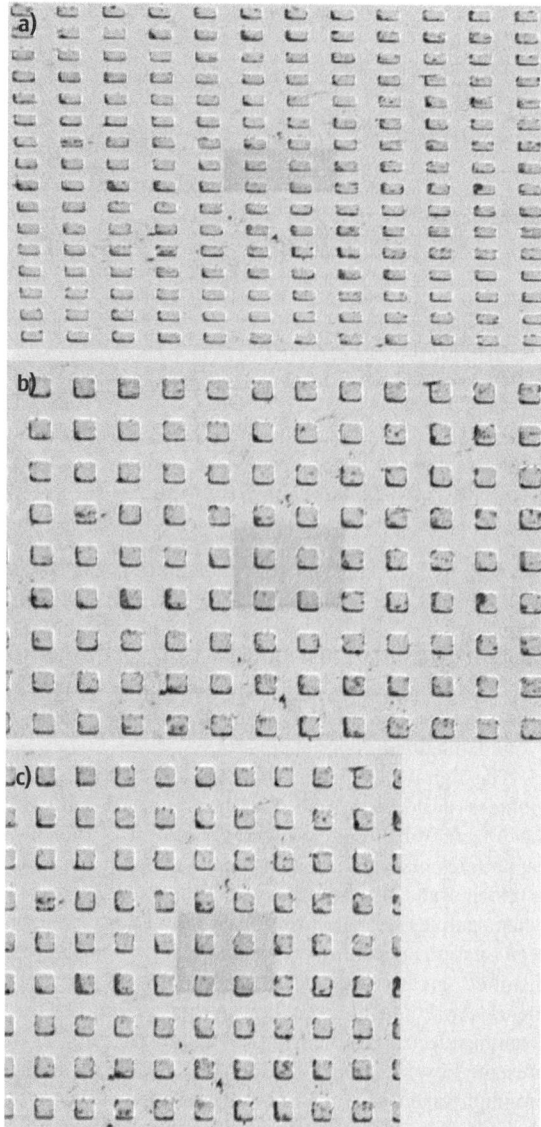

Fig. 51.9a–c SEMs of standard grid pattern taken at 60° sample tilt. (**a**) no correction, (**b**) SEM auto-tilt correction, (**c**) correction for all factors. (After [51.42])

lium, to create a focused particle beam. The ion beam is raster-scanned across the surface of the sample just as in the SEM. The impact of the gallium ions causes material of the sample to be sputtered or milled away. At low beam currents, the FIB system can be used to image a sample by collecting secondary electrons emitted from ion bombardment, as the sputter rate is relatively low. As the beam current is increased, the sputter rate increases as well. The precision of the electron optics allows for removal of material from a particular area in the shape of the raster-scan. The depth of removal is a function of the beam current and the milling time (dose). FIB is an excellent tool for performing cross sections of microstructures. It is a destructive technique, but is often the only way to access the exact area of interest. A thorough overview of FIB systems is given by *Reyntjens* [51.45]. His discussion covers additional FIB capabilities including using the ion beam for selective deposition of conductors and dielectrics to repair devices on the microscale. Also, some FIB systems have a second electron optics column for an electron gun mounted at a fixed solid angle to the ion beam column. Such dual-beam systems allow imaging and milling to be done simultaneously for greater process control.

A typical cross section of a microfabricated device [51.43] is shown in Fig. 51.10. The FIB system has been used to remove a section of material across the middle of the gear to expose the composite nature of its structure. The techniques discussed in the previous section on SEM image analysis can now be used to measure

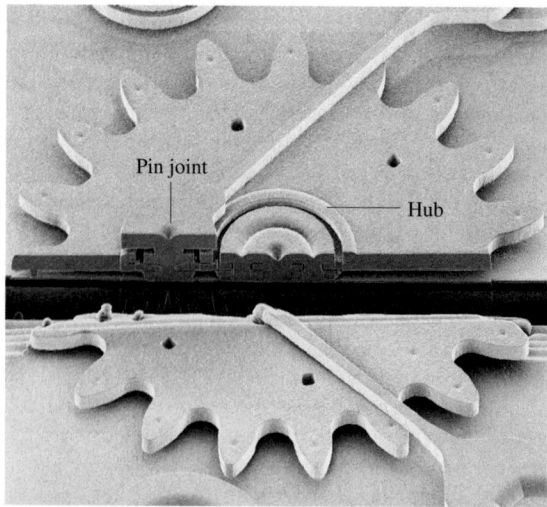

Fig. 51.10 FIB cross section through a micro-gear revealing the lower levels of a pin join and hub. (After [51.43])

parameters such as stiffness and mass, the accuracy of the results will only be as good as the accuracy of the measurements.

FIB Cross-Sectioning. An FIB system is a powerful tool for MEMS characterization [51.43, 44]. With a column of electron optics similar to an SEM, an FIB system instead uses a field-emitting ion source, such as gal-

dimensions for use in device characterization, subject to a similar warning about distortions.

FIB is a powerful analysis technique, although it always altering the sample to some degree. Gallium contamination, sputtering of all surfaces even during low-current imaging, and redeposition of milled material should be carefully considered before performing FIB.

Optical Inspection and Metrology. Optical microscopy is an obvious tool for MEMS analysis. Direct imaging of microstructures is possible during fabrication and also on completed devices that are not encapsulated by a protective cap or electronic package. The wide range of objective lenses available, with magnifications from 2× to 250×, allow fields of view to image individual device elements to entire chips. A range of cameras, including high-resolution digital still and video cameras, provide many options for data collection. Imaging software can be used to stitch multiple images together to form a large effective field of view at a given magnification.

The resolution limit is defined by the wavelength of light used for imaging divided by two times the numerical aperture (NA) of the objective lens. Typical limits fall in the hundreds of nanometer range, making optical microscopy well suited for most MEMS imaging. NEMS structures can easily fall below this resolution, though an optical micrograph can still be useful for imaging the larger-area chips.

Device measurements are made by using a reticle scale mounted in the eyepiece, a stage with a digital encoder system for measuring stage movements, or with off-line measurements of micrographs taking with a microscope imaging system. The reticle and micrograph techniques require that a calibration is known for each objective lens on the microscope.

The focus knob on most optical microscopes controlling the sample-to-objective distance is often calibrated and marked in micrometers. This feature can be used for out-of-plane device measurements, such as etch depths of reactive ion etching (RIE) or KOH-based etching, by focusing at two different levels of the device and subtracting the readings from the focus knob. By opening the aperture stop on the light source, thereby reducing the depth of focus of the image, precision can be increased.

Looking at a cross section of a device, for example to check etch depth and profile in a deep reactive-ion ion etching (DRIE) experiment can be a time-consuming process. Often it involves mounting the sample and preparing it for SEM to take advantage of orthographic viewing. Optical microscopy generally has the reso-

lution required for these features if only the sample could be imaged on its side. By cross-sectioning a wafer twice to form a strip of silicon thinner than the sample-thickness limit of your microscope, the silicon can be placed with the cross section of interest in the imaging plane of the microscope.

Profilometers. Stylus-based profilometers are a standard fixture in many cleanroom facilities for measuring the thickness of opaque films and the depth of etches. Typical systems include the Veeco Dektak and Tencor Alphastep product lines. A sharp tip stylus is dragged over the surface of a sample and surface-height variations are recorded for the line scan. The maximum resolution is several nanometers vertically and about one micrometer horizontally. This performance makes a profilometer well suited for measuring dimensions of micro- and nanostructures.

Besides obvious film thickness and etch depth measurements, the profilometer can be used to determine the pitch of a grating or comb drive and the height of a released beam. Care must be taken to account for stylus shape and size when trying to measure high-aspect-ratio features, but this is well documented in the system manual. Microdevices broken through rough handling become excellent measurement targets. A broken high-aspect-ratio flexure, laying on its side on a substrate, can now be precisely measured for thickness and height as these critical dimensions are better aligned with the sensitive measurement axes of the profilometer. This is demonstrated in Fig. 51.11.

Profilometers have also been used with released mechanical structures to determine material properties. Young's modulus and residual stress for materials used in microstructures have been characterized by dragging the stylus over structures made of the material of interest [51.46, 47]. Knowing the dimensions and boundary conditions of the test structures, the output of the profilometer can be used to extract material properties of the device.

Metallurgical Style Mounting for Cross-Sectioning. Cross-sectioning microstructures to look at internal features can be difficult for fragile structures. FIB is a good technique for small features, but it can be too slow when the structure to be analyzed is large (hundreds of micrometers). Standard techniques from the field of material science for grain structure analysis can be employed for microdevices.

The sample is embedded in a block of material such as epoxy or bakelite and allowed to cure. The polymer

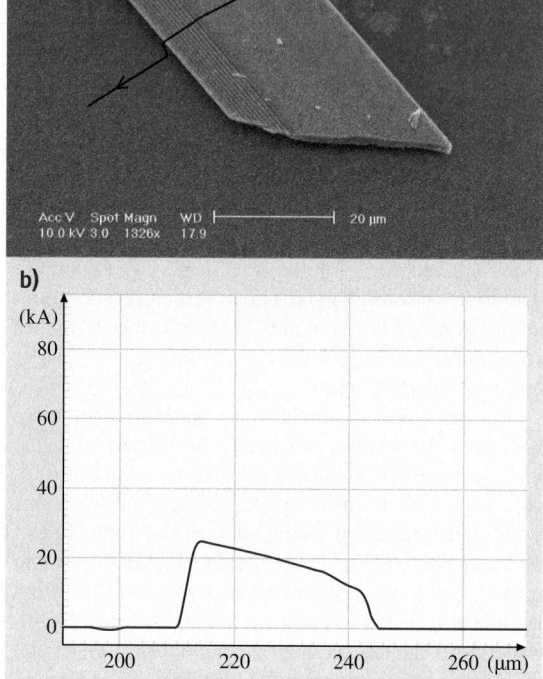

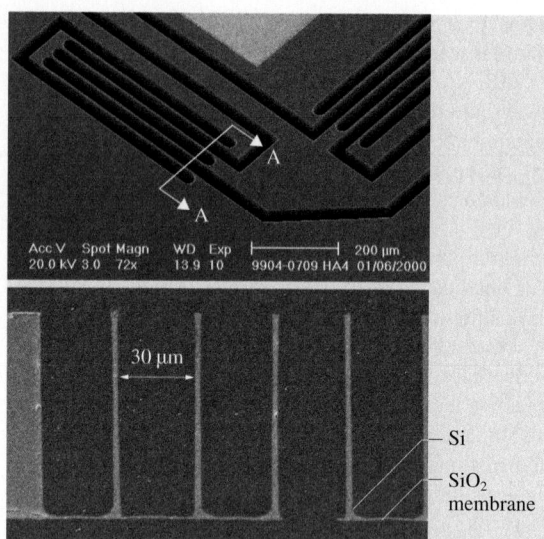

Fig. 51.12 SEM of silicon flexure device and cross section taken after embedding in epoxy and polishing. (After [51.13])

Fig. 51.11 (**a**) Broken silicon beam on substrate. The *black line* indicates the path of the stylus over the beam. (**b**) Profilometer scan of silicon beam showing 2.5-μm thickness and 35-μm height

holder is now thinned on a grinding wheel to expose the microdevice inside. Coarse grinding and fine polishing can be used to optimize time, surface quality, and exposing the exact region of interest. The advantage of polishing to a feature is that the position of the cross section can be precisely controlled. Also, after imaging, the polishing can be resumed to expose other slices of the microdevice for further analysis.

Figure 51.12 shows a microdevice with a flexure system and the corresponding cross section. The cross section was taken near the end of the first, third and fifth trench, and some silicon residue is visible at the base of these flexures. Although this residue was typically observed only near the flexure ends, the base of each fabricated flexure was either slightly narrower or slightly wider than the as-drawn line width, depending on the location of the beam on the wafer. This fine flexure structure and gaps would have been hard to preserve

during sample preparation if the device had not been embedded in epoxy.

In Situ Quasi-Static Device Characterization

Measurement of device motion and parameters at the nanoscale can be done using microinstrumentation fabricated on the chip with test devices. There have been many attempts to build in situ test structures for measuring spring constants of devices, the force generated by actuators, and the material properties of thin films [51.1, 48–56]. These test devices often either apply a known force, or a known displacement, and this allows information on the beam stiffness to be extracted.

On–Chip Micromechancial Loading Devices. Measurement of motion and parameters at the nanoscale can be accomplished using microinstrumentation fabricated on the chip with test devices and variable capacitor motion sensors. The benefits of in situ loading devices include self-attachment. If actuators are built of the same material as the test structure, clamping forces are not an issue as the sample is already attached to the loading mechanism. In addition, different types of forces can be applied. *Saif* describes a device with micron-scale buckling beams that are used to calibrate and characterize the spring constants of MEMS [51.54]. By monitoring the point at which a clamped–clamped, or clamped–free beam buckles, the force generated by a micromechanical actuator can be extracted. Once the force is known, the mechanical stiffnesses (linear and nonlinear) can be extracted. Torsional loading devices have also been reported. Figure 51.13 shows a schematic and SEM micrograph of such a loading device used to calibrate forces and spring constants of MEMS structures by applying a torsional load [51.52, 53]. The deflections of the microdevice or sample can be monitored in the SEM or using one of the optical techniques discussed previously.

In-situ micro-loading devices that apply relevant forces (typically milli-Newton or larger) are still very challenging to build using microfabrication techniques. In addition, the actuators are not often independently characterized, so the exact forces being applied are often not known with significant accuracy. This is a problem when relying on precise applied forces to characterize device behavior. An in-situ device should integrate a calibration mechanism to improve the accuracy in the extracted parameters.

Nanoindentation/Nanohardness. The nanoindenter is a commercial instrument that applies a vertical load and records the load versus displacement. In this way, similar to a microhardness tester, material properties can be extracted from indents made with controlled tips. This allows material properties of thin films to be determined. This instrument, being a controlled closed-loop force transducer, can be utilized for MEMS characterization. Slow, controlled measurements of force versus displacement can significantly enhance the techniques already discussed for the characterization of MEMS. One of the drawbacks of other techniques is that they apply force using the built-in MEMS actuators, which are not always independently characterized and thus the force being applied is not measured directly. This introduces significant error into calculations of quantities like spring constants, as the models depend on knowing the actuation force applied to the device. The nanoindenter is a tool that provides direct force input, independently of the actuator. This will allow direct extraction of mechanical spring constants, both linear and nonlinear. Figure 51.14 shows one experimental set-up for testing mechanical force versus displacement with a nanoindenter [51.57]. This particular experiment was developed to measure the contact resistance in a switch under repeated load. A standard load–displacement measurement would not need to monitor resistance externally.

The nanoindenter (available from Hysitron, MTS, Nano) has been used by a number of groups for MEMS

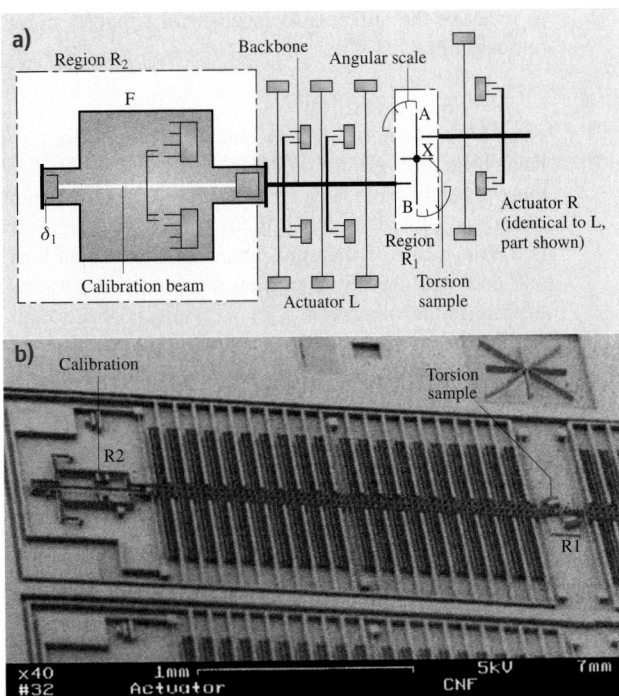

Fig. 51.13 (a) Schematic of an integrated, capacitively actuated micro-loading device capable of applying mN forces. **(b)** SEM of fabricated silicon micro-loading device. (After [51.53])

characterization. In [51.58], *Begley* and colleagues utilized an MTS nanoindenter to apply point loads to cantilever beams for adhesion experiments. In these experiments a nanoindenter with a Berkovich tip was used to apply loads necessary for adhesion of a polysilicon cantilever to the substrate. This technique was also used

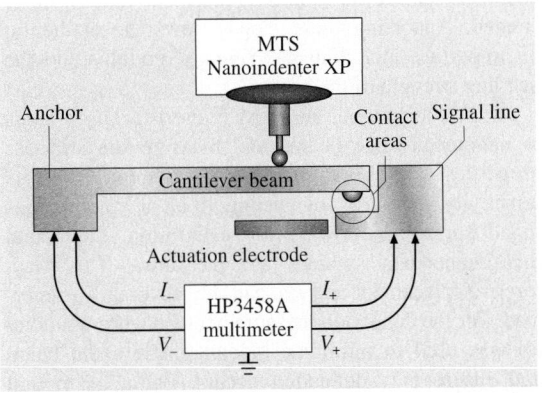

Fig. 51.14 Experimental set-up for the nanoindentation test on a cantilever beam. (After [51.57])

to measure the stiffness of mechanical supports in an optical switch [51.59].

AFM–Based Testing

Force Measurements. Micro- and nanoscale cantilevers have been utilized for imaging for many years, but more recently have been important for measuring and quantifying forces in many fields. A typical cantilever is a thin beam of material, connected to the bulk at one end, and usually having an atomically sharp protrusion at the other end. One example is the use of cantilevers to apply force to study the adhesion of a gecko *setal stalk* [51.60]. With specially designed AFM cantilevers, Autumn et al. were able to prove experimentally that the adhesion and friction that allows geckos and other animals to walk on walls and ceilings was dominated by van der Waals forces caused by significant surface area contact [51.60]. In addition, measurements were also done to determine forces produced by walking insects, including ants and cockroaches [51.61].

For force measurements using microcantilevers, the key is in microcantilever design and characterization. The cantilevers can be characterized using some of the techniques discussed above. It is essential to design a cantilever that has the correct stiffness for the desired application. Once characterized, integrated AFM cantilevers, and/or the complete AFM tool can be used to calibrate and measure the quasi-static behavior of MEMS and NEMS.

AFMs have been used for high-resolution characterization of quasi-static microbeams. The main technique utilizes a standard AFM (piezoelectric stack with a cantilever mounted on it) to apply forces at a particular location on a nanobeam. From this and the optical transducer on the AFM, force–displacement maps can be created. Analyzing these maps allows the extraction of properties, including the Young's modulus and the bending strength of the beams.

In this technique, an AFM is used to apply loads to nanometer-scale beams and measure the stiffness properties. The AFM cantilever is calibrated and its sensitivity measured by testing it on a smooth diamond surface. Following the calibration, the actual micro/nanobeam samples are measured. This way, force–displacement curves can be accurately generated. For the deflection measurements, a worn diamond tip was used to minimize indentation into the beam and ensure only deflection occurs. For a beam that is fixed at both ends and loaded in the center, force-displacement curves can be used to extract material

properties, as well as measure the stiffness of the beams directly.

AFM-based techniques can be extremely versatile when characterizing micro- and nanoscale devices. The versatility comes from the large design space of AFM cantilevers. The key to successful AFM measurements is to design or purchase a cantilever which is appropriate for the measurement of interest. With a good calibration technique, much information on quasi-static behavior and material properties can be gained from AFM-based measurements.

51.3.3 SEM–Based Testing

The scanning electron microscope can be a very useful tool for visualizing and quantifying motions in micro- and nanoscale systems. Testing in an SEM requires hardware to feed electrical signals into the vacuum chamber and to the device under test. If such connections are made, resistive contrast imaging [51.43] can be used to verify electrical continuity to various portions of the microstructure. Applying a voltage to a test pad will change the contrast of the image as the electron beam of the SEM interacts with the charge on the structure. Abrupt changes in contrast can be used to locate defects in a device.

With electrical connection to the device in the SEM, it can be actuated and the motion will appear as a *blurred* image. Burring is caused by aliasing between the SEM scan rate and the actuation frequency of the device or by signal averaging in the SEM image processing routines. Nanoscale resolution can be achieved with this technique. Using videotaped SEM and subsequent video analysis, the motions can be quantified. The techniques for handling distortions, as discussed in the section on measuring with an SEM, should be utilized here as well. A typical SEM image of a moving microstructure is shown in Fig. 51.15. This technique has been used very successfully by *Ruoff* [51.14, 62] to image nanowire vibration, and observe some very subtle features of nanostructures.

Stroboscopic, or time-resolved SEM offers excellent spatial resolution for imaging topography. By strobing the signal at the resonant frequency of the device, a moving device will appear still. Time resolution of less than 1 ns has been demonstrated using beam-blanking plates [51.63]. Also, gating the electron multiprinter provides time resolution of the order of 100 ns [51.64, 65]. The strengths of time-resolved SEM include high spatial resolution, large depth of focus and better than 100-ns time resolution for spatial detector systems.

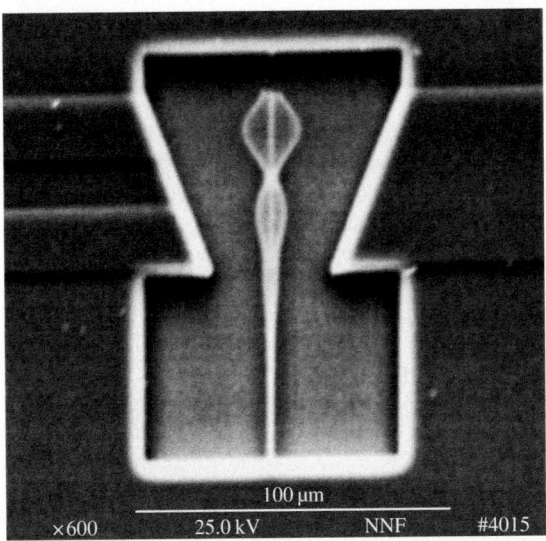

Fig. 51.15 SEM image of vibrating cantilever. The *blur* represents the envelope of motion. (After [51.64])

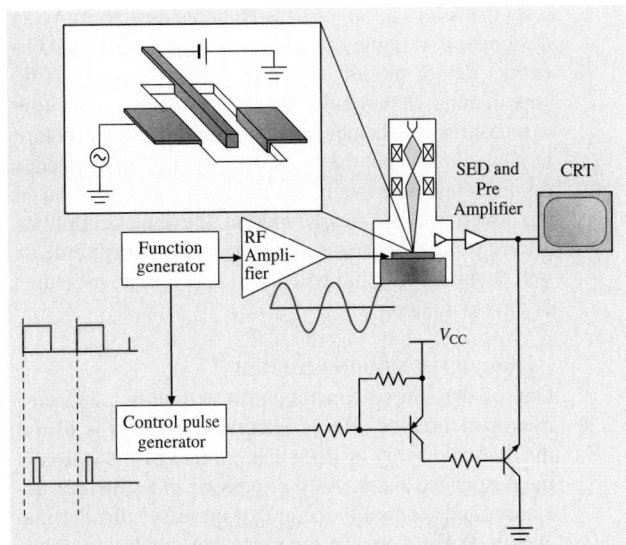

Fig. 51.16 Schematic for detector-modulated time-resolved SEM. (After [51.64])

The SEM can be used to characterize the smallest MEMS devices with excellent resolution [51.64–66]. The disadvantages of time-resolved SEM include the requirement for vacuum operation of the MEMS, charging effects, and difficulties in calibrating motion and displacements on a nanometer-scale. The addition of a laser-interferometer-based SEM stage or use of a calibrated linear standard can provide nanometer-scale calibration for MEMS measurements. A schematic for detector-modulated SEM is shown in Fig. 51.16. Figure 51.17 shows images from differently biased cantilevers obtained from this time-resolved SEM. The mode shapes and amplitudes are easily visible using this technique.

Fig. 51.17 The sequential time-resolved SEM images of microcantilevers. The silicon dioxidecantilever beam coated entirely by aluminium. (After [51.64])

51.3.4 Actuation Methods

Measuring the mechanical parameters of a device requires an actuation method to produce a response measurable by one of the previously described techniques. Integrated, or on-chip, capacitance-based sensors and actuators are most common. However, there are several methods for actuating or exciting mechanical structures using off-chip means. This can be very helpful when integrating an actuator, such as an electrostatic drive, for the purpose of testing would add complexity not required in the final application. These methods include Brownian motion, shakers, flowing gas streams, magnetic fields and impulses.

This section will cover various on- and off-chip actuation methods with examples using off-chip measurement techniques such as optical methods. The actuation methods will apply equally well to devices with integrated sensors. For example capacitance-change-based MEMS accelerometers, which are often tested using mechanical shakers or piezoresistive cantilevers, could easily be tested with the wind method.

On-Chip Capacitive Sensors/Actuators
Utilizing integrated variable capacitors (such as interdigitated comb actuators) Sect. 48.1.4 to measure motion and the resonant behavior of MEMS [51.67] is quite common. This technique can give good dynamic-

response information, provided that the devices move as designed. If variable capacitive sensing is to be used to extract device motion, however, a few things must be kept in mind. In particular, the use of variable capacitors to measure displacement assumes the capacitor structure is stable throughout the measurement and the electrodes do not rotate or move in the direction perpendicular to the direction of the desired motion. If these assumptions are not valid, then the resulting capacitive measurements will be misleading and will not be appropriate measures of device motion.

Brownian Motion Actuation

One of the simplest methods for actuating a released microstructure without an integrated actuator is to use the thermal energy or Brownian motion of the atoms in the suspended mass. As the mass of the structure decreases and the quality factor (Q) increases, the thermal energy of the atoms in the mass can create a measurable motion in the structure. Using one of the sensitive measurement techniques such as LDV or capacitance detection discussed elsewhere in this chapter, this *positional noise* can be determined. Examining the signal in the frequency domain, using a spectrum analyzer, will reveal the natural frequencies of the device. The quality factor of the device can be enhanced by performing the testing at a reduced pressure to remove damping from air.

For example, the Brownian motion of a 500-nanogram surface micromachined polysilicon device is $1\,\mu g/(Hz)^{1/2}$, in vacuum, as the Q value is $50\,000$. This corresponds to a position noise of $10^{-4}\,\text{Å}/(Hz)^{1/2}$ [51.68]. Additionally, (51.1), a statement of the equipartition theorem, can be used to determine the stiffness of a device, provided it is relatively linear [51.69].

$$\frac{1}{2}k\,\langle x\rangle^2 = \frac{1}{2}k_{\mathrm{B}}T \,, \tag{51.1}$$

where k_{B} is Boltzmann's constant, T is the temperature, and k is the linear stiffness.

Shakers

Many companies provide shakers or shake tables utilizing voice coil, piezoelectric, or other actuation methods for the purpose of evaluating typical macroscopic mechanical systems. One vendor, Bruel and Kjaer, has units in different sizes such as the 6502 with large-mass large-displacement capability but therefore limited frequency range. B&K's 6510 model is

a micro-shaker capable of 20-kHz actuation if the device fixture mass can be limited. With even smaller fixtures, a piezo scanner can extend the frequency range further.

Resonance for a microstructure can be found using a dual-fiber vibrometer, as described in the laser vibrometry section. One beam of the vibrometer should be aimed at the device while the other beam is directed onto the fixture. This differential technique will allow vibration amplitude to be measured as a function of frequency, finding the natural frequency as the peak amplitude [51.13]. Remember that the vibrometer requires motion parallel to the measurement beam so this is ideal for out-of-plane measurements. In-plane measurements can be done with creative fixtures, as shown in Fig. 51.18.

It is often helpful to mount a reference accelerometer onto the shaker to measure the shaker motion directly. A calibrated accelerometer, such as the B&K 4371 miniature piezo-accelerometer can be used without significantly increasing the fixture mass. Measuring the response of the shaker to the applied signal allows excitation levels to be known and transfer functions to be determined.

Scanning the vibrometer beam to different points on the released structure while it is being shaken can be done to measure the vibrational mode shape, as described in the laser vibrometry section.

Wind

Even the smallest shakers will have bandwidth limits, making them only useful to larger or softer microstructures with vibrational modes in the tens of kilohertz range. Higher-frequency excitation can still be done with off-chip actuation using a directed stream of gas. Using

Fig. 51.18 Shaker set-up for in-plane testing. Two optical fibers from the LDV are at the *right* of the picture; one is focused on the test fixture, the other on the sample mounted on top of the fixture. Microscope objective lenses are seen in the top of the image to aid in alignment

Fig. 51.19 Wind tester. High-pressure nitrogen is coming in from the tubing on the *left*. A microscope objective is seen at the top of the picture for focusing the LDV laser spot on the cantilever chip

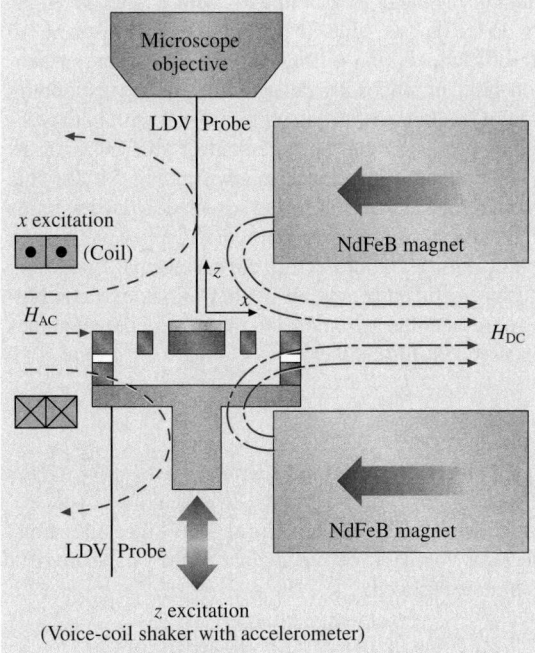

Fig. 51.20 Off-chip magnetic actuation of microstructures and integrated shaker. (After [51.13])

hypodermic, stainless-steel tubing with a 100-um inner diameter, Smithwas able to direct a turbulent stream of ultra-pure nitrogen at a cantilever structure and excite a 2.5-MHz vibrational mode. The vibration was measured using a vibrometer under a microscope and is illustrated in Fig. 51.19.

Magnetic Fields

Another method for actuating microstructures using off-chip forces is magnetic fields. *Horsley* [51.13] deposited a 2-μm permalloy (Ni − Fe) film using physical vapor deposition (PVD) on a released platform. Positioning a coil near the platform, as shown in Fig. 51.20, a frequency sweep was performed on the driving current in the coil and the vibration amplitude was observed with an optical microscope to determine the in-plane resonant frequency. Additionally, by using a permanent magnet, the platform could be moved statically away from its rest position. Performing frequency sweeps with the coil while the device was at a given distance allowed the authors to measure the in-plane resonant frequency as a function of displacement.

Actuation of high-frequency fixed–fixed silicon beams is report by *Cleland* [51.70] using Lorentz forces. Passing an AC current through a conducting beam with a network analyzer in the presence of a strong magnetic field perpendicular to the beam creates a Lorentz force capable of actuating the beam transversely, as shown in Fig. 51.21. By sweeping the drive frequency, the response of the beam is measured from the induced electromotive force (EMF), or voltage. Cleland used a vacuum chamber to get a high quality factor and cooled the sample to 4.2 K to use a superconducting solenoid to generate fields up to 7 T. The technique is useful to determine natural frequencies of resonators. Precise displacements cannot be determined, however, without knowing the coupling parameters of the test set-up to the microstructure.

Impulse Excitation

Another method of off-chip excitation that can be used is to provide an impulse to the device or the fixturing containing the device. This can be as simple as tapping on the test apparatus. If the device has a sufficiently high quality factor, it will vibrate at its resonant frequency for a finite time from the impulse. In some systems, quality factor can be increased by performing the test in a vacuum chamber at less than 100 torr [51.36, 71]. The impulse can be provided by putting a current pulse into a shaker or simply hitting the fixturing with a hard object. While this technique seems overly simple, it is often the quickest and easiest way to find the resonant frequency of a device. This technique works well for low-frequency devices, as the Fourier transform of a hammer blow consists primarily of low-frequency components.

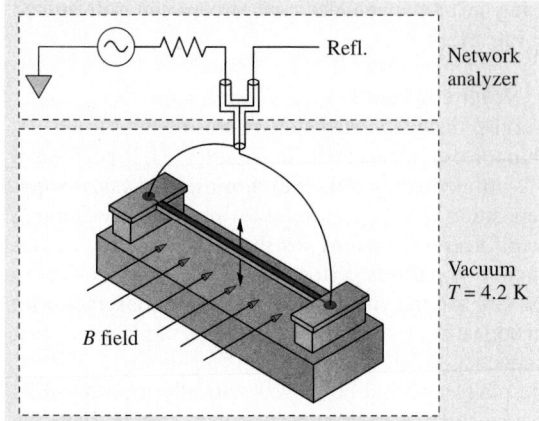

Fig. 51.21 Lorentz force for actuating nanowires. (After [51.70])

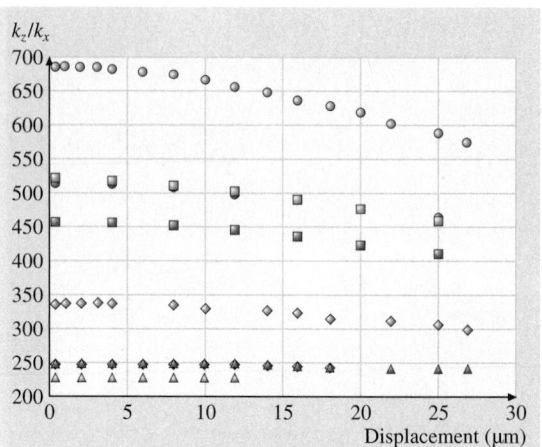

Fig. 51.22 Stiffness ratio of a suspension system as a function of displacement. After [51.13]

1-g Gravitational Reference

One should not overlook one of the handiest off-chip actuation methods—the Earth. Most microdevices with a DC-capable transduction method will be sensitive to the gravitational force from the mass of the earth. As such, a 1-g or 9.8-m/s^2 acceleration standard is available for use in any test facility. Typical local variations of the value of g are less than 0.05% except near the poles. The total variation on Earth is from 9.78 to 9.83 m/s^2 [51.72]. By orienting the sensitive axis of a device normal to the earth, a 1-g acceleration is applied. The response can be measured and used to calibrate the device. Rotating the device by 180° will now apply a −1-g reference standard and give a 2-g total shift in the signal and can be a quick check for the linearity of the device response.

Combinations

Doing in-plane measurements with a magnetic coil and out-of-plane measurements with a shaker, *Horsley* [51.13] was able to determine the in-plane to out-of-plane (k_z/k_x) stiffness ratio of a x–y stage suspension built in silicon by determining the corresponding natural frequencies. Using a DC magnet with both techniques produced the measurement of stiffness ratio as a function of displacement shown in Fig. 51.22. This characterizes the change in the suspension due to strains in the flexures resulting from the displacement from equilibrium as would occur during normal operation. This powerful and complete result was made by combining several of the actuation and measurement techniques discussed in this chapter.

51.4 Example: Characterizing an In-Plane MEMS Actuator

Once a suitable testing method has been utilized to collect dynamic information, device parameters can be extracted. For many structures, a simple device model can be used. As the device complexity increases, choosing an appropriate model can be difficult. In this section, a technique is outlined for extracting the stiffness information from a simple MEMS device consisting of a proof mass, spring supports, and an actuator. The electrostatically actuated device, shown in Fig. 51.7, is used to demonstrate how to extract the mechanical stiffness parameters (linear and nonlinear) and the quality factor from dynamic measurements. Although the dynamic

information was obtained from LDV measurements, the data could be obtained using other measurement techniques as well.

51.4.1 Determining the Quality Factor

The quality factor for the in-plane vibrations of a MEMS device can be extracted from motion measurements. Using a measurement technique not limited to repeatable motion, and using an integrated or external actuator, the device is set in motion by the application of a step function, similar to the impulse technique described in

Sect. 51.3.4. Often, a step-down response, while giving similar information as the impulse response, is more controllable and easily programmed in most function/voltage generators. Figure 51.23 shows the device response measured at atmospheric pressure. Observed optically, the device appears to jump to the new position. The true motion, as revealed by the measurement, shows behavior typical of an underdamped oscillator; the device oscillates as it settles to a new position.

By observing the motion of the device as it settles, the quality factor Q of the device in air can be determined. Q is defined as 2π times the number of cycles for the energy in the oscillator to diminish by the factor $1/e$ [51.73]. Since energy is proportional to the amplitude squared for a simple oscillator, the number of cycles for the ringing amplitude to decrease to $e^{-1/2}$ of the starting amplitude is defined as $Q/2\pi$. A DC bias is applied, and the device is allowed to settle. The bias is then removed and the ringing of the device as it returns to equilibrium is monitored by capturing the output. Signal analysis is performed on a computer by counting the cycles needed for the amplitude to drop by a factor of $e^{-1/2}$. At atmospheric pressure, the Q of this device is 16 ± 1.

By utilizing a variable pressure chamber, Q can be measured under different vacuum levels. For testing, the device was placed in a lidless package and mounted in the small vacuum chamber. The measurement of Q was repeated at a series of pressures. An example of repeated Q measurements (from the device shown in Fig. 51.7) is shown in Fig. 51.24. Consistent with previously reported studies on pressure-versus-Q relationships [51.71,74,75], there are three distinct regions. In the center region, Q is most dependent on pressure. This region is best modeled by momentum damping due to collisions with individual gas molecules, leaving Q inversely proportional to pressure. At high pressure, viscous damping is the dominant source of dissipation, and there is minimal pressure dependence. At low pressure, there is also a limit for Q, which depends on the device materials and the fabrication process [51.75].

For test techniques that do not allow for transient dynamic capture, Q can be monitored by measuring the following formula

$$Q = \frac{\omega_0}{\Delta\omega}, \tag{51.2}$$

where $\Delta\omega$ is the frequency band at the half-power points of the frequency response through the first fundamental mode. This is known as the full width at half maximum (FWHM) value of the response peak plotted in

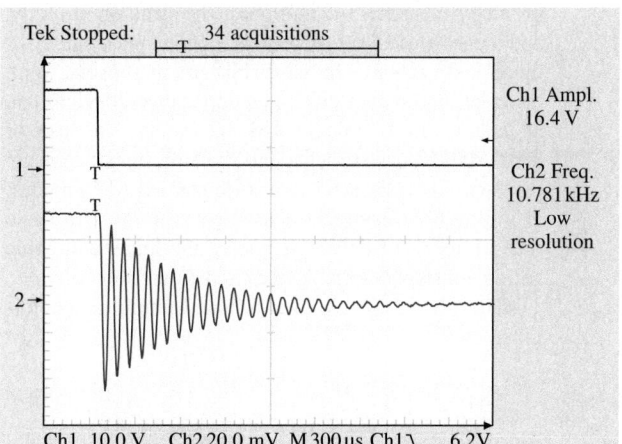

Fig. 51.23 Response of the device (*bottom*) to a step-down function (*top*)

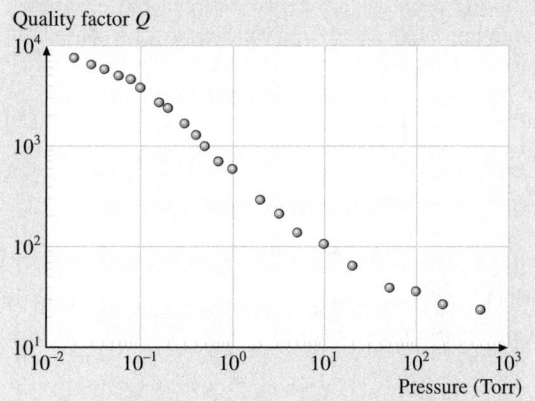

Fig. 51.24 Experimental plot of Q versus pressure

the frequency domain. For this method to be accurate, the device has to be behaving linearly when driven. For many MEMS devices, this transition to nonlinear behavior occurs at a rather low excitation, so the linearity should be verified before relying on this measurement to extract Q.

51.4.2 Determining Spring Constants

The measurement techniques discussed here can also be used to characterize the stiffness parameters of MEMS devices. Many MEMS use integrated variable capacitors to sense motion. However, motion sensing with variable comb-style capacitors requires that the capacitor electrodes remain rigid and do not move perpendicularly to the direction of the measured motion axis. The laser vibrometer measures the motion of the structure directly, and it can be used to verify the precision and accuracy of integrated capacitor-based displacement sensors. With the laser vibrometer technique, unpackaged devices can be characterized quickly and accurately. The use of quasi-static measurements and dynamic measurements can be used to characterize a nonlinear MEMS actuator. Using the vibrometer measurements, along with basic measurements of device geometry, the linear and cubic spring constants of the device can be determined.

The force from an interdigitated comb actuator is given by

$$F_{comb} = N\varepsilon_0 \frac{b}{d} V^2 , \tag{51.3}$$

where N is the number of movable fingers, b is the depth of the comb fingers, d is the gap between the movable and fixed electrodes, ε_0 is the dielectric constant of the material between the comb fingers (air, or vacuum), and V is the potential difference between the electrodes. From Fig. 51.25 and (51.3), we can extract a constant Ψ for the voltage–distance relationship

$$\Psi = \frac{x}{V^2} . \tag{51.4}$$

Combining (51.4) with (51.3), and Hooke's law, $F = kx$, the linear stiffness is found to be

$$k = \frac{N\varepsilon_0}{\Psi} \frac{b}{d} . \tag{51.5}$$

The device shown in Fig. 51.8 has 12 arms each with 29 comb fingers for a total of $N = 348$. The fingers are $d =$

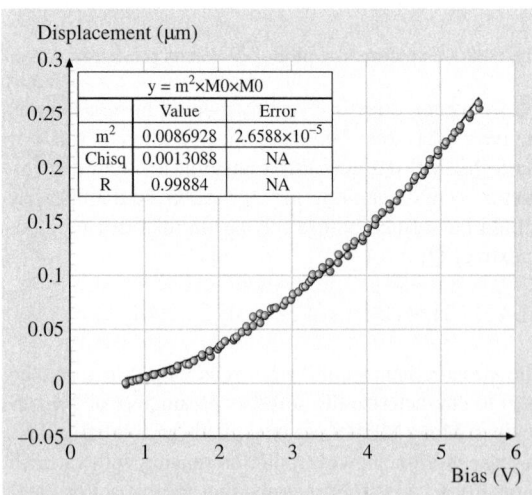

Fig. 51.25 Quasi-static voltage versus position obtained from a slow linear ramp input

20.5 μm tall and have a gape of $b = 2.0$ μm. Using $\Psi = 8.69 \times 10^{-9}$ m/V^2, found from the curve fit in Fig. 51.25, $k = 3.63$ N/m. The natural frequency of the device f_0 can be independently determined by applying an impulse voltage to the device and watching the response. For this sample experiment, a 16-V pulse was applied for a duration of 500 ns. Being an underdamped system, the device will *ring* at its natural frequency as a result of the pulse excitation. This can be seen in Fig. 51.23. Since k has been determined above and $\omega_0^2 = k/m$, the value of the mass m can be solved for f_0 is measured in Fig. 51.23 to be 10 781 Hz. With $\omega_0 ='' \pi f_0$ and k found earlier, the mass is calculated to be 7.92×10^{-10} kg.

A device designed with fixed–fixed springs, such as in this example, exhibits a nonlinear dynamic response. This is due to the cubic nonlinearity of the springs as the motion exceeds the linear range. Although higher order nonlinearities are present, their effects are small. Thus for sinusoidal forcing, the equation of motion can be approximated as:

$$mx'' + cx' + kx + k3x^3 = F\cos(\omega_1)t \tag{51.6}$$

where c is the damping coefficient, k is the linear stiffness, k^3 is the cubic stiffness, F is the amplitude of the driving force, and ω_1 is the driving frequency. This equation of motion is of the form known as the "Duffing Equation". In normalized form, this equation is:

$$x'' + cx' + \omega_0 x + hx^3 = G\cos\omega_1 t , \tag{51.7}$$

where ω_0 is the undamped resonant frequency, $G\cos\omega_1 t$ is the driving function, and h is the nonlinear stiffness term; h is positive for a hard spring (meaning the effective spring constant increases with applied signal) and negative (decreases with applied signal) for a soft spring. Duffing's equation has been applied to tunable MEMS resonators [51.76–80].

The frequency response for an undamped ($c = 0$) system of this type can be analytically determined [51.81]. The frequency–amplitude relationship is

$$\omega_1^2 = \omega_0^2 + \frac{3hA^2}{4} , \tag{51.8}$$

where A is amplitude of the displacement, ω_1 is drive frequency and ω_0 is the natural frequency of the corresponding linear system ($h = 0$). This equation describes the spline about which response curves are centered. This equation also gives the response for free vibrations of (51.5) with $G = 0$.

In systems with dissipation, the dissipation effects are similar to those in linear systems. Notice that

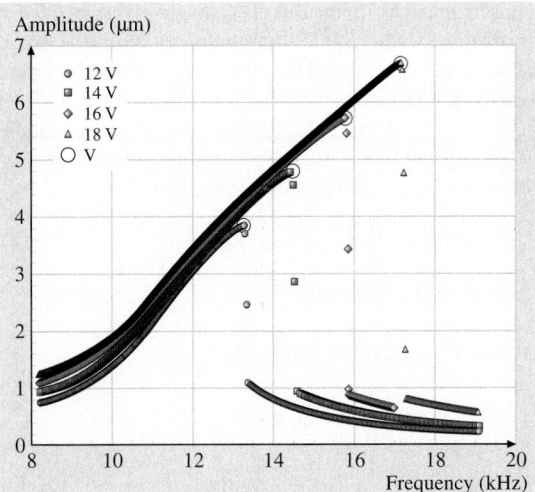

Fig. 51.26 Frequency-response curves for the resonator shown in Fig. 51.7

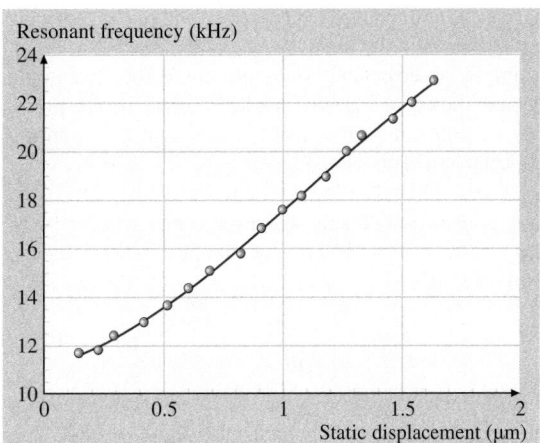

Fig. 51.27 Resonant-frequency shift as a device is statically displaced from equilibrium

(51.8) additionally describes the spline around which the response curves with dissipation are centered. Dissipation will limit the amplitude of the response. An undamped system will keep growing in amplitude, whereas a damped system will have a maximum amplitude, and show hysteresis in the response. Using one of the dynamic characterization techniques described above, the frequency response of the sample actuator can be measured. As seen in (51.8), the resonant frequency increases with amplitude. Figure 51.26 shows experimental frequency-response curves for the resonator (see Fig. 51.7 for a schematic) which has fixed–fixed springs. The curves are obtained by sweeping the frequency for varying amplitudes. The curves clearly show that the resonant frequency increases with increasing driving amplitude and is characteristic of a device with a hard non-linear stiffness. The large resonant-frequency shifts are due to the cubic nonlinearity, shown in (51.8).

The frequency response data can be used to obtain the linear and nonlinear stiffness coefficients of the device. In Fig. 51.26, the point on each curve where maximum amplitude is achieved is shown with a large circle. These points are described by (51.8). The curve described by the highlighted points (there are more data points in the actual curve fit; only four of the points are shown) can be fitted to (51.8), therefore extracting the cubic nonlinear stiffness, k_3/m. By also measuring k as described previously, $k/m = \omega_0^2$ and $k_3/m = h$ are used to solve for m and k_3 independently. The four circled points on Fig. 51.26 are used with (51.8) to determine

an average $h = 2.16\,\text{N}/(\text{kgm}^2)$. Using the previously determined mass, $k_3 = 1.71\,\text{N}/\text{m}^3$. Thus we can quantitatively determine the linear and nonlinear stiffness parameters and the mass, effectively characterizing the dynamic response of the device.

The change in resonant frequency with amplitude can be explored in more depth with an additional measurement. To see the change in resonant frequency, a DC signal is applied to the device to hold it at a specified offset displacement. Using the laser vibrometer and characterization suite, the response of the device is observed as it settles at each offset displacement. The frequency at which the device *rings* is its resonant frequency for that amplitude. This measurement was repeated for a variety of DC amplitudes or offset displacements. Figure 51.27 shows the resonant frequency versus offset amplitude of a device with fixed–fixed springs. As shown in the figure, the device exhibits a significant resonant frequency shift over the range of operation. A linear device would not show any appreciable frequency shift. This measurement can be useful in understanding the effects of the nonlinearities.

51.4.3 Error Analysis

When utilizing any measurement technique, it is important to quantify the error. For the previous example, an LDV system [51.28] was utilized to collect the measurement data. In this example, sources of error are due to SEM measurement of device parameters and the resolution limits of the vibrometer. Errors due to accuracy of the spectrum analyzer and oscilloscope are usually negligible, assuming that appropriate ranges for the tools are utilized and the units have a valid calibration. As an example, an error analysis is presented for the sample device shown in Fig. 51.7 and discussed in the previous section. The errors in the linear and cubic spring constants are determined below.

From (51.5), $k \propto b$, ε_0, N, and $k \propto 1/\psi$, $1/d$; ε_0 and N are known constants, however b and d have some error associated with the way they are measured. The error in k, dk, is:

$$dk = \left| \frac{N\varepsilon_0}{\Psi d} \, db \right| + \left| \frac{N\varepsilon_0 b}{\Psi d^2} \, dd \right| + \left| \frac{N\varepsilon_0 b}{\Psi^2 d} \, d\Psi \right| , \quad (51.9)$$

where db and dd are errors due to measurement of the dimensions d and b using the SEM or FIB; $d\Psi$ comes from the static-displacement equation, $x = \Psi V^2$, so $d\Psi = \frac{1}{V^2} \, dx$, where dx is the resolution error of the interferometric measurement. There is also a source of error in $d\Psi$ which is due to the curve fit of the data in Fig. 51.25. This will add an additional term to (51.9), which is the error in the curve fit. The total error in the linear spring constant is

$$dk = \left| \frac{N\varepsilon_0}{\Psi d} \, db \right| + \left| \frac{N\varepsilon_0 b}{\Psi d^2} \, dd \right| + \left| \frac{N\varepsilon_0 b}{\Psi d^2} \left(\frac{1}{V^2} \right) dx \right|$$
$$+ \left| \frac{N\varepsilon_0 b}{\Psi^2 d} \, d\Psi_{\text{curvefit}} \right| . \quad (51.10)$$

Different range settings on the vibrometer have different resolution limits. The range setting will change the dx error term.

A second source of error in the interferometric measurement involves the angular alignment of the micromachined mirror and the device axis of motion. If the mirror is placed or machined at an angle that is not parallel to the axis of the moving structure, errors can occur. These errors are generally known as cosine errors. If we consider a mirror that is out of alignment by an angle θ, the error in the length measured will be:

$$dx = x_0 \left(\frac{1}{\cos \theta} - 1 \right) , \quad (51.11)$$

where x_0 is the actual distance traveled, and dx is the error in length due to the misalignment. This error can be caused by improper placement of a mirror, or by rotation misalignment in the moving structure. A rotation of 0.5° is a reasonable estimate of errors incurred in machining mirrors as was done here, or in rotations of MEMS actuators due to instabilities. Using a 0.5° maximum rotation, the cosine error has a maximum value of 0.003 nm per μm of device motion. This error can also

occur in an out-of-plane direction due to errors in mirror fabrication. Using a 0.5° maximum rotation, the error in this term is also 0.003 nm, for a maximum error from cosines of 0.006 nm.

The error in the cubic term is developed in much the same way as the linear spring constant error. From (51.8), the cubic spring constant is defined as

$$k_3 = mh = \frac{4}{3A^2} \left(\omega_1^2 - \omega_0^2 \right) m . \quad (51.12)$$

The error in k_3 is therefore

$$|dk_3| = \left| \frac{-8}{3A^3} \left(\omega_1^2 - \omega_0^2 \right) m \, dA \right|$$
$$+ \left| \frac{4}{3A^2} \left(2\omega_0 \right) m \, d\omega_0 \right|$$
$$+ \left| \frac{4}{3A^2} \left(\omega_1^2 - \omega_0^2 \right) dm \right| , \quad (51.13)$$

where dm depends on the error in k which was calculated previously, and the error in ω_0.

$$|dm| = \left| \frac{dk}{\omega_0^2} + \frac{k}{\omega_0^3} \, d\omega_0 \right| . \quad (51.14)$$

The error in ω_0 is due to the effect of the vibrometer resolution to time errors. If the vibrometer error in displacement is dx and the velocity of the device as it crosses its equilibrium position is $v_{\max}$, then the error in time for measuring n cycles is $t_{\text{err}} = 2 \, dx/(n v_{\max})$. Refering again to Fig. 51.23, for the sinusoidal motion, $V_{\max} = 0.054$ m/s is found by multiplying the peak displacement and the frequency of oscillation (ω_0). The time for the measurement t is 8 divisions on the chart (2.4 ms) and includes $n = 29$ cycles. This can be converted to a frequency error in the following way

$$|d\omega_0| = \left| \frac{1}{t} - \frac{1}{t + t_{\text{err}}} \right| . \quad (51.15)$$

Using the above error formulas the error can be calculated for the sample device. Assuming that we can measure the device geometry in the SEM to an accuracy of < 50 nm; $db = dd = 50$ nm. dx is based on the total interferometric accuracy, which includes the error in the vibrometer decoder and cosine alignment errors. For the sample experiment, the vibrometer was set to a range of 20 μm/V, which has an error of ±40 nm. So $dx = 40$ nm (vibrometer error) $+0.006$ nm (cosine error). Based on these errors, $dk = 0.114$ N/m, which corresponds to an error in k of 3.13%.

The error in the cubic term dk_3 depends on A, which is the amplitude of the excitation. This error is also determined by the range setting of the vibrometer during the measurement. For this measurement, the range was also $20\,\mu m/V$, corresponding to $dA = 40\,nm$. Based on that

error, $dk_3 = 9.77 \times 10^9$, corresponding to a maximum error of 5.7%.

51.5 Design for Test

How are you going to actuate it: on chip or external forces? If needed, how are you going to connect to it electrically? How are you going to measure the response?

These questions may seem straightforward, but in the development stage they can have a very different answer than during the production stage. Measuring the motion of a mechanical structure using the optical methods discussed in this chapter may be necessary to verify the design and performance of on-chip sensors. This can be difficult to do if the MEMS part has been encapsulated to facilitate packaging.

A finished device will be bonded and packaged. During development, the cost of packaging for testing can be prohibitive if device yield is low. Wafer-level or die-level probing is a good alternative but may require design changes to be effective. Simple designs requiring a few connections can be measued on a manual probe station. More complex designs may require a custom probe card, for example from Kulicke and Soffa, to make and break dozens of connections. If probing is done before en-

capsulation, additional test points may be added due to additional die area being available that would otherwise be covered by the encapsulation die.

These considerations can force a device to have a different fabrication process in the development stage than the production stage. Wafer lots may be split out during production for analysis, process, or performance verification. The split is necessary as the actuation or measurement techniques may not be possible on the finished device.

MEMS testing is a new and dynamic field. The integrated-circuit (IC) industry is standardized to the level of fabless companies and fabrication foundries. Design, fabrication, and test procedures have been standardized and now follow a roadmap. As difficult as it has been to establish foundries in the MEMS community, testing standards are equally challenging to standardize. With a few exceptions, the testing methods discussed here have been developed by each group and are not available from a vendor. Surely there are also many more to be developed.

References

51.1 M. Haque, M. Saif: Strain gradient effect in nanoscale thin films, Acta Materialia **51**(11), 3053–61 (2003)

51.2 D. Peroulis, L. P. B. Katehi: Spring-loaded DC-contact RF MEMS switches, Internat. J. RF Microwave Comp.-Aid. Eng. **14**(4), 345–354 (2004)

51.3 H. V. Jansen, J. G. E. Gardeniers, J. Elders, H. A. C. Tilmans, M. Elwenspoek: Applications of fluorocarbon polymers in micromechanics and micromachining, Sensors Actuators A **41**(1–3), 136–40 (1994)

51.4 J. Bouchaud, H. Wicht: RF MEMS: status of the industry in 2004, application roadmap and market forecasts, 34th European Microwave Conference, London, 1569–1570 (2004)

51.5 B. Bircumshaw, G. Liub, H. Takeuchib, T.-J. Kingb, R. Howe, O. O'Reillya, A. Pisano: The Radial bulk annular resonator: towards a 50 Omega RF MEMS filter, Transducers 03: 12th International Conference on Solid-State Sensors, Actuators and Microsystems, Boston **2**, 876–880 (2003)

51.6 L. Linand R. Howe, A. Pisano: Microelectromechanical filters for signal processing, J. Microelectromech. Sys. **7**(3), 286–294 (1998)

51.7 S. Acar, A. Shkel: Nonresonant micromachined gyroscopes with structural-mode decoupling, IEEE Sensors J. **3**(4), 497–506 (2003)

51.8 G. Fedder andXie: Fabrication, characterization, and analysis of a DRIE CMOS MEMS Gyroscope, IEEE Sensors J. **3**(5), 622–631 (2003)

51.9 N. V. Lavrik, P. G. Datskos: Femtogram mass detection using photothermally actuated nanomechanical resonators, Appl. Phys. Lett. **82**(16), 2697–9 (2003)

51.10 N. V. Lavrik, M. Sepaniak, P. G. Datskos: Cantilever transducers as a platform for chemical and biological sensors, Rev. Sci. Instrum. **75**(7), 2229–2253 (2004)

51.11 T. Thundat, P. I. Oden, R. J. Warmack: Chemical, Physical and biological detection using microcantilevers. In: *Third international symposium on microstructures and microfabricated systems* (Electrochemical Society, Pennington 1997)

51.12 D. A. Horsley, N. Wongkomet, R. Horowitz, A. P. Pisano: Precision positioning using a microfabricated electrostatic actuator, IEEE Trans. Magnetics **35**(2), 993–999 (1999)

51.13 D. A. Horsley, P. G. Hartwell, R. G. Walmsley, J. Brandt, U. Yoon, S. Hoen: Multi-degree of freedom dynamic characterization of deep-etched silicon suspensions. In: *Solid State Sensor and Actuator Workshop* (IEEE, Hilton Head 2000) pp. 81–84

51.14 M.-F. Yu, G. J. Wagner, R. S. Ruoff, M. J. Dyer: Realization of parametric resonances in a nanowire mechanical system with nanomanipulation inside a scanning electron microscope, Phys. Rev. B Condens. Matter **66**(7), 073406/1–4 (2002)

51.15 X. Chen, M.-F. Yu, G. J. Wagner, R. S. Ruoff, M. J. Dyer: Mechanical Resonance of quartz microfibers and boundary condition effects, J. Appl. Phys. **95**(9), 4823–4828 (2004)

51.16 W. Zhang, K. L. Turner: A mass sensor based on parametric resonance. In: *Hilton Head 2004: A Solid State Sensor, Actuator and Microsystems Workshop*, Vol. 1 (IEEE, Hilton Head Island 2004) pp. 49–52

51.17 D. M. Freeman, A. J. Aranyosi, M. J. Gordon, S. S. Hong: Multidimensional motion analysis of mems using computer microvision. In: *Solid State Sensor and Actuator Workshop* (IEEE, Hilton Head 1998) pp. 150–155

51.18 D. M. Freeman, C. Q. Davis: Using video microscopy to characterize micromechanical systems. In: *1998 IEEE/LEOS Summer Topical Meeting Broadband Optical Networks and Technologies: An emerging reality.* (IEEE, Piscataway 1998)

51.19 M. Hart, R. A. Conant, K. Y. Lau, R. S. Muller: Stroboscopic Interferometer system for dynamic MEMS characterization, J. Microelectromechan. Syst. **9**(4), 409–415 (2000)

51.20 C. Q. Davis, D. M. Freeman: Using a light microscope to measure motions with nanometer accuracy, Opt. Eng. **37**, 1299–1304 (1998)

51.21 MMA G2 MEMS Motion Analyzer, Generation 2 0290 MMA G2 Principles and Performance 2000 MMA G2 Product sheet

51.22 OFV-3001, OFV-501, Polytec, GmbH

51.23 M. Hart, R. A. Conant, K. Y. Lau, R. S. Muller: Stroboscopic Phase-Shifting interferometry for dynamic characterization of optical MEMS, Proc. SPIE **3749**, 468–469 (1999)

51.24 C. Rembe, L. Muller, R. S. Muller, R. T. Howe: Full Three-Dimensional Motion Characterization of a gimballed Electrostatic Microactuator. In: *39th annual international reliability physics symposium, Orlando* (2001)

51.25 M. Zalalutdinov, A. Olkhovets, A. Zehnder, B. Ilic, D. Czaplewcki, J. M. Parpia, H. G. Craighead: Light Induced Parametric Amplification in MEMS Oscilla-

tions. In: *Design, Test, Integration and Packaging of a MEMS/MOEMS 2001* (SPIE, Bellingham WA 2001)

51.26 M. Zalalutdinov, A. Zehnder, A. Olkhovets, S. Turner, L. Sekaric, B. Ilic, D. Czaplewski, J. M. Parpia, H. G. Craighead: Autoparametric optical drive for micromechanical oscillators, Appl. Phys. Lett. **79**(5), 695–7 (2001)

51.27 L. Sekaric, M. Zalalutdinov, R. B. Bhiladvala, A. T. Zehnder, J. M. Parpia, H. G. Craighead: Operation of nanomechanical resonant structures in air, Appl. Phys. Lett. **81**(14), 2641–2643 (2002)

51.28 K. L. Turner: Multi-dimensional MEMS motion characterization using laser vibrometry. In: *Transducers '99. The 10th International conference on solid-state Sensors and Actuators, Digest of Technical Papers, Sendai* (1999)

51.29 D. Oliver: Scanning laser vibrometers, Automot. Eng. **104**, 71–75 (1996)

51.30 P. Castellini, G. M. Revel, E. P. Tomasini: Laser Doppler vibrometry: A review of advance and applications, Shock Vib. Dig. **30**(6), 443–456 (1998)

51.31 J. S. Burdess, A. J. Harris, D. Wood, R. J. Pitcher, D. Glennie: A system for the dynamic characterization of microstructures, J. Microelectromechan. Syst. **6**(4), 332–328 (1997)

51.32 A. Biswas, T. Weller, L. P. B. Katehi: Stress determination of micromembranes using laser vibrometry, Rev. Sci. Instrum. **67**(5), 1965–1969 (1996)

51.33 D. Eichner, W. von Muench: A two-step electrochemical etch-stop to produce freestanding bulk-micromachined structures, Sensors Actuators **60**, 103–107 (1997)

51.34 C. H. Liu, A. Barzilai, O. Ajakaiye, H. K. Rockstad, T. W. Kenny: Performance enhancements for the micromachined tunneling accelerometer, Proceedings 10th International Conference on Solid State, Sensors and Actuators, 1290 (1999)

51.35 T. W. Kenny, W. J. Kaiser, H. K. Rockstad, J. K. Reynolds, J. A. Podosek, E. C. Vote: Widebandwidth electromechanical actuators for tunneling displacement transducers, J. Microelectromech. Syst. **3**(3), 97–103 (1994)

51.36 D. W. Burns, J. D. Zook, R. D. Horning, W. R. Herb, H. Guckel: Sealed-cavity resonant microbeam pressure sensor, Sensors Actuators **48**, 179–186 (1995)

51.37 R. Lawton, M. Abraham, E. Lawrence: Characterization of non-planar motion in MEMS involving scanning laser interferometry, Proc. SPIE **3880**, 46–50 (1999)

51.38 Polytec: MSV-400 Microscope Scanning Vibrometer, Datasheet (Polytec, Tustin 2004)

51.39 Polytec: Vibrometer User's Manual 1999

51.40 Polytec: PMA-400 Planar Motion Analyzer, Data Sheet

51.41 D. E. Aspnes, A. A. Studna: Dielectric functions and optical parameters in Si, Ge, GaP, GaAs, GaSb, InP,

InAs, and InSb from 1.5 – 6.0 eV, Phys. Rev. B **27**(2), 985 (1983)

51.42 J. P. Kapur, D. Casasent: Geometric correction of SEM images, Hybrid Image and Signal Processing VII, Proc. SPIE **4387**, 204–214 (2001)

51.43 J. Walraven: Tools and Techniques for failure analysis and qualification of MEMS, Tech. Dig. **834** (2003)

51.44 K. A. Peterson, P. Tangyunyong, A. Pimental: *Materials and device characterization in micromachining symposium* (SPIE, Bellingham 1998)

51.45 S. Reyntjens, R. Puers: A review of focused ion beam applications in microsystem technology, J. Micromech. Microeng. **11**, 287–300 (2001)

51.46 Y. Tai, R. S. Muller: Measurement of Young's modulus on microfabricated structures using a surface profiler. In: *IEEE microelectromechanical systems workshop* (IEEE, Piscataway 1990) pp. 147–152

51.47 M. W. Denhoff: A measurement of Young's modulus and residual stress in MEMS bridges using a surface profiler, J. Micromech. Microeng. **13**, 686–692 (2003)

51.48 M. Haque, M. Saif: Application of MEMS force sensors for in situ mechanical characterization of nano-scale thin films in SEM and TEM. In: *Proceedings of 11th International Conference on Solid State Sensors and Actuators Transducers '01/Eurosensors XV* (Springer, Berlin, Heidelberg 2001)

51.49 M. Haque, M. Saif: In-situ tensile testing of nanoscale specimens in SEM and TEM, Exp. Mech. **42**(1), 123–8 (2002)

51.50 M. Haque, M. Saif: Microscale materials testing using MEMS actuators, J. Microelectromech. Syst. **10**(1), 146–52 (2001)

51.51 M. T. A. Saif, N. C. MacDonald: Planarity of large MEMS, J. Microelectromech. Syst. **5**(2), 79–97 (1996)

51.52 M. T. A. Saif, N. C. MacDonald: Measurement of forces and spring constants of microinstruments, Rev. Sci. Instrum. **69**(3), 1410–1422 (1998)

51.53 M. T. A. Saif, N. C. MacDonald: Microinstruments for submicron material studies, J. Mater. Res. **13**(12), 3353–3356 (1998)

51.54 M. T. A. Saif, N. C. MacDonald: A millinewton microloading device, Sensors Actuators **A52**, 65 (1996)

51.55 R. Grow, S. C. Minne, S. R. Manalis, C. F. Quate: Silicon nitride cantilevers with oxidation sharpened silicon tips for atomic force microscopy, J. Microelectromech. Syst. **11**(4), 317–321 (2002)

51.56 P. F. Indermuhle, G. Schurmann, G. A. Racine, N. F. de Rooij: Fabrication and characterization of cantilevers with integrated sharp tips and piezoelectric elements for actuation and detection for parallel AFM applications, Sensors Actuators **A60**, 186–90 (1997)

51.57 H. Lee, R. A. Coutu Jr., S. Mall, P. E Kladitis: Nanoindentation technique for characterizing cantilever beam style RF microelectromechanical systems

(MEMS) switches, J. Micromech. Microeng. **15**, 1230–1235 (2005)

51.58 E. Jones, K. Murphy, M. Begley: Mechanical measurements of adhesion in microcantilevers: transitions in geometry and cyclic energy changes, Exp. Mech. **43**, 280–288 (2003)

51.59 G. D. Cole, J. E. Bowers, K. L. Turner, N. C. MacDonald: Dynamic characterization of MEMS-tunable vertical-cavity SOAs, Optical MEMS and Their Applications Conference, 2005. IEEE/LEOS International Conference (Aug. 1-4, 2005) 99-100, Oulu, Finland

51.60 Y. A. Liang, K. Autumn, S. T. Hsieh, W. Zesch, W.-P. Chan, R. Fearing, R. J. Full, T. W. Kenny: Technical Digest of the 2000 Solid-State Sensor and Actuator Workshop, Hilton Head Island, SC (Transducers Research Foundation, Cleveland) 33-38

51.61 M. S. Bartsch, W. Federle, R. J. Full, T. W. Kenny: Small insect measurements using a custom MEMS force sensor TRANSDUCERS, Solid-State Sensors, Actuators and Microsystems, 12th International Conference, Vol. 2 (Boston, MA, 8-12 June 2003) 1039-1042

51.62 L. Shaoning, A. D. Dmitriy, Z. Sulin, F. T. Fisher, J. Lee, R. S. Ruoff: Realization of nanoscale resolution with a micromachined thermally actuated testing stage, Rev. Sci. Instrum. **75**(6), 2154–62 (2004)

51.63 N. C. MacDonald, C. Y. Robinson, R. M. White: Time-resolved scanning electron microscopy and its application to bulk-effect oscillators, J. Appl. Phys **40**, 4516 (1969)

51.64 I. Ogo, N. C. MacDonald: Applications of time-resolved scanning electron microscopy to the analysis of the motion of microelectromechanical structures, **B14**(3), 1630–1634 (1996)

51.65 J. J. Yao, N. C. MacDonald: Time-resolved scanning electron microscopy analysis of nanodynamical structures, Scanning Microsc. **6**(4), 939–942 (1992)

51.66 R. E. Mihailovich, N. C. MacDonald: Dissipation measurements of vacuum-operated single crystal silicon microresonators, Sensors Actuators A **50**, 199–207 (1995)

51.67 S. Pourkamali, R. Abdolvand, F. Ayazi: A 600 kHz Electrically Coupled MEMS Bandpass Filter, Proc. IEEE International Micro Electro Mechanical Systems Conference (MEMS'03), Kyoto, Japan, Jan. 2003 (IEEE, Piscataway 2003) pp. 702-705

51.68 R. Howe: Polysilicon integrated microsystems: technologies and applications. In: *Proceedings of the International Solid-State Sensors and Actuators Conference - TRANSDUCERS '95* (IEEE, Piscataway 1995)

51.69 T. B. Gabrielson: Mechanical-thermal noise in micromachined acoustic and vibration sensors, IEEE Trans. Electron Devices **40**(5), 903–9 (1993)

51.70 A. N. Cleland, M. L. Roukes: Fabrication of high frequency nanometer scale mechanical resonators from bulk Si crystals, Appl. Phys. Lett. **69**(18), 2653–5 (1996)

51.71   J. D. Zook, D. W. Burns, H. Guckel, J. J. Sniegowski, R. L. Engelstad, Z. Feng: Characteristics of polysilicon resonant microbeams, Sensors Actuators A **35**, 51–59 (1992)

51.72   M. B. Dobrin, C. H. Savit: *Introduction to Geophysical Prospecting*, 4th edn. (McGraw–Hill, New York 1988)

51.73   E. Purcell: *Electricity and Magnetism*, Vol. 2, 2nd edn. (McGraw–Hill, New York 1970)

51.74   K. Y. Yasumura, T. Stowe, E. Chow, T. Pfafman, T. W. Kenny, D. Rugar: A study of microcantilever quality factor, Proc. 1998 Hilton Head Workshop on Sensors and Actuators (IEEE, Piscataway 1998) p. 65

51.75   K. Y. Yasumura: Quality factors in micron- and submicron-thick cantilevers, J. Microelectromech. Syst. **9**(1), 117–25 (2000)

51.76   S. G. Adams: Design of Electrostatic Actuators to Tune the Effective Stiffness of Micro-Electro-Mechanical Systems. Ph.D. Thesis (Cornell University, Ithaca 1996)

51.77   S. G. Adams: Independent tuning of linear and nonlinear stiffness coefficients [actuators], J. Microelectromecha. Syst. **7**(2), 172–80 (1998)

51.78   W. Kun, W. Ark-Chew, C.-C. Nguyen: VHF free-free beam high-Q micromechanical resonators, J. Microelectromech. Syst. **9**(3), 347–60 (2000)

51.79   C. T. C. Nguyen: *Micromechanical Signal Processors* (University of California, Berkeley 1994)

51.80   C. T.-C. Nguyen: Micromechanical filters for miniaturized low-power communications (invited), Proceedings of SPIE: Smart Structures and Materials; Smart Electronics and MEMS (SPIE, Bellingham 1999)

51.81   Cunningham: *Introduction to Nonlinear Analysis* (McGraw–Hill, New York 1958)

52. Failure Mechanisms in MEMS/NEMS Devices

The commercialization of MEMS/NEMS devices is proceeding slower than expected, because the reliability problems of microscopic components differ from macroscopically known behavior. In this chapter, we provide an overview of the state of the art in MEMS/NEMS reliability. We discuss the specific, MEMS-related problems caused by stiction due to surface forces and electric charge. Materials issues such as creep and fatigue are treated as well. Nanoscale wear is covered briefly. MEMS packaging is also discussed, because the reliability of MEMS/NEMS components critically depends on the available protection from the environment.

Part G | 52

With the rapid spread of micro-electromechanical systems (MEMS) technology in the 1990s, it has become clear that the reliability of microscopic mechanical devices is a large concern [52.1]. It is one of the main reasons that much of the micromechanical technology available all over the world in laboratories has not been commercialized yet. The route from a working prototype to a commercial MEMS component should not be underestimated: the pioneering effort of Texas Instruments to bring the digital micromirror device (DMD) MEMS chip to the market took more than 10 years after its invention [52.2]. The same holds for gyroscopes and accelerometers: many years of hard work were spent on the conversion of laboratory prototypes to commercial products [52.3].

The failure mechanisms of MEMS/NEMS [in the rest of this chapter we will use MEMS as a generic name that also includes devices with nanometer-scale features, which are increasingly being referred to as nano-electromechanical systems (NEMS)] components are widely differing from those of their macroscopic counterparts, and this is the main reason that this tra-

jectory takes so long. Many MEMS designers still do not take into account the implications that the scale difference has on reliability when designing new devices, although the paradigmatic approach of *design for reliability*, i.e. taking into account reliability issues from the earliest design stage, is gaining more acceptance. A second point is that, even if people are willing to take into account all the available information on the failure modes and mechanisms of MEMS, they soon discover that not so much is known at all. Still, over the last few years, the information available on MEMS failure mechanisms has been steadily growing [52.4]. At present, we are not yet in a position to predict quantitatively when or where most of the different failure mechanisms will occur, but we can describe them in some detail and give rules of thumb to avoid many of them.

In this chapter we will introduce the basic MEMS failure mechanisms and point out when they are important and in which respects they differ from macroscopic failure mechanisms. We will study the most important of them in some detail and give recommendations for procedures that are known to counteract them.

52.1 Failure Modes and Failure Mechanisms

We have to start with a clear distinction between failure modes and failure mechanisms. The failure mode is the apparent failure. It might be that e.g. a MEMS beam is stuck to the substrate. If this causes a de-

Fig. 52.1 The failure mode of this resonator is broken beams; the failure mechanism is rough handling, causing an overload stress in the beams and support

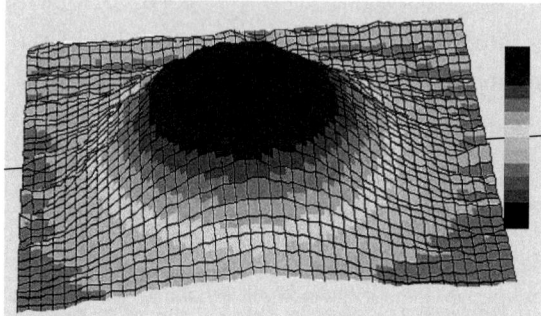

Fig. 52.2 Example of optical assessment of dynamic MEMS behavior: laser television (TV) holography measurement image of the first resonance mode of a MEMS pressure sensor membrane [52.5]

Table 52.1 Common MEMS failure mechanisms

Fracture	overload fracture
	fatigue fracture
Creep	applied stress
	intrinsic stress
	thermal stress
Stiction	capillary forces
	van der Waals molecular forces
	electrostatic forces
	solid bridging
Electromigration	
Wear	adhesive
	abrasive
	corrosive
Degradation of dielectrics	leakage
	charging
	breakdown
Delamination	
Contamination	
Pitting of contacting surfaces	
Electrostatic discharge (ESD)	

vice to malfunction, the failure mode of the device is a stuck beam. This failure mode may be caused by different failure mechanisms, such as stiction of the beam (stiction is a generic term for surfaces being stuck due to surface forces), it may be a fatigue fracture failure or caused by an overload condition, and so on. The failure mechanism is hence the physical cause of the failure mode. In Fig. 52.1 the failure mode is broken beams, but the failure mechanism is rough handling, causing a mechanical overload condition.

This distinction is important because often the failure mode is reported when something is wrong, while the problem can only be accurately addressed by investigating the failure mechanism. In Table 52.1, the most commonly encountered failure mechanisms are given (most of them taken from [52.6]).

Dedicated equipment to assess failure mechanisms is definitely needed, but only a few set ups have been reported to date. Sandia National Laboratories is famous for their Sandia high-volume measurement of micromachine reliability (SHiMMeR) system, where they can test a significant number of rotating micromachines simultaneously under various environmental conditions [52.7]. The operation of the micromotors is monitored with advanced software tools [52.8]. Inter-University Microelectronics Center (IMEC) has a dedicated system to assess radio-frequency (RF) MEMS switch reliability, where environmental conditions can be altered with in situ optical access and various stimuli can be applied [52.5]. Interferometry- and vibrometry-based optical systems of various kinds are in use to monitor the dynamic behavior of MEMS devices [52.9–12] (Fig. 52.2). Some commercial laboratories also have their dedicated systems but do not disclose their exact set ups.

Other equipment that is used for failure mechanism assessment has not been developed exclusively for MEMS, but the MEMS reliability community is relying heavily on them: scanning electron microscopy (SEM), focused ion beam (FIB) cross-sectioning, atomic force microscopy (AFM), optical microscopy, and various electronic actuation and detection systems are the most widespread among them.

52.2 Stiction and Charge–Related Failure Mechanisms

52.2.1 Stiction Due to Surface Forces

One of the most important problems in MEMS/NEMS devices is stiction, i. e. the unintentional sticking of contacting surfaces (Fig. 52.3). Because forces intentionally generated in MEMS by e.g. electrostatic actuation are generally very small, surface forces can dominate the device behavior. The fact that MEMS surfaces are so smooth that a large part of the surfaces is within the range of the atomic forces of the other surface aggravates the matter further. The most important surfaces forces commonly listed are: forces due to capillary condensation, van der Waals molecular forces, and chemical and hydrogen bonds between the surfaces. Surfaces tend to stick together when they are dried after the release etch, and a good overview of what happens is given by *Mastrangelo* and *Hsu* [52.14, 15]. From a reliability point of view, in-use stiction is more important.

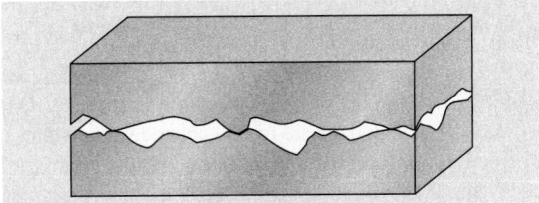

Fig. 52.4 Two rough surfaces in contact

The capillary condensation, van der Waals forces, and bonding force effects mentioned have in common that they are true interface forces. The force depends in this case on the situation at the interface, and not on effects far away from it. All these forces have in common that they are strong for surfaces very close together but fall off relatively fast when the distance is larger than a few nanometers.

As a result, surface roughness is an important parameter governing the stiction force, because the surface roughness of MEMS devices is commonly found to be of the same order of magnitude as the range of the surface forces (Fig. 52.4). This was demonstrated for the first time by *De Boer* et al. [52.13]. They monitored one-side-clamped free-standing cantilever MEMS structures that were stuck over a considerable part of their length, using an interferometer (Fig. 52.5). Fringes in the image indicated which part of the beams was free-standing and which part was stuck. Because the restoring spring force of the beams was known, the length over which the beams were not stuck gave

Fig. 52.3 Stiction failure of a comb drive

10 μm

Table 52.2 Deflection and surface interaction energy of S- and arc-shaped beams [52.13]

	S-shaped beam	Arc-shaped beam
Deflection $u(x)$	$u(x) = h \left(\dfrac{x}{s}\right)^2 \left(3 - 2\dfrac{x}{s}\right)$	$u(x) = h \left(\dfrac{x}{s}\right)^2 \left(3 - \dfrac{x}{s}\right)$
Surface interaction energy Γ	$\Gamma = \dfrac{3}{2} E \dfrac{t^3 h^2}{s^4}$	$\Gamma = \dfrac{3}{8} E \dfrac{t^3 h^2}{s^4}$

[a] E = Young's modulus of the beam material, the other parameters are given in Fig. 52.6

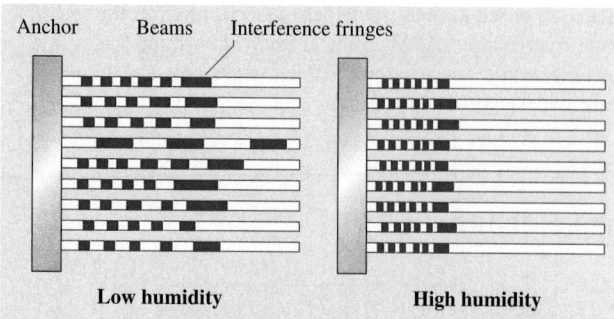

Low humidity   **High humidity**

Fig. 52.5 Schematic top view of the measurement of adhesion as a function of relative humidity using interference imaging by *De Boer* et al. [52.13]. The dark parts are fringes created with an interferometric objective. The fringe pattern reveals where the beams are bending, and where they are flat, because fringes are only observed where the beams are not running perfectly in plane

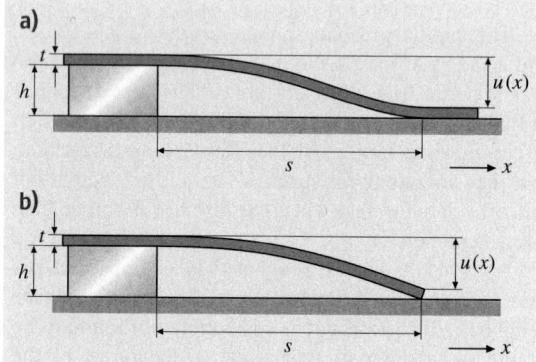

Fig. 52.6 (a) S-shaped beam; **(b)** arc-shaped beam

a measure for the surface interaction energy (Fig. 52.6 and Table 52.2).

We have to distinguish for this calculation between beams that adhere only at the tip (*arc-shaped*) and beams that adhere over a considerable length (*S-shaped*). It was argued by De Boer and coworkers that the S-shaped beam gives a more precise measure for the surface interaction energy because the arc-shaped beam introduces a large statistical error; only the properties of the surface at the very small contacting tip produce the measured result. Arc-shaped beams can very easily be used to find the surface interaction energy to first order however, because we only have to look at which beams are stuck and which beams are free-standing.

De Boer et al. [52.13] found that the surface interaction energy varied considerably with the relative humidity and attributed this effect to capillary condensation. Small liquid bridges are supposed to form when two

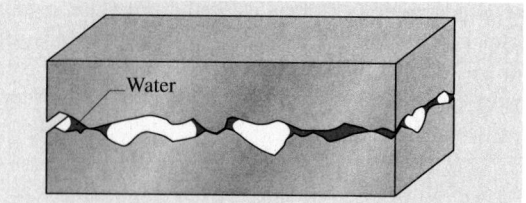

Fig. 52.7 Capillary condensation occurs on all places where the two contacting surfaces are closer together than a characteristic distance

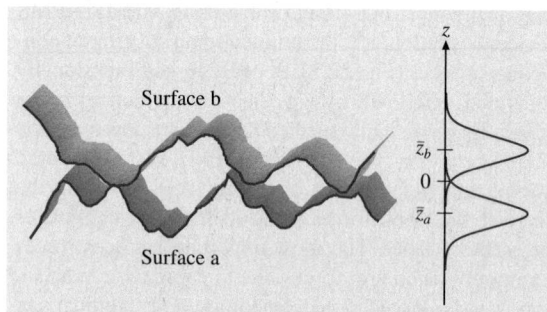

Fig. 52.8 The two surface-height distribution functions give rise to a distance distribution function $h(z)$ describing how large the part of the surface is that has an interaction with the other surface at a certain distance

hydrophilic surfaces are in close contact (Fig. 52.7). This effect had already been reported by *Bowden* and *Tabor* in 1950 [52.16] with a macroscopic measuring apparatus, but only with the development of MEMS technology has capillary condensation become a big technological issue. Because only the surface area that is closer to the other surface than a characteristic distance will be bridged by capillary condensed water, generally only part of the surface area is in contact in this sense.

Maboudian and *Howe* [52.17], *Legtenberg* et al. [52.18] and *Tas* et al. [52.19] provided the first quantitative models of stiction, making use of the work of *Israelachvilli* [52.20] on more generic systems. They described the surface forces as a function of the distance between two flat surfaces. The relation between surface interaction energy and distance for capillary forces (a function of humidity) and of van der Waals molecular forces (a function of the Hamaker constant of the molecules at the surface) was presented with very general applicability.

The roughness of the surfaces (some parts are closer together and contribute more to the stiction force than others) was discussed briefly by *Komvopoulos* et al. [52.21] and more quantitatively by *van Spengen* et al. [52.22]. The procedure boils down to the determination of a distance distribution function $h(z)$, giving the amount of surface at every distance z from the other surface (Fig. 52.8). The total surface interaction energy Γ_i due to a certain i-th force $e_i(z)$, e.g. the van der Waals force, is then given by

$$\Gamma_i = \int_0^\infty e_i(z)h(z)\,\mathrm{d}z\,, \tag{52.1}$$

This principle is generally applicable. As long as the forces do not influence each other, they will be reasonably well described by (52.1) and can be added together.

It is not easy to predict the distance distribution function $h(z)$, as the two surfaces may deform each other at the contacting points, giving rise to both plastic and elastic deformation. If we know the height distribution of the contacting surfaces independently, a rough estimate can be obtained by assuming only plastic deformation. This gives rise to the *bearing area* representation [52.22], from which the distance distribution can be obtained. Unfortunately, the surface interaction energy value obtained in this way is expected to be too low, as elastic deformation is expected to bring larger parts of the surfaces closer together than in the fully plastic case. Purely elastic deformation can be treated by the Greenwood–Williamson model [52.23], but has been found to be useful from an experimental point of view only for very soft, elastic materials like rubber. The technologically more interesting cases (silicon on silicon, metals etc.) are best described with a mixed plastic/elastic model, but unfortunately only numerical simulations can be employed to describe this [52.24, 25].

A more experimental approach to characterize the rough contact between two MEMS surfaces is followed by *Kogut* et al. [52.26], who use electrical current measurements through the contacting points to estimate the real contact area.

Whether the resulting surface interaction energies from the different force contributions result in a stiction failure or not depends on the restoring force. The peel number N_p (first coined by *Mastrangelo* and *Hsu* [52.15] and very well explained by *Zhao* [52.27]) can be used to find out whether a structure will suffer from stiction. For $N_p < 1$ the surfaces will adhere, while for $N_p > 1$ the surfaces will separate again if they are brought into contact.

The most cumbersome stiction force is that caused by capillary condensation. Water will only capillary condense onto hydrophilic surfaces, so a hydrophobic coating can be applied to prevent this. These layers are not unlike the boundary lubricants that have been used for decades to lower friction. Different ways of creating hydrophobic coatings have been applied but there is no industry standard yet. Anti-stiction coatings are generally made by either plasma deposition or self-assembled monolayer (SAM) deposition. The standard film quality test method is to bring a small water droplet to the surface and monitor the contact angle.

The gas used for the plasma deposition is usually a low-molecular-weight fluorinated organic compound, such as CHF_3 [52.28] or C_4F_{10} [52.29]. The MEMS device is bombarded with all kinds of free radicals and other reactive species generated from these molecules and forms a well-adhered water-repelling coating on the surfaces of the device. A big problem with plasma deposition is obtaining conformal coatings. Tricks such as having a Faraday cage inside the chamber are advantageous [52.29], but even then the conformity is usually not better than $2:1$ for the top and bottom surfaces of a MEMS device.

The durability of plasma-deposited coatings is relatively good, although they tend to come off in boiling water. *Man* et al. [52.29] tested their films for water contact-angle changes at elevated temperatures. By determining the lifetime at different temperatures and using an Arrhenius relation for the mean time to failure (MTTF)

$$\mathrm{MTTF} \propto \exp\left(\frac{E_a}{kT}\right), \tag{52.2}$$

they calculated a lifetime of more than 10 years at 150 °C. Knowing the Boltzmann constant k, one can obtain the activation energy E_a by performing experiments at different temperatures T. Impact wear was also tested.

To overcome the conformity problems associated with plasma deposition, and also because very thin layers are possible in this way, SAM coatings have been investigated since the middle of the 1990s. *Srinivasan* et al. [52.30] reported extensively on the subject. Silicon microstructures are released in HF, and while still in the liquid phase, the HF solution is replaced by a trichlorosilane solution. The most common are octadecyltrichlorosilane (OTS, $CH_3(CH_2)_{17}SiCl_3$) and 1H,1H,2H,2H-perfluorodecyltrichlorosilane (FDTS, $CF_3(CF_2)_7(CH_2)SiCl_3$). Figure 52.9 shows how such a film is formed on the surface of a MEMS device. A processing problem is that the reaction depends on trace amounts of water in the reaction environment. If the processing conditions

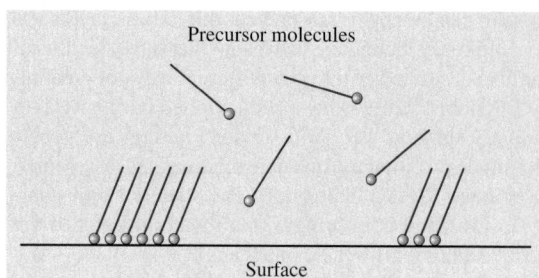

Fig. 52.9 SAM formation on a MEMS surface

are not extremely well controlled, micelles are formed as well as highly interconnected surface films that are easily peeled off [52.31].

Thermal stability tests show that, in air, the fluorocarbon films are able to stand 400 °C while the hydrocarbon films will degrade above 100 °C. In nitrogen no difference was observed [52.31]. *Kim* et al. [52.32] have suggested that the use of dimethyldichlorosilane (DDMS, $(CH_3)_2SiCl_2$) results in a process that is less sensitive to temperature, humidity and coating time. The films thus created withstood temperatures of above 450 °C without degradation.

De Boer et al. [52.13] found that FDTS films were susceptible to *water island formation* in a 90% relative humidity environment after only a few hours, while OTS films did not show such a dramatic effect. Whether this was a real difference or merely an effect of a difference in surface film quality has not been elucidated.

Another method besides anti-stiction coatings to reduce stiction problems is roughening the surfaces so that only small parts of the surfaces are in intimate contact. *Maboudian* et al. [52.22] show how this can be accomplished with different etchants (this also decreases van der Waals contributions). An absolutely water-free environment will also prevent capillary condensation problems altogether. The hermeticity of packages to accomplish this is discussed in the packaging section of this chapter (Sect. 52.3.4). The best solution to stiction due to surface interaction energies is the use of very stiff structures and high forces, which will separate the surfaces even under high adhesive contact forces. A typical higher limit for the surface interaction energy due to capillary forces is 0.1 J/m² [52.17], which corresponds to water everywhere between the surfaces. Surface interaction energies in rough devices and with anti-stiction coatings are typically 10–1000 times lower. If these numbers are taken into account, finite element modeling of the structures before they are produced can be used to predict whether stic-

tion due to surface forces will be a reliability threat or not.

52.2.2 Stiction Due to Electrostatic Attraction

Stiction due to dielectric charging is different from the previously discussed forces in the sense that it is not a true interface force. Charges may be present at the interface but may equally well reside in the bulk of a dielectric material. Electrostatic forces are very important on the microscale and hence an in-depth investigation of charge effects is important.

Charges may be generated in microsystems in different ways. It has been found that in devices with rubbing surfaces tribocharging may occur [52.33]. A rotating silicon micromotor was tested until it remained stuck due to the accumulated charge. Irradiation with a focused ion beam (FIB) neutralized the charges, after which the device resumed operation. Another situation where charge may be generated is where sparks in a small gap spray charge onto a surface [52.34]. Contact potentials of materials are usually < 0.5 V and are not expected to give any trouble in normal applications.

Charges generated by irradiation with ionizing radiation constitute a considerable challenge in MEMS devices. Accelerometers subjected to nuclear radiation were found to drift with increased dose because the dielectric under the moving mass of such a device was

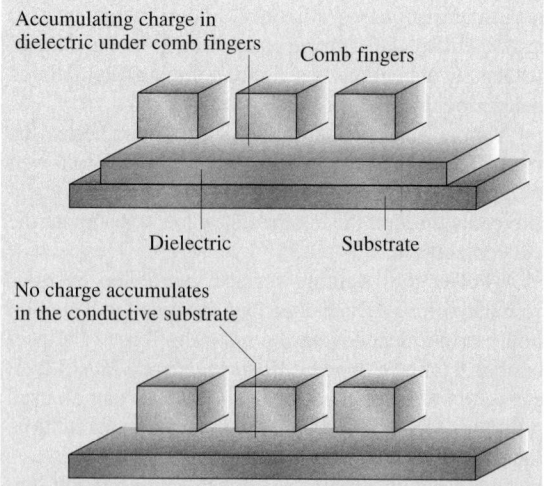

Fig. 52.10 Ionizing radiation can cause charging of the dielectric under the cantilevers (that are part of the moving mass) of an accelerometer and hence cause drift and in extreme cases stiction of the moving part

charged [52.35, 36]. Accelerometers without a dielectric under the moving mass were found to be much less susceptible, because no charge can accumulate (Fig. 52.10). One can imagine that excessive radiation would cause the moving mass to snap in and constitute a stiction failure, but this has not yet been reported.

The largest problems with stiction due to electrostatic attraction are found in devices where a large voltage is applied across a thin dielectric, such as in RF MEMS switches [52.37]. The high field causes charge injection into the dielectric by Schottky- or Poole–Frenkel-type conduction and is experimentally found to depend exponentially on the actuation voltage [52.38]. The situation resembles that of the high-k-dielectric gate oxides of the newest generations of metal–oxide–semiconductor (MOS) transistors, where similar charging effects threaten the stability of the threshold voltage [52.39].

The exact charging properties of a dielectric thin film depend on the deposition conditions. A *leaky* dielectric with a low resistivity may have many trapping sites, but can still be good from a technological point of view if the charge can flow away easily [52.40]. The films have to be empirically optimized because no consistent theory to describe good films has been developed yet.

Much more is known about how parasitic charges, when they are present, influence device behavior. For a free-sanding structure with small-scale deflections, *Wibbeler* et al. [52.34] have presented a theory that describes the deflection as a function of the accumulated charge. *Reid* [52.41] has shown finite element modeling (FEM) simulations for large deflections, pull-in and pull-out effects and the effect of parasitic charge on them.

The most comprehensive efforts, which are very useful from the point of view of failure due to stiction, are the treatises of *van Spengen* et al. [52.42] for uniform charge densities in the dielectric and *Rottenberg* et al. [52.43] for inhomogeneous charge densities. These two papers consider RF MEMS switches, but the principles are generic.

In these models, elementary electrostatics is used to describe the peculiar dynamic effects in MEMS with a structure like that in Fig. 52.11. A free-standing structure (the *bridge*) is located above a dielectric with a lower conductor. The whole assembly behaves as a capacitor. When a voltage is applied between the bridge conductor and the lower conductor, the bridge is attracted. If the bridge is moved over more than one third of the gap, the system becomes unstable and the bridge lands on the dielectric (pull-in). At this point the capacitance of the system becomes much higher and this is the reason that

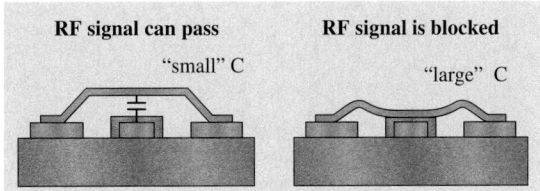

Fig. 52.11 RF MEMS switch cross section

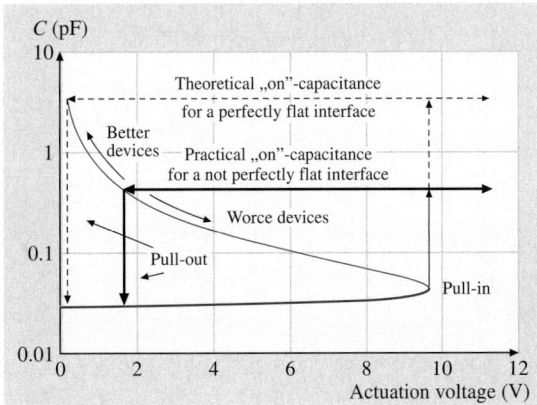

Fig. 52.12 C–V curve

it can be used to affect RF electronic signals. A much lower voltage is required to keep it there, so we have to lower the actuation voltage considerably to let it *pull out* again. This hysteresis is well known from RF MEMS C–V curves (Fig. 52.12).

Both positive and negative actuation voltages can cause the bridge to pull in (the C–V curve is symmetric for positive and negative voltages). For a uniform parasitic charge, the curve starts to shift. Negative charge causes the curve to shift to the right, while positive charge shifts the whole curve to the left on the voltage axis.

The effect of positive charge is easily understood: when the curve shifts so much to the left that the pull-out voltage crosses the 0 V line, the bridge will remain stuck even if the actuation voltage is removed. A negative charge results in a more tricky behavior. For small charge values, the pull-in and pull-out voltages will shift slightly upward so no stiction is expected. If too much charge accumulates, the pull out of the negative-going actuation will shift through 0 V (Fig. 52.13). For slow, positive-going voltages, this makes no difference. When we increase the voltage on an initially free-standing bridge, it will pull in at the shifted positive pull-in voltage, and return to its upward position when the voltage

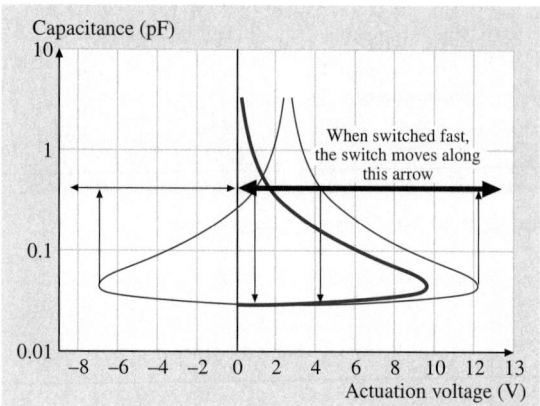

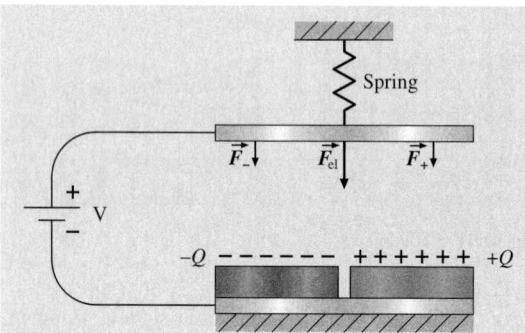

Fig. 52.14 A nonuniform charge can generate attraction of the upper conductor even when the net charge is zero [52.43]

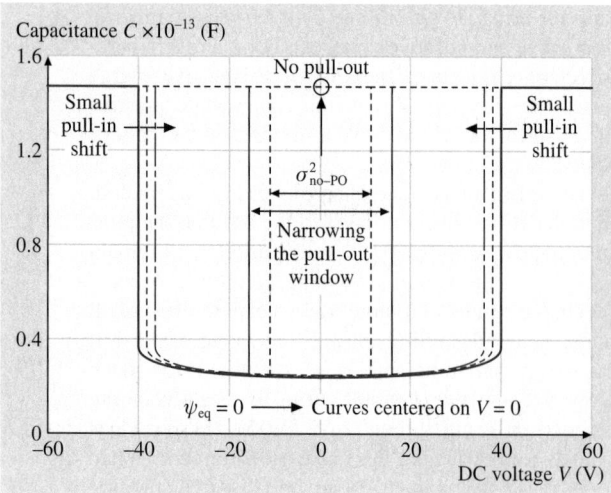

Fig. 52.15 The nonuniform charge causes the pull-out voltage to move to 0 V; in extreme cases the pull-out voltage disappears [52.43]

Fig. 52.13 With square-wave actuation, after the negative pull-out voltage has crossed 0 V due to a parasitic charge, the device moves only along the bold arrow and remains stuck◄

is again slowly decreased to below the pull-out voltage. Only when we go negative in actuation voltage will the bridge be pulled in again at the shifted negative pull-in voltage.

But when we change the actuation voltage quickly, as is often the case when devices are actuated, e.g. by a square wave, the behavior is completely different. We start in the same condition as before, with the bridge up, and increase the voltage quickly to above the pull-in voltage, the bridge will be pulled in after a characteristic time that is related to the mechanical resonance frequency of the device. When we quickly reduce the voltage to 0 V, the bridge stays behind again and is still down when the voltage is 0 V. However, now we are in a completely different situation. Because the pull-out voltage of the negative branch of the curve is positive, the bridge will remain stuck. The bridge does not know whether it came from a negative voltage or from a positive voltage. For a negative voltage it would only pull out above 0 V at the negative-branch pull-out voltage, so it will remain stuck. In this way both positive and negative charges can cause a MEMS device to remain stuck. It is even possible to calculate the approximate magnitude of the charge that is required to cause stiction in either case. The important case in which the negative pull-out voltage crosses zero and causes a device failure has also been verified experimentally [52.42].

Not only charge injection due to high voltages across the dielectric may cause this kind of reliability problems to emerge; ionizing radiation can also play a similar role. It has to be kept in mind however that, if the bridge or other moving structure does not intentionally or accidentally hit the surface below, stiction can only take place when the charge causes a shift that is so large that the much higher pull-in voltage crosses 0 V from either side.

The previous explanation is valid only for uniform charge distributions (either bulk or surface charge). *Rottenberg* et al. [52.43] have extended the theory to deal with nonuniform charge distributions that give rise to effects that have also been experimentally observed. The nonuniform distribution can cause a nonzero force even when the net charge on the dielectric is zero (Fig. 52.14). This causes a narrowing of the C–V curve so that the positive and negative pull-out and pull-in are closer together (Fig. 52.15).

In conclusion, even relatively small charge values of both polarities that are present in or on a dielectric may affect the device reliability when the moving part intentionally or accidentally hits the lower surface. If this is not the case, stiction can only occur when the C–V curve is shifted so much that the pull-in voltage crosses 0 V.

Reduction of the sensitivity for charging phenomena in MEMS devices can be accomplished in different ways. The most obvious way is the reduction of charging by optimizing the dielectrics for low charging. This can be done either by making a very-high-quality dielectric with few trapping sites, or by increasing the trap density but making sure that the average time a charge remains in a trap is very short (the *leaky dielectric*). Another option is to design the dielectrics that are prone to

charging out of the device concept. In RF MEMS this can be done with separate actuation pads where no dielectric is located so that no voltage acts across the high-*k* dielectric. In the accelerometer example of Fig. 52.10, this was done by removing the dielectric altogether. The third option, but probably the least satisfactory because of its added complexity and device cost, is to use advanced actuation schemes to reduce charging. *Goldsmith* et al. [52.38] use an advanced actuation waveform with a high initial voltage to pull the bridge down and a lower *holding* voltage to keep it down, so that charging is reduced. Polarity reversal is also used to sweep out the charges that are accumulating [52.44]. Both of these are useful with actuation signals that are known in advance and do not depend on the particular application in mind.

52.3 Creep, Fatigue, Wear, and Packaging-Related Failures

52.3.1 Creep

Creep is the permanent and irreversible deformation of a ductile material occurring below the yield strength. Its importance depends on temperature, mechanical stress, time, and material composition. It is easy to think of a solid metal as having a well-defined yield strength below which it does not flow and above which flow is rapid. This is unfortunately true only at absolute zero. Above this temperature, plastic flow is a dynamic process, where the strength of solids depends on strain and temperature. The irreversible time-dependent part of a strain ε is called creep.

The strain rate $\partial \varepsilon / \partial t$ (elongation speed) is extremely temperature-sensitive. It can often be approximated by a rate equation

$$\frac{\partial \varepsilon}{\partial t} = A \exp\left(-\frac{q}{kT}\right) , \tag{52.3}$$

where A and q are the fit parameters, k is Boltzmann's constant and T is the absolute temperature.

Thermal vibrations aid the applied stress in overcoming specific barriers (e.g. obstacles) to induce plastic flow in the metal. Slip, climb, shear on grain boundaries, and vacancy diffusion inside the grain (the so-called Nabarro–Herring effect) are the mechanisms responsible for deformation in metals [52.46]. The last of these is significant only at very high temperatures and under very small stresses. The normal operating temperatures in most practical engineering applications are

too low and usually the operating stresses are too high for the Nabarro–Herring mechanism to occur, so we can conclude that creep deformation in metals depends primarily on dislocation movement (slip, climb and grain-boundary shearing).

Three creep stages can be distinguished: primary, secondary (so-called steady-state creep) and tertiary creep, which ends up with fracture (Fig. 52.16). Especially devices made out of low-melting-point metals are sensitive to creep. This is the case because the melting temperature of a material has a major influence on its creep behavior. There exists a threshold temperature,

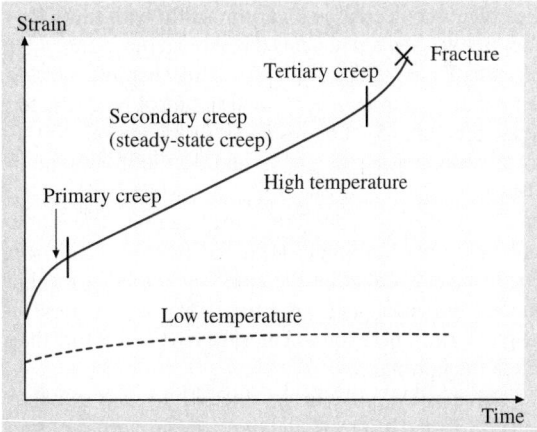

Fig. 52.16 Low-temperature and high-temperature creep under a constant stress [52.45]

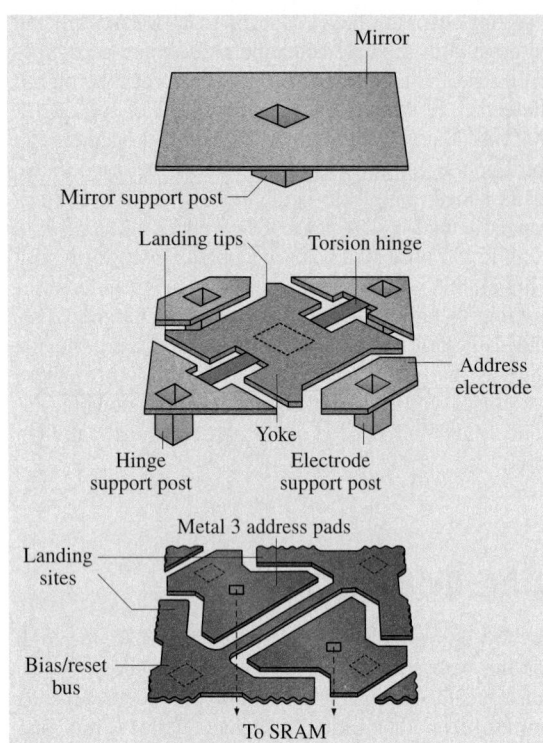

Fig. 52.17 Digital micromirror device for projection displays (courtesy of Texas Instruments)

Creep in MEMS

Creep is well characterized at the macroscopic level (usually under constant-stress/constant-temperature conditions), but has not been studied extensively in microsystems and therefore it can represent a reliability problem in MEMS. Many microdevices are fabricated by adopting low-melting-point metals (in particular aluminium). In most recent research work, long-term or accelerated reliability tests (i.e. creep tests) seldom attract much attention. In fact, feasibility studies are rarely conducted when starting research work (e.g. academic research). However, reliability became an issue in applications, where durability and reproducibility are a key element for commercial success. Creep can form a danger for commercial devices fabricated from low-melting-point metals, especially when operating at higher temperatures and stresses. Therefore, at an early stage of the design of metal MEMS, creep should be taken into account.

An important source of information on creep in MEMS is the work done by Texas Instruments on the digital micromirror device (DMD). *Van Kassel* et al. [52.48] have shown that creep was a problem in their micromirrors. This reliability problem was termed the hinge memory effect. Figure 52.18 depicts the sketch of a DMD. A single mirror is mounted on hinges and electrostatic attraction between the mirror structure and a semiconductor memory cell below is responsible for a $\pm 10°$ tilt. Hinge memory occurs when a mirror is operated in the same direction for a long period of time (high

which divides the deformation of a material by creep into two parts: high-temperature and low-temperature creep (Fig. 52.17). In the regime of high temperature, creep is a rather fast deformation and depends strongly on temperature, whereas in the low-temperature regime, the creep rate decreases logarithmically with time. It is commonly accepted that this threshold temperature T_c is given by $T_c = 0.5T_m$, where T_m is the absolute melting temperature, although each material must be evaluated separately [52.45]. For instance, T_c is around 200 °C for aluminium and its alloys and 315 °C for titanium alloys.

Shape changes caused by creep can be undesirable and represent a limiting factor in the lifetime of a device. For instance, blades on the spinning rotors in turbine engines are endangered by creep. They slowly grow in length during operation and must be replaced before they touch the housing.

Some MEMS, like for instance RF MEMS switches can be affected by creep, and the creep properties of thin metal films are not necessarily the same as those of the bulk material.

Fig. 52.18 SEM image of a test specimen [52.47]

duty cycle). A residual tilt when the voltage is removed appears over time. The lifetime of the mirror is reached when the residual tilt becomes excessive (dead pixels). *Van Kassel* et al. [52.48] related this effect to creep of the hinge material. The lifetime was increased by selecting more creep-resistant materials and by improving the fabrication process. The hinge memory effect was thermally activated. At 65 °C and high duty cycle, mirrors could operate for about 5000 h continuously, while at

45 °C they remained functional for more than 100 000 h even at high duty cycles.

Hoo-Jeong Lee et al. [52.47] presented a test specimen for measuring fatigue and creep in microscale structures. They patterned wafers in the shape of common (macroscopic) tensile test specimens with 50-µm-wide 500-µm-long metal beams (Fig. 52.19). Al and Al (1.5 at. %Ti) beams were tested in a piezoactuator-driven test apparatus. In stress relaxation tests (the samples were quickly loaded to a certain stress and then held at this strain while monitoring the change of stress), *Hoo-Jeong Lee* et al. [52.47] observed a load drop of 56% and 15% for Al and AlTi beams, respectively. The measurements were carried out for 10 min at room temperature. Figure 52.20 shows these results. *Hoo-Jeong Lee* et al. believe that the difference in the amount of relaxation was from Al_3Ti precipitates that are formed at the Al grain boundaries in the alloy samples. They showed that the microstructure accounted for the relaxation difference in the investigated films.

Cho et al. [52.49] presented creep data at elevated temperatures of electrodeposited lithography, electroforming and molding (LIGA, from the German initials) nickel dog-bone-shaped microsamples. Figure 52.21 presents creep tests conducted at 265 °C and 290 °C at stresses significantly below the yield strength. The results clearly show the primary-creep regime, where the creep rate decreases with time, and a steady-state regime, where strain rate is constant in time. *Cho* et al. showed that the time needed to attain 0.5% creep strain (considered as being a significant deformation) was more that 6 h at 265 °C, 1 h at 290 °C and only about 4 min at 400 °C.

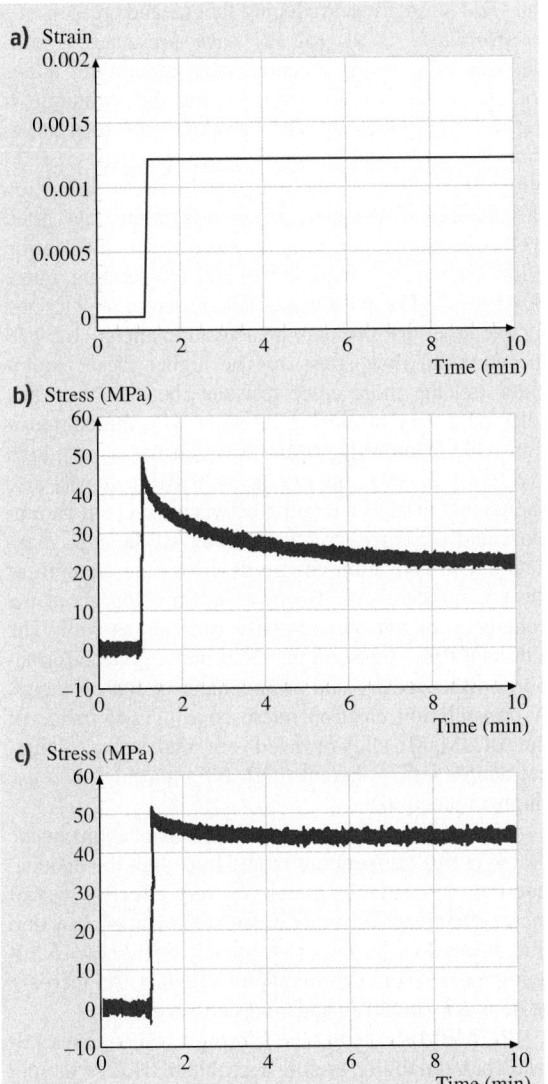

Fig. 52.19a-c Strain change with time (**a**) and corresponding stress change with time for Al beams (**b**) and AlTi beams (**c**) [52.47]

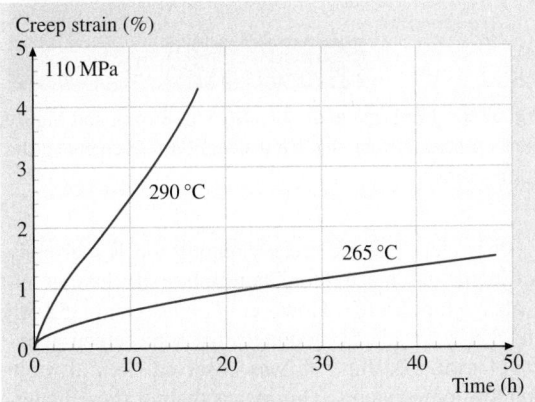

Fig. 52.20 Creep curves at 265 °C and 290 °C under the stress of 110 MPa [52.49]

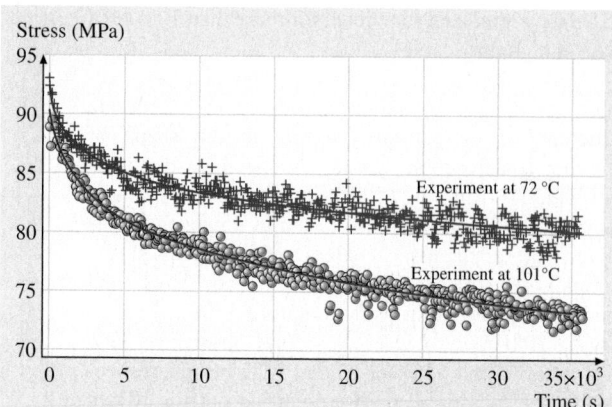

Fig. 52.21 Stress relaxations for Al(0.4 at. %Cu) 5-μm-thick film at 72 and 101 °C. *Solid lines* represent the fitted creep mechanism: dislocation glide limited by obstacles [52.52]

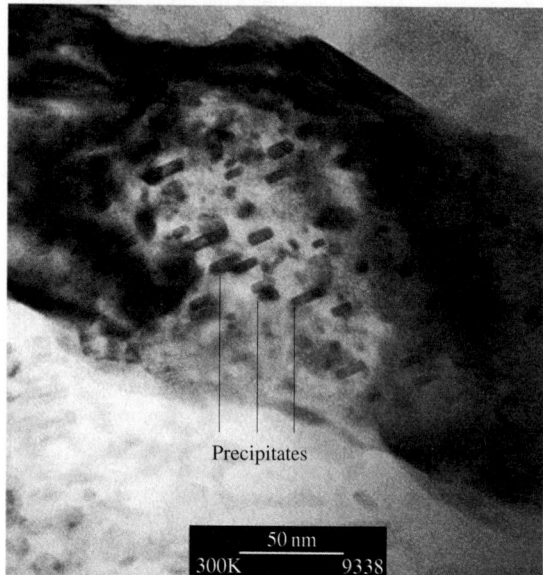

Fig. 52.22 TEM image of AlCuMgMn. Strong and highly dense precipitates hinder the dislocation movement in the grain interior

It is clear that the creep phenomenon is thermally activated and much more harmful to devices at elevated temperatures. However, *Larson* et al. [52.50] and *Yin* et al. [52.51] reported that some creep activity of nanocrystalline Ni was observed even at room and low temperatures. This means that for those materials care must be taken under almost any temperature conditions.

Hardened Pt alloys for MEMS have been presented by *Brazzle* et al. [52.53] to serve as creep-resistant materials. Solution-hardened Pt alloys for improved reliability of electromagnetically actuated micromirror devices exhibited precision angular displacement without observable fatigue and creep when used in flexures. *Brazzle* et al. [52.53] found that alloy 851 (79%Pt–15%Rh – 6%Ru) and Pt(10%Ir) were very hard flexure materials. The alloying components (rhodium, ruthenium and iridium) impede dislocation movement, thereby increasing the yield strength and reducing fatigue and creep.

Modlinski et al. [52.52] have presented a simple way of selecting creep-resistant aluminium alloys for use in RF MEMS switches, but the technique is generically applicable. They propose substrate curvature measurements to study creep properties of Al alloy films using isothermal tensile stress relaxation (Fig. 52.22). The tensile stress relaxation data fitted well to a known deformation mechanism (dislocation glide limited by obstacles) in the temperature range 60–110 °C. The Al alloys studied were characterized by two creep parameters: the activation energy ΔF and the athermal flow stress τ. The higher these parameters are, the more creep-resistant the alloy is. They selected a very promising Al alloy with the composition Al93.5Cu4.4Mg1.5Mn0.6 which has a very high ΔF (6.4 ± 2.5 eV) and τ (920 ± 420 MPa). It was also shown that there is a relation between the creep parameters and the microstructure of the Al alloys in more detail. The activation energy describes the strength of the precipitates, and also gives information about the coherency of the second-phase particles as well. The athermal flow stress, on the other hand, gives information on the density and arrangement of the obstacles. A transmission electron microscope (TEM) image of the AlCuMgMn alloy revealed very small, strong, coherent and/or semi-coherent (high-ΔF) and highly dense (high-τ) precipitates.

Creep is caused by the motion of dislocations inside grains in this temperature range. Impeding the dislocation motion inside the grains is a very effective way of increasing creep resistivity of an Al99.6Cu0.4 alloy thin film. It was done by a macroscopically well-known hardening mechanism: precipitation hardening. An increase of both ΔF (by 50%) and τ was observed.

RF MEMS switches are starting to be commercialized, but reliability is still a problem. Hot switching and the impact of RF power on the reliability of the MEMS switches has been extensively investigated, mainly by simulations. *Jensen* et al. [52.54], *Rottenberg* et al. [52.55] and many others reported that temperatures due to RF power present in some parts of a movable metal bridge are expected to be as high as 250 °C [52.54] and can easily exceed 100 °C. The high residual stresses (50–200 MPa) present in an RF MEMS bridge structure and the high temperatures caused by RF power may lead to creep deformation, deteriorating the functionality of a switch over time.

Creep conclusion

At the macroscale level, creep tests are performed using tensile specimens subjected to constant stress and temperature. These tests are rather easy to prepare and perform. In microscale experiments, the situation is very different. There are no standardized specimens, and there is no standard way of investigating creep in microdevices. Some perform the tests on real MEMS devices (DMD); some try to mimic macroscopic tensile test specimens and use them in the microworld. A very good idea, although not often used, is to implement well-known macroscopic techniques to render a material more creep-resistant. For example, hardening mechanisms such as solution hardening or precipitation hardening are interesting in this respect. These treatments produce stronger obstacles and hence decrease dislocation motion.

The microstructure (grain sizes, precipitates in case of alloys) of deposited films is very process-dependent. There are no standard deposited films. For instance, Al alloy films will exhibit different creep behavior when sputtered under different conditions. Their properties depend on the deposition methods as well as conditions. This variation is a considerable challenge for microscopic creep research.

A human factor is certainly an issue in understanding the state of the art of creep in MEMS. Most engineers and scientists dealing with MEMS are silicon- or, in general semiconductor-oriented. Creep is not really an issue in monocrystalline silicon, poly-silicon or silicon–germanium. There is probably a lack of know-how on plastic deformation in the MEMS society.

Research work in a field of metal plasticity in MEMS has been started. Increasing numbers of publications on understanding and investigating creep phenomenon on the microscale are being issued. MEMS engineers and scientists are facing new problems and challenges in metal MEMS, but fundamental research programs, discussions and exchange of views and knowledge of the plastic deformation of metals between specialists will speed up research in this field.

52.3.2 Fatigue

Fatigue of metals

The only large source of information on fatigue of metal MEMS devices is again the research carried out by Texas Instruments of the digital micromirror device (DMD). Macroscopically, the lifetime of many metal systems is limited by fatigue. A large, repetitive mechanical stress causes the accumulation of dislocation defects at the surface, resulting in fatigue cracks. In microscopic systems, the structural parts are only one, or a few grains thick, with very little dislocations. Therefore, not enough damage can accumulate at the surface to cause fatigue failures. It has been found indeed that MEMS devices suffer much less from fatigue than their macroscopic counterparts [52.56].

Fatigue behavior of materials is often shown in a so-called S-curve, in which the number of stress cycles to fracture is shown as a function of the applied maximum stress level. We usually find that at very high stresses only a small number of cycles is required to fracture the device, the low-cycle fatigue regime (Fig. 52.23). For lower stresses, the number of cycles increases. For very long low-stress cycling times, we find that different materials have different properties. Most steels e.g. have a fatigue limit (Fig. 52.23a). Below this limit, the structure survives forever. Other materials, like aluminium, do not have such a threshold. The curve decreases slowly, even for very large numbers of cycles (Fig. 52.23b). Brittle materials such as ceramics do not have a cycling fatigue effect; below the fracture strength the lifetime is infinite, above it, they immediately break without any cyclic stress effect (Fig. 52.23c).

It has been found that high-cycle fatigue is thin films is very unlikely, as explained above. Aluminium RF MEMS switches have likewise never been reported to suffer from high-cycle fatigue [52.57]. An interesting feature of microscopic structures however is that low-cycle fatigue has been found in aluminium [52.58, 59].

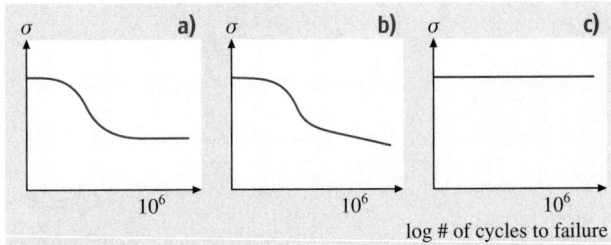

Fig. 52.23 (**a**) S-curve with threshold; (**b**) S-curve without threshold; (**c**) S-curve for ceramic

This raises the question whether in thin films at very high cyclic stresses new defects are being formed, while at lower stresses only existing defects can move to the surface (and there are too few of them to do any harm). More research is definitely needed to figure out how the process really behaves, and where the turning point is between microscopic and macroscopic fatigue behavior. For the moment we can only say qualitatively that metals that macroscopically tend to have no threshold for fatigue do have such a threshold when used as thin films. Therefore the S-curve for thin-film aluminium resembles that of bulk steel (Fig. 52.23a) instead of bulk aluminium (Fig. 52.23b).

Fatigue of silicon

From the early days of MEMS technology on, silicon in its mono- and polycrystalline form has been extensively used as a structural material. Silicon is a brittle material, and as such, should not intrinsically suffer degradation from fatigue (Fig. 52.23c). However, in air it is always covered with a native oxide, and concern was raised that this oxide might be susceptible to a stress corrosion cracking (SCC) phenomenon that is well known in glass technology [52.61]. SCC is the stress-assisted hydrolysis of the SiO_2 bonds at the crack tip under the influence of tensile stress and ambient humidity, causing crack propagation because atomic bonds are broken.

The phenomenon is thought to happen in the following way (Fig. 52.24). A small crack at the surface in the SiO_2 native oxide grows slightly due to SCC under the influence of a tensile stress, so that the silicon below the crack is slightly closer to the surface and oxidizes. Now the crack can propagate through the extended oxide layer, causing a new oxidation step in the silicon below. This continues until the whole structure fractures.

Connally et al. [52.62–66] initiated research into SCC in silicon microstructures, and indeed found a large humidity influence on the crack propagation rate. *Muhlstein* et al. [52.67] measured a more realistic configuration in which they did not only monitor crack propagation but also took into account crack initiation. Another advantage over the earlier measurements was that the stress intensity factor at the tip was not a required parameter to have a meaningful description. Similar measurements were also reported by *Kapels* et al. [52.68] and *Kahn* et al. [52.69].

It is likely that there is a critical stress in the silicon device, above which SCC may occur. The native oxide of silicon is normally under a compressive stress [52.6] and the applied external stress should be so large that this stress has to become tensile for SCC to occur at all. If the stress in the silicon is kept below this transition point, where the oxide surface layer turns from its intrinsic compressive stress to tensile, no problems due to SCC are expected.

Bagdahn and *Sharpe* [52.60] put all the data available, including their own, in one table and showed that SCC cannot explain all the properties of the behavior that has been found. Being a chemical, stress-assisted process, SCC should not have a pronounced cycling-speed dependency; only the absolute time that the stress is applied counts towards the reaction rate. *Bagdahn* and *Sharpe* [52.60] showed that the data available provide a definite trend of the number of cycles to failure as a function of the applied stress, which contradicts the SCC assumption. We see in Fig. 52.25 that these effects only show up at very high stresses, near the fracture strength of silicon. Therefore in practical MEMS devices the fatigue of silicon may not be a big issue because it is good design practice to keep some margin to the fracture strength for brittle materials anyway. However, in harsh environments it may turn out to be a problem.

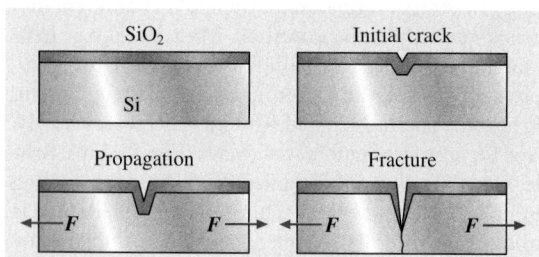

Fig. 52.24 SSC failure mechanism in silicon

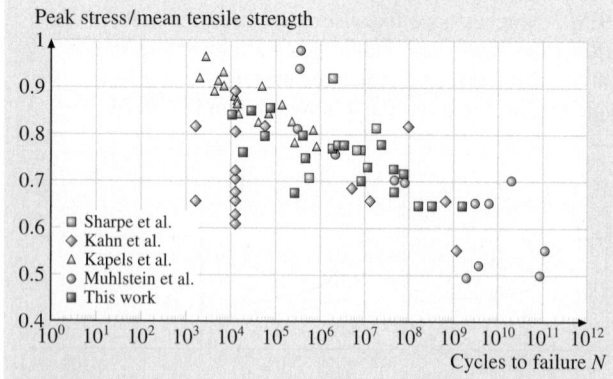

Fig. 52.25 *Bagdahn* and *Sharpe* [52.60] show that a definite trend appears if the applied stress is plotted as a function of the number of cycles to failure, contradicting the basic assumption of SCC

52.3.3 Wear

Wear is one of the major limiting factors in MEMS technology. After its occurrence had been shown by *Gabriel* et al. [52.71] and *Mehregany* et al. [52.72] in silicon micromotors in 1990, it was quickly recognized that wear is a severe limiting factor for almost all MEMS devices incorporating rubbing/sliding surfaces. At Sandia National Laboratories, a large effort was spent on understanding this effect, with experiments on a dedicated micromotor test vehicle (Fig. 52.26). This resulted in a quantitative empirical model to predict wear in certain types of micromotors [52.73]. Beautiful focused ion beam (FIB) cross sections in this work show that wear is indeed severe (Fig. 52.27). The wearing mechanism proposed is a kind of *adhesive wear*, where the surfaces adhere at their contacting points and shear off parts of the other surface (Fig. 52.28). Abrasive and corrosive wear take place at higher contact forces, or when a chemical reaction

assists the wear process. A comprehensive overview of their work can be found in their MEMS reliability report [52.74].

A good overview from the middle of the 1990s is given by *Komvopoulos* [52.75]. Studying wear at a more fundamental level is difficult, because typical analytical measurement methods do not cover the whole range of speeds and contact forces in MEMS devices (Fig. 52.29) [52.70]. Atomic force microscopy/friction force microscopy (AFM/FFM) systems are very accurate in mimicking a single asperity, but only at, for MEMS devices, very low speeds [52.76]. A dedicated *tribo-apparatus* measurement system [52.77]

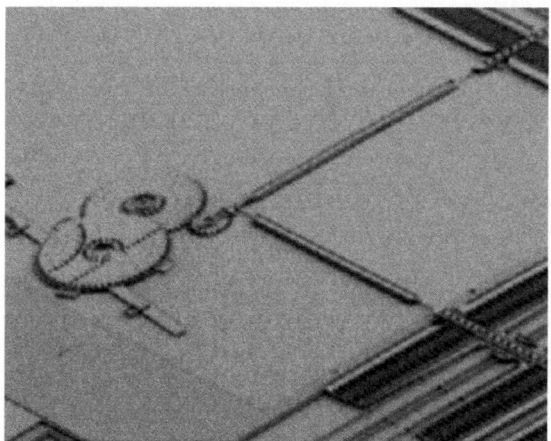

Fig. 52.26 Sandia's micromotor test vehicle to study wear

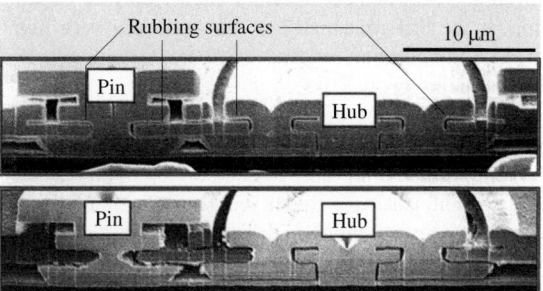

Fig. 52.27 FIB cross section of a new and a worn micromotor

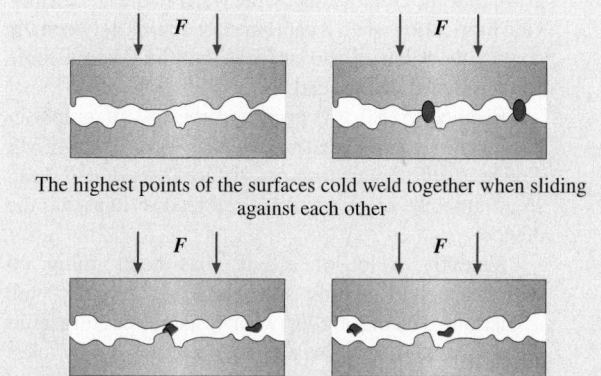

The highest points of the surfaces cold weld together when sliding against each other

Material is transferred between the surfaces. If the augmented asperities become too large, they break off, and debris accumulates between and around the worn surfaces

Fig. 52.28 Principle of adhesive wear

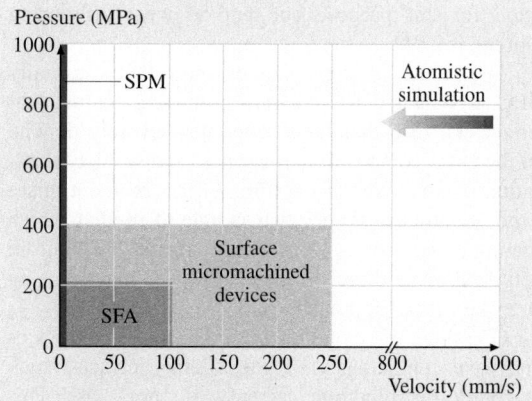

Fig. 52.29 MEMS surfaces have sliding speeds and contact forces not easily assessed by the available analysis methods: scanning probe microscopy (SPM) and surface force apparatus (SFA) [52.70]

fills part of the gap, but a large area remains outside the scope of the different (microscopic) measurement methods. Atomistic simulations can usually only be applied at much higher speeds, because otherwise computing time becomes prohibitively long. As a result, wear is often studied using MEMS devices themselves as the measurement vehicle [52.78–81], but, of course, this is generally not as accurate as one would like.

To counteract wear in MEMS devices, one can follow different routes. Common lubricants used in macroscopic systems (oil) cannot be applied, because the viscous drag of these materials is so large that it prevents high speeds on the microscale, if the devices work at all. In some circumstances, the water in the air can provide lubrication when it capillary condenses between the contacting points of the surfaces, but the friction force is not reduced dramatically.

Rymuza [52.82, 83] proposed the use of graphite-like materials embedded in the wear layer of a MEMS device. As the device loses its surface layer due to wear, small amounts of graphite are released to lubricate the device.

Recently, a lot of activity has been going on with respect to coating silicon MEMS devices with a friction-reducing and/or wear-resistant coating after the device has been released. This coating usually takes the form of a hard, chemically inert layer with a thickness of nanometers. Experiments have been performed using tungsten [52.84], SiN_x [52.85], diamond-like carbon (DLC) [52.86–89] ultra-nanocrystalline diamond (UNCD) [52.90], and SiC [52.91–93], among others. SAM coatings (Sect. 52.2.1) have also been investigated for this purpose but tend to wear off during rubbing [52.94].

The results of these studies show that especially DLC, UNCD and SiC are very promising coating materials, but also that their properties severely depend on the exact composition, roughness and surface termination (passivation) of the film. At this moment, there is no quantitative theory that is able to predict which coating compositions/properties are the best. Therefore optimization of these films is performed mainly using an experimental approach.

One finds that commercial applications of devices that have continuously rubbing/sliding surfaces, such as rotary micromotors, currently do not exist. This means that wear problems in MEMS devices are still so large that designers of commercial products avoid designs where rubbing surfaces have to survive more than a couple of million cycles.

52.3.4 Packaging

Introduction

Many MEMS contain movable and fragile parts. A specific packaging approach is required for protection during fabrication as well as during operation. For instance, standard wafer dicing will, if not destroy the MEMS device, at least introduce some contamination and particles that can lead to failure. Hence packaging must be carried out on the wafer during wafer processing, prior to die singulation. This packaging step is referred to as wafer-level or 0-level packaging. It defines the first protective interface for the MEMS device, achieved through on-wafer encapsulation of the movable parts in a sealed cavity. The cavity must be strong, for many applications hermetic, and equipped with electrical signal feedthroughs. An example of a 0-level packaging scheme for RF MEMS switches, which relies on wafer or chip stacking techniques using benzo-cyclobutene (BCB) as the bonding and sealing material, is shown in Fig. 52.30 [52.95].

Failure modes and failure mode investigation

Important requirements of MEMS packaging are resistance (against shear stress, pull stress, mechanical, thermal loads, etc.), and hermeticity against moisture, liquids, particles or gas contaminants, but also hermeticity to keep a controlled ambient (gas and pressure) inside the package. The most common failure modes that can occur are mainly related to the sealing material itself. For instance, a package can lose its shear or pull resistance in the case of bad adhesion between the sealing material and the MEMS die, although this is rarely the case and is process-related. More important is the loss of hermeticity in the case of an hermetic package. This can happen with the appearance of cracks in the sealing layer or at the bonding interfaces when the package device is subject to thermal shocks, for instance. The difference between the thermal coefficients of expansion of the different materials can induce a very high stress at the bonding interfaces and create some cracks, thus inducing gross leaks.

The importance of this can be seen in the following example. The package of Fig. 52.31b is made of a Si capping chip flip-chip bonded on top of an RF tunable capacitor using a solder seal (SnPb). Figure 52.32 shows the tuning ratio of the capacitor as a function of the actuation voltage, right after packaging and laser dicing, after storage in air, and after drying in dry N_2 at 75 °C, respectively. Compared to the curve obtained right after dicing, the tuning curve after storage in air

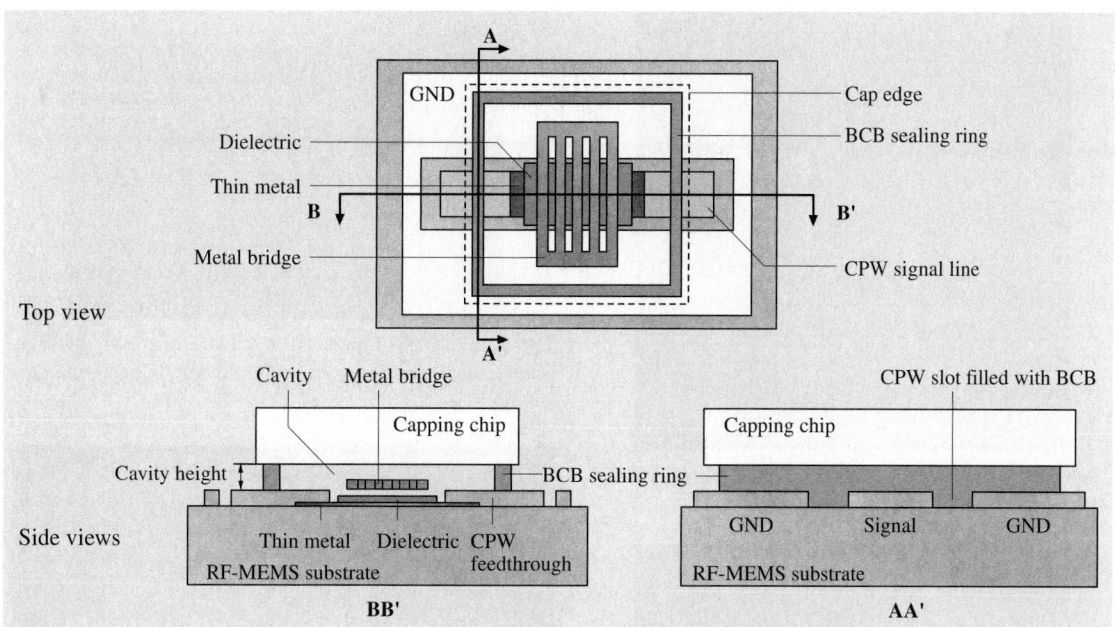

Fig. 52.30 Illustration of a wafer or 0-level packaged RF MEMS switch. The fragile MEMS part is housed in a cavity formed by a capping chip, bonded to the RF MEMS substrate using a layer of BCB

is shifted towards lower actuation voltages and shows nonreversible tuning. This phenomenon could be explained by a moisture-related charging effect of the RF capacitor when the package is stored in ambient air. This is confirmed by the tuning curve obtained after drying the packaged device in dry N_2 at 75 °C: the initial reversible tuning curve is restored. This clearly shows that, in this example, the package is not hermetic to moisture and that this influences the electrical performance of the MEMS device. Possible causes of packaging failure here could be microcracks in the solder sealing ring, bad adhesion between the solder and the MEMS die, particle contamination in the solder, or a local scarcity of solder.

For the same reasons, in the case of a (supposedly) hermetic package containing a controlled ambient (pressure and gas), the loss of hermeticity due to a leak in the sealing material will fill the package with air after a determined period of time, depending on the leak rate. For instance, a package of 100 nl (a few tens of nl are typically encountered in MEMS packaging) with an initial internal pressure of 1 mbar and displaying a leak rate as

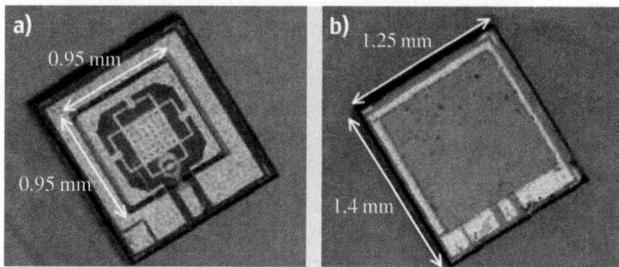

Fig. 52.31a,b Picture of a laser-diced *naked* (**a**) and packaged (**b**) RF tunable capacitor using solder (SnPb) flip-chip assembly

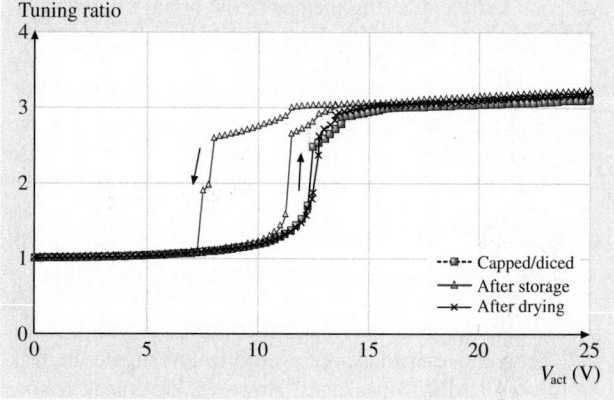

Fig. 52.32 Tuning ratio of the packaged tunable capacitor of Fig. 52.20 right after dicing (*squares*), after storage in ambient air (*triangles*) and after drying in dry N_2 at 75 °C (*crosses*)

Part G | 52.3

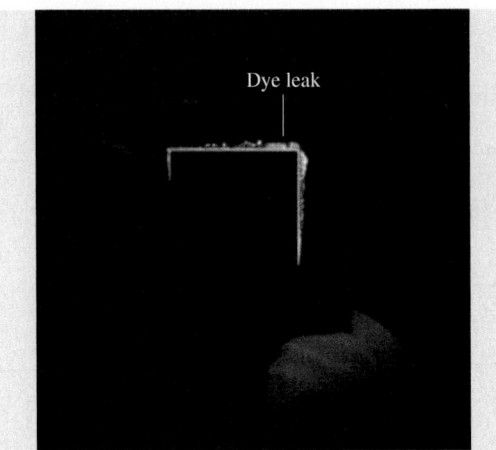

Fig. 52.33 UV inspection of a leaky package using the dye-penetrant method

Fig. 52.34 SAM picture of a Si cap bonded to a Si wafer using polymer. The black areas indicate positions where the bond failed (*arrows*) [52.96]

low as 10^{-10} millibar-liters/s will be filled with air after only 11 days.

Different testing methods exist or have been developed to first test the package (e.g. shear and pull strength, hermeticity, etc.) and second, to investigate the failure mechanisms. One can always refer to the MIL-STD-883 procedure to obtain the testing conditions and the rejection limits for the test [52.97]. This is the case for instance for the pull and shear testing (MIL-STD-883, method 2019.5). Hermeticity (fine leak) testing as specified by the MIL-STD-883 (method 1014.9) is still a problem because the fine-leak test conditions and specifications are clearly not applicable to MEMS packages due to the small cavity volumes they deal with [52.98]. Alternative methods are needed to investigate the leak rate of MEMS packages. However, gross-leak testing

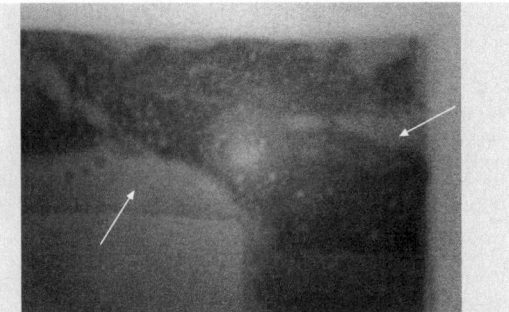

Fig. 52.35 X-ray image of a solder-sealed Si package. White areas are assumed to be solder oxide (*unbonded areas*)

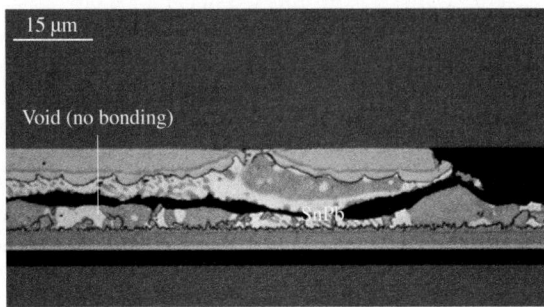

Fig. 52.36 Optical microscopy image of a cross section of a solder-sealed Si package

using fluorocarbon fluids is still applicable and is a good method to localize a leak path (e.g. a crack or a bad bonding) in the sealing material or the package itself. Similar to gross-leak testing is the dye penetrant method, as illustrated in Fig. 52.33. The package is immersed in a pressurized yellow dye (type Zyglo ZL-56 from Magnaflux). Next, the sample is rinsed with water, dried, and sprayed with a developer (type Zyglo ZP-9F). This developer has the property of sucking out the dye from the package if a leak exists, and some yellow fringes are then clearly visible under UV light.

Another method to localize a gross leak in a package is scanning acoustic microscopy (SAM). A picture of a polymer-sealed leaky package is shown in Fig. 52.34 [52.96]. The black areas indicate that there is no adhesion between the sealing layer and the Si chip.

Other methods to look at the bonding interface are X-ray imaging and cross sectioning. Figure 52.35 is an X-ray image of a solder-sealed Si package. Irregularities in the solder ring are visible, and the lighter areas are assumed to be oxidized solder areas, and thus potential leaky paths.

Cross sectioning is of course destructive for the package, but it can also be used to investigate the bond quality, as shown in the example of Fig. 52.36. In this picture, voids in the sealing layer are clearly visible, and are potential source of gross leaks.

Conclusion

Investigating the failure mechanisms of the package is at least as important as for the MEMS itself, because without a good and robust package, most MEMS devices cannot survive operating or storage conditions. Understanding why a package failed will help to improve the next generation of MEMS. A lot of development still has to be done in packaging reliability testing, and this requires dedicated methods of investigation, which is now slowly emerging.

52.4 Conclusions

We have seen that a lot about the reliability of MEMS/NEMS is already known, and that in some cases even macroscopic principles can be applied to improve the reliability of micromachines over the state of the art. The fact that reliability is such a problem is not because MEMS/NEMS are intrinsically unreliable. They are not, and in fact – due to the low mass and high strength of materials on a microscale – the reliability that can be attained is even higher than in many macroscopic situations. The reason that reliability is hampering the commercialization of MEMS has already briefly been addressed in the creep section: to produce any MEMS at all is already a considerable feat, so that reliability issues are often not taken into account in the early stages of development. In the macroscopic world, instead, reliability is one of the key design issues of almost any product.

This sense still has to pervade the MEMS society, although things are changing. Especially in commercial laboratories, *design for reliability* is the new mantra, and independent research and development labs have set up dedicated groups to study fundamental issues in microscale reliability, such as stiction, charging, creep, fatigue, wear and packaging. If this trend continues over the coming years, MEMS may become just as pervasive as microelectronics is now.

References

52.1 W. M. Miller, D. M. Tanner, S. L. Miller, K. A. Peterson: MEMS Reliability. The Challenge and the Promise. In: *Proc. 4th Annual "The Reliability Challenge"* (Dublin, Ireland 1998) pp. 4.1–4.7

52.2 M. R. Douglass: DMD reliability. A MEMS success story, Proc. SPIE **4980**, 1–11 (2003)

52.3 J. Bienstman: From product to production in automotive MEMS: In: *Proc. MicroMechanics Europe*, ed. by R. Puers (Leuven, Belgium 2004) pp. 107–108

52.4 W. M. van Spengen: MEMS reliability from a failure mechanisms perspective, Microelectron. Reliab. **43**, 1049–1060 (2003)

52.5 W. M. van Spengen, R. Puers, R. Mertens, I. De Wolf: High resolution optical investigation of small out-of-plane movements and fast vibrations; characterization and failure analysis of MEMS, Microsyst. Technol. **10**, 89–96 (2004)

52.6 B. Stark (ed): *MEMS Reliability Assurance Guidelines for Space Applications* (National Aeronautics Space Administration (NASA) and Jet Propulsion Laboratory (JPL), California Institute of Technology, Pasadena 1999)

52.7 D. M. Tanner, N. F. Smith, D. J. Bowman, W. P. Eaton, K. A. Peterson: First reliability test of a surface micromachined microengine using SHiMMeR, Proc. SPIE **3224**, 14–23 (1997)

52.8 G. F. LaVigne, S. L. Miller: A Performance Analysis System for MEMS using Automated Imaging Methods. In: *Proc. IEEE Int. Test Conference* (IEEE, Washington DC 1998) pp. 442–447

52.9 P. Krehl, S. Engemann, C. Rembe, E. P. Hofer: High-speed visualization, a powerful diagnostic tool for microactuators – retrospect and prospect, Microsyst. Technol. **5**, 113–132 (1999)

52.10 C. Rembe, L. Muller, R. S. Muller, R. T. Howe: Full three-dimensional motion characterization of a gimballed electrostatic microactuator. In: *Proc. IEEE Annual International Reliability Physics Symposium (IRPS)*, ed. by E. S. Sneyder (IEEE, Orlando, Florida 2001) pp. 91–98

52.11 J. S. Burdess, A. J. Harris, D. Wood, R. J. Pitcher, D. Glennie: A system for the dynamic characterization of microstructures, J. MEMS **6**, 322–328 (1997)

52.12 M. R. Hart, R. A. Conant, K. Y. Lau, R. S. Muller: Stroboscopic interferometer system for dynamic MEMS characterization, J. MEMS **9**, 409–418 (2000)

52.13 M. P. de Boer, J. A. Knapp, T. M. Mayer, T. A. Michalske: The role of interfacial properties on MEMS performance and reliability, Proc. SPIE **3825**, 2–15 (1999)

52.14 C. H. Mastrangelo, C. H. Hsu: Mechanical stability and adhesion of microstructures under capillary forces – part I: Basic theory, J. MEMS **2**, 33–43 (1993)

52.15 C. H. Mastrangelo, C. H. Hsu: Mechanical stability and adhesion of microstructures under capillary forces – part II: Experiments, J. MEMS **2**, 44–55 (1993)

52.16 F. P. Bowden, D. Tabor: *The Friction and Lubrication of Solids* (Clarendon, Oxford 1950)

52.17 R. Maboudian, R. T. Howe: Critical review: Adhesion in surface micromechanical structures, J. Vac. Sci. Technol. B **15**, 1–20 (1997)

52.18 R. Legtenberg, H. A. C. Tilmans, J. Elders, M. Elwenspoek: Stiction of surface micromachined structures after rinsing and drying: Model and investigation of adhesion mechanisms, Sensors Actuat. A **43**, 230–238 (1994)

52.19 N. Tas, T. Sonnenberg, H. Jansen, R. Legtenberg, M. Elwenspoek: Stiction in surface micromachining, J. Micromech. Microeng. **6**, 385–397 (1996)

52.20 J. Israelachvili: *Intermolecular and Surface Forces* (Academic, London 1991)

52.21 K. Komvopoulos: Surface engineering and microtribology for microelectro-mechanical systems, Wear **200**, 305–327 (1996)

52.22 W. M. van Spengen, W. M. R. Puers, I. De Wolf: A physical model to predict stiction in MEMS, J. Micromech. Microeng. **12**, 702–713 (2002)

52.23 J. A. Greenwood, J. B. P. Williamson: Contact of nominally flat surfaces, Proc. R. Soc. A **295**, 300–319 (1966)

52.24 B. Bhushan: Methodology for roughness measurement and contact analysis for optimization of interface roughness, IEEE Trans. Mag. **32**, 1819–1825 (1996)

52.25 B. Bhushan: *Principles and Applications of Tribology* (Wiley, New York 1999)

52.26 L. Kogut, K. Komvopoulos: Analysis of interfacial adhesion based on electrical contact resistance, J. Appl. Phys. **94**, 6386–6390 (2003)

52.27 Y.-P. Zhao, L. S. Wang, T. X. Yu: Mechanics of adhesion in MEMS – a review, J. Adhes. Sci. Technol. **17**, 519–546 (2003)

52.28 J. Elders, H. V. Jansen, M. Elwenspoek: Materials analysis of fluorocarbon films for MEMS applications. In: *Proc. An Investigation of Micro Structures, Sensors, Actuators, Machines and Robotic Systems* (New York, USA 1994) pp. 170–175

52.29 P. F. Man, B. P. Gogoi, H. Mastrangelo: Elimination of post-release adhesion in microstructures using

52.30 U. Srinivasan, M. R. Houston, R. T. Howe, R. Maboudian: Alkyltrichlorosilane-based self-assembled monolayer films for stiction reduction in silicon micromachines, J. MEMS **7**, 252–260 (1998)

52.31 R. Maboudian, W. R. Ashurst, C. Carraro: Self-assembled monoayers as anti-stiction coatings for MEMS: Characteristics and recent developments, Sensors Actuat. A **82**, 219–223 (2000)

52.32 B.-H. Kim, C.-H. Oh, K. Chun, T.-D. Chung, J.-W. Byun, Y.-S. Lee: A new class of surface modifiers for stiction reduction. In: *Proc. 12th International Conference on Micro Electro Mechanical Systems* (Piscataway, USA 1999) pp. 189–193

52.33 K. A. Peterson, P. Tangyunyong, A. A. Pimentel: Failure analysis of surface micromachined microengines, Proc. SPIE **3512**, 190–200 (1998)

52.34 J. Wibbeler, G. Pfeifer, M. Hietschold: Parasitic charging of dielectric surfaces in capacitive microelectromechanical systems (MEMS), Sensors Actuat. A **71**, 74–80 (1998)

52.35 A. R. Knudson, S. Buchner, P. McDonald, W. J. Stapor, A. B. Campbell, K. S. Grabowski, D. L. Knies: The effects of radiation on MEMS accelerometers, IEEE Trans. Nucl. Sci. **43**, 3122–3126 (1996)

52.36 C. I. Lee, A. H. Johnston, W. C. Tang, C. E. Barnes: Total dose effects on microelectromechanical systems (MEMS): Accelerometers, IEEE Trans. Nucl. Sci. **43**, 3127–3132 (1996)

52.37 G. M. Rebeiz: *RF MEMS: Theory, Design and Technology* (Wiley, Hoboken 2003)

52.38 C. L. Goldsmith, J. Ehmke, A. Malczewski, B. Pillans, S. Eshelman, Z. Yao, J. Brank, M. Eberly: Lifetime characterization of capacitive RF MEMS switches. In: *Proc. IEEE MTT-S International Microwave Symposium* (IEEE, New York 2001) pp. 227–230

52.39 S. Zafar, A. Callegari, E. Gusev, M. V. P. Fischetti: Charge trapping in high k gate dielectric stacks. In: *Proc. International Electron Devices Meeting (IEDM)*, ed. by S. Ikeda (IEEE, San Francisco, California 2002) pp. 517–520

52.40 J. C. Ehmke, C. L. Goldsmith, Z. J. Yao, S. M. Eshelman: Method and apparatus for switching high frequency signals, Texas Instruments Inc., US Patent 6391675 (1999)

52.41 J. R. Reid: Simulation and measurement of dielectric charging in electrostatically actuated capacitive microwave switches. In: *Proc. Modeling and Simulation of Microsystems (MSM)*, ed. by M. Laudon, B. Romanowicz (NSTI, San Juan, Puerto Rico 2002) pp. 250–253

52.42 W. M. van Spengen, R. Puers, R. Mertens, I. De Wolf: A comprehensive model to predict the charging and reliability of capacitive RF MEMS switches, Micromech. Microeng. **14**, 514–521 (2004)

52.43 X. Rottenberg, B. Nauwelaers, W. De Raedt, H. A. C. Tilmans: Distributed dielectric charging and

conformal fluorocarbon coatings, J. MEMS **6**, 25–34 (1997)

its impact on RF MEMS devices. In: *Proc. 34th European Microwave Conference* (Artech house, Amsterdam, The Netherlands 2004) pp. 77–80

52.44 W. M. van Spengen, R. Puers, I. De Wolf: RF MEMS Reliability – The Challenge, the Physics, and the Reward. In: *Proc. MME*, ed. by R. Puers (Leuven, Belgium 2004) pp. 319–325

52.45 H. E. Boyer: *Atlas of Creep and Stress-Rupture Curves* (ASM International, Metals Park 1988)

52.46 F. N. R. Nabarro, H. L. de Villiers: *The Physics of Creep* (Taylor Francis, New York 1995)

52.47 Hoo-Jeong Lee, G. Cornella, J. C. Bravman: Stress relaxation of free-standing aluminum beams for microelectromechanical systems applications, Appl. Phys. Lett. **76**, 3415–3417 (2000)

52.48 P. F. Van Kessel, L. J. Hornbeck, R. E. Meier, M. R. Douglass: A MEMS-based projection display, Proc. IEEE **86**, 1687–1704 (1998)

52.49 H. S. Cho, K. J. Hemker, K. Lian, J. Goettert, G. Dirras: Measured mechanical properties of LIGA Ni structures, Sensors Actuat. A **103**, 59–63 (2003)

52.50 K. P. Larsen, A. A. Rasmussen, J. T. Ravnkilde, M. Ginnerup, O. Hansen: MEMS devices for bending test: Measurements of fatigue and creep of electroplated nickel, Sensors Actuat. A **103**, 156–164 (2003)

52.51 W. M. Yin, S. H. Whang, R. Morshams, C. H. P. Xiao: Creep behavior of nanocrystalline nickel at 290 and 373 K, Mater. Sci. Eng. A **2301**, 18–22 (2001)

52.52 R. Modlinski, A. Witvrouw, P. Ratchev, V. Simons, A. Jourdain, H. A. C. Tilmans, R. Puers, J. den Toonder, I. De Wolf: Creep as a reliability problem in MEMS, J. Microelectron. Reliab. **44**, 1733–1738 (2004)

52.53 J. D. Brazzle, W. P. Taylor, B. Ganesh, J. J. Price, J. J. Bernstein: Solution hardened platinum alloy flexure materials for improved performance and reliability of MEMS devices, J. Micromech. Microeng. **15**, 43–48 (2005)

52.54 B. D. Jensen, J. L. Volakis, K. Saitou, K. Kurabayashi: Impact of skin effect on thermal behavior of RF-MEMS switches. In: *The 6th ASME-JSME Conf.*, ed. by S. Nishio, A. Lavine (Kona, Hawaii 2003) TED-AJ03-420

52.55 X. Rottenberg, B. Nauwelaers, W. De Raedt, H. A. C. Tilmans: RF current and power handling of RF-MEMS shunt switches. In: *Proceedings of MEMSWAVE* (Sweden 2004) pp. C1–C4

52.56 M. R. Douglass: Lifetime estimates and unique failure mechanisms of the digital micromirror device. In: *36th International Reliability Physics Symposium (IRPS)*, ed. by A. N. Campbell (IEEE, Reno, Nevada 1998) pp. 9–16

52.57 J. J. Yao: RF MEMS from a device perspective, J. Micromech. Microeng. **10**, R9–R38 (2000)

52.58 D. T. Read, J. W. Dally: Fatigue of microlithographically patterned free-standing aluminum thin film under axial stresses, J. Electron. Packaging **117**, 1–6 (1995)

52.59 G. Cornella, R. P. Vinci, R. Suryanarayanan Iyer, R. H. Dauskardt, J. C. Bravman: Observations of low-cycle fatigue of Al thin films for MEMS applications, Mater. Res. Soc. Symp. Proc. **518**, 81–86 (1998)

52.60 J. Bagdahn, W. N. Sharpe Jr.: Fatigue of polysilicon silicon under long-term cyclic loading, Sensors Actuat. A **103**, 9–15 (2003)

52.61 S. M. Wiederhorn, E. R. Fuller Jr., R. Thomson: Micromechanisms of crack growth in ceramics and glasses in corrosive environments, Met. Sci. **14**, 450–458 (1980)

52.62 W. W. van Arsdell, S. B. Brown: Subcritical crack growth in silicon MEMS, J. MEMS **8**, 319–327 (1999)

52.63 J. A. Connally, S. B. Brown: Micromechanical fatigue testing. In: *TRANSDUCERS '91, International Conference on Solid-State Sensors and Actuators Digest* (New York, USA 1991) pp. 953–956

52.64 S. B. Brown, W. van Arsdell, C. L. Muhlstein: Materials reliability in MEMS devices. In: *TRANSDUCERS '97, International Conference on Solid-State Sensors and Actuators Digest*, Vol. 1 (New York, USA 1997) pp. 591–594

52.65 S. B. Brown, E. Jansen: Reliability and long term stability of MEMS. In: *Summer Topical Meetings Digest, Optical MEMS and their Applications* (New York, USA 1996) pp. 9–10

52.66 S. B. Brown, G. Povirk, J. Connally: Measurement of slow crack growth in silicon and nickel micromechanical devices. In: *Proc. Micro Electro Mechanical Systems. An Investigation of Micro Structures, Sensors, Actuators, Machines and Systems* (IEEE, New York 1993) pp. 99–102

52.67 C. L. Muhlstein, S. B. Brown, R. O. Ritchie: High cycle fatigue and durability of polycrystalline silicon thin films in ambient air, Sensors Actuat. A **94**, 177–188 (2001)

52.68 H. Kapels, R. Aigner, J. Binder: Fracture strength and fatigue of polysilicon determined by a novel thermal actuator, Trans. Electron. Dev. **47**, 1522–1528 (2000)

52.69 H. Kahn, N. Tayebi, R. Ballerini, R. L. Mullen, A. H. Heuer: Fracture toughness of polysilicon MEMS devices, Sensors Actuat. A **82**, 274–280 (2000)

52.70 A. D. Romig Jr., M. T. Dugger, P. J. McWorther: Materials issues in microelectromechanical devices: Science, engineering, manufacturability and reliability, Acta Materialia **51**, 5837–5866 (2003)

52.71 K. J. Gabriel, F. Behi, R. Mahadevan: In-situ friction and wear measurements in integrated polysilicon mechanisms, Sensors Actuat. A **A21-A23**, 184–188 (1990)

52.72 M. Mehregany, S. D. Senturia, J. H. Lang: Friction and wear in microfabricated harmonic side-drive motors. In: *Technical Digest IEEE Solid-State Sensor and Actuator Workshop* (IEEE 1990) pp. 17–22

52.73 D. M. Tanner, W. M. Miller, W. P. Eaton, L. W. Irwin, K. A. Peterson, M. T. Dugger, D. C. Senft, N. F. Smith,

P. Tanyunyong, S. L. Miller: The effect of frequency on the lifetime of a surface micromachined microengine driving a load. In: *International Reliability Physics Symposium Proceedings (IRPS)*, ed. by A. N. Campbell (IEEE, Reno, Nevada 1998) pp. 26–35

52.74   D. M. Tanner: *MEMS Reliability: Infrastructure, Test Structures, Experiments and Failure Modes*, Sandia Rep. (Sandia National Laboratories, Livermore 2000), available from National Technical Information Service, U.S. Department of Commerce, Springfield VA 22161, USA or http://www.sandia.gov/mstc/technologies/micromachines/tech-info/bibliography/biblog_char.html

52.75   K. Komvopoulos: Surface engineering and microtribology for microelectro-mechanical systems, Wear **200**, 305–327 (1996)

52.76   J. A. Ruan, B. Bhushan: Atomic-scale friction measurement using friction force microscopy. 1. General – Principles and new measurement techniques, J. Tribol. **116**, 378–388 (1994)

52.77   H. Lui, B. Bhushan: Adhesion and friction studies of microelectromechanical systems/nanoelectromechanical systems materials using a novel triboapparatus, J. Vac. Sci. Technol. A **21**, 1528–1538 (2003)

52.78   M. P. de Boer: A hinged-pad test structure for sliding friction measurement in micromachining, Proc. SPIE **3512**, 241–250 (1998)

52.79   S. L. Miller, J. J. Sniegowski, G. LaVigne, P. J. McWorther: Friction in surface micromachined microengines, Proc. SPIE **2722**, 197–204 (1996)

52.80   D. C. Senft, M. T. Dugger: Friction and wear in surface micromachined tribological test devices, Proc. SPIE **3224**, 31–38 (1997)

52.81   M. P. de Boer, T. M. Mayer: Tribology of MEMS, MSR Bull. **26**, 302–304 (2001)

52.82   Z. Rymuza: Control tribological and mechanical properties of MEMS surfaces. Part 1: Critical review, Microsyst. Technol. **5**, 173–180 (1999)

52.83   Z. Rymuza, M. Misiak, L. Kuhn, K. Schmidt-Szalowski, Z. Ranek-Boroch: Control tribological and mechanical properties of MEMS surfaces. Part 2: Nanomechanical behavior of self-lubricating ultrathin films, Microsyst. Technol. **5**, 181–188 (1999)

52.84   J. G. Fleming, S. S. Mani, J. J. Sniegowski, R. S. Blewer: Tungsten coating for improved wear resistance and reliability of microelectromechanical devices, US Patent 6290859 (2001)

52.85   U. Beerschwinger, D. Mathieson, R. L. Reuben, S. J. Yang: A study of wear on MEMS contact morphologies, J. Micromech. Microeng. **4**, 95–105 (1994)

52.86   R. Bandorf, H. Lüthje, T. Staedler: Influencing factors on microtribology of DLC films for MEMS and microactuators, Diamond Rel. Mater. **13**, 1491–1493 (2004)

52.87   A. P. Musinho, R. D. Mansano, M. Massi, J. M. Jaramillo: Micro-machine fabrication using diamond-like carbon films, Diamond Rel. Mater. **12**, 1041–1044 (2003)

52.88   A. Erdemir: Superlubricity and wearless sliding in diamondlike carbon films, Proc. Mater. Res. Soc. **697**, 391–403 (2002)

52.89   A. Erdemir: Design criteria for superlubricity in carbon films and related microstructures, Tribol. Int. **37**, 577–583 (2004)

52.90   A. R. Krauss, O. Auciello, D. M. Gruen, A. Jayatissa, A. Sumant, J. Tucek, D. C. Mancini, N. Moldovan, A. Erdemir, D. Ersoy, M. N. Gardos, H. G. Busmann, E. M. Meyer, M. Q. Ding: Ultrananocrystalline diamond thin films for MEMS and moving mechanical assembly devices, Diamond Rel. Mater. **10**, 1952–1962 (2001)

52.91   W. R. Ashurst, M. B. J. Wijesundra, C. Carraro, R. Maboudian: Tribological impact of SiC encapsulation of released polycrystalline silicon microstructures, Tribol. Lett. **71**, 195–198 (2004)

52.92   S. Sundararajan, B. Bhushan: Micro/nanotribological studies of polysilicon and SiC films for MEMS applications, Wear **217**, 251–261 (1998)

52.93   X. Li, B. Bhushan: Micro/nanotribological characterization of ceramic films for microdevices, Thin Solid Films **340**, 210–217 (1999)

52.94   R. Maboudian, W. R. Ashurst, C. Carraro: Tribological challenges in microelectromechanical systems, Tribol. Lett. **12**, 95–100 (2002)

52.95   H. A. C. Tilmans, H. Ziad, H. Jansen, O. Di Monaco, A. Jourdain, W. De Raedt, X. Rottenberg, E. De Backer, A. De Caussemaeker, K. Baert: Wafer-level packaged RF-MEMS switches fabricated in a CMOS fab. In: *Proc. IEDM 2001* (IEEE, Washington, DC December 3-5 2001) pp. 921–924

52.96   I. De Wolf, W. Merlijn van Spengen, R. Modlinski, A. Jourdain, A. Witvrouw, P. Fiorini, H. A. C. Tilmans: Reliability and failure analysis of RF MEMS switches. In: *Proc. ISTFA 2002* (ASM, Phoenix, AZ 2002) pp. 275–281

52.97   Military Standard (MIL-STD-883): Test Methods and Procedures for Microelectronics

52.98   A. Jourdain, P. De Moor, S. Pamidighantam, H. A. C. Tilmans: Investigation of the hermeticity of BCB-sealed cavities for housing (RF-)MEMS devices. In: *Proc. MEMS 2002* (IEEE, Las Vegas, USA 2002) pp. 677–680

53. Mechanical Properties of Micromachined Structures

To be able to accurately design structures and make reliability predictions in any field, it is first necessary to know the mechanical properties of the materials that make up the structural components. The devices encountered in the fields of microelectromechanical systems (MEMS) and nanoelectromechanical systems (NEMS), are necessarily very small, and so the processing techniques and the microstructures of the materials used in these devices may differ significantly from bulk structures. Also, the surface-area-to-volume ratios in such structures are much higher than in bulk samples, and so surface properties become much more important. In short, it cannot be assumed that the mechanical properties measured for a bulk specimen of a material will apply when the same material is used in MEMS and NEMS. This chapter will review the techniques that have been used to determine the mechanical properties of micromachined structures, especially residual stress, strength and Young's modulus. The

experimental measurements that have been performed will then be summarized, in particular the values obtained for polycrystalline silicon (polysilicon).

53.1 Measuring Mechanical Properties of Films on Substrates

In order to accurately determine the mechanical properties of very small structures, it is necessary to test specimens made from the same materials, processed in the same way, and of the same approximate size. Not surprisingly it is often difficult to handle specimens this small. One solution is to test the properties of films that remain on substrates. Micro- and nanomachined structures are typically fabricated from films that are initially deposited onto a substrate, are subsequently patterned and etched into the appropriate shapes, and are then finally released from the substrate. If the testing is performed on the continuous film, before patterning and release, the substrate can be used as an effective "handle" for the specimen (in this case, the film). Of course, since the films are attached to the substrate, the types of tests possible are severely limited.

53.1.1 Residual Stress Measurements

One common measurement easily performed on films attached to substrates is residual film stress. The curvature of the substrate is measured before and after film deposition. Curvature can be measured in a number of ways. The most common technique is to scan a laser across the surface (or scan the substrate beneath the laser) and detect the angle of the reflected signal. Alternatively, profilometry, optical interferometry or even atomic force microscopy can be used. As expected, tools that map a surface or perform multiple linear scans can give more accurate readings than tools that measure only a single scan.

Assuming that the film is thin compared to the substrate, the average residual stress in the film, σ_f, is given

by the Stoney equation,

$$\sigma_f = \frac{1}{6} \frac{E_s}{(1 - \nu_s)} \frac{t_s^2}{t_f} \left(\frac{1}{R_1} - \frac{1}{R_2} \right) , \qquad (53.1)$$

where the subscripts f and s refer to the film and substrate, respectively; t is thickness, E is Young's modulus, ν is Poisson's ratio, and R is the radius of curvature before (R_1) and after (R_2) film deposition [53.2]. For a typical (100)-oriented silicon substrate, $E/(1 - \nu)$ (also known as the biaxial modulus) is equal to 180.5 GPa, independent of in-plane rotation [53.3]. This investigation can be performed on the as-deposited film or after any subsequent annealing step, provided no changes occur to the substrate.

This measurement will reveal the average residual stress of the film. Typically, however, the residual stresses of deposited films will vary throughout the thickness of the film. One way to detect this, using substrate curvature techniques, is to etch away a fraction of the film and repeat the curvature measurement. This can be iterated any number of times to obtain a residual stress profile for the film [53.4]. Alternatively, tools have been designed that can measure the substrate curvature during the deposition process itself, in order to obtain information on how the stresses evolve [53.5].

An additional feature of some of these tools is the ability to heat the substrates while performing the stress measurement. An example of the results obtained in such an experiment is shown in Fig. 53.1 [53.1], for an aluminium film on a silicon substrate. The slope of the heating curve gives the difference in thermal expansion between the film and the substrate. When the heating curve changes slope and becomes nearly horizontal, the yield strength of the film has been reached.

53.1.2 Mechanical Measurements Using Nanoindentation

Aside from residual stress, it is difficult to measure the mechanical properties of films attached to substrates without the measurement being affected by the presence of the substrate. Recent developments in nanoindentation equipment have allowed this technique to be used

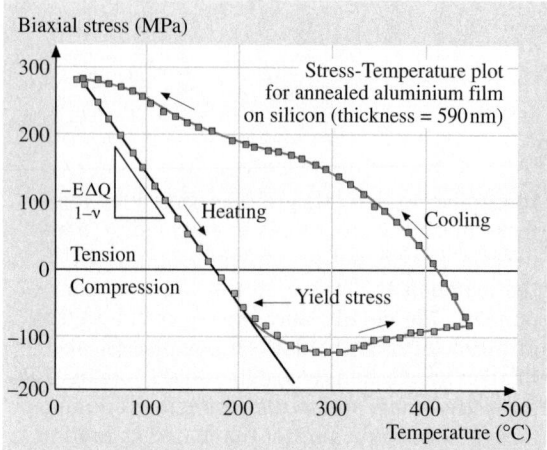

Fig. 53.1 Typical results for residual stress as a function of temperature for an aluminum film on a silicon substrate [53.1]. The stresses were determined by measuring the curvature of the substrate before and after film deposition, using the reflected signal of a laser scanned across the substrate surface

in some cases. With specially designed tools, indentation can be performed using very low loads. If the films being investigated are thick and rigid enough, measurements can be made that are not influenced by the presence of the substrate. Of course, this can be verified by depositing the same film onto different substrates. By continuously monitoring the displacement as well as the load during indentation, a variety of properties can be measured, including hardness and Young's modulus [53.6]. This area is covered in more detail in a separate chapter.

For brittle materials, cracks can be generated by indentation, and strength information can be gathered. But the exact stress fields created during the indentation process are not known exactly, and therefore quantitative values for strength are difficult to determine. Anisotropic etching of single-crystal silicon has been performed to create 30-μm-tall structures that were then indented to examine fracture toughness [53.7], but this is not possible with most materials.

53.2 Micromachined Structures for Measuring Mechanical Properties

Certainly the most direct way to measure the mechanical properties of small structures is to fabricate structures that would be conducive to such tests. Fabrication techniques are sufficiently advanced that virtually

any design can be realized, at least in two dimensions. Two basic types of devices are used for mechanical property testing: "passive" structures and "active" structures.

53.2.1 Passive Structures

As mentioned previously, the main difficulty encountered when testing very small specimens is handling. One way to circumvent this problem is to use passive structures. These structures are designed to act as soon as they are released from the substrate and to provide whatever information they are designed to supply without further manipulation. For all of these passive structures, the forces acting on them come from the internal residual stresses of the structural material. For devices on the micron scale or smaller, gravitational forces can be neglected, and therefore internal stresses are the only source of actuation force.

Stress Measurements

Since internal residual stresses act upon the passive devices when they are released, it is natural to design a device that can be used to measure residual stresses. One such device, a rotating microstrain gage, is shown in Fig. 53.2. There are many different microstrain gauge designs, but all operate via the same principle. In Fig. 53.2a, the large pads, labeled A, will remain anchored to the substrate when the rest of the device is released. Upon release, the device will expand or contract in order to relieve its internal residual stresses. A structure under tension will contract, and a structure under compression will expand. For the structure in Fig. 53.2, compressive stress will cause the legs to lengthen. Since the two opposing legs are not attached to the central beam at the same point – they are offset, they will cause the central beam to rotate when they expand. The device in Fig. 53.2 contains two independent gauges that point to one another. At the ends of the two central beams are two parts of a Vernier scale. By observing this scale, one can measure the rotation of the beams.

If the connections between the legs and the central beams were simple pin connections, the strain, ε, of the legs (the fraction of expansion or contraction) could be determined simply by the measured rotation and the geometry of the device, namely

$$\varepsilon = \frac{d_{\mathrm{beam}} d_{\mathrm{offset}}}{2 L_{\mathrm{central}} L_{\mathrm{leg}}} , \qquad (53.2)$$

where d_{beam} is the lateral deflection of the end of one central beam, d_{offset} is the distance between the connections of the opposing legs, L_{central} is the length of the central beam (measured to the center point between the leg connections), and L_{leg} is the length of the leg. However, since the entire device was fabricated from a single polysilicon film, this cannot be the case; some bending must occur at the connections. As a result, to get an accurate determination of the strain relieved upon release, finite element analysis (FEA) of the structure must be performed. This is a common situation for mi-

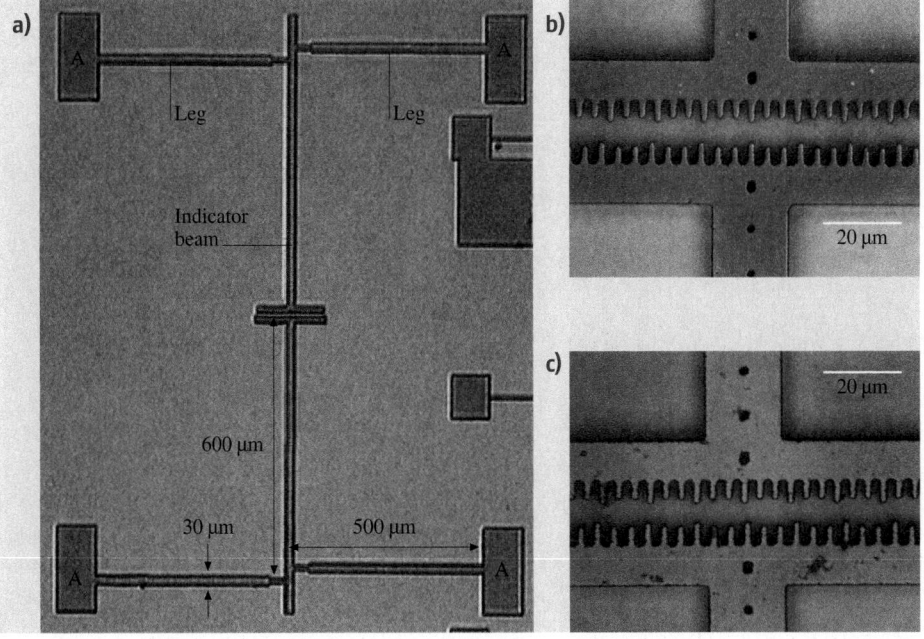

Fig. 53.2
(a) Microstrain gauge fabricated from polysilicon; (b) shows a close-up of the Vernier scale before release, and (c) shows the same area after release

Part G | 53.2

crodevices. FEA is a powerful tool for determining the displacements and stresses of nonideal geometries. One drawback is that the Young's modulus of the material must be known in order to do the FEA as well as to convert the measured strain into a stress value. But Young's moduli are known for many micromachined materials or they can be measured using other techniques.

Other devices besides rotating strain gauges have been designed that can measure residual stresses. One of the simplest is a doubly clamped beam, a long, narrow beam of constant width and thickness that is anchored to the substrate at both ends. If the beam contains a tensile stress it will remain straight, but if the beam contains a compressive stress it will buckle if its length exceeds a critical value, l_{cr}, according to the Euler buckling criterion [53.8],

$$\varepsilon_r = -\frac{\pi^2}{3}\left(\frac{h}{l_{cr}}\right)^2 , \tag{53.3}$$

where ε_r is the residual strain in the beam and h is the width or thickness of the beam, whichever is less. To determine the residual strain, a series of doubly clamped beams of varying lengths are fabricated. In this way, the critical length for buckling, l_{cr}, can be deduced after release. One problem with this technique is that during the release process, any turbulence in the solution will lead to enhanced buckling of the beams, and a low value for l_{cr} will be obtained.

For films with tensile stresses, a similar analysis can be performed using ring-and-beam structures, also called Guckel rings after their inventor, Henry Guckel. A schematic of this design is shown in Fig. 53.3 [53.8]. Tensile stress in the outer ring will cause it to contract.

This will lead to compressive stress in the central beam, even though the material was originally tensile before release. The amount of compression in the central beam can be determined analytically from the geometry of the device and the residual strain of the material. Again, by changing the length of the central beam it is possible to determine l_{cr}, and then the residual strain can be deduced.

Stress Gradient Measurements

For structures fabricated from thin deposited films, the stress gradient can be just as important as the stress itself. Figure 53.4 shows a portion of a silicon microactuator. The device is designed to be completely planar; however, stress gradients in the film cause the structures to bend. This figure illustrates the importance of characterizing and controlling stress gradients, and it also demonstrates that stress gradients are most easily measured for a simple cantilever beam. By measuring the end deflection δ of a cantilever beam of length l and thickness t, the stress gradient, $d\sigma/dt$ is determined by [53.9]

$$\frac{d\sigma}{dt} = \frac{2\delta}{l^2}\frac{E}{1-\nu} . \tag{53.4}$$

The magnitude of the end deflection can be measured by microscopy, optical interferometry, or any other technique.

Another useful structure for measuring stress gradients is a spiral. For this structure, the end of the spiral not anchored to the substrate will move out-of-plane. The diameter of the spiral will also contract, and the free end of the spiral will rotate when released [53.10].

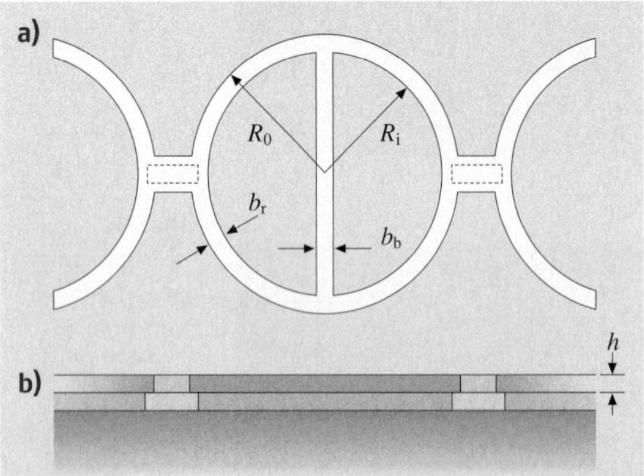

a)

b)

Fig. 53.3a,b Schematic showing (**a**) top view and (**b**) side view of Guckel ring structures [53.8]. The *dashed lines* in (**a**) indicate the anchors

50 μm

Fig. 53.4 Scanning electron micrograph (SEM) of a portion of a silicon microactuator. Residual stress gradients in the silicon cause the structure to bend

Strength and Fracture Toughness Measurements

As mentioned above, if a doubly clamped beam contains a tensile stress, it will remain taut when released because it cannot relieve any of its stress by contracting. This tensile stress can be thought of as a tensile load being applied at the ends of the beam. If this tensile load exceeds the tensile strength of the material, the beam will break. Since the tensile stress can be measured, as discussed in the Sect. 53.2.1, this technique can be used to gather information on the strength of materials. Figure 53.5 shows two different beam designs that have been used to measure strength. The device shown in Fig. 53.5a was fabricated from a tensile polysilicon film [53.11]. Different beams were designed with varying lengths of the wider regions (marked l_1 in the figure). In this manner, the load applied to the narrow center beam was varied, even though the entire film contained a uniform residual tensile stress. For l_1 greater than a critical value, the narrow center beam fractured, giving a measurement for the tensile strength of polysilicon.

The design shown in Fig. 53.5b was fabricated from a tensile Si_xN_y film [53.12]. As seen in the figure, a stress concentration was included in the beam, to ensure the fracture strength would be exceeded. In this case, a notch was etched into one side of the beam. Since the stress concentration is not symmetric with regard to the beam axis, this results in a large bending moment at that position, and the test measures the bend strength of the material. Again, like the beams shown in Fig. 53.5a, the geometry of various beams fabricated from the same film were varied, to vary the maximum stress seen at the notch. By seeing which beams fracture at the stress concentration after release, the strength can be determined.

The fracture toughness of a material can be determined with a similar technique, but an atomically sharp pre-crack is used instead of a stress concentration. Sharp pre-cracks can be introduced into micromachined structures before release by adding a Vickers indent onto the substrate, near the device; the radial crack formed by the indent will propagate into the overlying structure [53.14]. Accordingly, the beam with a sharp pre-crack, shown in Fig. 53.6, was fabricated using polysilicon [53.13]. Due to the stochastic nature of indentation, the initial pre-crack length varies from beam to beam. Because of this, even though the geometry of the beam remains identical, the stress intensity, K, at the pre-crack tip will vary. Upon release, only those pre-cracks whose K exceeds the fracture toughness of the material, K_{Ic}, will propagate, and in this way upper and lower bounds for K_{Ic} can be determined for the material.

For all of the beams discussed in this section, finite element analysis is required to determine the stress concentrations and stress intensities. Even though approximate analytical solutions may exist for these designs, the actual fabricated structure will not have idealized geometries. For example, corners will never be perfectly sharp, and cracks will never be perfectly straight. This reinforces the idea that FEA is a powerful tool when determining mechanical properties of very small structures.

Fig. 53.5a,b Schematic designs of doubly clamped beams with stress concentrations used for measuring strength. (**a**) was fabricated from polysilicon [53.11] and (**b**) was fabricated from Si_xN_y [53.12]

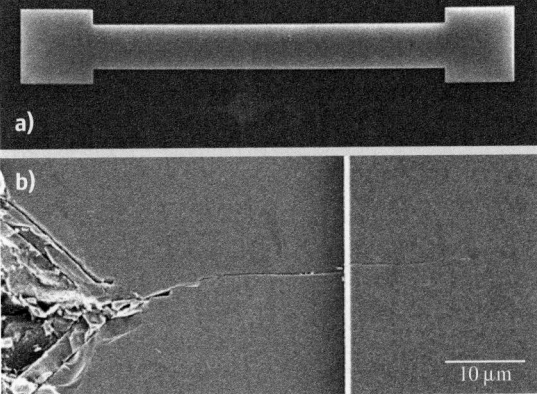

Fig. 53.6 (**a**) SEM of a 500 μm-long polysilicon beam with a Vickers indent placed near its center; (**b**) higher magnification SEM of the area near the indent showing the pre-crack traveling from the substrate into the beam [53.13]

53.2.2 Active Structures

As discussed above, it is very convenient to design structures that act upon release to provide information on the mechanical properties of the structural materials. This is not always possible, however. For example, those passive devices just discussed rely on residual stresses to create the changes (rotation or fracture) that occur upon release, but many materials do not contain high residual stresses as-deposited, or the processing scheme of the device precludes the generation of residual stresses. Also, some mechanical properties, such as fatigue resistance, require motion before they can be studied. Active devices are therefore used. These are acted upon by a force (the source of this force can be integrated into the device itself or can be external to the device) in order to create a change, and the mechanical properties are studied via the response to the force.

Young's Modulus Measurements

Young's modulus, E, is a material property critical to any structural device design. It describes the elastic response of a material and relates stress, σ, and strain, ε, by

$$\sigma = E\varepsilon \, . \tag{53.5}$$

In bulk samples, E is often measured by loading a specimen under tension and measuring displacement as a function of stress for a given length. While this is far more difficult for small structures, such as those fabricated from thin deposited films, it can be achieved with careful experimental techniques. Figure 53.7 shows a schematic of one such measurement system [53.15]. The fringe detectors in the figure detect the reflected laser signal from two gold lines deposited onto the polysilicon specimen, which act as gauge markers. This enables the strain in the specimen during loading to be monitored.

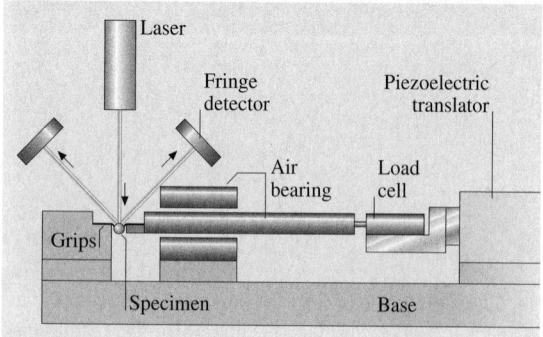

Fig. 53.7 Schematic of a measurement system for tensile loading of micromachined specimens [53.15]

Besides gold lines, Vickers indents placed in a nickel specimen can also serve as gauge markers [53.16], or a speckle interferometry technique [53.17] can be used to determine strain in the specimen. Once the stress-versus-strain behavior is measured, the slope of the curve is equal to E. By using a constant load, such as a dead weight, and resistive heating, high-temperature creep can also be investigated with this method [53.18].

In addition to the tensile test, Young's modulus can be determined by other measures of stress–strain behavior. As seen in Fig. 53.8, a cantilever beam can be bent by pushing on the free end with a nanoindenter [53.19]. The nanoindenter can monitor the force applied and the displacement, and simple beam theory can convert the displacement into strain in order to obtain E. A similar technique, shown schematically in Fig. 53.9 [53.20], involves pulling downward on a cantilever beam by means of an electrostatic force. An electrode is fabricated into the substrate beneath the cantilever beam, and a voltage is applied between the beam and the bottom electrode. The force acting on the beam is equal to the electrostatic force corrected to include the effects of fringing fields acting on the sides of the beam, namely

$$F(x) = \frac{\varepsilon_0}{2} \left(\frac{V}{g + z(x)} \right)^2 \left(1 + \frac{0.65[g + z(x)]}{w} \right) ,$$
$$\tag{53.6}$$

where $F(x)$ is the electrostatic force at x, ε_0 is the dielectric constant of air, g is the gap between the beam and the bottom electrode, $z(x)$ is the out-of-plane deflection of the beam, w is the beam width, and V is the applied voltage [53.20]. In this work, the deflection of the beam as a function of position is measured using optical interferometry. These measurements combine to give stress–strain behavior for the cantilever beam. An extension of this technique uses doubly clamped beams instead of cantilever beams. In this case, the deflection of the beam at a given electrostatic force depends on the residual stress in the material as well as Young's modulus. This method can therefore also be used to measure residual stresses in doubly clamped beams.

Another device that can be fabricated from a thin film and used to investigate stress–strain behavior is a suspended membrane, as shown in Fig. 53.10 [53.21]. As depicted in the schematic figure, the membrane is exposed to an elevated pressure on one side, causing it to bulge in the opposite direction. The deflection of the membrane is measured by optical or other techniques and related to the strain in the membrane. These membranes can be fabricated in any shape, typically square or circular. Both analytical solutions and finite element

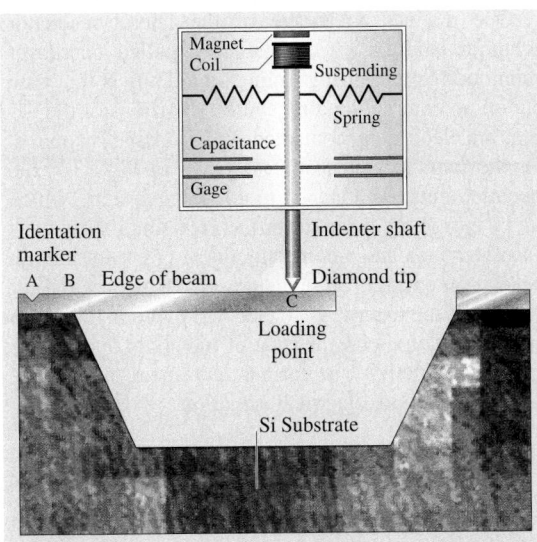

Fig. 53.8 Schematic of a nanoindenter loading mechanism pushing on the end of a cantilever beam [53.19]

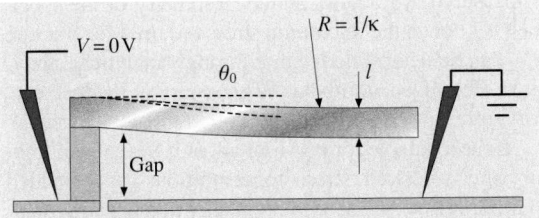

Fig. 53.9 Schematic of a cantilever beam bending test using an electrostatic voltage to pull the beam toward the substrate [53.20]

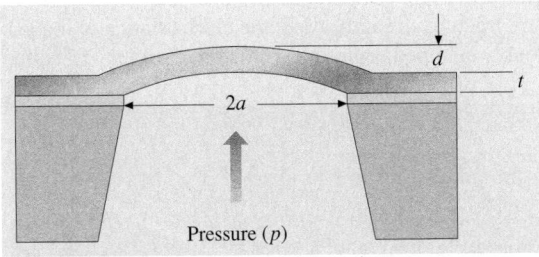

Fig. 53.10 Schematic cross-section of a microfabricated membrane [53.21]

analyses have been performed to relate the deflection to the strain. Like the doubly clamped beams, both Young's modulus and residual stress play a role in the deflected shape. Both of these mechanical properties can there-

fore be determined by the pressure-versus-deflection performance of the membrane.

Another measurement besides stress–strain behavior that can reveal the Young's modulus of a material is the determination of the natural resonance frequency. For a cantilever, the resonance frequency, f_r, for free undamped vibration is given by

$$f_r = \frac{\lambda_i^2 t}{4\pi l^2} \left(\frac{E}{3\rho} \right)^{1/2} , \qquad (53.7)$$

where ρ, l and t are the density, length and thickness of the cantilever; λ_i is the eigenvalue, where i is an integer that describes the resonance mode number; for the first mode $\lambda_1 = 1.875$ [53.23]. Given the geometry and density, measuring f_r allows E to be determined. The cantilever can be vibrated by a number of techniques, including a laser, loudspeaker or piezoelectric shaker. The frequency that produces the highest amplitude of vibration is the resonance frequency.

A micromachined device that uses an electrostatic comb drive and an AC signal to generate the vibration of the structure is known as a lateral resonator [53.24]. One example is shown in Fig. 53.11 [53.22]. When a voltage is applied across either set of the interdigitated comb fingers shown in Fig. 53.11, an electrostatic attraction is generated due to the increase in capacitance as the overlap between the comb fingers increases. The force, F, generated by the comb-drive is given by

$$F = \frac{1}{2} \frac{\partial C}{\partial x} V^2 = n\varepsilon \frac{h}{g} V^2 , \qquad (53.8)$$

Fig. 53.11 SEM of a polysilicon lateral resonator [53.22]

where C is capacitance, x is the distance traveled by one comb-drive toward the other, n is the number of pairs of comb fingers in one drive, ε is the permittivity of the fluid between the fingers, h is the height of the fingers, g is the gap spacing between the fingers, and V is the applied voltage [53.24]. When an AC voltage at the resonance frequency is applied across either of the two comb drives, the central portion of the device will vibrate. In fact, since force depends on the square of the voltage for electrostatic actuation, for a time t, a dependent drive voltage $v_D(t)$ (given by

$$v_D(t) = V_P + v_d \sin(\omega t) ,\qquad (53.9)$$

where V_P is the DC bias and v_d is the AC drive amplitude), the time-dependent portion of the force will scale with

$$2\omega V_P v_d \cos(\omega t) + \omega v_d^2 \sin(2\omega t) \qquad (53.10)$$

[53.24]. Therefore, if an AC drive signal is used with no DC bias, at resonance, the frequency of the AC drive signal will be one half of the resonance frequency. For this device, the resonance frequency f_r will be

$$f_r = \frac{1}{2\pi} \left(\frac{k_{sys}}{M} \right)^{1/2} ,\qquad (53.11)$$

where k_{sys} is the spring constant of the support beams and M is the mass of the portion of the device that vibrates. The spring constant is given by

$$k_{sys} = 24EI/L^3 ,\qquad (53.12)$$

$$I = \frac{hw^3}{12} ,\qquad (53.13)$$

where I is the moment of inertia of the beams, and L, h and w are the length, thickness, and width of each beam. Therefore, by combining these equations and measuring f_r, it is possible to determine E.

One distinct advantage of the lateral resonator technique and the electrostatically pulled cantilever technique for measuring Young's modulus is that they require no external loading sources. Portions of the devices are electrically contacted, and a voltage is applied. For the pure tension tests, as shown in Fig. 53.7, the specimen must be attached to a loading system, which can be extremely difficult for the very small specimens discussed here, and any misalignment or eccentricity in the test could lead to unreliable results. However, the advantage of the externally loaded technique is that there are no limitations on the type of materials that can be tested. Conductivity is not a requirement, nor is any compatibility with electrical actuation.

Strength and Fracture Toughness Measurements

As one might expect, any of the techniques discussed in the previous section that strain specimens in order to measure Young's modulus can also be used to measure fracture strength. The load is simply increased until the specimen breaks. As long as either the load or the strain is measured at fracture, and the geometry of the specimen is known, the maximum stress required for fracture σ_{crit} can be determined, either through analytical analysis or FEA. Depending on the geometry of the test, σ_{crit} will represent the tensile or bend strength of the material.

If the available force is limited, or if a localized fracture site is desired, stress concentrations can be added to the specimens. These are typically notches micromachined into the edges of specimens. Focused ion beams have also been used to carve stress concentrations into fracture specimens.

All of the external loading schemes, such as those shown in Figs. 53.7 and 53.8, have been used to measure fracture strength. Also, the electrostatically loaded doubly clamped beams can be pulled until they fracture.

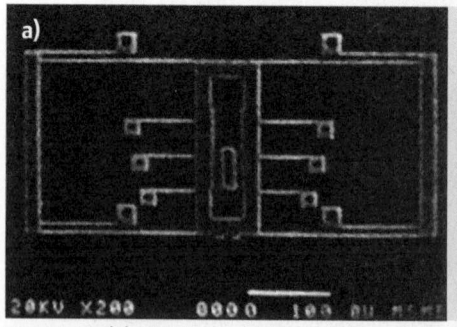 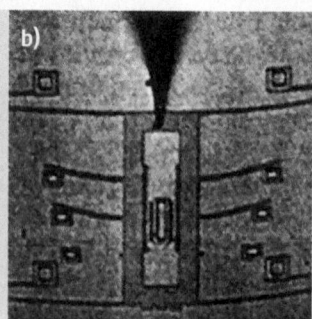 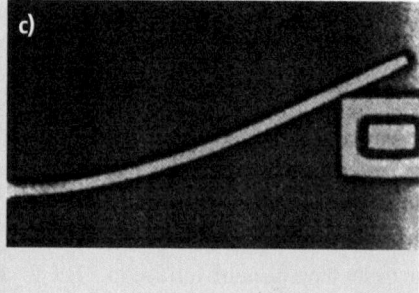

Fig. 53.12 (a) SEM of a device for measuring bend strength of polysilicon beams; (b) image of a test in process; (c) higher magnification view of one beam shortly before breaking [53.25]

In this case, there is one complication. The electrostatic force is inversely proportional to the distance between the electrodes, and at a certain voltage, called the "pull-in voltage," the attraction between the beam and the substrate will become so great that the beam will immediately be pulled into contact with the bottom electrode. As long as the fracture takes place before the pull-in voltage is reached, the experiment will give valid results.

Other loading techniques have been used to generate fracture of microspecimens. Figure 53.12 [53.25] shows one device designed to be pushed by the end of a micromanipulated needle. The long beams that extend from the sides of the central shuttle come into contact with anchored posts, and, at a critical degree of bending, the beams will break off. Since the applied force cannot be measured in this technique, the experiment is continuously optically monitored during the test, and the image of the beams just before fracture is analyzed to determine σ_{crit}.

Another loading scheme that has been demonstrated for micromachined specimens utilizes scratch drive actuators to load the specimens [53.26]. These types of actuators work like inchworms, traveling across a substrate in discrete advances as an electrostatic force is repeatedly applied between the actuator and the substrate. The stepping motion can be made on the nanometer scale, depending on the frequency of the applied voltage, and so it can be an acceptable approximation to continuous loading. One advantage of this scheme is that very large forces can be generated by relatively small devices. The exact forces generated cannot be measured, so (like the technique that used micromanipulated pushing) the test is continuously observed to determine the strain at fracture. Another advantage of this technique is that, like the lateral resonator and the electrically pulled cantilever, the loading takes place on-chip, and therefore the difficulties associated with attaching and aligning an external loading source are eliminated.

Another on-chip actuator used to load microspecimens is shown in Fig. 53.13, along with three different microspecimens [53.14]. Devices have been fabricated with each of the three microspecimens integrated with the same electrostatic comb-drive actuator. In all three cases, when a DC voltage is applied to the actuator, it moves downward, as oriented in Fig. 53.13. This pulls down on the left end of each of the three microspecimens, which are anchored on the right. The actuator contains 1486 pairs of comb fingers. The maximum voltage that can be applied is limited by the breakdown voltage of the medium in which the test takes place. In

air, this limits the voltage to less than 200 V. As a result, given a finger height of 4 μm and a gap of 2 μm, and using (53.8), the maximum force generated by this actuator is limited to about 1 mN. Standard optical photolithography has a minimum feature size of about 2 μm. As

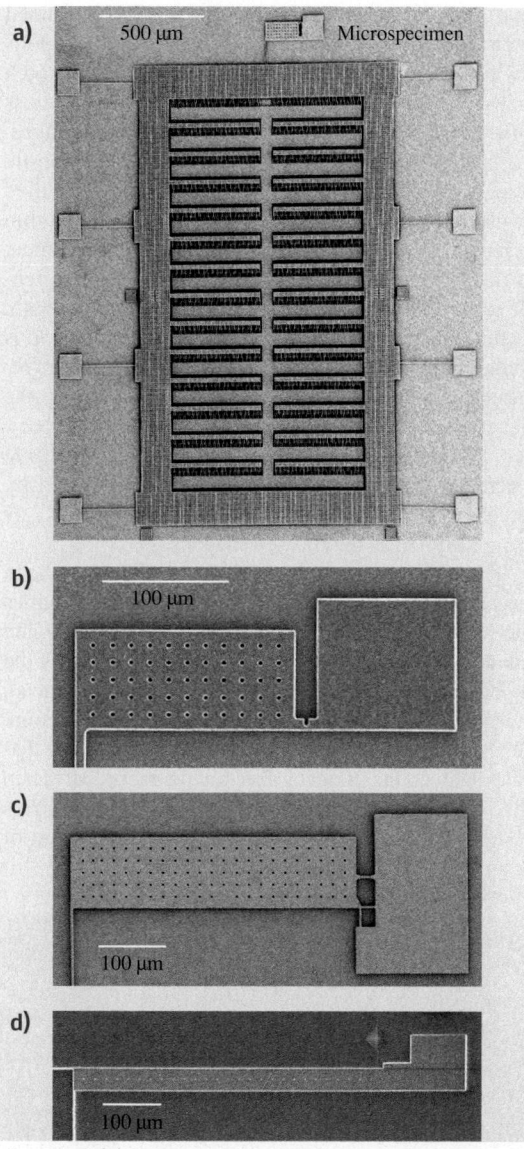

Fig. 53.13 (a) SEM of a micromachined device for conducting strength tests; the device consists of a large comb-drive electrostatic actuator integrated with a microspecimen; (b)–(d) SEMs of various microspecimens for testing bend strength, tensile strength and fracture toughness, respectively [53.14]

a result, the electrostatic actuator cannot generate sufficient force to perform a standard tensile test on MEMS structural materials such as polysilicon. The microspecimens shown in Fig. 53.13 are therefore designed such that the stress is amplified.

The specimen shown in Fig. 53.13b is designed to measure bend strength. It contains a micromachined notch with a root radius of 1 μm. When the actuator pulls downward on the left end of this specimen, the notch serves as a stress concentration, and when the stress at the notch root exceeds σ_{crit}, the specimen fractures. The specimen in Fig. 53.13c is designed to test tensile strength. When the left end of this specimen is pulled downward, a tensile stress is generated in the upper thin horizontal beam near the right end of the specimen. As the actuator continues to move downward, the tensile stress in this beam will exceed the tensile strength, causing fracture. Finally, the specimen in Fig. 53.13d is similar to that in Fig. 53.13b, except that the notch is replaced by a sharp pre-crack that was produced by the Vickers indent placed on the substrate near the specimen. When this specimen is loaded, a stress intensity K is generated at the crack tip. When the stress intensity exceeds a critical value K_{Ic}, the crack propagates. K_{Ic} is also referred to as the fracture toughness.

The force generated by the electrostatic actuator can be calculated using (53.8). However, (53.8) assumes a perfectly planar, two-dimensional device. In fact, when actuated, the electric fields extend out of the plane of the device, and so (53.8) is just an approximation. Instead, like many of the techniques discussed in this section, the test is continuously monitored, and the actuator displacement at the time of fracture is recorded. Then FEA is used to determine the magnitude of the stress or stress intensity seen by the specimen at the point of fracture.

In order to generate sufficient force to conduct tensile tests, a similar device to that shown in Fig. 53.13 has been designed which uses an array of parallel plate capacitors to provide the force, instead of comb-drives [53.28]. In this way the available force is increased but the maximum stroke is severely limited.

Fatigue Measurements

A benefit of the electrostatic actuator shown in Fig. 53.13 is that, besides monotonic loading, it can generate cyclic loading. This allows the fatigue resistance of materials to be studied. Simply by using an AC signal instead of a DC voltage, the device can be driven at its resonant frequency. The amplitude of the resonance depends on the magnitude of the AC signal. This amplitude can be increased until the specimen breaks; this will investigate the low-cycle fatigue resistance. Otherwise, the amplitude of resonance can be left constant at a level below that required for fast fracture, and the device will resonate indefinitely until the specimen breaks; this will investigate high-cycle fatigue. It should be noted that the resonance frequency of such a device is about 10 kHz. Therefore, it is possible to stress a specimen for over 10^9 cycles in less than a day. In addition to simple cyclic loading, a mean stress can be superimposed on the cyclic load if a DC bias is added to the AC signal. In this way, nonsymmetric cyclic loading (with a large tensile stress alternating with a small compressive stress, or vice versa) can be studied.

Another device that can be used to investigate fatigue resistance in MEMS materials is shown in Fig. 53.14 [53.27]. In this case, a large mass is attached to the end of a notched cantilever beam. The mass contains two comb drives on opposite ends. When an AC signal is applied to one comb drive, the device will resonate, cyclically loading the notch. The comb drive on

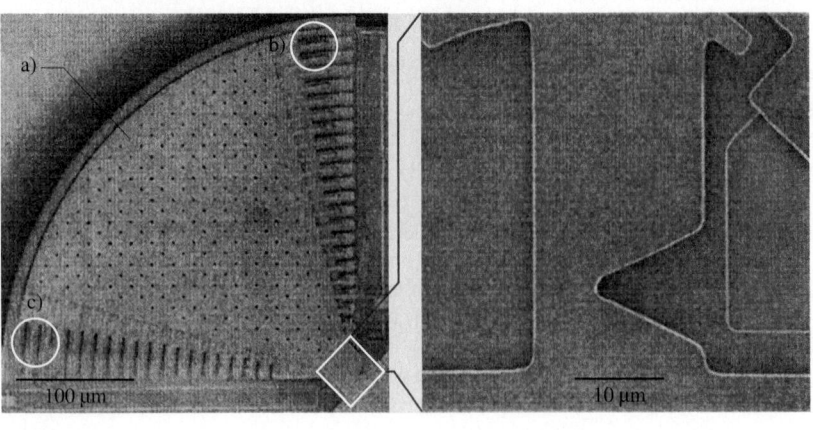

Fig. 53.14 SEM of a device used to investigate fatigue; the image on the right is a higher magnification view of the notch near the base of the moving part of the structure [53.27]: a) mass, b) comb drive actuator, c) capacitive displacement sensor)

the opposite side is used as a capacitive displacement sensor. This device contains many fewer comb fingers than the device shown in Fig. 53.13. As a result, it can apply cyclic loads by exploiting the resonance frequency of the device, but it cannot supply sufficient force to achieve monotonic loading.

Fatigue loading has also been studied using the same external loading techniques shown in Fig. 53.7. In this case, the frequency of the cyclic load is considerably lower, since the resonance frequency of the device is not being utilized. This leads to longer high-cycle testing times. Since the force is essentially unlimited, however, this technique allows a variety of frequencies to be studied to determine their effect on the fatigue behavior.

Friction Measurements

Friction is another property that has been studied in micromachined structures. To study friction, of course, two surfaces must be brought into contact with each other. This is usually avoided at all costs for these devices because of the risk of stiction. (Stiction is the term used when two surfaces that come into contact adhere so strongly that they cannot be separated.) Even so, a few devices have been designed that can investigate friction. One of these is shown schematically in Fig. 53.15 [53.29]. It consists of a movable structure with a comb-drive on one end and a cantilever beam on the other. Beneath the cantilever, on the substrate, is a planar electrode. The device is moved to one side

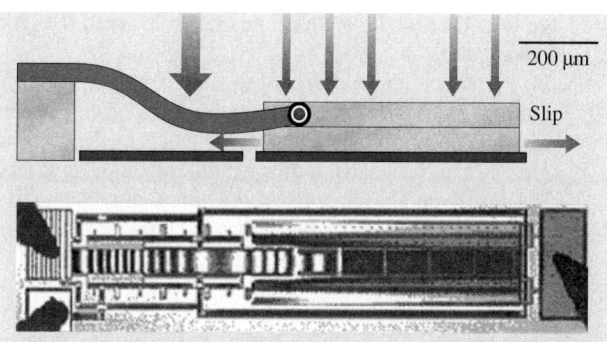

Fig. 53.16 Schematic cross-section and top-view optical micrograph of a hinged-cantilever test structure for measuring friction in micromachined devices [53.30]

using the comb-drive. Then a voltage is applied between the cantilever beam and the substrate electrode. The voltage on the comb-drive is then released. The device would normally return to its original position, to relax the deflection in the truss suspensions, but the friction between the cantilever and the substrate electrode holds it in place. The voltage to the substrate electrode is slowly decreased until the device starts to slide. Given the electrostatic force generated by the substrate electrode and the stiffness of the truss suspensions, it is possible to determine the static friction. For this device, bumps were fabricated on the bottom of each cantilever beam. This limited the surface area that came

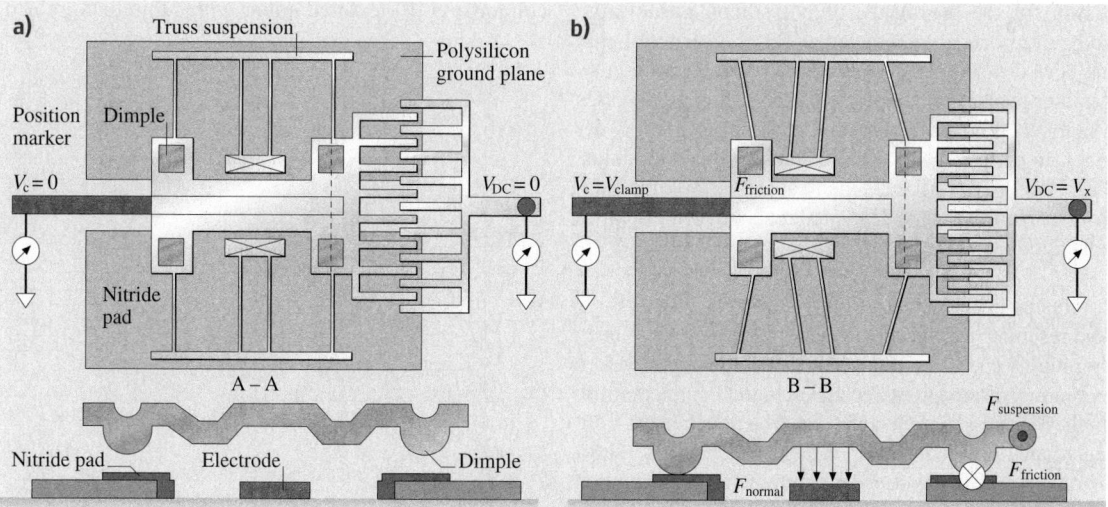

Fig. 53.15a,b Schematics depicting a device used to study friction. (a) shows top and side views of the device in its original position, and (b) shows views of the device after it has been displaced using the comb-drive and clamped using the substrate electrode [53.29]

into contact with the substrate and so lowered the risk of stiction.

Another device designed to study friction is shown in Fig. 53.16 [53.30]. This technique uses a hinged cantilever. The portion near the free end acts as the friction test structure, and the portion near the anchored end acts as the driver. The friction test structure is attracted to the substrate by means of electrostatic actuation, and when a second electrostatic actuator pulls down the driver, the friction test structure slips forward by a length proportional to the forces involved, including the frictional force. This distance, however, has a maximum of 30 nm, so all measurements must be exceedingly accurate in order to investigate a range of forces. This test structure can be used to determine the friction coefficients for surfaces with and without lubricating coatings.

53.3 Measurements of Mechanical Properties

All of the techniques discussed in Sects. 53.1 and 53.2 have been used to measure the mechanical properties of MEMS and NEMS materials. As a general rule, the results from the various techniques have agreed well with each other, and the argument becomes which of the measurement techniques is easiest and most reliable to perform. It is crucial to bear in mind, however, that certain properties (such as strength) are process-dependent, and so the results taken at one laboratory will not necessarily match those taken from another. This will be discussed in more detail in Sect. 53.3.1.

53.3.1 Mechanical Properties of Polysilicon

In current MEMS technology, the most widely used structural material is polysilicon deposited by low-pressure chemical vapor deposition (LPCVD). One reason for the prevalence of polysilicon is the large body of processing knowledge for this material that has been developed by the integrated circuit community. Another reason, of course, is that polysilicon possesses a number of qualities that are beneficial to MEMS devices, in particular high strength and Young's modulus. Therefore, most of the mechanical properties investigations performed on MEMS materials have focused on polysilicon.

Residual Stresses in Polysilicon
The residual stresses of LPCVD polysilicon have been thoroughly characterized using both the wafer curvature technique, discussed in Sect. 53.1.1, and the microstrain gauges, discussed in Sect. 53.2.1. The results from both techniques give consistent values. Figure 53.17 summarizes the residual stress measurements as a function of deposition temperature taken from five different investigations at five different laboratories [53.31]. All five sets of data show the same trend. The stresses change from compressive at the lowest deposition tem-

peratures to tensile at intermediate temperatures and back to compressive at the highest temperatures. The exact transition temperatures vary somewhat between the different investigations, probably due to differences in the deposition conditions: the silane or dichlorosilane pressure, the gas flow rate, the geometry of the deposition system, and the temperature uniformity. However, in each data set the transitions are easily discernible. The origin of these residual stress changes lies with the microstructure of the LPCVD polysilicon films.

As with all deposited films, the microstructure of the LPCVD polysilicon film is dependent on the deposition conditions. In general, the films are amorphous at the lowest growth temperatures (lower than $\sim 570\,°C$), display fine ($\sim 0.1\,\mu m$ diameter) grains at intermediate temperatures ($\sim 570\,°C$ to $\sim 610\,°C$), and contain columnar (110)-textured grains with a thin fine-grained

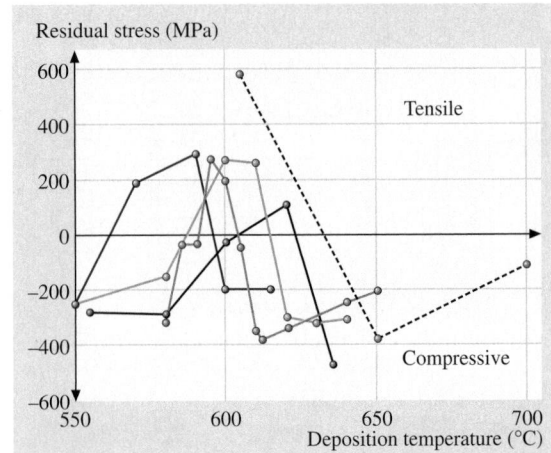

Fig. 53.17 Results for residual stress of LPCVD polysilicon films taken from five different investigations [53.31]. Data from the same investgation are connected by a line

nucleation layer at the substrate interface at higher temperatures ($\sim 610\,^\circ$C to $\sim 700\,^\circ$C) [53.31]. The fine-grained microstructure results from the homogeneous nucleation and growth of silicon grains within an as-deposited amorphous silicon film. In this regime, the deposition rate is just slightly faster than the crystallization rate. The as-deposited films will be crystalline near the substrate interface and amorphous at the free surface. (The amorphous fraction can be quickly crystallized by annealing above $610\,^\circ$C.) The columnar microstructure seen at higher growth temperatures results from the formation of crystalline silicon films as-deposited, with growth being fastest in the $\langle 110 \rangle$ directions.

The origin of the tensile stress in the fine-grained polysilicon arises from the volume decrease that accompanies the crystallization of the as-deposited amorphous material. The origins of the compressive stresses in the amorphous and columnar films are less well understood. One proposed explanation for compressive stress generation during thin film growth postulates that an increase in the surface chemical potential is caused by the deposition of atoms from the vapor; the increase in surface chemical potential induces atoms to flow into newly formed grain boundaries, creating a compressive stress in the film [53.32].

Stress gradients are also typical of LPCVD polysilicon films. The partially amorphous films contain large stress gradients since they are essentially bilayers of compressive amorphous silicon on top of tensile fine-grained polysilicon. The fully crystalline films also exhibit stress gradients. The columnar compressive films are most highly stressed at the film–substrate interface, with the compressive stresses decreasing as the film thickness increases; the fine-grained films are less tensile at the film–substrate interface, with the tensile stresses increasing as the film thickness increases [53.31]. Both stress gradients are associated with microstructural variations. For the columnar films, the initial nucleation layer corresponds to a very high compressive stress, which decreases as the columnar morphology develops. For the fine-grained films, the region near the film-substrate interface has a slightly smaller average grain size, due to heterogeneous nucleation at the interface. This region displays a slightly lower tensile stress than the rest of the film, since the increased grain boundary area reduces the local density.

Young's Modulus of Polysilicon
The Young's modulus of polysilicon films has been measured using all of the techniques discussed in Sect. 53.2.2. A good review of the experimental results taken from bulge testing, tensile testing, beam bending and lateral resonators are contained in [53.33]. All of the reported results are in reasonable agreement, varying from 130 to 175 GPa, though many values are reported with a relatively high experimental scatter. The main origin of the error in these results is the uncertainties involving the geometries of the small specimens used to make the measurements. For example, from (53.13), the Young's modulus determined by the lateral resonators depends on the cube of the tether beam width, typically about 2 μm. In general, the beam width and other dimensions can be measured via scanning electron microscopy to within about 0.1 μm; however, the width of the beam is not perfectly constant along the entire length or even throughout the thickness of the beam. These uncertainties in geometry lead to uncertainties in modulus.

In addition, the various experimental measurements lie close to the Voigt and Reuss bounds for Young's modulus calculated using the elastic stiffnesses and compliances for single-crystal silicon [53.33]. This strongly implies that Young's moduli of micro- and nanomachined polysilicon structures will be the same as for bulk samples made from polysilicon. This is not unexpected, since Young's modulus is a material property. It is related to interatomic interactions and should have no dependence on the geometry of the sample. It should be noted that polysilicon can display a preferred crystallographic orientation depending on the deposition conditions, and that this could affect the Young's modulus of the material, since the Young's modulus of silicon is not isotropic. However, the anisotropy is fairly small for cubic silicon.

A more recent investigation that utilized electrostatically actuated cantilevers and interferometric deflection detection yielded a Young's modulus of 164 GPa [53.20]. They found the grains in their polysilicon films to be randomly oriented, and calculated the Voigt and Reuss bounds to be 163.4–164.4 GPa. This appears to be a very reliable value for randomly oriented polysilicon.

Fracture Toughness and Strength of Polysilicon
Using the device shown in Fig. 53.13a and the specimen shown in Fig. 53.13d, the fracture toughness, K_{Ic}, of polysilicon has been shown to be 1.0 ± 0.1 MPa m$^{1/2}$ [53.34]. Several different polysilicon microstructures were tested, including fine-grained, columnar and multilayered. Amorphous silicon was also investigated. All of the microstructures displayed the same K_{Ic}. This indicates that, like Young's modulus,

fracture toughness is a material property, independent of the material microstructure or the geometry of the sample.

A tensile test, such as that shown in Fig. 53.7 but using a sample with sharp indentation-induced pre-cracks, yields a K_{Ic} of 0.86 MPa m$^{1/2}$ [53.35]. The passive, residual stress loaded beams with sharp pre-cracks shown in Fig. 53.6 gave a K_{Ic} of 0.81 MPa m$^{1/2}$ [53.13].

Given that K_{Ic} is a material property for polysilicon, the measured fracture strength σ_{crit} is related to K_{Ic} by

$$K_{Ic} = c\sigma_{crit}(\pi a)^{1/2} , \qquad (53.14)$$

where a is the crack-initiating flaw size, and c is a constant of order unity. The value for c will depend on the exact size, shape and orientation of the flaw; for a semicircular flaw, c is equal to 0.71 [53.36]. Therefore, any differences in the reported fracture strength of polysilicon will be the result of changes in a.

A good review of the experimental results available in the literature for polysilicon strength is contained in [53.37]. The tensile strength data vary from about 0.5 to 5 GPa. Like many brittle materials, the measured strength of polysilicon is found to obey Weibull statistics. This implies that the polysilicon samples contain a random distribution of flaws of various sizes, and that the failure of any particular specimen will occur at the largest flaw that experiences the highest stress. One consequence of this behavior is that, since larger specimens have a greater probability of containing larger flaws, they will exhibit decreased strengths. More specifically, it was found that the most important geometrical parameter is the surface area of the sidewalls of a polysilicon specimen [53.37]. The sidewalls, as opposed to the top and bottom surfaces, are those surfaces created by etching the polysilicon film. This is not surprising since LPCVD polysilicon films contain essentially no flaws within the bulk, and the top and bottom surfaces are typically very smooth.

As a result, the etching techniques used to create the structures will have a strong impact on the fracture strength of the material. For single-crystal silicon specimens it was found that the choice of etchant could change the observed tensile strength by a factor of two [53.38]. In addition, the bend strength of amorphous silicon was measured to be twice that of polysilicon for specimens processed identically [53.34]. It was found that the reactive ion etching used to fabricate the specimens produced much rougher sidewalls on the polysilicon than on the amorphous silicon.

The tensile strength of single crystal silicon has also been measured using a technique similar to that shown in Fig. 53.8. Sharp notches were introduced into the beams using a focused ion beam, and the apparent fracture toughness was measured for a variety of planes parallel to the notch front, along which the crack propagated. For the {110} notch plane, the fracture toughness was about 1 MPa m$^{1/2}$, and for the {100} notch plane it was about 2 MPa m$^{1/2}$ [53.39].

Fatigue of Polysilicon

Fatigue failure involves fracture after a number of load cycles, when each individual load is not sufficient by itself to generate catastrophic cracking in the material. For ductile materials, such as metals, fatigue occurs due to accumulated damage at the site of maximum stress and involves local plasticity. As a brittle material, polysilicon would not be expected to be susceptible to cyclic fatigue. However, fatigue has been observed for polysilicon tensile samples [53.35], polysilicon bend specimens with notches [53.27, 40], and polysilicon bend specimens with sharp cracks [53.41]. The exact origins of the fatigue behavior are still subject to debate, but some aspects of the experimental data are that the fatigue lifetime does not depend on the loading frequency [53.35], the fatigue behavior is affected by the ambient [53.13, 41], and the fatigue depends on the ratio of compressive to tensile stresses seen in the load cycle [53.13].

Friction of Polysilicon

The friction of polysilicon structures has been measured using the techniques described in Sect. 53.2.2. The measured coefficient of friction was found to vary from 4.9 [53.29] to 7.8 [53.30].

The static and dynamic friction of polysilicon coated with monolayer lubricants has been measured with a device similar to that shown in Fig. 53.15, but using a scratch-drive actuator instead of a comb-drive [53.42]. The dynamic friction at 0.2 m/s was approximately 80% of the static friction value; the static friction at zero applied load was due to an adhesive force of 0.95 nN/μm^2.

53.3.2 Mechanical Properties of Other Materials

As discussed above, of all the materials used for MEMS and NEMS, polysilicon has generated the most interest as well as the most research in mechanical properties characterization. However, measurements have been taken on other materials, and these are summarized in this section.

As discussed in Sect. 53.2.2, one advantage of the externally loaded tension test, as shown in Fig. 53.7, is that essentially any material can be tested using this technique. As such, tensile strengths have been measured to be 0.6 to 1.9 GPa for SiO_2 [53.43] and 0.7 to 1.1 GPa for titanium [53.44]. The yield strength for electrodeposited nickel was found to vary from 370 to 900 MPa, depending on the annealing temperature [53.16]. In addition, the yield strength was strongly affected by the current density during the electrodeposition process. Both the annealing and current density effects were correlated to changes in the microstructure of the material. Young's moduli were determined to be 100 GPa for titanium [53.44] and 215 GPa for electrodeposited nickel [53.16].

The tensile strength, Young's modulus, and Poisson's ratio of silicon nitride were measured to be 5.9 GPa, 0.23 and 257 GPa, respectively [53.45], and the same properties of amorphous diamond-like carbon were found to be 7.3 GPa, 0.17 and 759 GPa, respectively [53.46].

The technique of bending cantilever beams, shown in Fig. 53.8, can also be performed on a variety of materials. The yield strength and Young's modulus of gold were found to vary from 260 MPa and 57 GPa, respectively [53.19], to 300 MPa and 120 GPa, respectively [53.47], using this method. The same properties in Al were measured to be 150 MPa and 80 GPa, respectively [53.47], and in silicon nitride to be 6.9 GPa and 260 GPa, respectively [53.48]. Using a technique similar to that shown in Fig. 53.8, except that a doubly-clamped beam was used instead of a cantilever beam, the fracture toughness of ultrananocrystalline diamond was measured to be 4.5 MPa m$^{1/2}$ [53.49]. Resonating cantilever beams revealed a Young's modulus for gold of 47 GPa [53.50], and Young's moduli of 1.8 GPa and 14.4 GPa for silica and alumina aerogel thin films, respectively [53.51].

Another technique that can be used with a number of materials is the membrane deflection method, shown in Fig. 53.10. Silicon nitride measured with this technique revealed a Young's modulus of 258 [53.45] to 325 GPa and a burst strength of 7.1 GPa [53.52]. A polyimide membrane gave a residual stress of 32 MPa, a Young's modulus of 3.0 GPa, and an ultimate strain of about four percent [53.21]. Membranes were also fabricated from polycrystalline SiC films with two different grain structures [53.53]. The film with (110)-texture columnar grains had a residual stress of 434 MPa and a Young's modulus of 349 GPa. The film with equiaxed (110)- and (111)-textured grains had a residual stress of 446 MPa and a Young's modulus of 456 GPa.

Other devices that are used to measure mechanical properties require more complicated micromachining, namely patterning, etching and release, in order to operate. These devices are more difficult to fabricate with materials that are not commonly used as MEMS structural materials. However, the following examples demonstrate work in this area. The structure shown in Fig. 53.5b was fabricated from Si_xN_y and revealed a apparent fracture toughness of 1.8 MPa m$^{1/2}$ [53.12]. The devices shown in Fig. 53.6 revealed a fracture toughness of SiC of 3.1 MPa m$^{1/2}$ [53.54]. Lateral resonators of the type shown in Fig. 53.11 were processed using polycrystalline SiC, and the Young's modulus was determined to be 426 GPa [53.55]. The device shown in Fig. 53.12 was fabricated from polycrystalline germanium, and used to measure a bend strength of 1.5 GPa for unannealed Ge and 2.2 GPa for annealed Ge [53.56]. The same device was also fabricated from SiC and revealed a bend strength of 23 GPa [53.57]. Devices similar to that shown in Fig. 53.13 revealed the tensile strength of silicon nitride to be 6.4 GPa [53.52], and the Young's modulus and yield strength of aluminium to be 74.6 GPa and 330 MPa, respectively [53.58].

References

53.1 W. Nix: Mechanical properties of thin films, Metall. Trans. A **20**, 2217–2245 (1989)

53.2 G. G. Stoney: The tension of metallic films deposited by electrolysis, Proc. R. Soc. Lond. A **82**, 172–175 (1909)

53.3 W. Brantley: Calculated elastic constants for stress problems associated with semiconductor devices, J. Appl. Phys. **44**, 534–535 (1973)

53.4 A. Ni, D. Sherman, R. Ballarini, H. Kahn, B. Mi, S. M. Phillips, A. H. Heuer: Optimal design of multilayered polysilicon films for prescribed curvature, J. Mater. Sci., **38**, 4169–4173 (2003)

53.5 J. A. Floro, E. Chason, S. R. Lee, R. D. Twesten, R. Q. Hwang, L. B. Freund: Real-time stress evolution during $Si_{1-x}Ge_x$ heteroepitaxy: Dislocations, islanding, and segregation, J. Electron. Mater. **26**, 969–979 (1997)

53.6 X. Li, B. Bhusan: Micro/nanomechanical characterization of ceramic films for microdevices, Thin Solid Films **340**, 210–217 (1999)

53.7   M. P. de Boer, H. Huang, J. C. Nelson, Z. P. Jiang, W. W. Gerberich: Fracture toughness of silicon and thin film micro-structures by wedge indentation, Mater. Res. Soc. Symp. Proc. **308**, 647–652 (1993)

53.8   H. Guckel, D. Burns, C. Rutigliano, E. Lovell, B. Choi: Diagnostic microstructures for the measurement of intrinsic strain in thin films, J. Micromech. Microeng. **2**, 86–95 (1992)

53.9   F. Ericson, S. Greek, J. Soderkvist, J.-A. Schweitz: High sensitivity surface micromachined structures for internal stress and stress gradient evaluation, J. Micromech. Microeng. **7**, 30–36 (1997)

53.10   L. S. Fan, R. S. Muller, W. Yun, R. T. Howe, J. Huang: Spiral microstructures for the measurement of average strain gradients in thin films, Proc. IEEE Micro Electro Mechanical Systems Workshop, Napa Valley 1990 (IEEE, New York 1990) 177–182

53.11   M. Biebl, H. von Philipsborn: Fracture strength of doped and undoped polysilicon, Proc. Intl. Conf. Solid-State Sensors and Actuators, Stockholm 1995, ed. by S. Middelhoek, K. Cammann (Royal Swedish Academy of Engineering Sciences, Stockholm 1995) 72–75

53.12   L. S. Fan, R. T. Howe, R. S. Muller: Fracture toughness characterization of brittle films, Sens. Actuators A **21–23**, 872–874 (1990)

53.13   H. Kahn, R. Ballarini, J. J. Bellante, A. H. Heuer: Fatigue failure in polysilicon is not due to simple stress corrosion cracking, Science **298**, 1215–1218 (2002)

53.14   H. Kahn, N. Tayebi, R. Ballarini, R. L. Mullen, A. H. Heuer: Wafer-level strength and fracture toughness testing of surface-micromachined MEMS devices, Mater. Res. Soc. Symp. Proc. **605**, 25–30 (2000)

53.15   W. N. Sharpe Jr., B. Yuan, R. L. Edwards: A new technique for measuring the mechanical properties of thin films, J. Microelectromech. Syst. **6**, 193–199 (1997)

53.16   H. S. Cho, W. G. Babcock, H. Last, K. J. Hemker: Annealing effects on the microstructure and mechanical properties of LIGA nickel for MEMS, Mater. Res. Soc. Symp. Proc. **657** (2001) EE5.23.1–EE5.23.6

53.17   W. Suwito, M. L. Dunn, S. J. Cunningham, D. T. Read: Elastic moduli, strength, and fracture initiation at sharp notches in etched single crystal silicon microstructures, J. Appl. Phys. **85**, 3519–3534 (1999)

53.18   C.-S. Oh, W. N. Sharpe: Techniques for measuring thermal expansion and creep of polysilicon, Sens. Actuators A **112**, 66–73 (2004)

53.19   T. P. Weihs, S. Hong, J. C. Bravman, W. D. Nix: Mechanical deflection of cantilever microbeams: A new technique for testing the mechanical properties of thin films, J. Mater. Res. **3**, 931–942 (1988)

53.20   B. D. Jensen, M. P. de Boer, N. D. Masters, F. Bitsie, D. A. La Van: Interferometry of actuated micro-cantilevers to determine material properties and test structure nonidealities in MEMS, J. Microelectromech. Syst. **10**, 336–346 (2001)

53.21   M. G. Allen, M. Mehregany, R. T. Howe, S. D. Senturia: Microfabricated structures for the in situ measurement of residual stress, Young's modulus, and ultimate strain of thin films, Appl. Phys. Lett. **51**, 241–243 (1987)

53.22   H. Kahn, S. Stemmer, K. Nandakumar, A. H. Heuer, R. L. Mullen, R. Ballarini, M. A. Huff: Mechanical properties of thick, surface micromachined polysilicon films, Proc. IEEE Micro Electro Mechanical Systems Workshop, San Diego 1996, ed. by M. G. Allen, M. L. Redd (IEEE, New York 1996) 343–348

53.23   L. Kiesewetter, J.-M. Zhang, D. Houdeau, A. Steckenborn: Determination of Young's moduli of micromechanical thin films using the resonance method, Sens. Actuators A **35**, 153–159 (1992)

53.24   W. C. Tang, T.-C. H. Nguyen, R. T. Howe: Laterally driven polysilicon resonant microstructures, Sens. Actuators A **20**, 25–32 (1989)

53.25   P. T. Jones, G. C. Johnson, R. T. Howe: Fracture strength of polycrystalline silicon, Mater. Res. Soc. Symp. Proc. **518**, 197–202 (1998)

53.26   P. Minotti, R. Le Moal, E. Joseph, G. Bourbon: Toward standard method for microelectromechanical systems material measurement through on-chip electrostatic probing of micrometer size polysilicon tensile specimens, Jpn. J. Appl. Phys. **40**, L120–L122 (2001)

53.27   C. L. Muhlstein, E. A. Stach, R. O. Ritchie: A reaction-layer mechanism for the delayed failure of micron-scale polycrystalline silicon structural films subjected to high-cycle fatigue loading, Acta Mater. **50**, 3579–3595 (2002)

53.28   A. Corigliano, B. De Masi, A. Frangi, C. Comi, A. Villa, M. Marchi: Mechanical characterization of polysilicon through on-chip tensile tests, J. Microelectromech. Syst. **13**, 200–219 (2004)

53.29   M. G. Lim, J. C. Chang, D. P. Schultz, R. T. Howe, R. M. White: Polysilicon microstructures to characterize static friction, Proc. IEEE Micro Electro Mechanical Systems Workshop, Napa Valley 1990 (IEEE, New York 1990) 82–88

53.30   B. T. Crozier, M. P. de Boer, J. M. Redmond, D. F. Bahr, T. A. Michalske: Friction measurement in MEMS using a new test structure, Mater. Res. Soc. Symp. Proc. **605**, 129–134 (2000)

53.31   J. Yang, H. Kahn, A. Q. He, S. M. Phillips, A. H. Heuer: A new technique for producing large-area as-deposited zero-stress LPCVD polysilicon films: The MultiPoly process, J. Microelectromech. Syst. **9**, 485–494 (2000)

53.32   E. Chason, B. W. Sheldon, L. B. Freund, J. A. Floro, S. J. Hearne: Origin of compressive residual stress in polycrystalline thin films, Phys. Rev. Lett. **88** (2002) 156103-1–156103-4

53.33   S. Jayaraman, R. L. Edwards, K. J. Hemker: Relating mechanical testing and microstructural features of polysilicon thin films, J. Mater. Res. **14**, 688–697 (1999)

53.34   R. Ballarini, H. Kahn, N. Tayebi, A. H. Heuer: Effects of microstructure on the strength and fracture toughness of polysilicon: A wafer level testing approach, ASTM STP **1413**, 37–51 (2001)

53.35   J. Bagdahn, J. Schischka, M. Petzold, W. N. Sharpe Jr.: Fracture toughness and fatigue investigations of polycrystalline silicon, Proc. SPIE **4558**, 159–168 (2001)

53.36   I. S. Raju, J. C. Newman Jr.: Stress intensity factors for a wide range of semi-elliptical surface cracks in finite-thickness plates, Eng. Fract. Mech. **11**, 817–829 (1979)

53.37   J. Bagdahn, W. N. Sharpe Jr., O. Jadaan: Fracture strength of polysilicon at stress concentrations, J. Microelectromech. Syst., 302–312 (2003)

53.38   T. Yi, L. Li, C.-J. Kim: Microscale material testing of single crystalline silicon: Process effects on surface morphology and tensile strength, Sens. Actuators A **83**, 172–178 (2000)

53.39   X. Li, T. Kasai, S. Nakao, T. Ando, M. Shikida, K. Sato, H. Tanaka: Anisotropy in fracture of single crystal silicon film characterized under uniaxial tensile condition, Sens. Actuators A **117**, 143–150 (2005)

53.40   H. Kahn, R. Ballarini, R. L. Mullen, A. H. Heuer: Electrostatically actuated failure of microfabricated polysilicon fracture mechanics specimens, Proc. R. Soc. Lond. A **455**, 3807–3923 (1999)

53.41   W. W. Van Arsdell, S. B. Brown: Subcritical crack growth in silicon MEMS, J. Microelectromech. Syst. **8**, 319–327 (1999)

53.42   A. Corwin, M. P. de Boer: Effect of adhesion on dynamic and static friction in surface micromachining, Appl. Phys. Lett. **84**, 2451–2453 (2004)

53.43   T. Tsuchiya, A. Inoue, J. Sakata: Tensile testing of insulating thin films; humidity effect on tensile strength of SiO_2 films, Sens. Actuators A **82**, 286–290 (2000)

53.44   H. Ogawa, K. Suzuki, S. Kaneko, Y. Nakano, Y. Ishikawa, T. Kitahara: Measurements of mechanical properties of microfabricated thin films, Proc. IEEE Micro Electro Mechanical Systems Workshop (IEEE, New York 1997) 430–435

53.45   R. L. Edwards, G. Coles, W. N. Sharpe: Comparison of tensile and bulge tests for thin-film silicon nitride, Exp. Mech. **44**, 49–54 (2004)

53.46   S. Cho, I. Chasiotis, T. A. Friedmann, J. P. Sullivan: Young's modulus, Poisson's ratio and failure properties of tetrahedral amorphous diamond-like carbon for MEMS devices, J. Micromech. Microeng. **15**, 728–735 (2005)

53.47   D. Son, J.-H. Jeong, D. Kwon: Film-thickness considerations in microcantilever-beam test in measuring mechanical properties of metal thin film, Thin Solid Films **437**, 182–187 (2003)

53.48   W.-H. Chuang, T. Luger, R. K. Fettig, R. Ghodssi: Mechanical property characterization of LPCVD silicon nitride thin films at cryogenic temperatures, J. Microelectromech. Syst. **13**, 870–879 (2004)

53.49   H. D. Espinosa, B. Peng: A new methodology to investigate fracture toughness of freestanding MEMS and advanced materials in thin film form, J. Microelectromech. Syst. **14**, 153–159 (2005)

53.50   C.-W. Baek, Y.-K. Kim, Y. Ahn, Y.-H. Kim: Measurement of the mechanical properties of electroplated gold thin films using micromachined beam structures, Sens. Actuators A **117**, 17–27 (2005)

53.51   R. Yokokawa, J.-A. Paik, B. Dunn, N. Kitazawa, H. Kotera, C.-J. Kim: Mechanical properties of aerogel-like thin films used for MEMS, J. Micromech. Microeng. **14**, 681–686 (2004)

53.52   A. Kaushik, H. Kahn, A. H. Heuer: Wafer-level mechanical characterization of silicon nitride MEMS, J. Microelectromech. Syst. **14**, 359–367 (2005)

53.53   S. Roy, C. A. Zorman, M. Mehregany: The mechanical properties of polycrystalline silicon carbide films determined using bulk micromachined diaphragms, Mater. Res. Soc. Symp. Proc. **657** (2001) EE9.5.1–EE9.5.6

53.54   J. J. Bellante, H. Kahn, R. Ballarini, C. A. Zorman, M. Mehregany, A. H. Heuer: Fracture toughness of polycrystalline silicon carbide thin films, Appl. Phys. Lett. **86**, 071920–1–071920–3 (2005)

53.55   A. J. Fleischman, X. Wei, C. A. Zorman, M. Mehregany: Surface micromachining of polycrystalline SiC deposited on SiO_2 by APCVD, Mater. Sci. Forum **264-268**, 885–888 (1998)

53.56   A. E. Franke, E. Bilic, D. T. Chang, P. T. Jones, T.-J. King, R. T. Howe, G. C. Johnson: Post-CMOS integration of germanium microstructures, Proc. Int. Conf. Solid-State Sensors and Actuators (IEE Japan, Tokyo 1999) 630–637

53.57   D. Gao, C. Carraro, V. Radmilovic, R. T. Howe, R. Maboudian: Fracture of polycrystalline 3C-SiC films in microelectromechanical systems, J. Microelectromech. Syst. **13**, 972–976 (2004)

53.58   M. A. Haque, M. T. A. Saif: A review of MEMS-based microscale and nanoscale tensile and bending testing, Exp. Mech. **43**, 248–255 (2003)

Part G | 53

54. Thermo- and Electromechanical Behavior of Thin-Film Micro and Nanostructures

Applications using thin-film micro- and nanomechanical structures for actuation and sensing require the coupling of energy between various physical domains. This chapter focuses on two important couplings: thermomechanics and electromechanics. Thermomechanical phenomena is considered in Sect. 54.1, where we describe broad aspects of the deformation characteristics and stress states that arise when dealing with a large class of thin-film microstructures. These include the origin of stresses in multilayer films and their qualitative evolution through processing and release from the substrate. A basic framework is described for the analysis of the thermomechanics of multilayer films, emphasizing the *linear* response. Issues of *geometric* and *material* nonlinearity are then taken up, and equal emphasis is put on the generality of the analysis approach and specific applications. As much as possible, we show comparisons between theoretical predictions and companion experimental results.

A common use of electromechanics in microsystems involves the application of an electric potential between two electrodes where one is fixed and the other is connected to a deformable elastic structure. The electric potential produces an electric field and an associated electrostatic force that deforms the structure, and in turn alters the electrostatic force, resulting in fully-coupled nonlinear behavior. At some point an instability can occur where the deformable structure snaps into contact with the fixed electrode. This phenomena, called pull-in, is often used for switching applications. In Sect. 54.2 we describe the basic electromechanical phenomena using a parallel-plate electrostatic actuator as a reference. We discuss many important phenomena including pull-in, external forcing, stabilization, time response, the effects of dielectric charging, and breakdown of gases in small gaps. We address

these phenomena for a wide range of micromechanical structures including cantilevered beams and plates, torsionally suspended plates, and zipper actuators with curved electrodes.

Part G | 54

Microsystems rely heavily on thin-film technology – the deposition, patterning, etching, etc. of multiple film layers to yield micromechanical, and more recently, nanomechanical structures. In this chapter we do not discuss fabrication techniques, but instead refer the interested reader to many excellent references that describe the fabrication of structures by means of thin-film technology [54.1, 2] and, in particular, surface micromachining [54.3]. Regardless of the fabrication process flow, the end result is a thin-film microstructure that is connected to a substrate through one or more *sacrificial* film layers, which must be released to render it freestanding and thus useful. Figure 54.1a illustrates a simple, yet typical in terms of the microstructure and the materials involved, case: a cantilevered beam made of polycrystalline silicon (polysilicon) anchored to a silicon substrate and encased in an SiO_2 sacrificial layer. Also shown is an electrode used for electrostatic actuation of the cantilever after release. This is the state of affairs after the thin-film fabrication steps; we refer to it as the as-processed but unreleased condition. To render the beam freestanding, although anchored to the substrate at one end, the sacrificial material must be removed (Fig. 54.1b). For SiO_2, this is usually accomplished by etching in an HF solution, followed by drying. The latter step is important because during wet etching, strong (at the microscale) capillary forces can pull the beam into contact with the substrate. If the microstructure is too compliant, adhesive forces between the microstructure and the substrate can pin the beam to the substrate, rendering it useless. This phenomena is often referred to as stiction, and many recent references describe the phenomena, its impact on reliability, and practical ways to overcome it [54.4, 5]. For our purposes, we assume that the microstructure can be successfully released so that it is free from the substrate where desired and anchored where desired.

High stresses can develop in the film layers during processing, and these can vary significantly from layer to layer and even within a single layer. When the sacrificial layers are removed, the stresses redistribute, but at the expense of deformation. This can be detri-

mental in applications in which planarity is essential, or it can be used advantageously for actuation purposes. Indeed many applications are based on the use of thermally induced bending of bilayer beams and plates that arises due to the difference in thermal expansion coefficients between the film layers. An example is shown in Fig. 54.2 where a micromirror is suspended above the substrate by 2.0-μm-thick gold/polysilicon bilayer beams that position the mirror (or any desired device) and can then be electrostatically actuated to control the position of the micromirror. In addition to the stresses in individual film layers, each free surface (or interface between two materials) is subjected to a surface or interface stress [54.6]. This stress results because atoms at the surface or interface experience a local environment that differs from that in the interior of the film; the latter being that associated with a bulk material. For micrometer-scale structures, this surface stress is of little practical significance with regard to the overall deformation of the structure because it is associated with a region penetrating only a few lattice spacings from the surface. However, as characteristic sizes of structures decrease from the micrometer to the nanometer scale, this surface stress becomes significant and can dramatically impact atomistic structure, elasticity, and inelasticity of nanostructural elements such as rods, beams, and plates [54.7]. Upon release, micro- and nanostructures are typically used for actuation and/or sensing functions. This requires the coupling of the mechanical behavior of the structure with various energy domains. These

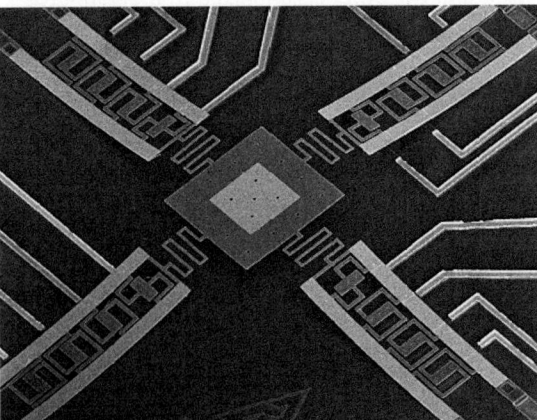

Fig. 54.2 Scanning electron microscope image of a micromirror (300 μm square) supported above the substrate by gold/polysilicon bilayer beams that can be individually actuated electrostatically to control the position of the micromirror

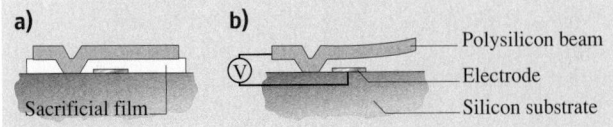

Fig. 54.1a,b Schematic of thin-film microstructures (**a**) attached to a substrate layer, and (**b**) free-standing after release from the substrate by etching of the sacrificial film

couplings include, among others, thermomechanics, electromechanics, electro-thermomechanics, and magnetomechanics. The most common, and perhaps the most easily realizable, is electromechanics, specifically the coupling of electrostatics and mechanical behavior. This coupling has been successfully demonstrated in many applications in the contexts of both sensors and actuators.

This chapter focuses on the behavior of thin-film structures under the action of thermomechanical and electromechanical loadings. Section 54.1 is devoted to the development of stresses and deformation in multilayer thin-film structures when subjected to thermomechanical loading. It focuses heavily on linear thermoelastic behavior, but describes important issues regarding geometric and material nonlinearity. It concludes with a discussion of the effects of surface stresses on nanostructures, and illustrates how the general formalism for analysis can be applied to nanostructures. Section 54.2 is devoted to electromechanics, emphasizing general approaches to modeling and then focusing on the behavior of common classes of electromechanical structures. The treatment is certainly not exhaustive, but we hope it is detailed enough to provide the interested reader with an entry-level understanding of the basic phenomena and some guidance regarding where to turn to obtain more in-depth information.

54.1 Thermomechanics of Multilayer Thin-Film Structures

In this section we describe aspects of the deformation characteristics and stress states in multilayer thin-film microstructures. Deformation and stress states will be considered for microstructures that are both attached to a substrate and freestanding after release from the substrate. In the modeling formalism that we will describe, the substrate simply serves as another layer. We will refer to microstructures as beams and/or plates, depending on the relevant dimensions. At times we will use the terms loosely, but we will describe the difference of their deformation characteristics in detail. The remainder of this section will be organized as follows: Sect. 54.1.1 provides an overview of the basic phenomena including the origin of stresses in multilayer films and their qualitative evolution through processing and then release from the substrate. Section 54.1.2 outlines a general framework for the analysis of the thermomechanical behavior of multilayer thin films, emphasizing *linear* response. Issues of *geometric* and *material* nonlinearity are then briefly taken up in Sects. 54.1.3 and 54.1.5, respectively. Section 54.1.6 touches on a number of related and important issues that were not covered in the other sections.

The objectives of this section are to describe the thermomechanical phenomena that arise when dealing with a large class of thin-film microstructures. A basic framework is described for analysis, and equal emphasis is put on the generality of the approach and specific applications. When possible, we show comparisons between theoretical predictions and companion experimental results. To this end, a specific material and geometric system must be chosen, and we use a gold/polysilicon multilayer for most of these cases. We note, though, that the phenomena discussed are not restricted to the gold/polysilicon system but arise in many multilayer film systems.

54.1.1 Basic Phenomena

During the multiple thin-film deposition and etching processes, many complicated mechanisms occur that result in straining of the film layers, and these generally differ from layer to layer. The requirement that the film layers maintain coherency at the interfaces between them means that these strains cannot occur freely, and thus they are constrained. This constraint leads to stresses in the layers and curvature of the multilayer film system; both can have significant practical implications. The mechanisms that lead to internal stresses include chemical reactions, lattice mismatch, and grain growth, among others, and are often called *intrinsic* stresses (see, for example [54.8]). The details of intrinsic-stress development in thin films have been the subject of extensive study for both scientific and technological reasons, but they are beyond the scope of our study. In addition, stresses may arise due to thermal expansion mismatch between layers when subjected to a temperature change. Such stresses are often called *extrinsic* stresses. Now, in the unreleased configuration of Fig. 54.1a, a complicated stress distribution exists as a result of the intrinsic stresses. The release process to yield the configuration of Fig. 54.1b obviously alters the stress distribution as well as the state of deformation. In order to gain physical insight, the remainder of this section discusses rather qualitatively the nature of the stress states in the unreleased and released configurations. In subsequent sections details are given regarding analysis of these thermomechanical phenomena.

Transformation Strains, Misfit Strains, and Thin-Film Stresses

The important concepts here are most easily described in terms of a bilayer film system in which both layers are isotropic. Furthermore, they are easier to grasp in the context of a specific example, of which many would suffice. Once this understanding is in place, the generalization of the concepts to multilayer film systems follows naturally. To begin, we define a couple of terms. A *transformation strain* ε^* is a strain that occurs in a solid, but with no accompanying stress. It is thus an inelastic strain. In the literature a number of other terms are used to describe this concept including *stress-free strain* and *eigenstrain*. It can result, for example, from thermal expansion, a crystallographic phase transformation, or a variety of sources. When thermal expansion is the source, ε^* can be specified as $\varepsilon^* = \alpha T$ where α is the thermal expansion coefficient of the material and T is the temperature change from a reference (usually stress-free) configuration. In general ε^* and α are second-order tensors. Stresses and deformation develop in bilayer film/substrate systems when, due to external or internal sources, each layer wants to undergo a *different* transformation strain, but the requirement of bonding of the films at the interface constrains each film to some degree. The source of the stress and deformation is then the difference in the transformation strains between the layers, i. e., $\Delta\varepsilon = \varepsilon_1^* - \varepsilon_2^*$ where ε_1^* and ε_2^* are the transformation strains of the two layers. For example, if a bilayer is subjected to a uniform temperature change T, then $\Delta\varepsilon = (\alpha_1 - \alpha_2) T = \Delta\alpha T$ where α_1 and α_2 are the thermal expansion coefficients of the two layers. $\Delta\varepsilon = \varepsilon_1^* - \varepsilon_2^*$ is termed the *misfit strain*. Its role as the direct source of deformation is obvious and clear for two-layer systems. The concept of misfit strain is not as direct for multilayer systems, and so we simply characterize the behavior in terms of the individual transformation strains of each layer.

It is important to understand the character of the stress distribution and deformation. To make the ideas concrete, consider the bilayer film system with planar dimensions (normal to the film thickness direction) that far exceed the total thickness. The two layers, denoted by 1 and 2, are perfectly bonded along the interface and have different isotropic thermal expansion coefficients and elastic moduli. The bilayer is initially flat, and each

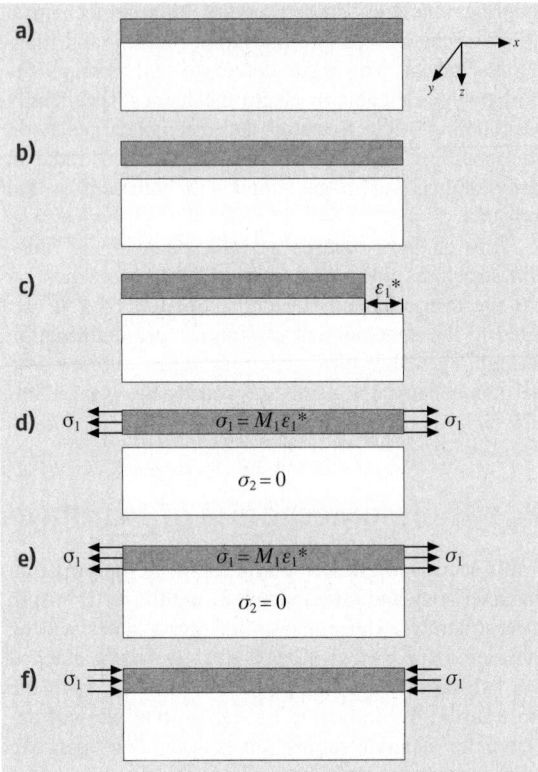

Fig. 54.3 Illustration showing the development of stresses and deformation in a bilayer film/substrate system

layer is stress-free. Now assume the bilayer is subjected to a uniform temperature change T, which would lead to equibiaxial ($\varepsilon^* = \varepsilon_{xx}^* = \varepsilon_{yy}^*$, $\gamma_{xy}^* = 0$) transformation strains $\varepsilon_1^* = \alpha_1 T$ and $\varepsilon_2^* = \alpha_2 T$ in the two layers if they were not bonded. Because they are bonded, this results in the development of equibiaxial stresses ($\sigma = \sigma_{xx} = \sigma_{yy}$, $\sigma_{xy} = 0$) in the layers, a change of length in the planar directions (in-plane straining) and bending of the bilayer (curvature). This can be understood by the following thought experiment, which is often used in the calculation of film stresses. We will only use it, though, to explain the basics of the resulting deformation and will leave the actual calculations to the next section.

Consider the bilayer system shown schematically in Fig. 54.3a. The film layers are stress-free and perfectly bonded. Our discussion will be in the context of biaxial transformation strains and stresses, but the ideas can be immediately applied to situations of uniaxial stress/strain behavior or plane-stress or plane-strain situations. We are interested in obtaining the deformation and stress state when the bilayer is subject to a uniform temperature change T. Since the films are stress free in Fig. 54.3a, one can cut the bonds that connect the two layers and separate the films (Fig. 54.3b) without generating any stresses or deforming either layer. We neglect the effects of interface and surface stresses in this description, but will take them up in Sect. 54.1.4. Now we (Fig. 54.3c) subject the layers to a uniform temperature change T; each layer will undergo its stress-free biaxial strain, $\varepsilon_1^* = \alpha_1 T$ and $\varepsilon_2^* = \alpha_2 T$ (for simplicity, Fig. 54.3c illustrates the deformation for the case where $\varepsilon_2^* = 0$, $T < 0$, and $\alpha_1 > 0$, but the ideas are the same for more general situations). Now the interfaces, which were originally in perfect registry, are no longer. To bring them into registry, a uniform stress must be applied to the film (Fig. 54.3d) by means of the application of forces along the film edge. At this stage the film is subjected to the uniform biaxial stress $\sigma_1 = M_1 \varepsilon_1^*$, where M_1 is the biaxial modulus of the film (see Sect. 54.1.2), the substrate is still stress-free, $\sigma_2 = 0$, and the layers are in perfect registry. They can then be reconnected (imagine reattaching the original bonds that were cut in Fig. 54.3b), without generating additional stresses or deforming the layers (Fig. 54.3e). The stress state in Fig. 54.3e is the same as that in Fig. 54.3d, the only difference is that the films are now bonded. Now, to recover the solution for the bilayer subjected to a uniform temperature change T, we only have to remove the edge forces. This can be accomplished by applying forces of equal magnitude but opposite sign to the edges (Fig. 54.3f). The solution is then obtained as the superposition of the problems in Fig. 54.3e, which is trivial, and Fig. 54.3f, which is difficult in general. Instead of now trying to solve the problem in Fig. 54.3f, we will simply describe the character of the resulting deformation and stress state. Two aspects are important: the behavior near the free edge where the stress in Fig. 54.3f is applied, and the behavior far away from the free edge. One can appeal to St. Venant's principle to simplify the latter, but more involved analysis is required for the former (see for example [54.9]).

At a distance of ten or so film thicknesses away from the edges, the solution to the problem in Fig. 54.3f can be decomposed, as shown in Fig. 54.4. The uniform force distribution on the edges of the film (Fig. 54.4a) can be replaced by a concentrated force acting at the middle z-coordinate of the film thickness (Fig. 54.4b). The solution to this problem can then be obtained by the superposition of the two problems in Fig. 54.4c and 54.4d, the loadings of which are statically equivalent to that in Fig. 54.4b. That in Fig. 54.4c is a concentrated force applied at the as-yet-unknown neutral surface of the bilayer; it will lead to a uniform in-plane (x–y plane) straining of the bilayer. That in Fig. 54.4d is a moment of magnitude Pd, where d is the distance between the z-coordinates of the middle of the film and the neutral surface of the bilayer; it will lead to bending of the bilayer. If there were also a film on the bottom of the substrate so that the layered film was symmetric in material properties and geometry (film thicknesses) in the z-direction, then no moment would result, and the layered system would not bend but would only strain in-plane.

It is probably obvious that very near the edge, the stress state is quite complicated. In fact, in the context of linear elasticity, a stress singularity generally exists at the intersection of the free edge and the interface between the two layers. This singular stress state is similar to that which exists at a crack tip, but the strength of the singularity differs depending on the mismatch in elastic constants of the materials [54.10, 11]. The region within which this stress state exists depends on the film thickness, and for very thin films it is often so small that it is insignificant. It has recently been shown, however, that in some cases it can be appreciable and in fact can be the primary source of failure [54.12]. As described in detail by *Hui* et al. [54.9], aside from the singular stresses, the stress state near the edges contains highly localized shear and normal stress (perpendicular to the interface) components. By St. Venant's principle these stress com-

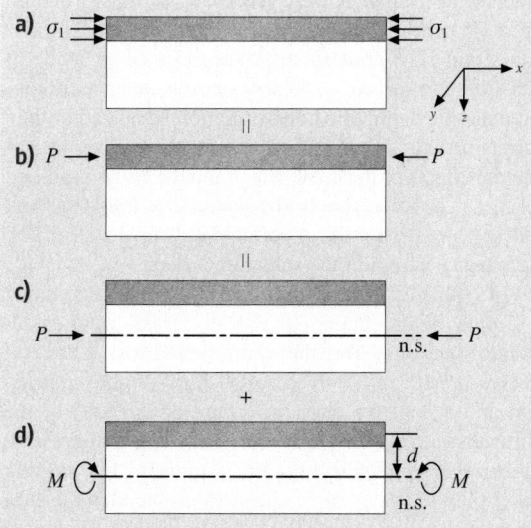

Fig. 54.4 Decomposition of the edge-loading problem into the superposition of a force and moment applied to the multilayer. Away from the edge, the force gives rise to uniform straining, and the moment gives rise to bending

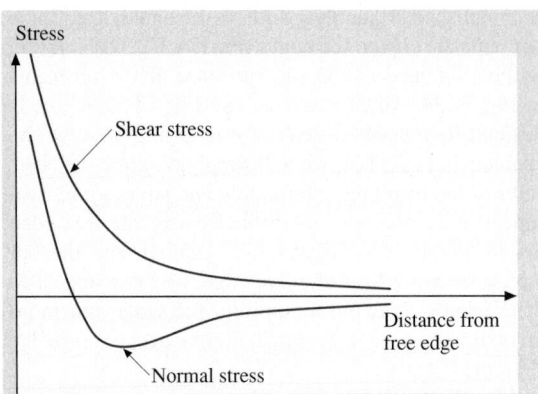

Fig. 54.5 Character of the normal and shear stress distribution at the film/substrate interface near the free edge

ponents decay over a rather short distance from the edge, but they can play a significant role in the durability of the film/substrate system as they can lead to delamination of the interface or cracking into one of the layers. The character of these stress components at the interface is shown qualitatively in Fig. 54.5. Note that, in order to satisfy equilibrium, the normal stress (often called the peel stress) changes sign from tensile to compressive. Furthermore, since these stress components are highly localized at the edge, in the interior of the system the interface is free of normal or shear stresses. In this chapter we will not further consider these localized stresses but instead will concern ourselves with the stresses away from the edges and the companion deformation.

Our discussion has been based on the assumption that transformation strains are constant throughout the cross section of each film. However, in microsystems applications, films will often have transformation strains that vary through the thickness. This will lead to bending of the film after it is released from the substrate. Transformation strains that vary through the thickness, i.e., $\varepsilon^*(z)$, can easily be accommodated in the description of Fig. 54.3. It is altered in that now in Fig. 54.3c an in-plane strain results, the magnitude of which is given by the through-thickness average of $\varepsilon^*(z)$, but in addition, the film bends with constant curvature. Then, both a force (uniform stress) and moment (linearly varying stress) must be applied in Fig. 54.3d to flatten the film and lengthen it so that it is again in perfect registry with the substrate and can then be rebonded [54.13].

This description provides the starting point for understanding the effects of transformation strains on stresses and deformation in a multilayer thin-film system. It is an accurate picture of the state of affairs before

release of a film from the substrate to yield a freestanding microstructure. In the next sections we will discuss the changes in the deformation that occur upon release as the constraint of the substrate is removed.

Single-Layer Beams

To keep things simple, in this and the following section we focus on *beams*, which have one nonzero normal stress component, that along the beam axis. In subsequent sections we will discuss the limitations of approximating such structures as beams and proceed with a more appropriate development in terms of *plates*. Consider the beam shown in Fig. 54.1. For simplicity we assume that the sacrificial material is much thinner than the substrate, has the same elastic moduli as the substrate, and so before release is mechanically equivalent to the substrate. In this way we can describe the stress distribution before release as that of a two-layer beam. To make the discussion more tangible, consider a specific but realistic case of a polysilicon beam on a silicon substrate. The top layer of thickness $1.5\,\mu m$ is the polysilicon film, and the bottom layer of thickness $500\,\mu m$ is the silicon substrate (including the very thin sacrificial layer that will be removed). Assume that the polysilicon film has an intrinsic mean stress S and a linear stress gradient β (we can easily consider a nonlinear stress gradient, but the simple linear gradient suffices to demonstrate the concepts involved). This intrinsic stress is interpreted as the stress state in the film if it were free, i.e., a transformation stress. It could be expressed in terms of a transformation strain $\varepsilon^*(z)$, but since reference is often made to thin-film stresses we will carry out this and the following example using this transformation stress rather than a transformation strain. We consider three cases to isolate the effects of the mean stress and the stress gradient: (a) $S = -10\,\text{MPa}$, $\beta = 0$, (b) $S = 0$, $\beta = 0.8\,\text{MPa}/\mu m$, and (c) $S = -10\,\text{MPa}$, $\beta = 0.8\,\text{MPa}/\mu m$. In all cases, the intrinsic stress in the substrate is zero.

In Table 54.1 we give calculated results using a simple beam theory (to be discussed in the following sections) for the three scenarios. Case (a) ($S = -10\,\text{MPa}$, $\beta = 0$) results in a slight contraction and a negative curvature (bent downward) of the film/substrate system prior to release. The compressive stress in the polysilicon beam is uniform but slightly lower than the value of S. The stress induced in the substrate is small and varies linearly through its thickness. It changes sign through the substrate, a necessary requirement to satisfy equilibrium. After release the stress is zero throughout the beam. To relax the stress, the beam extends (the midplane strain ε^0 is positive), but

Table 54.1 Stresses, curvature, and strains in a film/substrate system consisting of a 1.5-μm-thick polysilicon film and a 500-μm-thick silicon film. The elastic moduli of the film and the substrate are assumed equal, and the film has an intrinsic mean stress and stress gradient. The sign convention for curvature is that ($-$) indicates bending downward and ($+$) indicates bending upward

| | | (a) $S=-10\,\mathrm{MPa}$ $\beta=0\,\mathrm{MPa/\mu m}$ | | (b) $S=0\,\mathrm{MPa}$ $\beta=0.8\,\mathrm{MPa/\mu m}$ | | (c) $S=-10\,\mathrm{MPa}$ $\beta=0.8\,\mathrm{MPa/\mu m}$ | |
		Unreleased	Released	Unreleased	Released	Unreleased	Released
	ε^0 (με)	0.0624	62.5	0	0	0.0624	62.5
	κ (m^{-1})	-7.5×10^{-4}	0	5.0×10^{-9}	5.0	-7.5×10^{-4}	5.0
σ (MPa)	Top	-9.96	0	0.2	0	-9.76	0
Polysilicon film	Bottom	-9.96	0	-0.2	0	-10.16	0
σ (MPa)	Top	0.04	0	-2.0×10^{-7}	0	0.04	0
Silicon substrate	Bottom	-0.02	0	2.0×10^{-7}	0	-0.02	0

because $\beta=0$ it does not bend. Case (b) ($S=0\,\mathrm{MPa}$, $\beta=0.8\,\mathrm{MPa/\mu m}$) results in a slight positive curvature (bent upward) but no extension or contraction of the film/substrate system prior to release. The mean compressive stress in the polysilicon beam is zero, and the gradient of the stress distribution through the film is decreased from the value of β. The stress induced in the substrate is again small and varies linearly through its thickness, changing sign. After release the stress is again zero throughout the beam. The beam does not extend or contract since $S=0$, but it bends, assuming a curvature $\kappa=\beta/E$ where E is the Young's modulus of polysilicon (see Sect. 54.1.2 for details regarding this result). Careful examination of Table 54.1 shows that the strain, curvature, and stress for case (c) ($S=-10\,\mathrm{MPa}$, $\beta=0.8\,\mathrm{MPa/\mu m}$) can be obtained as the superposition of the results of cases (a) and (b). Specifically, after release the beam extends and bends.

These results illustrate the behavior of beams before and after release when they contain a mean stress and/or a linear stress gradient. The results also suggest natural means to attempt to measure the mean stress and the gradient. Test structures that are sensitive to length changes of beams after release can be used to characterize mean stress. For example, fracture of arrays of beams of varying width upon release can be used to estimate tensile stress, and fixed-fixed beams that buckle can be used to estimate compressive stress. The stress gradient in a beam can be determined by measuring its curvature after release. A survey of test structures suitable for these purposes can be found in [54.14].

Bilayer Beams
While stresses in single-layer microstructures are important for many applications, multilayer film micro-

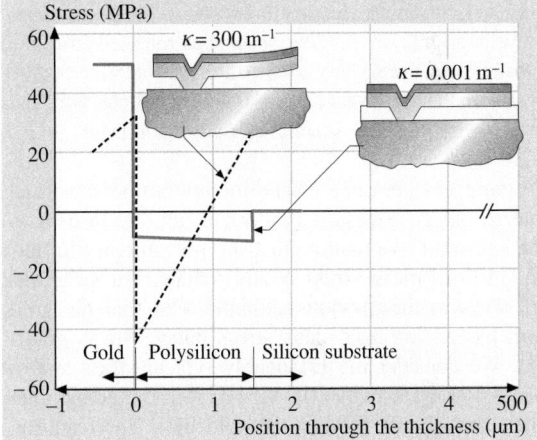

Fig. 54.6 Stress distribution through the thickness for a bilayer film consisting of a 0.5-μm-thick gold film on a 1.5-μm-thick polysilicon film subjected to an internal stresses and stress gradients. Stress distributions are shown before and after release from 500-μm-thick silicon

structures abound in microsystems technology, and the development of stresses and deformation in them is also important. In order to convey the basic ideas, we consider an example of a two-layer beam fixed to a substrate in the same manner as that in the previous section. Specifically we consider the layered structure shown in the inset of Fig. 54.6: a polysilicon beam of thickness 1.5 μm attached to a silicon substrate of thickness 500 μm, covered with a gold film of thickness 0.5 μm. Photos of actual beams of this form are shown in Fig. 54.7 where the length of the longest beam shown is 600 μm, and the beams are 50 μm wide. Again we assume that the sacrificial material is very

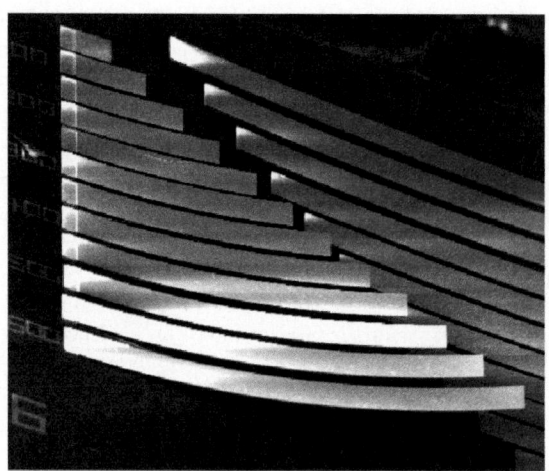

Fig. 54.7 Scanning electron microscope image of gold/polysilicon bilayer cantilever beams as released from the substrate. All beams have a 0.5-μm-thick gold film, but the beams on the *left* have a 1.5-μm-thick polysilicon film, and those on the *right* a 3.5-μm-thick polysilicon layer

thin and has the same elastic moduli as the substrate, and so before release it is mechanically equivalent to the substrate. We assume that the polysilicon film has an intrinsic mean stress S and a linear stress gradient β as in the previous example, and that the gold film has an intrinsic mean stress but no stress gradient. We consider the reasonably typical stress states: $S = -10\,\text{MPa}$ and $\beta = 0.8\,\text{MPa}/\mu\text{m}$ in the polysilicon beam and $S = 50\,\text{MPa}$ in the gold film. The resulting curvature and stress distribution before and after release are shown in Fig. 54.6. They show that before release the stresses in the polysilicon and gold layer are nearly uniform, compared to the range of stresses throughout the layers. The multilayer experiences a positive curvature; recall that the system consisting of just the polysilicon film and the substrate experienced a negative curvature of about the same magnitude. Thus the tensile stress in the gold is significant enough to drive the curvature from negative (bending down) to positive (bending up).

After release from the substrate, the gold/polysilicon beam contracts and bends substantially to a curvature of about $300\,\text{m}^{-1}$. This should be compared with $5\,\text{m}^{-1}$ for the polysilicon beam with a stress gradient. For a beam $300\,\mu\text{m}$ long cantilevered from one end as shown in Fig. 54.6, this amounts to a tip deflection of about $13\,\mu\text{m}$. The stress states in both the gold and polysilicon are significantly altered upon release. Perhaps most significantly, large stress gradients exist through the thickness of each layer. Starting at the top of the gold film, the

tensile stress increases through the gold thickness. Due to the discontinuity in elastic modulus and the residual stress and stress gradients, there is a jump in the stress of about $-77\,\text{MPa}$ at the gold/polysilicon interface. The stress is compressive at the top of the polysilicon film layer and increases linearly through the thickness of the polysilicon film, taking on its maximum tensile value at the bottom of the two-layer beam. In summary, the release process results in a reduction of the stress in the gold, an increase of the stress in the polysilicon, stress gradients in both layers, a small contraction, and a large curvature change.

These examples are meant to convey the basic phenomena and give a feel for the magnitudes of the quantities involved. The actual numerical results are based on calculations performed using a relatively simple multilayer beam theory and are reasonable estimates. Details regarding the analysis will be given in the following section, including a discussion of the suitability of the simple beam theory.

54.1.2 A General Framework for the Thermomechanics of Multilayer Films

Numerous studies have elucidated the basic thermomechanical response of multilayer material systems when subjected to temperature changes or other sources of transformation strains between the layers. These have come in the context of many technological applications, the most common being structural composite materials [54.15, 16] and thin film/substrate systems for microelectronics [54.17–24], which are directly applicable to the analysis of multilayer films.

The General Multilayer Film

When a layered film is subjected to transformation strains, two aspects of deformation play a primary role: straining of the midplane and bending. When the transverse deflections due to bending are of prime importance, as is often the case, one way to broadly characterize the deformation response, especially for plates with relatively large in-plane dimensions compared to their thickness, is in terms of the average curvature developed as a function of the transformation strains. Formally the curvature is a second-rank tensor, and for the type of layered film problems considered here it can be wholly described by the two principal curvature components, e.g., in the x- and y-directions, κ_x and κ_y. The curvature is a pointwise quantity, meaning it varies from point to point over the in-plane dimensions

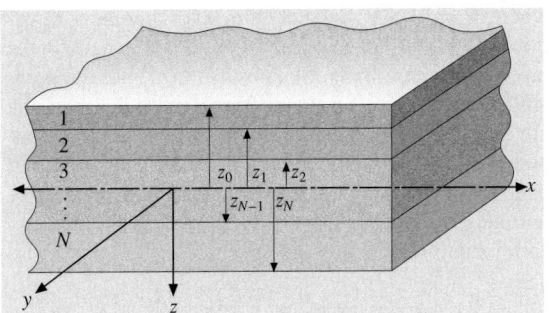

Fig. 54.8 Multilayer film system with a definition of the parameters involved in the analysis

of the plate. In terms of the average curvature variation as a function of misfit strain, three deformation regimes have been identified [54.18–21, 24, 25]. The first regime is a linear relation between the average curvature and temperature change where $\kappa_x = \kappa_y$, i. e., the average curvature is spherically symmetric (when the materials are isotropic). This symmetric deformation would not exist if the material properties were anisotropic. This deformation regime is characterized by both small transverse displacements and rotations, and so conventional thin-plate theory adequately describes the deformation. It is the subject of the analysis in this section. The second and third regimes, which arise due to geometric nonlinearity, will be taken up in Section 54.1.3.

Figure 54.8 shows a schematic of the geometry considered in this study; a multilayer film microstructure with lateral dimensions that far exceed the total thickness. As such we ignore the effects of the free edges and focus on a representative point in the layered film that lies at least a distance of a few film thicknesses from the edge. The layers are numbered $1, 2, \ldots, N$ where 1 is the top layer and N is the bottom layer. Each layer has a thickness t_i $(i = 1 \cdots N)$, and the total thickness of the microstructure is t. Each of the film layers consists of an isotropic material with a Young's modulus E_i, Poisson's ratio, ν_i, and may undergo a transformation strain $\varepsilon_i^*(z)$, which may vary with position z through the film thickness. The source of $\varepsilon_i^*(z)$ may be arbitrary, and the effect of multiple sources can be considered as the sum of the individual ones. A common source, and one that we will focus on, is thermal expansion upon a uniform temperature change. In this case $\varepsilon_i^*(z) = \varepsilon_i^* = \alpha_i T$, where α_i is the thermal expansion coefficient and T is the temperature change from a reference state. In addition, each layer may contain an internal stress $\hat{\sigma}(z)$, which also may vary with position z through the thickness. For simplicity we focus our re-

sults on isotropic layers, but the framework is applicable to anisotropic behavior of each layer. As a result of the difference in stress-free strains and/or internal stresses between each layer, stresses are developed in the layers and the structure deforms; our intent is to compute these.

Analysis

The analysis presented here is somewhat standard in the treatment of laminated composite plates but has not often been used in the analysis of multilayer thin-film problems for microelectronics and/or microelectromechanical systems. Detailed accounts of parts of the theory can be found in texts on composite materials (see for example [54.26, 27]), and so only a brief account is presented here.

The displacements u and v at any point z through the thickness of the film are assumed to be described by the classical Kirchoff hypothesis. As illustrated in Fig. 54.9, the Kirchoff hypothesis states that, after loads are applied to the layered film, the line a–a', which is initially straight and normal to the geometric midplane, remains straight, normal to the deformed geometric midplane, and its length is unchanged. In other words, the normal does not deform but only translates and rotates. As a consequence of this assumption, the displacements of any point in the layered film are given by:

$$
\begin{aligned}
u &= u^0 - z \frac{\partial w^0}{\partial x} \\
v &= v^0 - z \frac{\partial w^0}{\partial y} \\
w &= w^0 ,
\end{aligned}
\tag{54.1}
$$

where u^0, v^0, and w^0 are the displacements of the geometric midplane at $z = 0$. Note that, in addition to this explicit dependence of the displacements on z, they are generally functions of x and y.

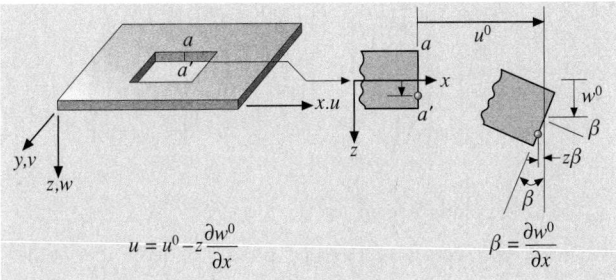

Fig. 54.9 Kinematics of the deformation of the thin-film multilayer

The nonzero normal and shear strains are then:

$$\varepsilon_x = \frac{\partial u}{\partial x} = \frac{\partial u^0}{\partial x} - z\frac{\partial^2 w^0}{\partial x^2} = \varepsilon_x^0 + z\kappa_x$$

$$\varepsilon_y = \frac{\partial v}{\partial y} = \frac{\partial v^0}{\partial x} - z\frac{\partial^2 w^0}{\partial x^2} = \varepsilon_y^0 + z\kappa_y$$

$$\gamma_{xy} = \frac{\partial u}{\partial y} + \frac{\partial v}{\partial x} = \frac{\partial u^0}{\partial y} + \frac{\partial v^0}{\partial x} - 2z\frac{\partial^2 w^0}{\partial x \partial y}$$

$$= \gamma_{xy}^0 + z\kappa_{xy} \tag{54.2a}$$

or in matrix form:

$$\{\varepsilon\} = \left\{\varepsilon^0\right\} + z\{\kappa\} \,, \tag{54.2b}$$

where:

$$\{\varepsilon\} = \left\{ \begin{array}{c} \varepsilon_x \\ \varepsilon_y \\ \gamma_{xy} \end{array} \right\} \,,$$

$$\left\{\varepsilon^0\right\} = \left\{ \begin{array}{c} \varepsilon_x^0 \\ \varepsilon_y^0 \\ \gamma_{xy}^0 \end{array} \right\} = \left\{ \begin{array}{c} \dfrac{\partial u^0}{\partial x} \\[6pt] \dfrac{\partial v^0}{\partial y} \\[6pt] \dfrac{\partial u^0}{\partial y} + \dfrac{\partial v^0}{\partial x} \end{array} \right\}$$

$$\{\kappa\} = \left\{ \begin{array}{c} \kappa_x \\ \kappa_y \\ \kappa_{xy} \end{array} \right\} = \left\{ \begin{array}{c} -z\dfrac{\partial^2 w^0}{\partial x^2} \\[6pt] -z\dfrac{\partial^2 w^0}{\partial y^2} \\[6pt] -2z\dfrac{\partial^2 w^0}{\partial x \partial y} \end{array} \right\} \,. \tag{54.3}$$

In (54.3), $\{\varepsilon^0\}$ is the midplane strain and $\{\kappa\}$ is the midplane curvature. As with the displacements, $\{\varepsilon^0\}$ and $\{\kappa\}$ are generally functions of x and y.

The stress–strain relations for each layer are:

$$\left\{ \begin{array}{c} \sigma_x(z) \\ \sigma_y(z) \\ \sigma_{xy}(z) \end{array} \right\}_k = \left[\begin{array}{ccc} Q_{11} & Q_{12} & Q_{16} \\ Q_{12} & Q_{22} & Q_{26} \\ Q_{16} & Q_{26} & Q_{66} \end{array} \right]_k$$

$$\times \left[\left\{ \begin{array}{c} \varepsilon_x \\ \varepsilon_y \\ \gamma_{xy} \end{array} \right\} - \left\{ \begin{array}{c} \varepsilon_x^*(z) \\ \varepsilon_y^*(z) \\ \gamma_{xy}^*(z) \end{array} \right\}_k \right] + \left\{ \begin{array}{c} \hat{\sigma}_x(z) \\ \hat{\sigma}_y(z) \\ \hat{\sigma}_{xy}(z) \end{array} \right\}_k \tag{54.4a}$$

or in compact form:

$$\{\sigma(z)\}_k = \{\overline{Q}\}_k \left[\{\varepsilon\} - \{\varepsilon(z)\}_k^*\right] + \{\hat{\sigma}(z)\}_k \,, \tag{54.4b}$$

where $k = 1 \cdots N$ denotes the layer number. Equation (54.4) is written in a global coordinate system for the layered structure, (x, y, z). As such, $[\overline{Q}]_k$ are the (generally anisotropic) elastic stiffness coefficients of the kth layer in the global coordinate system. They are obtained by transformation (see [54.26, 27], for example) of the stiffnesses, $[Q]_k$, from a natural coordinate system (x_1, x_2, x_3). For orthotropic materials the nonzero components of $[Q]$ for each layer are,

$$[Q] = \left[\begin{array}{ccc} \dfrac{E_1}{1-\nu_{12}\nu_{21}} & \dfrac{\nu_{12}E_2}{1-\nu_{12}\nu_{21}} & 0 \\[8pt] \dfrac{\nu_{21}E_1}{1-\nu_{12}\nu_{21}} & \dfrac{E_2}{1-\nu_{12}\nu_{21}} & 0 \\[8pt] 0 & 0 & G_{12} \end{array} \right] \,, \tag{54.5a}$$

where E_i, ν_{ij}, and G_{ij} are the orthotropic Young's moduli, Poisson's ratios, and the shear modulus, respectively, of the layer, in the local coordinate system. Note that $Q_{12} = Q_{21}$, and so the equality of these expressions in (54.5a) shows the reciprocal relationship between E_i and ν_{ij}; only three of the four are independent. For isotropic materials $[\overline{Q}] = [Q]$, and the nonzero components for each layer are:

$$[\overline{Q}] = [Q] = \left[\begin{array}{ccc} \dfrac{E}{1-\nu^2} & \dfrac{\nu E}{1-\nu^2} & 0 \\[8pt] \dfrac{\nu E}{1-\nu^2} & \dfrac{E}{1-\nu^2} & 0 \\[8pt] 0 & 0 & G \end{array} \right] \tag{54.5b}$$

In (54.4) $\{\hat{\sigma}(z)\}_k$ is the residual internal stress in the layer and may be an arbitrary function of z, i.e., a stress gradient may exist within each layer; $\{\varepsilon\}$ is the total strain, $\{\varepsilon^*\}_k$ is the inelastic transformation strain, and the term in rectangular brackets is the elastic strain. The stresses and strains are all defined in the (x, y, z) coordinate system. Substituting (54.2) into (54.4) yields the stress distribution through the thickness:

$$\{\sigma(z)\}_k = \{\overline{Q}\}_k \left[\{\varepsilon^0\} + z\{\kappa\} - \{\varepsilon^*(z)\}_k\right]$$
$$+ \{\hat{\sigma}(z)\}_k \,. \tag{54.6}$$

For the layered film system we define a resultant force $\{N\}$ and moment $\{M\}$ per unit length as:

$$\left\{ \begin{array}{c} N_x \\ N_y \\ N_{xy} \end{array} \right\} = \{N\} = \int_{-t/2}^{t/2} \{\sigma(z)\}_k \, dz$$

$$\left\{ \begin{array}{c} M_x \\ M_y \\ M_{xy} \end{array} \right\} = \{M\} = \int_{-t/2}^{t/2} \{\sigma(z)\}_k \, z \, dz \,. \tag{54.7}$$

For example, N_x is the force in the x-direction per unit length in the y-direction. $\{N\}$ and $\{M\}$ have units of $[F/L]$, $[F \cdot L/L]$, respectively.

Substituting (54.6) into (54.7) and breaking the integrals through the thickness into sums of integrals over each layer yields:

$$\{N\} = [A]\left\{\varepsilon^0\right\} + [B]\{\kappa\} - \left\{N^\varepsilon\right\} + \left\{N^\sigma\right\}$$

$$\{M\} = [B]\left\{\varepsilon^0\right\} + [D]\{\kappa\} - \left\{M^\varepsilon\right\} + \left\{M^\sigma\right\} \, , \quad (54.8)$$

where:

$$[A] = \int_{-t/2}^{t/2} \{\overline{Q}\}_k \, \mathrm{d}z = \sum_{k=1}^{N} \{\overline{Q}\}_k (z_k - z_{k-1})$$

$$[B] = \int_{-t/2}^{t/2} \{\overline{Q}\}_k \, z \, \mathrm{d}z = \frac{1}{2}\sum_{k=1}^{N} \{\overline{Q}\}_k \left(z_k^2 - z_{k-1}^2\right)$$

$$[D] = \int_{-t/2}^{t/2} \{\overline{Q}\}_k \, z^2 \, \mathrm{d}z = \frac{1}{3}\sum_{k=1}^{N} \{\overline{Q}\}_k \left(z_k^3 - z_{k-1}^3\right)$$

$$\{N^\varepsilon\} = \int_{-t/2}^{t/2} \{\overline{Q}\}_k \{\varepsilon^*(z)\}_k \, \mathrm{d}z$$

$$\{N^\sigma\} = \int_{-t/2}^{t/2} \{\hat{\sigma}(z)\}_k \, \mathrm{d}z$$

$$\{M^\varepsilon\} = \int_{-t/2}^{t/2} \{\overline{Q}\}_k \{\varepsilon^*(z)\}_k \, z \, \mathrm{d}z$$

$$\{M^\sigma\} = \int_{-t/2}^{t/2} \{\hat{\sigma}(z)\}_k \, z \, \mathrm{d}z \, . \quad (54.9)$$

In the terminology of laminated composite materials, $[A]$, $[B]$, and $[D]$ are the extensional, coupling, and bending constants and are functions of the elastic moduli of each layer and the arrangement of the layers through the thickness. $[A]$, $[B]$, and $[D]$ have units of $[F/L]$, $[F]$, and $[F \cdot L]$, respectively. Physically, $[A]$ describes the connection between in-plane forces and straining of the midplane, $[D]$ describes the connection between moments and curvature, and $[B]$ connects moments to midplane strain and forces to curvature. If both the geometry and material properties, including $\varepsilon^*(z)$ and $\hat{\sigma}(z)$, of the layered film system are symmetric about

$z = 0$, then $[B] = [0]$ and there is no coupling between in-plane straining and bending. Note that $[A]$, $[B]$, and $[D]$ are 3×3 matrices, as are $\left[\overline{Q}\right]_k$, while all other terms are 3×1 column vectors of the form of (54.3) and (54.4).

Equation (54.8) can be written in compact form as:

$$\left\{ \begin{array}{c} \{N\} \\ \{M\} \end{array} \right\} = \left[\begin{array}{cc} [A] & [B] \\ [B] & [D] \end{array} \right] \left\{ \begin{array}{c} \{\varepsilon^0\} \\ \{\kappa\} \end{array} \right\}$$

$$- \left\{ \begin{array}{c} \{N^\varepsilon\} \\ \{M^\varepsilon\} \end{array} \right\} + \left\{ \begin{array}{c} \{N^\sigma\} \\ \{M^\sigma\} \end{array} \right\} \, . \quad (54.10)$$

Equation (54.10) is most easily implemented in matrix form as a 6×6 system of equations that hold at each (x, y) coordinate. These are the constitutive equations for the layered film, incorporating the Kirchoff hypothesis and the strain–displacement relations. In order to completely describe the response of a layered film due to external or internal loads, (54.10) must be supplemented with the stress equilibrium equations and appropriate boundary conditions. In general the solution of these equations for a specified multilayer film geometry and loading is a complex undertaking and recourse is often taken to numerical methods such as the finite element method (FEM). Alternatively, the equations can be formulated using an energy approach that will be advantageous when we consider geometric nonlinearity in Section 54.1.3.

Fortunately, using this formulation we can solve a number of important problems simply using (54.10). For example, if there are no externally applied loads then $\{N\} = \{M\} = \{0\}$ and (54.10) can be inverted to yield:

$$\left\{ \begin{array}{c} \{\varepsilon^0\} \\ \{\kappa\} \end{array} \right\} = \left[\begin{array}{cc} [A] & [B] \\ [B] & [D] \end{array} \right]^{-1}$$

$$\times \left\{ \left\{ \begin{array}{c} \{N^\varepsilon\} \\ \{M^\varepsilon\} \end{array} \right\} - \left\{ \begin{array}{c} \{N^\sigma\} \\ \{M^\sigma\} \end{array} \right\} \right\} \, . \quad (54.11)$$

Recall that all terms on the right-hand side are functions of the elastic moduli, the known transformation strains and stresses, and the geometrical arrangement of the layers. Once the midplane strain and curvature are computed using (54.11), the stress distribution in each layer can be computed using (54.6).

Special Cases

Here we apply (54.11) to yield some simple, yet important, results. Before doing so, it is useful to explicitly express the form of the transformation stress distribution $\{\hat{\sigma}(z)\}_k$. To this end we restrict $\{\hat{\sigma}(z)\}_k$ to be a linearly varying function through each layer. This can be written as:

$$\{\hat{\sigma}(z)\}_k = \{S\}_k - \{\beta\}_k \left(z - \frac{z_{k-1}+z_k}{2}\right) , \quad (54.12)$$

where $\{S\}_k$ is the average stress in layer k and $\{\beta\}_k$ is the stress gradient as shown in Fig. 54.10. For this form of the stress, $\{N^\sigma\}$ and $\{M^\sigma\}$ can be expressed as:

$$\{N^\sigma\} = \sum_{k=1}^{N} \{S\}_k (z_k - z_{k-1})$$

$$\{M^\sigma\} = \frac{1}{2}\sum_{k=1}^{N} \{S\}_k \left(z_k^2 - z_{k-1}^2\right)$$
$$- \frac{1}{12}\sum_{k=1}^{N} \{\beta\}_k \left(z_k^3 - z_{k-1}^3\right) . \quad (54.13)$$

Furthermore, it is useful to specify the form of $\{\varepsilon^*(z)\}$ when its source is thermal expansion during a uniform temperature change T from a reference temperature. In this case:

$$\{\varepsilon^*(z)\}_k = \{\varepsilon^*\}_k = \{\alpha\}_k T . \quad (54.14)$$

For this form of the stress-free strain, $\{N^\varepsilon\}$ and $\{M^\varepsilon\}$ can be expressed as:

$$\{N^\varepsilon\} = \int_{-t/2}^{t/2} \{\overline{Q}\}_k \{\alpha\}_k T \, dz$$

$$= T\sum_{k=1}^{N} \{\overline{Q}\}_k \{\alpha\}_k (z_k - z_{k-1})$$

$$\{M^\varepsilon\} = \int_{-t/2}^{t/2} \{\overline{Q}\}_k \{\alpha\}_k Tz \, dz$$

$$= \frac{T}{2}\sum_{k=1}^{N} \{\overline{Q}\}_k \{\alpha\}_k \left(z_k^2 - z_{k-1}^2\right) . \quad (54.15)$$

In all of the examples described in this section we will consider only cases in which all layers have elastic constants, thermal expansion properties, and misfit strains that are isotropic. As a result, $A_{16} = A_{26} = B_{16} = B_{26} = D_{16} = D_{26} = 0$. The ramifications of this are that the midplane shear strains $\gamma_{xy}^0 = 0$ and the twist curvatures $\kappa_{xy} = 0$. In addition, the midplane strains and curvatures are equibiaxial, i.e., $\varepsilon_x^0 = \varepsilon_y^0 = \varepsilon^0$ and $\kappa_x = \kappa_y = \kappa$. So in the remainder of this section this equibiaxial deformation state is understood, and we refer to it simply by ε^0 and κ. While this is the case in the examples presented in this section, it is not the case when material anisotropy exists, the film is patterned in lines instead of as a blanket, or large deflections occur, which result in geometric nonlinearity. These issues will be discussed briefly in later sections.

Single-Layer Plate. Consider a single-layer film with an internal biaxial mean stress $S = S_{11} = S_{22}$ and stress gradient $\beta = \beta_{11} = \beta_{22}$ that is subjected to a uniform temperature change T. This scenario models the deformation of a film after release from a substrate when subjected to a temperature change where before release the film had an internal stress S and stress gradient β. In this case, after significant manipulation (54.11), with (54.12) and (54.13), yields:

$$\kappa = \frac{\beta}{M} \quad (54.16)$$

$$\varepsilon^0 = -\frac{S}{M} + \alpha T , \quad (54.17)$$

where $M = \frac{E}{1-\nu}$ is the biaxial modulus. As previously discussed, after release there is straining due to S and $\varepsilon^* = \alpha T$, and the plate bends with a biaxial curvature κ. The result of (54.16), and especially the simplification that results for beams (simply obtained by setting $\nu = 0$ and interpreting κ and ε^0 as the uniaxial curvature and midplane strain in the direction of the beam length) is commonly used in conjunction with measurements of the curvature after release to determine the stress gradient in a film.

Bilayer Plate. Another simple yet practically important special case is that of two isotropic layers of thicknesses t_1 and t_2, subject only to a uniform temperature change T. In this case, (54.11), with (54.12) and (54.13), yields:

$$\kappa = \frac{6\Delta\alpha T}{t_2} hm \left(\frac{1+h}{1+2hm(2+3h+2h^2)+h^4m^2}\right) \quad (54.18)$$

$$\varepsilon^0 = T\left(\frac{\alpha_1 mh (1+3h+3h^2+mh^3)}{1+2hm(2+3h+2h^2)+h^4m^2}\right.$$
$$\left. + \frac{\alpha_2 \left[1+mh(3+3h+h^2)\right]}{1+2hm(2+3h+2h^2)+h^4m^2}\right) , \quad (54.19)$$

where $h = \frac{t_1}{t_2}$, $m = \frac{M_1}{M_2}$, and $M_k = \frac{E_k}{1-\nu_k}$ (here $k = 1, 2$ corresponds to layers 1 and 2) and $\Delta\alpha = \alpha_2 - \alpha_1$.

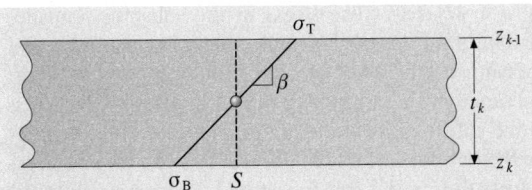

Fig. 54.10 Schematic showing a linear stress distribution characterized by a mean stress S and linear through-thickness stress gradient β

It is often convenient to define a nondimensional curvature $\hat{\kappa}$:

$$\hat{\kappa} = \frac{\kappa\, t_2}{6\Delta\alpha T} \tag{54.20}$$

Equation (54.18) can then be written as

$$\hat{\kappa} = \frac{hm\,(1+h)}{1+2hm\,\left(2+3h+2h^2\right)+h^4 m^2} \,. \tag{54.21}$$

Two limiting values of (54.21) are of interest. The first is the case where there is no elastic mismatch between the layers, i. e., $m = 1$. In this case (54.21) simplifies to:

$$\hat{\kappa} = \frac{h}{(1+h)^3} \,. \tag{54.22}$$

The other is the *thin-film limit* where $t_1 \ll t_2$. Expanding (54.18) in powers of h and retaining only the lowest-order term recovers *Stoney's* [54.28] well-known result:

$$\kappa = \frac{6\Delta\alpha T}{t_2} hm$$
$$\hat{\kappa} = hm \,. \tag{54.23}$$

In the thin-film limit, (54.19) reduces to $\varepsilon^0 = \alpha_2 T$.

Figure 54.11 shows the normalized curvature $\hat{\kappa}$ as a function of the layer thickness ratio h for various ratios of the biaxial moduli, m. The curves from top to bottom are for $m = 10, 0.65$, and 0.1. $m = 10$ and 0.1 represent a stiff film on a compliant substrate and a compliant film on a stiff substrate, respectively. $m = 0.67$ represents a gold film on a polysilicon substrate. The solid curves are the exact results from (54.21) and the dashed curves are the Stoney's result for the thin-film limit, (54.23). In the latter case, $\hat{\kappa}$ varies linearly with h, with slope m. The results in Fig. 54.11 show how the range of applicability of Stoney's result depends on both h and m; the range decreases for increasing h and decreasing m. For most microstructure applications where the layer thicknesses are comparable ($h > 0.1$), Stoney's result is not accurate

unless the film is much more compliant than the substrate. Thus physically, Stoney's result is accurate when the substrate stiffness controls the deformation, either because it is much thicker or has much higher elastic moduli than the film. Over the years many papers have discussed *corrections* to Stoney's result to increase the region of validity in terms of layer thicknesses and modulus mismatch. We emphasize, though, that (54.18) is valid for arbitrary layer thicknesses and modulus mismatch and is quite simple to use itself. We refer the reader interested in more details to [54.9, 20, 24].

Equally important from a technological viewpoint is the equibiaxial stress distribution through the thickness $\sigma_x(z) = \sigma_y(z) = \sigma(z)$ obtained from (54.6):

$$\sigma(z)_k = M_k\left(\varepsilon^0 + z\kappa - \alpha_k T\right) \,, \tag{54.24}$$

where $k = 1, 2$ denotes the layer. In (54.24), κ and ε^0 are obtained from (54.18) and (54.19), respectively, and M_k and α_k are the values for the kth layer. In the thin-film limit the equibiaxial stress varies so little through the film thickness that it can be taken as constant, i. e., $\sigma_x = \sigma_y = \sigma$. This is the reason one often refers to the *stress* in a thin-film by a single number. In the thin-film limit, the substrate (layer 2) thermal expansion dominates the deformation process, and so the film strains by an amount $\Delta\alpha T$, and the biaxial stress in the film is

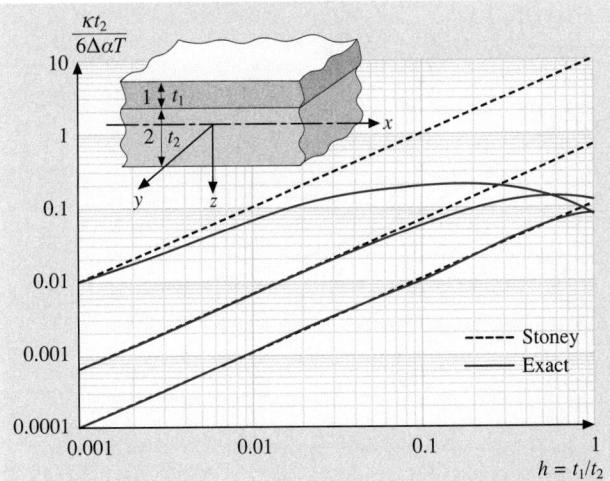

Fig. 54.11 Normalized curvature $\hat{\kappa}$ as a function of the film thickness ratio $h = t_1/t_2$ for three values of the biaxial modulus ratio, $m = M_1/M_2$. From *top* to *bottom*, the curves are for $m = 10$, 0.65 (gold/polysilicon), and 0.1. The *solid lines* are the exact solution, (54.21) and the *dashed lines* are Stoney's result in the thin-film limit, (54.23)

$\sigma_1 = M_1 \Delta\alpha T$, or in terms of the curvature:

$$\sigma_1 = \frac{M_2}{6} \frac{t_2^2}{t_1} \kappa .\qquad(54.25)$$

Note that, as written, the thermoelastic properties of the film do not enter (54.25). We will discuss the practical ramifications of this shortly.

Figure 54.12 shows the stress distribution through the thickness of both layers for a negative unit temperature change, (54.24), as a function of $h = t_1/t_2$ for $m = 0.65$, and $t_1 = 0.5\,\mu\text{m}$, representative of a gold film on a silicon substrate. Equation (54.24) shows, and we have previously discussed, that the biaxial stress varies linearly through each material. Thus we can completely specify the stress distribution by plotting it at four points: the top of material 1, the interface in material 1, the interface in material 2, and the bottom of material 2. Referring to Fig. 54.8, these positions are $z = z_0$, z_1, and z_2. Because of the perfect bonding between the two films, the total strain must be continuous across the interface. The elastic moduli are different for the two films, though, and so the stresses will be discontinuous at the interface. Hence we interpret $z = z_1$ as a position just above, or just below, the interface, and use M_k and α_k in (54.24) as appropriate. As $h \to 0$ the stresses at the top and at the interface in the film (layer 1) converge and take on a maximum value given

by $\sigma = M_1 \Delta\alpha T$. The stress in the substrate vanishes in this limit. As the substrate thickness increases (increasing h) the stress in the film decreases with the maximum value occurring at the interface. The stress in the substrate increases at the interface but decreases on the bottom surface. In general this results in a stress distribution in the substrate that is compressive on the bottom surface and tensile at the interface. Depending on h, the stresses in the film are either tensile throughout or tensile on the interface and compressive on the surface. The difference in stress between the interface and the surface is a measure of the stress gradient in each layer; it increases with increasing h. For different values of m, the behavior with varying h is similar to that in Fig. 54.12, with the stress in the film increasing with increasing m.

Equation (54.25) is the theoretical basis of what is perhaps the most common technique to measure thin-film stresses – the wafer curvature method [54.8]. It involves measuring the curvature of a thick (compared to the film thickness) wafer with and without a film, and then using the curvature difference between the two as in (54.25) to determine the film stress. It is particularly convenient because the film stress is obtained from the measured curvature regardless of the source of the stress and the thermoelastic moduli of the film. It is necessary, though, to know accurately the thickness of the film and substrate, as well as the elastic properties (biaxial modulus, $E/(1-\nu)$) of the substrate. An appealing feature of the method is that the elastic properties of the film, typically difficult to measure themselves, do not need to be known. This is because the much thicker substrate controls the deformation response.

Three-Layer Symmetric Plate. Considerable simplification also results for the technologically important case of a thin-film multilayer subject to a uniform temperature change T that consists of three layers, in which the top and bottom layer are identical in terms of thickness ($t_3 = t_1$) and material properties ($M_3 = M_1, \alpha_3 = \alpha_1$). In this case, from (54.9) $[B] = [0]$, and thus the plate does not bend, i. e., $\kappa = 0$. Equation (54.11), with (54.12) and (54.13), then yields:

$$\varepsilon^0 = \frac{(2\alpha_1 mh + \alpha_2)\,T}{2mh + 1} .\qquad(54.26)$$

In the thin-film limit $h \to 0$, $\varepsilon^0 = \alpha_2 T$.

The equibiaxial stress distribution is such that it does not vary through the thickness within each layer, i. e., $\sigma(z)_3 = \sigma(z)_1 = \sigma_1$, and $\sigma(z)_2 = \sigma_2$. From (54.6) with

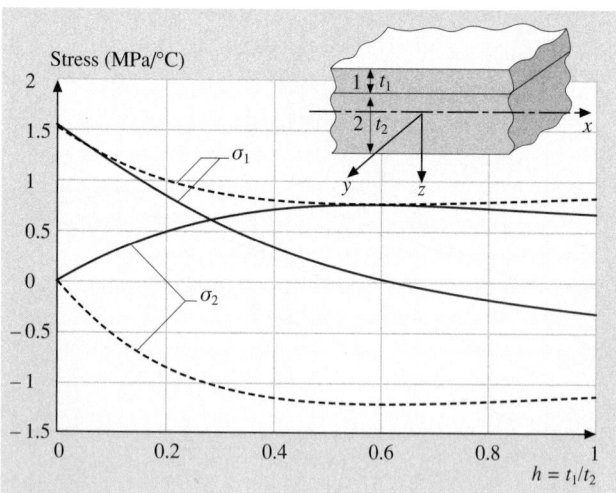

Fig. 54.12 Biaxial stress per unit temperature change, (54.24), for a gold/polysilicon bilayer ($m = 0.65$) for a 0.5-μm-thick gold film as a function of the polysilicon thickness $h = t_1/t_2$. Biaxial stresses are shown at four locations: the interface on the gold and polysilicon sides (*dashed lines*), and the surfaces of the gold and polysilicon (*solid lines*)

$\kappa = 0$ and ε^0 from (54.26):

$$\sigma_1 = \frac{M_1 \Delta \alpha T}{2mh + 1}$$

$$\sigma_2 = \frac{-2mh M_2 \Delta \alpha T}{2mh + 1} . \qquad (54.27)$$

In the thin-film limit, the stresses of (54.27) reduce to $\sigma_1 = M_1 \Delta \alpha T$ and $\sigma_2 = -2M_1 h \Delta \alpha T$.

Figure 54.13 shows the uniform biaxial stress in each material for a negative unit temperaturechange as a function of $h = t_1/t_2$ for $m = 0.65$, and $t_1 = 0.5 \, \mu\text{m}$, again representative of a gold film on a silicon substrate. As $h \to 0$, the stresses in the films (layers 1 and 3) are given by $\sigma_1 = M_1 \Delta \alpha T$, and that in the substrate (layer 2) approaches zero. The stress in the films is tensile and decreases with increasing h while that in the substrate is compressive, increasing in magnitude with increasing h.

Special Case – Beams
Our analysis thus far has focused on multilayer plates where both in-plane dimensions far exceed the thickness and where the transformation strains, and thus the stresses, are equibiaxial. When one in-plane dimension, say x, is much larger than the other, the structure is often referred to as a beam. In the classical treatment of the flexure of beams, it is assumed that only one nonzero normal stress component exists, σ_x. This greatly simplifies the analysis for homogeneous beams, but the situation is not as straightforward for multilayers with biaxial transformation strains. From the results in the previous section, one can obtain results for a beam theory by simply setting $\nu_1 = \nu_2 = 0$ in the formulae developed for biaxial stresses, but these results should be used with caution as the transverse component of the moment that results from the equibiaxial transformation strains cannot be neglected, as shown by three-dimensional finite element calculations [54.29, 30]. *Swanson* [54.31] describes how to apply the results from a multilayer plate theory to beams and specifically focuses on the effect of the beam width, giving easily usable results for narrow and wide beams. The main results are an effective structural stiffness $(EI)_{\text{eff}}$ that can be used to replace EI (E is the Young's modulus and I is the cross-sectional moment of inertia) in existing beam theory results, such as those that will be presented in Sect. 54.2 for electromechanics analysis. Insight into the behavior can be seen in the results in Table 54.2, which show measurements (using interferometry as described by *Dunn* et al. [54.32]) and predictions based on (54.18)

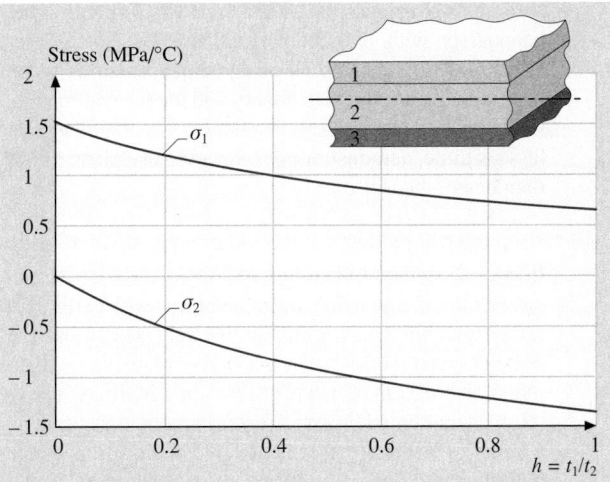

Fig. 54.13 Biaxial stress per unit temperature change for a gold/polysilicon/gold ($m = M_1/M_2 = 0.65$) three-layer film system. The calculations, (54.27), are for a 0.5-μm-thick gold film as a function of the polysilicon thickness $h = t_1/t_2$. Biaxial stresses are shown in the gold (*top curve*) and polysilicon (*bottom curve*)

for plates and the simplified version for beams obtained by setting $\nu_1 = \nu_2 = 0$. In all calculations both the gold and polysilicon are modeled as linear thermoelastic with isotropic material properties. Input parameters to the finite element calculations are $E_2 = 163$ GPa, $\nu_2 = 0.22$, $E_1 = 78$ GPa, $\nu_1 = 0.42$ [54.33]. The thermal expansion coefficients of the materials were assumed to vary linearly with temperature, and values at $100(24)°$C used are $\alpha_2 = 3.1(2.6) \times 10^{-6} \, /°$C, and $\alpha_1 = 14.6(14.2) \times 10^{-6} \, /°$C [54.33]. Although some uncertainty exists in the values of these material properties for the gold and polysilicon films, we think they are sufficiently accurate for the purpose of modeling the observed phenomena. The Young's modulus and Poisson's ratio of the polysilicon are in line with many mea-

Table 54.2 Predicted and measured curvature per unit negative temperature change for gold/polysilicon microstructures as a function of the polysilicon thickness. In all cases the gold thickness is 0.5 μm. The microstructures are 300 μm long and 50 μm wide and cantilevered from one end

Polysilicon thickness (μm)	$d\kappa/dT$ (m^{-1}/°C)		
	Measured	Predicted (plate)	Predicted (beam)
1.5	−5.36	−5.56	−4.86
3.5	−1.42	−1.44	−1.16

surements over many MUMPS runs [54.34] and agree adequately with bulk polycrystal averages of single-crystal elastic constants. Good agreement exists between the predictions using plate theory and the measurements, while predictions based on beam theory are significantly less accurate, demonstrating the need to use plate, rather than beam, theory.

Patterned Films

In microsystems applications, additional design freedom can be obtained by using multilayer films with patterned, rather than blanketed, films. In general one can use the finite element method with multilayer plate elements to model the deformation and stresses in a multilayer with arbitrary patterned layers. A useful practical example is a beam of width w_2 that is covered with a film in the pattern of a stripe of width w_1 as shown in Fig. 54.14. A simple modification of (54.18) yields the curvature along the length of the patterned beam as a function of the nondimensional line width, $c = w_1/w_2$:

$$\kappa_{\text{patterned}} = \frac{6\Delta\alpha T}{t_2} hcm \qquad (54.28)$$
$$\times \left[\frac{1+h}{1+2hcm\left(2+3h+2h^2\right)+h^4c^2m^2} \right].$$

Predictions from (54.28), normalized as $\eta = \kappa_{\text{patterned}}/T$, are shown in Fig. 54.14 for $m = 0.65$ (the gold/polysilicon bilayer) along with predictions from detailed finite element calculations and measurements over the entire range of $c = w_1/w_2$ [54.35]. Equation (54.28) accurately describes the effect of line width on curvature development and can be used to design bilayer beams.

Deflections

Thus far our discussion of deformation has been cast in terms of curvature. An important quantity for many applications where control of the deformation is important is the deflection of the multilayer. Formally this can be obtained by integrating the curvature–displacement relationship and applying the relevant boundary conditions. For example, assuming constant curvature, the tip deflection $w(x = L)$ for a cantilever beam of length L, is:

$$w(x = L) = \frac{\kappa L^2}{2}. \qquad (54.29)$$

54.1.3 Nonlinear Geometry

In the previous section we noted that in general the deformation of a layered film microstructure subjected to transformation strains consists of three regimes but focused on the first, linear regime. Here we focus on the second and third regimes. To illustrate, consider the seemingly simple case of a plate with total thickness much less than the in-plane dimensions, composed of two isotropic layers with different material properties (elastic moduli and thermal expansion coefficients) subjected to a temperature change. Although we use a temperature change for demonstration purposes, the results hold for any type of transformation strains. In terms of the average curvature variation as a function of temperature change, three deformation regimes have been identified, as illustrated in Fig. 54.15 [54.18–21, 24, 25]. The first, regime I, consists of a linear relation between the average curvature and temperature change where $\kappa_x = \kappa_y$, i.e., the average curvature is spherically symmetric. This deformation regime is characterized by both small transverse displacements and rotations, and so conventional thin-plate theory adequately describes the deformation. In Sect. 54.1.2 we outlined the analysis in this regime. The second, regime II, consists of a nonlinear relation between the average curvature and temperature, but again $\kappa_x = \kappa_y$. The behavior is due to *geometric nonlinearity* that results when the deflections become excessively large relative to the plate thickness, and they contribute significantly to the in-plane strains. It has been shown [54.17–21, 24, 25] that in these two regimes the symmetric deformation modes are stable. The second regime ends at the point when the deforma-

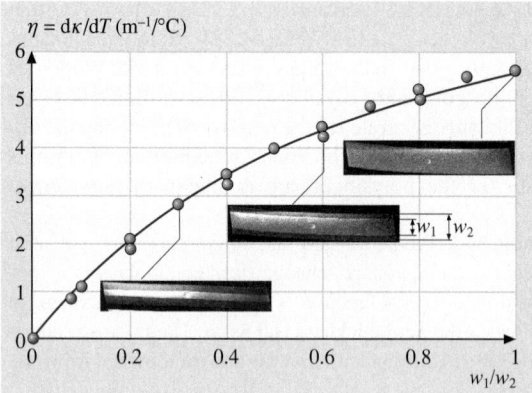

Fig. 54.14 Curvature per unit temperature change as a function of the ratio of gold to polysilicon film width w_1/w_2 for gold (0.5-μm-thick)/polysilicon (1.5-μm-thick) strip-patterned beams. *Grey circles* are finite element simulation results, the *solid line* is the analytical result of (54.28), and the *brown circles* are measurements

tion response bifurcates from a spherical to ellipsoidal deformation, i.e., $\kappa_x \neq \kappa_y$. At this point, the beginning of regime III, it becomes energetically favorable for the plate to assume the ellipsoidal shape because to retain the spherical deformation under an increasing temperature change requires increased midplane straining. After the bifurcation, the curvature in one direction increases while that perpendicular to it decreases; the plate tends toward a state of cylindrical curvature. This observation helps to explain the energetic argument, as unlimited cylindrical curvature can be obtained with no midplane straining, while spherical curvature cannot. This discussion has been cast in the context of linear material behavior. Additional deformation regimes result if material nonlinearity is present, for example, yielding [54.20].

Analysis

The understanding described above derives from a number of studies with different technological motivations, primarily structural laminated composites and thin films for microelectronics. Most of these studies are analytical [54.17–21, 24, 25, 36] and build upon the original work of *Hyer* [54.15,16]. To illustrate the computational approach in a reasonably simple setting, here we consider a two-layer plate with layer thicknesses t_1 and t_2. Each layer is isotropic and characterized by the Young's modulus E_i, Poisson's ratio ν_i, and thermal expansion coefficient α_i ($i = 1, 2$). We consider a square plate with side length L; a similar analysis can be carried out for circular [54.24] or other plate shapes. The plate is subject to a uniform temperature change, T. In other words, the stress-free strains $\{\varepsilon^*\}_k$ are due to thermal expansion as given by (54.14).

In order to compute the deformed shape when the plate is subject to a temperature change we use the approach of *Hyer* [54.16] as applied by *Masters* and *Salomon* [54.18]. The basic idea is to assume an admissible displacement field $w(x, y)$ in terms of unknown parameters (d_i) that are suitably chosen to be consistent with observed deformation modes. Values of the parameters d_i are then determined via a Ritz procedure so as to minimize the total potential energy of the system. Different choices of the assumed displacement field have been considered, and details of the procedures are given in the above references. Such analyses are sufficient to qualitatively, and in many cases quantitatively, explain the three regimes of deformation shown in Fig. 54.15. In fact, quite simple closed-form expressions result for special cases that provide illuminating descriptions of observed phenomena (see for exam-

ple [54.24]). A disadvantage of the analytical approaches is that for simplicity a displacement field consistent with a spatially constant curvature deformation mode is usually chosen. Additionally, these formulations are only useful for simple plate shapes. While this may be adequate for the structures demonstrated here, it is not for more complex in-plane shapes, of either or both layers, that arise in microsystems applications. In this more general case the best approach is probably to use the finite element method for the approximate analysis.

To proceed, we assume the transverse midplane displacement is of the form:

$$w^0(x, y) = d_1 x^2 + d_2 y^2 . \tag{54.30}$$

d_1 and d_2 are immediately recognized as one half of the curvature in the x- and y-directions, and the deformation is seen to be a constant-curvature mode. The midplane displacements u^0 and v^0 are assumed to be described by third-order polynomials in x and y, also with unknown constant coefficients d_3–d_8 to be determined:

$$u^0(x, y) = d_3 x + \frac{d_5}{3} x^3 + d_7 x y^2 - \frac{d_1^2}{6} x^3$$

$$v^0(x, y) = d_4 y + \frac{d_6}{3} y^3 + d_8 x^2 y - \frac{d_2^2}{6} y^3 . \tag{54.31}$$

The midplane strains are then computed from the nonlinear strain–displacement relations of the von Karman

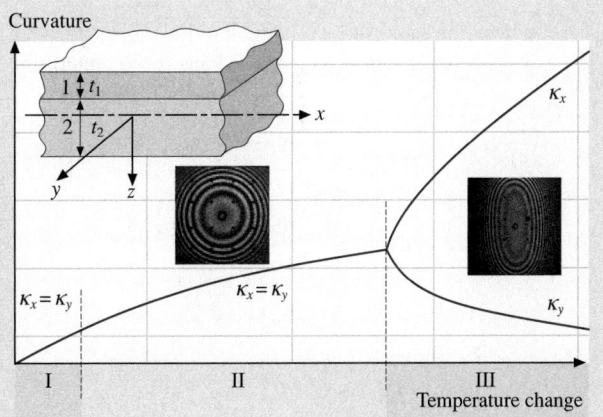

Fig. 54.15 General characteristics of the average curvature–temperature change of a two-layer plate microstructure showing the three deformation regimes. The interferograms in the *inset* show measured displacement contours for gold/polysilicon plates in regimes II and III

plate theory:

$$\varepsilon_x^0 = \frac{\partial u^0}{\partial x} + \frac{1}{2}\left(\frac{\partial w^0}{\partial x}\right)^2$$

$$\varepsilon_y^0 = \frac{\partial v^0}{\partial y} + \frac{1}{2}\left(\frac{\partial w^0}{\partial y}\right)^2$$

$$\gamma_{xy}^0 = \frac{\partial u^0}{\partial y} + \frac{\partial v^0}{\partial x} + \left(\frac{\partial w^0}{\partial x}\right)\left(\frac{\partial w^0}{\partial y}\right) . \tag{54.32}$$

The strains at any point through the thickness are then computed using the standard kinematic relations for thin plates, (54.2), and the stresses are computed from the strains using the conventional linear thermoelastic constitutive relations for each layer, (54.6) with $\{\varepsilon^*\}_k = \{\alpha\}_k T$. The first term in (54.32) is recognized as the conventional small deformation, and the second term arises from the large transverse deflections w^0, which result in straining of the midplane. This is shown schematically in Fig. 54.16 for the strain component ε_x^0; similar results are obtained for ε_y^0 and γ_{xy}^0.

The potential energy density of each layer is computed from the stress and strain in each layer, and the total potential energy of the plate is then computed by integrating it over the volume of the plate. This yields an expression for the potential energy of the plate in terms of the unknown coefficients d_i:

$$U(d_i) = \int_{-L/2}^{L/2}\int_{-L/2}^{L/2}\int_{-t/2}^{t/2} \{\varepsilon\}^T \{\sigma\}_k \; \mathrm{d}z \, \mathrm{d}x \, \mathrm{d}y , \tag{54.33}$$

where the superscript T denotes the transpose of the 3×1 column vector. The coefficients d_i are determined

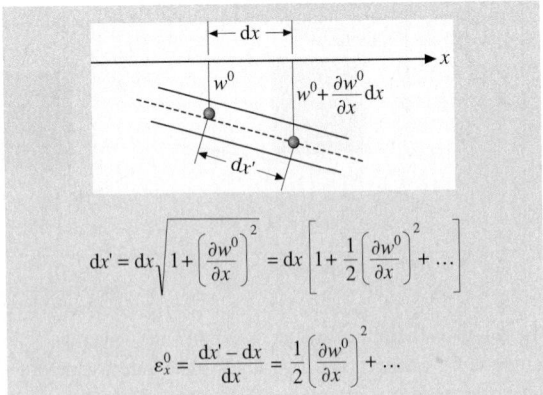

$$\mathrm{d}x' = \mathrm{d}x\sqrt{1 + \left[\frac{\partial w^0}{\partial x}\right]^2} = \mathrm{d}x\left[1 + \frac{1}{2}\left(\frac{\partial w^0}{\partial x}\right)^2 + \dots\right]$$

$$\varepsilon_x^0 = \frac{\mathrm{d}x' - \mathrm{d}x}{\mathrm{d}x} = \frac{1}{2}\left(\frac{\partial w^0}{\partial x}\right)^2 + \dots$$

Fig. 54.16 Illustration of the development of midplane strain due to bending when nonlinear geometry effects arise

by application of the principle of minimum potential energy:

$$\frac{\partial U(d_i)}{\partial d_i} = 0 . \tag{54.34}$$

This yields eight equations for the eight unknown d_i and thus solutions for the deformation response of Fig. 54.15. Complete details regarding this analysis can be found elsewhere [54.16, 18]; here we present only the pertinent results in what we hope is an accessible form. When the displacements are small, the solution of (54.34) and (54.35) recovers (54.18) and (54.19).

In the nonlinear but symmetric deformation regime II, the relationship between both the curvature and midplane strain, and the thermal expansion mismatch is too complex to present here. This is also the case in the nonlinear regime III after bifurcation. An important difference between the linear and nonlinear response, though, is that the nonlinear response depends on the plate size. The critical curvature at which bifurcation occurs, though, can be obtained explicitly and is given by:

$$\kappa_{cr} = \frac{12\sqrt{2}}{L^2}\sqrt{\frac{6 + A_{12}/A_{11}}{1 + A_{12}/A_{11}}}\frac{\sqrt{A_{66}D_{66} - B_{66}^2}}{A_{66}^2} \tag{54.35}$$

for a square plate of side length t, where A_{ij}, B_{ij}, and D_{ij} are the composite moduli of (54.8). In the simplified case where there is no elastic mismatch between the layers and the elastic response can be expressed in terms of the Young's modulus E and Poisson's ratio ν, (54.35) reduces to:

$$\kappa_{cr} = 2\sqrt{6}\sqrt{\frac{6 + \nu}{1 + \nu}}\frac{t}{L^2} , \tag{54.36}$$

where $t = t_1 + t_2$. These results have also been obtained by *Finot* and *Suresh* [54.20]; similar results exist for a circular plate [54.24, 32].

The analysis just discussed is made tractable by assuming a displacement field consistent with a constant curvature. In the nonlinear case, this assumption is questionable. While the assumed displacement field used in the Ritz procedure could be modified to incorporate the dependence of curvature on position, perhaps the simplest approach to tackle these more general problems is to use the finite element method to solve the geometrically nonlinear equations over an arbitrary spatial domain. This is also the most viable approach for more complicated in-plane geometries, including patterned films. Complete details regarding the use of the finite element method to carry out such calculations can be found in *Dunn* et al. [54.32]. Here we will simply show the result of some of these calculations. The input parameters used are the same as those given in Sect. 54.1.2.

Deformation Behavior

In order to illustrate the nature of the deformation phenomena in the nonlinear regimes, Fig. 54.17 shows contour plots of measured (using an interferometric microscope as described by *Dunn* et al. [54.32]) and predicted displacement fields $w(x, y)$ for square plates of four plate sizes that have been cooled from about $100\,°C$, where they are flat, to room temperature. Due to the thermal expansion mismatch between the polysilicon and gold, the $L = 150\,\mu m$ samples deform in a spherically symmetric manner; contours of constant transverse displacement $w(x, y)$ are nearly circles. This is also the case as the size increases to $L = 200$ and $250\,\mu m$, although the displacements increase as the plate size increases. At $L = 300\,\mu m$, though, the transverse displacement contours are not circular but elliptical, indicating that the

deformation is no longer spherically symmetric. Thus, when subjected to the same temperature change, both the magnitude and deformation mode depend on the plate size. As the in-plane dimension of the plate increases with the thickness held constant, the deformation mode changes from one of spherical symmetry to one more like cylindrical symmetry.

A complete picture of the deformation as a function of plate size and temperature change could be obtained by acquiring full-field displacements like those in Fig. 54.17 as a function of temperature. But a reasonable picture of the deformation behavior can be obtained in much simpler form by considering the *average* curvature in the x- and y-directions as a function of temperature. This is shown in Fig. 54.18, in which the average curvature in the x- and y-directions is plotted as a function of the magnitude of the temperature change during cooling. The temperature change is actually negative according to our convention, but its magnitude is plotted for convenience. The average curvatures are determined from the measured and computed $w(x, y)$ by averaging $\kappa_x = -\partial^2 w(x, 0)/\partial x^2$ and $\kappa_y = -\partial^2 w(0, y)/\partial y^2$ along

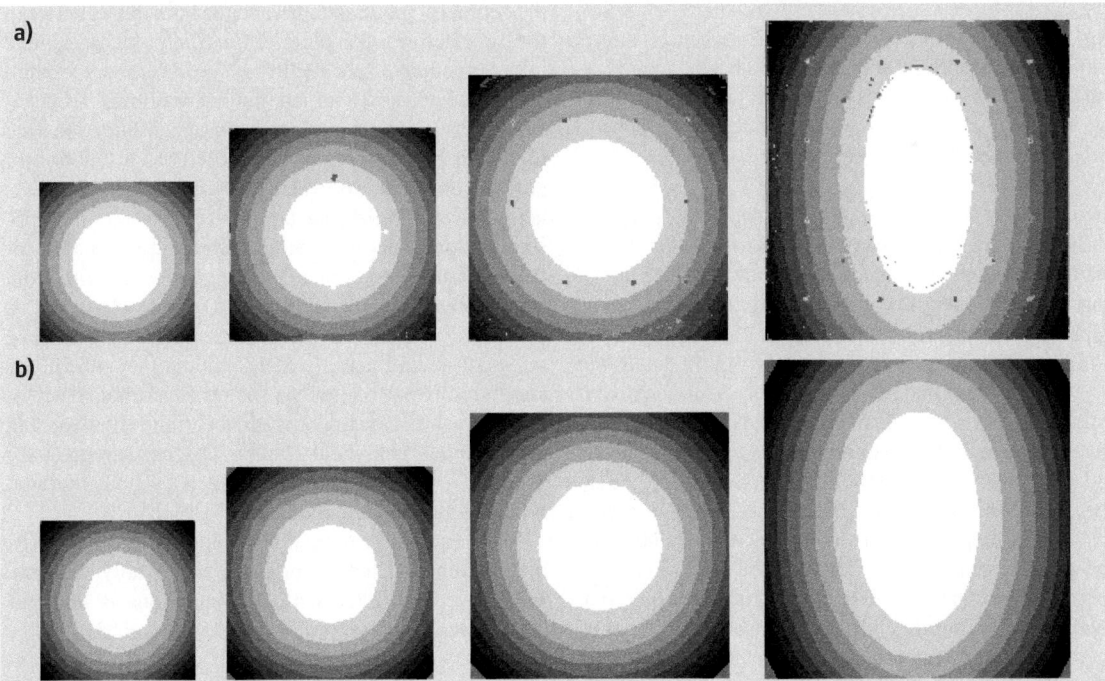

Fig. 54.17a,b Contour plots of the (**a**) measured, and (**b**) predicted transverse displacements $w(x, y)$ at room temperature following cooling from $100\,°C$ for the four gold/polysilicon square plates: $L = 150, 200, 250,$ and $300\,\mu m$, from *left* to *right*. Each contour band represents a displacement of $0.23, 0.35, 0.45,$ and $0.6\,\mu m$ for the $L = 150, 200, 250,$ and $300\,\mu m$ plates, respectively

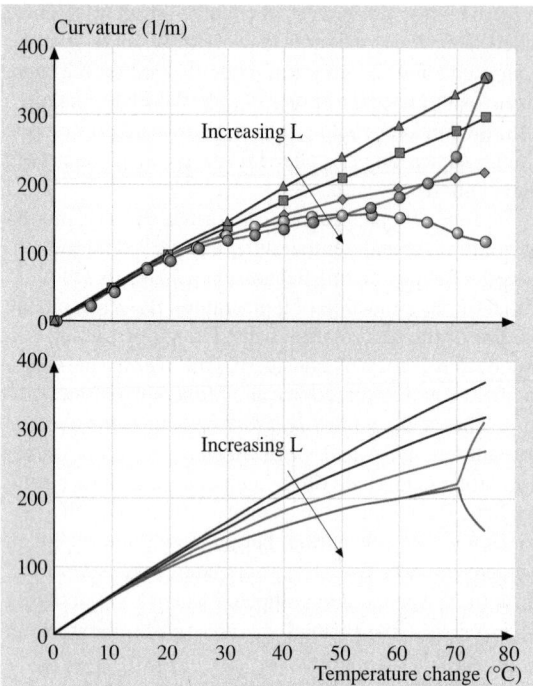

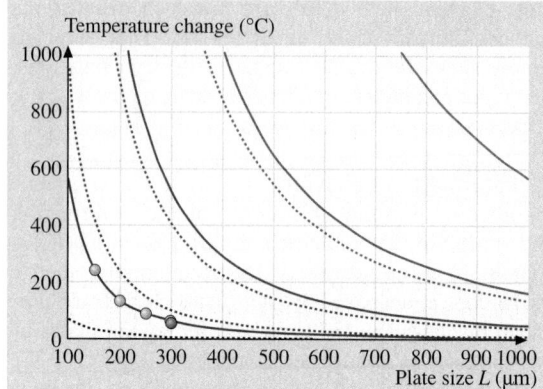

Fig. 54.19 Temperature change required for the initiation of nonlinear geometry effects (*dotted lines*) and critical temperature change for bifurcation (*solid lines*) as a function of square plate size and the thickness of polysilicon for the gold/polysilicon microstructures. Lines in each set from *left* to *right* represent 1.5-μm-, 3.5-μm-, 5.5-μm-, and 8.5-μm-thick polysilicon, respectively. *Grey circles* are finite element calculations, and the *brown circle* is a measurement

Fig. 54.18 Average measured (*top*) and predicted (*bottom*) curvature as a function of temperature change upon cooling from 100 °C to room temperature. The curves from *top* to *bottom* are for the $L = 150, 200, 250,$ and $300\,\mu m$ structures, respectively

the paths $y = 0$ and $x = 0$, respectively [54.32]. The x- and y-directions are taken to be aligned with the principal curvatures after bifurcation. The use of the average curvature as a measure of the plate deformation seems appropriate if the curvature is, or is close to, spatially uniform. This aspect will not be taken up in detail, but we refer the interested reader to [54.24,32] for details. It is apparent from both the measurements and predictions that in regime I, the curvature–temperature response is independent of plate size and shape. In regime II, though, there is a strong dependence on plate size, and this is also the case in regime III. The major discrepancy between the measurements and predictions is the sharpness of the bifurcation for the $L = 300\,\mu m$ plate; it is quite sharp in the predictions but much more gradual in the measurements. To understand this we note that the source of the bifurcation is an *imperfection* of some sort that breaks the ideal symmetry, and in general it is difficult to model the detailed imperfection accurately.

Figure 54.19 demonstrates the connection between the thermomechanical loading (the temperature change),

the geometry (plate size, L), and the boundaries between the three deformation regimes for gold/polysilicon plate microstructures. Specifically, it shows the temperature change necessary to initiate nonlinear effects (the transition between regions I and II), and bifurcation (the transition between regions II and III) as a function of polysilicon thickness when the gold film thickness is kept constant at 0.5 μm. Despite the fact that the constant curvature approximation becomes questionable for larger plate sizes, (54.35) is a good approximation, as seen by the agreement with the finite element calculations, at least for the elastic mismatch and plate sizes considered here. In fact, although not shown, the simplified result of (54.36) for no elastic mismatch is in reasonable agreement with the finite element and the complete analytical results. The measurement for $L = 300\,\mu m$ is accurately described by both the analytical and finite element results.

In the analytical treatments discussed previously, it is assumed that the curvature is spatially uniform. The power of finite element calculations is that this requirement is relaxed and the spatial variation of the curvature can be studied theoretically. Full-field interferometry measurements allow one to study this experimentally as well. We discuss the resulting behavior for gold/polysilicon plate microstructures briefly but refer the interested reader to [54.24, 32] for more details. In the linear regime, measurements and predictions both show that curvature is essentially uniform across the plate. In the nonlinear regime, though, the curvature varies appreciably with position, increasing by about a factor of two from the center to the periphery of the plate. The spatial nonuniformity of the curvature increases as the plate size increases. The spatial variation of the curvature raises concern regarding the suitableness of an analysis based on constant curvature. As mentioned previously, such an analysis may be adequate to describe the general deformation behavior but not for finer details.

54.1.4 Surface Stress: Scaling From Micro- to Nanostructures

As the size of a structural element decreases from micrometer to nanometer dimensions, the surface-area-to-volume ratio increases. Associated with the free surfaces, or interfaces in a multilayer, are surface energy and surface stress [54.6]. While these are often insignificant for micrometer-scale structures, they can play a significant role in the atomic structure, and mechanical and physical properties for nanostructures.

The surface energy γ is the excess free energy per unit area due to the existence of a free surface; it is equal to the reversible work per unit area needed to create a new surface [54.37]. Surface stress is the strain derivative of the surface energy [54.38]:

$$f_{ij} = \left(\partial\left(A\gamma\right)/\partial\varepsilon_{ij}\right)/A = \gamma\delta_{ij} + \partial\gamma/\partial\varepsilon_{ij} ,$$
$$i, j = 1, 2 \tag{54.37}$$

where A is the surface area, γ is the surface energy and ε_{ij} is the elastic strain in the plane of the surface. The surface stress is a two-dimensional second-rank tensor, while the surface energy is a scalar. The change of the total surface energy is equal to the work done by the surface stresses, owing to a strain $d\varepsilon_{ij}$ in a reversible process.

In metals surface stresses are typically tensile and result in contraction of surface atoms relative to the bulk atoms. This results in an *intrinsic* compressive stresses within the core. By core, we mean atoms that are sufficiently far from the surface so that they are in a local environment representative of that in an infinite lattice; this usually occurs a few lattice spacings from the surface. The surface-stress-induced intrinsic compressive stress state depends on the geometry of the nanostructure, i. e., whether the structure is a particle, rod, plate, or something more complex.

Fig. 54.20 illustrates the importance of the size scale by plotting the surface stress-induced intrinsic compressive stress in a gold nanowire. These results are based on a simple model of intrinsic-stress development [54.7] with a reasonable estimate of surface stress of $1\,\text{J/m}^2$. One sees that as the structural dimension (wire width) decreases from a micrometer to a nanometer, the intrinsic stress increases by about three orders of magnitude. These stresses have been shown to dramatically effect the atomistic structure, resulting in phase transformation and reorientation of the lattice [54.39]. In addition, they can dramatically effect the effective elastic [54.6, 7, 40–42] and inelastic [54.43] behavior of the nanostructure.

The analysis formalism developed in the previous section for multilayer films can be adapted to the study of various aspects of the deformation of nanometer-scale films with surface stress effects. For example, one can consider a nanofilm to consist of three distinct layers. The top and bottom layers are taken to represent the characteristics of the region near the surface, and the middle is taken to represent the core of the film. The surface stress effects can be modeled by appropriate modifications of the elastic constants and specification of a suitable transformation strain and/or internal srtess.

54.1.5 Nonlinear Material Behavior

So far we have discussed in some detail the linear thermoelastic response of multilayer films and to a lesser degree, the effects of geometric nonlinearity. An additional complication is material nonlinearity. It can arise

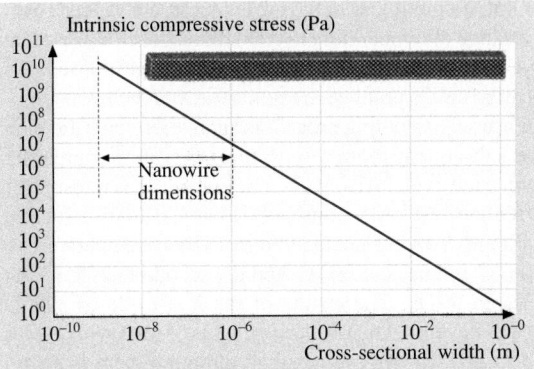

Fig. 54.20 Intrinsic compressive stress in the core of a gold nanowire as a function of the width of the square cross section. The wire is oriented in the [100] direction and it is assumed that a surface stress of $1.0\,\text{J/m}^2$ acts on each free surface

in numerous forms including plasticity, creep, stress relaxation, and evolution of the material microstructure (densification, grain growth, defect annihilation, etc.) during thermomechanical loading. In this section we briefly discuss nonlinear material behavior of multilayer films, focusing on the phenomena most relevant for the realization of reliable devices. The discussion in this section is concerned primarily with multilayer films where one of the films is a metal.

Nonlinear Material Behavior During Initial Thermal Cycles

As deposited, the material microstructure of a metal film is generally not stable. As the film is heated, microstructural evolution leads to the development of internal stress, which then can lead to changes in the curvature of the multilayer film. The general behavior is common to films made of many different metals, such as aluminum, copper, nickel, and gold. This has been studied extensively in the context of microelectronics applications (see for example [54.8, 44, 45]) where the substrate is much thicker than the film and the thin-film-limit results, which simplifies the interpretation of the results.

Figure 54.21 shows the general behavior of the curvature with temperature for a bilayer film consisting of a 0.5-μm gold film that was evaporated onto a 1.5-μm polysilicon film. During the initial heating the deformation is thermoelastic, characterized by a decrease in

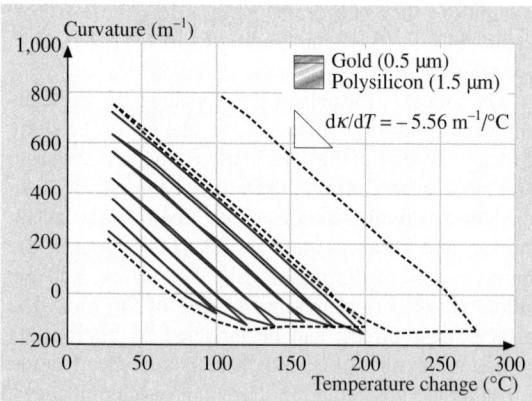

Fig. 54.21 Curvature versus temperature for gold (0.5-μm-thick)/polysilicon (1.5-μm-thick) 300 μm × 50 μm beam microstructures. The *solid lines* are tests with six cycles from room temperature to 200 °C with the maximum temperature in each successive cycle increasing by 20 °C. The *dashed lines* are tests with two cycles: the first from room temperature to 200 °C and then back to 30 °C, and the second to 275 °C and then back to 30 °C

curvature with a constant $d\kappa/dT$, until the temperature reaches about 70 °C (the theoretical thermoelastic slope from (54.18) is shown on the figure). Between 70 °C and 100 °C, the curvature decreases at a much lower rate; in fact the curve flattens, i.e., $d\kappa/dT \approx 0$. Interestingly, between room temperature and 100 °C, the curvature changes from positive to negative, our convention being that a positive curvature means a beam with the gold on top is curved upward. The strong departure of the curvature from the thermoelastic behavior results because the microstructure of the film is not stable in the as-deposited condition. The temperature increase during the first cycle promotes microstructural changes in the film. These result in tensile straining of the film, which competes with the thermoelastic deformation (see for example [54.8, 45]). Upon cooling from 100 °C, the response is again thermoelastic, this time throughout the entire cooling process. Although it does not occur here, if the temperature change is large enough, yielding of the metal film can occur upon cooling. The curvature upon return to room temperature is greater than that initially at room temperature. It is important to understand this behavior because many post-fabrication and packaging processes expose a microstructure to elevated temperature excursions. During these, the film is susceptible to this nonlinear deformation. In the tests in Fig. 54.21, the material has not been taken to a temperature high enough to stabilize fully the gold microstructure in the first cycle. As a result, subsequent cycles going to successively higher temperatures appear to have the effect of continuing to stabilize the microstructure. The response during the second cycle is similar to the first with the exception that the temperature change required to initiate nonlinear behavior is increased. Indeed, over the range of temperature cycles studied, each subsequent cycle is similar, exhibiting thermoelastic response upon heating until a point where nonlinear $\kappa–T$ behavior commences. The cooling process is again thermoelastic with the subsequent room-temperature curvature increasing with each cycle. Also shown in Fig. 54.21 is a result for a second test on a nominally identical microstructure with only two cycles: the first from room temperature to 200 °C and back to 30 °C, and the second to 275 °C and back to 30 °C. The results of the first cycle to 200 °C neatly envelope the results of the set of tests with cycles up to 200 °C. This shows that, although there is a path dependence of the deformation behavior, it is controlled by the maximum temperature reached during cycling, independent of how many increments are carried out to reach that temperature. The second cycle to 275 °C shows the same behavior observed in all cycles. The

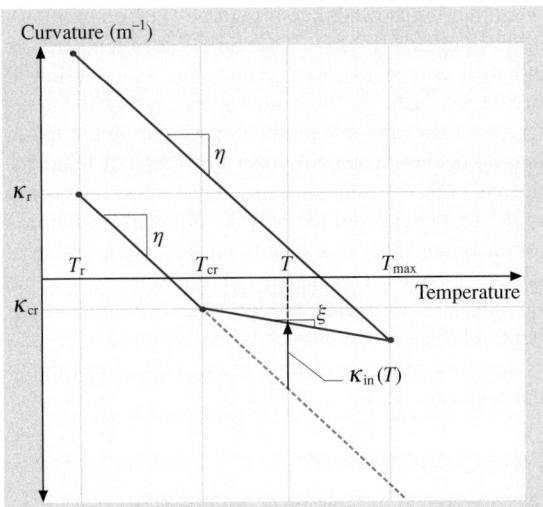

Fig. 54.22 Schematic diagram showing the parameters in the simple model that describes the curvature–temperature behavior of the bilayer beam during thermal cycling after release

results in Fig. 54.21 have been obtained at a constant heating/cooling rate. In terms of the stabilization of the material microstructure that results in nonlinear material behavior, the rate, along with the time held at the elevated temperature, plays a significant role.

Figure 54.21 suggests that a simple description of the evolution of curvature with temperature can be obtained with parameters as shown in Fig. 54.22: the as-released curvature κ_r (and associated temperature, T_r), a critical temperature at which inelastic mechanisms are activated, T_{cr} (and the associated curvature, κ_{cr}), the slope of the inelastic κ–T behavior (assuming simple linear behavior) for $T > T_{cr}$, ξ, and the thermoelastic slope $\eta = d\kappa/dT$. The as-released curvature is determined primarily by the intrinsic-stress development upon film deposition, cooling to room temperature, and release. These also influence T_{cr}. The parameter ξ is physically related to the curvature per unit temperature change due to inelastic mechanism(s) activated beyond T_{cr}. The situation in Fig. 54.21 is much simplified in that $\xi = 0$. With knowledge of these parameters, one can compute the curvature following a defined thermal loading history. The qualitative behavior in Fig. 54.22 also results for a substrate with patterned lines (Sect. 54.1.2), and the same formalism can be used with κ_r, η, and ξ depending on line width [54.35]. For example, (54.28) gives the dependence of η on line width. We also note that the recent microscopy work has focused on understanding the mechanisms of microstructural evolution that lead to this observed behavior, and thus to a mechanistic understanding of κ_r, η, and ξ [54.46].

Plasticity
Plasticity in thin-film multilayers has been studied in depth for thin metal films on thick substrates (see, for example [54.8, 47, 48]), but little work has been directed toward freestanding multilayers. This is probably because, at least when subject to thermal loading, the stresses are typically smaller in freestanding multilayer films than in systems with thin films on thick substrates (see Fig. 54.12). Many interesting open issues exist with regard to the general issue of plasticity in microstructures including the understanding of size effects and strain-gradient effects (see for example [54.49, 50]).

Creep and Stress Relaxation
For many microsystems applications, it is important to control accurately the deformation of thin-film microstructures over a significant period of time in order to meet performance and reliability requirements. This is especially important for microstructures subjected to thermal loading and/or operated at elevated temperatures. When one or more of the layers consists of a metal or polymer film, creep and stress relaxation in the film can significantly influence deformation and compromise device performance, and so their effects must be carefully considered. For example, *Miller* et al. [54.51] designed and fabricated microrelay switch arrays for radio-frequency (RF) communications applications using prestressed gold/polysilicon bimaterial beams as electrostatically actuatable switches. They observed a change in the switch shape and position over time and attributed it to stress relaxation in the gold. *Vickers-Kirby* et al. [54.52] report that creep in gold and nickel cantilever beams leads to voltage drops in micromachined tunneling accelerometers over time.

Creep and stress-relaxation phenomena have been investigated in some detail for thin film on thick substrate systems, motivated primarily by microelectronics applications (see for example [54.13, 44, 53–56]). These studies have focused on measuring, understanding, and modeling the stress–temperature behavior that occurs when a metal film–substrate system is heated from room temperature, then cooled back to room temperature, over one or a few cycles. The stress–temperature curves in the experimental works are typically obtained using the wafer curvature method that is now widely used to measure stress in thin films. During a typical test thermoelastic and inelastic mechanisms contribute to produce a complex nonlinear stress–temperature curve like that in Fig. 54.21. In most of these studies, the stress–temperature response is studied using wafer curvature measurements at a fixed heating or cooling rate so that rate-dependent and -independent phenomena are coupled in the response. Stress relaxation during an isothermal hold has been studied by [54.44, 53, 55, 57], among others. These studies all show that significant stress relaxation can occur over periods of only a few hours at modest temperatures of only about 100 °C for many metal thin-film systems. Due to the thin-film/thick-substrate system, the stress in the metal films is quite high, on the order of hundreds of MPa. Models incorporating power-law creep of the metal film were successfully able to describe the observed response qualitatively and, to a large degree, quantitatively. The latter, however, sometimes required modification of the power-law exponent. *Shen* and *Suresh* [54.44] showed experimentally that a thin passivation layer on a metal film on thick silicon substrates can significantly reduce creep and stress relaxation. *Thouless* et al. [54.55] also studied the effect of a thin passivation layer on the stress relaxation of thin films with a focus on the deformation mechanisms. Their experimental results suggest that the presence of a passivation layer on the surface of a film can have a substantial effect on relaxation rates likely by suppressing mechanisms associated with diffusion and dislocation climb. More recently, *Zhang* and *Dunn* [54.58] have carried out experiments and performed simulations of the creep/stress relaxation behavior of gold/polysilicon beams and plates like those of Fig. 54.21. This work illustrates the interesting coupling that can occur between material and geometric nonlinearity in thin-film microstructures.

Thermomechanical Fatigue

It is also important to understand the development of damage and its effect on deformation during cyclic thermomechanical loading. To date, only very limited attention has been directed toward this issue. *Zhang* and *Dunn* [54.35] showed that, if gold/polysilicon multilayers were cycled (Fig. 54.21) to an elevated temperature and then cooled to room temperature, they followed the thermoelastic path on subsequent cycling to a temperature below the maximum reached during the initial thermal cycle. Presumably over this range of temperature and time, the gold microstructure has been stabilized by the first cycle, and the polysilicon microstructure is not changing. This was confirmed for only a few, perhaps ten, cycles. *Gall* et al. [54.59] have shown that for thousands of cycles, though, a gradual shift of the thermoelastic curve downward is observed, possibly due to creep and stress relaxation in the gold, although the thermoelastic slope is maintained.

54.1.6 Other Issues

There are a number of additional issues regarding the thermomechanical behavior of thin-film microstructures that warrant discussion. Due to space limitations, however, we will only mention them and refer the reader to appropriate references.

In addition to the deformation itself, the stiffness characteristics of a microstructure, such as for example a beam or plate, can depend significantly on the stress in the beam. *Senturia* [54.60] describes the phenomena and discusses implications for microstructure design. This includes tensile stresses that tend to stiffen beams and plates as well as compressive stresses that can lead to buckling.

An important strategy has emerged to characterize the effects of processing on the residual-stress state and thermomechanical behavior of films: the use of on-chip test structures to extract material properties including film stresses. Many test structures, both passive and active, have been developed to measure residual stresses (tensile and compressive), stress gradients, elastic moduli, strength, toughness, thermal expansion, and many others. Nice surveys of many test structures are given by *Masters* et al. [54.14] and *de Boer* et al. [54.61] where further references to more extensive details can be found.

54.2 Electromechanics of Thin-Film Structures

Electromechanics, the coupling of electrostatics and mechanics, is perhaps the most common of the energy couplings used in microsystems technology for sensing and actuation. In this section we summarize the contem-

porary application of electromechanics, the analysis of fundamental electromechanics problems, and the practical issues that arise in the design of an electromechanical microsystem.

54.2.1 Applications of Electromechanics

Our reference to electromechanics will be limited to the coupling of electrostatics and the mechanics of constrained elastic media such as a cantilever beam. In this case, the electrical energy is transformed into strain energy of the elastic media but with typically negligible energy dissipated due to material damping. An alternative application of electromechanics is the coupling of electrostatics with rigid-body mechanics. In this case the mechanical structure is not constrained by a compliant support to the substrate but by a kinematic constraint, which is a pin joint between multiple structural layers or anchor between the substrate and a structural element. An example of these rigid-body constraints is the anchor in the electrostatic micromotors. In this case, the electrostatic energy is dissipated by friction (Coulomb-type friction between structural elements and pin joints, other structural elements, and/or the substrate) and is stored in the kinetic energy of the rigid body. In this case, the energy stored as strain energy in the structural element is, by design, negligible. In both electromechanics cases, the energy is dissipated by viscous forces under dynamic conditions. Finally the coupling of electrostatics and mechanics can more broadly include the coupling of electrostatics and fluids or electro-thermomechanics.

Microsystems have enabled many different commercial applications of electromechanics. These applications have included pressure sensors, accelerometers, gyros, resonators, micropumps, optical mirrors, optical shutters/VOAs (variable optical actuators), and direct-current (DC) and RF switches. In these examples, electrostatic forces are produced between a stationary electrode and a movable electrode, which is often the proof mass for an accelerometer or the diaphragm for a pump. The electrostatic forces are imparted by a voltage difference between the movable and stationary electrode. The electromechanics is typically implemented in two ways, with the first characterized by a stable, analog behavior, and the second characterized by an unstable, digital behavior. The stable behavior is implemented as a self-test capability for accelerometers such that a specific, stable displacement is produced with the application of the voltage. The unstable behavior is implemented with a switch so that when the instability point (snap-in or pull-in) is reached, the switch will instantaneously close.

Electrostatic actuation has been used for switches and microrelays [54.62–64], accelerometers [54.65,66], micropumps [54.67,68], and micromirrors [54.69–76]. The specific case of switches and relays involves multi-ple regions of electromechanical behavior and represents one of the richer electromechanic design problems. The general behavior is shown in Fig. 54.23. The first region is the stable actuation region. In this region, the switch is actuated to an instability point in the voltage–displacement curve called the pull-in or snap-in voltage. Up to the instability, the displacement increases monotonically until the pull-in voltage is reached. At the pull-in voltage, the switch instantaneously closes the gap because of the instability and establishes contact with the substrate. In the case of the switch, this means the closure of a contact. The electromechanics problem continues with a new set of boundary conditions in the next region because the switch is now *simply constrained* at the contact and fixed at its anchors to the substrate. In Fig. 54.23, the switch's tip displacement is shown, which is the reason the displacement is unchanged as the voltage increases. Region 3 is then characterized by a constant displacement for increasing voltage. The shape of the beam continues to change as the voltage is increased to the point that the electrodes could short if necessary precautions are not taken.

If the first consideration of the switch design is pull-in voltage, the second consideration of the switch design is the development of contact force, which is directly related to the contact resistance. The electromechanical analysis is continued beyond the pull-in voltage once contact is established to develop an increasing contact force. The electrostatic force is distributed between the anchor and the contacts, assuming other portions of the switch are not in contact. The electrostatic force increases as the square of the voltage difference, so the force acting on the contact will increase in proportion to this. Of course other physical limitations must be considered such as gas breakdown or a secondary pull-in that cause shorting between electrodes. The contact areas of the switch that are not the contacts should be minimized, so they do not detract from the contact force.

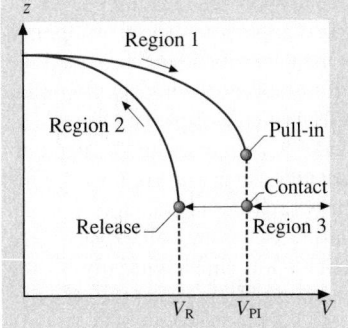

Fig. 54.23 Characteristic electromechanical hysteresis curve showing pull-in, contact, and release voltages

A third design consideration is the overdrive capability, defined as the maximum voltage the switch control electrodes can support before they touch and short. In some cases, the electrodes may be isolated from each other by a thin dielectric film, whose implications will be discussed in a later section. As suggested in the later section, the insulation material can be patterned to reduce the contact area between the electrodes and the insulating material, in order to reduce charge accumulation. With patterned isolation, the switch could be driven by a large enough voltage to short the uninsulated regions. This failure mode could be evaluated by the same electromechanics analysis.

A fourth design issue would be the switch's release voltage, which is the voltage when the contact force reaches zero and the switch opens. This is shown as the beginning of region 2, where the deformation is reduced as the voltage is decreased. The release voltage is lower than the pull-in voltage because the electrostatic forces are much greater for the reduced gap than for the larger gap just before the pull-in instability. A fifth design consideration is the switch's self-actuation voltage, which is that which occurs across open contact and unintentionally causes the switch to close. By design, the switch's self-actuation voltage should be maximized in the electromechanical analysis. A final design issue is the determination of breakdown limits across the switch's control electrodes. This is not limited to the voltage supported across the equilibrium gap of the control electrodes and needs to be considered for different cases. The breakdown will be described by the Paschen curve or by vacuum breakdown proportional to the gap.

54.2.2 Electromechanical Analysis

An electromechanical analysis requires modeling of the structural domain, the electrostatic domain, and the coupling between the electrostatic fields and the structural elements. It has been approached at three levels of complexity. The simplest approach is the development of lumped parameter models, which have provided the basic understanding of instability, the impact of nonlinear springs, and the pull-in voltage–gap relationship. The lumped parameter model is comprised of the parallel-plate capacitor and a spring (linear or nonlinear). One plate is constrained in space, by a substrate for example. The second plate is elastically constrained to one degree of freedom, typically, of motion, which is usually closing the air gap. The spring element represents the effective spring stiffness K of the structural element, which could

be a beam, a plate, a membrane. The parallel-plate capacitor represents the electrostatic coupling between the fixed and moving elements.

Increasing in complexity, the second level of analysis considers continuum structural elements with a distributed electrostatic load. For example, the solution for an isotropic homogeneous beam would satisfy the Euler–Bernoulli beam equation:

$$EI\frac{\mathrm{d}^4 w}{\mathrm{d}x^4} = q(x)\,, \tag{54.38}$$

where E is the Young's modulus, I is the structure's moment of inertia, $w(x)$ is the transverse deflection of the beam, x is the coordinate along the length of the beam, and $q(x)$ represents the distributed electrostatic load. For a multilayer film, an effective EI can be used, as mentioned in Sect. 54.1.2. The third level of analysis is to model the fully coupled problem as continuum solids. This approach has used different implementations of the finite difference (FD), finite element (FE), and boundary element (BE) methods. A typical example would be to implement a finite element code to solve the continuum mechanics problem and a boundary element code to solve the electrostatics problem. Another implementation would use finite element methods to solve both the mechanics and the electrostatics problems. In each case, a similar algorithm is followed for developing a self-consistent solution to the coupled electromechanics analysis.

The coupled electromechanics problem can be described simply as a pair of conductors, where one conductor has a fixed constraint and the second conductor is attached to an elastic structure, such as a beam, a plate, a membrane, or an elastic substrate. The conductors support a voltage difference between them that causes charge to be induced on the surface of the conductors. This charge produces an electrostatic force that attracts the two conductors together. As the elastically supported conductor deforms towards the fixed conductor, the charge redistributes and modifies the field, thereby the force distribution. The process continues until a redistribution of charge is no longer necessary to maintain equilibrium. The typical algorithm begins by calculating the displacements of the undeformed geometry, which is updated before proceeding to the electrostatic analysis. In the electrostatic analysis, the surface charge density distribution is calculated and then used to calculate the electrostatic forces. The electrostatic force distribution is the important parameter because it is passed to the mechanical analysis as the loading condition on the conductors in the undeformed state to deform the structure until a state of equilibrium is reached.

Electromechanical Systems Energy Balance

A general analytical methodology can be applied to many different structures including parallel-plate actuators with linear or nonlinear springs and plates supported by torsional suspensions [54.77]. They consider a single-input system with charge control or voltage control. To begin, a generalized actuator is considered that consists of two conductors. One conductor is fixed. The second conductor is constrained by an elastic support.

In the charge-controlled actuator, the total energy for the generalized actuator is written in terms of the mechanical energy as a function of the generalized coordinate χ and of the electrical energy stored in the capacitance of the actuator. The total energy is:

$$U_T(\chi, Q) = U_M(\chi) + \frac{Q^2}{2C(\chi)}, \qquad (54.39)$$

where U_T is the total energy, U_M is the mechanical energy, and the electrical energy is determined by the charge Q, and the capacitance $C(\chi)$. The first derivative of the total energy with respect to the generalized coordinate set equal to zero determines the equilibrium. The second derivative of the total energy with respect to the generalized coordinate set equal to zero determines the stability. The first and second derivatives are determined and then combined to determine what is referred to as the charge-controlled pull-in equation:

$$\frac{\partial U_M(\chi_{PI})}{\partial \chi} \frac{\partial^2}{\partial \chi^2}\left[\frac{1}{C(\chi_{PI})}\right]$$
$$-\frac{\partial^2 U_M(\chi_{PI})}{\partial \chi^2}\frac{\partial}{\partial \chi}\left[\frac{1}{C(\chi_{PI})}\right] = 0 \qquad (54.40)$$

and the pull-in charge Q_{PI}:

$$Q_{PI} = \left[\frac{-2\dfrac{\partial U_M(\chi_{PI})}{\partial \chi}}{\dfrac{\partial C(\chi_{PI})}{\partial \chi}}\right]^{1/2}. \qquad (54.41)$$

In the voltage-controlled case, the total co-energy for the generalized actuator is used because the capacitor is nonlinear, and the actuator is voltage-controlled. The co-energy is written in terms of the capacitance between the generalized conductors, the voltage across the conductors, and the mechanical energy. The total co-energy

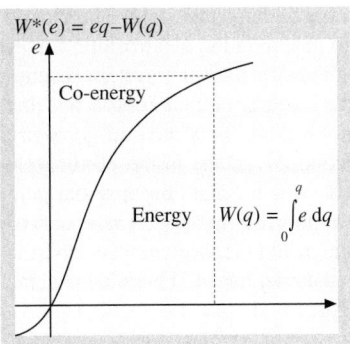

Fig. 54.24 Effort–displacement curve showing the relationship between energy and co-energy

is:

$$U_T^*(\chi, Q) = \frac{1}{2}C(\chi)V^2 - U_M(\chi). \qquad (54.42)$$

The relationship between energy and co-energy is shown in Fig. 54.24.

In a similar process to the charge-controlled system, the voltage-controlled pull-in equation is determined by the first derivative of the total co-energy with respect to the generalized coordinate and by the second derivative of the total co-energy with respect to the generalized coordinate. The equations for the first and second derivative are combined to determine the voltage-controlled pull-in equation

$$\frac{\partial U_M(\chi_{PI})}{\partial \chi}\frac{\partial^2 C(\chi_{PI})}{\partial \chi^2}$$
$$-\frac{\partial^2 U_M(\chi_{PI})}{\partial \chi^2}\frac{\partial C(\chi_{PI})}{\partial \chi} = 0 \qquad (54.43)$$

The pull-in voltage V_{PI} is:

$$V_{PI} = \left[\frac{2\dfrac{\partial U_M(\chi_{PI})}{\partial \chi}}{\dfrac{\partial C(\chi_{PI})}{\partial \chi}}\right]^{1/2}. \qquad (54.44)$$

The application of this approach to either the voltage-controlled or the charge-controlled case involves determining the mechanical energy and the capacitance. From the mechanical energy and the capacitance, the pull-in parameters and voltage are determined.

Lagrangian Approach to Electromechanics

Li and *Aluru* [54.78] and *Aluru* and *White* [54.79] have proposed a Lagrangian approach for both the electrostatic analysis and the mechanical analysis. Their algorithm begins by calculating the structural displacements for the undeformed geometry. They proceed to the electrostatic analysis to calculate the charge distribution on the conductors in the undeformed state. The charge distribution is used to calculate the electrostatic forces on the undeformed geometry until the system reaches equilibrium. The first advantage of their approach is that the algorithm does not require the structural geometry to be updated. A second advantage is the elimination of the integration error, which occurs because flat panels are used in other approaches to approximate curved surfaces. The overall algorithm is shortened because the interpolation functions do not have to be updated as the structure shape changes.

54.2.3 Electromechanics – Parallel-Plate Capacitor

The classic element for understanding electromechanics is the simple parallel-plate capacitor model with one plate fixed and the second, moveable, plate suspended by a spring. This basic electromechanics element is shown in Fig. 54.25. The capacitor is described by an area A; an initial gap g_0; and a dielectric constant ε in the air gap (where ε is the product of the relative dielectric of the media filling the gap, ε_r, and the permittivity of free space $\varepsilon_0 = 8.854 \times 10^{-12}$ F/m). The moveable plate is shown suspended by the spring, with spring constant K, and a single degree of freedom represented by the coordinate z. In this case, the equilibrium position is represented by having no charge on the plates (i.e. no electrostatic force) and no displacement of the spring (i.e. no spring force). It should be noted that, as the spring length grows by increasing z, the gap decreases. The behavior of this simple model is described in detail in many sources including [54.60, 77].

The parallel-plate capacitor can be analyzed with either a voltage source or a current source. With a voltage source, the charge on the capacitor plates is

$$Q = \frac{\varepsilon A}{g} V ,\qquad (54.45)$$

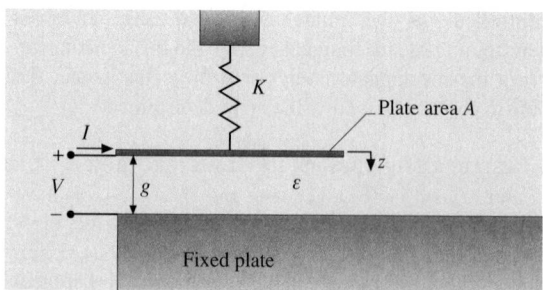

Fig. 54.25 Lumped parameter parallel plate capacitor electromechanical model

where Q is the charge on the capacitor plates, ε is the dielectric constant in the gap, A is the capacitor plate area, g is the instantaneous gap between plates, and V is the voltage across the plates.

The electrostatic force on the capacitor plates is determined by the dielectric constant, the capacitor plate area, the instantaneous gap, and the voltage across the capacitor:

$$F = \frac{\varepsilon A V^2}{2g^2} .\qquad (54.46)$$

The electrostatic force scales as the voltage squared and inversely decreases as the instantaneous gap squared. This is the force that acts on the movable plate; hence the spring stretches (or is relaxed if the voltage is lowered). The amount of stretch in the spring is:

$$z = \frac{F}{K} .\qquad (54.47)$$

The instantaneous gap is

$$g = g_0 - z = g_0 - \frac{\varepsilon A V^2}{2Kg^2} .\qquad (54.48)$$

**Electromechanics –
Parallel-Plate Capacitor with External Forcing**

Nemirovsky et al. [54.77] examined the case of a constant applied force acting on the suspended capacitor plate of a parallel-plate electrostatic actuator. This situation represents a force-balanced system, in which the electrostatic force is used to balance a pressure load acting on a membrane or an inertial load (e.g. acceleration-induced) on a suspended proof mass. With an applied mechanical force, the mechanical energy is:

$$U_M (z) = \frac{1}{2} K z^2 - F_{ext} z ,\qquad (54.49)$$

where F_{ext} is the externally applied force. The solution to the pull-in equation is

$$\zeta_{PI} = \frac{z}{z_{max}} = \frac{1}{3} \left(1 + 2 \frac{F_{ext}/K}{z_{max}} \right) ,\qquad (54.50)$$

where ζ_{PI} is the ratio z/z_{max}. It can be seen when the applied force is zero. Equation (54.50) reduces to the familiar result of $z = z_{max}/3$. The pull-in voltage is

$$V_{PI} = \left[\frac{8K \left(z_{max} - \dfrac{F_{ext}}{K} \right)^3}{27\varepsilon_0 A} \right]^{1/2} .\qquad (54.51)$$

Electromechanics –
Parallel-Plate Stabilization

An important practical consideration is the stabilization of the parallel-plate electrostatic actuator. *Seeger* et al. [54.80], *Hung* and *Senturia* [54.81], and *Nemirovsky* et al. [54.77] showed that the addition of a series capacitance (feedback capacitance) stabilizes the actuator such that it does not pull in to the substrate. The series capacitance and the variable actuator capacitance combine to form a voltage divider that provides passive control of the voltage across the actuator. In this case, *Nemirovsky* et al. [54.77] identified three regimes, defined in terms of δ, which is the ratio of the initial actuator capacitance to the series capacitance (*Nemirovsky* et al. [54.77] had defined the series capacitance in terms of a dielectric layer on one capacitor plate). The first regime is defined for $0 < \delta < 0.5$, when the series capacitance is less than half the nominal actuator capacitance. In this case, the actuator is stable so the pull-in effect is eliminated. The second regime is defined by $\delta > 0.5$ and the pull-in occurs between $1/3$ of the gap and the full distance of the gap. In the third regime $\delta \gg 1$, which means the series capacitance is much greater than the nominal capacitance. The pull-in parameter is given by:

$$\zeta_{\text{PI}} = \frac{1}{3}\left(1 + \frac{1}{\delta}\right), \tag{54.52}$$

which reduces to $\zeta_{\text{PI}} = \zeta_{\text{max}}/3$ when the series capacitance is eliminated. The pull-in voltage is:

$$V_{\text{PI}} = \left[\frac{8Kz_{\text{max}}^3}{27\varepsilon_0 A}\left(1 + \frac{1}{\delta}\right)^3\right]^{1/2}. \tag{54.53}$$

Electromechanics –
Extending the Range of Motion

Electromechanical systems have two obvious characteristics: nonlinearity and instability. If one desires digital operation or fast switching times, the inherent electromechanical instability is a significant benefit. On the other hand if analog control is desired, the control is faced with both the nonlinearity and the instability. In both cases, the range of motion of the electromechanical device is limited. The range of motion to the instability was extended by the addition of a series capacitor, but this did not address the linearity. *Toshiyoshi* et al. [54.70] addressed the linearity of electrostatically actuated mirrors in an optical scanner. Theirs was a two-degree-of-freedom scanner, so the increased

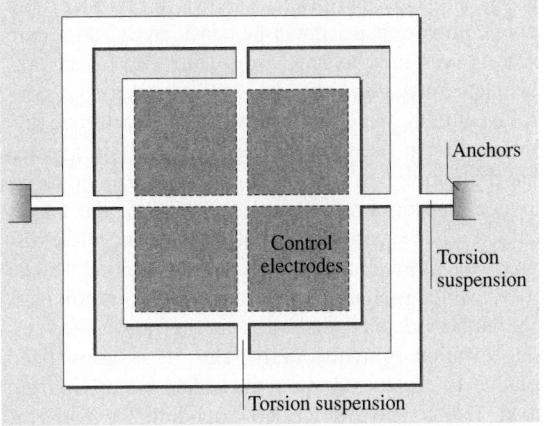

Fig. 54.26 Two-degree-of-freedom torsional electromechanical actuator with a rectangular plate and a stationary electrode divided into four sections

linearity was necessary to improve the crosstalk between the two different scanner modes. The scanner in question is a square plate suspended by torsion beams within a frame that is itself suspended by another pair of torsion beams. The plate and frame are suspended above the substrate, which has four stationary electrodes defined in the four quadrants of the plate, as shown in Fig. 54.26. To actuate the mirror with increased linearity, a bias voltage V_{bias} was applied between each electrode and the plate. Independent differential control voltages are superimposed on the bias voltage supplied between each electrode and the plate. The driving voltages are described by two independent differential voltages such that $V_{\text{diff1}} = (V_x + V_y)/2$, $V_{\text{diff2}} = (-V_x + V_y)/2$, $V_{\text{diff3}} = (-V_x + -V_y)/2$ and $V_{\text{diff4}} = (V_x + -V_y)/2$. A second approach taken to improve the linearity was to consider the shape of the mirror, because of the geometric relationship that determines the net electrostatic torque. They considered such simple geometries as square, diamond (square rotated

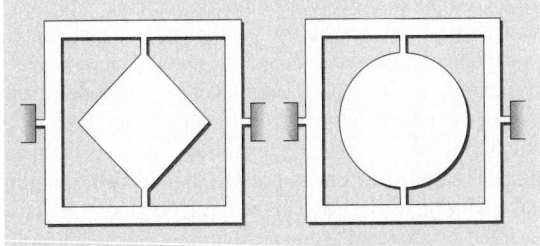

Fig. 54.27 Alternate two-degree-of-freedom torsional electromechanical actuators with diamond and circular plates

by 45), and a circular plate, as shown in Fig. 54.27, but more complex shapes could be considered. The square plate is presumed to have worse linearity because as the plate rotates the corners deflect toward the control electrodes, which increases the applied torque and further distorts the shape of the mirror. This suggested that the removal of the corners could contribute to increased linearity and reduced distortion of the mirror surface. They found the circular plates reduced the distortion by 50% over the rectangular and diamond-shaped plates, which is significant for the optical performance. The authors defined distortion as the normalized angle deviation from the ideal value. It is normalized relative to the maximum scan angle. *Nadal-Guardia* et al. [54.82] have described extending the range of travel by using current-drive methods instead of voltage drive.

54.2.4 Electromechanics of Beams and Plates

The basic understanding of electromechanics and pull-in phenomena has been developed around the simple parallel-plate capacitor model developed above. Many micro-electromechanical system (MEMS) designers innately understand that the electrostatic force is proportional to the square of the voltage, is inversely proportional to the square of the gap, and that the plate will pull in after displacing $1/3$ of the gap. While this is good to first order, it is not adequate for the electromechanical design of other structures such as cantilevered beams/plates, beams/plates with multiple supports, plates with torsional supports, and other structures.

O'Brien et al. [54.83] and *Choi* and *Lovell* [54.84] have studied the electromechanics of cantilever beams by presenting two improved models for cantilever beam electrostatic actuators. In both cases, the cantilever beam is a homogeneous material anchored to the substrate at one end and otherwise freely suspended along its length, which is supported above a stationary electrode. The new models are compared to the parallel-plate capacitance model and a simple extension of the parallel-plate model to a cantilever beam. The cantilever beam model uses a simple cantilever of width W, length L, thickness T, and initial air gap Z_0. The parallel-plate capacitance model is based on a capacitance determined by the length and width of the beam. The beam deflection was represented by the displacement function of a beam with a force applied at the beam tip. The beam spring constant was determined by the ratio of the force at the

beam tip to the displacement at the beam tip (maximum displacement). The capacitance was determined by integrating the parallel-plate capacitance for a differential length element along the length of the beam, where the beam gap is defined by the initial gap in the mechanical displacement function. This showed the beam tip displaced 45% of the initial gap before pull-in occurred. In this case the pull-in voltage can be expressed as a function of the beam's Young's modulus E, cross-sectional moment of inertia I, length L, and width W, the initial gap Z_0, and the dielectric constant in the air gap as

$$V_1 = \sqrt{\frac{18EIZ_0^3}{5\varepsilon_0 L^4 W}} = 0.5477\sqrt{\frac{ET^3 Z_0^3}{\varepsilon_0 L^4}} \quad \text{(Volts)}.$$

(54.54)

In the first improved model, the cantilever beam is defined by the same characteristics, but the electrostatic force acting along the length of the beam is replaced by an equivalent bending moment M applied at the beam tip. The beam displacement was determined for a uniform, homogeneous beam with a bending moment applied at its end. The beam's spring constant was determined by the ratio of the force at the beam tip and the displacement at the beam tip. The capacitance was determined by integrating the parallel-plate capacitance for a differential length element along the length of the beam, where the beam gap is defined by the initial gap in the mechanical displacement function. This model showed the beam tip displacement was approximately 46% of the initial gap before pull-in occurred. In this case the pull-in voltage is

$$V_2 = \frac{43}{50}\sqrt{\frac{18EIZ_0^3}{5\varepsilon_0 L^4 W}} = 0.471\sqrt{\frac{ET^3 Z_0^3}{\varepsilon_0 L^4}} \quad \text{(Volts)}.$$

(54.55)

In the second improved model, the cantilever beam was defined by a rigid beam, whose anchor was replaced by a simple hinge and continuum elasticity replaced by an equivalent spring at the beam's free end. Three new variables were introduced to describe the behavior in terms of radial coordinates centered at the hinge. The two radial coordinates describe the beginning of the electrode at r_0 and the end of the beam/electrode at r_1. The angle θ was introduced to describe the angular displacement of the beam. In this model, the tip of the beam displaced 44% of the initial gap before pulling into the substrate. The pull-in voltage is estimated by (54.56) or

(54.57) where r_0 and r_1 have been replaced by 0 and L:

$$V_3 = \frac{11}{25r_1^2} \sqrt{\frac{77ET^3Z_0^3}{25\varepsilon_0 \left[\frac{11(r_1 - r_0)}{r_1 - \frac{11}{25}r_0} - 14\ln\left(\frac{25r_1 - 11r_0}{14r_1}\right) \right]}} \,,$$

(54.56)

$$V_3 = \frac{11}{25L^2} \sqrt{\frac{77ET^3Z_0^3}{25\varepsilon_0}} = 0.4548 \sqrt{\frac{ET^3Z_0^3}{\varepsilon_0 L^4}} \text{ (Volts)} \,.$$

(54.57)

As expected, the dependence on parameters, such as Young's modulus, beam thickness, beam length, initial air gap, and the air-gap dielectric constant is the same for all cases presented. The difference in the three models is the coefficient of the pull-in voltage function and the tip deflection at pull-in. The coefficients of the pull-in voltage predictions are 0.5479, 0.471 and 0.4548, respectively. The beam tip deflected 45%, 44%, and 46% of the initial gap before pull-in occurred. This is in contrast to the parallel-plate model that predicts the plate deflection will be 33% of the initial gap before pull-in. By comparison to experimental test results, the models presented in (54.55) and (54.57) were more accurate at predicting the pull-in voltage for cantilever beams. The parallel-plate electromechanics model had greater error and under-predicted the pull-in voltage. The model represented by (54.54) over-predicted the pull-in voltage with greater error than either of (54.55) or (54.56). The beam models predicted more accurately for wider beams, which means they are better applied to beams with dimensions representative of a parallel-plate model used as a basis of the capacitance estimate. In all three cases, the beam models presented in (54.54–54.56) poorly predicted the pull-in voltage for narrow beams, which are dominated by the effects of nonuniform fringing fields and not the uniform parallel-plate fields. The typical application of the parallel-plate capacitor model is to geometries with a width W and length L that are much greater than the capacitor gap Z_0 ($W \gg Z_0$, $L \gg Z_0$; where greater than a factor of ten is a common rule of thumb). In Fig. 54.28, two capacitors are shown such that in one case the plate dimensions are much greater than the gap, and in the second case the plate dimensions are much smaller than the gap. When the plate dimensions are much larger than the gap, the uniform fields in the gap dominate the capacitive coupling. A small contribution to the capacitive coupling comes from the *fringing* fields at the edge of the plate (or the perimeter of the plate in three dimensions). As

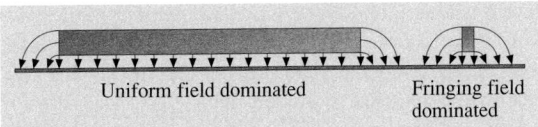

Fig. 54.28 Cross sections of wide and narrow electrodes showing uniform electric fields in the gap and fringing fields at the edges (perimeter)

shown in Fig. 54.28, the fringing fields could depend on the thickness of the plate. When the plate dimensions are of the same order as the gap, the fringing fields may dominate or be of the same order as the contribution of the uniform field in the gap. This is the nature of the narrow cantilever beams that were not predicted well by a parallel plate capacitor model. A correction can be introduced to the parallel plate model to approximate the contribution of the fringing fields. This approximation is often referred to as the Love approximation.

It is important to consider the simplifications that have been made for electromechanical systems. *Elata* et al. [54.85] have reconsidered these simplifications and added secondary degrees of freedom. In the simple models, bending or torsion of mechanical elements may have been ignored to first order. The secondary degree of freedom can significantly affect the pull-in voltage instability of devices especially relative to process variations or misalignment between stationary electrodes and moving structures. A simple example is the cantilever beam, where the pull-in voltage is primarily determined by displacement, z, and curvature, ρ. A secondary degree of freedom could be a twist angle, ϕ that could be associated with asymmetric fabrication.

Segmented Control of Electromechanical Membrane Deformation

Wang and *Hadaegh* [54.73] describe the electrostatic actuation of a deformable membrane mirror with segmented electrodes on the substrate. The membrane mirror, in this case silicon nitride with a conductive film, was suspended above the substrate containing the segmented electrode. The segmented electrodes are defined along an angular and radial pattern. A specified voltage can be applied between the membrane and each electrode segment to develop a specific deformation for the mirror. This increases the complexity of the control system but also the flexibility of controlling the shape of the mirror surface. The optical properties could be optimized by individual characterization of membranes or by feedback from the optical output. This basic concept was also applied by *Zhang* and *Dunn* [54.86], who were able to linearize the voltage–deflection relationship by tailoring the shape of the electrodes and the beam structure used to support them.

54.2.5 Electromechanics of Torsional Plates

Small–Angular–Deflection Approximation

The plate suspended by a torsional suspension instead of bending is the final configuration to be considered in terms of electromechanics. The electrostatic torsional plate has many applications, such as optical mirrors [54.69,74] and accelerometers [54.66]. The plate will define the optical surface of the mirror, the moveable plate of a varactor, or the proof mass of an accelerometer. It will be conductive or have a conductive film so that it can couple electrostatically to a stationary plate on the substrate. The moveable plate will be suspended by a torsional suspension such that the bending modes are suppressed and the primary kinematic mode is rotation of the plate. The variable capacitive coupling is used in two ways. First, it is used for actuation by applying a voltage between the moveable plate and a stationary plate. Second, it is used for sensing or use of the variable capacitance. Both the actuation and variable-capacitance functions can be present in a device. The key elements of interest are the voltage–displacement function and the pull-in voltage.

Degani et al. [54.87], *Xiao* et al. [54.88], and Nemirovsky et al. [54.77] have considered the torsional electrostatic actuator from an analytical perspective with comparison to experimental results and other modeling results. The effective torsional spring constant is determined from the torque–twist angle relationship, where the torque is the electrostatic torque. The electrostatic torque is developed for both small- and large-angle cases. The pull-in angle and pull-in voltage are determined from the pull-in equations for voltage control (54.43, 44) and charge control (54.40, 41). Their method is an improved approach for modeling torsional electrostatic actuators over approximations involving parallel-plate models and effective spring constants, as they have shown by comparison to experimental measurements. The torsional plate considered is suspended by a torsional suspension, whose axis of rotation defines a gap of d. It is assumed that the plates extend into the page (z-axis) to a depth b and have a total length a. The torsional plate electrostatic actuator is shown in Fig. 54.29. The solution assumes that the tilted plates are semi-infinite, ignoring the effects of fringing fields, hence their corrections, and are tilted by angles that sat-

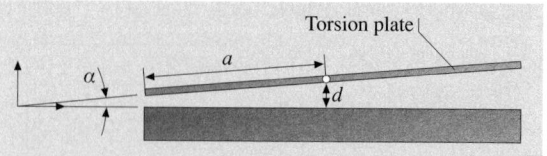

Fig. 54.29 Cross section of a torsional actuator with a uniform cross section into the page showing the angular degree of freedom α, plate length a; and initial gap d

isfy the small-angle approximation ($\tan\alpha \approx \sin\alpha \approx \alpha$ and $\cos\alpha \approx 1$). The capacitance is:

$$C(\Theta) = \frac{\varepsilon_0 b}{\alpha_{max}} \frac{1}{\Theta} \ln\left(\frac{1}{1-\Theta}\right), \tag{54.58}$$

where $\alpha_{max} = d/a$ and $\Theta = \alpha/\alpha_{max}$.

The capacitance function and the linear relationship for the elastic restoring force ($F_M = K\,a$) are used to develop the electrostatic and mechanical energy function to substitute in the pull-in equations. For voltage-controlled actuation, the pull-in parameter is developed from (54.43) to yield:

$$4 + \frac{5\Theta_{PI} - 4}{(1-\Theta_{PI})^2} - 3\ln(1-\Theta_{PI}) = 0. \tag{54.59}$$

Equation (54.59) is solved numerically to show that the voltage-controlled pull-in angle is $\Theta_{PI} \approx 0.44$. The pull-in angle is substituted in (54.44) to determine the voltage-controlled pull-in voltage:

$$V_{PI} = \left(\frac{0.827 K\alpha_{max}^3}{\varepsilon_0 b}\right)^{1/2}. \tag{54.60}$$

The charge-controlled pull-in parameter relationship is determined by substituting the electrostatic energy and mechanical energy functions into the charge-controlled pull-in parameter (54.40). This provides the nonlinear relationship for the charge-controlled pull-in angle:

$$\ln(1-\Theta_{PI}) - \frac{\Theta_{PI}}{(1-\Theta_{PI})^2}$$
$$- \frac{2\Theta_{PI}^2}{(1-\Theta_{PI})^2 \ln(1-\Theta_{PI})} = 0. \tag{54.61}$$

The charge-controlled pull-in angle is given by $\Theta_{PI} \approx 0.71$. The pull-in angle is used in the charge-controlled pull-in equation (54.41) to determine the pull-in charge relationship:

$$Q_{PI} = (1.798 K\varepsilon_0 b\alpha_{max})^{1/2}. \tag{54.62}$$

Large Angular Deflection and Plate Geometry

Nemirovsky and *Bochobza-Degani* [54.77] have extended the work to the case of large angle deflections, which included using the trigonometric relationship in the capacitance function and a nonlinear relationship for the mechanical spring constant. These nonlinear relationships were used in the voltage-controlled pull-in equation. In addition, their methodology was extended to planar plates with different shapes (Fig. 54.30). They specifically considered triangular plates (the apex of the triangle is near the axis of rotation), a reversed triangular plate (the triangle apex is at the distal end of the plate), a square plate (with the rotational axis on the edge of the plate), and a circular plate (with the axis of rotation passing through a point on the perimeter and the opposite point on the perimeter defining the most distal point of the plate). Only simple geometries were considered, and their methodology can be applied to other more-general geometries. The rectangular plate torsional, electrostatic actuator will provide the lowest actuation voltage relative to a fixed plate length. The reversed triangular plate has proved beneficial with regard to pixel area in mirror application and has been able to provide a higher torque for a specific plate length.

54.2.6 Leveraged Bending

As discussed, when the actuation electrode spans the entire length of the actuator, the stable controllable travel distance of the actuator plate is severely limited due to the pull-in instability. One can get around the problem of pull-in by applying an electrostatic force to only part of the structure and using the rest of the structure as a lever to move the structure through a larger range of motion. This concept, referred to as leveraged bending [54.89, 90], is shown schematically in Fig. 54.31 for a cantilever beam. The key idea is that the portions of the beam above the electrodes do not deflect far enough to violate the pull-in limit. The concept can be applied to other microstructure shapes and support conditions, but for fixed–fixed beams, the effect can be diminished by tensile residual stresses [54.89, 90].

54.2.7 Electromechanics of Zipper Actuators

The design requirements for electromechanical actuators usually require the minimization of the actuation voltage but maximizing the largest displacement. With the parallel-plate actuator, be it a plate, beam, or

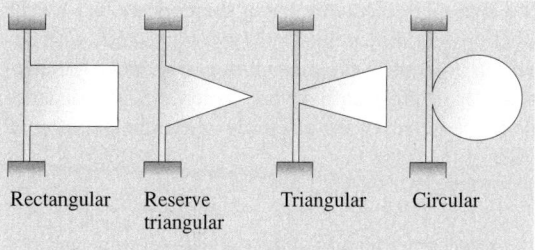

Fig. 54.30 Rectangular, triangular, reverse triangular, and circular plates are shown suspended by torsion beams for application to electromechanical mirrors or other actuators

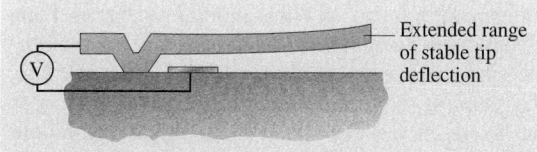

Fig. 54.31 Illustration of the concept of leveraged bending to increase the range of stable motion

torsional plate, the pull-in voltage is reduced by minimizing the air gap, but this limits the maximum displacement. Actuators with a zippering behavior have been implemented to achieve greater deflections for a lower actuation voltage [54.91]. In one example [54.92, 93] a plate is suspended by a torsional suspension. A long, slender electrode extends from the plate and terminates with a plate-like area. This electrode is supported above the fixed substrate electrode by a largely uniform gap. A voltage is applied between the substrate electrode and the extended electrode, such that the plate-like area is first pulled into the substrate. As the actuation voltage is increased, the long, slender electrode continues to collapse against the substrate electrode until a desired maximum deflection is achieved or until the drive voltage becomes impractical. This is the so-called zippering effect, which has been described for a motion perpendicular to the substrate. This uniform-gap scenario is not the best suited for minimizing the pull-in voltage, so other configurations such as the curved-electrode actuator have been proposed and applied to such devices as a switch and optical shutters as described by *Jin* et al. [54.94].

Legtenberg et al. [54.95, 96] described the basic operating features of the curved-electrode actuator. These actuators provide large deflections in the plane of the substrate. The curved-electrode actuator is composed of a stationary electrode and a compliant moving electrode.

The stationary electrode has a shape described by the function $s(x)$ shown in (54.63) and Fig. 54.32, which is written in terms of the maximum gap between the electrodes δ_{max}, the length of the electrode L, the distance along the length of the electrode x, and the polynomial order of the curve n.

$$s(x) = \delta_{max} \left(\frac{x}{L} \right)^n .$$ (54.63)

A necessary feature of the curved-electrode and zipper actuators, in general, is the contact between the moving and stationary electrodes. Without due consideration, these electrodes will short upon contact and damage the actuator. Two isolation solutions are implemented to prevent the shorting scenario. First, a continuous insulation film can be added between the electrodes to prevent them from shorting. In this case, the simplest solution is to add the insulation layer to the fixed, curved electrode. The alternative solutions include adding the insulating film to only the moving electrode or to both the moving and stationary electrodes. If the insulation is added to the moving electrode, it will become part of the stiffness of the beam. A second solution would be to add discrete, isolated contact bumpers that are made from the same conducting material as the moving electrode. As the gap closes, the moving electrode makes contact with the isolated bumpers but not the actuation electrode.

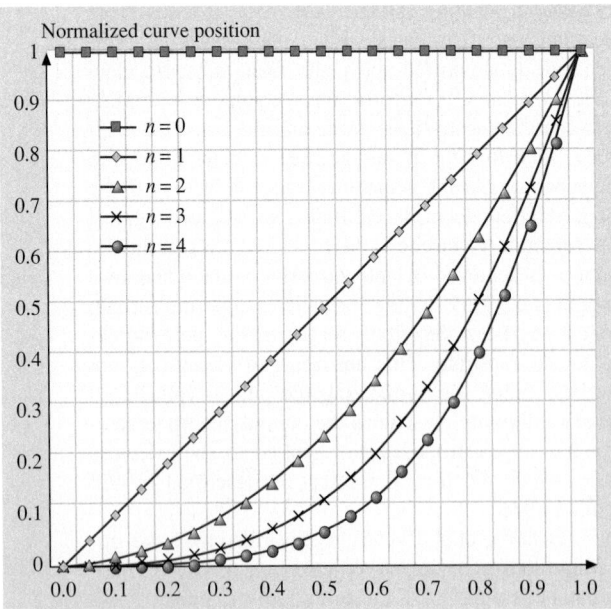

Fig. 54.32 Curved electrode shape curve shown as a function of order of the curve n

The operational differences are described by *Legtenberg* et al. [54.96]. Three other issues that must be considered during the design of a curved-electrode or zipper actuator are dielectric charging, stiction, and dielectric breakdown.

Legtenberg et al. [54.96] found that, as the polynomial order increased for the electrode curve, the pull-in voltage decreased for about the same maximum displacement. This means larger displacements can be achieved for lower voltages when compared to the parallel-plate actuator. Once the moving electrode is pulled in to the curved electrode, the moving electrode will collapse to the maximum displacement. It was shown for polynomial curves of order $n > 2$ that the maximum tip deflection becomes stable because of the geometric constraints provided by the curved electrode. This means that, to achieve the maximum displacement with the lowest pull-in voltage, a polynomial curve of order $n = 2$ should be specified. For polynomial curves with $n > 2$, stable continuous deflection along the curved electrode can be implemented, with the maximum deflection occurring at lower voltages for polynomial orders just above two. If an approximate linear range of operation is desired (i. e. the tip deflection is approximately linear in voltage), higher-order curves ($n > 2$) are desired. *Legtenberg* et al. [54.96] analyzed the curved-electrode problem by the application of analytical energy methods, specifically the Rayleigh–Ritz method and three-dimensional self-consistent electromechanical numerical simulations using CoSolve-EM in MEMCAD (now integrated into CoventorWare). From an energy perspective, the total potential energy U_T is determined in terms of the mechanical energy associated with the strain energy of bending and the electrostatic energy stored in the electric field. For the curved electrode of Fig. 54.32, the total potential energy is:

$$U_T = U_M + U_E$$
$$= \frac{1}{2} \int_0^L EI \left(\frac{d^2 w(x)}{dx^2} \right)^2 dx$$
$$- \frac{1}{2} \int_0^L \frac{\varepsilon_0 h V^2}{\frac{d}{\varepsilon_r} + s(x) - w(x)} dx ,$$ (54.64)

where L is the beam length, E is the beam's Young's modulus, I is the beam's cross-sectional moment of inertia, h is the thickness of the beam, V is the voltage applied between the beam and electrode, d is the initial gap at the start of the electrode, ε_0 is the permittivity of free space, and $w(x)$ is the beam's deflection curve.

The next step in the solution of this problem is to determine an admissible trial function that satisfies the beam boundary conditions and can adequately approximate the solution. In this case the deflection curve for a uniformly loaded cantilever beam is chosen as the trial function:

$$\omega(x) = cx^2 \left(6L^2 - 4Lx + x^2\right) = cf(x), \quad (54.65)$$

where c is a constant to be determined for specific cases.

The approach of *Nemirovsky* et al. [54.77] can be taken by substituting the potential-energy functions in the voltage-controlled pull-in equations. The simultaneous solution of the pull-in equations will result in the pull-in voltage and an implicit equation in terms of the constant c at pull-in. The pull-in voltage relationship is

$$V_{PI}^2 = \frac{EI}{\varepsilon_0 h} \frac{\displaystyle\int_0^L \left(\frac{d^2 f(x)}{dx^2}\right)^2 dx}{\displaystyle\int_0^L \frac{f^2(x)}{[d/\varepsilon_r + s(x) - c_{PI} f(x)]^3} dx} . \quad (54.66)$$

The implicit equation to determine the coefficient of the beam deformation function is

$$\int_0^L \frac{c_{PI} f^2(x)\,dx}{[d/\varepsilon_r + s(x) - c_{PI} f(x)]^3}$$

$$= \int_0^L \frac{f(x)\,dx}{[d/\varepsilon_r + s(x) - c_{PI} f(x)]^2} . \quad (54.67)$$

DeReus [54.97] designed a curved-electrode actuator to convert the curvilinear motion into a translational motion at the tip of the moving electrode. The transformation is accomplished by a folded spring that connects the electrode to the body that will be translating. A solid model of this actuator design is shown in Fig. 54.33. A pull-in analysis was performed with the results before

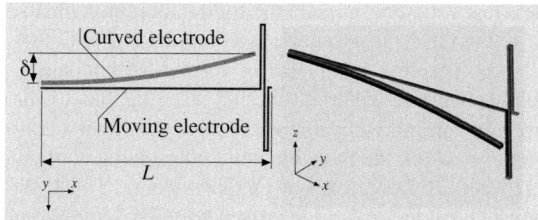

Fig. 54.33 Solid model of a curved electrode actuator with a folded suspension showing the curved electrode δ and the actuator beam length L

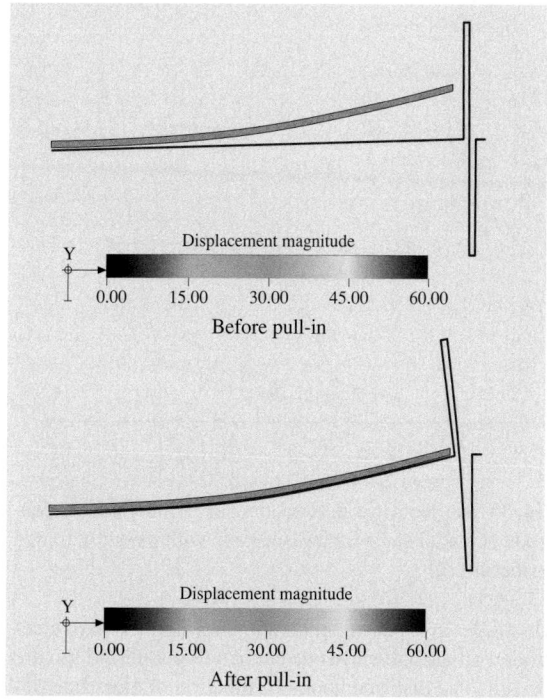

Fig. 54.34 Results of electromechanical pull-in analysis showing a curved-electrode actuator before and after pull-in has occurred

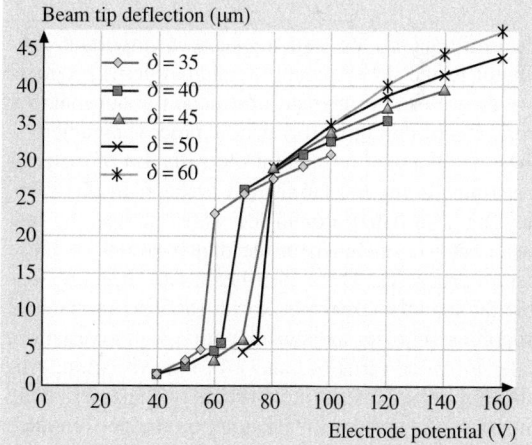

Fig. 54.35 Beam tip deformation as a function of electrode potential and δ for a suspension with a single fold

and after pull-in presented in Fig. 54.34. This shows that a 60-μm translation was accomplished with this particular design, which has a twofold flexure and could have application as a optical shutter or VOA. In Fig. 54.35,

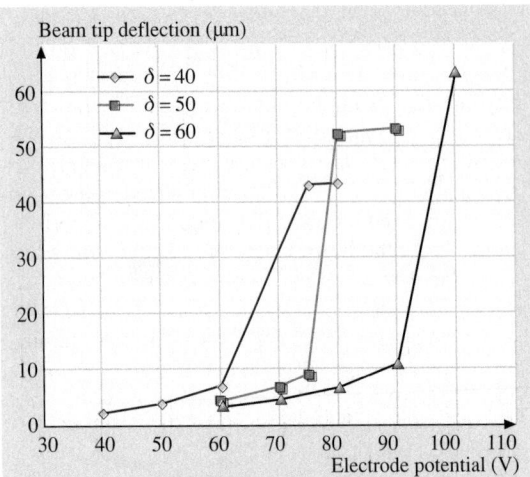

Fig. 54.36 Beam tip deformation as a function of electrode potential and δ for a suspension with a doubly folded suspension

electrode tip displacement is shown as a function of electrode voltage and electrode curve δ for a onefold flexure design. The displacement as a function of electrode potential and curve δ is shown for a twofold flexure in Fig. 54.36.

54.2.8 Electromechanics for Test Structures

Electromechanics test structures are an important contributor to the development of microsystems products as they are used to extract information about material properties of thin films for specific process parameters. They are also used to monitor the variation of material properties during the fabrication process. In this case, they are distributed around the wafer on specific test die or may be included in the dicing streets of more mature processes. These distributed test structures enable the manufacturer to track the variation in material properties with its associated within-wafer variation, wafer-to-wafer variation, and lot-to-lot variation. Another application of test structures is to characterize the impact of the first-level packaging on the performance of the product. With all the critical information being extracted from the test structures, it is important for the information to be easy to interpret. Test structures have been developed that involve electrical, optical, and mechanical interrogation of the structure, but these are not easily implemented at all levels of characterization. The purely electrical test structures offer simple evaluation at the wafer, die, and package level, so manufacturers often

prefer them. The most common electrical test structures used for extracting information on mechanical properties involve electromechanics, of course. This section has provided a basis for interpreting and modeling the behavior of specific test structures, such as a homogeneous cantilever beam, and can be extended to other structures.

Osterberg and *Senturia* [54.98] describe in detail the application of electromechanics to the extraction of mechanical properties for microsystems design. They labeled their test structures M-TEST for mechanical test structure. Their test structures comprised arrays of cantilever beams, fixed–fixed beams, and clamped, circular plates of various lengths, widths, or diameters. With these test structures, they demonstrated the extraction of important mechanical properties such as Young's modulus, plate modulus, and the residual film stress. Why choose an electromechanics test structure unless for the breadth and ease of application? An additional reason is the inherent instability that has been described as the pull-in of an electromechanical structure. The pull-in voltage is very distinct and can be measured in terms of a single parameter, the pull-in voltage. Given a control range of pull-in voltages for different structural layers in the manufacturing process, the pull-in parameter can be used as a simple control parameter.

Osterberg and *Senturia* [54.98] describe the methodology for extracting the mechanical properties. Their methodology includes the measurement of plan-view dimensions (i. e., width and length), out-of-plane curvature, pull-in voltage, film thickness, and air gap. From these measured parameters, the mechanical properties can be extracted. Their extraction methodology includes general closed-form expressions for pull-in voltage that depend on the material properties and geometry of a homogeneous, isotropic material.

Osterberg and *Senturia* [54.98] developed closed-form models (ultimately an empirical fit to an empirical expression) and two-dimensional (2-D) finite difference solutions based on an ideal test structure. The ideal test structure satisfied the following requirements. First, the ideal test structure is comprised of two conductors. The first conductor is a fixed ground plane. The second conductor is initially flat, parallel to the ground plane, and movable with respect to the first conductor. Second, the movable conductor operates in the small-deflection regime until pull-in occurs. Third, the movable conductor has ideal fixed boundary conditions such that all six degrees of freedom at each boundary are clamped. [This contrasts with actual test structures or devices where the fixed support (anchor) will re-

Table 54.3 Governing differential equations for three mechanical test structures: cantilever beams (CBs), fixed-fixed beams (FBs), and fixed, circular diaphragms (CDs)

Test structure	Governing differential equations
Cantilever beam (CB)	$\tilde{E}I\frac{d^4g}{dx^4} = -\frac{\varepsilon_0 V^2 w}{2g^2}(1+0.65\frac{g}{w})$
Fixed-fixed Beam (FB)	$\tilde{E}I\frac{d^4g}{dx^4} - T_b\frac{d^2g}{dx^2} = -\frac{\varepsilon_0 V^2 w}{2g^2}(1+0.65\frac{g}{w})$
Fixed circular diaphragm (CD)	$\frac{\tilde{E}t^3}{12}\nabla^4 g - T_d\nabla^2 g = -\frac{\varepsilon_0 V^2}{2g^2}$

Table 54.4 The calculated fitting parameters for the closed-form pull-in voltage expressions presented in (54.70), (54.71), and (54.72)

Fitting parameter	Test structure		
	CBs ($n=1$)	FBs ($n=2$)	CDs ($n=3$)
γ_{1n}	0.07	2.79	1.55
γ_{2n}	1.00	0.97	1.65
γ_{3n}	0.42	0.42	0.00

lax due to the residual stress.] Fourth, the ideal test structures have prismatic cross sections such that the effects of undercutting, over-etching, or lift-off profiles are not considered. Finally, the possible stress gradient is ignored in the ideal test structure. This means the cantilevers are not initially curved and the fixed–fixed beams and diaphragms are not buckled or bent out of plane.

The governing differential equations for the cantilever beam (CB), fixed–fixed beam (FB), and the fixed, circular diaphragm (CD) are shown in Table 54.3. In Table 54.3, $I = (1/12)wt^3$ is the cross-sectional moment of inertia, t is the thickness, w is the beam width, $g = g(x)$ is the gap between the moving and fixed conductors, $T_b = \sigma wt$, $T_d = \sigma t$, ε_0 is the permittivity of free space, $\tilde{E}$ is the effective modulus, σ is the effective residual stress, and V is the voltage. For the diaphragm, the effective modulus $\tilde{E}$ is $E/(1-v^2)$, where E is the Young's modulus and v is the Poisson ratio. For the cantilever beam and the fixed–fixed beams, the effective modulus will depend on the width of the beam, w. A wide beam ($w > 5t$) will exhibit plane-strain conditions so that the effective modulus will be the plate modulus $E/(1-v^2)$. A narrow beam ($w < 5t$) will have an effective modulus that is simply the Young's modulus E. The effective residual stress for a diaphragm is the biaxial stress σ_0. The effective residual stress for a cantilever beam is 0. For a fixed–fixed beam, the effective residual stress is $\sigma_0(1-v)$.

The next step is to identify the functional form of the pull-in equation. Recalling (54.52), the pull-in voltage is given by (54.68). In (54.68), K_{eff} is the effective spring constant given in (54.69) and γ_{3n} is a first-order correction for fringing electrostatic fields. In γ_{3n}, the index n refers to the test structure element where $n = 1$ is the cantilever, $n = 2$ is the fixed–fixed beam, and $n = 3$ is the fixed, circular diaphragm. K_{eff} is written in terms of L, the beam length and diaphragm radius and the factor $k = \sqrt{12S/B}$. The factor k is a measure of the relative strength of stress versus bending where $S = \sigma t g_0^3$ and $B = \tilde{E}t^3 g_0^3$. It is in the stress-limited regime when $k_L \geq 5$. The bending-limited regime is defined by $k_L \ll 5$.

$$V_{PI} = \sqrt{\frac{8K_{eff}g_0^3}{27\varepsilon_0\left[1+\gamma_{3n}\frac{g_0}{w}\right]}} \quad (54.68)$$

$$K_{eff} = \frac{S}{L^2\left[1+\frac{2\left(1-\cosh\left\{\frac{kL}{2}\right\}\right)}{\left(\frac{kL}{2}\right)\sinh\left\{\frac{kL}{2}\right\}}\right]} \quad (54.69)$$

Osterberg and *Senturia* [54.98] determined the general closed-form expressions for the pull-in voltage, as shown (54.70). Equation (54.71) shows the bending-dominated ($S \to 0$) limit for pull-in voltage. The stress-dominated ($B \to 0$) limit of pull-in voltage is shown in (54.72). In (54.70),

$$D_n = 1 + \frac{2\{1-\cosh(\gamma_{2n}kL/2)\}}{(\gamma_{2n}kL/2)\sinh(\gamma_{2n}kL/2)}$$

$$V_{PI} = \sqrt{\frac{\gamma_{1n}S}{\varepsilon_0 L^2 D_n(\gamma_{2n},k,L)\left[1+\gamma_{3n}\frac{g_0}{w}\right]}} \quad (54.70)$$

$$V_{PI}(S \to 0) = \sqrt{\frac{4\gamma_{1n}B}{\varepsilon_0 L^2\gamma_{2n}^2\left[1+\gamma_{3n}\frac{g_0}{w}\right]}} \quad (54.71)$$

$$V_{PI}(B \to 0) = \sqrt{\frac{\gamma_{1n}S}{\varepsilon_0 L^2\left[1+\gamma_{3n}\frac{g_0}{w}\right]}} \quad (54.72)$$

The closed-form expressions for the pull-in voltage are fitted to the 2-D finite difference solutions using a nonlinear least-squares analysis. The fitting parameters γ_{1n}, γ_{2n}, and γ_{3n} are shown in Table 54.4. With the expressions for pull-in voltage and the three test structures, the Young's modulus, plate modulus, and residual stress of the thin films can be determined.

Part G | 54.2

54.2.9 Electromechanical Dynamics: Switching Time

Switching time is an important factor in the design of such electromechanical devices as electrical or optical switches. The switching-time analysis requires consideration of the dynamic equations representing the device and the electrical source. *Castaner* et al. [54.99–101] have considered the optimization of the speed–energy product. In this case, a parallel-plate electrostatic actuator is described by an equivalent mechanical circuit that includes the mass m, the damping b, the spring k, and a variable gap g. The variable gap depends on the initial gap g_0 and the displacement of the moving electrode x, such that $g = g_0 - x$. In the parallel-plate variable-gap model the damping is described by squeeze film damping. (Other electrostatic actuators move parallel to the substrate with variable gaps defined by groups of moving and stationary fingers.) In these cases, the damping model will need to include squeeze film damping and shear flow damping (Couette or Stokes flow models). The electrical source is modeled as a voltage source and a source series resistance. A schematic of the model is shown in Fig. 54.37.

The equivalent electrical circuit of the parallel-plate electrostatic actuator is modeled by (54.73), where V_s is the source voltage, V is the voltage across the variable capacitance, R is the series source resistance, and Q is the charge on the capacitor plates:

$$\frac{V_s}{R} = \frac{V}{R} + \frac{dQ}{dT} .$$ (54.73)

The equivalent mechanical circuit is described by:

$$m\frac{d^2 X}{dT^2} + b\frac{dX}{dT} + kX = \frac{Q^2}{2C_0 G_0} .$$ (54.74)

Equation (54.74) includes the sum of the inertial force, the damping force, the spring force, and the electrostatic force. Equation (54.74) is written in terms of the mass of the moving capacitor plate m, the damping coefficient due to squeeze film damping b, the elastic constant of the spring by which the capacitor plate is suspended k, the charge on the capacitor plate Q, the initial gap G_0, and the initial capacitance C_0. The initial capacitance is written in terms of the area of the capacitor A, the dielectric constant of the capacitor air gap ε, and the initial gap:

$$C_0 = \frac{\varepsilon A}{G_0} .$$ (54.75)

The dynamics of the parallel-plate electrostatic actuation were divided into four events. The first occurrence

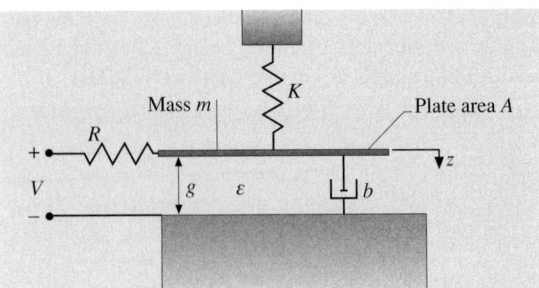

Fig. 54.37 Lumped parameter model of the parallel-plate electromechanical actuator including damping and series resistance for dynamic analysis

is a short charging time for the capacitor. The second phase is the acceleration phase, which defines the rapidly increasing velocity of the mass, but the position is still close to the initial equilibrium position. The third phase is the damping phase, defined by the velocity reaching a maximum, then decreasing due to increased damping, and then finally increasing due to the increasing electrostatic forces as the gap closes. The fourth and final phase is the pull-in phase, characterized by a rapidly increasing velocity. *Hung* and *Senturia* [54.89] have described the generation of efficient dynamical models from a minimal set of finite element simulations. *Pons-Nin* et al. [54.102] have described the behavior of the pull-in voltage and pull-in time for electromechanical actuators with current drives.

54.2.10 Electromechanics Issues: Dielectric Charging

Electrostatic actuators are operated in three different regimes depending on the application. In the first regime, the electrostatic actuator is operated over the voltage range that does not include the pull-in voltage. In this case, the actuated element does not contact the opposing substrate. In the second regime, the electrostatic actuator is operated through a voltage range that does include the pull-in voltage ($V_{act} \geq V_{PI}$). In this case, the desired response is to establish contact with or without some overdrive to achieve greater contact force or another response parameter. In the third regime, the electrostatic actuator is operated through a voltage range beyond the pull-in voltage. This is characteristic of electrostatic zipper actuators. In this case, the operating voltage is programmed to quickly establish pull-in, and thereby contact, and then the actuator operates in a range above the pull-in voltage ($V_{PI} < V_{act}$). In all three regimes, dielectric isolation layers can be considered to prevent shorting between actuation electrodes. For the first and second cases, the use of the insulating material is a preventative measure to protect against electrode shorting during an unintentional overdrive of the actuator. It is preventative because the actuation electrodes are not intended to touch. In the third case, the gap separating the moving and stationary actuation electrodes is reduced to zero so that an insulating layer is required. The insulating layer may be patterned or unpatterned.

The third regime defines touch-mode electrostatic actuators. *Cabuz* et al. [54.103] have described many of the failure modes that can be observed with touch-mode electrostatic actuators and that should be addressed during the design of the actuator and process. The observations and failure modes included: (1) a significant difference between the experimental actuation voltage (DC) and the theoretical actuation voltage, (2) a closed-state electrostatic pressure lower than predicted, (3) an actuation voltage that increases with each subsequent cycle until electrical breakdown occurs, (4) actuator vibration observed between an open and a closed state, (5) the actual release voltage is higher than theoretical, but the electrostatic forces should be much higher in the closed state such that the release voltage is always lower than the pull-in voltage, (6) the actuator remains closed temporarily after shorting, (7) permanent stiction will occur when in the closed state for a long period of time, (8) permanent stiction occurs after many actuation cycles.

Wibbeler et al. [54.104] have analyzed this problem in terms of one of the contributing factors, the accumulation and storage of localized charge in dielectrics. The common dielectrics in microsystems devices are SiO_2 and Si_3N_4, which show low mobility for surface charges and provide trap sites for positive or negative charge within their volume and at interfaces. *Kubo* et al. [54.105] showed that this charge can be produced by the fabrication process by contact electrification, and by gas discharge in the air gap (the case focused on by *Wibbeler* et al. [54.104]), and static electrification during handling (ESD). Contact electrification creates electrical charge through the mechanical contact of two materials with different work functions and the separation of the two materials.

To understand the impact of accumulated charge, *Wibbler* et al. [54.104] considered the impact of stored surface charge on the electrostatic force acting on a simple parallel-plate capacitor. In Fig. 54.38, a simple capacitor is considered with a dielectric (ε_d) filling part of the air gap (ε). The dielectric has a surface charge of $\sigma_s(x, y)$. The capacitor plates carry charge shown as σ_1 and σ_2, which are provided by the battery, of voltage V, connected across the capacitor plates. The thicknesses of the dielectric and air gap are d_d and d_g, respectively. The force was estimated from the applied voltage and from the offset voltage due to the accumulated surface charge. The force acting on the capacitor plate as a function of the applied voltage V, the offset voltage V_s, and the capacitance C is:

$$F = \frac{dC}{dz} \int_A \frac{[V - V_s(x, y)]^2}{2} \, dA .$$ (54.76)

The offset voltage is shown in (54.77) as a function of the surface charge, the thickness of the dielectric d_d, the dielectric constant ε_d, or the dielectric's contribution to the total capacitance C_d. The capacitance of the dielectric layer, per unit area, is defined as $C_d = \varepsilon_d/d_d$.

$$V_s = -\sigma_s(x, y) \frac{d_d}{\varepsilon_d} = -\frac{\sigma_s(x, y)}{C_d} .$$ (54.77)

Electromechanical Pull-In with Dielectric Charging

Nemirovsky et al. [54.77] and *Chan* et al. [54.106] studied the effect of residual charge at the interface of a dielectric and the air in the capacitor gap. With the dielectric in the capacitor gap and the residual charge, the total energy for the actuator is:

$$U_T = \frac{1}{2} K z^2 + \frac{Q_{res}^2}{2(C(z) + C_d)} - \frac{C(z)C_d}{2(C(z) + C_d)} V^2 .$$ (54.78)

In (54.78), Q_{res} is the residual charge, $C(z)$ is the air-gap capacitance, and C_d is the dielectric capacitance. From the total energy, the pull-in parameter and pull-in equations (54.43) and (54.44) can be solved for the pull-in parameter ζ_{PI} and pull-in voltage V_{PI}:

$$\varsigma_{PI} = \frac{1}{3} \left(1 + \frac{1}{\delta} \right) ,$$ (54.79)

$$V_{PI} = \left[\frac{8 K z_{max}^3}{27 \varepsilon_0 A} \left(1 + \frac{1}{\delta} \right)^3 - \frac{Q_{res}^2}{C_d^2} \right]^{1/2} .$$ (54.80)

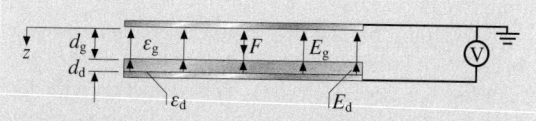

Fig. 54.38 Parallel-plate capacitor with dielectric and parasitic charges

In (54.79), ζ_{PI} is the ratio z/z_{max} and δ is the ratio $\varepsilon_d z_{max}/z_d$. Equation (54.80) shows that the dielectric layer will increase the pull-in voltage and the residual charge will lower the pull-in voltage. It can be seen that the residual charge does not impact on the normalized pull-in coordinate shown in (54.79).

Test structures were fabricated to verify the shift predicted by (54.76) that would result from a dielectric in the air gap. Test structures with a dielectric did show a shift in the voltage–deflection curve, whereas a shift was not observed when a dielectric was not present. A shift of -25 V was observed. Following electrostatic discharge events, offset voltages of -210 V and 145 V were observed during different phases of a charge–discharge cycle. This demonstrates the significant bias that the accumulated surface charge can create. As a result, *Wibbler* et al. [54.104] did not recommend trying to fix shorting by adding dielectric films over the electrodes. They did recommend estimating the influence of accumulated charge on the response of the device being designed. Instead of complete dielectric film coverage of an electrode, they recommended patterning the dielectric such that it was limited to the edges of the electrodes or patterned at locations that will touch first. By patterning small areas of insulation, the surface area available to store charge is smaller and the distance to the electrode is shorter for more timely dissipation of stored charge.

Cabuz et al. [54.103] identified the effects of humidity, DC or alternating-current (AC) driving voltage, and self-assembled monolayer (SAM) coatings on the performance of electromechanical actuators. With a DC operating voltage, the devices operated as expected below 35% RH except in dry air or dry N_2. In the dry environment the actuator would temporarily remain closed, would remain closed after high fields are applied, or would remain closed after a high number of cycles. When operated above 35% RH, the DC voltage was higher than expected, the closed-state electrostatic pressure was lower, the actuator did not release when expected, the actuator vibrated, and the voltage increased with subsequent cycles. When an AC actuation voltage is used the actuators operated up to 55% RH. At higher RH, the actuation voltage increased. The dry-condition operation was improved with AC driving compared to DC operation. AC voltage control is recommended for eliminating DC effects but cannot overcome the effects of humidity. To avoid the effects of humidity, two SAM coatings [octadecyltrichlorosilane (OTS) and perfluorodecyltrichlorosilane (PFTS)] were tried to influence the hydration of the surfaces.

OTS eliminated the effects observed for uncoated samples up to 95% RH for both DC and AC voltages. The PTFS improved operation up to 95% RH for AC voltages and 70% RH for DC control voltages. The residual in-use stiction was determined to be associated with accumulated and trapped charge on the surfaces or at interfaces.

54.2.11 Electromechanics Issues: Gas Discharge

As described in previous sections, the electromechanical actuation of a microstructure is performed by applying a voltage across two plates. The two plates can be defined by two conductors (e.g. Al to Al), by semiconductors (e.g. polysilicon), or by a combination of a conductor and a semiconductor (e.g. Al and polysilicon). When these two plates (or other similarly described structures) are fabricated in proximity to each other, they are usually fabricated with an air gap of less than 10 μm. With electrostatic actuators, a small gap is desirable to minimize the voltage required for actuation.

The presence of small gaps in microsystems devices has prompted the further investigation of electrical breakdown in air and other ambient conditions at small gaps (i. e. less than 50 μm). *Slade* and *Taylor* [54.107] have demonstrated that breakdown in small gaps is defined by three regimes: less than 4 μm, 4–6 μm, and greater than 6 μm. In the first case, breakdown events follows the behavior for a vacuum breakdown process at larger gaps. In the last case, the breakdown follows the behavior of Paschen's law for breakdown in air. In the middle case of gaps of 4–6 μm the breakdown process shows a transition from the vacuum breakdown process to the Paschen curve. In this case, the breakdown voltage is lower than predicted in either case, which indicates significant contributions from factors in both cases. The deviation from Paschen's curve was observed by *Torres* and *Dhariwal* [54.108] for gaps less than 4 μm.

The goal of this section is to describe some basic breakdown principles that can be used to determine whether or not a particular design will break down during an actuation event. Different processes will be considered for different gap regimes so that breakdown estimates can be made.

The vacuum breakdown region is characterized by a breakdown voltage that is a function of the gap separating the two electrodes. This contrasts with the breakdown behavior that follows the Paschen curve, which shows dependence on the product between electrode gap and pressure. The breakdown voltage in vacuum is

$$V_B = K_v t_g . \qquad (54.81)$$

Equation (54.81) shows that the breakdown voltage V_B is proportional to the gap t_g where the proportionality constant is $K_v = 97 \, \text{V}/\mu\text{m}$ for data from many sources, as indicated by *Slade* and *Taylor* [54.107]. The breakdown data in vacuum was determined for electrode gaps of $35-200 \, \mu\text{m}$. For electrode gaps of $0.2-40 \, \mu\text{m}$ in air, the proportionality constant was $65-110 \, \text{V}/\mu\text{m}$.

The electric field at the surface of the electrode is a critical parameter for determining breakdown in vacuum. This is estimated by the ratio of the voltage between the electrodes V and the gap between the electrodes t_g with a modification factor β_g. β_g is a geometrical enhancement factor that captures the field enhancement due to the macroscopic geometry:

$$E_g = \beta_g V / t_g . \qquad (54.82)$$

The geometrical enhancement factor can be determined by performing finite element analysis to determine the macroscopic electric field as a function of the electrode gap for the electrode geometry under consideration. The geometric enhancement factor is determined from the ratio of the macroscopic electric field E_g, from finite element simulations, to the electric field estimated by V/t_g [54.107, 109]. The estimated electric field, V/t_g, can be used as a reasonable approximation when the electrode gap is much smaller than the radius of curvature of the cathode electrode. If this is not the case, the geometric enhancement factor needs to be estimated from numerical analysis such as finite element or boundary element methods. The geometric enhancement factor is shown, by *Slade* and *Taylor* [54.107], to be between 1.01 and 1.57 for gaps ranging from 0.2 to $40 \, \mu\text{m}$, respectively, for the geometry of a needle cathode and a plate anode.

A second enhancement factor is attributed to the microscopic field enhancement due to the surface roughness of the electrodes. The electric field including the effects of the geometric enhancement and the microscopic enhancement is:

$$E = \beta_m \beta_g V / t_g . \qquad (54.83)$$

In (54.83) β_m is the microscopic enhancement factor. The total enhancement factor is $\beta_g \beta_m$, which can have a range of 100–250 for polished electrodes [54.110].

Following the estimate of the electric field at the surface of the electrodes, the current density for the field emission of electrons is calculated. This current density j_{FE} is described by the Fowler–Nordheim equation, which depends on the electric field E, the work function of the electrode material ϕ, and the dimensionless parameters of $t(y)$ and $v(y)$, where y is a parameter defined as $3.79 \times 10^{-5} \sqrt{E}/\phi$. Typically, $t(y)$ is one. $v(y)$ is given by $0.956 - 1.06 y^2$. The Fowler–Nordheim equation is presented in (54.84) with j_{FE} in A/m^2, E in V/m, and ϕ in eV.

$$j_{FE} = \frac{1.54 \times 10^{-6} E^2}{\phi t(y)^2}$$
$$\times \exp\left(\frac{-6.83 \times 10^9 \phi^{3/2} v(y)}{E}\right) . \qquad (54.84)$$

Equation (54.84) can be rewritten with E substituted from (54.83) and the current density replaced by an emission current I_e and the area A_e of a microprojection where the emission occurs. Equation (54.84) is shown rewritten in (54.85), such that the slope from the Fowler–Nordheim plot will provide the field enhancement factor as shown in (54.86) where m is the slope. The Fowler–Nordheim plot is the plot of I_e/V^2 versus $1/V$, which is constructed by measuring the voltage across the electrode gap and measuring I_e.

$$\text{Log}_{10}\left(\frac{I_e}{V^2}\right)$$
$$= \text{Log}_{10}\left(\frac{1.54 \times 10^{-6} A_e \beta^2}{\phi \, t(y)^2 t_g^2}\right)$$
$$- \frac{1}{2.303}\left(\frac{6.83 \times 10^9 \phi^{3/2} t_g v(y)}{\beta}\right)\frac{1}{V} , \qquad (54.85)$$

$$\beta = -\frac{2.303 m}{6.83 \times 10^9 \phi^{3/2} t_g v(y)} . \qquad (54.86)$$

For gaps larger than $6 \, \mu\text{m}$, the electrical breakdown as a function of gap does follow the Paschen curve for air and does not follow the vacuum breakdown curve for these larger gaps. The electrode gap is now larger than the mean free path of the air molecules so it is expected that emitted electrons will collide with the air molecules in the gap. For these larger electrode gaps, it is assumed that the electrical breakdown will follow the Townsend electron avalanche theory. A similar process is followed to estimate the breakdown. First, the electric field at the surface of the cathode electrode is estimated, including the enhancement factors for geometry and surface microstructure. An estimate of the total field enhancement factor is required for this. The electric field is substituted into the Fowler–Nordheim equation (54.84) to estimate the field emission current. In this case, the emitted electrons will interact with the background gas molecules, causing an avalanche in the current. The current avalanche process is described by (54.85) and (54.86), where t_g is the electrode gap, α is the first Townsend coefficient, and γ is the second Townsend coefficient. The first Townsend coefficient represents the number of electrons produced per unit distance along the direction of the electric field. The second Townsend coefficient is the number of electrons generated by secondary processes per primary avalanche. Breakdown will occur when the denominator of the expression for current approaches zero: $\gamma \exp(\alpha t_g)$ approaches unity.

In the transition region, the breakdown voltage can be lower than predicted by the Paschen curve or vacuum breakdown. It is suggested that in this regime, the partial electron avalanche process enhances the vacuum breakdown process to lead to a lower breakdown voltage [54.107].

The choice of materials is another consideration for the design of an electrostatic microactuator. As mentioned, the electrode pairs could include the following combinations: conductor–conductor, conductor–semiconductor, or semiconductor–semiconductor. In *Torres* and *Dhariwal* [54.108], the breakdown voltage was measured for electrode pairs of nickel, brass, or aluminum. A significant difference was not observed between the breakdown voltages of the different conductors or for the size of the electrode. At larger gaps, the breakdown voltage was different for different electrode shapes (cylindrical versus spherical) but this can be explained by the relative difference in the geometric enhancement factor for the electric field for the different electrodes. *Ono* et al. [54.111] made a comparison between electrode combinations where both a semiconductor (silicon) and a silicon–metal combination were used. In the case of a silicon electrode

with a metal electrode, the breakdown voltage deviated from the Paschen curve at approximately $6\,\mu m$. It was lower than the breakdown voltage predicted by the Paschen curve and trailed off with an approximate linear dependence on the electrode gap. With the silicon as one electrode, Au/Cr or Pt/Cr was used as the second electrode but little difference was found for the different metals. With the silicon–silicon electrode combination, the breakdown voltage followed the Paschen curve for air. For similar gaps, the breakdown voltage was significantly higher for the silicon–silicon electrodes than for the silicon–metal electrode. This was true for a thin silicon film over a Cr/Pt/Cr electrode.

Another option for reducing electric breakdown in the air gap is to introduce an insulating material over the electrode. However, *Wibbeler* et al. [54.104] have demonstrated this does not solve the problem of breakdown by itself and can lead to parasitic charging of the insulator. An electric breakdown will produce a large current, as described previously, that is destructive to electrodes without insulation. The dielectric limits the current so that the electrical breakdown event is not so destructive. *Wibbeler* et al. [54.104] demonstrated this through highly chaotic behavior in the voltage–displacement response of an electrostatic actuator. It is reasoned that the local field strength at the electrode reach a magnitude that caused an avalanche of free electrons and ions, as described here. These free charges travel the length of the air gap, but they are stopped and neutralized at the dielectric surface. This reduces the local electric field at the location of the surface charge so that the force on the cantilever is reduced and the process repeats itself. Throughout this process, a large amount of charge can be introduced onto the dielectric surface, which will typically have a very long dissipation time constant.

54.3 Summary and Topics not Covered

This chapter has focused on the thermomechanics and electromechanics of thin-film microstructures. Emphasis has been placed on the basic physical phenomena, approaches to the analysis of the phenomena, and the presentation of accessible results for special cases of practical importance in a variety of applications. We have not discussed many other energy-domain couplings that play major

roles in microsystems technology such as electrothermomechanics and magnetomechanics. We have also not discussed the role and behavior of materials such as piezoelectrics or shape-memory alloys, which play important roles in electromechanics and thermomechanics, respectively. We refer the interested reader to *Senturia* [54.60] for an introduction and further references.

References

54.1 M. Madou: *Fundamentals of Microfabrication* (CRC, Boca Raton 1997)

54.2 M. Gad-el-Hak: *The MEMS Handbook* (CRC, Boca Raton 2001)

54.3 J. J. Sniegowski, M. P. de Boer: IC-compatible polysilicon surface micromachining, Annu. Rev. Mater. Sci. **30**, 299–333 (2000)

54.4 C. H. Mastrangelo: Adhesion-related failure mechanisms in microelectromechanical devices, Tribol. Lett. **3**, 223–238 (1997)

54.5 M. P. de Boer, T. M. Mayer: Tribology of MEMS, MRS Bull. **26**, 302–304 (2001)

54.6 R. C. Cammarata, K. Sieradzki: Surface and interface stresses, Ann. Rev. Mater. Sci. **24**, 215–234 (1994)

54.7 J. Diao, K. Gall, M. L. Dunn: Atomistic simulation of the structure and elastic properties of gold nanowires, J. Mech. Phys. Solids **52**, 1935–1962 (2004)

54.8 W. D. Nix: Mechanical properties of thin films, Metall. Trans. A **20A**, 2217–2245 (1989)

54.9 C. Y. Hui, H. D. Conway, Y. Y. Lin: A reexamination of residual stresses in thin films and of the validity of Stoney's estimate, J. Electron. Packaging **122**, 267–273 (2000)

54.10 D. B. Bogy, K. C. Wang: Stress singularities at interface corners in bonded dissimilar materials, Int. J. Solids Struct. **7**, 993–1005 (1971)

54.11 V. L. Hein, F. Erdogan: Stress singularities in a two material wedge, Int. J. Fract. Mech. **7**, 317–330 (1971)

54.12 E. D. Reedy, T. R. Guess: Nucleation and propagation of an edge crack in a uniformly cooled epoxy/glass bimaterial, Int. J. Solids Struct. **39**, 325–340 (2002)

54.13 M. D. Thouless: Modeling the development and relaxation of stresses in films, Annu. Rev. Mater. Sci. **25**, 69–96 (1995)

54.14 N. D. Masters, M. P. de Boer, B. D. Jensen, M. S. Baker, D. Koester: Side-by-side comparison of passive MEMS residual strain test structures under residual compression. In: *Mechanical Properties of Structural Films*. In: *Mechanical Properties of Structural Films*, Vol. 1413, ed. by C. L. Muhlstein, S. B. Brown (ASTM STP, West Conshohocken 2001) pp. 168–200

54.15 M. W. Hyer: Calculation of the room-temperature shapes of unsymmetric laminates, J. Compos. Mater. **15**, 296–310 (1981)

54.16 M. W. Hyer: The room-temperature shape of four-layer unsymmetric cross-ply laminates, J. Compos. Mater. **16**, 318–340 (1982)

54.17 D. E. Fahnline, C. B. Masters, N. J. Salamon: Thin film stress from nonspherical substrate bending measurements, J. Vac. Sci. Technol. A **9**, 2483–2487 (1991)

54.18 C. B. Masters, N. J. Salamon: Geometrically nonlinear stress-deflection relations for thin film/substrate systems, Int. J. Eng. Sci. **31**, 915–925 (1993)

54.19 C. B. Masters, N. J. Salamon: Geometrically nonlinear stress-deflection relations for thin film/substrate systems with a finite element comparison, J. Appl. Mech. **61**, 872–878 (1994)

54.20 M. Finot, S. Suresh: Small and large deformation of thick and thin film multilayers: effects of layer geometry, plasticity and compositional gradients, J. Mech. Phys. Solids **44**, 683–721 (1996)

54.21 M. Finot, I. A. Blech, S. Suresh, H. Fujimoto: Large deformation and geometric instability of substrates with thin film deposits, J. Appl. Phys. **81**, 3457–3464 (1997)

54.22 L. B. Freund: The stress distribution and curvature of a general compositionally graded semiconductor layer, J. Cryst. Growth **132**, 341–344 (1993)

54.23 L. B. Freund: Some elementary connections between curvature and mismatch strain in compositionally graded thin films, J. Mech. Phys. Solids **44**, 723–736 (1996)

54.24 L. B. Freund: Substrate curvature due to thin film mismatch strain in the nonlinear deformation range, J. Mech. Phys. Solids **48**, 1159–1174 (2000)

54.25 N. J. Salamon, C. B. Masters: Bifurcation in isotropic thin film/substrate plates, Int. J. Solids Struct. **32**, 473–481 (1995)

54.26 R. M. Jones: *Mechanics of Composite Materials*, 2nd edn. (Taylor & Francis, London 1999)

54.27 M. W. Hyer: *Stress Analysis of Fiber-Reinforced Composite Materials* (McGraw-Hill, Boston 1998)

54.28 G. G. Stoney: The tension of metallic films deposited by electrolysis, Proc. R. Soc. Lond. A **82**, 172–175 (1909)

54.29 C. D. Pionke, G. Wempner: The various approximations of the bimetallic thermostatic strip, J. Appl. Mech. **58**, 1015–1020 (1991)

54.30 P. Krulevitch, G. C. Johnson: Curvature of a cantilever beam subjected to an equibiaxial bending moment, Mater. Res. Symp. Proc. **518**, 67–72 (1998)

54.31 S. R. Swanson: *Introduction to Design and Analysis with Advanced Composite Materials* (Prentice-Hall, Upper Siddel River 1997)

54.32 M. L. Dunn, Y. Zhang, V. Bright: Deformation and structural stability of layered plate microstructures subjected to thermal loading, J. Microelectromech. Syst. **11**, 372–383 (2002)

54.33 J. A. King: *Materials Handbook for Hybrid Microelectronics* (Teledyne Microelectronics, Los Angeles 1988)

54.34 W. N. Sharpe: Mechanical properties of MEMS materials. In: *The CRC Handbook of MEMS*, ed. by M. Gad-el-Hak (CRC, Boca Raton 2001) Chap. 3

54.35   Y. Zhang, M. L. Dunn: Deformation of blanketed and patterned bilayer thin film microstructures during post-release thermal and cyclic thermal loading, J. Microelectromech. Syst. **12**, 788–796 (2003)

54.36   B. D. Harper, C.-P. Wu: A geometrically nonlinear model for predicting the intrinsic film stress by the bending plate method, Int. J. Solids Struct. **26**, 511–525 (1990)

54.37   J. W. Gibbs: *The Collected Works of J. Willard Gibbs* (Longmans, New York 1928)

54.38   R. Shuttleworth: The Surface Tension of Solids, Proc. R. Soc. London A **63**, 444–457 (1950)

54.39   J. Diao, K. Gall, M. L. Dunn: Surface and stress induced phase transformation in metal nanowires, Nature Mater. **977**, 1–5 (2003)

54.40   L. G. Zhou, H. Huang: Are surfaces elastically softer or stiffer?, Appl. Phys. Lett. **84**, 1940–1942 (2004)

54.41   R. Dingreville, J. Qu, M. Cherkaoui: Surface free energy and its effect on the elastic behavior of nano-sized particles, wires, and films, J. Mech. Phys. Solids, in press (2005)

54.42   R. E. Miller, V. B. Shenoy: Size dependent elastic properties of nanosized structural elements, Nanotechnology **11**, 139–147 (2000)

54.43   K. Gall, J. Diao, M. L. Dunn: The strength of gold nanowires, Nanoletters **4**, 2431–2436 (2004)

54.44   Y. L. Shen, S. Suresh: Thermal cycling and stress relaxation response of Si-Al and Si-Al-SiO$_2$ layered thin films, Acta. Metall. Mater. **43**, 3915–3926 (1995)

54.45   S. P. Baker, A. Kretschmann, E. Arzt: Thermomechanical behavior of different texture components in Cu thin films, Acta Mater. **49**, 2145–2160 (2001)

54.46   D. C. Miller, C. F. Herrmann, H. J. Maier, S. M. George, C. R. Stoldt, K. Gall: Intrinsic stress development and microstructure evolution of Au/Cr/Si multilayer thin films subject to annealing, Scripta Materialia **52**, 873–879 (2005)

54.47   W. D. Nix: Elastic and plastic properties of thin films on substrates, Mater. Sci. Eng. A **234–236**, 37–44 (1997)

54.48   C. Thompson: The yield stress of polycrystalline thin films, J. Mater. Res. **8**, 237–238 (1993)

54.49   H. Gao, Y. Huang, W. D. Nix, J. W. Hutchinson: Mechanism-based strain gradient plasticity – I. Theory, J. Mech. Phys. Solids **47**, 1239–1263 (1999)

54.50   H. D. Espinosa, B. C. Prorok, M. Fischer: A novel method for measuring elasticity, plasticity, and fracture of thin films and MEMS materials, J. Mech. Phys. Solids **51**, 47–67 (2003)

54.51   D. C. Miller, M. L. Dunn, V. M. Bright: Thermally induced change in deformation of multimorph MEMS structures, Proc. SPIE **4558**, 32–44 (2001)

54.52   D. J. Vickers-Kirby, R. L. Kubena, F. P. Stratton, R. J. Joyce, D. T. Chang, J. Kim: Anelastic creep phenomena in thin film metal plated cantilevers for MEMS, Mater. Res. Soc. Symp. **657**, EE2.5.1–EE2.5.6 (2001)

54.53   R. M. Keller, S. P. Baker, E. Arzt: Stress–temperature behavior of unpassivated thin copper films, Acta Mater. **47**, 415–426 (1999)

54.54   M. D. Thouless, J. Cupta, J. M. E. Harper: Stress development and relaxation in copper films during thermal cycling, J. Mater. Res. **8**, 1845–1852 (1993)

54.55   M. D. Thouless, K. P. Rodbell, C. Cabral Jr.: Effect of a surface layer on the stress relaxation of thin films, J. Vac. Sci. Technol. **14**, 2454–2461 (1996)

54.56   R. P. Vinci, E. M. Zielinski, J. C. Bravman: Thermal stress and strain in copper thin films, Thin Solid Films **262**, 142–153 (1995)

54.57   J. Koike, S. Utsunomiya, Y. Shimoyama, K. Maruyama, H. Oikawa: Thermal cycling fatigue and deformation mechanism in aluminum alloy thin films on silicon, J. Mater. Res. **13**, 3256–3264 (1998)

54.58   Y. Zhang, M. L. Dunn: Geometric and Material Nonlinearity During the Deformation of Micron-Scale Thin-Film Bilayers Subject to Thermal Loading, J. Mech. Phys. Solids **52**, 2101–2126 (2004)

54.59   K. Gall, M. L. Dunn, Y. Zhang, B. Corff: Thermal cycling response of layered gold/polysilicon MEMS structures, Mech. Mater. **36**, 45–55 (2004)

54.60   S. D. Senturia: *Microsystems Design* (Kluwer Academic, Dordrecht 2001)

54.61   M. P. de Boer, N. F. Smith, N. D. Masters, M. B. Sinclair, E. J. Pryputniewicz: Integrated platform for testing MEMS mechanical properties at the wafer scale by the IMaP methodology. In: *Mechanical Properties of Structural Films*, ASTM STP **1413**, 85–95 (2001)

54.62   Z. J. Yao, S. Chen, S. Eshelman, D. Denniston, C. Goldsmith: Micromachined low-loss microwave switches, J. Microelectromech. Syst. **8**, 129–134 (1999)

54.63   B. McCarthy, G. G. Adams, N. E. McGruer, D. Potter: A dynamic model, including contact bounce of an electrostatically actuated microswitch, J. Microelectromech. Syst. **11**, 276–283 (2002)

54.64   H.-S. Lee, C. H. Leung, J. Shih, S.-C. Chang, S. Lorincz, I. Nedelescu: Integrated microrelays: Concept and initial results, J. Microelectromech. Syst. **11**, 147–153 (2002)

54.65   S. J. Cunningham, S. Tatic-Lucic, J. Carper, J. Lindsey, L. Spangler: A high aspect ratio accelerometer fabricated using anodic bonding, dissolved wafer, and deep RIE processes, Proc. Transducers '99, the 10th Intl. Conf. Solid-State Sensors and Actuators 1999, ed. by M. Esashi (IEE Jpn., Sendai 1999) 1522–1525

54.66   L. Spangler, C. Kemp: ISAAC: integrated silicon automotive acceleromter, Sens. Actuators A **54**, 523–529 (1996)

54.67   M. T. A. Saif, B. E. Alaca, H. Sehitoglu: Analytical modeling of electrostatic membrane actuator for micropumps, J. Microelectromech. Syst. **8**, 335–345 (1999)

54.68 C. Huang, C. Christophorou, K. Najafi, A. Naguib, H. M. Naguib: An electrostatic microactuator system for application in high-speed jets, J. Microelectromech. Syst. **11**, 222–235 (2002)

54.69 H. Toshiyoshi, H. Fujita: Electrostatic micro torsion mirrors for optical switch matrix, J. Microelectromech. Syst. **5**, 231–237 (1996)

54.70 H. Toshiyoshi, W. Piyawattanametha, C.-T. Chan, M. C. Wu: Linearization of electrostatically actuated surface micromachined 2-D optical scanner, J. Microelectromech. Syst. **10**, 205–214 (2001)

54.71 M. Fischer, M. Giousouf, J. Schaepperle, D. Eichner, M. Weinmann, W. von Münch, F. Assmus: Electrostatically deflectable polysilicon micromirrors – dynamic behaviour and comparison with the results from FEM modelling with Ansys, Sens. Actuators A **67**, 89–95 (1998)

54.72 P. K. C. Wang, R. C. Gutierrez, R. K. Bartman: A method for designing electrostatic-actuator electrode pattern in micromachined deformable mirrors, Sens. Actuators A **55**, 211–217 (1996)

54.73 P. K. C. Wang, F. Y. Hadaegh: Computation of static shapes and voltages for micromachined deformable mirrors with nonlinear electrostatic actuators, J. Microelectromech. Syst. **5**, 205–220 (1996)

54.74 J. Bühler, J. Funk, J. G. Korvink, F.-P. Steiner, P. M. Sarro, H. Baltes: Electrostatic aluminum micromirrors using double-pass metallization, J. Microelectromech. Syst. **6**, 126–135 (1997)

54.75 M. Fischer, H. Graef, W. von Münch: Electrostatically deflectable polysilicon torsional mirrors, Sens. Actuators A **44**, 83–89 (1994)

54.76 H. Schenk, P. Dürr, D. Kunze, H. Lakner, H. Kück: An electrostatically excited 2D-micro-scanning-mirror with an in-plane configuration of the driving electrodes, Proc. IEEE MEMS 2000, The 13th Ann. Intl. Conf. Micro Electro Mechanical Systems, Miyazaki 2000, ed. by I. Shimoyama, H. Kuwano (IEEE, Piscataway 2000)

54.77 Y. Nemirovsky, O. Bochobza-Degani: Methodology and model for the pull-in parameters of electrostatic actuators, J. Microelectromech. Syst. **10**, 601–615 (2001)

54.78 G. Li, N. R. Aluru: A Lagrangian approach for electrostatics analysis of deformable conductors, J. Microelectromech. Syst. **11**, 245–254 (2002)

54.79 N. R. Aluru, J. White: A multilevel Newton method for mixed-energy domain simulation of MEMS, J. Microelectromech. Syst. **8**, 299–308 (1999)

54.80 J. I. Seeger, S. B. Crary: Stabilization of electrostatically actuated mechanical devices, Proc. Transducers '97, 1997 Intl. Conf. Solid-State Sensors and Actuators, Chicago 1997, ed. by K. Wise (IEEE, Piscataway 1997) 1133–1136

54.81 E. S. Hung, S. D. Senturia: Extending the travel range of analog-tuned electrostatic actuators, J. Microelectromech. Syst. **8**, 497–505 (1999)

54.82 R. Nadal-Guardia, A. Dehe, R. Aigner, L. M. Castaner: Current drive methods to extend the range of travel of electrostatic microactuators beyond the voltage pull-in point, J. Microelectromech. Syst. **11**, 255–263 (2002)

54.83 G. J. O'Brien, D. J. Monk, L. Lin: Electrostatic latch and release; a theoretical and empirical study, Proc. Micro-Electro-Mechanical Syst. (MEMS) 2000, MEMS-Vol. 2, The 2000 ASME Intl. Mech. Eng. Cong. and Expo., Orlando 2000, ed. by A. J. Malshe, Q. Tan, A. P. Lee, F. R. Forster, R. S. Kenten (ASME, New York 2000) 19–26

54.84 B. Choi, E. G. Lovell: Improved analysis of microbeams under mechanical and electrostatic loads, J. Micromech. Microeng. **7**, 24–29 (1997)

54.85 D. Elata, O. Bochobza-Degani, S. Feldman, Y. Nemirovski: *Secondary DOF and Their Effect on the Instability of Electrostatic MEMS Devices* (Proc. MEMS 2003, Kyoto, Japan 2003) pp. 177–181

54.86 Y. Zhang, M. L. Dunn: A vertical electrostatic actuator with extended digital range via tailored topology. In: *Smart Structures and Materials 2002: Smart Electronics, MEMS, and Nanotechnology*, Vol. 4700, ed. by V. K. Varadan (Proceedings of SPIE – The International Society for Optical Engineering, Bellingham, Washington 2002) pp. 147–156

54.87 O. Degani, E. Socher, A. Lipson, T. Leitner, D. J. Setter, S. Kaldor, Y. Nemirovsky: Pull-in study of an electrostatic torsion microactuator, J. Microelectromech. Syst. **7**, 373–379 (1998)

54.88 Z. Xiao, X. Wu, W. Peng, K. R. Farmer: An angled-based design approach for rectangular electrostatic actuators, J. Microelectromech. Syst. **10**, 561–568 (2001)

54.89 E. S. Hung, S. D. Senturia: Generating efficient dynamical models for microelectromechanical systems from a few finite-element simulation runs, J. Microelectromech. Syst. **8**, 280–289 (1999)

54.90 E. S. Hung, S. D. Senturia: Leveraged bending for full-gap positioning with electrostatic actuation, Tech. Dig. Solid-State Sensors and Actuators Hilton Head '98, Hilton Head Island 1998, ed. by A. J. Ricco (Transducers Research Foundation, Cleveland Heights 1998) 83–86

54.91 J. R. Gilbert, S. D. Senturia: Two-phase actuators: Stable zipping devices without fabrication of curved structures, Tech. Dig. Solid-State Sensor and Actuator Workshop, Hilton Head Island, SC, USA 1996, ed. by R. T. Howe (Transducers Research Foundation, Cleveland Heights 1996) 98–100

54.92 J. R. Gilbert, R. Legtenberg, S. D. Senturia: 3D coupled electro-mechanics for MEMS: Application of CoSolve-EM, Proc. IEEE MEMS Conference, Amsterdam 1995, ed. by M. Elwenspoek, N. de Rooij (IEEE, Piscataway 1995) 122–127

54.93 J. R. Gilbert, G. K. Ananthasuresh, S. D. Senturia: 3D modeling of contact problems and hysteresis in coupled electro-mechanics, Proc. IEEE MEMS

Part G | 54

1996, The ninth Ann. Intl. Workshop on Micro Electro Mechanical Syst., San Diego 1996, ed. by M. G. Allen, M. L. Reed (IEEE, Piscataway 1996) 127–132

54.94   Y.-H. Jin, K.-S. Seo, Y.-H. Cho, S.-S. Lee, K.-C. Song, J.-U. Bu: An integrated SOI optical microswitch using electrostatic micromirror actuators with insulated touch-down beams and curved electrodes, Proc. Micro-Electro-Mechanical Syst. (MEMS) 2000, MEMS-Vol. 2, The 2000 ASME Intl. Mech. Eng. Cong. and Expo., Orlando 2000, ed. by A. J. Malshe, Q. Tan, A. P. Lee, F. R. Forster, R. S. Kenten (ASME, New York 2000) 177–181

54.95   R. Legtenberg, E. Berenschot, M. Elwenspoeke, J. Fluitman: Electrostatic curved electrode actuators, Proc. IEEE MEMS Conference, Amsterdam 1995 (IEEE, Piscataway 1995) 37–42

54.96   R. Legtenberg, J. Gilbert, S. D. Senturia, M. Elwenspoek: Electrostatic curved electrode actuators, J. Microelectromech. Syst. **6**, 257–265 (1997)

54.97   D. DeReus: Personal communications and internal reports (2002)

54.98   P. M. Osterberg, S. D. Senturia: M-TEST: A test chip for MEMS material property measurement using electrostatically actuated test structures, J. Microelectromech. Syst. **6**, 107–118 (1997)

54.99   L. M. Castaner, S. D. Senturia: Speed-energy optimization of electrostatic actuators based on pull-in, IEEE J. Microelectromech. Syst. **8**, 290–298 (1999)

54.100 L. Castaner, A. Rodriguez, J. Pons, S. D. Senturia: Measurement of power-speed product of electrostatic actuators, Proc. Transducers '99, the 10th Intl. Conf. Solid-State Sensors and Actuators, Sendai 1999, ed. by M. Esashi (IEE Jpn., Tokyo 1999) 1772–1775

54.101 L. Castaner, A. Rodriguez, J. Pons, S. D. Senturia: Pull-in time-energy product of electrostatic actuators: Comparison of experiments and simulation, Sens. Actuators **83**, 263–269 (2000)

54.102 J. Pons-Nin, A. Rodriguez, L. M. Castaner: Voltage and pull-in time in current drive of electrostatic

actuators, J. Microelectromech. Syst. **11**, 196–205 (2002)

54.103 C. Cabuz, E. I. Cabuz, T. R. Ohnstein, J. Neus, R. Maboudian: Factors enhancing the reliability of touch-mode electrostatic actuators, Sens. Actuators **79**, 245–250 (2000)

54.104 J. Wibbeler, G. Pfeifer, M. Hietschold: Parasitic charging of dielectric surfaces in capacitive microelectromechanical systems (MEMS), Sens. Actuators A **71**, 74–80 (1998)

54.105 H. Kubo, T. Namura, K. Yoneda, H. Ohishi, Y. Todokoro: Evaluation of charge build-up in wafer processing by using MOS capacitors with charge collecting electrodes, ICMTS 1995 Proc. IEEE Intl. Conf. Microelectronic Test Structures, Vol. 8, Nara 1995, ed. by T. Sugano, K. Asada (IEEE, Piscataway 1995) 5–9

54.106 E. K. Chan, K. Garikipati, R. W. Dutton: Characterization of contact electromechanics through capacitance–voltage measurement and simulations, IEEE J. Microelectromech. Syst. **8**, 208–217 (1999)

54.107 P. G. Slade, E. D. Taylor: Electrical breakdown in atmospheric air between closely spaced (0.2 μm–40 μm) electrical contacts, Proc. 47th IEEE Holm Conf. Electrical Contacts, Montreal 2001, ed. by K. Leung (IEEE, Piscataway 2001) 245–250

54.108 J-M. Torres, R. S. Dhariwal: Electric field breakdown at micrometer separations, Nanotechnology **10**, 102–107 (1999)

54.109 R. Longwitz, H. van Lintel, R. Carr, C. Hollenstein, P. Renaud: Study of gas ionization schemes for microdevices, Proc. Transducers '01, Eurosensors XV, The 11th Intl. Conf. Solid-State Sensors and Actuators, Munich 2001, ed. by O. Obermeier (Springer, Berlin Heidelberg New York 2001) 1258–1261

54.110 D. K. Davies, M. F. Biondi: Vacuum breakdown between plane-parallel copper plates, J. Appl. Phys. **37**, 2969–2977 (1966)

54.111 T. Ono, D. Y. Sim, M. Esashi: Micro-discharge and electric breakdown in a micro-gap, J. Micromech. Microeng. **10**, 445–451 (2000)

55. High Volume Manufacturing and Field Stability of MEMS Products

Low volume MEMS/NEMS production is practical when an attractive concept is implemented with business, manufacturing, packaging, and test support. Moving beyond this to high volume production adds requirements on design, process control, quality, product stability, market size, market maturity, capital investment, and business systems. In a broad sense, this chapter uses a case study approach: It describes and compares the silicon–based MEMS accelerometers, pressure sensors, image projection systems, and gyroscopes that are in high volume production. Although they serve several markets, these businesses have common characteristics. For example, the manufacturing lines use automated semiconductor equipment and standard material sets to make consistent products in large quantities. Standard, well controlled processes are sometimes modified for a MEMS product. However, novel processes that cannot run with standard equipment and material sets are avoided when possible. This reliance on semiconductor tools, as well as the organizational practices required to manufacture clean, particle-free products partially explains why the MEMS market leaders are integrated circuit manufacturers. There are other factors. MEMS and NEMS are enabling technologies, so it can take several years for high volume applications to develop. Indeed, market size is usually a strong function of price. This becomes a vicious circle, because low price requires low cost – a result that is normally achieved only after a product is in high volume production. During the early years, IC companies reduced cost and financial risk by using existing facilities for low volume MEMS production. As a result, product architectures are partially determined by capabilities developed for previous products. This chapter includes a discussion of MEMS product architecture with particular attention to the impact of electronic integration, packaging, and surfaces. Packaging and testing are critical, because they are significant factors in MEMS product cost.

These devices have extremely high surface/volume ratios, so performance and stability may depend on the control of surface characteristics *after packaging*. Looking into the future, the competitive advantage of IC suppliers will decrease as small companies learn to integrate MEMS/NEMS devices on CMOS foundry wafers. Packaging challenges still remain, because most MEMS/NEMS products must interact with the environment without degrading stability or reliability. Generic packaging solutions are unlikely. However, packaging subcontractors recognize that MEMS/NEMS is a growth opportunity. They will spread the overhead burden of high-capital-cost-facilities by developing flexible processes in order to package several types of moderate volume integrated MEMS/NEMS products on the same equipment.

Solid-state pressure sensors were first reported in the 1960s and commercialized by a number of start-up companies in the 1970s. By 1983, a Scientific American cover story [55.1] described pressure sensors, accelerometers, and ink-jet print heads and featured a gas chromatograph that combined an injection valve, a detector, and a chromatographic column on a silicon wafer. The future of the budding micromechanical device industry looked bright but commercial reality has seriously lagged the rosy market projections. The reason was simple: market forecasters fail to recognize the difference between lab prototypes and high volume production of stable products. Many companies have demonstrated that small quantities can be produced for niche markets. However, routine high volume production of reliable packaged devices is much more challenging. It requires the production prototype skills plus the manufacturing disciplines that are critical to long-term stable production.

Successful companies understand and apply these lessons. Indeed, over a hundred million MEMS pressure sensor, accelerometer, gyro, gas flow, and optical projection devices are shipped to customers annually. There is no universally accepted definition of MEMS. This chapter focuses on silicon-based products with movable micromechanical elements. It excludes disk drive heads, ink jet print heads, hearing aids, microscale plastic, cer-

amic, quartz and metal components, and test strips for in vitro diagnostics that are sometimes included within the definition of MEMS. Many MEMS products do not integrate support electronics with the MEMS element on the chip even though they are produced with IC equipment, materials, and processes. However, integrated MEMS products are becoming ubiquitous. For example, it took over nine years for Analog Devices to ship a hundred million surface micromachined MEMS accelerometers with intgrated electronics to customers. The second hundred million took less than three years, and by 2004, shipments were exceeding a million units a week. Texas Instruments experienced similar exponential growth for its DLP (Digital Light Processing) products. Initial sales were low because it took five years (1996–2001) to ship one million units. However, three million had been shipped by early 2004 and five million by the end of 2004 [55.2].

The literature related to high volume production of MEMS products is quite uneven. Some suppliers have published detailed process and design descriptions while others limit disclosures to market-focused publications. The two major suppliers of integrated MEMS products have discussed their designs and processes in detail. *Core* et al. [55.3], *Kuehnel* et al. [55.4], *Chau* et al. [55.5], and *Sulouff* [55.6] describe the Analog Devices *i*MEMS accelerometers and gyros. *Mignardi* et al. [55.7] give an in-depth review of the Texas Instruments DLP (Digital Light Processing) products.

MEMS devices are components in larger systems, so every MEMS product must define its system interface in a way that adds value from the perspective of the system designer. Some MEMS pressure sensor products are simply Wheatstone bridge elements in a circuit (Fig. 55.1). The image projection systems that use the Texas Instruments Digital Light Processing Technology™ could never be designed in this manner because up to two million mirrors are driven on each chip (Fig. 55.2). Such a system would be impractical unless electronics are integrated with the MEMS mirrors on the chip. Air bag sensors also illustrate how integration adds value. Non-MEMS ball-and-spring air bag sensors were often used in the early 1990s. Upon impact, a ball would be released and detected as it passed through the sensor tube. Unfortunately, these binary sensors could not always distinguish a frontal collision from the jolt of a pothole or a side collision. Therefore, multiple ball-and-spring sensors were placed at different locations in a vehicle and wired together in order to reliably identify accidents that should result in air bag deployment. High cost limited the use of these distributed systems. Analog

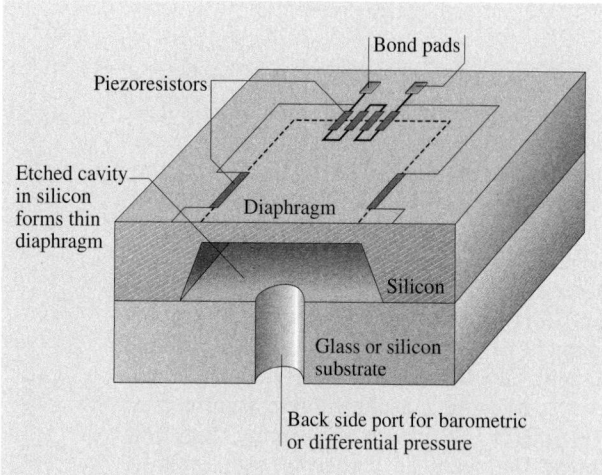

Fig. 55.1 Cross section of a piezoresistive silicon sensor after it is bonded to a silicon or glass substrate. When pressure is applied, the thin silicon diaphragm deforms. This causes changes in four implanted resistors that form a Wheatstone bridge. The resistors are located and oriented to cause resistance to increase in two resistors and decrease in the other two

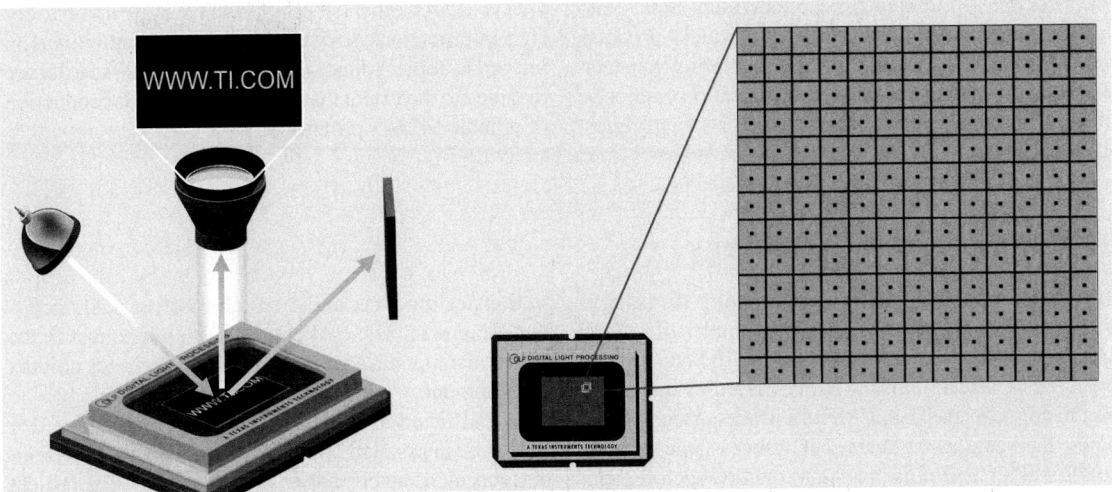

Fig. 55.2 Digital Light Processing technology[TM] is based on Digital Micromirror Device (DMD) chips that address and drive a "sea of mirrors." The DMD is the key element in small computer-based projection systems. Each of the 14 or 17 μm mirrors represents one pixel. The diagram on the *left* represents light reflecting from a mirror to the system lens. The illustration on the *right* is an exploded view of a small region of mirrors in a packaged DMD. (Courtesy of Digital Light Processing[TM], Texas Instruments, Inc.)

Devices changed the air bag market when it introduced the ADXL50 in 1993 (Fig. 55.3). The ADXL50 integrated electronics and an accelerometer on one chip to provide an analog output of deceleration versus time when an impact event occurred. This allowed automobile manufacturers to design air bag modules that compared the ADXL50 output to the known signature of a frontal collision. The result was greater reliability and lower cost, which led to the installation of air bags in almost every new automobile.

This chapter will not recite detailed descriptions of particular commercial products. Instead, brief descriptions will be used to illustrate basic principles and concepts. Part of the chapter is organized according to MEMS-specific issues like wafer singulation. This is useful because some designs and unit processes are well suited for lab scale or low volume production, but impractical in a high volume manufacturing environment. Complete reliance on such an organization would be artificial, because partitioning omits the great-

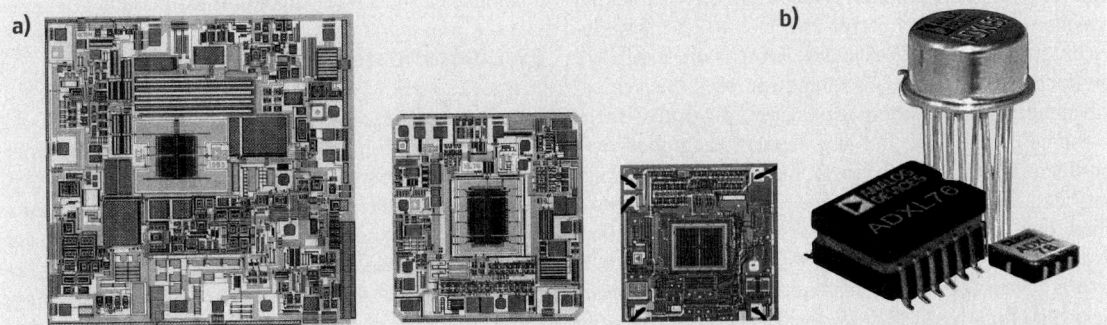

Fig. 55.3a,b Three generations of integrated accelerometers designed for single-axis air bag applications. (**a**) Die size is reduced by a factor of two about every four years. (**b**) Package technology also evolved to meet size, cost, and automated handling requirements. The ADXL50 was packaged in TO-100 seam-sealed metal headers. ADXL76 cerpacs (*bottom left*) are being replaced by the ADXL78 in 5 mm×5 mm LCC packages (*bottom right*). (Courtesy of Micromachined Products Division, Analog Devices, Inc.)

est challenge – combining these individual topics into a manufacturing flow that routinely produces products that meet cost, performance, and reliability expectations. The importance of process integration cannot be overemphasized, because changes in one step almost always affect other steps. Solutions may involve design modifications, so design-for-manufacturability is an important topic. Manufacturing and business skills are included to the extent that they apply to stable production of reliable MEMS products.

55.1 Manufacturing Strategy

MEMS manufacturing is based on existing IC technology. However, there are only a few examples of routine MEMS and IC production intermixed on the same equipment (or of MEMS devices with electronics integrated on the chip). Even when a MEMS process step is nominally the same as a standard IC process, the specific conditions or end point or some quality such as film stress will differ. Technologists from Analog Devices and Motorola have discussed these thin film issues in depth [55.8–11].

Manufacturers prefer to use dry etch processes when possible in order to achieve better control of features, minimize waste disposal and utilize automated equipment developed for the IC industry. However, anisotropic wet etching provides unique value in the production of some MEMS products like pressure, mass flow, and yaw rate sensors [55.12].

Although MEMS production economics is closely tied to the semiconductor model, useful lessons can be drawn from many industries. Some of the key characteristics that affect success are as follows:

55.1.1 Volume

Semiconductor economics is rooted in the principle that expensive processes are cost-effective when many devices are made simultaneously. Obviously, a MEMS product that is sold in low quantities loses this critical advantage. MEMS components are "disruptive technologies" – they are the key element in higher-level products that were previously impractical. The Analog Devices and Texas Instruments data cited above show that it can take several years to develop a higher-level products and build a customer base. Therefore, it is prudent to include an incubation phase in the business plan for a novel MEMS device. Low sales volume during this phase causes high unit costs – batch process cost benefits are achieved only after the market develops.

Startup companies often do not understand the implications of design-for-volume. It is a major challenge to transfer products made on lab tools to manufacturing equipment. The challenge is sufficient to justify the use of manufacturing people and equipment for product development.

Realistic price-volume estimates are also essential to a good business plan. The highest volume MEMS products (accelerometers and pressure sensors) all represent a small part of the total system cost, even though the functions they provide are absolutely critical to system performance.

Market-pull business models that respond to the needs of a specific large application in a particular industry are usually more successful than technology-driven models that seek applications for new devices. Given this reality, it is prudent to thoroughly understand the target industry and its quirks. Biotech, wireless communication, image projection, computer and consumer markets offer tremendous growth potential. However, the past and present growth in high volume MEMS products have been driven primarily by the automotive industry, with pressure sensors also benefiting from medical applications. *Sulouff* [55.13], *Weinberg* [55.14], *Marek* et al. [55.15], *Eddy* et al. [55.16], and *Verma* et al. [55.17] discuss automotive applications and the challenges a successful supplier must master.

55.1.2 Standardization

Mature industries develop standards in response to equipment vendors, material suppliers, and customer interface requirements. The MEMS industry is fragmented, so even after 30 years it is not mature. Deviations from semiconductor industry standards are inevitable because MEMS products impose unique requirements on wafer processing, packaging, handling, and testing. For example, KOH wet etching raises mobile ion concerns. Gold used in some optical products is incompatible with CMOS. Any deviation from standard process and material sets is serious, because the price-volume relationship is a steep curve in high volume applications.

55.1.3 Production Facilities

The semiconductor industry is divided into front-end (wafer fabrication) and back-end (product assembly and test) operations. IC design rules are based on pre-qualified fab, assembly, and test capabilities at specific sites. Consider how each of these functions matches the needs of MEMS production:

1. An IC product might be produced at an internal wafer fab, or designed to run on a standard foundry process. In contrast, production challenges and "know-how" have caused MEMS suppliers to retain wafer fab production within their internal facilities. High volume production in MEMS foundries is limited to bulk micromachined piezoresistive pressure sensors. To illustrate this point, Freescale (formerly Motorola) produced piezoresistive pressure sensors and accelerometers in an international Arizona wafer fab from the early 1980s until 2003. Production of the mature pressure sensors was transferred to *Dalsa* in 2003 [55.18]. However, production of surface micromachined accelerometers and tire pressure monitors was retained within Freescale; these products were transferred to Sendai, a wholly-owned Freescale subsidiary in Japan. Neither TI nor Analog Devices use MEMS foundries. These observations suggest that MEMS foundry growth is likely to be slow.
2. IC suppliers have shifted assembly operations to subcontractor facilities located in Southeast Asia. Some MEMS suppliers have followed this model. It is common to assemble low volume products in North America, Europe, or Japan. However, semiconductor assembly and package technology are evolving rapidly and are driven by subcontractors based in Southeast Asia. MEMS companies that do not utilize these resources pay a significant cost penalty. Obviously, the break point at which offshore assembly is attractive differs from one product to the next. Products that have unique assembly requirements and support high market prices may never reach this point.
3. IC testing is often carried out at the assembly subcontractor site. MEMS package and testing comprise a major portion of the total product cost, so a high volume manufacturing strategy should include Southeast Asia packaging and testing. This is complicated by the fact that MEMS testing requires unique stimuli that are not in the standard IC portfolio.

55.1.4 Quality

The goal in lab and low volume production environments is to make functioning devices. As quantities increase, other factors become more important: lot-to-lot repeatability, yield, device performance that meets well-defined specifications, process stability to ensure predictable on-time delivery, documentation and procedures that enable traceability and corrective actions when problems arise, etc. Methodologies that promote stable, high-quality manufacturing practices have been established in several industries. These standard procedures such as ISO/TS16949 [55.19], ISO9000, QS9000, and cGMP overlap to a large extent. QS9000:1998 [55.20] extended ISO9000:1994. It was developed to help automotive suppliers implement well-controlled processes, an essential requirement for long-term stable production of reliable products. Many of the QS9000:1998 features were incorporated into ISO9000:2000. ISO/TS16949 broadens this standard to more explicitly include business systems, management ownership and customer needs. It is written for automotive suppliers and shifts the emphasis from procedures to processes. QS9000 and ISO9000:2000 are used as the model in this chapter, but ISO/TS16949 has elements that should be considered by any ISO9000:2000 compliant organization.

55.1.5 Environmental Shield

IC wafers are completely passivated before packaging in order to meet product stability and reliability requirements. Aside from products like ink-jet print heads, the active regions of all MEMS devices currently in high volume production replicate this hermetic barrier in some manner. This passivation requirement is noteworthy because it has been a primary limitation on the growth of the MEMS industry. For example, it limits electrochemical sensors to benign or "throw-away" applications.

55.2 Robust Manufacturing

Stable production requires a manufacturing flow that is well controlled when measured against the product performance specifications. This has several implications:

55.2.1 Design for Manufacturability

Low maintenance processes must be designed to run on standard equipment. The semiconductor industry has invested billions of dollars to develop equipment that is automated, maintainable, reliable, and capable of supporting well controlled processes. Processes that are implemented on custom-designed equipment put this experience base aside and invariably have a long learning curve.

Equipment

Microstructure release illustrates this issue, and how it changes over time. Virtually all semiconductor processes operate in vacuum or at atmospheric pressure. In a brief, unpublished 1992 study related to ADXL-50 development, the author demonstrated the use of supercritical CO_2 mixtures for microstructure release and particle cleaning. Unfortunately, supercritical processes were incompatible with the industry infrastructure, because they require pressures in excess of 7.38 MPa (72.8 atmospheres; 1070 psi). Implementation would have required process and equipment development, so a release process that ran on existing equipment was selected.

Supercritical equipment designed for MEMS release was introduced a few years later. However, maintenance and cycle time concerns as well as single wafer capabilities limited its adoption to university and low volume manufacturing.

The intrinsic limitations of non-standard equipment have caused high volume MEMS suppliers to avoid supercritical processes. However supercritical equipment will continue to evolve and become competitive in applications where the design requirements make alternative processes unusable.

Materials

Process control can be no better than the materials used in the process. SAM (*self-assembled monolayer*) coatings illustrate this limitation. These materials suppress stiction following aqueous release of microstructures by treating the wafers in a SAM solution before drying. Although straightforward in principle, high volume manufacturers have not adopted SAM processes

(Sect. 55.2.3 reviews the processes that are used commercially). There are several reasons for this reticence. Classical SAM materials are chlorosilanes that have at least one organic substituent. Organosilanes have been used to treat the surface of inorganic materials for many years. Applications include coupling agents on the surface of fillers and reinforcing agents in polymers, as well as agglomeration control of particles. The chlorine sites hydrolyze when dissolved in solutions that contain a small amount of water. The resulting hydroxyl groups react with the microstructure surface oxide to produce a chemically bonded layer that reduces stiction due to the low surface energy and hydrophobic nature of the organic-rich surface. The concerns that cause manufacturers to use alternative stiction solutions include:

1. High variability due to the SAM chemical reactivity.
2. Particles are common by-products of the reactivity.
3. Many SAM process flows require organic baths. This raises health, safety, and waste disposal concerns.
4. Possibility of chloride corrosion on aluminum interconnects.

Maboudian et al. [55.21], *Srinivasan* et al. [55.22], *Kim* et al. [55.23], and *Pamidighantam* et al. [55.24] describe the use of SAM solutions to suppress stiction. There is also ongoing research focused on solving the manufacturing concerns [55.25–29].

Note that stiction also arises in packaged MEMS products. This yield and reliability problem is discussed in Sect. 55.3.1.

55.2.2 Process Flow and its Interaction with Product Architecture

MEMS products have structures and functions that do not exist in standard IC devices, so it is unrealistic to expect that every fab, assembly, and test step will reapply a previously qualified IC process. The challenge is to maximize utilization of available IC technology within the constraints of the product function and cost requirements. This leads to fundamental choices in product architecture, product design, and process flow, as discussed below.

Integration of MEMS and Circuits

The integration of MEMS and electronics on one chip has proven difficult. It adds little value in some MEMS products. However, as noted earlier, integration is es-

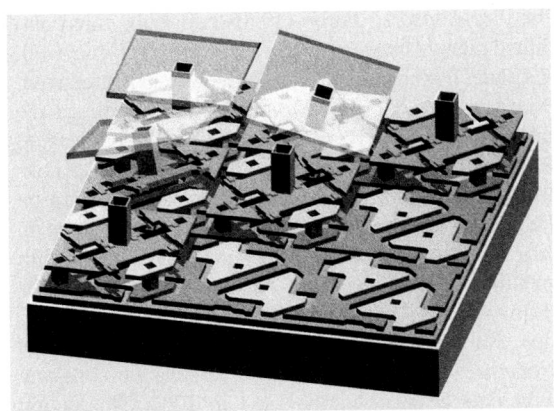

Fig. 55.4 DMD pixel array with tilted and non-tilted mirrors. (Courtesy of Digital Light Processing™, Texas Instruments, Inc.)

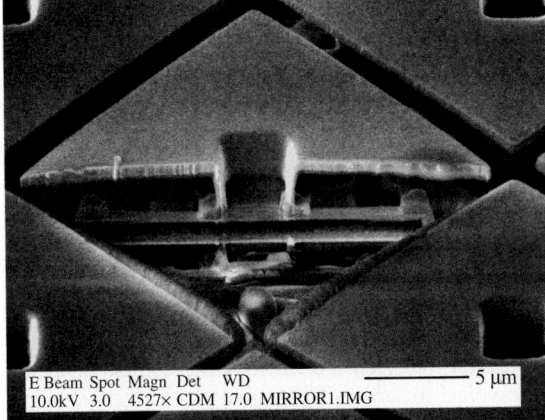

Fig. 55.5 Ion milled section of DMD pixel showing one mirror on its center support and its relation to the underlying layers. Note the close mirror spacing. (Courtesy of Digital Light Processing™, Texas Instruments, Inc.)

sential to products based on the Texas Instruments DLP (Digital Light Processing™). The third (and largest) category consists of applications where integrated and non-integrated products compete for market share. Pressure sensors and accelerometers are examples of this group.

Image Projection. The first projectors based on the DLP technology were shipped in 1996. DLP chips are produced on CMOS wafers in a mature TI process. The circuitry addresses and drives aluminum mirror arrays that function as on-off pixels – two million mirrors in the larger arrays (Fig. 55.2). *Mignardi* et al. [55.7] describe

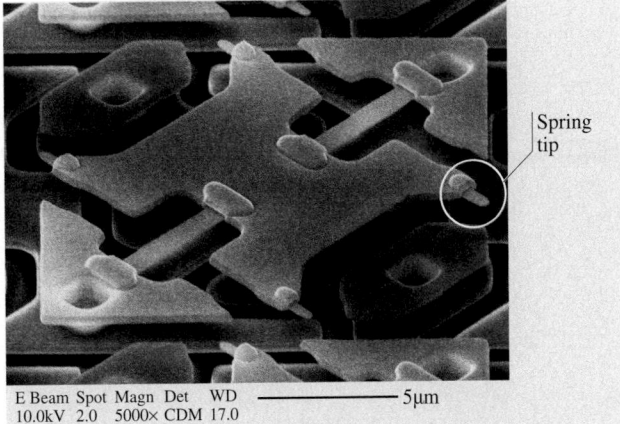

Fig. 55.6 DMD yoke and hinge layer. The spring tip touches landing pads when the drive electrode tilts the mirror. Elastic energy stored in the deformed spring is part of the restoring force that overcomes stiction when the drive electrode voltage is removed. (Courtesy of Digital Light Processing™, Texas Instruments, Inc.)

the DMD (Digital Mirror Device) fabrication process and illustrate some of the factors that must be considered when manufacturing flows are changed. The manufacturing flow patterns three aluminum depositions over SRAM cells that are positioned with 14 or 17 μm center-to-center spacing (Figs. 55.4 and 55.5). The first of these metal films form electrostatic drive electrodes and landing pads. The yoke and hinge layer has spring tips that land on the first metal layer when mirrors are rotated (Fig. 55.6). Mirrors are mechanically connected to the yokes through center pedestals. Sacrificial organic films separate the metal layers. These organics, along with a protective organic cover film, remain on the device until after the wafer is sawn and the chips are mounted in a ceramic package. They are removed by dry etching. Following surface passivation, inspection, and testing, an optical glass sub-assembly is hermetically sealed to the ceramic package (Fig. 55.7). At this point, the device

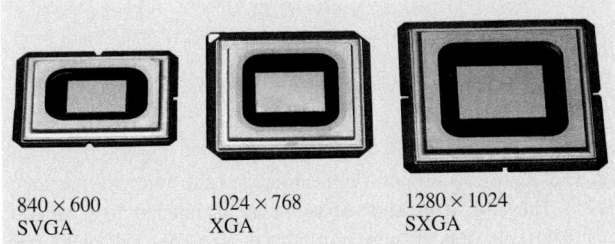

| 840 × 600 | 1024 × 768 | 1280 × 1024 |
| SVGA | XGA | SXGA |

Fig. 55.7 DMD products in hermetic packages. (Courtesy of Digital Light Processing™, Texas Instruments, Inc.)

Part G | 55.2

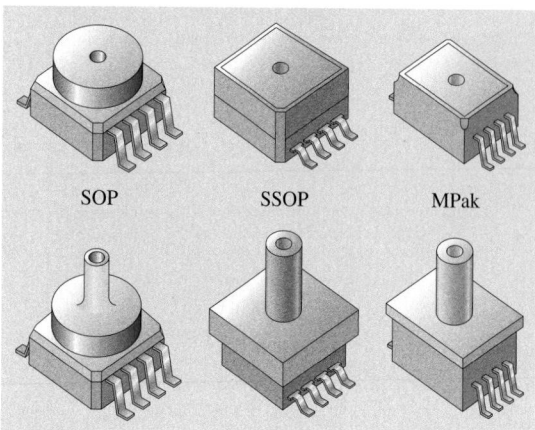

Fig. 55.8 Pressure sensor surface mount packages. (Courtesy of Semiconductor Products Sector, Motorola, Inc.)

is ready for burn-in and final test – a significant task because each mirror is actuated on chips with as many as two million mirrors.

In operation, a (nominal) 26 V electrode bias generates an electrostatic force that causes the mirror/yoke assembly to rotate either plus or minus 10 degrees around the hinge.

Unlike the pressure sensors and accelerometers discussed below, TI had no choice but to integrate. It would be difficult, if not impossible, to devise a cost-effective multi-chip solution to control and drive a matrix of over a million mirrors. Successful integration of circuits and mirrors was only one step in the commercialization process. To remain competitive against lower cost LCD and plasma products, TI continues to develop their package and test technologies, because package and test comprise a major part of the total product cost. This is discussed further in Sect. 55.2.8.

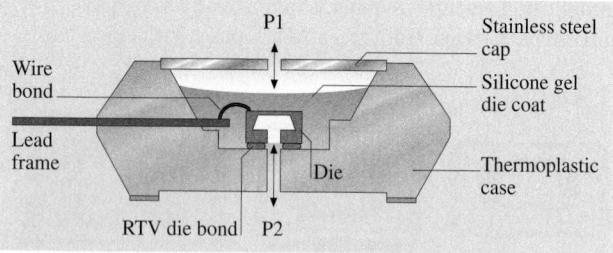

Fig. 55.9 Cross section of a typical gauge or differential pressure sensor. The silicon pressure sensor chip is bonded to a silicon substrate. This bonded unit is mounted in a pre-molded plastic package and protected from the ambient environment by silicone gel. (Courtesy of Semiconductor Products Sector, Motorola, Inc.)

Pressure Sensors. Early (1970s-era) bulk micromachined piezoresistive pressure sensors from Honeywell, ICT (later Foxboro-ICT), and Kulite were not integrated. However, in the 1980s, Motorola commercialized the first of its MPX5100 series piezoresistive pressure sensor products. These products had bipolar signal conditioning electronics and temperature compensation on the sensor chip. Current MPX products are offered in various plastic packages (Figs. 55.8 and 55.9) with maximum pressure ranges between 10 and 300 kPa (1.45–44 psi). Medical and tire pressure products are also available. Note that Motorola's Semiconductor Products Sector was spun off as a separate company (Freescale Semiconductor, Inc.) in 2004. This action split Motorola's MEMS business. Motorola retained the automotive module pressure sensor business. Freescale has the device level pressure sensors such as those sold for medical applications as well as tire pressure monitors. They also retained the accelerometer business. Over a 25 year period, Motorola/Freescale shipped more than 300 million pressure sensors [55.30].

Integrated Motorola piezoresistive pressure sensor products are cited here for illustrative purposes, but capacitive and piezoresistive silicon pressure sensors are available from many companies. Capacitive designs are less temperature sensitive and require less power than piezoresistive designs. However, die size tends to be larger. The fabrication process and interface electronics are also more involved. Piezoresistive products (sensor plus signal processing) are less expensive, smaller, easier to manufacture and suitable for most applications. They rely on the fact that the resistance of silicon changes in response to strain. This effect is very temperature sensitive. However, configuring four piezoresistors as a Wheatstone bridge, judicious selection of the doping level and junction depth and the use of a temperature compensating resistor network minimizes temperature error.

Piezoresistive pressure sensors (Fig. 55.1) require double-sided polished wafers and front-to-back alignment. The piezoresistors are implanted into the front side of (100) wafers and precisely located with respect to the edges of thin diaphragms. The diaphragms are produced by anisotropically wet etching cavities into the back surface of the wafers. KOH etch solutions are usually used to avoid the safety, toxicity, and waste disposal concerns that arise when large quantities of organic etchants like EDP (Ethylenediamine-pyrocatechol-water) are used. Front-to-back side wafer alignment is critical in order to achieve precise resistor location with respect to the cavity edges. Alignment to the wafer crystal axis is also important, because the slow-etching (111) planes determine the cavity wall, and thus the location of the diaphragm edges. The piezoresistors are commonly placed near these edges to maximize the signal (the effect of pressure on diaphragm stress is greatest at the edges). An etch stop may be used at additional cost because diaphragm thickness variations also have a large effect on sensitivity.

Control of mount stress is critical to achieving predictable performance. After wafer fab, some suppliers hermetically bond pressure-sensor wafers to a backup silicon wafer using glass frit. Others anodically bond them to a borosilicate glass wafer that has a thermal expansion coefficient close to silicon (Pyrex 7740 and Schott 8330 are two glasses used for this purpose). Absolute pressure sensors are produced when this bonding step is done in vacuum. As shown in Fig. 55.1 and Fig. 55.9, differential pressure and gauge pressure products incorporate through holes in the backup wafers. These ports allow fluid pressure to be applied through the cavities to the back side of the diaphragms.

After wafer singulation, the chips are mounted and sealed in cavity packages. Soft die attach materials, often silicone-based, are used to decouple the sensor from package and substrate stresses. In some products, the sensor chip is directly mounted to a substrate with a soft die attach without the intermediate backup wafer. Many package variations of MEMS pressure sensors are available, as summarized in Sect. 55.2.8.

The MEMS pressure sensor market is believed to exceed a hundred million pressure sensors annually [55.31]. Automotive applications like manifold absolute pressure (MAP) sensors, barometric absolute pressure (BAP) sensors and a wide range of gas and liquid pressure sensors form the largest market segment. Motorola, Bosch and Denso are the largest automotive market suppliers [55.32]. Automotive applications are the primary market for integrated pressure sensors. Health care and medical uses such as disposable blood pressure transducers and sensors to measure pressure in angioplasty catheters, infusion pumps, and intrauterine products are the second largest market. Industrial products such as process control pressure and differential pressure transmitters, household appliance, and aeronautical products are also significant. *Maudie* et al. [55.33] review the performance and reliability issues related to appliance applications like household washing machines.

Accelerometers. Most MEMS accelerometer designs apply some variant of Newton's Law, $F = ma$, to sense

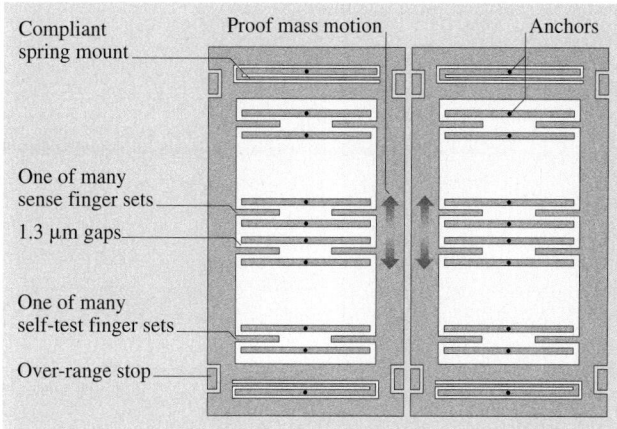

Fig. 55.10 Block diagram of the MEMS element in an ADXL78 single-axis accelerometer. The detailed design applies design-for-manufacturability principles to ensure close control of critical dimensions and residue-free removal of the sacrificial layer. (Courtesy of Micromachined Products Division, Analog Devices, Inc.)

the response of a proof mass. Many sensing principles, including piezoresistive, capacitive, piezoelectric, and resonant, have been examined [55.34]. Figure 55.10 has a conceptual view of the capacitive design used in Analog Devices ADXL accelerometers.

SensorNor was an early pioneer in silicon accelerometer manufacturing. Their product had a mass bonded to the end of a silicon piezoresistive element in an oil-filled package. However, the first MEMS accelerometer product to achieve large-scale market acceptance was the integrated ADXL50 air bag sensor from Analog Devices. When it was fully qualified in 1993, the ADXL50 was the first integrated surface-micromachined MEMS device of any type in production. The MEMS element in the ADXL50 was part of a closed loop differential capacitance circuit, but was designed to allow routine self-testing after the air bag module was installed in vehicles. This self-test capability eased concerns related to the adoption of a new technology in a critical safety application. The analog output feature also allowed the industry to design single point sensors – a significant system cost savings. *Core* et al. [55.3] and *Sulouff* [55.6] describe the ADXL wafer fabrication process. In essence, circuits are fabricated with a well-established BiCMOS process. The sensor polysilicon is deposited after these high temperature steps, followed by the lower temperature metal and passivation processes. The in situ doped polysilicon is connected to the circuit through N+ doped runners. Following wafer fabrication, thin film resistors are laser trimmed in an automated

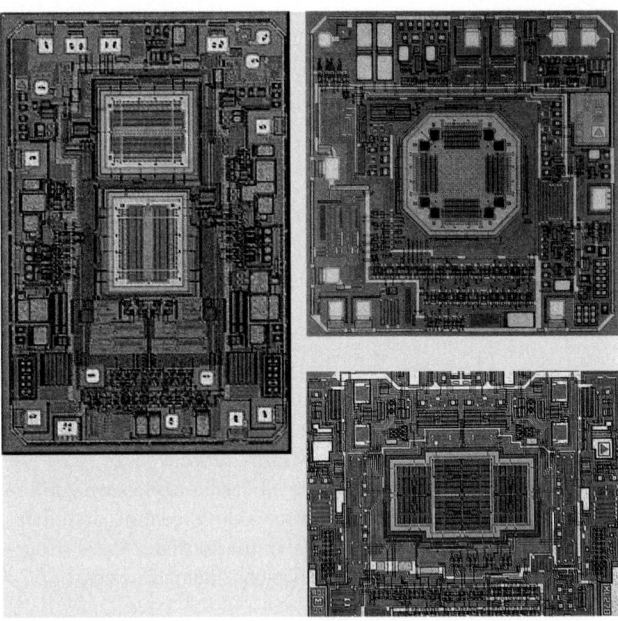

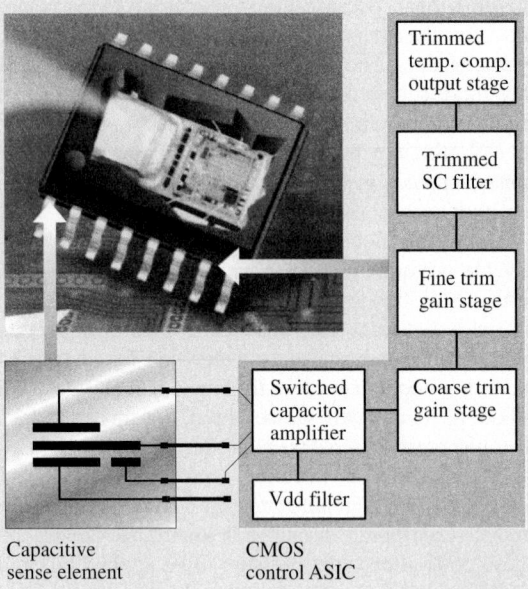

Fig. 55.11 Three two-axis accelerometers. *Top left:* ADXL276 was the first two-axis air bag accelerometer. The ADXL278 family (*bottom right*) is replacing it. *Top right:* 2-g ADXL202. The ADXL202 and ADXL278 are packaged in the LCC shown in Fig. 55.3. (Courtesy of Micromachined Products Division, Analog Devices, Inc.)

Fig. 55.12 Two-chip accelerometer in 16 lead plastic SOIC package. (Courtesy of Semiconductor Products Sector, Motorola, Inc.)

trim system to meet the specific end user's requirements. After wafer singulation (Sect. 55.2.5), the chips are assembled in hermetic cavity packages and screened in an automated test system to ensure compliance to the performance specification (Sect. 55.2.8).

Front air bag sensors have an output of 2 V at a full scale acceleration that is specified by the user (typically 35–50 g). Satellite air bag sensors are located in door pillars and near the front of the automobile and typically have higher ranges (250 g). Regardless of range, all air bag sensors must meet stringent cross-axis specifications, i.e., a front air bag sensor must not react to a side collision or to potholes in the road. This cross-axis requirement also applies to products that are designed to sense in two axes (Fig. 55.11).

Progressive reductions in chip and package sizes (Fig. 55.3) led to cost reductions and penetration into new markets. One example is the ADXL202 (Fig. 55.11). This two-axis, low-g accelerometer has a full scale of 2 g in each axis. It is widely used in industrial, consumer, and automotive applications. Production of integrated MEMS accelerometers at Analog Devices has grown rapidly in recent years (several million devices per month by 2001). It is worth noting that these

products are developed using an organizational philosophy that emphasizes design-for-manufacturability. For example, development is conducted on production equipment and directly involves production personnel. This may appear to be expensive and unwieldy, however, it ensures that product designs remain within the bounds of practical manufacturing. It also minimizes the difficulties that are normally encountered when new products are transferred to production.

Prior to the introduction of the ADXL50, most of the MEMS air bag candidates were nonintegrated, bulk-micromachined piezoresistive sensors [55.4]. However, Motorola and Bosch soon followed with an integrate-in-the-package strategy. Denso, Delphi, and VTI Hamlin are also active in some market sectors. The Motorola and Bosch products are nonintegrated capacitive MEMS products that are hermetically sealed at the wafer level (Sect. 55.2.4). By interconnecting a separate circuit chip and a sealed accelerometer chip in one package, they are able to use low cost CMOS for signal processing and near-standard plastic packaging (Sect. 55.2.8).

In contrast to the ADXL sensors, Motorola's MMA air bag products measure acceleration normal to the plane of the chip (z-axis). These differential capacitance sensors are formed from three layers of polysilicon. The middle layer is a "proof mass" that is suspended with ligaments and is free to move in response to an accelerating force. Ranges and outputs are similar to the ADXL products. Recent Motorola expansions add x-axis air bag products and a new line of low-g z-axis products (Fig. 55.12).

The Bosch x-axis sensor has interdigitated fingers that are patterned in a thick (11 μm) polysilicon film that is deposited in an epi reactor. Tight control of stress and stress gradients in these films is essential in order to meet sensor performance requirements [55.35, 36]. After patterning, the polysilicon is dry etched using fluorine-based chemistry [55.37] to form the differential capacitance sensor. An HF vapor process is used to remove the sacrificial oxide and release the microstructures. These thick structures have working capacitances near 1 pF [55.38]. *Laermer* et al. [55.39] explain how Bosch applies deep silicon etch processing to several production and experimental inertial sensors. Better known as the "Bosch etch," this deep reactive ion etching technique has pioneered the field of high-aspect-ratio silicon-based MEMS.

The Future of Integration. The high growth rate of integrated MEMS products demonstrates the value of integration in MEMS applications. To date, integration has been limited to large semiconductor companies that have in-house signal processing expertise and the willingness to invest substantial resources over a number of years. Each company identified a large volume market that could support a profitable integrated product. This "large semiconductor company" barrier-to-entry for integrated MEMS products will diminish in the future due to new SOI wafer technology (Sect. 55.2.4), new flip-chip equipment, and the development of low temperature MEMS processes. With these capabilities, MEMS-only companies will be able to integrate microstructures on circuit wafers that are preprocessed in CMOS foundries. To be successful, they will also have to develop cost-effective manufacturing and quality expertise (Sect. 55.2.9).

Risk

High risk is normal in research programs. However, the tremendous financial investments required to manufacture new products make the risk of failure a key consideration. This section discusses how the perception of manufacturing risk changes with time and is influenced by available knowledge and expertise.

Every development program makes fundamental product architecture decisions. These decisions can lead competing companies down different paths. For example, Analog Devices pursued chip-level integration, while Bosch and Motorola chose package-level integration. Why did these companies choose different paths? The answer is related to available resources, cost, time-to-market, and technical maturity. Analog Devices chose to integrate because of its expertise in process innovation and signal processing. This path did have risks. When the ADXL50 was designed, the fatigue life of polysilicon MEMS devices was unknown. Therefore, a closed loop circuit architecture was chosen in order to keep the proof mass motion under 100 Å. Polysilicon actually has excellent fatigue life. Once this was established, Analog changed their ADXL products to an open loop architecture that can detect capacitance changes as low as 20 zF (10^{-21} F). Most companies do not have the expertise to work in this regime, but the decision to integrate on one chip allowed Analog to make full use of its signal processing knowledge.

The two-chip Motorola and Bosch designs allow the circuit chip to be fabricated on any IC process. This product concept also removed the concern that integrating MEMS and circuits onto one chip might reduce overall product yield. However, it does require higher output signals from the MEMS chip in order to overcome parasitics that arise in the bond wires between the circuit and sensor chips. This requirement led Motorola toward large-area z-axis designs, while Bosch utilized its deep-etch process capability. To minimize cost and ensure reliability, both companies seal the sensors with glass at the wafer level. Sensor and circuit chips are then molded together in one plastic package. Each company built on expertise that it already had in order to minimize risk and cost. Motorola had in-house manufacturing expertise in wafer-level glass frit sealing of pressure sensors that are subsequently molded in plastic packages. Bosch had deep-etch process capability and the world's largest hybrid manufacturing plant, so screen printing and glass sealing were well-established. These considerations plus the concern that MEMS yield loss would undermine product yield contributed to their two-chip decision.

Semiconductor companies continually drive down cost and serve evolving markets by shrinking product size. Analog Devices believes that this long-term roadmap requires an integrated product. They chose to deal with one-chip risks, rather than package uncertainties. ADXL50 chips had been capped and molded in plastic long before the first qualified products were shipped in 1993 [55.40]. However, production-worthy capping equipment did not exist at that time. The effects of plastic package stress on yield and long-term device parametrics were also uncertain. By introducing the initial product in TO-100 hermetic metal packages, the risks associated with custom equipment and package stress were removed.

Reusable Engineering and Facilities to Achieve Economies of Scale

The IC industry produces many products on each process flow. Foundries and packaging subcontractors routinely apply this "economy-of-scale" principle. Only basic items like mask sets and bond pad diagrams are changed. The MEMS industry has been less successful in this area, although techniques like wafer-level laser trim and blowing poly fuses allow accelerometer suppliers to offer "specialty" products without losing the advantages of high volume production.

If only a few devices are produced in a capital-intensive facility, unit costs are high. This issue is particularly critical during the first years of product life when sales volume is normally low. One solution is to build the market by manufacturing the new product in an existing facility. The two-axis low-g ADXL202 (Fig. 55.11) was built using a new mask set with an existing air bag sensor process. Many ADXL202 consumer, industrial, and automotive applications are too small to justify the capital investment risk required to develop and produce a two-axis accelerometer, but use of an existing design and manufacturing base removed this constraint.

Market Dynamics Drive New Product Opportunities

In the 1980s, low cost and high reliability allowed MEMS pressure sensors to penetrate automotive applications like the measurement of manifold vacuum (MAP sensors). Once a technology is accepted, new opportunities usually develop. For example, tire pressure monitors with RF communication capability are now available in some car models. In the future, this type of monitor will be a safety feature expected by all consumers. Indeed, wireless communication capabilities will also be a common MEMS product requirement.

This example illustrates the fact that MEMS products are merely system components. They must support seamless integration into the larger system. MEMS opportunities in automobile safety systems will be affected by both communication protocols and interconnect technology (copper, fiber optic, and wireless). The growth spurt of the 1990s started with front-air-bag crash sensors located in the center module of automobiles. It has evolved to include satellite crash sensors behind bumpers, side-impact sensors in the B pillar (the vertical post between doors), impact sensors inside doors, rollover sensors, and vehicle dynamic control sensors to improve ride and handling. Each of these applications has unique characteristics that can favor one technology over another. For example, side-impact sensing systems must activate the air bags faster than front-impact systems. This response time characteristic can favor the use of pressure sensors that monitor air pressure inside door cavities – a good solution as long as the cavity remains sealed. An alternative is to place an accelerometer in the B pillar. These thin structural members have very little space, so z-axis accelerometers are sometimes easier to implement. Delphi and Motorola products have this characteristic. Analog's small x-axis accelerometers also meet this space requirement.

Gyros – New Products Produced with Preexisting Manufacturing Capabilities

Gyroscopes are transducers for rotational motion. They have a wide range of uses in platform stabilization and robotics, as well as automotive applications (dead reckoning backup for GPS navigation systems when satellite

Fig. 55.13 (a) ADXRS150 gyro with integrated electronics in 7 mm × 7 mm BGA package. **(b)** ADXRS150 chip. **(c)** MEMS portion of the ADXRS150 chip. (Courtesy of Micromachined Products Division, Analog Devices, Inc.)

communication links are lost, skid control systems, and rollover detection). Unfortunately, the non-MEMS and resonant quartz gyros used in military and aerospace applications are rather expensive for many high volume applications. BEI Systron Donner tailored their high performance quartz gyro to meet automotive needs and has shipped several million gyros. Other lower cost gyro and angular rate products, like the Murata vibrating ceramic bimorph products, are used in cameras. However, as a group, these low cost products do not meet the performance required for many applications.

Analog Devices and Bosch reapplied their accelerometer capabilities to develop and manufacture low cost gyroscopes. *Geen* et al. [55.41] describe the Analog Devices product. *Funk* et al. [55.42] and *Lutz* et al. [55.43] describe the Bosch product. These three publications reference previous work and discuss the challenges of producing gyros that meet performance, reliability, and cost targets.

The Analog Devices ADXRS150 gyro (Fig. 55.13) illustrates the technology. It has a resonant microstructure that is integrated with two electronic systems on one chip. One of the electronic systems drives the MEMS resonator. When the ADXRS150 is rotated, Coriolis acceleration is generated in a direction that is perpendicular to the vibration axis of the resonator. This motion is detected by the second accelerometer system. Full scale is rated at 150 deg/s, although it can sustain overloads up to 10 000 deg/s. Integration of electronics on the chip, microstructure modeling, and predictable manufacturing were essential to success, because the full-scale Coriolis motion is only about 1 Å. The combination of differential capacitance detection, area averaging, and correlation techniques allow this motion to be resolved from thermal noise down to about 0.00016 Å [55.41].

Resonant gyros are not new. However, as noted above, their high cost restricts them to specialty applications. Many are non-MEMS, or utilize the piezoelectric properties of quartz, so chip-level integration is not an option. Most require expensive vacuum packaging to achieve high mechanical Q at resonance. Unfortunately, vacuum packaging also makes them susceptible to mechanical damage. Viscous damping in gas-filled devices like the ADXRS150 suppresses the destructive impact events that occur in vacuum-packaged MEMS products during shipping and handling and in-use conditions. Equally important are ADXRS150 design features that allow the measurement to reject mechanical shock, vibration, and other environmental noise sources. As a result, under-the-hood location is practical in auto-

motive applications, an unmatched characteristic that significantly reduces system cost.

Analog Devices and Bosch both needed several years to develop their gyro products. Use of existing accelerometer infrastructure during the prototype and early production phases significantly reduced development costs and manufacturing risks. Early ADXRS150 gyros were sold as engineering grade products in socketable side braze packages. This allowed the development team and customers to understand the design and process parametrics that affect performance, reliability, and yield. Side braze packages are unacceptable in high volume applications, so the ADXRS150 is sold in a hermetic, ceramic, surface mount package.

55.2.3 Microstructure Release

Release stiction occurs when microstructures stick together after the sacrificial material is removed in a wet etch. It can cause considerable yield loss and must be considered in devices like capacitive pressure sensors and accelerometers that have closely spaced, mechanically compliant microstructures. Piezoresistive pressure sensors are not susceptible to release stiction, because these sensing diaphragms are not near other surfaces.

Release stiction is most frequently observed in silicon MEMS devices that use a sacrificial oxide. The oxide is removed by etching in wet HF, rinsed in deionized water, and then dried. Water promotes growth of a hydrophilic surface oxide. It also has a high surface tension. To minimize energy, the high surface tension in the shrinking water droplets causes the microstructures to be pulled together. When they touch, clean oxide surfaces stick, thus destroying device functionality. *Maboudian* et al. [55.21] review release stiction mechanisms, as well as the solutions that have been reported in the literature.

Some manufacturers do not publicly discuss their fab process. Those that do avoid release stiction by using gas phase processes to remove the sacrificial material. Texas Instruments uses photoresist, rather than silicon dioxide as the sacrificial material in their DMD manufacturing process. Two photoresist layers are under the aluminum mirrors. The wafer is also covered with a third organic layer to provide mechanical damage protection and allow for particle cleaning after the wafer is singulated into individual devices. By removing these organics in a dry etch process, the stiction problems that characterize wet etch processes are avoided.

Analog Devices avoids release stiction by dividing its accelerometer process flow into several steps, each

Part G | 55.2

of which is well controlled with standard manufacturing equipment. After the microstructure is formed, a few small channels are etched in the sacrificial oxide and filled with photoresist. These photoresist pedestals hold the microstructures in place when the remaining sacrificial oxide is etched, rinsed, and dried. The pedestals are then removed in a dry etch process. *Core* et al. [55.44] and *Sulouff* [55.6] describe this technique in more detail.

Bosch avoids release stiction by holding accelerometer wafers above an HF solution. HF vapors from the solution react with the sacrificial oxide, converting it to gaseous SiF_4 and water. The wafers are heated to prevent water from condensing. This process is noted in *Offenberg* et al. [55.45]. *Anguita* et al. [55.46] describe a similar process.

55.2.4 Wafer Bonding

To date, the growth of MEMS wafer bonding has been driven by three factors:

- Sealing microstructures by bonding wafers together addresses two sources of yield loss in MEMS manufacturing: wafer singulation and particles.
- Wafer-level mounting can be a cost-effective way to control stress in products that are sensitive to variations in mount stress.
- Fabrication of 3-D microstructures from elements that are formed on multiple wafers.

Piezoresistive pressure sensor wafers have been anodically bonded to borosilicate glass wafers for about 25 years. The earliest products were manufactured using little more than a hot plate and a high-voltage power supply. The era of custom built manual equipment has passed, because both Electronic Visions and Karl Suss now offer automated production-worthy tools for anodic, glass frit, organic, and silicon-silicon wafer bonding. *Schmidt* [55.47] reviews wafer-wafer bonding processes, while *Mirza* [55.48, 49] addresses MEMS wafer-level bonding applications and equipment.

Anodic Bonding

Anodic bonding is commonly used to seal glass wafers to the back side of bulk micromachined pressure sensor wafers. The process requires that a silicon wafer be placed in intimate contact with a glass wafer that contains a mobile ion at elevated temperature. An applied electric field causes mobile ions (usually sodium) in the glass to move away from the silicon interface toward the cathode at the other side of the glass. Bound negative charges remain in the glass near the silicon interface. These charges produce an electric field that pulls the wafers together and drives oxygen across the interface to anodically oxidize the silicon surface. Hermetic seals are routinely produced between flat wafers when particles are rigorously excluded. Process conditions depend on the glass composition and thickness, but 500–1000 V at 400 °C with 10 min cycle times is typical. Borosilicate glasses with thermal expansion coefficients close to silicon are used to minimize stress. Anodic bonding promotes surface conformation so hermetic glass-silicon seals can be formed with surface grooves as deep as 50 nm [55.50]. This process has been demonstrated to work with many material combinations [55.51].

Glass Frit Bonding

Glass frit bonding reapplies techniques and materials developed for hybrid processes and cerdip sealing. It is often used to seal pressure sensor wafers to silicon substrate wafers. A second common application is to glass-frit seal bulk-micromachined silicon cap wafers over the microstructures in accelerometer wafers. In this process, fine glass powder is dispersed in an organic binder to form a paste. This paste is screen printed or stenciled in the desired pattern on a wafer. The wafer is then passed through a furnace (typically 430–500 °C). This causes the organics to burn off and the individual glass particles to coalesce into the desired pattern.

The wafer bonding process starts with a wafer prepared as described in the preceding paragraph. This wafer is aligned to a second wafer, where they are held in intimate contact and heated until the glass softens. Liquid glass readily wets the surface of the adjacent wafer when at least trace amounts of oxygen are present. The liquid-wetting mechanism produces hermetic seals even when the wafer surfaces are relatively rough. However, seal widths less than 100 μm are difficult to achieve. Fixtures that maintain wafer alignment through the bonding process are also required. Low-melting lead oxide and boron oxide glasses are commonly used in processes that typically run at 450 °C. Glass wet-out occurs quickly, but process cycle times are limited by the cooling rate (rapid cooling can cause stress gradients and cracking in the glass if proper design practices are not followed). Glass frit bonding to semiconductor wafers requires a good understanding of glass composition effects and process conditions [55.52].

Silicon–Silicon Wafer Bonding

Silicon-silicon bonding has been demonstrated in many IC and MEMS applications over the last 30 years. These

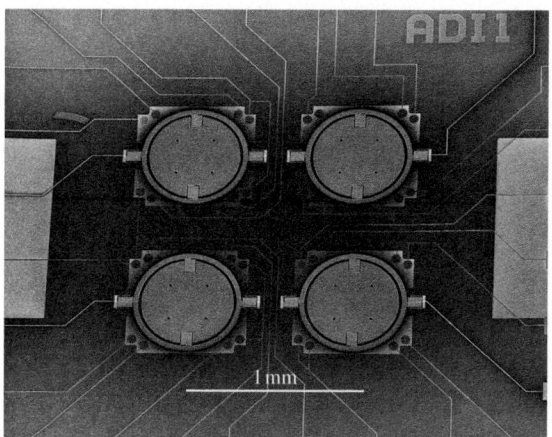

Fig. 55.14 Array of steerable mirrors designed for fiber optic network switching applications. The gimbal-mounted mirrors in this experimental product are electrostatically driven. Closed loop control based on capacitive sense electrodes ensures that the proper angle is maintained. High and low voltage circuits are integrated with mirrors on one chip. (Courtesy of Micromachined Products Division, Analog Devices, Inc.)

processes bring highly polished wafers into intimate contact and are often promoted by a thin hydrophilic oxide that is left on the surfaces by the aqueous cleaning process. Clean, well polished wafers will bond at room temperature. However, heating (800–1200 °C) increases bond strength by an order of magnitude. A fundamental difference between MEMS silicon–silicon bonding

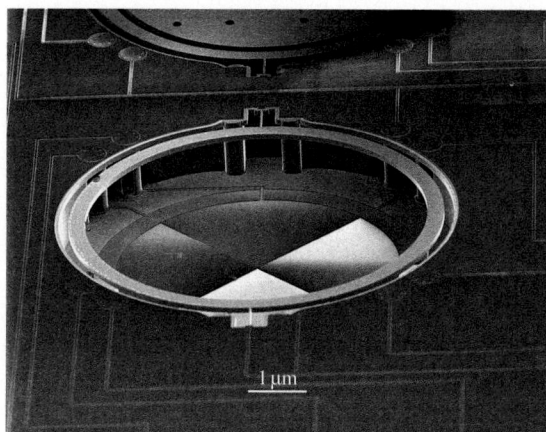

Fig. 55.15 View of the deep etched trenches, interlayer connections, and electrodes under a mirror of the SOI MEMS device illustrated in Fig. 55.14. (Courtesy of Micromachined Products Division, Analog Devices, Inc.)

and either anodic or glass frit bonding is that it is used to produce complex microstructures that cannot be made in one wafer. In contrast, most anodic and glass frit bonding applications address assembly and packaging issues.

Recent silicon-silicon advances have solved technical factors like stress control that had limited MEMS applications. For example, Analog Devices balances the stresses at oxide interfaces and the handle wafer to produce flat, three- and four-layer SOI wafers [55.53, 54]. The thickness of each layer is precisely controlled using grinding and CMP (chemical-mechanical polishing). Intermediate layers can be patterned, electrically isolated, electrically interconnected and locally etched to produce single-crystal silicon microstructures that are hundreds of microns thick on CMOS compatible wafers. With this new capability, several organizations are fabricating MEMS devices for biotech, automotive, and communications applications. One example is the array of gimbal-mounted mirrors shown in Figs. 55.14 and 55.15. These mirrors can be tilted in any axis to steer light between fibers in optical fiber bundles. The mirrors are driven by electrostatic force applied from electrodes that are patterned under the mirrors. Capacitance measurements based on a second set of electrodes are also part of a closed loop control system that maintains position control. Integration of high voltage drive electronics, low voltage control electronics, and MEMS mirrors would be impractical without this SOI capability and related MEMS processes.

55.2.5 Wafer Singulation

The cooling water and the particles generated in a standard IC diamond saw will destroy MEMS wafers unless some form of protection is used. Manufacturers solve this problem in several ways. Bosch and Motorola bulk micromachine cap wafers that are glass frit bonded to surface micromachined wafers. The caps protect the microstructures from water, particles, and mechanical damage so the wafers can be sawn with standard equipment.

The optical function of DMD chips required Texas Instruments to find a different solution for DMD wafers. Originally, they used a partial saw process on wafers that were protected with an organic film. This allowed wafer-level testing, but caused particle contamination and die loss when wafers were broken along the partial saw cuts. They now use a standard saw process, but do it before the mirrors are released. The wafers have the protective organic film over the mirrors, so normal cleaning processes are used. The singulated chips are

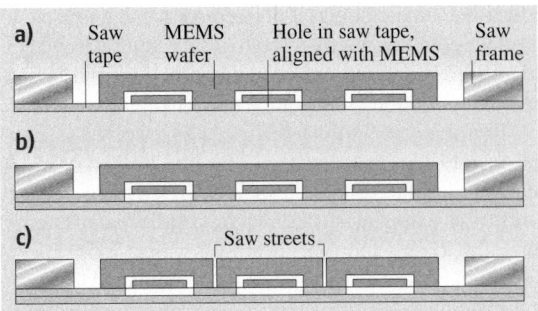

Fig. 55.16a–c Analog Devices "upside-down" saw process flow. (**a**) Saw tape is mounted on a saw frame. Holes are punched in the tape to match the microstructures on the wafer. The wafer is mounted-upside down on the saw tape and aligned so that the microstructures are in the holes. (**b**) A second layer of tape is placed over the first layer to form watertight pockets. (**c**) Wafer is aligned and sawn using standard equipment (adapted from[55.55])

mounted in ceramic packages before the organic layers are dry etched. This flow eliminates a primary source of particle contamination and problems associated with handling partially sawn wafers [55.7].

Analog Devices singulates uncapped MEMS wafers using an upside-down saw process and standard equipment and fixtures that are slightly modified (Fig. 55.16) (*Roberts* et al. [55.55]). A major attraction is that it allows wafer-level test and trim, so only chips that meet the product specification are assembled. It also avoids the added cost and yield loss associated with wafer capping. Although originally developed for accelerometers, this technology was extended to gyro wafers and to wafers that contain optical mirror arrays (see Figs. 55.13 and 55.14).

55.2.6 Particles

Particles cause yield loss in every product that has closely spaced movable microstructures. They also cause large area defects in anodic and silicon-silicon wafer bonded processes. Commercial MEMS suppliers use several solutions:

1. Cap the MEMS wafers before they leave the wafer fab clean area. Bosch and Motorola use this approach to produce accelerometer products (Fig. 55.12). It solves the wafer saw dilemma faced by most MEMS suppliers and allows the use of plastic packaging. Disadvantages include yield loss due to cap misalignment and stress, as well as the cost of the added

capping steps. Hermeticity testing is also difficult because capped devices require sensitivity levels beyond the capabilities of standard analytical test equipment.
   Capping protects MEMS from particle contamination during assembly, but does not prevent wafer fab contamination. A greater concern is that glass-frit seal processes and equipment do not usually meet fab cleanliness standards. Once capped, inspection is essentially impossible.

2. Analog Devices and Texas Instruments assemble their MEMS products in clean rooms. This option increases capital costs. TI invested in environmental and process control, rather than develop optical quality capping for the DMD. Analog Devices had solved the wafer saw dilemma, so environmental and process control was the low risk path for particle control.

Particle control is central to high yield and reliability. The first control level is to minimize particle contamination through proper handling procedures, equipment maintenance, environment controls, and process design. The DMD protective organic film illustrates how process design can be used to control particles. This film allows the wafer to be cleaned after wafer saw and die attach.

Even with the best of controls, some particle contamination occurs in every clean room. Electrical tests in the automated wafer- and package-level test programs are designed to detect particles in Analog Devices products. Texas Instruments also finds particles in DMD products by driving each mirror in an array to ensure proper operation. Such screens are good but imperfect. Attempts have been made to replace visual inspection with automated particle inspection based on pattern recognition software. To date, however, machine vision systems have not been able to match the human eye and mind aided by a high quality microscope.

55.2.7 Electrostatic Discharge and Static Charges

Susceptibility to electrostatic discharges (ESD) led semiconductor manufacturers to make ESD avoidance a central criterion in equipment and fixture design, as well as handling procedures. Control of ESD events on the manufacturing floor is only the first step. Analog Devices and Texas Instruments incorporate ESD protective circuitry into their MEMS products as a standard practice to avoid failures in their customer assembly lines, as well as in the final system.

Most MEMS products use electrostatic force to actuate or control suspended microstructures. This design characteristic introduces performance parasitics and failure mechanisms that are not encountered in normal IC devices. For example, charges that build up on a dielectric surface can deflect unshielded microstructures, even when they are several millimeters apart. Such effects are insidious, because closed loop control systems do not always correct for electrostatic-induced errors.

Section 55.3.1 discusses surface treatments designed to suppress stiction caused by surface forces. These treatments often produce a dielectric surface. If the dielectric properties of the treatment and the device design cause the surface to hold a charge, that anti-stiction coating may actually promote electrostatic stiction! Thus, any "solution" must be critically evaluated in order to identify and remove undesirable side effects.

55.2.8 Package and Test

IC packages are environmental (mechanical, chemical, electromagnetic, optical) barriers that protect the chips from surrounding media. Only power and electrical signals and inertial forces pass unimpeded, so most MEMS sensors and actuators require that this barrier be selectively penetrated. Package and test functions are unique to each company's design and MEMS product type. Standard semiconductor packaging equipment and processes are modified when possible. However, even the line of demarcation that separates wafer fab from packaging is unique to each product.

High volume MEMS testing is even more challenging than MEMS packaging. In essence, the task is to measure and trim the response of products mounted in non-standard packages to calibrated stimuli over a range of temperatures. Automated IC test equipment does not have the calibrated pressure, acceleration, or optical test functionality required for this task. *Maudie* et al. [55.56] detail the test system elements that are required to support volume production of MEMS products.

Pressure Sensors

Pressure sensors are extremely susceptible to stress, so die-attach stress variations broaden performance distributions. The uniformity produced by wafer-level bonding is a primary reason why these processes are used to make the first-level package in many pressure

sensor products. Soft silicone and fluorosilicone rubber products are often used to attach the sensor chip to the package to maximize isolation from mechanical stresses transmitted through the mount.

Most semiconductor products are assembled in plastic packages. The automated plastic presses transfer viscous epoxy compounds into multi-cavity molds at 175 °C and pressures near 6900 kPa (1000 psi). The hot liquid plastic creates high shear stress as it flows around devices in the mold cavities. Further stresses develop as the plastic hardens and cools. Even if this technology were adapted to create pressure ports in packages, the mold stresses would substantially affect pressure sensor performance.

Pre-molded plastic packages avoid the stress problem and incorporate pressure ports (Figs. 55.8 and 55.9). Polyphenylene sulfide (a high temperature thermoplastic) and medical-grade polysulfone are commonly used to make pre-molded pressure sensor packages.

The liquid or gas that is being measured can cause corrosion of interconnects and chip metallurgy, as well as parasitic leakage paths in the sensor. For that reason, a barrier such as silicone or fluorosilicone gel is often applied over the sensor. The soft gel transmits pressure with high fidelity over a wide temperature range and passivates the chip against many types of chemical attack. Parylene is also used for this purpose. This vapor-deposited organic coating is much stiffer than the gels, so thickness is typically controlled near 1 μm.

Petrovic et al. [55.57] tested the ability of fluorosilicone gels and several Parylene C thicknesses to protect powered pressure sensors against automotive and white goods benchmark liquids. One purpose of their study was to propose and demonstrate a formal media compatibility test protocol for pressure sensors similar to the IC industry standard tests. They found that the lifetime of sensors coated with both gel and parylene was considerably longer than sensors coated with either gel or parylene alone. *Petrovic* [55.58] later summarized the advantages and limitations of gels, parylene, and other techniques.

Figure 55.8 has examples of commercial plastic packaged pressure sensors. Pressure calibration equipment and software must be custom designed and built for these products. However, most pressure products are dimensioned to match IC packages in order to maximize compatibility with standard IC test handlers.

The harsh environment of some under-the-hood applications has been used to justify the high cost of metal cavity packages. For example, Bosch introduced a piezoresistive MAP sensor with signal conditioning that was hermetically sealed in an evacuated T08 header [55.59]. In-package trim was based on planar thyristors that were zapped as needed to bring the sensor within spec. The design avoided chip metallurgy corrosion and surface electrical parasitics by using the back surface of the silicon diaphragm as the fluid interface.

Much more expensive packaging is used in the pressure and differential pressure transmitters that are designed for industrial process control. These products seal the sensor in silicone oil behind thin metal diaphragms. Such products can measure pressure differences of 20 kPa superimposed on a pressure of 20 000 kPa (3000 psi) over wide temperature ranges with high accuracy and stability. Soft die attach materials decouple the sensor from thermal stresses that arise when they are mounted in metal housings. Note, however, that silicone die attach materials cannot be used if the transmitter is filled with silicone oil. Fluorosilicones are one alternative. Piezoresistive sensors have been used in pressure and differential pressure transmitters since the 1980s. *Fung* et al. [55.60] describe a relatively new version. This product uses a piezoresistive polysilicon sensor to measure both absolute and differential pressure on the same chip. See *Chau* et al. [55.61] for early work on a multirange version of this technology.

Image Projection

The Texas Instruments DMD poses unusual challenges because the package lid must be an optical quality glass with anti-reflection coatings to improve optical performance and reduce heat load. The glass lid has opaque borders around the image area to minimize stray light effects and create a sharp edge on the projected image. It is fused to Kovar frames to produce lid sub-assemblies. The package base is a multilayer alumina substrate with co-fired tungsten to provide the electrical interconnects. The package sidewalls are formed by brazing a Kovar seal ring to the substrate. Heat dissipation through this substrate is a significant consideration because the service life of DMD mirror hinges (Fig. 55.6) is largely determined by operating temperature [55.62]. Excessive high temperature creep results in "hinge memory" and would cause a gradual drift in mirror orientation.

The DMD chips are die attached in the package cavity before the mirrors are released. A dry etch process is used to remove the organic sacrificial layers. The die are then passivated and tested. Getter strips are attached on the inside surface of the glass at the sides to control vapor composition in the package cavity before the lid is aligned and seam sealed to the package seal ring. Image quality requirements place stringent requirements on handling, alignment, and spacing tolerances. Thermal stresses also arise when dissimilar materials are joined in high temperature processes to produce the final hermetic package. The DMD packages (Fig. 55.7) and the related processes are discussed further in *O'Connor* [55.63], *Mignardi* et al. [55.7], *Bang* et al. [55.64], and *Poradish* et al. [55.65].

DMD mirror release does not occur until after die attach. This has advantages with respect to wafer saw and particle suppression, but it sacrifices the economic attraction of wafer-level testing. There is no commercial test system capable of combining CMOS electrical testing with 100% testing of optical mirrors. Therefore, a custom electro-optic test system was built around an x-y-θ translation stage with a CCD camera, light source, control hardware and software, and test programs [55.7]. A characteristic of the papers published by the DMD group is their effective use of this test system to examine problems and statistically validate the solutions.

Assembly of the DMD on the projector electronics board requires alignment with the system optics, in addition to electrical connections. Initially, the DMD was held with a plastic clamp and electrically connected through elastomer pads that had alternating layers of conductive and nonconductive material. This allowed easy replacement if a DMD was damaged during assembly. However, the impedance of this connector system was too high for new, higher speed products. Electrical intermittents were also observed. Therefore, it was replaced [55.66] by a grid of c-shaped springs (cLGA[TM], Intercon Systems, Inc., Harrisburg, PA).

The DMD has about 40% of the business and entertainment image projection market [55.67]. Products based on liquid crystal technology serve the balance of the market. Market share is largely driven by price, so considerable effort has been placed on reducing the cost of DMD package and board-level assembly. *Migl* [55.66] reports on the assembly benefits achieved by replacing the epoxied heat sink with a mechanically attached heat sink. *Jacobs* et al. [55.67] describes the effort to replace the seam sealed window mount with a lower cost epoxy-bonded design. In theory, such a bond is not hermetic. However, the team realized that proper material selection, design, analysis, and use of moisture getters would produce a low humidity package through the product life. This change was not released to production. If it is implemented, success will require that adhesive bond integrity also be maintained, because the adhesive joint sustains thermal expansion and mechanical clamp stress cycling each time the projector is used.

Accelerometers and Gyros

The Bosch and Motorola capping processes are a first-level package that is applied at the end of wafer fab. After capping, the wafers can be sawn using standard equipment and do not require clean room assembly conditions. In principle, capped devices are compatible with standard plastic packaging. However, molding stresses can be a serious problem. To minimize stress, capped sensor chips are mounted on lead frames with a soft elastomer and often coated with a silicone gel to isolate the sensors from the package stress. Motorola products are calibrated by burning EPROMS after molding. Their early air bag accelerometers were molded in DIP and SIP packages for assembly on through-hole circuit boards. The shift to smaller surface mount SOIC packages reduces the quantity of plastic, so package stresses are reduced. To maintain compatibility with standard tooling in small thin plastic packages, either the cap or the MEMS wafers must be backlapped.

Analog Devices uses a near-standard saw process (Sect. 55.2.5 and Fig. 55.16), so capping is not required if hermetic cavity packages are used. Such packages are standard, but more expensive than molded plastic. A major attraction of cavity packages is that they eliminate plastic package stress (see page 1765), because the only mechanical connection between the chip and the package is through the die attach and bond wires. As a result, ADXL accelerometers and ADXRS gyro products are fully tested and trimmed on automated systems before the wafers are sawn. A lot-tracking system transfers this data to automated die attach systems. These systems are programmed to pick only chips that meet performance specifications from the wafer saw film frames. Thus, reject die are not assembled into packages. After packaging, devices are tested to ensure conformance to specification. However, package-level trim is not required, because the assembly process and cavity packages do not appreciably shift device parametrics.

The evolution of ADXL cavity packages is illustrated in Fig. 55.3. Initial products were packaged in TO-100 metal packages. This seam-sealed package is useful for development purposes. However, it is expensive and incompatible with the automated equipment used to assemble electronic circuit boards. Therefore, the early air bag sensors were soon switched to cerdips. Cerdips are made from low cost, molded, ceramic bases that have glass seal surfaces. The high process temperature requires use of a silver-glass die attach product. The full assembly flow includes two or three furnace passes near 450 °C to produce hermetic cavity packages.

IC products have been packaged in cerdips for decades, but through-hole circuit boards are seldom used today. A simple change in lead forming allows cerdips to be fully compatible with standard surface mount boards ("cerpacs"), so cerdips and cerpacs are assembled on the same equipment. Most of the ADXL products shipped in the late 1990s were packaged in cerpacs. However, small LCCs have become the package of choice for one- and two-axis accelerometers. The LCC (Leadless Chip Carrier) uses solder-sealed ceramic bases and metal lids. Organic die attach materials are practical, because the furnace gas is nitrogen, rather than air, and temperatures are about a hundred degrees lower than cerpac furnace temperatures. ADXRS gyros are assembled in a solder-sealed 7 mm × 7 mm ceramic ball grid array package (Fig. 55.13).

The fundamental message of this section is that IC package technology – both plastic and hermetic – is evolving very rapidly. Cost-competitive suppliers must remain cognizant of these trends in order to use them to best advantage. MEMS products that are packaged in a way that is not compatible with standard IC equipment and interfacing standards bear a significant cost premium. High package cost seriously limits market size because the price-volume curve is steep for most products. The package size trend is also critical to new market penetration. In summary, customer interface and new application requirements will continue to drive down both the cost and the size of most MEMS products.

The ADXL and ADXRS automated testers and handlers measure multiaxial linear and rotational acceleration. Analog Devices has a division that designs and manufactures automated test systems (many of the corporation's IC products have unique test requirements), so it avoided some of the problems associated with the procurement of custom-automated test equipment. MEMS suppliers have relied on custom-designed test equipment, but this situation is gradually changing. For example, Multitest GmbH now makes handlers with integrated shakers that are specifically designed for accelerometer testing.

55.2.9 Quality Systems

Quality systems are an intrinsic part of stable MEMS production. Management must drive a systemic approach to quality and set continuous improvement as a high priority goal.

Several quality systems define continuous improvement methodologies. In general, they formalize the process used to minimize defects and variations in products and in the processes used to manufacture them. Automotive supply companies must implement quality systems that meet QS-9000 [55.68] before product volume ramps up and continue them through the product life cycle. Four elements that are worthy of note:

1. *Failure Mode Effect Analysis* (*FMEA*). Early in the development phase, a team with representatives from several disciplines reviews the design (or process) in order to identify possible causes of failure. Each potential failure mode is given three numerical rankings. One ranking represents the likelihood of an occurrence, while the others rank the severity of that result and detectivity. These numerical scores are combined for each potential failure mode in order to identify the issues that merit the most attention before they become problems. FMEA spreadsheets are periodically updated as the product or process moves into production. More targeted versions are used in products or processes that are in stable production.
2. *Process Control*. Every significant process must have a "short loop" monitor to ensure stability. This measurement is tracked and statistically analyzed with respect to the control limits to identify changes before the process strays beyond the control limits.
3. *Review Boards*. Every change, unusual occurrence, and proposed solution is assessed by boards that meet on a regular basis to ensure that it does not put product quality at risk. Affected production material is put aside until the appropriate board approves its release. This may appear to be expensive and bureaucratic. However, the cost of a scraped wafer lot is insignificant when compared to the cost of a field replacement program.
4. *Procedures and Specifications*. Each step must be fully specified and identified in the process flow, along with the appropriate metrics for each lot. The compilation of this data in a retrievable form is an essential part of every continuous improvement program. By combining these product and process databases with lot tracking software, Pareto charts linking yield loss and test failures to process variations can be identified and eliminated. Experience has shown that unexpected second-order effects are present in every production line – but data-driven decisions on indirect effects cannot be made on small test populations.

Implementation of a continuous improvement program requires well-informed failure analysis teams, methodologies, and programs. Such analyses start with gathering facts and making relatively simple tests. Often this is sufficient. However, MEMS products are susceptible to uncommon failure modes, so *Walraven* et al. [55.69] gathered examples of more powerful analytical techniques and showed how they are applied to MEMS devices.

Each supplier uses the information and insights gathered from product performance, as well as control and yield data at different points in the process flow to refine their operations. *Douglass* [55.70], for example, outlined the yield loss and failure mechanisms observed in early Texas Instruments DMD products. Concerns like hinge memory, hinge fatigue, particles, stiction, and environmental robustness were each addressed and mitigated by focused teams. For example, *Mignardi* et al. [55.7] describe the partial wafer saw process used in the original DMD manufacturing flow process. Breaking these delicate wafers into individual product chips generated particles that caused yield loss and were potential sources of field failures. A new flow based on a standard full saw process eliminates this particle source. The result was an increase in both yield and reliability. Hinge fatigue characterization showed that bulk metal fatigue models do not properly describe thin film behavior. The hinge memory effort required an understanding of how thin film metal creep is affected by alloy composition and the environment [55.62]. Stiction control involves surface passivation, spring design, mirror dynamics, and moisture level [55.70, 71]. These publications give an insight into the quantities of data and the time required to bring robust products to market. They also illustrate how data-based evaluations can uncover unexpected effects like the acceleration of hinge creep by adsorbed moisture [55.71].

The second continuous improvement example applies to MEMS integration. High yield loss is perceived to be a serious risk when electronics and MEMS are integrated on one chip. Indeed, the initial yields on the Analog Devices integrated MEMS accelerometers were not impressive. However, it increased each year due to the work of many mission-oriented teams. Yield has reached defect density limited levels because the teams eliminated all significant failure mechanisms.

Coincident with the annual yield increases were reductions in customer failure rates. Failure rates in Analog's accelerometer products are in the low single digit ppm range. Quality does pay.

55.3 Stable Field Performance

Some topics discussed in Sect. 55.2 like particles and ESD are equally relevant to long-term stability. Most microstructure products have elements that are in close proximity, so stability is also affected by surface characteristics and mechanical shock.

55.3.1 Surface Passivation

Early, 1970s-era piezoresistive pressure sensors were not passivated. Like early ICs, these wafers only had oxide over the piezoresistors. Performance was inconsistent and drifted over time, because surface interactions with moisture and other atmospheric gases created a variety of shunts, parasitics, and charging issues.

Electrical Surface Passivation
The introduction of silicon nitride passivation and conductive field plates led to stable products that were hitherto unobtainable. The ability to achieve measurement stability was absolutely critical to growth of the MEMS pressure sensor industry.

Aside from pressure sensors, few MEMS products are passivated. Indeed, standard plasma nitride processes are usually impractical, because they produce dielectric coatings that support static charges. Such charges can cause electrical drift, or be the source of electrostatic forces that cause stiction.

There have been published reports attributing instability to the lack of passivation on microstructures. For example, Analog Devices implemented a special process in ADXL50 accelerometers before the product was released in order to suppress high temperature electrical drift [55.72]. This was not a moisture effect, because it occurred in dry nitrogen packages. Later ADXL designs eliminated the root cause.

Scientists at Lucent Technologies [55.73] observed anodic oxidation in polysilicon electrodes used to electrostatically drive mirrors in optical cross-connect products. If allowed to occur, this corrosion mechanism would cause the mirror position to drift and the product to eventually fail. The study concluded that, within the limits of the test, maintaining low moisture in these optical packages eliminates anodic oxidation.

The preceding paragraphs suggest that lack of MEMS passivation can affect long-term product stability. This issue is very design-related and can be driven by factors other than humidity. Even when low package humidity is determined to be adequate, design reviews should consider how normal manufacturing variations, outgassing and diffusion over the product life affect the moisture level in the gas adjacent to the microstructures.

Mechanical – Stiction and Wear
In-use stiction is difficult to predict and may not become an issue until manufacturing volumes increase, or when endusers handle the part in ways that are not anticipated [55.74]. The resulting product liability and field replacement programs carry substantial financial costs and have caused MEMS suppliers to withdraw from the market. End users have little tolerance for failure. Since in-use stiction is a failure mechanism in MEMS devices, a company that solves this problem creates an effective market barrier against competitors.

The scale of this problem varies between product areas. For example, the Texas Instruments DMD has moving mirrors that are designed to touch down on a substrate and later release. This product must address both stiction and wear. In contrast, the accelerometer products from Analog Devices, Bosch, and Motorola have proof masses that are suspended on compliant springs. By design, an automobile collision only displaces the proof mass by a few percent of the gap that separates it from adjacent surfaces. It is difficult to generate a shock wave in an air bag module that is sufficient to cause these proof masses to contact adjacent surfaces (acceleration levels of several thousand g are required). However, handling of the discrete packaged parts during test, shipment, and module assembly frequently cause shock events that bring MEMS surfaces into contact. For example, *Li* et al. [55.75] used both math modeling and tests to show that dropping a packaged MEMS accelerometer from the height of a table top generates several tens of thousands of g in the device when it lands on the floor. Thus, noncontacting MEMS products must be designed to withstand at least transient stiction.

Design. When shock, vibration, or functional operation causes MEMS elements to touch, the mounting springs are designed to pull them apart (Fig. 55.10). High stiffness springs are desirable to suppress stiction, but low stiffness springs increase measurement sensitivity and signal-to-noise ratio. Thus stiction is a fundamental MEMS design constraint that often limits product performance. Each manufacturer has proprietary design practices to address this performance versus reliability trade-off. In general, designers suppress stiction by minimizing the sources and applying supplementary techniques to ensure recovery when contact occurs:

1. Thorough analysis and testing to move harmful resonances beyond the range where they might be excited. Note that MEMS component-level analysis is insufficient – this is a system issue based on the packaged part as it is handled through manufacturing, system assembly, and end use.
2. Surface modification (discussed in the next section).
3. Minimizing contact area by integrating bumps and "stoppers" into the design [55.76].
4. Elimination of dielectric surfaces that may accumulate surface charges and result in electrostatic attraction.
5. Release of DMD mirrors from the touchdown position is assisted by pulsing the reset voltage to excite a mirror resonant frequency [55.70, 77].
6. Use of gas-filled packages to reduce contact velocity, as discussed later in this chapter.
7. ESD-protected designs that prevent voltage transients that may cause electrostatic attraction.

The only product in significant production that contains MEMS elements that are designed to touch is the Texas Instruments DMD. The proprietary aluminum alloy springs used in the mirror contact points touch and release millions of times during the product life. Considerable development was required to overcome metal creep that occurred in early designs.

Surface Modification. In addition to the touchdown springs, early DMD die were treated with perfluorodecanoic acid vapor before the glass lid was sealed onto the package. This treatment [55.78] created a low energy monolayer coating that suppressed initial stiction. However, improvements were required to achieve longer wear life, higher production rates, and lower contamination levels. A new process [55.79] places capsules inside DMD packages. After cleaning, the packages are immediately sealed. An oven bake releases anti-stiction vapor from the capsules to create a monolayer organic coating on surfaces inside these sealed packages. Excess anti-stiction material in the capsules maintains a low vapor pressure in the package cavities throughout the product life. Therefore, wear damage at the mirror contact springs is continuously repaired to prevent stiction failures before they occur.

Maintaining a low level of organic vapor inside the DMD reapplies a concept used with other organics to prevent corrosion of electronic metallurgy and contact fretting of separable connectors (see, for example,

[55.80, 81]). These treatments have been shown to add years to the useful life of electronic systems that are in enclosures but also exposed to aggressive environments.

Stiction can be minimized by design and process spin-offs – stiff springs, control of contact area, etch residues, etc. Some suppliers design within these limits; others seek to modify the MEMS surfaces. The requirements of any surface modification technology are quite substantial. In addition to normal manufacturing requirements (stable, scalable process and materials), the process must be conformal and must uniformly treat all areas, including microstructure bottom and side surfaces. If a conductive material is used, it must somehow be patterned in order to avoid electrical shorting. If a dielectric surface is formed, it must be extremely thin because thick dielectrics support surface charges. Such charges cause performance shifts and induce electrostatic stiction.

Since devices with movable microstructures are particularly susceptible to stiction, most suppliers who develop production-worthy surface modification processes consider this information a trade secret. Analog Devices chose to patent portions of its technology. The following paragraphs give an overview of the published Analog Devices surface treatment technologies. *Martin* [55.72] describes the evolution of these technologies from 1992 to 1999. Further information is contained in *Martin* et al. [55.82] and *Martin* [55.83].

The earliest of the published Analog surface treatment technologies was implemented after stiction was observed during automated testing and module assembly of air bag sensors that were packaged in cerdips. Cerdip assembly typically includes at least two furnace processes in air at 440–500 °C, so they seldom contain organic materials. The high temperature furnace air thermo-oxidizes adsorbed materials and removes them from the MEMS surface. The resulting surfaces are extremely clean. Unfortunately, clean inorganics have high surface energies. Therefore, adjacent microstructures readily stick together if shock and vibration during handling cause them to touch. Cerdips (and the surface mount cerpacs) have very low moisture levels (Mil specs allow up to 5000 ppm moisture, but < 100 ppm moisture is common in a well controlled manufacturing line). Deliberately raising the moisture level was found to be a potential solution because, within limits, moisture adsorption reduces surface energy. However, the solution adopted by Analog Devices in the mid-1990s was the anti-stiction agent (ASA) process. This process involved dispensing a controlled amount of a pure siloxane liquid (ASA) into each package immediately before sealing. ASA selection was based on thermo-oxidative stability, low volatility, purity, and the requirement that it be a liquid at the dispensing temperature. As packages heat in the furnace, adsorbed material on the microstructures is removed, leaving chemically active, high energy surfaces. Further heating volatilizes the ASA. Before escaping into the furnace, the ASA vapor surrounds the reactive MEMS surfaces and chemically bonds to them. The result is infusion of organic groups into the native oxide surface. Organic-rich surfaces are unreactive and have low surface energy. Thus the treatment is self-limiting after the first vapor molecules react. Obviously, ASA volatilization must be complete before the glass softens and seals the package cavity. Some ASA degrades in this process. However, silicone thermo-oxidation simply adds a few angstroms of silicon oxide to the native oxide, so the ASA process, by design, is non-contaminating.

Although variations and refinements of the ASA process were developed, this technology required that each individual package be treated. Automated dispense equipment made ASA acceptable when product volumes were moderate. However, as production approached several million per month, it became evident that a wafer-level anti-stiction process was desirable.

Assessment of candidate processes and materials led to a vapor treatment based on a custom-synthesized solid polymeric siloxane. This polymer is volatilized in a standard CVD furnace that treats up to a hundred wafers at a time. Repeatability is extraordinary because thickness variation within a furnace run, or from run to run, is only about one angstrom.

Package Environment Effects. The viscosity and density of the gas that contacts MEMS resonators limit the device Q, so resonant devices are often designed to operate in vacuum. Unfortunately, this makes them susceptible to handling damage. Viscous fill gases reduce destructive excursions of high-Q microstructures when they are mechanically shocked. They also provide squeeze film damping to cushion the contact event as surfaces approach each other. None of the high volume MEMS products on the market are evacuated. Presumably, this is due to the high cost of vacuum packaging and the susceptibility to handling damage.

Many investigators believe that suppression of in-use stiction requires low humidity in the package. This view is not surprising because most MEMS research groups have observed stiction caused by aqueous surface tension during microstructure release. However, in-use stiction and wear test results published by several investigators lead to a different conclusion [55.82, 84–89]. In general, they find high stiction in both dry and moist environments. However, stiction is substantially suppressed at intermediate humidity levels (approximately 15–40 % relative humidity at room temperature). This is equivalent to a moisture level of 3000–10 000 ppm. Note that the limits cited by different authors vary considerably due to the use of different materials and test conditions. Stiction at low humidity is probably caused by high surface energy on inorganic oxides that do not have adsorbed surface films. The low stiction and wear rates observed at intermediate humidity are attributed to the passivating and lubricating effects of adsorbed water. Capillary forces caused by the high surface tension of water become dominant at high humidity.

55.3.2 System Interface

MEMS products are mounted in systems that have much greater mass than the device itself. For this reason, it is difficult to transmit a mechanical shock wave to a MEMS sensor that is sufficient to cause failure after it is installed in a module. However, failures are possible when the devices are handled during module assembly. Improper functional response is also possible if the module mount system has a mechanical resonance that amplifies or reduces mechanical transmission under different conditions. This is a particular concern in automotive safety applications because every automobile platform has a different mechanical signature. Proper module mount design is required to ensure that the impact signal from an automobile crash is properly transmitted to air bag modules.

Suppliers of large IC die closely examine board-level stresses in order to minimize solder joint fatigue. Such stresses are more serious in MEMS products because the device, as well as the package are susceptible to mounting conditions. Furthermore, customer processes impose these stresses, so a supplier may have little information with respect to their origin or existence. One solution is to define and qualify a product-specific mount system, as Texas Instruments has done with the DMD. The ideal, however, is to devise products that are compatible with standard board footprint and assembly processes. This challenge became more difficult with the shift from through-hole to surface mount technology, because the mechanical isolation provided by the package pins is no longer present. Small package size, in-package, and within-chip isolation techniques, as well as thorough package-board stress analysis are used to address this concern.

References

55.1   J. B. Angell, S. C. Terry, P. W. Barth: Silicon micromechanical devices, Sci. Am. **248**, 44–50 (1983)

55.2   Texas Instruments, Inc.: Texas Instruments DLP™ Products Announce 5 Million Units Shipped. Press release, Dec. 13, 2004, www.dlp.com/about-dlp/about-dlp-press-release.asp?id=1222&bhcp=1

55.3   T. A. Core, W. K. Tsang, S. J. Sherman: Fabrication technology for an integrated surface-micromachined sensor, Solid State Technol. **36**(10), 39–47 (1993)

55.4   W. Kuehnel, S. Sherman: A surface micromachined silicon accelerometer with on–chip detection circuitry, Sens. Actuators A **45**, 7–16 (1994)

55.5   K. H.-L. Chau, R. E. Sulouff Jr.: Technology for the high-volume manufacturing of integrated surface-micromachined accelerometer products, Microelectron. J. **29**, 579–586 (1998)

55.6   B. Sulouff: Integrated surface micromachined technology. In: *Sensors for Automotive Technology*, Sensors Applications, Vol. 4, ed. by J. Marek, H.-P. Trah, Y. Suzuki, I. Yokomori (Wiley-VCH, Weinheim 2003) Chap. 5.2

55.7   M. A. Mignardi, R. O. Gale, D. J. Dawson, J. C. Smith: The digital micromirror device – A micro-optical electromechanical device of display applications. In: *MEMS and MOEMS Technology and Applications*, ed. by P. Rai-Choudhury (SPIE, Bellingham 2000) Chap. 4

55.8   K. Nunan, G. Ready, J. Sledziewski: LPCVD & PECVD operations designed for iMEMS sensor devices, Vac. Technol. Coat. **2**(1), 26–37 (2001)

55.9   M. Williams, J. Smith, J. Mark, G. Matamis, B. Gogoi: Development of low stress, silicon-rich nitride film for micromachined sensor applications, *Micromachining and Microfabrication Process Technology VI*, Proc. SPIE **4174**, ed. by J. Karam, J. Yasaitis, Proc. SPIE **4174** (2000) 436–442

55.10   Z. Zhang, K. Eskes: Elimination of wafer edge die yield loss for accelerometers, *Micromachining and Microfabrication Process Technology VI*, ed. by J. Karam, J. Yasaitis, Proc. SPIE **4174** (2000) 477–484

55.11   G. Bitko, A. C. McNeil, D. J. Monk: Effect of inorganic thin film material processing and properties on stress in silicon piezoresistive pressure sensors, Proc. Mat. Res. Soc. Symp. **444**, 221–226 (1997)

55.12   A. Hein, S. Finkbeiner, J. Marek, E. Obermeier: Material related effects on wet chemical micromachining of smart MEMS devices, *Micromachined Devices and Components V*, ed. by P. French, E. Peeters, Proc. SPIE **3876** (1999) 29–36

55.13   B. Sulouff: Commercialization of MEMS automotive accelerometers, 7th International Conference on the Commercialization of Micro and Nano Systems (COMS), Ypsilanti 2002 (MANCEF, Albuquerque 2002) 267–270

55.14   H. Weinberg: MEMS sensors are driving the automotive industry, Sensors **19**(2), 36–41 (2002)

55.15   J. Marek, M. Illing: Microsystems for the automotive industry, Electron Devices Meeting, IEDM Technical Digest International, San Francisco 2000 (IEEE, New York 2000) 3–8

55.16   D. S. Eddy, D. R. Sparks: Application of MEMS technology in automotive sensors and actuators, Proc. IEEE **86**(8), 1747–1755 (1998)

55.17   R. Verma, I. Baskett, B. Loggins: Micromachined electromechanical sensors for automotive applications, SAE Special Pub. **1312**, 55–59 (1998)

55.18   P. Adrian: Sensor Companies Can Use Foundries to Efficiently Boost Their Ability to Serve High-Volume Markets, Sensor Business Digest(Oct. 2002) www.sensorsmag.com / resources / businessdigest / sbd1002.shtml

55.19   ISO16949:2002: *Quality Management Systems – Particular Requirements for the Application of ISO 9001:2000 for Automotive Production and Relevant Service Part Organizations* (International Organization for Standardization, Geneva 2002)

55.20   *Quality Systems Requirements QS-9000*, 3rd edn. (Automotive Industry Action Group of the American Society for Quality, Milwaukee 1998)

55.21   R. Maboudian, R. T. Howe: Critical review: Adhesion in surface micromechanical structures, J. Vac. Sci. Technol. B **15**(1), 1–20 (1997)

55.22   U. Srinivasan, M. R. Houston, R. T. Howe, R. Maboudian: Alkyltrichlorosilane-based self-assembled monolayer films for stiction reduction in silicon micromachines, J. Microelectromech. Syst. **7**(2), 252–260 (1998)

55.23   B. H. Kim, T. D. Chung, C. H. Oh, K. Chun: A new organic modifier for anti-stiction, J. Microelectromech. Syst. **10**(1), 33–40 (2001)

55.24   S. Pamidighantam, W. Laureyn, A. Salah, A. Verbist, H. Tilmans: A novel process for fabricating slender and compliant suspended poly-Si micromechanical structures with sub-micron gap spacing, 15th IEEE 2002 Micro Electro Mechanical Systems (MEMS) Conf. (IEEE, New York 2002) 661–664

55.25   B. C. Bunker, R. W. Carpick, R. A. Assink, M. L. Thomas, M. G. Hankins, J. A. Voigt, D. Sipola, M. P. de Boer, G. L. Gulley: Impact of solution agglomeration on the deposition of self-assembled monolayers, Langmuir **16**, 7742–7751 (2000)

55.26   Y. Jun, V. Boiadjiev, R. Major, X.-Y. Zhu: Novel chemistry for surface engineering in MEMS, *Materials and Devices Characterization in Micromachining III*, ed. by Y. Vladimirsky, P. Coane, Proc. SPIE **4175** (2000) 113–120

55.27 R. Maboudian, W.R. Ashurst, C. Carraro: Self-assembled monolayers as anti-stiction coatings for MEMS: Characteristics and recent developments, Sens. Actuators A **82**(1), 219–223 (2000)

55.28 W.R. Ashurst, C. Yau, C. Carraro, R. Maboudian, M.T. Dugger: Dichlorodimethylsilane as an anti-stiction monolayer for MEMS: A comparison to the octadecyltrichlorosilane self-assembled monolayer, J. Microelectromech. Syst. **10**, 41–49 (2001)

55.29 W.R. Ashurst, C. Yau, C. Carraro, C. Lee, G.J. Kluth, R.T. Howe, R. Maboudian: Alkene based monolayer films as anti-stiction coatings for polysilicon MEMS, Sens. Actuators A **91**(3), 239–248 (2001)

55.30 Freescale Semiconductor: Sensor Solutions (2005), www.freescale.com / files% hack / sensors / doc / fact_sheet/SNSRSOLUTNTMFS.pdf

55.31 Nexus Task Force Report: *Market Analysis for Microsystems 1996-2002* (NEXUS, Grenoble 1998) www.nexus-mems.com

55.32 G. Dahlmann, G. Holzer, S. Hering, U. Schwarz: A Modular CMOS Foundry Process for Integrated Piezoresistive Pressure Sensors. In: *Advanced Microsystems for Automotive Applications 2005*, ed. by J. Valldorf, W. Gessner (Springer, Berlin 2005) pp. 413–424

55.33 T. Maudie, J. Wertz: Pressure sensor performance and reliability, IEEE Industry Appl. Mag. **3**(3), 37–43 (1997)

55.34 Lj. Ristic, R. Gutteridge, B. Dunn, D. Mietus, P. Bennett: Surface micromachined polysilicon accelerometer, IEEE 1992 Solid State Sensor and Actuator Workshop, IEEE 5th Technical Digest, Hilton Head 1992 (IEEE, New York 1992) 118–121

55.35 M. Furtsch, M. Offenberg, H. Munzel, J.R. Morante: Influence of anneals in oxygen ambient on stress of thick polysilicon layers, Sens. Actuators **76**, 335–342 (1999)

55.36 P. Lange, M. Kirsten, W. Riethmuller, B. Wenk, G. Zwicker, J.R. Morante, F. Ericson, J.A. Schweitz: Thick polycristalline silicon for surface microme-chanical applications: Deposition, structuring and mechanical characterization, Proc. 8th International Conf. on Solid State Sensors and Actuators and Eurosensors IX, Transducers '95, Vol.1 (IEEE, New York 1995) 202–205

55.37 F. Laermer, A. Schilp: Method of anisotropically etching silicon, US Patent 5 501 893 (1996)

55.38 M. Offenberg, H. Munzel, D. Schubert, O. Schatz, F. Laermer, E. Muller, B. Maihofer, J. Marek: Acceleration sensor in surface micromachining for airbag applications with high signal/noise ratio, SAE Special Pub. **1133**, 35–41 (1996)

55.39 F. Laermer, A. Schilp, K. Funk, M. Offenberg: Bosch deep silicon etching: Improving uniformity and etch rate for advanced MEMS applications, Proc 12th International Conf. on Micro Electro Mechanical Systems MEMS (IEEE, New York 1999) 211–216

55.40 J.R. Martin, C.M. Roberts Jr.: Package for sealing an integrated circuit die, US Patent 6 323 550 (2001)

55.41 J.A. Geen, S.J. Sherman, J.F. Chang, S.R. Lewis: Single chip surface micromachined integrated gyroscope with 50 deg/hour allan deviation, IEEE J. Solid-State Circuits **37**(12), 1860–1866 (2002)

55.42 K. Funk, H. Emmerich, A. Schilp, M. Offenberg, R. Neul, F. Larmer: A surface micromachined silicon gyroscope using a thick polysilicon layer, Proc. 12th International Conf. on Micro Electro Mechanical Systems MEMS (IEEE, New York 1999) 57–60

55.43 M. Lutz, W. Golderer, J. Gerstenmeier, J. Marek, B. Maihofer, S. Mahler, H. Munzel, U. Bischof: A precision yaw rate sensor in silicon micromachin-ing, 11th International Conf. on Solid State Sensors and Actuators, Transducers '97 (IEEE, New York 1997) 847–850

55.44 T.A. Core, R.T. Howe: Method for fabricating mi-crostructures, US Patent 5314572 (1994)

55.45 M. Offenberg, F. Laermer, B. Elsner, H. Munzel, W. Reithmuller: Novel process for a monolithic integrated accelerometer, Proc. 8th International Conf. on Solid State Sensors and Actuators and Eurosensors IX, Transducers '95 (IEEE, New York 1995) 589–592

55.46 J. Anguita, F. Briones: HF/H$_2$0 vapor etching of SiO$_2$ sacrificial layer for large-area surface-micromachined membranes, Sens. Actuators A **64**, 247–251 (1998)

55.47 M.A. Schmidt: Wafer-to-wafer bonding for mi-crostructure formation, Proc. IEEE **86**(8), 1575–1585 (1998)

55.48 A.R. Mirza: Wafer-level bonding technology for MEMS, Proc. 7th Intersociety Conf. on Thermal and Thermomechanical Phenomena in Electronic Sys-tems, ITHERM 2000, Vol.1 (IEEE, New York 2000) 113–119

55.49 A.R. Mirza: One micron precision, wafer-level aligned bonding for interconnect, MEMS and packaging applications, Proc. 50th Electronic Com-ponents & Technology Conf. (IEEE, New York 2000) 676–680

55.50 S. Mack, H. Baumann, U. Goesele: Gas tightness of cavities sealed by silicon wafer bonding, Proc. 10th Annual International Workshop on Micro Elec-tro Mechanical Systems, MEMS; IEEE Micro Electro Mechanical Systems (MEMS) (IEEE, New York 1997) 488–493

55.51 G. Wallis: Field assisted glass sealing, SAE Auto-motive Engineering Congress, Detroit 1971 (Society of Automotive Engineers, New York 1971) Paper 71023

55.52 S.A. Audet, K.M. Edenfeld: Integrated sensor wafer-level packaging, Proc. International Conf. on Solid State Sensors and Actuators, Trans-ducers '97, Vol.1 (IEEE, New York 1997) 287–289

Part G | 55

55.53 C. Gormley, A. Boyle, V. Srigengan, S. Black-stone: HARM processing techniques for MEMS and MOEMS devices using bonded SOI substrates and DRIE, *Micromachining and Microfabrication Process Technology VI*, ed. by J. Karam, J. Yasaitis, Proc. SPIE **4174** (2000) 98–110

55.54 K. Somasundram, D. Cole, C. McNamara, A. Boyle, P. McCann, C. Devine, A. Nevin: Fusion-bonded multilayered SOI for MEMS applications, Smart Sensors, Actuators and MEMS, Proc. SPIE **5116**, ed. by J.-C. Chiao, V. Varadan, C. Cané (2003) 12–19

55.55 C. M. Roberts Jr., L. H. Long, P. A. Ruggerio: Method for separating circuit dies from a wafer, US Patent 5 362 681 (1994)

55.56 T. Maudie, T. Miller, R. Nielsen, D. Wallace, T. Ruehs, D. Zehrbach: Challenges of MEMS device characterization in engineering development and final manufacturing, Proc. 1998 IEEE AUTOTESTCON (IEEE, New York 1998) 164–170

55.57 S. Petrovic, A. Ramirez, T. Maudie, D. Stanerson, J. Wertz, G. Bitko, J. Matkin, D. J. Monk: Reliability test methods for media-compatible pressure sensors, IEEE Trans. Industrial Electron. **45**(6), 877–885 (1998)

55.58 S. Petrovic: Progress in media compatible pressure sensors, Proc. of InterPACK'01, the Pacific Rim/International Intersociety Electronic Packaging Technical/Business Conf. & Exhibition (ASME, New York 2001) IPACK2001-15517

55.59 H.-J. Kress, J. Marek, M. Mast, O. Schatz, J. Muchow: Integrated pressure sensors with electronic trimming, Automotive Eng. **103**(4), 65–68 (1995)

55.60 C. Fung, R. Harris, T. Zhu: Multifunction polysilicon pressure sensors for process control, Sensors **16**(10), 75–79 (1999)

55.61 K. H.-L. Chau, C. D. Fung, P. R. Harris, J. G. Panagou: High-stress and overrange behavior of sealed cavity polysilicon pressure sensors, IEEE 4th Technical Digest, Solid State Sensor and Actuator Workshop (IEEE, New York 1990) 181–183

55.62 A. B. Sontheimer: Digital micromirror device (DMD) hinge memory lifetime reliability modeling, Proc. 40th Annual IEEE International Reliability Physics Symp. (IEEE, New York 2002) 118–121

55.63 J. P. O'Connor: Packaging design considerations and guidelines for the digital micromirror device™, Proc. of InterPACK'01, the Pacific Rim/International Intersociety Electronic Packaging Technical/Business Conf. & Exhibition (ASME, New York 2001) IPACK2001-15526

55.64 C. Bang, V. Bright, M. A. Mignardi, D. J. Monk: Assembly and test for MEMS and optical MEMS. In: *MEMS and MOEMS Technology and Applications*, ed. by P. Rai-Choudhury (SPIE, Bellingham 2000) Chap. 7

55.65 F. Poradish, J. T. McKinley: Package for a semiconductor device, US Patent 5 293 511 (1994)

55.66 T. W. Migl: Interfacing to the digital micromirror device for home entertainment applications, Proc. of InterPACK'01, the Pacific Rim/International Intersociety Electronic Packaging Technical/Business Conf. & Exhibition (ASME, New York 2001) IPACK2001-15712

55.67 S. J. Jacobs, J. J. Malone, S. A. Miller, A. Gonzalez, R. Robbins, V. C. Lopes, D. Doane: Challenges in DMD™ assembly and test, Proc. Mater. Res. Soc. **657** (2001)EE6.1.1–EE6.1.12

55.68 Automotive Industry Action Group: *Quality Systems Requirements QS-9000*, 3rd edn. (Automotive Industry Action Group of the American Society for Quality, Milwaukee 1998)

55.69 J. A. Walraven, B. A. Waterson, I. De Wolf: Failure analysis of micromechanical systems (MEMS). In: *Microelectronic Failure Analysis*, 4th edn. (ASM, Materials Park 2002) 2002 Suppl.

55.70 M. R. Douglass: Lifetime estimates and unique failure mechanisms of the digital micromirror device (DMD), Proc. 1998 36th Annual IEEE International Reliability Physics Symp. (IEEE, New York 1998) 9–16

55.71 S. J. Jacobs, S. A. Miller, J. J. Malone, W. C. McDonald, V. C. Lopes, L. K. Magel: Hermeticity and stiction in MEMS packaging, 40th Annual IEEE International Reliability Physics Symp., Dallas 2002 (IEEE, New York 2002) 136–139

55.72 J. R. Martin: Surface characteristics of integrated MEMS in high volume production. In: *Nanotribology: Critical Assessment and Research Needs*, ed. by S. M. Hsu, Z. C. Ying (Kluwer, Dordrecht 2002) Chap. 14

55.73 H. R. Shea, A. Gasparyan, C. D. White, R. B. Comizzoli, D. Abusch-Magder, S. Arney: Anodic oxidation and reliability of MEMS poly-silicon electrodes at high relative humidity and high voltages, *MEMS Reliability for Critical Applications*; Proc. SPIE **4180**, 117–122 (2000)

55.74 J. Martin: Stiction suppression in high volume MEMS products, Proc. 2003 STLE/ASME Joint International Tribology Conf. (ASME, New York 2003) 2003TRIB-266

55.75 G. X. Li, F. A. Shemansky Jr.: Drop Test and analysis on micro-machined structures, Sens. Actuators **85**, 280–286 (2000)

55.76 R. T. Howe, H. J. Barber, M. Judy: Apparatus to minimize stiction in micromachined structures, (1996)U.S. patent 5,542,295

55.77 L. J. Hornbeck, W. E. Nelson: Spatial light modulator and method, US Patent 5 096 279 (1992)

55.78 L. J. Hornbeck: Low reset voltage process for DMD, US Patent 5 331 454 (1994)

55.79 E. C. Fisher, R. Jascott, R. O. Gale: Method of passivating a micromechanical device within a hermetic package, US Patent 5 936 758 (1999)

55.80 B. A. Miksic: Use of vapor phase inhibitors for corrosion protection of metal products, Proc. 1983

NACE Annual Conf., Corrosion 83 (National Assoc. Corrosion Engineers, Houston 1983) Paper 308

55.81 D. Vanderpool, S. Akin, P. Hassett: Corrosion inhibitors in the electronics industry: Organic copper corrosion inhibitors, Proc. 1986 NACE Annual Conf., Corrosion 86 (National Assoc. Corrosion Engineers, Houston 1986) Paper 1

55.82 J. R. Martin, Y. Zhao: Micromachined device packaged to reduce stiction, US Patent 5 694 740 (1997)

55.83 J. R. Martin: Process for wafer level treatment to reduce stiction and passivate micromachined surfaces and compounds used therefor, (2001) International Patent Application WO 01/57920 A1

55.84 S. T. Patton, J. S. Zabinski: Failure mechanisms of a MEMS actuator in very high vacuum, Tribol. Int. **35**(6), 373–379 (2002)

55.85 S. T. Patton, K. C. Eapen, J. S. Zabinski: Effects of adsorbed water and sample aging in air on the µN level adhesion force between Si(100) and silicon nitride, Tribol. Int. **34**(7), 481–491 (2001)

55.86 S. T. Patton, W. D. Cowan, K. C. Eapen, J. S. Zabinski: Effect of surface chemistry on the tribological performance of a MEMS electrostatic lateral output motor, Tribol. Lett. **9**, 199–209 (2000)

55.87 S. T. Patton, W. D. Cowan, J. S. Zabinski: Performance and reliability of a new MEMS electrostatic lateral output motor, Proc. 1999 37th Annual IEEE International Reliability Physics Symp. (IEEE, New York 1999) 179–188

55.88 D. M. Tanner, J. A. Walraven, L. W. Irwin, M. T. Dugger, N. F. Smith, W. P. Eaton, W. M. Miller, S. L. Miller: The effect of humidity on the reliability of a surface micromachined microengine, Proc. 1999 37th Annual IEEE International Reliability Physics Symp. (IEEE, New York 1999) 189–197

55.89 M. P. de Boer, P. J. Clews, B. K. Smith, T. A. Michalske: Adhesion of polysilicon microbeams in controlled humidity ambients, Proc. Mater. Res. Soc. Symp. **518**, 131–136 (1998)

56. Packaging and Reliability Issues in Micro/Nano Systems

The potential of MEMS/NEMS technologies has been viewed as a revolution comparable or even bigger than that of microelectronics. These scientific and engineering advancements in micro-/nano-electromechanical systems (MEMS)/(NEMS) could bring previously unthinkable applications to reality, from space systems, environmental instruments, to daily-life appliances. As presented in previous chapters, the development of core MEMS/NEMS processes has already demonstrated a lot of commercial applications as well as future potentials with elaborate functionalities. However, creating a low-cost reliable package for the protection of these MEMS/NEMS products is still a very difficult task. Without addressing these packaging and reliability issues, no commercial products can be sold on the market. Packaging design and modeling, packaging material selection, packaging process integration, and packaging cost are the main issues to be considered. In this chapter, we will present the fundamentals of MEMS/NEMS packaging technology, including packaging processes, hermetic and vacuum encapsulations, thermal issues, packaging reliability, and future packaging trends. The future development of MEMS packaging will rely on the success of the implementation of several unique techniques, such as packaging design kits for system and circuit designer,

low-cost wafer-level and chip-scale packaging techniques, effective testing techniques, and reliable fabrication of an interposer [56.1] with vertical through-interconnects for device integrations.

56.1 Introduction to Micro-/Nano-Electromech -anical (MEMS)/(NEMS) Packaging

MEMS/NEMS are miniaturized systems that may have mechanical, chemical, or biomedical features with or without integrated circuits (IC) for sensing or actuation applications [56.2]. For example, pressure [56.3], temperature, flow [56.4], accelerometers [56.5], gyroscopes [56.6] and chemical sensors [56.7] can be fabricated by modern technologies for sensing applications. Fluidic valves [56.8], pumps [56.9], and inkjet printer heads are examples of actuation devices for medical, environmental, office and industrial applications.

Silicon is typically used as the primary substrate material for MEMS fabrication because it can provide unique electrical, thermal, and mechanical properties but can also be easily micromachined in a form of batch processing and be incorporated with microelectronic circuit by using most of the conventional semiconductor manufacturing processes and tools. As a result, smaller size, lighter weight, lower power consumption and cheaper fabrication cost are advantages of MEMS/NEMS devices. For example, with the advances

Part G | 56

of microfabrication technologies in the past decades, the MEMS market at the component level is currently in excess of 5 billion and is driving end-product markets larger than 100 billion [56.10].

Nevertheless, the road to the commercialization of MEMS/NEMS does not look as promising as expected. Many industrial companies took advantage of MEMS/NEMS technologies for product integrations and cost efficiency is the key toward commercialization. Several MEMS devices have been developed for and applied in the automotive industry and information technology and they dominate the market due to their high production volumes. On the other hand, most custom-designed MEMS/NEMS products are still very diverse, aiming for different applications with small- to medium-scale production. Based on past experience of the IC industry, the cost of packaging processes is about 30%, and sometimes can be more than 70%, of the total production expenses. MEMS/NEMS packaging processes are expected to be even more costly because of the stringent packaging requirements for MEMS components, in addition to the microelectronic circuitry, in a typical MEMS product [56.11].

56.1.1 MEMS/NEMS Packaging Fundamentals

Sealing or encapsulation is an important step in either IC or MEMS/NEMS packaging process to protect working devices. In traditional IC packaging procedures, the overall packaging steps often involve: (1) wafer dicing, (2) pick-and-place; (3) electrical connections, such as wire bonding, and (4) plastic molding or housing for sealing [56.12, 13]. With the increasing needs of high-performance multifunctional consumer electronic products, IC packaging processes have incorporated more-complex designs and advanced fabrication technologies, such as Cu interconnects [56.14], flip-chip bonding [56.15], ball grid arrays [56.16], wafer-level chip-scale packaging [56.17], and three-dimensional (3-D) packaging [56.18] to satisfy the necessity for high input/output (I/O) density, large die area, and high clock frequencies. The functions of conventional IC packaging are to protect, power, and cool the microelectronic chips or components and provide electrical and mechanical connection between microelectronic parts and the outside world. Unlike regular ICs, the diversity of MEMS/NEMS products complicates the sealing issues. MEMS/NEMS packaging processes cannot directly follow the procedures set by the IC packaging industry due to possible free-standing physical micro/nanostructures or chemical substances which cannot

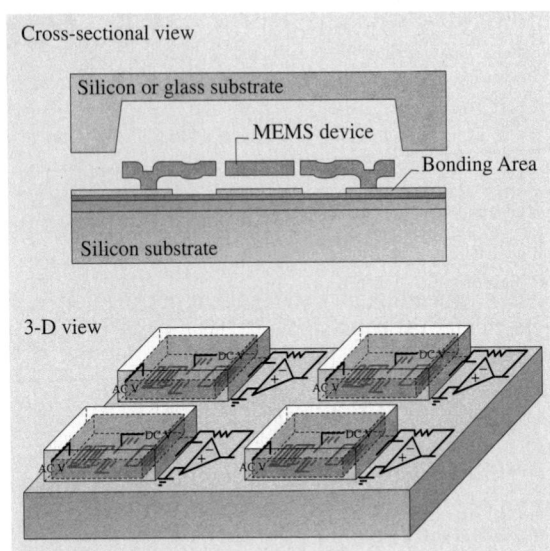

Fig. 56.1 A typical MEMS packaging illustration where a MEMS structure is encapsulated and protected by the packaging cap

survive the dicing or pick-and-place steps. Moreover, MEMS/NEMS components may need to interface with the outside environment (for example, fluidic interconnectors [56.19]), some other components may need to be hermetically sealed in vacuum (for example, inertial sensors [56.5]) in addition to the needs of electrical interconnects. Therefore, MEMS/NEMS packaging processes have to provide more functionalities, including better mechanical protection, thermal management, hermetic sealing and complex electrical and signal distribution.

It has been suggested that MEMS/NEMS packaging should be incorporated in the device fabrication stage as part of the micromachining process. Although this approach solves the packaging need for some specific devices, it does not solve the packaging need for general microsystems. In particular, many MEMS/NEMS devices are now fabricated by various foundry services [56.20, 21] and there is a tremendous need for a uniform packaging process. Figure 56.1 shows a typical MEMS device being encapsulated by a packaging cap. The most fragile part on this device is the suspended mechanical sensor, which is a freestanding mechanical, mass–spring microstructure. It is desirable to protect this mechanical structure during the packaging and handling processes. Moreover, vacuum encapsulation may be required for these microstructures in applications such as resonant accelerometers [56.5] or gyroscopes [56.6, 22]. A *packaging cap* with a properly designed microcavity is to be fabricated to encapsulate and protect the fragile MEMS/NEMS structure as the first-level post-packaging process. The wafer can be diced afterwards and the well-established packaging technology in IC industry can be followed to finish the packaging steps. A hermetic seal may be required to assure that no moisture or contamination can enter the package and affect the functionality of the micro/nanostructures. This increases the difficulty of common IC packaging processes tremendously. Although some MEMS chips can employ typical IC packaging techniques, such as die-attached processes and wiring interconnects for packaging [56.12], advanced packaging techniques, especially wafer-level packaging, are required for easy integration for multifunctional applications. For example, if chemical or biomedical substances are present [56.23], the sealing process must be carried out at a low processing temperature. If there are optical devices [56.24], the sealing process should provide good optical paths. If there are mechanical resonators [56.25], vacuum sealing might be required for better device performance and the desirable vacuum level depends on the specification of the device.

Before the state-of-the-art MEMS/NEMS packaging processes are discussed, several primary microfabrication processes for packaging applications are briefly summarized, including the flip-chip (FC) technique, ball grid arrays (BGA), through-wafer etching, and plating. Other silicon-based processes such as thin-film deposition, wet and dry chemical etching, lithography, lift-off, and wire bonding can be found in the textbooks [56.26].

Flip-Chip Technique (FC)

This technique is commonly used in the assembly process between a chip with microelectronics and a package substrate [56.15]. The microelectronic chip is *flip-joined* with the packaging substrate, and metal solder bumps are used as both the bonding agents and electrical paths between the microelectronic chip and the package substrate. Because the vertical bonding space can be very small, as controlled by the heights of the solder bumps, and the lateral distributions of bond pads can be on the whole chip instead of being only on the edge, this technique can provide a high density of input/output (I/O) connections. In the FC technique, solder bumps are generally fabricated by means of electroplating. Before the bumping process, multiple metal layers, such as TiW–Cu, Cr–Cu, Cr–Ni, TaN/Ta/Ni, have to be deposited as

a seed layer for electroplating and as a diffusion barrier to prevent the diffusion of solder into the electrical interconnects underneath.

Ball Grid Arrays (BGA)

This technology is very similar to the FC technique. An area array of solder balls on a single- or multi-chip module are used in the packaging process as electrical, thermal and mechanical connects to join the module with the next-level package, usually a printed circuit board [56.16]. The major difference between typical BGA and FC chips is the size of the solder bump. In BGA chips, the bumps are on the order of 750 μm in diameter, that is 10 times larger than those commonly used in FC chips.

Through-Wafer Etching

This is a chemical etching process to make through-wafer channels on a silicon substrate for the fabrication of vertical through-wafer interconnects. The chemical etching process can be either a wet or dry process. Anisotropic or isotropic etching solutions can be used in the wet etching process. The dry etch process is based on plasma and ion-assisted chemical reactions, which can cause either isotropic or anisotropic etching. In order to create high-density high-aspect-ratio through-wafer vias, deep reactive-ion etching (DRIE) is typically used. Two popular DRIE approaches, Bosch, and Cyro, are well described in the literature [56.27].

Electroplating

Electroplating is another common microfabrication process. It can be conducted for the deposition of an adherent metallic layer onto a conductive or nonconductive substrate. The process on a conductive substrate is called electrolytic plating that utilizes a seed layer as the anode to transfer metal ions onto the cathode surface when a direct current (DC) is passed through the plating solution. Plating processes that occur without applying an electrical current are called electroless plating, and can be used for both conducive and nonconductive surfaces. Electroless plating processes required a layer of a noble metal such as Pd, Pt, or Ru on the substrate to act as the catalyst to trigger the self-decomposition reaction in the plating solution. These electroplating processes are very important for electrical interconnect and solder-bump fabrication for packaging applications because of the low process temperature and cost.

These processes have generally been developed to provide electrical and thermal paths for various IC/MEMS packaging approaches.

56.1.2 Contemporary MEMS/NEMS Packaging Approaches

Several MEMS packaging issues and approaches before 1985 were discussed in the book, "Micromachining and Micropackaging of Transducers" [56.28] and researchers have been working on MEMS packaging approaches continuously since. For example, *Senturia* and *Smith* [56.29] discussed packaging and partitioning issues for microsystems. *Smith* and *Collins* [56.23] used epoxy to bond glass and silicon for chemical sensors. Several multi-chip module (MCM) methods have been proposed. *Butler* et al. [56.30] proposed adapting multi-chip module foundries using the chip-on-flex (COF) process. *Schuenemann* et al. [56.31] introduced a 3D stackable packaging concept for top–bottom ball grid arrays (TB-BGA) that includes electric, fluidic, optic, and communication interfaces. *Lee* et al. [56.32] and *Ok* et al. [56.10] presented a direct-chip-attach MEMS packaging method using through-wafer electrical interconnects. *Laskar* and *Blythe* [56.33] developed an MCM-type packaging process using epoxy. *Reichl* [56.34] discussed different materials for bonding and interconnection. *Grisel* et al. [56.35] designed a special process to package microchemical sensors. Special processes have also been developed for MEMS packaging, such as packaging for microelectrodes [56.36], packaging for biomedical systems [56.37] and packaging for space systems [56.38]. These specially designed, device-oriented packaging methods are aimed at individual systems. There is no reliable method yet that would qualify as a versatile post-packaging process for MEMS/NEMS with the rigorous process requirements of low temperature, hermetic sealing and long-term stability.

Previously, an integrated process using surface-micromachined microshells has been developed [56.39]. This process applies the concepts of sacrificial layer and low pressure chemical vapor deposition (LPCVD) sealing to achieve wafer-level post-packaging. Similar processes have been demonstrated. For example, *Guckle* et al. [56.40] and *Sniegowski* et al. [56.41] developed a reactive sealing method to seal vibratory micromachined beams. *Ikeda* et al. [56.42] adopted epitaxial silicon to seal microstructures. *Mastrangelo* et al. [56.43] used silicon nitride to seal mechanical beams as light sources. *Smith* et al. [56.44] accomplished a new fabrication technology by embedding microstructures and complementary metal–oxide–semiconductor (CMOS) circuitry. All of these methods have integrated the MEMS process with the post-packaging process such

that no extra bonding process is required. However, these schemes are highly process-dependent and not suitable for prefabricated circuitry.

New efforts for MEMS/NEMS post-packaging processes have been reported. *Butler* et al. [56.30] demonstrated an advanced MCM packaging scheme. It adopts the high-density interconnect (HDI) process consisting of embedding bare die into pre-milled substrates. Because the MEMS structures have to be released after the packaging process, this approach is undesirable for general microsystems. *Van der Groen* et al. [56.45] reported a transfer technique for CMOS circuits based on epoxy bonding. This process overcomes the surface roughness problem but epoxy is not a good material for hermetic sealing. In 1996, *Cohn* et al. demonstrated a wafer-to-wafer vacuum packaging process by using silicon–gold eutectic bonding with a 2-μm-thick polysilicon microcap. However, experimental results showed substantial leakage after a period of 50 days. *Cheng* et al. [56.46] developed a vacuum packaging technology using localized aluminium/silicon-to-glass bonding. In 2002, *Chiao* and *Lin* [56.47] demonstrated vacuum packaging of microresonators by rapid thermal processing. These recent and ongoing research efforts indicate the strong need for a versatile MEMS/NEMS post-packaging process.

56.1.3 Bonding Processes for MEMS/NEMS Packaging Applications

Previously, silicon-bonding technologies have been used in many fabrication and packaging applications, where two types of bonding processes are commonly used: (1) direct bonding processes such as anodic bonding and fusion bonding, and (2) bonding processes with intermediate layers such as epoxy bonding, eutectic bonding and solder bonding. Direct wafer-bonding processes are procedures that facilitate permanent attachments between two wafers without any intermediate layer. A permanent bond between two wafers can also be accomplished by using intermediate layers. Joining processes using intermediate layers have been used extensively by the ceramic industry to form metal-to-metal and metal-to-ceramic joints [56.48, 49] and can be characterized as [56.50]: (1) fusion or melting of two materials to form a stable intermediate compound which facilitates the bond; (2) diffusion, in which pressurized joint parts are heated to 70% of the material's melting temperature and a stable intermediate compound is formed at the interface and (3) brazing, in which a filler material is fitted into the two parts to be joined and, upon heating, a stable intermediate compound is formed. These processes are commonly used when lower bonding temperatures or a stronger bonding interface is required but cannot be achieved by the direct bonding process. Furthermore, the intermediate layers may reflow during the bonding process and fill the gaps between two bonding surfaces to overcome the surface roughness problem commonly encountered during the direct wafer-bonding processes. As such, the requirement for fine surface roughness for the direct wafer-bonding processes is greatly relieved for wafer-bonding processes with intermediate layers.

There have been many MEMS/NEMS applications of both direct bonding and bonding processes with the assistance of intermediate layers. For example, devices such as pressure sensors, micropumps, biomedical sensors or chemical sensors require mechanical interconnectors to be bonded on the substrate (see for example [56.7, 19, 51]). Glass has commonly been used as the bonding material, using anodic bonding at a temperature of about 300–450 °C (see, for example, [56.52,53]). *Klaassen* et al. [56.54] and *Hsu* et al. [56.55] have demonstrated different types of silicon fusion bonding and Si–SiO$_2$ bonding processes at very high temperatures of over 1000 °C. *Ko* et al. [56.28], *Tiensuu* et al. [56.56], *Lee* et al. [56.57] and *Cohn* et al. [56.58] have used eutectic bonding for different applications. All of these bonding techniques have different mechanisms that determine the individual bonding characteristics and process parameters. This section discusses the details of these processes.

Fusion Bonding for MEMS/NEMS Packaging

Silicon fusion bonding is an important fabrication technique for silicon on insulator (SOI) devices. The bonding is based on Si–O, Si–N, or Si–Si strong covalent bonds. However, very high bonding temperature (higher than 1000 °C), flat bonding surfaces (less than 6 nm) and intimate contact are the three basic requirements for strong, uniform, and hermetic bonding. The common silicon-to-silicon fusion bonding process starts with wafer hydration (soaking the wafers in a H$_2$O$_2$–H$_2$SO$_4$ mixture, diluted H$_2$SO$_4$, boiling nitric acid or oxygen plasma) to create a hydrophilic top layer consisting of O–H bonds [56.59]. Pre-bonding is accomplished when the two wafers are brought into intimate contact and Van der Waals forces creates bonds between the two wafers. An annealing step at elevated temperature is required to strengthen the bond. Although hydrophilic surface treatment can lower the bonding temperature, an annealing step higher than 800 °C is still needed to remove possible bubble formation at the bonding interface. *Bower* et al. [56.60] proposed that low-temperature Si$_3$N$_4$ fu-

sion bonding could be achieved at less than 300 °C. *Takagi* et al. [56.61] proposed that silicon fusion bonding could be carrier out at room temperature by using Ar$^+$-beam treatment on the wafer surface with a bond strength comparable to that of conventional fusion bonding.

Anodic Bonding
for MEMS/NEMS Packaging Applications

The invention of anodic bonding dates back to 1969 when *Wallis* and *Pomerantz* [56.62] found that glass and metal could be bonded together at about 200–400 °C below the melting point of glass with the aid of a high electrical field. This technology has been widely used for protecting onboard electronics in biosensors (see for example [56.63–65]) and sealing cavities in pressure sensors (see for example, [56.66]). Many reports have also discussed the possibility of lowering the bonding temperature by different mechanisms [56.67, 68]. Anodic bonding forms Si–O or Si–Si covalent bonds and is one of the strongest chemical bonds available for silicon-based systems. The bonding process can be accomplished on a hot plate with temperature of 180–500 °C in atmosphere or a vacuum environment. When a static electrical field is built up within the Pyrex glass and silicon, the sodium ions in the glass migrate away from the silicon–glass interface, creating a locally high electrical field and a bond is formed by electrochemical effects [56.62]. In order to create high electrical fields, a flat bonding surface with less than 50 nm roughness is required. In addition, the electrical field required for bonding is larger than 3×10^6 V/cm [56.28]. Such a high electrical field is generated by a power supply of 200–1000 V. Figure 56.2 shows the set up for anodic bonding where two bonding wafers are brought together and heated to an elevated temperature to supply the bonding energy. If there are freestanding, conductive micromechanical structures on any bonding wafer, care must be taken as the high voltage tends to pull the micromechanical structure and damage

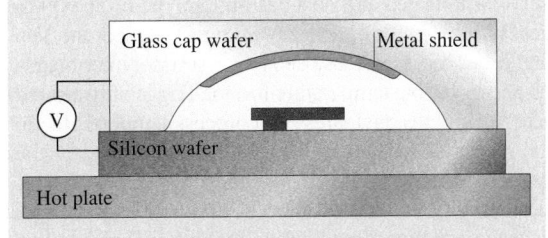

Fig. 56.2 Schematic diagram of the set up for the silicon-to-glass anodic bonding process

may occur. A thin-film metal pattern on the glass cap can be formed to provide shielding to solve this problem, as shown in Fig. 56.2. Furthermore, Corning 7740 Pyrex^TM is commonly used in the silicon-to-glass bonding system because it has a thermal expansion coefficient close to that of single-crystalline silicon in the range 200–300 °C. The induced residual-stress problem can be minimized in that temperature range. *Hanneborg* et al. [56.69] have successfully bonded silicon to other thin solid films, such as silicon dioxide, nitride, and polysilicon together with an intermediate glass layer using anodic bonding technique. *Chavan* and *Wise* [56.70] have reported on absolute pressure sensors fabricated by using the anodic bonding technique. In the process, a silicon cap with a thin heavily doped boron layer and a recess cavity was bonded in a vacuum environment to a glass substrate with prefabricated interconnection lines. The problem of oxygen outgassing due to the high electrical field in the anodic bonding process presents a challenge for the vacuum sealing process [56.71]. A thin Ti/Pt layer predeposited onto the glass surface has been shown to provide a good diffusion barrier, and the resulting pressure in the cavity can reach 200 mTorr [56.70]. In another example, microgyroscopes have been fabricated by *Hara* et al. [56.72] using the anodic bonding technique.

In practice, electrostatic bonding has become widely accepted in MEMS fabrication and packaging applications as described above. Unfortunately, possible contamination due to excess alkali metal in the glass; possible damage to microelectronics due to the high electrical field; and the requirement for flat surfaces for bonding limit the application of anodic bonding to MEMS post-packaging applications [56.73].

Epoxy Bonding (Adhesive Bonding)

Epoxy comprises four major components: epoxy resin, a filler such as silver slake, solvent or reactive epoxy diluent, and additives such as hardeners and catalysts [56.74, 75]. The bonding mechanism for epoxy is very complicated and depends on the type of epoxy. In general, the main source of bonding strength is the Van der Waals force. Because epoxy is a soft polymer material and its curing temperature for bonding is only around 150 °C, low residual stress and process temperature are the major advantages of epoxy bonding. However, the properties of epoxy can easily change with environmental humidity and temperature so the bonding strength decays over time. In addition, epoxy bonding has low moisture resistance and is a dirty process due to its additives. These disadvantages have made epoxy un-

favorable for the special hermetic or vacuum sealing MEMS/NEMS packaging requirements.

Eutectic Bonding

In many binary systems, there is a eutectic point corresponding to the alloy composition with the lowest melting temperature. If the environmental temperature is kept higher than the eutectic point, two contacted surfaces containing two elements with the eutectic composition can form a liquid-phase alloy. The solidification of the eutectic alloy forms *eutectic bonding* at a temperature lower than the melting temperature of either element in the alloy. Eutectic bonding can be a strong metal bonding. For example, in the case of the Au/Si alloy system, the eutectic temperature is only 363 °C when the composition is at the atomic ratio of 81.4% Au to 18.6% Si and bonding strength is higher than 5.5 GPa [56.76]. Because other alloy systems may have lower eutectic temperatures than the Al/Si system, they present great potential for MEMS packaging applications. In addition to the Au/Si system, the Al/Ge/Si, Au/SnSi, and Au/Ge/Si systems have been applied for MEMS packaging.

Solder Bonding

Solder bonding has been widely applied in microelectronic packaging [56.77]. Its low bonding temperature and high bonding strength are good characteristics for packaging. Furthermore, there are a variety of choices of solder material for specific applications. *Singh* et al. [56.78] have successfully applied solder bump bonding in the integration of electronic components and mechanical devices for MEMS fabrication [56.79]. In this case, indium metal was used to bond two separated silicon surfaces together by applying 350 MPa pressure; the bonding strength was as strong as 10 MPa. Glass frits can also be treated as a solder material and have been extensively used for vacuum encapsulation in MEMS industry. Glass frits are ceramic materials that can provide strong bonding strength with silicon with good hermeticity. Its bonding temperature is lower than 400 °C and is suitable for electronic components. However, a bonding area more than 200 μm wide is required to achieve good bonding results and this may become a drawback because area is the measure of manufacturing cost in IC industry. Nevertheless, glass frit is the most popular bonding process used in current products.

Localized Heating and Bonding

Low bonding temperature and short process time are desirable process parameters in MEMS packaging fabrication to provide a lower thermal budget and high

Table 56.1 Summary of bonding mechanisms

Bonding methods	Temperature	Roughness	Hermeticity	Post-packaging	Reliability
Fusion bonding	very high	highly sensitive	yes	yes by LH	good
Anodic bonding	medium	highly sensitive	yes	difficult	good
Epoxy bonding	low	low	no	yes	???
Integrated process	high	medium	yes	no	good
Low-temp. bonding	low	highly sensitive	???	no	???
Eutectic bonding	medium	low	yes	yes by LH	???
Brazing	very high	low	yes	yes by LH	good

???: no conlusive data

LH: localized heating

throughput. However, most chemical bonding reactions require a minimum and sufficient thermal energy to overcome the reaction energy barrier, also called the activation energy, to start the reaction and to form a strong bond. As a result, a high bonding temperature generally results in a shorter processing time to reach the same bonding quality at a lower bonding temperature [56.80]. The common limitations for the above bonding techniques are their individual bonding characteristics and temperature requirements. In general, MEMS packaging requires good bonding for hermetic sealing while the processing temperature must be kept low at the wafer level to reduce thermal effects on the existing devices. For example, a MEMS device may have prefabricated circuitry, biomaterial or other temperature-sensitive materials such as organic polymers, magnetic metal alloys, or piezoceramics. Since the packaging step comes after the device fabrication processes, the bonding temperature should be kept low to avoid the effects of high temperature on the system. Possible temperature effects include residual stress due to the mismatch of thermal expansion coefficient of bonding materials and substrates, electrical contact failure due to atomic interdiffusion at the interface, and contamination due to

the outgassing or evaporation of materials. In addition to the control of bonding temperature, the magnitude of the applied force required to create intimate contact for bonding and atmospheric environment control are other factors that should be considered. Based on heat-transfer simulation studies [56.81], it is possible to confine the high-temperature area to a small region by localized heating without heating the whole substrate. Therefore assembly steps can always be processed after device fabrication without having detrimental effects. As such, localized heating and bonding techniques have been introduced for postprocessing approaches [56.80, 82].

Table 56.1 summarizes these MEMS/NEMS packaging technologies and their limitations, including the localized heating and bonding approach. The localized heating approach introduces several new opportunities. First, better and faster temperature control can be achieved. Second, higher temperature can be applied to improve the bonding quality. Third, new bonding mechanisms that require high temperature such as brazing [56.83] may now be explored. As such, it has potential applications for a wide range of MEMS/NEMS devices and is expected to advance the field of MEMS/NEMS packaging.

56.2 Hermetic and Vacuum Packaging and Applications

Hermetic packaging is important because it provides a moisture-free environment to avoid charge separation in capacitive devices, corrosion in metallization, or electrolytic conduction in order to prolong the lifetime of electronic circuitry. Especially for MEMS/NEMS packaging, hermeticity of the packaging is desirable in most cases since one of the main failure mechanisms is humidity. Furthermore, the surface tension of wa-

ter could cause stiction of micromechanical structures, leading to malfunction. In several device applications, vacuum encapsulation is necessary but can be costly. Many resonant devices, such as comb-shaped μ-resonators and ring-type μ-gyroscopes that have very large surface-to-volume ratios and vibrate in a very tight space [56.22, 42], need a vacuum to improve their performance. Two important approaches to hermetic and vacuum packaging of MEMS/NEMS devices have been demonstrated: (1) the integrated encapsulation approach and (2) the postprocess packaging approach, as discussed in this section. Moreover, vacuum encapsulation by means of localized heating and bonding is discussed separately as a packaging example.

56.2.1 Integrated Micromachining Processes

Several MEMS hermetic and vacuum packaging processes have been demonstrated based on integrated micromachining processes where the construction of sealing or protection caps is integrated with device manufacturing. The integrated approach has the advantage of sealing mechanical components in situ prior to the chip dicing and handling steps to avoid contamination. An

integrated vacuum sealing process by LPCVD is presented here as an illustration. This integrated process can encapsulate comb-shaped microresonators [56.84] in vacuum at the wafer level. Figure 56.3 illustrates the cross-sectional view of the manufacturing process. First, a standard surface-micromachining process [56.85] is conducted by using four masks to define the first polysilicon layer, the anchors to the substrate, the dimples and the second polysilicon layer, as shown in Fig. 56.3a. The process so far is similar to the multi-user MEMS processes (MUMPs) [56.21] and comb-shaped microstructures are fabricated at the end of these steps. In the standard surface-micromachining process, the sacrificial layer (oxide) is etched away to release the microstructures. In the post-packaging process, a thick (7-μm) phosphorus-doped glass (PSG) is deposited to cover the microstructure and patterned by using 5:1 buffered HF (BHF) to define the microshell area, as shown in Fig. 56.3b. A thin (1-μm) PSG layer is then deposited and defined to form etch channels, as illustrated in Fig. 56.3c. The microshell material, low-stress silicon nitride, is now deposited with a thickness of 1 μm. Etch holes are defined and opened on the silicon nitride layer by using a plasma etcher. Silicon dioxide inside the packaging shell is now etched away by concentrated HF and the wafer is dried by using the supercritical CO_2 drying process [56.86]. After these steps, Fig. 56.3d applies. A 2-μm-thick LPCVD low-stress nitride is now deposited at a deposition pressure of 300 mTorr to seal the shell in the vacuum condition. Finally, the contact pads are opened, as shown in

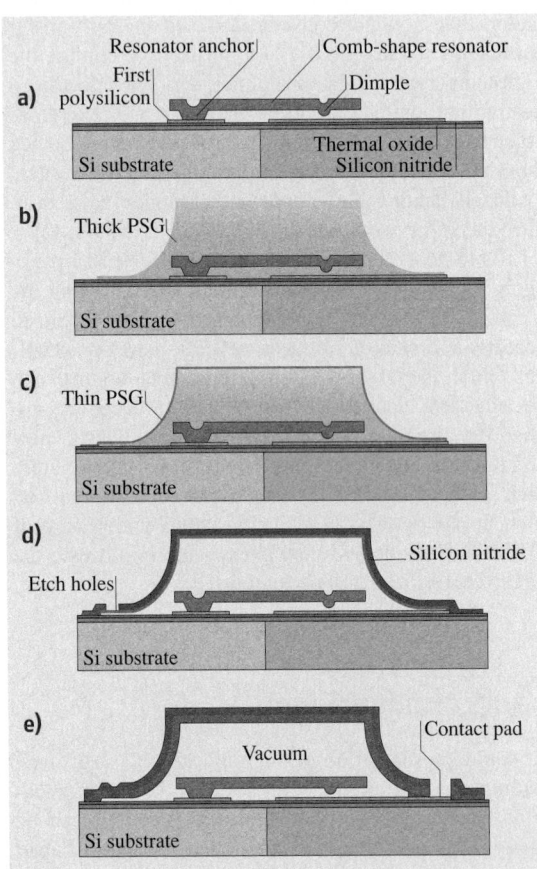

Fig. 56.3 An integrated vacuum encapsulation process using LPCVD nitride sealing to package micromechanical resonators [56.81]

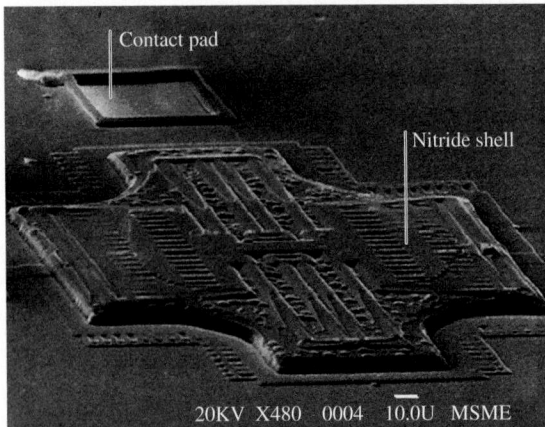

Fig. 56.4 SEM micrograph showing a vacuum packaged MEMS mechanical comb-shaped resonator packaged using an the integrated LPCVD sealing process as depicted in Fig. 56.3 [56.81]

Fig. 56.3e. Figure 56.4 is the scanning electron microscope (SEM) micrograph of a finished device with the protected microshell on top. The total packaging area (microshell) is about $400 \times 400 \, \mu m^2$. A contact pad is shown with the covering nitride layer removed. The shape of the microresonator, with beams $150 \, \mu m$ long and $2 \, \mu m$ wide is reflected on the surface of the microshell due to the integrated packaging process. The total height of the nitride shell is $12 \, \mu m$, as seen standing above the substrate. Spectral measurement of the comb resonator inside the packaging reveals that a vacuum level of about 200 mTorr has been accomplished [56.87].

Similarly, *Aigner* et al. [56.88] reported a Bi-CMOS-compatible integrated vacuum sealing process to package a polysilicon micro-accelerometer. The protecting shell was a polysilicon layer with supporting pillars anchored on the structural polysilicon. The release process was done in a HF gas-phase etching process to remove the sacrificial oxide layer and the release holes were sealed in a vacuum environment. The device was then injection-molded into a plastic package at a pressure of 100 bar; the supporting pillars were strong enough to hold the polysilicon shell under the high-pressure molding process.

An integrated sealing process using evaporation of aluminium has also been reported [56.89], as shown in Figs. 56.3 and 56.4. A silicon substrate was deposited with an n-type, 4-μm-thick epitaxial silicon layer. A controlled plasma etch and oxidation process formed a sharp tip and a layer of boro-phosphosilicate glass (BPSG) was used to fill the trench as the sacrificial layer. The nitride sealing cap was deposited and patterned and a 290-nm-thick PSG sacrificial release via was deposited and patterned, followed by the deposition and patterning of the polysilicon anode. After the release etching process, aluminium evaporation was done in a 2×10^{-6}-Torr vacuum chamber and an 800-nm-thick aluminium layer was deposited to seal the release via. The resulting pressure was estimated as 1 mPa by measuring the vacuum diode characteristics.

Other similar processes have been demonstrated based on the integrated encapsulation concept. For example, *Sniegowski* et al. [56.41] developed a reactive sealing method to seal vibratory micromachined beams; *Ikeda* et al. [56.42] used epitaxial silicon to seal microstructures; *Mastranglo* et al. [56.90] used silicon nitride to seal mechanical beams as light sources; *Smith* et al. [56.44] used the technology of embedding microstructures and CMOS circuitry. All of these approaches integrated the encapsulation process within the MEMS fabrication processes. The typical advantage of

this approach is that these devices could be ready for standard IC packaging processes such as dicing, pick-and-place etc. once the wafer-level integrated sealing processes are established.

Although this vacuum sealing processes successfully achieves hermetic and vacuum packaging, it has drawbacks. For example, these post-packaging processes are highly process-dependent and are not suitable for generic MEMS/NEMS post-packaging processes. Companies or researchers have to adapt these post-packaging processes for their own device manufacturing process. Currently, standard foundry services do not support any of these integrated processes. Also, integrated encapsulation does not allow the cavity pressure to be controlled, although it can achieve low pressure by wafer-level fabrication and provide lower manufacturing cost.

56.2.2 Post-Packaging Processes

The second approach is defined as post-packaging process. The packaging process starts when the device fabrication processes are completed, so this approach has great flexibility for various microsystems. For example, Fig. 56.5 shows a common industrial post-hermetic packaging called dual-in-line packaging (DIP) [56.91, 92]. A die is placed inside a ceramic holder covered by a sealing lid. Solder or ceramic joining is generally used to assemble the lid and holder under a pressure-controlled environment. Its high cost is the major drawback of this method because of the expensive ceramic holder and the low fabrication throughput. Another example of the post-packaging method is based on wafer-bonding techniques combined with microshell encapsulation. Devices are sealed by stacking another micromachined silicon or glass substrate, as illustrated in Fig. 56.1. Integrated microsystems and protection

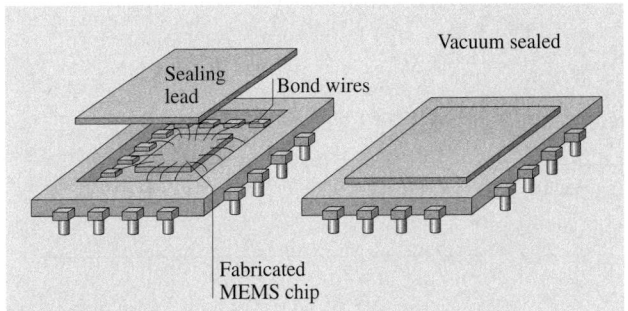

Fig. 56.5 A schematic diagram of industrial post-packaging (DIPS) using a ceramic holder to be covered by a sealing lid

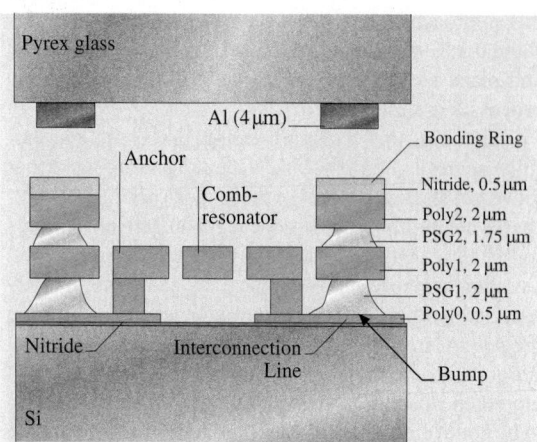

Fig. 56.6a,b Schematic diagrams of RTP bonding experiment. (**a**) The concept of aluminium-to-glass bonding. (**b**) Aluminum-to-nitride bonding with comb resonators. Drawn not to scale

shells are fabricated on different wafers, made of either silicon or glass, at the same time. After the two substrates are assembled together using silicon fusion, anodic, or low-temperature solder bonding to achieve the final encapsulation, these microshells will provide mechanical support, thermal path, or electrical contact for the MEMS/NEMS devices. Low packaging cost can be expected due to wafer-level processing.

A special heating method using rapid thermal processing (RTP) for wafer bonding applications is ex-

plained here to illustrate the roles of various control parameters such as temperature, time and intermediate bonding materials. *Chiao* and *Lin* [56.93] reported a wafer-bonding process based on the melting of an intermediate filler material to facilitate sealing of micromechanical structures. Figure 56.6a shows the concept of the bonding and sealing scheme using the aluminium-to-glass bonding system while Fig. 56.6b shows the aluminium-to-nitride bonding experimental set up with integrated comb resonators inside. Aluminium with a thickness of 3–4 μm was used and patterned to form sealing rings that surround the micromechanical structures on the device wafer. The width of a typical aluminium sealing ring was approximately 100–200 μm and the sealing area was $600 \times 600\,\mu m^2$. A glass (Pyrex^{TM}, Corning 7740) wafer was used as the cap to cover the MEMS devices. The heating and bonding energy was provided by RTP and the typical heating history is shown in Fig. 56.7 where the overall heating process can be completed in one minute, during which the temperature rose from room temperature to 990 °C and cooled down to 350 °C. The bonding and joining process of aluminium to glass was accomplished by heating at 990 °C for 2 s in the RTP chamber. It was demonstrated that aluminium could extract oxygen to form aluminium oxide to assist the bonding process [56.94]. Figure 56.8 shows a micro heatuator that has been successfully packaged using this bonding process and the surrounded liquid [in this case, isopropanol alcohol (IPA)] was sealed from penetrating inside the package.

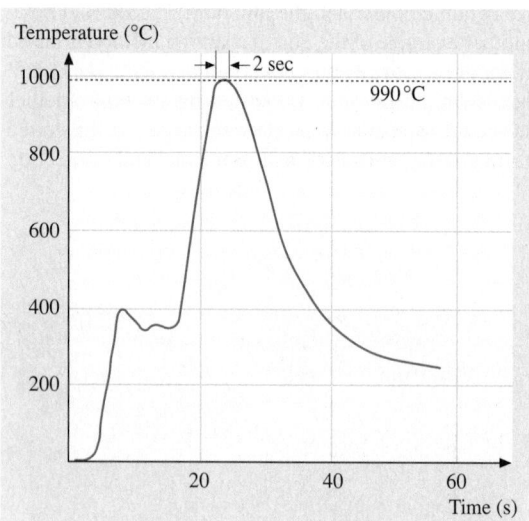

Fig. 56.7 Temperature history in an RTP bonding experiment

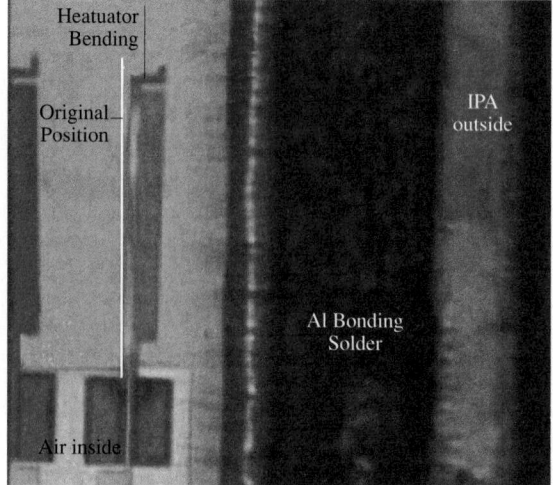

Fig. 56.8 A micro heatuator that was hermetically packaged and was operational when the package was immersed in liquid

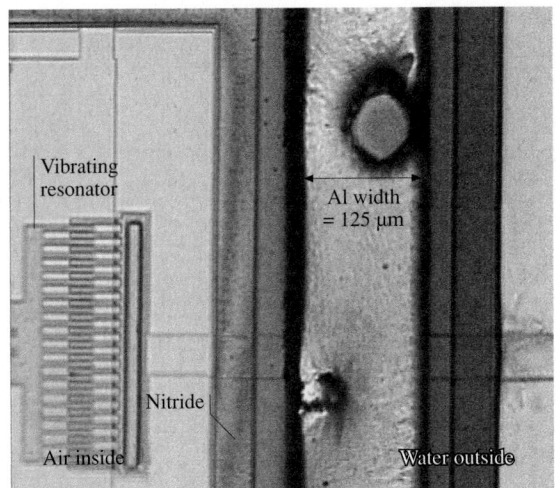

Fig. 56.9 A resonating comb-drive resonator that was sealed in the package chip and was operational after the package is immersed in water

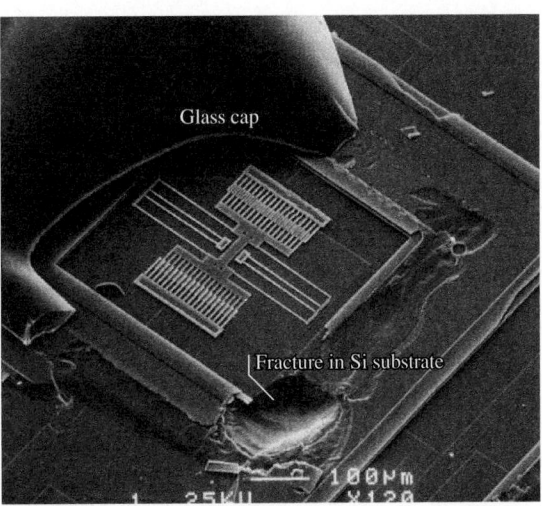

Fig. 56.10 SEM micrograph of the silicon substrate after forcefully breaking the aluminium-to-nitride bond. Glass debris is found attached to the silicon substrate

Other material systems have also been bonded by the RTP bonding process such as the aluminium-to-silicon nitride joining shown in Fig. 56.6(b) [56.95]. In this case, a 5000-Å-thick LPCVD silicon nitride layer was deposited and patterned on top of sealing ring structures that encompassed surface-micromachined comb-shaped resonators [56.85]. Using a process based on 10 s at the peak temperature of 750 °C achieved by RTP, a stable bond was formed at the aluminium–nitride interface. Figure 56.9 shows the packaged comb resonator that was resonating at 19.6 kHz when immersed in deionized (DI) water as seen under an optical microscope. The aluminium-to-nitride seal successfully blocked water from entering the package. In order to examine the bond strength, the package was forcefully broken, as shown in Fig. 56.10. The glass debris was attached to the sealing ring surrounding the comb-drive resonator on the silicon substrate. This shows that the bonding strength in the aluminium-to-nitride system is greater than the glass fracture strength, which is estimated to be around 270 MPa [56.96].

A vacuum sealing process by means of RTP bonding is discussed here in detail to address the technical issues in vacuum sealing processes. *Chiao* et al. [56.97] has reported a vacuum sealing process by RTP aluminium-to-nitride bonding. The RTP bonding process was conducted in a vacuum quartz tube, as shown in Fig. 56.11. Both device and cap wafers must be baked in vacuum at 300 °C for at least 4 h to drive out water and gas species that may adhere to the wafer sur-

face [56.98]. This pre-baking process in vacuum was necessary to minimize the outgassing effect during the bonding process in order to achieve a high-quality vacuum. Afterwards, the device and cap wafers were flip-chip assembled immediately, loaded into a sample holder and put inside a quartz chamber, as shown in Fig. 56.11. The system was then placed inside the RTP equipment and the base pressure was pumped down to about 1 mTorr by using a turbo pump. The vacuum was held steady for 4 h to drive out trapped gas inside the package cavity [56.99]. The bonding and vacuum sealing process was done by RTP heating for 10 s at 750 °C to complete the bonding process.

Figure 56.12 shows the measured spectrum of a vacuum-packaged, double-folded beam comb-drive resonator by using a microstroboscope [56.100]. The

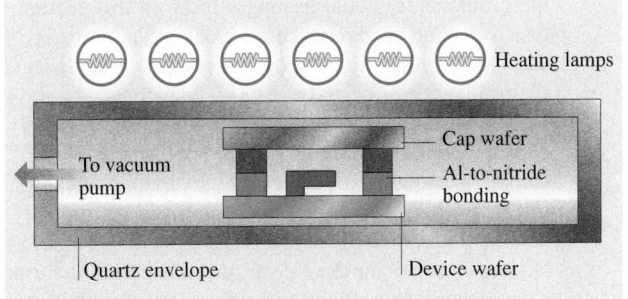

Fig. 56.11 Vacuum packaging apparatus using RTP aluminium-to-nitride bonding. Drawn not to scale

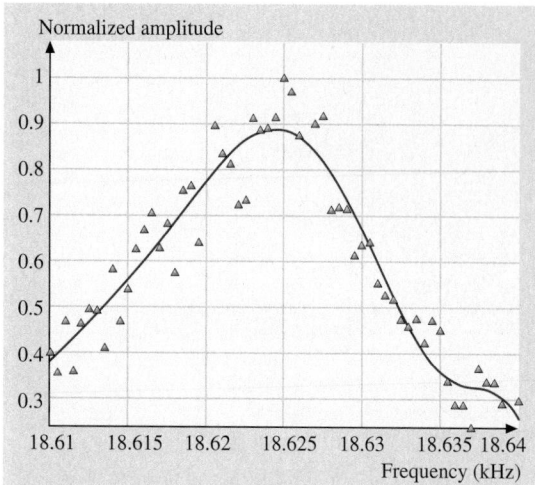

Fig. 56.12 Spectrum measurement results of a vacuum encapsulated comb-shaped resonator by using the RTP aluminium-to-nitride bonding method [56.47]

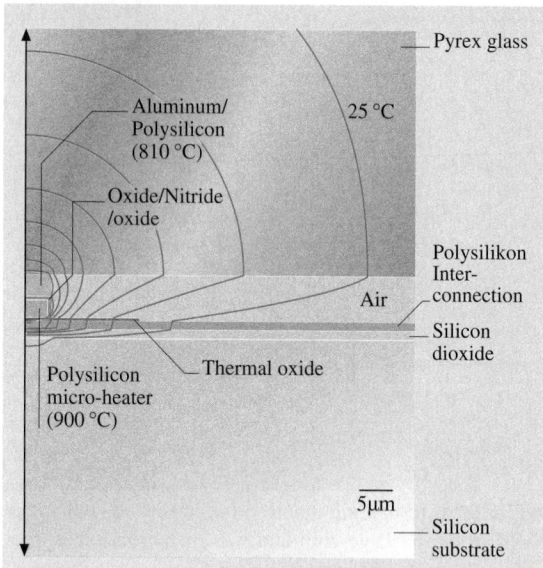

Fig. 56.13 A schematic diagram of the 2-D heat-transfer model, geometry and boundary conditions

central resonant frequency is at about 18 625 Hz and the quality factor is extracted as 1800 ± 200, corresponding to a pressure level of about 200 mTorr inside the package [56.99]. This type of post-packaging process at the wafer level has become the favorite approach to fabricate a hermetic encapsulation because it can provide lower cost and more process flexibility. However, this packaging process relies on good bonding techniques. A strong and reliable bonding between two substrates should be provided and this bonding procedure should be compatible with the other microsystem fabrication processes.

56.2.3 Localized Heating and Bonding

The approach to MEMS post-packaging based on localized heating and bonding has been proposed to address the problems of global heating effects. In this section, resistive microheaters are used as an example to provide localized heating, although several other means of localized heating have been demonstrated, including laser welding [56.101], inductive heating [56.102] and ultrasonic bonding [56.103]. The principle of localized heating is to achieve high temperatures for bonding while maintaining low temperatures globally at the wafer level. Resistive heating by using microheaters on top of the device substrate is applied to form a strong bond with silicon or the glass cap. According to the results of a two-dimensional (2-D) heat-conduction finite-element analysis, as shown in Fig. 56.13, the

steady-state heating region of a 5-μm-wide polysilicon microheater capped with a Pyrex glass substrate can be confined locally as long as the bottom of the silicon substrate is constrained to the ambient temperature. The physics of localized heating behind this design can be understood by solving the governing heat-conduction equations of the device structure without a cap [56.99]. As long as the width of the microheater and the thickness of the silicon substrate are much smaller than the die size and a good heat sink is placed underneath the silicon substrate, the heating can be confined locally. The temperature of the silicon substrate can be kept low or close to room temperature. Several localized resistive heating and bonding techniques have been successfully developed for packaging applications, including localized silicon-to-glass fusion bonding, gold-to-silicon eutectic bonding and localized solder bonding. Several solder materials have been successfully tested, including PSG, indium, and aluminium alloy [56.104].

The vacuum packaging example presented here is based on the localized aluminium/silicon-to-glass solder bonding technique. Built-in folded-beam comb-drive μ-resonators are used to monitor the pressure of the package. Figure 56.14 shows the fabrication process of the package and resonators. Thermal oxide (2 μm) and LPCVD Si$_3$N$_4$ (3000 Å) are first deposited on a silicon substrate for electrical insulation followed by the deposition of 3000 Å of LPCVD polysilicon. This

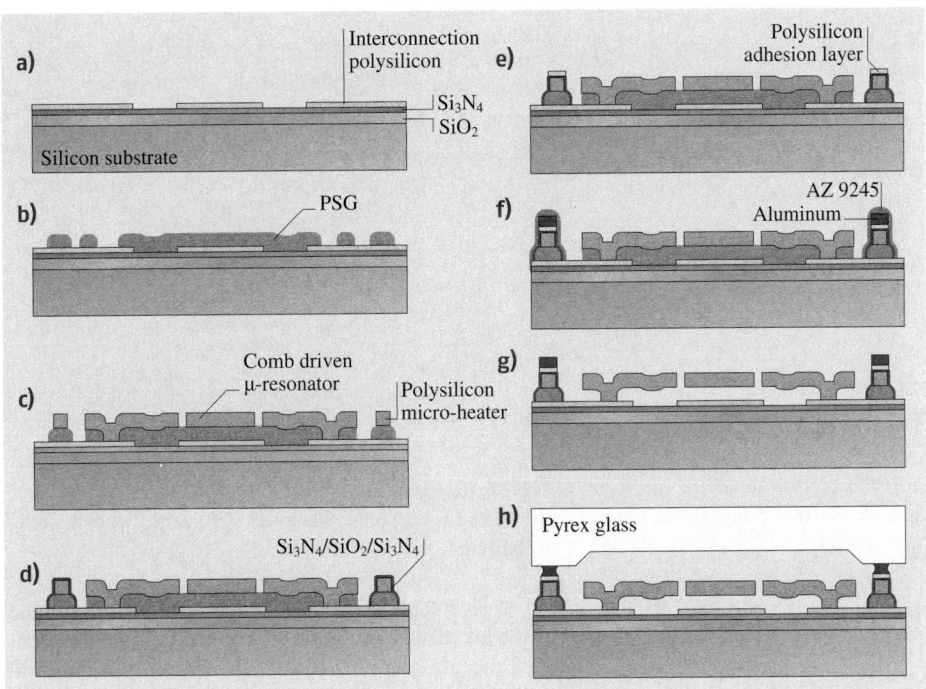

Fig. 56.14 The schematic process flow of vacuum encapsulation using localized aluminium/silicon-to-glass bonding

polysilicon is used as both the ground plane and the electrical interconnect to the μ-resonators, as shown in Fig. 56.14a. Figure 56.14b shows a 2-μm LPCVD SiO$_2$ layer that is deposited and patterned as a sacrificial layer for the fabrication of polysilicon μ-resonators using a standard surface-micromachining process. A 2-μm-thick phosphorus-doped polysilicon is used for both the structural layer of the microresonators and the on-chip microheaters. This layer is formed over the sacrificial oxide in two steps to achieve a uniform doping profile. Lower input power and better process compatibility are two major advantages to using the on-chip microheater in the glass package. The resonators are separated from the heater by a short distance (30 μm) to effectively prevent their exposure to the high heater temperature, as shown in Fig. 56.14c. This concludes the fabrication of the μ-resonators.

In order to prevent the current supplied to the microheater from leaking into the aluminium solder during bonding, an LPCVD Si$_3$N$_4$ (750 Å)/SiO$_2$ (1000 Å)/Si$_3$N$_4$ (750 Å) sandwich layer is grown and patterned on top of the microheater, as shown in Fig. 56.14d. Figure 56.14e and Fig. 56.14f show that aluminium (2.5 μm) and polysilicon (5000 Å) bonding materials are deposited and patterned. The sacrificial release is the final step to form freestanding μ-resonators.

Figure 56.14f shows that a thick AZ 9245 photoresist is applied to cover the aluminium/silicon-to-glass bonding system to ensure that the system withstands the attack from concentrated hydrofluoric acid. After 20 min of sacrificial release in concentrated HF, the system as shown in Fig. 56.14g is ready for vacuum packaging. A Pyrex glass cap with a 10-μm-deep recess is then placed on top with an applied pressure of ≈ 0.2 MPa

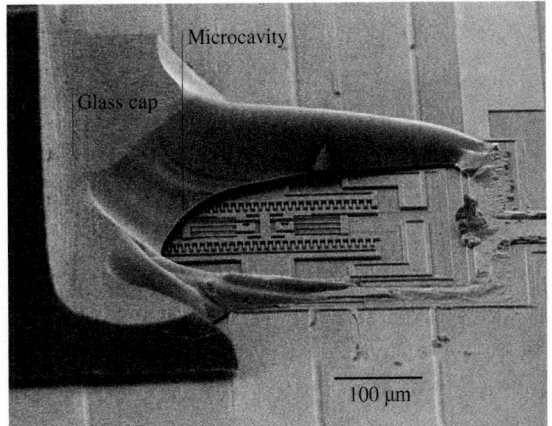

Fig. 56.15 The SEM micrograph of encapsulated microresonators after the glass cap is forcefully broken

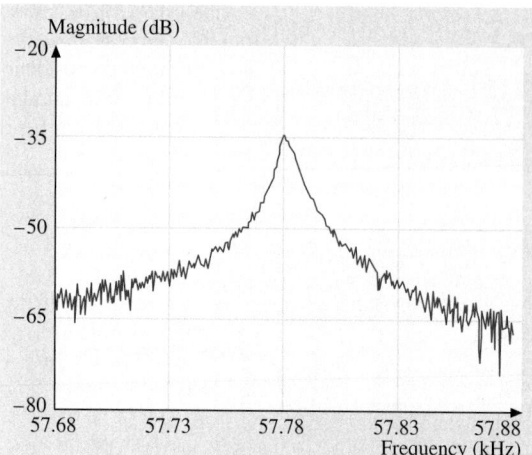

Fig. 56.16 The transmission spectrum of a glass-encapsulated μ-resonator after 120 min pump-down time in vacuum environment ($Q = 9600$)

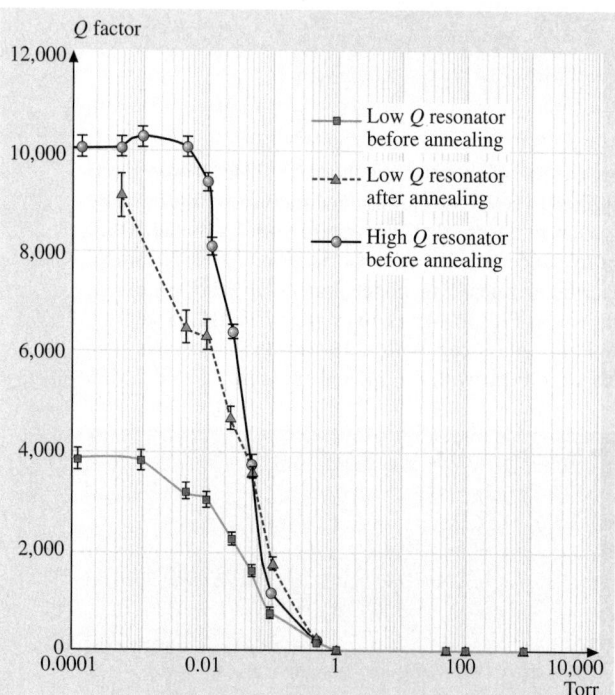

Fig. 56.17 Measured Q-factor versus pressure of un-packaged μ-resonators

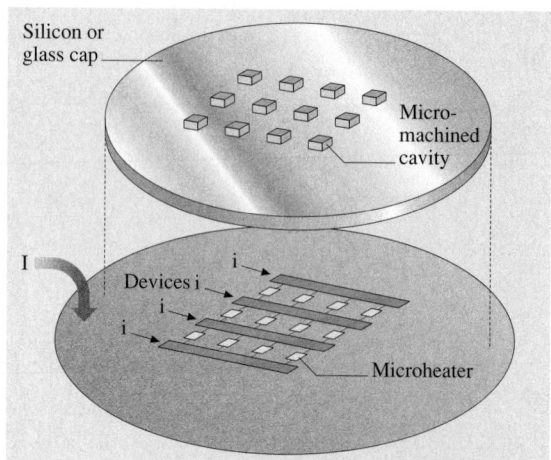

Fig. 56.18 Illustration of wafer-level vacuum packaging at the wafer level using the localized heating and bonding technique

To evaluate the integrity of the resonators packaged using localized aluminium/silicon-to-glass solder bonding, the glass cap is forcefully broken and removed from the substrate. It is observed that no damage is found on the μ-resonator and a part of the microheater is stripped away, as shown in Fig. 56.15, demonstrating that a strong uniform bond can be achieved without detrimental effects on the encapsulated device. Figure 56.16 shows a vacuum-encapsulated unannealed μ-resonator (≈ 57 kHz) after 120 min of wait time. The measured Q-factor after packaging is 9600. Based on the measurement of Q versus pressure of a high-Q unpackaged μ-resonator as shown in Fig. 56.17, it is demonstrated that the pressure inside the packaging is comparable to the vacuum level of th epackaging chamber.

Postprocess packaging using the localized heating and bonding technique includes four basic components: (1) an electrical and thermal insulation layer such as silicon dioxide or silicon nitride should be used for localized heating, (2) resistive microheaters are fabricated to provide the heating source for localized bonding, (3) materials, including metal and polysilicon, which can provide good bonding and hermeticity with silicon or glass substrates are considered as the bonding materials, and (4) a good heat sink under the device substrate for localized heating is provided during the bonding experiments. MEMS/NEMS devices will be fabricated on the device chip and hermetically sealed in the cavity formed by the device chip, resistive microheaters, and protection cap. The process can be either die level or wafer level. The schematic design of the

under a 25 mTorr vacuum, and the heater is heated using 3.4 Watts input power (the exact amount depends on the design of the microheaters) for 10 min to complete the vacuum packaging process, as shown in Fig. 56.14h.

wafer-level packaging process is shown in Fig. 56.18. The resistive microheaters are parallel to each other and connected together in order to ensure that identical current density is applied to individual packages at the same time. These heaters can either be fabri-cated on the chip or protection cap and can be built into a larger wafer for current inputs. The interconnections for these packaging cavities can be built into the dicing area such that no extra space is required for the packaging process.

56.3 Thermal Issues and Packaging Reliability

56.3.1 Thermal Issues in Packaging

The two key thermal issues related to MEMS packaging are: (1) heat dissipation from actuators and integrated circuitry components, and (2) thermal stress generated during the packaging process. These two topics are discussed separately.

Heat Dissipation Issues
In microelectronic chips, heat dissipation is a serious problem as the size of transistors continues to shrink and the density of transistors on a chip continues to increase with advances in IC fabrication technology. The trend of power packing into smaller packages has created increasing thermal management challenges [56.105]. Since the electrical characteristics of transistors change with working temperature, inefficient heat dissipation could raise the working temperature and affect device performance. Present MEMS/NEMS devices do not need high-power high-performance microprocessors, so power dissipation is not a problem. Nevertheless, some functional components in packaged MEMS/NEMS are very sensitive to temperature variation, such as biomaterials or laser diodes. Several chemical sensors and other applications such as micro polymerase chain reaction (PCR) chambers for deoxyribonucleic acid (DNA) replication actually require elevated temperature for operation and micro-thermal platforms are built for these devices. Thermal management to maintain the working temperature on these chips for stable operation is still an essential issue for packaging considerations. The geometrical complexity of MEMS/NEMS resulting from packing various functional components in a tight space increases the difficulty of thermal management. As packaging integration process becomes more complex, the fabrication constraint in the packaging process will have a great impact on the heterogeneous integration process in the front-end MEMS/NEMS and IC processes. For example, the requirement for low temperatures in the packaging process generally limits the possible choices of materials in the back-end process. In general, conventional IC packaging employs a heat sink attached to the chip to remove heat. The heat sink is generally made of a copper or stainless-steel bar with an array of fin structures on one side for better natural or forced heat convection to dissipate heat into the environment. In addition to heat sinks, thermal vias, heat pipes, immersion cooling, and thermoelectric cooling can also be used for effective heat removal. Because most MEMS/NEMS packages still follow the typical IC packaging architecture, one promising thermal management method, the heat pipe, is discussed for possible MEMS/NEMS packaging applications.

A heat pipe is a sealed slender tube containing a wick structure and a working fluid, typically water in electronics cooling. It is composed of three sections, the evaporator section at one end, the condenser section at the other end and the adiabatic section in the middle. In the evaporator section, heat is absorbed by the working fluid via a phase transformation from liquid to vapor. In the condenser section, heat dissipates into the outside environment. Thus, the fluid goes back to the liquid phase. The vapor phase is in a high-pressure high-temperature state that forces the vapor to flow into the condenser section at a lower temperature. Once the vapor condenses and gives up its latent heat, the condensed fluid is then pumped back to the evaporator section by the capillary force developed in the wick structure. Therefore, the middle adiabatic section contains two phases, the vapor phase in the core region and the liquid phase in the wick, flowing in opposite directions to each other and with no significant heat transfer between the fluid and the surrounding medium. Silicon has good thermal conductivity ($1.41 \, \text{W/cm}^\circ\text{C}$) and is easily micromachined to fabricate the heat pipe. Therefore, there is a great potential for the implementation of the silicon micro heat pipe in IC and MEM/NEMS packaging and several approaches have been proposed on this topic [56.106–108].

Packaging-Induced Thermal Stresses
Thermal-based bonding processes have been used in MEMS/NEMS packaging applications for many years, as discussed in this chapter. Thermal management is ex-

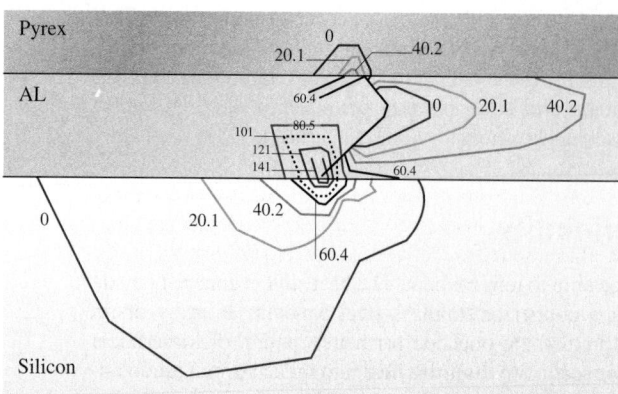

Fig. 56.19 Residual stress (GPa) for an aluminium solder width of 100 μm in the RTP silicon–aluminium–glass bonding system

tremely important during the bonding process to avoid fracture in the substrate or MEMS/NEMS device itself. Extremely high temperatures or rapid cooling conditions are may cause damage and should be carefully evaluated both analytically and experimentally. There are many ways to provide heating energy including electrical resistive heating, oven heating or induction heating [56.102]. These bonding processes may be put into two categories: (1) localized bonding, where the heat is directly applied only to the adhesive material used to bond the package to the device, and (2) global heating, where the entire system (MEMS/NEMS device, adhesive, and packaging material) is heated to produce bonding of the materials, which is the most common approach. Therefore, this section focuses on the thermal stress effects during the heating and cooling procedures in MEMS/NEMS packaging. The RTP aluminium-to-glass bonding process is used as a specific example for the discussion of thermal stresses [56.93]. The bonding process heats up the packaging system to 750 °C for 10 s, and then cools it back to room temperature. To simulate this process, an ANSYS program [56.109] was established to examine the shear stress due to coefficient of thermal expansion (CTE) variations in the bonding system as a result of temperature changes. The shear stress was recorded from the ANSYS analysis on the aluminium/Pyrex glass interface and the aluminium/silicon interface.

Two different models were analyzed. The first was the quartz–aluminium–silicon bonding system and the second was the Pyrex glass–aluminium–silicon bonding system [56.110]. The results of the ANSYS calculation were then analyzed and compared with experimental observations. Figure 56.19 shows the ANSYS results for a Pyrex glass bonding system, where the width of the aluminium solder is 100 μm and the maximum residual stress is 60 MPa in glass, which is slightly lower than the fracture strength of Pyrex glass at 70 MPa. It was discovered that increasing the aluminium width can lower these residual stresses. For example, the maximum residual stress analyzed from ANSYS in the Pyrex glass bonding systems is 74.5 Gpa, 58 Gpa, and 60 GPa for aluminium widths of 30 μm, 50 μm, and 100 μm, respectively [56.110]. Pyrex glass has a documented strength of around 69 GPa [56.111]. Fracture should always occur with an aluminium width of 30 μm or less, according to the ANSYS analysis. Fracture may occur sporadically at widths of 50 μm or 100 μm, depending on the amount and magnitude of the flaws in the Pyrex glass. Experiments were done on the Pyrex glass bonding systems with a width of 100 μm. The samples were heated up to 750 °C and then cooled down by taking them out of the oven. In all four experimental cases, small cracks were observed in the Pyrex glass, as shown in Fig. 56.20. These cracks may have occurred consistently for several reasons. Firstly, they may be a result of handling of the Pyrex glass before bonding. The Pyrex glass samples were kept in containers with each other, which may have resulted in abrasive contact. This may have caused flaws in the materials, which could have reduced the fracture strength. Secondly, it was observed that the cracks are small, only occurring tens of microns away from the aluminium and not propagating completely through the Pyrex glass. These cracks could be caused by the high stress applied, but the crack has not reached a critical size, and therefore has not propagated completely through the Pyrex glass. Therefore, the strength remains at the theoretical strength of 69 GPa, and the Pyrex glass is only partially cracked. Experimental analysis done by *Chaio* and *Lin* [56.93] shows that fracture was not observed when using aluminium

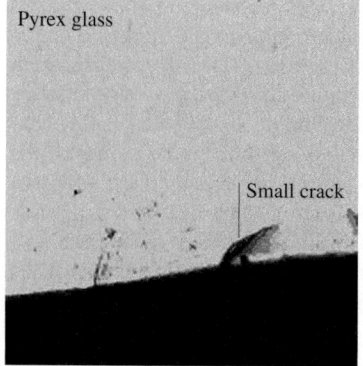

Fig. 56.20 Micrograph of the experimental result on the Pyrex glass–aluminum–silicon system. Small cracks can be observed

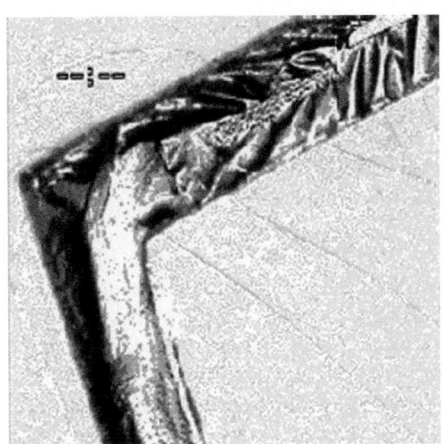

Fig. 56.21 Bonding result of the quartz–aluminum–silicon system where fracture can be observed

widths greater than 150 μm. This is consistent with the results of the ANSYS analysis that showed that, as the width of the aluminium is increased, the residual stress decreases.

The ANSYS stress predictions for the quartz bonding system yielded a maximum stress of 207 GPA, 117 GPa, and 100 GPa, for aluminium widths of 30 μm, 50 μm, and 100 μm, respectively. All three of these stresses are much larger than the theoretical strength of quartz at 48 GPa, and therefore fracture should always occur. Quartz has a much larger coefficient of thermal expansion (CTE) than silicon, which is why this was predicted. Experimentally, a quartz substrate was tested for the silicon–aluminium–quartz bonding test and the result is shown in Fig. 56.21. It is observed that cracks occur all over and cause serious damage on the quartz wafer. These cracks could be the failure mechanism of the hermetic package. Therefore, Pyrex glass is identified as a better bonding substrate than quartz.

The thermal stresses generated in the packaging process in quartz are much larger than the stresses in Pyrex glass, because of the difference between the CTE in the two systems. Quartz has a low CTE (0.54×10^{-6} K^{-1}) compared to the aluminium CTE (23×10^{-6} K^{-1}) and silicon CTE (3.5×10^{-6} K^{-1}). On the other hand Pyrex glass has a much closer CTE (3.2×10^{-6} K^{-1}) to silicon and aluminium, resulting in smaller stresses. The practical implications of the ANSYS results and the preceding information are that materials must be chosen carefully when carrying out the bonding process. Materials with a much higher or lower CTE than silicon should not be used to ensure fracture will not occur.

56.3.2 Packaging Reliability

Packaging is one of the key issues to be addressed for the evaluation of the reliability of MEMS/NEMS products. Any defects created during the sealing and packaging process may cause immediate device failure or may degrade the device performance over time. For example, micro-accelerometers that are used to deploy air bags in automobile safety applications require excellent reliability. If any leakage path is created during the sealing process between the two bonding interfaces, moisture may enter the sealed microcavity and cause device failure over time. Furthermore, thermal stress induced by the CTE mismatch is one of the main factors that affect the packaging reliability. In fact, the formation of the stress can happen not only during packaging process but also during the operation of devices. In particular, during device operation, the package will go through various temperature cycles because of environmental change. Such temperature variation causes the expansion of packaging materials when they are constrained by the packaged assembly. As a result of such thermal mismatch, significant stresses are induced in the package and may cause the device to fail. In addition to thermal mismatch, corrosion, creep, fracture, fatigue crack initiation and propagation, and delamination of thin films are all possible factors that may cause the failure of packaged devices [56.112]. These failure mechanisms could be prevented or deferred by using proper packaging designs. For instance, the thermally induced strain inside the packaging material is generally below the tolerance of the material and cannot cause immediate catastrophic damage. However, cyclic loading can generate and accumulate stresses and eventually cause failure. Several common designs in IC packaging have been used to prolong the lifetime of devices. For example, the strain in solder interconnects of BGA or flip-chip packaging can be effectively reduced by introducing a polymer underfill material between the chip and the substrate to distribute well the thermal stress induced by CTE mismatch [56.113]. The strain can be further reduced if excellent thermal paths are built around interconnects to alleviate thermal stress originating from the temperature gradient between the ambient and operation temperatures. On the other hand, delamination phenomena occur in the interface of adjacent material layers such as components made of dissimilar materials that are subsequently bonded together. Delamination can result in electrical or mechanical failures of devices such as mechanically cracking through the electrical via wall to make an electrical open because of the propagation of the delamination of metal line from the dielectric layer or overheating of the die because of delamination of the underside of the die, causing openings in the heat dissipation path. Because of the stress and thermal loading, the geometry, and the material properties are complex in MEMS/NEMS, the development of the packaging designs to increase the reliability is very important and requires extensive investigations.

Reliability testing is required before a new device can be delivered into the market. The test results can provide information for the improvement of packaging design and fabrication. Hence, how to analyze the failure data, which is called the reliability metrology, is very important in the packaging industry. The analysis method is to use the mathematical tools of probability and statistical distributions to evaluate data to understand the patterns of failure and to identify the sources of failure. For example, a failure density function is defined as the time derivative of the cumulative failure function

$$f(t) = \frac{\mathrm{d}F(t)}{\mathrm{d}t} \,, \tag{56.1}$$

$$F(t) = \int_0^t f(s)\,\mathrm{d}s \,. \tag{56.2}$$

The cumulative failure function, $F(t)$, is the fraction of a group of original devices that has failed at time t. The Weibull distribution function is one of the analytic mathematic models commonly used in the packaging reliability evaluation to represent the failure density function [56.12].

$$f(t) = \frac{\beta}{\lambda} \left(\frac{t}{\lambda} \right)^{\beta - 1} \exp\left[-\left(\frac{t}{\lambda} \right)^{\beta} \right] \,, \tag{56.3}$$

where β and λ are the Weibull parameters. The parameters β is called the shape factor and measures how the failure frequency is distributed around the average lifetime. The parameter λ is called the lifetime parameter and indicates the time at which 63.2% of the devices failure. By integrating both sides of the equation (56.3), $F(t)$ becomes

$$F(t) = 1 - \exp\left[-\left(\frac{t}{\lambda} \right)^{\beta} \right] \,. \tag{56.4}$$

Using the Weibull distribution function with the two parameters extrapolated by experimental data, one can estimate the number of failures at any time during the test. Moreover, knowing the meaning and values of the parameters, one can compare two sets of test data. For example, greater value of λ indicate that samples have a longer lifetime. Because all of the mathematical models are statistical approximations based on real experimental data, more testing samples provide better estimation accuracy.

56.3.3 Long-Term and Accelerated MEMS Packaging Tests

The capability to estimate the reliability or lifetime of a device provides valuable information for the manufacturer to maximize the profit margin by balancing the cost and the quality of the product. Moreover, a warranty period given by the manufacturer has to be built on the reliability information of the product. The reliability of MEMS/NEMS packages is best characterized using long-term tests with statistical data analyses. However, it is very difficult to measure the reliability or lifetime of a device in real time because testing over a prolonged time period may be required to prompt many devices to fail. In order to evaluate the reliability of a device in a timely fashion, accelerated testing is normally conducted in order to speed up the device aging process and thus shorten the required testing time. Accelerated testing, from the packaging and sealing point of view, is a testing method that emphasizes failure of the seal when foreign elements that may affect device performance leak into the microcavity. For a hermetic package, the lifetime of a MEMS/NEMS device is essentially an estimation of the time required for water to penetrate into the package. For vacuum encapsulated MEMS/NEMS devices, in addition to water penetration through the seal, gas penetration or outgassing from within the packaging materials such as the substrate, cap and sealing materials over time can degrade the vacuum level and thus the device performance. Therefore, the lifetime of a vacuum encapsulated MEMS package can be evaluated by the time that gas evolves into the package from either the seal or within the device materials, whichever comes first. Unfortunately, there are not many research publications that deal with long-term and accelerated test issues for MEMS/NEMS packaging reliability. In the conventional IC packaging industry, reliability estimation is carried out by accelerated tests and statistical predictions [56.12]. Accelerated tests often utilize high temperatures and humidities, such as autoclave tests [56.114] to speed up corrosion against the sealing boundary to accelerate the failure of the packages. The industry could use very similar accelerated tests to estimate the lifetime of a package because the basic assumptions of the failure mode and humidity issues are similar to those for conventional IC packages.

Several research groups have reported reliability studies for MEMS packages formed by different bonding methods and materials [56.94, 95, 115]. In this section, two packaging examples that aim to address long-term and accelerated tests are discussed.

Figure 56.22 shows long-term measurements of the Q-factor of a vacuum packaged μ-resonators using localized aluminium/silicon-to-glass bonding [56.94]. The vacuum encapsulation process is described in detail in Fig. 56.8. It was found that vacuum packaging by means of localized heating and bonding provides stable vacuum environments for the μ-resonator, and a quality factor of 9600 has been achieved with no degradation for at least one year. Since the performance of high-Q μ-resonators is very sensitive to environmental pressure, as shown in Fig. 56.17, any leakage can be easily detected. The fact that this high Q value can be maintained for one year indicates that the packaging process has been well performed and that both aluminium and Pyrex glass are suitable materials for vacuum packaging applications. According to a previous study of hermeticity in different materials, metal has a lower permeability to moisture than other materials such as glass, epoxy, and silicon. With a width of 1 μm, metal can effectively block moisture for more than 10 years [56.12]. In this vacuum package system, the bonding width is 30 μm such that it can sufficiently block the diffusion process of moisture. On the other hand, the diffusion effects of air molecules into these tiny cavities have not been studied extensively and the design guidelines for vacuum encapsulations are not clearly defined. Further investigations will be needed in this area and the example presented here serves as a good starting point.

On the other hand, accelerated testing puts a large number of samples in harsh environment, such as elevated temperature, elevated pressure and 100% humidity, to accelerate the corrosion process. The statistical failure data are gathered and analyzed to predict the lifetime of packages under a normal usage environment. As a result, the long-term reliability of the package can be predicted without going through true long-term tests. Unfortunately, accelerated tests have been an area that has not been addressed in MEMS/NEMS research papers. Although the industry must have done some extensive reliability tests, they do not publish their results, probably due to liability concerns. Among the very limited publications, this section uses a specific MEMS packaging system that has gone through accelerated tests as an illustrative example [56.47].

The package is accomplished by means of rapid thermal processing (RTP) bonding as described previously

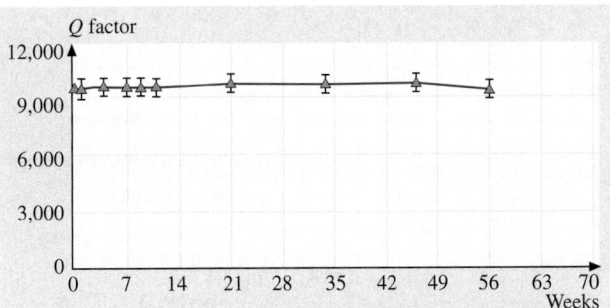

Fig. 56.22 Long-term measurement of encapsulated μ-resonators. No degradation of Q-factors is found after 56 weeks

in this chapter. The goal of the accelerated test is to examine the failure rate at the bonding interface. The accelerated tests start by putting the packaged samples into an autoclave chamber filled with high-temperature pressurized steam at 130 °C, 2.7 atm and 100% relative humidity for accelerated tests. The pressurized steam can penetrate small crevasses if there is a defect at the bonding interface [56.65]. Elevated temperature and humid environment speed up the corrosion process. A package is considered as a failure if water condenses or diffuses into the package. The statistical data gathered from accelerated tests in this case has been categorized as *right-censored* data [56.116]. Statistical failure data are gathered every 24 h under optical examination for a period of 864 h when new failure is seldom observed (therefore, right-censored on the time axis). In practice, this method was easier and more economical to implement than other methods. Owing to the robustness of the sample, it is difficult to conduct the tests to the point where all packages fail. The cumulative failure function

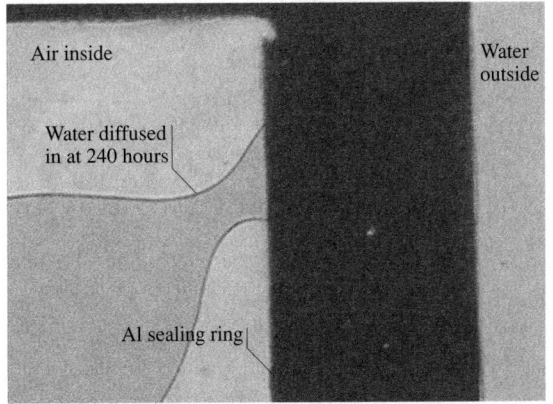

Fig. 56.23 A particular device failed at 240 h of testing time

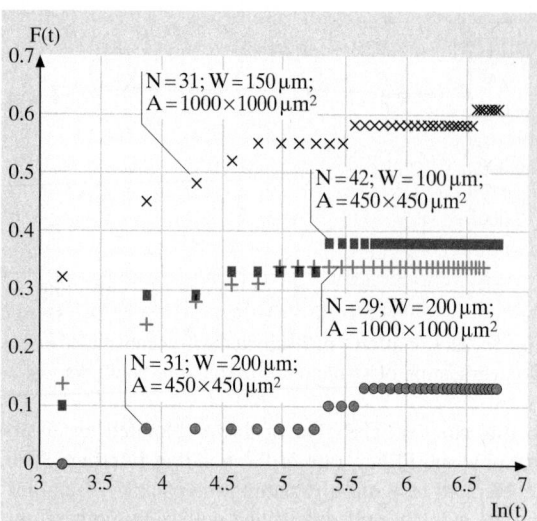

Fig. 56.24 Cumulative failure data

$F(t)$ is defined as:

$$F(t) = \frac{\text{Number of cumulative failures}}{\text{Number of samples } N} ,\qquad (56.5)$$

where N is the sample size at the beginning of the test. A package was considered as a failure if water condensed inside or diffused into the package. For example, water was found to diffuse into the cavity after

240 h into the test, as shown in Fig. 56.23. However, no leakage path could be identified under an optical microscope in this case. Figure 56.24 shows that the function $F(t)$ (in %) is plotted versus the logarithm of time. In general, most of the failures occurred in the first 96 h [$\ln(t) \approx 4.56$] and such a high number of early failures reflects the yielding issue of the sealing process. Moreover, packages with a smaller bonding width and larger bonding areas showed a higher percentage of failure. Both Weibull and lognormal statistical models [56.116] were used and compared to analyze the collected data to predict the lifetime of the packages, and the least-square fit method was used to determine the best fitting model. It was found that R^2, the *coefficient of determination* [56.116], values were generally in the range of 0.8 using the lognormal model, compared to values of 0.5 using the Weibull model. Therefore, the lognormal model was used to predict the life time of packages.

Figure 56.25 shows the inverse standard normal distribution function versus ln(time) and the maximum likelihood estimator (MLE) is then used to predict the mean, standard deviation and the mean time to failure (MTTF). Table 56.2 shows the MLE calculation results for the MTTF. The wide confidence level interval comes from the fact that only a small number of samples failed at the end of the test. It is also observed that packages with larger bonding widths and smaller bonding areas have larger MTTF values. The lower bound on the MTTF

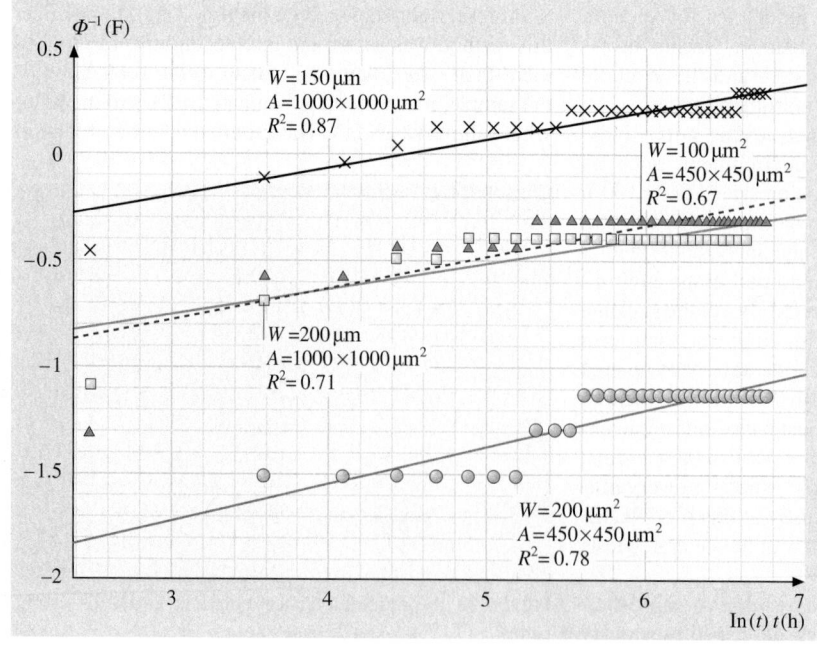

Fig. 56.25 Life data fitted by the lognormal distribution. R^2 is the coefficient of determination

Table 56.2 The maximum likelihood estimation for mean time to failure (MTTF)

MLE calculation resuls of MMTF[a]			
Bonding width W (μm)	Area A (μm^2)	MTTF		Worst cases in jungle condition (years)
		UB (years)	LB (years)	
200	450×450	1.8×10^7	0.57	1700
100	450×450	5.3	0.10	300
200	1000×1000	6.5×10^3	0.09	270
150	1000×1000	0.50	0.017	50

[a] UB is the upper bound, and LB is the lower bound of the 90% confidence interval, respectively. The MTTF LB times AF is the worst-case MTTF used in the jungle condition

provides the worst-case scenario. For example, only 4 out of 31 samples failed by the end of the test in the case of a bonding ring width of 200 μm and a sealing area of 450 μm×450 μm^2. The MTTF predicts, in the worst-case scenario, that there is a 90% chance that a package will fail in 0.57 years in the autoclave environment.

It is widely accepted that the acceleration factor (AF) for autoclave tests follows the Arrhenius equation [56.12] and can be modeled as:

$$ \text{AF} = \frac{(\text{RH}^{-n}\, e^{\Delta E_a/kT})_{\text{normal}}}{(\text{RH}^{-n}\, e^{\Delta E_a/kT})_{\text{accelerated}}} \quad (56.6) $$

Where RH is the relative humidity (85%, RH = 85), k is the Boltzmann constant and T is the absolute temperature. The recommended value for n, an empirical constant, is 3.0 [56.117], and ΔE_a, the activation energy, is 0.9 eV for a plastic dip package and 0.997 eV for an anodically bonded glass-to-silicon package [56.65].

If $\Delta E_a = 0.9$ is used to estimate the AF for the accelerated testing conditions compared with the jungle conditions (35 °C, 1 atm and 95% RH), the AF is about 3000 and the worst-case lifetime values in the jungle conditions are also listed in Table 56.2. The high values of estimated MTTF in the jungle condition could be a result of overestimation of the AF because plastic dip package may have smaller AFs than those of glass packages. Nevertheless, these data and analyses provide important guidelines in the area of accelerated tests for MEMS packages.

For vacuum packaged MEMS devices, the lifetime can be evaluated by monitoring the quality factor of microresonators inside their sealed cavities. Again, vacuum packaged MEMS resonators by using RTP aluminium-to-nitride bonding are discussed in detail here for better illustration of the various factors relevant to reliability. It was found that, under normal

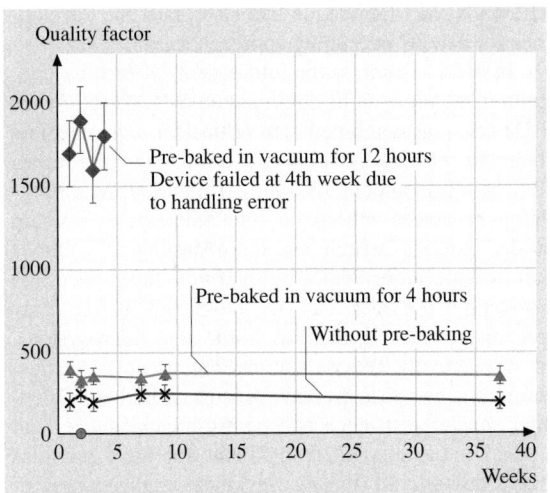

Fig. 56.26 Long-term stability tests to 37 weeks. The Q-value increases with the pre-baking time

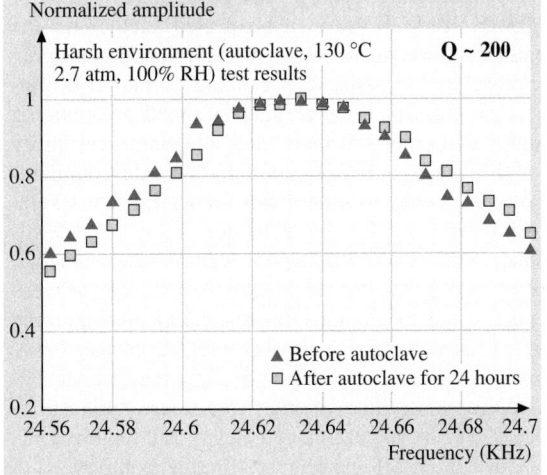

Fig. 56.27 Spectrum measured before and after accelerated testing for 24 h

condition at room-temperature storage, the quality factor of resonators remained constant after 37 weeks, as shown in Fig. 56.26. Furthermore, the vacuum quality under a harsh environment was characterized by putting a vacuum-encapsulated comb-resonator into an autoclave chamber (130 °C, 2.7 atm and 100% RH) for accelerated testing. The result is shown in Fig. 56.27 and the quality factor stayed at 200 after 24 h in the autoclave testing chamber. Since the slight differences between the two spectra are within the normal experimental errors, a conclusion may be drawn that this harsh environment test did not affect the vacuum seal.

In order to characterize the vacuum lifetime of the packages, two vacuum packaged comb-resonators were put in the harsh environment for continuous testing for up to 1008 h and the results are summarized in Table 56.3. The quality factor of each package has been measured in every 24 h interval and they were found to be maintained at 400 and 200, respectively, before failure. The first packaged resonator with a Q-value of 200 had an aluminium sealing ring with a width of 75 μm and a sealing area of 650 μm × 650 μm. The second packaged microresonator had a Q-value of 400 with an aluminium sealing ring width of 200 μm and a sealing area of 550 μm × 550 μm. If the accelerated

Table 56.3 Summary of the accelerated testing results of two vacuum packaged resonators

	Package 1	Package 2
Q-value	400	200
Hours under testing (h)	> 1008	576
MTTF, lower bond (h)	769	149

testing results on water penetration as discussed previously [56.95] were applied here for gas penetration as a measure of the vacuum sealing characteristics, the accelerated lifetime of the first and second packaged resonators would be 0.017–0.1 years (149–876 h) and 0.09–0.57 years (769–4993 h), respectively. Experimentally, the first packaged resonator failed at 576 h into the test in the autoclave chamber. This corresponds well to the lifetime prediction from the vapor-penetration work [56.95]. The second packaged resonator survived in the autoclave chamber for more than 1008 h (the device did not fail) and this result also verified the prediction made from the vapor-penetration work. However, it is noted that these results are preliminary data and more packaged devices and tests should be conducted to establish meaningful statistical analyses.

56.4 Future Trends and Summary

In the past, the development of MEMS/NEMS packaging mainly originated from IC packaging advancements because existing packaging techniques can significantly reduce the development cost of MEMS/NEMS. However, it is expected that the situation will change very soon so that MEMS/NEMS packaging approaches will assist IC packaging development. Recent progress in IC packaging is aimed at providing high I/O density and more chip integration capability for the needs of high-speed high-data-rate communications. In order to satisfy those requirements, several packaging concepts and techniques have been developed, such as 3-D packaging, wafer-level packaging, BGA, and the flip-chip technique. Although all of these concepts and methods can provide a package with more I/O density, flexibility in chip integration, and lower manufacturing cost for IC fabrication, they are still insufficient to provide solutions for future applications because of the increasing complexity and requirements of MEMS/NEMS packaging. On the contrary, with the progress of MEMS/NEMS fabrication technologies, several key processes such as deep reactive-ion

etching (DRIE), wafer bonding, and thick photoresist processes [56.118] have been utilized for IC packaging fabrication. Therefore, technologies developed in MEMS/NEMS fabrication can also assist the development of new IC packaging approaches.

In order to address the future needs of process integration, adaptive multi-chip module (MCM) [56.31] or 3-D packaging combined with vertical through-substrate interconnects [56.10, 119] are promising approaches for the development of future MEMS/NEMS packaging processes. Based on low-temperature flip-chip solder bonding techniques, these packaging methods can provide more flexibility in device fabrication and packaging. Devices can be fabricated before they are integrated together to form the microsystems to dramatically reduce the packaging cost. Vertical through-substrate interconnects can have higher I/O density, smaller resistance and parasitic capacitance, and mutual inductance. Although this approach provides many possible advantages, technical challenges exist. For instance, metal is commonly used as the fill material inside the vertical vias to form electrical interconnects.

This may introduce a large thermal mismatch with respect to the silicon substrate and generate huge thermal stress that could cause packaging reliability problems. Moreover, it will be an interesting engineering challenge to fill materials into those high-aspect-ratio vias.

The future development of MEMS/NEMS packaging depends on the successful implementation of unique techniques as described in the following:

1. Development of mechanical, thermal, and electrical models for packaging designs and fabrication processes;
2. Wafer-level chip-scale packaging with low packaging cost and high yields;
3. Effective testing techniques at the wafer level to reduce testing costs;
4. Device integration by vertical through-interconnects as an interposer [56.1] to avoid thermal mismatch problems.

In addition to these approaches and challenges, there are many other possibilities that have not been listed but also require dedicated investigation. For example, several key nanotechnologies have been introduced in the previous chapters but packaging solutions for these NEMS devices have not been addressed. Because it is feasible to use MEMS as a platform for NEMS fabrication, all the packaging issues discussed in the chapter can be directly applied to NEMS devices. On the other hand, nanotechnology may introduce new opportunities for MEMS/NEMS packaging applications by providing superior electrical, mechanical and thermal properties [56.120–123]. For example, carbon nanotubes have very high thermal conductivity [56.122] and may enable improved cooling effects for IC/MEMS/NEMS packaging applications.

In summary, MEMS/NEMS packaging issues have been introduced in the areas of fabrication, application, reliability and future development. Packaging design and modeling, packaging material selection, packaging process integration, and packaging cost are the main issues to be considered when developing a new MEMS/NEMS packaging process.

References

56.1 M. Matsuo, N. Hayasaka, K. Okumura, E. Hosomi, C. Takubo: Silicon interposer technology for high-density package, IEEE, ECTC, 1455–1459 (2000)

56.2 K. E. Peterson: Silicon as a mechanical material, Proc. IEEE **70**, 420–457 (1982)

56.3 L. Lin, H.-C. Chu, Y.-W. Lu: A simulation program for the sensitivity and linearity of piezoresistive pressure sensors, IEEE/ASME J. Microelectromech. Syst. **8**, 514–522 (1999)

56.4 Y. C. Tai, R. S. Muller: Lightly-doped polysilicon bridge as a flow meter, Sensors Actuators **15**, 63–75 (1988)

56.5 D. Hicks, S.-C. Chang, M. W. Putty, D. S. Eddy: Piezoelectrically activated resonant bridge microaccelerometer. In: *Solid-State Sensors and Actuators Workshop* (Transducers Research Foundation, Inc., Hilton Head Island 1994) pp. 225–228

56.6 J. Berstein, S. Cho, A. King, A. Kourepenis, P. Maciel, M. Weinberg: A micromachined comb-drive tuning fork rate gyroscope. In: *6th IEEE, Int. Conf. on MEMS* (IEEE, Fort Lauderdale, FL 1993) pp. 143–148

56.7 M. Madou: Compatibility and incompatibility of chemical sensors and analytical equipment with micromachining. In: *Solid-State Sensor and Actuator Workshop* (Transducers Research Foundation, Inc., Hilton Head 1994) pp. 164–171

56.8 M. J. Zdeblick, J. B. Angell: A Microminiature Electric-to-Fluidic Valve. In: *Proceedings of Trand-

sucers'87, the 4th International Conference on Solid-State Transducers and Actautors* (IEEE, Tokyo 1987) pp. 827–829

56.9 J. H. Tsai, L. Lin: A thermal bubble actuated micro nozzle-diffuser pump, IEEE/ASME J. Microelectromech. Syst. **11**, 665–671 (2002)

56.10 S. J. Ok, D. Baldwin: High density, aspect ratio through-wafer electrical interconnect vias for low cost, generic modular MEMS packaging. In: *8th IEEE, Int. Sym. Adv. Pack. Mat.* (IEEE, New York 2002) pp. 8–11

56.11 H. Reichl, V. Grosser: Overview and development trends in the field of MEMS packaging. In: *14th IEEE, Int. Conf. on MEMS* (IEEE, Interlaken, Switzerland 2001) pp. 1–5

56.12 R. R. Tummala, E. J. Rymaszewski, A. G. Klopfenstein: *Microelectronics Packaging Handbook, Semiconductor Packaging* (Chapman Hall, New York 1997)

56.13 C. A. Harper (Ed.): *Electronic Packaging and Interconnection Handbook.* (McGraw–Hill, New York 1991)

56.14 P. Kapur, P. M. McVittie, K. Saraswat: Technology and reliability constrained future copper interconnects, part II: performance implication, IEEE Trans. Elec. Dev. **49**, 598–604 (2000)

56.15 J. Lau: *Flip Chip Technologies* (McGraw–Hill, New York 1996)

56.16　R. Prasad: *Surface mount technology: principles and practice*, 2nd edn. (Chapman Hall, New York 1989)

56.17　M. Töpper, J. Auersperg, V. Glaw, K. Kaskoun, E. Prack, B. Keser, P. Coskina, D. Jäger, D. Petter, O. Ehrmann, K. Samulewicz, C. Meinherz, S. Fehlberg, C. Karduck, H. Reichl: Fab integrated packaging (FIP) a new concept for high reliability wafer-level chip size packaging, IEEE, ECTC (Las Vegas, NV 2000) 74–81

56.18　S. Savastiouk, O. Siniaguine, E. Korczynski: 3D wafer level packaging, Int. Conf. High-Density Interconnect and Systems packaging **9**, 26–31 (2000)

56.19　J. H. Tsai, L. Lin: Micro-to-macro fluidic interconnectors with an integrated polymer sealant, J. Micromech. Microeng. **11**, 577–581 (2001)

56.20　T. A. Core, W. K. Tsang, S. Sherman: Fabrication technology for an integrated surface-micromachined sensor, Solid State Technol., 39–47 (1993)

56.21　K. Koester, R. Majedevan, A. Shishkoff, K. Marcus: Multi-user MEMS processes (MUMPS) introduction and design rules, rev. 4, MCNC MEMS Technology Applications Center, Research Triangle Park, NC 27709, July (1996)

56.22　R. Lengtenberg, H. A. C. Tilmans: Electrically driven vacuum-encapsulated polysilicon resonantor, part I: design and fabrication, Sensors Actuators A **45**, 57–66 (1994)

56.23　R. L. Smith, S. D. Collins: Micromachined packaging for chemical microsensors, IEEE Trans. Electr. Dev. **ED 35**, 787–792 (1988)

56.24　M. C. Wu: Micromachining for Optical and Optoelectronic Systems, Proc. IEEE **85**(11), 1833–1856 (1997)

56.25　M. Putty, K. Najafi: A Micromachined Gyroscope. In: *Solid-State Sensors and Actuators Workshop* (Transducers Research Foundation, Inc., Hilton Head Island, SC 1994) pp. 212–220

56.26　S. Wolf: *Silicon Processing for the VLSI Era, Vol. I: Process Technology* (Lattice, Sunset Beach, CA 1995)

56.27　J. K. Bhardway, H. Ashraf: Advanced silicon etching using high density plasmas, SPIE Micromach. Fab. Technol. **2639**, 224–233 (1995)

56.28　W. H. Ko, J. T. Suminto, G. J. Yeh: *Bonding techniques for microsensors, Micromachining and Micropackaging for Transducers* (Elsevier Science, New York 1985)

56.29　S. D. Senturia, R. L. Smith: Microsensor packaging and system partitioning, Sensors Actuators **15**, 221–234 (1988)

56.30　J. T. Butler, V. M. Bright, P. B. Chu, R. J. Saia: Adapting multichip module foundries for MEMS packaging. In: *IEEE, Int. Conf. on Multi. Mod. and High Den. Pack.* (IEEE, Denver, CO 1998) pp. 106–111

56.31　M. Schuenemann, A. J. Kourosh, V. Grosser, R. Leutenbauer, G. Bauer, W. Schaefer, H. Reichl: MEM modular packaging and interfaces, IEEE, ECTC (2000) 681–688

56.32　D. W. Lee, T. Ono, T. Abe, M. Esashi: Fabrication of microprobe array with sub-100 nm nano-heater for nanometric thermal imaging and data storage. In: *Proceedings of IEEE Micro Electro Mechanical Systems Conference* (IEEE, Interlaken, Switzerland 2001) pp. 204–207

56.33　A. S. Laskar, B. Blythe: Epoxy multichip modules, a solution to the problem of packaging and interconnection of sensors and signal-processing chips, Sensors Actuators A **36**, 1–27 (1993)

56.34　R. Reichl: Packaging and interconnection of sensors, Sensors Actuators A **25-27**, 63–71 (1991)

56.35　A. Grisel, C. Francis, E. Verney, G. Mondin: Packaging technologies for integrated electrochemical sensors, Sensors Actuators **17**, 285–295 (1989)

56.36　J. L. Lund, K. D. Wise: Chip-level encapsulation of implantable CMOS microelectrode arrays. In: *Solid-State Sensor and Actuator Workshop* (Transducers Research Foundation, Inc., Hilton Head Island, SC 1994) pp. 29–32

56.37　T. Akin, B. Siaie, K. Najafi: Modular micromachined high-density connector for implantable biomedical systems. In: *Micro Electro Mechanical Systems Workshop* (IEEE, San Diego, CA 1996) pp. 497–502

56.38　L. Muller, M. H. Hecht: Packaging Qualification for MEMS-Based Space Systems. In: *Micro Electro Mechanical Systems Workshop* (IEEE, San Diego, CA 1996) pp. 503–508

56.39　L. Lin, K. McNair, R. T. Howe, A. P. Pisano: Vacuum encapsulated lateral microresoantors. In: *7th Int. Conference on Solid State Sensors and Actuators* (IEEE, Yokohama 1993) pp. 270–273

56.40　H. Guckel: Surface micromachined pressure transducers, Sensors Actuators A **28**, 133–146 (1991)

56.41　J. J. Sniegowski, H. Guckle, R. T. Christenson: Performance characteristics of second generation polysilicon resonating beam force transducers. In: *IEEE Solid-State Sensor and Actuator Workshop* (Transducers Research Foundation, Inc., Hilton Head Island 1990) pp. 9–12

56.42　K. Ikeda, H. Kuwayama, T. Kobayashi, T. Watanabe, T. Nishikawa, T. Oshida, K. Harada: Three dimensional micromachining of silicon pressure sensor integrating resonant strain gauge on diaphragm, Sensors Actuators A **21-23**, 1001–1010 (1990)

56.43　C. H. Mastrangelo, R. S. Muller: Vacuum-sealed silicon micromachined incandescent light source, IEEE, IEDM (1989) 503–506

56.44　J. Smith, S. Montague, J. Sniegowski, R. Manginell, P. McWhorter, R. Huber: Characterization of the embedded micromechanical device approach to the monolithic integration of MEMS with CMOS, SPIE (1996) 2879

56.45　S. Van der Groen, M. Rosmeulen, P. Jansen, K. Baert, L. Deferm: CMOS Compatible Wafer Scale Adhesive Bonding for Circuit Transfer. In: *International Conference on Solid-State Sensors and*

Actuators, Transducers'97 (IEEE, Chicago, IL 1997) pp. 629–632

56.46 Y. T. Cheng, Y. T. Hsu, L. Lin, C. T. Nguyen, K. Najafi: Vacuum packaging using localized aluminium/silicon-to-glass bonding using localized aluminium/silicon-to-glass bonding. In: *IEEE, Int. Conf. on MEMS* (IEEE, Interlaken, Switzerland 2001) pp. 18–21

56.47 M. Chiao, L. Lin: Vacuum packaging of microresonators by rapid thermal processing. In: *Proceedings of SPIE on Smart Electronics, MEMS, and Nanotechnology* (SPIE, San Diego 2002) pp. 17–21

56.48 J. H. Partridge: *Glass-to-Metal Seals* (The Society of Glass Technology, Sheffield 1949)

56.49 P. Kumar, V. A. Greenhut: *Metal-to-ceramic joining* (The Minerals, Metals and Materials Society, Warrendale 1991)

56.50 M. G. Nicholas, D. A. Mortimer: Ceramic/metal joining for structural applications, Mater. Sci. Technol. **1**, 657–665 (1985)

56.51 M. Esashi, S. Shoji, A. Nakano: Normally closed microvalve and micropump fabricated on a silicon wafer, Sensors Actuators **20**, 163–169 (1989)

56.52 M. E. Poplawski, R. W. Hower, R. B. Brown: A simple packaging process for chemical sensors. In: *Solid-State Sensor and Actuator Workshop* (Transducers Research Foundation, Inc., Hilton Head 1994) pp. 25–28

56.53 S. F. Trautweiler, O. Paul, J. Stahl, H. Baltes: Anodically bonded silicon membranes for sealed and flush mounted microsensors. In: *Micro Electro Mechanical Systems Workshop* (IEEE, San Diego, CA 1996) pp. 61–66

56.54 E. H. Klaassen, K. Petersen, J. M. Noworolski, J. Logan, N. I. Malfu, J. Brown, C. Storment, W. McCulley, G. T. A. Kovac: Silicon fusion bonding and deep reactive ion etching: A new technology for microstructures, Sensors Actuators A **52**, 132–139 (1996)

56.55 C. H. Hsu, M. A. Schmidt: Micromachined structures fabricated using a wafer-bonded sealed cavity process. In: *Solid State Sensor and Actuator Workshop* (Transducers Research Foundation, Inc., Hilton Head 1994) pp. 151–155

56.56 A. L. Tiensuu, J. A. Schweitz, S. Johansson: In situ investigation of precise high strength micro assembly using Au-Si eutectic bonding. In: *Int. Conf. on Solid-State Sensors and Actuators, and Eurosensors IX* (IEEE, Stockholm, Sweden 1995) pp. 236–239

56.57 A. P. Lee, D. R. Ciarlo, P. A. Krulevitch, S. Lehew, J. Trevino, M. A. Northrup: Practical microgripper by fine alignment, eutectic bonding and SMA actuation. In: *Int. Conf. on Solid-State Sensors and Actuators, and Eurosensors IX* (IEEE, Stockholm, Sweden 1995) pp. 368–371

56.58 M. B. Cohn, Y. Liang, R. Howe, A. P. Pisano: Wafer to wafer transfer of microstructures for vacuum packaging. In: *Solid State Sensor and Actuator Workshop* (Transducers Research Foundation, Inc., Hilton Head 1996) pp. 32–35

56.59 Q.-Y. Tong, U. Gosele: *Semiconductor Wafer Bonding, Science and Technology* (Wiley, New York 1999)

56.60 R. W. Bower, M. S. Ismail, B. E. Roberds: Low temperature Si_3N_4 direct bonding, Appl. Phys. Lett. **62**, 3485–3487 (1993)

56.61 H. Takagi, R. Maeda, T. R. Chung, T. Suga: Low temperature direct bonding of silicon and silicon dioxide by the surface activation method. In: *Int. Conf. on Solid-State Sensors and Actuators, Transducer 97* (IEEE, Chicago, IL 1997) pp. 657–660

56.62 G. Wallis, D. Pomerantz: Filed assisted glass-metal sealing, J. Appl. Phys. **40**, 3946–3949 (1969)

56.63 L. Bowman, J. Meindl: The packaging of implantable integrated sensors, IEEE Trans. Biomed. Eng. **BME-33**, 248–255 (1986)

56.64 M. Esashi: Encapsulated micro mechanical sensors, Microsyst. Technol. **1**, 2–9 (1994)

56.65 B. Ziaie, J. Von Arx, M. Dokmeci, K. Najafi: A hermetic glass-silicon micropackages with high-density on-chip feedthroughs for sensors and actuators, J. Microelectromech. Syst. **5**, 166–179 (1996)

56.66 Y. Lee, K. Wise: A batch-fabricated silicon capacitive pressure transducer with low temperature sensitivity, IEEE Trans. Electron. Dev. **ED-29**, 42–48 (1982)

56.67 S. Shoji, H. Kicuchi, H. Torigoe: Anodic bonding below 180 degree C for packaging and assembling of MEMS using lithium aluminosilicate-beta-quartz glass-ceramic. In: *Proceedings of the 1997 10th Annual International Workshop on Micro Electro Mechanical Systems* (IEEE, Nagoay 1997) pp. 482–487

56.68 M. Esashi, N. Akira, S. Shoji, H. Hebiguchi: Low-temperature silicon-to-silicon anodic bonding with intermediate low melting point glass, Sensors Actuators A **23**, 931–934 (1990)

56.69 A. Hanneborg, M. Nese, H. Jakobsen, R. Holm: Silicon-to-thin film anodic bonding, J. Micromech. Microeng. **2**, 117–121 (1992)

56.70 A. V. Chavan, K. D. Wise: Batch-processed vacuum-sealed capacitive pressure sensors, ASME/IEEE J. Microelectromech. Syst. **10**(4), 580–588 (2001)

56.71 H. Henmi, S. Shoji, Y. Shoji, K. Yoshimi, M. Esashi: Vacuum Packaging for Microsensors by Glass-Silicon Anodic Bonding, Sensors Actuators A Phys. **43**(1-3), 243–248 (1994)

56.72 T. Hara, S. Kobayashi, K. Ohwada: A New Fabrication Mmethod for Low-Pressure Package with Glass-Silicon-Glass Structure and Its Stability. In: *The 10th International Conference on Solid-State Sensors and Actuators, Transducers'99 Digest of technical papers*, Vol. 2 (IEEE, Sendai, Japan 1999) pp. 1316–1319

56.73 S. A. Audet, K. M. Edenfeld: Integrated sensor wafer-level packaging. In: *International Conference on Solid-State Sensors and Actuators, Transducers'97* (IEEE, Chicago, IL 1997) pp. 287–289

56.74 R. C. Benson, N. deHaas, P. Goodwin, T. E. Phillips: Epoxy adhesives in microelectronic hybrid applications, Johns Hopkins APL Tech. Dig. **13**, 400–406 (1992)

56.75 M. Shimbo, J. Yoshikawa: New silicon bonding method, J. Electrochem. Soc. **143**, 2371–2377 (1996)

56.76 P. M. Zavracky, B. Vu: Patterned eutectic bonding with Al/Ge thin film for MEMS, SPIE **2639**, 46–52 (1995)

56.77 G. Humpston, D. M. Jacobson: Principles of soldering and brazing, ASM Int. (1993) 241–244

56.78 A. Singh, D. Horsely, M. B. Cohn, R. Howe: Batch transfer of microstructures using flip-chip solder bump bonding. In: *International Conference on Solid State Sensors and Actuators, Transducer 97*, Vol. 1 (IEEE, Chicago, IL 1997) pp. 265–268

56.79 M. M. Maharbiz, M. B. Cohn, R. T. Howe, R. Horowitz, A. P. Pisano: Batch micropackaging by compression-bonded wafer-wafer transfer. In: *12th International Conference on MEMS* (IEEE, Orlando, FL 1999) pp. 482–489

56.80 Y. T. Cheng, L. Lin, K. Najafi: Localized silicon fusion and eutectic bonding for MEMS fabrication and packaging, IEEE/ASME, J. Microelectromech. Syst. **9**, 3–8 (2000)

56.81 L. Lin: Selective encapsulations of MEMS: micro channels, needles, resonators, and electromechanical filters, Ph.D. Thesis, UC Berkeley (1993)

56.82 Y. C. Su, L. Lin: Localized plastic bonding for micro assembly, packaging and liquid encapsulation. In: *Proceedings of IEEE Micro Electro Mechanical Systems Conference*, ed. by Interlaken (IEEE, Interlaken 2001) pp. 50–53

56.83 M. Schwartz: *Brazing* (Chapman Hall, London 1995)

56.84 L. Lin, R. T. Howe, A. P. Pisano: Microelectromechanical filters for signal processing, IEEE/ASME, J. Microelectromech. Syst. **7**, 286–294 (1998)

56.85 W. C. Tang, C. T.-C. Nguyen, R. T. Howe: Laterally driven polysilicon resonant microstructures, Sensors Actuators A **20**, 25–32 (1989)

56.86 G. T. Mulhern, D. S. Soane, R. T. Howe: Supercritical carbon dioxide drying of microstructures. In: *7th Int. Conference on Solid State Sensors and Actuators* (IEEE, Yokohama 1993) pp. 296–299

56.87 M. Judy: Micromechanisms Using Sidewall Beams, Ph.D dissertation, EECS Department, University of California at Berkeley, p. 162, (1994)

56.88 R. Aigner, K.-G. Oppermann, H. Kapels, S. Kolb: Cavity-Micromachining technology: zero-package solution for inertial sensors, TRANSDUCERS '01. EUROSENSORS XV. In: *11th International Conference on Solid-State Sensors and Actuators. Digest of Technical Papers*, Vol. 1 (IEEE, Munich, Germany 2001) pp. 186–189

56.89 M. Bartek, J. A. Foerster, R. F. Wolffenbuttel: Vacuum Sealing of Mcirocavities Using Metal Evaporation, Sensors Actuators A **61**, 364–368 (1997)

56.90 C. H. Mastrangelo, R. S. Muller, S. Kumar: Microfabricated Incandescent Lamps, Appl. Opt. **30**, 868–873 (1993)

56.91 A. M. Leung, J. Jones, E. Czyzewska, J. Chen, B. Woods: Micromachined accelerometer based on convection heat transfer. In: *11th International Conference on MEMS* (IEEE, Heidelberg, Germany 1998) pp. 627–630

56.92 D. R. Spark, L. Jordan, J. H. Frazee: Flexible vacuum-packaging method for resonating micromachines, Sensors Actuators A **55**, 179–183 (1996)

56.93 M. Chiao, L. Lin: Hermetic wafer bonding based on rapid thermal processing, Sensors Actuators A **91**, 398–402 (2001)

56.94 Y. T. Cheng, L. Lin, K. Najafi: Fabrication and hermeticity testing of a glass-silicon packaging formed using localized aluminium/silicon-to-glass bonding, IEEE/ASME, J. Microelectromech. Syst. **10**, 392–399 (2001)

56.95 M. Chiao, L. Lin: Accelerated Hermeticity Testing of a Glass-Silicon Package Formed by Rapid Thermal Processing Aluminum-to-Silicon Nitride Bonding, Sensors Actuators A Phys. **97–98**, 405–409 (2002)

56.96 M. K. Keshavan, G. A. Sargent, H. Conrad: Statistical Analysis of the Hertzian Fracture of Pyrex Glass Using the Weibull Distribution Function, J. Mater. Sci. **15**, 839–844 (1980)

56.97 M. Chiao, L. Lin: A Wafer-Level Vacuum Packaging Process by RTP Aluminum-to-Nitride Bonding. In: *Technical Digest of Solid-State Sensors, Actuator and Microsystems Workshop* (Transducers Research Foundation, Inc., Hilton Head 2002) pp. 81–85

56.98 F. Rosebury: *Handbook of Electron Tube and Vacuum Techniques* (Addison-Wesley, New York 1965)

56.99 Y. T. Cheng, W. T. Hsu, K. Najafi, C. T. Nguyen, L. Lin: Vacuum packaging technology using localized aluminium/silicon-to-glass bonding, J. Microelectromech. Syst. (2002) 556–565

56.100 SensArray Corporation, 47451 Fremont Blvd. Fremont, CA 94538

56.101 C. Luo, L. Lin: The application of nanosecond-pulsed laser welding technology in MEMS packaging with a shadow mask, Sensors Actuators A **97–98**, 398–404 (2002)

56.102 A. Cao, M. Chiao, L. Lin: Selective and localized wafer bonding using induction heating. In: *Technical Digest of Solid-State Sensors and Actuators Workshop* (Transducers Research Foundation, Inc., Hilton Head Island 2002) pp. 153–156

56.103 J. B. Kim, M. Chiao, L. Lin: Ultrasonic Bonding of In/Au and Al/Al for Hermetic Sealing of MEMS Packaging. In: *Proceedings of IEEE Micro Electro*

Mechanical Systems Conference (IEEE, Las Vegas 2002) pp. 415–418

56.104 Y. T. Cheng, L. Lin, K. Najafi: Localized bonding with PSG or indium solder as intermediate layer, 12th International Conference on MEMS (1999) 285–289

56.105 G. Thyrum, E. Cruse: Heat pipe simulation, a simplified technique for modeling heat pipe assisted heat sinks, Adv. Packag. **115**, 23–27 (2001)

56.106 G. P. Peterson, A. B. Duncan, M. H. Weichold: Experimental investigation of micro heat pipes fabricated in silicon wafers, J. Heat Transfer **115**, 751–756 (1993)

56.107 L. Jiang, M. Wong, Y. Zohar: Forced convection boiling in a microchannel heat sink, IEEE/ASME, J. Microelectromech. Syst. **10**, 80–87 (2001)

56.108 F. Arias, S. R. J. Oliver, B. Xu, E. Holmlin, G. M. Whitesides: Fabrication of metallic exchangers using sacrificial polymer mandrils, IEEE/ASME, J. Microelectromech. Syst. **10**, 107–112 (2001)

56.109 *ANSYS Modeling and Meshing Guide*, 3rd edn. (SAS IP, Inc., Canousburg 2002)

56.110 D. Bystrom, L. Lin: Residual stress analysis of silicon-aluminum-glass bonding processes. In: *ASME International Mechanical Engineering Congress and Exposition, MEMS Symposium, New Orleans, Louisiana, November* (ASME, New Orleans, Louisiana 2002)

56.111 H. Scholze: *Glass: Nature, Structure, and Properties*, 1st edn. (Springer, Berlin Heidelberg New York 1991)

56.112 R. R. Tummala: *Fundamentals of Microsystems Packaging* (McGraw-Hill, New York 2001)

56.113 S. J. Adamson: BGA, CSP, and flip chip, Adv. Packag. (2002) 21–24

56.114 JESD22-A102C: Accelerated Moisture Resistance-Unbiased Autoclave, JEDEC Solid State Technology Association, 2000

56.115 M. Dokmeci, K. Najafi: A high-sensitivity polyimide capacitive relative humidity sensor for monitoring anodically bonded hermetic micropackages, IEEE/ASME J. Microelectromech. Syst. **10**(2), 197–204 (2001)

56.116 E. E. Lewis: *Introduction to Reliability Engineering*, 2nd edn. (Wiley, New York 1996)

56.117 W. D. Brown: *Advanced Electronic Packaging* (IEEE, New York 1999)

56.118 F. Niklaus, P. Znoksson, E. Käluesten, G. Stemme: Void free full wafer adhesion bonding. In: *IEEE, Proceedings of IEEE Micro Electro Mechanical Systems Conference* (IEEE, Miyazaki, Japan 2000) pp. 241–252

56.119 C. H. Cheng, A. S. Ergun, B. T. Khuri-Yakub: Electrical through-wafer interconnects with sub-picofarad parasitic capacitance, IEEE, ECTC (2002) 18–21

56.120 D. Routkevitch, A. A. Tager, J. Haruyama, D. Al-mawlawi, M. Moskovits, J. M. Xu: Nonlithographic nano-array arrays: fabrication, physics, and device applications, IEEE, Trans. Electron. Dev. **43**, 1646–1658 (1996)

56.121 J. Gou, M. Lundstrom, S. Datta: Performance projections for ballistic carbon nanotube fielddefect transistors, Appl. Phys. Lett. **80**, 3192–3194 (2002)

56.122 S. U. S. Choi, Z. G. Zhang, W. Yu, F. E. Lockwood, E. A. Grulke: Anomalous thermal conductivity enhancement in nanotube suspensions, Appl. Phys. Lett **79**, 2252–2254 (2001)

56.123 K. Velikov, A. Moroz, A. Blaaderen: Photonic crystals of core-shell colloidal particles, Appl. Phys. Lett. **80**, 49–51 (2002)

Part H Technology

Part H Technological Convergence and Governing Nanotechnology

57. Technological Convergence from the Nanoscale

A series of scientific conferences and book-length publications predict that nanoscience will have its greatest impact through the convergence of four fields where research progress and engineering applications are expected to be especially significant. These are the so-called NBIC fields of nanotechnology, biotechnology, information technology, and new technologies based on cognitive science. This chapter is a first sociological reconnaissance of the convergenist movement in science and technology, based on the unity of nature at the nanoscale.

Nanotechnology may transform the world, but probably not through any single application such as nanostructured materials, nanoscale devices, or even the mythical nanobots. Although nanoscience and nanotechnology are typically conceptualized as a distinct field, their primary historical significance is likely to be the central role they play in the unification of most branches of science and technology into a single realm, united by a shared set of concepts, theories, research tools, and design principles. Already, nanotechnology has formed strong partnerships with biotechnology and with information technology [57.1]. The next and perhaps most decisive step will be nano-enabled unification of this triad with cognitive science to develop a host of new technologies to supplement and even enhance the human mind [57.2–4]. This, in turn, will give humans the ability to progress more rapidly in any scientific endeavor. These realizations spring from a series of converging technologies conferences and the associated emergence of what might be called a *convergenist movement* across science and technology but historically rooted in nanoscience and nanotechnology [57.5].

57.1 Nanoscience Synergy

There is nothing new about interdisciplinary work in science and engineering. Astrophysics unites astronomy and physics, biochemistry mediates between biology and chemistry, and social psychology bridges between sociology and psychology. A number of specialized technological fields connect a science with a branch of engineering. For example, earthquake engineers draw upon both mechanical engineering and geology, whereas information technologists tend to be trained either in computer science or electrical engineering. These local convergences were created to address particular empirical problems, and each has developed concepts and methods that are distinctive unto itself. The wider, nano-enabled convergence that has recently begun not only covers a vast territory, but is remarkable for developing universal concepts and methods that have immense potential to revolutionize science and technology more generally.

57.1.1 Organizational Background

Many individuals have commented in recent years about the progressive fusion of science and technology. Entomologist and sociobiologist *Edward O. Wilson* [57.6] has called the unification of scientific knowledge *consilience*. The term most often used in the context of technology is *convergence*. In his influential book *The Rise of the Network Society*, *Manuel Castells* ([57.7], p. 72) wrote,

> *Technological convergence increasingly extends to growing interdependence between the biological and micro-electronics revolutions, both materially and methodologically ... Nanotechnology may allow sending tiny microprocessors into the systems of living organisms, including humans.*

Only with the dawn of the new millennium has it become possible to envision what consilience and convergence will accomplish.

Perhaps the initial awareness emerged at the first major conference on the societal implications of nanoscience and nanotechnology, convened at the National Science Foundation (NSF) in Arlington, Virginia, September 28–29, 2000. The NSF organized the meeting at the request of the subcommittee on nanoscale science, engineering, and technology (NSET) of the national science and technology council (NSTC). The first paragraph of the introduction to the published, book-length report announces the fundamental insight, explicitly using the word *converge*:

> *A revolution is occurring in science and technology, based on the recently developed ability to measure, manipulate and organize matter on the nanoscale – 1 to 100 billionths of a meter. At the nanoscale, physics, chemistry, biology, materials science, and engineering converge toward the same principles and tools. As a result, progress in nanoscience will have very far-reaching impact ([57.8], p. 1.)*

A quickly produced version of the first societal implications volume was circulated among scientists and within the US government in March 2001, and on May 11 of that year a small group of scientists and engineers gathered at NSF to plan how to explore convergence further. The first converging technologies conference was held on December 3–4, 2001, with the sponsorship of both the National Science Foundation and the US Department of Commerce. A prompt government version of the report was circulated in June 2002, and the book was published early in 2003. The opening sentence of the report proclaimed:

> *In the first decades of the 21st century, concentrated efforts can unify science based on the unity of nature, thereby advancing the combination of nanotechnology, biotechnology, information technology, and new technologies based in cognitive science ([57.9], p. ix).*

Subsequently, the organizers of the 2000 and 2001 conferences also organized three annual converging technologies conferences: in February 2003 at Los Angeles, in February 2004 in New York City, and in February 2005 near Kona, Hawaii. The 2003 conference led to a volume in the series published by the New York Academy of Sciences [57.10], and a volume based on the 2004 conference was published by Springer of Berlin [57.11]. A similar book based on the 2005 conference is in progress. The world significance of the convergenist movement was recognized by the European Union when it convened its own conference on September 14–15, 2004, and issued the report, *Converging Technologies – Shaping the Future of European Societies*. In March 2000, the European Union had agreed to the Lisbon agenda, which was fundamentally the goal of becoming "the most competitive and dynamic knowledge-driven economy". Thus, the report from the 2004 EU conference declared:

> *The European Commission and Member States are called upon to recognize the novel potential of Converging Technologies (CTs) to advance the Lisbon Agenda. Wise investment in CTs stimulates science and technology research, strengthens economic competitiveness, and addresses the needs of European societies and their citizens. Preparatory action should be taken to implement CT as a thematic research priority, to develop Converging Technologies for the European Knowledge Society (CTEKS) as a specifically European approach to CTs, and to establish a CTEKS research community. ([57.12], p. 2)*

In the American context, some scientists saw converging technologies as a worthy successor to the information technology research (ITR) initiative that had completed its initial, five-year run in 2004, and to the national nanotechnology initiative (NNI) that continued later. Although a long-term goal is the unification of all fields of science and engineering, the initial discussions primarily involved the four NBIC realms of science and engineering: nanotechnology, biotechnology, information technology, and cognitive science (new technologies based on the convergence of computer science, psychology, neuroscience, philosophy, anthropology, economics, sociology, etc.).

Thus conceptualized, convergence is based on:

- material unity of nature at the nanoscale
- technology integration from the nanoscale
- key transforming tools for NBIC research and production
- the concept of reality as closely coupled, complex, hierarchical systems
- the goal to improve human performance
- the necessity for reformation of science and technology education, based on shared concepts across fields

57.1.2 Goal of Improving Human Performance

A certain amount of confusion has centered on the expressed goal of convergence to improve human performance. People who have not been present at the meetings have wondered what this word *performance* could mean? The problem is not semantic but historical, because our technological abilities will achieve very different levels over the coming decades, and thus the meanings of words will also change. In the near term, the convergence of NBIC will give people a number of very specific new tools to allow them to achieve their personal goals in science, industry, and everyday life. Some of these have been examined in great detail by convergenists.

The original converging technologies report listed 20 beneficial technological breakthroughs, illustrative of what NBIC might accomplish for human beings over the first half of the 21st century. Two years later, 26 of the contributors to the first three reports rated these developments in terms of how soon each was likely to be achieved to a significant degree. Predictions are always hazardous, of course, but their professional judgments provide a sense of how much work is required to achieve each of the 20. Here they are, arranged from the one expected soonest, to one that might not be achieved until the middle of the century:

- Anywhere in the world, an individual will have instantaneous access to needed information, whether practical or scientific in nature, in a form tailored for most effective use by the particular individual.
- New organizational structures and management principles based on fast, reliable communication of needed information will vastly increase the effectiveness of administrators in business, education, and government.
- Comfortable, wearable sensors and computers will enhance every person's awareness of his or her health condition, environment, chemical pollutants, potential hazards, and information of interest about local businesses, natural resources, and the like.
- People from all backgrounds and of all ranges of ability will learn valuable new knowledge and skills more reliably and quickly, whether in school, on the job, or at home.
- Individuals and teams will be able to communicate and cooperate profitably across traditional barriers of culture, language, distance, and professional specialization, thus greatly increasing the effectiveness of groups, organizations, and multinational partnerships.
- National security will be greatly strengthened by lightweight, information-rich war-fighting systems, capable uninhabited combat vehicles, adaptable smart materials, invulnerable data networks, superior intelligence-gathering systems, and effective measures against biological, chemical, radiological, and nuclear attacks.
- Engineers, artists, architects, and designers will experience tremendously expanded creative abilities, both with a variety of new tools and through improved understanding of the wellsprings of human creativity.
- Average people, as well as policymakers, will have a vastly improved awareness of the cognitive, social, and biological forces operating their lives, enabling far better adjustment, creativity, and daily decision making.
- Factories of tomorrow will be organized around converging technologies and increased human–machine capabilities as intelligent environments that achieve the maximum benefits of both mass production and custom design.
- Agriculture and the food industry will greatly increase yields and reduce spoilage through networks of cheap, smart sensors that constantly monitor the condition and needs of plants, animals, and farm products.
- The work of scientists will be revolutionized by importing approaches pioneered in other sciences,

for example, genetic research employing principles from natural language processing and cultural research employing principles from genetics.

- Robots and software agents will be far more useful for human beings, because they will operate on principles compatible with human goals, awareness, and personality.
- The human body will be more durable, healthier, more energetic, easier to repair, and more resistant to many kinds of stress, biological threats, and aging processes.
- A combination of technologies and treatments will compensate for many physical and mental disabilities and will eradicate altogether some handicaps that have plagued the lives of millions of people.
- Fast, broadband interfaces between the human brain and machines will transform work in factories, control automobiles, ensure military superiority, and enable new sports, art forms and modes of interaction between people.
- Machines and structures of all kinds, from homes to aircraft, will be constructed of materials that have exactly the desired properties, including the ability to adapt to changing situa-

tions, high energy efficiency, and environmental friendliness.

- The ability to control the genetics of humans, animals, and agricultural plants will greatly benefit human welfare; widespread consensus about ethical, legal, and moral issues will be built in the process.
- Transportation will be safe, cheap, and fast, due to ubiquitous real-time information systems, extremely high-efficiency vehicle designs, and the use of synthetic materials and machines fabricated from the nanoscale for optimum performance.
- Formal education will be transformed by a unified but diverse curriculum based on a comprehensive, hierarchical intellectual paradigm for understanding the architecture of the physical world from the nanoscale through the cosmic scale.
- The vast promise of outer space will finally be realized by means of efficient launch vehicles, robotic construction of extraterrestrial bases, and profitable exploitation of the resources of the Moon, Mars, or near-Earth approaching asteroids.

Each of these application areas requires the synthesis of at least two of the NBIC fields, and usually all four will be involved to some extent.

57.2 Dynamics of Convergence from the Nanoscale

One way to conceptualize the process of convergence is to concentrate on examples where two of the four realms are merging, while keeping an awareness that many future examples will involve three of the realms or even all four. From the standpoint of nanotechnology, these binary convergences are of two kinds: *primary nanoconvergence* and *secondary nanoconvergence*.

In primary nanoconvergence, one of the two realms in a binary convergence is nanotechnology itself: nano–bio, nano–info, and nano–cogno. Each of these areas will offer many examples in which nano contributes to the other technology, or where that technology contributes significantly to nano, as well as numerous applications requiring both. In secondary nanoconvergence, nanotechnology is not one of the two fields, but may be implicated indirectly in the convergence: bio-info, bio-cogno, and info–cogno. Developments in each of these areas may be facilitated by nanotechnology, for example through sensors or other instrumentation incorporating nanoscale components. Also, each of the binary convergences that do not involve nanotechnology directly

may nonetheless have implications for nano, for example creating new industries that will demand new nanoscale applications of many kinds.

A slightly different way of conceptualizing the binary convergences is to call them *regional*, as opposed to the smaller-scale *local* convergences already taking place between very specific disciplines, and in contrast to the *global* convergence of all four NBIC fields.

The nano–bio binary convergence is quite prominent already [57.13, 14]. For years, a number of the papers at major nanotechnology conferences have involved biology, and in 2005 a special bio-nanotechnology conference was held in conjunction with the Nanotech 2005 conference and trade show in Anaheim, California. *John H. Marburger*, director of the US government's office of science and technology policy, likes to call genetic engineering and other work involving complex molecules inside livings cells *wet nanotechnology* [57.15]. Nanoscience tools, including the atomic force microscope, can be used to study biological phenomena, such as cell membranes [57.16].

In return, biological process can be used to assemble artificial nanostructures [57.17]. Medicine is one of the application areas where the nano–bio convergence is likely to be especially useful, through engineered nanoscale structures employed in medication delivery systems, sensor nets placed inside the human body for diagnosis and monitoring of chronic conditions, and potentially in microscopic surgery [57.18–21]. Environmental protection is another very important nano–bio area [57.22].

Nano–info convergence is most familiar in the context of microelectronics, where the size of components on computer chips has shrunk into the nanoscale. Small size implies less waste of power but greater speed, and it has perhaps fortuitously been associated with lower costs. To sustain progress with information technology, essential for the well-being of civilization, requires a broad range of scientific research projects establishing the basis for radical new kinds of nanoelectronics. Another area of recent interest for defense and medical applications is nano-enabled microscale sensors, including devices that can recognize molecules. In addition, a field is emerging that we might call *computational nanoscience*, following one or both of two distinct approaches:

1. *computer simulation* to model the behavior of nanoscale structures, and
2. *nanoinformatics* of dynamic databases to store, manage, and retrieve information relevant to nanoscience and nanotechnology [57.23].

Nano–cogno convergence will primarily be a challenge for the future, but we see some foundations for it even today. Of course, many of the processes essential for the functioning of the human nervous system take place at the nanoscale, and thus need to be studied at that scale. The gap between neurons in the brain, across which neurotransmitters must flow, and the vesicles that store the neurotransmitters, are both nanoscale in size. Pigments in the eye, notably rhodopsin, are complex molecules whose nanoscale structural changes start the process that leads to perception of visual stimuli by the brain [57.24].

An intriguing example of future directions is the work done by *Rodolfo Llinas* and his colleagues, developing a new nano-enabled method of monitoring and stimulating the neurons in the brain [57.25]. Fine wires, intended to be a profuse bundle of nanowires in the operational system, would be inserted into the living brain through the blood vessels, until each is in contact with a different neuron and able to detect its interactions with adjacent neurons. Already, tests with platinum wires have been promising, and research is under way employing polymer wires that can be steered by applying minute electric currents that cause them to bend. Whether this particular approach proves successful or not, nanotechnology clearly has the significant potential to contribute instrumentation for fundamental cognitive science research. In return, cognitive science may help design effective educational programs to teach students about the structure and functions of molecules, based on a correct understanding of how children and young adults learn.

With respect to nanotechnology, the triad of secondary convergences (bio–info, bio–cogno, info–cogno) can be both enabled and enabler. We have already mentioned the crucial role for nanoelectronic components (sensors, computer chips, and the like) to support research in these regions of science. Among the most important bio–info areas is bioinformatics of nanoscale structures, such as proteins. An especially creative and perhaps risky convergent research project is the innovative attempt to apply methods from computer analysis of human language to decoding genome sequences [57.26, 27]. The info–cogno area includes the computer-science fields of human–computer interaction and artificial intelligence, both of which are fundamental to effective human use of the future science and engineering tools that will be needed to carry out work at the nanoscale [57.28, 29].

57.3 Ethical, Legal and Social Implications

The human implications of technological convergence were explored by one of the topical task forces of the second major conference on the societal implications of nanoscience and nanotechnology, held at the National Science Foundation on December 2–3, 2003. Human implications were also considered more fully by some of the scientists and scholars involved in the four converging technologies conferences.

57.3.1 Motivating Principles

The conference task force began with the prediction that converging technologies will in fact be very powerful shapers of the human future. Among the beneficial developments anticipated were:

- curing fatal diseases that have resisted previous treatment approaches
- improving agricultural productivity both in terms of higher crop yields and greater nutritional value
- overcoming disabilities such as blindness, deafness, and restricted mobility
- efficiently producing clean water from sources that are heavily polluted with natural or industrial contaminants
- cheap energy production without polluting the environment or depleting natural resources
- new manufacturing processes that are energy-efficient and nonpolluting
- a range of remedies for past environmental damage
- vastly improved portable computers and communication devices,
- a general acceleration of technological innovation

While the task force was enthusiastic about NBIC's potential for achieving these benefits, members also recognized that the same power could become destructive, either through accident or hostile action. The dangers are probably magnified by globalization. If one nation establishes reasonable precautions for research and deployment of powerful NBIC innovations, another may not, and the consequences may cross borders. At the very least, globalization complicates evaluation of new technologies, because the social implications are likely to be different depending upon the economic level, cultural values, and social structure of the countries involved [57.30]. NBIC may often be disruptive, by causing fundamental change through:

- making unexpected scientific discoveries
- creating radically novel engineering applications
- establishing entirely new industries, in services as well as manufacturing
- requiring new skills, knowledge, and associated educational programs
- breaking down the walls between academic disciplines, industrial sectors, and government research and regulatory agencies

In general, such disruptions reduce the value of many older skills and industries, even as they stimulate the emergence of new ones. Several economists at the second societal implications conference argued that the net result is increased opportunity and prosperity, although the social scientists reminded them that along the way some people in particular sectors of the economy will

suffer. For individuals, reskilling through adult education programs may be a partial answer, but governments will also need to contend with the dislocations when entire regions of their countries decline or segments of the population experience problems associated with unusually high rates of mobility.

The converging technologies task force of the societal implications conference argued for an holistic approach that would bring together a variety of stakeholders and experts in the natural and social sciences "to establish a framework that protects our rights as humans, while allowing for the technological developments that can help us achieve our full human potential".

57.3.2 Research Approaches

In general, participants in the debate about future implications of converging technologies believe that outright prediction of major developments will be difficult if not impossible, although research to identify the likely immediate impacts of narrowly defined innovations is possible and could often be valuable. For example, it is certainly desirable to do the necessary work in chemistry and biology to understand the possible toxicity of a new material, but it may not be possible to predict the social and economic impacts of introducing a new material that makes possible a range of new products. Nonetheless, flexible roadmaps that sketch possible future trends and constantly update themselves to take account of new insights and information are also valuable [57.31].

Therefore, it is important to develop a social-organizational system to monitor the results of innovation as they occur, to prepare contingency plans, and to respond quickly to new realities. The aim would not be to burden the innovative process with many extra hurdles, but rather to incorporate a broader range of well-grounded inputs into the decision processes, both at the level of specific innovation projects and the level of more general policy setting. Doing so could accelerate technological progress in many areas, by focusing scientific and engineering talent on applications that would be of especially great human benefit, and thus that could attract both government stimulus and private investment.

Another way of putting this is to say that ethical and social issues need to be addressed during the research process, rather than only after commercialization of already developed products. If industry develops something significantly new without incorporating socio-ethical inputs, and (as the expression goes) merely *throws it over the wall to see who picks it up*, the public may react negatively. The result could be rejection of a potentially beneficial innovation merely because its design and presentation were poorly adapted to the real needs and perceptions of customers. The negative reaction in Europe to foods based on genetically modified organisms is frequently mentioned as an apt but unfortunate example.

These insights immediately raise the problem that we do not currently know very well how to incorporate a wide range of societal inputs into the planning, doing, and application of research in science and engineering. Participants in the conference were practically unanimous in the view that innovation on balance was good, and NBIC research should be promoted vigorously. We therefore need scientific research on several social phenomena to understand best how to design organizations for rapid but responsible innovation [57.32, 33], indeed to design an agile society that is best able to exploit the tremendous possible benefits without the harms of the union of nanotechnology, biotechnology, information technologies, and cognitive technologies.

An especially crucial but also difficult area for research is communication with the public. Perhaps in a few decades, when convergence has progressed further, people may share much fundamental knowledge, many overarching concepts that united different fields, and a uniform language for discussing scientific issues. But at present, this is far from the case. Additionally, scientists who wish to communicate with the general public need to contend with the fact that there are many different *publics*, each with its own needs, assumptions, interests, hopes and fears.

NBIC must cope with the challenge that public perceptions of one of its fields may contaminate perceptions of others. For example, the stigma associated with biotechnologies such as genetic engineering, cloning and stem-cell research may stigmatize nanotechnology through guilt by association. Public understanding, therefore, will require a fine awareness that very real differences between fields remain even after convergence has progressed to a significant degree. This, in turn, will require a well-designed language for communicating both across NBIC fields and between them and the wider society, optimized simultaneously for facilitating cooperation between scientists in different fields and for helping policy makers and the public maintain whatever conceptual distinctions between technologies are required for sound evaluations of them.

Another area where social and behavioral research may be needed is the expanded scope for risk analysis that NBIC will demand. It is one thing to assess the impact of a single, well-defined technology, but something very different to understand the manifold and often contradictory effects of a phase change in a complex system. The task force hoped that improved quantitative models could be developed for risk characterization, risk analysis, and risk-based cost–benefit analysis. Advances in such methodologies would require related advances in the effective dissemination of information to the people responsible for risk management and public policy.

57.3.3 Regulatory Challenges

Currently, regulatory organizations – including professional organizations as well as government agencies – tend to be separated from each other and structured internally in terms of narrow areas of technical expertise. As technological convergence erodes the distinctions between industries and sciences, the task force believed there would be an increasing need for cross-training of regulators, and the development of multidisciplinary teams assisted by new information and cognitive technologies, which would require substantial investment. It may be necessary to rethink the missions and structures of the agencies, which in turn would demand legislative action. The complexity and velocity of technological change may require a simultaneous streamlining and strengthening of the processes by which new products are tested and approved.

At the same time, the NBIC technologies will give regulatory agencies new tools with which to handle their increasingly difficult jobs. These will include nano-enabled sensor systems that can measure chemical and biological byproducts in real time, massive but convenient information systems that can be accessed remotely from any location, and cognitive tools such as conceptual languages, data visualization systems, and communication systems for use by distributed work teams. At the same time, members of the task force argued that it was both necessary and desirable to explore nonregulatory approaches that emphasize cooperation among stakeholders rather than confrontation, building ethical principles directly into the processes of research, development, and deployment.

In several ways, the task force suggested, technological convergence will pose significant challenges for the legal profession, judiciary, and government agencies that administer law [57.34]:

- many legal professions will need considerable technical training to be able to understand NBIC issues
- the role of experts in science and engineering will increase, thus perhaps requiring new credentialing systems
- jurors will face ever more difficult challenges making sense of the evidence presented to them
- the ambiguity of undefined risks will undermine current standards of proof, for example in liability cases
- rapid but uneven change will make it difficult to settle international disputes, because laws, standards, and definitions will vary greatly
- patent disputes may increase, because it will be difficult to assess the scope of any patent application's prior art, the relationship to prior scientific publications, and the overlap with other patents

In addition, the fusion of nanotechnology with information technology may undercut traditional protections for privacy, through distributed and portable sensor systems, the ability of data fusion and data mining to extract unexpected knowledge from existing data, and the possible diversion of any of the NBIC technologies to new forms of criminal activity. Any of these developments might require innovations in legislation and enforcement.

The task force judged that technological convergence can be of great benefit for humanity, "if it does not become ensnared in unnecessary complexity, uncertainty and public alienation." To achieve this goal, members urged research on organizational effectiveness, design of a new regulatory environment, development of improved methods to assess risk, innovation to ensure greater public understanding and involvement, and an effort to achieve a global framework for research and regulation.

57.4 Transformative Synthesis

Observers may take a passive stance toward convergence, noting that it is occurring and merely commenting on its socio-cultural dynamics. However, a number of factors exist that may retard, wrongly direct, or even halt convergence, so it is necessary to do more than merely observe if the benefits of NBIC are to be achieved. Institutional inertia is especially problematic. Specialized industries, academic departments, and government agencies became established during one historical period, following one often-arbitrary category scheme, and then become a barrier to any progress that requires cooperation across categories. A well-established finding of social science is that such institutional sectors as companies in a particular industry became significantly more uniform than the technical nature of their real work requires, through imitation, government regulation, and the emergence of a shared conservative culture [57.35]. Breaking down barriers between fields, escaping the iron cage of bureaucracy, and eliminating *stovepipes* takes foresight and effort. To achieve what *James Canton* calls the innovation economy through convergence will require fundamental cultural and institutional changes [57.36, 37].

57.4.1 Common Languages and Communication Tools

The first three NBIC conferences discussed at length the new ways of communicating that would be both required and created by technological convergence. Creating a new language of science will be extremely difficult and time-consuming, not least because scientific discovery and technological innovation constantly generate new terms that initially catch on in one, narrow area and traditionally require years for diffusion to other areas [57.38, 39]. Even within a specific field, old language can be a barrier to progress, because it tends to make people think in obsolete concepts as well as terms. Until science has completed its job of understanding reality, which may require an additional several centuries of research if indeed it is ever accomplished, terminology will remain unstable.

Thus, it may be necessary to create *trading zones* at the intersections of fields, where a number of human, cultural, and technological resources help specialists communicate with each other [57.40–42]. The human resources may include *interactional experts* who have some technical training in all the relevant sciences but are also adept at communicating across cultures. The cultural resources will include *vocabularies* that blend terms from the different fields, what linguists call pidgins, creoles, or cajuns. The technological resources will include computer-based *collaboratories* where scientists and engineers from different fields may share, analyze, and visualize multimodal data [57.43].

57.4.2 The Example of Statistical Variation

To the extent that a unifying language already exists that can communicate across sciences, it is mathematics. Notably, the same descriptive and analytical statistics can be applied to numerous fields. The same correlation and regression coefficients can be calculated for crop yields in agriculture as for economic return on investments. The same normal curve or Gaussian distribution can be used to estimate the statistical significance of sample data in chemistry as in sociology. Even at this late date in the history of mathematics, it can be profitable to seek methods used exclusively in one area of science and see where else they might usefully be applied.

One of the most common patterns of variation, across many fields, is the *Zipf* [57.44] distribution, named after George Kingsley Zipf who believed it reflected a fundamental natural principle of least effort. Zipf's examples primarily concerned human language; for instance an analysis of the frequency of different words in the novel *Ulysses* by James Joyce. After counting how many times a word was used (its frequency), the words were ranked in terms of their frequencies, assigning the number 1 to the most common word, 2 to the second-most common word, and so on. Zipf postulated that the product of each word's frequency by its rank was approximately constant, across all words, for both common and rare words alike.

In support of his view that the underlying principle applied across many realms, Zipf also cited the example of city size. Take the populations of all the cities, towns and villages in a major nation and arrange them in descending order of size, then graph their populations. Or, I suggest, catalog all the objects in the solar system and arrange their data from the largest to smallest (Sun, Jupiter, Saturn ... grains of dust). Graphed on log–log scales, such tabulations produce an approximately straight line with negative slope. The distribution also appears to fit the sizes of corporations [57.45], and the sizes of religious congregations generated from computer simulations of group self-assembly through social interaction [57.46]. At the nanoscale, Zipf's so-called law also appears to describe the size distributions of physical structures that result from self-assembly [57.47]. There is considerable debate whether Zipf's discovery really reflects a meaningful natural law, or indeed what it means [57.48], but the distribution certainly qualifies as a pattern of variation or scaling law relevant to multiple sciences and technologies.

Metaphors drawn from one realm and applied to another can carry their mathematics with them. There already exists at least one example in which concepts developed to explain nanoscale phenomena have been applied along with their associated mathematics to the realm of human behavior. One of the most influential ideas in sociological research on social networks is the *vacancy chain* [57.49]. In a hierarchical organization, if an executive retires, someone else is promoted into the position, thereby opening that person's former position, into which somebody else will be promoted. The concept of vacancy chains was not original to sociology, however, but brought from semi-conductor physics by Harrison White who had earned a Ph.D. In a semiconductor material, when an electron moves step by nanoscale step across the lattice of molecules, a hole moves in the opposite direction. Applied to the sociology of formal organizations, this model gave rise to the concept of vacancy chains.

57.4.3 Proposed Higher-Level Concepts

It could be valuable to identify high-level concepts that categorize and thus that might inspire specific conceptual transfers from one field of science to another. One set of eight such higher-level concepts follows:

- **Conservation** – by analogy with the law of conservation of mass–energy, scientists can look for properties that are conserved in other realms, including symmetries, parity laws, and feedback-regulated stabilities in complex adaptive systems.
- **Indecision** – inconsistency, undecidability, uncertainty, chance, deterministic chaos and similar concepts are fundamental principles in the dynamics of systems over time; the tension between indecision and conservation is the fundamental fact of existence.
- **Configuration** – exemplified by the structure of complex molecules (responsible for the unity of nature at the nanoscale, where atoms assemble into complex structures), the detailed, dynamic structures of objects determine their properties.

- **Interaction** – elements of a system, (such as molecules in a solution, components in an electronic circuit, individual people in a community, or ideas in a mind) influence each other in terms of their separate configurations and their mode of combination, generating higher-level dynamics and other emergent phenomena.
- **Variation** – described by statistical analysis and caused by the combination of chance and divergent processes of interaction, this is essential for evolution and the emergence of complex systems of behavior based on the interaction of different configurations.
- **Evolution** – marked by drift, natural selection, and a competing trend toward greater complexity, evolution exploits variation to develop new configurations that compete through interactions.
- **Information** – scientific laws can be analyzed in terms of information content and flow, while the doing of any science today relies heavily upon information technology.
- **Cognition** – not to be confused with subjectivity, cognitive process are the dynamic aspect of information, involving potentially rigorous measurement, categorization, selection, and communication.

While all eight of these concepts deserve extensive explication, the last three require brief special discussion here, for two reasons. First, they relate directly to three of the four NBIC fields, namely biotechnology, information technology and cognitive technology. Second, their larger implications are controversial but may lead to a major reformulation of science.

Evolution is essential for the emergence of life and intelligence, and it can be harnessed for engineering creativity in many fields such as genetic algorithms in computing, evolutionary design, and the conscious development of new culture [57.50–52].

It is important to note that cognition is not the same as subjectivity, wishful thinking, or solipsism. Changing our minds about a physical phenomenon does not make it behave differently, although a fresh cognitive paradigm may offer insights we would not otherwise have about its characteristics, implications, and reality. Cognition is not merely a human activity, but also can be carried out by animals and computers, and it is closely tied to the other seven principles, notably to information.

Any violation of a conservation law or any indecision requires the input of information, and movement can be modeled as the transfer of information, so the laws of physics can be reconceptualized as principles of informatics. Classic information theory already employed concepts similar to those in thermodynamics [57.53], and there is a symmetrical relation between bits and atoms, as the name of the Center for Bits and Atoms at the Massachusetts Institute of Technology suggests. There is an information physics group at the University of New Mexico, and principles concerning communication of information constantly arise in work related to quantum theory [57.54, 55] and even appear in sober discussions of nanoscale molecular machines and nanostructured materials [57.56, 57]. Physicist *John Archibald Wheeler* is famous for having invented the slogan "it from bit", expressing the view that information is ontologically primary over matter, and he has offered a cognitive model of the origin of the universe [57.58, 59]. The point here is *not* to assert that we know the impact that future developments in evolutionary, information and cognitive sciences will have on nanoscience or natural science more generally, but merely to suggest that surprising intellectual developments may occur from NBIC convergence.

57.5 Cultural Implications of Convergence

The emphasis in NBIC discussions has naturally focused on the engineering applications that are likely to result in the coming decades, but we should not lose sight of the fundamental realignment of science and its role in society that is also likely. Despite high levels of public confidence in science, public understanding of its principles tends to be very weak [57.60]. This situation may change, when large sectors of science have consolidated intellectually and appropriate new educational curricula have been developed.

57.5.1 Education

Convergence in science and engineering education will require the development of vast new curriculum resources. A major effort in this direction has been NSF's program to build a national science digital library (NSDL) by creating and assembling curricular materials and then sharing them widely [57.61]. As the nsdl.org website explains, "The NSDL mission is to both deepen and extend science literacy through access to materials and methods that reveal the nature of the physical universe and the intellectual means by which we discover and understand it." Perhaps coincidentally, when the word *nanoscience* was entered into the NSDL's search engine, the first reference that came up connected nanoscience to genomics [57.62], thus illustrating convergence of nanotechnology, biotechnology, and – through the digital library itself – information technology.

This proliferation of curricular material will be bewildering, unless it is matched by a reorganization of the structure of classes taught in schools, reflecting the progressive convergence of the fields of professional science themselves. One excellent model that might be emulated is that of the science, mathematics, and computer science magnet program at Montgomery Blair High School in Maryland [57.63]. For example, the incoming student takes physics the first semester, chemistry the second, and the lessons closely connect the two subjects, even with an assignment at the beginning of chemistry relating to the quantum physics of electron orbits. The following paragraph from the program's website explains the intensely convergent design of the freshman curriculum:

> In the ninth grade, the magnet courses are strongly linked to each other. Students may learn a concept first semester in Physics, or second semester in Chemistry and bring that conceptual knowledge to their ninth grade R&E research and experimentation class. In R&E, they apply the knowledge they've gained from their science classes, and learn problem solving skills, not to mention a few engineering techniques. The data students have collected in their science and R&E classes is brought to their Fundamentals of Computer Science course, where the data is analyzed and further modeling of the experiment is completed on the computer. The Fundamentals of Computer Science course is not your typical computer science programming course at all. In fact, very little programming is done in this class – that comes sophomore year! Instead, students learn some of the ways scientists and mathematicians use a computer. They use modeling programs to perform virtual experiments and simulations. And they learn some of the fundamentals of computer science such as Boolean algebra and circuitry. [57.64]

Founded in 1985, this remarkable public school program admits about 100 highly gifted students each year, with an acceptance rate of only one out of every six students who apply, so we cannot be sure how widely its

paradigm can be applied. However, it would be shameful to assume that only a tiny elite can learn the fundamentals of science, if they are taught in a clear and well-organized manner. Furthermore, we can hope that convergence will in fact simplify the fundamental principles over the coming years, even as increasingly vast information-technology databases of facts are employed by professional scientists and engineers to apply these principles to specific applications. The result of even partial success in convergent science education of future generations will produce intellectual leaders capable of transforming human culture utterly.

57.5.2 Weltanschauung

Human understanding of the universe has been a difficult struggle, and each insight has been won at great cost. In prehistoric times, humans psychologically supplemented their limited technologies with magic, and their limited science with superstition [57.65]. Some of these superstitions became institutionalized, not only in religious organizations, but also in forms of political organization, the mass media, and numerous other institutions of society. The rise of science and science-based engineering both serves those institutions and threatens them.

At present, nanotechnology does not pose nearly as sharp a challenge to popular superstitions as do the three other fields. Biotechnology not only includes a number of activities that are anathema to traditional religion – artificial birth-control techniques, stem-cell research, and perhaps soon even human reproductive cloning – but also connects directly to the unscriptural theory of evolution by natural selection from random variation. Information technology erodes orthodoxy by disseminating information and ideas widely, notably via the Internet, and it poses a special challenge to traditional beliefs about personhood through its incomplete but growing successes in artificial intelligence.

Paul Bloom has argued that cognitive science will be the arena of religious conflict in the 21st century, because its powerful theories and growing findings contradict traditional notions of the unitary, transcendent, immortal soul [57.66]. Other cognitive scientists – who combine psychology, anthropology, and evolution theory – argue that the human species evolved with an innate bias to assume that complex events were caused by conscious agents [57.67–69]. This may have been adaptive thousands of years ago, because it helped people predict the behavior of the animals they hunted, the predators that hunted them, and the other people who cooperated to form the hunting and gathering bands in which ancient humanity lived. But one consequence of this mental bias to perceive an agent behind every action may have been a readiness to believe in supernatural beings, postulated to be the willful causes of natural phenomena. To the extent that nanotechnology assists cognitive science in unlocking the secrets of the human brain, it may therefore become entangled in the great struggle against religion that Bloom believes may be just over the horizon for science.

Convergence changes the rules of the game in several ways. Throughout the 20th century, scientists observed a truce with popular superstitions, in part because conflict would endanger public financial support for science, but in part also simply because any one scientific development was likely to be irrelevant to religious faith. This was the case because science was fragmented. With convergence, every part of science becomes connected to dangerous ideas like evolution, artificial intelligence, and research on human cognition [57.70]. As an ivory-tower activity, pure science may have little impact on the popular mind. But embodied in advanced technology, science is changing the fundamental conditions of human life, and thus cannot be ignored.

57.5.3 Policy Implications

Medical ethicist *James Hughes* [57.71, 72] has analyzed the competing ideologies that are likely to shape policy responses to new technologies in the 21st century, with special emphasis upon NBIC, as well as on biotechnology where much of the public debate currently rages. In the 20th century, he argues, it was possible to map political ideologies in terms of two dimensions: *economics* and *culture*. With respect to economic issues, progressives differed from conservatives in wanting to use government programs to reduce inequality. Among the cultural issues that divided progressives from conservatives were nationalism, religion and gender roles. Today, however, a third dimension of ideological variation may be emerging, describing attitudes toward *technology*.

In *techno-politics*, Hughes says, some individuals and groups will press for more rapid scientific and technical progress, and others will warn about potential dangers of technology and call for caution in its development. This ideological dimension is orthogonal to the other two. For example, *techno-conservatives* are not necessarily economic conservatives, but may be a mixture of people from the left and right ends of the politico-economic spectrum who share a suspicion of developments in such fields as stem cell research or

nanotechnology. Among *techno-progressives*, a key issue may be the extent to which policy should favor technological advances that could fundamentally change human nature, for example by enhancing the powers of mind or body, or by considerably extending the human lifespan.

Philosopher *George Khushf* [57.73] suggests that the scientists and engineers working on NBIC need to discuss more fully what it means to improve human performance. He notes that the title of the original NBIC report included the phrase *for improving human performance*. The title of the second book-length NBIC report refers to *human potential*, and the third talks more generally about *human progress* and *innovations*. Some readers have jumped to the conclusion that NBIC involved an attempt to transform human nature in some significant way, but most contributors to the reports have certainly not seen it that way.

Khushf notes that many of the applications described in the first NBIC report improve human performance only in the most general sense of providing valuable benefits, such as more efficient energy production and economic sustainability. They allow humans to achieve many of their traditional goals more cheaply, comfortably, or effectively, in areas that could be described as economic or as having to do with the day-to-day quality of life.

A second, and highly controversial meaning of *human enhancement* is physical transcendence of our inherited genetic, environmental, and cognitive condition. In a manner reminiscent of Hughes's analysis, Khushf suggests that a culture clash is fueling the controversy over human enhancement. Some people, perhaps comparable to Hughes's techno-conservatives, believe that there is such a thing as human nature and that technology should not be used to change it. In contrast, people like Hughes's techno-progressives see nothing wrong in using technology to increase native human abilities beyond what they have traditionally been. One way to describe this debate is in terms of NBIC applications that seek to provide *therapy* versus *enhancement* [57.74].

Among the scientists and engineers who are actually doing the NBIC discovery and innovation, the focus is on the technically exciting next steps, and there is great optimism that NBIC can benefit human beings. Interested outsiders, however, are making ethical distinctions about how NBIC might affect humans. Some ethicists, critics and other nontechnical observers especially distinguish benefits that help humans overcome hazards such as diseases and environmental threats from enhancements that would give humans powers they never possessed before.

The NBIC scientists and engineers do not think in terms of this distinction between therapy and enhancement. Khushf argues that NBIC proponents need to address this issue explicitly.

One approach would be for the convergenist movement to explicitly adopt an ethic of helping people overcome disadvantages but not changing their human natures, promoting therapy but not enhancement. Khushf argues instead that convergenists should promote enhancement as well, but only after clarifying its meaning and their own cultural values.

The distinction between therapy and enhancement is most clear in the general area of biotechnology, where medicine has traditionally sought to cure diseases and technology has long been used to overcome disabilities. Enhancements, such as steroid use by athletes, have been decried for violating the *rules of the game* in human competition. Biotechnology is also the area where the ethical debates are most acute at the present time. The feelings are not entirely logical, however. Many people would be outraged if some individuals started inserting computer chips in their brains to improve their mental functions, but nobody is offended that thousands of people carry pocket Internet-linked computers that enhance memory without placing any hardware within their skulls.

The whole distinction between therapy and enhancement may be erroneous, based on naive assumptions that are largely false. In one of the most influential books on the history of technology, *Man Makes Himself*, *V. Gordon Childe* [57.75] pointed out that the human species has not been *natural* for at least ten thousand years, because we rely upon major technologies that have transformed us. The invention of agriculture is his chief example, because it transformed nutrition, vastly increased population size, made sedentary life and cities possible, and permitted the invention of writing and formal education. Humans were designed by evolution to live in small bands of hunters and gatherers, having a very low population density and possessing cultural traits adapted to roaming a largely uninhabited wilderness. The ability to read and write is not part of original human nature, one could argue. Nobody knows how many people lived in the world 100 000 years ago, but less than 1 000 000 is quite plausible. Thus there may be as many as 10 000 times as many people in the world today than there were when most of our *natural* traits evolved.

57.6 Conclusion

People outside of science and engineering sometimes have unreasonable images of what technology can accomplish in the near future. Another way of putting this is to say that we may still be only half way through the historical transformation of the world that began with the European renaissance. Some date that shift from the so-called high renaissance of Leonardo da Vinci (1452–1519), in which case we have another 500 years until technological progress runs out of new innovations to make. Historians today are more likely to start the clock for modern civilization closer to the year 1200 or even earlier. In any case, while we cannot predict with any confidence, it would seem that some of the imaginable consequences of convergence – human transcendence among them – may be only dreams (or nightmares) for the very distant future.

Over the next few decades, we can anticipate progress across the NBIC fields of science and technology, that could be comparable to a renaissance. Like the original renaissance, convergence will be driven by two ideals. First, its central assumption is that nature can be understood intellectually as a unified system, operating according to rational principles of very broad scope. Second, its prime value is the humanistic goal of enhancing the lives of living people, rather than sacrificing them to abstract, inhuman ideals as may have been the case in the Middle Ages or in the darkness of 20th ideological regimes. Nanotechnology may seem remote from the vast tempests of human history, operating as it does on a scale a few billionths of the size of a human being. Through convergence, however, nanotechnology will become connected to all the movements related to science of the coming years.

References

57.1   E. S. Michelson: Measuring the merger: Examining the onset of converging technologies. In: *Managing Nano-Bio-Info-Cogno Innovations*, ed. by W. S. Bainbridge, M. C. Roco (Springer, Berlin Heidelberg New York 2006) pp. 47–69

57.2   W. S. Bainbridge: Cognitive technologies. In: *Managing Nano-Bio-Info-Cogno Innovations*, ed. by W. S. Bainbridge, M. C. Roco (Springer, Berlin Heidelberg New York 2006) pp. 203–226

57.3   W. Sententia: Cognitive enhancement and the neuroethics of memory drugs. In: *Managing Nano-Bio-Info-Cogno Innovations*, ed. by W. S. Bainbridge, M. C. Roco (Springer, Berlin Heidelberg New York 2006) pp. 153–171

57.4   Z. Lynch: Neuropolicy (2005–2035): Converging technologies enables neurotechnology creating new ethical dilemmas. In: *Managing Nano-Bio-Info-Cogno Innovations*, ed. by W. S. Bainbridge, M. C. Roco (Springer, Berlin Heidelberg New York 2006) pp. 173–191

57.5   M. C. Roco, W. S. Bainbridge: Converging technologies for improving human performance: Integrating from the nanoscale, J. Nanoparticle Res. **4**, 281–295 (2002)

57.6   E. O. Wilson: *Consilience: The Unity of Knowledge* (Knopf, New York 1998)

57.7   M. Castells: *The Rise of the Network Society* (Blackwell, Oxford 2000)

57.8   *Societal Implications of Nanoscience and Nanotechnology*, ed. by M. C. Roco, W. S. Bainbridge (Kluwer, Dordrecht 2001)

57.9   *Converging Technologies for Improving Human Performance*, ed. by M. C. Roco, W. S. Bainbridge (Kluwer, Dordrecht 2003)

57.10   *The Coevolution of Human Potential and Converging Technologies*, Vol. 1013, ed. by M. C. Roco, C. D. Montemagno (Ann. NY Acad. Sci., New York 2004)

57.11   *Managing Nano-Bio-Info-Cogno Innovations*, ed. by W. S. Bainbridge, M. C. Roco (Springer, Berlin Heidelberg New York 2006)

57.12   *Converging Technologies – Shaping the Future of European Societies*, ed. by A. Nordmann (European Commission, Brussels 2004)

57.13   C. D. Montemagno: Integrative technology for the twenty-first century. In: *The Coevolution of Human Potential and Converging Technologies*, Vol. 1013, ed. by M. C. Roco, C. D. Montemagno (Ann. NY Acad. Sci., New York 2004) pp. 38–49

57.14   C. D. Montemagno: Nanotechnology's implications for the quality of life. In: *Societal Implications of Nanoscience and Nanotechnology II: Maximizing Human Benefit*, ed. by M. C. Roco, W. S. Bainbridge (Springer, Berlin Heidelberg New York 2005)

57.15   J. H. Marburger: The future of nanotechnology. In: *Societal Implications of Nanoscience and Nanotechnology II: Maximizing Human Benefit*, ed. by M. C. Roco, W. S. Bainbridge (Springer, Berlin Heidelberg New York 2005)

57.16   T.-H. Fan, A. G. Federov: Electrohydrodynamic interactions of an AFM tip and a biological membrane, Proc. NSTI Nanotechnol., (2003) pp. 376–379

57.17   R. Bashir: Biologically mediated assembly of artificial nanostructures and microstructures. In: *Handbook of Nanoscience Engineering and Technology*, ed. by W. A. Goddard, D. W. Brenner, S. E. Lyshevski, G. J. Iafrate (CRC, Boca Raton 2004) pp. 15.1–15.32

57.18   J. Watson: Biomedicine eyes 2020. In: *Converging Technologies for Improving Human Performance*, ed. by M. C. Roco, W. S. Bainbridge (Kluwer, Dordrecht 2003) pp. 60–67

57.19   P. Connolly: Nanobiotechnology and life extension. In: *Converging Technologies for Improving Human Performance*, ed. by M. C. Roco, W. S. Bainbridge (Kluwer, Dordrecht 2003) pp. 182–190

57.20   M. Heller: The nano-bio connection and its implication for human performance. In: *Converging Technologies for Improving Human Performance*, ed. by M. C. Roco, W. S. Bainbridge (Kluwer, Dordrecht 2003) pp. 191–193

57.21   L. Goldenberg: The "integration/penetration model:" Social impacts of nanobiotechnology issues. In: *Societal Implications of Nanoscience and Nanotechnology II: Maximizing Human Benefit*, ed. by M. C. Roco, W. S. Bainbridge (Springer, Berlin Heidelberg New York 2005)

57.22   N. Savage: Converging technologies and their societal implications. In: *Societal Implications of Nanoscience and Nanotechnology II: Maximizing Human Benefit*, ed. by M. C. Roco, W. S. Bainbridge (Springer, Berlin Heidelberg New York 2005)

57.23   W. S. Bainbridge: Information technology for convergence. In: *Managing Nano-Bio-Info-Cogno Innovations*, ed. by W. S. Bainbridge, M. C. Roco (Springer, Berlin Heidelberg New York 2006)

57.24   J. Saam, E. Tajkhorshid, S. Hayashi, K. Schulten: Molecular dynamics investigation of primary photoinduced events in the activation of rhodopsin, Biophys. J. **83**, 3097–3112 (2002)

57.25   R. Llinas: Brain-machine interface via a neurovascular approach. In: *Converging Technologies for Improving Human Performance*, ed. by M. C. Roco, W. S. Bainbridge (Kluwer, Dordrecht 2003) pp. 244–251

57.26   J. Klein-Seetharaman, R. Reddy: Biological language modeling: Convergence of computational linguistics and biological chemistry. In: *Converging Technologies for Improving Human Performance*, ed. by M. C. Roco, W. S. Bainbridge (Kluwer, Dordrecht 2003) pp. 428–437

57.27   J. Klein-Seetharaman: The use of analogies for interdisciplinary research in the convergence of nano-, bio-, and information technology. In: *Societal Implications of Nanoscience and Nanotechnology II: Maximizing Human Benefit*, ed. by M. C. Roco, W. S. Bainbridge (Springer, Berlin Heidelberg New York 2005)

57.28   *Berkshire Encyclopedia of Human-Computer Inter-action*, ed. by W. S. Bainbridge (Berkshire, Great Barrington 2004)

57.29   R. Amant: Information technology and cognitive systems. In: *Managing Nano-Bio-Info-Cogno Innovations*, ed. by W. S. Bainbridge, M. C. Roco (Springer, Berlin Heidelberg New York 2006)

57.30   J. Hurd: Converging technologies in developing countries: Passionate voices, fruitful actions. In: *Managing Nano-Bio-Info-Cogno Innovations*, ed. by W. S. Bainbridge, M. C. Roco (Springer, Berlin Heidelberg New York 2006)

57.31   R. E. Albright: Roadmapping convergence. In: *Managing Nano-Bio-Info-Cogno Innovations*, ed. by W. S. Bainbridge, M. C. Roco (Springer, Berlin Heidelberg New York 2006)

57.32   E. Arkilick: Management of innovation for convergent technologies. In: *Societal Implications of Nanoscience and Nanotechnology II: Maximizing Human Benefit*, ed. by M. C. Roco, W. S. Bainbridge (Springer, Berlin Heidelberg New York 2005)

57.33   J. Spohrer, D. McDavid, P. P. Maglio, J. W. Cortada: NBIC convergence and technology-business coevolution: Towards a services science to increase productive capacity. In: *Managing Nano-Bio-Info-Cogno Innovations*, ed. by W. S. Bainbridge, M. C. Roco (Springer, Berlin Heidelberg New York 2006)

57.34   S. E. Miller: Converging technologies: Innovation, legal risks, and society. In: *Societal Implications of Nanoscience and Nanotechnology II: Maximizing Human Benefit*, ed. by M. C. Roco, W. S. Bainbridge (Springer, Berlin Heidelberg New York 2005)

57.35   P. J. DiMaggio, W. W. Powell: The iron cage revisited: Institutional isomorphism and collective rationality in organizational fields, Am. Sociol. Rev. **48**, 147–160 (1983)

57.36   J. Canton: NBIC convergent technologies and the innovation economy: Challenges and opportunities for the 21st century. In: *Managing Nano-Bio-Info-Cogno Innovations*, ed. by W. S. Bainbridge, M. C. Roco (Springer, Berlin Heidelberg New York 2006)

57.37   C. Mody: Short-term implications of convergence for scientific and engineering disciplines. In: *Societal Implications of Nanoscience and Nanotechnology II: Maximizing Human Benefit*, ed. by M. C. Roco, W. S. Bainbridge (Springer, Berlin Heidelberg New York 2005)

57.38   W. S. Bainbridge: Scientific nomenclature. In: *The Encyclopedia of Language and Linguistics*, ed. by R. E. Asher, J. M. Y. Simpson (Pergamon, Oxford 1994) pp. 3685–3690

57.39   W. S. Bainbridge: Technological nomenclature. In: *The Encyclopedia of Language and Linguistics*, ed. by R. E. Asher, J. M. Y. Simpson (Pergamon, Oxford 1994) pp. 4536–4541

57.40   M. E. Gorman: Combining the social and the nano-technology: A model for converging technologies. In: *Converging Technologies for Improving Human Performance*, ed. by M. C. Roco, W. S. Bainbridge (Kluwer, Dordrecht 2003) pp. 367–373

57.41   M. E. Gorman: Collaborating on convergent technologies. In: *The Coevolution of Human Potential and Converging Technologies*, Vol. 1013, ed. by M. C. Roco, C. D. Montemagno (Ann. NY Acad. Sci., New York 2004) pp. 25–37

57.42   M. E. Gorman, J. Groves: Collaboration on converging technologies: Education and practice. In: *Managing Nano-Bio-Info-Cogno Innovations*, ed. by W. S. Bainbridge, M. C. Roco (Springer, Berlin Heidelberg New York 2006)

57.43   T. A. Finholt, J. P. Birnholtz: If we build it, will they come? The cultural challenges of cyberinfrastructure development. In: *Managing Nano-Bio-Info-Cogno Innovations*, ed. by W. S. Bainbridge, M. C. Roco (Springer, Berlin Heidelberg New York 2006)

57.44   G. K. Zipf: *Human Behavior and the Principle of Least Effort* (Addison–Wesley, Cambridge 1949)

57.45   R. L. Axtell: Zipf distribution of U.S. firm sizes, Science **293**, 1818–1820 (2001)

57.46   W. Bainbridge: *God From the Machine: Artificial Intelligence Models of Religious Cognition* (AltaMira, Walnut Creek 2005)

57.47   A. T. Skjeltorp, J. Akselvoll, K. L. Kristiansen, G. Helgesen, R. Toussaint, E. G. Flekkoy, J. Cernak: Self-assembly and dynamics of magnetic holes. In: *Forces, Growth and Form in Soft Condensed Matter: At the Interface between Physics and Biology*, ed. by A. T. Skjeltorp, A. V. Belushkin (Kluwer, Dordrecht 2004) pp. 165–179

57.48   R. K. Belew: Weighting and matching against indices. In: *Finding Out About: A Cognitive Perspective on Search Engine Technology and the WWW* (Cambridge Univ. Press, New York 2000) pp. 60–104

57.49   H. C. White: *Chains of Opportunity* (Harvard Univ. Press, Cambridge 1970)

57.50   J. Pollack: Breaking the limits on design complexity. In: *Converging Technologies for Improving Human Performance*, ed. by M. C. Roco, W. S. Bainbridge (Kluwer, Dordrecht 2003) pp. 161–164

57.51   G. W. Strong, W. S. Bainbridge: A potential new science. In: *Converging Technologies for Improving Human Performance*, ed. by M. C. Roco, W. S. Bainbridge (Kluwer, Dordrecht 2003) pp. 318–325

57.52   W. S. Bainbridge: The evolution of semantic systems. In: *The Coevolution of Human Potential and Converging Technologies*, Vol. 1013, ed. by M. C. Roco, C. D. Montemagno (Ann. NY Acad. Sci., New York 2004) pp. 150–177

57.53   C. E. Shannon: A mathematical theory of communication, Bell Syst. Technic. J. **27**, 623–656, 379–423 (1948)

57.54   H. Ollivier, D. Poulin, W. H. Zurek: Objective properties from subjective quantum states: Environment
as a witness, Phys. Rev. Lett. **93**, 220401.1–220401.4
(2004)

57.55   C. A. Fuchs, A. Peres: Quantum theory needs no
"interpretation", Phys. Today **53**, 70–71 (2000)

57.56   T. D. Schneider: Sequence logos, machine/channel
capacity, Maxwell's demon, and molecular computers: A review of the theory of molecular
machines, Nanotechnology **5**, 1–18 (1994)

57.57   R. Pappu, B. Recht, J. Taylor, N. Gershenfeld: Physical one-way functions, Science **297**, 2026–2030
(2002)

57.58   J. A. Wheeler: Beyond the black hole. In: *Some
Strangeness in the Proportion: A Centennial Symposium to Celebrate the Achievements of Albert
Einstein*, ed. by H. Woolf (Addison-Wesley, Reading
1980) pp. 341–375

57.59   R. Matthews: I is the law, New Scientist **161**, 24–28
(January 1999)

57.60   National Science Board: Science and engineering indicators. In: *Division of Science Resources
Statistics* (National Science Foundation, Arlington
2004)

57.61   Center for Science and Engineering Education,
National Research Council: *Developing a Digital
National Library for Undergraduate Science, Math-
ematics, Engineering, and Technology Education:
Report of a Workshop* (National Academy, Wash-
ington, D.C. 1998)

57.62   J. Politz, A. Pombo: Genomics meets nanoscience:
Probing genes and the cell nucleus at 10^{-9} meters,
Genome Biology **3**, 4007.1–4007.3 (2002)

57.63   http://www.mbhs.edu/departments/magnet/

57.64   http://www.mbhs.edu/departments/magnet/
coursesandlife/

57.65   R. Stark, W. S. Bainbridge: *A Theory of Religion*
(Lang, New York 1987)

57.66   P. Bloom: *Descartes' Baby: How the Science of Child
Development Explains what Makes Us Human* (Ba-
sic Books, New York 2004)

57.67   P. Boyer: *Religion Explained: The Evolutionary Origins of Religious Thought* (Basic Books, New York
2001)

57.68   S. Atran: *In Gods We Trust: The Evolutionary Landscape of Religion* (Oxford Univ. Press, New York
2002)

57.69   J. L. Barrett: *Why Would Anyone Believe in God?*
(AltaMira, Walnut Creek 2004)

57.70   D. C. Dennett: *Darwin's Dangerous Idea: Evolution
and the Meanings of Life* (Simon & Schuster, New
York 1995)

57.71   J. J. Hughes: *Citizen Cyborg* (Westview, Cambridge
2004)

57.72   J. J. Hughes: Technopolitics of the 21st century. In:
Managing Nano-Bio-Info-Cogno Innovations, ed.
by W. S. Bainbridge, M. C. Roco (Springer, Berlin,
Heidelberg 2006)

57.73   G. Khushf: An ethic for enhancing human performance through integrative technologies. In:
Managing Nano-Bio-Info-Cogno Innovations, ed.
by W. S. Bainbridge, M. C. Roco (Springer, Berlin
Heidelberg New York 2006)

57.74   President's Council on Bioethics (U.S.): *Beyond
Therapy: Biotechnology and the Pursuit of Hap-
piness* (ReganBooks, New York 2003)

57.75   V. G. Childe: *Man Makes Himself* (New American Library, New York 1951)

58. Governing Nanotechnology: Social, Ethical and Human Issues

This chapter is a human-centered survey of nanotechnology's broader implications, reporting on the early phase of work by social scientists, philosophers, and other scholars. It begins with the social science agenda developed by governments, and the heritage of research on technology and organizations that social science brings to this mission. It then outlines current thinking about nanotechnology's economic impacts, health or environmental impacts, and social contributions. It discusses how technology can be regulated by a combination of informal ethics and formal law, then concludes by considering the shape of popular nanotechnology culture, as reflected in science fiction, public perceptions, and education.

58.1 Social Science Background

The role that social sciences will play in the future management of nanotechnology will depend both upon the tasks that society wishes to assign to it and upon the histories of social science itself, which have shaped their current scopes, theories, and research methodologies. Very recent governmental and professional publications suggest the tasks that social science will be asked to undertake.

58.1.1 The Scope of Societal Implications Research

In the United States, the President's budget request for the fiscal year 2006 called for significant efforts to understand the "practical implications and cultural context of nanotechnology research and development" ([58.1], p. 28), listing three subtopics:

1. Research directed at environmental, health, and safety (EHS) impacts of nanotechnology development, and risk assessment of such impacts

2. Education-related activities such as development of materials for schools, undergraduate programs, technical training, and public outreach

3. Research directed at identifying and quantifying the broad implications of nanotechnology for society, including social, economic, workforce, educational, ethical, and legal implications

Social and behavioral science research is potentially relevant to all three of these topics, but primarily for the third. Much of the research concerning EHS impacts will, of necessity, involve chemical and biological studies of the toxicity of nanostructured materials, and research on how such materials may travel in the natural environment, for example in groundwater. However, health and environmental protection are also socio-economic issues, involving studies of how human behavior affects the impact that such materials may have over the lifecycle of manufactured products, from production to ultimate disposal.

The nanoscience curricula developed for schools and colleges must, of course, be based on the findings and theories of physics, chemistry, materials science, and nanoscale biology. However, education will be more effective to the extent that it is also designed on the basis of a correct understanding of human cognition and the learning process. Phenomena distinctive to the nanoscale are remote from people's daily experience, and thus it will be important to develop effective means to help them to visualize these processes in a way that is not only accurate but also intelligible to students at a range of levels of sophistication.

The list in the third subtopic – social, economic, workforce, educational, ethical, legal – is a set of research areas, but it also implies a roster of scientific and scholarly approaches. The social topics naturally belong to the traditional social sciences: sociology, political science, cultural anthropology, social psychology, and linguistics. Economic issues naturally belong to economics, with some input from other sciences, notably the psychology of decision-making and the sociology of organizations. The workforce topic is shared primarily by sociology and economics. Studies in education draw upon all of the fields, notably cognitive science to understand the learning process itself, and the social sciences to understand the functioning of schools.

There is some controversy over which fields of science and scholarship have the most to contribute to an understanding of ethics. Within philosophy, generally considered one of the humanities rather than one of the sciences, ethics is an established discipline, and some philosophers – called *ethicists* – specialize in analyzing real-world problems. The processes by which norms, customs, and ethical principles become established in society, and their variations across societies, are studied by sociologists, social psychologists, cultural anthropologists, and historians. Throughout human existence, religion has been the primary institution of society establishing ethical standards, but its role in modern society is hotly debated.

Some would say that modern society has replaced the informality of traditional ethics with the formality of law and government regulations. Legal implications of nanotechnology are the domain of political science, socio-legal studies, jurisprudence, and criminology. However, government does not monopolize decisions about right and wrong, so the field of ethics remains largely autonomous, rather than being subsumed by law. The professional ethics of nanoscientists and nanotechnologists is important in its own right, as are ethical standards developed by corporations and business groups interacting with the general public.

Another starting point is the categorization of fields suggested by the organization of research programs at the National Science Foundation (NSF), which has taken a leading role in the U.S. National Nanotechnology Initiative. The chief home for social research is NSF's Directorate for Social, Behavioral and Economic Research, which contains two divisions whose missions are to support university-based scientific research: the Division of Behavioral and Cognitive Sciences, and the Division of Social and Economic Sciences. The Directorate for Engineering, which has played a leading role in the NNI, supports some socio-economic research on industrial management. The Directorate for Computer and Information Science and Engineering supports research on social implications of information technology, which will increasingly be affected by nanotechnology in such areas as sensors, mobile computing, and high-density data storage. The Directorate for Education and Human Resources has a responsibility for research as well as curriculum development and specific educational programs across science, engineering, and mathematics. One difference between the NSF and its counterparts in other countries is that the humanities are not included, except for a little support for the history and philosophy of science.

A European report, *The Social and Economic Challenges of Nanotechnology*, issued by the Economic and Social Research Council in the UK, suggests a rather different three-part research agenda for the social sciences [58.2]:

1. The governance of technological change;
2. Social learning and the evaluation of risk and opportunity under uncertainty;
3. The role of new technology in ameliorating or accentuating inequity and economic divides.

The first item suggests a more aggressive, centralized role of government in setting research priorities, imposing regulations (concerning health, safety and intellectual property), and supporting education than generally envisioned in the US. The second item suggests the challenges faced by corporations and individuals as they learn from experience about the costs and benefits of various technical possibilities. The third suggests the larger political and economic context that mediates between the inventions and their ultimate human impact.

The following pages will consider such issues from the perspective of social science, especially informed by a pair of major conferences on the societal implications of nanoscience and nanotechnology that the author helped organize and edit. The first major gathering to examine the societal implications of nanoscience and nanotechnology was held at NSF, from the 28th to the 29th of September 2000. At that time, the social sciences had not yet been mobilized to address nano issues, so the emphasis was on a somewhat pragmatic level: how nanotechnology could contribute specific innovations to a range of applications meeting human needs. However, the resulting book-length report – which we can call *Societal Implications I* – was able to identify a large number of social science hypotheses and the beginnings of a rich research agenda.

Over the following five years, a large number of smaller conferences, publications, and individual efforts have helped build a new field dedicated to the study of the societal implications of nano, but the most significant single event was a second NSF conference held from the 3rd to the 5th of December 2003. For the sake of concision, we can call its report *Societal Implications II*. By the time it was prepared, some real social science research had been performed on the topic, a variety of new theory-based hypotheses had been framed, and data collection activities had been launched. To provide a context for understanding recent developments, it is necessary to survey at least a small part of the historical background that formed the consciousness of contemporary social scientists.

58.1.2 Technological Determinism Theory

Social science has a very long and fruitful history of research into the societal implications of technology, and it has also addressed the social wellsprings of ethics and legal institutions. This background will be useful when addressing new issues concerning nanotechnology, but it is essential to recognize that a diversity of viewpoints have survived a century of debate, so a consensus should not be expected soon, if ever. Nonetheless, many insights, theories, research methodologies, and historical analogies have been developed that can be bought to bear on nanotechnology, as we attempt to understand and deal with its implications for human society.

The classic sociological theory about the role of technology in human history is *technological determinism*, the view that technological innovation is essentially self-causing and largely determines social trends. Although this perspective is certainly not fully correct, it still has

much to teach us and is a benchmark against which other theories are measured. The best statement of technological determinism is probably still the one that *William F. Ogburn* expressed the better part of a century ago in his 1922 textbook, *Social Change* [58.3]. For Ogburn, there were four primary steps in the process of development:

1. invention
2. accumulation
3. diffusion
4. adjustment

Invention is the process by which new technologies are created. Many people have the romantic notion that inventions are the products of unusually creative individuals, but Ogburn said instead that it was a largely impersonal process. He noted that a very large fraction of all worthwhile inventions are developed almost simultaneously by two or more different people, and suggested that inventions were the natural result of the particular level of science and technology at that moment. If all of the pieces exist, somebody is bound to put them together. For example, the telegraph is a combination of existing elements: electricity, coils, batteries, signaling, and alphabet codes. Once all of these were in place, several individuals simultaneously invented variants of the telegraph, and it joined the existing cultural base that would make still further developments possible, like the telephone, the Internet, and the World Wide Web. There is no single inventor for any major branch of nanotechnology.

Accumulation is the growth of technical culture, as new inventions are conceived at a faster rate than old ones are forgotten. Today, humans invest great effort into preserving the technical expertise required to develop and apply inventions, especially through a myriad of educational programs in such fields as engineering, computer science, mathematics, and the natural sciences. In ancient times, this was accomplished through apprenticeship relationships, which were one of the most significant societal institutions of the Renaissance. Today, information about nanotechnology accumulates in specialist journals, and is also communicated at conferences and within major research centers.

Diffusion is the transmission of an invention from one context to another. This can occur from one field of activity to another, from one geographic location to another, or even across cultural boundaries. Another, related term is *technology transfer*, which sometimes has the connotation of applying a scientific discovery to a practical problem or commercializing it, or it may mean applying an engineering approach from one industry to another. Much diffusion takes place when people move, for example when a student trained in some branch of nanoscience takes a job in industry.

Adjustment is the process by which the non-technical aspects of a culture respond to invention. For Ogburn, this was the one step where social factors really mattered, and he felt that this adjustment was often carried out by social movements. Today, government-sponsored activities like the *Societal Implications* conferences anticipate nanotechnology's impacts, in order to prepare a swift and hopefully painless adjustment.

Like many other American sociologists of the middle of the twentieth century, but hardly any twenty-first century sociologists, Ogburn was a *functionalist*. He believed that societies naturally developed harmonious cultures, in which the major institutions (including beliefs and values as well as formal organizations) were well-adapted to each other and to surrounding conditions such as the economy or the physical environment. Occasionally, a new invention would be so powerful that it would upset this equilibrium, and there might be an uncomfortable period of *cultural lag* before the balance was re-established. Foresight minimizes cultural lag in the case of nanotechnology.

Ogburn's four-step model is remarkably clear and compelling, but a few moments of thought and the research of eight decades suggest modifications. First of all, there would logically be feedback loops between some of the four steps. The accumulation of inventions makes more inventions possible, because many new ideas are fresh combinations of old ones. Likewise, diffusion promotes more innovation, because it brings together many ideas that may join to form new ideas. Furthermore, some inventions promote accumulation and diffusion – such as the development of writing in ancient times and the creation of the Internet today – which then, in turn, stimulate more invention. Taken together, these ideas suggest that under favorable circumstances, the innovation process can be self-reinforcing, leading to exponential rates of change.

However, Ogburn's notion that invention is self-generating is questionable. The sociologist *Robert K. Merton* argued that different areas of endeavor, including fields of science, receive greater or lesser attention as society's values change, thus increasing or decreasing the rate of invention in a particular field [58.4]. Several other sociologists argued that the variable structure of social relationships in society retards or advances the diffusion that can enable further invention [58.5–7]. Not surprisingly, economists like *Jacob Schmookler* argued that invention is an economic activity which will vary depending upon how much the market invests in it [58.8]. Using the specific example of the space program, *William Bainbridge* suggested that some major technological developments are the results of social movements that push their goals despite the indifference of most scientists or investors [58.9], while *Paul Freiberger* and *Michael Swaine* described the birth of the personal computer in similar terms [58.10]. Perhaps nanotechnology can be understood as a social movement, too [58.11].

At the present time, nanotechnology seems to be riding the crest of several enthusiasm waves at once. It integrates innovations from several traditional fields (chemistry, physics, materials science), bringing together ideas and methods that can be combined in various ways to produce many new inventions. Research is going on worldwide, with rapid diffusion of innovations across national boundaries and technology transfer from one area to another. Nanotechnology concepts have started to affect the wider culture, and heavy investment indicates that the field has significant value.

It is clear that Ogburn's model must be modified to take account of the roles that markets, government agencies, social movements, and the wider culture will play in accelerating or decelerating progress in particular areas, by investing in research or discouraging innovations of particular kinds through regulation. This observation raises questions regarding both policy and research, specifically directed at the twenty-first century's premier technological revolution, nanotechnology.

58.1.3 Organization Theory

Among all the many well-established social science approaches relevant to societal implications of technology, one deserves to be mentioned here: the study of formal organizations. Nanoscience and nanotechnology are being developed within organizations, notably universities and corporations. Today, we understand that an organization is a complex system of social roles (notably leadership and labor), commitments (including formal rules), positions (statuses, jobs), and information (not only technical knowledge but also economic data and information about customers and competitors). In an extremely influential 1981 textbook, *W. Richard Scott* [58.12] described three different *system* perspectives on organizations that had guided both scholars and managers alike, that emerged in the following order over the preceding decades: *rational*, *natural*, and *open*.

The *rational system* perspective assumes that corporations and other formal organizations are dedicated to achieving explicit goals and are formally structured in a manner that should help the organization achieve its goals efficiently and effectively. Among the most important principles of rational management are the following:

1. All of the organization's personnel should be linked into a single hierarchy of authority in the form of a pyramid.
2. Each person should receive orders from only one superior.
3. No superior should have more subordinates than he or she can effectively control.
4. All routine business should be handled by subordinates according to a set of well-established rules, and superiors should concentrate on dealing with exceptions that fall outside the routine.

Another way of putting this is to say that top management makes decisions, perhaps based on information passed up from below, and it issues commands about the work to be done. Many managers find this a very comfortable way of thinking about their organization, and it presents a relatively simple context for setting and following policies concerning nanotechnology. Management decides which nanotechnologies to invest in, which lines of inquiry their nanoscientists are to explore, and which safeguards should be in place to ensure an adequate level of safety, health, and environmental protection. By the middle of the twentieth century, however, social scientists had discovered that this "idealized" picture of organizations did not fit reality well.

The *natural system* perspective argued that organizations are collections of people who are affected by formal goals and rules but not entirely dominated by them. These people have their own goals and needs, which must be met reasonably well if the organization is to function. Indeed, the fundamental goal of the organization and its parts is simply survival, and specific goals set by management are only secondary. To understand an organization, one should study the informal relationships among the individual people, and the formal organizational chart may be quite misleading. Attempting to impose greater rationalization may be dysfunctional, and a better way to manage is to build leadership at all levels, motivate workers, and strengthen human relationships. This suggests that an understanding of social and ethical issues concerning nanotechnology must be widely shared by people throughout the organization, rather than merely possessed by top management.

The *open system* perspective challenged both of the earlier perspectives for wrongly imagining that organizations are stable structures with consistent long-term goals and programs. Instead, it says, organizations are shifting coalitions of interest groups, strongly influenced by external factors that also shift unpredictably. To use the terminology of the early twenty-first century, organizations are *complex systems* that may or may not be *adaptive* and that are susceptible to *chaotic behavior*. A open system organization does not chart a fixed course toward a solidly established goal, but employs feedback to adjust course constantly toward moving goals. In such a variegated and unstable context, there is no single best organizational form or management style, and different methods work better in different environments.

To the extent that the open system model correctly describes organizations in the modern world, it may be impossible to establish firm policies about how to exploit nanotechnologies for maximum human benefit, but it will be necessary to constantly re-examine situations and respond flexibly. However, all three perspectives probably describe modern organizations to some extent. Thus the best overall strategy will be for management to establish firm goals and policies to the extent that this is possible, invite all members of the organization to participate in the social and ethical debates, and to follow flexible strategies guided by constant feedback from both inside and outside the organization.

58.2 Human Impacts of Nanotechnology

Researchers in a number of fields have begun to monitor the economic, environmental, and social effects of nanotechnology. Fundamental to this effort are data regarding how extensively and in what areas nanoscience is being pursued and nanotechnology is being developed. For example, a team at the University of California, Los Angeles, combining expertise in both economics and sociology, is creating NanoBank [58.13]. This will be a massive database for use both by specialists and by members of the public, beginning with information about nano research publications, patents, industries, and eventually social impacts. Meanwhile, a number of participants in a growing community of scientists have begun to theorize and collect data about the economic, health, and social impacts.

58.2.1 Economic Impacts

Most participants in the *Societal Implications* conferences were confident that nanotechnology could enable a wide range of economically profitable applications progressively over the coming decades as techniques are devised and investments are made in the infrastructure required for research, development, and production [58.14–18]. At first, nanotechnology would enable improvements in existing industries. Radical innovations that launch entirely new industries would be likely further in the future.

Among the contributors to *Societal Implications I* was economist *Irwin Feller*, who was at the time a professor at Pennsylvania State University and chairman of the advisory committee for the NSF Directorate for Social, Behavioral and Economic Sciences. He argued that nanotechnology, in all its many forms, is likely to enter the marketplace first in applications where performance is especially important and people are willing to pay somewhat higher prices for incremental improvements [58.19]. An interesting example is the announced use of carbon nanotubes in the heads of golf clubs manufactured by the Wilson Golf company; here they are used to achieve the most effective balance by separating strength from weight, supposedly allowing golfers to hit balls farther. Golf and other sports greatly reward small improvements in performance. Other examples of fields where there is a premium on performance include medicine, aviation and space technology, as well as military technology.

Richard Freeman, of Harvard University, has offered a set of three predictions about the economic impact of nanotechnology, based on an analysis of its nature as well as on economic theory [58.20]. First, the chief beneficiary will be labor, in the form of increased wages for workers. Second, nano will gradually increase labor demand and thus wages across the entire economy, rather than only in a few high-tech sectors. Third, the effect on the economy will be smooth, rather than causing sudden changes in production and employment. In great measure, these conclusions rest on the observation that the forms of nanotechnology we can envision over the next decade or two include a very large number and variety of developments that can reinvigorate many different industries to some degree, rather than a single revolutionary innovation. We should note, however, that all analyses and predictions at this early point are uncertain, and nothing can substitute for careful, empirical research.

Among the questions that need to be studied concerning nanotechnology's economic impact, the following stand out:

1. Can the technologically most advanced nations capture the profits from their own innovations, or will nanotechnology chiefly enrich the front rank of developing nations (notably China) that may be able to manufacture quality products at lower cost?
2. Can nanotechnology reassert the importance of manufacturing industries (over service and information industries), or will the declining relative significance of manufacturing reduce nano's potential economic impact?
3. Will the economic benefits of nanotechnology be shared widely, or will they primarily accrue to owners and investors of large corporations?
4. Will the great cost of retooling prevent the introduction of potentially important nanotechnology innovations, such as the replacement of current "silicon chip" microelectronics by molecular electronics based (for example) on carbon nanotubes?

58.2.2 Health and Environment

Beginning in 2003, evidence began to be published that some nanomaterials could be toxic or otherwise harmful to health under some circumstances [58.21–24]. This would not be particularly surprising, of course, because many materials can be harmful, depending upon the dosage and the organism's ability to clean small quantities out of its system. Risk, in nanotechnology as in other fields, is a factor of both the harm an event might cause and the probability that the event would actually occur [58.25]. A toxic substance that is safely locked away in a durable container, such as perhaps carbon nanotubes inside a composite material, is not harmful. Also crucial to an understanding of these effects is research into how the substance is transported in the natural environment, for example in the atmosphere [58.26]. Thus, knowledge about the potential danger to human health of a nanostructured material is a good first step toward taking the precautions necessary to ensure that it has no opportunity to cause that harm [58.27]. Organizations with a responsibility in this area, such as the U.S. Environmental Protection Agency, have begun exploring such issues [58.28, 29].

Already in *Societal Implications I*, Lester Lave noted that it was important to consider the costs and benefits of a particular kind of nanotechnology in terms of its entire lifecycle – from manufacture, through performance in use, to disposal – and in terms of the entire technical system of which it is a part. For example, the use of carbon nanotubes in parts of vehicles may entail some health hazards in the workplace before they are safely embedded in composite materials, but it could reduce health hazards in use by reducing the weight of the vehicle, thereby also reducing fuel use and the associated air pollution. The savings and other benefits in use could pay for the extra costs of manufacture, and some costs involved in safe disposal or recycling once the vehicle was worn out. Even with all the costs figured in, *Lave* argued that the net benefit of nanotechnology could be enormous:

> *It promises to reduce by orders of magnitudes the inputs of energy and materials and associated environmental discharges required to produce a device that can perform a particular task. The result could be perhaps an order of magnitude increase in real income for the current world population without requiring more energy, materials, or resulting in additional discharges. Thus, nanotechnology offers the prospect of giving poor nations much higher standards of living and making the world economy sustainable. [58.30]*

Nanotechnology can reduce the use of energy and other resources, thus simultaneously promoting conservation of these resources while reducing the pollution that is caused when they are used. The economic gains achieved by nano can not only be used partly to mitigate or prevent any harmful health or environmental consequences of the technology itself, but also can be invested directly into improving physical quality of life. The other side of this argument is that economic growth is likely to be very costly to the environment if industry is unable to employ nanotechnology. Naturally, such claims are open to challenge, and the only way to be certain is to carry out extensive, rigorous research, not only into the possible toxicity of some nanomaterials, but also on the socio-technical systems that may employ them, to identify dangerous points in the lifecycle where precautions would be advisable.

58.2.3 Social Scenarios

Participants in the *Societal Implications II* task force on the "quality of life" predicted that nanotechnology could contribute significantly with respect to food, water, energy, and the environment [58.31]:

- *food*: improving inventory storage; growing a diversity of high-yield crops
- *water*: low-energy purification and desalination; reducing water waste in manufacturing and farming
- *energy*: reduced dependence on fossil fuels; solar photovoltaic energy production; renewable energy systems; energy distribution with hydrogen
- *environment*: remediating waste and pollution; permitting systems and materials that use resources most efficiently; recycling pollution into raw materials; ensuring safety and sustainability of new materials

Another task force, with the responsibility to frame alternative social scenarios, pointed out that we do not really know how adequate our current institutions are to manage rapid technological change in a graceful and maximally beneficial manner. Indeed, the open systems model of organizations, discussed above, suggests that institutions will need to change in order to adapt to the new technological possibilities. Given our lack of knowledge about how well our institutions will perform over the short run, we cannot be sure about which of two very different scenarios will actually happen:

1. the transition from current technologies to much more capable nano-enabled technologies will be smooth and benign, bringing benefits quickly to the majority of people, with relatively few unwanted consequences
2. the transition will be rough, with many bad decisions that either fail to take advantage of good technologies or apply hazardous technologies unthinkingly, causing many conflicts among societal institutions and triggering an anti-nano backlash

With this second, chilling scenario in mind, the group recommended research into how our current institutions function when dealing with questions about new technologies, with the aim to use the results of such studies to redesign faulty institutions. One starting point would be an inventory of existing institutions, evaluating how they cope with uncertainty, change, and conflict.

Looking further into the future, *Robin Hanson* [58.32] has argued that nanoscience may ultimately transform the structure of manufacturing industries in a series of five steps:

1. Methods are successfully developed for efficient manufacturing atom-by-atom
2. General-purpose factories are established, using methods analogous to three-dimensional printing

Part H | 58.2

(stereolithography) to make a great variety of products from common raw and recycled materials, following software instructions

3. A large number of small factories of this kind are built everywhere, near their customers, dominating manufacturing to the exclusion of today's large corporations and huge centralized factories
4. These local, generalized factories are usually idle, because they are cheap and efficient
5. These small factories can copy themselves

Note that this is not the infamous "gray goo" scenario, because the factories copy themselves only when human labor is invested to run off and assemble the necessary parts for a new factory. The net result would be not only increased prosperity but also a further reduction of the economic significance of manufacturing, and the same might be the case for the transportation industry and many kinds of raw material extraction and production as well. This would further increase the relative importance of service and information industries, which themselves may be local (in the case of services) or highly distributed (in the case of information). If all this occurred, as *Bruce Tonn* [58.33] has pointed out, local communities would become much more important socially, and large governmental units such as nations could weaken.

58.3 Regulating Nanotechnology

Given the alternative models of formal organizations developed by social scientists, deciding how to "govern" nanotechnology could present problems. In the rational system model, we need only decide which policies to follow, and organizations will carry them out. However, that model is faulty, and both the natural system and open system models predict that control will never be perfect. Thus, it will be important to instill a proper sense of ethics into engineers and decision-makers. The challenges for law and government are only beginning, but will require much work in the coming decades.

58.3.1 Ethics

A long-standing debate in philosophy concerns the extent to which ethical rules can be objectively defined, versus being judgments made by human beings from their particular standpoints and interests [58.34–36]. A modern view is that morality is negotiated between people, in order to enhance cooperation and to serve both their mutual interests and their enlightened self-interests [58.37, 38]. At a joint US-European workshop on the societal implications of nanotechnology, held in 2002, *Jesus Mosterin* noted that ethical standards are constantly changing as society and our technical capabilities change [58.39]. Thus, we cannot expect to settle nano-related ethical issues today, once and for all. Instead, each generation will need to engage these issues afresh.

Such observations cast doubt on the role of the philosophy of ethics itself, as at least three participants in *Societal Implications II* noted. Writing on "The Ethics of Ethics," *James R. Von Ehr* argued that professional ethicists who comment on nanotechnology may merely be expressing their personal, political opinions as if they were the result of objective philosophical analysis [58.40]. Similarly, *Robert E. McGinn* doubted that ethicists were sufficiently knowledgeable about nanotechnology to be able to apply philosophical analysis to it, and *John Trumpbour* speculated that whatever one wants to do with nano, it would be possible to find some ethicist or other who would endorse it [58.41, 42].

To see our way around this impasse, we need to recall the origins of philosophy in ancient Greece. Socrates did not pretend to know the exact truth himself, and with respect to ethics he ultimately deferred to his society when he drank that cup of hemlock that ended his own life. The role of the philosopher is not to proclaim truth but to help other people seek it, perhaps through that method of exploratory discussion that even today is called *Socratic dialog*. This is what philosopher *Vivian Weil* did when she contributed to both *Societal Implications* conferences [58.43, 44]. In a similar spirit, in *Societal Implications II*, philosopher *Rosalyn W. Berne* explored the meanings of the concept of quality of life in order to better understand how nanotechnology might affect it [58.45].

58.3.2 Law and Governance

Relatively little research has been carried out up to this point regarding legal issues that might distinctively concern nanotechnology, but one area of fairly active investigation is patents. The US Patent and Trademark office has established patent class 977 for nanotechnology:

> related to research and technology development at the atomic, molecular or macromolecular levels, in the length of scale of approximately 1–100 nm range in at least one dimension, and that provides a fundamental understanding of phenomena and materials at the nanoscale and to create and use structures, devices, and systems that have novel properties and functions because of their small and/or intermediate size [58.46].

One line of research charts the rising significance of nanotechnology in different industries and nations, by means of patent counts [58.47, 48]. *E. Jennings Taylor* notes that a primary purpose of patents is to encourage innovation by ensuring that inventors and the companies that support them will be able to profit, but he also points out that a poorly designed patent system can have the opposite effect of stifling innovation in nanotechnology [58.49]. Among the ways that this could happen is that nanotechnology patents could limit developments in nanoscience, unless special exemptions are made for research uses of patented techniques. While this issue affects many areas of science and technology, its is especially acute here because of the difficulty involved with drawing a conceptual line between nanotechnology and nanoscience.

The attorney *Sonia E. Miller* has suggested that nanotechnology might have implications for a number of legal specialties [58.50]. Criminal law might be affected through nanoscale forensics, which already employs DNA analysis and may in future exploit a very wide range of kinds of evidence collected or analyzed at the nanoscale, including surveillance information from nano-enabled microscale sensors. Health and environmental law might be complicated by the difficulty of defining nanostructured substances, the challenge of measuring their health effects, and the possibility that it could be hard to track the origins of uncontrolled releases of nanoparticles into the environment. In general, Miller argues that scientific expertise is going to play a wider role in legal cases, and she points out that it will be essential to arrive at consistent definitions of nano-related terms when writing legislation and applying the law.

Clearly, some objectively hazardous nanostructured materials will need to be covered by government regulations, and thus they will need to be unambiguously defined so that their presence or absence can be established for any particular case. There is a danger if vague and overly broad definitions of nanotechnologies are en-

shrined in laws that inhibit the development and use of safe and beneficial applications. Given the uncertainties and ambiguities in this area of rapid scientific and engineering progress, it may not be feasible or desirable to establish a comprehensive set of formal regulations. The alternative is to embed appropriate values into the organizations that are doing the research, development and deployment of new nanotechnologies [58.51]. That is, the organizations themselves should be dedicated to using the technology for human benefit – which after all is what their customers want – and naturally include effective review procedures and safeguards in their standard operating procedures.

Undoubtedly, in the future many nanostructured materials will be precisely defined and covered by government health and environmental regulations, but the potential societal impacts of nanotechnology go very far beyond adding a few exotic materials to the very long list of substances that can be implicated in pollution. Thus, a very important role for government will be supporting research into the wider implications, research that is often called *science and technology studies* (STS) [58.52]. The very real but secondary effects include the way that government funding initiatives such as the National Nanotechnology Initiative influence the nature of modern universities [58.53].

One area of intense government-supported nanotechnology activity is the development of military and national security applications, such as lightweight armor, sensors, battlefield medical equipment, and a host of other possibilities [58.54–56]. Reasonable people disagree about who has the responsibility to make decisions about developing technologies that are intended to harm human beings, or to protect fighters who harm other human beings. If one has a great deal of confidence in the political system of one's nation, then it is plausible to assert that the government itself has the effective moral responsibility to make proper decisions, and individuals should not be expected to second-guess these decisions. It can also be argued, however, that individuals can never abdicate their own moral responsibilities to their government, in which case nanoscientists, nanotechnologists, and those who support them must inform themselves and consider carefully the likely uses of their innovations.

If a government does not deserve its citizens' confidence, then scientists and engineers must consider carefully before developing technologies with military applications. In addition, they should pay attention to the international context. For example, a dual-use nanotechnology may be developed by an ethically responsible nation, then diffuse to an irresponsible one.

International discussions about the societal implications of nanotechnology have begun, notably at a conference held in Arlington Virginia, from the 17th to the 18th of June 2004, where representatives of 25 nations contributed [58.57]. At the very least, international communication will help all of the nations inform themselves about issues and the alternative responses to them. It would seem likely that international definitions of terms will be developed. In a linguistic sense,

nanotechnology was born in 1960, when the General Conference on Weights and Measures included the word "nanometer" in the internationally recognized definition of the metric system. Perhaps regulations governing international trade of some potentially hazardous nanostructured materials might be developed, and one could also imagine international partnerships to help develop especially beneficial nanotechnologies and distribute their benefits most widely around the globe.

58.4 The Cultural Context for Nanotechnology

For better or worse, our society's image of nanotechnology has largely been defined by fantasy stories, rather than by scientific research. However, the mass media are beginning to familiarize the public with the concept, and social scientists are beginning to explore public understanding and opinion about the topic. This is a crucial moment in history, when educational institutions need to examine how to incorporate nanoscience into curricula, and how to begin to produce the technically trained workforce that society will need.

58.4.1 Science Fiction

An old if minor tradition exists in science fiction where tales are set in very tiny environments. In 1919, the pulp magazine *All-Story Weekly* carried "The Girl in the Golden Atom" by Ray Cummings. Later expanded into a novel, this story concerned a scientist who was able to shrink himself down far below the nanoscale to visit an electron, which was depicted like a planet circling an atomic nucleus as its sun. This adventure begins when the scientist focuses a powerful new microscope on the wedding ring of his mother, and sees an attractive female inhabitant of the atomic planet in great danger. He has the technology to zoom down and rescue her, but if his instruments ever lose track of that particular atom, he will never find it again.

Cummings was actually well-informed about science and technology, because he was an assistant of the great electrical genius, Thomas Alva Edison. In 1911, the physicist Ernest Rutherford had discovered that atoms consisted of a nucleus surrounded by electrons, and his "solar system" model of the atom received considerable publicity. He had used gold foil as the target of his experiments, which naturally suggested a gold wedding ring for Cummings' romance. Even in 1911, it was clear that atomic structure was very different from

that of our solar system, however. For example, electron orbits are not limited roughly to a plane as is true for the planets, and there can be several identical electrons in the same orbit. In 1913, Niels Bohr suggested that electron orbits are possible only at certain energies, defined by the quantum theory that developed over the next few years, and the modern scientific picture of the atom is very different from a solar system. Importantly, electrons and other fundamental constituents of matter cannot function as worlds in which intelligent creatures could live and have adventures or romances, because below a certain size (called the Planck length) or energy (called the quantum) there can be no stable structures or entities [58.58].

This is not to say that Cummings should not have written his entertaining story. It probably encouraged readers who happened to be teenage boys to feel that atoms were interesting, and when the novel came out in 1923 its science was only a decade out of date. Later writers borrowed the idea that electrons could be the settings for stories [58.59]. In the critically acclaimed story "Surface Tension," *James Blish* imagined aquatic people only a millimeter in height, whose cells would of necessity be nanoscale, struggling to break through the conservatism of their water-bound culture, for which surface tension is a nice metaphor [58.60]. Perhaps the best-known novel about adventures at the nanoscale is *Fantastic Voyage*, novelized by *Isaac Asimov* from a movie with the same title starring Raquel Welch and Stephen Boyd, in which a submarine and its crew are miniaturized and injected into the blood stream of a human being to destroy a dangerous blood clot in his brain [58.61].

Quite separately, two of the contributors to *Societal Implications II*, *George M. Whitesides* and *Carol Lynn Alpert*, quoted one of the fundamental principles of science fiction, Arthur C. Clarke's third law: "Any sufficiently advanced technology is indistinguishable from magic" [58.62, 63]. Alpert also quoted the variant offered by science fiction writer Gregory Benford: "Any technology distinguishable from magic is insufficiently advanced" ([58.64] p. 5). The implications are two-fold: 1) to the uninitiated, nanotechnology seems like magic; 2) at the nanoscale, it is hard for nontechnical people to tell the difference between fact and fiction. Clarke's first and second laws also reveal the fact that science fiction writers wish reality were more magical than real scientists find it to be:

1. When a distinguished but elderly scientist states that something is possible, he is almost certainly right. When he states that something is impossible, he is very probably wrong.
2. The only way of discovering the limits of the possible is to venture a little way past them into the impossible [58.65, 66].

The historian of science fiction *Sam Moskowitz* explains: "Science fiction is a branch of fantasy identifiable by the fact that its eases the 'willing suspension of disbelief' on the part of its readers by utilizing an atmosphere of scientific credibility for its imaginative speculation in physical science, space, time, social science and philosophy" ([58.67], p. 11). Of concern to scientists and engineers is the possibility that the general public will either adopt an excessively cautious attitude toward nanotechnology, on the basis of false assumptions derived from horror fiction, or that they will fail to appreciate real developments because they are not so wondrous as the impossible ideas in science fiction.

Fully nine of the essays in the *Societal Implications II* report refer to *Michael Crichton's* technophobic novel, *Prey*. Crichton's introduction warns: "Sometime in the twenty-first century, our self-deluded recklessness will collide with our growing technological power. One area where this will occur is in the meeting point of nanotechnology, biotechnology, and computer technology. What all three have in common is the ability to release self-replicating entities into the environment" ([58.68], p. X). Practicing chemists and engineers in the relevant fields doubt that engineered nanoscale entities could become self-reproducing in the natural environment, unless they were genetically designed on the basis of the living things that have co-evolved with the Earth, but this is a standard assumption of science fiction about nanotech. Crichton's novel postulates that a monster is created by a corporate research project gone awry, consisting of biologically launched nanoparticles that evolve rapidly and develop a swarm intelligence. In

the first of two climaxes, the monster kills people, and in the second, it takes control over their bodies.

Aficionados of science fiction literature note that Crichton's novel is similar to a 1983 story by *Greg Bear*, that was later expanded into the novel, *Blood Music*, describing the fantastic transformation of all humanity by intelligent nanobots [58.69, 70]. Among the most highly acclaimed science fiction stories about nanotechnology is Neal Stephenson's novel, *The Diamond Age*, that imagines nano in terms of computer science. In a premise framed like a sociological theory, *Stephenson* explains, "Now nanotechnology had made nearly anything possible, and so the cultural role in deciding what should be done with it had become far more important that imagining what could be done with it" ([58.71], p. 37). In her nanotechnology quartet of novels, *Kathleen Ann Goonan* describes a world transformed by nanotechnology, but in contrast to Stephanson's computer science model, she imagines that future nanotechnology will be more like biological genetic engineering [58.72–75].

Our civilization will benefit from artistic creativity that takes its metaphors from nanotechnology. Science fiction can inspire young people to enter technical fields, and it can communicate to a wide audience the excitement that real scientists and engineers experience in their work. Thus, we should applaud the individual writers who have begun promulgating visionary images of the future of the field, and we cannot expect them to limit their imaginations to the proven nanoscale techniques of today. However, their stories are not an appropriate basis for deciding policy, and they may misinform the public about both investment opportunities and potential hazards. The unfortunate results could be nanocrazes and nanopanics.

Future historians will need to determine how much of the current interest from people who are neither nanoscientists nor nanotechnologists was inspired by science fiction, either directly or indirectly. One reason science fiction was especially ready to embrace nano was that it had suffered great disappointment about its primary earlier obsession, spaceflight. By the end of the 1970s, when manned spaceflight lagged after the end of the Apollo Program and space probes revealed what inhospitable places the other planets are, the plausibility of colonizing other worlds had declined significantly. Small organizations that sought to advance space colonization, like the L-5 Society, thus fell on hard times, and some members were ready for something new. Among them was Eric Drexler.

Drexler is undoubtedly very important in the history of popular nano-culture, and he would be an excellent subject for an in-depth biography. In the absence of deep study, we can note just a few facts. In the mid-1970s, Drexler was an assistant to *Gerard K. O'Neill*, an engineer who wanted to build a space city at the Lagrange-5 (L-5) point in the moon's orbit [58.76, 77]. Starting in 1977, *Drexler* published in the *L-5 News* some essays about the exploitation of extraterrestrial resources, a novel way of manufacturing space structures, and the legality of space development [58.78–80]. In 1987, the L-5 Society merged with the more conservative National Space Institute to form the National Space Society. Perhaps not entirely coincidentally, that was about the time Drexler published his visionary first book about nanotechnology, *Engines of Creation* [58.81], and the leading magazine of the science fiction field, *Analog*, first mentioned Drexler's nano ideas in 1987. Soon, that magazine was regularly publishing imaginative stories that included nanotechnology.

Thus, part of the significance of nanotechnology for popular culture is that it provides a set of metaphors to sustain a sense of wondrous possibility for the future, to some extent replacing the dream of spaceflight, even as it renders actual space travel more feasible technically and economically [58.82].

58.4.2 Public Perceptions

It is likely that only a small fraction of the general public in economically advanced nations still has much idea what nanotechnology is, although the mass media are beginning to popularize it [58.83, 84]. Certainly, readers of science fiction literature are a small minority of the general population, with a range of unusual attitudes toward science and technology [58.85], although they may informally share their views with many non-readers. In 2002, more than half of the respondents to a Eurobarometer survey could not answer an attitudinal question about how nanotechnology will impact their way of life, which was far more nonresponsive than for any of the other technologies listed [58.86]. A British survey found that only 29% of the general public claimed to have heard of nanotechnology, and only 19% were willing to try to define it ([58.87], p. 6).

It turns out to be rather difficult to frame simple questions that can evaluate how much the general public knows with any accuracy without eliciting quibbles from people who are very knowledgeable. Some questionnaire items have assumed nanotechnology deals only with things that are invisible because they are the size of atoms. However, surface coatings a few nanometers thick and bulk quantities of nanoparticles certainly can

be visible to the unaided human eye, and atoms are actually less than a nanometer in diameter. To measure knowledge accurately, one would need a several-item quiz of somewhat sophisticated questions, and this would be very costly in the expensive surveys of random samples of the general population; and then one would predict that almost everyone would flunk the test.

A non-random-sample study of members of the general public likely to be both interested and slightly knowledgeable was *Survey2001*, sponsored by the National Geographic Society [58.88]. This was an online questionnaire with an opportunistic set of respondents, and thus the percentages cannot be extrapolated to the general public. However, National Geographic readers are constantly bombarded with information about environmental degradation, and should be alert to the possible negative impacts of nano. Thus, it is interesting to see that fully 57.5% of 3909 English-speaking respondents agreed with the following statement: "Human beings will benefit greatly from nanotechnology, which works at the molecular level atom by atom to build new structures, materials, and machines." In contrast, only 9.0% agreed that: "Our most powerful 21st-century technologies – robotics, genetic engineering, and nanotechnology – are threatening to make humans an endangered species."

Given the sampling limitations of this study, statistical correlations within the data are probably more reliable than response frequencies [58.89, 90]. The questionnaire included 16 environmental issues, and attitudes toward nano correlated strongly with only one of them: "the introduction of genetically modified food into our food supply." People who were worried about so-called Frankenfoods also tended to be worried about nanotechnology, an indication that nanotechnology could become publicly stigmatized if it become associated in the public mind with controversial nanoscale biotechnologies. Also less affected by the sampling limitations was an open-ended question asking respondents to comment on the two statements that named nanotechnology. Interestingly, respondents gave very few negative answers, and none mentioned a concern about the hypothetical notion of self-reproducing nanoscale robots. In contrast, they were able to mention 82 distinct possible benefits of nanotechnology.

Michael D. Cobb and *Jane Macoubrie* carried out a random-sample telephone survey of 1536 Americans that asked them to rank five benefits and five risks of nanotechnology [58.91]. By far the most highly ranked benefit was "new ways to detect and treat human diseases," while "new ways to clean the environment" came in second. Among the possible risks, "losing personal privacy" was rated most important – presumably through nano-enabled microelectronic sensors – whereas "uncontrollable spread of nanorobots" was in last place. Interestingly, the risk that scientists currently take most seriously, "breathing in nanoparticles that accumulate in the body" was ranked in the middle of the five risks. While surveys like this are important, at the present time they probably say more about people's general concerns, rather than anything about nanotechnology itself.

58.4.3 Education

Education must be a high priority in any government nanotechnology initiative [58.92]. We can concern ourselves with two very different kinds of nanotechnology education: 1) preparing students to become professional nanoscientists, nanotechnology engineers, or technicians in nano-based industries; 2) including fundamental concepts about nanoscale phenomena in a liberal education and in the informal education that citizens receive after graduation.

To address issues in both areas, educators have begun looking at how to incorporate nano into school curricula, even in fairly early grades [58.93]. One idea of how

to do so is the creation of immersive, virtual reality environments to teach students at all levels about phenomena at the nanoscale, allowing them subjectively to amble among the atoms [58.94]. For nano professionals, training in mathematics will remain crucially important [58.95].

In the case of the United States, native-born citizens have recently shown a lack of interest in technical occupations, and reliance upon foreign nanotechnologists raises both security and sustainability concerns, especially as other countries upgrade both their educational institutions and career opportunities [58.96]. Discussions have begun about whether a distinctive curriculum and degree programs in nanoengineering need to be established, as opposed to continuing to draw upon graduates from many traditional fields [58.97, 98]. For future engineers and technicians to gain experience with the real materials associated with their future work, universities and corporations may need to set up nanofabrication facilities that are shared by both academic and industrial organizations in their particular geographic areas [58.99, 100]. Whatever form their technical education takes, nanoengineers also need to learn about the social and ethical implications of the work that they will be doing [58.101].

58.5 Conclusions

Scholars and social scientists in a variety of traditions are beginning to study the societal and social implications of nanotechnology, and to develop the knowledge base necessary for proper governance. A fundamental issue is how much we can rely upon formal, government regulation, versus informal guidance of the technology by the people involved in developing it. The latter may seem more difficult, but given the fallibility of formal organizations it may be more feasible.

The challenge is to enable humanity to take advantage of nanotechnology at the most rapid rate compatible with wise choices in those (hopefully few) areas where hazards or inequities could exist. Research will be

needed to identify nanostructured materials that are actually hazardous, leading to narrowly drawn regulation of these substances within the existing regulatory context for other hazardous substances. Social-scientific research will also be essential, not only to monitor nanotechnology development and identify problems promptly so they can be solved, but also to explore new avenues where nanotechnology can beneficially serve humanity. Equally important will be the public dialog and careful analysis needed to establish ethical principles and the factual knowledge with which to implement them among nanoscientists, nanoengineers, and decision-makers.

References

58.1   1. Nanoscale Science, Engineering, and Technology Subcommittee, National Science and Technology Council: *The National Nanotechnology Initiative: Research and Development Leading to a Revolution in Technology and Industry* (National Nanotechnology Coordination Office, Arlington 2005)

58.2   S. Wood, R. Jones, A. Geldart: *The Social and Economic Challenges of Nanotechnology* (Economic and Social Research Council, Swindon 2003)

58.3   W. F. Ogburn: *Social Change* (Huebsch, New York 1922)

Part H | 58

58.4   R. K. Merton: *Science, Technology and Society in Seventeenth-Century England* (Harper Row, New York 1970)

58.5   E. Katz, P. F. Lazarsfeld: *Personal Influence: The Part Played by People in the Flow of Mass Communications* (Free Press, Glencoe 1955)

58.6   D. Crane: *Invisible Colleges: Diffusion of Knowledge in Scientific Communities* (Univ. Chicago Press, Chicago 1972)

58.7   E. M. Rogers: *Diffusion of Innovations* (Free Press, New York 1995)

58.8   J. Schmookler: *Innovation and Economic Growth* (Harvard Univ. Press, Cambridge 1966)

58.9   W. S. Bainbridge: *The Spaceflight Revolution* (Wiley-Interscience, New York 1976)

58.10   P. Freiberger, M. Swaine: *Fire in the Valley: The Making of the Personal Computer* (McGraw-Hill, New York 2000)

58.11   T. A. Ten Eyck: Communication streams and nanotechnology: The (re)interpretation of a new technology. In: *Nanotechnology: Societal Implications – Individual Perspectives*, ed. by M. C. Roco, W. S. Bainbridge (Springer, Berlin, Heidelberg 2006) pp. 280–282

58.12   W. R. Scott: *Rational, Natural, and Open Systems* (Prentice-Hall, Englewood Cliffs 1981)

58.13   L. G. Zucker, M. R. Darby: Socio-economic impact of nanoscale science: Initial results and NanoBank. In: *Nanotechnology: Societal Implications – Individual Perspectives*, ed. by M. C. Roco, W. S. Bainbridge (Springer, Berlin, Heidelberg 2006) pp. 2–23

58.14   W. R. Boulton: Managing the nanotechnology revolution: Consider the Malcolm Baldrige national quality criteria. In: *Nanotechnology: Societal Implications – Individual Perspectives*, ed. by M. C. Roco, W. S. Bainbridge (Springer, Berlin, Heidelberg 2006) pp. 24–32

58.15   J. Canton: The emerging nanoeconomy: Key drivers, challenges, and opportunities. In: *Nanotechnology: SocietalImplications – Individual Perspectives*, ed. by M. C. Roco, W. S. Bainbridge (Springer, Berlin, Heidelberg 2006) pp. 32–43

58.16   S. Jurvetson: Transcending Moore's law with molecular electronics and nanotechnology. In: *Nanotechnology: SocietalImplications – Individual Perspectives*, ed. by M. C. Roco, W. S. Bainbridge (Springer, Berlin, Heidelberg 2006) pp. 43–56

58.17   G. Thompson: Semiconductor scaling as a model for nanotechnology commercialization. In: *Nanotechnology: SocietalImplications – Individual Perspectives*, ed. by M. C. Roco, W. S. Bainbridge (Springer, Berlin, Heidelberg 2006) pp. 56–61

58.18   L. Hornyak: Sustaining the impact of nanotechnology on productivity, sustainability, and equity. In: *Nanotechnology: SocietalImplications – Individual Perspectives*, ed. by M. C. Roco, W. S. Bainbridge (Springer, Berlin, Heidelberg 2006) pp. 64–67

58.19   I. Feller: An economist's approach to analyzing the societal impacts of nanoscience and nanotechnology. In: *Societal Implications of Nanoscience and Nanotechnology*, ed. by M. C. Roco, W. S. Bainbridge (Kluwer, Dordrecht 2001) pp. 108–113

58.20   R. Freeman: Non-nano effects of nanotechnology on the economy. In: *Nanotechnology: SocietalImplications – Individual Perspectives*, ed. by M. C. Roco, W. S. Bainbridge (Springer, Berlin, Heidelberg 2006) pp. 68–74

58.21   V. I. Colvin: The potential environmental impact of engineered nanomaterials, Nature Biotechnol. **21**, 1166–1170 (2003)

58.22   C.-W. Lam, J. T. James, R. McCluskey, R. L. Hunter: Pulmonary toxicity of single-wall carbon nanotubes in mice 7 and 90 days after intratracheal instillation, Toxicol. Sci. **77**, 126–134 (2004)

58.23   D. Warheit: Nanoparticles: Health impacts?, Mater. Today **7**, 32–35 (February 2004)

58.24   G. Oberdörster, E. Oberdörster, J. Oberdörster: Nanotoxicology: An emerging discipline evolving from studies of ultrafine particles, Environ. Health Persp. **113**(7), 823–839 (2005)

58.25   D. M. Berube: Communicating nanotechnological risks. In: *Nanotechnology: SocietalImplications – Individual Perspectives*, ed. by M. C. Roco, W. S. Bainbridge (Springer, Berlin, Heidelberg 2006) pp. 245–251

58.26   S. K. Friedlander, D. Y. H. Pui (eds): *Emerging Issues in Nanoparticle Aerosol Science and Technology* (Univ. California, Los Angeles 2003)

58.27   W. Luther (ed): *Industrial Applications of Nanomaterials – Chances and Risks* (Future Technologies Division, VDI Technologiezentrum, Düsseldorf 2004)

58.28   N. Savage: Converging technologies and their societal implications. In: *Nanotechnology: SocietalImplications – Individual Perspectives*, ed. by M. C. Roco, W. S. Bainbridge (Springer, Berlin, Heidelberg 2006) pp. 164–168

58.29   M. C. Roco: Environmentally responsible development of nanotechnology, Environ. Sci. Technol. **39**, 106A–112A (2005)

58.30   L. B. Lave: Lifecycle/sustainability implications of nanotechnology. In: *Societal Implications of Nanoscience and Nanotechnology*, ed. by M. C. Roco, W. S. Bainbridge (Kluwer, Dordrecht 2001) pp. 205–212

58.31   M. C. Roco, W. S. Bainbridge: Societal implications of nanoscience and nanotechnology: Maximizing human benefit, J. Nanopart. Res. **7**, 1–1 (2005)

58.32   R. Hanson: Five nanotech social scenarios. In: *Nanotechnology: SocietalImplications – Individual Perspectives*, ed. by M. C. Roco, W. S. Bainbridge (Springer, Berlin, Heidelberg 2006) pp. 109–113

58.33   B. Tonn: Co-evolution of social science and emerging technologies. In: *Managing Nano-Bio-Info-Cogno Innovations: Converging Technologies in Society*, ed. by W. S. Bainbridge, M. C. Roco (Springer, Berlin, Heidelberg 2006) pp. 309–335

58.34   I. Kant: *A Critique of Pure Reason* (Wiley, New York 1900 [1787])

58.35   G. E. Moore: *Principia Ethica* (Cambridge Univ. Press, Cambridge 1951)

58.36   J. Rawls: *A Theory of Justice* (Harvard Univ. Press, Cambridge 1971)

58.37   G. C. Homans: *Social Behavior: Its Elementary Forms* (Harcourt, Brace Jovanovich, New York 1974)

58.38   D. Gauthier: *Morals by Agreement* (Oxford Univ. Press, Oxford 1986)

58.39   J. Mosterin: Ethical implications of nanotechnology. In: *Nanotechnology: Revolutionary Opportunities and Societal Implications*, ed. by M. Roco, R. Tomellini (European Commission, Luxembourg 2002) pp. 91–94

58.40   J. R. Von Ehr: The ethics of ethics. In: *Nanotechnology: SocietalImplications – Individual Perspectives*, ed. by M. C. Roco, W. S. Bainbridge (Springer, Berlin, Heidelberg 2006) pp. 195–198

58.41   R. E. McGinn: Ethical issues in nanoscience and nanotechnology: reflections and suggestions. In: *Nanotechnology: SocietalImplications – Individual Perspectives*, ed. by M. C. Roco, W. S. Bainbridge (Springer, Berlin, Heidelberg 2006) pp. 169–172

58.42   J. Trumpbour: Technological revolutions and the limits of ethics in an age of commercialization. In: *Nanotechnology: SocietalImplications – Individual Perspectives*, ed. by M. C. Roco, W. S. Bainbridge (Springer, Berlin, Heidelberg 2006) pp. 113–121

58.43   V. Weil: Ethical issues in nanotechnology. In: *Societal Implications of Nanoscience and Nanotechnology*, ed. by M. C. Roco, W. S. Bainbridge (Kluwer, Dordrecht 2001) pp. 245–251

58.44   V. Weil: Ethics and nano: A survey. In: *Nanotechnology: SocietalImplications – Individual Perspectives*, ed. by M. C. Roco, W. S. Bainbridge (Springer, Berlin, Heidelberg 2006) pp. 172–182

58.45   R. W. Berne: Negotiations over quality of life in the nanotechnology initiative. In: *Nanotechnology: SocietalImplications – Individual Perspectives*, ed. by M. C. Roco, W. S. Bainbridge (Springer, Berlin, Heidelberg 2006) pp. 198–205

58.46   US Patent Trademark Office: Nanotechnology Classification 977 (USPTO, Classification Operations, Washington 2002), www.uspto.gov/go/classification/uspc977/defs977.htm

58.47   Z. Huang, A. Yip, G. Ng, F. Guo, Z. Chen, M. C. Roco: Longitudinal patent analysis for nanoscale science and engineering: Country, institution and technology field, J. Nanopart. Res. **5**, 333–363 (2003)

58.48   Z. Huang, H. Chen, Z. Chen, M. C. Roco: International nanotechnology development in 2003: Country, institution, and technology field analysis based on USPTO patent database, J. Nanopart. Res. **6**, 325–354 (2004)

58.49   E. J. Taylor: An exploration of patent matters associated with nanotechnology. In: *Nanotechnology: SocietalImplications – Individual Perspectives*, ed. by M. C. Roco, W. S. Bainbridge (Springer, Berlin, Heidelberg 2006) pp. 187–195

58.50   Sonia E. Miller: Law in a new frontier. In: *Nanotechnology: SocietalImplications – Individual Perspectives*, ed. by M. C. Roco, W. S. Bainbridge (Springer, Berlin, Heidelberg 2006) pp. 182–187

58.51   F. N. Laird: Problems of governance of nanotechnology. In: *Nanotechnology: SocietalImplications – Individual Perspectives*, ed. by M. C. Roco, W. S. Bainbridge (Springer, Berlin, Heidelberg 2006) pp. 207–211

58.52   B. E. Seely: Societal implications of emerging science and technologies: A research agenda for science and technology studies (STS). In: *Nanotechnology: SocietalImplications – Individual Perspectives*, ed. by M. C. Roco, W. S. Bainbridge (Springer, Berlin, Heidelberg 2006) pp. 211–223

58.53   T. L. Smith: Institutional impacts of government science initiatives. In: *Nanotechnology: SocietalImplications – Individual Perspectives*, ed. by M. C. Roco, W. S. Bainbridge (Springer, Berlin, Heidelberg 2006) pp. 223–232

58.54   W. M. Tolles: National security aspects of nanotechnology. In: *Societal Implications of Nanoscience and Nanotechnology*, ed. by M. C. Roco, W. S. Bainbridge (Kluwer, Dordrecht 2001) pp. 218–236

58.55   W. M. Tolles: In defense of nanotechnology in defense. In: *Nanotechnology: SocietalImplications – Individual Perspectives*, ed. by M. C. Roco, W. S. Bainbridge (Springer, Berlin, Heidelberg 2006) pp. 236–240

58.56   J. Reppy: Nanotechnology for national security. In: *Nanotechnology: SocietalImplications – Individual Perspectives*, ed. by M. C. Roco, W. S. Bainbridge (Springer, Berlin, Heidelberg 2006) pp. 232–236

58.57   Meridian Institute: *International Dialogue on Responsible Research and Development of Nanotechnology* (Meridian Institute, Alexandria 2004)

58.58   S. Weinberg: *The Discovery of Subatomic Particles* (Freeman, New York 1990)

58.59   F. Pragnell: *The Green Man of Graypec* (Greenberg, New York 1950)

58.60   J. Blish: *The Seedling Stars* (Gnome, New York 1957)

58.61   I. Asimov: *Fantastic Voyage* (Bantam, New York 1966)

58.62   G. M. Whitesides: Science and education for nanoscience and nanotechnology. In: *Nanotechnology: Societal Implications – Maximizing Benefit for Humanity*, ed. by M. C. Roco, W. S. Bainbridge (Springer, Berlin 2006) pp. 42–51

58.63   C. L. Alpert: Public engagement with nanoscale science and engineering. In: *Nanotechnology: SocietalImplications – Individual Perspectives*, ed. by M. C. Roco, W. S. Bainbridge (Springer, Berlin, Heidelberg 2006) pp. 265–274

58.64 T. Pratchett, I. Stewart, J. Cohen: *The Science of Discworld* (Ebury, New York 1999)

58.65 A. C. Clarke: Hazards of prophecy. In: *The Futurists*, ed. by A. Toffler (Random House, New York 1972) p. 144

58.66 A. C. Clarke: *Profiles of the Future* (Bantam, New York 1963)

58.67 S. Moskowitz: *Explorers of the Infinite* (Meridian, Cleveland 1963)

58.68 M. Crichton: *Prey* (Harper Collins, New York 2002)

58.69 G. Bear: Blood music, Analog **106**(6), 12–36 (1983)

58.70 G. Bear: *Blood Music* (Simon Schuster, New York 2002)

58.71 N. Stephenson: *The Diamond Age* (Bantam, New York 1995)

58.72 K. A. Goonan: *Queen City Jazz* (Tor, New York 1994)

58.73 K. A. Goonan: *Mississippi Blues* (Tor, New York 1997)

58.74 K. A. Goonan: *Crescent City Rhapsody* (Eos, New York 2000)

58.75 K. A. Goonan: *Light Music* (Eos, New York 2002)

58.76 G. K. O'Neill: *The High Frontier: Human Colonies in Space* (Morrow, New York 1976)

58.77 M. A. G. Michaud: *Reaching for the High Frontier: The American Pro-Space Movement, 1972–84* (Praeger, New York 1986)

58.78 E. Drexler: Non-terrestrial resources, L-5 News **2**(3), 4–5 (March 1977)

58.79 K. Henson, E. Drexler: Vapor phase fabrication of structures in space, L-5 News **2**(3), 6–7 (March 1977)

58.80 E. Drexler: Space mines, space law, and the third world, L-5 News **3**(4), 7–8 (April 1978)

58.81 K. E. Drexler: *Engines of Creation* (Anchor Press/Doubleday, New York 1986)

58.82 S. L. Venneri: Implications of nanotechnology for space exploration. In: *Societal Implications of Nanoscience and Nanotechnology*, ed. by M. C. Roco, W. S. Bainbridge (Kluwer, Dordrecht 2001) pp. 213–218

58.83 B. V. Lewenstein, J. Radin, J. Diels: Nanotechnology in the media: A preliminary analysis. In: *Nanotechnology: SocietalImplications – Individual Perspectives*, ed. by M. C. Roco, W. S. Bainbridge (Springer, Berlin, Heidelberg 2006) pp. 258–265

58.84 D. S. Hope, P. E. Petersen: A proposal to advance understanding of nanotechnology's social impacts. In: *Nanotechnology: SocietalImplications – Individual Perspectives*, ed. by M. C. Roco, W. S. Bainbridge (Springer, Berlin, Heidelberg 2006) pp. 251–258

58.85 W. S. Bainbridge: *Dimensions of Science Fiction* (Harvard Univ. Press, Cambridge 1986)

58.86 G. Gaskell, N. Allum, S. Stares: *Europeans and Biotechnology in 2002: Eurobarometer 58.0* europa.eu.int/comm/public_opinion/archives/ebl ebs _177_en.pdf (2003)

58.87 Royal Academy of Engineering: *Nanoscience and Nanotechnologies: Opportunities and Uncertainties* (The Royal Society, London 2003)

58.88 W. S. Bainbridge: Public attitudes toward nanotechnology, J. Nanopart. Res. **4**, 561–570 (2002)

58.89 W. S. Bainbridge: Validity of web-based surveys. In: *Computing in the Social Sciences and Humanities*, ed. by O. V. Burton (Univ. Illinois Press, Urbana 2002) pp. 51–66

58.90 W. S. Bainbridge: Sociocultural meanings of nanotechnology: Research methodologies, J. Nanopart. Res. **6**, 285–299 (2004)

58.91 M. D. Cobb, J. Macoubrie: Public perceptions about nanotechnology: Risks, benefits and trust, J. Nanopart. Res. **6**, 395–405 (2004)

58.92 G. M. Whitesides, J. C. Love: Implications of nanoscience for knowledge and understanding. In: *Societal Implications of Nanoscience and Nanotechnology*, ed. by M. C. Roco, W. S. Bainbridge (Kluwer, Dordrecht 2001) pp. 129–145

58.93 K. M. Kulinowski: Incorporating nanotechnology into K-12 education. In: *Nanotechnology: SocietalImplications – Individual Perspectives*, ed. by M. C. Roco, W. S. Bainbridge (Springer, Berlin, Heidelberg 2006) pp. 322–327

58.94 J. Klein-Seetharaman: Interactive, entertaining, virtual learning environments. In: *Nanotechnology: SocietalImplications – Individual Perspectives*, ed. by M. C. Roco, W. S. Bainbridge (Springer, Berlin, Heidelberg 2006) pp. 152–158

58.95 M. G. Forest: Mathematical challenges in nanoscience and nanotechnology: An essay on nanotechnology implications. In: *Societal Implications of Nanoscience and Nanotechnology*, ed. by M. C. Roco, W. S. Bainbridge (Kluwer, Dordrecht 2001) pp. 146–173

58.96 G. C. Black: Human resource implications of nanotechnology on national security and space exploration. In: *Nanotechnology: SocietalImplications – Individual Perspectives*, ed. by M. C. Roco, W. S. Bainbridge (Springer, Berlin, Heidelberg 2006) pp. 297–300

58.97 T. Chang: Educating undergraduate nanoengineers. In: *Nanotechnology: SocietalImplications – Individual Perspectives*, ed. by M. C. Roco, W. S. Bainbridge (Springer, Berlin, Heidelberg 2006) pp. 305–317

58.98 P. E. Stephan: Human resources for nanotechnology. In: *Nanotechnology: SocietalImplications – Individual Perspectives*, ed. by M. C. Roco, W. S. Bainbridge (Springer, Berlin, Heidelberg 2006) pp. 331–335

58.99 S. J. Fonash: Implications of nanotechnology for the workforce. In: *Societal Implications of Nanoscience and Nanotechnology*, ed. by M. C. Roco, W. S. Bainbridge (Kluwer, Dordrecht 2001) pp. 173–180

58.100 J. L. Merz: Technological and educational implications of nanotechnology – infrastructural and educational needs. In: *Societal Implications of Nanoscience and Nanotechnology*, ed. by

M. C. Roco, W. S. Bainbridge (Kluwer, Dordrecht 2001) pp. 186–196

58.101 B. E. Seely: Educational opportunities related to the societal implications of nanotechnology. In: *Nanotechnology: SocietalImplications – Individual Perspectives*, ed. by M. C. Roco, W. S. Bainbridge (Springer, Berlin, Heidelberg 2006) pp. 327–331

Acknowledgements

A.3 Introduction to Carbon Nanotubes
by Marc Monthioux, Philippe Serp,
Emmanuel Flahaut, Manitra Razafinimanana,
Christophe Laurent, Alain Peigney,
Wolfgang Bacsa, Jean-Marc Broto

The authors wish to acknowledge their membership to the European Research Group "Sciences and Applications of Nanotubes" (contact: annick.loiseau@onera.fr) and the French Carbon Group (contact: beguin@cnrs-orleans.fr).

A.4 Nanowires
by Mildred S. Dresselhaus, Yu-Ming Lin,
Oded Rabin, Marcie R. Black, Jing Kong,
Gene Dresselhaus

The authors gratefully acknowledge the stimulating discussions with Professors Charles Lieber, Gang Chen, S. T. Lee, Arun Majumdar, Peidong Yang, and Jean-Paul Issi, Dr. Joseph Heremans and Ted Harman. The authors are grateful for support for this work by the ONR Grant #000140-21-0865, the MURI program subcontract PO #0205-G-7A114-01 through UCLA, and DARPA contract #N66001-00-1-8603.

A.9 Stamping Techniques for Micro- and Nanofabrication
by Etienne Menard, John A. Rogers

The authors extend their deepest thanks to all of the collaborators who contributed the work described here.

A.12 Nanometer-Scale Thermoelectric Materials
by Joseph P. Heremans

The author is very grateful to Dr D. T. Morelli, for reviewing the manuscript and enlightening discussions.

A.13 Nano- and Microstructured Semiconductor Materials for Macroelectronics
by Yugang Sun, Seung-Hyun Hur, John A. Rogers

This work was supported by DARPA-funded AFRL-managed Macroelectronics Program (FA8650-04-C-7101). Funding is also partially provided by the U.S. Department of Energy under grant DEFG02-91-ER45439 and NSF through grant NIRT-04-3489. Devices fabrication and characterizations were carried out by using the Microfabrication and Crystal Growth Facility and Facilities of the Center for Materials Microanalysis in Frederick Seitz Materials Research Laboratory, University of Illinois, which is partially supported by the U.S. Department of Energy under Grant No. DEFG02-91-ER45439. The authors thank M. A. Mathew, S. Mack, K. Lee, K. Hurley, E. Menard and Dr. Z.-T. Zhu for providing some original images and curves.

B.15 MEMS/NEMS Devices and Applications
by Darrin J. Young, Christian A. Zorman,
Mehran Mehregany

The authors wish to thank Wen H. Ko for the helpful discussions and suggestions, Peng Cong for updating the references, Michael Suster and Joseph Seeger for preparing the figures.

B.16 Nanomechanical Cantilever Array Sensors
by Hans Peter Lang, Martin Hegner,
Christoph Gerber

We thank R. McKendry (University College, London, UK), J. Zhang, A. Bietsch, V. Barwich, M. Ghatkesar, Th. Braun, F. Huber, N. Backmann, J.-P. Ramseyer, A. Tonin, H. R. Hidber, E. Meyer and H.-J. Güntherodt (University of Basel, Basel, Switzerland) for valuable contributions and discussions, as well as U. Drechsler, M. Despont, H. Schmid, E. Delamarche, H. Wolf, R. Stutz, R. Allenspach, and P.F. Seidler (IBM Research, Zurich Research Laboratory, Rüschlikon, Switzerland). This project is funded partially by the IBM Zurich Research Laboratory (Rüschlikon, Switzerland), the National Center of Competence in Research in Nanoscience (Basel, Switzerland), the Swiss National Science Foundation and the Commission for Technology and Innovation (Bern, Switzerland).

B.17 Therapeutic Nanodevices
by Stephen C. Lee, Mark Ruegsegger,
Philip D. Barnes, Bryan R. Smith, Mauro Ferrari

The authors gratefully acknowledge many friends and colleagues whose input helped shape this article. Particularly, we acknowledge Beth S. Lee for many, many helpful discussions and for unflagging support. We also acknowledge Phil Streeter for many of the same services, performed in the office of friend rather than spouse. We acknowledge the support services of Anita Bratcher in manuscript preparation and the artistic stylings of Vladimir Marukhlenko for the figures incorporated in the manuscript. We also acknowledge Carol Bozarth for her kind and spontaneous support of the physical process of manuscript editing. This work is dedicated to the memories of Mildred A. Lee, Antonio Ferrari and Marialuisa Ferrari and the multitude of others whose lives have been tragically shortened by cancer, with the determination that therapeutic nanotechnology be used to help abolish the terrible power of the disease over human life.

D.30 Surface Forces and Nanorheology of Molecularly Thin Films
by Marina Ruths, Jacob N. Israelachvili

This work was supported by ONR grant N00014-00-1–0214. M. Ruths thanks the Academy of Finland for financial support.

D.31 Interfacial Forces and Spectroscopic Study of Confined Fluids
by Y. Elaine Zhu, Ashis Mukhopadhyay, Steve Granick

YZ gratefully acknowledges the financial support from the faculty research program at the University of Notre Dame under Award Number 370857. SG appreciates financial support from the NSF (Surface Engineering Programm) and also from the NSF (Polymers Program, Award DMR-0605947).

D.32 Scanning Probe Studies of Nanoscale Adhesion Between Solids in the Presence of Liquids and Monolayer Films
by Robert W. Carpick, James Batteas, Maarten P. de Boer

We gratefully acknowledge the help of Ms. Erin Flater who provided valuable assistance and insights into the literature on capillary formation. RWC acknowledges support from the National Science Foundation CAREER Program, grant #CMS-0134571, from the Army Research Office, grant #DAAD19-03-1-0102, and from the Air Force Office of Scientific Research, grant #FA9550-05-1-0204. Sandia is a multiprogram laboratory operated by Sandia Corporation, a Lockheed Martin Company, for the United States Department of Energy's National Nuclear Security Administration under contract DE-AC04-94AL85000.

D.35 Computer Simulations of Nanometer-Scale Indentation and Friction
by Susan B. Sinnott, Seong-Jun Heo, Donald W. Brenner, Judith A. Harrison

S.B.S. and S.-J.H. acknowledge support from the Air Force through grant FA9550-04-1-0367 and from the National Science Foundation supported Network for Computational Nanotechnology (EEC-0228390). D.W.B. acknowledges support from the Office of Naval Research through grant N00014-04-2006 and from the National Science Foundation through grant DMR-0304299. J.A.H. acknowledges support from the Air Force through grants F1ATA04295G001 and F1ATA04295G002 and from the Office of Naval Research via grant N0001405WX20129.

D.39 Mechanics of Biological Nanotechnology
by Rob Phillips, Prashant K. Purohit, Jané Kondev

We happily acknowledge useful discussions with Kai Zinn, Jon Widom, Bill Gelbart, Andy Spakowitz, Zhen-Gang Wang, Ken Dill, Carlos Bustamante, Tom Powers, Larry Friedman, Jack Johnson, Pamela Bjorkman, Paul Wiggins, Steve Williams, Wayne Falk, Adrian Parsegian, Alasdair Steven and Steve Quake. RP and PP acknowledge support of the NSF through grant number 9971922, the NSF supported CIMMS center, and the support of the Keck Foundation. JK is supported by the NSF under grant number DMR-9984471 and is a Cottrell Scholar of Research Corporation.

F.46 The "Millipede" – A Nanotechnology-Based AFM Data-Storage System
by Gerd K. Binnig, G. Cherubini, M. Despont, Urs T. Dürig, Evangelos Eleftheriou, H. Pozidis, Peter Vettiger

It is our pleasure to acknowledge our colleagues T. Albrecht, T. Antonakopoulos, P. Bächtold, A. Dholakia, U. Drechsler, B. Gotsmann, W. Häberle, D. Jubin, M.A. Lantz, T. Loeliger, H.E. Rothuizen, R. Stutz, and D. Wiesmann for their invaluable contributions to the millipede project.

In addition, thanks and appreciation go to H. Rohrer for his contribution to the initial millipede vision and concept and to our former collaborators, J. Brugger, now at the Swiss Federal Institute of Technology, Lausanne (Switzerland), M.I. Lutwyche, now at Seagate, Pittsburg, IL, and W.P. King, now at Georgia Tech, Atlanta, GA, as well as to K. Goodson, T.W. Kenny, and C.F. Quate of Stanford University, CA.

We are also pleased to acknowledge stimulating discussions with and encouraging support from our colleagues W. Bux and P.F. Seidler of the IBM Zurich Research Laboratory, J. Mamin, D. Rugar, and B.D. Terris of the IBM Almaden Research Center, San Jose, CA, and G. Hefferon of IBM, East Fishkill, NY.

Special thanks go to J. Frommer, C. Hawker, J. Mamin, and R. Miller of the IBM Almaden Research Center for their enthusiastic support in identifying and synthesizing alternative polymer media materials, and to H. Dang, A. Sharma, and S. Sri-Jayantha of the IBM T.J. Watson Research Center, Yorktown Heights, NY, for their contributions to the work on servo control.

F.47 Nanotechnology for Data Storage Applications
by Dror Sarid, Brendan McCarthy, Ghassan E. Jabbour

The authors would like to thank Digital Instruments (Veeco) for contributing their Multi-Mode Nanoscope III, the Department of Energy (DE-FG-3-02ER46013/A001) for a generous grant, EMC for their generous gift, and the Vice President for Research, University of Arizona, for equipment support.

G.54 Thermo- and Electromechanical Behavior of Thin-Film Micro and Nanostructures
by Martin L. Dunn, Shawn J. Cunningham

We are most grateful for the assistance of Dr. Yanhang Zhang of Boston University, for her help with the preparation of figures. Many of the calculations and measurements also draw from her work as a Ph.D. student at the University of Colorado. MLD acknowledges support from DARPA, Sandia National Laboratories, the National Science Foundation, and the AFOSR for support of aspects of his research that appear in this work.

About the Authors

Chong H. Ahn

University of Cincinnati
Department of Electrical
and Computer Engineering
and Computer Science
Cincinnati, OH, USA
chong.ahn@uc.edu

Chapter B.19

Dr. Chong Ahn is a Professor of Electrical and Computer Engineering at the University of Cincinnati. He obtained his Ph.D. degree in Electrical Engineering from the Georgia Institute of Technology in 1993 and then worked as a postdoctoral fellow at IBM T.J. Watson Research Center. His research interests include all aspects of design, fabrication, and characterization of magnetic MEMS devices, microfluidic devices, protein chips, lab-on-a-chips, nano biosensors, point-of-care testing and BioMEMS systems. He is an associate editor of the IEEE Sensors Journal.

Boris Anczykowski

nanoAnalytics GmbH
Münster, Germany
anczykowski@nanoanalytics.com

Chapter C.27

Dr. Boris Anczykowski is a physicist with an extensive research background in the field of dynamic Scanning Force Microscopy. He co-invented the Q-Control technique and received the Innovation Award Münsterland for Science and Economy in 2001 for this achievement. He is a managing director and co-founder of nanoAnalytics GmbH, a company specialized in the characterization of surfaces and interfaces on the micro- and nanometer scale.

Massood Z. Atashbar

Western Michigan University
Department of Electrical
and Computer Engineering
Kalamazoo, MI, USA
massood.atashbar@wmich.ed

Chapter A.7

Professor Massood Z. Atashbar received the B.Sc. degree in electrical engineering from the Isfahan University of Technology, Tehran, Iran, the M.Sc. degree in electrical engineering from the Sharif University of Technology, Tehran, and the Ph.D. degree from the Department of Communication and Electronic Engineering, RMIT University, Melbourne, Australia, in 1998. From 1998 to 1999, he was a Postdoctoral Fellow at the Center for Electronic Engineering and Acoustic Materials, The Pennsylvania State University, University Park. He is an Assistant Professor with the Electrical and Computer Engineering Department, Western Michigan University, Kalamazoo. His research interests include physical and chemical microsensors development, wireless sensors, and applications of nanotechnology in sensors, digital electronics, advanced signal processing, and engineering education.

Wolfgang Bacsa

University of Toulouse III (Paul Sabatier)
Laboratoire de Physique des Solides
(LPST), UMR 5477 CNRS
Toulouse Cedex, France
bacsa@ramansco.ups-tlse.fr;
bacsa@lpst.ups-tlse.fr

Chapter A.3

Professor Wolfgang Bacsa is an expert in the emerging field of nanooptics and carbon nanotubes. He has a Ph.D. from the Swiss Federal Institute of Technology (ETH) Zurich in Physics and has extensive experience in condensed matter physics, optics, microscopy, synthesis of ultrathin films and nanostructured carbon. Professor Bacsa worked at the ETH Zürich, PennState University and EPFL Lausanne.

Authors

William Sims Bainbridge   Chapters H.57, H.58

Division of Information, Science
and Engineering
National Science Foundation
Arlington, VA, USA
wbainbri@nsf.gov

William Sims Bainbridge earned his doctorate from Harvard University. He is the author of 11 books, 4 textbook-software packages, and about 180 shorter publications in information science, social science of technology, and the sociology of culture. His software employed innovative techniques to teach theory and methodology: Experiments in Psychology, Sociology Laboratory, Survey Research, and Social Research Methods and Statistics. He is the editor of the Berkshire Encyclopedia of Human-Computer Interaction and author of the forthcoming God from the Machine, a study using artificial intelligence techniques to understand religious belief. Most recently, he co-edited Converging Technologies for Improving Human Performance, which explores the combination of nanotechnology, biotechnology, information technology, and cognitive science, and he is currently editing the next book in the Converging Technologies series. At the National Science Foundation since 1992, he represented the social and behavioral sciences on five advanced technology initiatives: High Performance Computing and Communications, Knowledge and Distributed Intelligence, Digital Libraries, Information Technology Research, and Nanotechnology, before joining the staff of the Directorate for Computer and Information Science and Engineering. Currently, he is director of NSF's Science and Engineering Informatics program, after having directed the Sociology, Human Computer Interaction, and Artificial Intelligence programs, and a member of the faculty of George Mason University.

Antonio Baldi   Chapter A.7

Centro National Microelectrónica
Institut de Microelectronica
de Barcelona (IMB)
(CNM–CSIC)
Barcelona, Spain
antoni.baldi@cnm.es

Professor Antonio Baldi is a Ramon y Cajal researcher at the CNM-IMB, Spain. He received his BS degree in telecommunication engineering from the Universitat Politècnica de Catalunya, (1996) and his Ph.D. degree on electronics engineering from the Universitat Autònoma de Barcelona in 2001. From 2001 to 2003 he was at the University of Minnesota working in the field of bioMEMS as a postdoctoral fellow. In the summer of 2003 he joined the Chemical Transducers Group (GTQ), at the Centro Nacional de Microelectrónica, where he works on the development of microsystems and instrumentation for chemical and biochemical sensing.

Philip D. Barnes   Chapter B.17

Columbus, OH, USA
barnes.156@osu.edu

Phillip D. Barnes is completing his doctorate in biomedical engineering at The Ohio State University. He received the B.Sc. degree in electrical engineering and M.Sc. degree in biomedical engineering from The Ohio State University in 2001 and 2004, respectively.

James Batteas   Chapter D.32

Texas A&M University
Department of Chemistry
College Station, TX, USA
batteas@mail.chem.tamu.edu

James Batteas is an Associate Professor of Chemistry at Texas A&M University. He received his PhD in Chemistry in 1995 from UC Berkeley. Research in his group includes studies of charge transport in organic molecules on surfaces, tribological properties of oxide surfaces and self-assembled monolayers, self-organizing nanoscale materials for device applications, plant biopolymers, and nanofabrication approaches for the development of electronic and sensing architectures.

Roland Bennewitz   Chapter D.33

McGill University
Physics Department
Montreal, QC, Canada
roland@physics.mcgill.ca

Roland Bennewitz studied physics in Freiburg and Berlin where he received his PhD for work on defects at surfaces of insulators. He is now assistant at the University of Basel where his research activities focus on high-resolution force microscopy as a tool in nanotribology and surface science.

Bharat Bhushan

The Ohio State University
Nanotribology Laboratory
for Information Storage
and MEMS/NEMS
Columbus, OH, USA
bhushan.2@osu.edu

Chapters 1, C.22, D.29, D.34, D.38, D.40, D.41, E.42, E.43, E.44, G.50

Dr. Bharat Bhushan is an Ohio Eminent Scholar and The Howard D. Winbigler
Professor in the Department of Mechanical Engineering, a Graduate Research Faculty
Advisor in the Department of Materials Science and Engineering, and the Director of
the Nanotribology Laboratory for Information Storage and MEMS/NEMS (NLIM) at
the Ohio State University, Columbus, Ohio. He holds two M.Sc., a Ph.D. in
mechanical engineering/mechanics, an MBA, and three semi-honorary and honorary
doctorates. His research interests are in micro/nanotribology and its applications to
magnetic storage devices and MEMS/NEMS (Nanotechnology). He has authored 5
technical books, 45 handbook chapters, more than 450 technical papers in referred
journals, and more than 60 technical reports, edited more than 25 books, and holds 14
U.S. patents. Dr. Bhushan has previously worked for the R and D Division of
Mechanical Technology Inc., Latham, NY; the Technology Services Division of SKF
Industries Inc., King of Prussia, PA; the General Products Division Laboratory of IBM
Corporation, Tucson, AZ; and the Almaden Research Center of IBM Corporation, San
Jose, CA.

Gerd K. Binnig

IBM Zurich Research Laboratory
Micro-/Nanomechanics
Rüschlikon, Switzerland
gbi@zurich.ibm.com

Chapter F.46

Gerd Binnig obtained his Ph.D. from the Johann Wolfgang Goethe
University, Frankfurt, Germany, and joined IBM Research in 1978. He
was corecipient of the 1986 Nobel Prize in Physics for the invention of
the scanning tunnelling microscope, and he also invented the atomic
force microscope. His current research interests are micro- and
nanosystem techniques and "Fractal Darwinism", a theory he developed
to describe complex systems.

Marcie R. Black

Massachusetts Institute
of Technology
Department of Electrical Engineering
and Computer Science
Cambridge, MA, USA
marcie@alum.mit.edu

Chapter A.4

Marcie Black recently received her Ph.D. from Prof. Dresselhaus's
research group at MIT studying the optical properties of nanowires. In
particular, she identified the dominant optical absorption mechanism in
the IR of bismuth nanowires as an indirect interband transition that is
enhanced over bulk bismuth. Currently she is studying organic
opto-electronics with an emphasis on photovoltaics.

Maarten P. de Boer

Sandia National Laboratories
MEMS Devices and Reliability Physics
Department
Albuquerque, NM, USA
mpdebo@sandia.gov

Chapter D.32

Maarten P. de Boer received his Ph.D. in Materials Science from the University of
Minnesota in 1996 studying thin film adhesion. He is now a Principal Member of the
MEMS Devices and Reliability Physics Department, Sandia National Laboratories.
His interests are in the area of developing test structures, metrologies, and mechanics
to study mechanical and tribological properties in MEMS.

Donald W. Brenner

North Carolina State University
Department of Materials Science
and Engineering
Raleigh, NC, USA
brenner@ncsu.edu

Chapter D.35

Dr. Donald W. Brenner is currently a Professor in the Department of Materials Science
and Engineering at North Carolina State University, Raleigh. He received his B.Sc.
from the State University of New York in 1982 and his Ph.D. from Penn. State
University in 1987, both in Chemistry. He joined the Theoretical Chemistry Section at
the United States Naval Research Laboratory as a staff scientist in 1987, and joined
the North Carolina State University faculty in 1994. His research interests focus on
using atomic and mesoscale simulation and theory to understand technologically
important processes and materials. Professor Brenner's awards include the 2002
Feynman prize for research achievement in nanotechnology (theory), the Alcoa
Foundation Engineering Research Achievement Award (2000), Co-author of a
Veridian Medal Paper (1999), and an Outstanding Teacher Award from the North
Carolina State College of Engineering (1999). He is a member of the North Carolina
State University Academy of Outstanding Teachers.

Authors

Jean-Marc Broto

Institut National des Sciences
Appliquées of Toulouse
Laboratoire National des Champs
Magnétiques Pulsés (LNCMP)
Toulouse Cedex 4, France
broto@insa-tlse.fr; broto@lncmp.or

Chapter A.3

Jean-Marc Broto is Professor at the Université Toulouse III, France. He is a specialist in electronic transport and magnetization properties under high magnetic fields and contributed to the discovery of the giant magnetoresistance in 1988.

Guozhong Cao

University of Washington
Seattle, WA, USA
gzcao@u.washington.edu

Chapter A.5

Dr. Guozhong Cao is an Associate Professor of Materials Science and Engineering and Mechanical Engineering at the University of Washington (UW). He received his Ph.D. from Eindhoven University of Technology, The Netherlands, in 1991. Before joining UW faculty in 1996, Dr. Cao worked briefly at University of Twente, University of Nijmegen, and at the Advanced Materials Labs of the University of New Mexico and Sandia National Lab. His major awards include the European Union Fellowship in 1993, the college Outstanding Educator Award in 1999, and the university Distinguished Teaching Award in 2000. He has published over 160 refereed papers in a wide range of materials science field, edited three conference proceedings, and authored a book, "Nanostructures and Nanomaterials: Synthesis, Properties and Applications" in 2004. Dr. Cao's current research is focused mainly on processing, characterization, and applications of dielectrics, ferroelectrics, piezoelectrics, scintillation oxides, intercalation compounds, organic-inorganic hybrids, and nanostructured materials and coatings. Processing techniques include sol-gel processing, hydrothermal growth, electrochemical and electrophoretic deposition, and self-assembly. The research emphasis is to achieve novel properties for various applications through control of micro and nanostructure, and atomic engineering of materials through processing and composition design.

Robert W. Carpick

University of Wisconsin-Madison
Department of Engineering Physics
Madison, WI, USA
carpick@engr.wisc.edu

Chapter D.32

Robert Carpick has a Ph.D. in Physics (1997) from the University of California at Berkeley. He has been an Assistant Professor at the University of Wisconsin-Madison since 2000. He carries out research and publishes in the areas of nanotribology, nanomechanics, nanostructured materials, and scanning probe development. He serves on the editorial board of Review of Scientific Instruments. In 2002 he received a Faculty Early Career Development award from the U.S. National Science Foundation.

Tsung-Lin Chen

National Chiao Tung University
Department of Mechanical Engineering
HsinChu, Taiwan
tsunglin@mail.nctu.edu.tw

Chapter F.48

Tsung-Lin Chen received his B.Sc. and M.A. degrees in mechanical engineering from the National Tsing-Hua University, Taiwan in 1990 and 1992, respectively. He received Ph.D. degree in mechanical engineering from University of California at Berkeley, USA in 2001. He is currently an assistant professor in National Chiao Tung University in Taiwan. His research interests include MEMS devices design and MEMS fabrication process development.

Yu-Ting Cheng

National Chiao Tung University
Electronics Engineering Department
HsinChu, Taiwan
ytcheng@faculty.nctu.edu.tw

Chapter G.56

Professor Yu-Ting Cheng received his Ph.D. degree in Electrical Engineering from the University of Michigan, Ann Arbor, in 2000. After his graduation, he worked for IBM Watson Research Center, Yorktown Heights, as a research staff member. He is a member of IEEE, IOP, and Phi Tau Phi. Currently he is Assistant Professor at the National Chiao Tung University, Taiwan. His research interests include the development of novel materials and fabrication technologies for MEMS/NEMS applications and microsystems integration.

Giovanni Cherubini

IBM Zurich Research Laboratory
Storage Technologies
Rüschlikon, Switzerland
cbi@zurich.ibm.com

Chapter F.46

Dr. Giovanni Cherubini received a Ph.D. degree in electrical engineering from the University of California, San Diego, in 1986, and joined IBM Research in 1987. His interests include high-speed data transmission and data storage systems. He is Editor for CDMA systems, for IEEE Transactions on Communications, and served as Guest Editor for the IEEE Journal on Selected Areas in Communications issues on access technologies (1995) and on multiuser detection techniques (2001–2002). He is co-author of the book Algorithms for Communications Systems and their Applications.

Mu Chiao

The University of British Columbia
Mechanical Engineering Department
Vancouver, BC, Canada
muchiao@mech.ubc.ca

Chapter G.56

Professor Mu Chiao received a Ph.D. degree in Mechanical Engineering and Berkeley Sensor and Actuator Center from University of California at Berkeley, Ca, USA, in 2002. From August 2002 to Feb 2003, he had been with Berkeley Sensor and Actuator Center, University of California at Berkeley as a post-doctoral research fellow. He is currently with the Department of Mechanical Engineering at The University of British Columbia as an Assistant Professor. His research interests include MEMS packaging, MEMS power sources and novel nanowire/tube syntheses.

Jin-Woo Choi

Louisiana State University
Department of Electrical
and Computer Engineering
Baton Rouge, LA, USA
choi@ece.lsu.edu

Chapter B.19

Jin-Woo Choi received his B.S. and M.S. degree in Electrical Engineering from Seoul National University in Korea in 1994 and 1996, respectively. He received his Ph.D. degree in Electrical Engineering from the University of Cincinnati in 2000. Now he is an Assistant Professor at Louisiana State University, Baton Rouge, Louisiana. His current research activities include magnetic particle separators, microfluidic systems for biochemical detection, micro total analysis systems (μ-TAS), bioelectronics, and BioMEMS components and systems.

Shawn J. Cunningham

WiSpry, Inc.
Colorado Springs Design Center
Colorado Springs, CO, USA
shawn.cunningham@wispry.com

Chapter G.54

Shawn Cunningham is working on the development of RF MEMS switch and associated processes with wiSpry, Inc. His interests include materials characterization, reliability, and Design for MEMS Manufacturability. Prior to joining wiSpry, Shawn pursued MEMS research and product development at Coventor, Ford Microelectronics, and the University of Utah's Center for Engineering Design and in collaboration with the Univerisity of Colorado.

Dietrich Dehlinger

University of California, San Diego
Dept. of Electrical
and Computer Engineering
La Jolla, CA, USA
ddehling@ucsd.edu

Chapter B.14

Dietrich Dehlinger is a graduate student in the Department of Electrical and Computer Engineering at the University of California, San Diego. His background is in optics and MEMS and his current interest are in nanofabrication techniques.

Michel Despont

IBM Zurich Research Laboratory
Micro-/Nanomechanics
Rüschlikon, Switzerland
dpt@zurich.ibm.com

Chapter F.46

Michel Despont received his Ph.D. in physics from the University of Neuchâtel, Switzerland, in 1996. After a postdoctoral fellowship at IBM's Zurich Research Laboratory, he was visiting scientist at the Seiko Instrument Research Laboratory in Japan in 1997. His current research at IBM focuses on the development of micro- and nanomechanical devices and of processes to fabricate so-called system-on-chip.

Lixin Dong

Swiss Federal Institute of Technology (ETH), Zürich
Institute of Robotics and Intelligent Systems (IRIS)
Zürich, Switzerland
ldong@ethz.ch

Chapter F.49

Professor Lixin Dong is research scientist in IRIS of ETH Zürich. He received the B. Sc. and M.Sc. degrees in Mechanical Engineering from Xi'an University of Technology (XUT), China in 1989 and 1992, respectively. He became Research Associate in 1992, Lecturer in 1995, and Associate Professor in 1998 at XUT. He received his Ph.D. degree in Micro- and Nanosystem Engineering from Nagoya University in 2003, and became Assistant Professor there in the same year. His main research interests include nanorobotics and related topics including carbon nanotubes, nanorobotic manipulation, nanoassembly, nanofabrication, NEMS, and nanomechanochemistry.

Gene Dresselhaus

Massachusetts Institute
of Technology
Francis Bitter Magnet Laboratory
Cambridge, MA, USA
gene@mgm.mit.edu

Chapter A.4

Gene Dresselhaus received his Ph.D. in physics from the University of California in 1955. He was a faculty member at the University of Chicago, and assistant professor at Cornell before joining MIT Lincoln Laboratory in 1960 as a staff member. In 1976 he assumed his current position at the MIT Francis Bitter Magnet Laboratory. His area of interest is the electronic structure of nanomaterials and he has co-authored with
M.S. Dresselhaus several books on fullerenes, nanowires, and nanotubes.

Mildred S. Dresselhaus

Massachusetts Institute
of Technology
Department
of Electrical Engineering
and Computer Science
and Department of Physics
Cambridge, MA, USA
millie@mgm.mit.edu

Chapter A.4

Mildred Dresselhaus received her Ph.D. in physics from the University of Chicago in 1958. She joined the MIT faculty in 1967. She has been active in research across broad areas of solid state physics, especially in carbon science. Her present research activities focus on carbon nanotubes, bismuth nanowires, low dimensional thermoelectricity, and novel forms of carbon. She is the recipient of the National Medal of Science and 17 honorary degrees.

Martin L. Dunn

University of Colorado at Boulder
Department of Mechanical Engineering
Boulder, CO, USA
martin.dunn@colorado.edu

Chapter G.54

Martin L. Dunn received the Ph.D. in Mechanical Engineering from the University of Washington and was a postdoctoral appointee at Sandia National Laboratories. His research focuses on the micromechanical behavior of materials and structures. He has published over 75 articles in archival journals, and his research has been sponsored by NSF, DOE, NIST, DARPA, AFOSR, and Sandia National Laboratories.

Urs T. Dürig

IBM Zurich Research Laboratory
Micro-/Nanomechanics
Rüschlikon, Switzerland
drg@zurich.ibm.com

Chapter F.46

Urs Dürig received a Ph.D. degree from the Swiss Federal Institute of Technology, Zurich, in 1984. He joined IBM as a post-doc working on near-field optical microscopy. He is Research Staff Member since 1986: He worked in the field of scanning tunnelling and dynamic force microscopy. In 1997, he joined the Micro/Nanomechanics group focusing on polymer material issues and thermal modelling.

Evangelos Eleftheriou

IBM Zurich Research Laboratory
Storage Technologies
Rüschlikon, Switzerland
ele@zurich.ibm.com

Chapter F.46

He received a Ph.D. in Electrical Engineering from Carleton University, Ottawa, Canada, in 1985, and joined IBM Research in 1986. His research focuses on signal processing and coding for recording and transmission systems. His current research interests include nanotechnology, in particular probe-storage techniques. Dr. Eleftheriou was elected IEEE Fellow in 2002.

Mauro Ferrari

The University of Texas
Health Science Center at Houston
Houston, TX, USA
Mauro.Ferrari@uth.tmc.edu

Chapter B.17

Dr. Ferrari is a Professor at the Brown Institute of Molecular Medicine and Chairman of the Department of Biomedical Engineering, University of Texas Health Science Center at Houston. He is Professor of Experimental Therapeutics at the University of Texas M.D. Anderson Cancer Center in Houston, TX. and Professor of Bioengineering, Rice University, Houston, TX. as well as Professor of Biochemistry and Molecular Biology, University of Texas Medical Branch, Galveston, TX.~ Currently he is President of the Texas Alliance for NanoHealth in Houston, TX.

Emmanuel Flahaut

Université Paul Sabatier
CIRIMAT (Centre Interuniversitaire
de Recherche et d'Ingénierie
des Matériaux)
Toulouse, France
flahaut@chimie.ups-tlse.fr

Chapter A.3

Emmanuel Flahaut obtained his Ph.D in Materials Science in Toulouse working on CCVD synthesis of carbon nanotubes (CNTs) and dense ceramic-based composites including CNTs. He spent then more than one year as a post-doctoral researcher in Malcolm Green's Group in Oxford to work mainly on the filling of CNTs. He is now a permanent CNRS researcher at the University of Toulouse.

László Forró

Swiss Federal Institute
of Technology (EPFL)
Institute of Physics
of Complex Matter
Lausanne, Switzerland
laszlo.forro@epfl.ch

Chapter D.36

Professor László Forró is working on the synthesis, physical properties and manipulation of carbon nanostuctures and nanostuctured arrays, as well as, mechanical properties of carbon nanotubes, carbon onions, biological tubular systems. Transport and electron spin resonance studies of molecular materials, quasi-one-dimensional organic metals, organic superconductors, cuprates, manganates and fullerenes – up to high pressures. Tunneling spectroscopy in cuprate and fullerene superconductors. Optical properties of strongly correlated systems and biomaterials.

Jane Frommer

IBM Almaden Research Center
Department of Science
and Technology
San Jose, CA, USA
frommer@Almaden.ibm.com

Chapter E.45

Following a Ph.D. in Organometallic Chemistry from Caltech, Dr. Jane Frommer has been involved in a diverse set of research areas, including electronically conducting polymers and scanning probe microscopy. Her present AFM lab at IBM collaborates with a wide variety of laboratories involved in materials research, including lithography, chromatography, and storage. Common to all these studies is Frommer's interest in the properties and structure of molecules in confined geometries.

Authors

Authors

Harald Fuchs

Chapter C.27

Universität Münster
Physikalisches Institut
Münster, Germany
fuchsh@uni-muenster.de

1984 Ph.D. Universität des Saarlandes with Prof. H. Prof Gleiter (nano crystalline Systems), 1984–1985, Post doc with IBM Research Lab. Zurich in the group of G. Binnig and H. Rohrer, 1985–1993 Project manager 'Ultrathin Organic Films' with BASF AG, Ludwigshafen, Germany. Since 1993 he is full Professor and Director at the Physical Institute of the University of Münster, 2000: Cofounder and Scientific Director of the Center for Nanotechnology (CeNTech).

Christoph Gerber

Chapter B.16

Institute of Physics, University of Basel
National Competence Center for Research
in Nanoscale Science (NCCR) Basel
Basel, Switzerland
Christoph.Gerber@unibas.ch

Professor Christoph Gerber is the Director for scientific communication of the National Center of Competence for Nanoscale Science (NCCR) at the Institute of Physics, University of Basel, Switzerland and is a Research Staff Member emeritus in Nanoscale Science at the IBM Research Laboratory in Rüschlikon, Switzerland. He has served as a project leader in various programs of the Swiss National Science Foundation and currently in the European 6th Framework. During the past 25 years, his research has been focused on nanoscale science. He is a veteran in scanning probe microscopy and made major contributions to the invention of the scanning tunneling microscope and the atomic force microscope (AFM) and is a co-inventor of biochemical sensors based on AFM technology. He is an author and co-author of more than one hundred scientific papers that appeared in peer-reviewed journals with more than 12 000 citations in cross disciplinary fields. He belongs worldwide to the one hundred most cited researchers in Physical Sciences. He gave numerous plenary and invited talks at international conferences.

Franz J. Giessibl

Chapters C.24, C.26

Universität Regensburg
Naturwissenschaftliche Fakultät II –
Physik
Regensburg, Germany
*Franz.Giessibl@physik.uni-regensburg.
de*

Professor Giessibl is working in atomic force microscopy and scanning tunnelling microscopy in ultrahigh vacuum at room temperature and low temperatures. He is a Steering Committee Member of the International Conference on Noncontact Atomic Force Microscopy. He received the R and D 100 Award Chicago 1994, the German Nanoscience Award 2000, and the Rudolf Kaiser Price in 2001. Since May 2006 he is head of the Department of Experimental and Applied Physics at the University of Regensburg, Germany.

Enrico Gnecco

Chapter D.33

University of Basel
Department of Physics
Basel, Switzerland
Enrico.Gnecco@unibas.ch

Enrico Gnecco studied physics in Genoa, where he received his PhD for work on the mechanism of growth of nanostructured carbon films. He is now assistant at the University of Basel, where his research activities focus on friction force microscopy and molecular machinery.

Steve Granick

University of Illinois
at Urbana–Champaign
Department of Chemistry
Urbana, IL, USA
sgranick@uiuc.edu

Chapter D.31

Steve Granick is a Founder Professor of Materials Science and Engineering and also a Professor of Chemistry, Physics, Chemical and Biomolecular Engineering at the University of Illinois at Urbana-Champaign. He received his B.A. from Princeton University in 1978 and his Ph.D. from the University of Wisconsin-Madison in 1982. He joined the faculty of the University of Illinois in 1985 following postdoctoral research at the Collège de France with P.-G. de Gennes and at the University of Minnesota with Matthew Tirrell. His honors include the Beckman Research Board Award (1985), election to ASME Research Committee on Tribology (1989), Fellow in the Center for Advanced Study of UIUC (1990), Fellow American Physical Society (1992), Senior Xerox Award, UIUC (1993), NSF Award for Special Creativity (1993), Sabbatical Professor, Kyoto University, Japan (1994), University Scholar, UIUC (1997), named as Founder Professor of Engineering (1999), Paris-Sciences Professor (2002). His research interests are in the areas of polymers, complex fluids, and biomaterials.

Gérard Gremaud

Swiss Federal Institute
of Technology (EPFL)
Institute of Physics
of Complex Matter
Lausanne, Switzerland
gremaud@epfl.ch

Chapter D.36

Dr. Gérard Gremaud is a physicist and senior lecturer at the Ecole Polytechnique Fédérale de Lausanne (EPFL). He is active in the research fields of dislocation dynamic, acoustic and atomic force microscopy and granular physics. He is also responsible for the teaching of metrology and practical works to the physics students.

Jason H. Hafner

Rice University
Department of Physics
& Astronomy
Houston, TX, USA
hafner@rice.edu

Chapter C.23

Jason Hafner earned his Ph.D. in physics from Rice University in 1998. He then held an NIH postdoctoral fellowship at Harvard University working on nanotube probes for high resolution biological atomic force microscopy. In 2001 he returned to Rice as an assistant professor where his group is pursuing various biophysical applications of scanned probe microscopy and nanomaterials. Dr. Hafner received a Beckman Young Investigator award in 2002.

Judith A. Harrison

U.S. Naval Academy
Department of Chemistry
Annapolis, MD, USA
jah@usna.edu

Chapter D.35

Dr. Judith A. Harrison is currently a Professor of Chemistry at the United States Naval Academy. She received her B. A. in chemistry from Saint Anselm College in 1984 and her Ph.D. in chemistry from the University of New Hampshire in 1989. Her graduate work was in the area of gas-phase reaction dynamics. Before joining the faculty of the Naval Academy in 1993, she was an Office of Naval Research postdoctoral associate at the Naval Research Laboratory in Washington, DC. Her current research focuses on the theoretical examination of nanometer-scale processes, such as indentation, friction, wear, and tribochemistry of hydrocarbon systems. She has won several awards including the Naval Academy's Research Excellence Award and the Department of the Navy's Meritorious Civilian Service Award both in 2000. She is a member of Sigma Xi, the Society of Tribologists and Lubrication Engineers, the American Chemical Society, and the Materials Research Society.

Authors

Peter G. Hartwell   Chapter G.51

Hewlett Packard Laboratories
Palo Alto, CA, USA
peter.hartwell@hp.com

Peter G. Hartwell is currently a researcher at Hewlett-Packard Laboratories in Palo Alto, California. As a member of the Emerging Technologies Lab, his research focuses on design, fabrication and testing of microelectromechanical systems (MEMS). He has extensive experience in commercializing silicon MEMS products working on advanced sensors and actuators. He graduated cum laude from the University of Michigan in 1992 with a B.Sc.E in Materials Science and Engineering. He then joined NASA's Jet Propulsion Laboratory in Pasadena, California working in electronic packaging. He left JPL in 1993 and went to Cornell University where he received a Ph.D. in Electrical Engineering in 1999 after doing extensive research on MEMS testing. He did brief post doctoral work at HP Labs before joining the staff at Hewlett-Packard Laboratories in 2000.

Martin Hegner   Chapter B.16

Institute of Physics, University of Basel
National Competence Center for Research
in Nanoscale Science (NCCR) Basel
Basel, Switzerland
Martin.Hegner@unibas.ch

PD Martin Hegner received the M.Sc. degree in life sciences and his Ph.D. from the Swiss Federal Institute of Technology, Zurich, Switzerland. In 1994 he was awarded a Ciba Geigy postdoc fellowship for projects at the Physics Institute of University of Basel and worked on single biomolecule experiments involving scanning probe microscopy. In 1996 he received a SNF and a Human Frontier Science postdoctoral fellowship and joined for three years the research group of C. Bustamante at University of Oregon and thereafter at University of California at Berkeley, where he developed functional interfaces for optical tweezers single molecule experiments. On joining again the Physics Institute at the University of Basel in 1999, he built up a biophysically oriented research group focusing s on nanomechanical investigations on single biomolecules and lately on multifunctional cantilever arrays for biomedical applications.

Michael J. Heller   Chapter B.14

University of California San Diego
Dept. of Bioengineering,
Dept. of Electrical
and Computer Engineering
La Jolla, CA, USA
mheller@bioeng.ucsd.edu

Professor Michael J. Heller received his Ph.D. in Biochemistry of Colorado State University in 1973. Further steps in his career: NIH Postdoctoral Fellow, Northwestern University; Supervisor DNA Technology, Amoco, 1976–1984; Director Molecular Biology, Molecular Biosystems, Inc., 1984–1987; Co-founder and President, Integrated DNA Technologies, 1987–1989; Co-founder and CTO, Nanogen, 1993–2001. Dr. Heller is now Professor in the Departments of Bioengineering and Electrical and Computer Engineering at the University California, San Diego.

Stefan Hengsberger   Chapter D.36

Ecole d'ingénieur de Fribourg
Fribourg, Switzerland
stefan.hengsberger@eif.ch

Professor Stefan Hengsberger received his diploma in physics from the University of Saarbrücken in 1997. He started as a research scientist at Fraunhofer/Miami (Florida) where he worked until May 1998. In July 1998 he joined the Swiss Federal Institute of Technology (EPFL) where he earned his Ph.D. in biomechanics in spring 2002. He stayed for another year at EPFL as a postdoc in biomechanics and physics until summer 2003. Since October 2003 he is Professor of Physics at the University of Applied Science in Fribourg, Switzerland.

Seong-Jun Heo

Univeristy of Florida
Dept. of Materials Science and
Engineering
Gainesville, FL, USA
heogyver@ufl.edu

Chapter D.35

Mr. Seong-Jun Heo is a graduate student in the Department of Materials Science and Engineering at the University of Florida. He received his B. Sc. and M.D. from Hanyang University in Seoul, Korea in 1997 and 1999, respectively. He was a research engineer in the semiconductor division of Samsumg Electronics from 1999 to 2003. He joined Dr. Sinnott's group in 2003 and is working towards his Ph.D. He is interested in the computational simulation of the mechanical and electronic properties of carbon nanotubes.

Joseph P. Heremans

The Ohio State University
Dept. of Physics,
Dept. of Mechanical Engineering
Columbus, OH, USA
heremans.1@osu.edu

Chapter A.12

Dr. Joseph P. Heremans holds undergraduate (Ingénieur Civil Electricien, 1975) and graduate (Docteur en Sciences Appliquées, 1978) degrees from the Catholic University of Louvain, in Belgium. After appointments as a visiting scientist at the H. C. Ørsted Institute in Copenhagen, the Massachusetts Institute of Technology, and the Institute for Solid State Physics of the University of Tokyo, he joined the General Motors Research Labs in 1984, where he was a the group leader of electro-optical physics and later the manager of the semiconductor physics section of the Physics and Physical Chemistry Department. He joined the Delphi Research Labs, in Shelby Township, Michigan, in 1999 as a Research Fellow. He has co-authored 162 refereed publications, was the co-editor of two books, and holds 29 US patents. His expertise is in transport properties of semiconductors and low-dimensional materials, particularly in thermal transport. He is a fellow of the American Physical Society since 1987.

Peter Hinterdorfer

Johannes Kepler University of Linz
Institute for Biophysics
Linz, Austria
peter.hinterdorfer@jku.at

Chapter C.28

Peter Hinterdorfer earned a Dr. tech. from the University of Linz, Austria, Institute for Biophysics in 1992. He was a postdoctoral fellow at the University of Virginia, Department of Molecular Physiology and Biological Physics (1992/1993). Since then he is at the University of Linz, Institute for Biophysics, where he holds a position as Associate Professor. His current and ongoing research includes single molecule force spectroscopy and high resolution topography and recognition imaging of biological samples.

Dean Ho

University of California Los Angeles
Department of Bioengineering
Los Angeles, CA, USA
deanh@seas.ucla.edu

Chapter A.11

Dean Ho received his B.Sc. in Physiological Science from UCLA in 2001 and his Ph.D. in 2005 from the UCLA Department of Bioengineering. He is currently a postdoctoral research fellow in the UCLA Department of Mechanical and Aerospace Engineering, as well as Department of Bioengineering. His research has covered emerging areas in bionanotechnology to interface membrane-bound proteins with block copolymeric biomimetic membranes to fabricate biomolecular solar cells, as well as protein-based thin film devices to enhance existing fuel cell technologies. He has published over 20 peer-reviewed papers in the areas of biochemical energetics, polymeric biomembrane characterization, as well as developing novel power sources based on the harnessing of protein functionality. Dean is a member of Sigma Xi.

Dalibor Hodko

Nanogen, Inc.
San Diego, CA, USA
dhodko@nanogen.com

Chapter B.14

Dr. Hodko received his B.Sc., M.Sc. and Ph.D. degrees in chemical engineering and chemistry from the University of Zagreb, Croatia. He has more than 20 years of industrial experience in the development of innovative technologies (17 issued patents) and products in the areas of bioanalytical and micro- and nanofluidics devices as well as large systems with applications in physical and environmental chemistry. He has intensive experience in managing Government supported project and was a Principal Investigator on more than 25 federally funded projects. Dr. Hodko is presently a Director of Advanced Technology at Nanogen, Inc.

Roberto Horowitz

University of California at Berkeley
Department
of Mechanical Engineering
Berkeley, CA, USA
horowitz@me.berkeley.edu

Chapter F.48

Roberto Horowitz joined the Department of Mechanical Engineering at the University of California at Berkeley in 1982, where he is currently a Professor. Dr. Horowitz teaches and conducts research in the areas of adaptive, learning, nonlinear and optimal control, with applications to Micro-Electromechanical Systems (MEMS), computer disk file systems, robotics, mechatronics and intelligent vehicle and highway systems (IVHS).

Hirotaka Hosoi

Japan Science and Technology
Corporation
Sapporo, Japan
hosoi@sapporo.jst-plaza.jp

Chapter C.24

Hirotaka Hosoi received the D.E. degree in electronic enginnering from Hokkaido University in 1999. Since 2002 he is at Innovation Plaza Hokkaido, Japan Science and Technology Corporation (JST). His main research focus is in high-resolution magnetic imaging of magnetic materials surfaces using a scanning force microscope. His current research interests includes magnetism on metal-oxide surfaces.

Xinghui Huang

University of California at Berkeley
Department of Mechanical Engineering
Berkeley, CA, USA
xhhuang@me.berkeley.edu

Chapter F.48

Xinghui Huang received his B.Sc. and M.Sc. from Tsinghua University, Beijing, China, in 1994 and 1999, respectively. He is currently a Ph.D. candidate in Mechanical Engineering at the University of California at Berkeley. He is currently researching servo and vibration control design for dual-stage computer hard disk drives. His research interests include servo control, vibration control, robust control, and multi-rate control, with applications to MEMS devices, computer disk file systems, and mechatronics. He received the Kornie and Herta Otto Fellowship in Mechanical Engineering in 2005.

Seung-Hyun Hur

University of Illinois
at Urbana-Champaign
Department of Materials Science
and Engineering
Urbana, IL, USA
shur@uiuc.edu

Chapter A.13

Seung-Hyun Hur is currently a research scholar of Prof. John A. Roger's group in the department of Materials Science and Engineering at University of Illinois at Urbana-Champaign. He is also Ph.D. candidate in the department of Chemical and Biomolecular engineering of KAIST in Korea. His interests include nanofabrication, flexible electronics, single-walled carbon nanotube TFTs.

Jacob N. Israelachvili

University of California
Department of Chemical Engineering and
Materials Department
Santa Barbara, CA, USA
Jacob@engineering.ucsb.edu

Chapter D.30

Jacob Israelachvili earned his Ph.D. 1971 at the Cavendish Laboratory, University of Cambridge, UK. He held various positions at the Department of Applied Mathematics at the Australian National University (1974–1986), including those of Professional Fellow and Head of Department. In 1986 he joined the faculty at University of California, Santa Barbara as Professor in the Department of Chemical Engineering and Materials Department. In 1988 he was elected a Fellow of the Royal Society of London, and in 1991 he was awarded the Alpha Chi Sigma Award for Chemical Engineering Research by the AIChE. He was elected as a foreign associate of the US National Academy of Engineering in 1996.

Ghassan E. Jabbour

University of Arizona
Optical Sciences Center
Tucson, AZ, USA
gej@optics.arizona.edu

Chapter F.47

Ghassan E. Jabbour is the head of the research group Organic Optoelectronic Materials and Devices at the Optical Sciences Center working on organic and hybrid materials and their applications to light emitting dvices, solar cells, memory storage, solid state lighting, and other areas. Professor Jabbour is a SPIE Fellow, Track Chair of the Nanotechnology Program for SPIE, and Associate Editor of the Journal of the Society for Information Display (JSID). He has over 200 publications, invited talks, and conference proceedings.

Guangyao Jia

University of California, Irvine
Dept. of Mechanical
and Aerospace Engineering
Irvine, CA, USA
gjia@uci.edu

Chapter B.20

Guangyao Jia received his B.E. degree in Mechanical Engineering from Beijing University of Chemical Technology in China in 1990 and his M. Sc. degree in Chemical Engineering from Florida Institute of Technology in 2001. Currently he is working with Professor Marc Madou on his Ph.D. degree in Mechanical and Aerospace Engineering at the University of California, Irvine. His research includes centrifuge-based microfluidics for flow-through DNA hybridization and microfluidic PCR chips for rapid DNA amplification.

Anne Jourdain

Interuniversity Microelectronics Center
(IMEC)
Leuven, Belgium
jourdain@imec.be

Chapter G.52

Anne Jourdain received the M.Sc. degree in Opto-Microelectronics from the Ecole Nationale Supérieure d'Ingénieurs de Caën, France in 1994. She did her Ph.D. in LETI-CEA, Grenoble, France, and received the Ph. D. degree in 1998 from the University Joseph Fourier of Grenoble, France. In 1999, she joined the Interuniversity Microelectronics Center (IMEC), Leuven, Belgium, where she was involved in the MEMS processing development. In 2000, she joined the RF-MEMS team where she is currently in charge of the packaging of RF-MEMS devices.

Harold Kahn

Case Western Reserve University
Department of Materials Science
and Engineering
Cleveland, OH, USA
kahn@cwru.edu

Chapter G.53

Harold Kahn is Researcher Associate Professor of Materials Science and Engineering at Case Western Reserve University, Cleveland, Ohio. His research is focused on MEMS device processing and testing, particularly wafer-level mechanical testing and shape-memory actuated microfluidics. He received a B.S. in metallurgical engineering from Lafayette College and a Ph.D. in electronic materials from the Massachusetts Institute of Technology.

Horacio Kido

University of California at Irvine
Mechanical and Aerospace Engineering
Irvine, CA, USA
hkido@uci.edu

Chapter B.20

Horacio Kido holds B.Sc. and M.Sc. degrees from Stanford University in Biological Sciences as well as a Ph.D. from the University of California at Davis in Agricultural and Environmental Chemistry. He works on microfluidic systems for biological sample preparation.

Authors

Jitae Kim

University of California, Irvine
Dept. of Mechanical
and Aerospace Engineering
Irvine, CA, USA
jitaekim@uci.edu

Chapter B.20

Jitae Kim received the B.Sc. degree from Hanyang University, South Korea and MS from University of Southern California in 1998 and 2001, respectively. He is currently a doctoral student in mechanical engineering (BioMEMS) at UC, Irvine. His major research interest is CD (compact disk)-based microfluidics for sample preparation in nucleic acid analysis.

Jongbaeg Kim

Dicon fiber Optics, Inc.
Richmond, CA, USA
jongbaeg@gmail.com

Chapter G.56

Jongbaeg Kim received his Ph.D. degree in Mechanical Engineering from Berkeley Sensor and Actuator Center at the University of California, Berkeley, in 2004. Since 2004 he has been working in the MEMS group at DiCon Fiberoptics, Inc. in Richmond, CA, where he designs optical MEMS components. His research interests are system dynamics, mechatronic system modeling, design and fabrication of MEMS and nanotechnology.

Nahui Kim

University of California, Irvine
Department of Electrical Engineering
and Computer Science
Irvine, CA, USA
nahuik@uci.edu

Chapter B.20

Nahui Kim received the master degree in Electrical Engineering and Computer Science from the University of California at Irvine now pursuing her Ph.D. in EECS at UCI. She has four years of industrial experience as a circuit engineer. Her recent research work involves a compact disc-based microfluidic platform for genetic, molecular, behavioural, and pharmacological analysis of organismal stress response and antidepressants action.

Andras Kis

University of California
Physics Department
Berkeley, CA, USA
akis@lbl.gov

Chapter D.36

András Kis received his Ph.D. in physics from the Swiss Federal Institute of Technology, Lausanne, Switzerland, in 2003. He is currently working as a postdoctoral researcher at the University of California, Berkeley. His research is focused on mechanical properties of nanoscale objects and development of nanoelectromechanical devices based on nanotubes. András Kis was awarded the Latsis Foundation University prize in 2004.

Shih-Chung Kon

University of California at Berkeley
Department of Mechanical Engineering
Berkeley, CA, USA
stankon@me.berkeley.edu

Chapter F.48

Shih-Chung Stanley Kon is a Ph. D student in Mechanical Engineering at University of California at Berkeley. He received his B.Sc. from National Taiwan University in 1999. His research interests include design and fabrication of MEMS devices, piezoelectric materials, and vibration control.

Jané Kondev

Brandeis University
Physics Department
Waltham, MA, USA
kondev@brandeis.edu

Chapter D.39

Professor Kondev graduated with a PhD from Cornell University and did postdoctoral work at Brown University and Princeton University before joining the faculty at Brandeis University. He is a recipient of the CAREER award from the National Science Foundation and is a Cotrell Scholar of the Research Corporation.

Jing Kong

Massachusetts Institute of Technology
Department of Electrical Engineering
and Computer Science
Cambridge, MA, USA
jingkong@mit.edu

Chapter A.4

Jing Kong is Assistant Professor in the Department of Electrical Engineering and Computer Science at MIT. Her research interests lie in the synthesis, characterization, fundamental property studies and application with carbon nanotubes. She obtained her BS degree in Chemistry from Peking University in China, and Ph.D. in Chemistry from Stanford University.

Anders Kristensen

Technical University of Denmark (DTU)
MIC – Department of Micro
and Nanotechnology
Kgs. Lyngby, Denmark
ak@mic.dtu.dk

Chapter A.8

Professor Anders Kristensen, born in Denmark, was educated at University of Copenhagen, where he received Ph.D. in physics in 1994. He has been employed at Royal Holloway and Bedford New College, University of London (1994–96), Niels Bohr Institute fAFG, University of Copenhagen (1996–2001). He joined MIC – Department of Micro and Nanotechnology, Technical University of Denmark (DTU) as an Associate Professor in 2001. Before joining MIC his research was concerned with experimental investigations of mesoscopic electron transport in III–V semiconductor devices. At MIC his research interests are nanoimprint lithography and polymer based micro and nanostructures for lab-on-a-chip applications.

Andrzej J. Kulik

Swiss Federal Institute
of Technology (EPFL)
Institute of Physics of Complex Matter
Lausanne, Switzerland
andrzej.kulik@epfl.ch

Chapter D.36

Andrzej Kulik is the Head of the Biostructures and Nanomechanics Laboratory of the Swiss Federal Institute of Technology (EPFL) in Lausanne, Switzerland. His research concentrates on quantitative nanoscale materials properties, nanotribology. scanning probe microscopy, contact mechanics, nanoindentation, near-field ultrasonics, optical tweezers, nanolithography, and nanomanipulation.

Hans Peter Lang

National Competence Center for Research
in Nanoscale Science (NCCR) Basel
Institute of Physics, University of Basel
Basel, Switzerland
Hans-Peter.Lang@unibas.ch

Chapter B.16

Dr. Hans Peter Lang received a Ph.D. in physics from the University of Basel in 1994. As a postdoc, he directed research in the pulsed laser deposition and low temperature scanning tunneling microscopy groups at the Institute of Physics in Basel. Since 1996, he is working as a postdoc at IBM Zurich Research Laboratory in the field of cantilever array sensors. Since 2000, he is a project leader in a project focused on biochemical applications of cantilever array sensors. He has given more than 100 scientific presentations and has published about 100 scientific articles in renowned journals.

Carmen LaTorre

Insulating Systems Business
Owens Corning
Granville, OH, USA
carmen.latorre@owenscorning.com

Chapter D.40

Carmen LaTorre holds B.Sc. and M.Sc. degrees in mechanical engineering from The Ohio State University. He is currently an engineer at the Owens Corning Science and Technology Center in Granville, Ohio, where his primary focus is on fibreglass insulation product/process development.

Christophe Laurent

Université Paul Sabatier
CIRIMAT (Centre Interuniversitaire
de Recherche et d'Ingénierie
des Matériaux)
Toulouse, France
laurent@chimie.ups-tlse.fr

Chapter A.3

Dr. Ch. Laurent is Professor of Materials Chemistry at University Paul Sabatier and is the head of the Nanocomposites and Carbon Nanotubes group of CIRIMAT. His research include in the synthesis, characterization and mechanical properties of ceramic-matrix nanocomposites, and since 1994, carbon nanotubes (synthesis of single- and double-walled CNTs, formation mechanisms, characterization, localized growth, hydrogen storage, ceramic-matrix composites).

Abraham Lee

University of California at Irvine
Department of Biomedical Engineering
Irvine, CA, USA
aplee@uci.edu

Chapter B.21

Abraham Lee is Professor in Biomedical Engineering and Mechanical and Aerospace Engineering at the University of California, Irvine. Before UCI he held positions at the National Cancer Institute, DARPA, and Lawrence Livermore National Laboratory (LLNL). He has extensive experience in developing micro and nanotechnology for biomedical and biotech applications, including the development of a micromechanical release mechanism for the deployment of embolic coils into brain aneurysms. In the area of microfluidics, Dr. Lee and his student demonstrated the first AC magnetohydrodynamic (MHD) micropump for micro total analysis systems. Both technologies have been licensed to companies. Professor Lee's current research is focused on the development of integrated "digital" micro/nano fluidic chips for the following applications: point-of-care diagnostics, "smart" nanomedicine for early detection and treatment, automated cell sorting based on electrical signatures, tissue engineering and stem cells, the synthesis of ultra-pure materials, and biosensors to detect environmental and terrorism threats. He has published over 40 peer reviewed articles and owns 32 issued patents.

Stephen C. Lee

Ohio State University
Biomedical Engineering Center
Columbus, OH, USA
Lee@bme.ohio-state.edu

Chapter B.17

Stephen C. Lee is a pioneer in the field of semi-biological nanodevices (nanobiological devices), having published the first monograph devoted to the topic in 1998. His interests are in enabling technologies for the incorporation of functional proteins and nucleic acids into nanodevices, particularly for application in oncology and cardiovascular disease. He is currently Associate Professor of Cellular and Molecular Biochemistry, Chemical Engineering and Biomedical Engineering at the Ohio State University.

Wayne R. Leifert

CSIRO Commonwealth Scientific
and Industrial Research Organisation
Health Sciences and Nutrition
Adelaide, BC South Australia, Australia
Wayne.Leifert@csiro.au

Chapter B.18

Dr. Wayne Leifert received his Ph.D. in membrane physiology from the University of Adelaide, South Australia. His areas of interests are in cell and membrane physiology, biochemistry and molecular biology of G-protein coupled receptors. His recent research has focused on the G-protein coupled receptor signaling processes particularly for construction of cell-free biosensors utilizing nano- and biotechnologies to adapt for multiple applications including medical diagnostics and high throughput drug screening.

Yunfeng Li

University of California at Berkeley
Department
of Mechanical Engineering
Berkeley, CA, USA
yunfeng@me.Berkeley.edu

Chapter F.48

Yunfeng Li received the B.S. and M.S. degrees from Beijing University of Aeronautics and Astronautics, Beijing, China, in 1992 and 1995, respectively. He is currently working towards the Ph.D. degree in the Department of Mechanical Engineering, University of California, Berkeley, CA, USA. His research interests include motion control, vibration control, control of MEMS devices, and mechatronics.

Liwei Lin

University of California at Berkeley
Mechanical Engineering Department
Berkeley, CA, USA
lwlin@me.berkeley.edu

Chapter G.56

Liwei Lin received his Ph.D. from the University of California, Berkeley, in 1993. He was an Associate Professor in the Institute of Applied Mechanics, National Taiwan University, Taiwan (1994–1996) and an Assistant Professor in Mechanical Engineering Department at the University of Michigan (1996–1999). He joined UC-Berkeley in 1999 and is now a Professor at the Mechanical Engineering Department and Co-Director at Berkeley Sensor and Actuator Center.

Yu-Ming Lin   Chapter A.4

Massachusetts Institute
of Technology
Department of Electrical Engineering
and Computer Science
Cambridge, MA, USA
yming@mgm.mit.edu

Lin Yu-Ming performed experimental and theoretical studies on
Bi-based nanowires for next-generation thermoelectric materials. The
electronic transport properties of these systems are studied to
investigate quantum size effects. He received the Masterworks Award of
the Department of EECS, MIT (2000) and the Gold Medal, MRS
Graduate Student Award (2002)

Huiwen Liu   Chapter E.44

Eden Prairie, MN, USA
Huiwen.Liu@seagate.com

Dr. Huiwen Liu was the associate director of the Nanotribology Laboratory for
Information Storage and MEMS/NEMS at The Ohio State University. His research
interests are study of mechanical, tribological, and physical properties of advanced
materials and MEMS/NEMS devices on micro-nanoscale. He obtained an Alexander
von Humboldt fellowship and the Japanese Science and Technology Agency
Fellowship in 1997 and 1998, respectively.

Gustavo S. Luengo   Chapter D.36

Department of Physics Knowledge
L'Oréal Recherche 1
Aulnay sous Bois, France
gluengo@rd.loreal.com

Gustavo Luengo received his Ph.D. from the Physical Chemistry Department at the
Complutense University, Madrid, Spain, in 1993, His industrial experience started as a
lubricant researcher at REPSOL Oil Co. (Spain) followed by a postdoctoral research
at the Department of Chemical Engineering, University of California, Santa Barbara.
He is since 2000 in charge of the Nanophysics and Electron Microscopy laboratory at
L'Oréal Recherche in Aulnay sous Bois, France. His research focus is on the
nanotribology and local properties of cosmetic substrates (hair and skin) and materials
(polymers), and applications of nanotechnology.

Marc J. Madou   Chapter B.20

University of California
Dept. of Mechanical
and Aerospace Engineering
Irvine, CA, USA
mmadou@uci.edu

Dr. Madou is a Chancellor's Professor at UC Irvine in the Department
of Mechanical and Aerospace Engineering. His research focuses on
miniaturization science (MEMS and NEMS) with emphasis on
chemical and biological applications. Current projects include polymer
actuators (for drug delivery), C-MEMS and CD based fluidics. Dr.
Madou received his Ph.D. from the Rijksuniversiteit, Ghent, Belgium,
Solid-State Physics Laboratory in 1978.

Adrian B. Mann   Chapter D.37

Rutgers University
Department of Ceramics
and Materials Engineering
Piscataway, NJ, USA
abmann@rci.rutgers.edu

Dr. Mann's research focuses on the nanomechanics of materials and the
fabrication of nanostructured materials. His research is predominantly
on biomedical materials, but also includes ceramics, polymers and
metals. He is currently an Assistant Professor at Rutgers University,
New Jersey. Prior to this he was a lecturer at the University of
Manchester, England and a Fulbright scholar at The Johns Hopkins
University, Maryland.

Othmar Marti   Chapter C.22

University of Ulm
Department of Experimental Physics
Ulm, Germany
Othmar.Marti@physik.uni-ulm.de

Profesor Othmar Marti is Head of the Department of Experimental Physics at the
University of Ulm. His main research topics are polymers, scanning force microscopy,
friction, near-field optics, and the optics of nanoparticles. Studies: diploma from ETH
Zürich, Dr. sc. nat. ETH Zürich, Habilitation University Konstanz. He worked at IBM
Research Zurich, Switzerland, University of California, USA, ETH Zurich,
Switzerland and University of Konstanz, Germany.

Jack Martin

Analog Devices, Inc.
Micromachined Products Division
Cambridge, MA, USA
jack.martin@analog.com

Chapter G.55

Jack Martin has been a technologist and manager in the design, development and manufacture of industrial MEMS products for 25 years. His accomplishments include development of wafer fab and packaging processes that are used by Analog Devices to produce iMEMS integrated accelerometers and gyroscopes. He has a Ph.D. in Materials Science, a BS and MS in Chemical Engineering, and is a Licensed Professional Engineer.

Shinji Matsui

University of Hyogo
Laboratory of Advanced Science
and Technology for Industry
Hyogo, Japan
matsui@lasti.u-hyogo.ac.jp

Chapter A.6

Dr. Shinji Matsui is a Professor at University of Hyogo. He obtained his Ph. D. degree in Electrical Engineering from Osaka University in 1981. Prior to joining University of Hyogo, he worked at NEC Corporation. His significant works were the demonstration of electron-beam induced deposition and atomic-beam holography. His current research is focused on the three-dimensional nanofabrication by focused-ion-beam chemical-vapour-deposition and on room-temperature nanoimprint.

Brendan McCarthy

University of Arizona
Optical Sciences Center
Tucson, AZ, USA
bmccarthy@optics.arizona.edu

Chapter F.47

Brendan Mc Carthy is a postdoctoral researcher in the Optical Sciences Center, University of Arizona. He received his PhD in 2001, from Trinity College, University of Dublin, Ireland. His thesis topic was the characterisation of composite materials based on carbon nanotubes and conjugated polymers, using optical and vibrational spectroscopy, scanning tunnelling microscopy, and transmission electron microscopy.

Edward J. McMurchie

CSIRO Commonwealth Scientific
and Industrial Research Organisation
Health Sciences and Nutrition
Adelaide BC, SA, Australia
Ted.McMurchie@csiro.au

Chapter B.18

Ted McMurchie is Associate Professor and leads the Biosensors and Biochips Project at CSIRO Health Sciences and Nutrition. He has adjunct appointments at Flinders University and The University of Adelaide. His research interests span the area of cell membrane signaling processes in relation to human health and disease. Recent focus has been on the study of G-protein coupled receptors and their future application in diagnostics and drug screening technologies. This work is aimed at the construction of cell-free, high-throughput microarray platforms with molecular biology, nano- and biotechnology approaches.

Mehran Mehregany

Case Western Reserve University
Department of Electrical Engineering
and Computer Science
Cleveland, OH, USA
mxm31@cwru.edu

Chapters A.10, B.15

Professor Mehran Mehregany received his B.Sc. in Electrical Engineering from the University of Missouri in 1984, and his M.Sc. and Ph.D. in Electrical Engineering from Massachusetts Institute of Technology in 1986 and 1990, respectively. From 1986 to 1990 he was a consultant to the Robotic Systems Research Department at ATandT Bell Laboratories, where he was a key contributor to ground-breaking research in microelectromechanical systems (MEMS). In 1990 he joined the Department of Electrical Engineering and Applied Physics at Case Western Reserve University as an Assistant Professor. He was awarded the Nord Assistant Professorship in 1991 and was promoted to Full Professor in July 1997. He held the George S. Dively Professor of Engineering endowed chair from January 1998 until July 2000, when he was appointed the Goodrich Professor of Engineering Innovation. He served as the Director of the MEMS Research Center at Case from July 1995 until December 2002. Since January 2003, Professor Mehregany has been serving as Chairman of the Electrical Engineering and Computer Science Department. Professor Mehregany is well known for his research in the area of MEMS, and his work has been widely covered by domestic and foreign media. His research interests are in micro- and nano-electromechanical systems (MEMS and NEMS), including sensors, actuators, micromachining and microfabrication technologies. His additional specialized interest centers on developing silicon carbide as an enabling material for MEMS and NEMS, in particular for applications in harsh environments. Professor Mehregany is the founder/co-founder of several technology companies, including Advanced Micromachines Incorporated (now part of The Goodrich Corporation), FLX Micro, Inc. and NineSigma, Inc.

Etienne Menard

3714 Beckman Institute
Department of Materials Science
and Engineering
Urbana, IL, USA
emenard@uiuc.edu

Chapter A.9

Etienne Menard received an engineering diploma in Electronics from the National Polytechnic Institute of Engineering in Electrotechnology, Electronics, Computer Science and Hydraulics (ENSEEIHT) at Toulouse, France, in 2002. He is currently pursuing a Ph.D. in chemistry at the University Pierre et Marie Curie, Paris, France, and the Department of Material Science and Engineering at the University of Illinois at Urbana-Champaign (UIUC) under the direction of Denis Fichou at the Laboratoire des Semi-Conducteurs Organiques, CEA Saclay, France, and Professor John A. Rogers at UIUC. His current research focuses on semi-conducting materials for flexible 'macroelectronic' circuits, the development of soft lithography techniques and organic materials for molecular electronics. These efforts are highly multidisciplinary and combine expertise from nearly every traditional field of technical study: chemistry and chemical engineering, physics, electrical engineering, materials science, optics and mechanics. His research has been featured on the covers of recent issues of Applied Physics Letters, Physics Status Solidi A and other publications.

Ernst Meyer

University of Basel
Institute of Physics
Basel, Switzerland
Ernst.Meyer@unibas.ch

Chapter D.33

Ernst Meyer is professor of physics at the University of Basel. He is interested in friction force and dynamic force microscopy with true atomic resolution. He is also active in the field of sensors based upon micromechanics and magnetic spin resonance detection with force microscopy. Awarded from the Swiss Physical Society, he is member of the Swiss and American Physical Society, of the Editorial Board of Tribology Letters, and co-editor of books on atomic force microscopy.

Authors

Robert Modlinski

Interuniversity Microelectronics Center
(IMEC)
MCP/REMO
Leuven, Belgium
modlinsk@imec.be

Chapter G.52

Robert Modlinski received the M.Sc. degree in microsystems from the Wrocław University of Technology, Poland, in 2001. Currently he is working towards the Ph.D. at the Katholieke Universiteit Leuven, Belgium while staying at the Interuniversity Microelectronics Center IMEC, in the Microsystems, Components and Packaging Division, Leuven, Belgium. His main research interests are in the areas of MEMS with focus on reliability, testing and characterization. He is focusing on mechanical characterization of metals and alloys for use in MEMS, including the investigation of creep and fatigue phenomena and the identification of processing-microstructure-temperature relations.

Carlo D. Montemagno

University of California Los Angeles
School of Engineering
and Applied Science
Los Angeles, CA, USA
cdm@seas.ucla.edu

Chapter A.11

Dr. Carlo Montemagno is currently the Carol and Roy Doumani Professor of Biomedical Engineering, the Chairman of UCLA's Department of Bioengineering and Department of Biomedical Engineering IDP. He is also a Professor of Mechanical and Aerospace Engineering. After receiving his B.Sc. degree from Cornell in biological engineering, he spent eight years in the U.S. Navy as a Civil Engineering Corp Officer. In addition, during his service with U. S. Navy Dr. Montemagno earned a M.Sc. in Petroleum and Natural Gas Engineering at Pennsylvania State University. In 1988, Dr. Montemagno joined Argonne National Laboratory and went on to join the Biological and Environmental Engineering faculty of Cornell University in 1995. Joining UCLA in 2001, Professor Montemagno has published over 60 papers and holds several patents. For his cumulative work Professor Montemagno was awarded the Feynman Prize in Nanotechnology in 2003. Professor Montemagno's research is focused on the application of nanotechnology to biological systems. He is well known for having engineered and fabricated the first nanobiomechanical motor system. His current research projects are directed at the development of biomolecular motor powered nanoelectromechanical devices, muscle powered MEMS devices, micro-robotics, and the engineering of on-chip detectors for pathogens.

Marc Monthioux

UPR A-8011 CNRS
Centre d'Elaboration des Matériaux et
d'Etudes Structurales (CEMES)
Toulouse, France
monthiou@cemes.fr

Chapter A.3

Marc Monthioux has been working on carbon materials for more than 20 years. He is involved in research on carbon nanotubes since 1998, discovered the ability of single-wall nanotubes in being filled by foreign molecules the same year, associated with B. Smith and Prof. D.E. Luzzi from University of Pennsylvania. He is currently Director of Research at the French National Center for Scientific Research and European Associate Editor of CARBON Journal.

Markus Morgenstern

University of Hamburg
Institute of Applied Physics
Hamburg, Germany
mmorgens@physnet.uni-hamburg.de

Chapter C.25

Markus Morgenstern earned his Ph.D. from the Institute of Interface Research and Vacuum Physics of the Forschungszentrum Jülich, Germany in 1996. After one year of research at the University of Paris VII he joined the Group of Prof. Dr. R. Wiesendanger in 1997 as a Senior Scientist. In 2002 he completed his Habilitation at the University of Hamburg with the subject Scanning tunnelling spectroscopy on semiconductor systems and nanostructures

Seizo Morita

Osaka University
Department of Electronic Engineering
Suita-City, Osaka, Japan
smorita@ele.eng.osaka-u.ac.jp

Chapter C.24

Prof. Seizo Morita works in the atomic force microscopy (AFM). He has discovered two-dimensional friction with a lattice periodicity, and two-dimensional solid phase of densely contact-electrified electrons on SiO_2 thin films under ambient conditions. He has achieved mapping, discrimination and control of atomic force and atom with atomic resolution, and also atom manipulation based on a mechanical method using the noncontact AFM apparatus.

Koichi Mukasa

Hokkaido University
Nanoelectronics Laboratory
Sapporo, Japan
mukasa@nano.eng.hokudai.ac.jp

Chapter C.24

K. Mukasa is a Professor of electronics at Hokkaido University, Sapporo, Japan. In 1980 he joined Alps Electric Co.Ltd, where he worked on the magnetic thin film heads and materials. In 1987 he moved to the university. His research interests include spin-polarized STM, exchange force microscopy, magnetic force microscopy, Mott spin detectors, nanostructure concerning electron spin and molecular/biological materials and devices.

Ashis Mukhopadhyay

Wayne State University
Department of Physics
Detroit, MI, USA
ashis@physics.wayne.edu

Chapter D.31

Dr. Ashis Mukhopadhyay is an Assistant Professor of Physics at Wayne State University, Detroit. He received his Ph.D from Kansas State University, where he studied the structure of liquid-liquid interfaces of complex fluid mixtures using optical techniques. He did his postdoctoral study at the University of Illinois Urbana-Champaign. There he integrated force-based measurements of confined fluids with spectroscopy. This study made an important contribution to understand friction and lubrication on the nanoscale. His research interests include soft matter and complex fluids, polymer and glassy systems, nanotribology and nanorheology, adsorption and wetting phenomena. He uses ellipsometry, ultrafast laser spectroscopy and imaging techniques to understand the interfacial structure and dynamics at the molecular level.

Ryozo Nagamune

University of California at Berkeley
Department of Mechanical Engineering
Berkeley, CA, USA
ryozo@me.berkeley.edu

Chapter F.48

Dr. Ryozo Nagamune received his Ph.D. from the Division of Optimization and Systems Theory, the Department of Mathematics, the Royal Institute of Technology in Sweden in 2002. He was a postdoctoral researcher at the Mittag-Leffler Institute in Djursholm, Sweden, in 2003. He is currently a postdoctoral researcher at the University of California, Berkeley. His research interests include multivariable robust control and its application to dual-stage servo systems in hard disk drives.

Bradley J. Nelson

Swiss Federal Institute of Technology (ETH)
Institute of Robotics and Intelligent
Systems (IRIS)
Zürich, Switzerland
bnelson@ethz.ch

Chapter F.49

Brad Nelson is the Professor of Robotics and Intelligent Systems at ETH-Zürich and is the director of the Institute of Robotics and Intelligent Systems (IRIS). His primary research direction lies in extending robotics research into emerging areas of science and engineering. His most recent scientific contributions have been in the area of microrobotics, biomicrorobotics, and nanorobotics, including efforts in robotic micromanipulation, microassembly, MEMS (sensors and actuators), mechanical manipulation of biological cells and tissue, and NEMS. He has also contributed to the fields of visual servoing, force control, sensor integration, and web-based control and programming of robots. Professor Nelson received a B.Sc. (Mechanical Engineering) from the University of Illinois at Urbana-Champaign in 1984, an M.Sc. (Mechanical Engineering) from the University of Minnesota in 1987, and the Ph.D. degree in Robotics (School of Computer Science) from Carnegie Mellon University in 1995. During these years he also worked as an engineer at Honeywell and Motorola, and served as a United States Peace Corps Volunteer in Botswana, Africa. In 1995 he became Assistant Professor at the University of Illinois at Chicago, Associate Professor at the University of Minnesota in 1998, and Professor at ETH in 2002.

Michael Nosonovsky

National Institute of Standards
and Technology
Gaithersburg, MD, USA
Michael.Nosonovsky@nist.gov

Chapter D.38

Michael Nosonovsky got his Ph.D. degree from Northeastern University (Boston). He conducted research in nanotribology and nanomechanics as a Visiting Scholar at Ohio State University and as a NRC Postdoctoral Scholar at the National Institute of Standards and Technology, Gaithersburg, MD.

Kenn Oldham

University of California at Berkeley
Department
of Mechanical Engineering
Berkeley, CA, USA
oldham@newton.berkeley.edu

Chapter F.48

Kenn Oldham is a graduate student in Mechanical Engineering at the University of California at Berkeley, pursuing M.S. and Ph.D. through National Science Foundation Graduate Fellowship. He received his B.S. from Carnegie Mellon University, Pittsburgh, PA. He is currently researching microdevices for hard disk drives. Research interests include MEMS design for control and reliability, optimal and robust control, and materials for microdevices.

Hiroshi Onishi

Kanagawa Academy
of Science and Technology
Surface Chemistry Laboratory
Kanagawa, Japan
oni@net.ksp.or.jp

Chapter C.24

Dr. Onishi Hiroshi is an experimental chemist at the Kanagawa Academy of Science and Technology interested in molecule-scale reaction kinetics at interfaces. He likes to observe molecules moving and reacting over metal oxide surfaces by time-lapse imaging with scanning probe microscopes. Domestic societies encouraged him with awards to further develop his research towards nano-scale chemistry.

René M. Overney

University of Washington
Department of Chemical Engineering
Seattle, WA, USA
roverney@u.Washington.edu

Chapter E.45

René Overney received his Ph.D. in Physics at the University of Basel in 1992. After his postdoctoral years in Japan and at Exxon, CR, Annandale (NJ) he joined the University of Washington in 1996. His research interests are in mesoscale sciences involving nanorheological interpretations of processes that are as diverse as lubrication, membrane transport, or quantum yield in optoelectronic devices.

Alain Peigney

Maître de Conférences at University
of Toulouse III (Paul Sabatier)
Centre Inter-universitaire de Recherche
sur l'Industrialisation des Matériaux
(CIRIMAT)
Toulouse Cedex 04, France
peigney@chimie.ups-tlse.fr

Chapter A.3

Dr. Alain Peigney is ceramic engineer and has a Doctor in physical chemistry. He is Associate Professor of materials chemistry at the Paul Sabatier University, Toulouse, France. His research encompasses the synthesis, sintering and microstructural characterization of ceramics and ceramic matrix nanocomposites. Since 1994 he concentrates on the synthesis of single- and double-walled carbon nanotubes and the preparation of nanocomposites containing carbon nanotubes.

Oliver Pfeiffer

University of Basel
Institute of Physics
Basel, Switzerland
Oliver.Pfeiffer@stud.unibas.ch

Chapter D.33

Oliver Pfeiffer is PhD student at the University of Basel. The main topic of his work is the energy dissipation of oscillating cantilevers in non-contact AFM. Related to this research field is the examination of damping of torsional oscillations of cantilevers when approaching the sample.

Rob Phillips

California Institute of Technology
Mechanical Engineering
and Applied Physics
Pasadena, CA, USA
phillips@aero.caltech.edu

Chapter D.39

Rob Phillips is Professor of Mechanical Engineering and Applied Physics at CALTECH, Pasadena. Phillips' work aims to examine nanomechanics of both crystalline solids and biological molecules and their assemblies. Recent efforts have been aimed at investigating mechanics of DNA packing in viruses, the tension-induced gating of ion channels and the mechanical response of bio-functionalized cantilevers.

Haralampos Pozidis

IBM Zurich Research Laboratory
Storage Technologies
Rüschlikon, Switzerland
hap@zurich.ibm.com

Chapter F.46

Dr. Pozidis received the Ph.D. in electrical engineering from Drexel University, Philadelphia, PA, in 1998. After working with Philips Research, Eindhoven, the Netherlands, on signal processing and coding for optical storage technologies, with focus on DVD and Blue-Ray-Disc, Dr. Pozidis joined IBM Research in 2001. His research focuses on receiver design for alternative storage technologies, particularly scanning probe microscopy-based techniques.

Robert Puers

Katholieke Universiteit Leuven
ESAT/MICAS
Leuven, Belgium
bob.puers@esat.kuleuven.ac.be

Chapter G.52

Professor Robert Puers recieved his B.Sc. degree in electrical engineering in Ghent in 1974, and his M.Sc. degree at the Katholieke Universiteit te Leuven in 1977, where he also obtained his Ph.D. in 1986. From 1980 he was employed as a Research Assistant at the Laboratory ESAT at K.U.Leuven. In 1986 he became Director (NFWO) of the clean room facilities for silicon and hybrid circuit technology at the ESAT-MICAS laboratories of the same university. He was a pioneer in the European research efforts in silicon micromachined sensors, MEMS and packaging techniques, for biomedical implantable systems as well as for industrial devices. At present, he is as a full professor at the K.U.Leuven. He is the author or co-author of more than 300 papers on biotelemetry, sensors, MEMS and packaging in reviewed journals or international conferences. He is a Fellow of the Institute of Physics (UK), council member of the International Microelectronics and Packaging Society (IMAPS), senior member of the Institute for Electric and Electronic Engineers (IEEE).

Prashant K. Purohit

California Institute of Technology
Mechanical Engineering
Pasadena, CA, USA
prashant@caltech.edu

Chapter D.39

Purohit Prashant is currently a Postdoctoral Scholar in Applied Mechanics in the Mechanical Engineering department at Caltech. The overall theme of his research is to develop systematic methods to understand the physics operative at the nanometer scales. His research falls broadly into two categories. The first concerns the subject of coarse-graining methods for crystalline solids and the second mechanics problems in biology. His group hopes to extend the coarse-graining methods in solids to the arena of biological macromolecules.

Calvin F. Quate

Stanford University
Ginzton Laboratory
Stanford, CA, USA
quate@ee.stanford.edu

Chapter C.26

Professor Calvin F. Quate received his B. S. from the University of Utah in 1944 and the Ph.D. degree from Stanford University, Stanford, CA, in 1950. He held positions at Bell Laboratories and Sandia Corporation before joining the faculty at Stanford University, where he has been since 1961. Currently, he is a Faculty Member with the Department of Electrical Engineering and the Department of Applied Physics at Stanford University. His research interests revolve around the scanning probe microscopes. Dr. Quate is a Member of the National Academy of Engineering and the National Academy of Sciences, an Honorary Fellow of the Royal Microscopical Society, and a Foreign Member of the Royal Society London. He has received the IEEE Morris N. Liebmann Award in 1981, the Rank Prize for Optoelectronics in 1982, the IEEE Medal of Honor in 1988, the President's National Medal of Science in 1992, and the American Physical Society Keithley Award in 2000.

Oded Rabin

Massachusetts Institute
of Technology
Department of Chemistry
Cambridge, MA, USA
oded@mgm.mit.edu

Chapter A.4

Oded Rabin is a Ph.D. candidate in Prof. Mildred S. Dresselhaus' group working on the thermoelectric properties of bismuth-antimony nanowire systems, and on electrochemistry-based nanowire synthesis methods. He earned a B.A. degree in Chemistry from the Technion – Israel Institute of Technology, Haifa, Israel and an M. A. degree in Chemistry from the Weizmann Institute of Science, Rehovot, Israel.

Françisco M. Raymo

University of Miami
Department of Chemistry
Coral Gables, FL, USA
fraymo@miami.edu

Chapter A.2

Françisco M. Raymo is Assistant Professor in the Department of Chemistry at the University of Miami. His research interests lie at the interface of chemistry and materials science. In particular, he is exploring innovative strategies to process optical signals with molecular switches, design fluorescent probes for chemical sensing and assemble nanostructured films from electroactive building blocks.

Manitra Razafinimanana

University of Toulouse III (Paul Sabatier)
Centre de Physique des Plasmas
et leurs Applications (CPPAT)
Toulouse Cedex, France
razafinimanana@cpat.ups-tlse.fr

Chapter A.3

Professor Manitra Razafinimanana was born in Analalava, Madagascar. He received the 3rd cycle degree, and Doctorat d'Etat des Sciences Physiques degree from the Université Paul Sabatier, Toulouse, France, in 1982 and 1986, respectively. Since 2000, he has held the position of professor. He has worked on plasma diagnostics, arc-electrode interaction, transport coefficients and thermodynamical properties calculation.

Ziv Reich

Weizmann Institute of Science
Department of Biological Chemistry
Rehovot, Israel
ziv.reich@weizmann.ac.il

Chapter C.28

Professor Ziv Reich received his Ph.D. at the Department of Organic Chemistry, Weizmann Institute of Science. He was postdoc at the Howard Hughes Medical Institute (HHMI), Stanford University. His research concentrates on transport phenomena in biological systems, protein energy landscapes and single-molecule techniques in applications and development. Dr. Reich received the EMBO Young Investigator, the Teva Pharmaceuticals Award, and the Yigal Allon Fellowship for Outstanding Young Faculty, Israel.

John A. Rogers

University of Illinois
Department of Materials Science
and Engineering
Urbana, IL, USA
jrogers@uiuc.edu

Chapters A.9, A.13

Professor John A. Rogers obtained BA and BS degrees in chemistry and in physics from the University of Texas, Austin, in 1989. From MIT, he received SM degrees in physics and in chemistry in 1992 and the Ph.D. degree in physical chemistry in 1995. From 1995 to 1997, Rogers was a Junior Fellow in the Harvard University Society of Fellows. He joined Bell Laboratories as a Member of Technical Staff in 1997 and served as a Research Director from 2000-2002. He is currently Founder Professor of Engineering at University of Illinois at Urbana-Champaign, where he pursues his research interests in unconventional methods for micro/ nanofabrication, plastic and flexible electronics and unusual photonic systems.

Mark Ruegsegger

Ohio State University
Biomedical Engineering Center
Columbus, OH, USA
mark@bme.ohio-state.edu

Chapter B.17

Mark Ruegsegger received his PhD in biomedical engineering at Case Western Reserve Unviersity and is currently an Assistant Professor of Biomedical Engineering at The Ohio State University. His research focus is the development of superparamagnetic particles that can target specific cell or tissue types, in particular atherosclerotic plaque. Other projects include biomimetic surface coatings on biomaterials and characterization of biomolecular flow in nanochannels.

Authors

Marina Ruths

University of Massachusetts Lowell
Department of Chemistry
Lowell, MA, USA
Marina_Ruths@uml.edu

Chapter D.30

Dr. Marina Ruths is Assistant Professor at the University of Massachusetts at Lowell. She received her Ph.D. from the University of California, Santa Barbara in 1996 followed by postdoctoral research at the University of Illinois at Urbana-Champaign, at the Max-Planck-Institute for Polymer Research, Germany and at Åbo Akademi University. Her current research includes adhesion, friction and nanorheology of surfactant, polymer, and liquid crystal systems. She received an ASLA-Fulbright grant in 1991 and an Alexander von Humboldt fellowship in 1998.

Ozgur Sahin

Harvard University
Rowland Institute at Harvard
Nanomechanical Sensing
Cambridge, MA, USA
sahin@rowland.harvard.edu

Chapter C.26

Dr. Ozgur Sahin received B.Sc. degree in electrical engineering from Bilkent University, Ankara, Turkey in 2001 and M.Sc. and the Ph.D. degree in Electrical Engineering from Stanford University, CA. He is currently working at Nanomechanical Sensing Group at Harvard University, Cambridge. His research interests include developing tools for sensing physical and biological phenomena on the nanoscale.

Dror Sarid

University of Arizona
Optical Sciences Center
Tucson, AZ, USA
sarid@optics.arizona.edu

Chapter F.47

Dror Sarid is Professor of Optical Sciences and Director of the Optical Data Storage Center. He is conducting research in scanning tunneling microcopy, atomic force microscopy and related systems and in particular probe storage, nano-optics and nano-technology. He published three books and more than 150 papers, mostly on optics and nanotechnology related topics.

Akira Sasahara

Kanagawa Academy
of Science and Technology
Surface Chemistry Laboratory
Kanagawa, Japan
ryo@net.ksp.or.jp

Chapter C.24

Dr. Sasahara Akira is interested in local structures formed on solid surfaces. His current research focuses on elucidation of chemical and physical properties of nano-scale structures on metal oxide surfaces and their effect on chemical reaction.

Helmut Schift

Paul Scherrer Institute
Laboratory for Micro-
and Nanotechnology
Villigen PSI, Switzerland
helmut.schift@psi.ch

Chapter A.8

Dr.-Ing. Helmut Schift is head of the Micro- and Nanostructuring Group in the Laboratory for Micro- and Nanotechnology at the Paul Scherrer Institut (PSI) in Villigen, Switzerland. After studying electrical engineering at University of Karlsruhe, Germany, and the École Nationale Supérieure de Physique de Strasbourg (ENSPS) in Strasbourg, he performed his Ph.D. studies at the Institute of Microtechnology Mainz (IMM), Germany. After his graduation in 1994, he joined the Laboratory for Micro- and Nanotechnology at the Paul Scherrer Institut as a research staff member. The focus of his work is the development of new replication technologies, including nanoimprint lithography and injection molding. Apart from his research he is giving lectures on "Nanotechnology for Engineers" at the University of Applied Sciences Aargau, Switzerland, and seminar talks on nanotechnology for a non-scientific audience.

Authors

André Schirmeisen

University of Münster
Institute of Physics
Münster, Germany
schira@uni-muenster.de

Chapter C.27

Dr. André Schirmeisen is currently working with his research group on nanoscale mechanical phenomena at Münster University. First to combine field ion microscopy with force microscopy to investigate atomically defined nanocontacts. He spent several years in Canada at McGill University to earn his PhD degree in physics. Before he worked as a strategic business consultant and is always interested in connecting nanoscience research with business applications.

Alexander Schwarz

University of Hamburg
Institute of Applied Physics
Hamburg, Germany
aschwarz@physnet.uni-hamburg.de

Chapter C.25

Dr. Alexander Schwarz belongs to the scientific staff of the Center of Microstructure Research at the Institute of Applied Physics at the University of Hamburg, Germany. The group has 10 years (since 1993) experience in scientific research in the field of force microscopy and spectroscopy at cryogenic temperatures in ultrahigh vacuum and high magnetic fields. He is a Senior Scientist and works on Magnetic Force Microscopy (MFM) at low temperatures

Udo D. Schwarz

Yale University
Department
of Mechanical Engineering
New Haven, CT, USA
udo.schwarz@yale.edu

Chapter C.25

Udo D. Schwarz received his Ph.D. from the University of Basel in 1993 already using scanning force microscopy. Subsequently he moved to the University of Hamburg where he specialised on low-temperature scanning force microscopy and nanotribology. After spending a year at the Lawrence Berkeley National Laboratory he accepted a position as Associate Professor of Mechanical Engineering at Yale University in 2002.

Philippe Serp

Maître de Conférences, Institut National
Polytechnique de Toulouse
Laboratoire de Catalyse, Chimie Fine
et Polymères (LCCFP), Ecole Nationale
Supérieure d'Ingénieurs en Arts
Chimiques et Technologiques
Toulouse, France
pserp@ensct.fr;
Philippe.Serp@ensiacet.fr

Chapter A.3

Dr. Philippe Serp is Associate Professor in the LCCFP at ENSIACET. After receiving is Ph.D. from Paul Sabatier University, Toulouse, France, in 1994 with a work on the preparation of supported catalyst, he moved to Universidade do Porto to carry out post-doctoral research on catalytic CVD to prepare carbon fibers. His current research interests include CVD preparation of nanostructured materials and catalysis.

Huamei (Mary) Shang

University of Washington
Seattle, WA, USA
hmshang@u.washington.edu

Chapter A.5

Huamei (Mary) Shang is a Ph.D. candidate in Materials Science and Engineering at the University of Washington. Her research interests are mainly on the synthesis of optically transparent super-hydrophobic films and coatings, nanostructured and columnar structured scintillation oxide films for neutron detection, and single crystal [001] ZnO nanorod arrays for optical applications. She is the author of over 20 publications, and the major awards she held are the JIN fellowship funded by the Pacific North-west National laboratory and the University of Washington in 2004 and the Ford fellowship funded by Ford Company in 2005.

Susan B. Sinnott

University of Florida
Department of Materials Science
and Engineering
Gainesville, FL, USA
sinnott@mse.ufl.edu

Chapter D.35

Dr. Susan B. Sinnott is an Associate Professor of Materials Science and Engineering at the University of Florida. She received her B.Sc. in Chemistry from the University of Texas at Austin in 1987 and her Ph.D. in Physical Chemistry from Iowa State University in 1993. She was a National Research Council Postdoctoral Associate at the Naval Research Laboratory from 1993 to 1995 and was on the faculty at the University of Kentucky prior to joining the University of Florida in 2000. Her research is focused on investigating the properties and processing of materials using theoretical and computational approaches, including electronic structure calculations and atomistic simulations. Professor Sinnott's awards include University of Florida Materials Science and Engineering Faculty Excellence Awards in 2005, 2004, 2003, and 2002, a Japan Society for the Promotion of Science Fellowship in 2001, and an Oak Ridge Associated Universities Ralph E. Powe Junior Faculty Enhancement Award in Engineering in 1996.

Bryan R. Smith

Ohio State University
Biomedical Engineering Center
Columbus, OH, USA
bryan@bme.ohio-state.edu

Chapter B.17

Bryan Smith is a Ph.D. candidate in biomedical engineering at The Ohio State University, specializing in the microfabrication and bioconjugation of micro- and nanoparticles for proteomics as well as imaging and drug delivery in breast cancer and atherosclerosis.

Anisoara Socoliuc

University of Basel
Institute of Physics
Basel, Switzerland
A.Socoliuc@unibas.ch

Chapter D.33

Anisoara Socoliuc studied physics in Iasi, Romania. She is now PhD student at the University of Basel. The main topic of her work is the study of wear processes on the nanometer scale by friction force microscopy.

Olav Solgaard

E.L. Ginzton Laboratory
Stanford, CA, USA
solgaard@stanford.edu

Chapter C.26

Professor Olav Solgaard received the B.Sc. degree in Electrical engineering from the Norwegian Institute of Technology and the M.Sc. and Ph.D. degrees in Electrical Engineering from Stanford University, California. He was a postdoctoral researcher at the University of California at Berkeley, before joining the University of California at Davis as an Assistant Professor in 1995. In 1999 he joined Stanford University where he is now an Associate Professor of Electrical Engineering. His research interests are optical devices and systems for communication and measurements with an emphasis on semiconductor fabrication and MEMS technology. ~He has authored more than 150 technical publications, and holds 18 patents. He is a co-founder Silicon Light Machines, Sunnyvale, CA, and an active consultant in the MEMS industry.

W. Merlijn van Spengen

Leiden University
Kamerlingh Onnes Laboratory
Leiden, The Netherlands
spengen@physics.leidenuniv.nl

Chapter G.52

Merlijn van Spengen was born in Kasama, Zambia and has the Dutch nationality. He received the M.Sc. degree in electronic engineering (microelectronics failure analysis) from Eindhoven University of Technology, The Netherlands, in 1999. In 2004 he obtained the Ph.D. degree from the Catholic University of Leuven, Belgium, while staying at the independent microelectronics research institute IMEC. Here he studied MEMS reliability and characterization with a focus on stiction and charge-induced defects. Currently, he is at Leiden University, The Netherlands, working on the application of nanotribological principles to MEMS.

Kazuhisa Sueoka   Chapter C.24

Graduate School of Information Science
and Technology, Hokkaido University
Nanoelectronics Laboratory
Sapporo, Japan
sueoka@nano.isthokudai.ac.jp

Kazuhisa Sueoka is currently Professor at the Graduate School of Information Science and Technology, Hokkaido University. His research interests include spin polarized scanning tunnelling microscopy and local spin injection, non-contact atomic force microscopy and surface spin imaging, development of electron spin analyzer and application to scanning electron microscopy, ferromagnetic/semiconductor heterostructure, and fabrication of cantilevers equipped with sensor devices such as magnetoresistance sensors, and single electron transistor.

Yasuhiro Sugawara   Chapter C.24

Osaka University
Department of Applied Physics
Suita, Japan
sugawara@ap.eng.osaka-u.ac.jp

Yasuhiro Sugawara received his Ph.D. in 1988 from Tohoku University and is Professor in the Department of Applied Physics of the Graduate School of Engineering at Osaka University since 2002. His research focuses on the further development of scanning probe microscopes and their applications, especially the noncontact atomic force microscope for the observation of solid surfaces at the atomic and molecular level. His aim is also to develop new nanomaterials and nanodevices by manipulation of single atoms and molecules using the atomic force microscope.

Benjamin Sullivan   Chapter B.14

University of California, San Diego
Department of Bioengineering
La Jolla, CA, USA
bdsulliv@ucsd.edu

Benjamin Sullivan is a biomedical engineer with a research focus on nanoscale optoentropic transduction mechanisms, electrophoretic transport, and tear film biophysics. He is currently a graduate student at the University of California, San Diego.

Yugang Sun   Chapter A.13

University of Illinois at Urbana Champaign
Department of Materials Science
and Engineering
Urbana, IL, USA
ygsun@uiuc.edu

Yugang Sun received his B.Sc. and Ph.D. degrees in chemistry from the University of Science and Technology of China (USTC) in 1996 and 2001, respectively. After graduation from USTC, he came to the University of Washington for a postdoctoral research associate in Prof. Younan Xia's group until December 2003. In 2004, he moved to the University of Illinois at Urbana-Champaign for a postdoctoral appointment with Prof. John Rogers. His research interests include synthesis and characterization of nanostructures, micro/nanofabrication, bioanalysis, and devices for optics and electronics.

Paul Swanson   Chapter B.14

Nanogen, Inc.
Department Advanced Technology
San Diego, CA, USA
pswanson@nanogen.com

Dr. Swanson is a Principal Engineer at Nanogen, Inc. He received his Ph.D. degree in Electrical Engineering from the University of Illinois. He has designed generations of microelectrode arrays, with and without integrated active circuitry, for the electrophoretic manipulation of charged nanoscaled materials. He is an inventor on patents ranging from devices for molecular biological analysis to laser logic devices.

Nikhil S. Tambe   Chapter D.34

GE Global Research, Bangalore # 122
EPIP
Material Systems Technologies
Bangalore, India
nikhil.tambe@ge.com

Dr. Nikhil S. Tambe obtained his Ph.D. from The Ohio State University working on nanotribology of materials and coatings for nanotechnology applications. His research has been highlighted in Nature Materials and rated amongst the best in 2005 by the Institute of Physics. At GE Global Research, his current work is on the mechanics of biomimetically inspired nanoengineered surfaces for industrial applications.

Mike Yung-Chieh Tan

University of California at Irvine
Department of Biomedical Engineering
Irvine, CA, USA
ytan@uci.edu

Chapter B.21

Dr. Yung Chieh Tan received a B.Sc. in Bioengineering Premedical from the University of California San Diego in 2001. He received his M. Sc. and Ph.D. degrees in Biomedical Engineering from University of California Irvine in 2002 and 2005, respectively. His research interests include the control of droplets in microfluidic systems, drug encapsulation, and the design and formation of artificial cells.

Kimberly L. Turner

University of California, Santa Barbara
Department of Mechanical
and Environmental Engineering
Santa Barbara, CA, USA
turner@engineering.ucsb.edu

Chapter G.51

Dr. Kimberly L. Turner has been a member of the Mechanical Engineering faculty at University of California Santa Barbara since 1999, where she is an Associate Professor, a position she has held since 2004. She received her B.Sc. in Mechanical Engineering from Michigan Technological University, and her Ph.D. in Theoretical and Applied Mechanics from Cornell University, in 1994 and 1999, respectively. Dr. Turner's interests focus on micro-and nanoscale science, focusing on dynamics and characterization, and applications utilizing nonlinear dynamics. Dr. Turner is the recipient of the 1997 Varian fellowship award given by the American Vacuum Society, as well as an NSF graduate research fellowship, the NSF CAREER award, and an NSF/MEXT Young Scientist Exchange Fellowship. She was also recently named a UCSB Academic Senate Distinguished Teacher. She is co-director of the Sensors Task in the UCSB/ARMY Institute for Collaborative Biotechnologies, and a member of the California Nanosystems institute. She is an active member of ASME and IEEE, as well as the ASEE, the American Vacuum Society, Tau Beta Pi, Pi Tau Sigma, and the Cornell Society of Engineers.

George W. Tyndall

IBM Almaden Research Center
Science and Technology
San Jose, CA, USA
tyndallgw@netscape.net

Chapter E.45

Dr. George Tyndall's current research focuses on the friction and adhesion in boundary lubricants especially as it pertains to the tribology of the magnetic recording head-disk interface in hard-disk drive applications.

Peter Vettiger

IBM Zurich Research Laboratory
Manager Micro-/Nanomechanics
Rüschlikon, Switzerland
pv@zurich.ibm.com

Chapter F.46

In 1963 Peter Vettiger joined IBM Research (Rüschlikon). He established and headed micro/nanoscale fabrication activities for superconducting, electronic and opto-electronic devices. Together with G.K. Binnig, he initiated Millipede probe-storage activities in 1995. His research interests are micro/nanomechanical devices and systems for probe-storage and biological applications. He is an IEEE Fellow and received a Doctor honoris causa from the University of Basel, Switzerland.

Guohua Wei

Ohio State University
Department of Mechanical Engineering
Nanotribology Laboratory for Information
Storage and MEMS/NEMS
Columbus, OH, USA
guohua_wei1113@yahoo.com

Chapter D.40

Dr. Guohua Wei received his B.Sc. and M.Sc. degree in materials science and engineering from University of Science and technology Beijing, China, in 1996 and 1999, respectively, He received his Ph.D. degree from University of Alabama, Tuscaloosa in 2003. His research interests include nanoindentation, thin films, adhesion, surface analysis, MEMS/NEMS.

Authors

David Wendell   Chapter A.11

University of California Los Angeles
Department of Bioengineering
Los Angeles, CA, USA
dwendell@ucla.edu

David Wendell received his B.Sc. in Bioengineering from the department of Biological and Environmental Engineering at Cornell University in 2000. Subsequently, he worked under Dr. Greg Martin of the Boyce Thompson Institute for Plant Research mapping genetic pathogen responses in Tomato. After receiving his M. Sc. in Biomedical Engineering from Cornell University in 2002 for his work in atomic force spectroscopy, he relocated to UCLA's Bioengineering program. His current research focuses on building primitive neural networks using membrane-bound proteins in lipid and block copolymeric membranes to study emergent systems. His most recent publication focuses on measuring actin polymerization forces using a mixture of purified cytoskeleton proteins.

Darrin J. Young   Chapter B.15

Case Western Reserve University
Electrical Engineering
and Computer Science
Cleveland, OH, USA
djy@po.cwru.edu

Darrin J. Young received his BS, MS, and Ph.D. degrees from the EECS Department at University of California at Berkeley in 1991, 1993, and 1999, respectively. He pioneered RF MEMS high-Q tunable passive devices for wireless communications. He joined the EECS Department at Case Western Reserve University as an assistant professor in 1999. His main research interests include MEMS device design and fabrication.

Y. Elaine Zhu   Chapter D.31

University of Notre Dame
Dept. of Chemical and Biomolecular
Engineering
Notre Dame, IN, USA
yzhu3@nd.edu

Dr. Yingxi Elaine Zhu currently is an assistant professor of Chemical and Biomolecular Engineering at the University of Notre Dame, IN. She received her B.Sc. in Chemical Engineering and Polymer Science from Tsinghua University, China in 1997, and her Ph.D. in Materials Science and Engineering from the University of Illinois at Urbana-Champaign in 2001. Before she joined Notre Dame in November, 2004, she had worked at General Motors R&D as a senior research scientist in 2002–2003 and in the Division of Engineering and Applied Science and Physics at Harvard University as a postdoc research fellow in 2003–2004. Dr. Zhu's research interests include polymer, colloids and other soft matter, molecular structure and dynamics at interface, surface and interfacial phenomena, fluid dynamics and material science in micro- or nanofluidic devices, fuel cells, rheology and tribology.

Babak Ziaie   Chapter A.7

Purdue University
Department of Electrical and Computer
Engineering
West Lafayette, IN, USA
bziaie@purdue.edu

Dr. Babak Ziaie is an associate professor in the School of Electrical and Computer Engineering at Purdue University. Prior to joining Purdue in January 2005, he was with the Electrical and Computer Engineering Department of the University of Minnesota (1999–2004). He received his doctoral degree in Electrical Engineering from the University of Michigan in 1994. His research interests are mostly related to biomedical applications of MEMS and microsystems. Dr. Ziaie is the recipient of the NSF Career award in Biomedical Engineering (2001) and McKnight Endowment Fund Award for Technological Innovations in Neuroscience (2002).

Christian A. Zorman

Case Western Reserve University
Department of Electrical Engineering
and Computer Science
Cleveland, OH, USA
caz@po.cwru.edu

Chapters A.10, B.15

Christian A. Zorman, Ph.D. received his B.Sc. cum laude in physics and B.A. cum laude in economics from the Ohio State University in 1988, and M.Sc. and Ph.D. in physics from Case Western Reserve University in 1991 and 1994, respectively. His doctoral research involved an investigation of the secondary electron emission properties of CVD diamond films for vacuum electronics. Dr. Zorman joined the MEMS program at CWRU in 1994 as a Research Associate and immediately began working in the SiC MEMS area. He was promoted to Senior Research Associate in 1997 and Researcher in 2000. He currently is an Associate Professor in EECS at CWRU. He has been instrumental in the construction of AP- and LPCVD reactors for SiC thin films, and has led the development of recipes for the growth of single and polycrystalline 3C-SiC films for micromachined sensors and actuators. In addition to the development of novel bulk and surface micromachining techniques for SiC, Dr. Zorman was a key contributor in the development of novel polishing, wafer bonding, and low defect density growth processes for SiC. His current research interests include the development of SiC for NEMS. He has published over 120 technical papers, five book chapters, and has taught several short courses on SiC for MEMS. Professor Zorman is a past chairman of the MEMS Technical Group in the American Vacuum Society and is currently serving as co-chairman.

Jim Zoval

University of California Irvine
Department of Mechanical and Aerospace
Engineering
Irvine, CA, USA
jzoval@uci.edu

Chapter B.20

Dr. Zoval is a Research Specialist in the Department of Mechanical and Aerospace Engineering at UC Irvine. He has industrial experience in clinical and diagnostic platform development. His research interests are in nanotechnology, sensors, microscopy, and spectroscopy. He is currently researching and developing an automated medical diagnostic platform for detection of human bacterial and viral infection. Dr. Zoval received his Ph.D. in Physical Chemistry from UC Irvine in 1996.

Philippe K. Zysset

Vienna University of Technology
Institute of Lightweight Design
and Structural Biomechanics
Vienna, Austria
philippe.zysset@ilsb.tuwien.ac.at

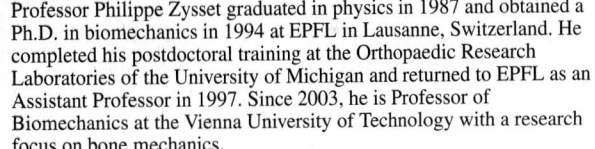

Chapter D.36

Professor Philippe Zysset graduated in physics in 1987 and obtained a Ph.D. in biomechanics in 1994 at EPFL in Lausanne, Switzerland. He completed his postdoctoral training at the Orthopaedic Research Laboratories of the University of Michigan and returned to EPFL as an Assistant Professor in 1997. Since 2003, he is Professor of Biomechanics at the Vienna University of Technology with a research focus on bone mechanics.

Authors

Subject Index

Subject Index

Subject Index

Subject Index

Subject Index